Jane's

All The World's Aircraft 2005-2006

Editor-in-Chief Paul Jackson FRAeS
Deputy Editor Kenneth Munson AMRAeS
Assistant Editor Lindsay Peacock

Ninety-sixth Edition

Total number of entries 1,532 New and updated entries 1,460
Total number of images 2,002 New images 1,017

Visit jawa.janes.com and view the list of latest updates that have been added to the online version of *Jane's All The World's Aircraft* subsequent to this print edition.

Bookmark jawa.janes.com today!
Jane's All The World's Aircraft online site gives you details of the additional information that is unique to online subscribers and the many benefits of upgrading to an online subscription. Don't delay, visit jawa.janes.com today and view the list of latest updates to this online service.

ISBN 0 7106 2684 3

In the US and its dependencies
Jane's Information Group Inc, 110 N. Royal Street, Suite 200, Alexandria, Virginia 22314, US

Printed in in the United Kingdom

Contents

This Edition has been compiled by:

Susan Bushell	AIRCRAFT: LIGHT AVIATION
Bill Gunston, OBE, FRAeS	GLOSSARY; AERO-ENGINES TABLES
Paul Jackson, FRAeS	AIRCRAFT: FRANCE (part), INTERNATIONAL (part), RUSSIAN FEDERATION (part), UKRAINE, UNITED KINGDOM (part) and UZBEKISTAN
Mike Jerram	AIRCRAFT: GENERAL AVIATION
Kenneth Munson, AMRAeS, ARHistS	AIRCRAFT: CIVIL AND MILITARY (part); LIGHTER THAN AIR
Lindsay Peacock	AIRCRAFT: INTERNATIONAL (military) and UNITED STATES OF AMERICA (military)

Jane's All the World's Aircraft website: jawa.janes.com

How to use *Jane's All the World's Aircraft*

Described on the following pages are all known powered aircraft, of which details have been received, currently in, or anticipating, commercial production in all countries of the World, apart from those built at home, entirely from plans; and rapidly dismantled, ultralight recreational machines. Exceptions to these criteria are made in the cases of 'one off' aircraft of technical interest – for example those undertaking pioneering work with NASA or designed as a practical exercise by aeronautical universities. Many other of the World's aircraft remaining in service, but no longer being built, will be found in *Jane's Aircraft Upgrades*.

Entries in this book are in alphabetical order (a) by country (including International) and then (b) by manufacturer's name. In the cases of manufacturers producing a diversity of aircraft, those of obvious military potential are presented first. However, it should be noted that a few of the larger constructors are divided into operating divisions, within which individual aircraft types are arranged. In the case of the multinational EADS, aircraft descriptions are retained in the French, German and Spanish sections.

Company entries begin with a brief introduction, including postal address, telephone/fax numbers, e-mail and website addresses and the names of some significant executives. The last-mentioned listing is by no means exhaustive and further details will be found in *Jane's International ABC Aerospace Directory* and *Jane's International Defence Directory*. Subcontractors and suppliers to the aerospace industry appear in *Jane's Aircraft Component Manufacturers,* which also provides rapid references to the subcontractors involved in certain prominent aircraft programmes.

For ease of access to information, entries on individual types of aircraft are subdivided under the following headings:

TYPE: A brief description of the aircraft's function. A full list of categories appears under the heading Type Classifications.
PROGRAMME: A record of key events in an aircraft's production history. In the case of significant aircraft, which may have a long and detailed development phase, a short summary lists the key dates in that phase in an easily retrievable format.
CURRENT VERSIONS: Where applicable, details of available models (marks) and cross-reference to earlier versions now out of production.
CUSTOMERS: Present total on order and produced, often in tabular form, for military aircraft and those civil types for which such a list would not be of prohibitive length.
COSTS: Price per unit or programme price, plus any other disclosed information on R&D expenditure.
DESIGN FEATURES: Where appropriate, opens with a broad statement of design objectives and the means by which they were achieved. This is followed by details such as aerofoil section and helicopter rotor speeds.
FLYING CONTROLS: Here is described the method of controlling the aircraft, it being assumed that the reader has a basic understanding of the function of ailerons, flaps, rudders, trim tabs and the other conventional manoeuvring surfaces (refer to 'conventional and manual' in the Glossary). Descriptions are concerned with the method by which the controls are operated (manual/powered) and appropriate control inputs determined (autopilot/fly-by-wire, for example).
STRUCTURE: Configuration, materials and any special manufacturing methods; details of subcontractors or partners producing significant elements of the airframe.
LANDING GEAR: Includes tyre sizes and pressures for wheeled aircraft, as well as ground turning circle. Braking parachutes, where fitted.
POWER PLANT: Number and power of engines; helicopter transmission ratings; fuel capacity. Brief additional details are provided in the Aero-Engines listing; fuller descriptions of turbine power plants are in *Jane's Aero-Engines*.
ACCOMMODATION: Seating arrangements, access, environmental control and, for transport aircraft, cargo loading capacity; type of ejection seat, if fitted.
SYSTEMS: Power generation provisions, de-/anti-icing equipment, pressurisation/air conditioning and similar equipment.
AVIONICS: The entry is subdivided into communications, radar, flight aids, instruments, mission equipment (mostly military or law-enforcement) and self-defence (military).
EQUIPMENT: Cargo-handling aids, spraying/firefighting apparatus, lighting, ballistic recovery parachutes and similar items.
ARMAMENT: Fixed and air-dropped/launched weapons listed by the manufacturer as actual or potential armament. Not all items may have been cleared for carriage and not all operators will use those which have. Refer also to the Missiles listing, where appropriate.
DIMENSIONS, EXTERNAL: Includes door sizes and certain ground clearances.
DIMENSIONS, INTERNAL: Includes areas and volumes where relevant.
AREAS: Wings, fixed tail surfaces and control surfaces.
WEIGHTS AND LOADINGS: As supplied by the manufacturer; individual aircraft may vary.
PERFORMANCE: Observations as above; all speeds assumed TAS unless stated otherwise.
OPERATIONAL NOISE LEVELS: Internationally recognised measurements of landing and take-off sound at airports.

Lighter than air craft are now included in the main section, along with aeroplanes, rotorcraft and a few specialised air vehicles. Inclusion is restricted to commercially and militarily operated airships, and vehicles of special technical interest, but not recreational balloons.

Measurements of all types are given in both SI (metric) and Imperial units, the more common conversion factors for which are in the Glossary. Performance details are quoted in good faith and certain critical conversions 'rounded' to give a margin of safety, although *Jane's* does not purport to be an alternative to the manufacturer's operating notes.

In addition to the main section on aeroplanes, helicopters and lighter than air craft are others detailing

Air-Launched Missiles, Aero-Engines and Propellers. Air-Launched Missiles is a rapid reference which seeks not to duplicate the separate and vastly more detailed *Jane's Air-Launched Weapons,* indicating instead how the potential of aircraft described is augmented by the weapons they carry. Aero-Engines are arranged in alphabetical order of manufacturer, irrespective of their country of origin, within the classifications of piston, turboprop, turboshaft and jet engine. Propellers lists some of the main airscrew manufacturers and provides explanations of their designation systems.

Indexes are arranged to speed reference to both current and past aircraft. The first index deals with aircraft in this volume only and is followed by a second covering the past 10 editions of the book. However, the Air-Launched Missiles and Aero-Engines listings are self-indexing, with multiple cross-references.

Each entry for a company or aircraft is annotated to indicate whether or not it has been altered since the last edition was published.

To help users evaluate the data of this edition, the following identifiers have been used:

● ***VERIFIED*** The editor has made a detailed examination of the entry's content and checked it's relevancy and accuracy for publication in the new edition to the best of his ability.

● ***UPDATED*** During the verification process, significant changes to content have been made to reflect the latest position known to Jane's at the time of publication.

● ***NEW ENTRY*** Information on new equipment and/or appearing for the first time in the title.

● ***NEW*** New images are identified as ***NEW*** and most are followed by a seven digit number for ease of identification by our image library.

A full list of all entries indicating their current status is provided in the index.

Total number of entries 1,532 New and updated entries 1,460
Total number of images 2,002 New images 1,017

Visit jawa.janes.com and view the list of latest updates that have been added to the online version of *Jane's All the World's Aircraft* subsequent to this print edition.

British Library Cataloguing-in-Publication Data.
A catalogue record for this book is available from the British Library.

Printed and bound in Great Britain by Thanet

Jane's Libraries

To assist your information gathering and to save you money, Jane's has grouped some related subject matter together to form 'ready-made' libraries, which you can access in whichever way suits you best – online, on CD-ROM, via Jane's EIS or through Jane's Data Service.

The entire contents of each library can be cross-searched, to ensure you find every reference to the subjects you are looking for. All Jane's libraries are updated according to the delivery service you choose and can stand alone or be networked throughout your organisation.

www.janes.com

Jane's Defence Equipment Library

Aero-Engines
Air-Launched Weapons
Aircraft Upgrades
All the World's Aircraft
Ammunition Handbook
Armour and Artillery
Armour and Artillery Upgrades
Avionics
C4I Systems
Electro-Optic Systems
Explosive Ordnance Disposal
Fighting Ships
Infantry Weapons
Land-Based Air Defence
Military Communications
Military Vehicles and Logistics
Mines and Mine Clearance
Naval Weapon Systems
Nuclear, Biological and Chemical Defence
Radar and Electronic Warfare Systems
Strategic Weapon Systems
Underwater Warfare Systems
Unmanned Aerial Vehicles and Targets

Jane's Defence Magazines Library

Defence Industry
Defence Weekly
Foreign Report
Intelligence Digest
Intelligence Review
International Defence Review
Islamic Affairs Analyst
Missiles and Rockets
Navy International
Terrorism and Security Monitor

Jane's Market Intelligence Library

Aircraft Component Manufacturers
All the World's Aircraft
Defence Industry
Defence Weekly
Electronic Mission Aircraft
Fighting Ships
Helicopter Markets and Systems
International ABC Aerospace Directory
International Defence Directory
Marine Propulsion
Naval Construction and Retrofit Markets
Police and Security Equipment
Simulation and Training Systems
Space Directory
Underwater Technology
World Armies
World Defence Industry

Jane's Security Library

Amphibious and Special Forces
Chemical-Biological Defense Guidebook
Facility Security
Fighting Ships
Intelligence Digest
Intelligence Review
Intelligence Watch Report
Islamic Affairs Analyst
Police and Security Equipment
Police Review
Terrorism & Security Monitor
Terrorism Watch Report
World Air Forces
World Armies
World Insurgency and Terrorism

Jane's Sentinel Library

Central Africa
Central America and the Caribbean
Central Europe and the Baltic States
China and Northeast Asia
Eastern Mediterranean
North Africa
North America
Oceania
Russia and the CIS
South America
South Asia
Southeast Asia
Southern Africa
The Balkans
The Gulf States
West Africa
Western Europe

Jane's Transport Library

Aero-Engines
Air Traffic Control
Aircraft Component Manufacturers
Aircraft Upgrades
Airport Review
Airports and Handling Agents –
- Central and Latin America (inc. the Caribbean)
- Europe
- Far East, Asia and Australasia
- Middle East and Africa
- United States and Canada

Airports, Equipment and Services
All the World's Aircraft
Avionics
High-Speed Marine Transportation
Marine Propulsion
Merchant Ships
Naval Construction and Retrofit Markets
Simulation and Training Systems
Transport Finance
Urban Transport Systems
World Airlines
World Railways

Jane's
Intelligence and Insight You Can Trust

104100405

EDITORIAL AND ADMINISTRATION

Director: Ian Kay, e-mail: Ian.Kay@janes.com

New Media Publishing Director: Sean Howe, e-mail: Sean.Howe@janes.com

Content Services Director: Anita Slade, e-mail: Anita.Slade@janes.com

Content Systems Manager: Jo Agius, e-mail: Jo.Agius@janes.com

Pre-Press Manager: Christopher Morris, e-mail: Christopher.Morris@janes.com

Team Leader: Melanie Rovery, e-mail: Melanie.Rovery@janes.com

Content Editor: Tracy Johnson, e-mail: Tracy.Johnson@janes.com

Production Controller: Laura-Jane Walker, e-mail: Laura-JaneWalker@janes.com

Content Update: Jacqui Beard, Information Collection Team Leader
Tel: +44 (0) 20 8700 3808 Fax: +44 (0) 20 8700 3959
e-mail: yearbook@janes.com

Jane's Information Group Limited, Sentinel House, 163 Brighton Road, Coulsdon, Surrey CR5 2YH, UK
Tel: +44 (0) 20 8700 3700 Fax: +44 (0) 20 8700 3900

SALES OFFICES

Send Europe and Africa enquiries to: *Mike Gwynn – Head of Information Sales*
Jane's Information Group Limited, Sentinel House, 163 Brighton Road, Coulsdon, Surrey CR5 2YH, UK
Tel: +44 (0) 20 8700 3750 Fax: +44 (0) 20 8700 3751
e-mail: customerservices.uk@janes.com

Send US/Latin America/Canada enquiries to: *Greg Wallis – VP Information Sales*
Jane's Information Group Inc, 110 N Royal Street, Suite 200, Alexandria, Virginia 22314, US
Tel: (+1 703) 683 37 00 Fax: (+1 703) 836 02 97 Telex: 6819193
Tel: (+1 800) 824 07 68 Fax: (+1 800) 836 02 97
e-mail: customerservices.us@janes.com

Send Asia enquiries to: *David Fisher – Group Business Manager*
Jane's Information Group Asia, 78 Shenton Way, #10-02, Singapore 079120, Singapore
Tel: (+65) 63 25 08 66 Fax: (+65) 62 26 11 85
e-mail: asiapacific@janes.com

Send Australia/New Zealand enquiries to: *Russel Smith – Business Manager*
Jane's Information Group, PO Box 3502, Rozelle Delivery Centre, New South Wales 2039, Australia
Tel: (+61 2) 85 87 79 00 Fax: (+61 2) 85 87 79 01
e-mail: oceania@janes.com

Send Middle East enquiries to: *Ali Abdellatif Siali – Regional Sales Manager*
Jane's Information Group, PO Box 502138, Dubai, United Arab Emirates
Tel: (+971 4) 390 23 36 Fax: (+971 4) 390 88 48
e-mail: mideast@janes.com

Send Japan enquiries to: *Norihisa Fukuyama – Information Consultant*
Jane's Information Group, Palaceside Building, 5F, 1-1-1, Hitotsubashi, Chiyoda-ku, Tokyo 100-0003, Japan
Tel: (+81 3) 52 18 76 82 Fax: (+81 3) 52 22 12 80
e-mail: japan@janes.com

Send India enquiries to: *T C Martin – Information Consultant*
Jane's Information Group, PO Box 3806, New Delhi 110049, India
Tel/Fax: (+91 11) 26 51 61 05
e-mail: india@janes.com

ADVERTISEMENT SALES OFFICES

(Head Office)
Jane's Information Group
Sentinel House, 163 Brighton Road,
Coulsdon, Surrey CR5 2YH, UK
Tel: (+44 20) 87 00 37 00
Fax: (+44 20) 87 00 38 59/37 44
e-mail: defadsales@janes.com

Richard West, Senior Key Accounts Manager
Tel: (+44 1892) 72 55 80 Fax: (+44 1892) 72 55 81
e-mail: richard.west@janes.com

Nicky Eakins, Advertising Sales Executive
Tel: (+44 20) 87 00 38 53 Fax: (+44 20) 87 00 38 59/37 44
e-mail: nicky.eakins@janes.com

(US/Canada office)
Jane's Information Group
110 N Royal Street, Suite 200,
Alexandria, Virginia 22314, USA
Tel: (+1 703) 683 37 00
Fax: (+1 703) 836 55 37
e-mail: defadsales@janes.com

USA and Canada
Katie Taplett, US Advertising Sales Director
Tel: (+1 703) 683 37 00 Fax: (+1 703) 836 55 37
e-mail: katie.taplett@janes.com

Northern US and Eastern Canada
Linda Hewish, Northeast Region Advertising Sales Manager
Tel: (+1 703) 683 37 00 Fax: (+1 703) 836 55 37
e-mail: linda.hewish@janes.com

Southeastern US
Kristin D Schulze, Advertising Sales Manager
PO Box 270190, Tampa, Florida 33688-0190, USA
Tel: (+1 813) 961 81 32 Fax: (+1 813) 961 96 42
e-mail: Kristin.Schulze@janes.com

Western US and Western Canada
Richard L Ayer
127 Avenida del Mar, Suite 2A, San Clemente, California 92672, USA
Tel: (+1 949) 366 84 55 Fax: (+1 949) 366 92 89
e-mail: ayercomm@earthlink.com

Australia: *Richard West* (see UK Head Office)

Benelux: *Nicky Eakins* (see UK Head Office)

Brazil: *Katie Taplett* (see US address)

Eastern Europe (excl. Poland): MCW Media & Consulting Wehrstedt
Dr Uwe H Wehrstedt
Hagenbreite 9, D-06463 Ermsleben, Germany
Tel: (+49) 07 00/WEHRSTEDT / (+49 03) 47 43/620 90
Fax: (+49 03) 47 43/620 91
e-mail: info@Wehrstedt.org

France: Patrice Février
BP 418, 35 avenue MacMahon,
F-75824 Paris Cedex 17, France
Tel: (+33 1) 45 72 33 11 Fax: (+33 1) 45 72 17 95
e-mail: patrice.fevrier@wanadoo.fr

Germany and Austria: *MCW Media & Consulting Wehrstedt* (see Eastern Europe)

Greece: *Nicky Eakins* (see UK Head Office)

Hong Kong: *Nicky Eakins* (see UK Head Office)

India: *Nicky Eakins* (see UK Head Office)

Iran: Eideh Info Company
Ali Jahangard
19 4th Street, Ghaem Magham Avenue, Tehran, Iran
Tel: (+98 21) 873 59 23
e-mail: eideh@mavara.com

Israel: Oreet – International Media
15 Kinneret Street, IL-51201 Bene Berak, Israel
Tel: (+972 3) 570 65 27 Fax: (+972 3) 570 65 27
e-mail: admin@oreet-marcom.com
Defence: Liat Shaham
e-mail: liat_s@oreet-marcom.com

Italy and Switzerland: Ediconsult Internazionale Srl
Piazza Fontane Marose 3, I-16123 Genoa, Italy
Tel: (+39 010) 58 36 84 Fax: (+39 010) 56 65 78
e-mail: genova@ediconsult.com

Middle East: *Nicky Eakins* (see UK Head Office)

Pakistan: *Nicky Eakins* (see UK Head Office)

Poland: *Nicky Eakins* (see UK Head Office)

Russia: Vladimir N Usov, PO Box 98, 622018 Nizhniy Tagil,
Sverdlovsk Region, Russian Federation
Tel/Fax: (+7 3435) 23 02 68
e-mail: uvn125@uraltelecom.ru

Scandinavia: The Falsten Partnership
PO Box 27, Portslade, East Sussex BN41 2XA, UK
Tel: (+44 1273) 77 10 20 Fax: (+ 44 1273) 77 00 70
e-mail: sales@falsten.com

Singapore: *Richard West/Nicky Eakins* (see UK Head Office)

South Africa: *Richard West* (see UK Head Office)

South Korea: Infonet Group Inc
Sanbu Rennaissance Tower 902, 456 Gongdukdong, Mapogu, Seoul, South Korea
Contact: Mr Jongseog Lee
Tel: (+82 2) 716 99 22
Fax: (+82 2) 716 95 31
e-mail: jslee@infonetgroup.co.kr

Spain: Via Exclusivas SL
e-mail: viaexclusivas@viaexclusivas.com

Turkey: *Richard West* (see UK Head Office)

ADVERTISING COPY
Linda Letori (Jane's UK Head Office)
Tel: (+44 20) 87 00 37 42 Fax: (+44 20) 87 00 38 59/37 44
e-mail: linda.letori@janes.com

For North America, South America and Caribbean only:
Lia Johns (Jane's US/Canada Office)
Tel: (+1 703) 683 37 00 Fax: (+1 703) 836 55 37
e-mail: lia.johns@janes.com

FREE ENTRY/CONTENT IN THIS PUBLICATION

Having your products and services represented in our titles means that they are being seen by the professionals who matter – both by those involved in procurement and by those working for the companies that are likely to affect your business. We therefore feel that it is very much in the interests of your organisation, as well as Jane's, to ensure your data is current and accurate.

- **Don't forget –** You may be missing out on business if your entry in a Jane's product is incorrect because you have not supplied the latest information to us.
- **Ask yourself –** Can you afford not to be represented in Jane's printed and electronic products? And if you are listed, can you afford for your information to be out of date?
- **And most importantly –** The best part of all is that your entries in Jane's products are TOTALLY FREE OF CHARGE.

Please provide (using a photocopy of this form) the information on the following categories where appropriate:

1. Organisation name: ____________
2. Division name: ____________
3. Location address: ____________
4. Mailing address if different: ____________
5. Telephone (please include switchboard and main department contact numbers, for example Public Relations, Sales, and so on): ____________
6. Facsimile: ____________
7. e-mail: ____________
8. Web sites: ____________
9. Contact name and job title: ____________
10. A brief description of your organisation's activities, products and services: ____________
11. Jane's publications in which you would like to be included: ____________

Please send this information to:
Jacqui Beard, Information Collection, Jane's Information Group,
Sentinel House, 163 Brighton Road, Coulsdon, Surrey, CR5 2YH, UK
Tel: (+44 20) 87 00 38 08
Fax: (+44 20) 87 00 39 59
e-mail: yearbook@janes.com

Copyright enquiries:
Contact: Keith Faulkner
Tel/Fax: (+44 1342) 30 50 32
e-mail: copyright@janes.com

Please tick this box if you do not wish your organisation's staff to be included in Jane's mailing lists ☐

JAWA

Fresh from its triumph in 2004, when it secured the prize for the first civilian spacecraft, Scaled Composites turned its attentions in 2005 to one of the last 'terrestrial' air records. The Scaled 311 Capricorn Virgin Atlantic GlobalFlyer (here accompanied by an Extra 300 chase-plane during a proving flight over the Sierra Nevada mountains) was designed to allow pilot and 'serial record-breaker' Steve Fossett to circumnavigate the Earth non-stop and, uniquely, alone. This he accomplished successfully between 1-3 March 2005, completing the 19,880 (36,818 km) journey in 67 hours, 1 minutes 46 seconds.
NEW/1042466

Executive Overview

In Cold War days, the battle of technology and production numbers between East and West would have dominated such annual reviews of the aerospace world as this. Since the turn of the 21st Century, *Jane's* executive overviews have, increasingly, been dominated by accounts of another non-shooting battle between West and Further West: Airbus and Boeing. Future historians of air transport will have to reserve a disproportionate number of column-inches to describe the events and their ramifications of the last 12 months.

Both major battles — orders and deliveries — have now been won for a second year by Airbus, leaving Boeing with no "Yes, but" figleaf to cover its embarrassment:

	Orders 2004	Net orders 2004	Deliveries 2004	Backlog	Orders 2003	Deliveries 2003
Airbus	370	366	320	1,500	254	305
Boeing	277	272	285	1,097	239	281

It will be noted that both companies have improved their order intake and levelled or increased their delivery totals over 2003, while suffering nothing like the number of cancellations recorded in 2002 and 2003. The latter figure amounted to a mere nine aircraft, although this does not reflect the usual trading up or down between the various types and subvariants offered by individual manufacturers.

Furthermore, this favourable year-end picture seemed far from certain when the half-year sums were done in July 2004:

	Orders	Net orders	Deliveries	Backlog
Airbus (1H04)	104	100	161	1,393
Boeing (1H04)	82	78	151	1,037

Projection of these figures, indicating a one-third fall in comparison with 2003, to the year-end would have prompted far gloomier predictions for the future and possibly even a return to slump. What the second-half rally does show, however, is that the airliner business can still be capricious.

To continue the Cold War analogy, we have recently witnessed the start of a new 'arms race' between the two airliner superpowers. Old systems have been retired and new weapons developed.

Major changes in the Airbus line-up have included:

- Announcement of the A350 rival to Boeing's 787. With this A330 derivative, Airbus has begun what it frequently criticised Boeing for doing: resorting to derivation as a substitute for innovation. The A350 is unlikely to match the 787's efficiency in range. Interestingly, although it will be added to the A330's type certificate, it is likely to be the first aircraft certified by the new European Aviation Safety Agency.
- Rollout of the A380. Hardly a surprise, the prototype 'officially' and figuratively emerged from its hangar 18 days into 2005. Nominally a 555-seater, at least one customer (Emirates) is to outfit its aircraft for 644

Boeing's announcements have not all been positive:

- Rollout in October 2004 of the 1,050th and final B757
- Abandonment of B717 marketing and, eventually, production; announced January 2005
- Rollout in February 2005 of the -200LR version of B777, a challenger to the A340-500 and -600 and critical to Boeing's attempt to wrest back market-share. A freighter is also planned
- Launch in January 2005 of the B787 Dreamliner, formerly designated 7E7.

While not announced, the list could be swelled by reports from within the industry that both the B747 and B767 are dangerously close to the abyss of a non-viable backlog, having (at 1 January 2005) just 32 and 25 unfulfilled orders, respectively. Balance is partly restored by *Jane's* deletion of the Airbus A310 from the printed version of *All the World's Aircraft* in recognition of the fact that, although never officially killed off, it quietly 'ceased to be' some years ago.

Detailed breakdowns of orders and production are in the Airbus, Boeing and individual aircraft sections of *All the World's Aircraft*.

Further on the theme of only publishing news when it is good, Airbus recently released its 'annual' 20-year airliner global forecast for the first time since 2000 (although the 2003 totals have been available to the industry).

	Seats	2003	2004	% change
Single-aisle	100	584	834	+43
	125	1,888	2,459	+30
	150	2,363	3,123	+32
	175	2,798	2,463	–12
	210	2,442	2,023	–17
Subtotal		**10,075**	**10,902**	**+8**
Small twin-aisle	250	1,782	1,800	+1
	300	1,219	1,319	+8
	Freight	176	187	+6
Subtotal		**3,177**	**3,306**	**+4**
Mid twin-aisle	350	744	797	+7
	400	649	533	–18
	Freight	174	142	–18
Subtotal		**1,567**	**1,472**	**–6**
Large	450	203	239	+18
	500	537	517	–4
	600	292	283	–3
	800	101	166	+64
	1,000	30	45	+50
	Freight	30	45	+50
Subtotal		**1,535**	**1,648**	**+7**
Total		**16,354**	**17,328**	**+6**

As ever, the difference between the major manufacturers' predictions is most marked in the 450-plus seat range (and equivalent freighters). Boeing's lack of an A380 challenger is justified by its predictions of a market for somewhat less than 400 of these leviathans, while Airbus foresees 1,650. It is predictable that at least one example of a freighter of this size will be required — to carry the eggs to be deposited on the face of whichever company got it so spectacularly wrong.

At the smaller, regional end of the airliner market, recent developments have included:

- First flight of the Antonov An-148 regional jet
- Launch of Bombardier's CSeries twinjet family
- First flights of Embraer's 190 and 195, completing the upper level of its 170 family

The last-mentioned move would appear to have been a wise one in view of recent indications that the 50-seat market – in the US, at least – is on the point of collapse. Nevertheless, Embraer's 135, 140 and 145 have been excellent sellers since launched, as demonstrated:

	Orders	Options	Deliveries Total	Deliveries 2004	Backlog
ERJ135	123	2	106	1	17
ERJ140	94	20	74		20
ERJ145	684	314	618	88	66
Subtotal	**901**	**336**	**798**	**89**	**103**
Embraer170	158	177	46	46	112
Embraer175	15				15
Embraer190	155	230			155
Embraer195	15	20			15
Subtotal	**343**	**427**	**46**	**46**	**297**
Total	**1,244**	**763**	**844**	**135**	**407**

Main rival, Bombardier has seen its fortunes rally slightly for the year 2004. Its lower-end aircraft is the twin-turboprop Dash 8Q, rather than a twin-jet, although the CSeries could see the company with an all-jet portfolio if it can gain enough support for a launch decision.

	Orders	Deliveries Total	Deliveries 2004	Backlog
Q100	299	299		0
Q200	96	95	1	1
Q300	231	211	8	20
Q400	114	88	10	26
Subtotal	**740**	**693**	**19**	**47**
CRJ100	226	226		0
CRJ200	793	670	75	123
CRJ440	75	74	33	1
CRJ700	261	176	52	85
CRJ900	45	25	15	20
Subtotal	**1,400**	**1,171**	**175**	**229**
Total	**2,140**	**1,864**	**194**	**276**

While the Dash 8Q order book fails to impress, it does appear from the experience of its European rival, the ATR series, that the turboprop regional market is continuing the modest rally begun in 2003. A mere couple of years before then, it appeared that the turboprop regional would be extinct by now. ATR reports a 25 per cent boost in revenues for 2004 and predicts viability for the near term, despite anticipated production at only half the annual rate of the late 1990s:

Year	1997	1998	1999	2000	2001	2002	2003	2004	2005	2006	2007
ATRs delivered *target	39	31	35	22	20	19	9	13	15*	18*	18*

General aviation trends are mixed, with business aircraft looking buoyant while manufacturers of lighter and private aircraft expect 2005 to be less hectic on the sales front than its predecessor. In fact, such is business jet optimism that the 2004 NBAA Convention at Las Vegas saw two new companies taking the gamble with their own versions of a supersonic business jet.

The words 'two', 'new' and 'own' are not entirely accurate, for both the SAI and Aerion companies are enlisting the support of big players such as Lockheed Martin, Boeing and Raytheon to do the sums, draw the blueprints and cut the metal – and put up some of the money. The logic of this novel form of matchmaking is not immediately apparent, for such giant firms as these are quite capable of arranging their own partnerships and designing their own high-tech aircraft, should they perceive the need. Full marks to Aerion and SAI for 'rattling the cage', but it may be that they become breakfast for the awakened occupant.

Leading makers of subsonic jet business aircraft have been cautiously optimistic of late, with Cessna reporting a "solid year" in 2004; Raytheon (Beech/Hawker) taking "strong orders" for all except its slow-moving Horizon; and Gulfstream starting 2005 almost sold-out and with "a nice backlog of orders for 2006". A slight mutter from the last-mentioned that it had accepted "perhaps more than we wanted" of orders for the G350 indicates the sanguinity to tempt providence which one hopes will not prove to be looking a gift turbofan in the intake.

Production during 2004 of the major business jets (with 2003 figures in parentheses) was:

Bombardier	
Learjet 40	17 (0)
Learjet 45/XR	22 (17)
Learjet 60	9 (12)
Challenger 300	28 (1)
Challenger 604	29 (24)
GlobalExpress	20 (14)
Global 5000	4(0)
Total	**129 (68)**
Cessna	
CitationJet 1	20 (22)
CitationJet 2	27 (56)
CitationJet 3	6 (0)
Citation Bravo	25 (31)
Citation Encore	24 (21)
Citation Excel	23 (48)
Citation XLS	32 (0)
Sovereign	9 (0)
Citation X	15(18)
Total	**181(196)**
Dassault	
Falcon 50EX	5 (8)
Falcon 2000	11 (12)
Falcon 900C	3 (3)
Falcon 900EX	15 (20)
Falcon2000EX	29(16)
Total	**63 (59)**
Gulfstream	
100/200	22 (24)
300-550	56(50)
Total	**78 (74)**
Raytheon	
Premier	37 (29)
Hawker 400XP	28 (24)
Hawker 800XP	50(47)
Total	**115(100)**
Grand total	**566(497)**

A regular check of the business aircraft manufacturers' predictions is the annual forecast produced by one of the major suppliers of engines and avionics, Honeywell. The company anticipates a market for 3,450 business jets over the coming five years, of which

- 1,195 will be long- and ultra-long range aircraft, exemplified by the Falcon 7X, Global 5000 and Gulfstream G450/550
- 760 large jets, such as Challenger 604, Embraer Legacy, Falcon 2000 and Gulfstream G350
- 2,460 medium- to medium-large types (Challenger 300, Sovereign, Horizon and Gulfstream G150)
- 2,350 light- to light-medium jets like Cessna's CJ3, the Learjet 40 and 45XR
- 1,560 very light jets, such as the Citation CJ1/CJ2, Mustang, Premier 1 and long-awaited Sino-Swearingen SJ30
- 125 or so commercial jets equipped for business travel, principally the Boeing BBJ and Airbus ACJ

As in previous years, the crystal ball contains only the cloudiest of images of the likely market for ultra-small jets. This is perhaps understandable, for while other segments of the market are micro-stratified into the classifications just noted, companies associated with the last-mentioned can be described as: manufacturers, possible manufacturers, hopeful manufacturers and unlikely manufacturers – to which we are annually adding a growing number of build-it-in-your-own-garage manufacturers.

It would be depressing to rehearse 2004's crop of light jet bankruptcies, Chapter 11 stays of execution, re-financings and stagnations, so it is more encouraging to mention one that has kept its targets for the past year. The Eclipse 500 was due to fly in pre-production form in 2004 and duly did so. Admittedly, this was on 31 December, but in a field in which years can take on an elasticity understood only by Albert Einstein, it is cheering to see a project – and one from a newcomer, too — keeping its deadlines.

The piston and turboprop market's projected slowing in 2005 should not be confused with impending recession. With the US having the lion's share of the market, its domestic trends are disproportionately reflected in the World arena and a distinct 'blip' on the graph for 2004 has resulted from the welcome effect of legislation offering tax breaks to customers buying GA aircraft before 31 December. That provided more than a 10 per cent increase in sales over 2003 and, while *Jane's* does not purport to be an economics textbook, it probably more than paid for the notional tax revenue losses through increased economic activity and consequent taxation at other points. Would that other governments could grasp the advantages of this reciprocal of 'spend to save'.

Increased sales of light singles have also been favourably influenced by the US General Aviation Revitalisation Act, signed 10 years ago as a move to limit product liability to 10 years. It resulted in the return of Cessna's previously discontinued product range and expansion at Piper, although Mooney failed to thrive under the new regime and required regeneration under new ownership. As 2004 production figures for four-seaters show, however, the traditional metal-built lightplanes are being challenged by newcomers to the ranks of the aircraft manufacturers, producing their products in composites.

Manufacturer	Type	Metal	Composites
Cessna			
	172	236	
	182	329	
Subtotal		**565**	
Cirrus			
	SR20		91
	SR22		459
	SRV		3
Subtotal			**553**
Diamond	DA40		**203**
Lancair	Columbia		**78**
Mooney			
	Bravo	9	
	Ovation	28	
Subtotal		**37**	
Raytheon/Beech	Bonanza	**62**	
Socata	TB series	**5**	
Piper			
	Warrior	18	
	Archer	19	
	Arrow	12	
Subtotal		**49**	
Tiger	AG-5B	**19**	
Totals		**737**	**834**

Low numbers for Socata reflect the fact that manufacture of the TB series has ended, except for special orders. The Czech Republic's Zlin – another famous name — has also suffered setbacks because of insolvency. Not mentioned is OMF aircraft, whose composites Symphony is intended to return to production after financial difficulties.

Still to be felt in the US is the stimulative effect of the much-delayed Light Sport Airplane legislation, which finally came into force on 1 September 2004 to create a new class of private pilot and certified aeroplane below 558 kg (1,232 lb) max take-off weight.

Military aviation has also enjoyed some government patronage over the past year, most notably in the US. Here, the falling dollar has assisted exports, although the government's 15 per cent boost (to $46 billion) of aircraft, engines and related services was a major factor, even before missiles (up 10 per cent) and space (up 5 per cent) are considered. Exports from the US of all aerospace material showed a 6.5 per cent rise to $56.8 billion and imports fell by almost 3 per cent.

Russia is making much out of its recent return to combat aircraft purchasing after more than a decade of stagnation. Deliveries in 2005 of 20 aircraft is hardly likely to transform the strategic situation – even less so when one reads the small print: all 10 of the Su-27 'Flankers' and one of the two Tu-160 'Blackjack' bombers are merely upgrades. However, other targets for this year include service-entry of the Mil Mi-24PN (a 'Hind' upgrade) and completion of flight tests of a modified Su-24 interdictor, both these aircraft having long disappeared from the pages of *All the World's Aircraft* on account of the end of new production.

The scarcely audible trickling of the rouble stream continues to stimulate muses within the defence ministry in Moscow. Current dreams are of a new strategic bomber, perhaps to be based on the Tu-160, and an increase in the aerial refuelling fleet. Latter aspirations are based on the stretched Ilyushin Il-76MF transport, except that Russia still has not been able to afford any of the MF variant for transport or any other duties, as yet.

Another tanker-transport requirement is exercising minds in both Europe and the US, as the USAF prepares to replace some of its older Boeing KC-135s. A previous proposal to obtain Boeing 767s — and, purely coincidentally, keep the production line open despite falling civil sales — was abandoned and the competition re-opened. Notionally, the Airbus A330 has a chance of competing and winning. It would be interesting to know even half the pro and con arguments now being fired backwards and forwards within offices in Washington and between there and the Seattle headquarters of Boeing.

Proving to the Europeans that it can be done is the recent maximum-profile sale of a fleet of helicopters to carry the US President. The rotorcraft in question is what the rest of the world calls an AgustaWestland EH101, but is known in North America as the Lockheed Martin US101 and accompanied at all times by a prominent list of all the subcontracting US companies involved. Could the A330 be dressed in similar clothes? Would the Europeans stoop to concocting a transvestite tanker? There is no evidence that Airbus is *that* keen to get a USAF sale, but it would be a source of innocent merriment to see it try.

A more detailed analysis of military markets and trends will be found in *Jane's World Air Forces*.

The US101 purchase merely serves to demonstrate how Europe is penetrating the US helicopter market. And Honeywell's predictions of a 10 per cent increase in the worldwide market for turbine-powered helicopters over the next five years indicate that the numbers will grow. In all, there will be 2,400 new sales between 2005 and 2009, growing to 5,550 by 2015. Market share is predicted to be:

- 25 per cent for corporate operators
- 22 per cent for emergency medical services
- 13 per cent for law enforcement
- 15 per cent utility use (over half in the US)
- 10 per cent offshore services
- 15 per cent other uses

As for type, the mix is expected to comprise:

- 48 per cent light singles
- 20 per cent light twins
- 25 per cent intermediate twins
- 7 per cent heavy

Output by the major manufacturers of turbine-powered civil helicopters during 2004 was:

Manufacturer	Europe	USA
AgustaWestland		
A109/A119	75	
Eurocopter (orders)		
EC 120 Colibri	39	
EC 130/Ecureuil, Fennec	146	
EC 135	67	
EC 145/BK 117	13	
EC 155 Dauphin/Panther	32	
EC 225/725 Super Puma	1	
Bell		
206 JetRanger		7
206L LongRanger		18
407		40
427		9
430		4
412		31
Sikorsky (estimated)		
S-76		25
S-92		4
Totals	**373**	**138**

Further, may be mentioned the USA's Robinson, whose piston-powered products have achieved new heights. During 2004, aided by the US dollar's weakness, it produced 234 two-seat R22s and 456 four-seat R44s, leaving Enstrom and Sikorsky's newly-acquired Schweizer far behind.

Notes and Acknowledgements

As ever, production of this edition relies heavily upon others who contribute in official and unofficial ways to its completeness. Susan Bushell, Mike Jerram, Kenneth Munson and Lindsay Peacock have researched and compiled the entries on the diverse flying machines which grace the following pages. James Goulding and Mike Badrocke have been responsible for many of the line drawings; Bill Gunston contributed to the completeness of the Glossary and Aero Engines tables.

Within the *Jane's* offices, Tracy Johnson has processed all the information for both computer screen and paper presentation, ably assisted by Carol Offer, plus Diana Burns as indexer. The editor's thanks, too, are extended to former Managing Editor, Simon Michell, who departed the company in December 2004.

It is our unfortunate duty to record a more permanent departure from the world of aerospace. Paul Duffy, who died on 16 January 2005, began contributing to aviation journals in 1960, when aged 14, and was soon supplying material to John W R Taylor for *Jane's* and the enthusuiasts' books he produced for Ian Allan. This association continued, and Paul's later move from his native Ireland to Russia was of great benefit to the present editor when bureaucracy threatened to keep *Jane's* out of the Moscow Air Show. Indeed, he contributed so much to the improvement of liaison between East and West that representatives of the Russian airline industry took the unprecedented step of placing a paid obituary in a UK aerospace journal. It would be hard to think of a higher professional honour.

Others still, happily, contributing include our current Russian specialist, Yefim Gordon, without whose regular inputs these pages would be much the poorer. Robert Hewson, Ryszard Jaxa-Malachowski and Sebastian Zacharias have been their usual, helpful selves, especially on the photographic side. In the same context we welcome the illustrations of Mitsuhiro Kadota which have much improved our coverage of Japan's aircraft. Despite Japanese domination of the camera manufacturing business, it seems that these devices seldom fall into the hands of local aircraft producers, making us, largely, dependent upon the enthusiastic private individual.

Thanks are also extended to Peter Cooper (Falcon Aviation), Kensuke Ebata (a second Japanese exception to the rule), John Fricker, Jean-Louis Gaynecoetche, Jamie Hunter (*Jane's Aircraft Upgrades*), Geoffrey P Jones, Arnold Nayler, Rod Simpson, Martin Streetly (*Jane's Radar & EW Systems* ; and *Electronic Mission Aircraft*); Jim Thorn (now retired from editorship of *Australian Aviation*), Mike Vines (Photo Link) and Simon Watson.

Paul Jackson

Pulham Market, Norfolk, March 2005

Editor Biography

Paul Jackson MRAeS, Editor-in-Chief, Jane's All the World's Aircraft

The chance gift of an aircraft book at the age of six was the Genesis of a passion for all matters aeronautical that has absorbed Paul Jackson for almost half a century. Editor of an aviation society newsletter in his native East Yorkshire shortly after leaving school, he contributed to amateur and professional aviation journals, on both historical and current matters, for a further decade before becoming a freelance aerospace writer and editor. In 1987, he was invited to join the compiling team of *Jane's All the World's Aircraft*, being appointed its editor in 1994. Other duties at *Jane's* include the editorship of *World Air Forces*, which was launched a year later. Inevitably, his hobbies are aviation related, including flying and maintaining a veteran Aeronca Chief lightplane; chairing a branch of the Popular Flying Association; and aviation photography.

Paul Jackson **NEW**/0567909

Type Classifications

To assist comparison between different makes of aircraft undertaking broadly similar roles, the 'TYPE' classifications which introduce each aircraft description are standardised and arranged into 13 classes.

Classification is according to *Jane's* own criteria; in some instances this may differ from a manufacturer's description. Exact capacities and engine types and numbers for large aircraft are omitted in these shorthand descriptions. However, as size diminishes, such aspects assume greater importance and, accordingly, are quoted.

Except where the type description immediately and obviously precludes it (for example, 'Strategic transport'), aircraft are assumed to be monoplanes powered by a single piston engine and propeller. Multiple engines, turboprop and jet propulsion are specifically mentioned. In deference to common usage, 'jet' includes turbojets and turbofans, as more properly described in the 'POWER PLANT' paragraph.

All aircraft are Class A (or equivalent) certified/certifiable and factory-assembled; where clarification is necessary, the term 'lightplane' is employed. Ultralights and/or aircraft built from kits are specifically noted as such. However, in view of widely differing national legislation on ultralights, it must be noted that the term has several interpretations: the most liberal (in terms of permissible maximum weight) is used in this book.

The more versatile makes of aircraft cannot be given justice in deliberately short type descriptions. Readers should be aware that the ability to undertake surveillance, geological survey, VIP transport, water bombing and many other duties can be easily bestowed on medium transports and some other types of aircraft. Crop sprayers can be reconfigured in moments for firefighting and oil pollution dispersal. The full story will be found under the 'CURRENT VERSIONS' heading.

As an aid to analysis, totals for each category are given in parentheses.

Class 1: Bomber and surveillance (24)

Northrop Grumman E-2C Hawkeye *NEW*/0589328

These are military or paramilitary aircraft of widely differing size and performance.

Strategic bomber (1)
Tupolev Tu-160 (Russian Federation)
Maritime reconnaissance four-jet (2)
BAE Systems Nimrod MRA. Mk 4 (United Kingdom)
Kawasaki P-X (Japan)
Maritime surveillance twin-jet (2)
Airbus MPA (International)
Boeing 737 MMA (United States)
Maritime surveillance twin-turboprop (3)
Airtech CN-235 MP Persuader and CN-235 MPA (International)
ATR 42 Surveyor (International)
CASA C-212 Patrullero (Spain)
Airborne early-warning and control system (3)
Airbus AEW&C (International)
Boeing 737 AEW&C (United States)
Northrop Grumman E-2C Hawkeye (United States)
Airborne ground surveillance system (4)
Airbus A321 AGS (International)
Boeing 767 Military Versions (United States)
Northrop Grumman E-8 Joint STARS (United States)
Raytheon Sentinel (United States)
Airborne multisensor command and control system (1)
Northrop Grumman E-10 (United States)
Multisensor surveillance twin-turboprop (1)
BNG BN2T-4S Defender 4000 (United Kingdom)
Multisensor surveillance turboprop (1)
Pilatus PC-12M and Spectre (Switzerland)
Multisensor surveillance twin-prop (1)
Vulcanair P.68 Observer and P.68 Diesel (Italy)
Multisensor surveillance lightplane (5)
BLA Dragonfly (Optica) (United Kingdom)
Diamond HK36TTC-MPX (Austria)
SAI G97V Spotter (Italy)
Schweizer SA 2-37 (United States)
Schweizer SA 2-38 (United States)

Class 2: Fighter and trainer (61)

Eurofighter Typhoon *NEW*/0589268

Air superiority fighter (3)
SAC (Sukhoi Su-27) J-11 (China)
Sukhoi Su-27 (Russian Federation)
Sukhoi Su-30 (Su-27PU) (Russian Federation)
Multirole fighter (20)
ADA Tejas (LCA) (India)
Boeing F-15E Eagle (United States)
Boeing F/A-18E/F Super Hornet (United States)
CAC J-7 (China)
CAC J-10 (China)
Dassault Mirage 2000 (France)
Dassault Rafale (France)
Eurofighter Typhoon (International)
KAI (Lockheed Martin) F-16C/D Fighting Falcon (Korea, South)
Lockheed Martin F-16 Fighting Falcon (United States)
Lockheed Martin (645) F-22 Raptor (United States)
Lockheed Martin F-35 Joint Strike Fighter (United States)
MiG-29 (Russian Federation)
Saab JAS 39 Gripen (Sweden)
SAC J-8B (China)
SAC F-8 IIM (China)
Sukhoi Su-30M (Russian Federation)
Sukhoi Su-33UB (Su-27KUB) (Russian Federation)
Sukhoi Su-35 and Su-37 (Su-27M) (Russian Federation)
Sukhoi PAKFA (Russian Federation)
Attack fighter (7)
AMX AMX (International)
CAC FC-1 Xiaolong (China)
HAL (SEPECAT) Jaguar International (India)
IRIAF Saeghe (Iran)
Mitsubishi F-2 (Japan)
Sukhoi Su-34 (Russian Federation)
XAC JH-7 (China)
Light fighter/advanced jet trainer (2)
Aero L 159 (Czech Republic)
EADS Mako (Germany)
Advanced jet trainer (3)
Boeing/BAE Systems T-45 Goshawk (International)
GAIC JL-9 and CY-1 (China)
Kawasaki T-4 (Japan)
Advanced jet trainer/light attack jet (8)
Aermacchi M-346 (Italy)
Avioane IAR-99 Soim (Romania)
BAE Systems Hawk 50, 60 and 100 Series (United Kingdom)
HAIG L-15 (China)
IACI (SAHA) Shafagh (Iran)
KAI T-50 and A-50 Golden Eagle (Korea, South)
MiG-AT (Russian Federation)
Yakovlev Yak-130 (Russian Federation)
Basic jet trainer (2)
IRIAF Tazarve (Iran)
ZRiPSL Bielik and Fenix (Poland)
Basic jet trainer/light attack jet (5)
Aermacchi MB-339 (Italy)
Aermacchi M-311 (Italy)
HAL HJT-36 Sitara (India)
LMAASA AT-63 Pampa NG (Argentina)
NAMC JL-8 (China)
Basic turboprop trainer/attack lightplane (1)
Embraer EMB-314 Super Tucano (Brazil)
Basic turboprop trainer (6)
Beech (3000) T-6A Texan II (United States)
Fuji T-3Kai/T-7 (Japan)
IAMI (HESA) HT-80 (Iran)
KAI KT-1 Woong-Bee (Korea, South)
Pilatus PC-9M Advanced Turbo Trainer (Switzerland)
Pilatus PC-21 (Switzerland)
Basic prop trainer/attack lightplane (1)
Aermacchi SF-260 (Italy)
Basic prop trainer (2)
ENAER (ECH-51) T-35 Pillan (Chile)
IAMI (HESA) Parastu (Iran)

Class 3: Miscellaneous and/or government (13)

Scaed White Knight/SpaceShipOne *NEW*/0567784

Aircraft of diverse or multiple duties employed generally, but not exclusively, by the state.

Missile defence system (1)
Boeing AL-1A (United States)
Multirole twin-jet (1)
KAC (Raytheon) Hawker 800 (Japan)
High-altitude platform (1)
Scaled 318 White Knight (United States)
Technology demonstrator (8)
AMV 211 (United States)
FanWing FW8 (United Kingdom)
Hunt Fuel-Less Aircraft (United States)
Northrop Grumman XSP (United States)
Piccard Solar Impulse (Switzerland)
Scaled 311 Capricorn (United States)
Scaled 316 SpaceShipOne (United States)
Tupolev Tu-156 (Russian Federation)
Ornithopter (1)
UTIAS 'Big Flapper' (Canada)

Class 4: Transport (30)

EADS-CASA C-295 *NEW*/0525768

Generally of a military nature, often with rear loading ramps. The larger aircraft are usually, but not exclusively, jet-powered. Light transports are those not exceeding 5,670 kg (12,500 lb).

Strategic transport (3)
Airbus A400M (International)
Antonov-70 (Ukraine)
Boeing C-17A Globemaster III (USA)
Tanker-transport (2)
Airbus MultiRole Tanker-Transport (MRTT) (International)
Boeing 767 Military Versions (USA)
Medium transport/multirole (6)
Boeing Advanced Tactical Transport (USA)
Ilyushin/Irkut/HAL Il-214 (International)
Kawasaki C-X (Japan)
Lockheed Martin 382U/V Hercules (USA)
Lockheed Martin Advanced Mobility Aircraft (AMA) (USA)
SAC Y-8 (China)
Twin-turboprop transport (16)
Airtech CN-235 (International)
Alenia/Lockheed Martin C-27J Spartan (Italy)
Antonov An-38 (Ukraine)
CASA C-212 Series 400 (Spain)
CASA C-295 (Spain)
CSIR Saras (India)
Frati F.2500 (Italy)
HAL (Dornier) 228 (India)
Ilyushin Il-112 (Russian Federation)
LZ L 610 G (Czech Republic)
MiG-110 (Russian Federation)
PZL M28 Skytruck (Poland)
PZL M28 Skytruck Plus (Poland)
PZL M28B Bryza (Poland)
Sukhoi Su-80 (Russian Federation)
TAI (Airtech) CN-235M (Turkey)

Turboprop light transport (3)
Beriev Be-132 (Russian Federation)
HAI Y-12 (China)
Skydesign SB100 Skylander (France)

Class 5: Airliner and freighter (58)

Antonov AN-140 *NEW*/0589976

Civilian passenger and cargo aircraft. Jet power is implied for all large airliners; number of engines given only for medium-size aircraft; regional jets are all twins.

Outsize freighter (1)
Antonov An-124 (Ukraine)
High-capacity airliner (1)
Airbus A380 (International)
Wide-bodied airliner (9)
Airbus A300-600 (International)
Airbus A310 (International)
Airbus A330 (International)
Airbus A340 (International)
Boeing 747 (USA)
Boeing 767 (USA)
Boeing 777 (USA)
Boeing 7E7 Dreamliner (USA)
Ilyushin Il-96-300 (Russian Federation)
New concept airliner (1)
Boeing Blended Wing Body (USA)
Four-jet freighter (1)
Boeing 747-400F (USA)
Twin-jet transport (2)
Antonov An-72 and An-74 (Ukraine)
Antonov An-74-300 and An-174 (Ukraine)
Twin-jet airliner (10)
Airbus A318 (International)
Airbus A319 (International)
Airbus A320 (International)
Airbus A321 (International)
Boeing 717 (USA)
Boeing 737 (USA)
Boeing 757 (USA)
Boeing 757-300 (USA)
Tupolev Tu-204 and Tu-214 (Russian Federation)
Tupolev Tu-334, Tu-336 and Tu-354 (Russian Federation)
Twin-jet freighter (5)
Airbus A300-600 Convertible and A300-600 Freighter (International)
Boeing 767-300 General Market Freighter (USA)
Tupolev Tu-154M (Russian Federation)
Tupolev Tu-156 (Russian Federation)
Tupolev Tu-330 (Russian Federation)
Regional jet airliner (16)
ACAC ARJ21 (China)
Antonov !An-148 (Ukraine)
Avcraft Dornier 328 (Germany)
Bombardier CRJ200 and Challenger 800 (Canada)
Bombardier CRJ700 (Canada)
Bombardier CRJ900 (Canada)
Embraer ERJ-145 (Brazil)
Embraer ERJ-135 (Brazil)
Embraer ERJ-140 (Brazil)
Embraer 170 and 190 (Brazil)
Irkut (Aviastep) 111 (Russian Federation)
JADC YSX (Japan)
Myasishchev M-70 (Russian Federation)
Sukhoi RRJ (Russian Federation)
Tupolev Tu-324 and Tu-414 (Russian Federation)
Yakovlev MS-21 (Russian Federation)
Twin-turboprop airliner (10)
Antonov An-140 (Russian Federation)
ATR 42 (International)
ATR 72 (International)
Beech 1900D (USA)
Bombardier Dash 8 Q100 and Q200 (Canada)
Bombardier Dash 8 Q300 (Canada)
Bombardier Dash 8 Q400 (Canada)
Embraer EMB-120 Brasilia (Brazil)
Ilyushin Il-114 (Russian Federation)
XAC MA60 (China)
Twin-turboprop freighter (2)
AUC FF-1080-200 Freight Feeder (USA)
IAI C-5WA Airtruck (Israel)

Class 6: Business (60)

ADAM A500 CarbonAero *NEW*/0589399

The established configuration of a business jet being a twin, only single-and tri-jet configurations are specifically noted. Detailed strata devised by participants in the business jet market are not reproduced here.

Instead, the main category is subdivided into small business jets with accommodation for six or fewer passengers; long-range (sub-airliner size) business jets with the 4,000 n mile (7,400 km; 4,600 mile) range necessary to fly from the US eastern seaboard to Western Europe; and the central core of twin-jets with upwards of seven seats.

Supersonic business jet (1)
Gulfstream SBJ (USA)
Long-range business jet (5)
Bombardier BD-700 Global Express (Canada)
Bombardier Global 5000 (Canada)
Gulfstream G300 and G400 (USA)
Gulfstream G450 (USA)
Gulfstream G500 and G550(USA)
Large business jet (2)
Boeing Business Jet (BBJ) (USA)
Boeing Business Jet 2 (BBJ 2) (USA)
Long-range business tri-jet (2)
Dassault Falcon 900 (France)
Dasault Falcon 7X (France)
Business tri-jet (1)
Dassault Falcon 50 (France)
Business jet (21)
Aerostar FJ-100 (USA)
Bombardier BD-100 Challenger 300 (Canada)
Bombardier CL-600 Challenger (Canada)
Cessna 550 Citation Bravo (USA)
Cessna 560 Citation Encore (USA)
Cessna 560XL Citation Excel and XLS (USA)
Cessna 680 Citation Sovereign (USA)
Cessna 750 Citation X (USA)
Dassault Falcon 2000 (France)
Embraer Legacy (Brazil)
Hawker 400 (USA)
Hawker 800 (USA)
Hawker Horizon (USA)
Gulfstream G100 (USA)
Gulfstream G150 (USA)
Gulfstream G200 (USA)
Learjet 31 (USA)
Learjet 40 (USA)
Learjet 45 (USA)
Learjet 60 (USA)
Smartfish Business Jet (Switzerland)
Light business jet (12)
Adam A700 (USA)
Ameur Altajet (France)
Avocet ProJet (USA)
Beech Premier I (USA)
Cessna Citation Mustang (USA)
Cessna 525 Citation CJ1 (USA)
Cessna 525A Citation CJ2 (USA)
Cessna 525B Citation CJ3 (USA)
Eclipse Eclipse 500 (USA)
Honda HA-420 HondaJet (USA)
Safire Jet (USA)
Sino Swearingen SJ30-2 (USA)
Light business jet/kitbuilt (1)
Maverick Leader (USA)
Business mono-jet (1)
Diamond D-Jet (Austria)
Business mono-jet kitbuilt (1)
Aerocomp CA-J Comp Air Jet (USA)
Business twin-turboprop (2)
Beech King Air C90B (USA)
Piaggio P.180 Avanti (Italy)
Business turboprop (6)
AIR Epic LT (USA)
Extra EA-500 (Germany)
FACL F1 (United Kingdom)
Grob G 140TP (Germany)
Piper PA-46-500TP Malibu Meridian (USA)
Socata TBM 700 (France)
Business turboprop kitbuilt (1)
Comp Air 12 (USA)
Business twin-prop (2)
Adam A500 CarbonAero (USA)
High Performance TT62 Alekto (Germany)
Business prop (2)
Extra EA-400 (Germany)
Piper PA-46-350P Malibu Mirage (USA)

Class 7: Utility (77)

PAC 750X *NEW*/0580568

Restricted to single- and twin-engine passenger or passenger/light freight aircraft with accommodation, typically, for four or six persons. An intermediate category bridging commercial and private ownership. Crop sprayers are included.

Utility turboprop twin (7)
Atlas Liftmaster (USA)
Beech King Air B200 (USA)
Beech King Air 350 (USA)
BNG BN2T Turbine Islander and Defender (United Kingdom)
Ilyushin Il-100 (Russian Federation)
Piaggio P.166 (Italy)
Reims F406 Caravan II (France)
Light utility twin-prop transport (8)
Avia Accord-201 (Russian Federation)
BNG BN2B Islander (United Kingdom)
MiG-201 (Russian Federation)
TAAL (Partenavia) P.68 Series (India)
TKEF 44 Angel (USA)
VulcanAir P.68, Observer and P.68 Diesel (Italy)
VulcanAir VA 300 (Italy)
Wolfsberg Raven 257 (Czech Republic)
Six-seat utility twin (2)
Beech Baron 58 (USA)
Piper PA-34-220T Seneca V (USA)
Four-seat utility twin (4)
Diamond DA42 Twin Star (Austria)
Dubna-4 (Russian Federation)
Piper PA-44-180 Seminole (USA)
ZRiPSL EM-11 Orka (Poland)
Light utility turboprop (15)
Aerocourier Aerocourier (USA)
Aeroprogress/ROKS-Aero T-101 Grach (Russian Federation)
Cessna 208 Caravan (USA)
Explorer Explorer (USA)
Ibis Ae 270 Spirit (International)
Integrity Integrity (Australia)
Intracom GM-17 Viper (Switzerland)
Intracom GM-19 (Switzerland)
Khrunichev T-507 Skvorets (Russian Federation)
Myasishchev M-101T Gzhel (Russian Federation)
PAC 750XL (New Zealand)
Pilatus PC-12 (Switzerland)
Smolensk SMG-92 Turbo Finist (Russian Federation)
Smolensk (Technoavia) SM-2000 (Russian Federation)
VulcanAir VF 600W Mission (Italy)
Light utility transport (4)
Gippsland GA-8 Airvan (Australia)
Sherpa Sherpa (USA)
Smolensk SM-92 Finist (Russian Federation)
Zlin Z 400 Rhino (Czech Republic)
Utility biplane (2)
SAMC Y-5 (China)
Six-seat utility transport (11)
Beech Bonanza A36 (USA)
Beech Turbo Bonanza B36TC (USA)
Cessna 206 Stationair and T206 Turbo Stationair (USA)
Found FBA-2C Bush Hawk (Canada)
Khrunichev T-415 Snegir (Russian Federation)
Piper PA-32R-301 Saratoga II HP (USA)
Piper PA-32R-301FT 6X (USA)
Piper PA-32R-301T Saratoga II TC (USA)
Piper PA-32R-301XTC 6XT (USA)
Smolensk (Technoavia) SM-2000P (Russian Federation)
Yak Alacon Yak-58 (Kazakhstan)
Agricultural sprayer (23)
Agrolot PZL-126P Mrówka 2001 (Poland)
AII AVA-303 (Iran)
Air Tractor AT-401Air Tractor (USA)
Air Tractor AT-402 Turbo Air Tractor (USA)
Air Tractor AT-502 Turbo (USA)
Air Tractor AT-602 (USA)
Air Tractor AT-802 (USA)
CZAW (Zenair) CH 701-AG STOL (Czech Republic)
EADS PZL PZL-106BT Turbo-Kruk (Poland)
Gippsland GA200 Fatman (Australia)

Khrunichev T-419 Strezoka (Russian Federation)
Kulon-2 SKhS (Russian Federation)
MVEN-1 Fermer (Russian Federation)
NEIVA EMB-202 Ipanema (Brazil)
PAC Cresco (New Zealand)
PZL M18 Dromader (Poland)
Sukhoi Su-38L (Russia)
Thrush Turbo-Thrush S2R (USA)
Thrush 660 Turbo-Thrush (USA)
Unikomtranso Don (Russian Federation)
Vitek Ezhik (Russian Federation)
Weatherly 620 (USA)
Yakovlev Yak-SKh (Russian Federation)

Agricultural sprayer ultralight (1)
Aerolites Aeromaster (USA)

Class 8: Amphibian (22)

Wust Seahawk *NEW*/0561696

All such aircraft (including the occasional flying boat) are included here for ease of reference, even those assembled from kits or classified as ultralights.

Four-turboprop amphibian (1)
ShinMaywa US-1A (Japan)

Twin-turboprop amphibian (1)
Bombardier 415 (Canada)

Twin-jet amphibian (1)
Beriev (Betair) Be-200 Altair (Russian Federation)

Utility Amphibian (1)
Aero-Volga LA-8 Flagman (Russian Federation)

Six seat amphibian (4)
Aero-Volga L-6 (Russian Federation)
Beriev Be-103 Bekas (Russian Federation)
LanShe Lake LA-250 (USA)
Warrior Centaur (United Kingdom)

Six-seat amphibian/kitbuilt (1)
Seastar Seastar (USA)

Five-Seat amphibian (1)
Seawind Seawind 300C (USA)

Four-seat amphibian (2)
Aquastar TA16 Seafire (USA)
Chernov Che-25 (Russian Federation)

Four-seat amphibian kitbuilt (2)
Cobra Landseair (Australia)
SG Sea Storm (Italy)

Three-seat amphibian/kitbuilt (1)
Gidroplan Che-22 Korvet (Russian Federation)

Three-seat amphibian kitbuilt (1)
Aeroprakt-24 (Ukraine)

Two-seat amphibian kitbuilt (2)
SG Sea Storm (Italy)
Quikkit Glass Goose (USA)

Two-seat amphibian /kitbuilt (1)
EDRA Helicentro Paturi (Brazil)

Two-seat amphibian ultralight kitbuilt (3)
Aerolites Aeroskiff (USA)
Colyaer Mascato (Spain)
Wüst Seahawk (Germany)

Class 9: Lightplane (factory built) (127)

Extra EA-300L *NEW*/0573324

Single-prop aircraft seating up to five persons and available only in complete form, unless otherwise stated.

Two-seat jet sportplane (4)
ATG-1 Javelin (USA)
Frati F.1000 (Italy)
Messerschmitt Me 262 (USA)
Smartfish Sportplane (Switzerland)

Five-seat lightplane/turboprop (1)
Maule M-7 and M-9 (USA)

Five-seat lightplane (1)
NLA AC-500 Aircar (China)

Four-seat lightplane (35)
Cessna 172 Skyhawk (USA)
Cessna 182 Skylane (USA)
Cessna T182T Turbo Skylane (USA)
Cirrus SR20 and SRV (USA)
Cirrus SR22 (USA)
Commander 115 (USA)
Diamond DA40-TDI Star (Austria)
Diamond DA40-180 Star (Canada)
EADS PZL PZL-104M Wilga 2000 (Poland)
EADS PZL PZL-110 Koliber 160 (Poland)
FACI F-3 (Iran)
III F.300 (Italy)
KARI Bora (Korea, South)
PZL I-23 Manager (Poland)
Ilyushin Il-103 (Russian Federation)
Lancair Columbia (USA)
Luscombe 11E (USA)
Mooney M20M Bravo (USA)
Mooney M20R Ovation 2 and M20S Eagle (USA)
OMF Symphony 4 (Germany)
Piper PA-28-161 Warrior III (USA)
Piper PA-28-181 Archer III (USA)
Piper PA-28-201 Arrow (USA)
Robin DR 400 Dauphin (France)
Robin DR 400/160 Major (France)
Robin DR 400/180 Régent (France)
Robin DR 500 Président (France)
SAIC LE-500 Little Eagle (China)
Scaled TAA-1 (USA)
Solaris Sigma and Alpha (USA)
Socata TB 9 Tampico and TB10 and TB 200 Tobago (France)
Socata TB 20 and TB 21 Trinidad (France)
Tiger AG-5B Tiger (USA)
Yakovlev Yak-18T (Russian Federation)
Zlin Z 143 (Czech Republic)

Three-seat lightplane (1)
PAC (AMF) Mushshak and Super Mushshak (Pakistan)

Three-seat biplane (1)
Waco Classic YMF Super (USA)

Primary prop trainer/sportplane (11)
Aerostar (Yakovlev) Iak-52 (Romania)
AII AVA-202 (Iran)
EADS PZL PZL-112 Koliber Junior (Poland)
III D-130 and D-200 (Italy)
NAL/TAAL Hansa-3 (India)
PAC Airtrainer CT-4E (New Zealand)
PZL M26 Iskierka (Poland)
Robin R 2120 Alpha (France)
Slingsby T67 Firefly (United Kingdom)
Zlin Z 142 C (Czech Republic)
Zlin Z 242 L (Czech Republic)

Aerobatic two-seat turboprop sportplane (1)
Endeavour Phoenix (Australia)

Aerobatic two-seat sportplane (7)
CAP 10 (France)
Extra EA-200 (Germany)
Extra EA-300 and EA-330 (Germany)
Sukhoi Su-29 (Russian Federation)
Wingtip to Wingtip Panzl S-330 (USA)
Yakovlev Yak-54 (Russian Federation)
Zivko Edge 540T (USA)

Aerobatic two-seat sportplane/kitbuilt (1)
Dyn'Aero CR100, CR110 and CR120 (France)

Aerobatic two-seat lightplane (4)
Aeromot AMT-600 Guri (Brazil)
Grob G 115 (Germany)
Grob G 120A (Germany)
LanShe SP20 and SP26 (USA)

Two-seat lightplane (44)
3Xtrim 550 Trener (Poland)
AMD Zenith CH 2000 Alarus (USA)
American Champion 7ACA Champ (USA)
American Champion 7ECA Citabria Aurora (USA)
American Champion 7GCAA Citabria Adventure (USA)
American Champion 7GCBC Citabria Explorer (USA)
American Champion 8GCBC Scout (USA)
American Champion 8KCAB Super Decathlon (USA)
Aquila A210 (Germany)
Association Zephyr Alizé (France)
Aviat Husky A-1 (USA)
Aviat Husky Pup (USA)
Aviatika-MAI-910 Interfly (Russian Federation)
CAG Toxo (Spain)
Cub Crafters PA-18-180 Top Cub (USA)
Cub Crafters Cub Light (USA)
Diamond DA20-W (Austria)
Diamond DA20-C1 Katana (Canada)
Dorna D139-PT1 Blue Bird (Iran)
Dubna-2 Osa (Russian Federation)
Eagle Eagle 150 (Malaysia)
Elitar IE-201 Senator (Russian Federation)
Elitar 202 (Russian Federation)
EV-AT Harmony (Czech Republic)
Flight Design CT (Germany)
Glass STOLGlass SG70 (Colombia)
Glass SpeedGlass (Colombia)
IAR IAR-46 (Romania)
III Sky Arrow (Italy)
II Speed Arrow (Italy)
Impulse Impulse (Germany)
Issoire APM-20 Lionceau and APM-21 Lion (France)
Khrunichev T-21 Sapsan (Russian Federation)
Liberty XL-2 (USA)
Lightwing AC4 (Switzerland)
MAI-217 (Russian Federation)
Maule M-4 (USA)
Montagne Mountain Goat (USA)
OMF-100 Symphony (Germany)
Renaissance 8F (USA)
Robin R 2160 (France)
Sauper Papango (France)
SSVOBB Impuls (Netherlands)
Taylorcraft F22 (USA)
Motor glider (6)
AMS Carat (Slovenia)
Aeromot Ximango (Brazil)
Diamond HK36 (Austria)
Scheibe SF25C Falke (Germany)
Stemme S10 and S15 (Germany)
Stemme S6, S8 and S15-8 (Germany)

Aerobatic two-seat biplane (2)
Aviat Pitts S-2C (USA)
SSH (Bücker) T-131 Jungmann (Poland)

Aerobatic single-seat sportplane (5)
CAP 232 (France)
Sukhoi Su-31 (Russian Federation)
Smolensk SP-55M (Russia)
Yakovlev Yak-3M and Yak-9U-M (Russian Federation)
Zivko Edge 540 (USA)

Aerobatic single-seat biplane (1)
SSH (Bücker) Jungmeister (Poland)

Glider tug (2)
Robin DR 400/180R Remo 180 (France)
Robin DR 400/200R Remo 200 (France)

Class 10: Utility kitbuilt (48)

Aerocomp Comp Air 7SX *NEW*/0589470

This class reflects the growing number of kitbuilt aircraft intended for more than recreational use. The largest has 10 seats and turboprop power.

Two-seat jet sportplane kitbuilt (1)
Viper ViperJet (USA)

Two-seat kitbuilt twin (4)
Leza AirCam (USA)
SlipStream SkyBlaster (USA)
SlipStream SkyQuest (USA)
VSTOL SST 2000 Pairadigm (USA)

Utility kitbuilt (5)
Aerocomp Comp Air 7 (USA)
Aerocomp Comp Air 8 (USA)
Aerocomp Comp Air 10 (USA)
Khrunichev T-411 Aist (Russian Federation)
KLS SUA-7 (USA)

Six-seat kitbuilt (5)
ADI Super Stallion (USA)
Aerocomp Comp Air 6 (USA)
Aviabellanca Skyrocket III (USA)
Barr BarrSix (USA)
Murphy Super Rebel and Moose (Canada)

Four-seat kitbuilt twin (3)
Aeroprakt-28 (Ukraine)
Creative Flight Aerocat (Canada)
Hensley H-1 Wolf (USA)

Four-seat kitbuilt (27)
Aerocomp Comp Air 4 (USA)
Aeronix Airelle (France)
AMD Zodiac CH 640 (USA)
Aviation Development Alaskan Bushmaster (USA)
AviPro Bearhawk (USA)
CLASS Bush Caddy (Canada)
Culp's Monoculp (USA)
Dream Tundra (Canada)
Dyn'Aero MCR4S (France)
Express Express (USA)
Four Winds 192, 210 and 250T (USA)
Jabiru J200 and J400 (Australia)
KARI (Velocity) Firefly (Korea, South)
Lambert Mission M212 (Belgium)
Lammer Geyer Jupiter (South Africa)
Lancair Lancair IV, IVP and Sentry (USA)
Lancair Lancair ES and Super ES (USA)
Pottier P 200 Series (France)

Pulsar Cruiser and Super Cruiser (USA)
Ravin Ravin 500 (South Africa)
St Just Super Cyclone (Canada)
Tapanee Levitation 4 (Canada)
Team Tango Foxtrot (USA)
Ullman Panther (USA)
Vans RV-10 (USA)
Wag-Aero 2+2 Sportsman (USA)
Zenith Super STOL CH 801 (USA)
Three-seat kitbuilt (2)
Aerocomp Comp Air 3 (USA)
Murphy Rebel (Canada)
Three-seat biplane kitbuilt (1)
Waco UMF (USA)

Aerobatic three-seat biplane/kitbuilt (1)
Culp's Special (USA)

Class 11: Kitplanes and/or ultralights (238)

Supermarine Spitfire Mk 26 ***NEW***/0580305

One- or two-seat, single-prop machines intended for private ownership. 'Kitbuilt' aircraft are in the lightplane category, unless described as ultralights (and flyable as such in at least some countries). Some aircraft are available only in complete form, while kit manufacturers may offer the option of fly-away aircraft to those requiring the aeroplane, but not the assembly work. These are indicated as follows:

Single-seat sportplane kitbuilt (4)
Bede BD-17 Nugget (USA)
Flug Werk Fw 190 (Germany)
Mini-Imp Mini-Imp (USA)
Supermarine Spitfire (Australia)

Single-seat ultralight/motor glider (1)
Noin Exel (France)

Single-seat ultralight (1)
Kolb Firefly (USA)

Single-seat ultralight kitbuilt (13)
Aerolites Bearcat (USA)
Blue Yonder E-Z KingCobra (Canada)
Fisher Avenger (USA)
Flightstar Spyder and Formula (USA)
Fly Synthesis Wallaby (Italy)
Joplin 1/2 Tun (USA)
Kappa 4.3 Baron (Czech Republic)
Kolb Firefly (USA)
Murphy JDM-8 (Canada)
Mecklenburger Me-109 (Germany)
Sky Raider Sky Raider (USA)
Slipstream Scepter (USA)
Tecnico SupaPup Mk 4 (Australia)
Single-seat ultralight/kitbuilt (5)
Interplane Griffon 103 (Czech Republic)
Rans S-17 Stinger (USA)
Sapphire LSA Mk II (Australia)
Silence Silence (Germany)
Skystar Litfox Lite (USA)

Single-seat ultralight biplane (1)
Aviatika-MAI-890 (Russian Federation)

Single-seat ultralight biplane kitbuilt (2)
Fisher Youngster and Youngster V (USA)
Yesteryear MiFyter (USA)

Single-seat motor glider/kitbuilt (1)
TeST TST-9 Junior (Czech Republic)

Tandem-seat kitbuilt (11)
AeroKuhlmann SCUB (France)
American Homebuilts' Vaquero John Doe (USA)
Blue Yonder E-Z Flyer (Canada)
Carlson CA6 Criquet (USA)
Cub Crafters Cub Light (USA)
Custom Flight North Star (Canada)
Partenair S45 Mystère (Canada)
Rans S-7 Courier (USA)
Van's RV-4 (USA)
Van's RV-8 and RV-8A (USA)
Wag-Aero Sport Trainer (USA)

Tandem-seat sportplane kitbuilt (13)
Aceair Aeriks A-200 (Switzerland)
Bede BD-18 Nugget (USA)
Flug Werk Ar 96 (Germany)
HB Flugtechnik HB 204 Tornado (Austria)
Kimball Model 15 Raptor A S (USA)
Legend Legend (USA)
MSW VOTEC 322 (Switzerland)
Super-Chipmunk Super-Chipmunk (Canada)
Supermarine Spitfire (Australia)
TBG Thunder Mustang (USA)
Titan T51 Mustang (USA)
Team Rocket F1 (USA)
Warner Sportster (USA)

Tandem-seat turboprop sportplane kitbuilt (2)
Legend Turbine Legend (USA)
Turbine Design TD-2 Tempest (USA)

Tandem-seat turboprop sportplane/kitbuilt (1)
Cameron P-51G (USA)

Tandem-seat ultralight (9)
Albatross AS-3A (Russian Federation)
DAR-21 Vector (Bulgaria)
III Sky Arrow (Italy)
Kolb Sling Shot (USA)
Let-Mont UL Piper (Czech Republic)
Louit Cougar (France)
Sauper J.300 Series 3 (France)
TeST-12 (Czech Republic)
VMA Ultralight (Vietnam)

Tandem-seat ultralight kitbuilt (10)
Sky Raider Sky Raider II (USA)
Joplin Tundra (USA)
Kolb Kolbra (USA)
Legend Lite Skywatch SS-11 (Canada)
Rans S-18 Stinger II (USA)
Reality Easy Raider (United Kingdom)
Reality (Just Plane) Escapade (United Kingdom)
RMT 03 Bateleur (Germany)
Rocky Mountain Ridge Runner (USA)
US Light Aircraft Hornet (USA)

Tandem-seat ultralight/kitbuilt (10)
Aeroprakt-20 (Ukraine)
Aeroprakt-20R-912 Sky Cruiser (Ukraine)
CFM Shadow Series D and E (United Kingdom)
CFM Streak (United Kingdom)
CFM SA-II Star Streak (United Kingdom)
Flaming Air Trener Baby (Czech Republic)
HB Flygtechnik Cubby (Austria)
Morava Zlin (Russo) Savage (Czech Republic)
Storch Storch SS Mk 4 (Australia)
TL-232 Condor (Czech Republic)

Tandem-seat biplane kitbuilt (3)
Aviat (Christen) Eagle II (USA)
Culp's Sopwith Pup (USA)
Steen Pitts 14 (USA)

Tandem-seat biplane/kitbuilt (1)
Kimball (Pitts) Model 12 (USA)

Tandem-seat ultralight biplane (1)
B&F FK 12 Comet (Germany)

Tandem-seat ultralight biplane kitbuilt (4)
Fisher R-80 and RS-80 Tiger Moth (USA)
Grosso Easy Eagle (USA)
Murphy Renegade II (Canada)
Murphy Renegade Spirit (Canada)

Tandem-seat ultralight twin/kitbuilt (2)
Aeroprakt-26 (Ukraine)
Aeroprakt-36 (Ukraine)

Side-by-side kitbuilt (26)
Aerocomp Merlin GT (USA)
Aeronix Airelle (France)
AirStart Griffin (Canada)
Avid Magnum (USA)
Creative Flight Aerocat (Canada)
DAC Ranger (Netherlands)
Dyn'Aero MCR01(France)
Glasair Super II (USA)
Glasair III (USA)
GlaStar GlaStar (USA)
HB Flugtechnik HB 207 Alfa (Austria)
HPA M1 Speedcruiser (USA)
IndUs Sport E (USA)
Lancair Legacy (USA)
Midwest Aerosport Formula GT (USA)
NuVenture Venture (USA)
Prosport Freebird Classic (USA)
Pulsar Super Pulsar 100 (USA)
Pulsar Sport 150 (USA)
Rans S-16 Shekari (USA)
Skystar Kitfox (USA)
Team Tango Tango (USA)
Ultravia Pelican (Canada)
Van's RV-7 and RV-7A (USA)
Van's RV-9 and RV-9A (USA)
Wag-Aero Wag-A-Bond

Side-by-side sportplane kitbuilt (1)
Kimball McCullocoupe (USA)

Side-by-side lightplane/kitbuilt (7)
Aero AT-3 (Poland)
Aeropract-27 (Russian Federation)
Aeropract-33 (Russian Federation)
Goair GT-1 Trainer (Australia)
Sequoia Falco F.8L (USA)
SG Storm (Italy)
Tecnam P92 Echo (Italy)
Side-by-side lightplane/motor glider kitbuilt (1)
Europa Europa (United Kingdom)

Side-by-side ultralight (42)
3Xtrim 450 Ultra (Poland)
Aeros UL-2000 Flamingo (Czech Republic)
Aero Services Guepard and Guepy (France)
Aerospool WT-9 Dynamic (Slovak Republic)
Aerostyle Breezer (Germany)
Aviatika-MAI-223 Kityonok (Russian Federation)
B&F FK 9 Mark 3 (Germany)
B&F FK 14 Polaris (Germany)
Bussard Raptor (Germany)
BUL Zùlù (France)
C+C Flying CC/03 (Germany)
Colyaer Martin 3 (Spain)
DAR-23 Aksent (Bulgaria)
DAR-25 Impuls (Bulgaria)
Delta Pegass (Czech Republic)
Ekolot JK-04 Albatros (Poland)
Ekolot JK-05 Beetle (Poland)
Ekoflug MXP-740 (Slovak Republic)
Elitar IE-101 Elitar (Russian Federation)
Elitar Sigma-4 (Russian Federation)
Espace Liberté Guerin G1 (France)
Euro ALA Jet Fox 97 (Italy)
Flight Design CT (Germany)
Fly Synthesis Texan (Italy)
Halley Apollo Fox (Hungary)
IAMI (HESA) Sanjaghak (Iran)
ICP Savannah (Italy)
Ikar Ai-10 Ikar (Ukraine)
Impulse Impulse (Germany)
Istasis01 (France)
Larus Larus (Czech Republic)
Pipistrel Virus (Slovenia)
PJB Vega (France)
PJB Spectra (France)
PJB Marauder (France)
Remos G-3 Mirage (Germany)
Remos G-4 Velocity (Germany)
Sauper Papango (France)
SG Rally (Italy)
Tecnam P96 Golf (Italy)
Tecnam P2002 Sierra (Italy)
TL-96 Star and TL-200 Sting Carbon (Czech Republic)

Side-by-side ultralight kitbuilt (29)
Alpi Pioneer 200 (Italy)
Alpi Pioneer 300 (Italy)
Avid Flyer Mark IV (USA)
Best Off Sky Ranger (France)
CAG Toxo (Spain)
Cobra Explorer (Australia)
DEA Yuma (Italy)
Dyn'Aero MCR01(France)
Fisher Dakota Hawk (USA)
Flightstar IISL and IISC (USA)
Kolb Mark III Extra (USA)
Murphy Maverick (Canada)
Noin Choucas (France)
Nordic 8 Mini Explorer (Canada)
Precision Tech Ferguson F-IIB Fergy (USA)
Raj Hamsa X-Air (India)
Raj Hamsa Hanuman (India)
Rans S-12XL Airaile and S-12S Super Airaile (USA)
SAI G97 Spotter (Italy)
Skygear Comet (Korea, South)
Skystar Kitfox Lite² (USA)
SlipStream Dragonfly (USA)
SlipStream Genesis and Revelation (USA)
Sonex Sonex (USA)
Teknico AeroPup (Australia)
Vol Mediterrani VM-1 Esqual (Spain)
Zenith Zodiac CH 601HD (USA)
Zenith Zodiac CH 601XL (USA)
Zenith STOL CH 701 (USA)

Side-by-side ultralight/kitbuilt (30)
Aeronix Airelle (France)
Aeroprakt-22 (Ukraine)
Aeropro Eurofox (Slovak Republic)
Aerostar 01 (Romania)
Ameur Altania (France)
ATEC Zephyr and Faeta (Czech Republic)
CLASS Kestrel (Canada)
EV-AT teamEurostar (Czech Republic)
Fantasy Air Allegro and Arius (Czech Republic)
Fläming Air FSU Saphir (Germany)
Fly Synthesis Storch (Italy)
HB Flugtechnik Amigo and Dandy (Austria)
Hughes Australian LightWing GR-912 (Australia)
Hughes Australian LightWing Sport 2000 (Australia)
Hughes Australian Lightwing Speed (Australia)
Humbert Tétras (France)
Ikarus C42 (Germany)
Interplane Skyboy (Czech Republic)
Jabiru Jabiru (Australia)
Kappa KP-2U Sova (Czech Republic)
Let-Mont UL Tulák (Czech Republic)
PCH Flight Pretty Flight (Germany)
Raj Hamsa Hanuman (India)
TeST TST-5 Variant (Czech Republic)
TeST TST-6 Duo (Czech Republic)
ULBI Wild Thing (Germany)
Urban UFM-10 Samba (Czech Republic)
Urban UFM-13 Lambáda (Czech Republic)
WD D4 BK Fascination (Germany)
WD D5 Evolution (Germany)
Side-by-side ultralight biplane (1)
Aviatika-MAI-890 (Russian Federation)

Side-by-side ultralight/motor glider (2)
Noin Excel (France)
Pipistrel Sinus (Slovenia)
Side-by-side ultralight motor glider (1)
Protoplane Massai (France)
Side-by-side ultralight motor glider kitbuilt (1)
Sonex Xenos (USA)
Side-by-side ultralight/motor glider kitbuilt (1)
Cobra Arrow (Australia)
Side-by-side ultralight twin (1)
Airox F-31 (Czech Republic)

Class 12: Rotary wing (149)

AgustaWestand A129 Mangusta *NEW*/0576301

Includes tiltrotors, autogyros and lifting platforms. Light utility helicopters are those not exceeding 5,000 kg (11,023 lb) maximum take-off weight.

Attack helicopter (13)
Agusta A 129 Mangusta (Italy)
Agusta A 129 International (Italy)
Bell 446 SuperCobra and King Cobra (USA)
Boeing AH-64 Apache (USA)
Boeing Sikorsky RAH-66 Comanche (USA)
Denel AH-2 Rooivalk (South Africa)
Eurocopter 665 Tiger/Tigre (International)
Fuji (Boeing) AH-64D (Japan)
HAL LCH (India)
Kamov Ka-50 Chernaya Akula (Russian Federation)
Mil Mi-28 (Russian Federation)
TAI (Bell) AH-1Z King Cobra (Turkey)
Westland WAH-64D Apache Longbow (United Kingdom)
Armed observation helicopter (2)
HAL Lancer (India)
Kawasaki OH-1 (Japan)
Naval combat helicopter (4)
CHAIG Z-8 (China)
Kaman (K894) SH-2G Super Seasprite (USA)
Mitsubishi (Sikorsky) SH-60 (Japan)
Sikorsky S-70B (USA)
Light observation helicopter (1)
HAL Light Observation Helicopter (LOH) (India)
Heavy lift helicopter (1)
Mil Mi-26 (Russian Federation)
Medium lift helicopter (3)
Boeing 114 and 414 (USA)
Kawasaki (Boeing) CH-47J Chinook (Japan)
Light lift helicopter (1)
Kaman K-1200 K-MAX (USA)
Medium transport helicopter (7)
Euromil Mi-38 (International)
Kamov Ka-29 (Russian Federation)
Kamov Ka-60 Kasatka (Russian Federation)
Kazan/Mil Mi-17 (Russian Federation)
Mil Mi-38 (Russian Federation)
Mitsubishi (Sikorsky) UH-60 (Japan)
Sikorsky S/H-92 (USA)
Multirole medium helicopter (18)
Agusta A149 (Italy)
Agusta-Bell 412 and Griffon (Italy)
Bell 412 (Canada)
CHRDI Z-10 (China)
Dirgantara (Eurocopter) NAS-332 Super Puma (Indonesia)
EH Industries EH101 (International)
Eurocopter Super Puma Mk I and Cougar Mk I (International)
Eurocopter Super Puma Mk II and Cougar Mk II (International)
IRGC Transport Helicopter (Iran)
KAI KMH (Korea, South)
KAL KMH (Korea, South)
Kamov Ka-32 (Russian Federation)
Kamov Ka-62 (Russian Federation)
Mil Mi-17 (Mi-8M), Mi-19, Mi-171 and Mi-172 (Russian Federation)
NH Industries NH90 (International)
PZL W-3 Sokól (Poland)
Sikorsky S-70A (USA)
Sikorsky S-76C+ (USA)
TAI (Eurocopter) Cougar Mk 1 (Turkey)
Light utility helicopter (38)
AVA-505 Thunder (Iran)
Agusta A 109 (Italy)
Agusta A 119 Koala (Italy)
Bell 206B-3 JetRanger III (Canada)
Bell 206L-4 LongRanger IV (Canada)
Bell 407 (Canada)
Bell 427 (Canada)
Bell 430 (Canada)
Bell/Agusta AB 139 (International)
CHAIG Z-11 (China)
CHRDI Z-X (China)
Enstrom 480 and TH-28 (USA)
Eurocopter AS 350 Ecureuil/AStar and AS 550 Fennec (International)
Eurocopter AS 355 Ecureuil 2/TwinStar and AS 555 Fennec (International)
Eurocopter AS 365N Dauphin 2 and AS 565 Panther (France)
Eurocopter EC 130 (International)
Eurocopter EC 135 and EC 635 (France)
Eurocopter EC 155B (International)
Eurocopter EC 120 B Colibri (International)
Eurocopter BK 117 (International)
Eurocopter EC 145 (International)
FH-1100 FHoenix (USA)
HAI (Eurocopter) Z-9 Haitun (China)
HAL Dhruv (India)
HAL Cheetah (India)
HAL Chetak (India)
IHSRC (PANHA) 287 (Iran)
IRGC Shahed 274 (Iran)
KAI-Bell SB 427 (Korea, South)
Kamov Ka-226A Sergei (Russian Federation)
Kazan Ansat (Russian Federation)
MD 500 and MD 530 (USA)
MD 520N (USA)
MD 600N (USA)
MD Explorer (USA)
Mitsubishi MH2000 (Japan)
PZL-Swidnik SW-4 (Poland)
Westland Lynx (United Kingdom)
AEW helicopter (1)
Kamov Ka-31 (Russian Federation)
Four-seat helicopter (3)
Mil Mi-34 (Russian Federation)
Robinson R44 (USA)
Schweizer 330 and 333 (USA)
Three-seat helicopter (6)
Aviampex Yanhol (Ukraine)
Enstrom F28 and 280 (USA)
Kazan Aktai (Russian Federation)
MK Helicopter MK III (Germany)
Mil Mi-60 (Russian Federation)
Schweizer 300C (USA)
Two-seat helicopter (7)
Aerokopter ZA-6 San'ka (Ukraine)
Aerocopter AK-1 (Ukraine)
Brantly B-2B (USA)
DFH Dragon 334 (Italy)
IL IS-2 (Poland)
IRHSC (Panha) Sanjaqak (Iran)
Robinson R22 (USA)
Four-seat helicopter kitbuilt (2)
Hillberg 1-04 Baby Huey (USA)
VAT Hummingbird (USA)
Two-seat helicopter kitbuilt (3)
Hillberg EH 1-02 TandemMouse (USA)
JAG JAG (USA)
Rotorway Exec 162F (USA)
Two-seat ultralight helicopter (1)
NA Design NA 40 Bongo (Czech Republic)
Two-seat ultralight helicopter kitbuilt (3)
ASII Ultrasport 496 (USA)
Helisport CH-7 Kompress (Italy)
Masquito M80 (Belgium)
Single-seat helicopter kitbuilt (1)
Hillberg EH 1-01 RotorMouse (USA)
Single-seat ultralight helicopter kitbuilt (4)
ASII Ultrasport 254 and 331 (USA)
Eagle Helicycle (USA)
Innovator Mosquito (Canada)
Yoshine EzyCopter (Taiwan)
Utility autogyro (1)
Groen RevCon 6 (USA)
Four-seat autogyro (1)
Groen Hawk 4T (USA)
Three-seat autogyro (1)
Irkut A-002 (Russian Federation)
Two-seat autogyro ultralight kitbuilt (3)
Air Command Commander Elite (USA)
Little Wing Roto-Pup (USA)
RAF 2000 (Canada)
Two-seat autogyro ultralight (4)
Aerocopter Futura (Spain)
Magni M16 Tandem Trainer (Italy)
Magni M 19 Shark (Italy)
Magni M21 (Italy)
Two-seat autogyro ultralight/kitbuilt (1)
AAI SparrowHawk (USA)
Single-seat autogyro ultralight (2)
Magni M 20 Talon (Italy)
MAI-205 (Russian Federation)
Single-seat autogyro ultralight kitbuilt (1)
Little Wing Roto-Pup (USA)
New-concept rotorcraft (2)
Boeing X-50A Dragonfly Canard Rotor/Wing (USA)
Piasecki Vectored Thrust Ducted Propeller (USA)
Lifting vehicle (4)
AD & D Hummingbird (Israel)
PAM 100B Individual Lifting Vehicle (USA)
Urban Aero CityHawk (Israel)
Urban Aero X-Hawk and TurboHawk (Israel)
Exoskelitor flying vehicle (2)
Polyot Yula (Russian Federation)
Yanagisawa Gen H-4 (Japan)
Light utility tiltrotor (1)
Bell/Agusta BA609 (International)
Multimission tiltrotor (3)
Bell Quad Tiltrotor (USA)
Bell Boeing V-22 Osprey (USA)
Eurotilt (International)
Convertiplane (4)
Carter CarterCopter CC1 (USA)
Carter Heliplane Transport (USA)
Groen Gyrolifter (USA)
Groen Gyroliner (USA)

Class 13: Lighter than air (33)

ATG AT-10 *NEW*/0589387

All are (at least broadly) cylindrical dirigibles, unless otherwise specified.

Helium rigid (1)
Aeros Aeroscraft ML (USA)
Helium semi-rigid (9)
AMSAT Mantha (France)
Blenntec UPship 50 Metre (USA)
EADS First 1A (France)
RosAeroSystems PD-300 (Russian Federation)
RosAeroSystems MD-900 (Russian Federation)
RosAeroSystems DPD-5000 (Russian Federation)
RosAeroSystems DTs-N1 (Russian Federation)
Voliris 900 (France)
Zeppelin NT (Germany)
Helium non-rigid (13)
ABC Lightship A-60+ (USA)
ABC Lightship/Spector Series (USA)
AERALL AVEA (France)
Aerostatica-300 (Russian Federation)
ATG AT-10 (United Kingdom)
Evolution FS 200 Friendship 200 (Germany)
GSI Skyship 500HL (USA)
GSI Skyship 600 (USA)
Pan Atlantic Cargo Airship System (Canada)
Parabounce Propbike (USA)
RosAeroSystems Au-11 (Russian Federation)
RosAeroSystems Au-12 (Russian Federation)
SVAM CA-80 (China)
Helium non-rigid special shape (1)
21st Century SPAS-R1 (Canada)
Helium non-rigid hybrid (2)
AHA Hornet (USA)
AHA Light Utility (USA)
Hybrid non-rigid air vehicle (1)
ATG SkyCat (United Kingdom)
Hybrid rigid air vehicle (1)
Ohio DynaLifter (USA)
Hot air non-rigid (4)
Cameron DP Series (United Kingdom)
Gefa-Flug AS 105 GD (Germany)
Kubicek AV-2 (Czech Republic)
Skymedia SMA (USA)
Helium balloon (1)
QinetiQ QinetiQ 1 (UK)

First Flights

Some of the first flights made during 2004

Having flown the Embraer 170 and 175 during the previous year, Brazil's leading aircraft manufacturer attended to the top end of its latest airliner range in 2004, with maiden sorties of both the Embraer 190 and 195. It ended the year by moutning this pectacular formation flight of the new family, led by the 195, with 190, 175 and 170 in close attendance *NEW*/1042423

Australia 10 Jul
Hughes PR-Bipe (19-4080)

Austria
Diamond DA42 Twin Star (Lycoming conversion of DA42-V1) (OE-VPS) 16 Apr
Diamond DA42 Twin Star, fourth (third flying) prototype (OE-VDB) 29 Apr

Belgium
Lambert Mission M212-100 (G-XFLY) 13 Jul

Brazil

190-100 *NEW*/1042422

Embraer 190-100, prototype (PP-XMA) 12 Mar
Embraer 190-100, second prototype (PP-XMB) 9 May

195-100 *NEW*/1042421

Embraer 195-100, prototype (PP-XMJ) 7 Dec

Canada
Bombardier Global 5000, second prototype (C-GLRM) 8 Jan
Innovator Mosquito XE .. Mar
Dream Tundra, second prototype (C-GAGH) .. Apr

SPAS-70-1A *NEW*/1111647

21st Century SPAS-70-1A (alias TSI SA-60) spherical airship 24 Jun
Bell 407 MAPL (TailFan) demonstrator (N41835) 15 Jul
Boeing 720, P&WC testbed with (one) PW610F engine (C-FETB) 16 Dec

China
CAC FC-1 Xiaolong, second prototype (03) 9 Apr
NLA AC-500 Air Car .. May
Changhe Z-8F 28 Aug
NLA AC-500 Air Car, second prototype (c/n 004) ('official' flight on 7th) 3 Dec
CHAIG Z-11W prototype 27 Dec

J-10B *NEW*/1042419

Czech Republic
CZAW Lake Sport (OK-JUR 99) 11 Mar

EV-AT *NEW*/1042418

EV-AT VUT-100-131i Cobra (OK-EVE) ('official' flight on 16th) 11 Nov

The second Bombardier Global 5000 lifted off from a chilly Montreal on 8 January 2004 *NEW*/1042420

LD-1/MFI-9UL *NEW*/1042417

LD Aviation LD-1/MFI-9UL (MFI-9HB replica) (unmarked) 21 Nov

Germany

160 Ranger *NEW*/1042416

Grob G 160 Ranger (D-FTBG) 29 Mar
Flug Werk FW 190A-8/N, first flyable replica (D-FWWC) 22 Jul

India

HJT-36 Sitara *NEW*/1042415

HAL HJT-36 Sitara, second prototype (S3474) 26 Mar

CSIR Saras *NEW*/1136950

CSIR Saras, prototype (VT-XSD) 29 May
Sukhoi Su-30MKI, first assembled by HAL 27 Nov

International

AS 665 *NEW*/1042414

Eurocopter AS 665 Tiger/Tigre, first ARH for Australian Army (F-ZVLH/A38-001) 20 Feb
Eurocopter Typhoon, first Spanish-built production single-seat (IPA4/PS001) (C.16-20) 27 Feb

Airbus A310-300 MRTT, first conversion for German Air Force (1027) 12 Mar

NH 90) *NEW*/1042458

NH Industries NH 90, first production TTH for German Army (TGEA01/9891) ('official' flight on 11th) 4 May
Airbus A340-300 testbed, first flight with (one) Rolls-Royce Trent 900 engine for A380 (F-WWAI) 17 May

Eurofighter Typhoon *NEW*/1042457

Eurofighter Typhoon F. Mk 2, first UK-built production single-seat (IPA5/PS002) (ZJ700) 7 Jun
Eurofighter EC 225, first production (F-WQDJ) 24 Jun
NH Industries NH 90, first French production and first TTT for Finland 15 Sep
NH Industries NH 90, first TTH for Italian Army (ITAR 001/CS.X81517) 15 Sep
AgustaWestland Merlin VX-X (US101), first flight with uprated CT7-8E engines (G-17-510) 5 Oct

Eurofighter Typhoon *NEW*/1042456

Eurofighter Typhoon, first production single-seat for Luftwaffe service (GS002/9839) 22 Oct
Airbus A318-100, first flight with PW6124A engines (F-WWIA) 9 Dec
Eurofighter Typhoon, first production single-seat for Italian Air Force (IS002) 20 Dec

Iran

IRIAF Saeghe 8 Jul

Italy

Aermacchi M-346, first prototype (MM.CS.X615) 15 Jul
Tecnam P2004 Bravo 9 Sep

SF-260EA *NEW*/1042454

Aermacchi SF-260EA, first for Italian Air Force (unmarked) 21 Oct

C-27J Spartan *NEW*/1042453

Alenia/Lockheed Martin C-27J Spartan, first for Greek Air Force (HA001) 15 Dec

Lithuania

Termikas Yak-52TD (Aerostar Iak-52 modified as 'taildragger') (LY-TTD) 23 Apr

Russian Federation

Sukhoi Su-27M first 'production' conversion 5 Mar
Rostvertol (Mil) Mi-28N, first production (02) 25 Mar
Yakovlev Yak-52M (01) 16 Apr
Yakovlev Yak-130, first production (01) 30 Apr

Slovenia

Taurus 503 *NEW*/1042452

Pipistrel Taurus 503 (S5-P1102) 15 May

South Africa

A 109 LUH *NEW*/1042451

Agusta A 109 LUH, first Denel-assembled for South African Air Force (4006) 9 Sep

During 2004, Eurofighter flew the last of its five 'Instrumented Production Aircraft' and the first single-seat machines for regular squadron service. IPA4 took to the air in Spain on 27 February *NEW*/1042455

Sweden
Saab JAS 39D Gripen, first production (39815) — 2 Jun

JAS 39C *NEW*/1042450

Saab JAS 39C Gripen, first for lease to Czech Air Force (9237) — 18 Oct

Switzerland
Pilatus PC-21, second prototype (HB-HZB) — 7 Jun
Pilatus PC-9M Advanced Turbo Trainer, first for Bulgarian Air Force (663) — 20 Sep

Ukraine
Antonov An-124-100M-150 upgrade (UR-82008) — .. Aug

Aeroprakt-36 *NEW*/1042449

Aeroprakt-36 Vulcan (unmarked) — 18 Sep

An-148 *NEW*/1135023

Antonov An-148 (UR-NTA) — 17 Dec
Antonov An-74TK-200A (UR-CES) — 24 Dec

UK
AgustaWestland Super Lynx 300, first for Royal Thai Navy (ZJ987/2313) — 26 Jul

Nimrod MRA *NEW*/0578168

BAE Systems Nimrod MRA. Mk 4 prototype (PA1/ZJ516, converted MR Mk 2 XV247) — 26 Aug
BAE Systems Nimrod MRA. Mk 4, second aircraft (PA2/ZJ518, converted MR Mk 2 XV234) — 21 Dec

United States
Sikorsky S-70B-6 Aegean Hawk, first for Greek Navy (PN-59) — 10 Feb
American Autogyro SparrowHawk (N4132R) — 26 Feb
Four Winds 210C (N250FW) — 27 Feb
Boeing AH-64D Sharaf, first for Israel Defence Force/Air Force — .. Feb
Kimball McCullocoupe (N511) — 1 Mar

311 Capricorn *NEW*/1042448

Scaled Composites 311 Capricorn *Virgin Atlantic GlobalFlyer* (N277SF) — 5 Mar
NASA/Boeing X-43A Hyper-X second vehicle; first successful flight — 27 Mar
Cadcor Chanute (N1903N) — 23 Apr

T-6B *NEW*/1042447

Beech T-6B (N2841B/N3000B) — 12 May
Boeing 737-700AE&C 'Wedgetail', first for Royal Australian Air Force (N378BC/A30-001) — 20 May
Bombardier/Raytheon Sentinel R. Mk 1, first production, first flight in fully equipped configuration (ZJ690) — 26 May
Sikorsky S-92A, first production (N192PH) — 14 Jun
Northrop Grumman E-2C Hawkeye 2000, first upgraded aircraft for JASDF — 14 Jul

Epic Air *NEW*/1042446

Epic Air Epic LT (N370JP) — 17 Jul
AeroComp CA-J CompAir Jet (N41861) — 17 Jul
Maule M-4-180V (N799ZZ) — Jul

Citation Jet *NEW*/1042445

Cessna 525B Citation Jet CJ1+ (N525AD) — 8 Oct
Quest Kodiak (N490KQ) — 16 Oct
Boeing EF-18A Hornet, first EADS upgrade for Spain (C.15-62) — 23 Nov
Boeing 747-121, GE testbed with Engine Alliance GP 7200 engine on port inboard pylon (N747GE) — 3 Dec
Boeing YAL-1A, first flight with integrated battle management and beam/fire control systems installed (00-0001) — 3 Dec
Boeing 720-023B, P&W testbed with one PW610F engine attached to forward fuselage, starboard (C-FETB) — 16 Dec

Eclipse Eclipse 500 *NEW*/1042467

Eclipse Eclipse 500, second (first 'conforming') prototype (N503EA) — 31 Dec

Some aircraft are destined to appear in 'first flights' more than once. The unique AL-1A airborne laser platform has flown as a plain Boeing 747, but took to the air once again on 3 December 2004 with its fire control systems installed. Later, it will again find a place in this section, having finally gained its large and complex laser 'gun' ***(James Heseltine, USAF)***
NEW/1042444

Aerospace Calendar

Some significant aerospace events in 2004

Returning to the Dryden Flight Research Center with an F/A-18 Hornet escort after its final mission – the launch of the X-43 scramjet test vehicle on 16 November – NASA's venerable Boeing NB-52B Stratofortress was formally retired on 17 December 2004, the 101st anniversary of the Wrights' powered flight. This was the eighth production B-52 and, remarkably, it had been in service for almost half the time since the Brothers' first sortie, having first flown in 1955 and transferred to NASA in June 1959 to assist an NB-52A with carrying the North American X-15. It has, itself, been replaced by another of the breed, a B-52H from the final Stratofortress batch, delivered in July 2001 and subsequently modified to testbed configuration. The NB-52B is structurally still youthful, having flown 2,443.8 hours in 1,051 sorties *(NASA/Tony Landis)*
NEW/1042437

2004	Military aviation	Civil aviation
8 Jan		First flight of second development and test Bombardier Global 5000

NEW/1042436

2004	Military aviation	Civil aviation
19 Jan	The first Austrian pilot flew in a Eurofighter Typhoon (at Manching, Germany) following his country's confirmation of purchase	
20 Jan	India signed a contract for 16 MiG-29K combat aircraft for service with naval air arm	

NEW/1042435

2004	Military aviation	Civil aviation
9 Feb	Lockheed Martin F/A-22 Raptor 4003 achieved the type's 5,000th hour of accident-free flight testing	

NEW/1042434

2004	Military aviation	Civil aviation
9 Feb		Roll-out of the prototype Embraer 190 regional airliner
13 Feb		First production Cessna 525B Citation CJ3 rolled out
23 Feb	Cancellation of Boeing-Sikorsky RAH-66 Comanche attack helicopter announced by US Department of Defense	Gulfstream Aerospace announced Gulfstream G350 model on eve of Asian Aerospace show in Singapore
24 Feb		FAA certification awarded to Angel 44 light utility transport aircraft
5 Mar		First flight of Scaled 311 Capricorn *(Virgin Atlantic GlobalFlyer)*
16 Mar	Italian Air Force received first squadron-assigned Eurofighter for 4° Stormo	
23 Mar		FAA certification awarded to Cessna 560 Citation XLS
24 Mar	Prototype ShinMaywa US-1AKai handed over to Japan Maritime Self Defence Force	
28 Mar		First flight of VAM VAM-1 ultralight aircraft at Phuoc Long Airport, Vietnam
29 Mar		Czech certification awarded to Kappa KP-5 ASA ultralight
31 Mar		Delivery of 50th Eurocopter EC 145 (to Securité Civile)
Mar	Request for proposals issued for South Korean KMH multirole medium helicopter	
17 Apr		First production Cessna Citation XLS rolled out
26 Apr		Formal launch of Boeing 7E7 (now 787) Dreamliner; first order (for 50 from All Nippon Airways) received on same date

2003	Military aviation	Civil aviation

NEW/1042433

2003	Military aviation	Civil aviation
28 Apr	Aermacchi M-346 advanced jet trainer began taxying trials at Venegono	
30 Apr	German Air Force wing JG 73 moved to operational base at Laage with five Eurofighter 2000s after instructor training at Manching	
Apr		Leteckê Závody of Czech Republic declared bankrupt
4 May	First flight of first production NH Industries NH 90 at Donauwörth (for German Army)	
13 May	RAF Eurofighter Typhoon issued with Release to Service	
15 May		First flight of Pipistrel Taurus ultralight sailplane
20 May	First flight (at Warton) of Eurofighter Typhoon in No. 29 Squadron colours	

NEW/1042432

2003	Military aviation	Civil aviation
26 May	First flight of first fully equipped UK RAF Raytheon Sentinel R. Mk 1 (Global Express)	GKN announced agreement concerning sale of its 50 per cent share of AgustaWestland to Finmeccanica
27 May	First three Spanish Air Force Eurofighter 2000s delivered to Ala 11 at Moron after instructor training at Getafe	
7 Jun	First flight of first production single-seat RAF Eurofighter Typhoon	
11 Jun		French and Chinese politicians in Paris signed an agreement under which Hafei Aviation Industry will build Eurocopter HC 120s under licence in China
16 Jun	Russian and Ukraine Air Force Commanders-in-chief signed agreement to resume military trials of Antonov An-70 tactical transport, ending suspension that began 25 Feb 2003	First flight of first production Sikorsky S-92
18 Jun		Moravan Aeroplanes (Zlin) declared bankrupt
21 Jun		Beriev Be-200 Altair jet water bomber detachment of Russian disaster relief agency declared operational at Zhukovsky
22 Jun		Transport Canada certification presented to Learjet 40
23 Jun		Interstate Aviation Committee awarded airworthiness certificates to Tupolev for Tu-334-100 and Tu-204-120CE freighter

2003	Military aviation	Civil aviation

NEW/1042431

2003	Military aviation	Civil aviation
24 Jun		Eurocopter flew the 600th member of the Super Puma/Cougar family - this also the first production EC 225
Jun		Fairchild Dornier Aeroindustries filed for insolvency after failing to obtain financial backing to re-launch Fairchild Dornier 728/928 regional jet airliner
6 Jul		Europa Management (International) Ltd declared bankrupt. New company formed to support Europa lightplane two months later
15 Jul	First flight of prototype Aermacchi M-346 advanced/lead-in fighter trainer	First flight of Bell 407 Modular Affordable Product Line 'TaiLFan' demonstrator
19 Jul		Sibir became launch customer for Sukhoi Russian Regional Jet, announcing order for 10 aircraft at Farnborough International
		Bombardier unveiled proposed CSeries airliner at Farnborough International
20 Jul		AgustaWestland unveiled A 109S Grand light twin helicopter at Farnborough International
21 Jul	Hand-over to UK Ministry of Defence of 67th and last AgustaWestland-built WAH-64D Apache	
27 Jul		Chinese CAAC certification awarded to Diamond DA40D Diamond Star
Jul	Oman armed forces placed order for 20 NH Industries NH 90 TTH helicopters	
20 Aug		Bella Aviation in UK announced suspension of operations two months after taking over the former CFM company and its Shadow series of light kitplanes
Aug		Taiwanese company AIDC completed the initial batch of 10 wing sets for the Ae 270 business turboprop and asked Czech partner Aero Vodochody to assume responsibility for the entire programme
1 Sep		US Sport Pilot rule and associated Light Sport Aircraft certification regulations came into effect
6 Sep		Europa Aircraft (2004) Ltd began operations, having taken over assets and premises of Europa Management (International) at Kirkbymoorside, UK
9 Sep	Royal Thai Navy formally accepted its first AgustaWestland Lynx at Yeovil	First flight of Tecnam P2004 Bravo
16 Sep		Bell/Agusta AB139 deliveries began with hand-over of first of two VIP-configured helicopters to Namibian Government

NEW/1042430

NEW/1042429

2003	Military aviation	Civil aviation
22 Sep		Hand-over of the first Embraer 170 regional airliner to Republic Airways Holdings
23 Sep		Unofficial roll-out of Antonov An-148 regional jet airliner
27 Sep		Kazan announced launch of production of Ansat light utility helicopter
29 Sep		Cessna handed over first two production Sovereign business jets to customers in Wisconsin and Mexico
6 Oct	First flight announced of US101 demonstrator version of AgustaWestland International EH101 for US Marine Corps VXX presidential airlift competition	
11 Oct		Aerion Supersonic Business Jet project announced at NBAA Convention in Las Vegas. Bombardier announced 'hot and high' Learjet 40 XR at NBAA
		Formal roll-out of Antonov An-148 regional jet.
15 Oct		FAA certification awarded to Cessna 525B Citation CJ3
19 Oct		Brazilian manufacturer Neiva, now an Embraer company and manufacturer of the Ipanema sprayer, celebrated 50 years in aviation
25 Oct	Danish version of AgustaWestland International EH101 formally named Merlin Joint Supporter	
26 Oct		Chinese CAAC certification awarded to Diamond DA20-C1

NEW/1042428

2003	Military aviation	Civil aviation
27 Oct	Lockheed Martin rolled out 41st F/A-22A Raptor at Marietta, this destined to be first for an operational squadron (at Langley AFB)	
28 Oct		Boeing rolled out the last of 1,050 Model 757s at Renton, ending a 26-year saga
29 Oct		Thielert Aircraft Engines received type certification for the 228 kW (310 hp) Centurion 4.0 Jetfuel engine from EASA
Oct		First flight of four-engined conversion of Dewald Sunny; claimed to be world's first four-motor ultralight

NEW/0590178

2003	Military aviation	Civil aviation
1 Nov	Model of LFC-16 new version of CY-1 advanced jet trainer exhibited at Airshow China at Zhuhai	Model of XAC WJ twin-turbofan regional airliner exhibited at Airshow China
11 Nov		First flight of EV-AT VUT 100 Cobra lightplane in Czech Republic

NEW/1042427

2003	Military aviation	Civil aviation
15 Nov	Polish air Force accepted first PZL SW-4 light helicopter at Swidnik	

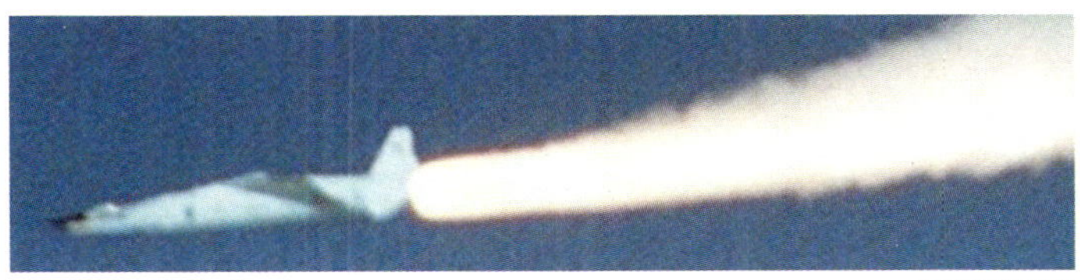

NEW/1042426

2003	Military aviation	Civil aviation
16 Nov	Third (second successful; final in series) flight of X-43 Hyper-X scramjet-powered UAV, mounted on a Pegasus booster and dropped from a Boeing B-52, achieved Mach 9.8	
23 Nov	Canada signed contract with Sikorsky Aircraft Corporation covering supply of 28 H-92 maritime helicopters to Canadian Forces from late 2008	
29 Nov		First Iranian-built Ir.An-140 entered revenue service with Safiran airline
3 Dec	Boeing YAL-1A Airborne Laser development aircraft returned to flight status after hiatus of almost two years	Ceremonial roll-out of 500th Boeing 777 at Everett, Washington; aircraft subsequently delivered to International Lease Finance Corp for operation by Air France

NEW/0590390

2003	Military aviation	Civil aviation
6 Dec	HAL rolled out first of 140 Sukhoi Su-30MKIs to be assembled in India	
8 Dec	Contract for start of development work on HAD (support) version of Eurocopter Tiger signed at Marignane by France and Spain	

2003	Military aviation	Civil aviation
11 Dec		Roll-out of first production-conforming Eclipse Aviation Eclipse 500 very light business jet
15 Dec	First two Eurocopter Tiger combat helicopters delivered to Australian Army South Africa signed declaration of intent to become partner in Airbus A400M military transport aircraft programme	First Mooney aircraft to be equipped with a Garmin G1000 glass cockpit delivered: an Ovation2 GX
17 Dec	Announced that UK had become the fourth and last (after Germany, Italy and Spain) to sign Eurofighter Typhoon production Tranche 2 agreement, comprising 236 aircraft; additional 18 are for Austria	First flight of Antonov An-148 regional airliner at Kiev
20 Dec	Australia signed contract for supply of five Airbus A330-200 multirole tanker/transport aircraft from 2008	
21 Dec		Air Europa of Spain became first airline to commit to the Airbus A350-800, signing MoU for supply of 10 aircraft with further two subject to option
21 Dec	First Australian-built Eurocopter Tiger ARH5 made its maiden flight at Brisbane	
22 Dec	First AgustaWestland International EH101 for Portuguese Air Force handed over at Vergiate	First series production Eurocopter EC 225 delivered to Algeria for use as VIP transport. Sikorsky announced signature of contract for VIP-configured S-92 helicopter to be supplied to Turkish Government
23 Dec	Swiss Federal Office for Civil Aviation awarded type certificate to Pilatus PC-21	Embraer announced that the 175 had been awarded CTA (Brazilian) certification

NEW/1042425

2003	Military aviation	Civil aviation
		Provisional certification awarded to Hawker Horizon; coincident with delivery of first production aircraft, which subsequently bailed back by Raytheon for further testing
29 Dec	First full-production-standard, single-seat Eurofighter Typhoon delivered to customer at Getafe, Spain	High Performance TT62 Alekto business turboprop twin began taxi trials
31 Dec		First flight of Eclipse Aviation Eclipse 500 very light business jet with PW610F engines

NEW/1042424

Official Records

(Corrected to 1 January 2005)

Reinvigorated with three Pratt & Whitney PT6 turboprops, vintage Dornier Do 24ATT RP-C2403 set two Sub-class C-3g Group 2 records on 3 September 2004 for times to heights of 3,000 m and 6,000 m; respectively 13 min 42 s and 47 min 22 s *(Paul Jackson)* ***NEW/***1042461

The Fédération Aéronautique Internationale (FAI) is the world authority for verification of aviation records. Each category is further broken down by type of power plant, weight band and sex of pilot, while time-to-height and height-with-payload records are subdivided into further bands. Additionally, point-to-point records can be set between an almost unlimited number of city pairs. Consequently, only the more significant records are quoted here.

The full range of categories is:
Class A Free balloons
Class B Airships
Class C Aeroplanes
Class D Gliders and motor gliders
Class E Rotorcraft
Class F Aeromodelling
Class G Parachuting
Class H VTOL aeroplanes
Class I Human-powered aircraft
Class K Spacecraft
Class M Tilting wing/engine aircraft
Class N STOL aeroplanes
Class O Hang gliders and paragliders
Class P Aerospacecraft
Class R Microlights
Class S Space models
Class U Unmanned Aerial Vehicles

ABSOLUTE WORLD RECORDS

CLASS A (Free Balloons)

Four records are classed as Absolute World Records for free balloons by the FAI, as follows:

Duration (Switzerland)
Bertrand Piccard (Switzerland) and Brian Jones (UK) in the Cameron R-650 combination hot air/helium balloon HB-BRA *Breitling Orbiter 3*, from Château d'Oex, Switzerland, to near Dâkhla, Egypt, between 1 and 21 March 1999. 477 hours 47 minutes.

Distance (Switzerland)
Bertrand Piccard (Switzerland) and Brian Jones (UK) in the Cameron R-650 combination hot air/helium balloon HB-BRA *Breitling Orbiter 3*, from Château d'Oex, Switzerland to near Dâkhla, Egypt, between 1 and 21 March 1999. 22,037.8 n miles (40,814.0 km; 25,361.34 miles).

Altitude (USA)
Cdr M D Ross and Lt Cdr V A Prather in Winzen Research ONXR290WRP-1 *Lee Lewis Memorial*, from USS *Antietam*, Gulf of Mexico, on 4 May 1961. 34,668 m (113,740 ft).

Shortest time around the world (USA)
Steve Fossett (USA) in Cameron/Cole R-550 combination hot air/helium balloon N277SF *Bud Light Spirit of Freedom* from Northam, Australia to meridian 116° 42′ 16″ East, over Australia, between 19 June and 2 July 2002. 320 hours 33 minutes.

Two airship records set by the Graf Zeppelin (here over Wembley Stadium, London) in 1928 have yet to be bettered 0543361

CLASS B (Airships)

Four records are classed as Absolute World Records for airships by the FAI, as follows:

Duration (Germany)
Kapt Hugo Eckener and crew in Zeppelin D-LZ127 *Graf Zeppelin* from Lakehurst, New Jersey, to Friedrichshafen, Germany, between 29 October and 1 November 1928. 71 hours 7 minutes.

Distance in a straight line (Germany)
Kapt Hugo Eckener and crew in Zeppelin D-LZ127 *Graf Zeppelin* from Lakehurst, New Jersey, to Friedrichshafen, Germany, between 29 October and 1 November 1928. 3,447.35 n miles (6,384.50 km; 3,967.14 miles).

Altitude (USA)
Holkan Colting and Tim Buss in Colting SPAS 62.5 C-FZRY between Frumheller, Alberta and Gull Lake, Saskatchewan, Canada, 12 June 2003. 6,234 m (20,453 ft).

Speed in a straight line (UK)
James Dexter and Michael Kendrick in ABC Lightship A-60+ N606LG at Halfpenny Green, UK, 19 January 2000. 50.11 kt (92.8 km/h; 57.66 mph).

CLASS C (Aeroplanes)

Seven records are classed as Absolute World Records for aeroplanes by the FAI, as follows:

Distance in a straight line (USA), and **Distance in a closed circuit** (USA)
Richard 'Dick' Rutan and Jeana Yeager in the Voyager (N269VA), between 14 and 23 December 1986.

Circumnavigation of the World, starting and finishing at Edwards AFB, California. 21,712.816 n miles (40,212.139 km; 24,986.664 miles).

Altitude (USSR)
Alexander Fedotov in an E-266M (MiG-25) on 31 August 1977. 37,650 m (123,523 ft).

Altitude in sustained horizontal flight (USA)
Captain Robert C Helt and Major Larry A Elliott, USAF, in an unidentified Lockheed SR-71A on 28 July 1976 at Beale AFB, California. 25,929.031 m (85,069 ft).

Altitude, after launch from a 'mother-plane' (USA)
Major R White, USAF, in the North American X-15A-3 (56-6672) on 17 July 1962, at Edwards AFB, California. 95,935.99 m (314,750 ft).

Speed on a straight course (USA)
Captain Eldon W Joersz and Major George T Morgan Jr, USAF, in an unidentified Lockheed SR-71A on 28 July 1976 over a 15/25 km course at Beale AFB, California. 1,905.81 kt (3,529.56 km/h; 2,193.17 mph).

Speed in a closed circuit (USA)
Major Adolphus H Bledsoe Jr and Major John T Fuller, USAF, in Lockheed SR-71A 64-17958, on 27 July 1976, over a 1,000 km closed circuit from Beale AFB, California. 1,818.154 kt (3,367.221 km/h; 2,092.294 mph).

WORLD CLASS RECORDS

Following are details of some of the more important world class records confirmed by the FAI:

CLASS C, GROUP 1 (Landplanes with piston engines):

Great circle distance without landing and **Distance in a closed circuit**
See Absolute World Records.

Altitude (Italy)
Mario Pezzi, in a Caproni Ca 161*bis*, on 22 October 1938. 17,083 m (56,046 ft).

Speed in a straight line (USA)
Lyle Shelton in the modified Grumman F8F Bearcat *Rare Bear* (N777L), with a 2,833 kW (3,800 hp) Wright R-3350 engine, on 21 August 1989, over a 3 km course at Las Vegas, New Mexico. 459.10 kt (850.25 km/h; 528.33 mph).

CLASS C, GROUP 2 (Landplanes with turboprop engines):

Great circle distance without landing (USA)
Lt Col E L Allison and crew in a Lockheed HC-130H Hercules (65-0972), on 20 February 1972. 7,587.99 n miles (14,052.95 km; 8,732.098 miles).

Distance in a closed circuit (USA)
Cdr Philip R Hite and crew in the Lockheed RP-3D Orion (158227), on 4 November 1972. 5,455.46 n miles (10,103.51 km; 6,278.03 miles).

Altitude (USA)
Einar Enevoldson in Grob Egrett-1, N14ES, on 1 September 1988, at Majors Field, Greenville, Texas. 16,329 m (53,573 ft).

Speed on a straight course (USA)
Cdr Donald H Lilienthal and crew in a Lockheed P-3C Orion, over a 15/25 km course on 27 January 1971. 435.26 kt (806.10 km/h; 500.89 mph).

Speed in a closed circuit (USSR)
Ivan Sukhomlin and crew in Tupolev Tu-114 SSSR-76459, on 9 April 1960, carrying a 25,000 kg payload over a 5,000 km circuit. 473.66 kt (877.212 km/h; 545.07 mph).

Andrew Keech exceeded two autogyro records in 2004, flying his Little Wing LW-5 Previous holders were a Rotorflight Dominator (altitude) and Wallis WA-116 (distance) *NEW/*1042462

CLASS C, GROUP 3 (Landplanes with jet engines):

Great circle distance without landing (USA)
Major Clyde P Evely, USAF, in a Boeing B-52H Stratofortress, between 10 and 11 January 1962, from Okinawa to Madrid, Spain. 10,890.27 n miles (20,168.78 km; 12,532.3 miles).

Distance in a closed circuit (USSR)
Vladimir Terski and crew in Antonov An-124 SSSR-82009, between 6 and 7 May 1987. Moscow-Astrakhan-Tashkent-Lake Baikal-Petropavlovsk-Chukot Peninsula-Murmansk-Zhdanov-Moscow. 10,880.625 n miles (20,150.921 km; 12,521.201 miles).

Altitude, speed on straight course and **speed in 1,000 km closed circuit**
See Absolute World Records.

Speed over a 3 km course at restricted altitude (USA)
Darryl Greenamyer in the modified Red Baron F-104RB Starfighter (N104RB), on 24 October 1977, at Mud Lake, Tonopah, Nevada. 858.77 kt (1,590.45 km/h; 988.26 mph).

Speed in a 100 km closed circuit (USSR)
Alexander Fedotov in an unidentified Mikoyan E-266 (MiG-25) on 8 April 1973. 1,406.641 kt (2,605.1 km/h; 1,618.734 mph).

Speed in a 500 km closed circuit (USSR)

Designed in 1967 and flown five years later, Dennis Polen's Special still holds the Sub-class C-1b speed record of 263.64 kt (488.26 km/h; 303.309 mph) set by current owner, Richard C Keyt, at AirVenture, Oshkosh, on 26 July 2001, making it the world's fastest four-cylinder aircraft
(Paul Jackson) *NEW/*1042463

M Komarov in a Mikoyan E-266 (MiG-25), on 5 October 1967, near Moscow. 1,609.88 kt (2,981.5 km/h; 1,852.62 mph).

Speed around the world (Eastbound; without aerial refuelling) (France)
Michel Dupont and Claude Hetru in Concorde F-BTSD on 16 August 1995, 704.68 kt (1,305.93 km/h; 811.46 mph). (Note: Exceeds record for circumnavigation with aerial refuelling and speed Westbound).

Greatest payload carried to a height of 2,000 m (Ukraine)
Aleksandr V Galunenko in Antonov An-225 Mriya UR-82060, on 11 September 2001. 253,820 kg (559,577 lb).

CLASS C.2, ALL GROUPS (Seaplanes):

Great circle distance without landing (UK)
Capt D C T Bennett and First Officer I Harvey, in the Short-Mayo S.20 *Mercury*, G-ADHJ, between 6 and 8 October 1938, from Dundee, Scotland, to the Orange River, South Africa. 5,211.66 n miles (9,652 km; 5,997.5 miles).

Altitude (USSR)
Georgi Buryanov and crew of two in Beriev M-10 '40', on 9 September 1961, over the Sea of Azov. 14,962 m (49,088 ft).

Speed on a straight course (USSR)
Nikolai Andrievsky and crew of two in Beriev M-10 '40', on 7 August 1961, at Joukovski-Petrovskoe, over a 15/25 km course. 492.44 kt (912 km/h; 566.69 mph). **Note:** Class C.3 Amphibians record is held by Sergei Andreev and Nikolay Shlykov in Beriev M-12 Tchaika '05' on 30 October 1987. 309.54 kt (573.27 km/h; 356.21 mph).

CLASS E.1 (Helicopters):

Great circle distance without landing (USA)
R G Ferry in a Hughes YOH-6A, between 6 and 7 April 1966, from Culver City, California, to Ormond Beach, Florida. 1,923.08 n miles (3,561.55 km; 2,213 miles).

Altitude (France)
Jean Boulet in an Aerospatiale SA 315B Lama on 21 June 1972, at Istres, France. 12,442 m (40,820 ft).

Speed on a straight course (UK)
Trevor Egginton and Derek Clews in Westland Lynx G-LYNX, on 11 August 1986, over a 15/25 km course, at Glastonbury, Somerset. 216.45 kt (400.87 km/h; 249.09 mph).

Speed in a 100 km closed circuit (USSR)
Boris Galitsky and crew of five in a Mil Mi-6, on 26 August 1964, near Moscow. 183.67 kt (340.15 km/h; 211.36 mph).

Speed in a 500 km closed circuit (USA)
Thomas F Doyle Jr in Sikorsky S-76A N5445J, at West Palm Beach, Florida, on 8 February 1982. 186.68 kt (345.74 km/h; 214.83 mph).

CLASS E.3 (Autogyros):

Altitude without payload (USA)
Andrew C Keech, in Little Wing LW-5 N100MK on 20 April 2004 at Frederick, Maryland. 8,049 m (17,785 ft).

Great circle distance without landing (USA)
Andrew C Keech, in Little Wing LW-5 N100MK from Little Rock, Arkansas, to Hickory, North Carolina, on 22 February 2004. 536.32 n miles (993.27 km; 617.20 miles).

Distance in a closed circuit (UK)
Wing Cdr K H Wallis, in Wallis WA-116/F/S G-BLIK, on 5 August 1988, at Waterbeach Airfield, Cambridge. 541.44 n miles (1,002.75 km; 623.08 miles).

Speed on a straight course (UK)
Wing Cdr K H Wallis, in a Wallis WA-121/Mc G-BAHH, over a 3 km course, at Shipdam, Norfolk, on 16 November 2002. 112.1 kt (207.7 km/h; 129.1 mph).

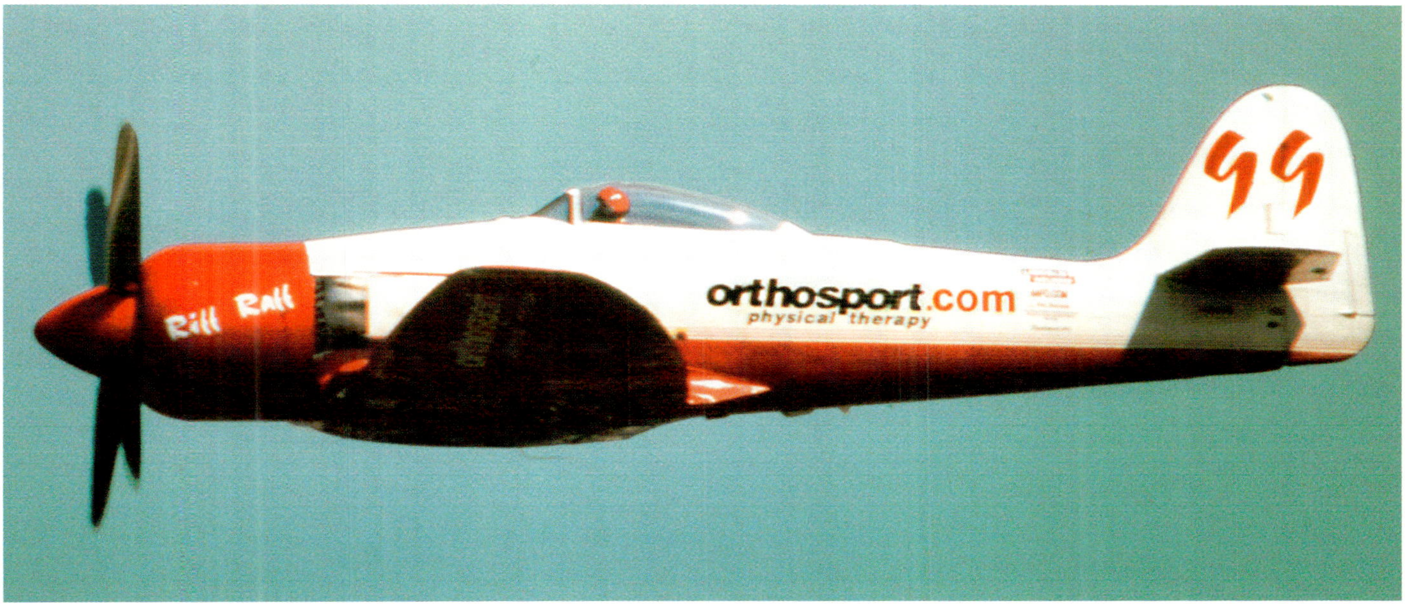

Hawker Sea Fury N62143 "Riff Raff", flown by Kike Keenum, set a Sub-class C-1e Group 1 piston-engined landplane record on 4 October 2004 at 285.4 kt (528.6 km/h; 328.5 mph). Refitted with a Wright R-3350 radial of 2,237 kW (3,000 hp), the aircraft broke a record set in 1950 *(Paul Jackson)* ***NEW/***1042460

ABC A-I-50 N155LG remained airborne for 24 hours 39 minutes 55 seconds on 13-14 September 2004, breaking six classes of endurance record for non-pressurised airships upwards of 3,000m3. Pilots were Carl Harbuck and Douglas McFadden *(ABC)* ***NEW/***1042459

International Aircraft Registration Prefixes

The inventiveness of those applying for special registrations knows no bounds. This particular thingumbob, whatsit or gizmo is an American Champion Super Decathlon on the UK civil aircraft register *(Paul Jackson)* ***NEW***1042443

Registration Requirements
An international system for the identification of civil aircraft was devised in 1919 and enshrined (with much else concerning aerial transport) in the Convention Relating to the Regulation of Air Navigation, which was signed in Paris on 13 October that year. This built upon embryo national systems, most notably that of the UK which had adopted the letter "G" as a national prefix in May 1919 following the issue, on 30 April, of what are believed to have been the world's first airworthiness and registration requirements.

Further conferences and conventions built upon this basis, the allocation of marks passing to the International Commission for Air Navigation (CINA). Today, Annex 7 of the International Civil Aviation Organisation regulations describes the application of aircraft registrations in broad terms. National prefixes have changed through the years, not only as the consequence of the birth of new countries, but for other reasons. Examples from the 1920s include CH- (for Switzerland); R- (Argentina); and UL- (Luxembourg). Lithuania, when recently returned to autonomy, was assigned LY- and not the RY- previously used.

Implementation of Annex 7 is by national governments, whose aviation authorities often have slightly different interpretations. Variations might include the wearing of Italic characters (not exceeding 30° slope) and dispensation for helicopters to carry smaller registrations. In addition, New Zealand allows serifed, variable thickness letters similar to Helvetica typeface and the omission of the national prefix for aircraft not intending to leave the islands.

The 'regulations' which follow must be taken, therefore, as only a guide.

(A) Permanent marking of aircraft nationality and registration shall (a) have no ornamentation; (b) contrast in colour with the background; and (c) be legible.

(B) Display of Marks: Each aircraft shall display marks consisting of the Roman capital letter denoting nationality, followed by the registration letters or number of the aircraft in Arabic numerals or Roman capital letters.

(C) Size of Marks: The character marks on vertical surfaces shall be at least 30 centimetres high (25 cm on helicopters in some countries, with dispensations down to 15 cm where space is unavailable); and 50 centimetres on lighter than air craft. When required to be carried, underside registrations (port wing of aeroplanes; beneath helicopter cabins) should be 50 cm in height.

Festering German Cessna 172 *(Paul Jackson)* ***NEW***1042440

Characters shall be formed by solid lines one-sixth as thick as the character is high. Characters must be two-thirds as wide as they are high, except the number "1", which must be one-sixth as wide as it is high, and the letters "M"and "W" which may be as wide as they are high. The space between each character may not be less than one quarter of the character width.

(D) Location of marks: These shall be on either the vertical tail surfaces or the sides of the fuselage between wing and empennage. Rotorcraft shall display marks horizontally on both surfaces of the cabin, fuselage, boom, or tail.

Airship marks are to be located lengthwise on each side of the hull and on its upper surface on the line of symmetry; or on the horizontal and vertical stabilisers. For the horizontal stabiliser, located on the right half of the upper surface and on the left half of the lower surface, with the tops of the letters and numbers toward the leading edge; and for the vertical stabiliser, located on each side of the bottom half stabiliser, with the letters and numbers placed horizontally.

The following two tables list (by alphanumeric order of prefix and alphabetical order of country, respectively) the markings which identify the national registrations of civil aircraft. The tables include some *de facto* prefixes which do not conform to ICAO rules. Mongolia, for example, has standardised on JU-, replacing both MT- and BNMAU-. Not all allocated markings are in current use, examples including Gibraltar and Vatican City. Brazil used only PP- and PT- from its allocation until PR- was brought into use in April 1999. In an unprecedented move in early 2002, Liberia responded to widespread abuse of its register by uncertified aircraft carrying illegal identities and revoked all EL- registrations. The new prefix A8- was introduced in 2003. At the same time, the first use was recorded in the Russian autonomous region of Karelia of the prefix KT-, while the Russian Federation began issuing privately owned aircraft with new registrations comprising four numbers and one letter, replacing five numbers.

Letter/number prefixes were introduced in 1948, while some countries which follow the national identifier with numbers omit the hyphen. In the case of some small members of the British Commonwealth, one or two letters after the hyphen are required to complete the national identity, leaving scope for as few as 26 aircraft on the national register.

Within national authorities, rules for allocation vary considerably. Not all letters of the alphabet need be employed. Owners may be permitted to request personal suffixes outside the strict alphabetical progression of the register, or certain classes of aircraft (by weight band or number of engines, for example) might be assigned specific groups of registrations. Civil aircraft flown under permit, sailplanes, ultralights, microlights and aircraft undergoing manufacturer's testing are in some countries allocated registration suffixes of an entirely different character (for example, numbers instead of letters).

It should be borne in mind that some suffix batches are reserved for military use. These call-signs may – notably in Africa – be applied externally to aircraft. Equally, some countries have a second, internal civil registration system, not mentioned here, for aircraft which do not normally cross international borders. Examples include UK sailplanes and French microlights.

A Stampe SV.4C (SNCAN-built) is no blushing debutante, having been designed in the 1930s, but this aircraft's status as a German (D-), light single (E) gets it almost half-way there *(Paul Jackson)* ***NEW***1042442

Flying under 'B Conditions' in the UK is this AugstaWestland EH101 G-17-510. The third element of the marking is at the manufacturer's discretion, in this instance indicating the Model number *(AgustaWestland)* ***NEW***1042441

Robinson R22 worth waiting for: G-ODOT *(Paul Jackson)* ***NEW***1042438

Registration prefixes: by prefix

AP-	Pakistan
A2-	Botswana
A3-	Tonga
A4O-	Oman
A5-	Bhutan
A6-	United Arab Emirates
A7-	Qatar
A8-	Liberia
A9C-	Bahrain
B-	China
B-H	Hong Kong
B-M	Macao
B-	Taiwan
C-, CF-	Canada
CC-	Chile
CN-	Morocco
CP-	Bolivia
CR-, CS-	Portugal
CU-	Cuba
CX-	Uruguay
C2-	Nauru
C3-	Andorra
C5-	Gambia
C6-	Bahamas
C9-	Mozambique
D-	Germany
DQ-	Fiji
D2-	Angola
D4-	Cape Verde
D6-	Comoros
EC-	Spain
EI-, EJ-	Ireland
EK-	Armenia
EL-	revoked 2002 (was Liberia)
EP-	Iran
ER-	Moldova
ES-	Estonia
ET-	Ethiopia
EW-	Belarus
EX-	Kyrgyzstan
EY-	Tajikistan
EZ-	Turkmenistan
E3-	Eritrea
F-	France and colonies
G-	United Kingdom
HA-	Hungary
HB-	Switzerland & Liechtenstein
HC-	Ecuador
HE-	Greece (ultralights)
HH-	Haiti
HI-	Dominican Republic
HK-	Colombia
HL-	Korea, South
HP-	Panama
HR-	Honduras
HS-	Thailand
HV-	Vatican City
HZ-	Saudi Arabia
H4-	Solomon Islands
I-	Italy
JA-	Japan
JU-	Mongolia
JY-	Jordan
J2-	Djibouti
J3-	Grenada
J5-	Guinea-Bissau
J6-	St Lucia
J7-	Dominica
J8-	St Vincent & Grenadines
LN-	Norway
LQ-, LV-	Argentina
LX-	Luxembourg
LY-	Lithuania
LZ-	Bulgaria
N	USA
OB-	Peru
OD-	Lebanon
OE-	Austria
OH-	Finland
OK-	Czech Republic
OM-	Slovak Republic
OO-, OQ-	Belgium
OY-	Denmark
P-	Korea, North
PH-	Netherlands
PJ-	Netherlands Antilles
PK-	Indonesia
PP- to PU-	Brazil
PZ-	Suriname
P2-	Papua New Guinea
P4-	Aruba
RA-, RF-	Russia
RDPL-	Laos
RP-	Philippines
SE-	Sweden
SP-	Poland
ST-	Sudan
SU-	Egypt
SU-YA	Palestine
SX-	Greece
S2-	Bangladesh
S5-	Slovenia
S7-	Seychelles
S9-	São Tomé e Principé
TC-	Turkey
TF-	Iceland
TG-	Guatemala
TI-	Costa Rica
TJ-	Cameroon
TL-	Central African Republic
TN-	Congo (Republic)
TR-	Gabon
TS-	Tunisia
TT-	Chad
TU-	Côte d'Ivoire
TY-	Benin
TZ-	Mali
T2-	Tuvalu
T3-	Kiribati
T7-	San Marino
T8A-	Palau
T9-	Bosnia-Herzegovina
UK-	Uzbekistan
UN-	Kazakhstan
UR-	Ukraine
VH-	Australia
VN-	Vietnam
VP-A	Anguila
VP-B	Bermuda
VP-C	Cayman Islands
VP-F	Falkland Islands
VP-G	Gibraltar
VP-LM	Montserrat
VP-LV	British Virgin Islands
VQ-H	St Helena & Ascension
VQ-T	Turks & Caicos Islands
VT-	India
V2-	Antigua & Barbuda
V3-	Belize
V4-	St Kitts & Nevis
V5-	Namibia
V6-	Micronesia
V7-	Marshall Islands
V8-	Brunei
XA- to XC-	Mexico
XT-	Burkina Faso
XU-	Cambodia
XY-, XZ-	Myanmar
YA-	Afghanistan
YI-	Iraq
YJ-	Vanuatu
YK-	Syria
YL-	Latvia
YN-	Nicaragua
YR-	Romania
YS-	El Salvador
YU-	Yugoslavia
YV-	Venezuela
Z-	Zimbabwe
ZA-	Albania
ZK- to ZM-	New Zealand
ZP-	Paraguay
ZS- to ZU-	South Africa
Z3-	Macedonia
3A-	Monaco
3B-	Mauritius
3C-	Equatorial Guinea
3D-	Swaziland
3X-	Guinea
4K-	Azerbaijan
4L-	Georgia
4R-	Sri Lanka
4X-	Israel
5A-	Libya
5B-	Cyprus
5H-	Tanzania
5N-	Nigeria
5R-	Madagascar
5T-	Mauritania
5U-	Niger
5V-	Togo
5W-	Samoa
5X-	Uganda
5Y-	Kenya
6O-	Somalia
6V-, 6W-	Senegal
6Y-	Jamaica
7O-	Yemen
7P-	Lesotho
7Q-	Malawi
7T-	Algeria
	Barbados
8P-	
8Q-	Maldives
8R-	Guyana
9A-	Croatia
9G-	Ghana
9H-	Malta
9J-	Zambia
9K-	Kuwait
9L-	Sierra Leone
9M-	Malaysia
9N-	Nepal
9Q-	Congo (Democratic Republic)
9U-	Burundi
9V-	Singapore
9XR-	Rwanda
9Y-	Trinidad & Tobago

United Kingdom 'B Conditions' Markings

Many countries' registration systems include provision for aircraft undergoing special testing, or being prepared for export, to have temporary markings in a reserved area of the register. In the UK, the system takes the form of the national identifying letter (G-) followed by a number to indicate the authorised manufacturer or modification centre, followed by an individual number, usually beginning at one (for example G-85-25). The last-mentioned numbers were frequently re-used, but reallocation is now rare.

In recent years, aircraft capable of carrying armament, which would have used B Conditions, have been required to carry a military serial number issued by the Ministry of Defence in the normal sequence of allocations.

Markings currently authorised (although not necessarily used on a regular basis) are detailed in the following table. These are not 'registrations' in the strict sense of the term, as there is no central record of allocations.

G-03	BAE Systems
G-04	BAE Systems
G-5	Raytheon Services Ltd (ex-de Havilland)
G-08	BAE Systems
G-9	BAE Systems (ex-Hawker)
G-11	BAE Systems (ex-Avro)
G-14	Short Brothers
G-16	BAE Systems (ex-Vickers)
G-17	AgustaWestland
G-31	BAE Systems (ex-Scottish Aviation)
G-36	Cranfield Aerospace
G-51	B-N Group
G-52	Marshall Aerospace
G-54	Cameron Balloons
G-61	Aviation Enterprises
G-63	Thunder & Colt
G-65	Solar Wings
G-67	Atlantic Aeroengineering
G-69	Cyclone Airsports
G-70	FLS Aerospace
G-71	FR Aviation
G-72	Lindstrand Balloons
G-75	Chichester-Miles Consultants
G-76	Police Aviation Services
G-77	Thruster Air Services
G-78	Bristow Helicopters
G-79	McAlpine Helicopters
G-80	British Microlight Aircraft Association
G-81	Cooper Aerial Services
G-82	European Helicopters
G-83	Mann Aviation Group
G-84	Intora-Firebird
G-85	CFM Aircraft
G-86	Advanced Technologies Group
G-87	CHC Scotia
G-88	Air Hanson

Registration prefixes: by country

Afghanistan	YA-
Albania	ZA-
Algeria	7T-
Andorra	C3-
Angola	D2-
Anguilla	VP-LA
Antigua & Barbuda	V2-
Argentina	LQ-, LV-
Armenia	EK-
Aruba	P4-
Australia	VH-
Austria	OE-
Azerbaijan	4K-
Bahamas	C6-
Bahrain	A9C-
Bangladesh	S2-
Barbados	8P-
Belarus	EW-
Belgium	OO-, OQ-
Belize	V3-
Benin	TY-
Bermuda	VP-B
Bhutan	A5-
Bolivia	CP-
Bosnia-Herzegovina	T9-
Botswana	A2-
Brazil	PP- to PU-
Brunei	V8-
Bulgaria	LZ-
Burkina Faso	XT-
Burundi	9U-
Cambodia	XU-
Cameroon	TJ-
Canada	C-, CF-
Cape Verde	D4-
Cayman Islands	VP-C
Central African Republic	TL-
Chad	TT-
Chile	CC-
China	B-
(Hong Kong	B-H)
(Macao	B-M)
Colombia	HK-
Comoros	D6-
Congo (Democratic Republic)	9Q-
Congo (Republic)	TN-
Costa Rica	TI-
Côte d'Ivoire	TU-
Croatia	9A-
Cuba	CU-
Cyprus	5B-
Czech Republic	OK-
Denmark	OY-
Djibouti	J2-
Dominica	J7-
Dominican Republic	HI-
Ecuador	HC-
Egypt	SU-
Equatorial Guinea	3C-
Eritrea	E3-
Estonia	ES-
Ethiopia	ET-
Falkland Islands	VP-F
Fiji	DQ-
Finland	OH-
France and colonies	F-
Gabon	TR-
Gambia	C5-
Georgia	4L-
Germany	D-
Ghana	9G-
Gibraltar	VP-G
Greece	SX-
(ultralights)	HE-
Grenada	J3-
Guatemala	TG-
Guinea	3X-
Guinea-Bissau	J5-
Guyana	8R-
Haiti	HH-

The sole remaining flyable de Havilland Sea Vixen naval fighter is now civilian-owned and carries the UK prefix G-, followed by the specially chosen individual letters CVIX *(Paul Jackson)* 0569557

Although flex-wings do not qualify for inclusion in the Aircraft section, this Italian-designed Ramphos Amphibious Trike illustrates the United States 'ultralight vehicle' registration sequences. As 10GJM, it is identified as coming within the series managed by the US Ultralight Association. Parallel bodies are the Aero Sports Connection (A, two numbers, three letters) and Experimental Aircraft Association (E, two numbers, three letters) *(Paul Jackson)* 0569558

Honduras	HR-
Hungary	HA-
Iceland	TF-
India	VT-
Indonesia	PK-
Iran	EP-
Iraq	YI-
Ireland	EI-, EJ-
Israel	4X-
Italy	I-
Jamaica	6Y-
Japan	JA
Jordan	JY-
Kazakhstan	UN-
Kenya	5Y-
Kiribati	T3-
Korea, North	P-
Korea, South	HL-
Kuwait	9K-
Kyrgyzstan	EX-
Laos	RDPL-
Latvia	YL-
Lebanon	OD-
Lesotho	7P-
Liberia	A8-
Libya	5A-
Liechtenstein	HB-
Lithuania	LY-
Luxembourg	LX-
Macedonia	Z3-
Madagascar	5R-
Malawi	7Q-
Malaysia	9M-
Maldives	8Q-
Mali	TZ-
Malta	9H-
Marshall Islands	V7-
Mauritania	5T-
Mauritius	3B-
Mexico	XA- to XC-
Micronesia	V6-
Moldova	ER-
Monaco	3A-
Mongolia	JU-
Montserrat	VP-LM
Morocco	CN-
Mozambique	C9-
Myanmar	XY-, XZ-
Namibia	V5-
Nauru	C2-
Nepal	9N-

Angry Italian VulcanAir P.68 *(Paul Jackson)*
NEW1042439

Netherlands	PH-
Netherlands Antilles	PJ-
New Zealand	ZK- to ZM-
Nicaragua	YN-
Niger	5U-
Nigeria	5N
Norway	LN-
Oman	A4O-
Pakistan	AP-
Palau	T8A-
Palestine	SU-YA
Panama	HP-
Papua New Guinea	P2-
Paraguay	ZP-
Peru	OB-
Philippines	RP-
Poland	SP-
Portugal	CR-, CS-
Qatar	A7-
Romania	YR-
Russian Federation	RA-, RF-
Rwanda	9XR-
El Salvador	YS-
Samoa	5W-
San Marino	T7-
São Tomé e Principé	S9-
Saudi Arabia	HZ-
Senegal	6V-, 6W-
Seychelles	S7-
Sierra Leone	9L-
Singapore	9V-
Slovak Republic	OM-
Slovenia	S5-
Solomon Islands	H4-
Somalia	6O-
South Africa	ZS- to ZU-
Spain	EC-
Sri Lanka	4R-
St Helena & Ascension	VQ-H
St Kitts & Nevis	V4-
St Lucia	J6-
St Vincent & Grenadines	J8-
Sudan	ST-
Suriname	PZ-
Swaziland	3D-
Sweden	SE-
Switzerland	HB-
Syria	YK-
Taiwan	B-
Tajikistan	EY-
Tanzania	5H-
Thailand	HS-
Togo	5V-
Tonga	A3-
Trinidad & Tobago	9Y-
Tunisia	TS-
Turkey	TC-
Turkmenistan	EZ-
Turks & Caicos	VQ-T
Tuvalu	T2-
Uganda	5X-
Ukraine	UR-
United Arab Emirates	A6-
United Kingdom	G-
United States	N
Uruguay	CX-
Uzbekistan	UK-
Vanuatu	YJ-
Vatican City	HV-
Venezuela	YV-
Vietnam	VN-
Virgin Islands	VP-L
Yemen	7O-
Yugoslavia	YU-
Zambia	9J-
Zimbabwe	Z-

US civilian registration previously carried a second letter following the national N to indicate the aircraft's ownership status. Dispensations are granted for historic aircraft to retain the marking, such as this painstakingly restored Boeing Stratoliner, now retired to the National Air & Space Museum in Washington, DC *(Paul Jackson)* 0569559

Glossary of Aerospace Terms in Jane's All the World's Aircraft

Ejection seat. Requiring no further elaboration, an assisted departure from a USAF Thunderbirds demonstration team Lockheed Martin F-16 Fighting Falcon dramatically illustrates the value of this invention (US Air Force) *NEW*/1042465

AAM Air-to-air missile.
AAR Air-to-air refuelling.
AB Aktiebolag (Swedish company constitution).
absolute ceiling Greatest altitude attainable by aircraft in level flight.
AC Alternating current.
ACC Air Combat Command (US).
ACLS Automatic carrier landing system.
ACMI Air combat manoeuvring instrumentation.
ACN Aircraft classification number (ICAO system for aircraft pavements).
ADC Air data computer.
ADF Medium-frequency automatic direction-finding (equipment).
ADI Attitude/director indicator.
aerofoil Any solid body so shaped that, as a fluid medium (air or hot gas) moves past it, it experiences a useful force perpendicular to the direction of relative motion; thus, a wing generates lift, while a turbine blade generates torque on a shaft.
aeroplane (N America, airplane) Heavier-than-air aircraft with propulsion and a wing that does not rotate in order to generate lift.
AEW Airborne early warning.
AFB Air Force Base (USAF).
AFCS Automatic flight control system.
AFRC Air Force Reserve Command (US).
AFRP Aramid fibre-reinforced plastics.
afterburning Temporarily augmenting the thrust of a turbofan or turbojet by burning additional fuel in the jetpipe.
AGM Air-to-ground missile.
Ah Ampère-hours.
AHRS Attitude/heading reference system.
airbrake Passive device extended from aircraft to increase drag. Most common form is hinged flap(s) or plate(s), mounted in locations where operation causes no significant deterioration in stability and control at any attainable airspeed.
aircraft All manmade vehicles for off-surface navigation within the atmosphere, including helicopters and balloons. For practical purposes, air-cushion vehicles and wing-in-ground-effect vehicles are excluded from the classification.
airship Power-driven lighter-than-air aircraft. Traditional classes are: blimp, a small non-rigid; non-rigid, in which envelope is essentially devoid of rigid members and maintains shape by inflation pressure; semi-rigid, non-rigid with strong axial keel acting as beam to support load; and rigid, in which envelope is itself stiff in local bending or supported within or around rigid framework.
airstair Retractable stairway built into aircraft.
ALARM Air-launched anti-radiation missile.
ALCM Air-launched cruise missile.
Allithium Aluminium-lithium alloy.
ALLTV All-light level television.
AM Amplitude modulation.
AMC Air Mobility Command (US).
AMRAAM Advanced Medium-Range AAM.
ANG Air National Guard (US).
anhedral Downward slope of wing or tailplane from root to tip.
anti-balance tab Hinged surface on trailing-edge of stabilator and operating in same direction, so as to dampen its movement.
ANVIS Aviator's night vision system.
AO Aktsionernoye Obshchestvo (Co Ltd; Russian company constitution).
AoA Angle of attack (see 'attack' below).
AOOT Aktsionernoye Obshchestvo Oktrytogo Tipa (Russian company constitution).
approach noise Measured 1 n mile from downwind end of runway with aircraft passing overhead at 113 m (370 ft).
APR Auxiliary power reserve.
APU Auxiliary power unit (part of aircraft).
ARINC Aeronautical Radio Inc, US company whose electronic box sizes (racking sizes) are the international standard.
ARM Anti-radiation missile.
ArNG Army National Guard (US).
ASE (1) Automatic stabilisation equipment; (2) Aircraft survivability equipment.
ASI Airspeed indicator.
ASM Air-to-surface missile.
aspect ratio Measure of wing (or other aerofoil) slenderness seen in plan view, usually defined as the square of the span divided by gross area.
AST Air Staff Target (UK).
ASTOVL Advanced STOVL.
ASUW Anti-surface unit warfare.
ASV Anti-surface vessel.
ASW Anti-submarine warfare.
ATC Air traffic control.

ATR Airline Transport Radio ARINC 404 black box racking standards.

attack, angle of (alpha) Angle at which airstream meets aerofoil (angle between mean chord and free-stream direction). Not to be confused with angle of incidence (which see).
augmented Boosted by afterburning (turbofan).

NEW/1042464

autogyro Rotary-wing aircraft propelled by a propeller (or other thrusting device) and lifted by a freely running autorotating rotor. Compare gyroplane.
AUW All-up weight (term meaning total weight of aircraft under defined conditions, or at a specific time during flight). Not to be confused with MTOW (which see).
avionics Aviation electronics.
AWACS Airborne warning and control system (aircraft category).
axisymmetric intakes Twin, circular engine air intakes mounted astride the spinner of New Piper light aircraft.
axisymmetric nozzle Circular jet-engine nozzle capable of unrestricted vectoring movement (within a cone specified by mechanical limitation) to enhance aircraft manoeuvrability.

ballistic parachute Emergency recovery parachute installed in (generally light) aircraft and capable of supporting both machine and occupants.
band See radar frequency.
bar Non-SI unit of pressure adopted by this yearbook pending wider acceptance of Pa. 1 bar = 10^5 Pa. ISA pressure at S/L is 1013.2 mb or just over 1 bar. ICAO has standardised hectopascal for atmospheric pressure, in which ISA S/L pressure is 101.32 hPa.
basic operating weight MTOW minus payload (thus, including crew, fuel and oil, bar stocks, cutlery and so on).
bearingless rotor Rotor in which flapping, lead/lag and pitch change movements are provided by the flexibility of the structural material and not by bearings. No rotor is truly rigid.
BITE Built-in test equipment.
bladder tank Fuel (or other fluid) tank of flexible material.
BLC Boundary-layer control.
bleed air Hot high-pressure air extracted from gas turbine engine compressor or combustor and taken through valves and pipes to perform useful work such as pressurisation, driving machinery or anti-icing by heating surfaces.
blown flap Flap across which bleed air is discharged at high (often supersonic) speed to prevent flow breakaway.
BOW Basic operating weight (which see).
BPR Bypass ratio.
BTU Non-SI energy unit (British Thermal Unit) = 0.9478 J.
bulk cargo All cargo not packed in containers or on pallets.
bus Busbar, main terminal in electrical system to which battery or generator power is supplied.
BV Besloten Vennootschap (Netherlands company constitution).
BVR Beyond visual range.
BWB Blended wing/body.
bypass ratio Air flow through fan duct (not passing through core) divided by airflow through core.

byte Group of bits of information forming unit in computer processing.

C³ Command, control and communications.
CAA Civil Aviation Authority (UK).

cabane Structure, usually of braced struts, to support load above fuselage or wing. May carry parasol wing, engine nacelle or upper wing of most biplanes.
cabin altitude Height above S/L at which ambient pressure is same as inside cabin.
CAD/CAM Computer-assisted design/computer-assisted manufacture.

0569565

CAHI Central Aero and Hydrodynamics Institute of the Russian Federation; also transliterated as TsAGI.
canards Foreplanes, fixed or controllable aerodynamic surfaces ahead of CG.
capacity The volume swept out on each stroke by the pistons of a piston engine. It is expressed in cc (cubic centimetres) for small engines and in litres (1 litre = 1,000 cc) for larger ones. Also known as displacement or swept volume.
carbon fibre Fine filament of carbon/graphite used as strength element in composites.
CAS (1) Calibrated airspeed, ASI calibrated to allow for air compressibility according to ISA S/L; (2) close air support.
casevac Casualty evacuation.
Cat Category. Meanings include runway visibility and decision height minima for ILS. See diagram.
CATIA Computer-aided three-dimensional interactive analysis; Anglicised form of French CAD proprietary system *(Conception assistée tridimensionelle interactive d'applications)*.
CBU Cluster bomb unit.
CEAM Centre d'Expériences Aériennes Militaires.
CEAT Centre d'Essais Aéronautiques de Toulouse.
Ceconite Manmade covering material for light aircraft; trade name.
CEO Chief executive officer.
CEV Centre d'Essais en Vol.
CFE Conventional Forces in Europe.
CFRP Carbon fibre-reinforced plastics.
CG Centre of gravity.
chaff Thin slivers of radar-reflective material cut to length appropriate to wavelengths of hostile radars and scattered in bundles to protect friendly aircraft.
chord Distance from leading-edge to trailing-edge measured parallel to longitudinal axis.
CIS Commonwealth of Independent [ex-USSR] States. See also RFAS.
CKD Component knocked down, for assembly elsewhere.
clean (1) In flight configuration with landing gear, flaps, slats and so on retracted;
(2) Without any optional external stores.
c/n Constructor's number; manufacturer's serial number.
C of A Certificate of Airworthiness; awarded to each individual aircraft (compare Type Certificate).
COIN Counter-insurgency.
collective pitch Controls pitch of all blades of helicopter main rotor in unison.
combi Civil aircraft carrying both freight and passengers on main deck.
comint Communications intelligence.
composite Material made of two constituents, such as filaments or short whiskers plus adhesive forming binding matrix.
constant-speed Variable-pitch propeller governed by a CSU so that its rotational speed is held constant.
contrarotating Propellers on same axis turning in opposite directions (compare C/R).
conventional and manual Aeroplane manoeuvring surfaces mechanically linked to pilot's hand and foot controls, unassisted (except, optionally, by aerodynamic or mass balances) and comprising ailerons on the outboard wing, rudder(s) to the rear of fixed tailfin(s) and elevators to the rear of a fixed (but optionally, incidence angle trimmable) tailplane. The description optionally includes leading-edge slats, flaps on inboard trailing-edges and trim tabs, all of which are mentioned separately, if installed. Ailerons which droop in unison with flaps (and thus are not the primary means of lowering stalling speed) are regarded as conventional. Control systems not conforming to the above – in that they have foreplanes, one or more all-moving tail surfaces, or flaperons, and those with mechanical/electronic assistance or interception of the pilot's movements – are described in appropriate detail.

convertible Transport aircraft able to be equipped to carry passengers or cargo, but not both simultaneously.
COO Chief operating officer.
core Gas generator portion of turbofan comprising compressor(s), combustion chamber and turbine(s).
C/R Counter-rotating; propellers of multi-engined aircraft turning in opposite directions on different axes (compare contrarotating).
CRT Cathode-ray tube.
cruising speed Flight speed on less than full engine power; maximum is normally at 75%, if not otherwise specified, but some manufacturers use higher throttle settings.
CSAS Command and stability augmentation system (part of AFCS).
CTOL Conventional take-off and landing (compare V/STOL).
CVR Cockpit voice recorder.
CY Calendar year; 1 January to 31 December. Compare FY.
cyclic pitch Controls variation of pitch as helicopter rotor blade makes each revolution.

Dacron Artificial fabric for light aircraft covering; trade name.
DADC Digital air data computer.
DADS Digital air data system.
DARPA Defense Advanced Research Projects Agency (US) (briefly ARPA before February 1996).
databus Electronic highway for passing digital data between aircraft sensors and system processors, usually MIL-STD-1553B or ARINC 419 (one-way) and 619 (two-way) systems.
dB Decibel.
DC Direct current.
DECU Digital engine (or electronic) control unit.
dem/val Demonstration/validation.
derated Engine restricted to power less than potential maximum (usually such engine is flat rated, which see).
design weight Different authorities have different definitions; weight chosen as typical of mission but usually much less than MTOW.
DF Direction-finder, or direction-finding.
DGAC Direction Générale à l'Aviation Civile. French certification authority.

0569566

dihedral Upward slope of wing from root (or intermediate point) to tip.
disposable load Sum of masses that can be loaded or unloaded, including payload, crew, removable equipment, usable fuel and other consumables; MTOW minus OWE.
DME UHF distance-measuring equipment; gives slant distance to a beacon; DME element of Tacan.
DoD Department of Defense.
dog-tooth A sharp discontinuity in the leading-edge of a wing or tail surface resulting from an increase in chord (see also sawtooth).
Doppler Short for Doppler radar – radar using fact that received frequency is a function of relative velocity between transmitter or reflecting surface and receiver; used for measuring speed over ground or for detecting aircraft or moving vehicles against static ground or sea.
double-slotted flap One having an auxiliary aerofoil ahead of main surface to increase maximum lift.

EAA Experimental Aircraft Association (divided into local branches called Chapters).
EAS Equivalent airspeed, RAS minus correction for compressibility.
ECCM Electronic counter-countermeasures.
ECM Electronic countermeasures.
ECS Environmental control system.
EEZ Economic exclusion (or exclusive-economic) zone.
EFIS Electronic flight instrument(ation) system, in which large multifunction CRT displays replace traditional instruments.
EGPWS Enhanced ground proximity warning system
EGT Exhaust gas temperature.
ehp Equivalent horsepower, measure of propulsive power of turboprop made up of shp plus addition due to residual thrust from jet.

EICAS Engine indication and crew alerting system.
ekW Equivalent kilowatts, SI measure of propulsive power of turboprop (see ehp).
elevon Wing trailing-edge control surface combining functions of aileron and elevator.
elint electronics intelligence.
ELT Emergency locator transmitter, to help rescuers home on to a disabled or crashed aircraft.
EMD Engineering and manufacturing development.
EO Electro-optical.
EPNdB Effective perceived noise decibel, SI unit of EPNL.
ERU Ejector release unit
ESM (1) Electronic surveillance (or support) measures; (2) Electronic signal monitoring.
ETOPS Extended-range twin (engine) operations (thus sometimes given as EROPS), routeing not more than a given flight time (120, 180 or 240 minutes) from a usable alternative airfield.
EW Electronic warfare.

FAA Federal Aviation Administration.
FAC Forward air control (or controller).
factored Multiplied by an agreed number to take account of extreme adverse conditions, errors, design deficiencies or other inaccuracies.
FADEC Full-authority digital engine (or electronic) control.
FAI Fédération Aéronautique Internationale.
fail-operational System which continues to function after any single fault has occurred.
fail-safe Structure or system which survives failure (in case of system, may no longer function normally).
FAR Federal Aviation Regulations.
FAR Pt 23 Defines the airworthiness of private and air taxi aeroplanes of 5,670 kg (12,500 lb) MTOW and below.
FAR Pt 25 Defines the airworthiness of public transport aeroplanes exceeding 5,670 kg (12,500 lb) MTOW.
FBL Fly-by-light (which see).
FBW Fly-by-wire (which see).
FCS Flight control system.
FDR Flight data recorder (which see).
FDS Flight director system.
feathering Setting propeller blades at pitch aligned with slipstream to minimise drag.
fence A chordwise projection on the surface of a wing, used to modify the distribution of pressure.
Fenestron Helicopter tail rotor with many slender blades rotating in short duct (registered name).
ferry range Extreme safe range with zero payload.
FFAR Folding-fin (or free-flight) aircraft rocket.
field length Measure of distance needed to land and/or take off; many different measures for particular purposes, each precisely defined
fixed-pitch Propeller with blades fixed to the hub.
FL Flight level. Notional altitude for air traffic control purposes which assumes ISA pressure (1013.25 mb; 29.92 in Hg) at S/L. Expressed in hundreds of feet; thus FL255 indicates approximately 25,500 ft
flap A surface carried on the leading- or trailing-edge of a wing and able to move relative to it. The simplest leading-edge flap and so-called plain (trailing-edge) flap is formed by hinging the entire edge of the wing. The Krueger is a leading-edge flap forming part of the wing undersurface, swung down and forwards on arms to give a bluff leading-edge. A split flap is formed by hinging only the undersurface of the trailing-edge. A slotted flap is a hinged trailing-edge which moves aft as well as down on tracks to leave a narrow slot ahead of it; hence double- and triple-slotted. A Fowler flap is a complete auxiliary aerofoil mounted on tracks under a fixed trailing-edge; initially it moves aft, to emerge behind the fixed part of the wing, and at the end of its travel it rotates down. A Gouge flap has an upper surface forming part of a cylinder, and rotates (on rails or brackets) about that cylinder's centre.
flaperon Wing trailing-edge surface combining functions of flap and aileron.
FLAR Federatsii Lyubitelei Aviatsii Rossii, Russian PFA.
flat-four Piston engine having four horizontally opposed cylinders; thus, flat-twin, flat-six and so on.
flight-adjustable pitch Propeller with blades that can be changed in pitch during flight to a limited extent (eg one way only). Compare variable pitch.
flat rated Propulsion engine capable of giving full thrust or power for take-off at an airfield well above S/L and/or at high ambient temperature (thus, probably derated at S/L).
flight data recorder Crash-protected recorder of dynamic/static pressure, air temperature, control-surface and slat/flap positions, 3-axis accelerations, engine parameters and possibly other variables.
FLIR Forward-looking infra-red.
fly-by-light Flight control system in which signals pass between computers and actuators along fibre-optic leads.
fly-by-wire Flight control system with electrical signalling, without mechanical interconnection between cockpit flying controls and control surfaces.
FM Frequency modulation.

FMS (1) Foreign military sales (US DoD);
(2) Flight management system.
footprint (1) A precisely delineated boundary on the surface around an airfield, inside which the perceived noise of an aircraft exceeds a specified level during take-off and/or landing;
(2) Dispersion of weapon or submunition impact points.

0569567

foreplanes Pivoted canard surfaces forming part of the primary flight control system with authority in pitch and possibly also in roll. See also canards.
FOV Field of view.
Fowler flap See flap.
frequency See radar frequency.
frequency agile (frequency hopping) Making a transmission harder to detect by switching automatically to a succession of frequencies.
FSD Full-scale development.
FSED Full-scale engineering development.
FY Fiscal year; in US government affairs, runs from 1 October to 30 September (FY05 begins 1 October 2004); in Japan, from 1 April (FY16 or FY04 began 1 April 2004).

g Acceleration due to mean Earth gravity, that is of a body in free-fall; or acceleration due to rapid change of direction of flight path.
gallons Non-SI measure; 1 Imp gallon (UK) = 4.546 litres, 1 US gallon = 3.785 litres.
GFRP Glass fibre-reinforced plastics.
'glass cockpit' Cockpit in which dial instruments are replaced by multifunction electronic displays.
glass fibre Spun molten glass; see GFRP.
glideslope Element giving vertical (height) guidance in ILS.
glove (1) Fixed portion of wing inboard of variable sweep wing;
(2) additional aerofoil profile added around normal wing for test purposes.
GmbH Gesellschaft mit beschränkter Haftpflicht (or Haftung) (German company constitution).
GPS Global Positioning System, US military/civil satellite-based precision navaid.
GPSS GPS Steering: integration of GPS with aircraft's autopilot to provide active navigation and smoother course changes.
GPU Ground power unit (not part of aircraft).
GPWS Ground-proximity warning system.
green aircraft Aircraft flyable but unpainted, unfurnished and basically equipped.
gross wing area See wing area.
ground-adjustable pitch Propeller with blades that can be adjusted in pitch by an engineer on the ground. Compare flight adjustable and variable pitch.
GS See glideslope.
gunship Aircraft designed for battlefield attack; helicopter gunships normally with slim body carrying pilot and weapon operator only.
GUP Gosudarstvennoye Unitarnoye Predpriyatie (Russian State Unitary Enterprise).
gyroplane Alternative term for autogyro favoured by some certification authorities, including FAA.

h Hour(s).
handed Rotating in opposite directions.
hardened Protected as far as possible against nuclear explosion.
hardpoint Reinforced part of aircraft to which external load can be attached, for example weapon or tank pylon.
HDU Hose-drum unit.
head-down display On the cockpit instrument panel (as distinct from a HUD).
head-level display Immediately below HUD.
helicopter Rotary-wing aircraft both lifted and propelled by one or more power-driven rotors turning about substantially vertical axes.
HF High frequency.
HIFR Helicopter in-flight refuelling.
HIRF High-intensity radiated field(s).
HMD Helmet-mounted display; hence HMS = sight.
HOCAC Hands on cyclic and collective.
homebuilt Aircraft built/assembled from plans or kits.
hot-and-high Adverse combination of airfield height and high ambient temperature, which lengthens required take-off distance.

0543354

HOTAS Hands on throttle and stick.
hot refuelling Replenishment of fuel while engine(s) running.
hovering ceiling Ceiling of helicopter (corresponding to air density at which maximum rate of climb is zero), either IGE or OGE.
HP High pressure (HPC, compressor; HPT, turbine).
hp Horsepower, non-SI unit of power.
HSI Horizontal situation indicator.
HUD Head-up display (bright numbers and symbols projected on pilot's aiming sight glass and focused on infinity so that pilot can simultaneously read display and look ahead). The term is increasingly rendered in the USA as ''heads up'', which is incorrect.
Hz Hertz, cycles per second.

IAS Indicated airspeed, airspeed indicator reading corrected for instrument error.
IATA International Air Transport Association.
ICAO International Civil Aviation Organisation.
IFF Identification friend or foe.
IFR (1) Instrument flight rules (compare VFR);
(2) in-flight refuelling.
IGE In ground effect: helicopter performance with theoretical flat horizontal surface just below it (for example mountain).
IIR Imaging infra-red.
ILS Instrument landing system. See Cat.
Imperial gallon 1.20095 US gallons; 4.546 litres.
INAS Integrated nav/attack system.
Inc Incorporated (company constitution).
incidence The angle at which the wing is set in relation to the fore/aft axis. Often wrongly used to mean angle of attack (which see).
inertial navigation Measuring all accelerations imparted to a vehicle and, by integrating these with respect to time, calculating speed at every instant (in all three planes) and, by integrating a second time, calculating total change of position in relation to starting point.
INS Inertial navigation system.
integral construction Machined from solid instead of assembled from separate parts.
integral tank Fuel (or other liquid) tank formed by sealing part of structure.
intercom Wired telephone system for communication within aircraft.
inverter Electric or electronic device for inverting (reversing polarity of) alternate waves in AC power to produce DC.
IOC Initial operational capability.
IR Infra-red.
IRCM Infra-red countermeasures.
IRLS Infra-red linescan (builds TV-type picture showing cool and hot regions as contrasting shades).
IRS Inertial reference system.
IRST Infra-red search and track.
ISA International Standard Atmosphere (1013.25 mb, 1,225 g/m³ and 15°C at mean sea level; lapse rate 1.98°C per 1,000 ft up to –56.5°C at 36,090 ft).

JAA Joint Aviation Authorities.
JAR Joint Aviation Requirements, agreed by all major EC countries (JAR 25 equivalent to FAR Pt 25).
JAR-VLA JAR classification for Very Light Aircraft (MTOW limit of 750 kg; 1,653 lb).
JASDF Japan Air Self-Defence Force.
JATO Jet-assisted take-off (actually means rocket-assisted).
JCAB Japan Civil Airworthiness Board.
JDA Japan Defence Agency.
JGSDF Japan Ground Self-Defence Force.
JMSA Japan Maritime Safety Agency.
JMSDF Japan Maritime Self-Defence Force.
joined wing Tandem wing layout in which forward and aft wings are swept so that the outer sections meet.

JPATS Joint Primary Aircraft Training System (Beech T-6A Texan II).
JSC Joint stock company.
JSF Joint Strike Fighter.
J-STARS US Air Force/Navy Joint Surveillance and Target Attack Radar System in Northrop Grumman E-8C.
JTIDS Joint Tactical Information Distribution System (NATO Link 16).

NEW/0569568

Junkers aileron Control surface (sometimes flaperon) suspended from mountings to the rear of wing trailing edge.

kbit One thousand bits of memory.
Kevlar Aramid fibre used as basis of high-strength composites material.
kg Kilogramme (2.20462 lb).
kitbuilt Prefabricated aircraft for amateur assembly.
KK Kabushiki Kaisha (Japanese company constitution).
km/h Kilometres per hour.
kN KiloNewtons (N×10^3). See N.
knot 1 n mile per hour (1.852 km/h; 1.15078 mph).
Krueger flap Hinges down and then forward from below the leading-edge.

kVA Kilovolt-ampères.
kW Kilowatt, SI measure of all forms of power (not just electrical).

LAMPS Light airborne multipurpose system.
LANTIRN Low-altitude navigation and targeting infra-red, night.
LAPES Low-altitude parachute extraction system.
LBA Luftfahrtbundesamt (German civil aviation authority).
lb Pound, non-SI unit of weight: 0.453592 kg.
lb st Pounds of static thrust.
LCD Liquid crystal display, used for showing instrument information.
LCN Load classification number, measure of 'flotation' of aircraft landing gear linking aircraft weight, weight distribution, tyre numbers, pressures and disposition.
LED Light-emitting diode.
lift dumper Spoiler designed to open on landing to reduce lift and thus increase effectiveness of wheel braking.
LINS Laser inertial navigation system.
litre SI unit of volume (0.264177 US gallon; 0.219975 Imp gallon).
LLTV Low-light TV (thus, LLLTV, low-light level); see ALLTV.
LO Low-observables, which see (stealth).
load factor (1) Percentage of maximum payload; (2) design factor (*g* limit) for airframe.
LOC Localiser (which see).
localiser Element giving steering guidance in ILS.
LOH Light observation helicopter.
loiter Fly for maximum endurance, at much less than normal cruise speed.
longerons Principal fore-and-aft structural members (for example in fuselage).
Loran Long-range navigation; family of hyperbolic navaids based on ground radio emissions, now mainly Loran C.
LOROP Long-range oblique photography.
LOS Line of sight.
low-observables Materials, structures and techniques designed to minimise aircraft signatures of all kinds.
lox Liquid oxygen.
LP Low pressure (LPC, compressor; LPT, turbine).
LRIP Low-rate initial production.
LRMTS Laser ranger and marked-target seeker.
LRU Line-replaceable unit.
Ltd Limited (company constitution).

m Metre(s), SI unit of length (3.28084 feet).
M or Mach number The ratio of the speed of a body to the speed of sound (340 m; 1,116 ft/s in air at 15°C) under the same ambient conditions.

NEW/0569569

MAD Magnetic anomaly detector.
mass balance Mass attached to flight control surface, typically ahead of hinge axis, internally or externally, to reduce or eliminate coupling with airframe flutter modes.
mass flow Mass of air passing per second (usually at T-O, S/L).
MAWS Missile-approach warning system.
mb Millibar, bar × 10^{-3}.
MCM Mine countermeasures.
medevac Medical evacuation.
MFD Multifunction (electronic) display.
MHz Megahertz: 1 million (10^6) Hertz.
microlight See ultralight.
MIDS Multifunction information distribution system.
MKR Marker beacon receiver.
MLS Microwave landing system.
MLU Mid-life update.
MLW Maximum landing weight.
mm Millimetres, metres × 10^{-3}.
M_{MO} Maximum operating Mach number.
MMS Mast-mounted sight.
MoD Ministry of Defence.
monocoque Structure with strength in outer shell, devoid of internal bracing (semi-monocoque, with some internal supporting structure).
MoU Memorandum of Understanding.
MPA Maritime patrol aircraft.
mph Miles per hour.
MSIP Multistaged improvement program (US).
MTBF Mean time between failures.
MTI Moving-target indication (radar).
MTOW Maximum take-off weight (minus taxi/run-up fuel).
MYP Multiyear procurement (US).
MZFW Maximum zero-fuel weight.

N Newton, SI unit of force, = 0.22480455 lb force.
NACES Navy aircrew common ejection seat (US).
NAS Naval Air Station (US).
NASA National Aeronautics and Space Administration (US).
NASC Naval Air Systems Command (also several other aerospace meanings) (US).
NATC Naval Air Training Command or Test Center (also several other aerospace meanings) (US).
NATO North Atlantic Treaty Organisation.
nav/com Navigation and communications receiver.
NBAA National Business Aircraft Association (US).
NBC Nuclear, biological, chemical (warfare).
NDT Non-destructive testing.
Newton See N.
NFO Naval flight officer; second crew member in US Navy aircraft; compare WSO.
Ni/Cd Nickel/cadmium.
Nib Forward-pointing extension at inner end of fixed glove on VG aircraft or leading-edge root extension on light aircraft.
n mile nautical mile, 1.852 km, 1.15078 miles.
NOE Nap-of-the-Earth (low flying in military aircraft, using natural cover of hills, trees and so on).
NV Naamloze Vennootschap (Belgian/Netherlands company constitution).
NVG Night vision goggles.
NVS Noise vibration suppression.

OAO Otkrytolye Aktsionernoye Obshchestvo (JSC; Russian company constitution).
OAT Outside air temperature.
OBIGGS Onboard inert gas generating system.
OBOGS Onboard oxygen generating system.
OCU (1) Operational Conversion Unit; (2) operational capabilities upgrade.
OEI One engine inoperative.
OEU Operational Evaluation Unit.
offset Workshare granted to a customer nation to offset the cost of an imported system.
OGE Out of ground effect; helicopter hovering, far above nearest surface.
OKB Opytnyi Konstruktorskoye Byuro (Russian experimental design bureau)
Omega Long-range hyperbolic radio navaid.
OOO Obshchestvo Ogranichennoye Otvetstvennostyu (Russian company constitution).
opeval Operational evaluation.
OTH Over-the-horizon (OTHT adds targeting).

OTPI On-top position indicator (indicates overhead of submarine in ASW).

The Electromagnetic Spectrum

Frequency	Wavelength	General Designation	NATO Band	US Band
30-3 kHz	10,000-100 km	ELF		
3-30 kHz	100-10 km	VLF		
30-300 kHz	10-1 km	LF		
300 kHz-3 MHz	1,000-100 m	MF		
3-30 MHz	100-10 m	HF	A	
30-230 MHz	10-1.3 m	VHF	A	
230-250 MHz	1.3-1.2 m	VHF	A	P
250-300 MHz	1.2-1 m	VHF	B	P
300-500 MHz	100-60 cm	UHF	B	P
500-1,000 MHz	60-30 cm	UHF	C	P
1-2 GHz	30-13 cm	UHF	D	L
2-3 GHz	15-10 cm	UHF	E	S
3-4 GHz	10-7.5 cm	SHF	F	S
4-6 GHz	7-5.5 cm	SHF	G	C
6-8 GHz	5-7.5 cm	SHF	H	C
8-10 GHz	3.75-3 cm	SHF	I	X
10-12.5 GHz	3-2.5 cm	SHF	J	X
12.5-18 GHz	2.5-1.6 cm	SHF	J	Ku
18-20 GHz	1.6-1.5 cm	SHF	J	K
20-26.5 GHz	1.5-1.1 cm	SHF	K	K
26.5-30 GHz	1.1-1 cm	SHF	K	Ka
30-40 GHz	10-7.5 mm	EHF	K	Ka
40-60 GHz	7.5-5 mm	EHF	L	mm
60-100 GHz	5-3 mm	EHF	M	mm
100-300 GHz	3-1 mm	EHF		

Notes: Three overlapping descriptive systems are used in the West.
General designations are Extremely Low, Very Low, Low, Medium, High, Very High, Ultra High, Super High and Extremely High Frequency.
Frequencies are measured in kilo (1,000), mega (1,000,000) and giga (1,000,000,000) cycles per second (Hertz); wavelengths measured in kilometres, metres, centimetres and millimetres.
'NATO' bands describe radar and electronic warfare equipment; 'US' bands are used for radar and satellite communications. The latter's bounds are slightly 'elastic'.
Aircraft-to-ground voice communications for air traffic control and similar purposes (including ground ration beacons) uses 108-136 MHz in the VHF band and 225-400 MHz in the V/UHF bands, the latter principally military, and not entirely accurately termed 'UHF'.

OWE Operating weight empty. MTOW minus payload, usable fuel and oil and other consumables (thus, includes crew).

pallet (1) for freight, rigid platform for handling by forklift or conveyor;
(2) for missile, interface mounting and electronics box outside aircraft.
Pascal SI unit of pressure =1 Nm^{-2} (one Newton per square metre).
payload Disposable load generating revenue (passengers, cargo, mail and other paid items); in military aircraft, loosely used to mean total load carried of weapons, cargo or other mission equipment.
Performance Aircraft capabilities after S/L take-off at MTOW in ISA with normal full tankage, except as otherwise specified, and with landing data at MLW (where different).
PFA Popular Flying Association (UK).
PGM Precision-guided munition.
phased array Radar in which the beam is scanned electronically in one or both axes without moving the antenna.
Pirate Passive infra-red airborne tracking equipment.
PLA Prelaunch activities.
plane A lifting surface (for example wing, tailplane).
plc Public limited company (company constitution).
plug door Door larger than its frame in pressurised fuselage, either opening inwards or arranged to retract parts before opening outwards.
plume The region of hot air and gas emitted by a helicopter jetpipe.
ply Indication (ply rating) of tyre strength in a specific application; not necessarily the actual number of carcass plies in the tyre.
pneumatic de-icing Covered with flexible surfaces alternately pumped up and deflated to throw off ice.
port Left side, looking forward.
power-by-wire Using electric power alone (not electro-hydraulic) to drive control surfaces and perform other mechanical tasks.
power loading Aircraft weight (usually MTOW) divided by total propulsive power or thrust at T-O. For helicopters, based on transmission rating rather than total engine power.
power train A complete mechanical drive system, for example the sequence of gearwheels, clutches and shafts transmitting power from one or more engines to the rotors of a helicopter.
PPV Pre-production verification.
prepreg Glass fibre cloth or rovings pre-impregnated with resin to simplify layup.
pressure fuelling Fuelling via a leakproof connection through which fuel passes at high rate under pressure.
primary flight controls Those used to control trajectory of aircraft (thus, not trimmers, tabs, flaps, slats, airbrakes or lift dumpers, and so on).
primary flight display Single screen bearing all data for aircraft flight-path control.
propfan A family of new-technology propellers characterised by multiple scimitar-shaped blades with thin sharp-edged profile. Single and contrarotating examples promise to extend propeller efficiency up to an aircraft Mach number of about 0.8.
proprotor Large propeller, tilting for forward or vertical flight.
PT Perusahaan Terbatas (Indonesian company constitution).
Pty Proprietary (company constitution).
pulse Doppler Radar sending out pulses and measuring frequency-shift to detect returns only from moving target (s) seen against background clutter.

NEW/0569570

pylon Structure linking aircraft to external load (engine nacelle, drop tank, bomb, and so on).

Radar frequency Operating bands of airborne radars are given according to frequency. That part of the electromagnetic spectrum appropriate to above-surface short-range communication and radar (but not OTH) used in aviation is given in the adjacent table with an approximate cross-reference to previously used wavelength bands.

radius The approximate distance an aircraft can fly from base and return without intermediate landing.
RAI Registro Aeronautico Italiano (Italian civil aviation authority).
RAM Radar absorbent material.
ramp weight Maximum weight at start of flight (MTOW plus taxi/run-up fuel).
range Too many definitions to list, but essentially the distance an aircraft can fly (or is permitted to fly) with specified load and usually while making allowance for specified additional manoeuvres (diversions, standoff, go-around and so on).
RAS Rectified airspeed, IAS corrected for position error.
raster Generation of large-area display, for example TV screen, by close-spaced horizontal lines scanned either alternately or in sequence.
RAT Ram air turbine.
rating Any of several values of thrust or shaft power which an engine is qualified (usually also guaranteed) to develop under specified conditions.
RCS Radar cross-section; apparent size of echo.
redundant Provided with spare capacity or data channels and thus able to survive failures.
reversion Ability to switch to manual control following failure of a powered system.
RFAS Russian Federation and Associated States (CIS).
RFP Request(s) for proposals.
rigid rotor See bearingless rotor.
RMI Radio magnetic indicator; combines compass and navaid bearings.
R/Nav Calculates position, distance and time from groups of airways beacons.
RON Research octane number of fuel.
roving Multiple strands of fibre, as in a rope (but usually not twisted).
rpm Revolutions per minute.
RPV Remotely piloted vehicle (pilot in other aircraft or on ground); contrast UAV.
RSA Réseau du Sport de l'Air.
ruddervators Flying control surfaces, usually a V tail, that control both yaw and pitch attitude.
RVSM Reduced vertical separation minimum. Halved (1,000 ft) air traffic control separation between FL290 and FL410.
RWR Radar warning receiver.

s Second(s)
SA Société Anonyme (France, Romania), Sociedad Anónima (Brazil, Spain) or Spółka Akeyjna (Poland) (company constitution).
safe-life A term denoting that a component has proved by testing that it can be expected to continue to function safely for a precisely defined period before replacement.
SAM Surface-to-air missile.
SAR (1) Search and rescue;
(2) synthetic aperture radar.
SAS Stability augmentation system.
satcom Satellite communications.
sawtooth Same as dog-tooth.
Sdn Bhd Sendirian Berhad (Malaysian company constitution).
SEAD Suppression of enemy air defence(s).
semi-active Homing on to radiation reflected from target illuminated by radar or laser energy beamed from elsewhere (for example, from launch aircraft).
sensitive altimeter Altitude indicator of mechanical type, having acute sensitivity.
service ceiling Usually height equivalent to air density at which maximum attainable rate of climb is 100 ft/min.
servo A device which acts as a relay, usually augmenting the pilot's efforts to move a control surface, or the like.
sfc Specific fuel consumption (which see).
shaft Connection between gas-turbine and compressor or other driven unit. Two-shaft engine has second shaft, rotating at different speed, surrounding the first (thus, HP surrounds inner LP or fanshaft).
shipment One item or consignment delivered (by any means of transport) to customer.
shp Shaft horsepower, measure of power transmitted via rotating shaft.
shroud Many meanings, including: (1) a fixed circular duct surrounding a fan or propfan;
(2) a ring formed by lateral projections on a rotor (for example fan) blade (part-span or at the tip);
(3) a portion of a wing or other fixed aerofoil projecting aft over the leading-edge of a hinged or otherwise movable surface such as a flap, aileron or elevator.
sideline noise EPNdB measure of aircraft landing and taking off, at point 0.25 n mile (2- or 3-engined) or 0.35 n mile (4-engined) from runway centreline.
sidestick Control column in the form of a short handgrip beside the pilot.
sigint Signals intelligence.
signature Characteristic 'fingerprint' of all acoustic or electromagnetic radiation (radar, IR, and so on).
single-aisle Passenger cabin has seats on each side of a single aisle along or near the centre.

single-shaft Gas-turbine in which all compressors and turbines are fixed to common shaft.
S/L Sea level.
SLAR Side-looking airborne radar.

NEW/0569560

slat Auxiliary curved or mini-aerofoil surface designed to prevent flow breakaway from a wing or tail. On a tail leading-edge it may be fixed, leaving a narrow slot. On a wing it is almost always retractable, normally flush with the wing profile but extended (under power or by aerodynamic lift) to leave a narrow slot for take-off, low-speed loiter or landing.
slot, slotted See slat.
snap-down Air-to-air interception of low-flying aircraft by AAM fired from fighter at a higher altitude.
SONAR, sonar Sound navigation and ranging.
SpA Società per Azioni (Italian company constitution).
specific fuel consumption Rate at which fuel is consumed divided by power or thrust developed, and thus a measure of engine efficiency. For jet engines (air-breathing, not rockets) unit is mg/Ns, milligrams per Newton-second; for shaft engines unit is µg/J, micrograms (millionths of a gram) per Joule (SI unit of work or energy).
spoiler Plank-like surface normally recessed into top of wing, hinged up under power to reduce (spoil) lift and increase drag. Used asymmetrically for lateral control.

spoileron Small spoiler augmenting ailerons.
sportplane Light aircraft design in which performance takes precedence over utility.
Sp. z o.o Spółka z ograniczoną odpowiedzialnością (Polish company constitution)
Srl Società Reponsibilita Limitata (Italian company constitution).
SSB Single-sideband (radio).
SSR Secondary surveillance radar.
SST Supersonic transport.
st Static thrust.
stabilator One-piece, all-moving horizontal tail, combining functions of horizontal stabiliser and elevator.
stabilizer Tailplane (US); vertical stabilizer = fin.
stall Sudden near-total loss of lift of a wing because AoA has exceeded a critical value.
stall strips Sharp-edged strips on wing leading-edge to induce stall to initiate at that point.
stalling speed Airspeed at which aircraft stalls at 1*g*.
starboard Right side, looking forward.
static inverter Solid-state (not rotary machine) inverter of alternating wave-form to produce DC from AC.
STC Supplementary Type Certificate.
stealth See low-observables.
stick-pusher Stall-protection device that forces pilot's control column forward as stalling angle of attack is neared.
stick-shaker Stall-warning device that noisily shakes pilot's control column as stalling angle of attack is neared.
STOL Short take-off and landing. (Several definitions, stipulating allowable horizontal distance to clear screen height of 35 or 50 ft or various SI measures.)
store Object carried as part of payload on external attachment (for example bomb, drop tank).
STOVL Short take-off, vertical landing.
strobe light High-intensity flashing beacon.
supercritical wing Wing of relatively deep, flat-topped profile generating lift right across upper surface instead of concentrated close behind leading-edge.
sweepback Backwards inclination of wing or other aerofoil, seen from above, measured relative to fuselage or other reference axis, usually measured at quarterchord (25 per cent) or at leading-edge.

t Tonne, 1 Megagram, 1,000 kg.
tab Small auxiliary surface hinged (flight-adjustable) or attached in a fixed position (ground-adjustable) to trailing-edge of control surface for trimming, balancing (reducing hinge moment: force needed to operate main surface) or in other way assisting pilot. Compare anti-balance tab.
tabbed flap Fitted with narrow-chord tab along trailing-edge which deflects to greater angle than main surface.

Tacan Tactical air navigation, UHF navaid giving bearing and distance to ground beacons; distance element (see DME) can be paired with civil VOR.
taileron Left and right tailplanes used as primary control surfaces in both pitch and roll.
tailplane Horizontal stabiliser; main horizontal tail surface, originally fixed and carrying hinged elevator (s) but today often a single 'slab' serving as control surface (see stabiliser, stabilator).
TANS Tactical air navigation system; Decca Navigator or Doppler-based computer, control and display unit.
TAS True airspeed, EAS corrected for density (often very large factor) appropriate to aircraft altitude.
TBO Time between overhauls.
t/c ratio Ratio of the thickness (aerodynamic depth) of a wing or other surface to its chord, both measured at the same place parallel to the fore-and-aft axis.
TCAS Traffic-alert and collision-avoidance system.
Tercom Terrain-comparison (or contour-matching), navigation aid which compares relief of terrain with profile stored in memory.
TFR Terrain-following radar (for low-level attack).
thickness Depth of wing or other aerofoil; maximum perpendicular distance between upper and lower surfaces.
thrust vectoring Rotation of a vehicle's thrust axis to control its trajectory or support its weight.
TIALD Thermal imaging and laser designation (pod).
tiltrotor Aircraft with fixed wing and rotors that tilt up for hovering and forward for fast flight.

NEW/0569563

T-O Take-off.

T-O noise EPNdB measure of aircraft taking off, at point directly under flight path 3.5 n miles from brakes-release.
TOGW Take-off gross weight (not necessarily MTOW)
ton Imperial (long) ton = 1.016 t or 2,240 lb, US (short) ton = 0.9072 t or 2,000 lb.
track Distance between centres of contact areas of main landing wheels measured left/right across aircraft (with bogies, distance between centres of contact areas of each bogie).
transceiver Radio transmitter/receiver.
transformer-rectifier Device for converting AC to DC at a different voltage.
transponder Radio transmitter triggered automatically by a particular received signal, as in secondary surveillance radar (SSR).
TsENTROSPAS (in Russian Federation) Ministry for Civil Defence, Emergencies and Elimination of the Consequences of Natural Disasters.
turbofan Gas-turbine jet engine generating most thrust by a large-diameter cowled fan, with small part added by jet from core.
turbojet Simplest form of gas turbine comprising compressor, combustion chamber, turbine and propulsive nozzle.
turboprop Gas turbine in which as much energy as possible is taken from gas jet and used to drive reduction gearbox and propeller.
turboshaft Gas turbine in which as much energy as possible is taken from gas jet and used to drive high-speed shaft (which in turn drives external load such as helicopter transmission).
twist Progressive change of angle of incidence of a wing, rotor blade or other aerofoil from root to tip.

Tyre classification systems

Classification name	Example	Nominal diameter	Nominal section width	Nominal rim diameter
Type III	8.50-10		8½ in	10 in
Type VII	49×17	49 in	17 in	
Three Part	49×19.0-20	49 in	19.0 in	20 in
Radial	32×8.8R16	32 in	8.8 in	16 in
Metric	670×210-12	670 mm	210 mm	12 in

Example microlight/ultralight maxima

Country	Name	Seat(s)	Empty weight	MTOW	Fuel	Vso
Australia	Ultralight	1-2		540 kg		
France	ULM[1]	1		300 kg[2]		35 kt[3]
	ULM[1]	2		450 kg[2]		35 kt[3]
Germany	Ultralicht	1-2		472.5 kg		
UK	Microlight	1-2		450 kg		35 kt
USA	Ultralight	1	254 lb		5 USg	
	LSA	1-2	1,232 lb			

[1] Ultra leger motorise; criteria accepted by several European countries
[2] Plus 10 per cent for seaplanes and amphibians
[3] Plus 5 per cent for seaplanes and amphibians

Type Certificate Airworthiness licence granted to enable a manufacturer to produce and market a specified type of aircraft (compare C of A).
tyre sizes Five systems of classification are in current use; Type I, consisting of a single figure indicating nominal diameter in inches, is obsolete. See adjacent table and also 'ply'

UAV Unmanned (or uninhabited) aerial vehicle; contrast RPV.
UHCA Ultra-high capacity airliner.
UHF Ultra-high frequency.
ultralight Light aircraft with parameters below specified national limits, qualifying for less rigorous licensing; also known as microlight. See table.
upper surface blowing Turbofan jet expelled over upper surface of wing to increase lift.
usable fuel Total mass of fuel consumable in flight, usually 95 to 98 per cent of system capacity.
useful load Usable fuel and other consumables plus payload.
US gallon 0.83267 Imperial gallon; 3.785 litres.
UV Ultra-violet.

'V-speeds' Shorthand notation of significant speeds within an aircraft's flight envelope (see table).
variable geometry Capable of grossly changing shape in flight, especially by varying sweep of wings.
variable pitch Propeller with its blades held in rotary bearings in the hub, so that pitch (of all blades in unison) can be altered in flight. See constant speed; compare ground- and flight-adjustable pitch.
VDU Video (or visual) display unit.
vectored Capable of being pointed in different directions.
vertrep Vertical replenishment.
VFR Visual flight rules.
VHF Very high frequency.
VLF Very low frequency (area-coverage navaid).
VMS Vehicle management system.
VOR VHF omnidirectional range (network of VHF radio beacons each providing to/from bearing).
vortex generators Small blades attached to wing and tail surfaces to energise local airflow and improve control.
vortillon Short-chord fence (particularly on MD-80 series) ahead of and below leading-edge.
VSI Vertical speed (climb/descent) indicator.
V/STOL Vertical/short take-off and landing.

washout Inbuilt twist of wing or rotor blade reducing angle of incidence towards the tip.
watt SI unit of power, equal to 1 Js^{-1} (one Joule per second).
WDNS Weapon delivery and navigation system.
wet Housing fuel; wet wing often has extra connotation of integral tankage. Wet pylon can accommodate external fuel tank.
wheelbase Minimum distance from nosewheel or tailwheel (centre of contact area) to line joining mainwheels (centres of contact areas).
white tail Aircraft completed but unsold, thus not bearing any operator's insignia.
wide-body Passenger aircraft with cabin wide enough to have two longitudinal aisles between seats.
wing area Total projected area of clean wing (no projecting flaps, slats and so on) including all control surfaces and area of fuselage bounded by leading- and trailing-edges projected to centreline (inapplicable to slender-delta aircraft with extremely large leading-edge sweep angle). Described in *Jane's* as gross wing area; net area excludes projected areas of fuselage, nacelles, and so on.
wing loading Aircraft weight (usually MTOW) divided by wing area.
winglet Small auxiliary aerofoil, usually sharply upturned and often sweptback, at tip of wing.
WSO Weapon(s) system(s) officer.

yoke Pilot's flight control interface for pitch and roll axes in the form of a stick (control column) to the top of which is laterally pivoted a pair of handgrips in the form of a Y.

ZAO Zakrytoe Aktsionernoye Obshchestvo (Russian company constitution).
zero-fuel weight MTOW minus usable fuel and other consumables, in most aircraft imposing severest stress on wing and defining limit on payload.
zero/zero seat Ejection seat designed for use even at zero speed on ground.
ZFW Zero-fuel weight.

V-speeds definitions

V_1 Decision speed, up to which it should be possible to abort a take-off after failure of the critical engine and stop safely within the remaining runway length. After reaching V_1 the take-off must be continued.
V_2 Minimum take-off safety speed.
V_A Design manoeuvring speed. The speed below which abrupt and extreme control movements are possible (though not advised) without exceeding the airframe's limiting load factors.
V_B Design speed for maximum gust intensity.
V_C Design cruising speed.
V_D Design diving speed.
V_{DF} Maximum demonstrated diving speed. Also **M_{DF}**, maximum demonstrated Mach No. in a dive.
V_E Maximum speed at which landing gear (or other item) may be extended or retracted (cycled).
V_{FE} Maximum flap extension speed (top of white arc on ASI).
V_H Maximum level-flight speed with maximum continuous power.
V_{LE} Maximum speed with landing gear extended.
V_{MCA} Minimum control speed (air). Minimum speed at which directional control of a multi-engined aircraft can be maintained after failure of critical engine (in effect, the lowest speed at which the aircraft possesses sufficient rudder authority to counteract the yaw induced by asymmetric thrust).
V_{MU} Minimum unstick speed.
V_{NE} Never-exceed speed.
V_{MO} Maximum operating speed. Also **M_{MO}**, maximum operating Mach No.
V_{NO} Normal operating speed. The maximum structural cruising speed allowable for normal operating conditions.
V_R Rotation speed, at which to raise the nose for take-off.
V_{RA} Rough-air speed. Maximum recommended airspeed for penetrating turbulent air.
V_{REF} Any reference or 'bug' speed, typically quoted for approach speeds.
V_{S0} Stalling speed at maximum take-off weight, in landing configuration with flaps and landing gear down, at sea level, ISA conditions (bottom of white arc on ASI). Also **V_S**, stalling speed 'clean', and **V_{S1}**, stalling speed for a given configuration other than 'clean'.
V_{SSE} Minimum speed for deliberate shutting down of one engine for purposes of asymmetric flight training.
V_X Best angle of climb speed on all engines. Sometimes (UK usage) **$V_{\%}$**.
V_{XSE} Best engine-out angle of climb speed.
V_Y Best rate of climb speed on all engines.
V_{YSE} Best engine-out rate of climb speed. Sometimes **$V_{\%SE}$**.
V_{ZRC} Zero rate of climb speed (on one engine, where drag of inoperative engine reduces climb gradient to zero).

Argentina

LMAASA

LOCKHEED MARTIN AIRCRAFT ARGENTINA SA

Avenida Fuerza Aérea Argentina Km 5500, X5010JMN Córdoba
Tel: (+54 351) 466 87 33
Fax: (+54 351) 466 87 34
e-mail: busdev@lmaasa.com
PRESIDENT: James A Taylor
GENERAL MANAGER, PROGRAMS AND MANUFACTURE: Alberto O Buthet
INTERNATIONAL BUSINESS DEVELOPMENT DIRECTOR: Bernard R Kelleher

Original FMA (Military Aircraft Factory) came into operation 10 October 1927 as central organisation for aeronautical research and production; underwent several name changes before reverting, in 1968, to original title as component of Aérea de Material Córdoba (AMC) of Argentine Air Force. FMA converted into a joint stock company from April 1992, Air Force buying 30 per cent of the shares to establish itself as holding and management authority; control of FMA handed over to Planning Secretary of MoD on 20 December 1993. MoD and Lockheed Aircraft Argentina SA signed concession agreement on 15 December 1994, allocating management of FMA to Lockheed Martin, now Lockheed Martin Aircraft Argentina SA, from 1 July 1995.

Principal activities are divided into three market segments; current tasks (1) aircraft maintenance, modification and upgrade, including A-4AR Fighting Hawk C-130 Hercules, Fokker F.27 and F.28, IA 58 Pucará, T-34 Mentor, Boeing 707 and IA 63 Pampa; (2) manufacturing of parts, subassemblies, assemblies and composites components; and (3) engine maintenance for T56, Atar 09C, TFE731, J52, T53 and JT8D power plants. Argentine Air Force maintenance contract renewed in January 2001 for five years at estimated value of US$230 million, but national financial crisis has severely curtailed military contracts.

Laboratories, factories and other aeronautical division buildings occupy total covered area of 220,000 m² (2,368,075 sq ft); had workforce of 900 in early 2003. Córdoba facility also accommodates Centro de Ensayos en Vuelo (Flight Test Centre), a separate division also controlled by Argentine Air Force, at which all aircraft produced in Argentina undergo certification testing.

UPDATED

LMAASA AT-63 PAMPA

TYPE: Basic jet trainer/light attack jet.

PROGRAMME: First-generation IA 63 Pampa initiated by Fuerza Aérea Argentina (FAA; Argentine Air Force) 1979, eventual configuration being selected over six other designs early 1980, with Dornier of Germany providing technical assistance (including manufacture of prototypes' wings and tailplanes); two static/fatigue test airframes and three flying prototypes built (first flight, by EX-01, 6 October 1984); first flight of production Pampa October 1987; 14 of initial batch of 18 (including three prototypes) delivered to FAA (10 survivors currently serving I Escuadron of 4 Grupo de Caza at El Plumerillo, Mendoza) from 1988; expected follow-on order for 46 did not materialise, and Argentine Navy requirement for 12 long remained in abeyance, but one new aircraft (E-816), assembled from existing and new components, delivered to FAA on 28 September 1999.

'New-generation' Pampa NG revealed late 1997 and offered to Argentine Air Force and Navy.

Resumption of production announced 29 June 2000 for a further batch of 12 Pampas in new AT-63 (with hyphen) configuration, deliveries then scheduled to begin in 2003 and end in June 2005; contract (2001) includes option for further 12. This reaffirmed in January 2001 to be part of five-year FAA support contract placed with LMAASA. US$230 million contract also includes upgrade of existing Pampas. Argentine Navy interest renewed in 2000; initial batch of eight in prospect.

Unofficial report of mid-2002 alleged Colombian interest in 24 Pampas.

Work was suspended in 2002 because of national financial crisis; no further announcements were made in 2003.

CURRENT VERSIONS: **IA 63 Pampa:** Standard Argentine Air Force version. Production complete.

Naval version: Would have strengthened landing gear, uprated engine and some changed avionics.

Pampa 2000 International: Not proceeded with.

Pampa NG A: Proposed advanced trainer with updated avionics. Essentially became AT-63 (initial version).

Pampa NG B: Proposed combat-capable version; uprated engine and avionics. Features to be incorporated in later upgrade of AT-63.

AT-63: Attack/trainer. Announced 2001 as new standard version for Argentine Air Force and potential export. Two prototypes modified from existing IA 63s in 2001–02; first flight planned for November 2002, but delayed following funding problems of FAA. Initial version (Pampa Phase II) features new processor for DECU, Elbit avionics suite (MIL-STD-1553B databus, mission computer, RLG INS/GPS, integrated weapons system, liquid crystal MFD in each cockpit, and front cockpit HUD). Advanced version (Pampa Phase III) proposed for 2005 with structurally enhanced (+7/–3 *g*) wing, 18.9 kN (4,250 lb

IA 63 Pampa of 4 Grupo de Caza, Argentine Air Force 0118292

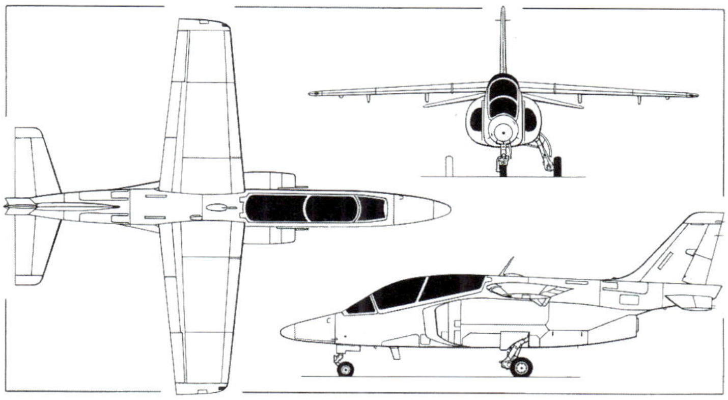

LMAASA AT-63 Pampa two-seat basic and advanced jet trainer *(Jane's/Dennis Punnett)* 0507117

st) TFE731-40R engine, missionised avionics with enhanced tactical capabilities, nose-mounted laser range-finder, fin-mounted RWR, conformal chaff/flare dispensers, two additional hardpoints (total seven) for AAMs and strengthened landing gear for increased MTOW.

AT-63 version also proposed with Lockheed Martin AN/APG-67(V)4 multimode radar in response to Colombian interest; announced at FIDAE, Chile, April 2002 that Lockheed Martin Naval Electronics & Surveillance Systems selected to integrate APG-67 in Pampa. Also reported that Singapore evaluated AT-63 in early 2003 as S.211 replacement.

Description applies to Phase II Pampa.

COSTS: US$8 million to US$9 million, including radar (estimated, 2002).

DESIGN FEATURES: Intended for cost-effective pilot training in mission management techniques, advanced fighter lead-in training and extended-range anti-drug patrol missions. High degree of commonality with the original IA 63 Pampa, providing a customised low life-cycle cost fleet. Fail-safe service life 8,000 hours.

Non-swept shoulder-mounted wings and anhedral tailplane, sweptback fin and rudder; single engine with twin lateral air intakes. Wing section Dornier DoA-7/-8 advanced transonic; leading-edge sweep 5° 24'; thickness/chord ratio 14.5 per cent at root, 12.5 per cent at tip; anhedral 3°.

FLYING CONTROLS: Conventional hydraulically powered ailerons, rudder, all-moving tailplane, single-slotted Fowler flaps, and door-type airbrake, deployable at all speeds, on each side of upper rear fuselage; primary surfaces have Liebherr tandem actuators and electromechanical trim.

STRUCTURE: Conventional all-metal semi-monocoque/stressed skin; two-spar wing box forms integral fuel tank. Modified fuselage nosecone, tailcone and fintip in AT-63.

LANDING GEAR: SHL (Israel) retractable tricycle type, with hydraulic extension/retraction and emergency free-fall extension. Oleo-pneumatic shock-absorbers. Single Messier-Bugatti wheel on each unit with Goodrich (main) or Continental (nose) low-pressure tyre; nosewheel offset 10 cm (3.9 in) to starboard. Tyre sizes 6.50–10 (10 ply) on mainwheels, 380×150 (4/6 ply) on nosewheel, with respective pressures of 6.55 bar (95 lb/sq in) and 4.00 bar (58 lb/sq in). Nosewheel retracts rearward, mainwheels inward into underside of engine air intake trunks. Messier-Bugatti mainwheel hydraulic disc brakes incorporate anti-skid device; nosewheel steering (±47°). Gear designed for operation from unprepared surfaces.

POWER PLANT: One 15.57 kN (3,500 lb st) Honeywell TFE731-2C-2N turbofan installed in rear fuselage. Single-point pressure refuelling, plus gravity point in upper surface of each wing. Standard internal fuel capacity of 968 litres (255 US gallons; 213 Imp gallons) in integral wing tank of 550 litres (145 US gallons; 121 Imp gallons) and 418 litre (110 US gallon; 92.0 Imp gallon) flexible fuselage tank with a negative *g* chamber permitting up to 10 seconds of inverted flight. Additional 415 litres (109 US gallons; 91.0 Imp gallons) can be carried in auxiliary tanks inside outer wing panels, to give a maximum internal capacity of 1,383 litres (364 US gallons; 304 Imp gallons). Provision on centre underwing stations for two external drop tanks, each of 317 litres (83.7 US gallons; 69.7 Imp gallons). Total fuel capacity 2,017 litres (533 US gallons; 444 Imp gallons).

ACCOMMODATION: Tandem, rear seat elevated, on UPC (Stencel) S-III-S3IA63 zero/zero ejection seats. Ejection procedure can be preselected for separate single ejections, or for both seats to be fired from front or rear cockpit. HOTAS operation; dual controls standard. One-piece wraparound windscreen. One-piece canopy, with internal screen, is hinged at rear and opens upward. Entire accommodation pressurised and air conditioned.

SYSTEMS: Honeywell environmental control system, maximum differential 0.30 bar (4.4 lb/sq in), supplied by high- or low-pressure engine bleed air, provides a 1,980 m (6,500 ft) cockpit environment up to flight level 5,730 m (18,800 ft) and also provides ram air for negative *g* system and canopy seal. Oxygen system supplied by 10 litre (0.35 cu ft) lox converter. Engine air intakes anti-iced by engine bleed air.

Two independent hydraulic systems, each at pressure of 207 bar (3,000 lb/sq in), each supplied by engine-driven pump. Each system incorporates a bootstrap reservoir pressurised at 4 bar (58 lb/sq in). No.1 system, with flow rate of 16 litres (4.2 US gallons; 3.5 Imp gallons)/min, actuates primary flight controls, airbrakes, landing gear and wheel brakes; No. 2 system, with flow rate of 8 litres (2.1 US gallons; 1.75 Imp gallons)/min, actuates primary flight controls, wing flaps, emergency and parking brakes, and nosewheel steering. Honeywell ram air turbine provides emergency hydraulic power for No. 2 system if engine shuts down in flight and pressure in this system drops below minimum.

Electrical system (28 V DC) supplied by Lear Siegler 400 A 11.5 kW engine-driven starter/generator; secondary supply (115/26 V AC power at 400 Hz) from two Flite-Tronics 450 VA static inverters and two SAFT 27 Ah sealed lead batteries. Thirty minutes of emergency electrical power available in case of in-flight engine shutdown.

AVIONICS: *Comms:* Two VHF/UHF transceivers and direction-finder, intercom and IFF or ATC transponder.

Radar: Lockheed Martin AN/APG-67(V)4 multimode radar optional (not fitted in Argentine Air Force aircraft).

Flight: VOR/ILS with marker beacon receiver; DME or Tacan optional; autonomous ESIS/air data computer, HSI, ADF, ADI; Honeywell HG 764 laser INS with GPS; radar altimeter; air data computer.

Instrumentation: Single 127 mm (5 in) HUD (multimode: UFCP, PDU, PSVS and camera); colour HUD camera/airborne videotape recorder; video repeater; 12.7 × 17.8 cm (5 × 7 in) liquid crystal MFD in each cockpit for Argentine Air Force; second, similar MFD in each cockpit (for EICAS) for export version; Multirole central computer and MIL-STD-1553B digital databus.

Mission: New integrated weapon delivery system. Mission computer/symbol generator/integrated comms; weapon management system/data transfer unit; steerable laser ring finder optional; radar warning receiver optional; chaff/flare dispenser.

ARMAMENT: Five stations for external stores, stressed for 440 kg (970 lb) on centre fuselage and each inboard underwing station, 290 kg (639 lb) on each outboard underwing station, all at +5.5/−2 *g*. Phase III aircraft will have further pair of pylons further outboard, each rated at 170 kg (375 lb), for total of seven, plus uprated inboard wing pylons. Several external stores configurations including Mk 81 and Mk 82 bombs; LAU-32, LAU-51 and LAU-10 rocket pods; 30 mm gun pod (centreline), twin machine gun pods; and CBLS 200 practice bomb carriers.

DIMENSIONS, EXTERNAL:
Wing span ... 9.69 m (31 ft 9½ in)
Wing aspect ratio ... 6.0
Length overall ... 10.93 m (35 ft 10¼ in)
Height overall ... 4.29 m (14 ft 1 in)
Tailplane span ... 4.58 m (15 ft 0¼ in)
Wheel track ... 2.66 m (8 ft 8¾ in)
Wheelbase ... 4.42 m (14 ft 6 in)

AREAS:
Wings, gross ... 15.63 m² (168.3 sq ft)
Ailerons (total) ... 0.89 m² (9.58 sq ft)
Trailing-edge flaps (total) ... 2.93 m² (31.54 sq ft)
Fin ... 1.86 m² (20.02 sq ft)
Rudder ... 0.655 m² (7.05 sq ft)
Tailplane ... 4.35 m² (46.87 sq ft)

WEIGHTS AND LOADINGS (estimated):
Weight empty ... 2,820 kg (6,217 lb)
Max fuel weight:
fuselage ... 327 kg (721 lb)
wings, internal inboard ... 430 kg (948 lb)
wings, internal outboard ... 325 kg (717 lb)
underwing drop tanks ... 496 kg (1,093 lb)
Max external stores load ... 1,900 kg (4,189 lb)
Max T-O weight ... 5,000 kg (11,023 lb)
Max wing loading ... 319.9 kg/m² (65.52 lb/sq ft)
Max power loading ... 321 kg/kN (3.15 lb/lb st)

PERFORMANCE (estimated at 3,764 kg; 8,300 lb clean T-O weight with normal internal fuel, except where indicated):
Max level speed at 7,985 m (25,900 ft) ... 440 kt (814 km/h; 506 mph)
Max operating speed (VMO) ... M0.8
Econ cruising speed at 9,145 m (30,000 ft) ... 350 kt (648 km/h; 402 mph)
Stalling speed at S/L, 50% normal internal fuel:
flaps up ... 104 kt (193 km/h; 120 mph)
flaps down ... 82 kt (152 km/h; 95 mph)
Max rate of climb at S/L ... 1,560 m (5,118 ft)/min
Service ceiling ... 12,900 m (42,320 ft)
T-O run ... 430 m (1,410 ft)
Landing run at 3,497 kg (7,710 lb) ... 460 m (1,510 ft)
Radius of action:
air-to-air (hi-hi), T-O weight of 4,300 kg (9,480 lb) with 254 kg (560 lb) external load, 5 min allowance for dogfight, normal internal fuel, 30 min reserves ... 380 n miles (703 km, 437 miles)
air-to-ground (hi-lo-lo-hi), 30 n mile dash out/in, T-O weight of 5,000 kg (11,023 lb) with 1,000 kg (2,205 lb) external load, max internal fuel, 5 min allowance for weapon delivery, plus 30 min reserves ... 236 n miles (127 km, 205 miles)
Ferry range 9,145 m (30,000 ft), ISA with 15 min reserves ... 1,140 n miles (2,111 km; 1,311 miles)
g limits ... +6/−3

UPDATED

Australia

COBRA

COBRA AVIATION

40 Industry Place, Bayswater, Victoria 3153
Tel: (+61 3) 97 20 69 66
Fax: (+61 3) 97 20 33 49
e-mail: sales@cobraaviation.com.au
Web: http://www.cobraaviation.com.au
MANAGING DIRECTOR: Tom Wickers

Company founded in 2000 as Tomair Aircraft Sales; subsequently formed Cobra Aviation, for marketing of Arrow and Explorer.

UPDATED

COBRA ARROW

TYPE: Side-by-side ultralight kitbuilt; motor glider kitbuilt.

PROGRAMME: Non-flying exhibition airframe, then incomplete, displayed at Avalon Air Show, February 2001; prototype (19-3700) was removed from its moulds in August 2001 and was fitted out in late 2001/early 2002. First flight 18 May 2002. Conventional tailplane and elevators replaced in July 2002 by stabilator. First tailwheel version under construction (Kim Penglaze, Victoria) in mid-2003; was due to fly in January 2004.

CURRENT VERSIONS: **Arrow:** Standard version, *as described*. Available to Ultralight or Certified rules.

Arrow M/G: Motor glider version with 3.66 m (12 ft) increase in wing span.

Explorer: *Described separately.*

CUSTOMERS: Total of 10 kits sold by February 2003; first production kit shipped to New Zealand customer in March 2003.

COSTS: Arrow Sport kit, less engine, instruments and retractable landing gear, A$35,000, plus tax (2003).

DESIGN FEATURES: Derived from US Polliwagen homebuilt, described in 1983–84 and earlier editions of *Jane's*. Designed to meet AUF 51 per cent amateur-built rules. Conventional low-wing monoplane; sweptback T tail. Kit comprises prejoined fuselage mouldings with cowling, canopy and main spar carry-through installed; tail surfaces; wing skins with main spar installed; laser-cut profiles for rib shapes; carbon fibre instrument panel; plus seats, fuel cells, control columns, rudder pedals and hardware. Quoted average build time 350 hours.

FLYING CONTROLS: Manual. Ailerons, rudder and stabilator. Flaps.

STRUCTURE: Mostly of pre-moulded composites.

LANDING GEAR: Option of tailwheel type, fixed; or tricycle type, fixed or retractable. From mid-2003, wheel sizes 6.00×6 main; 5.00×6 nose.

POWER PLANT: Prototype has one 74.6 kW (100 hp) Subaru EA81 and Bolly Optima three-blade propeller. Other suitable engines include JPX 4TX75, Rotax 582, 912 and 914, Jabiru 2200 and 3300 and Textron Lycoming O-200 and O-360. Fuel capacity 71.9 litres (19.0 US gallons; 15.8 Imp gallons); optionally 117 litres (30.9 US gallons; 25.7 Imp gallons).

ACCOMMODATION: Two persons side by side under enclosed canopy. Dual controls (sticks) standard.

DIMENSIONS, EXTERNAL:
Wing span: Arrow ... 8.53 m (28 ft 0 in)
Arrow M/G ... 12.19 m (40 ft 0 in)
Length overall ... 5.79 m (19 ft 0 in)
Height overall (nosewheel) ... 1.83 m (6 ft 0 in)
Wheelbase ... 1.83 m (6 ft 0 in)

WEIGHTS AND LOADINGS:
Weight empty, typical ... 280 kg (617 lb)
Max T-O weight: Ultralight ... 544 kg (1,200 lb)
Certified ... 700 kg (1,543 lb)
Max power loading (prototype) ... 6.63 kg/kN (10.90 lb/hp)

PERFORMANCE (Subaru engine):
Max level speed ... more than 200 kt (370 km/h; 230 mph)
Cruising speed at FL85 ... 132 kt (244 km/h; 152 mph)
Stalling speed, flaps down ... 38 kt (71 km/h; 44 mph)
T-O and landing run ... 122 m (400 ft)
Range ... 782 n miles (1,448 km; 900 miles)
g limits ... +10/−6

UPDATED

COBRA EXPLORER

TYPE: Side-by-side ultralight kitbuilt.

PROGRAMME: In February 2003, Cobra revealed it was building a prototype of a high-wing version of the Arrow, due to fly in the middle of that year. Strut-braced wings save 40 kg (88 lb) of structural weight. No further details revealed, although prototype structurally complete by mid-2003 and due to be publicly displayed at Narrowmine in April 2004.

UPDATED

Cobra Arrow prototype in current form ***(Paul Jackson)*** ***NEW***/0589390

Artist's impression of Cobra Landseair ***NEW***/0554412

COBRA LANDSEAIR

TYPE: Four-seat amphibian kitbuilt.

PROGRAMME: Designed by Martin Gischus; announced by Cobra in May 2003. Manufacture of tooling for prototype began in 2000. No further news by early 2004.

COSTS: Kit A$58,000 (2003).

POWER PLANT: One pusher piston engine. Options include Subaru EA82, Rotax 914 and Jabiru 3300. Fuel capacity 100 litres (26.4 US gallons; 22.0 Imp gallons).

DIMENSIONS, EXTERNAL:
Wing span ... 10.06 m (33 ft 0 in)
Length overall ... 6.10 m (20 ft 0 in)

DIMENSIONS, INTERNAL:
Cabin max width ... 1.12 m (3 ft 8 in)

AREAS:
Wings, gross ... 15.79 m² (171.0 sq ft)

WEIGHTS AND LOADINGS:
Weight empty ... 350 kg (772 lb)
Max T-O weight ... 620 kg (1,366 lb)

PERFORMANCE:
Cruising speed ... 105 kt (194 km/h; 120 mph)
Stalling speed ... 40 kt (75 km/h; 46 mph)
T-O run on water ... 213 m (700 ft)
Landing run on water ... 244 m (800 ft)

UPDATED

ENDEAVOUR

ENDEAVOUR AEROSPACE

PO Box 582, Werribee, Victoria 3030
Tel: (+61 0402) 29 37 35
Fax: (+61 397) 34 78 87
e-mail: endaero@optusnet.com.au
MANAGING DIRECTOR: Mark Graham
ENGINEERING DIRECTOR: Richard Bartholomeusz

Company formed in 2000 to create an Australian general aviation export business. Acquired rights to complete development of, and export, Hawker de Havilland Wamira turboprop trainer; aircraft will be offered at a competitive price, with emphasis on US private market. Financial backing being obtained in 2002–03. By the end of 2003, there had been no further announcements.

UPDATED

ENDEAVOUR PHOENIX

TYPE: Aerobatic two-seat turboprop sportplane.

PROGRAMME: HDH A10 Wamira programme begun in 1982 to meet Royal Australian Air Force requirement for turboprop pilot trainer; redefined as A10B in May 1985 but rejected in December 1985 in favour of Pilatus PC-9. Prototype then substantially complete and HDH continued development with aim of late 1986 first flight. However, venture was cancelled and A$80+ million costs written off.

Rights acquired by Endeavour, which plans to produce sportplane and touring versions for export market, while retaining military potential. Following implementation of funding plan, estimated two-year programme to complete prototype/demonstrator under Experimental certification. No further reports up to end of 2003.

CURRENT VERSIONS: **Tandem:** Sportplane or intermediate military trainer conforming to current preference.

Side-by-side: Original Wamira configuration. Tourer or basic military trainer.

Both to be available with choice of engine power and would be suitable for pilot training up to first 100 hours of a military syllabus. Version commonality is 80 per cent.

DESIGN FEATURES: Low-wing, high-performance, aerobatic turboprop; mid-mounted tailplane. Tapered wing planform with root nibs and dihedral. Enlarged rudder on higher-powered version. Sweptback fin with small fillet; tapered tailplane.

FLYING CONTROLS: Conventional and manual. Mass-balanced ailerons with tab each side, port tab electrically actuated. Electric trim tabs in elevators (two) and rudder. Pushrod actuation throughout. Electrically actuated flaps.

STRUCTURE: Aluminium 2024 semi-monocoque with two-spar wing; single-spar rudder; tailplane as wings; some composites fairings. Comprehensive corrosion-proofing.

LANDING GEAR: Tricycle type; retractable. Mainwheels, with 7.00–8 tyres, retract inwards electromechanically and have no doors; nosewheel, 5.00–5, retracts rearward. All tyre pressures 2.9 bar (42 lb/sq in). Manual emergency extension. Oleo-pneumatic shock-absorber on each leg.

POWER PLANT: One Pratt and Whitney Canada PT6A-25C turboprop, flat-rated to either 559 kW (750 shp) or 634 kW (850 shp), driving a Hartzell HC-B3TN-T10173-11R constant-speed, fully feathering, reversible-pitch, three-blade propeller. Single fuel tank in lower centre fuselage capacity 580 litres (153 US gallons; 128 Imp gallons).

ACCOMMODATION: Two pilots in tandem (single-piece canopy) or side-by-side (two-piece canopy) with raised rear seat. Separate windscreen; canopies slide rearwards. Partial pressurisation. Baggage space behind seats, with external access.

SYSTEMS: Hydraulic system for brakes; no pneumatic system. 28 V DC electrical system incorporating 28 V, 200A starter/generator, with two 125 VA static invertors for AC supply. Crew oxygen system. Environmental control system.

DIMENSIONS, EXTERNAL:
Wing span ... 11.00 m (36 ft 1 in)
Wing chord: at root ... 3.03 m (9 ft 11¼ in)
at tip ... 1.21 m (3 ft 11½ in)
Wing aspect ratio ... 6.1

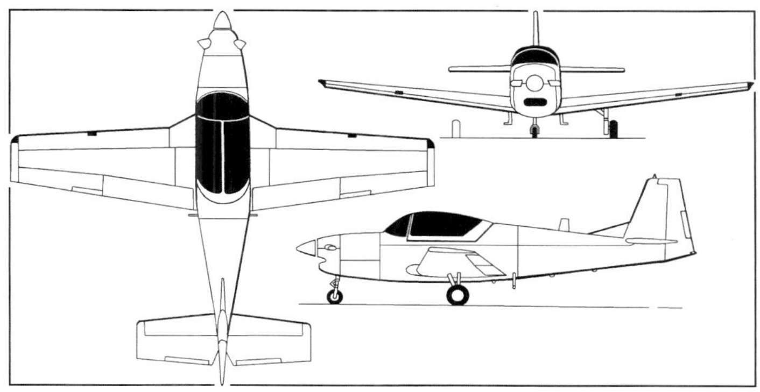

Side-by-side version of Phoenix more closely resembles the HDH Wamira ***NEW***/0552991

Length overall ... 10.20 m (33 ft 5½ in)
Height overall ... 3.51 m (11 ft 6¼ in)
Tailplane span ... 4.50 m (14 ft 9¼ in)
Wheel track ... 4.00 m (13 ft 1½ in)
Wheelbase ... 3.10 m (10 ft 2 in)
Propeller diameter ... 2.29 m (7 ft 6 in)

AREAS:
Wings, gross reference ... 20.00 m² (215.3 sq ft)

WEIGHTS AND LOADINGS (A: 559 kW; 750 shp engine, B: 634 kW; 850 shp engine):
Weight empty: A ... 1,636 kg (3,607 lb)
B ... 1,698 kg (3,744 lb)
Max ramp weight: A ... 2,213 kg (4,878 lb)
B ... 2,303 kg (5,077 lb)

PERFORMANCE (estimated; A, B as above):
Never-exceed speed (VNE) ... M0.575 (320 kt; 593 km/h; 368 mph) EAS
Max level speed:
at FL150:
A ... 238 kt (441 km/h; 274 mph)
B ... 248 kt (459 km/h; 285 mph)
at S/L: A ... 230 kt (426 km/h; 265 mph)
B ... 238 kt (441 km/h; 274 mph)
Max cruising speed:
at FL150:
A ... 229 kt (424 km/h; 264 mph)
B ... 244 kt (452 km/h; 281 mph)
at S/L: A ... 225 kt (417 km/h; 259 mph)
B ... 227 kt (420 km/h; 261 mph)
Stalling speed:
flaps up: A ... 69 kt (128 km/h; 80 mph)
B ... 71 kt (132 km/h; 82 mph)
flaps down: A ... 59 kt (110 km/h; 68 mph)
B ... 60 kt (112 km/h; 69 mph)
Max rate of climb at S/L: A ... 885 m (2,905 ft)/min
B ... 1,012 m (3,320 ft)/min
Time to FL150: A ... 6 min 40 s
B ... 6 min 5 s
Max certified ceiling ... 9,570 m (31,400 ft)
T-O to 15 m (50 ft): A ... 363 m (1,190 ft)
B ... 397 m (1,300 ft)
g limits ... +7.0/–3.5

UPDATED

GIPPSLAND

GIPPSLAND AERONAUTICS PTY LTD

Latrobe Regional Airport, Airfield Road, Traralgon, PO Box 881, Morwell, Victoria 3840
Tel: (+61 3) 51 72 12 00
Fax: (+61 3) 51 72 12 01
e-mail: sales@gippsaero.com
Web: http://www.gippsaero.com
DIRECTORS:
George Morgan
Peter Furlong
SALES CO-ORDINATOR: Marguerite Morgan

US MARKETING OFFICE:
AIRVAN USA
16644 Roscoe Boulevard, Hangar 12, Van Nuys Airport, California 91406
Tel: (+1 818) 908 99 69
e-mail: airvanusa@airvanusa.com
Web: http://www.airvanusa.com
PRESIDENT: Joe Desjardins

Involved since 1971 in design and modification programmes for a wide range of aircraft, from wooden homebuilts to pressurised turboprops. Conversion of five Piper Pawnees to two-seat configuration in second half of 1980s led eventually to a totally new design, the GA200 crop-sprayer. Second Gippsland aircraft, GA8 Airvan, first flew in 1995 and was delivered to customers from December 2000 onwards. First export Airvan flew in 2001.

Other activities include main spar life extension of Piper Navajo/Chieftain, of which 62 had been refurbished by October 2003.

In October 2003, Gippsland occupied 4,180 m² (45,000 sq ft) of factory area; personnel totalled 100.

UPDATED

GIPPSLAND GA200 FATMAN

TYPE: Agricultural sprayer.

PROGRAMME: Precursor was two-seat conversion of five Piper PA-25-235 Pawnees to -235/A8 or -235/A9 Fatman standard in 1986–90. Prototype GA200 (VH-BCE) registered 1991; this and second aircraft used airframes from damaged Pawnees; No 3 (VH-SKG) was first new-build, in 1992. Designation derived from hopper capacity in US gallons. Full Australian CAA certification in both Normal and Agricultural categories (to CAO 101.16 and 101.22 and FAR Pt 23 Amendment 23.36) awarded 1 March 1991; FAR Pt 23 certification awarded 15 October 1997 in Restricted category. Certification also achieved in Brazil, Canada and elsewhere. Production at Traralgon, Victoria, since 1993.

CURRENT VERSIONS: **GA200:** Initial version; 194 kW (260 hp) Textron Lycoming O-540-H2A5 engine. Hopper capacity 776 litres (205 US gallons; 170 Imp gallons). Replaced by GA200C.

GA200 Ag-trainer: Training version, with dual controls, dual rudder pedals and smaller hopper. Total of three built, plus unspecified number of conversions from GA200s.

GA200B: Unofficial designation of GA200 with extended wingtips. Most built up to 1998 were to this standard.

GA200C: *As described*. Uprated (224 kW; 300 hp IO-540-K1A5) engine beneath cowling of revised shape; constant-speed propeller; 1,050 litre (277.4 US gallon; 231 Imp gallon) hopper capacity. In early 1998, 21st production (23rd overall) aircraft completed as prototype GA200C; all subsequent aircraft to this standard. Textron Lycoming AEIO-580 of 246 kW (330 hp) installed in new-built 45th aircraft and first flown 11 February 2003 for engine evaluation.

CUSTOMERS: Delivery of 44th (including prototypes) effected in November 2001; 45th flown February 2003; none further by end of that year. Exports to Brazil (one), Canada (one), China (nine), New Zealand (eight, beginning 1994), South Africa (two) and USA (five).

DESIGN FEATURES: Purpose-designed agricultural aircraft. Braced low wing, with large integral hopper forward of cockpit; crash-resistant, corrosion-proofed structure; gap-sealed ailerons; detachable wingtips. Although initially based on Piper Pawnee, GA200 in current form is substantially different. Wing dihedral 7° from roots. Flaps and ailerons non-handed. Horn-balanced elevators and rudder.

FLYING CONTROLS: Conventional and manual, cable actuated. Single-slotted trailing-edge wing flaps can be deployed to tighten turning radius during agricultural operations; T-O setting 15°, maximum 38°. Interconnect system applies bias to elevator trim spring when flaps extended, to avoid pitch trim changes. Fixed tab on rudder.

STRUCTURE: Fuselage of welded SAE 4130 chromoly steel tube with removable metal side and top panels; wings (braced by overwing inverted V strut of aluminium each side) and wire-braced tail surfaces conventional all-metal, but wing spars constructed from sheet metal to expedite repairs; wingtips detachable. Rear fuselage and tail surfaces are fabric covered.

LANDING GEAR: Non-retractable, with 6 in diameter Cleveland P/N 40-84A mainwheels mounted on tubular steel side Vs, hinged to lower longerons, plus half-axles under centreline, each with bungee cord shock-absorber, and hydraulic dampers; wire-cutter on leading tube of landing gear legs; Cleveland heavy duty hydraulic disc brakes. Mainwheel tyres size 8.50×6 (6 ply). Scott 3200 steerable/castoring tailwheel, size 5.00-5 (4 ply) mounted on multileaf flat springs.

POWER PLANT: One 224 kW (300 hp) Textron Lycoming IO-540-K1A5 flat-six engine, driving a Hartzell HC-C2YR-1BF/F8475R constant-speed, two-blade metal propeller. AEIO-580 installation under test in 2003, offering additional 22 kW (30 hp). Fuel in integral tank in each wing, combined usable capacity 200 litres (52.8 US gallons; 44.0 Imp gallons), plus small 14 litre (3.7 US gallon; 3.1 Imp gallon) header tank in upper front fuselage. Oil capacity 11.4 litres (3.0 US gallons; 2.5 Imp gallons).

ACCOMMODATION: Two energy-absorbing seats, side by side (smaller, right-hand seat for loader/driver); dual controls and second set of rudder pedals for right-hand seat in Ag-trainer. Four-point 25 *g* restraint harness(es). Cockpit doors open upward, each with bulged window to improve shoulder room and outward view. Pressurisation system minimises dust and chemical ingress.

SYSTEMS: 14 V 95 A automotive alternator and automotive or R-35 aviation battery for electrical power; 50 A circuit breaker switch serves as master switch.

AVIONICS: *Instrumentation:* Basic VFR instruments and VHF com only.

EQUIPMENT: One 1,050 litre (277 US gallon; 231 Imp gallon) capacity hopper in forward fuselage (approximately 76 litres; 20.0 US gallons; 16.7 Imp gallons less in Ag-trainer), with minimum load dump time of five seconds. Multirole door/hopper outlet, eliminating need to change outlet when changing between solids and liquids, can also be used for fire bombing or laying of fire retardants. Spreader vanes can be added to increase swath width. Cable cutters on landing gear legs and windscreen. 100 W landing light in each wingtip; 28 V night working lights system (two retractable underwing 600 W lights, powered by separate 28 V 55 A alternator) also available.

DIMENSIONS, EXTERNAL:

Wing span, standard version	11.985 m (39 ft 3¾ in)
Wing chord, constant	1.60 m (5 ft 3 in)
Wing aspect ratio	7.3
Length overall (flying attitude)	7.48 m (24 ft 6½ in)
Height (static) over cockpit canopy	2.33 m (7 ft 7¾ in)
Tailplane span	2.90 m (9 ft 6¼ in)
Wheel track	2.335 m (7 ft 8 in)
Propeller diameter	2.13 m (7 ft 0 in)

AREAS:

Wings, gross	19.59 m² (210.9 sq ft)
Ailerons (total)	0.79 m² (8.49 sq ft)
Flaps (total)	0.71 m² (7.63 sq ft)
Fin	0.48 m² (5.155 sq ft)
Rudder, incl tab	0.63 m² (6.76 sq ft)
Tailplane	1.295 m² (13.92 sq ft)
Elevators (total)	1.37 m² (14.725 sq ft)

WEIGHTS AND LOADINGS:

Weight empty	758 kg (1,671 lb)
Operating weight empty	868 kg (1,914 lb)
Max hopper capacity, Normal category	544 kg (1,199 lb)
Max T-O weight: Normal category	1,524 kg (3,359 lb)
Agricultural category	1,995 kg (4,398 lb)
Max landing weight	1,448 kg (3,192 lb)
Max wing loading	101.8 kg/m² (20.86 lb/sq ft)
Max power loading	8.91 kg/kW (14.63 lb/hp)

PERFORMANCE:

Never-exceed speed (VNE)	144 kt (266 km/h; 165 mph)
Max cruising speed at 75% power	115 kt (213 km/h; 132 mph)
Econ cruising speed	100 kt (185 km/h; 115 mph)
Stalling speed: flaps up	58 kt (108 km/h; 67 mph)
flaps down	53 kt (99 km/h; 61 mph)
Max rate of climb at S/L	290 m (950 ft)/min
T-O run, normal category	350 m (1,148 ft)
T-O to 15 m (50 ft), normal category	475 m (1,558 ft)
Landing from 15 m (50 ft)	360 m (1,181 ft)
Max range	400 n miles (740 km; 460 miles)
g limits	+3.8/−1.52

UPDATED

Gippsland GA200 Fatman agricultural sprayer ***NEW***/1047216

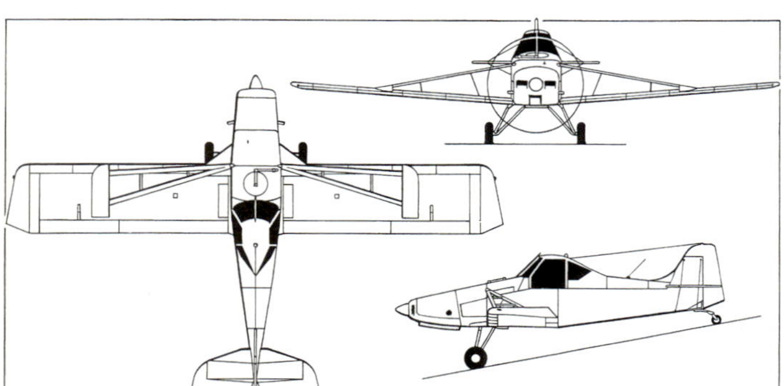

Gippsland GA200C Fatman sprayer *(Jane's/Mike Keep)* 0110575

GIPPSLAND GA8 AIRVAN

TYPE: Light utility transport.

PROGRAMME: Design completed and proof of concept prototype construction started early 1994; first flight 3 March 1995 (aircraft then marked only as 'GA-8' on the fin but allocated company test registrations VH-GAV and later VH-PTR); 186 kW (250 hp) Textron Lycoming O-540; prototype destroyed 7 February 1996 during spinning trials; second airframe completed to major component stage for static testing; third airframe (initially flown without test registration VH-2GI but with 'AIRVAN' on fin was second flying prototype, first flown in August 1996 with an interim 224 kW (300 hp) Textron Lycoming IO-540K-1A5 engine and three-blade propeller, pending installation of intended (but temporarily discontinued) production standard IO-580 in 1997; total of 350 hours flight testing completed by two prototypes by November 1998; second prototype re-registered VH-XGA in January 1999.

Provisional type certification obtained 10 March 1999, with full CASA certification to FAR Pt 23 Amendment 48 achieved 10 October 2000; subsequently upgraded to FAR Pt23

Gippsland GA8 Airvan of the US Civil Air Patrol ***NEW***/0569592

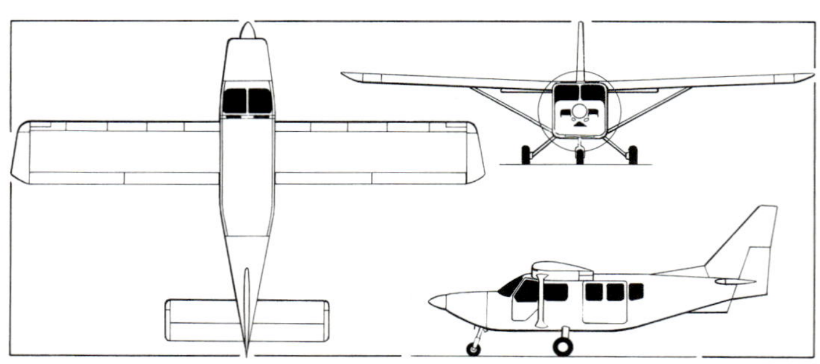

Gippsland GA8 Airvan (Textron Lycoming IO-540 engine) *(James Goulding)* 0110574

Amendment 54 on 14 March 2003; this applicable to c/n 026 (VH-HQE) and upwards, but earlier aircraft are upgradable. First delivery (VH-RYT) to Air Fraser Island Air 22 December 2000. Initial export aircraft (seventh overall, c/n 008/V3-HGI) first flew 24 November 2001, and delivered to Maya Island Air of Belize in following month. International debut (VH-BJY) at Asian Aerospace show in Singapore February 2002 before delivery to distributor PT Airvan Dirgantara of Indonesia. Transport Canada certification achieved in April 2003, followed by FAA FAR Pt23 Amendment 54 approval 30 May 2003; UK certification being sought in late 2003. Total of 36 built by October 2003, when production rate was 20 per year, rising to 50 in 2005 and 400 by 2007.

CURRENT VERSIONS: **GA8 Airvan:** Initial production version; IO-540 engine. Other engine options, including IO-580, turbocharged TIO-540 for hot-and-high operations and various aero diesel engines, under study in 2003.

GA10: Projected stretched, 10-seat turboprop version to be powered by a 313 kW (420 shp) Rolls-Royce 250-B17F under study in early 2002.

CUSTOMERS: Total of 33 delivered by October 2003, including exports to Belize (Maya Island Air, two), Botswana (two), Canada, Indonesia (PT Airvan Dirgantara, three), Mozambique (two), New Zealand (three), South Africa (two) and USA (five). Australian customers include Air Fraser Island (three), Alligator Airways (two), Shine Aviation Services (one), Skydive Oz (one), Slingair (one), TGS Air Charter (two), and Wrightsair (one). In January 2003, the US Civil Air Patrol appointed Gippsland to supply Airvans on an Indefinite Quantity Indefinite Delivery (IQID) basis over a three-year period. The first CAP Airvan, c/n 030 (VH-AUP/N605CP), was delivered in August 2003 and will be used for SAR and Homeland Security missions.

COSTS: US$500,000, VFR equipped (2003).

DESIGN FEATURES: Strut-braced, high-wing monoplane with sweptback vertical tail and fixed tricycle landing gear; designed to operate from unprepared strips. Fin and rudder modified, and ventral finlet added on second prototype, late 1996.

Wing aerofoil modified V-35; dihedral 2° 30'; incidence 2°; twist 1.6°.

FLYING CONTROLS: Conventional and manual. Trimmable tailplane, deflections +2°/–5°; slotted flaps, deflections 14 and 38°. Elevator deflections +15°/–19°; ailerons +17°/–16°; rudder ±21°.

STRUCTURE: Light alloy two-spar wing based on GA200 Fatman unit; composites cowlings and fairings; entire structure designed for easy manufacture, maintenance and repair.

LANDING GEAR: Non-retractable tricycle type; single-piece tubular spring main gear, steerable steel spring/oleo nose leg; Parker hydraulic disc brakes; mainwheels 8.50-6 (6 ply); nosewheel 6.00-6 (6 ply). Wipline float installation under development in Canada.

POWER PLANT: One 224 kW (300 hp) Textron Lycoming IO-540-K1A5 flat-six engine driving a two-blade Hartzell HC-2YR-1BF/F8475R constant-speed propeller. Fuel capacity 340 litres (89.8 US gallons; 74.8 Imp gallons) in two wing tanks, of which 332 litres (87.7 US gallons; 73.0 Imp gallons) usable. Oil capacity 11 litres (3.0 US gallons; 2.5 Imp gallons).

ACCOMMODATION: Pilot and up to seven passengers or equivalent cargo; full FAR Pt 23.562 compliant crashworthy seats. Amsafe seatbelt airbags under development in early 2003. Crew door each side of flight deck; flight openable forward-sliding cargo door on port side aft of wing. Cabin is heated and ventilated. Baggage compartment, with security net, at rear of cabin. Ventral cargo pod, volume 0.51 m³ (18.0 cu ft) under development in 2003.

SYSTEMS: Split bus electrical system powered by 14 V 95 A engine-driven alternator. Maintenance-free sealed battery and external power receptacle with overvolt protection standard.

AVIONICS: Standard Bendix/King avionics suite.

Comms: Dual KY 97A-61 VHF; KT 76A-12 transponder; KMA 24H71 audio panel and TX4400 intercom; pilot's yoke-mounted PTT switch; pilot's and co-pilot's headsets and microphone jacks.

Flight: KR 87 ADF; altitude encoder.

Instrumentation: KVD 150 moving map with 102 × 76 mm (4 × 3 in) display, airspeed indicator, artificial horizon, altimeter with hectoPascal subscale, electric turn co-ordinator, directional gyro, vertical speed indicator, OAT gauge, magnetic compass, electronic digital fuel flow and totaliser gauge, EGT gauge, tachometer, fuel supply gauge, oil temperature/pressure gauges, CHT gauge, and manifold pressure gauge.

EQUIPMENT: Standard equipment includes automatic fuel management system, windscreen demisting, instrument panel lighting, avionics dimming, map and cabin reading lights, navigation lights, dual landing/taxiing lights, low-profile red beacon lights; pilot's and co-pilot's four-point inertia reel shoulder harnesses; six three-point inertia reel harnesses for passengers; cabin soundproofing; and entrance step.

DIMENSIONS, EXTERNAL:
- Wing span: 12.41 m (40 ft 8¾ in)
- Wing chord, constant: 1.60 m (5 ft 3 in)
- Wing aspect ratio: 7.9
- Length overall: 8.95 m (29 ft 4¼ in)
- Fuselage max width: 1.37 m (4 ft 6 in)
- Height overall: 3.89 m (12 ft 9 in)
- Tailplane span: 4.17 m (13 ft 8 in)
- Wheel track: 2.79 m (9 ft 2 in)
- Wheelbase: 2.30 m (7 ft 6½ in)
- Cargo/passenger door: Height: 1.08 m (3 ft 6½ in)
  - Width: 1.07 m (3 ft 6 in)
  - Height to sill: 0.85 m (2 ft 9½ in)
- Propeller diameter: 2.13 m (7 ft 0 in)

DIMENSIONS, INTERNAL:
- Cabin: Length, incl flight deck: 4.01 m (13 ft 2 in)
  - Max width: 1.27 m (4 ft 2 in)
  - Max height: 1.19 m (3 ft 11 in)
  - Floor area: incl flight deck: 5.02 m² (54.0 sq ft)
    - excl flight deck: 3.44 m² (37.0 sq ft)
  - Volume, incl flight deck: 5.1 m³ (180 cu ft)

AREAS:
- Wings, gross: 19.32 m² (208.0 sq ft)
- Ailerons (total): 1.60 m² (17.20 sq ft)
- Trailing-edge flaps (total): 1.60 m² (17.20 sq ft)
- Fin: 1.90 m² (20.50 sq ft)
- Rudder: 0.77 m² (8.30 sq ft)
- Tailplane: 4.40 m² (47.40 sq ft)
- Elevators (total): 1.76 m² (18.90 sq ft)

WEIGHTS AND LOADINGS:
- Weight empty: 997 kg (2,198 lb)
- Baggage capacity: shelf: 113 kg (250 lb)
  - aft bin: 23 kg (50 lb)
- Cargo pod capacity: 150 kg (330 lb)
- Max fuel: 248 kg (547 lb)
- Max T-O and landing weight:
  - initial certification: 1,814 kg (4,000 lb)
  - target: 1,905 kg (4,200 lb)
- Max wing loading:
  - initial certification: 93.9 kg/m² (19.23 lb/sq ft)
  - target: 98.6 kg/m² (20.19 lb/sq ft)
- Max power loading:
  - first prototype: 9.74 kg/kW (16.00 lb/hp)
  - initial certification: 8.12 kg/kW (13.33 lb/hp)
  - target: 8.52 kg/kW (14.00 lb/hp)

PERFORMANCE:
- Never-exceed speed (VNE): 185 kt (342 km/h; 212 mph)
- Normal cruising speed: 121 kt (224 km/h; 139 mph)
- Econ cruising speed: 104 kt (193 km/h; 120 mph)
- Stalling speed: flaps up: 60 kt (112 km/h; 69 mph)
  - flaps down: 52 kt (97 km/h; 60 mph)
- Max rate of climb at S/L: 240 m (787 ft)/min
- Service ceiling, estimated: 6,100 m (20,000 ft)
- T-O run: 347 m (1,138 ft)
- T-O to 15 m (50 ft): 554 m (1,818 ft)
- Landing from 15 m (50 ft): 371 m (1,218 ft)
- Landing run: 155 m (509 ft)
- Range: with max fuel:
  - normal cruise: 730 n miles (1,352 km; 840 miles)
  - econ cruise: 930 n miles (1,722 km; 1,070 miles)
- Endurance, no reserves:
  - at normal cruising speed: 6 h
  - at econ cruising speed: 9 h

UPDATED

GOAIR

GOAIR PRODUCTS

675 Drover Road, Bankstown Airport, New South Wales 2200
Tel:(+61 2) 97 96 34 25
Fax: (+61 2) 97 91 03 54
e-mail: goair@bigpond.com
MANAGING DIRECTOR: Phil Goard

Company machines, welds and manufactures aircraft components. During the 1990s it designed, built and flight-tested the Goair Trainer, production of which is now under way at a low rate.

In 2001, Goair completed a replica of the 1920s Westland Widgeon.

VERIFIED

GOAIR GT-1 TRAINER

TYPE: Side-by-side lightplane/kitbuilt.

PROGRAMME: Four years in design; public debut (VH-BBR, then unflown) at Avalon Air Show, March 1995; first flight July 1995; some 120 hours of flight testing completed by November 1998; second aircraft (preproduction) displayed without markings at Aviex 2000 at Sydney/Bankstown and subsequently registered (VH-AYS) on 22 January 2001; had flown 30 hours by late 2001; certification to JAR-VLA was expected in late 2001 but had slipped; will also be marketed as a kitbuilt.

CUSTOMERS: Five aircraft built or under construction by early 2003, of which two had flown.

COSTS: Estimated US$75,000 for certified aircraft (2003).

DESIGN FEATURES: Conventional low-wing, fixed landing gear monoplane; design goals strength, simplicity, low costs, docile stall characteristics and ease of maintenance. Production aircraft differs from prototype in having a forward-sliding canopy; cabin 7.6 cm (3 in) wider; larger ailerons and slotted flaps; and leaf spring main landing gear legs.

FLYING CONTROLS: Conventional and manual; elevators and ailerons are pushrod operated, rudder is cable operated. Flaps range 0–40°.

STRUCTURE: All metal.

LANDING GEAR: Non-retractable tricycle type with spring steel main legs and rubber-in-compression nose leg. Wheel size (all) 5.00×5. Hydraulic toe brakes standard.

POWER PLANT: One 88 kW (118 hp) Textron Lycoming O-235-C2A (in first prototype) or -C1 (in preproduction aircraft) flat-four engine, driving a Sensenich two-blade metal propeller. Fuel contained in two integral tanks in leading-edges of wings, combined capacity 130 litres (34.3 US gallons; 28.6 Imp gallons).

ACCOMMODATION: Two persons, side by side under forward-sliding bubble canopy; dual controls (sticks) standard, control wheels optional; seats adjust fore and aft; four-point harnesses standard. Baggage compartment behind seats.

Prototype Goair Trainer *(Gerard Frawley/Australian Aviation)* 0507480

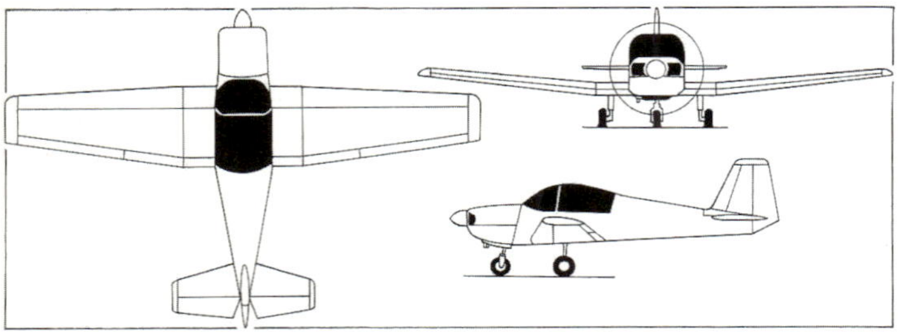

Goair Trainer (115 hp Textron Lycoming O-235) in prototype configuration *(Paul Jackson)* 0507481

DIMENSIONS, EXTERNAL:
Wing span 8.76 m (28 ft 9 in)
Wing chord at root 1.52 m (5 ft 0 in)
Wing aspect ratio: prototype 6.9
production 7.3
Length overall: prototype 6.25 m (20 ft 6 in)
production 7.16 m (23 ft 6 in)
Height overall: prototype 2.03 m (6 ft 8 in)
production 2.29 m (7 ft 6 in)

DIMENSIONS, INTERNAL:
Cabin max width 1.14 m (3 ft 9 in)
Baggage compartment volume 0.17 m³ (6.0 cu ft)

AREAS:
Wings, gross 10.50 m² (113.0 sq ft)

WEIGHTS AND LOADINGS:
Weight empty 470 kg (1,036 lb)
Max T-O weight 748 kg (1,650 lb)
Max wing loading 71.3 kg/m² (14.60 lb/sq ft)
Max power loading 8.51 kg/kW (13.98 lb/hp)

PERFORMANCE:
Max level speed 115 kt (213 km/h; 132 mph)
Cruising speed 100 kt (185 km/h; 115 mph)
Stalling speed: flaps up 50 kt (93 km/h; 58 mph)
flaps down 42 kt (78 km/h; 49 mph)
Max rate of climb at S/L 229 m (750 ft)/min
T-O run 366 m (1,200 ft)
Landing run 198 m (650 ft)
Endurance 5 h

VERIFIED

HUGHES

HOWARD HUGHES ENGINEERING PTY LTD

PO Box 89, Lot 8, Southern Cross Drive, Ballina, New South Wales 2478
Tel: (+61 2) 66 86 86 58
Fax: (+61 2) 66 86 83 43
e-mail: al@spot.com.au
Web: http://www.lightwing.com.au
MANAGING DIRECTOR: Howie Hughes

Company formed 1984 to design and produce the Australian LightWing series which has developed from GR-582 via GA-55 to GR-912 and the associated Sport 2000. Most production, undertaken in a 2,000 m² (21,525 sq ft) factory at Ballina Airport, is for local use and either full certification or registry under AUF rules; around 10 aircraft per year are built. Aircraft for export are issued with Australian certificates of airworthiness. By end 2003, over 150 aircraft had been produced.

In 2003, the company announced the LightWing Speed, a low-wing development of the GR-912 family. In early 2003, the workforce numbered 10.

UPDATED

HUGHES AUSTRALIAN LIGHTWING GR-912

TYPE: Side-by-side ultralight/kitbuilt.

PROGRAMME: Prototype Australian LightWing (ALW) made first flight June 1986; initially with 47.8 kW (64.1 hp) Rotax 582 UL; many sold conforming to CAO 95 Pt 25 requirements; also produced under homebuilt or CAO 101 Pt 28 regulations in non-retractable tailwheel and twin-float configurations; first amphibian model delivered in 1990. Type certificate issued 25 September 1998.

CURRENT VERSIONS: **GR-912:** Tailwheel version. *As described.*

GR-912-T: Tricycle landing gear version; marketing designation Sport 2000; see following entry.

LightWing Truck: Added to range in 2002; uses same wing married to 45 cm (1 ft 5¾ in) longer fuselage with extra space behind seats able to accommodate two further seats; reinforcing has obviated need for wire bracing on tail unit. Cabin width increased by 15 cm (6 in). Powered by 84.6 kW (113.4 hp) Rotax 914; will accept eight-cylinder 149 kW (200 hp) Jabiru once engine development completed. Initially offered in tailwheel configuration, with nosewheel version due to enter service in 2003.

CUSTOMERS: Over 150 aircraft (all models) delivered by end 2002; most registered under AUF rules. Claimed 29 per cent share of Australian two-seat light trainer market with GR-912.

COSTS: GR-912 A$85,900 factory built (2003).

DESIGN FEATURES: Modified Clark Y wing section.

FLYING CONTROLS: Conventional and manual. Optional flaps.

STRUCTURE: Strut-braced metal wings; welded steel tube fuselage and tail; tail surfaces are wire-braced; composites and polyfibre covering.

LANDING GEAR: Non-retractable tailwheel type; optional amphibious floats with retractable mainwheels and non-retractable tailwheels (latter doubling as water rudders).

POWER PLANT: One 73.5 kW (98.6 hp) Rotax 912 ULS flat-four driving a two- or three-blade propeller; Aerofiber wooden two-blade fixed-pitch propeller recommended. Fuel capacity 60 litres (15.8 US gallons; 13.2 Imp gallons) of which 59 litres (15.6 US gallons; 13.0 Imp gallons) are usable.

DIMENSIONS, EXTERNAL:
Wing span 9.50 m (31 ft 2 in)
Length overall 5.68 m (18 ft 7½ in)
Height overall 1.90 m (6 ft 2¼ in)

Hughes GR-912 tailwheel version *(Paul Jackson)* ***NEW***/0547733

AREAS:
Wings, gross 15.35 m² (165.2 sq ft)

WEIGHTS AND LOADINGS:
Weight empty 295 kg (650 lb)
Max T-O weight 544 kg (1,199 lb)

PERFORMANCE:
Max level speed 95 kt (176 km/h; 109 mph)
Max cruising speed 80 kt (148 km/h; 92 mph)
Loiter speed 50 kt (93 km/h; 58 mph)
Stalling speed, full flaps, power off 40 kt (74 km/h; 46 mph)
Max rate of climb at S/L 244 m (800 ft)/min
T-O run 100 m (330 ft)
Landing run 80 m (263 ft)
Range with max fuel 216 n miles (400 km; 248 miles)

UPDATED

HUGHES AUSTRALIAN LIGHTWING SPORT 2000

TYPE: Side-by-side ultralight/kitbuilt.

PROGRAMME: Developed version of GR-912 (which see); certified in Australian Primary category. Uses original LightWing type certificate.

CUSTOMERS: 10 built by end 2002.

COSTS: A$88,900 factory built; A$45,900 kit, including alternative 59.6 kW (79.9 hp) Werner engine (2003).

FLYING CONTROLS: Conventional and manual. Horn-balanced tail surfaces; elevator pushrod-operated; aileron and rudder cable-operated.

STRUCTURE: Welded steel tube fuselage; riveted aluminium alloy wings; mainly fabric covered, with non-structural glass fibre fairings.

LANDING GEAR: Non-retractable tricycle type.

POWER PLANT: One 59.6 kW (79.9 hp) Rotax 912 UL flat-four driving an Aerofiber two-blade fixed-pitch wooden propeller. Fuel capacity 62 litres (16.4 US gallons; 13.6 Imp gallons).

DIMENSIONS, EXTERNAL:
Wing span 9.50 m (31 ft 2 in)
Length overall 5.68 m (18 ft 7½ in)
Height overall 2.20 m (7 ft 2½ in)

AREAS:
Wings, gross 15.14 m² (163.0 sq ft)

WEIGHTS AND LOADINGS:
Weight empty 300 kg (661 lb)
Max T-O weight 480 kg (1,058 lb)

PERFORMANCE:
Cruising speed at 75% power 72 kt (133 km/h; 83 mph)
Stalling speed 35 kt (65 km/h; 41 mph)
T-O and landing from 15 m (50 ft) 300 m (985 ft)
Range with max fuel 300 n miles (555 km; 345 miles)
g limits +6/−4

VERIFIED

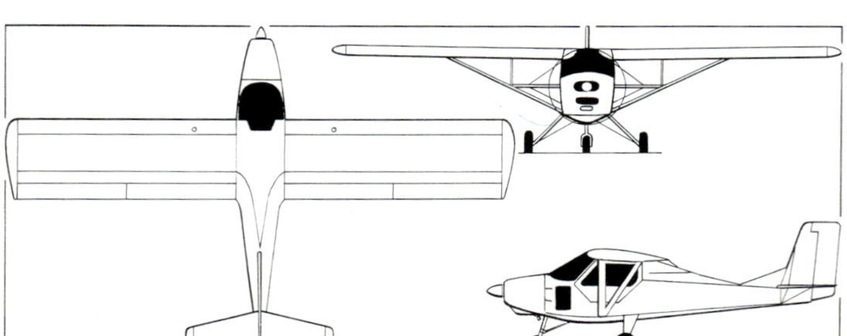

Hughes Australian LightWing Sport 2000, showing optional 'Heliview' cockpit with side window at floor height (Rotax 912 UL flat-four) *(James Goulding)* 0010201

HUGHES AUSTRALIAN LIGHTWING SPEED

TYPE: Side-by-side ultralight/kitbuilt.

PROGRAMME: Development of version announced in 2003. First flight date had not been announced by end of 2003. Will be available only as a kit for AUF or Experimental categories.

DESIGN FEATURES: Based on company's existing LightWing GR-912 family.

FLYING CONTROLS: As LightWing GR-912 but with flaps as standard.

STRUCTURE: As LightWing GR-912, but with low wing. Fuselage has integral roll-cage.

LANDING GEAR: Initially offered as a non-retractable tailwheel type.

POWER PLANT: One 84.6 kW (113.4 hp) Rotax 914 driving an in-flight adjustable propeller. Fuel tanks in glass-fibre wingtips.

ACCOMMODATION: Gull-wing doors each side of fuselage.

All data are provisional.

DIMENSIONS, EXTERNAL:

Wing span ... 7.40 m (24 ft 3¼ in)

Length overall ... 6.00 m (19 ft 8¼ in)

WEIGHTS AND LOADINGS:

Max T-O weight ... 544 kg (1,199 lb)

PERFORMANCE:

Normal cruising speed ... 150 kt (278 km/h; 173 mph)

Stalling speed, flaps down ... 45 kt (84 km/h; 52 mph)

Service ceiling ... 3,050 m (10,000 ft)

UPDATED

HUGHES PR-BIPE

TYPE: Tandem-seat, ultralight biplane/kitbuilt.

PROGRAMME: Name reflects biplane version of Hughes Pocket Rocket. Offered from 2003; kit priced at AU$34,990 (2004).

NEW ENTRY

Partly sectioned drawing of proposed Hughes PR-Bipe 0109141

INTEGRITY

INTEGRITY AIRCRAFT

PO Box 569, Wauchope, New South Wales 2446
Tel: (+61 409) 46 36 13
e-mail: lhwatson@ceinternet.com.au
OWNER AND MANAGER: Lance H Watson

In May 2003, Mr Watson gave evidence to the Australian House of Representatives Transport & Regional Services Committee hearings, which included details of his company's intention to develop an aircraft able to serve remote communities at "bus fare prices". Manufacture was due to begin in early 2005.

UPDATED

INTEGRITY INTEGRITY

TYPE: Light utility turboprop.

PROGRAMME: Announced 2003. Version of Britten-Norman Trislander powered by a single 746 kW (1,000 hp) Honeywell TPE331-12 turboprop in the tail position; centre of gravity maintained by 76 cm (30 in) extension to forward fuselage, increasing seating capacity to 20 including pilot(s). Design advantages include ability to keep engine running while reloading and refuelling. Typical route length 324 n miles (600 km; 373 miles). Received 20 expressions of interest by May 2003; estimated cost US$1.7 million.

Preliminary data include cruising speed of 162 kt (300 km/h; 186 mph); 2,358 kg (5,200 lb) MTOW; ferry range 971 n miles (1,800 km; 1,118 miles).

Integrity single-engined version of Britten-Norman Trislander ***NEW***/0567230

Integrity owns 12 uncompleted Trislander kits which would be used as prototypes and to launch production. Further demand would be met by new production at Romaero in Romania. Type certification to be undertaken in UK; design patent pending in USA.

UPDATED

JABIRU

JABIRU AIRCRAFT PTY LTD

PO Box 5186, Bundaberg West, Queensland 4670
Tel: (+61 7) 41 55 17 78
Fax: (+61 7) 41 55 26 69
e-mail: jabiru@tpgi.com.au
Web: http://www.jabiru.net.au
JOINT MANAGING DIRECTORS:
Phil Ainsworth
Rodney Stiff

EUROPEAN AGENT:
ST Aviation Ltd
Technology House, High Street, Downham Market, Norfolk PE38 9HH, UK
Tel: (+44 1366) 38 55 58
Fax: (+44 1366) 38 55 59
Web: http://www.jabiru.co.uk
MANAGING DIRECTOR: Kevin Pearce

Jabiru formed 1988 by Rodney Stiff and Phil Ainsworth to produce Jabiru LSA 55/2K, but branched into engine design and manufacture when Italian-built KFM engine, which powered this variant, was withdrawn from market. Now produces 1,600 and 2,200 cc flat-four and 3,300 cc flat-six engines for its own and other manufacturers' aircraft. A factory-built SLA version of Jabiru, the UL 450, was launched in 2000. The company and aircraft are named for Australia's sole indigenous stork. Production averages 100 aircraft and 360 engines per year. A prototype of a four-seat Jabiru, the J400, first flew in March 2001 and customer deliveries began almost immediately. By December 2003, 33 Jabirus were registered under Australian CAA regulations.

UPDATED

JABIRU JABIRU

TYPE: Side-by-side ultralight/kitbuilt.

PROGRAMME: Design of original Jabiru LSA, started 1987; prototype (first of two, VH-JCX and VH-JQX) made first flight August 1989; first customer delivery April 1991; factory-built version certified to CAO 101 Pt 55 on 1 October 1991; 20 built with KFM engine before change to 44.7 kW (60 hp) Jabiru 1600 flat-four, many of these subsequently re-engined; kitplane construction certification achieved in Australia, New Zealand, South Korea, UK and USA by early 1997; marketed in US as amateur-built kit, achieving FAR Pt 21.191 (g) certification on 8 February 1996 (as well as Australian CAO 19 equivalent); marketing office opened at Aiken, South Carolina, in anticipation of substantial US sales of at least 100 per year but later closed in favour of network of dealerships. Total of 54 built with 1600 engine between April 1993 and March 1996. More powerful ST version introduced in 1994.

Joint venture announced in 1998 with CDE Aviation Company of Sri Lanka, subsidiary of Lionair, to establish Jabiru Asia for manufacture of Jabiru ST3 for Asian markets, especially India. Jabiru factory at Koggala, in southern Sri Lanka; intended to have production capacity for 40 aircraft per year, using engines and avionics imported from Australia, and airframes manufactured locally. Nothing further had been heard of this by 2003.

CURRENT VERSIONS: **Jabiru LSA:** Light Sport Aircraft; factory-built; registered under AUF (Australian Ultralight Federation) rules (Pt 101.55). 'Short span; short fuselage' version: span 8.03 m (26 ft 4 in), length 5.03 m (16 ft 6 in), wing area 7.90 m^2 (85.0 sq ft).

Jabiru ST: Certified version; in factory production from mid-1994; first was 50th production Jabiru. In mid-1998, the 17th production ST was registered VH-JRU to Jabiru Aircraft as the first **ST3** short span and fuselage.

Factory-built Jabiru LSA ***(Paul Jackson)*** ***NEW***/0569585

Tailwheel Jabiru SP-T *(Paul Jackson)* ***NEW***/0569584

Jabiru Calypso registered in Canada *(Paul Jackson)* ***NEW***/0569583

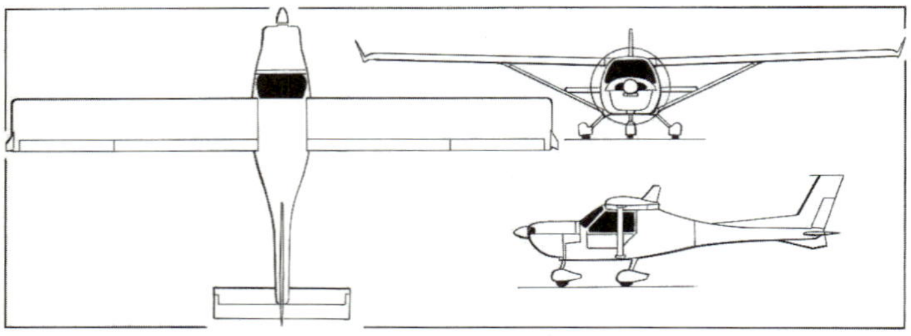

Jabiru Calypso two-seat ultralight with short rudder *(James Goulding)* 0143319

Jabiru SK: Quick-build kit version of ST (components are manufactured alongside production aircraft); first was 66th production Jabiru; quoted home-build time is 600 hours.

Jabiru SP4: Kit, combining longer fuselage of UL 450 with Jabiru 2200 engine and short-span wings of SK; originally designated SP. MTOW 470 kg (1,036 lb); stalling speed 39 kt (73 km/h; 45 mph); optimised for Canadian, Chilean and South African markets, among others. Fuel capacity 70 litres (18.5 US gallons; 15.4 Imp gallons).

Jabiru SP6: Company designation of SP4 kit with 89.5 kW (120 hp) Jabiru 3300 engine. Also known as **Jabiru SP 3300**. MTOW 500 kg (1,102 lb). Cruising speed 117 kt (217 km/h; 135 mph). Fuel capacity as SP4, or 85 litres (22.5 US gallons; 18.7 Imp gallons) optional.

Jabiru SP-T: Tailwheel version of SP. Enlarged rudder; mainwheel legs moved forward; steerable tailwheel and differential braking. Prototype built by Peter Kayne of Narrogin, Western Australia.

Jabiru SP 480: Proposed version of SP for JAR-VLA certification, with maximum take-off weight of 480 kg (1,058 lb); under development in late 1999.

Jabiru Calypso: Initially factory-built; now available only as kit; previously known as **UL 450**; SLA (Small Light Aeroplane to UK's now-replaced BCAR Section S), first aircraft was built in Ireland in 1998. Launched in UK in 2000 and meeting 35 kt (65 km/h; 41 mph) stalling speed requirement. Differences from SK include wing span increased by 1.37 m (4 ft 6 in); fuselage by 0.61 m (2 ft 0 in); flaps by total of 1.83 m (6 ft 0 in); and reduced ventral fin. Also available with tailwheel as **Calypso T**. Meets FAA LSA. 150 plus sold by end 2001.

J200 : and J400: Stretched version; described separately.

CUSTOMERS: Production of complete and kit aircraft totalled some 535 by December 2002. Sold to 22 countries. Manufacture in 2003 was mainly of four-seat versions.

COSTS: Calypso: US$20,900; SP US$19,900 (2003) both minus engine. Flyaway Calypso 2200 US$59,995 (April 2002).

DESIGN FEATURES: Designed to Australian AUF Pt 101.55 and British BCAR Section S, with reference to FAR Pt 21 and JAR-VLA. Ultralight or certified aircraft, suitable for factory or amateur assembly. Unswept wing (braced) and tailplane, swept fin; dorsal and ventral fins; all flying surfaces square-tipped; wings detachable for storage and transportation.

Wing section NASA 4412; drooped wingtips; dihedral 1° 15'; incidence 2° 30'; no twist.

FLYING CONTROLS: Conventional and manual. In-flight adjustable pitch trim; wide-span slotted flaps. Original short rudder supplanted by full-height rudder (also available as retrofit) on all versions. From 2002, enlarged fin version (¾ height rudder) adopted, entire vertical tail being moved further forward.

STRUCTURE: All-GFRP except metal wing struts, nosewheel assembly and engine mount.

LANDING GEAR: Non-retractable tricycle type, with 15° steerable nosewheel coupled to rudder. Cantilever self-sprung GFRP mainwheel legs; rubber block nosewheel shock-absorption; wheel fairings standard over 4.00–6 or 5.00–6 tyres; Bigfoot 6.00–6 tyres optional. Nosewheel 2.60–4, 4.00–4 or 5.00–6; tailwheel 250×50.

POWER PLANT: One 59.7 kW (80 hp) Jabiru 2200A/J flat-four engine, driving a two-blade, fixed-pitch, wood/composites propeller. Fuel capacity 65 litres (17.2 US gallons; 14.3 Imp gallons).

SYSTEMS: 12 V DC electrical system with 100 W capacity; 20 Ah battery.

AVIONICS: *Comms:* VHF antenna (inside fin) and intercom standard.

Instrumentation: Basic VFR; vacuum instruments optional.

Data below refer to Jabiru Calypso with Jabiru 2200 engine.

DIMENSIONS, EXTERNAL:

Wing span: normal	9.40 m (30 ft 10 in)
with winglets	9.56 m (31 ft 4½ in)
Length overall	5.64 m (18 ft 6 in)
Height overall	2.01 m (6 ft 7 in)

DIMENSIONS, INTERNAL:

Cockpit max width at shoulder	1.07 m (3 ft 6 in)

AREAS:

Wings, gross	9.29 m² (100.0 sq ft)

WEIGHTS AND LOADINGS:

Weight empty, equipped	242 kg (534 lb)
Max T-O weight	450 kg (992 lb)

PERFORMANCE:

Never-exceed speed (VNE)	120 kt (222 km/h; 138 mph)
Max level speed	110 kt (204 km/h; 127 mph)
Cruising speed at 75% power	100 kt (185 km/h; 115 mph)
Stalling speed, flaps down	35 kt (65 km/h; 41 mph)
Max rate of climb at S/L	305 m (1,000 ft)/min
T-O run	100 m (330 ft)
Landing run	168 m (555 ft)
Range with max fuel	430 n miles (796 km; 494 miles)

UPDATED

JABIRU J200 AND J400

TYPE: Four-seat kitbuilt.

PROGRAMME: Stretched development of Jabiru, described in previous entry. Prototype (19-3512), in ultralight configuration, first flew 16 March 2001. Maiden flight of first customer-built J200, 13 December 2001. Meets FAA 51% rule.

CURRENT VERSIONS: **J200:** Two-seat ultralight with same dimensions as J400. **200A** has maximum allowable T-O weight; **200B** limited to 544 kg (1,200 lb) to meet AUF ultralight rules.

J250: Announced at Sun 'n' Fun, April 2002; meets FAA LSA kitbuilt rules. Wing span increased 1.65 m (5 ft 5 in) to 9.75 m (32 ft 0 in) and area 3.89 m²(41.9 sq ft) to 11.89 m² (128.0 sq ft) to meet stall requirements of 44 kt (82 km/h; 51 mph) 'clean' and 39 kt (73 km/h; 45 mph) flaps down. Empty weight 326 kg (720 lb); MTOW 559 kg (1,232 lb).

J400: Four-seat lightplane.

J450: US market version (also sold in South Africa). Increased wing span (9.14 m; 30 ft 0 in) and area (11.15 m²; 120.0 sq ft); fuel 139 litres (36.8 US gallons; 30.7 Imp gallons).

CUSTOMERS: Total 150 sold by end of 2003, some being J400/J450 exports to New Zealand, South Africa and the United States.

COSTS: Kit US$41,500 without engine; US$55,400 with engine; J400 ready-to-fly US$65,000 (2004).

Data generally as for Jabiru Jabiru, except that below.

DESIGN FEATURES: Stretched fuselage with second pair of side windows. Wings of Jabiru LSA, with fuel in wings.

LANDING GEAR: Larger 5.00–6 wheels standard on J400 with 4.00–4 wheels optional.

POWER PLANT: One 89.5 kW (120 hp) Jabiru 3300 flat-six, driving two-blade fixed-pitch propeller. Fuel capacity 100 litres (26.4 US gallons; 22.0 Imp gallons) in wings.

ACCOMMODATION: Four persons; or two, if flown to ultralight regulations. Three doors on J400; rear port door operable only when flaps retracted.

DIMENSIONS, EXTERNAL:

Wing span	8.10 m (26 ft 7 in)
Wing aspect ratio	8.2
Length overall	6.55 m (21 ft 5¾ in)
Height overall	2.20 m (7 ft 2½ in)
Tailplane span	2.66 m (8 ft 8¾ in)
Wheel track	1.80 m (5 ft 10¾ in)
Propeller diameter	1.52 m (5 ft 0 in)

DIMENSIONS, INTERNAL:

Cabin: Max width	1.12 m (3 ft 8 in)
Max height	1.09 m (3 ft 7 in)

AREAS:

Wings, gross	8.00 m² (86.1 sq ft)

WEIGHTS AND LOADINGS:

Weight empty: J200A	300 kg (661 lb)
J200B/J400	310 kg (683 lb)
J450	313 kg (690 lb)
Max T-O weight: J200B	544 kg (1,200 lb)
J200A/J400, J450	700 kg (1,543 lb)

Jabiru J450 assembled in the US *(Paul Jackson)* ***NEW***/0580306

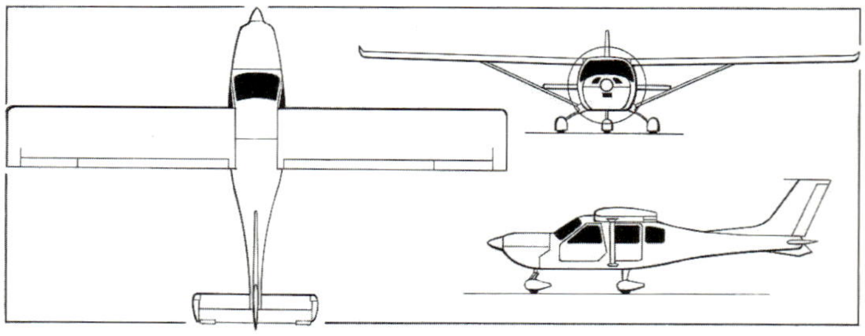

Two/four-seat Jabiru J200/J400 *(James Goulding)* 0143320

Max wing loading: J200B .... 68.0 kg/m² (13.92 lb/sq ft)
J200A/J400 .... 87.5 kg/m² (17.92 lb/sq ft)
Max power loading: J200B .... 6.08 kg/kW (10.00 lb/hp)
J200A/J400 .... 7.82 kg/kW (12.86 lb/hp)

PERFORMANCE:
Never-exceed speed (VNE) .... 138 kt (255 km/h; 158 mph)
Max level speed .... 130 kt (240 km/h; 150 mph)
Cruising speed .... 120 kt (222 km/h; 138 mph)
Manoeuvring speed .... 90 kt (166 km/h; 103 mph)
Stalling speed, power off:
flaps up:
J200B .... 51 kt (94 km/h; 58 mph)
J200A/J400 .... 60 kt (111 km/h; 69 mph)
flaps down: J200B .... 45 kt (83 km/h; 51 mph)
J200A/J400 .... 48 kt (88 km/h; 55 mph)
Max rate of climb at S/L: J200 .... 457 m (1,500 ft)/min
J450 .... 305 m (1,000 ft)/min
Service ceiling: all .... 4,570 m (15,000 ft)
T-O and landing run .... 200 m (660 ft)
Range:
except J450 .... 520 n miles (963 km; 598 miles)
J450 .... 660 n miles (1,222 km; 759 miles)
Endurance: except J450 .... 4 h 25 min
J450 .... 5 h 30 min
g limits .... +3.8/–1.9

UPDATED

SAPPHIRE

SAPPHIRE AIRCRAFT AUSTRALIA PTY LTD

175 Lloyd Street, Strathdale, Bendigo, Victoria 3550
Tel/Fax: (+61 3) 54 44 48 21
e-mail: sapphireaircraft@bigpond.com
Web: http://www.users.bigpond.com/stevendumesny
SALES MANAGER: Steven Dumesny

Sapphire markets an updated version of the Winton Sapphire, designed to CAO 19 by Scott Winton. A two-seat version, nominally designated Sapphire Scenic, and a self-launching glider were under development but have been cancelled.

UPDATED

Kitbuilt Sapphire LSA Mk II *(Paul Jackson)* ***NEW***/0547734

SAPPHIRE LSA MK II

TYPE: Single-seat ultralight/kitbuilt.

PROGRAMME: Introduced in 1998. First kit delivered in May 1999 and flown in September 1999. Company aims to produce one factory-built aircraft per year to evaluate production methods and develop kit manual; remainder of production is of kit version.

CURRENT VERSIONS: **Sapphire LSA:** Available as factory-built version meeting Australian CAO 95 Pt 25 certification standards, or as kitbuilt to CAO 95 Pt 10. Three kits available: Basic, comprising main airframe components; Intermediate, with control systems completed; and Advanced, including engine, propeller and instruments. All kits qualify for 51 per cent amateur-built kit regulations.

Sapphire Scenic: Proposed two-seat version under development will have MTOW of 500 kg (1,102 lb), 94 kt (174 km/h; 108 mph) cruise. Cost approximately A$35,000, plus tax (2002); project cancelled late 2003.

CUSTOMERS: First kit sold May 1999. Total of 42 flying by January 2003, including two factory-built.

COSTS: Factory-built flyaway A$35,266; Basic kit A$16,390; Intermediate kit A$23,650; Advanced kit A$29,414, all including tax (2003). Operating cost around A$24 per hour.

DESIGN FEATURES: Pod-and-boom fuselage with strut-braced shoulder-mounted wings; open cockpit with large windscreen. Baggage box, volume 25 litres (0.88 cu ft), in fuselage pod.

FLYING CONTROLS: Conventional and manual. Ailerons and trimmable all-flying tailplane operated by pushrods; rudder spring-coupled to tailwheel; quarter-span flaps; deflections 5° up (for fast cruising) and 20 and 40° down.

STRUCTURE: Mostly composites; GFRP fuselage pod reinforced with internal bulkheads; aluminium tailboom; wings of GFRP with foam ribs; fin is of GFRP; horizontal tail and rudder of aluminium tube with fabric covering. Wings and horizontal tail detach for storage or transport; quoted disassembly/assembly times 10 minutes and 5 minutes respectively.

LANDING GEAR: Tailwheel type; fixed. One-piece laminated GFRP main legs. Mainwheel size 11–4.00×5. Cable-operated brakes. Wheel fairings optional. At least one nosewheel-equipped Sapphire conversion exists.

POWER PLANT: One 29.5 kW (39.6 hp) Rotax 447 UL-1V or 34.0 kW (45.6 hp) Rotax 503 UL-1V two-cylinder two-stroke, driving a Sweetapple two- or three-blade wooden pusher propeller via 2.24:1 reduction gearing. Integral fuel tanks in wings, standard capacity 38 litres (10.0 US gallons; 8.4 Imp gallons), optional 58 litres (15.3 US gallons; 12.8 Imp gallons).

EQUIPMENT: Fully enclosed canopy and strut fairings optional.

DIMENSIONS, EXTERNAL:
Wing span .... 8.84 m (29 ft 0 in)
Length overall .... 5.00 m (16 ft 4¾ in)
Height overall .... 1.25 m (4 ft 1¼ in)
Propeller diameter .... 1.27 m (4 ft 2 in)

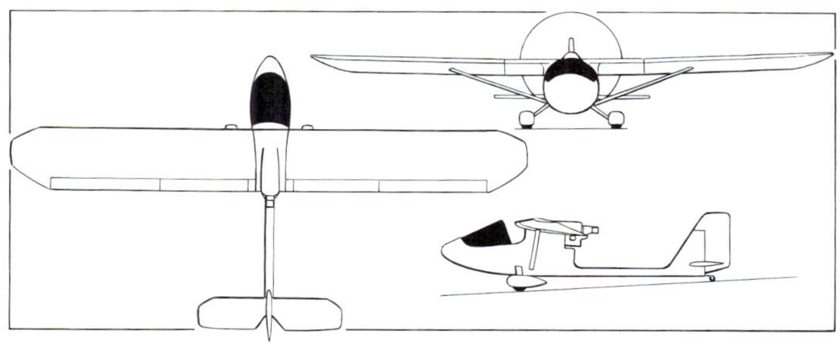

Sapphire LSA Mk II *(James Goulding)* ***NEW***/0561622

DIMENSIONS, INTERNAL:
Cabin max width .... 0.61 m (2 ft 0 in)

AREAS:
Wings, gross .... 9.13 m² (98.3 sq ft)

WEIGHTS AND LOADINGS:
Weight empty .... 158 kg (348 lb)
Max T-O weight .... 350 kg (771 lb)

PERFORMANCE:
Never-exceed speed (VNE) .... 98 kt (181 km/h; 112 mph)
Cruising speed at 75% power:
without fairings .... 75 kt (139 km/h; 86 mph)
with fairings: two-blade propeller .... 75 to 85 kt (139 to 157 km/h; 86 to 98 mph)
three-blade propeller .... 87 to 89 kt (161 to 165 km/h; 100 to 102 mph)
Stalling speed: power on .... 34 kt (63 km/h; 40 mph)
power off .... 36 kt (67 km/h; 42 mph)
Max rate of climb at S/L .... 305 m (1,000 ft)/min
Service ceiling .... 3,660 m (12,000 ft)
T-O run .... 55 m (180 ft)
Landing run .... 100 m (330 ft)
Range .... 328 n miles (607 km; 377 miles)
Endurance .... 3 h 30 min
g limits .... +8/–4

UPDATED

SEABIRD

SEABIRD AVIATION AUSTRALIA PTY LTD

Hervey Bay Airport, PO Box 618, Pialba, Hervey Bay, Queensland 4655
Tel: (+61 7) 41 25 31 44
Fax:(+61 7) 41 25 31 23
e-mail: seabirdaust@optusnet.com.au
CHAIRMAN AND MANAGING DIRECTOR: Donald C Adams
PRODUCTION DIRECTOR: Peter Adams

Seabird Aviation was founded in 1983 to develop the Seeker observation aircraft. Despite having built Seekers for certification, Seabird's business plan was always to franchise Seeker production capability, rather than sell individual aircraft; entire manufacturing equipment for a Seeker production line can be forwarded in two standard-size ISO shipping containers.

On 25 July 2000, Seabird signed an agreement with Evektor-Aerotechnik (EV-AT) of the Czech Republic leading to the formation of a joint venture company, Evektor-Seabird Pty Ltd, to manufacture and market the Seeker. This had produced no aircraft by mid-2003, when Seabird reached a new agreement with KADDB of Jordan for local assembly in quantity. Accordingly, the aircraft is described in the Jordanian section.

STORCH

STORCH AVIATION AUSTRALIA PTY LTD

113 Koree Island Road, Beechwood, New South Wales 2446
Tel/Fax: (+61 2) 65 85 64 58
Fax: (+61 2) 65 85 66 22
e-mail: info@storch.com.au
Web: http://www.storch.com.au
DESIGNER: Nestor Slepcev

Nestor Slepcev designed, built and flew two earlier aircraft before producing the Storch Mk 4. By 2003, 140 of the type had been produced. In August 1996, Mr Slepcev landed a Storch Mk 4 on Gran Sasso, Italy, in re-enactment of the rescue of Benito Mussolini, 53 years earlier.

UPDATED

STORCH STORCH SS MK 4

TYPE: Tandem-seat ultralight/kitbuilt.

PROGRAMME: First flight (VH-ZOR) 1991, then with single seat and powered by 59.6 kW (79.9 hp) Rotax 912 engine; two-seat version first flown 1994; JAR-VLA certification achieved in Australia, 14 October 1999.

CURRENT VERSIONS: **Storch 'Ultra':** Ultralight; Australian Primary Category; CAO 19 (amateur build).

Storch 'Muster': Utility version; envisaged uses include cattle mustering. Australian CAO 24 or certified categories.

Storch 'Criquet': Powered by Rotec Fireball 7 seven-cylinder radial engine of 82.0 kW (110 hp), thus resembling Morane-Saulnier MS 502 Criquet. Prototype VH-AJH public debut at Avalon Air Show, February 2001.

Super Storch Mk 4: Higher-powered version with 119 kW (160 hp) Textron Lycoming O-320 engine and top speed of 110 kt (203 km/h; 126 mph). 13 in Australia by April 2002.

Storch SS-FP: Floatplane; prototype, with full Lotus floats and Rotax 912S2 engine, registered VH-ANB in November 2001.

Storch Moose: M-14P radial engine (265 kW; 355 hp) installed in prototype VH-AYQ (100th Storch), which registered in November 2001. Fuel capacity 400 litres (106 US gallons; 88.0 Imp gallons). Length 7.62 m (25 ft 0 in); wing span 10.67 m (35 ft 0 in), max speed 105 kt (194 km/h; 120 mph); take-off run 61 m (200 ft).

CUSTOMERS: More than 140 sold by early 2003 (including some 50 factory-built) to customers in 20 countries in Africa, Australasia, Europe, North America and Pacific Rim. Approximately 60 per cent of production is factory-built.

COSTS: US$58,000, factory-complete (2000).

DESIGN FEATURES: Three-quarters scale version of Second World War German Fieseler Fi 156 Storch STOL liaison and observation aircraft. Wings are removable for transport or storage. Quoted build time for kitbuilt version approximately 600 hours. Aerofoil derived from Clark Y.

FLYING CONTROLS: Conventional and manual. Actuation by pushrods and cables; ground-adjustable tab on each aileron. Flaps and full span leading-edge slats. Horn-balanced tail surfaces. Electronic flight-adjustable tailplane incidence for pitch trim in Ultralight; VLA has fixed tailplane with trim tab in port elevator.

STRUCTURE: Welded 4130 chromoly steel tube fabric-covered fuselage and tail surfaces; metal engine cowling. Strut-braced 6061-T6 aluminium wings with stamped ribs and D-section leading-edge; latest versions aluminium-covered, except fabric flaps and ailerons. Composites wheel fairings.

LANDING GEAR: Tailwheel type; fixed. Centreline hinged, streamlined tube Vs attached to telescoping, bungee cord-sprung struts, upper ends of which fixed to apexes of tube pyramids on fuselage sides. Mainwheel size 8.00×6-6; Maule tailwheel, diameter 15 cm (6 in); hydraulic disc brakes. Float installation under development in late 1999.

POWER PLANT: Standard factory-built VLA version has one 73.5 kW (98.6 hp) Rotax 912 ULS flat-four driving two-blade Bruce de Chastel or Storch Aviation PROP-004 wooden propeller; 84.6 kW (113.4 hp) turbocharged Rotax 914 optional. Fuel in two wing tanks and a belly tank, combined capacity 90 litres (23.8 US gallons; 19.8 Imp gallons); a larger belly tank, capacity 100 litres (26.4 US gallons; 22.0 Imp gallons) optional.

ACCOMMODATION: Two, in tandem, with dual controls.

Storch SS Mk 4 'Muster' registered in the Australian CAO 24 series *(Paul Jackson)* ***NEW***/0572239

DIMENSIONS, EXTERNAL:

Wing span	10.00 m (32 ft 9¾ in)
Length overall	6.80 m (22 ft 3¾ in)
Height overall	2.40 m (7 ft 10½ in)
Propeller diameter	1.88 m (6 ft 2 in)
DIMENSIONS, INTERNAL:	
Cabin: Length	1.50 m (4 ft 11 in)
Max width	0.80 m (2 ft 7½ in)
Max height	1.00 m (3 ft 3¼ in)
AREAS:	
Wings, gross	16.00 m^2 (172.2 sq ft)
WEIGHTS AND LOADINGS (Storch 'Ultra'):	
Weight empty	350 kg (772 lb)
Max T-O weight	544 kg (1,200 lb)
PERFORMANCE:	
Never-exceed speed (VNE)	98 kt (181 km/h; 112 mph)
Max level speed	78 kt (144 km/h; 90 mph)
Cruising speed	70 kt (130 km/h; 81 mph)
Stalling speed, flaps down, power on	22 kt (41 km/h; 26 mph)
Max rate of climb at S/L	213 m (700 ft)/min
Service ceiling	4,575 m (15,000 ft)
T-O and landing run	9 to 15 m (30 to 50 ft)
Endurance: standard fuel	3 h 0 min
optional fuel	7 h 0 min
g limits	+6/–3

UPDATED

SUPERMARINE

SUPERMARINE AIRCRAFT PVT LTD

388 Hawkesbury Road, Moggill, Queensland 4070
Tel: (+61 7) 32 02 96 19
e-mail: sales@supermarineaircraft.com
Web: http://www.supermarineaircraft.com
MANAGER: Michael O'Sullivan
ASSISTANT MANAGER: John de Villiers

Michael (Mike) O'Sullivan founded Supermarine Aircraft Pty Ltd in 1992 to develop a kitbuilt scale replica of the Second World War Supermarine Spitfire. Factory area 700 m^2 (7,535 sq ft); workforce stood at 15 in October 2003.

UPDATED

SUPERMARINE SPITFIRE

TYPE: Tandem-seat sportplane kitbuilt.

PROGRAMME: Prototype Mk 25 first flew in November 1996. Designations follow from Mk 24 version produced by original Supermarine company.

CURRENT VERSIONS: **Spitfire Mk 25:** Single-seat version to 75 per cent scale. Discontinued by 2003.

Spitfire Mk 26: Two-seat and sole production version, with accommodation for pilot and passenger in tandem under single canopy, or pilot and additional fuel or baggage. *Data below refer to this version as approved for UK construction by the PFA.*

CUSTOMERS: Total of 35 of both versions sold by early 2003. Five under construction in UK by mid-2003. US marketing began at Sun 'n' Fun in April 2004.

COSTS: Complete basic kit from A$95,500 not including engine, propeller, instruments, shipping or taxes (2003).

DESIGN FEATURES: Scale replica of Supermarine Spitfire. Wings de-rig for storage or road transport. Supplied as complete kit; main airframe components are jig-assembled at factory for completion by customer; quoted build time 1,000 to 1,200 hours. Complies with FAA and other authorities' 51 per cent homebuilding rules.

Wing section similar to original; dihedral 5°; incidence 2° 10'; washout 2°.

FLYING CONTROLS: Conventional and manual; surfaces mass-balanced, with horn-balanced elevators; electric split flaps, max deflection 65°; pushrod actuation, except; cable actuated rudder and pitch trim.

STRUCTURE: Aluminium throughout, except for glass fibre cowling and non-structural fairings. Two-spar wing. Flush riveting.

LANDING GEAR: Tailwheel type; electrically retractable mainwheels. Main tyres 6.00. Hydraulic disc brakes.

POWER PLANT: One 134 kW (180 hp) Jabiru 5100 eight-cylinder four-stroke, driving a two- or four-blade constant-speed propeller. Fuel capacity 95 litres (25.1 US gallons; 20.9 Imp gallons).

Prototype Supermarine Spitfire Mk 26, wearing 'squadron code letters' to honour the original's designer, Reginald J Mitchell *(Paul Jackson)* ***NEW***/0580305

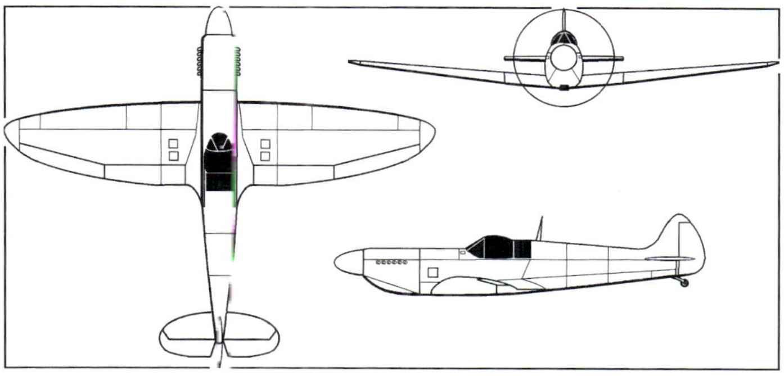

Authentic lines of the Supermarine Spitfire Mk 26 *(Paul Jackson)* ***NEW***/0569577

DIMENSIONS, EXTERNAL:
Wing span ... 8.44 m (27 ft 8¼ in)
Length overall ... 7.205 m (23 ft 7¾ in)
Height overall ... 1.98 m (6 ft 6 in)
Wheel track ... 1.68 m (5 ft 6¼ in)
AREAS:
Wings, gross ... 10.87 m^2 (117.0 sq ft)
WEIGHTS AND LOADINGS:
Weight empty ... 380 kg (837 lb)
Max T-O weight ... 650 kg (1,433 lb)
Max wing loading ... 59.81 kg/m^2 (12.25 lb/sq ft)
Max power loading (Jabiru) ... 5.59 kg/kW (9.18 lb/hp)

Supermarine Spitfire Mk 25 *(Paul Jackson)* ***NEW***/0554404

PERFORMANCE:
Never-exceed speed (VNE) ... 190 kt (351 km/h; 218 mph)
Cruising speed at 75% power ... 150 kt (278 km/h; 173 mph)
Manoeuvring speed ... 125 kt (232 km/h; 144 mph)
Stalling speed: flaps up ... 44 kt (82 km/h; 51 mph)
landing configuration ... 42 kt (78 km/h; 49 mph)
Max rate of climb at S/L ... 762 m (2,500 ft)/min
Service ceiling ... 5,490 m (18,000 ft)
T-O run ... 150 m (495 ft)
Landing run ... 350 m (1,150 ft)
Max range ... 434 n miles (804 km; 500 miles)
Endurance at econ cruising speed, with reserves ... 3 h 0 min
g limits ... +6/−3

UPDATED

TEKNICO

TEKNICO PTY LTD

PO Box 630, Oakbank, South Australia 5243
Tel: (+61 8) 83 88 43 49
Fax: (+61 8) 83 88 46 44
e-mail: supapup@arcom.com.au
Web: http://www.supapup.com
MANAGING DIRECTOR: John Cotton

Company previously named Aerosport Pty Ltd and Supapup Aircraft. Early versions (Mks 1, 2 and 3) of SupaPup hand-built at Hahndorf, South Australia, until cost of compliance with local airworthiness directives forced closure of company. Aerosport redesigned aircraft as Mk 4 in 1992 and returned to business with relaunched production line for kits. Most are operated under ultralight rules (homebuilt, uncertified; CAO 95 Pt 10). Two-seat AeroPup launched in 2002.

UPDATED

TEKNICO AEROPUP

TYPE: Side-by-side ultralight kitbuilt.
PROGRAMME: Kit production under way by November 2002.
COSTS: Kit, less engine, instrument and covering A$26,000 (2004).
The description for the SupaPup Mk 4 applies also to the AeroPup except as follows:
DESIGN FEATURES: Two-seat version of SupaPup Mk 4. Quoted build time 300 hours.
FLYING CONTROLS: Dual controls standard, with centrally mounted control column.
POWER PLANT: For engines in 48.5 kW (65 hp) to 74.6 kW (100 hp) range according to customer choice; Hirth, Jabiru, Rotax or Subaru recommended. Auxiliary belly tank optional.
ACCOMMODATION: Optional bulged windows increase cabin width to 1.14 m (3ft 9 in).
DIMENSIONS, EXTERNAL:
Wing span ... 8.20 m (26 ft 10¾ in)
Length overall ... 6.03 m (19 ft 9½ in)
DIMENSIONS, INTERNAL:
Cabin max width ... 1.07 m (3 ft 6 in)
AREAS:
Wings, gross ... 10.06 m^2 (108.3 sq ft)
WEIGHTS AND LOADINGS:
Weight empty, typical ... 240 kg (529 lb)
Baggage capacity ... 40 kg (88 lb)
Max T-O weight ... 544 kg (1,199 lb)
PERFORMANCE:
Stalling speed ... 42 kt (78 km/h; 49 mph)
T-O to 15 m (50 ft) ... 150 m (492 ft)
Endurance with aux tank and 10% reserves ... more than 4 h
g limits ... +5/−3

UPDATED

TEKNICO SUPAPUP MK 4

Teknico SupaPup Mk 4 ***NEW***/0102396

TYPE: Single-seat ultralight kitbuilt.
PROGRAMME: Original Aerosport company developed and produced three earlier versions of the SupaPup from the mid-1980s onwards; SupaPup Mk 4 first flown in early 1998. Some registered as lightplanes; most as ultralights with 300 kg (661 lb) MTOW limitation under CAO 95 Pt 10.
COSTS: Kit, less engine, instruments and covering A$18,000 (2004).
DESIGN FEATURES: Strut-braced, high-wing cabin monoplane; wings fold rearwards for storage or transportation without disconnection of fuel lines or control runs. Quoted build time 250 hours.
Wing dihedral 2°.
FLYING CONTROLS: Conventional and manual via torque tubes and cables.
STRUCTURE: Welded steel tube fuselage covered in Stitts fabric; glass fibre engine cowling and wingtips.
LANDING GEAR: Tailwheel type; fixed. Two faired-in side Vs hinged to lower longerons, plus half-axles, each with shock absorber, attached to compression frame. Steerable tailwheel.
POWER PLANT: One 34.0 kW (45.6 hp) Rotax 503 UL-1V two-cylinder two-stroke or 40.3 kW (54 hp) Jabiru 1600 four-cylinder four-stroke. Jabiru 2200 and Rotax 582 and 618 also suitable. Fuel contained in wing tanks, combined capacity 54 litres (14.3 US gallons; 11.9 Imp gallons).
ACCOMMODATION: Optional bulged windows can be fitted to increase cabin width to 0.71 m (2 ft 4 in).
EQUIPMENT: Optional cargo pod under development.
DIMENSIONS, EXTERNAL:
Wing span ... 7.74 m (25 ft 4¾ in)
Length overall ... 5.70 m (18 ft 8½ in)
DIMENSIONS, INTERNAL:
Cabin max width ... 0.61 m (2 ft 0 in)
AREAS:
Wings, gross ... 9.50 m^2 (102.3 sq ft)
WEIGHTS AND LOADINGS:
Weight empty ... 180 to 230 kg (397 to 507 lb)
Baggage capacity ... 20 kg (44 lb)
Max T-O weight ... 340 kg (750 lb)
PERFORMANCE:
Never-exceed speed (VNE) ... 130 kt (240 km/h; 149 mph)
Max level speed ... 100 kt (185 km/h; 115 mph)
Cruising speed ... 95 kt (176 km/h; 109 mph)
Stalling speed ... 36 kt (67 km/h; 42 mph)
Max rate of climb at S/L ... 335 m (1,100 ft)/min
T-O to 15 m (50 ft) ... 200 m (660 ft)
Endurance with 10% reserves ... 4 h 30 min

UPDATED

Austria

DIAMOND

DIAMOND AIRCRAFT INDUSTRIES GMBH

N A Otto-Strasse 5, A-2700 Wiener Neustadt
Tel: (+43 2622) 267 00
Fax: (+43 2622) 267 80
e-mail: office@diamond-air.at
Web: http://www.diamondair.at
PRESIDENT: Christian Dries
MANAGING DIRECTORS:
Michael Feinig
Christian Trieb
MARKETING DIRECTOR: Sylvia Mandl
PURCHASING DIRECTOR: Franz Schermann
SALES DIRECTOR: Joerg Schwarzmeier
TECHNICAL DIRECTOR: Martin Volck

Company formed 1981 in Friesach, Carinthia, southern Austria, as Hoffman Flugzeugbau GmbH; re-formed after August 1984 bankruptcy as Hoffman Aircraft Ltd, subsidiary of Simmering-Graz-Pauker AG; relocated to new facility at Wiener Neustadt 1987; name HOAC Austria Flugzeugwerk Wiener Neustadt GmbH adopted 1990 after 1989 management buyout; renamed Diamond Aircraft Industries GmbH March 1996 and achieved JAR 21 design approval in 2001.

Factory expanded to 5,200 m^2 (56,000 sq ft) of floor space in 1996 to accommodate wider model range, increase production levels and provide new research and maintenance facilities. Ground broken for further 10,900 m^2 (117,325 sq ft) composites plant in May 2003. Workforce 220. Current products include refurbished DA20 Katana. Diamond Aircraft Corporation produces the two-seat DA20 in Canada; the four-seat DA40 Star became available in 2000 and deliveries of production aircraft accelerated in 2001. DA40-TDI certified and DA42 Twin Star launched in 2002, and D-JET announced in early 2003.

Having achieved sales of 1,700 aircraft up to the end of 2000, Diamond gained a further 176 (including Canada) in 2001, 201 in 2002, 228 in 2003 and was projecting 391 and 540 in the years up to 2005.

Diamond acquired UK rotary engine manufacturer MidWest Engines in 1994, but acquisition not announced until Aero '03 show at Friedrichshafen in April 2003, under new company name Diamond Engines. Plans then announced for manufacture of GAE 50R and GIAE 110R Wankel-type rotary engines, last-named to be installed in new DA20-W variant of the DA20 Katana.

Diamond opened a new 10,900 m^2 (117,000 sq ft) composites production centre at Wiener Neustadt on 23 January 2004 where components will be manufactured to supply assembly lines in Austria and Canada, with potential to increase production to up to 600 aircraft per year.

UPDATED

DIAMOND DA 20-W

TYPE: Two-seat lightplane.
PROGRAMME: Fuselage mockup displayed at Aero '03 at Friedrichshafen in April 2003; to be manufactured in Austria.
COSTS: €84,900 (2003).

The description of the DA20C1, built by Diamond Aircraft Corporation in Canada, applies also to the DA 20-W.

DESIGN FEATURES: Based on DA 20C1 airframe with Diamond Engines GIAE 110R rotary engine.
POWER PLANT: One 82.0 kW (110 hp) Diamond Engines GIAE 110R Wankel-type rotary engine. Fuel capacity 80 litres (21 US gallons; 17.6 Imp gallons).
DIMENSIONS, EXTERNAL:
Length overall ... 7.23 m (23 ft 8½ in)
WEIGHTS AND LOADINGS:
Weight empty ... 510 kg (1,124 lb)
Max T-O weight ... 730 kg (1,609 lb)
Max wing loading ... 62.87 kg/m^2 (12.88 lb/sq ft)
Max power loading ... 9.32 kg/kW (15.31 lb/hp)
PERFORMANCE (estimated):
Max cruising speed ... 110 kt (204 km/h; 127 mph)

UPDATED

Trial installation of GIAE 110R rotary engine in a DA 20 Katana airframe *(Paul Jackson)* ***NEW***/0552262

DIAMOND HK36

European marketing name: Super Dimona
North American marketing name: Katana Xtreme
TYPE: Motor glider.
PROGRAMME: Original Hoffman H-36 Dimona (Diamond) was designed by Austrian-German team, first flight by first of three prototypes taking place in Germany 9 October 1980; production, in Austria, started May 1981; Dimona Mk II introduced May 1985 (smaller wingtip fairings, modified cowling, better propeller pitch control, stronger main gear, sprung steerable tailwheel); production of H-36 totalled over 275 by 1989.

Diamond HK36TTC Super Dimona motor glider *(Paul Jackson)* ***NEW***/0569582

Development of HK36R Super Dimona (considerably redesigned by Dieter Kohler — hence K in designation) started March 1987; first flight, with 67.1 kW (90 hp) Limbach L 2400 engine, October 1989, and one or two more completed with Limbach engines, but Rotax 912 adopted as standard power plant. Production began April 1990 and Austrian certification to JAR 22 awarded 15 May 1990. Manufacture totalled 114 by 1995, when supplanted by current versions. 'Short-wing' LF 2000 was proof-of-concept for DA20 Katana; TS and TTS versions introduced 1995, but discontinued two years later; these and current TC/TTC featured winglets; prototypes (OE-9415 and OE-9416) built 1995. In 1997, export Dimonas for the North American market were renamed Katana Xtreme.

Diamond Aircraft was selected in July 2003 to supply a Katana Xtreme airframe for the Boeing Fuel Cell Demonstrator project led by Boeing's Madrid-based Research and Technology Center. Work to integrate Proton Exchange Membrane fuel cells supplied by Intelligent Energy (UK) was scheduled to begin in the summer of 2003, with first flight expected in late 2004 or early 2005.

CURRENT VERSIONS: **HK36TC 80:** Tricycle landing gear version of now discontinued HK 36TS. Some 20 TS/TC built in 1996, four in 1998 and eight for Indian Air Force in early 2000.

HK36TC 100: As HK 36TC 80, but with Rotax 912S-3 engine.

HK36TTC 115: As HK 36TC 80 but with Rotax 914. Majority of production is this version. **Eco** version described separately.

HK36TTC-MPX: Described separately.

CUSTOMERS: Two prototypes and 234 of T/TT series built by mid-2003 but production line then reportedly in process of transfer to Pasewalk, Germany. Most are TTCs and TCs. Sold to customers in Austria, Canada, Denmark, France, Germany, Netherlands, UK, USA and elsewhere. First of five for French Air Force delivered to Ecole de l'Air on 1 June 2002. Production totals for previous versions given under Programme. Sales of all H/HK 36 versions totalled 937 by end of 2003, of which some 890 then delivered.
COSTS: TC 80 €112,900; TC 100 €118,900; TTC 115 €123,700, ex-works, excluding taxes (all 2003).
DESIGN FEATURES: Conventional motor glider, with T tail and low/mid wing. Wings can be detached and folded back alongside fuselage for transportation and storage. Improvements of T/TT series (from 1996) include carbon fibre main spar. Compared with H-36 Dimona, HK 36 has modified inboard leading-edge, sweptback wingtips and Schempp-Hirth (instead of DFS) type airbrakes; better access to engine and control system via completely removable cowlings; improved spin characteristics; stronger main landing gear and improved tailwheel springing; larger canopy with new mechanism; wingroot fairings and rear fuselage redesigned.

Wortmann wing sections: FX-63-137 at root, FX-71 at tip fairings

FLYING CONTROLS: Conventional and manual. Longitudinal stability improved in Super Dimona over original by increasing elevator chord and adding trim tab; airbrakes in wing upper surface.
STRUCTURE: Mainly of GFRP; carbon fibre wing spar and main bulkhead. GFRP mainwheel legs and fairings.
LANDING GEAR: Non-retractable tricycle type, with free-castoring nosewheel, ±30°. Wheel fairings on all three legs.
POWER PLANT: *TC 80:* One 59.6 kW (79.9 hp) Rotax 912A-3 flat-four with 2.27:1 reduction drive to an MT-Propeller composites, two-blade, three-position constant-speed and feathering propeller.

TC 100: One 73.5 kW (98.6 hp) Rotax 912S-3.

TTC 115: One 84.6 kW (113.4 hp) Rotax 914F-3.

Fuel tank in fuselage, capacity 77 litres (20.3 US gallons; 16.9 Imp gallons).

ACCOMMODATION: Two seats side by side. Cockpit canopy hinged at rear to open upward. Baggage space aft of seats.
AVIONICS: Optional avionics suites from Bendix/King and Garmin.

Comms: Bendix/King KX 125 nav/com, KY 97A com, and KT 76A or KT 76C transponder; KMD 150 GPS, Garmin GNS 420 nav/com, GNX 150XL GPS/com, GTX 327 transponder and GMA 340 audio panel.

DIMENSIONS, EXTERNAL:
Wing span ... 16.33 m (53 ft 7 in)
Wing aspect ratio ... 17.4
Length overall ... 7.28 m (23 ft 10½ in)
Width, wings folded ... 2.20 m (7 ft 2½ in)
Height over tail ... 2.40 m (7 ft 10½ in)
AREAS:
Wings, gross ... 15.30 m^2 (164.7 sq ft)
WEIGHTS AND LOADINGS:
Weight empty: TC 80, TC 100 ... 555 kg (1,224 lb)
TTC 115 ... 563 kg (1,241 lb)
Max T-O weight: all ... 770 kg (1,697 lb)
Max towing load: TC 80 ... 370 kg (815 lb)
TC 100 ... 500 kg (1,102 lb)
TTC 115 ... 600 kg (1,322 lb)
Max wing loading: all ... 50.3 kg/m^2 (10.31 lb/sq ft)

Max power loading: TC 80 12.92 kg/kW (21.23 lb/hp)
TC 100 10.48 kg/kW (17.21 lb/hp)
TTC 115 9.10 kg/kW (14.95 lb/hp)

PERFORMANCE, POWERED:
Never-exceed speed (VNE):
all 141 kt (261 km/h; 162 mph)
Max cruising speed: TC 80 108 kt (200 km/h; 124 mph)
TC 100 116 kt (215 km/h; 134 mph)
TTC 115 119 kt (220 km/h; 137 mph)
Cruising speed at 65% power:
TC 80 90 kt (167 km/h; 104 mph)
TC 100 100 kt (185 km/h; 115 mph)
TTC 115 106 kt (196 km/h; 122 mph)
Max rate of climb at S/L: TC 80 246 m (807 ft)/min
TC 100 294 m (965 ft)/min
TTC 115 324 m (1,063 ft)/min
T-O run: TC 80 201 m (659 ft)
TC 100 193 m (633 ft)
TTC 115 182 m (600 ft)
T-O to 15 m (50 ft): TC 80 338 m (1,110 ft)
TC 100 308 m (1,010 ft)
TTC 115 274 m (900 ft)
g limits: all +5.3/–2.65

PERFORMANCE, UNPOWERED:
Best glide ratio: all 27
Min sink rate: all 1.27 m (4.17 ft)/s

UPDATED

DIAMOND HK36TTC-MPX

TYPE: Multisensor surveillance lightplane.

CURRENT VERSIONS: **HK36TTC-MPX:** Generic designation for sensor-equipped Dimona. Developed by Diamond Aircraft Corporation in association with L-3 Wescam Inc of Ontario; multimission utility aircraft version of Diamond HK 36TTC Katana Xtreme. MPX (Multipurpose Xtreme) incorporates hardpoint under each wing at approximately mid-span for quick-release sensor pods. Each pod can accommodate up to 45 kg (100 lb) of equipment, including gyrostabilised video surveillance cameras (such as L-3 Wescam Model 16 ENG broadcast camera system) and IR sensors, with provision for onboard data and image collection, or downlink telemetry. An independent 28 V 40 A DC power supply is provided for surveillance equipment, and a cockpit equipment bay, capacity 0.4 m^3 (14 cu ft) or 30 kg (66 lb), accommodates rack-mounted electronics or instruments for in-flight monitoring. Standard fuel capacity of 110 litres (29.1 US gallons; 24.2 Imp gallons) provides endurance up to 7 hours.

HK36TTC-Eco: Specific designation for aircraft currently in service. Certified in FAA Utility category 29 March 1999 and Restricted category (aerial photography only) 21 December 2000; latter permits increased MTOW of 929 kg (2,050 lb).

Prototype OE-9481 currently operated (since 1998) in Switzerland by Met Air atmospheric sampling company as HB-2335. Five conversions undertaken in Canada for US market: N842WS, N845WS and N846WS operated by Diamond Aircraft; and N843WS and N844WS of Wescam Air Operations, Van Nuys, California.

Of these, N844WS became OE-9437 in Diamond (Austria) ownership and exhibited at ILA, Berlin, May 2004 with a 55 kg (1,213 lb) L-3 Wescam 16DB750 turret.

UPDATED

Diamond HK36TTC-MPX equipped with L-3 Wescam sensor equipment ***NEW***/0569644

L-3 Wescam 16DB750 turret ***(Paul Jackson)*** ***NEW***/0580302

DIAMOND DA40-TDI DIAMOND STAR

Engineering designation: DA40D

TYPE: Four-seat lightplane.

PROGRAMME: Formally launched 23 April 1997 at the Aero 97 show at Friedrichshafen, Germany, when mockup displayed; initially retained name of Katana; proof-of-concept prototype DA40-V1 (OE-VPC) first flew on 5 November 1997, powered by a Rotax 914 engine; a Teledyne Continental IO-240-engined DA40-V2 (OE-VPE) followed shortly thereafter; neither event was publicised; a third prototype, DA40-V3, with Textron Lycoming IO-360 engine, flew in late 1998; fourth, near-production standard, prototype (OE-VPM) joined the test programme in mid-1999 followed by three more prototypes comprising DA40-P5 (c/n 40005/OE-VPQ), -P6 (40006/OE-VPB), -P7 (40007/OE-VPW), these having enlarged ventral strake and cutaway rudder.

First production aircraft (40008/OE-KPN) registered June 2000; second exhibited at Farnborough in following month. Lycoming-powered variant was initial production version, for which JAR 23 certification was achieved 25 October 2000; FAA certification, 9 April 2001, followed by JAA IFR certification on 27 April 2001, and FAA IFR certification on 23 August 2001. First deliveries to US customers (seven aircraft) took place during EAA AirVenture 2001 at Oshkosh in July 2001.

Austrian production temporarily halted in early 2002 upon transfer of Lycoming-engined variant to Canada. Thielert-engined version now sole European-built DA40.

CURRENT VERSIONS: **DA40-180:** Original version with Lycoming IO-360 engine; now produced by Diamond Canada (which see).

DA40 TDI: Powered by Thielert Centurion 1.7 turbo-diesel; converted prototype (OE-KPP) first flew 22 November 2001; JAR 23 N-VFR certification achieved 22 November 2002 followed by IFR approval on 18 June 2003. Development aircraft

Diamond DA40-TDI prototype ***NEW***/0532053

Instrument panel of DA40-TDI ***NEW***/0532054

OE-VTA (80th Austrian-built). Initial batch comprised 75 aircraft; some 100 produced by mid-2004 for operators in Australia, Austria, Denmark, France, Germany, New Zealand, Norway, Sweden and Switzerland. *As described.*

CUSTOMERS: Austrian production of petrol-engined DA 40 totalled 84 up to 2002, including prototypes and two conversions to TDI; customers in Austria, Germany, Netherlands, Sweden, Switzerland, UK, USA and elsewhere.

Total of 140 DA40 TDIs ordered by mid-2003, including 17 for Lufthansa's pilot training school; 60th aircraft registered in October 2003; Embry-Riddle Aeronautical University of Florida has indicated that it may convert an order for 10 DA40-180s to DA40-TDIs. Cabair of UK ordered 16, of which eight received by late 2003. The Amaury de la Grange pilot school at I'Epag de Merville, France, ordered six DA40 TDIs for delivery in February 2004.

COSTS: DA40 TDI €205,400 basically (N-VFR) equipped; full standard avionics (IFR) €38,500 extra; all excluding tax (2003).

DESIGN FEATURES: Four-seat development of DA20; to be certified to FAR/JAR 23 or equivalents in Europe, North America, Russian Federation, South Africa and Turkey. DA 20 wings attach to new, wider centre-section to increase span; downturned tips to horizontal tail; single strake/tail bumper. Enlarged cabin increases fuselage length by 0.67 m (2 ft 2¼ in).

Wing section is Diamond-modified Wortmann FX 63-137/20; washout 1°.

FLYING CONTROLS: Conventional and manual. Ailerons and elevators operated by pushrods with low-friction bearings, rudder by cables; manual pitch trim standard, electric pitch trim optional via control stick-mounted switch; ground-adjustable trim tab on rudder; electrically actuated slotted flaps.

STRUCTURE: Fuselage, with integral vertical fin, is mostly of GFRP construction with local CFRP reinforcement in high-stress areas, comprising two half-shells bonded together. Two-spar wings with GFRP/PVC foam/GFRP sandwich skins; horizontal tail and rudder surfaces similarly covered.

LANDING GEAR: Fixed tricycle type, with cantilever spring steel leg on each main unit and elastomeric suspension on nose leg. Mainwheel size 6.00–6; nosewheel 5.00–5. Steering is provided by differential braking of mainwheel and friction-damped castoring nosewheel. Speed fairings on all three wheels.

POWER PLANT: One 99 kW (133 hp) Thielert Centurion 1.7 turbo-diesel, driving an MTV-12-B/180-17 three-blade constant-speed propeller. Fuel in two wing tanks, standard capacity 110 litres (29.0 US gallons; 24.2 Imp gallons); 155 litres (40.9 US gallons; 34.1 Imp gallons) optional.

ACCOMMODATION: Four persons in two side-by-side pairs. Single-piece canopy lifts up and forwards for access to front seats; rear occupants board through an upward-opening door to port. Cabin has 26 *g* composites self-adapting seats with three-point automatic safety harnesses and roll-over bar. Baggage compartment behind rear seats includes tubular extension into rear fuselage for extra-long items. Rigid baggage compartment extension for long objects such as skis, optional.

SYSTEMS: 14 V electrical system includes 90 A alternator; 12 V 35 Ah battery.

AVIONICS: *Comms:* Baseline N-VFR suite includes Garmin GNS 430 com/nav/GPS, GTX 327 transponder and intercom. IFR suite adds second GNS 430 and GMA 340 audio selector panel with marker beacon receiver. GNS 530 com/nav/GPS (exchange for GNS 430) and Ack E-01 or Artex 406 MHz ELT optional.

Flight: N-VFR: Garmin GI 106A CDI. IFR: Second GI 106A CDI, IFR: Second GI 106A, Bendix/King KAP 140 dual-axis autopilot with optional altitude preselect and Sigma Tek DG with heading bug, standard. Further options include Bendix/King KCS 55A slaved compass system, KR 87 ADF, KN 62A DME and Goodrich WX-500 Stormscope.

Instrumentation: N-VFR fit includes TAE engine indicating system, compass, ASI, altimeter (second optional), VSI, artificial horizon, turn co-ordinator, DG, OAT gauge, chronometer and alternate static port.

EQUIPMENT: Landing light, taxying light, position lights, instrument lighting system, overhead cabin light, power socket for hand-held GPS and heated pitot/static system, all standard. Options include Operational Package comprising gust lock, towbar, tools and pitot cover; fire extinguisher, and first aid kit.

DIMENSIONS, EXTERNAL:
Wing span ... 11.94 m (39 ft 2 in)
Wing aspect ratio ... 10.6
Length overall ... 8.01 m (26 ft 3¼ in)
Height overall ... 2.00 m (6 ft 6¾ in)

DIMENSIONS, INTERNAL:
Baggage compartment: Length ... 1.08 m (3 ft 6½ in)
Max width ... 0.35 m (1 ft 1¾ in)
Max height ... 0.80 m (2 ft 7½ in)

AREAS:
Wings, gross ... 13.50 m² (145.3 sq ft)

WEIGHTS AND LOADINGS:
Weight empty ... 750 kg (1,653 lb)
Baggage capacity ... 35 kg (77 lb)
Max T-O weight ... 1,150 kg (2,535 lb)
Max landing weight ... 1,092 kg (2,407 lb)
Max wing loading ... 85.2 kg/m² (17.45 lb/sq ft)
Max power loading ... 11.60 kg/kW (19.06 lb/hp)

PERFORMANCE:
Never-exceed speed (VNE) ... 178 kt (329 km/h; 204 mph) IAS
Max level speed at FL100 ... 154 kt (285 km/h; 177 mph)
Cruising speed at FL100:
at 70% power ... 132 kt (244 km/h; 152 mph)
at 50% power ... 110 kt (204 km/h; 127 mph)
at 40% power ... 89 kt (165 km/h; 102 mph)
Stalling speed ... 49 kt (91 km/h; 57 mph)
Max rate of climb at S/L ... 238 m (780 ft)/min
Max certified altitude ... 5,000 m (16,400 ft)
T-O run ... 335 m (1,100 ft)
T-O to 15 m (50 ft) ... 635 m (2,085 ft)
Range, 45 min reserves:
standard fuel ... 750 n miles (1,389 km; 863 miles)
optional fuel ... 1,100 n miles (2,037 km; 1,265 miles)

UPDATED

DIAMOND DA42 TWIN STAR

TYPE: Four-seat utility twin.

PROGRAMME: Prototype (then wearing D-GENI but not officially registered as such) shown statically at ILA Berlin, 6 to 12 May 2002. Although long reported, company decision to build was only made in September 2001, and assembly began in March 2002; DA42-V1 first flight (as OE-VPS) 9 December 2002; total of 45 hours logged by late April 2003; second prototype DA42-V2, OE-VDA, flew October 2003; third used for static testing only;

Diamond DA42-V2 second prototype *(Paul Jackson)* ***NEW***/0580299

DA42-V1 prototype, following its conversion to Lycoming power *(Paul Jackson)* ***NEW***/0580296

DA42-P4 OE-VDB was shown at ILA, Berlin, May 2004, prior to first flight. EASA certification presented 13 May 2004; FAA FAR Pt 23 approval expected in early 2005; first deliveries 2004.

CURRENT VERSIONS: **DA42 Twin Star:** Standard version, *as described.*

DA42 Twin Star (Lycoming): Conversion of DA42-V1 with 134 kW (180 hp) Textron Lycoming IO-360-M1A flat-sixes and Hartzell FC7495S two-blade propellers first flew on 16 April 2004. Max level speed 180 kt (333 km/h; 207 mph); normal cruising speed 145 kt (269 km/h; 167 mph). Further production conditional upon market response.

DA42 MPP Observer: Multipurpose platform version, equipped with a nose-mounted Wescam 40.6 cm (16 in) camera; auxiliary fuel tank taking total capacity to 280 litres (73.9 US gallons; 61.6 Imp gallons) optional, extending endurance to more than 24 hours at 27 per cent power. Launch customer, Royal Canadian Mounted Police, has ordered 12, plus three options.

Unmanned UAV version of the DA42 MPP under study in early 2002, in collaboration with an unnamed defence company. This possibly related to study for installation of SON 112 observation system from Russia's UOMZ, which announced in May 2003. Similar in size and position to L-3 Wescam turret; contains three cameras.

CUSTOMERS: Total of 317 orders by May 2004, including 10 for North American launch customer Embry-Riddle Aeronautical University for its Aviation & Space Technology Academy (ASTA) graduate first-officer training programme, an unspecified number for CTC McAlpine of the UK and one for Atlantic Flight Training of the UK.

Manufacturer intended to have built 50th aircraft by end of 2004 and be producing 16 per month by mid-2005.

COSTS: €440,000 basically equipped (2004).

DESIGN FEATURES: Twin-engined version of DA40 Diamond Star; integral wing centre section replaced by a bolt-on centre section for wing of increased span; deeper floor provides increased headroom; front bulkhead (formerly firewall) moved forward. DA40 outboard wing panels with winglets; T tail; engine cowlings of large frontal area. Downturned tailplane tips. Enlarged rudder and fin dorsal fillet; underfin faired into rudder.

FLYING CONTROLS: Conventional and manual. Inset horn balance in ailerons. Flight-adjustable tabs on elevator (central) and rudder; ground-adjustable tab on port aileron Slotted flaps outboard of engines; split flaps inboard; both electrically actuated.

STRUCTURE: Composites throughout. Generally as for DA40, but with increased use of carbon fibre for additional strength.

LANDING GEAR: Tricycle type; electrohydraulically retractable. Nosewheel, with 5.00-5 tyre, retracts rearward and covered by single, starboard-hinged door; mainwheels, with 15×6.00–6 tyres and trailing-link configuration, retract inboard, having single part-door on legs, leaving wheel exposed after retraction. Oleo-pneumatic shock-absorbers on all legs. Hydraulic brakes. Floatplane version planned.

POWER PLANT: Two 101 kW (135 hp) Thielert Centurion 1.7 turbocharged diesel engines, each driving an MT-Propeller MTV-6-A/187-129 three-blade, variable-pitch (hydraulic) propeller. Single-lever control, with FADEC and automatic engine-out feathering. Fuel tanks in wings, usable capacity 195 litres (51.5 US gallons; 42.9 Imp gallons); optional long-range capacity 280 litres (74.0 US gallons; 61.6 Imp gallons).

ACCOMMODATION: Four persons in two side-by-side pairs. Single-piece canopy lifts up and forwards for access to front seats; rear occupants board through an upward-opening door to

Third flying Diamond Twin Star, the DA42-P4 OE-VDB *(Paul Jackson)* ***NEW***/0580298

Computer-generated image of DA42 MPP surveillance aircraft with UOMZ SON 112 turret ***NEW***/0576827

Engine nacelles of Lycoming-powered Diamond DA42 *(Paul Jackson)* ***NEW***/0580297

port. Cabin has deluxe leather interior with 26 *g* composites self-adapting seats with three-point inertia reel safety harnesses, and roll-over bar. Baggage compartment behind seats includes tubular extension into rear fuselage for extra-long items. Rigid baggage compartment extension for long objects such as skis, optional. Forward baggage compartment in nose with doors both sides.

SYSTEMS: Optional pulse-controlled oxygen system. Electrical system 28 V with dual paralleling alternators (one per engine), each providing up to 60 A. Battery in nose; 13.6 Ah. Optional TKS fluid de-icing system for wing, tailplane, propellers and windscreen; weight 54 kg (118 lb); typical effectiveness 4 hours. Optional oxygen and air conditioning systems. Bendix/King KAP 140 autopilot.

AVIONICS: Garmin G1000 integrated avionics system as standard.

Comms: Dual Garmin GIA 63 com/nav/GPS/ILS, single GMA 1347 digital audio selector panel with marker beacon receiver and GTX 35 Mode S transponder with traffic information system (where available).

Flight: Bendix/King GRS 77, GMU 44 AHRS and terrain awareness system. Options include Bendix/King KN 63 DME, Becker RA 3502 ADF and Goodrich WX-500 Stormscope.

Instrumentation: Two 254 mm (10 in) GDU 1040A (1,024 × 768 resolution) flat panel, sunlight readable, landscape format colour TFT screens.

EQUIPMENT: Landing light, taxying light, position lights, instrument lighting system, overhead cabin light, heated pitot/static system, ANR headsets, gust lock, towbar, tools and pitot cover; fire extinguisher and first aid kit all standard.

DIMENSIONS, EXTERNAL:
Wing span ... 13.42 m (44 ft 0¼ in)
Wing aspect ratio ... 10.9
Length overall ... 8.50 m (27 ft 10¾ in)
Height overall ... 2.60 m (8 ft 6¼ in)
Propeller diameter ... 1.87 m (6 ft 1½ in)

AREAS:
Wings, gross ... 16.46 m² (177.2 sq ft)
Ailerons (total) ... 0.72 m² (7.75 sq ft)
Trailing edge flaps (total):
slotted ... 1.34 m² (14.42 sq ft)
split ... 0.84 m² (9.04 sq ft)

WEIGHTS AND LOADINGS:
Weight empty ... 1,089 kg (2,400 lb)
Max T-O weight ... 1,650 kg (3,637 lb)
Max wing loading ... 100.2 kg/m² (20.53 lb/sq ft)
Max power loading ... 8.20 kg/kW (13.47 lb/hp)

PERFORMANCE:
Never-exceed speed (V_{NE}) ... 202 kt (374 km/h; 232 mph) IAS
Cruising speed: max at 80% power at FL125 ... 181 kt (335 km/h; 208 mph)
econ at 60% power at FL25 ... 157 kt (291 km/h; 181 mph)
Max rate of climb: at S/L ... 527 m (1,730 ft)/min
at FL100 ... 472 m (1,549 ft)/min

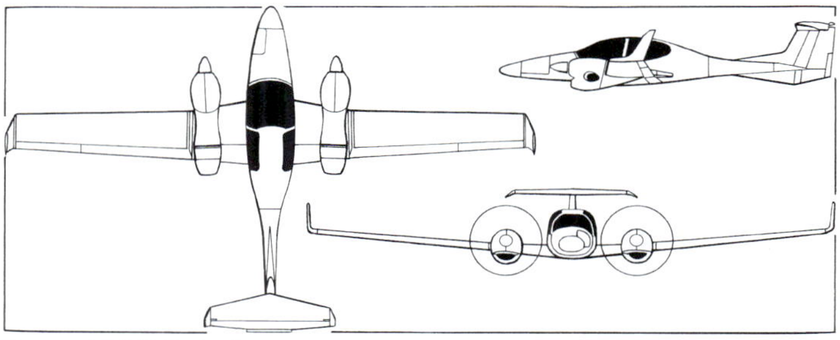

Diamond DA42 Twin Star *(James Goulding)* ***NEW***/1047849

Prototype Diamond DA42 Twin Star *(Paul Jackson)* ***NEW***/0576632

Max certified altitude ... 5,486 m (18,000 ft)
T-O run ... 290 m (955 ft)
T-O to 15 m (50 ft) ... 470 m (1,542 ft)
Landing from 15 m (50 ft) ... 500 m (1,640 ft)
Landing run ... 290 m (950 ft)
Range at econ cruising height and speed:
normal fuel ... 1,003 n miles (1,857 km; 1,154 miles)
optional fuel ... 1,442 n miles (2,670 km; 1,659 miles)

UPDATED

DIAMOND D-JET

TYPE: Business mono-jet.

PROGRAMME: Development first revealed in late 2002; first details released 14 January 2003; first flight scheduled for March or April 2005, with EASA certification and deliveries planned from second quarter 2006. To be certified in standard and high maximum take-off weight versions for European and US markets, respectively. Seven-seat, twin-engine derivative under study in early 2003, for availability from 2008.

CUSTOMERS: Sales by July 2004 totalled more than 180.

COSTS: Unit cost €870,000; programme cost €30 million (both 2003).

DESIGN FEATURES: Low-wing of laminar flow aerofoil section with moderate sweepback and small winglets; T tail; engine inlets in wingroots.

STRUCTURE: Composites throughout.

POWER PLANT: One 6.23 kN (1,400 lb st) Williams International FJ33-4 turbofan, with FADEC. Fuel capacity 725 litres (191.5 US gallons; 159.5 Imp gallons).

ACCOMMODATION: Five in two-plus-three seating configuration; pressurised cabin. Two cabin windows per side.

SYSTEMS: Pressurisation system, maximum differential 0.36 bar (5.2 lb/sq in) maintaining 2,440 m (8,000 ft) cabin environment to 7,620 m (25,000 ft). Ballistic parachute recovery system under consideration in mid-2004.

DIMENSIONS, EXTERNAL:
Wing span ... 11.99 m (39 ft 4 in)
Length overall ... 12.04 m (39 ft 6 in)
Height overall ... 3.10 m (10 ft 2 in)

DIMENSIONS, INTERNAL:
Cabin: Length ... 3.53 m (11 ft 7 in)
Max width ... 1.47 m (4 ft 10 in)
Max height ... 1.47 m (4 ft 10 in)

WEIGHTS AND LOADINGS (A: JAA certification; B: FAA certification):
Weight empty: A, B ... 1,175 kg (2,590 lb)
Max fuel weight ... 580 kg (1,279 lb)
Payload with max fuel: A ... 244 kg (538 lb)
B ... 400 kg (882 lb)
Max T-O weight: A ... 1,999 kg (4,407 lb)
B ... 2,155 kg (4,751 lb)

PERFORMANCE (estimated):
Max operating speed ... 295 kt (546 km/h; 339 mph)
Max cruising speed at FL250 ... 315 kt (583 km/h; 363 mph)
Stalling speed: A ... 61 kt (113 km/h; 71 mph)
B ... 63 kt (117 km/h; 73 mph)
Max rate of climb at S/L: A ... 823 kt (2,700 ft)/min
B ... 792 kt (2,600 ft)/min
Max certified altitude ... 7,620 m (25,000 ft)
Time to FL250 ... 8 min
T-O to 15 m (50 ft): A ... 620 m (2,035 ft)
B ... 723 m (2,375 ft)
Max range, IFR reserves at 240 kt (km/h; 276 mph) ... 1,320 n miles (2,444 km; 1,519 miles)

UPDATED

HB FLUGTECHNIK

HB FLUGTECHNIK GMBH

Dr Adolf Schärf Strasse 42, Postfach 74, A-4053 Haid-Ansfelden
Tel: (+43 7229) 79 10 40
Fax: (+43 7229) 791 04 15
e-mail: office@hb-flugtechnik.at
Web: http://www.hb-flugtechnik.at
MANAGING DIRECTOR: Ing Heino Brditschka
TECHNICAL DIRECTOR: Georg Passenbrunner

FACTORY AIRFIELD:
Flugpltatz Hofkirchen
A-4491 Hofkirchen

Company name indicates Heino Brditschka, designer of motor gliders, including the HB 23 and HB 202. Both had the unconventional mid-fuselage propeller, to which HB Flugtechnik has returned for some of its recent projects.

From its 2,400 m^2 (25,830 sq ft) factory, HB provides a wide variety of other aviation services, including overhaul and repair, aerial photography and advertising, flying training, aircraft charter and support of amateur constructors. In 1998, distribution of HB 207 kits was assigned to Aero-Service Thüringen of Erfurt, Germany, but this has now ceased.

In 2002, HB announced that it is working on more projects, including the **HB 401** high-wing pusher; **HB 402** high-wing design resembling Cessna 172; and **HB 403**, based around the company's Alfa.

The company also markets the Amigo, Cubby and Dandy ultralights produced by Let-Mont in the Czech Republic.

UPDATED

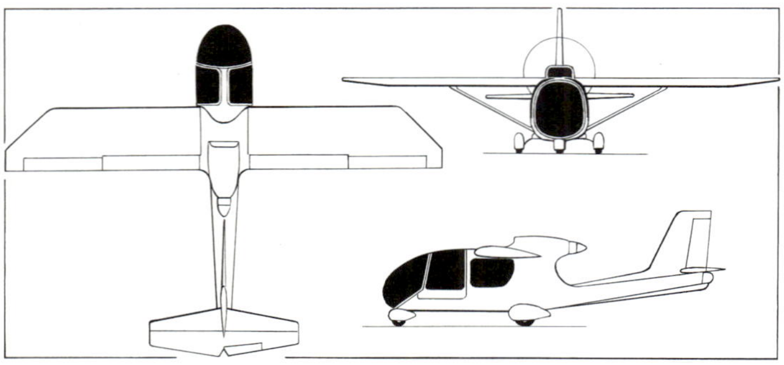

HB Flugtechnik HB 401 project, currently under development *(James Goulding)* ***NEW***/0547519

HB FLUGTECHNIK HB 204 TORNADO

TYPE: Tandem-seat sportplane kitbuilt.

PROGRAMME: Developed from previous, unbuilt, design; unflown airframe, based on components from first two aircraft, exhibited at Aero '99 at Friedrichshafen in April 1999; prototype completed by early 2001 and painted with spurious registration OE-VPP for Aero '01; first flight repeatedly postponed and had not been achieved by end of 2003. Tornado was not promoted at Aero '03, although development is reported to be continuing.

UPDATED

HB FLUGTECHNIK HB 207 ALFA

TYPE: Side-by-side kitbuilt.

PROGRAMME: First flight (HB 207RG OE-CHC) 14 March 1995. Second prototype (HB 207 OE-CAA) made public debut, unmarked and then unflown, at Aero '97 at Friedrichshafen, April 1997. Certified to JAR-VLA.

CURRENT VERSIONS: **HB 207:** Fixed landing gear version.

HB 207RG: Retractable landing gear version. Empty weight 498 kg (1,098 lb), MTOW 750 kg (1,653 lb); cruising speed at 75 per cent power 140 kt (259 km/h; 161 mph); service ceiling 4,570 m (15,000 ft); T-O run 340 m (1,115 ft), or 440 m (1,445 ft) to 15 m (50 ft); range 809 n miles (1,500 km; 932 miles).

Retractable landing gear (207RG) version of HB Alfa *(Paul Jackson)* ***NEW***/0572293

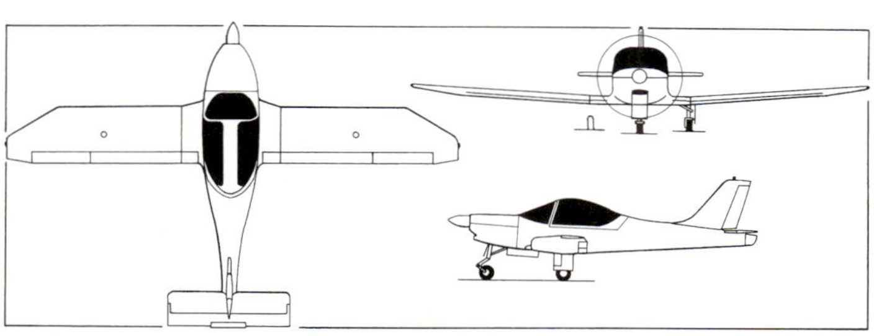

HB Flugtechnik HB-207RG Alfa (Volkswagen engine) *(Paul Jackson)* 0092124

CUSTOMERS: Total 85 sold in Austria, Brazil, Czech Republic, France, Germany, Lithuania, Sweden, Switzerland, United Kingdom and USA by April 2001; 70 completed by late 2003.

COSTS: Standard aircraft fast-build kit, no engine, propeller or avionics €22,500 (2003).

DESIGN FEATURES: Quoted build time 1,000 hours. Constant chord low wings with raked tips; slightly sweptback fin and rudder; one-piece elevator aft of vertical tail. Airframe features derigging and rigging guide points for road transport or storage.

FLYING CONTROLS: Conventional and manual. Ailerons and elevator actuated by pushrods; rudder by cables. Large central trim tab in elevator. Manual (optionally electric) operation for flaps.

STRUCTURE: Mainly metal primary structure with GFRP skin; moving surfaces have Ceconite covering.

LANDING GEAR: HB 207RG has retractable tricycle type, with size 4.00–4 tyre on each unit. Main units have brakes and rubber-in-compression shock-absorption, and retract inward; nosewheel retracts rearward. HB 207 has fixed tricycle type with streamlined fairings and spats on each unit.

POWER PLANT: Standard engine is an 80.9 kW (108.5 hp) VW-Porsche HB 2400 G/2 flat-four based on that of Porsche 911 sports car; alternatives include 59.6 kW (79.9 hp) Rotax 912 A, 73.5 kW (98.6 hp) Rotax 912S and 84.5 kW (113.3 hp) Rotax 914, or Limbach or Textron Lycoming engines up to 74.6 kW (100 hp). Geared drive to choice of propellers: two-blade wooden fixed-pitch, three-blade adjustable- or variable-pitch, or (as on prototypes) five-blade adjustable- or variable-pitch. Fuel tank in each wing, combined capacity 100 litres (26.4 US gallons; 22.0 Imp gallons).

ACCOMMODATION: Rearward-sliding fully transparent canopy.

SYSTEMS: Electrical system: 200 A 15 Ah (optionally 20 Ah) battery; 14 V 75 A alternator.

AVIONICS: *Comms:* Honeywell 760-channel KX 155 transceiver, KT 76A transponder and ELT optional.

Flight: Honeywell KX 125 VOR with KI 208 indicator optional.

Instrumentation: VFR flight and engine transmission instrumentation standard.

DIMENSIONS, EXTERNAL:	
Wing span	9.00 m (29 ft 6¼ in)
Wing aspect ratio	8.5
Length overall	5.95 m (19 ft 6¼ in)
Height overall	1.95 m (6 ft 4¼ in)
DIMENSIONS, INTERNAL:	
Cabin: Length	1.94 m (6 ft 4½ in)
Max width	1.10 m (3 ft 7¼ in)
Max height	0.97 m (3 ft 2¼ in)
AREAS:	
Wings, gross	9.50 m^2 (102.3 sq ft)
Vertical tail surfaces (total)	0.65 m^2 (7.00 sq ft)
Horizontal tail surfaces (total)	1.68 m^2 (18.08 sq ft)
WEIGHTS AND LOADINGS (U: Utility, N: Normal category):	
Weight empty: U, N	450 kg (992 lb)
Max T-O weight: U	640 kg (1,411 lb)
N	700 kg (1,543 lb)
Max wing loading: U	67.4 kg/m^2 (13.80 lb/sq ft)
N	73.7 kg/m^2 (15.09 lb/sq ft)
Max power loading:	
VW-Porsche HB 2400 G/2	8.66 kg/kW (14.22 lb/hp)
Rotax 912 A	11.75 kg/kW (19.31 lb/hp)
PERFORMANCE (HB 2400 engine, U and N as above):	
Never-exceed speed (VNE): U, N	166 kt (309 km/h; 192 mph)
Cruising speed: U	132 kt (245 km/h; 152 mph)
N	130 kt (240 km/h; 149 mph)
Stalling speed: flaps up: U	51 kt (93 km/h; 58 mph)
N	53 kt (98 km/h; 61 mph)
30° flap: U	46 kt (84 km/h; 53 mph)
N	47 kt (87 km/h; 55 mph)
Max rate of climb at S/L: U	330 m (1,082 ft)/min
N	300 m (984 ft)/min
Service ceiling	4,265 m (14,000 ft)
T-O run: U	190 m (625 ft)
N	210 m (690 ft)
T-O to 15 m (50 ft): U	300 m (985 ft)
N	320 m (1,050 ft)
Landing run: U	153 m (500 ft)
Range: U, N	647 n miles (1,200 km; 745 miles)
g limits: U	+4.4/–2.2
N	+3.8/–1.9

UPDATED

HB FLUGTECHNIK AMIGO, CUBBY AND DANDY

TYPE: Tandem-seat ultralight/kitbuilt; side-by-side ultralight/kitbuilt.

PROGRAMME: Introduced to HB range in 2002 as modular range, based on Let-Mont UL series of Piper Cub lookalikes produced in Czech Republic. Differences include the following:

CURRENT VERSIONS: **Amigo:** Side-by-side version of modern appearance, with nosewheel, short-span wings, square-topped fin and cut-down rear decking. Option of increased engine power. Shown at Aero '03, Friedrichshafen, 24–27 April 2003.

Cubby: Tandem-seat, tailwheel layout, most resembling classic J3 Cub.

HB Flugtechnik Amigo side-by-side ultralight *(Paul Jackson)* ***NEW***/0572300

HB Flugtechnik Dandy ***NEW***/1029721

Dandy: As Cubby, but side-by-side seating.

COSTS: Amigo kit €25,900 without engine, €39,900 with Rotax 912 (2004).

FLYING CONTROLS: Conventional and manual. All models have plain flaps and flight-adjustable trimmer in each elevator.

STRUCTURE: Amigo has metal tube, fabric-covered lower fuselage and strut-braced wings, latter with composites tips and leading edge; metal fin fillet, braced tailplane. Cubby and Dandy have metal tube fuselage with fabric covering aft of cabin and metal sheet forward, fabric-covered wooden wings.

LANDING GEAR: *Amigo:* Nosewheel type; fixed. Fuselage-mounted cantilever mainwheel legs. 4.00–6 mainwheels with cable-operated drum brakes; 4.00–4 nosewheel. Tail bumper.

Cubby/Dandy: Tailwheel type; fixed. Two faired-in side Vs hinged to lower longerons, plus half-axles, each with shock-absorber, attached to compression frame. 15×6.00–6 mainwheels and solid tailwheel.

POWER PLANT: One 59.6 kW (79.9 hp) Rotax 912 UL with two-blade propeller standard. Amigo option of 73.5 kW (98.6 hp) Rotax 912 ULS and two-blade AirPlast variable-pitch propeller. Fuel capacity 60 litres (15.8 US gallons; 13.2 Imp gallons) in all versions.

EQUIPMENT: Amigo has optional ballistic parachute.

DIMENSIONS, EXTERNAL (A: Amigo, C: Cubby, D:Dandy):

Wing span: A ..... 9.00 m (29 ft 6¼ in)
C ..... 9.90 m (32 ft 5¾ in)
D ..... 9.60 m (31 ft 6 in)
Length overall: A ..... 5.70 m (18 ft 8½ in)
C, D ..... 5.60 m (18 ft 4½ in)
Height overall: A ..... 2.20 m (7 ft 2½ in)
C, D ..... 2.00 m (6 ft 6¾ in)

AREAS:

Wings, gross: A ..... 11.80 m^2 (127.0 sq ft)
C ..... 13.00 m^2 (139.9 sq ft)
D ..... 12.80 m^2 (137.8 sq ft)

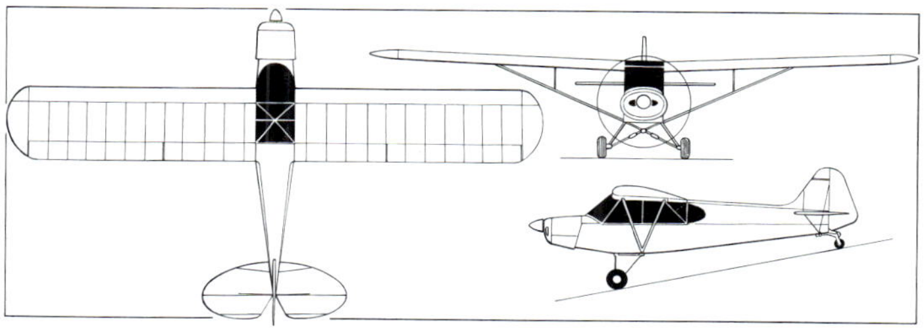

Cubby representative of Piper J3 Cub *(James Goulding)* ***NEW***/0547521

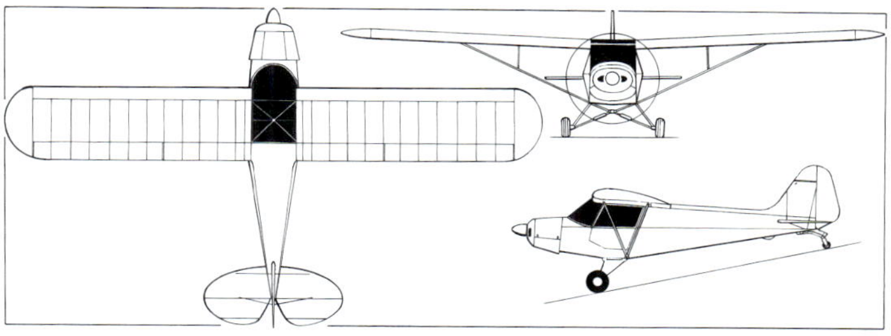

HB Flugtechnik Dandy side-by-side ultralight *(James Goulding)* ***NEW***/0547520

WEIGHTS AND LOADINGS:

Weight empty: A, C ..... 265 kg (584 lb)
D ..... 260 kg (573 lb)
Max T-O weight: all ..... 450 kg (992 lb)

PERFORMANCE:

Normal cruising speed: A ..... 86 kt (160 km/h; 99 mph)
C, D ..... 81 kt (150 km/h; 93 mph)
Stalling speed: A ..... 30 kt (55 km/h; 35 mph)
C, D, flaps down ..... 27 kt (50 km/h; 31 mph)
Max rate of climb at S/L ..... 300 m (984 ft)/min
T-O run: A ..... 60 m (200 ft)
C, D ..... 70 m (230 ft)
g limits ..... +6/–4

UPDATED

Belgium

LAMBERT

LAMBERT AIRCRAFT ENGINEERING BVBA

Lupinestraat 5, B-8500 Kortrijk
Tel/Fax: (+32 56) 21 33 47
e-mail: info@lambert-aircraft.com
Web: http://www.lambert-aircraft.com
DIRECTOR: Filip Lambert

In late 1995 Filip Lambert, a student at the College of Aeronautics, Cranfield, UK, was named one of three joint winners of the Royal Aeronautical Society Light Aircraft Design Competition with his Mission M212-100. The prototype was fabricated in Belgium with the assistance of Steven Lambert and the original intention of undertaking test flying at Koksijde air force base, where the company planned to build a 3,000 m^2 (32,290 sq ft) production facility during the first half of 2004; production scheduled to start in the second half of 2004, when extra staff employed, although a maiden flight did not take place until mid-2004. In September 2003 the workforce numbered three.

UPDATED

LAMBERT MISSION M212

TYPE: Four-seat kitbuilt.

PROGRAMME: Design started November 1992; fuselage mockup completed July 1993; construction of proof-of-concept prototype began at Geluveld, Belgium, January 1996; structural test programme completed 23 December 1998; prototype, M212-100 registered G-XFLY in February 2000, was structurally complete by March 2001; all composites work and landing gear completed November 2001; power plant and fuel system work began March 2002 in UK. Public debut in UK at PFA Rally, Cranfield, 21–23 June 2002. First flight planned for December 2002, but not achieved until 13 July 2004 at Cranfield.

CURRENT VERSIONS: Family of aircraft planned around core four-seat version.

M212-100: Prototype. One only to be built.

Lambert Mission M212 displayed at PFA Rally, Kemble, the week before its maiden flight *(Paul Jackson)* ***NEW***/0589265

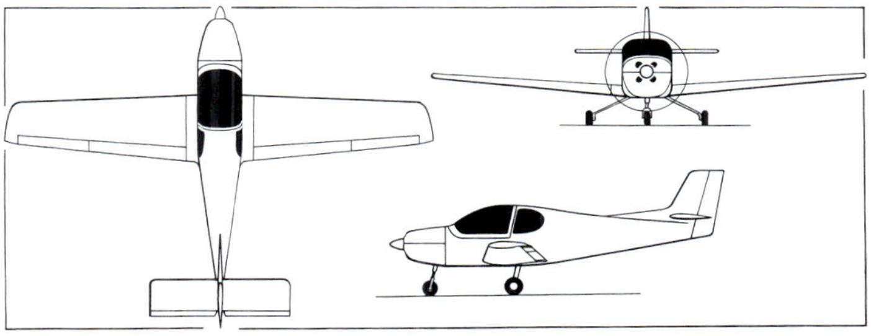

Provisional drawing of Lambert Mission M212–100 (Zoche ZO-01A diesel engine) *(James Goulding)* 0044443

M212-200: Basic four-seat version, will be available factory-built or as kit; certification to JAR/FAR Pt 23 planned.

M212-300: Proposed two-seat aerobatic trainer; since discontinued.

M212-400: Four-seat tourer with increased fuel and baggage capacity, optional retractable landing gear, maximum take-off weight 1,100 kg (2,425 lb) and range of over 1,000 n miles (1,852 km; 1,151 miles).

CUSTOMERS: Six orders for M212–200 and options on 20 M212–200 and –400 kits held in September 2003; customers in the Benelux countries, France and, mostly, the UK.

COSTS: Kit, including engine, propeller and VFR avionics, €110,000 (2003).

DESIGN FEATURES: Designed using CAD and wind tunnel research facilities; will meet FAR Pt 23 and JAR 23 certification criteria. Low-wing monoplane; wings, which have laminar flow, moderate taper and upturned tips. Wing dihedral 5°; incidence at root 2°; twist 2°. Quoted kit build time 800 hours.

FLYING CONTROLS: Conventional and manual. Cable and pushrod actuation. Electrically actuated single-slotted flaps.

STRUCTURE: All-composites, with extensive (95 per cent by weight) use of prepregs; glass and carbon in epoxy resin used for spars and longerons. Fuselage of monocoque construction; wing has dual main spars plus auxiliary spar for further strengthening.

LANDING GEAR: Fixed tricycle type. Mainwheels size 6.00–6, nosewheel 5.00–5 or 6.00–6; Cleveland brakes. Steerable nosewheel, maximum steering angle ±50°. Potential for development of retractable version (tailwheel option withdrawn).

POWER PLANT: One 112 kW (150 hp) Zoche ZO-01A radial four-cylinder, two-stroke diesel engine with multifuel capability, driving a three-blade constant-speed propeller. However, prototype fitted with second-hand 112 kW (150 hp) Textron Lycoming O-320-E2D flat-four; MT-Propeller MTV-18-C/175-17d three-blade, constant-speed (electric) propeller; and Gomolzig twin-muffler exhaust. Fuel capacity 120 litres (31.7 US gallons; 26.4 Imp gallons) in single fuselage tank in M212–100 prototype; production examples will have two wing tanks and total capacity of 280 litres (74.0 US gallons; 61.6 Imp gallons). In developed form as four-seat tourer, suitable for engines in 172 to 194 kW (230 to 260 hp) power range.

ACCOMMODATION: Four persons in 2+2 configuration, under one-piece, forward-hinged canopy. Seats and rudder pedals adjustable; dual controls standard.

All data are provisional.

DIMENSIONS, EXTERNAL:

Wing span	9.80 m (32 ft 1¾ in)
Wing aspect ratio	8.0
Length overall	7.40 m (24 ft 3½ in)
Height overall	2.90 m (9 ft 6 in)
Tailplane span	3.20 m (10 ft 6 in)
Wheel track	2.80 m (9 ft 2¼ in)

DIMENSIONS, INTERNAL:

Cabin: Length	2.50 m (8 ft 2½ in)
Max width	1.12 m (3 ft 8 in)
Max height	1.25 m (4 ft 1¼ in)
Baggage compartment volume: M212–100	0.50 m^3 (17.7 cu ft)

AREAS:

Wings, gross	12.00 m^2 (129.2 sq ft)
Ailerons (total)	0.60 m^2 (6.46 sq ft)
Flaps (total)	1.08 m^2 (11.62 sq ft)
Elevators (total)	2.65 m^2 (28.52 sq ft)

WEIGHTS AND LOADINGS:

Weight empty	485 kg (1,069 lb)
Max T-O weight	900 kg (1,984 lb)
Max wing loading	75.0 kg/m^2 (15.36 lb/sq ft)
Max power loading	8.04 kg/kW (13.20 lb/hp)

PERFORMANCE:

Never-exceed speed (VNE)	183 kt (338 km/h; 210 mph)
Max level speed at S/L	136 kt (252 km/h; 157 mph)
Max cruising speed at FL80: at 75% power	131 kt (243 km/h; 151 mph)
at 60% power	118 kt (219 km/h; 136 mph)
Stalling speed, flaps down	49 kt (91 km/h; 57 mph)
Max rate of climb at S/L	293 m (960 ft)/min
T-O run	224 m (735 ft)
Range with max fuel: at 75% power	640 n miles (1,185 km; 736 miles)
at 60% power	800 n miles (1,481 km; 920 miles)

UPDATED

MASQUITO

MASQUITO AIRCRAFT N.V.

Reigersbaan 31, B-1760 Roosdaal-Strijtem
Tel: (+32 54) 34 30 08
Fax: (+32 54) 34 30 09
e-mail: info@masquito.be
Web: http://www.masquito.be
DIRECTORS:
Paul Maschelein
Stefaan Maschelein John Pescod

This company was established to produce the Masquito M80 ultralight helicopter, four prototypes of which were completed. Production deliveries were due to begin in early 2004, but were delayed because of a change of licencing authority from the UK to France. Difficulties persisted, however, and in mid-2004 Masquito Aircraft announced that it had "ceased all activities on its very light helicopter project and associated engine development".

UPDATED

Masquito M80 ultralight helicopter, now terminated *(Paul Jackson)* ***NEW***/0572292

WOLFSBERG

WOLFSBERG AIRCRAFT CORPORATION NV

Woudstraat 23, B-3600 Genk
Tel: (+32 89) 38 08 31
Fax: (+32 89) 38 61 41
MANAGING DIRECTOR: Alec N Clark

Formerly Triloader Aircraft Corporation NV, which at one time intended to develop the Clark-Norman Triloader. Wolfsberg Aircraft Corporation NV is now developing the Raven 257 small multirole aircraft.

The prototype programme was managed in the Czech Republic by Wolfsberg-Evektor, construction being Letov Air for the wing and Evektor-Aerotechnik for the remainder. The prototype flew on 28 July 2000. In September 2000 the whole programme was transferred by Wolfsberg Aircraft, Belgium, to its wholly owned subsidiary Letov Air (now called Wolfsberg Letecká Tovarna) in Prague. After major management reorganisation at Wolfsberg Letecká Tovarna, development continues. Construction and certification is carried out by a subsidiary company in Czech Republic; marketing is from Belgium.

Brazil

AEROMOT

AEROMOT INDÚSTRIA MECÂNICO-METALÚRGICA LTDA

Caixa Postale 8031, Avenida das Indústrias 1210, 90200-290 Porto Alegre, RS
Tel: (+55 51) 33 71 16 44
Fax: (+55 51) 33 71 16 55
e-mail: industria@aeromot.com.br
Web (1): http://www.ximango.com.br
Web (2): http://www.aeromot.com.br
PRESIDENT: Claudio B Viana
MANAGING DIRECTOR: João Claudio Jotz
MARKETING CONTACT: Fabiane Dorneles

Aeromot Indústria is part of Aeromot Group with Aeromot Aeronaves e Motores SA (parent company, founded 1967, which provides maintenance).

Aeromot Indústria initially designed, certified and built seats for aircraft built by Embraer; it later produced several structural parts for Embraer aircraft, and designed and certified seats for Airbus, Boeing, Fokker and McDonnell Douglas (now Boeing) commercial transports. In 1985 it purchased assets of Fournier motor glider factory in France, including sole manufacturing rights for RF-10; has since incorporated several improvements to basic model, including different engines and propellers. A lightplane version, the AMT-600 Guri, with reduced wing span, first flew in 1999 and was certified in December 2001.

Factory has shop floor area of 3,100 m^2 (33,375 sq ft) and workforce of approximately 150.

In late 2003, Aeromot was reported to be in the final stages of negotiation with GAIC of China with regard to the latter's licensed production of the Guri and, subject to a change of Chinese air licencing law, the Super Ximango. Joint development and production of a four-seat aircraft is also under consideration.

UPDATED

AEROMOT XIMANGO

US Air Force designation: TG-14

TYPE: Motor glider.

PROGRAMME: Brazilian production version of French Aérostructure (Fournier) RF-10; first flight (French prototype) 6 March 1981; see Sailplanes section of 1990–91 *Jane's* for French production history. All production rights sold to Aeromot July 1985; Brazilian CTA certification of AMT-100 granted 5 June 1986 and French (DGAC) 10 October 1990; AMT-200, developed by Aeromot, made first flight July 1992; was certified 3 February 1993 in Brazil, on 29 December 1993 in USA (to FAR Pt 22), during 1995 in UK and on 24 November 1998 in Canada. Certification of AMT-200 also achieved in Australia, Colombia, France, Germany, Japan, Netherlands and New Zealand.

CURRENT VERSIONS: **AMT-100 Ximango:** Initial production version; powered by one 59.7 kW (80 hp) Limbach L 2000 EO1 flat-four engine driving a Hoffmann HO-V62RL/160BT two-blade three-position variable-pitch propeller. Total 43 built, 1988–93.

AMT-100P Ximango: Military/police and observation version. Two side windows below canopy; underfuselage pod for 100 kg (220 lb) of weapons and avionics or surveillance equipment. In service with Brazilian military police and other law enforcement and environment protection agencies. Production ceased; replaced by AMT-200P, AMT-200SP and AMT-300P with similar optional equipment.

AMT-200 Super Ximango: Generally similar to AMT-100 except for Rotax 912A power plant. Main production variant between 1995 and 2000. Improvements introduced in 1997 include redesigned instrument panel; new canopy locking mechanism; fully enclosing main landing gear doors; tailwheel fairing; and metal fuel tanks.

AMT-200P: Military, police and observation version.

AMT-200S Super Ximango S: Version optimised for pilot training and soaring (S indicating 'sport'). Generally similar to AMT-200 except for Rotax 912 S4 engine and optional removable winglets (span as AMT-300). Prototype first flew in September 1999;

Aeromot TG-14 (AMT-200) motor glider of the USAF *(Paul Jackson)* ***NEW***/0569604

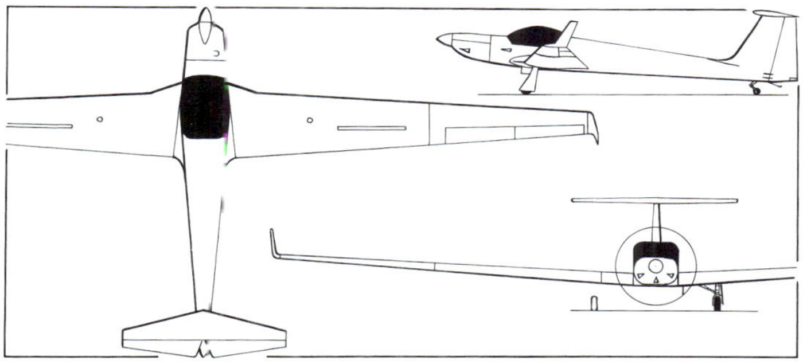

Aeromot AMT-200 Super Ximango S *(James Goulding)* ***NEW***/0589198

deliveries from December 1999, to customers in Australia (one), Brazil (one), Dominican Republic (one), South Africa (two), UK (two) and USA (22). Main production variant from 2000.

Selected from at least 13 types evaluated for the USAF's Introductory Flight Training Program (IFTP), replacing grounded Slingsby T-3A Fireflies in flight screening/primary training role; initial order for 14 aircraft announced October 2001; first (01-0136/N136XS) delivered to 94th Flight Training Squadron of US Air Force Academy at Colorado Springs on 24 June 2002, all operating with civilian registrations.

AMT-200SP: Military, police and observation version.

Detailed description applies to the AMT-200 except where indicated.

AMT-300 Turbo Ximango Shark: Prototype (PT-ZAM, converted from an early AMT-100) flew July 1997; based on AMT-200, but with Rotax 914F turbocharged engine, redesigned cockpit, new engine cowling and winglets. Brazilian certification achieved 31 March 1999, followed by FAA approval 19 July 1999; deliveries began in May 1999 with two aircraft for US market to Ximango US at Spruce Creek, Florida; believed none further by late 2004, however.

AMT-300P: Military, police and observation version.

AMT-300R Reboque: Glider tug version of AMT-300.

AMT-600 Guri: Trainer and sport aerobatic version. Described separately.

CUSTOMERS: Initial order for 50 from Brazilian Civil Aeronautical Department; also produced for military/paramilitary roles such as observation, patrol and counter-insurgency. Total of 163 AMT-100s, AMT-200s, AMT-200Ss and AMT-300s produced by December 2004, including deliveries to customers in Argentina (two), Australia (seven), Belgium (one), Brazil (63), Canada (one), Colombia (one), Dominican Republic (one), France (seven), Germany (two), Japan (four), South Africa (two), Mexico, UK (six), USA (49) and undisclosed (5). Further airframe (c/n 112) completed as prototype Guri. AMT-300 production began at No. 106, but continues in parallel with AMT-200.

COSTS: US$135,000 basic for AMT-200; US$143,000 basic for AMT-200S; US$154,000 basic for AMT-300 (all 2004).

DESIGN FEATURES: Typical motor glider; low, tapered wing with fixed incidence T tail. Wings detachable for transportation and storage; outboard half of wing can be folded inward without disconnecting aileron controls.

Wing section NACA 64_3-618; dihedral 2° 30'.

FLYING CONTROLS: Conventional and manual; ailerons and elevators operated by pushrods, rudder by cables; trim tab actuated by cable on AMT-100/200; trim control on AMT-200S/300 effected by spring device on elevator pushrod. Schempp-Hirth airbrakes in wing upper surface

STRUCTURE: All-GFRP/foam except for carbon fibre main spar and light alloy airbrakes.

LANDING GEAR: Mechanically retractable mainwheels (tyre size 330×130), with hydraulic suspension and Cleveland hydraulic disc brakes with differential action; steerable tailwheel with size 210×65 tyre.

POWER PLANT: One 59.6 kW (79.9 hp) Rotax 912A flat-four with Hoffmann HO-V62R/170FA propeller. Fuel in two main tanks in wings, combined capacity 90 litres (23.8 US gallons; 19.75 Imp gallons). AMT-200S has one 73.5 kW (98.6 hp) Rotax 914 S4, driving a three-position variable-pitch Hoffmann propeller. AMT-300 has one 84.6 kW (113.4 hp) Rotax 914 F3 turbocharged flat-four driving an MT/MTV-21 constant-speed feathering propeller; fuel capacity 90 litres (23.8 US gallons; 19.8 Imp gallons).

ACCOMMODATION: Two seats side by side. One-piece canopy hinged at rear to open upward. Dual controls standard.

SYSTEMS: Electric starter and 12 V 30 A alternator.

DIMENSIONS, EXTERNAL:

Wing span: 200, 200S ... 17.47 m (57 ft 3¾ in)
 300 ... 17.67 m (57 ft 11½ in)
Wing aspect ratio: 200, 200S ... 16.3
 300 ... 16.7
Width, wings folded ... 10.15 m (33 ft 3½ in)
Length overall ... 8.08 m (26 ft 6 in)
Height overall: 200, 200S, 300 ... 1.93 m (6 ft 4 in)
Tailplane span ... 3.68 m (12 ft 0¾ in)
Wheel track ... 2.80 m (9 ft 2¼ in)
Wheelbase ... 5.03 m (16 ft 6 in)
Propeller diameter: 200, 200S ... 1.70 m (5 ft 7 in)
 300 ... 1.65 m (5 ft 5 in)

DIMENSIONS, INTERNAL:

Cockpit: Length ... 1.85 m (6 ft 0¾ in)
 Max width ... 1.15 m (3 ft 9¼ in)
 Max height ... 1.00 m (3 ft 3¼ in)

AREAS:

Wings, gross: 200, 200S ... 18.70 m² (201.3 sq ft)
 300 ... 18.75 m² (201.8 sq ft)
Ailerons (total) ... 2.72 m² (29.28 sq ft)
Airbrakes (total) ... 1.16 m² (12.49 sq ft)
Fin ... 1.44 m² (15.50 sq ft)
Rudder ... 0.69 m² (7.43 sq ft)
Tailplane ... 1.95 m² (20.99 sq ft)
Elevator ... 0.95 m² (10.23 sq ft)

WEIGHTS AND LOADINGS:

Weight empty: 200 ... 620 kg (1,367 lb)
 200S ... 625 kg (1,378 lb)
 300 ... 630 kg (1,389 lb)
 300R ... 620 kg (1,367 lb)
Max T-O weight: 200, 200S, 300 ... 850 kg (1,874 lb)
Max baggage: 300 ... 10 kg (22 lb)
Max wing loading: 200, 200S ... 45.5 kg/m² (9.31 lb/sq ft)
 300 ... 45.3 kg/m² (9.28 lb/sq ft)
Max power loading: 200 ... 14.26 kg/kW (23.42 lb/hp)
 200S ... 11.56 kg/kW (19.00 lb/hp)
 300 ... 9.92 kg/kW (16.29 lb/hp)

PERFORMANCE, POWERED:

Never-exceed speed (VNE):
 200, 200S, 300 ... 133 kt (245 km/h; 153 mph)
Max cruising speed: 200 ... 110 kt (205 km/h; 127 mph)
 200S ... 119 kt (220 km/h; 137 mph)
 300 ... 124 kt (230 km/h; 143 mph)
Normal cruising speed: 300 ... 122 kt (225 km/h; 140 mph)
Stalling speed: 300 ... 42 kt (78 km/h; 48 mph)
Max rate of climb at S/L: 200 ... 156 m (512 ft)/min
 200S ... 180 m (591 ft)/min
 300 ... 198 m (650 ft)/min
Service ceiling: 200, 200S ... more than 4,900 m (16,076 ft)
 300 ... more than 8,700 m (28,543 ft)
T-O run, hard runway: 200 ... 226 m (745 ft)
 200S ... 210 m (689 ft)
 300 ... 174 m (570 ft)
T-O to 15 m (50 ft), hard runway:
 200 ... 305 m (1,001 ft)
 300 ... 296 m (970 ft)
Landing from 15 m (50 ft) ... less than 300 m (984 ft)
Landing run ... less than 150 m (492 ft)
Range with max fuel:
 200: best power ... 594 n miles (1,100 km; 683 miles)
 best economy ... 756 n miles (1,400 km; 870 miles)
 200S: best power ... 575 n miles (1,065 km; 661 miles)
 best economy ... 783 n miles (1,450 km; 901 miles)
 300: best power ... 548 n miles (1,015 km; 630 miles)
 best economy ... 810 n miles (1,500 km; 932 miles)
Max endurance:
 200: best power ... 5 h 30 min
 best economy ... 12 h
 200S: best power ... 4 h 54 min
 best economy ... 12 h
 300: best power ... 4 h 30 min
 best economy ... 12 h
g limits: 200, 200S, 300 ... +5.3/–2.65

PERFORMANCE, UNPOWERED:

Best glide ratio at 58 kt (108 km/h; 67 mph):
 200, 200S ... 31.1
 300 ... 32.1
Min rate of sink at 52 kt (97 km/h; 60 mph):
 200, 200S ... 0.93 m (3.05 ft)/s
 300 ... 0.96 m (3.15 ft)/s
Stalling speed: 200, 200S, 300 ... 42 kt (78 km/h; 48 mph)

UPDATED

AEROMOT AMT-600 GURI

English name: Boy

TYPE: Aerobatic two-seat lightplane.

PROGRAMME: Design began July 1998; prototype (PP-XBS), employing a Ximango airframe first flown 13 July 1999; Brazilian CTA certification was achieved 19 December 2001. First production aircraft (PR-AMT) registered in August 2003.

Negotiations under way in late 2003 for a retractable landing gear version to be built under licence by GIAC in China.

CURRENT VERSIONS: **AMT-600 Guri:** Initial production version, as described.

Data below are provisional, and refer to prototype.

Advanced Trainer: Projected derivative with 134 kW (180 hp) Textron Lycoming engine, retractable landing gear and IFR avionics.

CUSTOMERS: Brazilian DAC (civil aeronautics department) signed order for 25 in October 2003 to be distributed to several aero clubs.

DESIGN FEATURES: Generally as Ximango, but shortened wings (of same aerofoil section) and tricycle landing gear. Wings can be removed for storage.

FLYING CONTROLS: Conventional and manual; ailerons and elevator operated by pushrods, rudder by cables; ground-adjustable trim tab on rudder and starboard aileron. Control surface maximum deflections: elevator +22/–25°, ailerons +15/–27°, rudder 28°, flaps 45°.

STRUCTURE: Generally as for Ximango. Wing includes main-spar, plus rear-spar of Z shape to accommodate ailerons and flaps; two ribs in each panel, at centre-section break and tip, but three ribs per side at centre-section to strengthen walkways. Ailerons and flaps have leading- and trailing-edge spars, each with six glass fibre ribs stiffened with foam and aluminium skirt. Two fuselage bulkheads only: one horizontal, one vertical, for attachment of tail surfaces. Plywood for local strengthening in areas of equipment installations.

LANDING GEAR: Non-retractable tricycle type, with hydropneumatic trailing-link suspension on main units and rubber-in-compression suspension on nose leg. Oldi wheels and brakes. Mainwheel tyre size 6.00×6, nosewheel 5.00×5.

POWER PLANT: One 85.8 kW (115 hp) Textron Lycoming O-235-NBR flat-four, driving a two-blade Sensenich 72CK-0-50 aluminium propeller. Fuel contained in two wing tanks, combined capacity 90 litres (23.8 US gallons; 19.8 Imp gallons), with filler port in upper surface of each wing.

Prototype Aeromot AMT-600 Guri 0131751

Aeromot AMT-600 Guri with raised canopy intended for production version *(James Goulding)* 0121661

ACCOMMODATION: Two persons, side by side, under single-piece, upward- and rearward-hinged canopy. Dual controls standard. Boarding step on each side forward of wing on prototype will be relocated aft of wing on production aircraft.

AVIONICS: VFR avionics, including Garmin GPS 100; IFR avionics optional.

DIMENSIONS, EXTERNAL:

Wing span	10.50 m (34 ft 5½ in)
Wing aspect ratio	8.0
Length overall	8.07 m (26 ft 5¾ in)
Height overall	2.65 m (8 ft 8¼ in)
Tailplane span	2.60 m (8 ft 6¼ in)
Aileron span	1.99 m (6 ft 6½ in)
Flap span	2.44 m (8 ft 0 in)
Wheel track	2.99 m (9 ft 9¾ in)
Wheelbase	1.75 m (5 ft 9 in)
Propeller diameter	1.83 m (6 ft 0 in)

DIMENSIONS, INTERNAL:

Cabin max width	0.86 m (2 ft 10 in)

AREAS:

Wings, gross	13.79 m² (148.4 sq ft)
Ailerons (total)	0.90 m² (9.69 sq ft)
Flaps (total)	1.40 m² (15.07 sq ft)
Rudder	0.63 m² (6.78 sq ft)
Tailplane	1.41 m² (15.18 sq ft)
Elevator	0.76 m² (8.18 sq ft)

WEIGHTS AND LOADINGS:

Weight empty	675 kg (1,488 lb)
Max T-O weight	900 kg (1,984 lb)
Max wing loading	65.3 kg/m² (13.37 lb/sq ft)
Max power loading	10.49 kg/kW (17.23 lb/hp)

PERFORMANCE:

Never-exceed speed (VNE)	135 kt (250 km/h; 155 mph)
Cruising speed	108 kt (200 km/h; 124 mph)
Stalling speed	49 kt (90 km/h; 56 mph)
Service ceiling	5,300 m (17,380 ft)
Max rate of climb at S/L	216 m (709 ft)/min
T-O run	130 m (427 ft)
Max range	783 n miles (1,450 km; 901 miles)
Endurance	6 h 0 min
g limits	+3.8/−1.52

UPDATED

EDRA HELICENTRO

EDRA HELICENTRO, PEÇAS E MANUTENÇAO LTDA

Rodovia Estadual SP-191 Km 87.5, 13537-000 Ipeúna, Saõ Paulo
Tel: (+55 19) 576 12 92
Fax: (+55 19) 576 13 92
e-mail: mnd@edraescola.com.br
Web: http://www.edraescola.com.br

WORLD DISTRIBUTOR:
Amphibian Airplanes of Canada Ltd
Box 1100, Squamish, British Columbia, V0N 3G0, Canada
Tel: (+1 604) 898 53 27 and 32 00
Fax: (+1 604) 898 20 09 and (+1 775) 890 08 83
e-mail: harusch@uniserve.com
Web: http://www.seastaramphibian.com

Following cessation of manufacture by the Billie company in France, the Pétrel amphibian is being produced in Brazil for local and export markets. Marketing outside South America is undertaken by Amphibian Airplanes of Canada, which built the first North American SeaStar in 1998. Components are imported and augmented by North American standard items before sale as kits. Exports have been made as far afield as Australia.

EDRA Helicentro is one of three components of the EDRA group, founded in 1994 and with interests in pilot training, aircraft maintenance, air taxi services and aircraft dealership (Schweizer helicopters for Brazil).

UPDATED

EDRA HELICENTRO PATURI

English name: Masked Duck
Export marketing name: SeaStar
Canadian name: Sao Paolo Seabird

TYPE: Two-seat amphibian/kitbuilt.

PROGRAMME: Derived from Claude Tisserand's Hydroplum II, prototype of which first flew 1 November 1986; SMAN bought design rights in 1987; construction of Pétrel prototype began January 1989; first flight July 1989. At least 70 built in France; rights passed to Billie Aero Marine. Production transferred to Brazil by 1997.

CUSTOMERS: Total of 135 produced by EDRA by late 1999; in addition, over 100 kits sold by December 2003. Customers in Europe, Africa, Canada, Australia, USA and Brazil.

COSTS: Kit, less engine, US$18,500 with basic items only, or US$27,000 for complete components set; fly-away US$43,000 (2004).

DESIGN FEATURES: Experimental category pusher-engined biplane; road-towable on custom-designed trailer with assembly/disassembly time of 10 minutes. Quoted kit build time 500 hours (fast-build kit).

Equal-span, constant-chord wings, upper unit mounted on cabane; V-type interplane struts, with diagonal strut brace to fuselage from upper plane. Single-step hull. Boom-mounted empennage, wire-braced. Floats at mid-span of lower plane. (French-built Pétrels have shorter lower span with floats at tips.)

Wing section NACA 2412; dihedral 2° 13' on upper wings, 3° 26' on lower; sweepback 4° upper, 2° lower.

FLYING CONTROLS: Conventional and manual. Ailerons on upper wing only. No flaps. Actuation by wires (rudder) and pushrods. Electric trim.

STRUCTURE: Moulded monocoque single-step hull of epoxy/carbon fibre foam with carbon fibre tailboom; wings have 2014-T6 aluminium alloy tubular main spar with PVC foam ribs and glass fibre/epoxy leading-edge and tips covered with fabric and braced by single diagonal strut and V interplane struts of 6061-T6 aluminium. Wings fold for ground

EDRA Helicentro Paturi, marketed abroad as the SeaStar *(Paul Jackson)* 0084581

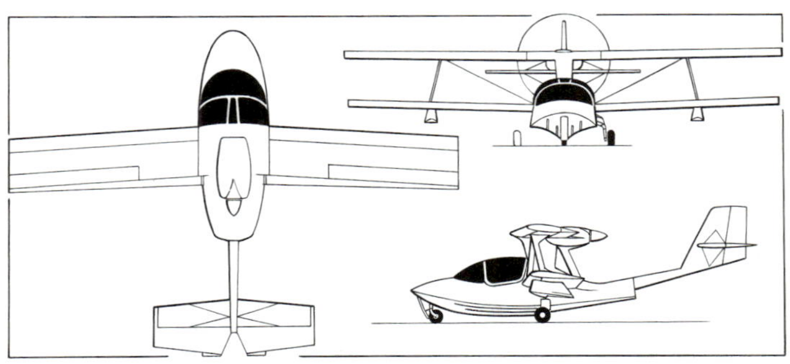

Paturi amphibious two-seat lightplane *(James Goulding)* 0100404

transportation. Tail surfaces of glass fibre spars and PVC foam ribs, with fabric and glass fibre covering, stainless steel wire braced.

LANDING GEAR: Retractable tricycle type; based on Mooney design, with Johnson bar actuation; main units retract upwards into hull and undersurface of lower wing; nosewheel retracts upwards and forwards, tyre remaining partially exposed to serve as docking bumper when operating on water. Nosewheel maximum steering angle 80°. Hydraulic disc brakes on main units. No water rudders.

POWER PLANT: One 59.6 kW (79.9 hp) Rotax 912 UL or 73.5 kW (98.6 hp) Rotax 912 ULS four-stroke piston engine driving an Airplast 175 three-blade pusher propeller. Optional Airplast PV 50 in-flight adjustable pitch propeller with electric or manual control. Fuel capacity 50 litres (13.2 US gallons; 11.0 Imp gallons), of which 45 litres (11.9 US gallons; 9.9 Imp gallons) usable. Optional auxiliary tank, capacity 106 litres (28.0 US gallons; 23.3 Imp gallons).

ACCOMMODATION: Two, side by side in open or enclosed cockpit, with windscreen or single-piece forward-hinging canopy. Dual controls.

EQUIPMENT: Optional BRS ballistic parachute. Bilge pump.

DIMENSIONS, EXTERNAL:

Wing span (both)	8.25 m (27 ft 8 in)
Length overall	6.47 m (21 ft 2¾ in)
Height overall	2.26 m (7 ft 5 in)

Wheel track .... 1.78 m (5 ft 10 in)
Wheelbase .... 2.10 m (6 ft 10¾ in)
Propeller diameter .... 1.65 m (5 ft 5 in)

DIMENSIONS, INTERNAL:
Cabin max width .... 1.13 m (3 ft 8½ in)

AREAS:
Wings, gross .... 16.50 m^2 (177.6 sq ft)

WEIGHTS AND LOADINGS:
Weight empty .... 300 kg (661 lb)
Max T-O weight .... 600 kg (1,323 lb)
Max wing loading .... 36.4 kg/m^2 (7.45 lb/sq ft)
Max power loading .... 10.08 kg/kW (16.56 lb/hp)

PERFORMANCE (Rotax 912 ULS):
Never-exceed speed (VNE) .... 98 kt (180 km/h; 112 mph)
Max level speed .... 92 kt (170 km/h; 106 mph)
Max cruising speed .... 86 kt (160 km/h; 99 mph)
Manoeuvring speed .... 81 kt (150 km/h; 92 mph)
Stalling speed .... 35 kt (65 km/h; 41 mph)
Max rate of climb at S/L .... 152 m (500 ft)/min
Service ceiling .... 3,660 m (12,000 ft)
T-O run: on land .... 200 m (660 ft)
on water .... 250 m (820 ft)
Landing run: on land .... 250 m (820 ft)
on water .... 150 m (495 ft)
Range with max fuel .... 600 n miles (1,111 km; 690 miles)
Radius of action, auxiliary fuel .... 405 n miles (750 km; 466 miles)
Endurance .... 2 h 30 min
g limits .... +4/−2

UPDATED

EMBRAER

EMPRESA BRASILEIRA DE AERONÁUTICA SA

Av Brig Faria Lima 2170, Caixa Postal 343, 12227-901 São José dos Campos, SP
Tel: (+55 12) 39 27 10 00
Fax: (+55 12) 39 21 23 54
Web: http://www.embraer.com

PRESIDENT AND CEO: Maurício Novis Botelho
EXECUTIVE VICE-PRESIDENT, INDUSTRIAL: Satoshi Yokota
EXECUTIVE VICE-PRESIDENT, COMMERCIAL: Frederico Pinheiro Fleury Curado
EXECUTIVE VICE-PRESIDENT, PLANNING AND ORGANISATIONAL DEVELOPMENT: Horácio Aragonés Forjaz
EXECUTIVE VICE-PRESIDENT, FINANCE: Antonio Luis Pizarro Manso
VICE-PRESIDENT, COMMERCIAL (MILITARY MARKETING): Romualdo Monteiro de Barros
VICE-PRESIDENT, CORPORATE JETS: Sam Hill
PRESS OFFICER: Marcia Benevides

EUROPEAN OFFICE:
Embraer Europe Corporate Jets
Aéroport du Bourget Zone d'Aviation d'Affaires, BP 74,
F-93352 Le Bourget Cedex, France
Tel: (+33 1) 49 38 44 44
Fax: (+33 1) 49 38 44 45
e-mail: corporatejets@embraer.com
MANAGER: Neil Patton

EMBRAER DELIVERIES
(at September 2003)

Type	Deliveries
EMB-110 Bandeirante	469
EMB-111 Bandeirante	
Patrulha	31
EMB-120 Brasilia	354
EMB-121 Xingu	105
ERJ-135	104
ERJ-140	74
ERJ-145	509
EMB-200/201/202	
Ipanema	860
EMB-312 Tucano	650
EMB-326GB Xavante	182
AMX	56
Light aircraft (incl Neiva)	2,470
Total	**5,864**

Created 19 August 1969, Embraer began operating on 2 January 1970. Its 250,000 m^2 (2,691,000 sq ft) factory at Faria Lima was officially joined on 15 January 2001 by the Eugênio de Mello plant, comprising 147,650 m^2 (1,589,300 sq ft) of covered space. Both facilities are near São José dos Campos. Fifth company plant, at Gavião Peixoto (300 km; 186 miles northwest of São Paulo) opened 11 June 2002; to expand over five years to 3,000,000 m^2 (32 million sq ft) and undertake final assembly of Legacy, Super Tucano, AMX-T and special missions versions of EMB-145.

Total Embraer workforce was 12,348 on 7 October 2003. Neiva (which see) became a subsidiary in March 1980. Embraer and subsidiaries have delivered nearly 5,500 aircraft.

Privatisation of Embraer was undertaken on 7 December 1994 with the auction of 55.4 per cent of voting stock. A consortium, led by Bozano Simonsen bank, acquired 45.44 per cent of auctioned stock, assumed a controlling interest in the company, and provided an extra US$36 million in capitalisation. A further 10 per cent of voting stock was offered to the public within 60 days, with Embraer employees and the Brazilian government retaining 10 and 18.4 per cent respectively. In October 1999 a consortium of French aerospace companies, comprising Aerospatiale Matra (now EADS France), Dassault, SNECMA and Thomson-CSF (now Thales), acquired 20 per cent of Embraer's voting shares. Embraer also embarked on a joint venture with Liebherr International of Germany to establish Embraer-Liebherr Equipamentos do Brasil SA (ELEB) to create additional opportunities for the company's landing gear and hydraulics components business. Embraer's controlling group — with 60 per cent of voting shares — comprises pension funds Previ and Sistel, plus investment house Companhia Bozano Simonsen.

Principal current own-design manufacturing programmes are EMB-120 Brasilia commuter transport, ERJ-135/ERJ-140/ERJ-145 and ERJ-170/190 regional jet families and EMB-312/EMB-314 Tucano turboprop military trainer. Subsidiary Neiva manufactures EMB-202 Ipanema agricultural aircraft and holds licences from New Piper in USA. Main production of International (with Italy) AMX attack fighter is complete, but Embraer holds outstanding contract for Venezuelan order.

In subcontract field, first deliveries made in 1994 of wingtips and vertical fin fairings for Boeing 777 under 1991 contract.

Embraer is a risk-sharing partner in the Sikorsky S-92 Helibus programme; under a June 1995 contract valued at US$170 million, it will supply 730 sets of S-92 fuel sponsons. In 1995 Embraer signed a co-operation agreement with PZL Warszawa-Okecie of Poland under which Embraer subsidiary Neiva SA would market PZL light aircraft in Brazil and PZL subsidiary Skypol would acquire Brasilias for its freight and charter services. This appears to have lapsed.

On 3 April 2002 Embraer, Dassault Aviation, SNECMA Moteurs and Thales Airborne Systems formed the Mirage 2000BR Consortium to develop a version of the Dassault Mirage 2000–5 Mk 2 as a contender for the Brazilian Air Force's F-X BR programme. If selected, FAB Mirage 2000BRs would be assembled at a new Embraer facility at Gavião Peixoto.

On 2 December 2002 Embraer signed an agreement with Harbin Aircraft Industry (Group) Co Ltd and Hafei Aviation Industry Co Ltd to form a new joint venture company, Harbin Embraer Aircraft Company Ltd, which will produce under licence the ERJ-135, ERJ-140 and ERJ-145, primarily for customers in the People's Republic of China. Launch customer China Southern Airlines ordered six ERJ-145s on 2 February 2004. A 24,000 m^2 (847,550 sq ft) facility has been established in Harbin, Heilongjiang Province, employing up to 220 staff. The first Harbin Embraer-built aircraft, an ERJ-145, was rolled out and made its first flight on 16 December 2003.

Embraer delivered 177 aircraft (45 ERJ-135s, 112 ERJ-145s, three EMB-135s and 17 lightplanes) in 2000, 174 aircraft (two EMB-120s, 27 ERJ-135s, 22 ERJ-140s, 104 ERJ-145, seven EMB-135, one EMB-145 and 11 light aircraft) in 2001. Totals for 2002 were 82 ERJ-145s, 36 ERJ-140s, three ERJ-135s, seven Legacy Executives, one Legacy Shuttle, one EMB-135 and one EMB-145 for government use, or 131 in all. Totals for 200 were 14 ERJ-135s, 16 ERJ-140s, 57 ERJ-145s, 13 Legacys, and one EMB-145 for government use, or 101 in all. Planned target for 2004 is 160, rising to 170 in 2005. Firm order backlog valued at US$10.9 billion at 31 March 2004 for airline, corporate and defence markets.

UPDATED

EMBRAER EMB-314 SUPER TUCANO

English name: Super Toucan
Brazilian Air Force designations: A-29 and AT-29
TYPE: Basic turboprop trainer/attack lightplane.

DEVELOPMENT MILESTONES

Development (as EMB-312H)	Jan 91
Announced	Jun 91
First flight	15 May 93
Certification	Aug 94
Selected by Brazilian Air Force	8 Aug 2001
Production started	Feb 2002
First delivery	18 Dec 2003

PROGRAMME: Design of original EMB-312 Tucano started January 1978. Ministry of Aeronautics contract received 6 December that year for two flying prototypes and two static/fatigue test airframes; first prototype (Brazilian Air Force serial number 1300) made first flight 16 August 1980, second (1301) on 10 December 1980; third to fly (PP-ZDK, on 16 August 1982) was to production standard. Total of 650 built by 1998; further details in 2000–01 and previous *Jane's* and in *Jane's Aircraft Upgrades*.

Development of EMB-314 began January 1991; announced (as EMB-312H) at Paris Air Show June 1991; Embraer development aircraft PT-ZTW (c/n 312161, previously used as prototype for TPE331-powered Tucano adopted by Royal Air Force) modified as Tucano H proof-of-concept (POC) prototype, making first flight in this form 9 September 1991. This aircraft toured US Air Force/Navy bases August and September 1992 as preliminary to Super Tucano entry in JPATS competition; Embraer teamed with Northrop May 1992 to bid Super Tucano for JPATS, but was unsuccessful. Provisional Brazilian type certification granted August 1994 after 500 hour, 396 sortie test and certification programme.

CURRENT VERSIONS: **EMB-314:** Two EMB-312H prototypes (PT[later PP]-ZTV, c/n 312454, first flight 15 May 1993, and PP-ZTF, c/n 312455, first flight 14 October 1993) tailored to US JPATS requirements as EMB-312HJ. Designation subsequently changed to EMB-314 to reflect extensive modifications to structure and systems.

ALX (EMB-314M): Brazilian Air Force (FAB) version, for border patrol missions under its SIVAM (*SIstema de Vigilancia da AMazonia*) programme. FAB finalised specification in early 1994; trials to validate projected flight characteristics, using POC aircraft and both Super Tucano prototypes, completed 1994. US$50 million development contract signed 18 August 1995 for two prototypes (one single-seat) to be modified from Super Tucano prototypes. The first flew in May 1996 and is being used for external stores compatibility and handling qualities testing; the second, which flew in early 1997, for testing the advanced

Embraer YA-29 Super Tucano ***NEW***/0569600

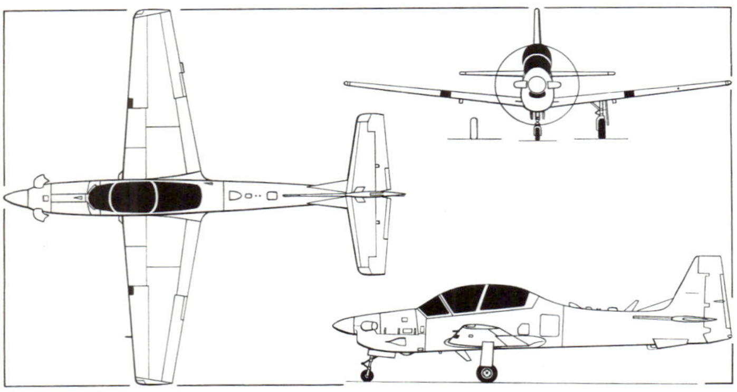

AT-29 advanced trainer version of Embraer Super Tucano *(James Goulding)* 0056316

weapons systems. Total 650 hours accumulated by early 2002, with up to 300 more required for completion.

FAB commitment to purchase 99 ALXs, of which 50 will be two-seat **AT-29**, 30 of these replacing AT-26 Xavante with 2°/5° Grupo of Training Command at Natal AFB, remainder configured for night intruder role and expected to serve with 1°/3° Grupo at Boa Vista and 2°/3° Grupo at Porto Velho; 49 will be single-seat **A-29**. Prototype 5700 (the former EMB-314 PP-ZTV), designated **YA-29**, rolled out 28 May 1999; IOC originally intended in May 2001 but FAB formal order only placed (76 firm, plus 23 options) on 8 August 2001; production started February 2002; deliveries began 18 December 2003. Elbit selected December 1996 to supply mission avionics, including ventral FLIR Systems AN-AAQ-22 turret, GPS/INS, radalt, Mode S transponder, DME, ILS, ADF, VOR, RWR, MAWS and chaff/flare dispenser. Export variants of both versions will be offered for border patrol/COIN missions and for basic/advanced pilot training. Can be flown as single-seat attack aircraft with fuel tank in rear cockpit.

Dominican Republic Air Force ordered 10 Super Tucanos on 20 August 2001.

Estimated market for 500 aircraft over 10 years.

DESIGN FEATURES: Meets requirements of FAR Pt 23 Appendix A, and MIL and CAA Section K specifications. Low-mounted wings, stepped cockpits in tandem, fully aerobatic. Small fillet forward of tailplane root each side.

EMB-314 differs from EMB-312 mainly in having more powerful engine, reprofiled wing and plugs of 0.37 m (1 ft 2½ in) forward and 1.00 m (3 ft 3¼ in) aft of cockpit to accommodate longer engine and retain CG and stability. Other changes include strengthened airframe for higher *g* loads and longer fatigue life, potentially 18,000 hours for typical training missions, or 12,000 hours in operational environments, depending on mission loads and utilisation; ventral strakes; five weapons hardpoints; NVG-compatible 'glass cockpit' with HOTAS controls. Able to cover whole primary and half of advanced pilot training syllabus, and fly precision weapons delivery and target towing missions.

Wing section NACA 63-415 at root; NACA 63A-212 at tip.

FLYING CONTROLS: Conventional and manual. Primary surfaces internally balanced; electrically actuated trim tab in, and small geared tab on, each Frise aileron; electromechanically actuated spring tab in rudder and port elevator. Electrically actuated single-slotted Fowler flaps on wing trailing-edges. Fixed incidence tailplane. Ventral airbrake.

STRUCTURE: Conventional all-metal construction from 2024 series aluminium alloys; continuous three-spar wing box forms integral fuel tankage. Steel flap tracks.

LANDING GEAR: Hydraulically retractable tricycle type, with single wheel and Piper oleo-pneumatic shock-absorber on each unit. Accumulator for emergency extension in the event of hydraulic system failure. Hydraulic steering for nose unit. Rearward-retracting steerable nose unit; main units retract inward into wings. Parker Hannifin 40-130 mainwheels, Oldi-DI-1.555-02-OL nosewheel. Tyre sizes 6.50-10 (8 ply) tubeless on mainwheels, 5.00-5 (6 ply) tubeless on nosewheel. Tyre pressures (±0.21 bar; 3 lb/ sq in in each case) are 5.17 bar (75 lb/sq in) on mainwheels, 4.48 bar (65 lb/sq in) on nosewheel. Parker Hannifin 30-95A hydraulic mainwheel brakes.

POWER PLANT: *ALX:* One 1,193 kW (1,600 shp) Pratt & Whitney Canada PT6A-68-3 turboprop, with FADEC, driving a Hartzell five-blade, constant-speed, fully feathering, reversible-pitch propeller.

EMB-314: One 969 kW (1,300 shp) PT6A-68A.

Single-lever combined control for engine throttling and propeller pitch adjustment. Two integral fuel tanks in each wing, total capacity 694 litres (183.3 US gallons; 152.7 Imp gallons). Fuel tanks lined with anti-detonation plastics foam. Optional self-sealing 303 litre (80.0 US gallon; 66.6 Imp gallon) tank in rear cockpit. Single-point pressure refuelling. Fuel system allows nominally for up to 30 seconds of inverted flight. Provision for three ferry fuel tanks (centreline and inboard wing pylons), each total capacity 330 litres (87.1 US gallons; 72.6 Imp gallons).

ACCOMMODATION: Two in tandem, on Martin-Baker Mk 10 LCX zero-zero ejection seats, in air conditioned and pressurised cockpit. One-piece fully transparent vacuum-formed canopy, opening sideways to starboard, with internal and external jettison provisions. Rear seat elevated 25 cm (9.9 in). Dual controls standard. Baggage compartment in rear fuselage, with access via door on port side.

SYSTEMS: Two-axis autopilot with embedded mission planning capability. Freon air cycle conditioning system, with engine-driven compressor. Single hydraulic system, consisting basically of (a) control unit, including reservoir with usable capacity of 1.9 litres (0.5 US gallon; 0.42 Imp gallon); (b) an engine-driven pump with nominal pressure of 131 bar (1,900 lb/sq in) and nominal flow rate of 4.6 litres (1.22 US gallons; 1.01 Imp gallons)/min at 3,800 rpm; (c) landing gear and gear door actuators; (d) filter; (e) shutoff valve; and (f) hydraulic fluid to MIL-H-5606. Under normal operation, hydraulic system actuates landing gear extension/retraction and control of gear doors. Landing gear extension can be performed under emergency operation; emergency retraction also possible during landing and T-O with engine running. Reservoir and system are suitable for aerobatics. No pneumatic system. 28 V DC electrical power provided by a 6 kW starter/generator, 26 Ah battery and, for 115 V and 26 V AC power at 400 Hz, a 250 VA inverter. Onboard oxygen generation system. Canopy and propeller de-icing. Integrated Data Acquisition Recorder (IDAR) includes VCR and FDR functions.

AVIONICS: *Comms:* Standard Rockwell Collins equipment. Integrated nav/com. Tactical VHF/UHF with datalink provisions; transponder.

Flight: Laser INS/dual GPS; twin VOR; ILS; DME; ADF; autopilot.

Instrumentation: Two 152 × 203 mm (6 × 8 in) CMFDs in each cockpit; HUD with up-front control panel in front cockpit. NVG Gen III compatible internal/external lighting. Optional HMD.

Mission: Provision for datalink; video camera and recorder. Embedded mission planning capability. Virtual radar/armament training systems. FLIR. Chaff/flare dispensers; MAWS and RWR.

EQUIPMENT: Landing light in each wing leading-edge; taxying lights on nosewheel unit. Optional Kevlar cockpit armour.

ARMAMENT: One 12.7 mm machine gun and 200 rounds mounted in each wing. Provision for a variety of ordnance including two Giat NC621 20 mm cannon pods, Mk 81/82 bombs, MAA-1 Piranha AAMs, BLG-252 cluster bombs, SBAT-70/19 or LAU-68A/G rocket pods or MLBs on underwing stations.

DIMENSIONS, EXTERNAL:
Wing span .... 11.14 m (36 ft 6½ in)
Wing chord: at root .... 2.30 m (7 ft 6½ in)
at tip .... 1.07 m (3 ft 6⅛ in)
Wing aspect ratio 6.4
Length overall .... 11.42 m (37 ft 5¾ in)
Fuselage: Length (excl rudder) .... 10.53 m (34 ft 6½ in)
Max depth .... 1.86 m (6 ft 1¼ in)
Height overall (static) .... 3.90 m (12 ft 9½ in)
Tailplane span .... 4.66 m (15 ft 3½ in)
Wheel track .... 3.76 m (12 ft 4 in)
Wheelbase .... 3.36 m (11 ft 0¼ in)
Propeller ground clearance (static) .... 0.345 m (1 ft 1½ in)
Baggage compartment door:
Height .... 0.60 m (1 ft 11½ in)
Width .... 0.54 m (1 ft 9¼ in)
Height to sill .... 1.25 m (4 ft 1¼ in)

DIMENSIONS, INTERNAL:
Cockpits: Combined length .... 2.90 m (9 ft 6⅛ in)
Max height .... 1.55 m (5 ft 1 in)
Max width .... 0.85 m (2 ft 9½ in)
Baggage compartment volume .... 0.17 m³ (6.0 cu ft)

AREAS:
Wings, gross .... 19.40 m² (208.8 sq ft)
Ailerons (total) .... 1.97 m² (21.20 sq ft)
Trailing-edge flaps (total) .... 2.58 m² (27.77 sq ft)
Fin, incl dorsal fin .... 2.29 m² (24.65 sq ft)
Rudder, incl tab .... 1.38 m² (14.85 sq ft)
Tailplane, incl fillets .... 4.77 m² (51.34 sq ft)
Elevators, incl tab .... 2.00 m² (21.53 sq ft)

WEIGHTS AND LOADINGS (EMB-314, except where indicated):
Basic weight empty .... 2,420 kg (5,335 lb)
Max external load .... 1,500 kg (3,307 lb)
Max internal fuel load (usable) .... 538 kg (1,186 lb)
Max T-O weight: EMB-314, clean .... 3,190 kg (7,033 lb)
ALX .... 3,600 kg (7,936 lb)
Max ramp weight .... 3,210 kg (7,077 lb)
Max zero-fuel weight: EMB-314 .... 2,670 kg (5,886 lb)
ALX .... 3,150 kg (6,945 lb)
Max wing loading .... 164.4 kg/m² (33.68 lb/sq ft)
Max power loading .... 3.29 kg/kW (5.41 lb/shp)

PERFORMANCE (EMB-314, at max clean T-O weight except where indicated):
Max level speed: EMB-314 at FL200 .... 301 kt (557 km/h; 346 mph)
ALX with external stores .... 245 kt (454 km/h; 282 mph)
Max cruising speed at FL200 .... 286 kt (530 km/h; 329 mph)
Econ cruising speed at FL200 .... 228 kt (422 km/h; 262 mph)
Stalling speed, power off:
flaps and landing gear up .... 85 kt (157 km/h; 98 mph) EAS
flaps and landing gear down .... 78 kt (145 km/h; 90 mph) EAS
Max rate of climb at S/L .... 895 m (2,925 ft)/min
Service ceiling .... 10,670 m (35,000 ft)
T-O run .... 350 m (1,150 ft)
T-O to 15 m (50 ft) .... 550 m (1,805 ft)
Landing from 15 m (50 ft) .... 860 m (2,820 ft)
Landing run .... 550 m (1,805 ft)
Range at FL300 with max fuel, 30 min reserves .... 847 n miles (1,568 km; 974 miles)
Ferry range at FL250 with underwing tanks and 30 min reserves
1,495 n miles (2,768 km; 1,720 miles)
Endurance on internal fuel at econ cruising speed at FL250, 30 min reserves . 6 h 30 min
g limits: fully Aerobatic category at 2,770 kg (6,107 lb) .... +7/-3.5
at 2,770 kg (6,107 lb) with external stores .... +4/-2.2

UPDATED

EMBRAER ERJ-145

Brazilian Air Force designations: R-99A, R-99B and P-99

TYPE: Regional jet airliner.

PROGRAMME: Development plans revealed 12 June 1989, aimed at first flight late 1991 and first deliveries mid-1993; programme delayed by company cutbacks, complete redesign of wing and other changes, as described in 2000-01 and earlier *Jane's*.

First metal cut for prototype, and tooling fabrication, in second quarter 1993; assembly

Embraer ERJ-145 of PB Air, Thailand, one of two leased from the manufacturer

NEW/0569599

DEVELOPMENT MILESTONES

Announced	12 June 89
First prototype material cut	2Q 93
First flight	11 Aug 95
Officially rolled out	18 Aug 95
Certification	10 Dec 96
First delivery	19 Dec 96

of prototype (PT-ZJA) began October 1994; fuselage sections mated January 1995; first flight 11 August 1995 ahead of formal roll-out and 'official' first flight a week later; first of three pre-series aircraft (PT-ZJB) first flown 17 November 1995; second ('ZJC) flew on 14 February and third ('ZJD) 2 April 1996; FAA and Brazilian CTA certification (to FAR/JAR 25, FAR Pt 36, ICAO Annex 16 and FAR Pt 121) achieved 10 December 1996. Single prototype and three pre-series aircraft undertook a 1,600 hour, 13 month development flight testing and certification programme. Deliveries began on 19 December 1996 with two aircraft (N15925 and N15926) to US launch customer Continental Express. Designation changed from EMB-145 to ERJ-145 in October 1997 to reflect 'Regional Jet' terminology, although former is retained for corporate and military variants. Certified by the aviation authorities of 27 countries by September 1998.

In addition to its suitability for executive transport and corporate shuttle roles, Embraer foresees military potential for the ERJ-145 as a tanker aircraft for small combat units, and as an AEW, elint, comint, sigint or battlefield surveillance platform. See Current Versions.

Two static test airframes: first (802) completed trials on 30 August 1996; second (803) began 10 year programme in December 1996.

CURRENT VERSIONS: **ERJ-145:** Initial version. Certified (by FAA) 10 December 1996 at 19,200 kg (42,328 lb) MTOW.

ERJ-145ER : (Extended Range) FAA co-certification 10 December 1996 at 20,600 kg (45,415 lb) MTOW.

ERJ-145MR: Certified by FAA 7 May 1998 at 22,000 kg (48,501 lb) MTOW. Two only, subsequently converted to ERJ-145LR standard. See ERJ-145MP, below.

ERJ-145LR : (Long Range) FAA co-certification 7 May 1998 for full passenger payload range of 1,640 n miles (3,037 km; 1,887 miles); increases in fuel capacity and all operating weights; 22,000 kg (48,501 lb) MTOW; and uprated AE 3007A1 turbofans providing 15 per cent more thermodynamic power, but flat rated to 33.1 kN (7,430 lb st), for improved climb and hot weather cruise performance.

ERJ-145MP: Combines ERJ-145LR's fuselage, zero-fuel weight and maximum landing weight with ERJ-145ER's landing gear, fuel tanks and range. Maximum take off weight 20,990 kg (46,275 lb) Available from 1999 as replacement for ERJ-145MR.

ERJ-145XR : (Extra Long Range) Announced at Farnborough International Air Show 25 July 2000 with launch order for 75, plus 100 options, from Continental Express subsequently increased to 104, plus 100 options. Features strengthened fuselage, wings and horizontal stabiliser, winglets, uprated AE 3007A1E turbofans providing lower specific fuel consumption, improved hot-and-high operation and higher single-engine ceiling; and auxiliary fuel tank in wing/fuselage fairing. Prototype (PT-ZJB, modified from first pre-series ERJ-145) first flown 29 June 2001; first production aircraft (c/n 590) flew early April 2002; two aircraft took part in a 400 hour test programme. Brazilian CTA granted 3 September 2002, followed by FAA approval 22 October 2002, and first deliveries to Continental Express/ExpressJet, which was scheduled to receive 18 ERJ-145XRs by the end of 2002.

ERJ-145EP: FAA certification 14 October 2003. MTOW 20,990 kg (46,275 lb); max ramp weight 21,090 kg (46,495 lb); MLW 18,700 kg (41,226 lb); MZFW 17,100 (37,699 lb); fuel as ER/MR versions.

ERJ-135: Short-fuselage, 37-seat version; *described separately*.

ERJ-140: Mid-size, 44-seat version; *described separately*.

EMB-145AEW&C: Formerly EMB-145SA; Brazilian Air Force designation **R-99A**. Airborne early warning and remote sensing version of ERJ-145LR developed for Brazilian government's *SIstema de Vigilancia de AMazonia* (SIVAM) programme for which Raytheon is prime contractor; initial requirement for five; contract signature March 1997. Selected on 15 December 1998 by Greek Air Force for its four-aircraft AEW requirement, with delivery from 2002; contract, signed 1 July 1999, valued at US$500 million. Announced late 1996 and features a strengthened fuselage, ventral strakes, more powerful APU, increased fuel capacity (three extra tanks at extreme rear of cabin, plus jettison capability), extending endurance to more than 8 hours; enhanced electrical system, five seats for relief crew and four (with provision for additional two) operators' consoles, including tactical co-ordinator. Flight crew of two.

Mission systems comprise civil version of Ericsson PS-890 Erieye side-looking airborne radar with antenna housed in long overfuselage fairing, optimised for lower-speed targets typically encountered in border incursions; five Erieye systems purchased at cost of US$145 million in 1997 for installation in EMB-145SAs, with first system delivery scheduled for 1999. Onboard command and control system, and BAE Systems North America Comms-Non Comms system. Radar is pulse Doppler type, operating in E/F-band, offering coverage from very low level up to about 25,000 m (82,000 ft) and at ranges exceeding 162 n miles (300 km; 186 miles). Datalink; GPS; secure communications. Second airframe (PP-XSA/6702) was first to fly, on 22 May 1999, ahead of formal roll-out on 28 May. Following systems integration by Raytheon in USA, first two R-99As and one R-99B handed over to 2°/6° Grupo at Anapolis on 24 July 2002; third and fourth R-99As followed in December 2002 and last on 12 December 2003.

Greek selection of Erieye-equipped EMB-145 announced 1 July 1999; four aircraft ordered, of which first handed over on 16 October 2003 to begin equipment outfitting.

Mexican government order for one EMB-145AEW&C announced 1 March 2001; equipment includes comint system installed by Raytheon.

EMB-145RS: Brazilian Air Force designation **R-99B**. Remote sensing version, of which three ordered for FAB's SIVAM programme for delivery commencing first quarter 2001. Similar to AEW variant, and with ventral strakes, but different mission systems for primary roles in natural resources exploitation, environmental and river pollution control, economic activities, ground occupation monitoring and illegal activities surveillance. Main sensor is version of Canadian MacDonald Dettwiler IRIS (Integrated Radar Imaging System) synthetic aperture radar, installed in underfuselage bulge with auxiliary antennas beneath wingroots, operating in D-band interferometric mode and capable of generating 3-D imagery. Other main sensors include Star Safire FLIR mounted behind nosewheel bay, Daedalus ultraviolet/visible/infra-red linescanner and BAE Systems North America Comms-Non Comms system. Roll-out (PP-XRT/6751) November 1999, with first delivery to 2°/6° Grupo at Anapolis AFB on 24 July 2002; all received by late 2003.

Greek Air Force Embraer EMB- 145AEW&C

NEW/0569598

Embraer EMB145AEW&C, cutaway drawing key

1 Radome
2 Weather radar scanner
3 ILS glideslope antenna
4 Radar scanner mounting
5 Avionics cooling air ram intake
6 Nosewheel doors
7 Nose avionics equipment compartment
8 Access hatch through nosewheel bay
9 Pitot heads, port and starboard
10 Battery bay
11 Nosewheel leg-mounted taxying light
12 Twin nosewheels, forward retracting
13 Hydraulic steering jack
14 Ground power socket
15 Cooling air blower and exhaust duct
16 Front pressure bulkhead
17 Temperature probe
18 Incidence vane
19 Rudder pedals
20 Control column
21 Instrument panel, five multifunction CRT displays
22 Instrument panel shroud
23 Windscreen wipers
24 Electrically heated windscreen panels
25 Overhead systems switch panel
26 Cockpit doorway
27 First Of cer's seat
28 Centre control pedestal
29 Captain's seat
30 Side console with nosewheel steering tiller and document stowage
31 Main entry door with airstairs
32 Folding handrail
33 Entry lobby
34 Starboard side lavatory and galley modules
35 Door surround structure
36 GPS antenna
37 TCAS antenna
38 VHF antenna
39 Cabin wall trim panelling
40 Folding tables
41 Fuselage frame and stringer structure
42 Crew rest area, five-seats
43 Composite cabin floor panels with seat mounting rails
44 Lower VHF antenna
45 Conditioned air risers from underfloor distribution ducts
46 Cabin window panels with pull-down sun blinds
47 Position of pressure refuelling connection on starboard side
48 Wing inspection light
49 Composites wing/fuselage fairing
50 Ventral navigation radome
51 Landing light
52 Air conditioning pack, dual system port and starboard
53 Operator's seats (four)
54 Operator's consoles with multifunction, full-colour CRT displays
55 Dorsal radome forward mounting struts
56 Cooling air ram intake
57 Starboard wing integral fuel tank
58 Vent tank
59 Starboard aileron tandem hydraulic actuator
60 Vortex generators
61 Outer wing panel dry bay
62 Starboard navigation and strobe lights
63 Winglet
64 Starboard aileron
65 Flap operating screw jack, torque shaft driven
66 Starboard outboard double-slotted flap segment, extended
67 Dorsal radome
68 Ericsson Microwave Systems PS-890 'Erieye' active phased array doppler surveillance radar antenna, port and starboard
69 Radar transmission equipment
70 Internal cooling air ducts
71 Starboard spoiler panels, open
72 ADF antenna
73 Anti-collision beacon
74 Communications equipment cabinet
75 Starboard side escape hatch
76 Erieye radar process units
77 Wing centre-section carry-through structure
78 Wing front spar/fuselage attachment joint
79 Emergency external lighting
80 Wing root multibolt rib joint
81 Rear spar/fuselage attachment joint
82 Main landing gear wheel bay
83 Central flap drive motor
84 Erieye radar processing equipment cabinet
85 ESM equipment racks
86 Dorsal radome aft support struts
87 Support strut attachment joint and fairing
88 Fuselage fuel tanks (six)
89 Cabin rear bulkhead
90 Electrical equipment compartment
91 Starboard engine nacelle
92 Dorsal radome cooling air exhaust
93 Starboard engine installation
94 Engine pylon
95 Engine pylon mounting fuselage main frames
96 Aft equipment compartment, mission equipment conditioning system
97 Up-and-over door
98 Composites n root fairing
99 Tailplane de-icing air duct
100 Rear avionics equipment rack, access hatch to starboard
101 Rear pressure bulkhead
102 Sloping n-spar mounting frames
103 Engine re suppression bottles (two)
104 Multispar n structure
105 Rudder tandem hydraulic actuators
106 VOD localiser antenna
107 Leading edge HF antenna
108 Tailplane trim actuator, tandem screw jacks
109 Starboard tailplane
110 Starboard elevator
111 Elevator servo tab
112 Inboard spring tab
113 Tail navigation light
114 Auxiliary fins, above and below
115 Port elevator rib structure
116 Static dischargers
117 Elevator horn balance
118 Tailplane leading edge de-icing
119 Tailplane two-spar torsion box structure
120 Tailplane pivot mounting
121 Two-segment rudder rib structure
122 Fore rudder panel
123 Training rudder
124 APU exhaust
125 Auxiliary Power Unit (APU)
126 APU mounting bulkhead and firewall
127 Ventral fins, port and starboard
128 Port engine nacelle exhaust duct
129 Rear engine mounting
130 Pylon-mounted bleed-air pre-cooler
131 Forward engine mounting
132 Equipment bay doorway
133 Hinged engine cowling panels
134 Rolls-Royce AE3007A turbofan
135 Multilobe exhaust mixer
136 FADEC engine controller
137 Engine accessory equipment gearbox
138 Generator cooling air intakes
139 Intake lip bleed-air de-icing
140 Electrical load centre
141 Aft window panel
142 Composites trailing edge root fairing
143 Hydraulic reservoir, dual system port and starboard
144 Port inboard double-slotted flap segment
145 Inboard ground spoiler/lift dumper
146 Outboard spoiler/speedbrake
147 Flap panel CFC composites structure
148 Outboard double-slotted flap segment
149 CFC composites aileron structure
150 Port winglet
151 Static dischargers
152 Aft strobe light
153 Port navigation and strobe lights
154 Leading edge vortillons
155 Port aileron tandem hydraulic actuators
156 Wing panel rib structure
157 Port wing integral fuel tankage
158 Gravity fuel filler
159 Corrugated inner leading edge skin de-icing air ducting
160 Twin mainwheels
161 Trailing axle mainwheel suspension
162 Mainwheel leg strut
163 Main landing gear leg pivot mounting
164 Hydraulic retraction jack
165 Side breaker strut
166 Fuel collector tank and pumps
167 Aileron control cable run
168 Leading edge de-icing air supply duct

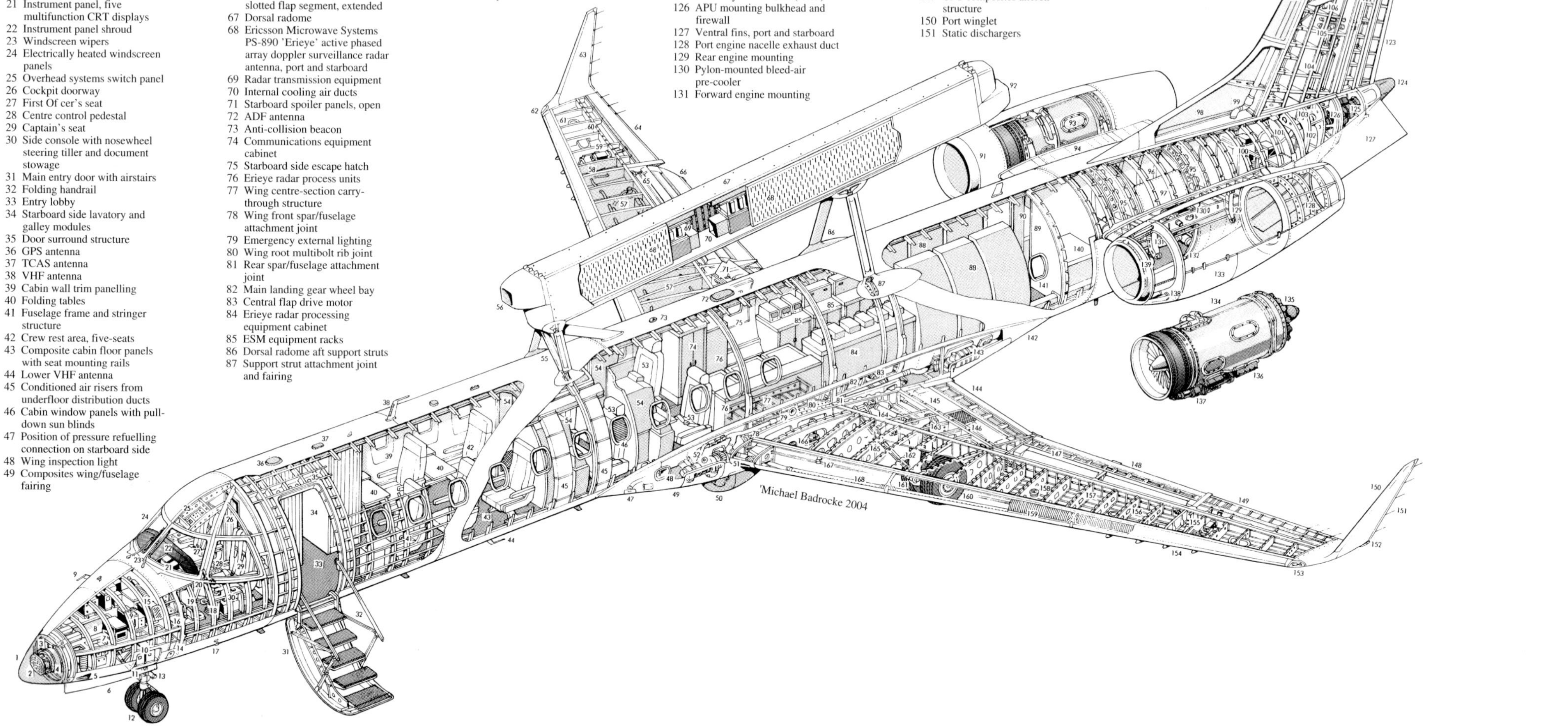

Extra long range Embraer ERJ-145XR *NEW*/0569597

EMB-145AGS: Airborne ground sensor version, under study during 2000; equipped with a mission package comprising Airborne Platform Subsystem (APS), Airborne Mission Equipment Subsystem (AMES) and Ground Exploitation Station Subsystem (GESS) including HF, UHF, VHF, ELINT and IMINT equipment, providing a self-deployable and cost-effective surface reconnaissance system.

EMB-145 airframe selected as aerial platform for Lockheed Martin's and Harris Corporation's bid for US Army Aerial Common Sensor (ACS) requirement, designed to replace Guardrail Common Sensor and Airborne Reconnaissance Low intelligence, surveillance and reconnaissance systems.

EMB-145MP : and EMB-145MP/ASW: Brazilian Air Force designation **P-99**. Maritime patrol and anti-submarine warfare versions, under development by 2000; equipped with surveillance radar with multiple target track-while-scan mode, autodetection, FLIR interface, digital map, incorporated tactical aids, SAR/ISAR mode allowing real-time imaging, adaptive processing for different sea states, and simultaneous side and range views; high-altitude and resolution FLIR; ESM suite; COMINT/ELINT; MAD; IFF/SSR and acoustics.

Mexican government order for two EMB-145MPs announced 1 March 2001. Equipment includes SeaVue radar and AN/APX-114 IFF interrogator to be installed by Raytheon at Greenville. FAA Type Certificate granted 3 July 2003; AE3007A, A1/1, A1/2, A1 or A1P engines; MTOW 20,990 kg (46,275 lb); max ramp weight 21,090 kg (46,495 lb); MLW 19,300 kg (42,549 lb); MZFW 17,900 kg (39,462 lb).

CUSTOMERS: Total of 697 firm orders and 440 options for ERJ-145 commercial variant by 31 March 2004. Four hundredth ERJ series delivery was an ERJ-145 (HB-JAL) to Crossair on 22 March 2001; 500th was an ERJ-145 (N2933K) delivered to Chatauqua Airlines of Indiana on 21 September 2001; 600th to Swiss (as HB-JAY) on 28 May 2002, and 700th to Alitalia Express (as I-EXML) on 9 May 2003. Total of 112 ERJ-145s delivered in 2000, and 104 in 2001. See table. Belgian Air Force took delivery of two ERJ-145LRs in VIP configuration (CE-03 and CE-04) on 11 December 2001 and 21 January 2002 respectively.

COSTS: Estimated development costs US$300 million.

Following description applies to ERJ-145ER except where indicated.

DESIGN FEATURES: Stretched EMB-120 Brasilia fuselage (with tailcone adapted for rear-mounted engine installation), allied to new-design wing with Embraer supercritical section; CBA-123 nose and cabin; T tailplane.

Wing sweepback 22° 43' 48" at quarter-chord.

FLYING CONTROLS: Conventional and assisted. Ailerons and two-section rudder hydraulically actuated, with artificial feel; mechanically actuated elevator with automatic and spring tab. Four-segment in-flight and ground spoilers; two pairs of electrically actuated double-slotted flaps. Control surface movements: ailerons +25/–15°, elevators +27/–14°, rudder ±15° (first segment) ±11° 15' (second segment), flaps +7° 30'/–45° (inboard) +9° 10'/–45° (outboard), spoilers +52° (inboard) +30° (outboard).

STRUCTURE: Fuselage as for Brasilia; two-spar wing with integral fuel tanks, plus auxiliary third spar supporting landing gear; T tail unit with aluminium main boxes; wing and tailplane leading-edges aluminium, fin leading-edge composites sandwich. Gamesa (Spain) builds wings, wing/body fairings, main landing gear doors and engine nacelles; rear fuselage section 1, including engine pylons and passenger/service/baggage doors, plus centre-fuselage section 1, including doors, by Sonaca (Belgium); fin, tailplane and elevators by ENAER (Chile); engine nacelles and thrust reversers by Hurel Hispano UK; nose radome by Norton; passenger cabin and baggage compartment interiors by C & D Interiors (USA). Structure designed for an economical service life of 60,000 flights.

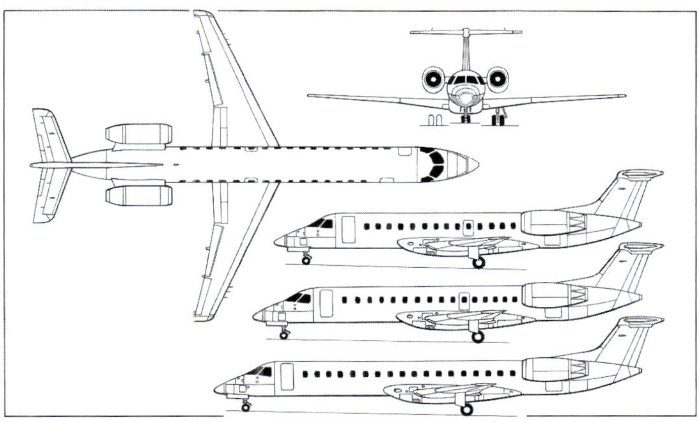

ERJ-145 rear-engined twin-turbofan regional airliner with additional side views of ERJ-135 (top) and ERJ-140 (centre) *(James Goulding)* 0143318

EMBRAER ERJ-135/140/145 PRODUCTION
(at 31 March 2004)

Customer	Variant	Ordered	Del'd	Backlog
Aerolitoral	145	5	-	5
Air Caraibes	145	2	2	
Alitalia Express	145	14	14	
American Eagle	135	40	40	
	140	59	59	
	145	136	71	65
Axon Airlines	145	3	3	
bmi (British Midland)	135	4	2	2
	145	11	9	2
British Regional	145	23	23	
Brymon Airways	145	7	7	
China Southern	145	6	-	6
Cirrus Airlines	145	1	1	
City Air Sweden	135	2	2	
ERA Spain	145	2	2	
Express Jet	135	30	30	
	145	245	200	45
	135	3	3	-
Flandre Air	145	7	5	2
GECAS	145	2	2	
Jet Magic	135	1	1	-
KLM Excel	145	2	2	
LOT Polish	145	14	14	
Luxair	145	9	9	
Mesa	145	36	36	
Midwest	140	20	0	20
Pan Européenne	135	1	1	
Portugalia	145	8	8	
Proteus	135	3	3	
	145	11	8	3
Regional Airlines France	135	3	3	
	145	17	14	3
Republic (Wexford Management)	135	15	15	
	140	15	15	
	145	68	47	21
Rheintalflug	145	3	3	
Rio-Sul	145	16	16	
Sichuan Airlines	145	5	5	
Skyways AB	135	2	2	
	145	4	4	
South African Airlink	135	20	5	15
Swiss	145	25	25	
Trans States Airlines	145	15	15	
Totals		**913**	**724**	**189**

Notes: Total excludes military versions, EMB-135/145s in corporate service and Legacys.

LANDING GEAR: Twin-wheel main legs retract inward into wing/fuselage fairings; twin-wheel nose unit retracts forward. EDE/Liebherr landing gear system, with EDE responsible for whole system and Liebherr for development and production of nose unit. Goodrich wheels and carbon brakes. Tyre sizes 30×9.5–14 (16 ply) tubeless (main), 19.5×6.75–8 (8 ply) tubeless (nose); tyre pressure 8.60 to 9.00 bar (125 to 130 lb/sq in). Minimum ground turning radius at nosewheel 12.51 m (41 ft 0 in). Minimum turning circle 29.22 m (95 ft 10½ in).

POWER PLANT: Two turbofans pylon-mounted on rear cone of ERJ-145ER, 145MR and 145LR have 33.7 kN (7,580 lb st), Rolls-Royce AE 3007A, 3007A1/1 or 3007A1/2 as standard; 37.1 kN (8,338 lb st) AE 3007A1P optional, all with FADEC. ERJ-145XR has two 39.6 kN (8,895 lb st) AE 3007A1Es. Clamshell-type thrust reversers optional.

Parker Hannifin fuel system. Fuel capacity of ER and MR 5,201 litres (1,374 US gallons; 1,144 Imp gallons) in two wing tanks; usable fuel 5,091 litres (1,345 US gallons; 1,120 Imp gallons). ERJ-145LR capacity increased to 6,439 litres (1,701 US gallons; 1,418 Imp gallons; 6,352 litres (1,678 US gallons; 1,397 Imp gallons) usable. ERJ-145XR fuel comprises two wing tanks, each of 3,199 litres (845 US gallons; 704 Imp gallons), plus 1,037 litre (274 US gallon; 228 Imp gallon) ventral tank, for total of 7,435 litres (1,964 US gallons; 1,635 Imp gallons), of which 7,382 litres 1,950 US gallons; 1,624 Imp gallons) usable.

ACCOMMODATION: Two pilots, flight observer and cabin attendant. Standard accommodation for 50 passengers, three-abreast at seat pitch of 79 cm (31 in). Carry-on baggage wardrobe,

Embraer ERJ-145 of Axon Airlines, Greece *NEW*/0532060

galley and cabin attendant's seat at front of cabin; lavatory and main baggage compartment at rear of cabin. Cabinet plus overhead bins carry-on baggage capacity 358 kg (789 lb); underseat capacity 450 kg (992 lb); main baggage compartment capacity 1,200 kg (2,646 lb). Additional baggage cabinet or galley capacity can be provided by removing one or two single forward passenger seats. Outward-opening plug-type door, incorporating airstair, at front on port side, identical to that of EMB-120; upward-sliding baggage door at rear on port side; sideways-opening service door at front on starboard side; inward-opening emergency exit above wing on each side. Entire accommodation, including baggage compartments, pressurised and air conditioned.

SYSTEMS: Liebherr Aerospace pressurisation system (maximum differential 0.54 bar; 7.8 lb/sq in) maintains 2,440 m (8,000 ft) cabin altitude to 11,275 m (37,000 ft). Hamilton Sundstrand air conditioning and bleed air systems (wing and tailplane leading-edges and engine intakes anti-iced by engine bleed air); electric anti-icing system for windscreen and static and pitot tubes and sensors. Lucas electrical power generation system. Hamilton Sundstrand T-62T-40C11 or C14 APU. Honeywell air turbine starter. Parker Hannifin flight control and steering systems. Hydro-Aire brake-by-wire control system. EROS oxygen system.

AVIONICS: Honeywell Primus 1000 as core system.

Comms: Dual Primus II radios and radio management units.

Radar: Primus 1000 colour weather radar.

Flight: Dual digital air data computers, dual AHRS, TCAS and GPWS standard. FMS/GPS optional. Flight Dynamics HUD selected April 1998 for certification in 2000, providing Cat. III landing capability.

Instrumentation: EFIS panel comprising five 280 × 180 mm (11 × 7 in) displays, two PFDs, two MFDs and IECAS.

DIMENSIONS, EXTERNAL:
- Wing span: ERJ-145ER/LR ... 20.04 m (65 ft 9 in)
  - ERJ-145XR over winglets ... 21.01 m (68 ft 11 in)
- Wing chord: at root ... 4.09 m (13 ft 5 in)
  - at tip ... 1.04 m (3 ft 5 in)
- Wing aspect ratio ... 7.9
- Length overall ... 29.87 m (98 ft 0 in)
- Fuselage: Length ... 27.93 m (91 ft 7½ in)
  - Max diameter ... 2.28 m (7 ft 5¾ in)
- Height overall ... 6.76 m (22 ft 2 in)
- Tailplane span ... 7.55 m (24 ft 9 in)
- Wheel track (c/l of shock-struts) ... 4.10 m (13 ft 5½ in)
- Wheelbase ... 14.45 m (47 ft 5 in)
- Passenger door (fwd, port): Height ... 1.70 m (5 ft 7 in)
  - Width ... 0.71 m (2 ft 4 in)
  - Height to sill (max) ... 1.63 m (5 ft 4 in)
- Baggage door (rear, port): Height ... 1.12 m (3 ft 8 in)
  - Width ... 1.00 m (3 ft 3¼ in)
  - Height to sill (max) ... 1.76 m (5 ft 9¼ in)
- Service door (rear, stbd): Height ... 1.42 m (4 ft 8 in)
  - Width ... 0.62 m (2 ft 0½ in)
  - Height to sill ... 1.60 m (5 ft 3 in)
- Emergency exits (overwing, each):
  - Height ... 0.92 m (3 ft 0¼ in)
  - Width ... 0.51 m (1 ft 8 in)

DIMENSIONS, INTERNAL:
- Cabin (excl flight deck and baggage compartment, incl lavatory):
  - Length ... 16.49 m (54 ft 1¼ in)
  - Max width ... 2.10 m (6 ft 10¾ in)
  - Max height ... 1.83 m (6 ft 0 in)
  - Max aisle width ... 0.52 m (1 ft 8½ in)
  - Floor area ... 25.7 m^2 (277 sq ft)
  - Volume ... 53.0 m^3 (1,872 cu ft)
- Baggage compartment: Length ... 3.26 m (10 ft 8¼ in)
- Baggage volume:
  - wardrobe and stowage compartment ... 1.4 m^3 (49 cu ft)
  - overhead bins ... 1.9 m^3 (67 cu ft)
  - underseat ... 2.3 m^3 (80 cu ft)
  - baggage compartment ... 9.2 m^3 (325 cu ft)

AREAS:
- Wings, gross ... 51.18 m^2 (550.9 sq ft)
- Ailerons (total) ... 1.70 m^2 (18.30 sq ft)
- Trailing-edge flaps (total) ... 8.36 m^2 (89.99 sq ft)
- Spoilers (total) ... 2.32 m^2 (24.97 sq ft)
- Fin ... 5.07 m^2 (54.57 sq ft)
- Rudder ... 2.13 m^2 (22.93 sq ft)
- Tailplane ... 11.20 m^2 (120.55 sq ft)
- Elevators (total, incl tabs) ... 3.34 m^2 (35.95 sq ft)

WEIGHTS AND LOADINGS:
- Operating weight empty:
  - ERJ-145ER ... 11,940 kg (26,323 lb)
  - ERJ-145LR ... 12,100 kg (26,676 lb)
  - ERJ-145XR ... 12,580 kg (27,734 lb)
- Baggage capacity ... 1,200 kg (2,646 lb)
- Max payload: ERJ-145ER ... 5,160 kg (11,376 lb)
  - ERJ-145LR ... 5,800 kg (12,787 lb)
  - ERJ-145XR ... 5,920 kg (13,051 lb)
- Max fuel: ERJ-145ER ... 4,173 kg (9,200 lb)
  - ERJ-145LR ... 5,187 kg (11,435 lb)
  - ERJ-145XR ... 6,032 kg (13,298 lb)
- Max T-O weight: ERJ-145ER ... 20,600 kg (45,415 lb)
  - ERJ-145LR ... 22,000 kg (48,500 lb)
  - ERJ-145XR ... 24,100 kg (53,131 lb)
- Max ramp weight: ERJ-145ER ... 20,700 kg (45,635 lb)
  - ERJ-145LR ... 22,100 kg (48,721 lb)
  - ERJ-145XR ... 24,200 kg (53,351 lb)
- Max landing weight: ERJ-145ER ... 18,700 kg (41,226 lb)
  - ERJ-145LR ... 19,300 kg (42,549 lb)
  - ERJ-145XR ... 20,000 kg (44,092 lb)
- Max zero-fuel weight:
  - ERJ-145ER ... 17,100 kg (37,698 lb)
  - ERJ-145LR ... 17,900 kg (39,463 lb)
  - ERJ-145XR ... 18,500 kg (40,785 lb)
- Max wing loading:
  - ERJ-145ER ... 402.5 kg/m^2 (82.44 lb/sq ft)
  - ERJ-145LR ... 429.8 kg/m^2 (88.04 lb/sq ft)
  - ERJ-145XR ... 470.9 kg/m^2 (96.44 lb/sq ft)
- Max power loading:
  - ERJ-145ER, ERJ-145XR ... 305 kg/kN (3.00 lb/lb st)
  - ERJ-145LR ... 326 kg/kN (3.20 lb/lb st)

PERFORMANCE:
- High cruising speed: ERJ-145ER, ERJ-145LR ... M0.78 (450 kt; 833 km/h; 518 mph)
  - ERJ-145XR ... M0.80
- Time to climb to FL350 ... 20 min
- Max certified altitude: all ... 11,275 m (37,000 ft)
- Service ceiling, OEI: ERJ-145ER/LR ... 6,100 m (20,000 ft)
- FAR T-O field length at S/L:
  - ERJ-145ER/LR ... 1,970 m (6,465 ft)
  - ERJ-145XR ... 2,089 m (6,855 ft)
- FAR landing field length, S/L, at typical landing weight:
  - ERJ-145ER/LR ... 1,390 m (4,560 ft)
  - ERJ-145XR ... 1,430 m (4,692 ft)
- Range, 50 passengers, 100 n miles (185 km; 115 mile) diversion, 45 min reserves
  - ERJ-145LR ... 1,620 n miles (3,000 km; 1,864 miles)
  - ERJ-145XR ... 2,000 n miles (3,704 km; 2,301 miles)

UPDATED

EMBRAER ERJ-135

TYPE: Regional jet airliner.

DEVELOPMENT MILESTONES

Programme launched	16 Sep 97
Rolled out	12 May 98
First flight	4 Jul 98
Certification	Jun 99
First delivery	23 Jul 99

PROGRAMME: Launched 16 September 1997; two pre-series ERJ-145s (001/PT-ZJA and 002/PT-ZJC) modified to create two prototype ERJ-135s; roll-out (PT-ZJA) 12 May 1998, followed by first flight 4 July 1998; public debut at Farnborough Air Show September 1998; second aircraft (PT-ZJC), flown 24 September 1998, for systems testing before conversion to production standard in March 1999. Brazilian Centro Técnico Aerospacial (CTA) certification achieved in June 1999; FAA certification (as EMB-135ER and LR) 15 July 1999. First delivery 23 July 1999 to Continental Express; other early aircraft to American Eagle. CTA and JAA approval granted 10 August 2003 for steep approach (5.5° glideslope) operations at London-City Airport.

CURRENT VERSIONS: **ERJ-135:** Regional airliner, *as described*. Available from the outset in 135ER and 135LR versions.

EMB-135BJ Legacy: Corporate version; *described separately*.

CUSTOMERS: Total of 122 firm commercial orders and two options by 31 March 2004. See table. Additionally, one VIP-configured ERJ-135LR handed over to Greek Air Force on 7 January 2000 and two, also in VIP configuration, to the Belgian Air Force on 4 June and in August 2001, for operation by No. 21 Squadron at Melsbroek with two similarly configured ERJ-145s. Deliveries have been 16 in 1999, 45 in 2000, 27 in 2001, three in 2002, 14 in 2003 and none in the first three months of 2004.

Programme based on estimates of 500 sales.

COSTS: Development cost US$100 million, of which 40 per cent provided by risk-sharing partners. Unit cost US$11.8 million.

DESIGN FEATURES: Shares 96 per cent commonality with ERJ-145 including engines, wings, tail surfaces, flight deck and main systems; fuselage shortened by 3.53 m (11 ft 7 in) by removal of two frames (4.84 m; 15 ft 10½ in ahead of wing and 3.07 m; 10 ft 0¾ in at rear) and substitution of two shorter frames (2.85 m; 9 ft 4¼ in and 1.53 m; 5 ft 0¼ in, respectively).

BMI (British Midland) Embraer ERJ-135 ***NEW***/0569596

Flight deck of Embraer Legacy and ERJ-135 0131839

FLYING CONTROLS: As for ERJ-145.
STRUCTURE: As for ERJ-145.
LANDING GEAR: As for ERJ-145.
POWER PLANT: Two 33.7 kN (7,580 lb st) Rolls-Royce AE 3007A3 or AE 3007A1/3 turbofans. Fuel as ERJ-145ER and ERJ-145LR, respectively.
ACCOMMODATION: Standard accommodation for 37 passengers in three-abreast configuration.
SYSTEMS: As for ERJ-145.
DIMENSIONS, EXTERNAL: As for ERJ-145 except:
Length: overall ... 26.33 m (86 ft 4½ in)
fuselage ... 24.39 m (80 ft 0¼ in)
Wheelbase ... 12.43 m (40 ft 9¼ in)
DIMENSIONS, INTERNAL:
Cabin (excl flight deck and baggage compartment, incl lavatory): Length ... 12.95 m (42 ft 5¾ in)
Baggage compartment: Length ... 3.34 m (10 ft 11½ in)
Baggage volume:
wardrobe and stowage compartment ... 1.0 m^3 (35 cu ft)
overhead bins ... 1.4 m^3 (49 cu ft)
underseat ... 1.7 m^3 (60 cu ft)
Galley volume ... 0.99 m^3 (35 cu ft)
WEIGHTS AND LOADINGS:
Operating weight empty:
ERJ-135ER ... 11,390 kg (25,111 lb)
ERJ-135LR ... 11,490 kg (25,331 lb)
Baggage capacity ... 1,000 kg (2,205 lb)
Max payload: ERJ-135ER ... 4,270 kg (9,414 lb)
ERJ-135LR ... 4,510 kg (9,943 lb)
Max fuel weight: ERJ-135ER ... 4,173 kg (9,200 lb)
ERJ-135LR ... 5,187 kg (11,435 lb)
Max T-O weight: ERJ-135ER ... 19,000 kg (41,888 lb)
ERJ-135LR ... 20,000 kg (44,092 lb)
Max landing weight:
ERJ-135ER, ERJ-135LR ... 18,500 kg (40,785 lb)
Max ramp weight: ERJ-135ER ... 19,100 kg (42,108 lb)
ERJ-135LR ... 20,100 kg (44,313 lb)
Max zero-fuel weight: ERJ-135ER ... 15,600 kg (34,392 lb)
ERJ-135LR ... 16,000 kg (35,274 lb)
Max wing loading:
ERJ-135ER ... 371.2 kg/m^2 (76.04 lb/sq ft)
ERJ-135LR ... 390.8 kg/m^2 (80.04 lb/sq ft)
Max power loading:
ERJ-135ER ... 282 kg/kN (2.76 lb/lb st)
ERJ-135LR ... 297 kg/kN (2.91 lb/lb st)
PERFORMANCE:
Max cruising speed ... M0.78 (450 kt; 833 km/h; 518 mph)
Time to FL350 ... 20 min
Max certified altitude ... 11,275 m (37,000 ft)
T-O field length at S/L ... 1,700 m (5,577 ft)
Landing field length, S/L, at typical landing weight ... 1,360 m (4,460 ft)
Range with 37 passengers, 100 n mile (185 km; 115 mile) diversion, 45 min reserves ... 1,700 n miles (3,148 km; 1,956 miles)

UPDATED

EMBRAER ERJ-140

TYPE: Regional jet airliner.

DEVELOPMENT MILESTONES

Programme launched	20 Sep 99
First flight	27 Jun 00
Certification	Jun 01
First delivery	Jul 01

PROGRAMME: Launched 30 September 1999 at the European Regional Airline Association annual meeting in Paris; first flight of prototype, modified from prototype ERJ-135 c/n 801/PT-ZJA, 27 June 2000; public debut at Farnborough International Air Show July 2000; Brazilian CTA and FAA certification (for both ER and LR versions) achieved in June and 26 July 2001 respectively; first delivery (PP-XGF/N800AE) to American Eagle late July 2001.
CURRENT VERSIONS: **ERJ-140ER:** Standard version, *as described*. Engineering designation EMB-135KE.
ERJ-140LR: Long-range version. Engineering designation EMB-135KL.
CUSTOMERS: Launch customer American Eagle announced order for 66 on 27 September 2000, subsequently reduced to 59 through conversion of orders and options to ERJ-145s. Total of 94 firm orders and 20 options by 31 March 2004. See table.
COSTS: US$15.2 million (1999).

Embraer ERJ-140 regional jet airliner ***NEW***/0569639

Descriptions for the ERJ-135 and ERJ-145 apply also to the ERJ-140 except as follows:
DESIGN FEATURES: Shares 98 per cent commonality with ERJ-135/145, including engines, wings, tail surfaces, flight deck and main systems; ERJ-135 fuselage stretched by 2.30 m (7 ft 6½ in) by removal of two frames (2.85 m; 9 ft 4¼ in ahead of wing and 1.35 m; 4 ft 5 in at rear) and substitution of two longer frames (3.94 m; 12 ft 11 in and 2.56 m; 8 ft 4¾ in respectively).
POWER PLANT: Rolls-Royce AE 3007A1/3 turbofans, each rated at 33.7 kN (7,580 lb st); fuel as ERJ-145ER/MR and 145LR, respectively.
ACCOMMODATION: Standard accommodation for 44 passengers, three-abreast at seat pitch of 39 cm (31 in). Flight attendant seat on port side immediately aft of flight deck standard; attendant seat in centre of aisle at rear of cabin optional. Wardrobe/carry-on baggage cabinet and galley at front of cabin, lavatory at rear.
DIMENSIONS, EXTERNAL:
Length: overall ... 28.47 m (93 ft 5 in)
fuselage ... 26.52 m (87 ft 0 in)
Wheelbase ... 13.51 m (44 ft 4 in)
DIMENSIONS, INTERNAL:
Baggage volume:
Wardrobe and stowage compartment ... 0.93 m^3 (32.84 cu ft)
WEIGHTS AND LOADINGS:
Basic operating weight: ER, LR ... 11,800 kg (26,014 lb)
Baggage capacity ... 1,200 kg (2,646 lb)
Max fuel weight: ER ... 4,173 kg (9,200 lb)
LR ... 5,187 kg (11,435 lb)
Max T-O weight: ER ... 20,100 kg (44,312 lb)
LR ... 21,100 kg (46,517 lb)
Max landing weight: ER, LR ... 18,700 kg (41,226 lb)
Max ramp weight: ER ... 20,200 kg (44,533 lb)
LR ... 21,200 kg (46,738 lb)
Max zero-fuel weight: ER, LR ... 17,100 kg (37,699 lb)
Max wing loading: ER ... 392.7 kg/m^2 (80.44 lb/sq ft)
LR ... 412.3 kg/m^2 (84.44 lb/sq ft)
Max power loading: ER ... 298 kg/kN (2.92 lb/lb st)
LR ... 313 kg/kN (3.07 lb/lb st)
PERFORMANCE:
Max cruising speed: ER, LR ... Mach 0.78 (450 kt; 833 km/h; 518 mph)
T-O field length: ER ... 1,720 m (5,643 ft)
LR ... 1,320 m (4,331 ft)
Time to climb to 10,670 m (35,000 ft): ER, LR ... 16 min
Service ceiling: ER, LR ... 11,278 m (37,000 ft)
Range with 44 passengers at long-range cruising speed 100 n mile (185 km; 115 mile) alternate and 45 min reserves:
ER ... 1,230 n miles (2,278 km; 1,415 miles)
LR ... 1,630 n miles (3,018 km; 1,875 miles)

UPDATED

EMBRAER EMB-135BJ LEGACY

TYPE: Business jet.

DEVELOPMENT MILESTONES

Announced	23 Jul 00
First flight	31 Mar 01
Certification	10 Dec 01

PROGRAMME: Announced on eve of Farnborough International Air Show 23 July 2000; first flight of prototype (PP-XJO), converted from second ERJ-135 prototype (PT-ZJC), 31 March 2001; Brazilian CTA certification achieved 10 December 2001, followed by JAA certification on 9 July, FAA approval on 23 August 2002, and Russian Interstate Aviation Committee approval on 18 November 2003. Early aircraft limited to MTOW of 22,200 kg (48,942 lb); increased by 300 kg (661 lb) from c/n 145625).
Improvements introduced in 2003 included removal of windscreen wipers, redesign of fairings and air inlets, polishing of wing leading-edges and surface smoothing, resulting in 50 n mile (93 km; 58 mile increase in maximum range in Executive configuration.
CURRENT VERSIONS: Offered in **Executive, Corporate Shuttle** and **Regional** variants.
CUSTOMERS: Total of 22 delivered by 30 September 2003 to customers in Europe, Africa, Middle East, Latin America and North America. Launch customer Swift Aviation of Phoenix, Arizona, ordered 25, with 25 options in July 2000; other announced customers include an unnamed major energy company based in Houston, Texas, which ordered one in corporate shuttle configuration in April 2001; the Greek Air Force, which ordered one in Executive configuration for delivery in December 2001; Skeikh Fahad Al Athel of Saudi Arabia, who took delivery of one on 16 December 2002; the Indian Government, which ordered five in Executive configuration on 19 September 2003, four of which will replace HAL/Avro 748s with the Indian Air Force's Air HQ Communications Squadron at Palam, and one operated by the Border Security force; and fractional ownership operator flight options, which signed an MoU on 27 October 2003 for up to four in Executive configuration. Market estimated at 240 aircraft over a ten-year period.

Embraer Legacy business jet ***NEW***/0569643

COSTS: Executive US$20.8 million, outfitted (2003); Shuttle US$16.4 million. Direct operating cost, Executive US$1,161 per hour; Shuttle US$1,020 per hour (2002).

Description for ERJ-135 applies also to the Legacy, except the following.

DESIGN FEATURES: Compared with ERJ-135 Legacy Executive has belly and aft fuel tanks in extended underwing fairing, plus winglets.

STRUCTURE: Airframe manufactured as ERJ-135 and modified to Legacy configuration; Embraer performs interior completion, using components supplied by Nordam or Duncan Aviation.

POWER PLANT: Legacy Executive initially had two 37.1 kN (8,338 lb st) Rolls-Royce AE 3007A1P turbofans; from c/n 145625, two AE 3007A1Es, each 39.7 kN (8,917 lb st). Legacy Shuttle has two 33.7 kN (7,580 lb st) AE 3007 A1/3s. Legacy Executive initially 10,152 litres (2,682 US gallons; 2,233 Imp gallons). Increased from c/n 145625 to 10,266 litre (2,712 US gallons; 2,258 Imp gallons) in two wing- and two forward fuselage tanks, usable capacity 10,160 litres (2,684 US gallons; 2,235 Imp gallons). Legacy Shuttle has 6,397 litres (1,690 US gallons; 1,407 Imp gallons) of usable fuel.

ACCOMMODATION: Typical accommodation for 13 passengers in Legacy Executive configuration. Legacy Shuttle can accommodate 16 to 19 in two-abreast arrangement, or 30 to 37 in three-abreast configuration. Baggage compartment and lavatory at rear of cabin.

SYSTEMS: Hamilton Sundstrand T-62T-40C14 APU.

AVIONICS: Honeywell Primus 1000 as core system.

DIMENSIONS, INTERNAL (A: Executive, B: Shuttle):

Cabin volume: B .... 40 m^3 (1,413 cu ft)
Baggage compartment volume: A .... 6.80 m^3 (240 cu ft)
B .... 9.20 m^3 (325 cu ft)

WEIGHTS AND LOADINGS:

Basic operating weight: A .... 13,600 kg (29,983 lb)
B .... 11,500 kg (25,353 lb)
Baggage capacity: A .... 454 kg (1,001 lb)
B .... 1,000 kg (2,205 lb)
Max payload: A .... 2,400 kg (5,291 lb)
B .... 4,500 kg (9,921 lb)
Max fuel weight: A .... 8,242 kg (1,817 lb)
B .... 5,135 kg (11,321 lb)
Payload with max fuel: A .... 728 kg (1,605 lb)
B .... 3,465 kg (7,639 lb)
Max T-O weight: A .... 22,500 kg (49,604 lb)
B .... 20,000 kg (44,092 lb)
Max landing weight: A, B .... 18,500 kg (40,785 lb)
Max ramp weight: A .... 22,570 kg (49,758 lb)
B .... 20,100 kg (44,313 lb)
Max zero-fuel weight: A, B .... 16,000 kg (35,274 lb)
Max wing loading: A .... 439.6 kg/m^2 (90.04 lb/sq ft)
B .... 390.8 kg/m^2 (88.04 lb/sq ft)
Max power loading: A .... 283.5 kg/kN (2.78 lb/lb st)
LR .... 296.7 kg/kN (2.91 lb/lb st)

PERFORMANCE:

Cruising speed:
A at FL390 .... M 0.80 (459 kt; 850 km/h; 528 mph)
B at FL370 .... M 0.78 (447 kt; 828 km/h; 514 mph)
Max certified altitude: A .... 11,885 m (39,000 ft)
B .... 11,278 m (37,000 ft)
T-O field length: A .... 1,759 m (5,770 ft)
B .... 1,707 m (5,600 ft)
Landing field length: A, B .... 817 m (2,681 ft)
Range with NBAA IFR reserves:
A with 8 pax .... 3,250 n miles (6,019 km; 3,740 miles)
B with 16 pax .... 1,840 n miles (3,408 km; 2,117 miles)

UPDATED

EMBRAER 170 AND 190

TYPE: Regional jet airliner.

DEVELOPMENT MILESTONES

Announced	Feb 99
Selected by Crossair and Regional Airlines	14 Jun 99
First prototype material cut	14 Jul 00
Rolled out	29 Oct 01
First flight	19 Feb 02
Certification	13 Nov 03
First delivery	8 Mar 04

PROGRAMME: Announced (officially 'pre-launch') in February 1999; designations then ERJ-170 and ERJ-190; engine selection May 1999; first orders announced 14 June 1999; risk-sharing partners (see 'Structure', below) revealed at European Regional Airline Association annual meeting in Paris, 30 September 1999. Joint definition phase leading to ERJ-170 design freeze, completed in April 2000; first metal cut for first of six pre-series ERJ-170s 14 July 2000; ERJ-170 (c/n 0001/PP-XJE) first rolled out 29 October 2001, first flight 19 February 2002; followed by c/n 0002/PP-XJC (to be HB-JCA) in Crossair (to become Swiss) colours on 9 April 2002, by which time the first aircraft had completed 40 hours of test flying; c/n 0003/PP-XJB (first equipped with full Honeywell Primus Epic avionics suite) on 25 May 2002; c/n 0004/PP-XJF on 19 June 2002; c/n 0005/PP-XJA on 14 July 2002; and c/n 0006/PP-XJS on 28 December 2002. Announced at roll-out that ERJ prefix was discontinued and that ERJ-190–200 to be known as Embraer 195; intermediate Embraer 175 simultaneously revealed.

Public debut (PP-XJE) at Regional Airline Association convention at Nashville, Tennessee 12 to 15 May 2002, followed by European debut (PP-XJA) at Farnborough International Air Show 22 July 2002, followed by European demonstration tour comprising 18 flights totalling more than 100 hours.

Six flying pre-series aircraft and one production aircraft conducted 3,000 hour flight test programme, culminating in provisional type certification by CTA on 13 November 2003, and by FAA on 8 December 2003, followed by full certification by CTA (19 February 2004), FAA (20 February 2004) and JAA in first quarter 2004 following completion of flight control system certification documentation. First deliveries, to LOT Polish and Alitalia, on 8 and 10 March 2004 respectively. Further two static test airframes (801 and 802) conducted and static fatigue test programme with aim of completing 5,000 simulated flights to clear aircraft for 2,500 actual flight hours by time of certification, and eventual demonstration of 80,000-cycle economic service life. Initial production of four per month in 2003, rising to six per month in 2004 and maximum capacity of 10 per month. Certification and first delivery of Embraer 175 scheduled for second quarter of 2004 (but see table, below). First flight of first of two pre-series Embraer 195s scheduled for third quarter of 2004, followed by certification in second quarter 2006 and first delivery (to Swiss, formerly Crossair) in third quarter 2006. First of three pre-series Embraer 190s rolled out with certification and deliveries in third quarter of 2005.

CURRENT VERSIONS: **Embraer 170:** Baseline version with 70 to 78 seats; available in Standard and Long-Range versions.

Embraer 175: Longer version with 1.77 m (5 ft 9¾ in) fuselage stretch by means of two plugs to accommodate 78 to 86 passengers; offered in Standard and Long-Range versions. First flight (c/n 0014-PP-XJD) 14 June 2003, followed by second aircraft on 7 August 2003.

Embraer 190: Further stretch by 6.34 m (20 ft 9½ in) to accommodate up to 104 passengers; wing span increased by 2.72 m (8 ft 11 in); GE CF34-8E-10E engines; strengthened landing gear; available in Standard and Long-Range versions. Launch customer jetBlue ordered 100, plus 100 options, on 10 June 2003. Prototype rolled out 9 February 2004; first flight 12 March 2004. First delivery expected in third quarter of 2005.

Embraer 195: Formerly ERJ-190–200. Further stretch of Embraer 190–200 by 2.41 m (7 ft 11 in) to accommodate up to 110 passengers; available in Standard and Long-Range versions; first metal cut 23 August 2002; first flight anticipated in third quarter 2004.

Embraer 170 sixth prototype *(Paul Jackson)* ***NEW***/0569595

Common flight deck of Embraer 170/190 0131838

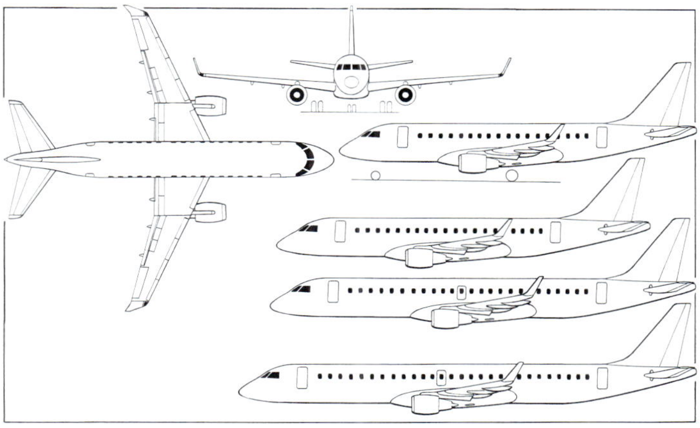

Embraer 170 regional jet, with additional side views of Embraer 175, Embraer 190 and Embraer 195 (bottom) *(James Goulding)* ***NEW***/0526909

Embraer 175 prototype during its maiden flight on 14 June 2003 ***NEW***/0552827

Artist's impression of Embraer 195, longest of the family ***NEW***/0576633

Corporate: Formerly ECJ-170. Proposed variant of Embraer 170 Long-Range with additional fuel tanks in baggage compartment area to extend range to more than 4,000 n miles (7,408 km; 4,603 miles). Launch decision expected following certification of Embraer 170.

CUSTOMERS: See table. Launch customers Crossair and Regional Airlines of France announced at Paris Air Show 14 June 1999; Crossair (now Swiss) ordered 30 Embraer 170s and 30 Embraer 195s, with options on a further 100 Embraer 170/195s, for intended delivery (Embraer 170) from December 2002, but subsequently reduced on 25 March 2003 to 15 of each model, with first delivery (Embraer 170) in August 2004 (but see table, below), followed by first Embraer 195 in 2006. North American launch customer US Airways ordered 85 Embraer 170s, plus 50 options, on 12 May 2003, for delivery commencing 2004.

COSTS: Embraer 170 US$24.8 million; Embraer 190, US$29.6 million (both 2003). US Airways order for 85 Embraer 170s valued at US$2.1 billion; jetBlue order for 100 Embraer 190s valued at US$3 billion (both 2003). Development US$850 million for whole family (2001).

DESIGN FEATURES: Design goals include low weight, simplicity of operation, common crew type rating, high reliability, ease and economy of maintenance and ability to operate from same airports as ERJ-135/145. Low-wing airliner of conventional appearance, with podded engine below each wing and (on Embraer 170) winglets; airframe designed for an economic life of 60,000 to 80,000 cycles.

EMBRAER 170/190 ORDERS
(at 31 March 2004)

Customer	Type	Orders
Air Caraibes	170	2
Alitalia	170	6
Cirrus Germany	170	1
GECAS	170	5
jetBlue	190	100
LOT Polish	170	6
Republic Airways	170	13
Swiss	170	15
	195	15
US Airways	170	85
Undisclosed	190	10
Total		**258**

[1] Further 342 options held. In October 2003 Swiss requested postponement of its firm orders for Embraer 170s and 190s.

FLYING CONTROLS: Fly-by-wire. Ailerons, rudder and all-moving tailplane. Double-slotted flaps, four-section leading-edge slats and five-section spoilers on each wing, of which three outboard panels are multifunctional and inboard pair operate as ground spoilers. Movable horizontal stabiliser for pitch trim.

Ventral airbrake optional to meet steep glideslope requirements for airports such as London City and Lugano.

STRUCTURE: Embraer is responsible for, forward fuselage, centre fuselage II, fuselage plugs, wing-to-fuselage fairing and final assembly; risk-sharing partners are C & D Interiors (cabin interior); Gamesa (rear fuselage and horizontal and vertical tail surfaces); General Electric (power plant and nacelles); Hamilton Sundstrand (tailcone, APU and air management and electrical systems); Honeywell (avionics); Kawasaki (wing stub, fixed leading- and trailing-edge assemblies, flaps, spoilers, control surfaces and engine pylons); Latécoère (centre fuselage I and III); Liebherr (landing gear); Parker Hannifin (hydraulic, flight control and fuel systems); and Sonaca (wing slats).

LANDING GEAR: Retractable tricycle type. Twin wheels on each unit; tyre size H38 × 13.0–18. Aircraft Braking Systems Corporation carbon brakes.

POWER PLANT: Embraer 170/175 series has two 63.2 kN (14,200 lb st) General Electric CF34-8E turbofans; 190/195 series has two 88.9 kN (20,000 lb st) CF34-10E turbofans; both engines have FADEC.

ACCOMMODATION: Total of 70 to 78 (Embraer 170), 78 to 86 (Embraer 175), 98 to 108 (Embraer 190) or 108 to 118 (Embraer 195) passengers, four abreast at 81 cm (32 in) seat pitch. Optional first class cabin on Embraer 170 and Embraer 190 with three-abreast seating.

AVIONICS: Honeywell Primus Epic five-tube EFIS. Thales Avionics integrated electronic standby instrument (IESI) system standard.

DIMENSIONS, EXTERNAL (all versions, except where stated):
Wing span over winglets:
 170, 175 .... 26.00 m (85 ft 3½ in)
 190, 195 .... 28.72 m (94 ft 2¾ in)
Length overall: 170 .... 29.90 m (98 ft 1¼ in)
 175 .... 31.68 m (103 ft 11¼ in)
 190 .... 36.24 m (118 ft 10¾ in)
 195 .... 38.65 m (126 ft 9½ in)
Height overall: 170 .... 9.67 m (31 ft 8¾ in)
 175 .... 9.73 m (31 ft 11 in)
 190 .... 10.28 m (33 ft 8¾ in)
 195 .... 10.52 m (34 ft 6 in)
Fuselage height: all .... 3.35 m (11 ft 0 in)
Fuselage width: all .... 3.01 m (9 ft 10½ in)
Tailplane span: all .... 10.00 m (32 ft 9¾ in)

DIMENSIONS, INTERNAL:
Cabin (excl flight deck):
 Length: 170 .... 19.43 m (63 ft 9 in)
  175 .... 21.20 m (69 ft 6¾ in)
  190 .... 25.76 m (84 ft 6 in)
  195 .... 28.17 m (92 ft 5 in)
 Max width: all .... 2.74 m (8 ft 11¾ in)
 Max height: all .... 2.00 m (6 ft 6¾ in)
 Aisle width: all .... 0.50 m (1 ft 7¾ in)

WEIGHTS AND LOADINGS:
Basic operating weight:
 170 .... 20,940 kg (46,165 lb)
 175 .... 21,810 kg (48,083 lb)
 190 .... 28,080 kg (61,906 lb)
 195 .... 28,970 kg (63,868 lb)
Max payload: 170 .... 8,660 kg (19,092 lb)
 175 .... 9,890 kg (21,804 lb)
 190 .... 12,720 kg (28,043 lb)
 195 .... 13,530 kg (29,828 lb)
Max fuel: 170, 175 .... 9,312 kg (20,529 lb)
 190, 195 .... 12,872 kg (28,378 lb)
Max T-O weight:
 170 Standard .... 35,990 kg (79,344 lb)
 170 Long-Range .... 37,200 kg (82,012 lb)
 175 Standard .... 37,500 kg (82,673 lb)
 175 Long-Range .... 38,790 kg (85,517 lb)
 190 Standard .... 47,790 kg (105,359 lb)
 190 Long-Range .... 50,300 kg (110,892 lb)
 195 Standard .... 48,790 kg (107,563 lb)
 190 Long-Range .... 50,790 kg (111,793 lb)
Max landing weight:
 170 .... 32,800 kg (72,312 lb)
 175 .... 34,000 kg (74,957 lb)
 190 .... 43,000 kg (94,799 lb)
 195 .... 45,000 kg (99,208 lb)
Max power loading:
 170 Standard .... 285 kg/kN (2.79 lb/lb st)
 170 Long-Range .... 294 kg/kN (2.89 lb/lb st)

175 Standard ... 297 kg/kN (2.91 lb/lb st)
175 Long-Range ... 307 kg/kN (3.01 lb/lb st)
190 Standard ... 268 kg/kN (2.63 lb/lb st)
190 Long-Range ... 282 kg/kN (2.77 lb/lb st)
195 Standard ... 274 kg/kN (2.69 lb/lb st)
195 Long-Range ... 285 kg/kN (2.80 lb/lb st)

PERFORMANCE (estimated):
Max operating speed: all ... M0.82
Max cruising speed at FL100:
all ... 320 kt (593 km/h; 368 mph)
Time to climb to FL350:
170, 195 ... 16 min
175 ... 19 min
190 ... 15 min
T-O field length:
170 Long-Range ... 1,689 m (5,545 ft)
175 Long-Range ... 1,910 m (6,270 ft)
190 Long-Range ... 1,983 m (6,510 ft)
195 Long-Range ... 2,179 m (7,150 ft)
Landing field length:
170 Long-Range ... 1,316 m (4,320 ft)
175 Long-Range ... 1,352 m (4,440 ft)
190 Long-Range ... 1,379 m (4,525 ft)
195 Long-Range ... 1,428 m (4,685 ft)
Range with max passengers at long-range cruising speed:
170 Standard ... 1,700 n miles (3,148 km; 1,956 miles)
170 Long-Range ... 2,000 n miles (3,704 km; 2,301 miles)
175 Standard ... 1,600 n miles (2,963 km; 1,841 miles)
175 Long-Range, 195 Long-Range ... 1,800 n miles (3,333 km; 2,071 miles)
190 Standard ... 1,700 n miles (3,148 km; 1,956 miles)
190 Long-Range ... 2,200 n miles (4,074 km; 2,531 miles)
195 Standard ... 1,500 n miles (2,778 km; 1,726 miles)

UPDATED

HELIBRAS

HELICÓPTEROS DO BRASIL SA (SUBSIDIARY OF EUROCOPTER)

Distritio Industrial de Itajubá, Rua Santos Dumont 200, Caixa Postal 184, 37500-000 Itajubá, Minas Gerais
Tel: (+55 35) 36 23 20 00
Fax: (+55 35) 36 23 20 01
Web: http://www.helibras.com.br
COMMERCIAL DEPARTMENT:
Aeroporto Campo de Marte, Avenida Santos-Dumont 1979, CEP 02012-010 Saõ Paolo, SP
Tel: (+55 11) 69 90 37 00
Fax: (+55 11) 62 21 55 35
PRESIDENT: Jean Raquin
COMMERCIAL AND CUSTOMER SUPPORT DIRECTOR: Vincent Kieffer
MARKETING MANAGER: Luis Henrique Testa

Formed 1978; owned jointly by Grupo Bueninvest (30 per cent), MGI Participações (25 per cent) and Eurocopter France (45 per cent). First assembly hall inaugurated 28 March 1980; facility occupies 12,000 m^2 (129,200 sq ft); new facilities, totalling 3,800 m^2 (40,900 sq ft) opened at São Paulo in second quarter of 1998 for maintenance, spares, training, sales and marketing. Total of 280 personnel employed in 2003.

Assembly, marketing and overhaul, under Eurocopter licence, of single- and twin-engined Ecureuil/Fennec helicopters and twin-engined Dauphin/Panther; markets in Brazil and overhauls civil Eurocopter AS 350/355 Esquilo (Fennec), EC 130, EC 135, AS 332 Super Puma; Eurocopter/Kawasaki BK 117C-1 (EC 145); and Eurocopter/CATIC/ST Aero EC 120 (see International section).

Total of 460 helicopters, including some 310 Esquilos, sold by 2003, representing 47 per cent of the turbine-powered helicopter fleet in Brazil); to more than 90 customers, including 206 to Brazilian police forces, Army, Navy and Air Force; nearly 10 per cent of overall total exported to Argentina, Bolivia, Chile, Paraguay and Venezuela. Supply of new EC 120B Colibri to local owners began in 1999.

UPDATED

NEIVA

INDÚSTRIA AERONÁUTICA NEIVA SA (SUBSIDIARY OF EMBRAER)

Caixa Postal 1011, Avenida Alcides Cagliari 2281, 18606-900 Botucatu-SP
Tel: (+55 14) 38 11 20 46
Fax: (+55 14) 38 11 19 36
e-mail: neiva@laser.com.br
PRESIDENT AND DIRECTOR: Mauricio Novis Botelho
MANAGING DIRECTOR: Paolo Urbanavicius
SALES MANAGERS:
Luiz Fabiano
Zaccarelli Cunha

Formed 12 October 1954 by José Carlos de Barros Neiva; produced Paulistinha, Regente, Universal and related lightplanes of own design. Became wholly owned Embraer subsidiary 11 March 1980; factory area 35,700 m^2 (384,275 sq ft), workforce 1,057 in December 2003, following October 2001 cut of 270 staff. Previously played major role in Embraer general aviation programmes, including complete production of EMB-711ST Corisco Turbo, EMB-710 Carioca, EMB-720D Minuano (of which some 300 delivered), EMB-721 Sertanejo and EMB-810D Seneca IV (licence-built Piper models); also responsible since 1980 for total production of EMB-202 and now discontinued EMB-201A agricultural sprayers, and for components manufacture for ERJ-145 and Embraer 170/190 programmes. Low-volume production of EMB-120 Brasilia was transferred from Embraer in 2001. Built 17 light aircraft in 1999; same in 2000; and 11 in 2001. Total of more than 3,500 aircraft built by 2003.

UPDATED

EMBRAER EMB-120 BRASILIA

Brazilian Air Force designation: VC-97

TYPE: Twin-turboprop airliner.

PROGRAMME: Design began September 1979; three flying prototypes, two static/fatigue test aircraft and one pre-series demonstrator built (first flight, by PT-ZBA, 27 July 1983); Brazilian CTA certification with original PW115 engines on 10 May 1985 followed by FAA

Embraer EMB-120 Brasilia ***NEW***/0569640

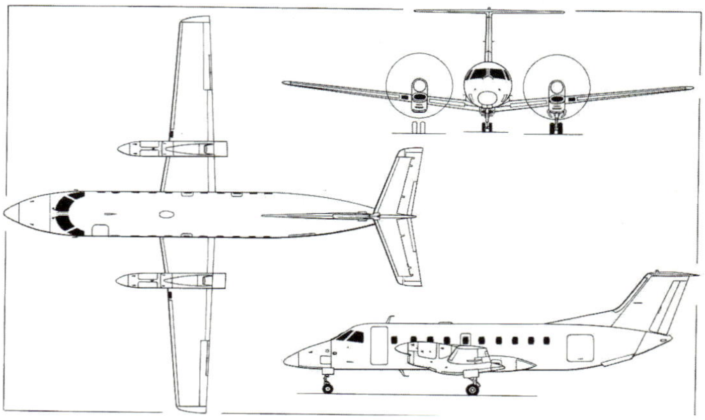

Embraer EMB-120 Brasilia twin-turboprop transport ***(Dennis Punnett)*** 0507118

(FAR Pt 25) type approval 9 July 1985, British/French/German approval 1986 and Australian in April 1990; deliveries began June 1985 (to Atlantic Southeast Airlines, USA, entering service that October); first order for corporate version (United Technologies Corporation, USA) received August 1986, delivered following month.

From October 1986 (c/n 120028), all Brasilias delivered incorporate composites materials equivalent to 10 per cent of aircraft basic empty weight; also since late 1986 has been available in hot-and-high version (certified 26 August 1986) with PW118A engines which maintain maximum output up to ISA + 15°C at S/L (first customer Skywest of USA); on 4 January 1989, first prototype began flight trials with Honeywell TPE331-12B turboprop on port side of rear fuselage, as testbed for engine installation of subsequently cancelled Embraer/FMA CBA-123; 300th Brasilia delivered 4 September 1995; more than five million flying hours by September 1998, at which time Brasilias had carried more than 60 million passengers.

Extended-range EMB-120ER (now standard version) announced June 1991, certified by CTA February 1992.

CURRENT VERSIONS: **EMB-120:** Launch production version, with 1,118.5 kW (1,500 shp) PW115 engines and Hamilton Sundstrand 14RF four-blade propellers (replaced at early stage by higher output PW115s of 1,193 kW; 1,600 shp). Details in 1996–97 and earlier *Jane's*.

EMB-120RT (Reduced Take-off): Initial production version; 1,342 kW (1,800 shp) PW118 engines for better field performance; most early aircraft now retrofitted to RT standard. Also, from late 1986, in hot/high version with PW118As (see Programme above). See 1996-97 and earlier editions for details.

EMB-120FC: 'Full Cargo' conversion of EMB-120RT developed in USA by IASG and approved by Embraer. Prototype (N453UE) converted June 2000 and exhibited at Farnborough in following month. Cargo volume 31.0 m^3 (1,095 cu ft); maximum payload 3,700 kg (8,157 lb); empty weight 7,200 kg (15,873 lb); maximum T-O weight 12,000 kg (26,455 lb). Initial 10 kits ordered by North South Airways.

EMB-120 Combi: Mixed configuration version of ER with quick-release seats, 9 *g* movable rear bulkhead and cargo restraint net; retains lavatory and galley; typical capacity 30 passengers and 700 kg (1,543 lb) of cargo or 19 passengers and 1,100 kg (2,425 lb) cargo.

EMB-120QC (Quick-Change): Convertible in 40 minutes from 30-passenger layout to 3,500 kg (7,716 lb) all-cargo configuration with floor and sidewall protection, fire detection system, smoke curtain aft of flight deck, 9 *g* movable rear bulkhead and cargo restraint net; conversion from cargo to passenger interior takes 50 minutes. First customer (14 May 1993 delivery), Total Linhas Aereas of Brazil.

EMB-120ER Brasilia Advanced: First delivery to Skywest Airlines in May 1993; standard production version from 1994, previously referred to as EMB-120X or Improved Brasilia. Additional range obtained by increasing maximum T-O weight without major structural change; this also allows for increased standard passenger-plus-baggage weight (97.5 kg; 215 lb per person instead of 91 kg; 200 lb) and obligatory fitting of TCAS and flight data recorder. Incorporates several modifications and some redesign aimed at maximising both passenger comfort and dispatchability while simultaneously reducing operational and maintenance costs. These include redesigned, interchangeable leading-edges for all flying surfaces; deletion of the fin de-icing boot; improved door seals and redesigned interior panel joints to reduce passenger cabin noise; more comfortable Sicma-built pilot seats; redesigned flight deck door; and new-design overhead bins with wider doors and increased capacity. Other features include new passenger cabin lighting; redesigned flight deck and cabin ventilation systems; new windscreen frame fairings to improve resistance to condensation; new flight deck sun visors; improved flap system; and an increase in baggage/cargo compartment capacity to 700 kg (1,543 lb).

Existing Brasilias can be retrofitted to ER standard (first two customers for retrofit Delta Air Transport and Luxair); initial production deliveries (two) to Great Lakes Aviation in USA, December 1994.

EMB-120 Cargo: All-cargo version of ER with 4,000 kg (8,818 lb) payload capacity; floor and sidewall protection, fire detection system, smoke curtain between flight deck and cargo cabin, and cargo restraint net.

Detailed description applies to EMB-120ER except where indicated.

EMB-120K: Proposed maritime surveillance or ASW version for Brazilian Air Force; would be equipped with in-flight refuelling, giving 15 hour endurance; under study in 1999.

VC-97: VIP transport model for Brazilian Air Force's 6° Esquadrao de Transporte Aérea and Grupo de Transporte Especial, both at Brasilia.

CUSTOMERS: See table. Embraer delivered 352nd (including prototypes) aircraft to SkyWest (as N586SW) in September 1999. Manufacture then transferred to Neiva, which delivered two in 2001 (none in 2000) to Avior Express of Venezuela. No further deliveries reported in 2002 or 2003.

DESIGN FEATURES: Low-mounted unswept wings, circular-section pressurised fuselage, all-swept T tail unit. Fixed incidence tailplane.

Wing section NACA 23018 (modified) at root, NACA 23012 at tip; incidence 2°; at 66 per cent chord, wings have 6° 30' dihedral.

FLYING CONTROLS: Conventional and assisted. Internally balanced ailerons, and horn-balanced elevators, actuated mechanically (ailerons by dual irreversible actuators); serially hinged two-segment rudder actuated hydraulically by Bertea CSD unit; trim tabs in ailerons (one port, two starboard) and each elevator. Hydraulically actuated, electrically controlled, double-slotted Fowler trailing-edge flap inboard and outboard of each engine nacelle, with small plain flap beneath each nacelle; no slats, slots, spoilers or airbrakes. Small fence on each outer wing between outer flap and aileron; twin ventral strakes under rear fuselage.

STRUCTURE: Kevlar-reinforced glass fibre for wing and tailplane leading-edges and tips, wingroot fairings, dorsal fin, fuselage nosecone and (when no APU fitted) tailcone; Fowler flaps of carbon fibre. Remainder conventional semi-monocoque/stressed skin structure of 2024/7050/7475 aluminium alloys with chemically milled skins. Fuselage pressurised between flat bulkhead forward of flight deck and hemispherical bulkhead aft of baggage compartment, and meets damage tolerance requirements of FAR Pt 25 (Transport category) up to Amendment 25–54. Wing is single continuous three-spar fail-safe structure, attached to underfuselage frames; tail surfaces also three-spar.

LANDING GEAR: Retractable tricycle type, with Goodrich twin wheels and oleo-pneumatic shock-absorber on each unit (main units 12 in, nose unit 8 in). Hydraulic actuation; all units retract forward (main units into engine nacelles). Hydraulically powered nosewheel steering. Mainwheel tyres 24×7.25–12 (12 ply) tubeless; nose tyre 18×5.5 (8 ply) tubeless; pressure 6.90 to 7.58 bar (100 to 110 lb/sq in) on main units, 4.14 to 4.83 bar (60 to 70 lb/sq in) on nose unit. Goodrich carbon brakes standard (steel optional). Hydro Aire anti-skid system standard; autobrake optional. Minimum ground turning radius 15.76 m (51 ft 8½ in). Nosewheel guard optional for operation from unpaved surfaces.

EMBRAER EMB-120 BRASILIA CUSTOMERS
(at 24 September 2003)

Customer	Qty
Air Aruba	1
Air Littoral	10
Air Midwest	14
ASA	62
Avior	2
Bop Air	2
Brazilian Air Force (VIP)	4
(SIVAM)	1
Comair	40
CSE Aviation	2
DAT	10
DLT	12
Ematec	1
Esquel	4
Flight West	2
Great Lakes	5
Interbrasil	3
Luxair	3
Mesa Air	2
Nordeste	1
Norsk Air	4
Ontario	5
Pantanal	2
Passaredo	3
Penta	1
Rico	1
Rio Sul	12
Skywest	70
TAT	2
Texas Air	34
Total	1
UTC	1
Westair	35
Embraer (retained)	2
Total	**354**

POWER PLANT: Two Pratt & Whitney Canada PW118 or PW118A turboprops, each rated at 1,342 kW (1,800 shp) for T-O and maximum continuous power, and driving a Hamilton Sundstrand 14RF-9 four-blade constant-speed reversible-pitch autofeathering propeller with glass fibre blades containing aluminium spars. Fuel in two-cell 1,670 litre (441 US gallon; 367.2 Imp gallon) integral tank in each wing; total capacity 3,340 litres (882 US gallons; 734.4 Imp gallons), of which 3,308 litres (874 US gallons; 728 Imp gallons) are usable. Single-point pressure refuelling (beneath outer starboard wing), plus gravity point in upper surface of each wing. Oil capacity 9 litres (2.4 US gallons; 2.0 Imp gallons).

ACCOMMODATION: Two-pilot flight deck. Main cabin accommodates attendant and 30 passengers in three-abreast seating at 79 cm (31 in) pitch, with overhead lockable baggage racks, in pressurised and air conditioned environment. Passenger seats of carbon fibre and Kevlar, floor and partitions of carbon fibre and Nomex sandwich, side panels and ceiling of glass fibre/Kevlar/Nomex/carbon fibre sandwich. Provisions for wardrobe, galley and lavatory. Quick-change interior available optionally (first customer, Total Linhas Aéreas in 1993), for 30 passengers or 3,500 kg (7,716 lb) of cargo. Downward-opening main passenger door, with airstairs, forward of wing on port side. Type II emergency exit on starboard side at rear. Overwing Type III emergency exit on each side. Pressurised baggage compartment aft of passenger cabin, with large door on port side.

Also available with all-cargo interior; executive or military transport interior; or in mixed-traffic version with 24 or 26 passengers (lavatory omitted in latter case), and 900 kg (1,984 lb) of cargo in enlarged rear baggage compartment.

SYSTEMS: Honeywell air conditioning/pressurisation system (differential 0.48 bar; 7 lb/sq in), with dual packs of recirculation equipment. Duplicated hydraulic systems (pressure 207 bar; 3,000 lb/sq in), each powered by an engine-driven pump, for landing gear, flap, rudder and brake actuation, and nosewheel steering. Emergency standby electric pumps on each system, plus single standby hand pump, for landing gear extension. Main electrical power supplied by two 28 V 400 A DC starter/generators; two 28 V 100 A DC auxiliary brushless generators for secondary and/or emergency power; one 24 V 40 Ah Ni/Cd battery for assisted starting and emergency power. Main and standby 450 VA static inverters for 26/115 V AC power at 400 Hz. Single high-pressure (127.5 bars; 1,850 lb/sq in) oxygen cylinder for crew; individual chemical oxygen generators for passengers. Pneumatic de-icing for wing and tail leading-edges; electrically heated windscreens, propellers and pitot tubes; bleed air de-icing of engine air intakes. Optional Honeywell GTCP36-150(A) APU in tailcone, for electrical and pneumatic power supply.

AVIONICS: Rockwell Collins Pro Line II digital package as core system.

Comms: Dual VHF-22 com transceivers, one TDR-90 transponder, Dorne & Margolin DMELT-81 emergency locator transmitter, Fairchild voice recorder, dual Avtech audio/interphones and Avtech PA and cabin interphone all standard. Third VHF com, second transponder, Motorola Selcal and flight entertainment music all optional. Alternative Bendix/King avionics package available to special order.

Radar: Rockwell Collins WXR-270 weather radar standard; WXR-300 optional.

Flight: Dual VIR-32 VHF nav receivers, one ADF-60A, DME-41 and CLT-22/32/62/92 control heads. Optional single/dual JET RNS-8000 3D or Racal Avionics RN 5000 nav and second DME. Optional equipment includes APS-65 digital autopilot, single/dual FCS-65 digital flight director, single/dual BAE Systems Canada CMA-771 Alpha VLF/Omega, MLS, GPWS, flight recorder.

Instrumentation: Dual AHRS-85 digital strapdown AHRS, dual ADI-84, dual EHSI-74, dual RMI-36 and JET standby altitude indicator standard. Optional equipment includes dual EFIS-86 electronic flight instrumentation systems, MFD-85 multifunction display, dual ALT-55 radio altimeters and altitude alerter/preselect.

DIMENSIONS, EXTERNAL:

Wing span	19.78 m (64 ft 10¾ in)
Wing chord: at root	2.81 m (9 ft 2¾ in)
at tip	1.40 m (4 ft 7 in)
Wing aspect ratio	9.9
Length overall	20.07 m (65 ft 10¼ in)
Fuselage: Length	18.73 m (61 ft 5½ in)
Max diameter	2.28 m (7 ft 5¾ in)
Height overall	6.35 m (20 ft 10 in)
Elevator span	6.94 m (22 ft 9¼ in)
Wheel track (c/l of shock-struts)	6.58 m (21 ft 7 in)
Wheelbase	6.97 m (22 ft 10½ in)
Propeller diameter	3.20 m (10 ft 6 in)
Propeller ground clearance (min)	0.52 m (1 ft 8½ in)
Passenger door (fwd, port): Height	1.70 m (5 ft 7 in)
Width	0.77 m (2 ft 6¼ in)
Height to sill	1.47 m (4 ft 10 in)
Cargo door (rear, port): Height	1.36 m (4 ft 5½ in)
Width	1.30 m (4 ft 3¼ in)
Height to sill	1.67 m (5 ft 5¾ in)
Emergency exit (rear, stbd): Height	1.37 m (4 ft 6 in)
Width	0.51 m (1 ft 8 in)
Height to sill	1.56 m (5 ft 1½ in)
Emergency exits (overwing, each):	
Height	0.91 m (3 ft 0 in)
Width	0.51 m (1 ft 8 in)
Emergency exits (flight deck side windows, each):	
Min height	0.48 m (1 ft 7 in)
Min width	0.51 m (1 ft 8 in)

DIMENSIONS, INTERNAL:

Cabin, excl flight deck and baggage compartment:	
Length	9.38 m (30 ft 9¼ in)
Max width	2.10 m (6 ft 10¾ in)
Max height	1.76 m (5 ft 9¼ in)
Floor area	15.0 m² (161 sq ft)
Volume	27.4 m² (968 cu ft)
Rear baggage compartment volume:	
30-passenger version	6.4 m² (226 cu ft)
all-cargo version	2.7 m² (95 cu ft)
Passenger/cargo version	11.0 m² (388 cu ft)
Cabin, incl flight deck and baggage compartment:	
Total volume	approx 41.8 m³ (1,476 cu ft)
Max available cabin volume (all-cargo version)	31.1 m³ (1,098 cu ft)

AREAS:

Wings, gross	39.43 m² (424.4 sq ft)
Ailerons (total)	2.88 m² (31.00 sq ft)
Trailing-edge flaps (total)	3.23 m² (34.77 sq ft)
Fin, incl dorsal fin	5.74 m² (61.78 sq ft)

Rudder 2.59 m^2 (27.88 sq ft)
Tailplane 6.10 m^2 (65.66 sq ft)
Elevators, (total, incl tabs) 3.90 m^2 (41.98 sq ft)

WEIGHTS AND LOADINGS:
Weight empty, equipped 7,150 kg (15,763 lb)[1]
Operating weight empty 7,560 kg (16,667 lb)
Max usable fuel 2,660 kg (5,864 lb)
Max payload 3,340 kg (7,363 lb)[1]
Max T-O weight 11,990 kg (26,433 lb)
Max ramp weight 12,080 kg (26,632 lb)
Max landing weight 11,700 kg (25,794 lb)
Max zero-fuel weight 10,900 kg (24,030 lb)
Max wing loading 304.1 kg/m^2 (62.28 lb/sq ft)
Max power loading 4.47 kg/kW (7.34 lb/shp)

[1] *4 kg (8.8 lb) payload decrease/empty weight increase with PW118A engines*

PERFORMANCE:
Max operating speed (VMO) 272 kt (504 km/h; 313 mph) EAS
Max level speed at FL200 327 kt (606 km/h; 377 mph)
Max cruising speed at FL250
PW118 300 kt (555 km/h; 345 mph)
PW118A 313 kt (580 km/h; 360 mph)
Long-range cruising speed at FL250 270 kt (500 km/h; 311 mph)
Stalling speed, power off:
flaps up 120 kt (223 km/h; 138 mph) CAS
flaps down 89 kt (165 km/h; 103 mph) CAS
Max rate of climb at S/L 762 m (2,500 ft)/min
Rate of climb at S/L, OEI 168 m (550 ft)/min
Service ceiling: PW118 8,840 m (29,000 ft)
PW118A 9,750 m (32,000 ft)
Service ceiling, OEI, AUW of 11,200 kg (24,690 lb):
PW118 or PW118A 4,600 m (15,100 ft)
FAR Pt 25 T-O field length 1,550 m (5,085 ft)
FAR Pt 135 landing field length, MLW at S/L 1,390 m (4,560 ft)
Range at FL300, reserves for 100 n mile (185 km; 115 mile) diversion and 45 min hold:
with max (30-passenger) payload (2,721 kg; 6,000 lb):
PW118 850 n miles (1,575 km; 979 miles)
PW118A 800 n miles (1,482 km; 921 miles)
with max fuel:
PW118, 20 passengers (1,785 kg; 3,935 lb payload) 1,629 n miles (3,017 km; 1,874 miles)
PW118A, 20 passengers (1,785 kg; 3,935 lb payload) 1,570 n miles (2,908 km; 1,806 miles)

OPERATIONAL NOISE LEVELS (FAR Pt 36, BCAR-N and ICAO Annex 16):
T-O 81.2 EPNdB
Approach 92.3 EPNdB
Sideline 83.5 EPNdB

UPDATED

NEIVA EMB-202 IPANEMA

TYPE: Agricultural sprayer.

PROGRAMME: Embraer design; first flight (EMB-200 Ipanema prototype PP-ZIP) 31 July 1970; certified 14 December 1971; see 1977–78 and earlier editions for EMB-200/200A (49/24 built), EMB-201 (200 built) and EMB-201R (three built); and 1992–93 edition for EMB-201A (402 built). Manufacture transferred to Neiva in March 1980; many detail improvements (listed in 1991–92 and earlier *Jane's*) introduced late 1988. Available from 2002 onwards as airframe only, giving option of customer engine installation.

On 10 October 2002, Embraer revealed the prototype (PP-XHQ) of an alcohol-fuelled (Lycoming) Ipanema.

CURRENT VERSIONS: **EMB-202 Ipanema:** Current version from 1992; generally similar to EMB-201A except for increased wing span, larger hopper and choice of Lycoming or Continental engine. First delivery 2 October 1992.

Experimental alcohol-fuelled version announced 10 October 2002 as joint development project between Neiva and CTA, with technical support from Textron Lycoming and Hartzell, aimed at reducing emissions and extending engine maintenance cycle; certification then expected within 18 months.

Prototype was conversion of standard EMB-202 PT-ULA which flew on trials as PP-XHO before reverting to original markings.

CUSTOMERS: Delivery of 900th Ipanema (including 678 of earlier versions) took place on 4 December 2003. Sales totalled 29 in 2002 and 56 in 2003, most customers reserving option of change to alcohol fuel.

COSTS: US$200,000; alcohol fuel retrofit US$20,000 (both 2002).

DESIGN FEATURES: Designed to Brazilian Ministry of Agriculture specifications. Typical agricultural aircraft features, including hopper ahead of pilot and high cockpit for good view over nose; low-mounted unbraced wings, with cambered leading-edges (detachable) and tips; rectangular-section fuselage; slightly sweptback vertical tail.

Prototype alcohol-fuelled Neiva EMB-202 Ipanema ***NEW***/0569621

Latest style of Ipanema wingtip ***NEW***/0531970

Wing section NACA 23015 (modified); incidence 3°; dihedral 7° from roots.

FLYING CONTROLS: Conventional and manual. Frise ailerons; trim tab in starboard elevator, ground-adjustable tab on rudder. Slotted flaps on wing trailing-edges.

STRUCTURE: All-metal safe-life fuselage frame of welded 4130 steel tube, with removable skin panels of 2024 aluminium alloy and glass fibre and specially treated against chemical corrosion. All-metal wings (single-spar) and tail surfaces (two-spar).

LANDING GEAR: Non-retractable mainwheels and tailwheel, with rubber shock-absorbers in main units. Tailwheel has tapered spring shock-absorber. Mainwheels and tyres size 8.50–10. Tailwheel diameter 250 mm (10 in). Tyre pressures: main, 2.07 to 2.41 bar (30 to 35 lb/sq in); tailwheel, 3.79 bar (55 lb/sq in). Hydraulic disc brakes on mainwheels.

POWER PLANT: One 224 kW (300 hp) Textron Lycoming IO-540-K1J5D flat-six engine, driving a Hartzell two-blade (optionally three-blade) constant-speed metal propeller. Optional engine is Teledyne Continental 224 kW (300 hp) IO-550-D with McCauley two-blade constant-speed propeller. Integral fuel tanks in each wing leading-edge, with total capacity of 292 litres (77.1 US gallons; 64.2 Imp gallons), of which 264 litres (69.75 US gallons; 58.1 Imp gallons) are usable. Refuelling point on top of each tank. Oil capacity 12 litres (3.2 US gallons; 2.6 Imp gallons).

ACCOMMODATION: Single horizontally/vertically adjustable seat in fully enclosed cabin with bottom-hinged, triple-lock, jettisonable window/door on each side and two overhead windows. Ventilation airscoop at top of canopy front edge; second airscoop on fin leading-edge pressurises interior of rear fuselage. Air conditioning and inertia reel shoulder harness standard.

SYSTEMS: 28 V DC electrical system supplied by 24 Ah batteries and a Bosch KI 28 V 35 A alternator. Power receptacle for external battery (AN-2552-3A type) on port side of forward fuselage.

AVIONICS: *Comms:* Optional portable VHF transceiver.
Flight: Optional portable Garmin GPS.

EQUIPMENT: Hopper for agricultural chemicals, capacity 950 litres (251 US gallons; 209 Imp gallons) liquid or 750 kg (1,653 lb) dry. Dusting system below centre of fuselage. Spraybooms or Micronair atomisers aft of or above wing trailing-edges respectively. Options include ram air pressure generator for use with liquid spray system; improved lightweight spraybooms; smaller/lighter Micronair AU5000 rotary atomisers; trapezoidal spreader with adjustable inlet to improve application of dry chemicals; and, from 2002, electrostatic spray system.

DIMENSIONS, EXTERNAL:
Wing span 11.69 m (38 ft 4¼ in)
Wing chord, constant 1.71 m (5 ft 7½ in)
Wing aspect ratio 6.9
Length overall (tail up) 7.43 m (24 ft 4½ in)
Height overall (tail down) 2.20 m (7 ft 2½ in)
Fuselage: Max width 0.93 m (3 ft 0½ in)
Tailplane span 3.66 m (12 ft 0 in)
Wheel track 2.20 m (7 ft 2½ in)
Wheelbase 5.20 m (17 ft 7¼ in)
Propeller diameter:
Hartzell two- and three-blade 2.13 m (7 ft 0 in)
McCauley two-blade 2.18 m (7 ft 2 in)

DIMENSIONS, INTERNAL:
Cockpit: Length 1.20 m (3 ft 11¼ in)
Max width 0.85 m (2 ft 9½ in)
Max height 1.34 m (4 ft 4¾ in)

AREAS:
Wings, gross 19.94 m^2 (214.6 sq ft)
Ailerons (total) 1.60 m^2 (17.22 sq ft)
Trailing-edge flaps (total) 2.30 m^2 (24.76 sq ft)

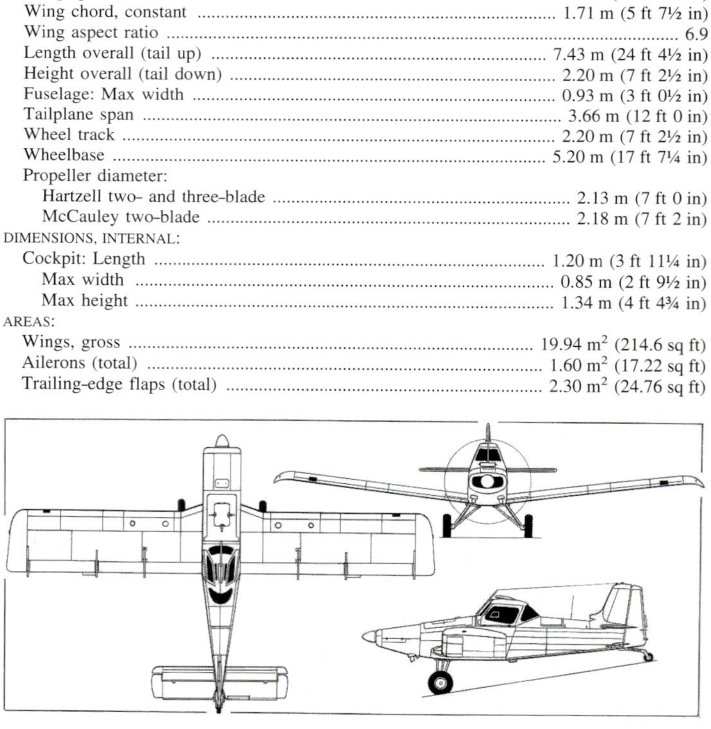

Neiva EMB-202 Ipanema (one Textron Lycoming IO-540) *(Jane's/Dennis Punnett)* 0507120

Fin .... 0.58 m² (6.24 sq ft)
Rudder .... 0.63 m² (6.78 sq ft)
Tailplane .... 3.17 m² (34.12 sq ft)
Elevators (total, incl tab) .... 1.50 m² (16.15 sq ft)

WEIGHTS AND LOADINGS (N: Normal, R: Restricted category):
Weight empty: N, R .... 1,020 kg (2,249 lb)
Max payload: N, R .... 750 kg (1,653 lb)
Max T-O and landing weight: N .... 1,550 kg (3,417 lb)
R .... 1,800 kg (3,968 lb)
Max wing loading: N .... 77.7 kg/m² (15.92 lb/sq ft)
R .... 90.3 kg/m² (18.49 lb/sq ft)
Max power loading: N .... 6.92 kg/kW (11.39 lb/hp)
R .... 8.03 kg/kW (13.23 lb/hp)

PERFORMANCE (clean):
Never-exceed speed (VNE):
N .... 147 kt (272 km/h; 169 mph)
Max level speed at S/L:
N .... 124 kt (230 km/h; 143 mph)
R .... 121 kt (225 km/h; 140 mph)
Max cruising speed at 75% power at FL60:
N .... 115 kt (213 km/h; 132 mph)
R .... 111 kt (206 km/h; 128 mph)
Stalling speed, power off (N):
flaps up .... 56 kt (103 km/h; 64 mph)
8° flap .... 54 kt (100 km/h; 62 mph)
30° flap .... 50 kt (92 km/h; 57 mph)
Stalling speed, power off (R):
flaps up .... 60 kt (110 km/h; 68 mph)
8° flap .... 58 kt (107 km/h; 66 mph)
30° flap .... 53 kt (99 km/h; 61 mph)
Max rate of climb at S/L, 8° flap:
N .... 283 m (930 ft)/min
R .... 201 m (660 ft)/min
Service ceiling, 8° flap: R .... 3,470 m (11,380 ft)
T-O run at S/L, asphalt runway:
N .... 200 m (655 ft)
R .... 354 m (1,160 ft)
T-O to 15 m (50 ft), conditions as above:
N .... 332 m (1,090 ft)
R .... 564 m (1,850 ft)
Landing from 15 m (50 ft) at S/L, 30° flap, asphalt runway: N .... 412 m (1,355 ft)
R .... 476 m (1,565 ft)
Landing run, conditions as above: N .... 153 m (505 ft)
R .... 170 m (560 ft)
Range at FL60, no reserves:
N .... 506 n miles (938 km; 583 miles)
R .... 474 n miles (878 km; 545 miles)

UPDATED

Bulgaria

DAR

DAR SAMOLETI OOD (DAR AIRCRAFT LTD)

Mladost 103-A-49, Sofia 1797
Tel/Fax: (+35 927) 571 65
e-mail: dar-info@dir.bg
Web: http://dar.dir.bg
MANAGER AND CO-OWNER: Tony Ilief
CHIEF DESIGNER AND CO-OWNER: Vesselin Valkanov

DAR's product line was expanded with the Impuls in 2003, although marketing of that aircraft had not begun by the end of the year.

Company name revives initials of former Darjavna Aeroplanna Rabotilnitza (State Aircraft Factory), which produced trainers and a prototype light attack aircraft in late 1930s/early 1940s.

UPDATED

DAR-21 VECTOR

TYPE: Tandem-seat ultralight.

PROGRAMME: Designed by Vesselin Valkanov. Prototype (LZ-DAR) first flew 20 May 2000, initially powered by a 39.0 kW (52.3 hp) Hirth 2704. No known production. Agricultural version marketed only in Bulgaria.

COSTS: US$50,000 (agricultural version, 2002).

DESIGN FEATURES: Conventional high-wing monoplane of angular appearance; constant-chord wing and tailplane, both strut-braced; sweptback fin; good all-round view from cockpit. Horizontal wingtip vanes and leading-edge slots outboard.

FLYING CONTROLS: Conventional and manual. Flaps (0, 15 and 30°). Trim tabs on starboard aileron and rudder; flight-adjustable (electric) tab on single-piece elevator. Aileron and rudder actuation by cables; elevator by pushrods. No aerodynamic balances.

STRUCTURE: Generally of 1050, 2024, 3130 and 6164 aluminium alloy sheet and type with milled and turned steel elements in landing gear. Composites engine cowling. Single-spar wing.

LANDING GEAR: Tailwheel type; fixed. Bungee shock-absorbers. Mainwheel size 6.00 standard; 8.00 optional. Steerable tailwheel.

POWER PLANT: One 47.8 kW (64.1 hp) Rotax 582 UL two-stroke with 2.58:1 reduction gear driving a Powerfin three-blade, composites propeller. Fuel tank in each wing and fuselage, each 30 litres (7.9 US gallons; 6.6 Imp gallons); total fuel 90 litres (23.8 US gallons; 19.8 Imp gallons).

ACCOMMODATION: Two pilots in tandem in enclosed, extensively glazed cockpit, with single door, starboard, to seats and baggage area to rear.

EQUIPMENT: Optional Spray Miser spraying bars mounted at mainwheel leg roots.

DIMENSIONS, EXTERNAL:
Wing span .... 8.00 m (26 ft 3 in)
Length overall .... 5.75 m (18 ft 10½ in)
Height overall .... 1.80 m (5 ft 10¾ in)

DIMENSIONS, INTERNAL:
Cockpit max width .... 0.63 m (2 ft 0¾ in)

AREAS:
Wings, gross .... 10.40 m² (111.9 sq ft)

Prototype DAR-21 Vector 0527113

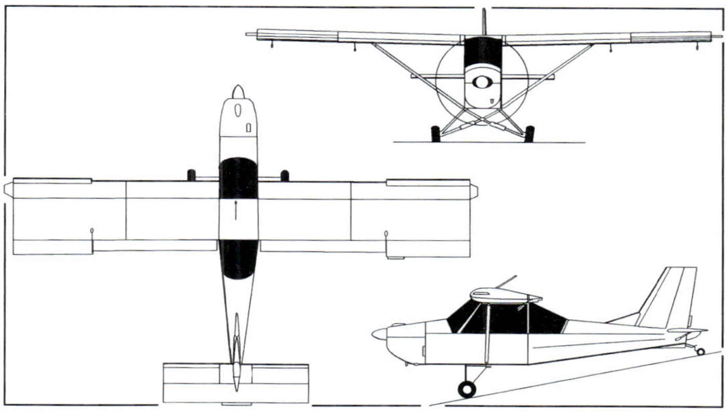
DAR-21 Vector tandem-seat ultralight *(Paul Jackson)* ***NEW***/0552994

WEIGHTS AND LOADINGS:
Weight empty .... 230 kg (507 lb)
Max T-O weight .... 450 kg (992 lb)

PERFORMANCE:
Max level speed .... 76 kt (140 km/h; 87 mph)
Max cruising speed .... 67 kt (125 km/h; 78 mph)
Stalling speed, flaps down .... 33 kt (60 km/h; 38 mph)
Endurance .... 9 h

UPDATED

DAR-23 AKTSENT

English name: Accent

TYPE: Side-by-side ultralight.

PROGRAMME: Prototype (LZ-DAS) flying by 2002; joined by second aircraft (LZ-DAF) early 2003. No known production.

COSTS: Available on application.

DESIGN FEATURES: Cabin pod underslung from boom carrying uncowled engine, wings and empennage. Wings braced to cabin; cabin braced to boom ahead of windscreen.

FLYING CONTROLS: Conventional and manual. Plain flaps. Flight-adjustable elevator tab. No aerodynamic balances. Ailerons, elevator and flaps actuated by pushrods; rudder by cables.

STRUCTURE: Metal boom and composites cabin. Metal wings fabric-covered.

LANDING GEAR: Tricycle type; fixed. Tubular steel legs. Mainwheels Azusa 6.00. Cable mainwheel brakes.

POWER PLANT: One 34.0 kW (45.6 hp) Rotax 503 UL two-stroke driving a Powerfin two-blade, composites propeller via 2.58:1 reduction gear. Fuel capacity 30 litres (7.9 US gallons; 6.6 Imp gallons).

ACCOMMODATION: Optional cover for cockpit.

DIMENSIONS, EXTERNAL:
Wing span .... 8.00 m (26 ft 3 in)
Length overall .... 5.00 m (16 ft 4¾ in)

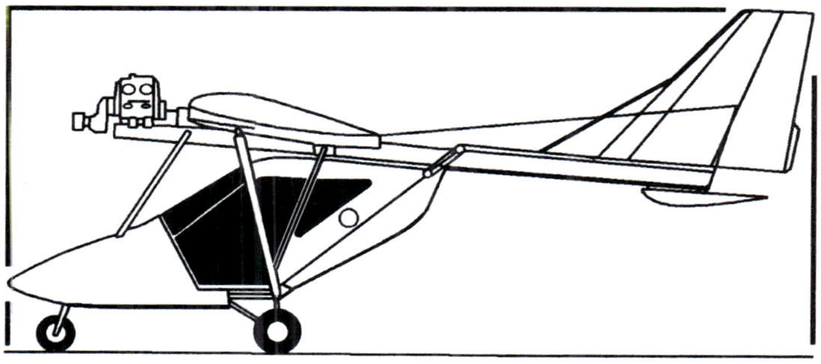
Profile of DAR-23 Aktsent *(Paul Jackson)* ***NEW***/0552054

Second prototype DAR-23 Accent ***NEW***/0552053

Height at wing .... 1.80 m (5 ft 10¾ in)

DIMENSIONS, INTERNAL:

Cockpit max width .... 0.90 m (2 ft 11½ in)

AREAS:

Wings, gross .... 10.40 m^2 (111.9 sq ft)

WEIGHTS AND LOADINGS:

Weight empty .... 180 kg (397 lb)

Max T-O weight .... 380 kg (837 lb)

PERFORMANCE:

Max level speed .... 67 kt (125 km/h; 78 mph)

Max cruising speed .... 51 kt (95 km/h; 59 mph)

Stalling speed, flaps down .... 27 kt (50 km/h; 32 mph)

Endurance .... 3 h

UPDATED

DAR-25 IMPULS

English name: Impulse

TYPE: Side-by-side ultralight.

PROGRAMME: Announced early 2003; prototype LZ-DAH then complete. Similar to DAR-23, but with enlarged cockpit. No further details disclosed by end of 2003.

UPDATED

Prototype DAR-25 Impuls in early 2003 ***NEW***/0552061

Canada

21ST CENTURY

21ST CENTURY AIRSHIPS INC

PO Box 177, Main Station, Newmarket, Ontario L3Y 4X1
Tel: (+1 905) 898 62 74
Fax: (+1 905) 898 72 45
e-mail: info@21stcenturyairships.com
Web: http://www.21stcenturyairships.com
CHAIRMAN AND CEO: Hokan Colting

SPAS-70-B internal cabin, looking rearwards from flight deck ***NEW***/0593265

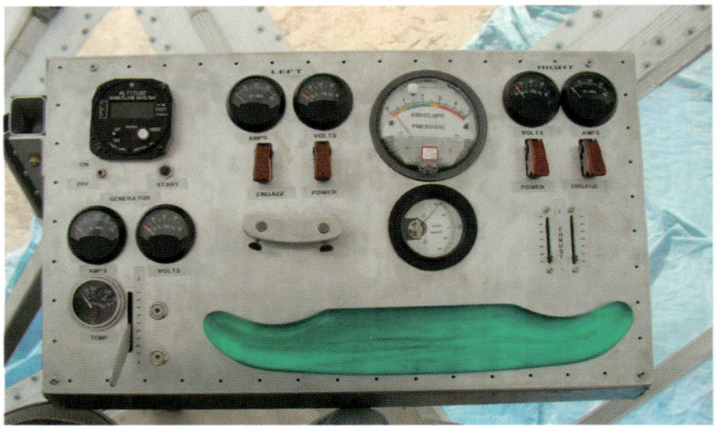

Hybrid electric propulsion control panel of the SPAS-70-B ***NEW***/0593267

SPAS-70-1A, built for Techsphere Systems as N8041X ***NEW***/1111647

Model	Reg'n	Seats	Engines (hp)	Diam (ft)	Vol (cu ft)	MTOW (lb)	Remarks
SPAS-1	C-FJHH	1	2× 24	35	23,000	1,280	First manned spherical prototype; first flight 8 June 1991
SPAS-2	C-FMKK	1	4×2.5	33	18,700	1,166	First flight 15 March 1992; later re-engined as SPAS-3
SPAS-3	C-FMKK	1	2× 5+1×24	33	18,700	1,166	SPAS-2 with new engine and control configuration; world altitude records in classes BA-2 to BA-10 on 8 November 1992
SPAS-13	C-FRLM	2	2× 50	43	41,500	2,400	Semi-rigid with internal frame for structure and engine support; internal flight deck. First flight 10 June 1994
SPAS-13-B	C-FRLM	2	2× 50	43	41,500	2,400	Re-engineered without internal frame but with internal flight deck
SPAS-70	C-FYOK	2	4×100	56	91,000	5,300	Four piston engines; first flight 5 September 1996
SPAS-70-B	C-FYOK	2	2× 90	56	91,000	5,300	Re-engined with two turbo-diesel engines driving 13 ft diameter propeller; stub wings redesigned. First flight 5 August 1997
SPAS-R1	C-GJKI	4	2× 90	60	112,000	5,830	First flight 30 June 2002
SPAS-62.5	C-FZRY	2	2× 80	62.5	126,700	7,480	Absolute world altitude record for airships, 12 June 2003. As of September 2004, was being converted to a hybrid-electric propulsion system with propellers driven by turbo-diesel diesel-generators
SPAS-70-1A	N8041X	2	2×80+1×60	62.5	126,700	7,480	Built for TSI as SA-60. First flight 24 June 2004 with temporary propulsion system. Converted to hybrid-electric propulsion system in September 2004
SPAS-396	C-FAUP	1	2× 5	30	13,860	825	Used solely for testing a new buoyancy control system for long-duration flights

Formed 1988 as a privately owned R&D company aimed at improving airship technology, and has developed a new type of airship which is spherical and finless. Current airships have an internal cabin for crew and passenger(s); a hybrid Diesel-electric propulsion system with a low noise profile; amphibious ability to land on and take off from water; internal illumination for spectacular visuals during night flights; and a basic design suitable for a variety of tasks.

Prototypes built and flown during development are listed in the accompanying table. In 2004, the basic design was being refined for: (1) a high-altitude platform for telecommunications (cellular phones, microwave and broadband) and national security applications; (2) sightseeing flights; (3) aerial advertising; and (4) heavy lifting.

The planned high-altitude platform, designated **SA-260**, will typically have a diameter of 79.25 m (260 ft), volume of 260,000 m³ (9.18 million cu ft), and will be able to operate at altitudes of approximately 18,300 to 20,700 m (60,000 to 68,000 ft) for extended periods. A smaller, piloted, prototype was the SPAS-62.5, registered as C-FZRY in June 2003. On 12 June 2003, crewed by Hokan Colting and Tim Buss, it set an absolute altitude record (ratified by the FAI on 6 January 2004) of 6,234 m (20,453 ft) in a flight from Drumheller, Alberta, to Gull Lake, Saskatchewan.

In 2003, manufacturing and sales rights to certain types of 21st Century spherical airship designs were sold to Techsphere Systems International of Atlanta, Georgia, USA (TSI, which see), for whom a SPAS-70-1A was built as a prototype under the new name AeroSphere SA-60.

UPDATED

SPAS-70-B during test flights in 2002 ***NEW***/0593258

AIRSTART

AIRSTART CORPORATION

Hangar 11, 11760 109th Street, City Center Airport, Edmonton, Alberta T5G 2T8
Tel:(+1 780) 944 92 10
Fax:(+1 780) 461 05 84
*e-mail:*griffin@airstart.com
*Web:*http://www.airstart.com
POSTAL ADDRESS: 11760 109th Street, Edmonton, Alberta T5G 2T8
CEO: Ron Wunder
VICE-PRESIDENT, SALES AND MARKETING: William (Bill) Crampton

Following refinancing, the former Canada Air RV company was reorganised. Manufacturing is now undertaken by AC Millennium Corporation, and Sales and Marketing by AirStart Corporation. Research and development is undertaken by ARV Development Corporation. The Griffin is at present available only in North America; research and development continues under the aegis of ARV Development Corporation.

VERIFIED

AIRSTART GRIFFIN

TYPE: Side-by-side kitbuilt.

PROGRAMME: Design work started January 1989; construction of first prototype Griffin I began September 1989; first flight 1993. Construction of Mark II began 1995 with first flight that year; Mark III first flown 1997 followed by Mark IV first flight October 1998. First kit delivered 4 January 1996.

CURRENT VERSIONS: **Griffin III:** Lower-powered version; no longer available.

Griffin IV: Baseline version, *as described.*

CUSTOMERS: 72 ordered and eight flying by mid-2002 including three in Canada and two in USA.

COSTS: Kit US$18,500 excluding engine, propeller and instruments (2003). Estimated design/programme cost US$3.5 million to date.

DESIGN FEATURES: High-wing monoplane. Good all-round visibility, including roof windows, enhanced by instrument panel positioned low in cockpit and convex side windows. Quoted build time 800 hours.

FLYING CONTROLS: Manual. Full-span continuously positionable flaperons deflect to 20° and also reflex slightly for cruise drag reduction. Electrically operated full-span trim on port elevator.

STRUCTURE: All-aluminium semi-monocoque fuselage and wings; elevator and rudder are fabric-covered. Single-strut laminar flow high-aspect ratio wing; thickness/chord ratio 19 per cent. Wing dihedral 1° 30'. Low-mounted tailplane.

LANDING GEAR: Choice of tricycle or tailwheel layout. Sprung aluminium mainwheel legs. Tricycle model has oleo-pneumatic nosewheel suspension. Tailwheel version has steerable Matco 11 cm (4½ in) tailwheel. Matco disc brakes. Mainwheels and nosewheel Matco 6.00–6, all tyres pressurised to 2.9 bar (42 lb/sq in).

POWER PLANT: *Griffin IV:* One 119 kW (160 hp) Textron Lycoming O-320 flat-four engine, driving a two-blade fixed-pitch wooden propeller in demonstrator; engines in range 74.6 to 119 kW (100 to 160 hp) can be fitted including 59.6 kW (79.9 hp) Rotax 912 and 74.6 kW (100 hp) Teledyne Continental O-200A.

Griffin III: 74.6 kW (100 hp) CAM 100 piston engine.

Fuel capacity 136 litres (35.9 US gallons; 29.9 Imp gallons) in two wing tanks. Oil capacity 8 litres (2.1 US gallons; 1.8 Imp gallons).

ACCOMMODATION: Pilot and passenger in side-by-side seating with dual controls. Upward-opening door on each side of fuselage. Baggage stowage behind seats.

DIMENSIONS, EXTERNAL:
Wing span .... 10.82 m (35 ft 6 in)
Wing chord: at root .... 1.23 m (4 ft 0½ in)
at tip .... 0.74 m (2 ft 5 in)
Wing aspect ratio 11.0
Length overall .... 6.25 m (20 ft 6 in)
Fuselage max width .... 1.09 m (3 ft 7 in)
Height overall: tricycle .... 2.44 m (8 ft 0 in)
Tailplane span .... 2.54 m (8 ft 4 in)
Wheel track .... 2.44 m (8 ft 0 in)
Wheelbase: tricycle .... 2.44 m (8 ft 0 in)
Propeller diameter .... 1.83 m (6 ft 0 in)

Propeller ground clearance: tricycle ... 0.26 m (10¼ in)
Passenger doors (each): Height ... 0.76 m (2 ft 6 in)
Width ... 0.76 m (2 ft 6 in)

DIMENSIONS, INTERNAL:
Cabin: Length ... 1.07 m (3 ft 6 in)
Max width ... 1.09 m (3 ft 7 in)
Max height ... 1.07 m (3 ft 6 in)
Floor area ... 1.28 m^2 (13.8 sq ft)
Baggage hold volume ... 0.51 m^3 (18.0 cu ft)

AREAS:
Wings, gross ... 10.68 m^2 (115.0 sq ft)
Rudder ... 0.55 m^2 (5.90 sq ft)
Tailplane ... 1.89 m^2 (20.34 sq ft)
Elevator ... 1.01 m^2 (10.87 sq ft)

WEIGHTS AND LOADINGS (With 119 kW; 160 hp engine):
Weight empty ... 490 kg (1,080 lb)
Baggage capacity ... 91 kg (200 lb)
Max T-O and landing weight:
Griffin IV ... 785 kg (1,731 lb)
Max zero-fuel weight ... 682 kg (1,503 lb)
Max wing loading: ... 73.7 kg/m^2 (15.09 lb/sq ft)
Max power loading: ... 6.60 kg/kW (10.84 lb/hp)

PERFORMANCE (With 119 kW; 160 hp engine):
Never-exceed speed (VNE):
Griffin IV ... 144 kt (266 km/h; 165 mph)
Max operating speed (VMO):
Griffin IV ... 141 kt (261 km/h; 162 mph)
Max cruising speed at 75% power:
Griffin IV ... 130 kt (241 km/h; 150 mph)
Econ cruising speed ... 114 kt (211 km/h; 131 mph)
Stalling speed ... 40 kt (73 km/h; 45 mph)
Max rate of climb at S/L ... 396 m (1,300 ft)/min
Service ceiling ... 5,485 m (18,000 ft)
T-O run ... 107 m (350 ft)
Landing run ... 183 m (600 ft)
Range with max fuel ... 565 n miles (1,046 km; 650 miles)

VERIFIED

AWT

ADVANCED WING TECHNOLOGIES CORPORATION

No evidence is available of recent trading by AWT.

UPDATED

BELL

BELL HELICOPTER TEXTRON CANADA (DIVISION OF TEXTRON CANADA LTD)

12800 Rue de l'Avenir, Mirabel, Québec J7J 1R4
Tel: (+1 514) 437 34 00
Fax: (+1 514) 437 60 10
e-mail: bhtchr@bellhelicopter.textron.com
PRESIDENT: Paul Costanzo

Memorandum of Understanding to start helicopter industry in Canada signed 7 October 1983; 38,900 m^2 (418,725 sq ft) factory opened late 1985 and employs 1,700 people; US civil production of 206B JetRanger transferred to Canada in 1986, 206L LongRanger by early 1987; then 212 in 1988 and 412 in 1989. The 230 was introduced and certified in Canada in 1992, but has been replaced by the 430. Newest products are the 407 and 427. About half of each helicopter made in Canada (dynamic systems supplied by Bell Fort Worth). Total of 233 commercial helicopters delivered in 1997, followed by 197 in 1998, 146 in 1999, 145 in 2000, 122 in 2001 and 92 in 2002.

UPDATED

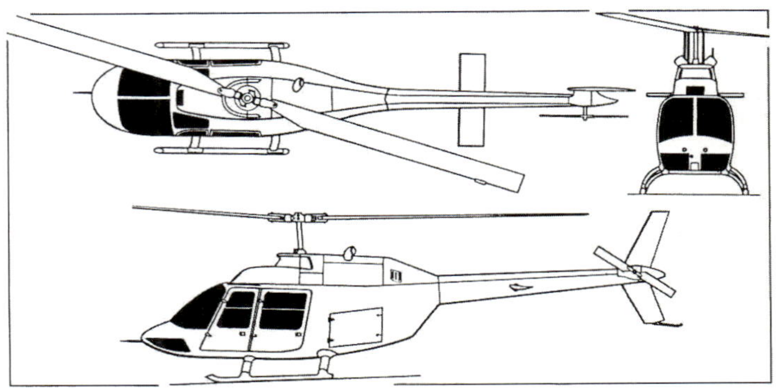

Bell 206B-3 JetRanger five-seat helicopter *(Paul Jackson)* 0507574

BELL 206B-3 JETRANGER III

US Army designation: TH-67 Creek

TYPE: Light utility helicopter.

PROGRAMME: Model 206 first flew (N73999) 8 December 1962 and unsuccessfully offered to US Army as four-seat OH-4 with Allison 250-C10 turboshaft and 1,134 kg (2,500 lb) MTOW; certified by FAA on 28 April 1964, but only five YOH-4As built. Redesigned as five-seat Model 206A Jet Ranger, with 236 kW (317 hp) Allison 250-C18 engine and 3,000 lb MTOW; first flown (N8560F) 10 January 1966; certified 20 October 1966. Model 206A-1, with longer main blades and re-shaped rear windows, certified 6 May 1969 and built as OH-58A/B Kiowa (of which several civilianised under former designation) and TH-57A Sea Ranger.

Model 206B Jet Ranger II certified 19 August 1971, with 236 kW (317 hp) Allison 250-C20 engine and 3,200 lb MTOW (some 206As retrofitted); Model 206B-1 certified 10 November 1971 but produced only for Australia; Model 206B-2 introduced 1976 and 206B-3 in mid-1977, latter with Allison (later Rolls Royce) 250-C20J power plant which originally certified for Bell 206L; also built for US Navy as TH-57B/C. Production transferred from Fort Worth, Texas, to Mirabel, Canada, in 1986, beginning with c/n 3959. Increase of 45 kg (100 lb) in payload under development in late 2003.

Bell 206B JetRanger III ***NEW***/0569620

CURRENT VERSIONS: **206B-3 JetRanger III:** Current civil production version.

Following description refers to JetRanger III.

TH-67 Creek (Bell designation TH-206): Selected March 1993 as US Army NTH (New Training Helicopter) choice to replace UH-1 at pilot training school, Fort Rucker, Alabama (223 Aviation Regiment at Cairns AAF and 212 Aviation Regiment at Lowe AHP). Instructor and pupil in front seats; plans abandoned for second pupil to be seated at the rear, observing flight instruments by closed-circuit TV screen mounted on back of right-hand front seat. Powered by Rolls-Royce (formerly Allison) 250-C20JN engine. First batch included nine cockpit procedures trainers outfitted by Frasca International; three configurations: VFR, IFR, and VFR with IFR provision.

TH-67 features also include dual controls, crashworthy seats, five-point seat restraints, heavy-duty battery, particle separator, bleed air heater, heavy-duty skid shoes and enlarged instrument panel. IFR version additionally possesses force trim system, auxiliary electrical system and is FAA certified for dual pilot operation. Also supplied to Taiwan Army, based at Gwe-Ren.

CUSTOMERS: TH-206 (TH-67 Creek) declared winner of US Army NTH competition March 1993; initial 102 ordered in IFR configuration; second batch of 35 VFR helicopters ordered February 1994; deliveries began 15 October 1993 with N67001 and N67014 (all TH-67s have civilian registrations); 45 aircraft (and six procedures trainers) delivered in time for first training course to open 5 May 1994; initial orders 137 TH-67s and nine cockpit procedures trainers; all delivered by late 1995. Additional orders placed on behalf of Taiwanese Army, which received 30 TH-67s in 1997–99. Deliveries in 1999 included six to the Bulgarian Air Force. Further 50 ordered for US Army for delivery between September 2001 and November 2003.

Production of Model 206 (not including four-blade Model 406s) up to January 2004 has comprised five YOH-4As, 713 Model 206As, 2,202 Model 206A-1/OH-58s, 74 Canadian CH-136/OH-58As, 56 Australian CA-32/OH-58As, 40 TH-57As, 50 OH-58Bs, 217 TH-67As, 1,177 Model 206Bs, 321 Model 206B-2s and 2,364 Model 206B-3s, while further 911 Jet Rangers built by Agusta in Italy. Total 8,130. Bell Canada delivered 28 206Bs in 1999; 14 in both 2000 and 2001; and 20 in 2002 (including 10 TH-67s).

COSTS: US Army initial NTH contract US$84.9 million.

DESIGN FEATURES: Typical light utility helicopter with skid landing gear, high-mounted tailboom and horizontal stabiliser. Two-blade teetering main rotor with preconed and underslung bearings; blades retained by grip, pitch change bearing and torsion-tension strap assembly; two-blade tail rotor; main rotor rpm 374 to 394.

FLYING CONTROLS: Hydraulic fully powered cyclic and collective controls and foot-powered tail rotor control; tailplane with highly cambered inverted aerofoil section and stall strip produces appropriate nose-up and nose-down attitude during climb and descent; optional autostabiliser, autopilot and IFR systems.

STRUCTURE: Conventional light alloy structure with two floor beams and bonded honeycomb sandwich floor; transmission mounted on two beams and deck joined to floor by three fuselage frames; main rotor blades have extruded aluminium D-section leading-edge with honeycomb core behind, covered by bonded skin; tail rotor blades have bonded skin without honeycomb core.

LANDING GEAR: Aluminium alloy tubular skids bolted to extruded cross-tubes. Tubular steel skid on ventral fin to protect tail rotor in tail-down landing. Special high skid gear (0.25 m; 10 in greater ground clearance) available for use in areas with high brush. Pontoons or stowed floats, capable of in-flight inflation, available as optional kits.

POWER PLANT: One Rolls-Royce 250-C20J turboshaft, rated at 313 kW (420 shp) for T-O; 276 kW (370 shp) max continuous. Optionally, one Rolls-Royce 250-C20R/4 rated at 336 kW (450 shp) for T-O. Transmission rating 236 kW (317 shp) for T-O; 201 kW (270 shp) continuous. Rupture-resistant fuel tank below and behind rear passenger seat, usable capacity 344 litres (91.0 US gallons; 75.75 Imp gallons). Refuelling point on starboard side of fuselage, aft of cabin. Oil capacity 5.68 litres (1.5 US gallons; 1.25 Imp gallons).

ACCOMMODATION: Two seats side by side in front and three-seat rear bench. Dual controls optional. Two forward-hinged doors on each side, made of formed aluminium alloy with transparent panels (bulged on rear pair). Baggage compartment aft of rear seats, capacity 113 kg (250 lb), with external door on port side.

SYSTEMS: Hydraulic system, pressure 41.5 bar (600 lb/sq in), for cyclic, collective and directional controls. Maximum flow rate 7.57 litres (2.0 US gallons; 1.65 Imp gallons)/min. Open reservoir. Electrical supply from 150 A starter/generator. One 28 V 17 Ah Ni/Cd battery; auxiliary 13 Ah battery optional. Optional ECS.

AVIONICS: *Comms:* VHF communications, intercom and speaker system.

Flight: VOR, ADF, DME and R/Nav optional.

EQUIPMENT: Standard equipment includes cabin fire extinguisher, first aid kit, door locks, night lighting, and dynamic flapping restraints. Optional items include clock, engine hour meter, turn and slip indicator, custom seating, internal stretcher kit, rescue hoist, cabin heater, camera access door, high-intensity night lights, engine fire detection system, and external cargo hook of 680 kg (1,500 lb) capacity.

DIMENSIONS, EXTERNAL:
- Main rotor diameter: 10.16 m (33 ft 4 in)
- Main rotor blade chord: 0.33 m (1 ft 1 in)
- Tail rotor diameter: 1.65 m (5 ft 5 in)
- Tail rotor blade chord: 0.12 m (4¾ in)
- Distance between rotor centres: 5.96 m (19 ft 6½ in)
- Length: overall, rotors turning: 11.82 m (38 ft 9½ in)
  - fuselage, incl tailskid: 9.50 m (31 ft 2 in)
- Height: over tailfin: 2.54 m (8 ft 4 in)
  - to top of rotor head:
    - standard skids: 2.89 m (9 ft 6 in)
    - high skids: 3.17 m (10 ft 4¾ in)
    - emergency floats: 3.20 m (10 ft 6 in)
- Stabiliser span: 1.97 m (6 ft 5¾ in)
- Width over landing gear:
  - standard skids: 1.95 m (6 ft 4¾ in)
  - high skids: 2.07 m (6 ft 9½ in)
  - emergency floats: 2.68 m (8 ft 9½ in)
- Forward cabin doors (each): Height: 1.02 m (3 ft 4 in)
  - Width: 0.61 m (2 ft 0 in)
- Rear cabin doors (each): Height: 1.02 m (3 ft 4 in)
  - Width: 0.91 m (3 ft 0 in)

DIMENSIONS, INTERNAL:
- Cabin: Length of seating area: 2.13 m (7 ft 0 in)
  - Max width: 1.27 m (4 ft 2 in)
  - Max height: 1.28 m (4 ft 3 in)
  - Volume: 1.1 m³ (40 cu ft)
- Baggage compartment:
  - Max width: 1.10 m (3 ft 7¼ in)
  - Max height: 0.55 m (1 ft 9½ in)
  - Max length: 0.94 m (3 ft 1¼ in)
  - Volume: approx 0.45 m³ (16 cu ft)

AREAS:
- Main rotor blades (each): 1.68 m² (18.05 sq ft)
- Tail rotor blades (each): 0.11 m² (1.18 sq ft)
- Main rotor disc: 81.07 m² (872.7 sq ft)
- Tail rotor disc: 2.14 m² (23.05 sq ft)
- Stabiliser: 0.90 m² (9.65 sq ft)

WEIGHTS AND LOADINGS:
- Weight empty: standard civil: 772 kg (1,702 lb)
  - TH-67: VFR: 852 kg (1,879 lb)
    - IFR: 911 kg (2,009 lb)
- Operating weight, TH-67: VFR: 1,369 kg (3,019 lb)
  - IFR: 1,428 kg (3,149 lb)
- Max payload: internal: 635 kg (1,400 lb)
  - external: 680 kg (1,500 lb)
- Max T-O weight: internal load: 1,451 kg (3,200 lb)
  - external load: 1,519 kg (3,350 lb)
- Max disc loading:
  - internal load: 17.9 kg/m² (3.67 lb/sq ft)
  - external load: 18.7 kg/m² (3.84 lb/sq ft)
- Transmission loading at max T-O weight and power:
  - internal load: 6.15 kg/kW (10.09 lb/shp)
  - external load: 6.43 kg/kW (10.57 lb/shp)

PERFORMANCE (at internal load max T-O weight, ISA):
- Never-exceed speed (VNE) at S/L: 122 kt (225 km/h; 140 mph)
- Max and econ cruising speed at S/L: 115 kt (214 km/h; 133 mph)
- Max rate of climb at S/L: 390 m (1,280 ft)/min
- Vertical rate of climb at S/L: 91 m (300 ft)/min
- Service ceiling: 4,115 m (13,500 ft)
- Hovering ceiling: IGE: 4,025 m (13,200 ft)
  - OGE: 1,615 m (5,300 ft)
- Range with max fuel, no reserves:
  - at S/L: 374 n miles (692 km; 430 miles)
  - at FL50: 395 n miles (732 km; 455 miles)
- Range, TH-67 VFR with normal 314 litres (83.0 US gallons; 69.0 Imp gallons) fuel: 327 n miles (605 km; 376 miles)
- Endurance: 4 h 30 min

OPERATIONAL NOISE LEVELS (FAR Pt 36):
- T-O: 88.7 EPNdB
- Flyover: 85.4 EPNdB
- Approach: 90.6 EPNdB

UPDATED

BELL 206L-4 LONGRANGER IV

TYPE: Light utility helicopter.

PROGRAMME: Stretched JetRanger. LongRanger announced 25 September 1973; first flight (N206L) 11 September 1974; initial versions were 206L (certified 22 September 1975, 1,814 kg; 4,000 lb MTOW) and 206L-1 LongRanger II (certified 17 May 1978, 1,837 kg; 4,050 lb (internal) MTOW); plus limited production of 206L-2 (derated 186 kW; 250 shp engine). Model 206L-3 LongRanger III certified 10 December 1981 with -C30P engine derated to 324 kW (435 shp) and 1,882 kg (4,150 lb) MTOW. Production having transferred to Canada in January 1987.

CURRENT VERSIONS: **LongRanger IV:** Announced March 1992 as new current standard model; transmission uprated to absorb 365 kW (490 shp) instead of 324 kW (435 shp) from same engine; gross weight raised from 1,882 kg (4,150 lb) to 2,018 kg (4,450 lb); certified 2 October 1992 and delivered from December that year.

TwinRanger: Twin-engined version; model **206LT;** 13 built on LongRanger IV line between 1993 and 1997. No longer available.

Gemini ST: Twin-engined rebuild of LongRanger III/IV developed by Tridair and Soloy in USA: see *Jane's Aircraft Upgrades*.

CUSTOMERS: Production up to late 2003 totalled 153 Model 206Ls, 637 Model 206L-1s, 17 Model 206L-2s, 612 Model 206L-3s (final 398 at Mirabel) and 281 Model 206L-4s (and LTs) or 1,700 in all. Total of 12 delivered in 1999, 27 in 2000, 10 in 2001 and 12 in 2002. Recent customers included Air Logistics of New Iberia, Louisiana, which ordered 14 in March 2003 for delivery between second quarter 2003 and 2005, and the Tulsa Police Air Support Unit of Oklahoma, which took delivery of one in May 2003.

DESIGN FEATURES: As JetRanger, but cabin length increased to make room for club seating and extra window; Bell Noda-Matic transmission to reduce vibration; vertical stabilisers added to horizontal tail surfaces. Improvements introduced on LongRanger II include new freewheel unit, modified shafting and increased-thrust tail rotor; main rotor rpm 394; rotor brake optional.

FLYING CONTROLS: As JetRanger, but with endplate fins on tailplane; single-pilot IFR with Rockwell Collins AP-107H autopilot; optional SFENA autopilot with stabilisation and holds for heading, height and approach.

STRUCTURE: As JetRanger.

LANDING GEAR: As JetRanger.

POWER PLANT: One 485 kW (650 shp) Rolls-Royce 250-C30P turboshaft (maximum continuous rating 415 kW; 557 shp). Transmission rated at 365 kW (490 shp) for take-off, with a continuous rating of 276 kW (370 shp); 340 kW (456 shp) transmission optional. Rupture-resistant fuel system, comprising three interconnected cells, total usable capacity 419 litres (111 US gallons; 92.0 Imp gallons).

ACCOMMODATION: Redesigned rear cabin, more spacious than JetRanger. With a crew of two, standard cabin layout accommodates five passengers in two canted rearward-facing seats and three forward-facing seats. Optional DeLuxe and Custom DeLuxe interiors with fabric, fabric/vinyl, all vinyl or fabric/leather seats. Port forward passenger seat has folding back to allow loading of a 2.44 × 0.91 × 0.30 m (8 × 3 × 1 ft) container, making possible carriage of such items as survey equipment, skis and other long components. Double doors on port side of cabin provide opening 1.55 m (5 ft 1 in) wide, for straight-in loading of stretcher patients or utility cargo; in ambulance or rescue role two stretcher patients and two ambulatory patients/attendants can be carried. Dual controls optional.

SYSTEMS: Hydraulic system; 28 V DC electrical power from 180 A starter/generator and 17 Ah battery. Engine bleed air ECS optional.

AVIONICS: *Comms:* Bendix/King suite includes dual nav/com and transponder.

Flight: ADF, DME and marker beacon receiver. Honeywell R/Nav, radio altimeter and encoding altimeter optional.

EQUIPMENT: Optional kits include emergency flotation gear, 907 kg (2,000 lb) cargo hook, rescue hoist, Nightsun searchlight (requires high skid gear).

DIMENSIONS, EXTERNAL:
- Main rotor diameter: 11.28 m (37 ft 0 in)
- Main rotor blade chord: 0.33 m (1 ft 1 in)
- Tail rotor diameter: 1.65 m (5 ft 5 in)
- Tail rotor blade chord: 0.135 m (5¼ in)

Bell 206L LongRanger IV ***NEW***/0569617

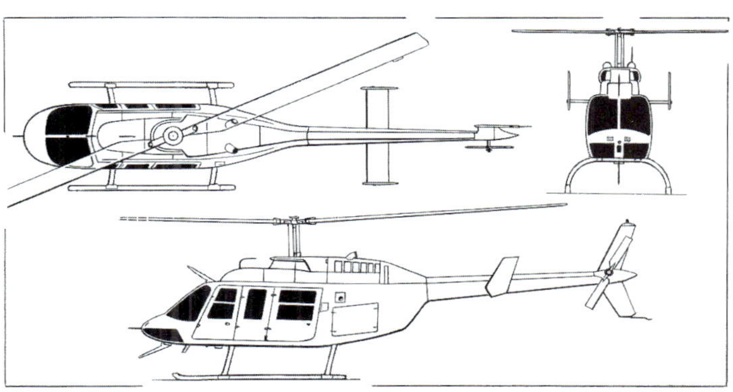

Bell 206L-4 LongRanger IV general purpose light helicopter *(Jane's/Dennis Punnett)* 0507575

Length: overall, rotors turning ... 13.02 m (42 ft 8½ in)
fuselage, incl tailskid ... 9.82 m (32 ft 2½ in)
Height: over tailfin ... 3.12 m (10 ft 2¾ in)
to top of rotor head ... 3.14 m (10 ft 3¾ in)
Fuselage: Max width ... 1.32 m (4 ft 4 in)
Stabiliser span ... 1.98 m (6 ft 6 in)
Width over skids ... 2.34 m (7 ft 8 in)
Forward cabin doors (each): Height ... 1.04 m (3 ft 5 in)
Max width ... 0.61 m (2 ft 0 in)
Centre cabin door (port): Height ... 1.04 m (3 ft 5 in)
Width ... 0.64 m (2 ft 1¼ in)
Rear cabin doors (each): Height ... 1.04 m (3 ft 5 in)
Width ... 0.91 m (3 ft 0 in)
Baggage door: Height ... 0.55 m (1 ft 9½ in)
Width ... 0.94 m (3 ft 1¼ in)

DIMENSIONS, INTERNAL:
Cabin: Length ... 2.74 m (9 ft 0 in)
Max width and height ... as JetRanger
Volume ... 2.4 m³ (83 cu ft)
Cabin cargo volume (all passenger seats removed) ... 2.83 m³ (100 cu ft)
Baggage compartment volume ... 0.45 m³ (16.0 cu ft)

AREAS:
Main rotor disc ... 99.89 m² (1,075.2 sq ft)
Tail rotor disc ... 2.13 m² (22.97 sq ft)

WEIGHTS AND LOADINGS:
Weight empty, standard ... 1,047 kg (2,309 lb)
Max external load ... 907 kg (2,000 lb)
Max T-O weight: normal ... 2,018 kg (4,450 lb)
external load ... 2,064 kg (4,550 lb)
Max disc loading: normal ... 20.7 kg/m² (4.23 lb/sq ft)
external load ... 19.3 kg/m² (3.95 lb/sq ft)
Transmission loading at max T-O weight and power:
internal load ... 5.52 kg/kW (9.08 lb/shp)
external load ... 5.65 kg/kW (9.29 lb/shp)

PERFORMANCE (at max normal T-O weight, ISA):
Never-exceed speed (VNE):
at S/L ... 130 kt (241 km/h; 150 mph)
at FL50 ... 133 kt (246 km/h; 153 mph)
Max cruising speed: at S/L ... 110 kt (204 km/h; 127 mph)
at FL50 ... 111 kt (206 km/h; 128 mph)
Max rate of climb at S/L ... 408 m (1,340 ft)/min
Service ceiling at max cruise power ... 3,050 m (10,000 ft)
Hovering ceiling: IGE ... 3,050 m (10,000 ft)
OGE ... 1,980 m (6,500 ft)
Range with max fuel, no reserves:
at S/L ... 324 n miles (600 km; 372 miles)
at FL50 ... 357 n miles (661 km; 411 miles)
Endurance ... 3 h 42 min

OPERATIONAL NOISE LEVELS (FAR Pt 36):
T-O ... 88.4 EPNdB
Flyover ... 85.2 EPNdB
Approach ... 90.7 EPNdB

UPDATED

BELL 407

TYPE: Light utility helicopter.

PROGRAMME: Design definition launched in 1993 as Bell Light Helicopter to supplement, and eventually replace, JetRanger and LongRanger; concept demonstrator Model 407 (N407LR) first flown 21 April 1994 (standard Bell 206L-3 modified with tailboom and dynamic system of military OH-58D, plus sidewall fairings to simulate broader fuselage); programme first revealed at Heli-Expo '95, Las Vegas, January 1995. Two prototype/preproduction 407s (C-GFOS and C-FORS) first flown on 29 June and 13 July 1995, respectively; first production airframe (C-FWQY/N407BT) flown 10 November 1995. Transport Canada certification 9 February 1996, with FAA certification same day, as extension of Bell 206 certificate; first customer delivery at Heli-Expo '96, Dallas, in February. MoU of June 1996 provided for licensed assembly and marketing by Dirgantara (formerly IPTN) of Indonesia.

Law Enforcement Demonstrator being promoted from 2002 features bubble windows, NightSun searchlights and FLIR, equipped by Edwards and Associates of Bristol, Tennessee.

Improvements under consideration in late 2003 included night vision goggles (NVG) qualification, and extended maintenance intervals aimed at reducing operating costs.

CUSTOMERS: Launch customers Petroleum Helicopters, Niagara Helicopters and Greenland Air. The 500th production Bell 407 was delivered in October 2001 to Pabst Air, Germany;

Bell 407 disembarking armed police ***NEW***/0569616

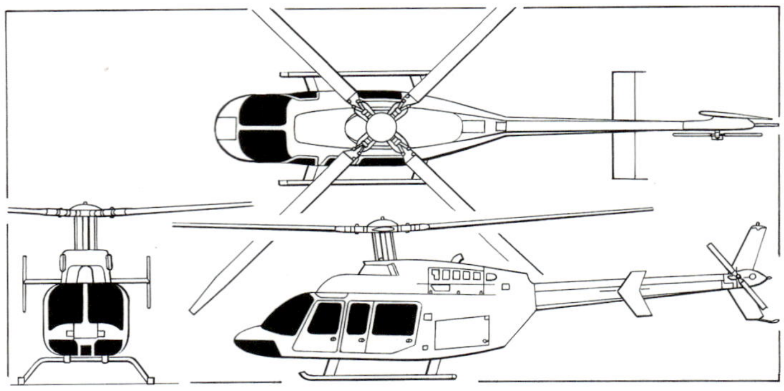

Bell 407 seven-seat light helicopter *(James Goulding)* 0103545

some 574 delivered to operators in 45 countries by late 2003; fleet time then totalled more than 745,000 hours. Law enforcement operators include Tampa Police Department, Florida; Palm Beach County Sheriff's Office; and the State Police Departments of Connecticut, Louisiana, New York and Virginia. Recent customers include Air Methods Corporation of Englewood, Colorado, which took delivery of two in June 2003, bringing its fleet total to 20; Anne Arundel County Police Department (one); The Flitner Ranch of Wyoming (one); L E Jones Production Company of Duncan, Oklahoma (one); Life Flight Eagle of Kansas City, Missouri (two); Med-Trans Corporation of Bismarck, North Dakota (five, supplementing existing fleet of 14 Bell 407s and two JetRangers); and Palm Beach County Sheriff's Office (one). Total of 62 delivered in 1999, 62 in 2000, 47 in 2001 and 33 in 2002.

COSTS: US$1.37 million (1999) flyaway. Average direct operating cost US$307 per hour (1999). Development cost estimated as US$50 million, of which US$9 million provided by Canadian government.

DESIGN FEATURES: As for JetRanger and LongRanger. Based on Bell 206L-4 LongRanger fuselage with cabin widened by 17.8 cm (7 in); 35 per cent larger cabin window area; Litton LCD active matrix cockpit displays; all-composites four-blade main rotor based on that of OH-58D, with soft-mounted pylon isolation system; single Rolls-Royce 250-C47 turboshaft; enlarged vertical stabilisers. 'Ring-fin' shrouded rotor tested in 1995 as possible future option. Rotor speed 413 rpm.

STRUCTURE: Generally as for LongRanger, but with carbon fibre tailboom and flush-fitting cabin door windows and skylights.

LANDING GEAR: LongRanger unit modified to match new rotor's ground-resonance characteristics, and with rear cross-tube pivoted in centre between outboard damping pads. High skids, low skid fairings and pop-out floats, optional.

POWER PLANT: One Rolls-Royce 250-C47B turboshaft rated at 606 kW (813 shp) for T-O, 523 kW (701 shp) maximum continuous; transmission rating 503 kW (674 shp) for T-O, 470 kW (630 shp) continuous operation. Single-channel FADEC standard. Standard usable fuel capacity 484 litres (128 US gallons; 106.6 Imp gallons); optional auxiliary fuel tank in aft baggage compartment, usable capacity 72 litres (19 US gallons; 15.8 Imp gallons); oil capacity 5.7 litres (1.5 US gallons; 1.25 Imp gallons). 'Quiet Cruise' system, in development in early 1997, reduces flyover noise level by using FADEC to reduce rotor rpm to 90 to 93 per cent in the cruise, with accompanying reduction in VNE to 110 kt (204 km/h; 126 mph).

ACCOMMODATION: With a crew of two, standard cabin layout accommodates five passengers in two rearward-facing seats with centre armrest/console, and three forward-facing seats, all fabric-covered. Optional utility cabin has vinyl-covered seats; corporate interior features two extra-wide forward-facing seats in the outboard positions, an occasional-use seat in the centre and additional soundproofing.

SYSTEMS: Electrical system, includes 28 V 17 Ah Ni/Cd battery and 180 A starter generator

AVIONICS: Bendix/King suite options to customer's choice include KX 155 and KX 165 nav/com, KY 196A VHF, KT 70 and KY 76A transponder, KR 87 ADF, KLN 89B GPS, KCS 55A compass system and ELT.

DIMENSIONS, EXTERNAL:
Main rotor diameter ... 10.67 m (35 ft 0 in)
Main rotor blade chord ... 0.27 m (10¾ in)
Tail rotor diameter ... 1.65 m (5 ft 5 in)
Tail rotor blade chord ... 0.16 m (6½ in)
Length: overall, rotors turning ... 12.74 m (41 ft 9½ in)
rotor in X configuration ... 11.16 m (36 ft 7¼ in)[1]
fuselage ... 10.58 m (34 ft 8½ in)
Stabiliser span over endplate fins ... 2.22 m (7 ft 3½ in)
Height: over tailfin ... 3.10 m (10 ft 2 in)
overall: low skids ... 3.11 m (10 ft 2½ in)
high skids ... 3.32 m (10 ft 10¾ in)
Ground clearance: ... 3.10 m (10 ft 2 in)
low skids ... 0.41 m (1 ft 4 in)
high skids ... 0.62 m (2 ft 0½ in)
Width: over skids ... 2.29 m (7 ft 6 in)
rotor in X configuration ... 7.65 m (25 ft 1¼ in)
Rear cabin door:
Width: port ... 1.55 m (5 ft 1 in)s
starboard ... 0.91 m (3 ft 0 in)
Height to sill (standard skids) ... 0.51 m (1 ft 8 in)

[1] *Blades at 45° to fuselage centreline*

DIMENSIONS, INTERNAL:
Cabin: Max width ... 1.37 m (4 ft 6 in)
Max height ... 1.00 m (3 ft 3¼ in)
Baggage compartment volume ... 0.45 m³ (16.0 cu ft)

AREAS:
Main rotor disc ... 89.38 m² (962.1 sq ft)
Tail rotor disc ... 2.08 m² (22.34 sq ft)

WEIGHTS AND LOADINGS:
Weight empty, equipped ... 1,203 kg (2,653 lb)
Max payload ... 1,178 kg (2,597 lb)
Max hook capacity ... 1,200 kg (2,646 lb)
Max T-O weight: internal load:
standard ... 2,268 kg (5,000 lb)
optional ... 2,381 kg (5,250 lb)
external load ... 2,721 kg (6,000 lb)
Max disc loading ... 27.9 kg/m² (5.72 lb/sq ft)

Max disc loading: internal load:
standard .... 25.4 kg/m² (5.20 lb/sq ft)
optional .... 26.6 kg/m² (5.46 lb/sq ft)
external load .... 30.4 kg/m² (6.24 lb/sq ft)
Transmission loading at max T-O weight and power:
internal load: standard .... 4.52 kg/kW (7.42 lb/shp)
optional .... 4.74 kg/kW (7.79 lb/shp)
external load .... 5.42 kg/kW (8.90 lb/shp)
PERFORMANCE (at internal load MTOW, ISA):
Never-exceed speed (VNE) .... 140 kt (259 km/h; 161 mph)
Max cruising speed:
at S/L .... 133 kt (246 km/h; 153 mph)
at 1,220 m (4,000 ft) .... 135 kt (250 km/h; 155 mph)
Long-range cruising speed:
at S/L .... 112 kt (207 km/h; 129 mph)
at 1,220 m (4,000 ft) .... 115 kt (213 km/h; 132 mph)
Max certified T-O height .... 5,180 m (17,000 ft)
Max certified altitude .... 6,100 m (20,000 ft)
Hovering ceiling: IGE .... 3,720 m (12,200 ft)
OGE .... 3,170 m (10,400 ft)
Max range .... 330 n miles (611 km; 379 miles)
Endurance .... 3 h 42 min

UPDATED

BELL 427

TYPE: Light utility helicopter.

PROGRAMME: Launched as New Light Twin (NLT) in February 1996 on signature of collaborative agreement with Samsung Aerospace Industries of South Korea. Prototype assembly began early 1997; first flight (C-GBLL) 11 December 1997; second prototype (C-FCSS) completed February 1998; two prototypes undertook flight test programme, gaining Transport Canada certification on 19 November 1999; FAA VFR certification achieved 24 January 2000, followed by Dual Pilot IFR (DPIFR) Category A certification on 24 May 2000. JAA certification was expected in early 2002. First production aircraft (C-GDEJ) flown June 1998; compared with prototypes, production 427 has longer exhausts and revised upper surface contours. Alternative FAR Pt 29 version, with increased T-O weight, will also be available.

CURRENT VERSIONS: **Bell 427:** *As described.*

SB427: South Korean-assembled version; see KAI entry in that national section.

Super Kiowa: Proposed armed scout offered by Bell.

CUSTOMERS: Orders for 22 placed during the mockup's first public display at Farnborough Air Show 1996; 85 on order by May 2000 from 50 customers, including five sold by Samsung to CitiAir in South Korea. Deliveries began 2000, but only 38 built by early 2003 and none further registered by December 2003. Total of five delivered during 2000, 15 in 2001 and five in 2002. Recent customers include Khalifa Airways of Algeria, which took delivery of one in VIP configuration in February 2002. MoU signed in 1999 with Elbit Systems Inc for joint pre-design phase of military light helicopter using Bell 427 as baseline platform.

DESIGN FEATURES: Similar in appearance to Bell 407, with cabin stretch of 33 cm (13 in), but is all-new design, incorporating twin-engine safety margins. Flight dynamics based on four-blade rotor system of Bell OH-58D Kiowa (see US section), allied to tail rotor of Bell 407; folding main blades. Main rotor rpm 395; tail rotor rpm 2,375. Purpose-designed 'flat pack' main transmission, with direct input from both engines, has only four gear meshes to simplify design and operation. Transmission attached to airframe by four liquid-inertia vibration eliminators. First Bell helicopter designed entirely with use of computer (Dassault CATIA programme).

FLYING CONTROLS: Sextant AFDS 95-1 AFCS, with two- to four-axis autopilot computer, flight director computer and associated equipment, to be certified by 2000.

STRUCTURE: Generally as for Bell 206, but extensive use of carbon/epoxy composites reduces airframe parts count by some 33 per cent. Cabin floor and roof are flat panels for ease of manufacture; minimal use of curved panels elsewhere. Composites main and tail rotors; main blades have nickel-plated stainless steel leading-edges. Soft-in-plane hub of main rotor employs a composites flexbeam yoke and elastomeric joints, eliminating lubrication and maintenance requirements. Brake and main rotor blade folding optional. Composites cabins and rolled aluminium tailbooms built by Samsung; assembly in Canada, except for sales to Korea and China; Hexcel honeycomb as stiffener.

LANDING GEAR: Twin skids with dynamically tuned cross tubes to reduce ground resonance. Low skids standard; optional high skids and emergency floats.

POWER PLANT: Two Pratt & Whitney Canada PW207D turboshafts with FADEC, each rated at 529 kW (710 shp) for T-O (5 minutes) or 466 kW (625 shp) maximum continuous; OEI ratings 611 kW (820 shp) for 30 seconds, 582 kW (780 shp) for 2 minutes, 559 kW (750 shp) for 30 minutes or 529 kW (710 shp) maximum continuous. Twin-engine transmission rating, T-O and maximum continuous, 597 kW (800 shp). OEI transmission rating, 485 kW (650 shp) for 30 seconds; 451 kW (605 shp) for 2 minutes; 343 kW (460 shp) maximum continuous.

Fuel contained in three crash-resistant tanks; two forward, one aft, total usable capacity 770 litres (203.5 US gallons; 169 Imp gallons). One forward fuel tank can be removed in EMS configurations to provide additional stretcher space in cabin or to permit stretcher to extend into port side of cockpit. Oil capacity (total, both engines) 10.2 litres (2.69 US gallons; 2.24 Imp gallons).

Bell 427 (P&WC PW207D turboshafts) ***NEW***/0569615

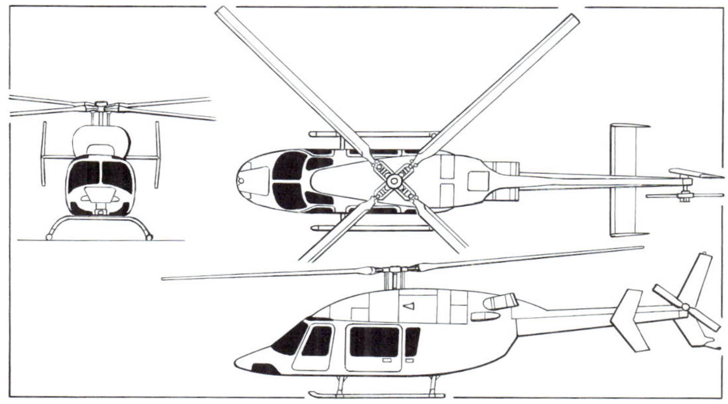

Bell 427 light utility helicopter *(James Goulding)* ***NEW***/0552086

ACCOMMODATION: Standard accommodation is for two crew in cockpit, on 20 *g* energy-attenuating seats, and six passengers in cabin on two rows of three seats in club configuration (all-forward-facing seats optional); all seats equipped with inertia-reel shoulder harnesses. Alternative configurations include corporate club-four seating with refreshment/entertainment console between each pair of seats, club-five with console between rearmost seats, or club-six, without consoles. Optional EMS interiors provide for carriage of one or two stretchers with up to two medical attendants, affording either full patient or head-only in-flight access, with single- or two-person crew. In cargo configuration, with all passenger seats removed, an optional removable flat cargo floor can be installed, equipped with integral tie-downs. Two forward-hinged doors each side; cabin doors, both sides, are forward-hinged, but port unit can be replaced by optional rearward-sliding door for cargo handling. External door on starboard side to rear baggage hold.

SYSTEMS: Hydraulic system, operating pressure 86 bar (1,250 lb/sq in), provides boost power for main and tail rotor controls; 28 V DC electrical power from 17 Ah Ni/Cd battery and two engine-mounted 17 Ah starter/generators; 28 Ah battery and 200 A starter/generator optional. Air conditioning optional.

AVIONICS: Rogerson-Kratos NeoAV two-screen LCD integrated instrument display system (IIDS) for monitoring engine instruments, fuel quantity, hydraulic and electrical systems and weight and balance functions. Rogerson Kratos NeoAV EFIS optional, featuring GPS interface, area navigation map, non-precision approach capability and weather radar display. Bendix/King nav/com avionics suite to customer's choice.

EQUIPMENT: Optional kits include engine air particle separator, cargo hook, cargo floor, sliding cabin door, external rescue hoist, NightSun searchlight, EMS installation and wire strike protection system.

DIMENSIONS, EXTERNAL:
Main rotor diameter .... 11.28 m (37 ft 0 in)
Main rotor blade chord .... 0.34 m (1 ft 1¼ in)
Tail rotor diameter .... 1.73 m (5 ft 8 in)
Tail rotor blade chord .... 0.18 m (7¼ in)
Length: overall, rotors turning .... 12.99 m (42 ft 7¼ in)
overall, rotor in X configuration .... 11.55 m (37 ft 10¾ in)[1]
fuselage, incl tailskid .... 10.94 m (35 ft 10¾ in)
Height over tailfin .... 3.49 m (11 ft 5¼ in)
Ground clearance: low skids .... 0.41 m (1 ft 4¼ in)
high skids .... 0.67 m (2 ft 2½ in)
Width: overall, rotor in X configuration .... 8.24 m (27 ft 0½ in)[1]
over skids .... 2.29 m (7 ft 6 in)
over tailfins .... 2.89 m (9 ft 5¾ in)
Cabin door: Height .... 1.07 m (3 ft 6 in)
Width .... 1.24 m (4 ft 0¾ in)

[1] *Blades at 45° to fuselage centreline*

DIMENSIONS, INTERNAL:
Cabin: Max height .... 1.30 m (4 ft 3 in)
Baggage hold volume .... 0.76 m³ (27.0 cu ft)

AREAS:
Main rotor disc .... 99.89 m² (1,075.2 sq ft)
Tail rotor disc .... 2.34 m² (25.2 sq ft)

WEIGHTS AND LOADINGS (provisional):
Weight empty .... 1,751 kg (3,860 lb)
Max payload .... 1,220 kg (2,690 lb)
Max baggage weight .... 113 kg (250 lb)
Max T-O weight: internal load: .... 2,880 kg (6,350 lb)
optional and external load .... 2,971 kg (6,550 lb)
Cargo hook capacity .... 1,361 kg (3,000 lb)
Max cargo floor loading .... 366.2 kg/m² (75 lb/sq ft)
Max disc loading .... 29.74 kg/m² (6.09 lb/sq ft)
Transmission loading at max T-O weight and power:
internal load .... 4.56 kg/kW (7.50 lb/shp)
external load .... 4.98 kg/kW (8.19 lb/shp)

PERFORMANCE:
Never-exceed speed (VNE) .... 140 kt (259 km/h; 161 mph)
Max cruising speed at S/L .... 136 kt (252 km/h; 157 mph)
Econ cruising speed at S/L .... 133 kt (246 km/h; 153 mph)
Service ceiling:
max continuous power .... 3,050 m (10,000 ft)
OEI (30 min) .... 1,220 m (4,600 ft)
Hovering ceiling: IGE .... 2,745 m (9,000 ft)
OGE .... 533 m (1,750 ft)
Range with max fuel, econ cruising speed, no reserves .... 387 n miles (716 km; 445 miles)
Endurance .... 4 h 0 min

UPDATED

BELL 412

Canadian Forces designation: CH-146 Griffon
UK armed forces designations: Griffin HT. Mk 1 and HAR. Mk 2

TYPE: Multirole medium helicopter.

PROGRAMME: Original 412 announced 8 September 1978 (see earlier editions of *Jane's*); FAR Pt 29 VFR approval received 9 January 1981, IFR 13 February 1981; 213 built in USA; production (SP version) transferred to Canada February 1989; first delivery (civil) 18 January 1981. Production licences obtained by Dirgantara of Indonesia and Agusta of Italy (which see).

Increase in payload of 91 kg (200 lb) in development in late 2003, through use of composites in engine access doors, chemical milling of landing gear skids and removal of mountings and avionics boxes for redundant equipment.

CURRENT VERSIONS: **412SP:** Special Performance version with increased maximum T-O weight, new seating options and 55 per cent greater standard fuel capacity. Superseded by 412HP early 1991. Details in 1991–92 *Jane's*.

Military 412: Announced by Bell June 1986; fitted with Lucas Aerospace chin turret and Honeywell Head Tracker helmet sight similar to that in AH-1S; turret carries 875 rounds, weighs 188 kg (414 lb) and can be removed in under 30 minutes; firing arcs ±110° in azimuth, +15° and –45° in elevation; other armament includes twin dual FN Herstal 7.62 mm gun pods, single FN Herstal 0.50 in pod, pods of seven or nineteen 2.75 in rockets, M240E1 pintle-mounted door guns, FN Herstal four-round 70 mm rocket launcher and a 0.50 in gun or two Giat M621 20 mm cannon pods.

412HP: Improved transmission giving better OGE hover; FAR Pt 29 certification 5 February 1991, first delivery (c/n 36020) later that month.

412EP (Enhanced Performance): PT6T-3D engine, dual digital automatic flight control system (DDAFCS), three-axis in basic aircraft but customer option for four-axis and EFIS. Category A certification was imminent in late 1998. Also customer option for SAR fit. Now standard current model.

Detailed description applies to Bell 412EP.

412CF (CH-146) Griffon: Canadian Forces C$700 million contract for 100 CH-146s (modified Bell 412EP) placed in 1992. Duties include armed support, troop/cargo transport, medevac, ASW, SAR and patrol; first flight (146000) 30 April 1994; deliveries began 14 October 1994; completed early 1998. Generally as commercial Bell 412EP except for avionics and mission equipment. Empty weight 3,402 kg (7,500 lb); maximum weight as civil version.

412EP Sentinel: First of two modified in 1998 by Heli-Dyne Systems with quick-change ASV and ASW mission packages; intended for Ecuadorean Navy, but order cancelled and aircraft became demonstrator. ASV equipment comprises Honeywell RDR-1500B chin radar, Hughes Starburst searchlight, radar warning receiver, Wescam sensor turret and possibly Penguin Mk 2 Mod 7 ASMs; ASW fit is L3 Ocean Systems AN/AQS-18A dipping sonar and Raytheon Mk 46 torpedo.

412SA: First three of 16 ordered by Royal Saudi Air Force built in 2001 for manufacturer's trials. Equipment standard not disclosed, but sufficiently different from 412EP to warrant separate c/n sequence beginning 33501. Production continued in 2002, although only eight confirmed built by late 2003.

412 Plus: Projected improved version under study in 1999 with MTOW increased to 5,647 kg (12,450 lb), uprated PT6C engines, new dynamic components and Rogerson-Kratos avionics. Development terminated in early 2001.

NBell-412: Indonesia's Dirgantara (which see) has licence to produce up to 100 Model 412SPs.

CUSTOMERS: Some 651 Bell 412s of all versions built in North America by early 2004, including 26 delivered in 1999 and 24 in 2000; 22 in 2001 (including five to El Salvador) and 25 in 2002 (including five to Saudi Arabia).

Military deliveries include Venezuelan Air Force (two), Botswana Defence Force (three), Public Security Flying Wing of Bahrain Defence Force (two), Sri Lankan armed forces (four), Nigerian Police Air Wing (two), Mexican government (two VIP transports), South Korean Coast Guard (one), Honduras (10), Royal Norwegian Air Force (19, of which 18 assembled by Helikopter Service, Stavanger, to replace UH-1Bs of 339 Squadron at Bardufoss and 720 Squadron at Rygge). Three 412EPs delivered to Slovenian Territorial Forces in 1995, for border patrol and rescue duties; four ordered by Philippine Air Force late in 1996, comprising two for VVIP transport and two SAR; first of nine 412EPs entered service in April 1997 as HT. Mk 1s with civilian-operated Defence Helicopter Flying School at RAF Shawbury, UK, within which they constitute No. 60 (Reserve) Squadron, RAF; two more ordered in May 2002, together with four HAR. Mk 2s, latter replacing Wessex SAR helicopters of No. 84 Squadron at Akrotiri, Cyprus, from April 2003. Four in SAR/utility fit delivered to Venezuelan Navy, 1999. Recent customers include the National Defence Secretariat of Mexico, which took delivery of four in mid-2002; Venezuelan Navy, which ordered four in SAR configuration in early 2002; Khalifa Airways of Algeria, which took delivery of one 412EP in June 2002, and the New York City Police Aviation Department, which took delivery of one 412EP in September 2003.

Royal Saudi Air Force Bell 412SA (Peter J Cooper/Falcon Aviation) ***NEW***/0552085

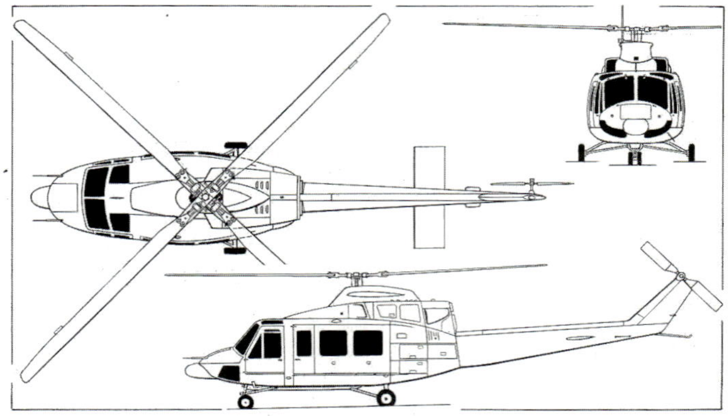

Bell 412EP (P&WC PT6T turboshaft) *(Jane's/Dennis Punnett)* 0507122

New York Police Department Bell 412EP ***NEW***/0569614

Bell 412EP used by the fire rescue service of Dade County, Miami ***NEW***/0569613

COSTS: Bell 412EP, VFR-equipped US$4.895 million (1999); Bell 412EP, IFR-equipped US$5.12 million (1999). Average direct operating cost US$746 per hour (1997).

DESIGN FEATURES: Four-blade main rotor with blades retained within central metal star fitting by single elastomeric bearings; shorter rotor mast than 212; blades can be folded; rotor brake standard; two-blade tail rotor; main rotor rpm 314.

FLYING CONTROLS: Fully powered hydraulic controls; gyroscopic stabiliser bar above main rotor; automatic tailplane incidence control.

STRUCTURE: Generally of conventional light metal. Main rotor blade spar unidirectional glass fibre with 45° wound torque casing of glass fibre cloth; Nomex rear section core with trailing-edge of unidirectional glass fibre; leading-edge protected by titanium abrasion strip and replaceable stainless steel cap at tip; lightning protection mesh embedded; provision for electric de-icing heater elements; main rotor hub of steel and light alloy; all-metal tail rotor.

LANDING GEAR: High skid, emergency pop-out float or non-retractable tricycle gear optional. Spats optional for last-named.

POWER PLANT: Pratt & Whitney Canada PT6T-3D Turbo Twin-Pac, rated at 1,342 kW (1,800 shp) for T-O and 1,193 kW (1,600 shp) maximum continuous. OEI ratings 850 kW (1,140 shp) for 2½ minutes, or 723 kW (970 shp) for 30 minutes. Transmission rating 1,022 kW (1,370 shp) for T-O, 828 kW (1,110 shp) maximum continuous; OEI rating 850 kW (1,140 shp). Optional 30 kW (40 shp) for accessory drives from main gearbox.

Seven interconnected rupture-resistant fuel cells, with automatic shutoff valves (breakaway fittings), have a combined usable capacity of 1,249 litres (330 US gallons; 275 Imp gallons). Two 76 or 310.5 litre (20.0 or 82.0 US gallon; 16.7 or 68.3 Imp gallon) auxiliary fuel tanks, in any combination, can increase maximum total capacity to 1,870 litres (494 US gallons; 411.6 Imp gallons). Single-point refuelling on starboard side of cabin.

ACCOMMODATION: Pilot and up to 14 passengers: one in front port seat and 13 in cabin. Dual controls optional. Accommodation heated and ventilated.

SYSTEMS: Dual hydraulic systems, pressure 69 bar (1,000 lb/ sq in) each. 28 V DC electrical system supplied by two completely independent 450 VA inverters. 40 Ah Ni/Cd battery.

AVIONICS: *Comms:* Optional IFR avionics include dual Bendix/King Gold Crown III.

Radar: Weather radar optional.

Flight: Dual KNR 660A VOR/LOC/RMI receivers, KDF 800 ADF, KMD 700A DME, KXP 750A transponder and KGM 690 marker beacon/glideslope receiver. Optional Honeywell AFCS.

EQUIPMENT: Optional equipment includes cargo sling and rescue hoist.

DIMENSIONS, EXTERNAL:

Main rotor diameter	14.02 m (46 ft 0 in)
Main rotor blade chord: at root	0.40 m (1 ft 4 in)
at tip	0.22 m (8½ in)
Tail rotor diameter	2.62 m (8 ft 7 in)

Tail rotor blade chord .... 0.29 m (11½ in)
Length: overall, rotors turning .... 17.12 m (56 ft 2 in)
fuselage, excl rotors .... 12.70 m (41 ft 8 in)
Height: to top of rotor head .... 3.48 m (11 ft 5 in)
overall, tail rotor turning .... 4.57 m (15 ft 0 in)
Stabiliser: span .... 2.87 m (9 ft 5 in)
chord .... 0.79 m (2 ft 7 in)
Width over skids .... 2.84 m (9 ft 4 in)
Rear sliding doors (each): Height .... 1.24 m (4 ft 1 in)
Width .... 1.88 m (6 ft 2 in)
Height to sill .... 0.76 m (2 ft 6 in)
Baggage compartment door: Height .... 0.53 m (1 ft 9 in)
Width .... 1.71 m (2 ft 4 in)
Emergency exits (centre cabin windows, each):
Height .... 0.76 m (2 ft 6 in)
Width .... 0.97 m (3 ft 2 in)

DIMENSIONS, INTERNAL:
Baggage compartment volume .... 0.79 m^3 (28.0 cu ft)

AREAS:
Main rotor disc .... 154.40 m^2 (1,661.9 sq ft)
Tail rotor disc .... 5.38 m^2 (57.86 sq ft)

WEIGHTS AND LOADINGS:
Weight empty, standard equipped .... 3,079 kg (6,789 lb)
Max external hook load .... 2,041 kg (4,500 lb)
Max T-O and landing weight, internal or external load .... 5,397 kg (11,900 lb)
Max disc loading .... 35.0 kg/m^2 (7.16 lb/sq ft)
Transmission loading at max T-O weight and power .... 5.29 kg/kW (8.69 lb/shp)

PERFORMANCE:
Never-exceed speed (VNE) .... 140 kt (259 km/h; 161 mph)
Max cruising speed: at S/L .... 122 kt (226 km/h; 140 mph)
at FL50 .... 124 kt (230 km/h; 143 mph)
Long-range cruising speed at FL50 .... 130 kt (241 km/h; 150 mph)
Service ceiling, OEI, 30 min power rating .... 1,920 m (6,300 ft)
Hovering ceiling: IGE .... 3,110 m (10,200 ft)
OGE .... 1,585 m (5,200 ft)
Range at FL50, long-range cruising speed, standard fuel, no reserves .... 402 n miles (744 km; 462 miles)
Endurance .... 3 h 42 min

OPERATIONAL NOISE LEVELS (FAR Pt 36) (412 EP):
T-O .... 92.8 EPNdB
Flyover .... 93.4 EPNdB
Approach .... 95.6 EPNdB

UPDATED

BELL 430

TYPE: Light utility helicopter.

PROGRAMME: Preliminary design 1991; four-blade rotor, higher-powered and stretched variant of Bell 230; programme launched February 1992; two prototypes, modified from Bell 230 airframes; first prototype (C-GBLL; wheel-equipped) flown 25 October 1994; second prototype (C-GEXP; skid-equipped, with complete avionics suite) flown 19 December 1994; first flight of production 430 (C-GRND) in 1995; deliveries began 25 June 1996 after Canadian type approval on 23 February. MoU of June 1996 with Dirgantara (formerly IPTN) for licensed assembly and marketing in Indonesia. Under an agreement signed in September 2003, Hafei Aviation Industry Company Ltd of China will become sole supplier of Bell 430 airframes to the manufacturer.

Second production aircraft, N4300 circumnavigated the world in a record time of 17 days 6 hours 14 minutes, landing back at Fairoaks, UK, on 3 September 1996.

CUSTOMERS: First delivery 25 June 1996, when sixth production aircraft (N6282X) handed over to IPTN (now Dirgantara) of Indonesia in eight-seat executive configuration. Thirteen delivered in 1996; followed by eight, 15 and 18 in 1997–99, 11 in 2000, 14 in 2001 and seven in 2002. Recent customers include Life flight Network of Portland, Oregon, which took delivery of one in August 2003, and the University of Tennessee's Medical Center, which took delivery of two in 2003, replacing Bell 412s in its Lifestar EMS operation. Total 98 built by late 2003.

COSTS: Programme US$ 8 million, 35 per cent financed by Canadian Defence Industry Productivity Program (DIPP) and repayable as royalty on each sale.

DESIGN FEATURES: Optimised for high cruising speed with (retractable) wheel landing gear, although traditional skids optional; inclined towards executive transport market. Bell 230 fuselage lengthened by 0.46 m (1 ft 6 in) plug; Bell 680 all-composites four-blade bearingless, hingeless main rotor; approximately 10 per cent power increase over Bell 230; uprated transmission, and optional EFIS. Short-span sponson each side of fuselage houses mainwheel units and fuel tanks, and serves as work platform.

Bell 430 used for news-gathering in Japan ***NEW***/0569637

Instrument panel of Bell 430 *(Paul Jackson)* ***NEW***/0569636

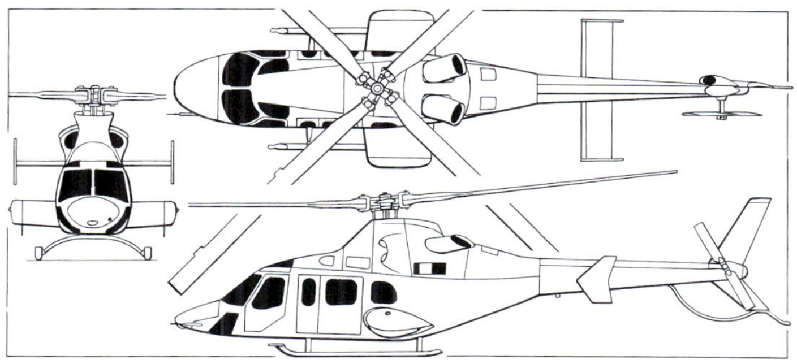

Skid version of Bell 430 nine-seat helicopter (two Rolls-Royce 250 turboshafts) *(James Goulding)* 0103543

FLYING CONTROLS: Fully powered hydraulic, with elastomeric pitch change and flapping bearings; fixed tailplane with leading-edge slats and endplate fins; strakes under sponsons; single-pilot IFR system without auto-stabilisation.

STRUCTURE: Semi-monocoque fuselage of light alloy, with limited use of light alloy honeycomb panels. Fail-safe structure in critical areas. One-piece nosecone tilts forward and down for access to avionics and equipment bay. Short span cantilever sponson set low on each side of fuselage, serving as main landing gear housings, fuel tanks and work platforms. Section NACA 0035. Dihedral 3° 12'. Incidence 5°. Sweepback at quarter-chord 3° 30'.

Fixed vertical fin in sweptback upper and lower sections. Tailplane, with slotted leading-edge and endplate fins, mounted midway along rear fuselage. Small skid below ventral fin for protection in tail-down landing. Four-blade main rotor with stainless steel spars and leading-edges, Nomex honeycomb trailing-edge with glass fibre skin, and glass fibre safety straps; tail rotor blades stainless steel. Rotors shaft-driven through gearbox with two spiral bevel reductions and one planetary reduction. Main blade and hub life, 10,000 hours.

LANDING GEAR: Tubular skid type on Utility version. Executive version has hydraulically retractable tricycle gear, single mainwheels retracting forward into sponsons; forward-retracting nosewheel fully castoring and self-centring; hydraulic disc brakes on main units. Mainwheel tyre size 18×5.5, nosewheel 5.00–5. Emergency floats optional.

POWER PLANT: Two Rolls-Royce 250-C40B turboshafts, each rated at 603 kW (808 shp) for T-O and 518 kW (695 shp) maximum continuous. OEI ratings 701 kW (940 shp) for 30 seconds, 656 kW (880 shp) for 2 minutes, 623 kW (835 shp) for 30 minutes and 602 kW (808 shp) continuous. Chandler Evans FADEC. Transmission rating 779 kW (1,045 shp) for 5 minutes for T-O, 737 kW (989 shp) maximum continuous. Power train TBO, 5,000 hours. Usable fuel capacity 935 litres (247 US gallons; 206 Imp gallons) in skid version, 710 litres (187.5 US gallons; 156 Imp gallons) in wheeled version; provision in both versions for 182 litre (48 US gallon; 40 Imp gallon) auxiliary tank. Fuel system is rupture resistant, with self-sealing breakaway fittings.

ACCOMMODATION: Standard layout has forward-facing seats for nine persons (two-two-two-three) including pilot. Options include 10-seat layout (two-two-three-three); eight-seat executive (rear six in club layout), six-seat executive (rear four in club layout with console between each pair); and five- and four-seat executive with one or two refreshment cabinets; seat pitches vary between 86 cm (34 in) and 91 cm (36 in). Pilots on crashworthy (energy attenuating) seats, which are optional for passengers. Customised emergency medical service (EMS) versions also available, configured for pilot-only operation plus one or two pivotable stretchers and four or three medical attendants/sitting casualties respectively. Two forward-opening doors each side; EMS version has optional stretcher door between forward and rear doors on port side. Entire interior ram air ventilated and soundproofed. Dual controls optional.

SYSTEMS: Dual hydraulic systems (dual for main rotor collective and cyclic, single for tail rotor). Dual 28 V DC electrical systems, powered by two 30 V 200 A engine-mounted starter/generators (derated to 180 A) and a 24 V 28 Ah Ni/Cd battery.

AVIONICS: *Comms:* Bendix/King Gold Crown III.

Flight: Honeywell KFC 500 AFCS. GPS optional.

Instrumentation: Rogerson-Kratos LCD integrated instrument display system (IIDS) comprising two active matrix LCDs displaying engine and system parameters; optional Rogerson-Kratos EFIS.

EQUIPMENT: Standard equipment includes rotor and cargo tiedowns, ground handling wheels for skid version, retractable 450 W search/landing light. Options include dual controls, auxiliary fuel tankage, force/feel trim system, more comprehensive nav/com avionics, 272 kg (600 lb) capacity rescue hoist, 1,587 kg (3,500 lb) capacity cargo hook, emergency flotation gear, heated windscreen, particle separator and snow baffles.

DIMENSIONS, EXTERNAL:
Main rotor diameter — 12.80 m (42 ft 0 in)
Main rotor blade chord — 0.36 m (1 ft 2¼ in)
Tail rotor diameter — 2.10 m (6 ft 10½ in)
Tail rotor blade chord — 0.25 m (10 in)
Length: fuselage (incl tailskid) — 13.44 m (44 ft 1¼ in)
overall, rotors turning — 15.30 m (50 ft 2½ in)
rotor in X configuration — 13.60 m (44 ft 7¼ in)
Fuselage max width over sponsons — 3.45 m (11 ft 4 in)
Height to top of rotor head:
standard skids — 4.03 m (13 ft 2½ in)
high skids — 4.33 m (14 ft 2½ in)
wheels — 3.72 m (12 ft 2½ in)
Skid track — 2.54 m (8 ft 4 in)
Wheel track — 2.78 m (9 ft 1½ in)
Wheelbase — 4.17 m (13 ft 8¼ in)
Passenger doors: forward:
Height — 1.34 m (4 ft 4¾ in)
Width — 0.88 m (2 ft 10¾ in)
aft:
Height — 1.22 m (4 ft 0 in)
Width — 0.91 m (3 ft 0 in)
Baggage door: Height — 0.60 m (1 ft 11½ in)
Width — 0.85 m (2 ft 9½ in)

DIMENSIONS, INTERNAL:
Cabin: Length, excl cockpit — 2.77 m (9 ft 1 in)
Max width — 1.37 m (4 ft 6 in)
Max height — 1.30 m (4 ft 3 in)
Volume — 4.5 m^3 (158 cu ft)
Baggage compartment volume — 1.05 m^3 (37.2 cu ft)

AREAS:
Main rotor disc — 128.71 m^2 (1,385.4 sq ft)
Tail rotor disc — 3.45 m^2 (37.12 sq ft)

WEIGHTS AND LOADINGS (A: wheeled version; B: skid version):
Weight empty: A — 2,430 kg (5,358 lb)
B — 2,415 kg (5,325 lb)
Useful load: A — 1,788 kg (3,942 lb
B — 1,803 kg (3,975 lb)
Max external load: A, B — 1,587 kg (3,500 lb)
Max T-O weight, all conditions — 4,218 kg (9,300 lb)
Max disc loading — 32.8 kg/m^2 (6.71 lb/sq ft)
Transmission loading at max T-O weight and power — 5.42 kg/kW (8.90 lb/shp)

PERFORMANCE:
Never-exceed speed (VNE):
A, B — 150 kt (277 km/h; 172 mph)
Max level speed: A — 143 kt (265 km/h; 165 mph)
B — 139 kt (257 km/h; 160 mph)
Long-range cruising speed:
A — 135 kt (250 km/h; 155 mph)
B — 131 kt (243 km/h; 151 mph)
Service ceiling: A, B — 5,590 m (18,340 ft)
Hovering ceiling, IGE: A, B — 3,078 m (10,100 ft)
OGE: A, B — 1,890 m (6,200 ft)
Max range, standard fuel, no reserves:
A — 275 n miles (509 km; 316 miles)
B — 353 n miles (653 km; 406 miles)
Max endurance: A, B — 2 h 48 min

OPERATIONAL NOISE LEVELS (FAR Pt 36, Stage 2):
T-O — 92.4 EPNdB
Sideline — 91.6 EPNdB
Approach — 93.8 EPNdB

UPDATED

BELL MODULAR AFFORDABLE PRODUCT LINE

At the NBAA Convention in Orlando, Florida, in October 2003, Bell announced brief details of Modular Affordable Product Line (MAPL) programme that is expected to result in three all-new helicopters sharing common cockpits, avionics and tailbooms, mated to fuselage of common configuration but varying in size. Initial certification target is for the end of the decade, with first MAPL model available in 2008, and all three variants achieving service entry by 2012. Further data released in February 2004 mentioned the five-seat, single-engined *Bell 351* and eight-seat *Bell 382* twin, both with four-blade rotors and 'fantail' anti-torque rotor. Design targets include 20 per cent improvements in speed and useful load, 20 per cent reduction in operating costs, and 10dB reduction in noise over existing helicopters in the same class, with 99 per cent despatch reliability. Price range expected to be between US$500,000 and US$1 million (2004).

MAPL 'TailFan' demonstrator, Bell 407 (N41835) modified with a 1.02 m (3 ft 4 in) diameter fan and duct replacing the standard tail rotor, made its first flight at Bell's XworX research center in Arlington, Texas, on 15 July 2004. It will be used for evaluation of the performance and acoustics of various duct configurations in hover and forward flight. High altitude flight testing will also take place at Bell's facility in Leadville, Colorado.

UPDATED

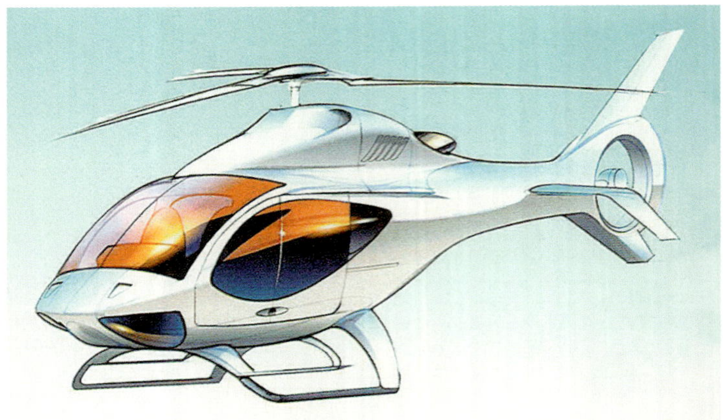

Artist's impression of Bell MAPL ***NEW***/0580704

Bell TailFan demonstrator making its first flight on 15 July 2004 ***NEW***/0583650

BLUE YONDER

BLUE YONDER AVIATION

Box 12, Suite 9, RR5, Calgary, Alberta T2P 2G6
Tel: (+1 403) 936 57 67
Fax: (+1 403) 936 51 08
e-mail: mailto:ezflyer@sprint.ca
Web: http://www.ezflyer.com
PRESIDENT: Wayne Winters

Blue Yonder designed the E-Z Flyer in 1991; built under licence by Merlin in USA until bankruptcy of December 1995; production tooling transferred to Canada and manufacture of it (and Merlin GT) resumed; Aerocomp is US agent, although Blue Yonder responsible for global sales. The company has added the E-Z KingCobra and Harvard to its product line.

UPDATED

BLUE YONDER E-Z FLYER

TYPE: Tandem-seat kitbuilt.

PROGRAMME: First kitbuilt E-Zs delivered by Merlin mid-1995. Programme transferred in 1998 to Blue Yonder, which has also produced a twin-engined prototype, designated E-Z Flyer Twin.

CURRENT VERSIONS: **E-Z Flyer:** *Basic version; as described.*

E-Z Flyer Twin: As E-Z Flyer, but with two Rotax engines in range 37.0 to 73.5 kW (49.6 to 98.6 hp). Wing span 10.82 m (35 ft 6 in); maximum T-O weight 635 kg (1,400 lb). First flown 1999; prototype (C-ITEZ) only to date with second aircraft under construction late 2003.

CUSTOMERS: Total 47 kits sold by late 2003.

COSTS: E-Z Flyer US$12,300 including engine; E-Z Flyer Twin US$20,670 (2004).

Blue Yonder E-Z Flyer two-seat kitbuilt 0507562

DESIGN FEATURES: Open tube fuselage married to high-mounted Aerocomp Merlin GT wing and tail sections. Quoted build time 300 hours for E-Z Flyer and 450 hours for E-Z Flyer Twin.

FLYING CONTROLS: Manual. Non-drooping Junkers-type ailerons.

STRUCTURE: Fuselage is welded 4130 chromoly. Wings are aluminium D-cell construction with aluminium and Styrofoam constructed ribs; Stits Poly-Fiber covering.

Prototype E-Z Flyer Twin ***NEW***/0572291

LANDING GEAR: Tricycle type; fixed. Fuselage-mounted cantilever mainwheel legs. Hegar rims with brakes; solid spring steel legs; 18 in tundra tyres on mainwheels.
POWER PLANT: One 37.0 kW (49.6 hp) Rotax 503 or 47.8 kW (64.1 hp) Rotax 582 engine, driving a two- or three-blade pusher propeller. Fuel capacity 34 litres (9.0 US gallons; 7.5 Imp gallons), optionally expandable to 83 litres (22.0 US gallons; 18.3 Imp gallons).
DIMENSIONS, EXTERNAL:
Wing span .... 9.45 m (31 ft 0 in)
Length overall .... 6.40 m (21 ft 0 in)
Height overall .... 1.83 m (6 ft 0 in)
AREAS:
Wings, gross, incl extended tips .... 16.35 m² (176.0 sq ft)
WEIGHTS AND LOADINGS:
Weight empty .... 225 kg (495 lb)
Max T-O weight .... 607 kg (1,340 lb)
PERFORMANCE:
Max level speed .... 87 kt (161 km/h; 100 mph)
Max cruising speed .... 56 kt (105 km/h; 65 mph)
Stalling speed .... 35 kt (65 km/h; 40 mph)
Max rate of climb at S/L .... 137 m (450 ft)/min
T-O run .... 31 m (100 ft)
T-O to 15 m (50 ft) .... 99 m (325 ft)
Landing from 15 m (50 ft) .... 61 m (200 ft)
Landing run .... 53 m (175 ft)
Range .... 330 n miles (611 km; 379 miles)

UPDATED

BLUE YONDER E-Z KINGCOBRA

TYPE: Single-seat ultralight kitbuilt.
PROGRAMME: Designed by Blue Yonder for Dr Jack Barlass; prototype (C-IFWW) first flown late 1998. Aircraft and design rights later purchased from Barlass estate.
CURRENT VERSIONS: **King Cobra:** *As described.*
Harvard: Adaptation resembling North American AT-6 Texan/Harvard.
CUSTOMERS: Two flying by end 2002.
COSTS: Kit US$17,895 including engine; firewall back US$13,450 (2004).
DESIGN FEATURES: Low-wing monoplane; uses (shortened) wing from Merlin and E-Z Flyer allied to new tail unit. Options include electric starter. Quoted build time 400 hours.
FLYING CONTROLS: Manual. Junkers-type ailerons.
STRUCTURE: Fabric-covered 4130 chromoly welded steel tube fuselage and fabric-covered aluminium D-cell cantilever wings.
LANDING GEAR: Tailwheel type. Hegar 46 cm (18 in) wheels and tundra tyres standard. Hydraulic brakes.
POWER PLANT: One 47.8 kW (64.1 hp) Rotax 582 UL driving two-blade, wooden propeller; design will accept engines up to 73.5 kW (98.6 hp). Fuel capacity 57 litres (15.0 US gallons; 12.5 Imp gallons).
DIMENSIONS, EXTERNAL:
Wing span .... 8.23 m (27 ft 0 in)
Length overall .... 6.40 m (21 ft 0 in)
Height overall .... 1.83 m (6 ft 0 in)
AREAS:
Wings, gross .... 14.68 m² (158.0 sq ft)
WEIGHTS AND LOADINGS:
Weight empty .... 246 kg (543 lb)
Max T-O weight .... 544 kg (1,200 lb)
PERFORMANCE:
Never-exceed speed (VNE) .... 104 kt (193 km/h; 120 mph)
Normal cruising speed .... 78 kt (145 km/h; 90 mph)
Stalling speed .... 27 kt (49 km/h; 30 mph)
Max rate of climb .... 305 m (1,000 ft)/min
Service ceiling .... 4,265 m (14,000 ft)
T-O run .... 19 m (60 ft)
Landing run .... 46 m (150 ft)
Range .... 400 n miles (740 km; 460 miles)

UPDATED

Blue Yonder E-Z KingCobra ***NEW***/0572304

Blue Yonder Harvard ***NEW***/0572303

BOMBARDIER

BOMBARDIER AEROSPACE

400 Chemin de la Côte Vertu West, Dorval, Québec H4S 1Y9
Tel: (+1 514) 855 50 00
Fax: (+1 514) 855 79 03
Web: http://www.aero.bombardier.com
DESIGN DIVISIONS
Bombardier Aerospace Montreal: formerly Canadair; follows Bombardier entry
Bombardier Aerospace Toronto: formerly de Havilland; follows Bombardier Aerospace Montreal entry
Bombardier Aerospace Learjet: see under US under Learjet
Bombardier Aerospace Belfast: formerly Shorts
PRESIDENT AND COO: Pierre Beaudoin
EXECUTIVE VICE-PRESIDENT, ENGINEERING AND PRODUCT DEVELOPMENT: John Holding
EXECUTIVE VICE-PRESIDENT, PROGRAMS AND STRATEGIC PLANNING: Steven A Ridolfi
VICE-PRESIDENT, OPERATIONS: Ken Brunrle
VICE-PRESIDENT, FINANCE: Claude Ferland
PRESIDENT, REGIONAL AIRCRAFT: John Giraudy
PRESIDENT, BUSINESS AIRCRAFT: Peter Edwards
PRESIDENT, AMPHIBIOUS AIRCRAFT: Michael Bourgeois

Bombardier Inc, a diversified Canadian corporation with 79,000 employees, formed Bombardier Aerospace in 1986, subsequently combining design and manufacturing activities of Canadair (1986), de Havilland (1992), Learjet (1990) and Shorts (1989). Sales, marketing and support are conducted by the Amphibious Aircraft, Regional Aircraft and Business Aircraft units.

Bombardier Aerospace also designs and manufactures components for Airbus and Boeing.

Bombardier Aerospace's revenue for the year ended 31 January 2003 totalled C$11.3 billion. Employment stood at 26,000 worldwide in late 2004.

UPDATED

BOMBARDIER AEROSPACE DELIVERIES

FY (ending 31 January)	Regional	Business	Amphibian	Total
1999–2000	104	183	5	292
2000–01	157	203	10	370
2001–02	206	162	2	370
2002–03	220	77	1	298
2003–04	232	89	3	324

BOMBARDIER CSERIES

TYPE: Regional jet airliner.
PROGRAMME: Formerly known as New Commercial Aircraft Program (NCAP). Announced, and provisional specification released, at Farnborough International Air Show, 19 July 2004. Service entry expected in 2010.
CURRENT VERSIONS: **CSeries 110-STD:** Baseline version.
CSeries 110-ER: Long-range version; strengthened fuselage.
CSeries 135-STD: Stretched version; fuselage plugs fore and aft of wing; strengthened structure.
CSeries 135-ER: Long-range version.
CUSTOMERS: Market for aircraft in CSeries class estimated at 5,800, valued at US$250 billion, between 2004 and 2024.
DESIGN FEATURES: Swept low wing, with winglets common to all versions; sweptback tail surfaces; wing-mounted, podded engines. Design goals include operational flexibility on short- and long-haul routes; total life-cycle cost improvement over existing airliners; 15 per cent improvement in direct operating costs, compared with contemporary in-service designs and 20 per cent improvement over out-of-service designs; 99 per cent despatch reliability, and high passenger appeal through comfort, cabin spaciousness and in flight entertainment systems.

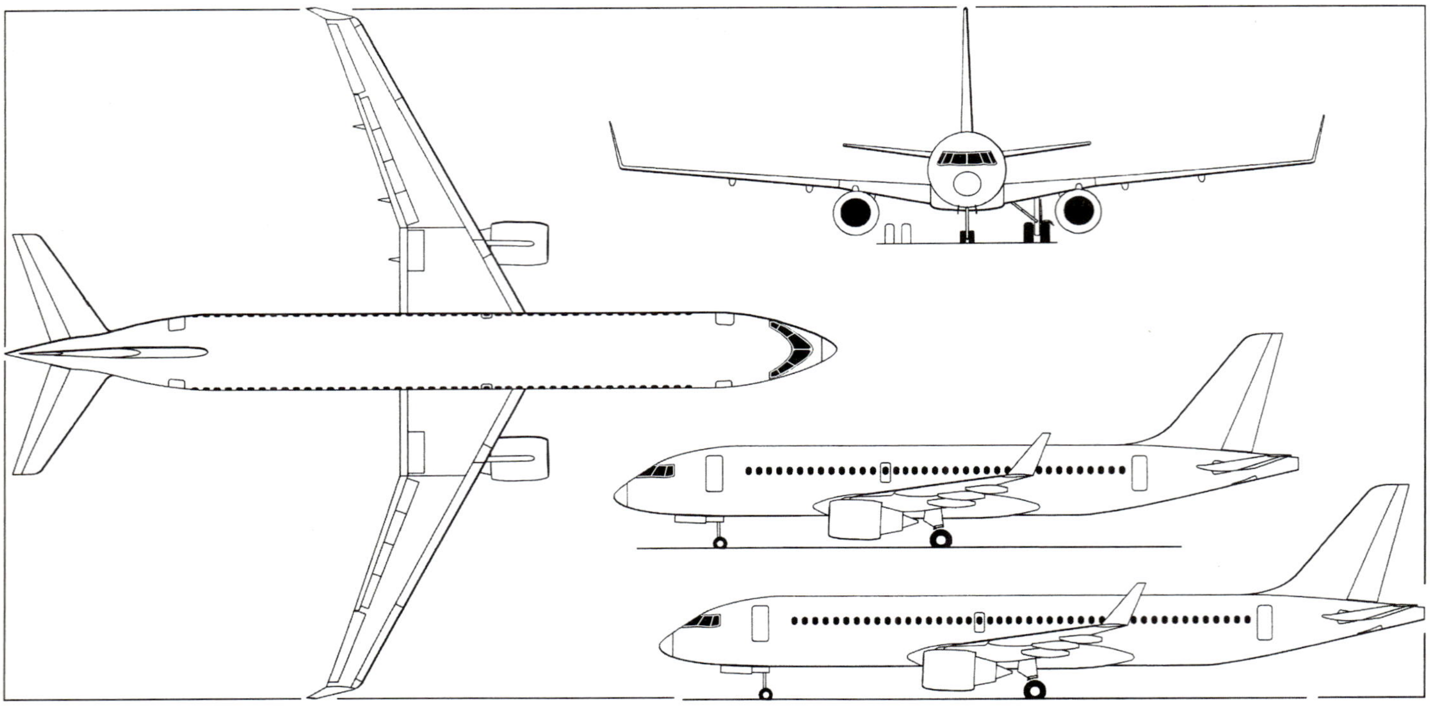

Bombardier CSeries 135, with additional side view (upper) of CSeries 110 *(Paul Jackson)* ***NEW***/0589194

Artist's impression of Bombardier CSeries 110 regional airliner ***NEW***/0589238

FLYING CONTROLS: Fly-by-wire, with sidestick controllers and advanced systems architecture with flight envelope protection. Fowler flaps; three-section leading-edge slats and five-section spoilers on each wing.

STRUCTURE: Common forgings for all versions; composites control surfaces, leading-edge slats and Fowler flaps, tail surfaces, rear fuselage and tailcone.

LANDING GEAR: Retractable tricycle type. Twin wheels on each unit. Main units retract inwards, nosewheel forward. Carbon brakes.

POWER PLANT: *CSeries 110-STD:* Two 87.6 kN (19,700 lb st) turbofans.

CSeries 110-ER and CSeries 135-STD: Two 99 kN (22,250 lb st) turbofans.

CSeries 135-ER: Two 108.8 kN (24,450 lb st) turbofans.

Selected power plant will offer highest bypass ratio for class and targeted at lowest life-cycle costs and Stage 5 noise compliance.

ACCOMMODATION: Two crew on flight deck and three or four cabin attendants. CSeries 110 accommodates 110 passengers at 81 cm (32 in) seat pitch in single class in 3+2 configuration; ATLAS standard galley fore and aft, each with dedicated service door on starboard side; two lavatories, one with access for handicapped passengers, and separate crew and passenger closets all standard. Optional two-class arrangement accommodates 12 passengers at 91 cm (36 in) seat pitch in forward business class cabin, and 85 at 81 cm (32 in) pitch in economy class. CSeries 135 accommodates 135 passengers at 81 cm (32 in) pitch in single class configuration, or, in two-class arrangement, 12 passengers at 91 cm (36 in) pitch in forward business class cabin, and 125 at 81 cm (32 in) pitch in economy class. Passenger doors on port side at front and rear of cabin. Overwing emergency exit on each side. Cabin design aims to provide at least one large window per row of seats on each side to maximise natural lighting and outside view.

SYSTEMS: Emphasis is on integrated aircraft/electrical systems.

DIMENSIONS, EXTERNAL:
Wing span: all ... 33.65 m (110 ft 5 in)

DIMENSIONS, INTERNAL:
Cabin:
Max width, all ... 3.45 m (11 ft 4 in)
Max height ... 2.21 m (7 ft 3 in)
Aisle width: business ... 0.61 m (2 ft 0 in)
economy ... 0.58 m (1 ft 11 in)
Underfloor baggage/cargo hold:
Width: at roof ... 2.21 m (7 ft 3 in)
at floor ... 1.09 m (3 ft 7 in)
Max height ... 1.01 m (3 ft 4 in)
Overhead storage bins:
Length ... 0.61 m (2 ft 0 in)
Height ... 0.28 m (0 ft 11 in)
Depth ... 0.36 m (1 ft 2 in)

WEIGHTS AND LOADINGS (A: CSeries 110-STD, B: CSeries 110-ER, C: CSeries 135-STD, D: CSeries 135-ER)::
Operating weight empty: A, B ... 34,813 kg (76,750 lb)
C ... 37,898 kg (83,550 lb)
D ... 38,328 kg (84,500 lb)
Max payload: A, B ... 14,470 kg (31,900 lb)
C, D ... 17,758 kg (39,150 lb)
Max T-O weight: A ... 55,928 kg (123,300 lb)
B, C ... 62,686 kg (138,200 lb)
D ... 68,628 kg (151,300 lb)
Max landing weight: A ... 50,848 kg (112,100 lb)
B ... 51,619 kg (113,800 lb)
C ... 57,924 kg (127,700 lb)
D ... 58,513 kg (129,000 lb)
Max zero-fuel weight: A, B ... 49,283 kg (108,650 lb)
C ... 55,656 kg (122,700 lb)
D ... 56,087 kg (123,650 lb)
Max power loading: A ... 319 kg/kN (3.13 lb/lb st)
B, C ... 317 kg/kN (3.11 lb/lb st)
D ... 316 kg/kN (3.09 lb/lb st)

PERFORMANCE(estimated: A, B, C, D, as above):
Max cruising speed: A, B, C ... M0.82
D ... M0.81
Initial cruising altitude: A, B, C ... FL390
D ... FL370
T-O balanced field length: A ... 1,677 m (5,500 ft)
B, C ... 1,860 m (6,100 ft)
D ... 2,043 m (6,700 ft)
Landing field length: A ... 1,509 m (4,950 ft)
B ... 1,540 m (5,050 ft)
C ... 1,723 m (5,650 ft)
D ... 1,738 m (5,700 ft)
Max range: A, C ... 1,800 n miles (3,333 km; 2,071 miles)
B ... 3,250 n miles (6,019 km; 3,740 miles)
D ... 2,800 n miles (5,185 km; 3,222 miles)

UPDATED

BOMBARDIER AEROSPACE MONTREAL OPERATIONS

400 Chemin de la Côte Vertu West, Dorval, Québec H4S 1Y9
POSTAL ADDRESS: PO Box 6087, Station Centreville,
Montréal , Québec H3C 3G9
Tel:(+1 514) 855 50 00
Fax:(+1 514) 855 79 03
Web: http://www.aero.bombardier.com
VICE-PRESIDENT AND GENERAL MANAGER, OPERATIONS: Serge Perron
VICE-PRESIDENT, OPERATIONS, ST LAURENT PLANT: Jean Séguin
VICE-PRESIDENT, OPERATIONS, DORVAL PLANT: Réal Gervais
VICE-PRESEIDENT, OPERATIONS, MIRABEL PLANT: Alain Dugas

Acquired by Bombardier Inc 23 December, 1986, Bombardier Aerospace Montreal (formerly Canadair) has manufactured more than 4,500 aircraft since 1944. Mirabel plant, adjacent to Montréal International Airport, until recently comprised 27,870 m^2 (300,000 sq ft) of floor space for manufacture and assembly of the Challenger and Regional Jet. However, 27,870 m^2 (300,000 sq ft) final assembly building for CRJ700 and CRJ900 formally opened on 22 October 2001, having begun operations in previous August. An 18,580 m^2 (200,000 sq ft) interior outfitting, flight testing, inspection and delivery centre is also part of the Mirabel facility. Parts, components and spare parts for various aircraft, including Challenger, Regional Jet series, Global Express and Bombardier 415, are manufactured at St Laurent plant, 214,350 m^2 (2,307,275 sq ft) in Québec. Structural components for other aircraft builders, such as Boeing and Aerospatiale Matra, are also manufactured in this plant.

The Challenger series and Global Express are completed at the 38,590 m^2 (415,400 sq ft) Bombardier Completion Center, Montréal, beside the Bombardier Aerospace headquarters at Dorval. Total workforce in the Montréal area is more than 15,000.

UPDATED

BOMBARDIER CRJ200 AND CHALLENGER 800

Engineering designation: CL-600-2B19

TYPE: Regional jet airliner.

PROGRAMME: Design studies began in third quarter of 1987; basic configuration frozen June 1988; formal programme go-ahead given 31 March 1989; extended-range CRJ100ER announced September 1990. Three development aircraft built (c/n 7001–7003), plus static test airframe (c/n 7991) and forward fuselage test article (7992); first flight of 7001 (C-FCRJ) 10 May 1991; 7002 (C-FNRJ) first flew 2 August 1991 and 7003 on 17 November 1991; all three in 1,400-hour flight test programme in Wichita, USA. CF34-3A1 engine obtained its US type certificate 24 July 1991. Transport Canada type approval (CRJ100 and CRJ100ER) 31 July 1992. Japanese Civil Aviation Bureau certification 23 May 2000.

First delivery aircraft (c/n 7004) flew 4 July 1992, and to Lufthansa CityLine of Germany (as D-ARJA) 29 October 1992; European JAA and US FAA certification 14 and 21 January 1993 respectively; long-range CRJ100LR certified 29 April 1994; CRJ200 with CF34-3B1 engines announced in 1995. Replaced CRJ100 after 226 of the latter had been delivered. Total fleet time at October 2003 (not including corporate aircraft) was 7,414,853 flight hours and 6,459,897 cycles, with 99.5 per cent despatch reliability rate. 200th aircraft delivered (to Lufthansa) 24 October 1997; 300th to Atlantic Coast Airlines in April 1999. 400th to Delta Connection/SkyWest in July 2000. 500th to Atlantic Coast Airliners 26 April 2001, and 600th to Atlantic Southeast Airlines 29 January 2002, and 700th to Air Nostrum 30 October 2002, and 1,000th aircraft in CRJ series, a CRJ 700, to Conair on 9 December 2003. Production of CRJ200 running at 9.5 per month in 2000, rising to 12.5 per month by late 2001, and 14.5 per month by 2003, with annual targets of 165 in 2003 and 174 in 2004. Achieved 155 deliveries in 2003, comprising 132 CRJ200s and 23 CRJ440s.

CURRENT VERSIONS: **CRJ100:** Original standard aircraft. Engineering designation CL-600-2B19.

CRJ100ER: Replaced by CRJ200ER.

CRJ100LR: Announced March 1994; launch customer, Lauda Air of Austria; replaced by CRJ200LR.

CRJ200: Standard aircraft; designed to carry 50 passengers over 985 n mile (1,824 km; 1,133 mile) range; CF34-3B1 engines with 2.8 per cent lower specific fuel consumption than CF34-3A1 of CRJ100, increasing initial cruise altitude by 213 m (700 ft), cruising speed by 2.5 kt (4.5 km/h; 3 mph), and range typically by 1.5 per cent; Class C baggage compartment as standard. First delivery, to Tyrolean Airways as OE-LCF, 15 January 1996. Further improvements in development for introduction on CRJ200 variants during early 1996 included 3 kt (5.5 km/h; 3.5 mph) reduction in V_2 speed to provide 91 m (300 ft) reduction in T-O run at maximum T-O weight; 1 kt (1.8 km/h; 1.2 mph) reduction in V_{REF} to provide 15 m (50 ft) reduction in landing run at typical landing weights; new 8° flap setting to improve second-segment climb performance; and GPS integrated with an upgraded FMS.

CRJ200ER: Extended-range capability with optional increase in maximum T-O weight to 23,133 kg (51,000 lb) and optional additional fuel capacity, for range of 1,645 n miles (3,046 km; 1,893 miles).

CRJ200LR: Longer-range version of CRJ200ER (more than 2,005 n miles; 3,713 km; 2,307 miles); maximum T-O weight increased by 907 kg (2,000 lb) to 24,040 kg (53,000 lb).

CRJ200B, CRJ200B ER and CRJ200B LR: As above, but with optional hot-and-high CF34-3B1 engines providing normal T-O thrust up to ISA+7.8°C (ISA+6.1°C for standard engines), and APR thrust up to ISA+15°C (ISA+6.1°C for standard engines).

CRJ440: Engineering designation CL-600–2B19. Version seating 44 passengers in standard configuration. Launch customer Northwest Airlines has ordered 75.

CRJ700: *Described separately.*

Bombardier CRJ200 operated by Air Wisconsin on behalf of AirTran ***NEW***/0572288

Air Canada Bombardier CRJ200 operating services for Jazz ***NEW***/0572287

Bombardier Challenger 800 business jet *(Paul Jackson)* ***NEW***/0572286

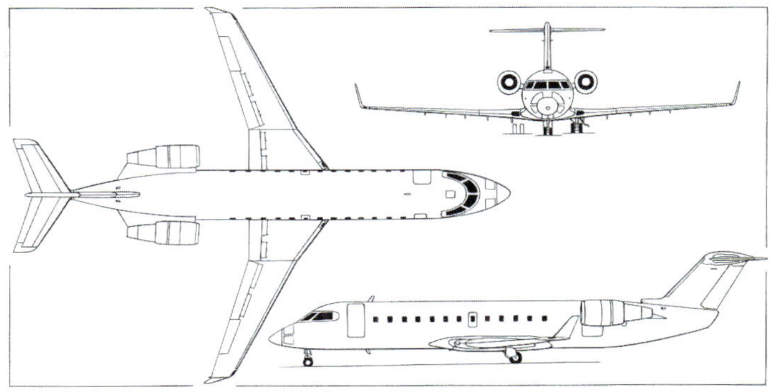

Bombardier CRJ200 (two General Electric CF34-3B1 turbofans) 0507125

Corporate Jetliner: Company shuttle version with more spacious cabin accommodation for 18 to 30 passengers. One delivered June 1993 to Xerox Corporation. Five ordered by the People's Republic of China in January 1997; operated on behalf of PRC government by China United Airlines' crews; contract value C$116 million, including outfitting, pilot and maintenance staff training and spares. Supplanted from September 2002 by corporate version of Challenger 800 (see below).

Challenger 800: Corporate version developed in consultation with launch customer TAG Aeronautics Ltd to meet requirement for non-stop flights, London to Jeddah or equivalent, with three crew and five passengers; or between Middle East city pairs with 15 passengers. First flown 26 May 1995 and formally announced at Paris Air Show in the following month; initially designated **Canadair Special Edition**; first delivery (N877SE) to TAG during Dubai International Aerospace Show in November 1995; second TAG aircraft delivered November 1997. Accommodation for up to 19 passengers in customised cabin; additional 1,814 kg (4,000 lb) of fuel carried in two auxiliary tanks behind main cabin, extending range to more than 3,000 n miles (5,556 km; 3,452 miles); maximum T-O weight 24,040 kg (53,000 lb); first aircraft powered by standard CF34-3A1 turbofans, but subsequent examples are equipped with CF34-3B1s increasing range to 3,120 n miles (5,778 km; 3,590 miles); Rockwell Collins Pro Line 4 avionics as on RJ, but with third FMS, third VHF, dual Collins HF and Selcal. Manufactured to special order only. Recent customers include Poly Technologies Inc, which ordered two on 16 August 2001 for operation by China Ocean Aviation Group, with deliveries scheduled for 2002. Renamed Challenger 800 on eve of NBAA Convention at Orlando, Florida, 8 September 2002.

CUSTOMERS: See table. Recent customers include Penske Jet, Inc, which ordered one in 42-seat corporate shuttle configuration for service entry in January 2004.

COSTS: Programme development costs C$275 million. Atlantic Coast Airlines order for 10 CRJ200ERs valued at US$200 million (September 1998). Order for two Special Editions valued at US$54.2 million (August 2001).

DESIGN FEATURES: Evolved from Challenger (which see), designed expressly for regional airline operating environment. Advanced transonic wing design, with winglets for high-speed operations; fuel-efficient GE turbofans; options include higher design weights, additional fuel capacity, more comprehensive avionics, and maximum certified altitude raised to 12,500 m (41,000 ft).

Wings, designed with computational fluid dynamics (CFD), have 13.2 per cent (root) and 10 per cent (tip) thickness/chord ratios, 2° 20' dihedral, 3° 25' root incidence and 24° 45' quarter-chord sweepback.

FLYING CONTROLS: Conventional and power-assisted. Primary controls with cables and push/pull rods for multiple redundancy; hydraulically actuated ailerons, elevators and rudder with at least two hydraulic power control unit actuators per surface (three on rudder and elevator); ailerons and elevators fitted with flutter dampers (dual on elevators); rudder with dual-channel control yaw damping; artificial feel and electric trim for roll and yaw; electronically controlled, variable incidence T tailplane for pitch trim and electronically controlled artificial pitch feel. Double-slotted electromechanical flaps with electronically controlled Datron electric motors; BAE fly-by-wire spoiler and spoileron system, four spoilers each side, with inner two functioning as ground spoilers, outer two comprising one flight spoiler and one spoileron, both also providing lift dumping on touchdown. Avionics suite includes engine indication and crew alerting system (EICAS).

STRUCTURE: Semi-monocoque fuselage is damage tolerant FAR/JAR 25 certified airframe with chemically milled skins; flat pressure bulkheads forward of flight deck and aft of baggage compartment; extensive use of advanced composites in secondary structures (passenger compartment floor, wing/fuselage fairings, nacelle doors, wing access door covers, winglets, tailcone, avionics access doors and landing gear doors); comprehensive anti-corrosion treatment and drainage. Wing is one-piece unit mounted to underside of fuselage; two-spar box joined by ribs, covered top and bottom with integrally stiffened skin panels (three upper and three lower each side) for smooth flow; machined or built-up spars and shearweb-type ribs. Short Brothers (UK) manufactures fuselage central section, fore and aft fuselage plugs, wing flaps, ailerons, spoilerons and inboard spoilers.

LANDING GEAR: Hydraulically retractable tricycle type, manufactured by Dowty. Inward-retracting main units each have 15 in Aircraft Braking System (ABS) wheels with H29×9.0-15 (16 ply) Goodyear tubeless tyres, pressure 11.17 bar (162 lb/sq in) unladen. Nose unit has 18×4.4 (12 ply) tyres (deflector type) and Dowty Canada steer-by-wire steering; unladen tyre pressure 8.62 bar (125 lb/sq in). Aircraft Braking System steel

BOMBARDIER CRJ ORDERS AND DELIVERY
(at 31 March 2004)

Customer	Variant	Ordered	Delivered	Backlog
Adria Airways	200	4	4	
Air Canada	100	24	24	
	200	2	2	
Air Dolomiti	200	5	5	
Air Littoral	100	19	19	
Air Nostrum	200	31	24	7
Air Wisconsin	200	64	63	1
American Eagle	701	25	20	5
Atlantic Coast Airlines	200	121	87	34
Atlantic Southeast (ASA)	200	45	45	
	701	12	12	
Austrian Arrows[1]	200	13	13	
Brit Air	100	20	20	
	701	12	10	2
British European	200	4	4	
China Eastern Yunnan	200	6	6	
Cimber Air	200	2	2	
COMAIR	100	110	110	
	701	20	19	1
DAC Air	200	2	2	
Delta Connection	200	116	84	32
	701	26	9	17
GECAS	200	5	5	
Horizon Air	701	28	18	10
Japan Air Lines (J-AIR)	200	6	6	
Kendell/Ansett	200	12	12	
Lauda Air	100	8	8	
Lufthansa CityLine	100	35	35	
	200	10	10	
	701	20	20	
Lufthansa/Eurowings	200	15	12	3
Maersk	200	11	11	
	701	5	5	
Malev	200	4	4	
Mesa Air	200	32	32	
	701	21	15	6
	900	39	16	23
Midway	200	24	24	
Northwest Airlines	200	54	35	19
	440	75	52	23
Saeaga Airlines	200	1	1	
Shandong Airlines	200	5	5	
	701	2	2	
Shanghai Airlines	200	3	3	
SkyWest	100	10	10	
	200	100	90	10
	701	32	3	29
South African Express	200	6	6	
Southern Winds	200	2	2	
The Fair Inc	200	2	2	
US Airways	200	60	16	44
	705	25	2	23
Challenger 800	SE	19	19[2]	
Totals		1,354	1,065	289

[1] Formerly Tyrolean
[2] Comprising 18 Series 200 and one Series 700.
Does not include Air Canada memorandum of understanding signed 19 December 2003 for 15 CRJ200s and 30 CRJ705s, plus options on 15 CRJ200s and 30 CRJ 705s, for delivery commencing third quarter 2004 for CRJ200s and second quarter 2005 for CRJ700s.

multidisc brakes and fully modulated Hydro Aire Mk III anti-skid system. Minimum taxiway width for 180° turn (with 3.35 m; 11 ft 0 in safety margin) is 22.86 m (75 ft 0 in).

POWER PLANT: Two General Electric CF34-3B1 turbofans, each rated at 41.0 kN (9,220 lb st) with APR and 38.8 kN (8,729 lb st) without. Nacelles produced by Short Brothers. Pneumatically actuated thrust reversers. Fuel in two integral wing tanks, combined capacity 5,300 litres (1,400 US gallons; 1,166 Imp gallons); increasable to 8,080 litres (2,135 US gallons; 1,778 Imp gallons) with optional centre-wing tank. Pressure refuelling point in starboard leading-edge wingroot; transfer rate 474 litres (125 US gallons; 104 Imp gallons)/min at 3.45 bar (50 lb/sq in); two gravity points on starboard wing (one for centre tank) and one on port wing.

ACCOMMODATION: Two-pilot flight deck; one or two cabin attendants. Main cabin seats up to 50 passengers in standard configuration, four-abreast at 79 cm (31 in) pitch, with centre aisle; maximum capacity 52 seats. Various configurations, from 15 to 50 seats, available for corporate version. Downward-opening front passenger door with integral airstairs on port side; plug-type forward emergency exit/service door opposite on starboard side (inoperative on Challenger 800). Inward-opening baggage door on port side at rear. Overwing Type III emergency exit each side (port side door inoperative on Challenger 800). Entire accommodation pressurised, including rear baggage compartment.

SYSTEMS: Cabin pressurisation and air conditioning system (maximum differential 0.57 bar; 8.3 lb/sq in). Primary flight control systems powered by hydraulic servo-actuators with distinct, alternate paths cable and pushrod systems. Electric trim and dual yaw dampers. Three fully independent 207 bar (3,000 lb/sq in) hydraulic systems. Three-phase 115 V AC electrical primary power at 400 Hz supplied by two 30 kVA engine-driven generators; alternative power provided by APU and air-driven generator. Conversion to 28 V DC by five transformer-rectifier units. Main (Ni/Cd) battery 17 Ah, APU battery 43 Ah. Honeywell GTCP 36-150 (RJ) APU and two-pack air conditioning system in rear of fuselage. Wing leading-edges and engine intake cowls anti-iced by engine bleed air. Electric anti-icing of windscreen and cockpit side windows, pitot heads, air data vanes, static sources and sensors. Ice detection system standard.

AVIONICS: *Comms:* Dual VHF nav/com radios. Options include HF radio, single Selcal and 8.33 kHz VHF.

Radar: Rockwell Collins digital weather radar system; split-scan weather radar and radar with turbulence mode optional.

Flight: Dual flight management systems optional. GPWS, windshear detection system and TCAS. EGPWS optional in place of GPWS. L3 flight data recorder and CNE. Dual FMS 4200 and dual IRS in Corporate Jetliner and Challenger 800.

Instrumentation: Rockwell Collins Pro Line 4 integrated all-digital suite, including dual primary flight displays, dual multifunction displays, dual EICAS, dual AFCS, dual AHRS, dual air data system and Cat. II capability with Cat. IIIa optional using head-up guidance system. Dual inertial reference system optional in lieu of AHRS. Flight Dynamics Inc HGS 2100 HUD approved by Transport Canada November 1995, permitting Cat. IIIa operation.

DIMENSIONS, EXTERNAL: As for Challenger 604 except:

Wing span over winglets ... 21.21 m (69 ft 7 in)
Wing chord: at fuselage c/l ... 5.13 m (16 ft 10 in)
at tip ... 1.27 m (4 ft 2 in)
Wing aspect ratio (excl winglets) ... 8.9
Length: overall ... 26.77 m (87 ft 10 in)
fuselage ... 24.38 m (80 ft 0 in)
Height overall ... 6.22 m (20 ft 5 in)
Fuselage max diameter ... 2.69 m (8 ft 10 in)
Wheel track ... 3.17 m (10 ft 5 in)
Wheelbase ... 11.41 m (37 ft 5 in)
Passenger/crew door (fwd, port):
Height ... 1.78 m (5 ft 10 in)
Width ... 0.91 m (3 ft 0 in)
Service door (stbd, fwd): Height ... 1.22 m (4 ft 0 in)
Width ... 0.61 m (2 ft 0 in)
Height to sill (crew/service) ... 1.63 m (5 ft 4 in)
Baggage door (port, rear): Height ... 1.09 m (3 ft 7 in)
Width ... 0.84 m (2 ft 9 in)
Height to sill ... 1.63 m (5 ft 4 in)
Emergency exit (overwing, stbd):
Height ... 0.96 m (3 ft 2 in)
Width ... 0.51 m (1 ft 8 in)

DIMENSIONS, INTERNAL: As for Challenger 604 except: Cabin (incl baggage compartment, excl flight deck):

Length ... 14.76 m (48 ft 5 in)
Max height ... 1.85 m (6 ft 1 in)
Width: at centreline: except 800 ... 2.57 m (8 ft 5 in)
800 ... 2.49 m (8 ft 2 in)
at floor level ... 2.18 m (7 ft 2 in)
Floor area: except 800 ... 32.1 m^2 (346 sq ft)
800 ... 30.3 m^2 (326 sq ft)
Volume: except 800 ... 57.1 m^3 (2,015 cu ft)
800 ... 53.8 m^3 (1,900 cu ft)
Stowage volume:
main (rear) baggage compartment ... 9.0 m^3 (318 cu ft)
wardrobes/bins/underseat (total) ... 4.7 m^3 (165 cu ft)

AREAS:

Wings: gross (excl winglets) ... 54.54 m^2 (587.1 sq ft)
net ... 48.35 m^2 (520.4 sq ft)
Ailerons (total) ... 1.93 m^2 (20.8 sq ft)
Trailing-edge flaps (total) ... 10.60 m^2 (114.1 sq ft)
Spoilers (total) ... 2.26 m^2 (24.3 sq ft)
Winglets (total) ... 1.38 m^2 (14.9 sq ft)
Fin ... 9.18 m^2 (98.8 sq ft)
Rudder ... 2.03 m^2 (21.9 sq ft)
Tailplane ... 9.44 m^2 (101.6 sq ft)
Elevators (total) ... 2.84 m^2 (30.52 sq ft)

WEIGHTS AND LOADINGS:

Manufacturer's weight empty:
200 ... 13,236 kg (29,180 lb)
200ER, 200LR ... 13,243 kg (29,195 lb)
800 Corporate ... 11,703 kg (25,800 lb)
Operating weight empty: 200 ... 13,730 kg (30,270 lb)
200ER, 200LR ... 13,835 kg (30,500 lb)
800 Corporate ... 14,424 kg (31,800 lb)
800 ... 15,377 kg (33,900 lb)
Max payload (structural): 200 ... 5,411 kg (11,930 lb)
200ER, 200LR ... 6,124 kg (13,500 lb)
800 Corporate ... 5,533 kg (12,200 lb)
800 ... 2,540 kg (5,600 lb)
Max fuel: 200 ... 4,254 kg (9,380 lb)
200ER, 200LR, 800 Corporate ... 6,489 kg (14,305 lb)
800 ... 8,303 kg (18,305 lb)
Payload with max fuel: 200 ... 3,651 kg (8,050 lb)
200ER ... 2,923 kg (6,445 lb)
200LR ... 3,831 kg (8,445 lb)
800 Corporate ... 2,333 kg (5,145 lb)
800 ... 474 kg (1,045 lb)
Fuel with max payload: 800 ... 6,237 kg (13,750 lb)
Max T-O weight: 200 ... 21,523 kg (47,450 lb)
200ER, 800 Corporate ... 23,133 kg (51,000 lb)
200LR, 800 Corporate (optional), 800 ... 24,040 kg (53,000 lb)
Max ramp weight: 200 ... 21,636 kg (47,700 lb)
200ER ... 23,246 kg (51,250 lb)
200LR, 800 ... 24,154 kg (53,250 lb)
Max zero-fuel weight: 800 ... 17,917 kg (39,500 lb)
200 ... 19,141 kg (42,200 lb)
200ER, 200LR, 800 Corporate ... 19,958 kg (44,000 lb)
Max landing weight: 200 ... 20,275 kg (44,700 lb)
200ER, 200LR, 800 Corporate, 800 ... 21,319 kg (47,000 lb)
Max wing loading: 200 ... 394.6 kg/m^2 (80.82 lb/sq ft)
200ER ... 424.1 kg/m^2 (86.87 lb/sq ft)
200LR, 800 Corporate (optional), 800 ... 440.8 kg/m^2 (90.27 lb/sq ft)
Max power loading (APR rating):
200 ... 263 kg/kN (2.57 lb/lb st)
200ER, 800 Corporate ... 282 kg/kN (2.77 lb/lb st)
200LR, 800 Corporate (optional), 800 ... 293 kg/kN (2.87 lb/lb st)

PERFORMANCE:

Max operating speed:
above FL314 ... M0.85
below FL254 ... 335 kt (621 km/h; 386 mph)

High-speed cruising speed: CRJ200 at FL370 ... M0.81 or 464 kt (859 km/h; 534 mph)
800 Corporate, 800 ... M0.80 or 459 kt (850 km/h; 528 mph)
Normal cruising speed: CRJ200 at FL370 ... M0.74 or 424 kt (785 km/h; 488 mph)
800 Corporate, 800 ... M0.77 or 442 kt (819 km/h; 509 mph)
Long-range cruising speed, 800 Corporate, 800 ... M0.74 or 424 kt (785 km/h; 488 mph)
Approach speed, 45° flap, AUW of 19,504 kg (43,000 lb) ... 135 kt (250 km/h; 155 mph)
Max rate of climb at FL15, 250 kt CAS/M0.74 climb
schedule: 200 ... 1,128 m (3,700 ft)/min
200LR ... 1,036 m (3,400 ft)/min
800 ... 1,034 m (3,395 ft/min)
Max certified altitude ... 12,500 m (41,000 ft)
FAR T-O field length at S/L, ISA:
200 ... 1,527 m (5,010 ft)
200ER ... 1,768 m (5,800 ft)
200LR ... 1,917 m (6,290 ft)
800 Corporate ... 1,765 m (5,790 ft)
800 ... 1,918 m (6,295 ft)
FAR landing field length at S/L, ISA, at max landing weight: 200 ... 1,423 m (4,670 ft)
200ER, 200LR ... 1,478 m (4,850 ft)
800 Corporate, SE ... 887 m (2,910 ft)
Range with max payload at long-range cruising speed, FAR Pt 121 reserves:
200 ... 965 n miles (1,787 km; 1,110 miles)
200ER ... 1,645 n miles (3,046 km; 1,893 miles)
200LR ... 2,005 n miles (3,713 km; 2,307 miles)
Corporate (30 seats), NBAA IFR reserves ... 2,017 n miles (3,735 km; 2,321 miles)
800 with 3,674 kg (8 100 lb) payload ... 1,541 n miles (2,853 km; 1,773 miles)
Range with max fuel:
800 Corporate ... 2,250 n miles (4,167 km; 2,589 miles)
800 ... 3,120 n miles (5,778 km; 3,590 miles)

OPERATIONAL NOISE LEVELS: (CRJ200, FAR Pt 36):
CRJ200:
T-O ... 77.6 EPNdB
Approach ... 92.1 EPNdB
Sideline ... 82.4 EPNdB
800 Corporate, 800:
T-O ... 78.7 EPNdB
Approach ... 92.1 EPNdB
Sideline ... 82.2 EPNdB

UPDATED

BOMBARDIER CRJ700

Engineering designation: CL-600-2C10

TYPE: Regional jet airliner.

PROGRAMME: Design and market evaluation began in 1995 in consultation with 15-member advisory panel of airline operators; stretched derivative of Regional Jet; originally designated CRJ-X. GE CF34-8C1 engines selected February 1995; low-speed wind tunnel testing began early 1995 at Institute of Aeronautical Research, Ottawa; high-speed wind tunnel testing began November 1995 at Rockwell facilities in California. More than 650 hours of wind tunnel testing completed by September 1996. Bombardier Board gave approval for launch on 21 January 1997, at which time orders included four from launch customer Brit Air of France. Aerodynamic configuration frozen 14 March 1997; design frozen 17 July 1998. New version of CF34 engine first ground run in February 1998, followed by flight tests on GE-owned Boeing 747 testbed in first quarter of 1999 and certification in November 1999. First flight (C-FRJX c/n 10001) 27 May 1999; official roll-out 28 May 1999; public debut (c/n 10004) at Farnborough International Air Show 23 July 2000; Transport Canada certification achieved 22 December 2000; first delivery, to Brit Air of France, in February 2001, followed by deliveries to Horizon Air and Lufthansa CityLine in May.

Four flying aircraft and two static test airframes participated in the certification programme; one static test airframe was used for Complete Aircraft Static Test (CAST), while the other underwent Durability and Damage Tolerance Test (DDTT), and was subjected to the equivalent of 160,000 flight cycles. Test programme was conducted at the Bombardier Flight Test Centre at Wichita, Kansas, totalling 1,600 flight hours on four aircraft: C-FRJX (handling and performance evaluation), 10002/C-FJFC (systems), 10003/C-FBKA (avionics) and 10004/C-FCRJ, the first to be fully furnished (function and reliability testing), which initially flew on 16 December 1999. In December 2002 Bombardier signed a tentative agreement with China Aviation Corporation I (AVIC I) that could lead to final assembly of the CRJ700 and CRJ900 by Shanghai Aviation Industrial Corporation.

Total fleet time, including CRJ900s, at January 2004 stood at 366,196 hours and 298,328 cycles, with a 98.9 per cent despatch reliability rate.

CURRENT VERSIONS: **CRJ701:** Formerly CRJ700 68-seat version in standard and extended range (ER) weight options.

CRJ705: Formerly CRJ701. 70-seat version in standard and extended range (ER) weight options.

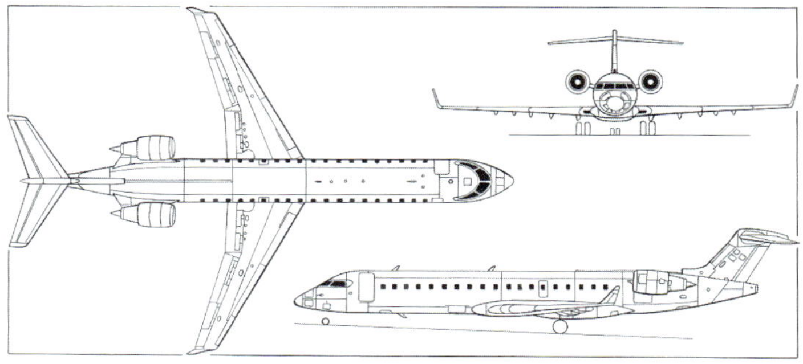

Bombardier CRJ700 series regional jet airliner *(James Goulding)* ***NEW***/0131848

CRJ900: further stretch, *described separately*.

CUSTOMERS: Refer to table in CRJ200 entry. By 31 March 2004 had gained 228 orders, of which 135 had then been delivered.

COSTS: Development cost C$645 million, of which C$440 million provided by Bombardier and balance by risk-sharing partners. Unit cost US$26.8 million (1998). Break-even point is at 200th of 400 aircraft production run.

DESIGN FEATURES: Commonality (including crew training and type rating) with CRJ100/200; direct operating cost per seat-mile to be 20 per cent lower than that of CRJ100/200 and lowest of any airliner in its class. Fuselage stretched 4.72 m (15 ft 6 in) by plugs fore and aft of centre-section to seat 70 passengers; rear pressure bulkhead moved aft by 1.29 m (4 ft 3 in); cabin 6.02 m (19 ft 9 in) longer; APU moved to tailcone; cabin floor lowered by 2.5 cm (1 in) and ceiling raised by 1.3 cm (½ in) to provide 1.89 m (6 ft 2¼ in) headroom; cabin windows raised 11.5 cm (4½ in); new underfloor baggage compartment, volume 3.09 m^3 (109 cu ft) to facilitate ramp check-in; wing span increased by 1.83 m (6 ft 0 in) by wingroot plug; wing leading-edge extended and equipped with high-lift devices; larger horizontal tail surfaces; new pitch control system; main landing gear lengthened; new wheels, tyres, brakes; air conditioning and anti-icing systems upgraded; new engines; new underfloor baggage compartment on forward port side; overhead stowage bins redesigned.

Incorporation of CRJ900's strengthened wing and avionics software upgrades under consideration in mid-2001 to increase commonality between CRJ700 and CRJ900, affording a 200 n mile (370 km; 230 mile) increase in range for CRJ700.

FLYING CONTROLS: Sextant and Menasco primary and secondary flight control system; flying controls are actuated via a dual network of cables, pulleys and pushrods which operate hydraulic power drive units; secondary controls are fly-by-wire.

STRUCTURE: Programme participants include Bombardier Canadair operations (wing, cockpit, rudder and doors, electrical system, primary flight controls, plus final assembly and interior completion), C & D Interiors (cabin interior), Gamesa (vertical and horizontal stabilisers), General Electric (power plant), GKN Westland (tailcone and doors), Hella Aerospace GmbH (lighting system), Honeywell (APU), Intertechnique (fuel system), Liebherr Aerospace Toulouse (air management system), Menasco Aerospace (landing gear), Mitsubishi Heavy Industries (aft fuselage), Parker Abex (hydraulic system), Rockwell Collins (avionics), Sextant Avionique (flight control system), Shorts (nacelles and thrust reversers) and Hamilton Sundstrand (flaps, leading-edge slats and electrical system).

LANDING GEAR: Hydraulically retractable tricycle type by Menasco with twin wheels on each unit. Mainwheel tyre size H36×12.0–18, pressure 10.55 bar (153 lb/sq in); nosewheel tyre size H20.5×6.75–10, pressure 9.24 bar (134 lb/sq in).

POWER PLANT: Two General Electric CF34-8C1 turbofans with dual-channel FADEC. Engine rating 56.4 kN (12,670 lb st), or 61.3 kN (13,790 lb st) with automatic power reserve, flat rated to ISA + 15°C. Intertechnique fuel management system with Ratier-Figeac controls. Fuel capacity 10,989 litres (2,903 US gallons; 2,417 Imp gallons).

ACCOMMODATION: Two-pilot flight deck. Main cabin seats 70 passengers, four-abreast at 79 cm (31 in) pitch. Baggage compartment and lavatory at rear of cabin; underfloor baggage compartment; various combinations of galleys, wardrobes and lavatory at front of cabin according to seating capacity. Passenger door, forward emergency exit/service door, overwing emergency exits and baggage door as for Regional Jet.

SYSTEMS: Hamilton Sundstrand electrical generation system comprising two 40 kVA integrated drive generators; tailcone-mounted Honeywell APU approved for operation up to 12,500 m (41,000 ft); Liebherr air management system; Intertechnique fuel system; Walter Kidde fire detection system; Goodrich potable water system; Hella lighting system; Parker/Abex Hydraulics hydraulic systems, and Teleflex anti-icing system.

AVIONICS: *Radar:* Rockwell Collins digital weather radar.

Flight: Rockwell Collins AHRS and TCAS.

Instrumentation: Rockwell Collins Pro Line 4 EFIS with six 127 × 178 mm (5 × 7 in) CRT displays, including dual PFD, dual MFD and dual EICAS; Flight Dynamics HGS 2000 head-up guidance system. Autopilot, FMS and centralised avionics maintenance functions are provided by integrated avionics processing system (IAPS). Windshear detection and recovery system standard. Equipped for Cat. II landings.

Following data are provisional.

Bombardier CRJ700 delivered to Comair on 9 December 2003 as 1,000th of CRJ line ***NEW***/0572284

DIMENSIONS, EXTERNAL:
Wing span 23.24 m (76 ft 3 in)
Wing aspect ratio 7.4
Length overall 32.51 m (106 ft 8 in)
Max diameter of fuselage 2.69 m (8 ft 10 in)
Height overall 7.57 m (24 ft 10 in)
Tailplane span 8.53 m (28 ft 0 in)
Wheelbase 13.67 m (44 ft 10 in)
Turning circle 22.86 m (75 ft 0 in)
Passenger door (port, fwd):
Height 1.78 m (5 ft 10 in)
Width 0.91 m (3 ft 0 in)
Height to sill 1.73 m (5 ft 8 in)
Baggage door (port, aft): Height 0.84 m (2 ft 9 in)
Width 1.09 m (3 ft 7 in)
Height to sill 2.31 m (7 ft 7 in)
Baggage door (port, fwd): Height 0.51 m (1 ft 8 in)
Width 1.07 m (3 ft 6 in)
Height to sill 1.28 m (4 ft 2½ in)
Service door (starboard, fwd): Height 1.22 m (4 ft 0 in)
Width 0.61 m (2 ft 0 in)
Height to sill 1.73 m (5 ft 8 in)

DIMENSIONS, INTERNAL:
Cabin (excl flight deck):
Length 17.27 m (56 ft 8 in)
Max width: at centreline 2.55 m (8 ft 4½ in)
at floor level 2.13 m (7 ft 0 in)
Max height 1.89 m (6 ft 2¼ in)
Floor area, excl cockpit 36.88 m^2 (397 sq ft)
Volume 75.95 m^3 (2,682 cu ft)
Baggage volume (total) 23.3 m^3 (824 cu ft)

AREAS:
Wings, net 68.63 m^2 (738.7 sq ft)
Horizontal tail surfaces (total) 20.74 m^2 (223.3 sq ft)
Vertical tail surfaces (total) 13.36 m^2 (143.8 sq ft)

WEIGHTS AND LOADINGS:
Operating weight empty 19,731 kg (43,500 lb)
Max payload 8,528 kg (18,800 lb)
Payload with max fuel: standard 4,364 kg (9,620 lb)
ER 5,384 kg (11,870 lb)
Max fuel weight 9,017 kg (19,880 lb)
Max T-O weight: standard 32,999 kg (72,750 lb)
ER 34,019 kg (75,000 lb)
Max ramp weight: standard 33,112 kg (73,000 lb)
ER 34,132 kg (75,250 lb)
Max landing weight 30,390 kg (67,000 lb)
Max zero-fuel weight 28,259 kg (62,300 lb)
Max wing loading: standard 480.8 kg/m^2 (98.48 lb/sq ft)
ER 495.7 kg/m^2 (101.53 lb/sq ft)
Max power loading: standard 293 kg/kN (2.87 lb/lb st)
ER 302 kg/kN (2.96 lb/lb st)

PERFORMANCE:
Cruising Mach No.:
high speed 0.825 (470 kt; 870 km/h; 541 mph)
normal 0.78 (444 kt; 822 km/h; 511 mph)
Max certified altitude 12,500 m (41,000 ft)
T-O field length at S/L, ISA:
standard 1,564 m (5,130 ft)
ER 1,676 m (5,500 ft)
Landing field length at S/L, ISA, at max landing weight:
standard and ER 1,551 m (5,090 ft)
Range, 70 passengers, at M0.77, IFR reserves:
standard 1,685 n miles (3,120 km; 1,939 miles)
ER 1,985 n miles (3,676 km; 2,284 miles)

OPERATIONAL NOISE LEVELS (FAR Pt 36):
T-O at max T-O weight 83.2 EPNdB
Approach at max landing weight 92.5 EPNdB
Sideline 89.6 EPNdB

UPDATED

BOMBARDIER CRJ900

Engineering designation: CL-600-2D24

TYPE: Regional jet airliner.

PROGRAMME: Announced October 1999; stretched derivative of CRJ700; wind tunnel testing began November 1999; partner/supplier selection finalised second quarter 2000; interior mockup completed March 2000; formal launch at Farnborough International Air Show 24 July 2000; prototype, modified from CRJ700 prototype C-FRJX, with fuselage plugs but retaining CRJ700 wings, landing gear and engines, first flown 21 February 2001; public debut at Paris Air Show 14 June 2001; first production aircraft, c/n 15001/C-GRNH, first flown 20 October 2001; Transport Canada certification achieved 9 September 2002, followed by FAA approval 25 October 2002 and JAA certification 23 December 2002; first customer delivery (c/n 15002) to Mesa Air on 3 February 2003, followed by service entry on 26 April.

In December 2002 Bombardier signed a tentative agreement with China Aviation Industry Corporation I (AVIC I) that could lead to final assembly of the CRJ700 and CRJ900 by Shanghai Aviation Industrial Corporation.

CURRENT VERSIONS: **900:** Standard version.

900ER: Extended-range version.

900ER European: As 900ER but with maximum T-O weight limited to 36,995 kg (81,560 lb) to minimise weight-related charges when operating in European airspace.

900LR: Long-range version.

900LR European: As 900LR but with maximum T-O weight limited to 37,995 kg (83,764 lb).

CUSTOMERS: Firm orders for 39 by 31 March 2004, from Mesa Air, of which 6 then delivered. Estimated market for 800 aircraft in CRJ900 class over 20-year period.

COSTS: Development cost C$200 million; unit cost C$30 million (2002).

DESIGN FEATURES: Compared to CRJ700, has fuselage stretched — by means of 2.29 m (7 ft 6 in) plug forward of centre-section and 1.57 m (5 ft 2 in) plug aft of centre-section — to

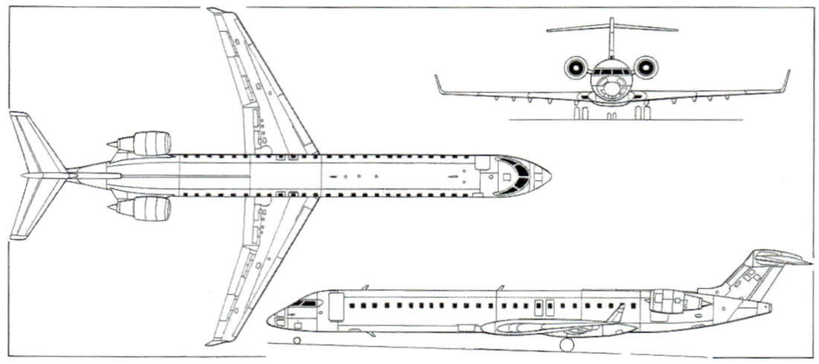

Bombardier CRJ900 *(James Goulding)* 0131847

First CRJ900 operator is Mesa Air ***NEW***/0573605

accommodate 86 to 90 passengers in four-abreast configuration with fore and aft lavatories; 5 to 10 per cent higher thrust engines; strengthened main landing gear with upgraded wheels and brakes; strengthened wing; two additional overwing emergency exits; increased volume in forward underfloor baggage hold, and an additional underfloor baggage door and aft service door on starboard side. Common crew qualification with CRJ200/700 series.

STRUCTURE: Programme partners as for CRJ700. Final assembly and completion by Bombardier at new CRJ700/900 facility at Montréal-Mirabel Airport.

LANDING GEAR: As for CRJ700.

POWER PLANT: Two General Electric CF34-8C5 turbofans. Engine rating 58.4 kN (13,123 lb st) or 63.4 kN (14,255 lb st) with automatic power reserve, flat rated to ISA + 15°C. Fuel capacity 10,989 litres (2,903 US gallons; 2,417 Imp gallons).

ACCOMMODATION: Standard dual-class accommodation for 86 passengers in four-abreast configuration at 79 cm (31 in) seat pitch with fore and aft lavatories and forward galley; alternative configurations include high density with accommodation for 90 passengers at 79 cm (31 in) seat pitch, dual-class with 15 business class seats three-abreast at 86 cm (34 in) seat pitch in forward section and 60 economy class four-abreast at 79 cm (31 in) seat pitch at rear, and dual class with 55 business class seats in four-abreast configuration at 84 cm (33 in) seat pitch in forward section and 24 economy class at 79 cm (31 in) seat pitch at rear. Standard additional floor beam facilitates offset seat rail for three-abreast seating throughout cabin.

SYSTEMS: Honeywell RE220 APU.

AVIONICS: As for CRJ700.

DIMENSIONS, EXTERNAL: As for CRJ700 except
Wing span 23.24 m (76 ft 3 in)
Length overall 36.19 m (118 ft 9 in)
Height overall 7.49 m (24 ft 7 in)
Wheelbase 14.73 m (48 ft 4 in)
Aft service door (stbd):
Height 1.22 m (4 ft 0 in)
Width 0.61 m (2 ft 0 in)
Height to sill 2.39 m (7 ft 10 in)
Emergency exits (four, overwing):
Height 0.91 m (3 ft 0 in)
Width 0.51 m (1 ft 8 in)

DIMENSIONS, INTERNAL:
Cabin (excl flight deck):
Baggage volume: checked 16.81 m^3 (593.5 cu ft)
total 25.57 m^3 (903 cu ft)

WEIGHTS AND LOADINGS:
Operating weight empty 21,546 kg (47,500 lb)
Max payload 10,206 kg (22,500 lb)
Payload with max fuel: 900 6,178 kg (13,620 lb)
900ER 6,972 kg (15,370 lb)
Max fuel weight 8,822 kg (19,450 lb)
Max T-O weight: 900 36,514 kg (80,500 lb)
900ER 37,421 kg (82,500 lb)
900ER European 36,995 kg (81,560 lb)
900LR 38,328 kg (84,500 lb)
900LR European 39,808 kg (87,763 lb)
Max ramp weight: 900 36,627 kg (80,750 lb)
900ER 37,535 kg (82,750 lb)
900LR 38,442 kg (84,750 lb)
Max landing weight 33,339 kg (73,500 lb)
Max zero-fuel weight 31,751 kg (70,000 lb)
Max wing loading: 900 532.1 kg/m^2 (108.98 lb/sq ft)
900ER 545.3 kg/m^2 (111.68 lb/sq ft)
900ER European 539.1 kg/m^2 (110.41 lb/sq ft)
900LR 558.5 kg/m^2 (114.39 lb/sq ft)
900LR European 580.1 kg/m^2 (118.81 lb/sq ft)
Max power loading: 900 313 kg/kN (3.07 lb/lb st)
900ER 320 kg/kN (3.14 lb/lb st)
900ER European 317 kg/kN (3.11 lb/lb st)
900LR 328 kg/kN (3.22 lb/lb st)
900LR European 325 kg/kN (3.19 lb/lb st)

PERFORMANCE:
Cruising Mach No:
high speed 0.83 (476 kt; 882 km/h; 548 mph)

for max range .... 0.78 (444 kt; 822 km/h; 511 mph)
Max certified altitude: 900, 900ER .... 12,500 m (41,000 ft)
Service ceiling, OEI: 900 .... 4,968 m (16,300 ft)
900ER .... 4,755 m (15,600 ft)
FAR T-O field length:
900 .... 1,871 m (6,140 ft)
900ER .... 1,969 m (6,460 ft)
900LR .... 2,068 m (6,785 ft)
FAR landing field length at S/L, ISA, at max landing weight:
900, 900ER, 900LR .... 1,660 m (5,445 ft)
Range, 86 passengers, at M0.77:
900 .... 1,540 n miles (2,852 km; 1,772 miles)
900ER .... 1,777 n miles (3,291 km; 2,045 miles)
900LR .... 1,941 n miles (3,594 km; 2,233 miles)

UPDATED

Bombardier Challenger 300 fifth prototype *(Paul Jackson)* ***NEW***/0573302

Bombardier Challenger 300 flight deck ***NEW***/0573243

BOMBARDIER BD-100 CHALLENGER 300

Engineering designation: BD-100-1A10

TYPE: Business jet.

PROGRAMME: Design study, then known as 'Bombardier Model 70', revealed at the Paris Air Show in June 1997; formally announced at NBAA Convention at Las Vegas 18 October 1998; launched at Paris Air Show 13 June 1999; initially named Continental; first metal cut 21 October 1999 following completion of joint definition phase; AS907 engine first flown 29 January 2000, engine certification achieved 25 June 2002; wing/fuselage mating of first aircraft achieved 19 November 2000; first flight (c/n 20001/C-GJCJ) from the Bombardier Flight Test Center at Wichita's Mid-Continent Airport 14 August 2001, followed by second aircraft (c/n 20002/C-GJCF) on 9 October. These and three further aircraft (c/n 20003/C-GIPX, dedicated to avionics test and flown 6 December 2001; c/n 20004/C-GJCV for systems testing and the first to be fully outfitted with standard interior, flew 5 April 2002; and c/n 20005/C-GIPZ, for function and reliability testing (including cabin systems), originally due to fly in May 2002 but delayed until 8 March 2003; participated in the flight test and certification programme culminating in Transport Canada 525 approval, FAA FAR Pt 25 and JAA JAR 25 certification, with RVSM approval, FAR Pt 36 Stage 3 noise compliance. Transport Canada certification achieved 31 May 2003, followed by FAA approval on 4 June 2003 and JAA certification on 31 July 2003, first delivery (N505FX c/n 20006) to Bombardier Business Jet Solutions (Flexjet) 23 December 2003; first revenue flight 8 January 2004. Total of seven delivered by 31 January 2004.

Public debut at NBAA Convention, New Orleans, 11 December 2001 (formal presentation 12 December). European debut (C-GJCV) at EBACE 2003 at Geneva 5 May 2003. Re-named Challenger 300 on 8 September 2002, immediately prior to NBAA Convention at Orlando, Florida. Target production up to 15 in 2003, rising to 40 in 2004 and the maximum planned rate of 60 per year from 2005.

CUSTOMERS: Two orders signed at time of launch, by customers in Germany and United Arab Emirates; total of 125 firm orders received by 5 May 2003, including 25 for Bombardier's Flexjet fractional ownership programme, 15 for European customers and five for its Middle East and Arab nations distributor TAG Aeronautics. Bombardier anticipates gaining 30 per cent of estimated 1,200-aircraft market in this class by 2012, with fractional ownership operations especially targeted.

COSTS: Development cost C$500 million (1998); break-even estimated at 300th aircraft. Unit cost US$16.29 million typically equipped (2002). Direct operating cost estimated at US$1,329 per hour (2002).

DESIGN FEATURES: Design goals included coast-to-coast range across USA with eight passengers in cabin with stand-up headroom and take-off field length less than 1,525 m (5,000 ft). General configuration is as shown in the accompanying illustrations; supercritical wing with winglets, sweepback 27° at quarter-chord.

FLYING CONTROLS: Conventional. Ailerons manually actuated via cables, pulleys and pushrods, each with a geared tab and fixed tab, plus trim tab on port aileron only; maximum aileron deflections ±18°. Horn-balanced elevators, maximum deflections +24/–18°, and single rudder panel, maximum deflection ±30°, hydraulically actuated by cables and pulleys with manual reversion, each with dual PCUs; variable incidence tailplane for pitch trim, maximum travel +2/–12°. Hydraulically actuated Fowler flaps, maximum deflection 30°; each wing has two-segment multifunction spoiler outboard, maximum deflection 45°, and two-segment ground spoiler/lift dumper inboard, maximum deflection 60°; yaw damper standard.

STRUCTURE: Primarily light alloy, with composites for some non-structural fairings; fuselage of semi-monocoque construction with frames and stringers; two-spar wing; three-spar fin. Programme suppliers include: AIDC Taiwan (rear fuselage and tail unit); Canadair (cockpit, forward fuselage and primary flight controls); DeCrane Aircraft (cabin interior) ECE (electrical system and cockpit lighting); Fischer Austria (wing-to-fuselage fairings); GKN Westland (engine nacelles); Goodrich (wheels and brakes); Hawker de Havilland Australia (tailcone and APU installation kit); Hella (lighting); Honeywell (power plant and APU); Hurel-Dubois (thrust reversers); Intertechnique (fuel system); Liebherr Aerospace-Toulouse (environmental control and anti-icing systems); Liebherr Aerospace Lindenberg (flap control system); Messier-Dowty (landing gear); Mitsubishi Heavy Industries (wing); Moog (secondary flight controls); NLX (flight training device and level C/D flight simulator); Parker Aerospace (hydraulic system); PPG Industries (cockpit windscreens and cabin windows); Rockwell Collins (avionics); Scott Aviation (oxygen system); Shorts (centre fuselage), and Walter Kidde (fire detection and suppression system). Final assembly will be at Bombardier's Learjet facility in Wichita, with interior completion in Tucson.

LANDING GEAR: Hydraulically retractable tricycle type by Messier-Dowty, with two wheels on each unit; trailing-link-type main units retract inwards, nosewheel forwards. Steerable nosewheel, maximum deflection ±65°. Mainwheel tyre size 26.5×8.0-18, nosewheel tyre size 18×5.5-10. Goodrich carbon composites multiple disc brakes. Turning radius 17.68 m (58 ft 0 in).

POWER PLANT: Two Honeywell HTF7000 (formerly AS907) turbofans with FADEC, each with thermodynamic rating of 35.81 kN (8,050 lb st), flat-rated to 28.91 kN (6,500 lb st) with APR at ISA+15°C. All fuel contained in two integral wing tanks, combined capacity 7,934 litres (2,096 US gallons; 1,745 Imp gallons). Oil capacity both engines, 10.40 litres (2.75 US gallons; 2.29 Imp gallons). Gravity fuelling point in top of each wing, near leading-edge, plus single-point pressure refuelling/defuelling port in starboard wingroot near leading-edge. Target-type reversers standard.

ACCOMMODATION: Two crew flight deck; cabin, with flat floor, accommodates eight passengers in standard 'double club' arrangement on tracking, swivelling and reclining 16 *g* seats with retractable headrests, cabin management controls, cupholders and shoulder harnesses; three-seat 16 *g* take-off and landing-certified divan with extending backrest optional as interchange for two club seats. Standard cabin equipment includes fold-out work tables; one 110 V electrical outlet per club seat group; hot drinks dispensers; DVD/CD player with hi-fi grade audio speakers and two 381 mm (15 in) flat screen monitors; Airshow 400 system; Magnastar 2000 in-flight telephone with two handsets and switchable locations; extended-life LED lighting; forward galley with microwave oven; forward passenger wardrobe and crew coat closet; aft lavatory and vanity unit with hot and cold water and removable waste tank, and flight-accessible baggage compartment. Free-fall opening/power-assisted-closure, semi-plug-type airstair cabin door, on port side

Bombardier Challenger 300 fourth prototype ***NEW***/0573303

Bombardier Challenger 300 cabin interior ***NEW***/0573244

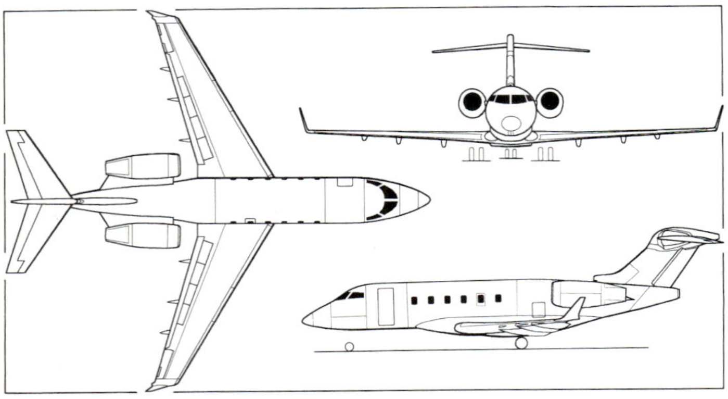

Bombardier Challenger 300 ***(Paul Jackson)*** ***NEW***/0580554

immediately aft of flight deck, also serves as Type I emergency exit; single plug-type overwing Type III emergency exit on starboard side, between rearmost pair of club seats. External baggage door aft of port wing trailing-edge. Cabin and baggage compartment pressurised, air-conditioned and heated.

SYSTEMS: Two independent phosphate-ester hydraulic systems with one engine-driven pump and one DC motor pump per system, pressure 207 bar (3,000 lb/sq in), plus one auxiliary system powered by an accumulator. Pressurisation system, differential 0.60 bar (8.78 lb/sq in), with auxiliary system providing pressurisation up to 10,670 m (35,000 ft). 28 V DC electrical system comprises three 400 Ah DC brushless generators (one each on the engines and one on the APU) and two 24 V 44 Ah Ni/Cd batteries which provide power for APU starting, in-flight emergency power and ground power. APU generator can carry load of a failed engine generator, and one battery can supply power for APU starting. Oxygen system, capacity to suit customer requirements, with demand-type masks for crew and drop-down masks for passengers.

Engine bleed air automatically controlled anti-icing for wing leading-edges and nacelle lips; electrically anti-iced windscreen; heated angle-of-attack vanes and pitot probes. Honeywell tailcone-mounted RE220 APU, with FADEC, will be certified for operation up to 11,280 m (37,000 ft) and in-flight starting to 9,150 m (30,000 ft).

AVIONICS: Rockwell Collins Pro Line 21 as core system.

Comms: Dual VHF com with 8.33 kHz frequency spacing capability; dual integrated radio control and display units; dual transponders, all standard. Third VHF com; dual HF com, satcom, VHF/satcom datalink capability, Selcal and ELT optional.

Radar: Dual-scan digital weather radar with optional turbulence detection.

Flight: Standard equipment includes dual ILS/VOR/markers, AHRS and air data computers; single ADF, DME, FMS/CDU, GPS sensor, EGPWS, TCAS II, EICAS, radio altimeter, CVR and flight deck aural warning system. Second ADF, DME, FMS/CDU and GPS, three-dimensional flight plan maps, FDR and lightning sensor optional.

Instrumentation: EFIS with four 305 × 254 mm (12 × 10 in) colour LCDs providing liquid PFD and MFD functions for pilot and co-pilot.

DIMENSIONS, EXTERNAL:

Wing span over winglets ... 19.46 m (63 ft 10 in)
Length overall ... 20.93 m (68 ft 8 in)
Height overall ... 6.17 m (20 ft 3 in)
Fuselage max diameter ... 2.34 m (7 ft 8 in)
Tailplane span ... 7.23 m (23 ft 8½ in)
Wheel track ... 3.20 m (10 ft 6 in)
Wheelbase ... 8.46 m (27 ft 9 in)
Passenger door: Height ... 1.89 m (6 ft 2½ in)
  Width ... 0.76 m (2 ft 6 in)
Baggage door: Height ... 0.76 m (2 ft 6 in)
  Width ... 0.61 m (2 ft 0 in)
  Height to sill ... 1.63 m (5 ft 4 in)
Emergency exit: Height ... 0.91 m (3 ft 0 in)
  Width ... 0.51 m (1 ft 8 in)

DIMENSIONS, INTERNAL:

Cabin (excl cockpit):
  Length ... 8.71 m (28 ft 7 in)
  Width: at centreline ... 2.18 m (7 ft 2 in)
    at floor ... 1.55 m (5 ft 1 in)
  Max height ... 1.85 m (6 ft 1 in)
  Floor area ... 13.5 m^2 (146 sq ft)
  Volume ... 24.35 m^3 (860 cu ft)
Baggage compartment volume ... 2.99 m^3 (105.50 cu ft)

AREAS:

Wings, net ... 48.49 m^2 (522.0 sq ft)
Ailerons, total ... 0.93 m^2 (10.00 sq ft)

Fifth Bombardier Challenger 300 taking off for the first time on 8 March 2003 0547101

Challenger 300 wearing its new name for the first time at NBAA Convention, Orlando, Florida, September 2002 ***(Paul Jackson)*** 0533392

Trailing-edge flaps, total ... 7.56 m^2 (81.40 sq ft)
Spoilers, total ... 3.55 m^2 (38.26 sq ft)
Rudder, incl tabs ... 1.89 m^2 (20.40 sq ft)
Tailplane ... 11.39 m^2 (122.55 sq ft)
Elevators ... 4.22 m^2 (45.40 sq ft)

WEIGHTS AND LOADINGS (provisional):

Operating weight empty ... 10,591 kg (23,350 lb)
Payload: max ... 1,247 kg (2,750 lb)
  with max fuel ... 522 kg (1,150 lb)
Fuel weight:max ... 6,418 kg (14,150 lb)
  with max payload ... 5,693 kg (12,550 lb)
Max T-O weight ... 17,463 kg (38,500 lb)
Max ramp weight ... 17,531 kg (38,650 lb)
Max landing weight ... 15,308 kg (33,750 lb)
Max zero-fuel weight ... 11,839 kg (26,100 lb)
Max wing loading ... 360.1 kg/m^2 (73.75 lb/sq ft)
Max power loading ... 244 kg/kN (2.39 lb/lb st)

PERFORMANCE (estimated):

Max level speed ... 476 kt (882 km/h; 548 mph)
Max operating speed:
  S/L to FL80 ... 300 kt (556 km/h; 345 mph)
  FL80 to 8,984 m (29,475 ft) ... 320 kt (593 km/h;368 mph)
  Above 8,984 m (29,475 ft) ... M0.83
High cruising speed ... M0.82 or 470 kt (870 km/h; 541 mph)
Normal cruising speed ... M0.80 or 459 kt (850 km/h; 528 mph)
Max rate of climb at S/L ... 1,097 m (3,600 ft)/min
Rate of climb at S/L, OEI ... 205 m (673 ft)/min
Initial cruising altitude ... 12,500 m (41,000 ft)
Max certified altitude ... 13,715 m (45,000 ft)
T-O balanced field length ... 1,439 m (4,720 ft)
Landing run at max landing weight ... 792 m (2,600 ft)
Range with eight passengers, NBAA IFR reserves ... 3,100 n miles (5,741 km; 3,567 miles)

UPDATED

BOMBARDIER CL-600 CHALLENGER 604

Canadian Forces designations: CC-144, CC-144B and CE-144A
Engineering designation: CL-600-2B16

TYPE: Business jet.

DEVELOPMENT MILESTONES

Challenger 600	
First flight	8 Nov 78
Certification	10 Aug 80
First flight, production	21 Sep 79
First delivery	30 Dec 80
Subsequent versions:	
Challenger 601	
First flight	10 Apr 82
First flight, production	17 Sep 82
Certification	25 Feb 83
First delivery	6 May 83
Challenger 604	
First flight	18 Sep 94
Certification	20 Sep 95

PROGRAMME: First flight of prototype (C-GCGR-X) 8 November 1978; first flight production Challenger 600 with AlliedSignal ALF 502L-2 turbofans 21 September 1979; first customer delivery 30 December 1980; first flight Challenger 601 with GE CF34s 10 April 1982; first 601-1A delivered 6 May 1983; first 601-3A 6 May 1987 and first 601-3A/ER 19 May 1989; first 601-3R 14 July 1993; first 604 25 January 1996. Challenger certified for operation in

Bombardier Challenger 604 *(Paul Jackson)* **NEW**/0573297

40 countries by 1998. By 28 February 2003 the Challenger fleet had flown 2.36 million hours, with a despatch reliability in excess of 99.7 per cent. 500th Challenger rolled out 'green' 25 May 2000 and handed over (as N816CC) 1 September 2000; 600th for customer delivery (c/n 5557) delivered for completion 6 February 2003 and (as C-GAWH) formally handed over to Clearwater Fine Foods of Nova Scotia at NBAA Convention, Orlando, Florida, on 8 October 2003.

CURRENT VERSIONS: **Challenger 600:** Total 84 (excluding one prototype) built (76 since retrofitted with winglets); 12 delivered to Canadian Department of National Defence as CC-144 (three) and CE-144A (three), plus three for coastal patrol, two for general transport and one test aircraft. Production completed with final delivery on 22 June 1983.

Challenger 601-1A: First production version to have CF34 engines; first flight 17 September 1982. Deliveries (66, including four CC-144Bs) between 6 May 1983 and 29 May 1987.

Challenger 601-3A: Version with 'glass' cockpit and CF34-3A engines; first flight 28 September 1986; Canadian and US certification 21 and 30 April 1987; also certified for Cat. II and in 22 other countries; improvements include CF34-3A engines flat rated to 21°C, and fully integrated digital flight guidance and flight management systems. Total of 134 delivered between 6 May 1987 and 29 October 1993.

Challenger 601-3R Extended-range option available on new 601-3As since 1989 (c/n 5135 and onwards) and as retrofit to 601-1As and 601-3As; range increased to 3,585 n miles (6,639 km; 4,125 miles) with NBAA IFR reserves; first flight 8 November 1988; Canadian certification 16 March 1989; tail fairing replaced with conformal tailcone fuel tank which extends fuselage length by 46 cm (1 ft 6 in) and adds 118 kg (260 lb) to operating weight empty; maximum ramp weight increased by 680 kg (1,500 lb). Optional gross weight increase of 227 kg (500 lb). Total of 92 modification kits supplied between March 1989 and October 1993. Challenger 601-3ER, incorporating extended-range modifications, CF34-3A1 engines and 20,457 kg (45,100 lb) max T-O weight, was standard production version from 14 July 1993 (first delivery); last of 59 new-build aircraft delivered by early 1996.

Challenger 604: Has range of 4,077 n miles (7,550 km; 4,691 miles) at M0.74 and is powered by General Electric CF34-3B engines each rated at 38.8 kN (8,729 lb st) T-O power at ISA + 15°C. Prototype (C-FTBZ) modified on the production line from a Challenger 601-3R; engineering designation CL-600-2B16; first flight (with CF34-3A engines) 18 September 1994; first flight with definitive CF34-3B engines 17 March 1995. Exploits systems developed in Regional Jet programme. Rockwell Collins Pro Line 4 EFIS; extra 1,242 litres (328 US gallons; 273 Imp gallons) of fuel in aft equipment bay, forward fuselage tank and tail tank. Automatic aft-CG control to reduce trim drag for longer range. New landing gear, carbon brakes and anti-skid system; strengthened tail unit; new wing-to-fuselage and underbelly fairings. Maximum T-O weight 21,863 kg (48,200 lb). Transport Canada certification achieved 20 September 1995; FAA certification 2 November 1995; 100th delivery to a customer was made in mid-1999.

From June 2001, Challenger 604s have been delivered with upgraded PrecisionPlus Collins Pro Line 4 avionics, intended to reduce pilot workload and make the aircraft more compatible with future air traffic environments. Standard PrecisionPlus features include automatic look-up and display of take-off, approach, landing and missed-approach speeds, eliminating the need to refer to manual charts; automatic look-up and display of thrust setting (N_1) for take-off, climb, cruise and go-around; blending of actual observed wind and entered wind to improve the prediction of flight time and fuel requirements; position reporting in non-radar environments such as the North Atlantic; improved polar navigation, enabling the crew to navigate and steer the aircraft at latitudes over 89°; full-time DME reporting on the pilot's MFD; EICAS improvements including the addition of metric fuel indication capability, logic enhancements and FMS performance enhancements; and full integration with the Flight Dynamics HUD and Safe Flight AutoPower autothrottle system.

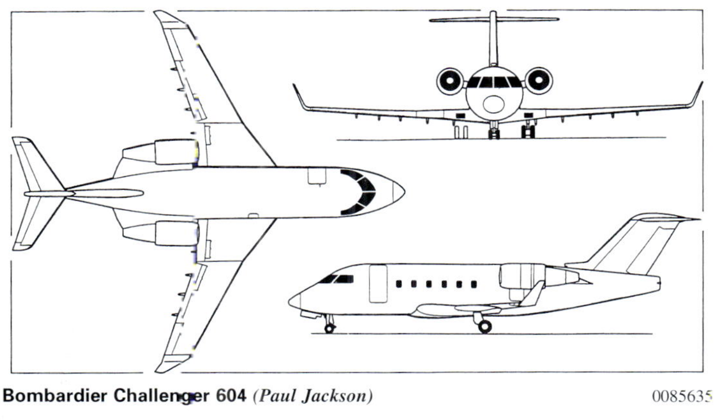

Bombardier Challenger 604 *(Paul Jackson)* 0085635

Optional features include flight plan map feature providing an intuitive, three-dimensional graphic representation of the programmed flight plan and predicted flight path for the pilot's and co-pilot's MFDs; long-range cruise feature allowing pilots to select a cruise speed computed by the FMS for either maximum range or maximum speed; search pattern feature offering automatic generation of waypoints; and expanded FDR to meet FAA FAR Pt 135.152 requirements. The PrecisionPlus avionics upgrade is also available for retrofit to earlier Challenger 604s.

Max-Viz EVS-1000 enhanced vision system received FAA certification on 13 March 2003 for installation on Challengers.

Recent customers include the US Federal Aviation Administration, which ordered one, with two options, in January 2004.

Detailed description applies to Challenger 604.

Special Missions: One Challenger 604 delivered in late 2000 to (South) Korean National Maritime Police with unspecified sensor and communications suite. First of three maritime surveillance 604s entered service with Royal Danish Air Force in late 2002 although first flight with full modifications (ventral radome and retractable EO sensor installed by Field Aviation) not achieved until 28 February 2003.

CUSTOMERS: See under individual headings in Current Versions. Total 630 Challengers of all versions delivered (including to completion centres) by 14 May 2004, including 285 Challenger 604s. Annual deliveries have included 33 in 1997, 36 in 1998, 40 in 1999, 38 in 2000, 41 in 2001, 31 in 2002, 24 in 2003, and six in the first three months of 2004.

COSTS: Unit cost (604), US$4 million, typically equipped (2002).

DESIGN FEATURES: Advanced wing section; quarter-chord sweep 25°; thickness/chord ratio 14 per cent at root, 12 per cent at leading-edge sweep break and 10 per cent at tip; dihedral 2° 33'; incidence at root 3° 30'; fuselage circular cross-section, pressurised.

FLYING CONTROLS: Conventional, fully powered hydraulic controls; electrically actuated variable incidence tailplane; two-segment spoilers (outboard airbrake panels, inboard lift dumpers); two-segment double-slotted flaps.

STRUCTURE: Two-spar wing torsion box; chemically milled fuselage skin panels with riveted frames and stringers form damage-tolerant structure; multispar fin and tailplane.

LANDING GEAR: Hydraulically retractable tricycle type, with twin wheels and Dowty oleo-pneumatic shock-absorber on each unit. Mainwheels retract inward into wing centre-section, nose unit forward. Nose unit steerable and self-centring. Mainwheels have H27×8.5-14 (16 ply) tubeless tyres, pressure 12.07 bar (175 lb/sq in); nosewheels have Goodrich 18×4.4 (12 ply) tubeless (deflector-type) tyres, pressure 10.00 bar (145 lb/sq in). ABS (Aircraft Braking Systems) hydraulically operated multiple-disc carbon brakes with fully modulated anti-skid system. Minimum ground turning radius 12.19 m (40 ft 0 in).

POWER PLANT: Two General Electric CF34-3B1 turbofans, each rated at 41.0 kN (9,220 lb st) with automatic power reserve, or 38.8 kN (8,729 lb st) without APR, pylon-mounted on rear fuselage and fitted with cascade-type fan-air thrust reversers. Nacelles and thrust reversers by Shorts. Integral fuel tank in centre-section, capacity 2,839 litres (750 US gallons; 624 Imp gallons), one in each wing (each 2,725 litres; 720 US gallons; 600 Imp gallons) and auxiliary tanks (combined capacity 1,181 litres; 312 US gallons; 260 Imp gallons) beneath cabin floor. Saddle tanks, total capacity 999 litres (264 US gallons; 220 Imp gallons); tank in tailcone, capacity 745 litres (197 US gallons; 164 Imp gallons). Total fuel capacity 11,214 litres (2,963 US gallons; 2,468 Imp gallons). Pressure and gravity fuelling and defuelling. Oil capacity 13.6 litres (3.6 US gallons; 3.0 Imp gallons).

ACCOMMODATION: Two-pilot flight deck with dual controls. Blind-flying instrumentation standard. Cabin interiors to customer's specifications; maximum of 19 passenger seats and three crew approved. Typical installations include lavatory, buffet, bar and wardrobe. Medevac version can carry up to seven stretcher patients, infant incubator, full complement of medical staff and comprehensive intensive care equipment. Baggage compartment, with own loading door, accessible in flight. Downward-opening, power-assisted door on port side, forward of wing. Overwing emergency exit on starboard side. Entire accommodation heated, pressurised and air conditioned. Optional extended cabin interior increases cabin length by 0.51 m (1 ft 8 in) and provides two additional cabin windows by removing rear closet and moving the lavatory and baggage compartment bulkheads, with corresponding 0.20 m^3 (7 cu ft) reductions in baggage capacity. Ultra Electronics active noise vibration control (ANVC) system optional.

SYSTEMS: Honeywell pressurisation and air conditioning systems, maximum pressure differential 0.63 bar (9.1 lb/sq in). Three independent hydraulic systems, each of 207 bar (3,000 lb/sq in). No. 1 system powers flight controls (via servo-actuators positioned by cables and pushrods); No. 2 system for flight controls and brakes; No. 3 system for flight controls, landing gear extension/retraction, brakes and nosewheel steering. Nos. 1 and 2 systems each powered by an engine-driven pump, supplemented by an AC electric pump; No. 3 system by two AC pumps. Two 30 kVA engine-driven generators supply primary 115/200 V three-phase AC electric power at 400 Hz. Four transformer-rectifiers to convert AC power to 28 V DC; one primary 24 V 17 Ah Ni/Cd battery and one auxiliary 24 V 43 Ah battery. Alternative primary power provided by APU and/or an air-driven generator, latter deployed automatically in flight if engine-driven generators and APU are inoperative. Stall warning system, with stick shakers and stick pusher. Honeywell GTCP-100E gas-turbine APU for engine start, ground air conditioning and other services. Electric anti-icing of windscreen, flight deck side windows and pitot heads; Hamilton Sundstrand bleed air anti-icing of wing leading-edges, engine intake cowls and guide vanes. Gaseous oxygen system, pressure 127.5 bar (1,850 lb/sq in). Continuous-element fire detectors in each engine nacelle, APU and main landing gear bays; two-shot extinguishing system for engines, single-shot system for APU.

AVIONICS: Rockwell Collins Pro Line 4 with PrecisionPlus upgrade.

Comms: Dual VHF; dual ATC transponders; dual HF; cockpit voice recorder.

Radar: Rockwell Collins WXP-4220 colour digital weather radar with turbulence detection.

Flight: Dual VHF nav with provision for third; dual DME; dual ADF; dual Litton LTN-101 laser inertial reference systems (LIRS) with full provision for third; dual flight management system with provision for third; digital automatic flight control system, with dual-channel autopilot and flight director; Mach trim and auto trim; dual digital air data system. Flight Dynamics HGS 2150 HUD received FAA approval on 4 April 2000 and is optional. Space provisions for flight data recorder, ELT, dual GPS, EGPWS, AFIS, TCAS II and Safe Flight Instrument Corporation AutoPower enhanced autothrottle system.

Instrumentation: Rockwell Collins digital avionics include Pro Line 4 six-tube EFIS with 184 × 184 mm (7¼ × 7¼ in) CRT displays which include two-tube EICAS display (MFD); standby instruments (artificial horizon, airspeed indicator, compass and altimeter). Systems certified for Cat. II operations.

EQUIPMENT (MEDEVAC VERSION): Includes cardiopulmonary resuscitation unit; physio control lifepack comprising heart defibrillator, ECG and cardioscope; ophthalmoscope; respirators and resuscitators; infant monitor; X-ray viewer; cardiostimulator; foetal heart monitor; and anti-shock suit.

DIMENSIONS, EXTERNAL:
Wing span over winglets ... 19.61 m (64 ft 4 in)
Wing chord: at root ... 3.99 m (13 ft 1 in)
at tip ... 1.27 m (4 ft 2 in)
Wing aspect ratio (excl winglets) ... 8.0
Length overall ... 20.85 m (68 ft 5 in)
Fuselage: Length ... 18.77 m (61 ft 7 in)
Max diameter ... 2.69 m (8 ft 10 in)
Height overall ... 6.30 m (20 ft 8 in)
Tailplane span ... 6.20 m (20 ft 4 in)
Wheel track (c/l of shock-struts) ... 3.18 m (10 ft 5 in)
Wheelbase ... 7.99 m (26 ft 2½ in)
Passenger door (port, fwd): Height ... 1.78 m (5 ft 10 in)
Width ... 0.94 m (3 ft 1 in)
Height to sill ... 1.63 m (5 ft 4 in)
Baggage door (port, rear): Height ... 0.84 m (2 ft 9 in)
Width ... 0.71 m (2 ft 4 in)
Height to sill ... 1.73 m (5 ft 8 in)
Overwing emergency exit (stbd):
Height ... 0.91 m (3 ft 0 in)
Width ... 0.51 m (1 ft 8 in)

DIMENSIONS, INTERNAL:
Cabin: Length, incl galley, lavatory and baggage area, excl flight deck ... 8.66 m (28 ft 5 in)
Max width ... 2.49 m (8 ft 2 in)
Width at floor level ... 2.18 m (7 ft 2 in)
Max height ... 1.85 m (6 ft 1 in)
Floor area ... 18.8 m^2 (202 sq ft)
Volume ... 32.6 m^3 (1,150 cu ft)

AREAS:
Wings, gross (excl winglets) ... 48.31 m^2 (520.0 sq ft)
Ailerons (total) ... 1.39 m^2 (15.0 sq ft)
Trailing-edge flaps (total) ... 7.80 m^2 (84.0 sq ft)
Fin ... 9.18 m^2 (98.8 sq ft)
Rudder ... 2.03 m^2 (21.9 sq ft)
Tailplane ... 6.45 m^2 (69.4 sq ft)
Elevators (total) ... 2.15 m^2 (23.1 sq ft)

WEIGHTS AND LOADINGS:
Manufacturer's weight empty ... 9,806 kg (21,620 lb)
Operating weight empty ... 12,331 kg (27,185 lb)
Max fuel ... 9,072 kg (20,000 lb)
Max payload ... 2,184 kg (4,815 lb)
Payload with max fuel ... 506 kg (1,115 lb)
Fuel with max payload ... 7,393 kg (16,300 lb)
Max T-O weight: standard ... 21,591 kg (47,600 lb)
optional ... 21,863 kg (48,200 lb)
Max ramp weight: standard ... 21,636 kg (47,700 lb)
optional ... 21,908 kg (48,300 lb)
Max landing weight ... 17,236 kg (38,000 lb)
Max zero-fuel weight ... 14,515 kg (32,000 lb)
Max wing loading:
standard ... 446.9 kg/m^2 (91.54 lb/sq ft)
optional ... 452.6 kg/m^2 (92.69 lb/sq ft)
Max power loading:
standard ... 263 kg/kN (2.58 lb/lb st)
optional ... 267 kg/kN (2.61 lb/lb st)

PERFORMANCE (at standard max T-O weight, except where indicated):
High-speed cruising speed ... M0.82 (470 kt; 870 km/h; 541 mph)
Normal cruising speed ... M0.80 (459 kt; 851 km/h; 529 mph)
Long-range cruising speed ... M0.74 (425 kt; 787 km/h; 489 mph)
Time to initial cruising altitude ... 22 min
Initial cruising altitude ... 11,460 m (37,600 ft)
Max certified altitude ... 12,500 m (41,000 ft)
Service ceiling, OEI: at mid-cruise weight 17,373 kg (38,300 lb) ... 6,920 m (22,700 ft)
at max T-O weight ... 5,170 m (16,960 ft)
Balanced T-O field length (ISA at S/L) ... 1,780 m (5,840 ft)
Landing distance at S/L at max landing weight ... 846 m (2,777 ft)
Range with max fuel and five passengers, NBAA IFR reserves:
long-range cruising speed ... 4,027 n miles (7,458 km; 4,634 miles)
normal cruising speed ... 3,714 n miles (6,878 km; 4,274 miles)
Design *g* limit ... +2.5

OPERATIONAL NOISE LEVELS:
T-O ... 81.2 EPNdB
Sideline ... 86.2 EPNdB
Approach ... 90.3 EPNdB

UPDATED

BOMBARDIER BD-700 GLOBAL EXPRESS

Engineering designation: BD-700-1A10

TYPE: Long-range business jet.

DEVELOPMENT MILESTONES

Announced	28 Oct 91
Rolled out	26 Aug 96
First flight	13 Oct 96
Public debut	Nov 96
Certification	31 Jul 98
First delivery	8 Jul 99

PROGRAMME: Announced 28 October 1991 at NBAA Convention; full-scale cabin mockup exhibited at NBAA Convention September 1992; conceptual design started early 1993. Programme launched 20 December 1993; high-speed configuration frozen June 1994; low-speed configuration established August 1994.

Ground test programme using static test airframe c/n 0001 began August 1996; prototype C-FBGX (engineering designation BD-700-1A10) rolled out 26 August 1996; first flight 13 October 1996; public debut at NBAA Convention at Orlando, Florida, November 1996; prototype and three other aircraft undertook 2,000-hour, 18-month flight test programme based at Bombardier's flight test centre in Wichita, Kansas; second aircraft, (C-FHGX), which is used for systems evaluation and testing, flew 3 February 1997, third (C-FJGX), which is used for avionics and autopilot testing, 22 April 1997; fourth (C-FKGX), first flown 8 September 1997 and the first to be fully outfitted, was used for function and reliability testing.

Transport Canada certification 31 July 1998; FAA certification 13 November 1998; JAA certification 7 May 1999; German LBA certification 26 May 1999; first customer delivery of completed aircraft 8 July 1999 to AirFlite Inc of Long Beach, California, which operates the aircraft on behalf of Toyota Motor Sales USA; 50th 'green' airframe delivered to Montréal completion centre 7 June 2000. Transport Canada and JAA RVSM approval granted 16 January 2001, followed by FAA RVSM approval on 29 January. Thales Avionics head-up flight display system (HFDS) achieved Transport Canada certification on 14 September 2001 and FAA approval on 4 October 2001. Following exploratory discussions with potential suppliers, in-house development of Bombardier Enhanced Vision System (BEVS) began in 2002 and scheduled for certification in first quarter 2005. At 28 February 2003 total fleet tiome stood at 73,000 flying hours with a despatch rate of 99.25 per cent. 100th aircraft (c/n 9100/N1SA) delivered 18 December 2002 to Stanford Financial Group of Houston, Texas.

CURRENT VERSIONS: **Global Express:** *As described.*

Bombardier Global Express long-range business jet ***NEW***/0546838

Computer-generated image of Global Express XRS long-range business jet *NEW*/0568991

Impression of Global Express AGS, offered to NATO *NEW*/0548618

Global Express XRS: Improved version, announced at NBAA Convention in Orlando, Florida, 6 October 2003. Design goals include increased range at high speed, improved take-off performance and new fast-refuelling technology. Additional forward fuselage tank in wing/fuselage fairing adds 674 kg (1,486 lb) of usable fuel; zero-flap take-off capability enhances hot-and-high performance with increased fuel load; software upgrades to the fuel computer, coupled with structural changes, reduce refuelling time by 15 minutes. Bombardier enhanced vision system (BEVS) is standard. Increased pressure differential maintains a 1,372 m (4,500 ft) cabin environment at FL450, and 1,737 m (5,700 ft) environment at FL510; upgraded humidification system optional. Cabin features two additional windows (forward starboard, opposite door; and rear port; adjacent to flaps), providing 40 per cent more natural light in forward vestibule area; redesigned floor plan with full galley on port side; 12-hour non-stop-flight-approved crew area with overhead storage on starboard side; 0-42 m³ (15 cu ft) of additional aft storage volume; increased overhead stowage in crew area; and LED lighting. Service entry scheduled for early 2006. Unit cost US$45.5 million, typically equipped (2003).

Sentinel R. Mk 1: Ground surveillance version for UK; prime contractor Raytheon Systems.

Sentinel AGS: Airborne ground surveillance. The Global Express has been chosen as the airborne platform for the Raytheon-led Co-operative Transatlantic AGS Solution (CTAS) bid for the NATO Alliance Ground Surveillance (AGS) system, for which a development contract was expected to be awarded in December 2004. If selected, the AGS-modified Global Express would have reduced service ceiling and range compared with the business jet, estimated at 13 715 m (45,000 ft) and 5,000 n miles (9,260 km; 5,754 miles), respectively.

CUSTOMERS: Delivery of 125th aircraft (excluding two Global 5000s, but including prototypes) was effected (to a completion centre) on 23 December 2003. Recent customers include Qatar Airways which ordered two, with two options, on 8 December 2003, for delivery in mid- and late 2004. Estimated market for 500 to 800 long-range business jets over 15 years; Bombardier anticipates capturing 50 per cent of the market, breaking even at approximately 100; target production rate 34 per year; total of 35 delivered in 2000, 21 in 2001, 16 in 2002, 17 in 2003, and nine in the first three months of 2004. Total of 134 in service by 14 May 2004.

COSTS: Development costs C$800 million; half carried by Bombardier, balance by risk-sharing partners. Unit cost US$43.35 million (2002).

DESIGN FEATURES: Design goal was longest possible range at highest speed from short runway with 99.5 per cent despatch reliability; wide-body fuselage, combining Challenger cabin cross-section with cabin length of Regional Jet; all-new, 'third-generation supercritical' wings with leading-edge slats and winglets.

Wing sweep 35° at quarter-chord, thickness/chord ratio 11 per cent, dihedral 2° 30', root incidence 2° 30'. Wing, high-lift devices and wing/fuselage interface and area-ruled rear fuselage/engine pylon junction contours developed with extensive use of computational fluid dynamics (CFD). Rear-mounted engines. Sweptback T tail with 38° sweep and 5° anhedral on tailplane. 45° sweep on fin.

FLYING CONTROLS: Conventional and mechanical. Fully powered primary flying controls with variable artificial feel and emergency back-up via ram air turbine following triple hydraulic failure; dual sidestick controllers; duplicated cable runs with automatic disconnect in the event of control surface jamming; dual power control units on ailerons (maximum deflections +26.5/−23° and elevators (maximum deflections +24/−19°), triple units on rudder (maximum deflection 37° left/right). Eight-section (total) leading-edge slats (maximum deflection 20°) and six-section (total) single-slotted Fowler flaps (maximum deflection 30°) are signalled by dual electronic control units and operated by dual-motor power units connected by rigid driveshafts to ball-screw actuators. Electrically signalled, hydraulically actuated multifunction spoilers (four per side, outboard, operating differentially to assist ailerons and improve roll response, and symmetrically for speed brake or lift dump functions) and ground spoilers (two per side, inboard), maximum deflection +40°. Horizontal stabiliser incidence adjustable for pitch trim (+13/−2°) via dual-channel electrically driven screw actuator; roll trim accomplished by electric trim actuator located at aileron feel simulator unit; yaw trim accomplished by electric trim actuator at summing unit in fin. Dual yaw damper stability augmentation system and stick shaker/pusher stall protection system standard.

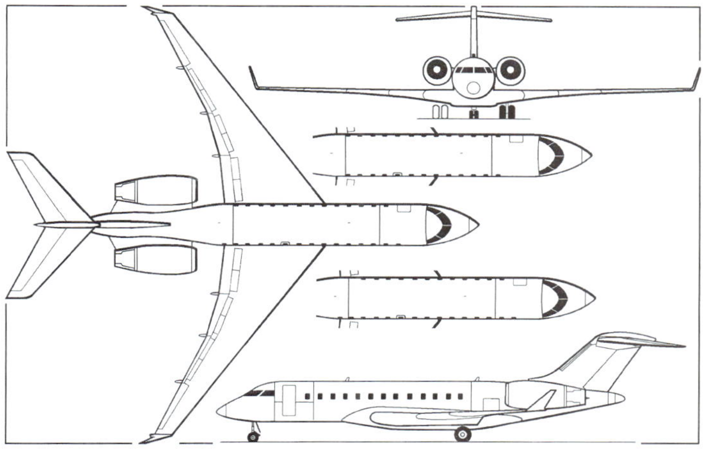

Bombardier BD-700 Global Express with maximum option of 15 standard cabin windows, plus scrap views of (upper) standard cabin glazing and (lower) Global Express XRS *(Paul Jackson)* *NEW*/0567062

STRUCTURE: Semi-monocoque fuselage with chemically milled C-188 A1 aluminium alloy skin riveted over alloy frames and stringers to form damage-tolerant structure; main two-spar torsion box wing structure mostly of alloy construction, with machined alloy spars and ribs and polyurethane-coated machined alloy skin panels; two-spar winglets of mixed alloy/composites construction; multispar fin is alloy; ailerons, flaps, spoilers, rudder, two-spar tailplane, elevators, wing/fuselage fairings, flap track fairings, main landing gear bay, upper and lower engine nacelle doors and cabin floor panels are of composites construction.

Bombardier's Canadair division is design authority and manufactures nose section; de Havilland manufactures rear fuselage, engine pylons and vertical stabiliser and is responsible for final assembly at Downsview; Mitsubishi supplies wings and centre fuselage; Short Brothers designed and manufactures forward fuselage, engine nacelles, horizontal stabiliser and other composites components; Bombardier is responsible for interior completions at its Montréal facility. Other participants in the programme are Honeywell Aerospace (APU), Ametek Aerospace (data acquisition unit, engine vibration monitoring system, fuel flow transmitters and engine thermocouples), Rolls-Royce Deutschland (power plant), Hella (lighting systems), Honeywell (avionics), Liebherr-Aerospace Toulouse (air management system), Lucas Aerospace (electrical systems), Messier-Dowty International (landing gear), Parker Bertea Aerospace (flight controls and fuel and hydraulic systems), Raytheon E-Systems (pitch feel systems), Thales Avionics (flight control system) and Hamilton Sundstrand (slat/flap actuation system and ram air turbine).

LANDING GEAR: Hydraulically retractable tricycle type with Messier-Dowty oleo-pneumatic shock-absorber and twin wheels on each unit; main units retract inward into wing, nosewheel forwards. Goodyear tyres, mainwheels tyre size H38×12.0-19 (20 ply) tubeless, maximum pressure 11.45 bar (166 lb/sq in); nosewheel tyres 21×7.25-10 (12 ply) tubeless (deflector-type), maximum pressure 9.93 bar (144 lb/sq in). Carbon brakes with dual Goodrich/HydroAire hydraulic digital brake-by-wire/modulated anti-skid system providing pilot-selectable, three-level autobrake capability. Steerable nosewheel, maximum steering angle ±75°; minimum ground turning radius 20.73 m (68 ft 0 in).

POWER PLANT: Two rear-mounted 65.6 kN (14,750 lb st) Rolls-Royce Deutschland BR710A2-20 turbofans, flat rated to ISA + 20°C, with FADEC. International Nacelle Systems (Shorts/Hurel-Dubois joint venture) hydraulically actuated two-petal target-type thrust reversers.

Fuel contained in two integral wing tanks, each of 8,479 litres (2,240 US gallons; 1,865 Imp gallons) capacity, centre-section tank, capacity 6,117 litres (1,616 US gallons; 1,346 Imp gallons), and auxiliary tank in aft fuselage, capacity 1,234 litres (326 US gallons; 271 Imp gallons), giving total standard capacity of 24,310 litres (6,422 US gallons; 5,347 Imp gallons). Fuel from centre-section and auxiliary tanks is transferred to wing tanks from where two AC main pumps and DC back-up pump feed to engines; automatic fuel management system balances quantities in port and starboard wing tanks. Gravity and pressure refuelling; single-point pressure fuelling/defuelling coupling in starboard wing/fuselage fairing. Oil capacity 20 litres (5.3 US gallons; 4.4 Imp gallons) with oil replenishment tank, capacity 5.7 litres (1.5 US gallons; 1.2 Imp gallons) permitting remote oil servicing from the cockpit.

ACCOMMODATION: Crew of three or four (including cabin attendant) and eight to 19 passengers depending on interior fit. Customised cabin interior according to customer requirements. Typical arrangement comprises three-compartment cabin with lavatory at rear, crew rest area, galley, small lavatory and wardrobe forward, and provision for 'office in the sky', stateroom or conference area. Flight-accessible baggage compartment at rear of cabin with external plug-type door forward of port engine intake. Accommodation is heated, air conditioned and pressurised; predicted cabin noise level 52 dB. Thirteen windows on port side of cabin; standard 13 to starboard, with option of up to two extra windows, one forward, one rear (for totals of 13, 14 or 15), each 40 cm (1 ft 3¾ in) high × 27.4 cm (10¾ in) wide; one window over wing on starboard side doubles as plug-type emergency exit. Electrically operated airstair door at front of cabin on port side.

SYSTEMS: Integrated air management system by Liebherr-Aerospace Toulouse provides engine bleed, wing anti-ice, air conditioning, cabin pressurisation and avionics ventilation. Digitally controlled dual cooling pack system with ozone converters and bleed air filters provides cabin air circulation at standard rate 1.81 m³ (64 cu ft)/min/person, maximum rate 2.29 m³ (81 cu ft)/min/person with crew-selectable 100 per cent fresh air or recirculation and three air sources for cabin pressure control; maximum pressure differential 0.66 bar (9.64 lb/sq in) maintains a 1,525 m (5,000 ft) cabin altitude to 12,500 m (41,000 ft) and a 2,200 m (7,220 ft) cabin altitude to maximum operating altitude of 15,545 m (51,000 ft). Engine bleed air anti-icing for wing leading-edge fixed surfaces and slats; tail surfaces unprotected; bleed management system automatically switches between low-and high-pressure compressor air to improve engine efficiency. Oxygen system comprises four 1,417 litre (50 cu ft) oxygen cylinders pressurised to 127.6 bar (1,850 lb/sq in) for passenger and crew use.

Lucas/Leach electrical power generation and distribution system comprises two 40 kVA variable frequency generators on each engine, supplying primary 115/200 V three-phase AC electrical power at 324 to 596 Hz; alternative AC power provided by 45 kVA APU-mounted generator and emergency power by 9 kVA air-driven generator, the latter automatically deployed in the event of power loss; electrical management system automatically performs priority-based load-shedding and reconfiguration in event of failure. Four 150 A TRUs convert AC to 28 V DC; emergency DC provided by 25 Ah and 42 Ah low-maintenance Ni/Cd batteries. Provision for external AC and DC power connection. Triple logic-controlled AC power centre performs primary AC power distribution and high-power secondary distribution via solid-state switches and 'smart' -contactors; low-power AC distributed through thermal circuit breakers in the cockpit. Triple logic-controlled DC power centre provides non-interruptible primary DC power distribution, emergency bus supplies and normal DC supplies to four secondary power distribution assemblies (SPDAs) throughout the aircraft to provide remote logic-controlled power to all DC loads. Two CDUs in the cockpit allow for remote sensing/setting and resetting of circuit breakers.

Tailcone-mounted Honeywell RE220(GX) APU provides electrical power (45 kVA ground; 40 kVA flight), as well as bleed air and main engine starting; APU is certified for operation up to 13,715 m (45,000 ft), in-flight starting up to 11,280 m (37,000 ft) and engine starting up to 9,145 m (30,000 ft).

Triple-redundant hydraulic systems at pressure of 207 bar (3,000 lb/sq in), with bootstrap reservoirs.

Walter Kidde Aerospace integrated aircraft fire detection and extinguishing system provides continuous fire detection monitoring in engine nacelles, APU compartment, main landing gear bays and cabin; dual extinguishers provide two-shot fire suppression in main engine and APU bays. Aircraft serviceability monitored by CAIMS (central aircraft information and maintenance system) with facilities including in-flight display.

AVIONICS: Honeywell Primus 2000XP as core system.

Comms: Dual VHF (third optional); dual Rockwell Collins HF; dual transponders; dual radio management systems; Coltech five-channel Selcal; Honeywell digital FDR and CVR; ELT; satcom optional, with antenna mounted in fin cap; Teledyne Magnastar Office in the Sky datalink optional.

Radar: Colour weather radar with dual controllers.

Flight: Dual flight management systems (third optional) with dual Cat. II autopilots and triple digital air data computers providing fail-safe AFCS; triple laser gyro inertial reference systems; GPS with option for second sensor; ADF; VOR/ILS; DME; TCAS II; Honeywell EGPWS with terrain database integrated into Primus 2000XP system for EFIS display.

Instrumentation: Dual EFIS comprising six 203 × 178 mm (8 × 7 in) CRT multifunction displays, for PFD and EICAS functions; dual Rockwell Collins digital radio altimeter; combined standby airspeed/altimeter, standby artificial horizon and stowable standby heading indicator. Thales HFDS with Cat. II landing capability and lightning sensor system optional. Bombardier Enhanced Vision System (BEVS) standard from first quarter 2005.

Data below apply to Global Express and Global Express XRS, except where noted.

DIMENSIONS, EXTERNAL:
Wing span over winglets ... 28.65 m (94 ft 0 in)
Wing chord: at root ... 6.43 m (21 ft 1 in)
at tip ... 1.24 m (4 ft 1 in)
Wing aspect ratio ... 8.6
Length: overall ... 30.30 m (99 ft 5 in)
fuselage ... 26.31 m (86 ft 4 in)
Diameter of fuselage, constant portion ... 2.69 m (8 ft 10 in)
Height overall ... 7.57 m (24 ft 10 in)
Tailplane span ... 9.68 m (31 ft 9 in)
Wheel track (c/l of shock-absorbers) ... 4.06 m (13 ft 4 in)
Wheelbase ... 12.78 m (41 ft 11 in)
Passenger door: Height ... 1.83 m (6 ft 0 in)
Width: ... 0.91 m (3 ft 0 in)
Baggage door: Height ... 0.84 m (2 ft 9 in)
Width ... 1.09 m (3 ft 7 in)
Emergency exit: Height ... 0.99 m (3 ft 3 in)
Width ... 0.51 m (1 ft 8 in)

DIMENSIONS, INTERNAL:
Flight deck volume ... 3.99 m^3 (141.0 cu ft)
Cabin (excl flight deck): Length ... 14.73 m (48 ft 4 in)
Width at floor ... 2.11 m (6 ft 11 in)
Max width ... 2.49 m (8 ft 2 in)
Max height ... 1.90 m (6 ft 3 in)
Floor area ... 31.1 m^2 (335 sq ft)
Volume, incl baggage compartment ... 60.6 m^3 (2,140 cu ft)

AREAS:
Wings, basic ... 94.95 m^2 (1,022.0 sq ft)
Horizontal tail surfaces (total) ... 22.76 m^2 (245.0 sq ft)
Vertical tail surfaces (total) ... 17.28 m^2 (186.0 sq ft)

WEIGHTS AND LOADINGS (A: Global Express; B: Global Express XRS):
Operating weight empty: A ... 22,816 kg (50,300 lb)
B ... 23,360 kg (51,500 lb)
Max payload: A ... 2,585 kg (5,700 lb)
Payload with max fuel: A ... 725 kg (1,600 lb)
B ... 805 kg (1,775 lb)
Fuel with max payload ... 17,804 kg (39,250 lb)
Max fuel weight: A ... 19,663 kg (43,350 lb)
B ... 20,400 kg (44,975 lb)
Max ramp weight: A standard ... 43,205 kg (95,250 lb)
A optional ... 43,658 kg (96,250 lb)
B ... 44,565 kg (98,250 lb)
Max T-O weight: A standard ... 43,091 kg (95,000 lb)
A optional ... 43,545 kg (96,000 lb)
B ... 44,452 kg (98,000 lb)
Max landing weight: A, B ... 35,652 kg (78,600 lb)
Max zero-fuel weight: A, B ... 25,401 kg (56,000 lb)
Max wing loading:
A standard ... 453.8 kg/m^2 (92.95 lb/sq ft)
A optional ... 458.6 kg/m^2 (93.93 lb/sq ft)
B ... 468.2 kg/m^2 (95.89 lb/sq ft)
Max power loading:
A standard ... 328 kg/kN (3.22 lb/lb st)
A optional ... 332 kg/kN (3.25 lb/lb st)
B ... 339 kg/kN (3.32 lb/lb st)

PERFORMANCE:
Max level speed (VMO):
S/L to FL80 ... 300 kt (555 km/h; 345 mph) CAS
FL80 to FL309 ... 340 kt (629 km/h; 391 mph) CAS
above FL309 ... M0.89
High cruising speed: A, B ... M0.89 (515 kt; 954 km/h; 593 mph)
Normal cruising speed: A, B ... M0.85 (488 kt; 904 km/h; 562 mph)
Long-range cruising speed at FL450: A ... M0.80 (459 kt; 850 km/h; 528 mph)
Initial cruising altitude: A ... 13,105 m (43,000 ft)
Time to climb to initial cruising altitude:
A, B ... 30 min
Max certified altitude: A, B ... 15,545 m (51,000 ft)
T-O balanced field length: A ... 1,774 m (5,820 ft)
B ... 1,887 m (6,190 ft)
Landing distance, A, B ... 814 m (2,670 ft)
Runway LCN ... 55
Range with max fuel and eight passengers, NBAA IFR reserves:
at M0.85: A ... 6,010 n miles (11,130 km; 6,916 miles)
B ... 6,150 n miles (11,390 km; 7,077 miles)
at M0.87: A ... 5,276 n miles (9,771 km; 6,071 miles)
B ... 5,450 n miles (10,093 km; 6,272 miles)
Range with max payload:
A at M0.85 ... 5,187 n miles (9,606 km; 5,969 miles)
A at M0.87 ... 4,596 n miles (8,511 km; 5,289 miles)
Design *g* limit: A, B ... +2.5

OPERATIONAL NOISE LEVELS (all):
T-O ... 82.4 EPNdB
Approach ... 89.8 EPNdB
Sideline ... 88.6 EPNdB

UPDATED

BOMBARDIER GLOBAL 5000

Engineering designation: BD-700-1A11

TYPE: Long-range business jet.

DEVELOPMENT MILESTONES

Design began	99
Announced	25 Oct 01
Marketing began	5 Feb 02
First order (Sino Private Aviation)	28 Feb 02
First flight	7 March 03

PROGRAMME: Market and design studies began 1999; announced 25 October 2001; formal launch 5 February 2002; first flight (c/n 9127/C-GERS) 7 March 2003; public debut at Paris Air Show in June 2003; second test aircraft (c/n 9130/C-GLRM) first flew 8 January 2004. Certification by Transport Canada, achieved 12 March 2004 following two aircraft flight test programme; deliveries of 'green' aircraft to completion centres scheduled to begin in second quarter 2004; EASA certification expected in June 2004; service entry (C-GERS, as demonstrator) fourth quarter 2004; corporate service entry in early 2005.

CUSTOMERS: Launch customer TAG Aeronautics ordered five on 5 November 2001; letters of intent for 15 aircraft at time of launch; Sino Private Aviation of Hong Kong ordered one aircraft during Asian Aerospace 2002 in Singapore on 28 February 2002. Potential market for 750 aircraft in class by 2010.

COSTS: US$33 million 'green' (2002).

DESIGN FEATURES: Based on Global Express, with 1.83 m (6 ft 0 in) reduction in fuselage length, and 1,200 n mile (2,222 km; 1,381 mile) reduction in maximum range. Up to 19 passengers.

Data generally as for Global Express, except the following.

DIMENSIONS, EXTERNAL:
Length overall ... 29.49 m (96 ft 9 in)

DIMENSIONS, INTERNAL:
Cabin (excl flight deck):
Length ... 12.95 m (42 ft 6 in)
Volume ... 53.29 m^3 (1,882 cu ft)

WEIGHTS AND LOADINGS:
Basic operating weight ... 22,838 kg (50,350 lb)

Maiden flight of the second Global 5000, 8 January 2004 ***NEW***/0573427

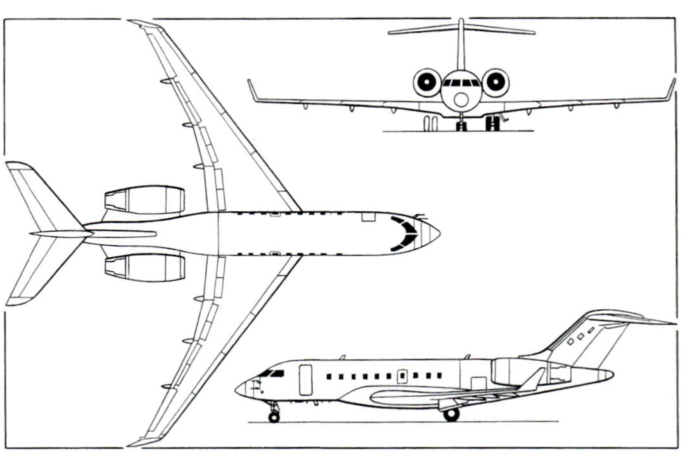

Bombardier Global 5000 general arrangement *(James Goulding)* 0126930

Prototype Bombardier Global 5000 ***NEW***/0573456

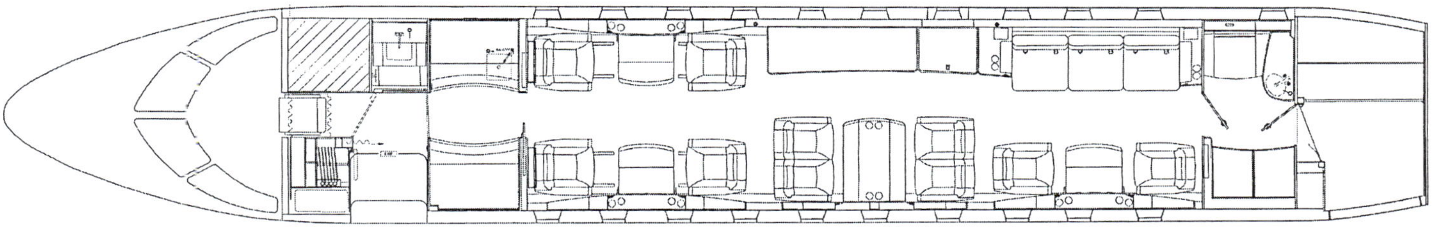

Floor plan of typical Global 5000 ***NEW***/0573242

Bombardier Global 5000 cabin ***NEW***/0573241

Payload: max	2,563 kg (5,650 lb)
with max fuel	726 kg (1,600 lb)
Max fuel weight	16,329 kg (36,000 lb)
Fuel with max payload	14,492 kg (31,950 lb)
Max T-O weight	39,780 kg (87,700 lb)
Max ramp weight	39,893 kg (87,950 lb)
Max landing weight	35,652 kg (78,600 lb)
Max zero-fuel weight	25,401 kg (56,000 lb)
Max wing loading	419.0 kg/m² (85.81 lb/sq ft)
Max power loading	303 kg/kN (2.97 lb/lb st)

PERFORMANCE:

Cruising speed: max	M0.89 (510 kt; 945 km/h; 587 mph)
normal	M0.85 (488 kt; 904 km/h; 562 mph)
Initial cruising altitude	13,105 m (43,000 ft)
Time to climb to initial cruising altitude	25 min
Max certified altitude	15,545 m (51,000 ft)
T-O balanced field length	1,525 m (5,000 ft)
Landing run	825 m (2,700 ft)
Range, NBAA IFR reserves, three crew, eight passengers:	
at M0.85	4,800 n miles (8,889 km; 5,523 miles)
at M0.88	3,700 n miles (6,852 km; 4,257 miles)

UPDATED

BOMBARDIER AEROSPACE TORONTO OPERATIONS

Garratt Boulevard, Downsview, Ontario M3K 1Y5
Tel: (+1 416) 633 73 10
Fax: (+1 416) 375 45 46

VICE-PRESIDENT, CANADIAN OPERATIONS: Serge Perron
VICE-PRESIDENT AND GENERAL MANAGER, OPERATIONS: Alain Dugas
VICE-PRESIDENT, DASH 8 SERIES 100/200/300 OPERATIONS: Rob Comand
VICE-PRESIDENT, DASH 8 SERIES 400 OPERATIONS: Bruce Vannus
VICE-PRESIDENT, GLOBAL EXPRESS OPERATIONS: Trevor Anderson
VICE-PRESIDENT, FINANCE: Colin Fernie
VICE-PRESIDENT, HUMAN RESOURCES: Bernard Cormier
MANAGER, PUBLIC RELATIONS: Colin Fisher

Bombardier Dash 8 Q200 of the Mexican Navy ***NEW***/0143488

Established 1928 as The de Havilland Aircraft of Canada Ltd, subsidiary of The de Havilland Aircraft Company Ltd, both absorbed 1961 by Hawker Siddeley Group; ownership transferred to Canadian government 26 June 1974; purchased by Boeing Company 31 January 1986 and made a division of Boeing of Canada Ltd; Boeing's intention to sell announced July 1990.

Sale to Bombardier Inc (51 per cent) and government of Ontario (49 per cent), signed 22 January 1992, supported by help from Ontario and federal governments, with Canadian Export Development Corporation to provide sales financing for Dash 8; all government support conditionally repayable. Bombardier acquired remaining 49 per cent of de Havilland in January 1997.

At its Downsview facility, Bombardier Aerospace Toronto manufactures Dash 8Q series, including spare parts and components, and also some components of the Bombardier Global Express; performs final assembly of Dash 8Q, Global Express and Global 5000; designed and builds wings for Learjet 45. Bombardier 415 assembly is in a 4,750 m² (51,000 sq ft) plant at North Bay, Ontario. Bombardier Aerospace Toronto operations' employment totalled more than 3,000 in late 2002.

VERIFIED

BOMBARDIER DASH 8 Q100 AND Q200

TYPE: Twin-turboprop airliner.

DEVELOPMENT MILESTONES

Announced	1980
First flight	20 Jun 83
Certification	28 Sep 84
First flight, production	3 Apr 84
First delivery (norOntair)	23 Oct 84
Entered service (norOntair)	Dec 84

PROGRAMME: Launched 1980; first flight of first prototype (C-GDNK) 20 June 1983, second prototype (C-GGMP) 26 October 1983, third 22 December 1983; fourth aircraft, first with production P&WC PW120 engines, 3 April 1984. Certified to Canadian DoT, FAR Pts 25 and 36 and SFAR No. 27 on 28 September 1984, followed by FAA type approval; also certified in Australia, Austria, Brazil, Cameroon, China, Colombia, Germany, Ireland, Italy, Maldives, Netherlands, New Zealand, Norway, Papua New Guinea, South Africa, Taiwan, UK and United Arab Emirates. First delivery (norOntair) 23 October 1984 followed by service entry in December; 500th Dash 8 (a Series 200, N355PH) delivered 21 November 1997 to Horizon Air. 600th of type delivered 6 March 2001 (see Q300). Total Dash 8 fleet time at October 2003 (not including Q400) was 13,105,842 hours and 15,891,727 cycles, with 99.4 per cent despatch reliability rate.

CURRENT VERSIONS: **Dash 8 Q:** Redesigned interior and noise and vibration suppression system (NVS) standard from second quarter of 1996, reducing cabin noise levels by 12 dB; all current production aircraft are equipped with NVS and known as Dash 8 Qs.

Dash 8 Series 100: Initial version, with choice of PW120A or PW121 engines; discontinued.

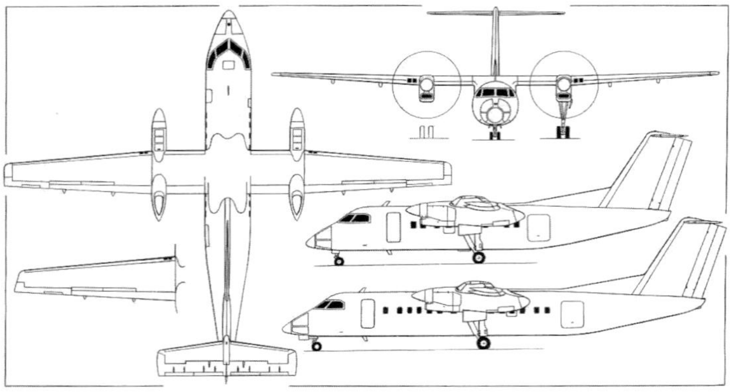

Bombardier Dash 8 Q100, with additional side view (bottom) and wingtip of Q300 *(Jane's/Dennis Punnett)* 0507404

Dash 8 Series 100A: Introduced 1990; PW120A (or optional PW121) engines and restyled interior with 6.35 cm (2.5 in) more headroom in aisle; first delivery to Pennsylvania Airlines July 1990; discontinued.

Dash 8 Series 100B: Improved version from 1992; PW121 engines enhance airfield and climb performance; discontinued.

Dash 8 Q100: Introduced 1998; PW 120A or optional PW 121 in basic weight version, PW 121 standard in high gross weight (HGW) version.

Dash 8 Series 200A: Increased speed/payload version of Series 100A with PW 123C engines. Transport Canada certification March 1995; first delivery 19 April 1995; discontinued.

Dash 8 Series 200B: As 200A, but with PW123D engines for full power at higher ambient temperatures; discontinued.

Dash 8 Q200: Introduced in1998; increased speed/payload version of Q100; increased OEI capability and greater commonality with Series 300. Same airframe as Q100, but PW 123C/D engines give 30 kt (56 km/h; 35 mph) increase in cruising speed, allowing airlines to increase frequencies or operational radius. PW 123D engine offers full power at higher ambient temperatures for improved hot-and-high airfield performance.

Detailed description applies to Q200 except where indicated.

Dash 8 Q300 and Q400: *described separately.*

Dash 8M and Triton: Special missions versions, previously promoted, including Canadian **CC-142** and US Air Force **E-9A**.

CUSTOMERS: By 31 March 2004, orders received for 299 Srs 100 subvariants and 96 Srs 200s, of which 299 and 94 respectively had then been delivered (including military Dash 8Ms). Refer to table in Dash 8 Q400 entry.

COSTS: Unit cost approximately US$12 million (2000).

DESIGN FEATURES: T tail, swept fin with large dorsal fin. Wing has constant chord inboard section and tapered outer panels; thickness/chord ratio 18 per cent at root, 13 per cent at tip; dihedral 2° 30' on outer panels; inboard leading-edges drooped; 3° washout at wingtips.

FLYING CONTROLS: Conventional and power-assisted. Fixed tailplane; horn-balanced elevator with four tabs; mechanically actuated horn-balanced ailerons with inset tabs; hydraulically actuated roll spoilers/lift dumpers forward of each outer flap; two-segment serially hinged rudder, hydraulically actuated; yaw damper; stall strips on leading-edges outboard of engines; two-section slotted Fowler flaps. Digital AFCS.

STRUCTURE: Fuselage near-circular section, flush riveted and pressurised; adhesive bonded stringers and cutout reinforcements; wing leading-edge, radome, nose bay, wing/fuselage and wingtip fairings, dorsal fin, fin leading-edge, fin/tailplane fairings, tailplane leading-edges, elevator tips, flap shrouds, flap trailing-edges and other components of Kevlar and Nomex; wing has tip-to-tip torsion box. Wheel doors of Kevlar and other composites.

LANDING GEAR: Retractable tricycle type, by Dowty Aerospace, with twin wheels on each unit. Steer-by-wire nose unit retracts forward, main units rearward into engine nacelles. Goodrich mainwheels and brakes; Hydro-Aire Mk 3 anti-skid system. Tyres 26.5×8.0-13 (12 ply) or high flotation H31×9.75-13 (10 ply) on main units; 18×5.5 (10 ply) or 22×6.50-10 (12 ply) on nose unit. Standard tyre pressures: main 9.03 bar (131 lb/sq in), nose 5.52 bar (80 lb/sq in). Low-pressure tyres optional, pressure 5.31 bar (77 lb/sq in) on main units, 3.31 bar (48 lb/sq in) on nose unit.

POWER PLANT: *Q100:* Two 1,491 kW (2,000 shp) Pratt & Whitney Canada PW 120A turboprops, driving Hamilton Sundstrand 14SF-7 four-blade constant-speed fully feathering aluminium/glass fibre propellers with reverse pitch, standard on basic weight version. Two 1,603 kW (2,150 shp) PW 121s optional on this version and standard on High Gross Weight version.

Q200: Two 1,603 kW (2,150 shp) PW 123C/D turboprops, driving Hamilton Sundstrand 14SF-23 propellers; PW 123C is flat-rated for full power at up to 26°C; PW 123D maintains same power at up to 45°C.

Standard usable fuel capacity (in-wing tanks) 3,160 litres (835 US gallons; 695 Imp gallons); optional auxiliary tank system increases this to 5,700 litres (1,506 US gallons; 1,254 Imp gallons). Pressure refuelling point in rear of starboard engine nacelle; overwing gravity point in each outer wing panel. Oil capacity 21 litres (5.5 US gallons; 4.6 Imp gallons) per engine.

ACCOMMODATION: Crew of two on flight deck, plus cabin attendant. Dual controls standard. Standard commuter layout provides four-abreast seating, with central aisle, for 37 passengers at 79 cm (31 in) pitch, plus buffet, lavatory and large rear baggage compartment. Wardrobe at front of passenger cabin, in addition to overhead lockers and underseat stowage, provides additional carry-on baggage capacity. Alternative 39-passenger, passenger/cargo or corporate layouts available at customer's option. Movable bulkhead to facilitate conversion to mixed traffic. Port side airstair door at front for crew and passengers; large inward-opening port side door aft of wing for cargo loading. Emergency exit each side, in line with wing leading-edge, and opposite passenger door on starboard side. Entire accommodation pressurised and air conditioned.

SYSTEMS: Pressurisation system with maximum differential 0.38 bar (5.5 lb/sq in). Normal hydraulic installation comprises two independent systems, each having an engine-driven variable displacement pump and an electrically driven standby pump; accumulator and hand pump for emergency use. Electrical system DC power provided by two starter/generators, two transformer-rectifier units, and two Ni/Cd batteries. Variable frequency AC power provided by two engine-driven AC generators and three static inverters. Ground power receptacles in port side of nose (DC) and rear of starboard nacelle (AC). Rubber-boot de-icing of wing, tailplane and fin leading-edges and nacelle intakes by pneumatic system; electric de-icing of propeller blades, pitot static ports, stall warning transducer, engine intake adaptor and elevator horn leading-edge. APU optional. Simmonds fuel monitoring system.

AVIONICS: Rockwell Collins and Honeywell com/nav.

Comms: Dual Rockwell Collins VHF-22; single Mode C transponder. Avtech audio integrating system. Telephonics PA system.

Radar: Primus P660 colour weather radar.

Flight: Rockwell Collins DME-42, ADF-60A, Honeywell Mk VIII EAPNS, Honeywell SPZ-8000 dual-channel digital AFCS with integrated fail-operational flight director/autopilot system, dual digital air data system. Optional FMS and GPS.

Instrumentation: Honeywell EFIS. Rockwell Collins/Flight Dynamics HGS 2000 head-up guidance system for Cat. IIIa operation optional.

DIMENSIONS, EXTERNAL:

Wing span ... 25.91 m (85 ft 0 in)
Wing aspect ratio ... 12.4
Length overall ... 22.25 m (73 ft 0 in)
Fuselage: Max diameter ... 2.69 m (8 ft 10 in)
Height overall ... 7.49 m (24 ft 7 in)
Elevator span ... 7.92 m (26 ft 0 in)
Wheel track (c/l of shock-struts) ... 7.87 m (25 ft 10 in)
Wheelbase ... 7.95 m (26 ft 1 in)
Propeller diameter ... 3.96 m (13 ft 0 in)
Propeller ground clearance ... 0.94 m (3 ft 1 in)
Propeller fuselage clearance ... 0.76 m (2 ft 6 in)
Passenger/crew door (fwd, port):
  Height ... 1.66 m (5 ft 5½ in)
  Width ... 0.76 m (2 ft 6 in)
  Height to sill ... 1.09 m (3 ft 7 in)
Baggage door (rear, port): Height ... 1.52 m (5 ft 0 in)
  Width ... 1.27 m (4 ft 2 in)
  Height to sill ... 1.09 m (3 ft 7 in)

DIMENSIONS, INTERNAL:

Cabin (excl flight deck): Length ... 9.14 m (30 ft 0 in)
  Max width ... 2.49 m (8 ft 2 in)
  Width at floor ... 2.03 m (6 ft 8 in)
  Max height ... 1.96 m (6 ft 5 in)
  Floor area ... 23.60 m^2 (254.0 sq ft)
  Volume ... 37.6 m^3 (1,328 cu ft)
Baggage compartment volume ... 8.5 m^3 (300 cu ft)

AREAS:

Wings, gross ... 54.35 m^2 (585.0 sq ft)
Ailerons (total) ... 1.12 m^2 (12.1 sq ft)
Fin ... 9.81 m^2 (105.6 sq ft)
Rudder ... 4.31 m^2 (46.4 sq ft)
Tailplane ... 8.97 m^2 (96.5 sq ft)
Elevators (total) ... 4.97 m^2 (53.5 sq ft)

WEIGHTS AND LOADINGS:

Operating weight empty: Q100 ... 10,433 kg (23,000 lb)
  Q200 ... 10,501 kg (23,151 lb)
Max usable fuel: standard ... 2,576 kg (5,678 lb)
  optional ... 4,647 kg (10,244 lb)
Max payload: Q100 ... 4,082 kg (9,000 lb)
  Q200 ... 4,195 kg (9,249 lb)
Payload with max fuel: Q100 ... 3,475 kg (7,662 lb)
  Q200 ... 3,389 kg (7,471 lb)
Max T-O weight: Q100 ... 15,649 kg (34,500 lb)
  Q100 HGW, Q200 ... 16,465 kg (36,300 lb)
Max landing weight: Q100 ... 15,377 kg (33,900 lb)
  Q200 ... 15,649 kg (34,500 lb)
Max zero-fuel weight:
  Q100, Q200 JAA ... 14,515 kg (32,000 lb)
  Q200 ... 14,696 kg (32,400 lb)
Max wing loading: Q100 ... 287.9 kg/m^2 (58.97 lb/sq ft)
  Q100 HGW, Q200 ... 303.0 kg/m^2 (62.05 lb/sq ft)
Max power loading: Q100 ... 5.25 kg/kW (8.63 lb/shp)
  Q100 HGW, Q200 ... 5.14 kg/kW (8.44 lb/shp)

PERFORMANCE (at 95% standard MTOW, except where indicated):

Max cruising speed: Q100 ... 265 kt (491 km/h; 305 mph)
  Q100 HGW ... 270 kt (500 km/h; 311 mph)
  Q200 ... 290 kt (537 km/h; 334 mph)
Stalling speed, flaps down: all ... 72 kt (134 km/h; 83 mph)
Max rate of climb at S/L: all ... 450 m (1,475 ft)/min
Max certified altitude: all ... 7,620 m (25,000 ft)
Service ceiling: Q100 ... 4,503 m (14,775 ft)
  Q100 HGW ... 5,105 m (16,750 ft)
  Q200 ... 4,938 m (16,200 ft)
FAR Pt 25 T-O field length:
  Q100 (PW 121) ... 991 m (3,250 ft)
  Q200 ... 1,000 m (3,280 ft)
FAR Pt 25 landing field length at max landing weight:
  Q100 ... 785 m (2,575 ft)
  Q200 ... 780 m (2,560 ft)
Range with 37 passengers:
  Q100 ... 1,020 n miles (1,889 km; 1,173 miles)
  Q200 ... 925 n miles (1,713 km; 1,064 miles)

OPERATIONAL NOISE LEVELS: (FAR Pt 36 Stage 3 and ICAO Annex 16):

T-O ... 80.5 EPNdB
Sideline ... 85.6 EPNdB
Approach ... 94.7 EPNdB

UPDATED

BOMBARDIER DASH 8 Q300

TYPE: Twin-turboprop airliner.

DEVELOPMENT MILESTONES

Announced	mid-85
Programme launched	Mar 86
First flight	15 May 87
Certification	14 Feb 89
First delivery (TimeAir)	27 Feb 89

PROGRAMME: Dash 8 Series 300 announced mid-1985 as stretch of Series 200; launched March 1986; first flight (modified Series 100 prototype C-GDNK) 15 May 1987; Canadian DoT certification 14 February 1989; first delivery (Time Air) 27 February 1989; FAA type approval 8 June 1989; now also certified in Antigua, Argentina, Australia, Austria, Bahamas, Brazil, Chile, China, Colombia, Egypt, Germany, Ireland, India, Indonesia, Italy, Jordan, Malaysia, Maldives, Mexico, Netherlands, New Zealand, Norway, Romania, Senegal, South Africa, Spain, Taiwan, Thailand, United Arab Emirates and Zambia. Low-noise Dash 8 Q (see Series 200 for details) became standard version from 1996. A Q300 delivered to Air Nippon on 6 March 2001 was the 600th Dash 8.

CURRENT VERSIONS: **Series 300, 300A, 300B and 300E:** Initial versions, described in 1998–99 and earlier editions.

Q300: Introduced 1998; differs from Q200 in having extended wingtips; 3.43 m (11 ft 3 in) two-plug fuselage extension giving standard seating for 50 at 81 cm (32 in) pitch or 56 at 74 cm (29 in) pitch, plus second cabin attendant; also larger galley, galley service door, additional wardrobe, larger lavatory, dual air conditioning packs and optional Turbomach T-40 APU; powered by 1,775 kW (2,380 shp) P&WC PW123s driving Hamilton Sundstrand 14SF-23 four-blade propellers standard in basic version; 1,864 kW (2,500 shp) PW 123Bs optional in basic and standard in high gross weight (HGW) versions provide increased mechanical power for improved take-off performance in low and cold conditions; optional 1,775 kW (2,380 shp) PW 123Es provide 5 per cent increase in thermodynamic power up to 40°C (96°F) for improved hot-and-high performance; fuel capacity as Q200; tyre pressures increased (mainwheels 6.69 bar; 97 lb/sq in, nosewheels 4.14 bar; 60 lb/sq in). NVS system standard on all aircraft produced from second quarter 1996.

CUSTOMERS: Firm orders for 211 by 31 March 2004, of which 207 then delivered. Refer to table in Dash 8 Q400 entry.

COSTS: US$14.3 million (2000).

DIMENSIONS, EXTERNAL: As for Q200 except:
- Wing span .... 27.43 m (90 ft 0 in)
- Wing aspect ratio 13.4
- Length overall .... 25.68 m (84 ft 3 in)
- Wheelbase .... 10.01 m (32 ft 10 in)

DIMENSIONS, INTERNAL: As for Q200 except:
- Cabin (excl flight deck): Length .... 12.65 m (41 ft 6 in)
  - Floor area .... 30.57 m² (329.0 sq ft)
  - Volume .... 52.0 m³ (1,838 cu ft)
- Baggage compartment volume:
  - with 50 passengers .... 9.1 m³ (320 cu ft)
  - with 56 passengers .... 7.9 m³ (280 cu ft)

AREAS:
- Wings, gross .... 56.21 m² (605.0 sq ft)
- Ailerons (total) .... 1.87 m² (20.18 sq ft)
- Tail surfaces .... as for Series 100A/200A

WEIGHTS AND LOADINGS:
- Operating weight empty: basic .... 11,812 kg (26,042 lb)
  - HGW .... 11,823 kg (26,065 lb)
- Max usable fuel: standard .... 2,576 kg (5,679 lb)
  - optional .... 4,646 kg (10,243 lb)
- Max payload: basic .... 5,061 kg (11,158 lb)
  - HGW .... 6,094 kg (13,435 lb)
- Payload with max fuel .... 5,106 kg (11,257 lb)
- Max T-O weight: basic .... 18,642 kg (41,100 lb)
  - HGW .... 19,504 kg (43,000 lb)
- Max landing weight: basic .... 18,144 kg (40,000 lb)
  - HGW .... 19,050 kg (42,000 lb)
- Max zero-fuel weight: basic .... 16,873 kg (37,200 lb)
  - HGW .... 17,917 kg (39,500 lb)
- Max wing loading: basic .... 331.7 kg/m² (67.93 lb/sq ft)
  - HGW .... 347.0 kg/m² (71.07 lb/sq ft)
- Max power loading: basic .... 5.25 kg/kW (8.63 lb/shp)
  - HGW .... 5.49 kg/kW (9.03 lb/shp)

PERFORMANCE (at 95% standard MTOW except where indicated):
- Max cruising speed: basic .... 287 kt (532 km/h; 330 mph)
  - HGW .... 285 kt (528 km/h; 328 mph)
- Stalling speed, flaps down .... 77 kt (141 km/h; 88 mph)
- Max rate of climb at S/L .... 549 m (1,800 ft)/min
- Rate of climb at S/L, OEI .... 137 m (450 ft)/min
- Max certified altitude .... 7,620 m (25,000 ft)
- Service ceiling, OEI: basic .... 4,145 m (13,600 ft)
  - HGW .... 3,734 m (12,250 ft)
- FAR Pt 25 T-O field length at S/L, ISA, 15° flap:
  - basic .... 1,097 m (3,600 ft)
  - HGW .... 1,180 m (3,870 ft)
- FAR Pt 25 landing field length at max landing weight:
  - basic .... 1,010 m (3,315 ft)
  - HGW .... 1,040 m (3,415 ft)
- Range, HGW with 50 passengers:
  - standard fuel .... 841 n miles (1,557 km; 967 miles)
  - auxiliary fuel .... 1,098 n miles (2,033 km; 1,263 miles)

OPERATIONAL NOISE LEVELS: (FAR Pt 36):
- T-O .... 79.5 EPNdB
- Sideline .... 87 EPNdB
- Approach .... 93.3 EPNdB

UPDATED

Bombardier Dash 8 Q300 twin-turboprop airliner ***NEW***/0546836

BOMBARDIER DASH 8 Q400

TYPE: Twin-turboprop airliner.

DEVELOPMENT MILESTONES

Programme launched	Jun 96
First prototype material cut	Dec 96
Rolled out	21 Nov 97
First flight	31 Jan 98
Certification	14 Jun 99
First delivery (SAS Commuter)	20 Jan 00
Entered service (SAS Commuter)	7 Feb 00

PROGRAMME: Launched June 1995 as stretch of Series 300; component manufacture began December 1996; engine first test flown in January 1997, mounted on nose of Pratt & Whitney Canada Boeing 720B testbed; roll-out 21 November 1997, first flight (C-FJJA) 31 January 1998; five aircraft participated in the 1,900 hour, 1,400 sortie flight test programme, based at the Bombardier Flight Test Centre in Wichita, Kansas, leading to Transport Canada CAR 525 Amendment 86 certification on 14 June 1999, JAA approval in December 1999 and FAA FAR Pt 25 approval on 8 February 2000. JAA approval for steep approach (5.5° glideslope) operations at London City Airport achieved 11 October 2001. First delivery (OY-KCA) to SAS Commuter 20 January 2000, followed by service entry 7 February on Copenhagen, Denmark, to Poznan, Poland route. Q400 fleet flight time totalled 154,800 hours and 177,000 cycles by November 2001. Mitsubishi joined programme as risk-sharing partner in October 1995, with responsibility for design, development and manufacture of fuselage and tail sections.

CUSTOMERS: Total of 103 firm orders by 31 March 2004, at which time 83 had been delivered; refer to table. Launch customer was Great China Airlines (now UNI Airways), which ordered six (since cancelled) in February 1996. US launch customer Horizon Air, which ordered 15 with 15 options, in June 1999, took delivery of its first aircraft on 8 January 2001. First for Changan handed over 27 October 2000. British European (now FlyBE) received first on 23 November 2001; Widerøe deliveries began 16 November 2001. Recent customers include FlyBE, which ordered 17, plus 20 options, on 23 April 2003, for delivery from second quarter 2003. All Nippon Airways, which ordered two in February 2004 and four in April 2004, bringing its firm order total to 12. Total fleet time at January 2004 stood at 330,178 hours and 365,328 cycles, with 99.4 per cent despatch reliability rate.

Bombardier Dash 8 Q400 of All Nippon Airways ***NEW***/0572283

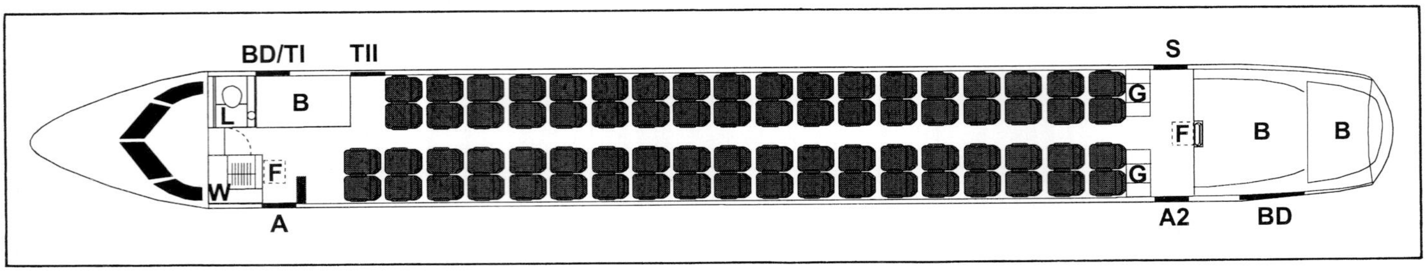

Dash 8 Q400 in 74-seat configuration *A: airstair/Type I exit, A2: passenger door/Type I exit, B: baggage, BD: baggage door, F: folding seat, G: galley, L: lavatory, S: service door/Type I exit, TI/TII: Type I/Type II exits, W: wardrobe* 0044509

COSTS: FlyBE order for 17 aircraft, valued at US$362 million (2003). Direct operating cost in US currency, based on 70 passengers over 200 n mile (370 km; 230 mile) stage length: in USA 7.40 cents per seat-mile; in Europe 7.50 cents per seat-kilometre (both 2003).

DESIGN FEATURES: Compared with Q300, fuselage stretched 6.83 m (22 ft 5 in) to seat up to 78. Other new features include engines; propeller; tapered inboard wing section increasing propeller/fuselage clearance to 1.09 m (3 ft 7 in); revised wing/fuselage fairings; ailerons, elevators and fin cap fairing; landing gear; baggage/service doors; upgraded avionics. Airframe designed for crack-free life of 40,000 flight hours/80,000 flight cycles and an economic life of 80,000 flight hours/160,000 flight cycles. Common type rating with Q100/200/300.

STRUCTURE: Generally as for Q100 and Q200. De Havilland Canada manufactures cockpit section and wing and performs final assembly; other participants in programme include AlliedSignal (electrical power system); Goodrich (brakes); Dowty Aerospace (propellers); Menasco (landing gear); Micro Technica (flap system); Mitsubishi (fuselage and tail sections); Parker Bertea Aerospace (hydraulics and fuel system); Pratt & Whitney Canada (engines); Sextant Avionique (avionics system); and Shorts (engine nacelles).

POWER PLANT: Two Pratt & Whitney Canada PW150A turboprops with FADEC, each flat rated to 3,781 kW (5,071 shp) at 37.4°C, driving Dowty R408 six-blade, slow-turning (1,020 rpm for T-O; 850 rpm cruise) composites propellers giving 15 per cent reduction in T-O rpm and 6 to 19 per cent reduction in cruise rpm over Q100/300. Fuel capacity 6,526 litres (1,724 US gallons; 1,436 Imp gallons).

ACCOMMODATION: Variety of cabin configurations providing four-abreast seating, with central aisle, for 70 passengers at 84 cm (33 in) pitch. 74 passengers at 79 cm (31 in) pitch, 78 passengers at 76 cm (30 in), or two-class layout for 10 business class passengers at 86 cm (34 in) pitch and 62 economy class passengers at 79 cm (31 in); NVS system standard.

AVIONICS: Prime contractor, Thales.

Comms: Dual VHF nav/com and mode S transponder; solid-state FDR and cockpit voice recorder; ELT antenna.

Radar: Weather radar.

Flight: GPWS, radar altimeter, ADF and DME. Provision for FMS, GPS and TCAS II.

Instrumentation: EFIS employs five 152 × 203 mm (6 × 8 in) LCDs. Optional Flight Dynamics HGS4100 (later HGS4200) head-up guidance system.

DIMENSIONS, EXTERNAL:
- Wing span ... 28.42 m (93 ft 3 in)
- Wing aspect ratio ... 12.8
- Length overall ... 32.84 m (107 ft 9 in)
- Fuselage max diameter ... 2.69 m (8 ft 10 in)
- Height overall ... 8.36 m (27 ft 5 in)
- Tailplane span ... 9.27 m (30 ft 5 in)
- Wheel track ... 8.79 m (28 ft 10 in)
- Wheelbase ... 13.94 m (45 ft 9 in)
- Propeller diameter ... 4.11 m (13 ft 6 in)
- Propeller ground clearance ... 0.97 m (3 ft 2 in)
- Propeller fuselage clearance ... 1.10 m (3 ft 7¼ in)
- Passenger/crew door (port, fwd):
  - Height ... 1.66 m (5 ft 5½ in)
  - Width ... 0.76 m (2 ft 6 in)
  - Height to sill ... 1.22 m (4 ft 0 in)
- Passenger/crew door (port, rear):
  - Height ... 1.73 m (5 ft 8 in)
  - Width ... 0.71 m (2 ft 4 in)
  - Height to sill ... 1.52 m (5 ft 0 in)
- Baggage door (port, rear): Height ... 1.55 m (5 ft 1 in)
  - Width ... 1.27 m (4 ft 2 in)
  - Height to sill ... 1.55 m (5 ft 1 in)
- Baggage door (stbd, fwd): Height ... 1.45 m (4 ft 9 in)
  - Width ... 0.71 m (2 ft 4 in)
  - Height to sill ... 1.17 m (3 ft 10 in)
- Service door (stbd, aft): Height ... 1.45 m (4 ft 9 in)
  - Width ... 0.71 m (2 ft 4 in)
  - Height to sill ... 1.52 m (5 ft 0 in)

DIMENSIONS, INTERNAL:
- Cabin (excl flight deck): Length ... 18.80 m (61 ft 8 in)
  - Max width ... 2.49 m (8 ft 2 in)
  - Max height ... 1.95 m (6 ft 5 in)
  - Floor area ... 43.6 m^2 (469 sq ft)
  - Volume ... 77.6 m^3 (2,740 cu ft)
- Baggage volume: fwd ... 2.58 m^3 (91.0 cu ft)
  - aft ... 11.6 m^3 (411 cu ft)
  - total ... 14.2 m^3 (502 cu ft)

AREAS:
- Wings, gross ... 63.08 m^2 (679.0 sq ft)
- Ailerons (total) ... 1.87 m^2 (20.18 sq ft)
- Vertical tail surfaces (total) ... 14.12 m^2 (152.0 sq ft)
- Horizontal tail surfaces (total) ... 16.72 m^2 (180.0 sq ft)

WEIGHTS AND LOADINGS:
- Operating weight empty ... 17,108 kg (37,717 lb)
- Max payload ... 8,747 kg (19,283 lb)
- Payload with max fuel ... 6,831 kg (15,059 lb)
- Max T-O weight: basic ... 27,986 kg (61,700 lb)
  - intermediate ... 28,998 kg (63,930 lb)
  - HGW ... 29,256 kg (64,500 lb)
- Max landing weight ... 28,009 kg (61,750 lb)
- Max zero-fuel weight ... 25,855 kg (57,000 lb)
- Max wing loading: basic ... 443.7 kg/m^2 (90.87 lb/sq ft)
  - intermediate ... 459.7 kg/m^2 (94.15 lb/sq ft)
  - HGW ... 463.8 kg/m^2 (94.99 lb/sq ft)
- Max power loading: basic ... 3.70 kg/kW (6.08 lb/shp)
  - intermediate ... 3.84 kg/kW (6.30 lb/shp)
  - HGW ... 3.87 kg/kW (6.36 lb/shp)

PERFORMANCE (at 95% of standard MTOW, except where indicated):
- Max cruising speed ... 360 kt (667 km/h; 414 mph)
- Max certified altitude ... 7,620 m (25,000 ft)
- Service ceiling, OEI ... 5,334 m (17,500 ft)
- FAR T-O field length at S/L, MTOW ... 1,404 m (4,605 ft)
- FAR landing field length at S/L, MLW ... 1,288 m (4,225 ft)
- Max range, 70 passengers, IFR reserves ... 1,360 n miles (2,518 km; 1,565 miles)

OPERATIONAL NOISE LEVELS (estimated, FAR Pt 36):
- T-O ... 78.3 EPNdB
- Sideline ... 84 EPNdB
- Approach ... 94.3 EPNdB

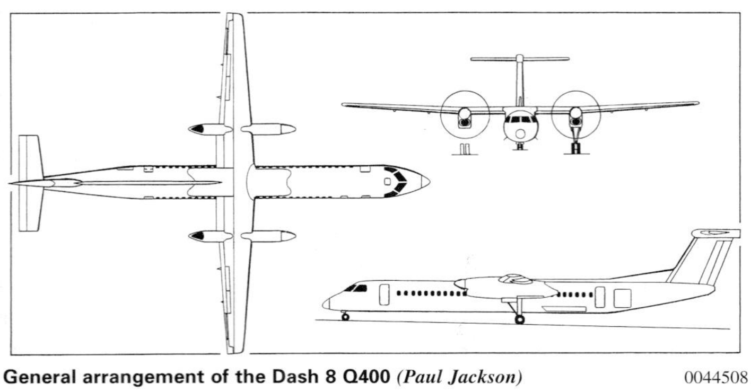

General arrangement of the Dash 8 Q400 *(Paul Jackson)* 0044508

BOMBARDIER Q SERIES ORDERS AND DELIVERIES
(at 31 March 2004)

Customer	Variant	Ordered	Delivered	Backlog
Abu Dhabi Aviation	200	2	2	
AGES/Worldwide	300	6	6	
Air Atlantic	100	15	15	
AirBC	100	12	12	
	300	6	6	
Air Creebec	100	1	1	
Air Dolomiti	300	3	3	
Air Maldives	200	1	1	
Air Manitoba	100	1	1	
Air Nippon Network/ ANK	300	5	5	
Air Niugini	200	2	2	
Air Nostrum	300	19	19	
Air Nova	100	6	6	
Air Ontario	100	30	30	
	300	6	6	
Air Senegal International	300	1	1	
Air Wisconsin	100	5	5	
	300	7	7	
Alberta Government	100	1	1	
All Nippon Airways	400	8	3	5
ALM	300	2	2	
Amakusa	100	1	1	
America West	100	12	12	
Ansett New Zealand	100	2	2	
Augsburg Airways	200	2	2	
	300	7	7	
	400	5	5	
Australian Airlines	100	4	4	
Austrian arrows[1]	100	11	11	
	300	19	19	
	400	10	8	2
Aviaco/Furlong	300	4	4	
Avline	300	4	4	
BPX Colombia	200	1	1	
Bahamasair	300	3	3	
British European	200	3	3	
	300	4	4	
	400	4	4	
Brymon Airways	300	10	10	
BWIA	300	4	4	
Canadian DND	100	6	6	

Customer	Variant	Ordered	Delivered	Backlog
CCAir	100	2	2	
Changan	400	3	3	
City Express	100	4	4	
Contact Air	100	2	2	
	300	4	4	
DAC Air	300	1	1	
Eastern Australia	200	1	1	
Eastern Metro Express	100	8	8	
Elveden Investments	100	3	3	
	300	1	1	
Fairways Corp	100	1	1	
FlyBE[3]	400	17	6	11
GPA Jetprop	100	16	16	
	300	14	14	
Hamburg Airlines	100	2	2	
	300	2	2	
Horizon Air	100	21	21	
	200	28	28	
	400	17	17	
HydroQuebec	400	2	2	
Interot	100	2	2	
	300	1	1	
JAS/Japan Air Commuter	400	5	3	2
LADS	200	1	1	
LIAT	100	5	5	
	300	3	3	
MarkAir	300	2	2	
Mesa Air	200	12	12	
Mexican Navy	200	1	1	
Midroc Aviation	200	2	2	
Mobil Oil	100	1	1	
National Jet	200	6	6	
	300	2	2	
Norfolk Airlines	100	1	1	
norOntair	100	2	2	
Northwest Airlines	100	25	25	
Norwegian CAA	100	1	1	
Oriental Air Bridge	200	2	2	
Palestinian	300	2	2	
Petroleum Air Service	300	3	3	
Qantas Airways	200	1	1	
	300	16	12	4
Rheintalflug	100	3	3	
	300	3	3	
Royal Wings	300	1	1	
Ryukyu Air Commuter	100	4	4	
South African Express	300	12	12	
SAS Commuter	400	28	28	
Saeaga Airlines	200	1	1	
	300	1	1	
Saudi Aramco	200	3	3	
Schreiner Airways	100	1	1	
	300	2	2	
Talair	100	2	2	
TAVAJ	200	2	2	
Time Air	100	4	4	
	300	10	10	
Transport Canada	100	2	2	
Tyco	400	1	1	
UNI Airways[2]	100	4	4	
	200	1	1	
	300	14	14	
USAF/Sierra	100	2	2	
USAir Express	100	56	56	
	200	19	19	
Wideroe's	100	15	15	
	300	4	4	
	400	3	3	
Zhejiang Airlines	300	2	2	
Undisclosed	100	3	3	
	200	5	3	
	300	1	1	
Totals		**709**	**683**	**26**

Notes: Operators with leased Dash 8 aircraft include, but are not necessarily restricted to, the following: Air Alliance, ERA, Lloyd Aviation, Lufthansa CityLine, Sabena, TABA.
US Airways operators include Piedmont Airlines and Allegheny Commuter.

[1] Formerly Tyrolean.
[2] UNI Airways received initial aircraft when named Great China Airlines.
[3] British European (now FlyBE) received initial aircraft when named Jersey European Airways.

SUMMARY

	Ordered	Delivered	Backlog
Series 100	299	299	
Series 200	96	94	2
Series 300	211	207	4
Series 400	103	83	20
Totals	**709**	**683**	**26**

UPDATED

BOMBARDIER 415

US marketing name: SuperScooper
Engineering designation: CL-215-6B11

TYPE: Twin-turboprop amphibian.

PROGRAMME: Introduced as product follow-on to piston-engined Canadair CL-215; given new designation Canadair 415 in 1991 to distinguish new production turboprop model from CL-215T retrofit, but engineering designation retained and new-build turboprop versions are CL-215-6B11. Launched officially 16 October 1991 with firm orders from France and (August 1992) Québec; first flight (C-GSCT) 6 December 1993, although preceded by initial CL-215T conversion (C-FASE), flown on 8 June 1989. Canadian certification 24 June 1994 in Restricted and Utility categories, FAA approval 14 October 1994 in Restricted category. RAI (Italy) approval 27 October 1994 in Restricted category. Fleet had achieved 50,000 flying hours and 190,000 scooping sorties by 31 March 2001. Final assembly relocated to North Bay, Ontario, in November 1998. Production suspended in October 2001 pending new orders (but see 415MP entry below). New designation Bombardier 415 adopted in February 2002.

CURRENT VERSIONS: Standard **415** and first production units are in **firefighting** configuration

415MP: Multipurpose version incorporating FLIR, SLAR and nose-mounted search radar for missions such as search and rescue, coastal and border patrol and environmental monitoring, while retaining firefighting capability. First flight (C-GHVX; 55th aircraft, destined for Greece) 6 March 2002, preceding a 200-hour flight test programme, and was delivered in late March 2004.

415MP can be modified for **maritime**, **SAR** and **special missions** .

415GR: Ordered by Greece, January 1999; based on 415MP; increased weights; boat handling and cargo hoisting provisions. Two (plus an optional third) will be configured for combat search-and-rescue (C-SAR) role, for which, during 2000, SAAB Nyge Aero of Sweden was awarded a three-year contract to install MSS 5000 mission equipment including SLAR (each side), wing-mounted FLIR Systems SeaFLIR, nose-mounted Honeywell Primus WX 660 weather/search radar, digital cameras, autopilot and provision for Have Quick secure radios and rescue beacon receivers. The aircraft will also have an enlarged cargo door to facilitate deployment of an inflatable rescue boat. The Hellenic Air Force C-SAR aircraft will be based at Elefsis AB; delivery was due in May 2003.

CUSTOMERS: See table. First French Canadair 415 delivered to CEV experimental unit 8 February 1995, but trials revealed need for modifications and acceptance delayed until 13 June 1995; deliveries completed June 1997. Ontario provincial government announced order for nine on 2 April 1998. Government of Malaysia expressed interest in acquiring two following demonstration flights in February 2002. Total 59 built by October 2001 production suspension; 60th registered in January 2002, and 62nd in August 2003.

COSTS: Approximately US$23 million in firefighting configuration (2000).

DESIGN FEATURES: Retains well-proven basic airframe of piston-engined CL-215 (thick wing with zero dihedral and 2° incidence; row of vortex generators on each wing outboard of

BOMBARDIER 415 CUSTOMERS
(at February 2004)

Customer	Qty	First aircraft	Delivered
Canada: Ontario govmt	9	C-GOGD	29 Apr 98
Québec govmt	8	C-GQBA	Jul 95
Croatia[1]	4	9A-CAG	Feb 97
France: Sécurité Civile	12	F-ZBFS	8 Feb 95
Greek govmt[2]	10	2039	Jan 99
Italy: Protezone Civile[3]	16	I-DPCD	27 Jan 95
Total	**59**		

[1] Requirement for six
[2] Plus five options; order total includes two 415MP
[3] Order for two in February 2004 includes option to convert one to 415MP

Bombardier 415 (two P&WC PW123 turboprops) ***NEW***/0547109

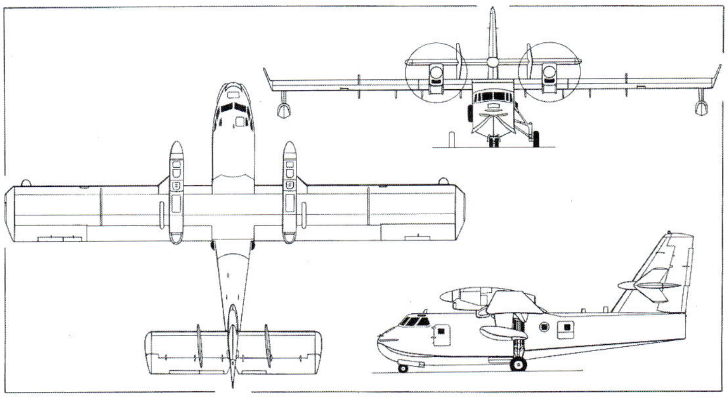

Canadair 415 SuperScooper twin-turboprop general purpose amphibian ***(Jane's/Dennis Punnett)*** 0507126

fence; long stall strip inboard of starboard fence; leading-edge strakes and fences beside engine nacelles; water scoops behind planing step; anti-spray channels in planing bottom chine) but incorporates upgrading modifications and improvements including higher operating weights for increased firefighting productivity; pressure refuelling; wing endplates for lateral stability; finlets and tailplane/fin bullet to recover longitudinal and directional stability affected by relocated thrust line, increased power and new propellers; powered rudder, ailerons and elevators; new electrical system; new 'glass' cockpit with air conditioning; enlarged four-tank firefighting drop system.

FLYING CONTROLS: Conventional and power-assisted. Hydraulically actuated ailerons, elevators and rudder, standard; manual reversion in event of hydraulic failure; geared tab in each aileron, spring tab in rudder and each elevator, plus trim tab in port aileron and port elevator. Hydraulically operated single-slotted flaps, each supported by four external hinges.

STRUCTURE: No-dihedral, no-twist high wing, of constant chord; one-piece structure with two conventional spars, extruded spanwise stringers and interspar ribs and aluminium alloy skins. All-metal, fail-safe, single-step boat-hull fuselage with numerous watertight compartments. Tail surfaces of aluminium alloy sheet and extrusions, with honeycomb panels on control surfaces.

LANDING GEAR: Hydraulically retractable tricycle type. Self-centring twin-wheel nose unit retracts rearward into hull and is fully enclosed by conformal doors. Nosewheel steering standard. Main gear legs retract into wells in sides of hull. Plate mounted on each main gear assembly encloses bottom of wheel well. Mainwheel tyres 15.00-16 (16 ply) tubeless, pressure 5.31 bar (77 lb/sq in); nosewheel tyres 6.50-10 (10 ply) tubed, pressure 6.55 bar (95 lb/sq in). Hydraulic disc brakes. Non-retractable stabilising floats, each carried near wingtip on pylon cantilevered from wing box structure, with breakaway provision.

POWER PLANT: Two 1,775 kW (2,380 shp) Pratt & Whitney Canada PW123AF turboprops, on damage-tolerant mounts capable of withstanding a breach of the compressor/turbine casing, each driving a Hamilton Sundstrand 14SF-19 four-blade constant-speed fully feathering reversible-pitch propeller. Two fuel tanks, each of eight identical flexible cells, in wing spar box, with total usable capacity of 5,796 litres (1,531 US gallons; 1,275 Imp gallons). Single-point pressure refuelling (rear fuselage, starboard side), plus gravity points in wing upper surface.

ACCOMMODATION: Normal crew of two side by side on flight deck, with dual controls. Additional station in maritime patrol/SAR versions for third cockpit member, mission specialist and two observers. For water bomber cabin installation, see Equipment paragraph. Combi layout offers cargo at front, full firefighting capability, plus 11 seats at rear. Other quick-change interiors available for utility/paratroop (up to 14 troop-type folding canvas seats in cabin) or other special missions according to customer's requirements. Flush doors to main cabin on port side of fuselage forward and aft of wings. Optional aft cargo door, height 1.33 m (4 ft 4½ in), width 1.46 m (4 ft 9½ in). Emergency exit on starboard side aft of wing trailing-edge. Crew emergency hatch in flight deck roof on starboard side. Mooring hatch in upper surface of nose. Provision for additional cabin windows.

SYSTEMS: Vapour cycle air conditioning system and combustion heater. Hydraulic system, pressure 207 bar (3,000 lb/sq in), utilises two engine-driven pumps (maximum flow rate 45.5 litres; 12.0 US gallons; 10.0 Imp gallons/min) to actuate nosewheel steering, landing gear, flaps, water drop doors, pickup probes, flight controls, main gear unlocking and wheel brakes. Hydraulic fluid (MIL-H-83282) in air/oil reservoir slightly pressurised by engine bleed air. Electrically driven third pump provides hydraulic power for emergency actuation of landing gear and brakes and closure of water doors. Electrical system includes two 800 VA 115 V 400 Hz static inverters, two 28 V 400 A DC engine-driven starter/generators and two 40 Ah Ni/Cd batteries. Pneumatic/electric intake de-icing system; airframe ice protection system optional.

AVIONICS: Dual Honeywell Primus 2 digital integrated VHF nav/com.

Comms: Global Wulfsberg VHF/UHF/AM/FM and Rockwell Collins HF radios with central control heads, ELT and dual transponders.

Radar: Search/weather radar optional.

Flight: Dual ADF, VOR/ILS, marker beacon receivers and single DME.

Instrumentation: Honeywell EDZ-605 EFIS with three-tube Integrated Instrument Display System for EADI and EHSI; dual Litef/Honeywell AHRS, dual air data computers, Honeywell radio altimeter.

EQUIPMENT (FIREFIGHTER): Four integral water tanks in main fuselage compartment, near CG (combined capacity 6,137 litres; 1,621 US gallons; 1,350 Imp gallons), plus eight inward-facing seats in forward cabin. Tanks filled by two hydraulically actuated scoops aft of hull step, fillable also on ground by hose adaptor on each side of fuselage. Four independently openable water doors in hull bottom. Onboard foam concentrate reservoirs (capacity 680 kg; 1,500 lb) and mixing system. Improved drop pattern and drop door sequencing compared with CL-215. Optional spray kit can be coupled with firefighting tanks for large-scale spraying of oil dispersants and insecticides. In a typical firefighting mission, with a water source 6 n miles (11 km; 7 miles) from the fire, aircraft can remain on station for 3 hours, dropping 55,267 litres (14,600 US gallons; 12,157 Imp gallons)/h. Water tanks can be scoop-filled completely (ISA at S/L, zero wind) in 12 seconds over a water distance of 1,341 m (4,400 ft); partial water loads can be scooped on smaller bodies of water. Minimum safe water depth for scooping is only 1.40 m (4 ft 7 in).

EQUIPMENT (OTHER VERSIONS): Stretcher kits, passenger or troop seats, cargo tiedowns, searchlight and other equipment according to mission and customer requirements. Bombardier 415 can be equipped with maritime surveillance radar and electro-optical sensors, precision navigation and communications equipment and autopilot.

DIMENSIONS, EXTERNAL:
Wing span ... 28.63 m (93 ft 11 in)
Wing chord, constant ... 3.54 m (11 ft 7½ in)
Wing aspect ratio ... 8.2
Length overall ... 19.82 m (65 ft 0½ in)
Beam (max) ... 2.59 m (8 ft 6 in)
Length/beam ratio ... 7.5
Height overall: on land ... 9.07 m (29 ft 9 in)
on water ... 6.96 m (22 ft 10 in)
Draught: wheels up ... 1.12 m (3 ft 8 in)
wheels down ... 2.03 m (6 ft 8 in)
Tailplane span ... 10.97 m (36 ft 0 in)
Wheel track ... 5.25 m (17 ft 2¾ in)
Wheelbase ... 7.23 m (23 ft 8½ in)
Propeller diameter ... 3.97 m (13 ft 0¼ in)
Propeller fuselage clearance ... 0.59 m (1 ft 11¼ in)
Propeller water clearance ... 1.16 m (3 ft 9¾ in)
Propeller ground clearance ... 2.77 m (9 ft 1 in)
Forward door: Height ... 1.37 m (4 ft 6 in)[1]
Width ... 1.03 m (3 ft 4 in)
Height to sill ... 1.68 m (5 ft 6 in)
Rear door: Height ... 1.12 m (3 ft 8 in)
Width ... 1.02 m (3 ft 4 in)
Height to sill ... 1.83 m (6 ft 0 in)
Optional large rear cargo door:
Height ... 1.12 m (3 ft 8 in)
Width ... 1.47 m (4 ft 10 in)
Height to sill ... 1.83 m (6 ft 0 in)
Water drop door: Length ... 1.60 m (5 ft 3 in)
Width ... 0.28 m (0 ft 11 in)
Emergency exit: Height ... 0.91 m (3 ft 0 in)
Width ... 0.51 m (1 ft 8 in)

[1] *incl 25 cm (10 in) removable sill*

DIMENSIONS, INTERNAL:
Cabin, excl flight deck: Length ... 9.38 m (30 ft 9½ in)
Max width ... 2.39 m (7 ft 10 in)
Max height ... 1.90 m (6 ft 3 in)
Floor area ... 19.7 m^2 (212 sq ft)
Volume ... 35.6 m^3 (1,257 cu ft)

AREAS:
Wings, gross ... 100.33 m^2 (1,080.0 sq ft)
Ailerons (total) ... 8.05 m^2 (86.60 sq ft)
Flaps (total) ... 22.39 m^2 (241.00 sq ft)
Fin ... 11.22 m^2 (120.75 sq ft)
Rudder, incl tabs ... 6.02 m^2 (64.75 sq ft)
Tailplane ... 20.55 m^2 (221.20 sq ft)
Elevators (total, incl tabs) ... 7.88 m^2 (84.80 sq ft)

WEIGHTS AND LOADINGS (A: firefighter, B: utility, land or water based):
Typical operating weight empty:
A ... 12,882 kg (28,400 lb)
B ... 12,733 kg (28,072 lb)
Max internal fuel weight: A, B ... 4,649 kg (10,250 lb)
Max payload: A (disposable) ... 6,123 kg (13,500 lb)
B ... 3,777 kg (8,328 lb)
Max ramp weight: A (land) ... 19,958 kg (44,000 lb)
A (water), B ... 17,236 kg (38,000 lb)
Max T-O weight: A (land) ... 19,890 kg (43,850 lb)
B (land) ... 18,597 kg (41,000 lb)
A (water), B (land and water) ... 17,168 kg (37,850 lb)
Max touchdown weight for water scooping:
A ... 16,420 kg (36,200 lb)
Max flying weight after water scooping:
A ... 21,319 kg (47,000 lb)
Max landing weight:
A, B (land and water) ... 16,783 kg (37,000 lb)
Max zero-fuel weight: A ... 19,504 kg (43,000 lb)
B ... 16,511 kg (36,400 lb)
Max wing loading:
A (after scoop) ... 212.5 kg/m^2 (43.52 lb/sq ft)
A (land) ... 198.2 kg/m^2 (40.60 lb/sq ft)
B (land) ... 185.4 kg/m^2 (37.96 lb/sq ft)
A (water), B (land and water) ... 171.1 kg/m^2 (35.05 lb/sq ft)
Max power loading:
A (after scoop) ... 6.01 kg/kW (9.87 lb/shp)
A (land) ... 5.60 kg/kW (9.21 lb/shp)
B (land) ... 5.24 kg/kW (8.61 lb/shp)
A (water), B (land and water) ... 4.84 kg/kW (7.95 lb/shp)

PERFORMANCE (at weights shown):
Max cruising speed at FL100, AUW of
16,329 kg (36,000 lb) ... 203 kt (376 km/h; 234 mph)
Long-range cruising speed at FL100, AUW of
16,329 kg (36,000 lb) ... 150 kt (278 km/h; 173 mph)
Patrol speed at S/L, AUW of 14,741 kg (32,500 lb) ... 130 kt (241 km/h; 150 mph)
Stalling speed: 15° flap, AUW of 21,319 kg (47,000 lb) ... 80 kt (149 km/h; 93 mph)
Max rate of climb at S/L, AUW of 21,319 kg (47,000 lb) ... 396 m (1,300 ft)/min
T-O distance at S/L, ISA:
land, AUW of 19,890 kg (43,850 lb) ... 844 m (2,770 ft)
land, AUW of 18,597 kg (41,000 lb) ... 783 m (2,570 ft)
water, AUW of 17,168 kg (37,850 lb) ... 808 m (2,650 ft)
Landing distance at S/L, ISA:
land, AUW of 16,783 kg (37,000 lb) ... 674 m (2,210 ft)
water, AUW of 16,783 kg (37,000 lb) ... 665 m (2,180 ft)
Scooping distance at S/L, ISA (incl safe clearance heights) ... 1,341 m (4,400 ft)
Ferry range with 998 kg (2,200 lb) payload ... 1,250 n miles (2,315 km; 1,438 miles)
Design *g* limits (15° flap) ... +3.25/−1

UPDATED

CLASS

CANADIAN LIGHT AIRCRAFT SALES AND SERVICES INC

880 Chemin St-Féréol, Les Cèdres, Québec J7T 1N3
Tel: (+1 450) 452 47 72 and (+1 888) 977 14 47
Fax: (+1 450) 452 26 94
e-mail: marla@bushcaddy.com
Web (1): http://www.bushcaddy.com
Web (2): http://www.class-inc.ca

PARTNERS:
Sean Gilmore
Marlene Gill Ronald Strauss

EUROPEAN, MIDDLE EAST AND NORTH AFRICAN AGENTS:
AMA Aeromax Affairs, Aérodrome St Laurent, 4 route de Sémignan, F-33112 St Laurent Médoc, France

CLASS took over production of the previously named Cadi from designer Jean Edues Potvin and has recently expanded the range to provide six versions of the Bush Caddy as well as introducing the Kestrel in 2003.

UPDATED

CLASS BUSH CADDY

TYPE: Four-seat kitbuilt.

PROGRAMME: Designed by Jean Edues Potvin, the original Cadi was first flown in first half of 1993. Kits of the refined Cadi were marketed from 1994. The re-engineered Bush Caddy appeared in 2000. Several new versions have been added to the range since then, using both Rotax and Lycoming power plants. In 2003, the company was developing a 149 kW (200 hp) Jabiru 5100 installation.

CURRENT VERSIONS: **R80:** Smallest aircraft in range and classified as ultralight in USA, Canada and Australia; two seats; designed for engines up to 59.7 kW (80 hp). Can be purchased as a kit or as a factory-completed aircraft. Voyageur option can be converted from tailwheel to tricycle layout in a few hours, as hardpoints for both variants are provided as standard. 'R' in designation indicates Rotax engine.

Description below refers to R80, unless otherwise stated.

R60: Ultralight version of the above, intended for European market; due for launch during 2003.

R120: Engines in the 48.5 to 89.5 kW (65 to 120 hp) range; two seats. Introduced at AirVenture, Oshkosh, July 2001. Voyageur option available as for R80.

L160: Accepts engines up to 119 kW (160 hp). Seats two-plus-one. 'L' in designation indicates Lycoming engine.

L162: Two-plus-two version of L160.

L164: Four-seat version, with 134 kW (180 hp) engine. Displayed statically at AirVenture 2001. Prototype (C-GJDK) first flew 8 September 2001.

CUSTOMERS: Nearly 100 of all versions sold by end 2002 (latest information supplied), of which 66 were then flying.

COSTS: Basic kit costs (excluding engine, propeller and internal finishing): R80, C$20,596; R120, C$21,895; L160 C$25,401; L162, C$27,535; L164, C$30,895, quick build kit C$50,000 (2004). Factory-finished R80, C$72,000 (all 2004).

DESIGN FEATURES: Conventional lightplane; high, constant chord wing braced by V struts each side; sweptback cantilever tail surfaces.

Kit includes many preformed and welded parts. Quoted build time 1,000 hours. Fast-build kits are available in which wings and fuselage internal structures are complete and need only covering.

FLYING CONTROLS: Conventional and manual. No horn balances on control surfaces. Flaps standard on L160 and higher-powered versions and optional for R120.

STRUCTURE: All-metal construction throughout. Two-spar wing has a modified Piper Super Cub aerofoil with flat lower surface and has twin 7.6 cm (3 in) extruded aluminium bracing struts and composites wingtips. Monocoque fuselage contains 6061-T6 bulkheads covered with sheet aluminium and has a composites cowling.

LANDING GEAR: Tailwheel type, fixed; R120 has quick-change option from tailwheel to tricycle. Two (optionally faired-in) side Vs hinged to lower longerons, plus half-axles spring under centreline with bungee cord. Optional floats, amphibious floats and skis can be fitted. Cleveland mainwheels are 7.00-6, with McCreary tyres and Matco hydraulic brakes. Castoring tailwheel, size 2.50-4.

POWER PLANT: *R80:* One 59.5 kW (79.9 hp) Rotax 912 UL driving a two-blade Sensenich metal fixed-pitch propeller; options include 73.5 kW (98.6 hp) Rotax 912 ULS and 84.6 kW (113.4 hp) Rotax 914 UL and two- and three-blade Warp Drive propellers. Fuel capacity 91 litres (24.0 US gallons; 20.0 Imp gallons) in two wingroot tanks.

R120: 73.5 kW (98.6 hp) Rotax 912 ULS standard; engines up to 89.5 kW (120 hp) optional. Standard fuel capacity 100 litres (26.4 US gallons; 22.0 Imp gallons); optional increase to 155 litres (40.8 US gallons; 34.0 Imp gallons).

L160 and *L162:* One 119 kW (160 hp) Lycoming O-320 standard. Fuel capacity 173 litres (45.6 US gallons; 38.0 Imp gallons).

CLASS Bush Caddy L164 prototype showing external access to baggage compartment *(Paul Jackson)* ***NEW***/0528676

L164: One 134 kW (180 hp) Lycoming IO-360 standard. Fuel capacity as L160/L162.

ACCOMMODATION: Pilot and passenger in fully adjustable seats in R80 and R120; bench seat in rear of L160 and L162 for one and two extra passengers, respectively; L164 is full four-seat version. Wraparound Lexan windscreen. Baggage locker behind cabin, with door on port side.

DIMENSIONS, EXTERNAL:
Wing span: R80, R120 ... 9.75 m (32 ft 0 in)
L160, L162, L164 ... 10.97 m (36 ft 0 in)
Wing chord, constant: all ... 1.60 m (5 ft 3 in)
Wing aspect ratio: R80, R120 ... 6.1
L160, L162, L164 ... 6.9
Length overall: R80, R120 ... 6.73 m (22 ft 1 in)
L160, L162 ... 7.26 m (23 ft 10 in)
L164 ... 7.75 m (25 ft 5 in)
Height: R80, R120 ... 2.01 m (6 ft 7¼ in)
L160 ... 2.29 m (7 ft 6 in)
L164 ... 2.32 m (7 ft 7¼ in)
Tailplane span ... 2.62 m (8 ft 7 in)
Wheel track ... 2.13 m (7 ft 0 in)
Wheelbase ... 4.44 m (14 ft 7 in)

DIMENSIONS, INTERNAL:
Cabin: Max width: R80, R120, L160 ... 1.12 m (3 ft 8 in)
L164 ... 1.19 m (3 ft 11 in)
Max height: R80, R120 ... 1.12 m (3 ft 8 in)
L160 ... 1.17 m (3 ft 10 in)
L164 ... 1.33 m (4 ft 4½ in)
Baggage hold volume ... 0.42 m^3 (15.0 cu ft)

AREAS:
Wings, gross: R80, R120 ... 15.61 m^2 (168.0 sq ft)
L160, L162, L164 ... 17.56 m^2 (189.0 sq ft)

WEIGHTS AND LOADINGS:
Weight empty: R80 ... 294 kg (648 lb)
R120 ... 373 kg (823 lb)
L160 ... 498 kg (1,098 lb)
L164 ... 562 kg (1,239 lb)
Baggage capacity ... 34 kg (75 lb)
Max T-O weight: R80 ... 559 kg (1,232 lb)
R120 ... 680 kg (1,500 lb)
L160, L162 ... 998 kg (2,200 lb)
L164 ... 1,134 kg (2,500 lb)
Max wing loading: R80 ... 35.8 kg/m^2 (7.33 lb/sq ft)
R120 ... 43.6 kg/m^2 (8.93 lb/sq ft)
L160, L162 ... 56.8 kg/m^2 (11.64 lb/sq ft)
L164 ... 64.6 kg/m^2 (13.23 lb/sq ft)
Max power loading: R80 ... 9.38 kg/kW (15.42 lb/hp)
R120 ... 9.26 kg/kW (15.21 lb/hp)
L160, L162 ... 8.37 kg/kW (13.75 lb/hp)
L164 ... 8.45 kg/kW (13.89 lb/hp)

PERFORMANCE:
Never-exceed speed (VNE):
R80, R120 ... 113 kt (209 km/h; 130 mph)
L160 ... 130 kt (241 km/h; 150 mph)
L164 ... 139 kt (257 km/h; 160 mph)
Cruising speed at 75% power:
R80 ... 83 kt (153 km/h; 95 mph)
R120 ... 87 kt (161 km/h; 100 mph)
L160 ... 100 kt (185 km/h; 115 mph)
L164 ... 117 kt (217 km/h; 135 mph)
Econ cruising speed at 60% power:
R80, R120 ... 78 kt (145 km/h; 90 mph)
L160 ... 91 kt (169 km/h; 105 mph)
L164 ... 104 kt (193 km/h; 120 mph)
Stalling speed, power off: R80 ... 30 kt (55 km/h; 34 mph)
no flaps: R120, L160, L162, L164 ... 37 kt (68 km/h; 42 mph)
flaps down: R120 ... 28 kt (52 km/h; 32 mph)
L160, L162, L164 ... 33 kt (62 km/h; 38 mph)
Max rate of climb at S/L: R80, L160 ... 259 m (850 ft)/min
R120 ... 274 m (900 ft)/min
L164 ... 366 m (1,200 ft)/min
Service ceiling: all ... 3,960 m (13,000 ft)
T-O run: R80 ... 61 m (200 ft)
R120 ... 107 m (350 ft)
L160 ... 110 m (360 ft)
L164 ... 76 m (250 ft)
Landing run: R80 ... 107 m (350 ft)
R120, L160 ... 153 m (500 ft)
L164 ... 183 m (600 ft)
Range: R80 ... 450 n miles (833 km; 517 miles)
R120 ... 500 n miles (926 km; 575 miles)
L160 ... 625 n miles (1,157 km; 719 miles)
L164 ... 800 n miles (1,481 km; 921 miles)
Endurance ... more than 4 h

UPDATED

CLASS KESTREL

TYPE: Side-by-side ultralight/kitbuilt; two-seat lightplane.

PROGRAMME: Epervier Aviation SA of Gosselies, Belgium, flew prototype Epervier (Sparrowhawk) in September 1990, initially as ultralight (ULM), registered OO-964; then as lightplane (ATL) OO-EPV with 11.15 m (36 ft 7 in) wingspan. One production Epervier ULM (OO-EPB) built before project terminated.

CLASS Kestrel static demonstrator, the former Epervier ULM *(Paul Jackson)* ***NEW***/0572322

Rights acquired by CLASS in April 2003, together with OO-EPB and five incomplete airframes; renamed Kestrel and marketing began immediately to ensure early 2004 launch, but with changes dictated by feedback from potential customers; Rotax 912S or Jabiru 3300 engine; composites wing covering replacing fabric.

CURRENT VERSIONS: **Kestrel 60:** Baseline version.

Kestrel 150: Upgraded version, with more powerful engine and higher take-off weight. First deliveries due during second half of 2004.

COSTS: Kestrel 60 kit C$23,125 for first five aircraft, thereafter C$25,000; Kestrel 150 kit C$24,950 for first five aircraft, thereafter C$27,000 (2003).

DESIGN FEATURES: Intended for cross-country and training; to complement all-metal Bush Caddy range. Primarily aimed at kitbuilt market, but will also be available as factory-finished aircraft.

Wing section NACA 23012.

FLYING CONTROLS: Conventional and manual.

STRUCTURE: Fuselage of composite epoxy/carbon material on glass and gelcoat polyester reinforced with bonded framework.

LANDING GEAR: Fuselage-mounted cantilever mainwheel legs with wheel fairings. Choice of tricycle or tailwheel type. Steerable nosewheel. Mainwheel cable brakes.

POWER PLANT: Kestrel 60 will have one 47.8 kW (64.1 hp) Rotax 582 driving two- or three-bladed propeller; Kestrel 150 will accept engines up to 112 kW (150 hp).

ACCOMMODATION: Two, side by side; baggage space behind seats. Forward-hinged door each side.

AVIONICS: Design accepts full VFR panel for pilot training.

DIMENSIONS, EXTERNAL:

Wing span	9.76 m (32 ft 0¼ in)
Wing chord (constant)	1.60 m (5 ft 3 in)
Length overall	7.03 m (23 ft 0¾ in)

AREAS:

Wings, gross	17.52 m² (188.6 sq ft)

WEIGHTS AND LOADINGS:

Weight empty: Kestrel 150	450 kg (992 lb)
Max T-O weight: Kestrel 150	725 kg (1,600 lb)

PERFORMANCE (estimated):

Normal cruising speed: Kestrel 60	78 kt (145 km/h; 90 mph)
Kestrel 150	96 kt (177 km/h; 110 mph)
Stalling speed	33 kt (62 km/h; 38 mph)
Endurance	5 h

UPDATED

CREATIVE FLIGHT

CREATIVE FLIGHT INC

Elico House, 140 Industrial Park Road, PO Box 596, Haliburton, Ontario K0M 1S0
Tel: (+1 705) 457 38 44
Fax: (+1 705) 457 96 51
e-mail: info@creativeflight.com
Web: http://www.creativeflight.com
CHIEF DESIGNER: Kirk Creelman
INFORMATION: Terrance Gavan

NEW ZEALAND OFFICE:
PO Box 13697, Johnsonville, Wellington
Fax: (+64 4) 233 98 44

This company's first product was renamed from MPA to Aerocat in 2002. Kit production has begun.

UPDATED

CREATIVE FLIGHT AEROCAT

TYPE: Side-by-side kitbuilt; four-seat kitbuilt twin.

PROGRAMME: Project began 1997. Launched at AirVenture, Oshkosh, July 2000. Development prototype (C-GYCC) first flown in single-engine form 15 July 2001; exhibited at Oshkosh 24 to 30 July 2001. Reconfigured in twin-engine form and flown on 5 September 2002; first flight with floats attached (from land) 15 October 2002; water taxying trials begun 17 October 2002; total flight time stood at 40 hours in August 2003, at which time kit production had commenced.

CURRENT VERSIONS: **Aerocat 120:** Two-seat (or two-plus-two); single Jabiru 3300 engine. No amphibious option.

Aerocat Twin 240: Two Jabiru engines.

Seacat: Floatplane/amphibian version of Twin 240.

Aerocat 200: Single Lycoming IO-360 engine.

Aerocat Single Amphibian: Amphibious Aerocat 200.

COSTS: Aerocat 120 US$93,750, Aerocat Twin 240 US$116,250, Seacat US$131,050, Aerocat 200 US$131,750, Aerocat Single Amphibian US$147,450, complete kit including engine(s) and US$10,000 avionics and extras allowance (2004).

DESIGN FEATURES: Designation indicates 'multi platform aircraft'. Kitplane meeting '51 per cent' rule for amateur construction; easily completed in single- or twin-engine configuration and with amphibian and seaplane landing gear.

Cabin pod, with broad outward visibility, mounted on cranked wings of two-stage taper and with sweptback tips. Engine(s) on arched cabane with central, vertical reinforcement strut. Twin tailbooms and high tailplane; forward boom sections contain main landing gear and attachment points for floats with cutouts for mainwheels when in amphibian mode. Production version has improved cabin aerodynamics including blunter nose reducing length by 15 cm (6 in) from prototype's 8.23 m (27 ft 0 in).

Wing section NLF-0215 (mod); dihedral 3°; twist −4°.

FLYING CONTROLS: Conventional and manual. Horn balances on elevator and rudders; flight-adjustable tab on starboard rudder and port aileron. Fowler flaps, max deflection 40°. Third flap, between booms, is independently actuated; potential advantages during take-off and landing are being assessed during trials. Flaps actuated by wires, with electric control; rudders wire-actuated, with manual control.

STRUCTURE: All structures manufactured by 'open layup, vacuum-formed' process. Fuselage, tailbooms, rudders, ailerons, flaps and floats of glass fibre/foam core construction. Wing spars and cabane of carbon fibre/foam core; nosewheel assembly of carbon fibre.

LANDING GEAR: Retractable tricycle type; retractable. Mainwheels 6.00–6; castoring, forward retracting nosewheel, size 5.00–5; differential hydraulic mainwheel brakes. Adjustable air bag suspension. Optional Series 2500 composites floats, with or without wheeled gear. Float installation takes 45 minutes.

POWER PLANT: One or two piston engines, driving three-blade propeller of constant-speed, flight- or ground-adjustable pitch or fixed-pitch type. Prototype flown with single and twin 89 kW (120 hp) Jabiru 3300 flat-six(es) and ground-adjustable pitch propeller(s). Airmaster three-blade, constant-speed propeller on trial in 2002. Recommended options include

Creative Flight Aerocat prototype in 2003 *(Paul Jackson)* ***NEW***/0572312

157 kW (210 hp) Textron Lycoming IO-360, similarly rated Jabiru or 194 kW (260 hp) Subaru turbo, all for four-seat, single-engine version. Fuel capacity (usable) 258 litres (68.0 US gallons; 56.7 Imp gallons).

ACCOMMODATION: Two or four persons, according to version. Rear seats may be replaced by freight. Door each side of cabin. Dual controls.

SYSTEMS: Electrical power 12 V; electro-pneumatic pump 6.9 bar (100 lb/sq in) for landing gear actuation.

AVIONICS: To customer's requirements. Prototype has GPS, ELT and VHF radio.

DIMENSIONS, EXTERNAL:
Wing span ... 10.36 m (34 ft 0 in)
Wing aspect ratio ... 7.0
Length: fuselage ... 7.90 m (25 ft 11 in)
overall ... 8.08 m (26 ft 6 in)
Height overall ... 2.72 m (8 ft 11 in)

DIMENSIONS, INTERNAL:
Cabin: Length ... 2.74 m (9 ft 0 in)
Max width ... 1.48 m (4 ft 9 in)
Max height ... 1.17 m (3 ft 10 in)
Float baggage compartment: (each) ... 0.28 m³ (10.0 cu ft)

AREAS:
Wings, gross ... 15.33 m² (165.0 sq ft)

WEIGHTS AND LOADINGS (120: Aerocat 120, 240: Aerocat Twin 240, Seacat, 200: Aerocat 200, SA: Aerocat Single Amphibian):
Weight empty: 120 ... 592 kg (1,305 lb)
240 ... 703 kg (1,549 lb)
Seacat ... 802 kg (1,769 lb)
200 ... 689 kg (1,520 lb)
SA ... 791 kg (1,744 lb)
Max T-O weight: 120 ... 952 kg (2,100 lb)
except 120 ... 1,202 kg (2,650 lb)
Max wing loading: 120 ... 62.1 kg/m² (12.73 lb/sq ft)
except 120 ... 78.4 kg/m² (16.06 lb/sq ft)
Max power loading: 120 ... 10.65 kg/kW (17.5 lb/hp)
240, Seacat ... 6.72 kg/kW (11.04 lb/hp)
200, SA ... 7.68 kg/kW (12.62 lb/hp)

PERFORMANCE (120, 240, Seacat, 200, SA):
Cruising speed at 80% power:
120 ... 124 kt (230 km/h; 143 mph)
240 ... 154 kt (285 km/h; 177 mph)
Seacat ... 146 kt (270 km/h; 168 mph)
200 ... 156 kt (290 km/h; 180 mph)
SA ... 149 kt (277 km/h; 172 mph)
Stalling speed, flaps down: S, T ... 45 kt (83 km/h; 51 mph)
Max rate of climb at S/L:
120 ... 244 m (800 ft)/min
240 ... 366 m (1,200 ft)/min
Seacat ... 351 m (1,150 ft)/min
200 ... 335 m (1,100 ft)/min
SA ... 320 m (1,050 ft)/min
Range at 55% power:
120 ... 700 n miles (1,296 km; 806 miles)
240 ... 1,000 n miles (1,666 km; 1,035 miles)
Seacat, SA ... 850 n miles (1,574 k; 978 miles)
200 ... 950 n miles (1,759 km; 1,093 miles)

UPDATED

CUSTOM FLIGHT

CUSTOM FLIGHT COMPONENTS LTD

RR1, Perkinsfield, Ontario L0L 2J0
Tel: (+1 705) 526 96 26
Fax: (+1 705) 526 25 29
e-mail: nstar@csolve.net
Web: http://www.customflightltd.com
PRESIDENT: Morgan Williams
EXECUTIVE VICE-PRESIDENT: Jeff Owen

Custom Flight Components produces the North Star and more basic Bright Star two-seat kitplanes, both modelled on the Piper Super Cub. The North Star wing can also be bought as a separate kit and retrofitted to Piper Super Cub and Aeronca Champ and Bellanca Citabria aircraft. Completions continue at low rate.

Company plans include an eight-seat aircraft to be named Galaxie, as well as four- and six-seat versions of the North Star.

VERIFIED

CUSTOM FLIGHT NORTH STAR

TYPE: Tandem-seat kitbuilt.

PROGRAMME: Re-creation of Piper Super Cub, but re-engineered for additional structural strength. Prototype North Star (C-FGYD) first flew in 1991 and made its public debut that year. Offered in kit form from 1992 onwards.

CURRENT VERSIONS: **North Star**: Standard version, *as described below*.

Bright Star: Basic version with no flaps, single one-piece door, leaf spring steel main landing gear, square-topped fin and fuel capacity of 87 litres (23.0 US gallons; 19.2 Imp gallons); introduced in 1996, but first aircraft had not flown by late 2002; optional tricycle or tailwheel versions. Accepts engines from 74.6 kW to 134 kW (100 to 180 hp).

North Star 4: Four-seat version, to be powered by engine in 108 to 194 kW (145 to 260 hp) range with a maximum take-off weight of around 1,134 kg (2,500 lb). Seeking launch customers and project reported to be on hold until North Star 6 established in production.

North Star 6: Six-seat version; maximum take-off weight of around 1,810 kg (4,000 lb). Engines between 194 and 265 kW (260 and 355 hp) can be installed.

CUSTOMERS: At least 14 (of which seven completed) in Canada; five (four completed) in USA. Production aircraft Nos. 7, 8 and 16 completed in Ontario by D E Blackburn, A King and R M Harkness in 1998. Nos. 3 (Subaru engine) and 15, built by W Smith, Ontario, and J Y Gratton, Québec followed in the first half of 1999. Total of at least 24 kits sold by September 2002; in addition five of the six-seat version were under construction at the factory in late 2002.

COSTS: North Star: US$39,631; Bright Star: C$19,602; both excluding engine (2004).

DESIGN FEATURES: Strut-braced mainplanes; wire-braced tailplanes; raked wingtips. Easy maintenance/inspection covers. Quoted build time 1,200 hours.

Compared with Super Cub, has landing gear legs 8 cm (3 in) taller; flatter spine; squared wingtips, with ailerons moved outboard by 0.61 m (2 ft 0 in); aileron and flap chord increased 10 cm (4 in); flap span increased 0.60 m (2 ft 0 in); cockpit 6 cm (2½ in) wider.

Aerofoil USA 35B.

FLYING CONTROLS: Conventional and manual. Horn-balanced rudder and elevators. Four-position flaps.

STRUCTURE: Welded 4130 chromoly tube, fabric-covered fuselage. Tubes injected with oil as anti-corrosion measure. Wing of prototype constructed of wood, but kit aircraft have I-beam spar, aluminium ribs and 2024-T3 leading-edges.

LANDING GEAR: Tailwheel type; fixed. Two faired-in side Vs, hinged to lower longerons, plus half-axles, each with hydraulic/coil-spring shock-absorber, attached to compression frame. Steerable tailwheel. Optional floats and skis. Mainwheel tyres 6.00-6 or 8.50-6; tailwheel 2.80-4 in.

POWER PLANT: One 112 kW (150 hp) Textron Lycoming O-320-E3D flat-four driving a McCauley two-blade fixed-pitch propeller. Fuel capacity 197 litres (52.0 US gallons; 43.3 Imp gallons) in two wing tanks.

ACCOMMODATION: Pilot and passenger in tandem within enclosed cockpit. Door on each side — upper half Perspex and lower half fabric-covered steel tubing. Baggage compartment reached through separate door on starboard side of fuselage.

Prototype Custom Flight North Star two-seat lightplane 0131799

Data estimated for Bright Star.

DIMENSIONS, EXTERNAL:
Wing span ... 11.07 m (36 ft 4 in)
Wing chord, average ... 1.60 m (5 ft 3 in)
Length overall: North Star ... 6.86 m (22 ft 6 in)
Bright Star ... 7.16 m (23 ft 6 in)
Height overall: North Star ... 2.08 m (6 ft 10 in)
Bright Star ... 2.29 m (7 ft 6 in)
Propeller diameter ... 2.08 m (6 ft 10 in)
Cabin door (each): Height ... 0.89 m (2 ft 11 in)
Width ... 1.17 m (3 ft 10 in)

DIMENSIONS, INTERNAL:
Cockpit, max width ... 0.72 m (2 ft 4¼ in)

AREAS:
Wings, gross ... 17.65 m² (190.0 sq ft)

WEIGHTS AND LOADINGS:
Weight empty: North Star ... 531 kg (1,170 lb)
Bright Star ... 476 kg (1,050 lb)
Baggage capacity ... 41 kg (90 lb)
Max T-O weight: North Star ... 997 kg (2,200 lb)
Bright Star ... 816 kg (1,800 lb)
Max wing loading: North Star ... 56.3 kg/m² (11.53 lb/sq ft)
Bright Star ... 46.3 kg/m² (9.47 lb/sq ft)
Max power loading (112 kW; 150 hp engine):
North Star ... 8.93 kg/kW (14.67 lb/hp)
Bright Star ... 7.30 kg/kW (12.00 lb/hp)

PERFORMANCE:
Max level speed ... 122 kt (225 km/h; 140 mph)
Max cruising speed:
North Star: landplane ... 100 kt (185 km/h; 115 mph)
floatplane ... 85 kt (158 km/h; 98 mph)
Bright Star ... 87 kt (161 km/h; 100 mph)
Normal cruising speed:
North Star ... 85 kt (158 km/h; 98 mph)
Stalling speed, power off:
flaps up ... 35 kt (65 km/h; 40 mph)
flaps down ... 31 kt (57 km/h; 35 mph)
Stalling speed, full power:
flaps up ... 31 kt (57 km/h; 35 mph)
flaps down ... 22 kt (41 km/h; 25 mph)
Max rate of climb at S/L:
North Star ... 335 m (1,100 ft)/min
Bright Star ... 305 m (1,000 ft)/min
Service ceiling ... 4,880 m (16,000 ft)
T-O run (half fuel): North Star ... 55 m (180 ft)
Bright Star ... 92 m (300 ft)
Landing run: North Star ... 61 m (200 ft)
Bright Star ... 84 m (275 ft)
Range ... 600 n miles (1,111 km; 690 miles)

UPDATED

DELASTEK

DELASTEK INC

Suite 14, 2699 5e Avenue, CP 123, Grand Mère, Québec G9T 5K7
Tel: (+1 450) 545 17 10 and (+1 819) 533 57 88
Fax: (+1 819) 533 34 94
e-mail: info@delastek.com
Web: http://www.delastek.com
PRESIDENT: Claude Lessard
VICE-PRESIDENT: Lucie McCutcheon
OPERATIONS MANAGERS:
Sylvie Hamel
Stéphane Cormier

Former Partenair company was formed in 1998 to develop the S44 Mystère, designed by Saleem Saleh and Frédéric Amblard. This evolved into the S45. In late 2003, Partenair was bought by Delaslek, which intends to continue development of the S45.

NEW ENTRY

DELASTEK S45 MYSTERE

English name: Mystery
TYPE: Tandem-seat kitbuilt.
PROGRAMME: Design work on series began in 1991. Prototype (C-FZHP) first flew as Rotax 912-powered Partenair S44 on 16 November 1996, and as S45 in 1998; total 50 hours flown by July 2001, when second prototype (C-GGYY), with Textron Lycoming engine, exhibited complete, but unflown, at AirVenture, Oshkosh. First flight eventually accomplished 4 October 2001.
CURRENT VERSIONS: **Mystère Mark I:** Retrospective designation, since 2003, for original aircraft.
Mystère Mark II: Announced early 2003. Engine lin raised 10 cm (4 in) to allow fitment of 1.78 m (5 ft 10 in) propeller; modified engine cowling. *As described.*
Mystère Mark III: Proposed retractable landing gear version, announced mid-2003.
CUSTOMERS: By end 2003, seven aircraft under construction.
COSTS: Basic kit, excluding engine, propeller and avionics, US$27,900 (2004); can be purchased in three sub-kits.
DESIGN FEATURES: Configured to eliminate pitch-up associated with high thrust-line; horizontal stabiliser immediately above propeller slipstream counteracts thrust increases by exerting proportional pitch-down and additionally has increased effectiveness at lower airspeeds. Streamlined, pod-and-boom configuration, with T tail and sweptback fin with large fillet. Wings have constant-chord inboard sections and tapered outer panels. Engine

Partenair (now Delastek) S45 second prototype *(Paul Jackson)* ***NEW***/0572309

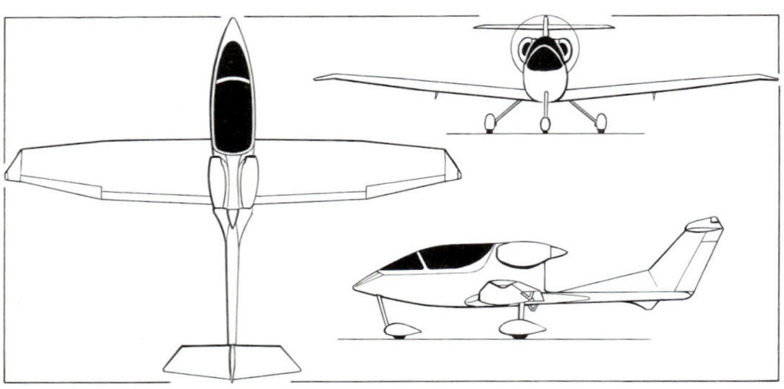

Delastek S45 Mystère tandem-seat kitbuilt *(James Goulding)* 0526897

Second (Mark II) Partenair S45 *(Paul Jackson)* ***NEW***/0561593

Partenair S45 Mystère in 2001 *(Paul Jackson)* ***NEW***/0110740

and pusher propeller behind rear seat. Quoted build time 1,000 hours. Provision for 4 or 6° wing dihedral to be incorporated during manufacture. Wing airfoil NACA65215 on S44 and modified NACA65218 on S45. Wing washout 2° on S45.
FLYING CONTROLS: Conventional and manual. Aileron movements +30/−15°. Double slotted flaps, electrically actuated; positions 10/25/45°. Spade-type balance below each aileron. Flight-adjustable tab in elevator.
STRUCTURE: Generally of composites. Wing has C-shaped main spars plus glass fibre secondary spar at 75% chord.
LANDING GEAR: Tricycle type; fixed. Cantilever sprung main legs of 2024-T3 aluminium; trailing link nosewheel. Speed fairings on all wheels. Hydraulic Cleveland brakes. Mainwheels 5.00–5; Azusa nosewheel 11×4.00–5.
POWER PLANT: One 119.3 kW (160 hp) Textron Lycoming IO-320 flat-four, driving a Prince two-blade metal propeller with Q tips. Fuel capacity 208 litres (55.0 US gallons; 45.8 Imp gallons) of which 197 litres (52.0 US gallons; 43.3 Imp gallons) usable.
ACCOMMODATION: Two persons in tandem, beneath two-piece canopy/windscreen. Dual controls.
EQUIPMENT: Prototype S45 fitted with BRS parachute.

DIMENSIONS, EXTERNAL:	
Wing span	7.13 m (28 ft 6 in)
Wing chord: at root	1.37 m (4 ft 6 in)
at tip	0.69 m (2 ft 3 in)
Wing aspect ratio	7.6
Length overall	7.32 m (24 ft 0 in)
Height overall	2.44 m (8 ft 0 in)
Wheel track	2.29 m (7 ft 6 in)
Tailplane span	2.62 m (8 ft 7 in)
Propeller diameter	1.83 m (6 ft 0 in)
DIMENSIONS, INTERNAL:	
Cabin max width	0.76 m (2 ft 6 in)
AREAS:	
Wings, gross	9.94 m^2 (107.0 sq ft)
WEIGHTS AND LOADINGS:	
Weight empty	522 kg (1,150 lb)
Baggage capacity	34 kg (75 lb)
Max T-O weight	861 kg (1,900 lb)
Max wing loading	86.70 kg/m^2 (17.76 lb/sq ft)
Max power loading	7.23 kg/kW (11.88 lb/hp)
PERFORMANCE:	
Never-exceed speed (VNE)	182 kt (338 km/h; 210 mph)
Max cruising speed	139 kt (257 km/h; 160 mph)
Stalling speed	48 kt (89 km/h; 55 mph)
Max rate of climb at S/L	427 m (1,400 ft)/min
T-O and landing run	183 m (600 ft)
Max range	1,000 n miles (1,852 km; 1,150 miles)
g limits	+6/−3

UPDATED

DIAMOND

DIAMOND AIRCRAFT CORPORATION

1560 Crumlin Sideroad, London, Ontario N5V 1S2
Tel: (+1 519) 457 40 00
Fax: (+1 519) 457 40 21
e-mail: sales@diamondair.com
Web: http://www.diamondair.com
CEO: Christian Dries
MARKETING DIRECTOR: John Gauch
TECHNICAL DIRECTOR: Peter Maurer

Diamond Aircraft Corporation incorporated 12 January 1992 and established 24,625 m^2 (265,000 sq ft) plant at London, Ontario in June 1994, to build DV 20 Katana light aircraft.
An experimental design, the DA 20-B1 (C-GKAT), was completed in June 1996, followed by the DA 20-C1 prototype in February 1997. Series production of the DA 20-C1 is currently under way, augmented from March 2002 onwards by the Diamond DA 40-180 Star (of which an Austrian version is in parallel production by Diamond Aircraft Industries). Workforce stood at 160 in late 2002.

UPDATED

DIAMOND DA 20-C1

English name: Samurai Sword
TYPE: Two-seat lightplane.
PROGRAMME: Original Austrian-built Katana revealed following first flight on 16 March 1991 of proof-of-concept LF 2000 (OE-VPX); followed by LF 2 Katana predecessor (OE-CPU, first flight December 1991); DV 20 prototype (OE-AKL, first flight 17 December 1992) developed from this by Dries and Volck and was representative of production aircraft; Austrian and German certification 26 April 1993; JAR-VLA certification received later in 1993; now also certified by Canada, UK and, on 9 December 1994, USA (FAA). Second

US-registered Diamond DA 20-C1 Eclipse *NEW*/0572281

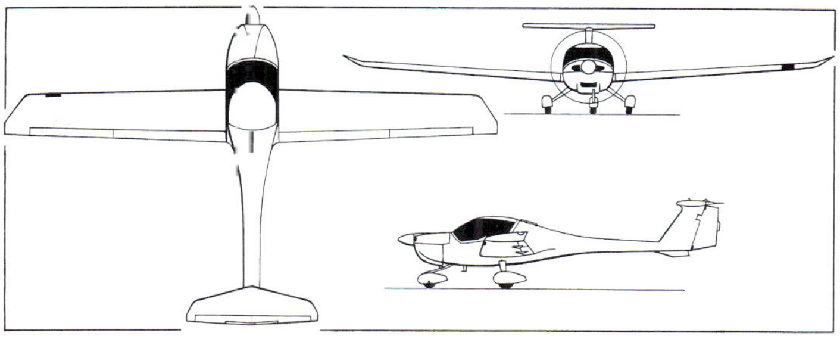

Diamond DA 20-C1 *(James Goulding)* 0044445

production line opened in Canada, initially for the DA 20-A1; European-built aircraft designated DV 20. European manufacture of the DV 20 ended in mid-1996.

The current C1 version, introducing minor airframe improvements, was announced at Aero '97 at Friedrichshafen, shortly after first flight of the Canadian prototype (C-FDVA); production of this version began in early 1998. Canadian variant introduced some 40 minor changes, including wider cabin, electric trim and doubled electrical capacity, but is unnamed. Exports of -C1s to USA began June 1998 with seventh production aircraft.

CURRENT VERSIONS: **DA 20-C1:** *As described.* Certified by FAA 6 April 1998. Relaunched in two versions from early 2001: **Evolution** basically equipped trainer, capable of 135 kt (250 km/h; 155 mph) cruise at 1,830 m (6,000 ft) and **Eclipse**, featuring carpets, leather seats and other interior features for private owners.

DA 20-A2: Intended to have DA 20-C's improved airframe, but retain DA 20-A's 59.7 kW (80 hp) Rotax 912 flat-four engine. Not launched.

DA 20-W: Based on DA 20C-1 airframe, but with 78.3 kW (105 hp) Diamond GIAE 110R rotary engine. To be built in Austria by Diamond Aircraft Industries GmbH.

CUSTOMERS: Total of more than 740 DV/DA 20s of all versions delivered by end 2003 from Austrian and Canadian production. European production of DV 20s terminated with the 160th aircraft in July 1996; last of 331 Canadian DA 20 A1s completed in late 1998. Total of 250 C1s registered up to December 2003. Most to USA, but some to Canada and one each to New Zealand (in March 1999) and UK (July 1999). Katanas of all versions operated by training schools in Australia (one), Austria (six), Canada (11), Czech Republic (one), Finland (one), Germany (six), New Zealand (one), Sweden (one), South Africa (one), Switzerland (one), UK (two) and USA (32).

Customers include Embry-Riddle Aeronautical University, which has leased 35 C1s to fulfil a contract to provide basic flight training of USAF cadet pilots at the Air Force Academy in Colorado Springs.

COSTS: Evolution US$123,900; Eclipse US$139,990 (2002).

DESIGN FEATURES: Conventional, mid-wing, T-tail monoplane of composites construction.

Wing section Wortmann FX-63-137/20; dihedral 4° at root, 27° at tip; sweepback 1°; incidence, 2° 30'. Fin sweepback 57°; tailplane incidence −2°.

FLYING CONTROLS: Conventional and manual. Ailerons and elevators are operated by push-pull tubes, rudder by cables; electrically actuated spring bias provides elevator trimming; electrically actuated flaps. Control surface movements: ailerons +16°/−13°; elevator +25°/−15°; rudder ±30°; flaps 15° for T-O, 45° landing.

STRUCTURE: Fuselage, with integral vertical fin, is mostly of GFRP construction with local CFRP reinforcement in high-stress areas, comprising two half-shells which are bonded together and incorporate transverse bulkheads in the cabin/centre-section area and three ring bulkheads in the tailcone. Wings comprise I-section spar of GFRP, with ribs providing mounting surfaces for control tubes and bellcranks; CFRP spar caps; centre section shear web is foam and fibre; wing skins are of GFRP/PVC foam/GFRP sandwich construction, as are the rudder and horizontal tail surfaces. Flaps and ailerons of GFRP.

LANDING GEAR: Fixed tricycle type, with fuselage-mounted cantilever self-sprung aluminium leg on each main unit and elastomeric suspension on nose leg; latter offset to starboard by 8 cm (3 in). Steering is provided by differential braking of the mainwheels and friction-damped castoring nosewheel. Speed fairings on all three wheels standard on Eclipse, optional on Evolution.

POWER PLANT: One 93 kW (125 hp) Teledyne Continental IO-240-B flat-four driving a Sensenich W69EK-63 two-blade wood and composites propeller. Fuel capacity 92.7 litres (24.5 US gallons; 20.4 Imp gallons), of which 91 litres (24.0 US gallons; 20.0 Imp gallons usable); oil capacity 5.7 litres (1.5 US gallons; 1.25 Imp gallons).

ACCOMMODATION: Two seats side by side, with baggage space to rear; leather seats and inertia reel safety harnesses optional. Canopy hinged at rear to open upward.

SYSTEMS: Electrical system includes 12 V 40 A alternator and 12 V 240 Ah battery.

AVIONICS: *Comms:* Eclipse has Garmin GMA 340 audio panel, GNS 430 comm/nav/GPS, GI 106A, and GTX 327 transponder, all as standard; other avionics to customer's choice include GNS 420, GPS 150XL, Bendix/King KMD 150 GPS and KI 208 VOR. Evolution has Bendix/King KX 155 nav/com, KI 209 VOR/LOC/GS indicator, KT 76C transponder and PM501 intercom as standard. AvPac 1 adds KLX 135A GPS/com and PMA 6000 audio panel/intercom with marker beacon receiver; AvPac 2 adds items as for AvPac 1, but with glideslope for KX 155

Flight: Optional S-TEC 30 two-axis autopilot and slaved HIS in Eclipse.

EQUIPMENT: Evolution options include sheepskin seats, deluxe carpet, inertia reel harnesses, integral instrument lighting, electro luminexsent panel floodlight, pop-out vent windows, engraved aluminium throttle quadrant cover plates, external power receptacle, engine heater system, and heavy duty battery.

DIMENSIONS, EXTERNAL:
- Wing span: 10.87 m (35 ft 8 in)
- Wing aspect ratio: 10.2
- Length overall: 7.165 m (23 ft 6 in)
- Height overall: 2.18 m (7 ft 2 in)
- Tailplane span: 2.65 m (8 ft 8¼ in)
- Wheel track: 1.90 m (6 ft 2¾ in)
- Wheelbase: 1.72 m (5 ft 7¾ in)
- Propeller diameter: 1.75 m (5 ft 9 in)

AREAS:
- Wing area, gross: 11.61 m^2 (125.0 sq ft)

WEIGHTS AND LOADINGS:
- Weight empty: 529 kg (1,166 lb)
- Baggage capacity: 20 kg (44 lb)
- Max T-O weight: JAA: 750 kg (1,653 lb)
  - FAA: 800 kg (1,763 lb)
- Max wing loading (JAA): 64.6 kg/m^2 (13.22 lb/sq ft)
- Max power loading (JAA): 8.03 kg/kW (13.20 lb/hp)

PERFORMANCE:
- Never-exceed speed (VNE): 164 kt (303 km/h; 188 mph)
- Cruising speed at 75% power at FL60: 140 kt (259 km/h; 161 mph)
- Manoeuvring speed: 106 kt (196 km/h; 122 mph) CAS
- Stalling speed: clean: 47 kt (87 km/h; 54 mph)
  - in landing configuration: 38 kt (71 km/h; 44 mph)
- Max rate of climb at S/L: 305 m (1,000 ft)/min
- Service ceiling: 5,365 m (17,600 ft)
- T-O run: 337 m (1,105 ft)
- T-O to 15 m (50 ft): 448 m (1,470 ft)
- Landing from 15 m (50 ft): 454 m (1,490 ft)
- Landing run: 390 m (1,280 ft)
- Range, 45 min reserves: Eclipse: 410 n miles (759 km; 471 miles)

UPDATED

DIAMOND DA40-180 DIAMOND STAR

TYPE: Four-seat lightplane.

PROGRAMME: Development by Diamond in Austria. Production of Lycoming-engined variant transferred to Canada in 2002.

CURRENT VERSIONS: **DA 40-180:** Baseline version with Lycoming IO-360 engine. Preparation for Canadian assembly (which see) began September 2001, and first from this source (C-GDFC, c/n 40201) delivered on 28 March 2002 to Academy Air Services, Ontario.

DA 40-TDI: Powered by 99 kW (133 hp) Thielert Centurion 1.7 turbo-diesel; Austrian-built.

DA 40-FP: Fixed-pitch propeller version; launched at Sun 'n' Fun, Lakeland, Florida, April 2004.

CUSTOMERS: Total of 130 produced in Canada by end of 2003.

COSTS: With standard Bendix/King avionics US$179,900; with standard Garmin avionics US$181,900; with Bendix/King or Garmin deluxe avionics packages, both US$189,900 (all 2002); with Avidyne FlightMax Entegra package US$228,800 (2003); with Garmin G1000 package estimated US$214,900 (2003).

The description of the DA 40-TDI (which see) applies also to the DA 40-180 except as follows:

POWER PLANT: One 134 kW (180 hp) Textron Lycoming IO-360-M1A flat-four with Lasar electronic ignition and self-adapting inlet cooling, driving an MTV-12-B/180-17 three-blade constant-speed (hydraulic) propeller. Fuel in two wing tanks, standard capacity 155 litres (40.9 US gallons; 34.1 Imp gallons); optional 200 litres (52.8 US gallons; 44.0 Imp gallons).

SYSTEMS: 28 V electrical system includes 70 A alternator.

AVIONICS: Choice of Bendix/King, Avidyne or Garmin packages.

Demonstrator DA 40-FP Diamond Star with fixed-pitch propeller *(Paul Jackson)* *NEW*/0576667

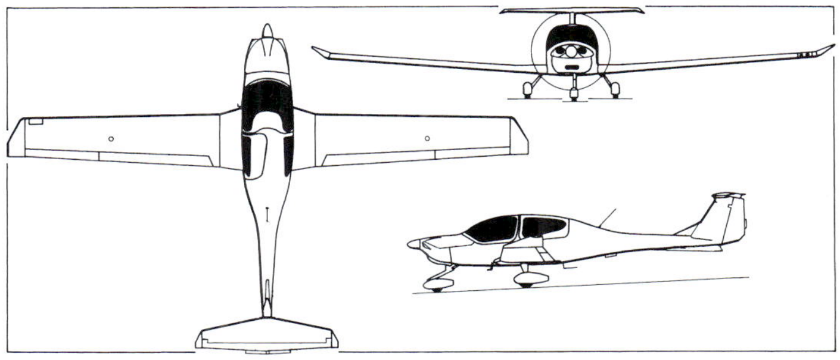

Diamond DA 40-180 Diamond Star (Lycoming IO-360) *(Paul Jackson)* 0092123

Canadian-built Diamond DA 40-180 Diamond Star *(Paul Jackson)* ***NEW***/0572305

Comms: Bendix/King Standard package includes KX 155 nav/com, KT 76A transponder with blind encoder, and PM 1000 II intercom; Deluxe package substitutes KX 155A for KX 155 and KT 76C for KT 76A, and adds KX 165A nav/com/glideslope, and KMA 28S audio panel with intercom and marker beacon receiver in place of PM 1000 II. Garmin Standard package includes GNS 430 nav/com/GPS/glideslope, GTX 327 transponder with blind encoder, and GMA audio panel with intercom. Deluxe package adds second GNS 430. GNS 530 nav/com/GPS optional in exchange for GNS 430.

Flight: Bendix/King Standard package includes KLN 94 GPS moving map and KI 208 VOR/LOC. Deluxe package substitutes KLN 94 GPS for VFR unit, and adds KI 209A VOR/GS. Further options include KAP 140 two-axis autopilot with altitude hold/preselect and electric pitch trim, KCS 55A slaved HSI (exchange for KI 209A) and MD-41 GPS/nav/annunciator. Garmin Standard package includes GI 106A VOR/LOC/GPS/GS indicator. Deluxe package adds second GI 106A. Further options include Bendix/King KAP 140 two-axis autopilot with altitude hold/preselect and electric pitch trim, KCS 55A slaved HSI (exchange for GI 106A), MD 41 GPS/nav/annunciator and Goodrich WX-500 Stormscope.

Instrumentation: Vision Microsystem VM 1000 engine control system comprising manifold pressure gauge, tachometer, engine hour meter, oil pressure gauge, oil/cylinder head temperature indicators, fuel flow indicator, ammeter and voltmeter; compass, ASI, altimeter, VSI, turn co-ordinator, artificial horizon, DG, OAT gauge, chronometer and alternate static port, all standard.

Avidyne FlightMax Entegra glass cockpit, with IFR-certified 26 cm (10.4 in) colour LCD PFD, EX5000 MFD, and E-max Total Engine Management System, introduced as an option in August 2003. Garmin G1000 integrated glass cockpit available as an option from first quarter 2004, comprising large format sunlight readable 25.4 cm (10 in) PFD and MFD; audio control system; dual integrated radio modules providing WAAS-capable IFR GPS, VHF/ILS and VHF comm functions with 8.33 kHz channel spacing; Mode S transponder with Traffic Information Service; solid state AHRS; digital air data computer; interface with Bendix/King KAP 140 two-axis autopilot; and optional weather and terrain data.

WEIGHTS AND LOADINGS:

Weight empty	740 kg (1,631 lb)
Baggage capacity	35 kg (77 lb)
Max T-O weight	1,150 kg (2,535 lb)
Max landing weight	1,092 kg (2,407 lb)
Max wing loading	85.2 kg/m² (17.45 lb/sq ft)
Max power loading	8.58 kg/kW (14.10 lb/hp)

PERFORMANCE:

Never-exceed speed (VNE)	178 kt (329 km/h; 204 mph) IAS
Max level speed	147 kt (272 km/h; 169 mph)
Cruising speed: at 75% power at 1,220 m (4,000 ft)	145 kt (269 km/h; 167 mph)
at 50% power at FL100	120 kt (222 km/h; 138 mph)
Stalling speed	49 kt (91 km/h; 57 mph)
Max rate of climb at S/L	326 m (1,070 ft)/min
Max certified altitude	5,000 m (16,400 ft)
T-O to 15 m (50 ft)	351 m (1,150 ft)
Range, 45 min reserves:	
standard fuel	570 n miles (1,055 km; 655 miles)
optional fuel	740 n miles (1,370 km; 851 miles)

UPDATED

DREAM

DREAM AIRCRAFT INC

565 Maisonneuve, Granby, Québec J2G 3H5
Tel: (+1 450) 372 99 29
Fax: (+1 450) 372 81 22
Web: http://www.dreamaircraft.com
VICE-PRESIDENT: Yvan Desmarais
SALES MANAGER: Luc Prémont

Dream's debut as an aircraft manufacturer was at the April 2003 Sun 'n' Fun festival at Lakeland, Florida. By the following year, it was able to show a second prototype of the Tundra lightplane more representative of the production version. The company is a division of specialist engineering firm D&G Inc.

UPDATED

Dream Tundra prototype *(Paul Jackson)* ***NEW***/0572321

DREAM TUNDRA

TYPE: Four-seat kitbuilt.

PROGRAMME: Construction of the prototype started in 1998 and first flew (C-GIPN) at Bromont on 12 May 2001; experimental certification 7 June 2001; first flight from floats 3 October 2001; displayed at Sun 'n' Fun in April 2003 (when it had accumulated 120 hours) and at AirVenture, Oshkosh, July 2003. Second prototype flown April 2004 (C-GAGH), displaying detail changes, including spring cantilever aluminium landing gear, double floor, thinner tailplane and rudder, redesigned engine cowling and higher-powered IO-360 engine.

CUSTOMERS: By April 2004, three sold and first set of wing and tail components had been delivered.

COSTS: Kit C$29,500 (tailwheel) or C$32,000 (nosewheel) (2004), excluding engine, propeller and avionics.

DESIGN FEATURES: Intended to carry four adults and baggage from short and semi-prepared runways. Designed on 3-D software and constructed of precision-drilled aluminium components. Wings braced by V struts; downturned tips. Quoted build time 1,000 hours.

FLYING CONTROLS: Conventional and manual. Large, horn-balanced elevators with flight-adjustable trim tab to port. Horn-balanced rudder. Fowler flaps.

STRUCTURE: Aluminium throughout, except composites wing-, fin- and tailplane tips.

LANDING GEAR: Tailwheel type; fixed. Cantilever spring aluminium mainwheel legs. (Prototype has two faired-in side Vs hinged to lower longerons, plus half-axles spring under centreline with bungee cord). Mainwheels 8.50×6 or 8.00–6; solid, steerable tailwheel. Optional float.

POWER PLANT: One 149 kW (200 hp) Textron Lycoming IO-360 flat-six (134 kW; 180 hp O-360-A in prototype), driving a Sensenich 76EM8-0-57 two-blade, metal propeller. Total fuel 208 litres (55.0 US gallons; 45.8 Imp gallons) in two wing tanks, of which 204 litres (53.9 US gallons; 44.9 Imp gallons) usable.

ACCOMMODATION: Four persons in two side-by-side pairs. Baggage space to rear. Upward-hinged door each side.

Second prototype of the Dream Tundra is representative of the production version *(Paul Jackson)* ***NEW***/0576668

DIMENSIONS, EXTERNAL:
- Wing span ... 10.97 m (36 ft 0 in)
- Wing chord (constant) ... 1.55 m (5 ft 1 in)
- Wing aspect ratio ... 7.1
- Length overall ... 7.14 m (23 ft 5 in)
- Height overall ... 3.05 m (10 ft 0 in)
- Tailplane span ... 3.23 m (10 ft 7 in)

DIMENSIONS, INTERNAL:
- Cabin: Length ... 1.98 m (6 ft 6 in)
- Max width ... 1.12 m (3 ft 8 in)

AREAS:
- Wings, gross ... 17.06 m² (183.6 sq ft)
- Horizontal tail surfaces ... 3.62 m² (39.0 sq ft)

WEIGHTS AND LOADINGS:
- Weight empty ... 669 kg (1,475 lb)
- Max payload ... 338 kg (745 lb)
- Max fuel weight ... 150 kg (330 lb)
- Max T-O weight ... 1,156 kg (2,550 lb)
- Max wing loading ... 67.8 kg/m² (13.89 lb/sq ft)
- Max power loading ... 8.62 kg/kW (14.17 lb/hp)

PERFORMANCE:
- Never-exceed speed (VNE) ... 130 kt (241 km/h; 150 mph)
- Cruising speed: max ... 115 kt (212 km/h; 132 mph)
- at 75% power ... 103 kt (190 km/h; 118 mph)
- Stalling speed: flaps up ... 46 kt (84 km/h; 52 mph)
- flaps down ... 37 kt (68 km/h; 42 mph)
- Max rate of climb at S/L ... 219 m (720 ft)/min
- Service ceiling ... 4,265 m (14,000 ft)
- T-O run ... 122 m (400 ft)
- Landing run ... 152 m (500 ft)
- Range ... 594 n miles (1,100 km; 684 miles)
- Endurance ... 5 h 50 min
- *g* limits ... +3.8/–1.7

UPDATED

EUROCOPTER CANADA

EUROCOPTER CANADA LTD

(Subsidiary of Eurocopter)

HEAD OFFICE: PO Box 250 1100 Gilmore Road, Fort Erie, Ontario L2A 5M9
Tel: (+1 905) 871 77 72
Fax: (+1 905) 871 33 20 and 35 99
e-mail: euro@eurocopter.ca
Web: http://www.eurocopter.ca
PRESIDENT AND CEO: Edouard Gaillat
VICE-PRESIDENT, FINANCE: Cyrille Fourny
VICE-PRESIDENT, SUPPORT AND SALES: Tony Brown
DIRECTOR, OPERATIONS: Benoit Marcoux
DIRECTOR, CUSTOMER SUPPORT AND SALES: Gordon Kay
MANAGER, MARKETING SERVICES AND CUSTOMER RELATIONS: Diane Bourdeau

Started as MBB Helicopter Canada on 30 March 1984, Eurocopter Canada Limited (ECL) now employs 150 people at its 9,290 m² (100,000 sq ft) facility at Fort Erie. The company holds the world product mandate and design authority for the BO 105 LS A-3 and provides products and services for the complete Eurocopter range, including the new EC 120 B Colibri, EC 130B4 and EC 135. Eurocopter Canada is certified for ISO 9001–2000; its production department is certified for AQA P1, JAR 145 and FAR Pt 45 requirements; the engineering department makes use of the CATIA design system and is responsible for certification of modifications and repairs for the Eurocopter range as well as construction and manufacture of optional equipment and spare parts. Pilot and engineer training is also carried out by the company and offered as part of the purchase package. Warehouses at Montréal and Vancouver provide parts support for over 340 Eurocopter helicopters at present operational in Canada. Previously produced BO 105 LS utility helicopter.

Current activities also involve assembly of the EC 120 B Colibri, AS 350/355 Ecureuil, EC 130B4, EC 135 and EC 155. Turnover in 1999 was C$62 million, in 2000, C$60 million, in 2001, C$60 million and in 2002, C$50 million.

UPDATED

FOUND

FOUND AIRCRAFT CANADA INC

RR# 2, Site 12, Box 10, Georgian Bay Airport, Parry Sound, Ontario P2A 2W8
Tel: (+1 705) 378 05 30
Fax: (+1 705) 378 05 94
e-mail: found@zeuter.com
Web: http://www.foundair.com

The present company of this name was establised to reinstate production of the FBA-2 Bush Hawk, beginning deliveries in mid-2000. An XP version was introduced in 2001 and at least 17 have now been built

UPDATED

FOUND FBA-2C BUSH HAWK

TYPE: Six-seat utility transport.

PROGRAMME: Original Found FBA-2C first flew on 9 May 1962 and was last described in the 1967–68 *Jane's*. Production numbered 27, plus five improved Centenniel 100s (1969–70 edition) before company ceased trading.

Prototype of current FBA-2C1 (C-FSVD), converted from an existing aircraft with IO-540-D engine, and other improvements, flew at Georgian Bay Airport, Ontario, in November 1996 and completed test flying, on wheels and floats, during 1998 and 1999. New production was due to start in July 1997 but programme slipped; preproduction aircraft C-GDWO (c/n 28; IO-540-L engine) made first flight October 1998 and certified 5 March 1999; was followed by C-GDWQ in 1999; initial production Bush Hawk C-GDWS (c/n 30) flew 2 March 2000 FAA certificate A7EA reinstated 10 March 2000, covering both FBA-2C and FBA-2C1; uprated (224 kW; 300 hp) version added 1 May 2001.

CURRENT VERSIONS: **FBA-2C Bush Hawk-260:** 194 kW (260 hp) version. Not currently available, due to Hawk-XP development taking priority.

FBA-2C1 Bush Hawk-300: 224 kW (300 hp) version.

Description and data apply to Bush Hawk-300 unless otherwise stated.

Bush Hawk-XP: Improved version; demonstrator C-GDWC (c/n 31) built 2000; available from 2001.

CUSTOMERS: By end 2003, 17 had been built. Customers include Reliance Airways, Northwest Territories; Nakina Outpost Camps; and Air Service, Ontario. Initial production aircraft exported to USA 24 May 2001.

COSTS: Bush Hawk-XP US$299,000 on wheels; US$326,000 on skis, US$369,000 on amphibious floats and US$341,000 on straight floats; hourly operating cost C$63.40 (2004).

DESIGN FEATURES: Fully corrosion-proofed, all-metal construction for operation in exceptionally poor weather over all terrain.

High wing with constant chord to mid-span; outer panels unswept, with sharply swept forward trailing-edges. Sweptback fin with two-stage fillet; mid-mounted tailplane. Floatplane directional stability enhanced by underfin and auxiliary fins above and below tailplane at three-quarters span.

FLYING CONTROLS: Conventional and manual. Horn-balanced rudder and elevators. Flaps.

STRUCTURE: Aluminium-covered steel tube front fuselage and semi-monocoque rear fuselage. Deep-section, low-drag, all-aluminium wing.

LANDING GEAR: Fixed. Prototype, preproduction and first production aircraft have tailwheel configuration; all following landplanes have tricycle layout with optional tailwheel layout. Sprung steel main legs with urethane shock-absorbers. Mainwheels 8.00–6, but 66 cm (26 in) 8.50–10 and 76 cm (30 in) Tundra tyres being evaluated. Steerable tailwheel. Other options include Wipline 3450 floats or amphibious floats or Aerocet 3500L floats and Federal C-3200 or C-3600 retractable skis.

POWER PLANT: One 194 kW (260 hp) Textron Lycoming IO-540-D4A5 or 224 kW (300 hp) IO-540-L1C5 flat-six engine, driving a Hartzell HC-C3YR-IRF constant-speed three-blade metal propeller. Fuel in two wing tanks, total capacity 379 litres (100 US gallons; 83.3 Imp gallons) of which 371 litres (98.0 US gallons; 81.6 Imp gallons) usable. Oil capacity 10 litres (2.75 US gallons; 2.3 Imp gallons).

Found FBA-2C Bush Hawk in its natural environment *(Paul Jackson)* ***NEW***/0576666

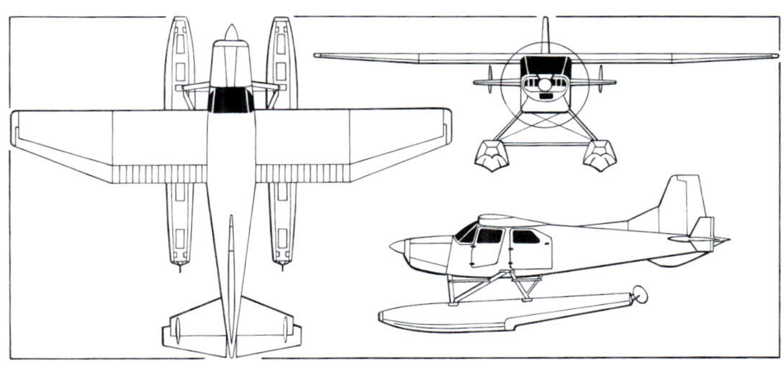

Found FBA-2C Bush Hawk six-seat utility transport *(James Goulding)* 0099158

ACCOMMODATION: Pilot and co-pilot, plus four passengers in cabin, with crew entrance through forward-hinged door on each side of fuselage. Rear hammock-style seat rolls up for storage in cabin roof when used in freight role; optional individual folding seats available. Rear cabin typically holds four 204 litre (54.0 US gallon; 45.0 Imp gallon) drums and has large, forward-hinged door on each side; door sills are flush with floor; doors swing 180° to allow easy loading. Alternatively, one stretcher and attendant in rear cabin. Camera hatch can be fitted in cabin floor.

SYSTEMS: 24 V 10 Ah battery, 28 V 70 A alternator.

AVIONICS: *Comms:* Optional Bendix/King KT-76A Mode C transponder.

Flight: KLX-135A GPS optional.

Data follow for Bush Hawk-300 landplane.

DIMENSIONS, EXTERNAL:
- Wing span ... 10.97 m (36 ft 0 in)
- Wing aspect ratio ... 7.2
- Length overall: landplane ... 8.08 m (26 ft 6 in)
- floatplane ... 8.74 m (28 ft 8 in)

Height overall: landplane .... 2.51 m (8 ft 3 in)
floatplane .... 4.17 m (13 ft 8 in)
Propeller diameter .... 2.13 m (7 ft 0 in)
Cabin door: Width .... 0.67 m (2 ft 2½ in)

DIMENSIONS, INTERNAL:
Cabin (including baggage compartment):
Length .... 4.01 m (13 ft 2 in)
Max width .... 1.12 m (3 ft 8¼ in)
Max height .... 1.27 m (4 ft 2 in)
Volume .... 3.40 m^3 (120 cu ft)

AREAS:
Wings, gross .... 16.72 m^2 (180.0 sq ft)

WEIGHTS AND LOADINGS:
Weight empty: landplane .... 862 kg (1,900 lb)
floatplane .... 1,021 kg (2,250 lb)
Baggage capacity .... 113 kg (250 lb)
Max T-O weight: landplane .... 1,587 kg (3,500 lb)
floatplane .... 1,678 kg (3,700 lb)
Max wing loading: landplane .... 94.9 kg/m^2 (19.44 lb/sq ft)
floatplane .... 100.4 kg/m^2 (20.56 lb/sq ft)
Max power loading: landplane .... 7.10 kg/kW (11.67 lb/hp)
floatplane .... 7.51 kg/kW (12.33 lb/hp)

PERFORMANCE:
Cruising speed at 75% power:
landplane .... 150 kt (278 km/h; 173 mph)
floatplane .... 130 kt (241 km/h; 150 mph)
Stalling speed, power off:
flaps up: landplane .... 57 kt (106 km/h; 66 mph)
foatplane .... 58 kt (108 km/h; 67 mph)
flaps down: landplane .... 52 kt (97 km/h; 60 mph)
floatplane .... 54 kt (100 km/h; 62 mph)
Max rate of climb at S/L:
landplane .... 341 m (1,120 ft)/min
floatplane .... 320 m (1,050 ft)/min
Service ceiling: landplane .... 5,485 m (18,000 ft)
floatplane .... 4,570 m (15,000 ft)
T-O run: landplane .... 229 m (750 ft)
floatplane .... 366 m (1,200 ft)
T-O to 15 m (50 ft): landplane .... 406 m (1,330 ft)
floatplane .... 580 m (1,900 ft)
Range with max fuel:
landplane .... 880 n miles (1,629 km; 1,012 miles)
floatplane .... 710 n miles (1,315 km; 817 miles)
Endurance .... 8 h 30 min

UPDATED

INNOVATOR

INNOVATOR TECHNOLOGIES

Box 17, Site 17, RR5, Calgary, Alberta T2P 2G6
Tel: (+1 403) 669 31 01, 936 54 23
e-mail: mosquito@innovatortech.ca
Web: http://www.innovatortech.ca
DIRECTOR/DESIGNER: John Uptigrove

Having originally been marketed as an open frame helicopter, the Innovator Mosquito was relaunched at Sun 'n' Fun, April 2004, with an enclosed cabin and structure.

UPDATED

INNOVATOR MOSQUITO

TYPE: Single-seat ultralight helicopter kitbuilt.

PROGRAMME: First flight 2000 in form of unenclosed frame with tripod landing gear; total 130 hours of airborne time by July 2003. Holds Canadian amateur-builder's certificate of airworthiness. In March 2004, a prototype of an enclosed version was flown. This retained the tricycle landing gear, but a production-representative machine with twin skids was shown at the public launch at Sun 'n' Fun, Lakeland, Florida in April 2004.

CURRENT VERSIONS: **Mosquito:** Original, unenclosed version.

Mosquito XE: Enclosed version; marketed from 2004.

Mosquito XEL: Float-equipped version of XE.

CUSTOMERS: By April 2004, 29 of the unenclosed version had been sold, of which four were flying.

COSTS: Kit US$26,000; or US$24,000 upon evidence of helicopter flying instruction; enclosed version (2004).

DESIGN FEATURES: To FAR Pt 103 regulations. Vertically mounted engine driving main rotor via toothed belt as second-stage reduction gear. Tailboom supported by V strut.

Semi-rigid main rotor. Unique control system: main rotor control from floor-mounted stick and collective through control mixer at base of main mast, then through push tubes in main mast to base of swashplate; last-mentioned is contained within mast and supported by a pushrod in rotor shaft. Control rods each side of pushrod transmit inputs through rotor shaft to control lever on top of main shaft and down to blade pitch horns through pitch links. Tail rotor controlled by foot pedals through Bowden cable to actuating lever on tail rotor shaft.

Main rotor speed 500 rpm; tail rotor 2,500 rpm. Quoted build time 200 hours.

STRUCTURE: Frame of 6061-T6 aluminium; glass fibre tripod landing gear replaced by twin skids from 2004; carbon fibre tailboom and V struts. Composites cabin. Main rotor blades constructed upon aluminium spar bonded to wrapped aluminium sheet skin; foam sealer plugs both ends. Titanium main gearbox; aluminium fuel tanks.

POWER PLANT: One 44.7 kW (60 hp) Zanzottera MZ202 two-cylinder two-stroke with electric starter and directly attached primary reduction gear. Centrifugal clutch and overspeed provision during autorotation. Flexible couplings and flexible driveshaft transmit power to tail rotor. Total fuel 19 litres (5.0 US gallons; 4.2 Imp gallons) in two tanks.

DIMENSIONS, EXTERNAL:
Main rotor diameter .... 5.49 m (18 ft 0 in)
Tail rotor diameter .... 1.02 m (3 ft 4 in)
Length: frame .... 4.88 m (16 ft 0 in)
overall (excl main rotor) .... 6.10 m (20 ft 0 in)
Max width .... 1.83 m (6 ft 0 in)
Height overall .... 2.11 m (6 ft 11 in)

Current, enclosed (XE) configuration of Innovator Mosquito *(Paul Jackson)* ***NEW***/0576669

Mr Uptigrove demonstrating his Innovator Mosquito in its original form *(Paul Jackson)* ***NEW***/0572320

WEIGHTS AND LOADINGS:
Weight empty .... 115 kg (254 lb)
Max T-O weight .... 240 kg (530 lb)

PERFORMANCE:
Max level speed .... 61 kt (113 km/h; 70 mph)
Cruising speed .... 52 kt (97 km/h; 60 mph)
Service ceiling: IGE .... 2,440 m (8,000 ft)
OGE .... 1,830 m (6,000 ft)

UPDATED

LEGEND LITE

LEGEND LITE INC

General Delivery, New Hamburg, Ontario N0B 2G0
Tel: (+1 519) 662 29 80
e-mail: teamskywatch@skywatchss11.on.ca
Web (1): http://www.skywatchss11.on.ca
Web (2): http://www.airsport.com/kits/skywtch.htm
CEO: Rob McIntosh

Freedom Lite company began production of the Skywatch SS-11 in 1996; changed name to Legend Lite during 2000. There have been no known completions of SS-11s since 2000, however.

UPDATED

MURPHY

MURPHY AIRCRAFT MANUFACTURING LTD

Unit 1, 8155 Aitken Road, Chilliwack, British Columbia V2R 4H5
Tel: (+1 604) 792 58 55
Fax: (+1 604) 792 70 06
e-mail: mursales@murphyair.com
Web: http://www.murphyair.com
PRESIDENT: Darryl Murphy
SALES MANAGER: Colleen Dyck
MARKETING MANAGER: Dean Mueller

Formed in 1985, the company moved from initial 745 m^2 (8,000 sq ft) plant to new 4,925 m^2 (53,000 sq ft) premises in 1995 to accommodate increase in business. In mid-2000, Murphy opened a fast-build factory in the Philippines, which employs 40 staff to produce Mooses and float kits; by end 2002, five aircraft per month were being delivered. Subassemblies are built from Canadian parts which are then shipped back.

In 2003, Murphy was chosen by Bombardier to fit its new 224 kW (300 hp) V-6 engine into a Moose, shown publicly at AirVenture, Oshkosh in July 2003; it has also announced that the JDM-8 has been developed further to enable it to be classified as a homebuilt rather than an ultralight.

In 2003 the Canadian workforce numbered 22.

UPDATED

MURPHY RENEGADE II

TYPE: Tandem-seat ultralight biplane kitbuilt.

PROGRAMME: Prototype Renegade II made first flight May 1985; first kits became available September that year. Ready assembled aircraft, plans and kits are available.

CUSTOMERS: By end-2003, 677 Renegade IIs and Renegade Spirits had been completed and flown.

COSTS: Quick-build kit US$14,040; full materials kit US$9,880; partial parts kit US$3,744; plans US$338 (2004 prices).

DESIGN FEATURES: Sporting biplane of classic configuration. Conforms to TP101.41 ultralight regulations in Canada and some European countries. Optional wingtip extensions (for countries such as UK requiring lower wing loadings) increase gross area to 15.83 m^2 (170.4 sq ft). Optional extended wings and rounded tips; optional floats and parachute.

NACA 23012 wing section; 10° sweepback on upper wings; 3° dihedral on non-swept lower wings. Angle of incidence for both wings of 4°.

Kits in four standards ranging from partial parts kit to 300/400 working hour quick-build kit with all parts premanufactured plus partly assembled fuselage, wings and engine mounting.

FLYING CONTROLS: Conventional and manual. Differentially rigged Frise ailerons.

STRUCTURE: Fabric-covered aluminium tubing; rear fuselage top-decking of preformed glass fibre; two-spar wings have aluminium sheet leading-edges; glass fibre wingtips optional; wire-braced tail assembly.

LANDING GEAR: Tailwheel type; fixed. Two faired-in side Vs hinged to lower longerons, plus half-axles sprung under centreline with bungee cord. Tailwheel steerable.

POWER PLANT: One 34.0 kW (45.6 hp) Rotax 503 UL-1V two-cylinder two-stroke engine standard; 37.0 kW (49.6 hp) twin-carburettor Rotax 503 UL-2V optional. Alternative engines up to 59.6 kW (80 hp) can be fitted, including Rotax 582 and Teledyne Continental O-65, O-75 and O-85. Fuel capacity 53 litres (14.0 US gallons; 11.6 Imp gallons).

DIMENSIONS, EXTERNAL:
Wing span (normal): upper ... 6.48 m (21 ft 3 in)
lower ... 6.05 m (19 ft 10 in)
Length overall ... 5.61 m (18 ft 5 in)
Height overall ... 2.08 m (6 ft 10 in)

AREAS:
Wings, gross ... 14.29 or 15.83 m^2 (153.8 or 170.4 sq ft)

WEIGHTS AND LOADINGS:
Weight empty ... 170–236 kg (375–520 lb)
Max T-O weight, Rotax 503 ... 431 kg (950 lb)

PERFORMANCE (Rotax 503 engine and two crew):
Never-exceed speed (VNE) ... 104 kt (193 km/h; 120 mph)
Max level speed at 305 m (1,000 ft) ... 70 kt (129 km/h; 80 mph)
Cruising speed at 75% power ... 56 kt (105 km/h; 65 mph)
Stalling speed, power off ... 33 kt (61 km/h; 38 mph)
Max rate of climb at S/L ... 152 m (500 ft)/min
T-O and landing run ... 92 m (300 ft)
Range at 75% power ... 197 n miles (365 km; 227 miles)

UPDATED

MURPHY RENEGADE SPIRIT

TYPE: Tandem-seat ultralight biplane kitbuilt.

PROGRAMME: First flight (C-IJLW) 6 May 1987; kit production began a month later.

DESIGN FEATURES: Variant of Renegade II. Alternative power plants; internal and external refinements.

COSTS: Fast-build kit US$15,600; materials kit US$10,920; partial parts kit US$4,160; plans US$338 (2004).

POWER PLANT: One 47.8 kW (64.1 hp) Rotax 582 in-line engine standard, in radial-type cowling; 59.7 kW (80 hp) Rotax 912 flat-four or Alvis rotary engine optional.

Murphy Renegade Spirit ***(Paul Jackson)*** ***NEW***/0572278

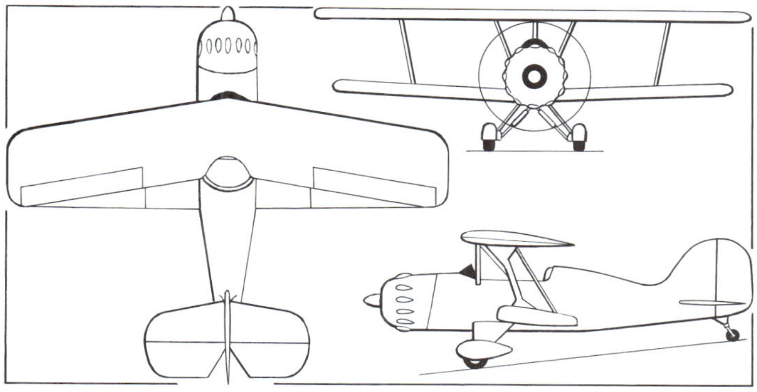

Murphy Renegade Spirit ***(James Goulding)*** ***NEW***/1047838

DIMENSIONS, EXTERNAL: As for Renegade II

AREAS: As for Renegade II

WEIGHTS AND LOADINGS:
Weight empty ... 190–227 kg (420–500 lb)
Max T-O weight ... 431 kg (950 lb)

PERFORMANCE (two crew and Rotax 582 engine):
As for Renegade II except:
Max level speed ... 78 kt (145 km/h; 90 mph)
Cruising speed at 75% power ... 63 kt (116 km/h; 72 mph)
Stalling speed ... 35 kt (64 km/h; 40 mph)
Max rate of climb at S/L ... 152 m (500 ft)/min
T-O and landing run ... 107 m (350 ft)
Range at 75% power ... 206 n miles (381 km; 237 miles)

UPDATED

MURPHY REBEL

TYPE: Three-seat kitbuilt.

PROGRAMME: Construction of prototype started October 1989; first flight 1990; building of production aircraft/kits began in October 1990.

CURRENT VERSIONS: **Rebel:** Standard version, *as described.*

Rebel Elite: Modified tricycle landing gear version, with metal-covered split flaps and fin of increased chord; O-320 or optional 134 kW (180 hp) Lycoming O-360 engine; and increased maximum T-O weight to 816 kg (1,800 lb). First flight (C-FWSF) 29 February 1996.

Super Rebel and Maverick: *Described separately.*

CUSTOMERS: Some 754 complete aircraft or kits (including 82 Elites) ordered by 1 September 2003; over 460 flying.

COSTS: Rebel kit without engine US$16,120; Elite kit without engine US$21,627 tailwheel, US$22,627 tricycle, fast-build extra US$14,000 (2004). Available in three sub-component kits.

DESIGN FEATURES: High-wing, cabin monoplane for recreational, training, cross-country, border patrol and other roles; designed to conform to FAR Pt 23 and JAR (Utility category); STOL performance. Optional fin fillet required for floatplane version. Wings braced by single streamline strut each side (no intermediate struts); wings detachable and horizontal tail folds upward for transportation and storage. Quoted build time 800 to 1,000 hours.

NACA 4415 (modified) wing section.

FLYING CONTROLS: Conventional and manual. Flaps. Elite has split flaps which droop up to 18° in 6° increments; metal-covered ailerons and cantilevered horizontal stabiliser.

STRUCTURE: Aluminium alloy, semi-monocoque fuselage; flaperons fabric covered; three-spar wings covered in aluminium sheet glass fibre wingtips. Tailplane also strut-braced. Rebel Elite wing leading-edge 60 per cent thicker than Rebel.

LANDING GEAR: Tailwheel type; fixed. Optional tricycle gear (Rebel Elite), skis, or Murphy 1500 or 1800 straight or wheeled or Montana 2100 wheeled floats. 6.00-6 wheels with 18 in high-profile tyres are standard. Standard mainwheel configuration of two faired-in side Vs hinged to lower longerons, plus half-axles, each with bungee cord shock-absorber, attached to compression frame; optionally, fuselage-mounted aluminium cantilever mainwheel legs with optional wheel fairings.

POWER PLANT: One 59.6 kW (79.9 hp) Rotax 912 UL, or 86.5 kW (116 hp) Textron Lycoming O-235-N2C engine; 2.27:1 reduction gear with Rotax engine and direct drive with O-235. Elite has 119 kW (160 hp) Lycoming O-320; or optional 134 kW (180 hp) Lycoming O-360; ground-adjustable wooden two-blade propeller; three-blade propeller optional. Standard fuel capacity 167 litres (44.0 US gallons; 36.6 Imp gallons) in wing tanks; optional extra 220 litre (58.0 US gallon; 48.3 Imp gallon) tank.

ACCOMMODATION: Pilot and two passengers. Door on each side of cabin, half- or fully glazed.

DIMENSIONS, EXTERNAL:
Wing span: Rebel ... 9.14 m (30 ft 0 in)
Elite ... 9.25 m (30 ft 4 in)
Wing chord, constant ... 1.52 m (5 ft 0 in)
Wing aspect ratio ... 6.0
Length overall: Rebel ... 6.50 m (21 ft 4 in)
Elite ... 6.78 m (22 ft 3 in)

Murphy Rebel on optional cantilever aluminium mainwheel legs ***NEW***/0559752

Customised Murphy Rebel with wheel fairings and larger fin fillet *(Paul Jackson)* ***NEW***/0572277

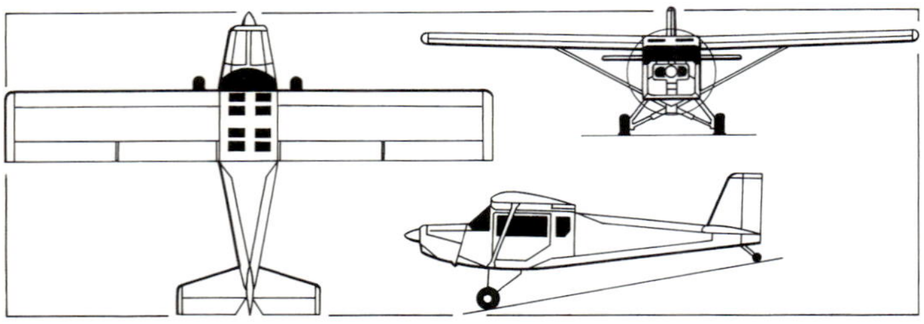

Murphy Rebel three-seat kitbuilt *(Paul Jackson)* 0064886

Fuselage max width ... 1.12 m (3 ft 8 in)
Height overall: Rebel, Elite (tailwheel) ... 2.03 m (6 ft 8 in)
Elite (tricycle) ... 2.39 m (7 ft 10 in)
Tailplane span ... 2.79 m (9 ft 2 in)
Wheel track: Rebel ... 2.18 m (7 ft 2 in)
Elite ... 2.32 m (7 ft 7½ in)
Wheelbase: Rebel, Elite (tailwheel) ... 4.93 m (16 ft 2 in)
Elite (tricycle) ... 1.70 m (5 ft 7 in)
Propeller diameter ... 1.78 m (5 ft 10 in)

AREAS:
Wings, gross: Rebel ... 13.94 m^2 (150.0 sq ft)
Elite ... 14.12 m^2(152.0 sq ft)

WEIGHTS AND LOADINGS (A: Rotax 912, B: O-235, E: Elite, O-360):
Weight empty: A ... 295–317 kg (650–700 lb)
B ... 374–408 kg (825–900 lb)
E ... 445 kg (980 lb)
Baggage capacity ... 45.5–68 kg (100–150 lb)
Max T-O weight: A ... 657 kg (1,450 lb)
B ... 748 kg (1,650 lb)
E ... 816 kg (1,800 lb)
Max wing loading: A ... 47.2 kg/m^2 (9.67 lb/sq ft)
B ... 53.7 kg/m^2 (11.00 lb/sq ft)
E ... 57.8 kg/m^2 (11.84 lb/sq ft)
Max power loading: A ... 11.03 kg/kW (18.13 lb/hp)
B ... 8.66 kg/kW (14.22 lb/hp)
E ... 6.09 kg/kW (10.00 lb/hp)

PERFORMANCE (A, B, E as above):
Never-exceed speed (VNE):
A ... 124 kt (230 km/h; 143 mph)
B ... 131 kt (243 km/h; 151 mph)
E ... 136 kt (252 km/h; 157 mph)
Max level speed: A ... 87 kt (162 km/h; 100 mph)
B ... 108 kt (201 km/h; 125 mph)
E ... 126 kt (233 km/h; 145 mph)
Cruising speed at 75% power:
A ... 78 kt (145 km/h; 90 mph)
B ... 100 kt (185 km/h; 115 mph)
E ... 115 kt (212 km/h; 132 mph)
Stalling speed, power off, flaps up:
A ... 35 kt (65 km/h; 40 mph)
B ... 39 kt (71 km/h; 44 mph)
E ... 40 kt (74 km/h; 46 mph)
Stalling speed, power off, flaps down:
A ... 32 kt (58 km/h; 36 mph)
B ... 35 kt (65 km/h; 40 mph)
E ... 37 kt (68 km/h; 42 mph)
Max rate of climb at S/L: A ... 164 m (500 ft)/min
B ... 244 m (800 ft)/min
E ... 457 m (1,500 ft)/min
Service ceiling: B ... 3,960 m (13,000 ft)
T-O run: A ... 137 m (450 ft)
B ... 122 m (400 ft)
E ... 143 m (470 ft)
T-O to 15 m (50 ft): A, B ... 244 m (800 ft)
Landing run: A ... 92 m (300 ft)
B, E ... 122 m (400 ft)
Range with max fuel:
A ... 764 n miles (1,416 km; 880 miles)
B ... 691 n miles (1,281 km; 796 miles)
E ... 593 n miles (1,099 km; 683 miles)
Endurance: B ... 7 h 36 min
E ... 5 h 12 min
g limits ... +3.8/−2.5
... (+5.7/−3.8 ultimate)

UPDATED

MURPHY SUPER REBEL AND MOOSE

TYPE: Six-seat kitbuilt.

PROGRAMME: Announced April 1995; prototype Super Rebel (then designated SR 2500) SR 2500 (C-GKSR), with initial MTOW of 1,134 kg (2,500 lb), first flew November 1995; parts shipments began in late 1995 with tail sections. Tricycle version first flew May 1997. Increased gross weight Moose (originally SR 3500) introduced at Sun 'n' Fun 2001; prototype converted from Super Rebel demonstrator C-GBZD.

CURRENT VERSIONS: **Super Rebel:** Standard version.

Moose: Increased gross weight version, first flown 2 April 2001; (VOKBM engine); choice of two power plants. Tailwheel version only.

CUSTOMERS: Some 228 of both versions sold by September 2003.

COSTS: *Super Rebel:* Basic kit less engine, propeller, instruments and avionics: tailwheel version US$26,500; tricycle version US$28,000 (2003).

Moose: Basic kit, less engine, propeller, instruments and avionics, US$33,920 (2004). Fast-build kit extra US$16,960.

DESIGN FEATURES: Development of Murphy Rebel (which see). All-metal construction. Quoted build time 1,400 to 1,800 hours. Fast-build kits have quoted build time of 800–1,000 hours.

FLYING CONTROLS: Conventional and manual. Horn-balanced tail surfaces. Electric trim and three-stage (0, 18, 33°) flaps; ailerons deflect 20° up and 12° down; ailerons droop 12° when 33° flap deployed.

STRUCTURE: Wing is identical in shape to that of Rebel and Elite; aerofoil modified NACA 4415. Three-spar wing; leading-edge of 0.032 sheet aluminium.

LANDING GEAR: Tailwheel configuration is standard; tricycle also available. Optional hardpoints enable rapid conversion. Main and nose tyres 8.00-6 in; tailwheel has 9 in tyre. Dual brakes. Optional 3600 series floats.

POWER PLANT: *Super Rebel:* Prototype has 186 kW (250 hp) Textron Lycoming O-540-A4A5 driving two-blade (three-blade optional) constant-speed Hartzell propeller, but design accommodates engines from 134 to 224 kW (180 to 300 hp). Fuel in two wing tanks, total capacity 227 litres (60.0 US gallons; 50.0 Imp gallons). Optional larger tanks increase capacity by further 303 litres (80.0 US gallons; 66.6 Imp gallons).

Moose: Choice of 186 kW (250 hp) Textron Lycoming IO-540 or 265 kW (355 hp) VOKBM M-14P radial, latter driving two-blade wooden propeller. Fuel capacity as SR 2500.

ACCOMMODATION: Pilot and up to five passengers in two pairs of side-by-side seats plus optional jump seat; dual controls. Rear bench seat can be removed to provide cargo capacity. Baggage compartment is reached by large cargo door on port side of fuselage, behind passenger door.

DIMENSIONS, EXTERNAL:
Wing span ... 10.97 m (36 ft 0 in)
Wing chord, constant ... 1.52 m (5 ft 0 in)
Wing aspect ratio ... 7.1
Length overall ... 7.01 m (23 ft 0 in)
Fuselage max width ... 1.12 m (3 ft 8 in)
Height overall: tailwheel ... 1.98 m (6 ft 6 in)
tricycle ... 2.67 m (8 ft 9 in)
Tailplane span ... 3.30 m (10 ft 10 in)
Wheel track ... 2.29 m (7 ft 6 in)
Wheelbase: tailwheel ... 5.94 m (19 ft 6 in)
tricycle ... 2.03 m (6 ft 8 in)
Propeller diameter: Super Rebel ... 2.13 m (7 ft 0 in)
Moose with M-14P engine ... 2.39 m (7 ft 10 in)
Passenger door: Height ... 1.07 m (3 ft 6 in)
Width ... 1.07 m (3 ft 6 in)
Baggage door: Height ... 0.83 m (2 ft 8½ in)
Width ... 0.93 m (3 ft 0½ in)

Murphy Super Rebel ***NEW***/0559750

Murphy Moose fitted with Murphy 3600 floats ***NEW***/0559749

Trial installation of Bombardier V300 engine in a Murphy Moose *(Paul Jackson)* ***NEW***/0572276

DIMENSIONS, INTERNAL:
- Cabin max width .... 1.12 m (3 ft 8 in)
- Height .... 1.30 m (4 ft 3 in)
- Length .... 3.05 m (10 ft 0 in)

AREAS:
- Wings, gross .... 16.91 m² (182.0 sq ft)
- Ailerons (total) .... 2.37 m² (25.50 sq ft)
- Flaps (total) .... 2.09 m² (22.50 sq ft)

WEIGHTS AND LOADINGS (Super Rebel with 186 kW; 250 hp engine, Moose with 265 kW; 355 hp engine):
- Weight empty: Super Rebel .... 748 kg (1,650 lb)
  - Moose .... 816 kg (1,800 lb)
- Baggage capacity .... 113 kg (250 lb)
- Max T-O weight: Super Rebel .... 1,361 kg (3,000 lb)
  - Moose .... 1,587 kg (3,500 lb)
- Max wing loading:
  - Super Rebel .... 80.5 kg/m² (16.48 lb/sq ft)
  - Moose .... 93.9 kg/m² (19.23 lb/sq ft)
- Max power loading:
  - Super Rebel .... 7.30 kg/kW (12.00 lb/hp)
  - Moose .... 6.00 kg/kW (9.86 lb/hp)

PERFORMANCE (engines as before):
- Never-exceed speed (VNE):
  - Super Rebel .... 153 kt (284 km/h; 177 mph)
  - Moose .... 164 kt (304 km/h; 189 mph)
- Max level speed:
  - Super Rebel .... 139 kt (257 km/h; 160 mph)
  - Moose .... 152 kt (282 km/h; 175 mph)
- Max cruising speed at 70% power
  - Super Rebel .... 126 kt (233 km/h; 145 mph)
  - Moose .... 135 kt (249 km/h; 155 mph)
- Stalling speed, power off:
  - flaps up: Super Rebel .... 46 kt (84 km/h; 52 mph)
    - Moose .... 49 kt (91 km/h; 56 mph)
  - flaps down: Super Rebel .... 40 kt (74 km/h; 46 mph)
    - Moose .... 44 kt (81 km/h; 50 mph)
- Max rate of climb at S/L:
  - Super Rebel .... 305 m (1,000 ft)/min
  - Moose .... 457 m (1,500 ft)/min
- Service ceiling .... 4,575 m (15,000 ft)
- T-O run: Super Rebel, Moose .... 183 m (600 ft)
- Landing run: Super Rebel .... 152 m (500 ft)
  - Moose .... 183 m (600 ft)
- Range: with max standard fuel:
  - Super Rebel .... 537 n miles (996 km; 619 miles)
  - Moose .... 521 n miles (965 km; 600 miles)
  - with optional larger tanks:
    - Super Rebel .... 729 n miles (1,350 km; 839 miles)
- Endurance: Super Rebel .... 4 h 15 min
  - Moose .... 4 h 0 min
- *g* limits (ultimate): both .... +5.7/–3.8

UPDATED

MURPHY MAVERICK

TYPE: Side-by-side ultralight kitbuilt.

PROGRAMME: Original design shelved in favour of Rebel, but revived in 1994 to meet Japanese ultralight restrictions; prototype/demonstrator G-MYSS used to gain UK BCAR Section S type approval in 1995.

CUSTOMERS: 49 completed and flown by late 2003, at which time 132 kits had been sold.

COSTS: Kit: US$14,040 without engine (2004).

DESIGN FEATURES: Derivative of Rebel, with 40 per cent commonality of parts, including optional wingtip extensions.

FLYING CONTROLS: As Rebel, except cable-operated ailerons only instead of full-span flaperons.

STRUCTURE: Similar to Rebel, but with weight-saving features. Wings omit full-span stringers and are fabric covered; glass fibre engine cowling; tail surfaces also fabric covered.

LANDING GEAR: As Rebel.

POWER PLANT: One two-cylinder two-stroke engine (39.5 kW; 53 hp Rotax 503 DC or 47.8 kW; 64.1 hp Rotax 582 UL); GSC two-blade propeller. Testing with Hpower HKS-700E four-stroke engine completed in November 1999. Fuel capacity 18.9 litres (5.0 US gallons; 4.2 Imp gallons) standard, 53 litres (14.0 US gallons; 11.7 Imp gallons) in optional wing tanks.

DIMENSIONS, EXTERNAL:
- Wing span: standard .... 8.97 m (29 ft 5 in)
  - with extended tips .... 9.88 m (32 ft 5 in)
- Length overall, tail up .... 6.30 m (20 ft 8 in)

Murphy Maverick built by a UK owner *(Paul Jackson)* ***NEW***/0589435

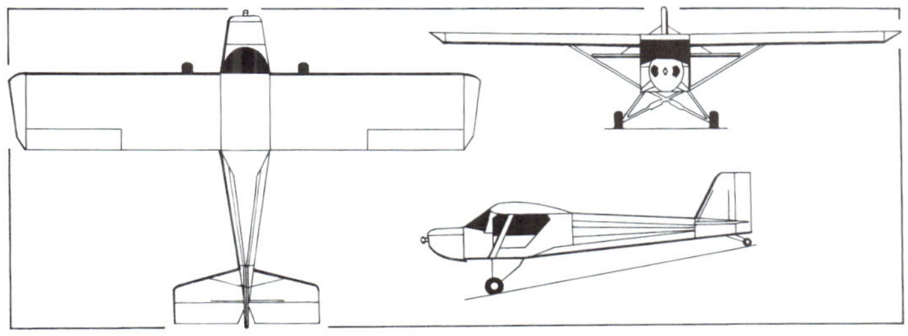

Murphy Maverick two-seat ultralight *(Paul Jackson)* 0064887

AREAS:
- Wings, gross: standard .... 13.66 m² (147.0 sq ft)
  - with extended tips .... 15.05 m² (162.0 sq ft)

WEIGHTS AND LOADINGS (standard span and Rotax 503 DC engine):
- Weight empty .... 179–191 kg (395–420 lb)
- Max T-O weight .... 431 kg (950 lb)

PERFORMANCE (standard span and Rotax 503 DC engine):
- Max level speed .... 82 kt (153 km/h; 95 mph)
- Cruising speed at 75% power .... 70 kt (129 km/h; 80 mph)
- Stalling speed .... 28 kt (52 km/h; 32 mph)
- Max rate of climb at S/L .... 183 m (600 ft)/min
- T-O run .... 46 m (150 ft)
- Landing run .... 61 m (200 ft)
- Range with max fuel at 75% power .... 243 n miles (450 km; 280 miles)
- *g* limits .... +5.7/–3.8

UPDATED

MURPHY JDM-8

TYPE: Single-seat ultralight kitbuilt.

PROGRAMME: Design commenced December 1998; initially shown in uncompleted form at Oshkosh in 2000. Prototype (C-IGTD) first flown 30 March 2001 and made public debut at Sun 'n' Fun the following month as 'conceptual' design to evaluate market response. Launched at AirVenture, Oshkosh, July 2001.

Following a customer survey undertaken in March 2003, the US legal ultralight production was superseded by the development of a homebuilt/advanced ultralight model.

CUSTOMERS: Two prototypes registered by September 2003: one FAR Pt 103 compliant ultralight (C-IGTD) and one homebuilt/advanced ultralight example (C-IFBZ).

COSTS: Projected cost US$8,495 excluding engine and chemical agents.

DESIGN FEATURES: Murphy's first low-wing design. Wings fold upwards for storage. Enclosed cockpit optional.

FLYING CONTROLS: Conventional and manual. No aerodynamic balancing. Frise ailerons.

STRUCTURE: Semi-monocoque fuselage and constant-chord wings covered with aluminium sheet; wings comprise three C-channel spars and have NACA 4415 section. Ailerons have

Murphy JDM-8 with optional spring steel landing gear ***NEW***/0559751

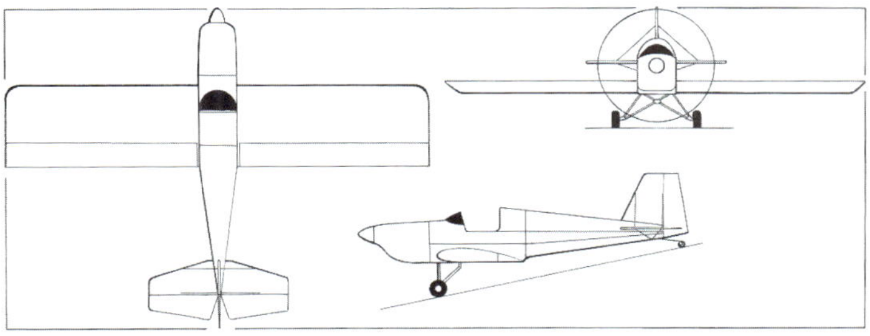

Murphy JDM-8 single-seat ultralight *(James Goulding)* 0132945

aluminium ribs; aluminium tube empennage with wire bracing; control surfaces fabric-covered.

LANDING GEAR: Tailwheel type; fixed. Standard two side Vs hinged to wingroots, plus half-axles, each with bungee cord shock-absorber; optional 4130 spring steel circular tube cantilever legs. 12.7 cm (5 in) mainwheels with 8.00-6 tyres and 8.8 cm (3½ in) tailwheel. Sealed ball bearings. Hydraulic brakes.

POWER PLANT: Design accepts most two- and four-stroke engines in the 37.3 to 59.7 kW (50 hp to 80 hp) range. Fuel capacity 75.7 litres (20.0 gallons; 16.7 Imp gallons). Optional 34 litre (9.0 US gallons; 7.5 Imp gallons) wing tanks.

DIMENSIONS, EXTERNAL:
- Wing span ... 6.10 m (20 ft 0 in)
- Width, wings folded ... 2.44 m (8 ft 0 in)
- Length overall ... 4.47 m (14 ft 8 in)
- Height overall ... 1.63 m (5 ft 4 in)

DIMENSIONS, INTERNAL:
- Cabin max width ... 0.66 m (2 ft 2 in)

AREAS:
- Wings, gross ... 9.29 m^2(100.0 sq ft)

WEIGHTS AND LOADINGS (44.7 kW; 60 hp HKS-700E):
- Weight empty ... 136–204 kg (300–450 lb)
- Max T-O weight ... 353 kg (780 lb)

PERFORMANCE:
- Never-exceed speed (VNE) ... 100 kt (185 km/h; 115 mph)
- Normal cruising speed ... 78 kt (145 km/h; 90 mph)
- Stalling speed ... 31 kt (57 km/h; 35 mph)
- Max rate of climb at S/L ... 305 m (1,000 ft)/min
- T-O run ... 53 m (175 ft)
- Landing run ... 91 m (300 ft)
- Endurance ... 5 h
- *g* limits ... ±3.8

UPDATED

NORDIC

NORMAN AVIATION INTERNATIONAL INC

1141 Chemin des Iles, St-Henri de Lévis, Québec G0R 3E0
Tel: (+1 418) 833 43 37
Fax: (+1 418) 833 70 57
e-mail: norman.aviation@sympatico.ca
Web: http://www3.sympatico.ca/norman.aviation
PRESIDENT Jacques Norman

EUROPEAN MARKETING:
Euro Fly Aero
Hubert Ferté, Aérodrome de Montardoise, F-10150
Montsuzain, France
Tel: (+33 3) 25 37 50 28
Fax: (+33 3) 25 37 50 27
e-mail: ferte.euro@wanadoo.fr

Previous products of Nordic have been the I, II, III, IV, V, VI and VII, of which II and VI versions remain available. Most recent addition is the Nordic 8 *(sic)*.

UPDATED

Nordic/Euro Fly Aero Mini Explorer registered in the French ultralight series *(Geoffrey P Jones)* ***NEW***/0572154

NORDIC 8 MINI EXPLORER

TYPE: Side-by-side ultralight kitbuilt.

PROGRAMME: Mini Explorer is third, and smallest, of a series of similar concept aircraft, begun with the twin-engined Explorer, produced by Hubert DeChevigny and Dean Wilson (the latter, designer of the Avid Flyer and other light aircraft). The single-engined Private Explorer (14.40 m; 47 ft 3 in span, 1,860 kg; 4,102 lb MTOW) flew in 1998 and was marketed as a kit from Idaho, USA. The sixth of the type was registered in Canada during September 2003.

First Mini Explorer (C-IGXF) was registered in November 2001; marketing began 2002. European marketing proceeding in parallel; first French-built aircraft was noted in 2003.

CUSTOMERS: At least three built in Canada and one in France by mid-2003.

COSTS: Kit C$34,500 less engine (2004).

DESIGN FEATURES: Rugged construction, with deep and wide fuselage to provide sleeping accommodation.

High, constant-chord wing with V bracing struts each side.

FLYING CONTROLS: Conventional and manual.

STRUCTURE: 4130 steel fuselage frame with fabric covering; wooden wings with geodesic ribs.

LANDING GEAR: Tricycle type; fixed. Fuselage-mounted cantilever mainwheel legs; levered nosewheel with helical spring shock-absorber.

POWER PLANT: One flat-four piston engine: 73.5 kW (98.6 hp) Rotax 912 ULS or 84.5 kW (113.3 hp) Rotax 914 UL. Fuel capacity 76 litres (20.0 US gallons; 16.7 Imp gallons).

ACCOMMODATION: Two persons, side by side. Floor to rear with space for (typically) utensil storage cupboard and hand-basin, plus bed.

DIMENSIONS, EXTERNAL:
- Wing span ... 10.97 m (36 ft 0 in)
- Wing chord (constant) ... 1.52 m (5 ft 0 in)
- Wing aspect ratio ... 7.2
- Length overall ... 6.40 m (21 ft 0 in)
- Height overall ... 1.47 m (4 ft 10 in)

DIMENSIONS, INTERNAL:
- Cabin max width ... 1.22 m (4 ft 0 in)
- Floor area ... 3.72 m^2 (40.0 sq ft)

AREAS:
- Wings, gross ... 16.72 m^2 (180.0 sq ft)

WEIGHTS AND LOADINGS:
- Weight empty ... 297 kg (654 lb)
- Max T-O weight ... 558 kg (1,232 lb)

PERFORMANCE:
- Never-exceed speed (VNE) ... 113 kt (209 km/h; 130 mph)
- Cruising speed at FL50 ... 74 kt (137 km/h; 85 mph)
- Stalling speed ... 33 kt (62 km/h; 38 mph)
- Max rate of climb at S/L (solo) ... 259 m (850 ft)/min
- Service ceiling ... 3,050 m (10,000 ft)
- T-O run ... 92 m (300 ft)
- Landing run ... 61 m (200 ft)
- Range ... 680 n miles (1,259 km; 782 miles)
- Endurance ... 7 h

UPDATED

PAN ATLANTIC

PAN ATLANTIC AEROSPACE CORPORATION (SUBSIDIARY OF AV-INTEL CORPORATION)

CHAIRMAN: Fredrick D Ferguson
VICE-PRESIDENT: Anthony Asterita
INFORMATION OFFICER: Ray Trudeau

Pan Atlantic's recent LTA-based projects included a high-altitude, satellite-like platform in partnership with Skysat Communications Networks Corporation of New York; the Cargo Airship System (CAS); the LEAP surveillance drone; and the Magnus advertising vehicle.

Patents for the CAS were purchased by an aerospace group in Texas in December 2002; an 80 ton payload airship development programme was then said to have started, with completion of a first article scheduled for late 2004. However, no further news had emerged by the beginning of that year.

UPDATED

PARTENAIR

PARTENAIR DESIGN INC

Absorbed by Delastek Inc.

UPDATED

RAF

ROTARY AIR FORCE MARKETING INC

PO Box 1236, 1107-9th Street W, Kindersley, Saskatchewan S0L 1S0
Tel: (+1 306) 463 60 30
Fax: (+1 306) 463 60 32
e-mail: info@raf2000.com
Web: http://www.raf2000.com
SALES MANAGER: Don LaFleur

Incorporated in 1987, Rotary Air Force employs 16 people.

In May 2001, RAF announced that it would be expanding its activities into commercial applications, including agricultural spraying and paramilitary functions.

UPDATED

RAF 2000

TYPE: Two-seat autogyro kitbuilt.
PROGRAMME: Introduced 1990; conforms to 51 per cent homebuilt rules.
CURRENT VERSIONS: **2000 STD-SE:** Basic version, *as described*.

2000 GTX-SE: Top-of-range model; kit includes carburated version of engine, rotor brake, heater, dual controls and adjustable pitch and roll trim assembly.

2000 GTX-SE FI: Fuel-injected version of GTX.

CUSTOMERS: RAF autogyros sold in Argentina, Australia, Austria, Brazil, Canada, Chile, China, Colombia, Czech Republic, Ecuador, Germany, Greece, Hungary, Ireland, Israel, Italy, Japan, Kazakhstan, Latvia, Lithuania, Mexico, Netherlands, New Caledonia, New Zealand, Norway, Poland, Portugal, Puerto Rico, Russia, Serbia & Montenegro, South Africa, Spain, Taiwan, Thailand, UK and USA. At least 550 believed completed and flown by end 2004.
COSTS: 2000 STD-SE kit price US$20,615, including Subaru engine (carb version); 2000 GTX-SE US$22,500, including carburated Subaru; 2000 GTX-SE FI US$25,500, including fuel-injected Subaru (all 2004).
DESIGN FEATURES: Conventional autogyro. Suitable for training, crop-spraying, power line inspection, predator control, stock mustering and aerial photography. Quoted build time 150–250 hours.
STRUCTURE: Composites RAF rotor blade features 6061-T6 aluminium spar and foam filler. Patented 5 × 10 cm (2 × 4 in) folding (two-part) rotor mast, rubber mast bushing and adjustable CG. Removable doors.
LANDING GEAR: Tricycle type; fixed; optional speed fairings.
POWER PLANT: One 97 kW (130 hp) Subaru EJ22 16-valve four-cylinder liquid-cooled engine driving a ground-adjustable three-blade Warp Drive composites propeller through RAF cog belt reduction gear, ratio 2.10:1. Fuel capacity 87 litres (23.0 US gallons; 19.2 Imp gallons) of premium unleaded Mogas, of which 79 litres (21.0 US gallons; 17.5 Imp gallons) are usable. Optional capacity 95 litres (25.0 US gallons; 20.8 Imp gallons), of which 87 litres (23.0 US gallons; 19.2 Imp gallons) usable. Fuel consumption carburated version 20.1 litres (5.3 US gallons; 4.4 Imp gallons)/h, injected version 18.2 litres (4.8 US gallons; 4.0 Imp gallons)/h, both at 80 per cent power.
ACCOMMODATION: Pilot and passenger side by side in enclosed cabin. Can be flown with doors removed.
SYSTEMS: 35 A alternator, electric starter.
EQUIPMENT: Agricultural version has 114 litre (30.0 US gallon; 25.0 Imp gallon) tank, moulded to fit into back and bottom of cabin enclosure, supplying a 24-nozzle spraybar fitted behind and below the engine. Spray width 6.7 to 7.3 m (22 to 24 ft).

RAF 2000 GTX-SE two-seat autogyro *(Paul Jackson)* ***NEW***/0572275

DIMENSIONS, EXTERNAL:
Rotor diameter ... 9.14 m (30 ft 0 in)
Rotor blade chord ... 0.22 m (8½ in)
Fuselage: Length ... 4.11 m (13 ft 6 in)
Max width ... 1.08 m (3 ft 6½ in)
Width over wheels ... 1.59 m (5 ft 2½ in)
Height overall: STD-SE ... 2.50 m (8 ft 2½ in)
GTX-SE ... 2.58 m (8 ft 5½ in)
Propeller diameter ... 1.73 m (5 ft 8 in)
Wheel track ... 1.55 m (5 ft 1 in)
Wheelbase ... 3.91 m (12 ft 10 in)
DIMENSIONS, INTERNAL:
Cabin: Length ... 1.50 m (4 ft 11 in)
Max width ... 1.09 m (3 ft 7 in)
Max freight ... 1.32 m (4 ft 4 in)
AREAS:
Rotor disc ... 65.67 m^2 (706.9 sq ft)
WEIGHTS AND LOADINGS:
Weight empty: STD-SE ... 345 kg (760 lb)
GTX-SE ... 363 kg (800 lb)
Max T-O weight ... 698 kg (1,540 lb)
Max disc loading ... 10.6 kg/m^2 (2.18 lb/sq ft)
Max power loading ... 7.21 kg/kW (11.85 lb/hp)
PERFORMANCE, POWERED (two occupants):
Max operating speed (VMO) ... 87 kt (161 km/h; 100 mph)
Cruising speed: STD-SE ... 61 kt (113 km/h; 70 mph)
GTX-SE ... 65 kt (121 km/h; 75 mph)
Min flying speed ... 17–26 kt (32–48 km/h; 20–30 mph)
Unstick speed ... 26–35 kt (48–64 km/h; 30–40 mph)
Max rate of climb at S/L ... 305 m (1,000 ft)/min
Service ceiling ... 3,050 m (10,000 ft)
T-O run ... 23–107 m (75–350 ft)
Landing run ... Up to 3 m (10 ft)
Range with max fuel:
STD-SE ... 243 n miles (450 km; 279 miles)
GTX-SE ... 261 n miles (483 km; 300 miles)
GTX-SE FI ... 313 n miles (579 km; 360 miles)
Endurance with 30 min reserves: standard fuel:
STD-SE, GTX-SE ... 3 h 30 min
Optional fuel, GTX-SE FI ... 4 h 18 min
PERFORMANCE, UNPOWERED:
Glide ratio ... 4:1

UPDATED

ST JUST

ST JUST AVIATION INC

Suite 204, 1310 Gay-Lussac, Boucherville, Québec J4B 7G4
Tel: (+1 450) 641 86 86
Fax: (+1 450) 641 84 21
e-mail: st-justaviation@sympatico.ca
Web: http://www.cyclone-aviation.com
PRESIDENT: Guy Cantin
DIRECTOR: Albert Beaudry

Original Avionerrie du Lac St John formed in 1991 to provide parts for Cessna 185 owners after that company removed the aircraft from its product line; later progressed to manufacturing kits of the Cyclone. Sold to St Just Aviation in 1997.

VERIFIED

ST JUST SUPER CYCLONE

TYPE: Four-seat kitbuilt.
PROGRAMME: Introduced in 1997; based on reverse engineering of the Cessna 180/185 (which see in *Jane's Aircraft Upgrades*).
CUSTOMERS: Total 69 sold and 38 flying by January 2004, including 20 in Canada.
COSTS: Basic kit US$39,900 (2002). Fast-build kits US$3,000 extra.
DESIGN FEATURES: Compared to Cessna 180/185, Super Cyclone has span increased by addition of 30 cm (1 ft) at each wingroot, which increases flap length, lowers stalling speed, enhances stability and increases maximum T-O weight. Quoted build time 2,000 hours.
FLYING CONTROLS: Conventional and manual. Flaps.
STRUCTURE: Single-strut braced, two-spar wing with constant-chord centre-section and tapered outer sections. Wing and empennage constructed of 2024 T-3 aluminium; composites engine cowling. Large fin fillet. Low-mounted tailplane.
LANDING GEAR: Tailwheel type, with sprung steel main legs; Federal wheel/skis and floats can be fitted. Mainwheels 6.00-6.
POWER PLANT: One 224 kW (300 hp) Teledyne Continental IO-520, flat-six engine, driving a three-blade, constant-speed McCauley propeller. Optional alternative engines between 149 and 261 kW (200 and 350 hp). Fuel capacity 341 litres (90.0 US gallons; 75.0 Imp gallons), of which 318 litres (84.0 US gallons; 70.0 Imp gallons) are usable.
ACCOMMODATION: Pilot and three passengers in two pairs of seats with upward-hinged door on each side of cabin. Additional baggage door on port side of fuselage behind cabin. Compared to Cessna 185, has a third cabin window each side.
AVIONICS: Full IFR panel available.

Seaplane version of St Just Super Cyclone 0137973

DIMENSIONS, EXTERNAL:
Wing span ... 11.58 m (38 ft 0 in)
Wing aspect ratio ... 7.2
Length overall ... 7.92 m (26 ft 0 in)
Height overall ... 2.41 m (7 ft 11 in)
Propeller diameter ... 2.18 m (7 ft 2 in)
DIMENSIONS, INTERNAL:
Cabin max width ... 1.07 m (3 ft 6 in)
AREAS:
Wings, gross ... 17.74 m^2 (191.0 sq ft)
WEIGHTS AND LOADINGS:
Weight empty ... 839 kg (1,850 lb)
Baggage capacity ... 68 kg (150 lb)
Max payload ... 680 kg (1,500 lb)
Max T-O weight ... 1,587 kg (3,500 lb)
Max wing loading ... 89.5 kg/m^2 (18.32 lb/sq ft)
Max power loading ... 7.10 kg/kW (11.67 lb/hp)
PERFORMANCE:
Max operating speed ... 152 kt (281 km/h; 175 mph)
Max cruising speed ... 142 kt (263 km/h; 163 mph)

Stalling speed: flaps up ..... 37 kt (68 km/h; 42 mph)
flaps down ..... 33 kt (62 km/h; 38 mph)
Max rate of climb at S/L ..... 488 m (1,600 ft)/min
Service ceiling ..... 4,570 m (15,000 ft)
T-O run ..... 69 m (225 ft)
Landing run ..... 122 m (400 ft)
Range with max fuel ..... 788 n miles (1,460 km; 907 miles)
g limits ..... +3.8/–1.5

UPDATED

SUPER-CHIPMUNK

SUPER-CHIPMUNK INC

109, Route 201, St Louis-de-Gonzague, Québec J0S 1T0
Tel: (+1 450) 373 07 06
Fax: (+1 450) 373 31 44
e-mail: info@super-chipmunk.com
Web: http://www.super-chipmunk.com
PRESIDENT: Gilles Leger

This company was launched to manufacture a modern reproduction of the de Havilland Canada DHC-1 Chipmunk, externally similar rear of the firewall, but with all-new structural design. Company owns Supplementary Type Certificate (STC) for original Super Chipmunk conversion of ex-military aircraft. Engine modification kits for existing Chipmunks are available. Promotion was continuing in 2002. The prototype is officially registered as a Leger Super Chipmunk.

Following the death of Gilles Leger in the crash of the prototype on 14 July 2002, work on the project appears to have ceased.

UPDATED

TAPANEE

TAPANEE AVIATION INC

437 Route 309 Nord, Mont St-Michel, Québec J0W 1PO
Tel: (+1 819) 586 20 59
Fax: (+1 819) 586 27 48
e-mail: info@tapanee.com
Web: http://www.tapanee.com
DESIGNER: Michel Lequin

The company's most recent known design is the Levitation 4, although the earlier Pégazair STOL bushplane remains available as kit or plans.

VERIFIED

TAPANEE LEVITATION 4

TYPE: Four-seat kitbuilt.

PROGRAMME: Prototype (C-GJUY) flown 2002. No other completions known.

COSTS: Kit US$29,600 (2004).

DESIGN FEATURES: Extrapolation of Pégazair, with deeper fuselage. High-mounted, parallel-chord wings with V-strut bracing and STOL optimisation including Handley Page leading-edge slats. Large rear surfaces for low-speed control, including sweptback fin with large ventral fillet and tailplane attached below. Quoted build time 1,500 hours.

FLYING CONTROLS: Manual. Junkers flaperons. All-moving fin and tailplane, latter having two anti-balance tabs.

STRUCTURE: Welded steel tube with aluminium skin. All metal wing and horizontal tail

LANDING GEAR: Tailwheel type; fixed. Two fared-in, hinged side Vs, plus half-axles sprung under centreline with rubber-in-compression shock-absorbers; steerable tailwheel. Mainwheels 8.00-6. Hydraulic brakes.

POWER PLANT: One 134 kW (180 hp) Textron Lycoming O-360 flat-six, driving a two-blade propeller. Fuel capacity 208 litres (55.0 US gallons; 45.81 Imp gallons).

ACCOMMODATION: Four persons in two side-by-side pairs. Upward-hinged door each side. Large baggage compartment to rear, with large, external (port) door.

DIMENSIONS, EXTERNAL:
Wing span ..... 10.21 m (33 ft 6 in)
Length overall ..... 7.16 m (23 ft 6 in)

Prototype Tapanee Levitation 4 *(Paul Jackson)* ***NEW***/0572317

DIMENSIONS, INTERNAL:
Cabin max width ..... 1.22 m (4 ft 0 in)
AREAS:
Wings, gross ..... 16.72 m² (180.0 sq ft)
WEIGHTS AND LOADINGS:
Weight empty ..... 621 kg (1,368 lb)
Max T-O weight ..... 1,133 kg (2,500 lb)
Max wing loading ..... 67.8 kg/m² (13.89 lb/sq ft)
Max power loading ..... 8.45 kg/kW (13.89 lb/hp)
PERFORMANCE:
Max operating speed ..... 113 kt (209 km/h; 130 mph)
Cruising speed ..... 100 kt (185 km/h; 115 mph)
Stalling speed, power on ..... 31 kt (67 km/h; 35 mph)
Max rate of climb ..... 213 m (700 ft)/min
T-O run ..... 122 m (400 ft)
Landing run ..... 91 m (300 ft)
Range ..... 575 n miles (1,064 km; 661 miles)

UPDATED

ULTRAVIA

ULTRAVIA AERO INTERNATIONAL INC

152-A Industrial, Gatineau, Québec J8P 7G7
Tel: (+1 819) 669 31 44
Fax: (+1 819) 669 84 06
e-mail: pelican@ultravia.ca
Web: http://www.ultravia.ca

The Pelican PL was added to the Ultravia line in 1991, followed by the Pelican Sport in 1998. In April 2002, New Kolb Aircraft of USA was appointed distributor. During 2003, the company announced a further model: the TuToR. Ultravia has a 1,486 m² (16,000 sq ft) factory near Ottawa-Gatineau Airport.

UPDATED

ULTRAVIA PELICAN

TYPE: Side-by-side kitbuilt.

PROGRAMME: Original Le Pelican first flown May 1982. The Pelican PL was introduced in 1991 and the Sport in 1998. Meets Transport Canada TP 10141 regulations. Pelican Sport also available as factory-finished aircraft.

CURRENT VERSIONS: **Pelican PL:** Standard ultralight version.

Pelican Sport: Redesigned wing with higher-lift aerofoil, longer span and STOL kit. Comes in three subvariants, depending on operating area: Sport 450 (Europe), Sport 500 (Brazil) and Sport 600 (US, Canada, Australia and Mexico).

Pelican TuToR: Derivative conforming to FAR Pt 23; certification under way during 2003 and expected to be completed in 2004; will also be certified to Canadian chapter 523-VLA standards.

Description applies to Pelican Sport 600 unless otherwise indicated.

CUSTOMERS: Over 350 Pelican PLs and 100 Pelican Sports completed and flown by December 2002. Ultravia had built nearly 700 Pelicans of all variants by January 2004, of which up to 40 exported to USA before association with Kolb.

COSTS: PL: standard US$21,500; fast-build US$27,900; Sport: standard US$22,025; fast-build US$28,575 (all 2003).

DESIGN FEATURES: High-wing of constant chord, single bracing strut each side; sweptback fin, with constant-chord tailplane at base. Cantilever landing gear.

Nosewheel version of Ultravia Pelican *(Paul Jackson)* ***NEW***/0572155

Fast-build kit, with 49 per cent of assembly completed, is also available. Quoted build time of standard kit 1,000 hours and of fast-build kit, 500 hours.

FLYING CONTROLS: Conventional and manual. Five position (0 to 45°) flaps. Pelican Sport ailerons droop 10° when full flap deployed. Mechanically operated elevator trim tab, ground-adjustable rudder and aileron tabs. Horn-balanced rudder.

STRUCTURE: Composites, vacuum-moulded fuselage with rigid PVC core. Single-strut, single-spar, all-metal wing with aluminium-covered control surfaces. Cantilever horizontal tail surfaces. Landing gear legs of 4130 steel.

LANDING GEAR: Fixed tricycle versions standard. Fuselage-mounted cantilever mainwheel legs with optional wheel fairings. Steerable nosewheel has bungee suspension and uses Azusa 15 cm (6 in) wheel. Optional tailwheel version uses a Matco WLT-6 or Kolb 10 cm (4 in) tailwheel. Cleveland 5.00–5 mainwheels with disc brakes on both versions. Optional floats and speed fairings.

POWER PLANT: Typically, one 73.5 kW (98.6 hp) Rotax 912 ULS four-cylinder four-stroke driving a three-blade Warp Drive ground-adjustable propeller. Other engines can be fitted,

Ultravia Pelican in US guise as Kolb Sport 600 *(Paul Jackson)* ***NEW***/0572301

including the 59.6 kW (79.9 hp) Rotax 912 UL and 84.6 kW (113.4 hp) Rotax 914 UL. Fuel capacity 91 litres (24.0 US gallons; 20.0 Imp gallons) in two wing tanks, of which 87 litres (23.0 US gallons; 19.2 Imp gallons) usable.

ACCOMMODATION: Pilot and passenger in side-by-side seating. Baggage area behind seats.

SYSTEMS: 12 V 250 W generator.

DIMENSIONS, EXTERNAL:

Wing span: PL ..... 8.99 m (29 ft 6 in)
Sport ..... 9.75 m (32 ft 0 in)
Length overall: PL ..... 6.02 m (19 ft 9 in)
Sport ..... 6.07 m (19 ft 11 in)
Height overall, nosewheel configuration
PL ..... 2.54 m (8 ft 4 in)
Sport ..... 2.59 m (8 ft 6 in)

AREAS:

Wings, gross: PL ..... 10.03 m² (108.0 sq ft)
Sport ..... 10.89 m² (117.3 sq ft)

WEIGHTS AND LOADINGS:

Weight empty: PL ..... 363–386 kg (800–850 lb)
Sport ..... 327–340 kg (720–750 lb)
Max T-O weight: PL ..... 635 kg (1,400 lb)
Sport ..... 599 kg (1,320 lb)

PERFORMANCE (Rotax 912 ULS engine):

Never-exceed speed (VNE):
PL ..... 142 kt (263 km/h;163 mph)
Sport ..... 135 kt (250 km/h;155 mph)
Normal cruising speed: both ..... 115 kt (213 km/h;132 mph)
Stalling speed, flaps down: PL ..... 44 kt (82 km/h;51 mph)
Sport ..... 39 kt (73 km/h;45 mph)
Max rate of climb at S/L: PL ..... 366 m (1,200 ft)/min
Sport ..... 411 m (1,350 ft)/min
Service ceiling: both ..... 4,875 m (16,000 ft)
T-O run: PL ..... 183 m (600 ft)
Sport ..... 91 m (300 ft)
Landing run: PL ..... 168 m (550 ft)
Sport ..... 152 m (500 ft)
Range with max fuel:
PL ..... 695 n miles (1,287 km;800 miles)
Sport ..... 608 n miles (1,126 km;700 miles)

UPDATED

UTIAS

UNIVERSITY OF TORONTO INSTITUTE FOR AEROSPACE STUDIES

4295 Dufferin Street, Toronto, Ontario M3H 5T6
Tel: (+1 416) 667 77 08
Fax: (+1 416) 667 77 99
e-mail: james.delaurier@rogers.com
Web (1): http://www.ornithopter.net
Web (2): http://www.ornithopter.ca
RESEARCHER: Prof James D DeLaurier

Although no further flight attempts had been possible by late 2003, some design improvements had been made to the University's Big Flapper ornithopter during the preceding year.

UPDATED

UTIAS 'BIG FLAPPER'

TYPE: Ornithopter

PROGRAMME: Early research resulted 1985 in quarter-scale, hand-launched, engine-powered and radio-controlled flying model; this progressively improved into more efficient model flown 4 September 1991 and recognised by FAI as first successful engine-powered ornithopter. Feasibility study 1993–94 for full-scale prototype, construction of which began 1995; this aircraft (C-GPTR) began taxi tests October 1996 and by 1999 had demonstrated ability to accelerate under own power to more than 43 kt (80 km/h; 50 mph) and make a number of brief hops. Further activity suspended in third quarter 2002 pending additional funding, but main landing gear redesigned in 2003 to improve shock absorption. New gear delivered 15 September 2003, but this too late for that year's test season, which UTIAS planned to resume in second quarter 2004. In addition to Toronto University, Project Ornithopter currently has some 16 sponsors including Canadian National Research Council, the Ken Molson Foundation and Toronto Aerospace Museum. Criterion for success is considered to be sustained flight of at least 11 seconds which, including take-off and landing runs, would cover the length of the runway.

COSTS: Development costs of full-size prototype £140,000 up to mid-2002.

DESIGN FEATURES: Conventional fuselage and tail unit. Flapping wings are hinged to a centre-section which is moved up and down by pylons connected to the drive train and pivot on the ends of vertical struts mounted on short outriggers attached to the base of the fuselage. Thus, when centre-section pushes up, wings flap down, and vice versa. The designed wing deflection is 53° 36', although compliance in the structure under flapping loads gives somewhat larger actual values. The wings are designed to flap 1.2 times per second.

Specially designed (by Prof Michael Selig) S 1020 wing aerofoil section.

UTIAS Big Flapper during taxi tests in August 2001 0527147

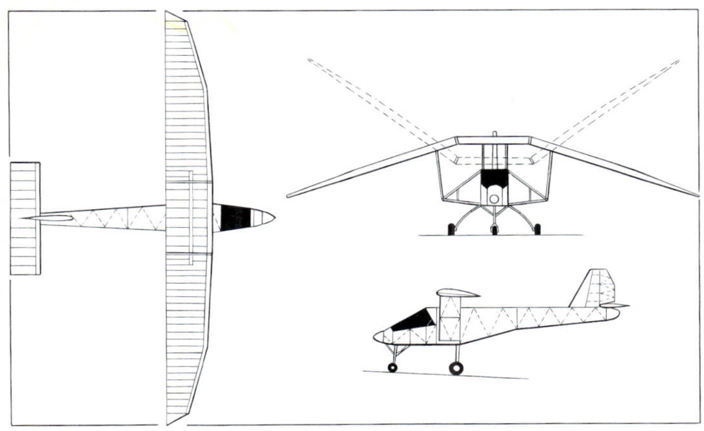

General arrangement of the University of Toronto ornithopter *(James Goulding)* 0526866

FLYING CONTROLS: All of thrust and most of lift created by mechanical flapping of wings. Thrust due mainly to low-pressure region around wing leading-edges, which creates suction. Wings also able to twist passively ('aeroelastic tailoring') to prevent airflow separation. Rudder provides control in roll as well as yaw, due to wing opposite direction of yaw gaining lift while that of other wing loses lift.

STRUCTURE: Fuselage frame of steel tube and aluminium; wings of Kevlar, carbon fibre and epoxy resin, plus wood and structural foam. Dacron overall covering. Original wing was optimised for 272 kg (600 lb) aircraft weight; to offset actual (higher) weight, a pair of small, fixed, auxiliary wings has been added as a temporary solution. Design study for larger replacement wing completed in mid-2003, but not yet funded.

LANDING GEAR: Tricycle type, fixed, with cantilever, self-sprung carbon fibre mainwheel bows; shock-absorption on nosewheel leg. Gear designed to give aircraft a nose-down attitude on ground, to minimise bounce during take-off run. New (2003) main gear has more 'bow-legged' appearance and is more robustly welded to fuselage.

POWER PLANT: One 17.9 kW (24 hp) König three-cylinder, two-stroke, fan-cooled engine, with 60:1 reduction gear drive to power wing centre-section. As latter requires more power on upstrokes than on downstrokes (outer wing panels operating in opposite direction), engine stores energy on flywheel between downstrokes.

ACCOMMODATION: Pilot only.

DIMENSIONS, EXTERNAL (approx):

Wing span ..... 12.3 m (40 ft 6 in)
Length overall ..... 7.3 m (24 ft 0 in)
Height overall ..... 2.7 m (9 ft 0 in)

WEIGHTS AND LOADINGS:

Weight empty ..... 363 kg (800 lb)

PERFORMANCE (estimated):

Unstick speed ..... 50 kt (92 km/h; 57 mph)
Cruising speed ..... 44 kt (82 km/h; 51 mph)

UPDATED

ZENAIR

ZENAIR LTD

Huronia Airport, Midland, Ontario L4R 4K8
Tel: (+1 705) 526 28 71
Fax: (+1 705) 526 80 22
e-mail: zenair@zenair.com
Web: http://www.zenair.com
PRESIDENT AND DESIGNER: Christophe Heintz
VICE-PRESIDENT: Mathieu Heintz
GENERAL MANAGER: Bruce Barker

Founded in 1974, Zenair designs and develops light aircraft. Christophe Heintz was formerly chief engineer of Avions Pierre Robin in France. In 1992 Zénith Aircraft Company was formed to manufacture and market all versions except the CH 2000 in the USA, and in September 1996 reached full production status. In June 1999, Zenith announced that it was transferring production of the CH 2000 to a new, purpose-built 2,800 m^2 (30,000 sq ft) factory at Eastman, Georgia; Aircraft Manufacturing and Development Company was formed in the US to produce the aircraft and its derivative, the kitbuilt CH 640; the transfer is now complete, although Zenair built three CH 2000s in 2002.

Zenair has licensed representatives in Africa, South America, Belgium, Czech Republic (see entry for CZAW), France, Germany, Israel, Italy (see ICP), Spain and the UK.

Zenith Aircraft Company (Mexico, Missouri, USA) licenses kit production of Zenair designs. In addition to light aircraft, company constructs metal floats and amphibious floats.

UPDATED

Chile

ENAER

EMPRESA NACIONAL DE AERONÁUTICA DE CHILE

Avenida José Miguel Carrera 11087, El Bosque, Santiago
Tel: (+56 2) 383 18 73
Fax: (+56 2) 528 26 99
e-mail: marketing@enaer.cl
Web: http://www.enaer.cl
PRESIDENT: Gen Patricio Rios
EXECUTIVE DIRECTOR: Carlos Traub Gainsborg
TECHNICAL DIRECTOR: Rodolfo Pinto
COMMERCIAL DIRECTOR: Ricardo Klima
OPERATIONS DIRECTOR: Maximo Bascuñan
SALES DIRECTOR: Manuel Vargas
MARKETING MANAGER: Felipe Contardo
COMMUNICATIONS MANAGER: Cecilia Sola

State-owned company, formed 1984 from IndAer industrial organisation set up 1980 by Chilean Air Force (FACh); aircraft manufacture started 1979 with assembly of 27 Piper PA-28 Dakota lightplanes for Chilean Air Force and flying clubs. Covered plant area 47,000 m^2 (505,900 sq ft); workforce of 1,615 in mid-2001 (latest information received).

Recent main programmes have mainly concerned various military upgrade programmes. Currently undertaking sole-source fin/tailplane manufacture of ERJ-135/140/145 (560 shipsets ordered) as Embraer risk-sharing partner, and subcontract manufacture of part of tail unit of EADS CASA CN-235 and C-295; is also manufacturing components for Dassault Falcon 900 and Falcon 2000.

Own-design programmes have comprised T-35 piston-engined Pillán, T-35DT Turbo Pillán military trainer and Ñamcu light aircraft. Last-named became centred in Netherlands under Euro-ENAER as the EE-10 Eaglet; ENAER was to have been South American distributor for this aircraft, but had not begun to supply structures by early 2002, when Euro-ENAER declared bankrupt.

UPDATED

ENAER (ECH-51) T-35 PILLÁN

English name: Devil
Spanish Air Force designation: E.26 Tamiz (Grader)

TYPE: Basic prop trainer.

PROGRAMME: First two prototypes (first flight 6 March 1981) developed by Piper; followed by three Piper kits for ENAER assembly (first flight, by FACh s/n 101, 30 January 1982); then slight redesign, as recorded in earlier editions; ENAER series production started September 1984; first flight of production T-35A, 28 December 1984. Deliveries to FACh began 31 July 1985; original export deliveries completed 1991. Increased-span Pillán 2000, proposed in 1998, did not materialise. Production resumed in November 1998 to build 12 for Dominican Republic and again in 2002 to meet an Ecuadorean Navy order for four. Promotion of T-35DT continued at FIDAE Air Show in April 2002, but no further orders by late 2003. Remains available.

CURRENT VERSIONS: **T-35A:** Primary trainer version for Chilean Air Force. In service with Escuela de Aviación 'Capitán Avalos' at El Bosque AB, Santiago.

Detailed description applies to the above version except where indicated.

T-35B: Chilean Air Force and Dominican Republic instrument trainer, with more comprehensive instrumentation.

T-35C: Primary trainer for Spanish Air Force (designation E.26 Tamiz), assembled by CASA from ENAER kits.

T-35D: Instrument trainer for Panama and Paraguay.

T-35S: Single-seat prototype (CC-PZB); not produced.

T-35DT Turbo Pillán: Turboprop version; first offered in early 1990s; no orders announced by April 2002, when shown at FIDAE Air Show after mid-March maiden flight with new 'glass cockpit' (203 × 254 mm; 8 × 10 in ARNAV ICDS 2000 display) and offered with IAI Tamam POP or FLIR Systems Ultra 7500 infra-red sensor in turret under port wing as border surveillance aircraft. Still no reported orders by November 2003, but promotion continuing for drug interdiction role; 90-hour flight test programme with Ultra 7500 FLIR completed some two months earlier.

T-35 TD Turbo Pillán with underwing Ultra 7500 FLIR turret ***NEW***/0562817

ENAER T-35 Pilláns of the Chilean Air Force *(FACh)* ***NEW***/0528743

CUSTOMERS: Total of 128 (excluding prototypes and trials aircraft) built for air forces of Chile (36 T-35A, 12 T-35B), Dominican Republic (12 T-35B), Panama (10 T-35D), Paraguay (13 T-35D) and Spain (41 T-35C/E.26), plus four T-35B delivered to Ecuadorean Navy on 2 April 2002. Surplus FACh T-35Bs also sold to El Salvador (five), Guatemala (five) and Panama (two).

COSTS: US$1.3 million, T-35DT (2002).

DESIGN FEATURES: Low-risk, low-cost and international marketability achieved by reliance on Piper for development, some structural components and FAA compliance. Based on Piper Cherokee series (utilising many components of PA-28 Dakota and PA-32 Saratoga); cleared to FAR Pt 23 (Aerobatic category) and military standards for basic, intermediate and instrument flying training. Low-wing, tandem-seat design; sweptback vertical tail, non-swept horizontal surfaces.

Wing section NACA 65$_2$-415 on constant chord inboard panels, NACA 65$_2$-415 (modified) at tips; incidence 2° at root, –0° 30' at tip; dihedral 7° from roots.

FLYING CONTROLS: Conventional and manual. Mass-balanced elevators and rudder; single-slotted wing flaps, aileron trim tab (port) and tailplane/elevator trim all electrically actuated; variable incidence tailplane.

STRUCTURE: Main structure of aluminium alloy and steel, with riveted skins, except for glass fibre engine cowling, wingtips and tailplane tips. Single-spar fail-safe wings, with components mainly from PA-28-236 Dakota (leading-edges) and PA-32R-301 Saratoga (trailing-edges), modified for shorter span; vertical tail virtually identical with Dakota; tailplane uses some standard components from Dakota and PA-31 (Navajo/Cheyenne); tailcone from Cherokee components, modified for narrower fuselage.

LANDING GEAR: Hydraulically retractable tricycle type, with single wheel on each unit. Main units retract inward, steerable nosewheel rearward. Piper oleo-pneumatic shock-absorber in each unit. Emergency free-fall extension. Cleveland mainwheels and McCreary tyres size 6.00-6 (8 ply), nosewheel and tyre size 5.00-5 (6 ply). Tyre pressures: 2.62 bar (38 lb/sq in) on mainwheels, 2.41 bar (35 lb/sq in) on nosewheel. Single-disc air-cooled hydraulic brake on each mainwheel. Parking brake. Minimum ground turning radius 6.20 m (20 ft 4 in).

POWER PLANT: One 224 kW (300 hp) Textron Lycoming IO-540-K1K5 flat-six engine, driving a Hartzell HC-C3YR-4BF/FC7663R three-blade constant-speed metal propeller. Fuel in two integral aluminium tanks in wing leading-edges, total capacity 291.5 litres (77.0 US gallons; 64.1 Imp gallons), of which 278 litres (73.4 US gallons; 61.1 Imp gallons) are usable. Overwing gravity refuelling point on each wing. Oil capacity 11.4 litres (3.0 US gallons; 2.5 Imp gallons). Fuel and oil systems permit unlimited inverted flight.

ACCOMMODATION: Two vertically adjustable seats, with seat belts and shoulder harnesses, in tandem beneath one-piece transparent jettisonable canopy which opens sideways to starboard. One-piece acrylic windscreen and one-piece window in glass fibre fairing aft of canopy. Rear (instructor's) seat 22 cm (8.7 in) higher. Dual controls standard. Baggage compartment aft of rear cockpit, with external access on port side. Cockpits ventilated; cockpit heating and canopy demisting by engine bleed air.

SYSTEMS: Electrically operated hydraulic system, at 124 bar (1,800 lb/sq in) pressure for landing gear retraction and 44.8 bar (650 lb/sq in) for gear extension; separate system at 20.7 bar (300 lb/sq in) for wheel brakes. Electrical system is 24 V DC, powered by 28 V 70 A engine-driven Prestolite alternator and 24 V 15.5 Ah lead-acid battery, with inverter for AC power at 400 Hz to operate RMIs and attitude indicators. External power socket. No oxygen or de-icing provisions.

AVIONICS: Standard basic avionics by Bendix/King.

Comms: Two VHF com transceivers, two audio selector panels, transponder and interphone standard; Bendix/King KX 165 VOR/ILS.

Flight: Bendix/King KR 87 ADF, KN 63 DME KR 21 marker beacon receiver and KI 525A HSI.

DIMENSIONS, EXTERNAL:

Wing span	8.84 m (29 ft 0 in)
Wing aspect ratio	5.7
Length overall	8.00 m (26 ft 3 in)
Height overall	2.64 m (8 ft 8 in)
Tailplane span	3.05 m (10 ft 0 in)
Wheel track	3.02 m (9 ft 11 in)
Wheelbase	2.09 m (6 ft 10¼ in)
Propeller diameter	1.93 m (6 ft 4 in)

DIMENSIONS,INTERNAL:

Cockpit: Length	3.24 m (10 ft 7½ in)

Max width .... 1.04 m (3 ft 5 in)
Max height .... 1.48 m (4 ft 10¼ in)

AREAS:
Wings, gross .... 13.69 m² (146.3 sq ft)

WEIGHTS AND LOADINGS:
Weight empty, equipped .... 930 kg (2,050 lb)
Fuel weight .... 210 kg (462 lb)
Max aerobatic T-O weight .... 1,315 kg (2,900 lb)
Max T-O and landing weight .... 1,338 kg (2,950 lb)
Max wing loading .... 97.73 kg/m² (20.03 lb/sq ft)
Max power loading .... 5.98 kg/kW (9.83 lb/hp)

PERFORMANCE:
Never-exceed speed (VNE) .... 241 kt (446 km/h; 277 mph)
Max level speed at S/L .... 168 kt (311 km/h; 193 mph)
Cruising speed:
at 75% power at 2,680 m (8,800 ft) .... 144 kt (266 km/h; 166 mph) IAS
at 55% power at 5,120 m (16,800 ft) .... 138 kt (255 km/h; 159 mph) IAS
Stalling speed: flaps up .... 67 kt (125 km/h; 78 mph)
flaps down .... 62 kt (115 km/h; 72 mph)
Max rate of climb at S/L .... 465 m (1,525 ft)/min
Time to: 1,830 m (6,000 ft) .... 4 min 42 s
3,050 m (10,000 ft) .... 8 min 48 s
Service ceiling .... 5,840 m (19,160 ft)
Absolute ceiling .... 6,250 m (20,500 ft)
T-O run .... 287 m (940 ft)
T-O to 15 m (50 ft) .... 494 m (1,620 ft)
Landing from 15 m (50 ft) .... 509 m (1,670 ft)
Landing run .... 238 m (780 ft)
Range with 45 min reserves:
at 75% power at 2,440 m (8,000 ft) .... 590 n miles (1,093 km; 679 miles)
at 55% power at 3,660 m (12,000 ft) .... 650 n miles (1,204 km; 748 miles)
Range, no reserves:
at 75% power at 2,440 m (8,000 ft) .... 680 n miles (1,260 km; 783 miles)
at 55% power at 3,660 m (12,000 ft) .... 735 n miles (1,362 km; 846 miles)
Endurance at S/L: at 75% power .... 4 h 24 min
at 55% power .... 5 h 36 min
g limits .... +6/–3

UPDATED

China

ACAC

AVIC I COMMERCIAL AIRCRAFT COMPANY

23/F, 1027 Changning Road, Shanghai 200050
Tel: (+86 21) 52 41 37 37
Fax: (+86 21) 52 41 37 71
Web: http://www.acac.com.cn

PARTICIPATING COMPANIES:
Chengdu Aircraft Industry Group (CAC)
Shanghai Aircraft Research Institute (SARI) Shanghai Aircraft Industry Company (SAIC) Shenyang Aircraft Corporation (SAC) Xian Aircraft Design and Research Institute (XADRI) Xian Aircraft Company (XAC)

CHAIRMAN: Yang Yuzhong
PRESIDENT AND VICE-CHAIRMAN: Tang Xiaoping
VICE-PRESIDENT: Tao Zhihai

Six Chinese organisations formed a consortium in 1998 to study the prospects of launching, with risk-sharing foreign industrial partners, a programme for a new regional jet in the 70-seat class. Initial concepts were focused on 58-seat (NRJ 58) and 76-seat (NRJ 76) variants. However, the latter project received the designation ARJ21 (Advanced Regional Jet, 21st Century) in 2001 and State Council approval in June 2002; ACAC was formed in October 2002 to manage this programme.

UPDATED

ACAC ARJ21

TYPE: Regional jet airliner.

PROGRAMME: Project officially launched by AVIC I on 7 November 2000 as US$700 million programme under the auspices of a newly created New Regional Jet Programme Management Company. Aircraft was renamed ARJ21 (Advanced Regional Jet, 21st century) in 2001 and allocated initial investments from AVIC I (20 million yuan); CAC, SAIC and XAC (each 5 million yuan); and SAC (3 million yuan). Further investment is being sought from other Chinese and foreign companies. The parent company is responsible for defining airframe configuration; defining and subcontracting work-shares; schedule and quality control; final assembly, certification, marketing and customer support. Programme go-ahead approved by Chinese government June 2002; announced October 2002 with formation of ACAC and allocation of 5.2 billion yuan (US$650 million) start-up funding; Boeing to provide engineering support. Manufacture at seven factories, including Xian and Shanghai; final assembly at Shanghai. Selection of General Electric CF34 announced 5 November 2002. Prototype construction began simultaneously at Chengdu, Shanghai and Xian plants on 20 December 2003. First flight targeted for mid-2006, certification for October 2007 and service entry by end of 2007.

CURRENT VERSIONS: **ARJ21-700**: Baseline version, seating 78 to 85 passengers.

ARJ21-900: Stretched version, with seats for 98 to 105 passengers.

Freighter and Business jet: Planned future variants.

CUSTOMERS: Total of 35 ordered by 1 October 2003 by Shandong Airlines (launch operator, 10), Shanghai Airlines (launch customer, five) and Shenzhen Financial Leasing Company (20). Estimated sale of 500 (350 domestic and 150 export) over 20 years.

DESIGN FEATURES: Low, swept wing (25° at quarter-chord) with winglets; engines pod-mounted to rear fuselage; all-swept T tail.

Model of ARJ21 airliner *(Paul Jackson)* ***NEW***/0572302

ARJ21 flight deck mockup ***NEW***/0558609

Supercritical wing section being designed and wind tunnel tested by Antonov design bureau. Wings 3° dihedral, tailplane 3° anhedral.

FLYING CONTROLS: Honeywell/Parker Hannifin electrically actuated (fly-by-wire) ailerons, elevators and rudder; no mechanical back-up. Three-segment leading-edge flaps. Two-segment trailing-edge flaps, with single spoiler/lift dumper forward of inboard flap and three-segment spoilers forward of each outer flap. Hamilton Sundstrand flap actuation.

STRUCTURE: Wings and centre-fuselage by Xian (XAC); tailplane and final assembly by Shanghai (SAIC).

LANDING GEAR: Liebherr Aerospace and Aircraft Braking Systems retractable tricycle type, with twin wheels on each unit.

POWER PLANT: Two 80.1 kN (18,000 lb st) class rear-mounted General Electric CF34-10A turbofans with FADEC. Smiths Aerospace thrust reverser actuators. Standard fuel capacity 12,719 litres (3,360 US gallons; 2,798 Imp gallons).

ACCOMMODATION: Flight crew of two, plus two cabin attendants. Five-abreast seating for up to 85 (in ARJ21-700) or 105 (ARJ21-900) passengers in all-economy basic layout at 81 cm (32 in) pitch. Alternative mixed class configuration offers eight-seat business section (two rows, four abreast, at 96.5 cm; 38 in pitch) at front, with 70 and 90 economy class, respectively, to rear. Galleys (two), wardrobe, storage area and lavatory at front of cabin; second storage area and lavatory at rear. Type I passenger door at front on port side, with Type I service door opposite; two Type III emergency exits over wing on each side. Underfloor baggage door forward and aft of wing on starboard side.

SYSTEMS: Parker Hannifin hydraulic and fuel systems. Hamilton Sundstrand APS 2300 APU, electrical power generation and distribution system (115 V AC from three 40 kVA

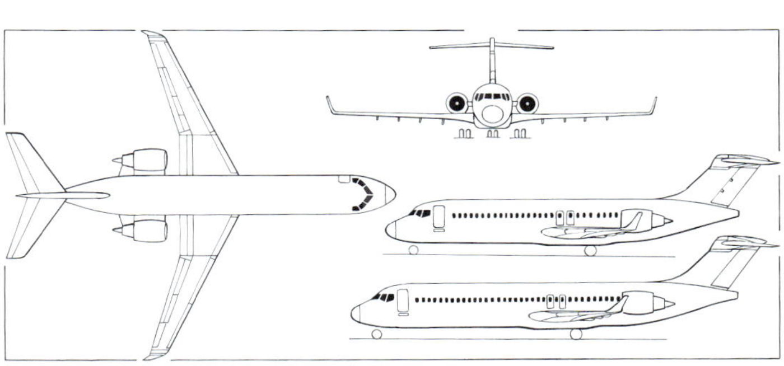

Baseline ARJ21-700, with additional side view of the stretched ARJ21-900 *(James Goulding)* ***NEW***/0561621

generators; two engine-driven and one APU-driven), and emergency ram-air turbine. Liebherr Aerospace air conditioning, de-icing and bleed air systems. Sagem cabin systems; Kidde fire protection system.

AVIONICS: Rockwell Collins Pro Line 21 selected April 2003; suite includes VHF-4000 voice/data receiver, RIU-4000, five 254 × 203 mm (10 × 8 in) LCDs, EICAS, solid-state weather radar, AHS-3000 and FMS-4200. Eaton Corporation flight deck panel assemblies and lighting controls.

DIMENSIONS, EXTERNAL (A: ARJ21-700, B: ARJ21-900):
- Wing span over winglets: A, B ... 27.67 m (90 ft 0¾ in)
- Length overall: A ... 32.68 m (107 ft 3½ in)
  - B ... 36.06 m (118 ft 4½ in)
- Fuselage length: A ... 29.465 m (96 ft 8 in)
  - B ... 32.84 m (107 ft 9 in)
- Height overall: A, B ... 8.49 m (27 ft 10¼ in)
- Tailplane span: A, B ... 10.51 m (34 ft 5¾ in)
- Wheel track (c/l of shock-struts):
  - A, B ... 4.63 m (15 ft 2¼ in)
- Wheelbase: A ... 14.87 m (48 ft 9½ in)
  - B ... 16.80 m (55 ft 1½ in)
- Passenger door (fwd, port): Height ... 1.83 m (6 ft 0 in)
  - Width ... 0.86 m (2 ft 10 in)
- Service door (fwd, stbd): Height ... 1.22 m (4 ft 0 in)
  - Width ... 0.69 m (2 ft 3 in)
- Overwing emergency exits (four,each):
  - Height ... 0.91 m (3 ft 0 in)
  - Width ... 0.51 m (1 ft 8 in)

DIMENSIONS, INTERNAL:
- Passenger cabin: Max width ... 3.145 m (10 ft 3¾ in)
  - Aisle width ... 0.48 m (1 ft 7 in)
  - Max height ... 2.06 m (6 ft 9¾ in)

WEIGHTS AND LOADINGS (A and B as above):
- Operating weight empty:
  - A, standard range ... 23,083 kg (50,889 lb)
  - A, extended range ... 23,183 kg (51,110 lb)
  - B, standard range ... 24,507 kg (54,029 lb)
  - B, extended range ... 24,607 kg (54,249 lb)
- Max standard fuel weight: A, B ... 10,386 kg (22,897 lb)
- Max payload: A ... 8,935 kg (19,698 lb)
  - B ... 11,246 kg (24,793 lb)
- Max T-O weight:
  - A, standard range ... 37,645 kg (82,993 lb)
  - A, extended range ... 40,737 kg (89,809 lb)
  - B, standard range ... 41,412 kg (91,297 lb)
  - B, extended range ... 43,951 kg (96,895 lb)
- Max landing weight:
  - A, standard range ... 35,010 kg (77,184 lb)
  - A, extended range ... 37,886 kg (83,524 lb)
  - B, standard range ... 38,513 kg (84,907 lb)
  - B, extended range ... 40,874 kg (90,112 lb)

PERFORMANCE (estimated):
- Max cruising Mach number ... 0.78
- Max cruising speed: A, B ... 450 kt (833 km/h; 518 mph)
- Max certified altitude: A, B ... 10,670 m (35,000 ft)
- Service ceiling, OEI, at T-O weight for 500 n mile (926 km; 575 mile) range:
  - A, standard range ... 6,720 m (22,050 ft)
  - A, extended range ... 6,693 m (21,960 ft)
  - B, standard range ... 6,482 m (21,260 ft)
  - B, extended range ... 6,461 m (21,200 ft)
- T-O field length at S/L:
  - A, standard range ... 1,472 m (4,830 ft)
  - A, extended range ... 1,728 m (5,670 ft)
  - B, standard range ... 1,616 m (5,300 ft)
  - B, extended range ... 1,822 m (5,980 ft)
- Landing field length at MLW at S/L:
  - A, standard range ... 1,436 m (4,710 ft)
  - A, extended range ... 1,508 m (4,950 ft)
  - B, standard range ... 1,514 m (4,970 ft)
  - B, extended range ... 1,573 m (5,160 ft)
- Range with max payload:
  - A and B, standard range ... 1,200 n miles (2,222 km; 1,380 miles)
  - A, extended range ... 2,000 n miles (3,704 km; 2,301 miles)
  - B, extended range ... 1,800 n miles (3,333 km; 2,071 miles)

UPDATED

AVIC

CHINA AVIATION INDUSTRY CORPORATION (ZHONGGUO HANGKONG GONGYE ZONGGONGSI)

67 Jiao Nan Street (PO Box 33), Beijing 100712
Tel: (+86 10) 64 01 33 32
Fax: (+86 10) 64 01 36 48
Web (1): http://www.avic1.com.cn
Web (2): http://www.avic2.com.cn

PRESIDENTS:
Liu Gaozhuo (AVIC I)
Zhang Yanzhong (AVIC II)

VICE-PRESIDENTS:
Yang Yuzhong (AVIC I)
Liang Zhenhe (AVIC II) Xu Zhanbin (AVIC II)

DIRECTORS-GENERAL OF MARKETING:
Tang Xiaoping (AVIC I)
Cui Degang (AVIC II)

INTERNATIONAL MARKETING:
CATIC (Zhongguo Hangkong Jishu Jinchukou Zonggongsi: China National Aero-Technology Import and Export Corporation)
CATIC Plaza, 18 Beichen Dong Street, Chaoyang District, Beijing 100101
Tel: (+86 10) 64 94 22 55, 64 94 03 70 and 64 94 10 90
Fax: (+86 10) 64 94 06 58 and 64 94 10 88
e-mail: webmaster@catic.com.cn
Web: http://www.catic.com.cn
PRESIDENT: Yang Chunshu
VICE-PRESIDENT, EXPORTS: Ding Shiqing
DIRECTOR, PUBLIC RELATIONS: Bi Jianfa

Present Chinese aviation industry created in 1951 and has since manufactured some 14,000 aircraft (including more than 10,000 military), more than 50,000 aero-engines and 10,000 air-to-air and tactical missiles. Some 700 aircraft have been exported, approximately 10 per cent of them civil types.

Former Ministry of Aero-Space Industry abolished 1993 and AVIC created on 26 June 1993 as economic entity to develop market economy and expand international collaboration in aviation programmes. CATIC Group formed 26 August 1993, with CATIC (founded January 1979) as its core company, to be responsible for import and export of aero and non-aero products, subcontract work and joint ventures.

Xian, Chengdu, Shanghai, Shenyang, Harbin and other factories carry out subcontract work on Airbus A300/310/318/320; ATR 72; Boeing 737/747/757; de Havilland Dash 8Q; and Bombardier 415. Licensed manufacture of Sukhoi Su-27s is undertaken at Shenyang. Co-development with UK/France/Italy and Singapore of AE-100 regional airliner abandoned in 1998 and replaced by AVIC I's ARJ21 project for 79/99-seat family of aircraft (managed by ACAC); AVIC II programme for smaller regional jet now being met by co-production of Embraer ERJ-135/140/145 family by Harbin Embraer Aircraft Industry (HEAI).

Total workforce of aerospace industry was reduced to about 500,000 in 1998, when about 34,000 workers were made redundant and some 14,000 others transferred to non-aerospace activities. AVIC's President announced plans at Airshow China in November 1998 to restructure aircraft industry into two "competing and co-operating" groups in the near future. Draft organisation plan submitted to State Council in February 1999, resulting in division of AVIC into two separate companies (AVIC I and II) with effect from July 1999. Workforces in 2002 were 240,000 and 210,000 respectively, with corresponding assets of 34.9 billion and 31.5 billion yuan. AVIC I comprises 54 industrial enterprises and 30 research institutes; corresponding figures for AVIC II are 54 and three. AVIC II also controls 22 other enterprises, institutes and specialised companies, including China Aviation Supply and Marketing Corporation (CASM); and (with AVIC I) has 50 per cent ownership of CATIC. Principal organisations in this structure, and their major indigenous programmes, are shown below. AVIC I announced intention in late 2002 to restructure itself, with future emphasis on its core aerospace businesses, which would operate independently of the parent organisation as specialised airframe, engine and equipment business units. AVIC II was stated in mid-2002 to have government approval to float its non-military business on the Hong Kong Stock Exchange; to this end, new company AviChina Industry and Technology created in 2003; in October 2003, a 5 per cent stake in this company was acquired by EADS. In same year, new MoU signed with Antonov, to pursue possibility of producing An-70 and An-124-300 in China.

AVIC I: AVIC I Commercial Aircraft Company (ARJ21)
- Beijing Aviation Simulator
- Chengdu Aircraft Industries Group (J-7/F-7, FC-1/JF-17 and J-10)
- China Air-to-Air Missile Research Institute
- Guizhou Aviation Industries Group (JJ-7/FT-7; JL-9; jet engines; missiles)
- Shanghai Aviation Industries Group (airliner subcontracts)
- Shenyang Aircraft Industries Group (J-8/F-8 and J-11/Su-27; civil subcontracts)
- Xian Aero-Engine Company (WP8 and WS9)
- Xian Aircraft Industries Group (JH-7/FBC-1 and MA60)

AVIC II: Changhe Aircraft Industries Group (Z-8 and Z-11)
- Chengdu Engine Company (WP7 and WP13)
- China Helicopter Design and Research Institute (Z-10)
- Dongan Engine Company (WJ5)
- Harbin Aircraft Industries Group (Hafei Z-9, 'Z-X' and Y-12)
- Hongdu Aviation Industries Group (NAMC JL-8/K-8)
- Shaanxi Aircraft Company (Y-8)
- Shijiazhuang Aircraft Industry Corporation (Y-5)
- South Aero-Engine Company (HS5, WJ6 and WZ8)

UPDATED

CAC

CHENGDU AIRCRAFT INDUSTRY GROUP (CHENGDU FEIJI GONGYE GONGSI) (SUBSIDIARY OF AVIC I)

Huangtianba (PO Box 800), Chengdu, Sichuan 610092
Tel: (+86 28) 740 10 33
Fax: (+86 28) 740 49 84
e-mail: cacoa@mail.cac.com.cn
Web: http://:www.cac.com.cn
CHAIRMAN: Yang Tingkuo
PRESIDENT: Yang Baoshu
DEPUTY GENERAL MANAGER: Li Shaoming
DIRECTOR, INTERNATIONAL CO-OPERATION DIVISION: Wang Yinggong

Major centre for fighter development and production, founded 1958; has since built over 2,000 fighters of more than 10 models or variants; current facility occupies site area of 462 ha (1,142 acres) and had 1995 workforce (latest figure known) of nearly 20,000. Production continues of some variants of J-7/F-7 fighter series; new J-10 and FC-1 fighters currently under development and/or early production. Chengdu also thought to be in concept stage of an advanced combat aircraft, possibly to the same outline requirement as the SAC (Shenyang)

'XXJ'. In addition, CAC was named in 2001 as a risk-sharing partner in the ACAC ARJ21 regional jet programme.

Subcontract work includes passenger doors for the Airbus A320; wing components for Boeing 737/747; production of fuselage components for the Dassault Falcon 2000EX was due to begin in 2003 under an agreement signed in early 2002. Non-aerospace products, which previously accounted for about 10 per cent of current output, were targeted to have reached 40 per cent by 2000; it is not known whether this was achieved.

UPDATED

CAC J-7

Chinese name: Jianjiji-7 (Fighter aircraft 7)
Westernised designation: F-7

TYPE: Multirole fighter.

PROGRAMME: Origins of MiG-21 licensed manufacture in China (1961–67) detailed in earlier editions. Chengdu production of J-7 I began June 1967 (first flight 16 June 1969 but not accepted in large numbers); development of J-7 II began 1975, followed by first flight 30 December 1978 and production approval September 1979; development of F-7M and J-7 III started 1981; J-7 III first flight 26 April 1984; F-7M (first flight 31 August 1983) revealed publicly October 1984, production go-ahead December 1984, named Airguard early 1986; first F-7P deliveries to Pakistan 1988; first F-7MPs to Pakistan mid-1989; F-7MG public debut November 1996; J-7FS revealed September 1998 and F-7MF November 2000.

CURRENT VERSIONS (DOMESTIC): **J-7B:** Modified and improved J-7 I, original designation J-7 II; WP7B turbojet of increased thrust (43.2 kN; 9,700 lb st dry, 59.8 kN; 13,450 lb st with afterburning); 720 litre (190 US gallon; 158 Imp gallon) centreline drop tank for increased range; brake-chute relocated at base of rudder to improve landing performance and shorten run; rear-hinged canopy, jettisoned before ejection seat deploys; new Chengdu Type II seat operable at zero height and speeds down to 135 kt (250 km/h; 155 mph); and new Lanzhou compass system. Small batch production (typically, 14 in 1989), notwithstanding advent of J-7 III and J-7E.

J-7C : (originally J-7 III) Chinese equivalent of MiG-21MF, with blown flaps and all-weather, day/night capability. Main improvements are change to WP13 engine with greater power; additional fuel in deeper dorsal spine; JL-7 (J-band) interception radar, with correspondingly larger nose intake and centrebody radome; sideways-opening (to starboard) canopy, with centrally located rearview mirror; improved HTY-4 low-speed/zero height ejection seat; more advanced fire-control system; twin-barrel 23 mm gun under fuselage (with HK-03D optical gunsight); broader-chord vertical tail surfaces, incorporating antennas for LJ-2 omnidirectional RWR in hemispherical fairing each side at base of rudder; increased weapon/stores capability (four underwing stations), similar to that of F-7M; and new or additional avionics (which see). Joint development by Chengdu and Guizhou (GAIC); entered AFPLA service from 1992, but reportedly limited production and use only (15th and 19th Divisions).

J-7D : (originally J-7 IV) Further (early 1990s) attempt to improve upon J-7 III; 71.6 kN (16,093 lb st) WP13F1 engine, JL-7A interception radar, RWR antennas atop vertical fin; HUD, Tacan and ADC; weapons upgrade to include PL-8 AAMs. Believed to equip 15th and 29th Divisions, including two night fighter regiments.

J-7E: Upgraded version of J-7B with modified, double-delta wing, retaining existing leading-edge sweep angle of 57° inboard but reduced sweep of only 42° outboard; span increased by 1.17 m (3 ft 10 in) and area by 1.88 m^2 (20.2 sq ft), giving 8.17 per cent more wing area; four underwing stations instead of two, outer pair each plumbed for 480 litre (127 US gallon; 106 Imp gallon) drop tank; new WP7F version of WP7 engine, rated at 44.1 kN (9,921 lb st) dry and 63.7 kN (14,330 lb st) with afterburning; armament generally as listed for F-7M, but capability extended to include PL-8 (Python 3) air-to-air missiles; *g* limits of 8 (up to M0.8) and 6.5 (above M0.8); avionics include HUD and ADC. Believed to have made first flight in April 1990 and entered service 1993. Operators include 1st, 2nd, 3rd, 4th, 14th and 37th AFPLA Divisions and Navy 4th Division. In production.

J-7EB: Version of J-7E equipping AFPLA 'August 1st' aerobatic team (nine in 1998 display season); fitted with smoke canisters for display purposes.

J-7FS: Technology demonstrator, modified from standard J-7 II, with new chin-mounted intake and central splitter plate under reconfigured ogival nosecone; more powerful (73.4 to 78.3 kN; 16,502 to 17,604 lb thrust class with afterburning) Liyang (LMC) WP13F II turbojet. Began 22 month flight test programme on 8 June 1998. Enlarged nose avionics bay able to accept 60 cm (23.6 in) diameter multimode pulse Doppler radar believed to be under development in China. Development work formed basis for new F-7MF export variant (which see). Planned future changes were to include wing modifications based on J-7E/F-7MG double-delta configuration.

J-7G: Reported as further upgrade of J-7E; maiden flight in mid-2002. New fire-control radar, PL-8 IR-guided AAMs and updated cockpit; intended for Chinese air force and naval units.

JJ-7: Tandem two-seat operational trainer, based on J-7 II and MiG-21US; developed at Guizhou by GAIC.

CURRENT VERSIONS (EXPORT): **F-7A:** Export counterpart of J-7 I/J-7A, supplied to Albania and Tanzania.

F-7B: Export version of J-7 II/J-7B, with R550 Magic missile capability; supplied to Egypt and Iraq in 1982–83 and also to Sudan. Some supplied to Air Force of Zimbabwe (known locally as F-7 II) appear to be of this version.

F-7BS: Hybrid version supplied to Sri Lanka 1991: has F-7B fuselage/tail and Chinese avionics (no HUD), combined with four-pylon wings of F-7M. Equips No. 5 Squadron of SLAF. Zimbabwe also has some four-pylon aircraft with its No. 5 Squadron; these known locally as the F-7 IIN.

CAC J-7Es of the AFPLA ***NEW***/0569633

CAC J-7D of the AFPLA's trials unit ***NEW***/0569632

F-7M Airguard: First upgraded export version, developed from J-7B; new avionics imported from May 1979 included Marconi HUDWAC (head-up display and weapon aiming computer); new ranging radar, air data computer, radar altimeter and IFF; more secure com radio; improved electrical power generation system for the new avionics; two additional underwing stores points; improved WP7B(BM) engine; birdproof windscreen; strengthened landing gear; ability to carry PL-7 air-to-air missiles; nose probe relocated from beneath intake to top lip of intake, offset to starboard. Exported to Iran and Myanmar.

Description applies to F-7M version, except where indicated.

F-7MB: Variant of F-7M; mentioned in 1996 F-7MG brochure. Customer believed to be Bangladesh (16); one planned to be equipped with reconnaissance pod.

F-7MF: Latest known variant; debut at Airshow China, November 2000; further development of J-7FS and F-7MG for export market. Larger 'solid' nose; shorter, rectangular intake located farther back under nose; small, shoulder-mounted canards just forward of wingroot leading-edge; WP13F engine; 1553B databus; avionics to include 43 n mile (80 km; 50 mile) range pulse Doppler radar (possibly IAI Elta EL/M-2032), single HUD and dual HDDs; 3,000 kg (6,614 lb) external stores load.

Wind tunnel testing completed; first flight then targeted for late 2001/early 2002, but most recently forecast for second half of 2003, though not reported by March 2004. Performance expectations include M1.8 top speed, 16,000 m (52,500 ft) ceiling, 650 m (2,135 ft) T-O run and 1,403 n mile (2,600 km; 1,615 mile) ferry range. Possibly testbed for some features of J-10.

F-7MG: Improved version of F-7M (G suffix indicates *gai:* modified), combining double-delta wings of J-7E with Grifo MG radar, other upgraded avionics, uprated (WP13F) engine and leading/trailing-edge manoeuvring flaps. Said to have 45 per cent better manoeuvrability than F-7M. Public debut (aircraft 0142 and 0144) at Airshow China, Zhuhai, November 1996. Pakistan (see F-7PG) only customer so far, but Bangladesh said to have requirement for up to 12 and Zimbabwe interested in enough for two squadrons.

F-7MP: Further modified variant of F-7P; improved cockpit layout and navigation system incorporating Rockwell Collins AN/ARN-147 VOR/ILS receiver, AN/ARN-149 ADF and Pro Line II digital DME-42. Avionics (contract for up to 100 sets) delivered to China from early 1989. FIAR (now Galileo) Grifo MG fire-control radar (range of more than 30 n miles; 55 km; 34 miles) for F-7P and MP ordered 1993, to replace Skyranger; flight trials began May 1996 and completed in 1997.

F-7N: Variant of F-7M; mentioned in 1996 F-7MG brochure, but no details given. Possibly an alternative designation for Zimbabwe F-7 IIN (see F-7BS paragraph above).

F-7P Airguard: Variant of F-7M (briefly called Skybolt), embodying 24 modifications to meet specific requirements of Pakistan Air Force, including ability to carry four air-to-air missiles (Sidewinders) instead of two and fitment of Martin-Baker Mk 10L ejection seat. Delivered 1988–91.

F-7PG: Pakistan Air Force designation of F-7MG; 57 (plus six FT-7PGs) ordered in late 2000. Deliveries to Nos. 17 and 23 Squadrons at Samungli, replacing Shenyang F-6s, completed by second quarter 2002; third squadron expected to be No. 2 at Masroor. Grifo 7PG radars (still awaiting firm contract at end 2002) would be produced by KARF factory of Pakistan Aircraft Complex. Possibility of follow-on order for up to 25 more (including FT-7PG trainers).

FT-7: Export designation of GAIC JJ-7 two-seat trainer.

CUSTOMERS: Several thousand built for Chinese air forces; Air Divisions currently equipped with J-7 variants comprise Nos. 1, 2, 3, 4, 9, 12, 14, 15, 18, 21, 24, 26, 29, 33, 35, 42 and 44, plus Naval 4th, each with two or three Regiments totalling up to 100 aircraft; Nos. 7, 17, 31 and 32 believed disbanded.

More than 400 exported to Albania (12 F-7A), Bangladesh (16 F-7MB), Egypt (approximately 90 F-7B?), Iran (18 or more F-7M), Iraq (approximately 90 F-7B?), Myanmar (24 F-7M), Pakistan (20 F-7P and 100 F-7MP, all designated F-7P by PAF; followed by 57 F-7PG), Sri Lanka (four F-7BS), Sudan (22 F-7B); Tanzania (16 F-7A) and Zimbabwe (approximately 12 F-7B/F-7IIN variants). Pakistan Air Force F-7P squadrons are No. 2 at Masroor, Nos. 18 and 20 at Rafiqui and No. 25 at Mianwali; No. 17 first with F-7PG (27 March 2002), followed by No. 23, both at Quetta.

DESIGN FEATURES: Typical mid-1950s design of fighter, incorporating diminutive delta wing (double-delta on J-7E/EB and F-7MG/PG), with clipped tips to mid-mounted wings, plus all-moving horizontal tail; circular-section fuselage with dorsal spine; nose intake with conical centrebody; swept tail, with large vertical surfaces and ventral fin.

'Red 139', the Chengdu J-7FS technology demonstrator prototype, following its repainting in camouflage ***NEW***/0547104

Prototype CAC J-7G on an early test flight ***NEW***/0573877

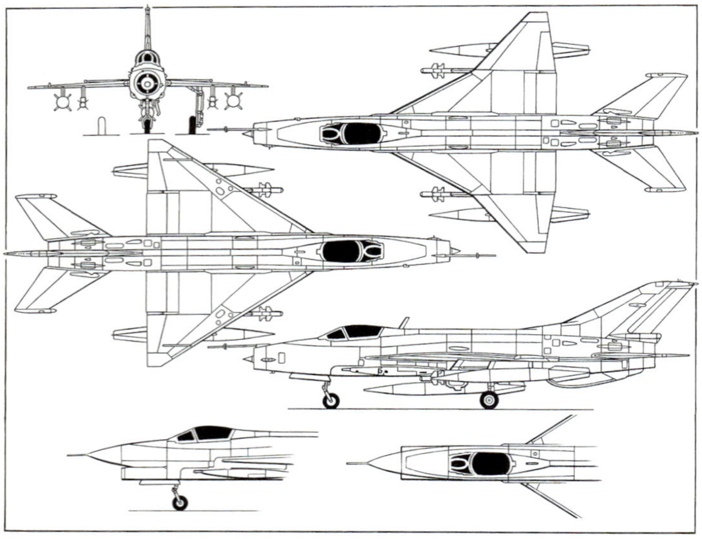

CAC F-7M Airguard single-seat fighter and close support aircraft; upper plan view shows modified outer wings of J-7E and F-7MG; lower scrap views of J-7FS nose in profile and plan *(Jane's/Mike Keep)* 0126688

Wing anhedral 2° from roots; incidence 0°; thickness/chord ratio approximately 5 per cent at root, 4.2 per cent at tip; quarter-chord sweepback 49° 6′ 36″ (reduced on J-7E/EB and F-7MG/PG outer panels); no wing leading-edge camber.

FLYING CONTROLS: Manual operation, with autostabilisation in pitch and roll; hydraulically boosted inset ailerons; plain trailing-edge flaps, actuated hydraulically; forward-hinged door-type airbrake each side of underfuselage below wing leading-edge; third, forward-hinged airbrake under fuselage forward of ventral fin; airbrakes actuated hydraulically; hydraulically boosted rudder and all-moving, trimmable tailplane. Leading/trailing-edge manoeuvring flaps on J-7E/EB and F-7MG/PG.

STRUCTURE: All-metal; wings have two primary spars and auxiliary spar; semi-monocoque fuselage, with spine housing control pushrods, avionics, single-point refuelling cap and fuel tank; blister fairings on fuselage above and below each wing to accommodate retracted mainwheels.

LANDING GEAR: Inward-retracting mainwheels, with 600×200 tyres (pressure 11.50 bar; 167 lb/sq in) and LS-16 disc brakes; forward-retracting nosewheel, with 500×180 tyre (pressure 7.00 bar; 102 lb/sq in) and LS-15 double-acting brake. Nosewheel steerable ±47°. Minimum ground turning radius 7.04 m (23 ft 1¼ in). Tail braking parachute at base of vertical tail.

POWER PLANT: *F-7M:* One LMC (Liyang) WP7B(BM) turbojet (43.2 kN; 9,700 lb st dry, 59.8 kN; 13,448 lb st with afterburning).

J-7C: LMC WP13 turbojet (40.2 kN; 9,039 lb st dry, 64.7 kN; 14,550 lb st with afterburning).

J-7E/EB: See Current Versions.

J-7FS: See Current Versions.

F-7MF/MG/PG: WP13F (44.1 kN; 9,921 lb st dry, 64.7 kN; 14,550 lb st with afterburning).

Total F-7M internal fuel capacity 2,385 litres (630 US gallons; 525 Imp gallons), contained in six flexible tanks in fuselage and two integral tanks in each wing. Provision for carrying a 500 or 800 litre (132 or 211 US gallon; 110 or 176 Imp gallon) centreline drop tank, and/or a 500 litre drop tank on each outboard underwing pylon. Maximum internal/external fuel capacity 4,185 litres (1,105 US gallons; 921 Imp gallons).

ACCOMMODATION: Pilot only, on CAC zero-height/low-speed ejection seat operable between 70 and 459 kt (130 and 850 km/h; 81 and 528 mph) IAS. Martin-Baker Mk 10L seat in F-7P/MP/PG. One-piece canopy, hinged at rear to open upward. J-7C canopy opens sideways to starboard.

SYSTEMS: Improved electrical system in F-7M, using three static inverters, to cater for additional avionics. Jianghuai YX-3 oxygen system.

AVIONICS: *Comms:* BAE Systems AD 3400 UHF/VHF multifunction com, Chinese Type 602 IFF transponder; Type 605A ('Odd Rods' type) IFF in J-7C.

Radar: BAE Systems Type 226 Skyranger ranging radar in F-7M; Chinese JL-7 fire-control radar in J-7C; Galileo (formerly FIAR) Grifo MG in F-7P/MP/MG/PG (look-down, shoot-down and track-while-scan capability).

Flight: Navigation function of BAE Systems HUDWAC includes approach mode. WL-7 radio compass, XS-6A marker beacon receiver, Type 0101 HR A/2 radar altimeter and BAE Systems air data computer in F-7M. Beijing Aeronautical Instruments Factory KJ-11 twin-channel autopilot and FJ-1 flight data recorder in J-7C. F-7MG/PG suite includes VOR/DME/INS and Tacan.

Instrumentation: BAE Systems Type 956 HUDWAC (head-up display and weapon aiming computer) in F-7M provides pilot with displays for instrument flying, with air-to-air and air-to-ground weapon aiming symbols integrated with flight-instrument symbology. It can store 32 weapon parameter functions, allowing for both current and future weapon variants. In air-to-air combat its four modes (missiles, conventional gunnery, snapshoot gunnery, dogfight) and standby aiming reticle allow for all eventualities. VCR and infra-red cockpit lighting in F-7MG/PG, for which licence-built Russian helmet sight, slaved to PL-9 AAM, is also in production.

Self-defence: Skyranger ECCM in F-7M. Chinese LJ-2 RWR and GT-4 ECM jammer in J-7C.

ARMAMENT (F-7M): Two 30 mm Type 30-1 belt-fed cannon, with 60 rds/gun, in fairings under front fuselage just forward of wingroot leading-edges. Two hardpoints under each wing, of which outer ones are wet for carriage of drop tanks. Centreline pylon used for drop tank only. Each inboard pylon capable of carrying a PL-2, -2A, -5B, -7 or -8 (Python 3) missile (and PL-9 on F-7MG/PG) or, at customer's option, an R550 Magic; one 18-tube pod of Type 57-2 (57 mm) air-to-air and air-to-ground rockets; one Type 90-1 (90 mm) seven-tube pod of air-to-ground rockets; or a 50, 150, 250 or 500 kg bomb. Each outboard pylon can carry one of above rocket pods, a 50 or 150 kg bomb, or a 500 litre drop tank.

ARMAMENT (J-7C): One 23 mm Type 23-3 twin-barrel gun in ventral pack. Five external stores stations can carry two to four PL-2 or PL-5B air-launched missiles; two or four Qingan HF-16B 12-round launchers for Type 57-2 or seven-round pods of Type 90-1 rockets; or two 500 kg, four 250 kg or ten 100 kg bombs, in various combinations with 500 litre (one centreline and/or one under each wing) or 800 litre (underfuselage station only) drop tanks.

DIMENSIONS, EXTERNAL:

Wing span: except J-7E/F-7MG	7.15 m (23 ft 5½ in)
J-7E/F-7MG	8.32 m (27 ft 3½ in)
Wing chord: at root	5.51 m (18 ft 0¾ in)
at tip (except J-7E/F-7MG)	0.46 m (1 ft 6¼ in)
Wing aspect ratio: except J-7E/F-7MG	2.2
J-7E/F-7MG	2.8
Length overall: excl nose probe	13.945 m (45 ft 9 in)
incl nose probe	14.885 m (48 ft 10 in)
Fuselage: Length	12.175 m (39 ft 11½ in)
Max diameter	1.34 m (4 ft 4¾ in)
Height overall	4.105 m (13 ft 5½ in)
Tailplane span	3.74 m (12 ft 3¼ in)
Wheel track	2.69 m (8 ft 10 in)
Wheelbase	4.805 m (15 ft 9¼ in)

AREAS:

Wings, gross: except J-7E/F-7MG	23.00 m² (247.6 sq ft)
J-7E/F-7MG	24.88 m² (267.8 sq ft)
Ailerons (total): except J-7E/F-7MG	1.18 m² (12.70 sq ft)
Trailing-edge flaps (total)	1.87 m² (20.13 sq ft)
Fin	3.48 m² (37.46 sq ft)
Rudder	0.97 m² (10.44 sq ft)
Tailplane	3.94 m² (42.41 sq ft)

WEIGHTS AND LOADINGS:

Weight empty: F-7M	5,275 kg (11,629 lb)
F-7MF	5,500 kg (12,125 lb)
F-7MG	5,292 kg (11,667 lb)
Normal T-O weight with two PL-2 or PL-7 air-to-air missiles:	
F-7M	7,531 kg (16,603 lb)
J-7C	8,150 kg (17,967 lb)
F-7MG	7,540 kg (16,623 lb)
Max T-O weight: F-7MG	9,100 kg (20,062 lb)
Wing loading at normal T-O weight:	
F-7M	327.4 kg/m² (67.06 lb/sq ft)
J-7C	354.3 kg/m² (72.56 lb/sq ft)
Max wing loading: F-7MG	365.8 kg/m² (74.91 lb/sq ft)
Power loading at normal T-O weight:	
F-7M, J-7C	126 kg/kN (1.23 lb/lb st)
Max power loading: F-7MG	141 kg/kN (1.38 lb/lb st)

PERFORMANCE (F-7M at normal T-O weight with two PL-2 or PL-7 air-to-air missiles, except where indicated):

Never-exceed speed (VNE) above 12,500 m (41,010 ft)	M2.35 (1,346 kt; 2,495 km/h; 1,550 mph)
Max level speed between 12,500 and 18,500 m (41,010–60,700 ft)	M2.05 (1,175 kt; 2,175 km/h; 1,350 mph)
Unstick speed	167–178 kt (310–330 km/h; 193–205 mph)
Touchdown speed	162–173 kt (300–320 km/h; 186–199 mph)
Max rate of climb at S/L	10,800 m (35,435 ft)/min
Acceleration from M0.9 to M1.2 at 5,000 m (16,400 ft)	35 s
Max sustained turn rate: M0.7 at S/L	14.7°/s
M0.8 at 5,000 m (16,400 ft)	9.5°/s
Service ceiling	18,200 m (59,720 ft)
Absolute ceiling	18,700 m (61,360 ft)
T-O run	700–950 m (2,300–3,120 ft)
Landing run with brake-chute	600–900 m (1,970–2,955 ft)
Typical mission profiles:	
combat air patrol at 11,000 m (36,080 ft) with two air-to-air missiles and three 500 litre drop tanks, incl 5 min combat	45 min
long-range interception at 11,000 m (36,080 ft) at 351 n miles (650 km; 404 miles) from base, incl M1.5 dash and 5 min combat, stores as above	
hi-lo-hi interdiction radius, out and back at 11,000 m (36,080 ft), with three 500 litre drop tanks and two 150 kg bombs	324 n miles (600 km; 373 miles)
lo-lo-lo close air support radius with four rocket pods, no external tanks	200 n miles (370 km; 230 miles)
Range: two PL-7 missiles and three 500 litre drop tanks	939 n miles (1,740 km; 1,081 miles)
self-ferry with one 800 litre and two 500 litre drop tanks, no missiles	1,203 n miles (2,230 km; 1,385 miles)
g limit	+8

PERFORMANCE (J-7C at normal T-O weight):

Max operating Mach No. (MMO)	2.1
Unstick speed with afterburning	173 kt (320 km/h; 199 mph)
Touchdown speed with flap blowing	146 kt (270 km/h; 168 mph)
Min level flight speed	140 kt (260 km/h; 162 mph)
Max rate of climb at S/L	9,000 m (29,525 ft)/min
Service ceiling	18,000 m (59,060 ft)
Acceleration from M1.2 to M1.9 at 13,000 m (42,660 ft)	3 min 27 s
Air turning radius at 5,000 m (16,400 ft) at M1.2	5,093 m (16,710 ft)
T-O run with afterburning	800 m (2,625 ft)
Landing run with flap blowing, drag-chute and brakes deployed	550 m (1,805 ft)
Range: on internal fuel	518 n miles (960 km; 596 miles)
with 800 litre belly tank	701 n miles (1,300 km; 807 miles)

with 800 litre belly tank and two 500 litre
underwing tanks ..... 1,025 n miles (1,900 km; 1,180 miles)
g limits: up to M0.8 ..... +8.5
above M0.8 ..... +7

PERFORMANCE (F-7MG):
Max operating Mach No. (MMO) 2.0
Max level speed ..... 648 kt (1,200 km/h; 745 mph) IAS
Min level speed ..... 114 kt (210 km/h; 131 mph) IAS
Max rate of climb at S/L ..... 11,700 m (38,386 ft)/min
Max instantaneous turn rate ..... 25.2°/s
Sustained turn rate: at 1,000 m (3,280 ft) ..... 16°/s
at 5,000 m (16,400 ft) ..... 11°/s
at 8,000 m (26,240 ft) ..... 8°/s
Service ceiling ..... 17,500 m (57,420 ft)
Theoretical ceiling ..... 18,000 m (59,060 ft)
T-O run, with afterburning, and landing run ..... 600–700 m (1,970–2,300 ft)
Operational radius:
air superiority (hi-hi-hi) with two AIM-9P AAMs and three 500 litre drop tanks, incl 5 min combat with afterburner ..... 459 n miles (850 km; 528 miles)
air-to-ground attack (lo-lo-hi) with two Mk 82 bombs and two 500 litre drop tanks ..... 297 n miles (550 km; 342 miles)
Ferry range ..... 1,187 n miles (2,200 km; 1,367 miles)
g limits ..... +8/−3

UPDATED

CAC FC-1 XIAOLONG

English name: Fierce Dragon
Pakistan Air Force designation: JF-17 Thunder

TYPE: Attack fighter.

PROGRAMME: Fighter China (FC) programme launched 1991 following cancellation of US participation in development of Chengdu Super-7. Some design assistance from MiG OKB, possibly based on (then-designated MiG-33) mid-1980s project for single-engined variant of the MiG-29. (Sources at MiG experimental bureau quoted as saying that FC-1 was designed there to a military specification as Izd (*Izdeliye*: article) 33 and later offered for Chinese production following cancellation of Russian requirement.)

Full-scale mockup completed 2001 and exhibited at Zhuhai November 2002. First five prototypes (including 02 for static and 05 for fatigue test) under construction by then, with oft-postponed maiden flight targeted for mid-2003. Formal approval for production, at CAC's No. 132 factory, reported in January 2003. Avionics competition still open in 2003–04; contenders include IAI Elta, SAGEM and Thales. Choice of a multimode pulse Doppler radar apparently still lies between Phazotron of Russia (Kopyo), Galileo (FIAR) of Italy (Grifo S-7), and Thales (RC 400). Selection had not been made by early 2004, due to Western misgivings about technology transfer, although Pakistan which renamed Super-7 as JF-17 (Joint Fighter) Thunder in 2003, expected to opt for Grifo; China may look elsewhere for domestic version if clearance is much further delayed.

Chinese domestic requirement for the FC-1 is said to be for several hundred aircraft, with Pakistan wanting about 150 to 200, but a Pakistani order is conditional upon AFPLA commitment, which was still awaited in late 2003; however, PLA Navy still said to be interested. Pakistan reportedly to receive 12 from 2004 for familiarisation training. A letter of intent for joint development by China and Pakistan was reportedly signed in February 1998, followed by signature of a joint development and production agreement in June 1999.

First FC-1 rolled out 31 May 2003 and began low-speed taxying trials 27 June, followed by eight-minute maiden flight (first of two that day) on 25 August, and first 'official' flight, by then also bearing JF-17 markings, 3 September 2003, all at Chengdu's Wenjiang Airport.

Prototype FC-1 following repainting ***NEW***/0576621

FC-1 prototype wearing its alternative designation ***NEW***/0576622

Second flying FC-1/J-17, the blue/grey-camouflaged '03' ***(Pakistan Air Force)*** ***NEW***/0563928

Pakistan news reports at that time indicated following timetable for that country's participation: finalise MoU September 2003; sign contract December 2003; low-rate initial production by Pakistan Aeronautical Complex from June 2004 to December 2005; begin full rate production January 2006.

Maiden flight of second flight test prototype (03) occurred on 9 April 2004. Pakistan Air Force test pilots flew first prototype for first time two days earlier.

CUSTOMERS: Envisaged for air forces of China and Pakistan initially, but costed to be competitive in wider export market; promoted to Bangladesh, Egypt and Nigeria in 2003. Seen as potential replacement for Shenyang J-6, Chengdu J-7, Nanchang Q-5, Northrop F-5 and Dassault Mirage III/5.

Preproduction batch of 16 to be built (eight for each partner country), of which PAC will assemble four of those for each customer.

COSTS: Development investment of some US$150 million by mid-2001.

DESIGN FEATURES: Agile light fighter. Mid-mounted delta wing with narrow wingroot strakes at leading-edge; leading-edge manoeuvring flaps; single turbofan engine; side-mounted twin intakes, with splitter plates; large intake trunks provide space for considerable internal fuel capacity. Large main fin with dorsal fairing; two smaller, uncanted ventral fins.

FLYING CONTROLS: Conventional hydraulic servo-operated control of ailerons, rudder and all-moving tailplane initially, with single analogue fly-by-wire system for back-up; provision for FBW to become primary system later. Trailing-edge and leading-edge flaps.

STRUCTURE: Primary structure conventional aluminium and steel alloy semi-monocoque. Some components may be manufactured in Pakistan.

LANDING GEAR: Retractable tricycle type, with single wheel and oleo shock-absorber on each unit. Mainwheels retract upward into engine intake trunks; nosewheel retracts rearward.

FC-1 prototype under construction ***NEW***/0576620

Underside of Chengdu FC-1 *(Pakistan Air Force)* ***NEW***/0563929

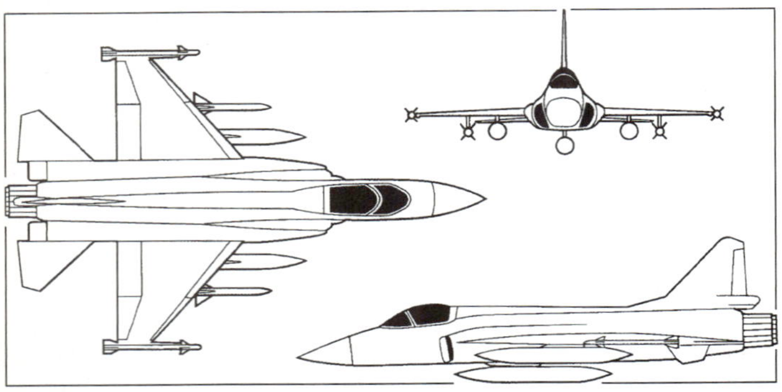

Chengdu FC-1 Xiaolong (Klimov RD-93 turbofan) ***NEW***/0576478

POWER PLANT: One Klimov RD-93 (RD-33 derivative) turbofan (81.4 kN; 18,298 lb st with afterburning), possibly to be licence built by Liyang Machinery Corporation (LMC) for production aircraft. Could have alternative Western engine at customer's option, as China reportedly does not have rights to re-export RD-93. Substantial internal fuel capacity. Provision for external fuel tanks.

ACCOMMODATION: Single Jianghan TY6 zero/zero ejection seat under one-piece canopy. Two-seat training version also planned.

AVIONICS: Chinese avionics in prototype include ring laser gyro INS/GPS, IR search and track system, HUD and NVG-capable instruments. Western avionics being sought for production aircraft. Prototype(s) reportedly to have IAI Elta EL/M-2032 fire-control radar; 04 will be avionics and weapon system testbed.

ARMAMENT: Underfuselage centreline station for 23 mm GSh-23-2 twin-barrel cannon or other store; two attachments under each wing and one at each wingtip. Weapons expected to include advanced AAMs, ASMs, bombs, gun and rocket pods, or other stores. Mockup seen in November 2002 mounting the AMRAAM-class LETRI (No. 607 Institute) SD-10 (PL-12) BVR, active radar homing AAM (lock-on range 20 km; 12.4 miles).

DIMENSIONS, EXTERNAL:
Wing span, to c/l of AAMs ... 8.50 m (27 ft 10¾ in)
Length overall ... 14.00 m (45 ft 11¼ in)
Height overall ... 5.10 m (16 ft 8¾ in)
Wheel track ... 2.30 m (7 ft 6½ in)
Wheelbase ... 5.14 m (16 ft 10¼ in)

AREAS:
Wings, gross ... 24.40 m²(262.6 sq ft)

WEIGHTS AND LOADINGS:
Max external stores load ... 3,720 kg (8,201 lb)
Max internal fuel weight ... 2,300 kg (5,071 lb)
T-O weight:
normal (wingtip AAMs only) ... 9,100 kg (20,062 lb)
max ... 12,700 kg (27,998 lb)
Max wing loading ... 520.5 kg/m² (106.61 lb/sq ft)
Max power loading ... 156 kg/kN (1.53 lb/lb st)

PERFORMANCE (estimated):
Max level speed at altitude, clean ... M1.6
Service ceiling ... 16,700 m (54,800 ft)
T-O run ... 500 m (1,640 ft)
Landing run ... 700 m (2,300 ft)
Combat radius:
fighter ... 648 n miles (1,200 km; 745 miles)
ground attack ... 378 n miles (700 km; 435 miles)
Max range on internal fuel ... 971 n miles (1,800 km; 1,118 miles)
Max ferry range ... 1,619 n miles (3,000 km; 1,864 miles)
g limit ... +8

UPDATED

CAC J-10

English name: Vigorous Dragon
Chinese name: Jianjiji-10 (Fighter aircraft 10)
Westernised designation: F-10

TYPE: Multirole fighter.

PROGRAMME: Reports of the existence of this new Chinese fighter began to emerge in October 1994, following the detection by a US intelligence satellite of a prototype at Chengdu. Said to be in the weight and performance class of the Eurofighter Typhoon and Dassault Rafale, the J-10 bears a close external resemblance to the cancelled IAI Lavi of the late 1980s, despite Israeli government statements that fears of unauthorised technology transfer are unfounded. A photograph, released in 1996 by the People's Liberation Army, of a wind tunnel model of the J-10 showed it to be outwardly identical to the Lavi in all essential respects, apart from slightly raised foreplanes and the addition of wingroot trailing-edge extensions.

This was confirmed in January 2001 when an unauthorised source posted a photograph of prototype 1001 on the Worldwide Web (although it was rapidly removed). The photograph showed the artist's impression first published in the 1997–98 edition to be accurate in all respects, except that it was larger, and fin chord was some 20 per cent greater from root to tip.

It now appears that go-ahead for what became the J-10 was given in September 1988, development by No. 611 Research Institute beginning in the following month. An all-metal mockup was completed by late 1993, but poor performance predictions, coupled with a requirement change from air superiority to multirole, caused some redesign and consequent delays in the programme.

In late 1995, Russian sources indicated that the first flight, then expected during the early part of 1996, would be powered by a single AL-31FN turbofan; first prototype 1001 is said to have achieved this milestone in mid-1996. Russian avionics manufacturer Phazotron offered its Zhuk (Beetle) radar (already selected for the F-8 IIM upgrade programme) and the more capable Zhemchug (Pearls) and RP-35 as alternatives to the Elta EL/M-2035 or a derivative.

In late 1997, the slightly modified prototype 1002 was lost in a fatal accident. Two further prototypes (1003 and 1004) had been completed, and designation J-10 bestowed, during that year; the officially announced 'first' flight date of 23 March 1998 apparently referred to resumption of flight testing by one of these aircraft. According to China-based sources, 10 AL-31FN engines had been imported by end 1997 and four J-10 prototypes completed. An unofficial CAC source in mid-1999 stated that two prototypes (evidently 1001 and 1003) were then flying, with four others undergoing static test, still being assembled or only just completed; of these, 1004 and/or 1005 may have been static/fatigue test aircraft. Chinese sources at Zhuhai in November 2000 stated that between five and eight J-10s then built and more than 140 test flights made; nine aircraft identified by mid-2002, of which 1007 to 1009 may be further prototypes or preproduction aircraft, one having made a 'first' flight on 28 June 2002. Service entry should be in about 2005. Meanwhile, unconfirmed reports in late 2002/early 2003 indicated that 10 J-10s had been deployed with the Nanjing Military Region in August 2002 for operational evaluation or familiarisation training; and that manufacture of the first two two-seat J-10B examples would begin in 2003 and would have enhanced air-to-ground and maritime attack capability. Maiden flight of two-seater reported as occurring on 26 December 2003, with photographs appearing on Internet at that time. First (preproduction or serial production) J-10As delivered to AFPLA training unit on 23 February 2003, according to Chinese Xinhua news agency. Serial numbers 1015 and 1016 seen during that year.

Prototype CAC J-10B two-seat version undergoing manufacturer's testing in 2004 ***NEW***/1043423

CAC J-10 with a full fit of external fuel tanks, apparently the 38th production aircraft pictured before delivery ***NEW***/1043422

Model of Chengdu's proposed twin-engined stealth development of the J-10 *(AVIC I via Y Chang)* ***NEW***/0531455

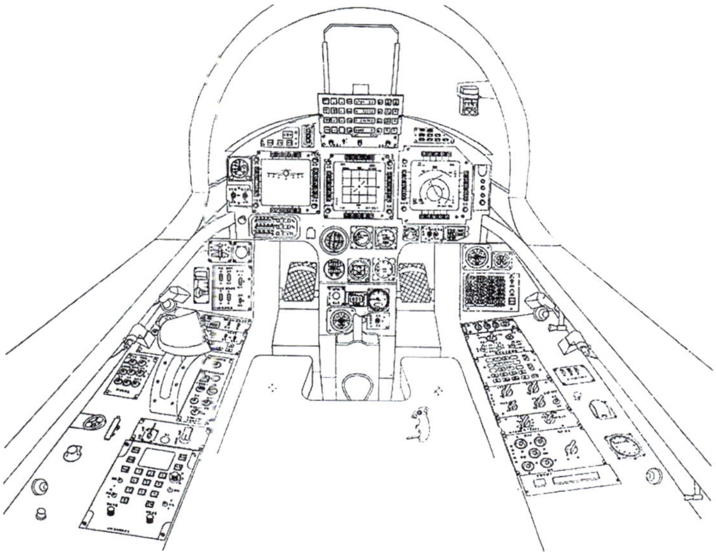

Uncorroborated drawing depicting possible J-10 cockpit layout 0096948

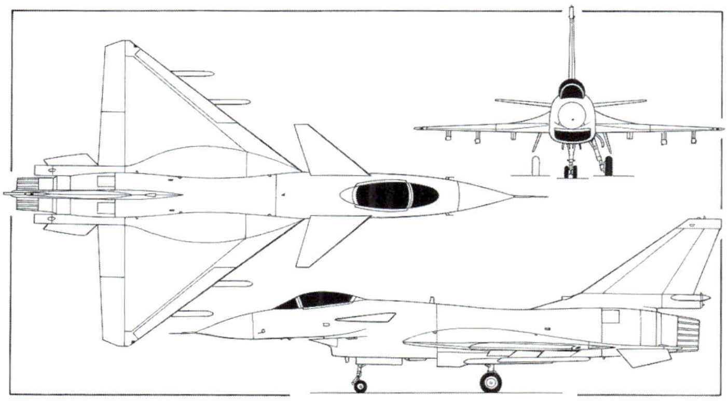

Provisional general arrangement of J-10 fighter *(Michael Badrocke)* 0137950

CAC J-10 '1013' ***NEW***/0573880

The J-10 failed to make its widely predicted public debut at Zhuhai in November 2002, but it emerged shortly after the show that CAC and the No. 611 Institute had completed conceptual design work on two advanced variants of the J-10: one single- and one twin-engined, both having twin vertical tailfins and embodying stealth characteristics. Both also featured a redesigned and more angular nose section. It has also been noted that in 2002 the *China People's Daily* referred to the J-10 as *Qiang Shi* (Attack 10), rather than the *Jian* (Fighter) title that might have been expected. This appears to support the belief that the aircraft is viewed as a J-7/Q-5 replacement rather than a rival to the Su-27/J-11 to replace the Shenyang J-8.

CUSTOMERS: Reports have suggested an AFPLA requirement for up to 300, but achievement of this may depend upon China's progress with licensed manufacture of the Sukhoi Su-27 (see entry for Shenyang (SAC) J-11).

DESIGN FEATURES: Tail-less delta wing and close-coupled foreplanes; single sweptback vertical tail with twin outward-canted ventral fins; single, rectangular ventral engine air intake.

FLYING CONTROLS: All-moving foreplanes; inboard and outboard elevons; single-piece rudder; wing leading-edge manoeuvring flaps.

LANDING GEAR: Retractable tricycle type. Main units retract forwards, twin-wheel nose unit rearwards.

POWER PLANT: One Saturn/Lyulka AL-31FN turbofan (79.4 kN; 17,857 lb st dry and 122.6 kN; 27,558 lb st with afterburning) in prototypes and initial production aircraft; total 54 engines delivered by MMPP Salyut 2001–2003, according to Moscow-based Jane's source; contract completed in January 2004. Non-approval of manufacturing licence for this engine indicates possibility of change to domestic (WS10A?) power plant in later aircraft.

Fuel capacity reported to be in region of 3,175 litres (839 US gallons; 698.5 Imp gallons) in wings and 1,775 litres (469 US gallons; 390.5 Imp gallons in fuselage), giving total internal capacity of 4,950 litres (1,308 US gallons; 1,089 Imp gallons). Provision for auxiliary fuel tanks on centreline and inboard underwing stations.

ACCOMMODATION: Pilot only, on zero/zero ejection seat (apparently Martin-Baker in prototypes; possibly domestic Jianghan TY6 planned later).

AVIONICS: Reportedly equipped with quadruplex digital fly-by-wire FCS, GPS/INS navigation, HOTAS controls, a wide field of view HUD, one colour and two monochrome liquid crystal MFDs (weapons, RWR/navigation and radar information), air data computer, 1553B-standard databus and a helmet-mounted weapon sight. FCS based on active control technology developed in F-8 II ACT testbed. Pulse Doppler radar apparently yet to be selected; quoted candidates include Phazotron Zhuk or Zhemchug, domestic (LETRI) JL-10A, Elta EL/M-2035 and FIAR Grifo 2000. Prototypes and initial production J-10As reported to have domestic Type 1473 radar.

ARMAMENT: Recently released photographs indicate 11 external stores points, including one on centreline, tandem pairs on fuselage sides and three under each wing, the outboard wing stations each carrying PL-8 (Python 3) or later AAMs such as PL-11 or PL-12. Other potential weapons could include Vympel R-73 and R-77 AAMs; C-801 or C-802 ASMs; YJ-8K (anti-ship) or YJ-9 (anti-radiation) missiles; and laser-guided or free-fall bombs. Internally mounted 23 mm cannon on port side of nosewheel; provision for a Chinese-developed (Luoyang Filat?) IR/laser navigation and targeting pod.

Following estimated data have not been confirmed.

DIMENSIONS, EXTERNAL:
Wing span .... 8.78 m (28 ft 9¾ in)
Length overall .... 14.57 m (47 ft 9½ in)
Height overall .... 4.78 m (15 ft 8¼ in)

AREAS:
Wings, gross .... 33.1 m² (356.3 sq ft)
Foreplanes (total) .... 5.45 m² (58.66 sq ft)

WEIGHTS AND LOADINGS:
Weight empty .... 9,750 kg (21,495 lb)
Max internal fuel .... 4,500 kg (9,921 lb)
Max external stores load .... 4,500 kg (9,921 lb)
Max T-O weight with external stores .... 18,500 kg (40,785 lb)

PERFORMANCE:
Max level speed: at altitude M1.85
at S/L M1.2
Service ceiling .... 18,000 m (59,050 ft)
T-O run, clean .... 350 m (1,150 ft)
Combat radius .... 250–300 n miles (463–555 km; 287–345 miles)
Ferry range with max internal and external fuel ... 1,000 n miles (1,850 km; 1,150 miles)
g limits .... +9/–3

UPDATED

CHAIG

CHANGHE AIRCRAFT INDUSTRIES GROUP (CHANGHE FEIJI GONGYE GONGSI) (SUBSIDIARY OF AVIC II)

PO Box 109, Jingdezhen, Jiangxi 333002
Tel: (+86 798) 844 20 19 and 844 24 30
Fax: (+86 798) 844 71 69 and 844 14 60
Web: http://www.changhe.com
PRESIDENT: Yang Jinhua

CHAIG (formerly Changhe Aircraft Factory), which occupies a 433 ha (1,070 acre) site at Jingdezhen, is one of the 500 largest industries in China. It has a workforce of more than 10,000 and began producing coaches, minivans and other commercial road vehicles in 1974. These and other automotive products still account for much of output, but batch-produced helicopters have included the Z-8 and Z-11.

CHAIG is responsible for manufacture of the tailcone, vertical fin and horizontal stabiliser of the Sikorsky S-92 helicopter. The tail for the first S-92 was delivered to Sikorsky in May 1997. Changhe is also thought likely to become the centre for Chinese licensed assembly and/or manufacture of the AgustaWestland A 109E Power, which was the subject of Sino-Italian negotiations in 2002–03.

UPDATED

CHAIG Z-8

Chinese name: Zhishengji-8 (Vertical take-off aircraft 8)

TYPE: Naval combat helicopter.

PROGRAMME: Design work begun 1976, but suspended from 1979 to mid-1984; initial flight of first prototype 11 December 1985, second prototype October 1987; domestic type approval awarded 8 April 1989; first Z-8 handed over to PLA Naval Aviation for service trials 5 August 1989; initial production approved; final design approval granted 12 November 1994. Applications include troop transport, ASW/ASV, search and rescue, minelaying/sweeping, aerial survey and firefighting.

CURRENT VERSIONS: **Z-8:** Standard version, *as described.*

Z-8A: Designation of upgraded tactical transport version, with Turbomeca Makila 2A engines matched to Turmo gearbox. Two reportedly delivered to PLA Army Aviation for evaluation in 2001; camouflaged; lack nose radome and side-mounted floats. Reports indicate that production deliveries began in November 2002, with about seven in service by late 2003.

Z-8F: Improved and more powerful version, for which the 1,380 kW (1,850 shp) P&WC PT6B-67A turboshaft was selected in November 2002. First flight expected in second quarter 2004; certification, as 30-passenger civil transport, targeted for 2007. Additional 671 kW (900 shp) of power expected to enhance 'hot-and-high' performance, including increase in service ceiling to 4,700 m (15,420 ft) at MTOW and increased payload capacity. Other improvements said to include new (composites) main rotor blades, with anti-icing; new avionics and mission equipment. Duties envisaged are search and rescue, general utility missions and logistics support.

CUSTOMERS: Up to 20 delivered to PLA Navy by end of 1999; upgraded versions still being promoted by CATIC in 2002–03. Sole operating naval unit is Shipborne Helicopter Group within East Sea Fleet at Dachang (Shanghai).

Changhe Z-8 in service with PLA Naval Aviation 0525833

DESIGN FEATURES: Chinese equivalent of Aerospatiale Super Frelon, of which 16 supplied to PLA Navy in 1977–78. Six-blade main rotor and five-blade tail rotor; boat-hull fuselage with watertight compartments inside planing bottom; stabilising float at rear each side, attached to small stub-wing; small, strut-braced fixed horizontal stabiliser on starboard side of tail rotor pylon. Search radar in nose 'thimble' on SAR version.

FLYING CONTROLS: Pitch control fitting at root of each main rotor blade; drag and flapping hinges for each blade mounted on rotor head starplates; each main blade also has a hydraulic drag damper. Fully redundant flight control system, with Dong Fang KJ-8 autopilot.

STRUCTURE: Stressed skin metal fuselage, with riveted watertight compartments; gearboxes manufactured by Zhongnan Transmission Machinery Factory.

LANDING GEAR: Non-retractable tricycle type, with twin wheels and low-pressure oleo-pneumatic shock-absorber on each unit. Small tripod tailskid under rear of tailboom. Boat hull and side floats permit emergency water landings and take-offs.

POWER PLANT: Three Changzhou (CLXMW) WZ6 turboshafts, each with maximum emergency rating of 1,156 kW (1,550 shp) and 20 per cent power reserve at S/L, ISA. Two engines side by side in front of main rotor shaft and one aft of shaft. Transmission rated at 3,072 kW (4,120 shp).

Standard internal fuel capacity 3,900 litres (1,030 US gallons; 858 Imp gallons), in flexible tanks under floor of centre-fuselage. Auxiliary fuel tanks can be carried inside cabin for extended-range or self-ferry missions, increasing total capacity to 5,800 litres (1,532 US gallons; 1,276 Imp gallons).

ACCOMMODATION: Crew of two or three on flight deck. Accommodation in main cabin for up to 27 fully armed troops, or 39 without equipment; up to 15 stretchers and a medical attendant in ambulance configuration; a BJ-212 Jeep-type vehicle and its crew; or other configurations according to mission. Entire accommodation heated, ventilated, soundproofed and vibration-proofed. Forward-opening crew door on each side of flight deck. Rearward-sliding door at front of cabin on starboard side. Hydraulically actuated rear-loading ramp/door.

EQUIPMENT: Equipment for SAR role can include 275 kg (606 lb) capacity hydraulic rescue hoist and two five-person liferafts. Can also be equipped with sonar, sonobuoys, search radar, or equipment for oceanography, geological survey and forest firefighting.

ARMAMENT: Can be equipped with Yu-7 torpedoes, anti-shipping missiles, or gear for minelaying (eight 250 kg mines) or minesweeping.

DIMENSIONS, EXTERNAL:
- Main rotor diameter ... 18.90 m (62 ft 0 in)
- Tail rotor diameter ... 4.00 m (13 ft 1½ in)
- Length overall, rotors turning ... 23.035 m (75 ft 7 in)
- Height overall, rotors turning ... 6.66 m (21 ft 10¼ in)
- Width over main gear sponsons ... 5.20 m (17 ft 0¾ in)

AREAS:
- Main rotor blades (each) ... 5.10 m^2 (54.90 sq ft)
- Tail rotor blades (each) ... 0.56 m^2 (6.03 sq ft)
- Main rotor disc ... 280.48 m^2 (3,019.1 sq ft)
- Tail rotor disc ... 12.57 m^2 (135.3 sq ft)

WEIGHTS AND LOADINGS:
- Manufacturer's weight empty ... 6,980 kg (15,388 lb)
- Weight empty, equipped ... 7,550 kg (16,645 lb)
- Max cargo payload: internal ... 4,000 kg (8,818 lb)
  - on external sling ... 5,000 kg (11,023 lb)
- Max T-O weight: standard fuel ... 10,592 kg (23,351 lb)
  - with auxiliary fuel ... 13,000 kg (28,660 lb)
- Max disc loading:
  - standard fuel ... 37.8 kg/m^2 (7.73 lb/sq ft)
  - auxiliary fuel ... 46.35 kg/m^2 (9.49 lb/sq ft)
- Transmission loading at max T-O weight and power:
  - standard fuel ... 3.45 kg/kW (5.67 lb/shp)
  - auxiliary fuel ... 4.23 kg/kW (6.96 lb/shp)

PERFORMANCE (A: at T-O weight of 9,000 kg; 19,841 lb, B: at 11,000 kg; 24,251 lb, C: at 13,000 kg; 28,660 lb):
- Never-exceed speed (VNE):
  - A ... 170 kt (315 km/h; 195 mph)
  - B ... 159 kt (296 km/h; 183 mph)
  - C ... 148 kt (275 km/h; 170 mph)
- Max cruising speed: A ... 143 kt (266 km/h; 165 mph)
  - B ... 140 kt (260 km/h; 161 mph)
  - C ... 134 kt (248 km/h; 154 mph)
- Econ cruising speed: A ... 137 kt (255 km/h; 158 mph)
  - B ... 132 kt (246 km/h; 153 mph)
  - C ... 125 kt (232 km/h; 144 mph)
- Rate of climb at S/L (15° 30' collective pitch, OEI):
  - A ... 690 m (2,263 ft)/min
  - B ... 552 m (1,811 ft)/min
  - C ... 396 m (1,299 ft)/min
- Service ceiling: A ... 6,000 m (19,680 ft)
  - B ... 4,900 m (16,080 ft)
  - C ... 3,050 m (10,000 ft)
- Hovering ceiling IGE: A ... 5,500 m (18,040 ft)
  - B ... 3,600 m (11,800 ft)
  - C ... 1,900 m (6,240 ft)
- Hovering ceiling OGE: A ... 4,400 m (14,440 ft)
  - B ... 2,300 m (7,540 ft)
- Range with max standard or auxiliary fuel, OEI, no reserves: A ... 232 n miles (430 km; 267 miles)
  - B ... 442 n miles (820 km; 509 miles)
  - C ... 431 n miles (800 km; 497 miles)
- Ferry range with auxiliary fuel tanks, OEI, no reserves:
  - C ... 755 n miles (1,400 km; 870 miles)
- Endurance with max standard or auxiliary fuel, OEI, no reserves: A ... 2 h 31 min
  - B ... 4 h 43 min
  - C ... 4 h 10 min

UPDATED

CHAIG Z-11

Chinese name: Zhishengji-11 (Vertical take-off aircraft 11)

TYPE: Light utility helicopter.

PROGRAMME: First details officially released at China Air Show in Zhuhai November 1996, together with photographs showing one or two Z-11s in flight. Development by Chinese Helicopter R&D Institute (CHRDI) began in 1991; quoted first flight date of 22 December 1994 thought to refer to re-engined modification of two second-hand ex-US AStars (Ecureuils) acquired earlier that year. In early 1997, a Chinese government agency announced that the Z-11 had flown for the first time on 26 December 1996.

The Z-11 appears identical externally to Eurocopter AS 350B Ecureuil except for nose contours, but CHAIG publicity in mid-2001 stated that "China owns its independent intellectual property rights". Eurocopter (which sold eight Ecureuils to China in 1996) declines to comment on its provenance. Project reportedly launched in 1989, with development work beginning in 1992. Technical appraisal was completed in 1996, and small batch production began in 1997, first customer deliveries being made in September 1998. Test programme included flights totalling 719 n miles (1,332 km; 827 miles) in temperatures from −43 to +6°C (−41.7 to 42.8°F), including a 2-hour sortie cruising at −38°C (−38.9°F). Design was finalised in December 2000, and the Z-11 received CAAC certification in April 2001. CAAC approval for series production for civilian use was announced on 23 December 2002. Trials of an armed version, equipped with cannon, ATMs, unguided rockets and a roof-mounted low-light/infra-red sight, were reported to be under small-batch evaluation in late 2003.

A twin-engined version is planned, for which the Rolls-Royce 250 and Turbomeca Arrius 1A were in contention in 2002–03.

CURRENT VERSIONS: **Civil:** Applications include powerline inspection and maintenance; geological survey; forest fire protection; emergency medical service; aerial seeding and crop-spraying; touring and sightseeing.

Changhe Z-11 (27th of the type) in service with a PLA Aviation flying school ***NEW***/0573881

Z-11MB1 (Arriel 2B1A turboshaft), first flown on 7 March 2003 ***(Turbomeca)*** ***NEW***/0552271

Model of Z-11 observation helicopter with roof-mounted sight ***(Robert Hewson)*** ***NEW***/0547105

Military: Quoted roles include reconnaissance; control and communications; frontier patrol; public security and law enforcement; executive transport; and crew training.

Z-11MB1: Latest version to be announced, which made its first flight on 7 March 2003 powered by a 632 kW (848 shp) Turbomeca Arriel 2B1A turboshaft. Changhe hoped to receive certification of this version by the end of 2003.

CUSTOMERS: Eight said to have been completed by late 1996, of which four had been delivered. All photographs released before late 1998 showed Z-11 in military camouflage with PLA insignia. One report from Aviation Expo China in October 1997 stated that the PLA had ordered 20 military Z-11s, these apparently going to Army Aviation's training school and 4th Army Aviation Regiment. Production amounted to 'several dozen' by mid-2001, at which time some 10,000 hours flown and 50,000 take-offs and landings. According to one estimate, about 59, including some armed examples, were in PLA Army Aviation service by late 2003.

Chongqing Three Gorges General Aviation Airlines (one ordered) reported as first civil customer in mid-2001; China Central Television of Jiangxi Province received one on 25 August 2002. One Z-11MB1 reportedly sold and delivered to (unidentified) Chinese customer by September 2003.

POWER PLANT: One 510 kW (684 shp) WZ8D turboshaft (licence-built Turbomeca Arriel 1D), reportedly produced by Liming engine factory.

DIMENSIONS, EXTERNAL:
Main rotor diameter ... 10.69 m (35 ft 0¾ in)
Tail rotor diameter ... 1.86 m (6 ft 1¼ in)
Length overall, main rotor turning ... 13.01 m (42 ft 8¼ in)
Fuselage: Length ... 11.24 m (36 ft 10½ in)
Max width ... 1.80 m (5 ft 10¾ in)
Height: over tailfin ... 3.02 m (9 ft 11 in)
to top of rotor hub ... 3.14 m (10 ft 3½ in)
Tailplane span ... 2.53 m (8 ft 3½ in)
Skid track ... 2.09 m (6 ft 10¼ in)
Fuselage/ground clearance beneath cabin ... 0.39 m (1 ft 3¼ in)

WEIGHTS AND LOADINGS:
Weight empty ... 1,120 kg (2,469 lb)
Fuel weight (standard) ... 423 kg (933 lb)
Max T-O weight ... 2,200 kg (4,850 lb)

PERFORMANCE:
Max level speed ... 150 kt (278 km/h; 172 mph)
Cruising speed ... 130 kt (240 km/h; 149 mph)
Hovering ceiling IGE ... 5,240 m (17,200 ft)
Hovering ceiling OGE:
Z-11 ... 4,500 m (14,760 ft)
Z-11MB1 ... 6,000 m (19,680 ft)
Max range ... 324 n miles (600 km; 372 miles)
Endurance ... 3 h 54 min

UPDATED

CHRDI

CHINESE HELICOPTER RESEARCH AND DEVELOPMENT INSTITUTE (ZHONGGUO ZHISHENGJI SHEJI YANGJIUSO) (SUBSIDIARY OF AVIC II)

Jingdezhen, Jiangxi
CHIEF DESIGNER: Wu Ximing

This institute (workforce approximately 2,000) is the overall design authority for Chinese indigenous helicopter programmes. Few details yet available of Z-10; other projects reported to include 10 tonne heavy-lift helicopter.

UPDATED

Computer graphic of medium transport helicopter believed to be the CMH/Z-10 *(Robert Hewson)* ***NEW***/0547112

CHRDI Z-10

Chinese name: Zhishengji-10 (Vertical take-off aircraft 10)

TYPE: Multirole medium helicopter.

PROGRAMME: Thought to have been initiated in about 1994; Nos. 602 and 608 Institutes reported to be taking part in airframe design. Eurocopter France became partner in programme 15 May 1997 with US$70 million to US$80 million, nine-year contract to assist in developing rotor system; joined 22 March 1999 by Agusta (now AgustaWestland) with contract (approximately Lit50 billion; US$30 million) to be responsible for transmission system and vibration analysis. Nomenclature of Chinese Medium Helicopter (CMH) introduced by AgustaWestland in July 2000; officially linked with Z-10 designation by Chinese sources by mid-2002. Reports suggest three transmission sets built in Italy and four, collaboratively, being assembled in China.

In size and weight class of Bell 412 and Sikorsky S-76 (MTOW variously reported as 5, 5.5 or 6 tonnes; 11,023 to 13,227 lb), with capacity for two crew and up to 14 passengers or troops. First flight said to have taken place in first half of 2003 (April quoted by one apparently reliable Chinese source), following start of ground tests in May 2002. Resumed flight testing early 2004 after unspecified modifications; second prototype also flown by that time. Flight testing planned to end in 2006; eventual production reportedly to be entrusted to Changhe.

DESIGN FEATURES: Intended to fulfil tactical military transport, attack helicopter and commercial transport roles. No design features yet officially confirmed, but indications suggest five-blade main and six-blade tail rotors. Intention of both military and civil use suggests that such aspects as composites construction, crashworthy airframe and seats, rotor system ballistic tolerance, engine run-dry capacity and ability to carry slung loads are likely, depending upon version.

POWER PLANT: Two turboshafts. Pratt & Whitney Canada 1,268 kW (1,700 shp) PT6C-67C selected in mid-2001 (10 ordered) to power prototypes; first set of these delivered by end of that year, and all 10 by early 2004. Production aircraft may be powered by domestic SAEC WZ9 turboshaft.

UPDATED

GAIC

GUIZHOU AVIATION INDUSTRY GROUP (GUIZHOU HANGKONG GONGYE GONGSI) (SUBSIDIARY OF AVIC I)

110 Jinjiang Road, Xiaohe District, Guiyang, Guizhou 550009
Tel: (+86 851) 831 72 31 and 831 72 12
Fax: (+86 851) 831 72 14
e-mail: gaig@gaic.com.cn
Web: http://www.gaic.com.cn
CHAIRMAN: Zhou Wanfeng
PRESIDENT: Zhang Jun
DIRECTOR OF PROGRAMMES: Hu Xupin

The Guizhou Aviation Industry Group incorporates many enterprises, factories and institutes engaged in various aerospace and non-aerospace activities; Group employment totals some 51,000 (2004), with assets of 10.6 billion yuan; aerospace workforce is about 6,000. Main aircraft manufacturing plants are named Longyan, Shuangyang and Yunma; Liyang is aero-engine producer. Aviation programmes include JL-9 fighter trainers, two series of turbojets, air-to-air missiles and rocket launchers, plus participation in Chengdu (CAC, which see) production of J-7/F-7. GAIC also manufactures maintenance jigs and tools for the Airbus airliner family.

In late 2003, GAIC was negotiating with Aeromot of Brazil with regard to licenced production of a retractable landing gear version of the AMT-600 Guri and possible later manufacture of AMT-200/300 Super Ximango motor gliders. A joint project for a four-seat lightplane was also mentioned.

UPDATED

Prototype JL-9 making its first flight on 13 December 2003 ***NEW***/0569775

GAIC JL-9 SHANYING AND CY-1

Export designation: FTC-2000
English name: Plateau Eagle

TYPE: Advanced jet trainer.

PROGRAMME: Revealed as FTC-2000, in model form at Airshow China, November 2000, and as full-size mockup two years later. Proposed as lead-in fighter trainer for CAC FC-1 (which see) and J-8D/F-8D, and as replacement for JJ-7/FT-7. Wind tunnel testing completed by late 2000. Three prototypes under construction by September 2003, first of which then nearing completion; maiden flight (c/n JL90001) 13 December 2003. Service entry targeted for 2005.

CURRENT VERSIONS: **JL-9:** *As described.*

CY-1/LFC-16: Modified version, designed by Beijing SuperWing Technology Research Institute and displayed in model form (marked CY-1) at November 2002 Airshow China in Zhuhai; would be produced by GAIC if go-ahead received. Features sweptback, shoulder-mounted foreplanes plus what were described as 'side-plate canards' comprising narrow rectangular surface each side of fuselage from just below foreplane trailing-edge to extreme tail, where there are small, rectangular, downward-canted tailfins with small leading-edge strakes. Same model shown in November 2004, but designation changed to **LFC-16**.

COSTS: Predicted at US$2.4 million, excluding radar.

DESIGN FEATURES: Upgraded derivative of JJ-7/FT-7, having J-7E double-delta wings, lateral intakes and 'solid' nose to permit installation of fire-control radar. Airframe otherwise similar to JJ-7/FT-7.

FLYING CONTROLS: Conventional mechanical (hydraulic actuation).

LANDING GEAR: Retractable tricycle type, actuated hydraulically.

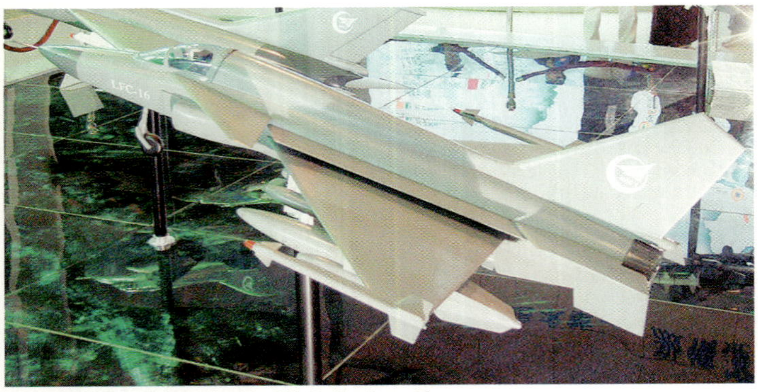

Model of LFC-16 (previously CY-1) displayed in 2004 *(Jane's/Robert Hewson)* ***NEW***/1047698

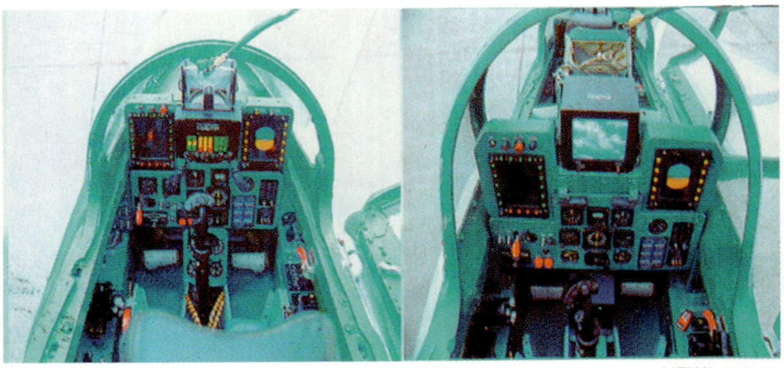

Front and rear cockpits of the GAIC JL-9 ***NEW***/0589391

GAIC JL-9 Shanying forward fuselage detail ***NEW***/0589392

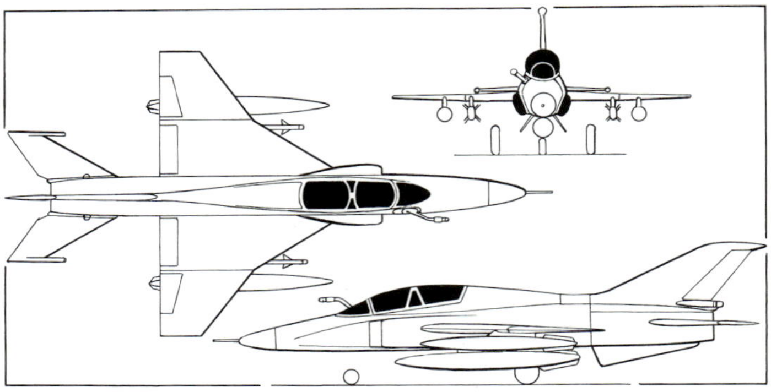

GAIC JL-9 Shanying *(James Goulding)* ***NEW***/0589197

Mainwheels retract inwards, nosewheel forwards.

POWER PLANT: One WP13F (C) turbojet (43.15 kN; 9,700 lb st dry, 63.25 kN; 14,220 lb st with afterburning). Gravity and pressure refuelling. In-flight refuelling probe (detachable) on starboard side of nose, common with that of J-8D/F-8D. Underfuselage and inboard underwing stations 'wet' for carriage of drop-tanks.

ACCOMMODATION: Crew of two in tandem, on TY6D zero/zero ejection seats. Individual starboard-hinged cockpit canopies; wraparound windscreen.

AVIONICS: *Comms:* Com radio, IFF and ATC transponder.

Radar: Pulse Doppler fire-control radar.

Flight: EFIS with HOTAS controls; GPS/INS.

Instrumentation: Two MFDs in each cockpit; HUD in front cockpit, with HUD repeater, rear; air data computer; mission computer; radio compass; radar altimeter; Tacan; marker beacon receiver; RWR; all linked via GJB289A databus.

Mission: Stores management system; video system.

ARMAMENT: Internal gun. Stores attachments under fuselage (one) and wings (two each side). Shown with PL-9 AAMs on outer wing stations and PL-8s inboard.

Following details all provisional:

DIMENSIONS, EXTERNAL:
Wing span ... 8.32 m (27 ft 3½ in)
Wing aspect ratio ... 2.6
Length overall ... 14.555 m (47 ft 9 in)
Height overall ... 4.105 m (13 ft 5½ in)
Wheel track ... 2.69 m (8 ft 10 in)
Wheelbase ... 4.87 m (15 ft 11¾ in)

AREAS:
Wings, gross ... 26.15 m² (281.5 sq ft)

WEIGHTS AND LOADINGS:
Fuel weight: internal ... 2,060 kg (4,542 lb)
external ... 1,302 kg (2,870 lb)
T-O weight: clean ... 7,800 kg (17,196 lb)
max ... 9,800 kg (21,605 lb)
Max wing loading ... 374.8 kg/m² (76.76 lb/sq ft)
Max power loading ... 155 kg/kN (1.52 lb/lb st)

PERFORMANCE (estimated):
Operating Mach number (MMO) ... 1.6
Max level speed ... 648 kt (1,200 km/h; 757 mph) IAS
Unstick speed ... 135–146 kt (250–270 km/h; 155–168 mph)
Touchdown speed ... 124–135 kt (230–250 km/h; 143–155 mph)
Min flying speed ... 114 kt (210 km/h; 131 mph)
Max rate of climb at S/L ... 15,600 m (51,181 ft)/min
Service ceiling ... 16,000 m (52,500 ft)
T-O distance ... 400–500 m (1,315–1,640 ft)
Landing distance ... 500–600 m (1,640–1,970 ft)
Range with max internal and external fuel ... 1,349 n miles (2,500 km; 1,553 miles)
Operational endurance ... 2 h

UPDATED

HAI

HAFEI AVIATION INDUSTRY COMPANY (HAFEI HANGKONG GONGYE GONGSI) (SUBSIDIARY OF AVIC II)

15 Youxie Street, Pingfang District, Harbin, Heilongjiang 150066
Tel: (+86 451) 650 11 22 and 653 07 36
Fax: (+86 451) 652 01 44
e-mail: hai@hafei.com
Web: http://www.hafei.com
CHAIRMAN: Wang Bin
PRESIDENT: Cui Xuewen
GENERAL MANAGER: Qu Jingwen
EXECUTIVE DEPUTY GENERAL MANAGER: Xu Zhanbin
PUBLIC RELATIONS OFFICER: Li Changjun

Established as Harbin Aircraft Manufacturing Corporation (Harbin Feiji Zhizao Gongsi: HAMC) 1952, subsequently producing H-5 light bomber (Soviet-designed Il-28) and Z-5 helicopter (Soviet-designed Mi-4) in large numbers, as well as smaller quantities of Chinese-designed SH-5 flying-boat and Y-11 agricultural light twin (NATO reporting name 'Chan-A'). Now a core company of Harbin Aircraft Industry Group (HAIG, which see), along with its sister company Hafei Motor. Occupies 514 ha (1,270 acre) site, including 350,000 m² (3,767,400 sq ft) of workshop space. Workforce in 1998 (latest figure provided) numbered approximately 18,000.

Currently producing own-design Y-12 utility light twin and licence-manufacturing Eurocopter Dauphin 2 as Z-9; also developing new 'Z-X' helicopter. Subcontract work includes Dauphin doors for Eurocopter.

HAI (24 per cent) is partnered by Eurocopter (61 per cent) and Singapore Technologies Aerospace (15 per cent) in the Eurocopter EC 120 Colibri programme, for which it was previously subcontracted to build the cabin module (200th shipset handed over 5 September 2000). This participation was enlarged by an agreement of 20 November 2003 under which HAI was to begin co-production and full local assembly of this helicopter, with a capability to complete 20 aircraft per year for the Chinese market. The official contract was signed in Paris on 11 June 2004, and the first Hafei-assembled EC 120 (Chinese designation **HC 120**) was expected to roll off the line by the end of that year; the Chinese government has reportedly ordered an initial batch of eight, with a further 50 on option. The agreement has potential to be extended to cover other Eurocopter types such as the AS 355 and EC 135.

In September 2003, HAI also concluded an agreement with Bell Helicopter Textron for sole-source manufacture of the cabin module, tailboom and other components for the Bell 430; these will be shipped to Canada for final assembly at Mirabel, Québec, starting in 2004.

UPDATED

HAI (EUROCOPTER) Z-9 HAITUN

Chinese name: Zhishengji-9 (Vertical take-off aircraft 9)
English name: Dolphin

TYPE: Light utility helicopter.

PROGRAMME: Licence-built Eurocopter AS 365N Dauphin 2 (which see in International section). Licence agreement (Aerospatiale/CATIC) signed 2 July 1980; first (French-built) example made initial acceptance flight in China 6 February 1982; Chinese parts manufacture began 1986; initial agreed batch of 50, last of which delivered January 1992. Production continuing under May 1988 domestic contract. Plans to introduce Arriel 2C announced mid-2001, following delivery of two of these engines in April; first flight achieved in September 2001.

CURRENT VERSIONS: **Z-9:** Initial Chinese version, equivalent to French AS 365 N1 and assembled from French-built kits; 28 completed. Used mainly as troop transport.

Z-9A: Follow-on kit-built version, to AS 365N2 standard. Final 22 of first 50 were of this version. Some equipped to carry machine gun pods or rocket launchers, but lack roof-mounted sight of more fully armed WZ-9 version.

HAI approached in early 2004 by Dirgantara with proposal to produce Z-9A in Indonesia to meet an Indonesian Army requirement.

Z-9A C²: Reported designation of command and control (command post) version for artillery fire direction.

Z-9A-100: Effectively, prototypes for domestic licence-built version, with WZ8A engines and much increased local manufacture (72.2 per cent of airframe and 91 per cent of engine). Two built; first flight 16 January 1992; flight test programme completed 20 November 1992 after almost 200 flight hours (408 flights); Chinese type approval received 30 December 1992.

Chinese Army Aviation WZ-9 armed with rocket pods and roof-mounted sight *(Jane's Robert Hewson)* ***NEW***/0105853

Z-9B: First indigenous production version, based on Z-9A-100. Modified Fenestron with 11 wider-chord, all-composites blades instead of 13 metal blades as in AS 365N1. Principal unarmed PLA version for SAR, artillery direction, EW, troop transport (accommodates eight), communications and other utility duties. Certified also for domestic commercial operations 19 April 2001.

Data apply to Z-9A except where indicated.

Z-9C: Version for PLA Naval Aviation (6th Naval Independent Regiment) for deployment aboard certain older classes of destroyers and frigates; known to be flying by late 2000. Believed to be equivalent to Arriel 2-engined Eurocopter AS 565MA Panther, but equipped with Thales HS-12 dipping sonar and KLC-1 I/J-band (Chinese version of Agrion 15) surface search radar; armament includes two Yu-7 torpedoes or TV-guided C-701 anti-surface vessel missiles. May have entered limited production by late 2003.

DZ-9: Reported designation of electronic warfare version (also reported as **Z-9EW**) for communications jamming.

WZ-9: Armed (*wuzhuang*) version (also reported as **Z-9W**); four Norinco HJ-8A (*Hongjian:* Red Arrow) wire-guided anti-tank missiles, twin 12.7 mm machine gun or 23 mm cannon pods, or twin 57 or 90 mm rocket pods, and gyrostabilised, roof-mounted optical sight. First flight thought to have been in early 1989; in production. Export designation **Z-9G**, available with or without roof-mounted sight.

A photograph shown on an Internet website has illustrated a model of a dedicated attack helicopter, clearly based on the Z-9 airframe but having a modified nose embodying a side-by-side two-person cockpit. Armament and equipment includes up to four HJ-8A anti-tank missiles or TY-90 IR-guided short-range AAMs, nose-mounted FLIR and roof-mounted optical sighting system.

H410A: Version with 635 kW (851 shp) Arriel 2C (WZ8C) turboshafts for improved 'hot-and-high' performance. (Designation signifies 4.10 tonne MTOW.) First flight September 2001; CAAC certification 10 July 2002. Initial orders for eight.

H425: Further-improved development of H410A; MTOW 4,250 kg (9,369 lb). Retains Arriel 2C engines, but features 200 mm (7¾ in) wider cabin, redesigned main rotor blade tips, Rockwell Collins 'glass cockpit' avionics, crashworthy fuel system and structure, and improved interior layout. First flight 30 December 2003; targeted to achieve certification by end of 2004.

H450: Projected (4.50 tonne MTOW) development of H425: further improvements to rotors, transmission and control system.

CUSTOMERS: Total 180 (all versions) reportedly built by late 2003; eight H410As ordered mid-2002 by Far Eastern Leasing Company, China State Oceanic Administration and Zhoushan Civil Aviation Development Company.

Most early production for Chinese armed services (People's Naval Aviation and Army Aviation) and, more recently, People's Armed Police (four delivered 12 September 2001). Entered service with two PLA army groups January and February 1988 (Beijing and Shenyang Military Regions respectively); units include 6th, 8th and 9th Army Aviation Regiments. Also with AFPLA unit in Hong Kong. According to one estimate in late 2003, Army then had some 61 Z-9/9As and 31 WZ-9s in service. People's Naval Aviation believed to use Z-9A for commando transport as well as shipboard communications. Ten based at former Royal Air Force airfield at Sek Kong in third quarter of 1997, following 1 July handover of Hong Kong by UK. First export made in late 2000 (two Z-9As to Mali Air Force). Mauritanian Air Force received two Z-9As in 2003; at least one other to Mali for military use.

Civil models used for various duties including offshore oil rig support and air ambulance (four stretchers/two seats or two stretchers/five seats). In 1992, Flying Dragon Aviation received two (late production Z-9s, augmenting an Aerospatiale Dauphin) which are operated on behalf of the Ministry of Forestry. Five civil Z-9s ordered by Shenzhen Financial Leasing Company (SFLC) in 2001. One Z-9A deployed to Arctic regions in mid-1999, on scientific survey ship.

STRUCTURE: Transmission manufactured by Dongan Engine Manufacturing Company at Harbin, hubs and tail rotor blades by Baoding Propeller Factory.

POWER PLANT: Two 526 kW (705 shp) Turbomeca Arriel 1C1 turboshafts, produced by SAEC at Zhuzhou as WZ8A. From 2001, H410A and H425 powered by Arriel 2Cs as WZ8C. Fuel capacity 1,140 litres (301 US gallons; 251 Imp gallons). Option for 180 litre (47.5 US gallon; 39.6 Imp gallon) auxiliary tank.

ARMAMENT (WZ-9): Up to four HJ-8A or HJ-8E ATMs (range 1.6 n miles; 3 km; 1.9 miles); twin 12.7 mm or 23 mm gun pods; or two pods of 57 or 90 mm rockets. Possible other weapons include TY-90 IR-guided AAM (already test-flown on Z-9 in 1998; range of 3.2 n miles; 6 km; 3.7 miles) and C-701 TV-guided anti-ship missile (8.1 n miles; 15 km; 9.3 miles).

WEIGHTS AND LOADINGS (A: Z-9A, B: Z-9G):

Weight empty, equipped: A ... 2,050 kg (4,519 lb)
B ... 2,462 kg (5,428 lb)
Standard fuel weight: A ... 900 kg (1,984 lb)
Max payload: A ... 2,038 kg (4,493 lb)
Max load on cargo sling: A ... 1,600 kg (3,527 lb)
Max T-O weight, internal or external load:
A, B ... 4,100 kg (9,039 lb)

PERFORMANCE (A and B as above):

Never-exceed speed (VNE) ... 170 kt (315 km/h; 195 mph)
Max cruising speed at S/L: A ... 151 kt (280 km/h; 174 mph)
B ... 138 kt (255 km/h; 158 mph)
Max vertical rate of climb at S/L:
A ... 252 m (827 ft)/min
B ... more than 240 m (787 ft)/min
Max forward rate of climb at S/L:
A ... 396 m (1,299 ft)/min
B ... more than 480 m (1,575 ft)/min
Service ceiling: Z-9A ... 4,570 m (15,000 ft)
WZ-9 ... 5,000 m (16,400 ft)
Hovering ceiling:
IGE: A ... 2,150 m (7,050 ft)
B ... 3,400 m (11,150 ft)
OGE: A ... 1,150 m (3,770 ft)
B ... 2,500 m (8,200 ft)
Max range at 135 kt (250 km/h; 155 mph) normal cruising speed, no reserves:
standard tanks:
A, B ... 464 n miles (860 km; 534 miles)
with auxiliary tank:
A ... 539 n miles (1,000 km; 621 miles)
Endurance: B ... 4 h 20 min

UPDATED

HAI 'Z-X'

TYPE: Light utility helicopter.

PROGRAMME: Hafei is understood to be developing a new light utility helicopter. No other details could be confirmed at the time of closing for press.

UPDATED

HAI Y-12

Chinese name: Yunshuji-12 (Transport aircraft 12)
Export marketing name: Twin Panda

TYPE: Twin-turboprop light transport.

PROGRAMME: Initiated as uprated version of Y-12 (I); first flight 16 August 1984. Y-12 (IV) (B-569L, first flight 30 August 1993) received production approval September 1995.

CURRENT VERSIONS: **Y-12 (I):** Initial version (first flight 14 July 1982), with PT6A-11 engines; three prototypes and approximately 30 production examples built.

Y-12 (II): Major production version; higher-rated engines, no leading-edge slats and smaller ventral fin. Certified by CAAC 25 December 1985 and UK CAA (BCAR Section K) 20 June 1990.

Y-12 (IV) (two P&WC PT6A-27 turboprops) 0106481

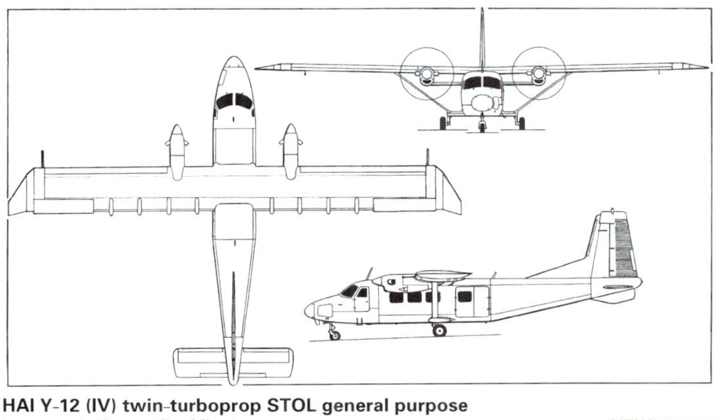

HAI Y-12 (IV) twin-turboprop STOL general purpose transport *(James Goulding)* ***NEW***/0589196

Y12 (I) AND Y-12 (II) DELIVERIES (AT MID-2002)

Country	Operator	Quantity	First Aircraft
Domestic (mostly Y-12 (I))			
China	AVIC (No. 630 Institute)	1	B-3820
	China General Aviation	5	B-3810
	China Southwest Airlines	4	B-3811
	Flying Dragon Aviation	9	B-3802
	Guizhou Aviation Corporation	1	B-3809
	State Oceanic Administration	1	?
Subtotal		**21**	
Export			
Australia	Aviex	1	VH-LLK
Bangladesh	Aero Bengal Airlines	2	?
Cambodia	Air Force	2	XU-016
Egypt	Air Force	2	?
Eritrea	Air Force	3	ER 800
	Red Sea Air[1]	1	E3-AAI
Fiji	Fiji Air	3	DQ-FHF
Guyana	Defence Force	1	8R-GDS
Iran	Air Force	9	15–2245
Kenya	Air Force	12	128?
Kiribati	Air Kiribati	1	T3-ATI
Laos	Lao Aviation	7	RDPL-34115
Malaysia	Berjaya Air Charter	2	9M-TAB
Mauritania	Air Force	2	5T-MAE
Mongolia	Mongolian Airlines	6	D-0064
Namibia	Defence Force	2	NDF97-600
Nepal	Nepal Airways	5	9N-ACD
Pakistan	Air Force	2	96–035
	Army	4	045
Peru	Air Force	6	333
	National Police	3	224
Philippines	Philippine Eagle Airlines	3	RP-C1518
Sri Lanka	Air Force	9	CR-851
Tanzania	Air Force	2	JW9029
Zambia	Air Force	4	AF214
Subtotal		**94**	
Total		**115**	

[1] Unclear if aircraft is ex-military

Detailed description applies to Y-12 (II) except where indicated.

Y-12 (IV): Improved version with sweptback wingtips; modifications to control surface actuation, main gear and brakes; redesigned seating for 18 to 19 passengers; starboard side rear baggage door; maximum payload and maximum T-O weight increased. Further changes include Rockwell Collins or Honeywell com/nav for both VFR and IFR, plus optional colour weather radar, GPS, Omega navigation, wing and tail de-icing, and oxygen system. Domestic certification received 3 July 1994 and FAR Pt 23 approval on 26 March 1995; Indonesian certification 16 November 2000. Sichuan Airlines order for 20 (7 November 2000) assumed to be of this version.

Y-12E Harbinger: Next-generation version for Chinese market; 559 kW (750 shp) PT6A-135A engines (derated to 462 kW; 620 shp) and Hartzell four-blade lightweight propellers for improved 'hot and high' performance, plus strengthened structure, reduced noise levels and improved avionics. Announced at Airshow China, November 2000; passed CAAC preliminary design review 16 November 2000. First flight (B-610L) August 2001; award of CAAC type certificate announced 26 February 2002; launch customer Sichuan Airlines (option for 10). Twelve engines for Y-12E delivered by February 2004; three aircraft said to be in service in China at that time.

Y-12AEW: Being promoted in 2004; bulged nose housing BAE Systems Skymaster AEW and maritime surveillance radar.

Y-12F: Concept study for version with pressurised fuselage and retractable landing gear.

Y-12G: In design stage 2002–03 as dedicated cargo version with large side door and capacity for three LD3 containers.

CUSTOMERS: About 120 built by Harbin by mid-2003: see table. (In February 2004, Pratt & Whitney Canada confirmed shipment of some 300 PT6A-27 engines since beginning of Y-12 programme.)

COSTS: Batch of 15 PT6-engined aircraft valued at US$58 million (1999).

DESIGN FEATURES: Designed to standards of FAR Pts 23 and 135 Appendix A and developed to improve upon modest payload/range of piston-engined Y-11. Constant-chord high braced wings, with small stub-wings at cabin floor level supporting mainwheel units; basically rectangular-section fuselage, upswept at rear; non-swept tail surfaces; large dorsal fin; ventral fin under tailcone.

Wing section LS(1)-0417; thickness/chord ratio 17 per cent; dihedral 1° 41'; incidence 4°.

FLYING CONTROLS: Conventional and manual. Drooping ailerons, horn-balanced elevators and rudder; trim tab in starboard aileron, rudder and each elevator; electrically actuated two-segment double-slotted flaps on each wing trailing-edge.

STRUCTURE: Conventional all-metal structure with two-spar fail-safe wings and stressed skin fuselage; Ziqiang-2 resin bonding on 70 per cent of wing structure and 40 per cent of fuselage; integral fuel tankage in wing spar box; bracing strut from each stub-wing out to approximately one-third span.

LANDING GEAR: Non-retractable tricycle type, with oleo-pneumatic shock-absorber in each unit. Single-wheel main units, attached to underside of stub-wings. Single trailing-link, non-steerable nosewheel. Mainwheel tyres size 640×230, pressure 5.50 bar (80 lb/sq in); nosewheel tyre size 480×200, pressure 3.50 bar (51 lb/sq in). Hydraulic brakes. Minimum ground turning radius 16.75 m (54 ft 11½ in).

POWER PLANT: Two Pratt & Whitney Canada PT6A-27 turboprops, each derated to 462 kW (620 shp) and driving a Hartzell HC-B3TN-3B/T10173B-3 three-blade constant-speed reversible-pitch propeller. All fuel in tanks in wing spar box, total capacity 1,616 litres (427 US gallons; 355.5 Imp gallons), with overwing gravity filling point each side. No recent news of domestic WJ9 turboprop (442 to 520 kW; 593 to 697 shp), said to have been under development for Y-12 and other aircraft.

ACCOMMODATION: Crew of two on flight deck, with access via forward-opening door on port side. Four-way adjustable crew seats. Dual controls. Main cabin can accommodate up to 19 passengers in commuter configuration, in three-abreast layout (with aisle), at seat pitch of 71 cm (28 in). Alternative layouts for up to 15 parachutists, or all-cargo configuration with 11 tiedown rings. Air ambulance (six stretchers plus four patient or medical attendant seats) and VIP layouts also available. Passenger/cargo double door on port side at rear, rear half of which opens outward and forward half inward; foldout steps in passenger entrance. Emergency exit on each side at front of cabin and opposite passenger door on starboard side at rear. Baggage compartment in nose and at rear of passenger cabin, for 100 kg (220 lb) and 260 kg (573 lb) respectively.

SYSTEMS: Hamilton Sundstrand R70-3WG environmental control system. Hydraulic system (operating pressure 118 to 147 bar; 1,711 to 2,132 lb/sq in) for mainwheel brakes. Two 6 kW DC starter/generators, two 600 VA 400 Hz single-phase static inverters and one 43 Ah Ni/Cd battery for electrical power. Goodrich Type 29S-7D 5178 anti-icing system optional for wing, tailplane and fin leading-edges. Oxygen system optional.

AVIONICS: *Comms:* VHF-251 and HF-230 radio, AUD-251A intercom and TDR-950 transponder.

Radar: Honeywell 1400C, RDS-81 or RDS-82 colour weather radar optional.

Flight: ALT-50 radio altimeter, DME-451, ADF-650A, MKR-350 marker beacon receiver, KLN 90 GPS receiver (optional), VIR-351 VOR, GLS-350 glideslope receiver and PN-101 pictorial navigation display. Doppler navigation with satellite responder (Y-12 (IV), Omega nav) optional (for example, in mineral detection role).

Instrumentation: 20025-11324 airspeed indicator, 510-8D10 horizon, 101420-11934 encoding altimeter, 30230-11101 vertical speed indicator, 9551-BN541 bank indicator, LC-2 magnetic compass, ZWH-1 outside air temperature indicator, and ZEY-1 flap position indicator; dual engine torquemeters, interturbine temperature indicators, gas generator tachometers, oil temperature and pressure indicators, fuel pressure and quantity indicators; 309W clock; and XDH-10B warning light box. Flight data recorder and GPWS optional on Y-12E.

EQUIPMENT: Hopper for 1,200 litres (317 US gallons; 264 Imp gallons) of dry or liquid chemical in agricultural version. Appropriate specialised equipment for firefighting, geophysical survey (for example, long, kinked sensor tailboom) and other missions.

DIMENSIONS, EXTERNAL:
Wing span: Y-12 (II) ... 17.235 m (56 ft 6½ in)
Y-12 (IV) ... 19.20 m (63 ft 0 in)
Wing chord, constant ... 2.00 m (6 ft 6¾ in)
Wing aspect ratio: Y-12 (II) ... 8.7
Length overall ... 14.86 m (48 ft 9 in)
Height overall ... 5.675 m (18 ft 7½ in)
Elevator span ... 5.365 m (17 ft 7¼ in)
Wheel track ... 3.61 m (11 ft 10¼ in)
Wheelbase ... 4.70 m (15 ft 5 in)
Propeller diameter ... 2.49 m (8 ft 2 in)
Propeller ground clearance ... 1.325 m (4 ft 4¼ in)
Distance between propeller centres ... 4.94 m (16 ft 2½ in)
Fuselage ground clearance ... 0.65 m (2 ft 1½ in)
Crew door: Height ... 1.12 m (3 ft 8 in)
Width ... 0.65 m (2 ft 1½ in)
Passenger/cargo door: Height ... 1.38 m (4 ft 6¼ in)
Width (passenger door only) ... 0.65 m (2 ft 1½ in)
Width (double door) ... 1.45 m (4 ft 9 in)
Emergency exits (three, each):
Height ... 0.68 m (2 ft 2¾ in)
Width ... 0.68 m (2 ft 2¾ in)
Baggage door (nose, port):
Max height ... 0.56 m (1 ft 10 in)
Width ... 0.75 m (2 ft 5½ in)

DIMENSIONS, INTERNAL:
Cabin, excl flight deck and rear baggage compartment:
Length ... 4.82 m (15 ft 9¾ in)

Max width 1.46 m (4 ft 9½ in)
Aisle width 0.23 m (9 in)
Max height 1.70 m (5 ft 7 in)
Floor area 7.0 m^2 (75.8 sq ft)
Volume 12.9 m^3 (455 cu ft)
Baggage compartment volume:
nose 0.77 m^3 (27.2 cu ft)
rear 1.89 m^3 (66.7 cu ft)

AREAS:
Wings, gross: Y-12 (II) 34.27 m^2 (368.9 sq ft)
Ailerons (total, incl tab) 2.88 m^2 (31.00 sq ft)
Trailing-edge flaps (total) 6.00 m^2 (64.58 sq ft)
Fin, incl dorsal fin 2.24 m^2 (24.11 sq ft)
Rudder, incl tab 3.34 m^2 (35.95 sq ft)
Tailplane 3.10 m^2 (33.37 sq ft)
Elevators (total, incl tabs) 4.06 m^2 (43.70 sq ft)

WEIGHTS AND LOADINGS (A: Y-12 (II), B: Y-12 (IV), C: Y-12E):
Max usable fuel weight: A, B, C 1,230 kg (2,712 lb)
Max payload: A 1,700 kg (3,748 lb)
B, C 1,984 kg (4,374 lb)
Max T-O weight: A 5,300 kg (11,684 lb)
B, C 5,670 kg (12,500 lb)
Max ramp weight: A 5,330 kg (11,750 lb)
B, C 5,700 kg (12,566 lb)
Max landing weight: A 5,300 kg (11,684 lb)
B, C 5,400 kg (11,905 lb)
Max zero-fuel weight: A 4,900 kg (10,803 lb)
B, C 5,188 kg (11,438 lb)
Max cabin floor loading (cargo):
A, B 750 kg/m^2 (154 lb/sq ft)
Max wing loading: A 145.9 kg/m^2 (29.90 lb/sq ft)
Max power loading: A 5.74 kg/kW (9.42 lb/shp)
B 6.14 kg/kW (10.08 lb/shp)

PERFORMANCE (A, B and C as above):
Max operating speed (VMO) at FL100:
A 177 kt (328 km/h; 204 mph)
B 175 kt (325 km/h; 201 mph)
C 186 kt (344 km/h; 214 mph)
Max cruising speed at FL100:
A 157 kt (292 km/h; 181 mph)
Econ cruising speed at FL100:
A 135 kt (250 km/h; 155 mph)
B 140 kt (260 km/h; 162 mph)
Max rate of climb at S/L: A 486 m (1,594 ft)/min
B, C 468 m (1,535 ft)/min
Rate of climb at S/L, OEI: A 84 m (275 ft)/min
B 90 m (295 ft)/min
C 87 m (285 ft)/min
Service ceiling: A, B, C 7,000 m (22,960 ft)
Service ceiling, OEI, 15 m (50 ft)/min rate of climb, max continuous power: A 3,000 m (9,840 ft)
T-O run, 15° flap:
normal: A 340 m (1,115 ft)
B 370 m (1,215 ft)
C 450 m (1,480 ft)
STOL: A 230 m (755 ft)
T-O to 15 m (50 ft), 15° flap:
normal: A 420 m (1,380 ft)
B 490 m (1,610 ft)
STOL: A 370 m (1,215 ft)
Landing from 15 m (50 ft), 20° flap: with braking and propeller reversal:
normal: A 480 m (1,575 ft)
STOL: A, B 370 m (1,215 ft)
C 320 m (1,050 ft)
with brakes only:
normal: A 620 m (2,035 ft)
B 630 m (2,070 ft)
C 590 m (1,935 ft)
STOL: A 200 m (656 ft)
Landing run, 20° flap:
with braking and propeller reversal: A 200 m (660 ft)
with brakes only: A, B 340 m (1,120 ft)
C 330 m (1,085 ft)
Range at econ cruising speed at FL100 with max fuel, 45 min reserves:
A 723 n miles (1,340 km; 832 miles)
B 707 n miles (1,310 km; 814 miles)
C 723 n miles (1,340 km; 832 miles)
Endurance, conditions as above: A 5 h 12 min
B 5 h 15 min

UPDATED

HAIG (HARBIN)

HARBIN AIRCRAFT INDUSTRIES GROUP (HARBIN FEIJI GONGYE GONGSI) (SUBSIDIARY OF AVIC II)

15 Youxie Street, Pingfang District, PO Box 201-29, Harbin, Heilongjiang 150066
Tel: (+86 451) 650 11 22
Fax: (+86 451) 650 22 73
PRESIDENT AND GENERAL MANAGER: Cui Xuewen
VICE-PRESIDENT: Xu Zhanbin
PUBLIC RELATIONS: Chen Xiaoyi

SUBSIDIARY:
Harbin Embraer Aircraft Industry Company
VICE-CHAIRMAN: Guan Dongyuan
GENERAL MANAGER: Marcelo Ramon Ferroni

HAIG at Harbin is the parent organisation of Hafei Aviation Industry Company. It has overall responsibility for HAI's participation in the Eurocopter EC 120 programme.

Contract, signed 2 December 2002, to create new company (Harbin Embraer Aircraft Industry Co Ltd: HEAI, which see), located in Harbin, for co-production of ERJ-145 regional jet; Embraer 51 per cent holding, HAIG and HAI 24.5 per cent each. Embraer investing US$25 million in new 22,000 m^2 (236,800 sq ft) production facility employing up to 150 people on ERJ-145 programme; ERJ-135/140 could also be built. Facility inaugurated August 2003; first locally built ERJ-145 rolled out and first flown 16 December 2003; seven planned in 2004; eventual capacity for 24 aircraft per year. First of six for China Airlines to be delivered in June 2004.

Official photograph issued to mark roll out and simultaneous first flight of the initial Harbin-built Embraer ERJ-145. The aircraft is in an exact replica of the Embraer company colours; has port engine cowlings removed; and the image shows obvious signs of retouching in the location normally occupied by Brazilian civilian registrations ***NEW***/0573891

UPDATED

HAIG (HONGDU)

HONGDU AVIATION INDUSTRY GROUP (HONGDU HANGKONG GONGYE GONGSI) (SUBSIDIARY OF AVIC II)

Xinxiqiao, PO Box 5001-506, Nanchang, Jiangxi 330024
Tel: (+86 791) 846 95 55 and 845 29 32
Fax: (+86 791) 846 82 24
e-mail: aerobiz@hongdu.com.cu
Web: http://www.hongdu.com.cn
PRESIDENT: Jiang Liang

Following the division of AVIC's responsibilities in 1999, Hongdu Aviation Industry Group became the parent organisation of the former Nanchang Aircraft Manufacturing Company (NAMC). HAIG, which had a 2003 workforce of 5,600, also manufactures motorcycles, textile machinery and a range of other non-aerospace products.

NAMC, which became a core unit of HAIG in 1998, was created in 1951 as the state-owned Hongdu Machinery Building Factory; it built 379 CJ-5s (licence Soviet Yak-18s) between 1954 and 1958, and in 1960s shared in large production programme for J-6 fighter (Chinese development of MiG-19); also built (1957–68) 727 Y-5 (Chinese An-2) biplanes. Current programme is JL-8/K-8 jet trainer. Agreements with and via Canadian Aerospace Group in 1998 provided for North American assembly and marketing of N-5A agricultural aircraft in return for option to build CAG Windeagle and Monitor Jet in China; however, project subsequently foundered. Also built prototype of NLA AC-500 AirCar.

Although manufacture of the NAMC Q-5 (A-5) single-seat attack fighter ended some years ago, Hongdu showed a model of a two-seat variant, designated JQ-5J and optimised for interdiction, at the Zhuhai air show in November 2004. At the same event, it continued promotion of the CJ-6A basic trainer.

Proposed JQ-5J two-seat interdictor shown by Hongdu (NAMC) in 2004; CJ-6A (PT-6) model in background ***(Jane's Robert Hewson)*** ***NEW***/1043424

The company occupies a 500 ha (1,235 acre) site, with 10,000 m^2 (107,600 sq ft) of covered space, and delivered its 4,000th aircraft in 1993.

UPDATED

HAIG JL-8

Chinese name: Jiaolianji 8: Training aircraft 8
Export designation: K-8 Karakoram
TYPE: Basic jet trainer/light attack jet.
PROGRAMME: Launched publicly (as NAMC L-8) at 1987 Paris Air Show as proposed export

Pakistan Air Force HAIG K-8 jet trainers ***NEW***/1047016

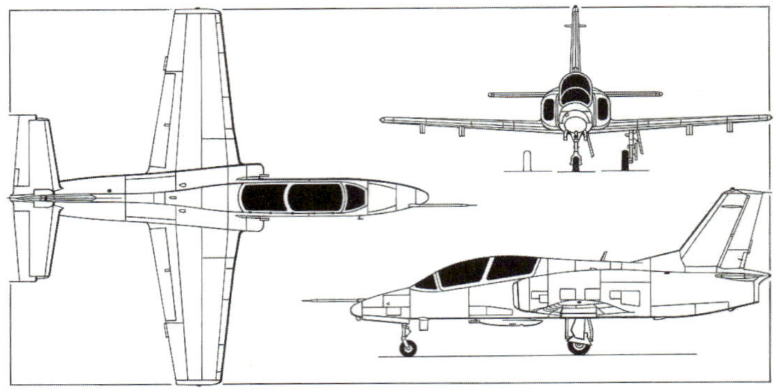

JL-8/K-8 jet trainer and light attack aircraft *(Jane's/Mike Keep)* 0507166

JL-8/K-8 upgraded front avionics panel, due for introduction in 2005 *(Jane's/Robert Karniol)* ***NEW***/0590235

aircraft to be developed jointly with international partner. Subsequently proposed to be co-developed with Pakistan as partner (25 per cent share); aircraft then redesignated K-8 and named after mountain range forming part of China/Pakistan border. Pakistan decided 1994 against own assembly line, but agreed to reconsider if subcontracting share (12 per cent initially, but being increased to 25 per cent in mid-2001 by addition of front fuselage) should later increase to 45 per cent.

Manufacture of four prototypes started January 1989; three flying prototypes: 001 (first flight 21 November 1990), 003 (first flight 18 October 1991) and 004; 002 is static and fatigue test aircraft.

First preproduction aircraft (1001/L8 320101) used as demonstrator. Preproduction batch of six for Pakistan Air Force evaluation (ordered 9 April 1994) handed over in China on 21 September 1994 and delivered to PAF on 10 November that year; PAF evaluation (approximately 1,200 hours) completed in August 1995; aircraft reported subsequently in use from Air Academy, Risalpur. China importing Progress AI-25 turbofans for use in domestic JL-8 version (see under Power Plant). One aircraft in use from 1997 by China Flight Test Establishment as variable stability testbed (K-8VSA) with fly-by-wire flight control system.

CURRENT VERSIONS: **K-8**: Initial version with TFE731 engine; *as described*.

K-8E: TFE731-powered version for Egyptian Air Force, 80 of which ordered under contract signed on 27 December 1999. First 10 Chinese-built (first flight 5 July 2000); Arab Organisation for Industrialisation (AOI) then assembling batches of 15 and 10 from CKD kits of medium-sized and smaller components respectively before progressing to 90 per cent local manufacture of final 45; first kit-assembled K-8E completed mid-2001 and some 35 achieved by end of 2002. Total of 33 items newly selected, developed or upgraded for this version include instrument panels and consoles; com/nav systems; fire-control system; fuel system; environmental control system; hydraulic system; and landing gear.

JL-8: Reported designation of Chinese domestic version when fitted with AI-25 engine.

K-8VSA: In-flight variable stability testbed (serial number 320203), in use from 1997 by China Flight Test Establishment (CFTE). Equipped with digital fly-by-wire AFCS, sidestick controller and data acquisition system.

CUSTOMERS: Total of 12 preproduction K-8s (six each to AFPLA and Pakistan Air Force) delivered by end of 1996. Original joint venture agreement reportedly involved up to 75 for Pakistan Air Force, but Pakistan Secretary for Defence Production quoted in early 1996 as saying up to 100 needed eventually to replace Cessna T-37; however, SLEP for T-37 has postponed main K-8 requirement to about 2005. AFPLA originally had requirement for several hundred, of which 25 to 30 delivered by late 1999 and at least 40 confirmed by late 2003; equip No. 4 Flying Training School at Shijiazhuang and No. 13 FTS; AFPLA now widely reported to have abandoned plans for full adoption. Myanmar Air Force last three of 12 delivered in September 1999. Zambian Air Force eight, and Namibian Air Force first four, also delivered in 1999. Three of Sri Lanka six (No. 14 Squadron) destroyed in Tamil Tiger attack on Katunayake AB, 24 July 2001. Egyptian Air Force had received about 45 (of 80) of K-8E version by late 2002, to replace Aero L-29. Second batch of six delivered to Pakistan Air Force in late 2003. Interest also reported from Bangladesh, Cambodia, Eritrea, Laos and Thailand.

K-8 EXPORT ORDERS(at March 2004):

Country:	
Colombia	6
Egypt	80
Morocco	?[3]
Myanmar	12[1]
Namibia	12
Pakistan	12
Sri Lanka	6
Zambia	8[2]
Total	**136+**

[1] *Reported total requirement for 30*
[2] *Reported option for further 8*
[3] *To replace approximately 14 Cessna T-37Bs*

COSTS: US$3 million to US$3.5 million flyaway (1996) with TFE731 and Western avionics. Egyptian Air Force (80 aircraft) contract quoted as US$345 million (1999).

DESIGN FEATURES: Intended for full basic flying training plus parts of primary and advanced syllabi, but capable also of light ground attack missions. Sweptback vertical/non-swept horizontal tail surfaces.

Tapered low wings, with NACA 64A-114 (mod) root and NACA 64A-412 tip sections; sweepback 2° 13' 8" at quarter-chord, 1° 30' incidence at root, 3° dihedral from roots; −2° twist.

FLYING CONTROLS: Conventional and power-assisted; ailerons have hydraulic boost and artificial feel; variable incidence tailplane; electrically operated trim tab in rudder and port elevator. Aileron travel ±18°, elevators 16° up/28° down, rudder ±28°. Two-position Fowler flaps (23° for T-O, 35° for landing), and split airbrake under fuselage just aft of mainwheel doors, are hydraulically actuated.

STRUCTURE: All-metal damage-tolerant main structure; ailerons of honeycomb, fin and rudder of composites. PAC share (initially only tailplane and elevators) increased to include fin, rudder, rear fuselage and engine cowling/access panels. First Pakistan-built subassemblies delivered to China in mid-1997; planned output by PAC of 12 tail units and three front fuselages in 2001; 24 and 12 respectively in 2002; 36 tails and undecided number of front fuselages in 2003. However, first PAC-built front fuselage not handed over until 5 November 2002, at which time two more under construction.

LANDING GEAR: Hydraulically retractable tricycle type, with single wheel and oleo-pneumatic shock-absorber on each unit. Main units retract inward into underside of fuselage; nosewheel, which has hydraulic steering, retracts forward. Mainwheel tyres size 561×169, pressure 6.90 bar (100 lb/sq in). Chinese hydraulic disc brakes. Anti-skid units. Minimum ground turning radius 6.69 m (21 ft 11½ in).

POWER PLANT: Including prototypes, all except JL-8 have one 16.01 kN (3,600 lb st) Honeywell TFE731-2A-2A turbofan with Lucas Aerospace FADEC, mounted in rear fuselage. Production JL-8 for China powered by 16.87 kN (3,792 lb st) ZMKB Progress AI-25TLK turbofan, ordered in 1997; initial batch of 58 imported; further 42 being negotiated in 2004.

Fuel in two flexible tanks and one inverted flight tank in fuselage and one integral tank in wing centre-section, combined capacity 1,000 litres (264 US gallons; 220 Imp gallons); single refuelling point in fuselage. Provision for carrying one 250 litre (66.0 US gallon; 55.0 Imp gallon) drop tank on outboard pylon under each wing. Oil capacity 4 kg (8.8 lb).

ACCOMMODATION: Instructor and pupil in tandem, on Jianghan TY7A zero/zero ejection seats (Martin-Baker Mk 10L in Pakistan aircraft); rear seat elevated 28 cm (11 in). One-piece wraparound windscreen; two-piece canopy opens sideways to starboard. Cockpits pressurised and air conditioned.

SYSTEMS: Honeywell ECS 51833 air conditioning and pressurisation system, with maximum differential of 0.27 bar (3.9 lb/sq in). Hydraulic system, pressure 207 bar (3,000 lb/sq in), for operation of landing gear extension/retraction, wing flaps, airbrake, aileron boost, nosewheel steering and wheel brakes. Flow rate 15 litres (3.96 US gallons; 3.30 Imp gallons)/min, with air-pressurised reservoir, plus emergency back-up hydraulic system. Abex AP09V-8-01 pump. Electrical systems powered by 12 kW, 28.5 V DC generator (primary) and 24 V DC (auxiliary), with 115/26 V single-phase AC and 36 V three-phase AC available from 40 Ah Ni/Cd battery and two static inverters, both at 400 Hz. Liquid oxygen system for occupants. Demisting of cockpit transparencies.

AVIONICS: *Comms:* Bendix/King KTR 908 VHF or KTR 909 UHF; intercom; FJ-20 flight data recorder.

Flight: Bendix/King KNR 634A VOR/glideslope/MKR with KA 26 beacon receiver, ADF 462, Type 265 radio altimeter, HZX-4A AHRS, SS/SC-5 air data computer and KTU 709 Tacan or WL-7A radio compass.

Instrumentation: Rockwell Collins EFIS-86T in first 100 aircraft, incorporating CRT primary flight and navigation displays for each crew member plus dual display processing units and selector panels for tandem operation. Blind-flying instrumentation standard. Standby flight instruments include ASI, rate of climb indicator, barometric altimeter, emergency horizon and standby compass.

Chinese (Sicong Group) upgrade for domestic and export aircraft expected to enter production in 2005, includes HUD, Multifunction Display System (MDS), digital video recorder, up-front control panel and data transfer card. Systems, which began testing in September 2004, said to include GPS, Tacan and radio compass navigation, and to function as a weapons fire-control system.

ARMAMENT (OPTIONAL): One 23 mm gun pod under centre-fuselage; self-computing optical gunsight in cockpit, plus gun camera. Two external stores points under each wing. Twin ejector racks on inboard stations can carry total of four 6, 11.5 or 50 kg practice bombs; single-store outboard stations can each carry a PL-5E air-to-air missile, a 12-round pod of 57 mm rockets, a 200 kg, 250 kg or BL755 bomb, or a drop fuel tank.

DIMENSIONS, EXTERNAL:
- Wing span ... 9.63 m (31 ft 7¼ in)
- Wing mean aerodynamic chord ... 1.85 m (6 ft 0¾ in)
- Wing aspect ratio ... 5.4
- Length overall: incl nose pitot ... 11.60 m (38 ft 0¾ in)
  - excl nose pitot ... 10.40 m (34 ft 1½ in)
- Height overall ... 4.21 m (13 ft 9¾ in)
- Elevator span ... 4.20 m (13 ft 9½ in)
- Wheel track ... 2.54 m (8 ft 4 in)
- Wheelbase ... 4.44 m (14 ft 6¾ in)

AREAS:
- Wings, gross ... 17.02 m^2 (183.2 sq ft)
- Ailerons (total, incl tab) ... 1.095 m^2 (11.80 sq ft)
- Trailing-edge flaps (total) ... 2.65 m^2 (28.55 sq ft)
- Fin ... 1.92 m^2 (20.67 sq ft)
- Rudder, incl tab ... 1.115 m^2 (12.02 sq ft)
- Tailplane ... 2.84 m^2 (30.57 sq ft)
- Elevators (total, incl tab) ... 1.32 m^2 (14.21 sq ft)

WEIGHTS AND LOADINGS:
- Weight empty, equipped ... 2,757 kg (6,078 lb)
- Max fuel: internal ... 780 kg (1,720 lb)
  - external (two drop tanks) ... 388 kg (855 lb)
- Max external stores load ... 943 kg (2,080 lb)
- T-O weight: clean ... 3,700 kg (8,157 lb)
  - with two 250 litre drop tanks ... 4,204 kg (9,268 lb)
- Max T-O weight with external stores ... 4,332 kg (9,550 lb)
- Max wing loading ... 254.5 kg/m^2 (52.13 lb/sq ft)
- Max power loading (TFE731) ... 271 kg/kN (2.65 lb/lb st)

PERFORMANCE (at clean T-O weight):
- Never-exceed speed (VNE) ... 512 kt (950 km/h; 590 mph) IAS
- Max level speed at S/L ... 432 kt (800 km/h; 497 mph)
- Unstick speed ... 100 kt (185 km/h; 115 mph)
- Approach speed ... 108 kt (200 km/h; 124 mph)
- Touchdown speed, 35° flap ... 86 kt (160 km/h; 99 mph)
- Stalling speed, 35° flap ... 81 kt (150 km/h; 94 mph)
- Max rate of climb at S/L ... 1,800 m (5,905 ft)/min
- Service ceiling ... 13,600 m (44,620 ft)
- T-O run ... 440 m (1,445 ft)
- T-O to 15 m (50 ft) ... 600 m (1,970 ft)
- Landing from 15 m (50 ft) ... 518 m (1,700 ft)
- Landing run ... 530 m (1,740 ft)
- Range:
  - max internal fuel ... 842 n miles (1,560 km; 969 miles)
  - max internal/external fuel ... 1,155 n miles (2,140 km; 1,329 miles)
- Endurance: max internal fuel ... 3 h 12 min
  - max internal/external fuel ... 4 h 12 min
- *g* limits (clean) ... +7.33/−3.0

UPDATED

HAIG JL-15

Chinese name: Jiaolianji 15 (Training aircraft 15)

TYPE: Advanced jet trainer/light attack jet.

PROGRAMME: Revealed (as L-15) as feasibility study at Aviation Expo 2001, Beijing, September 2001 and displayed in model form at Zhuhai November 2002. International partner(s) being sought. Parallel design study being undertaken by Yakovlev OKB, revealed at MAKS air show in August 2003 and due to be presented to AVIC II by end of that year. However, by September 2003, when details of the L-15 were again made available at an air show in Beijing the aircraft's appearance had changed significantly, becoming more like the Yak-130, especially in profile. Yakovlev confirmed in mid-2004 that collaboration on the JL-15 was continuing. First flight not expected before at least 2005.

Full-scale model of HAIG JL-15 shown at Zhuhai in November 2004 *(Robert Hewson)* ***NEW***/1047699

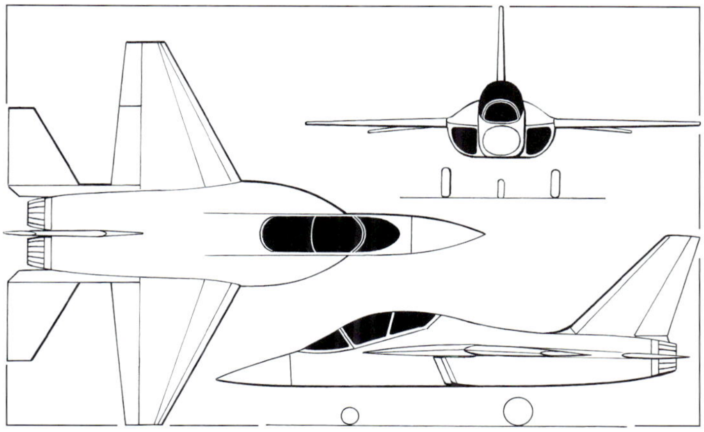

HAIG JL-15 general arrangement (2004 configuration) *(James Goulding)* ***NEW***/0589195

DESIGN FEATURES: Shoulder-mounted wings with curved, leading-edge wingroot strakes; leading-edge manoeuvring slats; tandem cockpits; twin engines; single vertical fin. Envisaged as supersonic trainer for J-10 and J-11. Service life 10,000 hours or 30 years.

Following details highly provisional:

FLYING CONTROLS: Digital fly-by-wire. Large, door-type dorsal airbrake.

POWER PLANT: Twin turbofans; 24.5 kN (5,512 lb st) Motor-Sich AI-222K-25Fs or 34.1 kN (7,660 lb st) class domestic WS11s originally proposed, but most recent literature refers to two engines, each of 41.2 kN (9,259 lb st).

ACCOMMODATION: Two zero/zero ejection seats in tandem.

AVIONICS: HOTAS controls, HUD(s) and MFDs; stores management system.

DIMENSIONS, EXTERNAL:
- Wing span ... 9.48 m (31 ft 1¼ in)
- Length overall ... 12.27 m (40 ft 3 in)
- Height overall ... 4.37 m (14 ft 4 in)

WEIGHTS AND LOADINGS:
- Normal T-O weight, clean ... 6,500 kg (14,330 lb)
- Max T-O weight ... 9,500 kg (20,950 lb)

PERFORMANCE:
- Max level speed ... M1.4
- Max rate of climb at S/L ... 9,000 m (29,530 ft)/min
- Service ceiling ... 16,000 m (52,490 ft)
- *g* limits ... +8/−3

UPDATED

HMDH

HONGDU MD HELICOPTERS

Nanchang, Jiangxi

New joint venture created in early 2003 by RDM Holdings (60 per cent) and Hongdu Aviation Industries Group (40 per cent), with initial US$10 million investment by RDM, under which HMDH assembles MD 500 and 600 series helicopters from US-supplied kits, first of which were delivered in March 2003. Programme launch planned for October 2003, with three MD 500s and one MD 600 to be completed by year-end. Initial commitments for about 30 helicopters reported by that time.

Sales to China in 2002 included an MD 500E and an Explorer to Guangdong General Aviation Company (which also operates an MD 600N), and one MD 500E to Xiongying Aero Club. At its formation, HMDH had received commitments from local law enforcement and tourist agencies. Future plans envisage gradual progression towards an increased share in the manufacturing process.

UPDATED

HTATC

HUZHOU TAIXIANG AVIATION TECHNOLOGY COMPANY

Huzhou, Zhejiang

Company formed 29 November 2001 as joint venture between Jinjiang Medicine Company and Aitas Corporation of USA with total investment of US$29.9 million and registered capital of US$20 million. Is assembling ASII (Sportscopter) Ultrasport, Rotorway Exec, a three-seat helicopter and an unspecified fixed-wing aircraft from imported kits. First deliveries (five aircraft, including one Ultrasport and one Exec) made in September 2002. HTATC aimed to achieve output of 150 in its first two years of operation and full local manufacture, except for engines, by end of third year.

VERIFIED

NLA

Room 102, Building D, 29 Yudao Street, Nanjing, Jiangsu 210016
Tel: (+86 25) 489 16 36 and 489 36 55
Fax: (+86 25) 489 16 37
e-mail: njqf@pub.jlonline.com
Web: http://www.pub.jlonline.com
PRESIDENT: Wo Dingzhu

NLA was created in 1998 by Nanjing University of Aeronautics and Astronautics (NUAA), the municipality of Nanjing, the provincial government of Jiangsu and other business interests. It has its headquarters on the NUAA campus and is responsible for the five-seat AC-500, which made its first flight in 2004.

NEW ENTRY

First public flight of second AC-500 Air Car ***NEW***/1043425

NLA AC-500 AIR CAR

TYPE: Five-seat lightplane.

PROGRAMME: Revealed at Airshow China in November 1998; joint development by NUAA and NLA. Prototype originally targeted to fly in 2000; built and rolled out at HAIG plant, Nanchang, in third quarter of 2000 and exhibited at Airshow China in November 2000. After a long period of apparent inactivity, the Chinese *People's Daily* news agency reported on 14 June 2004 that the aircraft had recently (May ?) made its maiden flight at the No. 320 Plant in Nanchang. Three examples completed at that time; a second aircraft (c/n0004) flew for the first time on 3 December 2004 (official first flight on 7 December), this being in trainer configuration, equipped with dual-controls. FAR Pt 23 certification is intended.

COSTS: 3.6 million yuan (2004).

DESIGN FEATURES: General appearance similar to that of Piper and Socata types in same weight/performance category. Constant-chord, low-mounted wings with upturned tips; sweptback vertical tail and small underfin.

Said to be intended mainly for business flights by senior government and industry officials; and for forestry protection, mail delivery, agricultural, environmental monitoring, remote sensing and other activities.

FLYING CONTROLS: Conventional and manual; flaps fitted. One-piece horn-balanced elevator; ground-adjustable tab on rudder.

STRUCTURE: All-metal main structure; some non-load-bearing components manufactured from composites.

LANDING GEAR: Tricycle type; fixed. Single wheel on each unit.

POWER PLANT: One 194 kW (260 hp) Textron Lycoming IO-540 flat-six engine, driving a three-blade propeller. (Teledyne Continental in prototype, according to one report.)

NLA AC-500 Air Car No. 4, second to fly ***NEW***/1043426

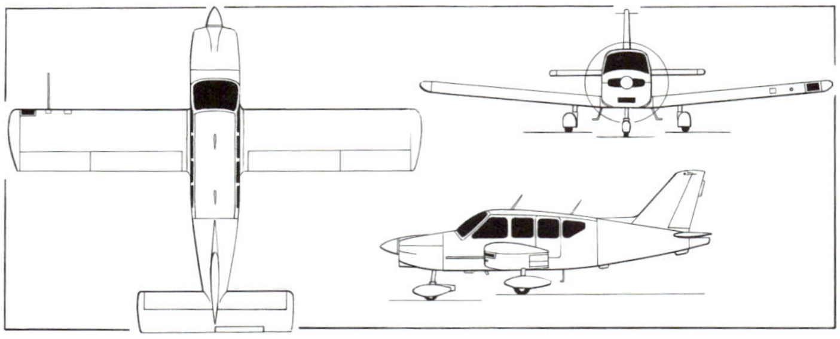

NLA AC-500 light aircraft ***(James Goulding)*** ***NEW***/0126689

ACCOMMODATION: Pilot and co-pilot or passenger in front, with two rearward-facing seats behind and a forward-facing single seat aft of these. Crew door each side of cockpit, plus passenger door aft of wing on port side.

DIMENSIONS, EXTERNAL:

Wing span	10.20 m (33 ft 5½ in)
Length overall	8.14 m (26 ft 8½ in)
Fuselage max width	1.20 m (3 ft 11¼ in)
Height overall	3.05 m (10 ft 0 in)
Wheel track	2.80 m (9 ft 2¼ in)
Wheelbase	2.18 m (7 ft 1½ in)
Propeller diameter	1.96 m (6 ft 5 in)

WEIGHTS AND LOADINGS:

Max baggage weight	65 kg (143 lb)
T-O weight	1,540 kg (3,395 lb)
Max power loading	7.95 kg/kW (13.06 lb/hp)

PERFORMANCE (estimated):

Max level speed	156 kt (289 km/h; 179 mph)
Max cruising speed	135 kt (250 km/h; 155 mph)
Stalling speed	60 kt (110 km/h; 69 mph)
Service ceiling	3,000 m (9,840 ft)
Range	540 n miles (1,000 km; 621 miles)
Endurance	4 h

NEW ENTRY

SAC (SHAANXI)

SHAANXI AIRCRAFT COMPANY (SHAANXI FEIJI GONGSI) (SUBSIDIARY OF AVIC II)

PO Box 34, Hanzhong, Shaanxi 723213
Tel: (+86 916) 288 62 71 and 288 52 72
Fax: (+86 916) 288 61 82 and 288 51 82
*e-mail:*sac@shanfei.com
*Web:*http://www.shanfei.com
PRESIDENT: Hu Xiaofeng
MARKETING MANAGER: Li Yousen

Founded early 1970s; occupies a 300 ha (741 acre) site and had 2003 workforce of more than 10,000; covered workspace includes largest final assembly building in China. Status raised to Industry Group in 2001. Main aircraft programme, recently receiving increased development/upgrading effort, is Y-8 transport; non-aerospace products include coaches and small trucks.

UPDATED

SAC Y-8

Chinese name: Yunshuji-8 (Transport aircraft 8)

TYPE: Medium transport/multirole.

PROGRAMME: Redesign, as Chinese development of Antonov An-12B, started at Xian March 1969; first flight of first (Xian-built) prototype 25 December 1974, followed by second (c/n 001802, first built by SAC) 29 December 1975; production go-ahead given 11 February 1980. Pressurised Y-8C made first flight 17 December 1990. Production rate was approximately five per year during first half of 1990s; new (mainly freighter) designations introduced with Y-8F100 and F200 from 1997; further F-prefixed variants revealed November 2000, presumably reflecting current manufacturing standard; three-person flight crews resulting from use of modernised avionics; apparently now seen to have more future in cargo and military roles, following reported difficulties with pressurisation system.

CURRENT VERSIONS: **Y-8:** Prototype and baseline military transport.

Y-8A: Helicopter carrier. Main cabin height increased by 120 mm (4.72 in) by deleting internal gantry; downward-opening rear ramp/door, as in Y-8C. Deliveries began 1987. In service.

Y-8B: Mainly unpressurised civil transport. First deliveries 1986; CAAC certification 1993. Military equipment deleted; empty weight reduced by 1,720 kg (3,792 lb); some avionics differ. In service.

Y-8C: First fully pressurised version, developed with Lockheed collaboration. Changes included redesigned (downward-opening ramp-type) cargo loading door and main landing gear; handling system for standard freight containers and pallets; improved com/nav/ATC equipment, air conditioning and oxygen systems; additional emergency exits. First flight 17 December 1990 (SAC 182, converted from first Shaanxi prototype); CAAC certification 1993. Five delivered by January 1994 (latest figure received); apparently superseded by Y-8F200/F400.

Y-8D: Export baseline version, with main avionics by Rockwell Collins and Litton. Later versions designated **Y-8D II**. Deliveries began 1987 (Y-8D) and 1992 (Y-8D II); eight delivered by early 1997; none reported since then.

Y-8E: Drone carrier version of baseline Y-8 for WZ-5 Chang Hong (Long Rainbow) UAV. Forward pressure cabin accommodates drone controller's console; carrier/launch trapeze for one UAV under each outer wing panel. First flight 1989; first deliveries later that year, replacing obsolete Tu-4s. In service.

Y-8F: Livestock carrier version of baseline Y-8, with cages to hold up to 500 sheep or goats. First flight early 1990; first deliveries later that year; CAAC type approval 25 January 1994. In service.

Y-8F100: Upgraded cargo version. Modernised avionics (colour weather radar, EFIS, VOR/ILS, DME, GPS navigation, FDR, HF and VHF radios and ATC transponder) and redesigned loading system. Launch customer China Postal Airlines (three); delivered 1996 with WJ6A engines. Crew of five. Now the baseline part-unpressurised version.

Y-8F200: Pressurised version of F100, certified by CAAC in October 1997. Increased internal passenger/cargo volume. Two delivered to Tanzanian Air Force 26 October 2003.

Detailed description applies to F-100/200 unless otherwise indicated.

Y-8F300: Freighter. Generally as Y-8F100, but with 'solid' nose and reduced (three-person) flight crew. Not currently being promoted in 2004.

Y-8F400: Pressurised freighter. Generally as Y-8F200, but with 'solid' nose and reduced (three-person) flight crew. Said to be scheduled for refit with PW150A engines; Maiden

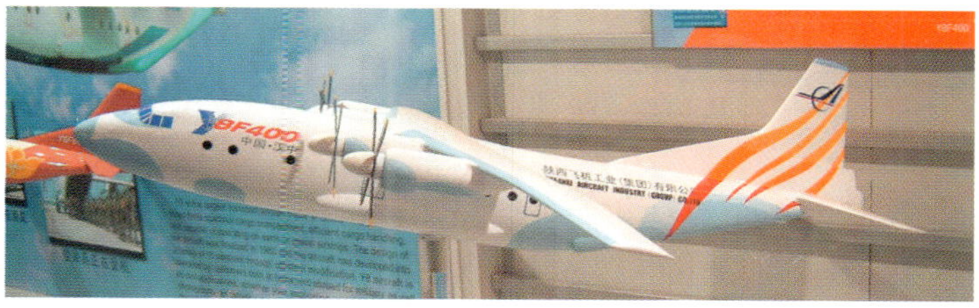

Model of the solid-nosed Y-8F400 freighter *(Robert Hewson)* ***NEW***/0547123

Canadian turboprops and six-blade propellers characterise the Y-8F600 *(Robert Hewson)* ***NEW***/0547117

November 2002 display model of projected new Y-8X *(Robert Hewson)* ***NEW***/0547124

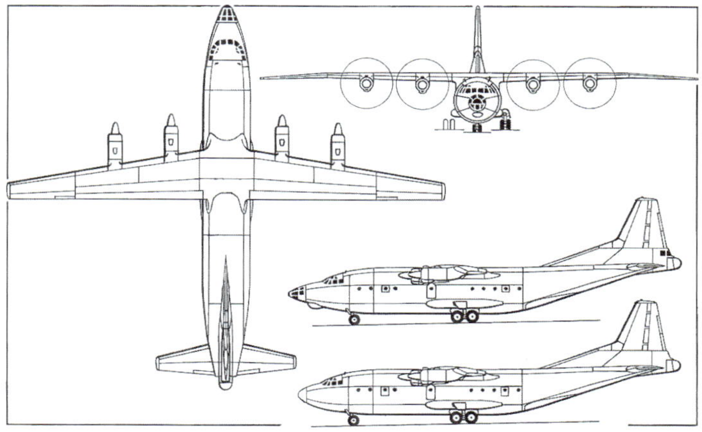

SAC Y-8 four-turboprop multipurpose transport, with additional side view (bottom) of F300/400 freighter *(Jane's/Mike Keep)* 0137951

flight 25 August 2001; certified by CAAC 26 August 2002; passed acceptance review 27 November 2002. China Postal Airlines suggested as possible launch customer.

Y-8F600: Further Westernised, 'solid'-nosed freighter version, with Pratt & Whitney Canada PW150B turboprops and Dowty R408 six-blade composites propellers, Honeywell Primus Epic avionics suite, two-person flight crew; otherwise generally similar to Y-8F400. Development approved by Defence and Industry Commission November 2001; co-development agreement with Antonov design bureau 4 November 2001. Two under construction late 2000; first flight originally expected by end of 2001, but first engine deliveries not due until mid-2004. According to SAC President at November 2004 Airshow China, was then 'on schedule' to fly in July 2005, with certification targeted for 2006.

Y-8F800: Proposed version: stretched, with redesigned wings and landing gear; payload/range target of 30,000 kg (66,139 lb) over 4,200 n miles (7,778 km; 4,833 miles).

Y-8H: Aerial survey version. No details yet known.

Y-8X: Designation applied originally to **maritime patrol prototype** (B-4101), with Western com/nav, radar, surveillance and search equipment; larger chin radome. Received type approval in September 1984; no further examples confirmed, although some reports suggest PLA Navy has eight. Details in earlier editions.

Designation revived in 2002 and applied to new **strategic airlift version**.

Y-8X: Designation (2002) for much-enhanced version with redesigned fuselage and tail unit, all-new wings and 4,847 kW (6,500 shp) class turboprops. Antonov ASTC collaborating in design As of mid-2004, estimated OWE of 39,000 kg (85,980 lb), normal payload 25,000 kg (55,115 lb), max payload 30,000 kg (66,140 lb), max T-O and landing weight 77,000 kg (169,755 lb), overload T-O weight 81,000 kg (178,575 lb). Performance targets include 324-351 kt) 600-650 km/h; 373-404 mph) cruise at 9,000 m (29,525 ft), max range 4,211 n miles (7,800 km; 4,846 miles) and max endurance 12 hours. Load capacity nine 2.24 × 2.74 m (88 × 108 in) or eight 2.44 × 3.18 m (96 × 125 in) pallets, with low-altitude airdrop of 20,000 kg (44,090 lb) single-item cargo or 132 parachutists. Lifetime expectancy 50,000 flight hours, 20,000 sorties or 30 years.

'Y-8 AEW': At least one aircraft (CFTE serial number 079) fitted with BAE Systems (formerly Racal) Skymaster AEW radar, six of which ordered under US$66 million contract in 1996. Reportedly first flew November 1998. Radar, installed in bulbous nose fairing, has 360° scan and more than 200 n mile (370 km; 230 mile) range. Observed in PLA Navy exercises, reportedly transmitting information via datalink to ship-based Z-9 helicopters. Chinese official designation not yet known. Further details in *Jane's Electronic Mission Aircraft*.

CUSTOMERS: Total of about 90, including exports, reportedly delivered (of which 50 can be confirmed) by late 2002. In service in China with commercial operators, China Postal service (three Y-8F100s from 1996) and PLA Air Force (AFPLA); eight military exports (Y-8D) to air forces of Myanmar (four from 1994), Sri Lanka (three from 1987) and Sudan (one in 1991). First two Sri Lankan aircraft modified locally for use as bombers (both since lost). Two (Y-8F100) leased to BonAir of Iran from March 1998; two (Y-8F200) exported to Tanzania 26 October 2003. Company reported receipt of letters of intent for 40 Y-8F600s from Chinese airlines by November 2004, at which time military exports also being negotiated.

DESIGN FEATURES: High-mounted wing; circular-section fuselage (forward section and tail turret pressurised), upswept at rear; angular tail surfaces with large dorsal fin. More pointed nose transparencies than An-12, probably from Chinese H-6 (Tu-16) production; shorter, 'solid' nose replaces this glazing in F300/F400.

Wing sections C-5-18 at root, C-3-16 at rib 15 and C-3-14 at tip, final two digits indicating thickness/chord ratio; incidence 4°; 1° dihedral on intermediate panels, 4° anhedral on outboard panels; 6° 50' sweepback at quarter-chord; fixed-incidence tailplane.

FLYING CONTROLS: Conventional and manual. Aerodynamically balanced differential ailerons, elevators and rudder, each of which has inset trim tab; two-segment, hydraulically actuated double-slotted Fowler flaps on each wing trailing-edge; comb-shaped spoilers forward of flaps.

STRUCTURE: All-metal (aluminium alloy) conventional semi-monocoque/stressed skin; wings, tailplane and fin are all two-spar box structures; landing gear and all hydraulic components manufactured by Shaanxi Aero-Hydraulic Component Factory (SAHCF).

LANDING GEAR: Hydraulically retractable tricycle type, with Shaanxi (SAHCF) nitrogen/oil shock-struts on all units. Four-wheel main bogie on each side retracts inward and upward into blister on side of fuselage. Twin-wheel nose unit, hydraulically steerable to ±35°, retracts rearward. Mainwheel tyres size 1,050×300, pressure 28.40 bar (412 lb/sq in); nosewheel tyres size 900×300, pressure 16.70 bar (242 lb/sq in). Hydraulic disc brakes and Xingping inertial anti-skid sensor. Minimum ground turning radius 13.75 m (45 ft 1½ in).

POWER PLANT: *Except F600:* Four 3,126 kW (4,192 shp) SAEC (Zhuzhou) WJ6A turboprops, each driving a Baoding four-blade J17-G13 constant-speed fully feathering propeller.

F600: Four 3,781 kW (5,071 shp) Pratt & Whitney Canada PW 150B turboprops, each driving a Dowty six-blade all-composites propeller.

All fuel (F100/F300) in two integral tanks and 29 bag-type tanks in wings (20,102 litres; 5,310.5 US gallons; 4,422 Imp gallons) and fuselage (10,075 litres; 2,661.5 US gallons; 2,216 Imp gallons), giving total capacity of 30,177 litres (7,971.5 US gallons; 6,638 Imp gallons). Reduced fuel load in F200/F400 (see under Weights and Loadings). Refuelling points in starboard side of fuselage (between frames 14 and 15), mainwheel fairing, and in wing upper surface.

ACCOMMODATION: Flight crew of five (pilot, co-pilot, navigator, engineer and radio operator) in early versions and F100/F200; three crew only in F300/F400/F600. Forward portion of fuselage in F100/F300 is pressurised, and can accommodate up to 14 passengers in addition to crew. Cargo compartment (between frames 13 and 43) is unpressurised in these versions. Pressurised volume increased in F200/F400; effective length of cargo hold extended internally by 2.00 m (6 ft 6¾ in). Maximum accommodation for up to 96 troops; or 80 paratroops; or up to 92 casualties with three medical attendants; or two 'Liberation' army trucks, plus Jeep-sized vehicle on loading ramp. Drive on/drive off vehicle loading, using auxilliary ramp.

Short-hold freighter versions (Y-8F100 and Y-8F300) can accommodate optimum eight 2.24 × 1.37 m (88 × 54 in), four 2.24 × 2.74 m (88 × 108 in), three 2.44 × 3.18 m (96 × 125 in), or three 88 × 125 in with one 88 × 108 in and one 88 × 54 in standard cargo pallets; or 17 LD3 containers. Larger (Y-8F200 and Y-8F400) hold can accept optimum four 2.24 × 3.18 m (88 × 125 in), four 2.44 × 3.18 m (96 × 125 in) pallets or 19 LD3s. F600 hold capacity, five 96 × 125 or seven 88 × 108 in pallets. Individual cargo items of up to 7,400 kg (16,315 lb) can be airdropped.

Crew door and two emergency exits in forward fuselage. Three additional emergency exits in cargo compartment, access to which is via a large rear-loading ramp/door in underside of rear fuselage.

SYSTEMS: Forward fuselage of F100/F200 pressurised to maintain a differential of 0.20 bar (2.8 lb/sq in) at altitudes above 4,300 m (14,100 ft). (F300/F400 fully pressurised.) Two independent hydraulic systems, with operating pressures of 152 bar (2,200 lb/sq in) (port) and 147 bar (2,130 lb/sq in) (starboard), plus hand and electrical standby pumps, for actuation of landing gear extension/retraction, nosewheel steering, flaps, brakes and rear ramp/door. Electrical DC power (28.5 V) supplied by eight 12 kW generators, an 18 kW (24 hp) Xian Aero Engine Company APU (mainly for engine starting) and four 28 Ah batteries. Four 12 kVA alternators provide 115 V AC power at 400 Hz. Gaseous oxygen system for crew. Electric de-icing of windscreen, propellers and fin/tailplane leading-edges; hot air de-icing for wing leading-edges. WDZ-1 APU.

AVIONICS (Y-8F300/400): *Comms:* Rockwell Collins VHF-42B and HF-9000 com radios and TDR-94 ATC transponder.

Radar: Rockwell Collins TWR-850 colour weather radar.

Flight: Universal Avionics UNS-1K flight management system; Rockwell Collins VOR-432, DME-442, AHS-85E AHRS and EFIS-86E; Honeywell ED-55 flight data recorder.

Instrumentation: Honeywell Mk V nav display.

EQUIPMENT (FREIGHTER VERSIONS): Electric winch, tow, and 2,300 kg (5,070 lb) capacity hoist for cargo loading and unloading. Roller system for containerised or palletised cargo handling. Cargo tiedowns and barrier net.

DIMENSIONS, EXTERNAL (all versions, except where indicated):

Wing span .... 38.00 m (124 ft 8 in)
Wing chord: at root .... 4.73 m (15 ft 6¼ in)
at tip .... 1.69 m (5 ft 6½ in)
mean aerodynamic .... 3.45 m (11 ft 3¾ in)
Wing aspect ratio 11.9
Length overall: F100, F200 .... 34.02 m (111 ft 7¼ in)
F300, F400 .... 32.93 m (108 ft 0½ in)
Fuselage:
Max diameter of circular section .... 4.10 m (13 ft 5½ in)
Height overall .... 11.16 m (36 ft 7½ in)
Tailplane span .... 12.195 m (40 ft 0¼ in)
Wheel track (c/l of shock-struts) .... 4.92 m (16 ft 1¾ in)
Wheelbase (c/l of main bogies) .... 9.575 m (31 ft 5 in)
Propeller diameter: except F600 .... 4.50 m (14 ft 9¼ in)
F600 .... 4.115 m (13 ft 6 in)
Propeller ground clearance: except F600 .... 1.89 m (6 ft 2½ in)
Crew door: Height .... 1.455 m (4 ft 9¼ in)
Width .... 0.80 m (2 ft 7½ in)

Rear-loading hatch: Length ... 7.67 m (25 ft 2 in)
Width: min ... 2.65 m (8 ft 8¼ in)
max ... 3.10 m (10 ft 2 in)
Emergency exits (each): Height ... 0.55 m (1 ft 9¾ in)
Width ... 0.60 m (1 ft 11½ in)

DIMENSIONS, INTERNAL:
Cabin (incl flight deck, galley and toilet):
Length: F100, F300 ... 13.70 m (44 ft 11¼ in)
F200, F400 ... 15.70 m (51 ft 6 in)
F600 ... 18.00 m (59 ft 0¾ in)
Width: min ... 3.00 m (9 ft 10 in)
max ... 3.50 m (11 ft 5¾ in)
Height: min ... 2.20 m (7 ft 2½ in)
max ... 2.60 m (8 ft 6½ in)
Floor area: F100, F300 ... 55.0 m^2 (592 sq ft)
Volume: F100, F300 ... 123.3 m^3 (4,354 cu ft)
F200, F400 ... 137.6 m^3 (4,859 cu ft)
F600 ... 170.0 m^3 (6,003 cu ft)

AREAS:
Wings, gross ... 121.86 m^2 (1,311.7 sq ft)
Ailerons (total) ... 7.84 m^2 (84.39 sq ft)
Trailing-edge flaps (total) ... 26.91 m^2 (289.66 sq ft)
Rudder ... 6.535 m^2 (70.34 sq ft)
Tailplane ... 27.05 m^2 (291.16 sq ft)
Elevators (total) ... 7.10 m^2 (76.42 sq ft)

WEIGHTS AND LOADINGS:
Weight empty, equipped: F100 ... 34,500 kg (76,060 lb)
F200 ... 34,760 kg (76,635 lb)
F400 ... 34,000 kg (74,955 lb)
F600 ... 33,500 kg (73,855 lb)
Max fuel load: F100, F300 ... 22,909 kg (50,505 lb)
F200, F400 ... 14,566 kg (32,115 lb)
Max payload:
containerised: all versions ... 15,000 kg (33,069 lb)
bulk cargo: F100, F200, F600 ... 20,000 kg (44,090 lb)
Max airdroppable cargo: total ... 13,200 kg (29,100 lb)
single piece ... 7,400 kg (16,315 lb)
Max T-O weight: F100, F200 ... 61,000 kg (134,480 lb)
F600 ... 65,000 kg (143,300 lb)
Max ramp weight: F100, F200 ... 61,500 kg (135,585 lb)
Max landing weight: F100, F200 ... 58,000 kg (127,870 lb)
F600 ... 65,000 kg (143,300 lb)
Max zero-fuel weight:
F100, F300 ... 55,278 kg (121,865 lb)
F200, F400 ... 55,538 kg (122,440 lb)
Max wing loading: F100, F200 ... 500.6 kg/m^2 (102.53 lb/sq ft)
F600 ... 533.4 kg/m^2 (109.25 lb/sq ft)
Max power loading: F100, F200 ... 4.88 kg/kW (8.02 lb/shp)
F600 ... 4.30 kg/kW (7.06 lb/shp)

PERFORMANCE:
Max level speed at FL230:
F100, F300 ... 357 kt (662 km/h; 411 mph)
F200, F400 ... 345 kt (640 km/h; 397 mph)
Max cruising speed at FL262:
except F600 ... 297 kt (550 km/h; 342 mph)
F600 ... 324 kt (600 km/h; 373 mph)
Econ cruising speed at FL262:
all versions ... 286 kt (530 km/h; 329 mph)
Unstick speed ... 129 kt (238 km/h; 148 mph)
Touchdown speed at MLW ... 130 kt (240 km/h; 150 mph)
Max rate of climb at S/L ... 473 m (1,552 ft)/min
Rate of climb at S/L, OEI ... 231 m (758 ft)/min
Service ceiling, AUW of 51,000 kg (112,435 lb):
F100, F300 ... 10,400 m (34,120 ft)
F200, F400 ... 10,050 m (32,970 ft)
Service ceiling, OEI, AUW of 51,000 kg (112,435 lb) ... 8,100 m (26,580 ft)
Runway ACN ... 15
T-O run: except F600 ... 1,270 m (4,170 ft)
F600 ... 1,540 m (5,050 ft)
FAR T-O field length ... 1,900 m (6,235 ft)
T-O to 15 m (50 ft) ... 3,007 m (9,870 ft)
Landing from 15 m (50 ft) at MLW ... 2,174 m (7,135 ft)
FAR landing field length ... 1,650 m (5,415 ft)
Landing run at MLW: except F600 ... 1,050 m (3,445 ft)
F600 ... 1,510 m (4,955 ft)
Range with max payload:
F100 ... 687 n miles (1,273 km; 791 miles)
Range with max fuel:
F100 ... 3,032 n miles (5,615 km; 3,489 miles)
F200, F400 ... 1,857 n miles (3,440 km; 2,137 miles)
F300 ... 2,861 n miles (5,300 km; 3,293 miles)

UPDATED

SAC (SHENYANG)

SHENYANG AIRCRAFT CORPORATION (SHENYANG FEIJI GONGSI) (SUBSIDIARY OF AVIC I)
1 Lingbei Street, Huanggu District, Shenyang, Liaoning 110034
Tel: (+86 24) 86 89 66 80 and 86 59 92 05
Fax: (+86 24) 86 89 66 89
Web: http://www.sac.com.cn
PRESIDENT: Li Fangyong

Pioneer Chinese fighter design centre; built 767 examples of J-5 (licence MiG-17F) from 1956–59 and from 1963 was major producer of several thousand of J-6 series (reverse-engineered MiG-19), including 634 JJ-6 tandem-seat fighter trainers; initiated development and early production of J-7 (see under CAC). On 29 June 1994, SAC became core enterprise in newly formed Shenyang Aircraft Industries Group (SAIG). Occupies site area of more than 800 ha (1,976 acres) and has workforce of 30,000; only some 30 per cent of current activities are in aerospace.

Principal programme is J-8 II/J-8B fighter, currently the subject of various upgrade programmes. Now also engaged in major assembly/co-production programme of Sukhoi Su-27 variants as J-11 for PLA Air Force (AFPLA).

Aerospace subcontract manufacture includes cargo doors for Boeing 757, wing ribs and emergency exits for Airbus A319 and A320, tailcone/landing gear door/pylon components for Lockheed Martin C-130, rear fuselage and tail components for Boeing 737, and other machined parts for BAE, Boeing and EADS Deutschland. Collaboration with Hellenic Aerospace Industry Ltd announced in February 1997 on formation of Shenyang Hellenic Aircraft Repair Company.

UPDATED

J-8B of PLA Navy's 22nd Fighter Regiment/9th Air Division at Lingshui ***NEW***/0572274

SAC J-8Bs of 9th Division, PLA Air Force ***NEW***/0572273

SAC J-8B

Chinese name: Jianjiji-8 (Fighter aircraft 8)
Westernised designation: F-8
NATO reporting name: Finback-B

TYPE: Multirole fighter.

PROGRAMME: Development of original J-8 started 1964; first flight 5 July 1969; initial production authorised July 1979, began December 1979. All-weather J-8 I made maiden flight 24 April 1981 and production approved July 1985; ended 1987 after between 100 and 150 J-8s and J-8 Is.

Present baseline configuration is improved J-8 II (now J-8B), on which design work began 1980; maiden flight of first of four prototypes took place 12 June 1984. Peace Pearl programme, to upgrade J-8 II with Western avionics, embargoed by US government mid-1989 and cancelled by China 1990; alternative (F-8 IIM) upgrade programme now in progress (see separate entry). Trials with some parts of structure covered in Xikai SF18 radar-absorbent material were reported in early 1999.

CURRENT VERSIONS: **J-8**: ('Finback-A'): Initial clear-weather day fighter. Single-piece forward-opening canopy; simple ranging radar in nose centrebody; single-barrelled gun on each side of nosewheel bay, with PL-2 AAMs on inboard wing pylons and provision for external tanks outboard.

J-8 I: ('Finback-A'): Improved (all-weather) version of J-8. Same power plant, but fitted from outset with Sichuan SR-4 fire-control radar in intake centrebody; single twin-barrelled 23 mm cannon on each side of lower front fuselage. Production completed. Approximately 54 in service in mid-2001, by which time apparently redesignated **J-8A**; figure includes some converted to reconnaissance role (see JZ-8 paragraph). Units have included 1st and 3rd Regiments of 1st Fighter Division at Anshan; some (including reconnaissance versions) remain in use with 70th Fighter Regiment at Yangkun.

J-8B: ('Finback-B'): All weather version (originally J-8 II), some 70 per cent redesigned compared with J-8 I. Main configuration change is to 'solid' nose and twin lateral air intakes, providing more nose space for fire-control radar and other avionics, plus increased airflow for more powerful WP13A II turbojets. In production and service, but manufactured in small economic batches rather than continuous production.

Late-production J-8Bs have an upgraded (KLJ-1 ?) fire-control radar with a lookdown/shootdown mode compatible with PL-8 (Python 3) IR-guided and PL-11 semi-active radar-guided AAMs, plus a new KJ-8602 RWR antenna on the fintip. New avionics, some possibly of Israeli design or origin, include an HK-13E HUD, 563B INS, JD-3II Tacan and an RKL 800A integrated ECM suite. In service with PLA Air Force and Navy.

Detailed description applies to J-8B.

J-8 II ACT: Active Control Technology (fly-by-wire) testbed, which first flew 29 December 1996 and completed its 49th and last sortie on 21 September 1999; shown in

Fintip RWR antenna identifies this 'Finback-B Mod' as a J-8D of the PLA Navy 9th Division's 2nd Regiment. Starboard refuelling probe appears to have been removed ***NEW***/0558581

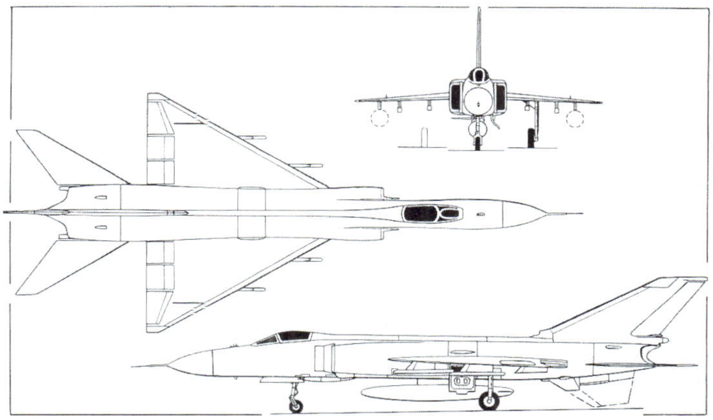

J-8B version of the 'Finback' twin-jet fighter ***(Jane's/Dennis Punnett)*** 0507133

model form at Airshow China in November 2000. Full-authority quad-redundant, three-axis digital AFCS; small canards mounted high on air intakes to induce instability; two 1553B-standard flight computers with databus interface; integrated servo actuators for all moving control surfaces.

J-8C: New production variant (originally J-8 III), apparently based on development work with J-8 II ACT. Features include fly-by-wire flight controls, canards, WP14 turbojets (73.6 kN; 16,535 lb st with afterburning), in-flight refuelling probe and new IAI Elta EL/M-2034 or similar fire-control radar. Prototype (8301) reportedly flew in 1993, but flight test not completed until late 2001; may now be in limited use.

J-8D: ('Finback-B Mod'): Designation (originally J-8 IV) of 12 or more J-8Bs built or modified for in-flight refuelling; non-retractable (but possibly removable) probe on starboard side of cockpit; combat radius increased to 648 n miles (1,200 km; 745 miles). In service with PLA Air Force and Navy.

JZ-8: Reconnaissance version (also reported as **J-8E** and **J-8 V**), believed to be converted from J-8A; at least six known. Retains gun armament; undernose sensor package similar to that fitted to some Su-17/20/22 variants; centreline pod similar in appearance to that carried by MiG-21R, incorporating large rectangular camera window or SLAR antenna. In service with PLA Air Force.

F-8 II: Originally proposed export version, subsequently modified to become present F-8 IIM.

F-8 IIM: Upgraded version; described separately.

CUSTOMERS: PLA Air Force and Navy.

DESIGN FEATURES: Extension of late 1950s Soviet heavy fighter theory. Thin-section, mid-mounted delta wings and all-sweptback tail surfaces; fuselage has area rule 'waisting', detachable rear portion for engine access, and dorsal spine fairing. Large ventral fin under rear fuselage, main portion of which folds sideways to starboard during take-off and landing, provides additional directional stability; small fence on each wing upper surface near tip; small airscoops at foot of fin leading-edge and at top of fuselage each side, above tailplane. Sweepback 60° on wing and tailplane leading-edges; wings have slight anhedral.

FLYING CONTROLS: Hydraulically boosted ailerons, rudder and low-set all-moving tailplane; two-segment single-slotted flaps on each wing trailing-edge, inboard of aileron; four door-type underfuselage airbrakes, one under each engine air intake trunk and one immediately aft of each mainwheel well.

STRUCTURE: Conventional aluminium alloy semi-monocoque/stressed skin construction, with high-tensile steel for high load-bearing areas of wings and fuselage and titanium in high-temperature fuselage areas; ailerons, rudder and rear portion of tailplane are of aluminium honeycomb with sheet aluminium skin; dielectric skins on nosecone, tip of main fin, and on non-folding portion of ventral fin leading-edge.

LANDING GEAR: Hydraulically retractable tricycle type, with single wheel and oleo-pneumatic shock-absorber on each unit. Steerable nose unit retracts forward, main units inward into centre-fuselage; mainwheels turn to stow vertically inside fuselage, resulting in slight overwing bulge. Brake-chute in bullet fairing at base of rudder.

POWER PLANT: Two LMC (Liyang) WP13A II turbojets, each rated at 42.7 kN (9,590 lb st) dry and 65.9 kN (14,815 lb st) with afterburning, mounted side by side in rear fuselage with pen-nib fairing above and between exhaust nozzles. Lateral, non-swept air intakes, with automatically regulated ramp angle and large splitter plates similar in shape to those of MiG-23. Internal fuel capacity (four integral wing tanks plus fuselage tanks) approximately 5,400 litres (1,426 US gallons; 1,188 Imp gallons). Single-point pressure refuelling. Provision for auxiliary fuel tanks on fuselage centreline and each outboard underwing pylon. J-8D retrofitted with probe for in-flight refuelling from Xian H-6 (Tu-16) bombers converted as aerial tankers.

ACCOMMODATION: Pilot only, on zero/zero ejection seat under one-piece canopy hinged at rear and opening upward. Cockpit pressurised, heated and air conditioned. Heated windscreen.

SYSTEMS: Two simple air-cycle environmental control systems, one for cockpit heating and air conditioning and one for radar cooling; cooling air bled from engine compressor. Two

CHINESE J-8 OPERATORS

Service	Air Division	Air Regiment	Base	Variant
PLA AF	1st	3rd	Anshan	J-8B
	9th	n/k	Foshan	J-8D
	18th	54th	Changsha	J-8D
	24th	70th[1]	Yangkun	J-8D, JZ-8
	24th	72nd	Zunhua	J-8
	26th	n/k	Chongming	J-8B
	30th	88th	Dandong	J-8
	30th	89th		J-8
	30th	90th		J-8
		4th	Yuhong	JZ-8
PLA Navy	5th	15th	Qingdao	J-8A
	9th	22nd	Lingshui	J-8B

[1] Previously equipped with J-8

207 bar (3,000 lb/sq in) independent hydraulic systems (main utility system plus one for flight control surfaces boost), powered by engine-driven pumps. Primary electrical power (28.5 V DC) from two 12 kW engine-driven starter/generators, with two 6 kVA alternators for 115/200 V three-phase AC at 400 Hz. Pneumatic bottles for emergency landing gear extension. Pop-out ram air emergency turbine under fuselage.

AVIONICS: *Comms:* VHF/UHF and HF/SSB radios; 'Odd Rods'-type IFF.

Radar: Obsolete Type 208 monopulse radar in nose. To be retrofitted with Phazotron Zhuk-8 II multimode fire-control radar (100 ordered in June 2001).

Flight: ILS, Tacan, marker beacon receiver, radio compass, radar altimeter, autopilot.

Mission: Gyro gunsight and gun camera.

Self-Defence: RWR (antenna in fintip); chaff/flare dispensers in tailcone.

Enlarged avionics bays in nose and fuselage provide room for modernised fire-control system and other upgraded avionics of later versions.

ARMAMENT: One 23 mm Type 23-3 twin-barrel cannon, with 200 rounds, in underfuselage pack immediately aft of nosewheel doors. Seven external stations (one under fuselage and three under each wing) for a variety of stores which can include PL-2B IR air-to-air missiles, PL-7 medium-range semi-active radar homing air-to-air missiles, Qingan HF-16B 12-round pods of 57 mm Type 57-2 unguided air-to-air rockets, launchers for 90 mm air-to-surface rockets, bombs, or (centreline and outboard underwing stations only) auxiliary fuel tanks.

DIMENSIONS, EXTERNAL:

Wing span ..... 9.345 m (30 ft 8 in)
Wing aspect ratio 2.1
Length overall: excl nose probe ..... 20.53 m (67 ft 4¼ in)
incl nose probe ..... 21.39 m (70 ft 2¼ in)
Height overall ..... 5.41 m (17 ft 9 in)
Wheel track ..... 3.74 m (12 ft 3¼ in)
Wheelbase ..... 7.335 m (24 ft 0¾ in)

AREAS:

Wings, gross ..... 42.20 m² (454.2 sq ft)

WEIGHTS AND LOADINGS:

Weight empty ..... 9,820 kg (21,649 lb)
Normal T-O weight ..... 14,300 kg (31,526 lb)
Max T-O weight ..... 18,879 kg (41,621 lb)
Wing loading:
at normal T-O weight ..... 338.9 kg/m² (69.40 lb/sq ft)
at max T-O weight ..... 447.4 kg/m² (91.63 lb/sq ft)
Power loading:
at normal T-O weight ..... 108 kg/kN (1.06 lb/lb st)
at max T-O weight ..... 143 kg/kN (1.40 lb/lb st)

PERFORMANCE:

Design max operating Mach No. (MMO) 2.2
Design max level speed ..... 701 kt (1,300 km/h; 808 mph) IAS
Unstick speed ..... 175 kt (325 km/h; 202 mph)
Touchdown speed ..... 156 kt (290 km/h; 180 mph)
Max rate of climb at S/L ..... 12,000 m (39,370 ft)/min
Acceleration from M0.6 to M1.25 at 5,000 m (16,400 ft) ..... 54 s
Service ceiling ..... 18,000 m (59,050 ft)
T-O run with afterburning ..... 670 m (2,200 ft)
Landing run, brake-chute deployed ..... 1,000 m (3,280 ft)
Combat radius ..... 432 n miles (800 km; 497 miles)
Ferry range ..... 1,025 n miles (1,900 km; 1,180 miles)
g limit in sustained turn at 5,000 m (16,400 ft) ..... +4.83

UPDATED

SAC F-8 IIM

TYPE: Multirole fighter.

PROGRAMME: Upgraded J-8B; outgrowth of earlier proposals for F-8 II export version. Developed jointly by SAC and Shenyang Aircraft Research Institute (SARI); first flight 31 March 1996, two years after delivery of drawings; flight testing of aircraft and WP13B engine completed 19 January 1998. Second F-8 IIM completed by late 1998. Additional proposed upgrades revealed by then included No. 607 Institute Blue Sky low-altitude navigation pod; Southwest China Electronic Equipment Research Institute KG 300G airborne self-protection jammer pod; No. 613 Institute FLIR/laser targeting pod; and a triple redundant digital fly-by-wire flight control system. These would presumably be similar to the systems quoted for the J-8 II ACT.

CUSTOMERS: Developed primarily for export, but none yet ordered; however, improvements now being adopted as upgrade for in-service AFPLA J-8Bs.

DESIGN FEATURES: Main differences from J-8B are more powerful engine, improved avionics and modernised cockpit with HOTAS controls.

POWER PLANT: Two LMC (Liyang) WP13B turbojets, each rated at 47.1 kN (10,582 lb st) dry and 68.7 kN (15,432 lb st) with afterburning.

SYSTEMS: Electrical system includes two 15 kVA starter/generators.

AVIONICS: *Comms:* Advanced com/nav radios; IFF.

Radar: Phazotron Zhuk-8 II multifunction, look-up/look-down pulse Doppler radar, with 38 n mile (70 km; 43.5 mile) detection range for approaching targets and 21.5 n mile (40 km; 25 mile) range for receding targets (both targets assumed to have 3 m²; 32.3 sq ft radar cross-section). N010 Zhuk-27 being flight tested by late 1998, presumably in second aircraft.

Flight: Include Tacan and datalink; locally developed HUD and INS/GPS navigation; MFDs; HOTAS controls; ARINC 429 databus.

F-8 IIM in attack mode, with bombs, rocket pods and air-to-air missiles 0010256

The modernised cockpit of the F-8 IIM 0507588

Self-defence: Omnidirectional RWR; rear hemisphere noise jammer against threat (including pulse Doppler) radars; chaff/flare dispenser.

ARMAMENT: Up to six Chinese PL-5 or PL-9 short-range AAMs on underwing stations, or two Russian R-27R1 (AA-10 'Alamo') medium-range AAMs; up to four seven-round pods of 90 mm Type 90-1 rockets; up to 10 anti-runway bombs or 10 Type 250-III or 250-IV low-drag bombs (four under wings and six under fuselage); or eight anti-tank bombs; or five 500 kg low-drag bombs. Internal cannon as for J-8 II.

DIMENSIONS, EXTERNAL, AND AREAS: As for J-8B

WEIGHTS AND LOADINGS:

Weight empty ... 10,371 kg (22,864 lb)
Normal fuel load ... 4,200 kg (9,259 lb)
Max external stores load ... 4,500 kg (9,921 lb)
Normal T-O weight ... 15,288 kg (33,704 lb)
Max T-O weight ... 18,879 kg (41,621 lb)
Max wing loading ... 447.4 kg/m² (91.63 lb/sq ft)
Max power loading ... 138 kg/kN (1.35 lb/lb st)

PERFORMANCE:

Max operating Mach No. (MMO) ... 2.2
Unstick speed ... 179 kt (330 km/h; 206 mph)
Touchdown speed ... 162 kt (300 km/h; 186 mph)
Level acceleration:
M0.7 to M1.0 at 1,000 m (3,280 ft) ... 21 s
M0.6 to M1.25 at 5,000 m (16,400 ft) ... 55 s
Max rate of climb at M0.9:
at 1,000 m (3,280 ft) ... 13,440 m (44,094 ft)/min
at 5,000 m (16,400 ft) ... 9,600 m (31,496 ft)/min
Service ceiling ... 18,000 m (59,060 ft)
T-O run with afterburning ... 630 m (2,070 ft)
Landing run with brake-chute ... 900 m (2,955 ft)
Typical mission radius:
air-to-air interception, out and back at M0.8:
at 500 m (1,640 ft) with 3 min combat ... 189 n miles (350 km; 217 miles)
at 11,000 m (36,080 ft) with 5 min combat ... 540 n miles (1,000 km; 621 miles)
combat air patrol, incl 10 min patrol and 5 min combat, out at M0.8 at 5,000 m (16,400 ft), back at M0.8 at 11,000 m (36,080 ft) .. 324 n miles (600 km; 372 miles)
air-to-ground attack, out and back at M0.8 at 10,000 m (32,800 ft), incl 5 min combat ... 486 n miles (900 km; 559 miles)
Ferry range ... 1,026 n miles (1,900 km; 1,180 miles)
g limits during sustained turn at M0.9:
at 1,000 m (3,280 ft) ... +6.9
at 5,000 m (16,400 ft) ... +4.7

UPDATED

SAC (SUKHOI SU-27SK) J-11

TYPE: Air superiority fighter.

PROGRAMME: In February 1996, the Russian military sales organisation Rosvooruzheniye (now Rosoboronexport) announced a contract under which China would be licensed to manufacture the Sukhoi Su-27 'Flanker' (up to 50 per year; total of 200 in all) at Shenyang.

An initial batch of 26 Russian-built Su-27s, delivered from 1992, comprised 24 Su-27SKs ('Flanker-B') and two two-seat Su-27UBK combat trainers ('Flanker-C'). These equip the AFPLA's 3rd Division, based at Wuhu, Anhui Province, in the Nanjing Military Region. They were followed in 1996 by a further 24 (14 and 10, respectively), mostly delivered to the 2nd Division in Shuixi, Guangdong Province, Guangzhou Military Region. In February 1997 Russian licence granted for Chinese manufacture at Shenyang, initially in the form of CKD kit assembly; in 1998 KnAAPO delivered the first two kits; both aircraft made their first flight in December 1998. However, it was reported in mid-2000 that substandard work had caused Russian technicians to rebuild first two aircraft, necessitating import, from 14 December 2000 onwards, of 28 additional Russian-built two-seat Su-27UBKs to offset shortfall in Chinese production and maintain pilot training schedule. Eight of these delivered by 31 December 2000, with 10 each following in 2001 and 2002. Additional units now include 1st, 2nd, 7th, 19th and 33rd Divisions.

As regards Chinese production, six or seven Su-27s were planned to be assembled annually during 1999–2001, increasing to 15 to 20 per year from 2002. Chinese-built Su-27s are designated **J-11** (single-seat) and **JJ-11** (two-seat), but none of the latter has been reported. By late 2003 all 95 parts kits (of diminishing completeness) representing the first batch had been delivered from Russia, but there was no sign of the follow-on order for 105, nor of a change of local manufacture to Su-30MKK variant, as once expected. Acquisition of MKK (for which local designation J-13 suggested) is now from Russian production. PLA Naval Aviation also received 24 Su-30MK2s in 2004, bringing total 'Flanker' family acquisition to 273.

According to a late 2002 report Shenyang is developing its own multirole version of the J-11, to incorporate an indigenous multimode radar and other avionics and be capable of deploying the domestic PL-12/SD-10 active radar-homing AAM and Chinese or Russian PGMs. One Su-27K/J-11 was also said to have been used in 2002 to flight test the domestic Liming WS10A turbofan engine. However, it is suggested that a Russian upgrade programme (Su-27SMK) may have China as principal customer.

UPDATED

SAC-built J-11 No. 80, one of the most recent reported (early 2004), serving 33rd Division ***NEW***/0589264

SAC (?) 'J-X'

The embryonic advanced combat aircraft to which the US Office of Naval Intelligence (ONI) originally allocated the provisional designation XXJ was last described in the late 1990s, at which time it was thought likely to be a Chengdu (CAC) design. Little was then heard of it for some years, but in the early months of 2001 further US-sourced information indicated that the design had been considerably modified, then revealing a configuration suggesting some Sukhoi influence. This appeared to make Shenyang the more likely developer, although both Chinese companies are thought to have (possibly competing) advanced combat aircraft projects in hand.

According to the 2001 information, the aircraft had a canard delta configuration, combined with clipped-delta horizontal tail surfaces and twin outward-canted fins and rudders. Conformal underfuselage intakes fed a pair of thrust-vectoring turbojets (possibly the new 116 kN (26,000 lb) class WP15) or turbofans. Empty weight was estimated at 20,000 kg (44,100 lb); stealth characteristics and FBW flight controls were assumed.

However, according to a senior AVIC I official in late 2002, this was one of a number of design concepts created by Shenyang in collaboration with No. 601 Research Institute, wind

Wind tunnel models of two possible 'J-X' designs *(AVIC I via Y Chang)* ***NEW***/0531381

tunnel testing of which was then under way. Two such configurations were illustrated in an AVIC I video shown at Airshow China in November 2002, both showing twin-engined, tailed-delta designs broadly similar in outline to the fuselage of the US F/A-22 and the wings and vertical tail of the F-16. Possible power plant is the domestically developed Liming WS10A turbofan, with thrust-vectoring nozzles. Multifunction fire-control radars under consideration are said to include the Chinese Type 1473, reported to have a search range of 81 n miles (150 km; 93 miles) and be able to track up to 15 targets of which up to eight could be attacked simultaneously. Russia's Phazotron Zhemchug radar is also thought to be under consideration. The designs appear to cater for internal weapons stowage.

UPDATED

SAIC (SHANGHAI)

SHANGHAI AVIATION INDUSTRY CORPORATION (SHANGHAI HANGKONG GONGYE GONGSI) (SUBSIDIARY) (SUBSIDIARY OF AVIC II)

2668 Zhongshan North Road, Shanghai 200063
Tel: (+86 21) 62 57 33 51
Fax: (+86 21) 62 57 33 50
e-mail: saiccn@sh163a.sta.net.cn
Web: http://www.saic-china.com
CHAIRMAN: Li Wanxin

SAIC is the core company of the Shanghai Aviation Industry Group. It has more than 20 subordinate enterprises, of which the principal ones are Shanghai Aircraft Manufacturing Factory (SAMF); Shanghai Aircraft Research Institute (SARI); Shanghai Aero-Engine Manufacturing Factory; and Shanghai International Aero Technology. SAIC and SARI are taking part in the development programme for the ACAC ARJ21 regional jet. Subcontract work includes components for the ATR 72 and Boeing 737/757.

UPDATED

SAIC (SHIJIAZHUANG)

SHIJIAZHUANG AIRCRAFT INDUSTRY CORPORATION (SHIJIAZHUANG FEIJI GONGYE GONGSI) (SUBSIDIARY OF AVIC II)

PO Box 164, 25 Beihuan-lu, Shijiazhuang, Hebei 050062
Tel: (+86 311) 777 52 83
Fax: (+86 311) 775 29 93
e-mail: samc@yeah-net
Web: http://www.samc.com.cn
GENERAL MANAGER: Cheng Binyou
CHIEF ENGINEER: Lang Zhidong

SAIC (formerly SAMC) has produced Chinese Y-5 versions of the An-2 general purpose biplane since 1970, concentrating since 1989 on Y-5B and C customised agricultural, forestry, tourist and parachutist versions; other products have included W-5 and W-6 ultralight series. With the effective cessation of Polish An-2 production, SAIC is the only company now regularly building new examples of this long-serving biplane. Became part of Xian Aircraft Industrial Group July 1992, but relocated within AVIC II in 1999 reorganisation. Latest venture is a TB20 Trinidad lookalike known as the LE-500 Little Eagle.

Occupies 46 ha (113.7 acre) site, including over 100,000 m² (1,076,400 sq ft) of covered space. Workforce of 2,805 in April 2001 included 461 engineers and technicians.

The 'tipsailed' Y-5C for the PLA Air Force 0507590

SAIC LE-500 LITTLE EAGLE

TYPE: Four-seat lightplane.

PROGRAMME: Mockup displayed at November 2002 Airshow China in Zhuhai; prototype then expected to fly in October 2003, but not reported by early 2004. Service entry targeted for 2005.

DESIGN FEATURES: Said to be a joint development by SAIC, an AVIC I design institute, and China Civil Aviation Flying Academy (CCAFA). Appears outwardly identical to Socata TB20 Trinidad, with which it shares Lycoming IO-560 power plant and similar MTOW and cruising speed, but developed without French company's assistance.

CUSTOMERS: Intention to purchase signified by CCAFA, but may face competition from Western manufacturers to replace existing fleet of French-built Trinidads.

Mockup of Shijiazhuang LE-500 shown at Zhuhai in November 2002 ***(Robert Hewson)*** ***NEW***/0552052

SAIC Y-5

Chinese name: Yunshuji-5 (Transport aircraft 5)

TYPE: Utility biplane.

PROGRAMME: Antonov An-2 has been built under licence in China since 1957, chiefly at Nanchang (727 produced up to 1968) and latterly by SAIC; production of Y-5N civil transport and general purpose version ended in 1986, at which time 221 had been completed. Y-5B dedicated agricultural and forestry version made first flight 2 June 1989; now in batch production; first nine Y-5Bs produced in 1990; total of 112 Y-5B/Cs produced by end of 2000 (latest total received). SAIC now only source of continuing production of this aircraft.

CURRENT VERSIONS: **Y-5B:** Dedicated agricultural and forestry version; certified to CCAR 23 Chinese equivalent of FAR Pt 23.

Following description applies to Y-5B, except where indicated.

Y5B-100: Designation of version displayed at Zhuhai in November 1998, fitted with wing 'tipsails' of the Y-5C and registered B-8448.

Y-5B(K): Tourist version, first flown 1993; certified to CCAR 23.

Y-5B(D): Multipurpose (agri-forest or tourist) version, first flown 1995; certified to CCAR 23.

Y-5C: Parachutist version, with wingtip vanes ('tipsails'); first flown 1996. Ordered by PLA Air Force (AFPLA).

CUSTOMERS: See table.

Y-5 B/C PRODUCTION (at mid-2002):

Customer:	
Military (Y-5C)	
PLA Air Force	48
Airlines (Y-5B)	
Changchun Airlines	4
Hebei Jihua Airlines	3
Jingmen United Corporation	11
Xinjiang Airlines	15
Other local operators	6
General (Y-5B)	
Anhui General Aviation	3
Chaoyang General Aviation	8
Guanghan Civil Aviation Flying School	5
Guangzhou General Aviation	3
Hebei General Aviation	3
Jiangnan General Aviation	3
Total	**112**

DESIGN FEATURES: Unequal span, single-bay biplane; strut-braced wings (single I strut each side) and tail; fuselage circular section forward, rectangular in cabin section, oval in tail section; fin integral with rear fuselage. RPS wing section, thickness/chord ratio 14 per cent (constant); dihedral, both wings, approximately 2° 48'.

FLYING CONTROLS: Conventional and manual. Differential ailerons, plus elevators and rudder, actuated by cables and push/pull rods; electric trim tab in port aileron, rudder and port elevator; full-span automatic leading-edge slots on upper wings; electrically actuated slotted trailing-edge flaps on both wings. Wingtip vanes on some variants (see Current Versions) for additional control at low speeds.

STRUCTURE: All-metal, with fabric covering on wings aft of main spar and on tailplane. Y-5B specially treated to resist corrosion; cabin doors sealed against chemical ingress; empty weight reduced.

LANDING GEAR: Non-retractable split axle type, with long-stroke oleo-pneumatic shock-absorbers. Mainwheel tyres size 800×260, pressure 2.25 bar (33 lb/sq in). Pneumatic shoe brakes on main units. Fully castoring and self-centring tailwheel, size 470×210, with electropneumatic lock. For rough field operation, shock-absorbers can be charged from compressed air cylinder in rear fuselage. Interchangeable ski landing gear optional.

POWER PLANT: One 735 kW (985 hp) SAEC (Zhuzhou) HS5 nine-cylinder radial engine, driving a Baoding J12B-G15 four-blade variable-pitch propeller. Fuel capacity 1,240 litres (328 US gallons; 273 Imp gallons).

ACCOMMODATION: Flight crew of one or two; dual controls; seats for 12 tourist class passengers (Y-5B(K)) or 10 parachutists (Y-5C). Emergency exit on starboard side at rear. Cabin heating and ventilation improved by new ECS.

AVIONICS: *Comms:* Honeywell KHF 950 HF and KY 196 VHF radios and KMA 24 audio control panel; some electrical and other instrument installations also improved.

Flight: GPS 155.

EQUIPMENT: Large hopper/tank with emergency jettison of contents; high flow rate, wind-driven pump; sprayers with various nozzle sizes, depending upon spray volume required.

DIMENSIONS, EXTERNAL:
Wing span: upper ... 18.18 m (59 ft 7¾ in)
lower ... 14.24 m (46 ft 8½ in)
Wing aspect ratio: upper ... 7.6
lower ... 7.3
Length overall, tail down ... 12.40 m (40 ft 8¼ in)
Height overall, tail down ... 4.01 m (13 ft 2 in)
Wheel track ... 3.36 m (11 ft 0¼ in)
Wheelbase ... 8.19 m (26 ft 10½ in)
Propeller diameter ... 3.40 m (11 ft 2 in)
Cargo door (port): Mean height ... 1.55 m (5 ft 1 in)
Mean width ... 1.39 m (4 ft 6¾ in)

DIMENSIONS, INTERNAL:
Cargo compartment: Length ... 4.10 m (13 ft 5½ in)
Max width ... 1.60 m (5 ft 3 in)
Max height ... 1.80 m (5 ft 10¾ in)

AREAS:
Wings, gross: upper ... 43.54 m² (468.7 sq ft)
lower ... 27.98 m² (301.2 sq ft)

WEIGHTS AND LOADINGS (A: Y-5B with dry chemical spreader, B: Y-5B with liquid spray system, C: Y-5C):
Max payload: A, B, C ... 1,500 kg (3,307 lb)
Max T-O weight: A, B, C ... 5,250 kg (11,574 lb)
Max wing loading: A, B, C ... 73.4 kg/m² (15.03 lb/sq ft)
Max power loading: A, B, C ... 7.15 kg/kW (11.75 lb/hp)

PERFORMANCE (A, B and C as for Weights):
Max level speed at S/L:
A ... 110 kt (205 km/h; 127 mph)
B ... 108 kt (200 km/h; 124 mph)
C ... 129 kt (239 km/h; 148 mph)
Max level speed at FL56:
A ... 119 kt (220 km/h; 137 mph)
B ... 116 kt (215 km/h; 133 mph)
C ... 138 kt (256 km/h; 159 mph)
Operating speed: A, B ... 86 kt (160 km/h; 99 mph)
Stalling speed: A, B, C ... 46 kt (85 km/h; 53 mph)
Max rate of climb at S/L: A ... 120 m (394 ft)/min
B ... 114 m (374 ft)/min
C ... 213 m (699 ft)/min
Rate of climb at FL52:
A ... 133 m (436 ft)/min
B ... 123 m (404 ft)/min
C ... 225 m (738 ft)/min
Service ceiling: A, B ... 4,500 m (14,760 ft)
C ... 5,000 m (16,400 ft)
Air turning radius: A, B, C ... 350 m (1,150 ft)
T-O run: A ... 170 m (560 ft)
B ... 180 m (595 ft)
C ... 150 m (495 ft)
Landing run: A ... 160 m (525 ft)
B ... 157 m (515 ft)
C ... 170 m (560 ft)
Range at S/L with 670 litres (177 US gallons; 147 Imp gallons) fuel:
A, B ... 456 n miles (845 km; 525 miles)
C ... 475 n miles (880 km; 546 miles)
Endurance, conditions as above: A, B ... 5 h 36 min
Swath width (A, B): HV, LV ... 40–50 m (130–165 ft)
ULV ... 60 m (200 ft)

UPDATED

SAMF

SHANGHAI AIRCRAFT MANUFACTURING FACTORY (SHANGHAI FEIJI ZHIZAO GONGCHANG) (SUBSIDIARY OF SAIC SHANGHAI)

3115 Chang Zhong Road, Shanghai 200436
Tel: (+86 21) 56 68 11 22
Fax: (+86 21) 56 68 43 36
PRESIDENT: Liu Qianyou

Shanghai Aircraft Manufacturing Factory (SAMF) created 1951; now part of Shanghai Aviation Industry Corporation (SAIC, which see); occupies site area of 135.5 ha (334.8 acres); SAMF workforce 4,630 in late 2000. Built small batch of McDonnell Douglas MD-82/83 airliners in mid-1990s, and produced main and landing gear doors for MD-80 series from 1979; produced cargo and service doors, avionics access doors and tailplanes for these aircraft. Current subcontract work includes tailplanes for Boeing 737 NG. Also built Q-2 ultralight.

VERIFIED

SSAC

SHANGHAI SIKORSKY AIRCRAFT COMPANY (SHANGHAI SIKORSKY FEIJI GONGSI)

3115 Chang Zhong Road, Shanghai 200436
Tel: (+86 21) 66 50 74 76 and 56 68 08 99
Fax: (+86 21) 66 50 74 67
e-mail (1): public@shanghaisikorsky.com
e-mail (2): caiyan@shanghaisikorsky.com
Web: http://www.shanghaisikorsky.com
CEO: Teng Wei
GENERAL MANAGER: Chris Jaran
SALES MANAGER: Cai Yan

Agreement to form this private joint venture company announced in February 2002 by Sikorsky Aircraft of USA and Shanghai Little Eagle Science and Technology Company (SLEC). Objective is Chinese kit assembly and/or local manufacture of Schweizer 300C, 300CB and 333 (which see), with certification and marketing assistance from Sikorsky; shareholding SLEC 51 per cent, Sikorsky 49 per cent. Reported capacity to produce up to 24 aircraft per year from 2004. New 3,000 m² (32,292 sq ft) assembly plant under construction; due to come into operation September 2004.

UPDATED

SSAC (SCHWEIZER) SHEN SERIES

PROGRAMME: First three (of five) kits delivered 26 August 2002; assembly in China by Schweizer personnel under way by November 2002. Target of 24 for completion in first full year of programme; plan to reach eventual peak of 48 per year. Chinese government approval to begin business operations received April 2003. Name 'Shen' derived from Shanghai's river.

CURRENT VERSIONS: **Shen2B:** Schweizer 300CBi (two seats, Lycoming HO-360-G1A engine); optimised for training and police patrol.

Shen3A: Schweizer 300C (three seats, Lycoming HIO-360-D1A engine); utility missions include agriculture and police or powerline patrol.

Shen4T: Schweizer 333 (four seats, R-R 250-C20W turboshaft); law enforcement, observation, powerline patrol and turbine training.

One of the first Shen2Bs delivered to China for service with Panyu Police ***NEW***/0552270

CUSTOMERS: First two Shen2Bs leased to Panyu Police by Guangdong Baiyun General Aviation Company 2002; two Shen2Bs and one Shen3A to Xinjin branch of China Civil Aviation Flying College in September 2002; one Shen3A to Sichuan Forestry Bureau in 2003; two Shen3A for April 2004 delivery to Pianjin Aviation University; one Shen4T for unnamed customer.

COSTS: Shen2B US$260,000, Shen3A US$280,000, Shen4T US$600,000 (approximate, 2003).

UPDATED

SVAM

SHANGHAI VANTAGE AIRSHIP MANUFACTURING COMPANY (SHANGHAI VANTAGE FEITING ZHIZAO GONGSI)

Shanghai
Tel: (+86 21) 39 51 01 81
Fax: (+86 21) 52 53 14 86
e-mail: info@vantageship.com
Web: http://www.vantageship.com
GENERAL MANAGER: Xu Shunli

Company formed 1991 and registered in May 2000; workforce of 28 are mainly former military personnel, many (including three pilots) wth helium airship experience; has CAAC approval. Completed its first airship (single-place CA-50) in September 1999; second is medium-sized, twin-engined CA-80 for advertising, aerial photography and surveillance; combined flying hours more than 220 by late 2002. Long-term plans include 60 tonne payload CA-800.

UPDATED

SVAM CA-80

TYPE: Helium non-rigid.
PROGRAMME: Completed September 2001. In service as advertising and aerial survey or surveillance platform. Certified to equivalent of FAA airworthiness standard. A CA-80B has also been reported, built in 2002.
DESIGN FEATURES: Conventional envelope shape with ventral gondola. Four tailfins, with rudders/elevators, in X configuration.
POWER PLANT: Twin, Teledyne Continental IO-240-B vectored-thrust piston engines, each 93 kW (125 hp), driving MT-Propeller MTV-7 variable-pitch, pusher propellers.
ACCOMMODATION: Pilot and up to five passengers.
AVIONICS: *Comms:* Bendix/King KY 96 VHF and KT 76A transponder.
Flight: Garmin 130C GPS.
DIMENSIONS, EXTERNAL:
Envelope: Length ... 43.00 m (141 ft 1 in)
Max diameter ... 10.60 m (34 ft 9¼ in)
Fineness ratio ... 4.1
Height overall ... 14.00 m (45 ft 11¼ in)
Banner area (each side):
Length ... 31.00 m (101 ft 8½ in)
Depth ... 9.71 m (31 ft 10¼ in)

The twin-engined SVAM CA-80 helium semi-rigid 0525362

Gondola: Length ... 3.90 m (12 ft 9½ in)
Max width ... 1.60 m (5 ft 3 in)
Max height ... 1.85 m (6 ft 0¾ in)
DIMENSIONS, INTERNAL:
Envelope volume ... 2,533.0 m^3 (89,450 cu ft)
Ballonet volume (25%) ... 633.25 m^3 (22,363 cu ft)
AREAS:
Advertising banner area (total) ... 602.0 m^2 (6,480 sq ft)
WEIGHTS AND LOADINGS:
Weight empty ... 1,800 kg (3,968 lb)
Max fuel weight ... 300 kg (661 lb)
Max payload ... 742 kg (1,636 lb)
Max buoyancy ... 2,570 kg (5,666 lb)
Max T-O weight ... 2,542 kg (5,604 lb)
PERFORMANCE:
Max level speed ... 51 kt (95 km/h; 59 mph)
Max cruising speed ... 36 kt (66 km/h; 41 mph)
Max rate of climb at S/L ... 486 m (1,594 ft)/min
Max rate of descent ... 426 m (1,398 ft)/min
Pressure ceiling ... 2,745 m (9,000 ft)
Range ... 412 n miles (763 km; 474 miles)
Endurance ... 12 h

UPDATED

WHIC

WUHAN HELICOPTER INDUSTRY COMPANY (WUHAN ZHISHENGJI GONGYE GONGSI)

9th Floor, 871 Jiefang Avenue, Wuhan 430013
Tel: (+86 27) 85 79 31 13
Fax: (+86 27) 85 78 38 23
e-mail: whic@public.whhb.cn
Web: http://www.whic.com.cn
PRESIDENT: Dinghe Zhang

Established August 1993, under auspices of Wuhan local government, and with ¥64.5 million capital, to co-produce Enstrom F28F, 280FX, TH-28 and 480 helicopters (which see); assembled first two (280FX and TH-28) from CKD kits October 1993; delivered to Wuhan Public Security Bureau and PLA respectively; at least one more US-built TH-28 supplied to China in May 1997. Two Enstrom 480s delivered for reassembly in June 1998. WHIC also operates an Enstrom 480 as demonstrator and for charter work. New factory completed on a 150 ha (370.7 acre) site and reportedly became operational in 1998, although no production had occurred by end of 2002. However, agreement revived by March 2003, when Enstrom announced expectation that component manufacture by WHIC expected to begin in third quarter of that year. Manufacture duly launched on 3 September 2003, coupled with news that

Wuhan Enstrom TH-28 in Chinese markings 0044499

plant had received Y202 million investment from Wuhan Helicopter General Airline Company. Factory is expected to have eventual capability for completing up to 100 helicopters per year. Most are expected to be used for forest fire patrol work.

XAC

XIAN AIRCRAFT INDUSTRY COMPANY (XIAN FEIJI GONGYE GONGSI) (SUBSIDIARY OF AVIC I)

PO Box 140-84, Xian, Shaanxi 710089
Tel: (+86 29) 684 56 65
Fax: (+86 29) 620 37 07
e-mail: zh-yang@xac.com.cn
Web: http://www.xac.com.cn
CHAIRMAN: Yang Zhong
PRESIDENT OF XIAN AIRCRAFT INDUSTRIAL GROUP, AND GENERAL MANAGER OF XAC:
Gao Dacheng
VICE-PRESIDENT, XAC: Nie Zhongliang
AIRCRAFT MARKETING MANAGER: Wang Zhigang

Aircraft factory established at Xian 1958; current company is core enterprise of Xian Aircraft Industry Group (XAIG); it has covered area of some 300 ha (741.5 acres); of 20,000 workforce, about 90 per cent are engaged in aircraft production.

Major current programmes concern JH-7 attack aircraft and MA60 transport (Y-7/An-24 derivative). XAIG and Xian Aircraft Design and Research Institute (XADRI) are participants in the ACAC ARJ21 regional jet programme.

Subcontract work includes glass fibre header tanks, water float pylons, ailerons and various doors for Bombardier 415 amphibian; fins and tailplanes for Boeing 737/757; and A320 wing components for Airbus Industrie. In 1997, subcontracting for ATR (which began with ATR 42 wingtips in 1986) was extended to include ATR 72 rear fuselage sections.

UPDATED

XAC JH-7

Chinese name: Jianjiji Hongzhaji-7 (Fighter-Bomber aircraft 7)
Westernised designation: FBC-1 (formerly B-7)
Export name: Flying Leopard

TYPE: Attack fighter.
PROGRAMME: Revealed publicly September 1988 as model at Farnborough International Air Show; first of four or five prototypes said to have been rolled out during previous month; first flight 14 December 1988 and first supersonic flight 17 November 1989; service entry originally scheduled for 1992–93, but delayed; first seen openly in TV broadcast of October 1995 naval exercise.

XAC JH-7 '69', believed to be the first production aircraft ***NEW***/0573890

Photograph showing a full-scale mockup of the JH-7A ***NEW***/0573882

JH-7s apparently of the second Regiment to equip, cockpit covers obscuring the obvious difference between JH-7 and JH-7A. Aircraft far left is in primer; second from left has the unit code 08 ***NEW***/0573889

First public appearance (by third prototype '083', wearing Flight Test Establishment colours) was made in November 1998 flypast at Airshow China in Zhuhai. This coincided with announcement of new export designation and name, and statement that aircraft was "being redesigned" for export market.

CURRENT VERSIONS: **JH-7:** Domestic version.

Following description applies to JH-7 except where indicated.

JH-7A: Various China-, Russia- and US-based sources have reported that No. 603 Institute still undertaking improvements that include No. 613 Institute FLIR/laser targeting pod and No. 607 Institute Blue Sky low-altitude navigation pod; modified LETRI JL-10A Shen Ying radar; digital (and possibly quadruple) fly-by-wire controls; a health monitoring system; INS/GPS; new databus; two additional hardpoints each side; one-piece wraparound windscreen; greater use of composites; and integration of additional Russian weapons such as Kh-31P (AS-7 'Krypton') standoff missiles and KAB-500 laser-guided bombs.

Development completed in 2001 and first flight achieved 1 July 2002. Possibility of 20 to 30 being produced for service entry in 2004.

FBC-1 Flying Leopard: Designation (from 1998) of proposed export version; would have customer-defined radar, avionics and armament. No orders yet announced.

FBC-1M Flying Leopard II: Upgraded export version, revealed in model form at Aviation Expo/China exhibition in Beijing, September 2003. Improvements generally parallel those described for JH-7A, plus possible further increase of external hardpoints to 11, enabling maximum stores load of up to 9,000 kg (19,842 lb).

CUSTOMERS: Earlier (1997) Chinese reports, of up to 24 in service with PLA Naval Air Force in operational evaluation role in mid-1997, were apparently exaggerated (stated at 1998 Airshow China that only seven prototype/preproduction aircraft then completed). By early 2001 20 aircraft, excluding perhaps five (four seen) prototypes, confirmed by serial numbers (69 to 88); this is compatible with known number (50) of Spey engines initially purchased, but it has been reported that a further 80 to 90 surplus or second-hand Speys were delivered in mid-2001; these possibly for JH-7A. First identified JH-7 unit is PLA Naval Air Force 16th Bomber Regiment, 6th Naval Air Division, at Dachang, Shanghai. By 2003, aircraft numbered between 04 and 10 had been noted with 6th Division, but probably equipping a second Regiment.

DESIGN FEATURES: In same role and configuration class as Russian Sukhoi Su-24 'Fencer'. High-mounted wings with compound sweepback, dog-tooth leading-edges and 7° anhedral; twin turbofans, with lateral air intakes; all-swept tail surfaces, comprising large main fin, single small ventral fin and low-set all-moving tailplane with anti-flutter weights at tips; small overwing fence at approximately two-thirds span. Wings have 47° 30' sweepback on leading-edges. Quarter-chord sweep angles approximately 45° on fin, 55° on tailplane.

STRUCTURE: Conventional all-metal, except for dielectric panels. Plans to fit SF18 radar-absorbent material revealed in early 1999.

LANDING GEAR: Retractable tricycle type, with twin wheels on each unit. Trailing-link main units retract inward, nose unit rearward.

POWER PLANT: All aircraft so far built are powered by two licence-built (Xian WS9) or UK-supplied Rolls-Royce Spey Mk 202 turbofans: each 91.2 kN; 20,515 lb st with afterburning). Centreline and outboard underwing stations plumbed for auxiliary fuel tanks.

Original intention to power production JH-7 with LM (Liming) WS6 turbofans of 71.1 kN (15,990 lb st) dry rating (122.1 kN; 27,445 lb st with afterburning) abandoned, as was licensed production of Russian alternative such as Saturn/Lyulka AL-31F or SNECMA M53-P2.

ACCOMMODATION: Crew of two in tandem (rear seat elevated). HTY-4 ejection seats, operable at speeds from zero to 540 kt (1,000 km/h; 621 mph) and altitudes from S/L to 20,000 m (65,600 ft). Individual canopies, hinged at rear and opening upward. One-piece wraparound windscreen on JH-7A.

AVIONICS(FBC-1): *Comms:* Short wave and ultra short wave transceivers of Italian origin.

Radar: LETRI JL-10A Shen Ying J-band pulse Doppler fire-control radar (search range 43 n miles; 80 km; 50 miles and tracking range 22 n miles; 40 km; 25 miles in look-up mode, or 29 n miles; 54 km; 34 miles and 17 n miles; 32 km; 20 miles, respectively, in look-down mode). Scans ±60° in azimuth and can track four targets simultaneously.

Flight: Automatic flight control system; GPS/INS navigation.

Instrumentation: Includes No. 603 Institute HUD and two other MFDs.

Chinese TV picture of the JH-7A production line ***NEW***/0573883

Chinese TV picture of JH-10A radar in a JH-7A ***NEW***/0573884

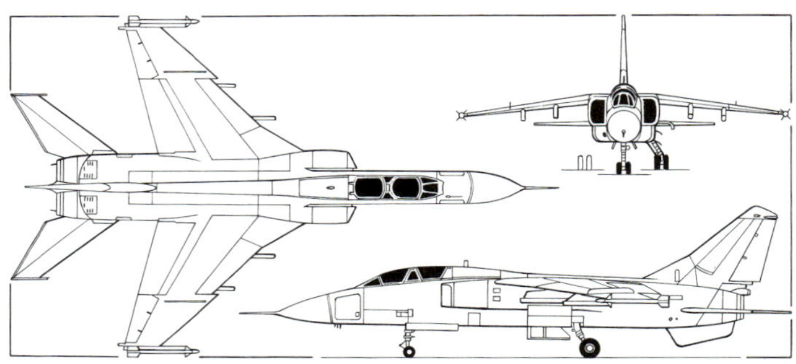

XAC JH-7 interdictor ***(James Goulding)*** 0105034

Mission: AFCS linked to fire-control system; No. 613 Institute helmet-mounted sight.

Self-defence: Chaff/flare dispenser at base of vertical tail; omnidirectional RWR; active and passive ECM.

ARMAMENT: Twin-barrel 23 mm gun, with 200 rounds, in starboard side of lower fuselage, just forward of mainwheel bay. JH-7 has fuselage centreline stores station, plus two under each wing and rail for PL-5B, PL-7 or similar close-range air-to-air missile at each wingtip; JH-7A has additional mid-point station under each wing, plus one on lower side of each engine intake trunk. Typical underwing load for maritime attack, two C-801K or C-802K sea-skimming anti-ship missiles (inboard) and two drop tanks (outboard); C-803 (YJ-83) test-launched in 2002. Other potential weapons may include C-701 TV-guided anti-ship missiles and 500 kg LGBs.

DIMENSIONS, EXTERNAL:	
Wing span	12.705 m (41 ft 8¼ in)
Wing aspect ratio	3.1
Length overall: excl probe	21.025 m (68 ft 11¾ in)
incl probe	22.325 m (73 ft 3 in)
Fuselage length (excl probe)	19.105 m (62 ft 8¼ in)
Height overall	6.575 m (21 ft 6¾ in)
Tailplane span	7.39 m (24 ft 3 in)
Wheel track	3.06 m (10 ft 0½ in)
Wheelbase	7.805 m (25 ft 7¼ in)
AREAS:	
Wings, gross	52.30 m^2 (563.0 sq ft)
WEIGHTS AND LOADINGS:	
Max fuel weight	10,050 kg (22,156 lb)
Max external stores load	6,500 kg (14,330 lb)
Max T-O weight	28,475 kg (62,776 lb)
Max landing weight	21,130 kg (46,583 lb)
Max wing loading	544.5 kg/m^2 (111.51 lb/sq ft)
Max power loading (WS9 engines)	156 kg/kN (1.53 lb/lb st)

PERFORMANCE:
Max level speed at 11,000 m (36,080 ft) (clean) ..... M1.7 (975 kt; 808 km/h; 1,122 mph)
Max operating speed .... 653 kt (1,210 km/h; 751 mph) IAS
Cruising speed .... M0.80-0.85 (459–487 kt; 850–903 km/h; 528–561 mph)
Service ceiling (clean) .... 15,600 m (51,180 ft)
T-O run .... 920 m (3,020 ft)
Landing run .... 1,050 m (3,445 ft)
Combat radius .... 891 n miles (1,650 km; 1,025 miles)
Ferry range .... 1,970 n miles (3,650 km; 2,268 miles)
g limit .... +7

UPDATED

XAC MA60

TYPE: Twin-turboprop airliner.

PROGRAMME: Model of MA60 as proposed Y7-200A (PW127 turboprop) variant first shown at Asian Aerospace, Singapore, February 2000. First flight (prototype B-995L) reportedly 12 March 2000; CAAC type certificate issued 22 June 2000; revenue service began August 2000; public debut (Airshow China) November 2000; production certificate issued 12 December 2000.

Designation indicates Modern Ark, 60 (high density) seats. To replace earlier Y-7 variants, which from 2001 no longer meet amended Chinese airworthiness requirements. XAC production capacity 12 to 15 a year; domestic operators will lease from central leasing agency SFLC, which will be responsible for sole-source marketing (first arrangement of this kind in China).

CURRENT VERSIONS: **MA60:** Initial production passenger version; *as described*. Current life-cycle of 25,000 landings and 30,000 flight hours.

MA60-100: Improvements planned include reduction in empty weight by about 400 kg (882 lb); increase in 'hot-and-high' MTOW of about 900 kg (1,984 lb); increase in OEI ceiling; 162 n mile (300 km; 186 mile) increase in range; airframe drag reduction; Rockwell Collins Pro Line 21 avionics upgrade. Not known (early 2004) whether improvements yet implemented.

MA60H-500: Updated MA60 equivalent of former Y7H, with rear-loading ramp/door inherited from An-26. Fully pressurised cargo compartment, with electric winch and hydraulic conveyor; cargo door can be retracted under belly to permit direct loading from truck or ground. Available from 2004. Demonstrator bears CFTE serial number 073.

MA60-MPA (Fearless Albatross): Projected maritime patrol version (previously Y7-200BF); nose-mounted search radar; raked wingtips; auxiliary fuel tanks scabbed to fuselage sides to extend range to estimated 1,350 n miles (2,500 km; 1,553 miles); four underwing stations for air-to-air and anti-ship missiles. Shown as model at Airshow China, November 2000.

MA40: Shortened (40-passenger) version; under consideration.

CUSTOMERS: Sole-source domestic marketing by newly formed Shenzhen Finance Leasing Corporation which, on 7 November 2000, signed contract for delivery of 60 MA60s over five-year period. Launch operator Sichuan Airlines (five ordered, plus five on option) introduced first leased example (B-3426) into revenue service August 2000. Other reported orders by March 2003 from Changan Airlines (two), China Northern (five), Wuhan Airlines (three) and Lao Aviation (three). Of these 18, six said to be in airline operation at that time. Most of first 60 expected to be of improved MA60-100 standard.

COSTS: Approximately US$11 million (2002).

DESIGN FEATURES: Further upgrade of Y-7 series, mainly through Western engines and avionics and improved passenger comfort; incorporates best features of earlier Y7-200A and B programmes. Airframe life 40,000 hours, 25,000 cycles or 25 years.

Non-swept high-mounted wings, with 2° 12' 12" anhedral on tapered outer panels; basically circular-section fuselage; sweptback fin and rudder; 9° dihedral on tailplane; twin ventral strakes under tailcone. Wing sweepback 6° 50' at quarter-chord of outer panels; incidence 3°.

FLYING CONTROLS: Conventional mechanical. Mass-balanced, servo-compensated ailerons with electrically actuated trim tabs; tab in each elevator; electric trim/servo tab in rudder. Hydraulically actuated single-slotted (inboard) and double-slotted (outboard) trailing-edge flaps; landing deflection 30°.

STRUCTURE: Conventional light alloy; two-spar wing, with spot-welded skins; bonded/welded semi-monocoque fuselage.

LANDING GEAR: Retractable tricycle type with twin wheels on all units. Hydraulic actuation, with emergency gravity extension. Multidisc carbon brakes on mainwheels; hydraulically steerable (±45°) and castoring nosewheel unit.

POWER PLANT: Two 2,051 kW (2,750 shp) Pratt & Whitney Canada PW127J (PW127G in cargo version) turboprops, each driving a Hamilton Sundstrand 247F-3 four-blade, composites, constant-speed, fully-reversible 'scimitar' propeller. Fuel in integral wing tanks immediately outboard of nacelles, and bag-type tanks in centre-section, total usable capacity 5,200 litres (1,374 US gallons; 1,144 Imp gallons). Provision for additional tanks in centre-section. Optional pressure refuelling point in starboard engine nacelle; gravity fuelling point above each tank.

MA60 demonstrator at Singapore in February 2002 ***(Paul Jackson)*** ***NEW***/0546841

Model of the projected 'Fearless Albatross' maritime patrol variant of the MA60 ***(Robert Hewson)*** ***NEW***/0547125

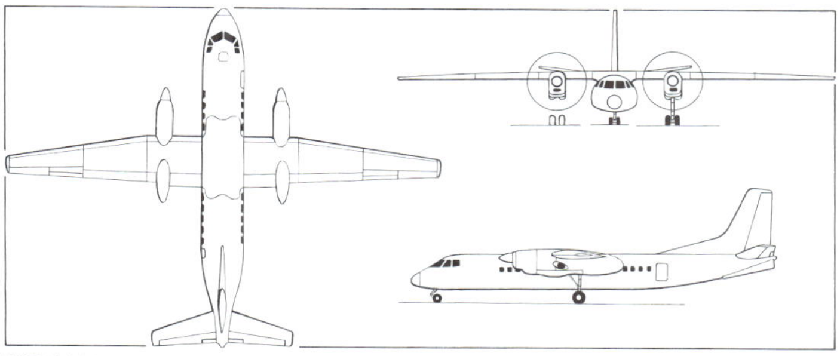

XAC MA60 twin-turboprop airliner ***(James Goulding)*** 0524604

ACCOMMODATION: Crew of two on flight deck, plus one or two cabin attendants. Standard layout has four-abreast seating, with centre aisle, for 56 or 60 (with forward wardrobe removed) passengers at 76 cm (30 in) pitch in air conditioned, soundproofed and pressurised cabin. Galley and lavatory at rear on starboard side. Wardrobes forward and aft of passenger cabin, plus overhead stowage bins in cabin. Passenger airstair door on port side at rear of cabin; emergency exit opposite. Freight door forward, starboard. All doors open inward.

SYSTEMS: Hamilton Sundstrand GTCP36-150CY APU. Hydraulic system, pressure 157 bar (2,280 lb/sq in), for actuation of flaps, landing gear extension/retraction, nosewheel steering and anti-skid braking system; emergency system for flaps and wheel brakes. Electrical power supplied by 12 kW, 28 V APU-driven AC generator and two 115/26 V, 400 Hz static inverters; two 24 V, 43 Ah Ni/Cd batteries for emergency power and engine starting. Pneumatic de-icing of wing and tail landing-edges; windscreen, propeller blades and sensors anti/de-iced electrically. Oxygen system for passengers and crew. Fire detection and extinguishing system.

AVIONICS: *Comms:* Bendix/King dual KTR 908 VHF, single KHF 950 HF and KXP 756 ATC transponder; Hamilton Sundstrand AV 557C CVR; Becker 3100 audio system; Chinese FJ-30A FDR.

Radar: Rockwell Collins WXR-350 weather radar; Honeywell CAS-67A collision avoidance system.

Flight: Dual VIR-32 VOR/ILS/marker, dual ADF-60A, dual DME-42 and ALT-55B low-altitude radio altimeter (all Rockwell Collins); Universal UNS-1M nav system; Sfena H321AKM1 standby horizon.

Instrumentation: Four-tube EFIS-85 (B14), ADS-85 air data system, APS-85 autopilot and dual AHS-85 attitude and heading systems (all Rockwell Collins); Hamilton Sundstrand Mk II GPWS.

ARMAMENT (MA60-MPA): Six external stores stations: two on fuselage sides, each for 1,000 kg (2,205 lb) load; two inboard (each 1,500 kg; 3,307 lb) and two outboard (each 500 kg; 1,102 lb) under wings. Fuselage stations can each support two torpedoes, a 1,000 kg bomb, sonobuoy container or gun pod. Inboard wing stations suitable for drop fuel tank, anti-shipping missile or AAM. Outer wing stations able to carry single torpedo, 500 kg bomb, searchlight, gun pod or additional AAM.

DIMENSIONS, EXTERNAL:
Wing span .... 29.20 m (95 ft 9½ in)
Wing chord: at root .... 3.50 m (11 ft 5¾ in)
at tip .... 1.095 m (3 ft 7 in)
Wing aspect ratio 11.4
Length overall .... 24.71 m (81 ft 0¾ in)
Fuselage max width .... 2.90 m (9 ft 6¼ in)
Height overall .... 8.855 m (29 ft 0½ in)
Wheel track (c/l of shock-struts) .... 7.90 m (25 ft 11 in)
Wheelbase .... 9.565 m (31 ft 4½ in)
Propeller diameter .... 3.96 m (13 ft 0 in)
Passenger door: Height .... 1.405 m (4 ft 7¼ in)
Width .... 0.75 m (2 ft 5½ in)
Forward freight door: Height .... 1.22 m (4 ft 0 in)
Width .... 1.19 m (3 ft 10¾ in)
Rear baggage door: Height .... 1.41 m (4 ft 7½ in)
Width .... 0.75 m (2 ft 5½ in)
Emergency exits (two, each):
Height .... 0.93 m (3 ft 0½ in)
Width .... 0.51 m (1 ft 8 in)

DIMENSIONS, INTERNAL:
Passenger cabin: Length .... 10.795 m (35 ft 5 in)
Max width .... 2.685 m (8 ft 9¾ in)

Max height .... 1.905 m (6 ft 3 in)
Aisle width .... 0.435 m (1 ft 5¼ in)
Width at floor .... 2.29 m (7 ft 6¼ in)
Baggage volume .... 9.5 m^3 (335 cu ft)
AREAS:
Wings, gross .... 74.98 m^2 (807.1 sq ft)
WEIGHTS AND LOADINGS (A: passenger, B: cargo, C: MPA):
Operating weight empty: A .... 13,700 kg (30,203 lb)
Max fuel: A, B .... 4,030 kg (8,885 lb)
C .... 7,500 kg (16,535 lb)
Baggage capacity .... 840 kg (1,852 lb)
Max payload: A, B .... 5,500 kg (12,125 lb)
Max T-O weight: A, B .... 21,800 kg (48,060 lb)
C .... 24,000 kg (52,910 lb)
Max ramp weight: A, B .... 21,900 kg (48,281 lb)
Max landing weight: A, B .... 21,200 kg (46,738 lb)
Max zero-fuel weight: A, B .... 19,500 kg (42,990 lb)
Max wing loading: A, B .... 290.7 kg/m^2 (59.55 lb/sq ft)
C .... 320.1 kg/m^2 (65.56 lb/sq ft)
Max power loading: A, B .... 5.32 kg/kW (8.74 lb/shp)
C .... 5.86 kg/kW (9.62 lb/shp)
PERFORMANCE (A, B and C as above):
Max cruising speed: A .... 292 kt (540 km/h; 335 mph)
Econ cruising speed: A .... 232 kt (430 km/h; 267 mph)
Max operating altitude: A .... 7,620 m (25,000 ft)
C .... 8,500 m (27,880 ft)
Service ceiling, OEI: A .... 3,920 m (12,860 ft)
B .... 4,000 m (13,120 ft)
T-O run: A .... 812 m (2,665 ft)
C .... 969 m (3,180 ft)
T-O field length: A .... 1,425 m (4,675 ft)
B .... 1,400 m (4,595 ft)
Landing field length: A .... 1,460 m (4,790 ft)
Landing run at MLW: A .... 875 m (2,870 ft)
C .... 650 m (2,135 ft)
Range:
with 56 passengers, 100 n miles + 45 min reserves
A .... 864 n miles (1,600 km; 994 miles)
B .... 567 n miles (1,050 km; 652 miles)
with max fuel:
A .... 1,322 n miles (2,450 km; 1,522 miles)
B .... 1,187 n miles (2,200 km; 1,367 miles)
C .... 2,591 n miles (4,800 km; 2,982 miles)
Max endurance: C .... 12 h

UPDATED

Colombia

GLASS

GLASS AIRCRAFT DE COLOMBIA

Aeródromo las Gaviotas, Vereda la Argentina, Km 5 al sur del Aerópuerto el Eden
Tel: (+157 011) 576 754 17 32
e-mail: stolglass@netxos.com.co

US DISTRIBUTOR:
CR Aviation, 14532 Southwest 127th Street, Hangar 109, Miami, Florida 33186
Tel: (+1 305) 255 42 22
Fax: (+1 305) 255 30 03

Three types of light aircraft are currently under development by this company which was formed in January 1999. One of these is a light utility twin with Rotax engines, to be known as the TwinGlass. No development timetable has been announced for this aircraft.

The company has a workforce numbering 72 and in November 2001 reported production at around four aircraft per year.

UPDATED

Projected TwinGlass utility twin ***NEW***/0561677

GLASS STOLGLASS SG70

TYPE: Two-seat lightplane.
PROGRAMME: North American marketing launched at AirVenture, Oshkosh, in July 2003 with a US-registered demonstrator.
CUSTOMERS: Venezuelan Yaracuy state government reported to have ordered 12 aircraft of which three delivered by November 2001. By May 2002, 27 aircraft had been built. Orderes for 2002 numbered 45.
DESIGN FEATURES: Short-field performance. Constant-chord, high-lift, high-mounted mainplanes with V-strut bracing resemble those of Zenith CH 701. Sweptback fin and root-mounted tailplane.
FLYING CONTROLS: Conventional and manual. Junkers ailerons and flaps. No aerodynamic balances on tail surfaces.
STRUCTURE: Generally of glass fibre.
LANDING GEAR: Tricycle type; fixed. Fuselage-mounted cantilever main legs with hydraulic brakes. Speed fairings on all wheels.
POWER PLANT: One flat-four piston engine; option of 73.5 kW (98.6 hp) Rotax 912 S or 84.5 kW (123.3 hp) Rotax 914, each driving a three-blade propeller. Fuel capacity 76 litres (20.0 US gallons; 16.7 Imp gallons).
ACCOMMODATION: Two persons, side by side. Baggage area behind seats. Door each side.
DIMENSIONS, EXTERNAL (Rotax 912 or 914 engine):
Wing span .... 9.15 m (30 ft 0 in)
Length overall .... 6.50 m (21 ft 4 in)
Height overall .... 2.57 m (8 ft 5¼ in)
WEIGHTS AND LOADINGS:
Weight empty .... 330 kg (727 lb)
Max T-O weight .... 600 kg (1,322 lb)
Max power loading: 912 .... 8.16 kg/kW (13.41 lb/hp)
PERFORMANCE (Rotax 912 or 914 engine):
Never-exceed speed (VNE) .... 121 kt (225 km/h; 140 mph)
Cruising speed at 75% power:
912 .... 100 kt (185 km/h; 115 mph)
914 .... 109 kt (201 km/h; 125 mph)
Stalling speed, power off:

STOLGlass SG70 employed as a US demonstrator *(Paul Jackson)* ***NEW***/0576735

flaps up .... 31 kt (57 km/h; 35 mph)
flaps down .... 26 kt (49 km/h; 30 mph)
Max rate of climb at S/L: 912 .... 335 m (1,100 ft)/min
914 .... 366 m (1,200 ft)/min
Service ceiling: 912 .... 4,265 m (14,000 ft)
T-O run: at S/L, ISA: 912 .... 50 m (165 ft)
914 .... 45 m (150 ft)
1,220 m (4,000 ft), ISA: 912 .... 125 m (410 ft)
914 .... 120 m (395 ft)
Landing run .... 130 m (430 ft)
Range: 912 .... 365 n miles (675 km; 420 miles)
914 .... 347 n miles (643 km; 400 miles)

UPDATED

GLASS SPEEDGLASS

TYPE: Two-seat lightplane.
PROGRAMME: First flown June 2003.
COSTS: Company estimated unit cost at US$45,000 in 2002.
DESIGN FEATURES: Low wing with winglets; sweptback fin with root-mounted tailplane.
FLYING CONTROLS: Conventional and manual. No aerodynamic balances on tail surfaces.
STRUCTURE: Generally of glass fibre, with carbon fibre and duralumin.
LANDING GEAR: Tricycle type. Mainwheels retract inwards; nosewheel rearwards.
POWER PLANT: Not specified. Three-blade propeller. Fuel capacity 114 litres (30.0 US gallons; 25.0 Imp gallons).
DIMENSIONS, EXTERNAL:
Wing span .... 8.75 m (28 ft 4½ in)
Length overall .... 7.50 m (24 ft 7¼ in)
Height overall .... 3.20 m (10 ft 6 in)
WEIGHTS AND LOADINGS:
Weight empty .... 450 kg (992 lb)
Max T-O weight .... 800 kg (1,763 lb)
PERFORMANCE (estimated):
Never-exceed speed (VNE) .... 156 kt (290 km/h; 180 mph)
Cruising speed at 75% power .... 130 kt (241 km/h; 150 mph)
Stalling speed, power off:
flaps up .... 53 kt (97 km/h; 60 mph)

Prototype Glass SpeedGlass ***NEW***/0561678

flaps down .... 44 kt (81 km/h; 50 mph)
Max rate of climb at S/L .... 304 m (1,000 ft)/min
Service ceiling .... 4,570 m (15,000 ft)
T-O run, ISA: at S/L .... 200 m (660 ft)
at 1,220 m (4,000 ft) .... 350 m (1,150 ft)
Landing run .... 400 m (1,315 ft)
Range .... 955 n miles (1,770 km; 1,100 miles)

UPDATED

Czech Republic

AERO

AERO VODOCHODY A.S

Uherské Letiště 374, CZ-250 70 Odolena Voda
Tel: (+420 255) 76 31 52
Fax: (+420 255) 76 32 25
e-mail: vitezslav.kulich@aero.cz
Web: http://www.aero.cz

PRESIDENT AND CHAIRMAN: Antonín Jakubše
CEO: Vladimir Nývlt
VICE-PRESIDENTS:
Vladimír Jaroš (Finance)
Jiří Fidranský (Research and Development) Viktor Kučera (Military Aircraft Programmes) Miloš Vališ (Civil Aircraft Programmes) František Bílek (Aero Structures) Martin Paloda (Marketing) Václav Pavlíček (Services)
CHIEF DESIGNERS:
Zdeněk Stucklík (Military Programmes)
Josef Jironč (Civil Programmes)
PRESS OFFICER: Vitezslav Kulich

Established on 1 July 1953, Aero Vodochody is the successor to the Aero Aircraft factory founded in Prague in 1919. Over the years it has produced more than 11,000 aircraft, most notably the L-29 Delfin (3,665 built) and more than 2,800 L-39 Albatros jet trainers. Major current programme concerns the L159.

The company underwent a change of ownership as part of a restructuring programme approved on 31 July 1996, when it had yet to recover fully from debts occurring in 1990–91. It became a joint-stock company in which the majority shareholder was the Czech government, through shares held by Letka a.s. (35.66 per cent) and the Konsolidační Banka (29.00 per cent). From 17 August 1998, a further 35.29 per cent was acquired by Boeing Česká s.r.o., a joint venture comprising Boeing and Czech Airlines (CSA), owned 90 per cent by Boeing and 10 per cent by CSA. The remaining 0.05 per cent of Aero was owned by other shareholders. However, Boeing decided to withdraw its participation in early 2004; terms agreed on 11 October 2004, whereby Boeing's Česká holding repurchased by Czech government for nominal Kcs2, 53 per cent being allocated to Letka and 47 per cent to the Konsolidačni Banka, making the company 99.95 per cent state-owned. Workforce at end of 2003 was about 2,000, but being reduced to 1,600 in May 2004.

Company is currently organised into four main business units: Military Aircraft, Civil Aircraft, Aero Structures and Sikorsky. Aircraft programmes are the L159 and the Ae 270 Ibis civil transport, the latter in a 50/50 venture with AIDC of Taiwan. Delivery of the first L159As to the Czech Air Force enabled the company to return to an operating profit. By early 2003, it had 67 firm orders and six options for the Ae 270, but no export orders have yet been obtained for the L159, a situation which led the Czech government in late 2003 to call for an audit of the company's affairs. It had agreed in July 2003 to cover existing bank loan repayments until 2008, but subsequently indicated that this support could be withdrawn if no sales of the L159 were obtained by the end of 2004. Meanwhile, Aero required to pay Czech government penalty of approximately Ckr750 million (US$28 million) due to late delivery of L159A. A profit of Kcs 313 million was made in 2003.

Under a June 2000 agreement with Sikorsky, Aero is undertaking fabrication and subassembly of S-76+ helicopter airframes (but not their dynamic systems). The agreement originally called for delivery of 15 airframes per year over a seven-year period; first set, assembled from US-built kits, was redelivered to Sikorsky at Stratford, Connecticut, in January 2001. In-country manufacture by Aero had begun by the end of that year, with the number of airframes rising annually since then; 21 completed in 2003 (bringing total to 44) and 28 scheduled for 2004.

UPDATED

AERO L159

TYPE: Light fighter/advanced jet trainer.

PROGRAMME: Design started late 1992; F124 turbofan selected as power plant mid-1994; Rockwell North American (now Boeing) awarded avionics contract late 1994. Development approved by Czech government 7 April 1995, followed by production commitment later the same month and 72-aircraft production contract 4 July 1997. Two prototypes (plus two airframes for static and dynamic test) built. First (two-seat) prototype/engine demonstrator (5831) rolled out 12 June and made first flight 2 August 1997.

Second (single-seat) prototype (5832), in Czech Air Force configuration (first flight 18 August 1998) named ALCA (Advanced Light Combat Aircraft); is first with full avionics and is avionics/weapons trials aircraft. Official first flight of 5832 was 20 August; public debut during Czech International Air Fest at Hradec Králové 29 August 1998. Gun firing tests in Czech Republic followed by air-launched weapon trials in Norway in April and May 1999; nav/attack system trials August/September 1999. Weapons qualification tests at Cazaux, France, from May to July 2000.

First flight of production L159A (6001) 20 October 1999. Deliveries to CzAF due to have begun late 1999 (five aircraft, followed by 16 in 2000, 26 in 2001 and final 25 in 2002);

Aero L159B tandem-seat advanced jet trainer, equipped with in-flight refuelling probe ***NEW***/0593109

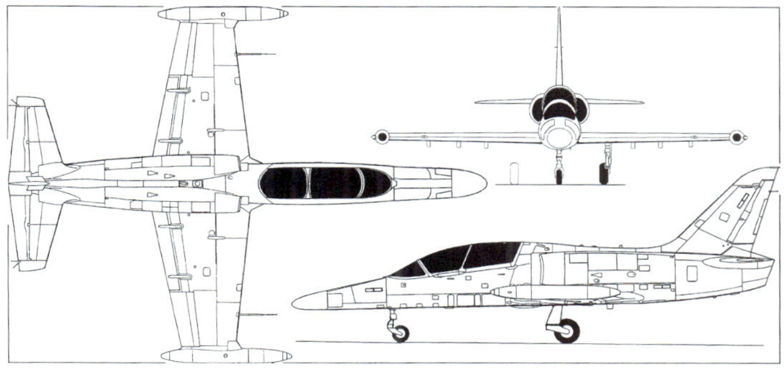

Aero L159 light attack aircraft ***NEW***/0572161

however, first two (6006 and 6007) not delivered (to 4 Tactical Fighter Base at Caslav) until 29 December 2000, with first pilot conversion training beginning in mid-March 2001. Deliveries then rescheduled as 37 by end of 2001, 21 during 2002 and final 14 during first half of 2003. However, in May 2001, during which month aircraft evaluated by US Navy Test Pilots' School, Czech defence minister stated that CzAF will operate only 37 aircraft.

The L159A fleet (then 25 aircraft) was grounded on 14 November 2001 following incidents to two aircraft apparently due to suspected avionics malfunctions. No other significant problems had been registered by the end of 2002, but in mid-2003 Czech government stated intention to reduce CzAF inventory further, to only 18 aircraft, with surplus being put up for sale. Final L159A was delivered to Coslav AB on 3 February 2004.

First production L159B (6073) achieved maiden flight on 1 June 2002 and made international debut at Farnborough Air Show in following month.

CURRENT VERSIONS: **L159A ALCA:** Single-seater.

Detailed description applies to L159A except where indicated.

L159B Albatros II: Tandem-seat trainer version, designed primarily for advanced and operational training; can also be configured to customer requirements from basic to lead-in fighter training, reconnaissance or ground attack. Further US$15 million invested by Boeing in July 2001 to develop software for this version.

CUSTOMERS: Production lots 1 and 2 comprised five and 10 aircraft respectively, of which 12 delivered by 31 March 2001. Total 58 delivered by May 2003, but only 14 of these were then operations-capable, reportedly due to avionics and hydraulic systems problems. Deliveries completed in February 2004. Primary roles intended to be close air support, counter-insurgency, anti-ship missions, tactical reconnaissance, point air defence, air defence against slow/low-flying targets and border patrol. Initial order for L159B still awaited in April 2004. Prospect of exports to other countries, including Poland, to whom a Maverick-armed version was offered in late 2000. Company said to be preparing offsets for sale to Slovak Air Force in July 2001, and to be negotiating with other 'very potential' customers, including Indian Air Force, but no orders for either version by April 2004.

COSTS: Czech Air Force order for 72 L159As valued at approximately US$1 billion. L159B (baseline configuration) US$9.5 million (2002).

DESIGN FEATURES: Configuration generally as L-39/L-59 predecessors except for larger, redesigned nose to accommodate radar; longer fuselage; single-seat armoured cockpit; HOTAS controls; Western avionics; 297 kg (655 lb) additional fuel tank in lieu of second seat; permanent wingtip fuel tanks; seven external stores stations.

Wing section NACA 64A-012; sweepback 6° 30' on leading-edges; dihedral 2° 30'; incidence 2°; twist 0°.

FLYING CONTROLS: Conventional primary surfaces; ailerons and elevators actuated hydraulically, rudder mechanically with electrically operated trim tab. Artificial feel standard. Twin underfuselage airbrakes and double-slotted trailing-edge flaps, also actuated hydraulically. Maximum flap deflection 44°.

STRUCTURE: Based on that of L-59, but substantially modified. Approximately 200 Czech subcontractors involved in programme.

LANDING GEAR: Hydraulically retractable tricycle type, with single wheel and oleo-pneumatic shock-absorber on each unit. Main units retract inward into wings, nosewheel forward. Moravan K36 wheels and 610×215 tyres (pressure 6.00 to 8.00 bar; 87 to 116 lb/sq in) on main units; nose unit has Moravan K37 wheel and 460×180 tyre (pressure 3.52 to 4.48 bar; 51 to 65 lb/sq in). Moravan K52 air-cooled hydraulic brakes with anti-skid system. Nose unit hydraulically steerable ±59°. Minimum ground turning radius 5.70 m (18 ft 8½ in).

POWER PLANT: One 28.0 kN (6,300 lb st) Honeywell/ITEC F124-GA-100 turbofan with dual-redundant FADEC. Internal fuel of L159A (total capacity 1,980 litres; 523 US gallons; 436 Imp gallons) is contained in six fuselage and two wingtip tanks and protected by OBIGGS (onboard inert gas generation system); L159B omits one 350 litre (92.5 US gallon; 77.0 Imp gallon) fuselage tank, internal capacity thus 1,630 litres (431 US gallons; 359 Imp gallons) and provision for OBIGGS. Up to four 500 litre (132 US gallon; 110 Imp gallon) external tanks can be carried underwing: two on inboard stations, two on centre stations. Pressure fuelling point in port side of lower front fuselage; gravity points in upper fuselage aft of canopy and in each wingtip tank. In-flight refuelling capability began flight testing on L159B in August 2002.

ACCOMMODATION: Single VS-2C ejection seat in armoured cockpit (L159A); two VS-2B seats and no armour in L159B. Canopy opens sideways manually in L159A, upwards hydraulically in L159B.

SYSTEMS: PBS simple cycle environmental control system, using engine bleed air; Honeywell cockpit pressure control system (maximum differential 0.28 bar; 4.1 lb/sq in). Dual hydraulic systems, each at pressure of 210 bar (3,045 lb/sq in), actuating ailerons, elevators, flaps, airbrakes, landing gear, wheel brakes and nosewheel steering. System maximum flow rate 35 litres (9.25 US gallons; 7.7 Imp gallons)/min. Hydraulic tank pressurised by nitrogen at 2.50 bar (36.3 lb/sq in). Pneumatic system (1.80 bar; 26.1 lb/sq in) for canopy sealing.

Hamilton Sundstrand 40 kVA integrated drive generator provides 200/115 V three-phase main AC power at 400 Hz; 28 V DC power from 6 kW VUES LUN-2134 back-up generator and Varta/Aero 25 Ah battery. Intertechnique OBOGS oxygen system. Windscreen and air

Aero L159B Albatros II prototype *(Paul Jackson)* ***NEW***/0572272

intakes de-iced by engine bleed air. PBS Safir 5F APU, operable up to 9,000 m (29,520 ft) altitude, for engine starting and emergency hydraulic or electrical power.

AVIONICS: Integration of complete avionics suite by Boeing.

Comms: Two Rockwell Collins ARC-210 UHF/VHF radios; Honeywell AN/APX-100 IFF.

Radar: Galileo Grifo L multimode pulse Doppler radar (L159A only).

Flight: Bendix/King KNR 634A VOR/glideslope/MKR, KDM 706A DME and air data computer; Honeywell H-746G ring laser gyro INS with embedded GPS.

Instrumentation: Flight Visions FV-3000 HUD; dual Honeywell 102 × 102 mm (4 × 4 in) liquid crystal colour head-down MFDs; Astronautics mechanical ADI and HSI. Optional helmet-mounted display/sight and NVG-compatible lighting.

Mission: Dual Flight Visions 6400–100 mission computers with HUD symbology generator; provision for reconnaissance and jamming pod. Hamilton Sundstrand weapon delivery system. Dynamic Control Corporation stores management system.

Self-defence: BAE Systems Sky Guardian 200 RWR; Thales (Vinten) Vicon 78 Series 455 countermeasures dispenser. Optional on L159B.

ARMAMENT: No built-in armament. Seven external stores stations: one under fuselage and three under each wing. Fuselage station stressed for 300 kg (661 lb) load; inboard, centre and outboard wing pylons stressed for 550, 320 and 150 kg (1,213, 705 and 331 lb) each respectively (L159B centre pylons 500 kg; 1,102 lb only). Primary armament comprises AIM-9 Sidewinder AAMs, AGM-65 Maverick ASMs, and SUU-20 or CRV7 unguided rocket pods. Has also been trialled with Brimstone missile and TIALD targeting pod. For typical weapon loads, see accompanying illustration.

DIMENSIONS, EXTERNAL:
Wing span: excl tip tanks ... 8.70 m (28 ft 6½ in)
incl tip tanks ... 9.54 m (31 ft 3½ in)
Wing chord: at root ... 2.80 m (9 ft 2¼ in)
at tip ... 1.40 m (4 ft 7 in)
Wing aspect ratio, incl tip tanks ... 4.8
Length: overall ... 12.72 m (41 ft 8¾ in)
fuselage ... 12.59 m (41 ft 3¾ in)
Height overall ... 4.87 m (15 ft 11¾ in)
Tailplane span ... 4.70 m (15 ft 5 in)
Wheel track ... 2.42 m (7 ft 11¼ in)
Wheelbase ... 4.56 m (14 ft 11½ in)

AREAS:
Wings, gross ... 18.80 m^2 (202.4 sq ft)
Ailerons (total) ... 1.69 m^2 (18.19 sq ft)
Trailing-edge flaps (total) ... 2.68 m^2 (28.85 sq ft)
Fin ... 2.59 m^2 (27.88 sq ft)
Rudder, incl tab ... 0.91 m^2 (9.80 sq ft)
Tailplane ... 3.90 m^2 (41.98 sq ft)
Elevators (total) ... 1.40 m^2 (15.07 sq ft)

WEIGHTS AND LOADINGS:
Weight empty, equipped ... 4,350 kg (9,590 lb)
Max fuel weight:
internal, L159A ... 1,547 kg (3,411 lb)
internal, L159B ... 1,320 kg (2,910 lb)
external: both ... 1,620 kg (3,571 lb)
Max external stores load: both ... 2,700 kg (5,952 lb)
Max T-O and ramp weight ... 8,000 kg (17,637 lb)
Max landing weight ... 6,000 kg (13,228 lb)
Max zero-fuel weight ... 6,810 kg (15,013 lb)
Max wing loading ... 425.5 kg/m^2 (87.15 lb/sq ft)
Power loading, L159A clean with 50% internal fuel ... 284 kg/kN (2.79 lb/lb st)

PERFORMANCE (C: clean, S: with 1,470 kg; 3,241 lb of external stores, except where indicated):
Never-exceed speed (VNE) ... 518 kt (960 km/h; 596 mph)
Max level speed at S/L: C ... 505 kt (936 km/h; 581 mph)
S ... 468 kt (867 km/h; 539 mph)
Stalling speed at 5,500 kg (12,125 lb) AUW, engine idling, flaps and gear down ... 100 kt (185 km/h; 539 mph)
Max rate of climb at S/L: C ... 3,726 m (12,220 ft)/min
S ... 2,280 m (7,480 ft)/min
Sustained turning radius at 5,000 m (16,400 ft):
C ... 612 m (2,008 ft)
S ... 882 m (2,894 ft)
Sustained turn rate at 5,000 m (16,400 ft):
C ... 11.1°/s
S ... 7.7°/s
Sustained load factor:
at S/L: C ... 5.7 *g*
S ... 3.8 *g*
at 5,000 m (16,400 ft): C ... 3.5 *g*
S ... 2.4 *g*
Service ceiling ... 13,200 m (43,300 ft)
T-O run at S/L:
C at 5,925 kg (13,062 lb) AUW ... 470 m (1,545 ft)
S at 7,190 kg (15,851 lb) AUW ... 853 m (2,800 ft)
T-O to 15 m (50 ft) at S/L:
C at 5,925 kg (13,062 lb) AUW ... 720 m (2,365 ft)
S at 7,190 kg (15,851 lb) AUW ... 1,276 m (4,185 ft)
Landing from 15 m (50 ft) at S/L:
C at 4,489 kg (9,897 lb) AUW ... 992 m (3,255 ft)
S at 5,257 kg (11,590 lb) AUW ... 1,254 m (4,115 ft)
Landing run at S/L:
C at 4,489 kg (9,897 lb) AUW ... 628 m (2,060 ft)
S at 5,257 kg (11,590 lb) AUW ... 816 m (2,675 ft)
Typical mission radius (A: L159A, B: L159B):
with gun pod, two Mk 82 bombs, two AIM-9 missiles and two 500 litre drop tanks:
A, lo-lo-lo ... 305 n miles (565 km; 351 miles)
A, hi-lo-hi ... 426 n miles (790 km; 490 miles)
B, lo-lo-lo ... 264 n miles (490 km; 304 miles)
B, hi-lo-hi ... 364 n miles (675 km; 419 miles)
with gun pod, two Mk 82 bombs and two 500 litre drop tanks:
A, lo-lo-lo ... 310 n miles (575 km; 357 miles)
A, hi-lo-hi ... 442 n miles (820 km; 509 miles)
B, lo-lo-lo ... 270 n miles (500 km; 310 miles)
B, hi-lo-hi ... 378 n miles (700 km; 435 miles)
with two 500 litre drop tanks only:
A, lo-lo-lo ... 345 n miles (640 km; 397 miles)
A, hi-lo-hi ... 526 n miles (975 km; 605 miles)
B, lo-lo-lo ... 299 n miles (555 km; 344 miles)
B, hi-lo-hi ... 450 n miles (835 km; 518 miles)
Range, reserves of 10% of internal fuel:
L159A, max internal fuel only ... 848 n miles (1,570 km; 975 miles)
L159B, max internal fuel only ... 659 n miles (1,220 km; 758 miles)
L159A, max internal and external fuel ... 1,366 n miles (2,530 km; 1,572 miles)
L159B, max internal and external fuel ... 1,242 n miles (2,300 km; 1,429 miles)
g limits (ultimate) ... +8/–4

UPDATED

AEROS

AEROS S.R.O.

Nádražní 211, CZ-284 01 Kutná Hora
Tel: (+420 327) 56 22 44
Fax: (+420 327) 56 25 58
e-mail: flamingo@aeros.cz or c.ilona@volny.cz
Web: http://www.aeros.cz
MANAGING DIRECTOR: Miroslav Staněk
INFORMATION OFFICER: Miss Ilona Černá

Known until 2002 as Lusus s.r.o., Aeros has 10 employees at its 700 m^2 (7,550 sq ft) plant.

UPDATED

AEROS UL-2000 FLAMINGO

TYPE: Side-by-side ultralight.

PROGRAMME: Entered production in 2000 and international debut at Friedrichshafen Air Show, April 2001.

CURRENT VERSIONS: **UL-2000 S Flamingo Sport:** Baseline version, with 48.5 kW (65 hp) Hirth 2706 engine and two-blade fixed-pitch propeller.

UL-2000 T Flamingo Trainer: As Sport, but with 74.6 kW (100 hp) Hirth F30 and two-blade ground-adjustable pitch propeller; GPS navigation; dual controls. No longer available.

UL-2000 A Flamingo Aerobatic: As Trainer except for 82.0 kW (110 hp) Hirth and three-blade ground-adjustable propeller; or 73.5 kW (98.6 hp) Rotax 912 ULS or 84.6 kW (113.4 hp) Rotax 914. Export-only version; no longer available.

UL-2000 Flamingo side-by-side ultralight *(Paul Jackosn)* 0131725

UL-2000 B Business Flamingo: Generally similar to Trainer, but without dual controls, no longer available.

UL-2000 D Flamingo De Luxe: Customised version, with modified 74.6 kW (100 hp) engine.

CUSTOMERS: Sales in Canada, Lithuania and the Czech Republic.

COSTS: UL-2000 S basic kit US$26,933; fast-build kit US$30,991 (2004).

DESIGN FEATURES: Unbraced high-wing cabin monoplane with sweptback vertical tail. Wings detach for transportation and storage.

FLYING CONTROLS: Manual. Non-balanced rudder; large one-piece elevator with inset, in-flight-adjustable trim tab. Three-position, electrically operated flaps.

STRUCTURE: All-composites fuselage; wings of sandwich construction with GRP main spar.

LANDING GEAR: Tricycle type; fixed; steerable nosewheel optional. Fuselage-mounted spring composites cantilever main legs with 400×100 tyres and mechanical drum brakes; hydraulic drum brakes and speed fairings optional, nosewheel tyre 12×4. Floats and skis can also be fitted.

POWER PLANT: Accepts a variety of engines, including 48.0 kW (64.4 hp) Rotax 582, 73.5 kW (98.6 hp) Rotax 912ULS, 84.6 kW (113.4 hp) Rotax 914, 48.5 hp (65 hp) Hirth 2706 and 74.6 kW (100 hp) Hirth F30. Fuel capacity 45 litres (11.9 US gallons; 9.9 Imp gallons) in UL-2000 S, T, A and D; 60 litres (15.9 US gallons; 13.2 Imp gallons) in B.

ACCOMMODATION: Upward-opening window/door each side; hatshelf behind seats; pilot's relief tube.

SYSTEMS: 12 V electrical system.

DIMENSIONS, EXTERNAL:
Wing span .... 8.60 m (28 ft 2½ in)
Length overall .... 6.40 m (21 ft 0 in)
Height overall .... 2.50 m (8 ft 2½ in)

DIMENSIONS, INTERNAL:
Cabin max width .... 1.10 m (3 ft 7¼ in)

AREAS:
Wings, gross .... 10.50 m² (113.0 sq ft)

WEIGHTS AND LOADINGS:
Weight empty .... 265–285 kg (584–628 lb)
Max T-O weight (all versions) .... 520 kg (1,146 lb)

PERFORMANCE (S: Sport, D: De Luxe):
Never-exceed speed (VNE):
both .... 145 kt (270 km/h; 167 mph)
Max level speed: both .... 129 kt (240 km/h; 149 mph)
Cruising speed: both .... 97 kt (180 km/h; 112 mph)
Stalling speed: all .... 36 kt (65 km/h; 41 mph)
Max rate of climb at S/L: S .... 270 m (886 ft)/min
D .... 420 m (1,378 ft)/min
T-O to 15 m (50 ft): both .... 180 m (590 ft)
Landing from 15 m (50 ft): both .... 250 m (820 ft)
Range with max fuel:
S .... 432 n miles (800 km; 497 miles)
D .... 540 n miles (1,000 km; 621 miles)
g limits: all .... ±6

UPDATED

AIROX

AIROX S.R.O.

Dolní Radová, CZ-463 03 Stráž nad Nisou
Tel/Fax: (+420 48) 515 92 24
e-mail: airox@airox.cz
Web: http://www.airox.cz
DIRECTOR: Ing. Miroslav Hajek
INFORMATION: Tomáše Valouška

Airox was formed in 1993 to build aircraft from composites materials, as well as parts for ships. In mid-2003 the company's workforce numbered 24.

VERIFIED

AIROX F-31

TYPE: Side-by-side ultralight twin.

PROGRAMME: Public debut (F-31A) at aviation exhibition in Prague, September 2000. Engines installed and ground running underway by late 2003.

CURRENT VERSIONS: **F-31A:** Tailwheel version; *as described*. Prototype is of this version.

F-31B: Tricycle landing gear; under development.

F-31C: Modified lifting and control surface geometry; under development.

F-31D: Multiseat, non-ultralight version; under development.

COSTS: Provisional cost US$44,538 with Rotax 618 engines (2004).

DESIGN FEATURES: Low-wing monoplane with dihedral from roots and upturned wingtips; cruciform tailplane.

FLYING CONTROLS: Conventional and manual; flaps fitted.

STRUCTURE: All-composites (CFRP sandwich shell); single-spar wing.

LANDING GEAR: Tailwheel type; fixed. Wing-mounted cantilever self-sprung main units.

POWER PLANT: Two 33.6 kW (45 hp) Hirth F23 flat-twin engines, each driving a SportProp 1360W two-blade, glass fibre propeller. Fuel capacity 75 litres (19.8 US gallons; 16.5 Imp gallons). Other suitable engines include Rotax 503 UL-2V, 582 DCDI-2V or 618 DCDI-2V.

AVIONICS: VFR instrumentation standard; IFR to customer's requirements, including Bendix/King KY 97 transceiver Garmin GPS.

Prototype Airox F-31A twin-engined ultralight ***NEW***/0528737

DIMENSIONS, EXTERNAL:
Wing span .... 9.60 m (31 ft 6 in)
Length overall .... 6.60 m (21 ft 7¾ in)

AREAS:
Wings, gross .... 12.00 m² (129.2 sq ft)

WEIGHTS AND LOADINGS:
Weight empty .... 278 kg (613 lb)
Max T-O weight .... 450 kg (992 lb)

PERFORMANCE:
Never-exceed speed (VNE) .... 152 kt (282 km/h; 175 mph)
Max level speed .... 135 kt (251 km/h; 156 mph)
Cruising speed .... 112 kt (208 km/h; 129 mph)
Stalling speed, flaps down .... 36 kt (65 km/h; 41 mph)
Max rate of climb at S/L .... 480 m (1,575 ft)/min

UPDATED

ATEC

ATEC V.O.S

Opolanská 171, CZ-289 07 Libice nad Cidlinou
Tel/Fax: (+420 325) 63 73 71
e-mail: sales@atecvos.cz
Web: http://www.atecvos.cz
PRODUCTION DIRECTOR: Petr Volejnik
SALES DIRECTOR: Jan Franěk

ATEC was formed in 1992 to develop, manufacture and sell ultralight aircraft.

UPDATED

ATEC ZEPHYR AND FAETA

TYPE: Side-by-side ultralight/kitbuilt.

PROGRAMME: Prototype Zephyr (OK-CUA 01) built 1997.

CURRENT VERSIONS: **Zephyr Standard:** Original version, no longer available; 10.60 m (34 ft 9¼ in) wing span and 59.6 kW (79.9 hp) Rotax 912 UL engine.

Zephyr 2000: Introduced April 2001; C suffix indicates change to new shape all-composites fuselage construction; shorter wing and tail spans.

ATEC Zephyr 2000 two-seat ultralight *(Paul Jackson)* ***NEW***/0576254

Description below applies to Zephyr 2000C.

Zephyr Solo: Single-seat version. Prototype (OK-FUH-01) made initial flight 29 October 2000. Not proceeded with.

Prototype ATEC 321 Faeta *(Paul Jackson)* ***NEW***/0576253

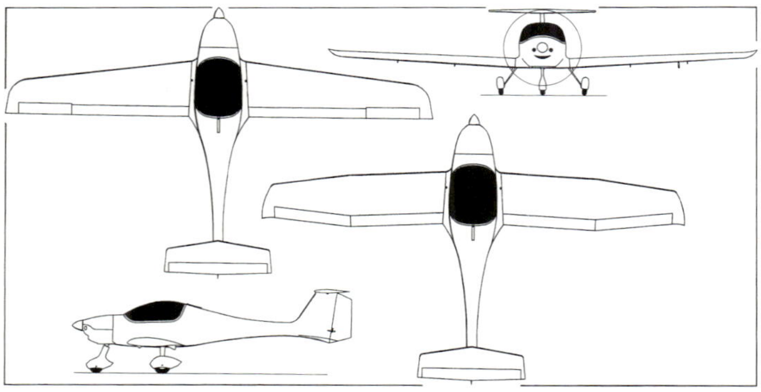

ATEC Faeta general arrangement, with extra plan view (lower) of Zephyr ***NEW***/0547522

ATEC 321 Faeta: Maiden (OK-IUG 15) flight in February 2003; introduced at Aero '03, Friedrichshafen, April 2003. As Zephyr, but has composites skinned carbon fibre wing with constant sweepback and differently proportioned ailerons and flaps. Two-blade ground adjustable Fiti propeller and ballistic parachute. Generally as Zephyr unless otherwise indicated.

CUSTOMERS: Over 90 (all versions) built and sold by October 2002. Exported to Australia, Canada, France, Germany, Greece, Ireland, Italy, Netherlands, Portugal, Spain and Sweden. Marketed in Germany by Fläming Air.

COSTS: Zephyr 2000: US$64,450 with Rotax 912 ULS (2004); Faeta: €49,900 with Rotax 912 UL; €52,000 with Rotax 912 ULS (2003) all flyaway, excluding tax. Zephyr kit (excluding engine) US$42,950 (2004).

DESIGN FEATURES: Low-wing monoplane with T tail. Wing section UA-2 laminar flow.

FLYING CONTROLS: Conventional and manual; spring trim. Electric flaps and elevator trim optional. Dual controls.

STRUCTURE: Fuselage of carbon fibre; wing leading-edge, horizontal tailplane, wingtips, elevator, rudder, mainwheel legs and engine cowling of carbon fibre composites. Wing main spar, wing trailing-edge, ailerons and flaps made of wood. Fabric-covered wings.

LANDING GEAR: Tricycle type, fixed. Fuselage-mounted cantilever mainwheel legs. Aerodynamic composites nosewheel leg and integral wheel fairing. Castoring nosewheel. Tyres sizes 380×100 (main), 300×100 (nose); Faeta has 350×120 main tyres. Hydraulic disc brakes. Speed fairings on all three wheels.

POWER PLANT: *Zephyr 2000:* One 59.6 kW (79.9 hp) Rotax 912 UL or 73.5 kW (98.6 hp) Rotax 912 ULS flat-four engine; FITI two- or three-blade ground-adjustable pitch propeller. Fuel capacity 60 litres (15.9 US gallons; 13.2 Imp gallons) standard, 80 litres (21.1 US gallons; 17.6 Imp gallons) optional.

Faeta: Power plant options as for Zephyr. Fuel capacity 70 litres (18.4 US gallons; 15.4 Imp gallons).

ACCOMMODATION: Pilot and passenger side by side under one-piece canopy.

EQUIPMENT: Optional ballistic parachute system.

DIMENSIONS, EXTERNAL:
Wing span ..... 9.60 m (31 ft 6 in)
Length overall ..... 6.20 m (20 ft 4 in)
Height overall ..... 2.00 m (6 ft 6¾ in)

AREAS:
Wings, gross ..... 10.10 m^2 (108.7 sq ft)

WEIGHTS AND LOADINGS:
Weight empty ..... 275 kg (606 lb)
Max T-O weight: standard ..... 450 kg (992 lb)
with ballistic parachute ..... 473 kg (1,043 lb)

PERFORMANCE (912 ULS engine; A: Zephyr, B: Faeta):
Max level speed: A ..... 132 kt (245 km/h; 152 mph)
B ..... 138 kt (255 km/h; 158 mph)
Max cruising speed: A ..... 116 kt (215 km/h; 133 mph)
B ..... 124 kt (230 km/h; 143 mph)
Stalling speed, flaps down: A ..... 36 kt (65 km/h; 41 mph)
B ..... 33 kt (60 km/h; 38 mph)
Max rate of climb at S/L: A ..... 390 m (1,280 ft)/min
B ..... 420 m (1,378 ft)/min
T-O run ..... 120 m (395 ft)
Landing run ..... 200 m (660 ft)
Range: standard fuel ..... 432 n miles (800 km; 497 miles)
optional fuel ..... 594 n miles (1,100 km; 683 miles)
g limits ..... +4/–2

UPDATED

CZAW

CZECH AIRCRAFT WORKS S.R.O.

Luční 1824, CZ-686 02 Staré Město
Tel: (+420 572) 54 34 56
Fax: (+420 572) 54 36 92
e-mail: aircraft@czaw.cz
Web: http://www.airplane.cz
CEO: Chip W Erwin

In 1997 Zenair Ltd of Canada formed a joint venture with CZAW, established earlier that year, for manufacture in the Czech Republic of its STOL CH 701 and Zodiac CH 601 series of light aircraft. First Czech components were manufactured in 1998. Factory output comprised 30 in 1997, 44 in 1998 and 54 in 1999 and has now exceeded 100 per year; by September 2003, over 500 aircraft had been produced. The STOL CH 801 and CH 2000 are now also available from CZAW. Workforce in 2003 was 115.

CZAW offers the CH 601, 701 and 801 in three different stages of completion: as a factory-jigged, fast-build kit; and in full flyaway form with Rotax 912 engine, avionics, instruments and two-tone paint finish. A total of 85 CH 701s, with options for 48 more, was ordered by the Indian National Cadet Corps in late 2000 for the training of Indian Air Force pilots; 25 of these had been delivered by 21 February 2001. Development was completed in late 1998 of a modified CH 701 for ULV agricultural applications. A Zodiac floatplane, flown by CZAW's CEO, won the 1999, 2001 and 2003 Schneider Cup seaplane races. Additionally, CZAW is European agent for the AMD CH 2000.

A new design, the Lake Sport, was unveiled at Sun 'n' Fun in April 2004.

CZAW produces a complete line of wheeled and non-wheeled aircraft floats, with displacements ranging from 250 to 1,000 kg (551 to 2,205 lb), suitable for such aircraft as the GlaStar, Kitfox, Zenair and Van's types. 'Jump-start' wings, ailerons, flaps and tail surfaces for the GlaStar kitplane are also manufactured.

In addition, CZAW refurbished the prototype Luscombe 8F under an agreement with Renaissance Aircraft LLC and Zenair Ltd; this programme was reported to be on hold in late 2003.

UPDATED

Zenair CH 701 on CZAW floats ***NEW***/0576136

CZAW (ZENAIR) CH 701-AG STOL

TYPE: Agricultural sprayer.

PROGRAMME: Developed by CZAW following cessation of Zenith production of a dedicated agricultural version. The CH 701-SP with 500 kg (1,100 lb) maximum take-off weight has recently been added to the range.

COSTS: CH 701 kit US$12,980, ready to fly US$40,500; CH 701-AG US$50,000 ready-to-fly (all 2004).

DESIGN FEATURES: STOL performance, plus ULV crop treatment system optimised for low cost and precise application.

Differences from basic, US-fabricated CH 701 as follows:

EQUIPMENT: Spray Miser ULV crop treatment system, comprising 113.6 litre (30.0 US gallon; 25.0 Imp gallon) belly tank and 12 wind-driven rotary atomisers with controlled droplet application. Hydraulic nozzles produce droplet sizes varying from 1 to 500 microns in size. Electrically powered pump (12 V DC), with remotely controllable on/off switch, can deliver liquid at up to 30.3 litres (8.0 US gallons; 6.7 Imp gallons)/min. Tank has emergency dump facility.

WEIGHTS AND LOADINGS:
Weight empty, equipped ..... 270 kg (595 lb)
Max T-O weight ..... 450 kg (992 lb)

PERFORMANCE:
Application speed range ..... 39–65 kt (72–121 km/h; 45–75 mph)
Max rate of climb at S/L ..... 305 m (1,000 ft)/min
Swath width ..... 9.1–30.5 m (30–100 ft)
Range ..... 217 n miles (402 km; 250 miles)

UPDATED

CZAW LAKE SPORT MERMAID

TYPE: Two-seat amphibian ultralight.

PROGRAMME: First flight (OK-JUR 99) 11 March 2004. Following preliminary testing, shipped to Lakeland, Florida, for public debut at Sun 'n' Fun, 13–19 April 2004, at which time a second prototype was under construction. Flight testing was due for completion by end of 2004.

CUSTOMERS: More than 30 sold by April 2004 and 98 production slots reserved by September 2004. Deliveries were scheduled to begin in first quarter of 2005.

COSTS: US$78,500 (2004).

DESIGN FEATURES: Intended as certifiable, roomy, low-cost amphibian meeting US LSA criteria and with useful range and dual controls.

Mid-wing, T-tail configuration; constant-chord wings with stabilising floats at half span. Wings removable to reduce width of airframe for storage or transport; folding wing option

Prototype CZAW Lake Sport Mermaid conducting early trials with wing floats removed ***NEW***/0576823

to be added. Pusher propeller above and behind cockpit; retractable water rudder at rear of single-step hull. Five hull compartments, each watertight and with bilge pump socket in port wing, adjacent to cockpit.

FLYING CONTROLS: Conventional and manual. Flight-adjustable trim tabs in elevator and starboard aileron. Fowler flaps; deflections 20 and 35°. Aileron trim optional.

STRUCTURE: Aluminium alloy throughout, except for rubber nosecone. Full zinc chromate coating.

LANDING GEAR: Tricycle type; retractable. Mainwheels retract hydraulically rearward into wing, tread protruding through port in upper wing surface and visible from cockpit rear quarterlights. Nosewheel retracts forward and is steerable. Mainwheels size 14×4 with hydraulic brakes; nosewheel 4.00×4. Rubber-in-compression shock-absorbers throughout, with twin units on each mainwheel leg. Tail bumper.

POWER PLANT: One 73.5 kW (98.6 hp) Rotax 912 ULS flat-four driving a Woodcomp three-blade, ground-adjustable pitch, scimitar profile, carbon fibre propeller. Optionally, one 84.5 kW (113.3 hp) Rotax 914.

Fuel tanks in wing leading-edge, total capacity 100 litres (26.4 US gallons; 22.0 Imp gallons).

ACCOMMODATION: Two persons, side by side, beneath forward-hinged, single-piece canopy; dual controls. Baggage space to rear, above and below main spar.

AVIONICS: VFR instrumentation standard.

EQUIPMENT: Portable, hand-operated bilge pump. Optional BRS ballistic recovery parachute and landing light in starboard wing leading-edge.

Lake Sport Mermaid amphibian prior to its US debut ***NEW***/0576824

DIMENSIONS, EXTERNAL:	
Wing span	9.50 m (31 ft 2 in)
Wing aspect ratio	7.8
Length overall	7.30 m (23 ft 11½ in)
DIMENSIONS, INTERNAL:	
Cabin max width	1.12 m (3 ft 8 in)
AREAS:	
Wings, gross	11.50 m² (123.8 sq ft)
WEIGHTS AND LOADINGS:	
Weight empty	330 kg (728 lb)
Max T-O weight	560 kg (1,234 lb)
PERFORMANCE (PROVISIONAL):	
Max level and never-exceed (VNE) speed	115 kt (213 km/h; 132 mph)
Cruising speed at 75% power	100 kt (185 km/h; 115 mph)
Stalling speed: flaps up	44 kt (80 km/h; 50 mph)
flaps down	35 kt (64 km/h; 40 mph)
Max rate of climb at S/L	300 m (984 ft)/min
T-O run: on land	137 m (450 ft)
on water	229 m (755 ft)
Landing run: on land, water	152 m (500 ft)
Range	432 n miles (800 km; 497 miles)
Endurance	4 h 30 min

NEW ENTRY

DELTA

DELTA SYSTEM-AIR AS

Bratří Štefanů 101, CZ-500 03 Hradec Králové
Tel: (+420 495) 40 74 05
Fax: (+420 495) 40 74 08
e-mail: office@dsa.cz
Web: http://www.dsa.cz
MANAGER, FLIGHT OPERATIONS: Jan Rusek

Promotion of the Pegass ultralight appears to have ceased, although Delta offers an overhaul service for the type among its varied aerial work and maintenance activities.

UPDATED

EV-AT

EVEKTOR-AEROTECHNIK AS

Letecká čp. 1384, CZ-686 04 Kunovice
Tel: (+420 572) 53 71 11 and 53 71 17
Fax: (+420 572) 53 79 10
e-mail: marketing@evektor.cz
Web: http://www.evektor.cz
CHAIRMAN: Jaroslav Růžička
MANAGING DIRECTOR: Jiří Habarta
COMMERCIAL DIRECTOR: Milan Mach
MARKETING MANAGER: Vik Kotek

The Evektor aircraft design office (workforce 150 in 1999) was formed in 1991, a major part of its initial work consisting of contribution to the designs of the L 159 and Ae 270 Ibis under contract to Aero Vodochody. On 2 February 1996 it acquired a 100 per cent shareholding in Aerotechnik CZ, and in 1996–97 the two companies (now a single joint-stock company) developed and delivered preseries and test front fuselage components of the L 159 to Aero.

Aerotechnik was formed in 1970, originally as an enterprise of the former Czechoslovak defence organisation Svazarm. Its first aircraft programmes involved single- and two-seat rotorcraft. Two facilities, at Kunovice and Moravská Třebová, have a combined floor space of 7,340 m² (79,000 sq ft). The combined Evektor-Aerotechnik workforce totalled 400 in mid-2003.

In recent years Aerotechnik produced kits for the Pottier P 220 S Koala all-metal lightplane, powered by its own Vikron III engine. This design was further developed via the P 220 UL into the EV-97 Eurostar with a Rotax 912 UL engine, and became the first aeroplane to bear the Evektor name. A VLA version of the Eurostar, the Harmony, followed in 2001.

Evektor-Aerotechnik assisted in early stages of the Wolfsberg Raven 257 programme, including assembly of the prototype; this aircraft was subsequently transferred to Wolfsberg Letov. A separate joint venture was formed in October 1999 with GAIC of China, leading to establishment in June 2000 of Guizhou Evektor Aircraft Corporation for local production, sales and support of EV-AT aircraft in that country. The first EV-97 CKD kit for Chinese assembly was delivered in August 2000. This arrangement superseded that previously agreed for Guizhou to become a licensed production centre for the Raven 257. However, there has been little evidence of progress with the Chinese EV-97.

On 26 July 2000, Evektor-Aerotechnik signed an agreement with Seabird Aviation Australia to manufacture and market the latter's Seeker SB7L-360 light observation aircraft. By 2002, nothing further had been heard of the venture, and it appears to have been discontinued.

In late 2003, Evektor headed a 17-company consortium of Czech aerospace companies to develop and market a twin-engined 10- to 15-seat turboprop business aircraft under the designation EV-55. Also under development is the four-seat VUT 100 Cobra lightplane.

Other activities include development and production of parts, jigs and tools for the aircraft industry; spares and components for light and ultralight aircraft; and the overhaul and re-engining of L-13SV Vivat and L-60 Brigadýr aircraft.

UPDATED

EV-AT EV-55

TYPE: Business twin-turboprop.

PROGRAMME: Announced in December 2003. Development being undertaken by Evektor as leader in a 17-company consortium. Funding provided by offset programme tied into Saab/BAE Systems Gripen purchase. Prototype expected to be flying by 2007.

DESIGN FEATURES: High-wing transport with T-tail. Tapered, high-aspect ratio wing.

FLYING CONTROLS: Conventional and manual. Horn-balanced rudder and elevators. Trim tab in each aileron and elevator. Flaps divided inboard and outboard.

STRUCTURE: All-metal.

Artists' impression of EV-AT EV-55 ***NEW***/0576219

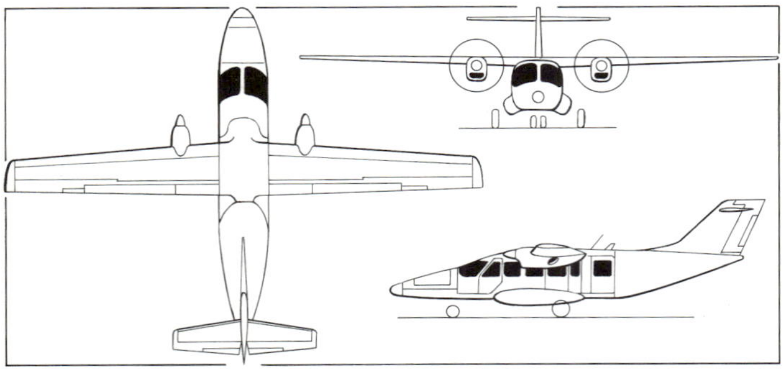

EV-55 provisional general arrangement *(James Goulding)* ***NEW***/0583032

LANDING GEAR: Retractable tricycle type; mainwheels retract into panniers on lower fuselage sides.

POWER PLANT: Two engines in the 336 to 373 kW (450 to 500 hp) class, driving constant-speed propellers.

ACCOMMODATION: Pilot and up to nine passengers standard; high-density option for 14 passengers; alternatively, cabin can be configured for freight. Passenger door on port side of rear fuselage incorporates forward extension for freight loading; cockpit door on port side forward of wing. Baggage compartment in nose accessed by upward-hinged doors.

All data are provisional.

DIMENSIONS, EXTERNAL:

Wing span	16.00 m (52 ft 6 in)
Length overall	13.04 m (42 ft 9½ in)
Height overall	4.23 m (13 ft 10½ in)
Tailplane span	5.00 m (16 ft 4¾ in)
Wheel track	3.00 m (9 ft 10 in)
Wheelbase	3.89 m (12 ft 9¼ in)
Passenger door: Width	1.10 m (3 ft 7¼ in)

DIMENSIONS, INTERNAL:

Cabin: Length	4.30 m (14 ft 1¼ in)
Max width	1.56 m (5 ft 1½ in)
Max height	1.48 m (4 ft 10¼ in)
Volume	9.0 m³ (317.8 cu ft)
Baggage hold volume (total)	3.5 m³ (123.6 cu ft)

WEIGHTS AND LOADINGS:

Weight empty	2,333 kg (5,143 lb)
Max payload	1,700 kg (3,748 lb)
Max T-O and landing weight	4,400 kg (9,700 lb)
Max power loading (373 kW; 500 hp engines)	5.90 kg/kW (9.70 lb/hp)

PERFORMANCE:

Normal cruising speed	more than 200 kt (370 km/h; 230 mph)
T-O run	341 m (1,120 ft)
Range: with max fuel, 45 min reserves	1,455 n miles (2,694 km; 1,674 miles)
with 1,400 kg (3,086 lb) payload	400 n miles (740 km; 460 miles)

NEW ENTRY

EV-AT EV-97 TEAMEUROSTAR

TYPE: Side-by-side ultralight/kitbuilt.

PROGRAMME: Developed as P 220 UL from Pottier P 220 S Koala; extensive redesign started in September 1996; prototype (OK-CUR 97), then known as Eurostar, made first flight May 1997; first customer delivery October 1997. Certified as ultralight in Belgium, Czech Republic, Finland, France, Germany, Italy, Slovak Republic and United Kingdom. First UK-built example (G-NIDG, by distributor Cosmik Aviation) flown in April 2000; local assembly in China by Guizhou Evektor Aircraft Corporation (GEAC); first flight by Chinese-assembled EV-97, 24 October 2000.

CURRENT VERSIONS: **Eurostar 2000:** Introduced late 1999, replacing Eurostar Model 99. Increased fuselage width (4 cm; 1½ in); modified wingtips and increased span; larger ailerons; steerable nosewheel; flush, oval nose air intake. Available factory-built or as kit.

teamEurostar 2001: Further-improved version, introduced in early 2001; choice of Rotax 912 UL or more powerful 912 ULS engine. Name changed to avoid conflict with Eurostar France-UK rail link.

teamEurostar 2002: Further improved version meeting Sport Pilot NRPM in USA (where marketed as **SportStar**), *as described.*

CUSTOMERS: More than 200 (all versions) ordered, including 36 Model 2000. Sales in 2001 totalled 56 and in 2002, 72. Customers in Belgium, Czech Republic, Estonia, Finland, France, Germany, Ireland, Italy, Netherlands, Poland, Portugal, Russia, Slovak Republic, Spain, Sweden, UK, Ukraine and USA.

COSTS: €49,900 excluding taxes (2003).

DESIGN FEATURES: Low-wing monoplane; modified NACA 4415 aerofoil section; dihedral 3°. Wing folding optional.

FLYING CONTROLS: Conventional and manual. Four-position (0/15/30/50°) split flaps. Elevators, starboard aileron and rudder have trim tabs.

STRUCTURE: Mostly of riveted metal; wingtips and mainwheel legs in GFRP.

LANDING GEAR: Tricycle type; fixed. Fuselage-mounted cantilever mainwheel legs. Hydraulic mainwheel disc brakes and (in current version) steerable nosewheel. Wheel sizes 350×140 (main), 300×100 (nose). Optional wheel fairings or mudguards. Optional floats.

Early production Eurostar employed as a US demonstrator *(Paul Jackson)* ***NEW***/0576139

teamEurostar 2002 version assembled in the UK *(Paul Jackson)* ***NEW***/0576168

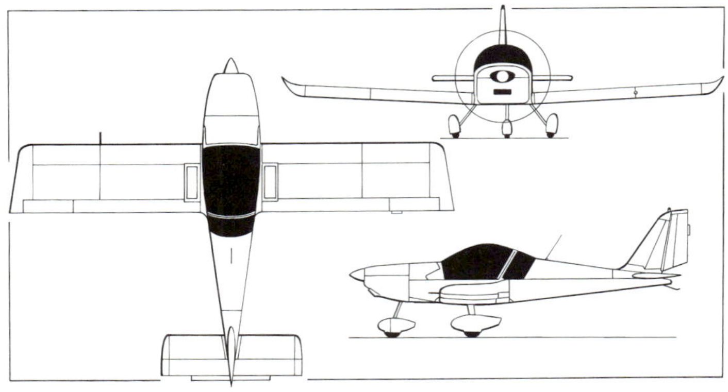

EV-AT EV-97 teamEurostar 2001 two-seat ultralight *(James Goulding)* 0059970

POWER PLANT: One 59.6 kW (79.9 hp) Rotax 912 UL flat-four engine standard, with VZLU V230C two-blade, fixed-pitch wooden propeller; 73.5 kW (98.6 hp) Rotax 912 ULS and V534BD-UL three-blade, constant-speed (hydraulic), wooden propeller optional. Jabiru 2200 also optional. Fuel capacity 70 litres (18.5 US gallons; 15.4 Imp gallons) of which 65 litres (17.2 US gallons; 14.3 Imp gallons) are usable; optionally 50 litres (13.2 US gallons; 11.0 Imp gallons).

ACCOMMODATION: Front-hinged, upward-opening canopy, with additional transparency aft of rear frame. Space for 15 kg (33 lb) of baggage aft of seats. Early versions had canopy/rear transparency break line further forward. Optional replacement of rear transparency by metal fairing and bulged cabin sides.

SYSTEMS: 12 V (battery) electrical system.

AVIONICS: To customer's specification.

EQUIPMENT: Optional ballistic parachute, wing-folding mechanism and aero-tow hook.

DIMENSIONS, EXTERNAL:

Wing span	8.10 m (26 ft 7 in)
Length overall	5.98 m (19 ft 7½ in)
Fuselage max width	1.04 m (3 ft 5 in)
Height overall	2.34 m (7 ft 8¼ in)
Wheel track	1.60 m (5 ft 3 in)
Wheelbase	1.40 m (4 ft 7 in)

DIMENSIONS, INTERNAL:

Cabin max width: standard	1.04 m (3 ft 5 in)
optional	1.18 m (3 ft 10½ in)

AREAS:

Wings, gross	9.84 m² (105.9 sq ft)

WEIGHTS AND LOADINGS:

Weight empty	262 kg (578 lb)
Max T-O weight	450 kg (992 lb)[1]

[1] *480 kg (1,058 lb) permitted in Slovak Republic*

PERFORMANCE (at 450 kg; 992 lb MTOW. A: Rotax 912 UL engine, B: Rotax 912 ULS):

Never-exceed speed (VNE):	
A, B	145 kt (270 km/h; 167 mph) IAS
Max level speed: A	121 kt (225 km/h; 139 mph) IAS
B	132 kt (245 km/h; 152 mph) IAS
Max cruising speed: A	97 kt (180 km/h; 112 mph) IAS
B	108 kt (200 km/h; 124 mph) IAS
Stalling speed, flaps down:	
A, B	36 kt (65 km/h; 41 mph) CAS
Max rate of climb at S/L: A	330 m (1,083 ft)/min
B	480 m (1,574 ft)/min
Service ceiling: A, B	5,000 m (16,400 ft)
T-O run: A	145 m (475 ft)
B	100 m (330 ft)
Landing run: A, B	200 m (660 ft)
Max range, standard fuel:	
A, B	432 n miles (800 km; 497 miles)
g limits: A, B	+6/−3

OPERATIONAL NOISE LEVELS:

To German LSL Chapter 6	60 dBA

UPDATED

EV-AT HARMONY

TYPE: Two-seat lightplane.

PROGRAMME: Revealed (then called Eurostar VLA) 1999; renamed Harmony and made first flight (OK-VLA) 6 April 2001; certification end 2002 and first deliveries early 2003. Initial batch of six, to be followed by further 12.

COSTS: €79,900 to €95,490 (excluding VAT), depending upon avionics fit (2003).

DESIGN FEATURES: Heavier/more powerful version of teamEurostar, conforming to JAR-VLA and BCAR Section S. Reinforced fuselage aft of cockpit; slightly wider wheel track. Split flaps deflect 0/15/40° and extend over two-thirds of wing length.

EV-AT Harmony prototype/demonstrator *(Paul Jackson)* ***NEW***/0576241

Description as for teamEurostar 2002 except as follows:

POWER PLANT: One 73.5 kW (98.6 hp) Rotax 912 ULS flat-four engine; VZLU V230E two-blade, fixed-pitch, wooden propeller. Standard fuel capacity 67 litres (17.7 US gallons; 14.7 Imp gallons).

ACCOMMODATION: Two seats side by side; dual controls. Front-hinged canopy, with larger size of rear transparency (refer EV-97). Baggage space aft of seats.

AVIONICS: VFR instrumentation only in basic aircraft. 'Effect' version adds VSI, turn co-ordinator, VHF com antenna, PM 1000 intercom, Bendix/King KT 76A transponder, ACK encoder, KY 97A VHF and DME/transponder antenna. Top-of-range 'VFR night' suite omits KY 97A and ACK encoder, but adds KLN 35A GPS with KA 92 antenna, KX 125 nav/com, KT 76A transponder, DME/transponder and VHF antennae, VOR/glideslope antenna, artificial horizon and directional gyro. Further options available.

DIMENSIONS: As for teamEurostar

AREAS:

Wings, gross	10.13 m² (109.0 sq ft)

WEIGHTS AND LOADINGS:

Weight empty	330 kg (728 lb)
Max baggage weight	15 kg (33 lb)
Max T-O weight	575 kg (1,267 lb)
Max wing loading	56.8 kg/m² (11.63 lb/sq ft)
Max power loading	7.82 kg/kW (12.85 lb/hp)

PERFORMANCE (preliminary):

Never-exceed speed (VNE)	145 kt (270 km/h; 167 mph) IAS
Max level speed	129 kt (240 km/h; 149 mph) IAS
Max cruising speed	111 kt (205 km/h; 127 mph) IAS
Stalling speed, flaps down	44 kt (80 km/h; 50 mph) CAS
Max rate of climb at S/L	300 m (984 ft)/min
Service ceiling	5,000 m (16,400 ft)
T-O run	130 m (425 ft)
T-O to 15 m (50 ft)	260 m (855 ft)
Landing from 15 m (50 ft)	480 m (1,575 ft)
Landing run	240 m (790 ft)
Range	388 n miles (720 km; 447 miles)
g limits	+5.7/–2.25

UPDATED

EV-AT VUT 100 COBRA

TYPE: Four-seat lightplane.

PROGRAMME: Project started in 2000 at Brno University; intended to be first of a family of four- to seven-seat aircraft. Development began in August 2001, with construction of first prototype beginning in February 2003; first flight 11 November 2004 (official first flight 16 November 2004). Serial production intended; will meet JAR 23/FAR 23 requirements; certification intended in early 2006.

CURRENT VERSIONS: **VUT 100-120i:** Baseline version, *as described.*

VUT 100-120TDi: Powered by 169 kW (227 hp) SMA SR 305 diesel engine driving an MT Propeller MTV-9 three-blade, constant-speed propeller, diameter 1.96 m (6 ft 5 in).

Prototype EV-AT VUT 100 Cobra on an early test flight ***NEW***/1043437

First Evektor VUT 100 Cobra ***NEW***/1043434

VUT 100-131i: Intended to be second version introduced; will have 231 kW (310 hp) Teledyne Continental IO-550N driving MT-Propeller MTV-9 three-blade, constant-speed propeller, diameter 1.98 m (6 ft 6 in). Fuel capacity as VUT 100-120i.

DESIGN FEATURES: Multipurpose, all-metal aircraft designed to meet a range of tasks including sport flying, basic and advanced training and aero-towing.

FLYING CONTROLS: Conventional and manual. Flaps.

STRUCTURE: Metal throughout.

LANDING GEAR: Tricycle type; retractable.

POWER PLANT: One 149 kW (200 hp) Textron Lycoming IO-360-A flat-four, driving an MT-Propeller MTV-12 three-blade, constant-speed, 1.95 m (6ft 4¾ in) propeller. Fuel capacity 340 litres (89.8 US gallons; 74.8 Imp gallons) in all versions.

ACCOMMODATION: Pilot and up to four passengers.

All data are provisional.

AVIONICS: *Instrumentation:* Two-screen EFIS; mechanical backup for primary instruments.

DIMENSIONS, EXTERNAL:

Wing span	10.20 m (33 ft 5½ in)
Length overall	8.00 m (26 ft 3 in)
Propeller diameter	1.83 m (6 ft 0 in)

DIMENSIONS, INTERNAL:

Cabin: Length	2.92 m (9 ft 7 in)
Max width	1.31 m (4 ft 3½ in)
Max height	1.22 m (4 ft 0 in)

WEIGHTS AND LOADINGS (A: 100-120i; B: 100-120TDi; C: 100-131i):

Weight empty: A	800 kg (1,763 lb)
B:	880 kg (1,940 lb)
Max T-O weight: A	1,330 kg (2,932 lb)
B	1,330 kg (2,932 lb)
C	1,451 kg (3,200 lb)

Two-screen EFIS of VUT 100 Cobra ***NEW***/1043436

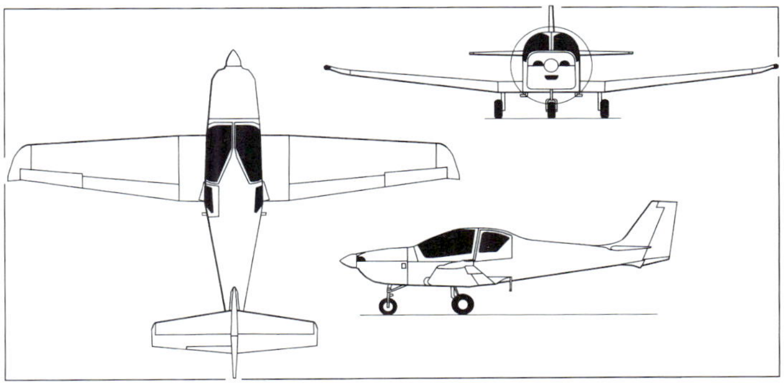

Evektor VUT 100 Cobra *(Paul Jackson)* ***NEW***/0587529

First Evektor VUT 100 Cobra ***NEW***/1043433

PERFORMANCE (A, B, C AS ABOVE, ESTIMATED):
Max level speed: A ... 160 kt (296 km/h; 184 mph)
C ... 180 kt (333 km/h; 207 mph)
Stalling speed, flaps down:
A, B ... 50 kt (93 km/h; 58 mph)
C ... 55 kt (102 km/h; 64 mph)
Max rate of climb at S/L: A, C ... 300 m (985 ft)/min
Range with max fuel:
A ... 1,080 n miles (2,000 km; 1,242 miles)
B ... 1,300 n miles (2,407 km; 1,496 miles)
C ... 1,000 n miles (1,852 km; 1,150 miles)
Endurance: A ... 8 h
B ... 10 h
C ... 6 h
g limits: Normal ... +3.8/–1.52
Utility ... +4.4/–1.76

NEW ENTRY

FANTASY AIR

FANTASY AIR GROUP

Kollàrova 511, CZ-397 01 Pisek
Tel: (+420 3) 82 21 02 46
Fax: (+420 3) 82 21 29 62 and 21 44 57
e-mail: sales@fantasyair.com
Web: http://www.fantasyair.com
MANAGING DIRECTOR: Josef Jestřáb

Fantasy Air established at Písek 1995; initial product launched as Cora, in Legato and Allegro versions. By mid-2000 was standardising on variants of Allegro. Fantasy Air employs 23 workers at its 750 m2 (8,073 sq ft) factory.

UPDATED

FANTASY AIR ALLEGRO AND ARIUS

TYPE: Side-by-side ultralight/kitbuilt.

PROGRAMME: Certified in Czech Republic, Finland, France and Germany; public debut at Friedrichshafen Air Show, May 1997. All are subvariants of Fantasy **Cora**. Legato subvariant no longer marketed. Production rate three per month from 2001 onwards.

A new wing profile was first tested in May 2002. Revealed in the same month were plans to fit a Diesel Air D280 twin-piston engine, derated to 60.0 kW (80 hp); these were in abeyance by October 2002.

CURRENT VERSIONS: **Allegro ST:** Baseline version; 47.8 kW (64.1 hp) Rotax 582 UL. Remains available.

Allegro 200: As ST, but with 55.0 kW (73.8 hp) Rotax 618 UL. Remains available.

Allegro 2000: Introduced from 1 March 2000 (kit version introduced September 2000); 59.6 kW (79.9 hp) Rotax 912 UL or (Allegro S) 73.5 kW (98.6 hp) 912 ULS. SportProp glass fibre or VZLU V331-3NC wooden three-blade propeller. Quoted build time 510 hours for standard kit and 190 hours for quick-build kit. Certified in Belgium, Czech Republic, Denmark, Finland and Germany.

Allegro SW: Short-span (Speed Wing) version, without taper on outer panels. New SM 701 aerofoil section wing and 47.8 kW (64.1 hp) Rotax 582 UL or 59.6 kW (79.9 hp) Rotax 912 UL. German marketing only with Rotax 912 UL and propeller options as for Allegro 2000. Four to France in September 1999, where certification subsequently awarded; also certified in Czech Republic and Spain. Kit version available from 1 March 2000.

Allegro 2000 F (formerly Arrius F): Float-equipped, standard-span wing (SM 701 section) and Rotax 912 UL or 912 ULS. Two to Finland in August 1999. First flight (converted Allegro, OK-DUU 03) 5 November 1999.

CUSTOMERS: Total 70 (51 kits and 19 factory-built, including a few Cora Legatos) registered in Czech Republic by November 2000. Exports to Finland and Hungary in 2000. Total sales by October 2002 exceeded 140, including 75 in Czech Republic, 27 in France and 17 in Germany; 32 built in 2000, 35 in 2001 and 34 in 2002.

COSTS: Allegro 2000 €55,445; Allegro SW €52,915; Allegro 2000 quick-build kit US$20,625 (all 2004, incl tax).

DESIGN FEATURES: Braced high-wing monoplane; T tail; tapered outer panels on 2000 and Arius, constant chord throughout on SW. UA-2 aerofoil section on ST and 200, improved SM 701 on 2000, SW and Arius; dihedral 1° 10' on Allegro 2000; 1° 30' on Allegro SW.

FLYING CONTROLS: Conventional and manual. Electrically actuated flaperons (settings 0/15/48°). In-flight-adjustable tab in port half of elevator; ground-adjustable rudder tab.

STRUCTURE: Fabric- and metal-covered metal wings with welded steel tube centre-section and full-span metal flaperons; laminated glass fibre fuselage and integral fin; all-metal rudder, tailplane and one-piece elevator.

LANDING GEAR: Tricycle type; fixed. Fuselage-mounted spring Kevlar cantilever mainwheel legs and 350 mm wheels, rubber-block shock-absorbers, hydraulic mainwheel brakes and steerable nosewheel; wheel spats. Full Lotus twin-float gear to be certified for Arius F; testing undertaken (with OK-DUU 03) in August 2001.

POWER PLANT: One Rotax engine (details under Current Versions); SportProp three-blade, glass fibre or Junkers Profly two-blade, ground-adjustable pitch, carbon fibre propeller. Fuel capacity 55 litres (14.5 US gallons; 12.1 Imp gallons).

EQUIPMENT: Junkers recovery parachute optional.

DIMENSIONS, EXTERNAL (A: 2000, B: 2000 F), C: SW):
Wing span: A, B ... 10.81 m (35 ft 5½ in)
C ... 9.63 m (31 ft 7 in)
Length overall: A ... 6.36 m (20 ft 10½ in)
B, C ... 6.10 m (20 ft 0¼ in)
Height overall: A ... 2.05 m (6 ft 8¾ in)
B, C ... 2.00 m (6 ft 6¾ in)
DIMENSIONS, INTERNAL:
Cabin max width: A, B, C ... 1.08 m (3 ft 6½ in)
AREAS:
Wings, gross: A, B ... 11.37 m² (122.4 sq ft)
C ... 11.55 m² (124.3 sq ft)
WEIGHTS AND LOADINGS:
Weight empty: A, B, C ... 285–295 kg (628–650 lb)
Max T-O weight: A, C ... 450 kg (992 lb)
B ... 520 kg (1,146 lb)
PERFORMANCE, POWERED (Rotax 912 UL):
Never-exceed speed (VNE):
A ... 118 kt (220 km/h; 136 mph)
B, C ... 108 kt (200 km/h; 124 mph)
Max cruising speed at 75% power:
A ... 89 kt (165 km/h; 103 mph)
B ... 97 kt (180 km/h; 112 mph)
Stalling speed: A, B, C ... 36 kt (65 km/h; 41 mph)
Max rate of climb at S/L: A ... 360 m (1,181 ft)/min
B ... 240 m (787 ft)/min
C ... 300 m (984 ft)/min
T-O run: A, C ... 150 m (495 ft)
B ... 220 m (722 ft)
Landing run: A, C ... 100 m (330 ft)
B ... 120 m (395 ft)
Max range: A, C ... 270 n miles (500 km; 310 miles)
B ... 297 n miles (550 km; 341 miles)
PERFORMANCE, UNPOWERED:
Best glide ratio: A, B ... 12
C ... 10

UPDATED

Short-span Fantasy Air Allegro SW *(Paul Jackson)* ***NEW***/0576164

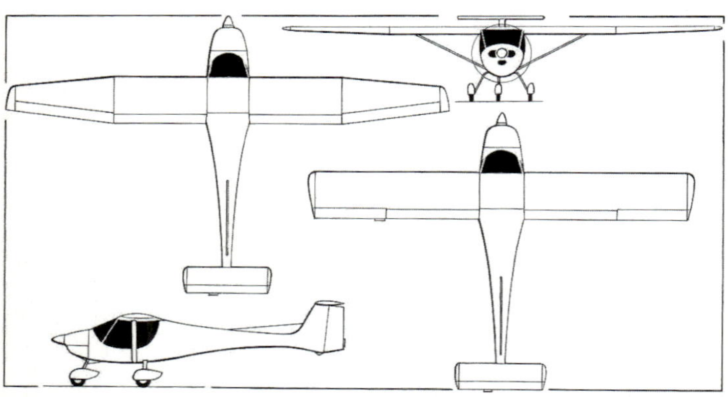

Fantasy Air Cora Allegro 2000, with additional plan view of short-span SW version *(James Goulding)* 0137210

INTERPLANE

INTERPLANE SPOL S.R.O.

Letiště Zbraslavice, CZ-285 21 Zbraslavice
Tel/Fax: (+420 327) 59 13 81
e-mail: interplane@cmail.cz
Web: http://www.InterplaneAircraft.com
SALES MANAGER AND EXECUTIVE: Andrea Greplová
DIRECTOR OF MANUFACTURING: Radek Hurt
CHIEF DESIGNER: Jaroslav Dostal

NORTH AND SOUTH AMERICAN SALES:
Interplane LLC
30 Lee Gate Lane, Grosse Point Farms, Michigan 48236, USA
Tel: (+1 313) 882 34 00
e-mail: interplane@hotmail.com
PRESIDENT: Ben Dawson

Interplane was created as an ultralight manufacturer in 1992; design and product development is by Gryf Development company at Hodonin. An early product, the Griffon, is no longer being promoted. Interplane also markets Alexander Dewald's **Sunny** forward-stagger, tail-less biplane in its single-seat form, powered by a 61.0 kW (82 hp) Hirth F30E flat-four, price US$28,900 (2004); and the Aerosport Qualt 200L low-wing two-seater as the **Skyboy 2000L** (or **Falcon 2000L**), price US$39,900.

UPDATED

Interplane Sunny tail-less biplane ***NEW***/0576130

INTERPLANE SKYBOY

TYPE: Side-by-side ultralight/kitbuilt.

PROGRAMME: Prototype designed by team led by Jaroslav Dostal; first flown 1993; entered production 1994; conforms to Czech UL-2 and German BFU-DULV standards. Certified in Czech Republic (1994), France (1996), Slovak Republic and Germany (excluding noise tests, 1999). Current EX and UL versions introduced in 1999. New canopy design permits operation without doors for hot-weather flying.

CURRENT VERSIONS: **Skyboy:** Initial version, with Rotax 503 or 582 engine as standard and cockpit width of 1.20 m (3 ft 11¼ in); no longer produced.

Skyboy S: Reinforced version for higher MTOW and speeds; Rotax 582 engine and choice of wing spans; no longer produced.

Skyboy EX: Standard version from 1999; for European customers; also for customers in Australia and USA, where it is classified in the FAA Experimental category.

Skyboy UL: Specially lightened (empty weight 225 kg; 496 lb) to meet US FAR Pt 103 ultralight trainer regulations.

Skyboy ZK-F: Version with wing flaps; MTOW 475 kg (1,047 lb), where local regulations allow.

Skyboy XJ: Tractor propeller version with WarpDrive three-blade, ground-adjustable pitch propeller. Composites Fuselage with fabric-covered wings and ailerons. Fuel capacity 56 litres (14.8 US gallons; 12.3 Imp gallons); prototype OK-IUR 21. Other features include fuselage-mounted spring cantilever composites mainwheel legs; built-up rear decking to pod; GRS parachute; hydraulic brakes and V-strut ZK-F wings. External mass balance on elevators; Flight-adjustable trim tabs attached to each elevator; ground-adjustable rudder tab. Length 6.86 m (22 ft 6 in); height 2.34 m (7 ft 8 in).

CUSTOMERS: Total of 78 registered by late 2003, including 16 on the FAA register. Scales to Australia, Austria, Belgium, Bulgaria, Canada, Czech Republic, Denmark, France, Germany, Luxembourg, Mexico, Netherlands, Poland, Portugal, Slovak Republic, South Korea and USA.

COSTS: *UL:* Flyaway US$29,900 (2004).

Description applies to Skyboy UL.

DESIGN FEATURES: High, constant-chord wing, boom-mounted empennage and pusher propeller. Available factory-built or (with or without power plant and instruments) as kit. Wing struts and tail surfaces can be folded for transportation and storage.

Wing aerofoil section NACA 4412; dihedral 1° 30'; incidence 4° 30'. Leading-edges swept forward 2°.

FLYING CONTROLS: Conventional and manual. Tabs on rudder and each elevator; flaps optional.

STRUCTURE: Aluminium tube and glass fibre; fabric covering.

LANDING GEAR: Tricycle type; fixed. Coil spring/hydraulic (motorcar-type) shock-absorbers on trailing-link mainwheels and castoring nosewheel; mechanical or hydraulic brakes on nosewheel only. Ski and float gear optional.

POWER PLANT: One 47.3 kW (64.1 hp) Rotax 582 UL two-cylinder in-line two-stroke engine standard; 59.6 kW (79.9 hp) Rotax 912 UL or 73.5 kW (98.6 hp) Rotax 912 ULS flat-four; Rotax 462, 503 or 618; Hirth 2706; or Verner SVS 1400 optional. Wide choice of two- to six-blade wooden or laminate pusher propellers (three-blade, carbon fibre Junkers Profly standard). Standard fuel capacity 38 litres (10.0 US gallons; 8.4 Imp gallons); optional fuel capacity 57 litres (15.0 US gallons; 12.5 Imp gallons).

ACCOMMODATION: Optional cabin doors.

EQUIPMENT: Optional Magnum 450 ballistic recovery parachute.

DIMENSIONS, EXTERNAL:
Wing span .... 9.12 m (29 ft 11 in)
Length overall .... 6.37 m (20 ft 10¾ in)
Height overall .... 2.13 m (6 ft 11¾ in)

DIMENSIONS, INTERNAL:
Cockpit max width .... 1.24 m (4 ft 1 in)

AREAS:
Wings, gross: without flaps .... 12.87 m² (138.5 sq ft)
with flaps .... 13.38 m² (144.0 sq ft)

WEIGHTS AND LOADINGS:
Weight empty, equipped .... 225 kg (496 lb)
Max T-O weight .... 450 kg (992 lb) or 558 kg (1,232 lb)

PERFORMANCE (Rotax 912, 558 kg; 1,232 lb MTOW):
Max level speed .... 75 kt (138 km/h; 86 mph)
Econ cruising speed .... 56 kt (105 km/h; 65 mph)
Stalling speed .... 34 kt (63 km/h; 40 mph)
Max rate of climb at S/L .... 306 m (1,004 ft)/min
T-O run .... 122 m (400 ft)
Landing run .... 152 m (500 ft)
Range with standard fuel .... 160 n miles (296 km; 184 miles)

UPDATED

Interplane Skyboy UL ultralight *(Paul Jackson)* ***NEW***/0532087

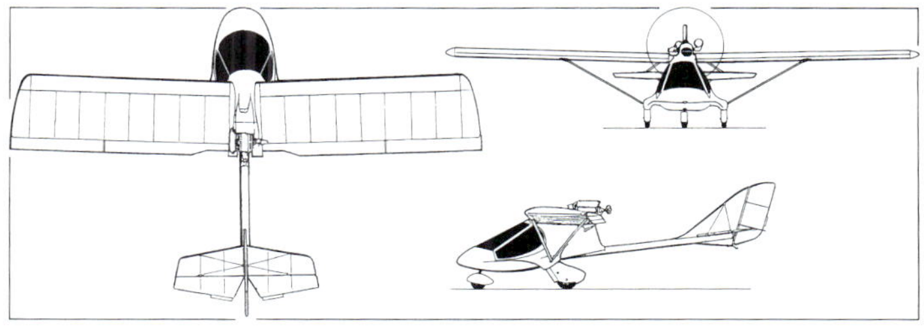

Interplane Skyboy, showing optional flaps *(Paul Jackson)* 0110846

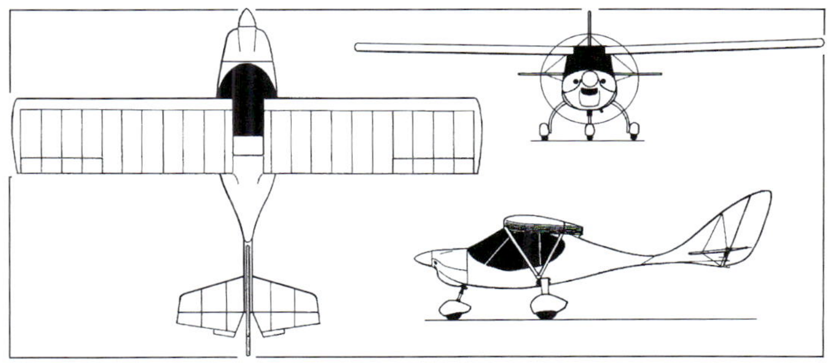

Interplane Skyboy XJ tractor propeller version *(Paul Jackson)* ***NEW***/0587528

Interplane Skyboy XJ tractor propeller version *(Paul Jackson)* ***NEW***/0583085

KAPPA

KAPPA 77 A.S.

Brtnická 21, CZ-586 01 Jihlava
Tel: (+420 567) 30 81 21 and 31 00 18
Fax: (+420 567) 30 81 22
e-mail: info@kappa77.cz and kappa77@mbox.vol.cz
Web: http://www.kappa77.cz
MANAGING DIRECTOR: Ing Miroslav Navrátil
CHIEF DESIGNER: Ing Thomas Prihoda
TECHNICAL DESIGNER: Ing Thomas Navrátil
MARKETING DIRECTOR AND PUBLIC RELATIONS MANAGER: Dr. Vit Tausch

Kappa established 1991, initially for machine manufacture; founders previously employed by Aero. Design of KP-2U began 1994; Kappa 77 a.s. created as subsidiary in June 1997; Sova production started 1998. Baron introduced 1 January 2004. The company employs a workforce of 140 in its 4,250 m^2 (45,750 sq ft) factory.

UPDATED

KAPPA KP-2U SOVA

English name: Owl

TYPE: Side-by-side ultralight/kitbuilt.

PROGRAMME: Designed by Kappa company and Institute of Aerospace Engineering at Brno Technical University and developed over a 22-month period. First of two flying prototypes (OK-BUU 23) first flown 26 May 1996; first public demonstration the following month at Jihlava Aero '96. Czech certification 23 September 1997; German certification, and of kit by FAA, in June 1999, followed by Polish certification in April 2001 and Portuguese certification in June 2002. Australian and Spanish certification under way at end 2002.

CURRENT VERSIONS: **KP-2U:** *As described.*

KP-5: Proposed version with fixed (optionally retractable) landing gear, square wingtips and option of Rotax 912, 912 S or 914 engine. Span and length as KP-2U; height overall 2.46 m (8 ft 0¾ in); wing area as KP-2U; weight empty 320 kg (705 lb); MTOW up to 544 kg (1,199 lb); cruising speed 113 kt (210 km/h; 131 mph); range 378 n miles (700 km; 435 miles); *g* limits as KP-2U.

CUSTOMERS: Over 140 sold by October 2003; customers in Australia, Belgium, Brazil, China, Czech Republic, Ecuador, Finland, France, Germany (known as KPD-2U), Italy, Netherlands, Poland, Portugal, Russia, Spain, South Africa and USA.

COSTS: €45,000 flyaway with Rotax 912 UL (2003).

DESIGN FEATURES: Low-wing monoplane conforming to BFU 95 regulations; JAR-VLA version being developed.

Wing sections NASA GA(W)-1 at root and GA(W)-2 at tip; upturned wingtips, dihedral 6°. Sweptback vertical tail.

FLYING CONTROLS: Conventional and manual. Electrically operated trim tab in port elevator; manually operated Fowler flaps (deflection 10 and 35°).

STRUCTURE: All-metal, except for CFRP cockpit frame.

LANDING GEAR: Electrically retractable tricycle type (all units rearward); rubber shock-absorption; mainwheels semi-exposed when retracted. Steerable nosewheel; mechanically operated brakes (hydraulic optional). Fixed gear optional.

POWER PLANT: One 59.6 kW (79.9 hp) Rotax 912 UL engine, with 4:1 reduction drive to a two- or three-blade, flight-adjustable or three-blade ground-adjustable pitch propeller; can alternatively accommodate Rotax 582 UL, 912 ULS or 914. Fuel capacity (two wing tanks) 64 litres (16.9 US gallons; 14.1 Imp gallons); 94 litres (24.8 US gallons; 20.7 Imp gallons) for Australian market. Optionally a 59.7 kW (80 hp) Jabiru 2200 can be fitted.

ACCOMMODATION: Forward-hinged canopy. Dual controls. Space behind seats for 14 kg (31 lb) of baggage.

EQUIPMENT: Optional emergency system under consideration.

DIMENSIONS, EXTERNAL:
Wing span .... 9.90 m (32 ft 5¾ in)
Length overall .... 7.17 m (23 ft 6¼ in)
Height overall .... 2.60 m (8 ft 6¼ in)
Propeller diameter .... 1.73 m (5 ft 8 in)

AREAS:
Wings, gross .... 11.85 m^2 (127.6 sq ft)

Kappa KP-2U Sova ***(Paul Jackson)*** ***NEW***/0576163

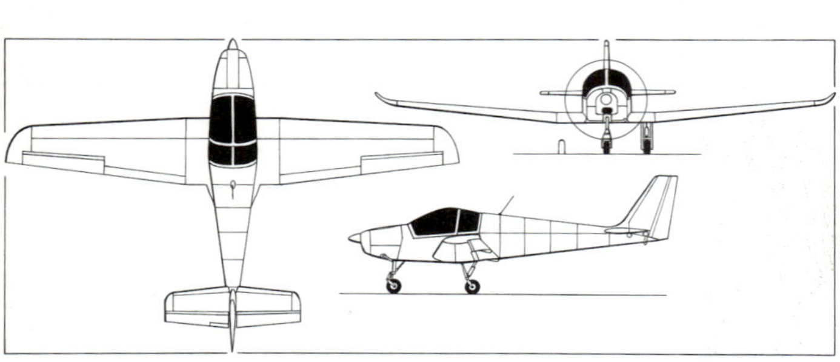

Kappa KP-2U Sova two-seat ultralight ***(James Goulding)*** 0062594

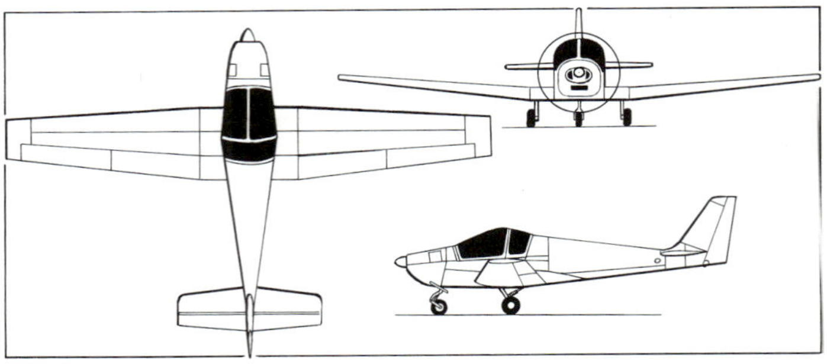

Proposed KP-5 version of Kappa Sova ***(James Goulding)*** ***NEW***/0583031

WEIGHTS AND LOADINGS:
Weight empty .... 285 kg (628 lb)
Max T-O weight: Czech Republic .... 450 kg (992 lb)
Canada .... 480 kg (1,058 lb)
Australia .... 544 kg (1,199 lb)

PERFORMANCE (Rotax 912 UL engine):
Max level speed .... 129 kt (240 km/h; 149 mph)
Max cruising speed at 75% power .... 108 kt (200 km/h; 124 mph)
Stalling speed: flaps up .... 39 kt (71 km/h; 45 mph)
flaps down .... 26 kt (48 km/h; 30 mph)
Max rate of climb at S/L .... 390 m (1,279 ft)/min
T-O run .... 100 m (330 ft)
Landing run .... 140 m (460 ft)
Range with max fuel .... 518 n miles (960 km; 596 miles)
g limits .... +4/−2

UPDATED

KAPPA 4.3 BARON

TYPE: Single-seat ultralight kitbuilt.

PROGRAMME: Promoted at Aero '03, Friedrichshafen, Germany, April 2003, when first deliveries were scheduled for August 2003; this subsequently revised, and marketing began in January 2004.

CURRENT VERSIONS: **Baron II:** *As described.*

COSTS: Kit €9,970 (2003).

DESIGN FEATURES: Simple design with constant-chord parasol wing; interchangeable tail surfaces; and circular-section fuselage. Quoted build time 250 hours.

Clark Y wing section.

FLYING CONTROLS: Conventional and manual.

STRUCTURE: General structure requires only Duralumin angle bars and T6 sheets.

LANDING GEAR: Tailwheel type; fixed.

POWER PLANT: One 29.8 kW (40.0 hp) piston engine to customer's preference. Fuel capacity 40 litres (10.6 US gallons; 8.8 Imp gallons).

DIMENSIONS, EXTERNAL:
Wing span .... 9.96 m (32 ft 8¼ in)
Length overall .... 9.58 m (31 ft 5¼ in)
Height overall .... 2.89 m (9 ft 5¾ in)

AREAS:
Wings, gross .... 13.50 m^2(145.3 sq ft)

WEIGHTS AND LOADINGS:
Weight empty .... 170 kg (375 lb)
Max T-O weight .... 300 kg (661 lb)

PERFORMANCE (estimated):
Normal cruising speed .... 70 kt (130 km/h; 81 mph)
Stalling speed, power off .... 29 kt (52 km/h; 33 mph)
Range .... 216 n miles (400 km; 248 miles)
g limits .... +4/−1.5

VERIFIED

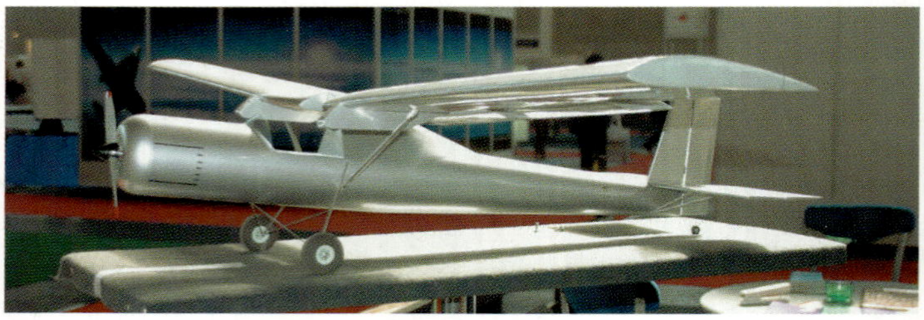

Model of Kappa 4.3 Baron II shown at Aero '03 ***(Paul Jackson)*** ***NEW***/0567753

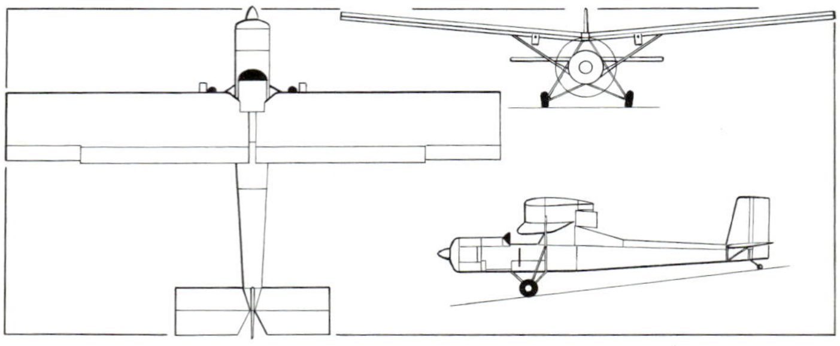

Kappa 4.3 Baron II single-seat ultralight kitbuilt ***(James Goulding)*** ***NEW***/0561620

KUBICEK

BALONY KUBICEK SPOL S.R.O. (KUBICEK BALLOONS LTD)

Francouzská 81, CR-602 00 Brno
Tel: (+420 5) 45 21 19 17 and 45 21 47 55
Fax: (+420 5) 45 21 30 74
e-mail: sales@kubicekballoons.cz
Web: http://www.kubicekballoons.cz
OWNER: Dipl Ing Aleš Kubíček
MANAGING DIRECTOR: Dipl Ing Radím Polaček
SALES MANAGER: Michael Suchý

First Kubíček hot-air balloon was manufactured in 1983 and was followed by about 50 more of Kubíček design by various state-owned companies such as Aviatik and Aerotechnik. Present company was formed as private enterprise in 1991 and began new range of hot-air balloons under BB series designations (total of 270 produced by October 2003). Currently markets standard, ready-to-fly balloons in 12 envelope sizes or types from 1,200 m^3 (42,380 cu ft) to 6,000 m^3 (211,900 cu ft); also special-shape balloons to order. Own polyester fabric for these is produced by sister company, Kubiček Textil. Workforce of 25 in early 2003, in factory area of 1,400 m^2 (15,069 sq ft).

Kubíček's first hot-air airship design was the AV-1. A new design, the AV-2, made its maiden flight in 1999. AV-1 development has been suspended by Kubíček to concentrate on balloon production, but the airship continues to be marketed in Russia by Avgur and RosAeroSystems. Development and marketing of AV-2 continues in Czech Republic.

UPDATED

KUBICEK AV-2

TYPE: Hot-air non-rigid.

PROGRAMME: First flight 26 May 1999; public debut (c/n 106/OK-8036) June 1999; then sold to Reprox Kosice in Slovak Republic as OM-ADS, though returned to factory in 2003 for resale to another customer.

COSTS: US$35,000 (2003).

STRUCTURE: Envelope of PES proprietary fabric, with rip panel and safety valves; unbraced, air-filled tail surfaces. All-metal (welded chromoly steel tube) frame, laminated plastics skin gondola.

LANDING GEAR: Tricycle type, with twin nosewheels; fixed.

POWER PLANT: One 34.0 kW (45.6 hp) Rotax 503 UL-2V engine and Avia V-230C two-blade wooden pusher propeller. Fuel tank beneath centre seat, capacity 30 litres (7.9 US gallons; 6.6 Imp gallons).

ACCOMMODATION: Pilot and two passengers.

AVIONICS: *Comms:* Icom A3E 760-channel VHF radio.

Kubicek AV-2 three-place hot-air airship ***NEW***/0573437

Flight: Garmin 12XL nav receiver.
Instrumentation: Flytec 3040 instrument pack.

DIMENSIONS, EXTERNAL:
- Envelope: Length ... 38.00 m (124 ft 8 in)
- Max diameter ... 14.20 m (46 ft 7 in)
- Gondola: Length ... 3.00 m (9 ft 10 in)
- Max width ... 1.70 m (5 ft 7 in)
- Max height ... 1.85 m (6 ft 0¾ in)

DIMENSIONS, INTERNAL:
- Envelope volume ... 3,500 m^3 (123,601 cu ft)

AREAS:
- Display area (each) side) ... 100.0 m^2 (1,076.4 sq ft)

WEIGHTS AND LOADINGS:
- Weight empty ... 440 kg (970 lb)
- Fuel weight ... 20 kg (44 lb)
- Max T-O weight ... 870 kg (1,918 lb)

PERFORMANCE (Rotax 503):
- Max level speed ... 10.5 kt (20 km/h; 12 mph)
- Pressure ceiling (estimated) ... 2,440 m (8,000 ft)
- Endurance ... 1 h 30 min to 2 h

UPDATED

LARUS

LARUS

Sadová 338, CZ-675 71 Náměšt nad Oslavou
Fax: (+420 5) 09 62 07 70
e-mail: Ludovit.packo@volny.cz; info@larus-czech.com
Web: http://www.larus-czech.cz
CHIEF DESIGNER: Ludovit Packo

The Larus is the company's first product.

UPDATED

LARUS LARUS

TYPE: Side-by-side ultralight.

PROGRAMME: Design work started 1991; prototype (OK-EUC 01) made first flight 22 October 1997. A second prototype has also flown. Will be certified in both ultralight and VLA categories.

DESIGN FEATURES: Typical small lightplane low-wing configuration. Wing sections NACA 2305 at root, NACA 2312 at tip.

FLYING CONTROLS: Conventional and manual. Horn-balanced rudder and one-piece elevator; central trim tab in elevator. Flaps fitted.

STRUCTURE: Mainly composites fuselage and wooden wings with single spar.

LANDING GEAR: Tricycle type; fixed. Fuselage-mounted spring cantilever main units; castoring nosewheel.

POWER PLANT: One 70.8 kW (95 hp) Aero Max 2400 piston engine; two-blade propeller. Fuel capacity 40 litres (10.6 US gallons; 8.8 Imp gallons).

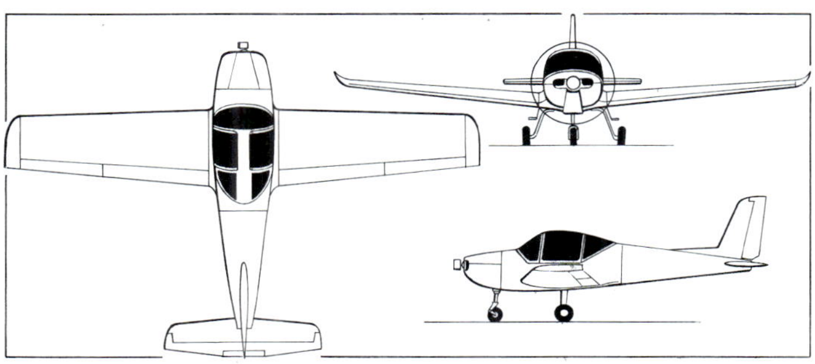

Larus two-seat ultralight *(James Goulding)* 0121274

Prototype Larus OK-EUC 01 ***NEW***/0576232

ACCOMMODATION: Pilot and passenger seated side by side. Doors open forwards. Baggage compartment behind seats.

DIMENSIONS, EXTERNAL:
- Wing span ... 9.80 m (32 ft 1¾ in)
- Length overall ... 6.30 m (20 ft 8 in)
- Height overall ... 2.80 m (9 ft 2¼ in)

AREAS:
- Wings, gross ... 12.86 m^2 (138.4 sq ft)

WEIGHTS AND LOADINGS:
- Weight empty ... 305 kg (672 lb)
- Max T-O weight ... 450 kg (992 lb)

PERFORMANCE:
- Max level speed ... 124 kt (230 km/h; 143 mph)
- Cruising speed ... 94 kt (175 km/h; 109 mph)
- Stalling speed, flaps down ... 38 kt (70 km/h; 44 mph)
- Max rate of climb at S/L ... 330 m (1,083 ft)/s
- Service ceiling ... 3,000 m (9,840 ft)
- T-O to 15 m (50 ft) ... 250 m (820 ft)
- Landing from 15 m (50 ft) ... 280 m (919 ft)

UPDATED

LET-MONT

LET-MONT S.R.O.

Vikýřovice 226, CZ-788 13 Šumperk
Tel/Fax: (+420 585) 03 51 32
DIRECTORS:
Petr Podešva
Jaroslav Podešva
BUSINESS MANAGER: Tomáš Podešva

The company's Tulak and Piper ultralights, together with a development of more streamlined appearance, are marketed in western Europe by Austrian aircraft manufacturer, HB Flugtechnik.

Tomáš Podešva also has his own company which produces the Trener Baby.

VERIFIED

LET-MONT UL TULÁK AND UL PIPER

English name: Rambler

TYPE: Side-by-side ultralight/kitbuilt; tandem-seat ultralight.

PROGRAMME: Tulák certified in 1996 and Piper in 1998. Variations of the design are marketed by HB Flugtechnik of Austria as the Amigo and Cubby.

CURRENT VERSIONS: **UL Tulák:** Side-by-side seats; factory-built or kit. Marketed by HB as Dandy.

UL Piper: Tandem seats; otherwise as Tulák. No kit version. German name **Tandem Tulák**. Span 9.82 m (32 ft 2½ in); length 5.96 m (19 ft 6½ in); maximum cruising speed 70 kt (130 km/h; 81 mph). Engine options include 58.8 kW (78.9 hp) or 73.5 kW (98.6 hp) BMW; 58.8 kW (78.9 hp) Verner; 62.5 kW (83.8 hp) Tatra; and 51.5 kW (69.0 hp) Limbach. Marketed by HB as Cubby.

Let-Mont UL Piper two-seat ultralight ***NEW***/0540532

CUSTOMERS: Total 55 kits and complete aircraft built by end of 2002; ninth factory-built machine was registered in Poland in 2002. Known sales in Czech Republic, France, Germany and Poland.

Data apply to both Tulák and Piper, except where indicated.

DESIGN FEATURES: High, constant-chord wing braced by V struts; optional tapered or downturned wingtips. Wings fold back for storage and transportation. Resemble Piper designs of 1940s/1950s.

Clark Y wing section, thickness/chord ratio 12.5 per cent.

FLYING CONTROLS: Conventional and manual. Flight-adjustable trim tab in each elevator. Plain flaps.

STRUCTURE: Fabric-covered metal tube fuselage; two-spar wooden wing; engine cowling metal; some composites.

LANDING GEAR: Tailwheel type; fixed. Two faired side Vs hinged to lower longerons, plus half-axles sprung under centreline with bungee cord. Steerable tailwheel; 420×150 mainwheels with bungee suspension and cable-operated brakes. Mainwheel spats optional.

POWER PLANT: One 37.0 kW (49.6 hp) Rotax 503 UL-2V engine; SportProp three-blade, ground-adjustable pitch, glass fibre propeller. Fuel capacity 50 litres (13.2 US gallons; 11.0 Imp gallons).

DIMENSIONS, EXTERNAL:
Wing span: tapered tips .... 9.60 m (31 ft 6 in)
downturned tips .... 9.40 m (30 ft 10 in)
Width, wings folded .... 3.10 m (10 ft 2 in)
Length overall .... 5.60 m (8 ft 4½ in)
Height overall .... 2.00 m (6 ft 6¾ in)

DIMENSIONS, INTERNAL:
Cabin max width: Tulák .... 1.14 m (3 ft 9 in)
Piper .... 0.70 m (2 ft 3½ in)

AREAS:
Wings, gross .... 13.00 m² (139.9 sq ft)

WEIGHTS AND LOADINGS:
Weight empty: Tulák .... 255 kg (562 lb)
Piper .... 265 kg (584 lb)
Max T-O weight: both .... 450 kg (992 lb)

PERFORMANCE (both):
Never-exceed speed (VNE) .... 89 kt (165 km/h; 102 mph)
Econ cruising speed .... 59 kt (110 km/h; 68 mph)
Stalling speed, flaps down .... 33 kt (60 km/h; 38 mph)
Max rate of climb at S/L .... 300 m (984 ft)/min
T-O run .... 65 m (215 ft)
Landing run .... 60 m (200 ft)
g limits .... +6/–4

UPDATED

LZ

LETECKÉ ZÁVODY A.S. (LZ AERONAUTICAL INDUSTRIES INC)

PO Box 1177, Uherské Hradiště, CZ-686 04 Kunovice
Tel: (+420 572) 81 60 00
Fax: (+420 572) 81 60 06
e-mail: let@let.cz
Web: http://www.let.cz
PRESIDENT: Libor Soska
EXECUTIVE VICE-PRESIDENT: Tomáš Štefánek
VICE-PRESIDENT, MARKETING AND SALES: Patrik Joachimczyk
TECHNICAL DIRECTOR: Vlastimil Janulík
COMMERCIAL DIRECTOR: Lubor Smericka
MANUFACTURING DIRECTOR: Pavel Zábrana
BUSINESS DEVELOPMENT DIRECTOR: Pavel Knotek
MAINTENANCE DIRECTOR: Andrej Ziarovsky
CHIEF DESIGNER: Miroslav Pešák

LZ Aeronautical Industries was part of a diversified consortium of companies primarily producing aircraft, aircraft components and automobile industry logistics systems. As Letecké Závody (formerly Let a.s), it was established in 1936 as a repair shop for Ba 33, B-534, Junkers W 34 and Arado Ar 96B aircraft. Since then, it has produced and delivered more than 8,000 aircraft of various types and categories, including (but not limited to) such well-known types as the Yak-11 trainer (as C-11), Aero Ae 45/145, L-200 Morava, L-29 Delfin, L-13 Blanik sailplane, Zlin Z 37 Cmelak crop-sprayer and, most recently, some 1,100 examples of the L 410 twin-engined commuter transport. Until being declared bankrupt on 24 October 2000, Let had focused on upgrading the L 410/420; repairing, modernising and selling used L 410s, especially low-time aircraft from the former USSR; continuing the L 610 G certification programme; production of L-13/23/33 series of Blanik sailplanes; and attracting subcontract work.

On 16 July 2001, Moravan Aeroplanes acquired the company for Kcs200 million (US$5 million).

Its subsequent activities were then OEM (original equipment manufacturer) aircraft production, as well as near- and long-term production subcontracting. It also supplied components for the aviation and other industries. Aircraft manufacture included L-13AC/L-23/L-33 Blanik/Solo Blanik sailplane series and role conversion of L 410; development of the L 610 G continued until the company was again declared bankrupt in early April 2004. Plans to resume aerospace activities have included talks with Rekkof Aircraft of the Netherlands with a view to returning the Fokker 70/100 regional jet to production.

UPDATED

MORAVA ZLIN

MORAVA ZLIN AERO SERVICE S.R.O.

Tel: (+39 347) 470 04 19
Fax: (+39 045) 896 12 62
e-mail: info@zlinaero.com and savage.ulm@tiscalinet.it
Web: http://www.zlinaero.com
PRESIDENT: Paquale Russo

Morava Zlin Aero Service set up in late 2000 to market the Russo Savage ultralight.

MORAVA ZLIN (RUSSO) SAVAGE

TYPE: Tandem-seat ultralight/kitbuilt.

PROGRAMME: First flown 1997. Initial development of Russo Savage was undertaken in Italy, with parts being manufactured in Czech Republic; entire operation now undertaken in Czech Republic; Czech type approval being sought in late 2003. Available in both ready to fly and kit form.

COSTS: Kit US$17,800; ready-to-fly US$38,918 (2003).

DESIGN FEATURES: Based on Piper J3 Cub; wings fold for storage and transportation. Quoted build time 350 hours.

FLYING CONTROLS: Conventional. Manually operated three-position flaps (maximum 35°). Manual trim in elevator.

STRUCTURE: Fuselage of 4130 welded steel; two-spar wings of 6061-T6 anodised aluminium; all fabric-covered.

Morava (Russo) Savage in the colour scheme of a Second World War British Army Auster ***NEW***/0567727

Wing section NACA 4412.

LANDING GEAR: Tailwheel type; fixed. Bungee suspension, castoring tailwheel. Hydraulic brakes. Tundra tyres can be fitted for rough-ground operations.

POWER PLANT: One 73.5 kW (98.6 hp) Rotax 912 ULS driving a two-blade wooden propeller. Fuel capacity 68 litres (18.0 US gallons; 15.0 Imp gallons) in two welded aluminium tanks in wingroots.

DIMENSIONS, EXTERNAL:
Wing span .... 9.31 m (30 ft 6½ in)
Length overall .... 6.39 m (20 ft 11½ in)
Height overall .... 2.03 m (6 ft 8 in)

AREAS:
Wings, gross .... 14.20 m^2 (152.8 sq ft)
WEIGHTS AND LOADINGS:
Weight empty .... 283 kg (624 lb)
Max T-O weight .... 450 kg (992 lb)
PERFORMANCE:
Never-exceed speed (VNE) .... 110 kt (205 km/h; 127 mph)
Max operating speed .... 104 kt (192 km/h; 119 mph)
Normal cruising speed .... 92 kt (170 km/h; 106 mph)
Stalling speed, flaps down .... 28 kt (52 km/h; 33 mph)
Max rate of climb at S/L .... 270 m (886 ft)/min
T-O run .... 90 m (295 ft)
Landing run .... 70 m (230 ft)
Range with max fuel .... 168 n miles (312 km; 193 miles)

NA DESIGN

NA DESIGN COMPANY INC

Brněnská 415, CZ-664 34 Kuřim
Tel: (+420 603) 45 44 94
Fax: (+420 5) 41 26 34 17 and 850 77 41
e-mail: nadc@nadc.cz
Web: http://www.bongo.cz and http://www.nadc.cz
CHAIRMAN AND CHIEF DESIGNER: Jan Námisňák
CUSTOMER SERVICE DIRECTOR: Jaroslav Najman
GENERAL MANAGER: Jan Opiatek

NA Design was created in third quarter of 1999, in co-operation with Moravan Aeroplanes Inc. Development of the (formerly UNIS) Bongo continued into early 2002, at which time NA Design was seeking a Western partner to assist with certification, production and marketing; the project was put on hold in late 2002 while finance was sought but, despite this, a new version, the four-seat NA50, was announced in late 2002. Some US$2.5 million said to be required to bring aircraft to production readiness; development is continuing towards a 2005 launch date. Designer Jan Námisňák had previously built the Caprice 21 tandem-seat, high-wing pusher lightplane.

UPDATED

NA 40 Bongo two-seat light utility helicopter 0044014

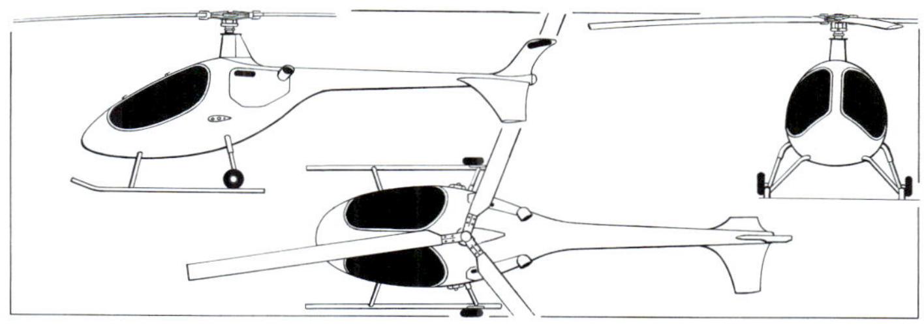

General arrangement of the NA 40 Bongo light helicopter *(James Goulding)* 0110505

NA DESIGN NA 40 BONGO

TYPE: Two-seat ultralight helicopter.
PROGRAMME: Design work 1992 followed by model tests 1993; construction of Bongo technology demonstrator started 1996; public debut at Brno International Machinery Fair 1997 (then called UNIS NA40, and with two-blade main rotor) by first prototype OK-CIU. Began 70 hours of tethered hover flights March 1998. Two further prototypes completed by mid-1998. NA Design plans Normal category certification to FAR Pts 27, 33, 34 and 36 by late 2003, subject to finding joint venture Western partner to share certification, manufacturing and marketing costs. However, by 2002, financial support of US$2.5 million was still being solicited, but nothing further had been heard by early 2003; by March 2003 the company stated that development was continuing and commercial launch was due for 2005.
CURRENT VERSIONS: **NA 40 Bongo:** Basic civil version; *as described.*
NA 42 Barracuda: Proposed military/police version. Power plant as NA 40 or single Rolls-Royce 250 engine; stub-wings for weapons carriage; undernose sensor turret; intercom. Otherwise generally as NA 40.
NA 44 Bion: Projected UAV version of NA 42 for various military and civil applications.
NA 50: Proposed four-seat version announced in late 2002 with estimated unit cost of US$1 million.
COSTS: Approximately US$750,000 (2001).
DESIGN FEATURES: Three-blade teetering rotor; pod and boom fuselage with inverted Y-tail unit. Cocomo patented anti-torque system eliminates need for tail rotor, reduces transmission complexity and operating noise level.
FLYING CONTROLS: Conventional cyclic and collective controls. Ducted air anti-torque system.
STRUCTURE: Mainly composites, including rotor blades; some aluminium in fuselage, otherwise double-curvature monocoque sandwich; laminated, vacuum-formed elastomeric rotor head.

Artist's impression of four-seat NA 50 ***NEW***/0576155

LANDING GEAR: Tubular twin-skid gear with ground handling wheels. Inflatable permanent or emergency floats optional.
POWER PLANT: Two 86 kW (115 shp) Prvni Brnenská Strojirna PBS Velká Bíteš TE 50B turboshafts, with FADEC and dual ignition. Engines mounted side by side behind cockpit; transmission via combining gearbox. Single self-sealing fuel tank beneath engines, capacity 210 litres (55.5 US gallons; 46.2 Imp gallons).
ACCOMMODATION: Adjustable, foldable and anatomically shaped seats for pilot and passenger. Baggage space aft of seats; provision for additional, aerodynamic baggage container under fuselage. Gull-wing window/door each side, hinged on centreline and opening upward.
SYSTEMS: Dual 27 V DC electrical systems; external power receptacle.
AVIONICS: *Instrumentation:* VFR standard; IFR to be offered later.
EQUIPMENT: Rocket-deployed emergency recovery parachute optional.
DIMENSIONS, EXTERNAL:
Rotor diameter .... 7.48 m (24 ft 6½ in)
Fuselage: Length .... 6.15 m (20 ft 2¼ in)
Max width .... 1.30 m (4 ft 3¼ in)
Height to top of rotor head .... 2.35 m (7 ft 8½ in)
Skid track .... 1.76 m (5 ft 9¼ in)
DIMENSIONS, INTERNAL:
Baggage space .... 0.23 m^3 (8.1 cu ft)
AREAS:
Rotor disc .... 43.94 m^2 (473.0 sq ft)
WEIGHTS AND LOADINGS:
Weight empty, equipped: NA 40 .... 480 kg (1,058 lb)
NA 42 .... 495 kg (1,091 lb)
Max load on external sling .... 250 kg (551 lb)
Max T-O weight .... 950 kg (2,094 lb)
Max disc loading .... 21.62 kg/m^2 (4.43 lb/sq ft)
PERFORMANCE:
Never-exceed speed (VNE) .... 151 kt (280 km/h; 174 mph)
Max cruising speed .... 135 kt (250 km/h; 155 mph)
Econ cruising speed .... 124 kt (230 km/h; 143 mph)
Max rate of climb at S/L .... 480 m (1,575 ft)/min
Hovering ceiling IGE .... 4,000 m (13,120 ft)
Max range .... 270 n miles (500 km; 310 miles)
Endurance: NA 42 .... 2 h 30 min

UPDATED

PODEŠVA

TOMÁŠ PODEŠVA

CZ-783 96 Újezd u Uničova 87
Tel: (+42 605) 55 07 08
Fax: (+42 585) 03 51 32
PRESIDENT: Tomáš Podešva

EUROPEAN DISTRIBUTOR:
Fläming Air GmbH
Flugplatz Oehna, D-14913 Zellendorf
Tel: (+49 337) 426 74 20
Fax: (+49 337) 426 74 27
e-mail: flugplatzoehna@t-online.de
Web: http://www.flaemingair.de

Tomáš Podešva is also responsible for the design of the Let-Mont UL Tulák and UL Piper ultralights.

UPDATED

PODEŠVA TRENER BABY

TYPE: Tandem-seat ultralight/kitbuilt.
PROGRAMME: Designed and built as ultralight prototype by Peter Podešva; scale replica of

Akro version of Podešva Trener Baby on show at Berlin in May 2004 *(Paul Jackson)* ***NEW***/0583087

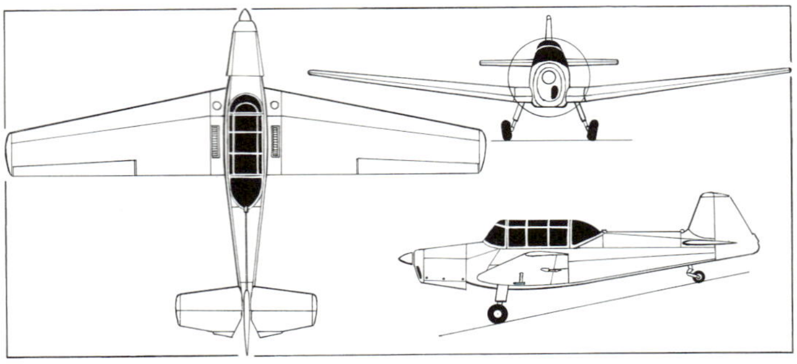

Podešva Trener Baby (Walter Mikron III engine) *(James Goulding)* ***NEW***/0558570

Podešva Trener Baby *(Paul Jackson)* ***NEW***/0576252

Czechoslovak Zlin 126 Trener (member of Zlin 26 family, which was produced between late 1940s and mid-1970s). Production version built by son, Tomáš Podešva at Hranice; certified and distributed by Fläming Air; second aircraft (OK-FUD 01) exhibited at ILA, Berlin, June 2000. Certification in mid-2001.

CURRENT VERSIONS: **Ultralight:** Wooden wings; MTOW limited to 450 kg (992 lb). Also available as kit, meeting '51 per cent rule'.

Experimental: Metal wings; 520 kg (1,146 lb) MTOW; kitbuilt.

Akro: Version with all-metal wing and tailplane. Fluted aileron skins and split flaps; fabric-covered elevator. Public debut (D-EETB) at Berlin, May 2004.

COSTS: Ultralight: kit €41,667; complete example with engine €59,524. Experimental: €52,530; with retractable landing gear €53,372. Akro kit €46,578 (all 2004).

DESIGN FEATURES: Low wing with sweptback leading-edge and unswept trailing-edge; mid-mounted, braced tailplane. Replica of Zlin 126 aerobatic trainer to 80 per cent scale, except full-scale cockpit.

FLYING CONTROLS: Conventional and manual. Split flaps. Flight-adjustable tabs on both elevators; ground-adjustable tab on rudder and port aileron.

STRUCTURE: Steel tube fuselage with fabric covering, except for sheet metal upper surface and engine cowling; fabric-covered empennage. Wings wood or metal, according to version, with aluminium leading-edge and fabric covering. Composites nose fairing.

LANDING GEAR: Tailwheel type; fixed. Wing-mounted, trousered, cantilever sprung main legs. Mainwheels 5.00-5; tailwheel 200×50. Hydraulic brakes.

POWER PLANT: One 55.1 kW (73.9 hp) Walter Mikron III four-cylinder, in-line piston engine driving a Kremen two-blade, variable-pitch (electric), wooden propeller or Woodcomp fixed-pitch propeller. (VZLÚ two-blade wooden propeller on prototype.) Fuel capacity 35 litres (9.2 US gallons; 7.7 Imp gallons).

AVIONICS: To customer's specification.

EQUIPMENT: Optional rescue system.

DIMENSIONS, EXTERNAL:
Wing span ... 8.68 m (28 ft 5¾ in)
Length overall ... 6.52 m (21 ft 4¾ in)
Height overall ... 1.80 m (5 ft 10¾ in)

AREAS:
Wings, gross ... 10.80 m² (116.3 sq ft)

WEIGHTS AND LOADINGS (UL: Ultralight, E: Experimental):
Weight empty: UL ... 296 kg (653 lb)
E ... 380 kg (837 lb)
Max T-O weight: UL ... 450 kg (992 lb)
UL where regulations permit ... 472.5 kg (1,041 lb)
E ... 600 kg (1,322 lb)

PERFORMANCE (UL, E, as above):
Never-exceed speed (VNE):
UL ... 121 kt (225 km/h; 139 mph)
E ... 178 kt (330 km/h; 205 mph)
Cruising speed at 75% power:
UL ... 94 kt (175 km/h; 109 mph)
E ... 124 kt (230 km/h; 143 mph)
Stalling speed: UL ... 36 kt (65 km/h; 41 mph)
E ... 45 kt (83 km/h; 52 mph)
Max rate of climb at S/L ... 270 m (886 ft)/min
Max range ... 432 n miles (800 km; 497 miles)
g limits: UL ... +4/−2
E ... +6/−3

UPDATED

TEST

TEST S.R.O.

Dobrovskeho 78, CZ-612 00 Brno
Tel/Fax: (+420 5) 49 24 90 73
e-mail: test@infoline.cz
Web: http://www.test.infoline.cz
PRODUCTION DIRECTOR: Zdeněk Teply
MARKETING DIRECTOR: Zbynek Jaros

Company was established 1992 to design, develop and produce ultralight sailplanes, powered sailplanes and aircraft; first product was TST-1 Alpin sailplane. Capacity increased to 10 aircraft per year in 1995 and to 40 per year from September 1998. Developed versions of Alpin (TST-3 and TST-8), available in both sailplane and motor glider variants) still form core of production, along with TST-5, TST-6 and TST-9 described in the following entries; all are available in plans-only, various kits or full flyaway form. Workforce was 15 in 2001.

In October 2002, a further sailplane - the TST-10 - was added to the product line, and the company announced that it had started development of the TST-12, a two-seat, low-wing, tailwheel ultralight.

UPDATED

TeST's first product was the Alpin sailplane *(Paul Jackson)* ***NEW***/0531977

TEST TST-5 VARIANT

TYPE: Side-by-side ultralight/kitbuilt.

PROGRAMME: Developed in mid-1990s.

DESIGN FEATURES: Braced high-wing monoplane; forward-swept, constant chord wings (wing section NACA 4415). Basically a short-wing version of TST-6 (see separate entry for TeST-6 Duo).

TeST TST-5 Variant ***NEW***/0531978

FLYING CONTROLS: Conventional and manual (ailerons and elevators rod-actuated, rudder by cables).

STRUCTURE: Two-spar wing, with plywood and fabric covering; glass fibre wingtips; V bracing struts. All-wood fuselage except for glass fibre panel behind cockpit.

LANDING GEAR: Tricycle; fixed. Fuselage-mounted spring cantilever mainwheel legs with wheel fairings and 400×100 wheels; 300 mm diameter steerable nosewheel. Mainwheel brakes.

POWER PLANT: One 47.8 kW (64.1 hp) Rotax 582 UL engine and two-blade propeller standard; or suitable equivalent. Fuel capacity 36 litres (9.5 US gallons; 7.9 Imp gallons).

DIMENSIONS, EXTERNAL:
Wing span ... 10.50 m (34 ft 5½ in)
Wing aspect ratio ... 8.0
Length overall ... 6.40 m (21 ft 0 in)
Height overall ... 2.30 m (7 ft 6½ in)

DIMENSIONS, INTERNAL:
Cabin max width ... 1.02 m (3 ft 4¼ in)

AREAS:
Wings, gross ... 13.70 m² (147.5 sq ft)

WEIGHTS AND LOADINGS:
Weight empty ... 260 kg (573 lb)
Max T-O weight ... 450 kg (992 lb)

PERFORMANCE:

Never-exceed speed (VNE)	91 kt (170 km/h; 105 mph)
Max cruising speed	81 kt (150 km/h; 93 mph)
Econ cruising speed	49–65 kt (90–120 km/h; 56–75 mph)
Stalling speed	36 kt (65 km/h; 41 mph)
Max rate of climb at S/L	210 m (689 ft)/min
Service ceiling	4,000 m (13,120 ft)
T-O and landing run	150 m (495 ft)
g limits	+4/–2

UPDATED

TEST TST-6 DUO

TYPE: Side-by-side ultralight/kitbuilt.

PROGRAMME: Developed in mid-1990s.

DESIGN FEATURES: Forward-swept shoulder-wing monoplane.

FLYING CONTROLS: Conventional and manual; primary surfaces as for TST-5. Two-position trailing-edge flaps; upper-surface spoilers.

STRUCTURE: As described for TST-5 except wing has Wortmann FX61-184 section, has one main and two auxiliary spars, is entirely plywood-covered and has glass fibre wingtips.

LANDING GEAR: Tricycle type; fixed. Fuselage-mounted spring cantilever mainwheel legs with wheel fairings and 400×100 wheels; 300 mm diameter steerable nosewheel. Mainwheel brakes.

POWER PLANT: One 47.8 kW (64.1 hp) Rotax 582 UL engine and three-blade propeller standard; or suitable equivalent. Fuel capacity 36 litres (9.5 US gallons; 7.9 Imp gallons).

DIMENSIONS, EXTERNAL (As TST-5 except):

Wing span	14.60 m (47 ft 10¾ in)
Wing aspect ratio	16.5

AREAS:

Wings, gross	12.90 m² (138.9 sq ft)

WEIGHTS AND LOADINGS:

Weight empty	265 kg (584 lb)
Max T-O weight	450 kg (992 lb)

PERFORMANCE, POWERED (As TST-5 except):

Max speed with spoilers extended	86 kt (160 km/h; 99 mph)
Econ cruising speed	49–59 kt (90–110 km/h; 56–68 mph)
Service ceiling	5,000 m (16,400 ft)

PERFORMANCE, UNPOWERED:

Best glide ratio at 54 kt (100 km/h; 62 mph)	16
Min rate of sink	2.00 m (6.56 ft)/s

UPDATED

TeST TST-6 Duo ***NEW***/0531979

TEST TST-9 JUNIOR 2000

TYPE: Single-seat motor glider/kitbuilt.

PROGRAMME: Developed in late 1990s.

DESIGN FEATURES: Forward-swept, high-wing monoplane. Improved version of earlier TST-7 Junior, from which it differs mainly in having circular section fuselage and T tail.

FLYING CONTROLS: Conventional and manual; similar to TST-6 except for one-piece elevator and absence of flaps.

STRUCTURE: Wings and fuselage as for TST-6; all-wood tail unit.

LANDING GEAR: Tricycle type; fixed. Fuselage-mounted spring cantilever mainwheel legs with wheel fairings and 300×100 wheels; 250 mm diameter steerable nosewheel.

POWER PLANT: One 31.0 kW (41.6 hp) Rotax 447 engine and two-blade SportProp 1600 propeller standard; or suitable equivalent. Fuel capacity 25 litres (6.6 US gallons; 5.5 Imp gallons).

DIMENSIONS, EXTERNAL:

Wing span	12.40 m (40 ft 8¼ in)
Wing aspect ratio	15.4
Length overall	5.90 m (19 ft 4¼ in)
Height overall	2.10 m (6 ft 10¾ in)

DIMENSIONS, INTERNAL:

Cabin max width	0.59 m (1 ft 11¼ in)

AREAS:

Wings, gross	10.00 m² (107.6 sq ft)

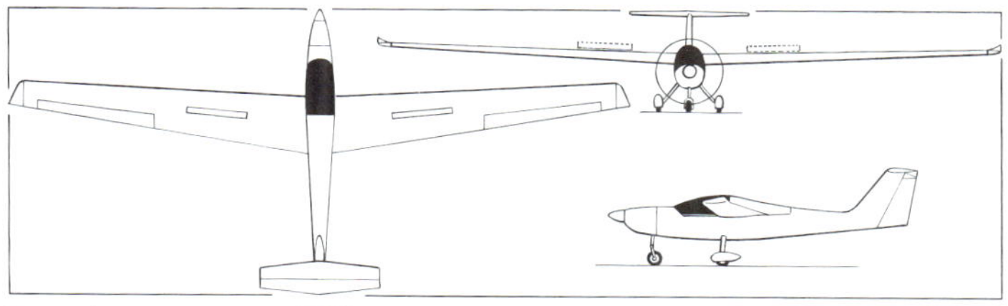

General arrangement of TST-9 Junior *(James Goulding)* ***NEW***/0552088

WEIGHTS AND LOADINGS:

Weight empty	175 kg (386 lb)
Max T-O weight	300 kg (661 lb)

PERFORMANCE, POWERED:

Never-exceed speed (VNE), and max speed with spoilers extended	97 kt (180 km/h; 111 mph)
Max cruising speed	89 kt (165 km/h; 103 mph)
Econ cruising speed	49–65 kt (90–120 km/h; 56–75 mph)
Stalling speed	35 kt (63 km/h; 40 mph)
Max rate of climb at S/L	270 m (886 ft)/min
Service ceiling	5,000 m (16,400 ft)
T-O run	100 m (330 ft)
Landing run	120 m (395 ft)
g limits	+4/–2

PERFORMANCE, UNPOWERED:

Best glide ratio at 49 kt (90 km/h; 56 mph)	20
Min rate of sink	1.50 m (4.92 ft)/s

UPDATED

TEST TST-12

TYPE: Tandem-seat ultralight.

PROGRAMME: Development being undertaken during late 2002; announced January 2003.

DESIGN FEATURES: Will meet ultralight criteria in most countries. Low wing with slight sweepback and upturned wingtips; sweptback fin; constant-chord tailplane. Wing has removable outer panels for storage and ground transportation.

FLYING CONTROLS: Conventional and manual.

Artist's impression of the TST-12 ***NEW***/0576223

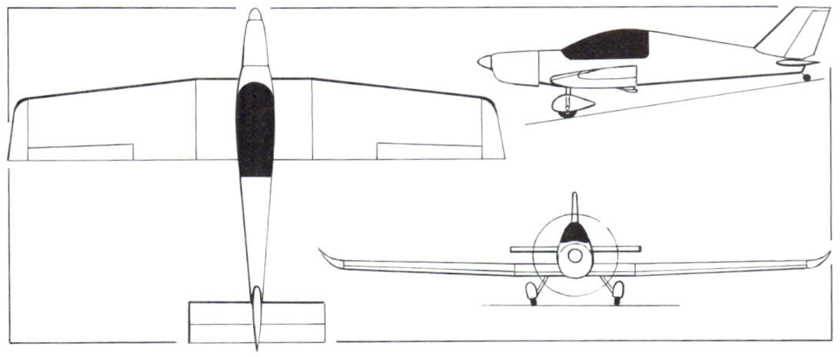

TST-12 two-seat ultralight general arrangement *(James Goulding)* ***NEW***/0552087

TeST TST-9 Junior single-seat kitbuilt ***NEW***/0531980

STRUCTURE: Fuselage of wood with elliptical cross-section and composites panels behind cockpit. Two-spar cantilever wooden wing has wooden leading-edges with fabric-covered rear section. Fabric-covered control surfaces.

LANDING GEAR: Tailwheel type; fixed. Wing-mounted, steel cantilever main legs with wheel fairings and 400×100 mainwheels and brakes. Steerable tailwheel.

POWER PLANT: One 47.8 kW (64.1 hp) Rotax 582 UL engine in prototype; options will include Rotax 912 and Parma-Technik Mikron engines. Fuel capacity 35 litres (9.2 US gallons; 7.7 Imp gallons).

DIMENSIONS, EXTERNAL:
Wing span ... 9.42 m (30 ft 10¾ in)
Length overall ... 6.68 m (21 ft 11 in)

WEIGHTS AND LOADINGS:
Weight empty ... 280 kg (617 lb)
Max T-O weight ... 450 kg (992 lb)

PERFORMANCE:
Never-exceed speed (VNE) ... 97 kt (180 km/h; 111 mph)
Stalling speed ... 34 kt (62 km/h; 39 mph)
g limits ... +4/–2

UPDATED

TL

TL ULTRALIGHT

Dobrovského 734, CZ-500 02 Hradec Králové
Tel: (+420 49) 521 17 53 and 521 89 10
Fax: (+420 49) 521 33 78
e-mail: info@tl-ultralight.cz
Web: http://www.tl-ultralight.cz
DIRECTOR: Ing Ivan Lelák

GERMAN DISTRIBUTOR:
Martin Wezel Flugzeugtechnik
Erlenbachstrasse 38, D-72768 Reutlingen
Tel: (+49 7121) 684 08
Fax: (+49 7121) 67 72 38
e-mail: wezel.martin@t-online.de
Web: http://www.Wezel-Flugzeugtechnik.de

TL also markets the TL-22 Duo/Eso Rogallo-wing trike. Formed in 1990, the company previously produced the TL-1, TL-2 and TL-32 ultralights, progressing then to the TL-132, predecessor of the current TL-232.

UPDATED

TL-96 STAR AND TL-2000 STING CARBON

TYPE: Side-by-side ultralight.

PROGRAMME: Prototype Star first flew November 1997; has German certification.

CURRENT VERSIONS: **TL-96 Star:** Baseline version; *as described.* Marketed in USA as StarSport.

TL-2000 Sting Carbon: All-CFRP version; prototype OK-GUU 10; announced 2000; type certificate issued, 19 December 2002. Marketed in USA as StingSport. Retractable gear version shown (prior to first flight) at Aero '03, Friedrichshafen, April 2003. Estimated 15 kt (28 km/h; 17 mph) speed advantage for 3 kg (7 lb) weight penalty US version has increased weight of 386 kg (851 lb) empty and 648 kg (1,430 lb) max T-O, with Rotax 914 engine.

CUSTOMERS: Exports include several registered in Netherlands; others to Cyprus, Hungary, Portugal and USA. Total of at least 121 TL-96s and 66 TL-2000s registered by April 2004.

COSTS: TL-96 €58,850; TL-2000 €65,850 (2004).

TL-2000 Sting Carbon RG version in the US registry ***(Paul Jackson)*** ***NEW***/0583086

TL-2000 Sting Carbon ***(Paul Jackson)*** ***NEW***/0576157

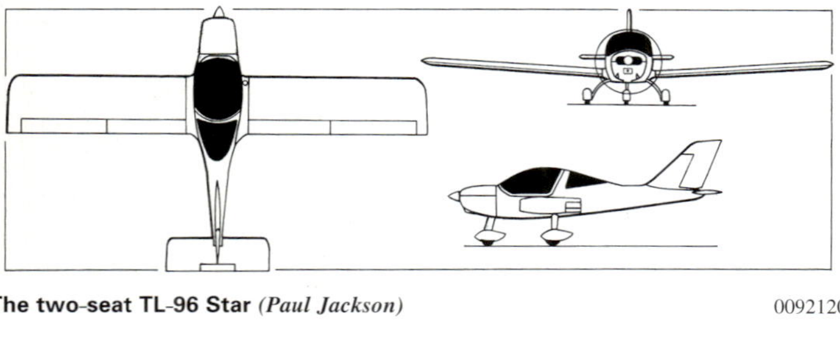

The two-seat TL-96 Star ***(Paul Jackson)*** 0092120

DESIGN FEATURES: Conventional, low-wing monoplane with constant-chord wings and tailplane, plus sweptback fin.

MS 313 wing aerofoil section. Optional tailcone extension for towing hook.

FLYING CONTROLS: Manual. All-moving tailplane with anti-balance tab. Two-position split flaps.

STRUCTURE: All-composites (GFRP and CFRP).

LANDING GEAR: Tricycle type; fixed. Fuselage-mounted spring GFRP cantilever legs and wheel fairings. Steerable nosewheel. Tyre sizes 4.00-8/400×100 (main), 4.00-4 (nose). Hydraulic mainwheel brakes.

POWER PLANT: One 59.6 kW (79.9 hp) Rotax 912 UL-DCDI or 84.6 kW (113.4 hp) Rotax 914 flat-four engine; three-blade Albastar propeller. Optional 67.1 kW (90 hp) Aero Prag AP-45 flat-four and two-blade wooden propeller. Fuel capacity (RON 95) 50 litres (13.2 US gallons; 11.0 Imp gallons) in TL-96 and 63 litres (16.6 US gallons; 13.9 Imp gallons)in TL-2000, optionally 100 litres (26.4 US gallons; 22.0 Imp gallons).

ACCOMMODATION: Heated as standard.

DIMENSIONS, EXTERNAL (A: TL-96, B: TL-2000):
Wing span: A ... 9.20 m (30 ft 2¼ in)
B ... 8.44 m (27 ft 8¼ in)
Length overall: A ... 6.50 m (21 ft 4 in)
B ... 5.93 m (19 ft 5½ in)
Height overall: A ... 2.20 m (7 ft 2½ in)
B ... 2.30 m (7 ft 6½ in)

DIMENSIONS, INTERNAL:
Cabin max width: A ... 1.15 m (3 ft 9¼ in)

AREAS:
Wings, gross: A ... 12.20 m² (131.3 sq ft)
B ... 9.80 m² (105.5 sq ft)

WEIGHTS AND LOADINGS:
Weight empty: A ... 285 kg (628 lb)
B ... 275 kg (606 lb)
Max T-O weight: A, B ... 450 kg (992 lb)

PERFORMANCE, POWERED (Rotax 912):
Never-exceed speed (VNE):
A ... 148 kt (275 km/h; 170 mph)
B ... 164 kt (305 km/h; 189 mph)
Max level speed: A ... 138 kt (255 km/h; 158 mph)
B ... 145 kt (270 km/h; 167 mph)
Max cruising speed: A ... 119 kt (220 km/h; 137 mph)
B ... 140 kt (260 km/h; 162 mph)
Stalling speed: A ... 34 kt (63 km/h; 40 mph)
B ... 36 kt (65 km/h; 41 mph)
Max rate of climb at S/L: A ... 300 m (984 ft)/min
B ... 420 m (1,378 ft)/min
Range: A ... 540 n miles (1,000 km; 621 miles)
B ... 810 n miles (1,500 km; 932 miles)

PERFORMANCE, UNPOWERED:
Best glide ratio: A ... 16

UPDATED

TL-232 CONDOR

TYPE: Tandem-seat ultralight/kitbuilt.

PROGRAMME: Side-by-side seat TL-132 (Rotax 503 engine) first flown 1993; entered production, and TL-232 introduced, in 1994. TL-132 subsequently discontinued.

CURRENT VERSIONS: **TL-232 Condor Plus:** Compared to TL-132, has modified wing profile; option of more powerful engine; cut-down rear fuselage and additional glazing for rear cockpit.

Power Condor: Version with 73.5 kW (98.6 hp) Rotax 912 ULS and glider-towing equipment; able to aero-tow sailplanes of up to 400 kg (882 lb) AUW.

Condor S: Aerotowing version, certified 23 October 2001. Prototype (D-MPAK) fitted with Hirth fuel-injected engine.

CUSTOMERS: Operators in Czech Republic, Germany, Netherlands, Poland and Sweden. Total of 51 (including nine kits) registered in Czech Republic by late 2000, at which time more than 100 flying in Europe.

COSTS: Kit only €21,010; with Rotax 582 UL-DCDI €38,000; Rotax 912 UL €47,700; Rotax 912S UL €58,800 (all 2003).

TL Ultralight TL-232 Condor Plus 0097996

DESIGN FEATURES: Constant-chord, high wing with V-strut bracing; mid-mounted tailplane; sweptback fin with long fillet. Wings and horizontal tail surfaces foldable for transportation and storage.

FLYING CONTROLS: Conventional and manual. In-flight-adjustable tab in starboard elevator. Two-position flaps. Dual controls.

STRUCTURE: Fabric-covered metal. Composites rear top-decking.

LANDING GEAR: Fixed; tricycle or tailwheel type. Fuselage-mounted spring cantilever mainwheel legs with wheel fairings. Tyre sizes 16×4 (main), 400×100 (nose). Cable brakes. Leg and wheel speed fairings optional.

POWER PLANT: One 47.8 kW (64.1 hp) Rotax 582 UL-DCDI two-cylinder two-stroke in-line engine in basic TL-232 driving a three-blade Albastar ground-adjustable pitch propeller. Optional alternatives include 59.6 kW (79.9 hp) Rotax 912 UL-DCDI or 73.5 kW (98.6 hp) Rotax 912 ULS flat-four with Ivoprop three-blade composites propeller. Fuel capacity 50 litres (13.2 US gallons; 11.0 Imp gallons) standard, 70 litres (18.5 US gallons; 15.4 Imp gallons) optional.

DIMENSIONS, EXTERNAL:
- Wing span .... 10.60 m (34 ft 9¼ in)
- Length overall: tailwheel .... approx 6.50 m (21 ft 4 in)
  - tricycle .... 6.08 m (19 ft 11¼ in)
- Height overall: tailwheel .... approx 1.85 m (6 ft 1 in)
  - tricycle .... 2.30 m (7 ft 6½ in)

AREAS:
- Wings, gross .... 14.84 m² (159.7 sq ft)

WEIGHTS AND LOADINGS:
- Weight empty .... 265–280 kg (584–617 lb)
- Max T-O weight .... 450 kg (992 lb)

PERFORMANCE (Rotax 912 UL-DCDI engine):
- Max level speed .... 97 kt (180 km/h; 111 mph)
- Cruising speed .... 81 kt (150 km/h; 93 mph)
- Stalling speed .... 33 kt (60 km/h; 38 mph)
- Max rate of climb at S/L .... 276 m (906 ft)/min

UPDATED

URBAN

URBAN AIR S.R.O.

CZ-561 61 Donlí Libchavy 83
Tel/Fax: (+420 465) 58 21 53
e-mail: urbanair@libchavy.cz
Web: http://www.aeroklubmo.macroware.cz/lambada
DIRECTOR: Pavel Urban
SALES MANAGER: Ing Milos Mladek

This company produces the UFM-10 Samba and UFM-13 Lambáda ultralight at its three factories in the area of Usti nad Orlici.

UPDATED

URBAN UFM-13 LAMBÁDA

TYPE: Side-by-side ultralight/kitbuilt.

PROGRAMME: Design started June 1994; first flight (UFM-11 prototype) 23 May 1996; production began October 1996; first customer delivery October 1997.

CURRENT VERSIONS: **UFM-11:** Short-span (11.80 m; 38 ft 8½ in) version. Production discontinued in 2000, but restarted in 2003. *Details in 2001–02 and earlier editions.*

UFM-13: Standard version, *as described.* Optional wingtips to extend span from 13 to 15 m (UFM-13/15). Also available as sailplane (**UFM-13W**) with single mainwheel.

A UFM-13/15 (D-MZGG) shown at Berlin in May 2002 featured a Hirth fuel-injected engine and Hoffmann HO-V62R two-blade, variable pitch, metal propeller.

CUSTOMERS: Total 16 factory-built UFM-11s and 42 UFM-13s built by mid-2004, exported to Brazil, Estonia, France, Germany, Ireland, Netherlands, South Africa, Spain, Switzerland and USA.

COSTS: Flyaway, excluding VAT, €47,300 (2004).

DESIGN FEATURES: Conforms to JAR-VLA. Forward-swept wings with cranked leading-edge and laminar profile (aerofoil section SM 701); incidence 3°. T tail. Wings and horizontal tail detach for storage and transportation.

FLYING CONTROLS: Conventional (manual) flaperons, rudder and one-piece elevator; upper surface Schempp-Hirth spoilers.

STRUCTURE: Laminated glass fibre and carbon fibre, with CFRP wing spar.

LANDING GEAR: Fixed; tricycle or tailwheel type. Fuselage-mounted spring cantilever mainwheel legs with wheel fairings and hydraulic brakes. Mainwheel tyres 4.00–8.

Urban UFM-13 Lambáda ***NEW***/0576125

Urban UFM-11, with additional plan view of UFM-13/UFM-15 (starboard) (centre) and three views of low-wing UFM-10 (bottom) 0131723

POWER PLANT: One 59.6 kW (79.9 hp) Rotax 912 UL, 73.5 kW (98.6 hp) Rotax 912 ULS or 59.7 kW (80.0 hp) Jabiru 2000 four-stroke engine; driving SportProp two-blade, glass fibre (Rotax) or Kremen three-blade, wooden (Jabiru) propeller. Fuel tank in port wing, capacity 50 litres (13.2 US gallons; 11.0 Imp gallons); second tank can be fitted into starboard wing.

DIMENSIONS, EXTERNAL:
- Wing span: standard .... 13.00 m (42 ft 7¾ in)
  - extended .... 15.00 m (49 ft 2½ in)
- Length overall .... 6.60 m (21 ft 7¾ in)
- Height overall .... 1.95 m (6 ft 4¾ in)

DIMENSIONS, INTERNAL:
- Cabin max width .... 1.06 m (3 ft 5¾ in)

AREAS:
- Wings, gross: standard span .... 12.16 m² (130.9 sq ft)
  - extended span .... 12.85 m² (138.3 sq ft)

WEIGHTS AND LOADINGS:
- Weight empty: standard span .... 265 kg (562 lb)
  - extended span .... 280 kg (617 lb)
- Max T-O weight: both .... 472 kg (1,040 lb)

PERFORMANCE, POWERED (Rotax 912 UL):
- Never-exceed speed (VNE) .... 108 kt (200 km/h; 124 mph)
- Max cruising speed .... 81 kt (150 km/h; 93 mph)
- Stalling speed .... 35 kt (63 km/h; 40 mph)
- Max rate of climb at S/L .... 300 m (984 ft)/min
- Range .... 432 n miles (800 km, 497 miles)

PERFORMANCE, UNPOWERED:
- Best glide ratio: standard wing 26
  - extended wing 30
- Min rate of sink .... 1.10 m (3.61 ft)/s

UPDATED

URBAN UFM-10 SAMBA

TYPE: Side-by-side ultralight/kitbuilt.

PROGRAMME: Previously known as Speed Lambáda. Prototype OK-EUU 38 rolled out July 1999; first production aircraft (OK-EUU 39) flew 21 July 1999.

CURRENT VERSIONS: **UFM-10 Samba:** *as described*

UFM-10 Samba XXL: Introduced in 2003. Superficially similar to standard Samba, but has redesigned carbon fibre fuselage and empennage. Cockpit section is from ATEC Zaphyr, and Samba XXL is thus broadly in parallel to FSU Saphir, marketed by Fläming Air of Germany. Wing maintains 10.00 m (32 ft 9¾ in) span. Fuselage length increased by 10 cm (4 in); height 2.20 m (7 ft 2½ in), due to redesigned, longer landing gear legs. Other differences from standard Samba include front-hinged canopy and fin fillet. Third prototype (OK-IUA 24) was first to fly - on 15 April 2003.

Urban UFM-10 Samba in South African ownership ***NEW***/0576124

Urban UFM-10 Samba XXL *(Paul Jackson)* ***NEW***/0576141

CUSTOMERS: Total of at least 40 Sambas and seven XXLs produced by mid-2004, including examples to Germany, Ireland, Netherlands, South Africa and USA.
COSTS: €47,300 excluding VAT (2004).
DESIGN FEATURES: Short-span version of UFM-13 (which see); low-mounted wing and tailplane; shorter wheelbase. Suitable as tug for sailplanes of up to 380 kg (837 lb) MTOW.
FLYING CONTROLS: Conventional and manual. Split flaps.
POWER PLANT: One 59.6 kW (79.9 hp) Rotax 912 UL or 73.5 kW (98.6 hp) Rotax 912 ULS flat-four, driving two- or three-blade, ground-adjustable pitch propeller as per UFM-13. Jabiru 2200 planned as future option. Fuel capacity 50 litres (13.2 US gallons; 11.0 Imp gallons) in port wing tank; second tank can be fitted to starboard wing.
EQUIPMENT: BRS recovery parachute optional.
DIMENSIONS, EXTERNAL:
Wing span ... 10.00 m (32 ft 9¾ in)
Length overall ... 5.90 m (19 ft 4¼ in)
Height overall ... 1.95 m (6 ft 4¾ in)
AREAS:
Wings, gross ... 8.90 m^2(95.8 sq ft)
WEIGHTS AND LOADINGS:
Weight empty: standard ... 250 kg (551 lb)
with parachute and hook ... 286 kg (630 lb)
Max T-O weight: standard ... 472 kg (1,040 lb)
with parachute and hook ... 520 kg (1,146 lb)
PERFORMANCE, POWERED (Rotax 912 UL):
Never-exceed speed (VNE) ... 140 kt (260 km/h; 161 mph)
Max cruising speed ... 135 kt (250 km/h; 155 mph)
Stalling speed: flaps down ... 36 kt (65 km/h; 41 mph)
Max rate of climb at S/L ... 300 m (984 ft)/min
PERFORMANCE, UNPOWERED:
Best glide ratio ... 19
Min rate of sink ... 1.50 m (4.92 ft)/s

UPDATED

WOLFSBERG LETOV

WOLFSBERG LETECKÁ TOVARNA S.R.O. (SUBSIDIARY OF WOLFSBERG AIRCRAFT CORPORATION NV)

Beranových 65, CZ-199 02 Praha 9-Letňany
CEO: J Vondrak
PRODUCTION MANAGER: J Turek
FINANCE MANAGER: T Kotlik
QUALITY MANAGER: V Matousek

Company in process of formation from part of former Letov, established in 1918 and privatised in mid-1990s as separate divisions, including Letov Letecká Výroba, Letov Simulator, Letov Nastrojana and Letov Air. Following 1998 bankruptcy of Letov group, Letov Letecká Výroba (Letov Aircraft Production) was sold by administrator on 1 June 2000 to Latécoère of France and continued with manufacture of aerostructures, while Letov Air became a wholly owned subsidiary of Wolfsberg Aircraft Corporation of Belgium as (initially) Letecká Tovarna ('Aircraft Factory', abbreviated to Letov), having discontinued production of ultralight aircraft (LK-2, LK-3 and ST-4).

On 21 November 2000, it was announced that Letecká Tovarna had been assigned design and production responsibility for the Wolfsberg Raven 257 cargo aircraft, taking over from Evektor-Aerotechnik, another Czech firm. Marketing and financing is the responsibility of the Belgian company. Following a major reorganisation of management in 2003, the company began restructuring, with the name of Wolsberg taking precedence.

WOLFSBERG RAVEN 257

TYPE: Light utility twin-prop transport.
PROGRAMME: Definitive design began in June 1997. Wings and tail surfaces of prototype made by Letov Air in Prague; rest of aircraft and flight testing by Evektor-Aerotechnik in Kunovice, Czech Republic. First flight (OK-RAV) 28 July 2000. Original date set for certification to FAR Pt 23 was April 2002, but delays caused this to be postponed. Design and manufacture transferred back to Letov Air in September 2000. Some major modifications were carried out to prototype, increases in wing span and MTOW; and forward port and starboard doors repositioned to provide direct access to flight deck. Production of the conforming prototype and structural test aircraft was proceeding in late 2003/early 2004 after reorganisation and changing the name of the company to Wolfsberg Letecká Tovarna s.r.o. Initial plan to customise 'green' aircraft at various Western European locations has been abandoned.

Wolfsberg-Letov Raven 257 prototype after modification to production configuration 0132901

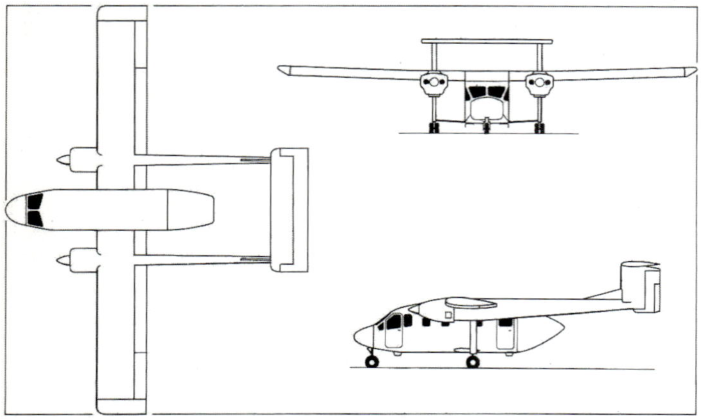
Wolfsberg-Letov Raven 257 utility aircraft *(James Goulding)* 0132944

Prototype Raven 257 in flight during mid-2002 ***NEW***/0540539

COSTS: Approximately €750,000 (2003).
DESIGN FEATURES: High-wing, twin-boom configuration with square section fuselage and high-set tailplane. Design goals include ability to operate from short unprepared strips, simple systems and easy field maintenance for operation in less-developed areas, combined with low acquisition and operating costs; seen as modern replacement for aircraft in the B-N Islander and Piper Aztec class, and as a 'step up' for operators of single-engined utility aircraft such as the Cessna 206, with applications in passenger, cargo, medevac and airdrop roles.

NACA 23018 section wings, with 2° 30' dihedral; 3° incidence; NACA 0012 section horizontal and vertical tail surfaces.
FLYING CONTROLS: Conventional and manual, via push-pull rods. Control surfaces statically balanced. Two-section elevator with independent controls and trim tab on port side. Maximum elevator deflection +27/–22°, aileron deflection +25/–15°; rudder deflection ±25°. Trim tab on starboard rudder. Four-section electrically actuated flaps, T-O setting 20°, landing 48°.
STRUCTURE: Conventional, mostly riveted metal, with glass fibre wingtips, nosecone, cargo hatch and engine cowlings.
LANDING GEAR: Non-retractable tricycle type, based on, and mostly interchangeable with, that of B-N Islander, with twin wheels on each main unit and single nosewheel; oleo strut suspension; mainwheel legs braced to bottom of fuselage. Mainwheel tyre size 7.00–6; nosewheel 6.00–6; Cleveland 40–90F brakes on mainwheels; 40–76D on nosewheel.
POWER PLANT: Two 224 kW (300 hp) Teledyne Continental IO-550-N8B flat-six engines, driving two-blade Hartzell HC-C2YF-2CUF/FC8475K-6/SM8 composites propellers. Fuel contained in two integral tanks in outer wings, combined capacity 536 litres (141.5 US gallons; 118 Imp gallons), with filler port in upper surface of each wing.
ACCOMMODATION: Two-crew cockpit (although aircraft will be certified for single-pilot operation); up to eight passengers, or combi configuration with four passengers and 500 kg (1,102 lb) of cargo. Crew door on each side; sliding semi-bulkhead separates flight deck from cabin/cargo area, which has flat floor and quick-change features, with passenger door at rear on port side and upward-hinged rear cargo door. In cargo configuration, fuselage can accommodate 304 × 139 cm pallets, 1.2 × 1.0 m Europallets and bulk cargo under netting. Baggage door on port side, emergency exit on starboard side.
SYSTEMS: 24 V electrical system includes two engine-driven alternators and 24 V 19 Ah accumulator.
AVIONICS: Dual VHF nav/com, ADF, transponder, GPS and DME.

Data refer to production version.

DIMENSIONS, EXTERNAL:
Wing span ... 15.40 m (50 ft 6¼ in)
Wing chord, constant ... 1.85 m (6 ft 0¾ in)
Wing aspect ratio ... 8.4
Length overall ... 11.28 m (37 ft 0 in)
Height overall ... 4.00 m (13 ft 1½ in)
Tailplane span ... 4.80 m (15 ft 9 in)
Wheel track ... 4.00 m (13 ft 1½ in)
Wheelbase ... 3.86 m (12 ft 8 in)
Propeller diameter ... 1.98 m (6 ft 6 in)
DIMENSIONS, INTERNAL:
Cabin, excl flight deck: Length ... 3.91 m (12 ft 10 in)
Max width ... 1.47 m (4 ft 10 in)
Max height ... 1.44 m (4 ft 8¾ in)
Volume ... 8.30 m^3(293 cu ft)
AREAS:
Wings, gross ... 28.40 m^2(305.7 sq ft)
Ailerons (total) ... 1.48 m^2(15.93 sq ft)
Trailing-edge flaps (total): inner ... 1.06 m^2(11.41 sq ft)
outer ... 1.60 m^2(17.22 sq ft)
Tailplane ... 4.54 m^2(48.87 sq ft)
Elevators, incl tabs ... 3.36 m^2(36.17 sq ft)

Fin (total) .... 3.84 m²(41.33 sq ft)
Rudders (total) .... 2.36 m²(25.40 sq ft)

WEIGHTS AND LOADINGS:
Weight empty .... 1,800 kg (3,968 lb)
Max T-O weight .... 3,100 kg (6,834 lb)
Max zero-fuel weight .... 3,000 kg (6,614 lb)
Max wing loading .... 109.2 kg/m²(22.36 lb/sq ft)
Max power loading .... 6.93 kg/kW (11.39 lb/hp)

PERFORMANCE (estimated):
Max level speed .... 145 kt (268 km/h; 167 mph)
Cruising speed, 75% power .... 138 kt (255 km/h; 158 mph)
Econ cruising speed .... 130 kt (240 km/h; 149 mph)
Stalling speed .... 53 kt (98 km/h; 61 mph)
Max rate of climb at S/L .... 444 m (1,457 ft)/min
Rate of climb at S/L, OEI .... 108 m (354 ft)/min
Range with max fuel .... 715 n miles (1,324 km; 822 miles)

ZLIN

MORAVAN AEROPLANES INC (A DIVISION OF MORAVAN INC)

Letiště 1578, CZ-765 81 Otrokovice
Tel: (+420 57) 608 30 00 and 609 39 00
Fax: (+420 57) 608 39 29
e-mail: m.aeroplanes@moravan.cz
Web: http://www.moravan.cz
PRESIDENT: Libor Soska
VICE-PRESIDENT: Patrik Joachimczyk
MANAGING DIRECTOR: Dalibor Jarolim
TECHNICAL DIRECTOR: Stepan Urban
SALES DIRECTOR: Michal Harnsik
MARKETING: Eva Buresova

Formed 18 September 1934 as Zlinská Letecká Akciová Společnost (Zlin Aviation Company Ltd) in Zlin; manufacture of Zlin aircraft started 1933 by Masarykova Letecká Liga (Masaryk League of Aviation); total of approximately 650 aircraft, including gliders and military transports, manufactured by end of Second World War.

Company then renamed Moravan; post-war aircraft production included 1,495 of the Z 26 family (162 Z 26, 170 Z 126, 366 Z 226, 436 Z 326, 331 Z 526 and 30 Z 726) built between 1949 and 1975. Production began in 1970 of Z 42/142/242 and Z 43/143 families; Algerian Z 43s (five supplied in 1991) designated **Safir 43**; Algerian Z 142s (17 complete and 29 kits in 1987–91) designated **Firnas 142**. In July 1997, the factory suffered serious flood damage, causing production to be suspended until the following year.

In June 1999, Moravan announced the sale of 100 Z 242s to Egypt, equivalent to a two-year backlog of work at then-current production rates. This order then represented approximately half of Moravan's backlog value of US$40 million, but was later cancelled. Moravan sold 22 aircraft in 2000, compared with only 15 in 1999. Aerolease, a joint venture with Dutch financial support, was formed in 2000 to lease Zlin aircraft to European flying clubs and flying training schools. The Z 142 C, out of production since 1995, was to be resumed as the result of a 2001 order; these to have been manufactured by LZ. Licensed manufacture of the Russian Technoavia Finist, with an Orenda OE-600 piston diesel engine, was planned under the designation Z 400 Rhino. Under a contract with EADS Socata, signed on 22 June 2001, Moravan Aeroplanes produced complete fuselages for the TB series; deliveries were to begin in August 2002, at a rate of six to eight per month. Also built Russo Savage kitplane for Italian designer Pasquale Russo.

It was announced on 16 July 2001 that approval had been given for Moravan Aeroplanes to acquire Let Kunovice, with the exception of the L 610 G programme, for the sum of Kcs200 million (US$5 million). Moravan intended to keep open the L 410/420 production line, as well as Let's L-13, L-23 and L-33 sailplanes, and planned to use the factory (renamed Letecké Zavody: LZ) also for its work on the Z 400 Rhino programme. L 610 G programme subsequently acquired by LZ in early 2002. Shortly after its acquisition of Let, Moravan Aeroplanes announced plans to establish a final assembly centre for its aircraft at Beeville, Texas, USA, in a joint venture with Pereira Internacional Aérea of Houston.

Moravan also manufactured ejection seats and wing movable surfaces for other aerospace companies.

On 18 June 2004, the company was declared bankrupt and a receiver called in to restructure and reduce the business in size.

UPDATED

Zlin Z 142 C aerobatic light trainer ***NEW***/0064437

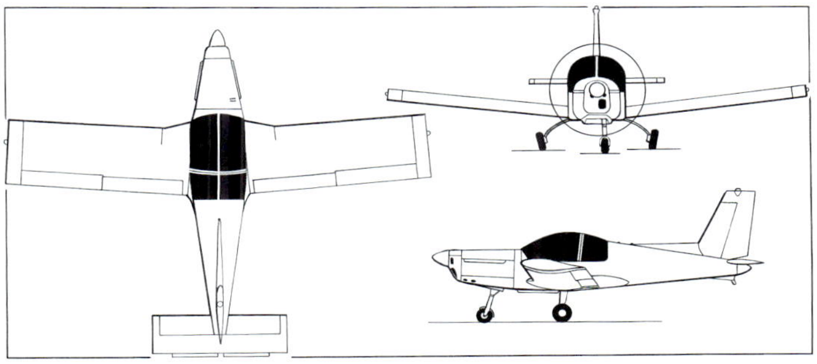

Zlin Z 142 C (LOM M 337 AK engine) *(James Goulding)* ***NEW***/0573876

ZLIN Z 142 C

TYPE: Primary prop trainer/sportplane.

PROGRAMME: Design of original Z 142 begun late 1977/early 1978, prototype (OK-078) construction April 1978; first flight 29 December 1978; FAR Pt 23 certification (Aerobatic, Utility and Normal) 1980; deliveries from April 1980; Canadian DoT certification 26 November 1991. Z 142 C received Czech certification on 18 July 1991; also certified in Australia (5 March 1996), Canada (4 August 1993) and Slovak Republic (18 June 1991). Full-rate production ended in 1995, but aircraft still available to order; standard delivery time, four months after contract signature.

CURRENT VERSIONS: **Z 142 C:** Current version, certified in A (aerobatic trainer), U (general trainer) and N (touring) categories. *Description applies to this version.*

Z 142 CAF: Military version; eight delivered to Czech Air Force.

Z 242 L: Lycoming-powered development of Z 142; *described separately.*

CUSTOMERS: Total 378 (all versions) built by 1995, including 10 Z 142 C for Canada/USA and eight Z 142 CAF for Czech Air Force. Further order(s) reported in 2001; four to Angola in 2003.

COSTS: Standard aircraft US$119,860 (VFR) or US$139,860 (IFR) (1999).

DESIGN FEATURES: Development of Zlin Z 142 M; basic, advanced and aerobatic training, glider towing and (with appropriate equipment) night flying and IFR training.

Wing section NACA $63_2416.5$; dihedral 6° from roots; sweep forward 4° 20' at quarter-chord.

POWER PLANT: One LOM M 337 AK inverted six-cylinder air-cooled in-line engine (157 kW; 210 hp at 2,750 rpm), with supercharger and low-pressure injection pump, driving a two-blade Avia V 500 A constant-speed metal propeller. Fuel tanks in each wing leading-edge, with combined capacity of 120 litres (31.7 US gallons; 26.4 Imp gallons). Normal category version has auxiliary 50 litre (13.2 US gallon; 11.0 Imp gallon) tank at each wingtip, increasing usable fuel capacity to 220 litres (58.1 US gallons; 48.4 Imp gallons). Fuel and oil systems permit inverted flying for up to 1½ minutes. Oil capacity 12 litres (3.2 US gallons; 2.6 Imp gallons).

ACCOMMODATION: As for Z 242 L.

DIMENSIONS, EXTERNAL:
Wing span .... 9.16 m (30 ft 0½ in)
Length overall .... 7.33 m (24 ft 0½ in)
Height overall .... 2.75 m (9 ft 0¼ in)
Elevator span .... 2.905 m (9 ft 6¼ in)
Wheel track .... 2.33 m (7 ft 7½ in)
Wheelbase .... 1.66 m (5 ft 5¼ in)
Propeller diameter .... 2.00 m (6 ft 6½ in)
Propeller ground clearance .... 0.40 m (1 ft 3¾ in)

DIMENSIONS, INTERNAL:
Cockpit max width .... 1.12 m (3 ft 8 in)

AREAS:
Wings, gross .... 13.15 m² (141.5 sq ft)

WEIGHTS AND LOADINGS (A: Aerobatic, U: Utility, N: Normal category):
Basic weight empty (all versions) .... 730 kg (1,609 lb)
Max T-O weight: A .... 970 kg (2,138 lb)
U .... 1,020 kg (2,248 lb)
N .... 1,090 kg (2,403 lb)
Max landing weight: A .... 970 kg (2,138 lb)
U .... 1,020 kg (2,248 lb)
N .... 1,050 kg (2,315 lb)
Max wing loading: A .... 73.8 kg/m² (15.11 lb/sq ft)
U .... 77.6 kg/m² (15.89 lb/sq ft)
N .... 82.9 kg/m² (16.98 lb/sq ft)
Max power loading: A .... 6.19 kg/kW (10.17 lb/hp)
U .... 6.51 kg/kW (10.69 lb/hp)
N .... 6.96 kg/kW (11.43 lb/hp)

PERFORMANCE (at relevant max T-O weight):
Never-exceed speed (VNE) (all versions):
at S/L ... 170 kt (316 km/h; 196 mph)
at 500 m (1,640 ft) ... 174 kt (323 km/h; 200 mph)
Max level speed at S/L:
A, U ... 125 kt (232 km/h; 144 mph)
N ... 123 kt (228 km/h; 142 mph)
Max cruising speed at S/L at 75% power:
A, U ... 112 kt (207 km/h; 129 mph)
N ... 109 kt (202 km/h; 126 mph)
Max cruising speed at FL656:
A ... 115 kt (213 km/h; 132 mph)
U ... 113 kt (210 km/h; 130 mph)
N ... 111 kt (206 km/h; 128 mph)
Econ cruising speed at S/L at 65% power:
A, U ... 104 kt (193 km/h; 120 mph)
N ... 100 kt (186 km/h; 116 mph)
Econ cruising speed at 500 m (1,640 ft):
A, U ... 97 kt (179 km/h; 111 mph)
N ... 92 kt (171 km/h; 106 mph)
Stalling speed at S/L, flaps up:
A ... 61 kt (113 km/h; 71 mph)
U ... 63 kt (115 km/h; 72 mph)
N ... 65 kt (120 km/h; 75 mph)
Stalling speed at S/L, T-O flap setting:
A ... 60 kt (110 km/h; 69 mph)
U ... 61 kt (112 km/h; 70 mph)
N ... 63 kt (116 km/h; 73 mph)
Stalling speed at S/L, flaps down:
A ... 53 kt (98 km/h; 61 mph)
N ... 56 kt (103 km/h; 64 mph)
Max rate of climb at S/L, ISA:
A ... 312 m (1,023 ft)/min
U ... 388 m (945 ft)/min
N ... 252 m (827 ft)/min
Service ceiling: A ... 5,000 m (16,400 ft)
N ... 4,200 m (14,180 ft)
T-O run: A ... 231 m (760 ft)
U ... 236 m (775 ft)
N ... 252 m (830 ft)
T-O to 15 m (50 ft): A ... 510 m (1,675 ft)
U ... 560 m (1,840 ft)
N ... 620 m (2,035 ft)
Landing from 15 m (50 ft): A ... 425 m (1,395 ft)
U ... 440 m (1,445 ft)
N ... 460 m (1,510 ft)
Landing run: A ... 190 m (625 ft)
U ... 200 m (660 ft)
N ... 220 m (725 ft)
Range at econ cruising speed:
A, U ... 316 n miles (586 km; 364 miles)
N ... 522 n miles (967 km; 601 miles)
g limits: A ... +6/–3.5
U ... +6/–3
N ... +3.8/–1.5

UPDATED

ZLIN Z 242 L

TYPE: Primary prop trainer/sportplane.

PROGRAMME: Lycoming-powered version of Z 142 (see 1997–98 and earlier *Jane's*). Design started December 1988; first flight (OK-076) 14 February 1990; second prototype (SE-KMS) followed, both converted on production line from Z 142; Czech certification to FAR Pt 23 (A, U and N categories) 22 March 1992; now certified in Argentina, Australia, Austria, Belgium, Canada, Israel, Luxembourg, Slovenia, Sweden, UK and USA.

CURRENT VERSIONS: **Z 242 L:** Aerobatics-capable trainer.

Description applies to this version.

Z 242 LA: Lower-powered, non-aerobatic version for primary, day/night and IFR training and glider/banner towing; *g* limits +4.4/–1.76. Prototype (OK-ANA) on test early 1996. No subsequent information received.

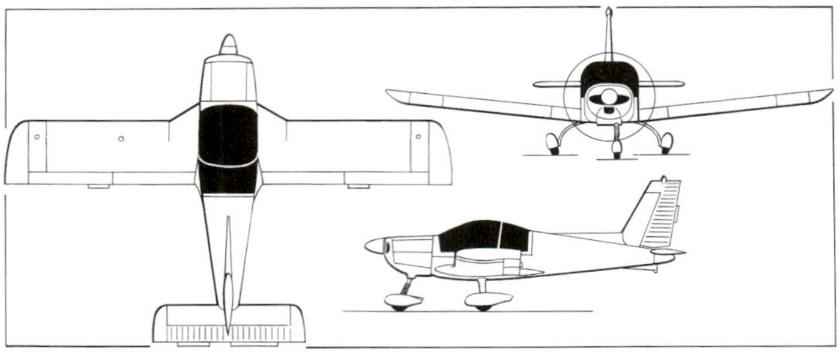

Zlin Z 242 L primary prop trainer *(James Goulding)* ***NEW***/0589210

CUSTOMERS: Total of 120 delivered by late 2002, for customers in Argentina (one), Australia (one), Canada (15), Czech Republic (four), Israel (three), Macedonia (four), Peru (Air Force, 18), Slovenia (Defence Force, seven), Sweden (two), UK (three), USA (38) and elsewhere. Yemen Air Force received 12 in 2002. Ten ordered by Mexican Navy for Escuela de Aviación at Baja; delivered 13 June 2002, with negotiations continuing for up to 10 more. Bolivia and Ecuador also discussing possible orders at that time. Company reported an order backlog of more than 100 for US flying schools in mid-2001; was also then bidding (unsuccessfully) for USAF T-3A Firefly replacement contract. Five to Yemen and one to Macedonia in 2003; additional (unspecified) contracts in that year from unnamed Asian countries.

COSTS: Mexican Navy contract reported as approximately US$2.6 million (2002).

DESIGN FEATURES: Conventional low-wing monoplane. Constant-chord horizontal surfaces with wingroot glove. Sweptback fin with fillet. Changes from Z 142, apart from new engine, include redesigned (and shorter) engine cowling and front fuselage, wing incidence, 0° sweep, wingroot glove, redesigned wing- and tailplane tips, redesigned fuel system and updated instruments. Spin recovery strake each side of cowling.

Wing section NACA $63_2416.5$.

FLYING CONTROLS: Conventional and manual. Slotted, mass-balanced surfaces used for ailerons and flaps; horn-balanced elevator with trim tab; ground-adjustable tabs in ailerons and rudder; ailerons and elevator operated by rods, rudder by cables.

STRUCTURE: Mainly metal; wing has main and auxiliary spars; duralumin skins, fluted on control surfaces; metal engine cowlings; centre-fuselage is steel tube cage with composites skin panels.

LANDING GEAR: Non-retractable tricycle type, with nosewheel offset 13 cm (5.1 in) to port. Oleo-pneumatic nosewheel shock-absorber. Mainwheels carried on flat spring steel legs. Nosewheel steered (±38°) by rudder pedals. Mainwheels and Barum tyres size 420×150 or Goodyear 6.00–6.5, pressure 1.90 bar (28 lb/sq in); nosewheel and Barum tyre size 350×135 or Goodyear 5.00–5, pressure 2.50 bar (36 lb/sq in). Hydraulic disc brakes on mainwheels can be operated from either seat. Parking brake standard.

POWER PLANT: One Textron Lycoming AEIO-360-A1B6 flat-four engine (149 kW; 200 hp at 2,700 rpm) driving an MTV-9-B-C/C-188-18a three-blade constant-speed wood/composites propeller or Hartzell HC-C3YR-4BF/FC 6890 three-blade constant-speed metal propeller. Fuel capacity 120 litres (31.7 US gallons; 26.4 Imp gallons); Normal category version wingtip tanks 55 litres (14.5 US gallons; 12.1 Imp gallons) each, bringing usable capacity to 230 litres (60.7 US gallons; 50.6 Imp gallons). Inverted flight limited to 1 minute. Oil capacity 8 litres (2.1 US gallons; 1.8 Imp gallons).

ACCOMMODATION: Side-by-side seats for two persons, instructor's seat to port. Both seats adjustable and permit use of back-type parachutes. Baggage space (20 kg; 44 lb) aft of seats. Cabin and windscreen heating and ventilation standard. Forward-sliding cockpit canopy. Dual controls standard.

SYSTEMS: Electrical system includes 1.6 kW 28 V engine-driven generator and 24 V 19 Ah Gill battery. External power source can be used for engine starting.

AVIONICS: To customer's specification, usually from Bendix/King Silver Crown or Garmin range.

EQUIPMENT: Standard equipment includes EGT gauge, fuel flow indicator, *g* meter and anti-collision beacon. Optional items include cockpit/instrument/cabin lights, landing/taxying lights, anti-collision lights, and towing gear for gliders of up to 500 kg (1,102 lb) weight.

DIMENSIONS, EXTERNAL:
Wing span ... 9.34 m (30 ft 7¾ in)

Appropriately registered Zlin Z 242 L US demonstrator *(Paul Jackson)* ***NEW***/0589262

Wing chord, constant portion 1.50 m (4 ft 11 in)
Wing aspect ratio 6.3
Length overall 6.94 m (22 ft 9¼ in)
Height overall 2.95 m (9 ft 8¼ in)
Elevator span 3.20 m (10 ft 6 in)
Wheel track 2.33 m (7 ft 7¾ in)
Wheelbase 1.76 m (5 ft 9¼ in)
Propeller diameter: MTV 1.88 m (6 ft 2 in)
Hartzell 1.78 m (5 ft 10 in)
Propeller ground clearance: MTV 0.33 m (1 ft 1 in)
Hartzell 0.38 m (1 ft 3 in)

DIMENSIONS, INTERNAL:
Cabin: Length 1.80 m (5 ft 10¾ in)
Max width 1.12 m (3 ft 8 in)
Max height 1.20 m (3 ft 11¼ in)
Baggage space 0.20 m^3 (7.1 cu ft)

AREAS:
Wings, gross 13.86 m^2 (149.2 sq ft)
Ailerons (total) 1.41 m^2 (15.18 sq ft)
Trailing-edge flaps (total) 1.41 m^2 (15.18 sq ft)
Fin 0.54 m^2 (5.81 sq ft)
Rudder, incl tab 0.81 m^2 (8.72 sq ft)
Tailplane 1.23 m^2 (13.24 sq ft)
Elevator, incl tabs 1.36 m^2 (14.64 sq ft)

WEIGHTS AND LOADINGS (A: Aerobatic, U: Utility, N: Normal category):
Basic weight empty: A, U, N 730 kg (1,609 lb)
Max T-O weight: A 970 kg (2,138 lb)
U 1,020 kg (2,248 lb)
N 1,090 kg (2,403 lb)
Max landing weight: A 970 kg (2,138 lb)
U 1,020 kg (2,248 lb)
N 1,050 kg (2,315 lb)
Max wing loading: A 70.3 kg/m^2 (14.40 lb/sq ft)
U 73.6 kg/m^2 (15.07 lb/sq ft)
N 78.6 kg/m^2 (16.11 lb/sq ft)
Max power loading: A 6.50 kg/kW (10.68 lb/hp)
U 6.84 kg/kW (11.24 lb/hp)
N 7.31 kg/kW (12.01 lb/hp)

PERFORMANCE:
Never-exceed speed (VNE):
all versions 170 kt (315 km/h; 195 mph)
Max level speed at S/L:
A 127 kt (236 km/h; 146 mph)
U 125 kt (233 km/h; 144 mph)
N 124 kt (231 km/h; 143 mph)
Cruising speed at FL65 at 75% power:
A 123 kt (227 km/h; 141 mph)
U 125 kt (233 km/h; 144 mph)
N 124 kt (231 km/h; 143 mph)
Max cruising speed at FL65 at 75% power:
A 123 kt (227 km/h; 141 mph)
U 121 kt (225 km/h; 140 mph)
N 120 kt (223 km/h; 139 mph)
Stalling speed at S/L:
flaps up: A 60 kt (111 km/h; 69 mph)
U 62 kt (114 km/h; 71 mph)
N 64 kt (118 km/h; 74 mph)
T-O flap setting: A 57 kt (105 km/h; 66 mph)
U 58 kt (107 km/h; 67 mph)
N 60 kt (111 km/h; 69 mph)
flaps down: A 51 kt (94 km/h; 59 mph)
U 53 kt (97 km/h; 61 mph)
N 54 kt (100 km/h; 63 mph)
Max rate of climb at S/L: A 336 m (1,102 ft)/min
U 300 m (984 ft)/min
N 270 m (886 ft)/min
Service ceiling: A 4,800 m (15,740 ft)
N 4,500 m (14,760 ft)
T-O run: A 242 m (795 ft)
U 233 m (765 ft)
N 306 m (1,005 ft)
T-O to 15 m (50 ft): A 560 m (1,840 ft)
U 495 m (1,625 ft)
N 706 m (2,320 ft)
Landing from 15 m (50 ft): A 516 m (1,695 ft)
U 525 m (1,725 ft)
N 560 m (1,840 ft)
Landing run: A 265 m (870 ft)
N 285 m (935 ft)
Range with max fuel:
A, U 267 n miles (495 km; 308 miles)
N 570 n miles (1,056 km; 656 miles)
Range at FL100, 65% power:
A 228 n miles (424 km; 263 miles)
N 504 n miles (935 km; 581 miles)
g limits: A +6/−3.5
U +5/−3
N +3.8/−1.5

UPDATED

ZLIN Z 143

Czech Air Force designation: Z 143 MAF

TYPE: Four-seat lightplane.

PROGRAMME: Launched May 1991, with prototype construction beginning mid-year and first flight (Z 143 L prototype OK-074) 24 April 1992. Czech certification of this version 10 June 1994; now also certified in Austria, Bulgaria, Germany, Slovak Republic, Switzerland and USA.

CURRENT VERSIONS: **Z 143 L**: Standard version, with Textron Lycoming engine.

Detailed description applies to Z 143 L.

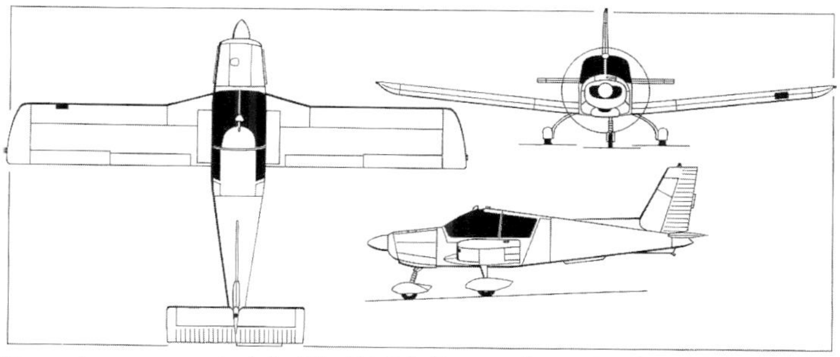

General arrangement of the Zlin Z 143 L (Textron Lycoming O-540-J3A5 engine) *(Jane's/Mike Keep)* 0137334

Z 143 L instrument panel 0507592

Z 143 LSi: Proposed version with 194 kW (260 hp) Textron Lycoming fuel-injection engine. No recent information.

CUSTOMERS: Total of 49 (excluding prototype and three testbeds/demonstrators) delivered by December 2003 to customers in Argentina (one), Austria (one), Bulgaria (two), Czech Republic (Air Force, four), Egypt (five), Germany (eight), Israel (two), Macedonia (one), Slovenia (two), Switzerland (four), Thailand (two) and USA (seven) and elsewhere.

COSTS: Standard aircraft approximately US$200,000, depending on avionics/equipment fit (2003).

DESIGN FEATURES: Generally as for Z 242. For basic aerobatics (except inverted flight), glider towing and (with equipment) night and IFR training. Non-swept wings, of similar planform to Z 242 L but with NACA $63_2416.5$ wing section and 6° dihedral.

FLYING CONTROLS: As for Z 242 L.

STRUCTURE: As for Z 242 L.

LANDING GEAR: As for Z 242 L.

POWER PLANT: One Textron Lycoming O-540-J3A5 flat-six engine (175 kW; 235 hp at 2,400 rpm); MTV-9-B/195-45a three-blade variable-pitch propeller. Fuel capacity 122 litres (32.2 US gallons; 26.8 Imp gallons) in two wing tanks; provision for two 51 litre (13.5 US gallon; 11.2 Imp gallon) auxiliary tanks in Normal category version, raising total to 224 litres (59.2 US gallons; 49.3 Imp gallons). Oil capacity 12 litres (3.2 US gallons; 2.6 Imp gallons).

ACCOMMODATION: Four seats in 2 + 2 configuration, with pilot in front left-hand seat. All seats have four-point safety belts. Baggage compartment (capacity 60 kg; 132 lb) aft of rear seats, with external door on port side. One-piece forward-sliding canopy. Dual controls standard. Cabin and windscreen heating and ventilation standard.

SYSTEMS: As for Z 242 L.

AVIONICS: As for Z 242 L.

DIMENSIONS, EXTERNAL:
Wing span 10.14 m (33 ft 3¼ in)
Wing chord, mean aerodynamic 1.49 m (4 ft 10¾ in)
Wing aspect ratio 7.0
Length overall 7.58 m (24 ft 10½ in)
Height overall 2.91 m (9 ft 6½ in)
Elevator span 3.01 m (9 ft 10½ in)
Wheel track 2.44 m (8 ft 0 in)
Wheelbase 1.75 m (5 ft 9 in)
Propeller diameter 1.95 m (6 ft 4¾ in)
Propeller ground clearance 0.28 m (11 in)

DIMENSIONS, INTERNAL:
Baggage compartment volume 0.25 m^3 (8.83 cu ft)

AREAS: As for Z 242 L except:
Wings, gross 14.78 m^2 (159.1 sq ft)

WEIGHTS AND LOADINGS (A: Aerobatic, N: Normal category):
Weight empty, equipped: A, N 850 kg (1,874 lb)
Max T-O weight: A 1,080 kg (2,381 lb)
N 1,350 kg (2,976 lb)
Max landing weight: A 1,080 kg (2,381 lb)
N 1,280 kg (2,822 lb)
Max wing loading: A 73.1 kg/m^2 (14.97 lb/sq ft)
N 91.4 kg/m^2 (18.72 lb/sq ft)
Max power loading: A 6.16 kg/kW (10.13 lb/hp)
N 7.70 kg/kW (12.66 lb/hp)

PERFORMANCE:
Never-exceed speed (VNE):
A, N 170 kt (315 km/h; 195 mph)
Max level speed at S/L: A 144 kt (267 km/h; 166 mph)
N 142 kt (264 km/h; 164 mph)

Zlin Z 143 L operated in Germany *(Paul Jackson)* ***NEW***/0532083

Max cruising speed at FL656 at 75% power:
A 127 kt (235 km/h; 146 mph)
N 125 kt (232 km/h; 144 mph)
Econ cruising speed at 60% power:
A 116 kt (216 km/h; 134 mph)
N 111 kt (205 km/h; 127 mph)
Stalling speed:
flaps up: A 58 kt (108 km/h; 67 mph)
N 63 kt (117 km/h; 73 mph)
flaps down: A 54 kt (100 km/h; 63 mph)
N 60 kt (110 km/h; 69 mph)
Max rate of climb at S/L: A 444 m (1,457 ft)/min
N 294 m (964 ft)/min
Service ceiling: A 5,700 m (18,700 ft)
N 4,170 m (13,680 ft)
T-O run: A 295 m (970 ft)
N 170 m (560 ft)
T-O to 15 m (50 ft): A 450 m (1,480 ft)
N 640 m (2,100 ft)
Landing from 15 m (50 ft): A 590 m (1,935 ft)
N 765 m (2,510 ft)
Landing run: A 305 m (1,000 ft)
N 380 m (1,250 ft)
Range at FL100 at 65% power:
A 631 n miles (1,170 km; 727 miles)
N 259 n miles (480 km; 298 miles)
g limits: A +4.4/−1.76
N +3.8/−1.52

UPDATED

ZLIN Z 400 RHINO

TYPE: Light utility transport.

PROGRAMME: Announced at Sun 'n' Fun, 8 April 2001, that Moravan was to manufacture under licence, as the Z 400, a re-engined (engineering work by NA Design) version of the Russian Technoavia SM-92 Finist. Prototype then under conversion; rolled out at Kunovice 21 January 2002 and made first flight (OK-HZL) on 22 May 2002. Production planned to start July 2002, with certification (JAR 23) targeted for late 2002 and deliveries from early 2003. However, programme in abeyance and no production undertaken by October 2003 as a consequence of Orenda's decision to terminate the OE-600 engine programme. Amphibious float version under study. Predicted market for 500 Orenda-powered Finists.

COSTS: Development costs approximately US$3 million by mid-2002; unit price Kcs26 million (US$700,000) at that time.

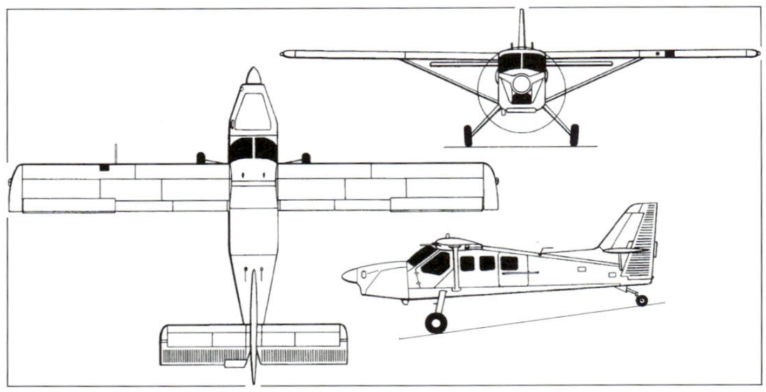

Zlin Z 400 Rhino (Orenda V-8) *(James Goulding)* ***NEW***/0535821

POWER PLANT: One 447 kW (600 hp) Orenda OE-600A eight-cylinder V-type piston diesel engine (derated to 362 kW; 485 hp); three-blade Avia AV 803 propeller. Fuel capacity 760 litres (201 US gallons; 167 Imp gallons). Provision for two auxiliary tanks internally, each 200 litres (52.8 US gallons; 44.0 Imp gallons).

DIMENSIONS, EXTERNAL (As for Finist, except):
Length overall 9.85 m (32 ft 3¾ in)
Propeller diameter 2.60 m (8 ft 6¼ in)

WEIGHTS AND LOADINGS:
Weight empty 1,540 kg (3,395 lb)
Max T-O weight 2,700 kg (5,952 lb)
Max wing loading 132.1 kg/m² (27.05 lb/sq ft)
Max power loading 7.47 kg/kW (12.27 lb/hp)

PERFORMANCE:
Never-exceed speed (VNE) 183 kt (340 km/h; 211 mph)
Max level speed 145 kt (270 km/h; 167 mph)
Service ceiling 6,000 m (19,680 ft)
T-O run 220 m (725 ft)
Landing run without propeller reversal 200 m (660 ft)
Range:
with max payload 863 n miles (1,600 km; 994 miles)
with max fuel 1,080 n miles (2,000 km; 1,242 miles)
with auxiliary fuel 1,295 miles (2,400 km; 1,491 miles)

UPDATED

Prototype Orenda-engined Zlin Rhino preparing for its maiden flight ***NEW***/0535819

Finland

PATRIA

PATRIA AVIATION OY (SUBSIDIARY OF PATRIA INDUSTRIES OY)

Naulakatu 3, FIN-33100 Tampere
Tel: (+358 20) 46 91
Fax: (+358 20) 469 26 90
e-mail: finavitec@patria.fi
Web: http://www.patria.fi
EXECUTIVE VICE-PRESIDENT, AVIATION: Aarne Nieminen

Patria Aviation (formerly Patria Finavitec) and its Valmet predecessors have built 30 types of aircraft, including 19 of Finnish design, since 1922; current activities include production of aircraft and aero-engine parts; and maintenance, overhaul, repair and modification of aircraft, aero-engines and their accessories. Workforce was approximately 1,300 in 2003, including engines facility at Linnavuori and Patria Ostermans Aero AB, a Swedish helicopter maintenance company acquired in 1999. Helsinki-based Pilot Aircraft Oy became a wholly owned subsidiary of Patria Aviation on 24 March 2004; activities include pilot and ATC training.

Patria Aviation currently manufactures rear and centre fuselage sections for the Embraer ERJ-135/145 under subcontract to Sonaca of Belgium, to which first shipsets were delivered in second half of 1998. MoUs signed in December 1999 by Patria Finavitec and Airbus Industrie for former to take part in conceptual definition of A380 and to explore other possible industrial co-operation. Agreed in September 2000 to support Sikorsky bid of S-92 for Nordic Standard Helicopter Programme (NSHP) requirement but, following selection of EHI and NHI as joint winners, Patria contracted on 19 October 2001 to assemble 50 NH90s and associated RTM322 engines at Jämsä and Linnavuori, respectively, deliveries taking place between 2005 and 2011. Work at new Jämsä factory began in September 2003.

In June 2001, EADS acquired a 26.8 per cent interest in parent company Patria Industries Oy.

UPDATED

France

AERALL

ASSOCIATION POUR L'ETUDE ET LA RECHERCHE SUR LES AERONEFS ALLEGES

2 bis, avenue Odette (BP 46), F-94130 Nogent-sur-Marne
Tel: (+33 1) 49 00 15 14
Fax: (+33 1) 47 75 09 56
PRESIDENT: Jean René Fontaine

AERALL is a non-profit organisation created in 1971 to promote lighter than air technology and new aeronautical concepts. Since that time it has organised a series of international symposia aimed at shaping industry and government opinion regarding the future use of airships, particularly for transporting heavy and indivisible loads, or as stratospheric communications, traffic and pollution surveillance platforms.

Due to a misunderstanding, AERALL was previously reported to be the development agency for a CNES project known as AVEA. This is now known to be incorrect: AERALL was merely asked in 2000 to report on the viability of AVEA for the French Ministry of Transport. Its purpose is not to indicate support (or otherwise) for individual projects.

UPDATED

AEROKUHLMANN

AEROKUHLMANN

Aérodrome de Cerny, F-91590 Cerny
Tel:(+33 1) 69 90 17 80
Fax:(+33 1) 69 23 34 78
e-mail: aerokuhlmann@magic.fr
Web: http://www.chez.com/scub
DIRECTOR: Hervé Kuhlmann
COMMERCIAL DIRECTOR: Daniel Feldzer

BUSINESS ADDRESS:
Immeuble Aéroclub de France, 6 rue Galilée, F-75782 Paris Cedex 16
Tel/Fax: (+33 1) 40 70 13 76

AeroKuhlmann's first product, built in the company's 600 m^2 (6,450 sq ft) factory on La Ferté-Alais airfield, near Cerny, is the SCUB.

UPDATED

AEROKUHLMANN SCUB

TYPE: Tandem-seat kitbuilt.

PROGRAMME: Prototype (91-KJ) first flew 5 May 1996. Certified to JAR-VLA standard in France.

CURRENT VERSIONS: **Ultralight:** Optional 450 kg (992 lb) MTOW in Europe or 544 kg (1,200 lb) in USA.

Certified: MTOW 598 kg (1,320 lb).

Description applies to certified version, except where otherwise stated.

CUSTOMERS: Production totalled 17 by June 1999. No more recent news has been received although promotion was continuing in 2002.

COSTS: FFr290,000, basically equipped, plus VAT (1997). Floats FFr36,000 extra.

DESIGN FEATURES: Braced high-wing monoplane, suitable for wide range of tasks including surveillance, mapping, cargo, medical and agricultural duties. Wings and tail fold for storage. Airframe design life of 10,000 hours.

Wing supercritical profile NASA LS1 0417 mod; incidence 4° 30', dihedral 2°.

FLYING CONTROLS: Conventional and manual. Horn-balanced rudder. Elevator movement ±24°, rudder movement ±21°, aileron movement +8/–15°. Flight-adjustable trim tab on starboard elevator. No flaps.

STRUCTURE: Dacron-covered chromium-molybdenum tubular steel fuselage. Wing main spar of three-ply birch with carbon fibre spar cap, glass fibre ribs and plywood-covered leading-edge, all covered with Dacron.

LANDING GEAR: Tailwheel type; fixed. Two faired side Vs hinged to lower longerons; outboard oleo-pneumatic shock-absorbers attached at upper ends to fuselage sides. Mainwheels 8.00–6; drum brakes. Optional float (Stéphane Morosini/Hydro Aero Concept two-step) or ski gear can be installed in quoted time of 2 hours; additional pick-up point for both forward of main legs. Tundra tyres can be fitted for rough field operations.

POWER PLANT: One 62.5 kW (83.8 hp) JPX 4TX75A flat-four with dual ignition, driving a ULX two-blade wooden propeller. Fuel capacity 110 litres (29.0 US gallons; 24.2 Imp gallons). Oil capacity 11.4 litres (3.0 US gallons; 2.5 Imp gallons).

ACCOMMODATION: Pilot and passenger in tandem, with dual controls.

SYSTEMS: All electric systems corrosion proof on floatplane version.

AVIONICS: *Flight:* GPS optional.

Instrumentation: ASI, VSI, compass, altimeter, oil temperature, oil pressure, tachometer, fuel gauges.

EQUIPMENT: For surveillance, video cameras can be fitted beneath wings; four Micronair 7000 spray atomisers can be fitted on underwing bar, with a 150 litre (39.6 US gallon; 33.0 Imp gallon) tank fitted under fuselage.

AeroKuhlmann SCUB *(Paul Jackson)* ***NEW***/0546842

DIMENSIONS, EXTERNAL:
- Wing span ... 11.00 m (36 ft 1 in)
- Wing aspect ratio ... 8.5
- Width, wings folded ... 1.83 m (6 ft 0 in)
- Length overall ... 7.01 m (23 ft 0 in)
- Height: overall ... 1.83 m (6 ft 0 in)
  - folded ... 2.29 m (7 ft 6 in)
- Tailplane span ... 2.74 m (9 ft 0 in)
- Tailplane chord ... 0.71 m (2 ft 4 in)

AREAS:
- Wings, gross ... 14.30 m^2 (153.9 sq ft)
- Ailerons (total) ... 0.72 m^2 (7.80 sq ft)
- Fin ... 0.37 m^2 (4.00 sq ft)
- Rudder, incl tab ... 0.56 m^2 (6.00 sq ft)
- Tailplane ... 1.21 m^2 (13.00 sq ft)
- Elevators, incl tab ... 1.21 m^2 (13.00 sq ft)

WEIGHTS AND LOADINGS:
- Weight empty: landplane ... 250 kg (551 lb)
  - floatplane ... 354 kg (780 lb)
- Max T-O weight: landplane ... 450 kg (992 lb)
  - floatplane ... 500 kg (1,102 lb)
- Max wing loading: landplane ... 31.5 kg/m^2 (6.45 lb/sq ft)
  - floatplane ... 35.0 kg/m^2 (7.16 lb/sq ft)
- Max power loading: landplane ... 7.20 kg/kW (11.83 lb/hp)
  - floatplane ... 8.00 kg/kW (13.14 lb/hp)

PERFORMANCE, POWERED:
- Never-exceed speed (VNE) ... 91 kt (169 km/h; 105 mph)
- Max operating speed ... 67 kt (124 km/h; 77 mph)
- Stalling speed ... 33 kt (60 km/h; 38 mph)
- Max rate of climb at S/L ... 240 m (787 ft)/min
- Service ceiling ... 4,570 m (15,000 ft)
- T-O and landing run: on land ... 55 m (180 ft)
  - on water ... 91 m (300 ft)
- T-O to 15 m (50 ft) ... 137 m (450 ft)
- Landing from 15 m (50 ft) ... 201 m (660 ft)
- Range with max fuel ... 543 n miles (1,005 km; 625 miles)
- Endurance ... 7 h 0 min
- *g* limits ... +6.6/–3.3

PERFORMANCE, UNPOWERED:
- Best glide ratio at 49 kt (90 km/h; 56 mph) ... 15

UPDATED

AERONIX

AERONIX SARL

2 rue Jean Monnet, F-78990 Elancourt
Tel: (+33 1) 30 16 17 70
Fax: (+33 1) 30 62 57 71
e-mail: contacts@aeronix.fr
Web: http://www.aeronix.fr
DIRECTOR GENERAL: Lionel de Mauduit du Plessis
TECHNICAL DIRECTOR: Rémi Cuvelier
COMMERCIAL DIRECTOR: Gérard Mosali

Formed in 1999, Aeronix has recently launched production of its first aircraft, the Airelle. The company also markets avionics and provides simulated training.

VERIFIED

AERONIX AIRELLE

TYPE: Side-by-side ultralight/kitbuilt; side-by-side kitbuilt; four-seat kitbuilt.
PROGRAMME: Development began in 1999; one-third scale model flew October 2000. Announced at Paris Air Show, June 2001; first flight February 2002 (unannounced); mockup shown at Paris, June 2003. Deliveries of kit version began in 2003; flyaway deliveries due to have begun in August 2003.
CURRENT VERSIONS: **Ultralight:** (*Ultra Léger*) Two 29.8 kW (40 hp) Zanzottera flat-twin engines.
Two-seat lightplane: (*Avion Biplace*) Two engines up to 74.6 kW (100 hp).
Four-seat lightplane: (*Avion Quadriplace*) Stretched version with two engines up to 74.6 kW (100 hp). Available from 2006.
CUSTOMERS: Seven sales and 80 options by June 2003.
COSTS: Kit €60,000 to €100,000; flyaway €100,000 to €150,000, according to version (2003).
DESIGN FEATURES: Unconventional 'four-wing, push-pull' design, featuring tapered, anhedral foreplane with downturned tips and sweptback (30°), dihedral mainplane with vertical fins at tips; plus flight control system lacking pilot's rudder pedals and incorporating 'glass cockpit'.
FLYING CONTROLS: Manual. Ailerons and flaps on mainplane, plus rudder in each fin; full-span elevators on foreplane. Rudders split as airbrakes.
STRUCTURE: Generally of glass fibre and carbon fibre.
LANDING GEAR: Tricycle type; fixed. Fuselage-mounted spring composites cantilever mainwheel legs. Mainwheels 4.00–6; steerable nosewheel, 11×4.00–5.
POWER PLANT: Two piston engines, power according to variant, in tractor/pusher arrangement. Fuel capacity 60 litres (15.9 US gallons; 13.2 Imp gallons) in ultralight; 160 litres (42.3 US gallons; 35.2 Imp gallons) in two- and four-seat versions.
ACCOMMODATION: Two or four persons, according to variant. Two centreline-hinged, upward-opening canopy/doors.
AVIONICS: *Comms:* Bendix/King KY 97 VHF.
Instrumentation: Single LCD.
EQUIPMENT: BRS-5 ballistic parachute.
DIMENSIONS, EXTERNAL (U: ultralight, B: two-seat, Q: four-seat):
Wing span: U, Q ... 8.90 m (29 ft 2½ in)

Aeronix Airelle's unconventional layout ***NEW***/0576222

B ... 7.00 m (22 ft 11½ in)
Length overall: U ... 5.93 m (19 ft 5½ in)
B ... 4.83 m (15 ft 10¼ in)
Q ... 5.80 m (19 ft 0¼ in)
DIMENSIONS, INTERNAL:
Cabin max width ... 1.14 m (3 ft 9 in)
AREAS:
Wings, gross: U, Q ... 15.00 m² (161.5 sq ft)
WEIGHTS AND LOADINGS (U, B, Q, as above):
Weight empty, equipped: U ... 250 kg (551 lb)
B ... 280 kg (617 lb)
Q ... 320 kg (705 lb)
Max T-O weight: U ... 450 kg (992 lb)
B ... 550 kg (1,212 lb)
Q ... 650 kg (1,433 lb)
Max wing loading: U ... 30.0 kg/m² (6.14 lb/sq ft)
B ... 36.7 kg/m² (7.51 lb/sq ft)
Q ... 43.3 kg/m² (8.88 lb/sq ft)
PERFORMANCE (estimated; U, B, Q, as above):
Max level speed: U ... 124 kt (230 km/h; 142 mph)
B ... 162 kt (300 km/h; 186 mph)
Q ... 151 kt (280 km/h; 174 mph)
Stalling speed: U ... 34 kt (62 km/h; 39 mph)
B, Q ... 50 kt (92 km/h; 58 mph)
Max rate of climb at S/L: U ... 600 m (1,968 ft)/min
B, Q ... 360 m (1,181 ft)/min
Range with max payload: U ... 378 n miles (700 km; 435 miles)
B ... 809 n miles (1,500 km; 932 miles)
Q ... 674 n miles (1,250 km; 776 miles)

UPDATED

AERO SERVICES

AERO SERVICES GUEPARD

F-12200 Toulonjac
Tel: (+33 5) 65 45 28 54
Fax: (+33 5) 65 45 24 64
e-mail: contact@aeroservices.fr
Web: http://www.aeroservices.fr
DIRECTOR: Jean-Daniel Roman

Aéro Services Guépard produces the Guépard and Guépy ultralights.

VERIFIED

AERO SERVICES GUEPARD AND GUEPY

TYPE: Side-by-side ultralight.
PROGRAMME: Certified by DGAC in ULM category, August 2000.
CURRENT VERSIONS: **Guépard:** Highest-powered version. *As described.* Flyaway or kit.
Guépy: Powered by 48.0 kW (64.4 hp) Rotax 582 or 59.7 kW (80 hp) Jabiru 2200. Metal wing with fabric covering and full-span flaperons. Weight empty 240 kg (529 lb) or 286 kg (630 lb) with optional extras. Flyaway or kit.
Guépy Club: Lowest-powered version, with 34.0 kW (45.6 hp) Rotax 503 or 48.0 kW (64.4 hp) Rotax 582. Weight empty (fabric-covered wing only) 210 kg (463 lb) or 286 kg (630 lb) with optional extras.
CUSTOMERS: More than 80 sold in France by 2003.
COSTS: Guépard, flyaway; fabric-covered wing: Rotax 582 €35,329, Rotax 912 €40,826, Rotax 912S €42,396; all-aluminium wing: Rotax 582 €40,040, Rotax 912 €46,865, Rotax 912S €48,678. Guépard kit (less engine) with fabric wing €19,261, with aluminium wing €24,823.
Guépy, flyaway: Rotax 582 €31,085, Jabiru €36,647; kit (less engine) €19,261. Guépy Club Kit €15,687. (All 2004.)
DESIGN FEATURES: Constant-chord, high wing with V strut bracing. Tailplane at base of sweptback fin.
FLYING CONTROLS: Conventional and manual. Flaps.
STRUCTURE: Airframe of 4130 aluminium tube, with Dacron covering; wings 2017 and 2024 aluminium with Dacron or aluminium covering.
LANDING GEAR: Fixed; option of nosewheel or tailwheel type. Brakes.
POWER PLANT: One piston engine: 48.0 kW (64.4 hp) Rotax 582, 73.5 kW (98.6 hp) Rotax 912 or 84.5 kW (113.3 hp) Rotax 912S. Two-blade (optional three-blade) propeller. Standard fuel 65 litres (17.2 US gallons; 14.3 Imp gallons).

Aéro Services Guépy ultralight ***NEW***/0561680

DIMENSIONS, EXTERNAL:
Wing span ... 9.60 m (31 ft 6 in)
AREAS:
Wings, gross ... 14.60 m² (157.2 sq ft)
WEIGHTS AND LOADINGS (Rotax 912S):
Weight empty: basic ... 262 kg (578 lb)
with all options ... 286 kg (631 lb)
Max T-O weight ... 472.5 kg (1,041 lb)
PERFORMANCE (Rotax 912S):
Max level speed ... 103 kt (190 km/h; 118 mph)
Normal cruising speed ... 97 kt (180 km/h; 112 mph)
Stalling speed, flaps down ... 30 kt (55 km/h; 35 mph)
T-O run ... 50 m (165 ft)
Landing run ... 70 m (230 ft)

UPDATED

AMEUR

AMEUR AVIATION SA

2 rue de la Serre, ZAC la Novialle, F-63670 La Roche Blanche
Tel: (+33 4) 73 78 62 24
Fax: (+33 4) 73 78 61 67
e-mail: ameur.aviation@wanadoo.fr
Web: http://www.slicom.fr/sc2a
PRESIDENT AND CHIEF DESIGNER: Boussad Ameur

Ameur Aviation (known until 2001 as Ameur Aviation Technologie - AAT) originated on the island of Corsica, but has moved to mainland France.

UPDATED

AMEUR ALTANIA

TYPE: Side-by-side ultralight/kitbuilt.

PROGRAMME: Prototype (F-WARA, later F-PARA), then known as AAT Balbuzard, first flew 9 July 1995, powered by 59.6 kW (79.9 hp) Rotax 912. Second prototype (also F-PARA, but different airframe) followed in July 1996, fitted with 88.0 kW (118 hp) Textron Lycoming O-235 and three-blade, fixed-pitch propeller; also featured lower positioned wing of increased span and with winglets; plus redesigned landing gear of greater height; flew 200 hours, achieving 180 kt (333 km/h; 207 mph) at 550 m (1,800 ft), but glass fibre construction judged too heavy; refitted with Airplast constant-speed propeller; lost in 1997, following propeller driveshaft break in flight.

Third prototype (F-PTCD) produced in July 1998; name changed to AAT Baljims 1A (acronym of designer, test pilot and family members) after Eurocopter found to have prior claim to Balbuzard name; flew 150 hours. Fourth prototype was 15 m (49 ft) motor glider (F-PBAL), which was discontinued after 30 hours of experimental flying.

Further redesign, most notably in pre-impregnated carbon fibre for weight reduction, plus double-slotted flaps and fail-safe wing structure, resulted in Ameur Altania prototype (registration F-WSTA reserved), scheduled to have flown in mid-2001, although appears not to have been proceeded with and replaced with fifth aircraft (F-WWMU), which first flew December 2001 and has since been fitted with upturned wingtips. Altania is initially in ultralight category, with first deliveries due in fourth quarter of 2002, while JAR-VLA and FAR Pt 23 versions will follow.

CURRENT VERSIONS: **Balbuzard/Baljims:** Earlier versions.

Altania RG 80 UL: Retractable gear, 80 hp ultralight; *as described.*

Altania Vista: Surveillance version, suitable for security/law enforcement, environmental protection and powerline/pipeline inspection; promoted by Thales Aerosurveillance; mockup shown at Paris in June 2001. Day/night sensor turret in starboard underside, with controls and operating mechanism replacing passenger's seat; datalink to ground station. Observation speed 60 to 90 kt (111 to 167 km/h; 69 to 104 mph); VNE 180 kt (333 km/h; 207 mph).

Altania Saphir : (formerly Altania 4) Four-seat version, with same fuselage, proposed for Experimental category; 89.5 kW (120 hp) Jabiru 3300 flat-six engine; available only as kit.

UAV MALE: Unmanned drone version, under development.

Iguidar: Two-seat trainer version, under development.

Altajet B1: *Described separately.*

COSTS: Kit €67,500; flyaway €89,500; both plus VAT (2001).

DESIGN FEATURES: Aerodynamically efficient configuration with 'buried' engine, pusher propeller at extreme rear and V tail with two ventral strakes. Air intake above cockpit. Driveshaft to propeller has universal joints and restraining rings in the event of joint failure.

FLYING CONTROLS: Manual. Frise ailerons, plus combined elevators and rudders. Electrically actuated double-slotted flaps. Optional electric trim.

STRUCTURE: Pre-impregnated carbon/epoxy and honeycomb, including single-piece fuselage and single-spar, fail-safe wing.

LANDING GEAR: Tricycle type; retractable. Mainwheels retract inward and rearward; nosewheel rearward; mainwheel size 5.00-5; nosewheel 11×4.00-5. Hydraulic brakes on mainwheels.

POWER PLANT: One 59.7 kW (80 hp) Jabiru 2200 flat-four driving a three-blade propeller; fuel in fuselage tank, total 85 litres (22.5 US gallons; 18.7 Imp gallons). Certified versions have two wing tanks, increasing total capacity to 328 litres (86.6 US gallons; 72.2 Imp gallons). Four-seat version fuel capacity is 120 litres (31.7 US gallons; 26.4 Imp gallons).

ACCOMMODATION: Pilot and passenger under one-piece blown Perspex canopy. Baggage locker.

AVIONICS: *Comms:* Transponder standard.
Flight: GPS standard.
Instrumentation: Optional 'glass cockpit'.

EQUIPMENT: Ballistic parachute standard.

DIMENSIONS, EXTERNAL:
Wing span ... 7.70 m (25 ft 3 in)
Wing aspect ratio ... 9.5
Length overall ... 6.00 m (19 ft 8¼ in)
Height overall ... 2.30 m (7 ft 6½ in)
Tailplane span ... 2.60 m (8 ft 6¼ in)
Wheel track ... 1.80 m (5 ft 11 in)
Wheelbase ... 2.35 m (7 ft 8½ in)

DIMENSIONS, INTERNAL:
Cabin: Length ... 2.70 m (8 ft 10¼ in)
Max width ... 1.22 m (4 ft 0 in)

AREAS:
Wings, gross ... 6.22 m² (67.0 sq ft)

WEIGHTS AND LOADINGS:
Weight empty: RG ... 265 kg (584 lb)
Saphir ... 315 kg (694 lb)
Max T-O weight: RG ... 472 kg (1,040 lb)
Saphir ... 675 kg (1,448 lb)
Max wing loading: RG ... 75.9 kg/m² (15.54 lb/sq ft)
Saphir ... 108.9 kg/m² (22.56 lb/sq ft)
Max power loading: RG ... 7.91 kg/kW (13.00 lb/hp)
Saphir ... 7.54 kg/kW (12.39 lb/hp)

PERFORMANCE, POWERED:
Never-exceed speed (VNE):
RG ... 178 kt (330 km/h; 205 mph)
Saphir ... 216 kt (400 km/h; 248 mph)
Max level speed: RG ... 155 kt (287 km/h; 178 mph)

Prototype Ameur Altania at Paris in 2003 *(Paul Jackson)* ***NEW***/0576235

Saphir ... 179 kt (332 km/h; 206 mph)
Max cruising speed at 2,500 m (5,520 ft):
RG ... 140 kt (260 km/h; 162 mph)
Saphir ... 175 kt (325 km/h; 202 mph)
Stalling speed: RG ... 36 kt (65 km/h; 41 mph)
Saphir ... 49 kt (90 km/h; 56 mph)
Max rate of climb at S/L: RG ... 420 m (1,378 ft)/min
Saphir ... 360 m (1,180 ft)/min
Service ceiling ... 5,500 m (18,040 ft)
T-O run: RG ... 150 m (495 ft)
Saphir ... 270 m (886 ft)
Landing run: RG ... 170 m (560 ft)
Saphir ... 350 m (1,148 ft)
Range: pilot only ... 1,443 n miles (2,673 km; 1,660 miles)
two persons ... 866 n miles (1,605 km; 997 miles)
Endurance: pilot only ... 9 h 25 min
two persons ... 3 h 35 min
Endurance: RG ... 5 h 30 min
Saphir ... 5 h
g limits ... +6/-4

PERFORMANCE, UNPOWERED:
Best glide ratio: RG ... 18.5
Stalling speed: RG ... 36 kt (65 km/h; 41 mph)
Saphir ... 49 kt (90 km/h; 56 mph)

UPDATED

AMEUR ALTAJET

TYPE: Light business jet.

PROGRAMME: Jet version of Ameur Altania; announced at Paris in June 2001, when known as Fanjet. Development contract signed 19 June 2001; first flight had been due 2003, but delayed to 2004. However, mockup revealed at Paris, 15-22 June 2003. Building of prototype was to have begun in September 2003.

CURRENT VERSION: **Altajet B1:** Four-seat version, *as described.*

Altania B2: Proposed two-seat version with maximum level speed of 270 kt (500 km/h; 311 mph).

Description generally as for Altania, except the following.

DESIGN FEATURES: Envisaged with single turbofan partly buried in lower portion of tailcone; curved intake duct drawing air from below fuselage, adjacent to wing trailing-edge. By 2004, had evolved into twin-engined aircraft, 335 kg (739 lb) heavier, with wingtip tanks and single-piece canopy replacing windscreen and two doors. Later variant envisaged with more powerful engines and max cruising speed of 400 kt (741 km/h; 460 mph). Both versions to be certified to FAR Pt 23.

COSTS: US$500,000 flyaway (2003 estimate).

POWER PLANT: Two 3.43 kN (770 lb st) Williams EJ22 turbofans.

ACCOMMODATION: Four persons, plus baggage, beneath single-piece canopy. Pressurised; S/L atmosphere to 6,550 m (21,000 ft); 2,440 m (8,000 ft) environment up to max altitude.

SYSTEMS: De-icing.

AVIONICS: IFR. 'Glass cockpit'.

DIMENSIONS, INTERNAL:
Cabin: Length ... 3.00 m (9 ft 10 in)
Max width ... 1.22 m (4 ft 0 in)
Max height ... 1.15 m (3 ft 9¼ in)
Baggage volume ... 0.90 m³ (31.8 cu ft)

WEIGHTS AND LOADINGS (estimated):
Max T-O weight ... 1,500 kg (3,306 lb)
Max power loading ... 219 kg/kN (2.15 lb/lb st)

PERFORMANCE (estimated):
Max cruising speed ... 300 kt (556 km/h; 345 mph)
Endurance ... 3 h

UPDATED

Ameur Altajet mockup *(Paul Jackson)* ***NEW***/0576234

APEX

APEX INTERNATIONAL

3 rue Troyon, F-75017 Paris
Tel: (+33 1) 40 72 61 10
Fax: (+33 1) 40 72 62 98
Web: http://www.capaviation.com
e-mail: info@apex-aircraft.com
CHAIRMAN AND CEO: Guy Pellissier
SALES MANAGER: Pierre Pelletier

Three French light aircraft manufacturers are included in the holding company Apex International (formerly known as Aéronautique Service); products of BUL, CAP and Robin are described where appropriate; related companies are detailed here. In total, the group had 200 employees in 2001.

In late 2001, the company announced that it was embarking on restructuring its component parts and building a new customer service centre at Darois, due to be operational in April 2002.

However, Apex entered voluntary receivership in September 2002 and underwent financial restructuring while aircraft production at its member companies was slowed or stopped; on 30 May 2003 a French court approved an extension of receivership until July, by which time Apex hoped to have completed restructuring, including downsizing of its workforce and focusing on FAA certification for CAP and Robin aircraft, although work on the CAP 222 will continue when finance allows; receivership lifted on 25 August 2003. Production will be at 25 to 30 per year, and in the year following imposition of receivership Apex delivered 23 aircraft of various types.

Constructions Aéronautiques de Bourgogne (CAB): Construction of BUL Zùlù, CAP 10 and (formerly) CAP 222 at Darois, and Robin aircraft at Bernay. President: Patrice Serignac.

CAP Industries: Construction of CAP 232 and after-sales service of all CAP aircraft (including former Mudry CAP designs) at Bernay. Director-General: Eric Willaert.

Aviation Service Center (ASCO): After-sales service and support. Director-General: Ludovic Paul.

Group administrative services are provided by **Aeronautique Financements & Services**.

UPDATED

ASSOCIATION ZEPHYR

ASSOCIATION ZEPHYR

34 rue Victor Hugo, F-92300 Levallois-Perret
Tel/Fax: (+33 1) 41 06 69 88
e-mail: assoc.zephyr@mail.meloo.com
Web: http://www.zephyr.fr.fm
PROJECT MANAGER: Guillaume Bullin

Association Zéphyr was founded in 1994 by fourth-year student engineers of the Ecole Supérieure des Techniques Aérotechniques et de Construction Automobile (ESTACA), who have designed the Alizé light aircraft, intended for amateur construction. Sponsorship and support is being provided by Dassault, EVRA and the RSA.

In 2002–03, students at ESTACA built a replica of the Wright Flyer for exhibition at the Paris Air Show of June 2003.

UPDATED

ASSOCIATION ZEPHYR ALIZE

English name: Tradewind

TYPE: Two-seat lightplane.

PROGRAMME: Design started 1996; construction of prototype began January 1998; first flight scheduled for 2000, but by mid-2001 only empennage had been completed and wings were under construction. Assembly undertaken at Creil aerodrome.

DESIGN FEATURES: Low wing; swept fin and rudder. Wing section NACA NLF 0215F. Drawings prepared by Autocad and CATIA.

STRUCTURE: All-wood airframe structure with glass fibre skin.

LANDING GEAR: Retractable tricycle type.

POWER PLANT: One 119 kW (160 hp) Textron Lycoming IO-320-B1E, driving an Evra propeller in prototype, with future provision for SMA MR 200 of 149 kW (200 hp). Fuel in two tanks, combined capacity 140 litres (37.0 US gallons; 30.8 Imp gallons).

ACCOMMODATION: Two persons, side by side, under bubble canopy.

DIMENSIONS, EXTERNAL:
Wing span (without wingtips) ... 7.53 m (25 ft 8½ in)
Wing aspect ratio ... 7.0
Length overall ... 6.40 m (21 ft 0 in)

Model of the Association Zéphyr Alizé *(Paul Jackson)* 0131729

DIMENSIONS, INTERNAL:
Cabin max width ... 1.10 m (3 ft 7¼ in)
Baggage volume ... 0.04 m^3 (1.41 cu ft)

AREAS:
Wings, gross ... 8.10 m^2 (87.2 sq ft)

WEIGHTS AND LOADINGS:
Weight empty, equipped ... 370 kg (816 lb)
Max T-O weight ... 645 kg (1,422 lb)
Max wing loading ... 79.6 kg/m^2 (16.31 lb/sq ft)
Max power loading (prototype) ... 5.41 kg/kW (8.89 lb/hp)

PERFORMANCE (estimated):
Cruising speed ... 173 kt (320 km/h; 199 mph)
Landing speed ... 59 kt (110 km/h; 68 mph)
Max rate of climb at S/L ... 390 m (1,280 ft)/min
Range ... 648 n miles (1,200 km; 745 miles)
g limits ... +4.4/–2.2

VERIFIED

BEST OFF

BEST OFF

Aérodrome, BP 943, F-82009 Montauban Cedex
Tel: (+33 5) 63 67 97 15
Fax: (+33 5) 63 67 97 16
e-mail: skyranger@free.fr
Web: http://www.skyranger.net

This company holds rights to the Sky Ranger ultralight, which is built under licence in Ukraine by Aeros and distributed by several dealerships throughout the world.

VERIFIED

BEST OFF SKY RANGER

TYPE: Side-by-side ultralight kitbuilt.

PROGRAMME: Designed in Toulouse, France, in 1990 by Philippe Prevot; winner of World FAI two-seat ultralight championships in 1996, 1999 and 2003, plus silver medal in European

Best Off Sky Ranger *(Paul Jackson)* ***NEW***/0576230

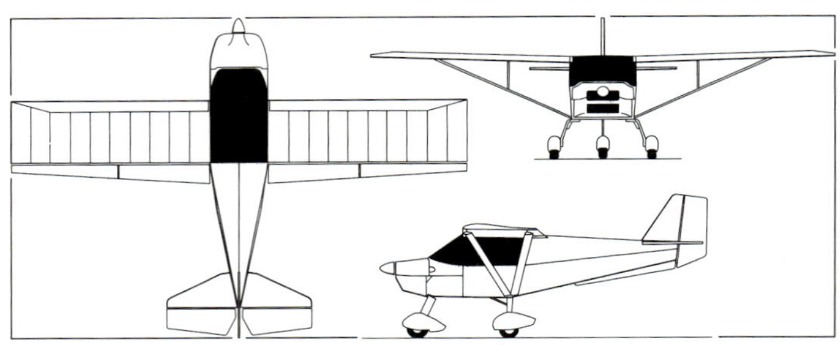

Best Off Sky Ranger side-by-side ultralight *(Paul Jackson)* ***NEW***/0583030

ultralight championship in 2002. After production of 200 by Synairgie in France, between 1992 and 1997, manufacture was transferred to Aeros in 1998; current version is Sky Ranger 2, offered in two versions: standard V.Fun and BCAR-compliant V.Max.

CUSTOMERS: Over 600 (including 200 of first series) delivered by early 2004.

COSTS: Kit €10,980 less engine; with engine €12,000 extra (2004).

DESIGN FEATURES: Intended to combine high performance with simple construction. Delivered as fast-build kit, assembled by two persons in two weeks. Wings foldable for storage by two persons in 20 minutes.

High wing with V-strut bracing and auxiliary struts; wire-braced empennage. Purpose-designed wing section. Directional stability enhanced from 2003 onwards by addition of large underfin, although this quickly changed to revised design comprising smaller underfin and flush, downward extension of rudder. US also having downward extension to rudder.

FLYING CONTROLS: Conventional and manual; cable operated. Three-position flaps; pushrod actuated.

STRUCTURE: Employs only straight 2024 aluminium tubing; no bends or welds; fabric covering, except for composites engine cowling and forward fuselage. No composites in structural elements. Other parts of stainless steel and plated mild steel. Optional composites wingtips.

LANDING GEAR: Tricycle type; fixed. Spring steel cantilever mainwheel legs with drag bracing tubes attached to lower longerons. All tyres 4.00–6. Cantilever-sprung main legs. Disc brakes. Optional wheel fairings and floats.

POWER PLANT: One 59.6 kW (79.9 hp) Rotax 912 UL with 1:2.27 gearing in highest-powered version. Rotax 503, 582, 912 UL-S (France only) and Jabiru 2200 are alternatives. Three-blade, ground-adjustable pitch propeller. Fuel capacity 50 litres (13.2 US gallons; 11.0 Imp gallons) in tank behind seats.

DIMENSIONS, EXTERNAL:

Wing span	9.50 m (31 ft 2 in)
Length overall	5.50 m (18 ft 0½ in)
Height overall	2.00 m (6 ft 6¾ in)

AREAS:

Wings, gross	14.10 m^2 (151.8 sq ft)

WEIGHTS AND LOADINGS (Rotax 912 engine):

Weight empty	250 kg (551 lb)
Max T-O weight	560 kg (1,234 lb)

PERFORMANCE (Rotax 912 engine):

Never-exceed speed (VNE)	105 kt (195 km/h; 121 mph)
Max cruising speed	81 kt (150 km/h; 93 mph)
Stalling speed, flaps down	35 kt (64 km/h; 40 mph)
Max rate of climb at S/L	274 m (900 ft)/min
T-O run	90 m (295 ft)
Landing run	120 m (394 ft)
g limits (ultimate)	+6/–4

UPDATED

BUL

BUL AVIATION

1 bis route de Troyes, F-21121 Darois
Tel: (+33 3) 80 44 20 75
Fax: (+33 3) 80 44 20 69 or 80 35 60 80
e-mail: bulaero@wanadoo.fr
PRESIDENT: Philippe Corne
CONTACT: David Morin

BUL (Bourgogne Ultra Léger) is one of three light aircraft manufacturers owned by Apex International (formerly Aéronautique Service), its specialist area being ultralights. Details of the Zùlù, its first product, were announced in 1998. Development has been protracted, but certification of a completely redesigned version was expected in 2002 or early 2003; Company's receivership in September 2002 has delayed this, but it is believed to still be active.

UPDATED

BUL ZÙLÙ

TYPE: Side-by-side ultralight.

PROGRAMME: Design began August 1996 and programme launched two months later; prototype (WBA-01) construction began November 1996; first flight 15 March 1998; French certification achieved July 1998. Assembly of first production aircraft began October 1998 for first flight in August 1999; however, flight testing of prototype had by then demonstrated insufficient margin for useful load within ultralight weight limits.

The design was then completely revised and a new Zùlù made its first flight, unannounced, in late 2001; certification of ultralight version intended to be complete by mid-2002 and of heavier version (to meet VLA flight conditions) in mid-2003. Neither target achieved due to company's financial position.

CUSTOMERS: First order placed 5 October 1998; four on order by December 1998.

No data have been released for current Zùlù.

UPDATED

CAP

CAP AVIATION

9 rue de l'Aviation, F-21121 Dijon-Darois
Tel:(+33 3) 80 35 65 10
Fax: (+33 3) 80 35 65 15
*e-mail:*info@capaviation.com
Web: http://www.capaviation.com
PRESIDENT, CEO AND HEAD OF DESIGN: Dominique Roland
SALES MANAGER, EUROPE: Alain Ruelloux

CAP Aviation (known as Akrotech Europe until January 1999) is owned by holding company Apex International (formerly known as Aéronautique Service — AES) (which see), also responsible for the Robin and BUL concerns. It formed in 1997, initially to market the Giles G-202 two-seat aerobatic aircraft, subsequently marketed in factory-completed form as the CAP 222.

Apex took over production of the Mudry series of aerobatic aircraft on 12 May 1997 following that company's bankruptcy. All aerobatic aircraft produced by Apex are marketed by CAP Aviation: the CAP 232 built at Bernay and CAP 10 and CAP 222 at Darois.

CAP Aviation's parent company, Apex Aircraft, emerged from voluntary receivership on 25 August 2003.

UPDATED

CAP 10

TYPE: Aerobatic two-seat sportplane.

PROGRAMME: Derived from Piel CP-30/CP-301 Emeraude, first flown 19 June 1954 and built by amateur constructors as well as by 11 commercial plants, under various names. Super Emeraude modified with taller fin and 119 kW (160 hp) engine as CP-100 prototype, flown 13 August 1966; with broader chord rudder and 134 kW (180 hp) engine, became prototype CAP 10 B, first flight (F-WOPX) 22 August 1968; further three pre-series aircraft; certified 4 September 1970; FAA certification for day and night VFR on 13 June 1974. Produced by Mudry until 1996.

Production transferred to Dijon, beginning with CAP 10 C in 2001.

CURRENT VERSIONS: **CAP 10 B:** Previous version; now out of production, but can be upgraded to CAP 10 C in one day's work.

CAP 10 C: Announced in early 1999; features carbon fibre-reinforced wing spar for reduced weight and increased speed and roll rate.

Improvements effected in CAP 10 C include substitution of pushrods for aileron control, replacing cables; larger (25 per cent) ailerons of increased chord with repositioned axis for faster roll rate, compensated by two spade-type servos per side; electric actuation of flaps; electric fuel gauges and flush filler caps; optional constant-speed propeller (in prospect); upgraded instrumentation with four standard options; redesigned landing gear to improve ground handling; local airframe reinforcement and landing gear attachments; and revised canopy design to reduce cockpit noise levels. Roll rate improved by 30 per cent. New wing available for retrofitting to existing CAP 10 Bs and is supplied complete with landing gear.

Production CAP 10 C No. 302 ***(Paul Jackson)*** ***NEW***/0572271

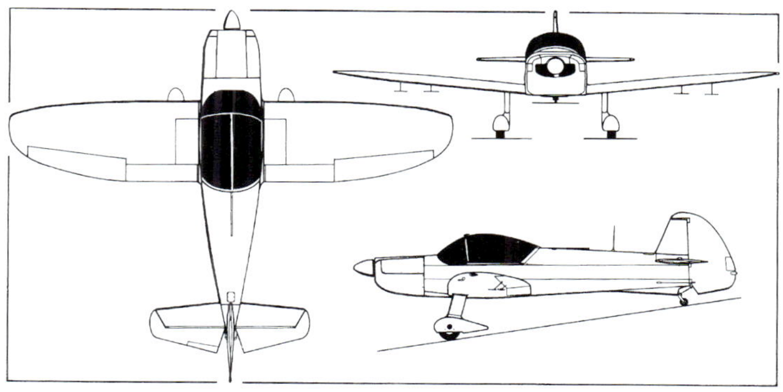

CAP 10 C, showing new ailerons ***(Paul Jackson)*** 0137207

Static test of wing undertaken 27 April 2000; prototype first flight (CAP 10 B HB-SAX retrofitted with new wing) 5 March 2001; first 'true' CAP 10 C (F-WWNN, c/n 300) first flown 15 May 2001 and following initial flight testing was delivered to CEV during July. DGAC certification achieved in early 2002. Reregistered F-GYKA on completion of trials. CAP 10Cs are registered under CAP 10B type certificate. FAA certification achieved 2 July 2004.

Description refers to CAP 10 C unless otherwise stated.

CUSTOMERS: Total 284 of earlier versions (including prototypes) built to mid-1999, of which 279 completed by Mudry before 1996 bankruptcy. Most recent five produced by Akrotech Europe in temporary factory at Bernay and include exports to UK (three) and USA (one); last, in 1999, was G-LORN (c/n 282) to UK. Production then paused, pending launch of CAP 10 C and transfer of production to Dijon. Previous major users, French Air Force (56) and Mexican Air Force (20) began disposals in 1994–95. Two more recent aircraft operated by 208 Squadron of South Korean Air Force; French Navy has eight; at end 2000, around 60 were still in military service and a further 180 in civil use.

First production aircraft (c/n 301; G-CPXC) registered in December 2001 to Cole Aviation in UK; 11th registered in January 2004; customers include SEFA of France (five). By mid-2004 an additional 45 sets of CAP 10 C wings had been sold to existing CAP 10 B operators.

COSTS: €148,966 to €170,138 depending on equipment level (2002).

DESIGN FEATURES: Simple, sporting lightplane with low, basically elliptical wing, mid-mounted tailplane and curved rudder.

Wing section NACA 23012; dihedral 5°; incidence 0°.

FLYING CONTROLS: Conventional and manual. Electric trim tabs on port elevator; balance tab on rudder; tailplane incidence adjustable on ground; electric three-position (0, 15, 40°) flaps. Control surface deflections: ailerons +25/–15°, elevators ±25°; rudder ±18°.

STRUCTURE: Main spar of carbon fibre caps and birch ply webs; wooden ribs; birch ply skin; carbon fibre wingtips; birch ply ailerons, flaps, fin/rudder and tailplane/elevator. Wooden fuselage, with okoumé ply upper decking and polyester fabric covering elsewhere.

LANDING GEAR: Tailwheel type; fixed. Wing-mounted, trousered, cantilever main legs of light alloy, with integral ERAM type 9 270 C oleo-pneumatic shock-absorbers. Single wheel on each main unit, tyre size 380×150/15×6.00–5. Solid tailwheel tyre, size 6×2.00. Tailwheel is steerable by rudder linkage but can be disengaged for ground manoeuvring. Hydraulically actuated mainwheel disc brakes (controllable from port seat) and parking brake. Dual toe brakes standard. Streamline fairings on mainwheels and legs.

POWER PLANT: One 134 kW (180 hp) Textron Lycoming AEIO-360-B2F flat-four engine, driving a Hoffmann HO29HM-189–170 two-blade, fixed-pitch wooden propeller; optional Evra 3.180-170-H5.F. Standard fuel tank aft of engine fireproof bulkhead, usable capacity 72 litres (19.0 US gallons; 15.8 Imp gallons); total 75 litres (19.8 US gallons; 16.5 Imp gallons). Auxiliary tank, usable capacity 78 litres (20.6 US gallons; 17.2 Imp gallons), total capacity 79 litres (20.8 US gallons; 17.3 Imp gallons), beneath baggage compartment, must be empty for aerobatics. Inverted fuel and oil (Aviat/Christen) systems permit continuous inverted flight.

ACCOMMODATION: Side-by-side adjustable seats for two persons, with provision for back parachutes, under rearward-sliding and jettisonable moulded transparent canopy. Five-point aerobatic shoulder harness standard. Leather upholstery optional. Space for 20 kg (44 lb) of baggage aft of seats in training and touring models.

SYSTEMS: Electrical system includes Delco-Rémy 40 A engine-driven alternator and STECO ET24 Ni/Cd battery.

AVIONICS: Bendix/King and Garmin suites optional.

Comms: Optional KY97A VHF com; GMA 340 audio panel and Sigtronics SPA400 intercom; GTX 320 or GTX 327 transponder; ACK 30 encoder; EBC 102A ELT.

Flight: Garmin GNC 250 XL, GNS 420 or GNS 430 GPS; GI 106A VOR/ILS; Sandel SN3308 EHSI.

Instrumentation: ASI (kt and mph); altimeter; VSI; recording *g*-meter; positive/negative slip indicator; compass; tachometer; fuel gauges; oil pressure/temperature, EGT and CHT gauges; and manifold pressure/fuel flow gauge, all standard. Artificial horizon, turn co-ordinator, and directional gyro optional or standard according to selected equipment fit level.

Equipment: Panel light, cockpit light, landing light, navigation and strobe lights, external power socket, aerobatic sighting device, and canopy cover, all optional.

DIMENSIONS, EXTERNAL:
Wing span ..... 8.08 m (26 ft 6 in)
Wing aspect ratio 6.0
Length overall ..... 7.16 m (23 ft 6 in)
Height overall ..... 2.55 m (8 ft 4½ in)
Tailplane span ..... 2.90 m (9 ft 6 in)
Wheelbase ..... 4.30 m (14 ft 1¼ in)
Wheel track ..... 2.06 m (6 ft 9 in)
Propeller diameter ..... 1.90 m (6 ft 2¾ in)

DIMENSIONS, INTERNAL:
Cabin: Length ..... 1.05 m (3 ft 5½ in)
Max width ..... 1.05 m (3 ft 5½ in)
Max height ..... 0.98 m (3 ft 2½ in)

AREAS:
Wings, gross ..... 10.85 m^2 (116.8 sq ft)
Vertical tail surfaces (total) ..... 1.32 m^2 (14.25 sq ft)
Horizontal tail surfaces (total) ..... 1.86 m^2 (20.02 sq ft)

WEIGHTS AND LOADINGS (A: Aerobatic, U: Utility):
Weight empty, equipped ..... 540 kg (1,190 lb)
Fuel weight: A ..... 54 kg (119 lb)
U ..... 108 kg (238 lb)
Max T-O weight: A ..... 780 kg (1,719 lb)
U ..... 830 kg (1,830 lb)
Max wing loading: A ..... 71.9 kg/m^2 (14.72 lb/sq ft)
U ..... 76.5 kg/m^2 (15.67 lb/sq ft)
Max power loading: A ..... 5.82 kg/kW (9.56 lb/hp)
U ..... 6.19 kg/kW (10.16 lb/hp)

PERFORMANCE:
Never-exceed speed (VNE) ..... 183 kt (340 km/h; 211 mph)
Max level speed at S/L and cruising speed at 75% power at FL80 ..... 148 kt (274 km/h; 170 mph)
Stalling speed: clean ..... 53 kt (99 km/h; 61 mph)
flaps down ..... 46 kt (85 km/h; 53 mph) IAS
Max rate of climb at S/L ..... 488 m (1,600 ft)/min
Service ceiling ..... 5,000 m (16,400 ft)
T-O run ..... 350 m (1,150 ft)
T-O to 15 m (50 ft) ..... 395 m (1,300 ft)
Landing from 15 m (50 ft) ..... 600 m (1,970 ft)
Landing run ..... 360 m (1,185 ft)
Range with max fuel ..... 521 n miles (965 km; 600 miles)
g limits: A ..... +6/−4.5
U ..... +4.4/−1.8

UPDATED

CAP 232

TYPE: Aerobatic single-seat sportplane.

PROGRAMME: Prototype (F-WZCH) first flew 7 July 1994; French certification and first sales March 1995. First place in 2000 World Aerobatic Championship in France was taken by Eric Vazeille in CAP 232 F-GKDX; Catherine Maunoury was fifth (women's first) in F-GXCM.

CUSTOMERS: Total 44 built by late 2004, including prototype, one destroyed before first flight and one crashed on test. Mainly to French civil register (19) and four to French Air Force, but others sold to Australia (one, 1996 and USA eight). Others employed in Switzerland and UK, but retaining French registry. Moroccan Air Force placed order for nine in mid-2000 to re-equip Marche Verte aerobatic team; deliveries between early 2001 and 2004.

COSTS: US$250,000 (2001).

DESIGN FEATURES: Optimised for world-class aerobatic competition. Similar to, and improvement on, CAP 231 EX. Strong, carbon fibre wing of broad root chord, sharply tapered for high roll rates; tapered tail surfaces.

Moroccan Air Force CAP 232 ***(Jean-Louis Gaynecoetche)*** ***NEW***/0572158

David Ellison's CAP 232 aerobatic sportplane ***(Paul Jackson)*** ***NEW***/0572269

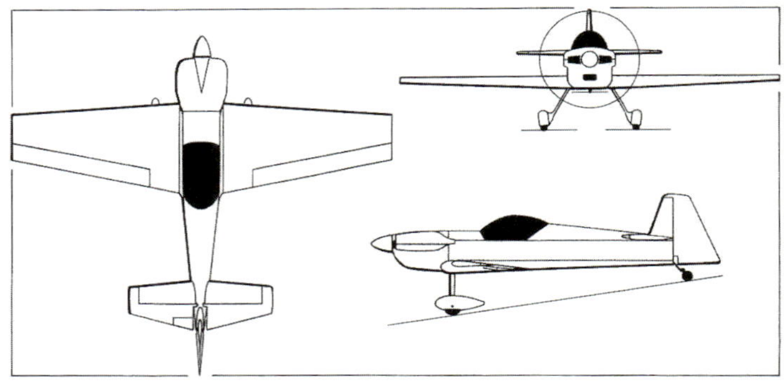

CAP Aviation CAP 232 ***(Paul Jackson)*** 0507681

CAP 232 ***(Paul Jackson)*** ***NEW***/0110946

FLYING CONTROLS: Conventional and manual. Electrically actuated elevator tab; elevator servo tab to reduce stick forces. Almost full-span ailerons for high roll rates.

STRUCTURE: Two-spar carbon fibre wings; 10 ribs.

LANDING GEAR: Fixed tailwheel type; fuselage-mounted cantilever mainwheel legs with wheel fairings. Mainwheel tyres 5.00–5.

POWER PLANT: One 224 kW (300 hp) Textron Lycoming AEIO-540-L1B5D flat-six engine, driving a variable-pitch (hydraulic) MT-Propeller MTV-14-BC-C190-17 four-blade propeller. Total fuel capacity 189.3 litres (50.0 US gallons; 41.6 Imp gallons) in one fuselage tank and two wing tanks. Fuel and oil system designed for prolonged inverted flight.

ACCOMMODATION: Single seat under rear-hinged canopy. Baggage space behind pilot.

SYSTEMS: Electrical system includes engine-driven alternator and 12 V 60 Ah battery.

AVIONICS: Customer specified.

Comms: Radio and transponder optional.

Flight: GPS optional.

EQUIPMENT: Sighting frame can be attached to wingtip for judging exact verticals in competition aerobatics.

DIMENSIONS, EXTERNAL:
Wing span ..... 7.39 m (24 ft 3 in)
Wing chord: at root ..... 1.83 m (6 ft 0 in)
at tip ..... 0.91 m (3 ft 0 in)
Wing aspect ratio 5.4
Length overall ..... 6.76 m (22 ft 2 in)
Tailplane span ..... 2.74 m (9 ft 0 in)
Wheel track ..... 1.75 m (5 ft 9 in)
Propeller diameter ..... 1.90 m (6 ft 2¾ in)
Propeller ground clearance ..... 0.30 m (1 ft 0 in)

AREAS:
Wings, gross ..... 10.15 m^2 (109.3 sq ft)
Ailerons (total) ..... 1.00 m^2 (10.76 sq ft)
Fin ..... 0.55 m^2 (5.92 sq ft)
Rudder ..... 0.77 m^2 (8.29 sq ft)
Tailplane ..... 1.02 m^2 (10.99 sq ft)
Elevators ..... 1.11 m^2 (11.93 sq ft)

WEIGHTS AND LOADINGS (A: Aerobatic, N: Normal category):
Weight empty ..... 585 kg (1,290 lb)
Max payload ..... 227 kg (500 lb)
Fuel weight ..... 123 kg (270 lb)
Max T-O and landing weight: N ..... 730 kg (1,610 lb)[1]
A ..... 821 kg (1,810 lb)[1]
Max wing loading ..... 80.9 kg/m^2 (16.56 lb/sq ft)

Max power loading .... 3.67 kg/kW (6.03 lb/hp)

[1] *As quoted by manufacturer*

PERFORMANCE:

Never-exceed speed (VNE)	219 kt (405 km/h; 252 mph)
Max level speed	189 kt (349 km/h; 217 mph)
Max cruising speed at 75% power	180 kt (333 km/h; 207 mph)
Manoeuvre speed (VA)	170 kt (314 km/h; 195 mph)
Econ cruising speed	145 kt (269 km/h; 167 mph)
Stalling speed, power off	56 kt (104 km/h; 65 mph)
Max rate of climb at S/L	1,003 m (3,290 ft)/min
Rate of roll at manoeuvre speed	420°/s
Service ceiling	4,575 m (15,000 ft)
T-O run	150 m (495 ft)
T-O to 15 m (50 ft)	180 m (595 ft)
Landing from 15 m (50 ft)	450 m (1,480 ft)
Range with max fuel, 45% power	650 n miles (1,203 km; 748 miles)
g limits	±10

UPDATED

DASSAULT

DASSAULT AVIATION

9 Rond-Point Champs Elysées-Marcel Dassault, F-75008 Paris
Tel: (+33 1) 53 76 93 00
Fax: (+33 1) 53 76 93 20
ADMINISTRATION AND COMMUNICATIONS OFFICE: 78 quai Marcel Dassault, Cedex 300, F-92552 St Cloud Cedex
Tel: (+33 1) 47 11 86 90
Fax: (+33 1) 47 11 87 40
Web: http://www.dassault-aviation.fr
WORKS:
F-92552 St Cloud, F-95100 Argenteuil,
F-33127 Martignas-sur-Jalles,
F-33701 Bordeaux-Mérignac,
F-33260 La Teste (Cazaux), F-64205 Biarritz,
F-13804 Istres, F-74370 Argonay,
F-59472 Lille-Seclin, F-86580 Biard (Poitiers)
HONORARY CHAIRMAN: Serge Dassault
CHAIRMAN AND CEO: Charles Edelstenne
VICE-CHAIRMAN: Bruno Revellin-Falcoz
SENIOR EXECUTIVE VICE-PRESIDENT, OPERATIONS: Christian Decaix
SENIOR EXECUTIVE VICE-PRESIDENT, INTERNATIONAL: Bruno Cotté
VICE-PRESIDENT, COMMUNICATIONS: Gérard David
INFORMATION OFFICER: Luc Berger

Former Avions Marcel Dassault-Breguet Aviation formed from merger of Dassault and Breguet aircraft companies in December 1971; French government acquired 20 per cent of stock in January 1979, raised to 45.76 per cent in November 1981; present company name adopted June 1990. Of government shareholding, 10.75 per cent held double voting rights so that French government had majority (55 per cent) control. On 17 September 1992, Dassault joined with Aerospatiale in state-owned joint holding company, SOGEPA (which see), which has 35 per cent holding in Dassault Aviation; Dassault and Aerospatiale then pooled resources and co-ordinated R&D strategy, although each remained separate entity.

French government intention to merge Aerospatiale and Dassault Aviation implemented after State relinquished control of Aerospatiale; agreement, ratified by shareholders on 23 December 1998, allocated government's shares in Dassault Aviation to Aerospatiale, but without double voting rights. Aerospatiale later became Aerospatiale Matra and then incorporated into EADS. Last-mentioned affirmed continued stake in Dassault on 27 June 2000 and this endorsed by shareholders on 4 September 2000.

Related measures include separation from Dassault Aviation of Dassault Systèmes, this becoming Dassault Participations, owned 45.94 per cent by EADS, 50.02 per cent by Groupe Industriel Marcel Dassault and 4.04 per cent privately. Dassault intention to separate civil and military businesses, with Falcon business jet activities concentrated at Biarritz, Martignas and Seclin, plus Mérignac for flight test, vetoed by Aerospatiale Matra on 30 June 1999. However, *de facto* partition implemented on 1 January 2000, on "internal and informal basis".

Employees total some 8,600 at 10 industrial sites: St Cloud (2,800), Argenteuil (1,350), Argonay (550), Biarritz (1,200), Mérignac (1,150), Martignas (400), Istres (800), Cazaux (50), Poitiers (150) and Seclin (250). Sales in 2002 totalled €3.44 billion, of which 65 per cent Falcon, 30 per cent home military, 5 per cent military exports, including sales of 72 Falcons. Consolidated orders in 2002 amounted to €3.52 billion.

Work began on 16 September 2002 on new manufacturing building at Bordeaux for Falcon 7X. Signed co-operation accord with Sukhoi of Russia in June 2003, collaborative opportunities including supersonic business jet, corporate shuttle aircraft and UAVs.

Dassault assembles and tests its civil and military aircraft in its own factories, but operates wide network of subcontractors. Other products include flight control system components, maintenance and support equipment. Related companies include Dassault Falcon Jet Corporation, Dassault Falcon Service and SOGITEC. All are subordinate to Groupe Industriel Marcel Dassault, previously including Dassault Electronique, which became part of Thomson-CSF Detexis (now Thales) on 1 January 1999.

Dassault Aviation shared in Atlantique programme with Belgium, Germany, Italy, Netherlands and UK; and Alpha Jet with Germany. Offset manufacture of Dassault aircraft components arranged with Belgium, Egypt, Greece, Spain and UAE. European Aerosystems Ltd established with BAE Systems as a joint venture military aircraft company. On 5 July 2002, Dassault, EADS CASA, EADS Military Aircraft and Saab signed industrial collaboration agreement for feasibility study of Advanced European Jet Pilot Training system.

Dassault has produced more than 1,600 executive jets for service in 60 countries and received orders for over 2,700 Mirages of all types. The company has delivered some 7,500 aircraft to 75 countries and these have accumulated 20 million flying hours, including 10 million achieved by Falcon series business jets up to 2002.

UPDATED

DASSAULT MIRAGE 2000

Indian Air Force name: Vajra (Divine Thunder)

TYPE: Multirole fighter.

PROGRAMME: Selected as main French Air Force combat aircraft 18 December 1975; first of four single-seat prototypes flew 10 March 1978, followed by two-seat version on 11 October 1980; initially developed as interceptor with SNECMA M53 power plant and Thomson-CSF (now Thales) RDM multimode pulse Doppler radar; M53-5 in early production aircraft succeeded by M53-P2; fitted with RDI radar from 38th French Air Force 2000C onwards; first flight of production 2000C 20 November 1982; first flight of production two-seat 2000B, 7 October 1983; first unit, EC 1/2 'Cigognes', formed at Dijon 2 July 1984. Subsequently developed for strike/attack as 2000N/D.

Second-generation Mirage 2000-5 first flown as prototype 24 October 1990; initial production aircraft flown in October 1995 and type qualification granted (for SF1 French production standard) by DGA procurement agency on 13 June 1997; initial export delivery in May 1997, followed by first to French Air Force in December 1997. By end of 2003, Mirage 2000s of eight air forces had accumulated 1 million flying hours.

Third-generation Mirage 2000-5 Mk 2 launched in April 1999; first flew December 2000; deliveries began April 2003.

Potential engine upgrade revealed in early 2000; M53 PX3 under study to provide additional 8 to 10 per cent thrust. PX3 offered to Brazil for Mirage 2000BR in 2002.

DEVELOPMENT MILESTONES

Mirage 2000C	
Official go-ahead	18 Dec 75
First flight	10 Mar 78
First flight, production	20 Nov 82
Entered service (EC 1/2)	2 Jul 84
Subsequent versions:	
Mirage 2000N	
First flight	3 Feb 83
First flight, production	3 Mar 83
First delivery	30 Mar 88
Entered service (EC 1/4)	1 Jul 88
Mirage 2000D	
First flight	19 Feb 91
First delivery (CEAM)	9 Apr 93
Entered service (EC 1/3)	29 Jul 93
Mirage 2000-5	
First flight	24 Oct 90
First flight, production	Oct 95
First delivery (Taiwan)	May 97
Mirage 2000-9	
First flight	14 Dec 00
First delivery (Abu Dhabi)	14 Apr 03

Dassault Mirage 2000-9DADs re-delivered to Abu Dhabi in 2003 ***(F Robineau-Dassault/Aviaplans)*** ***NEW***/0569628

CURRENT VERSIONS: **2000AT:** Advanced Trainer; announced November 2001. Based on 2000-5 Mk 2, including 'glass cockpit', but with operational equipment, such as radar and EW, removed. JTIDS-type (Link 16) datalink to provide realistic sensor inputs for training, obviating need for real air combat targets. Cost 20 to 25 per cent less than operational Mirage 2000; savings in training costs to be achieved by extending turboprop basic stage, then transferring student directly to 2000AT, thus eliminating intermediate jet trainer.

2000B: (*Biplace*: two-seat). Trainer counterpart of 2000C; Nos. 501 to 514 (Series) S3 with RDM radar and M53-5 power plant; Nos. 515 to 520, S4 with RDI J1-1 radar and M53-5, but Nos. 516 and 517 retrofitted with RDM; No. 521 S4-2 with RDI J2-4 and M53-5; No. 522 also S4-2, but M53-P2; Nos. 523 to 530 S5 with RDI J3-13 and M53-P2. Final delivery in December 1994. See also Mirage 2000DA below. One or two Mirage 2000Bs distributed to each 2000C operating squadron, but most operated by OCU, EC 2/5 at Orange, which assumed task from EC 2/2 at Dijon on 1 July 1998. S3 aircraft have been upgraded with RDI radar.

One aircraft, No. 504, converted to **2000BOB** (*Banc Optronique Biplace:* two-seat electro-optics testbed); flown 28 June 1989 after modification by CEV at

Dassault Mirage 2000D tactical attack fighters of EC 2/3 'Champagne' *(F Robineau-Dassault/Aviaplans)* ***NEW***/0569625

Dassualt Mirage 2000C of French Air Force unit EC 2/5 'Ile de France' *(Paul Jackson)* ***NEW***/0569627

Prototype Dassault Mirage 2000B posing as a 2000-5 Mk 2, with MICA AAMs under the fuselage, Magic 2s outboard underwing and two LGBs on the centreline, complete with laser designation pod *(F Robineau-Dassault/Aviaplans)* ***NEW***/0569626

Brétigny-sur-Orge; trials of Rubis FLIR, VEH-3020 holographic HUD, night vision goggles, helmet-mounted display and other electro-optical systems. Trials mainly in connection with the Dassault Rafale programme, but flew with Shehab/Nahar designator/FLIR system for Mirage 2000-9 programme. Fitted with OBOGS.

2000BR: Version of 2000-5 Mk 2 proposed for Brazil, 2001. Development and promotion agreement signed by Dassault, Embraer, Snecma and Thales 27 March 2002. Total 48 had been required, of which 32 to be assembled locally by Embraer, but Brazilian government cancelled requirement in January 2003.

2000C: (*Chasse:* fighter). Standard interceptor; Nos. 1 to 37 built as series S1, S2 and S3 with RDM radar and M53-5 power plants; since upgraded to S3 radar standard. Loosely called Mirage **2000RDM**; Mirage 2000B and C collectively known as Mirage **2000DA** *(Défense Aérienne).* Later aircraft (loosely **2000RDI**) have RDI radar and M53-P2 power plants: Nos. 38 to 48 Series S4, delivered from July 1987 and later upgraded to S4-1; Nos. 49 to 63 S4-1; Nos. 64 to 74 S4-2; Nos. 75 to 124 Series S5, delivered between late 1990 and June 1995. Equipment standards of Mirage 2000B/Cs are **S3**, incapable of launching Matra BAe 530D (530F only); **S4** RDI J1-1 radar; **S4-1** retrofit of all S4s with improved J1-2; **S4-2** further radar upgrade to J2-4; **S4-2A** retrofit of all S4-1 and S4-2 aircraft (Nos. 38 to 74, 515 and 518 to 522) with improved J2-5 radar; **S5** definitive standard with J3-13 radar and, from No. 93 onwards, Spirale chaff/flare dispenser; and **S5-2C** retrofit introduced 1995 (aircraft of EC 1/12) providing better anti-jamming protection for RDI radar. Conversion completed of 37 2000B/Cs to Mirage 2000-5F (which see). Withdrawal of S3 began in January 1998 for upgrading with RDI radar and final aircraft returned to service (at Orange) in March 1999; EC 2/2 gained late production (RDI) aircraft, with which it became operational in the air defence role on 1 August 1998.

Detailed description applies to 2000C except where indicated.

2000D: Two-seat conventional attack version of 2000N, lacking ASMP missile interface and nose pitot but with HOTAS controls, additional display screens for both crew members, Antilope 5-3D terrain-following/terrain-reference radar, GPS and improved (ICMS Mk 2) ECM; functions of pilot and navigator more clearly demarcated. First flight (D01, ex-N01) 19 February 1991; second prototype, D02 (ex-N02), flown 24 February 1992 ; first 2000D (No. 601) delivered CEAM Mont-de-Marsan for trials 9 April 1993; first squadron, EC 1/3 'Navarre', achieved limited IOC (six aircraft only) 29 July 1993 at CEAM. Combat debut was 5 September 1995, launching AS 30L ASMs in Bosnia. Last of 86 delivered February 2002.

Initial six aircraft built to 'interim baseline' configuration known as **R1N1L**, with ability to launch AS-30 laser-guided ASM and Magic AAM only. Further 18 (Nos. 607 to 624), designated **R1N1**, had only 'classic' (free-fall, non-laser) options. Later **R1** aircraft (No. 625 onwards) have full range of current French Air Force armament (BAP 100, BAT 120, Belouga, 68 mm rockets, AS 30L, GBU-12 (Mk 82), BGL 1000, Magic 2 and PDLCT designator). Deliveries of new aircraft ended in December 2001. PDLCT-S (*Synergie*), first employed over Yugoslavia in March 1999, is modified version with improved imagery contrast. All interim 2000Ds were modified to full R1 standard by June 1995. **R2** standard, operational July 2001 (as R2.1), introduced improved navigation system, Eclair fully integrated self-defence suite and Atlis II laser designator pods (from Jaguar force) for use with Paveway II and III bombs; APACHE capability was added in late 2003. First two R2 conversions completed early 2000 at Nancy; further 18 followed, last in 2002. All aircraft to receive Thales RSA NG improved electronic warfare system. Planned **R3** version, with SCALP SOM and a reconnaissance pod, was abandoned in June 1996. However, further enhancements in prospect, including SCALP-EG and MIDS/Link 16 datalink for NATO inter-operability. AASM GPS-guided bomb due in 2005. Full NATO interoperability is planned.

2000E: Multirole fighter for export; M53-P2 power plant throughout. Details in Customers subsection. Baseline version for India, Egypt and Peru, differences from 2000C including RDM radar with CW illumination for Super 530D AAM; two main computers, with expanded memory; ULISS 52 INS; improved ECM (integrated system with VCM-65 display or, alternatively, Remora and Caiman pods); VE-130 HUD; VMC-180 head-down; and expanded weapon options. Abu Dhabi and Greek 2000Es have extra computing power, further armament options and improved self-defence (SAMET system for Abu Dhabi; ICMS Mk 1 for Greece).

2000N: Two-seat low-altitude penetration version to deliver ASMP nuclear standoff missile; two prototypes; first flight 3 February 1983; one preseries aircraft (No. 301) built at Istres and first flown 3 March 1986; first 24 production aircraft (Nos. 302-325) were 2000N-K1 with ASMP capability only; from July 1988 remaining aircraft, designated 2000N-K2, have full conventional and ASMP capability; production ended 1993; all K1s (initially of squadrons 3/4 and part of 2/4) retrofitted to K2 for conventional attack, programme completed by late 1998. Equipment includes Antilope 5 terrain-following radar, two SAGEM inertial platforms, two improved AHV-12 radio altimeters, colour head-down CRT, two Magic self-defence missiles, and internal countermeasures system comprising

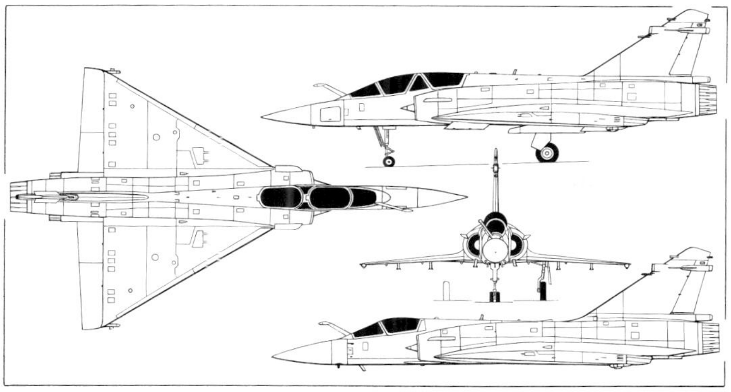

Dassault Mirage 2000-5DDA, with added side view (bottom) of Mirage 2000-5Ei *(Jane's/Dennis Punnett)* 0062696

Cameleon jamming system, Serval RWR and Spirale automatic chaff/flare dispenser system. Further improvement (K2-4C) in 2002 integrated self-defence systems with nav/attack system.

Follow-on upgrade, 2000N K3, launched in September 2003 for incorporation in 50 aircraft of EC 1/4, 2/4 and 3/4, gaining IOC in 2007. Thales Reco-NG reconnaissance pod; new self-protection avionics and upgraded ASMP-Ameliore (which authorised late 2000 for 2011 service entry).

2000N' (N Prime): Initial designation of 2000D.

2000R: Single-seat day/night reconnaissance export version of 2000E but with normal radar nose; various sensor pods possible (see Avionics paragraph).

2000-3: Private venture upgrade, begun 1986, with Rafale-type multifunction (five-CRT) cockpit displays known as APSI (advanced pilot system interface); prototype BY1/F-ZJTB (ex-No. B01) flew 10 March 1988. Later received RDY radar.

2000–4: Private venture integration of Matra BAe MICA AAM. First guided flight against target drone, 9 January 1992.

2000–5: Multirole upgrade incorporating -3 and -4 improvements, plus Thales RDY radar and new central processing unit, Thales VEM 130F HUD and ICMS Mk 2 countermeasures; laser-guided bombs and ASMs in air-to-ground role; MICA AAM. First Mirage flight of RDY radar in BY1 (later numbered BY2) May 1988; first flight of full 2000–5 (two-seat) 24 October 1990 (same aircraft; initially no serial number; later reverted to B01); first single-seater, 01 (conversion of trials aircraft CY1) flown 27 April 1991.

Export orders from Taiwan and Qatar; air-to-air firing trials with MICA completed July 1996; air-to-ground trials completed May 1997; operational use (export; Taiwan) from June 1997.

FFr4,600 million (US$830 million) conversion programme announced November 1992 for upgrading 37 French Air Force 2000Cs (34 S4-2As, Nos. 38 to 49, 51 to 59, 61 to 63 and 65 to 74 and three S5s, Nos. 77 to 79) mainly from EC 2/5 and 3/5 at Orange to **2000-5F** for continued service; contract awarded 25 November 1993; funding included only one reworked aircraft in 1994 defence budget; 10 more in 1995 funding; and 23 in 1996. Delivery schedule (not achieved) was one in 1997; 11 in 1998; 22 in 1999; and three in 2000; one lost on 16 March 1999. Rework (at Argenteuil, with reassembly at Bordeaux) required six months per aircraft and involved complete dismantling and return of fuselages to Argenteuil. RDI radars from these upgraded aircraft retrofitted in early production (RDM) Mirage 2000Cs. Prototypes were conversions of Nos. 51 and 77 (latter an S5); both modified at Istres; initial aircraft reflown 26 February 1996; second followed two months later.

First 'production' conversion, No. 38 handed over 30 December 1997 at Istres test centre before delivery in April 1998 to CEAM, Mont-de-Marsan, to begin conversion of pilots of EC 1/2 from Dijon. IOC at Dijon (12 aircraft) 31 March 1999; FOC 4 February 2000; re-equipment of EC 2/2 began late 1999; deliveries complete by 2001.

Identification features are additional (LAM) 'bullet' antenna on fin leading-edge, absence of nose pitot and four horizontal antenna bands on radome. Normal external configuration is two MICA and two Magic AAMs, plus two 2,000 litre (528 US gallon; 440 Imp gallon) and one 1,300 litre (343 US gallon; 286 Imp gallon) external tanks. Optional configuration

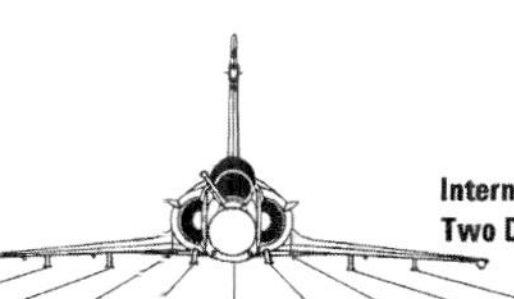

230	1830	450	450	1400	450	450	1830	230	Pylon max weight (kg)
		●	●		●	●			MICA AAM
●								●	MICA/Magic IR AAM
	●						●		Underwing fuel tank
				●					Underfuselage fuel tank
	●●●●	●	●	●●●●	●	●	●●●●		500 lb/250 kg bomb
	●			●			●		2,000 lb/1,000 kg LGB
		●	●	●●●●	●	●			Durandal bomb
				18					BAP 100 penetration bomb
	●	●	●	●	●	●	●		Beluga grenade dispenser
	●			●			●		F2 practice bomb launcher
●	●						●	●	F4 (18) rocket launcher
			●						EO or IR targeting pod
	●						●		AS 30 laser-guided missile
				●					APACHE/SCALP missile
	●						●		Anti-radar missile
	●						●		AM 39 anti-ship missile
				●●					GBU-12 LGB
			●						FLIR navigation pod
				●					Twin-gun pod
				●					Reconnaissance pod
				●					ECM/ELINT pod
				●					Buddy refuelling pod

Mirage 2000-5 potential weapons options *NEW*/0528742

FRENCH MIRAGE 2000 SQUADRONS

Squadron	Base	Type	Commissioned	Remarks
EC 1/2 'Cicognes'	Dijon	2000C (RDM)	2 Jul 84	Temp stood down Jan 98[4]
		2000-5F	31 Mar 99	Operational 4 Feb 00
EC 2/2 'Cote d'Or'	Dijon	2000B/C (RDM)	27 Jul 86	OCU until Jan 98
		2000C	1 Aug 98	
		2000-5F	2000	
EC 3/2 'Alsace'	Dijon	2000C (RDM)	Mar 86	Disbanded 31 Jul 93
EC 1/3 'Navarre'	Nancy	2000D	1 Apr 94[1]	
EC 2/3 'Champagne'	Nancy	2000N	30 Aug 91	Conventional attack only
		2000D	24 Apr 97[2]	Converted 1996–98
EC 3/3 'Ardennes'	Nancy	2000D	21 Aug 95	
EC 1/4 'Dauphine'	Luxeuil	2000N	1 Jul 88	
EC 2/4 'La Fayette'	Luxeuil	2000N	1 Jul 89	
EC 3/4 'Limousin'	Istres	2000N	1 Jul 90	
EC 1/5 'Vendée'	Orange	2000C	1 Apr 89	
EC 2/5 'Ile de France'	Orange	2000C	1 Apr 90	
		2000B/C	1 Jul 98	OCU
EC 3/5 'Comtat Venaissin'	Orange	2000C	3 Sep 90[3]	Disbanded 31 Aug 97
EC 1/12 'Cambrésis'	Cambrai	2000C	1 Jul 92	
EC 2/12 'Picardie'	Cambrai	2000C	1 Sep 93	
EC 4/33 'Vexin'	Djibouti	2000C/D[5]	5 Aug 2002	

[1] Limited IOC 29 Jul 93
[2] When only partly equipped with 2000D
[3] First delivery
[4] And equipped with Alpha Jets to maintain pilot currency
[5] Multirole unit with five Mirage 2000Cs and three 2000Ds

(Standard Kilo) is four MICAs and two Magics. MICA delivered (first 25 rounds) in radar mode from August 1999 followed by IR version in 2003. Early 2000-5F deliveries were to **SF1** standard; SF1C upgrade due to follow, improving radar range by 15 per cent and adding non-co-operative target mode which analyses returns from target's engines. Follow-on upgrades projected with MIDS terminal for Link 16 datalink capability for IR version of MICA and provision for helmet-mounted sight.

2000-5 Mk 2: Announced April 1999; purchased initially by Greece. Features in common with 2000-9 include modular avionics, larger LCDs, secure communications, tactical datalink, additional decoy dispenser, laser gyro INS, upgraded ECM, expanded aircraft-missile datalink, Damoclès laser-designation pod (known as Shehab on 2000-9), Nahar navigational FLIR in Damoclès pylon, upgraded version of RDY radar (RDY2, with multitarget air-to-sea search and track, high-resolution SAR mapping mode and search and track of mobile land targets), increased MTOW of 17,500 kg (38,580 lb), new multichannel digital recording system, new rear cockpit colour display repeater and, possibly, helmet-mounted sight. Can carry six MICA AAMs (active radar or IR) in addition to air-to-surface weapons.

2000-8: Mirage 2000EAD/RAD/DAD supplied to Abu Dhabi/UAE from 1989. Standard AD8.

2000-9: Version of 2000-5 for United Arab Emirates incorporating long-range air-to-ground capability with weapons including Black Shaheen SOM and Hakim. Configuration includes M53-P2 engines, RDY-2 radar with synthetic aperture and beam-sharpening modes, Thales Totem 3000 laser INS, Elettronica IMEWS (integrated modular electronic warfare system) and digital terrain system. First 2000-9s undertook flight testing and proving (three aircraft) at Istres from early 2001; first flight (DAD10; two-seat version) 14 December 2000; sole single-seat 2000-9RAD (RAD19) flew 25 January 2001. Further features as for 2000-5 Mk 2. Deliveries began in 2003.

CUSTOMERS: See table for rapid reference. **France** required seven prototypes and 372 production aircraft; reduced in late 1991 to 318 by abandonment of final 24 2000Cs and 30 2000Ds and transfer of some single-seat aircraft to trainer contract; no Mirages funded in 1992 or 1993, but one 2000B and 14 2000Cs cancelled, then re-ordered in 1994 defence budget as 2000Ds. French orders subsequently amended to 30 2000Bs, 124 2000Cs, 75 2000Ns and 86 2000Ds (total 315); last in 2001. Mirage 2000C equipped three squadrons of EC 2 at Dijon (1984–86), three of EC 5 at Orange (1988–90) and two of EC 12 at Cambrai.

Mirage 2000N deliveries began at Luxeuil 30 March 1988; EC 1/4 'Dauphine' formed 1 July 1988, now with full dual-role K2 series aircraft, followed by two more squadrons. Pending 2000D, EC 2/3 'Champagne' operational at Nancy 1 September 1991 with 2000N in conventional role. EC 1/3 'Navarre' fully operational with 2000D from 31 March 1994; EC 3/3 'Ardennes' began re-equipment, July 1994; EC 2/3 began converting to 2000D (ex-2000N) in mid-1996, completing conversion in 1999. Mirage 2000N depot servicing required every 900 hours or 36 months, whichever sooner.

Abu Dhabi (United Arab Emirates) ordered 18 aircraft on 16 May 1983 and took up 18 options in 1985 for a total of 22 2000EADs, eight 2000RADs and six 2000DADs; these to Standard AD8; deliveries delayed from 1986 to 1989 by provision for US weapons such as Sidewinder; deliveries to Maqatra/Al Dhafra completed November 1990 for Nos. 1 and 2 Shaheen (Warrior) Squadrons. Abu Dhabi 2000RADs carry COR2 multicamera pod, but alternatives include Raphaël-type SLAR 2000 or Harold pods; second 18 have Elettronica ELT/158 threat warning receivers and ELT/558 self-protection jammers; all with Spirale chaff/flare system. Self-defence suite code-named SAMET. Deliveries 7 November 1989 to November 1990. Weapons include BAE Systems PGM Hakim ASM.

Follow-on batch authorised in late 1996; reported 16 December 1997 as 30 new 2000-9s and upgrade to this standard of 33 earlier aircraft; contract signed 18 November 1998; new element later confirmed as 32 aircraft, while upgrades reduced by attrition to 30; total value estimated as US$3.4 billion for aircraft, or US$5.5 billion, including weapons and support. Weapons include Black Shaheen version of Storm Shadow/SCALP SOM, IR and active radar MICA AAMs; IMEWS, incorporating Spirale and Eclair ECM systems; Shebab target designator pod (export version of Damoclès/PDLCT-S) and Nahar navigation FLIR mounted in starboard shoulder pylon.

Deliveries began 14 April 2003 with new-build aircraft, of which first nine formally accepted on 26 April; first refurbished aircraft were received in June 2003. Early aircraft are Standard AD91 with MICA AAM capability; Standard AD92, with additional air-to-ground radar modes and weapons, available from 2004.

Egypt ordered 16 2000EMs and four BMs in December 1981; deliveries 30 June 1986 to January 1988; based at Beni Suef with 82 Squadron in interceptor role.

Greece ordered 36 2000EGs and four 2000BGs on 20 July 1985; handed over from

21 March 1988 and delivered from 27 April 1988 for 331 'Aegeas' and 332 'Geraki' Mire Pandos Kairou within 114 Pterix Mahis at Tanagra; deliveries suspended October 1989 at 28th aircraft; resumed 1992 and completed 18 November 1992. RDM3 radar; Spirale chaff/flare system installed, as part of ICMS self-defence suite; full ICMS system operational from August 1995. Some upgraded to 2000EG-SG3 standard with launch capability for AM 39 Exocet anti-ship missile. Further 15 Mirage 2000-5 Mk 2s confirmed 30 April 1999 and formally ordered on 21 August 2000, together with upgrade to this standard of 10 earlier aircraft by HAI. Deliveries from 2004, with equipment including Thales Damoclès laser designation pods and ICMS Mk 3. New weapons include MICA AAM (Standard EG 51) and SCALP EG ASM (EG 52). Option held on further three new aircraft and five upgrades.

India first ordered 36 2000Hs and four THs in October 1982; 26 Hs and four THs temporarily powered by M53-5; final 10 by M53-P2 from outset; first flight of 2000H (KF-101) 21 September 1984; first flight of 2000TH (KT-201) early 1985; No. 7 IAF Squadron 'Battle Axe' formed at Gwalior AB 29 June 1985, coincident with first arrivals in India. Named *Vajra* (Divine Thunder); second Indian order for six Hs and three THs signed March 1986 and delivered April 1987 to October 1988 to complete No. 1 'Tigers' Squadron. Third order placed 19 September 2000 for four 2000Hs and six 2000THs to be delivered in 2004. Despite grey-and-blue air defence camouflage, unconfirmed report claimed Indian Mirages optimised for attack, with Antilope 5 radar and twin INS. Role (at least) confirmed by 1996 selection (subject to operational trials) of the Rafael Litening laser-designator pod for IAF Mirage 2000s and Jaguars. By 1993 IAF aircraft appearing in experimental brown-and-green low-level colours, with Spirale chaff dispensers. By 1998, 38 remaining 2000Hs had been upgraded by HAL at Bangalore with local flare dispensers and were preparing for receipt of LGBs. In 2002, India planning acquisition of 126 Mirage 2000-5s to equip seven squadrons in strike role; final 108 would be assembled locally by HAL. Discussions continued into 2003, when India also announced intention of buying 12 aircraft from Qatar.

Peru ordered 24 2000Ps and two 2000DPs in December 1982, but reduced this to 10 2000Ps and two 2000DPs; first 2000DP handed over 7 June 1985; deliveries to Peru from December 1986; Escuadron de Caza-Bombardeo 412 of Grupo Aéreo de Caza 4 at La Joya inaugurated 14 August 1987.

Qatar ordered 12 Mirage 2000-5s under contract 'Falcon' on 31 July 1994, together with MICA and Magic 2 AAMs. Versions are 2000-5EDA (nine; single-seat) and 2000-5DDA (three; two-seat). Equipment includes GPS, ICMS Mk 2 and provision for Spirale. First flight late 1995; first three handed over at Bordeaux on 8 September 1997; four delivered to Qatar 18 December 1997; four more on 1 April 1998. All 12 offered for sale to India in mid-2003.

Taiwan ordered 60 Mirage 2000-5s with M53-P2 power plants on 18 November 1992. First export sale for -5; first flight late 1995; first handed over on 9 May 1996 for training in France; initial five arrived in Taiwan as sea freight on 5 May 1997; equips 41, 42 and 48 Squadrons of 2nd Tactical Fighter Wing at Hsinchu; IOC of initial squadron, November 1997. Final eight delivered late October 1998, at which time second squadron declared operational. Versions are 2000-5Ei (48 aircraft; single-seat) and 2000-5Di (12 aircraft; two-seat). Requirement revealed in 1997 for an additional 60. Armament includes MICA and Magic AAMs.

Additionally, Jordan ordered 10 2000EJs and two 2000DJs on 22 April 1988; all were cancelled in August 1991.

Total 601 firm orders (excluding seven prototypes and six company-owned trials and demonstrator aircraft) by 1 January 2001 (315 French; 286 exports); by late 2003, more than 570 had been delivered. France had received all 124 Cs, 30 Bs, and 75 Ns by 31 December 1995, only 2000Ds being delivered after that date (all 86 by December 2001). Provisional agreement signed with Pakistan in January 1992 for 44 Mirage 2000Es; under further discussion in 1994–95, quantity having been revised to 36, including nine two-seat aircraft, all to 2000-5 standard. No further progress made.

Demonstration and evaluation sorties by a Mirage 2000-5 two-seat version given to Czech, Hungarian and Polish air forces in August 1996. Possible Polish final assembly by PZL Mielec is covered by an MoU announced in June 1997. MoU for assembly by Embraer was agreed before postponement of Brazilian requirement.

MIRAGE 2000 ORDERS

Customer	Qty	Version	First aircraft
Abu Dhabi/UAE[1]	6	2000DAD	701
	8	2000RAD	711
	22	2000EAD	731
	20	2000-9RAD	719
	12	2000-9DAD	707
Egypt	16	2000EM	101
	4	2000BM	201
France	30	2000B	501
	124[2]	2000C	1
	86	2000D	601
	75	2000N	301
Greece	36[3]	2000EG	210
	4	2000BG	201
	15	2000-5 Mk 2	—
India	46	2000H	KF101
	13	2000TH	KF201
Peru	10	2000P	050
	2	2000DP	193
Qatar	9	2000-5EDA	QA90
	3	2000-5DDA	QA86
Taiwan	48	2000-5Ei	2001
	12	2000-5Di	2051
Subtotal	**601**		
Prototypes	7		
Development	6		
Total	**614**		

[1] Some 30 being upgraded to 2000-9DAD/RAD. All rebuilt EADs have reconnaissance pod capability and thus become RADs
[2] 37 upgraded to 2000-5F
[3] 10 to be upgraded to 2000-5 Mk 2

COSTS: Programme unit cost: Taiwan FFr 333 million (1997).

DESIGN FEATURES: Multirole combat aircraft; low-set, thin delta wing for high internal volume and low wave drag (as in Mirage III/5), but with delta's disadvantages in manoeuvrability and landing/take-off requirements offset by relaxed stability and leading-edge slats. Area-ruled fuselage.

Wing has cambered section, 58° leading-edge sweep and moderately blended root employing Karman fairings. Cleared for 9 *g* and 270°/s roll at sub- and supersonic speed carrying six air-to-air missiles.

FLYING CONTROLS: Full fly-by-wire control with Thales autopilot; two-section elevons on wing move up 16° and down 25°; inner leading-edge slat sections droop up to 17° 30' and outer sections up to 30°; fixed strakes on intake ducts create vortices at high angles of attack that help to correct yaw excursions; small airbrakes above and below wings.

STRUCTURE: Multispar metal wing; elevons have carbon fibre skins with AG5 light alloy honeycomb cores; carbon fibre/light alloy honeycomb panel covers avionics bay; most of fin and all rudder skinned with boron/epoxy/carbon; rudder has light alloy honeycomb core.

LANDING GEAR: Retractable tricycle type by Messier-Bugatti, with twin nosewheels; single wheel on each main unit. Hydraulic retraction, nosewheels rearward, main units inward. Oleo-pneumatic shock-absorbers. Electro-hydraulic nosewheel steering (±45°). Manual disconnect permits nosewheel unit to castor through 360° for ground towing. Light alloy wheels and tubeless tyres, size 360×135-6 or 360×135R6, pressure 9.5 bar (138 lb/sq in) on nosewheels; size 750×230R15, pressure 15.00 bar (217 lb/sq in) on mainwheels. Messier-Bugatti hydraulically actuated polycrystalline graphite disc brakes on mainwheels, with anti-skid units. Compartment in lower rear fuselage for brake parachute, arrester hook or additional chaff/flare dispenser.

POWER PLANT: One SNECMA M53-P2 turbofan, rated at 64.3 kN (14,462 lb st) dry and 95.1 kN (21,385 lb st) with afterburning. Alternative M53-P20, rated at 98.1 kN (22,046 lb st), is no longer offered. Movable half-cone centrebody in each air intake.

Internal wing fuel tank capacity 1,325 litres (350 US gallons; 291 Imp gallons); fuselage tank capacity 2,675 litres (707 US gallons; 588 Imp gallons) in single-seat aircraft, 2,595 litres (686 US gallons; 571 Imp gallons) in two-seat aircraft. Total internal fuel capacity 4,000 litres (1,057 US gallons; 880 Imp gallons) in 2000C and E, 3,920 litres (1,036 US gallons; 862 Imp gallons) in 2000B, N, D and S. Provision for one jettisonable 1,300 litre (343 US gallon; 286 Imp gallon) RPL-522 96 kg (212 lb) fuel tank under centre of fuselage, and a 1,700 litre (449 US gallon; 374 Imp gallon) RPL-501/502 210 kg (463 lb) or 2,000 litre (528 US gallon; 440 Imp gallon) drop tank under each wing. Total internal/external fuel capacity 8,526 litres (2,252 US gallons; 1,876 Imp gallons) in 2000C and E, 8,450 litres (2,233 US gallons; 1,859 Imp gallons) in 2000B.

Detachable flight refuelling probe forward of cockpit on starboard side, including export aircraft. Dassault type 541/542 tanks of 2,000 litres (528 US gallons; 440 Imp gallons) are available for the 2000-5, 2000-9, 2000N, D and S wing attachments (and optional on 2000B/C), empty weight 240 kg (529 lb) each, and may be complemented by an RPL-522 on centreline.

ACCOMMODATION: One or two occupants (see Current Versions) on SEMMB licence-built Martin-Baker Mk 10Q zero/zero ejection seat(s), in air conditioned and pressurised cockpit. Pilot-initiated automatic ejection in two-seat aircraft; 500 milliseconds delay between departures. Canopy/ies hinged at rear to open upward and, on Mirage 2000D, covered in gold film to reduce radar signature.

SYSTEMS: ABG-Semca air conditioning and pressurisation system. Two independent hydraulic systems, pressure 280 bar (4,000 lb/sq in) each, to actuate flying control servo units, landing gear and brakes. Hydraulic flow rate 110 litres (29.1 US gallons; 24.2 Imp gallons)/min. Electrical system includes two Auxilec 20110 air-cooled 20 kVA 400 Hz constant frequency alternators (25 kVA in Mirage 2000D and 2000-5), two Bronzavia DC transformers, a SAFT 40 Ah battery and ATEI static inverter. Intertechnique oxygen system.

AVIONICS: *Comms:* Thales ERA-7000 V/UHF com transceiver, ERA-7200 UHF (with optional Have Quick II) or SCP 5000 secure voice com; Thales NRAI-7A/NRAI-11 IFF transponder/interrogator (SC10/IDEE 1 on 2000-5) for Taiwan; SC10/TSA2535 on 2000-9; TSC2031/TSA2531 on 2000-5 Mk 2).

Radar: Thales RDM multimode radar or RDI pulse Doppler radar, each with operating range of 54 n miles (100 km; 62 miles). (Mirage 2000N/D have Thales Antilope terrain-following radar for automatic flight down to 61 m (200 ft) at speeds not exceeding 600 kt (1,112 km/h; 691 mph); Antilope 5TC in 2000N includes altitude-contrast updating of navigation system; Antilope 5-3C in 2000D has full terrain-reference navigation facility.) Thales RDY multimode, multitarget radar in 2000-5 has the ability to detect 24 targets while tracking eight.

Flight: SOCRAT 8900 solid-state VOR/ILS and IO-300-A marker beacon receiver, Thales radio altimeter (AHV-6 in 2000B and C, AHV-9 in 2000-5/5 Mk 2/-9 export aircraft, two AHV-12 in 2000N. Thales NC12 or Deltac Tacan. SAGEM Uliss 52 inertial platform (52D in 2000C and B; 52E for export; and two 52P in 2000N/D; plus integrated GPS in 2000D and 2000-5). Thales Totem 3000 RLG in 2000-9 and 2000-5 Mk 2, combined with GPS and modular data processing. Thales Type 2084 central digital computer and Digibus digital databus (2084-XR in 2000D; Xi in 2000-5. Thales AP 605 autopilot (606 in 2000N, 607 in 2000D, 608 in 2000-5), 609 in 2000-5 Mk 2 and 2000-9). Thales TLS 2020 MMR (multimode receiver) in 2000-5 Mk 2 and 2000-9, combining ILS, microwave landing and differential GPS). Undisclosed operator ordered BASE (BAE Systems) Terprom GPWS in 1996 for retrofit by Orbital Sciences Corporation. Digital terrain system (DTS) on 2000-9.

Instrumentation: Thales TMV 980 data display system (VE 130 head-up and VMC 180 head-down) (two ICC 55 head-down in 2000N/D). Mirage 2000-5 has Thales integrated multidisplay system featuring combined HUD/head-level display with Thales recording camera. Provision optional in 2000-5 and -9 for Thales Topsight E helmet-mounted sight/display system.

Mission: Mirage 2000-5 has LAM (*liaison avion-missile*) 'bullet' antenna on fin leading-edge for mid-course guidance of MICA AAMs. Sensors of strike/attack and export versions include 570 kg (1,257 lb) Thales Raphaël SLAR 2000 pod, 400 kg (882 lb) Thales COR2 multicamera pod or 680 kg (1,499 lb) Dassault AA-3-38 Harold long-range oblique photographic (Lorop) pod; 110 kg (243 lb) Thales/Intertechnique Rubis FLIR pod; Thales Atlis laser designator and marked target seeker (in pod on forward starboard underfuselage station); 340 kg (750 lb) Thales PDLCT day/night (TV/thermal imaging) laser designator pod on Mirage 2000D (or CLDP/Atlis II on export aircraft)(2000D squadrons also being issued with Atlis II designation pods — being refurbished Atlis systems from Jaguar force), while PDLCT-S available from mid-1999; one 550 kg (1,213 lb) Thales TMV 004 (CT51J) Caiman offensive or intelligence ECM pod and one 400 kg (882 lb) Thales Astac elint (interferometer) pod. Export PDLCT-S known as Damoclès (and Shehab in UAE). Mirage 2000-5 Mk 2 and 2000-9 have Thales Thomrad 5000/6000 secure communications and datalink.

Self-defence: Systems in 2000C and 2000N include Thales Serval radar warning receiver (antennas at each wingtip and on trailing-edge of fin, near tip, plus VCM-65 cockpit display); Thales Caméléon (2000N), Caméléon C2 (2000D) or Sabre (2000C) jammer at base of fin (detector on fin leading-edge); and MBDA Spirale, comprising chaff dispensers in Karman fairings at wing trailing-edge/fuselage intersection and flares in lower rear fuselage. French Air Force DDM (*Détecteur Départ Missile*) missile plume detector requirement satisfied by 1994 purchase of SAGEM SAMIR system for 1995 fitment in rear of Magic launch rails (2000D/N first, but also to 2000Cs patrolling Bosnia). Spirale fitted to 2000N-K2; retrofitted to 2000N-K1 and installed on 2000Cs from No. 93; earlier 2000Cs have Eclair system (Alkan LL5062 chaff and flare launcher) in place of braking parachute, lacking automatic operation. Spirale on some export 2000Es.

Upgrade planned of 2000E with ICMS Mk 1 (integrated countermeasures system; as in 2000EG) comprising RWR and SHR, Thales high-band jammer (leading-edge of fin and bullet fairing at base of rudder) and Spirale; automated ICMS Mk 2 of Mirage 2000-5 adds receiver/processor in nose to detect missile command links; extra pair of antennas near top of fin and additional DF antennas scabbed to existing wingtip pods (fin and secondary wingtip antennas also on Greek Mirage 2000s); ABD2000 export version of Sabre in some Mirage 2000Es. Mirage 2000-5 Mk 2 has ICMS Mk 3. ICMS Mk 3/IMEWS eventually to comprise fully integrated suite of RWR, jammer, Spirale (chaff/flare), additional Eclair (chaff/flare), ESM, tactical situation awareness and targeting. External equipment can include two 182 kg (401 lb) Thales DB 3141/3163 Remora self-defence ECM pods.

EQUIPMENT: Optional 250 kg (551 lb) Intertechnique 231-300 buddy-type in-flight refuelling pod.

ARMAMENT: Two 30 mm DEFA 554 guns in 2000C, 2000E and single-seat 2000-5 (not fitted in B, D or N), with 125 rds/gun. Nine attachments for external stores, five under fuselage and two under each wing. On 2000-5, fuselage centreline stressed for 1,400 kg (3,086 lb) loads; other four fuselage points for 450 kg (992 lb) each; inner wing pylons for 1,830 kg (4,034 lb) each; and outboard wing points for 230 kg (507 lb) each.

Typical interception weapons comprise two 275 kg (606 lb) Super 530D or (if RDM radar not modified with target illuminator) 250 kg (551 lb) 530F missiles (inboard) and two 90 kg (198 lb) 550 Magic or Magic 2 missiles (outboard) under wings. Alternatively, each of four underwing hardpoints can carry a Magic. MICA (active radar or — qualified 31 July 2002 — IR versions) AAM (110 kg; 243 lb) on Mirage 2000-5 Mk 2 and 2000-9. Primary weapon for 2000N is 900 kg (1,984 lb) ASMP tactical nuclear missile mounted on LM-770 centreline pylon.

In air-to-surface role, the Mirage 2000 can carry up to 6,200 kg (13,669 lb) of external stores, including MBDA 250 kg retarded bombs or 32.5 kg (72 lb) TDA BAP 100 anti-runway bombs; 16 MBDA Durandal 219 kg (483 lb) penetration bombs; one or two 990 kg (2,183 lb) MBDA BGL 1000 laser-guided bombs; five or six 305 kg (672 lb) MBDA Belouga cluster bombs or 400 kg (882 lb) TDA BM 400 modular bombs; up to three Rafaut F2 practice bomb launchers; US Mk 20, Mk 82, GBU-24 and GBU-12 bombs; two 520 kg (1,146 lb) AS 30L, Armat anti-radar, or 655 kg (1,444 lb) AM 39 Exocet anti-ship, air-to-surface missiles; four 185 kg (408 lb) MBDA LR F4 rocket launchers, each with eighteen 68 mm rockets; two packs of 100 mm rockets; or a 765 kg (1,687 lb) Dassault CC 630 gun pod, containing two 30 mm guns and total 600 rounds of ammunition.

Mirage 2000D to receive 1,200 kg (2,646 lb) APACHE-AP standoff (216 n mile; 400 km; 249 mile) weapons dispenser. APACHE-SCALP (*Système de Croisière Autonome à Longue Portée*) selected December 1994 to satisfy APTGD (*Arme de Précision Tirée à Grande Distance*) requirement for 216 to 324 n mile (400 to 600 km; 249 to 374 mile) range stealthy cruise missile; service entry in 2003. For air defence weapon training, a Cubic Corporation AIS (airborne instrumentation subsystem) pod, externally resembling a Magic missile, can replace Magic on launch rail, enabling pilot to simulate a firing without carrying actual missile.

DIMENSIONS, EXTERNAL:
Wing span .... 9.13 m (29 ft 11½ in)
Wing aspect ratio 2.0
Length overall: 2000C, E .... 14.36 m (47 ft 1¼ in)
–5 .... 14.33 m (47 ft 0¼ in)[1]
2000B, N .... 14.55 m (47 ft 9 in)[1]
Height overall: 2000C, E, –5 .... 5.14 m (16 ft 10¼ in)
2000B, N, D .... 5.10 m (16 ft 8¾ in)
Wheel track .... 3.50 m (11 ft 5¾ in)
Wheelbase .... 5.00 m (16 ft 4¾ in)

[1] *2000D, -5, -5 Mk 2 and -9 versions lack nose pitot*

AREAS:
Wings, gross .... 41.0 m² (441.3 sq ft)

WEIGHTS AND LOADINGS:
Weight empty: 2000C, E, –5 .... 7,500 kg (16,534 lb)
2000B, N, D .... 7,600 kg (16,755 lb)
2000-5 Mk 2; single-seat .... 7,920 kg (17,460 lb)
two-seat .... 7,990 kg (17,615 lb)
Max internal fuel: 2000C, –5 .... 3,160 kg (6,967 lb)
2000B, N, D .... 3,100 kg (6,834 lb)
Max external fuel .... 4,140 kg (9,127 lb)
Max external stores load .... 6,200 kg (13,669 lb)
Combat weight: 2000-5 .... 9,500 kg (20,944 lb)
T-O weight clean: 2000C, E, –5 .... 10,860 kg (23,940 lb)
2000B, N, D .... 10,960 kg (24,165 lb)
Max T-O weight: except –5, –9:
normal .... 14,000 kg (30,864 lb)
overload .... 17,000 kg (37,480 lb)
–5, –9 .... 17,500 kg (38,580 lb)
Max landing weight .... 16,700 kg (36,817 lb)
Max wing loading .... 414.6 kg/m² (84.92 lb/sq ft)
Max power loading (M53-P2) .... 179 kg/kN (1.75 lb/lb st)

PERFORMANCE (M53-P2 power plant):
Max level speed: at height M2.2
at S/L M1.2
Max continuous speed: 2000C, E M2.2
2000N, D M1.6
Min speed in stable flight .... 100 kt (185 km/h; 115 mph)
Authorised min flight speed .... zero
Approach speed .... 140 kt (259 km/h; 161 mph)
Landing speed .... 125 kt (232 km/h; 144 mph)
Max rate of climb at S/L:
M53-P2 .... 17,060 m (56,000 ft)/min
M53-P20 .... 16,765 m (55,000 ft)/min
Time to 11,000 m (36,080 ft) and M1.8:
2000-5 .... approx 5 min
Time from brake release to intercept target flying at M3.0 at 24,400 m (80,000 ft) .... less than 5 min
Service ceiling: 2000C .... 16,460 m (54,000 ft)
2000-5 .... 18,290 m (60,000 ft)
Range: hi-hi-hi .... 1,000 n miles (1,852 km; 1,151 miles)
interdiction, hi-lo-hi .... 800 n miles (1,480 km; 920 miles)
attack, hi-lo-hi .... 650 n miles (1,205 km; 748 miles)
attack, lo-lo-lo .... 500 n miles (925 km; 575 miles)
with one 1,300 litre and two 1,700 litre drop tanks .... 1,800 n miles (3,333 km; 2,071 miles)
Operational loiter (2000-5) at M0.8 at 7,620 m (25,000 ft) with three external tanks, four MICA and two Magic AAMs .... 2 h 30 min
Operational range (2000-5) for 5 min combat at M0.8 at 9,145 m (30,000 ft) with four MICA and two Magic AAMs, tanks jettisoned .... 780 n miles (1,445 km; 898 miles)
g limits: .... +9.0/–3.2 normal
.... +11.0/–4.5 ultimate

UPDATED

DASSAULT RAFALE

English name: Squall
TYPE: Multirole fighter.

DEVELOPMENT MILESTONES

Design began	June 82
Official go-ahead	late 82
First order (demonstrator)	13 Apr 83
Rolled out	14 Dec 85
First flight	4 July 86
First flight, pre-production	19 May 91
Production go-ahead	23 Dec 92
First flight, production	24 Nov 98
First delivery (Landivisiau)	Dec 00
Entered service (12 F)	18 May 01

PROGRAMME: Ordered (as *Avion de Combat Tactique; ACT*) to replace French Air Force Jaguars and (as *Avion de Combat Marine; ACM*) Navy Crusaders and Super Etendards; first flight of Rafale A prototype (F-ZJRE) 4 July 1986; first flight with SNECMA M88 replacing one GE F404, 27 February 1990 (was 461st flight overall); 867th and final sortie, 24 January 1994. ACE International (*Avion de Combat Européen*) GIE set up in 1987 by Dassault Aviation, SNECMA, Thomson-CSF (now Thales) and Dassault Electronique, partly to attract international partners; none found. Four preproduction aircraft, as described under Current Versions (specific). Production launch officially authorised, 23 December 1992 (and 31 December 1992 for M88-2 power plant). First Rafale B and Rafale M ordered 26 March 1993; first production aircraft (Rafale B No. 301) flew 24 November 1998 and made 'inaugural' (official) first flight 4 December; to CEV at Istres, early 1999, for development of F2 production standard. First production Rafale M (No. 1) flew at Bordeaux 7 July 1999. Rafale had by then accumulated over 4,000 sorties.

Test airframe, in Rafale M configuration, delivered to CEAT at Toulouse for ground trials 10 December 1991. Between 17 December 1991 and 2 March 1993, completed 10,000 simulated flights, including 3,000 catapult take-offs and 3,000 deck landings. Rafale structural validation achieved 15 December 1993.

French Air Force preference switched to operational two-seat (pilot and WSO) derivative of Rafale C in 1991; announced 1992 that 60 per cent of procurement to be two-seat, although 16 aircraft deleted from requirements at this time. Procurement target further reduced, as detailed in Customers paragraph. Two-seat version of naval Rafale announced September 2000.

Funding constraints and French government demands for cost reductions resulted in suspension of Rafale programme in November 1995 and temporary blocking of most 1996 funds. Plans were simultaneously abandoned for three progressively more sophisticated service standards of Rafale (*Standard Utilisateur* 0, 1 and 2) and replaced by a common French military standard and an export parallel, although three basic software standards (F1 to F3) will phase-in operational capabilities, as described under Current Versions (general). Work on production Rafales temporarily halted in April 1996. On 22 January 1997, Dassault and French defence ministry agreed on 48-aircraft multiyear procurement (1997–2002) in return for a 10 per cent cost reduction, effectively relaunching the programme. This initiative lapsed with the change of government after June 1997, but reinstated in January 1999, with firm orders for 28, plus 20 options, covering deliveries between 2002 and 2007. Contract awarded January 2000 for development of F2 Standard Rafale, representing first capability upgrade.

First production aircraft completed late 1998; second and third (No. 302 and M1) delivered to CEV trials unit late 1999/early 2000; M2 and M3 to Landivisiau naval air station, December 2000; first naval squadron, 12 F, formed 18 May 2001 and achieved planned strength of 10 Rafales in September 2002; first operational carrier deployment by

Rafale M launching from ***Charles de Gaulle*** ***(F Robineau-Dassault/Aviaplans)*** ***NEW***/0569629

Dassault Rafale C101, the first production single-seat aircraft for the French Air Force *(F Robineau-Dassault/Aviaplans)* ***NEW***/0569630

four aircraft (M2 to M5) aboard FS *Charles de Gaulle* for exercise Trident d'Or in Mediterranean, 21 to 29 May 2001. Rafale B No. 303 is first for air force (CEAM trials establishment), in early 2004; first air force squadron to be equipped with 20 aircraft during 2006; 80th aircraft due in 2008, 140th in 2011; final (294th) French delivery due in 2020.

Associated programmes include Thales electronic scanning RBE2 (*Radar à Bayalage Electronique deux plans*) multimode radar, ordered November 1989; test flights begun in Falcon 20 No. 104, 10 July 1992; first RBE2 flight in Rafale 7 July 1993 (B01); first production RBE2 flew on 16 October 1997 in a Falcon 20 before being refitted in Rafale B01 from November 1997. Development authorised in early 1999 of upgraded RBE2 version (full air-to-ground weapons capability) for 2003 delivery and installation in F2 standard Rafales.

Thales/MBDA defensive aids package named Spectra (*Système pour la Protection Electronique Contre Tous les Rayonnements Adverses*); wholly internal IR detection, laser warning, electromagnetic detection, missile approach warning, jamming and chaff/flare launching; nine prototypes ordered; total weight 250 kg (551 lb); Spectra trials begun on Mirage 2000 in 1992, while full suite installed in Falcon 20 No. 252 by Dassault at Istres between December 1992 and September 1994 before flight trials; Spectra flown in Rafale M02 on 20 September 1996 at launch of integration programme at CEV Istres. Development contract awarded 1991 for Thales OSF (*Optronique Secteur Frontal*) with IRST, FLIR and laser range-finder in two modules ahead of windscreen; surveillance, tracking and lock-on by port module; target identification, analysis and optical identification by starboard module; combined output in pilot's head-level display; initially tested in Mirage 2000BOB.

M88 engine, which has flown only on Rafales, achieved type certification on 22 March 1996. First of initial production batch of 42 M88-2 engines delivered by SNECMA 30 December 1996. Conformal fuel tanks flight-tested in April 2001. Integration of Matra BAe MICA AAM completed 5 July 2000, after 27 launches.

Rafale offered for export; Greek evaluation in January 2000; model displayed February 2000 carried two conformal fuel tanks, each of 1,250 litres (330 US gallons; 275 Imp gallons), increasing range with two SCALP missiles to 1,000 n miles (1,852 km; 1,150 miles), hi-lo-hi. Rafale B No. 302 equipped with supplementary software to enable demonstration of LGB capability; test separation of GBU-12 bombs in April 2000, followed by live drop in October 2000. Offered in South Korea's F-X competition (40 aircraft), but unsuccessful against Boeing F-15K, early 2002.

CURRENT VERSIONS (GENERAL): **Rafale B:** Originally planned two-seat, dual-control version for French Air Force; weight initially envisaged as 350 kg (772 lb) more than Rafale C; 3 to 5 per cent higher cost than Rafale C. Being developed into fully operational variant for either pilot/WSO or single-pilot combat capability. Serial numbers begin at 301. First two assigned to CEV.

Rafale C: Single-seat combat version for French Air Force. Serial numbers begin at 101, first flown 16 April 2003.

Rafale D: Original configuration from which production versions derived; now '*Rafale Discret*' (stealthy) generic name for French Air Force versions.

Rafale M: Single-seat carrierborne fighter; serial numbers begin at 1. Navalisation weight penalty, 610 kg (1,345 lb); has 80 per cent structural and equipment commonality with Rafale C, 95 per cent systems commonality. Navy's financial share of French programme cut in 1991 from 25 to 20 per cent.

Initial operational software standard for Rafale M, designated F1, permits air defence missions against multiple targets using Magic and radar-homing version of MICA; self-defence provided by Spectra system. F2 will apply to both naval and air force Rafales delivered from late 2004 (15 M/Ns, 22 Bs and 11 Cs) and combine F1 with air-to-ground radar modes and the ability to launch IR-guided MICA, SCALP and AASM weapons as well as OSF electro-optics suite and MIDS datalink; operational in 2006. Funding for F2 development granted 31 December 1998; first stage of F2, known as E12, is OSF, SCALP and standard air-to-surface ordnance. First production aircraft, M1, on loan to Dassault for F2 integration. Last of 35 single-seat naval Rafales to be delivered in 2008. In same year (operational 2009), F3 standard (for which initial, risk mitigation phase launched January 2003) provides full capabilities to naval and air force Rafales, including air-to-sea attack, AM39 and ASMP-A weapons, refuelling and reconnaissance pods, and helmet-mounted display. Unspecified and unfunded F4 envisaged for 2010, but early naval aircraft will all have been upgraded to F2 and F3 by 2008; Meteor AAM and associated electronically scanned radar antenna are key elements of F4.

Rafale N: Two-seat, dual control, naval version; also known briefly as Rafale BM; announced September 2000; development and production contracts placed December

No. 302, the second French Air Force trials Rafale B *(Jean-Louis/Gaynecoetche)* ***NEW***/0528668

Dassault Rafale M of 12 F, French Navy *(Paul Jackson)* ***NEW***/0569631

Rafale refuelling probe and OSF sensors *(Paul Jackson)* ***NEW***/0552891

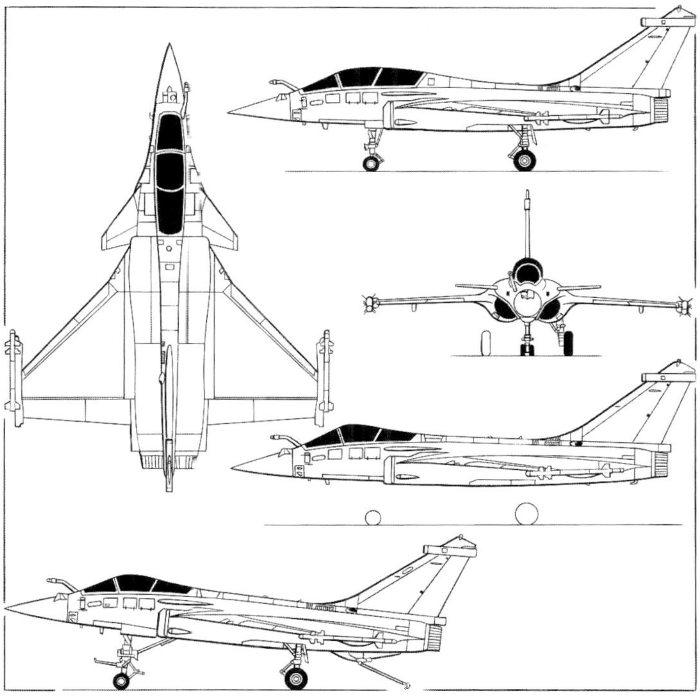

Dassault Rafale B, with additional side views of single-seat Rafale C (centre) and navalised Rafale M (bottom) *(James Goulding)* 0507496

2002;prototype, No. 16, for trials in 2006; requirement for 25 aircraft within overall Navy purchase; first production aircraft in 2008; final delivery in 2012. Cost 5 per cent more than single-seat M; 250 kg (551 lb) heavier, but 260 kg (573 lb) less internal fuel; 85 per cent commonality with three previous versions; deletion of internal canon replaces 200 litres (52.8 US gallons; 44.0 Imp gallons) of fuel lost with addition of second seat.

'Rafale Mk 2': Export version, under active consideration by 2000, featuring active antenna radar, M88-3 engines of 88.3 kN (19,850 lb st) each. Available from 2006; conformal tanks and Damoclès laser target designator. Development cost estimated in 2001 as €1.3 billion; joint venture agreed in January 2001 by Dassault, Thales and SNECMA; offered to South Korea.

CURRENT VERSIONS (SPECIFIC): **C01:** Single-seat Rafale C prototype, C01/F-ZWVR, ordered 21 April 1988; flown 19 May 1991; officially flight tested at CEV in October 1991, two months ahead of schedule; 100th sortie 12 May 1992. First Rafale gun firing, 5 March 1993; first Magic 2 AAM launch, 26 March 1993. Continued high AoA trials in 2000, having exceeded 30°. M88-2 Stage 4 engine trials in late 2000; continued engine trials into 2003.

Second Rafale C order (C02) not placed; abandoned 1991.

M01: First navalised prototype, F-ZWVM, ordered 6 December 1988; flown 12 December 1991. Assigned to structural qualification, FCS and aerodynamic trials. Catapult trials ashore at US Naval Air Warfare Center, Patuxent River, and Lakehurst, 13 July/23 August 1992; second series of US trials 15 January/18 February 1993 followed by deck trials on *Foch*; first deck landing 19 April 1993; first deck launch 20 April 1993, although first take-offs with 'jump strut' nosewheel leg began in following month; third US trials series 18 November/16 December 1993, carrying external loads; fourth series, October/December 1995, including dummy deck launch at maximum weight of 22,300 kg (49,163 lb). On 8 June 1995, M01 made the first launch of a MICA AAM against an aerial target acquired by RBE2 radar fitted in a Rafale. Employed on development trials for F2 production standard. Total 438 hours in 723 sorties by October 1997. In storage 1999; F1 standard avionics trials in 2000. Retired by 2003.

B01: Two-seat dual-control trainer Rafale B prototype, B01/F-ZWVS, ordered 19 July 1989 as first with RBE2 radar and Spectra defensive systems; first flight 30 April 1993; first flight with RBE2 7 July 1993. Longest Rafale sortie, Istres to Dubai, November 1995: approximately 3,020 n miles (5,600 km; 3,475 miles) in 6 hours 30 minutes with three aerial refuellings (including one precautionary). Heavy configuration trials (23,400 kg; 51,588 lb, including two APACHEs) completed in February 1997. Total 990 sorties by mid-1999. First Rafale to fly with conformal tanks, 18 April 2001.

M02: Second naval prototype, ordered 4 July 1990; first flight 8 November 1993; assigned to operational and maintenance testing aboard ship and navigation/weapons trials. Joint carrier trials with M01 aboard *Foch* (second series) 27 January/4 February 1994. Third series of deck trials (M02 only) aboard *Foch* begun 17 October 1994 for three weeks (total 28 launches including two at night); included maintenance, electromagnetic compatibility, RBE2 radar and Spectra ECM tests. Fitted with model of OSF in late 1994, for vibration tests. Flew Singapore to Istres (6,300 n miles; 11,668 km; 7,250 miles) in under 15 hours, February 1996. First launch of Magic 2 AAM at moving target, 4 April 1996. Total 297 hours in 348 sorties by October 1997. With B01, employed by 1998 on development of F2 and F3 avionics standards. On 6 July 1999 became only second jet fighter (following Super Etendard) to land on new carrier FS *Charles de Gaulle*. Continued Spectra trials in 2000.

CUSTOMERS: Anticipated worldwide market for 500 aircraft in addition to originally planned 250 for French Air Force (225 Cs and 25 Bs) and 86 for French Navy; former service announced revised requirement for 234, comprising 95 Rafale Cs and 139 two-seat (pilot and WSO) combat versions, in 1992. Defence economies in 1996 included reduction of requirements to 60 Ms and Ns.

Naval deliveries began with No. 2 to CEPA trials unit on 19 July 2000, followed by No.3 in September 2000; service familiarisation began 4 December 2000 when Nos. 2 and 3 delivered to Landivisiau naval air base; Landivisiau's 12 Flottille re-formed 18 May 2001 with four aircraft; first operational voyage began aboard *Charles de Gaulle* (commissioned 1999) on 1 December 2001, and Nos. 2 to 8 had embarked by 10 March 2002, when stationed in Arabian Gulf; Nos. 9 and 10 received mid-2002 and IOC then declared.

Balance of naval order replaces Super Etendards, 11 Flottille being first recipient, in 2006–07; third squadron to follow. Navy received 10 aircraft in F1 configuration (actually LF1, which upgraded to partial F1 in early 2002, incorporating some air-to-air capability, although full F1, with cannon and MICA AAM, not due until late 2002) 15 F2s and 35 F3s, last in 2012, all to be based at Landivisiau.

RAFALE PLANNED ORDERS AND DELIVERIES

Date	Orders		Deliveries	
	AF	Navy	AF	Navy
up to 2002	36	25	3	10
2003–2008	94	31	46	19
post 2008	104	4	185	31
Totals	**234**	**60**	**234**	**60**

RAFALE PROCUREMENT

Year	Rafale C	Rafale B	Rafale M/N	Total
1993		1	1	2
1994		1	2	3
1995			5	5
1996		1	2	3
1997				
1998				
1999[1]	7	14	7	28
2001	5	7	8	20
2002				
2003	46[2]		13	59
Totals	**58**	**24**	**38**	**120**

[1] start of F2 production version; earlier aircraft are F1; 62nd and subsequent will be F3
[2] May include some Rafale Bs

Notes: Orders as announced. Third Rafale B was subsequently exchanged with first production Rafale C.

Air Force deliveries originally planned for 1996–2009, including first 20 in interim configuration; two year postponement announced 1992; further slips in development funding delayed first receipts to 2002 and IOC to 2006, when first squadron to be established at St Dizier with 20 Rafale Bs. Only Air Force's first three aircraft 301, 302 and 101 (two Bs and one C) to F1 standard.

Official authorisation to launch production given 23 December 1992. Initial production contract in 1993 defence budget (formally awarded 26 March 1993), comprising one aircraft each for Air Force and Navy. Total 13 by 1996, while 1997–2002 plan envisaged 33 B/Cs and 15 naval versions (total 48) to be ordered (and two Bs and 12 Ms delivered), followed by orders for 15 B/Cs per year from 2003. In early 1997, French defence procurement agency, DGA, agreed with Dassault a multiyear procurement of the 48 aircraft in return for a 10 per cent cost reduction, but this later suspended. Authorisation to order the first 13 Rafales was only granted in May 1997, however. At the same time, separate plans were being formulated for acceleration into service of the first 10 air force Rafales to equip an export-promoting and operational trials (half) squadron, but those plans also soon abandoned. Eventually, go-ahead given on 14 January 1999 for 48 aircraft, including 20 options, these confirmed on 21 December 2001. In 2003, details finalised for further batch of 59 (46 Air Force and 13 Navy), making 120 (82 + 38) in all.

Deliveries in 2003–08 five-year plan will be 57 to Air Force and 19 to Navy. Export versions of naval variant were available to potential customers from 1999 onwards. China expressed interest in 1996–97. By 2015 French Air Force combat arm is expected to comprise 140 Rafales in front-line units.

COSTS: Programme estimated at FFr155 billion (1991), including FFr40 billion for R&D; revised to FFr178 billion in 1993, FFr198.4 billion in 1995 and FFr202.37 billion in 1996. Last-mentioned total comprises FFr48.62 billion (of which 25 per cent paid by industry) for development, FFr17.583 billion for industrialisation, FFr76.25 billion for 234 Rafale B/Cs, FFr20.89 billion for 60 Rafale Ms, FFr37.812 billion for spares and FFr1.215 billion for simulators. Early 1997 agreement on 10 per cent cost reduction resulted in flyaway price falling to FFr282 million for a Rafale C, FFr299 million for Rafale B and FFr315 million for Rafale M. Total of FFr30 billion spent by 1995. Second production order for eight aircraft (1994/1995 authorisations) estimated at FFr1.5 billion, excluding engines, radar and weapons system. In 1998, however, new cost estimate for 294 aircraft was FFr320 billion as a consequence of programme delays. Cost (1999) of 48 aircraft given as FFr17.2 billion. Introduction of Rafale N version increased programme cost by €274 million.

DESIGN FEATURES: Multirole combat aircraft, rivalling Eurofighter Typhoon; designated "omnirole" to describe simultaneous air-to-air and air-to-ground capabilities. Minimum weight and volume structure to hold costs to minimum; thin, mid-mounted delta wing with moving canard; individual fixed, kidney-shaped intakes without shock cones. HOTAS controls, with sidestick controller on starboard console and small-travel throttle lever.

Wing leading-edge sweepback approximately 48°.

FLYING CONTROLS: Fully digital fly-by-wire controls with fully modulated two-section leading-edge slats and two elevons per wing; canard incidence automatically increased to 20° when landing gear lowered. Specification includes 30° AoA in stable flight.

STRUCTURE: Most of wing components made of carbon fibre including elevons; slats in titanium; wingroot and tip fairings Kevlar; canard made mainly by superplastic forming and diffusion bonding of titanium; fuselage 50 per cent carbon fibre; fuselage side skins of aluminium-lithium alloy; wheel and engine doors carbon fibre; fin made primarily of carbon fibre with aluminium honeycomb core in rudder. Composites account for 25 per cent by weight of structure and 20 per cent of surface area; weight saving directly attributable to composites is 300 kg (661 lb) — equivalent to a 1 tonne reduction in empty weight.

LANDING GEAR: Hydraulically retractable tricycle type supplied by Messier-Dowty, with single 790×275-15 (20 ply) or 790×275R15 mainwheels and twin, hydraulically steerable, 360×135-6 or 360×135R6 nosewheels. All wheels retract forward. Designed for impact at vertical speed of 3 m (10 ft)/s, or 6.5 m (21 ft)/s in naval version, without flare-out. Rafale M has same mainwheels; but 520×140R10.5 nosewheels. Messier-Bugatti carbon brakes on all three units, controlled by fly-by-wire system.

Rafale M has 'jump strut' nosewheel leg which releases energy stored in shock-absorber at end of deck take-off run, changing aircraft's attitude for climb-out without need for ski-jump ramp. 'Jump strut' advantage equivalent to 9 kt (16 km/h; 10 mph) or 900 kg (1,984 lb) extra weapon load; not to be used aboard aircraft carrier *Foch*, which to have 1° 30' ramp giving 20 kt (37 km/h; 23 mph) or 2,000 kg (4,409 lb) advantage. Dowty Aerospace Yakima holdback fitting. Naval nosewheel steerable ±70°; or almost 360° under tow. Hydraulic (Rafale M) or tension-stored (Rafale B/C) arrester hook. Landing gear management (braking and steering) by Thales computer.

POWER PLANT: Two SNECMA M88-2 augmented turbofans, each rated at 48.7 kN (10,950 lb st) dry and 72.9 kN (16,400 lb st) with afterburning. Stage 1 standard engines in first 13 (F1) production Rafales were limited to 300 hour hot-section TBO; Stage 4, flight tested from late 2000 and certified in December 2001, achieves initial 600 hours and intended for eventual 1,000 hours, had been retrofitted in all aircraft by late 2000. M88-3 of 88.3 kN (19,840 lb st) maximum rating offered as follow-on, and being developed for export Rafale.

Internal tanks in single-seat versions for approximately 5,700 litres (1,506 US gallons; 1,254 Imp gallons) of fuel. Rafale B internal fuel confirmed as 5,300 litres (1,400 US gallons; 1,166 Imp gallons). Fuel system by Lucas Air Equipement, Lebozec and Zenith Aviation; equipment by Intertechnique. Five 'wet' hardpoints: centreline, two inboard wing and two centre wing; all able to accommodate a 1,250 litre (330 US gallon; 275 Imp gallon) external tank; alternative 2,000 litre (528 US gallon; 440 Imp gallon) centreline and inboard; alternative conformal tanks on spine, length 7.50 m (24 ft 7¼ in); total capacity 2,300 litres (608 US gallons; 506 Imp gallons), and able to accept in-flight replenishment. Pressure refuelling in 7 minutes, or 4 minutes for internal tanks only. Fixed (detachable) in-flight refuelling probe on all versions.

ACCOMMODATION: Pilot only, on SEMMB (Martin-Baker) Mk 16 zero/zero ejection seat, reclined at angle of 29°. One-piece Sully Produits Spéciaux blister windscreen/canopy, hinged to open sideways to starboard. Canopy gold-coated to reduce radar reflection.

SYSTEMS: Technofan cockpit air conditioning system; Cryotechnologies avionics cooling system. Dual hydraulic circuits, pressure 350 bar (5,075 lb/sq in), each with two Messier-Bugatti pumps and Bronzavia ancillaries. Auxilec electrical system, with two 30/40 kVA Auxilec variable frequency alternators. Triplex digital plus one dual analogue fly-by-wire flight control system, integrated with engine controls and linked with weapons system. Air Liquide OBOGS; EROS oxygen system; L'Hotellier fire detection system; Microturbo APU below fin fillet.

AVIONICS: Provision for more than 780 kg (1,720 lb) of avionics equipment and racks.

Comms: Thales V/UHF and Thales TRA 6032 SATURN UHF radios; Thales SB25A transponder/interrogator. TEAM intercom; Thales voice-activated radio controls and voice alarm warning system. Chelton aerials.

Radar: GIE Radar (Thales) RBE2 look-down/shoot-down radar, able to track up to eight targets simultaneously, with automatic threat assessment and allocation of priority. RBE2 AA (active array) version offered for export Rafales from 2006. Radar access via port-hinged radome.

Flight: Thales TLS-2020 integrated ILS/MLS, VOR/DME; SAGEM Sigma 95N (RL-90) RLG INS (SAGEM Telemir interface with carrier's navigation on Rafale M); Thales AHV-17 radio altimeter and SFIM/Thales ESPAR static memory flight recorder.

Instrumentation: Digital display of fuel, engine, hydraulic, electrical, oxygen and other systems information on two 127 × 127 mm (5 × 5 in) lateral multifunction touch-sensitive colour LCD displays by Thales. Third cockpit screen is 20 × 20° head-level tactical navigation/sensor display. Thales CTH3022 wide-angle, holographic HUD (30 × 22° field of view) incorporating Signaal USFA OTA-1320 CCD camera and recorder. Both displays collimated at infinity. Thales/Intertechnique Topsight E helmet-mounted sight.

Mission: Thales OSF electro-optical sensors and SAGEM OSF infra-red sensors. MIDS (Multifunctional Information Distribution System) datalink (equivalent to JTIDS/Link 16). Various reconnaissance, ECM, FLIR and laser designation pods, including Thales Damoclès target designator and projected Thales RECO-NG pod. Terrain-following system initially cleared (1999) to 152 m (500 ft); over land, reduced to 91 m (300 ft) by 2002, plus 30 m (100 ft) over sea; eventual goal is 30 m (100 ft) land and 15 m (50 ft) over water.

Self-defence: Spectra radar warning and ECM suite by Thales and MBDA. Thales DAL (*Détecteur d'Alerte Laser*) system. Flare launchers in rear wing/fuselage fairing.

EQUIPMENT: Integral, electrically operated, folding ladder in Rafale M.

ARMAMENT: One 30 mm Giat DEFA 791B cannon in side of starboard engine duct (except naval two-seat). Fourteen external stores attachments: two on fuselage centreline, two beneath engine intakes, two astride rear fuselage, six under wings and two at wingtips; of these, five stressed for heavy stores and fuel tanks. Forward centreline position deleted on Rafale M. Normal external load 6,000 kg (13,228 lb); maximum permissible, 9,500 kg (20,944 lb); see weapon options table. In strike role, one ASMP standoff nuclear weapon. In interception role, up to eight MICA AAMs (with IR or active homing) and two underwing fuel tanks; or six MICAs and three external fuel tanks. In air-to-ground role, typically sixteen 227 kg (500 lb) bombs, two MICAs and two 1,250 litre (330 US gallon; 275 Imp gallon) tanks; or two APACHE standoff weapon dispensers, two MICAs and three tanks; or FLIR pod, Atlis laser designator pod, two 1,000 kg (2,205 lb) laser-guided bombs, two AS.30L laser ASMs, four MICAs and single tank. In anti-ship role, two Exocet sea-skimming missiles, four MICAs and two external fuel tanks. Future weapons will include AASM (*Armement Air-Sol Modulaire*) powered LGB.

DIMENSIONS, EXTERNAL:
Wing span, incl wingtip missiles ... 10.80 m (35 ft 5¼ in)
Wing aspect ratio ... 2.6
Length overall ... 15.27 m (50 ft 1¼ in)
Height overall (Rafale D) ... 5.34 m (17 ft 6¼ in)

AREAS:
Wings, gross ... 45.70 m² (491.9 sq ft)

WEIGHTS AND LOADINGS (estimated):
Basic weight empty, equipped:
Rafale C ... 9,850 kg (21,716 lb)
Rafale B ... 10,450 kg (23,038 lb)
Rafale M ... 10,460 kg (23,060 lb)
Rafale N ... 10,710 kg (23,611 lb)
External load (incl fuel): normal ... 6,000 kg (13,228 lb)
max ... 9,500 kg (20,944 lb)
Max fuel weight: internal:
single seat ... 4,500 kg (9,921 lb)
two seat: Rafale B ... 4,350 kg (9,590 lb)
Rafale N ... 4,240 kg (9,348 lb)
underwing ... 7,500 kg (16,535 lb)
conformal ... 1,850 kg (4,079 lb)
Max T-O weight: early production ... 19,500 kg (42,990 lb)
subsequent production ... 22,500 kg (49,604 lb)
developed version ... 24,500 kg (54,013 lb)
Max landing weight ... 22,500 kg (49,605 lb)
Max wing loading:
initial version ... 426.7 kg/m² (87.39 lb/sq ft)
developed version ... 536.1 kg/m² (109.80 lb/sq ft)
Max power loading:
initial version ... 134 kg/kN (1.31 lb/lb st)
developed version (M88-3) ... 141 kg/kN (1.38 lb/lb st)

PERFORMANCE (estimated):
Max level speed: at altitude ... M1.8
at low level ... 750 kt (1,390 km/h; 864 mph)
Approach speed ... 120 kt (223 km/h; 139 mph)
Max rate of climb at S/L ... approx 18,290 m (60,000 ft)/min
Roll rate ... 270°/s
Max instantaneous turn rate ... up to 30°/s
Service ceiling ... 16,765 m (55,000 ft)
T-O distance: air defence ... 400 m (1,315 ft)
attack ... 600 m (1,970 ft)
Landing distance ... 450 m (1,480 ft)
Radius of action: low-level penetration with 12 × 250 kg bombs, four MICA AAMs and 4,000 litres (1,056 US gallons; 880 Imp gallons) of external fuel in three tanks ... 570 n miles (1,055 km; 655 miles)
air-to-air, long-range with eight MICA AAMs and 6,000 litres (1,585 US gallons; 1,320 Imp gallons) of external fuel in four tanks, 12,200 m (40,000 ft) transit ... 950 n miles (1,759 km; 1,093 miles)
Operational loiter ... up to 3 h
g limits ... +9.0/–3.2

UPDATED

DASSAULT FALCON 7X

TYPE: Long-range business tri-jet.

PROGRAMME: Announced, under temporary name/designation Falcon Next/NXT, at the Paris Air Show in June 2001; then became Falcon FNX; formal designation Falcon 7X announced 29 October 2001. By September 2002, high- and low-speed wind tunnel tests had been conducted at the ONERA facilities in Modane and Toulouse and in the European Transonic Wind Tunnel at Cologne, Germany; design frozen and first metal cut May 2003. First flight second quarter of 2005; mid-2005; certification scheduled for second half of 2006; first customer deliveries anticipated in late 2006.Initial production rate 15 per year.

CUSTOMERS: Market estimated at 400 aircraft over unspecified period. Total of 40 orders held by May 2004, split approximately evenly between US and other customers.

COSTS: Development cost estimated at US$600 million to US$700 million. Unit cost US$35 million to US$37.1 million (all 2003).

DESIGN FEATURES: Similar in configuration to Falcon 900C/900EX, but with cabin some 20 per cent longer, redesigned nose, double-curved windscreen panels and an entirely new high-subsonic section wing with 20 per cent fewer parts, 5° more sweepback, (34° on inboard section, 30° on outboard section), 40 per cent more area and featuring full-span, two-section leading-edge slats. The wing design will be employed on future developments within the Falcon range, expected to be designated Falcon 5X and 9X.

FLYING CONTROLS: Fly-by-wire controls, with sidestick controllers.

STRUCTURE: Total of 21 programme partner suppliers announced by May 2003, including: CASA (horizontal stabiliser); Stork-Fokker (trailing-edge control surfaces); Hurel Hispano/Aermacchi (engine nacelles and thrust reversers); Latécoère (T5 fuselage section); Socata (T34 upper fuselage section and body fairing); Sonaca (wing leading-edges); and Sully Saint Gobain (windscreen and cabin windows). Final assembly in new 23,226 m² (250,000 sq ft) Charles Lindbergh Building at Bordeaux, with capacity for four Falcon 7Xs per month.

POWER PLANT: Three Pratt & Whitney Canada PW307A turbofans, each flat-rated to 27.1 kN (6,100 lb st) at ISA +18°C. Usable fuel capacity 16,326 litres (4,313 US gallons; 3,591 Imp gallons).

LANDING GEAR: Retractable tricycle type by Messier-Dowty. Wheels and brakes by ABSC. Steerable nosewheel ± 60°. Minimum ground turning radius 8.80 m (61 ft 8 in).

ACCOMMODATION: Typical configuration will provide three lounge areas, berthing capability for six passengers, lavatories, galleys, crew rest area and a large flight-accessible baggage compartment. Pressurisation system will provide a 1,830 m (6,000 ft) cabin environment at high cruise altitudes.

SYSTEMS: Honeywell air management system. Pressurisation system maintains 1,830 m (6,000 ft) cabin environment at maximum operating altitude. Honeywell 35-150(FN) APU. Honeywell/Parker hydraulic power generation system. Parker hydraulic system; Intertechnique oxygen system; L'Hotelier fire detection and extinguishing; TRW electrical generation and distribution.

AVIONICS: Dassault EASy flight deck as core system, incorporating Honeywell Primus Epic platform.

DIMENSIONS, EXTERNAL:
Wing span ... 25.15 m (82 ft 6 in)

Computer generated image of Falcon 7X long-range business tri-jet ***NEW***/0573454

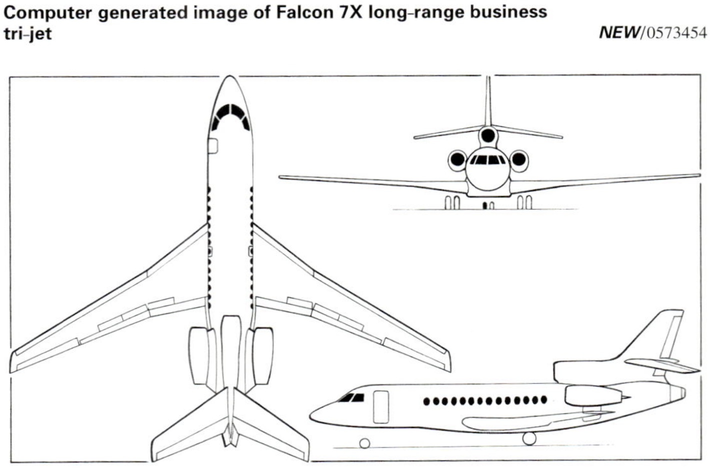

Provisional configuration of Dassault Falcon 7X *(James Goulding)* 0121273

Wing aspect ratio 8.9
Length overall .... 23.19 m (76 ft 1 in)
Height overall .... 7.77 m (25 ft 6 in)
Wheel track .... 9.75 m (31 ft 11¾ in)
Wheelbase .... 4.27 m (14 ft 0 in)
Passenger door: Height .... 1.72 m (5 ft 7¾ in)
Width .... 0.80 m (2 ft 7½ in)
Type III emergency exit: Height .... 0.91 m (3 ft 0 in)
Width .... 0.53 m (1 ft 9 in)

DIMENSIONS, INTERNAL:
Cabin
Length, third flight deck seat to baggage compartment .... 11.91 m (39 ft 1 in)
Width: max .... 2.64 m (8 ft 8 in)
at floor level .... 1.91 m (6 ft 3¼ in)
Max height .... 1.88 m (6 ft 2 in)
Volume .... 43.95 m^3 (1,552 cu ft)
Baggage compartment volume .... 4.45 m^3 (157 cu ft)

AREAS:
Wings, gross .... 70.70 m^2 (761.00 sq ft)

WEIGHTS AND LOADINGS:
Max fuel .... 13,109 kg (28,900 lb)
Max T-O weight .... 28,893 kg (63,700 lb)
Max landing weight .... 27,442 kg (60,500 lb)
Max ramp weight .... 28,985 kg (63,900 lb)
Max zero-fuel weight .... 17,735 kg (39,100 lb)
Max wing loading .... 408.7 kg/m^2 (83.71 lb/sq ft)
Max power loading .... 355 kg/kN (3.48 lb/lb st)

PERFORMANCE:
Max operating Mach No. (MMO) 0.90
Max operating speed (VMO) .... 370 kt (685 km/h; 425 mph)
Approach speed, landing weight of 16,828 kg (37,100 lb), 8 passengers, NBAA IFR reserves .... 104 kt (193 km/h; 120 mph)
Max operating altitude .... 15,545 m (51,000 ft)
FAR Pt 25 balanced T-O field length, S/L, ISA, with eight passengers and max fuel .... 1,585 m (5,200 ft)
FAR Pt 91 landing distance at landing weight of 16,828 kg (37,100 lb) with eight passengers and NBAA IFR reserves .... 716 m (2,350 ft)
Range: with three crew, eight passengers and NBAA IFR reserves:
at M0.80 .... 5,700 n miles (10,556 km; 6,559 miles)

UPDATED

Dassault Falcon 50EX long-range business jet ***NEW***/0573289

Dassault Falcon 50EX interior *(Bud Shannon/Dassault)* ***NEW***/0573314

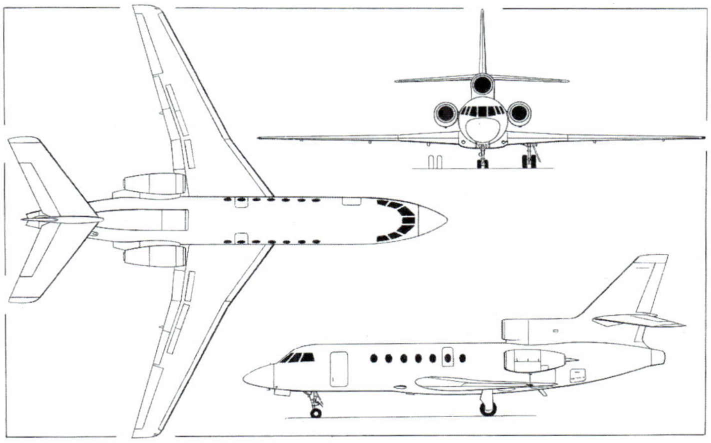

Dassault Falcon 50 long-range three-turbofan business transport *(Jane's/Dennis Punnett)* 0507137

DASSAULT FALCON 50

Spanish Air Force designation: T.16

TYPE: Business tri-jet.

DEVELOPMENT MILESTONES

Falcon 50	
First flight	7 Nov 76
First flight, pre-production	13 Jun 78
Certification	27 Feb 79
First delivery	Jul 79
Subsequent versions	
Falcon 50EX	
Announced	26 Apr 95
First flight	10 Apr 96
Certification	15 Nov 96
First delivery	Feb 97

PROGRAMME: First flight of prototype Falcon 50 (F-WAMD) 7 November 1976; second prototype 18 February 1978; first (and only) preproduction 13 June 1978. French certification 27 February 1979; FAA certification 7 March 1979; deliveries began July 1979. Available in sigint version from 1994.

Falcon 50EX announced 26 April 1995. Long-range variant; uprated turbofans provide a 7 per cent improvement in fuel consumption and improved initial cruising altitude of 12,500 m (41,000 ft). Production of initial batch of 40 Falcon 50EX airframes began 1995. First flight (F-WOND, c/n 251) 10 April 1996; new avionics suite installed in Falcon 50 c/n 252 for integration flight testing leading to avionics certification by FAA on 15 July 1996; DGAC certification achieved 15 November 1996, followed by FAA approval 20 December; initial production aircraft (F-WWHA, c/n 253) first flight October 1996; flown in 'green' condition Le Bourget, Paris, to Teterboro, New Jersey, 16 November, thence to Orlando, Florida, for US debut at National Business Aircraft Association Convention; first customer delivery (to Volkswagen of Germany) February 1997. Total of 80 EXs delivered by May 2003. Estimated market for 150 to 200 of EX version over unspecified period.

CURRENT VERSIONS: **Falcon 50:** Previous version, now superseded by Falcon 50EX.

Falcon 50-40: Retrofit of earlier version with Honeywell TFE731-40 engines, each 16.46 kN (3,700 lb). See *Jane's Aircraft Upgrades*.

Falcon 50EX: *As described.*

Falcon 50M Surmar: Maritime surveillance version.

First flight (No. 36/F-ZWTA) November 1998; first delivery in January 2000; operational with Flottille 24 at Lann-Bihoué from September 2000. Final (fourth) delivery in early 2002.

Falcon 50 Sigint: Model of a potential signals intelligence Falcon 50 displayed at Dubai Air Show in November 1997.

CUSTOMERS: Total 336 Falcon 50s and 50EXs delivered for outfitting by 1 April 2003, including 18 in 2000, 13 (all 50EXs) in 2001, 10 in 2002, eight in 2003, and none in the first three months of 2004. Adopted by governments of Burundi, Djibouti, France, Iraq (later Iran), Italy, Jordan, Libya, Morocco, Portugal, Rwanda, South Africa, Spain, Sudan and Yugoslavia; three of Italian Air Force convertible for medevac.

COSTS: 50EX US$20.6 million (2004).

DESIGN FEATURES: Three-engine layout permits overflight of oceans and desert areas within public transport regulations. Sharply waisted rear fuselage and engine pod designed by computational fluid dynamics; wing has compound leading-edge sweep (24° 50' to 29° at quarter-chord) and optimised section.

FLYING CONTROLS: Fully powered controls with pushrods, dual-barrel hydraulic actuators and artificial feel; variable incidence anhedral tailplane with dual electrical actuation by screwjack; drooped leading-edge inboard and slats outboard; double-slotted flaps; three two-position airbrake/spoiler panels on each wing.

STRUCTURE: All-metal, circular-section fuselage with rear baggage compartment inside pressure cabin; wing boxes are integral fuel tanks bolted to carry-through box. Carbon fibre horizontal tail surfaces introduced during 2001. Latécoère, Potez, Reims Aviation, SEFCA and Socata are subcontractors on the programme; other suppliers include Dassault Equipements (flight controls, flaps, slats and airbrakes), Liebherr-Aerospace Toulouse (engine bleed air system); SARMA (flight control actuating rods) and Sully Produits Spéciaux (windscreen panels).

LANDING GEAR: Retractable tricycle type by Messier-Dowty, with twin wheels on each unit. Hydraulic retraction, main units inward, nosewheels forward. Nosewheels steerable ±60° for taxying, ±180° for towing. ABS wheels, brakes and braking system. Mainwheel tyres size 26×6.6 (14 ply) or 26×6.6R14 tubeless, pressure 14.55 bar (211 lb/sq in). Nosewheel tyres size 14.5×5.5-6 (14 ply) tubeless, pressure 9.31 bar (135 lb/sq in). Four-disc brakes designed for 400 landings with normal energy braking. Minimum ground turning radius (about nosewheels) 13.54 m (44 ft 5 in).

POWER PLANT: Three Honeywell TFE731-40 turbofans, each rated at 16.46 kN (3,700 lb st) at ISA + 17°C in Falcon 50EX. Two engines pod-mounted on sides of rear fuselage, third attached by two top mounts. Thrust reverser on centre engine. Fuel in integral tanks, with capacity of 5,787 litres (1,529 US gallons; 1,273 Imp gallons) in wings and 2,976 litres (786 US gallons; 655 Imp gallons) in fuselage tanks. Total fuel capacity 8,763 litres (2,315 US gallons; 1,928 Imp gallons). Single-point pressure fuelling. Intertechnique fuel distribution and gauges.

ACCOMMODATION: Standard accommodation for two crew and nine passengers. In typical arrangement, cabin is divided into a forward section with four armchairs and two fold-out tables, and a rear section with three-seat sofa (convertible into a single bed) and two armchairs separated by a fold-out table; two galleys at forward end of cabin, toilet at rear. Alternative layouts to customer choice, with accommodation for a maximum of 19 passengers. Cabin and rear baggage compartment are pressurised and air conditioned; Barber-Colman temperature controls. Access is by separate door on port side.

SYSTEMS: Air conditioning system utilises bleed air from all three engines or APU. Maximum pressure differential 0.63 bar (9.14 lb/sq in). Pressurisation maintains a maximum cabin altitude of 2,440 m (8,000 ft) to a flight altitude of 14,935 m (49,000 ft). Two independent Messier-Bugatti and Vickers-Sterer hydraulic systems, pressure 207 bar (3,000 lb/sq in), with three engine-driven pumps and one emergency electric pump, actuate primary flying controls, flaps, slats, landing gear, wheel brakes, airbrakes and nosewheel steering. Plain reservoir, pressurised by bleed air at 1.47 bar (21 lb/sq in). 28 V DC electrical system, with a 9 kW 28 V DC Auxilec starter/generator on each engine and two 23 Ah batteries. Labinal electrical harnesses. Wing leading-edge, centre engine S-duct and engine nacelles have engine bleed air anti-icing. Automatic emergency oxygen system. Honeywell 36-150 APU standard.

AVIONICS: *Comms:* Dual Rockwell Collins VHF and mode S transponders; dual Rockwell Collins HF-9000 HF transceivers with Selcal; Teledyne Controls MagnaStar or Honeywell Flitefone 800 radiotelephones; cockpit voice recorder and ELT standard.

Radar: Rockwell Collins TWR 850 Doppler turbulence detection weather radar.

Flight: Dual Rockwell Collins VIR-432 VOR, ADF-462 and DME-442. Rockwell Collins APS-4000 autopilot, ADC-850C air data systems, dual Honeywell FHS-6100 with integrated GPS receiver, Honeywell EGPWS and dual Honeywell Laseref III laser gyro inertial reference systems standard. Dual Universal UNS-1C FMS optional, replacing GNS-XES.

Instrumentation: Rockwell Collins Pro Line 4 (EFIS-4000) four-tube EFIS; Thales three-tube LCD engine indicating electronic display (EIED).

DIMENSIONS, EXTERNAL:
Wing span ... 18.86 m (61 ft 10½ in)
Wing chord (mean) ... 2.84 m (9 ft 3¾ in)
Wing aspect ratio ... 7.6
Length: overall ... 18.52 m (60 ft 9¼ in)
fuselage ... 17.66 m (57 ft 11 in)
Height overall ... 6.98 m (22 ft 10¾ in)
Tailplane span ... 7.74 m (25 ft 4¾ in)
Wheel track ... 3.98 m (13 ft 0¾ in)
Wheelbase ... 7.24 m (23 ft 9 in)
Passenger door: Height ... 1.52 m (4 ft 11¾ in)
Width ... 0.80 m (2 ft 7½ in)
Height to sill ... 1.30 m (4 ft 3¼ in)
Emergency exits (each side, over wing):
Height ... 0.92 m (3 ft 0¼ in)
Width ... 0.51 m (1 ft 8 in)
Baggage door: Height ... 0.73 m (2 ft 4¾ in)
Width ... 0.99 m (3 ft 3 in)

DIMENSIONS, INTERNAL:
Cabin, incl forward baggage space and rear toilet:
Length ... 7.16 m (23 ft 6 in)
Max width ... 1.86 m (6 ft 1¼ in)
Max height ... 1.80 m (5 ft 10¾ in)
Volume ... 20.2 m^3 (712 cu ft)
Baggage volume ... 2.55 m^3 (90.0 cu ft)
Baggage compartment (rear) ... 3.26 m^3 (115 cu ft)

AREAS:
Wings, gross ... 46.83 m^2 (504.1 sq ft)
Horizontal tail surfaces (total) ... 13.35 m^2 (143.69 sq ft)
Vertical tail surfaces (total) ... 9.82 m^2 (105.70 sq ft)

WEIGHTS AND LOADINGS:
Weight empty, equipped ... 9,603 kg (21,170 lb)
Basic operating weight ... 9,888 kg (21,800 lb)
Baggage capacity (rear) ... 1,000 kg (2,205 lb)
Max payload: normal ... 1,710 kg (3,770 lb)
with max fuel ... 1,170 kg (2,579 lb)
Max fuel ... 7,040 kg (15,520 lb)
Max T-O weight: standard ... 18,007 kg (39,700 lb)
optional ... 18,497 kg (40,780 lb)
Max ramp weight: standard ... 18,098 kg (39,900 lb)
optional ... 18,497 kg (40,780 lb)
Max zero-fuel weight ... 9,888 kg (21,800 lb)
Max landing weight ... 16,200 kg (35,715 lb)
Max wing loading ... 384.5 kg/m^2 (78.75 lb/sq ft)
Max power loading ... 365 kg/kN (3.58 lb/lb st)

PERFORMANCE:
Max operating Mach No. (MMO) ... 0.86
Max operating speed (VMO):
at S/L ... 350 kt (648 km/h; 402 mph) IAS
at FL237 ... 370 kt (685 km/h; 425 mph) IAS
Max cruising speed ... M0.85 or 487 kt (902 km/h; 560 mph)
Normal cruising speed ... M0.80 or 459 kt (850 km/h; 528 mph)
Long-range cruising speed at FL350 ... M0.75 (430 kt; 797 km/h; 495 mph)
Approach speed, eight passengers and NBAA IFR reserves . 107 kt (198 km/h; 123 mph)
Initial cruising altitude ... 12,497 m (41,000 ft)
Max certified altitude ... 14,935 m (49,000 ft)
Time to climb to FL410 ... 23 min
FAR Pt 25 balanced T-O field length, S/L, ISA with eight passengers and max fuel ... 1,490 m (4,890 ft)
FAR Pt 91 landing distance at MLW with eight passengers and NBAA IFR reserves ... 666 m (2,185 ft)
Range with eight passengers and NBAA IFR reserves:
at M0.80 ... 3,025 n miles (5,602 km; 3,481 miles)
at M0.75 ... 3,285 n miles (6,083 km; 3,780 miles)

OPERATIONAL NOISE LEVELS:
T-O ... 83.8 EPNdB
Approach ... 95.2 EPNdB
Sideline ... 92.0 EPNdB

UPDATED

DASSAULT FALCON 900

Spanish Air Force designation: T.18

TYPE: Long-range business tri-jet.

PROGRAMME: Falcon 900 announced 27 May 1983; first flight of prototype (F-GIDE *Spirit of Lafayette*) 21 September 1984, second aircraft (F-GFJC) 30 August 1985; flew non-stop 4,305 n miles (7,973 km; 4,954 miles) Paris to Little Rock, Arkansas, September 1985; returned Teterboro, New Jersey, to Istres, France, at M0.84; French and US certification March 1986, including status close to FAR Pts 25 and 55 for damage tolerance of entire airframe.

Prototype Falcon 900 in use as testbed (first flight 12 April 1994) for new laminar flow wing section intended to provide significant reductions in drag. New section installed as sleeve on inner wing and designed to demonstrate hybrid laminar flow with boundary layer suction via seven channels in laser-drilled titanium skin over some 10 per cent chord of upper wing surface. Following test programme, modified aircraft has returned to Dassault Falcon Service to validate laminar flow capability under normal commercial operating conditions. The 250th Falcon 900 series aircraft was delivered in mid-2000.

CURRENT VERSIONS: **Falcon 900B:** French and UK certification received end 1991; complies with FAR Pt 36 Stage III and ICAO 16 noise requirements; approved for Cat. II approaches, for operations from unpaved fields; re-engined with TFE731-5BR-1C turbofans, to give 5.5 per cent power increase; initial cruising altitude 11,855 m (39,000 ft) and NBAA IFR range increased by 100 n miles (185 km; 115 miles); retrofit offered to existing operators. Supplanted by Falcon 900C.

Falcon 900C: Announced 26 June 1998. Combines airframe, engines and cabin of 900B with Honeywell Primus 2000 avionics of 900EX, but without autothrottles; 900B F-WWFP/F-GRDP (c/n 169) served as prototype for certification; first flown 17 December 1998; deliveries began in December 1999 (c/n 180 to Sony Aviation Corp in USA), replacing 900B on production line. French DGAC certification achieved 15 June 1999, with FAA certification following on 26 August 1999.

Detailed description applies to Falcon 900C, except where indicated.

Falcon 900DX: Announced at the European Business Aviation Convention and Exhibition at Geneva 24 May 2004 as future replacement for Falcon 900C. Features the

Standard flight deck of Dassault Falcon 900EX ***NEW***/0573311

Flight deck of Falcon 900EX *(Philippe Stroppa/Arovn Production/Dassault)* ***NEW***/0573313

DEVELOPMENT MILESTONES

Falcon 900	
Announced	27 May 83
First flight	21 Sep 84
Certification	Mar 86
Subsequent versions:	
Falcon 900EX	
Announced	Oct 94
Rolled out	13 Mar 95
First flight	1 Jun 95
Certification	31 May 96
First delivery	1 Nov 96
Falcon 900C	
Announced	Jun 98
First flight	17 Dec 98
Certification	15 Jun 99

Dassault Falcon 900EX EASy business jet *(Paul Bowen/Dassault)* ***NEW***/0573312

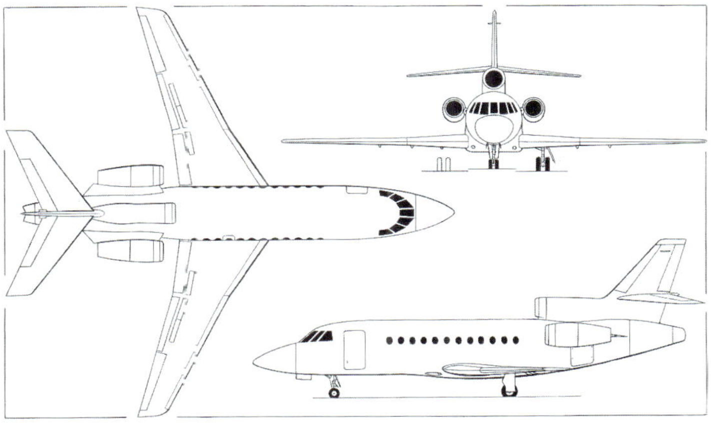

Dassault Falcon 900C (three Honeywell TFE731 turbofans) *(Jane's/Dennis Punnett)* 0507138

Falcon 900EX's Honeywell TFE731-60 engines and Dassault EASy flight deck, but with fuel capacity reduced to 8,541 kg (18,830 lb). Provisional data include: cost US$31.65 million (2004); equipped operating weight 11,099 kg (24,470 lb); maximum T-O weight 21,183 kg (46,700 lb); initial cruising altitude 12,497 m (41,000 ft); time to climb to FL370 17 minutes; T-O balanced field length 1,490 m (4,890 ft); maximum range 4,100 n miles (7,593 km; 4,718 miles). Certification and first deliveries are scheduled for December 2005.

Falcon 900EX: Long-range development of 900B, announced October 1994. Re-engined with 22.24 kN (5,000 lb st) (ISA+17°C) Honeywell TFE731-60 turbofans, to give 5.8 per cent increase in retained thrust at 12,200 m (40,000 ft) and more than 8 per cent improvement in cruise specific fuel consumption. Engine nacelles, pylons, thrust reversers and portions of centre engine S-duct redesigned; maximum fuel capacity increased to 11,865 litres (3,134 US gallons; 2,610 Imp gallons) by addition of centre-fuselage tank of 591 kg (1,303 lb) capacity and tank in rear fuselage, capacity 240 kg (530 lb).

Upgraded standard avionics comprise fully integrated Honeywell Primus 2000 suite with five-tube 20 × 17.75 cm (8 × 7 in) colour EFIS; one engine instrument display; three IC-800 integrated avionics computers; dual FMZ-2000 flight management systems with a third optional; dual fail-operational autopilots; T-O to landing autothrottle; Honeywell EGPWS; dual Laseref III inertial reference systems with third optional; Primus colour weather radar; optional single or dual 12-channel GPS, multichannel satcom, communications management unit (CMU) and Flight Dynamics HGS-2850 head-up display. Dassault EASy integrated flight deck with four 330 × 254 mm (13 × 10 in) active matrix liquid crystal displays (AMLCDs) to be introduced on 900EX from 2003. First flight with EASy cockpit was made by F-WNCD (c/n 97) on 22 February 2002.

Risk-sharing partners, representing 20 per cent of total development investment, are Honeywell (engines and primary avionics), SABCA (centre engine intake cowlings), Hellenic Aircraft Industries (rear fuselage fuel tank), Latécoère (T5 fuselage section and engine pylons), and Alenia (nacelles and centre engine thrust reverser).

Prototype (F-WREX) rolled out 13 March 1995; first flight 1 June 1995; flew Luton, England, to Las Vegas, Nevada, non-stop on 24 September 1995, completing the 4,700 n mile (8,704 km; 5,409 mile) flight in 11 hours 40 minutes including 30 minutes hold for air traffic delays; DGAC certification 31 May 1996; FAA approval granted 19 July after 350-hour flight test programme; first customer delivery to Anheuser-Busch Companies Inc (N200L) 1 November 1996; production aircraft delivered to Little Rock, Arkansas, for outfitting.

Japan MSA: Two Falcon 900s for long-range maritime surveillance entered service with the Japan Maritime Safety Agency September 1989; US search radar, special communications radio, operations control station, U-125A-style search windows and drop hatch for sonobuoys, markers and flares.

CUSTOMERS: Total 201 Falcon 900s and 124 Falcon 900EXs delivered to completion centres by 31 March 2003. Government/VIP versions operated by Algeria, Australia, Belgium, Equatorial Guinea, France, Gabon, Italy, Malaysia, Nigeria, Russia, Saudi Arabia, Spain, Syria and United Arab Emirates. Production totalled 20 in 1998, 24 in 1999, 29 (six 900C and 23 900EX) in 2000, 27 (six 900C and 21 900EC) in 2001, 21 (four 900C and 17 900EX) in 2002, 13 (three 900C, six 900EX and four 900EX EASy) in 2003, and five (all 900EX EASy) in the first three months of 2004.

COSTS: Standard equipped 900C US$31.3 million (2004); 900EX EASy US$34.75 million (2004).

DESIGN FEATURES: Larger cross-section and cabin length than Falcon 50; added economy and further power increase of engines achieved by mixer compound nozzle tailpipe, mixing cold and hot flows.

Wing adapted from Falcon 50 but increased span and area, and optimised for M0.84 cruise; compound leading-edge sweep (24° 50' to 29° at quarter-chord); dihedral 0° 30'.

FLYING CONTROLS: Fully powered flying controls with artificial feel and variable-incidence tailplane as for Falcon 50; full-span slats and double-slotted Fowler flaps; three-position airbrakes.

STRUCTURE: Design and manufacture computer assisted; damage-tolerant structure; extensive use of carbon fibre and aramid (Kevlar); Kevlar radome, wingroot fairings and tailcone; secondary rear cabin pressure bulkhead allows access to baggage in flight and additional protection against pressure loss. Nosewheel doors of Kevlar; mainwheel doors of carbon fibre. Kevlar air intake trunk for centre engine, and rear cowling for side engines. Carbon fibre central cowling around all three engines.

New horizontal tail surface featuring cast titanium central box with resin transfer moulding composite spars and carbon fibre skin panels, resulting in a 13.6 kg (30 lb) reduction in weight, certified by the DGAC in December 1999 and introduced as standard on Falcon 900C and 900EX from December 2000 deliveries.

LANDING GEAR: Retractable tricycle type by Messier-Bugatti, with twin wheels on each unit. Hydraulic retraction, main units inward, nosewheels forward. Oleo-pneumatic shock-absorbers. Mainwheels fitted with Michelin radial tyres size 29×7.7-15, pressure: 900C 13.60 bar (197 lb/sq in), 900EX 13.80 bar (200 lb/sq in). Nosewheel tyres size 17.5×5.75R8, pressure: 900C 10.20 bar (148 lb/sq in, 900EX 10.90 bar (158 lb/sq in). Hydraulic nosewheel steering (±60° for taxying, ±180° for towing). Messier-Bugatti triple-disc carbon brakes and anti-skid system. Minimum ground turning radius (about nosewheels) 14.55 m (47 ft 8¾ in).

POWER PLANT: Three Honeywell TFE731-5BR-1C turbofans, each rated at 21.13 kN (4,750 lb st) at ISA + 10°C. Thrust reverser on centre engine. Fuel in two integral tanks in wings, capacity 3,428 litres (906 US gallons; 754 Imp gallons) and 3,437 litres (908 US gallons; 756 Imp gallons; forward fuselage tank, capacity 2,061 litres (544 US gallons; 453 Imp gallons), and rear fuselage tank capacity 1,899 litres (502 US gallons; 418 Imp gallons), for total fuel capacity of 10,825 litres.

ACCOMMODATION: Type III emergency exit on starboard side of cabin permits wide range of layouts for up to 19 passengers. Flight deck for two pilots, with central jump-seat. Flight deck separated from cabin by door, with crew wardrobe and baggage locker on either side. Galley at front of main cabin, on starboard side opposite main cabin door. Passenger area is divided into three lounges. Forward zone has four 'sleeping' swivel chairs in facing pairs with tables. Centre zone is dining area, with two double seats facing a transverse table. On starboard side, storage cabinet contains foldaway bench, allowing five to six persons to be seated around table, while leaving emergency exit clear. In rear zone, inward-facing three-seat settee on starboard side converts into a bed. On port side, two armchairs are separated by a table. At rear of cabin, a door leads to toilet compartment, on starboard side, and a second structural plug door to large rear baggage area. Baggage door is electrically actuated.

Other interior configurations available. Alternative eight-passenger configuration has bedroom at rear and three personnel seats in forward zone. A 15-passenger layout divides a VIP area at rear from six (three-abreast) chairs forward; full fuel can still be carried with

15 passengers. The 18-passenger scheme has four rows of three-abreast airline-type seats forward, and VIP lounge with two chairs and settee aft. Many optional items, including additional windows, front toilet unit, video system with one or more monitors, 'Airshow 200' navigation display system, compact disc deck, aft cabin partition, one or two couches in aft cabin convertible to bed(s), storage cabinet in baggage hold, aft longitudinal table, individual listening devices for passengers, lifejackets and rafts.

SYSTEMS: Air conditioning system uses engine bleed air or air from Honeywell GTCP36-150 APU installed in rear fuselage. Softair pressurisation system, with maximum differential of 0.64 bar (9.3 lb/sq in), maintains sea level cabin environment to height of 7,620 m (25,000 ft), and cabin equivalent of 2,440 m (8,000 ft) at 15,550 m (51,000 ft). Cold air supply by single oversize air cycle unit. Two independent hydraulic systems, pressure 207 bar (3,000 lb/sq in), with three engine-driven pumps and one emergency electric pump, actuate primary flying controls, flaps, slats, landing gear retraction, wheel brakes, airbrakes, nosewheel steering and thrust reverser. Bootstrap hydraulic reservoirs. DC electrical system supplied by three 9 kW 28 V Auxilec starter/generators and two 23 Ah batteries. Heated bleed air anti-icing of wing leading-edges, intakes and centre engine duct; electrically heated windscreens. Eros (SFIM/Intertechnique) oxygen system.

AVIONICS: *Comms:* Honeywell Primus 2000 as core system.

Radar: Honeywell Primus 870 colour weather radar.

Flight: Dual autopilot. Honeywell FMZ-2000 FMS with two AZ-840 micro air data systems and two Laseref III LINS; Rockwell Collins dual VIR-432 VOR/ILS marker receiver, dual ADF-462 and DME-442; Honeywell AA-300 radar altimeter and IC-800 autopilot; and Honeywell EGPWS.

Instrumentation: Honeywell Primus 2000 five-tube EFIS, comprising two MFDs, two PFDS and one EIED, each measuring 203 × 178 mm (8 × 7 in). Flight Dynamics HGS-2850 head-up guidance system optional.

DIMENSIONS, EXTERNAL:
Wing span ... 19.33 m (63 ft 5 in)
Wing chord: at root ... 4.08 m (13 ft 4¾ in)
at tip ... 1.12 m (3 ft 8 in)
Wing aspect ratio ... 7.6
Length overall ... 20.21 m (66 ft 3¾ in)
Fuselage: Max diameter ... 2.50 m (8 ft 2½ in)
Height overall ... 7.55 m (24 ft 9¼ in)
Tailplane span ... 7.74 m (25 ft 4¾ in)
Wheel track ... 4.45 m (14 ft 7¼ in)
Wheelbase ... 7.90 m (25 ft 11 in)
Passenger door: Height ... 1.72 m (5 ft 7¾ in)
Width ... 0.80 m (2 ft 7½ in)
Height to sill ... 1.64 m (5 ft 4½ in)
Emergency exit (overwing, stbd):
Height ... 0.92 m (3 ft 0¼ in)
Width ... 0.53 m (1 ft 8¾ in)
Baggage door: Height ... 0.75 m (2 ft 5½ in)
Width ... 0.95 m (3 ft 1½ in)

DIMENSIONS, INTERNAL:
Cabin, excl flight deck, incl toilet and baggage compartments:
Length ... 10.11 m (33 ft 2 in)
Max width ... 2.34 m (7 ft 8¼ in)
Width at floor ... 1.91 m (6 ft 3¼ in)
Max height ... 1.88 m (6 ft 2 in)
Volume ... 35.8 m^3 (1,264 cu ft)
Rear baggage compartment volume ... 3.6 m^3 (127 cu ft)
Flight deck volume ... 3.8 m^3 (132 cu ft)

AREAS:
Wings, gross ... 49.00 m^2 (527.4 sq ft)
Vertical tail surfaces (total) ... 9.82 m^2 (105.7 sq ft)
Horizontal tail surfaces (total) ... 13.35 m^2 (143.7 sq ft)

WEIGHTS AND LOADINGS:
Weight empty, equipped (typical):
900EX ... 10,829 kg (23,875 lb)
Basic operating weight: 900C ... 10,977 kg (24,200 lb)
900EX ... 11,204 kg (24,700 lb)
Max payload: 900C ... 1,823 kg (4,020 lb)
900EX ... 2,796 kg (6,164 lb)
Payload with max fuel: 900C ... 1,059 kg (2,335 lb)
900EX ... 1,270 kg (2,800 lb)
Max fuel: 900C ... 8,693 kg (19,165 lb)
900EX ... 9,525 kg (21,000 lb)
Max ramp weight: 900C standard ... 20,729 kg (45,700 lb)
900C optional ... 21,183 kg (46,700 lb)
900EX standard ... 22,000 kg (48,500 lb)
900EX optional ... 22,317 kg (49,200 lb)
Max T-O weight: 900C standard ... 20,640 kg (45,500 lb)
900C optional ... 21,092 kg (46,500 lb)
900EX standard ... 21,909 kg (48,300 lb)
900EX optional ... 22,226 kg (49,000 lb)
Max landing weight:
900C, 900EX standard ... 19,050 kg (42,000 lb)
900EX optional ... 20,185 kg (44,500 lb)
Normal landing weight, eight passengers and fuel reserves: 900C .. 12,639 kg (27,865 lb)
900EX ... 12,846 kg (28,321 lb)
Max zero-fuel weight: 900C standard ... 12,800 kg (28,220 lb)
900C optional ... 14,000 kg (30,865 lb)
900EX ... 14,000 kg (30,865 lb)
Max wing loading:
900C standard ... 421.2 kg/m^2 (86.27 lb/sq ft)
900C optional ... 430.5 kg/m^2 (88.17 lb/sq ft)
900EX standard ... 447.1 kg/m^2 (91.58 lb/sq ft)
900EX optional ... 453.6 kg/m^2 (92.91 lb/sq ft)
Max power loading:
900C standard ... 326 kg/kN (3.19 lb/lb st)
900C optional ... 333 kg/kN (3.26 lb/lb st)
900EX standard ... 328 kg/kN (3.22 lb/lb st)
900EX optional ... 333 kg/kN (3.27 lb/lb st)

PERFORMANCE (at AUW of 12,250 kg; 27,000 lb, except where indicated):
Max operating speed (VMO): 900C, 900EX:
at S/L ... M0.87 (350 kt; 648 km/h; 403 mph IAS)
between FL100 and FL250 ... M0.84 (370 kt; 685 km/h; 425 mph IAS)
Max cruising speed:
900C, 900EX ... M0.84 or 481 kt (891 km/h; 554 mph)
Normal cruising speed:
900C, 900EX ... M0.80 or 459 kt (850 km/h; 528 mph)
Long-range cruising speed:
900C, 900EX ... M0.75 or 430 kt (796 km/h; 495 mph)
Approach speed, eight passengers and NBAA IFR fuel reserves:
900C ... 107 kt (199 km/h; 124 mph)
900EX ... 109 kt (202 km/h; 126 mph)
Stalling speed:
900C, 900EX, clean ... 106 kt (196 km/h; 122 mph)
900C, 900EX, landing configuration ... 85 kt (158 km/h; 98 mph)
Initial cruising altitude:
900C, 900EX ... 11,885 m (39,000 ft)
Max certified altitude:
900C, 900EX ... 15,550 m (51,000 ft)
FAR Pt 25 balanced T-O field length at standard max T-O weight: 900C ... 1,504 m (4,935 ft)
900EX ... 1,590 m (5,215 ft)
FAR Pt 91 landing distance:
with NBAA IFR reserves:
900C with eight passengers ... 716 m (2,350 ft)
900EX with eight passengers ... 724 m (2,375 ft)
Range with NBAA IFR reserves:
900C with five passengers at M0.75 ... 4,000 n miles (7,408 km; 4,603 miles)
at M0.80 ... 3,830 n miles (7,093 km; 4,407 miles)
900EX with eight passengers:
at M0.84 ... 3,810 n miles (7,056 km; 4,384 miles)
at M0.80 ... 4,335 n miles (8,028 km; 4,988 miles)
at M0.75 ... 4,500 n miles (8,334 km; 5,178 miles)

OPERATIONAL NOISE LEVELS:
T-O: 900C, 900EX ... 79.8 EPNdB
Approach: 900C ... 91.7 EPNdB
900EX ... 92.3 EPNdB
Sideline: 900C ... 91.2 EPNdB
900EX ... 90.5 EPNdB

UPDATED

DASSAULT FALCON 2000

TYPE: Business jet.

DEVELOPMENT MILESTONES

Falcon 2000	
Announced	Jun 89
Launched	4 Oct 89
First order	4 Oct 89
First flight	24 Mar 93
First flight, production	10 May 94
Certification	30 Nov 94
Subsequent versions:	
Falocn 200EX	
Announced	10 Oct 00
Rolled out	19 Jul 01
First flight	25 Oct 01
First flight, EASy flight deck	29 Jan 03
Certification	25 Mar 03
First delivery	2Q 03

PROGRAMME: Announced Paris Air Show 1989 as Falcon X; follow-on to Falcon 20/200; launched as Falcon 2000 on 4 October 1990, following first orders; Alenia joined as 25 per cent risk-sharing partner February 1991, with responsibility for rear fuselage section and engine nacelles; selection of CFE738 engine announced 2 April 1990; first flight (F-WNAV) 4 March 1993; one prototype only; third airframe (F-WWFA) was second to fly, on 10 May 1994; ferried 'green' to US subsidiary Dassault Falcon Jet Corporation at Little Rock, Arkansas, 13 July 1994 for completion as US demonstrator, and set two world speed records 31 October to 1 November 1994 flying Los Angeles (Chino) to Bangor, Maine, in 4 hours 36 minutes 27 seconds and Bangor to Paris in 5 hours 26 minutes 12 seconds; Dassault demonstrator, second airframe (F-WNEW), flew 11 July 1994.

JAA certification to JAR 25 obtained 30 November 1994, at which time five aircraft (prototype, two demonstrators and two customer aircraft) had accumulated 1,055.5 flight hours; FAR Pt 25 and FAR Pt 36 Stage 3 noise levels certification 2 February 1995; first delivery (F-WNEW/ZS-NNF to South African customer) 16 February 1995. Approved to operate into London City Airport during 1996. Commonwealth of Independent States certification 14 April 1997. HGS-2850 HUD first flown in prototype 8 December 1995; initial JAA certification granted 30 August 1996, followed by approval for low-visibility

Dassault Falcon 2000EX deliveries began in 2003 ***(François Robineau/Dassault)*** ***NEW***/0573309

take-offs (down to 91 m; 300 ft RVR) and hand-flown Cat. II and Cat. IIIa instrument approach approved by FAA 30 July 1997. Full JAA certification 16 December 1997; FAA certification 18 May 1998.

CURRENT VERSIONS: **Falcon 2000:** Initial production version. *As described.*

Falcon 2000EX: Extended-range version. Development began October 1999; announced at the NBAA Convention in New Orleans, 10 October 2000. Prototype (c/n 01/F-WMEX) rolled out 19 July 2001 and first flown 25 October 2001; two aircraft (F-WMEX and c/n 02/F-W66A). Completed 568 hours of flight testing in 242 flights, culminating in JAA and FAA certification on 25 March 2003; first three 'green' production airframes delivered to Little Rock completion centre in January 2003, with first customer deliveries of outfitted aircraft beginning in the second quarter of 2003 (Pro Line avionics). EASy flight deck available from second quarter of 2004. Falcon 2000EX manufactured in parallel with Falcon 2000.

New features include two Pratt & Whitney Canada PW308C turbofans, with FADEC, each rated at 31.1 kN (7,000 lb st) for take-off at ISA +15°C; Nordam advanced single pivot (ASP) thrust reversers; revised fuel system with 2,044 kg (4,506 lb) increase in usable fuel capacity to 9,414 litres (2,487 US gallons; 2,071 Imp gallons); strengthened main landing gear with heavy-duty braking system brake/steering control unit, and Falcon 900EX nose landing gear; and new bleed air system. The first 40 Falcon 2000EXs will retain the current Collins Pro Line 4 avionics suite, but aircraft delivered from March 2004 will be equipped with Dassault EASy flight deck with four-tube flat panel displays based on those of the Honeywell Primus Epic modular avionics system and designated **Falcon 2000EX EASy**; first flight with EASy cockpit F-WXEY) 29 January 2003.

CUSTOMERS: The 206th Falcon 2000 was delivered to the USA for outfitting on 10 December 2003. Production totalled 20 in 1997, 15 in 1998, 34 in 1999, 26 in 2000, 35 in 2001, 35 in 2002, 16 in 2003, and three in the first three months of 2004. Falcon 2000 No. 159/N259QS, delivered to Executive Jet on 11 September 2001, was also 1,500th Falcon/Mystère business jet produced by Dassault. Total of 72 Falcon 2000s, and 28 Falcon 2000EXs, plus 25 Falcon 2000EXs on option, ordered by Executive NetJets and NetJets Middle East fractional ownership programmes in Europe, Middle East and USA, of which 38 were in operation by October 2002. Royal Australian Air Force ordered three in 2000 to replace five leased Falcon 900s. Total market estimated at more than 300 in 10 years.

COSTS: Basic price 2000 US$23.2 million; 2000EX US$24.97 million (both 2004).

DESIGN FEATURES: Same fuselage cross-section as Falcon 900, but 1.98 m (6 ft 6 in) shorter. Falcon 900 wing with modified leading-edge and no inboard slats; sweepback at quarter-chord 24° 50' to 29°.

FLYING CONTROLS: Fully powered. Two-section flaps and three-section spoilers each side. Variable incidence tailplane. Leading-edge inboard slats. Maximum control surface movements: elevators +20/–16°; ailerons +25° 20'/–24° 50'; rudder ±29°; flaps –40°; airbrakes, inboard +68°, centre +50°, outboard +37°; leading-edge slats –30°.

STRUCTURE: Largely as for Falcon 900. Rear fuselage, including engine pods and pylons, by Alenia and Piaggio; thrust reversers by Dee Howard. Agreement reached in early 2002 for production of fuselage components of 2000EX by Chengdu Aircraft of China. New horizontal tail surface featuring cast titanium central box with resin transfer moulding composite spars and carbon fibre skin panels, resulting in a 13.6 kg (30 lb) reduction in weight, certified by the DGAC in December 1999 and introduced as standard from December 2000 deliveries.

LANDING GEAR: Retractable tricycle type; mainwheels retract inwards, nosewheel forwards. Main tyres 26×6.6R14 (14 ply) tubeless; nose 14.5×5.5R6 tubeless. Tyre pressures: 2000 main 13.6 bar (197 lb/sq in), nose 11.1 bar (161 lb/sq in); 2000EX main 15.1 bar (219 lb/sq in), nose 12.6 bar (183 lb/sq in). Steerable nosewheel (±60°). Minimum turning radius on ground 15.03 m (49 ft 3¾ in).

POWER PLANT: Two GE/Honeywell CFE738-1-1B turbofans, each rated at 26.3 kN (5,918 lb st) at ISA + 15°C. Fuel contained in six integral tanks (three each, outboard wing, inboard wing and centre wing box, port and starboard sides) plus two feeder tanks, combined capacity 6,927 litres (1,830 US gallons; 1,524 Imp gallons), of which 6,867 litres (1,814 US gallons; 1,510 Imp gallons) are usable. Honeywell GTCP36-150(F2M) APU, with start up approval to 10,668 m (35,000 ft). Dee Howard clamshell-type thrust reversers certified by FAA and JAA during 1995.

ACCOMMODATION: Up to 19 passengers and two flight crew; standard passenger accommodation is four seats in forward lounge and four seats and a two-person sofa in aft lounge. Pressurised, flight accessible baggage compartment at rear of cabin.

AVIONICS: Rockwell Collins Pro Line 4.

Comms: Dual com/nav; dual TRD-94D Mode-S transponders; dual RTU-4420 radio tuning units; Rockwell Collins HF-9000 transceiver.

Radar: Rockwell Collins WXR-840 colour weather radar standard; TWR-850 Doppler weather radar optional.

Flight: Rockwell Collins FMS-6100 flight management system and Cat. II autopilot standard. Rockwell Collins dual DME-442, dual ADF-462; dual AHRS and dual digital air data computers linked to dual-channel integrated avionics processor system (IAPS). Dual Honeywell Laseref III, Honeywell EGPWS, IRS, second FMS (with GPS), flight data recorder, traffic alert and collision avoidance system (TCAS) and GPWS all optional.

Instrumentation: Rockwell Collins Pro Line 4 four-tube EFIS-4000. Sextant Avionique three-tube engine indicating electronic display (EIED); Flight Dynamics HGS-2850 HUD.

Mission: Optional satcom antenna in fintip pod.

Data below apply to Falcon 2000 and 2000EX unless otherwise indicated.

DIMENSIONS, EXTERNAL:
Wing span 19.33 m (63 ft 5 in)
Wing chord (mean) 2.89 m (9 ft 5¾ in)
Wing aspect ratio 7.6
Length overall 20.22 m (66 ft 4 in)
Height overall 7.06 m (23 ft 2 in)
Wheel track 4.45 m (14 ft 7¼ in)
Wheelbase 7.39 m (24 ft 3 in)
Passenger door: Height 1.72 m (5 ft 7¾ in)
Width 0.80 m (2 ft 7½ in)
Baggage door: Height 0.775 m (2 ft 6½ in)
Width 0.75 m (2 ft 5½ in)
Emergency exits (Type III): Height 0.92 m (3 ft 0¼ in)
Width 0.53 m (1 ft 8¾ in)

DIMENSIONS, INTERNAL:
Cabin: Length 7.98 m (26 ft 2¼ in)
Width: max 2.34 m (7 ft 8¼ in)
at floor level 1.91 m (6 ft 3¼ in)
Max height 1.88 m (6 ft 2 in)
Volume 29.0 m³ (1,024 cu ft)

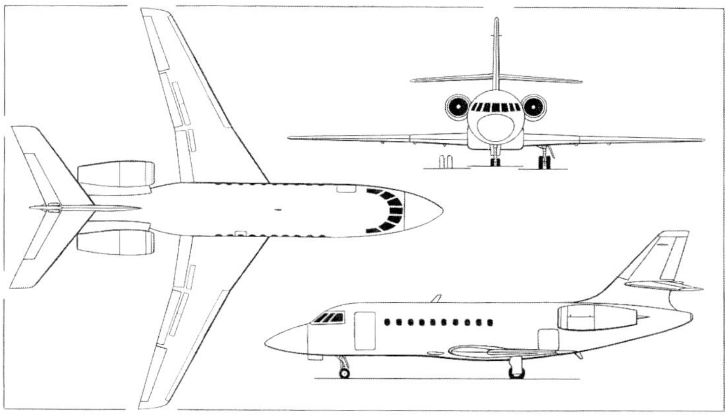

Dassault Falcon 2000 twin-engined business transport *(Jane's/Dennis Punnett)* 0507139

Baggage volume: standard 3.7 m³ (131 cu ft)
optional 5.4 m³ (191 cu ft)

AREAS:
Wings, gross 49.02 m² (527.6 sq ft)

WEIGHTS AND LOADINGS:
Weight empty, equipped: 2000 9,405 kg (20,735 lb)
2000EX 10,142 kg (22,360 lb)
Basic operating weight: 2000 9,798 kg (21,600 lb)
2000EX 10,519 kg (23,190 lb)
Max payload: 2000 3,202 kg (7,060 lb)
Max baggage: 2000, 2000EX 753 kg (1,660 lb)
Max fuel weight: 2000 5,513 kg (12,154 lb)
2000EX 7,557 kg (16,660 lb)
Max ramp weight: 2000: standard 16,329 kg (36,000 lb)
optional 16,556 kg (36,500 lb)
2000EX: standard 18,824 kg (41,500 lb)
optional 19,232 kg (42,400 lb)
Max T-O weight: 2000: standard 16,238 kg (35,800 lb)
optional 16,556 kg (36,500 lb)
2000EX: standard 18,733 kg (41,300 lb)
optional 19,232 kg (42,400 lb)
Max landing weight: 2000 14,970 kg (33,000 lb)
2000EX 17,826 kg (39,300 lb)
Max zero-fuel weight: 2000 13,000 kg (28,660 lb)
2000EX 13,472 kg (29,700 lb)
Max wing loading: 2000:
standard 331.3 kg/m² (67.85 lb/sq ft)
optional 337.8 kg/m² (69.18 lb/sq ft)
2000EX: standard 384.0 kg/m² (78.65 lb/sq ft)
optional 392.3 kg/m² (80.36 lb/sq ft)
Max power loading: 2000 standard 308 kg/kN (3.02 lb/lb st)
optional 314 kg/kN (3.08 lb/lb st)
2000EX: standard 302 kg/kN (3.49 lb/lb st)
optional 309 kg/kN (3.03 lb/lb st)

PERFORMANCE:
Max operating Mach No. (MMO) 0.85–0.86
Max operating speed (VMO):
2000 350–370 kt (648–685 km/h; 403–425 mph) IAS
Max cruising speed: 2000 M0.84 or 481 kt (891 km/h; 554 mph)
Normal cruising speed: 2000EX M0.85
2000 M0.80 or 459 kt (850 km/h; 528 mph)
Long-range cruising speed:
2000 M0.75 or 430 kt (796 km/h; 495 mph)
2000EX M0.77
Approach speed: 2000 109 kt (202 km/h; 125 mph)
2000EX 112 kt (207 km/h; 129 mph)
Initial cruising altitude: 2000 at M0.80 12,497 m (41,000 ft)
2000 at M0.75 13,106 m (43,000 ft)
2000EX at M0.77 12,497 m (41,000 ft)
Max certified altitude 14,330 m (47,000 ft)
Time to climb to FL410:
2000 24 min
2000EX 21 min
FAR Pt 25 balanced T-O field length, S/L, ISA:
2000: with eight passengers and max fuel 1,597 m (5,240 ft)
at max T-O weight 1,658 m (5,440 ft)
2000EX: at standard max T-O weight 1,806 m (5,925 ft)
at optional max T-O weight 1,884 m (6,180 ft)
Landing field length:
FAR Pt 91 with eight passengers and NBAA IFR reserves 780 m (2,560 ft)
at max landing weight 952 m (3,125 ft)
2000EX with six passengers and NBAA IFR reserves 803 m (2,633 ft)
Range with eight passengers and NBAA IFR reserves:
2000: at M0.75 3,120 n miles (5,778 km; 3,590 miles)
at M0.80 3,040 n miles (5,630 km; 3,498 miles)
2000EX with max fuel, six passengers, at
Mach 0.80, NBAA IFR reserves 3,800 n miles (7,037 km; 4,373 miles)
g limits +2.64/–1

OPERATIONAL NOISE LEVELS, 2000:
T-O 79.4 EPNdB
Approach 93.1 EPNdB
Sideline 86.4 EPNdB

UPDATED

DYN'AÉRO

SOCIÉTÉ DYN'AÉRO

19 rue de l'Aviation, F-21121 Darois
Tel: (+33 3) 80 35 60 62
Fax: (+33 3) 80 35 60 63
e-mail: dynaero@aol.com
Web: http://www.dynaero.com
CHAIRMAN: Christophe Robin
DESIGN ASSOCIATE: Michel Colomban

Dyn'Aéro formed September 1992 by Christophe Robin, son of the founder of Avions Pierre Robin; his initials (CR) appear in aircraft designations. Initial product was CR100, flown two weeks before formation of company. Two new designs derived from CR100 announced in March 1995: CR110 and CR120; simultaneously, MCR01 also announced. The new MCR01 Club flew in June 1998 and a four-seat version in June 2000. Production facilities at Darois aerodrome comprise 2,450 m² (26,375 sq ft) of workshops, storage and offices. Plans announced in February 2002 for establishment of a subsidiary, Dyn'Aéro Iberia at a new factory in Portugal, permitting an increase in annual production from 50 to 100 aircraft (initially of MCR ULC), with first airframe from new facility expected within 18 months.

By early 2004, more than 350 Dyn'Aéro kits had been sold.

UPDATED

DYN'AÉRO CR100, CR110 AND CR120

TYPE: Side-by-side sportplane/kitbuilt.

PROGRAMME: Prototype CR100 (F-PDYN) first flew 27 August 1992; third aircraft (F-PAFC) was designated CR100D); first five CR100s flew total of 600 hours of aerobatics in first year of operations. Available factory-built or as a kit.

CURRENT VERSIONS: **CR100:** Baseline version. Prototype (F-WASP).

CR100T: Tricycle landing gear version, first flown December 2000.

Description applies to CR100, except where otherwise stated.

CR110: Uprated version with 149 kW (200 hp) Textron Lycoming engine.

CR120: High-agility version. First flight (F-PPCR) September 1996; second aircraft (F-PTCR) flying by 1998. Engine as CR110, but airframe all carbon fibre; longer span ailerons (2.70 m; 8 ft 10¼ in), no flaps, shorter wing span (7.77 m; 25 ft 6 in), 270°/s rate of roll at 140 kt (260 km/h; 162 mph) compared with CR100's 195°/s, 149 kW (200 hp) Textron Lycoming AEIO-360 engine driving MTV-12-B-C/C-183-17e two-blade propeller, 85 litres (22.4 US gallons; 18.7 Imp gallons) aerobatic fuel in starboard wingroot and 105 litres (27.7 US gallons; 23.1 Imp gallons) in port wingroot, tyre pressure 2.00 bar (29.0 lb/sq in). Aircraft are registered under CR100 designation.

CUSTOMERS: Two CR100s ordered March 1995 by l'Equipe de Voltige de l'Armée de l'Air at Salon de Provence, of which first (No. 22 'CJ') delivered 21 July 1995; however, both withdrawn by 1997 and sold in 2000, one to UK owner Cole Aviation. Total of at least 37 CR100/110/120 kits sold; of these, 16 had been registered by early 2002. There were no known completions or new sales in 2003.

COSTS: €76,200 as a kit without engine; about €137,200 assembled, with engine and instruments.

DESIGN FEATURES: Single-engined low-wing monoplane of conventional layout.

Aerofoil section NACA 21015/12 modified; dihedral 0°.

FLYING CONTROLS: Conventional and manual. Elevator tab for pitch trim. Ground-adjustable tab on ailerons on rudder. Plain flaps.

STRUCTURE: Wood and fabric, but including carbon fibre main wing spar.

LANDING GEAR: Tailwheel type; fixed. Wing-mounted, trousered, cantilever main legs with integral oleo-pneumatic shock-absorbers. Wheel fairings on main legs. Disc brakes. Mainwheel tyre size 380×150; pressure 2.30 bar (33.0 lb/sq in).

POWER PLANT: One 134 kW (180 hp) Textron Lycoming AEIO-360-B2F flat-four engine, driving either Evra fixed-pitch or MT variable-pitch propeller. Compatible with alternative engines of 119 kW (160 hp) and above. Fixed-pitch or constant-speed propellers can be fitted. Fuel capacity for aerobatics 85 litres (22.4 US gallons; 18.7 Imp gallons), of which 82 litres (21.7 US gallons; 18.0 Imp gallons) are usable; max fuel capacity 160 litres (42.3 US gallons; 35.2 Imp gallons). Oil capacity 7.6 litres (2.0 US gallons; 1.7 Imp gallons).

Dyn'Aéro CR100 ***NEW***/0547496

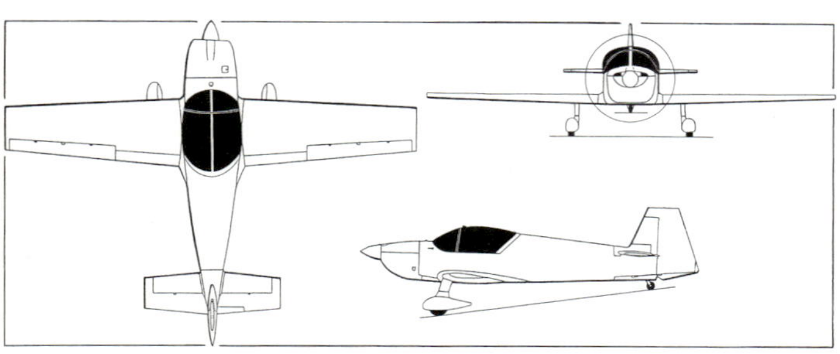

General arrangement of the Dyn'Aéro CR100 *(Paul Jackson)* 0044672

ACCOMMODATION: Two seats, side by side, with dual controls. Fixed windscreen and rearward-sliding canopy.

SYSTEMS: Electrical system: 12 V 30 Ah battery.

AVIONICS: Customer specified.

DIMENSIONS, EXTERNAL:
- Wing span ... 8.50 m (27 ft 10½ in)
- Wing aspect ratio ... 6.8
- Length overall ... 7.50 m (24 ft 7¼ in)
- Tailplane span ... 2.80 m (9 ft 2¼ in)
- Wheel track ... 2.37 m (7 ft 9¼ in)
- Propeller diameter ... 1.90 m (6 ft 3 in)

AREAS:
- Wings, gross ... 10.65 m² (114.6 sq ft)
- Ailerons (total) ... 1.215 m² (13.08 sq ft)
- Trailing-edge flaps (total) ... 1.21 m² (13.02 sq ft)
- Vertical tail surfaces (total) ... 1.71 m² (18.40 sq ft)
- Horizontal tail surfaces (total) ... 2.20 m² (23.68 sq ft)

WEIGHTS AND LOADINGS:
- Weight empty ... 550 kg (1,213 lb)
- Max T-O weight: Aerobatic ... 760 kg (1,676 lb)
  - Normal ... 850 kg (1,873 lb)
- Max wing loading: Aerobatic ... 71.5 kg/m² (14.64 lb/sq ft)
  - Normal ... 80.0 kg/m² (16.38 lb/sq ft)
- Max power loading: Aerobatic ... 5.66 kg/kW (9.31 lb/hp)
  - Normal ... 6.34 kg/kW (10.41 lb/hp)

PERFORMANCE:
- Never-exceed speed (VNE) ... 205 kt (379 km/h; 235 mph)
- Max level speed ... 165 kt (305 km/h; 190 mph)
- Max cruising speed ... 151 kt (280 km/h; 174 mph)
- Manoeuvring speed (VA) ... 140 kt (260 km/h; 161 mph)
- Max rate of climb at S/L ... 480 kt (1,575 ft)/min
- T-O run ... 153 m (500 ft)
- T-O to 15 m (50 ft) ... 351 m (1,150 ft)
- Range: normal fuel ... 302 n miles (560 km; 348 miles)
  - max fuel ... 604 n miles (1,120 km; 696 miles)
- *g* limits ... +8/–6

UPDATED

DYN'AÉRO MCR01

North American marketing names: Lafayette I Sportster; II Touring (Club); III Bushplane (ULM)

TYPE: Side-by-side kitbuilt; side-by-side ultralight kitbuilt.

PROGRAMME: Composites version of the metal MC100 Ban-Bi designed by Michel Colomban; fuselage first shown at RSA Rally in July 1995. Continues to be referred to by some sources as **Ban Bi**. MCR identifies Michel Colomban/Christophe Robin. Prototype (F-PECH) flew 1995.

CURRENT VERSIONS: **MCR01 Sportster:** Certified version meeting JAR-VLA requirements; **Sportster 912** or **Sportster 914**, according to engine. JAR-VLA certification awarded 26 June 2001 in France and 7 December 2001 in Peru.

MCR01 Club: Version of MCR01 Sportster intended for flight training. First flight (F-PCLB) June 1998. Certification to JAR-VLA was due in 2001.

MCR01 ULM: Ultralight version meeting FAI ULM requirements; 59.7 kW (80 hp) engine. **MCR01 ULC** has 490 kg (1,080 lb) MTOW and is known as the **MCR01 SLA** in the UK, where the first example (G-BZXG) made its first flight on 30 November 2001. DGAC certification received 1 March 2001.

Motor Glider: Under development. 'Winggrid' wingtips developed in conjunction with La Roche Consulting of Switzerland; 59.7 kW (80 hp) engine, best glide ratio 30:1.

MCR M: Tailwheel version. Certification issued 26 June 2002.

Electra-Plane: Being developed by Advanced Technology Products; stated to be world's first fuel cell-powered aircraft. Prototype is based on MCR01; second prototype will utilise a Hoffmann H-36 Dimona as base aircraft. Three stages planned: first, with lithium-ion batteries, will have 86 n mile (160 km; 100 mile) range; second, with combination of batteries and 10 to 15 kW (13 to 20 hp) fuel cell, will have 217 n mile (402 km; 250 mile) range. Final version will be powered solely with fuel cells and have 434 n mile (804 km; 500 mile) range.

Dyn'Aéro MCR ULC ultralight *(Paul Jackson)* ***NEW***/0576612

Dyn'Aéro MCR01 built by a UK home constructor *(Paul Jackson)* ***NEW***/0576618

Description applies to all versions except where indicated.

CUSTOMERS: Over 270 kits sold by January 2004, in Austria, Germany, Italy, Japan, Netherlands, New Zealand, Spain, Switzerland, UK and USA. More than 100 flying by early 2004.

COSTS: Kits, less engine and instruments: Sportster €28,300, ULC/ULM €33,100, Club €31,300 (all 2003).

DESIGN FEATURES: Single-engined low-wing monoplane with T tail; design goals included extremely low structural weight, fast kit production and high performance in comfort and safety.

FLYING CONTROLS: Manual. All-moving tailplane with trim tab, which is full span on ULM version; electrically actuated flaps and pitch trim; ULM and M versions have full-span double-slotted flaperons: Sportster has flaperons (both three-position: 0, 17, 45°); Club has separate double slotted flaps and ailerons.

STRUCTURE: Wing and control surfaces built on carbon fibre spars and cellular ribs, to which preformed aluminium skins are glued in partial vacuum; monocoque carbon fibre fuselage; quoted build time of 500 to 1,000 hours from three modular kits without special tools. Airframe disassembles for storage or road-towing, with recommended disassembly/reassembly time of 4 minutes, single-handed. Wing dihedral 3°.

LANDING GEAR: Tricycle type; fixed. Fuselage-mounted spring cantilever mainwheel legs with wheel fairings on all three units; nosewheel steerable. Large wheels optional for rough field operation. Tailwheel configuration optional on MCR01 Club and MCR01 ULC versions, which have further option of wing-mounted, trousered cantilever main legs with integral shock-absorbers. Skis tested on MCR M during winter 2002–03.

POWER PLANT: One 59.6 kW (79.9 hp) Rotax 912 UL; 63.4 kW (85 hp) JPX 4T75; or 73.5 kW (98.6 hp) normally aspirated Rotax 912 ULS or turbocharged Rotax 914 flat-four driving an EVRA 156-178-106 two-blade fixed-pitch propeller optimised for cruise; three-blade MT Propeller MTV-7-A/152-106 or MTV-6-A/152-106 constant-speed propeller optional. Jabiru-powered version first flown 19 April 2002. Standard fuel capacity 80 litres (21.1 US gallons; 17.6 Imp gallons) in all versions; optional additional 90 litre (23.75 US gallon; 19.8 Imp gallon) tank in Club and ULM or 264 litre (69.7 US gallon; 58.1 Imp gallon) tank in VLA.

MCR01 in its ULC/ULM/Sportster form *(Paul Jackson)* ***NEW***/0547128

Sportster (VLA) version of MCR *(Paul Jackson)* ***NEW***/0547121

Tailwheel MCR M version of MCR01 with fuselage-mounted spring cantilever main gear *(Paul Jackson)* ***NEW***/0547129

Dyn'Aéro MCR M tailwheel version, with wing-mounted landing gear option ***NEW***/0547069

AVIONICS: To customer's choice.

EQUIPMENT: Optional road-towing trailer/storage hangar. BRS ballistic parachute recovery system optional on ULM.

DIMENSIONS, EXTERNAL:
- Wing span: Sportster — 6.63 m (21 ft 9 in)
  - Club — 6.92 m (22 ft 8½ in)
  - ULM, M — 8.66 m (28 ft 5 in)
  - Motor Glider — 9.80 m (32 ft 1¾ in)
- Wing aspect ratio: Sportster — 8.5
  - Club — 7.0
  - ULM — 8.6
- Length overall: all — 5.53 m (18 ft 1¾ in)
- Height overall: Club, ULM — 1.53 m (5 ft ¼ in)

DIMENSIONS, INTERNAL:
- Cabin max width: all — 1.12 m (3 ft 8 in)

AREAS:
- Wings, gross: Sportster — 5.20 m² (56.0 sq ft)
  - Club — 6.45 m² (69.4 sq ft)
  - ULM, M — 8.31 m² (89.4 sq ft)
  - Motor Glider — 9.35 m² (100.6 sq ft)

WEIGHTS AND LOADINGS:
- Weight empty, equipped: Sportster — 235 kg (518 lb)
  - Club — 245 kg (540 lb)
  - ULM, M — 250 kg (551 lb)
  - Motor Glider — 262 kg (578 lb)
- Baggage capacity — 15 kg (33 lb)
- Max T-O weight: Sportster, ULM — 450 kg (992 lb)
  - Club — 490 kg (1,080 lb)
  - M — 544 kg (1,199 lb)

PERFORMANCE (59.7 kW; 80 hp engine):
- Never-exceed speed (VNE):
  - normal: Sportster — 172 kt (320 km/h; 198 mph)
- Club, ULM — 162 kt (300 km/h; 186 mph)
- M — 170 kt (315 km/h; 195 mph)
  - with BRS: all — 145 kt (270 km/h; 167 mph)
- Max level speed: Sportster 912 — 163 kt (302 km/h; 188 mph)
  - Sportster 914 — 197 kt (364 km/h; 226 mph)
  - Club — 151 kt (281 km/h; 174 mph)
  - ULM — 150 kt (278 km/h; 172 mph)
  - Motor Glider — 135 kt (250 km/h; 155 mph)
- Max cruising speed at 75% power:
  - at FL80:
    - Sportster 912 — 158 kt (292 km/h; 181 mph)
    - Club — 150 kt (278 km/h; 173 mph)
    - ULM — 146 kt (270 km/h; 168 mph)
    - Motor Glider — 130 kt (240 km/h; 149 mph)
    - M — 119 kt (221 km/h; 137 mph)
  - at FL110:
    - Sportster 914 — 154 kt (286 km/h; 178 mph)
    - ULM — 143 kt (265 km/h; 165 mph)
- Stalling speed, flaps down:
  - Sportster — 50 kt (91 km/h; 57 mph)
  - Club — 43 kt (79 km/h; 49 mph)
  - ULM — 34 kt (63 km/h; 40 mph)
  - Motor Glider — 30 kt (55 km/h; 35 mph)
  - M — 38 kt (70 km/h; 44 mph)
- Max rate of climb at S/L: Sportster, Club — 609 m (2,000 ft)/min
  - ULM — 426 m (1,400 ft)/min
- T-O run: Sportster — 285 m (935 ft)
  - Club — 210 m (689 ft)
  - ULM — 195 m (640 ft)
- T-O to 15 m (50 ft): Sportster, Club — 424 m (1,395 ft)
  - ULM — 203 m (666 ft)
- Range at econ cruising speed and altitude:
  - with normal fuel:
    - Sportster — 894 n miles (1,657 km; 1,029 miles)
    - Club — 829 n miles (1,536 km; 954 miles)
    - ULM — 834 n miles (1,545 km; 960 miles)
  - with max optional fuel:
    - Sportster — 3,452 n miles; (6,394 km; 3,973 miles)
    - Club — 1,623 n miles (3,006 km; 1,867 miles)
    - ULM — 1,611 n miles (2,985 km; 1,854 miles)
- *g* limits: all except Sportster — +4/–2
  - Sportster — +3.8/–1.9

UPDATED

DYN'AÉRO MCR4S

TYPE: Four-seat kitbuilt.

PROGRAMME: Mockup first shown at Aero '99, Friedrichshafen, April 1999. Kits available from late 1999. Prototype (FWWUZ) first flew 14 June 2000 and displayed at PFA International Air Rally, Cranfield, UK, 23–25 June 2000. First customer aircraft, built by noted French light aircraft designer/manufacturer Pierre Robin, flew in mid-2001 and DGAC certification achieved 26 June 2001. PFA approval was being sought in mid-2003.

CURRENT VERSIONS: **Three-seat:** With 59.6 kW (79.9 hp) Rotax 912 UL or JPX 4TX75 engine; 640 kg (1,411 lb) MTOW.

Four-seat: With 73.5 kW (98.6 hp) Rotax 912 UL-S normally aspirated engine and MT-Propeller MTV-7-A/156-122 fixed-pitch propeller; 750 kg (1,653 lb) MTOW.

Four-seat Performance: 84.6 kW (113.4 hp) Rotax 914 UL turbocharged engine and constant-speed propeller; optimised for cruise at 3,050 m (10,000 ft).

CITEC: On 29 July 2001, Dyn'Aero and Wilksch Airmotive of the UK announced a joint programme to develop the MCR4S CITEC, powered by the Wilksch WAM-120 diesel engine. Launch customer for this version was M. Serge Blanchard.

CUSTOMERS: 35 kits sold by mid-2003, of which 15 were then flying.

COSTS: Kit €48,100, less engine; €86,600, €91,900 or €131,700 with Rotax 912 ULS plus fixed propeller, 912 ULS plus c/s MT propeller or 914 plus c/s propeller (2003).

DESIGN FEATURES: Three/four-seat version of MCR01, having 75 per cent commonality. Dihedral 3°. Winglets introduced in 2001, and modified in 2002.

Description generally as for MCR01, except that below.

FLYING CONTROLS: Ailerons, plus two-segment flaps. Aileron max deflection −20°/+10°. Tailplane (all moving) max deflection +3.5°/−10°. Rudder deflection ±20°. Double-slotted flaps, deflections 0, 17 and 30°.

LANDING GEAR: Tricycle type; fixed. Wing-mounted, trousered, cantilever main legs with integral shock-absorbers. Fairing on all three wheels.

POWER PLANT: See Current Versions. Fuel in two wing tanks; standard capacity 120 litres (31.7 US gallons; 26.4 Imp gallons); optionally two 100 litre (26.4 US gallon; 22.0 Imp gallon) tanks can be fitted. In both cases, 2 litres (0.5 US gallons; 0.4 Imp gallons) is unusable.

Dyn'Aéro MCR4S built from 32nd kit *(Paul Jackson)* ***NEW***/0589433

Dyn'Aéro MCR4S four-seater with revised winglets, introduced in 2002 ***NEW***/0547481

Dyn'Aéro MCR4S three/four-seat tourer with current wingtip configuration *(Paul Jackson)* ***NEW***/0576479

ACCOMMODATION: Four persons, in two side-by-side pairs. Forward-hinged, two-piece canopy/windscreen, plus two fixed side windows.

EQUIPMENT: Ballistic parachute optional.

DIMENSIONS, EXTERNAL:
- Wing span ... 8.66 m (28 ft 5 in)
- Wing chord (mean) ... 0.96 m (3 ft 1¾ in)
- Wing aspect ratio ... 9.2
- Length overall ... 6.72 m (22 ft 0½ in)
- Height overall ... 1.95 m (6 ft 4¾ in)
- Tailplane span ... 2.50 m (8 ft 2½ in)

DIMENSIONS, INTERNAL:
- Cabin max width ... 1.17 m (3 ft 10 in)

AREAS:
- Wings, gross ... 8.15 m² (87.7 sq ft)

WEIGHTS AND LOADINGS:
- Weight: empty ... 325 kg (717 lb)
- equipped ... 350 kg (772 lb)
- Baggage capacity ... 40 kg (88 lb)
- Max T-O weight ... 640–750 kg (1,411–1,653 lb)

PERFORMANCE (Rotax 912, 912 ULS and 914 UL engines):
- Never-exceed speed (VNE) ... 165 kt (306 km/h; 190 mph)
- Max level speed: 912 ... 143 kt (265 km/h; 165 mph)
- 912 ULS ... 156 kt (288 km/h; 179 mph)
- 914 UL ... 173 kt (320 km/h; 199 mph)
- Manoeuvring speed ... 118 kt (219 km/h; 136 mph)
- Max cruising speed: 912 ... 136 kt (252 km/h; 157 mph)
- 912 ULS ... 150 kt (277 km/h; 172 mph)
- 914 UL ... 155 kt (287 km/h; 178 mph)
- Econ cruising speed: 912 ... 131 kt (242 km/h; 150 mph)
- 912 ULS, 914 UL ... 146 kt (270 km/h; 168 mph)
- Stalling speed: flaps up ... 54 kt (100 km/h; 63 mph)
- flaps down ... 45 kt (83 km/h; 52 mph)
- T-O run with c/s prop: 912 ... 317 m (1,040 ft)
- 912 ULS ... 203 m (670 ft)
- 914 UL ... 153 m (505 ft)
- T-O to 15 m (50 ft) with c/s prop: 912 ... 455 m (1,495 ft)
- 912 ULS ... 295 m (970 ft)
- 914 UL ... 221 m (725 ft)
- Landing run with c/s prop: 912 ULS ... 201 m (660 ft)
- Range with max fuel:
- 912 ... 1,034 n miles (1,916 km; 1,190 miles)
- 912 ULS, 914 UL ... 921 n miles (1,707 km; 1,060 miles)
- *g* limits ... +3.8/−1.5

UPDATED

EADS FRANCE

EADS FRANCE

37 boulevard Montmorency, F-75781 Paris Cedex 16
Tel: (+33 1) 42 24 24 24
Fax: (+33 1) 42 24 26 19
Web: http://www.eads.net

The Délégation Générale des Armements (DGA) is studying a number of domestic and foreign airship proposals for future strategic and tactical transport use by the French Army. EADS France is involved in the FIRST 1A helium semi-rigid.

UPDATED

EADS FIRST 1A

TYPE: Helium semi-rigid.

PROGRAMME: FIRST (*Force pour l'Innovation, la Recherche Scientifique et Technique*) is a current design study for an outsize transport airship.

DESIGN FEATURES: Not yet finalised (2004); ballonet pressure control.

STRUCTURE: Composites.

Following data are provisional.

POWER PLANT: Six 4,500 kW (6,035 shp) turboprops, each driving a 10-blade propeller.

DIMENSIONS, EXTERNAL:
- Length overall ... 300 m (984.3 ft)
- Envelope max diameter ... 73 m (239.5 ft)

DIMENSIONS, INTERNAL:
- Envelope volume ... 600,000 m³ (21.2 million cu ft)
- Cargo bay volume ... 5,760 m³ (203,400 cu ft)

WEIGHTS AND LOADINGS:
- Max payload ... 250,000 kg (551,155 lb)

PERFORMANCE (estimated):
- Cruising speed ... 86 kt (160 km/h; 99 mph)
- Range ... 3,239 n miles (6,000 km; 3,728 miles)

UPDATED

ESPACE LIBERTE

ESPACE LIBERTÉ SARL

Aéroport de Rouen/Boos, F-76520 Vallée de la Seine
Tel: (+33 2) 35 79 06 06
Fax: (+33 2) 35 79 26 40
DIRECTOR: Charles Guérin

MARKETING:
DKL Air Light SARL
rue de Luxembourg 24, L-8360 Goetzingen, Luxembourg
Tel: (+352) 26 30 02 55
e-mail: info@dklairlight.com
Web: http://www.dklairlight.com

Having worked on design of the DEA Yuma STOL lightplane in Italy, Charles Guérin returned to France to pursue development of a modified version, incorporating refinements including the option of three stages of high-lift wing.

UPDATED

ESPACE LIBERTE GUERIN G1

TYPE: Side-by-side ultralight.

PROGRAMME: Development began 1994 as DEA Yuma. Adaptations of design, leading to G1, began in 2002; several trials aircraft. Series production aircraft '00' shown at Aero '03, Friedrichshafen, in April 2003; No. 01 then intended to fly by mid-2003.

CURRENT VERSIONS: **V3:** Version 3, but baseline version. Empty weight of 282.5 kg (623 lb) to meet new ultralight limits. 'High-tech' wing with full fitment of Handley Page slats and slotted Fowler flaps. *As described.*

V2: Version 2. Empty weight 275 kg (606 lb). Fixed slats and slotted Fowler flaps.

V1: Version 1. Empty weight 265 kg (584 lb). Fixed slats and plain flaps.

CUSTOMERS: Total of 25 orders by early 2003; four more under construction early 2004.

COSTS: V3 €53,435; V2 €51,145; V1 €46,295; all plus tax (2004).

DESIGN FEATURES: Objectives included better performance than existing STOL light aircraft, but without sacrifice of cruising speed; effective and harmonised controls; foldable wings. Aircraft can be folded for transport by one person in 10 minutes.

High wings with V-struts; sweptback fin.

FLYING CONTROLS: Conventional and manual. Large, cable-actuated control surfaces for effectiveness of response at slow speeds. Optional Handley Page slats on wing leading-edge and/or slotted Fowler flaps, according to version.

STRUCTURE: Generally of heat-treated and corrosion-proofed aluminium alloy; skins bonded before riveting. Welded tubular 4130 steel frame cockpit structure; steel alloy wing struts. Control surfaces covered in Dacron fabric. Wing leading-edge slats of carbon fibre. Composites tips to wing and empennage. Composites engine cowling. Lexan windscreen; polycarbonate side windows.

LANDING GEAR: Tricycle type; fixed. Fuselage-mounted metal cantilever mainwheel legs with optional wheel fairings. Steerable, telescoping nosewheel leg. Mainwheels 8.00-6 (optionally larger); nosewheel 15×6.00-6. Hydraulic disc brakes. Mainwheel and parking brakes. Optional floats and skis.

POWER PLANT: One 73.5 kW (98.6 hp) Rotax 912 ULS flat-four. Fuel in wing tanks and header tank, total capacity 84 litres (22.2 US gallons; 18.5 Imp gallons). Duc two-blade propeller, ground- or flight-adjustable.

EQUIPMENT: Ballistic parachute. Optional agricultural spraying equipment.

DIMENSIONS, EXTERNAL:
Wing span .... 9.75 m (31 ft 11¾ in)

First production Guérin G1 two-seat STOL ultralight *(Paul Jackson)* ***NEW***/0576504

Length overall .... 6.45 m (21 ft 2 in)
Height overall .... 2.45 m (8 ft 0½ in)

DIMENSIONS, INTERNAL:
Cabin max width .... 1.22 m (4 ft 0 in)

AREAS:
Wings, gross .... 13.44 m² (144.7 sq ft)

WEIGHTS AND LOADINGS:
Weight empty .... 282.5 kg (623 lb)
Max T-O weight .... 450 kg (992 lb)

PERFORMANCE (estimated):
Never-exceed speed (VNE) .... 100 kt (185 km/h; 115 mph)
Cruising speed at 75% power .... 86 kt (160 km/h; 99 mph)
Stalling speed, power off, flaps down .... 22 kt (40 km/h; 25 mph)
Max rate of climb at S/L .... 480 m (1,575 ft)/min
T-O run .... 30 m (100 ft)
Landing run .... 55 m (180 ft)
Endurance .... 4 h 18 min

UPDATED

HUMBERT

HUMBERT AVIATION

1 rue du Menil, F-88160 Ramonchamp
Tel: (+33 3) 29 25 05 75
Fax: (+33 3) 29 25 98 97
e-mail: contact@humbert-aviation.com
Web: http://www.humbert-aviation.com
DESIGNER: Jean-Jacques Humbert

Formed by Jacques Humbert in 1994, the company's most recent design is the Tétras, although the earlier Moto du Ciel remains available.

UPDATED

HUMBERT TETRAS

TYPE: Side-by-side ultralight/kitbuilt.

PROGRAMME: Prototype unveiled at 1992 RSA Rally at Moulins. Named for a fish of the genus *Characidae*. Marketing began in 1994.

CUSTOMERS: At least 10 on French ultralight register by December 2001. Has also been evaluated by Centre d'Essais en Vol (CEV) for possible military applications, and supplied to Niger for insecticide spraying.

COSTS: Kit, with Rotax 912 UL engine, €15,000; flyaway €31,500 (2001).

DESIGN FEATURES: High wing with single bracing strut on each side; strut-braced tailplane. Wing section NACA 23012. Quoted build time 150 to 250 hours.

FLYING CONTROLS: Conventional and manual. Three-position plain flaps, deflections 0, 15 and 45°.

STRUCTURE: Steel tube fuselage with Dacron covering; composites wing; light alloy ailerons and flaps.

LANDING GEAR: Tailwheel type; fixed. Two faired-in side Vs hinged to lower longerons, plus half-axles sprung under centreline with bungee cord.

POWER PLANT: One 53.7 kW (72 hp) Humbert-Volkswagen HW 2000 or 59.6 kW (79.9 hp) Rotax 912 UL flat-four, driving a two-blade DUC composites propeller. Other options available, including JPX engine and 73.4 kW (98.4 hp) Rotax 912 ULS. Fuel in two wing tanks, combined capacity 60 litres (15.9 US gallons; 13.2 Imp gallons).

EQUIPMENT: Spray system optional.

Humbert Tétras two-seat ultralight ***NEW***/0552267

DIMENSIONS, EXTERNAL:
Wing span .... 10.10 m (33 ft 1¾ in)
Length overall .... 6.50 m (21 ft 4 in)
Height overall .... 2.06 m (6 ft 9 in)
Propeller diameter .... 1.71 m (5 ft 7¼ in)

AREAS:
Wings, gross .... 15.70 m² (169.0 sq ft)

WEIGHTS AND LOADINGS:
Weight empty .... 277 kg (611 lb)
Max T-O weight .... 450 kg (992 lb)

PERFORMANCE (Rotax 912 UL):
Max level speed .... 105 kt (194 km/h; 121 mph)
Cruising speed at 75% power at FL100 .... 101 kt (187 km/h; 116 mph)
Stalling speed, flaps down .... 30 kt (55 km/h; 35 mph)
Max rate of climb at S/L .... 381 m (1,250 ft)/min
T-O run .... 50-80 m (164-262 ft)
g limits .... +4/-2

UPDATED

ISATIS

ISATIS

BP 16, F-77211 Avon Cedex
Tel: (+33 6) 07 08 74 33
Fax: (+33 1) 60 71 19 37
e-mail: isatis01@wanadoo.fr
Web: http://www.persc.wanadoo.fr/isatis01
TECHNICAL DIRECTOR: Philippe Baffray
COMMERCIAL DIRECTOR: Philippe Gandolphe

The company's first product is the Isatis01 ultralight.

VERIFIED

ISATIS01

TYPE: Side-by-side ultralight.

PROGRAMME: Partly complete prototype first shown at Aero '03, Friedrichshafen, April 2003.

Incomplete Isatis01 prototype on show at Aero '03, Friedrichshafen, April 2003 *(Paul Jackson)* ***NEW***/0552066

Artist's impression of the completed Isatis01 ***NEW***/0552067

DESIGN FEATURES: Optimised for low noise and emissions; economy and ease of stowage. Mid-engined configuration with high, strut-braced, foldable wings. V tail with small underfin.

FLYING CONTROLS: Manual. Ailerons and twin ruddervators. Flaps.

STRUCTURE: Composites skin throughout; carbon fibre driveshaft (passing through cockpit from engine to propeller) and cantilever mainwheel legs. Metal tube cockpit frame. Carbon fibre wing structure with glass fibre skin.

LANDING GEAR: Tricycle type; fixed. Fuselage-mounted spring composites mainwheel legs with wheel fairings. All tyres 4.00-6; speed fairings over each.

POWER PLANT: One mid-mounted BMW R 1100 RS flat-four, rated at 66.2 kW (88.8 hp) at 7,200 rpm, with catalytic exhaust converter, driving a three-blade carbon fibre propeller.

EQUIPMENT: Ballistic parachute standard.

DIMENSIONS, EXTERNAL:
Wing span 9.80 m (32 ft 2 in)
Length overall 6.00 m (19 ft 8¼ in)

AREAS:
Wings, gross 12.10 m^2 (130.2 sq ft)

WEIGHTS AND LOADINGS:
Weight empty 258 kg (569 lb)
Max T-O weight 472.5 kg (1,041 lb)

PERFORMANCE (estimated):
Never-exceed speed (VNE) 183 kt (340 km/h; 211 mph)
Normal cruising speed 124 kt (230 km/h; 143 mph)
Stalling speed, flaps down 33 kt (60 km/h; 38 mph)
Landing from 15 m (50 ft) 188 m (617 ft)
Range 809 n miles (1,500 km; 932 miles)
g limits +4/−2

UPDATED

ISSOIRE

ISSOIRE AVIATION

ZA la Béchade, BP1, Aérodrome Issoire-Le Broc, F-63501 Issoire
Tel: (+33 4) 73 89 01 54 and 73 89 78 11
Fax: (+33 4) 73 89 54 59
e-mail: iav@issoire-aviation.com
Web (1): http://www.issoire-aviation.com
Web (2): http://www.apm20lionceau.com
PRESIDENT AND CEO: Philippe Moniot
DIRECTOR: Laurent Bourdier
SALES MANAGER: Yves Moyne-Bressand

Issoire Aviation was established in 1978 as a successor to Wassmer Aviation. Its subsidiary, Rex Composites, is responsible for carbon fibre airframes. Current product is the APM-20/21 lightplane, which appears to be in very low-rate production. An unmanned version is planned.

UPDATED

ISSOIRE APM-20 LIONCEAU AND APM-21 LION

English name: Lion Cub

TYPE: Two-seat lightplane.

PROGRAMME: Design (as Moniot APM-20) started 1992 by Les Industries de Composites d'Auvergne Réunités (ICAR); prototype (F-WWMP) exhibited at Paris Air Show 1995 before first flight on 21 November 1995; third (second flying) aircraft (F-WWXX) exhibited statically at Paris Air Show, June 1997, fitted with JPX flat-four engine, which is not offered on production aircraft. No. 4 (also F-WWXX) exhibited at Paris in June 1999 and No. 5 (F-GRRE) in 2001. Certified to JAR-VLA 17 May 1999 and to JAR-21 in 2000. First all-carbon fibre, single-engine aircraft to gain JAR-VLA certification. APM-21 will follow at unspecified date. Tailwheel and diesel-engined variants were under consideration in early 2003.

CURRENT VERSIONS: **APM-20:** Primary trainer *As described*.

APM-21 Lion: Aerial work version powered by 73.5 kW (98.6 hp) Rotax 912 ULS engine driving a two-blade, constant-speed Hoffmann propeller. Prototype (F-WWMP, modified from prototype Lionceau) demonstrated at Paris Air Show in June 1999.

Four-seat: Under development; no major structural changes.

APM-22 Liondo: Prototype APM-21 equipped with large winglets and exhibited at Paris, 15–22 June 2003, as representative of this UAV proposal, with 24-hour endurance and 556 kg (1,225 lb) MTOW.

CUSTOMERS: Three flying prototypes; one static test airframe (No. 02). Six production aircraft registered by late 2003; one other (No. 13) exhibited at Paris in June 2003 as APM-21. Break-even point estimated at 200 aircraft.

COSTS: Development cost around FFr10 million. Unit cost FFr620,000, minimally equipped or FFr700,000 with standard equipment (2001). Operating cost FFr350 per hour (1998).

DESIGN FEATURES: Low wing, NACA 63618 aerofoil, thickness/chord ratio 18 per cent, dihedral 3°, incidence 2°, twist 1°.

FLYING CONTROLS: Conventional and manual. Spring elevator tab for pitch trim. Electrically operated slotted flaps to about two-thirds span.

STRUCTURE: Carbon fibre/epoxy.

LANDING GEAR: Fixed tricycle type, with wheel fairings. Fuselage-mounted GFRP cantilever mainwheel legs with steerable oleo-sprung nosewheel. Mainwheels and nosewheel diameter 330 mm; maximum pressure 2.35 bar (34 lb/sq in).

POWER PLANT: One 59.6 kW (79.9 hp) Rotax 912A four-cylinder four-stroke driving an Evra AL1 two-blade, fixed-pitch wooden propeller. Fuel capacity 68 litres (18.0 US gallons; 15.0 Imp gallons), of which 65 litres (17.2 US gallons; 14.3 Imp gallons) are usable. Refuelling point on port side of fuselage. Oil capacity 4 litres (1.1 US gallons; 0.9 Imp gallon).

APM-21 prototype, converted from the first APM-20 *(Paul Jackson)* ***NEW***/0572268

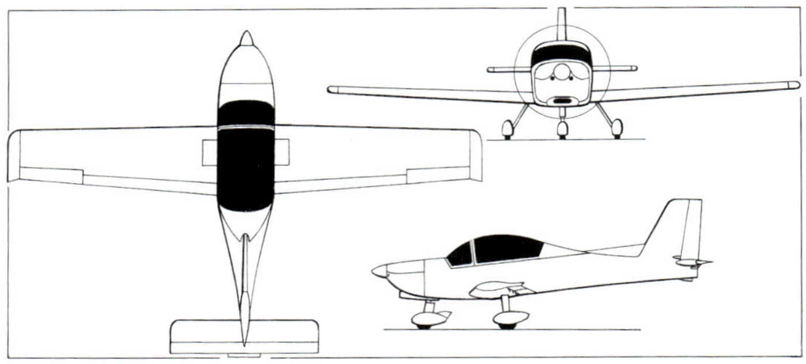

APM-20 Lionceau (Rotax 912A engine) *(James Goulding)* 0064443

ACCOMMODATION: Two, side by side under rearward-sliding tinted canopy. Dual controls standard. Baggage compartment at rear of seats. Fixed step forward of wing leading-edge on each side.

AVIONICS: *Instrumentation:* VFR panel standard.

DIMENSIONS, EXTERNAL:
Wing span 8.66 m (28 ft 5 in)
Wing chord: at root 1.10 m (3 ft 7¼ in)
at tip 0.84 m (2 ft 9 in)
Wing aspect ratio 7.9
Length overall 6.60 m (21 ft 7¾ in)
Max width of fuselage 1.15 m (3 ft 9¼ in)
Height overall 2.40 m (7 ft 10½ in)
Tailplane span 2.50 m (8 ft 2½ in)
Wheel track 1.92 m (6 ft 3½ in)
Propeller diameter 1.64 m (5 ft 4½ in)
Propeller ground clearance 0.31 m (1 ft 0¼ in)

AREAS:
Wings, gross 9.50 m^2 (102.3 sq ft)
Ailerons (total) 0.50 m^2 (5.38 sq ft)
Trailing-edge flaps (total) 1.40 m^2 (15.07 sq ft)
Fin 1.80 m^2 (19.37 sq ft)
Tailplane 2.00 m^2 (21.53 sq ft)
Elevator, incl tab 0.50 m^2 (5.38 sq ft)

WEIGHTS AND LOADINGS:
Weight empty 380 kg (838 lb)
Baggage capacity 20 kg (44 lb)
Max fuel 49 kg (108 lb)
Max T-O weight: current 620 kg (1,366 lb)
Planned 634 kg (1,397 lb)
Max wing loading 65.3 kg/m^2 (13.37 lb/sq ft)
Max power loading 10.39 kg/kW (17.06 lb/hp)

PERFORMANCE:
Never-exceed speed (VNE) 135 kt (250 km/h; 155 mph)
Cruising speed: at S/L 113 kt (210 km/h; 130 mph)
at FL55 124 kt (230 km/h; 143 mph)
Econ cruising speed 94 kt (175 km/h; 109 mph)

Fifth Issoire APM-20 Lionceau *(Paul Jackson)* ***NEW***/0546844

Manoeuvring speed	108 kt (200 km/h; 124 mph)
Stalling speed: flaps up	54 kt (100 km/h; 63 mph)
25° flap	44 kt (80 km/h; 50 mph)
Max rate of climb at S/L	204 m (669 ft)/min
T-O run	240 m (787 ft)
T-O to 15 m (50 ft)	460 m (1,509 ft)
Landing run	150 m (492 ft)
Endurance	4 to 5 h

UPDATED

LH AVIATION

Aérodrome de Merville, F-59660 Merville
Tel: (+33 6) 21 71 50 52
Web: http://www.lhaviation.com
e-mail: sebastien.lefebvre@lhaviation.com
DIRECTOR: Sébastien Lefebvre
ENGINEER Frédéric Hubschwerlen

MM Lefebvre and Hubschwerlen formed LH Aviation in early 2004 to market the LH-10 Ellipse, an aircraft they designed themselves.

NEW ENTRY

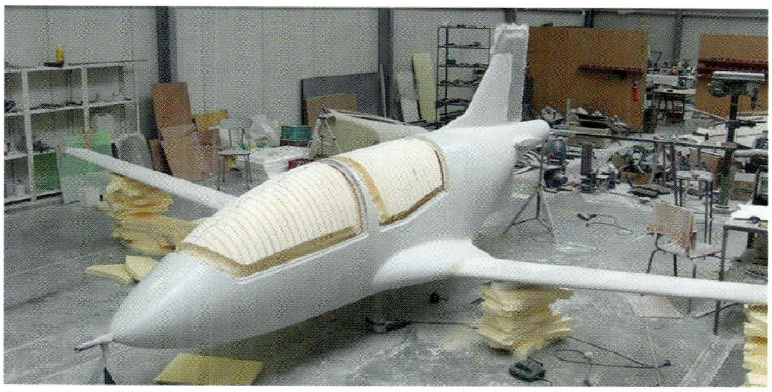

Assembly of LH-10 mockup at Merville ***NEW***/1047458

LH-10 ELLIPSE

TYPE: Tandem-seat kitbuilt.

PROGRAMME: Design began June 2003; prototype construction started September 2003; announced at AviaExpo, Lyon, France, 18 June 2004, when a full-scale fuselage mockup was displayed.

DESIGN FEATURES: Objectives included wide speed range, permitting long-range travel at 200 kt (370 km/h; 230 mph) or pleasure flying at 100 kt (185 km/h; 115 mph) using only 7 litres (1.8 US gallons; 1.5 Imp gallons) per hour; operation from small airfields through use of high-lift devices; and large and comfortable cockpit.

Mid-mounted engine and pusher propeller; high aspect ratio wings with thickened root gloves; broad underfin; extended wheelbase achieved by forward-raked nosewheel; dual sidestick controls. Designed to FAR Pt 23 criteria.

Marketed as quick-build kit, with most composites structural parts bonded and ready for assembly. Builder assistance programme available.

FLYING CONTROLS: Manual. Full-span flaperons; horn-balanced rudder; all-moving tailplane.

STRUCTURE: Generally of composites.

LANDING GEAR: Tricycle type; fixed and retractable (electrically actuated) options.

POWER PLANT: One 77.2 kW (103.5 hp) Masschi 105 liquid-cooled flat-four piston engine in mid-fuselage, driving a four-blade, automatically variable pitch, carbon fibre pusher propeller via a carbon fibre drive shaft. Fuel capacity 70 litres (18.5 US gallons; 15.4 Imp gallons).

DIMENSIONS, EXTERNAL:	
Wing span	8.00 m (26 ft 3 in)
Wing aspect ratio	14.2
Length overall	5.11 m (16 ft 9¼ in)
Height overall	2.14 m (7 ft 0¼ in)
DIMENSIONS, INTERNAL:	
Cockpit max width	0.68 m (2 ft 2¾ in)
AREAS:	
Wings, gross	4.50 m² (48.4 sq ft)
WEIGHTS AND LOADINGS	
Useful load	240 kg (529 lb)

Artist's impression of LH-10 ***NEW***/1047457

PERFORMANCE (estimated):	
Never-exceed speed (VNE)	270 kt (500 km/h; 310 mph)
Max level speed	200 kt (370 km/h; 230 mph)
Cruising speed at 75% power	100 kt (185 km/h; 115 mph)
Stalling speed	53 kt (99 km/h; 61 mph)
T-O to 15 m (50 ft)	250 m (820 ft)
Landing from 15 m (50 ft)	450 m (1,480 ft)
g limits	+4.4/–2.2

NEW ENTRY

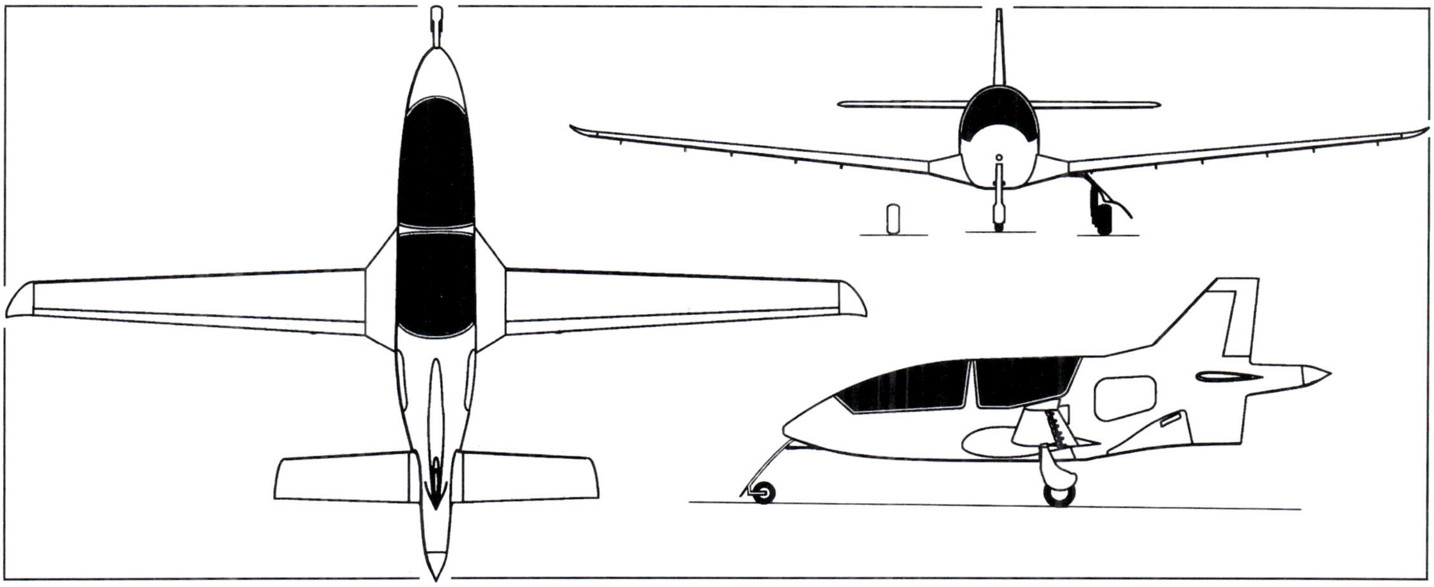

LH Aviation LH-10 Ellipse two-seat, high-performance kitbuilt *(Paul Jackson)* ***NEW***/1047839

LIFTIUM

LIFTIUM SA

135 avenue Victor Hugo, F-75116 Paris
MANAGING DIRECTOR: Didier Costes
PILOT: Thierry Garçon

Liftium's founder, Didier Costes, was a member in 1990–92 of a team which developed a human-powered airship known as the Zeppy 2 Continent. His Liftium 1 is a one-place ultralight airship of unusual design.

NEW ENTRY

LIFTIUM LIFTIUM 1

TYPE: Helium non-rigid.
PROGRAMME: First flight 7 November 2002 at Calvi, Corsica.
DESIGN FEATURES: 'Envelope' is in the form of a *flechette*, or dart, consisting of three inflated wings of narrow delta planform, arranged in inverted Y configuration and powered by a pusher engine. Four-wing version also possible.
FLYING CONTROLS: Elevator/rudder on each trailing-edge.
STRUCTURE: Envelope/wings of polyester and polyurethane.
POWER PLANT: Prototype powered by one 11.0 kW (14.8 hp) piston engine, driving a two-blade pusher propeller.
ACCOMMODATION: Single seat.
DIMENSIONS, EXTERNAL:
Length overall ... 17.50 m (57 ft 5 in)
DIMENSIONS, INTERNAL:
Envelope/wings volume, total ... 250.0 m^3 (8,828.7 cu ft)
PERFORMANCE:
Max level speed ... 27 kt (50 km/h; 31 mph)

NEW ENTRY

NOIN

NOIN AERONAUTIQUE ALPAERO

R N 85, Chateauvieux, F-05130 Tallard
Tel/Fax: (+33 4) 92 54 15 04
e-mail: info@alpaero.com
Web: http://www.alpaero.com
DIRECTOR: Claude Noin

Company formed 1976. The Exel is the fourth design produced by Claude Noin, whose father was a noted soaring pilot of the 1950s and was first to cross the Alps by glider. The company's earlier Sirius motor glider is no longer available, although it will return to active status if interest in design is increased.

UPDATED

NOIN CHOUCAS

English name: Jackdaw
TYPE: Side-by-side ultralight kitbuilt.
PROGRAMME: Design started 1994 by Claude Noin; construction of prototype started 1995; first flight (05-CO) February 1996; construction of production prototype started 1996, not been notified to *Jane's* by April 2001. Marketed as basic kit for advanced builders, quoted build time 1,200 hours. A special version for physically handicapped pilots is to be launched, if there is sufficient demand.
CUSTOMERS: One flying and two under construction by early 2004.
COSTS: Kit (glass fibre) €7,738, (carbon fibre) €8,756 (2004).
DESIGN FEATURES: Tail-less design, stated to be stallproof and spinproof; design aimed at meeting ultralight category certification requirements while matching performance of modern gliders and conventional three-axis ultralight aircraft. Dihedral 4°; zero twist.
FLYING CONTROLS: Unconventional flying wing; pitch control by elevators in inboard wing trailing-edge; ailerons outboard; large rudder. Trim tabs on elevators; spoilers/airbrakes on upper surface of wing.
STRUCTURE: Standard fuselage moulded in half-shells of carbon fibre/epoxy; glass fibre/epoxy fuselage available at lower cost for advanced builders, but with weight penalty; composites wing structure with carbon fibre main spar, glass fibre/epoxy-stiffened Styrofoam ribs and birch plywood D-section leading-edge, covered with heat-shrink Dacron; moulded carbon fibre/epoxy ailerons; Lexan/polycarbonate canopy with carbon fibre frame.
LANDING GEAR: Tailwheel type, fixed. Fuselage-mounted spring cantilever mainwheel legs with wheel fairings; drum brakes.
POWER PLANT: One 37.0 kW (49.6 hp) Rotax 503 UL-2V with 2.58:1 reduction gear driving a two-blade ground-adjustable DUC 20 carbon fibre propeller. Fuel capacity 36 litres (9.5 US gallons; 7.9 Imp gallons), of which 35 litres (9.25 US gallons; 7.7 Imp gallons) are usable.
ACCOMMODATION: Two, side by side beneath one-piece upward-hinging canopy; baggage shelf to rear of seats.
SYSTEMS: 14 V DC alternator provides electrical power.
AVIONICS (FACTORY-BUILT AIRCRAFT): *Comms:* VHF comm and intercom.
Instrumentation: ASI, altimeter, variometer, slip indicator, compass, tachometer, fuel gauge, chronometer, CHT and EGT gauges and voltmeter.
EQUIPMENT: BRS 5 ballistic parachute recovery system optional.
DIMENSIONS, EXTERNAL:
Wing span ... 14.35 m (47 ft 1 in)
Length overall ... 5.35 m (17 ft 6½ in)
Height overall ... 1.95 m (6 ft 4¾ in)
Propeller diameter ... 1.60 m (5 ft 3 in)
DIMENSIONS, INTERNAL:
Cabin max width ... 0.97 m (3 ft 2 in)
AREAS:
Wings, gross ... 21.30 m^2 (229.3 sq ft)
WEIGHTS AND LOADINGS:
Weight empty ... 260–290 kg (573–639 lb)
Max T-O weight:
FAI ultralight certification ... 450 kg (992 lb)
motor glider certification ... 510 kg (1,124 lb)
PERFORMANCE, POWERED:
Max level speed ... 97 kt (180 km/h; 111 mph)
Cruising speed at 75% power ... 70 kt (130 km/h; 81 mph)
Stalling speed ... 33 kt (60 km/h; 38 mph)
Max rate of climb at S/L ... 180 m (590 ft)/min
T-O run ... 100 m (330 ft)
Landing run ... 150 m (495 ft)
Range at 75% power, no reserves ... 175 n miles (325 km; 201 miles)
PERFORMANCE, UNPOWERED:
Best glide ratio ... 23
Min rate of sink ... less than 1.00 m (3.28 ft)/s

UPDATED

NOIN EXEL

TYPE: Single-seat ultralight/motor glider.
PROGRAMME: Design started 1997; first flight of prototype September 1998; construction of first production aircraft began 1999. Marketed as fast-build kit or ready to fly.
CUSTOMERS: Total of seven ordered and six built by February 2004, including one export to Chile.
COSTS: Kit less engine, propeller and instruments, €23,000, plus tax (2004).

First Noin Choucas ultralight kitbuilt ***NEW***/0576598

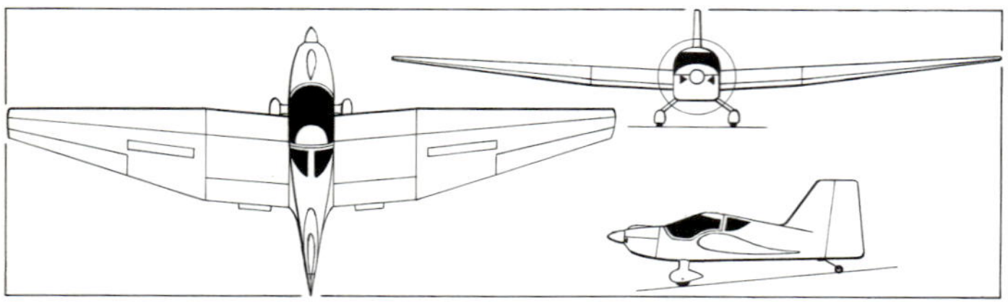

Noin Choucas tail-less light aircraft *(James Goulding)* 0044680

Noin Exel single-seat ultralight motor glider 0098690

DESIGN FEATURES: Pod-and-boom configuration, with T tail. Aimed at meeting FAI ultralight and JAR 22 certification criteria. Quoted construction time of fast-build kit, 300 hours.
FLYING CONTROLS: Conventional and manual; cable-operated elevators, flaperons and rudder; three-position flaps, deflections −5, 0 and +5°. Airbrakes on upper wing surfaces.
STRUCTURE: Fuselage moulded in half-shells of glass fibre/epoxy with plywood bulkheads. Wings have carbon fibre spar caps and glass fibre skins. Fabric-covered rudder. Optional winglets. Quoted build time 300 hours.
LANDING GEAR: Monowheel type; fixed. Tailwheel incorporated in base of rudder for directional control on the ground; small outrigger wheel at each wingtip. Drum brake on mainwheel.
POWER PLANT: One 13.4 kW (18 hp) JPX D-320 two-stroke flat twin with 2.38:1 reduction drive to a carbon fibre two-blade folding pusher propeller (three-blade propeller being tested October 2001). Fuel capacity 19 litres (5.0 US gallons; 4.2 Imp gallons).
EQUIPMENT: Optional ballistic parachute recovery system.
DIMENSIONS, EXTERNAL:
Wing span 13.74 m (45 ft 1 in)
Length overall 5.90 m (19 ft 4¼ in)
Propeller diameter 1.28 m (4 ft 2½ in)
AREAS:
Wings, gross 11.62 m² (125.1 sq ft)
WEIGHTS AND LOADINGS:
Weight empty 196 kg (432 lb)
Max T-O weight 310 kg (683 lb)
PERFORMANCE, POWERED:
Never-exceed speed (VNE) 97 kt (179 km/h; 111 mph)
Cruising speed at 75% power 65 kt (120 km/h; 75 mph)
Stalling speed 34 kt (63 km/h; 40 mph)
Max rate of climb at S/L 120 m (394 ft)/min
T-O run 150 m (492 ft)
Landing run 140 m (459 ft)
Endurance 3 h
g limits +4.4/−2.2
PERFORMANCE, UNPOWERED:
Best glide ratio 30
Min rate of sink 0.75 m (2.5 ft)/s

UPDATED

PJB

PJB AEROCOMPOSITE SARL

Zone Industrielle Aérodrome, Batiment 43, F-10500 St Leger sous Brienne
Tel: (+33 3) 25 27 55 54
Fax: (+33 3) 25 27 75 62
e-mail: contact@pjbaerocomposite.com
Web: http://www.pjb-aviation.net
SALES MANAGER: Pierre-Jean Back

PJB Aerocomposite, a division of spares supplier PJB Aviation Support, took over construction and development of the Aviakit Hermes, renaming it Vega. New 15,000 m² (161,460 sq ft) (total area) assembly plant at Brienne-le-Chateau inaugurated in March 2003. While the aviation support element of PJB remains in operation, there has been no recent mention of its lightplane division, which appears to have lapsed.

UPDATED

POTTIER

AVIONS POTTIER

4 rue de Poissy, F-78130 Les Mureaux
Tel: (+33 1) 30 99 13 85
Fax: (+33 1) 34 92 97 26
e-mail: 113015.217@compuserve.com
DIRECTOR: Jean Pottier

The prolific Jean Pottier designed numerous light aircraft for amateur construction or assembly. The P 200 series has been built commercially by several manufacturers in addition to being available to amateur builders. Most recent design was the **P 320UL**, the prototype of which (79-BC) was shown at the RSA Rally at Chambley in 2002.

Pottier was killed on 3 September 2003 when a P 320 apparently suffered a structural failure in the air.

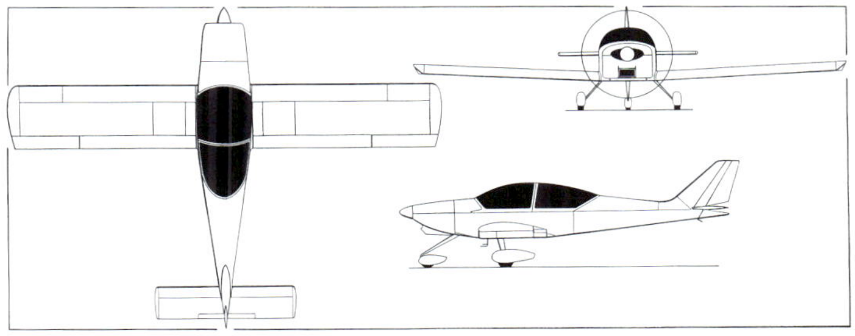

Pottier P 230 Panda three-seat kitbuilt ***(Paul Jackson)*** 0044683

P 320, Jean Pottier's last design ***(Geoffrey P Jones)*** ***NEW***/0567249

Pottier P 230 Panda built by a French amateur constructor ***(Paul Jackson)*** ***NEW***/0552273

POTTIER P 200 SERIES

TYPE: Four-seat kitbuilt.
PROGRAMME: First aircraft in P 200 series was single-seat P 210 S Coati with tailwheel-type landing gear. This does not appear to have entered production, but several two- to four-seat kitbuilt versions have appeared subsequently.
CURRENT VERSIONS: **P 220 S Koala:** Side-by-side two-seat version; 55.9 kW (75 hp) VW/Limbach engine standard. At least 20 kits delivered by early 1996; six flying and 20 building reported by early 1997. Modified version with Walter Mikron engine formerly produced in Czech Republic by Evektor-Aerotechnik (which see) and now marketed, with Limbach or Rotax engine, as the AT-3 by Aero Sp z.o.o of Poland. (see appropriate entry). Rotax-, Lycoming- and Continental-engined versions built in Czech Republic by EV-AT and in Italy by SG Aviation (both which see).

P 230 S Panda: Three-seater; 74.6 kW (100 hp) Continental engine standard. Prototype first flew 1990. Reportedly also available as **P230i** ultralight with 63.4 kW (85 hp) Rotax 912 engine.

P 240 S Saiga: Four-seat version (2 + 2); 134 kW (180 hp) Textron Lycoming engine. No specification data received for this version, and may no longer be current.

P 250 S Xerus: Two-seater; essentially tandem-seat version of Koala, but some dimensions differ.

P 270 S Amster Four-seater (2 + 2); similar to P 240 S but with 112 kW (150 hp) Textron Lycoming engine.

CUSTOMERS: Sold to customers in Czech Republic, Finland, France, Germany, Netherlands, Norway, Poland and elsewhere.
DESIGN FEATURES: Typical low-wing monoplane with sweptback vertical tail; wing aerofoil section NACA 4415 on all versions.
FLYING CONTROLS: Conventional and manual. Aileron deflection +20/−15° on all versions; all-moving tailplane with automatic balance/trim tab. All versions have flaps (settings 0, 10 and 30°).
STRUCTURE: Mainly metal; some components in composites.
LANDING GEAR: All versions have non-retractable tricycle gear. Mainwheel/nosewheel sizes 330 and 270 mm respectively on two-seaters, 350 and 300 mm on three- and four-seaters. Hydraulic mainwheel brakes. Wheel speed fairings optional.
POWER PLANT: See under Current Versions above.
ACCOMMODATION: See under Current Versions above. Dual controls standard. One-piece canopy on P 250 S, two-piece on other versions.
DIMENSIONS, EXTERNAL:
Wing span: 220 6.50 m (21 ft 4 in)
230, 270 8.10 m (26 ft 7 in)
250 6.40 m (21 ft 0 in)
Wing chord, constant: all 1.25 m (4 ft 1¼ in)
Wing aspect ratio: 220 5.3
230, 270 6.6
250 5.2
Length overall: 220 5.54 m (18 ft 2 in)
230 6.35 m (20 ft 10 in)
250 5.82 m (19 ft 1¼ in)
270 6.45 m (21 ft 2 in)
Height overall: 220, 250 1.95 m (6 ft 4¾ in)
230 2.05 m (6 ft 8¾ in)
270 2.15 m (7 ft 0¾ in)
Tailplane span: 220, 230, 250 2.60 m (8 ft 6½ in)
270 3.00 m (9 ft 10 in)
Wheel track: 220 1.70 m (5 ft 7 in)
230, 270 1.75 m (5 ft 9 in)
250 1.40 m (4 ft 7 in)
Wheelbase: 220 1.32 m (4 ft 4 in)
230, 270 1.45 m (4 ft 9 in)
250 1.55 m (5 ft 1 in)
Propeller diameter: 220 1.44 m (4 ft 8¾ in)
230 1.72 m (5 ft 7¾ in)
250 1.50 m (4 ft 11 in)

DIMENSIONS, INTERNAL:
Cabin: Length: 220 ... 1.64 m (5 ft 4½ in)
230, 250 ... 2.00 m (6 ft 6¾ in)
270 ... 2.10 m (6 ft 10¾ in)
Max width: 220 ... 1.04 m (3 ft 5 in)
230, 270 ... 1.10 m (3 ft 7¼ in)
250 ... 0.70 m (2 ft 3½ in)
Max height: 220 ... 1.05 m (3 ft 5¼ in)
230, 270 ... 1.10 m (3 ft 7¼ in)
250 ... 1.00 m (3 ft 3¼ in)

AREAS:
Wings, gross: 220 ... 8.00 m^2 (86.1 sq ft)
230, 270 ... 10.00 m^2 (107.6 sq ft)
250 ... 7.90 m^2 (85.0 sq ft)
Ailerons (total): 220, 250 ... 0.40 m^2 (4.31 sq ft)
230, 270 ... 0.60 m^2 (6.46 sq ft)
Flaps (total): 220, 250 ... 0.90 m^2 (9.69 sq ft)
230, 270 ... 1.10 m^2 (11.84 sq ft)
Vertical tail surfaces (total):
220, 230, 250 ... 0.80 m^2 (8.61 sq ft)
270 ... 1.00 m^2 (10.76 sq ft)
Horizontal tail surfaces (total):
220, 230, 250 ... 1.80 m^2 (19.38 sq ft)
270 ... 2.10 m^2 (22.60 sq ft)

WEIGHTS AND LOADINGS:
Weight empty: 220 ... 275 kg (606 lb)
230 ... 380 kg (838 lb)
250 ... 240 kg (529 lb)
270 ... 450 kg (992 lb)
Max T-O weight: 220 ... 500 kg (1,102 lb)
230 ... 700 kg (1,543 lb)
250 ... 470 kg (1,036 lb)
270 ... 870 kg (1,918 lb)
Max wing loading: 220 ... 62.5 kg/m^2 (12.80 lb/sq ft)
230 ... 70.0 kg/m^2 (14.34 lb/sq ft)
250 ... 59.5 kg/m^2 (12.19 lb/sq ft)
270 ... 87.0 kg/m^2 (17.82 lb/sq ft)
Max power loading: 220 ... 8.95 kg/kW (14.70 lb/hp)
230 ... 9.39 kg/kW (15.43 lb/hp)
250 ... 9.70 kg/kW (15.94 lb/hp)
270 ... 7.78 kg/kW (12.78 lb/hp)

PERFORMANCE:
Never-exceed speed (VNE):
220, 230, 250 ... 135 kt (250 km/h; 155 mph)
270 ... 156 kt (290 km/h; 180 mph)
Max level speed: 220 ... 113 kt (210 km/h; 130 mph)
230 ... 121 kt (225 km/h; 140 mph)
250 ... 119 kt (220 km/h; 137 mph)
270 ... 146 kt (270 km/h; 168 mph)
Max cruising speed: 220 ... 103 kt (190 km/h; 118 mph)
230 ... 113 kt (210 km/h; 130 mph)
250 ... 108 kt (200 km/h; 124 mph)
270 ... 130 kt (240 km/h; 149 mph)
Stalling speed, flaps up:
220, 230 ... 49 kt (90 km/h; 56 mph)
250 ... 48 kt (88 km/h; 55 mph)
270 ... 54 kt (100 km/h; 63 mph)
Stalling speed, flaps down:
220, 250 ... 44 kt (80 km/h; 50 mph)
270 ... 49 kt (90 km/h; 56 mph)
Rate of climb at S/L: 220 ... 210 m (689 ft)/min
230 ... 240 m (787 ft)/min
250 ... 222 m (728 ft)/min
270 ... 396 m (1,299 ft)/min
T-O run: 220 ... 200 m (660 ft)
230 ... 210 m (690 ft)
250 ... 180 m (590 ft)
270 ... 330 m (1,085 ft)
Landing run: 220 ... 220 m (725 ft)
230 ... 240 m (790 ft)
250 ... 200 m (660 ft)
270 ... 370 m (1,215 ft)
Range with max fuel:
220, 270 ... 378 n miles (700 km; 435 miles)
230 ... 405 n miles (750 km; 466 miles)
250 ... 351 n miles (650 km; 403 miles)
g limit: all ... +5.7

PROTOPLANE

PROTOPLANE AIRCRAFT DESIGN & PROTOTYPING

Tel: (+33 5) 62 95 36 31
e-mail: protopla@aol.com

In 2003, a mockup of the Massaï motor glider made its public debut at Paris.

UPDATED

PROTOPLANE MASSAÏ

TYPE: Side-by-side ultralight motor glider.

PROGRAMME: First reported in early 2002, when maiden flight said to be imminent. However, only a mockup was displayed at Paris in June 2003, since which time no further reports have been received.

DESIGN FEATURES: Low-wing motor glider with retractable landing gear and retractable propeller behind forward-sliding nosecone. Constant-chord inboard wing; outer panels with curved taper on leading- and trailing-edges. Constant-chord tailplane at base of sweptback fin. JAR 22 conformal.

Wing section FX66-17AII-182/26 at root; FX60-126.1/26 at tip.

Protoplane Massaï mockup ***(Paul Jackson)*** ***NEW***/0576839

FLYING CONTROLS: Conventional and manual. Two-section ailerons extend over tapered and constant-chord sections. Flaps, max deflection 35°. Schempp-Hirth airbrakes.

STRUCTURE: Composites throughout.

LANDING GEAR: Monowheel type; retractable, including outriggers at one-third span. Semi-buried tailwheel, mainwheel size 5.00–5; outriggers 125×45.

POWER PLANT: One 48.0 kW (64.5 hp) Rotax 582 piston engine driving a 1.68 m (5 ft 6¼ in) diameter two-blade Protoplane wood and glass fibre folding ('Z-type') propeller via a belt reduction system. Fuel capacity 70 litres (18.5 US gallons; 15.4 Imp gallons).

ACCOMMODATION: Two persons, side by side, beneath single piece, forward-hinged canopy.

EQUIPMENT: Provision for surveillance sensors.

DIMENSIONS, EXTERNAL:
Wing span ... 16.395 m (53 ft 9½ in)
Length overall, propeller retracted ... 7.42 m (24 ft 4¼ in)

AREAS:
Wings, gross ... 15.46 m^2(166.4 sq ft)

WEIGHTS AND LOADINGS:
Weight empty ... 250 kg (551 lb)
Max T-O weight ... 450 kg (992 lb)

PERFORMANCE, POWERED:
Cruising speed ... 119 kt (220 km/h 137 mph)
Max rate of climb at S/L ... 330 m (1,083 ft)/min
T-O run ... 73 m (240 ft)
Landing from 15 m (50 ft) ... 195 m (640 ft)
Range ... 449 n miles (832 km; 517 miles)
g limits ... +5.73/–73

PERFORMANCE, UNPOWERED:
Stalling speed, flaps down ... 30 kt (55 km/h; 35 mph)
Best glide ratio ... 36
Min rate of sink ... 0.61 m (2.00 ft)/s

UPDATED

REIMS

REIMS AVIATION INDUSTRIES

Aérodrome de Reims-Prunay, F-51360 Prunay
Tel: (+33 3) 26 48 46 65
Fax: (+33 3) 26 49 18 57
*e-mail:*reims.aviation@reims-aviation-industries.com
Web: http://www.reims-aviation-industries.com
CEO: Bernard Grouchko
DIRECTOR GENERAL: Philippe Denavit
VICE-PRESIDENT, MARKETING AND SALES: Philippe Denaoit
AREA SALES MANAGER: Laurent Mesmin
EXTERNAL RELATIONS: Max Boirame

Originally Avions Max Holste, founded 1933; Cessna acquired 49 per cent share in February 1960; name then changed to Reims Aviation, which licensed to manufacture Cessna aircraft for sale in Europe, Africa and Asia, but stopped making small piston-engined types when Cessna did; Reims had built 6,351 aircraft of all types by late 2001. Reims developed twin-turboprop F406 Caravan II (see below); but Cessna sold its share in Reims Aviation to Compagnie Française Chaufour Investissement (CFCI) in early 1989.

Filed for bankruptcy protection on 30 October 2002 following loss of aerostructures work in wake of Fairchild Dornier's bankruptcy. Company sold in two separate parts, with aerostructures and light aircraft maintenance becoming Reims Aerospace under ownership of Austrian group, Ventana.

Caravan production, marketing and maintenance activities purchased by Green Recovery venture capital group, becoming Reims Aviation Industrie. Employees in 2003 totalled 40. Expansion plans included manufacture of four aircraft in 2004.

UPDATED

REIMS F406 CARAVAN II

TYPE: Utility turboprop twin.

PROGRAMME: Announced mid-1982; first flight (F-WZLT) 22 September 1983; French certification 21 December 1984, FAA later; first flight production F406 (F-ZBEO) 20 April 1985.

CURRENT VERSIONS: **F406 Caravan II:** Initial production version, available in passenger, freight (with underbelly cargo pod), medevac, skydiving, aerial survey, training, navaid calibration and target towing variants. These retrospectively known as **F406 Mark I**.

F406 Mark II: Upgraded version announced (as F406 NG New Generation) in October 2000. Features include 473.5 kW (635 shp) PT6A-135A engines driving four-blade propellers; reduced structural weight; same MZFW; increase in maximum take-off weight to 4,700 kg (10,361 lb); quick-release access panels for easier maintenance; new lightweight instrument panel with liquid crystal displays; redesigned cabin interior by Air Esthetic featuring composites panels that can be painted or upholstered, improved acoustic and thermal insulation, new ergonomically designed seats with integral harnesses for improved comfort on long missions and integral window blinds; and standard Bendix/King Silver Crown avionics with Silver Crown Plus comms and two-tube EFIS display. Three standard cabin interiors offered commuter, special missions and VIP. Take-off, climb and OEI rate of climb performance are improved, and endurance increased by up to one hour.

Additionally, six versions of the basic **Special Missions** configuration, each with a ventral 360° radar:

Vigilant: Surveillance version (land or sea) with mission equipment to suit customer requirement, including (in latest configuration) Thales Airborne Maritime Situation Control System (AMASCOS) which comprises belly-mounted Telephonics AN/APS-143, Texas Instruments 1022 or Thales/DASA Ocean Master 100/200 360° surveillance radars, chin-mounted FLIR turret for FSI Ultra, Inframetrics Mk III, SAGEM Hesis, Star Safire, Thales Chlio or Wescam FLIR, mission management and communications systems and single operator's console in Srs 100 form; or with additional console for electronic surveillance operator in Srs 200. BAE Systems Seaspray 2000 radar for Scottish Fisheries; Texas Instruments AN/APS-134 radar for Australian Customs.

Vigilant Frontier: Border patrol and anti-drug surveillance version with mission equipment including surveillance radar in anti-drug configuration, FLIR and datalink.

Vigilant Polmar I: Pollution surveillance (maritime police) version with SAGEM Cyclope 2000 IR linescanner and Ericsson or Terma SLAR.

Vigilant Polmar II: As Polmar II, but with updated survey systems and LLTV camera. One ordered by French Customs in 2001 for delivery in 2004.

Vigilant Surmar: Surveillance (maritime) version with 360° radar, such as Texas Instruments AN/APS-134 or Telephonics AN/APS-143 and optional equipment including Litton night vision system. In 1996 Reims developed underwing hardpoints for the F406 which enable it to carry light weapons such as machine gun pods or rocket launchers, or camera pods or SAR liferafts.

Vigilant Surpolmar: Maritime surveillance version for Hellenic Coast Guard, combining Surmar and Polmar missions. Equipment specific to Greek aircraft comprises Surmar suite of Honeywell 1500B radar and FLIR. Systems Star Safire turret aided by Caledonian Airborne Systems track-while-scan system for 20-target tracking; and Polmar suite of Ericsson SLAR, Swedish Space Corporation MSS 5000 and SAGEM Cyclope 2000 IR linescan

Vigilant Comint/Imint: Dedicated communications and imaging intelligence version of the Caravan II, developed jointly with Thomson-CSF Communications, first shown at Paris '97. Equipped with Thomson TCC SAS airborne Elint system, an operator's console with MS Windows NT-based workstation and an array of direction-finding antennas housed in a ventral radome. Is also first Vigilant with a pair of underwing pylons for carriage of stores which can include gun or rocket pods and airdroppable SAR equipment.

CUSTOMERS: One prototype and 90 production aircraft built by late 2003. Current operators include but are not limited to: Air Charter Australia (one); Air Guyane, French Guyana (one); Air Namibia/Comav, South Africa (four); Air St Pierre, St Pierre & Miquelon (one); Aircraft Sales & Services, Pakistan (one); Airtech Servicios Aéreos-Tamarindo, Venezuela (one); Atlantic Express, UK (four); Aussie Air, Australia (one); Australian Coastwatch (three); Avirex, Gabon (one); Bay Air, Namibia (one); BenAir, Denmark (one); Bering Strait School, Alaska (one); Century Textiles & Industries, India (one); Clint Aero, Virgin Islands (one); Direct Flight, UK (four); Duijvestijn Aviation, The Netherlands (four); East African Air Charters, Kenya (one); Falcon Air, Zimbabwe (four); Fehlhaber Flugdienst, Germany (three); Fisheries Patrol of Namibia (one); French Army (two, used for target towing); French Customs Service (13, see below); Hageland Aviation, Alaska (three); Hellenic Coast Guard, Greece (three, see below); Homegrown Ltd, Kenya (one); Island Development, Seychelles (one); Lucorp Air Limo, Belgium (one); Mann Aviation, Jersey (one); Out of Blue Air Safari, South Africa (one); Republic of Korea Navy (five, see below); Sociedada de Aviçao Ligeira, Angola (three); Tanzanair, Tanzania (two); Vincent Aviation, Australia (one); Vincent Aviation, New Zealand (one); and Westair Aviation, Namibia (two).

French Customs Service received 10 in Surmar and two in Polmar I/II configuration. One in Polmar III configuration (No. 90), ordered in late 2000 and intended for delivery in 2002 but delayed until first quarter 2004. First Surmar (No. 74, although registered as a Vigilant) handed over 11 January 1995, followed by Nos. 75 and 77 on 12 September 1995, the latter two replacing Cessna 404s in Martinique, French Antilles.

Recent customers for the F406 include the Republic of Korea Navy, which ordered five in target tug configuration, the first delivered on 23 November 1998 and the last in mid-June 1999, total contract value being US$24 million (1997); and the Hellenic Coast Guard, which, in June 1999, ordered two for delivery in late 2000 as Surpolmar variants and additional one in November 1999; the first Surpolmar (F-WWSS/AC-21) for this contract made its maiden flight on 21 August 2000 and was delivered on 27 March 2001; remaining two received by December 2001, at which time a fourth was under negotiation. One to Air Saint-Pierre in mid-2003, fitted with TCAS II (first produced by Reims Aviation Industries). Total 92 sold by October 2003 (excluding prototype). An agreement with the Brazilian government signed in April 1998 provides for the purchase of three aircraft in multisensor configurations.

COSTS: Standard commuter aircraft US$2.4 million (2003). Maritime patrol versions typically US$4 million to US$6 million with mission equipment. Korean Navy contract for five aircraft valued at US$24 million, including technician training and one-year technical support (1997). Greek contract for two Surmars valued at FFr90 million (1999); single Polmar FFr33 million (1999). Direct operating cost US$338 per hour (1997).

DESIGN FEATURES: Extrapolated from Cessna Conquest airframe; wing section NACA 23018 at root and 23012 at tip; dihedral 3° 30' on centre-section, 4° 55' on outer panels; twist −3°; incidence 2° at root; fin offset 1° to port; tailplane dihedral 9°; cabin not pressurised.

FLYING CONTROLS: Conventional and manual; Trim tabs in elevators, port aileron and rudder; hydraulically operated Fowler flaps.

STRUCTURE: Conventional light metal with three-spar fail-safe wing centre-section to SFAR 41C; two-spar outer wings.

LANDING GEAR: Hydraulically retractable tricycle type with single wheel on each unit. Mainwheel tyre size 22×7.75–10, nosewheel 6.00-6. Main units retract inward into wing, nosewheel rearward. Emergency extension by means of a 138 bar (2,000 lb/sq in) rechargeable nitrogen bottle. Cessna oleo-pneumatic shock-absorbers. Main units of articulated (trailing link) type. Single-disc hydraulic brakes. Parking brake.

POWER PLANT: Two Pratt & Whitney Canada PT6A-112 turboprops (each 373 kW; 500 shp), each driving a McCauley 9910535-2 three-blade, reversible-pitch, and automatically feathering metal propeller. Fuel capacity 1,823 litres (481 US gallons; 401 Imp gallons) of which 1,798 litres (475 US gallons; 395.5 Imp gallons) usable. Oil capacity 17.2 litres (4.5 US gallons; 3.8 Imp gallons).

ACCOMMODATION: Crew of two and up to 12 passengers, in pairs facing forward, with centre aisle, except at rear of cabin in 12/14-seat versions. Alternative basic configurations for six VIP passengers in reclining seats in business version, and for operation in mixed passenger/freight role. Business version has partition between cabin and flight deck, and lavatory on starboard side at rear. Split main door immediately aft of wing, on port side, with built-in airstair in downward-hinged lower portion. Optional cargo door forward of this door to provide single large opening. Overwing emergency exit on each side. Passenger seats removable for cargo carrying, or for conversion to ambulance, air photography, maritime surveillance and other specialised roles. Baggage compartments in nose, with three doors, at rear of cabin and in rear of each engine nacelle. Ventral cargo pod optional.

SYSTEMS: Freon air conditioning system of 16,000 BTU capacity, plus engine bleed air and electric boost heating. Electrical system includes 28 V 250 A starter/generator on each engine and 39 Ah Ni/Cd battery. Hydraulic system, pressure 120 bar (1,750 lb/sq in), for operation of landing gear. Separate hydraulic system for brakes. Optional Goodrich pneumatic de-icing of wings and tail unit, and electric windscreen de-icing.

AVIONICS: Standard Bendix/King Silver Crown; Gold Crown optional.

Comms: Dual Bendix/King transceivers.

Radar: Honeywell RDR 2000 weather radar optional. Maritime surveillance radar as described under Current Versions.

Flight: Dual ADF and marker beacon receiver. Autopilot optional.

Instrumentation: Provision for equipment to FAR Pt 135A standards, including dual controls and instrumentation for co-pilot.

EQUIPMENT: Optional cargo interior includes heavy-duty sidewalls, utility floorboards, cabin floodlighting and cargo restraint nets. Optional pylons under each wing for carriage of stores including gun or rocket pods and airdroppable SAR equipment.

DIMENSIONS, EXTERNAL:

Wing span	15.08 m (49 ft 5½ in)
Wing aspect ratio	9.7
Length overall	11.89 m (39 ft 0¼ in)
Height overall	4.01 m (13 ft 2 in)
Tailplane span	5.87 m (19 ft 3 in)
Wheel track	4.28 m (14 ft 0½ in)
Wheelbase	3.81 m (12 ft 6 in)
Propeller diameter	2.36 m (7 ft 9 in)
Cabin door: Height	1.27 m (4 ft 2 in)
Width	0.58 m (1 ft 10¾ in)
Cargo double door (optional):	
Total width	1.24 m (4 ft 1 in)

DIMENSIONS, INTERNAL:

Cabin (incl flight deck): Length	5.71 m (18 ft 8¾ in)
Max width	1.42 m (4 ft 8 in)
Max height	1.31 m (4 ft 3¼ in)
Min height (at rear)	1.21 m (3 ft 11½ in)
Width of aisle	0.29 m (11½ in)
Volume	8.6 m^3 (305 cu ft)
Baggage compartment (nose):	
Length	2.00 m (6 ft 6¾ in)
Volume	0.74 m^3 (26.0 cu ft)
Nacelle lockers: Length	1.55 m (5 ft 1 in)
Width	0.73 m (2 ft 4¾ in)
Baggage volume: total, internal	2.2 m^3 (79 cu ft)
incl cargo pod	3.5 m^3 (124 cu ft)

Reims F406 Polmar 2 in service with the French Customs ***NEW***/1030726

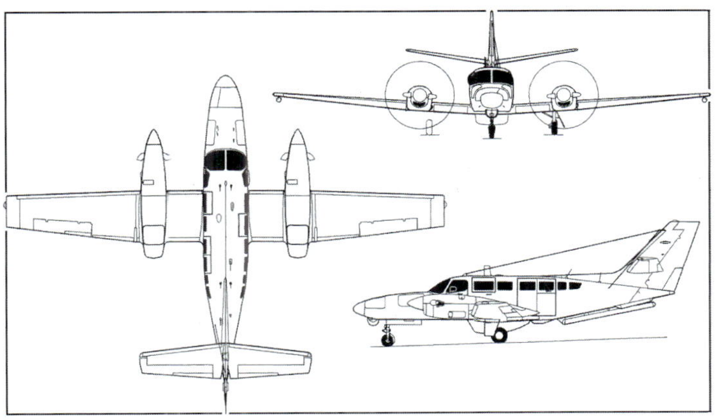

Reims F406 Vigilant Surpolmar maritime surveillance aircraft ***NEW***/0547130

AREAS:
- Wings, gross ... 23.48 m² (252.8 sq ft)
- Ailerons (total) ... 1.36 m² (14.64 sq ft)
- Trailing-edge flaps ... 3.98 m² (42.84 sq ft)
- Fin ... 4.05 m² (43.59 sq ft)
- Rudder, incl tab ... 1.50 m² (16.15 sq ft)
- Tailplane ... 5.81 m² (62.54 sq ft)
- Elevators, incl tabs ... 1.66 m² (17.87 sq ft)

WEIGHTS AND LOADINGS:
- Standard empty weight ... 2,283 kg (5,033 lb)
- Max payload ... 2,219 kg (4,892 lb)
- Max fuel ... 1,444 kg (3,183 lb)
- Max ramp weight ... 4,502 kg (9,925 lb)
- Max T-O and landing weight ... 4,468 kg (9,850 lb)
- Max zero-fuel weight ... 3,856 kg (8,500 lb)
- Max wing loading ... 190.3 kg/m² (38.97 lb/sq ft)
- Max power loading ... 5.99 kg/kW (9.85 lb/shp)

PERFORMANCE:
- Max operating Mach No. (MMO) ... 0.52
- Max operating speed (VMO) ... 229 kt (424 km/h; 263 mph) IAS
- Max cruising speed ... 246 kt (455 km/h; 283 mph)
- Econ cruising speed ... 200 kt (370 km/h; 230 mph)
- Stalling speed:
  - wheels and flaps up ... 94 kt (174 km/h; 108 mph) IAS
  - wheels and flaps down ... 81 kt (150 km/h; 93 mph) IAS
- Max rate of climb at S/L ... 564 m (1,850 ft)/min
- Rate of climb at S/L, OEI ... 121 m (397 ft)/min
- Service ceiling ... 9,145 m (30,000 ft)
- Service ceiling, OEI ... 4,935 m (16,200 ft)
- T-O run ... 526 m (1,725 ft)
- T-O to 15 m (50 ft) ... 803 m (2,635 ft)
- Landing from 15 m (50 ft), without reverse pitch ... 674 m (2,215 ft)
- Range with max fuel, at max cruising speed, 45 min reserves ... 1,153 n miles (2,135 km; 1,327 miles)
- *g* limits ... +3.6/−1.44

OPERATIONAL NOISE LEVELS:
- Flyover ... 72.0 EPNdB

UPDATED

ROBIN

ROBIN AVIATION

1 route de Troyes, F-21121 Darois
Tel: (+33 3) 80 44 20 50
Fax: (+33 3) 80 35 60 80
Web: http://www.robinaviation.com
NON-EXECUTIVE CHAIRMAN: Philippe Corne
CEO AND GENERAL MANAGER: Guy Pellissier
SALES MANAGER, EUROPE: Pierre Pelletier

Formed October 1957 as Centre Est Aéronautique; name changed to Avions Pierre Robin 1969; acquired July 1988 by Compagnie Française Chaufour Investissement (CFCI) and incorporated into Aéronautique Service group (now Apex International) with Robin SA (after-sales support company of Dijon Val-Suzon); Pierre Robin left company 1990. Total of 3,700 aircraft produced by early 2002 (excluding 35 Sperwer UAVs). Factory area (Constructions Aéronautiques de Bourgogne) 9,000 m² (96,875 sq ft).

Former Robin subdivisions, Constructions Aéronautiques de Bourgogne and Aéronautique Finance et Services, are now separate companies. The first-mentioned is responsible for the construction of Robin aircraft, as explained in the entry for Aéronautique Service.

Latest production addition to Robin range is R 2120 Alpha trainer, introduced 2001. However, on 25 June 2004 Alpha Aviation of New Zealand announced its plans to acquire the R 2120 programme and transfer manufacture from Europe, leaving Robin to concentrate on its range of wooden aircraft.

Robin's parent company, Apex Aircraft, emerged from voluntary receivership 25 August 2003.

UPDATED

ROBIN R 2120 ALPHA AND R 2160

Manufacturing rights being transferred to Alpha Aviation of New Zealand.

UPDATED

ROBIN DR 400 DAUPHIN

English name: Dolphin

TYPE: Four-seat lightplane.

PROGRAMME: First flight original DR 400 Petit Prince 15 May 1972; based on DR 300 (391, plus four prototypes), but with forward-sliding canopy; DR 400/100 Cadet (two-seat) and DR 400/108 Dauphin 80 included among early versions. French and UK certification 1977; DR 400 Dauphin introduced 1979; improvements introduced 1988 and 1993.

CURRENT VERSIONS: **DR 400/120 Dauphin 2+2:** Production version with 88.0 kW (118 hp) engine, to carry two adults and two children. Based on DR 400/120 Petit Prince, but with extra cabin windows.

DR 400/125i: Version of Dauphin 2+2 with fuel-injected 93.2 kW (125 hp) Teledyne Continental IO-240 driving three-bladed Mühlbauer constant-speed propeller, diameter 1.70 m (5 ft 7 in); prototype (F-WNNK) first flown 1995 and made its public debut at the Paris Air Show in June 1995; engine/propeller combination improves take-off, climb and cruising performance and reduces noise levels; one delivered in 1996. Weights and performance data in 2000–01 and earlier editions. No evidence of production.

DR 400/135 CDI: Version of Dauphin 2+2 with 99 kW (132.8 hp) Thielert Centurion 1.7 aero diesel engine; prototype (D-ECDI) completed ground running tests in August 2003 and first flew 10 October 2003. Deliveries due to have begun in October 2004.

DR 400/140B Dauphin 4: Full four-seater with 119 kW (160 hp) engine. Previously known as Major 80, which replaced DR 400/140 Earl.

Robin DR 400/140 Dauphin 4 *(Paul Jackson)* ***NEW***/0062940

Older Robin DR 400/120 Dauphin 2+2 *(Paul Jackson)* ***NEW***/0552272

DR 400/160 : and DR 400/180: *Described separately.*

CUSTOMERS: Some 1,850 of DR 400 series built by late 2003, including 30 in 2002, of which at least eight /120s and eight /140Bs and six in 2003 (two /120s, one 140B, one 160 and two 180s). Recent exports mainly to Germany, Netherlands, Norway, Switzerland and UK.

DESIGN FEATURES: Generally as for R 2120.

Wing section NACA 23013.5 modified with leading-edge droop; centre panels parallel chord, slight twist; outer panels tapered with dihedral 14°; twist −6°.

FLYING CONTROLS: Manual. Conventional ailerons and rudder. All-moving tailplane with trimmable anti-balance tab on each side; plain flaps.

STRUCTURE: All-wood; single box spar with ribs threaded over box; plywood-covered leading-edge box; polyester fabric covering elsewhere; fuselage plywood-covered; flaps all-metal and interchangeable; ailerons interchangeable.

LANDING GEAR: Non-retractable tricycle type, with long-stroke oleo-pneumatic shock-absorbers and hydraulically actuated disc brakes. All three wheels and tyres are size 380×150, pressure 1.57 bar (23 lb/sq in) on nose unit, 1.77 bar (26 lb/sq in) on main units. Nosewheel steerable via rudder bar. Fairings over all three legs and wheels. Tailskid with damper. Toe operated disc brakes and parking brake.

POWER PLANT: *Dauphin 2+2:* One 88.0 kW (118 hp) Textron Lycoming O-235-L2A flat-four engine, driving a Sensenich 72CKS6-0-56 two-blade, fixed-pitch aluminium propeller, or Hoffmann two-blade wooden propeller.

Dauphin 4: One Textron Lycoming O-320-D2A flat-four engine developing 104 kW (140 hp) at 2,300 rpm and 119 kW (160 hp) at 2,700 rpm. Sensenich 74DM6S5-2-64 propeller.

Dauphin 135: One TAE Centurion 1.7 diesel engine, rated at 99.0 kW (132.8 hp), driving an MT-Propeller, three-blade, variable-pitch metal propeller. Fuel tank in fuselage, capacity 110 litres (29.1 US gallons; 24.2 Imp gallons) with filler port on left side; Dauphin 4 has optional 51 litre (13.5 US gallon; 11.2 Imp gallon) auxiliary tank, with filler port on right side. Oil capacity 5.7 litres (1.5 US gallons; 1.25 Imp gallons).

ACCOMMODATION: Enclosed cabin, with fore-and-aft adjustable seats for three or four persons. Three-point inertia-reel harnesses standard. Maximum weight of 154 kg (340 lb) on front pair and 136 kg (300 lb), including baggage, at rear in Dauphin 2+2. Additional 55 kg (121 lb) of disposable load in Dauphin 4. Access via forward-sliding jettisonable transparent canopy. Dual controls standard. Cabin heated and ventilated. Baggage compartment with internal access.

SYSTEMS: Standard equipment includes a 12 V 50 A alternator, 12 V 32 Ah battery, electric starter, audible stall warning and windscreen de-icing.

AVIONICS: To customer's choice, typically Bendix/King KX 155 nav/com, KI 208 VOR/LOC indicator, KT 76C transponder with ACK 30 altitude encoder, Garmin 150 XL GPS, and SPA 400 intercom system, four microphone and headset jack plugs, and PTT switch on each control stick.

Instrumentation: Standard equipment includes ASI, sensitive altimeter, rate of climb indicator, artificial horizon, directional gyro, electric turn co-ordinator, magnetic compass, recording tachometer, eight-light annunciator/warning panel, oil pressure and temperature

Prototype Robin DR 400/135 CDI (TAE Centurion 1.7 diesel engine) *(Paul Jackson)* ***NEW***/0580293

gauges, fuel quantity gauges, fuel pressure gauge, and flight time recorder. Options include OAT gauge, CHT gauge, EGT gauge, and carburettor air temperature indicator.

EQUIPMENT: Standard equipment includes navigation lights, anti-collision strobe light, instrument panel lighting, and overall polyurethane paint with two-colour scheme. Options include pitot heat, external power socket, alternative static source, and additional exhaust silencer.

DIMENSIONS, EXTERNAL:
Wing span ... 8.72 m (28 ft 7¼ in)
Wing chord:
centre-section, constant ... 1.71 m (5 ft 7½ in)
at tip ... 0.90 m (2 ft 11½ in)
Wing aspect ratio ... 5.6
Length overall ... 6.96 m (22 ft 10 in)
Height overall ... 2.23 m (7 ft 3¾ in)
Tailplane span ... 3.20 m (10 ft 6 in)
Wheel track ... 2.60 m (8 ft 6¼ in)
Wheelbase ... 5.20 m (17 ft 0¾ in)
Propeller diameter: 2+2 ... 1.83 m (6 ft 0 in)
4 ... 1.88 m (6 ft 2 in)

DIMENSIONS, INTERNAL:
Cabin: Length ... 1.62 m (5 ft 3¾ in)
Max width ... 1.10 m (3 ft 7¼ in)
Max height ... 1.23 m (4 ft 0½ in)
Baggage volume ... 0.39 m^3 (13.8 cu ft)

AREAS:
Wings, gross ... 13.60 m^2 (146.4 sq ft)
Ailerons, total ... 1.15 m^2 (12.38 sq ft)
Flaps, total ... 0.70 m^2 (7.53 sq ft)
Fin ... 0.61 m^2 (6.57 sq ft)
Rudder ... 0.63 m^2 (6.78 sq ft)
Horizontal tail surfaces, total ... 2.88 m^2 (31.00 sq ft)

WEIGHTS AND LOADINGS:
Weight empty, equipped: 2+2 ... 550 kg (1,212 lb)
4 ... 580 kg (1,279 lb)
Max baggage: 2+2, 4 ... 40 kg (88 lb)
Max T-O and landing weight: 2+2 ... 900 kg (1,984 lb)
4 ... 1,000 kg (2,205 lb)
Max wing loading: 2+2 ... 66.2 kg/m^2 (13.56 lb/sq ft)
4 ... 73.5 kg/m^2 (15.05 lb/sq ft)
Max power loading: 2+2 ... 10.23 kg/kW (16.81 lb/hp)
4 ... 8.38 kg/kW (13.78 lb/hp)

PERFORMANCE:
Never-exceed speed (VNE):
2+2, 4 ... 166 kt (308 km/h; 191 mph)
Max level speed at S/L
2+2 ... 130 kt (241 km/h; 150 mph)
4 ... 143 kt (265 km/h; 165 mph)
Cruising speed at 75% power:
2+2 ... 116 kt (215 km/h; 133 mph)
4 ... 128 kt (237 km/h; 147 mph)
Stalling speed, flaps down:
2+2 ... 45 kt (82 km/h; 51 mph)
4 ... 47 kt (87 km/h; 54 mph)
Max rate of climb at S/L: 2+2 ... 183 m (600 ft)/min
4 ... 264 m (865 ft)/min
Service ceiling: 2+2 ... 3,660 m (12,000 ft)
4 ... 4,265 m (14,000 ft)
T-O run: 2+2 ... 235 m (775 ft)
4 ... 245 m (805 ft)
T-O to 15 m (50 ft): 2+2 ... 535 m (1,755 ft)
4 ... 485 m (1,595 ft)
Landing from 15 m (50 ft): 2+2 ... 460 m (1,510 ft)
4 ... 470 m (1,545 ft)
Landing run: 2+2 ... 200 m (660 ft)
4 ... 220 m (725 ft)
Range with standard fuel at 65%, no reserves: 2+2 ... 495 n miles (916 km; 569 miles)
4 ... 447 n miles (828 km; 514 miles)
Range, Dauphin 4 with optional fuel at max cruising speed, no reserves ... 645 n miles (1,195 km; 742 miles)

UPDATED

ROBIN DR 400/160 MAJOR

TYPE: Four-seat lightplane.

PROGRAMME: First flight of original DR 400 Chevalier 29 June 1972; certified France and UK same year; Major introduced 1980, being based on DR 400/140B Major 80 with additional cabin windows.

CUSTOMERS: See main DR 400 entry; total 140 delivered by December 1998. Six built in 2000, three in 2001, six in 2002 and one in 2003.

DESIGN FEATURES: Main differences from Dauphin are external baggage compartment door on port side and extended wingroot leading-edges to house additional fuel tanks.

Differences from Dauphin listed below:

POWER PLANT: One 119 kW (160 hp) Textron Lycoming O-320-D2A flat-four engine, driving a Sensenich 74DM6S5-2-64 two-blade, fixed-pitch aluminium propeller. Fuel tank in fuselage, capacity 110 litres (29.1 US gallons; 24.2 Imp gallons), and two tanks in wingroot leading-edges, giving total capacity of 190 litres (50.2 US gallons; 41.8 Imp gallons), of which 182 litres (48.1 US gallons; 40.0 Imp gallons) are usable. Provision for auxiliary tank, raising total capacity to 240 litres (63.4 US gallons; 52.8 Imp gallons). Oil capacity 7.6 litres (2.0 US gallons; 1.7 Imp gallons).

ACCOMMODATION: Seating for four persons, on adjustable front seats (maximum load 154 kg; 340 lb total) and rear bench seat (maximum load 154 kg; 340 lb total). Forward-sliding transparent canopy, but port rear window/door provides outside access to baggage area. Up to 40 kg (88 lb) of baggage can be stowed aft of rear seats when four occupants are carried.

AVIONICS: To customer's choice, typically Garmin GNS430 GPS/com/nav/ILS with GI 106 indicator, Bendix/King KX 155 nav/com 2, Bendix/King KI 203 indicator, and Garmin GTX 327 transponder.

DIMENSIONS, EXTERNAL:
Propeller diameter ... 1.83 m (6 ft 0 in)
Baggage door: Height ... 0.47 m (1 ft 6½ in)

Robin DR 400/160, showing additional rear transparencies and port rear door to baggage area *(Paul Jackson)* 0062942

Width ... 0.55 m (1 ft 9½ in)

AREAS:
Wings, gross ... 14.20 m^2 (152.8 sq ft)

WEIGHTS AND LOADINGS:
Weight empty, equipped ... 598 kg (1,318 lb)
Max T-O and landing weight ... 1,050 kg (2,315 lb)
Max wing loading ... 74.0 kg/m^2 (15.15 lb/sq ft)
Max power loading ... 8.81 kg/kW (14.47 lb/hp)

PERFORMANCE:
Never-exceed speed (VNE) ... 166 kt (308 km/h; 191 mph)
Max level speed at S/L ... 146 kt (271 km/h; 168 mph)
Max cruising speed at 75% power at 2,440 m (8,000 ft) ... 132 kt (245 km/h; 152 mph)
Econ cruising speed at 65% power at 3,200 m (10,500 ft) ... 130 kt (241 km/h; 150 mph)
Stalling speed: flaps up ... 56 kt (103 km/h; 64 mph)
flaps down ... 50 kt (93 km/h; 58 mph)
Max rate of climb at S/L ... 255 m (836 ft)/min
Service ceiling ... 4,115 m (13,500 ft)
T-O run ... 295 m (970 ft)
T-O to 15 m (50 ft) ... 590 m (1,940 ft)
Landing from 15 m (50 ft) ... 545 m (1,790 ft)
Landing run ... 250 m (820 ft)
Range at econ cruising speed, no reserves:
standard fuel ... 813 n miles (1,505 km; 935 miles)
optional fuel ... 1,026 n miles (1,900 km; 1,180 miles)

UPDATED

ROBIN DR 400/180 REGENT

TYPE: Four-seat lightplane.

PROGRAMME: First flight (F-WSQO) 27 March 1972; certified 10 May 1972. Later versions with additional cabin windows.

CUSTOMERS: See main DR 400 entry; total 363 delivered by early 1999. Some nine built in 2000, 11 in 2001, six in 2002 and one in 2003.

DESIGN FEATURES: Generally as for DR 400 series; two-seat rear bench.

Differences from DR 400/160 listed below:

POWER PLANT: One 134 kW (180 hp) Textron Lycoming O-360-A3A flat-four engine driving a Sensenich 76EM8S5-0-64 two-blade, fixed-pitch aluminium propeller. Fuel tankage unchanged.

ACCOMMODATION: Basically as for DR 400/160. Baggage capacity 60 kg (132 lb).

DIMENSIONS, EXTERNAL:
Propeller diameter ... 1.93 m (6 ft 4 in)

Robin DR 400/180 Régent *(Paul Jackson)* ***NEW***/0572265

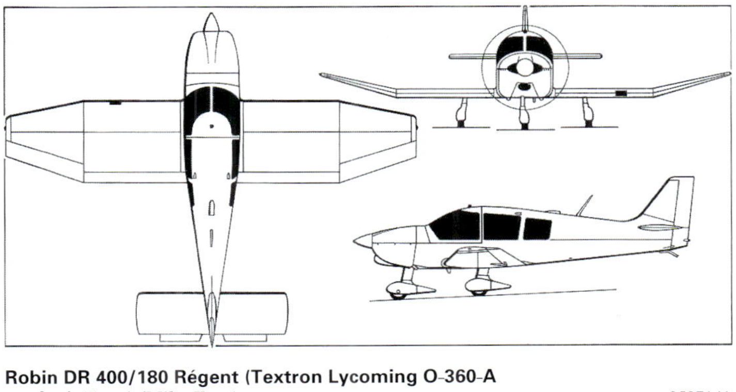

Robin DR 400/180 Régent (Textron Lycoming O-360-A engine) *(Jane's/Mike Keep)* 0507141

WEIGHTS AND LOADINGS:
Weight empty, equipped ... 610 kg (1,345 lb)
Max T-O and landing weight ... 1,100 kg (2,425 lb)
Max wing loading ... 77.48 kg/m² (15.87 lb/sq ft)
Max power loading ... 8.21 kg/kW (13.47 lb/hp)

PERFORMANCE (at max T-O weight):
Max level speed at S/L ... 150 kt (278 km/h; 173 mph)
Max cruising speed at 75% power at 2,285 m (7,500 ft) ... 140 kt (260 km/h; 162 mph)
Econ cruising speed at 60% power at 3,660 m (12,000 ft) ... 132 kt (245 km/h; 152 mph)
Stalling speed: flaps up ... 57 kt (105 km/h; 65 mph)
flaps down ... 52 kt (95 km/h; 59 mph)
Max rate of climb at S/L ... 252 m (825 ft)/min
Service ceiling ... 4,720 m (15,475 ft)
T-O run ... 315 m (1,035 ft)
T-O to 15 m (50 ft) ... 515 m (1,690 ft)
Landing from 15 m (50 ft) ... 530 m (1,740 ft)
Landing run ... 249 m (820 ft)
Range, no reserves:
standard fuel ... 753 n miles (1,395 km; 866 miles)
optional fuel ... 950 n miles (1,760 km; 1,094 miles)

UPDATED

ROBIN DR 400/180R REMO 180

TYPE: Glider tug.

PROGRAMME: First flight and certification 1972 as DR 400/180R (*Remorqueur*, abbreviated Remo); flown 1985 with Porsche PFM 3200 engine as DR 400RP or Remo 212; became first Porsche-powered aircraft to be certified; Remo 212 production ceased 1990 after 29 had been built.

CUSTOMERS: See main DR 400 entry; total 311 Remos delivered by December 1998. Eight 180Rs produced in 2000; one in 2001; none known in 2002; one (180RD) in 2003.

DESIGN FEATURES: Same as Régent, except no external baggage door, and baggage compartment covered in transparent Plexiglas to maximise rearward view; towing hook under tail; Dauphin wing (13.60 m²; 146.4 sq ft) without extended wingroot leading-edges.

Differences from DR 400/180 listed below:

POWER PLANT: One 134 kW (180 hp) Textron Lycoming O-360-A flat-four engine, driving (for glider towing) a Sensenich 76EM8S5-0-58 or Hoffmann HO27HM-180/138 two-blade, fixed-pitch metal propeller. For touring, a coarser pitch Sensenich 76EM8S5-0-64 propeller can be fitted. Fuel 110 litres (29.1 US gallons; 24.2 Imp gallons) normal; additional 50 litre (13.2 US gallon; 11.0 Imp gallon) tank optional.

WEIGHTS AND LOADINGS:
Weight empty, equipped ... 592 kg (1,305 lb)
Max T-O and landing weight ... 1,000 kg (2,205 lb)
Max wing loading ... 73.5 kg/m² (15.05 lb/sq ft)
Max power loading ... 7.46 kg/kW (12.25 lb/hp)

PERFORMANCE (A: with 300 kg; 661 lb glider, B: with 600 kg; 1,323 lb glider):
Max level speed ... 146 kt (270 km/h; 168 mph)
Cruising speed at 70% power at 2,440 m (8,000 ft) ... 124 kt (230 km/h; 143 mph)
Stalling speed, flaps down: A ... 47 kt (87 km/h; 54 mph)
B ... 45 kt (84 km/h; 52 mph)
Max rate of climb at S/L: A ... 282 m (925 ft)/min
B ... 210 m (690 ft)/min
Service ceiling ... 6,100 m (20,000 ft)
T-O run: A ... 205 m (675 ft)
B ... 300 m (985 ft)
T-O to 15 m (50 ft): A ... 400 m (1,315 ft)
B ... 470 m (1,542 ft)
Landing from 15 m (50 ft) ... 470 m (1,545 ft)
Range, no reserves:
max normal fuel ... 494 n miles (915 km; 568 miles)
with supplementary fuel ... 585 n miles (1,085 km; 674 miles)

UPDATED

Robin Remo 180 glider tug ***(Paul Jackson)*** ***NEW***/0572262

ROBIN DR 400/200R REMO 200

TYPE: Glider tug.

PROGRAMME: Prototype (F-ZWWX) flown 1993; further 12, plus one conversion, produced up to 2001, including one US export for Charles G Kalko of Honesdale, Pennsylvania. None reported built in 2002 or 2003.

Differences from DR 400/180 listed below:

POWER PLANT: One 149 kW (200 hp) Textron Lycoming IO-360 flat-four engine, driving a constant-speed propeller.

WEIGHTS AND LOADINGS:
Weight empty, equipped ... 650 kg (1,433 lb)
Max T-O and landing weight ... 1,100 kg (2,425 lb)
Max wing loading ... 77.5 kg/m² (15.87 lb/sq ft)
Max power loading ... 7.38 kg/kW (12.12 lb/hp)

PERFORMANCE (A: with 300 kg; 661 lb glider, B: with 600 kg; 1,323 lb glider):
Cruising speed at 75% power ... 135 kt (250 km/h; 155 mph)
Max rate of climb at S/L: A ... 312 m (1,024 ft)/min
B ... 258 m (846 ft)/min
T-O to 15 m (50 ft): A ... 400 m (1,312 ft)
B ... 415 m (1,362 ft)
Range, no reserves:
max normal fuel ... 447 n miles (828 km; 514 miles)
with supplementary fuel ... 610 n miles (1,130 km; 702 miles)

UPDATED

ROBIN DR 500 PRESIDENT

TYPE: Four-seat lightplane.

PROGRAMME: Prototype (F-WZZY), then known as DR 400/200i, first flew 5 June 1997; revealed at Paris Air Show, 13 June. Renamed as above but certified as **DR 400/500**.

CURRENT VERSIONS: **DR 500/200i Président:** *As described.*

DR 500 Super Régent: 180 hp engine, lacks fuel injection. None built.

DR 500C: Prototype, F-WZZY (c/n 27), flown in 2002 with 134 kW (180 hp) Textron Lycoming O-360 and Sensenich metal, fixed-pitch propeller of 1.63 m (5 ft 4 in) diameter — which may be reduced to 1.57 m (5 ft 4 in) on production version. Total of some 300 hours flown by mid-2004, although certification then still under way. Optimised for flying schools requiring upgrade from DR 400; has DR 500's wider fuselage, but economies and simplicity of carburettor and fixed-pitch propeller.

CUSTOMERS: Deliveries began in 1998 with eight aircraft. Total 40 built by mid-2002 for owners in France, Germany, New Zealand, Spain, Switzerland and UK; none further by end of 2003.

DESIGN FEATURES: Uprated version of DR 400/180 Régent, including that version's 14.20 m² (152.8 sq ft) wing; 149 kW (200 hp) Textron Lycoming IO-360-A1B6 engine driving a Hartzell F7 666A-2 constant-speed propeller; width of cabin increased by 10 cm (4 in) and headroom by same amount; IFR instrumentation as standard; Honeywell avionics, including KAP 140 autopilot. Fuel capacity increased. Flaps electrically operated.

Differences from DR 400/180 listed below:

LANDING GEAR: Nosewheel size 15 × 6.00-5, maximum pressure 1.8 bar (26 lb/sq in); mainwheel size 330×150, pressure 2.0 bar (29 lb/sq in).

POWER PLANT: One 149 kW (200 hp) Textron Lycoming IO-360-A1B6 driving a Hartzell two-blade, constant-speed metal propeller. Fuel contained in four tanks: two, each of capacity 40 litres (10.6 US gallons; 8.8 Imp gallons), in wingroots; main tank, capacity 105 litres (27.7 US gallons; 23.1 Imp gallons), below cabin floor; and fourth tank, capacity 90 litres (23.8 US gallons; 19.8 Imp gallons), below baggage compartment. Total capacity of 275 litres (72.6 US gallons; 60.5 Imp gallons).

ACCOMMODATION: Four adults, or two plus three children on fore-and-aft adjustable leather seats. Baggage shelf at rear of seats; upward-opening rear window on port side doubles as external access to baggage compartment.

AVIONICS: To customer's choice, typically Bendix/King KX 155 nav/com, KI 208 VOR/LOC indicator, KT 76C transponder with ACK 30 altitude encoder, Garmin 150 XL GPS, and SPA 400 intercom system, four microphone and headset jack plugs, and PTT switch on each control stick. Options include S-Tec 30 or S-Tec 55 autopilot and Stormscope WX 500 or WX 900.

Instrumentation: As for DR 400/180 except: dual altimeters, fuel flow/manifold pressure gauge, four fuel quantity gauges, and ammeter, all standard. Sandel Electronics HIS optional.

EQUIPMENT: Electric rudder trim, alternate static source, and three strobe lights (on upper fuselage and at wingtips), all standard.

DIMENSIONS, EXTERNAL:
Length overall ... 7.22 m (23 ft 8¼ in)
Propeller diameter ... 1.93 m (6 ft 4 in)

DIMENSIONS, INTERNAL:
Cabin max width ... 1.20 m (3 ft 11¼ in)

Prototype Robin DR 500C version of President, optimised for training ***(Paul Jackson)*** ***NEW***/0580295

Robin DR 500/200i Président ***(Paul Jackson)*** ***NEW***/0572259

WEIGHTS AND LOADINGS:
Weight empty ... 650 kg (1,433 lb)
Baggage capacity ... 60 kg (132 lb)
Max T-O weight ... 1,150 kg (2,535 lb)
Max wing loading ... 81.0 kg/m² (16.59 lb/sq ft)
Max power loading ... 7.72 kg/kW (12.68 lb/hp)

PERFORMANCE:
Max level speed ... 147 kt (272 km/h; 169 mph)
Cruising speed at 75% power ... 143 kt (265 km/h; 165 mph)
Max rate of climb at S/L ... 305 m (1,000 ft)/min
T-O to 15 m (50 ft) ... 463 m (1,520 ft)
Landing from 15 m (50 ft) ... 530 m (1,739 ft)
Range, standard fuel, no reserves:
at 75% power ... 995 n miles (1,842 km; 1,145 miles)
at 65% power ... 1,001 n miles (1,855 km; 1,152 miles)

UPDATED

SAUPER

SAUPER AVIATION SA

33 rue Fortuny, F-75017 Paris
Works: Aérodrome du Breuil, BP 1035, F-41010 Blois Cedex
Tel: (+33 2) 54 42 94 88
Fax: (+33 2) 54 42 00 11
e-mail: infosite@sauper-aviation.com
Web: http://www.sauper-aviation.com
MANAGING DIRECTOR: Didier Juery

Sauper formed in 1992 to build the Chéreau J.300 Joker ultralight, which is now marketed under the name of the parent company. A new aircraft, the Papango, flew in 2002. Personnel total 17 in a new 1,500 m² (16,150 sq ft) plant at Bois le Breuil.

UPDATED

SAUPER J.300 SERIES 3 JOKER

TYPE: Tandem-seat ultralight.

PROGRAMME: Designed by Tony Minguet in 1987. Series 3, introduced in 2000, has modified empennage, wider windscreen, lower door sills and new instrument panel.

CURRENT VERSIONS: **Civilian version** is used for private flying, crop-spraying and aerial photography. **Military version** features larger tyres and a full instrument panel including artificial horizon, UHF radio, GPS and transponder.

CUSTOMERS: Total 90 built by 2003, including at least 18 in Africa; most recent sales to Madagascar, Cameroon and the Arabian Peninsula. Two evaluated by the French Army in 1996.

COSTS: €38,200 (2003) excluding tax.

DESIGN FEATURES: Braced, high-wing monoplane with mid-mounted, braced tailplane; constant chord wing with tapered tips.

Wing section Clark Y; 2.2° anhedral.

FLYING CONTROLS: Conventional and manual. Self-centring (by helical spring) elevator. Horn-balanced rudder. Flight-adjustable tab in each elevator. No flaps.

STRUCTURE: Welded steel (25CD4S) tube airframe; Diatex 1500EV3 fabric covering; composites engine cowling.

LANDING GEAR: Tailwheel type; fixed. Two side Vs hinged to lower longerons, plus half-axles sprung under centreline with bungee cord. Mainwheels 8.00-6; speed fairings optional; hydraulic Grimeca brakes; tailwheel 2.60×85. Nosewheel version under development.

POWER PLANT: One 59.6 kW (79.9 hp) Rotax 912 or (optional) 73.5 kW (98.6 hp) Rotax 912 ULS flat-four driving a DUC Hélices three-blade propeller is standard; options include engines from Limbach. Two fuel tanks beneath seats, each 48.0 litres (12.7 US gallons; 10.6 Imp gallons) capacity. Oil capacity 3.0 litres (0.8 US gallon; 0.7 Imp gallon).

SYSTEMS: 12 V 17 Ah battery.

AVIONICS: *Comms:* IC A3 radio and intercom. Optional Bendix/King KY 97 VHF.

Flight: Garmin GPS 92 standard; GPS 295 optional.

EQUIPMENT: Optional ballistic parachute.

DIMENSIONS, EXTERNAL:
Wing span ... 9.60 m (31 ft 6 in)
Length overall ... 5.95 m (19 ft 6¼ in)
Height overall ... 2.00 m (6 ft 6¾ in)

AREAS:
Wings, gross ... 17.00 m² (183.0 sq ft)

WEIGHTS AND LOADINGS:
Weight empty ... 282 kg (622 lb)
Max T-O weight ... 450 kg (992 lb)

PERFORMANCE (Rotax 912 ULS):
Max operating speed (VMO) ... 102 kt (190 km/h; 118 mph)
Max cruising speed ... 97 kt (180 km/h; 112 mph)
Econ cruising speed ... 76 kt (140 km/h; 87 mph)
Stalling speed ... 30 kt (55 km/h; 35 mph)
Max rate of climb at S/L ... 300 m (984 ft)/min
T-O run ... 60 m (197 ft)
Landing run ... 70 m (230 ft)
Range ... 432 n miles (800 km; 497 miles)
Endurance ... 7 h

UPDATED

Sauper J.300 Joker ultralight ***NEW***/1029188

SAUPER PAPANGO

English name: Black Kite

TYPE: Side-by-side ultralight; two-seat lightplane.

PROGRAMME: First flown (F-WWNO) as VLA, 30 November 2002; had achieved 10 hours in 13 sorties by February 2003. Marketing began in 2003.

CURRENT VERSIONS: **VLA:** Lightplane. Features include certified Rotax 912 S engine, Evra propeller, electric elevator trim, Cleveland hydraulic brakes and tyres, and standard fit of artificial horizon, GPS, transponder and navigation lights. Marketing from early 2004.

ULM: Ultralight. Rotax 912 ULS engine, Duc propeller, 8.00×6 tyres with Grimeca hydraulic brakes and manual elevator trim. Empty weight 275 kg (606 lb). POC prototype flown third quarter of 2003.

Kit: Experimental category; engines up to 89.5 kW (120 hp) and 650 kg (1,433 lb) MTOW. Marketing from late 2004.

COSTS: VLA €88,000, including tax, complete. ULM €55,000, including tax, complete. Kit approximately €30,000, excluding engine (2003).

DESIGN FEATURES: Rugged lightplane with extensive glazing. Braced, high-wing monoplane; mid-mounted tailplane and sharply tapered fin; long mainwheel legs.

Wing section NACA 23015; tailplane section NACA 0010.

FLYING CONTROLS: Conventional and manual. All-moving tailplane with electrically actuated, full-span anti-balance tab. (Initially flown with full-span flaperons which replaced by mid-2003). Actuation by pushrods (elevator) and cables.

STRUCTURE: Fuselage structure of welded 4130 steel tube in cockpit area, and aluminium 5086. Wings aluminium 5086 with Diatex 1500EV3 fabric covering. Polyester engine cowling. Composites wingtips.

LANDING GEAR: Tailwheel type; fixed. Two streamlined side Vs hinged to cockpit frame and attached to oleo shock-absorbers anchored to cockpit sides. Cleveland 6.00-6 mainwheels (8.00-6 optional) and Cleveland hydraulic disc brakes. Solid tailwheel; steerable.

POWER PLANT: One 73.5 kW (98.6 hp) Rotax 912S or 912 ULS flat-four driving an Evra two-blade, fixed pitch, wooden or Duc propeller. Fuel tank behind seats; capacity 70 litres (18.5 US gallons; 15.4 Imp gallons).

SYSTEMS: 17 Ah battery and 20 A generator.

AVIONICS: *Comms:* Bendix/King KY 97 radio and Flightcom 403MC intercom. Kannad 406AF ELT

Flight: Garmin 295 GPS.

EQUIPMENT: BRS 1050 ballistic parachute optional.

DIMENSIONS, EXTERNAL:
Wing span ... 9.80 m (32 ft 1¾ in)
Wing chord, constant ... 1.63 m (5 ft 4¼ in)
Length overall ... 6.50 m (21 ft 4 in)
Tailplane span ... 2.40 m (7 ft 10½ in)
Tailplane chord, constant ... 0.80 m (2 ft 7½ in)
Wheel track ... 2.40 m (7 ft 10½ in)

DIMENSIONS, INTERNAL:
Cabin max width ... 1.21 m (3 ft 11¾ in)

AREAS:
Wings, gross ... 16.00 m² (172.2 sq ft)

WEIGHTS AND LOADINGS:
Max T-O weight: VLA, kit ... 650 kg (1,433 lb)
ULM ... 472.5 kg (1,042 lb)

PERFORMANCE (VLA; provisional/demonstrated):
Never-exceed speed (VNE) ... 126 kt (235 km/h; 146 mph)
Max level speed ... 92 kt (170 km/h; 106 mph)
Econ cruising speed ... 86 kt (160 km/h; 99 mph)
Stalling speed, clean ... 34 kt (63 km/h; 40 mph)
Max rate of climb ... 300 m (984 ft)/min
T-O run ... 180 m (591 ft)
Landing run ... 170 m (558 ft)

UPDATED

Sauper Papango VLA prototype ***NEW***/1029189

SKYDESIGN

GECI INTERNATIONAL

105 bis boulevard Malesherbes, F-75008 Paris
Tel: (+33 1) 53 53 00 53
Fax: (+33 1) 53 53 00 50
e-mail: marc.demontessus@geci.net
Web: http://www.geci.net
CHAIRMAN AND CEO: Serge Bitboul
GENERAL MANAGER: Marc de Montessus

ASIA BRANCH:
PT GECI Nusantara
Gedung BBU, 8th Floor, Jl Asia Africa No 141-149,
Bandung 40112, Indonesia
Tel: (+62 22) 424 19 06
Fax: (+62 22) 420 42 90
e-mail: alexander.supelli@geci.co.id

Formed in 1979, GECI International is a transport-orientated engineering and consulting company with a staff of 400. It has contributed to the Fairchild Dornier 728 and various business jet and Airbus airliner programmes. In 2001, it formed Skydesign SAS to develop the Skylander transport, formal launch of which was then intended to take place in the second half of 2003.

UPDATED

SKYDESIGN SB100 SKYLANDER

TYPE: Twin-turboprop light transport.

PROGRAMME: Announced 17 October 2001 at Seoul Air Show. Design supervision by the late Desmond Norman, co-designer of the BNG Islander. First metal scheduled to be cut in January 2004; first of two prototypes expected to fly in May 2005, leading to JAR/FAR Pt 23 certification and first delivery in mid-2006.

Programme structured for joint funding (initially US$120 million) by three risk-sharing partners, Korean Aerospace Industries being first, on 17 October 2001, to pledge US$30 million to secure responsibility for wing production. Assembly of complete aircraft will be by a French company under contract to GECI.

CURRENT VERSIONS: **Skylander:** *As described.* Freighter, passenger and maritime patrol versions.

Skylander L: Lightened version with 559 kW (750 shp) PT6A engines, 2,000 kg (4,400 lb) payload and reduced price.

COSTS: Estimated US$3.6 million unit cost for 'green' airframe; estimated direct operating cost US$435 per hour, based on annual utilisation of 1,500 flying hours (both 2003). Sales goal over 20 years is some 1,300, being equivalent to 30 per cent of estimated market for 4,485 aircraft in this class.

DESIGN FEATURES: Intended for operation from rough airfields with minimal maintenance facilities. Fixed landing gear and conventional alloy structure; unpressurised. Certifiable to FAR Pt 23.

Model of Skydesign Skylander with original single mainwheels *(Paul Jackson)* ***NEW***/0576224

Fuselage of square cross-section, with upswept rear; tailplane at base of fin. High wing with single bracing strut each side.

Wing section NACA 23015.

FLYING CONTROLS: Conventional and manual, by cable. Electrically actuated flaps. Horn-balanced rudder.

STRUCTURE: Principally of aluminium alloy.

LANDING GEAR: Tricycle type; fixed. Twin mainwheels mounted on sponsons. Single, steeerable nosewheel. Tyre size 8.50-10 (all round).

POWER PLANT: Two Pratt & Whitney Canada PT6A-65B turboprops, each 820 kW (1,100 shp), driving Hartzell five-blade propellers. Fuel contained in two integral wing tanks, combined capacity 3,285 litres (868 US gallons; 723 Imp gallons).

ACCOMMODATION: Two pilots; 18 passengers (two plus one abreast) and one cabin attendant. Alternative combi (nine to 12 passengers ahead of LD3 container and separated by 9 *g* bulkhead) and all-freight configurations. Latter include three LD3s plus 4.0 m^3 (141 cu ft) to rear; and four 1.83 × 1.40 m (72 × 55 in) pallets plus 5.4 m^3 (191 cu ft). Baggage hold beneath flight deck.

Crew door forward, port; double freight door (divided vertically, with forward- and rearward-opening elements) rear, port.

SYSTEMS: Electrical system includes 28V 3004 starter generator. Hydraulic system for brakes and nosewheel steering. Wing and tail surface leading-edges de-iced by pneumatic boots.

AVIONICS: Bendix /King avionics fit, with EFIS display.

DIMENSIONS, EXTERNAL:	
Wing span	21.60 m (70 ft 10½ in)
Wing chord, constant	2.00 m (6 ft 6¾ in)
Wing aspect ratio	10.9
Length overall	14.29 m (46 ft 10½ in)
Height overall	6.10 m (20 ft 0¼ in)
Tailplane span	6.92 m (22 ft 8½ in)
Tailplane chord, constant	1.46 m (4 ft 9½ in)
Wheel track	3.40 m (11 ft 1¾ in)
Wheelbase	5.405 m (17 ft 8¾ in)
Propeller diameter	2.82 m (9 ft 3 in)
Distance between propeller centres	6.77 m (22 ft 2½ in)
DIMENSIONS, INTERNAL:	
Hold: Length: parallel section	5.64 m (18 ft 6 in)
total flat floor	6.17 m (20 ft 3 in)
Width: max	1.94 m (6 ft 4½ in)
mean	1.88 m (6 ft 2 in)
at floor	1.83 m (6 ft 0 in)
Max height	1.83 m (6 ft 0 in)
Volume	28.0 m^3 (989 cu ft)
Baggage hold volume	0.85 m^3 (30.0 cu ft)
AREAS:	
Wings, gross	42.63 m^2 (458.9 sq ft)
WEIGHTS AND LOADINGS:	
Weight empty, freight version	4,320 kg (9,524 lb)
LD3 container version	4,500 kg (9,921 lb)
Passenger version	4,670 kg (10,296 lb)
Max fuel	2,450 kg (5,401 lb)
Max T-O and landing weight	8,390 kg (18,496 lb)
Max ramp weight	8,440 kg (18,607 lb)
Max wing loading	196.8 kg/m^2 (40.31 lb/sq ft)
Max power loading	5.12 kg/kW (8.41 lb/shp)
PERFORMANCE (estimated):	
Cruising speed: max	231 kt (428 km/h; 266 mph)
econ	190 kt (352 km/h; 219 mph)
Max rate of climb at S/L	610 m (2,000 ft)/min
Rate of climb at S/L, OEI	183 m (600 ft)/min
T-O run	420 m (1,378 ft)
T-O to 11 m (35 ft)	550 m (1,805 ft)
Landing run	385 m (1,263 ft)
Landing from 15 m (50 ft)	690 m (2,265 ft)
Range: with max fuel	1,200 n miles (2,222 km; 1,381 miles)
with max payload	150 n miles (277 km; 172 miles)

UPDATED

SOCATA

SOCATA SOCIÉTÉ DE CONSTRUCTIONS D'AVIONS DE TOURISME ET D'AFFAIRES (SUBSIDIARY OF EADS)

Aéroport de Tarbes Pyrénées Lourdes, F-65291 Tarbes Cedex 9
Tel: (+33 5) 62 41 73 00
Fax: (+33 5) 62 41 76 54
Web: http://www.socata.eads.net
CEO: Stéphane Mayer
CFO: Xavier Flory
VICE-PRESIDENT, SALES AND MARKETING, GENERAL AVIATION: Alain Brodin
VICE-PRESIDENT, GENERAL AVIATION: Jacques Lordon
VICE-PRESIDENT, INDUSTRY AND AEROSTRUCTURES: Albert Varenne
DEPUTY VICE-PRESIDENT, SALES AND MARKETING AND CEO SOCATA: Stéphane Bernard
DIRECTOR PUBLIC RELATIONS: Philippe de Segovia

US OPERATING AND SERVICE FACILITY:
Socata Aircraft, North Perry Airport, 7501 Pembroke Road, Pembroke Pines, Florida 33023
Tel: (+1 954) 893 14 00
Fax: (+1 954) 893 14 02

Formed 1966 as a subsidiary of Aerospatiale responsible for light aircraft. By mid-2002, Socata had produced over 2,100 TB series aircraft, excluding the TBM 700. In 1999, sold and delivered 40 TBs; sold 16 and delivered 21 TBM 700s. Reduced deliveries of 45 TBs and 33 TBMs in 2000; target for 2001 was 100+ and 33, respectively, but 63 and 33 actually achieved; 2002 deliveries totalled 70 TBs and 34 TBMs, and 2003 turned out 40 and 34. Revenue for 2002 totalled US$200 million.

Also makes components for Airbus A300/320/330/340/380/400M, Lockheed Martin C-130, ATR 42/72, Dassault Falcons, Eurocopter Super Puma, Dauphin, Ecureuil and EC 130, and Embraer 170 and 190, satellite structures. Covered floor area 77,675 m^2 (830,000 sq ft); workforce in early 2004 was 1,081.

Following formation of EADS on 10 July 2000, Socata is 100 per cent owned by that company.

UPDATED

SOCATA TB 9 TAMPICO, TB 10 AND TB 200 TOBAGO GT

TYPE: Four-seat lightplane.

PROGRAMME: Design of original TB series launched 1975; first flight of original TB 10 (F-WZJP), powered by 119 kW (160 hp) Textron Lycoming O-320, 23 February 1977; second prototype powered by 134 kW (180 hp) Textron Lycoming.

In June 1999, Socata announced its Nouvelle Génération (New Generation) series of single piston-engined models. The title Generation Two (GT) was subsequently adopted. These feature aerodynamic and other improvements that may subsequently be incorporated in the Socata MS diesel-engined light aircraft described elsewhere in this entry. Modifications include a curved dorsal fairing for the fin, upturned wingtips, raised cabin roofline with new single-piece carbon fibre/honeycomb roof, revised window pillar design, flush-mounted windows, larger baggage door, redesigned interior and new fuel filler door. From 201st TB in mid-2003 (TB 10 F-GOIH c/n 2200, displayed at Paris, 15 to 22 June 2003) all aircraft fitted with Moritz engine control panel.

The TB9C Tampico Club, TB 9 Sprint, TB 10 Tobago, TB 10 Tobago Privilège and TB 200 Tobago XL are now out of production, having been replaced in February 2000 by the GT series.

CURRENT VERSIONS: **TB 9 Tampico Sprint GT:** Baseline model. 119 kW (160 hp) Textron Lycoming O-320-D2A flat-four engine driving a Sensenich two-blade fixed-pitch

Socata TB 9 Tampico Sprint GT *NEW*/0546973

TAMPICO AND TOBAGO GT DELIVERIES

Year	TB 9 GT	TB 10 GT	TB 200 GT
2000	2[1]	1[1]	nil
2001	2	8	8
2002	3	7	2
2003	2	7	3
Total	**9**	**23**	**13**

[1] Estimated

metalpropeller. Fuel capacity 158 litres (41.7 US gallons; 34.8 Imp gallons), of which 152 litres (40.2 US gallons; 33.4 Imp gallons) are usable. Oil capacity 8 litres (2.1 US gallons; 1.8 Imp gallons).

TB 10 Tobago GT: 134 kW (180 hp) Textron Lycoming O-360-A1AD flat-four engine. *Detailed description applies to this version.*

TB 200 Tobago GT: Similar to TB 10 Tobago GT, but with 134 kW (180 hp) fuel-injected Textron Lycoming IO-360-A1B6 engine

CUSTOMERS: Total of 468 Tampicos and Tampico Clubs in service by June 1999 with customers in Africa and Middle East, Asia-Pacific, Australasia, Europe, and North America. Total of 637 Tobagos and 87 Tobago XLs delivered by June 1999, to customers in Africa and Middle East, Australasia, Europe, North America and elsewhere. Production of first-generation TBs (9, 10, 200, 20 and 21) totalled 1,928, including prototypes.

Deliveries of GT versions totalled some 45 by 1 January 2004.

Recent customers include the Indonesian Navy, which ordered five Tobago GTs in February 2004, of which two were scheduled for delivery in June 2004 and the remainder in early 2005.

DESIGN FEATURES: Conventional light tourer and trainer.

Wing section RA 16.3C3; thickness/chord ratio 16 per cent; dihedral 4° 30'.

FLYING CONTROLS: Manual, with pushrod actuation for ailerons, rudder and all-moving tailplane; cable-actuated anti-balance tab on tailplane, plus ground-adjustable tabs on ailerons and rudder; strakes on lower edges of fuselage just aft of wing control turbulence under rear fuselage; electrically actuated flaps.

STRUCTURE: Conventional light alloy; single-spar wing; GFRP tips, engine cowlings, cabin roof structure and dorsal strake. Triple anti-corrosion protection.

LANDING GEAR: Non-retractable tricycle type. Wing-mounted, trousered, cantilever main legs with integral oleo-pneumatic shock absorbers and trailing-link suspension; steerable nosewheel, oleo-pneumatic shock absorber; glass fibre wheel fairings on all three units. Mainwheel tyres size 6.00-6 (6 ply), pressure 2.30 bar (33 lb/sq in) on TB 10 GT/200 GT; 15×6.00–6 (6 ply) on TB 20 GT/21 GT. Nosewheel 5.00-5 on TB 10 GT/200 GT and 5.00-4 on TB 20 GT/21 GT. Hydraulic disc brakes. Parking brake.

POWER PLANT: One 134 kW (180 hp) Textron Lycoming O-360-A1AD flat-four engine, driving a Hartzell two-blade constant-speed propeller. Two integral fuel tanks in wing leading-edges; total capacity 210 litres (55.5 US gallons; 46.2 Imp gallons), of which 204 litres (53.9 US gallons; 44.9 Imp gallons) are usable. Oil capacity 8 litres (2.1 US gallons; 1.8 Imp gallons).

ACCOMMODATION: Four or five seats in enclosed cabin, with dual controls. Separate, adjustable front seats with inertia reel seat belts. Two separate rear seats or removable three-place bench seat with safety belts. Sharply inclined low-drag windscreen. Access via upward-hinged window/doors of glass fibre. Baggage compartment aft of cabin, with external door on port side. Cabin carpeted, soundproofed, heated and ventilated. Windscreen defrosting standard.

SYSTEMS: Electrical system includes 28 V 70 A alternator and 24 V 10 A battery. Hydraulic system for brakes only.

AVIONICS: Honeywell Silver Crown avionics suite to customer choice, including Bendix/King KLN 90 GPS.

Instrumentation: Typical IFR package includes dual altimeters; dual true airspeed

Socata TB 10 Tobago GT *NEW*/0580549

indicators; dual artificial horizons; vertical speed indicator, electric turn co-ordinator; directional gyro, tachometer; oil temperature, oil pressure, fuel pressure, fuel quantity, manifold pressure, CHT/EGT and OAT gauges; ammeter, voltmeter and compass.

EQUIPMENT: Includes armrests for all seats, map pockets, anti-glare visors, stall warning indicator, tiedown fittings and towbar, landing and navigation lights, four individual cabin lights and instrument panel lighting.

DIMENSIONS, EXTERNAL(TB 9 GT, TB 10 GT and TB 200 GT):

Wing span	10.01 m (32 ft 10 in)
Wing chord, constant	1.22 m (4 ft 0 in)
Wing aspect ratio	8.0
Length overall: TB 9 GT	7.72 m (25 ft 4 in)
TB 10 GT/200 GT	7.75 m (25 ft 5 in)
Height overall	3.02 m (9 ft 11 in)
Tailplane span	3.20 m (10 ft 6 in)
Wheel track	2.30 m (7 ft 6½ in)
Wheelbase	1.96 m (6 ft 5 in)
Propeller diameter	1.88 m (6 ft 2 in)
Propeller ground clearance	0.10 m (4 in)
Cabin doors (each): Width	0.90 m (2 ft 11½ in)
Height	0.76 m (2 ft 6 in)
Baggage door: Width	0.64 m (2 ft 1¼ in)
Max height	0.53 m (1 ft 9 in)

DIMENSIONS, INTERNAL(TB 9, TB 9C, TB 10 and TB 200):

Cabin:	
Length: firewall to rear bulkhead	2.53 m (8 ft 3½ in)
panel to rear bulkhead	2.00 m (6 ft 6¾ in)
Max width: at rear seats	1.28 m (4 ft 2¼ in)
at front seats	1.15 m (3 ft 9¼ in)
Max height	1.20 m (3 ft 11¼ in)

AREAS (TB 9 GT, TB 10 GT and TB 200 GT):

Wings, gross	11.90 m^2 (128.1 sq ft)
Ailerons (total)	0.91 m^2 (9.80 sq ft)
Trailing-edge flaps (total)	3.72 m^2 (40.04 sq ft)
Fin	0.88 m^2 (9.47 sq ft)
Rudder	0.63 m^2 (6.78 sq ft)
Horizontal tail surfaces (total)	2.56 m^2 (27.56 sq ft)

WEIGHTS AND LOADINGS:

Weight empty, with unusable fuel and oil:	
TB 9 GT	684 kg (1,508 lb)
TB 10 GT	730 kg (1,609 lb)
TB 200 GT	745 kg (1,642 lb)
Baggage: all	65 kg (143 lb)
Max T-O weight: TB 9 GT	1,060 kg (2,336 lb)
TB 10 GT, TB 200 GT	1,150 kg (2,535 lb)
Max wing loading: TB 9 GT	89.1 kg/m^2 (18.25 lb/sq ft)
TB 10 GT, TB 200 GT	96.6 kg/m^2 (19.79 lb/sq ft)
Max power loading: TB 9 GT	8.91 kg/kW (14.64 lb/hp)
TB 10 GT	8.57 kg/kW (14.08 lb/hp)
TB 200 GT	7.72 kg/kW (12.68 lb/hp)

PERFORMANCE:

Max cruising speed:	
TB 9 GT at 70% power at FL60	115 kt (213 km/h; 132 mph)
TB 10 GT at 75% power at FL60	127 kt (235 km/h; 146 mph)
TB 200 GT at 75% power at FL85	130 kt (240 km/h; 149 mph)
Econ cruising speed:	
TB 9 GT at 65% power at FL80	105 kt (194 km/h; 121 mph)
TB 10 GT at 60% power at FL80	109 kt (202 km/h; 125 mph)
TB 200 GT at 60% power at FL85	115 kt (213 km/h; 132 mph)
Stalling speed:	
flaps up: TB 9 GT	58 kt (107 km/h; 67 mph)
TB 10 GT	61 kt (112 km/h; 70 mph)
flaps down: TB 9 GT	48 kt (89 km/h; 56 mph)
TB 10 GT	52 kt (97 km/h; 60 mph)
TB 200 GT	53 kt (98 km/h; 61 mph)
Max rate of climb at S/L: TB 9 GT	203 m (665 ft)/min
TB 10 GT	240 m (787 ft)/min
TB 200 GT	286 m (937 ft)/min
Max certified altitude: TB 9 GT	3,350 m (11,000 ft)
TB 10 GT	3,965 m (13,000 ft)
TB 200 GT	4,875 m (16,000 ft)
T-O to 15 m (50 ft):	
TB 9 GT	570 m (1,870 ft)
TB 10 GT	505 m (1,657 ft)
TB 200 GT	475 m (1,558 ft)
Landing from 15 m (50 ft): TB 9 GT	420 m (1,378 ft)
TB 10 GT	460 m (1,509 ft)
TB 200 GT	449 m (1,473 ft)
Max range: TB 9 GT	564 n miles (1,046 km; 650 miles)
TB 10 GT	697 n miles (1,290 km; 802 miles)
TB 200 GT	637 n miles (1,179 km; 733 miles)

UPDATED

Socata TB 200 Tobago GT ***NEW***/0580551

SOCATA TB 20 AND TB 21 TRINIDAD

Israel Defence Force name: Pashosh (Lark)

TYPE: Four-seat lightplane.

PROGRAMME: First flight TB 20 (F-WDBA) 14 November 1980; French certification 18 December 1981; FAA certification 27 January 1984; first delivery (F-WDBB) 23 March 1982; first flight TB 21, 24 August 1984; French certification 23 May 1985; FAA certification 5 March 1986.

Earlier TB 20 Trinidad, TB 20 Trinidad Excellence, TB 20 C and TB 21 Trinidad TC variants (of which some 700 built) no longer in production.

CURRENT VERSIONS: **TB 20 Trinidad GT:** At the Paris Air Show in June 1999, Socata announced its Nouvelle Génération (New Generation) series of single piston-engined aircraft, the prototype of which, TB 20 NG F-WWRG, had first flown on 21 April 1999. The title Generation Two (GT) was subsequently adopted. The TB 20 GT received DGAC certification on 31 January 2000. The GT series was formally launched at Tarbes on 2 February 2000 with roll-out of the first production TB 20 GT, which became US demonstrator (N163GT c/n 2000) and was exhibited at Sun 'n' Fun, April 2000.

GT variants feature aerodynamic and other improvements that will be incorporated in the Socata Morane diesel-engined series described below. Identifying features include upturned wingtip, similar to those on the TBM 700 turboprop; a curved dorsal fillet; raised cabin roofline with new single-piece carbon fibre/honeycomb roof; revised cabin window/pillar

Socata TB 20 Trinidad GT *(Paul Jackson)* ***NEW***/0554410

Socata TB 21 Trinidad GT ***NEW***/0576132

Turbocharged Socata TB 21 Trinidad GT TC *(Paul Jackson)* ***NEW***/0576154

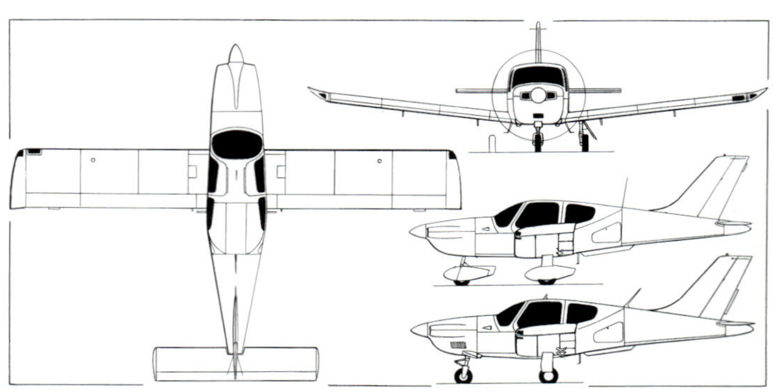

Socata TB 21 Trinidad GT, with additional side view (upper) of TB 10 Tobago GT *(Jane's/Dennis Punnett)* 0110866

TRINIDAD DELIVERIES

Year	TB 20 GT	TB 21 GT
2000	30[1]	12[1]
2001	33	12
2002	44	14
2003	19	9
Totals	**126**	**47**

[1] Approximately

design with flush-mounted windows; plus redesigned interior, larger baggage door, new fuel filler door, retractable footstep and optional three-blade Hartzell propeller.

In June 2003, Socata launched **Premium 2003** version of TB 20 Trinidad GT with additional features as standard, including Hartzell Scimitar propeller (with liquid de-icing), Garmin GNS 530 VHF/VOR/ILS/GPS, GNS 430, GTX 327 transponder and Bendix/King KFC 225 autopilot.

TB 21 Trinidad GT TC: As above, but with computer-controlled turbocharger. **Premium 2003** version launched at Paris in June 2003, with standard features as for concurrent TB 20.

CUSTOMERS: At time of launch of GT series, the Aeronautical Academy of Europe in Evora, Portugal, took options on four TB 20 GTs. By mid-2000, TB 20 GTs had been delivered to customers in France, UK and USA from planned first year production of 72. Orders in 2000 included 69 from New Avex, distributor for US West Coast; early customers included the Nigerian College of Aviation Technology (NCAT) of Zaria, Kaduna State, which ordered 10 Trinidad GTs for delivery beginning mid-2001 and French training school SEFA, which ordered 10 TB 20 GTs in June 2002, of which the first four were delivered in July 2002. Total of some 173 Trinidads delivered by 1 January 2004.

DESIGN FEATURES: Mainly as for Tobago; dihedral 6° 30'.

FLYING CONTROLS: As Tobago, but rudder trim and flap preselector added.

STRUCTURE: Largely as Tobago.

LANDING GEAR: Hydraulically retractable tricycle type, with single wheel on each unit. Free-fall emergency extension. Steerable nosewheel retracts rearward. Main units, with trailing-link suspension, retract inward into fuselage. Hydraulic disc brakes. Parking brake.

POWER PLANT: *TB 20 GT:* One 186 kW (250 hp) Textron Lycoming IO-540-C4D5D flat-six, driving a two-blade Hartzell constant-speed propeller.

TB 21 GT: One 186 kW (250 hp) Textron Lycoming TIO-540-AB1AD turbocharged flat-six. Two-blade Hartzell constant-speed propeller; three-blade propeller optional.

Fuel in two integral wing tanks, total capacity 336 litres (88.8 US gallons; 73.9 Imp gallons), of which 326 litres (86.1 US gallons; 71.7 Imp gallons) usable. Oil capacity 12.6 litres (3.3 US gallons; 2.8 Imp gallons).

ACCOMMODATION: Generally as for TB 9 Sprint GT, TB 10 Tobago GT and TB 200 XL GT; new headrests for all seats introduced on GT variants; rear seat can be removed for carrying 250 kg (551 lb) of cargo.

SYSTEMS: Electrical system comprises 28 V 70 A alternator and 24 V 10 Ah battery.

AVIONICS: TB 20/21 GT basic navigation package comprises single nav/com and VOR receiver. IFR packages and EFIS optional.

DIMENSIONS, EXTERNAL: As for TB 10 GT, except:
- Wing span .... 9.97 m (32 ft 8½ in)
- Wing aspect ratio 8.2
- Length overall .... 7.75 m (25 ft 5 in)
- Height overall .... 2.85 m (9 ft 4¼ in)
- Tailplane span .... 3.68 m (12 ft 0¾ in)
- Wheel track .... 2.17 m (7 ft 1½ in)
- Wheelbase .... 1.91 m (6 ft 3¼ in)
- Propeller diameter .... 2.03 m (6 ft 8 in)

DIMENSIONS, INTERNAL (TB 20 GT):
- Cabin: Width .... 1.28 m (4 ft 2½ in)
- Max height .... 1.20 m (3 ft 11¼ in)

AREAS: As for TB 10 GT, except:
- Horizontal tail surfaces (total) .... 3.06 m² (32.94 sq ft)

WEIGHTS AND LOADINGS:
- Operating weight empty: TB 20 GT .... 800 kg (1,764 lb)
- TB 21 GT .... 867 kg (1,911 lb)
- Max baggage: TB 20 GT, TB 21 GT .... 65 kg (143 lb)
- Max T-O and landing weight:
  - TB 20 GT, TB 21 GT .... 1,400 kg (3,086 lb)
- Max wing loading:
  - TB 20 GT, TB 21 GT .... 117.6 kg/m² (24.10 lb/sq ft)
- Max power loading:
  - TB 20 GT, TB 21 GT .... 7.51 kg/kW (12.35 lb/hp)

PERFORMANCE:
- Cruising speed at 75% power:
  - TB 20 GT at FL65 .... 163 kt (302 km/h; 188 mph)
  - TB 21 GT at FL250 .... 190 kt (352 km/h; 219 mph)
- Econ cruising speed at 65% power:
  - TB 20 GT at FL85 .... 150 kt (278 km/h; 173 mph)
  - TB 21 GT at FL250 .... 169 kt (313 km/h; 194 mph)
- Rate of climb at S/L: TB 20 GT .... 366 m (1,200 ft)/min
  - TB 21 GT .... 344 m (1,130 ft)/min
  - TB 21 GT at FL170 .... 244 m (800 ft)/min
- Certified ceiling: TB 20 GT .... 6,100 m (20,000 ft)
  - TB 21 GT .... 7,620 m (25,000 ft)
- T-O to 15 m (50 ft): TB 20 GT .... 595 m (1,955 ft)
  - TB 21 GT .... 595 m (1,955 ft)
- Landing from 15 m (50 ft):
  - TB 20 GT, TB 21 GT .... 540 m (1,775 ft)
- Max range:
  - TB 20 GT .... 1,100 n miles (2,037 km; 1,265 miles)
  - TB 21 GT .... 1,035 n miles (1,916 km; 1,191 miles)

UPDATED

SOCATA MS LIGHT AIRCRAFT

Plans for a family of Diesel-powered TB-type lightplanes were delayed by the priority development of the TBM 700C and have now been placed in abeyance, pending more favourable market conditions.

UPDATED

SOCATA TBM 700

TYPE: Business turboprop.

PROGRAMME: Three prototypes built: first flights 14 July 1988 (F-WTBM), 3 August 1989 (F-WKPG) and 11 October 1989 (F-WKDL); French certification received 31 January 1990; FAR Pt 23 type approval awarded 28 August 1990; first delivery 21 December 1990; Canadian public transport certification 1993. Wider, single-piece, upward-opening door replaced horizontally split door, port side, rear, from late 1998 and is standard. Aircraft used for freighting have optional crew door, port, forward. Increased weight version (TBM 700C2) announced late 2002 and achieved FAR 23 certification on 7 March 2003.

CURRENT VERSIONS: **TBM 700:** In addition to basic transport, Socata offers multimission versions, including military medevac, target towing, ECM, freight, maritime patrol, law enforcement, navaid calibration and vertical photography versions. Prototype of last-mentioned (F-GLBF) exhibited at the 1996 Farnborough Air Show.

TBM 700B Freighter: Cargo version with port side cargo door, 1.19 m (3 ft 10¾ in) × 1.07 m (3 ft 6 in), maker's reference MOD 70–091–52, and separate port side cockpit door; reinforced metal cargo floor (188 kg/m²; 38.5 lb/sq ft) with tiedown points in rails; optional port side cockpit door; maximum cargo capacity 825 kg (1,819 lb); volume 3.5 m³ (124 cu ft); target maximum T-O weight 3,300 kg (7,275 lb); quick conversion to six/seven-seat passenger configuration; announced June 1995; initially known as TBM 700C; French certification 23 November 1998. Launch customer Air Open Sky of France, which began TBM 700 freight operations in November 1999. Three handed over to French Army on 28 June 2000. Quest Diagnostics of the USA has ordered six, of which first three (N700QD, N701QD and N702QD) delivered 26 November 2001.

TBM 700C: First of two prototypes (c/n 205 F-WWRL/N778C) flew February 2002; officially announced 10 September 2002 at NBAA Convention, Orlando, Florida; second aircraft (c/n 240 F-WWRO/N811SW) followed as a 700C2. Changes include option of increased (**TBM 700C2**) or original (**TBM 700C1**) MTOW, according to local certification rules; strengthened spar box and wing attachments; stronger wheels and 10-ply tyres; additional (unpressurised) baggage compartment to rear of cabin; new interior with 20*g* seats; improved air conditioning. Sole variant from late 2002; first production 700C1 was c/n 244 (F-WWRI/N700DY); first production 700C2 c/n 245 (F-WWRM/N6750Y) completed November 2002. All aircraft have structural and interior modifications, but 700C1 retains original seats (not 20*g*) and landing gear; later upgrade to 700C2 is possible from c/n 299 onwards (278th production) all TBM 700Cs RVSM-approved on delivery. Total fleet time (270 aircraft) stood at more than 330,000 hours in May 2004.

Description and specification refer to TBM 700C1/2.

CUSTOMERS: Total 265 delivered by end of 2003 (see table). Deliveries to French Air Force began 27 May 1992 with first of initial six for liaison duties with Groupe Aérien d'Entrainement et de Liaison (GAEL) and ETE 43; further six supplied 1993–94 (also for ETE 41 and ETE 44, plus CEAM); total Air Force requirement is 22 (of which 17 funded by mid-1998); two officially handed over to French Army 13 January 1995 for 3 GHL at Rennes, followed by six more, last three of which delivered (in TBM 700B configuration) 28 June 2000. Two delivered to Indonesian Civil Aviation Academy in September 1996. Total of 24 delivered in 2000, 33 in 2001, 34 in 2002, and 34 in 2003. Total of 34 scheduled for delivery in 2004, including 24 for US customers.

COSTS: TBM 700C1 US$2,518,700; TBM 700C2 US$2,672,000; direct operating cost US$301.10 per hour (all 2003).

DESIGN FEATURES: High-speed, long-range, single-turboprop business transport. Low wing; twin strakes under rear fuselage; sweptback fin (with dorsal fin) and mass balanced rudder; non-swept tailplane with mass balanced elevators. Airframe life 12,000 cycles/16,000 hours from TBM 700C.

Wing of Aerospatiale RA 16–43 root section with 6° 30' dihedral from roots.

FLYING CONTROLS: Conventional and manual. Pushrod and cable actuated with electric trim tabs in port aileron, rudder and each elevator; 'scaled-down ATR' single-slotted Fowler flaps, also electrically actuated, along 71 per cent of each wing trailing-edge; slotted spoiler forward of each flap at outer end, linked mechanically to aileron; yaw damper.

STRUCTURE: Mainly of light alloy and steel except for control surfaces, flaps, most of tailplane and fin of Nomex honeycomb bonded to metal sheet; wing leading-edges and landing gear doors GFRP/CFRP; tailcone and wingtips GFRP; two-spar torsion box forms integral fuel tank in each wing.

LANDING GEAR: Hydraulically retractable tricycle type, with emergency manual operation. Inward-retracting main units of trailing-link type; rearward-retracting steerable nosewheel (±28°). Main tyres 18×5.5 (8 ply); 10 ply on 700C2) tubeless; nose 5.00–5 (10 ply) tubeless. Parker hydraulic disc brakes. Minimum ground turning radius (based on nosewheel) 23.98 m (78 ft 8 in).

POWER PLANT: One 1,178 kW (1,580 shp) Pratt & Whitney Canada PT6A-64 turboprop, flat rated at 522 kW (700 shp), driving a Hartzell HC-E4N-3/E9083S(K) four-blade constant-speed, fully feathering, reversible-pitch, metal propeller. Fuel in integral tank in each wing, combined capacity 1,100 litres (290.5 US gallons; 242 Imp gallons), of which 1,066 litres (282 US gallons; 234 Imp gallons) usable. Gravity filling point in top of each tank. Oil capacity 12 litres (3.2 US gallons; 2.6 Imp gallons).

ACCOMMODATION: Adjustable seats for one or two pilots at front. Dual controls standard. Four seats in club layout aft of these, with centre aisle, or five seats in high-density layout. TBM

FIRST 270 TBM 700 DELIVERIES

Country	Total
Australia	3
Austria	4
Canada	2
France	
Civil	3(prototypes)
	24
Air Force	19
Army	8
CEV	1
Germany	9
Indonesia	3
Ireland	1
Italy	1
Japan	3
Luxembourg	3
Netherlands	1
Switzerland	1
Thailand	1
USA	177
Unknown	6
Total	**270**

Notes: Based on initial permanent registration; transfers have since been effected

Socata TBM 700C2 business turboprop ***NEW***/0576151

700C1 has three-point safety harnesses, and front and centre seats have folding armrest, and headrest; TBM 700C2 has four-point harnesses on front seats, and folding armrests, and headrests, on all seats. Large upward-opening door on port side aft of wing; overwing emergency exit on starboard side. Oxygen system, comprising three under-seat bottles with individual emergency oxygen mask for each passenger and masks with integral microphones for crew. Pressurised baggage compartment at rear of cabin, with internal access only; additional unpressurised compartments in nose, between engine and firewall, and behind cabin, both with external access via doors on port side. Nose compartment of TBM 700C reduced in size by two-thirds, compared with 700B, due to installation of environmental control system. Optional crew door, port side, front is maker's OPT 70-52-002. Reinforced metal floor with tie-down points (OPT 70–25–027).

SYSTEMS: Engine bleed air pressurisation (to 0.43 bar; 6.2 lb/sq in) and Honeywell environmental control system/vapour control system. Hydraulic system for landing gear only. Electrical system powered by one 28 V 200 A (main) and one 28V 70 Ah engine-driven starter/generator, and a 28 V 43 Ah lead/acid (optionally Ni/Cd) battery. Pneumatic rubber-boot de-icing of wing/tailplane/fin leading-edges. Propeller blades, pitot tubes and stall warning sensor; anti-iced electrically heated engines inlet. Electric anti-icing and hot air demisting of windscreen. Gaseous oxygen system.

AVIONICS: *Comms:* Dual Garmin GNS 530 VHF com/VOR/ILS/GPS with GI 106A VOR/ILS indicator; dual GTX 327 Mode C transponders; GMA 340 stereo audio panel with marker beacon and music input; tri-band ELT; dual Bose X headsets with noise attenuation, all standard.

Radar: Honeywell RDR-2000 weather radar.

French Air Force Socata TBM 700 ***(Paul Jackson)*** ***NEW***/0576178

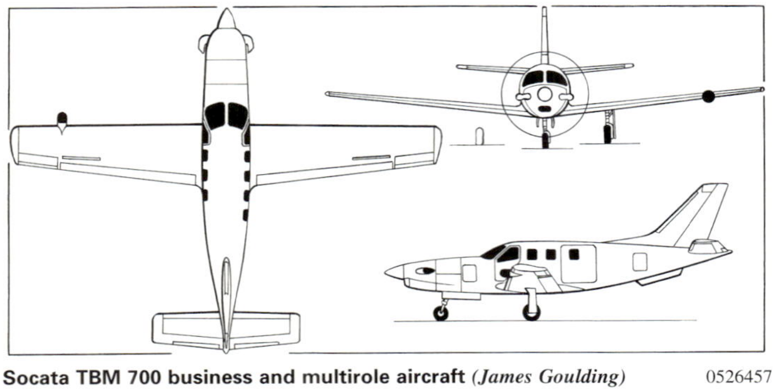

Socata TBM 700 business and multirole aircraft ***(James Goulding)*** 0526457

Flight: Bendix/King KCS 55 A heading and KI 525 A HSI indicators; KR 87 SC+ ADF; KRA 405 B radar altimeter with KNI 415 indicator; KN 63 DME; WX 500 Stormscope; and KMH 880 TCAS and TAWS, all standard.

Instrumentation: Bendix/King EFIS 40 two-tube EFIS; KMD 850 MFD; electric artificial horizon; standby vacuum artificial horizon; ASI; VSI; KEA 346 encoding drum altimeter; KNI 582 RMI indicator; KDI 574 DME indicator; rate of climb indicator; magnetic compass; torque indicator; propeller tachometer; ITT indicator; OAT gauge; electric generator control and monitoring on overhead panel; annunciator panel with master caution, master warning and aural warnings; cabin temperature gauge; cabin altitude and differential pressure indicator; flight time hours meter; and digital chronometer, all standard. Thommen needle chronometer (replaces digital chronometer), and engine hours meter (replaces flight time hours meter), optional.

EQUIPMENT: Standard cabin equipment includes three-way adjustable front seats; passenger seats with adjustable backrests, and individual lights and air vents; Alcantara upholstery; forward and aft 24 V power outlets; storable work table; refreshment cabinet; coat rack; and Halon cabin fire extinguisher. Optional cabin equipment includes CD player cabinet with four Bose X headsets. Platinum Edition interior features leather seats; wood veneered overhead panel, baggage compartment frame, retractable work table and cabinet doors; and black chrome or gold coloured metallic trim. Anti-collision strobe lights, navigation lights, taxy and landing lights, ice detection light, and ground power socket, all standard. Pulse light anti-collision system, parking protection kit, pearlescent paint, and customised paint schemes, all optional.

DIMENSIONS, EXTERNAL:

Wing span	12.68 m (41 ft 7¼ in)
Wing chord, mean aerodynamic	1.51 m (4 ft 11½ in)
Wing aspect ratio	8.9
Length overall	10.645 m (34 ft 11 in)
Height overall	4.355 m (14 ft 3¼ in)
Tailplane span	4.99 m (16 ft 4½ in)
Wheel track	3.87 m (12 ft 8¼ in)
Wheelbase	2.91 m (9 ft 6½ in)
Propeller diameter	2.31 m (7 ft 7 in)
Crew door (optional): Height	1.02 m (3 ft 4 in)
Width	0.78 m (2 ft 6¾ in)
Cabin door: Height	1.19 m (3 ft 10¾ in)
Width	1.08 m (3 ft 6½ in)
Rear baggage door: Height	0.39 m (1 ft 3¼ in)
Width	0.25 m (9¾ in)

DIMENSIONS, INTERNAL:

Cabin: Length	4.05 m (13 ft 3½ in)
Max width	1.21 m (3 ft 11¾ in)
Max height	1.22 m (4 ft 0 in)
Cargo floor area	3.15 m^2 (33.91 sq ft)
Volume	3.5 m^3 (124 cu ft)
Baggage compartment volume:	
front	0.08 m^3 (2.8 cu ft)
rear	0.10 m^3 (3.5 cu ft)

AREAS:

Wings, gross	18.00 m^2 (193.75 sq ft)
Vertical tail surfaces (total)	2.56 m^2 (27.55 sq ft)
Horizontal tail surfaces (total)	4.76 m^2 (51.24 sq ft)

WEIGHTS AND LOADINGS:

Weight empty, basic: TBM 700C1	2,075 kg (4,575 lb)
TBM 700C2	2,125 kg (4,685 lb)
Baggage capacity (all areas)	135 kg (298 lb)
Payload: max: TBM 700C1	647 kg (1,426 lb)
TBM 700C2	611 kg (1,347 lb)
with max fuel: TBM 700C1	205 kg (452 lb)
TBM 700C2	365 kg (805 lb)

Fuel weight (usable) .... 866 kg (1,910 lb)
Max T-O weight: TBM 700C1 .... 2,984 kg (6,578 lb)
TBM 700C2 .... 3,354 kg (7,394 lb)
Max ramp weight .... 3,000 kg (6,613 lb)
Max zero fuel weight .... 2,736 kg (6,032 lb)
Max landing weight: TBM 700C1 .... 2,835 kg (6,250 lb)
TBM 700C2 .... 3,186 kg (7,024 lb)
Max ramp weight: TBM 700C1 .... 3,000 kg (6,614 lb)
TBM 700C2 .... 3,370 kg (7,430 lb)
Max wing loading:
TBM 700C1 .... 165.8 kg/m² (33.95 lb/sq ft)
TBM 700C2 .... 186.3 kg/m² (38.16 lb/sq ft)
Max power loading:
TBM 700C1 .... 5.72 kg/kW (9.48 lb/shp)
TBM 700C2 .... 6.43 kg/kW (10.56 lb/shp)

PERFORMANCE:
Max cruising speed at FL260
.... 300 kt (555 km/h; 345 mph)
Econ cruising speed at FL310 .... 255 kt (472 km/h; 293 mph)
Stalling speed: clean .... 82 kt (152 km/h; 95 mph) IAS
flaps and landing gear down .... 65 kt (121 km/h; 75 mph) IAS
Max rate of climb at S/L .... 725 m (2,380 ft)/min
Time to FL260 .... 24 min
Max certified altitude .... 9,450 m (31,000 ft)
T-O to 15 m (50 ft): TBM 700C1 .... 650 m (2,135 ft)
TBM 700C2 .... 865 m (2,840 ft)
Landing from 15 m (50 ft) without reverse: TBM 700C1 .... 650 m (2,135 ft)
TBM 700C2 .... 741 m (2,435 ft)
Range at econ cruising speed, 45 min reserves:
with max payload: TBM 700C1 .... 330 n miles (611 km; 379 miles)
TBM 700C2 .... 1,074 n miles (1,989 km; 1,235 miles)
with max fuel: TBM 700C1 .... 1,637 n miles (3,031 km; 1,883 miles)
TBM 700C2 .... 1,565 n miles (2,898 km; 1,801 miles)
g limits .... +3.8/–1.5

UPDATED

VOLIRIS

VOLIRIS SARL

10 chemin des Heuleux, F-78240 Chambourcy
Tel: (+33 1) 39 79 39 88
Fax: (+33 1) 30 65 88 48
e-mail: info@voliris.com
Web: http://www.voliris.com

Voliris was formed in February 1999 and has since produced a number of small remotely controlled airships. A prototype of its first piloted product, the Voliris 900, flew on 26 June 2003.

UPDATED

VOLIRIS 900

TYPE: Helium semi-rigid.

PROGRAMME: Development process included 9.5 m (31.2 ft) long, 25 m³ (883 cu ft) radio-controlled model. Assembly of full-scale prototype completed in June 2002 and engine fitted early in following month. First helium inflation 21 June 2002; revealed publicly 26 June 2002 at Meudon. First flight (F-WAAN) was made at Clermont-Ferrand air base on 26 June 2003.

DESIGN FEATURES: Envelope is modified version of RosAeroSystems Au-12 (which see), and was delivered in May 2002. Gondola is French-built.

POWER PLANT: One (unspecified) 89.5 kW (120 hp) piston engine, driving pusher propeller at rear of gondola.

ACCOMMODATION: Pilot and one passenger.

DIMENSIONS, EXTERNAL:
Length overall .... 31.00 m (101 ft 8½ in)
Envelope max diameter .... 8.00 m (26 ft 3 in)

Gondola of Voliris 900 ***NEW***/0558611

Prototype Voliris 900 under construction ***NEW***/0558610

DIMENSIONS, INTERNAL:
Envelope volume .... 996 m³ (35,175 cu ft)
Helium volume .... 900 m³ (31,783 cu ft)

AREAS:
Envelope surface area .... 615.0 m² (6,620 sq ft)

PERFORMANCE (estimated):
Max level speed .... 45–48 kt (85–90 km/h; 52–55 mph)
Cruising speed .... 35–38 kt (65–70 km/h; 40–43 mph)
Max operating altitude .... 1,000 m (3,280 ft)
Endurance .... 5 h

UPDATED

Georgia

TBILAVIAMSHENI

TBILAVIAMSHENI JSC (TBILISI AEROSPACE MANUFACTURING JSC)

ulitsa Bogdan Khmelnitska 181B, 380036 Tbilisi
Tel: (+995 32) 70 84 12
Fax: (+995 32) 98 25 51
e-mail: tordia@gol.ge
Web: http://www.tam.ge
CHAIRMAN: Pantiko Tordia
DIRECTOR GENERAL: Nodar Beridze
FIRST DEPUTY GENERAL DIRECTOR AND CHIEF ENGINEER: Gocha Goguadze

Former Tbilisi Aviation State Association (TASA) became Tbilisi Aerospace Manufacturing (TAM) upon privatisation in 2000, adopting present name when incorporated as a joint stock company in 2002. TASA was established 15 December 1941 when No. 448 Engine Manufacturing plant was augmented by the evacuated No. 31 Aircraft Factory from Taganrog and No. 45 Aircraft Repair Plant from Sevastopol; began aircraft production with the LaGG-3 fighter, and progressed through La-5, Yak-3, Yak-15, Yak-17/17UTI, Yak-23, MiG-15 and MiG-17 to building 1,677 MiG-21U/UMs (for which it offers an upgrade programme) between 1957 and 1984. The plant also began Su-25/25U production in 1978 and has produced 875 basic single-seat versions, including 50 Su-25BM target-tugs. In total, it has built over 8,070 manned aircraft, 2,555 UAVs and over 36,000 AAMs (30,000 R-60/AA-8s and 6,000 R-73/AA-11s). Continues to support Kela ultralight, first built in 1970s.

Current manufacture, at low rate, is of the Su-25UB two-seat trainer; also produced small batch of its Su-25T (Su-39) derivative. Production is also stated to be under way of a batch of

20 Yak-58 light transports, although there is no evidence of recent sales. Upgrade activities include a 'Frogfoot' modernisation, the Su-25KM Scorpion, a prototype of which flew in 2001.

The 250,000 m^2 (2,691,000 sq ft) TAM plant employed 15,000 personnel at its peak, but currently has a workforce of about 2,000. A 1999 alliance with Kelowna Flightcraft of Canada provides for TAM to assist with production of materials for the Convair 5800 conversion programme and eventually undertake the entire programme, if demand warrants. TAM additionally plans to fabricate components for the US-designed ViperJet two-seat kitbuilt.

Also reported in late 2003 to have built two Leader light business jets, following agreement, with Maverick, also of US. However, TAM terminated the Maverick agreement in January 2004, 10 months after it had begun, and before the first aircraft could be flown. By August 2004 had been appointed by Aircraft Investor Resources of USA to build fuselage, wing and empennage for Epic LT.

Non-aviation products include ground power units, domestic space heaters, combined harvesters and roofing materials.

UPDATED

Germany

AEROSTYLE

AEROSTYLE ULTRALEICHT FLUGZEUGE GMBH

Norderreeg 2, D-25852 Bordelum
Tel: (+49 4671) 93 13 93
Fax: (49 4671) 93 13 94
Web: http://www.Aerostyle-GmbH.de
e-mail: info@Aerostyle-GmbH.de
DIRECTOR: Ralf Magnussen

The company's first product, the Breezer, is in production as a kit and is being marketed by Ikarus in complete form.

UPDATED

AEROSTYLE BREEZER

TYPE: Side-by-side ultralight kitbuilt.

PROGRAMME: Prototype (D-MOOV) exhibited unflown at Aero'99 at Friedrichshafen in April 1999; first flight was made in December 1999. Noise tests under way in mid-2000; German certification was achieved in second quarter of 2001. Prototype was exhibited at Friedrichshafen in April 2001 with modifications including steerable nosewheel replacing original trailing-link type. In 2002 an Experimental version with empty weight of 320 kg (705 lb) and max T-O weight of 580 kg (1,278 lb) was being offered.

CUSTOMERS: Ten kits and 25 complete aircraft sold by Aerostyle up to end 2002. Production of complete aircraft now undertaken by Grilz of Kamenz with marketing by Ikarus at Höhentengen.

COSTS: €19,500 excluding taxes (2003). Complete Ikarus Breezer €53,999 (2004).

DESIGN FEATURES: Low, constant-chord wing, with wingtips upturned at rear. Sweptback fin. Wing aerofoil section NACA 4414.

FLYING CONTROLS: Conventional and manual. Horn-balanced elevators and rudder; trim tab in port elevator; mechanical (electric optional) half-span Fowler flaps.

STRUCTURE: Mainly riveted aluminium, with glass fibre for some non-structural fairings, such as wingtips, cowling and rear fuselage.

LANDING GEAR: Tricycle type; fixed. Fuselage-mounted, spring composites cantilever mainwheel legs with wheel fairings; composites steerable nose leg; with wheel fairings.

Aerostyle Breezer as marketed by Ikarus in complete form *(Paul Jackson)* ***NEW***/0583329

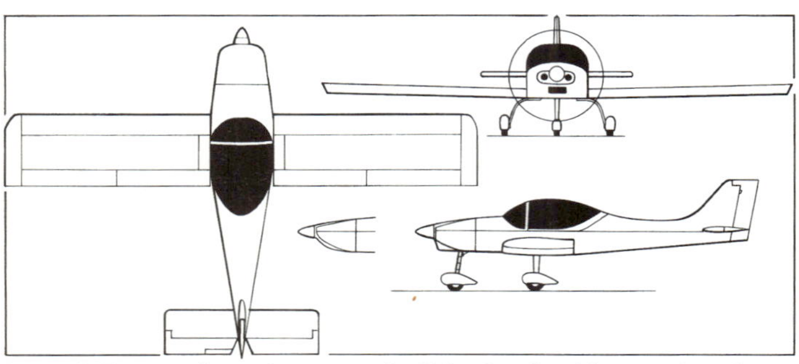

Aerostyle Breezer all-metal two-seat ultralight aircraft, with scrap view of Rotax engine nose *(James Goulding)* ***NEW***/0583331

Aerostyle Breezer prototype ***NEW***/0099161

Hydraulic brakes on mainwheels. Mainwheel size 14×4, nosewheel 4.00×4.

POWER PLANT: One 67.1 to 74.6 kW (90 to 100 hp) fuel-injected Take Off-BMW engine driving a two-blade (59.7 kW; 80 hp engine) or three-blade (74.6 kW; 100 hp engine) ground-adjustable Neuform T3 propeller; other optional engines include the 73.5 kW (98.6 hp) Rotax 912 ULS and 59.7 kW (80 hp) Jabiru 2200. Optionally a 73.5 kW (98.6 hp) Rotax 912 ULS can be fitted. Fuel capacity 70 litres (18.5 US gallons; 15.4 Imp gallons).

SYSTEMS: 12V battery.

EQUIPMENT: Optional Junters or BRS-5-UL ballistic parachute.

DIMENSIONS, EXTERNAL:
Wing span .... 8.71 m (28 ft 7 in)
Length overall .... 6.36 m (20 ft 10½ in)
Height overall .... 2.12 m (6 ft 11½ in)

AREAS:
Wings, gross .... 11.85 m^2 (127.6 sq ft)

WEIGHTS AND LOADINGS (estimated):
Weight empty .... 285–295 kg (628–650 lb)
Max T-O weight .... 472.5 kg (1,041 lb)

PERFORMANCE (estimated, A: 67.1 kW/90 hp engine; B: 73.5 kW/98.6 hp engine):
Max level speed: A .... 119 kt (220 km/h; 137 mph)
B .... 132 kt (245 km/h; 152 mph)
Cruising speed: A .... 105 kt (195 km/h; 121 mph)
B .... 113 kt (210 km/h; 130 mph)
Stalling speed: A, B .... 36 kt (65 km/h; 41 mph)
Max rate of climb at S/L: A .... 270 m (886 ft)/min
B .... 360 m (1,181 ft)/min
T-O run: A .... 135 m (395 ft)
B .... 95 m (312 ft)
Landing run: A, B .... 140 m (460 ft)
Range .... 459 n miles (850 km; 528 miles)
g limits .... +4/–2

UPDATED

Aerostyle Breezer *(Paul Jackson)* ***NEW***/0558598

AQUILA

AQUILA TECHNISCHE ENTWICKLUNGEN GMBH

Flugplatz, D-14959 Schönhagen
Tel: (+49 33) 731 70 70
Fax: (+49 33) 73 17 07 11
e-mail: info@aquila-aero.com
Web: http://www.aquila-aero.com
DIRECTORS:
Peter Grundhoff
Alfred Schmiderer
Markus Wagner
MARKETING MANAGER: Siegfried Dörfler

Company formed August 1996 to develop A210; employed 12 staff at Schönhagen aerodrome by April 2000; increased to 20 during 2001 and reached 38 by December 2003, and 43 by May 2004; following mid-2000 commissioning of new production hangar; capacity 50 aircraft per year. Distrubutors for France, the Netherlands, Scandinavia and Switzerland have been appointed.

UPDATED

AQUILA A210

TYPE: Two-seat lightplane.

PROGRAMME: Launched 1995; design, using CAD/CAM techniques, began 1997, assisted by Berlin Technical University; prototype (D-EQUI) displayed at Aero'99, Friedrichshafen, in April 1999; first flight 5 March 2000; initially flew without winglets; certification to JAR-VLA received September 2001; deliveries began July 2002 with first aircraft (D-ETHG) to Flugschule Hans Grade at Schönhagen. Certification in USA awarded 4 November 2003; Australia, France and Switzerland also completed during 2003; certification in Austria, Scandinavia and the UK was expected during 2004.

CUSTOMERS: Total 51 firm orders by May 2004, including pilot trainers for Lufthansa, and club aircraft for EADS and airline Swiss. Deliveries totalled 21 by May 2004.

COSTS: €119,500 excluding tax (2004).

DESIGN FEATURES: Design goals included crashworthy cabin structure, forgiving handling characteristics and low operating costs. Low wing with sharply tapered fuselage aft of cabin and sweptback fin; high aspect ratio wing with upturned tips.

Laminar flow wing section Horstmann/Quast HQ-42 modified. Dihedral 4° 30'.

FLYING CONTROLS: Conventional and manual. Actuation via pushrods for elevators and ailerons; cables for rudder. Integral fixed-angle rudder tab and balanced ailerons. Externally hinged, single-slotted Fowler flaps, deflections 15 and 35°.

STRUCTURE: Entirely CFRP/GFRP. Fuselage is a GFRP monocoque with CFRP stringers and GFRP/CFRP frames; wing shell structure and tail unit and control surfaces are GFRP/polyurethane foam sandwich with CFRP wingspar, load-bearing structures and local reinforcements.

LANDING GEAR: Tricycle type; fixed. Fuselage-mounted spring steel mainwheel legs; steerable nose leg with rubber-in-compression suspension; all wheels size 5.00-5; speed fairings on all wheels; hydraulic brakes; combined ventral fillet/tailskid.

POWER PLANT: One 73.5 kW (98.6 hp) Rotax 912S flat-four engine driving an MT-Propeller MTV-21-A/175-05 two-blade, variable-pitch (hydraulic) propeller. Fuel in two integral wing tanks, total capacity 120 litres (31.7 US gallons; 26.4 Imp gallons). Fuel filler in upper surface of each wing.

Aquila A210 delivered to France in 2004 *(Paul Jackson)* ***NEW***/0583327

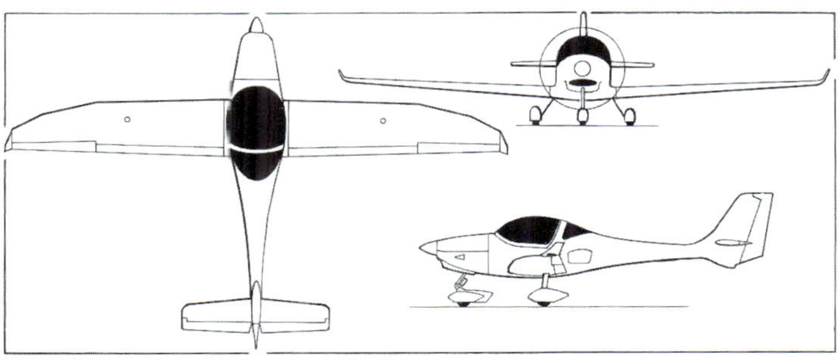

Aquila A210 all-composites light aircraft (Rotax 912 ULS flat-four) *(James Goulding)* 0137230

ACCOMMODATION: Two persons side by side under forward- and upward-opening windscreen/canopy; baggage door on port side of aft wing. Four-point safety harnesses for both occupants. Accommodation heated and ventilated. Choice of leather or cloth seating.

SYSTEMS: 14 V electrical system, including 40 Ah alternator and 12 V battery.

AVIONICS: *Comms:* Honeywell KX 125 com/nav; KT 76A transponder; encoding altimeter, PS Engineering PM 501 intercom, and ELT, all standard. Becker AR 4201 second com, Honeywell KT 76C digital transponder with encoder, Garmin KT 73 Mode S datalink, GTX 327 transponder, KMA 28 audio panel, GMA 340 audio panel, and 406 MHz ELT, all optional.

Flight: Honeywell KX 155 com/nav with GI 106A VOR/LOC/GS indicator and KMD 150 GPS with colour MFD, Garmin GNS 430 com/nav/GPS with GI 106A VOR/LOC/GS indicator, optional.

Instrumentation: Three-point altimeter, ASI, VSI, magnetic compass, artificial horizon, directional gyro, turn co-ordinator, CHT gauge, oil temperature/pressure gauges, fuel gauge, tachometer, manifold pressure gauge, OAT gauge, hour meter, Davtron M800 chronometer, all standard.

EQUIPMENT: Navigation and anti-collision lights and tiedown fittings, standard. Optional equipment includes auxiliary/external power receptacles, tinted canopy, fire extinguisher and aero-tow system.

DIMENSIONS, EXTERNAL:
Wing span, excl winglets ... 10.33 m (33 ft 10¾ in)
Wing aspect ratio ... 10.1
Length overall ... 7.33 m (24 ft 0½ in)
Height overall ... 2.30 m (7 ft 6½ in)
Fuselage max width ... 1.20 m (3 ft 11¼ in)
Tailplane span ... 3.00 m (9 ft 10 in)
Wheel track ... 1.94 m (6 ft 4½ in)
Wheelbase ... 1.66 m (5 ft 5¼ in)
Propeller diameter ... 1.75 m (5 ft 9 in)
Baggage door: Max width ... 0.51 m (1 ft 8 in)

DIMENSIONS, INTERNAL:
Cabin max width ... 1.20 m (3 ft 11¼ in)
Baggage volume ... 0.50 m^3 (17.66 cu ft)

AREAS:
Wings, gross ... 10.50 m^2(113.0 sq ft)

WEIGHTS AND LOADINGS:
Weight empty ... 490 kg (1,080 lb)
Baggage capacity ... 40 kg (97 lb)
Max T-O weight ... 750 kg (1,653 lb)
Max wing loading ... 71.4 kg/m^2(14.63 lb/sq ft)
Max power loading ... 10.20 kg/kW (16.76 lb/hp)

PERFORMANCE:
Max level speed ... 164 kt (305 km/h; 189 mph)
Cruising speed at 75% power ... 121 kt (224 km/h; 139 mph)
Stalling speed: flaps up ... 56 kt (102 km/h; 64 mph)
flaps down ... 44 kt (81 km/h; 51 mph)
Max rate of climb at S/L ... 229 m (750 ft)/min
Service ceiling ... 4,420 m (14,500 ft)
T-O run ... 250 m (820 ft)
T-O to 15 m (50 ft) ... 470 m (1,542 ft)
Landing from 15 m (50 ft) ... 500 m (1,640 ft)
Landing run ... 200 m (656 ft)
Range with max fuel at 55% power at FL500, 45 min reserves
620 n miles (1,148 km; 713 miles)
Endurance ... 6 h 5 min

UPDATED

AVCRAFT

AVCRAFT AEROSPACE GMBH

1005 Sycolin Road South East, Leesburg, Virginia 20175, USA
Tel: (+1 703) 669 08 89
Fax: (+1 703) 771 26 53
Web: http://www.avcraft.com
CEO: Ben Bartel
WORKS: Flugplatz Oberpfaffenhofen, PO Box 1103, D-88230 Wessling

On 20 December 2002, AvCraft agreed purchase of former Fairchild Dornier 328Jet and 428 programmes with bankruptcy administrators; assets included Dornier name, type and production certificates, tooling, spares stock, production parts and works production (five aircraft on assembly line and 18 'white tails'). Sale finalised in January 2003 and last of assets formally transferred in September 2003, at which time AvCraft had sold three-quarters of the 'white tail' stock and executed first delivery (Hainan airlines).

AVCRAFT DORNIER 328JET

TYPE: Regional jet airliner.

PROGRAMME: Development of Dornier 328 turboprop twin begun in mid-1980s, but then suspended; relaunched 3 August 1988; rolled out 13 October 1991; first flight (D-CHIC) 6 December 1991; first flight of first production 328-100 (D-CITI) 23 January 1993. JAA 25 certification 15 October 1993; FAA certification 10 November 1993; first delivery, to Air Engiadina, 21 October 1993. Production (including three prototypes) totalled 111; final aircraft (OE-LKC) delivered to Air Alps Aviation, Austria, 13 October 1999.

Twin turbofan version announced 5 February 1997 and formally launched at Paris Air Show in June 1997; marketed as 328JET, engineering designation being 328-300; known as 328Jet from 2003. High-speed wind tunnel tests, confirming cruise performance, and ground vibration tests completed by November 1997; prototype D-BJET, converted from second prototype 328 turboprop (D-CATI), rolled out 6 December 1997; first flight 20 January 1998; public debut 4 February 1998; three further prototypes built on the production line; second (D-BWAL) dedicated to performance certification testing, flew 20 May 1998; third (D-BEJR), used for avionics certification, flew 10 July 1998; fourth (D-BALL), used for function and reliability testing and simulated airline operation trials, flew on 15 October 1998; JAA certification was achieved on 8 July 1999 after completion of 1,560 flight hours in 950 sorties; FAA certification 15 July 1999; high gross weight certification achieved April 2000.

First delivery in July 1999 to Skyway Airlines (N351SK); total of 15 delivered in 1999 and 33 in 2000. By June 2001 the in-service fleet of aircraft had logged 100,000 hours, with a 99.23 per cent flight completion rate; high-time aircraft, operated by Hainan Airlines, had then logged nearly 4,000 hours. Production suspended by late 2002, following Fairchild Dornier's bankruptcy.

Programme purchased by AvCraft in January 2003; assets included 18 'white tail' aircraft in complete condition and five on production line; deliveries began on 11 September 2003 with B-3948 to Hainan Airlines; production relaunched in December 2003 16 'whitetails' sold by late January 2004; roll-out of first new production aircraft expected in fourth quarter 2004; initial target production rate of 18 per year; negotiations under way in early 2004 for manufacture of wing by European supplier.

CURRENT VERSIONS: **328Jet:** Regional airliner, *as described*; 32 passengers at 79 cm (31 in) seat pitch. Engineering designations **328-300** and **328-310**.

328Jet SMA: Proposed multirole special missions aircraft for troop transport, VIP missions, surveillance, SAR, AEW and medevac. Under study in 2002.

'White tail' AvCraft Dornier 328Jet employed as an Envoy envoy *(Paul Jackson)* ***NEW***/0567735

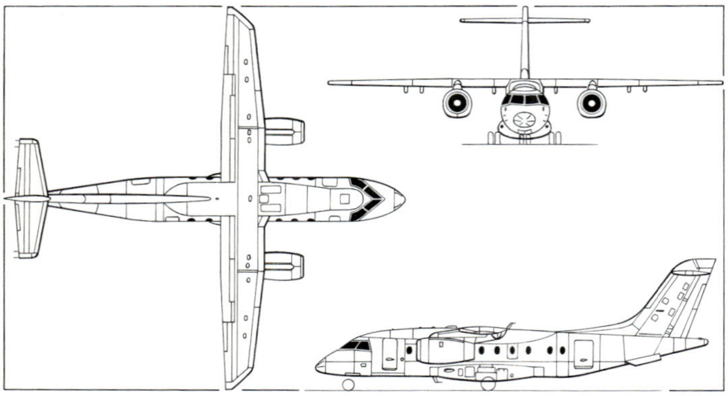

AvCraft Dornier 328Jet *(James Goulding)* 0084564

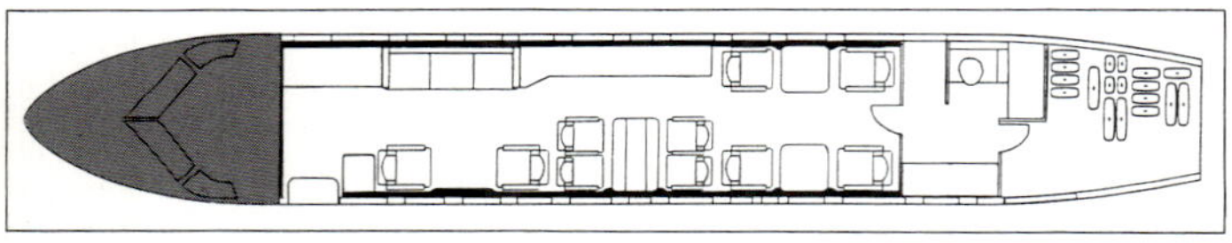

Typical Envoy internal arrangement 0044702

Cabin of Envoy Executive Shuttle 0132388

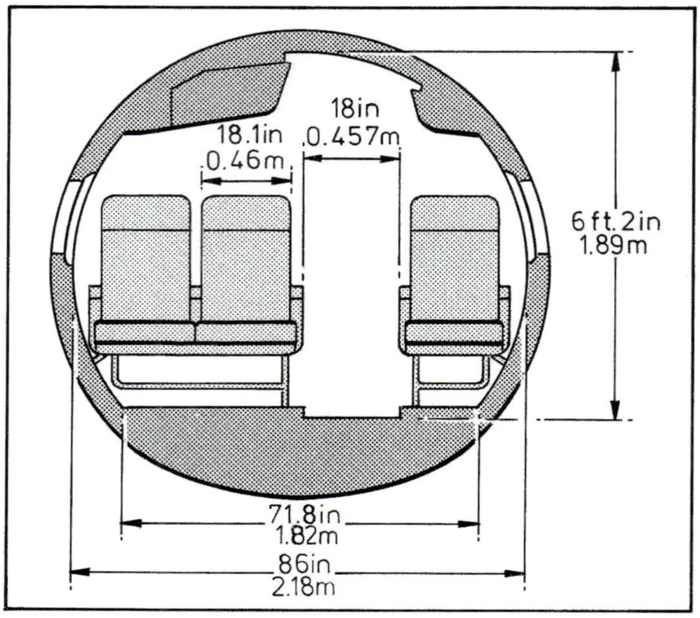

Cabin cross-section of AvCraft Dornier 328Jet *(Jane's/Mike Keep)* 0507367

Freighter: All-cargo version with large door in aft fuselage for loading of containers or palletised freight. No longer being promoted.

Envoy: Business jet and corporate shuttle version, launched at the 1997 National Business Aviation Association Convention in Dallas, Texas; known as Envoy 3 until 2003. Subvariants are Envoy **Executive** with accommodation for 12 passengers and forward service galley and wardrobe and aft galley and lavatory; Envoy **Executive Shuttle** accommodating 22 to 32 passengers; and Envoy **Convertible** for 14 and 18 passengers and with forward galley. Executive standard fuel weight 5,042 kg (11,116 lb); Shuttle and **Convertible** have standard 328Jet capacity, with option of extra tanks.

Total of 10 Envoy orders by December 2001. Announced customers include Aircraft Jetcharter, which has ordered a Corporate Shuttle variant for operation on behalf of an Ohio-based customer, and Grupo Protexa of Monterey, Mexico, which has ordered one aircraft in executive configuration. Aircraft Completions of Tyler, Texas, is US completions centre.

Dornier 428: Developed version; programme suspended on 8 August 2002.

CUSTOMERS: Total of 141 orders and 91 options quoted by early 2002; regular deliveries suspended after February 2002, with only one further aircraft supplied up to 31 December 2002, increasing total to 87. By October 2003, AvCraft had orders and options for 45, of which 13 represented firm sales due for delivery from stock before end of January 2004.

Notes: Excludes prototype converted from turboprop.

COSTS: Regional airliner US$13 million (2002); Corporate US$13.65 million, outfitted (2000). Airliner direct operating costs estimated at US$3.16 per n mile, based on 32 passengers and 300 n mile sectors.

DESIGN FEATURES: Combines basic TNT supercritical wing of Dornier 228 with new pressurised fuselage from NRT (Neue Rumpf Technologien) programme; internal volume designed to give passengers more seat width than in a Boeing 727 or 737 and stand-up headroom in aisle. Jet retains high commonality with turboprop model; major changes include new pylon/nacelle and centre wing attachments, strengthened landing gear and brakes, APU as standard, and modifications to Honeywell Primus 2000 avionics software, FADEC and environmental control systems.

FLYING CONTROLS: Conventional and manual. Optional lateral control (one) and ground (two) spoilers ahead of each aileron; horn-balanced elevators; trim tab in each elevator and rudder; single-slotted Fowler flaps.

STRUCTURE: Wing mainly light alloy structure; entire rear fuselage and tail surfaces of CFRP, except dorsal fin made of Kevlar/CFRP sandwich and aluminium alloy tailplane leading-edge; Kevlar/CFRP sandwich also used for wing trailing-edge structure, nosecone, tailcone and for long wing/fuselage fairing housing system components outside pressure hull; cabin doors of superplastic formed aluminium alloy; engine nacelles of superplastic formed titanium and carbon composites.

OGMA of Portugal manufactures fuselage shells that are assembled into complete fuselages by risk-sharing partner Aermacchi in Italy, which also manufactures flight deck structure; engine nacelles and doors by Westland Aerostructures; wing fairings by HAIG (Harbin) of China; rear fuselage and tail surfaces assembled by Fairchild Dornier from carbon fibre components; wing produced by Fairchild Dornier in San Antonio, USA; final assembly at Oberpfaffenhofen.

LANDING GEAR: ERAM (with SHL of Israel) retractable tricycle type, with twin Honeywell wheels on each unit; nose unit retracts forward, main units into Kevlar/CFRP sandwich unpressurised fairings on fuselage sides. Main tyres size 24×7.7 (14 ply), nose tyre 19.5×6.75–8 (10 ply); tyre pressures 4.40 bar (64 lb/sq in) on nose unit, 8.00 bar (116 lb/sq in) on main units; alternative 25.5×8.75–10 (14 ply) flotation main tyres; Honeywell brakes.

POWER PLANT: Two 26.9 kN (6,050 lb st) Pratt & Whitney Canada PW 306/9 turbofans with FADEC, pod-mounted under wings. Fuel capacity 4,268 litres (1,128 US gallons; 939 Imp gallons).

ACCOMMODATION: Flight crew of two and cabin attendant. Main cabin seats 32 to 34 passengers, three-abreast at 79 cm (31 in) or 76 cm (30 in) pitch, with single aisle; galley to rear of passenger seats; lavatory at rear of cabin, baggage compartment between passenger cabin and rear pressure bulkhead, with external access via baggage door in port side; crew/passenger airstair door at front on port side, with Type III emergency exit opposite; Type III emergency exit on port side at rear of cabin, with service door Type II exit at rear on starboard side.

SYSTEMS: Air conditioning and pressurisation systems standard (maximum differential 0.47 bar; 6.75 lb/sq in), with GKN Westland ECS. Hydraulic and two independent AC/DC electrical systems housed in main landing gear fairings. Tailcone-mounted Honeywell 36-150 (DD) APU as standard.

AVIONICS: Honeywell Primus 2000 suite.

Comms: Dual Primus II integrated radio system and Mode S transponder standard. HF comms optional.

Radar: Primus 650 weather radar standard; Primus 880 weather radar optional.

DORNIER 328JET DELIVERIES
(at 1 January 2003)

Customer	Variant	First delivery	Qty
Air Vallee	300	18 Apr 00	1
	310	12 Jun 01	1
Atlantic Coast Airlines	310	27 Apr 00	34
Gandalf Airlines	300	17 Nov 99	8
Grossman Air Service	Envoy	8 Feb 01	1
Hainan Airlines	300	3 Nov 99	20
Johnson Controls	Envoy	8 Oct 99	1
Philip Morris Management Corp	Envoy	26Apr 01	1
Skyway Airlines	300	30 Jul 99	9
	310	20 Feb 02	1
Great Plains (ex-Ozark)	300	27 Oct 99	2
Shell Nigeria (Bristow)	Envoy	3 Aug 99	3
Tyrolean Jet Service	Envoy	10 Sep 99	2
Ultimate Air Charters	Envoy	21 May 01	1
Wanair Tahiti	300	15 Dec 99	1
Welcome Air	300	24 Oct 02	1
Subtotal			**87**

Flight: AFCS, AHRS, dual integrated avionics computer, dual digital air data reference unit, TCAS, GPWS with windshear detection system standard. Optional GPS with GPS approach and head-up guidance system allowing Cat.II landings.

Instrumentation: Five-tube EFIS (203 × 178 mm; 8 × 7 in screens) with EICAS.

DIMENSIONS, EXTERNAL:
- Wing span ... 20.98 m (68 ft 10 in)
- Wing aspect ratio ... 11.0
- Length overall ... 21.23 m (69 ft 7¾ in)
- Fuselage: Length ... 20.92 m (68 ft 7¾ in)
  - Max width ... 2.415 m (7 ft 11 in)
  - Max depth ... 2.425 m (7 ft 11½ in)
- Height overall ... 7.05 m (23 ft 1½ in)
- Elevator span ... 6.70 m (21 ft 11¾ in)
- Wheel track (c/l of shock-struts) ... 3.22 m (10 ft 6¾ in)
- Wheelbase ... 7.42 m (24 ft 4¼ in)
- Passenger door (fwd, port): Height ... 1.70 m (5 ft 7 in)
  - Width ... 0.70 m (2 ft 3½ in)
- Service door (rear, stbd): Height ... 1.25 m (4 ft 1¼ in)
  - Width ... 0.51 m (1 ft 8 in)
- Baggage door (rear, port): Height ... 1.40 m (4 ft 7 in)
  - Width ... 0.92 m (3 ft 0¼ in)

DIMENSIONS, INTERNAL:
- Cabin (excl flight deck): Length ... 10.33 m (33 ft 10¾ in)
  - Max width ... 2.18 m (7 ft 2 in)
  - Width at floor ... 1.83 m (6 ft 0 in)
  - Max height in aisle ... 1.89 m (6 ft 2½ in)
- Baggage hold volume ... 6.4 m^3 (226 cu ft)

AREAS:
- Wings, gross ... 40.00 m^2 (430.6 sq ft)
- Ailerons (total) ... 2.42 m^2 (26.00 sq ft)
- Trailing-edge flaps (total) ... 7.61 m^2 (81.90 sq ft)
- Fin ... 11.06 m^2 (119.00 sq ft)
- Rudder ... 3.92 m^2 (42.20 sq ft)
- Tailplane ... 9.03 m^2 (97.20 sq ft)
- Elevators ... 3.08 m^2 (33.20 sq ft)

WEIGHTS AND LOADINGS (A: standard, B: high gross weight and Envoys except where otherwise stated):
- Operating weight empty: A ... 9,420 kg (20,600 lb)
  - B ... 9,394 kg (20,710 lb)
  - Envoy Executive ... 10,360 kg (22,840 lb)
  - Envoy Shuttle ... 9,543 kg (21,039 lb)
  - Envoy Convertible ... 10,118 kg (22,306 lb)
- Fuel weight: A, B ... 3,646 kg (8,039 lb)
  - Envoy Executive ... 5,042 kg (11,115 lb)
- Max payload: A ... 3,650 kg (7,200 lb)
  - B ... 3,676 kg (8,104 lb)
- Max T-O weight: A ... 15,200 kg (33,510 lb)
  - B ... 15,660 kg (34,524 lb)
- Max landing weight: A, B ... 14,390 kg (31,724 lb)
- Max ramp weight: A ... 15,350 kg (33,841 lb)
  - B ... 15,780 kg (34,789 lb)
- Max zero-fuel weight: A ... 12,610 kg (27,800 lb)
  - B ... 13,070 kg (28,814 lb)
- Max wing loading: A ... 380.0 kg/m^2 (77.83 lb/sq ft)
  - B ... 391.5 kg/m^2 (80.18 lb/sq ft)
- Max power loading: A ... 282.0 kg/kN (2.77 lb/lb st)
  - B ... 291.0 kg/kN (2.85 lb/lb st)

PERFORMANCE:
- Max cruising speed, MMO at 7,620 m (25,000 ft), ISA +10°:
  - A, B ... 400 kt (741 km/h; 460 mph)
- Service ceiling: A, B ... 10,670 m (35,000 ft)
- Service ceiling, OEI: A, B ... 7,528 m (24,700 ft)
- T-O required field length: A ... 1,382 m (4,535 ft)
  - B ... 1,422 m (4,665 ft)
- Landing required field length: A ... 1,306 m (4,285 ft)
  - B ... 1,387 m (4,550 ft)
- Range: 328Jet with 32 passengers, JAA reserves, and allowance for 100 n mile (185 km; 115 mile) diversion:
  - B ... 1,130 n miles (2,092 km; 1,300 miles)
  - Envoy Executive with 4 passengers, Envoy Shuttle with 22 passengers and NBAA reserves, standard fuel 1,200 n miles (2,222 km; 1,380 miles) and NBAA reserves ... 1,800 n miles (3,333 km; 2,071 miles)

B&F

B&F TECHNIK VERTRIEBS GMBH

Anton-Dengler-Straße 8, D-67346 Speyer
Tel: (+49 62) 327 20 76
Fax: (+49 62) 327 20 78
e-mail: info@fk-leichtflugzeuge.de
Web(1): http://www.fk-leichtflugzeuge.de
Web(2): http://www.fk-lightplanes.com
PRESIDENT: Peter Funk

The FK series, which began in 1959, is designed by father and son, Dipl Ing Otto Funk and Peter Funk. Latest products are the nostalgic FK 12 and FK 14 Polaris composites ultralights and a Mk IV version of FK 9.

Work on B&F's new 900 m^2 (9,700 sq ft) factory on the same airfield began in January 2002.

In 2003, B&F employed six people at its headquarters plus a further 45 at its Krosno, Poland, factory.

UPDATED

B&F FK 9

TYPE: Side-by-side ultralight.

PROGRAMME: The earlier versions of B&F's FK 9 were built at Krosno, Poland, employing metal tube fuselages and composites wings, both fabric-covered. The Mark 3 all-composites version was announced in 1997; 85 per cent of airframe is manufactured in Poland, for completion in Germany. Launched at Friedrichshafen Air Show, April 1997; UK sales promotion from July 1998. Following trials in earlier FK 9 (D-MCFK), 40.5 kW (54.2 hp) Suprex M160 three-cylinder engine, developed from that of DaimlerChrysler Smart motorcar, offered as alternative power plant from 2000 in version designated **FK 9 Smart**.

CURRENT VERSIONS: **FK 9 Mark 3:** Baseline version, now superceded by Mk IV.

FK 9 Mark 3B: Sold by Brazilian dealership as Alpha Bravo FK-9. Span 10.33 m (33 ft 10¾ in); wing area 12.22 m^2 (131.5 sq ft). At least one exported to Portugal.

FK 9 Mark 3 Club: Utility version; launched at Aero '03, Friedrichshafen, April 2003.

FK 9 Mark IV: Under development to replace Mk3; has wider (4 cm; 1½ in), longer (8 cm; 3¼ in) cabin and larger flaps and tailplane. Prototype (D-MBKA) first flown by October 2003. *Detail description applies to this version.*

FK 9 Mark IV Utility: Utility version for flight training and towing.

CUSTOMERS: More than 215 FK 9s of all series built by January 2004.

COSTS: €51,090 including tax (2004).

STRUCTURE: Composites wing and fuselage; metal tailplane; fabric-covered ailerons, slotted flaps and elevators.

B&F FK 9 Mk IV prototype ***NEW***/0576628

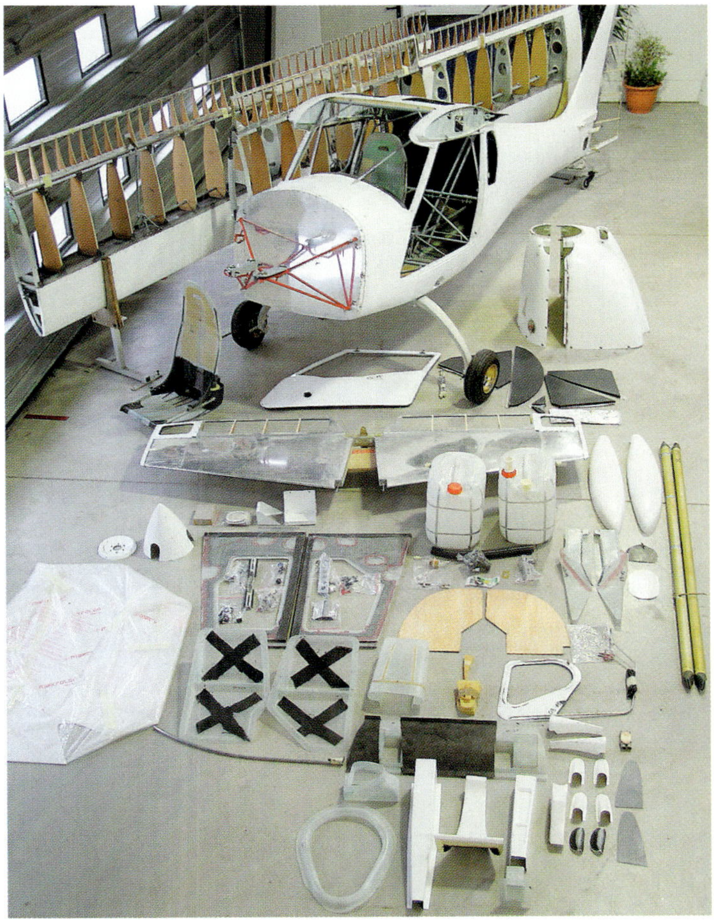
B&F FK 9 Mk IV kit, ready for assembly ***NEW***/0576630

LANDING GEAR: Fixed; tricycle type standard; tailwheel type optional. Fuselage-mounted spring cantilever mainwheel legs with wheel fairings. Trailing-link nosewheel with wheel fairing; steerable. Mainwheels 6.00-6.

POWER PLANT: One 59.6 kW (79.9 hp) Rotax 912 UL, 73.5 kW (98.6 hp) Rotax 912 ULS or 40.5 kW (54.2 hp) Suprex driving a Junkers three-blade, ground-adjustable pitch, carbon fibre propeller. Max fuel capacity 70 litres (18.5 US gallons; 15.4 Imp gallons), of which 65 litres (17.2 US gallons; 14.3 Imp gallons) are usable. Optional extra 20 litre (5.3 US gallon; 4.4 Imp gallon) tank can be installed.

DIMENSIONS, EXTERNAL:
- Wing span ... 9.85 m (32 ft 3¾ in)
- Length overall ... 5.85 m (19 ft 2¼ in)
- Height overall ... 2.15 m (7 ft 0¾ in)

B&F FK 9 tailwheel variant ***NEW***/0576631

AREAS:
Wings, gross ... 11.60 m^2 (124.9 sq ft)
WEIGHTS AND LOADINGS:
Weight empty ... 270 kg (595 lb)
Max T-O weight ... 520 kg (1,146 lb)
PERFORMANCE (Rotax 912 UL):
Never-exceed speed (VNE) ... 124 kt (230 km/h; 143 mph)
Cruising speed:
at 75% power ... 103 kt (190 km/h; 118 mph)
at 65% power ... 97 kt (180 km/h; 112 mph)
Stalling speed ... 35 kt (64 km/h; 40 mph)
Max rate of climb at S/L ... 330 m (1,080 ft)/min
T-O run ... 120 m (395 ft)
Landing run ... 220 m (722 ft)
Range ... 540 n miles (1,000 km; 621 miles)
g limits ... +6.1/−3

UPDATED

B&F FK 12 COMET

TYPE: Tandem-seat ultralight biplane/kitbuilt.
PROGRAMME: Design, by Peter Funk, started late 1994; construction of prototype began 1995; first flight (D-MPLI) March 1997; first flight of production aircraft May 1997; certification received in 1999.
CUSTOMERS: 57 Flying by January 2003.
COSTS: Kit €25,000, ready-to-fly €53,300, including taxes (2003).
DESIGN FEATURES: Wings, of laminar aerofoil section, have sweepback, and fold back for storage. Quoted build time 500 hours.
FLYING CONTROLS: Conventional and manual rudder and elevator; full-span flaperons on upper and lower wings.
STRUCTURE: Welded steel tube forward fuselage with riveted aluminium tube rear section; composites and GFRP laminate skinning; wings have carbon fibre main spar and leading-edge with aluminium secondary spar and flaperons, rear section fabric-covered.
LANDING GEAR: Tailwheel type; fixed. Fuselage-mounted spring cantilever mainwheel legs with wheel fairings. Mainwheels 6.00-6.
POWER PLANT: One 59.6 kW (79.9 hp) Rotax 912 UL, 73.5 kW (98.6 hp) Rotax 912 ULS or turbocharged 84.6 kW (113.4 hp) Rotax 914 driving an MT-Propeller two-blade wooden propeller; Junkers, Duc WarpDrive three-blade, carbon fibre propeller optional. Fuel capacity 58 litres (15.3 US gallons; 12.8 Imp gallons).
DIMENSIONS, EXTERNAL:
Wing span ... 6.70 m (21 ft 11¾ in)
Length overall ... 5.30 m (17 ft 4¾ in)
Height overall ... 1.98 m (6 ft 6 in)

Fiftieth B&F FK 12 Comet aerobatic biplane *(Paul Jackson)* ***NEW***/0576600

AREAS:
Wings, gross ... 13.40 m^2 (144.2 sq ft)
WEIGHTS AND LOADINGS:
Weight empty ... 265 kg (584 lb)
Max T-O weight ... 450 kg (992 lb)
PERFORMANCE (Rotax 912 ULS):
Never-exceed speed (VNE) ... 118 kt (220 km/h; 136 mph)
Max level speed ... 109 kt (202 km/h; 126 mph)
Cruising speed ... 89 kt (165 km/h; 103 mph)
Stalling speed, flaperons down ... 36 kt (65 km/h; 41 mph)
Max rate of climb at S/L ... 457 m (1,500 ft)/min
T-O run ... 140 m (460 ft)
Landing run ... 160 m (525 ft)
Range with max fuel ... 351 n miles (650 km; 403 miles)

UPDATED

B&F FK 14 POLARIS

TYPE: Side-by-side ultralight.
PROGRAMME: Development began early 1998; prototype (D-MVFK), then unflown, exhibited at Aero '99 at Friedrichshafen in April 1999 and first flown May 1999. In production by 2000.
CURRENT VERSIONS: **FK 14**: Baseline version; *as described.*
FK 14B: New tailwheel variant first shown at Aero '03, Friedrichshafen, at which time prototype (D-MVFK upgraded) not then flown. New carbon wing and 78 litre (20.6 US gallons; 17.2 Imp gallon) fuel tank. Mainwheels as baseline version, with solid tailwheel.
CUSTOMERS: 30 flying by January 2003.
COSTS: With Rotax 912, €72,000; with Rotax 912 ULS, €74,200; both including tax (2003).
DESIGN FEATURES: Low wing with upturned wingtips incorporating navigation lights; wing section designed with assistance of Stuttgart University wind tunnel.
FLYING CONTROLS: Conventional and manual; electrically operated Fowler flaps extend on rails to 20°; spring trim system; horn-balanced rudder.
STRUCTURE: Mainly composites, with metal ailerons, tailplane and flaps; elevators and rudder fabric-covered.
LANDING GEAR: Tricycle type; fixed. Fuselage-mounted spring composites cantilever mainwheel legs; steerable nosewheel; drum brakes; speed fairing on all wheels. Mainwheels 6.00-6.
POWER PLANT: One 59.6 kW (79.9 hp) Rotax 912 UL or 73.5 kW (98.6 hp) Rotax 912 ULS driving a Junkers three-blade, ground-adjustable pitch, carbon fibre propeller. Fuel contained in single tank behind cockpit, capacity 65 litres (17.2 US gallons; 14.3 Imp gallons).
EQUIPMENT: BRS 5 emergency parachute recovery system.
DIMENSIONS, EXTERNAL:
Wing span ... 9.10 m (29 ft 10¼ in)
Length overall ... 5.58 m (18 ft 3¾ in)
Height overall ... 1.88 m (6 ft 2 in)
AREAS:
Wings, gross ... 9.40 m^2 (101.2 sq ft)
WEIGHTS AND LOADINGS:
Weight empty ... 272 kg (600 lb)
Max T-O weight ... 475 kg (1,047 lb)
PERFORMANCE (estimated):
Never-exceed speed (VNE) ... 156 kt (290 km/h; 180 mph)
Cruising speed ... 130 kt (240 km/h; 149 mph)
Min flying speed ... 35 kt (65 km/h; 41 mph)
Max rate of climb at S/L ... 390 m (1,280 ft)/min
T-O run ... 122 m (400 ft)
Landing run ... 152 m (500 ft)
Range ... 350 n miles (648 km; 403 miles)

UPDATED

B&F FK 14 Polaris operated in France *(Pierre Gaillard)* ***NEW***/1047840

BUSSARD

BUSSARD DESIGN GMBH

Marie-Curiestrasse 6, D-85055 Ingolstadt
Tel: (+49 841) 901 44 60
Fax: (+49 841) 901 44 62
e-mail (1): info@bussard-design.de
e-mail (2): info@raptor-online.de
Web (1): http://www.bussard-design.de
Web (2): http://www.raptor-online.de
DEVELOPMENT DIRECTOR: Georg Heinrich

A design services subcontractor to the aerospace and automotive industries, Bussard planned to enter lightplane manufacturing with its own concept, the Raptor. However, no news of the programme has been received for more than a year.

UPDATED

C+C FLYING

C+C FLYING GMBH

Flugplatz Parchim-Schwerin, Ludwigsluster Strasse 5, D-19370 Parchim
Tel: (+49 3871) 45 14 54
Fax: (+49 3871) 45 14 56
e-mail: CundC_FlyingGmbH@web.de
Web: http://www.c-cflyinggmbh.com
CHIEF DESIGNER: Dr Ernst Alban

In 2003, C+C was building a prototype of its latest design, the CC/03; the earlier CC/01 and CC/02 remain available. A further aircraft, the CC/04, is under development..

UPDATED

C+C FLYING CC/03

TYPE: Side-by-side ultralight.
PROGRAMME: Incomplete prototype displayed at Aero '03 at Friedrichshafen in April 2003.
COST: €49,021 (2004).
DESIGN FEATURES: Low-wing, pod-and-boom configuration with ducted pusher propeller and T tail.
FLYING CONTROLS: Conventional and manual. Slotted flaps.
STRUCTURE: Glass fibre sandwich construction, with fabric-covered elevators, rudder and rear half of wings.
LANDING GEAR: Tricycle type; fixed. Fuselage-mounted, spring cantilever mainwheel legs.
POWER PLANT: One 73.5 kW (98.6 hp) Rotax 912S four-cylinder, air-cooled piston engine driving a ducted, three-blade, constant-speed propeller. Fuel capacity 70 litres (18.5 US gallons; 15.4 Imp gallons) in two wing tanks.
DIMENSIONS, EXTERNAL:
Wing span ... 9.40 m (30 ft 10 in)
Length overall ... 5.65 m (18 ft 6½ in)
Height overall ... 2.04 m (6 ft 8¼ in)
DIMENSIONS, INTERNAL:
Cabin max width ... 1.25 m (4 ft 1¼ in)
AREAS:
Wings, gross ... 10.85 m² (116.8 sq ft)
WEIGHTS AND LOADINGS:
Weight empty ... 250 kg (551 lb)
Max T-O weight ... 500 kg (1,102 lb)
PERFORMANCE (estimated):
Never-exceed speed (VNE) ... 124 kt (230 km/h; 143 mph)
Max level speed ... 110 kt (204 km/h; 127 mph)
Stalling speed, power off, flaps up ... 42 kt (78 km/h; 48 mph)
Max rate of climb at S/L ... 182 m (600 ft)/min
T-O run ... 250 m (820 ft)
Range ... 486 n miles (900 km; 559 miles)

UPDATED

C+C CC/03 prototype at its public debut *(Paul Jackson)* ***NEW***/0576503

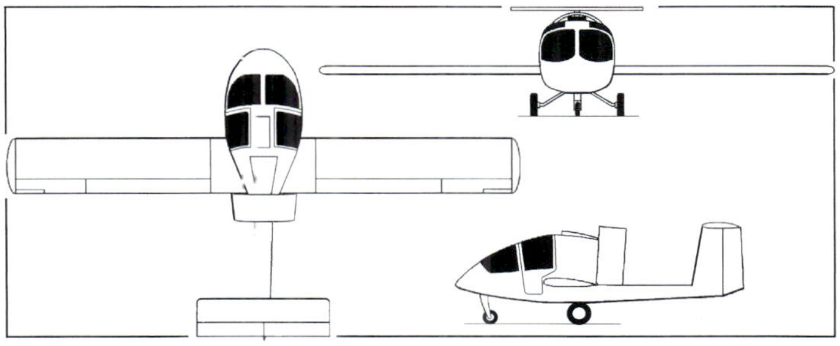

Provisional general arrangement of the C+C Flying CC/03 two-seat ultralight *(Paul Jackson)* ***NEW***/0554614

C+C FLYING CC/04

TYPE: Side-by-side ultralight.
PROGRAMME: Announced in 2004.
COST: €55,230 (2004).
DESIGN FEATURES: Mid-wing configuration based on CC/03 (which see). Wings fold for storage and transportation.
FLYING CONTROLS: Conventional and manual.
STRUCTURE: Glass fibre sandwich construction.
LANDING GEAR: Tricycle type; fixed.
POWER PLANT: Not yet selected; fuel capacity intended to be 65 litres (17.2 US gallons; 14.3 Imp gallons).

All data are provisional.

DIMENSIONS, EXTERNAL:
Wing span ... 9.40 m (30 ft 10 in)
Length overall ... 5.88 m (19 ft 3½ in)
DIMENSIONS, INTERNAL:
Cabin max width ... 1.13 m (3 ft 8½ in)
AREAS:
Wings, gross ... 10.85 m² (116.8 sq ft)
WEIGHTS AND LOADINGS:
Weight empty ... 250 kg (551 lb)
Max T-O weight ... 450 kg (992 lb)
PERFORMANCE:
Max operating speed ... 220 kt (407 km/h; 253 mph)
Normal cruising speed ... 180 kt (333 km/h; 207 mph)
Stalling speed, flap down ... 65 kt (121 km/h; 75 mph)
Max rate of climb at S/L ... 182 m (600 ft)/min
T-O run ... 250 m (820 ft)
Range ... 486 n miles (900 km; 559 miles)

NEW ENTRY

CARGOLIFTER

CARGOLIFTER AG

CHAIRMAN OF BOARD: Arnd Middelmann

CargoLifter was declared bankrupt in June 2002. Of its former assets, the subscale POC prototype 'Joey', which had made only 11 flights during its three-year existence, was severely damaged in a storm in the same year, as was the CL 75 AirCrane outsize transport balloon. In July 2003, the company's operator training vehicle, Skyship 600B-02 D-LCLA *Charly* (formerly N602CL), was sold to Skycruiser Corporation, and its giant hangar at Brand was acquired by an Anglo-Malaysian consortium, which has proposed converting it to an indoor tropical rainforest leisure centre. 'Joey' was sold by auction to a Malaysian purchaser for €13,500 in October 2003.

UPDATED

D'LONG

D'LONG AEROSPACE

Crusiusstrasse 30, D-80803 München
Tel: (+49 8921) 02 96 50
Fax: (+49 8921) 02 95 89
Web: http://www.d-long.com
PROJECT MANAGER: 728/928 Werner von Anhalt

D'Long Aerospace was formed in June 2003 as a division of D'Long International Strategic Investment Co Ltd of China following the latter's acquisition of the 728/928 twinjet airliner from the bankrupt Fairchild Dornier. The company employs 20 former Fairchild Dornier personnel. During the second half of 2003, D'Long began efforts to raise between US$300 million and US$400 million from private and public sources to re-launch the programme, with the aim of starting deliveries in 2006. The company intends to retain the Oberpfaffenhofen assembly line, but expects to subcontract some 40 per cent of manufacture to China to ensure a price competitive with the Embraer 170/190.

EVOLUTION

EVOLUTION SKYSERVICE

HEAD OFFICE: Jakobsgutweg 9, D-78333 Stockach
Tel: (+49 77 71) 87 34 34
Fax: (+49 77 71) 87 34 35
e-mail: sales@evo.aero
Web: http://www.evo.aero
WORKS: Himmelreichstrasse 6, D-78333 Stockach am Bodensee

Company makes a range of balloons, radio-controlled unmanned blimps and tethered aerostats. It describes its FS 200 as the world's first true ultralight airship.

UPDATED

EVOLUTION FS 200 FRIENDSHIP 200

TYPE: Helium non-rigid.
PROGRAMME: Revealed at Aero '03, Friedrichshafen, in April 2003, when said to be near to making first flight, though this had not been reported by early 2004. To be certified under ultralight aircraft regulations. Expected production of five per year. Two-seat version also planned.

FS 200 seat/engine unit *(Paul Jackson)* ***NEW***/0558617

Friendship 200 envelope on public display at Aero '03 *(Paul Jackson)* ***NEW***/0558616

CURRENT VERSIONS: Being marketed in ready-to-fly form and as homebuilt kit.
CUSTOMERS: Aimed at advertising market.
COSTS: €78,890 flyaway; kit €58,000; helium cost €1,000 for one week's supply (2003).
DESIGN FEATURES: Ovoid envelope; three wire-braced tailfins in inverted Y configuration, each with rudder/elevator; strake each side of lower envelope; suspended seat with engine mounted at rear.
STRUCTURE: Double envelope, outer one air-inflated, inner one helium-filled.
LANDING GEAR: Single wheel under pilot seat.
POWER PLANT: One 17.9 kW (24 hp) three-cylinder piston engine, driving three-blade pusher propeller.
ACCOMMODATION: Single seat. Ground handling crew of three or four.
DIMENSIONS, EXTERNAL:
Envelope: Length ..... 20.80 m (68 ft 3 in)
Max diameter ..... 3.80 m (12 ft 5½ in)
Display panels (each):
Length ..... 10.00 m (32 ft 9¾ in)
Depth ..... 3.00 m (9 ft 10 in)
DIMENSIONS, INTERNAL:
Envelope volume:
helium ..... 120 m^3 (4,240 cu ft)
total ..... 260 m^3 (9,180 cu ft)
WEIGHTS AND LOADINGS:
Weight empty ..... 70 kg (154 lb)
Max T-O weight ..... 160 kg (352 lb)
PERFORMANCE (estimated):
Max level speed ..... 43–52 kt (80–97 km/h; 50–60 mph)
Min flying speed ..... 5.5 kt (10 km/h; 6 mph)
Endurance ..... 4 h 30 min

UPDATED

EXTRA

EXTRA AIRCRAFT LLC

1935 Fruitville Pike, # 104, Lancaster, Pennsylvania 17601
Tel: (+1 717) 394 97 97
Fax: (+1 717) 394 51 06
e-mail: info@extraaircraft.com
Web: http://www.extraaircraft.com
FACTORY AND EUROPEAN HEADQUARTERS:
Extra GmbH, Flugplatz Dinslaken, Schwarze Heide 21
D-46569, Hünxe
Tel: (+49 2858) 913 70
Fax: (+49 2858) 91 37 30
PRESIDENT: Kenneth Weaver
CHIEF EXECUTIVE OFFICER: Kenneth L Keith
CHIEF DESIGNER: Walter Extra
GENERAL MANAGER: Extra GmbH: Corvin Huber
INFORMATION CONTACT: Bessy Donnelly

The former Extra Flugzeugbau GmbH, founded by Walter Extra, was forced to curtail activities in early 2003 as a consequence of insolvency. By mid-2003 the company had been restructured under US ownership as Extra Aircraft LLC. Research, development and manufacturing activities will remain in Germany, with establishment of a US completions centre under consideration after 2004.

UPDATED

EXTRA EA-200

TYPE: Aerobatic two-seat sportplane.
PROGRAMME: Prototype (D-ETEL) first flown 2 April 1996 and almost immediately shipped to USA for demonstrations and public debut at EAA Sun 'n' Fun in Lakeland, Florida, later that month; second prototype and principal flight test aircraft (D-EAZG) exhibited at ILA, Berlin, May 1996; German certification scheduled for 15 June 1996, followed by FAA certification (initially in Experimental/Exhibition Category) on 20 December 1996. Engineering designation **Extra EA-300/200**. May former the basis for future development of a new low-cost aerobatic aircraft.
CUSTOMERS: Prototype and five production aircraft delivered in 1996; 15 in 1997; five in 1998; further two in mid-1999; and one in mid-2000; further two completed in late 2001 for customers in USA and Ireland; total 31. None further in 2002 or 2003, but two under construction to special orders in 2004. Exports to USA account for 88 per cent of production.
COSTS: US$220,000 (late 2002).
Data generally as for Extra EA-300, except that below:
FLYING CONTROLS: Control surface movements: ailerons ±30°; elevators ±22°; rudder ±30°; trim tab ±15°.
POWER PLANT: One 149 kW (200 hp) Textron Lycoming AEIO-360-A1E flat-four engine with Gomolzig exhaust, driving an MT-Propeller MTV-12-B-C/C183-17e three-blade, constant-speed propeller. Standard usable fuel capacity 117 litres (30.9 US gallons; 25.7 Imp gallons); optional long-range tanks increase usable capacity to 185 litres (48.9 US gallons; 40.7 Imp gallons). Aerobatic usable fuel limited to 32 litres (8.5 US gallons; 7.0 Imp gallons). Unusable fuel 5 litres (1.3 US gallons; 1.1 Imp gallons) in all cases.
DIMENSIONS, EXTERNAL:
Wing span ..... 7.50 m (24 ft 7¼ in)

Extra EA-200 aerobatic/training aircraft ***NEW***/0573459

Wing aspect ratio 5.4
Length overall ..... 6.81 m (22 ft 4 in)
Height overall ..... 2.56 m (8 ft 4¾ in)
Propeller diameter ..... 1.83 m (6 ft 0 in)
AREAS:
Wings, gross ..... 10.40 m^2 (111.9 sq ft)
WEIGHTS AND LOADINGS:
Weight empty ..... 544 kg (1,199 lb)
Max T-O weight: Aerobatic, solo ..... 700 kg (1,543 lb)
Aerobatic, two-seat ..... 800 kg (1,763 lb)
Normal ..... 840 kg (1,851 lb)
Max wing loading:
Aerobatic, solo ..... 67.31 kg/m^2 (13.78 lb/sq ft)
Aerobatic, two-seat ..... 76.92 kg/m^2 (15.76 lb/sq ft)
Normal ..... 80.77 kg/m^2 (16.54 lb/sq ft)
Max power loading:
Aerobatic, solo ..... 4.69 kg/kW (7.71 lb/hp)
Aerobatic, two-seat ..... 5.36 kg/kW (8.81 lb/hp)
Normal ..... 5.63 kg/kW (9.25 lb/hp)
PERFORMANCE (N: at Normal category T-O weight, SA: at solo Aerobatic weight):
Never-exceed speed (VNE):
N ..... 220 kt (407 km/h; 253 mph)
Manoeuvring speed ..... 158 kt (293 km/h; 182 mph)
Cruising speed at 75% power:
N ..... 150 kt (278 km/h; 173 mph)
Stalling speed: N ..... 56 kt (104 km/h; 64 mph)
SA ..... 52 kt (96 km/h; 60 mph)
Max rate of climb at S/L: N ..... 488 m (1,600 ft)/min
SA ..... 686 m (2,250 ft)/min
Service ceiling: N ..... 4,575 m (15,000 ft)
Range with single pilot at 75% power at FL80, 45 min reserves
N: standard fuel ..... 450 n miles (833 km; 517 miles)
optional fuel ..... 600 n miles (1,111 km; 690 miles)
g limits: SA ..... ±10

UPDATED

EXTRA EA-300 AND EA-330

TYPE: Aerobatic two-seat sportplane.
PROGRAMME: Design began January 1987; first flight (D-EAEW) 6 May 1988; LBA certification 16 May 1990; certified FAR Pt 23 Amendment 33 Normal and Aerobatic categories for USA 26 February 1993; production started October 1988.
CURRENT VERSIONS: **Extra EA-300:** Baseline, mid-wing, two-seat aircraft.
Details apply to Extra EA-300 except where indicated.

Extra EA-300L: Low-wing, two-seat version with other structural changes; low wing enhances visibility during take-off and landing and eases installation of full IFR instrumentation; new ailerons give slightly higher roll rate (approximately 400°/s) and improved snap-rolling capability, both erect and inverted; shorter fuselage with all-composites side and belly access panels; cockpit interior modified for more comfort, with new reclined carbon fibre seats with optional leather trim; first aircraft (D-EUWH) delivered November 1994. FAA certification 31 March 1995. Current production version.

Extra EA-300L Touring Edition: First shown (D-EDGE, c/n 1175) at ILA, Berlin, May 2004. Cost US$245,000. Features include leather seats with three-way adjustment, heater, high quality paint, electric rudder pedal adjustment and option of upgraded instrument panel.

Extra EA-300S: Single-seat version of EA-300 with same power plant; wing shortened by 50 cm (19½ in) and more powerful ailerons; first flight (D-ESEW) 4 March 1992; certified March 1992; FAA certification 28 October 1993.

Extra EA-330LX: Strengthened version of EA-300L with CFRP fuselage panels, fin and rudder; wider-chord rudder and elevators, all with horn balances; 246 kW (330 hp) Textron Lycoming AEIO-580 engine. Prototype (D-ESEW, modified from prototype EA-300S) first flown January 1998 and since converted to EA-330L. First production aircraft (D-EPET) flew 11 May 1998 and delivered to Hungarian aerobatic pilot Peter Besenyei; normally flown with single-seat cockpit canopy; first public performance at North Weald, UK, 29 May 1998. Third, D-EDGE (marked '300LX') built 1999; none further known.

Extra EA-330XS: Designation applied to the former EA-300S (including D-EJKS of Aerobatics unlimited and N76PK of display pilot Gene Soucy) retrofitted with wide-chord rudder.

Extra 300L Touring Edition at its public debut, Berlin, May 2004 *(Paul Jackson)* ***NEW***/0580545

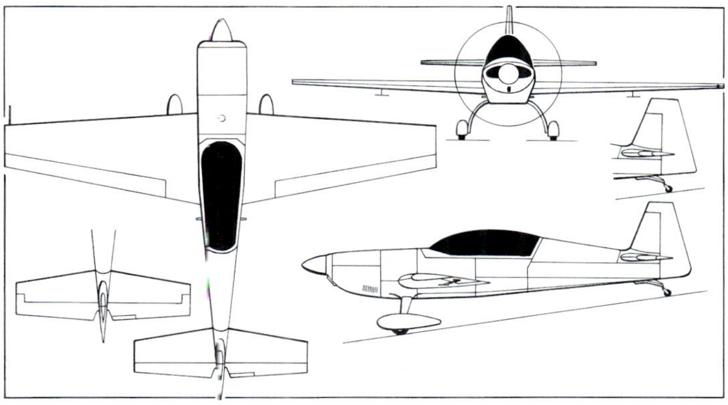

Extra EA-300L competition aerobatic aircraft, with scrap views of Extra 330 tail surfaces *(James Goulding)* 0064457

CUSTOMERS: Total 66 (including prototype) two-seat original ('mid-wing') EA 300s delivered by April 1997; additional one to USA in November 1998. Further 28 EA-300Ss built by August 1995, plus one in 1997, one in 1999 and one in 2002; total 31. EA-300L production totalled 172 by late 2003. EA-330L/LX production two (excluding prototype). Thus, 272 of all EA-300/330 by mid-2003. Total of eight were scheduled for production in 2003, then 36 in 2004, rising to maximum capacity of 48 in 2005. Sales of EA-300L totalled 189 by May 2004.

COSTS: Basic price (300L, excluding avionics) US$210,000 (2004).

DESIGN FEATURES: Designed for unlimited competition aerobatics; tapered, square-tipped mid-mounted wing with 4° leading-edge sweepback; aerofoil symmetrical MA-15S at root, MA-12S at tip; no twist or dihedral.

FLYING CONTROLS: Conventional and manual. Rod and cable operated; nearly full-span ailerons, assisted by spade-type suspended tabs; trim tab in starboard elevator. Controls surface movements: ailerons ±30°; elevators ±26°; rudder ±30°; trim tab ±15°. Dual controls standard.

STRUCTURE: Fuselage (excluding tail surfaces) of steel tube frame with part aluminium, part fabric covering; wing spars of carbon composites and shells of carbon composite sandwich; tail surfaces of carbon spars and glass fibre shells.

LANDING GEAR: Non-retractable tailwheel type. Fuselage-mounted spring cantilever GFRP mainwheel legs each with single polyamide-faired wheel; leaf-sprung steerable tailwheel. Cleveland heavy-duty disc brakes.

POWER PLANT: One 224 kW (300 hp) Textron Lycoming AEIO-540-L1B5 or AEIO-540-L1B5D flat-six engine, with Gomolzig exhaust, driving an MT-Propeller MTV-9-B-C/C200-15 three-blade, constant-speed propeller with Woodward governor (four-blade MTV-14-B-C/C190-17 optional). With either standard or optional propeller and Gomolzig System 3 silencer, Extra 300s meet German and US noise limits. Christen Industries inverted oil system. Fuel capacity: 300L standard capacity of 171 litres (45.1 US gallons; 37.6 Imp gallons), of which 165 litres (43.7 US gallons; 36.3 Imp gallons) usable; optional long-range tanks raise total fuel capacity to 208 litres (55.1 US gallons; 45.9 Imp gallons), of which 200 litres (52.8 US gallons; 44.0 Imp gallons) usable. Aerobatic category usable fuel restricted to 45 litres (12 US gallons; 10 Imp gallons). 300S usable fuel capacity 169 litres (44.6 US gallons; 37.2 Imp gallons).

ACCOMMODATION: Pilot and co-pilot/passenger in tandem composite seats under single-piece canopy opening to starboard; additional transparencies in lower sides of cockpit. Hooker harness with ratchet adjustment in rear seat. Electrically adjustable rudder pedals in rear cockpit.

SYSTEMS: 70 A alternator. Smoke system optional.

AVIONICS: To customer's choice, including Garmin GNS 430/530 GPS; Garmin 150XL or 250XL GPS/com; Becker AR 4201 com/intercom with PTT switch; Bendix/King KT 76 or Garmin GTX 327 transponder; ELT with remote panel control; and David Clark or Bose Series II headsets.

Instrumentation: Dual ASIs and altimeters. Digital *g*-meter in front cockpit. Rear cockpit features as standard magnetic compass; manifold pressure/fuel flow indicator; digital tachometer with recording time; hours meter; digital accelerometer/*g*-meter; oil temperature/pressure gauge; CHT/EGT gauge; ammeter; alternator warning light. Optional instrumentation includes: front cockpit: slip indicator and analogue accelerometer; rear cockpit: VSI; turn-and-bank indicator; slip indicator for erect/inverted flight; DG; electric artificial horizon; Shadin Mini-Flow L fuel computer; and digital clock.

EQUIPMENT: Standard equipment includes stall warning system; wingtip strobe lights; tiedown rings; and external paint scheme in three colours. Optional equipment includes titanium firewall; brake fairings; leather seats and cockpit trim; Hooker harness with ratchet for front seat; canopy lock; single-seat canopy; cabin heater; removable aerobatic sequence card holder, 45/90° wingtip sighting devices; audio/video recording and downlink systems; battery charging system; external power socket; glider towhook; polyurethane tape wing walk protection; fitted airframe cover; custom exterior paint schemes.

DIMENSIONS, EXTERNAL:
Wing span: 300, 300L ... 8.00 m (26 ft 3 in)
300S ... 7.50 m (24 ft 7¼ in)
Wing chord:
at root: 300, 300S ... 1.85 m (6 ft 0¾ in)
at tip: 300 ... 0.83 m (2 ft 8¾ in)
300S ... 0.93 m (3 ft 0½ in)
Wing aspect ratio: 300 ... 6.0
300S ... 5.4
Length overall: 300 ... 7.12 m (23 ft 4¼ in)
300L ... 6.94 m (22 ft 9¼ in)
300S ... 6.65 m (21 ft 9¾ in)
Height overall: 300, 300S ... 2.62 m (8 ft 7¼ in)
Tailplane span: 300 ... 3.20 m (10 ft 6 in)
Wheel track: 300 ... 1.80 m (5 ft 10¾ in)
Propeller diameter: MTV-9 ... 2.00 m (6 ft 6¾ in)
MTV-14 ... 1.90 m (6 ft 2¾ in)

AREAS:
Wings, gross: 300, 300L ... 10.70 m^2 (115.2 sq ft)
300S ... 10.44 m^2 (112.4 sq ft)
Ailerons, total: 300, 300L ... 1.71 m^2 (18.41 sq ft)

WEIGHTS AND LOADINGS:
Weight empty: 300 ... 630 kg (1,389 lb)
300L ... 667 kg (1,470 lb)
300S ... 609 kg (1,343 lb)
Max T-O weight:
300, Aerobatic, solo and 300S ... 820 kg (1,808 lb)
300, Aerobatic, two-seat ... 870 kg (1,918 lb)
300L Aerobatic, solo ... 820 kg (1,808 lb)
300L Aerobatic, two-seat ... 870 kg (1,918 lb)
300, 300L, Normal ... 950 kg (2,095 lb)
300S, Normal ... 920 kg (2,028 lb)
Max wing loading:
300, Aerobatic, solo ... 76.5 kg/m^2 (15.67 lb/sq ft)
300, 300L, Normal ... 88.8 kg/m^2 (18.19 lb/sq ft)
300S, Normal ... 88.12 kg/m^2 (18.05 lb/sq ft)
Max power loading:
300, Aerobatic, solo ... 3.72 kg/kW (6.12 lb/hp)
300, Normal ... 4.24 kg/kW (6.98 lb/hp)
300S ... 4.11 kg/kW (6.75 lb/hp)

PERFORMANCE:
Never-exceed speed (VNE):
300L, 300S ... 220 kt (407 km/h; 253 mph)
Max level speed: 300S ... 185 kt (343 km/h; 213 mph)
Max manoeuvring speed:
300L, 300S ... 158 kt (293 km/h; 182 mph)
Cruising speed, 45% power at 2,440 m (8,000 ft):
300L ... 170 kt (315 km/h; 196 mph)
Stalling speed:
300L, 300S at solo Aerobatic max T-O weight ... 55 kt (102 km/h; 64 mph)
300L, 300S at Normal max T-O weight ... 60 kt (111 km/h; 69 mph)
Max rate of climb at S/L:
300L, 300S ... 975 m (3,200 ft)/min
Service ceiling: 300L, 300S ... 4,875 m (16,000 ft)
T-O run: 300L ... 115 m (375 ft)
T-O to 15 m (50 ft): 300L ... 248 m (815 ft)
Landing run: 300L ... 177 m (580 ft)
Range at 65% power cruising speed at FL80, 45 min reserves:
300S ... 440 n miles (814 km; 506 miles)
Range at 45% power cruising speed, at FL80, 45 min reserves:
300L, standard fuel ... 414 n miles (766 km; 476 miles)
300L, long-range fuel ... 510 n miles (944 km; 586 miles)
Endurance: 300S ... 2h 30 min
g limits: Aerobatic, solo ... ±10
Two-seat Aerobatic ... ±8

UPDATED

EXTRA EA-400

TYPE: Business prop.

PROGRAMME: Announced February 1993; designed in collaboration with Delft University; first flight (D-EBEW) 4 April 1996; initial LBA certification to FAR Pt 23 received 23 April 1997. Second prototype (D-EGBU) flew April 1998, featuring downturned wingtips and ventral stake, both of which had been added to first aircraft before its retirement in 1997. This aircraft lost on delivery to first customer, 21 August 1998. FAA certification 15 April 1998.

CURRENT VERSIONS: **Extra EA-400**: *As described.*

Extra EA-500: Turboprop. Described separately.

CUSTOMERS: Two prototypes and 26 production aircraft completed by mid-2003; none further by May 2004, although two then nearing completion. Combined rate of one per month for EA-400/500 anticipated by December 2004, rising to 18 to 24 per year.

COSTS: Typically equipped, US$895,000 (2003).

DESIGN FEATURES: Cantilever high-wing NLF aerofoil section, T tail layout; spacious cabin; optimised for low internal and external noise.

FLYING CONTROLS: Conventional and manual. Frise ailerons, maximum deflections +27/–19°; two-section electrically actuated Fowler flaps approximately two-thirds of span, deflection 0, 15 and 30°; horn-balanced elevators, maximum deflection +33/–18°; trim tab in starboard elevator, maximum deflection +30/–20°; rudder maximum deflection ± 25°; ground-adjustable tab on rudder. Interconnected aileron and rudder controls, with override. Electric pitch trim.

STRUCTURE: Composites throughout. Airframe consists of skin with integral longerons and frames. Wing built on double front spar and rear spar, interconnected by ribs; tailplane has front and rear spar. Skins generally of carbon fibre facings and honeycomb; longerons, frames, spars and ribs of carbon fibre with foam core; wing leading-edge of glass fibre with honeycomb and glass fibre ribs. Service life 25 years/20,000 hours/30,000 cycles/22,500 pressurisation cycles.

LANDING GEAR: Hydraulically retractable tricycle type; designed and manufactured by Gomolzig; single wheels throughout; main units retract forward into fuselage sides; nosewheel retracts rearwards. Mainwheel size 15×6.0-6, nosewheel size 5.00-5. Minimum turning radius 20.4 m (67 ft). Independent hydraulic brakes on mainwheels. Nosewheel steerable ±30°.

POWER PLANT: One 261 kW (350 hp) Teledyne Continental Voyager TSIOL-550-C six-cylinder turbocharged and intercooled liquid-cooled engine, with tuned exhaust system, driving an MT-Propeller MTV-14-D/195-30a four-blade, constant-speed (hydraulic) propeller; three-blade propeller optional. Fuel in two integral wing tanks, combined capacity

Extra EA-400 (Teledyne Voyager flat-six) *(Paul Jackson)* ***NEW***/0573322

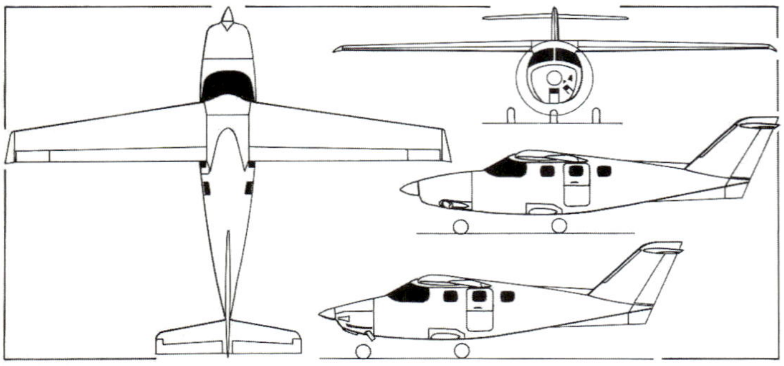

Extra EA-400 with additional side view (upper) of Extra EA-500 ***(Paul Jackson)*** 0137336

468 litres (123.6 US gallons; 102.9 Imp gallons), of which 404 litres (107 US gallons; 88.9 Imp gallons) are usable. Oil capacity 12.3 litres (3.2 US gallons; 2.7 Imp gallons).

ACCOMMODATION: Pilot and five passengers in pressurised cabin; four-way adjustable pilot's and co-pilot's seats; rear seats in club configuration; retractable shoulder harnesses for all accupants; single door on port side is horizontally split, with airstair in bottom half; top half hinges upwards. Emergency exit starboard side, opposite door. Cabin is heated and air conditioned.

SYSTEMS: Electrical system 24 V, with direct-driven 100 A alternator, belt-driven 85 A alternator and 24 V battery in tailcone. External power receptacle in tailcone. Pressurisation system, maximum differential 0.38 bar (5.5 lb/sq in). Wing and empennage leading-edges de-iced by Goodrich preumatic estane (silver) inflatable boots.

AVIONICS: Standard Garmin avionics suite, comprising GNS 530 and GNS 430 GPS/nav/coms, each with colour LCD MFD; GI-106A nav indicator with glideslope; GMA 340 audio panel with marker beacon receiver and six-place Bose X intercom interfaces; GTX 327 transponder; Honeywell EO461 EADI EFIS; Honeywell EFIS attitude indicator; Litef LCR-92/CCU EA-85511 laser gyro platform; S-Tec 55X autopilot with auto and command electric trim and electronic yaw trim; Pointer 3000 ELT; and Bose pilot's and co-pilot's headsets. Optional avionics include second GTX 327; GTX 330 Mode S transponder; remote-mounted Bendix/King KN 62or KDM 706A DME; Honeywell RDR-2000 colour weather radar in wingtip pod; Garmin GDL 49 weather uplink system; L-3 Communications Skywatch TCAS; and Aircell ACT02 voice/data system.

Instrumentation: Standard instrument includes dual ASIs, attitude indicators and directional gyros; pilot's Bendix/King KEA 130A encoding altimeter; co-pilot's standard altimeter; VSI; turn co-ordinator; magnetic compass; Moritz digital engine instrumentation system; Alcor CHT gauge; and master warning and annunciator panel. Options include EDM-800 graphic engine monitor with data recording facility, Thommen encoding drum altimeter; and Shadin air data computer.

EQUIPMENT: Standard equipment includes heated propeller, pilot's windscreen, pitot heads and stall vane; dual cockpit dome lights; multiple LED eyebrow panel lights; four overhead cabin reading lights; ice detection light; wingtip navigation and strobe lights; wing-mounted recognition lights; cowling-mounted landing light; grey leather seats and armrests; alcantara ultrasuede cabin headliner and sidewalls; grey chequerboard wool blends carpet; and stowaway writing desk. Options include video/audio entertainment cabinet (replaces one cabin seat); wood and carbon fibre cabin trim; cockpit/cabin privacy curtain; window shades; galley A-frame; cabin and engine preheat winter package; and customised exterior paint scheme.

DIMENSIONS, EXTERNAL:
Wing span ... 11.50 m (37 ft 8¾ in)
Wing aspect ratio ... 9.3
Length overall ... 9.56 m (31 ft 4½ in)
Max width of fuselage ... 1.46 m (4 ft 9½ in)
Height overall ... 3.09 m (10 ft 1¾ in)
Tailplane span ... 3.80 m (12 ft 5½ in)
Wheel track ... 2.20 m (7 ft 2½ in)
Wheelbase ... 2.56 m (8 ft 4¾ in)
Propeller diameter ... 1.95 m (6 ft 4¾ in)
Cabin door:
Width ... 0.68 m (2 ft 2¾ in)
Height ... 1.15 m (3 ft 9¼ in)

DIMENSIONS, INTERNAL:
Cabin (between forward and aft bulkheads):
Length ... 4.11 m (13 ft 6 in)
Max width ... 1.40 m (4 ft 7 in)
Max height ... 1.22 m (4 ft 0 in)

AREAS:
Wings, gross ... 14.26 m^2 (153.5 sq ft)
Horizontal tail surfaces ... 2.81 m^2 (30.25 sq ft)
Vertical tail surfaces ... 2.41 m^2 (25.94 sq ft)

WEIGHTS AND LOADINGS:
Basic operating weight empty ... 1,389 kg (3,062 lb)
Baggage capacity ... 90 kg (198 lb)
Useful load ... 522 kg (1,150 lb)
Max T-O and landing weight ... 1,999 kg (4,407 lb)
Max zero-fuel weight ... 1,772 kg (3,907 lb)
Max wing loading ... 140.2 kg/m^2 (28.71 lb/sq ft)
Max power loading ... 7.66 kg/kW (12.58 lb/hp)

PERFORMANCE:
Never-exceed speed (VNE) ... 221 kt (409 km/h; 254 mph) IAS
Manoeuvring speed ... 156 kt (289 km/h; 179 mph) IAS
Max cruising speed at 75% power at FL200 ... 212 kt (393 km/h; 244 mph)
Econ cruising speed at 65% power at FL200 ... 188 kt (348 km/h; 216 mph)
Stalling speed in landing configuration ... 59 kt (110 km/h; 68 mph)
Max rate of climb at S/L ... 326 m (1,070 ft)/min
Max certified altitude ... 7,620 m (25,000 ft)
T-O run ... 500 m (1,640 ft)
T-O to 15 m (50 ft) ... 860 m (2,820 ft)
Landing from 15 m (50 ft) ... 625 m (2,050 ft)
Landing run ... 275 m (900 ft)
Range, with reserves ... 1,060 n miles (1,963 km; 1,219 miles)
g limits ... +4.0/–1.6

UPDATED

EXTRA EA-500

TYPE: Business turboprop.

PROGRAMME: Turboprop version of Extra EA-400; prototype (D-EKEW), employing previously reserved airframe, first flew 26 April 2002; damaged in forced landing during third sortie, 30 April 2002; announced at, but unable to participate in, Berlin Air Show 6 to 12 May 2002. Repaired and resumed flight testing within three months. Public debut at the NBAA convention in Orlando, Florida, 7 October 2003, at which time the prototype had flown some 70 hours. JAA certification scheduled for first quarter 2004, followed by FAA approval in second quarter, and first deliveries in late 2004. Target of two to four deliveries by end of 2004.

CUSTOMERS: First year's production (unspecified) sold following NBAA debut in October 2003.

Data generally as for Extra EA-400, except that below.

COSTS: US$1,495,000 (2003).

POWER PLANT: One 336 kW (450 shp) Rolls-Royce 250-B17F/2 turboprop, driving a five-blade MT-Propeller composites, constant-speed, reversible-pitch propeller with Woodward governor. Fuel capacity, 687 litres (181 US gallons; 151 Imp gallons, of which 666 litres (176 US gallons; 146 Imp gallons) are usable.

ACCOMMODATION: Four-point pilot and co-pilot safety harnesses and three-point passenger harnesses, and baggage and cargo tiedowns, standard.

SYSTEMS: Electrical system has 28V 200A generator with long-life GCU automatic start and cut-off, and 25A back-up alternator with regulator. Cabin has bleed air heating and electrical defrosting system with ventilation fan, plus DC-powered air conditioning. Goodrich pneumatic de-icing and heated propeller.

AVIONICS: Honeywell APEX core system.

Comms: Garmin GMA 340 audio panel with MKR; Garmin GTX 327 transponder (Mode S).

Flight: Traffic advisory; EGPWS; Bendix/King KFC 225 flight director/autopilot. Optional TCAS, DME.

Instrumentation: Honeywell ED 461 EFIS and EADI attitude indicator. Two 264 mm (10.4 in) active matrix LCDs.

EQUIPMENT: Engine chip detector, and heated inlet ring, standard. Metallised exterior paint, optional.

DIMENSIONS, EXTERNAL:
Length overall ... 9.95 m (32 ft 7¾ in)
Propeller diameter ... 2.00 m (6 ft 6¾ in)

WEIGHTS AND LOADINGS:
Weight empty ... 1,400 kg (3,086 lb)
Max T-O weight: EASA ... 2,000 kg (4,409 lb)
FAA ... 2,130 kg (4,695 lb)
Max landing weight ... 2,000 kg (4,409 lb)
Max wing loading: EASA ... 140.3 kg/m^2 (28.73 lb/sq ft)
Max power loading: EASA ... 5.96 kg/kW (9.80 lb/shp)

PERFORMANCE:
Max operating speed ... 219 kt (406 km/h; 252 mph)
Manoeuvring speed ... 156 kt (289 km/h; 180 mph)
Cruising speed at mid-cruise weight:
90% power at FL140 ... 227 kt (420 km/h; 261 mph)
75% power at FL250 ... 200 kt (370 km/h; 230 mph)
Stalling speed ... 61 kt (113 km/h; 71 mph)
Max rate of climb at S/L ... 518 m (1,700 ft)/min
Service ceiling ... 7,620 m (25,000 ft)
T-O run ... 351 m (1,150 ft)
T-O to 15 m (50 ft) ... 427 m (1,400 ft)
Landing from 15 m (50 ft) ... 594 m (1,950 ft)
Landing run ... 366 m (1,200 ft)
Range, with reserves:
204 kg (450 lb) useful load ... 1,600 n miles (2,963 km; 1,841 miles)
363 kg (800 lb) useful load ... 1,000 n miles (1,852 km; 1,151 miles)
454 kg (1,000 lb) useful load ... 700 n miles (1,296 km; 805 miles)
Endurance:
at 90% power ... 4 h 6 min
at 75% power ... 7 h 36 min

UPDATED

Prototype Extra EA-500 business turboprop ***NEW***/0573448

FLÄMING AIR

FLÄMING AIR GMBH
Flugplatz Oehna-Zellendorf, D-14913 Zellendorf
Tel: (+49 33742) 61 70
Fax: (+49 33742) 617 21
e-mail: flugplatzoehna@t-online.de
Web: http://www.flaemingair.de

In addition to operating a flying school and air charter business, Fläming Air imports and sells Eastern European aircraft to the German market. These include the ATEC Zephyr, Urban Lambáda/Samba, Podešva Trener Baby, which is marketed under the Fläming Air name, although built in the Czech Republic, and a two-seater originally known as the FSU Saphir.

In late 2003, the company announced it was preparing to fly the FA 02 Smaragd, an ultralight motorglider based on the Saphir. By mid-2004, however, this had been renumbered FA 03, the Saphir being retrospectively designated FA 01 Smaragd and FA 02 Smaragd VLA.

UPDATED

FLÄMING AIR FA 03 SMARAGD TMG

English name: Emerald

TYPE: Side-by-side ultralight motor glider kitbuilt.

PROGRAMME: Announced late 2003; partly complete prototype exhibited at ILA, Berlin, May 2004, when first flight planned for two months later.

COSTS: Kit estimated €50 740, less engine (2004).

DESIGN FEATURES: Development of company's FA 01/02 Smaragd lightplane with stretched fuselage, tailwheel and increased wingspan. Laminar profile wing.

Data generally as for Saphir, except that below.

FLYING CONTROLS: Conventional and manual. Airbrakes.

STRUCTURE: Composites wing with carbon fibre reinforcement and landing edge; glass fibre and carbon fibre fuselage.

LANDING GEAR: Mainwheels 360×120; tailwheel 210×50.

POWER PLANT: Optimised for Rotax piston engine and Mühlbauer propeller.

Fuel in two wing tanks, combined capacity 120 litres (31.7 US gallons; 23.4 Imp gallons).

DIMENSIONS, EXTERNAL:
Wing span .... 16.00 m (52 ft 6 in)

WEIGHTS AND LOADINGS:
Weight empty .... 300 kg (661 lb)

NEW ENTRY

Prototype Fläming Air FA 03 Smaragd TMG on display at Berlin, May 2004 *(Paul Jackson)* ***NEW***/0583321

FLÄMING AIR FSU SMARAGD

English name: Emerald

TYPE: Side-by-side ultralight/kitbuilt.

PROGRAMME: Incomplete prototype of aircraft then known as Saphir (Sapphire) exhibited at Aero 2001 at Friedrichshafen in April 2001; first flight then scheduled for August or September 2001; D-MATR exhibited at Berlin, May 2002. Certification to LTF-UL regulations received in late 2003, at which time a second prototype was nearing completion.

Renamed Smaragd in mid-2004, coincident with upgrade including additional 0.20 m (8 in) wing span, adjustable seats, individual mainwheel brakes, new instrument panel and strengthened canopy with increased opening angle.

Prototype FA 02 Smaragd VLA version *(Paul Jackson)* ***NEW***/0583326

FA 01 Smaragd ultralight *(Paul Jackson)* ***NEW***/0583325

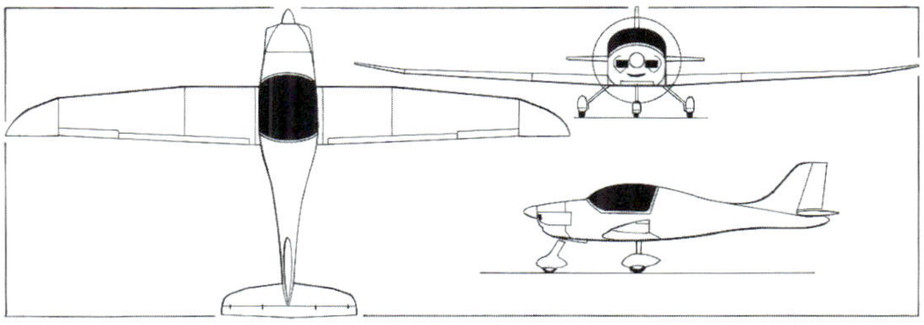

Fläming Air FA 01 Smaragd (Rotax 912 UL four-stroke) *(James Goulding)* 0137231

CURRENT VERSIONS: **FA 01 Smaragd 80:** Ultralight version, *as described.*

FA 01 Smaragd 100: As Saphir 80 but with 73.5 kW (98.6 hp) Rotax 912 ULS for glider- and banner-towing.

FA 01 Smaragd J: As Saphir 80 but with 59.7 kW (80 hp) Jabiru 2200.

FA 01 Smaragd D: Proposed version with 59.7 kW (80 hp) Diesel Air D-280.

FA 02 Smaragd VLA: Kitbuilt lightplane version to JAR-VLA; maximum T-O weight 600 kg (1,322 lb); span 9.00 m (29 ft 6¼ in). Prototype D-EERJ (with Diesel Air D-280 engine) exhibited incomplete at Berlin, May 2004, then known as Super Saphir.

COSTS: Kit: Saphir 80 €65,500; Saphir 100 €66,500; Saphir J €64,500, Saphir D €65,000 (including instruments and taxes) (all 2004), then known as Super Saphir.

DESIGN FEATURES: Low-wing monoplane with slightly sweptback wings and upturned wingtips. Combines elements of two Czech aircraft: wing and tail unit of Urban UFM 10 Samba with cockpit section of ATEC Zephyr, both of which are described in the Czech Republic section. Urban Samba XXL has same combination, but detail differences.

Wing section SM 701.

FLYING CONTROLS: Conventional and manual. One-piece elevator; plain flaps. Electrically actuated elevator trim tab and flaps.

STRUCTURE: All-composites.

LANDING GEAR: Tricycle type; fixed. Fuselage-mounted spring cantilever mainwheel legs. Mainwheel diameter 400 mm, nosewheel 300 mm. Brakes. Speed fairings on all three wheels.

POWER PLANT: One 59.6 kW (79.9 hp) Rotax 912 UL (Saphir 80), 73.5 kW (98.6 hp) Rotax 912 ULS (Saphir 100) or 59.7 kW (80 hp) Jabiru 2200, driving a Kremen SR 200 three-blade, ground-adjustable pitch wooden propeller; optional Kremen SR 2000 variable-pitch (electric), wooden propeller. Saphir D offered with 59.7 kW (80 hp) Diesel Air D-280 power plant. Fuel in two tanks, combined capacity 60 litres (15.9 US gallons; 13.2 Imp gallons) in standard versions and 120 litres (31.7 US gallons; 26.4 Imp gallons) in Super Saphir.

DIMENSIONS, EXTERNAL:
Wing span .... 10.30 m (33 ft 9½ in)
Length overall .... 5.90 m (19 ft 4¼ in)
Height overall .... 2.13 m (6 ft 11¾ in)
Propeller diameter .... 1.70 m (5 ft 7 in)

DIMENSIONS, INTERNAL:
Cabin max width .... 1.10 m (3 ft 7¼ in)

AREAS:
Wings, gross .... 9.25 m² (99.6 sq ft)

WEIGHTS AND LOADINGS:
Weight empty .... 270 to 290 kg (595 to 639 lb)
Max T-O weight .... 472.5 kg (1,042 lb)

PERFORMANCE, POWERED:
Max level speed .... 146 kt (270 km/h; 167 mph)
Cruising speed .... 119 kt (220 km/h; 137 mph)
Stalling speed .... 36 kt (65 km/h; 41 mph)

PERFORMANCE, UNPOWERED:
Best glide ratio 19
Min rate of sink .... 1.50 m (4.92 ft)/s

UPDATED

FLIGHT DESIGN

FLIGHT DESIGN GMBH
Sielminger Strasse 65, D-70771 Leinfelden-Echterdingen
Tel: (+49 711) 90 28 70
Fax: (+49 711) 902 87 99
e-mail: info@flightdesign.com
Web: http://www.flightdesign.com
FOUNDER: Matthias Betsch

PRODUCTION CENTRE:
Ost-Vest Konsalting SP
ulítsa Neftyanikov 15-A, 325000 Kherson, Ukraine
Tel: (+38 55) 223 51 35
Fax: (+38 55) 229 90 32
e-mail: owc@tlc.kherson.ua
DIRECTORS: Andrei Ya Prtygin, Aleksandr A Bondar

Flight Design's current product is the CT. Fabrication of parts is at Flight Design's Ukrainian factory producing composites to German specifications. At present Flight Design has a workforce exceeding 200 and is also involved in the production of hang gliders and paragliders.

UPDATED

FLIGHT DESIGN CT

TYPE: Two-seat lightplane; side-by-side ultralight.

PROGRAMME: Original CT version certified in Germany on 6 June 1997, having flown some 1,000 hours. CT and kitbuilt CTX version now discontinued.

CURRENT VERSIONS: **CT2K:** Standard version, *as described*. Lighter airframe; downturned wingtips; German certified maximum T-O weight increased to 472.5 kg (1,042 lb) in October 2003.

CTU: Longer span version to achieve 38 kg/m² (7.78 lb/sq ft) wing loading required in some countries; extra 1.00 m (3 ft 3¼ in) span and 1.20 m² (12.9 sq ft) increase in area; available to order only.

CTsw: Short span version (8.55 m; 28 ft ½ in), max T-O weight 472.5 kg 11.042 lb). Two aircraft exhibited at Berlin in May 2004.

CUSTOMERS: More than 250 of all versions built by September 2002, when production was running at six per month.

COSTS: Complete, with Rotax 912 UL-2, propeller and Junkers ballistic recovery system €53,987 excluding tax (2003). Kits no longer available.

DESIGN FEATURES: Optimised for safe handling, speed and range. High-mounted, cantilever, constant chord wing, sharply waisted fuselage and sweptback fin with large underfin. CT2K features new trim tab and optional downturned wingtips.

Wing dihedral 1° 40'.

FLYING CONTROLS: Conventional ailerons and horn-balanced rudder, manually actuated; all-moving tailplane with anti-balance tab. Externally-hinged slotted ailerons. Electrically actuated plain flaps, deflection −12° and +40°; internal (external on early aircraft) mass balances for ailerons and tailplane.

STRUCTURE: Principally of composites, including carbon/glass fibre and carbon/aramid sandwich construction.

LANDING GEAR: Tricycle type; fixed. Fuselage-mounted, forward-swept spring aluminium tube cantilever mainwheel and nosewheel legs; steerable nosewheel. Optional speed fairings on all wheels. Hydraulic mainwheel brakes.

POWER PLANT: One 59.6 kW (79.9 hp) Rotax 912 UL-2 or 73.5 kW (98.6 hp) Rotax 912 ULS flat-four, driving a Neuform two-blade C2-V variable-pitch or C2 fixed-pitch propeller. Fuel in two wing leading-edge tanks, combined capacity 130 litres (34.3 US gallons; 28.6 Imp gallons).

ACCOMMODATION: Two persons, side by side; dual controls; upward-hinged door with gas spring for each occupant. Baggage space behind seats, with external door on each side.

AVIONICS: Optional avionics include Becker Radio AR 4201 com, Honeywell KY97A com, KX 125 nav/com, KT 76A Mode A/C transponder, Garmin GNC 300 XL nav/com/GPS with Jeppesen database, Garmin Pilot III GPS, Flightcom headsets and ELT.

Instrumentation: Winter variometer and Rotax FlyDat engine management system optional.

EQUIPMENT: Optional equipment includes cabin comfort pack; anti-collision and position lights; and Junkers High Speed or BRS UL 1050 ballistic parachute recovery systems.

DIMENSIONS, EXTERNAL:
Wing span ... 9.31 m (30 ft 6½ in)
Wing chord, constant ... 1.17 m (3 ft 10 in)
Wing aspect ratio ... 8.0
Length overall ... 6.215 m (20 ft 4½ in)
Height overall ... 2.165 m (7 ft 1¼ in)
Tailplane span ... 2.38 m (7 ft 9¾ in)
Tailplane chord ... 0.70 m (2 ft 3½ in)
Wheel track ... 1.70 m (5 ft 7 in)
Wheelbase ... 1.45 m (4 ft 9 in)
Propeller diameter ... 1.73 m (5 ft 8 in)
Propeller ground clearance ... 0.28 m (11 in)

DIMENSIONS, INTERNAL:
Cabin max width ... 1.24 m (4 ft 0¾ in)

AREAS:
Wings, gross ... 10.80 m² (116.3 sq ft)
Ailerons (total) ... 0.98 m² (10.55 sq ft)
Fin and rudder (total) ... 1.14 m² (12.27 sq ft)
Tailplane (total incl elevators and tab) ... 1.65 m² (17.76 sq ft)
Flaps (total) ... 1.51 m² (16.25 sq ft)

WEIGHTS AND LOADINGS:
Weight empty ... 262 kg (578 lb)
Max T-O weight ... 600 kg (1,322 lb)[1]
Max baggage weight ... 60 kg (132 lb)
Max wing loading ... 55.6 kg/m² (11.38 lb/sq ft)
Max power loading: 912 UL ... 10.07 kg/kW (16.54 lb/hp)
912 ULS ... 8.16 kg/kW (13.41 lb/hp)

[1] *or 472.5 kg (1,042 lb) to meet national limits*

PERFORMANCE (downturned wingtips, A: 912 UL; B: 912 ULS):
Never-exceed speed (VNE):
A, B ... 167 kt (310 km/h; 192 mph) IAS
Max level speed: A, B ... 146 kt (270 km/h; 168 mph) IAS
Cruising speed at 75% power:
A ... 121 kt (225 km/h; 140 mph) IAS
B ... 127 kt (235 km/h; 146 mph) IAS
Stalling speed at 450 kg (992 lb) MTOW:
A, B ... 34 kt (62 km/h; 39 mph) IAS
Service ceiling: A, B ... 4,270 m 14,000 ft)
Max rate of climb at S/L: A ... 210 m (689 ft)/min
B ... 300 m (984 ft)/min
T-O run ... 90 m (295 ft)
T-O to 15 m (50 ft) ... 160 m (525 ft)
Range with max fuel, 30 min reserves ... 1,079 n miles (2,000 km; 1,242 miles)
g limits: A, B:
at 450 kg (992 lb) MTOW ... +6
at 600 kg (1,322 lb) MTOW ... +4.3

UPDATED

'Short Wing' Flight Design CTsw on display at ILA, Berlin, 2004 *(Paul Jackson)* ***NEW***/0583330

UK-registered Flight Design CT2K *(Paul Jackson)* ***NEW***/0576605

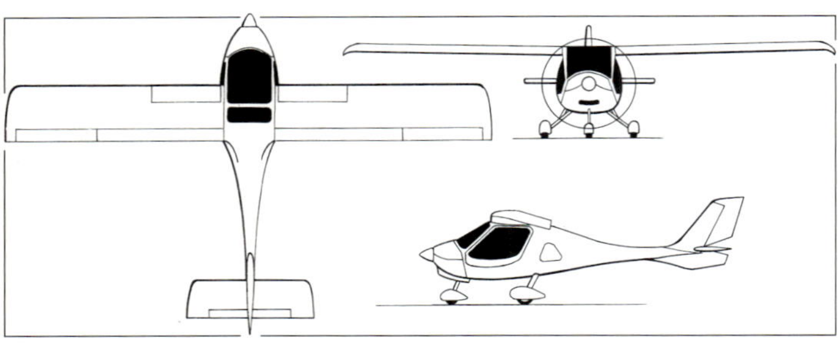

General arrangement of the Flight Design CT2K *(James Goulding)* 0110593

FLUG WERK

FLUG WERK GMBH

Kothingried 4, D-85408 Gammelsdorf
Tel: (+49 87) 66 93 98 78
Fax: (+49 87) 66 93 98 79
Web(1): http://www.flugwerk.com
Web(2): http://www.flugwerk.de
e-mail: colling@flugwerk.de
DIRECTOR: Claus Colling

Formed on 13 June 1996 and specialising in the reproduction of classic German aircraft, Flug Werk plans to augment its Fw 190 project with resumed manufacture of the Messerschmitt Bf 109 (for which it produces 'spare' parts for existing aircraft) and wings for the P-51 Mustang. A Heinkel He 112 replica was reported to be under consideration, but no further news has been received of this project or the planned Arado Ar 96B reproduction, the latter known to have been suspended in 2003. New 600 m² (6,450 sq ft) hangar inaugurated 31 March 2000; this has been outgrown and the company is looking to move to Ingoldstadt/Manching, where it has hired hangar space before constructing new facilities.

It was at Ingoldstat on 22 July 2004 that the first Fw 190 was flown. Other recent activities include repair of a genuine Bf 109G after a taxying accident, overhaul of a North American T-6 and construction of components for the P-51 Mustang. Company is capable of building complete replacement wings and fuselages for Mustangs and can supply major components for Bf 109 restorations. Earlier work on Fw 190s included refurbishment of WNr 173056, a genuine aircraft belonging to Don Hansen of Baton Rouge, Louisiana, for which an ASh-82 engine conversion kit was also engineered and supplied.

In late 2004, FW received airframes of one Bf 109E and one Bf 109F for restoration, the latter to be reverse-engineered for subsequent manufacture of further airframes with new DB 601 engines. A less complete Bf 109G-2 was also received in 2004 for restoration and resale on behalf of a third party. A North American T-28 Trojan was being overhauled in 2004-05.

Co-director Hans Günther Wildmoser died in September 2003.

UPDATED

FLUG WERK FW 190

TYPE: Single-seat sportplane kitbuilt.

PROGRAMME: Flug Werk GmbH started assembling a batch of 12 Focke-Wulf Fw 190 Second World War fighters in June 1996, these to be in FW 190A-8/N and FW 190D-9/N configurations, their /N designation indicating *nachbau*: replica. Components are being manufactured in seven countries (principally in Romania by Aerostar) from original drawings, resulting in airframes claimed to be 98 per cent faithful to the original.

Power plant of the 10 A-8s is a 1,400 kW (1,877 hp) ASh-82FN 14-cylinder twin-row radial engine, a Russian version of the original BMW 801D, driving a three-blade, 3.30 m (10 ft 10 in), constant-speed MT wood/composites propeller. The first aircraft (WNr 8900001, D-FWWC) shown statically, April 2000, but due to integration problems with the power plant, first engine runs were postponed to the third quarter of 2002; first flight was planned for August 2003, but not achieved until 22 July 2004. Three FW 190A-8/Ns will

First FW 190A-8/N built by Flug Werk *(EADS)* ***NEW***/1047378

be sold in Europe, the remaining seven going to the USA, South Africa and Australia; all had found firm buyers by early 2002. Price for kits (including engine, but not avionics, instruments, oxygen system, electrical wiring looms or paint) is €555,000 ex-works (2004).

FW-built aircraft are 450 kg (992 lb) lighter than the original, due to omission of military items, and carry 880 litres (232 US gallons; 194 Imp gallons) of fuel - an increase of 230 litres (61 US gallons; 51 Imp gallons) - although they are otherwise 98 per cent faithful to the original, including genuine Second World War tailwheel assemblies discovered in storage.

The remaining two aircraft, advertised in 2002, are the long-nose Fw 190D 'Dora', the original having a 1,402 kW (1,880 hp) Jumo 213.

Brief data for new-built FW 190A include:

DIMENSIONS, EXTERNAL:

Wing span	10.50 m (34 ft 5½ in)
Length overall	9.10 m (29 ft 10¼ in)
Height overall, fuselage	3.95 m (12 ft 3¼ in)

AREAS:

Wings, gross	18.30 m² (197.0 sq ft)

WEIGHTS AND LOADINGS:

Weight empty	2,900 kg (6,393 lb)
Max T-O weight	3,800 kg (8,377 lb)

DIMENSIONS, EXTERNAL:

Max level speed at FL200	343 kt (635 km/h; 395 mph)
Cruising speed at 2,300 rpm	316 kt (585 km/h; 364 mph)
Unstick speed	89 kt (165 km/h; 103 mph)
Approach speed	100 kt (185 km/h; 115 mph)
Initial rate of climb at S/L	1,219 m (4,000 ft)/min
Range, VFR reserves	534 n miles (990 km; 615 miles)

UPDATED

GEFA-FLUG

GESELLSCHAFT ZUR ENTWICKLUNG UND FÖRDERUNG AEROSTATISCHER FLUGSYSTEME GMBH (AEROSTATIC FLIGHT SYSTEMS DEVELOPMENT AND PROMOTION COMPANY)

Weststrasse 24C, D-52074 Aachen
Tel: (+49 241) 88 90 40
Fax: (+49 241) 889 04 20
e-mail: j.leisten@gefa-flug.de
Web: http://www.gefa-flug.de
MANAGING DIRECTOR: Karl Ludwig Busemeyer
SALES AND MARKETING DIRECTOR: Jürgen Leisten

GEFA-Flug was established 1975 to operate advertising and passenger-carrying hot-air balloons; has R&D department to develop airships for manned and unmanned civil and environmental research applications such as aerial photogrammetry and pollution monitoring.

Company currently operates more than six hot-air balloons, hot-air airships and a number of remotely controlled aerostats, and has sold about 100 systems (mostly unmanned and remotely controlled) worldwide. Its 27th hot-air airship was ordered in late 2002. Fleet includes three AS 105GD hot-air airships, including one four-seater; production of this model continues. Future market seen as survey projects for environmental research; GEFA-Flug also developing scientific measuring equipment in partnership with German Mining Technology, a partly government-funded body.

UPDATED

GEFA-FLUG AS 105 GD

TYPE: Hot-air non-rigid.

PROGRAMME: Modified version of AS 105 Mk II first flown 1990 and built until 1994 by Thunder & Colt in UK; German development began 1993. Certification (by UK CAA) and first production AS 105 GD completed in 1996; also certified by German LBA.

CUSTOMERS: Total of 26 built in Germany by early 2003, of which at least eight have German-certified four-seat gondolas; 27th, for Deutsche Post, was under construction in mid-2003. First sales to UK and Switzerland (latter include 0016/HB-QIQ and 0026/HB-QIP) effected in 2002.

COSTS: Development programme DM1.4 million; standard airship approximately €160,000 (2002).

DESIGN FEATURES: Longer and slimmer envelope than UK-built version; cruciform tail surfaces with rudder on each vertical fin; twin catenary load suspension system distributes loads of gondola weight and power plant forces evenly into envelope.

FLYING CONTROLS: Burners located inside envelope and operated electrically, with manual override; pilot lights fitted with electric spark and piezoelectric ignition, specially modified to operate directly underneath inflation fan without blowing out; steering by rudder on each vertical fin, operated by cables connecting them to gondola.

STRUCTURE: Envelope made from high-tenacity and high temperature-resistant fabric; engine drives generator supplying electric fan used to pressurise envelope; additional air for envelope pressurisation via scoop located in propeller slipstream. Gondola is a chromoly aircraft-grade steel tube spaceframe with Macrolon skin panels; windscreen, of polycarbonate sheet, forms partially enclosed cockpit.

LANDING GEAR: Non-retractable; two front and two rear wheels; spring shock-absorbers.

GEFA-Flug AS 105 GD in African surroundings ***NEW***/0540557

POWER PLANT: One 47.8 kW (64.1 hp) Rotax 582 two-cylinder water-cooled engine, driving a Helix four-blade carbon fibre pusher propeller. Fuel capacity 28 litres (7.4 US gallons; 6.2 Imp gallons).

ACCOMMODATION: Pilot, co-pilot and two passengers in pairs.

AVIONICS: *Comms:* Dittel FSG 71 720-channel VHF transceiver; transponder optional.
Instrumentation: Standard VFR.

DIMENSIONS, EXTERNAL:

Envelope: Length overall	41.00 m (134 ft 6¼ in)
Max diameter	12.80 m (42 ft 0 in)
Fineness ratio	3.2
Tail unit span	18.00 m (59 ft 0¾ in)
Propeller diameter	1.60 m (5 ft 3 in)
Propeller ground clearance	0.10 m (4 in)

DIMENSIONS, INTERNAL:

Envelope volume	2,973 m³ (105,000 cu ft)
Gondola cabin: Length	4.00 m (13 ft 1½ in)
Max width	1.60 m (5 ft 3 in)
Max height	1.80 m (5 ft 10¾ in)

WEIGHTS AND LOADINGS:

Weight empty	approx 450 kg (992 lb)
Max T-O weight: two-seat	850 kg (1,873 lb)
four-seat	900 kg (1,984 lb)

PERFORMANCE:

Max level speed	22 kt (40 km/h; 25 mph)
Max cruising speed	20 kt (37 km/h; 23 mph)
Econ cruising speed	15 kt (28 km/h; 17 mph)
Max rate of climb at S/L	183 m (600 ft)/min
Endurance with max payload	1 h

UPDATED

GROB

BURKHART GROB LUFT- UND RAUMFAHRT E K (ASSOCIATED WITH GROB-WERKE E K)

Lettenbachstrasse 9, D-86874 Tussenhausen-Mattsies
Tel: (+49 8268) 99 80
Fax: (+49 8268) 99 81 14
e-mail: sales@grob-aerospace.de
Web: http://www.grob-aerospace.de
OWNER: Dr hc Dipl Ing Burkhart Grob
CHAIRMAN: Dipl Ing Christian Grob
CEO: Dr Andreas Plesske
SALES MANAGER: Dipl Ing Hans Doll

Company founded 1928; began aviation activities in 1971, since when has built more than 3,500 aircraft; name changed in 1988 from Burkhart Grob Flugzeugbau (formed 1974) to Burkhart Grob Luft- und Raumfahrt GmbH with light and heavy sections; light section produces sailplanes, supports G 109 motor glider and the G 115, G 120, G 140 and G 160 powered aircraft; heavy section supports continued operations by earlier Egrett and Strato 2C and deals with space activities in support of Weltraum-Institut Berlin.

Factory covered area 28,000 m² (301,400 sq ft); workforce 100 in late 2002.

UPDATED

GROB G 115

Royal Air Force name: Tutor

TYPE: Aerobatic two-seat lightplane.

PROGRAMME: Developed from G 110 (first flown 6 February 1982) via G 112 (4 May 1984). First flight of Grob G 115 (D-EBGF) 15 November 1985; first flight of second prototype second quarter of 1986 with taller fin and rudder and relocated tailplane; LBA certification to FAR Pt 23 on 31 March 1987; British certification February 1988; FAA certification

21 December 1988; later gained full public transport certification and German spinning clearance; production terminated August 1990 after total of 103 (including prototypes) G 115/115As. Power plants and equipment updated and designations changed late 1992 for 1993 product line; total of 203 built by December 1999, comprising 103 early versions, one G115T, 14 G 115TAs, single G 115B and 84 G 115C/Ds. Five for Royal Navy pilot training designated **Heron**.

CURRENT VERSIONS: **G 115E:** Selected in June 1998 for UK Ministry of Defence requirement to replace BAe Bulldogs in RAF University Air Squadrons (UAS) and Air Cadet Air Experience Flights, plus a squadron of the Central Flying School (CFS). Total of 99 ordered for operation at 13 sites in civil markings under contract managed by Vosper Thornycroft Aerospace (formerly Bombardier Defence Services). Development aircraft (D-ERAF) first flown early 1999; first production aircraft (D-EUKB/G-BYUA) and three others delivered to UK 15 July 1999; official handover of first five aircraft (G-BYUB/C/D/E/F) at RAF Cranwell, Lincolnshire, 13 September 1999, for CFS; deliveries to Cambridge UAS began 14 September 1999; final aircraft (G-BYYB) registered in August 2001.

Description applies to G 115E RAF version.

G 115EG: Version based on G115E for the Egyptian Air Force, which ordered 74 in May 2000 for service at the Egyptian Air Force Academy at Bilbays and Bilbays 2 Air Bases. AEIO-360-B1B engine and MTV-12-B-C/C-183-17e propeller. First aircraft (D-EGYB/EAF 6801) formally handed over in Germany 9 October 2000; first batch of eight handed over by 1 December; deliveries continued at the rate of four per month until mid-2001, then eight per month until contract completion in February 2002. FAA certification 6 February 2001.

CUSTOMERS: By January 2002, production totalled 203 of early versions and 174 G 115E/EGs, or 377 in all, including prototypes and demonstrators, but excluding two G 110s and two G 112s. No further production by late 2003.

DESIGN FEATURES: Low-wing monoplane of composites construction optimised for flight training. Wing trailing-edge moderately swept forward; slightly tapered horizontal tail surfaces; sweptback fin. G 115E based on earlier G 115s, but with fuselage entirely of carbon fibre, saving more than 40 kg (88 lb) and thus allowing aerobatics to be performed at MTOW.

Wing section Eppler E696; dihedral 5°; incidence +2° at root, −2° at tip; tailplane section NACA 64010.

FLYING CONTROLS: Conventional and manual, via pushrods. Horn-balanced elevators with mechanically actuated trim tab in left unit; elevator deflection +34/−20°; mass-balanced ailerons with ground-adjustable trim tabs, aileron deflection +20° 25'/−18° 25'; horn-balanced rudder with ground adjustable tab, rudder deflection ±30°. Electrically actuated plain flaps, deflection 0, 15 and 60°; intermediate 45° setting optional.

STRUCTURE: Entirely of CFRP; semi-monocoque fuselage has rigid CFRP shell formed in two vertically split halves with frames and web members and integral fin; wing of CFRP/honeycomb sandwich with I-beam main spar and carbon fibre roving spar caps; CFRP/honeycomb horizontal tail surfaces of two-spar construction; rudder, ailerons and flaps are GFRP/rigid foam sandwich. Airframe is protected from lightning strike damage by means of aluminium fibres embedded in the outer layer of carbon fabric and copper mesh in the wing surfaces above and below the fuel tanks, all bonded to metal ground strips. Structural service life (aerobatic) 24,000 flight hours.

LANDING GEAR: Non-retractable tricycle type. Fuselage-mounted spring steel; cantilever main legs; gas-damped nosewheel steerable via rudder pedals, maximum deflection ±9° or ±47° with use of differential braking; mainwheel tyre size 6.00–6, nosewheel tyre size 5.00–5; wheel fairings on all three units; hydraulic disc brakes.

POWER PLANT: One 134 kW (180 hp) Textron Lycoming AEIO-360-B1F/B flat-four driving Hoffmann HO-V343K-V/183GY a three-blade, constant-speed (hydraulic), wood/composites propeller. Christen inverted oil system. Fuel in two integral wing tanks, combined capacity 150 litres (39.6 US gallons; 33.0 Imp gallons) of which 143 litres (37.8 US gallons; 31.5 Imp gallons) are usable, with collector tank, capacity 5.4 litres (1.4 US gallons; 1.2 Imp gallons) for up to three minutes' inverted flight; fuel filler port in top of each wing. Oil capacity 7.6 litres (2.0 US gallons; 1.7 Imp gallons).

ACCOMMODATION: Two seats side by side, under one-piece, rearward-sliding framed canopy; dual controls (sticks) standard; seats can accommodate backpack parachutes and have five-piece harnesses; baggage space behind seats. Cockpit is heated and ventilated.

SYSTEMS: Electrical system comprises 28 V DC 35 Ah engine-driven generator and 24 V high-capacity battery for engine starting and to provide 30 minutes' emergency power in the event of generator or main bus failure.

AVIONICS: Basic aircraft supplied without avionics. Optional minimum avionics pack includes Bendix/King KX 155A nav/com/glideslope, KI 203 VOR/LOC and KT 76A transponder. Typical customer-specified Bendix/King avionics fit for military trainer detailed below.

Royal Air Force Grob G 115E Tutor *(Paul Jackson)* ***NEW***/0573436

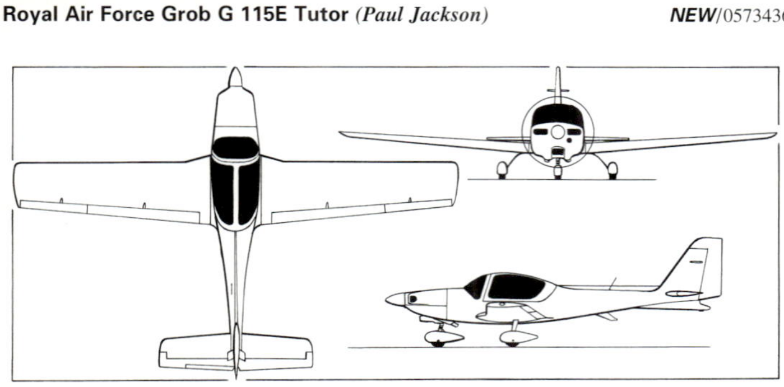

General arrangement of Grob G 115E Tutor *(James Goulding)* 0085599

Comms: KTR 909 UHF selector; with KFS 599A; dual KX 155A with KI 204 VOR/LOC/GS; KT 76A; KMA 28 audio panel; ELT; PTT switches on control columns and NATO-standard quick-release sockets for headsets.

Flight: KCS 55A compass with KG 102A slaved gyro unit; KI 525A HSI; KA 51B slaving control unit with KMT 112 magnetic azimuth transmitter; KN 63 DME with KDI 572 indicator; KG 102A slaved directional gyro; Filser LX500TR differential GPS.

Instrumentation: Standard ASI, VSI, turn and slip indicator, attitude gyro, directional gyro, tachometer, magnetic compass, CHT/fuel pressure gauge, manifold pressure/fuel flow gauge, oil pressure/temperature gauge, OAT/EGT gauge, fuel quantity gauge, voltage indicator, *g* meter and clock.

EQUIPMENT: Standard equipment includes navigation, anti-collision, landing and instrument panel lights; map light on flexible arm between seats optional.

DIMENSIONS, EXTERNAL:	
Wing span	10.00 m (32 ft 9¾ in)
Wing chord: at root	1.43 m (4 ft 8¼ in)
at tip	0.94 m (3 ft 1 in)
Wing aspect ratio	8.2
Length overall	7.79 m (25 ft 6¾ in)
Height overall	2.82 m (9 ft 3 in)
Tailplane span	3.50 m (11 ft 5¾ in)
Wheel track	2.56 m (8 ft 4¾ in)
Wheelbase	1.51 m (4 ft 11½ in)
Propeller diameter	1.83 m (6 ft 0 in)
Propeller ground clearance	0.19 m (0 ft 7½ in)
DIMENSIONS, INTERNAL:	
Cabin (incl baggage area):	
Length	2.14 m (7 ft 0¼ in)
Max width	1.03 m (3 ft 4½ in)
Max height	1.13 m (3 ft 8½ in)
Baggage volume	0.22 m^3 (7.8 cu ft)
AREAS:	
Wings, gross	12.21 m^2 (131.4 sq ft)
Ailerons (total)	1.12 m^2 (12.06 sq ft)
Fin	1.05 m^2 (11.30 sq ft)
Rudder	0.64 m^2 (6.89 sq ft)
Tailplane	1.86 m^2 (20.02 sq ft)
Elevators (total)	0.86 m^2 (9.26 sq ft)
WEIGHTS AND LOADINGS:	
Weight empty, basic	670 kg (1,477 lb)
Baggage capacity	55 kg (121 lb)
Max fuel	103 kg (227 lb)
Max T-O weight	990 kg (2,182 lb)
Max wing loading	81.08 kg/m^2(16.61 lb/sq ft)
Max power loading	7.38 kg/kW (12.12 lb/hp)
PERFORMANCE:	
Never-exceed speed (VNE)	184 kt (341 km/h; 212 mph)
Max level speed	135 kt (250 km/h; 155 mph)
Cruising speed at 75% power at FL50, ISA	124 kt (230 km/h; 143 mph)
Long-range cruising speed	96 kt (178 km/h; 110 mph)
Stalling speed: flaps up	52 kt (99 km/h; 61 mph)
60° flap	49 kt (91 km/h; 57 mph)
Max rate of climb at S/L	320 m (1,050 ft)/min
Max operating altitude	6,095 m (20,000 ft)
T-O run	248 m (815 ft)
T-O to 15 m (50 ft)	461 m (1,512 ft)
Landing from 15 m (50 ft)	457 m (1,500 ft)
Landing run	171 m (560 ft)
Range with 45 min reserves:	
75% power at FL40	446 n miles (826 km; 513 miles)
55% power at FL80	569 n miles (1,053 km; 654 miles)
45% power at FL80	620 n miles (1,148 km; 713 miles)
g limits	+6/-3

UPDATED

GROB G 120A

Israel Defence Force name: Snunit (cyclone).

TYPE: Aerobatic two-seat lightplane.

PROGRAMME: First flight (D-ELHU; unannounced) 1999; development of G 115 series to meet modern airline pilot and military training requirements, having advanced features including EFIS. With retractable landing gear and 194 kW (260 hp) Textron Lycoming AEIO-540 flat-six engine.

Announced January 2001, when Lufthansa ordered three (plus four options, of which three taken up in June 2001) for its US Airline Training Center Arizona (ATCA) operation at Goodyear, near Phoenix. German LBA provisional certification achieved 22 November 2001, followed by FAA FAR Pt 23 approval on 29 January 2002 and full LBA certification on 27 February 2002. Deliveries began (with N861AF) in February 2002.

CURRENT VERSIONS: **G 120A:** Baseline version; aerobatic MTOW 1,350 kg (2,976 lb).

G 120A-I Snunit: Israel Defence Force version. Certified by LBA on 27 September 2002 and by Israeli CAA on 14 October 2002. *As described.*

G 120A Observer: Long-range/endurance version; usable fuel capacity 378 litres (99.8 US gallons; 83.2 Imp gallons).

CUSTOMERS: Lufthansa, seven. On 19 February 2002, Cyclone Aviation, a subsidiary of Elbit Systems, contracted by Israel Defence Force/Air Force to supply pilot screening and primary training services over 10 years with company-owned and -operated fleet of up to 27 G 120A-Is; first three handed over at Hatzerim AFB on 27 October 2002; however, initial Israeli order covered only 17 aircraft, last of which delivered in July 2003. One further aircraft (25th production) as company demonstrator. Total 26, including prototype, built by mid-2004.

Description for G 120A generally as for G 115, except that below:

DESIGN FEATURES: Aerofoil Eppler E884; dihedral 2°; incidence 0°. Tailplane aerofoil NACA 64-010; fin 64-009. Airframe life 15,000 hours.

FLYING CONTROLS: Conventional and manual. Pushrod actuation for all primary surfaces. Slotted flaps electrically actuated; max deflection 60°. Servo tab and mass balance on each aileron; elevator trim tab starboard side with electric and manual actuation; rudder and elevator have mass balances, plus horn balance on rudder.

STRUCTURE: Semi-monocoque fuselage comprises a self-supporting carbon fibre-reinforced plastic shell with frame and web members. Cantilever wing of single-trapezoidal

Company demonstrator Grob G 120A D-ESAI *(Paul Jackson)* ***NEW***/0580546

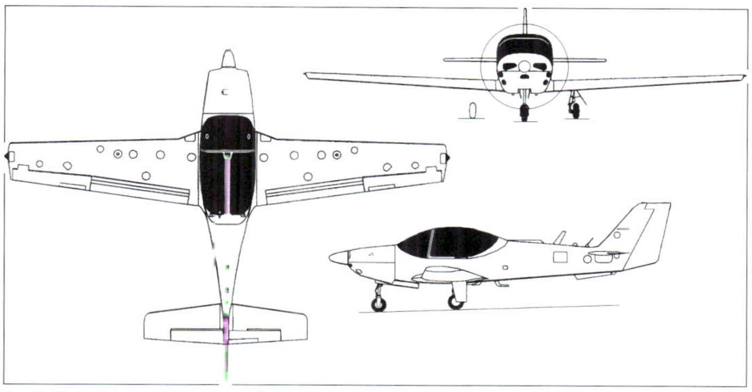

Grob G 120A-I aerobatic lightplane 0530193

cross-section has an I-beam main spar with caps of carbon fibre rovings. Wing shell of honeycomb sandwich, except tank section of PVC foam sandwich. Interconnection of wings is via the spar stubs, bolted together with two large bolts which connect the stub spars. Each wing is attached to the fuselage by three bolts. The main spar and strong webs hold mountings for the main gear leg. An auxiliary spar closes the wing trailing-edge and carries the flaps and ailerons.

Flaps with upper and lower CFRP skins in rigid PVC foam core. Aileron structure same as flaps. Fin, integrated with fuselage, comprises main and end spar of honeycomb sandwich design and a carbon-reinforced full laminate shell. Structural configuration of tailplane as for wings. Tailplane attached to the fuselage by four fittings. Both elevators have top and bottom CFRP shells; structure as for ailerons. Airframe is protected from moisture and ultra-violet radiation by UP gel-coat. A white filler on the outer surface of the aircraft conducts electricity, being part of lightning protection system. White two-component polyurethane lacquer covers the filler and completes the finish.

LANDING GEAR: Tricycle type; retractable. Mainwheels 6.00-6; nosewheel 5.00-5. Nosewheel steerable ±10°.

POWER PLANT: One 194 kW (260 hp) Textron Lycoming AEIO-540-D4D5 flat-six, driving a Hartzell HC-C3YR-1RF/F7663R three-blade, constant-speed propeller. Fuel capacity 262 litres (69.2 US gallons; 57.6 Imp gallons), of which 256 litres (67.6 US gallons; 56.3 Imp gallons) are usable. Max oil capacity 11.4 litres (3.0 US gallons; 2.5 Imp gallons).

SYSTEMS: Grob 120A-5700 landing gear hydraulic system, pressure 103 bar (1,500 lb/sq in); Grob ECS 115TA-7301 air conditioning system. Electrical system 28 V DC with 80 A alternator and Concorde RG-24-20 24 V 19 Ah battery. External power socket, port side, to rear of cockpit.

AVIONICS: *Comms:* Bendix/King KMA 24H-70 audio control/intercom, KY 196A VHF, KT 76C transponder.

Flight: Bendix/King KX 165A nav/com/glideslope, KN 62 DME, KLN. 94 GPS, KCS 55A compass, KR 87 ADF with KI 227 indicator.

DIMENSIONS, EXTERNAL:

Wing span	10.19 m (33 ft 5¼ in)
Length overall	8.065 m (26 ft 5½ in)
Height overall	2.57 m (8 ft 5¼ in)
Tailplane span	3.80 m (12 ft 5½ in)
Wheel track	2.42 m (7 ft 11¼ in)
Wheelbase	1.87 m (6 ft 1½ in)
Propeller diameter	1.98 m (6 ft 6 in)

AREAS:

Wings, gross	13.29 m² (143.1 sq ft)
Ailerons (total)	0.915 m² (9.85 sq ft)
Trailing-edge flaps (total)	1.61 m² (17.33 sq ft)
Fin	1.71 m² (18.41 sq ft)
Rudder	0.77 m² (8.33 sq ft)
Tailplane	3.04 m² (32.72 sq ft)
Elevator	0.70 m² (7.53 sq ft)

Instrument panel of Grob G 120A-I *(Paul Jackson)* ***NEW***/0573431

WEIGHTS AND LOADINGS:

Weight empty	960 kg (2,116 lb)
Baggage capacity	50 kg (110 lb)
Max T-O and landing weight	1,440 kg (3,174 lb)
Max zero-fuel weight	1,315 kg (2,899 lb)
Max wing loading	108.3 kg/m²(22.18 lb/sq ft)
Max power loading	7.42 kg/kW (12.21 lb/hp)

PERFORMANCE:

Never-exceed speed (VNE)	235 kt (435 km/h; 270 mph)
Max level speed	172 kt (319 km/h; 198 mph)
Manoeuvre speed	165 kt (306 km/h; 190 mph)
Cruising speed at 75% power at FL50	166 kt (307 km/h; 191 mph)
Stalling speed	55 kt (102 km/h; 64 mph)
Service ceiling, all	5,490 m (18,000 ft)
Max rate of climb at S/L	390 m (1,280 ft)/min
T-O to 15 m (50 ft)	656 m (2,150 ft)
Landing from 15 m (50 ft)	563 m (1,845 ft)
Range at 45% power, 30 min reserves:	
G 120A at FL80	830 n miles (1,537 km; 955 miles)
Observer at FL60	1,360 n miles (2,518 km; 1,565 miles)
Endurance at 45% power at FL20, 30 min reserves:	
Observer	11 h 40 min
g limits: G 120A Aerobatic	+6/–4
G 120A Utility, Observer	+4.4/–1.76

UPDATED

GROB G 140TP

TYPE: Business turboprop.

PROGRAMME: Development started early 2001. Prototype (D-ETPG), then unflown and previously unannounced, exhibited at Paris Air Show from 16 to 24 June 2001.

Design freeze completed in 2002 following incorporation of changes based on feedback from potential customers. First flight 12 December 2002, with EASA certification expected in late 2004.

CUSTOMERS: Potential market for 20 to 40 per year, for military and civilian markets. Interest reported from several air forces by early 2002.

COSTS: Unpressurised version US$1 million (2002).

DESIGN FEATURES: Generally similar to G 115/120 series. By 2004, a ventral strake had been added to the rear fuselage.

FLYING CONTROLS: Conventional and manual. Trim tab on port elevator and each aileron; horn-balanced rudder. Slotted flaps. Dual controls (sticks) standard.

STRUCTURE: Entirely CFRP. Structural service life (aerobatic) 15,000 flight hours.

LANDING GEAR: Retractable tricycle type with trailing-link suspension on main units. Main legs retract inwards into wing root area, nosewheel rearwards.

POWER PLANT: One 336 kW (450 shp) Rolls-Royce 250-B17F turboprop driving an MT-Propeller MTV-5 five-blade, constant-speed (hydraulic) propeller.

ACCOMMODATION: Four persons in individual seats. Separate upward-opening gullwing-type door for each seat, with circular 'porthole' window at rear of cabin on each side. Baggage

Grob G 140TP prototype *(Paul Jackson)* ***NEW***/0573430

Grob G 140TP in May 2004, following addition of a rear fuselage strake *(Paul Jackson)* ***NEW***/0580547

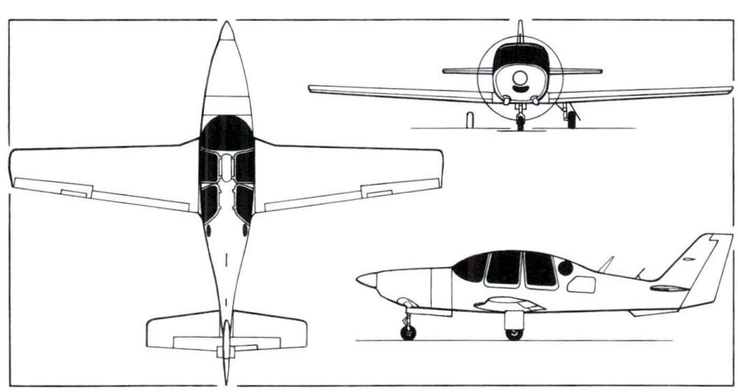

Grob G 140TP business turboprop *(James Goulding)* 0580557

Cabin of the Grob G 140TP *(Paul Jackson)* ***NEW***/0554431

door at rear of cabin on port side. Air-conditioning standard, pressurisation will be offered as an option.

AVIONICS: Bendix/King EFS 40 two-screen EFIS as core system, with Litef Lasergyro AHRS and provision for TCAS, weather radar and Stormscope.

DIMENSIONS, EXTERNAL:
- Wing span ... 10.30 m (33 ft 9½ in)
- Wing aspect ratio ... 8.0
- Length overall ... 8.90 m (29 ft 2½ in)
- Height overall ... 2.80 m (9 ft 2¼ in)

AREAS:
- Wings, gross ... 13.30 m^2 (143.2 sq ft)

WEIGHTS AND LOADINGS (A: Aerobatic; B: Utility):
- Weight empty ... 1,100 kg (2,425 lb)
- Max T-O and landing weight: A ... 1,500 kg (3,307 lb)
- B ... 1,800 kg (3,968 lb)
- Max wing loading ... 135.3 kg/m^2 (27.72 lb/sq ft)
- Max power loading ... 5.37 kg/kW (8.82 lb/shp)

PERFORMANCE (estimated):
- Cruising speed at FL100:
  - max ... 220 kt (407 km/h; 253 mph)
  - long range ... 170 kt (315 km/h; 196 mph)
- Stalling speed: A ... 55 kt (102 km/h; 64 mph)
- B ... 61 kt (113 km/h; 70 mph)
- Max rate of climb at S/L: A ... 853 m (2,800 ft)/min
- B ... 652 m (2,140 ft)/min
- T-O to 15 m (50 ft): A ... 374 m (1,225 ft)
- B ... 570 m (1,870 ft)
- Landing from 15 m (50 ft): A ... 573 m (1,880 ft)
- B ... 670 m (2,198 ft)
- Range at FL100, 30 min reserves:
  - with max payload ... 1,050 n miles (1,944 km; 1,208 miles)
  - with max fuel ... 1,150 n miles (2,129 km; 1,323 miles)
- *g* limits: U ... +4.4/−1.76
- A ... +6/−4

UPDATED

GROB G 160 RANGER

TYPE: Business turboprop.

PROGRAMME: Announced 5 May 2003; public debut of prototype (c/n 87000/D-FTGB), then unflown, at Paris Air Show 15 June 2003; first flight 29 March 2004; second prototype expected to fly in fourth quarter 2004; EASA certification anticipated in second quarter of 2005, followed by FAA approval in third quarter.

COSTS: Approximately US$2.8 million (2004).

DESIGN FEATURES: Low-wing monoplane; twin strakes under rear fuselage; sweptback fin; non-swept tailplane.

FLYING CONTROLS: Conventional and manual. Fowler flaps. Tabs in both ailerons; horn-balanced elevators with tab each side; horn-balanced rudder; inset horn on ailerons. Dual controls (yokes) standard.

STRUCTURE: Carbon fibre composites throughout.

LANDING GEAR: Retractable tricycle type with single wheel on each unit. Nosewheel retracts rearwards, main units retract inwards. Mainwheels 9.5×6.75–8 with hydraulic disc brakes; nosewheel 6.00–6, steerable.

POWER PLANT: One 634 kW (850 shp) Pratt & Whitney Canada PT6A-42A turboprop, flat rated at 41°C, driving a four-blade Hartzell D9383K fully reversible propeller. Usable fuel capacity 1,655 litres (437 US gallons; 364 Imp gallons).

Prototype Grob G 160 Ranger displayed at Berlin in May 2004 *(Paul Jackson)* ***NEW***/0580544

Proposed (2006) Grob G 160 instrument panel ***NEW***/0573464

ACCOMMODATION: Pilot and co-pilot or passenger in cockpit; four passengers in main cabin in club arrangement, with baggage area and lavatory in aft cabin separated by curtain; alternative layouts provide for one additional seat in rear cabin with reduced baggage space, or two additional seats without lavatory. Cabin door on port side aft of wing split horizontally; crew emergency exit, forward, port; cabin emergency exit, starboard, above wing. Four cabin windows per side. Fold-out tables, individual reading lights and 110V/220V power supply for laptop computers standard. Refreshment centre optional. Accommodation is pressurised and air conditioned.

SYSTEMS: Honeywell cabin pressure control system (CPCS) standard.

AVIONICS: Initial deliveries will be equipped with Bendix/King EFS 40 EFIS. From 2006 standard avionics fit will be Honeywell APEX integrated 'glass cockpit' comprising three-tube EFIS with 15 × 20 cm (6 × 8 in) displays for PFD, MFD and EICAS functions; Mode S transponder; and EGPWS. Options will include TCAS, radio altimeter, DME, ADF, Honeywell flight information system.

DIMENSIONS, EXTERNAL:
- Wing span ... 13.00 m (42 ft 7¾ in)
- Wing aspect ratio ... 8.2
- Length overall ... 11.50 m (37 ft 8¾ in)
- Height overall ... 3.40 m (11 ft 1¾ in)
- Tailplane span ... 5.00 m (16 ft 4¾ in)

DIMENSIONS, INTERNAL:
- Cabin: Length ... 4.96 m (16 ft 3¼ in)
  - Max width ... 1.58 m (5 ft 2¼ in)
  - Max height ... 1.42 m (4 ft 8 in)

AREAS:
- Wings, gross ... 20.50 m^2 (220.7 sq ft)

WEIGHTS AND LOADINGS:
- Max T-O and landing weight ... 3,300 kg (7,275 lb)
- Max payload ... 720 kg(1,587 lb)
- Max wing loading ... 161.0 kg/m^2 (32.97 lb/sq ft)
- Max power loading ... 5.21 kg/kW (8.55 lb/shp)

PERFORMANCE:
- Max cruising speed at FL250 ... 270 kt (500 km/h; 311 mph)
- Stalling speed ... 61 kt (113 km/h; 71 mph)
- Service ceiling ... 7,620 kt (25,000 ft)
- Max rate of climb at S/L ... 719 m (2,360 ft)/min
- T-O to 15 m (50 ft) ... 650 m (2,133 ft)
- Landing from 15 m (50 ft) without propeller reversal ... 650 m (2,133 ft)
- Range, five passengers, cruising at FL250,
  - 30 min reserves ... 1,800 n miles (3,333 km; 2,071 miles)
- Range with max fuel, cruising at FL250,
  - 30 min reserves ... 2,200 n miles (4,074 km; 2,531 miles)

UPDATED

HIGH PERFORMANCE

HIGH PERFORMANCE AIRCRAFT

Am Flugplatz 2, D-17419 Zirchow
Tel: (+49 38376) 29 90
Fax: (+49 38376) 299 99
e-mail: info@hp-aircraft.de
Web: http://www.hp-aircraft.de
CEO: Reiner Nochowitz
HEAD OF DESIGN: Christian Majunke

Company founded by the late Heiko Teegen on 10 April 2002 as joint venture by professional aviation magazine *Pilot und Flugzeug* and local government of Mecklenburg-Vorpommern. Intention is to produce TT62 Alekto light twin-prop, for which full funding was in place by time of public launch at Berlin Air Show, 6 to 12 May 2002. Secondary purpose is restoration of Rostock-Laage airfield as centre of aviation development with initial creation of 41 jobs at High Performance plant.

UPDATED

HIGH PERFORMANCE TT62 ALEKTO

TYPE: Business twin-prop.

PROGRAMME: Promotion begun at ILA Berlin, 6 May 2002. Fifth-scale model flown 23 November 2002. Fuselage mockup at Aero, Friedrichshafen, April 2003; complete, but unflown, prototype (D-IXTT) at ILA, Berlin May 2004. Taxy trials began 29 December 2004, with first flight expected in early January 2005.

CURRENT VERSIONS: Initial version *as described*. Projected developments include an air ambulance and aerial surveillance variant.

COSTS: €490,000 plus tax (2004).

Unflown TT62 Alekto prototype exhibited at Berlin in May 2004 *(Paul Jackson)* ***NEW***/0580558

High Performance TT62 Alekto port engine and propeller *(Paul Jackson)* ***NEW***/0580559

HP TT62 Alekto cooling radiators in rear of pylon *(Paul Jackson)* ***NEW***/0580560

DESIGN FEATURES: Design aims include speed and economy, achieved in part by internal mounting of engines for reduced frontal area and use of diesel power. Consumption 79 litres (20.9 US gallons; 17.4 Imp gallons)/h at normal cruise; 58 litres (15.3 US gallons; 12.8 Imp gallons)/h at econ cruise. Specific range, respectively, 3.84 n miles/kg and 4.80 n miles/kg. Mid-mounted wing. T tail. Two internal turbo-diesel engines connected by tubular drive-shafts to propellers mounted on upward-canted pylons at fuselage shoulders behind wing trailing-edge.

Wing section NASA LS(1)-0413 (alias GA(W)-2); dihedral 5° incidence 1°, sweepback at leading-edge 12°. Tailplane LWK 80-120/K25, no dihedral. Fin AH 85-L-120 at root; AH 85-L-250 at tip.

FLYING CONTROLS: Conventional and manual with sidesticks. Ailerons and elevator all 30 per cent chord with aerodynamic and static balance; rudder 25 per cent chord, similar; 20 per cent chord plain flaps. Large trim tab in each aileron and in elevator; upper quarter of rudder is separate trimming rudder. Cable actuated ailerons, elevator and rudder; electrically actuated flaps; electric trim.

STRUCTURE: Carbon fibre sandwich with epoxy resin matrix throughout. Wing has two spars and four main ribs; tailplane similar; fuselage has two bulkheads and one frame; two-spar tailplane; single-spar fin.

LANDING GEAR: Retractable tricycle type, electrohydraulically actuated. Can accommodate sink rates up to 3 m; 10 ft/s. Single wheel on each trailing-link main unit; single nosewheel. Main units of aluminium 7075 and 1.4548, retracting inward into wings; nosewheel retracts forward. Mainwheels 6.50–8, nosewheel 5.0–4 (6.00–6 on prototype). Parker Cleveland brakes.

POWER PLANT: Two 228 kW (306 hp) TAE Centurion 4.0 V-8 turbocharged, FADEC-controlled diesel engines, each driving an MT-Propeller MTV-25-1-DCF/CF175-512 five-blade constant-speed (2,300 rpm), fully feathering

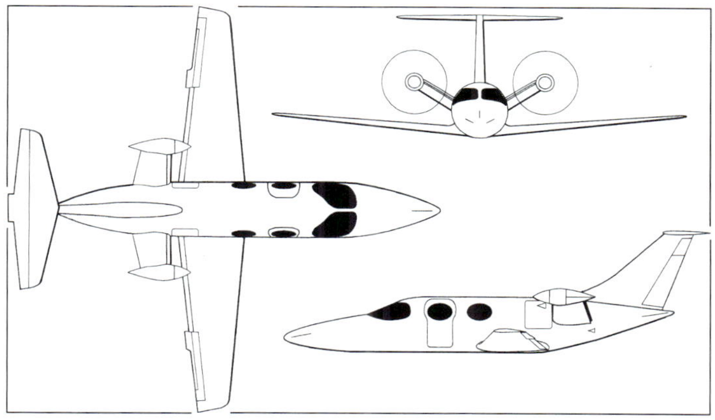

High Performance TT62 business twin-prop *(Paul Jackson)* ***NEW***/0580556

Mockup of TT62 fuselage shown at Friedrichshafen in April 2003 *(Paul Jackson)* ***NEW***/0561673

propeller. Cooling radiators in rear of propeller pylons. Large engine access panel each side. Fuel in wings, with gravity filler each side; total capacity 617 litres (163 US gallons; 136 Imp gallons).

ACCOMMODATION: Pilot and four passengers in two-one-two configuration (centre seat rearward-facing) in pressurised cabin with baggage space behind seats; differential 0.37 bar (5.5 lb/sq in); door forward of wing on port side; emergency exit for crew on lower starboard side; two cabin windows each side.

SYSTEMS: Hydraulic and electrical systems; liquid de-icing system standard. Optional air conditioning. Electrical system comprises two buses, each with separate battery. Liquid de-icing for wing, empennage and windscreen; electric propeller de-icing.

AVIONICS: Garmin G1000 suite, with two-screen EFIS. IFR standard.

DIMENSIONS, EXTERNAL:
Wing span ... 11.24 m (36 ft 10½ in)
Wing chord: at root ... 1.75 m (5 ft 9 in)
at tip ... 0.80 m (2 ft 7½ in)
Wing aspect ratio ... 9.0
Length overall ... 11.685 m (38 ft 4 in)
Fuselage: Length ... 10.665 m (35 ft 0 in)
Max diameter ... 1.55 m (5 ft 1 in)
Height without landing gear ... 3.435 m (11 ft 3¼ in)
Tailplane span ... 4.40 m (14 ft 5¼ in)
Propeller diameter ... 1.75 m (5 ft 9 in)
Cabin door: Max width (upper) ... 0.90 m (2 ft 11½ in)
Min width (lower) ... 0.68 m (2 ft 2¾ in)

DIMENSIONS, INTERNAL:
Cabin (between pressure bulkheads):
Length ... 4.50 m (14 ft 9¼ in)
Max width (diameter) ... 1.40 m (4 ft 7¼ in)
Baggage compartment:
Length ... 0.70 m (2 ft 3½ in)
Width ... 1.10 m (3 ft 7¼ in)

AREAS:
Wings, gross ... 15.46 m^2 (166.4 sq ft)
Ailerons (total) ... 0.67 m^2 (7.21 sq ft)
Flaps (total) ... 2.16 m^2 (23.25 sq ft)
Fin ... 2.40 m^2 (25.83 sq ft)
Rudder ... 0.60 m^2 (6.46 sq ft)
Tailplane ... 2.66 m^2 (28.63 sq ft)
Elevators (total) ... 1.14 m^2 (12.27 sq ft)

WEIGHTS AND LOADINGS:
Weight empty ... 1,780 kg (3,924 lb)
Baggage capacity ... 75 kg (165 lb)
Payload with max fuel ... 270 kg (595 lb)
Max fuel weight ... 500 kg (1,102 lb)
Max T-O weight ... 2,550 kg (5,621 lb)
Max wing loading ... 164.9 kg/m^2 (33.78 lb/sq ft)
Max power loading ... 5.59 kg/kW (9.19 lb/hp)

PERFORMANCE (estimated):
Max operating speed ... 250 kt (463 km/h; 288 mph) IAS
Max cruising speed at FL200 ... 240 kt (444 km/h; 276 mph) TAS
Normal cruising speed at FL200 ... 230 kt (426 km/h; 265 mph) TAS
Long-range cruising speed ... 190 kt (352 km/h; 219 mph) TAS
Stalling speed: flaps up ... 80 kt (149 km/h; 92 mph) IAS
flaps down ... 74 kt (137 km/h; 86 mph) IAS
Service ceiling ... 7,620 m (25,000 ft)
Service ceiling, OEI ... 4,420 m (14,500 ft)
Max rate of climb at S/L ... 549 m (1,800 ft)/min
Max rate of climb at S/l, OEI ... 152 m (500 ft)/min

Time to FL200 .... 15 min
T-O to 15 m (50 ft) .... 670 m (2,198 ft)
Landing from 15 m (50 ft) .... 660 m (2,165 ft)
Range, 45 min reserves:
with max payload:
normal cruising speed .... 890 n miles (1,648 km; 1,024 miles)
long-range cruising speed .... 1,290 n miles (2,389 km; 1,484 miles)
with max fuel:
normal cruising speed .... 1,440 n miles (2,666 km; 1,657 miles)
long-range cruising speed .... 1,930 n miles (3,574 km; 2,221 miles)

UPDATED

IKARUS

IKARUS DEUTSCHLAND (COMCO IKARUS GERÄTEBAU GMBH)

Am Flugplatz 11, Flugplatz Mengen, D-88367 Höhentengen
Tel: (+49 7572) 600 80
Fax: (+49 7572) 33 09
e-mail: post@comco-ikarus.de
Web: http://www.comco-ikarus.de
CEO: Ing Edward Grilz

Formed in 1976 to manufacture hang gliders, Ikarus undertakes sales and distribution of the C42 and Aerostyle Breezer ultralights in Europe as well as marketing completed examples of the latter. Ikarus also produces the C22 open cockpit ultralight and several hang glider designs. Over 1,500 Ikarus aircraft and 2,000 hang gliders have been sold.

UPDATED

IKARUS C42

TYPE: Side-by-side ultralight/kitbuilt.

PROGRAMME: Designed by Hans Gygax, and simultaneously launched in Europe (Friedrichshafen Air Show) and North America (Sun 'n' Fun) in second quarter of 1997. European deliveries began immediately. On 10 May 1998 an Ikarus C42 set a new ultralight world speed record over a closed circuit of 54 n miles (100 km; 62 miles) at 90.4 kt (167.5 km/h; 104.1 mph), and on 13 May 1998 established another world record over a 270 n mile (500 km; 311 mile) closed circuit at 71.78 kt (132.93 km/h; 82.60 mph). Upgraded version, introduced April 1999, offers optional winglets of GFRP and further option of GFRP ailerons, flaps, elevators and rudder.

CURRENT VERSIONS: **C42:** Production version until 2003.

C42 B: Introduced 2003. Compared to earlier version, has reprofiled nose with air intake below propeller displaced rearwards. *As described.*

Ikarus C42 B side-by-side seat ultralight *(Paul Jackson)* ***NEW***/0576606

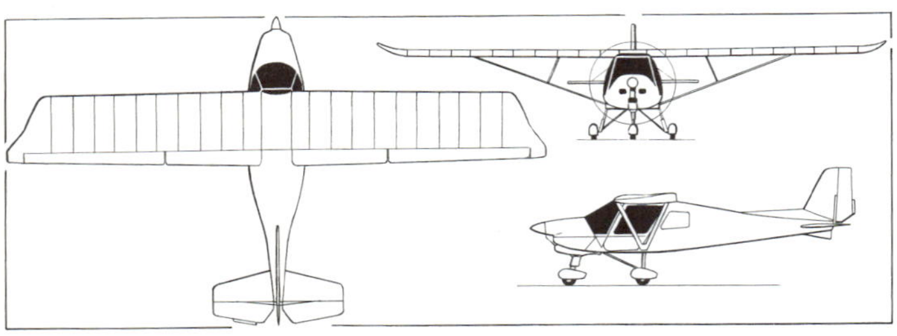

Ikarus C42 B current production version *(James Goulding)* ***NEW***/0583332

Floatplane version of Ikarus C42 B *(Paul Jackson)* ***NEW***/0558588

C42 Competition I/II: Kitbuilt versions with additional refinements.

CUSTOMERS: First production C42 to Netherlands owner (PH-2Y5) mid-1997. Others sold in Austria, Estonia, Finland, Germany, Norway and Sweden. United Kingdom PFA type clearance, and first deliveries, early 2002. By April 2004, over 650 had been sold and production was set at 80 per year.

COSTS: Ready-to-fly €44,950 plus tax (2004).

DESIGN FEATURES: Strut-braced high-wing monoplane; strut-braced tail unit. Wings fold for storage.

NACA 2412 wing section.

FLYING CONTROLS: Conventional and manual. Ailerons and elevators pushrod-operated, rudder cable-operated. Plain flaps standard; electrically actuated flaps optional; flight adjustable trim tab on port elevator.

STRUCTURE: Glass fibre fuselage assembled around full-length aluminium boom; tubular aluminium wings and tail surfaces, covered with Mylar/polyester laminate (GT-foil). Aluminium wings optional. Glass fibre-reinforced epoxy resin fairings; stainless steel fittings; Makrolon transparencies. Optional carbon fibre winglets.

LANDING GEAR: Tricycle type; fixed. Oleo-pneumatic suspension and speed fairings on all three units; hydraulic brakes on main units. Full Lotus inflatable floats optional.

POWER PLANT: One 59.6 kW (79.9 hp) Rotax 912 UL flat-four driving a Warp Drive two-blade, carbon fibre propeller via 1:2.27 reduction gearing or 73.5 kW (98.6 hp) Rotax 912 ULS driving a GSC three-blade, ground-adjustable pitch propeller via 1:2.43 reduction gearing. Standard fuel capacity 50 litres (13.2 US gallons; 11.0 Imp gallons). Further 50 litres (13.2 US gallons; 11.0 Imp gallons) optional.

SYSTEMS: 12 V battery.

EQUIPMENT: Optional BRS 50L4 or Magnum Speed S ballistic parachute, and ELT. Spraying equipment can be fitted for agricultural operations, and a towing hook for glider- and banner-towing purposes.

DIMENSIONS, EXTERNAL:
Wing span: standard .... 9.45 m (31 ft 0 in)
winglet option .... 9.50 m (31 ft 2 in)
Length overall .... 6.38 m (20 ft 11¼ in)
Height overall .... 2.24 m (7 ft 4¼ in)

DIMENSIONS, INTERNAL:
Cabin max width .... 1.22 m (4 ft 0 in)

AREAS:
Wings, gross .... 12.50 m^2 (134.5 sq ft)

WEIGHTS AND LOADINGS:
Weight empty: Standard .... 268 kg (591 lb)
Competition .... 275 kg (606 lb)
Max T-O weight, all .... 472.5 kg (1,042 lb)

PERFORMANCE (Rotax 912 UL):
Max level speed .... 100 kt (185 km/h; 115 mph)
Cruising speed at 75% power .... 94 kt (175 km/h; 109 mph)
Stalling speed .... 35 kt (65 km/h; 41 mph)
Max rate of climb .... 300 m (984 ft)/min
T-O run .... 90 to 100 m (295 to 330 ft)
Range with standard fuel .... 351 n miles (650 km; 403 miles)

UPDATED

IMPULSE

IMPULSE AIRCRAFT GMBH

Otto-Lilienthalstrasse 1, D-06188 Oppin/Halle
Tel: (+49 346) 04 25 10
Fax: (+49 346) 042 51 22
e-mail: info@impulse-aircraft.de
Web: http://www.impulse-aircraft.de

Impulse was established to produce the lightplane of the same name. This exists in Ultralight and Experimental category versions.

UPDATED

IMPULSE IMPULSE

TYPE: Two-seat lightplane; side-by-side ultralight.

PROGRAMME: Designed by Steinbach. First flight 28 July 2000. Tailwheel version first shown at Berlin in May 2002, having achieved certification on 25 April 2002. Belgian certification received 21 January 2004.

CURRENT VERSIONS: **100:** Restricted weight and power to meet 450 kg (992 lb) limit.

100TD: Tailwheel ('tail-dragger') version of 100.

130: Uprated version within US Experimental category limit. Prototype D-EKRK.

130TD: Tailwheel version introduced in 2002.

180: Higher-powered version; under development by 2002; to be certified under VLA requirements. Originally known as 360.

180TD: Tailwheel version of 180; under development. D-EEZZ shown unflown at Aero '03, Friedrichshafen in April 2003.

CUSTOMERS: Total 36 built and further six on order by May 2004, including nine in New Zealand.

COSTS: Kit: 100 €44,500 or 100TD €43,600 less engine, interior and optional factory assistance (€3,800 for three days; €14,900 for 35 days).

Flyaway: 100 Jabiru 2200 €65,900; 100 Rotax 912S €75,800; 100TD Jabiru 2200 €64,900; 100TD Jabiru 3300 €71,900 (2004), all plus tax (2004).

DESIGN FEATURES: Streamlined, low-wing configuration, with fuselage sharply waisted to rear of cockpit; laminar flow wing; sweptback fin with mid-mounted tailplane. Wide speed range. Available in Experimental and Ultralight weight categories. Quoted build time 450 hours; builder assistance programme available. Wings can be removed for storage and transportation.

FLYING CONTROLS: Conventional and manual. Horn-balanced rudder; flight-adjustable trim tab in port elevator. Actuation by pushrods (elevator and ailerons) and cables (rudder). Electrically operated three-quarter-span Fowler flaps.

STRUCTURE: Fuselage of carbon-honeycomb sandwich with 1.7734.4 chromoloy engine mount. Wings of carbon-honeycomb sandwich with single main spar passing through fuselage under occupants' knees. Tail unit also of sandwich with E-glass fin, into which radio antennas are embedded. Aluminium main landing gear legs.

LANDING GEAR: Fixed. Tricycle or tailwheeel (TD suffix) type with 5×5.00 mainwheels; tailwheel has solid 100 × 50 mm (4 × 2 in) tyre. Fuselage-mounted spring cantilever mainwheel legs with optional speed fairings. Foam cushion shock dampeners. Hydraulic brakes.

POWER PLANT: *100:* One piston engine: 59.7 kW (80 hp) Jabiru 2200 with two-blade, fixed-pitch propeller; or 73.5 kW (98.6 hp) Rotax 912 S flat-four, driving an MT 167R152-2M wooden three-blade, variable-pitch propeller; or 89 kW (120 hp) Jabiru 3300 flat-six with MT-Propeller two-blade, constant-speed propeller. Fuel in wing tanks, capacity 110 litres (29.1 US gallons; 24.2 Imp gallons). Option to fit 89.5 kW (120 hp) Jabiru 3300.

130 and 130TD: Options of 95 kW (128 hp) Limbach L2400ET flat-four and 118 kW (158 hp) LOM M332 in-line four. Fuel capacity 160 litres (42.3 US gallons; 35.2 Imp gallons).

180 and 180TD: One 134 kW (180 hp) Textron Lycoming IO-360-M1A driving an MT-Propeller MTV-12-B three-blade, variable-pitch propeller. Fuel capacity 200 litres (52.8 US gallons; 44.0 Imp gallons) in two integral wingtanks.

Optional power plants include BMW motorcycle engine and Thielert TAE 125 turbo-diesel.

ACCOMMODATION: Two persons, side by side, beneath single-piece hood; baggage space behind seats.

SYSTEMS: 12 V starter and generator.

EQUIPMENT: BRS parachute system.

DIMENSIONS, EXTERNAL (UL: Ultralight, E: Experimental):

Wing span	8.74 m (28 ft 8 in)
Wing aspect ratio	7.8
Length overall: 100	6.30 m (20 ft 8 in)
130, 130TD	6.10 m (20 ft 0¼ in)
180, 180TD	6.20 m (20 ft 4 in)
Height overall: 180	1.70 m (5 ft 7 in)
180TD	1.49 m (4 ft 10¾ in)
Tailplane span	3.29 m (10 ft 9½ in)
Wheel track	2.10 m (6 ft 10¾ in)
Wheelbase: 180TD	5.40 m (17 ft 8½ in)

DIMENSIONS, INTERNAL:

Cockpit max width	1.18 m (3 ft 10½ in)

AREAS:

Wings, gross	9.82 m^2 (105.7 sq ft)

WEIGHTS AND LOADINGS (100 with Rotax, 130 with LOM engine):

Weight empty: 100	260 kg (573 lb)
130	320 kg (705 lb)
130TD	350 kg (772 lb)
180, 180TD	400 kg (882 lb)
Baggage capacity: 100	20 kg (44 lb)
130	50 kg (110 lb)
180	80 kg (176 lb)
Max T-O weight: 100	472.5 kg (1,042 lb)
130, 130TD, 180, 180TD	720 kg (1,587 lb)
Max wing loading: 130, 130TD, 180, 180TD	73.3 kg/m^2 (15.01 lb/sq ft)
Max power loading: 130, 180, 180TD	6.11 kg/kW (10.04 lb/hp)

PERFORMANCE (100, 130 as above):

Never-exceed speed (VNE): 100	145 kt (270 km/h; 167 mph)
130, 130TD	226 kt (420 km/h; 261 mph)

Impulse 130 nosewheel version *(Paul Jackson)* ***NEW***/0580801

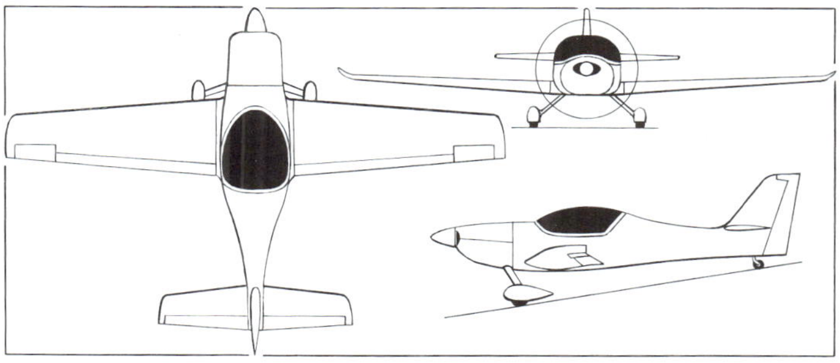

Impulse 100 side-by-side ultralight *(James Goulding)* ***NEW***/0558569

Max level speed: 180	200 kt (370 km/h; 230 mph)
Max cruising speed at 75% power: 100	130 kt (240 km/h; 149 mph)
130	162 kt (300 km/h; 186 mph)
130TD, 180	170 kt (315 km/h; 196 mph)
180TD	200 kt (370 km/h; 230 mph)
Stalling speed, flaps up: 100	35 kt (65 km/h; 41 mph)
130	41 kt (75 km/h; 47 mph)
130TD, 180, 180TD	44 kt (82 km/h; 51 mph)
Max rate of climb at S/L: 180	610 m (2,000 ft)/min
Service ceiling: all	5,490 m (18,000 ft)
T-O run: 100	60 m (197 ft)
130, 130TD	100 m (330 ft)
180	120 m (395 ft)
T-O to 15 m (50 ft): except 180	200 m (656 ft)
180	220 m (722 ft)
Landing from 15 m (50 ft): all except 180	210 m (690 ft)
180	260 m (855 ft)
Range: 100	647 n miles (1,200 km; 745 miles)
130, with auxiliary tank	3,509 n miles (6,500 km; 4,038 miles)
180	702 n miles (1,300 km; 807 miles)

UPDATED

KAISER

KAISER FLUGZEUGBAU GMBH

Flugplatz Schönhagen, D-14959 Schönhagen
Tel: (+49 3045) 02 28 53 or (+49 3329) 629 40
Fax: (+49 3329) 629 41
e-mail: service@kaiser-flugzeugbau.de
Web: http://www.Kaiser-Flugzeugbau.de
DIRECTOR: Dipl Ing Jörg Kaiser

Kaiser's first product, the Magic, had its debut as a static exhibit at the Berlin Air Show in May 1998. A second design, the Whisper, was later added to the company's portfolio. No confirmation has been received that either has flown, although the prototype Magic has undertaken taxying trials.

In early 2004, Kaiser announced its plans to acquire a Yakovlev Yak-54 aerobatic aircraft from Russia with a view to promoting and selling the aircraft in Western Europe, subject to satisfactory evaluation.

UPDATED

MECKLENBURGER

MECKLENBURGER ULTRALEICHT (FLUGZEUGBAU CHRISTIAN ENGELEN)

am Binning 5, D-17367 Eggesin
Tel: (+49 175) 811 62 13
Fax: (+49 3843) 33 48 89
e-mail: info@meck-ul.de
Web: http://www.meck-ul.de
CEO: Christian Engelen

Mecklenburger was established to develop ultralight versions of significant Second World War aircraft, principally fighters. A scale Messerschmitt 109 has been produced in composites, based on an original replica devised by Tassilo Bek.

UPDATED

MECKLENBURGER ME-109

TYPE: Single-seat ultralight kitbuilt.

PROGRAMME: Mecklenburger launched production of current version in late 2003, following construction of some eight of earlier (Bek) series, including c/n 007 with Rotax 912 UL flat-four of 59.6 kW (79.9 hp). By April 2004, the first production aircraft was 58 per cent complete, aircraft '9' (c/n 9) exhibited incomplete at Berlin, May 2004.

CURRENT VERSIONS: **Ultralight:** MTOW limited to 300 kg (661 lb).

Experimental: Short-span, increased weight version.

COSTS: *Ultralight:* Basic kit (minus engine and instruments) €23,780; flyaway €41,200 (with Rotax 582) or €42,350 (with Mikron IIIB) (all including tax, 2004).

Experimental: Kit €21,350 (all 2003).

DESIGN FEATURES: Approximately three-quarters scale replica of Messerschmitt Bf 109. Low, tapered wing; strut-braced tailplane. Quoted build time 600 hours.

FLYING CONTROLS: Conventional and manual. Horn-balanced rudder; plain ailerons. No flaps. Trim tab in starboard elevator.

STRUCTURE: Predominantly of glass and carbon fibre, with fabric-covered wooden control surfaces.

LANDING GEAR: Tailwheel type with manually retractable (outwards) mainwheels. Mainwheel doors of wood. Tailwheel 2.00–5; mainwheels 4.00–4 with cable brakes.

POWER PLANT: One piston engine. Options include 48.0 kW (64.4 hp) Rotax 582, 59.6 kW (79.9 hp) Rotax 912 and 55.0 kW (70 hp) Parma-Technik Mikron IIIB for Ultralight and 59.7 kW (80 hp) Subaru EA-71 for Experimental category. Fuel capacity 55 litres (14.5 US gallons; 12.1 Imp gallons).

DIMENSIONS, EXTERNAL (UL: Ultralight, EX: Experimental):

Wing span: UL	8.00 m (26 ft 3 in)
EX	6.70 m (21 ft 11¾ in)

Length overall: UL, EX ... 6.30 m (20 ft 8 in)
Height: propeller turning ... 2.10 m
to fin tip ... 1.17 m

AREAS:
Wings, gross: UL ... 10.10 m² (108.7 sq ft)

WEIGHTS AND LOADINGS:
Weight empty: UL ... 190 kg (441 lb)
EX ... 280 kg (617 lb)
Max T-O weight: UL ... 322 kg (661 lb)
EX ... 420 kg (926 lb)

PERFORMANCE (UL with Rotax 582, EX with Subaru):
Max level speed: UL ... 129 kt (220 km/h; 149 mph)
UL, Rotax ... 912 kt (240 km/h)
EX ... 145 kt (270 km/h; 167 mph)
Cruising speed at 65% power:
UL ... 97 kt (180 km/h; 112 mph)
EX ... 103 kt (190 km/h; 118 mph)
Stalling speed: UL ... 32 kt (59 km/h; 37 mph)
EX ... 36 kt (65 km/h; 41 mph)

UPDATED

Mecklenburger Me-109 kitbuilt version of Messerschmitt Bf 109 *(Paul Jackson)* ***NEW***/0583324

MK

MK HELICOPTER GMBH

Hans-Böcklerstrasse 1, D-65468 Trebur-Astheim
Tel: (+49 6147) 91 91 28
Fax: (+49 6147) 91 91 29
e-mail: MK-helicopter-Engineering@t-online.de
TECHNICAL MANAGER: Uwe Mathes

The MK II light helicopter has been developed into the MK III; work is continuing on the project.

UPDATED

MYLIUS

MYLIUS FLUGZEUGWERK GMBH & CO KG

Am Tower 10, D-54634 Bitburg
Tel: (+49 6561) 950 50
Fax: (+49 6561) 95 05 50
MANAGING DIRECTOR: Albert Mylius
FINANCIAL DIRECTOR: Sabine Palzer
PRODUCTION DIRECTOR: Ludwig Pflüger
SYSTEMS ENGINEER AND INFORMATION OFFICER: Jim Byrum

Mylius Flugzeugwerk was founded in 1996 by Albert Mylius, son of Hermann Mylius whose MHK-101 design was the proof-of-concept prototype for the MBB (Bölkow) BO 209 Monsun light aircraft. The company was developing a new range of light aircraft, based on Hermann Mylius' designs and similar in appearance to the Monsun, which employs modular, interchangeable structures for manufacturing efficiency, minimal spares stockage and low maintenance down-time. All versions were to be certified initially to JAA standards and later to FAR Pt 23, but development efforts most recently concentrated on the two-seat MY-103 at the expense of the four-seat MY-104.

No further developments were reported during 2003, and promotion appears to have been discontinued.

OMF

OSTMECHLENBURGISCHE FLUGZEUGBAU GMBH

Flughafenstrasse, D-17039 Trollenhagen
Tel: (+49 395) 42 56 00
Fax: (+49 395) 425 60 20
e-mail: info@omf-aircraft.com
Web: http://www.omf-aircraft.com
SALES DIRECTOR: Mario Böhme

CANADIAN COMPANY:
Symphony Aircraft Industries
3005 Lindberg
Trois-Rivières
Québec G9A 5E1
Tel: (+1 819) 377 39 79
PRESIDENT, OMF AIRCRAFT: Paul Costanzo
VICE-PRESIDENTS, OMF AIRCRAFT: Ian Boud Christian Ouimet
MARKETING DIRECTOR: Bill Sprague

OMF (then Ostmechlenburgische Flugzeugbau GmbH) was formed on 1 May 1998, having acquired aerodynamic technology for the GlaStar from Arlington Aircraft Developments of the USA in March 1998. Company's 1,500 m² (16,146 sq ft) factory on Trollenhagen airfield at Neubrandenburg was opened on 10 May 2000. In September 2001, OMF had 54 employees and by December 2003 this had reached 61; it had hoped to expand to 300 by 2004.

In April 2003, work started on a new C$25 million, 10,000 m² (107,650 sq ft) manufacturing facility for OMF Aircraft at Trois-Rivières, Québec, Canada; the first 2,500 m² (26,900 sq ft) was completed in August 2003.

OMF GmbH entered voluntary bankruptcy on 9 December 2003, but its Canadian branch, OMF Aircraft, 70 per cent owned by OMF GmbH, was initially unaffected, and expected to launch its own production in May 2004. However, production halted January 2004; its assets were put up for sale by its major creditor in April 2004.

In early 2004, OMF GmbH's assets were sold to German company SMW, while the management of OMF Aircraft was aiming to bid for the assets under the title Symphony Aircraft Industries. German company became OMF Flugzeugwerke GmbH in April 2004 and sold to Symphony Aircraft Industries.

UPDATED

Thielert Diesel of the Symphony 135, first flown on 3 June 2003 *(Thielert)* ***NEW***/0572341

OMF-100 SYMPHONY

TYPE: Two-seat lightplane; four-seat lightplane.

PROGRAMME: Redesign of GlaStar to JAR 23 certification standards began in May 1998. First prototype (D-ETCW, c/n P1) was essentially unchanged and first flew May 1999; second prototype D-EMVP (c/n 0001), with modifications to meet JAR 23, first flown 7 October 1999; third aircraft (D-ENVG) had not flown by June 2000, when it made type's public debut at Berlin Air Show. LBA evaluation began mid-July 2000 and certification and production certificate both awarded 29 August 2000; FAA certificate A43CE awarded 9 April 2001; Transport Canada certification awarded 30 June 2003. IFR certification issued June 2002. Customer deliveries began November 2001.

CURRENT VERSIONS: **Symphony 160:** Baseline two-seat version; *as described.*

Symphony 250: Four-seat version announced at Sun 'n' Fun, April 2003 (as Symphony 4); first flight then expected before end 2003, but delayed by bankruptcy. Data given are provisional. Has longer, tapered wing and longer cabin section with wider door.

OMF-100–160 Symphony operated in USA ***NEW***/0580694

OMF Symphony 135TD prototype *(Paul Jackson)* ***NEW***/0580787

Symphony 135TDI: First flown, as modification of D-EMVP, June 2003; powered by diesel engine.

CUSTOMERS: By December 2003 some 41 production aircraft delivered and 44 on order, of which 21 were Symphony 160 and remainder a mixture of 250 and 135TDI versions. Company had a target production rate of 300 aircraft per year by 2008 before bankruptcy intervened.

COSTS: Symphony 160 US$120,000 VFR equipped; US$140,000 IFR equipped; Symphony 250 under US$200,000 (estimated); Symphony 135TDI US$147,000 (all 2003).

DESIGN FEATURES: Intended to be an inexpensive trainer and lightplane. Completely redesigned landing gear and wing; retains only some 10 per cent commonality with GlaStar. Crashworthy cockpit and seats. Around 60 per cent of aircraft built by OMF; remainder sourced from abroad, including Czech Republic (aluminium flying surfaces), Poland, France, UK and USA. Service life of 18,000 hours. Braced, unswept high wing of constant chord; foldable for storage. Tailplane has curved, leading-edge root strakes; tall, sweptback fin. Centre-section of top wing thickened in late 2001, increasing maximum T-O weight.

FLYING CONTROLS: Conventional and manual. Mass-balanced Frise ailerons with three hinges each (compared to GlaStar's two), travel up 23°/down 17°; electrically operated, metal, three-position Fowler flaps (0, 20 and 40°). Horn-balanced, single-piece elevator with anti-balance tab, travel up 21°/down 20°; horn-balanced rudder, travels 21° in each direction. Two T-shape vortex generators added to wings in front of each aileron. Ailerons have end-fences: protruding above the surface at inboard end and below, outboard.

STRUCTURE: Fuselage of GFRP with 4130 steel tube frame strengthening in cockpit area. Metal wings, struts, flaps, ailerons, tailplane, elevator and rudder; composites wingtips, fin fillet, tailplane strakes and wheel fairings.

LANDING GEAR: Tricycle type; fixed. Fuselage-mounted spring steel cantilever mainwheel legs with wheel fairings. Castoring nosewheel with speed fairing. Cleveland hydraulic brakes. All wheels 5.00-5. Nosewheel turning circle 6.7 m (22 ft).

POWER PLANT: *Symphony 160:* One 119 kW (160 hp) Textron Lycoming O-320-D2A flat-four, driving a two-blade fixed-pitch MT-186R-140-3D P-244-3 propeller. Fuel in two main tanks, each 58.5 litres (15.45 US gallons; 12.9 Imp gallons), and one 6 litre (1.6 US gallon; 1.3 Imp gallon) feeder tank, total capacity 123 litres (32.5 US gallons; 27.1 Imp gallons) of which 110 litres (29.0 US gallons; 24.2 Imp gallons) usable. Oil capacity 7.6 litres (2.0 US gallons; 1.7 Imp gallons).

OMF Symphony 135TDI instrument panel *(Paul Jackson)* ***NEW***/0580785

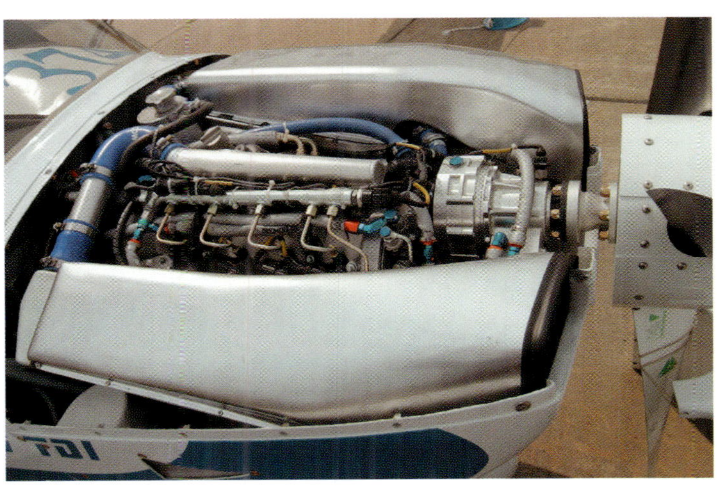

Thielert Centurion 1.7 installation in Symphony 135TDI *(Paul Jackson)* ***NEW***/0580784

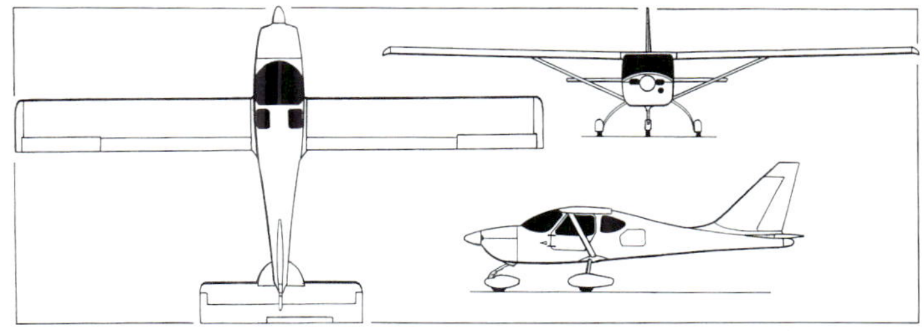

OMF-100-160 Symphony two-seat lightplane *(Paul Jackson)* 0093640

Symphony 250: One 186 kW (250 hp) Textron Lycoming IO-540-C flat-six. Fuel capacity 189 litres (50.0 US gallons; 41.6 Imp gallons).

Symphony 135TDI: One 99.3 kW (133 hp) Thielert Centurion 1.7 four-cylinder, four-stroke, turbocharged diesel engine, driving an MT-Propeller MTV-6-A/187-129 two-blade fixed-pitch propeller. Usable fuel capacity 114 litres (30.2 US gallons; 25.1 Imp gallons).

ACCOMMODATION: Two persons side by side on leather-covered glass fibre-reinforced composites material seats capable of withstanding 26 *g* forward and 19 *g* vertical crash; front-hinged door each side; baggage compartment behind crew with separate door on port side as alternative access.

SYSTEMS: Electric system 24 V 70 A DC. Vision Microsystems VM1000 engine monitoring system.

AVIONICS: Bendix/King KX 125 nav/com, KLX 135 GPS/com and KT 76 transponder as standard; Skyforce colour or black and white GPS optional.

EQUIPMENT: BRS parachute system being developed late 2003; will be standard on Symphony 250 and option for Symphony 135TDI.

DIMENSIONS, EXTERNAL:	
Wing span: 160	10.67 m (35 ft 0 in)
250	12.19 m (40 ft 0 in)
Wing chord, constant	1.12 m (3 ft 8 in)
Wing aspect ratio	9.5
Width, wings folded	2.43 m (7 ft 11¾ in)
Length overall	6.96 m (22 ft 10 in)
Max width of fuselage	1.22 m (4 ft 0 in)
Height overall	2.82 m (9 ft 3 in)
Tailplane span	3.08 m (10 ft 1¼ in)
Wheelbase	1.73 m (5 ft 8 in)
Propeller diameter	1.85 m (6 ft 1 in)
Propeller ground clearance	0.18 m (7 in)
Passenger door: Height	0.80 m (2 ft 7½ in)
Max width	0.81 m (2 ft 7¾ in)
Height to sill	0.84 m (2 ft 9 in)
Baggage door: Height	0.45 m (1 ft 5¾ in)
Max width	0.44 m (1 ft 5¼ in)
Height to sill	0.84 m (2 ft 9 in)
DIMENSIONS, INTERNAL:	
Cabin: Length	1.22 m (4 ft 0 in)
Max width: 160	1.10 m (3 ft 7¼ in)
250	1.17 m (3 ft 10 in)
Max height	1.14 m (3 ft 9 in)
Floor area	1.34 m^2 (14.4 sq ft)
Volume	1.53 m^3 (54 cu ft)
Baggage compartment: volume	0.06 m^3 (2.1 cu ft)
AREAS:	
Wings, gross	11.93 m^2 (128.4 sq ft)
Ailerons (total)	1.26 m^2 (13.56 sq ft)
Flaps (total)	1.56 m^2 (16.79 sq ft)
Tailplane	1.52 m^2 (16.36 sq ft)
Elevators (total, incl tabs)	1.29 m^2 (13.89 sq ft)
WEIGHTS AND LOADINGS:	
Weight empty: 160	658 kg (1,450 lb)
135TDI	680 kg (1,500 lb)
250	762 kg (1,680 lb)
Baggage capacity	75 kg (165 lb)
Max T-O weight: 160, 135TDI	975 kg (2,150 lb)
250	1,301 kg (2,870 lb)
Max wing loading: 160	81.8 kg/m^2 (16.74 lb/sq ft)
Max power loading: 160	8.18 kg/kW (13.44 lb/hp)
250	6.99 kg/kW (11.48 lb/hp)
135TDI	9.84 kg/kW (16.17 lb/hp)
PERFORMANCE:	
Never-exceed speed (VNE)	162 kt (300 km/h; 186 mph)
Max level speed	105 kt (194 km/h; 121 mph)
Max cruising speed at 75% power:	
160	131 kt (243 km/h; 151 mph)
135TDI	128 kt (237 km/h; 147 mph)
250	145 kt (269 km/h; 167 mph)
Econ cruising speed at: FL 80	130 kt (240 km/h; 149 mph)

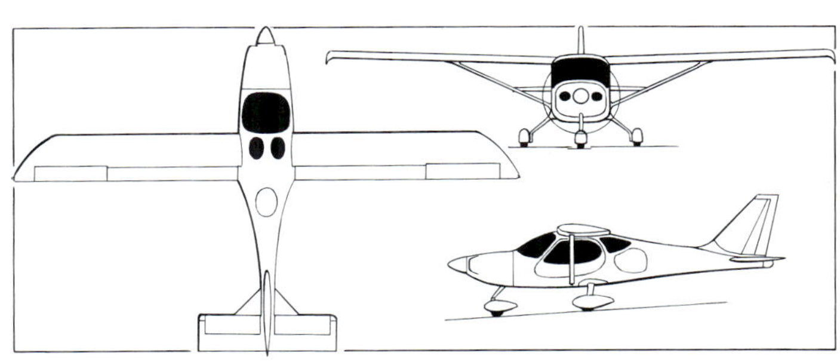

OMF Symphony 250 four-seat development of OMF-100-160 *(James Goulding)* ***NEW***/0561624

Stalling speed, power off:
 flaps up: 160, 135TDI ... 57 kt (106 km/h; 66 mph)
 250 ... 65 kt (121 km/h; 75 mph)
 flaps down: 160, 135TDI ... 48 kt (89 km/h; 56 mph)
 250 ... 55 kt (102 km/h; 64 mph)
Max rate of climb at S/L ... 259 m (850 ft)/min
Service ceiling ... 5,000 m (16,400 ft)
T-O run ... 280 m (919 ft)
T-O to 15 m (50 ft) ... 350 m (1,150 ft)
Landing from 15 m (50 ft): 160 ... 450 m (1,480 ft)
 135TDI ... 560 ft (1,840 ft)
Landing run ... 230 m (755 ft)
Range with max fuel:
 160 ... 522 n miles (966 km; 600 miles)
 250 ... 485 n miles (898 km; 558 miles)
 135TDI ... 700 n miles (1,296 km; 805 miles)
g limits: 160, 135 TDI ... +3.8/−1.5

UPDATED

PCH

PCH-FLUGZEUGBAU GMBH

e-mail: galogan@pch-flugzeugbau.de
Web (1): http://www.pch-flugzeugbau.de
Web (2): http://www.sam-aircrafts.de
PRESIDENT AND DESIGNER: Dipl Ing Calin Gologan

Originally known as PC-Flight, the company derived its name from the initials of its founders, designer Calin Gologan and test pilot Prof Dr Peter Maderitch. The first product, the Pretty Flight, was built in Romania from 1996 and assembled in Germany by Nitsche Flugzeugbau GmbH (now Aircraft Philipp GmbH).

PC-Flight ceased trading, but Dipl Ing Gologan then established PCH to continue marketing of the Pretty Flight.

VERIFIED

Pre-series PCH Pretty Flight ***NEW***/0567692

PCH PRETTY FLIGHT

TYPE: Side-by-side ultralight/kitbuilt.

PROGRAMME: Prototype (D-MNPF, c/n 01) first flew in November 1996; first flight of production aircraft (D-MOOI, c/n 001) 10 November 1997; German certification achieved 25 September 1998; at least three pre-series aircraft then completed; manufacture of first batch of 10 series production aircraft began in November 1998, but no further information received.

CURRENT VERSIONS: **Ultralight:** MTOW 450 kg (992 lb); *as described.*

Kit: Quoted build time 300 hours.

Experimental: Projected kit version in Experimental category with 290 kg (639 lb) empty weight and 560 kg (1,234 lb) MTOW.

STR: Stretched version with four-seat cabin. Under development.

COSTS: €52,000 with Rotax 912 UL and Smart; €54,000 with Rotax 912 ULS; €51,000 with Mikron, all flyaway (2004).

DESIGN FEATURES: Metal structure for high utilisation and reliability; available only in complete form. Designed using company's own CAD programme. High-wing braced monoplane with upturned wingtips, sweptback fin, and tailplane tip fences. Quick-build kit version also available, with quoted build time of 300 hours. Wing section GAW PC-1.

FLYING CONTROLS: Manual. Conventional horn-balanced rudder; 75 per cent span flaperons; and all-moving tailplane with anti-balance tabs.

STRUCTURE: Predominantly metal, with fabric-covered rear wing panels and control surfaces. Production in Romania at Star Tech Impex Srl, with final assembly by Nitsche Flugzeugbau GmbH (which see) in Germany.

LANDING GEAR: Tricycle type; fixed. Fuselage-mounted spring cantilever mainwheel legs; optional wheel fairings.

POWER PLANT: One 59.6 kW (79.9 hp) Rotax 912 UL or 73.5 kW (98.6 hp) Rotax 912 ULS flat-four driving a Fiti Speed three-blade fixed-pitch propeller. Alternatives are 55.0 kW (74 hp) Ecofly/Daimler Chrysler Smart three-cylinder and Parma-Technik Mikron IIIB four-cylinder of same power. Fuel in two tanks, combined capacity 100 litres (26.4 US gallons; 22.0 Imp gallons).

EQUIPMENT: BRS 1050 or USH 520 parachute recovery system.

DIMENSIONS, EXTERNAL:
 Wing span ... 10.00 m (32 ft 9¾ in)

PCH Flight Pretty Flight (Rotax 912 engine) *(James Goulding)* 0064484

 Length overall ... 6.25 m (20 ft 6 in)
 Height overall ... 2.57 m (8 ft 5¼ in)
DIMENSIONS, INTERNAL:
 Cabin max width ... 1.16 m (3 ft 9¾ in)
AREAS:
 Wings, gross ... 11.66 m^2 (125.5 sq ft)
WEIGHTS AND LOADINGS:
 Weight empty, including parachute ... 290 kg (639 lb)
 Max T-O weight ... 450 kg (992 lb)
PERFORMANCE:
 Max level speed ... 116 kt (215 km/h; 134 mph)
 Cruising speed at 75% power ... 100 kt (185 km/h; 115 mph)
 Stalling speed ... 34 kt (62 km/h; 39 mph)
 Max rate of climb at S/L ... 300 m (984 ft)/min
 T-O run ... 180 m (591 ft)
 Range ... 540 n miles (1,000 km; 621 miles)

UPDATED

REMOS

REMOS AIRCRAFT GMBH

Waldweg 1, D-85283 Eschelbach
Tel: (+49 8442) 96 77 77
Fax: (+49 8442) 96 77 96
e-mail: ac0312@remos.com
Web(1): http://www.remos.com
Web(2): http://www.rlsair.com
PRESIDENT: Lorenz Kreitmayr
HEAD OF DESIGN AND MARKETING: Hans Fuchs

PRODUCTION PLANT:
 Remos Sp zoo, Wytwornia Konstrukcji Lekkich, ulitsa Wapienicka 16, PL-43-300 Bielsko-Biała, Poland
 Tel: (+48 33) 818 30 80
 e-mail: remosbielsko@poczta.onet.pl

Remos company formed in 1983, and Remos Aircraft in 1990; currently markets the G-3 Mirage and announced the G-4 at Friedrichshafen in April 2001. Non-aviation activities of the parent company involve production of automotive components, including an enclosed road trailer for the Mirage.

In 2003, Remos appointed RLS Air of Bloomfield, Indiana as its US distributor.

UPDATED

REMOS G-3 MIRAGE

TYPE: Side-by-side ultralight/kitbuilt.

PROGRAMME: Prototype (D-MRAE) exhibited at Friedrichshafen Aero '97, April 1997. First flight 20 September 1997. First prize in 1998 German two-seat ultralight championships. Production version, shown in April 1999, featured rudder horn balance and minor changes to ailerons and elevator. **Mirage S**, certified 16 August 1999 with 74.6 kW (100 hp) Rotax 912 ULS, has 113 kt (210 km/h; 130 mph) cruising speed. In June 2000, D-MPCJ shown (then unflown) at ILA, Berlin, with 53.7 kW (72 hp) Swiss Auto SAB 430 turbocharged, two-cylinder, four-stroke motorcar engine, which installation weighs 30 kg (66 lb) less than Rotax. **Mirage RS**, with redesigned rudder and landing gear, plus addition of roof window and more extensive instrumentation, introduced at Aero '01, April 2001. At Aero '03, April 2003, Remos unveiled the Mirage RS/L (*Leicht*; light), which features more than 100 weight-saving modifications totalling 10 kg (22 lb), the main change being to a Jabiru 2200A engine. Prototype D-MPLK (c/n 137) flown shortly after exhibition and all subsequent aircraft to this standard.

In 2003, Remos announced **Mirage ARF** (almost-ready-to-fly) kit. **G-3/600** marketed from early 2004 to take advantage of LTF-2003 increased max T-O weight limit of 472.5 kg (1,042 lb) for German (and French) ultralights; prototype displayed at Sun 'n' Fun, April 2004, and had flown by end of that month; '600' indicates tested to 6 *g* load at 600 kg (1,322 lb).

G-3 RaLi: Version marketed by Jordan Aerospace Industries and announced at Dubai Air Show, 2004.

CUSTOMERS: Over 100 sold by 31 December 2001, when 73 on German ultralight register; had increased to 112 by May 2002; by April 2004, more than 130 were flying. Exports include Argentine police and undisclosed military agency in Romania; two to Netherlands in 1998–99 and one to US in 2002 for distribution, pending implementation of Sport Aircraft regulations.

COSTS: €69,950, excluding tax (2004). Upgrade to Rotax 912S extra €2,200.

DESIGN FEATURES: Meets FAR Pt 23. Intended for recreational and instructional use. Waisted rear fuselage, sweptback fin and braced wing with tapered outer panels. Wings fold for transportation and storage.

FLYING CONTROLS: Conventional and manual. Single-piece elevator with flight-adjustable inset tab. Horn-balanced rudder, ailerons and elevator. No rudder tab. Horn-balanced control surfaces on production version. Electrically operated flaps and trim.

STRUCTURE: Composites fuselage; wings of sandwich composites; leading edges and tips composites, remainder fabric-covered; all tail surfaces composites. Tail bumper. GFRP ailerons and flaps. Composites parts produced in Poland; metal parts in Germany.

Remos Mirage ultralight and optional kitplane *(Paul Jackson)* ***NEW***/0580789

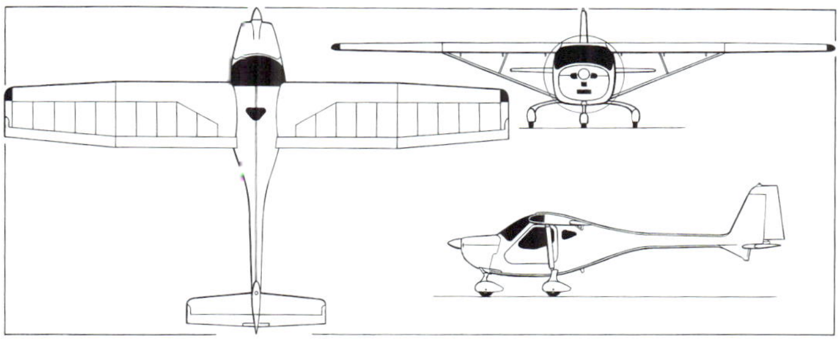

Remos G-3 Mirage *(James Goulding)* 0089515

LANDING GEAR: Tricycle type; fixed. Fuselage-mounted spring cantilever main gear with hydraulic drum brakes and integral speed fairings; rubber-in-compression nosewheel leg with speed fairing. Mainwheels 4.00–6; nosewheel 4.00–6.

POWER PLANT: Early models had one 59.6 kW (79.9 hp) Rotax 912 UL four-stroke engine, driving a GT-2/169.5 two-blade wooden propeller. Optionally, a certified 73.5 kW (98.6 hp) Rotax 912 ULS or /S2 can be fitted. Mirage RS/L has one 59.7 kW (80 hp) Jabiru 2200A flat-four driving a GT-2/166/VSR-FW101STRC two-blade, wood-and-composites, fixed-pitch propeller. Fuel capacity 70 litres (18.5 US gallons; 15.4 Imp gallons).

AVIONICS: To customer's specification.

EQUIPMENT: Junkers or BRS ballistic parachute.

DIMENSIONS, EXTERNAL:
Wing span ... 9.80 m (32 ft 1¾ in)
Length ... 6.47 m (21 ft 2¾ in)
Height overall ... 1.70 m (5 ft 7 in)

DIMENSIONS, INTERNAL:
Cockpit max width ... 1.17 m (3 ft 10 in)

AREAS:
Wings, gross ... 12.04 m² (129.6 sq ft)

WEIGHTS AND LOADINGS:
Weight empty ... 284 kg (626 lb)
Max T-O weight: BFU-95 ... 450 kg (992 lb)
LTF-2003 ... 472.5 kg (1,041 lb)
US LSA ... 558 kg (1,232 lb)

PERFORMANCE, POWERED (A: Rotax 912UL, B: Rotax 912 ULS):
Never-exceed speed (VNE) ... 121 kt (225 km/h; 139 mph)
Max level speed: ... 105 kt (195 km/h; 121 mph)
Normal cruising speed: A ... 105 kt (195 km/h; 121 mph)
B ... 111 kt (205 km/h; 127 mph)
Stalling speed ... 34 kt (63 km/h; 40 mph)
Max rate of climb at S/L: A ... 390 m (1,280 ft)/min
B ... 348 m (1,141 ft)/min
T-O run ... 60 m (197 ft)
T-O to 15 m (50 ft): A ... 150 m (495 ft)
B ... 180 m (590 ft)
Landing from 15 m (50 ft): both ... 200 m (660 ft)
Range ... more than 486 n miles (900 km; 559 miles)
Endurance ... 6 h
g limits ... +4/–2

PERFORMANCE, UNPOWERED:
Best glide ratio ... 17
Min rate of sink ... 1.80 m (5.91 ft)/s

UPDATED

REMOS G-4 VELOCITY

TYPE: Side-by-side ultralight.

PROGRAMME: Remos revealed artist's impression of its new G-4 high-performance, mid-wing ultralight at Aero '01, Friedrichshafen, in April 2001. The tricycle-gear aircraft, which is generally based on the existing G-3 Mirage, but with repositioned wings and tailplane, will have a 9.00 m (29 ft 6¼ in) wing span, be 6.50 m (21 ft 4 in) long and have an empty weight of 290 kg (639 lb). Fitted with a 73.5 kW (98.6 hp) Rotax 912 ULS, planned cruising speed is 126 kt (233 km/h; 145 mph) with a never-exceed speed (VNE) of 154 kt (285 km/h; 177 mph), stalling speed of 35 kt (65 km/h; 41 mph) and maximum sea level rate of climb of 395 m (1,300 ft)/min. By early 2002, Remos was predicting a maiden flight in November of that year but by April 2003 development had been halted pending September 2003 increase of German ultralight MTOW to 472.5 kg (1,041 lb). Although that was achieved, promotion of the G-4 had not resumed by mid-2004.

UPDATED

RMT

RMT AVIATION GMBH & CO KG

Dorfstrasse 16 A, D-15913 Siegadel
Tel: (+49 354) 718 08 05
Fax: (+49 354) 718 08 07
e-mail: rmtaviation@hotmail.com
Web: http://www.rmtaviation.com
MANAGING DIRECTOR: André von Schoenebeck

RMT Aviation markets the Bateleur kit, designed by its managing director, having moved to Germany from South Africa in 2001. In 2002 RMT was building a 2,500 m² (26,900 sq ft) plant in South Africa. A European facility will be located at Bremgarten aerodrome, Germany, while plans for a North American plant were then under negotiation. RMT envisages production of over 200 Bateleurs per year by 2006.

VERIFIED

RMT 03 BATELEUR

English name: Impostor

TYPE: Tandem-seat ultralight kitbuilt.

PROGRAMME: Derivative of Richter Delta Dart. Construction of prototype began in August 1993; this (ZU-AND; initially registered as Delta Dart) first flew on 11 September 1994; work on first production aircraft began in May 1996. Design refinement work undertaken by Pretoria University. South African ultralight certification 23 June 1997. On 3 November 2001, ZU-AND was damaged beyond repair following a forced landing and recovery the following day.

CURRENT VERSIONS: **Bateleur 100:** Baseline kit version, powered by 73.5 kW (98.6 hp) Rotax 912 ULS.

Bateleur 115T: Higher-powered version, *as described.*

VLA: Higher-weight version for VLA category. Engine 97 to 112 kW (130 to 150 hp).

Sport: Intended for US Sport class; fixed landing gear; 560 kg (1,234 lb) MTOW.

COSTS: Bateleur 100 kit: €78,800 (2004).

DESIGN FEATURES: Designed for ease of handling, manoeuvrability and quiet approach. Propeller and engine shielded for damage through positioning above the wing. Airframe life 25 years. Pusher layout with large, low-positioned delta wing, wingtip fins and close-coupled midships canard.

FLYING CONTROLS: Hydraulic. Ailerons; twin rudders (operating outwards only); elevators on wing. Electrically actuated flaps on canards also used for lateral trim.

STRUCTURE: Glass fibre, carbon fibre and Aromite.

LANDING GEAR: Retractable tricycle type; mainwheels retract inwards and nosewheel rearwards. Pneumatic actuation; emergency gravity deployment. Mainwheel tyres size 41 cm (16 in), nosewheel tyres size 33 cm (13 in). Hydraulic disc brakes.

RMT 03 Bateleur ***NEW***/0538545

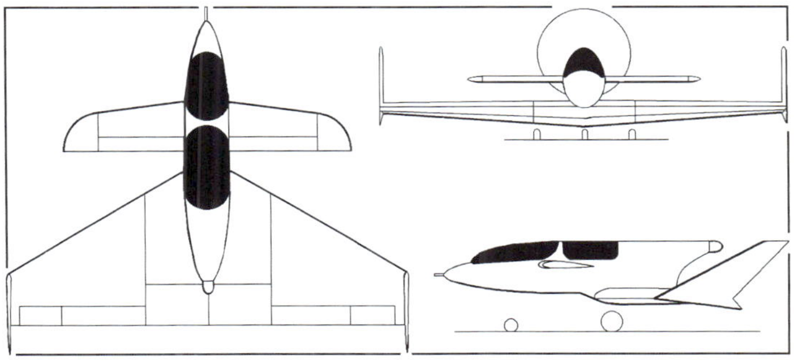

General arrangement of RMT 03 Bateleur *(Paul Jackson)* ***NEW***/0561660

POWER PLANT: One 84.6 kW (113.4 hp) Rotax 914S turbocharged engine, driving Rospeller three-blade variable-pitch propeller; one 73.5 kW (98.6 hp) Rotax 912ULS can optionally be fitted. Fuel capacity 76 litres (20.1 US gallons; 16.7 Imp gallons) of which 74 litres (19.5 US gallons; 16.3 Imp gallons) are usable. Additional, optional 35 litre (9.2 US gallon; 7.7 Imp gallon) long-range tank between wheel wells. Oil capacity 4.0 litres (1.1 US gallons; 0.9 Imp gallon).

ACCOMMODATION: Pilot and passenger in separate tandem cockpits with dual controls. Pilot's canopy opens forward; that of passenger sideways. Baggage compartment behind passenger seat.

SYSTEMS: 8 bar (116 lb/sq in) pneumatic system operates landing gear and canopy. 12 V 25 Ah battery.

EQUIPMENT: Optional ballistic parachute.

DIMENSIONS, EXTERNAL:

Wing span	6.25 m (20 ft 6 in)
Wing chord: at root	2.50 m (8 ft 2½ in)
at tip	0.93 m (3 ft 0½ in)
Canard span	4.765 m (15 ft 7½ in)
Length: fuselage	4.335 m (14 ft 2½ in)
except noseprobe	5.90 m (19 ft 4¼ in)
overall	6.02 m (19 ft 9 in)
Height overall	1.50 m (4 ft 11 in)
Wheelbase	1.70 m (5 ft 7 in)

AREAS:

Wings, gross	11.25 m^2 (121.1 sq ft)
Canard	3.75 m^2 (40.36 sq ft)

WEIGHTS AND LOADINGS:

Weight empty	250 kg (512 lb)
Baggage capacity	50 kg (110 lb)
Max T-O weight: ultralight	450 kg (992 lb)
VLA	600 kg (1,322 lb)

PERFORMANCE (Rotax 914 engine):

Never-exceed speed (VNE):	172 kt (320 km/h; 198 mph)
Max level speed	151 kt (280 km/h; 174 mph)
Cruising speed:	
at 75% power	140 kt (260 km/h; 162 mph)
econ	124 kt (230 km/h; 143 mph)
Stalling speed, flaps down	36 kt (65 km/h; 41 mph)[1]
Max rate of climb at S/L (pilot plus half fuel)	518 m (1,700 ft)/min
Service ceiling: Rotax 912	3,660 m (12,000 ft)
Rotax 914	5,485 (18,000 ft)
T-O to 15 m (50 ft)	160 m (525 ft)
Landing from 15 m (50 ft)	100 m (330 ft)
Range with max fuel/max payload	755 n miles (1,400 km; 869 miles)
g limits	±6

[1] *No conventional stall; enters controllable sink*

UPDATED

SCHEIBE

SCHEIBE FLUGZEUGBAU GMBH

August-Pfaltz-Strasse 23, D-85221 Dachau (PO Box 1829, D-85208 Dachau)
Tel: (+49 8131) 720 83 and 720 84
Fax: (+49 8131) 73 69 85
e-mail: sfflugzeug@t-online.de
Web: http://www.scheibe-flugzeugbau.de
CHAIRMAN AND MARKETING DIRECTOR: Dipl Ing Werner Hoffman
QS MANAGER: Heinz F Haferkorn

Scheibe has produced hundreds of gliders since it was founded on 31 October 1951 by the late Dipl Ing Egon Scheibe. Current motor glider production comprises the steel tube and wood SF 25C Falke and Rotax Falke. Its 2,000 m^2 (21,528 sq ft) factory had a workforce of 19 in late 2003.

Company has built (or licensed) almost 1,500 aircraft including 119 SF-28 Tandem Falke versions of current production aircraft.

UPDATED

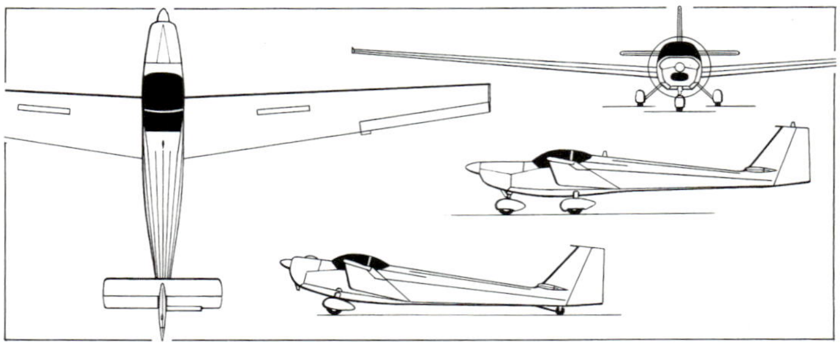

Tri-gear Scheibe SF 25C Falke powered by Rotax 912 A engine, with second side view (lower) of Falke 2000 with conventional landing gear *(Jane's/Mike Keep)* 0092142

SCHEIBE SF 25C FALKE

TYPE: Motor glider.

PROGRAMME: Bergfalke sailplane first flew 5 August 1951; motorised as SF-25A Motor Falke; D-KEDO first flew May 1963. Built in several versions, including under licence by Sportavia-Pützer (Germany) and Slingsby (UK, as T.61).

Current SF 25C version first flew (D-KBIK) in March 1971 and gained its type certificate in September 1972; FAA certrification April 2001. Falke 1700, with 48.5 kW (65 hp) Limbach 1700 engine discontinued. Falke 2100, with Sauer SS2100 engine, tested 1988 (D-KIAC), but only two others built.

CURRENT VERSIONS: **Falke 2000:** Limbach L 2000 EA flat-four engine; monowheel landing gear or conventional tailwheel with two mainwheels available.

Rotax Falke: Prototype D-KIAJ flown 1989. Has essentially the same performance as Falke 2000, but is powered by a water-cooled Rotax 912 A or, from 1999, Rotax 912 S flat-four engine; nosewheel landing gear optional. Most production now of this version; some earlier Falkes have been upgraded with Rotax engines. Tricycle landing gear first tested on Falke 2000 D-KBUG in 1987; deliveries from 1990.

Superschlepper: Glider tug version; has max T-O weight of 600 kg (1,323 lb), in this mode. Rotax 912 A or 912 S. Towing trials by D-KIEK in 1998–99.

CUSTOMERS: Total of 1,259 SF 25s of all types delivered by late 2003, including 698 SF-25Cs by the parent firm. More than 90 Rotax Falkes used as glider tugs.

COSTS: Falke 2000 with Limbach engine and monowheel landing gear, €90,060; nosewheel landing gear €3,130 extra; Rotax Falke €89,440 with 912 A, €91,380 with 912 S (nose-gear versions respectively €3,150 and €3,110 extra (all prices 2003).

DESIGN FEATURES: Docile aerofoil with moderate aspect ratio, designed for safe and simple handling; aircraft can be easily dismantled and transported by trailer. Mid-mounted wing with taper and slight forward sweep; constant chord tailplane; sweptback fin. Various landing gear and engine options.

FLYING CONTROLS: Conventional and manual. Door-type wing spoilers at one-third span; upper wing only. Trim tab on starboard aileron; 13 mm (0.5 in) endplates on wing.

STRUCTURE: Steel tube and fabric fuselage; wooden wing and empennage; wing fabric-covered rearwards from one-third chord; fabric-covered elevators and fin.

LANDING GEAR: Standard landing gear is central monowheel, outriggers under wings, and tailwheel; two-wheel, fuselage-mounted, sprung cantilever main landing gear with steerable tailwheel, and tricycle landing gear (only at 650 kg MTOW), are optional; mainwheels and nosewheel (where fitted) 5.00–4.

POWER PLANT: *Rotax Falke:* One 59.6 kW (79.9 hp) Rotax 912 A with water-cooled cylinder heads, running at 5,800 rpm and turning the propeller at 2,500 rpm; or a 73.5 kW (98.6 hp) Rotax 912 S running at 5,800 rpm and turning the propeller at 2,380 rpm.

Falke 2000: One 59.7 kW (80 hp) Limbach L 2000 EA driving the propeller at 3,450 rpm.

All engines have electric starter and alternator; propeller feathering optional. Glider tug has MTV-1-A/175-05 constant-speed (for Rotax 912 A) or MT 175R130-2A fixed-pitch (for Rotax 912 S) propeller.

Standard fuel tankage (all versions) is 55 litres (14.5 US gallons; 12.0 Imp gallons); optional fuel tankage 80 litres (21.1 US gallons; 17.6 Imp gallons).

ACCOMMODATION: Side-by-side seating under forward-hinged canopy; optional large cockpit opening. Dual controls standard; instrument panel space for radio.

SYSTEMS: 12 V electrical system.

AVIONICS: Choice of optional Bendix/King or Becker nav/com radios.

DIMENSIONS, EXTERNAL:

Wing span	15.30 m (50 ft 2¼ in)
Wing aspect ratio	12.9
Length overall	7.60 m (24 ft 11¼ in)
Height overall: tailwheel	1.68 m (5 ft 6¼ in)
nosewheel	2.50 m (8 ft 2½ in)

DIMENSIONS, INTERNAL:

Cockpit max width	1.00 m (3 ft 3¼ in)

AREAS:

Wings, gross	18.20 m^2 (195.9 sq ft)

WEIGHTS AND LOADINGS (Rotax 912 A):

Manufacturer's weight empty	approx 438 kg (966 lb)
Max T-O weight	650 kg (1,433 lb)
Max wing loading	35.7 kg/m^2 (7.31 lb/sq ft)
Max power loading	10.90 kg/kW (17.91 lb/hp)

PERFORMANCE, POWERED (L: Limbach, A: Rotax 912 A, S: Rotax 912 S, fixed-pitch (FP) or constant-speed (CS) propeller):

Max level speed (all)	102 kt (190 km/h; 118 mph)
Cruising speed: L	81 kt (150 km/h; 93 mph)
A	92 kt (170 km/h; 106 mph)
S	97 kt (180 km/h; 111 mph)
Stalling speed (all)	35 kt (65 km/h; 41 mph)
Max rate of climb at S/L: L	192 m (630 ft)/min
A: FP	240 m (787 ft)/min
CS	288 m (945 ft)/min

Scheibe SF 25C Rotax Falke *(Paul Jackson)* ***NEW***/0572254

S: FP ... 300 m (984 ft)/min
CS ... 330 m (1,082 ft)/min
T-O run: L ... 140 m (460 ft)
A ... 100 m (330 ft)
S ... 90 m (295 ft)
Range, standard fuel:
L, A ... 378 n miles (700 km; 435 miles)
S ... 324 n miles (600 km; 372 miles)

Endurance, standard fuel:
L ... 4–5 h
A ... 4 h 30 min
S ... 4 h
PERFORMANCE, UNPOWERED:
Best glide ratio ... 24
Min rate of sink ... 1.10 m (3.61 ft)/s

UPDATED

SILENCE

SILENCE AIRCRAFT

Otto Lilienthal Weg 2, Am Flugplatz, D-76646 Bruchsal
Tel: (+49 7251) 302 00
Fax: (+49 7251) 302 02 00
e-mail: info@silence-aircraft.de
Web: http://www.silence-aircraft.de
PARTNERS:
Matthias Strieker
Thomas Strieker
Michael Rudbach
Herbert Funke

Silence's first project is an ultralight kitbuilt with flying surfaces similar in shape to those of the Supermarine Spitfire. Until late 2003, the aircraft also was known as Silence. In July 2003, a new company, Silence Flugzeugbau, was established by sailplane manufacturer DG Flugzeugbau to build and market the Twister in conjunction with designers Matthias and Thomas Strieker. Manufacture is being undertaken at Verl.

UPDATED

SILENCE TWISTER

TYPE: Single-seat ultralight/kitbuilt.

PROGRAMME: Developed over a period of four years in a garden gazebo. Prototype (D-MTMH, incorporating partners' initials) made first flight 30 September 2000; LBA certification received May 2002; PFA VLA certification process under way in UK early 2004. Kits produced by DG Flugzeugbau.

CURRENT VERSIONS: **SA155:** Original version; with Diamond engine; *as described.*

SA180: As SA155, but with Jabiru engine. Prototype, D-MTMN (c/n 3), shown at Aero '03, Friedrichshafen, April 2003.

Akron: Experimental version under development for US market.

CUSTOMERS: Three flying and two more under construction by May 2003. Subsequent sales include two to UK.

COSTS: SA155 €71,000; SA180 kit €29,500; flyaway €69,500; Akron €29,500 (all kits less engine) (2003).

DESIGN FEATURES: Compact, low-wing configuration, with elliptical wings and tailplane. Extensive use of carbon and glass fibre honeycomb construction throughout to maintain strength and minimise weight. Wings detachable in 10 minutes for transportation and storage. Propeller spinner has three strakes and rotates freely in the airflow, thereby governing propeller angle. Quoted build time 500 hours.

FLYING CONTROLS: Conventional and manual. Plain flaps. Horn-balanced rudder.

STRUCTURE: Monocoque single-piece fuselage of composites, with honeycomb core. Kevlar sandwich centre-section, incorporating pilot's seat and attachments for wing spar, landing gear and ballistic parachute. Single-piece wing spar. Aerofoil Stuttgart LWK180/25.

LANDING GEAR: Tailwheel type. Electric retraction; carbon fibre main legs retract rearwards. Tailwheel non-retractable. Mainwheels 14×4. Hydraulic brakes.

POWER PLANT: Engines in the region of 70 kW (93.9 hp) can be fitted, to drive a three-blade, ground-adjustable pitch Helix carbon fibre propeller; prototype SA155 has 40 kW (53.6 hp) Wankel engine; second prototype SA155 (D-MTMO) has 37.3 kW (50 hp) Diamond (formerly Mid-West) AE 50 engine driving two-blade Silence carbon fibre propeller. SA180 has 59.7 kW (80 hp) Jabiru 2200 flat-four driving a three-blade Silence carbon fibre propeller. Fuel capacity 80 litres (21.1 US gallons; 17.6 Imp gallons) of which 75 litres (19.8 US gallons; 16.5 Imp gallons) is usable.

EQUIPMENT: Junkers ProFly ballistic parachute system.

DIMENSIONS, EXTERNAL:
Wing span ... 7.50 m (24 ft 7¼ in)
Length overall ... 6.18 m (20 ft 3¼ in)
Height overall ... 1.44 m (4 ft 8¾ in)
AREAS:
Wings, gross ... 8.72 m^2 (93.9 sq ft)
WEIGHTS AND LOADINGS:
Weight empty: SA155 ... 200 kg (441 lb)
SA180 ... 215 kg (474 lb)

Second prototype Silence SA155 Twister (Diamond engine) ***NEW***/0561658

Prototype Silence SA 180 Twister *(Paul Jackson)* ***NEW***/0576507

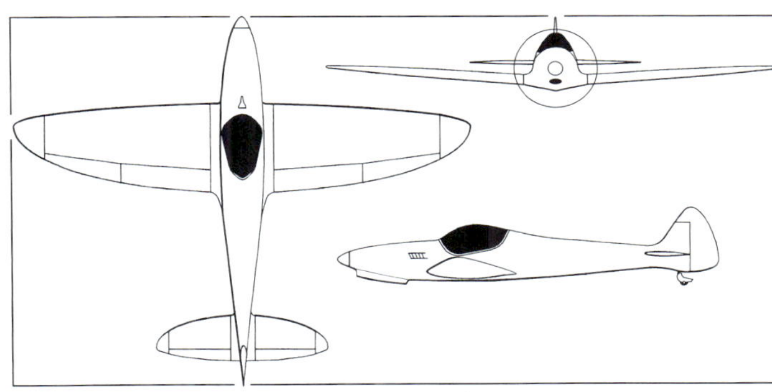

Silence Twister single-seat ultralight *(Paul Jackson)* ***NEW***/0561659

Max T-O weight ... 340 kg (749 lb)
PERFORMANCE:
Never-exceed speed (VNE) ... 162 kt (300 km/h; 186 mph)
Cruising speed: SA155 ... 108 kt (200 km/h; 124 mph)
SA180 ... 130 kt (240 km/h; 149 mph)
Stalling speed: both ... 35 kt (65 km/h; 41 mph)
Max rate of climb at S/L: SA155 ... 330 m (1,083 ft)/min
SA180 ... 510 m (1,673 ft)/min
T-O run: SA180 ... 90 m (295 ft)
Range with max fuel: both ... 702 n miles (1,300 km; 807 miles)
g limits ... +6/–4

UPDATED

STEMME

STEMME AG

7 Flugplatzstrasse F2, D-15344 Strausberg
Tel: (+49 3341) 36 12 13
Fax: (+49 3341) 36 12 30
e-mail: info@stemme.de
Web: http://www.stemme.de
CEO: Dr Reiner Stemme
TECHNICAL DIRECTOR: Dipl Ing Lothar Dalldorff
PRODUCTION DIRECTOR: Dr Jörg Suttan
MARKETING: Karen Stemme

NORTH AMERICAN REPRESENTATIVE:
Stemme USA Inc, 190 Carondelet Plaza, Suite 1111, Saint Louis, Missouri 63144
Tel: (+1 314) 721 59 04
Fax: (+1 314) 726 51 14
e-mail: info@stemme.com
Web: http://www.stemme.com

Founded in 1985, Stemme was originally based in Berlin but moved to nearby Strausberg airfield in 1991; current factory area is 3,000 m^2 (32,292 sq ft); workforce of 45 at beginning of 2003. The 100th aircraft was delivered in September 1998. Turnover was €4 million in 2001 and is forecast to increase to €20 million in 2005.

Stemme's first aircraft design, the S10 high-performance motor glider, introduced an unorthodox light aircraft concept by combining a modular airframe with the engine mounted amidships and driving a front-mounted propeller with retractable blades. Some 165 have been delivered, in three different versions. Current product range also includes the S2 (originally S7), a two-seat competition sailplane.

In 2004, S10 series was to be followed by new range of two-seat aircraft (TSA programme), a modular family of side-by-side sport aircraft comprising the S2 sailplane, S6 soaring motor glider and S8 touring motor glider. Subsidiary Stemme UMS (Umwelt-Messflug-Systeme) is developing the S15-8 entry-level surveillance version and the S8-Safari for semi-professional use.

UPDATED

STEMME S10 AND S15

US marketing name: Chrysalis
US Air Force designation: TG-11A

TYPE: Motor glider.

PROGRAMME: Prototype (D-KKST), with Glaser-Dirks DG500 wing, flown 6 July 1986; first flight of second prototype (D-KCHS) with definitive wing 2 June 1987; S10-VC surveillance and observation platform introduced in 1989; designed to JAR 22. Certified in Germany 31 December 1990, in UK 29 October 1991 and in USA, to FAR Pt 21, 8 July 1992. Production of S10 suspended March 1994; resumed late 1994 until 1997 with S10-V.

CURRENT VERSIONS: **S10:** Original version, with 69.4 kW (93 hp) Limbach L 2400 EB1.AD flat-four engine and retractable fixed-pitch propeller; 54 produced until March 1994, including prototypes.

S10-VC: Surveillance/observation conversion with underwing and small wingtip sensor pods for pollution control and resources investigation. One former S10 (D-KGCM/N600PL) used by Greenpeace since May 1994; one former S10-V (PH-1055) operated by Aerial Surveyance den Hollander, Netherlands, since 1995.

S10-V: Variable-pitch propeller version; power plant as for S10. Demonstrator (D-KGCX) completed mid-1994 as conversion of S10 but written off in Tanzania 10 March 1995; certified September 1994; total of 29 built between late 1994 and 1997; further five, or more, conversions from S10; production completed. (S10-V has engineering designation S14.)

S10-VT: Current production version with turbocharged Rotax 914 engine and variable-pitch propeller. Prototype (D-KGCR) built 1997; LBA certification August 1997, FAA certification September 1997 and CAA certification May 1998; total of 87 produced by mid-2004; of these, 39 to USA. (S10-VT has engineering designation S11.) Set world soaring record (D-KMTE) of 1,330 n miles (2,463 km; 1,530 miles) on 26 November 2000, piloted by Klaus Ohlmann.

Following description applies to S10-VT, except where indicated.

S15: Launched at the 1996 Berlin Air Show; Rotax 914 engine and a 20.00 m (65 ft 7½ in) wing span; two underwing hardpoints for sensor pods in law-enforcement or scientific research roles suitable for loads up to 65 kg (143 lb) each, including FLIR, Nightsun searchlights, video cameras, and environmental monitoring equipment being developed by Berlin Technical University's Geography Institute. Prototype (D-ESTE) registered 1996, but no record of completion. S15B, light surveillance version, based on S10-VT, was then due to enter service in 2002, followed by S15A advanced surveillance model in 2004; both sponsored by Deutsche Bundesstiftung Umwelt. First S10-VT (D-KGCR) shown at Berlin, June 2000 (and again in May 2002) as POC S15. Now superseded by S15-8 (see separate entry).

Condor: Medium-altitude long-endurance (MALE) manned or unmanned version proposed jointly in 2004 by OHB System (Germany) and Elop (Israel); would use Stemme S10-VT airframe, modified to V-tail configuration and possibly turboprop- or Diesel-powered, although version considered with podded turbofan above fuselage, immediately ahead of tailfins. Sensor-carrying capabilities generally similar to those of Stemme S15-8, with Thales Optronique Agile E-O/IR turret offered as standard mission payload.

Stemme S10-VT motor glider with propeller extended *(Paul Jackson)* ***NEW***/0572253

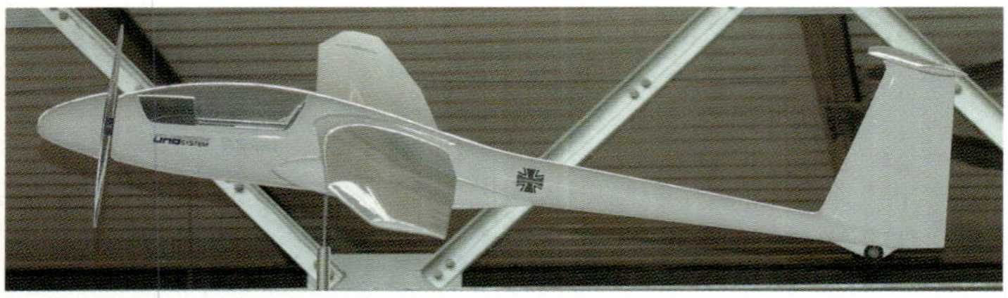

Model of Stemme S10 modified by OHB System for surveillance *(Paul Jackson)* ***NEW***/1047013

Artist's impression of proposed S10 derivative for Condor optionally piloted surveillance mission ***NEW***/1047031

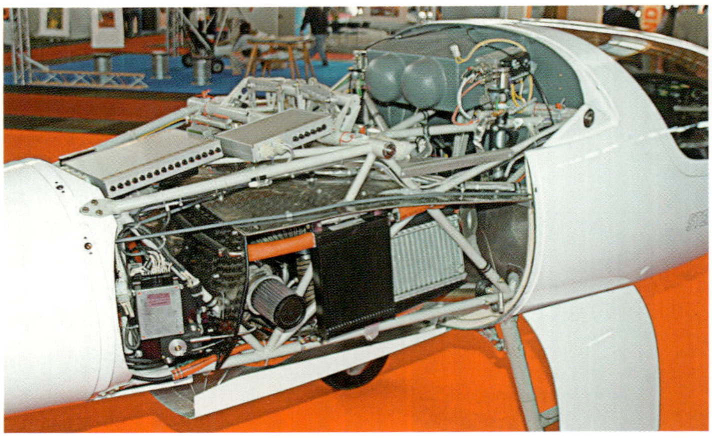

Engine installation of S10 *(Paul Jackson)* ***NEW***/0572252

S-UAV: Unpiloted surveillance conversion. One S10-V (N600V) used as testbed by Platforms International Corporation, Mojave, California.

CUSTOMERS: Owners in Australia, Austria, Belgium, Brazil, Canada, Czech Republic, Denmark, Finland, France, Germany, Italy, Mexico, Netherlands, New Zealand, South Africa, Spain, Switzerland, UK and USA. Two delivered in 1995 to 94th Air Training Squadron, USAF Academy, Colorado Springs, Colorado. Total of 165 S10 variants (including prototypes) built by late 2003, including three registered in that year.

COSTS: S 10-VT €168,000 basic (2003).

DESIGN FEATURES: Three-part shoulder-wing, T-tail motor glider with mid-mounted engine and completely retractable nose-mounted propeller. Transition time between soaring and powered flight is 5 seconds. Winglets option introduced from 1997. Outer wings can be folded by one person for ground handling and hangarage; wing sections carried in gantries inside trailer so that one person can fit centre and outer sections straight from trailer to fuselage.

Wing section Horstmann & Quast laminar flow HQ41/14.35; dihedral 1°; no twist.

FLYING CONTROLS: Conventional and manual. Full-span, six-position (−10/−5/0/+5/+10/+16°) trailing-edge flaps, outer segments of which function also as ailerons. Flap/aileron linkage for manoeuvrability and docility at low airspeeds; Schempp-Hirth airbrakes in outer ends of wing centre-section.

STRUCTURE: CFRP wings in three sections detachable from fuselage; CFRP rear fuselage and tail mounted to steel tube centre frame carrying wing, engine and landing gear; cockpit section, of CFRP structure with Kevlar safety lining, mounted at front of centre frame. Engine fully accessible and horizontal firewall separates engine from wing and flying controls. Winglets optional. Airframe made in Poland by Wytwornia Konstrukcji Kompozytowych, ulitsa Strumienska 829A, PL-43-385 Jasienica; *telephone* (+48 33) 815 33 31; *fax* (+48 33) 815 33 07; *e-mail* wkkpapiorek@pro.onet.pl.

LANDING GEAR: Tailwheel type; electrically actuated, inward-retracting, narrow track mainwheels; non-retractable tailwheel, steered by rudder. Mainwheel tyres 5.00-5 or 6.00-5, with electrically actuated, mechanical standby disc brakes. Tailwheel: 210×65 tyre; fairing standard.

POWER PLANT: One 84.6 kW (113.4 hp) Rotax 914 F2/S1 turbocharged liquid/air-cooled flat-four engine mounted in the central fuselage steel tube frame aft of cockpit and driving a two-blade, variable-pitch retractable propeller via a centrifugal clutch, a CFRP extension shaft and a reduction gear. Engine cooling by adjustable ram air intake. Nosecone is moved forward and spring-folded blades emerge through peripheral slot under centrifugal force.

Fuel in two 45 litre (11.9 US gallon; 9.9 Imp gallon) fuel tanks in outer ends of centre wing; 60 litre (15.8 US gallon; 13.2 Imp gallon) tanks optional. Total capacities 90 litres (23.8 US gallons; 19.8 Imp gallons) or 120 litres (31.7 US gallons; 26.4 Imp gallons).

ACCOMMODATION: Two pilots side by side; dual controls standard; seats adjustable for position and rake; one-piece canopy hinged at forward end and held open by gas struts.

SYSTEMS: 12 V electrical system; full night lighting, landing light and solar cells optional. Solar cells can be fitted behind cockpit to provide 12 V 37.5 W.

AVIONICS: *Comms:* Intercom, VHF radio and transponder optional, as is special cut-down panel for mountain soaring.

DIMENSIONS, EXTERNAL:	
Wing span, excl winglets	23.00 m (75 ft 5½ in)
Wing aspect ratio	28.3
Width, wings folded	11.40 m (37 ft 4¾ in)
Length overall	8.42 m (27 ft 7½ in)
Fuselage max width	1.18 m (3 ft 10½ in)
Height over tailplane	1.80 m (5 ft 10¾ in)
Wheel track	1.15 m (3 ft 9¼ in)
Wheelbase	5.42 m (17 ft 9½ in)
Propeller diameter	1.63 m (5 ft 4¼ in)
Fuselage ground clearance at mainwheels	0.72 m (2 ft 4¼ in)
DIMENSIONS, INTERNAL:	
Cockpit: Width	1.16 m (3 ft 9½ in)
Height	0.93 m (3 ft 0½ in)
AREAS:	
Wings, gross	18.70 m^2 (201.3 sq ft)

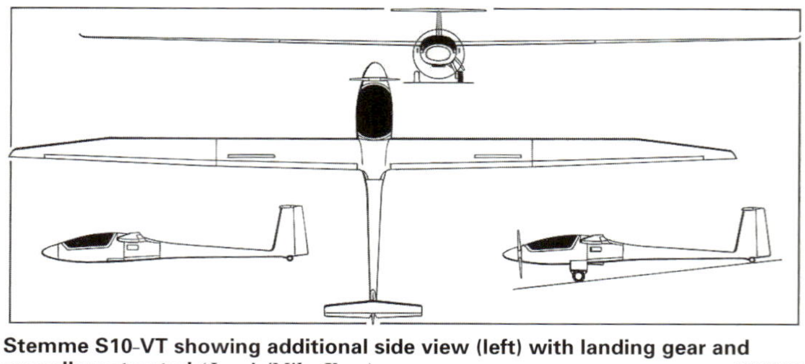

Stemme S10-VT showing additional side view (left) with landing gear and propeller retracted *(Jane's/Mike Keep)* 0507368

Vertical tail surfaces (total) .... 1.51 m² (16.25 sq ft)
Horizontal tail surfaces (total) .... 1.46 m² (15.72 sq ft)

WEIGHTS AND LOADINGS:
Weight empty .... 660 kg (1,455 lb)
Max T-O and landing weight .... 850 kg (1,874 lb)
Max wing loading .... 45.5 kg/m² (9.31 lb/sq ft)
Max power loading .... 10.06 kg/kW (16.53 lb/hp)

PERFORMANCE, POWERED:
Never-exceed speed (VNE), smooth air
.... 146 kt (270 km/h; 168 mph)
Manoeuvring speed .... 97 kt (180 km/h; 112 mph)
Max cruising speed:
at S/L .... 121 kt (225 km/h; 139 mph)
at FL100 .... 134 kt (248 km/h; 154 mph)
Stalling speed, flaps down .... 42 kt (78 km/h; 48 mph)
Max rate of climb at S/L .... 249 m (817 ft)/min
Service ceiling .... 9,140 m (30,000 ft)
T-O run .... 205 m (675 ft)
T-O to 15 m (50 ft) .... 447 m (1,465 ft)
Max range:
standard fuel .... 696 n miles (1,290 km; 801 miles)
max optional fuel
.... 928 n miles (1,720 km; 1,068 miles)

PERFORMANCE, UNPOWERED:
Best glide ratio at 57 kt (106 km/h; 66 mph) 50
Min rate of sink at 45 kt (83 km/h; 52 mph) .... 0.57 m (1.87 ft)/s
g limits .... +5.3/–2.65

OPERATIONAL NOISE LEVELS:
To German light aircraft rules (LSL Chapter X) .... 71.3 dBA

UPDATED

STEMME S6, S8 AND S15-8

TYPE: Motor glider.

PROGRAMME: Development nearing completion in late 2002 of modular family known as TSA (two-seat aircraft) and comprising S2 glider (two versions); S6 motor glider (five versions); and S8 motor glider (six versions). S8 announced at ILA, Berlin, May 2000; S6 at Aero '01, Friedrichshafen, April 2001. Deliveries of S6 were scheduled to begin in second quarter 2003, but revised schedule as at May 2004 forecast S6 maiden flight for late 2004; S2 to be next to fly, then S8.

CURRENT VERSIONS: **S6**: Soaring motor glider with large one-piece canopy, fixed tricycle landing gear and naturally aspirated Rotax 912 S engine. Suitable for towing gliders of up to 750 kg (1,653 lb) MTOW.

S6-L: As S6 except for lower MTOW.

S6-T: As S6, but turbocharged Rotax 914 F engine.

S6-R: As S6, but fully retractable tricycle landing gear.

S6-RT: As S6, but turbocharged Rotax 914 F and retractable tricycle landing gear.

S8: Touring motor glider with roomier cockpit and door on each side; otherwise as S6.

S8-T: As S8, but turbocharged Rotax 914 F.

S8-R: As S8, but retractable tricycle landing gear.

Impression of flood surveillance by an S15-8 ***NEW***/0538273

S8-RT: As S8, but turbocharged Rotax 914 F and retractable tricycle landing gear.

S8-Safari: Derivative of S8, S8-R, S8-T or S8-RT with hardpoints and underwing pods for semi-professional use.

S15-8: Light reconnaissance and surveillance aircraft for manned or MALE (medium-altitude, long-endurance) unmanned (UAV) missions. Derived from S8-T or S8-RT for governmental or commercial applications. Hardpoints for external loads; quick-change underwing mission equipment pods. Sensor payloads can include day/night E-O, SAR, lidar, imaging spectrometer and digital photogrammetric cameras; sensor fusion and integration with GPS/INS available.

DESIGN FEATURES: Based on platform strategy with S10 design principles, including proven propulsion system with mid-mounted engine, driveshaft and front propeller. New features include non-retractable constant-speed propeller with feathering for soaring mode, tricycle landing gear and task-adapted, low-drag aerodynamics. In area of wing/fuselage junction a geometric twist is combined with specially designed turbulent flow aerofoils, which minimise separations of boundary layer. Ease of maintenance through rapid change potential of power plant (engine, from drive flange to end of muffler, is mounted on a subframe), landing gear and propeller driveshaft group. Innovative engine installation offers upward exhaust emission for a noise level far below international limit.

FLYING CONTROLS: Primary controls (elevator, ailerons, rudder) conventional and manual; full dual controls. Secondary controls (flaps, trim, airbrakes). Electric trim. Manually actuated flaps. Advanced flap/aileron linkage for middle and outer wing control surfaces. Schempp-Hirth airbrakes in outer wings.

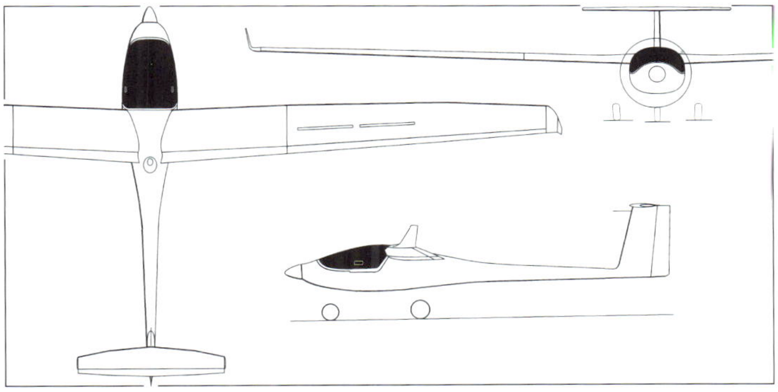

Stemme S6 general arrangement ***(Paul Jackson)*** ***NEW***/0535822

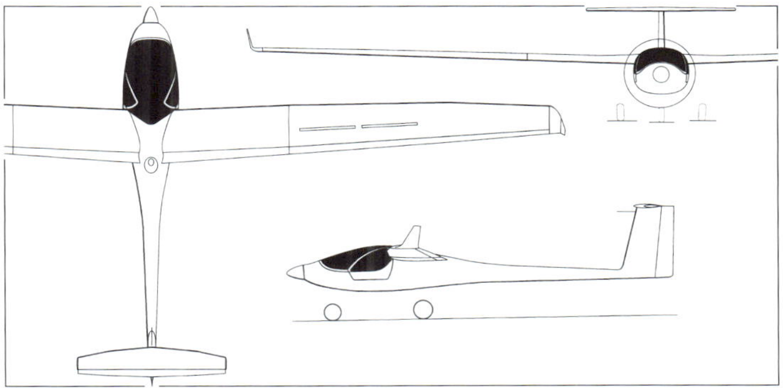

Stemme S8 motor glider ***(Paul Jackson)*** ***NEW***/0535823

Computer simulation of Stemme S6 ***NEW***/0538264

Enlarged cabin and three-blade propeller of S8 ***NEW***/0538266

STRUCTURE: CFRP wings in three sections, plus winglets. Wing detachable from fuselage; CFRP tailcone and tail unit mounted to centre steel framework accommodating engine and landing gear; safety cockpit with different types of composites.

LANDING GEAR: Tricycle type, fixed; or electrically actuated fully retractable (see Current Versions). Former has speed fairings on all wheels. Mainwheels size 380×50, nosewheel 300×100.

POWER PLANT: *Except T and RT versions:* One 73.5 kW (98.6 hp) Rotax 912 S flat-four engine, mounted on subframe aft of cockpit and driving a two-blade (optionally three-blade) constant-speed propeller via a CFRP extension shaft and a reduction gear.

T and RT versions and S15-8: One 84.5 kW (113.3 hp) turbocharged Rotax 914 F flat-four; five-blade propeller in S15-8; otherwise as described in preceding paragraph.

Standard fuel capacity (S6) 70 litres (18.5 US gallons; 15.4 Imp gallons); 140 litres (37.0 US gallons; 30.8 Imp gallons) optional for S6-RT, standard for S8 variants. Additional option in S15-8 for 230 litre (60.8 US gallon; 50.6 Imp gallon) long-range tank and/or two 65 litre (17.2 US gallon; 14.3 Imp gallon) underwing tanks.

ACCOMMODATION: Pilot and passenger/student side by side in S6, under single-piece, forward-hinged canopy. More spacious cockpit in S8 versions, with door each side.

SYSTEMS: As described for S10.

AVIONICS: *Comms:* VHF radio and intercom.

Flight: VOR; GPS; ATC transponder.

Instrumentation: Moving map; soaring computer.

Following data are as projected in late 2002:

DIMENSIONS, EXTERNAL (all):
- Wing span: 18.00 m (59 ft 0¾ in)
- Wing aspect ratio: 18.1
- Width, wings folded: 7.20 m (23 ft 7½ in)
- Length overall: 8.52 m (27 ft 11½ in)
- Height to top of tail: 2.45 m (8 ft 0½ in)
- Wheel track: 2.00 m (6 ft 6¾ in)
- Wheelbase: 2.00 m (6 ft 6¾ in)

AREAS:
- Wings, gross: 17.40 m² (187.3 sq ft)
- Vertical tail surfaces (total): 1.50 m² (16.15 sq ft)
- Horizontal tail surfaces (total): 1.74 m² (18.73 sq ft)

WEIGHTS AND LOADINGS:
- Weight empty: S6: 583 kg (1,285 lb)
  - S6-R: 587 kg (1,295 lb)
  - S6-RT: 599 kg (1,321 lb)
  - S8, S15-8: 594 kg (1,310 lb)
  - S8-R: 597 kg (1,317 lb)
  - S8-RT: 610 kg (1,345 lb)
- Max T-O weight: except S6-L: 850 kg (1,873 lb)*
  - S6-L: 800 kg (1,764 lb)
- Max wing loading:
  - S6-L: 48.8 kg/m² (10.00 lb/sq ft)
  - S6-L: 46.0 kg/m² (9.42 lb/sq ft)
- Max power loading:
  - Rotax 912 S (except S6-L): 11.56 kg/kW (19.00 lb/hp)
  - S6-L: 10.89 kg/kW (17.89 lb/hp)
  - Rotax 914 F: 10.06 kg/kW (16.53 lb/hp)

* *S15-8 950 kg (2,094 lb) with special C of A*

PERFORMANCE, POWERED (estimated, with two-blade propeller; A: S6-RT, B: S8-RT, C: S15-8):
- Max level speed at S/L:
  - A, B, C: 151 kt (280 km/h; 174 mph)
- Max level speed at FL160: C: 175 kt (324 km/h; 201 mph)
- Max cruising speed at S/L:
  - at 55% power: A: 130 kt (240 km/h; 149 mph)
  - B: 127 kt (235 km/h; 146 mph)
  - C: 114 kt (211 km/h; 131 mph)
- Max cruising speed at FL100:
  - A, B: 167 kt (310 km/h; 193 mph)
- Max rate of climb at S/L: A: 348 m (1,142 ft)/min
  - B: 330 m (1,083 ft)/min
  - C: 266 m (873 ft)/min
- Service ceiling: A, B: 8,535 m (28,000 ft)
  - C: 7,620 m (25,000 ft)
- Range at S/L, 140 litres fuel, 55% power:
  - A: 1,295 n miles (2,400 km; 1,491 miles)
  - B: 1,241 n miles (2,300 km; 1,429 miles)

PERFORMANCE, UNPOWERED:
- Best glide ratio:
  - S6, S6-T at 61 kt (113 km/h; 70 mph): 33
  - S6-R, S6-RT at 64 kt (119 km/h; 74 mph): 39
  - S8, S8-T at 63 kt (117 km/h; 72 mph): 32
  - S8-R, S8-RT at 66 kt (122 km/h; 76 mph): 38
- Min rate of sink at empty weight +70 kg (154 lb):
  - S6, S6-T: 0.85 m (2.79 ft)/s
  - S6-R, S6-RT: 0.73 m (2.40 ft)/s
  - S8, S8-T: 0.86 m (2.82 ft)/s
  - S8-R, S8-RT: 0.74 m (2.43 ft)/s

UPDATED

ULBI

ULTRALEICHTBAU INTERNATIONAL

Flugplatzstrasse 18, D-97437 Hassfurt
Tel: (+49 9521) 61 83 93
Fax: (+49 9521) 61 83 95
e-mail(1): info@ulbigmbh.de
e-mail(2): ulbi-gmbh@t-online.de
Web(1): http://www.ulbi-aircraft.de
Web(2): http://www.ulbi-gmbh.de
CEO: Angelika Kaiser
SALES MANAGER: Alfred Kaiser
CHIEF DESIGNER: Rainer Kurtz
CHIEF TEST PILOT: 'Biggi the Rottweiler'

The Romanian-developed Wild Thing lightplane is marketed by Ultraleichtbau, known until 2001 as Air Light. A nosewheel version was introduced in 1999.

UPDATED

ULBI WILD THING

TYPE: Side-by-side ultralight/kitbuilt.

PROGRAMME: Developed by Aerostar (which see) in Romania; prototype registered YR-6107. Launched at Friedrichshafen Air Show, April 1997; certification August 1997. Initial production aircraft had Jabiru 2200 engines; alternatives available from 1998. By early 2001, examples were operating in Austria, Czech Republic, France, Germany, Hungary and Netherlands.

CURRENT VERSIONS: **WT-01:** Baseline version; tailwheel.

WT-02: Nosewheel version introduced early 1999 (D-MWTA, 50th production aircraft, as demonstrator).

CUSTOMERS: Total of at least 85 built by 2003.

DESIGN FEATURES: Bears strong outward resemblance to Canadian Murphy Rebel. Built by Aerostar in Romania for distribution out of Germany. Wings fold alongside fuselage for storage or transportation on aluminium trailer.

FLYING CONTROLS: Conventional and manual. Frise ailerons. Elevators and ailerons pushrod-operated, rudder cable-operated. Horn-balanced tail surfaces; trim tab in port elevator; ground-adjustable tab on rudder. Flaps.

Nosewheel (WT-02) ULBI Wild Thing *(Paul Jackson)* ***NEW***/0580742

STRUCTURE: Main structure is metal monocoque, with fabric-covered ailerons and flaps.

LANDING GEAR: Tailwheel type; fixed. Fuselage-mounted. WT-01 has two faired-in side Ys plus half-axles, each with bungee cord shock-absorber, attached to compression frame two-blade, fixed-pitch. Mainwheel size 8.00-6, tailwheel size 200×50. Oversize Tundra wheels and tyres optional. Cable-operated disc brakes. WT-02 has spring cantilever mainwheel legs with attachment point further aft and nosewheel with compressed rubber shock-absorber; nosewheel size 140×6.

POWER PLANT: One 59.7 kW (80 hp) Jabiru 2200, 59.6 kW (79.9 hp) Rotax 912 UL, 78.3 kW (105 hp) Mid West Hawk, 89.5 kW (120 hp) Jabiru 3300 or 73.5 kW (98.6 hp) Rotax 912 ULS driving a two-blade Helix fixed-pitch carbon fibre ground-adjustable pitch, carbon fibre or Kremen SR 2000 XC three-blade, variable-pitch (electric) wooden propeller. Fuel in two wing tanks, combined capacity 80 litres (21.1 US gallons; 17.6 Imp gallons).

ACCOMMODATION: Prototype had third seat in rear of cabin, hence a spacious cockpit area.

AVIONICS: Becker, Filser and Icom avionics to customer's choice.

EQUIPMENT: Junkers Magnum speed parachute emergency rescue system and gliding towing equipment both optional.

DIMENSIONS, EXTERNAL:
- Wing span: 9.20 m (30 ft 2¼ in)
- Length overall: 6.90 m (22 ft 7¾ in)
- Height overall: 1.92 m (6 ft 3½ in)

AREAS:
- Wings, gross: 13.94 m² (150.0 sq ft)

ULBI WT-01 Wild Thing in the style of Equus Burchelli *(Paul Jackson)* ***NEW***/0589013

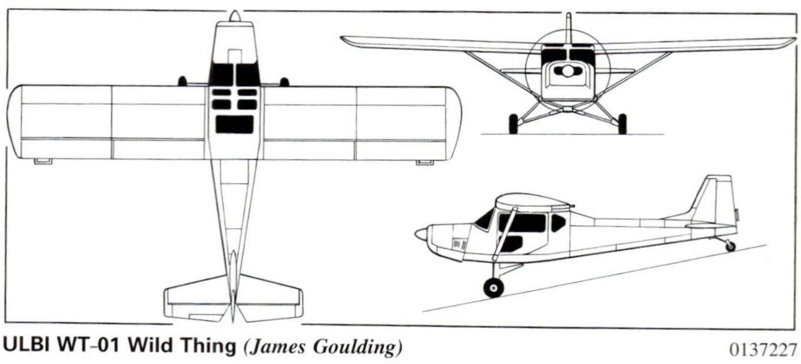

ULBI WT-01 Wild Thing *(James Goulding)* 0137227

WEIGHTS AND LOADINGS:
Weight empty (typical) ... 300 kg (661 lb)
Max T-O weight ... 450 kg (992 lb)
PERFORMANCE (Jabiru 2200 engine):
Never-exceed speed (VNE) ... 108 kt (200 km/h; 124 mph)
Max level speed ... 86 kt (160 km/h; 99 mph)
Cruising speed ... 76 kt (140 km/h; 87 mph)
Stalling speed ... 31 kt (56 km/h; 35 mph)
Max rate of climb at S/L ... 152 m (500 ft)/min
T-O run ... 98 m (320 ft)
Landing run ... 44 m (145 ft)
Range with max fuel ... 432 n miles (800 km; 497 miles)

UPDATED

WD

WD FLUGZEUGLEICHTBAU GMBH

Sudetenstrasse 57/2, D-73540 Heubach
Tel: (+49 7173) 92 99 90
Fax: (+49 7173) 92 99 99
e-mail: info@dallach.de
Web: http://www.dallach.de
PRESIDENT: Wolfgang Dallach

Formed in 1982, WD Flugzeugleichtbau markets the Fascination D4 BK and D5 Evolution, designed by Wolfgang Dallach. The earlier D3 Sunwheel is still available; WD is at present working on the **Revolution**, which will be powered by a 265 kW (355 hp) VOKBM M-14P. The company moved to its current premises in 1997.

In 1999, WD opened a new production facility in the Czech Republic, where kits and complete aircraft are manufactured.

UPDATED

WD D4 BK FASCINATION

TYPE: Side-by-side ultralight/kitbuilt.

PROGRAMME: Original D4, of steel, wood, fabric and composites construction, first flew in early 1996; public debut (D-MOVE and D-MBWH) at the Réseau du Sport de l'Air rally at Epinal, France, in July 1996. Prototype D4 BK (D-MPMM, carrying spurious registration D-MMMM) exhibited at Aero '99, Friedrichshafen, in April 1999, shortly before maiden flight; US debut at EAA Air Venture at Oshkosh, Wisconsin, in July 2000.

CURRENT VERSIONS: **D4:** Initial version, last described in 1999–2000 edition.

D4 BK: All-composites version, *as described*. BK indicates *bugrad, kraftstoff:* nosewheel, composites.

D4 Fascination VLA: Revealed at Aero '03 in mockup form; at which time certification was expected in third quarter of 2003; prototype (D-EVLA) first flew 4 October 2003. Dimensions as D4 BK except length overall 6.87 m (22 ft 6½ in).

D5: High-wing version; described separately.

CUSTOMERS: Total of 90 D4s sold (including 40 as kits), of which 50 were flying by April 1999, including 15 in Brazil. D4 BK sales totalled five by April 1999; in total, 190 (90 D4 and 100 D4 BK) sold by December 2003.

COSTS: Fully equipped €84,000 including tax (2003). Kit version €58,800.

DESIGN FEATURES: Wings and tail surfaces detach for storage/transport. Quoted kit build time 800 hours.

FLYING CONTROLS: Conventional and manual. Balance tabs suspended beneath each aileron; aileron gaps sealed. Horn-balanced rudder. Fowler flaps.

STRUCTURE: All composites.

LANDING GEAR: Standard fixed tricycle type, with nosewheel power steering. Fuselage-mounted spring cantilever mainwheel legs. Retractable optional; main units retract outwards, nosewheel rearwards. Cable-operated disc brakes.

POWER PLANT: One 59.6 kW (79.9 hp) Rotax 912 UL (standard), 73.5 kW (98.6 hp) Rotax 912 ULS (optional) or (on completion of development) 86.8 kW (116.3 hp) DZ 100 flat-four engine driving a Rospeller two-blade electric constant-speed propeller. Fuel capacity 80 litres (21.1 US gallons; 17.6 Imp gallons).

D4 Fascination VLA: One 73.5 kW (98.6 hp) Rotax 912 ULS driving a three-blade constant-speed (electric) propeller. Fuel capacity 180 litres (47.6 US gallons; 39.6 Imp gallons).

AVIONICS: GPS and transponder optional.

EQUIPMENT: Ballistic recovery parachute optional.

WD Fascination D4 BK composites kitbuilt *(Paul Jackson)* ***NEW***/0580746

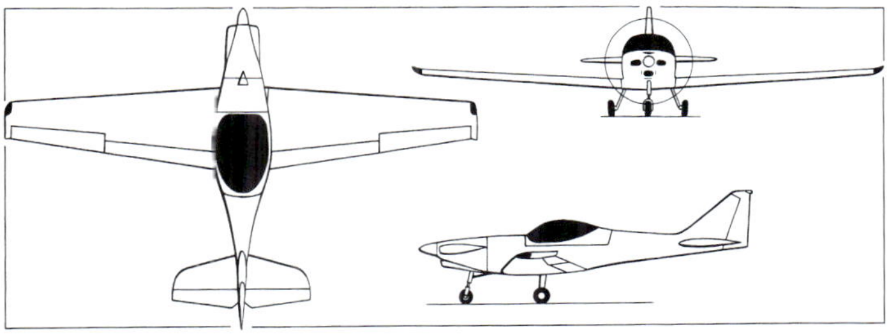

General arrangement of the WD Fascination D4 BK *(James Goulding)* ***NEW***/0580696

DIMENSIONS, EXTERNAL:
Wing span ... 9.00 m (29 ft 6¼ in)
Length overall ... 6.98 m (22 ft 10¾ in)
Height overall ... 1.85 m (6 ft 0¾ in)
AREAS:
Wings, gross ... 10.40 m² (111.9 sq ft)
WEIGHTS AND LOADINGS:
Weight empty: BK ... 290 kg (639 lb)
VLA ... 350 kg (772 lb)
Max T-O weight: BK ... 450 kg (992 lb)
VLA ... 650 kg (1,433 lb)
PERFORMANCE (Rotax 912 ULS):
Never-exceed speed: BK ... 157 kt (290 km/h; 180 mph)
VLA ... 151 kt (279 km/h; 173 mph)
Cruising speed: BK ... 145 kt (269 km/h; 167 mph)
VLA ... 140 kt (259 km/h; 161 mph)
Stalling speed: BK ... 34 kt (63 km/h; 40 mph)
VLA ... 45 kt (84 km/h; 52 mph)
Max rate of climb at S/L: BK ... 518 m (1,700 ft)/min
VLA ... 335 m (1,100 ft)/min
T-O run: BK ... 77 m (250 ft)
VLA ... 90 m (295 ft)
Landing run: BK ... 100 m (330 ft)
VLA ... 130 m (430 ft)
Range with max fuel: BK ... 700 n miles (1,296 km; 805 miles)
VLA ... 1,100 n miles (2,037 km; 1,265 miles)

UPDATED

WD D5 EVOLUTION

TYPE: Side-by-side, ultralight/kitbuilt.

PROGRAMME: Announced at Aero '01 at Friedrichshafen April 2001; first flight of prototype (D-MPMM) then expected within two months but not achieved until 16 February 2002. Series production began in 2002.

COSTS: Factory-built €91,600 including tax (2003).

DESIGN FEATURES: Utilises wings, Fowler flaps and tail surfaces of D4 Fascination. 'Glass cockpit'. Wings and tail surfaces detach for storage and transportation.

FLYING CONTROLS: Conventional and manual. Dual controls standard. Fowler flaps.

STRUCTURE: Primarily composites, with metal landing gear doors.

LANDING GEAR: Tricycle type; retractable. All three units retract rearwards, with main legs housed within lower fuselage.

POWER PLANT: One 59.6 kW (79.9 hp) Rotax 912 UL driving Rospeller three-blade constant-speed propeller via 1:2.27 reduction gearing, or 73.5 kW (98.6 hp) Rotax 912 ULS driving same propeller via 1:1.43 reduction gearing. Fuel capacity 120 litres (31.7 US gallons; 26.4 Imp gallons).

EQUIPMENT: Ballistic parachute recovery system optional.

DIMENSIONS, EXTERNAL:
Wing span ... 9.10 m (29 ft 10¼ in)
Length overall ... 6.62 m (21 ft 8½ in)
Height overall ... 2.04 m (6 ft 8¼ in)
DIMENSIONS, INTERNAL:
Cabin max width ... 1.25 m (4 ft 1¼ in)
AREAS:
Wings, gross ... 10.40 m² (111.9 sq ft)
WEIGHTS AND LOADINGS:
Weight empty ... 270 kg (595 lb)
Max T-O weight ... 450 kg (992 lb)
PERFORMANCE (Rotax 912 ULS):
Max level speed ... 157 kt (290 km/h; 180 mph)
Cruising speed ... 146 kt (270 km/h; 168 mph)
Stalling speed ... 35 kt (64 km/h; 40 mph)
Max rate of climb at S/L ... 510 m (1,673 ft)/min
T-O run: solo ... 75 m (246 ft)
dual ... 100 m (330 ft)
Range ... 918 n miles (1,700 km; 1,056 miles)

UPDATED

Prototype WD D5 Evolution demonstrating at Berlin in May 2004 *(Paul Jackson)* ***NEW***/0580808

WÜST

WÜST GMBH

PO Box 11027, 63718 Aschaffenburg
Tel: (+49 6021) 279 35
Fax: (+49 6021) 279 36
e-mail: mail@wuestenflug.de
Web: http://www.mark-wuest.de

Wüst GmbH was founded in October 2001 and in December 2001 acquired the assets of the former Mark GmbH Flugzeugbau. Wüst's development and production facilities are located in the Czech Republic.

VERIFIED

WÜST SEAHAWK II

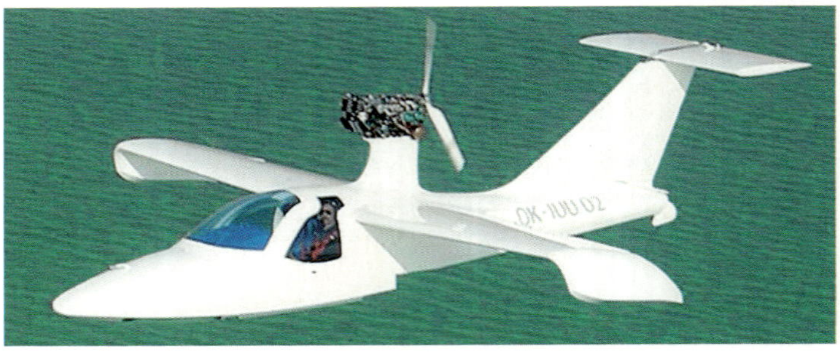

Wüst Seahawk II prototype during its first flight on 29 March 2003 ***NEW***/0561696

TYPE: Two-seat amphibian ultralight kitbuilt.

PROGRAMME: Scale model of Seahawk I built by Mark GmbH for preliminary tests; mockup displayed at Aero 2001 at Friedrichshafen April 2001; redesign by Wüst GmbH as Seahawk II began December 2001; prototype (OK-IUU 02) completed January 2003 and first flown 29 March 2003; manufacture of second prototype began September 2003.

DESIGN FEATURES: Single-step hull; parallel chord wing with turned-down wingtips forming narrow stabilising floats; sweptback fin with T tail configuration; upward-opening gullwing doors.

FLYING CONTROLS: Conventional and manual; one-piece horn-balanced elevator with trim tab; electrically actuated Fowler flaps. Dual controls standard.

STRUCTURE: Composites/honeycomb sandwich; single-spar wings.

LANDING GEAR: Retractable tricycle type; mainwheels retract upwards into sides of hull; nosewheel retracts forwards. Mainwheels on metal tube side Vs, hinged at top, with twin bracing struts mounted above and single oleo-pneumatic/helical spring (motorcar-type) shock absorber.

POWER PLANT: One 73.5 kW (98.6 hp) Rotax 912 ULS or 74.6 kW (100 hp) Textron Lycoming O-200 flat-four, pylon-mounted above wing in pusher configuration and driving a three-blade, fixed-pitch or constant-speed propeller. Fuel contained in two wing tanks, combined capacity 90 litres (23.8 US gallons; 19.8 Imp gallons).

DIMENSIONS, EXTERNAL:
- Wing span .... 8.38 m (27 ft 6 in)
- Length overall .... 7.35 m (24 ft 1¼ in)
- Height overall .... 2.65 m (8 ft 8½ in)

WEIGHTS AND LOADINGS:
- Weight empty .... 350 kg (772 lb)
- Max T-O weight .... 549 kg (1,212 lb)
- Max power loading (Rotax) .... 7.48 kg/kW (12.29 lb/hp)

UPDATED

ZEPPELIN

ZEPPELIN LUFTSCHIFFTECHNIK GMBH

Allmannsweilerstrasse 132, D-88046 Friedrichshafen
Tel: (+49 7541) 20 25 33
Fax: (+49 7541) 20 24 99
e-mail: info@zeppelin-nt.com
Web: http://www.zeppelin-nt.com
CHAIRMAN AND CEO: Dr Bernd Sträter
COO: Günter Schwenk
VICE-PRESIDENT, ENGINEERING: Robert Gritzbach
SALES AND MARKETING MANAGER: Dietmar Blasius
HEAD OF PUBLIC RELATIONS: Wolfgang Schröder

Zeppelin Luftschifftechnik (ZLT) was formed in September 1993 for the development and production of new technology (NT) airships. Luftschiffbau Zeppelin GmbH, of which ZLT is a subsidiary, was founded by Count Ferdinand von Zeppelin in 1908 and still exists under the auspices of the Zeppelin Foundation. It had a mid-2001 workforce of 75 full-time and 25 freelance personnel.

Corporate members (percentage shareholdings in parentheses) are ZF Friedrichshafen AG (44.8), Luftschiffbau Zeppelin GmbH (38.9), Zeppelin GmbH (11.5) and ZF-Lemförder Metallwaren AG (4.8). Deutsche Zeppelin Reederei (DZR) formed in January 2001 as 100 per cent subsidiary of Zeppelin Luftschifftechnik; will be developed as operator, pilot training school and service centre for latter's airships; first commercial service flown 15 August 2001 and over 23,000 passengers carried by August 2003. (First-generation Zeppelin undertook initial passenger service on 24 June 1910.)

UPDATED

ZEPPELIN NT

TYPE: Helium semi-rigid.

PROGRAMME: Study group for revival of modern semi-rigid airships formed 1989; concluded that combination of vectored thrust and new constructional approach offers best solution, resulting in NT programme; 10 m (32 ft 10 in) long, remotely controlled proof-of-concept model tested extensively in 1991. Design definition of N 07 series (originally designated N 05) completed late 1992; construction started October 1995, leading to completion in July 1997 and first flight (D-LZFN) on 18 September 1997 (69th anniversary of maiden flight(1928) of LZ 127 *Graf Zeppelin*) following public debut (April) at Friedrichshafen Air Show. Test flights undertaken from new airship base at Friedrichshafen Airport; damaged by gale in February 1999, but returned to test flying by following April; made 100th flight in August 1999; D-LZFN named *Friedrichshafen* 2 July 2000; flight test programme completed in December 2000 (more than 1,000 hours in approximately 250 flights); certified by LBA 26 April 2001. Being brought up to production standard in 2002; then to undertake mineseeking operations in Croatia; scheduled for traffic monitoring duties during 2004 Olympic Games in Athens.

First production airship (D-LZZR *Bodensee*, c/n 02) made maiden flight 19 May 2001; second (D-LZZF, c/n 03) entered construction in third quarter of 2000 and made its first flight 9 February 2003. First revenue-earning flight (by 02) 15 August 2001; had carried more than 23,000 fare-paying passengers by August 2003; sold to Japan March 2004, at which time had accumulated approximately 2,500 flying hours.

CURRENT VERSIONS: **Zeppelin NT:** *As described.*

Zeppelin ZET: Projected long-range, enlarged version, with overall length of 125 m (410.1 ft) and max diameter of 27 m (88.6 ft); envelope volume 50,000 m^3 (1,765,735 cu ft); 40-passenger gondola cabin capacity. Launch customer Zeppelin Europe Tours plans fleet of 10, with further two on option. Estimated five-year development programme for 2008 service entry; investment requirement quoted in 2003 as approximately €400 million (US$465 million).

CUSTOMERS: First three NT airships operated via DZR (Deutschen Zeppelin Reederei) subsidiary initially, to gain experience. D-LZZR purchased by Nippon Airship Corporation of Japan 2 March 2004; delivery then planned for June of that year.

COSTS: €8 million, including ground infrastructure (2002).

DESIGN FEATURES: Revives original design principles of Count Ferdinand von Zeppelin; first airship in world since 1940 with load-bearing internal structure. Intended as precursor of product line of commercial airships for variety of applications including scientific research, environmental monitoring, TV missions and tourism. Internal 'prism' primary structure modified and optimised from Stuttgart University proposal of 1975 and British patent of 1924; single helium cell; placement of vectored thrust units near CG and also at rear enhances manoeuvrability and control at low speeds, further improving safety and reducing ground crew requirements to three persons, thereby also reducing operating costs. Tailfins in inverted Y configuration.

FLYING CONTROLS: Computer-assisted flight control, propulsion and landing. Dual fly-by-wire flight control system; no mechanical back-up. Elevator or rudder on each tail surface. Engine thrust control (see under Power Plant) in pitch and yaw.

STRUCTURE: Internal primary structure consists of three rows of aluminium tubular longerons, interconnected with triangular carbon fibre frames and cross-braced with Kevlar cables,

Zeppelin NT No. 03 ***(Paul Jackson)*** ***NEW***/0573413

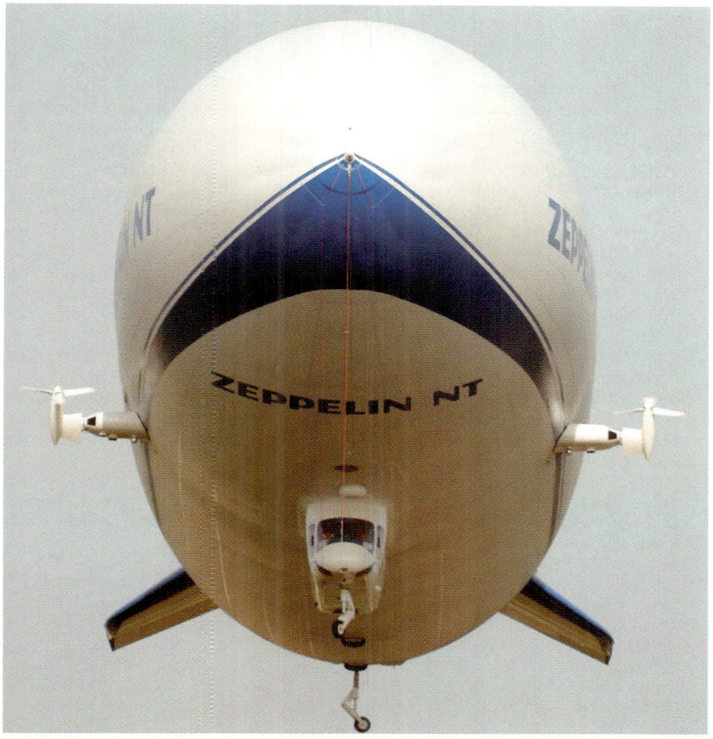

Zeppelin NT in hovering flight *(Paul Jackson)* ***NEW***/0573439

providing internal prism-shaped structure accounting for only 12 per cent of gross weight; this frame and pressurised envelope are both load-carrying structures to enhance safety factor (airship remains fully manoeuvrable even if internal pressure drops). Envelope, attached continuously to all longerons, is a high-strength, multilayer laminate of polyester and Tedlar; this material has low gas permeability and allows airship to be moored permanently in the open. Air ballonets in lower part of envelope afford protection to cabin and airframe in the event of a hard landing.

LANDING GEAR: Twin landing/ground handling wheels in tandem under gondola and rear of envelope. Airship can be moored to a fixed or mobile mast.

POWER PLANT: Three 149 kW (200 hp) Textron Lycoming IO-360 flat-four engines: one each side of hull above gondola in vectored-thrust propulsion unit, each driving a three-blade, variable-pitch, swivelling (up to 120°) tractor propeller; engine in tailcone drives, via a belt-driven gearbox, a lateral fan (for precise yaw control during low-speed and hovering manoeuvres) and a vectored propeller used as a pusher in forward flight and for precise pitch control in landing configuration. Combination of thrust vectoring and fuel trim tanks virtually obviates need for ballast management.

ACCOMMODATION: Crew of two, on ergonomically designed flight deck; gondola main cabin accommodates up to 12 passengers on individual seats, with galley, lavatory and wardrobe provision, or equivalent cargo or other payload.

AVIONICS: *Radar:* Weather radar.

Flight: Dual digital fly-by-wire aerodynamic steering; triple thrust vector steering; modern LCD flight deck avionics, including EFIS, EICAS, radio nav and GPS.

DIMENSIONS, EXTERNAL:
Envelope: Length overall .... 75.00 m (246 ft 0¾ in)
Max diameter .... 14.16 m (46 ft 5½ in)
Fineness ratio 5.3

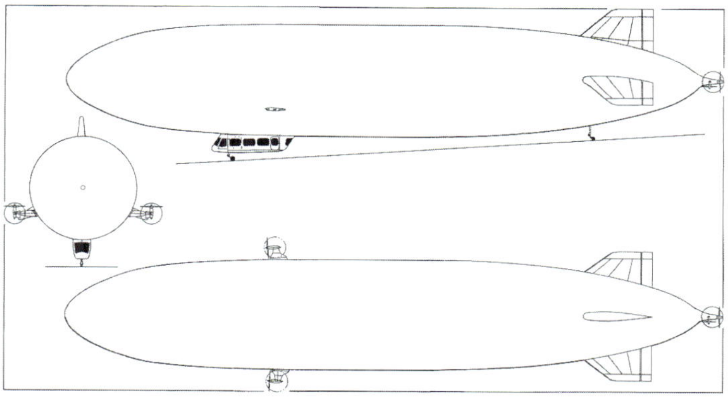

Zeppelin NT airship *(Paul Jackson)* 0143738

Max width over propeller ducts .... 19.50 m (63 ft 11¾ in)
Height overall .... 17.40 m (57 ft 1 in)

DIMENSIONS, INTERNAL:
Envelope volume .... 8,225 m³ (290,450 cu ft)
Ballonet volume (max) .... 2,200 m³ (77,700 cu ft)
Gondola: Length .... 10.70 m (35 ft 1¼ in)
Max width .... 2.25 m (7 ft 4½ in)
Volume .... 26.0 m³ (918.2 cu ft)

WEIGHTS AND LOADINGS:
Gondola: structural weight .... 440 kg (970 lb)
incl equipment and furniture .... 1,200 kg (2,646 lb)
Max payload .... 1,900 kg (4,189 lb)
Max T-O weight .... 10,690 kg (23,567 lb)

PERFORMANCE:
Max level speed .... 67 kt (125 km/h; 77 mph)
Cruising speed .... 62 kt (115 km/h; 71 mph)
Cruising altitude .... 1,000 m (3,280 ft)
Pressure ceiling .... 2,600 m (8,530 ft)
Range .... 486 n miles (900 km; 559 miles)
Endurance at 38 kt (70 km/h; 43 mph):
with max payload .... 14 h
with reduced payload .... 24 h

UPDATED

ZEPPELIN ZET

TYPE: Helium semi-rigid.

PROGRAMME: By 2004, and following the accumulation of experience in operating the Zeppelin NT airship on sightseeing flights, the separate operating company Zeppelin-Tourismus was proposing the construction of a larger craft with the ability to carry some 45 passengers, compared with 13 in the current airship. As yet, however, the manufacturer has not launched development, which will be conditional upon finance and accommodation.

The operating plan proposed by Förderverein Zeppelin-Tourismus eV, led by Wolfgang von Zeppelin, envisages operating bases in 10 European cities, with 10 airships flying tourist sorties between each leg every two days as well as circular flights from all the centres. Cost of the complete scheme will be €393 million. ZET will employ technology for the NT version.

DIMENSIONS, EXTERNAL:
Envelope: Length overall .... 125 m (410 ft)

DIMENSIONS, INTERNAL:
Envelope volume .... 50,000 m³ (1,766,000 sq ft)

NEW ENTRY

Model of enlarged Zeppelin NT, provisionally designated ZET, proposed by Zeppelin-Tourismus *(Paul Jackson)* ***NEW***/0576734

Hungary

DANEX

Apagyi u 29, 4552 Napkor
Tel: (+36 30) 624 29 45
e-mail: danex@mail.axelero.hu
Web: http://w1.171.telia.com/~u17112535/euro

Danex builds the EuroCub ultralight which is now marketed in the UK, having been sold in the Nordic countries for over a decade.

NORDIC SALES AGENT:
Edland International AB, SE-756 35 Bålsta, Sweden
Tel: (+46 171) 571 38
e-mail: edland@telia.com

UK SALES AGENT:
14 The Street, Oaksey, Wiltshire, SN16 9TG, UK
Tel/Fax: (+44 1666) 57 50 83
e-mail: info@eurocub.com
Web: http://www.eurocub.com

UPDATED

DANEX EUROCUB

TYPE: Side-by-side ultralight.

PROGRAMME: Derived from Al Jasman's Canadian-designed Aces High Cuby II (Cuby I being single-seat), now out of production following manufacture of some 20 kits. Swedish consortium acquired production licence in 1989; European prototype, SE-YLM with Rotax 503; others followed with Rotax 582 and Rotax 912. Sales to Denmark, Italy, Norway and Sweden with Rotax 503 and to Finland, Norway and Sweden with Rotax 912, all retaining Cuby II name.

Production transferred in 1994 to Nyiregyháza, Hungary, as Lamco (Light Aircraft Manufacturing Company KFT) EuroCub Mk I with Rotax 912 UL engine and Warp Drive IvoProp three-blade propeller, initially retaining 10.27 m (33 ft 8¼ in) wing span, although this reduced and wing endplates added. Mk I became Mk IV upon 30 per cent increase in size of tailplane, plus larger fin and rudder. Lamco became Danex Engineering KFT in 1998.

UK-based EuroCub Ltd formed to market aircraft as factory-built ultralight.

First production EuroCub Mk IV (Rotax 912 UL flat-four) ***(Paul Jackson)*** ***NEW***/0589243

CUSTOMERS: Some 10 European Cuby IIs, two EuroCub 912 ULs, 20 Mk Is and two Mk IVs (plus conversions) produced by 2004. Three operated by Swedish Ultraflyers from Frolunda, Stockholm.

DESIGN FEATURES: Classic high-wing monoplane with spacious cabin and wide-track landing gear. Constant-chord wing, with endplates, attached to upper longerons and braced to lower longerons by streamlined V struts with vertical intermediate struts. Tailplane braced below with struts and to fin with wires.

Wing section Clark Y (modified).

FLYING CONTROLS: Conventional and manual. Full-span, non-drooping ailerons (no flaps). Horn-balanced rudder and elevators. Flight-adjustable tab in port elevator.

STRUCTURE: Welded 4130 steel tube fuselage covered in doped Ceconite fabric. Wings of 6061-T6 aluminium with aluminium leading-edges and endplates and Ceconite to the rear. Glass fibre engine cowling. Circular section steel wing bracing struts have streamline wooden fairings to the rear.

LANDING GEAR: Tailwheel type; fixed. Two faired-in side Vs attached to lower longerons, plus half-axles, each with bungee cord shock-absorber, attached to compression frame. Wide track, with tundra tyres. Mainwheels 6.00-6; steerable tailwheel 200×50. Hydraulic brakes. AeroComp SF 1400 floats and amphibious floats optional.

POWER PLANT: One 59.6 kW (79.9 hp) Rotax 912 UL or 73.5 kW (98.6 hp) Rotax 912 ULS flat-four piston engine, driving a Warp Drive ground-adjustable pitch propeller. Fuel capacity 80 litres (21.1 US gallons; 17.6 Imp gallons) in wing tanks; optionally 100 litres (26.4 US gallons; 22.0 Imp gallons) in fuselage tank.

DIMENSIONS, EXTERNAL:
Wing span .... 9.55 m (31 ft 4 in)
Length overall .... 6.70 m (21 ft 11¾ in)
Height overall .... 1.80 m (5 ft 10¾ in)

AREAS:
Wings, gross .... 15.13 m² (162.9 sq ft)

WEIGHTS AND LOADINGS :
Weight empty: Ultralight .... 255 kg (562 lb)
Max T-O weight: Ultralight landplane .... 450 kg (992 lb)
floatplane .... 490 kg (1,080 lb)
amphibious floatplane .... 550 kg (1,212 lb)

PERFORMANCE (Rotax 912 ULS at 450 kg; 992 lb):
Normal cruising speed .... 95 kt (176 km/h; 109 mph)
Stalling speed .... 32 kt (60 km/h; 37 mph)
T-O run .... 50 m (165 ft)
Landing run .... 100 m (330 ft)
Endurance (absolute) .... 5 h 15 min
g limits .... +6/−4

UPDATED

HALLEY

HALLEY KFT

PO Box 425, Mester út 3, Eger H-3301
Fax: (+36 36) 32 02 08
e-mail: apollo@mail.datanet.hu
Web: http://www.halley.hu
PRESIDENT: Zoltán Molnár

Apollo Aircraft formed at Eger in 1980 by Mr Molnár. First product (1981) was P21 hang glider. Hungarian agent for Rotax engines in 1990. Renamed Halley Kft in 1992. Also produces Jet Star flex-wing. Had sold more than 1,000 flex-wings by 2003. Workforce over 20; factory area 2,500 m² (26,900 sq ft).

VERIFIED

HALLEY APOLLO FOX

TYPE: Side-by-side ultralight.

PROGRAMME: Derived from Apollo Classic prototype (HA-YNBA). First flight 1998. Certified in Belgium, Croatia, France, Hungary, Romania and Sweden. In 2003, an Apollo Fox flew from Hungary to Australia without the assistance of a ground support team.

CUSTOMERS: More than 50 sold by 2003.

COSTS: €25,000 with Rotax 582 engine; €31,000 with Rotax 912; both excluding tax (2003).

DESIGN FEATURES: Built to conform to British BCAR Section S regulations. High, constant-chord wing attached to upper longerons and braced to lower longerons by streamline V struts with vertical intermediate struts. Braced tailplane and sweptback fin. Wings foldable in 10 minutes.

FLYING CONTROLS: Manual. Junkers flaperons, tailplane and rudder.

STRUCTURE: Welded steel/chrome molybdenum fuselage with Ceconite fabric covering. Wing of two aluminium spars with 14 main and 13 auxiliary ribs; Ceconite covering.

Tailwheel version of Halley Apollo Fox ***NEW***/0567749

Halley Apollo Fox nosewheel variant ***NEW***/0567748

LANDING GEAR: Tricycle or tailwheel type: fixed. Two faired-in side Vs with half-axles sprung under centre fuselage. Nosewheel option has forward inclined leg with combined intermediate brace/compressed rubber shock-absorber and fuselage-mounted spring cantilever mainwheel legs. Optional mechanical drum or hydraulic disc brakes.

POWER PLANT: One 59.6 kW (79.9 hp) Rotax 912 flat-four; or 48.0 kW (64.4 hp) Rotax 582 two-cylinder, liquid-cooled piston engine; or 74.6 kW (100 hp) Saburu EA81 flat four. Wooden two-blade propeller or ground-adjustable pitch three- or four-blade propellers of glass fibre or carbon fibre. Fuel tanks in wings, combined capacity 62 litres (16.4 US gallons; 13.6 Imp gallons).

DIMENSIONS, EXTERNAL:
Wing span .... 9.15 m (30 ft 0 in)
Length overall .... 5.80 m (19 ft 0¼ in)
Height overall .... 1.78 m (5 ft 10 in)

DIMENSIONS, INTERNAL:
Cabin max width .... 1.10 m (3 ft 7¼ in)

AREAS:
Wings, gross .... 11.40 m² (122.7 sq ft)

WEIGHTS AND LOADINGS:
Weight empty .... 274 kg (604 lb)
Max T-O weight .... 450 kg (992 lb)

PERFORMANCE:
Cruising speed at 75% power .... 81 kt (150 km/h; 93 mph)
Stalling speed .... 35 kt (64 km/h; 40 mph)
Max rate of climb at S/L (solo) .... 1,325 m (4,347 ft)/min
T-O run .... 70 m (230 ft)
Landing run .... 94 m (310 ft)
Range .... 270 n miles (500 km; 310 miles)
g limits .... +4/−2

UPDATED

India

ADA

AERONAUTICAL DEVELOPMENT AGENCY

PO Box 1718, Vimanapura Post Office, Bangalore 560 017
Tel: (+91 80) 523 30 60 and 523 20 83
Fax: (+91 80) 523 84 93 and 523 44 93
e-mail: pai@ada.ernet.in
Web: http://www.ada.gov.in
LCA PROGRAMME DIRECTOR: Dr Kota Harinarayana
PROJECT DIRECTOR, TECHNOLOGY DEVELOPMENT AND LCA NAVY: Dr T G Pai

The ADA, an autonomous organisation under the Indian Ministry of Defence, was created in 1984 to be responsible for design and development of the next-generation Light Combat Aircraft (LCA), which made its maiden flight in early 2001 and was named Tejas in May 2003; its manufacturing partner is Hindustan Aeronautics Ltd (HAL, which see). Several laboratories of the Defence Research and Development Organisation (DRDO) have also contributed to the LCA programme, initial production of which was approved in 2002. Also participating in Multirole Transport Aircraft (MTA), an international programme in the form of the Ilyushin/Irkut/HAL Il-214.

UPDATED

ADA TEJAS

English name: Radiant
TYPE: Multirole fighter.

DEVELOPMENT MILESTONES

Feasibility study began	83
Project definition completed	late 88
Prototype construction started	mid-91
FSED Phase 1 authorised	Jun 93
Rolled out	17 Nov 95
First flight	4 Jan 01
FSED Phase 2 authorised	Nov 01
Initial production approved	Mar 02
Public debut	4 May 03
First supersonic flight	1 Aug 03

PROGRAMME: Development, as Light Combat Aircraft (LCA), approved by Indian government in 1983 as MiG-21 replacement; project definition begun second quarter 1987, completed late 1988, with first flight then predicted for April 1990 and service entry in about 1995. However, due to early slippages, basic design not finalised until 1990; construction by HAL started mid-1991. FSED Phase 1 sanctioned by Indian government in June 1993. First aircraft (TD1: technology demonstrator; serial number KH2001) was nine months behind revised schedule at roll-out on 17 November 1995; several subsequent postponements of new (June 1996) target date for first flight, which eventually took place on 4 January 2001; completed first phase of flight tests with 12th flight on 2 June 2001, reaching 8,000 m (26,250 ft) altitude, 18° AoA and M0.71 speed. FSED Phase 2 authorised November 2001. First supersonic flight targeted for January 2002, but eventually achieved (M1.08) on 1 August 2003.

TD2 (KH2002) rolled out on 14 August 1998 and scheduled to fly in September 2001, but did not do so until 6 June 2002. This has marginally lighter empty weight, increased usable fuel and Indian-developed HUD compared with TD1. Both aircraft powered by F404-GE-F2J3 afterburning turbofan, ground runs of which began on 9 April 1998; indigenous Kaveri engine being developed by Gas Turbine Research Establishment (GTRE) Bangalore and will undertake flight trials in Russia in Tu-16 testbed. Bench trials of this engine had reportedly totalled some 1,200 hours by early 2003 (of 8,000 required), but flight trials still not started by that time; now re-targeted for mid-2004. SNECMA and Eurojet reported to have offered technological assistance in developing Kaveri engine, which now considered unlikely to be ready until 2007. Meanwhile, to offset delays in Kaveri programme, prototypes PV1 to PV5 and at least first eight production aircraft will be fitted with F404 engine, further 40 of which received US export approval in 2002 and 17 more in February 2004. Latter batch will be F404-GE-IN20 and power first eight limited production aircraft and two naval prototypes. TD1 and TD2 had made 119 flights by 6 November 2003 (55 and 64 respectively); aircraft named Tejas (pronounced thay' jus) by Indian Prime Minister on 4 May 2003, at which time aircraft made its first public appearance (TD1 and TD2 flying) and third aircraft (PV-1) rolled out. Four of the PVs to be single-seaters; the fifth will be a two-seat operational trainer having commonality with the naval variant of LCA.

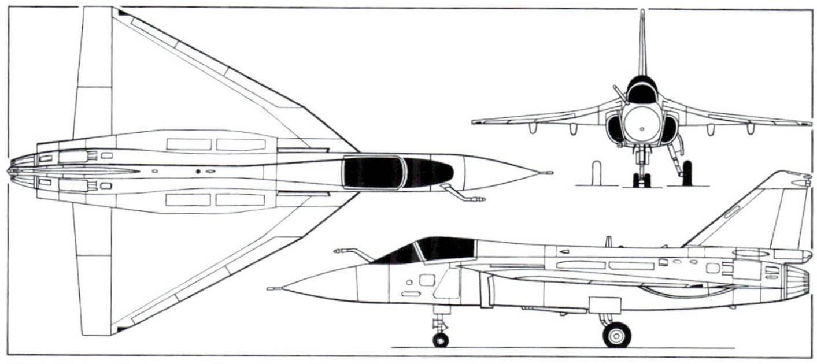
ADA Tejas light combat aircraft *(Jane's/Mike Keep)* 0110517

Programme delays further exacerbated by May 1998 US embargo on supplies and assistance following India's refusal to abandon nuclear weapons testing, necessitating all-Indian completion of development of integration of digital AFCS and some other systems (aircraft is inherently unstable and thus unable to fly without AFCS). However, sanctions lifted on 22 September 2001 and deliveries of air data and initial sensors by BAE Systems Controls (ex-Lockheed Martin) due to resume in mid-2003. Development of the indigenously developed radar is said to be on schedule; this is being flight tested in a HAL HS 748 and will then be installed in PV2. Additional US$125 million released in May 1999 to accelerate remaining development.

Phase 1 of flight test programme completed by April 2004, in which envelope expanded to M1.4 speed and 11,750 m (38,550 ft) altitude during approximately 200 hours by TD1/2 and PV-1.

Mid-2002 estimate indicated third (PV1) prototype (KH 2003) due to fly in 2002 (not achieved; eventually occurred on 25 November 2003); fourth and fifth (PV2 and 3) under construction in 2003 of which PV-2 nearing completion in March 2004; two-seater (PV5) to follow. IOC now expected to be declared with delivery of half a squadron at end of 2007; full operational clearance by December 2010.

CURRENT VERSIONS: **Tejas**: Single-seat version for Indian Air Force. *As described.*

Tejas Trainer: Tandem two-seat operational trainer; PV-5 prototype under development.

Tejas Navy: Carrierborne version; drooped nose, arrester hook, long-stroke landing gear with twin nosewheels, retractable canards and movable vortex controls at the wingroot. Design approved early 1999; development approved by Indian government in mid-2002. Two prototypes (two-seat NP-1 and single-seat NP-2); first flight (NP-1) targeted for 2007; F404-GE-IN20 engines; max T-O weight 12,500 kg (27,557 lb); max carrier landing weight 9,500 kg (20,944 lb).

CUSTOMERS: Indian Air Force (requirement for 200 single-seat and 20 trainers); Indian Navy may order up to 40 single-seaters. First eight for IAF approved in MoU March 2002; production by HAL at Bangalore.

COSTS: Phase 1 (TD1 and TD2) costs, including engine, estimated at approximately Rs30 billion (US$675 million) by late 2000. Estimated unit cost, based on 220 production total, US$17 million to US$20 million (2001).

DESIGN FEATURES: Tail-less delta planform with relaxed static stability; shoulder-mounted delta wings with compound sweep on leading-edges; large twist from inboard to outboard leading-edges; wing-shielded, side-mounted, fixed-geometry Y-duct air intakes. Advanced materials for minimum structural weight; fly-by-wire and HOTAS controls; high agility; supersonic at all altitudes; wide range of external stores.

FLYING CONTROLS: Hydraulically actuated two-segment trailing-edge elevons and three-section leading-edge slats; vortex-shedding inboard leading-edges with inboard slats to form vortices over wingroot and fin; airbrake in top of fuselage each side of vertical fin; these and rudder also actuated hydraulically.

STRUCTURE: Advanced materials include aluminium-lithium alloy, carbon composites and titanium alloys; CFRP wings (including elevons), fin (except GFRP tip) and rudder; Kevlar

Formation flight by TD1 and TD2 Tejas prototyps ***NEW***/0572251

radome. Wings manufactured with one-piece CFRP top and bottom skins, bolted on to wing box; majority of spars and ribs in composites. Fin, rudder, elevons, airbrakes and landing gear doors embody co-cured, co-bonded techniques. Carbon fibre composites to be increased from 30 per cent of airframe weight in TD1/2 to 42 per cent in PVs, with corresponding reduction of aluminium alloys from 57 to 43 per cent.

Hindustan Aeronautics (HAL) responsible for fuselage, communications system, electrical system, mechanical system LRUs and utility systems management; Aeronautical Development Establishment (ADE) flight control system, cockpit displays and display processors; National Aerospace Laboratories (NAL) responsible for development of fin and fabrication of rudder.

LANDING GEAR: Hydraulically retractable tricycle type. Single inward-retracting mainwheels; twin-wheel forward-retracting and steerable nose unit. DRDL carbon disc brakes; computer-controlled brake management system. ADRDE brake-chute in fairing at base of rudder.

POWER PLANT: *TD1, TD2 and PV-1 to -5:* One 80.1 kN (18,000 lb st) General Electric F404-GE-F2J3 afterburning turbofan; Indian GTRE GTX-35VS Kaveri turbofan (52.0 kN; 11,700 lb st dry, 80.5 kN; 18,100 lb st with afterburning), with digital engine control unit (KADECU), under development for later production aircraft. Multi-axis thrust vectoring nozzle planned.

NP-1, NP-2 and initial production aircraft: One 83.2 kN (18,700 lb st) General Electric F404-GE-IN20 afterburning turbofan, with FADEC.

Internal fuel in wing and fuselage integral tanks. Fixed in-flight refuelling probe on starboard side of front fuselage. Provision for up to three 1,200 or five 800 litre (317 or 211 US gallon; 264 or 176 Imp gallon) external fuel tanks on wing inboard, mid-board and underfuselage stations.

ACCOMMODATION: Pilot only, on Martin-Baker zero/zero ejection seat. Canopy opens sideways to starboard. Development will include two-seat training version.

SYSTEMS: Hydraulic system (by HAL) for powered flying controls, brakes and landing gear; electrical system for fly-by-wire and avionics power supply; Spectrum Infotech environmental control system for cockpit, radar and fuel tanks; lox system; HAL GTSU-110 gas turbine starter unit; aircraft-mounted accessory gearbox; USMS for aircraft utilities systems management. All utility systems (fuel, hydraulic, environmental control, electrical and brake management) are controlled by microprocessors.

AVIONICS (PRODUCTION VERSION): Avionics systems have capability to monitor health of main utility systems.

Comms: HAL V/UHF (with built-in ECCM) and standby UHF radios and audio management unit; HAL air-to-air/air-to-ground secure datalink; HAL IFF transponder/interrogator.

Radar: Electronics Research and Development Establishment (ERDE)/HAL multimode radar with multitarget search and track-while-scan and ground mapping functions; coherent pulse Doppler system, with look-up/look-down modes, Doppler beam-sharpening and moving target indication.

Flight: Quadruplex digital fly-by-wire AFCS. RLG-based INS; provision for GPS/INS; HAL RAM 1701A radio altimeter.

Instrumentation: NVG-compatible 'glass cockpit'; two Bharat Electronics active matrix, reconfigurable 76 × 76 mm (3 × 3 in) colour LCD MFDs; CSIO (India) collimated HUD; dedicated LCD 'get-you-home' panel; LCD multifunction keyboard.

Mission: Three multiplexed MIL-STD-1553B digital databusses; 32-bit mission computer operating in Ada, backed up by a second, equally powerful computer; IRST; laser range-finder/designator pod. Provision for reconnaissance, EW or other sensor pods. Helmet-mounted display/sight.

Self-defence: EW suite, by Advanced Systems Integration and Evaluation Organisation (ASIEO), includes radar warning receiver, self-protection jammer, laser warning system, missile approach warning system, countermeasures dispensing system and chaff/flare dispenser.

ARMAMENT: GSh-23 twin-barrel 23 mm gun with 220 rounds in blister beneath starboard engine intake trunk. Eight external stores stations (three under each wing, one on fuselage centreline and one beneath port intake trunk) for wide range of short/medium-range air-to-air missiles, PGMs, air-to-surface (including anti-ship) missiles, targeting or ECM pod, unguided rockets, conventional and retarded bombs, and cluster bomb dispensers. Indigenous Astra active radar-guided ASM under development by DRDL. Underwing stations stressed for loads of 1,200 kg (2,646 lb) inboard, 800 kg (1,764 lb) centre and 150 kg (331 lb) outboard; centreline station also 1,200 kg; port intake trunk station can carry 200 kg (441 lb) pod (FLIR, IRST, laser designator or reconnaissance). All except outboard underwing stations are wet for carriage of drop tanks.

DIMENSIONS, EXTERNAL:
- Wing span ... 8.20 m (26 ft 10¾ in)
- Wing aspect ratio ... 1.8
- Length overall ... 13.20 m (43 ft 3¾ in)
- Height overall ... 4.40 m (14 ft 5¼ in)
- Wheel track ... 2.20 m (7 ft 2½ in)
- Wheelbase ... 4.34 m (14 ft 2¾ in)

AREAS:
- Wings, gross ... approx 38.40 m^2 (413.3 sq ft)

WEIGHTS AND LOADINGS:
- Weight empty ... approx 5,500 kg (12,125 lb)
- Max external stores load ... more than 4,000 kg (8,818 lb)
- T-O weight (clean) ... approx 8,500 kg (18,740 lb)
- Wing loading (clean) ... approx 221.4 kg/m^2 (45.35 lb/sq ft)
- Power loading (TD1 and 2, clean) ... 106 kg/kN (1.04 lb/lb st)

PERFORMANCE (estimated):
- Max level speed at altitude ... M1.8
- Service ceiling ... above 15,240 m (50,000 ft)
- *g* limits ... +9/−3.5

UPDATED

ADA MEDIUM COMBAT AIRCRAFT (MCA)

The ADA is studying an advanced version of the Tejas under the above designation, as a potential replacement for Indian Air Force Jaguars and Mirage 2000s from about 2010. The single-seat MCA is intended to embody stealth technology, be powered by a non-afterburning variant of the GTRE Kaveri turbofan with thrust vectoring, carry laser-guided weapons and have a greater range than the Tejas. Other design features are planned to include the adoption of radar-absorbent materials; twin outward-canted tailfins; and overwing external fuel tanks. Design and development costs were provisionally estimated at US$2.3 billion in 1996 dollars.

WEIGHTS AND LOADINGS (approx):
- T-O weight: clean ... 12,000 kg (26,455 lb)
- max ... 18,000 kg (39,685 lb)

UPDATED

CSIR

COUNCIL OF SCIENTIFIC AND INDUSTRIAL RESEARCH

Centre for Civil Aircraft Design and Development, C-MMACS Campus, Belur, Bangalore 560 037
Tel: (+91 80) 526 32 19
Fax: (+91 80) 526 77 81 and 529 82 92
e-mail: narayan@ccadd.cmmacs.ernet.in
Web: http://www.cmmacs.ernet.in-/nal/pages/saraspg/sarashm.htm
HEAD OF CCADD AND SARAS PROJECT MANAGER: Dr K Yegna Narayan

Design and development of the Saras twin-turboprop business and commuter transport is nearing fruition as an Indian programme, having previously appeared in the International section under the Myasishchev/NAL heading and latterly under that of the Indian National Aerospace Laboratories (NAL). It has received Rs 1.58 billion (US$33 million) funding from various Indian government agencies and is currently under the leadership of CSIR, which has created the Centre of Civil Aircraft Design and Development (CCADD) to act as the nodal agency to monitor and manage the programme. Many Indian public and private sector industries are participating in the venture, with manufacture of production components being carried out in 35 work centres across the country. Final assembly will be by NAL.

VERIFIED

Saras prototype taxying for its maiden flight ***NEW***/0580550

Model of Saras-S, shown at Asian Aerospace, Singapore, February 2004 ***(Paul Jackson)*** ***NEW***/0580548

CSIR SARAS

English name: Crane

TYPE: Twin-turboprop transport.

PROGRAMME: Originated as six/nine-passenger light transport shown in model form at Moscow Aerospace '90 exhibition; revised design announced by Myasishchev 1993, with name Delphin; India had comparable project; general agreement concluded 1993 to combine Myasishchev and Indian programmes, Russian version being known as M-102 Duet and Indian version as Saras (species of Indian crane). Full-scale mockup shown at 1993 and 1995 Moscow Air Shows. Prototype construction (two in Russia, one in India originally) planned to begin in September 1994, but Indian share of private venture capital not secured until early 1996. Detailed engineering began in April 1996 following finalisation of design late in previous year; however, in 1997 Myasishchev was proceeding with unilateral development of the similar M-202, yet by 2001, was again promoting M-102 version; neither has progressed to the prototype stage in Russia. In early November 2001, Russia and India signed a protocol on construction of the M-102/Saras.

India elected to continue with the Saras as a national programme, although some collaboration with the former Russian partner was not ruled out. Government secured private backing of Taneja Aerospace and Kumaran Industries, plus financial support from the Indian Technology Development Board. Hindustan Aeronautics Ltd also became a major partner in the development programme and to build production aircraft at Nasik. Release of further government funding in mid-1999 enabled work to begin on manufacture (by HAL) of two flying prototypes and a structural test airframe. First prototype (VT-XSD) reported in February 2001 to be 80 per cent complete, with first flight then expected by end of 2001, but in following month, programme management board called for "rigorous weight control exercise". Roll-out 4 February 2003; first engine run conducted 30 October 2003; first flight 29 May 2004, followed by second flight on 7 June 2003. 'Inaugural' flight of 22 August 2004 was seventh sortie, performed for VIPs. Registration VT-XRM reserved for second prototype. Certification, will be to FAR Pt 25; initial deliveries anticipated no earlier than mid-2007.

CURRENT VERSIONS: **Saras:** Initial variant, *as described.*

Saras-S: Stretched 22- to 25-seat version, first publicly revealed in model form at Asian Aerospace 2004 in Singapore in February 2004 and expected to become standard production model. Provisional data includes: maximum T-O weight 7,900 kg (17,416 lb); maximum

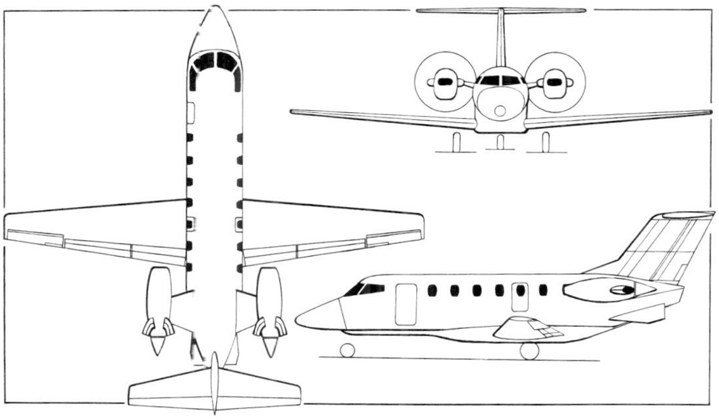

CSIR Saras twin-turboprop transport *(James Goulding)* 0130859

fuel 900 kg (1,984 lb); maximum speed at 9,144 m (30,000 ft) 249 kt (462 km/h; 287 mph); cruising speed 239 kt (442 km/h; 275 mph); T-O run 880 m (2,887 ft); service ceiling 9,144 m (30,000 ft); max rate of climb at S/L 826 m (2,170 ft)/min; range with max payload at FL300 540 n miles (1,000 km; 621 miles).

CUSTOMERS: At least 250 Indian orders forecast over 15 years. Indian Air Force has initial requirement for six, to replace Dornier Do 228s, by 2005; Indian coast guard also considering Saras as Do 228 replacement.

COSTS: Design and development cost quoted in February 1996 as US$40 million. Late in 1996, cost re-estimated as US$60 million, including flight test phase. Indian government approved Rs1.3 billion (US$30 million) in June 1999 to enable prototype manufacture to begin. Estimated unit price US$3.75 million (2001).

DESIGN FEATURES: Suitable for operations in hot-and-high conditions (typically, airfield at 2,000 m; 6,560 ft, at up to 45°C), and from semi-prepared runways. Unconventional twin pusher propeller configuration. High aspect ratio, low-mounted wings, with straight taper; pressurised circular-section fuselage; T tailplane on sweptback fin; engines pylon-mounted each side of rear fuselage.

Wing section GA(W)-2 (modified); leading-edge sweep 5°; thickness/chord ratio 15 per cent; 4° dihedral; 2° incidence; 2° twist.

FLYING CONTROLS: Conventional and manual. Tabs in all control surfaces; electrically operated pitch, roll and yaw trim; yaw damper; spoilers and 25 per cent chord, single-slotted Fowler flaps; dual-channel three-axis autopilot. Highly swept ventral fins provide nose-down pitching moment at high AoA.

STRUCTURE: Mixed construction of aluminium alloy (fuselage, engine pylons and fixed portion of wing) and composites (nosecone, wing/body fairing, wing moving surfaces and tips); designed for 30,000 hour life; fail-safe philosophy for all primary structures and major attachments; incorporates damage tolerance features. Two-spar wing with integrally milled skin riveted to stringers; two-spar flaps with honeycomb sandwich filler; single-spar ailerons with honeycomb filler; single-piece, two-spar metal tailplane and three-spar metal fin.

LANDING GEAR: Hydraulically retractable tricycle type, with single mainwheels and single steerable (±53°) nose unit. Mainwheels retract inward into wingroot fairings, nosewheel forward. Oleo-pneumatic shock-absorbers. Mainwheels 26×8.75, pressure 5.52 bar (80 lb/sq in); nosewheel 17.5×6.3, pressure 3.79 bar (55 lb/sq in); air-cooled carbon brakes. Minimum turning radius 8.10 m (26 ft 7 in) at nosewheel.

POWER PLANT: Two 634 kW (850 shp) Pratt & Whitney Canada PT6A-66 turboprops, pylon-mounted on sides of rear fuselage; Hartzell five-blade, constant-speed pusher propellers, rotating in opposite directions. Integral fuel tank in each wing, combined capacity 1,608 litres (425 US gallons; 354 Imp gallons), of which 1,595 litres (421 US gallons; 351 Imp gallons) are usable. Gravity refuelling point in port wing upper surface; optional single-point pressure refuelling point at starboard wingroot.

ACCOMMODATION: Two-person flight deck with dual controls, but to be certified also for one-pilot operation. Alternative layouts for 14 economy class passengers in single seats at 76 cm (30 in) pitch, with centre aisle, rear lavatory and total of 1.1 m^3 (39 cu ft) for baggage in forward and rear compartments; or 18 passengers in commuter layout. Typical executive interior could have eight seats, tables, wardrobe, baggage compartment, galley and lavatory. Ambulance version capable of carrying six stretchers, with seats for two medical attendants, medical supplies storage, baggage compartment and lavatory. Various passenger/cargo arrangements or special mission interiors optional in combi version. Door at front on port side, hinged downward, with integral airstairs; overwing Type III emergency exit on each side. Baggage compartment door on port side of tailcone.

SYSTEMS: Bootstrap environmental control system and pneumatic cabin pressure control system; cabin pressure differential 0.45 bar (6.5 lb/sq in). Hydraulic system (pressure 207 bar; 3,000 lb/sq in) for landing gear actuation, brakes and nosewheel steering; 4 litres (1.06 US gallons; 1.27 Imp gallons)/min electrically powered hydraulic pump; pneumatic system for emergency lowering of landing gear.

Primary electrical power is 28 V DC, supplied by two 12 kW starter/generators; 43 Ah Ni/Cd battery provides emergency power for essential loads for approximately 30 minutes, including three engine restarts. Two solid-state inverters supply 115/26/5 V AC power at 400 Hz for instrumentation; AC power for anti-icing and windscreen heating obtained via alternators. Emergency oxygen system for crew and passengers (two 1,400 litre; 49.4 cu ft bottles), plus two 122 litre (4.3 cu ft) portable bottles and masks for first aid. Engine fire and cabin smoke detection systems. Halon fire extinguishing system. De-icing boots on wing leading-edge nacelle intakes.

AVIONICS: Integrated digital system (ARINC 429 compatible).

Comms: VHF (two), optional HF, intercom/PA system, TCAS II transponder and CVR.

Radar: Weather radar.

Flight: Autopilot (dual-channel, three-axis), ADF, VOR/ILS, marker beacon receiver, DME, radar altimeter, air data sensor and computer, AHRS, flight control computer and flight director standard; GPS, satcom, Selcal and flight management system optional.

Instrumentation: Four-tube EFIS; 'return home' standby instrumentation.

DIMENSIONS, EXTERNAL:

Wing span	14.70 m (48 ft 2¾ in)
Wing chord: at c/l	2.65 m (8 ft 8¼ in)
at tip	0.85 m (2 ft 9½ in)
Wing aspect ratio	8.4
Length overall	15.02 m (49 ft 3¼ in)
Fuselage: Length	13.90 m (45 ft 7¼ in)
Max diameter	1.95 m (6 ft 4¾ in)
Height overall	5.20 m (17 ft 0¾ in)
Tailplane span	6.09 m (19 ft 11¾ in)
Wheel track	3.20 m (10 ft 6 in)
Wheelbase	6.465 m (21 ft 2½ in)
Propeller diameter	2.16 m (7 ft 1 in)
Propeller ground clearance	1.30 m (4 ft 3¼ in)
Distance between propeller centres	3.89 m (12 ft 9 in)
Passenger door (fwd, port): Height	1.48 m (4 ft 10¼ in)
Width	0.75 m (2 ft 5½ in)

DIMENSIONS, INTERNAL:

Cabin: Length	7.74 m (25 ft 4¾ in)
Max width	1.80 m (5 ft 10¾ in)
Aisle width at floor	0.54 m (1 ft 9¼ in)
Max height	1.72 m (5 ft 7¾ in)
Floor area	7.08 m^2 (76.2 sq ft)
Volume	21.3 m^3 (752 cu ft)
Baggage hold volume: fwd, stbd	0.64 m^3 (22.6 cu ft)
rear, port	0.46 m^3 (16.2 cu ft)

AREAS:

Wings, gross	25.70 m^2 (276.6 sq ft)
Ailerons (total)	1.27 m^2 (13.71 sq ft)
Trailing-edge flaps (total)	3.99 m^2 (42.93 sq ft)
Spoilers (total)	0.65 m^2 (7.00 sq ft)
Fin	5.70 m^2 (61.40 sq ft)
Rudder, incl tab	1.81 m^2 (19.45 sq ft)
Tailplane	7.00 m^2 (75.35 sq ft)
Elevators, incl tabs	2.05 m^2 (22.07 sq ft)

WEIGHTS AND LOADINGS:

Weight empty, equipped	4,116 kg (9,074 lb)
Max fuel weight	1,326 kg (2,923 lb)
Max payload	1,232 kg (2,716 lb)
Max T-O and landing weight	6,100 kg (13,448 lb)
Max zero-fuel weight	5,348 kg (11,790 lb)
Max wing loading	237.4 kg/m^2 (48.61 lb/sq ft)
Max power loading	4.81 kg/kW (7.91 lb/shp)

PERFORMANCE (estimated):

Never-exceed speed (VNE)	372 kt (690 km/h; 428 mph)
Max level speed at FL250	283 kt (524 km/h; 326 mph)
Econ cruising speed at FL250	216 kt (400 km/h; 249 mph)
Stalling speed at S/L, power off:	
flaps up	96 kt (178 km/h; 111 mph)
flaps down	78 kt (145 km/h; 90 mph)
Max rate of climb at S/L	700 m (2,297 ft)/min
Rate of climb at S/L, OEI	130 m (427 ft)/min
Max certified altitude	9,100 m (29,860 ft)
Service ceiling	11,000 m (36,090 ft)
Service ceiling, OEI	6,000 m (19,680 ft)
T-O run	720 m (2,362 ft)
T-O field length	1,000 m (3,280 ft)
Landing field length	1,200 m (3,937 ft)
Landing run	720 m (2,362 ft)
Runway LCN	10
Range, IFR reserves:	
with 8 passengers	756 n miles (1,400 km; 870 miles)
with 14 passengers	216 n miles (400 km; 248 miles)
self-ferry	1,038 n miles (1,924 km; 1,195 miles)
Endurance	5 h

UPDATED

HAL

HINDUSTAN AERONAUTICS LIMITED

CORPORATE OFFICE: PO Box 5150, 15/1 Cubbon Road, Bangalore 560 001
Tel: (+91 80) 286 46 36/7 and 286 46 39/43
Fax: (+91 80) 286 71 40 and 286 87 58
e-mail: marketing@hal-india.com
Web: http://www.hal-india.com
CHAIRMAN: N R Mohanty
EXECUTIVE DIRECTOR, PLANNING: B K Banerjee
DIRECTOR, CORPORATE PLANNING AND MARKETING: S C Das
GENERAL MANAGER, MARKETING: D Chowdhury
SENIOR MANAGER, PUBLIC RELATIONS: A Nageswara Rao

Hindustan Aircraft (Pvt) Ltd, established at Bangalore in 1940, was taken over by the Indian Ministry of Industry and Supplies in 1945, transferring to Ministry of Defence control in 1951. In August 1963, Hindustan Aeronautics Ltd (HAL) was incorporated as a wholly government-owned entity, adopting its present existence upon merging (with no change of name) with Aeronautics India Ltd (AIL) on 1 October 1964. HAL had manufactured more than 3,300 aircraft (12 types of in-house design and 13 types by licensed manufacture) and 3,400 aero-engines by February 2004. It has 14 manufacturing divisions at eight locations (seven at Bangalore, one each at Nasik, Koraput, Hyderabad, Barrackpore, Kanpur, Lucknow and Korwa), plus Design and Development Complex; nine R&D centres located with manufacturing divisions; all divisions have ISO 9000, 9001 and 9002 accreditation. Workforce just over 31,000 in early 2004. Additional production centres at Nasik and Koraput were due to open during 2002. Hyderabad Division manufactures avionics for all aircraft produced by HAL, plus air route surveillance and precision approach radars; Lucknow Division produces instruments and other accessories under licence from manufacturers in France, Russia and the UK; Korwa manufactures inertial navigation and nav/attack systems.

Former Multirole Transport Aircraft (MTA) project now an international joint venture (IRTA-21: Indo-Russian Transport Aircraft, 21st century) with Ilyushin and Irkut of Russian Federation.

Turnover during 2002–03 totalled Rs31.2 billion, including Rs1.03 million in exports; net profit during this period was Rs4.51 billion; turnover targeted to increase by 20 per cent in 2003–04, in which period HAL planned to invest Rs9 billion in improved facilities to produce Tejas, HJT-36 and Su-30MKI.

Major elements of HAL are:

DESIGN AND DEVELOPMENT COMPLEX Follows this entry
BANGALORE COMPLEX Follows Design and Development Complex
TRANSPORT AIRCRAFT DIVISION Follows Bangalore Complex
MIG COMPLEX Follows Transport Aircraft Division
ACCESSORIES COMPLEX
PO Box 215, Lucknow 226 016
Tel: (+91 522) 34 03 27
Fax: (+91 522) 34 03 42
MANAGING DIRECTOR: L M Bhardwaj

Comprises Lucknow (Accessories), Hyderabad (Avionics) and Korwa (Avionics) Divisions.

UPDATED

DESIGN AND DEVELOPMENT COMPLEX

PO Box 1789, Bangalore 560 017
Tel: (+91 80) 526 34 57
Fax: (+91 80) 526 30 96
Telex: 845 8083
DIRECTOR: Ashok K Baweja
GENERAL MANAGERS:
Yogesh Kumar (ARDC)
Fateh Singh (ARDC)
N H Venkatesh (RWRDC)

HAL currently has nine Research and Design (R&D) Centres, including one each for fixed-wing aircraft (ARDC) and rotary-wing (RWRDC) programmes. Former's main programmes are LCA Tejas (managed by ADA) and HJT-36 Sitara; latter concerned mainly with Dhruv ALH. Transport Aircraft R&D Centre was responsible for developing oversize cabin door, integration of Super Marec radar, gun pod, IR/UV scanner and flight data recorder for HAL-built Dornier 228 programme. Earlier designs have included HT-2, Pushpak, Krishak, Basant, Marut, Kiran, Deepak and HTT-34. Jet trainer successor to Kiran was revealed as HJT-36 in 1998; flight testing began in 2003. Complex workforce 1,901 in January 2003.

UPDATED

HAL LCH

TYPE: Attack helicopter.

PROGRAMME: Derivative of Dhruv (which see); displayed in model form at Paris Air Show, June 2001 as LAH (light attack helicopter) proposal, but was restyled LCH, signifying light combat. Original slimmed-down 'gunship' fuselage now discarded and basic Dhruv airframe retained except for forward fuselage modified to tandem crew seating. Officially launched 5 February 2003 and replaces earlier LOH programme, which now to be met by foreign design. LCH scheduled to fly within two years of government go-ahead, which still awaited in April 2004; intended to augment, and eventually replace, Indian Air Force Mi-35 fleet.

COSTS: Rs300 million earmarked for design and development (2003).

DESIGN FEATURES: Generally as for Dhruv except for modified nose section with tandem crew seating. Four-blade hingeless main rotor with swept blade tips. Intended for anti-tank, close air support, air-to-air combat and scout roles.

STRUCTURE: Extensive use of composites to reduce radar signature.

LANDING GEAR: Non-retractable and crashworthy tricycle type.

POWER PLANT: Two 895 kW (1,200 shp) Turbomeca/HAL TM333-2C2 Ardiden 1H turboshafts (Indian name Shakti) with FADEC, derated to 798 kW (1,070 shp). Crashworthy fuel system.

ACCOMMODATION: Crew of two in tandem; ergonomic cockpit.

SYSTEMS: Four-axis autostabilisation system; anti-resonance isolation system (ARIS). Dual-redundant electrical and hydraulic systems as for Dhruv.

AVIONICS: *Flight:* Digital AFCS and nav/com as for Dhruv.
Instrumentation: Dual night attack MFDs in each cockpit.
Mission: FLIR, TV and laser range-finder/designator; helmet-mounted sights.
Self-defence: Radar and laser warning receivers; MAWS; chaff/flare dispenser.

ARMAMENT: Undernose 20 mm cannon. Stub-wing hardpoints for ASMs, AAMs or rocket launchers.

DIMENSIONS, EXTERNAL:
Main rotor diameter .... 13.20 m (43 ft 3¾ in)
Length: fuselage .... 13.94 m (45 ft 8¾ in)
overall, rotors turning .... 15.86 m (52 ft 0½ in)
Height to top of rotor head .... 5.27 m (17 ft 3½ in)
Stub-wing span .... 4.43 m (14 ft 6½ in)
Wheel track .... 2.50 m (8 ft 2½ in)

WEIGHTS AND LOADINGS:
Max T-O weight .... 5,500 kg (12,125 lb) class

PERFORMANCE (estimated):
Never-exceed speed (VNE) .... 178 kt (330 km/h; 205 mph)
Max cruising speed .... 151 kt (280 km/h; 174 mph)
Service ceiling .... 6,500 m (21,320 ft)
Range with standard fuel .... 378 n miles (700 km; 435 mph)

UPDATED

Artist's impression of HAL Light Combat Helicopter ***NEW***/0576665

BANGALORE COMPLEX

Post Bag 1785, Bangalore 560 017
Tel: (+91 80) 522 82 30
Fax: (+91 80) 522 95 64
MANAGING DIRECTOR: A K Sakena

Bangalore Complex comprises Aircraft Division, Helicopter Division, Aerospace Division, Engine Division, Overhaul Division, Foundry and Forge Division, Industrial and Marine Gas Turbines Division, and Airport Services Centre. Programmes include subcontract work for leading aerospace companies such as BAE Systems, EADS, Boeing and Latécoère. Aircraft Division will complete 50 of 66 BAE Hawk Mk 132s ordered on 26 March 2004 for the Indian Air Force (eight from CKD kits, 42 by local manufacture, after delivery of 16 UK-built aircraft from BAE); first UK deliveries due in 2007.

Aerospace Division manufactures light alloy structures and assemblies for satellites and launch vehicles. Engine Division manufactures, overhauls and repairs Adour Mk 811, Artouste IIIB and TPE331-5 engines; it also overhauls and repairs Adour Mk 804E, Dart, Gnome, Orpheus and Avon engines. Overhaul of Jaguar, Kiran, Mirage and An-32 aircraft, Cheetah helicopters and Pratt & Whitney and Textron Lycoming piston engines is undertaken by Overhaul Division.

AIRCRAFT DIVISION:
Post Bag 1788, Bangalore 560 017
Tel: (+91 80) 522 89 69
Fax: (+91 80) 522 51 88
GENERAL MANAGER: G Parasuramireddy

Main programmes Jaguar International production and upgrade; composites and metal drop tanks; Dornier 228 landing gears; sheet metal items; forward passenger doors for Airbus A319/A320/A321 (600 shipsets); overwing emergency exit doors for Boeing 757 and 777. Floor area 125,100 m^2 (1,346,565 sq ft); workforce 2,102 in January 2003.

HELICOPTER DIVISION:
Post Bag 1790, Bangalore 560 017
Tel: (+91 80) 523 86 02
Fax: (+91 80) 523 47 17
GENERAL MANAGER: S M Kapoor

Manufacture and overhaul of Cheetah, Chetak, Lancer and Dhruv light helicopters. MoU in February 2004 for local production of rotorblades for Bell 206.

UPDATED

HAL (SEPECAT) JAGUAR INTERNATIONAL

Indian Air Force name: Shamsher (Assault Sword)

TYPE: Attack fighter.

Comprehensive details in Jane's Aircraft Upgrades*; following refers generally to Jaguar S, and particularly to Indian version.*

PROGRAMME: Forty (including five two-seat) Jaguar Internationals with Adour Mk 804 engines delivered from UK, beginning March 1981; 45 more with Adour Mk 811s and new DARIN nav/attack system assembled in India, making first flight (JS136) 31 March 1982; further 31 manufactured under licence in India (first delivery early 1988); deliveries of further 15, ordered 1993, were completed by March 1999 with delivery of last three of this batch. Indian government ordered additional batch of 17 two-seaters in March 1999. These being equipped with upgraded DARIN II nav/attack system and employed in a night attack role with laser-guided weapons. Deliveries to Indian Air Force to be made between 2003 and 2005. Letter of intent for further 20 single-seaters has been received, which would extend Jaguar production until 2007; these will have upgraded DARIN II nav/attack system and new LRMTS.

HAL Maritime Jaguar of No. 6 Squadron, Indian Air Force *(Simon Watson)* 0132398

INDIAN AIR FORCE JAGUARS

Mark/Variant	Serial Nos.	Assembled[1]	Remarks	Qty
GR. Mk 1	JI003 to 018	BAe	Loaned by/returned to RAF[2]	(16)
S(I)	JS101 to 135	BAe	Phase 2; Adour 804, NAVWASS	35
	JS136 to 170	HAL	Phase 3; Adour 811, DARIN	35
	JS171 to 194	HAL	Phase 4; Adour 811, DARIN	24[3]
	JS195 to 205	HAL	Phase 5; Adour 811, DARIN	11
Maritime	JM251 to 258	HAL	Phase 4; Adour 811, DARIN, Agave	8[3]
	JM259 to 262	HAL	Phase 5; Adour 811, DARIN, Agave	4
	JM263 to 264	HAL	Phase 6	2
T. Mk 2	JI001 to 002	BAe	Loaned by/returned to RAF[2]	(2)
B(I)	JT051 to 055	BAe	Phase 2; Adour 804, NAVWASS	5
	JT056 to 065	HAL	Phase 3; Adour 811, DARIN	10
	JT066 to 080	HAL	Phase 6; night attack, Adour 811, DARIN	15
Total				**149(18)**

[1] All co-built by (SEPECAT) BAe and Dassault
[2] Adour 804 engines and NAVWASS nav/attack avionics; two single-seat lost in Indian service
[3] Phase 4 originally planned as 56 aircraft of entirely local manufacture; amended to 32 kits

CURRENT VERSIONS: **Jaguar B(I):** Two-seat trainer version; 15 delivered; further 15 on order with DARIN II nav/attack system, laser designator pod and modified fin; first eight of these due for delivery by end March 2004.

Jaguar S(I): Standard single-seat strike version; 106 delivered; further 20 scheduled to follow.

Jaguar Maritime: Jaguars of IAF No. 6 Squadron, assigned to anti-shipping role, have Agave radar, interfaced with DARIN nav/attack system and Sea Eagle anti-shipping missiles; first modified aircraft delivered January 1986; 12 ordered, of which all delivered by end of 1999. To be upgraded by substitution of Elta EL/M-2032 for Agave, with which flight trials were under way in 2002. In 2001, deliveries began of a further two Jaguar M.

CUSTOMERS: Indian Air Force received 132, comprising 117 single-seat (including 12 Maritime) and 15 two-seat combat-capable trainers. Further 15 Jaguar Bs and two Ms (ordered 1999) in production for delivery from 2003; additional 20 single-seaters expected to follow. Basic strike version equips Nos. 5 ('Tuskers') and 14 ('Bulls') Squadrons at Ambala and Nos. 16 ('Cobras') and 27 ('Flaming Arrows') Squadrons at Gorakhpur; anti-shipping version equips No. 6 ('Dragons') Squadron at Pune.

DESIGN FEATURES: Purpose-designed attack aircraft. Shoulder-wing monoplane. Anhedral 3°. Sweepback 40° at quarter-chord. Outer leading-edges fitted with slat which also gives effect of extended chord. Tail unit sweepback at quarter-chord 40° on horizontal, 43° on vertical surfaces; tailplane anhedral 10°. Ventral fins beneath rear fuselage.

FLYING CONTROLS: Fairey Hydraulics powered flying controls. No ailerons: control by two-section spoilers, forward of outer flap on each wing, in association (at low speeds) with differential tailplane. Hydraulically operated (by screwjack) full-span double-slotted trailing-edge flaps. Leading-edge slats can be used in combat. All-moving slab-type tailplane, the two halves of which can operate differentially to supplement the spoilers.

STRUCTURE: Wing is an all-metal two-spar torsion box structure; skin machined from solid aluminium alloy, with integral stiffeners. Main portion built as single unit, with three-point attachment to each side of fuselage. Fuselage all-metal, mainly aluminium, built in three main units and making use of panels and, around the cockpit(s), honeycomb panels. Local use of titanium alloy in engine bay area. Two door-type airbrakes under rear fuselage, immediately aft of each mainwheel well. Structure and systems aft of cockpit(s), identical for single-seat and two-seat versions. The tail unit is a cantilever all-metal structure, covered with aluminium alloy sandwich panels. Rudder and outer panels and trailing-edge of tailplane have honeycomb core.

LANDING GEAR: Messier-Hispano-Bugatti retractable tricycle type, all units having Dunlop wheels and low-pressure tyres for rough field operation. Hydraulic retraction, with oleo-pneumatic shock-absorbers. Forward-retracting main units each have twin wheels, tyre size 615×225-10, pressure 5.8 bar (84 lb/sq in). Wheels pivot during retraction to stow horizontally in bottom of fuselage. Single rearward-retracting nosewheel, with tyre size 550×250-6 and pressure 3.9 bar (57 lb/sq in). Twin landing/taxying lights in nosewheel door. Dunlop hydraulic brakes. Anti-skid units and arrester hook standard. Irvin brake parachute of 5.5 m (18 ft 0 in) diameter in fuselage tailcone.

POWER PLANT: Two HAL-built Rolls-Royce Turbomeca Adour Mk 811 turbofans (Phase 3 aircraft onwards), each rated at 25.0 kN (5,620 lb st) dry and 37.4 kN (8,400 lb st) with afterburning. Fixed-geometry air intake on each side of fuselage aft of cockpit. Fuel in six tanks, one in each wing and four in fuselage. Total internal fuel capacity 4,200 litres (1,110 US gallons; 924 Imp gallons). Armour protection for critical fuel system components. Provision for carrying three auxiliary drop tanks, each of 1,200 litres (317 US gallons; 264 Imp gallons) capacity, on fuselage and inboard wing pylons. Provision for in-flight refuelling, with retractable probe forward of cockpit on starboard side.

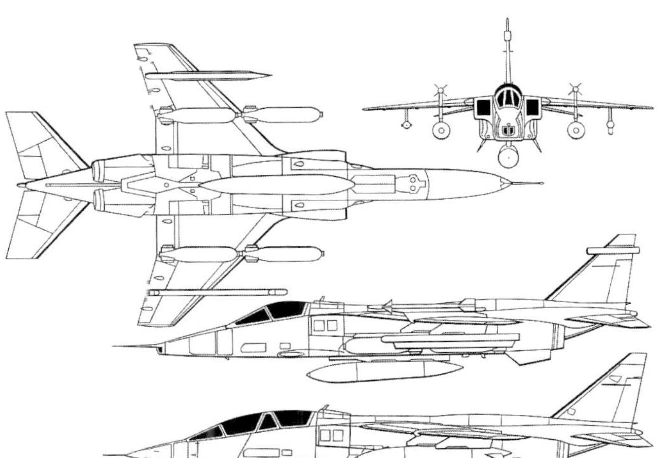

Jaguar single-seat tactical support aircraft and two-seat operational trainer *(Jane's/Mike Keep)* 0503190

ACCOMMODATION: Jaguar S(I) has enclosed cockpit for pilot, with rearward-hinged canopy and Martin-Baker Mk 9B zero/zero ejection seat. Jaguar B(I), crew of two in tandem on Martin-Baker Mk 9B Srs II zero/zero ejection seats. Individual rearward-hinged canopies. Rear seat 38 cm (15 in) higher than front seat. Windscreen bulletproof against 7.5 mm rifle fire.

SYSTEMS: Air conditioning and pressurisation systems for cockpit(s) also control temperature in certain equipment bays. Two independent hydraulic systems, powered by two engine-driven pumps. Hydraulic pressure 207 bar (3,000 lb/sq in). First system (port engine) supplies one channel of each actuator for flying controls, hydraulic motors which actuate flaps and slats, landing gear retraction and extension, brakes and anti-skid units. Second system supplies other half of each flying control actuator, two further hydraulic motors actuating slats and flaps, airbrake and landing gear emergency extension jacks, nosewheel steering and wheel brakes. In addition, there is a hydraulic power transfer unit and an emergency electrohydraulic pump.

Electrical power provided by two 15 kVA AC generators, either of which can sustain functional and operational equipment without load shedding. DC power provided by two 4 kW transformer-rectifiers. Emergency power for essential instruments provided by 40 Ah battery and static inverter. De-icing, rain clearance and demisting standard. Liquid oxygen system, which also pressurises pilot's anti-*g* trousers.

Jaguar is fully power-controlled in all three axes and is automatically stabilised as a weapons platform by gyros which sense disturbances and feed appropriate correcting data through a computer to the power control assemblies, in addition to human pilot manoeuvre demands. Power controls are all of duplex tandem arrangement, with mechanical and electrical servo-valves of Fairey platen design. Air-to-air combat capability can be enhanced by inclusion of roll/yaw dampers, to increase lateral stability, and by increasing slat and flap angles.

AVIONICS: HAL-manufactured DARIN (display attack and ranging inertial navigation) nav/attack system initially, incorporating INS, HUDWAC, COMED, interconnected MIL-STD-1553B dual-redundant databus and interfaced with LRMTS. The 15 new two-seaters and (when ordered) additional 20 strike versions will be fitted with new RLG/INS with GPS, new HUD and smart MFDs (see below); HOTAS upgrade postponed.

Comms: Main (20-channel) V/UHF; standby UHF and HAL COM 326A HF transceivers (being replaced by Incom integrated com system); IFF-400 AM transponder (single-seaters only).

Radar: Thales Avionics Agave originally in Maritime version, interfaced with DARIN system; planned to be replaced by Elta EL/M-2032.

Flight: SAGEM ULISS 82 INS (to be replaced by RLG/INGPS); HAL ADF and radar altimeter.

Instrumentation: Smiths HUDWAC (head-up display and weapon aiming computer) (to be replaced by Elop HUD with Smiths/HAL co-developed mission computer); BAE Systems COMED 2045 (combined map and electronic display) to be replaced by Thales (Sextant) digital autopilot and 152 × 152 mm (6 × 6 in) smart multifunction display (SMFD), ordered November 1999.

Mission: Laser ranger and marked target seeker (LRMTS). Provisional selection of Rafael Litening laser designation pod announced February 1996.

Self-defence: RWR; active and passive ECM.

ARMAMENT: Two 30 mm Aden guns in lower fuselage aft of cockpit in single-seater, with 150 rds/gun; single Aden on port side in two-seater (to be removed to accommodate self-protection jammer). One stores attachment on fuselage centreline and two under each wing. Centreline and inboard wing points can each carry up to 1,134 kg (2,500 lb) of weapons, outboard underwing points up to 567 kg (1,250 lb) each. Typical alternative loads include one air-to-surface missile and two 1,200 litre (317 US gallon; 264 Imp gallon) drop tanks; eight 1,000 lb bombs; various combinations of free-fall, laser-guided, retarded or cluster bombs; overwing R.550 Magic missiles; air-to-surface rockets; or a reconnaissance camera pack. Maritime Jaguars equipped with one or two Sea Eagle anti-shipping missiles.

DIMENSIONS, EXTERNAL:

Wing span ... 8.72 m (28 ft 7¼ in)
Wing aspect ratio ... 3.1
Length overall, incl probe:
  single-seat ... 16.955 m (55 ft 7½ in)
  two-seat ... 17.52 m (57 ft 6 in)
Height overall ... 4.89 m (16 ft 0½ in)
Wheel track ... 2.40 m (7 ft 10½ in)
Wheelbase ... 5.69 m (18 ft 8 in)

AREAS:

Wings, gross ... 24.03 m² (258.7 sq ft)

WEIGHTS AND LOADINGS:

Typical weight empty ... 7,000 kg (15,432 lb)
Max external stores load (incl overwing) ... 4,763 kg (10,500 lb)
Normal T-O weight (single-seater, with full internal fuel and ammunition for built-in cannon) ... 10,954 kg (24,149 lb)
Max T-O weight with external stores ... 15,700 kg (34,612 lb)
Max wing loading ... 653.3 kg/m² (133.79 lb/sq ft)
Max power loading ... 210 kg/kN (2.06 lb/lb st)

PERFORMANCE:

Max level speed at S/L ... M0.98 (648 kt; 1,200 km/h; 745 mph)
Max level speed above 6,000 m (19,680 ft) ... M1.5 (907 kt; 1,680 km/h; 1,044 mph)
Touchdown speed ... 115 kt (213 km/h; 132 mph)

Service ceiling ... 13,715 m (45,000 ft)
T-O run, clean: ... 565 m (1,855 ft)
  with four 1,000 lb bombs ... 880 m (2,890 ft)
  with eight 1,000 lb bombs ... 1,250 m (4,100 ft)
T-O to 15 m (50 ft) with typical tactical load ... 940 m (3,085 ft)
Landing from 15 m (50 ft) with typical tactical load ... 785 m (2,575 ft)
Landing run:
  normal weight, with brake-chute ... 470 m (1,540 ft)
  normal weight, without brake-chute ... 680 m (2,230 ft)
  overload weight, with brake-chute ... 670 m (2,200 ft)
Typical attack radius, internal fuel only:
  hi-lo-hi ... 460 n miles (852 km; 530 miles)
  lo-lo-lo ... 290 n miles (537 km; 334 miles)
Typical attack radius with external fuel:
  hi-lo-hi ... 760 n miles (1,408 km; 875 miles)
  lo-lo-lo ... 495 n miles (917 km; 570 miles)
Range with max external fuel ... 1,400 n miles (2,593 km; 1,611 miles)
g limits ... +8.6 (+12 ultimate)/−3

UPDATED

HAL DHRUV

English name: Polaris
TYPE: Light utility helicopter.

DEVELOPMENT MILESTONES

Agreement with MBB	Jul 84
Design started	Nov 84
Rolled out	29 Jun 92
First flight	20 Aug 92
First flight (military)	28 May 94
First flight (naval)	23 Dec 95
First flight (civil)	6 Mar 02
First deliveries	20 Mar 02
First civil order (Azal)	5 Feb 03
Certification (civil)	Oct 03

PROGRAMME: Agreement signed with MBB (Germany) July 1984 to support design, development and production of Advanced Light Helicopter (ALH); design started November 1984; ground test vehicle runs began April 1991; five flying prototypes (two basic, one air force/army, one naval and one civil); PT1 first prototype (Z3182) rolled out 29 June 1992, first flight 20 August and 'official' first flight 30 August 1992; PT2 second prototype (Z3183) made its first flight 18 April 1993; PT-A (army/air force prototype Z3268) on 28 May 1994; PT-N (naval prototype), with CTS 800 engines, flew for first time (IN901) on 23 December 1995. Total hours flown, including 'hot-and-high' trials in environments of 45°C (113°F) and more than 6,000 m (19,680 ft), were about 1,500. Military certification of air force/army, naval and coast guard versions completed in March 2002.

Naval trials by PT-N conducted in March 1998 aboard aircraft carrier INS *Viraat* and smaller decks of other Indian Navy vessels. May 1998 US trade embargo, imposed following India's refusal to sign nuclear test ban treaty, blocked import of CTS 800 engines (30 ordered) and delayed planned first flight of PTC-2 civil fifth prototype (VT-XLH) with this engine until 6 March 2002. Instead, all current variants now to be powered by TM 333, including retrofit of PT-N prototype; contract announced 7 February 2003 for HAL to co-develop and co-produce Turbomeca Ardiden 1H (Indian name Shakti) for future, higher-powered versions of Dhruv. Weight reduction programme initiated in mid-1998; RFPs issued later same year for cockpit display system. By the end of 1998, manufacture was well advanced of three preproduction aircraft (PPN-1, PPA-2 and PPA-3: one for each of the three armed services).

PT-A development aircraft for HAL Dhruv *(Paul Jackson)* ***NEW***/0572250

Deliveries (four each to Indian Air Force and Army, two each to Navy and Coast Guard) were due to begin in late 1999 but programme slipped; new schedule provided for first two (to Army) instead becoming due for delivery by end of 2001, to be followed by two each to Coast Guard, Navy and Air Force by the end of March 2002. Seven deliveries actually achieved by this date: Army two (IA-1101 and -1103) starting 20 March 2002, Air Force two (J-4041/4042 on 20 March), Navy two (IN-701/702 on 28 March) and Coast Guard (CG-851 on 18 March) one. However, Army's IA-1102 had been delivered for trials use earlier, on 4 January 2002. Indian Navy user trials conducted on board INS *Godavari* in November 2002. Eight more scheduled for delivery by 31 March 2003, of which Indian Navy received two on 24 March; total deliveries had increased to 18 by June 2003 (Army eight, Air Force four, Navy and Coast Guard three each). Initial batch of 30 TM 333-2B2 ordered in mid-1999 to power first 12 (including two civil) production Dhruvs; all then intended for delivery by 2002. Further 52 TM 333s ordered mid-2000 to power next 20; deliveries of these almost completed by February 2003. Additional contract at that time for over 300 TM 333s (some of which to re-engine Cheetahs) and Ardidens, former type for delivery from early 2004.

Development and marketing agreement between HAL and Israel Aircraft Industries announced in late 2002; involves both Dhruv and LCH derivative; IAI (Lahav) to concentrate on integrated avionics package and other internal systems. Second prototype equipped in 2003 as demonstrator, with suite including 'glass cockpit', MAWS and NVGs.

Indian Air Force Sarang (Peacock) display team formed 2003; made public debut at Aero India. Asian Aerospace air show, Singapore, February 2004.

CURRENT VERSIONS: **Air force/army:** Skid gear, crashworthy fuel tanks, bulletproof supply tanks, IR and flame suppression; night attack capability; roles to include attack and SAR.

Naval: Retractable tricycle gear, ADA SV-2000 surveillance radar, harpoon decklock, pressure refuelling; fairings on fuselage sides to house mainwheels, flotation gear and batteries.

Civil: Roles to include passenger and utility transport, commuter/offshore executive, rescue/emergency medical service and law enforcement. Wheel landing gear. Prototype targeted to fly in 2001, but this not achieved until 6 March 2002; DGCA certification received October 2003; to be followed by FAA/JAA type approval. Civil version entered production in 2003. Launch customer Azal India Helicopter (one ordered 5 February 2003, for delivery later that year).

Coast Guard: High commonality with naval version; nose-mounted surveillance radar; roof-mounted FLIR; starboard side, cabin-mounted 7.62 mm machine gun; radar console and operator's seat; liferaft, loudhailer.

LCH: Light attack development: described separately.

CUSTOMERS: Indian government requirement for armed forces and Coast Guard, to replace Chetaks/Cheetahs; letter of intent for 300 (Army 110, Air Force 150, Navy/Coast Guard 40)

IAF 'Sarong' demonstration team of HAL Dhruvs *(Paul Jackson)* ***NEW***/0576661

Dhruv naval and coast guard version with nose-mounted SV-2000 radar ***NEW***/0576662

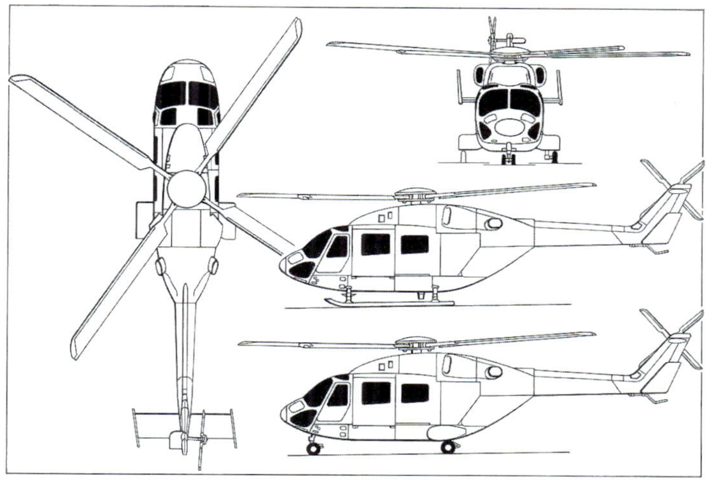

Naval version of the Dhruv Advanced Light Helicopter, with additional side view (centre) of air force/army variant ***(Jane's/Mike Keep)*** 0507145

Indian Army Aviation HAL Dhruv ***(Robert Hewson)*** ***NEW***/0552089

followed by contract for 100 in late 1996, but allocation revised by 2001 as Army 120, Navy 120, Air Force 60 and Coast Guard seven; all to be delivered by 2015. Second production lot contains 20. HAL predicts total military/civil domestic orders for about 650.

Initial Indian Army aircraft to 201 Squadron.

COSTS: Unit price of basic aircraft approximately Rs250 million (US$5.2 million) (2002). Total programme costs US$170 million by 1997 (latest information provided).

DESIGN FEATURES: First modern helicopter of local design and construction. Conventional layout, including high-mounted tailboom to accommodate rear-loading doors. Four-blade hingeless main rotor with advanced aerofoils and sweptback tips; Eurocopter FEL (fibre elastomer) rotor head, with blades held between pair of cruciform CFRP starplates; manual blade folding and rotor brake standard; integrated drive system transmission; four-blade bearingless crossbeam tail rotor on starboard side of fin; fixed tailplane; sweptback endplate fins offset to port; vibration damping by Lord AVCS (active vibration control system), comprising four isolator elements between main gearbox and fuselage.

Main rotor blade section DMH 4 (DMH 3 outboard); tail rotor blade section S 102C (S 102E at tip). Rotor speeds 314 rpm (main), 1,564 rpm (tail).

FLYING CONTROLS: Integrated dynamic management by four-axis AFCS (actuators have manual as well as AFCS input); constant-speed rpm control, assisted by collective anticipator (part of FADEC and stability augmentation system acting through AFCS).

STRUCTURE: Composites account for some 67 per cent of airframe. Main and tail rotor blades and rotor hub glass fibre/carbon fibre; Kevlar nosecone, crew/passenger doors, cowling, upper rear tailboom and most of tail unit; carbon fibre lower rear tailboom and fin centre panels; Kevlar/carbon fibre cockpit section; aluminium alloy sandwich centre cabin and remainder of tailboom.

LANDING GEAR: Non-retractable metal skid gear standard for air force/army version. Hydraulically retractable tricycle gear on naval and civil versions, with twin nosewheels and single mainwheels, latter retracting into fairings on fuselage sides which also (on naval version) house flotation gear and batteries; rearward-retracting nose unit; naval version has harpoon decklock system. Spring skid under rear of tailboom on all versions, to protect tail rotor. FPT Industries (UK) Kevlar inflatable flotation bags for prototypes, usable with both skid and wheel gear.

POWER PLANT: First three and fifth prototypes each powered by two Turbomeca TM 333-2B2 or -2C turboshafts, with FADEC, rated at 740 kW (993 shp) for T-O, 783 kW (1,050 shp) maximum contingency and 666 kW (893 shp) maximum continuous. LHTEC CTS 800-4H (998 kW; 1,338 shp) selected late 1994 and test-flown in fourth prototype, but subsequently embargoed; all now to have TM 333-2B2 until availability of 895 kW (1,200 shp) class Ardiden 1H (Shakti) in about 2006.

Transmission ratings (two engines) 1,280 kW (1,716 shp) for 30 minutes for T-O and 1,156 kW (1,550 shp) maximum continuous; OEI ratings 800 kW (1,073 shp) for 30 seconds (super contingency), 700 kW (939 shp) for 2½ minutes. Transmission input from both engines combined through spiral bevel gears to collector gear on stub-shaft. AVCS system (see Design Features) gives 6° of freedom damping. Power take-off from main and auxiliary gearboxes for transmission-driven accessories.

Total usable fuel, in self-sealing crashworthy underfloor tanks (three main and two supply), 1,400 litres (370 US gallons; 308 Imp gallons). Pressure refuelling in naval version. Crossfeed and fuel dump systems in all military versions.

ACCOMMODATION: Flight crew of two, on crashworthy seats in military/naval versions. Main cabin seats 12 persons as standard, 14 in high-density configuration. EMS interior (first flown by PT2/Z3183 in January 2001) can accommodate two stretchers and four medical attendants, or four stretchers and two medical personnel. Crew door and rearward-sliding door (military) or hinged door (civil) on each side; clamshell cargo doors at rear of passenger cabin.

SYSTEMS: DC electrical power from two independent subsystems, each with a 6 kW starter/generator, with battery back-up for 15 minutes of emergency operation; AC power, also from two independent subsystems, each with a 5/10 kVA alternator. Three hydraulic systems (pressure 207 bar; 3,000 lb/sq in, maximum flow rate 25 litres; 6.6 US gallons; 5.5 Imp gallons)/min: systems 1 and 2 for main and tail rotor flight control actuators, system 3 for landing gear, wheel brakes, decklock harpoon, rescue hoist (naval variant) and optional equipment. Oxygen system.

AVIONICS: *Comms:* V/UHF, HF/SSB and standby UHF com radio, IFF and intercom.

Radar: Weather radar optional. ERDE SuperVision-2000 360° I-band surveillance radar in Navy/Coast Guard version.

Flight: SFIM four-axis AFCS, Doppler navigation system, TAS system, ADF, radio altimeter, heading reference standard. GPS nav system in civil version, with additional VOR/ILS, DME and marker beacon.

Mission: Roof-mounted FLIR in Coast Guard version. EMS version equipped with navaids, patient monitoring, data recording systems, and datalink to transmit medical information to ground-based hospitals.

EQUIPMENT: Depending on mission, can include two to four stretchers, external rescue hoist, liferaft and 1,500 kg (3,307 lb) capacity cargo sling.

ARMAMENT: Cabin-side pylons for two torpedoes/depth charges or four anti-ship missiles on naval variant; on army/air force variant, stub-wings which can be fitted with eight anti-tank guided missiles, four pods of 68 mm or 70 mm rockets or two pairs of air-to-air missiles. Army/air force variant can also be equipped with ventral 20 mm gun turret or sling for carriage of land mines. Cabin-mounted 7.62 mm machine gun in Coast Guard version, firing from starboard side doorway.

DIMENSIONS, EXTERNAL:

Main rotor diameter ... 13.20 m (43 ft 3¾ in)
Main rotor blade chord: inboard ... 0.50 m (1 ft 7¾ in)
at tip ... 0.165 m (6½ in)
Tail rotor diameter ... 2.55 m (8 ft 4½ in)
Length:
overall, both rotors turning ... 15.87 m (52 ft 0¾ in)
fuselage (except Coast Guard version) ... 13.43 m (44 ft 0¾ in)
Height: overall, tail rotor turning:
army/air force version ... 4.98 m (16 ft 4 in)
naval version ... 4.91 m (16 ft 1¼ in)
to top of main rotor head:
army/air force version ... 3.93 m (12 ft 10¾ in)
naval version ... 3.76 m (12 ft 4 in)
Fuselage max width ... 2.00 m (6 ft 6¾ in)
Width over mainwheel sponsons (naval version) ... 3.15 m (10 ft 4 in)
Tail unit span (over fins) ... 3.19 m (10 ft 5½ in)
Wheel track (naval version) ... 2.80 m (9 ft 2¼ in)
Wheelbase (naval version) ... 4.37 m (14 ft 4 in)
Skid track (army/air force version) ... 2.60 m (8 ft 6¼ in)
Tail rotor ground clearance ... 2.34 m (7 ft 8¼ in)

DIMENSIONS, INTERNAL:

Cabin, excl flight deck: Max width ... 1.97 m (6 ft 5½ in)
Max height ... 1.42 m (4 ft 8 in)
Volume ... 7.33 m^3 (259 cu ft)
Cargo compartment volume ... 2.16 m^3 (76.3 cu ft)

AREAS:

Main rotor disc ... 136.85 m^2 (1,473.0 sq ft)
Tail rotor disc ... 5.11 m^2 (55.0 sq ft)
Main fin ... 2.126 m^2 (22.88 sq ft)
Endplate fins (total) ... 1.45 m^2 (15.61 sq ft)
Tailplane ... 2.40 m^2 (25.83 sq ft)

WEIGHTS AND LOADINGS (A: army/air force standard, B: naval version):

Weight empty, equipped: A ... 2,550 kg (5,622 lb)
B ... 2,685 kg (5,919 lb)
Max fuel weight: A ... 1,075 kg (2,370 lb)
B ... 1,140 kg (2,513 lb)
Max sling load: A ... 1,000 kg (2,205 lb)
B ... 1,500 kg (3,307 lb)
Max T-O weight: A ... 4,500 kg (9,920 lb)
B ... 5,500 kg (12,125 lb)
Max disc loading: A ... 32.9 kg/m^2 (6.74 lb/sq ft)
B ... 40.2 kg/m^2 (8.23 lb/sq ft)

Transmission loading at max T-O weight and power:
A .... 3.04 kg/kW (4.99 lb/shp)
B .... 3.72 kg/kW (6.10 lb/shp)

PERFORMANCE (A and B as above, at MTOW at S/L, ISA + 15°C):
Never-exceed speed (VNE):
A .... 178 kt (330 km/h; 205 mph)
B .... 162 kt (300 km/h; 186 mph)
Max cruising speed: A .... 143 kt (265 km/h; 165 mph)
B .... 135 kt (250 km/h; 155 mph)
Econ cruising speed: A .... 121 kt (225 km/h; 140 mph)
B .... 119 kt (220 km/h; 137 mph)
Max rate of climb at S/L: A .... 780 m (2,559 ft)/min
B .... 600 m (1,968 ft)/min
Service ceiling: A .... 6,500 m (21,320 ft)
B .... 4,580 m (15,020 ft)
Hovering ceiling:
IGE: A .... 4,400 m (14,435 ft)
B .... 1,900 m (6,235 ft)
OGE: A .... 3,800 m (12,470 ft)
B .... 1,200 m (3,940 ft)
Range with max fuel, 20 min reserves:
A .... 378 n miles (700 km; 435 miles)
B .... 351 n miles (650 km; 403 miles)
Endurance, conditions as above:
A .... 4 h 20 min
B .... 4 h 0 min

UPDATED

HAL HJT-36 SITARA

English name: Star

TYPE: Basic jet trainer/light attack jet.

DEVELOPMENT MILESTONES

Design started	97
Revealed	Feb 98
Mockup displayed	Dec 98
Development contract awarded	Jul 99
Production approval	Jan 01
First flight	7 Mar 03
Official first flight	21 Mar 03

PROGRAMME: Revealed at Singapore Air Show, February 1998; design started 1997. Conceived as successor to IAF HJT-16 Kiran. In design development stage; mockup, with Thales (Sextant) avionics suite, exhibited at Air India Show in December 1998, differed considerably in appearance from general arrangement released at Singapore. Indian government contract awarded in July 1999, covering completion, flight test (4,500 hours) and certification of prototypes; avionics contract placed June 2001. First metal cut July 2001; systems integration started December 2002; first taxi tests February 2003. Maiden flight of first prototype (S3466) 7 March 2003; 'inaugural' (official) first flight 21 March 2003 was seventh sortie; about 50 flights by 1 February 2004; second prototype (S3474), with full 'glass cockpit' and digital avionics, flew on 26 March 2004, and due to be followed by third later same year.

Surprise announcement by HAL in second quarter 2004 indicated that Larzac power plant to be replaced for production aircraft by still-in-development Saturn AL-55 turbofan, produced by Ufa Engine Building Association (UMPO). Sitara production start will be correspondingly delayed (from late 2004/early 2005), since delivery of first batch of Russian engines not expected before 2006 at the earliest.

CUSTOMERS: Indian Air Force and Navy as replacement for existing fleet of approximately 170 Kirans from about 2005. Government provisional approval for 211 awarded in January 2001, including 24 for Indian Navy; total production expected to be between 210 and 250.

First prototype HJT-36 Sitara intermediate jet trainer ***NEW***/0572156

Second HJT-36 Sitara prototype ***NEW***/1043417

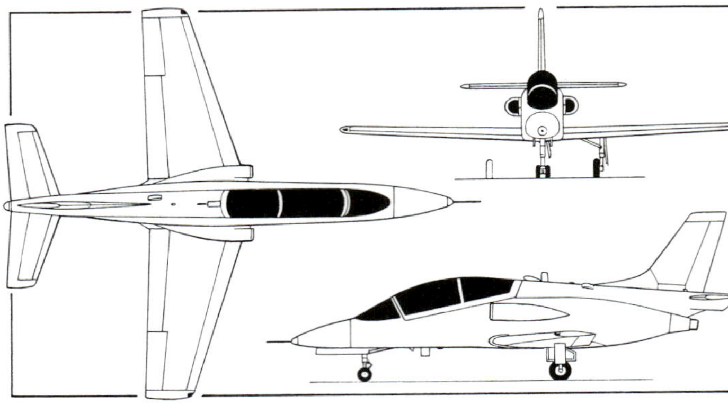
Provisional general arrangement of the HAL HJT-36 jet trainer *(Michael Badrocke)* ***NEW***/0572160

Announced 6 February 2003 that initial order would be 16 for IAF; however, by January 2004 HAL was still seeking authorisation to begin LSP (limited series production) batch of 12.

COSTS: Approximately US$15 million (2004).

DESIGN FEATURES: To meet intermediate jet trainer (IJT) requirement. Capable of high-speed training, but with simple handling at low speeds; cockpit layout compatible with current combat aircraft; some 25 per cent of LRUs are common with ADA Tejas. Conventional, low-wing CAD/CAM design with moderate (16°) wing leading-edge sweepback and 2° dihedral. NACA 64-series aerofoil (t/c ratio 15 per cent at root, 12 per cent at tip).

FLYING CONTROLS: Conventional and manual, with three-axis trim (rudder, tailplane and each aileron); horn-balanced elevators.

STRUCTURE: Light alloy and composites. Intended fatigue life of more than 7,500 hours, extendable to 10,000 hours.

LANDING GEAR: Retractable tricycle type. Inward-retracting main units, forward-retracting twin nosewheels.

POWER PLANT: Prototypes powered by one 14.12 kN (3,175 lb st) SNECMA Larzac 04-H20 non-afterburning turbofan in rear fuselage, fed by bifurcated air intake. Saturn 16.68 kN (3,750 lb st) AL-55 turbofan selected for production aircraft. Fuel tanks in fuselage and wings. Usable fuel capacity 1,150 litres (304 US gallons; 253 Imp gallons).

ACCOMMODATION: Crew of two in tandem, on Zvezda K-36CT lightweight zero/zero ejection seats; rear (instructor's) seat raised. One-piece canopy opens sideways to starboard. Forward vision over nose −8°.

AVIONICS: Smiths Aerospace integrated system, with units supplied by French, German and Indian manufacturers, comprising dual V/UHF comms, open systems architecture mission computer, HUD (Elop), HUD repeater, GPS, AHRS, active matrix LCDs (Thales), air data computers and rear cockpit data entry panel.

ARMAMENT: One underfuselage and four underwing weapons pylons for bombs, rocket pods and gun pods.

DIMENSIONS, EXTERNAL:
Wing span .... 9.80 m (32 ft 1¾ in)
Length overall .... 10.91 m (35 ft 9½ in)
Height overall .... 4.13 m (13 ft 6½ in)
Wheel track .... 2.51 m (8 ft 2¾ in)
Wheelbase .... 4,525 m (14 ft 10¼ in)

WEIGHTS AND LOADINGS (approx):
Max external stores load .... 1,000 kg (2,204 lb)
Max fuel weight (usable) .... 917 kg (2,022 lb)
T-O weight: clean .... 3,500 kg (7,716 lb)
max .... 4,500 kg (9,920 lb)

PERFORMANCE (estimated):
Max operating Mach No. 0.8
Max permissible diving speed .... 445 kt (824 km/h; 512 mph)
Service ceiling .... 12,000 m (39,370 ft)
Endurance .... 3 h
g limits (ultimate) .... +8/−3

UPDATED

HAL CHEETAH

TYPE: Light utility helicopter.

PROGRAMME: Licence-built Aerospatiale SA 315B Lama. Production by HAL for Indian armed forces started 1972. In early 2000, HAL received further orders for production for several years and has taken up possible employment of Turbomeca TM 333 turboshaft to enhance performance.

CUSTOMERS: HAL had delivered 260 by mid-2002, at which time manufacture of a batch of 10 for the Indian Air Force was under way, and order for further five anticipated. Nepal negotiating for further two in second quarter 2003. Re-engining with 740 kW (933 shp) TM 333-2M2 engines in programme to enhance performance; prototype (Z3455) flew for first

HAL Cheetah of the Indian Army *(Simon Watson)* ***NEW***/0532102

time on 1 February 2003 prospect of entire (Indian) fleet retrofit of up to 230 aircraft. Upgrade also includes integrated helmet, countermeasures dispensing system and oxygen system.

UPDATED

HAL LANCER

TYPE: Armed observation helicopter.

PROGRAMME: Upgraded, counter-insurgency version of Cheetah. Prototype, rebuilt from late production aircraft Z2867, unveiled late 1998: cabin has lightweight composites armour protection for pilot, control linkages and fuel tank; bulletproof flat-plate transparencies; twin gun/rocket pods. Can be rebuild or new production.

CUSTOMERS: Indian Army order for 12 (converted from Cheetah), of which five delivered March 2002; remainder due to follow during 2003. Ten ordered in November 2001 by Nepal for newly established Armed Police Force.

ACCOMMODATION: Crew of two, side by side; cleared for single-pilot operation.

ARMAMENT: Tubular metal outrigger each side of cabin, outboard of landing skids, each supporting jettisonable pod containing one 12.7 mm machine gun and three 70 mm unguided air-to-surface rockets. Pilot's gunsight.

WEIGHTS AND LOADINGS:

Basic weight empty: Cheetah	1,090 kg (2,403 lb)
Lancer, incl weapon pods	1,350 kg (2,976 lb)
Fuel weight	393 kg (866 lb)
Ammunition (500 rds) and six rockets	130 kg (287 lb)
Max T-O weight	1,950 kg (4,299 lb)

PERFORMANCE:

Never-exceed speed (VNE)	113 kt (210 km/h; 130 mph)
Cruising speed at S/L, ISA + 20°C	95 kt (176 km/h; 109 mph)
Max rate of climb at S/L	330 m (1,083 ft)/min
Service ceiling	5,400 m (17,720 ft)
Operational radius, incl 30 min over target and 20 min fuel reserves, ISA + 20°C	78 n miles (145 km; 90 miles)
Endurance	2 h 30 min

VERIFIED

HAL Lancer prototype, an armed and armoured version of the Cheetah
0099550

HAL CHETAK

TYPE: Light utility helicopter.

PROGRAMME: Licence-built Aerospatiale SA 316B Alouette III; remains in production only in India. HAL output had totalled at least 359 (of which 35 built from French kits) by mid-2002. Further orders, to prolong production for several years, received in early 2001. Turbomeca TM 333-2B2 under consideration as alternative power plant.

UPDATED

TRANSPORT AIRCRAFT DIVISION

PO Chakeri, Kanpur 208 008
Tel: (+91 512) 245 03 61 and 240 27 74
Fax: (+91 512) 245 05 05 and 245 00 85
e-mail: halknp@vsnl.net
GENERAL MANAGER: D K Mahajan

This Division established in 1960 to manufacture Hawker Siddeley (Avro) 748; is currently responsible for manufacture of the Dornier 228, and for overhaul, maintenance and repair of the Dornier 228, HS 748, HAL HPT-32 and civil aircraft. Co-operation agreement with Israel Aircraft Industries 21 February 2002 involves HAL in IAI (Bedek) Boeing 737 freighter conversion programme. Division assisting in design and development, and manufacturing prototypes, of the CSIR Saras (which see); series production at Kanpur is planned. Also plans co-production of ATR 42/72, for which launch order awaited in 2004. Co-design and development of Il-214 with Ilyushin and Irkut to fulfill IRTA-21 (Indo-Russian Transport Aircraft, 21st century) requirement.

Division holds ISO 9001 and ISO 14001 accreditation and is equipped with Transport Aircraft Research and Design Centre (TARDC) specialising in role modifications, sensor integration, aircraft upgrades and repair technology.

UPDATED

Civil-registered HAL Chetak *(Simon Watson)* ***NEW***/0572159

HAL (DORNIER) 228

TYPE: Twin-turboprop transport.

PROGRAMME: First flight of Dornier 228-100 prototype (D-IFNS) in Germany 28 March 1981; first flight 228-200 prototype (D-ICDO) 9 May 1981; British CAA certification 17 April 1984, FAR Pt 23 and Appendix A Pt 135 11 May 1984, Australian 11 October 1985. European production (including 11 for India), is not described here.

Contract for licensed manufacture of up to 150 Dornier 228s in India signed 29 November 1983; one pattern aircraft supplied complete from Germany; production planned in four phases (six, 10, 10 and 43 aircraft respectively), fourth phase representing full local manufacture after production from locally assembled kits; first Phase 1 Kanpur-assembled aircraft (VT-EJN, c/n 1002) handed over to Vayudoot on 1 November 1985 and made maiden flight on 31 January 1986. Full local manufacture from 18th onwards (CG758, first flight 7 March 1991). Indian production programme dormant from first quarter of 1997, but now continuing with current production (early 2003) of three for Indian Air Force; additional orders from Indian Navy and Coast Guard anticipated at that time. Following discontinuation of German production, worldwide sales and marketing rights held by HAL since November 1998; HAL pursuing export market and anticipates future orders to maintain production for further eight to 10 years. Power plant, landing gear and some avionics also manufactured by HAL to preserve self-sufficiency.

CURRENT VERSIONS: **Regional Airliner:** Ten **228-201**s (five each by Dornier and HAL) delivered to Vayudoot.

Maritime Surveillance: Thirty-six ordered for Indian Coast Guard (including three from Germany, delivered from July 1986); CG754 first flight 18 February 1988; first 18 from Indian production are **228-101s** and were delivered by March 1997. Go-ahead for next batch (seven **228-201**s) received in September 1999; these ordered under US$72 million contract in early 2000, of which all delivered in basic configuration by early 2003, with sensor integration then under way. One also to Mauritius Coast Guard in 1990.

In service with Coast Guard Air Squadrons at Daman and Chennai for coastal, environmental and anti-smuggling patrol; 360° Marec 2 search radar under fuselage (replaced in 15 aircraft by Super Marec and last seven to have 360° Elta radar); Matra infra-red/ultraviolet linescanner for pollution detection. Search and rescue liferafts, searchlight, hand-held aerial camera, bubble windows, side-mounted loudhailer, marine markers, and provision for two Micronair underwing spraypods to combat oil spills; in-flight-openable roller door for parajumping/paradropping. Normal crew two pilots, radar operator and observer. Optional armament includes two underwing 7.62 mm twin-gun pods.

Anti-ship: Indian Navy has acquired 25 specially equipped Dornier **228-201**s (ordered in batches of five, 10 and 10). Deliveries began (IN221) on 24 August 1991 for service with INAS 310 ('Cobras') at Dabolim; last batch of 10 completed in 2000. First two batches have Super Marec 360° radar; Elta EL/M-2022A search radar in second batch of 10 (integration of eight completed by early 2003). Apart from radar (360° scan, near-SAR/MTI and track-while-scan) under forward fuselage, current aircraft have IAI Tamam Airborne Multimission Optronic Stabilised Payload (AMOSP) (TV/FLIR/laser range-finder) ball turret in starboard main gear fairing; air/ground data downlink; and cabin consoles for four systems operators with MFDs and moving map display. Further Indian Navy orders expected.

Utility transport: Indian Air Force received 25 **228-201**s; deliveries started (HM667 and HM668) 15 December 1987 to Nos. 41 'Otters' and 59 'Hornbills' Squadrons; can carry 21 field-equipped troops on inward-facing composites folding seats; large HAL-developed

Indian Air Force HAL Dornier 228-201 *(Simon Watson)* ***NEW***/0532103

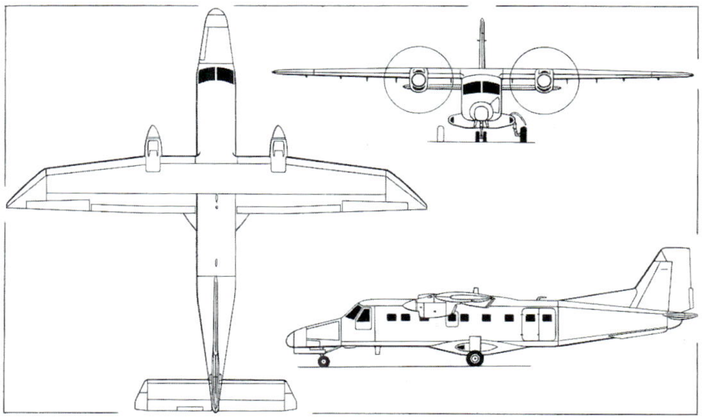

HAL (Dornier) 228-201 (two TPE331 turboprops) *(Jane's/Dennis Punnett)*
0507143

GERMAN/INDIAN DORNIER 228 DELIVERIES
(to January 2003)

Country	Operator	Series	Dornier	HAL/CKD	HAL	Total
India	Air Force	201		5	23	28
	Airports Authority	201	1	1		2
	Coast Guard	101	3	4	11	18
		201			7	7
	Navy	201	1		14	15
	Oil and Natural Gas Commission	101	1			1
	UB Air	101		1		1
	Vayudoot	201	5	5		10
Mauritius	Coast Guard	101			1	1
Totals			**11**	**16**	**56**	**83**

cargo double door, to meet IAF requirement, at rear, port side. Expectation of IAF follow-on order for further 18, but only three extra aircraft included in batch due for delivery by March 2003.

Executive/Air taxi: Various configurations including six- or 10-seat executive or 15-passenger air taxi, with customised interior, cabin attendant, work table and galley/wardrobe/lavatory; built-in APU for air conditioning and lighting in flight or on ground. One (VT-EIX) for Indian Oil and Natural Gas Commission.

CUSTOMERS: HAL deliveries are given in the accompanying table. The company anticipates building a further 56 aircraft beyond the 72 currently produced.

DESIGN FEATURES: Optimised for efficient cruising flight. High wing for propeller clearance, despite minimised fuselage ground clearance and pannier-mounted landing gear. Variable sweep wing leading-edge; unswept tailplane; sweptback fin with dorsal fillet.

Special Dornier wing with Do A-5 supercritical aerofoil; 8° leading-edge sweep on outer wing panels; raked tips; no dihedral or anhedral.

FLYING CONTROLS: Conventional and manual. Variable incidence tailplane with actuator switch on aileron wheel; horn-balanced elevators; rudder trim tab; single-slotted Fowler flaps augmented by drooping ailerons; two strakes under rear fuselage for low-speed stability.

STRUCTURE: Two-spar wing box; mainly light alloy structure, but with CFRP wingtips and tips of tailplane and elevators; GFRP nosecone, tips of rudder and fin; Kevlar landing gear fairings and in part of wing ribs; hybrid composites in fin leading-edge; fuselage unpressurised, built in five sections.

LANDING GEAR: Retractable tricycle type, with single mainwheels and twin-wheel nose unit; main units retract inward into fuselage fairings; hydraulically steerable nosewheels retract forward; Goodyear wheels and tyres, 25.5×8.75–10 (650×220-10) (10/12 ply) or 8.50–10 (10 ply) on mainwheels, 6.00-6 (8 ply), on nosewheels.

POWER PLANT: HAL-built Dornier 228-201s have two HAL-built Honeywell TPE331-5-252D turboprops, each flat rated at 533 kW (715 hp) to ISA +18°C and driving a Hartzell four-blade fully feathering propeller with reverse pitch. Post-2000 production aircraft have HAL-built TPE331-10 series engines; performance improvements of –5 engines through turbine blade enhancements also planned.

Primary wing box forms integral fuel tank with standard usable capacity 2,386 litres (630 US gallons; 525 Imp gallons); can be increased to 2,850 litres (753 US gallons; 627 Imp gallons). Oil capacity per engine 5.9 litres (1.56 US gallons; 1.30 Imp gallons).

ACCOMMODATION: Crew of one or two; pilots' seats adjustable fore and aft; two-abreast seating with central aisle; maximum capacity 19 (more in military versions); flight deck door on port side; combined two-section passenger and freight door, with integral steps, on port side of cabin at rear; one emergency exit on port side of cabin, two on starboard side; baggage compartment at rear of cabin, accessible externally and from cabin; capacity 210 kg (463 lb). Enlarged baggage door optional; additional baggage space in fuselage nose, with separate access; capacity 120 kg (265 lb); modular units using seat rails for rapid changes of role.

SYSTEMS: Entire accommodation heated (by engine bleed air) and ventilated. Hydraulic system, pressure 207 bar (3,000 lb/sq in), for landing gear, brakes and nosewheel steering; hand pump for emergency landing gear extension. Primary 28 V DC electrical system, supplied by two 28 V 250 A engine-driven starter/generators and two 24 V 25 Ah Ni/Cd batteries; two 350 VA inverters supply 115/26 V 400 Hz AC system. Air intake anti-icing standard; de-icing system optional for wing and tail unit leading-edges, fuselage, windscreen and propellers.

AVIONICS: *Comms:* Standard avionics include Rockwell Collins dual VHF-22A and single HF-230 com; Becker audio selector and intercom; Motorola Selcal; H R Smith ELT-503.

Radar: HAL-built Primus 500 weather radar; Elta or Super Marec 360° search radars.

Flight: Dual Rockwell Collins VIR-32A VOR/ILS; dual or single Rockwell Collins ADF-62A; dual or single Bendix/King KNI 582 RMI indicators; dual or single Rockwell Collins DME-42. Bendix/King KFC 250 autopilot optional to permit single-pilot IFR operation.

Instrumentation: IFR instrumentation standard, comprising dual Honeywell GH14B gyro horizons; dual ADIs; dual VSIs.

EQUIPMENT: Standard equipment includes complete internal and external lighting, hand fire extinguisher, first aid kit, gust control locks and tiedown kit. Stores attachment point beneath each wing outboard of engine nacelle.

ARMAMENT: Provision for gun pod, other weapons or mission pod under each wing.

DIMENSIONS, EXTERNAL:
- Wing span ... 16.97 m (55 ft 8 in)
- Wing aspect ratio ... 9.0
- Length overall ... 16.56 m (54 ft 4 in)
- Height overall ... 4.86 m (15 ft 11½ in)
- Tailplane span ... 6.45 m (21 ft 2 in)
- Wheel track ... 3.30 m (10 ft 10 in)
- Wheelbase ... 6.29 m (20 ft 7½ in)
- Propeller diameter ... 2.69 m (8 ft 10 in)
- Propeller ground clearance ... 1.08 m (3 ft 6½ in)
- Passenger door (port, rear): Height ... 1.35 m (4 ft 5 in)
  - Width ... 0.64 m (2 ft 1¼ in)
  - Height to sill ... 0.91 m (2 ft 11¾ in)
- Freight door (port, rear): Height ... 1.35 m (4 ft 5 in)
  - Width, incl passenger door ... 1.28 m (4 ft 2½ in)
- Emergency exits (each): Height ... 0.66 m (2 ft 2 in)
  - Width ... 0.48 m (1 ft 7 in)
- Baggage door (nose): Height ... 0.50 m (1 ft 7½ in)
  - Width ... 1.20 m (3 ft 11¼ in)
- Standard baggage door (rear):
  - Height ... 0.90 m (2 ft 11½ in)
  - Width ... 0.53 m (1 ft 9 in)

DIMENSIONS, INTERNAL:
- Cabin, excl flight deck and rear baggage compartment:
  - Length ... 7.08 m (23 ft 2¾ in)
  - Max width ... 1.346 m (4 ft 5 in)
  - Max height ... 1.55 m (5 ft 1 in)
  - Floor area ... 9.6 m^2 (103 sq ft)
  - Volume ... 14.7 m^3 (519.1 cu ft)
- Rear baggage compartment volume ... 2.6 m^3 (92 cu ft)
- Nose baggage compartment volume ... 0.89 m^3 (31.4 cu ft)

AREAS:
- Wings, gross ... 32.00 m^2 (344.3 sq ft)
- Ailerons (total) ... 2.71 m^2 (29.17 sq ft)
- Trailing-edge flaps (total) ... 5.87 m^2 (63.18 sq ft)
- Fin, incl dorsal fin ... 4.50 m^2 (48.44 sq ft)
- Rudder, incl tab ... 1.50 m^2 (16.15 sq ft)
- Horizontal tail surfaces (total) ... 8.33 m^2 (89.66 sq ft)

WEIGHTS AND LOADINGS (Indian-built 228-201):
- Operating weight empty ... 3,687 kg (8,128 lb)
- Max T-O weight: civil ... 6,200 kg (13,668 lb)
  - military ... 6,400 kg (14,110 lb)
- Max ramp weight: civil ... 6,230 kg (13,734 lb)
  - military ... 6,430 kg (14,175 lb)
- Max landing weight ... 6,100 kg (13,448 lb)
- Max zero-fuel weight ... 5,590 kg (12,324 lb)
- Max wing loading: civil ... 193.8 kg/m^2 (39.68 lb/sq ft)
  - military ... 200.0 kg/m^2 (40.96 lb/sq ft)
- Max power loading: civil ... 5.82 kg/kW (9.56 lb/shp)
  - military ... 6.00 kg/kW (9.87 lb/shp)

PERFORMANCE (228-201):
- Never-exceed speed (VNE) ... 255 kt (472 km/h; 293 mph) IAS
- Max operating speed (VMO) ... 223 kt (413 km/h; 256 mph) IAS
- Max cruising speed: at S/L ... 222 kt (411 km/h; 255 mph)
  - FL 100 ... 231 kt (428 km/h; 266 mph)
- Stalling speed: flaps up ... 79 kt (147 km/h; 91 mph) IAS
  - flaps down ... 63 kt (117 km/h; 73 mph) IAS
- Max rate of climb at S/L ... 582 m (1,909 ft)/min
- Service ceiling, 30.5 m (100 ft)/min rate of climb ... 8,535 m (28,000 ft)
- Service ceiling, OEI, 15 m (50 ft)/min rate of climb ... 3,870 m (12,700 ft)
- T-O run ... 442 m (1,450 ft)
- T-O to 15 m (50 ft) ... 580 m (1,905 ft)
- Accelerate/stop distance, with anti-skid ... 762 m (2,500 ft)
- Landing from 15 m (50 ft) at MLW ... 457 m (1,500 ft)
- Range at FL 100 with 19 passengers, reserves for 50 n mile (93 km; 57 mile) diversion, 45 min hold and 5% fuel remaining:
  - at max cruising speed ... 560 n miles (1,038 km; 645 miles)
  - at max range speed ... 630 n miles (1,167 km; 725 miles)
- Range with 775 kg (1,708 lb) payload, conditions
  - as above: at max cruising speed ... 1,160 n miles (2,148 km; 1,335 miles)
  - at max range speed ... 1,320 n miles (2,445 km; 1,519 miles)

UPDATED

MIG COMPLEX

Ojhar Township Post Office, Nasik, Maharashtra 422 207
Tel: (+91 2550) 27 51 00
Fax: (+91 2550) 27 58 21
MANAGING DIRECTOR: K P Puri
GENERAL MANAGERS:
M S Nadgir (Nasik Aircraft Division and Aircraft Upgrade Research & Design Centre)
BN Mishra (Nasik Aircraft Overhaul Division)
M Fakruddin (Su-30, Nasik)
J R Mehta (Koraput Engine Division)
D N Vyas (Su-30 Engine Division, Koraput)

This complex comprises the Nasik Aircraft and Koraput Engine Divisions of HAL.

Licensed manufacture of Sukhoi Su-30MKI is responsibility of this complex. Agreement signed at Nasik 28 December 2000 for licensed local manufacture of up to 140 following earlier order for initial batch of 50 Russian-built Su-30Ks (deliveries from 1997 and due to be completed in 2004); contract valued at US$3.7 billion; HAL investment in programme reportedly valued at US$650 million. HAL manufacture starting in 2004 and targeted for completion in 2017; first deliveries to Indian Air Force due 2004. Other present activities are MiG-21/27 upgrade, overhaul and repair.

UPDATED

IITB

INDIAN INSTITUTE OF TECHNOLOGY BOMBAY

Department of Aerospace Engineering, Mumbai 400 076
Tel: (+91 22) 25 72 25 45 extn 4125
Fax: (+91 22) 572 26 02
e-mail: rkpant@aero.iitb.ac.in
Web: http://www.aero.iitb.ac.in/~airships
PROGRAMME DIRECTOR, PADD: Professor Rajkumar S Pant

The IITB is heading a team of Indian national aerospace organisations, sponsored by a department of the Ministry of Science and Technology, to develop airship designs for various domestic applications under the generic title PADD (Programme on Airship Design and Development). As proposed, PADD will proceed through three phases (conceptual design, preliminary design, and detailed design and development, of six, 12 and 18 months duration respectively.

Initial Phase 1 activity has included a small unmanned prototype, the **PADD Micro**, which made its first flight on 23 January 2002; this craft is 5 m (16.4 ft) long and has an envelope volume of 6.8 m³ (240 cu ft). It is to be followed by a larger, twin-engined version (**PADD Demo**) with 100 kg (220 lb) payload capacity. Future plans include a twin-engined successor (**PADD PaxCargo**) with the ability to carry 15 passengers or 1,500 kg (3,307 lb) of cargo.

IITB Micro remotely controlled airship prototype ***NEW***/0573470

UPDATED

RAJ HAMSA

RAJ HAMSA ULTRALIGHTS PVT LTD

40 Goshala Road, Mahadevapura, Bangalore 560 048
Tel: (+91 80) 851 69 37
Fax: (+91 80) 560 80 25
e-mail: contact@rajhamsa@x-air.info
Web: http://www.x-air.info
MANAGING DIRECTOR: Joël Koechlin

PARTNER AND DISTRIBUTOR:
Rand Kar SA
Canal de la Martinière, F-44320 Frossay, France
Tel: (+33 2) 40 64 21 65
Fax: (+33 2) 40 64 15 22
e-mail: contact@randkar.fr
Web: http://www.randkar.fr

Company was founded in 1980, initially building hang gliders; powered hang gliders introduced 1983 and X-Air range 10 years later. Currently also markets Mosquito, Javelin and Racer hang-gliders and Clipper and Voyager powered hang-gliders.

VERIFIED

RAJ HAMSA X-AIR

TYPE: Side-by-side ultralight kitbuilt.

PROGRAMME: Launched 1993 as development of US Weedhopper; has BCAR Section S certification.

CURRENT VERSIONS: **X-Air**: Baseline version. Available as **X-Air 502T** (Rotax 503 engine '50 horsepower, two-stroke'), **X-Air 602T** (Rotax 582 '60 horsepower, two-stroke'); **X-Air 604T** (44.7 kW; 60 hp HKS 700E); and **X-Air 702T** (52.2 kW; 70 hp AMW 540L70 engine). Wing span 9.80 m (32 ft 1¾ in); area 16.00 m² (172.2 sq ft).

X-Air F Gumnam: Has more efficient wing, of higher aspect ratio (6.4 instead of 6.0) and fitted with flaps; deeper rear fuselage, providing space for 20 kg (44 lb) of baggage aft of seats; optional auxiliary fuel tank in wing; wire-braced tailplane. Known simply as X-Air F for European and US marketing; or in UK as **Falcon**, with Rotax 912 flat-four. Available as **X-Air F 502T**, **602T**, or **804 TJ** with Jabiru engine.

CUSTOMERS: More than 900 kits produced by January 2004.

Description applies to X-Air F, *except where indicated.*

COSTS: Basic X-Air F airframe kit €9,276 (or €10,261 optimised for Jabiru), plus €323 for engine instruments (€416 for Jabiru), €727 for flight instruments and either €3,675 (Rotax 503), €5,192 (Rotax 582), €7,191 (Rotax 582 with silencers and electric starter) or €10,534 (Jabiru 2200) for engine (all plus tax, 2003).

Raj Hamsa X-Air two-seat ultralight *(Paul Jackson)* 0137211

Original X-Air remains available at €7,900 for airframe kit (€8,204 for HKS-optimised version), or €12,625 (502T), €14,141 (602T), €16,138 (602T with silencers and electric starter) and €17,436 (604T) including engine and instruments (all plus tax, 2003).

DESIGN FEATURES: High-wing monoplane with sweptback leading-edges and wing-mounted engine. Quick-build kit (quoted build time only 40 hours). Potential applications include basic training, crop-spraying, aerial observation and surveillance.

Wing section NACA 4412. Sweepback 8°; dihedral 1° 12'; washout 4°.

FLYING CONTROLS: Conventional and manual. Differential ailerons; fixed tab on port elevator. All control surfaces non-balanced. Flaps; maximum deflection 25°.

STRUCTURE: Aluminium alloy subframe, with composites nose module; rear fuselage, wings and tail surfaces fabric-covered. Braced wings (V struts); tailplane strut-braced on X-Air, wire-braced on X-Air F/Gumnam.

LANDING GEAR: Tricycle type, fixed; two side Vs hinged to lower longerons, plus half-axles; shock-absorbers on all three units. Mainwheel drum brakes; nosewheel steerable by rudder pedals. All tyres 350×80. Optional mainwheel speed fairings (standard on Gumnam) and nosewheel guard. Small tailskid at base of fin. Can be equipped with Puddlejumper floats.

POWER PLANT: *X-Air F 502T:* One 37.0 kW (49.6 hp) Rotax 503 UL-2V air-cooled two-stroke driving a two-blade, wooden propeller.

X-Air F 602T: One 47.8 kW (64.1 hp) Rotax 582 UL water-cooled two-stroke driving a two-blade wooden, or three-blade composites, propeller.

Deeper rear fuselage characterises the Gumnam, or Falcon, version of X-Air F *(Paul Jackson)* ***NEW***/0528684

X-Air F 804TJ: One 59.7 kW (80 hp) Jabiru 2200 air-cooled four-stroke driving a two-blade, composites propeller.

Fuel in two tanks aft of seats, combined capacity 50 litres (13.2 US gallons; 11.0 Imp gallons); auxiliary wing tank optional.

EQUIPMENT: Ballistic recovery parachute optional.

DIMENSIONS, EXTERNAL:
- Wing span ... 9.40 m (30 ft 10 in)
- Length overall ... 5.65 m (18 ft 6½ in)
- Height overall ... 2.55 m (8 ft 4½ in)

DIMENSIONS, INTERNAL:
- Cabin max width ... 1.16 m (3 ft 9¾ in)

AREAS:
- Wings, gross ... 14.32 m^2 (154.1 sq ft)

WEIGHTS AND LOADINGS:
- Weight empty: 502T ... 230 kg (507 lb)
- 602T ... 237 kg (522 lb)
- 804TJ ... 268 kg (591 lb)
- Max T-O weight ... 450 kg (992 lb)

PERFORMANCE, POWERED (804TJ):
- Never-exceed speed (VNE) ... 83 kt (155 km/h; 96 mph)
- Max level speed at S/L ... 78 kt (145 km/h; 90 mph)
- Econ cruising speed at S/L ... 59 kt (110 km/h; 68 mph)
- Stalling speed: flaps up ... 27 kt (50 km/h; 32 mph)
- 25° flap ... 24 kt (43 km/h; 27 mph)
- Max rate of climb at S/L ... 288 m (945 ft)/min
- Service ceiling ... 4,000 m (13,120 ft)
- T-O run ... 80 m (265 ft)
- T-O to 15 m (50 ft) ... 220 m (725 ft)
- Landing from 15 m (50 ft) ... 110 m (360 ft)
- Landing run ... 70 m (230 ft)
- Endurance at cruising speed, standard fuel ... more than 3 h
- *g* limits ... +6/–3

PERFORMANCE, UNPOWERED (804TJ):
- Best glide ratio at 35 kt (64 km/h; 40 mph) ... 7
- Min sinking speed ... 2.50 m (8.20 ft)/s

UPDATED

RAJ HAMSA HANUMAN

US marketing name: X-Air H

TYPE: Side-by-side ultralight/kitbuilt.

PROGRAMME: Prototype (82.0 kW; 110 hp Hirth F30 engine and full-span flaperons) complete by July 2001 and immediately shipped to France for test flying by Rand Kar. Debut at Blois Air Show, September 2001. Launched at Paris, 15-22 June 2003, when second of planned six development and demonstrator aircraft was shown. Deliveries from first quarter 2004.

CURRENT VERSIONS: **X-Air H Hanuman 602T:** '60 horsepower, two-stroke' version with Rotax 582 engine.

X-Air H Hanuman 804TJ: Jabiru 2200 engine.

COSTS: In India, Rs1,400,000 flyaway, with Jabiru 2200. In Europe, basic kit €11,497; engine and instruments €8,239 extra for Rotax or €11,630 for Jabiru; flyaway price €22,104 with Rotax or €25,902 with Jabiru (all 2003, plus tax). In US (2004) basic kit US$13,995; engine and instruments, plus shipping from France, US$22,273 for Rotax or US$27,605 for Jabiru.

DESIGN FEATURES: Further development of X-Air/Gumnam, redesigned for even faster kit assembly. High wing with V strut bracing; upward-swept rear fuselage and sweptback fin.

FLYING CONTROLS: Conventional and manual. Plain, three-position flaps, maximum deflection 45°. Ailerons and flaps actuated by pushrods; rudder and elevators by cables. Flight adjustable trim tab on starboard elevator. No aerodynamic balances.

US demonstrator for the Raj Hamsa X-Air H, or Hanuman *(Paul Jackson)* ***NEW***/0576733

STRUCTURE: Aluminium tube fuselage and wings with Dacron covering, except composites forward fuselage. Wire-braced empennage.

LANDING GEAR: Tricycle type; fixed. Mainwheels and steerable nosewheel all 3.50-8. Bungee cord levered suspension. Cable-actuated mainwheel brakes. Mainwheel speed fairings standard.

POWER PLANT: One 48.0 kW (64.4 hp) water-cooled Rotax 582 two-stroke driving an Arplast EcoProp three-blade, ground-adjustable pitch, composites propeller; or one 59.7 kW (80 hp) air-cooled Jabiru 2200 four-stroke with two-blade composites propeller. Fuel capacity 80 litres (21.1 US gallons; 17.6 Imp gallons).

EQUIPMENT: BRS1050 ballistic recovery parachute optional.

DIMENSIONS, EXTERNAL:
- Wing span ... 10.05 m (32 ft 11¾ in)
- Length overall ... 6.09 m (19 ft 11¾ in)
- Height overall ... 2.30 m (7 ft 6½ in)

DIMENSIONS, INTERNAL:
- Cockpit max width ... 1.14 m (3 ft 9 in)

AREAS:
- Wings, gross ... 13.92 m^2 (149.8 sq ft)

WEIGHTS AND LOADINGS:
- Weight empty, equipped:
- 602T ... 248 kg (547 lb)
- 804TJ ... 263 kg (580 lb)
- Max T-O weight ... 450 kg (992 lb)

PERFORMANCE (804TJ):
- Never-exceed speed (VNE) ... 105 kt (195 km/h; 121 mph)
- Max level speed ... 89 kt (165 km/h; 102 mph)
- Econ cruising speed ... 70 kt (130 km/h; 81 mph)
- Stalling speed, power off:
- flaps up ... 34 kt (63 km/h; 40 mph)
- 25° flap ... 32 kt (58 km/h; 36 mph)
- 45° flap ... 29 kt (53 km/h; 33 mph)
- Max rate of climb at S/L ... 288 m (945 ft)/min
- Service ceiling ... 4,000 m (13,120 ft)
- T-O run ... 80 m (265 ft)
- T-O to 15 m (50 ft) ... 220 m (725 ft)
- Landing from 15 m (50 ft) ... 110 m (360 ft)
- Landing run ... 70 m (230 ft)
- *g* limits ... +6/–3

PERFORMANCE, UNPOWERED (804TJ):
- Best glide ratio at 49 kt (90 km/h; 56 mph) ... 9
- Min sinking speed ... 2.50 m (8.20 ft)/s

UPDATED

TAAL

TANEJA AEROSPACE AND AVIATION LTD

B Block, 2nd Floor, Akshaya Commercial Complex, 26 Victoria Road, Bangalore 560 047
Tel: (+91 80) 557 46 00
Fax: (+91 80) 557 46 17
e-mail: taal@vsnl.com
Web: http://www.taal.co.in
CHAIRMAN: Khushroo Rustumji
MANAGING DIRECTOR: Salil Taneja
JOINT MANAGING DIRECTOR: Arvind Goel
DEPUTY GENERAL MANAGER, AIRCRAFT SALES: Santosh Despande

TAAL is part of Indian Seamless Metal Tubes Group; has modern plant at Hosur, near Bangalore, with 1,300 m (4,265 ft) captive runway, hangars, laboratories, paint shops and other facilities for aircraft manufacture and overhaul. Entered into technical agreement second quarter 1992 with Partenavia of Italy to produce P.68C/TC, P.68 Observer and AP.68TP-600 Viator light twins in India. Second Viator delivered to South African customer in March 2003, but no other recent evidence of any output. More recently, has received kits of about 10 Sport E lightplanes (current incarnation of former Thorp T-211) from IndUS Aviation of Dallas, Texas, and had begun assembly of first two by end of 2003, with first delivery scheduled for March 2004. Indian parts manufacture expected to begin three months later.

TAAL's major domestic aircraft programme is the Hansa-3 light trainer. Also active in air taxi operations and manufacture of UAVs for Aeronautical Development Establishment and structural parts for space launchers.

UPDATED

NAL/TAAL HANSA-3

English name: Swan

TYPE: Primary prop trainer/sportplane.

PROGRAMME: Developed under agreement with Indian National Aerospace Laboratories and originally known as NALLA: NAL Light Aircraft. Design started May 1989; construction of Hansa-2 prototype began December 1991, and this aircraft (VT-XIW) made first flight 23 November 1993. Prototype received Indian DGCA Experimental category type certificate; re-engined 1995 as Hansa-2RE with more powerful IO-240 replacing original

Prototype NAL/TAAL Hansa-3 civil trainer ***NEW***/0528740

74.6 kW (100 hp) O-200 flat-four. Retired to HAL's museum after 128 hours of flying. Deliveries of production Hansa-3s began on 20 March 2001. Hansa manufacturing techniques being transferred to external agencies, beginning 28 August 2002 when Reinforced Plastics Industries, Bangalore, initiated manufacture of two composites fuselage shells.

In early 2002, NAL was in the early stages of designing a 'stretched' Hansa with between four and six seats. No further news of this had emerged by early 2004.

CURRENT VERSIONS: **Hansa-2:** Prototype in original configuration.

Hansa-2RE: Prototype (VT-XIW) re-engined with 93 kW (125 hp) IO-240B; span increased and flaps added; first flight 26 January 1996.

Hansa-3: Production version of Hansa-2RE, first flown (VT-XAL; designated Hansa-3I) on 25 November 1996. First flight with Rotax 914 F3 engine, 11 May 1998, made by second (and lighter) prototype VT-XBL (designated Hansa-3II); provisional JAR-VLA certification awarded to 'XBL in December 1998 in day VFR category. Current version (c/n 001 prototype VT-HNS) has lightning protection for night operations; made first flight 14 May 1999 and received DGCA certification on 1 February 2000 under FAR Pt 23, Amendment 23-42 (JAR-VLA night operation requirements).

Following description applies to this version.

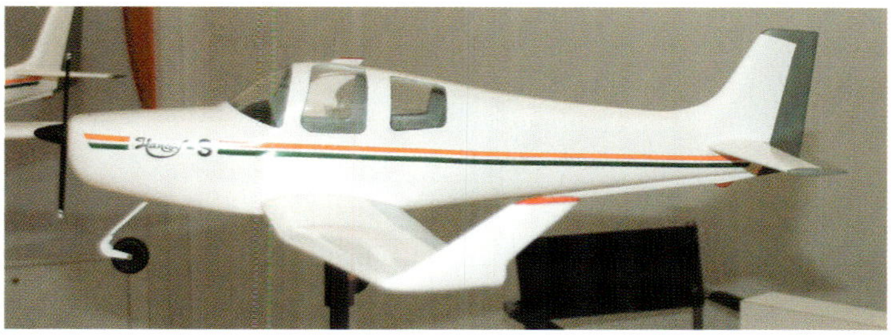

Model of the proposed Hansa-S *(Paul Jackson)* ***NEW***/0576732

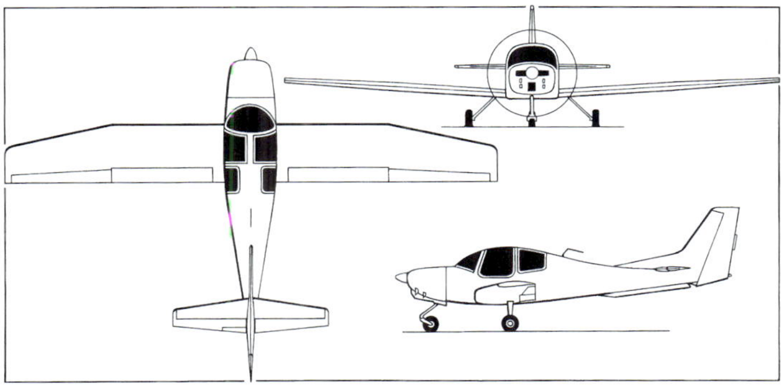

NAL/TAAL Hansa-3 *(Jane's/Mike Keep)* 0121113

Hansa-S: Four-seat version with raised rear decking and winglets; previously known as Hansa-4, powered by SMA SR305-230 diesel; model shown at Asian Aerospace, Singapore, February 2004; production start targeted for late 2004.

CUSTOMERS: Four prototypes, including Hansa-2, of which VT-HNS handed over on 22 March 2002 for use by Department of Aerospace Engineering, IIT Kanpur.

Three Hansa-3s ordered by DGCA for Indian flying clubs (c/n 002,003 and 004: VH-HNT, 'HNU and 'HNV), of which first handed over to DGCA on 20 March 2001, then presented to Andhra Pradesh Aviation Academy on 12 April 2001 and delivered to Hyderabad on 8 May 2001. Second and third to DGCA on 22 March 2002, of which 'NU to Kerala Aviation Training Centre and 'NV to Madhya Pradesh Flying Club on 12 June 2002. Fourth Hansa-3 (c/n 005: VH-HNW) flown 30 March 2003; to DGCA 14 July 2003 for Kerala Aviation Training Centre at Trivandrum. Fifth aircraft (c/n 006: VH-HNX) registered to National Aerospace Laboratory at Bangalore 11 May 2003.

COSTS: Programme development US$40 million (1997); standard aircraft Rs4.5 million (2001).

DESIGN FEATURES: Docile handling qualities for *ab initio* training; robust construction and low acquisition/operating cost. Low-wing monoplane with circular-section waisted fuselage; outer wings tapered on leading-edges, with unswept trailing-edge; sweptback fin and rudder; shallow ventral strake; conventional unswept, straight-tapered horizontal tail surfaces.

Laminar-flow wing, with NASA LS(1)-0415 aerofoil section on constant-chord inboard portion, linearly tapered to LS(1)-0413 from there to tip; sweepback 12° 18' 36 " on outer leading-edges; dihedral 4° from roots; incidence 0°; twist 0° between root and kink; washout –2° between kink and tip.

FLYING CONTROLS: Conventional and manual. Frise ailerons (100 per cent internal mass balance); horn-balanced plain elevators and large rudder, all actuated by pushrods; pitch trim by electrically operated tab in port elevator. Single-slotted Fowler flaps, deflection 20°.

STRUCTURE: Built entirely of composites (CFRP/GFRP reinforced epoxy) with hand-laid, vacuum-bagged, room temperature-cured sandwich shells; components are post-cured before assembly. Conventional rib and three-spar wings, with moulded sandwich shell and PVC foam core; two-spar tail surfaces similar. Fuselage is also a moulded sandwich shell with PVC foam core.

LANDING GEAR: Non-retractable tricycle type, with wing-mounted cantilever spring steel mainwheel legs and steerable (±33°) trailing-link nosewheel. Cleveland wheels and McCreary tyres (main 6.00–6, nose 5.00–5), pressure 2.07 bar (30 lb/sq in) on all units. Cleveland hydraulic disc brakes. Minimum ground turning radius 2.64 m (8 ft 8 in).

POWER PLANT: One 84.6 kW (113.4 hp at 5,800 rpm) Rotax 914 F3 turbocharged flat-four engine; Hoffmann NOV 352FQ+8 two-blade constant-speed propeller. Single composites fuel tank aft of cockpit, usable capacity 85 litres (22.5 US gallons; 18.7 Imp gallons). Single gravity refuelling point on top of fuselage, port side. Oil capacity 3 litres (0.8 US gallon; 0.7 Imp gallon).

ACCOMMODATION: Two seats side by side; dual controls standard. Upward-opening gullwing doors. Baggage compartment aft of pilot's seat.

SYSTEMS: Manual (toe-operated) hydraulic mainwheel brakes. Electrical (DC) system powered by 14 V, 40 A generator and 12 V, 18 Ah Gill G25M lead-acid battery.

AVIONICS: *Comms:* Bendix/King KLX 125 combined VHF/VOR com/nav unit with concealed foil antennas; ACK Technologies E-01 ELT; Clark intercom.

Flight: Optional GPS.

Instrumentation: Conventional VFR (IFR version planned). For day operations: ASI, VSI, altimeter, magnetic compass, slip/skid indicator and OAT gauge. For night operations: artificial horizon, directional gyro, turn co-ordinator, navigation lights, landing lights, anti-collision lights, panel and map lights.

DIMENSIONS, EXTERNAL:

Wing span	10.47 m (34 ft 4¼ in)
Wing chord: at root	1.30 m (4 ft 3¼ in)
at tip	0.80 m (2 ft 7½ in)
Wing aspect ratio	8.8
Length overall	7.66 m (25 ft 1½ in)
Fuselage max width	1.13 m (3 ft 8½ in)
Height overall	2.615 m (8 ft 7 in)
Tailplane span	3.60 m (11 ft 9¾ in)
Wheel track	2.20 m (7 ft 2½ in)
Wheelbase	1.95 m (6 ft 4¾ in)
Propeller diameter	1.73 m (5 ft 8 in)
Propeller ground clearance	0.33 m (1 ft 1 in)

DIMENSIONS, INTERNAL:

Cockpit: Length	2.15 m (7 ft 0¾ in)
Max width	1.07 m (3 ft 6¼ in)
Max height	1.00 m (3 ft 3¼ in)
Floor area	0.96 m^2 (10.3 sq ft)
Volume	1.4 m^3 (49.9 cu ft)
Baggage compartment volume	0.24 m^3 (8.5 cu ft)

AREAS:

Wings, gross	12.47 m^2 (134.2 sq ft)
Ailerons (total)	0.97 m^2 (10.44 sq ft)
Flaps (total)	1.515 m^2 (16.31 sq ft)
Fin	0.75 m^2 (8.07 sq ft)
Rudder	0.70 m^2 (7.53 sq ft)
Tailplane	1.19 m^2 (12.81 sq ft)
Elevators (total, incl tab)	0.85 m^2 (9.15 sq ft)

WEIGHTS AND LOADINGS:

Weight empty: normal	485 kg (1,069 lb)
night operations	545 kg (1,202 lb)
Max fuel weight	65 kg (143 lb)
Max T-O and landing weight	750 kg (1,653 lb)
Max wing loading	60.1 kg/m^2 (12.32 lb/sq ft)
Max power loading	8.87 kg/kW (14.57 lb/hp)

PERFORMANCE:

Max cruising speed at FL100	115 kt (213 km/h; 132 mph)
Stalling speed, flaps down	48 kt (89 km/h; 56 mph) IAS
Max rate of climb at S/L	198 m (650 ft)/min
T-O run	413 m (1,355 ft)
T-O to 15 m (50 ft)	450 m (1,480 ft)
Landing from 15 m (50 ft)	600 m (1,970 ft)
Landing run	540 m (1,775 ft)
Range with max fuel	455 n miles (842 km; 523 miles)
Endurance	4 h

UPDATED

Indonesia

DIRGANTARA

PT DIRGANTARA INDONESIA (INDONESIAN AEROSPACE)

HEAD OFFICE AND WORKS: PO Box 1562 BD, GPM 4th Floor, Jalan Pajajaran 154, Bandung 40174
Tel: (+62 22) 603 45 26, 601 07 54 and 601 07 59
Fax: (+62 22) 601 95 38, 607 56 71 and 603 16 96
e-mail: pub-rel@indonesian-aerospace.com
Web: http://www.indonesian-aerospace.com

PRESIDENT DIRECTOR AND CEO: Edwin Sudarmo
EXECUTIVE VICE-PRESIDENT: Rudy M Mokobombang
VICE-PRESIDENTS:
Iwan W Soemekto (Commerce and Business Development)
Mohamad Mochajan (Technology)
Budi Wuraskito (Production)
Muhammad Nuril Fuad (General Affairs)
Hidayat Hasan (Finance)
CORPORATE PUBLIC RELATIONS OFFICER: Soleh Affandi
PRESS OFFICER: Alex Lim

SALES OFFICE:
PO Box 3752 JKT, 14th and 15th Floor, BBD Plaza Building, Jalan Imam Bonjol 61, Jakarta 10310
Tel: (+62 21) 32 22 47
Fax: (+62 21) 310 00 81 and 32 53 19

SUBSIDIARIES:
PT Nusantara Turbin and Propulsi,
Jalan Pajajaran 154, Kawasan Pabrik IV, Bandung 40174
Tel: (+62 22) 603 19 85 and 603 18 51
e-mail: umc@bdg.centrin.net.id PT GE Nusantara Turbine Services,

Jalan Pajajaran 154, Kawasan Pabrik IV, Bandung 40174
Tel: (+62 22) 603 57 20
Fax: (+62 22) 603 64 41 PT GE Technology Indonesia,

Menara Batavia, 5th Floor,
Jalan KH Mas Mansyur Kav 126, Jakarta 10220
Tel: (+62 21) 574 71 23
Fax: (+62 21) 574 71 17 PT Nusantara Systems International,

Wisma Bisnis Indonesia, 14th Floor, Jalan S Parman Kav 12, Jakarta 11480
Tel: (+62 21) 530 72 22
Fax: (+62 21) 530 72 23
e-mail: info@nsi.co.id
Web: http://www.nsi.co.id

IPTN North America Inc,
1035 Andover Park West, Suite B, Tukwila, Seattle, Washington 98188-7681, USA
Tel: (+1 206) 575 65 07
Fax: (+1 206) 575 03 18
e-mail: iptn250@aol.com
PRESIDENT: M S Noer

IEU GmbH,
Papenstrasse 23, D-22089 Hamburg, Germany
Tel: (+49 40) 357 68 40
Fax: (+49 40) 35 76 84 44
e-mail: SJAM-IEU@t-online-de

Originally formed by Indonesian government as PT Industri Pesawat Terbang Nurtanio (Nurtanio Aircraft Industry Ltd) 23 August 1976 to centralise all aerospace facilities in one company; renamed PT Industri Pesawat Terbang Nusantara (IPTN) in late 1985. Company restructured in January 2000, with PT Bahana Pengelola Industri Strategis as major shareholder, becoming limited liability company with nine strategic business units, six (now four) profit centres, four resource centres and six subsidiaries/corporate functions. New name PT Dirgantara Indonesia (internationally, Indonesian Aerospace, or IAe) inaugurated 24 August 2000 (start of company's 25th year of existence). Workforce in July 2003 was 9,600, including 2,500 engineers and 5,000 technicians and operators, compared with some 9,500 in early 2002; further drastic reductions are planned. Site area is 79.3 ha (196 acres) including 600,000 m^2 (6,458,350 sq ft) covered area.

International Monetary Fund (IMF) bail-out of Indonesia's collapsed economy, beginning in 1997, demanded immediate suspension of further state funding for N-250 and N-2130 programmes. Alternative support for N-2130 was sought, but no new partners for either venture by early 2004; both programmes now moribund.

Manufacturing SBU co-manufactures Airtech CN-235 Series 200/220, produces NC-212 Series 200 under licence from CASA of Spain, and was responsible for indigenous N-250 and projected N-2130 regional airliners. Two CN-235s delivered to South Korea in December 2001 and six more by end of 2002; Pakistan Air Force ordered four (first of which handed over 28 January 2004) and Royal Malaysian Air Force two. Agreement 19 June 1991 with BAe (now BAE Systems) to collaborate on production and assembly of Hawks for Indonesian Air Force, although all were assembled in UK with Indonesia receiving offset work. Residual orders for nine NC-212-200s in late 2003 for Sabang Meruake Raya Air Charter (five) and Indonesian Police (four); production to end in 2004.

Helicopter SBU is responsible for Eurocopter BO 105 and Super Puma (as NBO-105 and NAS-332 respectively), and Bell 412 (as NBell-412). Deliveries resumed in late 1999; dedicated Eurocopter maintenance centre for NBO-105, NAS-332 and EC 120 officially opened 2 October 2002. Agreement with Bell, June 1996, to include Bell 407 and 430, has lapsed.

Weapon Systems SBU in Menang Tasikmalaya, West Java (plus smaller plant at Batu Poron, Madura), develops and produces weaponry for military aircraft.

Aerospace Systems SBU main business comprises satellite ground systems and components, application technology, offset and satellite engineering services.

Subcontract work includes production of components for Boeing 737 and 767, Airbus A380 (wing ribs) and Lockheed Martin F-16; first A380 deliveries 28 April 2003. Company's 1,400 m^2 (15,070 sq ft) gas-turbine maintenance centre can maintain, overhaul and repair

HELICOPTER DELIVERIES

Type	Army	Navy	Air Force	Civil	Total
NBell-412	9	5	—	15	29[1]
NBO-105	22	8	—	85	115[2]
NSA-330					
Puma	—	—	9	4	13
NAS-332					
Super Puma	—	4	3	7	14[3]
Totals	**31**	**17**	**12**	**111**	**171**

[1] Three for 2004 delivery to Army; production completed
[2] Five more remained for delivery in 2003–04; production then to end
[3] Nine more to be completed; production to end by mid-2004

Rolls-Royce 250, P&W JT8D, P&WC PT6T-3/3B, R-R Dart RDa.7, Honeywell LTS101 and TPE331 and General Electric CT7 engines, plus some components of Allison T56 and R-R Tay.

Indonesian government considering new five-year business plan for Dirgantara in late 2003. This envisaged drastic reduction of workforce to some 3,400; expansion of aerostructures business; continuing production of CN-235; and development, possibly as joint ventures with other Southeast Asian partners, of a 19-passenger transport (N-219) and a new light helicopter. Feasibility of former under discussion with CTRM (Malaysia) in early 2004, as well as a larger (up to 50 passengers) turboprop twin. At that time, licensed assembly of Polish SW-4 being examined as possible future helicopter venture.

UPDATED

DIRGANTARA (EUROCOPTER) NAS-332 SUPER PUMA

TYPE: Multirole medium helicopter.

PROGRAMME: Assembly of 11 SA 330J Pumas agreed 1977 and began 1981; switched to AS 332C and L Super Puma in early 1983. First NAS-332 for Pelita Air Service rolled out 22 April 1983.

CUSTOMERS: Deliveries by early 2004 included four as commando and general purpose transports for Indonesian Navy and seven civil. Sales of seven civil NAS-332s to Iran, reportedly for oil-rig duties, was approved in October 1996 but still in abeyance in early 2004; 16 requested by Indonesian Air Force as S-58T replacements, but only 12 funded. First two delivered in May and August 2001, but only one further example delivered by end of 2002; remaining nine due for delivery by mid-2004.

UPDATED

International

AGUSTAWESTLAND

AGUSTAWESTLAND

Web: http://www.agustawestland.com
CHAIRMAN: Kevin Smith
CEO: Amedeo Caporaletti
MANAGING DIRECTOR: Richard Case
COO:
A Gustapane (Italy)
A Johnson (UK)
MARKETING MANAGER: G Orsi
PUBLIC RELATIONS MANAGER:
Gianluca Grimaldi (Italy)
David Bath (UK)

MoU covering joint merger of helicopter divisions signed by Finmeccanica of Italy and GKN of UK on 16 April 1998; heads of agreement 18 March 1999; details finalised (subject to regulatory approval) and name AgustaWestland announced 26 July 2000; company declared fully operational 12 February 2000.

Westland also contributed its 50 per cent share in EH 101, GKN's aerospace transmissions business and 50 per cent share in Aviation Training International (joint venture with Boeing to support British Army WAH-64 Apaches). Agusta also contributed half of EH 101, transmissions and aerostructures business and own shares in NH90 (32 per cent) and Bell/Agusta (45 per cent). Workforce in mid-2004 was 8,000, including 4,000 in UK.

Together, the two companies have built over 7,100 helicopters for customers in more than 80 countries. Deliveries in 2003 totalled 118, including 75 for commercial use.

AgustaWestland owns 100 per cent of both Agusta SpA and Westland Holdings Ltd, latter comprising Westland Helicopters Ltd and Westland Transmissions Ltd. Subsidiaries are as follows.

Agusta SpA
AgustaWestland Inc (USA) (50 per cent)
Agusta Aerospace Corporation (USA)
Agusta Aerospace Services BV (Belgium)
AgustaWestland International Ltd (50 per cent)
Bell/Agusta Aerospace Company (USA) (50 per cent)
NHI sarl (32 per cent)
other subsidiaries and holdings

Westland Helicopters Ltd
AgustaWestland Inc (USA) (50 per cent)
AgustaWestland International Ltd (50 per cent)
Aviation Training International Ltd (50 per cent)
other subsidiaries and holdings.

On 31 October 2001, AgustaWestland initiated a joint marketing effort with Lockheed Martin, under which the latter will promote a US market version of the EHI EH101 (which see), known as the US101.

Agusta and Westland products are marketed under the AgustaWestland name, but helicopter type certificate ownership (and thus civil register entries) remains unaltered. The Bell/Agusta partnership does not involve Westland.

Fundamental change announced 26 May 2004 upon GKN's agreement to sell its 50 per cent share to Finmeccanica for £1,065 million, which transfer was then expected to be complete by September 2004, with AgustaWestland becoming wholly Italian-owned.

Agusta
Via Giovanni Agusta 520, I-21017 Cascina Costa di Samarate (VA)
Tel: (+39 0331) 22 97 69
Fax: (+39 0331) 22 99 84
Web: http://www.agusta.com
CHAIRMAN AND CEO: Amedeo Caporaletti
VICE-PRESIDENT, EXTERNAL RELATIONS AND COMMUNICATIONS: Gian Luigi Ghezzi
GENERAL MANAGER: Giuseppe Orsi

Original Agusta company established 1907 by Giovanni Agusta; acquired licence for Bell 47 in 1952; first flight of first Agusta example 22 May 1954; later produced Bell 204, 205, 206 and 212; Bell AB412 still in low-volume production; also produced various versions of Sikorsky S-61 under licence; participates in NH 90 programme. Own designs include A 109 multirole helicopter, A 119 Koala, A 129 anti-tank helicopter and AB139 joint venture with Bell Helicopter Textron.

Agusta group completely reorganised from 1 January 1981 under new holding company Agusta; became part of Italian public holding company EFIM, but integrated into Finmeccanica on 20 December 1996.

In November 1998 Agusta and Bell Helicopter Textron signed an agreement to form Bell/Agusta Aerospace Company, to share development, manufacture and marketing of the Bell/Agusta BA609 civil tiltrotor and AB139 utility helicopter.

On 14 January 2002, Agusta and Denel Group of South Africa signed an agreement for licensed production of the A 109 and A 119 in South Africa. Under the terms of the agreement Denel is manufacturing the helicopters at its Kempton Park facility for marketing in Africa, the Middle East, South America and Southeast Asia. Denel building 25 of the 30 A 109s ordered by the South African Air Force.

Westland
Lysander Road, Yeovil, Somerset BA20 2YB
Tel: (+44 1935) 47 52 22
Fax: (+44 1935) 70 21 31
*e-mail:*info@whl.co.uk
Web: http://www.agustawestland.com
CEO: Richard Case

Westland Aircraft Ltd (later expanded to Westland Group plc) formed July 1935, taking over aircraft branch of Petters Ltd (known previously as Westland Aircraft Works) that had designed/built aircraft since 1915; entered helicopter industry having acquired licence to build US Sikorsky S-51 as Dragonfly 1947.

GKN shareholding in Westland Group progressively increased until overall control achieved on 18 April 1994.

Westland Helicopters Ltd included GKN Westland Industrial Products Ltd at Weston-super-Mare, Somerset, and had 50 per cent interest in Aerosystems International Ltd at Yeovil (aerospace software; partnership with BAE Systems) and EH Industries Ltd.

Under June 1989 agreement with former McDonnell Douglas, Westland obtained co-production rights for Boeing AH-64 Apache; this selected for British Army Air Corps, July 1995; Westland is prime contractor to UK MoD. Production undertaken in newly built 5,200 m² (56,000 sq ft) assembly hall at Yeovil, opened 14 January 1999.

Aviation Training International (ATI) joint venture company formed with Boeing on 3 August 1998 to provide air, ground and maintenance crews for British Army Apaches under £650 million, 30 year contract; began operations 29 July 1999 at Sherborne HQ; is opening branches at Middle Wallop, Dishforth and Wattisham flying bases, plus engineering school at Arborfield.

Support of Sea King, Puma and Gazelle previously undertaken at Weston-super-Mare by Westland Industrial Products Ltd. Plant closure announced in January 2002.

AgustaWestland International Ltd (AWIL)

EH Industries (European Helicopter Industries) formed June 1980 by Westland Helicopters and Agusta (50 per cent each), to undertake joint development of new anti-submarine warfare helicopter for Royal Navy and Italian Navy. Programme handled on behalf of both governments by UK Ministry of Defence; Westland allocated design leadership for commercial version, Agusta for rear-loading military/utility version; naval version developed jointly for UK and Italian navies and export. IBM Federal Systems (now Lockheed Martin Aerospace Systems Integration Corporation) selected to manage Royal Navy programme in 1991, overseeing and taking responsibility for RN-specific development activity, systems integration and aircraft production and delivery.

Kawasaki of Japan, in conjunction with trading company Okura, signed an agreement in June 1995 for joint marketing and support of the EH101, this to include local manufacture, if warranted by orders.

Partnership between AgustaWestland, Bell Helicopter Textron, CAE and Boeing Canada (Team Cormorant) is promoting EH101 for the Canadian maritime helicopter programme; similarly, AgustaWestland, Bell Helicopter Textron and Lockheed Martin have teamed to promote EH101 as the US101 for US requirements including replacement of residential VH-3D helicopter.

On 19 March 2004, EHI Ltd was renamed AgustaWestland International Ltd upon official opening of new headquarters at Farnborough, UK.

UPDATED

AgustaWestland A 129C/EES-1 to current production standard ***NEW***/0576301

AGUSTAWESTLAND A 129 MANGUSTA

English name: Mongoose
Italian Army designations: EA-1, EES-1
TYPE: Attack helicopter.
PROGRAMME: Italian Army specification issued 1972; A 129 given go-ahead March 1978; final

Development Milestones

Requirement issued	72
Official go-ahead	March 78
Design submitted	30 Nov 82
First flight	11 Sept 83
First delivery	Oct 90

form settled 1980; detail design completed 30 November 1982; first flights of five development aircraft 11 September 1983, 1 July and 5 October 1984, 27 May 1985 and 1 March 1986; first delivery October 1990. In 1999, Italian government approved A 129 upgrade, which will raise first 45 army helicopters to multirole standard. Raytheon Stinger, in AAM version, selected in March 2000 to arm Italian A 129s; integration due to have been completed by 2003.

CURRENT VERSIONS: **Anti-tank**: Initial Italian Army version; local designation EA-1 (*Elicottero da Attacco*: Attack Helicopter). Also offered for export. Lot 2 aircraft (from 16th production onward — MM81391) equipped with secure communications, cockpit lighting compatible with third-generation NVGs, IR suppressors, IR jammers, laser warning system, improved main computer software, auxiliary fuel tanks and folding main rotor blades.

Multirole: Italian Army is receiving aircraft Nos. 46 to 60 in this configuration; originally designated **A 129 CBT** (*combattimento:* combat), but known to Agusta as **A 129C**. Army designation is EES-1 (*Elicottero da Esplorazione e Scorta*). Reconnaissance and Escort Helicopter). First aircraft, MM81421 'EI-951', handed over on 25 October 2002.

Main features are five-blade rotor; transmission uprated to 1,268 kW (1,700 shp); tail rotor speed increased by 7 per cent; maximum take-off weight increased to 4,600 kg (10,141 lb); Lockheed Martin/Otobreda TM 197B 20 mm cannon in nose turret; and Stinger AAMs. Fifth A 129 prototype (MMX598) converted to this configuration by 1997. Other features are G13 software standard, small strake along port side of tailboom; long, tray-shaped ammunition tank on port side of fuselage; revised cockpit layout, including HOCAC controls; updated Galileo Avionica 'HIRNS Plus' navigation/sighting system; integrated GPS; higher landing gear; improved LO paint scheme and AN/ALQ-144 IR jammer.

Similar upgrade of first 45 Mangustas began in 2002 and due for completion by 2007, but all will additionally receive G15 software in associated programme, permitting upgrade of defensive aids suite (*sistema integrato di autoprotezione:* integrated automatic protection system) with Marconi MILDS II missile approach warner and RALM-01(V)2 laser warner, Elettronica ELT-156X(V)4 RWR and MES ECDS-2 chaff/flare dispensers integrated with current AN/ALQ-144 IR jammer. First 45 helicopters receiving G15 first (2002 to 2007), followed by new-build A 129Cs. Italian Army also seeking follow-on to TOW missile and associated sight.

Shipborne: Proposed maritime anti-ship version; no orders received by early 2000.

A 129 International: *Described separately.*

CUSTOMERS: First five of planned 60 for Italian Army anti-tank squadrons (15 Lot 1 and 45 Lot 2) delivered October 1990 after delay of more than a year to allow fitting of Saab/ESCO HeliTOW system with nose-mounted sight; initially operated by 493° Squadrone Elicotteri da Attacco of 49° Gruppo Squadroni at Rimini/Miramare. First Lot 2 A 129, with Honeywell/OMI helicopter IR navigation system (HIRNS), entered service in August 1993; operated by 7th Attack Helicopter Regiment 'Vega' at Casarsa and Army Aviation Centre at Viterbo; 45 in service by end of 1996.

Last 15 aircraft being delivered as A 129 EES version from 2002 onwards. Upgrade contract for first 45 awarded December 2001. Current operating squadrons are 491° and 492° Squadroni Elicotteri da Attacco (49° Gruppo Squadroni 'Capricorno' of 5° Regimento at Casarsa) and 481° and 482° Sq EA (48° GS 'Pavone' of 7° Regimento at Rimini/Miramare). First operator of EES-1 is 48° GS.

DESIGN FEATURES: Fully articulated four-blade main rotor with blades retained by single elastomeric bearing and restrained by hydraulic drag damper and mechanical droop stop; main rotor blade folding on Lot 2 aircraft.

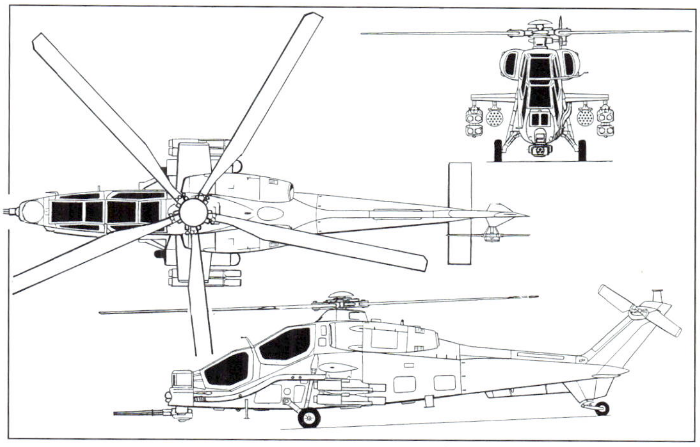

AgustaWestland A 129C Mangusta *(James Goulding)* ***NEW***/0561619

Main transmission has independent oil cooling system; intermediate and tail rotor gearboxes grease lubricated; all designed for at least 30 minutes' run dry; accessory gearbox can be run independently on ground without rotor engagement by No. 1 engine engaged by pilot-operated clutch.

FLYING CONTROLS: Full-time dual electronic flight controls, with full manual reversion, provide automatic heading hold, autohover, autopilot modes and autostabiliser modes, all selectable by pilot; gunner in front seat has cyclic side-arm controller, normal collective lever and pedals and has full access to AFCS; electrical inputs from AFCS integrated with hydraulic-powered control units.

STRUCTURE: Composites materials account for 45 per cent of fuselage weight (less engines) and 16.1 per cent of total empty weight; material used for fuselage panels, nosecone, tailboom, tail rotor pylon, engine nacelles, canopy frame and maintenance panels; each blade has CFRP and Nomex main spar, Nomex honeycomb leading- and trailing-edges, composites skins, stainless steel leading-edge abrasion strip and frangible tip; control linkage runs inside driveshaft to reduce radar signature, avoid icing and improve ballistic tolerance; blades tolerant to 12.7 mm hits, possibly also 23 mm; delta-hinged two-blade tail rotor with broad-chord blades for ballistic tolerance. Total 70 per cent of airframe surface is composites. Bulkhead in nose and A frame running up through fuselage to rotor pylon protect crew against roll-over; overall IR suppressing paint; airframe meets MIL-STD-1290 crashworthiness covering vertical velocity changes of 11.2 m (36 ft 9 in)/s and longitudinal changes of 13.1 m (43 ft 0 in)/s.

LANDING GEAR: Non-retractable tailwheel type, with single wheel on each unit. Two-stage hydraulic shock-strut in each main unit designed to withstand normal loads and hard landings at descent rates in excess of 10 m (32.8 ft)/s.

POWER PLANT: Two Rolls-Royce Gem 1004 turboshafts, each with a T-O rating of 657 kW (881 shp) for 30 minutes; maximum continuous rating of 615 kW (825 shp) for normal twin-engined operation; intermediate contingency rating of 657 kW (881 shp) for 30 minutes; maximum contingency rating of 704 kW (944 shp) for 2½ minutes; and emergency rating (S/L, ISA) of 759 kW (1,018 shp) for 20 seconds. Transmission rating (Lot 2) is 969 kW (1,300 shp) (two engines), 704 kW (944 shp) for single-engined operation, with emergency rating of 759 kW (1,018 shp); power input into transmission is at 27,000 rpm from the RR 1004. Standard transmission rating for Lot 3 (Multirole) aircraft is 1,268 kW (1,700 shp); 1,566 kW (2,100 shp) in emergency. Production engines licence-built in Italy by Piaggio. Fireproof engine compartment, with engines widely spaced to improve survivability from enemy fire.

Two separate fuel systems, with cross-feed capability; interchangeable self-sealing and crash-resistant tanks, self-sealing lines, and digital fuel feed control. Tanks can be foam-filled for fire protection. Single-point pressure refuelling. IR exhaust suppression system (from Lot 2) and low engine noise levels. Separate independent lubrication oil-cooling system for each engine. Provision (Lot 2 aircraft) for auxiliary (self-ferry) fuel tanks on outboard underwing stations.

ACCOMMODATION: Pilot and co-pilot/gunner in separate cockpits in tandem. Elevated rear (pilot's) cockpit. External crew field of view exceeds MIL-STD-850B. Each cockpit has a flat plate, low-glint canopy with upward-hinged door panels on starboard side, blow-out port side panel for exit in emergency, and Martin-Baker crashworthy seat with sliding side panels of composites armour. Landing gear design and crashworthy seats reduce impact from 50 *g* to 20 *g* in crash.

SYSTEMS: Hydraulic system includes two main circuits dedicated to flight controls and two independent circuits for rotor and wheel braking. Main system operates at pressure of 207 bar (3,000 lb/sq in) and is fed by two independent power groups integrated and driven mechanically by the main transmission. Tandem actuators are provided for main rotor flight controls. Hydraulic system flow rate 23.6 litres (6.2 US gallons; 5.2 Imp gallons)/min in each main group. Spring-type reservoirs, pressurised at 0.39 bar (5.6 lb/sq in).

AVIONICS: *Comms:* Dual Marconi SRT-651/A U/VHF and single Marconi SRT-170/EB4 HF/SSB radios; Italtel IFF (with Mode 4 encryption unit in Lot 2 aircraft). A 129C suite additionally comprises Have Quick V secure radios and Marconi SRT-651/S SINCGARS for communication with ground forces.

Flight: Fully integrated digital multiplex system (IMS) controls navigation, flight management, weapon control, autopilot, monitoring of transmission and engine condition, fuel/hydraulic/electrical systems, caution and warning systems; IMS managed by two Agusta Sistemi/Harris central computers, each capable of operating independently, backed by two interface units which pick up outputs from sensors and avionic equipment and transfer them, via redundant MIL-STD-1553B databusses, to main computers for real-time processing; processed information is presented to pilot and co-pilot/gunner on separate graphic/alphanumeric head-down multifunction displays (MFDs) with standard multifunction keyboards for easy access to information, including area navigation using up to 100 waypoints, weapons status and selection, radio tuning and mode selection, caution and warning, and display of aircraft performance; conventional instruments and dials are provided as back-up; IMS computer can store up to 100 preset frequencies for HF, VHF and UHF radio management; navigation is controlled by navigation computer of IMS coupled to Doppler radar and radar altimeter with low-airspeed indicator, normally used for rocket aiming, providing back-up velocity data when the Doppler is beyond limits; synthetic map presentation of waypoints, target areas and dangerous areas is shown on pilot's or co-pilot's MFD; Litton strapdown inertial reference for both flight control and navigation is integrated into the IMS; AFCS provides either three-axis stabilisation or full attitude and heading hold, automatic hover, downward transition to hover or holds for altitude, heading and airspeed or groundspeed and automatic track following. Litton LISA 4000 AHRS is integrated on A 129C with GPS; later variant also has Marconi ANV-351 Doppler velocity sensor.

Instrumentation: Full day/night operational capability, with equipment designed to give both crew members a view outside helicopter irrespective of light conditions; cockpit lighting compatible with night vision goggles.

Mission: Pilot's night vision system (HIRNS: helicopter IR night system) allows nap-of-earth (NOE) flight by night with outside view generated by BAE North America FLIR sensor mounted on a GEC/OMI steerable platform at nose of aircraft and presented to both crewmen through the monocle of the Honeywell integrated helmet and display sighting system (IHADSS), to which it is slaved by helmet position sensors; flight information symbology superimposed on to image, giving true head-up reference. HeliTOW sight gives co-pilot/gunner direct view optics and FLIR, plus laser for ranging; provision for mast-mounted sight (MMS). A 129C has Galileo Avionica HIRNS Plus and TEAC video recording system for helmet and display sighting, plus voice, allowing post-mission analysis.

Self-defence: Active and passive self-protection systems (ECCM and ECM) standard on Italian Army A 129; Have Quick II frequency-hopping radio in Lot 2; onboard nav/weapon system can connect directly or by datalink with Italian CATRIN C^3I combat information system; active and passive electronic warfare systems include Elettronica ELT-554 radar jammer and ELT-156 RWR (ELT-156–05 in Lot 2) radar warning receiver; BAE Italia RALM-101 laser warning receiver; Sanders AN/ALQ-144A IR jammer; provisions for chaff/flare dispenser. Upgraded SIALP defensive aids under development for Lot 3 and retrofit, including ELT-156X(V)4, Marconi MILDS II missile approach warner, MES ECDS-2 chaff/flare dispenser and BAE RALM-01(V)2 laser warner.

ARMAMENT: Four attachments beneath stub-wing stressed for loads of 200 kg (441 lb) outboard and 350 kg (772 lb) inboard or 350 kg (772 lb) outboard and 100 kg (220 lb) inboard; outboard stations incorporate articulation which allows pylon to be elevated 2° and depressed 10° to increase missile launch envelope; they are aligned with aircraft automatically, with no need for boresighting. Initial armament of up to eight TOW, ITOW or TOW 2/2A wire-guided anti-tank missiles (two, three or four in carriers suspended from each wingtip station), with Saab/ESCO HeliTOW aiming system; with these can be carried, on inboard stations, either two 7.62, 12.7 or 20 mm gun pods, or two launchers each for seven or 19 air-to-surface rockets. For general attack missions, rocket launchers can be carried on all four stations; Italian Army has specified SNIA-BPD 81 and 70 mm rockets. Multirole and International versions can carry, in addition to the above, a Lockheed Martin/Otobreda TM 197B (standard 350 or optional 500 rounds) or Giat M621 20 mm cannon mounted in a nose turret; and Stinger AAMs. Stinger to be integrated with Italian A 129s, following 2003 trials. Optional upgrades offered on export versions include an autotracking sight; provision for up to eight Hellfire anti-tank missiles with autonomous laser designation capability; or a mix of four Hellfires and four TOWs. Other armament options include up to eight Hot missiles; AIM-9L Sidewinder, Mistral or Javelin AAMs; or grenade launchers. Lucas 0.50 in self-contained gun turret qualified, but not used by Italian Army.

DIMENSIONS, EXTERNAL:
Main rotor diameter ... 11.90 m (39 ft 0½ in)
Tail rotor diameter ... 2.32 m (7 ft 7 in)
Wing span ... 3.20 m (10 ft 6 in)
Width over TOW pods ... 3.60 m (11 ft 9¾ in)
Length overall, both rotors turning ... 14.29 m (46 ft 10½ in)
Fuselage: Length ... 12.275 m (40 ft 3¼ in)
  Max width ... 0.95 m (3 ft 1½ in)
Height:
  over tailfin, tail rotor horizontal ... 2.75 m (9 ft 0¼ in)
  tail rotor turning ... 3.315 m (10 ft 10½ in)
  to top of rotor head ... 3.35 m (11 ft 0 in)
Tailplane span ... 2.50 m (8 ft 2¼ in)
Wheel track ... 2.23 m (7 ft 3¾ in)
Wheelbase ... 6.955 m (22 ft 9¾ in)

AREAS:
Main rotor disc ... 111.20 m^2 (1,197.0 sq ft)
Tail rotor disc ... 4.23 m^2 (45.53 sq ft)

WEIGHTS AND LOADINGS (Italian Army):
Weight empty, equipped ... 2,529 kg (5,575 lb)
Max internal fuel load ... 750 kg (1,653 lb)
Max external weapons load ... 1,200 kg (2,645 lb)
Max T-O weight: standard ... 4,100 kg (9,039 lb)
  multirole version ... 4,600 kg (10,141 lb)
T-O weight, normal mission ... 3,950 kg (8,708 lb)
Max disc loading: standard ... 36.9 kg/m^2 (7.55 lb/sq ft)
  multirole version ... 41.4 kg/m^2 (8.47 lb/sq ft)
Transmission loading at max T-O weight and power:
  standard ... 4.23 kg/kW (6.95 lb/shp)
  multirole version ... 3.63 kg/kW (5.97 lb/shp)

PERFORMANCE (Italian Army, with eight TOW, at mission T-O weight of 3,950 kg (8,708 lb), at 2,000 m (6,560 ft), ISA+20°C, except where indicated):
Dash speed ... 159 kt (294 km/h; 183 mph)
Max level speed at S/L ... 135 kt (250 km/h; 155 mph)
Max rate of climb at S/L ... 612 m (2,008 ft)/min
Service ceiling ... 4,725 m (15,500 ft)
Hovering ceiling: IGE ... 3,140 m (10,300 ft)
  OGE . 1,890 m (6,200 ft)Basic 2 h 30 min mission profile with eight TOW and 20 min fuel reserves:
  Fly 54 n miles (100 km; 62 miles) to battle area, mainly in NOE mode, 90 min loiter (incl 45 min hovering), and return to base
Max endurance, no reserves ... 3 h 5 min
g limits ... +3.5/–0.5

UPDATED

AGUSTAWESTLAND A 129 INTERNATIONAL

TYPE: Attack helicopter.

PROGRAMME: Upgraded export version of A 129 with increased power and military load, plus improved avionics. First flight of prototype powered by two T800 turboshafts October 1988; demonstrated in Arabian Gulf during 1990; first flight with production-standard five-blade rotor (converted prototype MMX592, 29002) 9 January 1995; tail rotor diameter slightly increased; T800 gives between 20 and 40 per cent more power than Gem 1004. First firing of AGM-114 Hellfire ATMs, November 1999; Rafael NT-D Dandy ATM integrated and launched in same test series. By early 2003 no orders had been received.

CURRENT VERSIONS: **International:** *As described.*

Scorpion: Offered to Australia in 1998 to meet Project Air 87 for 25 to 30 armed reconnaissance helicopters. Based on Multirole version, but some A 129 International features. Competition won in 2001 by Eurocopter Tiger.

A 129DE: Trials aircraft registered (I-INTR) to Agusta in February 1998.

Data generally as for Mangusta, except as stated.

DESIGN FEATURES: Five-blade main rotor; pannier on each side of forward fuselage.

POWER PLANT: Two LHTEC T800-LHT-800 turboshafts, each 996 kW (1,335 shp) for twin-engined operation or 1,047 kW (1,404 shp) maximum contingency OEI; transmission rating 1,268 kW (1,700 shp). Optional 33 per cent additional internal fuel.

AVIONICS: *Flight:* INS/GPS as main navigation sensor.

Instrumentation: Two 152 × 203 mm (6 × 8 in) MFDs in each crew position.

Mission: Upgraded observation and targeting sensors (CCD TV and laser range-finder, replacing original TOW telescopic sight), permitting autonomous laser designation and maximum range launch of Hellfire ATMs.

ARMAMENT: Hellfire and/or TOW anti-tank missile, 70 and 81 mm rockets; Lockheed Martin/Otobreda TM 197B 20 mm cannon in nose turret; Stinger or Mistral AAMs.

WEIGHTS AND LOADINGS:
Primary mission T-O weight ... 4,800 kg (10,582 lb)
Max T-O weight ... 5,100 kg (11,243 lb)

PERFORMANCE (mixed weapons at mission T-O weight):
Cruising speed ... 150 kt (278 km/h; 173 mph)
Max rate of climb at S/L ... 677 m (2,220 ft)/min
Rate of climb at S/L, OEI ... 274 m (900 ft)/min
Vertical rate of climb at S/L ... 326 m (1,070 ft)/min
Hovering ceiling: IGE ... 4,206 m (13,800 ft)
  OGE ... 3,292 m (10,800 ft)
Max range, internal fuel, no reserves ... 303 n miles (561 km; 348 miles)

UPDATED

Unique AgustaWestland A 129DE development aircraft (I-INTR) ***NEW***/0576298

AGUSTAWESTLAND WAH-64D APACHE AH.MK 1

TYPE: Attack helicopter.

PROGRAMME: Westland selected, 13 July 1995, to build McDonnell Douglas (now Boeing) AH-64D Apache for Army Air Corps (AAC) and Royal Marines. Total 67 ordered, although original requirement was for 91; contract H12b/400 finalised 25 March 1996; programme involved some 240 UK companies (including 60 direct subcontractors) that received over 50 per cent of UK Apache work. Value increased to some £3,100 million by 1999, in part through addition of Marconi Avionics (later part of BAE Systems) to provide HIDAS self-protection. UK version similar to US Apache, including Longbow radar, but powered by Rolls Royce Turbomeca RTM 322 turboshafts.

AgustaWestland-built WAH-64D Apache undergoing shipboard trials ***NEW***/0589309

Engine integration trials conducted on leased sixth preproduction AH-64D (85-25408) at Boeing's Mesa plant; first engine run 6 April 1998; first flight 29 May 1998. Eight WAH-64Ds built and flown at Mesa; ZJ166/N9219G first flew 25 September 1998 and handed over to GKN Westland 28 September; initial trials and pilot training conducted in US, where four WAH-64s were retained by January 2000; further four supplied, complete, to Westland; remainder followed in kit form from 30 August 1999; final kit (to become ZJ233) left Boeing at Mesa, Arizona, on 27 March 2003; first delivery to AAC 15 March 2000; IOC on delivery of ninth aircraft, 14 December 2000; initial military aircraft release (IMAR) signed 22 December 2000; formal in-service date and release-to-service achieved 16 January 2001; initial weapons clearance planned for July 2001; HIDAS clearance December 2001; cannon and CRV-7 clearance August 2002, allowing full deployment; shipboard clearance 2004, following one-month trial aboard HMS *Ocean* in March 2004; final delivery was formal hand-over of ZJ233 to MoD at Farnborough International on 21 July 2004; IOC with 656 Squadron, August 2004; 664 Squadron next; full environmental clearance in November 2004. First Apache with full suite of mission equipment (including HIDAS, HUMS and low-height warning system) delivered early 2003. Initial 37 aircraft built to interim specification and being upgraded.

Shorts contracted in September 1996 to supply engine nacelles, stub-wings and horizontal stabiliser; heat exchangers manufactured by IMI Marston under Hughes-Treitler licence. UK and Netherlands agreed on 5 September 1996 on joint support of their Apache fleets. US and UK governments signed co-operation agreement 24 May 2000 for future Apache development, operation and support. UK contract also includes 68 Longbow radars, 980 Hellfire missiles and 204 launchers.

CURRENT VERSIONS (SPECIFIC): **WAH1/ZJ166 : (N9219G)** US-built; first flight at Mesa 25 September 1998; handed over to Westland 28 September 1998 for RTM 322 engine trials in US. Arrived UK 11 October 2002.

CUSTOMERS: British Army. Service allocation comprises 19 for operational evaluation, training (of which eight at Middle Wallop) and attrition replacement, plus 16 for each of three Army Air Corps regiments: No. 9 at Dishforth (656 and 664 Squadrons, intended from third quarter of 2001) and Nos. 3 (654 and 669 Squadrons, intended from mid/late 2002) and 4 (662 and 663 Squadrons from early/mid-2003) at Wattisham. Units have additional responsibility for supporting Royal Marines with 'navalised' WAH-64s and integrated (Army/RM) crews. Re-equipment dates were not met because of crew shortage, and new aircraft delivered to storage at Fleetlands (beginning ZJ182 on 5 October 2001) and at Shawbury (beginning ZJ191 and ZJ192 on 2 July 2002). By mid-2002, Advanced Helicopter Trials Unit (AHTU) at Middle Wallop had some 12 Apaches for development work, including weapon firing by Attack Helicopter Training Unit. AHTU redesignated 673 Squadron on 1 April 2003. First operational unit, 656 Squadron, began conversion on 1 September 2003, following delivery of aircraft to Dishforth.

DESIGN FEATURES: Manual blade folding (windspeed limit 45 kt; 83 km/h; 52 mph).

FLYING CONTROLS: Manual back-up to FBW system.

POWER PLANT: Two 1,556 kW (2,100 shp) RRTI RTM 322-01/12 turboshafts.

SYSTEMS: Rotor blade de-icing system.

AVIONICS: *Comms:* Mode S IFF.

Instrumentation: Honeywell IHADSS helmet sight to be upgraded from about 2004.

Mission: Video recorder.

Self-defence: BAE Systems HIDAS helicopter integrated defensive aids system, including Sky Guardian 2000 RWR, Type 1223 LWR, Thales (Vinten) Vicon 78 Srs 455 chaff/flare dispenser and BAE Systems AN/AAR-57(V) common missile warning system (CMWS). Trial installation of three warning elements first flown (in USA) on UK Apache in July 2000. Standard US fit of AN/APR-48 RFI also included.

ARMAMENT: Additional provision for Bristol Aerospace CRV-7 70 mm rocket pods and Shorts Starstreak AAM.

UPDATED

AGUSTAWESTLAND LYNX

TYPE: Light utility helicopter.

PROGRAMME: Developed within Anglo-French helicopter agreement confirmed 2 April 1968; Westland given design leadership; first flight of first of 13 prototypes (XW835) 21 March 1971; first flight of fourth prototype (XW838) 9 March 1972, featuring production-type monobloc rotor head; first flights of British Army Lynx prototype (XX153) 12 April 1972, French Navy prototype (XX904) 6 July 1973, production Lynx (RN HAS. Mk 2 XZ229) 20 February 1976; first RN operational unit (No. 702 Squadron) formed on completion of intensive flight trials December 1977; AH. Mk 5 first flew (ZE375) 23 February 1985; other development details and records in 1975–76 and subsequent *Jane's*. Production shared 70 per cent Westland, 30 per cent Aerospatiale; for details of G-LYNX's 1986 world helicopter absolute speed record, and Lynx AH. Mk 7 XZ170's 1989 agility trials, see 1990–91 *Jane's*.

Second phase of development was Super Lynx (naval) and Battlefield Lynx, to which standards later UK military Lynx (Mks 8 and 9) were built; also exported. Battlefield Lynx

FLIR turret and defensive aids sensors on Omani Lynx Mk 120 ***(Paul Jackson)*** ***NEW***/0589335

Royal Air Force of Oman AgustaWestland Lynx Srs 300 Mk 120, accompanied by a Malaysian counterpart ***NEW***/0589357

mockup displayed at 1988 Farnborough Air Show (converted demonstrator G-LYNX), featuring wheeled landing gear, exhaust diffusers and provision for anti-helicopter missiles each side of fuselage; first flight of wheeled prototype (converted trials AH. Mk 7 XZ170) 29 November 1989; first flight of South Korean Super Lynx (90–0701, temporarily ZH219) 16 November 1989 (also first Lynx with Seaspray Mk 3 radar); first flight of Portuguese Super Lynx (9201, temporarily ZH580, ex-RN ZF559) 27 March 1992.

In September 1996, GKN Westland formally launched versions of Lynx powered by CTS800 turboshafts, with or without a six-screen EFIS. Current production versions were simultaneously redesignated, those now on offer being Srs 100, 200 or 300. Fuselages built from 1999 have potential for MTOW extension to 5,443 kg (12,000 lb) and beyond. First Super Lynx 300 (ZT800) flew 27 January 1999, although maiden flight with CTS800 engine was on 12 June 2001.

Potential life extension for Lynx possible following 30 January 2002 award to Westland of £20+ million, 18-month study contract expected to lead to upgrade of 80 British Army Mk 7 and Mk 9 Lynxes under Future Lynx Battlefield Light Utility Helicopter programme at total cost of £1 billion. This followed, on 22 July 2002, by announcement of assessment phase for Future Lynx to meet Royal Navy's Surface Combatant Maritime Rotorcraft requirement, requiring upgrade of 40 Lynx to enter service from 2008.

CURRENT VERSIONS: Early versions of Lynx for UK armed forces were last described in the 2002–03 *Jane's All the World's Aircraft*. Those for overseas operators appeared in the 1990–01 and earlier editions.

Super Lynx Series 100: Introduced September 1996; export version of HMA. Mk 8; conventional cockpit instrumentation, Rolls-Royce Gem 42-1 turboshafts and 360° radar. Applies retrospectively to aircraft sold to Brazil (Mk 21A), Portugal (Mk 95) and South Korea (Mk 99) as well as to seven Mk 88As sold to Germany, also in September 1996.

Super Lynx Series 200: Conventional cockpit, but electronic power system displays and two LHTEC CTS800-4N turboshafts with FADEC. Original Lynx demonstrator G-LYNX fitted with two 1,007 kW (1,350 shp) T800 turboshafts as **Battlefield Lynx 800** private venture (LHTEC funding power plants and gearboxes, Westland provided airframe for full flight demonstration programme); first flight 25 September 1991; programme terminated early 1992 after 17 hours. CTS800 is derived from the T800 (as in the Boeing Sikorsky RAH-66 Comanche) and offers an additional 30 per cent of power to operators using the Lynx in hot climates. Version for production Lynx is CTS800-4N

Super Lynx Series 300: Prototype first flown (ZT800) 27 January 1999, initially with Gem 42 engines; flew with CTS800-4N engines 12 June 2001. Six-screen EFIS cockpit (including four 158 mm; 6¼ in square integrated display units); dual redundant MIL-STD-1553B and ARINC 429 databusses; new navigation system and AHRS; revised communications suite; CTS800 engines. Ordered by Malaysia (with Sea Skua anti-ship missiles); Thailand confirmed contract for two in 2001; Oman announced intention to purchase on 17 March 2001 and signed contract for 16 on 19 January 2002; aircraft equipment includes Avitronics EW suite.

South Africa, announced intention to purchase on 18 November 1998, but later deferred expected contract for four to unspecified date; authorisation to proceed given February

Malaysian Navy AgustaWestland Lynx Srs 300 Mk 100 ***NEW***/0589355

First Thai Navy Lynx Srs 300 Mk 110 ***NEW***/0589434

FLIR turret and defensive aids sensors on Malaysian Lynx Mk 100 *(Paul Jackson)* ***NEW***/0589334

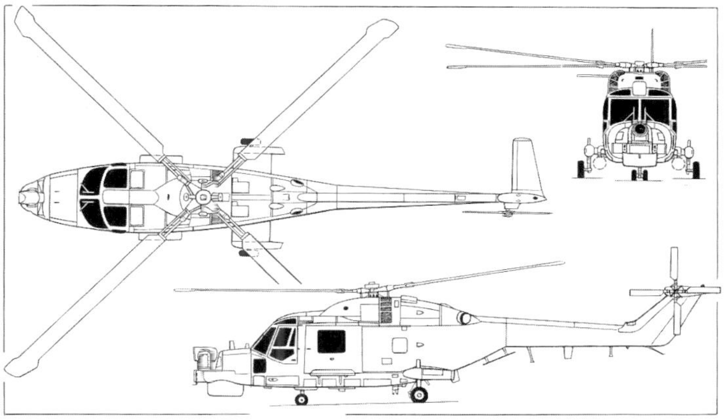

AgustaWestland Lynx HMA. Mk 8, based on the Super Lynx advanced export version *(Jane's/Dennis Punnett)* 0507240

2003; contract confirmation announced 14 August 2003; South African Srs 300s will be designated Mk 64 and equipped with Telephonics AN/APS-143(V)-3 radar, Sysdel Sea Raven 118 ESM, Kentron Cumulus LEO Mk 2 sensor turret with video datalink, Grintek GUS-1000 communications system (two Reutech ACR-500 V/UHF and one Grintek AT-2820 HF) and Telemat PT-200 IFF.

Future Lynx: Development of Series 300; offered to UK armed forces as retrofit meeting Army's Battlefield Light Utility Helicopter and Navy's Surface Combatant Maritime Rotorcraft requirements. Growth potential to MTOW of 6,250 kg (13,778 lb).

CUSTOMERS: Refer to table. Contracts for later versions include Super Lynx ordered by South Korea 1988 (12 **Mk 99** with Racal Avionics Doppler 71/TANS N nav system, Seaspray Mk 3 360° radar, AN/AQS-18 dipping sonar and Sea Skua), handed over between 26 July 1990 and May 1991 for 'Sumner' and 'Gearing' class destroyers; further 13 delivered 1999–2000 as Mk 99As to order confirmed in June 1997. Mk 99A has composites tailplane; first aircraft flown 3 July 1999 and departed Yeovil for surface delivery 1 September 1999. Portugal ordered five (first two ex-Royal Navy modified airframes) Super Lynx **Mk 95** 1990 (plus three options) with Racal RNS252 GPS-aided INS and Doppler 91 navigation systems and some US equipment including AN/AQS-18 sonar and Honeywell RDR 1500 radar; first two handed over 29 July 1993 for 'Vasco da Gama' class (MEKO 200) frigates; final two delivered 16 November 1993. Brazil ordered nine **Mk 21As** (plus five conversions from Mk 21). First for upgrade, N3027 delivered to Yeovil on 1 February 1995 and reflown as Mk 21A (N4010) on 22 December 1995; last two redelivered in late April 1998. First new-build helicopter (N4001) flew 12 June 1996; last delivered in August 1997. Mk 21A avionics include 360° Seaspray 3000 radar, RNS252 INS and Doppler 71 (but no FLIR or CTS); armament includes Sea Skua missiles. Mk 21A introduced 5,330 kg (11,750 lb) MTOW.

German Navy ordered seven Super Lynx **Mk 88As** in September 1996 and simultaneously took an option (confirmed on 25 June 1998) on upgrading (with new-build airframes) of its existing 17 (subsequently reduced by attrition to 15); new deliveries began with roll-out on 14 July 1999 (post first flight) of initial aircraft. Westland converted initial upgraded helicopter (8303), first flown 28 February 2001. First of 14 upgrades by Eurocopter at Donauwörth (of 8309) was redelivered on 7 March 2002. Mk 88A is a Srs 100 aircraft with 360° Seaspray Mk 3000 radar, FLIR turret (or nose fairing when not fitted), Rockwell Collins GPS and Racal Doppler 91 and RNS252; Sea Skua ASM armament.

Eight Danish aircraft upgraded to **Mk 90B** by 2004 with new airframes under contract announced 20 January 1998; all except one undertaken locally; initial conversion (S-191) by Westland and redelivered to Denmark on 1 November 2000; eighth entered re-work in June 2004. Super Lynx also offered to Australia and New Zealand (both unsuccessfully) and Malaysia, last-mentioned announcing order for six Mk 100s on 7 September 1999; deliveries began mid-2003; first flight in Malaysia was on 21 July 2003; final aircraft (No. 3) delayed in UK until mid-2004 for Sea Skua anti-ship missile trials and formally handed-over at Farnborough International on 19 July 2004.

Oman's first Lynx **Mk 120** flew 25 October 2003; first three delivered on 24 June 2004,

LYNX PRODUCTION

Variant	Customer	Qty	First aircraft	First flights	Operators
AH. Mk 1	Army Air Corps	113[13]	XZ170	11 Feb 1977-24 Jan 1984	
HAS. Mk 2	Fleet Air Arm	60	XZ227	20 Feb 1976-26 May 1981	Converted to Mk 3
HAS. Mk 2(FN)	French Navy	26	260	4 May 1977-5 Sep 1979	31F, 34F, ERCE
HAS. Mk 3	Fleet Air Arm	31	ZD249	4 Jan 1982-21 Oct 1988	815, 702 Sqdns
HAS. Mk 4(FN)	French Navy	14	801	1 Apr 1982-26 Aug 1983	31F, 34F, ERCE
Mk 5	MoD(PE)	3[13]	ZD285	21 Nov 1984-23 Feb 1985	DERA
AH. Mk 6	Royal Marines	-	-	None built	-
AH. Mk 7	Army Air Corps	11[13]	ZE376	23 Apr 1985-5 Jun 1987	
HMA. Mk 8	Fleet Air Arm	-		Conversions	815, 702 Sqdns
AH. Mk 9	Army Air Corps	16[13,14]	ZG884	20 Jul 1990-21 Jun 1992	653, 659 Sqdns
Mk 21	Brazilian Navy	9	N3020	30 Sep 1977-14 Apr 1978	Five upgraded to 21A
Mk 21A	Brazilian Navy	9[14]	N4001	12 Jun 1996-24 Apr 1997	1° EHEAA
Mk 22	Egyptian Navy	-	-	None built	-
Mk 23	Argentine Navy	2	0734	17 May 1978-23 Jun 1978	Withdrawn from service
Mk 24	Iraqi Army	-	-	None built	-
Mk 25[1]	Netherlands Navy	6	260	23 Aug 1976-16 Sep 1977	7/860 Sqdns
Mk 26	Iraqi Army (armed)	-	-	None built	-
Mk 27[2]	Netherlands Navy	10	266	6 Oct 1978-12 Nov 1979	7/860 Sqdns
Mk 28	Qatari Police	3[13]	QP-31	2 Dec 1977-12 Apr 1978	Withdrawn from service[7]
Mk 64	South African Navy[10]	4[14]	-	-	22 Squadron
Mk 80[12]	Danish Navy	8[11]	S-134	3 Feb 1980-15 Sep 1981	See Mk 90
Mk 81[3]	Netherlands Navy	8	276	9 Jul 1980-24 Mar 1981	7/860 Sqdns
Mk 82	Egyptian Army	-	-	None built	-
Mk 83	Saudi Army	-	-	None built	-
Mk 84	Qatari Army	-	-	None built	-
Mk 85	UAE Army	-	-	None built	-
Mk 86	Norwegian Coast Guard	6	207	23 Jan 1981-11 Sep 1981	Skv 337
Mk 87	Argentine Navy	-	-	Embargoed	-
Mk 88	German Navy	19[9]	8301	26 May 1981-10 Dec 1988	3/MFG 3
Mk 88A	German Navy	7[14]	8320	30 Apr 1999 — Mar 2000	3/MFG 3
Mk 89	Nigerian Navy	3	01-F89	29 Sep 1983-14 Mar 1984	101 Sqdn
Mk 90[12]	Danish Navy	1[4]	S-256	19 Apr 1988[6]	Søværnets Flyvetjeneste
Mk 95	Portuguese Navy	3[14,5]	9203	9 Jul 1993–1993	EHM[8]
Mk 99	South Korean Navy	12[14]	90–0701	16 Nov 1989-14 May 1991	627 Sqdn
Mk 99A	South Korean Navy	13[14]	99–0721	3 Jul 1999-Aug 2000	-
Srs 300/Mk 100	Malaysian Navy	6[14]	M501-1	28 Mar 2002	501 Sqdn
Srs 300	Thai Navy	2[14]	2313	26 Jul 04	-
Srs 300/Mk 120	Oman Air Force	16[14]	757	25 Oct 2003	-
Subtotal		**417**			
Lynx 3		1	ZE477	14 Jun 1984	
Demonstrators		3	G-LYNX	18 May 1979-27 Jan 1999	
Prototypes		13	XW835	21 Mar 1971-5 Mar 1975	
Total		**434**			

Notes: Note A: Army Air Corps 651, 652, 654, 655, 656, 657, 659, 661, 662, 663, 664, 665, 667, 669 and 671 Squadrons; 847 Squadron (Royal Marine Commando)

[1] Netherlands designation UH-14A; all to SH-14D
[2] Netherlands designation SH-14B; all to SH-14D
[3] Netherlands designation SH-14C; all to SH-14D
[4] Plus one conversion from demonstrator; these two upgraded to Super Lynx Mk 90B with new airframes
[5] Plus two conversions from Mk 3
[6] Completion and first flight at Vaerløse, Denmark
[7] Sold to Royal Navy for spares recovery and use as ground instructional airframes
[8] Esquadrilha de Helicópteros de Marinha
[9] Surviving 15 upgraded to Super Lynx with new airframes in 2001–2003
[10] Programme delay announced in 1999; go-ahead given February 2003
[11] Surviving six converted to Super Lynx Mk 90B with new airframes
[12] Converted during 1990s to Mk 80A/90A with Gem 42-1 engines
[13] Army version
[14] Super/Battlefield Lynx

although formal acceptance (sixth aircraft) not effected until 19 July 2004 in ceremony at Farnborough International. Are first Lynx with new-generation HUMS and related maintenance data transfer device.

COSTS: £100 million for seven Mk 88As (Germany), 1996. Proposed four Srs 300s for South Africa estimated to cost total of £80 million (1998). Two Thai Lynx 300s, plus logistic support and services, £25 million (2001). Six Malaysian Lynx 300s valued at RM700 million (US$184 million) (1999).

DESIGN FEATURES: Compact design suited to hunter-killer ASW and missile-armed anti-ship naval roles from frigates or larger ships (superseding ship-guided helicopters), armed/unarmed land roles with cabin large enough for squad, or other tasks; manually folding tail pylon on most (but not all) naval versions; single four-blade semi-rigid main rotor (foldable), each blade attached to main rotor hub by titanium root plates and flexible arm; rotor drives taken from front of engines into main gearbox mounted above cabin ahead of engines; in flight, accessory gears (at front of main gearbox) driven by one of two through shafts from first stage reduction gears; four-blade tail rotor, drive taken from main ring gear; single large window in each main cabin sliding door; provision for internally mounted armament, and for exterior universal flange mounting each side for other weapons/stores. Super Lynx has all-weather day/night capability; extended payload/range; advanced technology swept-tip (BERP) composites main rotor blades offering improved speed and aerodynamic efficiency and reduced vibration; and reversed direction tail rotor for improved control.

FLYING CONTROLS: Rotor head controls actuated by three identical tandem servojacks and powered by two independent hydraulic systems; control system incorporates simple stability augmentation system; each engine embodies independent control system providing full-authority rotor speed governing, pilot control being limited to selection of desired rotor speed range; in event of one engine failure, system restores power up to single-engine maximum contingency rating; main rotor can provide negative thrust to increase stability on deck after touchdown on naval versions; hydraulically operated rotor brake mounted on main gearbox; sweptback fin/tail rotor pylon, with starboard half-tailplane.

STRUCTURE: Conventional semi-monocoque pod and boom, mainly light alloy; glass fibre access panels, doors, fairings, pylon leading/trailing-edges, and bullet fairing over tail rotor gearbox; composites main rotor blades; main rotor hub and inboard flexible arm portions built as complete unit, as titanium monobloc forging; tail rotor blades have light alloy spar, stainless steel leading-edge sheath and rear section as for main blades.

LANDING GEAR (GENERAL PURPOSE MILITARY VERSION): Non-retractable tubular skid type. Provision for a pair of adjustable ground handling wheels on each skid. Flotation gear optional. Battlefield Lynx and AH. Mk 9 equivalent have non-retractable tricycle gear with twin nosewheels.

LANDING GEAR (NAVAL VERSIONS): Non-retractable oleo-pneumatic tricycle type. Single-wheel main units, carried on sponsons, fixed at 27° toe-out for deck landing; can be manually turned into line and locked fore and aft for movement of aircraft into and out of ship's hangar. Twin-wheel nose unit steered hydraulically through 90° by the pilot to facilitate independent take-off into wind. Sprag brakes (wheel locks) fitted to each wheel prevent rotation on landing or inadvertent deck roll. These locks disengaged hydraulically and re-engage automatically in event of hydraulic failure. Maximum vertical descent 1.83 m (6 ft)/s; with lateral drift 0.91 m (3 ft)/s for deck landing. Flotation gear, and deck securing system, optional. Latter include RDM Deadlock, Trigon 3, Trigon 5, Assist and PRISM, to customer's requirements.

POWER PLANT: Current option of two Rolls-Royce Gem 42-1 turboshafts, each rated at 746 kW (1,000 shp) for T-O or 664 kW (890 shp) max continuous; or two LHTEC CTS800-4Ns, each of 1,016 kW (1,362 shp) and 945 kW (1,267 shp), respectively. Transmission rating 1,372 kW (1,840 shp). Exhaust diffusers for IR suppression optional on Battlefield Lynx.

Fuel in five internal tanks; usable capacity 957 litres (253 US gallons; 210 Imp gallons) when gravity-refuelled; 985 litres (260 US gallons; 217 Imp gallons) when pressure-refuelled. Optional internal tank, replacing bench seat at rear of cabin, capacity 345 litres (91.0 US gallons; 75.9 Imp gallons). For ferrying, two tanks each of 441 litres (116 US gallons; 97.0 Imp gallons) in cabin, replacing bench tank. Maximum usable fuel 1,867 litres (493 US gallons; 411 Imp gallons). Engine oil tank capacity 6.8 litres (1.8 US gallons; 1.5 Imp gallons). Main rotor gearbox oil capacity 28 litres (7.4 US gallons; 6.2 Imp gallons).

ACCOMMODATION: Pilot and co-pilot or observer on side-by-side seats. Dual controls optional. Individual forward-hinged cockpit door and large rearward-sliding cabin door on each side; cockpit doors jettisonable; windows of cabin doors also jettisonable. Cockpit accessible from cabin area. Maximum high-density layout (military version) for one pilot and 10 armed troops or paratroops, on lightweight bench seats in soundproofed cabin. Alternative VIP layouts for four to seven passengers, with additional cabin soundproofing. Seats can be removed quickly to permit carriage of up to 907 kg (2,000 lb) of freight internally. Tiedown rings provided. In casualty evacuation role, with a crew of two, Lynx can accommodate up to six Alphin stretchers and a medical attendant. Both basic versions have secondary capability for search and rescue (up to nine survivors) and other roles.

SYSTEMS: Two independent hydraulic systems, pressure 141 bar (2,050 lb/sq in). Third hydraulic system provided in naval version when sonar equipment installed. No pneumatic system. 28 V DC electrical power supplied by two 6.3 kW engine-driven starter/generators and an alternator. External power sockets. 24 V 23 Ah (optionally 40 Ah) Ni/Cd battery fitted for essential services and emergency engine starting. 200 V three-phase AC power available at 400 Hz from two 25 kVA transmission-driven alternators. Cabin heating and ventilation system. Optional supplementary cockpit air conditioning system. Electric anti-icing and demisting of windscreen, and electrically operated windscreen wipers, standard; windscreen washing system.

AVIONICS: *Comms:* Typically Rockwell Collins VOR/ILS; DME; Rockwell Collins AN/ARN-118 Tacan; I-band transponder (naval version only); BAE PTR446, Rockwell Collins APX-72, DaimlerChrysler STR 700/375 or Italtel APX-77 IFF.

Flight: BAE duplex three-axis automatic stabilisation equipment; BAE GM9 Gyrosyn compass system; Racal tactical air navigation system (TANS); Racal 71 Doppler, E2C standby compass. BAE Mk 34 AFCS. Additional units fitted in naval version, when sonar is installed, to provide automatic transition to hover and automatic Doppler hold in hover.

Latest versions have Racal Doppler 91 and RNS 252 navigation; Honeywell/Smiths AN/APN-198 radar altimeter; Rockwell Collins 206A ADF; Rockwell Collins VIR 31A VOR/ILS.

Radar: Optional Seaspray Mk 3000 or Honeywell RDR 1500 360° scan radar in chin fairing. BAE Sea Owl thermal imaging equipment optional above nose.

Mission: Optional Honeywell AN/AQS-18 or Thomson HS-312 sonars in naval variants. Detection of submarines by dipping sonar or magnetic anomaly detector. Dipping sonar operated by hydraulically powered winch and cable hover mode facilities within the AFCS.

EQUIPMENT: All versions equipped as standard with navigation, cabin and cockpit lights; adjustable landing light under nose; and anti-collision beacon. For search and rescue, with three crew, both versions can have a waterproof floor and a 272 kg (600 lb) capacity clip-on hydraulic or electric hoist on starboard side of cabin; cable length 30 m (98 ft).

ARMAMENT: For armed escort, anti-tank or air-to-surface strike missions, army version can be equipped with two 20 mm cannon mounted externally so as to permit fitment of pintle-mounted 7.62 mm machine gun inside cabin. External pylon can be fitted on eachside of cabin for variety of stores, including two Minigun or other self-contained gun pods; two rocket pods; or up to eight HOT, Hellfire, TOW, or similar air-to-surface missiles. Additional six or eight reload missiles carried in cabin. For ASW role, armament includes two Mk 44, Mk 46, A244S or Sting Ray homing torpedoes, one each on an external pylon on each side of fuselage, and six marine markers; or two Mk 11 depth charges. Alternatively, up to four Sea Skua semi-active homing missiles; on French Navy Lynx, four AS.12 or similar wire-guided missiles.

Following data refer to Super Lynx.

DIMENSIONS, EXTERNAL:
Main rotor diameter ... 12.80 m (42 ft 0 in)
Tail rotor diameter ... 2.36 m (7 ft 9 in)
Distance between rotor centres ... 7.66 m (25 ft 1½ in)
Length of fuselage, tail rotor turning ... 13.34 m (43 ft 9¼ in)
Length overall:
main rotor turning ... 15.24 m (50 ft 0 in)
main rotor blades and tail folded ... 10.85 m (35 ft 7¼ in)
Width overall, main rotor blades folded ... 2.94 m (9 ft 7¾ in)
Height overall: tail rotor turning ... 3.67 m (12 ft 0½ in)
main rotor blades and tail folded ... 3.25 m (10 ft 8 in)
Tailplane half-span ... 1.32 m (4 ft 4 in)
Wheel track ... 2.94 m (9 ft 7¾ in)
Wheelbase ... 3.02 m (9 ft 11 in)

DIMENSIONS, INTERNAL:
Cabin, from back of pilots' seats:
Min length ... 2.055 m (6 ft 9 in)
Max width ... 1.78 m (5 ft 10 in)
Max height ... 1.42 m (4 ft 8 in)
Floor area ... 3.45 m^2 (37.1 sq ft)
Volume ... 4.9 m^3 (173 cu ft)
Cabin doorway: Width ... 1.37 m (4 ft 6 in)
Height ... 1.19 m (3 ft 11 in)

AREAS:
Main rotor disc ... 128.71 m^2 (1,385.4 sq ft)
Tail rotor disc ... 4.37 m (47.04 sq ft)

WEIGHTS AND LOADINGS:
Manufacturer's basic weight ... 3,291 kg (7,255 lb)
Operating weight empty (including crew and appropriate armament):
ASW (two torpedoes) ... 4,618 kg (10,181 lb)
ASV (four Sea Skuas) ... 4,373 kg (9,641 lb)
surveillance and targeting ... 3,708 kg (8,174 lb)
search and rescue ... 3,778 kg (8,329 lb)
Max underslung load ... 1,361 kg (3,000 lb)
Max fuel weight: normal ... 787 kg (1,735 lb)
bench seat ... 275 kg (606 lb)
ferry tanks, total two ... 696 kg (1,534 lb)
Max T-O weight ... 5,330 kg (11,750 lb)
Max disc loading ... 41.4 kg/m^2 (8.48 lb/sq ft)
Transmission loading at max T-O weight and power ... 3.89 kg/kW (6.39 lb/shp)

PERFORMANCE (CTS800 engines):
Max continuous cruising speed ... 132 kt (244 km/h; 152 mph)
Hovering ceiling IGE, ISA +20°C ... 2,105 m (6,900 ft)
Hovering ceiling OGE, ISA +20°C ... 1,445 m (4,740 ft)
Radius of action:
range with auxiliary fuel ... 540 n miles (1,000 km; 621 miles)
anti-submarine, 2 h on station, dipping sonar and one torpedo ... 20 n miles (37 km; 23 miles)
point attack with four Sea Skuas ... 125 n miles (232 km; 143 miles)
surveillance, 3 h 50 min on station ... 75 n miles (139 km; 86 miles)
Max endurance with auxiliary fuel ... 5 h 20 min

UPDATED

AGUSTAWESTLAND A149

TYPE: Multirole medium helicopter.

PROGRAMME: In October 2003, Agusta revealed preliminary plans for a 6.8 tonne multirole helicopter which it planned to offer to Australia as part of the AgustaWestland/BAE Systems combined bid for the Air 9000 requirement complementing, and in the long term possibly replacing, Sikorsky S-70A Black Hawks. The helicopter appears to be a reactivation of the A129 Utility project, which Agusta allowed to lapse in the mid-1990s. Australia was offered participation in design of the cabin and tailboom, but the helicopter was not selected for the RAAF.

CUSTOMERS: Predicted market for 2,000 helicopters of this class in civil and military employment outside the USA, of which a 20 per cent share is anticipated.

DESIGN FEATURES: Employs A129 technology, but with a new fuselage and drive train.

UPDATED

AGUSTAWESTLAND A 109

US Coast Guard designation: MH-68A Stingray
Swedish armed forces designation: Hkp 15

TYPE: Light utility helicopter.

PROGRAMME: First flight of original A 109 4 August 1971; deliveries of A 109A started early 1976; single-pilot IFR certification 20 January 1977; deliveries of uprated A 109 Mk II began September 1981; A 109C certified 1989; first deliveries February 1989; A 109 Max was medevac version; A 109CM and A 109EOA for military use.

First flight of A 109K April 1983; first flight of production representative second aircraft March 1984. A 109K2s of REGA air ambulance fitted with Sextant AFDS 95–1 AFCS, granted FAA single-pilot IFR certification late in 1996.

A 109 Power programme begun in 1993, intending to unify A 109C and A 109K2; initially known as A 109 Unified; launched November 1993; unique A 109D development aircraft with FADEC-equipped Allison 250-C22R9s flew 1 October 1994; prototype A 109E (I-EPWS) flew 8 February 1995. A 109D became second A 109E prototype.

CURRENT VERSIONS: **A 109KM:** Military version of A 109K2; roles include anti-tank/scout, escort, command and control, utility, ECM and SAR/medevac; fixed landing gear; sliding side doors.

A 109KN: Shipboard version with equivalent roles to A 109KM, including anti-ship, over-the-horizon surveillance and targeting and vertical replenishment.

AgustaWestland A 109 Power Elite NEW/0576296

AgustaWestland A 109LUH NEW/0576291

A 109K2: Special civil rescue version first sold to Swiss REGA non-profit rescue service; REGA equipment includes Spectrolab SX-16 searchlight, 1,000 kg (2,204 lb) cargo hook, GPS, Elbit moving map display and single-pilot IFR instrumentation; NVG compatible. Equipped with Thales AFDS 95–1 AFCS from 1996.

A 109K2 Law Enforcement: Dedicated police version; optional equipment includes 907 kg (2,000 lb) cargo hook, 204 kg (450 lb) capacity variable speed rescue hoist with 50 m (164 ft) of cable, rappeling kit, wire-strike protection, SX-16 searchlight, MA3 retractable light, external loudspeakers, emergency floats, GPS, FM tactical communications, weather radar, LLTV and FLIR.

A 109 Power: Revealed at 1995 Paris Air Show, by which time prototype had accumulated more than 60 hours of flight testing. Engineering designation is **A 109E**. Based on A 109K2 airframe with new, A 129-derived lightweight, low-maintenance titanium main rotor head connected to composites material grips via single elastomeric bearing on each blade; Pratt & Whitney Canada PW206C or Turbomeca Arrius 2K1 engines; new heavy-duty high-clearance landing gear in revised position below fuselage; cabin as for A 109K2, retaining quick-change facility from passenger to EMS operation. First production aircraft (I-PWER) completed late 1995; RAI IFR certification 31 May 1996; FAA IFR certification 26 August 1996. Aircraft with Rogerson IIDS (integrated instrument display system) cockpit (I-EAPW; first production) displayed at 1997 Paris Air Show. First 120 or so Powers had P&W engines; Turbomeca variant phased in with delivery of N7YL to Erie Lifestar in early September 2001. PZL-Świdnik of Poland contracted to build fuselages for A 109E between 1996 and 2002 as alternative production source.

US Coast Guard MH-68A (see Customers, below) is equipped with night vision device (NVD) cockpit, FLIR, weather radar and machine gun. Following trials aboard the USCG *Gallatin* (WHEC 721) in early 2001, the MH-68A has been cleared for deployment aboard all helicopter-capable cutters in the USCG fleet. MH-68A fleet had logged 10,667 hours by 28 April 2004.

A 109 Power Elite: Special edition of A 109E; improved interior and soundproofing; limited to 50 aircraft, c/ns 11151 to 11200, of which first (I-RAIB) exhibited at Paris in June 2001; initial export, fourth Elite, was delivered to Air Harrods in UK, May 2002; some 22 built by end of 2003, including first for Italian Carabinieri.

MH-68A: US Coast Guard version of A 109E (P & W version). Provision for MG240 machine gun in port doorway and rescue hoist over starboard door. Night vision compatible cockpit; encrypted radios.

A 109LUH: Formerly A109M. Military version of A 109E with PW207C or Arrius 2K2 engine. Swedish armed forces ordered 20, under designation Hkp 15, on 20 June 2001; first two handed over 10 September 2002. Hkp 15s will be used for crew training, utility, ASW, SAR and Medevac missions, from shore and shipboard bases. Requirement for 40 A 109LUHs announced by South African Air Force in November 1998 of which 30 ordered, with 10 options; these likely to be equipped with IST Dynamics turret mounting a Vektor multiple-calibre gun; first five manufactured in Italy, remainder by the Denel Group in South Africa. Initial Agusta-built aircraft rolled out 6 September 2002; Turbomeca engines. First Denel-built aircraft (4006) entered final assembly in February 2003 and made its public debut at the Africa Aerospace and Defence 2004 exhibition on 22 September 2004.

A 109LOH: Light observation helicopter variant. Selected by Malaysian Army Aviation in September 2003, with order for 11, plus initial support and training package, valued at US$70 million.

A 109S Grand: Described separately.

CUSTOMERS: Total of some 663 A 109s built by January 2004, comprising 336 A 109A series, 79 A 109C series, 41 A 109K2s and 207 A 109Es (including 22 Elites and 10 Turbomeca-powered). Recent customers include the US Coast Guard, which ordered four A 109Es (with four options, all taken up and delivered by late 2001, and two further options announced in February 2003) in interdiction configuration in April 2000 to equip Helicopter Interdiction Tactical Squadron 10 (HITRON TEN) based at Jacksonville, Florida, for armed anti-drug missions; Italy's Carabinieri, which took delivery of two A 109Es in 2000; Duke Life Flight of Durham, North Carolina, which took delivery of two A 109Es in EMS configuration in December 2001 and January 2002; the Italian government, which ordered one A 109E in VIP configuration for the Prime Minister's use; car manufacturer and Formula One racing team Ferrari, one A 109E; Air Harrods of the UK, one A 109E; Careflite, of Dallas-Fort Worth, Texas, which ordered four A 109Es in EMS configuration, the last of which was delivered on 30 December 2002, plus one ordered in February 2003, plus three options; Shenzhen Police of China, which ordered one A109E, plus two options, in January 2003; and the Slovenian Ministry of Defence, which ordered one, plus two options, in January 2003 for border patrol, civil protection, SAR and EMS duties. Greek government ordered five A 109Es in EMS configuration in 1999, and Dyfed/Powys Police Authority in the UK took delivery of an A 109E in late 1999, and ordered a replacement in March 2003. Three A 109K2s delivered to Dubai Police in 1995 and 1996. Switzerland ordered one A 109 Power, which was delivered in May 1998 to the Federal Office for Civil Aviation. Nigerian Navy ordered four A 109 Powers for anti-smuggling duties in the Niger delta, first two delivered mid-2002, followed by remaining two in September 2004. Deer Jet of China ordered two A 109E Powers for harbour pilot shuttle services. Japanese National Police Agency ordered three A 109Es in June 2003, for delivery by end of year, for operation in Hiroshima, Fukushima and Niigata. UK operator Euro SkyLink took delivery of an A109E Power in May 2004. Phoenix Police Department Air Patrol Unit and the Government of Rajasthan each ordered one A109 Power in July 2004; Airlift Northwest of Seattle ordered two A 109 Powers in October 2004; LifeFlight of Maine took delivery of two A 109 Powers in EMS configuration in December 2004. A 109E orders stood at nearly 300 by the end of 2002.

COSTS: A 109E US$3.2 million to US$3.5 million (2002). A 109K2 US$3.5 million (2001). A 109E direct operating cost US$380 per hour (2000). US Coast Guard contract for four A 109Es valued at US$18.6 million (2000). Swedish armed forces order for 20 A 109Ms valued at €130 million (2001).

DESIGN FEATURES: Fully articulated four-blade metal main rotor hub with tension/torsion blade attachment and elastomeric bearings; delta-hinged two-blade stainless steel tail rotor; manual blade folding and rotor brake optional. Main blade section NACA 23011 with drooped leading-edge; thickness/chord ratios 11.3 per cent at root, 6 per cent at tip. Tail rotor with Wortmann aerofoil and stainless steel skins; optional rotor brake. Compared with earlier models, A 109K has lengthened cabin to hold two stretchers fore and aft; modified fuel system; and smaller instrument panel.

FLYING CONTROLS: Fully powered hydraulic; IFR system with autopilot available. A 109KM has three-axis stability augmentation/attitude hold system; dual redundant IFR system and four-axis AFCS with flight path computer optional.

LANDING GEAR: A 109K series has non-retractable tricycle type, giving increased clearance between fuselage and ground. A 109E Power has tricycle type, with oleo-pneumatic shock-absorber in each unit. Single mainwheels and self-centring nosewheel castoring ±45°. Hydraulic retraction, nosewheel forward, mainwheels upward into fuselage. Hydraulic emergency extension and locking. Magnaghi disc brakes on mainwheels. All tyres are tubeless, of same size (360×135–6 or 380×135–6 or 14.5×5.5–6, 12/14 ply tubeless) and pressure (5.90 bar; 85 lb/sq in). Tailskid under ventral fin. Emergency pop-out flotation gear and fixed snow skis optional on all models.

POWER PLANT: *A 109K:* Two Turbomeca Arriel 1K1 turboshafts, each rated at 575 kW (771 shp) for 2½ minutes, 550 kW (737 shp) for take-off (30 minutes) and 471 kW (632 shp) maximum continuous power. Engine particle separator optional. Main transmission uprated to 671 kW (900 shp) for take-off and maximum continuous twin-engined operation; single-engine rating is 477 kW (640 shp) for 2½ minutes and 418 kW (560 shp) maximum continuous. Main rotor rpm 384, tail rotor 2,085. Standard usable fuel capacity 750 litres (198 US gallons; 165 Imp gallons), with optional 150 litre (39.6 US gallon; 33.0 Imp gallon) auxiliary tank for EMS operations, or 200 litre (52.8 US gallon; 44.0 Imp gallon) auxiliary tank in the A 109KM. Self-sealing fuel tanks optional. Independent fuel and oil system for each engine.

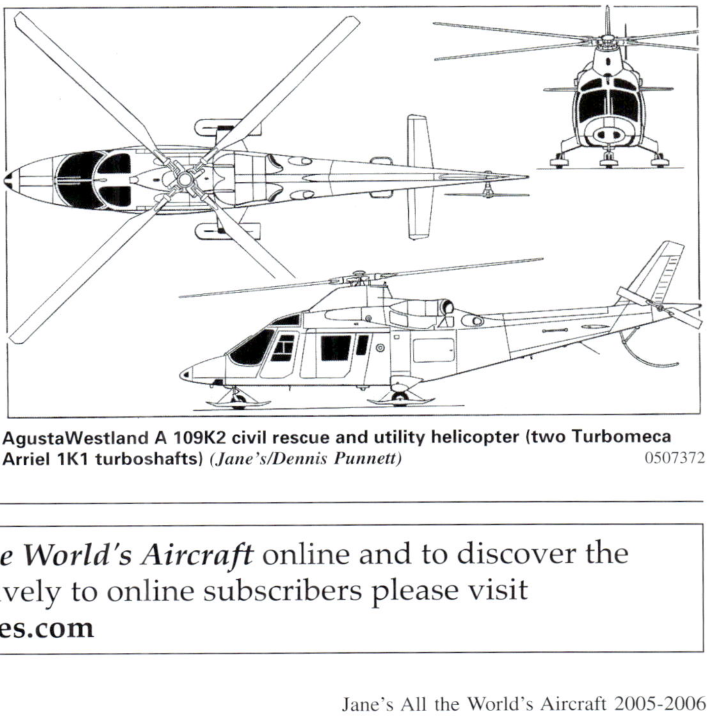

AgustaWestland A 109K2 civil rescue and utility helicopter (two Turbomeca Arriel 1K1 turboshafts) *(Jane's/Dennis Punnett)* 0507372

A 109E: Two Pratt & Whitney Canada PW206C engines, each rated at 477 kW (640 shp) for T-O, 423 kW (567 shp) for twin-engine operation, 546 kW (732 shp) for 2½ minutes' OEI contingency rating and 500 kW (670 shp) maximum continuous OEI; or two Turbomeca Arrius 2K1, each rated at 500 kW (670 shp) for T-O, 425 kW (570 shp) for twin-engine operation, 559 kW (750 shp) for 2½ minutes' OEI contingencies and 500 kW (670 shp) maximum continuous OEI; transmission rating 671 kW (900 shp) for T-O and maximum continuous for twin-engine operation, 418 kW (560 shp) maximum continuous and 477 kW (640 shp) for 2½ minutes' OEI; FADEC and liquid crystal multifunction displays for engine management. Standard fuel capacity (three cells) 605 litres (160 US gallons; 133 Imp gallons); optional capacity 710 litres (187 US gallons; 156 Imp gallons) with one extra cell or 870 litres (230 US gallons: 191 Imp gallons) with two extra cells.

SYSTEMS: Military versions have electrical system supplied by two 160 A 28 V DC self-cooled starter/generators and 27 Ah 28 V Ni/Cd battery. Optional AC electrical system comprises two 250 VA or two high-load 600 VA 115/26 V AC 400 Hz static inverters. Optional high-load AC system comprises one 6 kVA alternator and one standby 250 VA solid-state inverter. Dual independent hydraulic systems for flight controls, each capable of operating main actuators in the event of failure of the other system; utility hydraulic system with normal and emergency accumulators for operation of rotor brake, wheel brakes and nosewheel centring.

AVIONICS: A109E has Rockwell Collins Pro Line II or Bendix/King Silver Crown as core system. Military versions have ergonomic, NVG-compatible flight deck with provision for IFR instrumentation and role/mission dedicated displays.

Comms: VHF/AM, VHF/FM, UHF, HF, three-station intercom, IFF and ELT.

Radar: Colour weather radar optional.

Flight: ADF, VOR/GS/ILS, DME, GPS and VLF-Omega. Optional Rockwell Collins CMS80 cockpit management system with one or more centrally mounted CDUs and automatic target hand-off system compatibility.

Mission: FLIR.

Self-defence: Radar/laser warning receivers and chaff/flare/smoke dispensers.

EQUIPMENT: Optional equipment for military use includes windscreen wipers, rear view mirror, bleed air heater, particle separator, engine fire extinguisher, oxygen system, environmental control unit, one- or two-stretcher installation, air ambulance kit, external loudspeaker system, high-intensity searchlight, cargo platform, external cargo hook with maximum capacity 1,000 kg (2,205 lb), rescue hoist maximum capacity 270 kg (595 lb), snow skis and emergency floats.

ARMAMENT (OPTIONAL): Internal armament comprises pintle-mounted 7.62 mm machine gun and 12.7 mm machine gun in doorway. Provision for carriage of four or eight TOW, TOW 1, TOW 2 or TOW 2A missiles on external lateral pylons, each of 300 kg (661 lb) maximum capacity, with roof-mounted HeliTOW sight or APX M 334 or Helios gyrostabilised sights. Alternatively, pylons can accommodate seven- or 12-tube pods for 2.75 in or 81 mm rockets; rocket/machine gun (RPM) pods each with three 70 mm rockets and a 12.7 mm machine gun with 200 rounds; or machine gun (MG) pods with 12.7 mm gun (and 250 rounds) or 7.62 mm gun. A 109M has 12.7 mm or 20 mm nose-mounted turret gun.

DIMENSIONS, EXTERNAL:
- Main rotor diameter: 11.00 m (36 ft 1 in)
- Tail rotor diameter: 2.00 m (6 ft 6¾ in)
- Length overall, rotors turning: 13.04 m (42 ft 9 in)
- Fuselage length: 11.44 m (37 ft 6 in)
- Height over tailfin: 3.50 m (11 ft 5¾ in)
- Tailplane span: 2.88 m (9 ft 5½ in)
- Width over mainwheels: 2.45 m (8 ft 0½ in)
- Wheelbase: 3.535 m (11 ft 7¼ in)
- Main rotor tip ground clearance: 2.45 m (8 ft 0½ in)
- Fuselage ground clearance: 0.40 m (1 ft 3¾ in)
- Passenger doors (each): Height: 1.06 m (3 ft 5¾ in)
  - Width: 1.15 m (3 ft 9¼ in)
  - Height to sill: 0.65 m (2 ft 1½ in)
- Baggage door (port, rear): Height: 0.51 m (1 ft 8 in)
  - Width: 1.00 m (3 ft 3¼ in)

DIMENSIONS, INTERNAL:
- Cabin: Length: 2.10 m (6 ft 10¾ in)
  - Max width: KM, K2: 1.59 m (5 ft 2½ in)
  - E: 1.61 m (5 ft 3½ in)
  - Max height: 1.28 m (4 ft 2½ in)
  - Volume, incl flight deck: K2: 4.9 m^3 (173 cu ft)
  - E: 5.1 m^3 (180 cu ft)
- Baggage compartment volume: K2: 0.85 m^3 (30.0 cu ft)
  - E: 0.95 m^3 (33.6 cu ft)

AREAS:
- Main rotor disc: 95.03 m^2 (1,022.9 sq ft)
- Tail rotor disc: 3.14 m^2 (33.82 sq ft)

WEIGHTS AND LOADINGS:
- Weight empty: KM: 1,660 kg (3,660 lb)
  - K2: 1,650 kg (3,638 lb)
  - E: 1,570 kg (3,461 lb)
  - LUH: 1,639 kg (3,614 lb)
- Max slung load: 1,000 kg (2,204 lb)
- Max T-O weight: all: 2,850 kg (6,283 lb)
- Max T-O weight with slung load: all: 3,000 kg (6,613 lb)
- Max disc loading:
  - KM, K2, E: 30.0 kg/m^2 (6.14 lb/sq ft)
- Transmission loading at max T-O weight and power:
  - all: 4.24 kg/kW (6.97 lb/shp)

PERFORMANCE:
- Never-exceed speed (VNE):
  - KM, K2: 152 kt (281 km/h; 174 mph)
  - E, LUH: 168 kt (311 km/h; 193 mph)
- Max cruising speed at S/L, clean:
  - KM, K2: 143 kt (264 km/h; 164 mph)
  - E: 154 kt (285 km/h; 177 mph)
  - LUH: 151 kt (280 km/h; 174 mph)
- Max rate of climb at S/L: KM, K2: 594 m (1,950 ft)/min
  - E, LUH: 588 m (1,930 ft)/min
- Rate of climb at S/L, OEI:
  - KM, K2, LUH: 274 m (900 ft)/min
- Service ceiling: KM, K2: 6,100 m (20,000 ft)
  - E: 5,974 m (19,680 ft)
  - LUH: 5,029 m (16,500 ft)
- Service ceiling, OEI: KM, K2: 3,660 m (12,000 ft)
  - E, LUH: 3,990 m (13,100 ft)
- Hovering ceiling IGE: KM, K2: 5,305 m (17,400 ft)
  - E: 5,060 m (16,600 ft)
  - LUH: 4,328 m (14,200 ft)
- Hovering ceiling OGE: KM, K2: 3,900 m (12,800 ft)
  - E: 3,597 m (11,800 ft)
  - LUH: 2,957 m (9,700 ft)
- Max range, best height and speed, max optional (five cells) fuel: KM, K2: 434 n miles (805 km; 500 miles)
  - E: 521 n miles (964 km; 599 miles)
  - LUH: 447 n miles (827 km; 514 miles)
- Endurance: KM, K2: 4 h 0 min
  - E: 5 h 4 min
  - LUH: 4 h 29 min

UPDATED

AGUSTAWESTLAND A 109S GRAND

TYPE: Light utility helicopter.

PROGRAMME: Version of A 109 with enlarged cabin, unveiled in mockup form at Farnborough International, 20 July 2004. However, first of two prototypes flown in 2002 (unannounced) and by that time had accumulated combined 200 hours.

CUSTOMERS: Approximately 10 sold by time of unveiling, mid-2004. Deliveries planned from 2005 onwards, with 15 supplied to customers by end of 2006

COSTS: US$4.0 million to US$4.7 million (2004). Variable operating cost US$615.24 per hour (2004).

Data generally as for A 109, except that following.

DESIGN FEATURES: A 109 fuselage stretched by 0.21 m (8¼ in), allowing second window in cabin door. Quieter, scimitar-shape tail rotor. Certified to FAR/JAR 27, but with capabilities approaching FAR/JAR 29 standards

POWER PLANT: Two Pratt & Whitney Canada PW207C turboshafts with FADEC, each rated at 548 kW (735 shp) for T-O; 466 kW (625 shp) max continuous; 608 kW (815 shp) OEI max contingency for 2 min 30 s; and 548 kW (735 shp) OEI max continuous. Transmission rating 716 kW (960 shp) for T-O; 671 kW (900 shp) max continuous; 545 kW (730 shp) OEI max contingency; and 448 kW (600 shp) OEI max continuous. Fuel capacity (all minus 10 litres; 2.6 US gallons; 2.2 Imp gallons unusable): 576 litres (152 US gallons; 127 Imp gallons) with three-cell system; 671 litres (177 US gallons; 148 Imp gallons) with four-cell system; or 807 litres (213 US gallons; 178 Imp gallons) with five-cell system.

ACCOMMODATION: Eight persons, including one or two crew. Corporate/VIP configuration with five or six seats in club configuration. Baggage compartment, with external door, in tailboom.

AVIONICS: *Instrumentation:* Four flight/navigation LCDs, plus two data monitoring LCDs.

Modified image of AgustaWestland A 109 Grand ***NEW***/0589475

AgustaWestland A 109 Grand emergency services interior ***NEW***/0589462

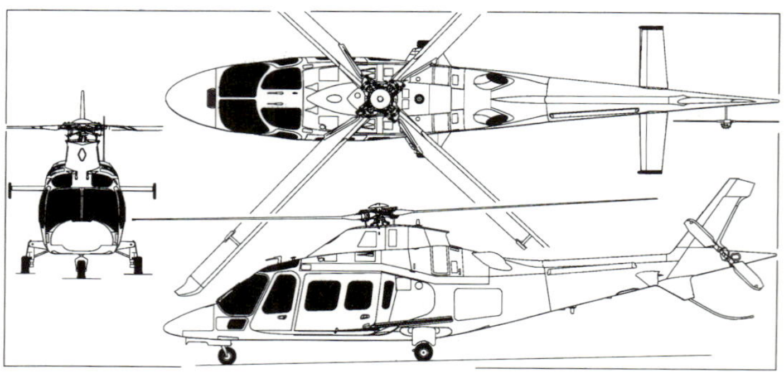

AgustaWestland A 109 Grand eight-seat utility helicopter ***(Paul Jackson)*** ***NEW***/0589208

DIMENSIONS, EXTERNAL:
Main rotor diameter ... 10.83 m (35 ft 6½ in)
Tail rotor diameter ... 1.94 m (6 ft 4½ in)
Length overall: rotors turning ... 12.96 m (42 ft 6¼ in)
Fuselage: Length ... 11.65 m (38 ft 2½ in)
Max width ... 1.64 m (5 ft 4½ in)
Height over tailfin ... 3.44 m (11 ft 3½ in)
Tailplane span ... 2.88 m (9 ft 5½ in)
Wheel track ... 2.15 m (7 ft 0½ in)
Wheelbase ... 3.74 m (12 ft 3¼ in)
Ground clearance: fuselage ... 0.40 m (1 ft 3¾ in)
tail rotor ... 1.00 m (3 ft 3¼ in)
Passenger doors (each): Width ... 1.40 m (4 ft 7 in)
DIMENSIONS, INTERNAL:
Cockpit volume ... 1.7 m² (58 cu ft)
Cabin (excl cockpit):
Length: at floor ... 2.30 m (7 ft 6½ in)
at roof ... 1.90 m (6 ft 2¾ in)
Max height ... 1.28 m (4 ft 2½ in)
Volume ... 3.9 m² (138 sq ft)
Baggage compartment: Length ... 2.30 m (7 ft 6½ in)
Max height ... 0.71 m (2 ft 4 in)
Volume ... 0.95 m² (33.5 cu ft)
AREAS:
Main rotor disc ... 92.12 m² (991.6 sq ft)
Tail rotor disc ... 2.96 m² (31.82 sq ft)
WEIGHTS AND LOADINGS:
Weight empty ... 1,655 kg (3,649 lb)
Max fuel weight: three cells ... 460 kg (1,014 lb)
four cells ... 535 kg (1,179 lb)
five cells ... 644 kg (1,420 lb)
Max T-O weight: internal load ... 3,175 kg (7,000 lb)
underslung load ... 3,200 kg (7,054 lb)
Max disc loading: internal load ... 34.5 kg/m² (7.06 lb/sq ft)
underslung load ... 34.7 kg/m² (7.11 lb/sq ft)
Transmission loading at max T-O weight and power:
internal load ... 4.44 kg/kW (7.29 lb/shp)
external load ... 4.77 kg/kW (7.84 lb/shp)
PERFORMANCE:
Never-exceed speed (VNE) ... 168 kt (311 km/h; 193 mph)
Max cruising speed at S/L ... 155 kt (287 km/h; 178 mph)
Max rate of climb at S/L ... 577 m (1,894 ft)/min
Max rate of climb at S/L, OEI ... 326 m (1,070 ft)/min
Hovering ceiling: IGE ... 4,345 m (14,260 ft)
OGE ... 2,845 m (9,340 ft)
Service ceiling ... 5,705 m (18,720 ft)
Service ceiling, OEI ... 3,710 m (12,160 ft)
Range with five fuel cells at FL50 ... 470 n miles (870 km; 540 miles)
Endurance with five fuel cells at FL50 ... 4 h 33 min
OPERATIONAL NOISE LEVELS:
T-O ... 89.3 EPNdB
Approach ... 89.4 EPNdB
Overflight ... 88.9 EPNdB

UPDATED

AGUSTAWESTLAND A 119 KOALA

TYPE: Light utility helicopter.

PROGRAMME: First flight (I-KNEW, second airframe) early 1995; public debut by static test first airframe (I-KOAL) at Paris Air Show June 1995; both initially with 597 kW (800 shp) Turbomeca Arriel 1 turboshaft; rebuilt with PT6B turboshafts early in 1997 as I-KOAL c/n 14003 and I-KNEW c/n 14004; demonstrator (I-KNOW) added in 1999 and US demonstrator in 2000; initial delivery (VH-FOX) to Linfox Holdings, Melbourne, Australia, 25 September 2000.

As part of South African purchase of 40 military A 109s, Denel (which see) is assembling and marketing the Koala in Africa and providing components for the Italian production line.

In February 2004 Agusta announced that assembly of the Koala would be transferred to Agusta Aerospace Corporation's Philadelphia, Pennsylvania facility during 2004. A new 3,716 m² (40,000 sq ft) assembly plant will have an initial capacity of 15 to 20 A 119s per year.

CURRENT VERSIONS: **A 119**: *as described.*

A 119 Military: Multirole military variant.

CUSTOMERS: Total of 46 produced by late-2004 (including prototypes), but not two rebuilds). Six ordered by Omniflight Helicopters in February 1996; orders and options totalled more than 45 by January 2003 from customers in Europe, North and South America and the Far East. Southeast Mississippi Air Ambulance District of Hattiesburg, Mississippi, took delivery of one in EMS configuration in February 2002. Interest shown by Carabinieri (Italian military police) as replacement for AB 206 and Guardia di Finanza (customs) as NH 500 replacement. Recent customers include the Pennsylvania State Police Department, which ordered two in January 2003, New York City Police's Department of Aviation, which ordered four in February 2003, of which the first was delivered in July 2004, Tex-Air Helicopters, which ordered 10 in June 2003, of which the first three were scheduled for

AgustaWestland A 119 Koala ***NEW***/0576287

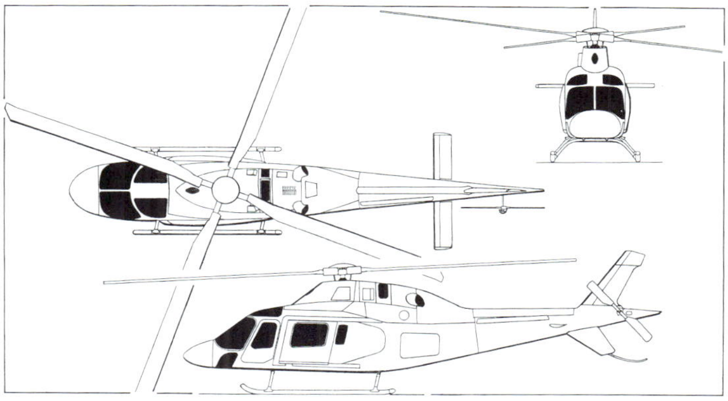

AgustaWestland A 119 Koala (one P&WC PT6B turboshaft) *(Paul Jackson)* 0137339

delivery in late 2003, for offshore operations in the Gulf of Mexico, and Zhengzhou Police of China, which has ordered one.

COSTS: US$1.85 million (2000).

DESIGN FEATURES: Fully articulated four-blade composites main rotor with titanium hub, composites material grips and elastomeric bearings; two-blade tail rotor; large underfin.

STRUCTURE: Aluminium alloy fuselage.

LANDING GEAR: Fixed skids.

POWER PLANT: One Pratt & Whitney Canada PT6B-37A turboshaft rated at 747 kW (1,002 shp) for T-O, and 650 kW (872 shp) maximum continuous. Transmission rating 671 kW (900 shp) for take-off and maximum continuous. Standard three-cell fuel system, combined capacity 606 litres (160 US gallons; 133 Imp gallons); optional four- and five-cell fuel tanks, combined capacities respectively 721 litres (188 US gallons; 156.5 Imp gallons) and 871 litres (230 US gallons; 191.5 Imp gallons).

ACCOMMODATION: Pilot and passenger in front; six passengers in main cabin in three-abreast club configuration; flight-accessible baggage compartment in cabin; main baggage compartment in rear fuselage with optional extensions; in EMS configuration can accommodate two stretchers and three medical attendants. Large sliding doors on each side of main cabin; forward-hinged door to cockpit on both sides. Cabin is heated, air conditioned and soundproofed.

AVIONICS: Bendix/King Silver Crown suite and integrated instrument display system (IIDS) standard.

Comms: Silver Crown com/nav and ELT standard.

Flight: Three-axis autopilot with duplex SAS and altitude hold mode standard; radar altimeter, Garmin GPS and moving map display optional.

Instrumentation: NVG compatibility optional.

EQUIPMENT: Optional equipment includes 500 kg (1,102 lb) or 1,000 kg (2,205 lb) fixed cargo hook; 450 kg (992 lb) rescue hoist; Simplex 323 fire attack system with 1,200 litre (317 US gallon; 263.9 Imp gallon) capacity; snow skids, external emergency floats, particle separator, Spectrolab SX-5 searchlight and FLIR/LLTV camera.

DIMENSIONS, EXTERNAL:
Main rotor diameter ... 10.83 m (35 ft 6½ in)
Tail rotor diameter ... 2.00 m (6 ft 6¾ in)
Length overall, rotors turning ... 13.01 m (42 ft 8¼ in)
Height overall ... 3.77 m (12 ft 4½ in)
Fuselage: max width ... 1.67 m (5 ft 5¾ in)
Width overall, rotor in 'X' ... 7.66 m (25 ft 1½ in)
Main rotor tip ground clearance ... 2.54 m (8 ft 4 in)
Tail rotor tip ground clearance ... 1.29 m (4 ft 2¾ in)
Fuselage ground clearance ... 0.54 m (1 ft 9¼ in)
DIMENSIONS, INTERNAL:
Cabin: Length ... 2.10 m (6 ft 10¾ in)
Max width ... 1.67 m (5 ft 5¾ in)
Max height ... 1.28 m (4 ft 2½ in)
Floor area ... 2.60 m² (28.0 sq ft)
Volume, incl flight deck ... 4.96 m³ (175.2 cu ft)
Baggage compartment: Length ... 2.30 m (7 ft 6½ in)
Volume ... 0.95 m³ (33.5 cu ft)
AREAS:
Main rotor disc ... 92.12 m² (991.6 sq ft)
Tail rotor disc ... 3.14 m²(33.82 sq ft)
WEIGHTS AND LOADINGS:
Basic weight empty ... 1,430 kg (3,153 lb)
Max T-O weight: internal load ... 2,720 kg (5,996 lb)
external load ... 3,150 kg (6,944 lb)
Max disc loading:
internal load ... 29.53 kg/m² (6.05 lb/sq ft)
external load ... 34.19 kg/m² (7.00 lb/sq ft)
Transmission loading at max T-O weight and power:
internal load ... 4.05 kg/kW (6.65 lb/shp)
external load ... 4.69 kg/kW (7.71 lb/shp)
PERFORMANCE (max internal load):
Never-exceed speed (VNE) ... 152 kt (281 km/h; 174 mph)
Max level speed ... 144 kt (267 km/h; 166 mph)
Hovering ceiling: IGE ... 4,450 m (14,600 ft)
OGE ... 3,261 m (10,700 ft)
Service ceiling ... 6,100 m (20,000 ft)
Range with max auxiliary fuel, at 1,525 m (5,000 ft),
no reserves ... 525 n miles (972 km; 604 miles)
Endurance ... 5 h 34 min

UPDATED

AGUSTA-BELL 412 AND GRIFFON

TYPE: Multirole medium helicopter.

PROGRAMME: First flight of Bell 412SP (in USA) August 1979; deliveries started January 1981; Agusta licensed production of civil version started 1981; first flight of military Griffon August 1982; deliveries began January 1983; Bell 412HP certified 29 June 1990; Bell 412EP is a civilian version currently manufactured under licence by AgustaWestland.

CURRENT VERSIONS: **Griffon:** Military derivative developed for direct fire support, scouting, assault transport, equipment transport, SAR and maritime surveillance.

Creso: Battlefield surveillance version; trial installation for Italian Army in AB 412 MM81196 as an element (with Mirach 26 UAVs and other sensors) of the CATRIN C^3I system. FIAR Creso E-band MTI radar in a circular radome below the nose, plus Elettronica emitter locators at the nose and on the tailboom. Galileo FLIR turret above pilot's seat. Maximum operational altitude of 1,500 m (4,920 ft) gives radar range of 39 to 49 n miles (72 to 91 km; 45 to 56 miles). First flight of operational system mid-1996; Creso is Italy's contender in the NATO Alliance Ground Surveillance programme.

CUSTOMERS: More than 260 produced by 2004, including 22 412EPs for local paramilitary use. Those in Italy include Army (24), Carabinieri (35), Coast Guard (nine), national fire service (14), national forest service (11, including two in 2003) and Guardia di Finanza (23, of which 18 received by 2004). Exports to Zimbabwe Air Force (10), Ugandan Army (two), Finnish Coast Guard (four); Royal Netherlands Air Force (three); Dubai Air Force (nine); Ghana Air Force (two, in EMS configuration); Royal Saudi Air Force (in parallel with Bell-built 412SAs); Swedish Army (five); and Dubai Police (two). Nine ordered for SAR by Turkish Coast Guard in late 1999.

COSTS: Five Turkish Coast Guard, total cost US$52 million (1999).

DESIGN FEATURES: Griffon has reinforced impact-absorbing landing gear, selective armour protection and differences noted below.

POWER PLANT: One Pratt & Whitney Canada PT6T-3D Twin-Pac rated at 1,342 kW (1,800 shp) for T-O and 1,194 kW (1,600 shp) maximum continuous (single-engine ratings 850 kW; 1,140 shp for 2½ minutes and 723 kW; 970 shp continuous). Transmission rating 1,181 kW (1,584 shp) for 5 minutes, 846 kW (1,135 shp) maximum continuous and 850 kW (1,140 shp) single engine. IR emission reduction devices optional. Fuel capacity 1,249 litres (330 US gallons; 275 Imp gallons). Two 75.7 or 341 litre (20.0 or 90.0 US gallon; 16.7 or 74.9 Imp gallon) auxiliary fuel tanks optional; single-point refuelling.

ACCOMMODATION: One or two pilots on flight deck, on energy-absorbing, armour-protected seats. Fourteen crash-attenuating troop seats in main cabin in personnel transport roles, six patients and two medical attendants in ambulance version, or up to 1,814 kg (4,000 lb) of cargo or other equipment. Space for 181 kg (400 lb) of baggage in tailboom. Total of 51 fittings in cabin floor for attachment of seats, stretchers, internal hoist or other special equipment.

SYSTEMS: Generally as for Bell 212/412.

AVIONICS: Optional four-screen EFIS.

EQUIPMENT: SAR and coastal surveillance versions equipped with 360° panoramic search radar integrated with FLIR and TV; video and still camera systems; datalink; dual digital four-axis AFCS with auto approach to hover, mark on target and search pattern; dual GPS; EFIS-LCD IFR cockpit instrumentation; operator's console in passenger compartment; gyrostabilised binoculars; searchlight and rescue light; electric rescue hoist; liferafts and lightweight emergency floats. BAE Systems MST-S multisensor turret system supplied for undisclosed AB 412EP customer in 1996, to operate in conjunction with Honeywell RDR-1500 maritime surveillance radar.

ARMAMENT: Wide variety of external weapon options for Griffon include swivelling turret for 12.7 mm gun, two 25 mm Oerlikon cannon, four or eight TOW anti-tank missiles, two launchers each with nineteen 2.75 in SNORA or twelve 81 mm rockets, 12.7 mm machine guns (in pods or door-mounted), four air-to-air or air defence suppression missiles, or, for attacking surface vessels, four Sea Skua or similar air-to-surface missiles.

DIMENSIONS: As for Bell 212/412

WEIGHTS AND LOADINGS:

Weight empty, equipped (standard configuration) ... 2,914 kg (6,425 lb)
Max T-O weight ... 5,398 kg (11,900 lb)
Transmission loading at max T-O weight and power ... 4.57 kg/kW (7.51 lb/shp)

PERFORMANCE:

Never-exceed speed (VNE) at S/L ... 140 kt (259 km/h; 161 mph)
Cruising speed: at S/L ... 122 kt (226 km/h; 140 mph)
at FL15 ... 125 kt (232 km/h; 144 mph)
at FL30 ... 123 kt (228 km/h; 142 mph)
Max rate of climb at S/L ... 542 m (1,780 ft)/min
Rate of climb at S/L, OEI ... 168 m (551 ft)/min
Service ceiling, 30.5 m (100 ft)/min climb rate ... 5,395 m (17,700 ft)
Service ceiling, OEI, 30.5 m (100 ft)/min climb rate ... 2,320 m (7,620 ft)
Hovering ceiling: IGE ... 3,110 m (10,200 ft)
OGE ... 1,585 m (5,200 ft)
Range with max standard fuel at appropriate cruising speed (see above), no reserves:
at S/L ... 354 n miles (656 km; 407 miles)
at FL15 ... 402 n miles (745 km; 463 miles)
at FL30 ... 434 n miles (804 km; 500 miles)
Max endurance: at S/L ... 3 h 36 min
at FL15 ... 4 h 12 min

UPDATED

Cabin of Agusta-Bell 412 ***NEW***/0589299

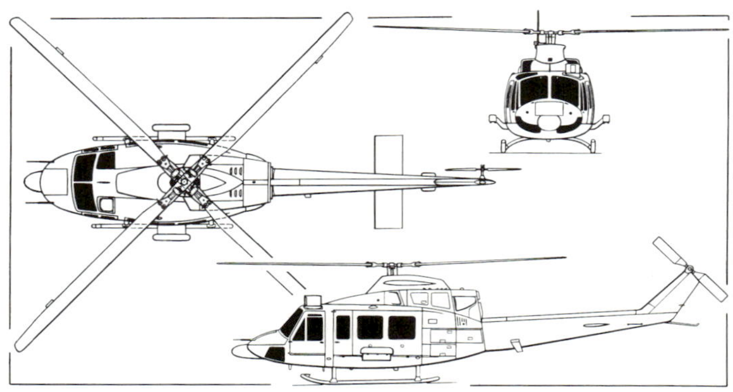

Agusta-Bell 412 Griffon military helicopter *(Jane's/Dennis Punnett)* 0052775

Italian Army Agusta-built Bell 412 *(Paul Jackson)* ***NEW***/0576281

Digital cockpit of Agusta-Bell 412, available as option ***NEW***/0589298

AGUSTAWESTLAND EH101

Royal Navy designation: Merlin HM. Mks 1 and 2
RAF designation: Merlin HC. Mk 3
Canadian Forces designation: CH-149 Cormorant

TYPE: Multirole medium helicopter.

PROGRAMME: Stems from Westland WG 34; selected in mid-1978 by UK MoD to meet SR(S) 6646 for a Sea King replacement; broadly similar requirement by Italian Navy led to 1980 joint venture with Agusta; subsequent market research confirmed compatibility of basic design with commercial payload/range and tactical transport/logistics requirements, resulting in decision to develop naval, civil and military variants based on common airframe.

Nine month project definition phase approved by UK/Italian governments 12 June 1981; full programme go-ahead announced 25 January 1984; design and development contract signed 7 March 1984; selected by Canadian government August 1987; Italian-built iron bird ground test airframe followed by nine preproduction aircraft (PP1-9: see Current Versions); RTM 322 engines selected for RN Merlin June 1990; T700-GE-T6A engines selected for Italian Navy version September 1990; fourth UK/Italian government MoU signed 30 September 1991 (starting industrialisation phase); UK MoD commitment to 44 Merlins 9 October 1991; Canadian order for 50 (35 CH-148 and 15 CH-149) announced 24 July 1992; UK and Italian civil certification of Srs 300 and 500 planned for late 1993, but eventually achieved on 24 November 1994; US FAA approval gained on following day.

Flight testing halted following January 1993 loss of PP2; resumed 24 June 1993. US$1 million study completed for US Marine Corps, 1993, assessing EH101 as 30-troop or cargo transport as fallback in event of Bell/Boeing MV-22 cancellation. First flight with RTM 322 engines, 6 July 1993; two aircraft to fly total of 260 hours for development. Canadian requirement cut to 43 by deletion of seven CH-148s in August 1993 as cost-saving measure, but incoming government cancelled entire programme on 4 November 1993, despite award of substantial offsets to Canadian firms. UK government confirmed EH101 as next RAF tactical transport helicopter on 1 December 1993; formal announcement of order for 22 made 9 March 1995; Italian order confirmed October 1995 for 16, plus eight options. First production aircraft, RN01/ZH821 for Royal Navy, first flew 6 December 1995 and officially rolled out on 6 March 1996, although lacking mission avionics. Military Aircraft Release trials began in May 1998, using RN02/ZH822 (first flight 14 January 1997) and the first iteration of the Release was issued in the last quarter of 1998, allowing the start of Intensive Flight Trials with the commissioning of No. 700M Squadron (the Merlin IFTU) at RNAS Culdrose on 1 December 1998, and of RN aircrew training. First Merlin transferred to RN charge was RN05/ZH825 on 12 November 1998; first to civil customer, March 1999; first production EH101 for Italian Navy flew 6 December 1999; first squadron (824 NAS, Royal Navy) commissioned 2 June 2000 at RNAS Culdrose.

Danish EH101 Mk 512 with operational fit ***NEW***/1047790

CURRENT VERSIONS: **Srs 100/Naval:** Primary roles ASW, ASV, anti-ship surveillance/tracking, amphibious operations and SAR; others include AEW, vertrep and ECM (deception, jamming and missile seduction); designed for autonomous all-weather operation from land bases, large and small vessels (including merchant ships) and oil rigs and, specifically, from a 3,500-tonne frigate, with dimensions tailored to frigate hangar size. Capabilities include Type 23 frigate launch and recovery in conditions up to Sea State 6 with ship on any heading and windspeed (from any direction) up to 50 kt (93 km/h; 57 mph); endurance and carrying capacity includes ability to operate for up to 5 hours with state-of-the-art equipment and weapons.

Italian Navy ASW/ASVW/Mk 110: Operates from shore bases and aircraft or helicopter carriers against surface and underwater targets with HELRAS dipping sonar; armed with two Marte Mk. 2/S ASMs. First of eight (MMX81480 '2-01') flew on 6 December 1999; first two delivered December 2000 for service trials; four more handed over by end of 2001.

Italian Navy AEW/Mk 112: Requirement revealed in 1994 for AEW helicopter. Four ordered in 1995, with first for delivery in 2003–04. Also known as **ASVW/E** (Anti-Submarine and Vessel Warfare, Enhanced). MM/APS-784-based Eliradar HEW-784 air and surface surveillance radar; radome almost doubled in size (from 1.80 m; 5 ft 10¾ in) to accommodate 3.00 m (9 ft 10 in) antenna, but other avionics similar to Italian Navy Mk 110. Initial aircraft, MM81488 '2-09', shown at Paris, 15–22 June 2003.

Merlin HM. Mk 1/Mk 111: Royal Navy ASW version for Staff Requirement (Sea) 6646; Lockheed Martin (formerly Loral) ASIC is prime contractor for systems integration; operates from Type 23 frigates, 'Invincible' class aircraft carriers, RFAs and other ships, and land bases. Initial production aircraft, RN01/ZH821, first flew 6 December 1995; second, RN02/ZH822, first with mission avionics, flew 14 January 1997. Of first seven production Merlins, RN01 and RN02 initially assigned to operational performance acceptance procedure (OPAP) trials at the fully instrumented Atlantic Underwater Test and Evaluation Center (AUTEC) in Bahamas, 1998–99, and to assistance in sea trials; RN03 to DTEO, Boscombe Down, in September 1997 for military aircraft release trials; joined by

AgustaWestland US101 demonstrator on its maiden flight with CT7-8E engines, October 2004 ***NEW***/1047791

Royal Air Force Merlin HC. Mk 3s ***NEW***/0573358

First Italian Navy EH101 Mk 410 utility variant on display at Farnborough in July 2004 *(Paul Jackson)* ***NEW***/0589258

RN02 for pre-IFTU stage of release, achieved November 1998. The fully instrumented RN04 followed in late 1997; RN05-RN08 formed Intensive Flight Trials Unit (No. 700M Squadron) on 1 December 1998; PP5, RN02 and RN03 underwent operational tests at AUTEC from February 1999 (Operation Pearly King). Subsequent deep water trials were undertaken from Benbecula in 1999. In April 2000, RN12 from Boscombe Down and RN17 from Culdrose undertook extended Ship Helicopter Operational Limit trials aboard RFA *Argus*, experiencing winds of up to 50 kt (93 km/h; 58 mph) and up to Sea State 10. Snow and icing trials accomplished by RN aircraft in Canada in first half of 2001.

RN02, RN03 and RN14 undertook ASuW trials from Benbecula from late June 2000, following 30-day series of ASW tests off Bahamas, during Operation 'Trial Wizard' in March-April 2000. These included dropping eight Sting Ray torpedoes and 800 sonobuoys.

By the end of 1998, the Royal Navy had seven Merlins including two with No. 700M Squadron, which received its third on 21 April 1999. 700M evolved from IFTU to OEU on 1 September 2001. Training role to No. 824 Squadron at Culdrose on 2 June 2000, with No. 814 Squadron forming as the first front-line squadron on 5 October 2001 for deployment aboard HMS *Ark Royal.* Embarked operational capability achieved in second quarter of 2002 with four helicopters, although complement will rise to six in late 2006. The aircraft also equips a small-ships squadron (No. 829, with six Merlin flights) and a second carrier squadron, No. 820, which officially reformed at Culdrose on 5 December 2003. The Royal Navy received its 25th Merlin in August 2000 and 35 had been delivered by September 2001, with the 44th and last example handed over shortly before the end of 2002.

Operational debut made in 'Telic' (UK contribution to Iraq conflict of 2003), when four helicopters of 814 Squadron deployed to Gulf aboard RFA *Fort Victoria;* primarily engaged in anti-surface warfare mission, but also undertook vertical replenishment, troop transport and long-range delivery taskings; lack of ASW requirement permitted removal of dipping sonar, making more space available in cabin. Over 800 hours accumulated in combat conditions, with just one sortie lost because of unserviceability.

In June 2003, Lockheed Martin UK Ltd selected to undertake two-year study programme and awarded £18 million contract for assessment phase of Merlin Capability Sustainment Plus (MCSP) project that is intended to address obsolescence issues and weapon system upgrades, including operational aspects arising from recent combat experience in the Gulf. Cost savings to figure prominently, with increased use of COTS (commercial off-the-shelf) equipment and open system computer architecture. Upgraded helicopters, which could be designated **Merlin HM. Mk 2**, anticipated to enter service in 2009.

Merlin AEW: UK MoD-funded study, submitted August 1998, of possible Sea King AEW. Mk 7 replacement to meet Royal Navy's FOAEW requirement. Ventral radome and stub-wings; latter would permit speeds up to 250 kt (463 km/h; 288 mph).

Heliliner (Srs 300): Commercial variant intended to offer 360 n mile (666 km; 414 mile) range, with full IFR reserves, carrying 30 passengers and baggage; flight crew of two, provision for cabin attendant, stand-up headroom, airline-style seating, overhead baggage stowage, full environmental control, passenger entertainment, and provision for lavatory and galley. Category A VTO performance, capable of offshore/oil rig operations or scheduled flights into city centres at high all-up weights under more rigorous future civil operating rules; rear-loading ramp optional. An executive version is also proposed.

Military Utility (Srs 400): Tactical or logistic transport variant with rear-loading ramp; able to airlift 6 tons or up to 30 combat-equipped troops. See Italian Navy and Merlin headings below. Naval equivalent **(Srs 200)** lacks ramp. Tail- and rotor-folding Utility version under consideration, with role options including mine countermeasures, towing EDO Mk 106 sled. Version will probably form the basis of the EHI bid to meet the UK's Support, Amphibious and Battlefield Rotorcraft (SABR) requirement to replace Royal Navy Sea King HC. Mk 4s, RAF Pumas and RAF SAR Sea King HAR. Mk. 3s.

Italian Navy Utility/Mk 410: Based on rear-ramp Srs 400, but with Galileo Avionica GaliFlir FLIR (also specified for other Italian versions), cargo hook, basic avionics and Elettronica ESM/ECM and self-defence suite similar to Italian ASW version. Weather radar. Four ordered in 1995. Serves in the commando support role with Nucleo Elicotteristico per la Lotta Anfiibia at Grottaglie. First aircraft, MM81492 '2-13', made public debut at Farnborough International, July 2004. Italian Air Force also has requirement for EH101s to replace two AS-61A-4 VIP helicopters.

Italian Navy Amphibious Support/Mk 413: Four ordered at beginning of 2002. Will have upgraded avionics for night missions (including NVG-compatible cockpit) as well as weather radar, self-protection systems, personnel locator systems and armour and machine guns; plumbing for in-flight refuelling equipment will be installed, although it is not intended to fit refuelling probes at present.

Merlin HC. Mk 3/Mk 411: Bid for RAF contract entered May 1994; revised cockpit layout for low-level operations; provision for pintle-mounted machine guns in side doors. Order for 22 announced 9 March 1995, in satisfaction of Staff Requirement (Air) 440. Each has provision for rapid installation of chin-mounted FLIR turret and (non-telescopic) refuelling probe beneath the nose, offset to starboard (though neither routinely fitted). Uprated engines. Other features include integrated defensive aids system (with Nemesis directional IR countermeasures, AN/AVR-2A(V) laser warning and Sky Guardian 200 RWR), NVG compatibility, crash-attenuating seats for all occupants (two pilots; loadmaster; optional fourth crewman; and 24 troops), active noise-reduction headsets for all passengers, non-folding main rotor blades and tailboom, improved navigation suite (compared with RN version), variable-speed cargo winch and roller conveyor for cargo handling, SAR hoist to starboard and cargo hook for external loads (installation in floor resulting in slightly reduced fuel capacity). Capable of carrying long wheelbase Land Rover;

Royal Navy Merlin HM. Mk 1 ***NEW***/0573349

an overload of 30 troops can board and strap-in within 2 minutes, and deplane within 40 seconds. Critical design review completed July 1997. Assembly of first (RAF01/ZJ117) began November 1997; rolled out 25 November 1998; first flew 24 December 1998. First aircraft delivered to DERA (now QinetiQ) at Boscombe Down 19 January 2000. Sixth aircraft handed over to Defence Procurement Agency on 27 June 2000, this date thereby becoming official acceptance into RAF inventory; formal release to service signed on 11 January 2001. Fifth and sixth aircraft used for conversion of first RAF instructors at Yeovil in June 2000.

Deliveries due between September 1999 (first to Boscombe Down for DTEO trials) and October 2001, but latter date not met (13 received by November 2001). Initial units, from April 2000, were Rotary Wing Operational Evaluation and Training Unit (originally at Boscombe Down) and the Operational Conversion Flight (actually an element of No. 28 Squadron at Benson) with a limited Release-to-Service for day training. Delivery of the first aircraft, due on 6 November 2000, was delayed by a worldwide EH101 grounding, following the loss of an HM. Mk 1. ZJ122 to Benson for technical familiarisation 11 December 2000; formal handover of five aircraft to RAF on 7 March 2001; by mid-June 2001, RAF had six Merlins at Benson, while further seven on flight trials with QinetiQ.

No. 28 Squadron officially re-formed 17 July 2001 and undertook first field deployment during 17–21 September 2001, when participated in exercise 'Pegasus Trial', which included operations from RAF Honington and first practice tactical troop-lift. A new Medium Support Helicopter Aircrew Training Facility, with two Merlin simulators, opened at Benson on 17 July 2000. Final aircraft handed over at Yeovil on 19 November 2002, with service allocation of 12 to Benson (No. 28 Squadron and Rotary Wing OEU); six to Aldergrove (No. 72 Squadron); and four in-use reserves (for support of 16 Air Assault Brigade). First unit declared operational was No. 28 in first quarter of 2003.

SAR/Utility (Srs 500): For commercial and paramilitary operators requiring rear-loading facility. First production **Mk 510** (I-AGWH) built for certification as a 'white tail'; rolled out in Italy 28 May 1997, and flew 17 June. Subsequently used for cold weather trials, from Fairbanks, Alaska, at temperatures from −5 to −32°C (23 to −26°F) until April 2000, and then for hot-and-high trials in USA. First order in November 1996 for a Mk 510 for Tokyo Metropolitan Police Agency; first flew September 1997; delivered from Italian production in January 1998 as first to civil customer; fitting-out at Kawasaki Heavy Industries was delayed by bankruptcy of original financiers; handed over at Gifu on 25 March 1999.

SAR aircraft for Canada (described separately), Denmark and Portugal among others, are Series 500s. Initial Danish Mk 512 first flew 12 December 2003 with conventional nose, but by public debut at Farnborough International in July 2004, had gained complex radome incorporating radar antenna and, below, fixed EGPWS window, plus chin FLIR turret and side-mounted ESM.

Danish helicopters have RTM322-02/08 turboshafts and dual-processor ASMC (aircraft system management computer) for higher level of systems integration. Avionics include Inmarsat Constellation secure voice satcom, Finding/Personnel Location System (F/PLS, with both global maritime distress safety system and COSPAS/SARSAT), TCAS, EGPWS and wireless intercom. Equipment includes dual electrical hoists, operating table and associated medical apparatus. For alternative troop transport role, Danish EH101s have 16 seats, DIRCM and customer-specific EW suite. Danish variant named **Merlin Joint Supporter** in ceremony, at Yeovil on 25 October 2004.

Portuguese EH101s fitted with ASMC, RTM322-02/08 engines, F/PLS and two hoists, plus ventral radar. SAR-specific equipment includes Nitesun searchlight, SAR Tech seats, provision for six stretchers and 11 troop seats. Additional CSAR equipment comprises EW suite, airborne interrogator and guidance system, folding main and tail rotors and machine gun attachment points at side-door and on ramp. Fishery protection (SIFICAP) variant has data management console with recording capability, plus external loudspeaker. First Portuguese EH101 handed-over at Vergiate on 22 December 2004.

'CH-X': Offered to USMC; chin gun turret; not adopted.

CH-148 Petrel and CH-149 Chimo: Intended Canadian ASW and SAR versions; cancelled after payment of US$353 million in compensation.

CH-149 Cormorant/Mk 511: Adaptation of civil EH101 (with ramp) offered to Canada in early 1995 as a CH-113 replacement (in place of the CH-149 Chimo) for SAR missions; selected late 1997. Deck landing capability; reduced cost achieved by reliance on mainly commercial avionics and by original proposed use of twin hoists in single door, rather than one per side. Since then, the hoist arrangement has been revised, with a primary winch in the starboard door and a secondary in the forward port door. Features Honeywell RDR-1400 radar, Litton LN-100G embedded laser INS/GPS, FLIR, Spectrolab SX-16 searchlight, crashworthy fuel tanks, Breeze-Eastern primary and secondary hoists and an underslung load hook. Planned to include provision for internal extended range tanks (ERTs) and hover in-flight refuelling from ships under way and wire-strike protection.

Total of 15 ordered from Italian production to replace CH-113 Labrador, with deliveries originally to begin in October 2000 and be complete by end of 2002; however, delays in certification and qualification resulted in deliveries being deferred. First Cormorant (149901) flew at Vergiate on 7 March 2000, with an official first flight on 31 May 2000. To equip 442 Squadron (19 Wing) at Comox, followed by 413 Squadron (14 Wing) at Greenwood, 424 Squadron (8 Wing) at Trenton and 103 Squadron (9 Wing) at Gander; first two helicopters (149904 and '905) handed over in Italy on 29 September 2001 and arrived Comox on 11 October 2001. Formal acceptance at Comox on 29 October 2001; CH-149 subsequently certified for operational duty on 15 July 2002, with SAR standby alert duty from 25 July; first operational mission involving successful SAR accomplished 30 July. 413 Squadron received its first CH-149 on 24 August 2002. Deliveries were completed in mid-2003.

Estimated to cost 40 per cent less than original CH-149 with 108 per cent offsets ('industrial regional benefits') totalling C$629 million. Associated industrial team includes Bombardier, Bristol Aerospace, Spar Aerospace and CAE, with Canadian Helicopters to provide a leasing and follow-on maintenance option. EHI has demonstrated a 15,500 kg (34,171 lb) MTOW (900 kg; 1,984 lb overweight) which would give a 1,080 n mile (2,000 km; 1,242 mile) SAR radius.

Profile of Danish EH101 radome ***(Paul Jackson)*** ***NEW***/0589256

Italian Navy AEW (Mk 112) version of EH101 at its public debut, Paris, June 2003 ***(Paul Jackson)*** ***NEW***/0552895

Canadian CH-149 Cormorant SAR helicopter ***NEW***/0573372

First Portuguese EH101 Mk 514 ***NEW***/0589255

Cormorant MHP: Tentative designation for EH101 subvariant offered by EHI-led 'Team Cormorant' to meet remaining Canadian requirement (Maritime Helicopter Program) for a shipborne ASW Sea King replacement. Sikorsky S-92 selected in July 2004.

US101: Versions optimised for US market, following announcement on 31 October 2001 of agreement between AgustaWestland and Lockheed Martin under which latter will promote the EH101 in the USA as the US101; Bell Helicopter Textron subsequently added to team in May 2003 and is expected to build the US101 in the event of major orders being secured. Targets include USAF's combat SAR requirement and other government agencies requiring EH101's range and payload combination. Initial purchase could originate from US Navy, which requires 23 helicopters, provisionally designated VXX, for presidential transport fleet; bids from Lockheed Martin and Sikorsky received during January 2004, with decision expected in May, but postponed. Earlier, in October 2003, Lockheed Martin took delivery of the ninth development EH101 at Owego, New York, for use as a demonstrator. First flight with CT7-8E engines and five 254 × 203 mm (10 × 8 in) LCDs announced 6 October 2004 (UK 'B Conditions' registration G-17-510).

On 28 January 2005, US101 selected as VXX, with 23 to be delivered to Marine Corps squadron HMX-1 at Quantico, Virginia, between 2009 and 2014.

CUSTOMERS: Orders totalled 142 at end of 2003; see table.

Total 66 for UK, comprising 44 Royal Navy Merlins and 22 RAF Merlins. Italian Navy ordered 16 (reduced from 36, and then from 24), comprising eight ASW (and ASV) versions, four AEW versions and four marines tactical transports with blade- and tail-folding; option on four of further eight converted to firm order at start of 2002; to appear as amphibious support helicopters. Westland produced one Merlin in 1995, three in 1997, six in 1998, 13 in 1999 and 12 in 2000; Agusta built two civil EH101s in 1997 and delivered first in 1998, the other being retained for trials and demonstration. Total of 90 delivered by beginning of October 2003. Long-range utility version for logistic and tactical support role also under consideration by Italy and proposed to US Marine Corps, with RAF Merlin HC. Mk 3 used for month-long demonstration tour of USA in May 2001. VVIP version offered to Saudi Arabia in 1995. EH101 selected in 2001 by Portugal for SAR, combat SAR and fisheries protection and by Denmark for SAR/Utility tasks; latter expects to receive first five helicopters in 2004; Portuguese EH101s built at Vergiate; those for Denmark at Yeovil; first Danish (ZJ990/M-501) and Portuguese helicopters both flew on 12 December 2003. Danish deliveries due late 2004 to 2006. Most recent customer is Japan, which committed in June 2003 to initial quantity of 14 for service with naval air arm and could undertake local production of as many as 67 more; service entry expected in 2007, initially to replace MH-53EJ Sea Dragon mine countermeasures and S-61A utility helicopters. Local assembly anticipated by Kawasaki Heavy Industries.

EH101 was offered to Australia for Air 9000 requirement in August 2003, but subsequently eliminated from contest in December on grounds of size; also in contention for orders from South Korea and Canada.

Prototypes assigned a 3,750 hour flight development programme, but had flown 5,000 hours by mid-1999. Additional 5,600 hours flown by PP8 and PP9 from Brindisi (Italy) and, from September 1998, Aberdeen (Scotland) in trial between March 1996 and June 2000, to improve reliability and prove extended (2,000 hour) overhaul intervals.

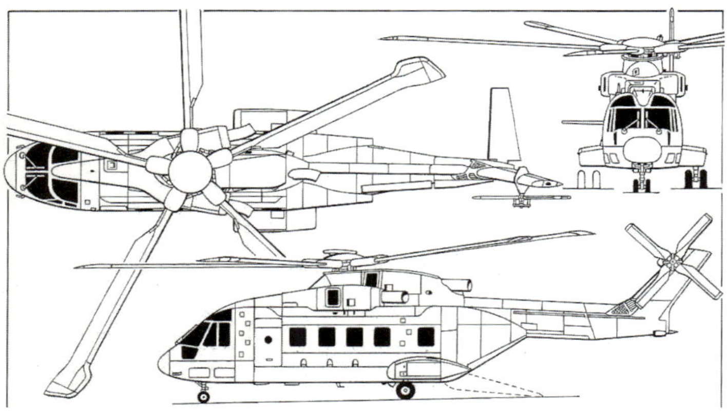

Utility version of the EH101, showing modified rear fuselage with rear-loading ramp/door *(Jane's/Mike Keep)* 0507157

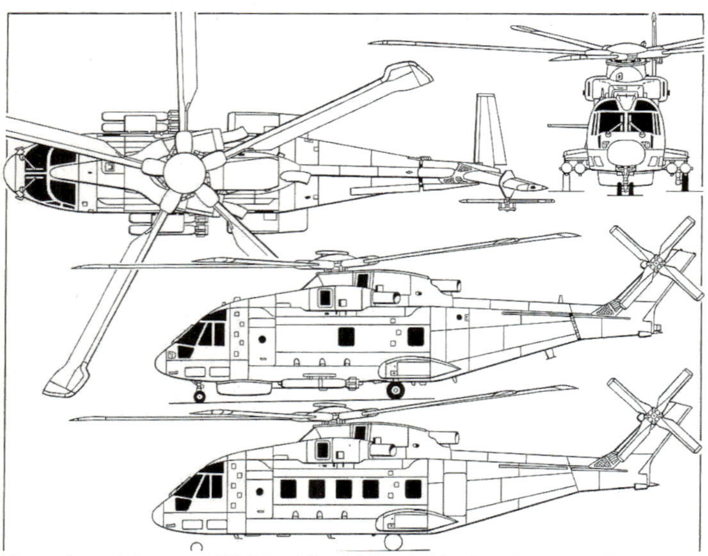

Three-view of the naval EH101, with additional side view (bottom) of the Heliliner *(Jane's/Mike Keep)* 0507156

First Danish EH101 on its maiden flight at Yeovil, temporarily fitted with standard nose radome ***NEW***/0573335

COSTS: Royal Navy £1.5 billion (1991) for 44 Merlin; RAF £500 million for 22 Merlin HC. Mk 3s; Canada C$4.4 billion (1992) for 50 CH-148/149, reduced to C$579 million for 15 CH-149 (AW320). Production investment phase (design, tooling, maturity and product support) valued at £200 million (1993). First Italian batch (16) valued at Lit1,250 billion (US$775 million), 1995, of which Westland's share is £150 million. UK official audit of early 1996 reported Merlin HM. Mk 1 as £351 million (36.1 per cent) over budget and 60 months late. Danish order for 14 reportedly worth US$343.3 million.

DESIGN FEATURES: Three-engine power margin with long-endurance 120 kt (222 km/h; 138 mph) cruise possible on two engines and good twin-engined hover performance. Fail-safe/damage-tolerant airframe and rotating components, high system redundancy, onboard monitoring of engines/transmission/avionics/utility systems; airframe/power plant/rotor and transmission systems/flight controls/utility systems common to all variants; five-blade main rotor with multiple load path hub and elastomeric bearings; blades of advanced aerofoil section with BERP-derived high-speed tips; four-blade teetering tail rotor; transmission has minimum 30 minutes (60 minutes demonstrated) run-dry capacity; fuselage in four main modules (front and centre ones common to all variants, modified rear fuselage and slimmer tailboom on military utility variant to accommodate rear-loading ramp); automatic power folding of main rotor blades and tail rotor pylon on naval variant, with emergency manual back-up (tail section folds forward/downward, stowing starboard half of tailplane beneath rear fuselage). Folding version of utility (rear ramp) EH101 has been designed. New active vibration cancelling system ACSR (active control of structural response) reduces vibration by 80 per cent at blade passing frequency.

Airframe overhaul interval 1,000 hours on service entry; eventual target 3,000 hours. Intended service life of 40,000 hours.

FLYING CONTROLS: Conventional cyclic, collective and yaw pedals for pilot and co-pilot. Dual digital AFCS with four-axis autopilot incorporating coupled flight director, SAR and ASW modes.

STRUCTURE: Rotor head of composites surrounding a metal core; composites blades; fuselage mainly aluminium alloy, with bonded honeycomb main panels; composites for such complex shapes as forward fuselage, upper cowling panels, tailfin, tailplane and windscreen. Engine air intakes of Kevlar reinforced with aero-web honeycomb. Tail unit of carbon epoxy and Kevlar epoxy skinned sandwich panels over central skeleton of metal- or foam-cored composites ribs and longerons; Kevlar-Nomex-Kevlar sandwich for leading-edge of tailfin. Single-sourced series production, with final assembly lines in Italy and UK.

LANDING GEAR: Hydraulically retractable tricycle type, with single mainwheels and steerable twin-wheel nose unit, designed and manufactured by AP Precision Hydraulics in association with Officine Meccaniche Aeronautiche. Main units retract into fairings on sides of fuselage. Goodrich wheels, tyres and brakes: main units have size 8.50–10 wheels with 24×7.7 tyres, unladen pressure 6.96 bar (101 lb/sq in); nosewheels have size 19.5×6.75 tyres, unladen pressure 8.83 bar (128 lb/sq in). Twin-mainwheel gear optional for all variants; adopted for all Italian Navy helicopters and RAF Merlin. FPT Industries emergency flotation bags.

POWER PLANT: Three Rolls-Royce Turbomeca RTM 322-01/8 turboshafts in Royal Navy Merlin (maximum contingency rating 1,724 kW; 2,312 shp, T-O rating 1,566 kW; 2,100 shp and maximum continuous rating 1,394 kW; 1,870 shp); RTM 322-02/8 in RAF version, T-O rating 1,670 kW (2,240 shp); RTM 322 Mk 250 specified for Japanese helicopters; version of this engine also chosen by Denmark and Portugal. General Electric T700-GE-T6A turboshafts in Italian naval variant, and T6A1 in Cormorant, rated at 1,521 kW (2,040 shp) for T-O and 1,327 kW (1,780 shp) maximum continuous. Engines for Italian naval variant assembled by Alfa Romeo Avio and Fiat; Cormorant engines assembled in Canada. Commercial and utility variants powered by three General Electric CT7-6 turboshafts (CT7-6A in PP3) with ratings of 1,491 kW (2,000 shp) for take-off and OEI and 1,282 kW (1,718 shp) maximum continuous.

Transmission rated at 4,161 kW (5,580 shp) for T-O, 3,715 kW (4,982 shp) maximum continuous and 2,769 kW (3,713 shp) OEI maximum continuous.

Standard fuel in three tanks, each of 1,074 litres (276 US gallons; 230 Imp gallons) capacity; total 3,222 litres (851 US gallons; 709 Imp gallons). Each tank feeds separate engine, except on selection of emergency cross-feed; self-sealing optional. Additional fourth or fifth tanks (all same size) optional; maximum fuel capacity 5,370 litres (1,417 US gallons; 1,181 Imp gallons). Merlin HC. Mk 3 capacity approximately 4,075 litres (1,077 US gallons; 896 Imp gallons), augmented by optional tank in cargo hold. Computerised fuel management system. Pressure refuelling point on starboard side; maximum transfer rate 682 litres (180 US gallons; 150 Imp gallons)/min; individual gravity refuelling positions on port side. Provision for detachable refuelling probe on RAF Merlins.

ACCOMMODATION: One or two pilots on flight deck (naval version will be capable of single-pilot operation, if required, and RN will operate with pilot, observer and crewman; commercial variant will be certified for two-pilot operation). ASW version will normally also carry observer and acoustic systems operator. Italian Navy ASW crew members are designated pilot/mission commander, co-pilot/tactical co-ordinator, and two sensor operators.

CH-149 Cormorant operates with pilot, co-pilot/navigator, flight engineer and two crewmen. Martin-Baker crew seats in naval version, able to withstand 10.7 m (35 ft)/s impact. Socea or Ipeco crew seats in commercial variant. Commercial version able to accommodate 30 passengers four-abreast at approximate seat pitch of 76 cm (30 in), plus cabin attendant, with lavatory, galley and baggage facilities (including overhead bins). Offshore variant offers 'club 4' grouped seating to facilitate rapid egress through windows in event of ditching. Military variant can accommodate up to 30 (seated) or 45 (non-seated) combat-equipped troops, 16 stretchers plus a medical team, palleted internal loads, or can carry externally slung loads of up to 5,443 kg (12,000 lb). SAR version can seat 28 survivors; or four seated patients, four stretchers and six medics; or 20 fully equipped Arctic rescuers with skis. It has also demonstrated an emergency evacuation capability with 55 in the cabin.

Main passenger door/emergency exit at front on port side with additional emergency exits on starboard side and on each side of cabin at rear, above main landing gear sponson. Large sliding door at mid-cabin position on starboard side, with inset emergency exit. Commercial variant has baggage bay aft of cabin, with external access via door on port side. Cargo loading ramp/door at rear of cabin on military and utility versions. Cabin floor loading 976 kg/m^2 (200 lb/sq ft) on PP1.

SYSTEMS: Hamilton Sundstrand/Microtecnica environmental control system. Dual-redundant integrated hydraulic system pressurised by three Vickers pumps each supplying fluid at 207 bar (3,000 lb/sq in) nominal working pressure, with flow rates of 55, 59 and 60 litres (14.5, 15.6 and 15.9 US gallons; 12.1, 13.0 and 13.2 Imp gallons)/min respectively. Hydraulic system reservoirs are of the piston load pressurised type, with a nominal pressure of 0.97 bar (14 lb/sq in).

Primary electrical system is 115/200 V three-phase AC, powered by two Lucas brushless, oilspray-cooled 45 kVA generators (90 kVA if Lucas Spraymat blade ice protection system fitted), with one driven by main gearbox and the other by accessory gearbox, plus a third, separately driven standby alternator. Sundstrand T-62 APU for main engine air-starting, and to provide electrical power, plus air for ECS, without running main engines or using external power supplies. Lucas Spraymat electric de-icing of main/tail blades standard on naval variant, optional on others; in February 2003, Smiths Aerospace received contract to supply main and tail rotor blade ice protection system with effect from July 2003, this incorporating electro-thermal heater mats on leading-edges of blades. Dunlop electric anti-icing of engine air intakes. Fire detection and suppression systems by Graviner and Walter Kidde respectively.

AVIONICS (MILITARY): Integrated systems based on two MIL-STD-1553B multiplex databusses that link basic aircraft management, avionics and mission systems.

Comms: Royal Navy version has BAE Systems communications subsystem including internal voice intercommunication to six positions, plus secure voice transmission via two AD3400 V/UHF radios, UHF and HF, plus M/A-COM Ltd ARI 5983 I-band transponder and Link 11. Racal RA 800 Light secure communications control system for RAF Merlin. Italian Navy equipment, by Elmer, includes three SRT-651 V/UHF, two SRT-170L HF, TD8503 Link 11, SP-1450 intercom and Italtel Mk 12 IFF. RAF Merlins have ADELT (automatically deployed emergency loader transmitter) containing FDR and CVR and locator beacon, scabbed on to starboard rear fuselage. Telephonics Corporation SDI (secure digital intercommunications) system to be installed on EH101s for Denmark and Portugal.

Radar: ASW version fitted with 360° search radar (pulse-compressed, frequency-agile BAE Systems Blue Kestrel 5000 in UK's Merlin; Eliradar MM/APS-784 in Italian helicopters). AEW version has Eliradar HEW-784 coherent pulse Doppler, 360° scan radar. Italian Navy utility helicopters have second-hand (from Agusta-Bell 212ASW) Officine Galileo MM/APS-705B search and weather radar which is removable, if demanded by certain missions. Telephonics Corporation RDR-1600 SAR Weather Avoidance Radar system to be used by Danish EH101s. Portuguese EH101 to be fitted with Galileo APS-717 radar.

Flight: Smiths Industries OMI SEP 20 dual-redundant digital AFCS is standard, providing fail-operational autostabilisation and four-axis autopilot modes (auto hover, auto transitions to/from hover standard on naval variants, optional on commercial and military variants). AFCS sensors on naval variant include BAE Systems LINS 300 ring laser gyro inertial reference unit (IRU) and Litton Italia LISA-4000 strapdown AHRS; IRU also provides self-contained navigation, with Racal Doppler 91E velocity sensor (Elmer system for Italy); Cossor Electronics GPS receiver selected for Royal Navy variant, Elmer GPS for Italian Navy aircraft. Other avionics on naval variants include Thomson-CSF AHV 16 radar altimeters (two) (Elettronica J-band units for Italian Navy), BAE Systems low-airspeed sensing and air data system, and Alenia/BAE Systems aircraft management computer. DRS Technologies subcontracted by Smiths Industries to provide flight control computer systems for Danish EH101s and any future sales.

Instrumentation: Litton EFIS with six 160 × 160 mm (6 × 6 in) Smiths multifunction screens. In mid-2003, consideration reportedly being given to introduction of three 330 × 200 mm (13 × 8 in) screens, utilising split images to present information that is currently displayed on two or more MFDs.

Mission: On naval variant, main processing element of management system is a dual-redundant aircraft management computer which carries out navigation, control and display management, performance computation and health and usage monitoring of principal systems (engines, drive systems, avionics and utilities); it also controls basic bus. Alenia/Racal cabin mission display unit. Surveillance radar (see above). Underwater detection in Royal Navy Merlin is by active/passive sonobuoys and Thomson Marconi Sonar TMS 118 ADS (active dipping sonar) system, incorporating same company's AQS-903 acoustic processor for LOFAR, DIFAR, VLAD, Barra, DICAS and CAMBS buoys and Thomson Sintra folding lightweight acoustic system for helicopters (FLASH) expandable array, which has 750 m (2,460 ft) of cable and is manoeuvred by a high-speed winch. Merlin has Racal Lightweight Common Control Unit as interface between crew and tactical navigation and communications systems. The HM. Mk 1 has four units (one each for pilot, co-pilot and two operators); HC. Mk 3 has one in the cabin primarily for maintenance management. Royal Navy Merlin also has Racal Orange Reaper ESM, Normalair-Garrett mission recorder and sonobuoy/flare dispenser, Chelton sonobuoy homing system and Ultra processor for Link 11 datalink. The TMS 118 sonar is to be upgraded from 2001 with new, commercial, off-the-shelf processors, adding to speed and capacity. Honeywell HELRAS Mod 2 dipping sonar, Alenia AYK-204 processors (four) and Alenia SL/ALR-735 ESM in Italian ASW version, which also includes two operators' consoles with 350 mm (14 in) displays. ASST (anti-ship surveillance and tracking) version will carry equipment for tactical surveillance and OTH (over the horizon) targeting, to locate and relay to a co-operating frigate the position of a target vessel, and for mid-course guidance of frigate's missiles. On missions involving patrol of an exclusive economic zone it can also, with suitable radar, monitor every hour all surface contacts within area of 77,700 km^2 (30,000 sq miles); can patrol an EEZ 400 × 200 n miles (740 × 370 km; 460 × 230 miles) twice in one sortie; and can effect boarding and inspection of surface vessels during fishing protection and anti-smuggling missions. FLIR specified for all Italian Navy versions; BAE Systems MST-S FLIR turret, including magnifying mode, can be rapidly installed on RAF Merlins.

First Danish EH101 Mk 512 in operational configuration *(Paul Jackson)* ***NEW***/0589257

First production Italian military EH101, a naval ASW variant ***NEW***/0535826

Self-defence: RAF Merlins have integrated defensive aids including Raytheon laser detection, BAE Systems Sky Guardian 2000 RWR, Doppler-based MAWS, Northrop Grumman AN/AAQ-24 Nemesis DIRCM and BAE Systems North America AN/ALE-47 chaff/flare dispensers. Chaff and flare dispensers on all Italian versions, co-ordinated by Elettronica ELT/156X(V2) RWR and BAE Systems RALM/1 laser warner.

AVIONICS (CIVIL): Integrated avionics system of commercial variant based on ARINC 429 data transfer bus.

Comms: Racal intercom system; Rockwell Collins or Honeywell communications system.

Radar: Honeywell weather radar.

Flight: BAE Systems Canada CMA-900 flight management system for fuel flow, fuel quantity and specific range computations; tuning of nav/com radios; interfaces with electronic instrument systems; two-dimensional multisensor navigation; and built-in navigational database with update service. AFCS sensors on commercial variant include two Litton Italia LISA-4000 strapdown AHRS. Standard avionics include Penny and Giles air data system.

Instrumentation: Smiths Industries/OMI electronic instrument system (EIS) providing colour flight instrument, navigation and power systems displays. CMA-900 includes colour CRT display with graphics and alphanumeric capability.

EQUIPMENT: ASW variants have two sonobuoy dispensers, external rescue hoist and Fairey Hydraulics (Merlin) decklock. BAJ Ltd four-float emergency flotation gear.

ARMAMENT (NAVAL AND MILITARY UTILITY VERSIONS): Naval version able to carry up to four homing torpedoes (BAE Systems Sting Ray on Merlin; Mk 46 or Eurotorp MU90 on Italian version) or other weapons, including Mk 11 Mod 3 depth charges. ASV version designed to carry two air-to-surface missiles (Marte Mk 2/S for Italian Navy) and other weapons, for use as appropriate, from strikes against major units using sea-skimming anti-ship missiles to small arms deterrence of smugglers. Armament optional on military utility versions; options include pintle-mounted machine guns in doorway/rear ramp, chin turret for 12.7 mm machine gun and stub-wings for rocket pods.

EH101 ORDERS

Country	Service	Quantity	Commitment	Mk/Version	First aircraft
Canada	Armed Forces	15	5 Jan 98	511 (SAR)	149901
Denmark	Air Force	14	7 Dec 01[1]	512 (SAR/Utility)	M-501
Italy	Navy	8	Oct 95	110 (ASW/ASVW)	MMX81480
	Navy	4	Oct 95	112 (AEW)	MM81488
	Navy	4	Oct 95	410 (Util)	MM81492
	Navy	4	Jan 02	413 (Amphib Support)	
Japan	Tokyo Police	1	Nov 96	510 (SAR)	JA01MP
	Navy	14	Jun 03		
Portugal	Air Force	6	Dec 01	514 (SAR)	19601
	Air Force	2	Dec 01	515 (SIFICAP)	
	Air Force	4	Dec 01	516 (CSAR)	
UK	Navy	44	9 Oct 91	111/HM. Mk 1	ZH821
	RAF	22	9 Mar 95	411/HC. Mk 3	ZJ117

[1] Contract signature

DIMENSIONS, EXTERNAL (A: naval variant, B: Heliliner, C: military/utility variant):
Main rotor diameter 18.59 m (61 ft 0 in)
Tail rotor diameter 4.01 m (13 ft 2 in)
Length:
overall, both rotors turning 22.80 m (74 ft 9¾ in)
fuselage 19.53 m (64 ft 1 in)
main rotor and tail pylon folded:
A 15.75 m (51 ft 8 in)
C 16.00 m (52 ft 6 in)
Width: cabin 2.80 m (9 ft 2¼ in)
fuselage overall (port sponson to starboard tailplane) 5.09 m (16 ft 8¼ in)
Over sponsons 4.61 m (15 ft 1½ in)
Main rotor and tail pylon folded:
A 5.20 m (17 ft 0¾ in)
C 5.60 m (18 ft 4½ in)
Crew door open 5.32 m (17 ft 5½ in)
Height: overall, both rotors turning 6.62 m (21 ft 8¾ in)
Main rotor and tail pylon folded:
A 5.20 m (17 ft 0¾ in)
C 5.30 m (17 ft 4¾ in)
Tailplane half-span 2.78 m (9 ft 1½ in)
Wheel track 4.55 m (14 ft 11¼ in)
Wheelbase 6.98 m (22 ft 11 in)
Passenger door (fwd, port): Height 1.70 m (5 ft 7 in)
Width 0.91 m (3 ft 0 in)
Sliding cargo door (mid-cabin, stbd):
Height 1.55 m (5 ft 1 in)
Width 1.83 m (6 ft 0 in)
Baggage compartment door (rear, port, B):
Height 1.38 m (4 ft 6 in)
Width 0.55 m (1 ft 10 in)
Rear-loading ramp/door (rear, military/utility variant):
Height 1.95 m (6 ft 4¾ in)
Width 2.26 m (7 ft 5 in)
Main rotor ground clearance (turning) 4.70 m (15 ft 5 in)

DIMENSIONS, INTERNAL:
Cabin:
Length: A 7.09 m (23 ft 3 in)
C 6.50 m (21 ft 4 in)
Max width 2.49 m (8 ft 2 in)
Width at floor 2.26 m (7 ft 5 in)
Max height: B 1.90 m (6 ft 2¾ in)
Volume: A 29.0 m^3 (1,024 cu ft)
B 27.5 m^3 (970 cu ft)
Baggage compartment volume (B) 3.8 m^3 (135 cu ft)
Rear ramp (C): Length 2.10 m (6 ft 10¾ in)
Width 1.80 m (5 ft 10¾ in)

AREAS:
Main rotor disc 271.51 m^2 (2,922.5 sq ft)
Tail rotor disc 12.65 m^2 (136.2 sq ft)

WEIGHTS AND LOADINGS (A, B, C, as above):
Operating weight empty (estimated):
A 10,500 kg (23,149 lb)
B (IFR, offshore equipped) 9,300 kg (20,503 lb)
C 9,350 kg (20,613 lb)
Merlin HC. Mk 3 10,250 kg (22,597 lb)
Max fuel weight (four internal tanks, total):
A (JP-1) 3,406 kg (7,509 lb)
B, C (JP-4) 3,360 kg (7,408 lb)
Merlin HC. Mk 3 3,200 kg (7,055 lb)
Max fuel weight (five internal tanks, total):
B, C (JP-4) 4,200 kg (9,259 lb)
Disposable load/payload:
A (four torpedoes) 960 kg (2,116 lb)
B (30 passengers plus baggage) 2,850 kg (6,283 lb)
C (24 combat-equipped troops) 3,120 kg (6,878 lb)
Max underslung load 5,443 kg (12,000 lb)
Max T-O weight: A, B, C: normal 14,600 kg (32,188 lb)
overload 15,600 kg (34,392 lb)
Max normal disc loading: A, B, C 53.8 kg/m^2 (11.01 lb/sq ft)
Transmission loading at normal max T-O weight and power
A, B, C 3.51 kg/kW (5.76 lb/shp)

PERFORMANCE:
Never-exceed speed (VNE) at S/L, ISA 167 kt (309 km/h; 192 mph) IAS
Average cruising speed 150 kt (278 km/h; 173 mph)
Best range cruising speed 125 kt (232 km/h; 144 mph)
Best endurance speed 80 kt (148 km/h; 92 mph)
Service ceiling 4,575 m (15,000 ft)
Hovering ceiling: IGE 2,225 m (7,300 ft)
OGE 1,128 m (3,700 ft)
Range (B):
four tanks, offshore IFR equipped, with reserves 610 n miles (1,129 km; 702 miles)
five tanks, offshore IFR equipped, with reserves 750 n miles (1,389 km; 863 miles)
Ferry range: C (four tanks plus internal auxiliary tank) 1,130 n miles (2,093 km; 1,300 miles)
SAR radius for 26 survivors, with two minute hover per survivor 350 n miles (648 km, 403 miles)
Endurance 5 h
g limit +3

UPDATED

AIRBUS

AIRBUS SAS
(AN EADS JOINT COMPANY WITH BAE SYSTEMS)

1 Rond Point Maurice Bellonte, F-31707 Blagnac Cedex, France
Tel: (+33 5) 61 93 33 33
Fax: (+33 5) 61 93 37 92
Web: http://www.airbus.com

CHAIRMAN OF SUPERVISORY BOARD: Manfred Bischoff
PRESIDENT AND CEO: Noël Forgeard
COO: Gustav Humbert
EXECUTIVE VICE-PRESIDENTS:
Alain Garcia (Engineering)
John Leahy (Commercial)
Patrick Gavin (Customer Services)
Gérard Blanc (Operations)
Gustav Humbert (Programmes)
Charles Champion (Large Aircraft Division)
SENIOR VICE-PRESIDENT, COMMUNICATIONS: Jean Claude Nicolas
VICE-PRESIDENT, CORPORATE COMMUNICATIONS: Michel Guérard
REGIONAL MANAGER, MEDIA RELATIONS: David Velupillai

AIRFRAME PRIME CONTRACTORS: **Airbus Deutschland** (see EADS)
Airbus España (see EADS CASA)
Airbus France (see EADS)
Airbus UK (see BAE Systems)

Airbus final assembly line in Hamburg ***NEW***/1044652

SUBSIDIARIES:
Airbus North America, Washington, DC, USA
Tel: (+1 703) 834 34 00
Fax: (+1 703) 834 33 41
Airbus China, Beijing
Tel: (+86 10) 804 861 61
Fax: (+86 10) 804 861 62
Airbus Japan, Tokyo
Tel: (+81 35) 220 02 41
Fax: (+81 35) 220 02 53

Airbus Industrie set up 18 December 1970 as Groupement d'Intéret Economique (GIE: a company which makes no profits or losses in its own right) and has been profitable since 1990, producing an operating surplus that is shared among its partners. Its first task was to manage development, manufacture, marketing and support of A300; this management now extends to A300-600, A310, A318, A319/ACJ, A320, A321, A330, A340 and A380. Sales reached the 1,000 mark in 1989; 2,000 in 1996; 3,000 in 1998; 4,000 in 2000; and 5,000 in August 2004. Turnover in 2003 was €19.2 billion. Large Airliner Division created in March 1996 to oversee A3XX (later A380) project. Airbus delivered its 3,500th aircraft, an A318 for Air France , on 2 April 2004. It is currently in concept pre-development of the A350, a long-range derivative of the A330 seen as a potential competitor to the Boeing 787.

A new single-aisle internal flight deck door, complying with pre-existing and new FAA anti-intrusion regulations, received JAA design approval on 21 May 2002 and is in production; it was certified for the A330/A340 family on 26 June 2002, and subsequently for all Airbus aircraft. In March 2003, Airbus selected Thales as its preferred supplier of HUDs for its entire range of FBW airliners, expected to appear first in the A320 family.

Plans for establishment of a single corporate entity (SCE) overtaken and modified by formation of European Aeronautic, Defense and Space Company (EADS, which see). Planned restructuring of the consortium into Airbus Integrated Company (AIC) was revealed in March 2000 and approved by the EC in late 2000; official starting date was 1 January 2001, although completion of formalities was delayed. EADS holds 80 per cent and BAE Systems 20 per cent of AIC, which incorporated in France on 11 July 2001 as an SAS (Société par Actions Simplifiées) following a formal decision on 23 June 2000.

Airbus is responsible for all work by partner companies and has some 45,000 employees, including workers at its spare parts centre in Hamburg and its US, Chinese and Japanese subsidiaries. Approximate European breakdown is Germany 16,000, France 14,000, UK 8,300 and Spain 2,400. Stork (formerly Fokker) is an associate in A300 and A310 and Belairbus (Belgian consortium) in A310, A320, A330 and A340. Alenia manufactures front fuselage plug for A321. An engineering centre was established in Moscow in a joint venture (registered in December 2002) with the Kaskol group, which has shares in Sokol, Hydromash, Rostvertol and other Russian aerospace companies; this was opened, with a workforce of 25, in late March 2003, and will develop and co-produce components for the A380 and other Airbus types; General Director is Vladimir Raschupkin.

Subsidiaries include Airbus North America, Airbus Finance Company (AFC), formed in 1994, Airbus Japan and Airbus China. Airbus's training centre in Toulouse, previously a subsidiary known as Aeroformation, and the main spares centre, Airbus Materiel Support, formerly Airspares Hamburg, are integrated within the consortium's Customer Services

Simulated formation flight by Airbus's most successful family: the A318, A319, A320 and A321 ***NEW***/1044679

TOTALS OF AIRBUS AIRLINERS
(at 1 January 2005)

	A300	A310	A318	A319	A320	A321	A330	A340	A380	Totals
Firm orders	591	260	61	1,030	1,864	416	517	374	139	5,252
Delivered	537	255	19	651	1,348	324	329	289	0	3,752
Operating	423	233	19	650	1,337	323	326	287	0	3,598

AIRBUS ORDERS and DELIVERIES 2004

	A300	A310	A318	A319	A320	A321	A330	A340	A380	Totals
Net orders	2	0	−18	105	185	7	48	27	10	366
Delivered	12	0	10	87	101	35	47	28	0	320
Backlog	54	5	42	379	516	92	188	85	139	1,500

Directorate; a new, purpose-built mockup centre was added to the Toulouse facilities during 1999 and was fully operational by end 2000. In July 1996, Airbus Training and Support Centre in Beijing was opened, becoming operational in October 1997. A Training Centre was opened in Miami, Florida, in October 1999.

UPDATED

AIRBUS A300-600

TYPE: Wide-bodied airliner.

PROGRAMME: Launched 29 May 1969; initial variants were A300B1 (first flight 28 October 1972, service entry November 1974, A300B2 (first flight June 1973, service entry 30 May 1974) and A300B4 (first flight December 1974, service entry June 1975; 248 built. A300-600 go-ahead 16 December 1980; first flight (F-WZLR) 8 July 1983; certified (with JT9D-7R4H1 engines) 9 March 1984; first delivery (to Saudia) 26 March 1984.

Airbus A300-600 twin-turbofan airliner in the insignia of Japan Air Lines *(JAL)* ***NEW***/0567736

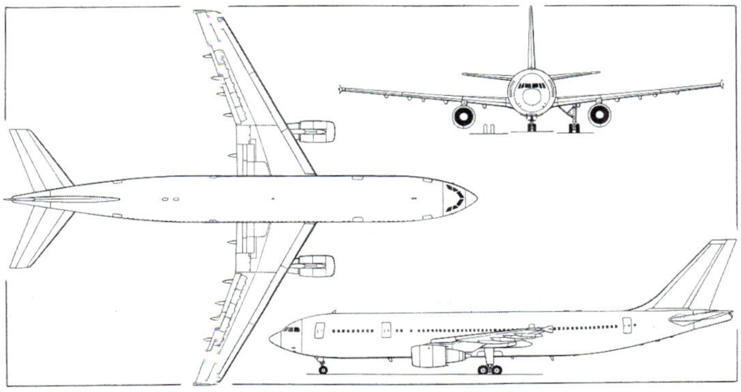

Airbus A300-600R wide-bodied transport (two GE CF6-80C2 turbofans) *(Jane's/Dennis Punnett)* 0507146

Improved version with CF6-80C2 engines and other changes (see Current Versions) made first flight 20 March 1985; French certification for Cat. IIIb take-offs and landings 26 March 1985; first delivery of improved version (to Thai Airways) 26 September 1985. Extended-range A300-600R (then known as -600ER) made first flight 9 December 1987, receiving European and FAA certification 10 and 28 March 1988 respectively, deliveries (to American Airlines) beginning 20 April 1988; A300-600 powered by GE CF6-80C2A5 with FADEC granted 180-minute ETOPS April 1994. CIS certification granted May 1996. Bulk of production backlog are A300F4-600R Freighters for United Parcel Service.

CURRENT VERSIONS: **A300-600:** Advanced version of A300B4-200; major A300 version since early 1984. Passenger and freight capacity increased by fitting rear fuselage of A310 with pressure bulkhead moved aft; wings have simple Fowler flaps and increased trailing-edge camber; forward-facing two-person flight deck with EFIS; new digital avionics; new braking control system; new APU; simplified systems; weight saving by use of composites for some secondary structural components; payload/range performance and fuel economy improved by comprehensive drag clean-up. Further improvements introduced in 1985 included CF6-80C2 or PW4000 as engine options, carbon brakes, wingtip fences and 'New World' flight deck; basic equipment of aircraft delivered from late 1991 further improved by incorporating standard options.

Cargo conversions of A300-600 and earlier A300B4 are offered; see *Jane's Aircraft Upgrades* for details.

Detailed description applies to current production A300-600/600R except where indicated.

A300-600R: Extended-range version of A300-600, differing mainly in having fuel trim tank in tailplane and higher maximum T-O weight.

A300-600 Convertible: Convertible passenger/cargo version, described separately.

A300-600 Freighter: Non-passenger version, described separately.

Airbus Super Transporter: A300-600R conversion as Super Guppy replacement.

CUSTOMERS: Total of 589 of all A300 versions ordered, of which 533 delivered, by 1 October 2004. There were six orders and eight deliveries in 2003. Outstanding contracts at 1 October 2004 comprised five for Air Hong Kong and 51 for United Parcel service.

AIRBUS A300 MARKS AND VARIANTS

Series	Mark	Power Plant	FAA Certification
B1	—	CF6-50A	not certified
B2	1A	CF6-50A	30 May 74
B2	1C	CF6-50C	19 Jun 75
B2K	3C	CF6-50C	30 Jun 76
B2	203	CF6-50C2	1 Oct 80
B4	2C	CF6-50C	30 Jun 76
B4	103	CF6-50C2	4 Oct 79
B4	203	CF6-50C2	2 Oct 81
B4	601	CF6-80C2A1	28 Mar 88
B4	603	CF6-80C2A3	19 Sep 88
B4	605R	CF6-80C2A5	28 Mar 88
B4	620	JT9D-7R4H1	19 Sep 88
B4	622R	PW4158	1 Aug 91
C4	605 (F)	CF6-80C2A5	21 Jun 02
F4	605R	CF6-80C2A5	27 Apr 94
F4	622R	PW4158	14 Jul 00

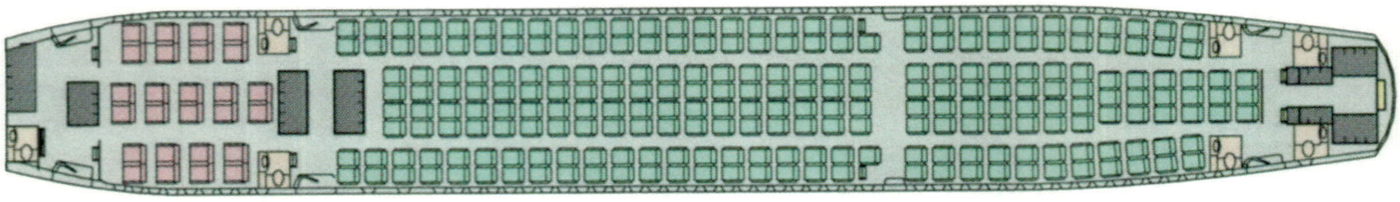

Two-class A300-600 interior for 26 first and 240 economy class passengers

0044877

COSTS: US$109.9 million (2001).

DESIGN FEATURES: Mid-mounted wings with 10.5 per cent thickness/chord ratio, 28° sweepback at quarter-chord, and (since 1985) tip fences; circular-section pressurised fuselage; all-swept tail unit.

FLYING CONTROLS: Power-assisted. Each wing has three-segment, two-position (T-O/landing) leading-edge slats (no cutout over engine pylon), small Krueger flap at leading-edge wingroot, three cambered tabless flaps on trailing-edge, all-speed aileron between inboard flap and outer pair, and seven spoilers forward of flaps on each wing; flaps occupy 84 per cent of trailing-edge, increasing wing chord by 25 per cent when fully extended; ailerons deflect 9° 2' downward automatically when flaps are deployed; all 14 spoilers used as lift dumpers: outboard 10 for roll control and inboard 10 as airbrakes; variable incidence tailplane. Ailerons/elevators/rudder fully powered by hydraulic servos (three per surface), controlled mechanically; secondary surfaces (spoilers/flaps/slats) fully powered hydraulically with electrical control, tailplane by two independent hydraulic motors electrically controlled with additional mechanical input; preselection of spoiler/lift dump lever permits automatic extension of lift dumpers on touchdown; flaps and slats have similar drive mechanisms, each powered by twin motors driving ball screwjacks on each surface with built-in protection against asymmetric operation.

STRUCTURE: Two-spar main wing box, integral with fuselage and incorporating fail-safe principles; third spar across inboard sections; semi-monocoque fuselage (frames and open Z-section stringers), with integrally machined skin panels in high-stress areas; primary structure is of high-strength, damage-tolerant aluminium alloy, with steel or titanium for some critical fuselage components, honeycomb panels or selected glass fibre laminates for secondary structures; metal slats, flaps and ailerons. CFRP fins replaced aluminium alloy unit from 1988; secondary structure composites include AFRP for flap track fairings, rear wing/body fairings, cooling air inlet fairings and radome; GFRP for wing upper surface panels above mainwheel bays, fin leading/trailing-edges, fintip, fin/fuselage fairings, tailplane trailing-edges, elevator leading-edges, tailplane and elevator tips and elevator actuator access panel; carbon-reinforced GFRP for elevators and rudder; CFRP for spoilers, outer flap deflector doors and fin box; all CFRP moving surfaces have aluminium or titanium trailing-edges. Nosewheel doors and mainwheel leg fairing doors also of CFRP. Nose gear is structurally identical to that of B2/B4/A310; main gear is generally reinforced, with a new hinge arm and a new pitch damper hydraulic and electrical installation. Nacelles have CFRP cowling panels and are subcontracted to Rohr (California); pylon fairings are of AFRP.

Airbus France builds nose (including flight deck), lower centre-fuselage, four inboard spoilers, wing/body fairings and engine pylons; Airbus Deutschland builds forward fuselage (flight deck to wing box), upper centre-fuselage, rear fuselage (including tailcone), vertical tail, 10 outboard spoilers and some cabin doors; it also equips wings and installs interiors and seats; Airbus UK (formerly BAe) designed wings and builds wing box; Airbus España manufactures horizontal tail, port and starboard forward passenger doors and mainwheel/nosewheel doors; Stork Aerospace produces wingtips, ailerons, flaps, slats and main gear leg fairings. Large, fully equipped and inspected airframe sections airlifted by Beluga to Airbus France at Toulouse for assembly and painting, aircraft then being flown to Hamburg for outfitting and returned to Toulouse for customer acceptance.

LANDING GEAR: Hydraulically retractable tricycle type, of Messier-Bugatti design, with Messier-Bugatti/Liebherr/Dowty shock-absorbers and wheels standard; twin-wheel nose unit retracts forward, main units inward into fuselage; free-fall extension; has four-wheel main bogies interchangeable left with right. Standard bogie size is 927 × 1,397 mm (36½ × 55 in); wider bogie of 978 × 1,524 mm (38½ × 60 in) is optional. Mainwheel tyres size 49×17–20 or 49×17.0R20 (30 ply) (standard) or 49×19–20 (30 ply) (wide bogie), with respective pressures of 12.41 and 11.10 bar (180 and 161 lb/sq in). Nosewheel tyres size 40×14 or 40×14.0R16 (22 ply), pressure 9.38 bar (136 lb/sq in). Steering angles 65°/95°. Messier-Bugatti/Liebherr/Dowty hydraulic disc brakes standard on all mainwheels. Normal braking powered by 'green' hydraulic system, controlled electrically through two master valves and monitored by a brake system control box to provide anti-skid protection. Standby braking (powered automatically by 'yellow' hydraulic system if normal 'green' system supply fails) controlled through a dual metering valve; anti-skid protection is ensured through same box as normal system, with emergency pressure supplied to brakes by accumulators charged from 'yellow' system. Automatic braking system optional. Bendix or Goodrich wheels and brakes available optionally. Minimum ground turning radius (effective, aft CG) 22.00 m (72 ft 2¼ in) about nosewheel, 34.75 m (114 ft 0 in) about wingtips.

POWER PLANT: Two turbofans in underwing pods. A300-600 was launched with 249 kN (56,000 lb st) Pratt & Whitney JT9D-7R4H1 and currently available with 249 kN (56,000 lb st) Pratt & Whitney PW4156 or 262 kN (59,000 lb st) General Electric CF6-80C2A1. A300-600R is offered with 274 kN (61,500 lb st) CF6-80C2A5 or 258 kN (58,000 lb st) PW4158. CF6-80C2A5 and PW4158 also available as options on A300-600.

Fuel in two integral tanks in each wing, and fifth integral tank in wing centre-section, giving standard usable capacity of 62,000 litres (16,379 US gallons; 13,638 Imp gallons). Additional 6,150 litre (1,625 US gallon; 1,353 Imp gallon) fuel/trim tank in tailplane (-600R only) increases this total to 68,150 litres (18,004 US gallons; 14,991 Imp gallons). Optional extra fuel cell in aft cargo hold can increase total to 73,000 litres (19,285 US gallons; 16,058 Imp gallons) in -600R. Two standard refuelling points beneath starboard wing; similar pair optional under port wing.

ACCOMMODATION: Crew of two on flight deck, plus two observers' seats. Passenger seating in main cabin in six-, seven-, eight- or nine-abreast layout with two aisles; typical mixed class layout has 266 seats (26 first class and 240 economy), six/eight-abreast at 96/86 cm (40/32 in) seat pitch with two galleys and one lavatory forward, one galley and two lavatories at Door 2 position, and one galley and four lavatories at rear; typical economy class layout for 285 passengers eight-abreast at 86 cm (34 in) pitch. Maximum capacity (subject to certification) 361 passengers. Closed overhead baggage lockers on each side

Series	Variant	MTOW (kg)
B2-1A	00	137,000
B2-1C	00	137,000
B2-1C	02	142,000
B2K-3C	00	142,000
B2-203	00	142,000
B4-2C	00	150,000
B4-2C	02,03,14	157,500
B4-103	00	150,000
B4-103	02,03,14	157,500
B4-203	00, 07	165,000
B4-601/03/20	00	165,000
B4-605R/22R	00	170,500
B4-605R/22R	01, 02	171,700
B4-605R/22R	03	167,800
C4-605R	00	170,500
F4-600R/22R	00	170,500
F4-600R/22R	06	165,100
F4-600R/22R	09	168,000

Notes: Variant parameters may include differences in data other than max T-O weight. Except as constrained by Series, Variants are applicable to several engine options

AIRBUS A300 ORDERS
(at 1 October 2004)

Customer	Qty
Air Afrique	3
Air France	23
Air Hong Kong	6
Air India	3
Air Inter Europe	8
Alitalia	8
American Airlines	35
Amiri Flight	2
Ansett	8
Australian Airlines	5
China Airlines	15
China Eastern Airlines	7
China Northern Airlines	6
China Northwest Airlines	3
CityBird	2
Continental Airlines	3
Cruzeiro	2
Eastern Airlines	34
Egyptair	17
Emirates	5
Federal Express	36
Finnair	2
Garuda	9
Hapag Lloyd	7
Iberia	6
Indian Airlines	10
ILFC	9
Iran Air	8
Japan Air System	32
Japan Fleet Service	2
Korean Air	32
Kuwait Airways	8
Laker	3
LaTur	2
Lufthansa	23
Malaysia Airlines	4
Monarch	4
Olympic	10
Pakistan International Airlines	4
Pan Am	12
Philippine Airlines	5
Polaris Aircraft Leasing	5
Saudi Arabian Airlines	11
SAS	4
Singapore Airlines	8
SOGERMA SOCEA	1
South African Airways	7
Thai Airways	33
Trans European Airways	1
Tunis Air	1
United Parcel Service	90
Varig	2
VASP	3
Total	**589**

(total capacity 10.5 m^3; 370 cu ft) and in double-sided central 'super-bin' installation (total capacity 14.5 m^3; 512 cu ft), giving 0.03 to 0.09 m^3 (1.2 to 3.2 cu ft) per passenger in typical economy layout.

Two outward parallel-opening Type A plug-type passenger doors ahead of wing on each side, and one on each side at rear. Type I emergency exit on each side aft of wing. Underfloor baggage/cargo holds fore and aft of wings, with doors on starboard side; forward hold can accommodate 12 LD3 containers, or four 2.24 × 3.17 m (88 × 125 in) pallets or, optionally, 2.43 × 3.17 m (96 × 125 in) pallets, or engine modules; rear hold can accommodate 10 LD3 containers; additional bulk loading of freight provided for in an extreme rear compartment with usable volume of 17.3 m^3 (611 cu ft); alternatively, rear hold can carry 11 LD3 containers, with bulk cargo capacity reduced to 9.0 m^3 (318 cu ft); bulk cargo compartment can be used to transport livestock. Entire accommodation is pressurised, including freight, baggage and avionics compartments.

SYSTEMS: Air supply for air conditioning system taken from engine bleed and/or APU via two high-pressure points; conditioned air can also be supplied direct to cabin by two low-pressure ground connections; ram air inlet for fresh air ventilation when packs not in use. Pressure control system (maximum differential 0.574 bar; 8.32 lb/sq in) consists of two identical, independent, automatic systems (one active, one standby); automatic switchover from one to other after each flight and in case of active system failure; in each system, pressure controlled by two electric outflow valves, function depending on preprogrammed cabin pressure altitude and rate of change of cabin pressure, aircraft altitude, and preselected landing airfield elevation. Automatic prepressurisation of cabin before take-off, to prevent noticeable pressure fluctuation during take-off. Modular box system provides passenger oxygen to all installation areas.

Hydraulic system comprises three fully independent circuits, operating simultaneously; each system includes reservoir of direct air/fluid contact type, pressurised at 3.52 bar (51 lb/sq in); fire-resistant phosphate ester-type fluid; nominal output flow 136 litres (35.9 US gallons; 30 Imp gallons)/min delivered at 207 bar (3,000 lb/sq in) pressure; 'blue' and 'yellow' systems have one pump each, 'green' system has two pumps. The three circuits provide triplex power for primary flying controls; if any circuit fails, full control of aircraft is retained without any necessity for action by crew. All three circuits supply ailerons, rudder and elevators; 'blue' circuit additionally supplies spoiler 7, spoiler/airbrake 4, airbrake 1, yaw damper and slats; 'green' circuit additionally supplies spoiler 6, flaps, Krueger flaps, slats, landing gear, wheel brakes, steering, tailplane trim, artificial feel, and roll/pitch/yaw autopilot; 'yellow' circuit additionally supplies spoiler 5, spoiler/airbrake 3, airbrake 2, flaps, wheel brakes, cargo doors, artificial feel, yaw damper, tailplane trim, and roll/pitch/yaw autopilot. Ram air turbine pump provides standby hydraulic power should both engines become inoperative.

Main electrical power supplied under normal flight conditions by two integrated drive generators, one on each engine; third (auxiliary) generator, driven by APU, can replace either of main generators, having same electromagnetic components but not constant-speed drive; each generator rated at 90 kVA, with overload ratings of 112.5 kVA for 5 minutes and 150 kVA for 5 seconds; APU generator driven at constant speed through gearbox. Three unregulated transformer-rectifier units (TRUs) supply 28 V DC power. Three 25 Ah Ni/Cd batteries used for emergency supply and APU starting; emergency electrical power taken from main aircraft batteries and emergency static inverter, providing single-phase 115 V 400 Hz output for flight instruments, navigation, communications and lighting when power not available from normal sources.

Hot air anti-icing of engines, engine air intakes, and outer segments of leading-edge slats; electrical heating for anti-icing flight deck front windscreens, demisting flight deck side windows, and for sensors, pitot probes and static ports, and waste water drain masts.

Honeywell 331-250F APU in tailcone, exhausting upward; installation incorporates APU noise attenuation. Self-contained fire protection system, and firewall panels protect main structure from an APU fire. APU provides bleed air to pneumatic system, and drives auxiliary AC generator during ground and in-flight operation; APU drives 90 kVA oilspray-cooled generator, and supplies bleed air for main engine start or air conditioning system. For current deliveries of A300-600, APU has improved relight capability, and can be started throughout flight envelope.

For new A300-600s and -600Rs, two optional modifications offered for compliance with full extended-range twin-engined operations (ETOPS) requirements: hydraulically driven fourth generator and increased cargo hold fire suppression capability. ETOPS kit qualified for aircraft with CF6-80C2 and JT9D-7R series engines and, since mid-1988, for those with PW4000 series.

AVIONICS: *Comms:* Standard communications radios include two VHF, with provision for a third, two HF, two transponders, one Selcal, interphone and passenger address systems, ground crew call system and cockpit voice recorder. Provision for Mode S transponders.

Radar: Weather radar standard, with provision for second.

Flight: Radio navigation avionics include two VOR, two ILS, two DME, one ADF, two marker beacon receivers and two radio altimeters; TCAS and GPWS. Most other avionics are to customer requirements, only those relating to the instrument landing system (Honeywell or Rockwell Collins ILS and Rockwell Collins or TRT radio altimeter) being selected and supplied by the manufacturer. Two Honeywell digital air data computers standard; basic digital AFCS has dual flight control computers (FCCs) for flight director and autopilot functions (for Cat. III automatic landings), single thrust control computer (TCC) for speed and thrust control, and two flight augmentation computers (FACs) to provide yaw damping, electric pitch trim, and flight envelope monitoring and protection. Options include second FCC (for Cat III automatic landing); second TCC; two flight management computers (FMCs) and two control display units for full flight management system. Basic aircraft also fitted with ARINC 717 data recording system with digital flight data acquisition unit, digital flight data recorder and three-axis linear accelerometer; optional additional level of windshear protection is available. Honeywell Enhanced GPWS available from March 1997.

Instrumentation: Six identical and interchangeable CRT electronic displays (four electronic flight instrument system and two electronic centralised aircraft monitor), plus digitised electromechanical instruments with liquid crystal displays.

DIMENSIONS, EXTERNAL:
Wing span 44.84 m (147 ft 1 in)
Wing chord at root 9.40 m (30 ft 10 in)
Wing aspect ratio 7.7
Length overall 54.08 m (177 ft 5 in)
Fuselage: Length 53.30 m (174 ft 10½ in)
Max diameter 5.64 m (18 ft 6 in)
Height overall 16.51 m (54 ft 2 in)
Tailplane span 16.26 m (53 ft 4 in)
Wheel track 9.60 m (31 ft 6 in)
Wheelbase (c/l of shock-absorbers) 18.62 m (61 ft 1 in)
Passenger doors (each): Height 1.93 m (6 ft 4 in)
Width 1.07 m (3 ft 6 in)
Height to sill: forward 4.60 m (15 ft 1 in)
centre 4.80 m (15 ft 9 in)
rear 5.50 m (18 ft 0½ in)
Emergency exits (each): Height 1.60 m (5 ft 3 in)
Width 0.61 m (2 ft 0 in)
Height to sill 4.87 m (15 ft 10 in)
Underfloor cargo door (forward):
Height 1.71 m (5 ft 7¼ in)
Width 2.69 m (8 ft 10 in)
Height to sill 3.07 m (10 ft 1 in)
Underfloor cargo door (rear):
Height 1.71 m (5 ft 7¼ in)
Width 1.81 m (5 ft 11¼ in)
Height to sill 3.41 m (11 ft 2¼ in)
Underfloor cargo door (extreme rear):
Height (projected) 0.95 m (3 ft 1 in)
Width 0.95 m (3 ft 1 in)
Height to sill 3.56 m (11 ft 8 in)

DIMENSIONS, INTERNAL:
Cabin, excl flight deck: Length 40.70 m (133 ft 6¼ in)
Max width 5.28 m (17 ft 4 in)
Max height 2.54 m (8 ft 4 in)
Underfloor cargo hold:
Length: forward 10.60 m (34 ft 9¼ in)
rear 7.95 m (26 ft 1 in)
extreme rear 3.40 m (11 ft 2 in)
Max height 1.76 m (5 ft 9 in)
Max width 4.20 m (13 ft 9¼ in)
Underfloor cargo hold volume:
forward 75.1 m^3 (2,652 cu ft)
rear 55.0 m^3 (1,942 cu ft)
extreme rear 17.3 m^3 (611 cu ft)

AREAS:
Wings, gross 260.00 m^2 (2,798.6 sq ft)
All-speed ailerons (total) 7.06 m^2 (75.99 sq ft)
Trailing-edge flaps (total) 47.30 m^2 (509.13 sq ft)
Leading-edge slats (total) 30.30 m^2 (326.15 sq ft)
Krueger flaps (total) 1.115 m^2 (12.00 sq ft)
Spoilers (total) 5.40 m^2 (58.13 sq ft)
Airbrakes (total) 12.59 m^2 (135.52 sq ft)
Fin 45.20 m^2 (486.53 sq ft)
Rudder 13.57 m^2 (146.07 sq ft)
Tailplane 44.80 m^2 (482.22 sq ft)
Elevators (total) 19.20 m^2 (206.67 sq ft)

WEIGHTS AND LOADINGS (A: CF6-80C2A1/A5 engines, B: PW4156/4158 engines, both in 266-seat configuration)[1]:
Manufacturer's weight empty:
A (600) 79,210 kg (174,630 lb)
A (600R) 80,070 kg (176,525 lb)
B (600) 79,151 kg (174,500 lb)
B (600R) 79,320 kg (174,870 lb)
Operating weight empty:
A (600) 90,115 kg (198,665 lb)
A (600R) 91,040 kg (200,700 lb)
B (600) 90,065 kg (198,565 lb)
B (600R) 90,965 kg (200,550 lb)
Max payload (structural): A (600) 39,885 kg (87,931 lb)
A (600R) 38,962 kg (85,896 lb)
B (600) 39,993 kg (88,169 lb)
B (600R) 39,037 kg (86,061 lb)
Max usable fuel:
600: standard 49,786 kg (109,760 lb)
600R: standard 54,721 kg (120,640 lb)
with optional cargo hold tank 58,618 kg (129,230 lb)
Max T-O weight (A and B):
600 165,000 kg (363,765 lb)
600R (basic) 170,500 kg (375,885 lb)
600R (option) 171,400 kg (377,870 lb)
Max ramp weight (A and B):
600 165,900 kg (365,745 lb)
600R (basic) 171,400 kg (377,870 lb)
600R (option) 172,600 kg (380,520 lb)
Max landing weight (A and B):
600 138,000 kg (304,240 lb)
600R (basic) 140,000 kg (308,645 lb)
Max zero-fuel weight (A and B):
600, 600R (basic) 130,000 kg (286,600 lb)
Max wing loading: 600 634.6 kg/m^2 (129.98 lb/sq ft)
600R (basic) 655.8 kg/m^2 (134.32 lb/sq ft)

[1] *Production aircraft from late 1996 onward. See 1989–90 and previous editions for original versions; 1996–97 and six previous editions for intermediate versions*

PERFORMANCE (A and B as for Weights and Loadings):
Max operating speed (VMO) from S/L to FL267
335 kt (621 km/h; 386 mph) CAS
Max operating Mach No. (MMO) above FL267 0.82
Max cruising speed:
at FL250 480 kt (890 km/h; 553 mph)
at FL300 M0.82 (484 kt; 897 km/h; 557 mph)
Typical long-range cruising speed at FL310 M0.80 (472 kt; 875 km/h; 543 mph)
Approach speed: 600 135 kt (249 km/h; 155 mph)
600R 136 kt (251 km/h; 156 mph)
Max operating altitude 12,200 m (40,000 ft)
Runway ACN for flexible runway, category B:
standard bogie and tyres: 600 56
600R 59
600R (option) 60
optional bogie and tyres: 600 52
600R 55
600R (option) 56
T-O field length at S/L, ISA + 15°C:
600: A 2,378 m (7,800 ft)
B 2,270 m (7,450 ft)
600R: A (C2A5 engines) 2,408 m (7,900 ft)

B (PW4158 engines) .... 2,362 m (7,750 ft)
Landing field length: 600 .... 1,536 m (5,040 ft)
600R .... 1,555 m (5,100 ft)
Range (1996 and subsequent deliveries) at typical airline OWE with 266 passengers and baggage, reserves for 200 n miles (370 km; 230 miles):
600, GE/PW engines .... 3,700 n miles (6,852 km; 4,257 miles)
600R, GE/PW engines, standard fuel .... 4,050 n miles (7,500 km; 4,660 miles)
600R, GE/PW engines, optional fuel .... 4,157 n miles (7,700 km; 4,784 miles)

OPERATIONAL NOISE LEVELS (A300-600R, ICAO Annex 16, Chapter 3):
T-O: A .... 91.1 EPNdB (96.3 limit)
B .... 92.2 EPNdB (96.3 limit)
Sideline: A .... 98.6 EPNdB (99.9 limit)
B .... 97.7 EPNdB (99.9 limit)
Approach: A .... 99.8 EPNdB (103.3 limit)
B .... 101.7 EPNdB (103.3 limit)

UPDATED

AIRBUS A300-600 CONVERTIBLE AND A300-600 FREIGHTER

TYPE: Twin-jet freighter.

PROGRAMME: Specialised versions of A300-600. First flight of A300-600F 2 December 1993; certified early April 1994, delivered to Federal Express 27 April and entered service in same month; A300-600F powered by GE CF6-80C2A5 with FADEC was first A300-600 version to operate with 180-minute ETOPS in May 1994.

CURRENT VERSIONS: **Convertible:** For all-passenger or all-cargo configuration. Typical options include accommodation (in mainly eight-abreast seating) for maximum 375 passengers (subject to certification) on the main deck; or up to twenty 2.24 × 3.17 m (88 × 125 in) pallets; or five 88 × 125 in plus nine 2.44 × 3.17 m (96 × 125 in) pallets. Three built by October 2004, including one each for Air France and Air Hong Kong.

Freighter: For freighting only; no passenger systems provided; various systems options give airlines ability to adapt basic aircraft to specific freight requirements; Airbus offers conversion with port-side forward freight door, or as new-production **A300F4-600R**, with payload capacity of 54,000 kg (119,050 lb). Freighter conversions, offered by Airbus UK and EADS EFW, are detailed in *Jane's Aircraft Upgrades*; over 65 under conversion or completed by June 2001; maiden flight 13 December 2001 (D-ASAE, c/n 477); certified by LBA and FAA eight days later.

General Freighter: Similar to A300-600F, but with side door and cargo loading system able to handle all sizes of freight from small packets to large containers. Launch customer Air Hong Kong, which announced order for six in January 2003 (two delivered by 1 October 2004), with further four on option.

CUSTOMERS: Federal Express became A300-600 Freighter launch customer July 1991 with order for 25 and commitments for 50 more, of which 11 confirmed as orders in September 1996; all now delivered. UPS placed order on 9 September 1998 for 30 PW4158 powered A300F4-600R Freighters plus options on a further 15 (later confirmed and a further 15 firm orders, to total 60, of which 45 delivered and in operation by 1 October 2004); 30 more on option. Deliveries began mid-2000; UPS has an option to convert some of the order to A380s if it wishes.

STRUCTURE: Generally similar to A300-600. Main differences are large port-side main deck cargo door, reinforced cabin floor, smoke detection system in main cabin; main deck cargo door is on opposite side to door of forward underfloor hold, allowing simultaneous loading or unloading at all positions.

POWER PLANT: Options as for A300-600R; first example was first Airbus aircraft powered by GE CF6-80C2A5 with FADEC.

DIMENSIONS, EXTERNAL: As A300-600R, plus:
Main deck cargo door (fwd, port):
Height (projected) .... 2.57 m (8 ft 5¼ in)
Width .... 3.58 m (11 ft 9 in)
Height to sill .... 4.91 m (16 ft 1 in)

DIMENSIONS, INTERNAL:
Cabin main deck usable for cargo:
Length .... 33.45 m (109 ft 9 in)
Min height .... 2.01 m (6 ft 7 in)

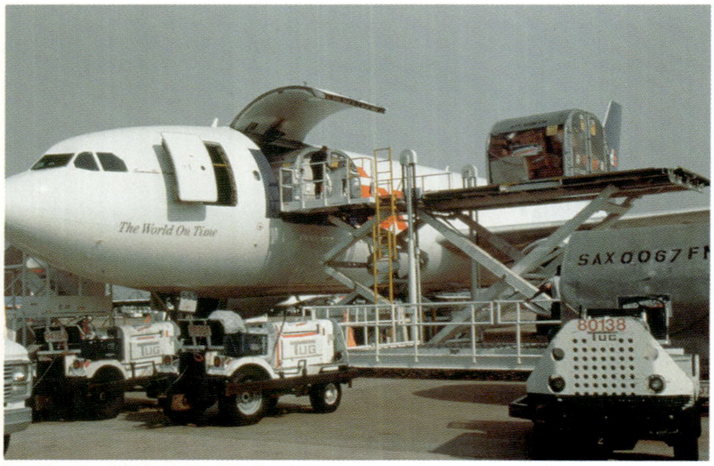

Loading an A300-600F of Federal Express ***NEW***/1044678

A300-600F in Air Hong Kong livery ***NEW***/1044677

Max height:
ceiling trim panels in place .... 2.22 m (7 ft 3½ in)
without ceiling trim panels .... 2.44 m (8 ft 0 in)
Volume .... 192.0–203.0 m³ (6,780–7,169 cu ft)

WEIGHTS AND LOADINGS (basic Convertible. A: with CF6-80C2A5 engines, B: with PW4158 engines):
Manufacturer's weight empty:
A, passenger mode .... 82,555 kg (182,000 lb)
B, passenger mode .... 82,470 kg (181,815 lb)
A, freight mode .... 80,345 kg (177,130 lb)
B, freight mode .... 80,260 kg (176,945 lb)
Operating weight empty:
A, passenger mode .... 93,550 kg (206,240 lb)
B, passenger mode .... 93,475 kg (206,075 lb)
A, freight mode .... 81,600 kg (179,895 lb)
B, freight mode .... 81,525 kg (179,730 lb)
Max payload (structural):
A, passenger mode .... 36,448 kg (80,354 lb)
B, passenger mode .... 36,523 kg (80,519 lb)
A, freight mode .... 48,400 kg (106,705 lb)
B, freight mode .... 48,475 kg (106,870 lb)
Max T-O weight: A, B .... 170,500 kg (375,900 lb)
Max landing weight: A, B .... 140,000 kg (308,650 lb)
Max zero-fuel weight: A, B .... 130,000 kg (286,600 lb)

WEIGHTS AND LOADINGS (basic Freighter variant of -600R):
Manufacturer's weight empty:
A .... 78,335 kg (172,700 lb)
B .... 78,250 kg (172,510 lb)
Operating weight empty: A .... 79,050 kg (174,275 lb)
B .... 78,980 kg (174,120 lb)
Max payload (structural):
A, range mode .... 50,950 kg (112,325 lb)
B, range mode .... 51,020 kg (112,480 lb)
A, payload mode .... 54,750 kg (120,705 lb)
B, payload mode .... 54,820 kg (120,855 lb)
Max T-O weight: A, B:
range mode .... 170,500 kg (375,900 lb)
payload mode .... 165,100 kg (363,980 lb)
Max landing weight: A, B:
range mode .... 140,000 kg (308,650 lb)
payload mode .... 140,600 kg (309,970 lb)
Max zero-fuel weight: A, B:
range mode .... 130,000 kg (286,600 lb)
payload mode .... 133,800 kg (294,980 lb)

PERFORMANCE:
Range with max (structural) payload, allowances for 30 min hold at 460 m (1,500 ft) and 200 n mile (370 km; 230 mile) diversion:
A, B, range mode .... 2,650 n miles (4,908 km; 3,050 miles)
A, B, payload mode .... 1,900 n miles (3,519 km; 2,186 miles)

UPDATED

AIRBUS A320

TYPE: Twin-jet airliner.

DEVELOPMENT MILESTONES

A320-100	
Launched	23 Mar 84
First flight	22 Feb 87
Certification	26 Feb 88
First delivery (Air France)	28 Mar 88
Subsequent versions	
A320-200	
Certification	8 Nov 88

PROGRAMME: Launched 23 March 1984; four-aircraft development programme (first flight 22 February 1987 by F-WWAI); JAA (UK/French/German/Dutch) certification of A320-100 with CFM56-5 engines, for two-crew operation, awarded 26 February 1988; first deliveries (Air France and British Airways) 28 and 31 March 1988 respectively; JAA certification of A320-200 with CFM56-5s received 8 November 1988, followed by FAA type approval for both models 15 December 1988; certification with V2500 engines (first flown 28 July 1988) received 20 April (JAA) and 6 July 1989 (FAA), deliveries with this power plant (to Adria Airways) beginning 18 May 1989; FAA approved common type rating on A320 and A321 without further training in early 1994; 500th A320 delivered, 20 January 1995, to United Airlines; 1,000th member of family (an A319) and 1,001st (A320 for United Airlines) were delivered 15 April 1999. Chosen by Thales as platform for its competitor in the South Korean Air Force's E-X AEW&C requirement. In 2000, Airbus agreed to transfer A320 assembly from Toulouse to Hamburg, where A319 and A321 already produced. EASA approval for 180-minute ETOPS granted in April 2004.

Cabin innovations unveiled in 2004 included a prototype cinema-style fold-up seat for the A320 family, designed to allow passengers to stow baggage in the overhead bins without blocking the aisle. It was hoped to certify the new design by the end of that year. In September 2004, Airbus announced successful completion, aboard an A320, if the first in-flight trial of GSM mobile telephones aboard an airliner, linking them to ground-based networks via the Globalstar satcom network. It hopes to offer passengers this facility from 2006.

CURRENT VERSIONS: **A320-100:** Initial version (21 ordered). Superseded by A320-200. Max T-O weight 68,000 kg (149,940 lb); fuel capacity (usable) 24,090 litres (6,364 US gallons; 5,299 Imp gallons).

A320-200: Now called simply **A320**. Basic version from third quarter 1988; differs from initial A320-100 in having wingtip fences, wing centre-section fuel tank and higher maximum T-O weights.

Detailed description applies to A320-200.

MPA 320: Proposed maritime patrol version as replacement for Dassault-Breguet Atlantic 1; see AMC entry.

A320 in the evocative colour scheme of Air New Zealand *NEW*/1044675

Airbus A320 of tour operator Thomas Cook *NEW*/1044674

A320 research: An A320 was used for riblet research 1989–91. In mid-1998 Airbus Industrie began testing an experimental laminar-flow fin on an A320; air is sucked through small holes in the leading-edge to reduce drag and save fuel.

A320 Freighter, Convertible and Quick-Convertible: Freight variants under consideration by Airbus and Hindustan Aeronautics; cargo capacities would be 20,000 kg (44,090 lb), 22,000 kg (48,500 lb) and 30,000 kg (66,140 lb), respectively. EADS EFW began feasibility studies in 2001 into producing freight versions of the A320 and A321; considering the start of a conversion line in 2008.

A318: Shortened version of A320; described separately.

A319: Shortened version of A320; described separately.

A321: Stretched version of A320; described separately.

AIRBUS A320 MARKS AND VARIANTS

Series	Mark	Power Plant	FAA Certification
100	111	CFM56-5A1	15 Dec 88
200	211	CFM56-5A1	15 Dec 88
200	212	CFM56-5A3	26 Nov 90
200	214	CFM56-5B4 or 5B4/P or 5B4/2P	12 Dec 96
200	231	V2500-A1	6 Jun 89
200	232	V2527-A5	12 Nov 93
200	233	V2527E-A5	17 Nov 95

Series	Variant	MTOW (kg)
100	000	68,000
200	000	73,500
200	001	68,000
200	003	75,500
200	007	77,000
200	008	73,500
200	009	75,500
200	010	77,000
200	011	75,500
200	012	77,000
200	013	71,500
200	014	73,500

Notes: Variant parameters may include differences in data other than max T-O weight. Except as constrained by Series, Variants are applicable to several engine options.

CUSTOMERS: Total 1,765 sold by 30 September 2004, of which 1,323 then delivered.

COSTS: Average cost of A320-200 is US$53.7 million, depending on choice of engines, customisation and design weights (2001).

DESIGN FEATURES: First subsonic commercial aircraft to have composites for major primary structures, and centralised maintenance system; advanced-technology wings have 25° sweepback at quarter-chord, 5° 6′ 36″ dihedral plus experience from A310 and significant commonality with other Airbus aircraft where cost-effective; 6° tailplane dihedral.

FLYING CONTROLS: A320 is first subsonic commercial aircraft equipped for fly-by-wire (FBW) control throughout entire normal flight regime, and first to have sidestick controller (one for each pilot) instead of control column and aileron wheel. Thales/SFENA digital FBW system features five main computers and operates, via electrical signalling and hydraulic jacks, all primary and secondary flight controls; pilot's pitch and roll commands are applied through sidestick controller via two different types of computer; these have redundant architecture to provide safety levels at least as high as those of mechanical systems they replace; system incorporates flight envelope protection features to a degree that cannot be achieved with

AIRBUS A320 ORDERS
(at 1 October 2004)

Customer	Qty
ACES	8
Adria Airways	5
Aer Lingus	11
Aero Lloyd	4
Aeroflot	1
Aerostar	12
Air Caledonie International	1
Air Canada	28
Air China Zhejiang Company	3
Air Deccan	2
Air France	30
Air Inter Europe	22
Air Jamaica	4
Air Malta	2
Air New Zealand	5
Alitalia	11
All Nippon Airways	28
America West Airlines	31
Ansett Australia	19
Asiana Airlines	2
Austrian Airlines	8
Boullioun Aviation Services	21
British Airways	32
British Midland Airways	6
Canadian Airlines International	2
China Aviation Supplies	6
China Eastern Airlines	30
China Northwest Airlines	13
China Southern Airlines	35
CIT Group	39
Condor Flugdienst	12
Croatia Airlines	2
Cyprus Airways	8
Debis AirFinance	16
Dragonair	5
Edelweiss Air AG	3
Egyptair	12
Finnair	9
First Choice Airways	4
Flightlease	9
GATX/CL AIR	35
GATX/Flightlease	2
GB Airways	7
GECAS	79
GPA	51
Gulf Air	14
Iberia	56
Iberworld	2
ILFC	200
Indian Airlines	31
JetBlue Airways	173
Kawasaki Leasing International	8
Kuwait Airways	4
LAN Chile	14
Lotus Airline	1
LTU	4
Lufthansa	37
Mexicana	16
Monarch Airlines	2
MyTravel Airways	6
Northwest Airlines	80
Nouvelair	2
ORIX Corporation	24
Philippine Airlines	4
Qantas	20
Qatar Airways (incl Amiri Flight)	11
RBS Aviation	5
Royal Jordanian	3
SALE	45
Sabena	3
Shorouk Air	2
Sichuan Airlines	2
Silkair	10
South African Airways	22
Spanair	11
SriLankan Airlines	2
Sudan Airways	1
Swissair	17
Syrian Arab Airlines	6
TACA	40
TAM	30
TAP-Air Portugal	5
Thomas Cook	2
TransAsia Airways	5
Tunis Air	12
United Airlines	117
US Airways	30
private customer	1
undisclosed	23
undisclosed cancellations	(−13)
Total	**1,765**

conventional mechanical control systems and its computers will not allow aircraft's structural and aerodynamic limitations to be exceeded; even if pilot holds sidestick fully forward, it is impossible to go beyond aircraft's maximum operating speed (VMO) for more than a few seconds; if pilot holds sidestick fully back, aircraft is controlled to an 'alpha

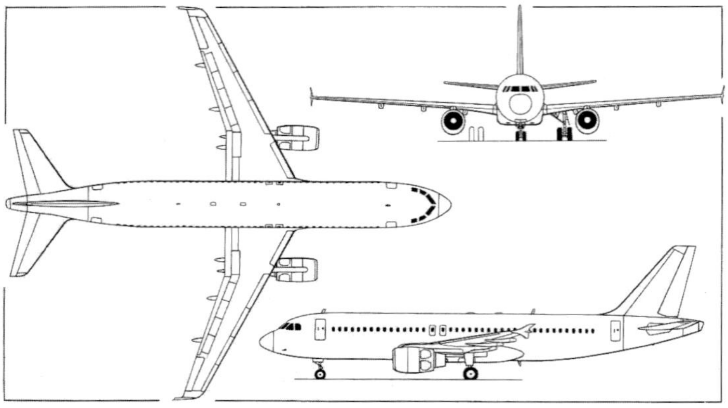

Airbus A320 twin-turbofan single-aisle 150/179-seat airliner *(Jane's/Dennis Punnett)* 0507148

floor' angle of attack, a safe airspeed above stall and throttles opened automatically to ensure positive climb. Nor is it possible to exceed *g* limits while manoeuvring. If a bank angle of more than 30° is commanded with the sidestick, the bank angle is automatically returned to 30° when pressure is released.

Fly-by-wire system controls ailerons, elevators, spoilers, flaps and leading-edge slats; rudder movement and tailplane trim connected to FBW system, but also signalled mechanically when used to provide final back-up pitch and yaw control, which suffices for basic instrument flying. Each wing has five-segment leading-edge slats (one inboard, four outboard of engine pylon), two-segment Fowler trailing-edge flaps, and five-segment spoilers forward of flaps; all 10 spoilers used as lift dumpers, inner six as airbrakes, outer eight and ailerons for roll control and outer four and ailerons for gust alleviation; slat and flap controls by Liebherr and Lucas.

STRUCTURE: Generally similar to A310, but with AFRP for fuselage belly fairing skins; GFRP for fin leading-edge and fin/fuselage fairing; CFRP for wing fixed leading/ trailing-edge bottom access panels and deflectors, trailing-edge flaps and flap track fairings, spoilers, ailerons, fin (except leading-edge), rudder, tailplane, elevators, nosewheel/mainwheel doors, and main gear leg fairing doors. A320 was first airliner to go into production with CFRP tailplane.

Airbus France builds entire front fuselage (forward of wing leading-edge), cabin rear doors, nosewheel doors, centre wing box and engine pylons, and is responsible for final assembly; centre and rear fuselage, tailcone, wing flaps, fin, rudder and commercial furnishing undertaken by Airbus Deutschland; Airbus UK builds main wings, including ailerons, spoilers and wingtips, and main landing gear leg fairings; Belgian consortium Belairbus produces leading-edge slats; Airbus España responsible for tailplane, elevators, mainwheel doors, and sheet metal work for parts of rear fuselage; Mitsubishi builds wingroot shroud box under Airbus UK subcontract; AVIC I of China provides wing components and signed an MoU in November 2000 to increase its participation, possibly leading to complete wing production by 2007; GKN Aerospace providing cargo door actuators from January 2002. Final assembly undertaken at Toulouse until 2000; now at Hamburg.

LANDING GEAR: Hydraulically retractable tricycle type, with twin wheels and oleo-pneumatic shock-absorber on each unit (four-wheel main-gear bogies, for low-strength runways, optional); Dowty main units retract inward into wing/body fairing; steerable Messier-Bugatti nose unit retracts forward; nosewheel steering angle ±75° (effective turning angle ±70°). Tyre size 46×16 or 46×17.0R20 (30 ply) on main gear and 30×8.8 or 30×8.8-R15 (16 ply) on nose gear; optional tyres for main gear are 49×17 or 49×17R20 or 49×19R20 or 46×16–20 or 49×19–20. Tyres for main-gear bogie option are 915×300R16 or 36×11 or 46×17.0R20. Carbon brakes standard. Minimum ground turning radius 13.80 m (45 ft 3¼ in) about nosewheels, 22.90 m (75 ft 1½ in) about mainwheels. Minimum width of pavement for 180° turn 23.1 m (75 ft 9½ in).

POWER PLANT: Two turbofans. Options comprise IAE V2500-A1, V2527-A5 or V2527E-A5, all 110.3 kN (24,800 lb st); CFM International CFM56-5A1 of 111.2 kN (25,000 lb st); CFM-56-5A3 of 117.9 kN (26,500 lb st); or CFM56-5B4, -5B4/P or -5B4/2P of 120.1 kN (27,000 lb st). Nacelles by Rohr Industries; thrust reversers by Hispano-Suiza (pivoting door type) for CFM56 engines, by IAE (cascade type) for V2500s. Dual-channel FADEC system standard on each engine.

Standard usable fuel capacity in wing and wing centre-section tanks is 23,860 litres (6,303 US gallons; 5,249 Imp gallons); further 82 litres (21.6 US gallons; 18.0 Imp gallons unusable. One or two additional centreline tanks (ACTs), each holding 2,900 litres (766 US gallons; 638 Imp gallons); can be fitted in rear underfloor baggage/cargo hold.

ACCOMMODATION: Standard crew of two on flight deck, with one (optionally two) forward-facing folding seats for additional crew members; seats for four cabin attendants. Single-aisle main cabin has seating for up to 179 (FAR) or 180 (JAR) passengers, depending upon layout, with locations at front and rear of cabin for galley(s) and lavatory(ies). Multiple customer choice of four-, five- and six-abreast layouts and standard or double-width aisle. Typical two-class layout has 12 seats four-abreast at 91.5 cm (36 in) pitch in 'super first' and 138 six-abreast at 81 cm (32 in) pitch economy class; alternative 152 six-abreast seats (84 business + 68 economy) at 86 and 78 cm (34 and 31 in) pitch respectively; single-class economy layout could offer 164 seats at 81 cm (32 in) pitch, or up to 179 in high-density configuration. Compared with existing single-aisle aircraft, fuselage cross-section is significantly increased, permitting use of wider triple seats to provide higher standards of passenger comfort; five-abreast business class seating provides standard equal to that offered as first class on major competitive aircraft. In addition, wider aisle permits quicker turnrounds. Overhead stowage space superior to that available on existing aircraft of similar capacity, and provides ample carry-on baggage space; best use of underseat space for baggage is provided by improved seat design and optimised positioning of seat rails.

Passenger doors at front and rear of cabin on port side, forward one having optional integral airstairs; service door opposite each of these on starboard side. Two overwing emergency exits each side. Fuselage double-bubble cross-section provides increased baggage/cargo hold volume and working height, and ability to carry seven containers derived from standard interline LD3 type. As base is same as that of LD3, all existing wide-body aircraft and ground handling equipment can accept these containers without modification. Forward and rear underfloor baggage/cargo holds, plus overhead lockers; with 164 seats, overhead stowage space per seat is 0.06 m^3 (2.0 cu ft). Mechanised cargo loading system will allow up to seven LD3-based containers to be carried in freight holds (three forward and four aft). Additional bulk cargo hold at rear, underfloor.

SYSTEMS: Liebherr/ABG-Semca air conditioning, Hamilton Sundstrand/Nord-Micro pressurisation. Honeywell 36-300 APU standard; Honeywell 131-9(A) or APIC APS 3200 available as standard options. All are interchangeable on A318/319/320/321 family. Primary electrical system powered by two Hamilton Sundstrand 90 kVA constant frequency generators, providing 115/200 V three-phase AC at 400 Hz; third generator of same type, directly driven at constant speed by APU, can be used during ground operations and, if required, during flight. Hydraulic system pressure 207 bar (3,000 lb/sq in).

AVIONICS: *Flight:* Fully equipped ARINC 700 digital avionics including advanced digital automatic flight control and flight management systems; AFCS integrates functions of SFENA autopilot and Honeywell FMS; Honeywell air data and inertial reference system.

Instrumentation: Each pilot has two Thales/VDO electronic flight instrumentation system (EFIS) displays (primary flight display and navigation display); PFD was first on an airliner to incorporate speed, altitude and heading. Between these two pairs of displays are two Thales/VDO electronic centralised aircraft monitor (ECAM) displays developed from the ECAM systems on A310 and A300-600; upper display incorporates engine performance and warning, lower display carries warning and system synoptic diagrams.

DIMENSIONS, EXTERNAL:
- Wing span ... 34.09 m (111 ft 10 in)
- Wing chord at root ... 6.10 m (20 ft 0 in)
- Wing aspect ratio ... 9.5
- Length overall ... 37.57 m (123 ft 3 in)
- Fuselage: Max width ... 3.94 m (12 ft 11 in)
  - Max depth ... 4.14 m (13 ft 7 in)
- Height overall ... 11.76 m (38 ft 7 in)
- Tailplane span ... 12.45 m (40 ft 10 in)
- Wheel track (c/l of shock-struts) ... 7.59 m (24 ft 11 in)
- Wheelbase ... 12.65 m (41 ft 6 in)
- Passenger doors (port, forward and rear), each:
  - Height ... 1.85 m (6 ft 1 in)
  - Width ... 0.81 m (2 ft 8 in)
  - Height to sill ... 3.415 m (11 ft 2½ in)
- Service doors (stbd, forward and rear), each ... as corresponding passenger doors
- Overwing emergency exits (two port and two stbd), each:
  - Height ... 1.02 m (3 ft 4¼ in)
  - Width ... 0.51 m (1 ft 8 in)
- Underfloor baggage/cargo hold doors (stbd, forward and rear),
  - each: Height ... 1.25 m (4 ft 1¼ in)
  - Width ... 1.82 m (5 ft 11½ in)

DIMENSIONS, INTERNAL:
- Cabin, excl flight deck: Length ... 27.50 m (90 ft 2¾ in)
  - Max width ... 3.68 m (12 ft 1 in)
  - Max height ... 2.16 m (7 ft 1 in)
- Aisle width: standard ... 0.48 m (1 ft 7 in)
  - option 1 ... 0.58 m (1 ft 11 in)
  - option 2 ... 0.635 m (2 ft 1 in)
  - option 3 ... 0.69 m (2 ft 3 in)
- Underfloor baggage/cargo holds:
  - Length: forward ... 4.85 m (15 ft 11 in)
    - rear ... 9.80 m (32 ft 2 in)
  - Max width ... 2.63 m (8 ft 7½ in)
  - Max height ... 1.24 m (4 ft 1 in)
  - Volume: forward ... 13.30 m^3 (469 cu ft)
    - rear ... 24.15 m^3 (853 cu ft)

AREAS:
- Wings, gross ... 122.40 m^2 (1,317.5 sq ft)
- Ailerons (total) ... 2.74 m^2 (29.49 sq ft)
- Trailing-edge flaps (total) ... 21.10 m^2 (227.12 sq ft)
- Leading-edge slats (total) ... 12.64 m^2 (136.06 sq ft)
- Spoilers (total) ... 8.64 m^2 (93.00 sq ft)
- Airbrakes (total) ... 2.35 m^2 (25.30 sq ft)
- Vertical tail surfaces (total) ... 21.50 m^2 (231.4 sq ft)
- Horizontal tail surfaces (total) ... 31.00 m^2 (333.7 sq ft)

WEIGHTS AND LOADINGS (typical 150-passenger configuration. A: CFM56-5B4/P engines, B: V2527-A5s):
- Operating weight empty: A ... 42,100 kg (92,815 lb)[1]
  - B ... 42,482 kg (93,657 lb)
- Baggage capacity:
  - forward hold ... 3,402 kg (7,500 lb)
  - rear hold ... 4,536 kg (10,000 lb)
  - bulk hold ... 1,497 kg (3,300 lb)
- Max payload: A ... 18,633 kg (41,079 lb)
  - B ... 18,518 kg (40,825 lb)
- Max fuel ... 19,159 kg (42,238 lb)
- Max T-O weight: basic ... 73,500 kg (162,040 lb)
  - option 1 ... 75,500 kg (166,445 lb)
  - option 2 ... 77,000 kg (169,755 lb)
- Max ramp weight: basic ... 73,900 kg (162,920 lb)
  - option 1 ... 75,900 kg (167,330 lb)
  - option 2 ... 77,400 kg (170,635 lb)
- Max landing weight: basic ... 64,500 kg (142,195 lb)
  - options 1 and 2 ... 66,000 kg (145,505 lb)
- Max zero-fuel weight: basic ... 61,000 kg (134,480 lb)
  - options 1 and 2 ... 62,500 kg (137,790 lb)
- Max wing loading:
  - basic ... 599.5 kg/m^2 (122.79 lb/sq ft)
  - option 1 ... 615.8 kg/m^2 (126.13 lb/sq ft)
  - option 2 ... 628.1 kg/m^2 (128.64 lb/sq ft)
- Max power loading (V2527 engines):
  - basic ... 312 kg/kN (3.06 lb/lb st)
  - option 1 ... 320 kg/kN (3.14 lb/lb st)
  - option 2 ... 327 kg/kN (3.20 lb/lb st)

[1] *Options raise OWE to a maximum of 43,000 kg (94,800 lb)*

PERFORMANCE (engines A and B as for Weights and Loadings, C: CFM56-5A3/5B4, D: 77,000 kg; 169,755 lb max T-O weight):
- Max operating Mach No. (MMO) ... 0.82
- Optimum cruising speed ... M0.78
- Max operating speed ... 350 kt (648 km/h; 403 mph) IAS
- Max rate of climb at S/L ... 152 m (500 ft)/min
- Initial cruise altitude ... 11,280 m (37,000 ft)
- Max certified altitude ... 12,130 m (39,800 ft)

Service ceiling, OEI .... 5,945 m (19,500 ft)
T-O distance at S/L, ISA + 15°C: A .... 1,960 m (6,430 ft)
B .... 1,950 m (6,400 ft)
C .... 2,180 m (7,155 ft)
D .... 2,250 m (7,385 ft)
Landing distance at max landing weight:
A, B, C, D .... 1,490 m (4,890 ft)
Runway ACN (flexible runway, category B):
twin-wheel, standard 45×16R20 tyres 41
four-wheel bogie option, 36×11-16 Type VII or 900×300-R16 22
Range with 150 passengers and baggage in two-class layout, FAR domestic reserves and 200 n mile (370 km; 230 mile) diversion:
basic: A .... 2,592 n miles (4,800 km; 2,982 miles)
B .... 2,596 n miles (4,807 km; 2,987 miles)
option 1: A .... 2,800 n miles (5,185 km; 3,222 miles)
B .... 2,871 n miles (5,317 km; 3,303 miles)
option 2:
A, C .... 3,045 n miles (5,639 km; 3,504 miles)
B .... 3,065 n miles (5,676 km; 3,527 miles)

OPERATIONAL NOISE LEVELS (ICAO Annex 16, Chapter 3; A, B and C as for performance):
T-O: A .... 88.0 EPNdB (91.5 limit)
B .... 84.8 EPNdB (91.5 limit)
C .... 85.4 EPNdB (91.5 limit)
Sideline: A .... 94.4 EPNdB (96.8 limit)
B .... 92.8 EPNdB (96.8 limit)
C .... 93.8 EPNdB (96.8 limit)
Approach: A .... 96.4 EPNdB (100.5 limit)
B .... 95.9 EPNdB (100.5 limit)
C .... 96.0 EPNdB (100.5 limit)

UPDATED

AIRBUS A318

TYPE: Twin-jet airliner.

PROGRAMME: Short-bodied version of A319 (itself a truncated A320). Formally announced at Farnborough Air Show, 7 September 1998, although known to be under consideration (as A319M5) since late 1997 when AVIC/AIA/STPL AE316/AE317 venture became uncertain. Smallest aircraft in Airbus family; programme launched 26 April 1999 with orders, commitments and options for 109 aircraft; launch customer Air France (via ILFC). Final assembly at Hamburg, resulting in some A319 production being moved to Toulouse. First metal cut November 2000; final assembly began 9 August 2001; three scheduled for completion in 2002.

Prototype (F-WWIA; 1,599th of A320 family) made maiden flight 15 January 2002, powered by PW6000 engines. Public debut at ILA, Berlin, 6 to 12 May 2002. Second prototype, F-WWIB, flew 3 June 2002; shown at Farnborough July 2002. Certification originally planned for October and service entry December 2002; however, delays in PW6000 programme caused revised schedule; as a result, F-WWIA re-engined with CFM56-5B/Ps, making maiden flight in this form on 29 August 2002 and receiving JAA certification 23 May 2003. FAA approval came in following month, preceding initial deliveries to Frontier Airlines (N80IFR, 22 July) and Air France (F-GUGA, 9 October 2003). Second prototype allocated to development of the PW6000-powered version, certification of which now anticipated in November 2005.

CURRENT VERSIONS: **A318-100:** Basic version.

First of the 15 A318s ordered by Air France ***NEW***/1044687

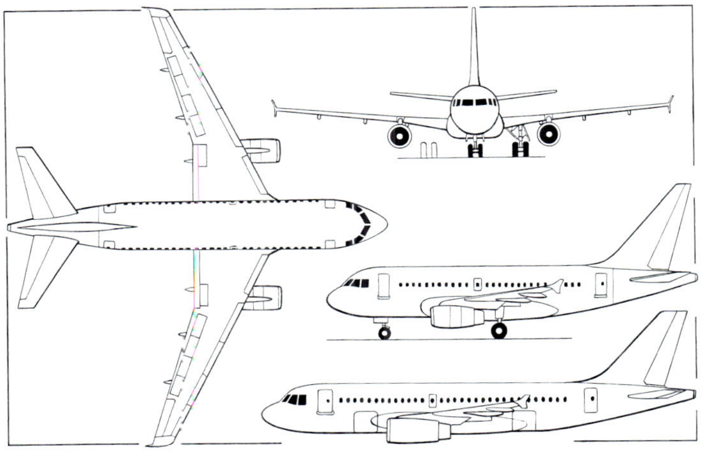
General arrangement of the Airbus A318 with additional side view (lower) of A319 *(James Goulding)* 0121271

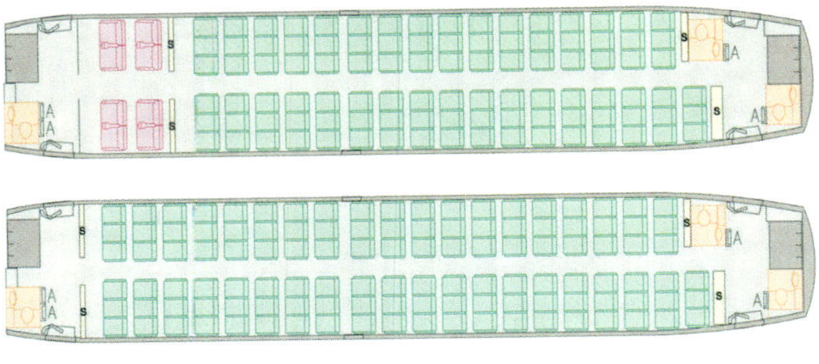
Airbus A318 representative cabin layouts for 107 seats (top) and 117 seats A: attendant's seat, S: screen 0044887

AIRBUS A318 MARKS AND VARIANTS

Series	Mark	Power Plant	FAA Certification
100	111	CFM56-5B8/P	4 Jun 03
100	112	CFM56-5B9/P	4 Jun 03

Series	Variant	MTOW (kg)
100	000	59,000
100	001	61,500
100	002	63,000
100	003	64,500
100	004	66,000
100	005	68,000
100	006	56,000
100	007	61,000
100	008	64,000

Notes: Variant parameters may include differences in data other than max T-O weight. Except as constrained by Series, Variants are applicable to several engine options.

CUSTOMERS: First customer, International Lease Finance Corporation (ILFC), signed MoU for up to 30 aircraft on 17 November 1998, subject to launch of project, followed by firm order 30 April 1999. First conditional airline customer, TWA, signed LoI on 9 December 1998 for 50, confirming 25 by firm order 14 December 1999 but cancelled when TWA taken over by American Airlines; other early customers include Egyptair (three on 17 July 1999 — first firm airline order, followed by two further examples, but subsequently all cancelled). Total orders 82, of which 12 delivered, at 1 October 2004. Despite slow start, in 2004 Airbus was predicting long-term sales of 300 to 400 A318s.

AIRBUS A318 ORDERS
(at 1 October 2004)

Customer	Engines	Qty
Air France	CFM	15
America West Airlines	PW	15
Frontier Airlines	CFM	5
GECAS	CFM	28
ILFC	CFM	15
Tarom	TBD	4
Total		**82**

COSTS: Programme cost estimated as US$300 million. Unit price US$41.7 million (2001).

DESIGN FEATURES: Shorter fuselage and taller fin than rest of A320 family, with which it has 95 per cent commonality, including A319 wing, pylon and interface. Laser welding (rather than riveting) used on lower fuselage to reduce costs and weight; first use of this technology on airliner. Common pilot type rating with A319, A320 and A321.

FLYING CONTROLS: As A320.

STRUCTURE: As A320, but fuselage 4½ frames (2.39 m; 7 ft 10 in) shorter than A319; 0.79 m (2 ft 7 in) removed forward and 1.60 m (5 ft 3 in) aft of wing. Redesigned cargo doors. Fin has tip extension. Assembled in Germany by Airbus Deutschland. AVIC of China will have share of production work to offset AE31X project cancellation.

LANDING GEAR: As A320. Minimum ground turning radius 11.30 m (37 ft 1 in) about nosewheels, 19.50 m (63 ft 11¾ in) about mainwheels.

POWER PLANT: Designed originally for two 98.3 kN (22,100 lb st) Pratt & Whitney PW6122 or 105.9 kN (23,800 lb st) PW6124 turbofans with clamshell-type thrust reversers; A318 is launch aircraft for PW6000 family. Alternative (and first in-service) power plant is 96.1 kN (21,600 lb st) CFM International CFM56-5B8/P or 103.6 kN (23,300 lb st) CFM International CFM56-5B9/P, officially announced 4 August 1999. Thrust reversing and deflection, fuel tank capacities and locations as A320.

ACCOMMODATION: Two flight crew plus three cabin crew. Typical passenger capacity eight first class and 99 second class with 96/81 cm (38/32 in) seat pitch, or 117 in single-class seating; high-density seating for 129 passengers; certified for up to 136. Front and rear passenger doors on port side; service door opposite each on starboard side. Overwing emergency exit each side. A318 will not carry containerised freight due to smaller baggage doors (of which size reduced to maintain same engine nacelle clearance for loading vehicles as A319).

SYSTEMS: Generally as A320; new-generation cabin intercom data system (CIDS).

AVIONICS: Generally as A320, but with new LCD screens.

DIMENSIONS, EXTERNAL: As A320 except:
Length overall .... 31.45 m (103 ft 2 in)
Height overall .... 12.55 m (41 ft 2 in)
Wheelbase .... 10.25 m (33 ft 7½ in)
Baggage doors (each): Height .... 1.24 m (4 ft 1 in)
Width .... 1.28 m (4 ft 2½ in)

DIMENSIONS, INTERNAL: As A320 except:
Cabin: Length .... 21.38 m (70 ft 1¾ in)
Underfloor baggage/cargo holds:
Length: forward .... 2.57 m (8 ft 5in)
rear .... 6.07 m (19 ft 11 in)
Volume: forward .... 6.51 m³ (230 cu ft)
rear .... 14.70 m³ (519 cu ft)

AREAS: As A320 except:
Fin and rudder (total, incl tab) ... 21.50 m^2 (231.4 sq ft)

WEIGHTS AND LOADINGS:
Typical operating weight empty ... 39,035 kg (86,057 lb)
Baggage capacity:
forward hold ... 1,614 kg (3,558 lb)
rear hold ... 2,131 kg (4,698 lb)
bulk hold ... 1,372 kg (3,025 lb)
Max payload ... 13,965 kg (30,788 lb)
Max T-O weight: basic ... 59,000 kg (130,075 lb)
option ... 68,000 kg (149,915 lb)
Max ramp weight: basic ... 59,400 kg (130,955 lb)
option ... 68,400 kg (150,800 lb)
Max landing weight: basic ... 56,000 kg (123,460 lb)
option ... 57,500 kg (126,765 lb)
Max zero-fuel weight: basic ... 53,000 kg (116,845 lb)
options ... 54,500 kg (120,150 lb)
Max wing loading: basic ... 481.2 kg/m^2 (98.57 lb/sq ft)
option ... 555.6 kg/m^2 (113.79 lb/sq ft)
Max power loading, CFM56-5B8/P engines:
basic ... 307 kg/kN (3.01 lb/lb st)
option ... 354 kg/kN (3.47 lb/lb st)

PERFORMANCE:
Max operating Mach No. (MMO) ... 0.82
Max operating speed ... 350 kt (648 km/h; 403 mph) IAS
Max certified altitude ... 12,200 m (40,000 ft)
Runway ACN: flexible Cat. B runway: basic MTOW ... 29
option 3 MTOW ... 34
T-O run at basic MTOW, 500 n mile (925 km; 575 mile) mission, elevation 610 m (2,000 ft), ISA +15°C:
PW6122 ... 1,670 m (5,479 ft)
CFM56 ... 1,630 m (5,348 ft)
Landing run at MLW, S/L, ISA:
PW6122 ... 1,332 m (4,370 ft)
CFM56 ... 1,355 m (4,446 ft)
Range with 107 passengers, FAR domestic reserves, 200 n miles (370 km; 230 miles) diversion, max payload:
basic MTOW:
PW6122 ... 1,462 n miles (2,707 km; 1,682 miles)
CFM56 ... 1,455 n miles (2,694 km; 1,674 miles)
option 1 MTOW:
PW6122 ... 1,960 n miles (3,630 km; 2,255 miles)
CFM56 ... 1,980 n miles (3,667 km; 2,278 miles)
option 2 MTOW:
PW6122 ... 2,820 n miles (5,222 km; 3,245 miles)
CFM56 ... 2,880 n miles (5,333 km; 3,314 miles)
option 3 MTOW ... 3,250 n miles (6,019 km; 3,740 miles)

OPERATIONAL NOISE LEVELS (estimated):
T-O with PW6122 at MTOW ... 79.7 EPNdB
Sideline ... 90.4 EPNdB
Approach ... 89.7 EPNdB

UPDATED

AIRBUS A319

TYPE: Twin-jet airliner.

PROGRAMME: Short-fuselage A320. Airbus Board officially authorised start of sales 22 May 1992; programme launched 10 June 1993. Final assembly of first aircraft (F-WWDB, the 546th of the A319/320/321 family) began 23 March 1995; rolled out at Hamburg 24 August; first flight (with CFM56-5A engines) on 29 August; second aircraft (F-WWAS, No. 572) flew (with CFM56-5A engines and commercial interior) 31 October 1995; 650 hour flight test programme resulted in initial certification (CFM56-5B) on 10 April 1996 (CFM56-5A and V2500-A5 followed); first delivery, HB-IPV to ILFC on 25 April 1996 and immediately to Swissair on 30 April, flying first service on 8 May; F-WWAS re-engined with IAE V2524 engines, first flight 22 May 1996 and certified on 18 December 1996; other 1996 first receipts included Air Inter (F-GPMA 21 June); Lufthansa (D-AILA 19 July); and Air Canada (C-FYIY 12 December). JAA 120 minute ETOPS approval granted 14 February 1997 and extended to ACJ in December 2000; 180-minute ETOPS (all versions) granted by EASA April 2004. 1,000th member of A320 family (an A319 for Air France via ILFC) delivered 15 April 1999.

A319 airliner of low-cost operator easyJet ***NEW***/1044684

A319 Long Range of German operator PrivatAir ***NEW***/1044682

CURRENT VERSIONS: **A319-100:** Baseline version.

Description applies to A319-100.

ACJ: Airbus Corporate Jetliner. Announced at 1997 Paris Air Show. Standard aircraft will carry up to 40 passengers over a range of 6,300 n miles (11,667 km; 7,250 miles), cruising at 12,500 m (41,000 ft) at speeds of up to M0.82. Certified as a commercial airliner to Cat. IIIb landing criteria and 180-minute ETOPS, and will convert easily to airliner configuration. First customer, announced on 18 December 1997, is Mohamed Abdulmohsin Al Kharafi of Kuwait. First aircraft (G-OMAK, 913th of A320 family), type A319-132, first flew 12 November 1998 and delivered to Jet Aviation, Switzerland, for outfitting on 31 December 1998; customer receipt 30 October 1999, operated by Twinjet Aircraft for Al Kharafi. JAR certification (as amendment of A319 certificate) received August 1999. Demonstrator F-WWIC first flew (as D-AVYB) 28 May 1999 and to Toulouse for trials 31 May; is 910th of A320 family, but followed c/n 913; eventually sold to UAE Air Force.

Orders and commitments totalled more than 30 by mid-2002, of which 14 then delivered; see accompanying table for latest customer list. FAA (FAR Pt 121) certification for both scheduled service and private operation in USA received October 2002.

Set world record non-stop 15 hour 13 minute flight of 6,918 n miles (12,812 km; 7,961 miles) from Santiago to Le Bourget on 16 June 1999.

Airbus announced in 2000 that production of the ACJ would be restricted to four per year until 2003; this was increased to six, due to demand. By mid-2002, six companies were recommended for outfitting: Lufthansa Technik, Hamburg; Jet Aviation, Basle; Air France Industries, Toulouse; Ozark Aircraft Systems, Bentonville, Arkansas; and EADS Sogerma, Toulouse.

A319LR: Long-range variant of ACJ: launched February 2003, when offered with 48-seat all business class interior and 35,400 litre (9,352 US gallon; 7,787 Imp gallon) fuel capacity, giving range of 4,500 n miles (8,334 km; 5,178 miles). Second aircraft for Qatar

Tunis Air Airbus A319 ***NEW***/1044683

Blue Mountain Aviation's A319 Executive is used to ferry major sports teams to important fixtures ***NEW***/1044711

AIRBUS A319 AND ACJ ORDERS
(at 1 October 2004)

Customer	Qty
Aero Services Executive	1[1]
Aeroflot	4
Air Bosna	2
Air Canada	37
Air China	17[1]
Air France	19[1]
Air Inter Europe	9
Air Mauritius	2
Al Kharafi Group	1[1]
Alitalia	12
America West Airlines	37
Austrian Airlines	7
Azerbaijan Hava Yollari	4[1]
Boullioun Aviation Services	5
Brazilian Air Force	1[1]
British Airways	36
Cebu Pacific	12
China Aviation Supplies	3
China Southern Airlines	6
CIT Group	23
Croatia Airlines	4
Cyprus Airways	2
DaimlerChrysler Aviation	1[1]
Debis AirFinance	10
Druk Air	2[1]
easyJet	120
Eurofly	1[1]
Finnair	5
French Air Force	2[1]
Frontier Airlines	27
GATX/CL AIR	1
GECAS	45
Germanwings	6
Iberia	3
ILFC	137
Independence Air	16
Italian Air Force	3[1]
JAT	8
LAN Chile	14
Lufthansa	20
Northwest Airlines	82
Qatar Airways	3[1]
Royal Thai Air Force	1[1]
SALE	1
Sabena	15
Sichuan Airlines	4
Silkair	6
South African Airways	11
Spirit Airlines	11
Swissair	6
TACA	9
TAM	13
TAP-Air Portugal	13
Tunis Air	3
UAE Air Force	1[1]
Ukrainian Air Force	1[1]
United Airlines	78
US Airways	66
Venezuelan Air Force	1[1]
Private customers	5
Total	**995**

[1] Indicates (or includes) ACJ

Airways (two-class layout for 110 passengers) is to this standard; entered service August 2003. Three more ordered by 30 September 2003: two for PrivatAir via CIT (48-seat all business class cabin) and one undisclosed.

A319 Executive: Variant of ACJ: 44-seat interior; lacks ACJ's auxiliary fuel tanks; 3,400 n mile (6,296 km; 3,912 mile) range. Three ordered by 30 September 2003: two for PrivatAir via CIT and one for Blue Moon Aviation (USA) via SALE.

CUSTOMERS: Total of 994 sold, of which 627 delivered, by 30 September 2004, including 25 ACJs.

COSTS: Total development cost estimated at US$275 million, entirely financed by Airbus Industrie. A319-100 cost approximately US$48.7 million depending on choice of engines and level of customisation (2001); ACJ cost US$36 million (2001), excluding outfitting of between US$4 million and US$10 million.

DESIGN FEATURES: Seven fuselage frames shorter than A320 (1.60 m; 5 ft 3 in forward of wing, 2.13 m; 7 ft 0 in aft); modified rear cargo hold; bulk hold door and forward overwing emergency exit deleted; derated engines; otherwise little changed. Seats 124 passengers in typical two-class layout, compared with 150 in A320 and 185 in A321; range of 3,550 n miles (6,574 km; 4,085 miles) said to be the longest in this category of airliner; common pilot type rating with A320 and A321.

FLYING CONTROLS: Same flight deck and flying control system as A320.

STRUCTURE: As A320. Assembled in Germany by EADS Deutschland Airbus, alongside the stretched A321; partner workshares rearranged to maintain overall workshare balance between France and Germany.

LANDING GEAR: As A320, but 46×16 or 46×17.0R20 main tyre options only. Minimum ground turning radius 12.10 m (39 ft 8½ in) about nosewheels, 20.60 m (67 ft 7 in) about mainwheels.

POWER PLANT: Two turbofans. Options comprise CFM56-5B5, -5B5/P or -5A4, all 97.9 kN (22,000 lb st); IAE V2522-A5 of 102.5 kN (23,040 lb st); CFM56-5B6, -5B6/P or -5B6/2P or -5A5, all 104.5 kN (23,500 lb st); IAE V2524-A5 of 108.9 kN (24,480 lb st), IAE V2527M-AS of 110.3 kN (24,800 lb st); or CFM56-5B7 or -5B7/P of 120.1 kN (27,000 lb st).

Standard (three tanks) fuel as for A320: 23,860 litres (6,303 US gallons; 5,249 Imp gallons) usable. Up to six additional centreline tanks (ACT) for maximum (nine tanks) of 40,640 litres (10,737 US gallons; 8,939 Imp gallons).

ACCOMMODATION: Typically 124 passengers (eight in 'super first' class plus 116 economy); maximum 145 passengers in all-economy configuration. Built-in airstairs in ACJ.

'Prestige' cabin option for ACJ configures forward and centre sections with dining and work areas complete with tables, lounge and bar; aft cabin has private office area and enclosed bedroom. Amenities include 64 kbyte communications links, large-screen video displays, external video cameras for security and flight progress monitoring, and additional aft cabin soundproofing.

AIRBUS A319 MARKS AND VARIANTS

Series	Mark	Power Plant	FAA Certification
100	111	CFM56-5B5 or -5B5/P	20 Jun 97
100	112	CFM56-5B6 or -5B6/P or -5B6/2P	30 Aug 96
100	113	CFM56-5A4	20 Jun 97
100	114	CFM56-5A5	20 Jun 97
100	115	CFM56-5B7 or -5B7/P	22 Oct 02
100	131	V2522-A5	20 Jun 97
100	132	V2524-A5	20 Jun 97
100	133	V2527M-A5	22 Oct 02

Series	Variant	MTOW (kg)
100	000	64,000
100	001	70,000
100	002	75,700

Notes: Variant parameters may include differences in data other than max T-O weight. Except as constrained by Series, Variants are applicable to several engine options.

DIMENSIONS, EXTERNAL: As A320 except:
Length overall .... 33.83 m (111 ft 0 in)
Wheelbase .... 11.05 m (36 ft 3 in)

DIMENSIONS, INTERNAL: As A320 except:
Cabin length (excl flight deck) .... 23.77 m (77 ft 11¾ in)
Underfloor baggage/cargo holds:
Length: forward .... 3.35 m (11 ft 0 in)
rear .... 7.67 m (25 ft 2 in)
Volume: bulk total .... 27.64 m^3 (976 cu ft)

WEIGHTS AND LOADINGS:
Typical operating weight empty:
standard .... 40,160 kg (88,537 lb)
option .... 41,203 kg (90,837 lb)
Baggage capacity: forward hold .... 2,268 kg (5,000 lb)
rear hold .... 3,020 kg (6,600 lb)
bulk hold .... 1,497 kg (3,300 lb)
Max payload .... 16,653 kg (36,714 lb)
Max T-O weight: standard .... 64,000 kg (141,095 lb)
option .... 75,500 kg (166,450 lb)
Max landing weight: standard .... 61,000 kg (134,480 lb)
option .... 62,500 kg (137,790 lb)
Max zero-fuel weight: standard .... 57,000 kg (125,665 lb)
option .... 58,500 kg (128,970 lb)
Max wing loading:
standard .... 522.9 kg/m^2 (101.09 lb/sq ft)
option .... 616.8 kg/m^2 (126.34 lb/sq ft)
Max power loading: standard, CFM56-5B5 .... 327 kg/kN (3.21 lb/lb st)
option, CFM56-5B7 .... 314 kg/kN (3.08 lb/lb st)

PERFORMANCE (A: with CFM56-5A4, CFM56-5B5 or V2522-A5; B: with CFM56-5A5, CFM56-5B6 or V2524-A5):

Max operating Mach No. (MMO) ... 0.82
Max operating speed ... 350 kt (648 km/h; 403 mph) IAS
Typical operating Mach No. ... 0.78
Max rate of climb ... 152 m (500 ft)/min
Max certified altitude: A319 ... 12,130 m (39,800 ft)
ACJ ... 12,500 m (41,000 ft)
Service ceiling, OEI ... 6,400 m (21,000 ft)
T-O distance, S/L ISA+15°C:
standard: A ... 1,720 m (5,643 ft)
option: B ... 2,640 m (8,661 ft)
Landing distance, S/L ISA: A, B ... 1,430 m (4,692 ft)
Range on normal fuel tankage with 124 passengers and baggage, FAR domestic reserves and 200 n mile (370 km; 230 mile) diversion:
A, at T-O weight 64,000 kg (141,095 lb) ... 1,813 n miles (3,357 km; 2,086 miles)
B, at T-O weight 75,500 kg (166,450 lb) ... 3,697 n miles (6,846 km; 4,254 miles)

OPERATIONAL NOISE LEVELS (A, B as above):

T-O: A ... 82.5 EPNdB
B ... 82.4 EPNdB
Sideline: A ... 92.6 EPNdB
B ... 92.2 EPNdB
Approach: A ... 93.9 EPNdB
B ... 94.2 EPNdB

UPDATED

AIRBUS A321

TYPE: Twin-jet airliner.

PROGRAMME: Stretched version of A320. Announced 22 May and launched 24 November 1989; four development aircraft; rolled out 3 March 1993, first flight with V2530 lead engine 11 March 1993 (F-WWIA), second aircraft with alternative CFM56-5B engine in May 1993; V2530-powered version received European JAA certification 17 December 1993; CFM56-5B2-powered version certified by JAA 15 February 1994; CFM56-5B1 certified by JAA 27 May 1994; first delivery (D-AIRA to Lufthansa) 27 January 1994; first service with Lufthansa 18 March 1994. A321 powered by CFM56-5Bs passed cold-weather trials in Kiruna, January 1994; first with alternative engine handed over to Alitalia 18 March 1994. JAA approval for Cat. III automatic landings achieved in December 1994. 120-minute ETOPS granted 29 May 1996 and upgraded to 180-minute approval in April 2004. The 100th A321 (Alitalia's I-BIXZ) flew 1 July 1998. A321 also recommended by Northrop Grumman as platform for Joint STARS, competing for NATO's Airborne Ground Surveillance requirement (later dropped) and by Raytheon for the Republic of Korea Air Force's E-X AEW&C competition.

CURRENT VERSIONS: **A321-100:** Initial version.

A321-200: Extended-range version, launched 12 April 1995; features reinforced structure, higher-thrust versions of existing engines and optional additional centre tank (ACT), capacity 2,900 litres (766 US gallons; 638 Imp gallons) which increases maximum T-O weight to 89,000 kg (196,210 lb) and range by 350 n miles (648 km; 402 miles). A321-200 expected to have increased market appeal on North American domestic routes; charter routes between northern and southern Europe; and on scheduled routes between Europe and Middle East. First aircraft flew 12 December 1996 at Hamburg; became G-OZBC of Monarch Airlines and delivered 24 April 1997. Higher-MTOW version, with option of second ACT, launched January 1999; first customer Spanair, for delivery September 2000.

A321CJ: Airbus Industrie reported to be considering a corporate version with additional fuel tanks; no formal plans to launch known by late 2004.

A321 Freighter: EADS EFW conducting feasibility studies into an A321 Freighter with capacity for 14 standard pallets.

CUSTOMERS: Total of 415 sold, of which 310 delivered by 1 October 2004.

COSTS: A321-2001 unit cost approximately US$65.6 million, depending on engine choice and customisation level (2001).

DESIGN FEATURES: Compared with A320, A321 has 4.27 m (14 ft 0 in) fuselage plug immediately forward of wing and 2.67 m (8 ft 9 in) plug immediately aft; pairs of wing fuel tanks unified and system simplified; other changes include local structural reinforcement of existing assemblies, slightly extended wing trailing-edge with double-slotted flaps, uprated landing gear and higher T-O weights.

STRUCTURE: As for A320 except for airframe changes noted under Design Features; front fuselage plug by Alenia and rear one by BAE Systems; final assembly and outfitting by Airbus Deutschland at Hamburg.

LANDING GEAR: Uprated, with 22 in wheel rims, 49×18.0-22 or 46×17.0R20 (30 ply) mainwheel tyres and increased energy brakes; wheels and brakes by Aircraft Braking Systems. Minimum ground turning radius 18.00 m (59 ft 0¾ in) about nosewheels, 27.60 m (90 ft 6½ in) about mainwheels.

Airbus A321 of Monarch Airlines ***NEW***/1044697

AIRBUS A321 MARKS AND VARIANTS

Series	Mark	Power Plant	FAA Certification
100	111	CFM56-5B1 or -5B1/P or -5B1/2P	20 Dec 95
100	112	CFM56-5B2 or -5B2/P	20 Dec 95
100	131	V2530-A5	20 Dec 95
200	211	CFM56-5B3/P o r-5B3/2P	18 Sep 97
200	231	V2533-A5	18 Sep 97

Series	Variant	MTOW (kg)
100	000	83,000
100	002	83,000
100	003	85,000
200	000	89,000
200	000	93,000

Notes: Variant parameters may include differences in data other than max T-O weight. Except as constrained by Series, Variants are applicable to several engine options.

POWER PLANT: Two turbofans. Options comprise: A321-100: IAE V2530-A5 of 133.0 kN (29,900 lb st); CFM56-5B1, -5B1/P or -5B1/2P of 133.4 kN (30,000 lb st); CFM56-5B2 or -5B2/P of 137.9 kN (31,000 lb st); and, for A321-200: IAE V2533-A5 of 140.6 kN (31,600 lb st); or CFM56-5B3/P or -5B3/2P of 142.3 kN (32,000 lb st).

Normal three-tank fuel capacity 23,700 litres (6,261 US gallons; 5,213 Imp gallons) usable. Fourth tank increases usable quantity to 26,600 litres (7,027 US gallons; 5,851 Imp gallons) in high pressure fuel system or 26,692 litres (7,051 US gallons 5,872 Imp gallons) in low pressure system. Fifth tank increases usable quantity to 29,684 litres (7,842 US gallons; 6,530 Imp gallons) low pressure (only).

ACCOMMODATION: Typically offers 24 per cent more seats and 40 per cent more hold volume than A320; examples are 185 passengers in two-class layout (16 'super first' class at 91 cm; 36 in seat pitch and 169 economy class at 81 cm; 32 in), or 220 passengers (certified limit) in all-economy high-density configuration. Each fuselage plug incorporates one pair of emergency exits, replacing single overwing pair of A320.

SYSTEMS: Choice of Honeywell 36-300 or APIC APS3200 APU; Honeywell 131-9(A) available from 1998; full commonality with A320 installation.

DIMENSIONS, EXTERNAL: As for A320 except:

Length overall ... 44.51 m (146 ft 0 in)
Height overall ... 11.81 m (38 ft 9 in)
Wheelbase ... 16.92 m (55 ft 6¼ in)
Emergency exits (forward stbd and rear port/stbd, each):
Height ... 1.52 m (5 ft 0 in)
Width ... 0.76 m (2 ft 6 in)
Emergency exit (forward port, usable also as passenger door): Height .. 1.85 m (6 ft 1 in)
Width ... 0.76 m (2 ft 6 in)

Airbus A321 of Alitalia ***NEW***/1044698

AIRBUS A321 ORDERS
(at 1 October 2004)

Customer	Qty
Aer Lingus	3
Aero Flight	5
Aeroflot	3
Air Canada	10
Air France	4
Air Inter Europe	7
Air Macau	1
Alitalia	23
All Nippon Airways	7
Ariana Afghan Airlines	10
Austrian Airlines	6
Boullioun Aviation Services	4
British Airways	12
British Midland Airways	6
China Eastern Airlines	4
China Northern Airlines	10
CIT Group	3
Debis AirFinance	5
Dragonair	2
Egyptair	4
Finnair	4
Flightlease	1
GATX/CL AIR	7
GB Airways	3
GECAS	12
Iberia	16
ILFC	85
Leisure International Airways	2
LTU	2
Lufthansa	26
Middle East Airlines	6
Monarch Airlines	7
MyTravel Airways	4
Onur Air	2
Qatar Airways	2
Royal Air Maroc	4
Sabena	3
SALE	5
Scandinavian Airlines System	12
Sichuan Airlines	2
Spanair	3
Spirit Airlines	4
Swissair	6
TAP-Air Portugal	3
TransAsia Airways	6
US Airways	41
Vietnam Airlines	5
undisclosed	13
Total	**415**

DIMENSIONS, INTERNAL: As A320 except:
Cabin, excl flight deck: Length ... 34.44 m (113 ft 0 in)
Underfloor baggage/cargo holds:
Length: forward ... 8.15 m (26 ft 9 in)
rear ... 11.40 m (37 ft 5 in)
Volume: forward ... 22.81 m^3 (806 cu ft)
rear ... 28.93 m^3 (1,022 cu ft)

WEIGHTS AND LOADINGS (A321-200):
Max fuel weight: three tanks ... 18,960 kg (41,800 lb)
four tanks ... 21,280 kg (46,914 lb)
five tanks ... 23,746 kg (52,350 lb)
Max T-O weight: basic ... 89,000 kg (196,210 lb)
option ... 93,000 kg (205,030 lb)
Max landing weight: basic ... 75,500 kg (166,450 ft)
option ... 77,800 kg (171,520 lb)
Max zero-fuel weight: basic ... 71,500 kg (157,630 lb)
option ... 73,800 kg (162,705 lb)

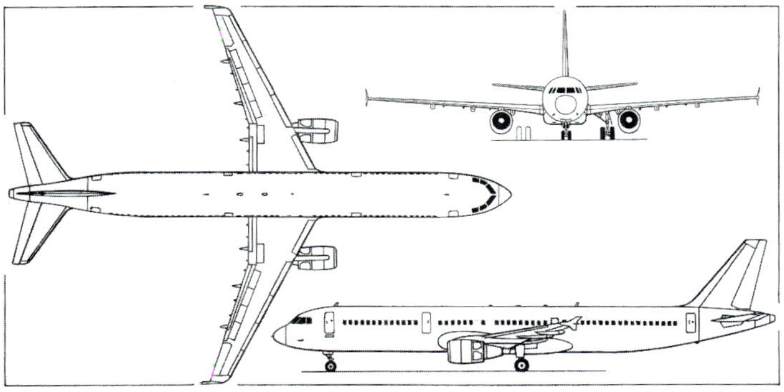

A321 stretched development of the Airbus A320 *(Jane's/Dennis Punnett)* 0507411

Max wing loading:
basic ... 727.1 kg/m^2 (148.93 lb/sq ft)
option ... 759.8 kg/m^2 (156.62 lb/sq ft)
Max power loading, CFM56-5B3:
basic ... 313 kg/kN (3.07 lb/lb st)
option ... 327 kg/kN (3.20 lb/lb st)

PERFORMANCE ((A321-100, except where indicated; typical 185-passenger layout. C1: CFM56-5B1, V: V2530-A5, C2: CFM56-5B2, C3: CFM56-5B3, V2: V2533-A5)::
Max operating Mach No. (MMO) ... 0.82
Max operating speed ... 350 kt (648 km/h; 403 mph) IAS
Typical operating Mach No. ... 0.78
Max rate of climb at S/L ... 152 m (500 ft)/min
Max certified altitude ... 12,130 m (39,800 ft)
Service ceiling, OEI: A321-100 ... 5,180 m (17,000 ft)
A321-200 ... 5,000 m (16,400 ft)
T-O distance at max T-O weight, S/L, ISA +15°C:
C1 ... 2,220 m (7,285 ft)
V ... 2,065 m (6,775 ft)
C2 ... 2,270 m (7,450 ft)
C3 ... 2,330 m (7,645 ft)
V2 ... 2,341 m (7,680 ft)
Landing distance at max landing weight:
C1, V ... 1,540 m (5,055 ft)
C2 ... 1,530 m (5,020 ft)
C3, V2 ... 1,577 m (5,175 ft)
Runway ACN (flexible runway, category B):
basic ... 48
Range on normal fuel tankage with 185 passengers and baggage at typical airline OWE, FAR domestic reserves and 200 n mile (370 km; 230 mile) diversion: A321-100:
C1, C2:
basic ... 2,138 n miles (3,959 km; 2,460 miles)
optional ... 2,250 n miles (4,167 km; 2,589 miles)
V: basic ... 2,253 n miles (4,172 km; 2,592 miles)
optional ... 2,386 n miles (4,418 km; 2,745 miles)
A321-200:
with one ACT ... 2,700 n miles (5,000 km; 3,107 miles)
with two ACT ... 3,000 n miles (5,556 km; 3,452 miles)
V2: basic ... 2,384 n miles (4,415 km; 2,743 miles)
with one ACT ... 2,714 n miles (5,026 km; 3,123 miles)

OPERATIONAL NOISE LEVELS (ICAO Annex 16, Chapter 3, estimated):
T-O: C1 ... 87.0 EPNdB (92.1 limit)
V ... 85.4 EPNdB (92.1 limit)
Sideline: C1 ... 96.2 EPNdB (97.2 limit)
V ... 94.5 EPNdB (97.2 limit)
Approach: C1 ... 96.1 EPNdB (100.9 limit)
V ... 95.4 EPNdB (100.9 limit)

UPDATED

AIRBUS A340

TYPE: Wide-bodied airliner.

PROGRAMME: Launched 5 June 1987 as parallel programme with A330, uniquely offering twin- and four-engined variants of same basic design. First flight of four-engined A340-300 25 October 1991; A340-200 and -300 certified simultaneously by 18 European joint airworthiness authorities (JAA) on 22 December 1992; first deliveries to Lufthansa on 29 January 1993 (-200) and Air France on 26 February 1993 (-300); both versions entered service March 1993; both received FAA certification 27 May 1993. A340/A330 certified by JAA for GPS satellite navigation January 1994; successful trials conducted at Toulouse in October 1994 using differential global positioning (DGPS) for fully automatic landings, including roll-out. The 200th aircraft from A330/340 assembly line flew in November 1997; 100th A340 delivered to Singapore Airlines in February 1997; 180-minute ETOPS for Rolls-Royce Trent 700 engines awarded May 1996.

A340-500 and -600 were launched on 8 December 1997, at an estimated investment cost of US$2.9 billion. First flights took place on 11 February 2002 and 23 April 2001 respectively, followed by certifications on 3 December 2002 and 29 April 2002. A340-600 entered service with Virgin Atlantic 1 August 2002. 2,000th Airbus was A340-300 for Lufthansa, delivered 18 May 1999. 400th A330/A340 family member first flown 23 March 2001.

Cathay Pacific took delivery of an A340-300 covered with 700 m^2 (7,535 sq ft) of plastic film riblets in October 1996 in a trial intended to reduce fuel consumption (average ¾ tonne per medium- or long-range flight) by reducing drag (see illustrations in 1998–99 edition).

CURRENT VERSIONS: **Initial A340-300:** Higher-capacity version, carrying up to 375 passengers (standard) or 440 (optional) and powered initially by CFM56-5C2 turbofans. Able to carry typical load of 295 passengers over distances of 7,300 n miles (13,519 km; 8,400 miles).

2004 record-setter: Singapore Airlines' A340-500 ***NEW***/1044721

A golden-topped A340-300 of Air Jamaica ***NEW***/1044722

Current A340-300: (engineering designation -300X) powered by 151.2 kN (34,000 lb st) CFM56-5C4 turbofans; maximum T-O weight 271,000 kg (597,450 lb) with optional MTOW of 275,000 kg (606,275 lb); stronger landing gear, and aerodynamic and engine refinements, plus optional additional centre tank (ACT) for increased fuel capacity compared with basic A340-300. First flight (F-WWJH; CFM56-5C4 engines) 25 August 1995; first delivery, to Singapore Airlines, 17 April 1996.

A340-300E: Enhanced version with upgraded CFM56-5C4/P engines (trial engine first flew on an A340 on 19 November 2002); optional increased MTOW of 276,500 kg (609,575 lb); new-style cabin from A340-600. Initial customer, South African Airways, ordered six in 2002; first delivery 9 March 2004.

A340-300 Awiator: Company testbed for Aircraft Wing with Advanced Technology Operation (Awiator) R & D programme to explore ways of reducing aircraft wake, drag, noise and fuel consumption, and other technologies, which could be applied to future Airbus designs. Two-year programme (to mid-2005) testing such features as enlarged winglets, modified inboard spoilers/airbrakes, turbulence sensors and various vortex devices.

Initial A340-200: Short-fuselage, longer-range version of A340-300, with same initial power plant; seating capacity 420 passengers, but more typically 303 in a two-class or 263 in three-class configuration; first flight (F-WWBA) 1 April 1992; entered service January 1993 with Lufthansa.

Longest Airbus of all: the A340-600, seen here in Iberia markings ***NEW***/1044720

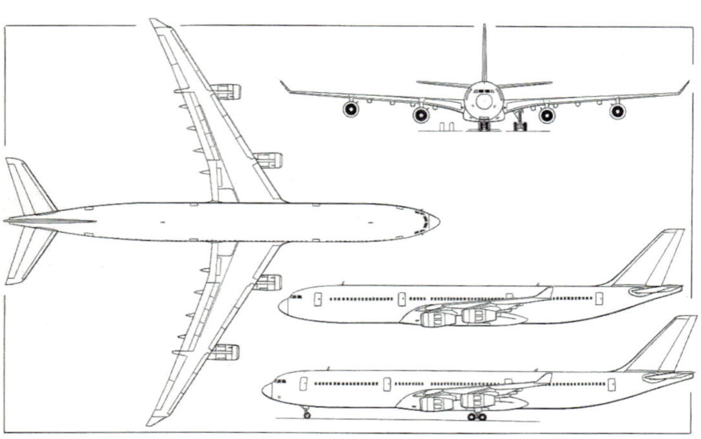

Airbus A340-300 four-turbofan long-range airliner, with additional side view (upper) of A340-200 ***(Jane's/Dennis Punnett)*** 0507149

Current A340-200: Advanced version (previously referred to as A340-8000) with additional fuel in two tanks in rear cargo hold, strengthened fuselage and wings, CFM56-5C4 engines and 275,000 kg (606,275 lb) maximum T-O weight; first order placed March 1996 by Prince Jefri of Brunei for one VVIP aircraft, which first flew 19 December 1997 and was retained by Airbus during 1998; stored at Schönefeld-Berlin in 'green' condition, in1999. Sold to Jordan for Royal Flight. New-build aircraft no longer marketed by Airbus.

A340-400: Series discontinued in favour of A340-600.

A340-500: Ultra-long-range variant of A340-600 able to carry (typically) 313 passengers in three classes across 8,650 n miles (16,020 km; 9,954 miles). Fuselage 3.20 m (10 ft 6 in) longer than A340-300, representing 14-frame shrink of A340-600, plus additional fuel in rear centre tank. Launched at 1997 Paris Air Show and received go-ahead December 1997. Commonality with existing A330/A340 variants allows cross-crewing and employment of common cargo containers and interiors to minimise training, staffing, provisioning and maintenance costs. Claims to be the world's longest-range airliner, justified twice in 2004 by Singapore Airlines with flights to Los Angeles and New York. Former, on 3 February, covered 7,609 n miles (14,093 km; 8,757 miles) in 14 hours 42 minutes; latter, distance 8,963 n miles (16,600 km; 10,314 miles), completed in 18 hours 18 minutes on 28 June. Development time of A340-500/-600 series reduced by 25 per cent by use of Airbus Concurrent Engineering (ACE) shared CADD-CAD/CAM systems.

Compared with -300, has 0.53 m (1 ft 9 in) fuselage plug ahead of wings and 1.07 m (3 ft 6 in) plug to the rear; wings of increased chord (incorporating a third fuselage plug of 1.60 m; 5 ft 3 in) with span stretched to 63.45 m (208 ft 2 in); 31° 6' sweepback at quarter-chord; taller vertical stabiliser is married to a new horizontal stabiliser to give increased chord fin with 0.50 m (1 ft 7¾ in) height extension. Max T-O weight 368,000 kg (811,300 lb). Landing gear adaptation involves replacement of central twin-wheel unit with forward-retracting brakeable double-bogie unit. Rolls-Royce Trent 553 of 236 kN (53,000 lb st) chosen as power plant; this combines fan diameter of Trent 700 with scaled IP and HP compressors and turbines of Trent 800, plus new high-lift LP turbine. Honeywell 331-600 APU chosen for A340-500/-600 in March 1999. A340-500 initial launch commitments received from Air Canada, Emirates, SIA and ILFC. Maiden flight (F-WWTE, c/n 394) took place 11 February 2002; certified 3 December 2002. Higher gross weight (HGW) version (380,000 kg; 837,755 lb) under consideration in 2004.

A340-600: Derivative of A340-300 with 20-frame fuselage stretch, 267 kN (60,000 lb st) class engines, increased horizontal tail area, full electrical control of rudder (replacing mechanical linkage between computer and actuator for both primary and secondary systems), additional fuel capacity, four-wheel central landing gear and 365,000 kg (804,690 lb) basic and 368,000 kg (811,300 lb) optional maximum T-O weights; can carry 380 three-class passengers or 475 in all-economy class up to 7,500 n miles (13,890 km; 8,630 miles) and designed as Boeing 747 replacement with significantly lower costs and full commonality with A330/A340 family. Compared with A340-300, has same wing chord/fuselage centre-section, wing span, tailplane and fin modifications as -500, but with further 5.87 m (19 ft 3 in) plug ahead of wing and 3.20 m (10 ft 6 in) to the rear. Airbus and GE Aircraft Engines agreed in April 1996 that latter should be sole power plant source for -600, but this accord dissolved in February 1997 and Rolls-Royce Trent 556 of 249 kN (56,000 lb st) chosen on non-exclusive basis. Ram-air turbine fitted in underwing pod between engines on starboard wing. Launch commitments from Aerolineas Argentinas, Air Canada, Egyptair, Lufthansa, Swissair and Virgin Atlantic; initial firm order placed by Virgin on 15 December 1997 for eight, plus options. First metal cut 27 July 1998. Final assembly began June 2000; airframe completed September 2000 and engines fitted November 2000. Prototype (c/n 360) rolled out 23 March 2001 and first flew (F-WWCA) 23 April 2001; public debut at Paris Air Show, June 2001; second aircraft (c/n 371) handed to test department 8 June 2001 and first flown (F-WWCB) 18 June 2001; third test example, (c/n 376) flew 24 September 2001 (F-WWCC/G-VATL); three-aircraft 1,600 hour test programme completed. Programme of weight reduction and engine improvements implemented as a result of test data; weight optimisation to be achieved gradually by structural modifications; optimum configuration reached in 2003. Certification by JAA awarded 29 April 2002; first delivery (fourth aircraft, G-VSHY, to Virgin Atlantic) 22 July and entered service 1 August 2002. Higher gross weight (HGW) version (376,000 kg; 828,935 lb) launched at Paris Air Show, June 2003; expected to become standard version from mid-2006, following maiden flight late 2005; Emirates to be launch customer.

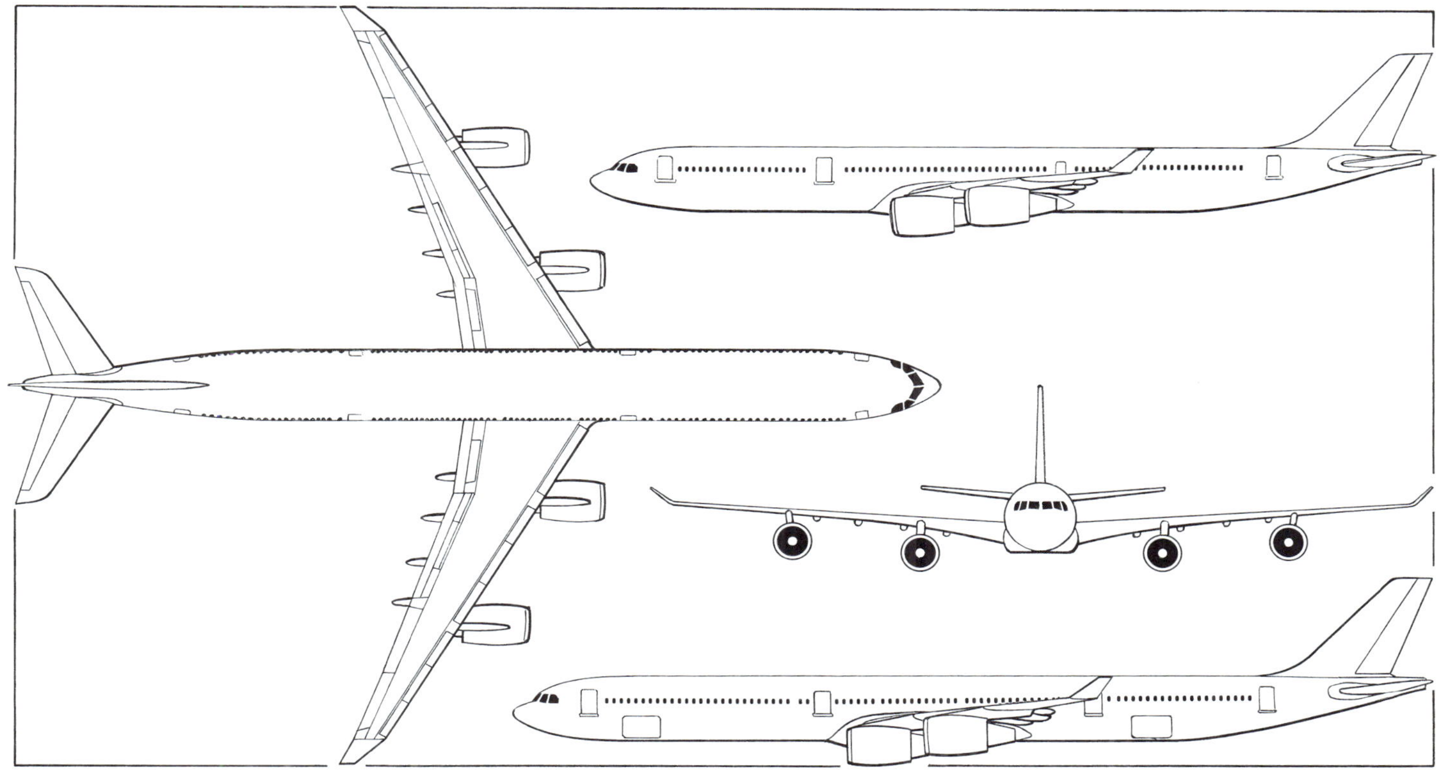

Airbus A340-600 four-turbofan ultra-long-range airliner, with additional side view (upper) of A340-500 *(James Goulding)* 0126690

A340-600F: Development study for freighter version; considered by UPS for order eventually placed for MD-11 freighters; airline has option to convert some of its A300-600Fs to A340-600Fs if required.

CUSTOMERS: Sales at 1 October 2004 totalled 352, of which 284 delivered. See table.

COSTS: Development cost of A340-500/-600 set at US$2.9 billion, 20 per cent lower than basic A340. Unit cost of A340-300 US$161.1 million; A340-500 US$177.8 million; A340-600 US$186.4 million (2002 prices).

DESIGN FEATURES: A340 capitalises on commonality with A330 (identical wing/cockpit/tail unit and same basic fuselage) to create aircraft for different markets, and also has much in common (for example, existing Airbus wide-body fuselage cross-sections, A310/A300-600 fin (except on -500/-600), advanced versions of A320 cockpit and systems) with rest of Airbus range; FAA has approved cross-crew qualification between A320 series, A330 and A340.

Airbus A340-600 flight deck 0527110

Series	Variant	MTOW (kg)
200	000	253,500
200	001	257,000
300	000	253,500
300	001	257,000
500	000	368,000
500	001	372,000
600	000	365,000
600	001	368,000

Notes: Variant parameters may include differences in data other than max T-O weight. Except as constrained by Series, Variants are applicable to several engine options.

AIRBUS A340 MARKS AND VARIANTS

Series	Mark	Power Plant	FAA Certification
200	211	CFM56-5C2 or -5C2/F or -5C2/G	27 May 93
200	212	CFM56-5C3/F or -5C3/G	7 Jul 94
200	213	CFM56-5C4	2 Oct 94
300	311	CFM56-5C2 or-5C2/F or -5C2/G	27 May 93
300	312	CFM56-5C3/F or -5C3/G	7 Jul 94
300	313	CFM56-5C4	2 Oct 97
500	541	Trent 553-61	27 Jan 03
600	642	Trent 556-61	22 Jul 02

AIRBUS A340 ORDERS
(at 1 October 2004)

Customer	Qty
Aerolineas Argentinas	6
Air Canada	13
Air China	3
Air China Southwest	3
Air France	14
Air Mauritius	3
Air Tahiti Nui	4
Austrian Airlines	4
Cathay Pacific	11
China Airlines	6
China Eastern Airlines	10
Egyptair	3
Emirates	28
Flightlease	2
Gulf Air	6
Iberia	29
ILFC	27
Kuwait Airways	4
LAN Chile	6
Lufthansa	45
Olympic Airways	4
Philippine Airlines	8
Qatar Airways	4
Sabena	5
SAS	7
Singapore Airlines	22
South African Airways	12
SriLankan Airlines	3
Swiss International Air Lines	12
TAP-Air Portugal	4
Thai Airways International	8
Turk Hava Yollari	7
UTA	7
Virgin Atlantic Airways	17
Private	4
Undisclosed	1
Total	**352**

New design wing (by BAE Systems), approximately 40 per cent larger than that of A300-600, has 30° sweepback at quarter-chord and winglets raked at 29° 42'; thickness/chord ratios 15.25 per cent at root, 11.27 per cent at inner kink, 9.86 per cent at outer kink and 10.60 per cent at tip. A340-500/600 wing 20 per cent larger than basic A340, has increased sweepback of 30° 6' and 1.60 m (5 ft 3 in) (removable) extension to each wingtip, rake angle 31° 30'.

FLYING CONTROLS: In A330/A340 electronic flight control system (EFCS), roll axis is controlled by two individual outboard ailerons and five outboard spoiler panels on each wing; pitch axis control is by trimmable tailplane and separate left and right elevators; tailplane can also be mechanically controlled from flight deck, but fly-by-wire computer inputs are superimposed; single rudder is directly linked to rudder pedals, with dual yaw damping inputs superimposed. High-lift devices consist of full-span slats, flaps and aileron droop; speed braking and lift dumping by raising all six spoilers on each wing and raising all ailerons. Slats and flaps controlled outside main fly-by-wire complex by duplicated slat and flap control computers (SFCC).

Control surface maximum deflections (A340-500 and 600 in parentheses, where different): ailerons ±25° (inner +20°/–30°; outer ±25°); aileron droop 10°; elevator +15°/–30° (+17°/–30°); rudder ±31° 35' (±35°); flaps 32° (33° 40'); No 1. spoiler, speed brake 25°, lift dumper 25°; Nos. 2 and 3 spoilers, roll 35°, speed brake 30° (35°), lift dumper 50°; Nos. 4 and 5 spoilers (and No. 6), roll 35° (40°), speed brake 30° (40°), lift dumper 50°; slat No. 1 (21°); Nos. 2 to 7 slats 24°; stabilisers +2°/–14°.

Control surface actuation by three hydraulic systems (green, yellow, blue); two powered control units (PCU) at each aileron and elevator are controlled either by primary or secondary flight control computers; single actuators at spoiler panels controlled by primary or secondary flight control computers; dual PCUs for fly-by-wire tailplane trimming, and for centrally located flap and slat actuators; three PCUs at rudder.

Fly-by-wire computers include three flight control primary computers (FCPC) and two flight control secondary computers (FCSC); each computer has two processors with different software; primary and secondary computers have different architecture and hardware; power supplies and signalling lanes are segregated; system provides stall protection, overspeed protection and manoeuvre protection as in the A320, but the A330/340 computer arrangement maintains the protections for longer in the face of failures of sensors and inputs; FCPC and FCSC all operate continuously and provide comparator function to active channels, but only one in active control at any one control surface; reconfiguration logic can provide alternative control after failures; different normal and alternative control laws apply fly-by-wire basically as a *g* demand in pitch and rate demand in roll, plus complex manoeuvre limitations; if all three inertial systems fail (removing attitude information), system reverts to direct mode in which control surface angle is directly related to sidestick position; ultimate control mode is direct control of rudder and tailplane angle from rudder pedals and manual trim wheel, which is sufficient for accurate basic instrument flight.

Pilots have sidestick controllers and normal rudder pedals; EFIS instrumentation consists of duplicated primary flight displays (PFD), navigation displays (ND) and electronic centralised aircraft monitors (ECAM); three display management computers, with separate EFIS and ECAM channels, can each control all six displays in their four possible formats; flight path control by duplicated flight management and guidance and envelope computers (FMGEC); they control every phase of flight including course, attitude, engine thrust and flight planning using information from GPS and inertial systems; point of no return calculations made automatically for long-range flights in A340/A330. Control system data are collected for maintenance purposes by two flight control data concentrator (FCDC) computers. Honeywell Pegasus flight management system evaluated 1999, before January 2000 certification.

In normal flight, bank angle limited to 33° hands-off (autopilot control) and 67° with full stick displacement; airspeeds limited to 305 kt (565 km/h; 350 mph) and M0.82 under automatic flight control; if stick is held fully forward, the nose is automatically raised when airspeed reaches VMO + 15 kt (28 km/h; 17 mph); if nose is raised, equivalent protections apply; 'alpha max' (13° clean and 19° with flap), slightly below maximum lift coefficient, is the greatest achievable with sidestick; 'alpha floor' is the angle of attack beyond which throttles progressively open to go-around power and airspeed is finally held steady just above stall, even if stick is continuously held back; alpha protection applied at 'normal' and 'hard' modes according to alpha rate; below protection speed (VPROT) of 142 kt (263 km/h; 163 mph) automatic and manual trimming stops; outer ailerons remain centred at over 200 kt (371 km/h; 230 mph); if a spoiler panel fails, the symmetrical opposite panel stops operating; if rudder yaw damping fails, pairs of spoiler panels are used instead; minimum-speed marker on PFD adjusts to changing aircraft configuration, airbrake and pitch rate; fuel automatically transferred between wing and tailplane tanks to minimise trim drag when cruising above 7,620 m (25,000 ft).

Engines controlled by setting throttle levers to marks on quadrant, such as climb (CLB), maximum continuous and flexible take-off (Flex T-O); digital engine control makes detailed settings appropriate to altitude and temperature.

During landing and take-off, nosewheel steering, by rudder pedals, automatically disengages above 100 kt (185 km/h; 115 mph); demands for more than 4°/s nose-up pitch restrained near ground; maximum airspeeds for flaps and slats signalled on airspeed scale; trim automatically cancels effect of flaps, landing gear and airspeed; ailerons droop with full flap selection and deflect 25° up with spoilers in lift dump after touchdown; thrust reverser failure automatically countered by cancelling symmetrical opposite reverser; voice warning demands throttle closure at 6 m (20 ft) during landing flare; autothrottle disengaged when throttles closed at touchdown; on touch-and-go landing, trim automatically reset for take-off when Flex T-O power selected; engine failure in flight compensated automatically, with wings held level, slight heading drift and spoiler panels sucked down to avoid unnecessary drag. A340-600 has forward-facing taxi-aid cameras mounted in fin and belly fairing to assist pilots during ground manoeuvres.

STRUCTURE: A330 and A340 wings almost identical except latter strengthened in area of outboard engine pylon with appropriate modification of leading-edge slats 4 and 5; main three-spar wing box and leading/trailing-edge ribs and fittings of aluminium alloy, with Al-Li for some secondary structures; steel or titanium slat supports; approximately 13 per cent (by weight) of wings is of CFRP, GFRP or QFRP, including outer flaps and flap track fairings, ailerons, spoilers, leading/trailing-edge fixed surface panels and winglets; common fuselage for all initial versions, except in overall length (A340-300 same size as A330-300; A340-200 and A330-200 respectively eight and ten frames shorter; A340-500 and A340-600 described above; see also dimensions, below); construction generally similar to that of A310 and A300-600 except centre-section to accept new wing; tail unit (common to all versions except A330-200 and A340-500/600, which have larger tailplane, fin and rudder) utilises same CFRP fin as A300-600 and A310; new tailplane incorporates trim fuel tank and has CFRP outer main boxes bridged by aluminium alloy centre-section. A340-500/600 have Aircelle nacelles constructed from carbon composites.

Work-sharing along lines similar to those for A310 and A300-600, with percentages similar to those held in the original consortium. Airbus France thus responsible for flight deck, engine pylons, part of centre-fuselage, centre-section cabin doors wing-to-body fairings and final assembly and outfitting at Toulouse; Airbus UK for wings and landing gear; Airbus Deutschland for most of fuselage, tailcone, fin, wing moving surfaces, cabin doors and interior; Airbus España for tailplane and forward starboard cabin doors; Belgian consortium Belairbus for leading-edge slats and slat tracks. Aircelle for engine nacelles. Japanese contributions from Bridgestone (tyres), Furukawa/Sky Aluminium (metal/plastics interfaces) and ShinMaywa (wing walkways). Airbus Deutschland (Bremen) developing lightweight spoiler for 2006 introduction. Before final assembly at Toulouse, Saint Nazaire (Airbus France) pre-assembles flight deck and three forward fuselage sections; and two (of three) rear fuselage sections.

LANDING GEAR: Main (four-wheel bogie) and twin-wheel nose units identical on all A330/340 versions. Main tyres size 54×21.0–23 or 46×17.0R20 or 1400×530R23 (32/36 ply); nose tyres 40.5×15.5-6 or 30×8.8R15 or 1050×395R16 (28 ply). A340-200/300 have additional twin-wheel auxiliary unit on fuselage centreline amidships, retracting rearward; A340-500 and -600 have four-wheel centre bogie, retracting forward. Goodyear tyres available on all units; Michelin and Bridgestone tyres also available for A340-500/600.

POWER PLANT: Four 138.8 kN (31,200 lb st) CFM56-5C2 turbofans initially; 144.6 kN (32,500 lb st) CFM56-5C3 and 151.2 kN (34,000 lb st) CFM56-5C4/P; CFM56 upgrade announced in July 2000, combining CFM56-5C with core of CFM56-5B/P. Rolls-Royce Trent 553-61 of 248 kN (55,780 lb st) in A340-500; Trent 556-61 of 260 kN (58,460 lb st) in A340-600.

A340-200/300: Maximum fuel capacity: (-200 and -300 until 1996) 138,600 litres (36,614 US gallons; 30,488 Imp gallons); (-200/300 1996–97) 140,640 litres (37,154 US gallons; 30,937 Imp gallons); (-200, 300, 1997 and subsequent deliveries) 141,500 litres (37,381 US gallons; 31,126 Imp gallons); (-200/300, 1997 and subsequent deliveries, including optional ACT) 148,700 litres (39,283 US gallons; 32,710 Imp gallons). Totals include 6,230 litres (1,646 US gallons; 1,370 Imp gallons) in tailplane trim tank.

A340-500: Total capacity 214,405 litres (56,641 US gallons; 47,164 Imp gallons) total, including 771 litres (204 US gallons; 170 Imp gallons) unusable. Centre and wing group tanks as for A340-600; plus Rear Centre tank of 19,883 litres (52,500 US gallons; 43,716 Imp gallons) and Trim tank of 8,011 litres (2,116 US gallons; 1,762 Imp gallons.

A340-600: Total capacity 194,897 litres (51,511 US gallons; 42,892 Imp gallons) total, including 761 litres (201 US gallons; 167 Imp gallons) unusable. Centre tank of 55,373 litres (14,732 US gallons; 12,181 Imp gallons), Trim tank of 8,386 litres (2,215 US gallons; 1,845 Imp gallons) and six wing tanks: Tank 1/4 of 49,070 litres (12,963 US gallons; 10,794 Imp gallons), Tank 2/3 of 69,744 litres (18,425 US gallons; 15,342 Imp gallons) and Outer tanks with combined 12,324 litres (3,256 US gallons; 2,711 Imp gallons).

ACCOMMODATION: Crew of two on flight deck (all versions); flight deck can be supplied with humidifier system. Passenger seating typically six-abreast in first class, six-abreast in business class and eight-abreast in economy (nine-abreast optional), all with twin aisles. A typical three-class layout seats 295 in A340-300, 239 in the A340-200, 313 in A340-500 and 380 in A340-600. Single-class seating capacities for A340-300/200/500/600 are 440, 420, 375 and 475 respectively. A340-300 is available with optional, removable, lower deck crew rest module, replacing pallet in rear cargo hold with a sleeper cabin for seven or eight cabin crew. Optional lower deck facilities in rear cargo hold of -600 include an area with stand-up headroom and lavatories plus up to eight crew rest bunks and galleys. A further facility is under study for -500 and -600 that could offer up to 10 full-length beds for passengers or a lower-deck lounge. A340-600 (except prototype) has two overwing Type III emergency exits in addition to standard eight Type A doors. Underfloor cargo holds house up to 32 LD3 containers or 11 standard 2.24 × 3.17 m (88 × 125 in) pallets in A340-300, 26 LD3s or nine pallets in A340-200, 30 LD3s or 10 pallets in A340-500, and 42 LD3s or 14 pallets in A340-600; front and rear cargo holds have doors wide enough to accept 2.44 × 3.17 m (96 × 125 in) pallets; all versions have a 19.7 m^3 (695 cu ft) bulk cargo hold aft of the rear cargo hold.

SYSTEMS: Hamilton Sundstrand GTCP 331-350C APU.

AVIONICS: Airbus Future Air Navigation System (FANS-A), comprising SmithsIndustries digital control and display system married to Honeywell FMS, underwent testing end 1999 and was certified July 2000; can be retrofitted to all A330/A340s. Tenzing Communications in-flight e-mail/Internet access system was successfully tested on an A340-600 during flight testing with a full passenger load.

DIMENSIONS, EXTERNAL:

Wing span: A340-200/300	60.30 m (197 ft 10 in)
A340-500/600	63.45 m (208 ft 2 in)
Wing chord at root: A340-200/300	10.60 m (34 ft 9¼ in)
A340-500/600	12.20 m (40 ft 0¼ in)
Wing aspect ratio: A340-200/300	10.1
A340-500/600	9.3
Length overall: A340-200	59.42 m (194 ft 11¼ in)
A340-300	63.68 m (208 ft 11 in)
A340-500	67.51 m (221 ft 6 in)
A340-600	74.96 m (245 ft 11 in)
Fuselage: Max diameter (all versions)	5.64 m (18 ft 6 in)
Height overall:	
A340-200/-300	16.84 m (55 ft 3 in)
A340-500/-600	17.75 m (58 ft 3 in)
Tailplane span: A340-300	19.41 m (63 ft 8 in)
A340-500/-600	22.96 m (75 ft 4 in)
Wheel track (all versions)	10.69 m (34 ft 5 in)
Wheelbase: A340-200	23.47 m (77 ft 0 in)
A340-300	25.40 m (83 ft 4 in)
A340-500	27.58 m (90 ft 6 in)
A340-600	32.89 m (107 ft 11 in)

DIMENSIONS, INTERNAL:

Cabin:	
Length (excl flight deck):	
A340-200	46.08 m (151 ft 2 in)
A340-300	50.34 m (165 ft 2 in)
A340-500	53.54 m (175 ft 8 in)
A340-600	60.99 m (200 ft 1 in)
Max width	5.28 m (17 ft 4 in)
Max height	2.40 m (7 ft 10½ in)
Underfloor baggage/cargo holds:	
Max height	1.70 m (5 ft 7 in)
Width at floor	3.175 m (10 ft 5 in)
Volume: forward hold:	
A340-300/-500 standard	69.1 m^3 (2,442 cu ft)
A340-300/-500 option	80.5 m^3 (2,844 cu ft)
A340-600 standard	92.2 m^3 (3,256 cu ft)
A340-600 option	107.4 m^3 (3,792 cu ft)
Volume: rear hold:	
A340-300 standard	55.6 m^3 (1,965 cu ft)
A340-300 option	62.6 m^3 (2,212 cu ft)

A340-500 standard .... 46.1 m^3 (1,628 cu ft)
A340-500 option .... 53.7 m^3 (1,896 cu ft)
A340-600 standard .... 69.1 m^3 (2,442 cu ft)
A340-600 option .... 80.5 m^3 (2,844 cu ft)
Volume: bulk hold (all versions) .... 19.7 m^3 (695 cu ft)

AREAS:
Wings, gross:
A340-200/-300 .... 361.60 m^2 (3,892.2 sq ft)
A340-500/-600 .... 437.00 m^2 (4,703.8 sq ft)

WEIGHTS AND LOADINGS:
Baggage capacity:
A340-200:
forward hold .... 18,507 kg (40,801 lb)
rear hold .... 15,241 kg (33,601 lb)
bulk hold .... 3,468 kg (7,646 lb)
A340-300:
forward hold .... 22,861 kg (50,400 lb)
rear hold .... 18,507 kg (40,801 lb)
bulk hold .... 3,468 kg (7,646 lb)
A340-500:
forward hold .... 24,494 kg (54,000 lb)
rear hold .... 16,330 kg (36,001 lb)
bulk hold .... 3,458 kg (7,624 lb)
A340-600:
forward hold .... 30,482 kg (67,201 lb)
rear hold .... 22,861 kg (50,400 lb)
bulk hold .... 3,468 kg (7,646 lb)
Operating weight empty (basic weight options):
A340-200 current .... 129,500 kg (285,500 lb)
A340-300 basic .... 130,000 kg (286,600 lb)
A340-500 .... 170,900 kg (376,770 lb)
A340-600 .... 177,700 kg (391,760 lb)
Max payload (maximum weight options):
A340-200: initial .... 44,000 kg (97,005 lb)
current .... 45,530 kg (100,375 lb)
A340-300 basic .... 50,900 kg (112,215 lb)
A340-500 .... 54,100 kg (119,270 lb)
A340-600 .... 67,200 kg (148,150 lb)
Max T-O weight:
A340-200, -300: basic .... 253,500 kg (558,875 lb)
option .... 257,000 kg (566,600 lb)
A340-500: basic .... 368,000 kg (811,300 lb)
option .... 372,000 kg (820,125 lb)
A340-600: basic .... 365,000 kg (804,675 lb)
option .... 243,000 kg (535,725 lb)
Max ramp weight:
A340-200, 300: basic .... 254,400 kg (560,850 lb)
option .... 257,900 kg (568,575 lb)
A340-500: basic .... 369,200 kg (813,950 lb)
option .... 373,200 kg (822,750 lb)
A340-600: basic .... 366,200 kg (807.325 lb)
option .... 369,200 kg (813,950 lb)
Max landing weight:
A340-200 .... 181,000 kg (399,025 lb)
A340-300 basic and option .... 186,600 kg (410,050 lb)
A340-500: basic .... 240,000 kg (529,105 lb)
option .... 243,000 kg (535,725 lb)
A340-600 basic .... 256,000 kg (564,380 lb)
option .... 259,000 kg (570,995 lb)
Max zero-fuel weight:
A340-200 .... 169,000 kg (372,575 lb)
A340-300 basic and option .... 174,000 kg (383,600 lb)
A340-500: basic .... 225,000 kg (496,040 lb)
option .... 230,000 kg (507,075 lb)
A340-600: basic .... 242,000 kg (533,520 lb)
option .... 245,000 kg (540,130 lb)
Max wing loading:
A340-200, -300 .... 760.5 kg/m^2 (155.76 lb/sq ft)
A340-500, -600 .... 835.2 kg/m^2 (171.07 lb/sq ft)

PERFORMANCE:
Max operating Mach No. (MMO) 0.86
Max operating speed .... 330 kt (611 km/h; 380 mph) IAS
Typical operating Mach No.:
A340-200/300 .... 0.82
A340-500/600 .... 0.83
Stalling speed:
A340-300, at 267,000 kg (588,625 lb), wheels up:
flaps up .... 161 kt (299 km/h; 186 mph)
flaps down .... 133 kt (247 km/h; 153 mph)
Max certified altitude: all .... 12,525 m (41,100 ft)
T-O field length at S/L, basic MTOW, ISA + 15°C:
A340-200 .... 3,017 m (9,900 ft)
A340-300 .... 3,125 m (10,250 ft)
A340-500 (estimated) .... 3,125 m (10,250 ft)
A340-600 .... 3,140 m (10,300 ft)
Range at typical OWE, international allowances and 200 n mile (370 km; 230 mile) diversion:
A340-200 with 239 passengers .... 8,000 n miles (14,816 km; 9,206 miles)
A340-300 with 295 passengers:
basic .... 7,200 n miles (13,334 km; 8,285 miles)
optional .... 7,400 n miles (13,704 km; 8,515 miles)
A340-500 with 313 passengers .... 8,650 n miles (16,019 km; 9,954 miles)
A340-600 with 380 passengers:
basic .... 7,500 n miles (13,890 km; 8,630 miles)

OPERATIONAL NOISE LEVELS (at basic MTOW):
T-O, fly-over: A340-300 .... 95.6 EPNdB
A340-600 .... 93.5 EPNdB
Sideline: A340-300 .... 96.1 EPNdB
A340-600 .... 95.5 EPNdB
Approach: A340-300 .... 96.9 EPNdB
A340-600 .... 99.9 EPNdB

UPDATED

AIRBUS A330

TYPE: Wide-bodied airliner.

PROGRAMME: Twin-engined A330 was developed simultaneously with four-engined A340; launched 5 June 1987; first flight (F-WWKA) with GE engines 2 November 1992; first R-R Trent 700-powered A330 flew 31 January 1994; simultaneous European and US certification with initial GE CF6-80E1 engines received 21 October 1993; first delivery (Air Inter) December 1993, entered service January 1994; certification with PW4164/4168 obtained 2 June 1994; A330 powered by GE CF6-80E1 with FADEC granted 120-minute ETOPS approval May 1994; Aer Lingus flew first ETOPS services across the Atlantic in May 1994; A330 powered by PW4164/4168 granted 90-minute ETOPS approval November 1994; certification with R-R Trent achieved 22 December 1994; 90-minute ETOPS approval granted before first delivery, to Cathay Pacific Airways, 24 February 1995. Further extensions of ETOPS to 180 minutes granted on 6 February 1995 (GE engines), 4 August 1995 (P&W) and 17 June 1996 (R-R); 180-minute ETOPS for PW4168A-powered A330-300 granted July 1999.

All main structural and systems information common to both A330 and A340 is listed in A340 entry. Differences in A330 given here.

CURRENT VERSIONS: **A330-300:** Baseline version; seating capacity 295 in three classes (12 first, 42 business, 241 economy). Payload increase of 7,000 kg (15,432 lb) offered for standard A330-300 in October 1993. Maximum T-O weight increased to 217,000 kg (478,400 lb) from November 1995 to allow typical 335 passengers to be carried 4,850 n miles (8,982 km; 5,581 miles), further increase to 230,000 kg (507,050 lb) with range increased to 5,600 n miles (10,370 km, 6,444 miles) offered in 1997; first aircraft to this

Airbus A330-300 in the insignia of Philippine Airlines

NEW/1044728

TAM Airbus A330-200 ***NEW***/1044730

standard ordered by Air Canada in 1997, but first in service was a Korean Air example in May 1999. Optional higher maximum T-O weight of 233,000 kg (513,675 lb) available from early 2001; earlier 230,000 kg (507,050 lb) versions can be retrofitted.

A330-200: Extended-range version; launched 24 November 1995; 10-frame reduction in fuselage length to 59.00 m (193 ft 6¾ in); maximum T-O weight 230,000 kg (507,050 lb); higher MTOW available from early 2001, as per A330-300. 253 passengers in three classes (12 first, 36 business, 205 economy) or 293 passengers in two classes; range with 253 passengers 6,650 n miles (12,315 km; 7,652 miles); engine choice as for A330-300; initial order, for 13, placed in March 1996 by ILFC. First flight (c/n 181, F-WWKA, with CF6-80E1 engines) 13 August 1997 was followed by public debut at Dubai Air Show in November 1997. Canada 3000 first user (on lease from ILFC) with first flight of production aircraft (C-GGWA) 20 January 1998 and delivery on 29 May 1998 following FAA/JAA/Transport Canada certification on 31 March 1998. First direct airline order from Korean Air, received second built in June 1998 (c/n 195, with PW4000 engines) following 4 December 1997 first flight and May 1998 certification. Version with Rolls-Royce Trents (re-engined first prototype) flew 24 June 1998 and following JAR and Transport Canada certification in January 1999 was delivered to Air Transat in February 1999. First with uprated CF6-80A1A3s delivered to Air France 17 December 2001.

AirTanker consortium bid for UK Future Strategic Tanker Aircraft (FSTA) contract offered A330-200 tankers, using new airframes converted by Cobham, which would enter service in 2007; selected as preferred platform January 2004.

A330-100: Proposed nine-frame reduction of A330-200; not built.

A330-200F: Proposed freighter version; possible DC-10/MD-11 replacement; planned launch at Paris Air Show in June 2001 was deferred to unspecified date. Maximum payload 63,500 kg (140,000 lb) over 4,200 n miles (7,778 km; 4,833 miles). Capacity for 20 2.24 × 3.17 × 2.44 m (88 × 125 × 96 in) pallets or containers on main deck and a combination of up to 14 LD3 plus up to 6 LD6 containers on the lower deck. Estimated market for 200 aircraft over 10 years.

A330-500: Proposed eight-frame reduction of A330-200, announced July 2000 and initially internally designated A330-M18; suspended in favour of A380 programme.

A350 Provisional designation of proposed A330 derivatives being studied in 2004; described separately.

CUSTOMERS: Total of 478 sold, of which 312 delivered, by 1 October 2004. See table.

Winglet detail of a Qantas A330-200 ***NEW***/1044729

COSTS: A330-200 US$135.3 million; A330-300 US$150.6 million (2002).

Flying Controls Control surface maximum deflections: ailerons ±25°; aileron droop 10°; elevator +15°/–30°; rudder ±30°; flaps 32°; No. 1 spoiler, speed brake 23°, lift dumper 35°; Nos. 2 to 6 spoilers, roll 35°, speed brake 30°, lift dumper 50°; slats 23°; stabilisers +2°/–14°.

POWER PLANT: Two turbofans. Initial application was 287 kN (64,530 lb st) GE CF6-80E1A2 turbofans (297 kN; 64,530 lb st E1A4 on longer-range version); alternative engines, using common pylon and mount, GE CF6-80E1A3 (each 305 kN; 68,530 lb st), P&W PW4168/A (305 kN; 68,600 lb st), R-R Trent 772-60/772B-60A (316 kN, 71,100 lb st), PW4164 (287 kN; 64,500 lb) and Trent 768-60 (300 kN; 67,500 lb st).

Fuel capacities: A330-200: Total 139,527 litres (36,860 US gallons; 30,693 Imp gallons), of which 437 litres (115.4 US gallons; 96.1 Imp gallons) unusable. Wing tanks total 91,648 litres (24,212 US gallons; 20,160 Imp gallons); centre tank 41,643 litres (11,001 US gallons; 9,160 Imp gallons); trim tank 6,236 litres (1,647 US gallons; 1,372 Imp gallons).

A330-301/321/322/341/342: Total 97,525 litres (25,764 US gallons; 21,453 Imp gallons), of which 354 litres (93.4 US gallons; 77.9 Imp gallons unusable. Wing tanks total 91,404 litres (24,147 US gallons; 20,107 Imp gallons); trim tank 6,121 litres (1,617 US gallons; 1,346 Imp gallons).

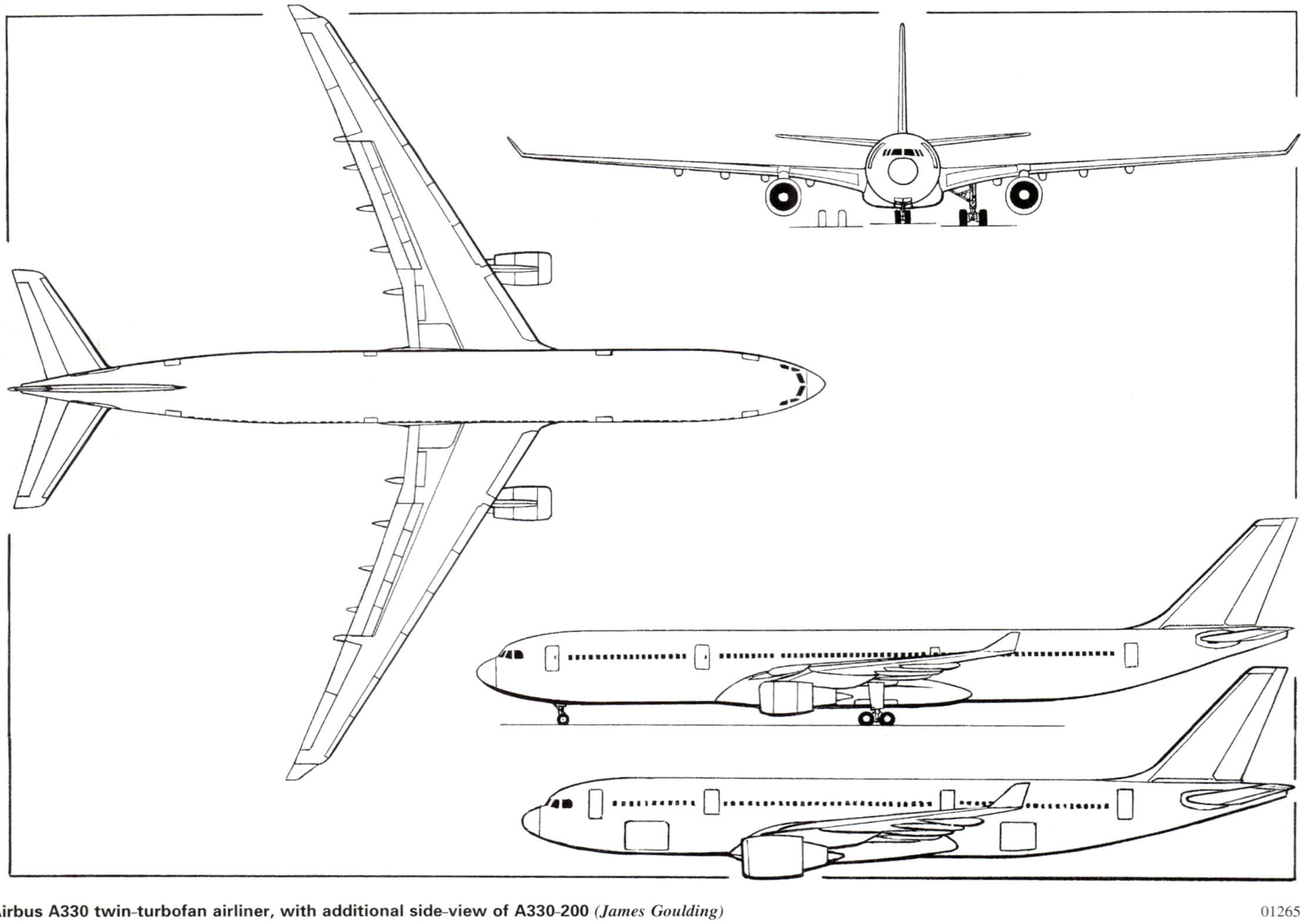

Airbus A330 twin-turbofan airliner, with additional side-view of A330-200 *(James Goulding)* 0126564

AIRBUS A330 MARKS AND VARIANTS

Series	Mark	Power Plant	FAA Certification
200	201	CF6-80E1A2	1 Apr 03
200	202	CF6-80E1A4	31 Mar 98
200	203	CF6-80E1A3	1 Nov 02
200	223	PW4168A	21 Jun 99
200	243	Trent 772B-60	21 Dec 00
300	301	CF6-80E1A2	21 Oct 93
300	321	PW4164	21 Jun 99
300	322	PW4168	21 Jun 99
300	323	PW4168A	8 Oct 99
300	341	Trent 768-60	21 Dec 00
300	342	Trent 772-60	21 Dec 00
300	343	Trent 772B-60	21 Dec 00

Series	Variant	MTOW (kg)
200	020	230,000
300	000	212,000
300	001	184,000
300	002	212,000
323/343 only	020	230,000
323 only	022	233,000
323 only	050	230,000
323 only	052	233,000

Notes: Variant parameters may include differences in data other than max T-O weight. Except as constrained by Series, Variants are applicable to several engine options.

A330-323/343: Total 98,239 litres (25,952 US gallons; 21,610 Imp gallons), of which 354 litres (93.4 US gallons; 77.9 Imp gallons) unusable. Wing tanks total 92,112 litres (24,334 US gallons; 20,262 Imp gallons); trim tank 6,127 litres (1,619 US gallons; 1,348 Imp gallons).

ACCOMMODATION: A330-200 and -300 certified for 375 passengers, with three pairs of Type A and one pair of Type 1 emergency exits, or 379 passengers with four pairs of Type A exits.

DIMENSIONS, EXTERNAL: As A340 except:
Length overall: A330-200 ... 59.00 (193 ft 7 in)
A330-300 ... 63.58 m (208 ft 7 in)
Height overall: A330-200 ... 17.88 m (58 ft 8 in)
A330-300 ... 16.84 m (55 ft 3 in)

DIMENSIONS, INTERNAL: As A340 except:
Underfloor baggage/cargo holds:
Volume: forward hold:
A330-200 basic ... 55.0 m³ (1,944 cu ft)
A330-200 option ... 62.6 m³ (2,212 cu ft)
A330-300 basic ... 69.1 m³ (2,442 cu ft)
A330-300 option ... 80.5 m³ (2,844 cu ft)
Volume: rear hold:
A330-200 basic ... 46.1 m³ (1,628 cu ft)

AIRBUS A330 ORDERS
(at 1 October 2004)

Customer	Qty
Aer Lingus	3
Air Algérie	5
Air Caledonie International	2
Air Canada	8
Air France	8
Air Inter Europe	6
Asiana Airlines	6
Austrian Airlines	3
British Midland Airways	1
Cathay Pacific Airways	26
China Airlines (Taiwan)	14
China Southern Airlines	4
CIT Group	14
Corsair	2
Dragonair	5
Egyptair	7
Emirates	28
EVA Air	2
Flightlease	9
Garuda Indonesia	9
GECAS	18
Gulf Air	6
ILFC	92
KLM	6
Korean Air	19
LTU	5
Lufthansa	10
Malaysia Airlines	10
Monarch Airlines	2
MyTravel Airways	7
Northwest Airlines	32
Philippine Airlines	8
Qantas	14
Qatar Airways	25
Sabena	3
SAS	4
SriLankan Airlines	6
Swiss International Airlines	4
TAM	5
Thai Airways International	12
US Airways	19
Undisclosed	9
Total	**478**

A330-200 option ... 53.7 m³ (1,896 cu ft)
A330-300 basic ... 55.6 m³ (1,965 cu ft)
A330-300 option ... 62.6 m³ (2,212 cu ft)
Volume: bulk hold (all versions) ... 19.7 m³ (695 cu ft)

WEIGHTS AND LOADINGS (C: with CF6-80E1A3/A4, P: with PW4168A, T: with Trent 772B):
Operating weight empty, basic versions, three-class layout: C ... 120,500 kg (265,655 lb)
P ... 121,100 kg (266,980 lb)
T ... 120,600 kg (265,875 lb)
A330-300: C ... 124,500 kg (274,475 lb)
P ... 125,100 kg (275,800 lb)
T ... 124,600 kg (274,695 lb)
Baggage capacity:
A330-200:
forward hold ... 18,869 kg (41,599 lb)
rear hold ... 15,241 kg (33,601 lb)
bulk hold ... 3,468 kg (7,646 lb)
A330-300:
forward hold ... 22,861 kg (50,400 lb)
rear hold ... 18,507 kg (40,800 lb)
bulk hold ... 3,468 kg (7,646 lb)
Max payload, basic versions: A330-200: C ... 47,500 kg (104,720 lb)
P ... 46,900 kg (103,395 lb)
T ... 47,400 kg (104,500 lb)
A330-300: C ... 48,500 kg (106,925 lb)
P ... 47,900 kg (105,600 lb)
T ... 48,400 kg (106,705 lb)
Max T-O weight:
A330-200/-300 basic ... 230,000 kg (507,060 lb)
-300 option ... 233,000 kg (513,675 lb)
Max landing weight:
A330-200 basic ... 180,000 kg (396,825 lb)
A330-200 option ... 182,000 kg (401,250 lb)
A330-300 basic ... 185,000 kg (407,855 lb)
A330-300 option ... 187,000 kg (412,265 lb)
Max zero-fuel weight:
A330-200 basic ... 168,000 kg (370,375 lb)
A330-200 option ... 170,000 kg (374,775 lb)
A330-300 basic ... 173,000 kg (381,400 lb)
A330-300 option ... 175,000 kg (385,810 lb)
Max wing loading:
A330-200/-300 basic ... 633.4 kg/m² (129.74 lb/sq ft)
A330-200/-300 option ... 644.4 kg/m² (131.97 lb/sq ft)

PERFORMANCE:
Max operating Mach No. (MMO) ... 0.86
Max operating speed ... 360 kt (666 km/h; 414 mph) IAS
Typical operating Mach No. ... 0.82
Max certified altitude ... 12,525 m (41,100 ft)
T-O field length at S/L, ISA + 15°C, Trent 772B engines:
A330-200, basic MTOW ... 2,530 m (8,300 ft)
A330-200, option MTOW ... 2,652 m (8,700 ft)
A330-300, basic MTOW ... 2,515 m (8,250 ft)
A330-300, option MTOW ... 2,606 m (8,550 ft)
Landing field length at S/L, ISA + 15°C, Trent 772B engines:
A330-200, option MLW ... 1,722 m (5,650 ft)
A330-200, option MLW ... 1,753 m (5,750 ft)
Range, typical OWE plus max passengers, international allowances and 200 n mile (370 km; 230 mile) diversion:
A330-200 at option MTOW ... 6,650 n miles (12,315 km; 7,652 miles)
A330-300 at option MTOW ... 5,600 n miles (10,371 km; 6,444 miles)

OPERATIONAL NOISE LEVELS (A330-300. A: 217,000 kg (478,400 lb) with Trent 768; B: 230,000 kg (507,050 lb) with Trent 772):
T-O, flyover: A ... 89.8 EPNdB
B ... 98.0 EPNdB
Sideline: A ... 96.5 EPNdB
B ... 101.0 EPNdB
Approach: A ... 96.8 EPNdB
B ... 104.0 EPNdB

UPDATED

AIRBUS A350

TYPE: Wide-bodied airliner.

PROGRAMME: Earlier hints of project, as counter to longer-range variants of Boeing 7E7 Dreamliner, gained credence at Farnborough Air Show, July 2004; designation confirmed two months later and formal launch anticipated before end of year. Delivery of major subassemblies required by mid-2007 to enable first flight in 2008 and service entry in 2009.

CURRENT VERSIONS(PROPOSED): **A350-800:** Based on A330-200. Passenger capacity about 250, range approximately 7,721 n miles (14,300 km; 8,885 miles).

A350-900: Based on A330-300. Passenger capacity about 350, range approximately 6,641 n miles(12,300 km; 7,642 miles).

DESIGN FEATURES: Lighter-weight, longer-range derivative of A330; objective 'maximum effect with minimum change'. Increased use of composites to reduce empty weight; new engines and redesigned wing. Flight control avionics based on those for A380.

STRUCTURE: Generally as for A330, but additional items in composites (keel, floor beams, centre-section wing box and rear pressure bulkhead) and use of 'Glare' fibre/metal laminate for parts of upper fuselage.

POWER PLANT: Twin 311.4 kN (70,000 lb st) class variants of General Electric GENX or Rolls-Royce Trent 1000 turbofans. Estimated fuel capacity 150,000 litres (39,627 US gallons; 32,996 Imp gallons).

WEIGHTS AND LOADINGS(estimated):
Max T-O weight ... 237,000 kg (522,500 lb)

NEW ENTRY

AIRBUS A380

TYPE: High-capacity airliner.

PROGRAMME: Engineering work began in early June 1994; known as A3XX until December 2000; separate A3XX directorate (Large Aircraft Division) formed within Airbus in March

Computer-generated image of the twin-deck Airbus A380-800 in the livery of Singapore Airlines ***NEW***/1044820

Airbus A340-300 Trent 900 testbed ***NEW***/1044818

1996; concept definition began April 1996; Airbus allocating approximately 40 per cent of programme to new partners. Full-size mockup of fuselage cross-section shown at Paris in 1997; completed, including partial concept interior, in 2000.

Airbus Industry Supervisory Board authorised programme go-ahead 8 December 1999; commercial launch authorised 23 June 2000; industrial launch and A380 designation confirmed 19 December 2000, on receipt of required 50th launch order. First metal and first carbon fibre lap were cut at Nantes on 23 January 2002. Test programme calls for 2,200 flying hours over 15 months. Trent 900 engine began flight testing in No. 2 position on A340-300 testbed (F-WWAI) 17 May 2004. Static testing at Toulouse began 25 May 2004; fatigue testing at Dresden from November 2005. Roll-out of first flying prototype due by end of 2004 and first flight early 2005, with certification and first deliveries to Emirates and Singapore Airlines by end of first quarter of 2006. Production intended to reach four per month by 2008.

First metal and/or CFRP cut at Nantes (France) January 2002; at Bremen (Germany) March 2002; at Varel (Germany) April 2002; at Filton (UK) August 2002; at Illescas (Spain) February 2003. First metal cut for wing centre-section 'bathtub' by EADS Military Aircraft at Augsburg 18 February 2003; first Saab-built wing leading-edge components delivered 24 April 2003; new Airbus UK factory at Broughton opened 4 July 2003. Six main subassemblies for static test aircraft (front, centre and rear fuselage, tailcone, tail unit and wings) arrived in Toulouse by April 2004 for final assembly. Followed in May (wings, fuselage) and June 2004 (horizontal and vertical tail surfaces) by major subassemblies for first flying prototype.

CURRENT VERSIONS (SPECIFIC): **001:** First prototype (F-WWOW), for airframe, systems and flight trials; R-R Trent 970 engines (311 kN; 70,000 lb st). Moved to final assembly 7 July 2004.

002: Second aircraft (second to be built); to fly third quarter 2005; fully equipped passenger cabin; early long-range flight trials, then eventual delivery to customer; R-R engines.

003: Third aircraft (fifth to be built): first for Singapore Airlines; R-R engines.

004: Fourth aircraft (third to be built): due to fly second quarter 2005; airframe and systems trials, then eventual delivery to customer; R-R Trent 970 engines. Major subassemblies reached Toulouse 1 July 2004.

005: Fifth aircraft (sixth to be built): second for Singapore Airlines; R-R engines.

006: Sixth aircraft (seventh to be built): third for customer airline; R-R engines.

007: Seventh aircraft (fourth to be built): to fly fourth quarter 2005, first to production standard; fully equipped cabin; route-proving flights in 2006, then eventual delivery to customer; R-R engines initially, then GP7200s.

008: Eighth aircraft; R-R engines.

009: Ninth aircraft; fifth to fly; first with (and used to test and certify) GP7270 engines.

CURRENT VERSIONS (GENERAL): **A380-700:** Potential short-fuselage version, more closely aligned to Boeing 747 replacement market.

A380-800: Baseline version; nominal 555 passengers in three-class layout and range of 8,000 n miles (14,816 km; 9,206 miles). Engine thrust 311 kN (70,000 lb), MTOW 560,000 kg (1,234,585 lb). Lower deck capacity 13 pallets or 38 LD3 containers. Deliveries to begin March 2006.

Following description applies to A380-800 except where indicated.

A380-800F: All-cargo version, carrying 150,000 kg (330,700 lb) of payload over 5,600 n miles (10,371 km; 6,444 miles). Standard layout is 25 pallets on upper deck, 33 pallets on main deck and 13 pallets in cargo hold/lower deck. Optional high-volume configuration offers more than 1,133 m^3 (40,000 cu ft) of capacity. Some composites to be replaced by aluminium-lithium (Al-Li). First flight planned for first half of 2007.

A380-900: Potential stretched version, initially known as A3XX-200; 656 seats in three-class layout and up to 990 at high density. Increased MTOW and fuel volume.

CUSTOMERS: Total of 129 firm orders by 1 October 2004 (see table). Potential market for 1,200 and over 300 cargo aircraft in A380 class up to 2020. Initial customer, on 24 July 2000, was Emirates, which committed to five A380s and two A380-800Fs, plus five options, subsequently increased to 20 passenger examples plus 10 options and 2 freighters in November 2001; further 21 of passenger version ordered 16 June 2003. First recipient will be Singapore Airlines. Federal Express became A380-800F launch customer in January 2001.

COSTS: Estimated development cost US$10.7 billion for whole A380 family given during June 2001, of which US$3.1 billion was sought from risk-sharing partners. Projected unit cost at launch was US$217 million for passenger version and US$233 million for freighter; revised to average unit price for A380-800 of US$265 million by mid-2002.

DESIGN FEATURES: First version, designated A380-800, is conventional in external appearance except for two rows of windows, but incorporates new developments in structures, materials, systems, landing gear design and aerodynamics. Dassault CATIA and IBM computer-aided design. Flight deck commonality with in-service Airbuses permits crew cross-qualification.

Fuselage is vertically orientated oval three-deck arrangement; this 'vertical ovoid' accommodates 10 passengers abreast on main deck and eight-abreast on upper deck, offering greater space per passenger than Boeing 747. Seating ranges from a nominal 555 (22 first class and 334 economy on main deck and 96 business and 103 economy on upper deck) in three-class layout to 840 in high-density, shorter-range applications. Typical launch customer layouts are for about 525 passengers. Dual-lane boarding stair allows four-aisle boarding and deplaning through main deck.

Lower deck can accommodate shop, bar, restaurant and/or normal range of 38 LD3 cargo containers or 13 pallets and 18.4 m^3 (650 cu ft) of bulk freight; main deck is large enough to accommodate two 2.44 × 2.44 m (8 ft 0 in × 8 ft 0 in) containers side by side in the freighter version.

Modifications to engines, nacelles and aerodynamics at customer request late in the launch phase have resulted in major reductions in noise levels.

Wing sweep 33° 30' at 25 per cent chord.

FLYING CONTROLS: Single-slotted flaps incorporate droop-nose device to improve climb performance. Two ailerons and two actuators on each wing, plus eight spoilers with

Airbus A380 No. 001 partly complete at Toulouse ***NEW***/1044819

Interior impression of the A380, showing the first class upper cabin ***NEW***/1044856

individual actuators. Elevators have two panels and actuators on each side; rudder also has two panels and actuators. Flaps, ailerons and engines have all been specifically positioned to keep wake vortex at a minimum.

STRUCTURE: Extensive use of CFRP for all flaps and spoilers, rear pressure bulkhead, centre wing box (first in CFRP on any Airbus), all tail surfaces, tailcone aft of fin leading-edge intersection with fuselage, and engine cowlings. New 'Glare' material, consisting of alternate layers of aluminium and glass fibre-reinforced adhesive which offers significant weight reduction and fatigue/damage resistance, developed by Stork Aerospace with Technical University of Delft and Netherlands National Aerospace Laboratories; tested on a German Air Force A310 from October 1999 and used for upper fuselage shell. Laser beam welding, which reduces cost and weight, used to attach stringers to lower fuselage panels. Wing leading-edge constructed of thermoplastics. Outer wings metal bonded.

Rear central area of the A380's main cabin 0527166

Simulation of the A380's flight deck ***NEW***/1044855

Computer-aided impression of A380-800F in the colours of Federal Express ***NEW***/1044854

Upper floor beams on A380-800 constructed of CFRP; those in A380-800F aluminium-lithium; throughout the structure, lighter 2524 aluminium alloys used in place of more traditional 2024.

Work allocation of major subassemblies is Airbus France (St Nazaire): flight deck and centre fuselage; Airbus Deutschland (Hamburg): forward centre fuselage, rear fuselage; Airbus Deutschland (Stade): fin and rudder; Airbus UK (Broughton): wing main panels; Airbus España: wing/fuselage fairings, belly fairing and fixed horizontal tail; Airbus France (Toulouse): engine pylons and final assembly. For more detailed breakdown, see table.

Transport of major components undertaken by purpose-built Chinese-built Ro-Ro ship (MV Ville de Bordeaux) ship from Hamburg via UK and St Nazaire (forward centre fuselage disembarked, joined to flight deck and re-embarked, accompanied by centre fuselage), the aircraft set then transferred to barge at Bordeaux and joined by Spanish-built elements for river transport to within 80 km (50 miles) of Toulouse, completing journey by road. New 228 × 120 × 23 m (748 × 394 × 75.5 ft) component assembly hall at Airbus

German-built A380 centre fuselage barrel ***NEW***/1044853

UK-built wing of the Airbus A380 ***NEW***/1044852

Airbus A380 main deck economy seating ***NEW***/1044851

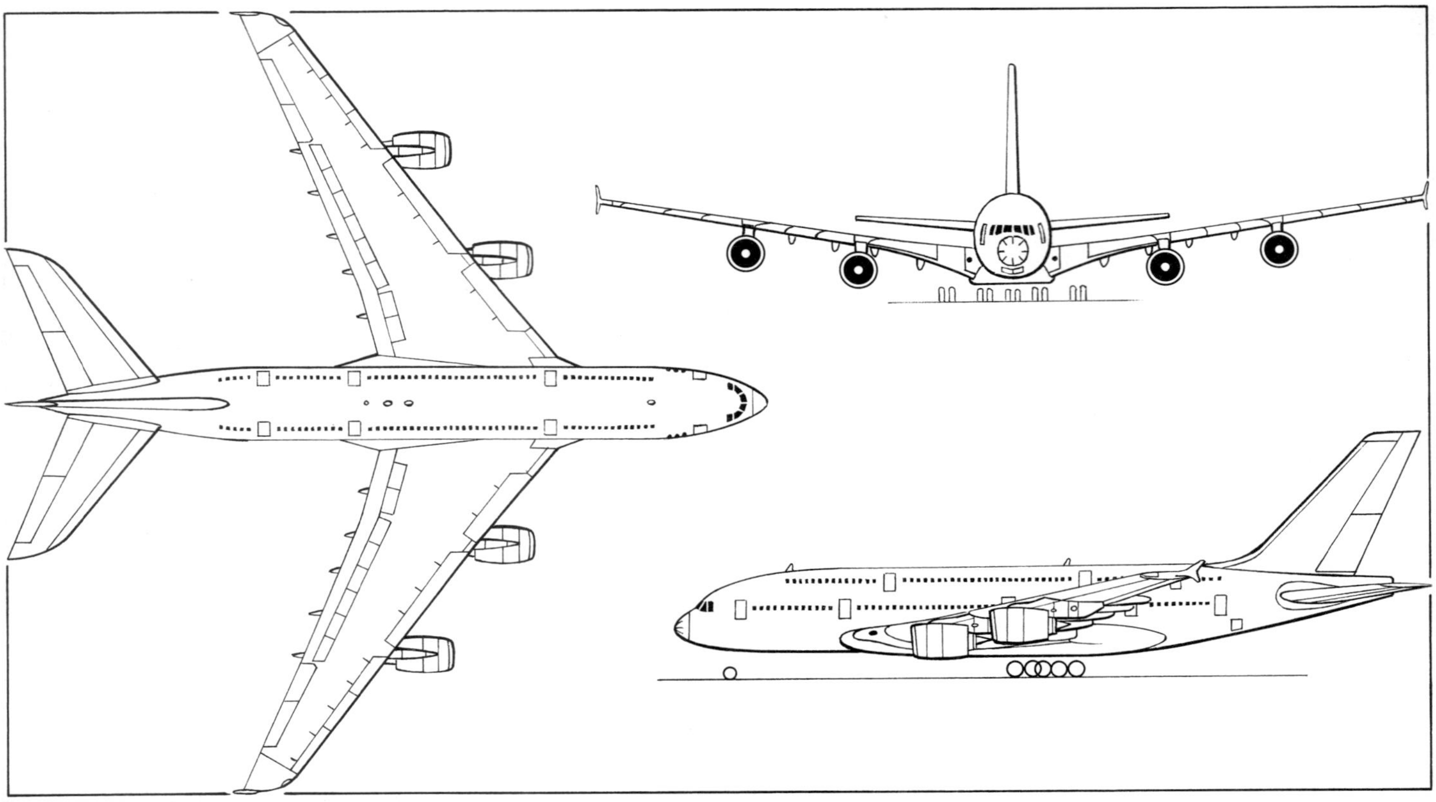

Airbus A380-800 555-seat airliner *(James Goulding)* 0103547

AIRBUS A380 CONFIRMED ORDERS
(at 1 October 2004)

Airline	Passenger	Freighter	Options	Engines	Order confirmed	First delivery
Air France	10	—	4	GP7200	18 Jun 01	Nov 2006
Emirates	41	2	10	GP7200	4 Nov 01	Mar 2006
Federal Express	—	10	20	GP7200	Jul 02	2008
ILFC	5	5	—	GP7200	17 Jun 01	2006
Korean Air	5	—	3	TBD	23 Oct 03	2007
Lufthansa	15	—	—	Trent 900	20 Dec 01	2007
Malaysia Airlines	6	—	—	TBD	11 Dec 03	2007
Qantas Airways	12	—	12	Trent 900	6 Mar 01	2006
Qatar Airways	2	—	—	TBD	9 Dec 03	2009
Singapore Airlines	10	—	15	Trent 900	16 Jul 01	Mar 2006
Virgin Atlantic Airways	6	—	6	Trent 900	25 Apr 01	Jul 2006
Total	**112**	**17**	**70**			

[1] Additional commitments from Etihad Airways (four) and Thai Airways International (six)

Deutschland's Hamburg site started 6 December 2001 and inaugurated 21 May 2003; Airbus France's 490 × 250 × 46 m (1,608 × 820 × 151 ft) final assembly hall at Toulouse inaugurated 7 May 2004.

Once completed, 'green' aircraft flown to Hamburg for internal fitting out. European and Middle Eastern aircraft delivered from Hamburg; those for the rest of the world return to Toulouse before despatch to customer.

LANDING GEAR: Goodrich main landing gear; each four-wheel wing-mounted unit weighs 2,310 kg (5,093 lb) and each six-wheel underfuselage unit weighs 4,080 kg (8,995 lb). Messier-Dowty twin-wheel nose landing gear. Michelin AIR X NZG tyres: 1400×530 R 23 (40 ply) on A380-800 main units, 1270×455 R 22 (32 ply) on nose unit; corresponding sizes for -800F are 56×22.0 R 24 (40 ply) and 1400×530 R 23 (40 ply). Bridgestone (Japan) is alternative tyre supplier. Underfuselage main gear set slightly aft of underwing units. Normal steering on nose gear and rear axle of body gear; back-up steering of nose gear in event of hydraulic power loss. Carbon brakes on all main group wheels. Manoeuvring compatible with 45 m (148 ft) wide runways and 23 m (75.5 ft) wide taxiways; U-turn possible on 60 m (200 ft) wide runways, by differential braking or asymmetrical thrust.

POWER PLANT: A380 is offered with a choice of Alliance (GE/P&W) GP7200 or Rolls-Royce Trent 900 series; Thrust reversers on inboard engines only. Airbus Industrie and Rolls-Royce signed an MoU specifying the Trent as favoured power plant in November 1996; due for certification in 2005. Trent 970 will be initially certified at 311 kN (70,000 lb st) but derated to 302 kN (68,000 lb st), with eventual growth to 374 kN (84,000 lb st). FADEC for GP7270 under development by BAE Systems Controls and Hispano-Suiza.

Alliance GP7200 uses same core as power plants for Boeing 747X and long-range 767. Detailed design work was due to have started in August 2001 but slipped to early 2003, with a first run achieved in March 2004, certification due at 363 kN (81,500 lb st) in July 2005 and first flight in November 2005.

Standard fuel capacity of both models is 310,000 litres (81,893 US gallons; 68,192 Imp gallons); extra fuel tanks would be fitted in wing centre box for longer-range models.

ACCOMMODATION: Flight deck crew of two, located at mezzanine level between main and upper deck, with access via main deck; rest areas are provided for crew in flight deck area; see Design Features for details of passenger cabins. Five main deck and three upper deck emergency exits on each side of fuselage; Goodrich evacuation slides, those for upper decks stored within the airframe rather than the door. Seat pitches: first class 173 cm (68 in), business class 122 cm (48 in), economy class 81 or 84 cm (32 or 33 in). Airbus 'Prestige' interior planned as VIP transport option.

SYSTEMS: Integrated modular avionics (IMA) system provided by Thales Avionics in conjunction with Diehl Avionik Systeme using computing modules slotted into cabinets throughout aircraft. Rockwell Collins supplying Ethernet avionics communications infrastructure at 100 Mbit/s speed with full duplex networking. Variable frequency AC electrical generating systems to be incorporated. Cameras fitted to top of fin and under fuselage for taxying assistance. Fuel management systems provided by Parker Aerospace including in-tank sensors and wiring, avionics and software to fit into IMA system.

A380 MAJOR INTERNAL SYSTEMS SUPPLIERS
(at January 2004)

System	Supplier
Air conditioning	Hamilton Sundstrand/ L'hotellier
Air data	Goodrich
APU	Pratt & Whitney Canada
Brake-by-wire	Messier-Bugatti
Cargo loading system	Goodrich
Electrical harnesses	Labinal
Electrical power distribution	ECE/Intertechnique
Electrical power generation	Aerolec (Thales Avionics/ TRW)
Environmental control	Nord-Micro
Evacuation	Goodrich/Diehl Avionik
Fire detection	Siemens-Cerberus/Meggitt
Flight deck displays	Thales Avionics/Diehl Avionik
Flight management	Honeywell
Fuel distribution	FR-HiTEMP/Intertechique
Fuel quantity and management	Parker Aerospace
Hydraulic	Eaton Corporation
Ice detection	Goodrich
Oxygen	Dräger/AirLiquide
Pneumatic	Honeywell
Primary flight controls	Goodrich (ailerons, elevators and rudder); Liebherr (spoilers)
Ram-air turbine (emergency electrical power generation)	Hamilton Sundstrand
Supplementary cooling	Fairchild Controls/ Microtecnica
Water and waste management	Monogram Systems

Airbus A380-800, cutaway drawing key

1 Radome
2 Weather radar antenna
3 Antenna mounting structure
4 Dual ILS antennas
5 Front pressure bulkhead
6 Nose landing gear wheel bay
7 Flight deck pressure floor
8 Rudder pedals
9 Side console with sidestick controller
10 Instrument panel with 10 full colour multifunction displays
11 Windscreen wipers
12 Electrically heated windscreen panels
13 Overhead systems switch panel
14 Two-pilot flight deck with central observer's seat
15 Flight deck bulkhead
16 Maintenance station
17 Folding supernumerary crew seats, port and starboard
18 Main avionics equipment bay
19 Nose landing gear pivot mounting
20 Nosewheel leg doors
21 Leg-mounted taxying lights
22 Twin nosewheels, forward retracting
23 Hydraulic nosewheel steering
24 Forward main door, all doors outward opening
25 Crew lavatory
26 Wardrobe
27 Crew rest compartment
28 Upper avionics equipment bay
29 Forward staircase
30 Lower deck lavatory
31 Fuselage lower lobe structure with welded skin/stringer panels
32 First class passenger cabin, 22 seats, six abreast
33 Upper deck toilet compartment
34 Overhead baggage lockers
35 Cabin roof trim/lighting panels
36 Passenger service units
37 Business class passenger cabin, 96 seats, six abreast
38 Cabin divider curtain
39 Main deck galley compartment
40 Wing and engine inspection lights
41 Forward cargo hold, door on starboard side
42 Conditioned air mixing and distribution units
43 Wing centre box carry-through structure with CFRP skins and web panels, dry bay on 800-series aircraft
44 Upper deck forward door, port and starboard
45 Upper deck window panels
46 Upper deck galley
47 Forward service trolley lift
48 Cabin wall trim panelling
49 Starboard wing inboard fuel tank
50 Starboard mid tank
51 Inboard drooped leading-edge, lowered
52 Starboard thrust reverser, open
53 Starboard engine nacelles
54 Nacelle pylons
55 Central leading-edge slats, extended
56 Pressure refuelling/defuelling connectors
57 Slat torque shaft and rack-and-pinion drive mechanism
58 Wing skin panelling
59 Wing stringers
60 Starboard feed tank
61 Fuel feed and vent piping
62 Surge tank
63 Outer fuel tank
64 Outboard leading-edge slats, extended
65 Wingtip vent tank
66 Starboard navigation (green) and strobe (white) lights
67 Starboard winglet
68 Obstruction light
69 Static dischargers
70 Starboard three-segment aileron
71 Aileron hydraulic actuators
72 Outboard two-segment single-slotted flap, extended
73 Flap aluminium alloy structure with honeycomb trailing-edge
74 Outboard spoiler panels (six)
75 Flap carriages and hinge links
76 Flap operating links, torque shaft driven
77 Twin anti-collision beacons
78 Fuselage frame and stringer structure
79 Floor beam structure above wing box centre-section
80 Forward end of aft T-section cargo hold
81 Starboard wing-mounted main landing gear, stowed position
82 Lower deck overwing door panel, port and starboard
83 Cabin wall insulation
84 Lower deck overhead baggage lockers
85 Starboard fuselage-mounted main landing gear, stowed position
86 Aft end of business class cabin
87 Curtained cabin divider
88 ADF antennae
89 Upper deck tourist class seating, eight abreast
90 Main deck tourist class seating, 10 abreast, total of 437 tourist class seats
91 Main deck window panels
92 Escape chute stowage, all upper deck door positions
93 Upper deck mid doors, port and starboard
94 Fuselage 'Glare' upper skin panels
95 Aft cargo hold door
96 Main deck cabin wall trim panelling
97 Upper deck aft door, port and starboard
98 Cabin attendant's folding seat, typical
99 Aft upper deck galley unit, crew rest area on port side
100 Aft service trolley lift
101 Aft staircase
102 Rear pressure bulkhead, aluminium alloy frame and CFRP dome structure
103 Machined fin support frames
104 Fin root bolted attachment joints
105 CFRP fin leading-edge structure
106 Starboard trimming tailplane
107 Tailplane integral trimming fuel tank
108 Tailplane vent tank, starboard side only
109 Starboard outboard elevator panel
110 Fin CFRP skin panels
111 Fin two-spar and rib all-CFRP torsion box structure
112 Upper rudder segment
113 Rudder hydraulic actuators
114 Lower rudder segment
115 Rudder CFRP rib and skin structure
116 Tailplane mounting bulkhead
117 Tailplane pivot mountings, port and starboard
118 APU bay fireproof bulkhead
119 APU intake
120 PW980A auxiliary power unit (APU)
121 Rear position light
122 APU exhaust
123 Port inboard elevator
124 Elevator hydraulic actuators
125 Port outboard elevator segment
126 Elevator CFRP rib and skin structure
127 Static dischargers
128 Aluminium alloy tailplane tip fairing
129 Tailplane two-spar and rib all-CFRP torsion box structure
130 Fin 'Logo' light
131 Tailplane sliding root seal
132 Tailplane screw jack trim actuator, hydraulic motor driven
133 Fuselage tailcone frame and stringer structure
134 Main deck aft door, port and starboard
135 Aft lavatories
136 Main deck tourist class cabin aft seat rows
137 Seat mounting rails
138 Bulk cargo hold, side door
139 Cargo hold bulkhead
140 Aft cargo hold floor with roller conveyors
141 Wing root trailing-edge fairing, Nomex honeycomb core and glass/CFRP skin panels on aluminium and titanium substructure
142 Port fuselage-mounted (BLG) main landing gear wheel bay
143 Wheel bay doors
144 Leg pivot mounting and hydraulic retraction jack
145 Wing-mounted (WLG) main landing gear wheel bay
146 Six-wheel BLG bogie with steerable aft wheel pair
147 WLG pivot mounting and hydraulic retraction jack
148 Side breaker strut
149 Shock-absorber leg strut
150 Four-wheel WLG bogie
151 Wing panel semi-span centre spar
152 Inboard spoiler panels (two)
153 Inboard single-slotted flap segment
154 Flap interconnection
155 Hinged flap track fairings
156 Outboard spoiler panels
157 Spoiler hydraulic jacks
158 Port two-segment outboard single-slotted flap
159 Fuel jettison, port and starboard
160 Port flap extended position
161 Aileron CFRP/Nomex honeycomb core structure
162 Port three-segment aileron
163 Fixed trailing-edge segment
164 Port winglet
165 Aft facing obstruction light
166 Port navigation (red) and strobe (white) lights
167 Wingtip vent tank
168 Port outboard three-segment leading-edge slat
169 Fixed leading-edge rib structure
170 Two-spar outer wing torsion box structure
171 Wing bottom skin/stringer panel with access manholes
172 Outer wing panel aluminium alloy ribs
173 Outboard engine pylon mounting
174 Engine pylon structure
175 Bleed air pre-cooler
176 Exhaust plug
177 Hot stream exhaust nozzle
178 Cold stream exhaust duct
179 Engine turbine section
180 Oil cooler
181 Ventral accessory equipment gearbox
182 Full-authority digital engine controller (FADEC)
183 Rolls-Royce Trent 900 turbofan
184 Fan case forward mounting
185 Acoustically lined engine intake
186 Nacelle pylon fairing
187 Wing three-segment centre leading-edge slat
188 Slat guide rails
189 Centre slat bleed-air de-icing
190 De-icing air telescopic duct
191 Port wing integral fuel tankage
192 CFRP wing ribs, typical
193 Port pressure refuelling/defuelling connectors
194 Inboard engine pylon mounting
195 Bleed air pre-cooler exhaust louvres
196 Translating rear engine cowling
197 Thrust reverser cascades, inboard engines only
198 Hinged cowling panels
199 Inboard Trent 900 turbofan engine
200 Engine starter air duct
201 Intake lip bleed-air de-icing
202 Wing inboard two-segment drooped leading-edge
203 Drooped leading-edge hinge links
204 Engine bleed-air ducting
205 Port wing inboard feed tank
206 Wingroot skin attachment cruciform member
207 Air conditioning system dual air generating units
208 Fixed leading-edge structure
209 Leading-edge taxying light
210 Landing and runway turn-off lights
211 Wingroot leading edge fairing

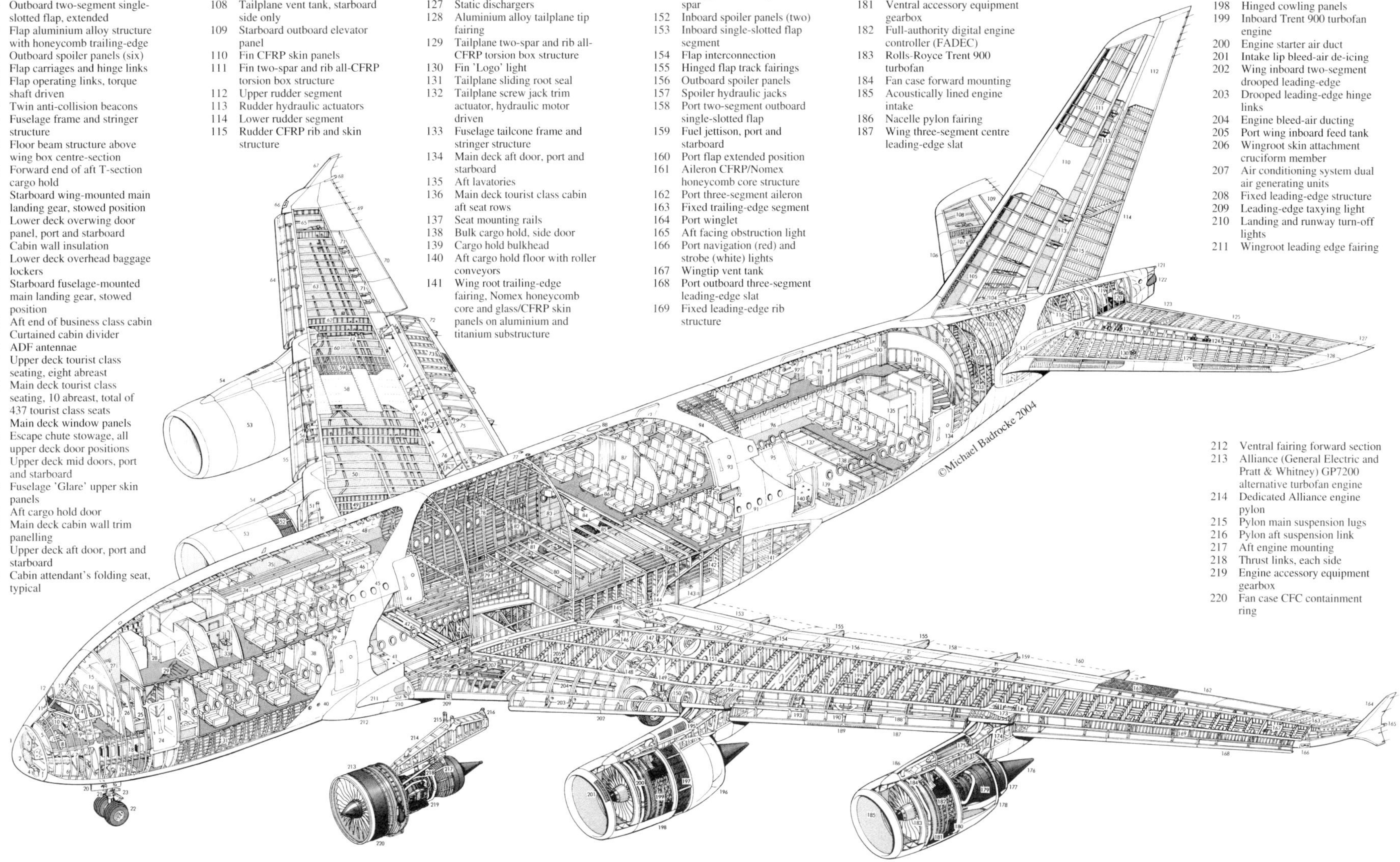

212 Ventral fairing forward section
213 Alliance (General Electric and Pratt & Whitney) GP7200 alternative turbofan engine
214 Dedicated Alliance engine pylon
215 Pylon main suspension lugs
216 Pylon aft suspension link
217 Aft engine mounting
218 Thrust links, each side
219 Engine accessory equipment gearbox
220 Fan case CFC containment ring

A380 MAJOR AIRFRAME INDUSTRIAL PARTNERS
(at January 2004)

Country	Company	Item
Australia	Hawker de Havilland	Wingtip fences
Austria	Fischer Advanced Composites	Flap track fairings; CFRP window frames; other components
Belgium	Belairbus[1]	Wing slat tracks and droop noses
	Sabca[1]	Centre rear lower fuselage shell
Finland	Finavitec[1]	Composites wing spoilers
France	EADS Socata[1]	Nose lower structure; nosewheel doors
	EADS Sogerma Services[1]	Centre rear main deck floor; flight deck seats
	Hurel-Hispano	Engine nacelles and thrust reversers
	Latécoère[1]	Upper deck passenger doors; lower nose section
	Mecachrome	Wing spars; upper and lower floor assemblies
	Messier-Dowty	Nose landing gear
	Michelin	Tyres
Germany	EADS Military Aircraft[1]	Wing inner inboard fixed leading-edges
	Eurocopter Deutschland[1]	Passenger and cargo doors
Italy	Alenia Aeronautica[1]	Centre upper and forward lower fuselage
Japan	Bridgestone	Tyres
	Fuji[1]	Vertical tail composites assemblies
	JAMCO	Upper floor deck crossbeams; vertical tail reinforcements
	Mitsubishi	Front and rear lower cargo doors
	Nippi[1]	Tailplane tips
	ShinMaywa	Wing root end fillet fairings
Korea, South	KAI	Wing bottom panels
Malaysia	CTRM[1]	Wing fixed leading-edge lower panels and inboard outer fixed leading-edges
Netherlands	Stork Aerospace[1]	Thermoplastic wing leading-edge 'J-noses'; selected forward and aft fuselage panels
Spain	Aries	Elevator and rudder components
	Gamesa[1]	Rear fuselage metal structures
	MASA	Tailplane leading/trailing-edges and ribs; wing ribs
	Sacesa	Belly fairing panels
Sweden	Saab Aerostructures[1]	Wing fixed leading-edge outboard of inner engine nacelles
Switzerland	RUAG Aerospace[1]	D-nose skins for wing inner fixed leading-edge and outer fixed trailing-edge
UK	BAE Systems Aerostructures[1]	Wing inboard outer fixed leading-edges
	Dunlop Aerospace	Nosewheels and brakes
	GKN Aerospace Services[1]	Wing fixed trailing-edge composites secondary structures; flap track beams
USA	Goodrich	Main landing gear
	Honeywell	Mainwheels and brakes
	Ralee Engineering	Wing top skin stringers

[1] Risk-sharing partner

Two 241 bar (3,500 lb/sq in) hydraulic systems (yellow and green) and two electrical systems (red and orange) for flight controls, each of the latter using at least two different systems in case of failure of any one; each system fully independent. Hydraulic systems use lighter, more compact pipework compared with earlier Airbus products. Wing-mounted landing gear powered by green system; underfuselage main gear runs on yellow system. Power generation systems with 180 kVA generators for each engine provided by Hamilton Sundstrand, which also selected (with P&WC) to provide APU. Eaton 345 bar (5,000 lb/sq in) hydraulic generators using eight engine-driven pumps and four electric pumps.

Onboard oxygen generator (OBOGS) optional.

AVIONICS: Honeywell flight management system, with Thales/Diehl displays; Rockwell Collins com/nav standard (includes VHF-920 and HF-900 data radios, multimode receiver, VOR-900 omnirange/marker beacon receiver, DME-900 and ADF-900). Elta ADT 406 ELT. Honeywell RDR-4000 colour weather radar. Dual Thales HUDs optional. Cockpit layout, by Airbus Toulouse division, compatible with other Airbus family members.

Eight new-generation 15 × 20 cm (6 × 8 in) LCDs form centre of package. Pilots can modify objects such as performance and navigation objects inside the screens using a cursor. The cursor is controlled by a new addition to the cockpit - a fixed, wireless, trackball integrated into the pedestal. Cursor control has been made possible by moving away from an 'MS-DOS-like' systems environment to a modern user-friendly 'Windows-like' system in the A380, using a cursor to control and command.

New features specific to the displays include a split-screen feature on the navigation screen. In addition to the conventional horizontal graphic display there is an additional vertical view, from which pilots can see a cut along the trajectory of the aircraft containing all data relevant to the vertical trajectory, such as safety altitudes, terrain, weather or vertical flight plan.

Onboard information system (OIS) offers flight crews easy access to non-avionics-related information and applications. This includes access to flight manuals, maintenance tools, cabin log books, passenger lists and more. As the system uses web technology, information can be passed in real time from ground to air and vice-versa. The OIS also provides passengers with web-browsing facilities. Flight crew access the OIS from display units, integrated at either side of the instrument panel, using a keyboard installed in the folding table. Onboard maintenance system (OMS) provides real-time information to ground crew. Honeywell terrain guidance and on-ground navigation will be integrated into FMS.

DIMENSIONS, EXTERNAL:
- Wing span ... 79.80 m (261 ft 9¾ in)
- Wing chord at root ... 16.60 m (54 ft 5½ in)
- Wing aspect ratio ... 7.5
- Length overall ... 72.75 m (238 ft 8 in)
- Fuselage max width ... 7.14 m (23 ft 5 in)
- Height overall ... 24.08 m (79 ft 0 in)
- Tailplane span ... 30.38 m (99 ft 8 in)
- Wheel track: A380-800 ... 14.33 m (47 ft 0¼ in)
  - A380-800F ... 14.39 m (47 ft 2½ in)
- Wheelbase: A380-800 ... 30.40 m (99 ft 8¾ in)
- Rear cargo door (A380-800F):
  - Max width ... 3.43 m (11 ft 3 in)
  - Max height ... 2.54 m (8 ft 4 in)

DIMENSIONS, INTERNAL:
- Cabin, A380-800: Length ... 50.68 m (166 ft 3¼ in)
  - Max width: main deck ... 6.58 m (21 ft 7 in)
    - upper deck ... 5.92 m (19 ft 5 in)
  - Width at floor: main deck ... 6.20 m (20 ft 4 in)
    - upper deck ... 5.33 m (17 ft 6 in)
- Bulk hold volume: A380-800 ... 18.4 m^3 (650 cu ft)
  - A380-800F ... 20.0 m^3 (706 cu ft)
- Total hold volume: A380-800 ... 171.0 m^3 (6,039 cu ft)
  - A380-800F ... 1,134 m^3 (40,047 cu ft)

AREAS:
- Wings, gross ... 845.0 m^2 (9,095.5 sq ft)

WEIGHTS AND LOADINGS:
- Operating weight empty:
  - A380-800 ... 277,000 kg (610,680 lb)
  - A380-800F ... 252,000 kg (555,565 lb)
- Max payload: A380-800 ... 84,000 kg (185,190 lb)
  - A380-800F ... 150,000 kg (330,700 lb)
- Max T-O weight: A380-800 ... 560,000 kg (1,234,580 lb)[1]
  - A380-800F ... 590,000 kg (1,300,720 lb)
- Max ramp weight: A380-800 ... 562,000 kg (1,239,000 lb)
  - A380-800F ... 592,000 kg (1,305,140 lb)
- Max landing weight:
  - A380-800 ... 386,000 kg (850,980 lb)[1]
  - A380-800F ... 427,000 kg (941,370 lb)
- Max zero-fuel weight:
  - A380-800 ... 361,000 kg (795,860 lb)[1]
  - A380-800F ... 402,000 kg (886,260 lb)
- Max wing loading:
  - A380-800 ... 662.7 kg/m^2 (135.74 lb/sq ft)
  - A380-800F ... 698.2 kg/m^2 (143.01 lb/sq ft)
- Max power loading:
  - A380-800 ... 463 kg/kN (4.54 lb/lb st)
  - A380-800F ... 433 kg/kN (4.25 lb/lb st)

[1] *As of December 2003, Airbus was considering offer of higher MTOW and MLW options on A380-800 from outset, of 569,000 kg (1,254,430 lb) and 391,000 kg (862,005 lb) respectively. New MZFW had not been defined at that time.*

PERFORMANCE (estimated, Trent 900 engines):
- Max operating speed ... M0.89 (340 kt; 630 km/h; 391 mph)
- Econ cruising Mach No. ... 0.85
- Initial cruising altitude ... 10,670 m (35,000 ft)
- Service ceiling ... 13,100 m (42,980 ft)
- T-O field length, ISA + 15°C:
  - A380-800 ... 2,987 m (9,800 ft)
  - A380-800F ... 3,009 m (9.870 ft)
- Runway ACN ... approx 68
- Time from brake release to FL350:
  - A380-800 ... 36 min
  - A380-800F ... 35 min
- Range:
  - A380-800 ... 8,000 n miles (14,816 km; 9,206 miles)
  - A380-800F ... 5,600 n miles (10,371 km; 6,444 miles)

OPERATIONAL NOISE LEVELS: Complies with London airports' 'Quota Count 2' category for departures (3 dB quieter than, or half the noise energy level of, most in-service Boeing 747s.

UPDATED

AIRBUS MULTIROLE TANKER-TRANSPORT (MRTT)

TYPE: Tanker-transport.

PROGRAMME: Belgian, Canadian, French, German and Thai air forces already operate A310s variously fitted for VIP, troop and/or freight transport. Airbus has delegated development and marketing of flight refuelling versions to its major partners, using either pre-owned or new aircraft. Demonstrator MRTT produced by conversion of former airline A310-324 N816PA; undertook compatibility trials with RAF aircraft, July 1995. Marketing efforts originally centred on the A310 version, which offered for the RAF's FSTA (future strategic tanker aircraft) requirement. This planned to entail start of replacement of VC10s in 2005, although more than one type of tanker is expected to be obtained, probably by lease rather than purchase. Exclusive marketing rights for MRTT assigned to Raytheon (which will also become Design Authority), June 1999. Four Luftwaffe A310-300s being converted into MRTTs, first of which (1027) flew 20 December 2003. Canada has also announced intent to modify two Polaris MRTs into MRTTs in same programme.

A330-200 is being developed as strategic tanker/transport and has been chosen by the AirTanker consortium as most suitable platform for UK FSTA programme. Consortium members EADS, Rolls-Royce, Thales and FRA are advocating a private finance initiative (PFI) solution with service entry in 2008 and programme life of 27 years; AirTanker bid selected on 26 January 2004 and will involve a mix of about 16 new-build and second-hand A330-200s; at same time, it was revealed that several key aspects of PFI solution still to be resolved before formal signature of contract, which could occur in second quarter of 2004. A330-based tanker also offered to Australia for AIR 5402 requirement for up to five aircraft; tenders submitted in October 2003, with selection and contract signature in first half of 2004, followed by service entry in 2007.

CURRENT VERSIONS: Interim MRT (MultiRole Transport) version of A310, without tanker capability, converted by Elbe Flugzeugwerke, Dresden and Lufthansa Technik, Hamburg, for Luftwaffe. Structural strengthening and 3.50 m × 2.50 m (11 ft 5¾ in × 8 ft 2½ in) cargo door added in port forward fuselage. Capacity of up to 214 passengers; or 36 tonnes of cargo/passengers; or 56 stretchers and six intensive care patients in the casevac role. First redelivered from Dresden in June 1999; four of seven Luftwaffe A310s redelivered as MRTs by early 2002. In service, Luftwaffe aircraft proved partially incompatible with military cargo handling equipment and upper deck therefore rarely used for freight; full conversion of two MRTs to MRTT, plus conversion of further pair of standard aircraft directly to MRTT, being undertaken between 2002 and 2005; first tanker kit delivered by EADS CASA in November 2002, with first flight following modification set to take place in fourth quarter of 2003. Remaining Luftwaffe A310s comprise two in VIP configuration and one passenger transport.

DESIGN FEATURES: Conversions offer greater refuelling and transport capability than earlier airliners in combination with modern aircraft with better lifetime costs and longer life expectancy. Possible roles include tanker with underwing HDUs and fuselage-mounted boom and/or hose transfer systems and carrying in excess of 111,270 kg (245,300 lb) of fuel (139,060 litres; 36,737 US gallons; 30,590 Imp gallons); and cargo and personnel transports which can be combined with refuelling, medevac, airborne command post and reconnaissance/airborne warning.

EADS has completed design of new aerial refuelling boom and this will undergo rig testing from November 2004, followed by flight trials on an Airbus A310-300 starting in late 2005; thereafter, it will be adapted for use by A330 aircraft. Maximum fuel flow rate is expected to be 4,550 litres (1,200 US gallons; 1,000 Imp gallons) per minute.

Airbus conversions offer payloads from 35,000 to 50,000 kg (77,161 to 110,231 lb), full payload transatlantic range, long on-station time, combined boom and hosereel transfer capability (fuel transfer rate at each refuelling point of 1,590 litres (420 US gallons; 350 Imp gallons) per minute), standard Airbus forward port-side freight door (projected height 2.57 m; 8 ft 5¼ in, width 3.58 m; 11 ft 9 in); quick-change main deck layout, probe or receptacle fuel receiver capability; commonality with existing airliners and same worldwide support resources, predictable spares requirements and longer remaining airframe life.

About 100 civil operators fly A300/310, and first-generation Airbus airliners now available on second-hand market; military rendezvous and self-protection systems can be fitted; main deck can be converted with palletised seating for up to 270 passengers in under 24 hours; up to 28,000 kg (61,729 lb) of additional fuel can be carried in tanks in underfloor cargo compartments.

DIMENSIONS, EXTERNAL (310: A310-300 MRTT, 330: A330-200 MRTT):
Length overall, incl probe: 310 ... 47.36 m (155 ft 4½ in)
330 ... 59.69 m (195 ft 10 in)

DIMENSIONS, INTERNAL:
Usable cabin length: 310 ... 43.90 m (144 ft 0 in)
330 ... 45.00 m (147 ft 7¾ in)
Cabin height: both ... 2.28 m (7 ft 5¾ in)
Max cabin width: both ... 5.29 m (17 ft 4¼ in)
Underfloor freight hold volume:
310 ... 80.0 m³ (2,825 cu ft)
330 ... 136.0 m³ (4,803 cu ft)

WEIGHTS AND LOADINGS:
Operating weight empty:
310 ... 80,830 kg (178,200 lb)
330 ... 120,500 kg (265,650 lb)
Max non-fuel payload: 310 ... 37,000 kg (81,571 lb)
330 ... 61,300 kg (135,140 lb)
Max normal fuel capacity:
310 ... 47,940 kg (105,690 lb)
330 ... 111,270 kg (245,300 lb)
Additional fuel: 310 ... 28,240 kg (62,258 lb)
Max T-O weight:
normal, 310 ... 157,000 kg (346,125 lb)
330 ... 230,000 kg (507,075 lb)
optional, 310 ... 164,000 kg (361,550 lb)
330 ... 233,000 kg (513,675 lb)
Max ramp weight: 310 ... 164,900 kg (363,550 lb)
330 ... 233,900 kg (515,650 lb)
Max landing weight: 310 ... 124,000 kg (273,375 lb)
330 ... 182,000 kg (401,250 lb)
Max zero-fuel weight: 310 ... 114,000 kg (251,325 lb)
330 ... 170,000 kg (374,775 lb)

PERFORMANCE:
Refuelling speed:
both, boom ... 240–320 kt (444–592 km/h; 276–368 mph)
both, hose and drogue:
S/L to FL275 ... 200–325 kt (370–602 km/h; 230–374 mph)
FL275 to FL350 ... M0.82
T-O run: 310 ... 2,347 m (7,700 ft)
Max range, standard fuel:
310 ... 4,800 n miles (8,889 km; 5,523 miles)
Max range, using transferable fuel:
310 ... 7,200 n miles (13,334 km; 8,285 miles)
330 ... 9,000 n miles (16,668 km; 10,357 miles)

UPDATED

AIRBUS AEW&C

TYPE: Airborne early warning and control system.

PROGRAMME: A310 originally selected by Raytheon Systems as airborne platform in unsuccessful bid for the Australian Project Wedgetail AEW&C requirement in which it competed against Boeing 737 and Lockheed Martin C-130J. Featured Elta Phalcon 360° phased-array radar in a fixed dorsal dome; formally announced in January 1997. Australia selected Boeing 737 in July 1999, but A310 AEW&C also offered to Turkey (again unsuccessfully) and South Korea, which due to select E-X platform in 2004 and award contract in 2005. However, latest proposals envisage using either A320 (Thales) or A321 (Raytheon).

UPDATED

Airbus A310 AEW&C proposal with non-rotating Elta radar above rear fuselage ***(Paul Jackson)*** 0062385

Standard production Airbus A330 and two German Panavia Tornados simulating a refuelling exercise at the 2003 Royal International Air Tattoo ***(Paul Jackson)*** ***NEW***/0569624

AIRBUS A321 AGS

TYPE: Airborne ground surveillance system.

PROGRAMME: Northrop Grumman chose the A321 as its preferred Joint STARS platform to compete for NATO's AGS (airborne ground surveillance) requirement. Considered in 1996 and rejected on cost grounds, the A321 was reconsidered in the light of NATO's unwillingness to accept a Boeing 707-based solution. Plans called for selection in 2002 and IOC, with six aircraft, by 2007, but implementation delayed by US refusal to release all technology. Northrop Grumman's AGS proposal features the US Army's AN/APY-X RTIP (Radar Technology Insertion Program) with an electronically scanned antenna array, giving spot/swath SAR, wide area MTI, ultra-high resolution SAR, inverse SAR and long-range high-resolution MTI modes. It would have 10 to 12 operator stations.

By 2002, Northrop Grumman, EADS and Galileo Avionica had teamed up to promote TIPS (Transatlantic Industrial Proposed Solution), which is based on A321 airframe and will involve participation of wide range of subcontractors throughout NATO countries. General Dynamics Canada, Indra and Thales had joined industrial team by 9 January 2004, when operating proposal submitted to NATO. National Armament Directors' conference approved proposal on 16 April 2004 (rejecting rival Raytheon bid) with intention of awarding contract in second quarter of 2005. By 2004, AGS envisaged with A321 "almost certainly" having sensor raised above rear fuselage (in "topknot" position) and with Northrop Grumman Global Hawk UAVs as 'gap-fillers'. Contract value estimated as US$4.8 billion (2004), of which initial design and development worth some US$300 million. IOC could be achieved by 2010, with four aircraft and 18 ground stations to provide minimum single-orbit coverage. Full requirement may be 12 and 48, respectively, although six and 24 would ensure two simultaneous missions.

UPDATED

Airbus A321 AGS configuration proposed in 2004 *(TIPS)* ***NEW***/0576826

AIRBUS MPA

TYPE: Maritime surveillance twin-jet.

PROGRAMME: Alenia Aeronautica/Finmeccanica and EADS Military Aircraft revealed at Farnborough Air Show in July 2002 that they had joined forces to submit proposal based on Airbus A320 to satisfy joint German/Italian MPA-R (Maritime Patrol Aircraft Replacement) requirement. Joint concept by EADS Deutschland (Ottobrunn) and EADS France (Toulouse). Ventral search radar; undernose EO/IR sensor; capacity for internal torpedo carriage in wing/body box; seven operator stations in cabin. Total development time expected to be five years. Request for proposals issued in November 2001 by German-Italian programme management team, with Airbus offer anticipating production of 24 MPA320 aircraft (10 German and 14 Italian), as well as provision of training simulators and a complete package of logistic and customer support services. Competing bids received by 26 July 2002 from Boeing, L-3 Communications and Lockheed Martin; outcome still awaited in mid-November 2003; if selected, entry into service now unlikely before beginning of 2009 at earliest.

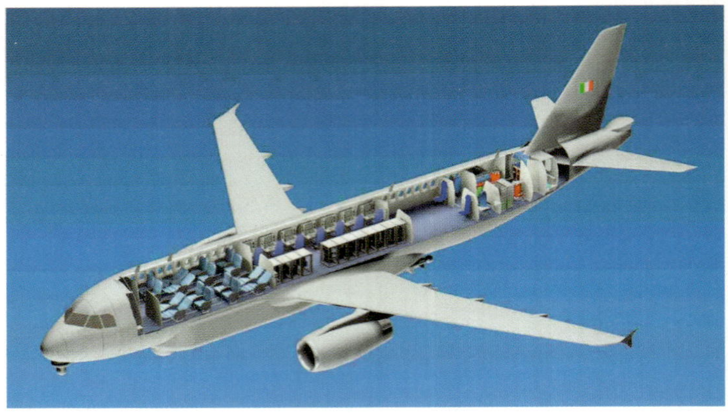

Computer-generated image of proposed Airbus MPA320 0527122

In second quarter of 2003, EADS Military Aircraft revealed MPA319. Based on commercial A319, the MPA319 is targeted at customers requiring a smaller and less costly maritime patrol aircraft than the MPA320. MPA319 concept features ventral 'canoes' ahead of and behind main landing gear in which weapons, including torpedoes and air-to-surface missiles, are contained. Mission system equipment expected to include retractable radar and electro-optical sensor turrets in nose section. MPA319 customers will be able to specify IAE V2500 or CFM International CFM 56 turbofan power plants.

Description for MPA 320 generally as for Airbus A320, except that below:

WEIGHTS AND LOADINGS:
- Weight empty: manufacturer's .... 35,400 kg (78,044 lb)
- operational .... 44,600 kg (98,326 lb)
- Max weapon load .... 4,900 kg (10,803 lb)
- Max fuel weight .... 27,200 kg (59,966 lb)
- Max T-O weight .... 77,000 kg (169,755 lb)
- Max landing weight .... 66,000 kg (145,505 lb)

PERFORMANCE (estimated):
- Turn radius, clean .... less than 1,850 m (6,070 ft)
- Ferry range .... 4,200 n miles (7,778 km; 4,833 miles)

UPDATED

AIRTECH

AIRCRAFT TECHNOLOGY INDUSTRIES

PARTICIPATING COMPANIES

EADS CASA: (Spain)
Dirgantara: (Indonesia)

Airtech was formed by CASA (now part of EADS) and IPTN (now known as Dirgantara) to develop the CN-235 twin-turboprop transport; responsibility for design and production was shared equally. The partnership applied only to the Series 10 and Series 100/110, with later versions being developed independently by CASA, according to a statement by that company.

UPDATED

AIRTECH CN-235

Spanish Air Force designations: T.19A and T.19B

TYPE: Twin-turboprop transport.

DEVELOPMENT MILESTONES

Design began	Jan 1980
First prototype material cut	May 1981
Rolled out	10 Sep 1983
First flight	11 Nov 1983
Certification	20 Jun 1986
First flight, production	19 Aug 1986
First delivery (Merpati)	15 Dec 1986
Entered service (Merpati)	1 Mar 1988

PROGRAMME: Launched as joint venture between CASA and Indonesian manufacturer IPTN (now Dirgantara, which see), which formed Airtech company to manage programme. Series 10 and Series 100/110 versions covered by this agreement; subsequent versions, notwithstanding Indonesian Series 220 and 330 equivalents, stated by CASA to be wholly Spanish.

Preliminary design began January 1980, prototype construction May 1981; one prototype completed in each country, with simultaneous roll-outs 10 September 1983; first flights 11 November 1983 (by CASA's ECT-100) and 30 December 1983 (IPTN's PK-XNC); Spanish and Indonesian certification 20 June 1986; first flight of production aircraft 19 August 1986; FAA type approval (FAR Pts 25 and 121) 3 December 1986; deliveries began 15 December 1986 from IPTN line and 4 February 1987 from CASA; entered service (with Merpati Nusantara Airlines) 1 March 1988; JAR 25 type approval October 1993.

Licence agreement with TAI (see Turkish section) announced January 1990, initially to assemble and later to manufacture locally 50 of 52 ordered; first flight of Turkish-assembled aircraft 24 September 1992; first delivery 13 November 1992; final air force delivery 10 August 1998, but TAI subsequently produced follow-on batch of nine maritime patrol variants and may build further 10 for maritime missions.

One of two CN-235 Srs 300s operated by Devon Holding & Leasing Inc of Lexington, North Carolina, pictured at Baghdad in late 2003 while delivering stocks of newly minted Iraqi dinars from the US to relaunch the local currency *(Patrick Allen)* ***NEW***/1033220

Artist's impression of a US Coast Guard CN-235 ***NEW***/0569651

In 1995, CASA unilaterally launched development of a stretched CN-235, as C-295; this is described under CASA heading in Spanish section.

CURRENT VERSIONS: **CN-235 Series 10:** Initial production version (15 built by each company), with CT7-7A engines.

CN-235 Series 100/110: Generally as Series 10, but CT7-9C engines in new composites nacelles; replaced Series 10 in 1988 from 31st production aircraft. Series 100 is Spanish-built and, following JAA certification, was certified by FAA in February 1992. Series 110 is Indonesian-built, with improved electrical, warning and environmental systems to comply with JAR 25; certification of this version achieved in Europe (JAA), July 1995.

Detailed description applies to the above version except where indicated.

CN-235 Series 200/220: Structural reinforcements to cater for higher operating weights, aerodynamic improvements to wing leading-edges and rudder, reduced field length requirements and much-increased range with maximum payload; Series 200 is Spanish-built and was certified by FAA March 1992. Series 220 is Indonesian-built, with improvements similar to Srs 110; prototype, flown early 1996, is converted from a company development aircraft (PK-XNV, the 20th production aircraft from the Indonesian line); orders include eight for Malaysian Air Force, of which six completed to Srs 220 standard (including three for maritime patrol) by early 1998 (34th to 39th Indonesian-built), with remaining two to be completed in VIP configuration. Revised leading-edge shape led to requirement to requalify pneumatic de-icer boots, delaying initial deliveries. Further orders for Series 220 from South Korea (eight, including one for VIP use and one for VVIP use) and Pakistan (four, including one for VIP use).

CN-235 Series 300/330: IPTN originally offered Series 330 **Phoenix** (with new Honeywell avionics, ARL-2002 EW system and 16,800 kg; 37,037 lb MTOW) to Royal Australian Air Force to meet Project Air 5190 tactical airlift requirement, but was forced by financial constraints to withdraw in 1998. Separately, CASA offered its own Series 300 to meet the same specification.

CN-235 Series 300 under certification in 2000 with an open-systems avionics architecture, based on MIL-STD-1553B and ARINC 429 digital databusses. Full NVG-compatible cockpit; four-dimensional navigation system with avionics suite, including Thales (Sextant) Topdeck colour weather radar, radios, solid-state flight data and cockpit voice recorders, enhanced TCAS, enhanced GPWS and four 152 × 203 mm (6 × 8 in) LCDs; twin HUDs and Totem 3000 ring laser gyro INS optional. Other features include in-flight refuelling capability, improved pressurisation (2,440 m; 8,000 ft cabin environment at 7,620 m; 25,000 ft) and provision for optional twin nosewheel installation to provide better soft-field taxying capability.

CN-235 AEW: Proposals were revealed in December 1995 for fitment of an Ericsson Erieye electronically scanned phased-array radar above the fuselage of a CN-235. Initial interest was from the Indonesian Air Force, but primarily in the ocean surveillance role; retrofit of three existing aircraft was considered, but has not been undertaken. Radar, three surveillance operators' positions and associated equipment increase aircraft weight by approximately 2,000 kg (4,409 lb). No recent news of this proposal has been received.

CN-235ER: Extended-range version (based on Series 300) originally selected by US Coast Guard in 2002 as fixed-wing element of Project Deepwater re-equipment programme, but subsequently shelved in favour of the basic CN-235 Series 300M MPA, described separately.

CN-235 M: Other military transport versions.

CN-235 MP Persuader and CN-235 MPA: Maritime patrol versions; described separately.

CN-235 QC: Quick-change cargo/passenger version; certified by Spanish DGAC May 1992.

CN-245: Indonesian stretched version; not built.

C-295: Spanish stretched version; described separately under CASA heading.

N2XXM: Project abandoned.

CUSTOMERS: See table. One (s/n 66049) acquired (presumably second-hand) by USAF Special Operations Command in 1998. Turkey signed a lease agreement on 16 April 1999 to allow a one-year renewable lease of two Turkish Air Force CN-235s to Jordan. Three Merpati aircraft leased to Air Venezuela from May 1999; three leased to Asian Spirit Airlines, Philippines, from March 2000, including two on lease-purchase. One CN-235-300 (first of this subvariant in service) leased by Austrian Ministry of Defence for six months from April 2000. CN-235 is contender in Taiwanese Air Force requirement for 18 light transports, and for the US Army Airborne Common Sensor platform requirement. Winner of US Coast Guard Project Deepwater competition, with total of 35 aircraft expected to be acquired.

COSTS: US$17.1 million (2002) programme unit cost, Malaysia.

DESIGN FEATURES: Optimised for short-haul operations, enabling it to fly four 860 n mile (1,593 km; 990 mile) stage lengths (with reserves) before refuelling and to operate from paved runways or unprepared strips; high-mounted wing; pressurised fuselage (including baggage compartment) of flattened circular cross-section, with upswept rear end incorporating cargo ramp/door; sweptback fin (with dorsal fin) and rudder; low-set non-swept fixed incidence tailplane and elevators; two small ventral fins; vortex generators on rudder and elevator leading-edges; optional extended nose radome.

NACA 65_3-218 aerofoil with no-dihedral/constant chord centre-section; tapered outer panels have 3° dihedral and 3° 51' 36" sweepback at quarter-chord.

FLYING CONTROLS: Conventional and manual. Ailerons, elevators and rudder statically and dynamically balanced (duplicated actuation for ailerons); mechanical servo tab and electric trim tab in each aileron, rudder and starboard elevator, trim tab only in port elevator; single-slotted inboard and outboard trailing-edge flaps (each pair interchangeable port/starboard), actuated hydraulically by Dowty irreversible jacks.

STRUCTURE: Conventional semi-monocoque, mainly of aluminium alloys with chemically milled skins; composites (mainly glass fibre or glass fibre/Nomex honeycomb sandwich, with some carbon fibre and Kevlar) for leading/trailing-edges of wing/tail moving surfaces, wing/fuselage and main landing gear fairings, wing/fin/tailplane tips, engine nacelles, ventral fins and nose radome. Propeller blades are of glass fibre, with metal spar and urethane foam core.

CASA builds wing centre-section, inboard flaps, forward and centre fuselage, engine nacelles; Dirgantara builds outer wings, outboard flaps, ailerons, rear fuselage and tail unit; both manufacturers use numerical control machinery extensively. Final assembly line in each country. Part of tail unit built by ENAER Chile under subcontract from CASA. TAI (Turkey) initially assembled under licence before progressing gradually to local manufacture.

LANDING GEAR: Messier-Bugatti retractable tricycle type with levered suspension, suitable for operation from semi-prepared runways. Electrically controlled hydraulic extension/retraction, with mechanical back-up for emergency extension. Oleo-pneumatic

Turkish Air Force Airtech CN-235 *NEW*/0569650

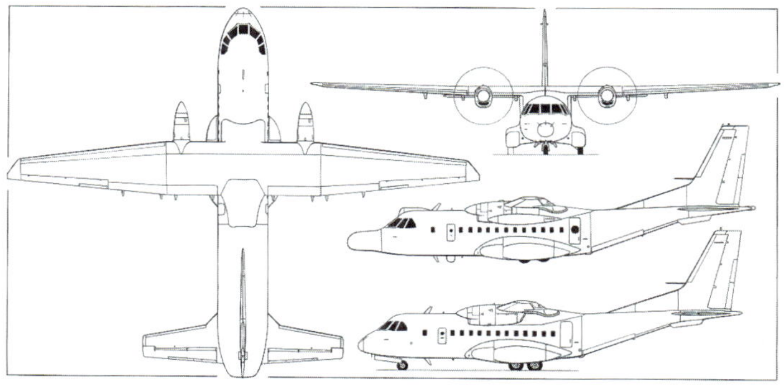

CASA CN-235 twin-turboprop multipurpose transport, with additional side view (centre) of a representative CN-235 MPA *(Dennis Punnett)* 0507150

shock-absorber in each unit. Each main unit comprises two wheels in tandem, retracting rearward into fairing on side of fuselage. Mainwheels semi-exposed when retracted. Single steerable nosewheel (±48°) retracts forward into unpressurised bay under flight deck. Dunlop 28×9.00–12 (12 ply) tubeless mainwheel tyres standard, pressure 5.17 bar (75 lb/sq in) on civil version, 5.58 bar (81 lb/sq in) on military version; low-pressure mainwheel tyres optional, size 11.00-12 (10 ply), pressure 3.45 bar (50 lb/sq in). Dunlop 24×7.7 (10/12 ply) tubeless nosewheel tyre, pressure 5.65 bar (82 lb/sq in) on civil version, 6.07 bar (88 lb/sq in) on military version; optional 8.50×10 (12 ply). Dunlop hydraulic differential disc brakes; Dunlop anti-skid units on main gear. Chilean Army aircraft used in Antarctic have wheel/ski gear. Minimum ground turning radius 9.50 m (31 ft 2 in) about nosewheel, 18.98 m (62 ft 3¼ in) about wingtip.

POWER PLANT: Two General Electric CT7-9C turboprops (CT7-9C3 in Srs 300), each flat rated at 1,305 kW (1,750 shp) (S/L, to 41°C) for take-off and 1,394.5 kW (1,870 shp) up to 31°C with automatic power reserve. Hamilton Sundstrand 14RF-21 (14RF-37 in Srs 300) four-blade constant-speed propellers, with full feathering and reverse-pitch capability. Fuel in two 1,042 litre (275 US gallon; 229 Imp gallon) integral main tanks in wing centre-section and two 1,592 litre (421 US gallon; 350 Imp gallon) integral outer-wing auxiliary tanks; total fuel capacity 5,264 litres (1,391 US gallons; 1,158 Imp gallons), of which 5,128 litres (1,355 US gallons; 1,128 Imp gallons) are usable. Single pressure refuelling point in starboard main landing gear fairing; gravity filling point in top of each tank. Propeller braking permits starboard engine to be used as on-ground APU. Oil capacity 14 litres (3.7 US gallons; 3.1 Imp gallons).

ACCOMMODATION: Crew of two on flight deck, plus cabin attendant (civil version) or third crew member (military version). Accommodation in commuter version for up to 44 passengers in four-abreast seating, at 76 cm (30 in) pitch, with 22 seats each side of central aisle. Lavatory, galley and overhead luggage bins standard. Pressurised baggage compartment at rear of cabin, aft of movable bulkhead; additional stowage in rear ramp area and in overhead lockers. Can also be equipped as mixed passenger/cargo combi (for example, 19 passengers and two LD3 containers), or for all-cargo operation, with roller loading system, carrying four standard LD3 containers, five LD2s, or two 2.24 × 3.18 m (88 × 125 in) and one 2.24 × 2.03 m (88 × 80 in) pallets; or for military duties, carrying up to 57 fully equipped troops or 46 paratroops (51 troops or paratroops on Srs 300). Other options include layouts for aeromedical airlift (18 stretchers and two medical attendants on Srs 300), electronic warfare, geophysical survey or aerial photographic duties.

Main passenger door, outward- and forward-opening with integral stairs, aft of wing on port side, serving also as a Type I emergency exit. Type III emergency exit facing this door on starboard side. Crew/service downward-opening door (forward, starboard) has built-in stairs, and serves also as a Type I emergency exit, or as passenger door in combi version; second Type III exit opposite this door on port side. Wide ventral door/cargo ramp in underside of upswept rear fuselage, for loading of bulky cargo. Accommodation fully air conditioned and pressurised.

SYSTEMS: Hamilton Sundstrand air conditioning system, using engine compressor bleed air. Honeywell electropneumatic pressurisation system (maximum differential 0.25 bar; 3.6 lb/sq in) giving cabin environment of 2,440 m (8,000 ft) up to operating altitude of 5,480 m (18,000 ft) on Srs 200; Srs 300 cabin pressurisation increased to 0.38 bar (5.5 lb/sq in), giving cabin environment of 2,350 m (7,700 ft) at altitude of 7,620 m (25,000 ft). Hydraulic system, operating at nominal pressure of 207 bar (3,000 lb/sq in), comprises two engine-driven, variable displacement axial electric pumps, a self-pressurising standby mechanical pump, and a modular unit incorporating connectors, filters and valves; system is employed for actuation of wing flaps, landing gear extension/retraction, wheel brakes, emergency and parking brakes, nosewheel steering, cargo ramp and door, and propeller braking. Accumulator for back-up braking system.

28 V DC primary electrical system powered by two 400 A Auxilec engine-driven starter/generators, with two 24 V 37 Ah Ni/Cd batteries for engine starting and 30 minutes' (minimum) emergency power for essential services. Constant frequency single-phase AC power (115/26 V) provided at 400 Hz by three 600 VA static inverters (two for normal operation plus one standby); two three-phase engine-driven alternators for 115/200 V variable frequency AC power. Fixed oxygen installation for crew of three (single cylinder at 124 bar; 1,800 lb/sq in pressure); three portable units and individual masks for passengers.

Pneumatic boot anti-icing of wing (outboard of engine nacelles), fin and tailplane leading-edges. Electric anti-icing of propellers, engine air intakes, flight deck windscreen, pitot tubes and angle of attack indicators. Engine fire detection and extinguishing system.

AVIONICS (CIVIL): *Comms:* Two Rockwell Collins VHF-22B com radios, one Avtech DADS crew interphone, Rockwell Collins TDR-90 ATC transponder. Fairchild A-100A cockpit voice recorder, Avtech PACIS PA system. Dorne & Margolin ELT 8-1 emergency transmitter. Optional second TDR-90; optional HF-230 radio.

Radar: Rockwell Collins WXR-300 weather radar.

Flight: Two VIR-32 VOR/ILS/marker beacon receivers; DME-42; ADF-60A; two 332D-11T vertical gyros; two MCS-65 directional gyros; two ADI-85A; two HSI-85; two RMI-36; APS-65 autopilot/flight director; ALT-55B radio altimeter; two 345A-7 rate of turn sensors (all by Rockwell Collins); SFENA H-301 APM standby attitude director indicator; Hamilton Sundstrand Mk II GPWS; and Fairchild/Teledyne flight data recorder. Options include second DME-42 and ADF-60A, Rockwell Collins RNS-325 radar nav, Litton LTN-72R inertial nav or Global GNS-500A Omega navigation system.

Instrumentation: Rockwell Collins EFIS-85B five-tube CRT system standard.

AVIONICS (MILITARY) (INDONESIAN AIRCRAFT): *Comms:* Rockwell Collins AN/ARC-182 VHF/UHF; Rockwell Collins HF 9000 HF; IFF.

Flight: Rockwell Collins VIR-32 VHF nav; Litton LTN92 GPS-aided INS; Rockwell Collins DF-206A ADF; Rockwell Collins AN/APS-65F autopilot; GPWS.

CN-235 PRODUCTION
(at end-2003)

Customer	Qty	Series	First order	First aircraft	First delivery	Delivered	Mfr
Civil version:							
Austral (Argentina)	2	100[1]	19 Dec 89	LV-VHM	1993	2	CASA
Binter Canarias (Spain)	4[2]	10	10 Jun 88	EC-EMO	22 Dec 88	4	CASA
Binter Mediterraneo (Spain)	4	100[1]	19 Dec 89	EC-FAD	4 Sep 90	3	CASA
Mandala Airlines (Indonesia)	3	110		—	—	—	Dirgantara
Merpati Nusantara (Indonesia)	15	10		PK-MNA	15 Dec 86	15	Dirgantara
Devon Holding & Leasing[13] (USA)	2	300	2002	N168D	2002	2	CASA
Military version:							
Abu Dhabi Air Force	7	110	—	810	31 Aug 93	7	Dirgantara
Botswana Defence Force	2	10	10 Jun 86	OG-1	21 Dec 87	2	CASA
Brunei Air Wing	3	MPA		—	—	—	Dirgantara
	1	110		ATU-501	1997	1	Dirgantara
Chilean Army	4	100	12 Feb 89	E-216	31 Aug 89	4	CASA
Colombian Air Force	3	200	Jul 97	1260	28 Jan 98	3	CASA
Columbian Navy	2	200	Dec 02	—	2003	2	CASA
Ecuadorean Army	1	100	6 Jun 89	AEE-502	6 Jun 89	1	CASA
Ecuadorean Navy	1	100	27 Jul 88	ANE-204	13 Jun 89	1	CASA
French Air Force	15[3]	200	11 Apr 90	043	28 Feb 91	15	CASA
	5	200	2002	152	2002	5	CASA
Gabon Air Forces	1	100	26 Feb 90	TR-KJE	19 Mar 91	1	CASA
Indonesian armed forces	24[4]	110		A-2301	12 Jan 93	7	Dirgantara
Irish Air Corps	1[5]	100	3 Apr 91	250	10 Apr 91	1	CASA
	2	MP	3 Apr 91	252	8 Dec 94	2	CASA
Malaysian Air Force	8[6]	220		M44-01	26 Aug 99	6	Dirgantara
Moroccan Air Force	7[7]	100	19 Sep 89	CNA-MA	27 Sep 90	7	CASA
Oman Police	2	100	15 Feb 92	A40-CU	14 Jan 93	2	CASA
Pakistan Air Force	4	220	29 Jun 01	—	2004		Dirgantara
Panama National Guard	1[8]	10	19 Mar 87	SAN-265	13 Sep 88	1	CASA
Papua New Guinea Defence Force	2	100	26 Oct 91	P2-0501	15 Nov 91	2	CASA
Saudi Air Force	4	10	5 Feb 84	118	9 Feb 87	4	CASA
South African Air Force (ex-Bophuthatswana Air Force)	1	10	29 May 90	8026	6 Jan 91	1	CASA
South Korean Air Force	12	100	19 Aug 92	078	13 Nov 93	12	CASA
	8	220	21 Oct 97	044	18 Dec 01	8	Dirgantara
Spanish Air Force	2[9]	10	16 Nov 88	T.19-01	7 Dec 88	2	CASA
	18	100	28 Dec 90	T.19-03	1 Feb 91	18	CASA
	4	MP		—	—	—	CASA
Thai Ministry of Agriculture and Co-operatives	2	220	Oct 96	2221	Apr 99	2	Dirgantara
Thai Police	0[10]	200	Apr 95	28053	4 Mar 96	0	CASA
Turkish Air Force	52[11]	100	11 Dec 90	051	25 Jan 92	52	TAI/CASA
Turkish Navy	6	MP	23 Sep 98	TCB-651	23 Dec 01	6	TAI/CASA
Turkish Coast Guard	3	MP	23 Sep 98	TCSG-551	23 Dec 01	3	TAI/CASA
US Coast Guard	2	MP	May 2003	—	2006	—	CASA
Subtotals	**240**					**205**	
Demo/trials	3[12]	(various)		EC-016		3	CASA
	5	(various)		PK-XNC		5	Dirgantara
Totals	**248**					**213**	

[1] Converted from Series 100 to Series 200
[2] Withdrawn 1998; three to Luftmeister, South Africa; one to Turkish Army (lost 16 May 2001)
[3] Including option on seven taken up in February 1996; first eight as Srs 100, but upgraded to Srs 200 from 1999
[4] Includes six maritime patrol of which three on firm order
[5] Withdrawn at end of lease, 1995
[6] Including two VIP versions ordered in 2002
[7] Includes one VIP version
[8] To Flight International (USA) 1995
[9] VIP version
[10] Former demonstrator
[11] 50 built in Turkey by TAI
[12] Includes one -100QC; plus one -200QC sold in 1996 to East Texas Aircraft Services Corporation
[13] Appears to be a US government company

Notes: CASA reported 189 military orders by mid-2000, which implies reduction in Indonesian requirement

Instrumentation: Rockwell Collins EFIS-85B(14) EFIS (four or five screens). IPTN developing cockpit lighting system compatible with night vision goggles.

AVIONICS (MILITARY): Series 300: Thales Avionics Topdeck suite (see Current Versions) as core system.

Flight: Twin ADU 3000 air data units, GPSs and AHRSs; radar altimeter; TCAS; GPWS; weather radar; optional Totem 3000 LINS, Cat. II landing capability, MLS and satcom.

Instrumentation: Four 152 × 203 mm (6 × 8 in) LCDs; optional HUDs. Optional electro-optical sensors display imagery on LCDs. NVG compatibility.

Mission: Four-dimensional navigation FMS calculates high-altitude and computed air release points for load-dropping.

EQUIPMENT: Navigation lights, anti-collision strobe lights, 600 W landing light in front end of each main landing gear fairing, taxying lights, ice inspection lights, emergency door lights, flight deck and flight deck emergency lights, cabin and baggage compartment lights, individual passenger reading lights, and instrument panel white lighting, all standard. Hand-type fire extinguishers on flight deck (one) and in passenger cabin (two); smoke detector in baggage compartment.

ARMAMENT (MILITARY VERSION): Three attachment points under each wing. Weapons can include AGM-84 Harpoon anti-ship missiles; Indonesian MPA version (which see) can be fitted with two Mk 46 torpedoes or AM 39 Exocet anti-shipping missiles.

Data follow for CASA-built Srs 300.

DIMENSIONS, EXTERNAL:
- Wing span ... 25.81 m (84 ft 8 in)
- Wing chord: at root ... 3.00 m (9 ft 10 in)
  - at tip ... 1.20 m (3 ft 11¼ in)
- Wing aspect ratio ... 10.2
- Length overall, standard nose ... 21.40 m (70 ft 2½ in)
- Fuselage: Max width ... 2.90 m (9 ft 6 in)
  - Max depth ... 2.615 m (8 ft 7 in)
- Height overall ... 8.18 m (26 ft 10 in)
- Tailplane span ... 10.60 m (34 ft 9¼ in)
- Wheel track (c/l of mainwheels) ... 3.90 m (12 ft 9½ in)
- Wheelbase ... 6.92 m (22 ft 8½ in)
- Propeller diameter: Srs 200 ... 3.35 m (11 ft 0 in)
  - Srs 300 ... 3.66 m (12 ft 0 in)
- Propeller ground clearance ... 1.66 m (5 ft 5¼ in)
- Distance between propeller centres ... 7.00 m (22 ft 11½ in)
- Passenger door (port, rear) and service door (stbd, fwd): Height ... 1.70 m (5 ft 7 in)
  - Width ... 0.73 m (2 ft 4¾ in)
  - Height to sill ... 1.22 m (4 ft 0 in)
- Paratroop doors (port and stbd, rear, each):
  - Height ... 1.75 m (5 ft 9 in)
  - Width ... 0.90 m (2 ft 11½ in)
  - Height to sill ... 1.22 m (4 ft 0 in)
- Ventral upper door (rear): Length ... 2.365 m (7 ft 9 in)
  - Width ... 2.35 m (7 ft 8½ in)
  - Height to sill ... 1.22 m (4 ft 0 in)
- Ventral ramp/door (rear): Length ... 3.04 m (9 ft 11¾ in)
  - Width ... 2.35 m (7 ft 8½ in)
  - Height to sill ... 1.22 m (4 ft 0 in)
- Type III emergency exits (port, fwd, and stbd, rear):
  - Height ... 0.92 m (3 ft 0¼ in)
  - Width ... 0.51 m (1 ft 8 in)

DIMENSIONS, INTERNAL:
- Cabin, excl flight deck: Length ... 9.65 m (31 ft 8 in)
  - Max width ... 2.70 m (8 ft 10¼ in)
  - Width at floor ... 2.365 m (7 ft 9 in)

Max height ..... 1.90 m (6 ft 2¾ in)
Floor area ..... 22.8 m^2 (246 sq ft)
Volume ..... 43.2 m^3 (1,527 cu ft)

AREAS:
Wings, gross ..... 59.10 m^2 (636.1 sq ft)
Ailerons (total, incl tabs) ..... 3.14 m^2 (33.80 sq ft)
Trailing-edge flaps (total) ..... 10.87 m^2 (117.00 sq ft)
Fin, incl dorsal fin ..... 11.11 m^2 (119.59 sq ft)
Rudder, incl tabs ..... 4.20 m^2 (45.21 sq ft)
Tailplane ..... 21.20 m^2 (228.2 sq ft)
Elevators (total, incl tabs) ..... 6.17 m^2 (66.41 sq ft)

WEIGHTS AND LOADINGS (Srs 300):
Operating weight empty ..... 9,909 kg (21,846 lb)
Max payload ..... 6,000 kg (13,228 lb)
Max fuel weight ..... 4,230 kg (9,326 lb)
Max T-O weight ..... 16,500 kg (36,376 lb)
Max ramp weight ..... 16,550 kg (36,486 lb)
Max landing weight ..... 16,500 kg (36,376 lb)
Max zero-fuel weight ..... 15,400 kg (33,951 lb)
Max wing loading ..... 279.2 kg/m^2 (57.19 lb/sq ft)
Max power loading (without APR) ..... 6.33 kg/kW (10.39 lb/shp)

PERFORMANCE (Srs 300):
Max cruising speed ..... 246 kt (455 km/h; 283 mph)
Max rate of climb at S/L ..... 183 m (600 ft)/min
Service ceiling ..... 9,145 m (30,000 ft)
Service ceiling, OEI ..... 4,275 m (14,020 ft)
T-O run ..... 398 m (1,305 ft)
T-O to 15 m (50 ft) ..... 754 m (2,474 ft)
Landing from 15 m (50 ft) ..... 603 m (1,978 ft)
Range:
with max fuel ..... 2,701 n miles (5,003 km; 3,108 miles)
with 4,000 kg (8,818 lb) payload ..... 1,549 n miles (2,870 km; 1,783 miles)
with max payload ..... 393 n miles (727 km; 452 miles)
g limits: at MTOW ..... +2.5/−1
below 14,100 kg (31,085 lb) ..... +3/−1

UPDATED

AIRTECH CN-235 MP PERSUADER AND CN-235 MPA

TYPE: Maritime surveillance twin-turboprop.

CURRENT VERSIONS: **CN-235 MP Persuader:** CASA version; different avionics from Indonesian MPA. In service with Irish Air Corps and ordered by Spain (four) and Turkey (nine: six for Navy, three for Coast Guard, assembled by TAI at Ankara). In mid-1999, Turkey sought proposals from at least seven potential integrators of surveillance systems to provide radar, FLIR and an acoustics suite for naval CN-235s; on 6 September 2002, contract signed with Thales covering supply and integration of maritime patrol mission equipment for these nine aircraft by 2006. Contract worth US$350 million and could be followed by further 10 systems to equip additional batch of aircraft.

CN-235 Series 300M MPA: CASA version for US Coast Guard, which ordered initial two of anticipated 35 aircraft in FY03 budget at cost of US$147 million. Delivery expected in 2006 and will feature EADS CASA Fully Integrated Tactical System (FITS), including

Maritime surveillance CN-235 MP Persuader of the Irish Air Corps *(Paul Jackson)* ***NEW***/0569649

multimode search radar; EO/IR sensors; VHF, UHF and HF communications, SATCOM terminal and large observation windows. Entire tactical system will be palletised, allowing aircraft to be easily reconfigured for use as personnel and/or cargo transport.

CN-235 MPA: Indonesian-developed version; available either with lengthened nose housing radar and IFF; or with normal CN-235 nose, plus belly radar; CN-235 prototype PK-XNC served as testbed. Maximum T-O weight 15,400 kg (33,951 lb), endurance more than 8 hours. Provision for quick-change configuration for general transport, communications or other duties. Required by Indonesian Navy (six included in national order for 24), Indonesian Air Force (three) and Brunei (three).

Indonesia confirmed initial three firm orders in May 2000, when Thomson-CSF (now Thales) selected to supply AMASCOS airborne maritime situation control system, comprising Elettronica ALR-733 RWR, T-CSF Ghlio thermal imager and Sextant Gemini navigation computer. Brunei chose Boeing as Argo Systems integrator for its three aircraft in late 1995, specifying individual sensors in October 1996 as AN/AAQ-21 FLIR, BAE Sky Guardian ESM, Cossor 3500 IFF and AN/APS-134 radar, plus two operators' consoles. BAE Systems Australia marketing CN-235 MPA in Asia-Pacific region under September 1997 agreement; BAE also to provide advanced systems development for proposed configurations.

AVIONICS (PERSUADER): *Radar:* Litton APS-504(V)5.

Mission: FLIR-2000HP undernose-mounted night vision system and Litton AN/ALR-85(V) ESM system, fully integrated via a central tactical processor with reconfigurable consoles.

AVIONICS (CN-235 MPA): *Radar:* BAE Systems Seaspray 4000, or Raytheon AN/APS-134 (LW) or Thales Ocean Master 100.

Flight: Litton LN92 ring laser gyro INS; Trimble TNL 7900 Omega/GPS.

Mission: Argo data processing and display system with multifunction consoles. BAE Systems Sky Guardian SG-300, or Argo Systems AR-700 or Litton AN/ALR-93(V)4 ESM. FLIR Systems AN/AAQ-21 Safire or BAE Systems MRT FLIR. Cossor 3500 IFF interrogator. (Trials aircraft originally equipped with APS-504 and Ocean Master; SG-300. Reconfigured by 1994 with AN/APS-134, MRT, AR-700, LN92 and TNL 7900. Further alternatives available at customer's option.)

UPDATED

AM

AIRBUS MILITARY SAS

404 Avenida de Aragon, E-28022 Madrid, Spain
Web: http://www.airbusmilitary.com
PRESIDENT: Francisco Fernández Sáinz
TECHNICAL DIRECTOR: Alain Cassier
COMMERCIAL DIRECTOR: Richard Thompson
INDUSTRIAL DIRECTOR: Angel Hurtado
HEAD OF MARKETING: David R Jennings

PARTICIPATING COMPANIES
Airbus SAS (France)
EADS CASA (Spain)
FLABEL (Belgium)
TAI (Turkey)

Airbus Military was legally established in January 1999 as a 'Société par Actions Simplifiées' as the prospective manufacturer of the Airbus A400M, formerly known as the Future Large Aircraft (FLA). Airbus is the major (63 per cent) shareholder in Airbus Military; TAI, OGMA and FLABEL are full risk-sharing partners. Airbus Military has assigned overall programme management, during development, to Airbus in Toulouse; as the programme reaches production, responsibility will progressively transfer to Spain.

Conceptual work was undertaken by the European FLA Group (Euroflag). Euroflag Srl originally formed 17 June 1991, with headquarters in Alenia head office in Rome, to manage European FLA development. Aerospatiale, Alenia, British Aerospace, CASA and Daimler-Benz Aerospace Airbus (DaimlerChrysler from 1998) had equal shares in Euroflag Srl; MoUs established 1992 with FLABEL (SABCA, SONACA, ASCO and BARCO) of Belgium, OGMA of Portugal and Turkish Aerospace Industries (TAI) of Turkey to allow integrated participation in FLA programme; BAE and FLABEL were industrial, not national, partners contributing their own funds, although the UK government announced in December 1994 that membership was to be upgraded to national participation.

The partners agreed in September 1994 to industrialise the programme by transferring it to their existing airliner production company; formal announcement was made on 14 June 1995 that Airbus Military Company would be established, replacing Euroflag, which then disbanded. Programme makes use of Airbus procedures and industrial infrastructure and takes advantage of technologies developed for Airbus airliners. The projected percentage shares for R&D financing at that time were Germany 25.7, France 17.2, UK 15.5, Italy 15.1, Spain 12.4, Turkey 6.9, Belgium 4.1 and Portugal 3.1.

By early 1997, the FLA programme had lost development sponsorship by the principal participating governments, although military commitments remained, subject to the aircraft being produced with commercial funding. Programme was weakened during 1997 by unilateral German negotiations with Ukraine over Antonov An-7X (Westernised An-70). German MoD attempts to involve Russia and Ukraine continued into 1999, but these were not supported by Airbus Industrie, which declined to become the prime contractor and assume the commercial risk of a programme based on the An-70; a study commissioned by the German MoD and carried out by DaimlerChrysler Aerospace reached a similar conclusion in September 1998. By mid-2000, senior German government sources were stressing the need for a European solution, prompting Airtruck to request assurances that its An-7X was still under consideration.

Airbus Military submitted responses to the seven-nation FLA RFP (request for proposals, dated September 1997) on 29 January 1999 and to the competitive Future Transport Aircraft (FTA) RFP issued to Boeing, Airbus and Lockheed Martin by Belgium, France, Spain and the UK on 31 July 1998. Acceptance of A400M was formally announced by all seven members on 27 July 2000; subsequently, on 19 June 2001, MoU signed by seven of nine participating nations (Belgium, France, Germany, Luxembourg, Spain, Turkey and the UK) concerning joint procurement through OCCAR, with Italy and Portugal expected to follow suit in the near future, although Italy announced intention to withdraw from project on 25 October 2001, with Portugal following suit in early 2003.

Programme encountered further difficulties in early 2002. Although OCCAR signed contract with AM on 18 December 2001 for 196 aircraft for eight countries, Germany failed to secure parliamentary approval for funding before agreement expired on 31 January 2002; Germany finally announced intention to go ahead with purchase at beginning of December 2002, although the number of aircraft involved had fallen from 73 to 60. Subsequently, on 27 May 2003, launch order for 180 aircraft signed in Bonn by OCCAR and Airbus Military, with industrial programme formally launched four days later, on 31 May; previously, on 6 May, Europrop International TP400-D6 engine chosen to power the A400M.

Clarification of the programme organisation led to Airbus Military being reconstituted as a Spanish company, Airbus Military S.L., with its registered offices in Madrid.

UPDATED

AIRBUS MILITARY A400M

TYPE: Strategic transport.

DEVELOPMENT MILESTONES

Requirement issued	Apr 89
Programme launched	May 96
Official go-ahead	May 03
First order (UK)	May 00
First flight (scheduled)	Jan 08
First delivery (France)	Nov 09

PROGRAMME: Original FIMA programme replaced April 1989 by five-nation industry MoU to develop four-turbofan, new technology transport to replace C-130 Hercules and C.160 Transall; Independent European Programme Group (IEPG) defined Outline European Staff Target (OEST) during 1991; initial studies undertaken by Euroflag organisation, which name reflected working designation Future Large Aircraft (FLA). Western European Union

Artist's impression of in-flight refuelling by an Airbus A400M ***NEW***/0576182

report in third quarter of 1991 concluded Euroflag FLA should form core of future European military transport capability to support Rapid Reaction Corps; national armament directors of Belgium, France, Germany, Italy, Portugal, Spain and Turkey affirmed support for 12 month prefeasibility study completed by Euroflag in late 1992; UK MoD declined involvement, but retained observer status; UK participation privately maintained by BAe and Shorts (10 per cent of BAe work); European Staff Target and intergovernmental MoU signed by seven nations in 1993; full feasibility programme officially started October 1993, by which time cargo hold width and height increased from original 3.66 m (12 ft 0 in) and 3.55 m (11 ft 7¼ in), respectively; study finished May 1995 and submitted to European defence ministries. Meanwhile, FLA underwent profound change in April 1994 when turbofans deemed incapable of providing desired performance; aircraft recast with four turboprops of new design. Discussion of a 'close association' between Euroflag and Airbus Industrie began third quarter of 1993 and formalised in June 1995.

Launch of the predevelopment phase (PDP) was postponed at least six months from early 1996 as a result of funding uncertainties. Original intention was for PDP to run from 1996 to 1998 and define a comprehensive specification for the aircraft and contractual forms and conditions against which the partner nations would make commitments. Full development and production phase (DPP) scheduled to follow directly on from PDP and terminate with first flight in 2002. Customer deliveries were then planned to begin in 2004.

France announced funding withdrawal from FLA development on 13 May 1996 and UK failed to rejoin the programme later that year, despite intention announced in December 1994 (when the RAF purchased Lockheed Martin C-130J Hercules). However, Germany became first to sign a European Staff Requirement, on 24 July 1996, although having terminated official funding for FLA development in previous month. AMC accordingly announced a 'single-phase' programme in May 1996.

The new programme schedule started in mid-1998 with a set of formal prelaunch activities (PLA), largely funded by industry and lasting 12 months, leading to a fully documented proposal for the Airbus A400M. This contained the technical proposal, including the aircraft specification and performance guarantees, and the commercial proposal with firm and fixed prices; a full set of contractual terms and conditions; and detailed planning of the single-phase programme. Strategic workshares (detailed enough to allow industry to provide the necessary resources to complete the proposal) were agreed at the start of PLA.

February 1999 delivery of the A400M proposal, initiated a 12-month period of negotiation of individual national requirements before planned official launch during early 2000 to meet an ESR now supported by Belgium, France, Germany, Italy, Spain, Turkey and the UK, but not Portugal. The 'PLA + single phase' programme provides industry with an uninterrupted development schedule and strong commitments from governments, while it also meets the customers' requirement that industry carries as much of the development risk as possible.

A meeting in March 2000 saw Belgium, France, Italy, Spain and Turkey identify a requirement for 131 A400Ms (37 fewer than expected) while, on 16 May, UK announced its intention to buy 25 aircraft (becoming first to fully commit). France and Germany followed on 9 June 2000. Seven participating nations announced selection of A400M on 27 July, committing to 225, including one for Luxembourg, although Turkey had reduced planned procurement to 10 at time of MoU signature in June 2001. Portugal announced requirement for four shortly thereafter and subsequently rejoined programme as a risk-sharing and industrial partner in early 2001. By June 2001, however, number of aircraft had fallen to three and Portugal subsequently withdrew in early 2003. MoU of 19 June 2001 on development and acquisition of A400M covered 196 aircraft, omitting 16 for Italy which in late October 2001 revealed it would not proceed with purchase. Formal launch was expected in early 2002, but delayed for more than a year as consequence of German failure to obtain parliamentary funding approval. However, Germany made commitment in December 2002 to 60 aircraft and programme officially launched on 27 May 2003, with signature of contract by OCCAR and Airbus Military. Programme then anticipated production of 180 aircraft for seven customers. Engine selection announced 6 May 2003; propeller on 24 July 2003; Milestone 1 (submission of quality plan) passed 29 August 2003, this accepted by OCCAR on 14 October 2003; Milestone 2 (launch of engine development) achieved 20 January 2004; landing gear selection 9 February 2004.

Flight testing is expected to be at EADS CASA's Seville plant and Airbus' Toulouse facility, with certification by a single authority; six development aircraft are to be built, of which five will be refurbished and sold on completion of test duties. EADS CASA at Seville will have sole production line, assembling components from Belgium, France, Germany, Spain, Turkey and the UK. Germany committed US$4.4 billion to the programme in November 2000, though this represented 60 per cent of the amount required to guarantee Germany's 73 aircraft and planned 33 per cent workshare. France committed US$2.6 billion at the same time.

A timetable of 56 months between contract and first flight has been agreed, with first delivery and IOC of a common configuration 'logistics' aircraft 77 months after contract signature. Maiden flight now scheduled for January 2008, with deliveries beginning in November 2009. France and Turkey will be first to accept aircraft, with Germany and the UK receiving their first A400Ms in 2010; Spain following in 2011, Luxembourg in 2017 and Belgium in 2018. Final deliveries in 2029.

CURRENT VERSIONS: Primarily for carriage of military personnel and outsize cargoes such as helicopters, armoured fighting vehicles, trucks and 40 ft ISO containers. Also designed to provide full aerial delivery and tactical air land capability, particularly into soft, natural surface airstrips. Typical strategic air transport capability will be 30,000 kg (66,139 lb) payload over 2,450 n miles (4,537 km; 2,819 miles) with full reserves, or 20,000 kg (44,092 lb) payload over 3,550 n miles (6,574 km; 4,085 miles). A400M can be modified to operate as a single-hose refueller (or, in conjunction with wing pods, as a three-point refueller); a pallet-mounted hose drum unit, secured to the closed rear ramp, will dispense fuel via a centreline hose passing through a resealable aperture in the ramp. Fuel load can be increased by installation of two additional tanks in cargo hold, up to total capacity of 11,500 kg (25,350 lb). With additional fuel tanks installed, an A400M could transfer 40,000 kg (88,185 lb) of fuel at a point 400 n miles (740 km; 460 miles) from base with loiter time of two hours. A400M speed envelope permits safe refuelling of fighter and large multi-engined aircraft, as well as helicopters.

CUSTOMERS: Procurement agency is Organisation Conjointe de Co-operation en matière d'Armement (OCCAR) in Bonn, acting for all prospective NATO purchasers. Maximum production rate 27½ per year to meet European orders. Exports expected, and attempts made in early 1995 to interest Japan; Australia briefed in February 2001 and again in November 2002 and may acquire A400M as eventual replacement for existing fleet of C-130H Hercules; export market estimated at about 480 aircraft over 25 years, with A400M expected to account for some 200 of this total.

A400M REQUIREMENTS
(at January 2004)

Country	Qty
Belgium	7
France	50
Germany	60
Luxembourg	1
Spain	27
Turkey	10
UK	25
Total	**180**

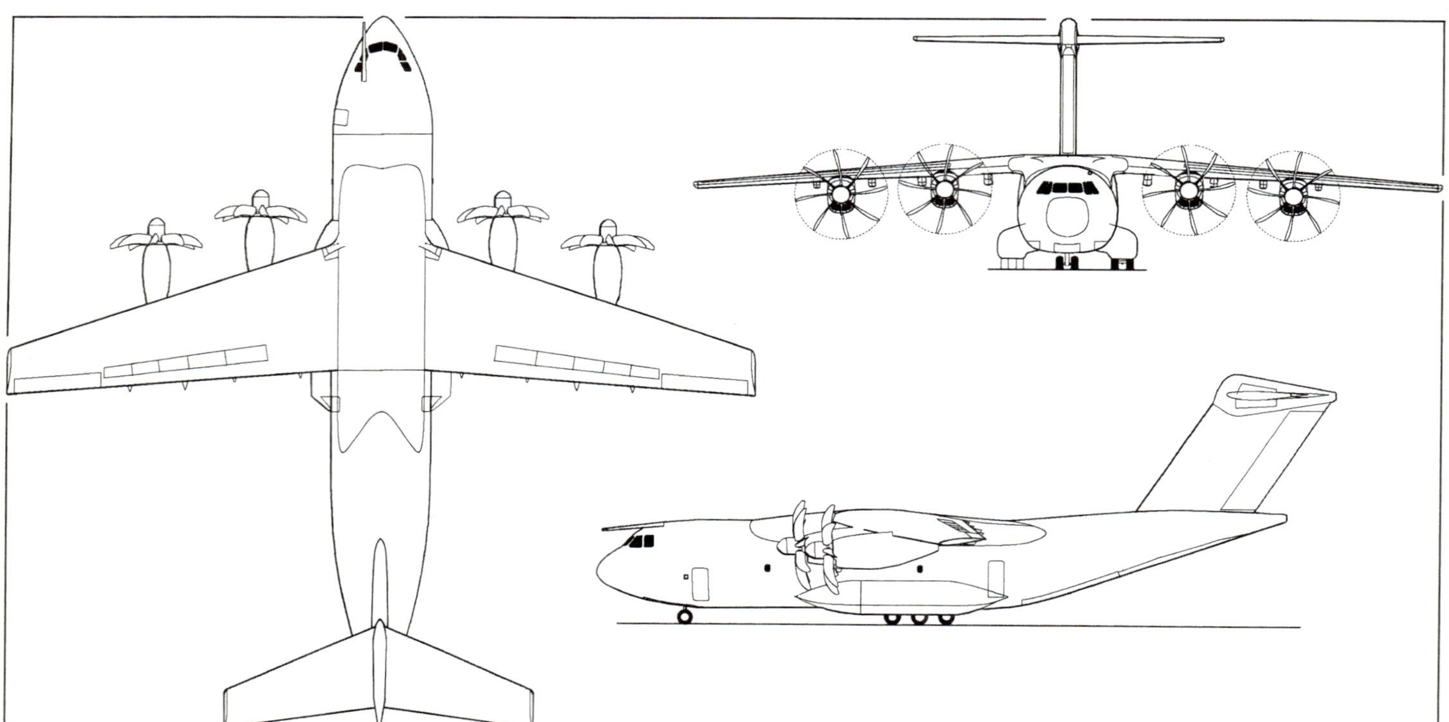

Airbus Military A400M military transport in production configuration *(Paul Jackson)* ***NEW***/0580228

COSTS: Total development cost expected to be €20 billion. Unit price estimated in early 2003 as €85 million, for basic aircraft. In December 2002, however, average flyaway price said to be €118.5 million, including 16 per cent VAT. Discrepancy probably explained by different equipment configuration.

DESIGN FEATURES: High-wing, T-tailed aircraft with rough-field landing gear and much larger cabin/hold floor area and cross-section than C-130/C.160, permitting high payload factors with low-density cargo vehicles or mixed passenger/cargo loads. Use of propellers felt to be essential for adequate thrust-reverse performance for taxying and short landing; for maximising power response; and for minimising FOD vulnerability. Long-range cruising speed of M0.68 to M0.72 up to 11,280 m (37,000 ft). Tactical mission parameters of 150 m (500 ft) AGL in IMC on predetermined route with civil standard of safety; 92 m (300 ft) night VMC; 46 m (150 ft) day VMC. Airbus Military has noted that its extensive use of new technology gives twice the volume and payload of the C-130J at the same life-cycle cost. Compared with the C-17, the A400M is said to be less than half the price and to have one third of the life-cycle cost. Minimum service life 30,000 hours, including allowance for low-level flight and short-field performance. Optimised for autonomous deployment; AMC offering 15-day away-from-base serviceability guarantee, all necessary support being within flight crew's capabilities.

Wing sweep 15° at 25 per cent chord; anhedral 4°; taper ratio 0.345; mean aerodynamic chord 5.671 m (18 ft 7¼ in). Tailplane sweep 20° at 25 per cent chord.

FLYING CONTROLS: Fly-by-wire, hydraulically powered; manually actuated electrohydrostatic back-up for ailerons and elevator; rudder actuated by two electrical back-up hydraulic actuators, each powered by one hydraulic circuit and one electrical circuit. Four spoilers and two-section double slotted fixed-vane flaps on each wing; tailplane trimmable by screw-jack. No slats. Spoilers used for roll control, lift dumping and as speed brakes.

STRUCTURE: Aluminium alloy, with titanium alloy in highly loaded areas (around windscreen, wing/fuselage joint and landing gear anchorage) and glass fibre or carbon fibre for lightly loaded components (landing gear doors and various fairings). Tailplane has aluminium alloy central joint and two outer composites box structures; elevator primary structure of carbon fibre. Fin has three-spar main box, trailing-edge shroud and single-piece rudder, all primarily of composites, plus metal/composites removable leading-edge. Rudder of carbon fibre, with aluminium, hinge-connecting ribs. Extensive use of carbon fibre composites in wing for skins, stringers, spars and moving surfaces; metal for ribs, engine mountings and fuselage pick-ups. Front spar at 15 per cent chord; composites rear spar at 62.5 per cent chord. Modern design and manufacturing techniques expected to afford major reductions in maintenance man-hour requirements and increases in aircraft availability/survivability.

LANDING GEAR: Retractable Messier-Dowty tricycle type with sufficient 'flotation' for semi-prepared and/or unsurfaced runways. Each main unit has six wheels in tandem pairs, retracting rearwards into fairings on fuselage sides. Each pair of mainwheels has independent, lever-type shock-absorbers. Twin nosewheels retract forwards. Emergency gravity extension of all units. Multidisc carbon brakes on mainwheels can operate differentially to assist steerable nosewheel in ground manoeuvring. Turning radius: landing gear 15 m (50 ft); wingtip 28.6 m (94 ft). Operable from CBR6 soft field. Mainwheels can 'kneel' for unloading of large cargoes. Hydraulic strut at rear of each sponson supports and stabilises aircraft during loading and unloading.

POWER PLANT: Initial candidate engines rated at approximately 6,898 kW (9,250 shp): M138 turboprop offered by Turboprop International SNECMA (33 per cent), MTU (33 per cent), Fiat Avio (22 per cent) and ITP (12 per cent) and based on SNECMA M88-2 core; Rolls-Royce Deutschland offered the 8,949 kW (12,000 shp) BR700-TP turboprop development of the BR715 turbofan; and Pratt & Whitney Canada proposed a 'Twinpac' version of the PW150. Required engine power, as defined by Airbus, was up to 7,457 kW (10,000 shp).

Choice initially settled on three-shaft 7,457 to 9,694 kW (10,000 to 13,000 shp) turboprop TP400 developed by Fiat Avio, ITP, MTU, Rolls-Royce, SNECMA and Techspace Aero, although this also rejected in February 2002 on basis of being too costly and too heavy as well as insufficiently powerful. Engine competition subsequently re-opened, with European and US manufacturers invited to submit proposals. Airbus announced intent to choose engine by September 2002, but decision delayed until 6 May 2003, when 7,979 kW (10,700 shp) EuroProp International (ITP, MTU, Rolls-Royce and Snecma) TP400-D6 selected in preference to Pratt & Whitney Canada PWC180. Engines are to be 'handed', with one of each wing pair rotating in opposite direction to the other, offering reduction in torque and elimination of asymmetric airflow over wing.

Eight-blade FH386 composites curved propellers to be supplied by Ratier-Figeac/Hamilton Sundstrand. Propeller tip speed 290 m/s (951 ft/s) at M0.68. Composites blades with integral de-icing blankets. Full reversal of pitch will allow aircraft to back up 2° slope at MTOW on concrete surface. Adjacent propellers turn in opposite directions to produce DBE (down between the engines) airflow. Viewed from rear Nos.1 (port outer) and 3 (stbd inner) turn clockwise; Nos. 2 and 4, anticlockwise. Advantages comprise more symmetric airflow, weight savings, reduced wing loading, increased (4 per cent) and better lift distribution, quieter cargo bay, better OEI handling, and reduced empennage area (fin 17 per cent; horizontal tail 20 per cent).

Fuel capacity 64,030 litres (16,915 US gallons; 14,085 Imp gallons) in seven tanks (one tank per engine, one transfer tank per wing, one centre tank); electric pumps and valves all mounted outside tanks. Detachable in-flight refuelling probe. Provision for inert gas system; provision for wing-mounted refuelling pods; optional hose drum unit (HDU) in cargo hold; and optional additional fuel tanks, totalling up to 14,400 litres (3,804 US gallons; 3,168 Imp gallons), in fuselage. Any standard A400M can be converted to a tanker in two hours. Pressure refuelling, with gravity back-up.

ACCOMMODATION: Two-man, NVG-compatible flight deck with dual sidestick controllers and additional optional forward-facing workstation for third 'mission crew member' to assist with tactical and special tasks, when required. View from flight deck exceeds JAR 25 and MIL-STD-850B. Provision for bulletproof flight deck windows, 68 mm (2¾ in) thick, and armour protection around crew's and loadmaster's seats. Loadmaster station forward of and overlooking cargo area. Two fixed, screened urinals and fixed hand-basin, starboard, rear. Flight crew rest area with two foldaway bunks; seat for fourth occupant on left, in front of bunks, with fifth seat as part of right-hand side of the bunk.

Two passenger doors forward; two rear. Forward, port, for normal access; forward, starboard, for emergency exit; rear doors for paratroop dropping. Three emergency exits in roof for flight crew and passengers. Cargo door, hinged at aft end, raised hydraulically to hold roof for loading via rear ramp. Optional closed-circuit TV surveillance of cargo hold, with imagery selectable on flight deck displays.

Cargo floor with 253 tiedown rings stressed to 4,536 kg (10,000 lb), 62 to 11,340 kg (25,000 lb) and 90 on ramp, also of 4,536 kg (10,000 lb). Typical loads include M113, Warrior, MRAV or LAV-III armoured vehicles; one Super Puma or NH90 or two Tiger or Apache helicopters; nine pallets (88 × 108 in military or 88 × 125 in civil); plus 57 troops and second loadmaster on permanent (tip-up) sidewall seats; two 20 ft ISO containers; Patriot SAM system; six Land Rovers, plus trailers; 66 stretchers and 10 medical attendants; or 120 armed troops on sidewall and removable centreline seats. Ramp stressed for 6,000 kg (13,228 lb) loads and has three hydraulically powered toes and 90 tiedowns stressed to 4,536 kg (10,000 lb).

SYSTEMS: FBW FCS derived from Airbus airliners, including sidestick controllers (left hand for captain, right hand for co-pilot, with conventional central power-lever throttle quadrant).

Electrical power provided by four engine-driven generators, each of 75 kVA. Additional power from three-phase generator on APU (90 kVA, A320 type, with in-flight restart capability) in right-hand side of rear wing fairing; three-phase generator on RAT; emergency battery; and external power receptacle. DC power from four 200 A transformer/rectifier units: two feed separated, main DC busbars; one feeds the 'flight essential' busbar and charges the emergency busbar and battery; and one feeds the APU starting system and charges the APU battery. Two Ni/Cd batteries are additional source for DC power.

Two hydraulic systems, Blue and Yellow, each operating at 207 bar (3,000 lb/sq in). Blue (driven by Nos. 1 and 4 engines) powers port aileron and elevator, No. 4 (outboard) and No. 2 spoiler on each wing, back-up brakes, cargo ramp and ground stabiliser struts. Yellow (Nos. 2 and 3 engines) responsible for starboard aileron and elevator, Nos. 1 and 3 spoilers on each side, landing gear kneeling, brakes and landing gear actuation. Both systems power flaps, rudder, trimmable horizontal stabiliser and steering. Each system has 140 litre (37.0 US gallon; 30.8 Imp gallon)/min engine-driven pump and 40 litre (10.6 US gallon; 8.8 Imp gallon)/min alternate current motor pump; power transfer unit between systems for emergency use.

Pneumatic system for air conditioning and pressurisation; wing and engine air intake anti-icing; engine starting; and pressurisation of other onboard systems. Two computers control four engine air bleed units. Interior divided into three air conditioning zones; one for flight deck and two in cargo hold. Cabin pressure altitude 2,440 m (8,000 ft) when flying at 11,280 m (37,000 ft).

AVIONICS: *Comms:* HF and V/UHF with COMSEC capability, SELCAL and optional Inmarsat SATCOM. Audio management system, wireless intercom system, cockpit voice recorder and passenger address system. ELT and IFF transponder.

Radar: Weather/navigation radar, with indication of turbulence and windshear; replaceable by optional military radar with ground mapping mode.

Flight: Three inertial platforms with embedded air data systems. VOR, DME, Tacan, ADF (optional), multimode receiver (ILS, MLS and GNSS), military GPS, two radar altimeters, Enhanced-GPWS and TCAS. Terrain-referenced navigation system, future air navigation system (FANS A), tactical ground collision avoidance system (T-GCAS), terrain masking low level flight system and formation keeping system as options.

Instrumentation: Two wide-angle HUDs; seven identical full-colour 152 × 203 mm (6 × 8 in) head-down displays (HDDs); two keyboard devices and two cursor control devices (CCDs); an eighth identical HDD, a third keyboard device and a third CCD as options for a third crew member.

Mission: Two flight management computers and two military mission management computers; optional MIDS tactical datalink.

Self-defence: Modular defensive aids subsystem (DASS) with optional elements including central computer, RWR, MWS (passive/active), LWR, expendables dispensing system (chaff/flare), direct energy infra-red countermeasures (DIRCM) and towed radar decoy.

EQUIPMENT: Winch at forward end of cargo hold. Optional 5-tonne power crane in roof above cargo ramp. Provision for one 1,200 kg (2,646 lb)/min refuelling pod under each wing and/or 1,800 kg (3,968 lb)/min HDU in rear of cargo hold.

DIMENSIONS, EXTERNAL:

Wing span	42.355 m (138 ft 11½ in)
Wing aspect ratio	8.1
Length: fuselage	39.09 m (128 ft 3 in)
overall	41.89 m (137 ft 5¼ in)
Height overall	14.63 m (48 ft 0 in)
Wheel track (mean)	6.22 m (20 ft 4¾ in)
Wheelbase (mean)	13.29 m (43 ft 7¼ in)
Propeller diameter	5.335 m (17 ft 6 in)
Cargo ramp: Length	5.80 m (19ft 0¼ in)
Width	4.00 m (13 ft 1½ in)
Height to sill	1.60 m (5 ft 3 in)
Cargo door: Length	7.76 m (25 ft 5½ in)

AREAS:

Wing area, gross	221.50 m² (2,384.2 sq ft)

DIMENSIONS, INTERNAL:

Hold: Length excl ramp	17.71 m (58 ft 1¼ in)
Length of ramp	5.40 m (17 ft 8½ in)
Width at floor, continuous	4.00 m (13 ft 1½ in)
Height: forward of wing box	3.85 m (12 ft 7½ in)
aft of wing box	4.00 m (13 ft 1½ in)
Floor area: incl ramp	92.4 m² (995 sq ft)
Volume: incl ramp (approx)	356 m³ (12,570 cu ft)
excl ramp (approx)	274 m³ (9,680 cu ft)

WEIGHTS AND LOADINGS: (A: logistic operation at max 2.5 *g*, B: logistic at 2.25 *g*, C: tactical operation at 2.5 *g*):

Operating weight, empty:	66,500 kg (146,605 lb)[1]
Max payload: A	34,000 kg (74,957 lb)
B	37,000 kg (81,570 lb)
C	29,500 kg (65,036 lb)
Max T-O weight: A	126,500 kg (278,885 lb)
B	130,000 kg (286,600 lb)
C	116,500 kg (256,835 lb)
Max landing weight: A, B	114,000 kg (251,325 lb)
C	106,500 kg (234,790 lb)
Max zero-fuel weight: A	100,500 kg (221,564 lb)
B	103,500 kg (228,180 lb)
C	96,000 kg (211,645 lb)
Max wing loading: A	571.1 kg/m² (116.97 lb/sq ft)
B	586.9 kg/m² (120.21 lb/sq ft)
C	526.0 kg/m² (107.72 lb/sq ft)
Max power loading: A	3.86 kg/kW (6.34 lb/shp)
B	3.96 kg/kW (6.51 lb/shp)
C	3.55 kg/kW (5.84 lb/shp)

[1] *including 1,000 kg (2,205 lb) allowance for optional equipment*

PERFORMANCE (estimated):

Max operating speed and Mach No. (VMO/MMO)	M0.72 (300 kt; 555 km/h; 345 mph)
Normal cruising Mach No.	0.68
Airdrop speed	130–200 kt (241–370 km/h; 150–230 mph)
Max rate of climb at S/L	1,524 m (5,000 ft)/min
Cruise ceiling: normal operations	11,280 m (37,000 ft)
special operations	12,192 m (40,000 ft)

T-O run at 116,500 kg (256,835 lb) Military critical field length, ISA, sea level, soft/dry runway ... 1,285 m (4,216 ft)
Landing run at 106,500 kg (234,792 lb), 152 m (500 ft) roll-out, ISA, sea level, soft/dry runway ... 615 m (2,018 ft)
Runway LCN: normal tyre pressure ... 24
40% deflation ... 18
Range with 5% reserves, missed approach, 200 n mile (370 km; 230 mile) diversion and 30 min hold at 445 m (1,500 ft), B:
with 30,000 kg (66,139 lb) payload ... 2,450 n miles (4,537 km; 2,819 miles)
with 20,000 kg (44,092 lb) payload ... 3,550 n miles (6,574 km; 4,085 miles)
Ferry range ... 4,900 n miles (9,074 km; 5,638 miles)

UPDATED

AMX

AMX INTERNATIONAL

c/o Alenia, Via Giulio Vincenzo Bona 85, I-00156 Rome, Italy
Tel: (+39 06) 41 72 31
Fax: (+39 06) 411 44 39
PRESIDENT: Dott Ing Giovanni Gazzaniga (Alenia)
VICE-PRESIDENTS:
F Grandi (Aermacchi)
R Pesce (Embraer)

PARTICIPATING COMPANIES
Alenia: see under Italy
Aermacchi: see under Italy
Embraer: see under Brazil

AMX production for the Italian and Brazilian air forces is now complete, but an order for 12 aircraft from Venezuela has still to be fulfilled.

UPDATED

AMX AMX

Brazilian Air Force designations: A-1 and A-1B
Italian Air Force name: Ghibli (Desert Wind)
TYPE: Attack fighter.

PROGRAMME: Resulted from June 1977 Italian Air Force specification for small tactical fighter-bomber; original Aeritalia/Aermacchi partnership joined by Embraer July 1980; seven single-seat prototypes built (first flight 15 May 1984): production of first 30 (Italy 21, Brazil nine), and design of two-seater, began mid-1986; first production aircraft rolled out at Turin 29 March 1988, making first flight 11 May; second contract (Italy 59, Brazil 25, including six and three two-seaters respectively) placed 1988.

Deliveries to Italian Air Force (six for Reparto Sperimentale di Volo at Pratica di Mare) began April 1989; production A-1 for Brazilian Air Force (s/n 5500) made first flight 12 August 1989, deliveries (two to Nucleo A-1 training nucleus at Santa Cruz) following from 17 October 1989; in-flight refuelling test programme completed (by Embraer) August/September 1989; first flight by first (of three) two-seat AMX-T prototypes 14 March 1990 (MM55024), followed by second on 16 July; first flight of Embraer two-seater (serial number 5650), 14 August 1991; third production batch authorised early 1992 (one year late); first two-seater for Brazilian Air Force (5650) delivered 7 May 1992.

Italian single-seater production temporarily halted following delivery on 1 February 1993 of 72nd aircraft (MM7160); resumed late 1994, with both AMX and first production batch of AMX-T. Final Italian single-seater delivered in 1997; total 110, comprising 74 built by Alenia and 36 by Aermacchi. Batch 4 (35 AMX and 16 AMX-T) and Batch 5 (42 AMX and 9 AMX-T) cancelled. Final Italian two-seater followed in 1998; 26 built: 17 Alenia, nine Aermacchi. Production continued in Brazil, where 50th was delivered on 1 December 1998 and last in 1999. No further manufacture until 2005 planned delivery of Venezuelan aircraft.

Istrana- and Amendola-based AMX squadrons flew 252 combat sorties (667 flying hours) during Operation Allied Force against Yugoslavia in 1999, dropping 39 Opher LGBs.

CURRENT VERSIONS: **AMX:** Replaced G91R/Y and some F-104G/S in Italian Air Force (eight squadrons originally planned) and some EMB-326GB Xavante in Brazilian Air Force for close support/interdiction/reconnaissance, sharing counter-air duties with IDS Tornado (Italy) and F-5E/Mirage 50 (Brazil); in service with five Italian Stormi (see table) and 10° and 16° Grupos (Brazil); Brazilian Air Force aircraft (designated **A-1**) differ primarily in avionics and weapon delivery systems, have two 30 mm guns instead of Italian version's single multibarrel 20 mm weapon and are usually fitted with in-flight refuelling probes.

AMX-MLU: Mid-life upgrade originally to be undertaken jointly by Brazil and Italy, but subsequently abandoned because of high cost and limited funding. Italy then opted to adopt more limited upgrade project, including GPS navigation, an improved EW suite, integration of Thales CLDP laser designator pod and ability to use newer weapons such as JDAM; consideration also reportedly given to installing new real-time reconnaissance pod. In this guise, Italy expects upgraded aircraft to be available for service in 2005.

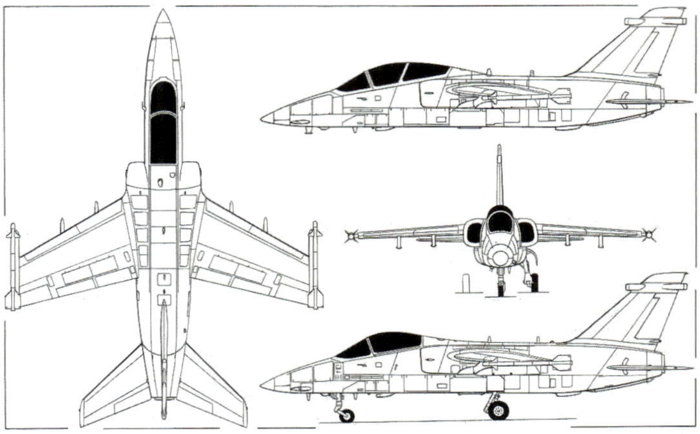

Alenia/Aermacchi/Embraer AMX; upper side view shows two-seater *(Jane's/Dennis Punnett)* 0507151

Super AMX: Designation of two-seater unsuccessfully offered to South Africa. Would have featured wide-angle HUD, 'glass cockpit', improved HOTAS, GPS, HMD, integrated defensive aids, new radar and new weapons.

Detailed description applies to single-seater except where indicated.

AMX-T: Second cockpit accommodated by removing forward fuselage fuel tank and relocating environmental control system; dual controls, canopy, integration of rear cockpit GEC-Marconi HUD monitor, and oxygen systems, designed/redesigned by Embraer; intended both as operational trainer and, suitably equipped, for such roles as EW, reconnaissance and maritime attack; most Italian AMX-Ts assigned to operational squadrons of 32° Stormo; others assigned, one per squadron, as trainers. Brazilian designation **A-1B.**

AMX-ATA: Venezuela announced intent to acquire eight two-seaters in September 1999, but contract not finalised until 18 December 2002, by which time total had increased to 12, comprising eight standard trainer aircraft and four electronic warfare aircraft. Venezuelan aircraft will have Elta EL/M-2032 radar and Elbit avionics.

AMX-ATA 2: Light combat aircraft and advanced trainer for export with AMX-MLU avionics and non-afterburning EJ200 engine.

AMX-E: Original proposal envisaged two-seat EW version for use as escort jammer and SEAD platform with AGM-88 HARM missiles. Flight controls removed from rear cockpit. MM55027 used for development. Feasibility study completed but further development cancelled. Subsequently, AMX-E version resurrected in different form for Venezuela, which to receive four aircraft outfitted with Elta Electronics EW suite. System able to undertake passive and active countermeasures and uses elements of EL/L-8222 Comprehensive Self Protection Pod, which includes RWR, MAWS, LWR, electro-magnetic jammers and expendables.

CUSTOMERS: Total of 192 (136 Italy/56 Brazil, including 26 and 11 two-seaters) delivered by March 2000 (see table) to original partners. Eight two-seaters selected by Venezuela on 9 September 1999; firm order for 12 aircraft for delivery from 2005 confirmed on 18 December 2002. Italy has placed all remaining Batch 1 aircraft in reserve for possible sale. By 1998, 23 Batch 2 aircraft had been upgraded to 'pre-FOC' (full operational capability) standard and plans were in hand to modify 15 to FOC. These 38 plus all 56 remaining from Batch 3 (FOC), formed baseline fleet of 94, though AMI active inventory quoted as 104 aircraft by May 1999.

COSTS: US$18.75 million (1999) AMX-T Venezuelan programme unit cost.

AMX of Italy's 51° Stormo *(Paul Jackson)* ***NEW***/0569646

DESIGN FEATURES: Intended for high-subsonic/very low-altitude day/night missions, in poor visibility and, if necessary, from poorly equipped or partially damaged runways. Required fatigue life of 16,000 hours being extended to 24,000 hours.

Wing sweepback 31° on leading-edges, 27° 30' at quarter-chord; thickness/chord ratio 12 per cent.

FLYING CONTROLS: Hydraulically actuated ailerons and elevators; leading-edge slats and Fowler double-slotted trailing-edge flaps (each two-segment on each wing, positions 0, 30 and 41°) actuated electrohydraulically; pair of hydraulically actuated spoilers forward of each flap pair, deployable separately in inboard and outboard pairs; fly-by-wire control of spoilers, rudder and variable incidence tailplane by Alenia/BAE Systems flight control computer; ailerons, elevators, rudder have manual reversion for fly-home capability, even with both hydraulic systems inoperative; spoilers serve also as airbrakes/lift dumpers.

STRUCTURE: Mainly aluminium alloy except for carbon fibre fin and elevators; shoulder-mounted wings, each with three-point attachment to fuselage, have three-spar torsion box with integrally stiffened skins; oval-section semi-monocoque fuselage, with rear portion (including tailplane) detachable for engine access.

Work split gives programme leader Alenia 46.7 per cent (centre-fuselage, nose radome, tail surfaces, ailerons and spoilers); Aermacchi has 23.6 per cent (forward fuselage including gun and avionics integration, canopy, tailcone) and Embraer 29.7 per cent (air intakes, wings, leading-edge slats, flaps, wing pylons, external fuel tanks and reconnaissance pallets). single-sourced production, with final assembly lines in Italy and Brazil.

LANDING GEAR: Hydraulically retractable tricycle type, of Messier-Bugatti levered suspension design, built in Italy by Magnaghi (nose unit) and in France by ERAM (main units). Single wheel and oleo-pneumatic shock-absorber on each unit. Nose unit retracts forward; main units retract forward and inward, turning through approximately 90° to lie almost flat in underside of engine air intake trunks. Nosewheel hydraulically steerable (±6° normal; ±45° with full pedal movement), self-centring, and fitted with anti-shimmy device. Towing travel ±90°. Mainwheel tyres size 670×210-12 (18 ply), pressure 9.65 bar (140 lb/sq in); nosewheel tyre size 18×5.5–8 (10 ply), pressure 10.70 bar (155 lb/sq in). Hydraulic brakes and fully modulated anti-skid system. No brake-chute. Runway arrester hook. Minimum ground turning radius 7.53 m (24 ft 8½ in).

POWER PLANT: One 49.1 kN (11,030 lb st) Rolls-Royce RB 168 Spey Mk 807 non-afterburning turbofan, built under licence in Italy by Fiat, Piaggio and Alfa Romeo Avio, in association with Companhia Eletro-Mecânica (CELMA) in Brazil. Self-sealing, compartmented, rubber fuselage bag tanks and two integral wing tanks with combined capacity of 3,500 litres (924.6 US gallons; 770 Imp gallons). Brazilian AMX carry 200 litres (52.8 US gallons; 44.0 Imp gallons) more internal fuel in additional saddle tank behind cockpit. Auxiliary fuel tanks of up to 1,100 litres (290 US gallons; 242 Imp gallons) capacity can be carried by Brazilian aircraft on each inboard underwing pylon, and up to 580 litres (153 US gallons; 128 Imp gallons) on each outboard pylon (Italian or Brazilian). Single-point pressure or gravity refuelling of internal and external tanks. Optional in-flight refuelling capability (probe and drogue system) is standard in Brazil.

ACCOMMODATION: Pilot only, on Martin-Baker Mk 10L zero/zero ejection seat; 18° downward view over nose. One-piece wraparound windscreen (reinforced on Brazilian aircraft); one-piece hinged canopy, opening sideways to starboard. Cockpit pressurised and air conditioned. Tandem two-seat combat trainer/special missions version also produced, with Mk 10LY-2 (front) and Mk 10LY-3 (rear) seats.

SYSTEMS: Microtecnica environmental control system (ECS) provides air conditioning of cockpit, avionics and reconnaissance pallets, cockpit pressurisation, air intake and inlet guide vane anti-icing, windscreen demisting and anti-*g* systems. Duplicated redundant hydraulic systems, driven by engine gearbox, operate at pressure of 207 bar (3,000 lb/sq in); both actuate primary flight control system (aileron, elevators and rudder) and secondary (flap and slat system); No. 1 circuit additionally supplies outboard spoilers, nosewheel steering and gun; No. 2 also supplies inboard spoilers and landing gear actuation. Primary electrical system AC power (115/200 V at fixed frequency of 400 Hz) supplied by two 30 kVA IDG generators, with two transformer-rectifier units for conversion to 28 V DC; 36 Ah Ni/Cd battery for emergency use, to provide power for essential systems in the event of primary and secondary electrical system failure. Aeroeletrônica (Brazil) external power control unit. Fiat FA 150 Argo APU for engine starting. APU-driven electrical generator for ground operation. Liquid oxygen system.

AVIONICS: All avionics/equipment packages pallet-mounted and positioned for rapid access. Modular design and space provisions within aircraft permit retrofitting of alternative avionics.

Comms: UHF and VHF com, and IFF.

Radar: Pointer ranging radar in Italian AMXs is I-band set modified from Elta (Israel) EL/M-2001B and built in Italy by FIAR. Brazilian aircraft were originally intended to have Tecnasa/SMA-built (Alenia) SCP-01 Scipio radar, but this was not installed at time of production due to local bankruptcies of two Brazilian companies selected to produce the radar. It will now be fitted as part of the mid-life upgrade. Venezuela specified Elta EL/M-2032 radar for its aircraft.

Flight: Litton Italia INS, with standby AHRS and Tacan, for Italian Air Force; VOR/ILS for Brazil. Data processing, with Microtecnica air data computer. BAE Systems MED 2067 video monitor display in rear cockpit of two-seater, for use by instructor/navigator as HUD monitor.

Instrumentation: Alenia computer-based weapon aiming and delivery, incorporating radar and Alenia stores management system; digital data displays (OMI/Alenia head-up, Alenia multifunction head-down, and weapons/nav selector). Provision for night vision goggles.

Mission: Italian aircraft of 3° Stormo equipped with Oude Delft Orpheus reconnaissance pods, and was an internal sensor suite being sought for deployment in 2001. This is believed to have been cancelled, but was to have comprised any one of three interchangeable Aeroeletrônica (Brazil) pallet-mounted photographic systems installed internally in forward fuselage, complementing external IR/EO pod on centreline pylon. Each system fully compatible with aircraft, and not affecting operational capability. Camera bay in lower starboard side of fuselage, forward of mainwheel bay.

Self-defence: Elettronica active and passive ECM, including fin-mounted radar warning receiver. AMX-E for Venezuela will incorporate EW suite by Elta Electronics.

ARMAMENT: One M61A1 multibarrel 20 mm cannon, with 350 rounds, in port side of lower forward fuselage of aircraft for Italian Air Force (one 30 mm DEFA 554 cannon on eachside in aircraft for Brazilian Air Force). Single stores attachment point on fuselage centreline, plus two attachments under each wing, and wingtip rails for two AIM-9L Sidewinder or similar IR air-to-air missiles (MAA-1 Piranha on Brazilian aircraft). Fuselage and inboard underwing points each stressed for loads of up to 907 kg (2,000 lb); outboard underwing points stressed for 454 kg (1,000 lb) each; wingtip stations stressed for 113 kg (250 lb) each. Twin carriers can be fitted to all five stations. Total external stores load 3,800 kg (8,377 lb). Attack weapons can include free-fall or retarded Mk 82/83/84 bombs, laser-guided bombs,

AMX PRODUCTION

	Italy				Brazil				
Batch	**AMX**	**First aircraft**	**AMX-T**	**First aircraft**	**AMX**	**First aircraft**	**AMX-T**	**First aircraft**	**Total**
1	19	MM7089	2	MM55024	8	5500	1	5650	30
2	53	MM7108	6	MM55026	22	5508	3	5651	84
3	38	MM7161	18	MM55034	15	5530	7	5654	78
Total	**110**		**26**		**45**		**11**		**192**

Notes: Brazil retains option on 23 aircraft. Excludes seven single-seat prototypes and 12 aircraft for Venezuela.

ITALIAN AMX UNITS

Squadron	Wing	Base	Type	First aircraft and acceptance	
13° Gruppo	32° Stormo	Amendola	AMX	MM7112	Nov 94
			AMX-T[1]	MM55029	by Apr 95
14° Gruppo	2° Stormo	Rivolto[2]	AMX	MM7130	10 Jul 91
			AMX-T[1]	MM55028	24 Nov 94
28° Gruppo[5]	3° Stormo	Villafranca	AMX	MM7129	Jun 93
			AMX-T[1]	MM55029	by Dec 94
101° Gruppo[3]	32° Stormo	Amendola	AMX	MM7091	31 Jul 95
			AMX-T	MM55030	21 Nov 94
103° Gruppo	51° Stormo	Istrana	AMX	MM7098	30 Sep 89
			AMX-T[1]	MM55027	by Apr 95
132° Gruppo	51° Stormo	Istrana[6]	AMX	MM7110	15 Nov 90[4]
			AMX-T[1]	MM55037	4 Dec 95

[1] One aircraft only
[2] At Istrana until February 1994
[3] Unit initially re-equipped as 201° Gruppo (until 31 July 1995); AMX OCU
[4] Initial working up at Istrana
[5] Disbanded 29 September 1997
[6] Villafranca until March 1999

BRAZILIAN AMX UNITS

Squadron	Wing	Base	First delivery
1° Esquadrão 'Adelfi'	16 Grupo de Aviação	Santa Cruz	Formed 7 November 1990[1]
1° Esquadrão 'Poker'	10 Grupo de Aviação	Santa Maria	1996
3° Esquadrão 'Centauro'	10 Grupo de Aviação	Santa Maria	1998

[1] Out of Nucleo A-1, originally established 7 November 1988

cluster bombs, air-to-surface missiles (including area denial, anti-radiation and anti-shipping weapons), electro-optical precision-guided munitions and rocket launchers.

Exocet firing trials conducted 1991; Marte trials 1994; carriage trials of GBU-16 Paveway II LGB on Italian AMX in 1995 and aircraft used Elbit Opher LGB system during Operation Allied Force over Kosovo, May-July 1999.

DIMENSIONS, EXTERNAL:
Wing span:
excl wingtip missiles and rails ... 8.875 m (29 ft 1½ in)
over missiles ... 9.97 m (32 ft 8½ in)
Wing aspect ratio ... 3.8
Wing taper ratio ... 0.5
Length: overall ... 13.23 m (43 ft 5 in)
fuselage ... 12.55 m (41 ft 2 in)
Height overall ... 4.55 m (14 ft 11¼ in)
Tailplane span ... 5.20 m (17 ft 0¾ in)
Wheel track ... 2.15 m (7 ft 0¾ in)
Wheelbase ... 4.70 m (15 ft 5 in)

AREAS:
Wings, gross ... 21.00 m^2 (226.0 sq ft)
Ailerons (total) ... 0.88 m^2 (9.47 sq ft)
Trailing-edge flaps (total) ... 3.86 m^2 (41.55 sq ft)
Leading-edge slats (total) ... 2.07 m^2 (22.28 sq ft)
Spoilers (total) ... 1.30 m^2 (13.99 sq ft)
Fin (exposed) ... 4.265 m^2 (45.91 sq ft)
Rudder ... 0.83 m^2 (8.93 sq ft)
Tailplane (total exposed) ... 5.10 m^2 (54.90 sq ft)
Elevators (total) ... 1.00 m^2 (10.76 sq ft)

WEIGHTS AND LOADINGS (all versions):
Operational weight empty ... 7,000 kg (15,432 lb)
Max fuel weight: internal ... 2,720 kg (5,997 lb)
external ... 1,726 kg (3,805 lb)
Max external stores load ... 3,800 kg (8,377 lb)
T-O weight (clean) ... 9,694 kg (21,371 lb)
Typical mission T-O weight ... 10,750 kg (23,700 lb)
Max T-O weight ... 13,000 kg (28,660 lb)
Normal landing weight ... 7,000 kg (15,432 lb)
Combat wing loading (clean) ... 457.1 kg/m^2 (93.62 lb/sq ft)
Max wing loading ... 619.1 kg/m^2 (126.79 lb/sq ft)
Max power loading ... 265 kg/kN (2.60 lb/lb st)

PERFORMANCE (A at typical mission weight of 10,750 kg; 23,700 lb with 907 kg; 2,000 lb of external stores, B at max T-O weight with 2,721 kg; 6,000 lb of external stores, ISA in both cases):
Max level speed: at S/L ... M0.83
at 9,140 m (30,000 ft) ... M0.86
Max rate of climb at S/L ... 3,124 m (10,250 ft)/min
Service ceiling ... 12,800 m (41,995 ft)
T-O run at S/L: A ... 631 m (2,070 ft)
B ... 982 m (3,220 ft)
T-O to 15 m (50 ft) at S/L: B ... 1,442 m (4,730 ft)
Landing from 15 m (50 ft) at S/L: B ... 753 m (2,470 ft)
Landing run at S/L ... 505 m (1,657 ft)
Attack radius, allowances for 5 min combat over target and 10% fuel reserves:
lo-lo-lo: A ... 300 n miles (556 km; 345 miles)
B ... 285 n miles (528 km; 328 miles)
hi-lo-hi: A ... 480 n miles (889 km; 553 miles)
B ... 500 n miles (926 km; 576 miles)
Ferry range with two 1,000 litre (264 US gallon; 220 Imp gallon) drop tanks, 10% reserves ... 1,800 n miles (3,333 km; 2,071 miles)
g limits ... +7.33/–3

UPDATED

ATR

AVIONS DE TRANSPORT REGIONAL INTEGRATED

1 allée Pierre Nadot, F-31712 Blagnac Cedex, France
Tel: (+33 5) 61 21 62 21
Fax: (+33 5) 61 21 63 18
Web: http://www.atraircraft.com

CHAIRMAN: Jean-Michel Léonard
CEO: Filippo Bagnato
SENIOR VICE-PRESIDENTS:
John Moore (Commercial)
Luigi Lombardi (Operations)
Jean-Pierre Cousserans (Customer Services)
Serge Queille (Finance)
DIRECTOR, COMMUNICATIONS AND PRESS RELATIONS: Frédéric Lahache

PARTICIPATING COMPANIES
Alenia Aeronautica: Italy
EADS: International

First Aerospatiale/Aeritalia (later Alenia) agreement July 1980; ATR programme started 4 November 1981; Groupement d'Intéret Economique (50:50 joint management company) formally established 5 February 1982 to develop ATR series of transport aircraft. Assembly or licensed production by Xian Aircraft in a new factory at Shenzhen, near Hong Kong, was discussed, but not pursued; Xian already produces rear fuselage components for ATR. Fuselage and wing production remain subcontracted to Alenia Aeronautica and EADS Sogerma (for Airbus France) respectively.

ATR marketing and support office opened in Washington 15 July 1986; ATR airline support centre in Singapore opened 18 November 1988; ATR Training Centre opened 1 July 1989; these functions taken over by AI(R) from 1996, but returned when this joint venture with BAe was dissolved on 1 July 1998. Beijing office opened in September 2003. Phased production lines introduced in 1999 have reduced delivery times to three months from one year. The 600th ATR was delivered to Air Dolomiti on 28 April 2000.

The ATR family logged its 10 millionth flight on 10 October 2000. Sales by 31 December 2003 totalled 380 ATR 42s and 296 ATR 72s, or 676 in all, of which 372 ATR 42s and 289 ATR 72s had been delivered to 115 operators in 73 countries, including 2003 deliveries of three and six respectively. Further 12 aircraft (one ATR 42-500 and 11 ATR 72-500s) ordered by October 2004. The company is planning to reach proposed ICAO Stage 4 noise levels

The ATR assembly line at Toulouse ***NEW***/1044742

which become effective on 1 January 2006. In 2002, ATR developed a modified internal door designed to resist unauthorised flight deck incursions, in anticipation of the FAA's 9 April 2003 deadline for all airliners to be retrofitted to this standard. By mid-2002, ATR had received 100 firm orders for the modified door from American Eagle (first customer), Finnair, Jet Airways and Trans States Airlines. The new door is resistant to forcible impacts of up to 300J and penetration of 9 mm and 44 Magnum ammunition. Total of 260 doors sold by October 2004.

ATR's Asset Management arm is responsible for second-hand sales.

ATR expected to have become a single corporate entity by early 1999 but this did not materialise until 1 June 2001 with the formal establishment of ATR Integrated combining the activities of the two partners, still with GIE status and around 500 employees. Holding is 50 per cent each. Workforce 510 in October 2004.

UPDATED

ATR 42

TYPE: Twin-turboprop airliner.

DEVELOPMENT MILESTONES

42	
Programme launched	4 Nov 81
First flight	16 Aug 84
Certification	24 Sep 85
First delivery	3 Dec 85
42–500	
Programme launched	14 Jun 93
First flight	16 Sep 94
Certification	28 Jul 95
First delivery	31 Oct 95

PROGRAMME: Joint launch by Aerospatiale (now in EADS) and Aeritalia (now Alenia) in November 1981, following June 1981 selection of P&WC PW120 turboprop as basic power plant; first flights of two prototypes 16 August 1984 (F-WEGA) and 31 October 1984 (F-WEGB); first flight production aircraft 30 April 1985; simultaneous certification to JAR 25 by France and Italy 24 September 1985, followed by USA (FAR Pt 25) 25 October 1985, Germany 12 February 1988, UK 31 October 1989; deliveries began 3 December 1985 to Air Littoral.

Series 500, with 'new look' interior, announced at Paris Air Show 14 June 1993; first flight F-WWEZ (c/n 443) 16 September 1994; French and UK certification 28 July 1995; first delivery (F-OHFF to Air Dolomiti) 31 October 1995. FAA certification 13 May 1996.

CURRENT VERSIONS **ATR 42–300:** Initial version; phased out of production in 1996. Two Pratt & Whitney Canada PW120 turboprops, each flat rated at 1,342 kW (1,800 shp) for normal operation and 1,492 kW (2,000 shp) OEI; Hamilton Sundstrand 14SF four-blade constant-speed fully feathering and reversible-pitch propellers.

ATR 42–320: Identical to 42–300 except for optional 1,566 kW (2,100 shp) PW121 engines for improved hot/high performance; OWE increased/payload decreased by 5 kg (11 lb). Entered service 1987; phased out in 1996.

ATR 42–400: P&WC PW121A engines with six-blade Hamilton Sundstrand 568F propellers; maximum cruising speed 266 kt (493 km/h; 306 mph); maximum range 825 n miles (1,527 km; 949 miles) with full payload. First flight 12 July 1995 (F-WWEF/OK-AFE); two Srs 420s ordered for CSA, and both delivered 14 March 1996, having received DGAC certification on 27 February. No further civil aircraft.

ATR 42–500: Principal ATR 42 version from 1996. Compared with Series 300, has more powerful engines, reinforced wings to allow greatly increased cruising speed and higher weights; all systems improvements of ATR 72, including flight management computers; cockpit, elevators and fin from ATR 72–210; strengthened landing gear; electrically operated main doors; reinforced fuselage and wing centre-section.

Description applies to ATR 42–500, except where indicated.

ATR 42 Quick-Change : (formerly Tube, previously Cargo QC:) Quick-change interior to hold nine containers. Available as new-build or retrofit. Cabin equipped with smoke detectors, air distribution shut-off valve and modifications necessary for certification for full freighter operations. Conversion takes about 30 minutes. Launch customer, DHL Aviation of South Africa, received first of two converted –300s in October 2000.

ATR 42 full freighter version: Fully dedicated freight transport version (bulk, containers or pallets). Various conversions available, including 'tube' and/or large cargo door (LCD). Full freighter version available with Alenia Aeronautica STC.

'Tube' package offered by Aeronavali includes E-class cabin compartment with floor reinforced to 400 kg/m$_2$ (82 lb/sq ft). Total volume for cargo is 56 m$_3$ (1,978 cu ft); max payload 5,883 kg (12,970lb). Parallel cabin section 10.25 m (33 ft 7½ in) long and tapered section 4.47 m (14 ft 8 in) long, with positions for up to four spider nets. MTOW in this configuration is 16,900 kg (37,258 lb); MLW 16,400 kg (36,156 lb); MZFW 15,540 kg (34,260 lb). Other 'tube' modifications by independent outfitters also available.

LCD installation is part of the STC, and provides a 2.94 × 1.80 m (9 ft 7¾ in × 5 ft 10¾ in) upward-opening cargo door on port side at front to permit loading of three 224 × 274 cm (88 × 108 in) pallets or five LD3 containers. Max payload for an ATR 42 fitted with both 'tube' and LCD is 5,300 kg (11,684 lb). First such retrofit, for Northern Air Cargo, took place in 2003.

ATR 42 F: Military/paramilitary freighter with modified interior, reinforced cabin floor, port-side cargo/airdrop door can be opened in flight; can carry 3,800 kg (8,377 lb) of cargo or 42 passengers over 1,250 n miles (2,315 km; 1,438 miles). One delivered to Gabon 1989.

ATR Calibration: Projected navaid calibration version.

ATR 42L: Projected freighter with lateral cargo door; available as ATR 42 Large Cargo Door (see above).

ATR 42 Surveyor: Maritime and rescue version; *described separately*.

CUSTOMERS: Total 381 firm orders, of which 376 delivered, by 31 August 2004. Four ordered and 10 delivered in 1998; 14 ordered and 12 delivered in 1999; six (plus four options) sold in 2000; five (all-500s) delivered in 2001 and five in 2002. Ten -500s ordered and three delivered in 2003; one by Air Tahiti in May 2004.

COSTS: ATR 42–500: development cost US$50 million; unit price US$14.2 million (2004).

DESIGN FEATURES: Designed to JAR 25/FAR Pt 25; high wing of medium aspect ratio, with constant-chord centre section and tapered outer panels; T-type tail with tapered tailplane and sweptback fin and fillet; pannier-mounted main landing gear.

Wing section Aerospatiale RA-XXX-43 (NACA 43 series derivative); thickness/chord ratio 18 per cent at root, 13 per cent at tip; constant-chord, no-dihedral centre-section with 2° incidence at root; outer panels 3° 6' sweepback at quarter-chord and 2° 30' dihedral.

FLYING CONTROLS: Conventional and manual. Lateral control assisted by single spoiler surface ahead of each outer flap; each aileron has electrically actuated trim tab; fixed incidence tailplane; horn-balanced rudder and elevators, each with electrically actuated trim tab; two-segment double-slotted flaps on offset hinges with Ratier-Figeac hydraulic actuators. Flap settings 0, 15, 25 and 35°.

STRUCTURE: Two-spar fail-safe wings, mainly of aluminium alloy, with leading-edges of Kevlar/Nomex sandwich; wing top skin panels aft of rear spar are of Kevlar/Nomex with carbon reinforcement; flaps and ailerons have aluminium ribs and spars, with skins of carbon fibre/Nomex and carbon/epoxy respectively; fuselage is fail-safe stressed skin, mainly of light alloy except for Kevlar/Nomex sandwich nosecone, tailcone, wing/body fairings, nosewheel doors and main landing gear fairings; fin (attached to rearmost fuselage frame) and tailplane carbon structure; CFRP/Nomex sandwich rudder and elevators; dorsal fin of Kevlar/Nomex and GFRP/Nomex sandwich; engine cowlings of CFRP/Nomex and Kevlar/Nomex sandwich, reinforced with CFRP in nose and underside; propeller blades have metal spars and GFRP/polyurethane skins.

EADS France originally responsible for design and construction of wings and engine nacelles, flight deck and cabin layout, installation of power plant, flying controls, electrical and de-icing systems, and final assembly and flight testing of civil passenger versions; wing manufacture and testing reallocated to EADS Sogerma at Bordeaux from late 2001. Alenia Aeronautica builds fuselage and tail unit, installs landing gear, hydraulic system, air conditioning and pressurisation systems. ATR 42/72 manufactured at St Nazaire and Nantes (France), Pomigliano d'Arco and Capodichino (Italy), and assembled in Toulouse.

LANDING GEAR: Hydraulically retractable tricycle type, of Messier-Dowty trailing-arm design, with twin wheels and oleo-pneumatic shock-absorber on each unit. Nose unit retracts forward, main units inward into fuselage and large underfuselage fairing. Goodrich wheels and multiple-disc mainwheel brakes and Hydro-Aire anti-skid units. Mainwheel tubeless tyres, size 32×8.8R16 (10/12 ply), pressure 8.69 bar (126 lb/sq in) or H34×10.0R16 (14 ply). Nosewheel tubeless tyres, size 450×190-5 (10 ply), pressure 4.34 bar (63 lb/sq in) or 450×190R5 (10 ply). Minimum ground turning radius: at nosewheel 10.14 m (33 ft 3½ in); at mainwheel 8.26 m (27 ft 1¼ in); at wingtip 17.40 m (57 ft 1 in).

POWER PLANT: Two 1,790 kW (2,400 shp) Pratt & Whitney Canada PW127E turboprops; ATR 72–210 nacelles; six-blade Ratier-Figeac/Hamilton Sundstrand 568F propellers with new electronic control giving faster response and better synchrophasing (as for ATR 72–500 from 1996). Propeller brake on starboard engine to enable engine to be used as auxiliary power unit for internal air conditioning.

ATR 42-500 in ACES livery ***NEW***/1044767

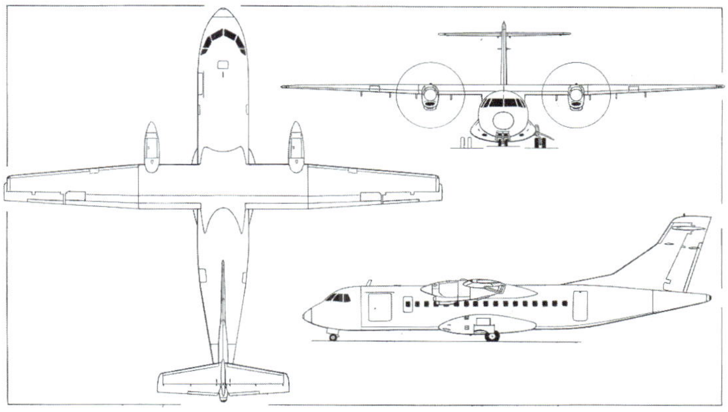
ATR 42 twin-turboprop regional transport *(Jane's/Dennis Punnett)* 0507152

Fuel in two integral tanks in spar box, total capacity 5,736 litres (1,515 US gallons; 1,262 Imp gallons). Single pressure refuelling point in starboard wing leading-edge. Gravity refuelling points in wing upper surface. Oil capacity 40 litres (10.6 US gallons; 8.8 Imp gallons).

ACCOMMODATION: Crew of two on flight deck; folding seat for observer. Seating for 42 passengers at 84 cm (33 in) pitch; or 46, 48 (standard) or 50 passengers at 76 cm (30 in) pitch; compared with Series 300, ATR 42–500 has completely new interior with new ceiling and sidewalls, indirect lighting, more sound damping; call buttons and reading lights relocated; overhead bins lengthened to 2 m (6 ft 6¾ in) to accommodate skis, golf clubs and fishing equipment carried as hand baggage. Baggage volume increased by 40 per cent. Corporate version also available; executive seating for eight to 24 passengers plus lounge area. Active noise control system, previously offered as an option, is no longer available. ATR 42–500s have structural acoustic treatment comprising reinforcement of seven fuselage frames adjacent propeller plane; dynamic vibration absorbers in this area; and internal aluminium skin damping material forward and aft of wing.

Passenger door, with integral steps, at rear of cabin on port side. Front baggage/cargo compartment between flight deck and passenger cabin, with access from inside cabin and separate loading door on port side; lavatory, galley, wardrobe and seat for cabin attendant at rear of passenger cabin, with service door on starboard side; rear baggage/cargo compartment aft of passenger cabin; additional baggage space provided by overhead bins and underseat stowage. Entire accommodation, including flight deck and baggage/cargo compartments, pressurised and air conditioned. Emergency exit via rear passenger and service doors, and by window exits on each side at front of cabin.

SYSTEMS: Improved in parallel with development of ATR 72; following refers to baseline aircraft. Honeywell air conditioning and Softair pressurisation systems, utilising engine bleed air. Pressurisation system (nominal differential 0.41 bar; 6.0 lb/sq in) provides cabin altitude of 2,040 m (7,000 ft) at flight altitudes of up to 7,620 m (25,000 ft).

Two independent hydraulic systems, each at pressure of 207 bar (3,000 lb/sq in), driven by electrically operated Abex pump and separated by interconnecting valve controlled from flight deck; system flow rate 7.9 litres (2.09 US gallons; 1.74 Imp gallons)/min; one system actuates wing flaps, spoilers, propeller brake, emergency wheel braking and nosewheel steering; second system for landing gear and normal braking. Kléber-Colombes pneumatic system for de-icing of wing leading-edges, tailplane leading-edges and engine air intakes; noses of aileron and elevator horns have full-time electric anti-icing.

Main electrical system is 28 V DC, supplied by two Auxilec 12 kW engine-driven starter/generators and two Ni/Cd batteries (43 Ah and 15 Ah), with two solid-state static inverters for 115/26 V single-phase AC supply; 115/200 V three-phase supply from two 20 kVA frequency-wild engine-driven alternators for anti-icing of windscreen, flight deck side windows, stall warning and airspeed indicator pitots, propeller blades and control surface horns. Eros/Puritan oxygen system. Instead of APU, starboard propeller braked and engine run to give DC and 400 Hz power, air conditioning and hydraulic pressure.

AVIONICS: Rockwell Collins com/nav equipment. Improved in parallel with development of ATR 72; following refers to baseline aircraft. For ATR 42-500, see ATR 72-500 description.

Comms: CVR, PA system.

Radar: Honeywell P-660 weather radar.

Flight: Honeywell DFZ 600 AFCS; AZ-800 ADCs; AH 600 AHRS with avionics standard communication bus; Hamilton Sundstrand GPWS; L-3 digital FDR; Rockwell Collins DME; Honeywell FMZ-800 flight management system and dual GPS receivers installed in four Continental Airlines ATR 42s to allow autonomous approaches.

Instrumentation: EZ-820 electronic flight instrument system.

DIMENSIONS, EXTERNAL:

Wing span .... 24.57 m (80 ft 7½ in)
Wing chord: at root .... 2.57 m (8 ft 5¼ in)
at tip .... 1.41 m (4 ft 7½ in)

ATR 42 flight deck ***NEW***/1044766

Wing aspect ratio 11.1
Length overall .... 22.67 m (74 ft 4½ in)
Fuselage max width .... 2.865 m (9 ft 4½ in)
Height overall .... 7.59 m (24 ft 10¾ in)
Elevator span .... 7.31 m (23 ft 11¾ in)
Wheel track (c/l of shock-struts) .... 4.10 m (13 ft 5½ in)
Wheelbase .... 8.78 m (28 ft 9¾ in)
Propeller diameter .... 3.96 m (13 ft 0 in)
Distance between propeller centres .... 8.10 m (26 ft 7 in)
Propeller fuselage clearance .... 0.82 m (2 ft 8¼ in)
Propeller ground clearance .... 1.17 m (3 ft 10 in)
Passenger door (rear, port): Height .... 1.75 m (5 ft 9 in)
Width .... 0.75 m (2 ft 5½ in)
Height to sill (at OWE) .... 1.46 m (4 ft 9½ in)
Service door (rear, stbd): Height .... 1.22 m (4 ft 0 in)
Width .... 0.61 m (2 ft 0 in)
Height to sill .... 1.375 m (4 ft 6¼ in)
Standard cargo/baggage door (fwd, port):
Height .... 1.575 m (5 ft 2 in)
Width .... 1.295 m (4 ft 3 in)
Height to sill (at OWE) .... 1.17 m (3 ft 9¼ in)
Large cargo door (fwd, port, optional)
Height .... 1.80 m (5 ft 11 in)
Width .... 2.95 m (9 ft 8 in)
Height to sill (at OWE) .... 1.05 m (3 ft 5¼ in)
Emergency exits (fwd, each): Height .... 0.91 m (3 ft 0 in)
Width .... 0.51 m (1 ft 8 in)
Crew emergency hatch (flight deck roof):
Length .... 0.51 m (1 ft 8 in)
Width .... 0.48 m (1 ft 7 in)

DIMENSIONS, INTERNAL:
Cabin: Length (excl flight deck, incl toilet and baggage compartments .... 14.72 m (48 ft 3½ in)
Max width .... 2.57 m (8 ft 5¼ in)
Max width at floor .... 2.26 m (7 ft 5 in)
Aisle width .... 0.46 m (1 ft 6 in)
Max height .... 1.90 m (6 ft 3 in)
Floor area .... 31.0 m^2 (334 sq ft)
Volume .... 58.0 m^3 (2,048 cu ft)
Baggage/cargo compartment volume
front (42–46 passengers) .... 6.0 m^3 (212 cu ft)
front (48 passengers) .... 4.8 m^3 (170 cu ft)
front (50 passengers) .... 3.6 m^3 (127 cu ft)
rear .... 4.8 m^3 (170 cu ft)
total front & rear (46 passengers) .... 10.8 m^3 (381 cu ft)
Containerised cargo volume
4 pallets .... 30.0 m^3 (1,059 cu ft)
5 LD3 .... 22.4 m^3 (791 cu ft)
Bulk cargo volume:
with 4 pallets .... 11.9 m^3 (420 cu ft)
with 5 LD3 .... 10.0 m^3 (353 cu ft)
Gross usable cargo volume .... 56.0 m^3 (1,978 cu ft)

AREAS:
Wings, gross .... 54.50 m^2 (586.6 sq ft)
Ailerons (total) .... 3.12 m^2 (33.58 sq ft)
Flaps (total) .... 11.00 m^2 (118.40 sq ft)
Spoilers (total) .... 1.12 m^2 (12.06 sq ft)
Fin, excl dorsal fin .... 12.48 m^2 (134.33 sq ft)
Rudder, incl tab .... 4.00 m^2 (43.05 sq ft)
Tailplane .... 11.73 m^2 (126.26 sq ft)
Elevators (total, incl tabs) .... 3.92 m^2 (42.19 sq ft)

WEIGHTS AND LOADINGS:
Operating weight empty .... 11,250 kg (24,802 lb)
Max fuel weight .... 4,500 kg (9,921 lb)
Max payload .... 5,450 kg (12,015 lb)[1]
Max ramp weight .... 18,770 kg (41,380 lb)
Max T-O weight .... 18,600 kg (41,005 lb)
Max landing weight .... 18,300 kg (40,345 lb)
Max zero-fuel weight .... 16,700 kg (36,817 lb)[1]
Max wing loading .... 341.3 kg/m^2 (69.90 lb/sq ft)
Max power loading .... 5.20 kg/kW (8.54 lb/shp)

[1] *Optional increase of 300 kg (661 lb)*

PERFORMANCE:
Max cruising speed at FL170 at 97% MTOW .... 300 kt (556 km/h; 345 mph)
Time to climb to FL170 .... 9.9 min
Service ceiling OEI, ISA +10°C, 97% MTOW .... 5,485 m (18,000 ft)
T-O distance: ISA, S/L .... 1,165 m (3,825 ft)
ISA +10°C at 915 m (3,000 ft) for 300 n mile (556 km; 345 mile) stage with 48 passengers .... 1,163 m (3,815 ft)
FAR landing field length:
S/L at landing weight with 48 passengers .... 1,040 m (3,415 ft)
S/L at MLW .... 1,126 m (3,695 ft)
Runway ACN for flexible runway, category B 10
Range with max fuel .... 1,600 n miles (2,963 km; 1,841 miles)
Max range with 48 passengers .... 840 n miles (1,555 km; 966 miles)

OPERATIONAL NOISE LEVELS:
Flyover .... 76.6 EPNdB
Sideline .... 80.7 EPNdB
Approach .... 92.4 EPNdB

UPDATED

ATR 42 MP SURVEYOR

TYPE: Maritime surveillance twin-turboprop.

PROGRAMME: Variant of ATR 42 airliner developed by Alenia Aeronautica. Exhibited in model form at Dubai Air Show, November 1995. Initially designated **SAR 42** and **ATR 42MP**. First airframe modified by Officine Aeronavali's Capodichino plant; maiden flight (CMX62166) in Surveyor configuration, but without equipment installed, 1 February 1999, was also delivery to Alenia at Caselle for systems integration. Italian civil certification received 24 October 1999; first delivery 14 December 1999.

CURRENT VERSIONS: Other missions include vessel search, identification and surveillance; maritime and coastal surveillance; pollution detection; search and rescue; medical evacuation; and VIP/troop/cargo/corporate/humanitarian transport.

ATR 42 MP Surveyor: Basic model, *as described.*

CUSTOMERS: Two ordered by Guardia di Finanza (Italian customs service) 1996 and first aircraft (MM62165) delivered that November but in normal transport configuration for training; converted to operational version in 2000; is ATR 42–400 for first two aircraft; third aircraft based on 42–500. All these now delivered.

Italian Guardia Costiera (Coast Guard) ordered one plus another option for unarmed version; first delivery (MM62170 '10-01', formerly of Italian Air Force) 30 May 2001; second aircraft delivered 2003.

Data as ATR 42–400, except particulars below.

POWER PLANT: Two 1,566 kW (2,100 shp) Pratt & Whitney Canada PW121A turboprops driving Ratier-Figeac/Hamilton Sundstrand 568F six-blade propellers.

ACCOMMODATION: Flight crew of two, plus three mission operators in modified cabin. Rest/debrief area towards rear of cabin; galley/lavatory facilities at rear. Observers' stations with bubble windows one each side of fuselage aft of wing. Rear door modified for in-flight opening; civil-style freight door in forward port side. SAR package behind rest area.

AVIONICS: *Comms:* VHF/UHF transceiver, AM-FM, VHF/FM/HF com transceiver, transponder/IFF, interphone and optional secure datalink.

Radar: Raytheon SV 2022 360° search radar; weather radar from ATR 42 retained.

Flight ADS, VOR, DME, ADF, optional Tacan, FMS, IRS, INS/GPS, radio altimeter, direction-finder.

Mission Galileo Avionica Airborne Tactical Observation and Surveillance system (ATOS) comprises mission management system; communications subsystem; L-3 Wescam turret beneath starboard landing gear pannier containing Galileo FLIR; daylight TV sensor; Spectrolab SX16E searchlight and Thiokol LUU-2B/B flare launcher all mounted in pod on starboard side of forward fuselage; Elettronica ALR-733 ESM. MIL-STD-1553B, RS-422 and ARINC 429 databus system. Two multifunction operator consoles with 48 cm (19 in) display, 25 cm (10 in) sensor display, keyboard, trackball, joystick and colour printer on starboard side; communications console. Provision for future growth of sensor suite.

EQUIPMENT: SAR equipment includes searchlight, loudspeakers and flare launcher. IR/UV scanner and optional SLAR and MVR for pollution detection.

ARMAMENT: Optional FN Herstal HPM twin machine gun pod on port side of forward fuselage.

WEIGHTS AND LOADINGS:
Typical operating weight empty .... 13,275 kg (29,266 lb)
Max payload .... 3,425 kg (7,551 lb)
Optional auxiliary fuel .... 650 kg (1,433 lb)
Max T-O weight .... 18,600 kg (41,005 lb)
Max landing weight .... 18,300 kg (40,345 lb)
Zero-fuel weight: typical .... 13,200 kg (29,101 lb)
max .... 16,700 kg (36,817 lb)

Italian Coast Guard ATR 42 Surveyor *(Paul Jackson)* ***NEW***/1044765

PERFORMANCE (at max T-O weight except where indicated):
- Max cruising speed at FL160 at 97% MTOW .... 281 kt (520 km/h; 323 mph)
- Typical patrol speed at FL20 at 90% MTOW
  - for max range .... 189 kt (350 km/h; 217 mph)
  - for max endurance .... 135 kt (250 km/h; 155 mph)
- Service ceiling, OEI, at 97% MTOW .... 3,960 m (13,000 ft)
- T-O distance at S/L ISA .... 1,050 m (3,445 ft)
- Landing distance at MLW, S/L, ISA .... 1,150 m (3,775 ft)
- Ferry range .... 2,019 n miles (3,740 km; 2,323 miles)
- Endurance on station at 200 n mile (370 km; 230 mile
  - radius .... 8 h
  - at 600 n mile (1,111 km; 690 mile) radius .... 3 h 30 min

UPDATED

ATR 72

TYPE: Twin-turboprop airliner.

DEVELOPMENT MILESTONES

72	
Programme launched	15 Jan 86
First flight	27 Oct 88
Certification	25 Sep 89
First delivery (Kar Air)	27 Oct 89
72-500	
First flight	19 Jan 96
Certification	14 Jan 97
First delivery (American Eagle)	31 Jul 97

PROGRAMME: Stretched version of ATR 42; announced at 1985 Paris Air Show; launched 15 January 1986; three development aircraft built: first flights 27 October 1988 (F-WWEY), 20 December 1988 (F-WWEZ, c/n 108) and 18 April 1989 (OH-KRA, c/n 126); French and US certification 25 September and 15 November 1989 respectively; deliveries, to Kar Air of Finland, began 27 October 1989 (OH-KRB); UK certification 30 July 1993.

CURRENT VERSIONS: **ATR 72-200:** Initial production version; two Pratt & Whitney Canada PW124B turboprops, each rated at 1,611 kW (2,160 shp) for normal take-off and 1,790 kW (2,400 shp) with ATPCS; Hamilton Sundstrand 14SF-11 four-blade propellers. Also cargo version, capable of carrying 13 small containers. Now discontinued.

ATR 72-210: Improved hot/high performance version with PW127 engines rated at 1,849 kW (2,480 shp) and Hamilton Sundstrand 247F propellers with composites blades on steel hubs; ATPCS power 2,059 kW (2,760 shp); carries 17 to 19 more passengers than standard ATR 72 in WAT-limited conditions; French and US certification 15 and 18 December 1992, German on 24 February 1993; first delivery December 1992.

ATR 72-500: Launched as ATR 72-210A. Improved hot/high performance version with PW127 engines, six-blade propellers and redesigned interior. First flight 19 January 1996; DGAC certification achieved 14 January 1997; first delivery (to American Eagle) 31 July 1997.

Description applies to ATR 72-500, except where indicated.

ATR 72 Quick-Change: Similar to ATR 42 Quick-Change, except capacity for 13 ATR containers; conversion time approx 45 minutes.

ATR 72 full freighter version: Remarks as for equivalent ATR 42. Total cargo volume 75.5 m^3 (2,666 cu ft); parallel cabin section 14.75 m (48 ft 4¾ in) long, with positions for up to six spider nets. MTOW in this configuration 22,000 kg (48,501 lb); MLW 21,350 kg (47,069 lb); MZFW 20,000 kg (44,092 lb). LCD as for ATR 42 freighter; permits loading of five 224 × 274 cm (88 × 108 in) pallets or seven LD3 containers. Max payload for first-generation aircraft fitted with 'tube' and LCD is 8,093 kg (17,842 lb). First retrofit (to c/n 108) completed for Farnair by Aeronavali (subsidiary of Alenia Aeronautica) in June 2002; second for same customer, in 2004.

ATR 72 ASW: Projected anti-submarine warfare version, based on ATR 42 but also offering 1,270 kg (2,800 lb) payload including torpedoes, depth charges and anti-ship missiles. None yet built.

CUSTOMERS: Total 307 ATR 72s ordered up to 31 August 2004, at which time 294 delivered; 300th (destined for Air New Zealand) completed manufacture on 23 November 2004. Deliveries in 1998 totalled 21 and in 1999 numbered 23. Orders in 2000 totalled 18 plus six options; 15 (all -500s) delivered in 2001, 14 in 2002 and six in 2003; further 11 (Binter Canarias six, Air Tahiti four, Air New Zealand one) ordered by July 2004.

COSTS: ATR 72-500, US$17.5 million (2004).

DESIGN FEATURES: As ATR 42 (which see), but with more power, more fuel, greater wing span/area, and longer fuselage for up to 74 passengers.

FLYING CONTROLS: As for ATR 42 but vortex generators ahead of ailerons and aileron horn balances shielded by wingtip extensions; vortex generators under leading-edge of elevators. Flap settings 0, 15 and 30°.

Binter Canarias ATR 72-500 ***NEW***/1044763

STRUCTURE: Generally as for ATR 42, but new wings outboard of engine nacelles have CFRP front and rear spars, self-stiffening CFRP skin panels and light alloy ribs, resulting in weight saving of 120 kg (265 lb); sweepback on outer panels 2° 18' at quarter-chord. Trials of an all-composites tail assembly were conducted in 1997 and the structure incorporated in all production aircraft from 1998. The major airframe inspection period for the ATR 72 was increased on 2 October 1997 from 24,000 to 36,000 cycles, with a corresponding reduction in maintenance cost, bringing it in line with the ATR 42 family.

LANDING GEAR: Messier-Hispano-Bugatti units with Dunlop wheels (tyres size H34×10.0R16 (14 ply), pressure 7.86 bar; 114 lb/sq in) and structural carbon brakes; nosewheel tyre as ATR 42. Minimum ground turning radius: at nosewheel 12.45 m (40 ft 10¼ in); at mainwheel 8.26 m (27 ft 1¼ in); at wingtip 19.74 m (64 ft 9¼ in).

POWER PLANT: *ATR 72-500:* Two Pratt & Whitney Canada PW127F turboprops, each rated at 1,864 kW (2,500 shp) for normal flight and 2,051 kW (2,750 shp) for take-off, driving Ratier-Figeac/Hamilton Sundstrand 568F six-blade, all-composites propellers. Fuel capacity 6,337 litres (1,674 US gallons; 1,394 Imp gallons), comprising ATR 42 tanks, plus additional 637 litres (168 US gallons; 140 Imp gallons) in outer wings; pressure refuelling point in starboard main landing gear fairing.

ACCOMMODATION: Basic 68 passengers at 79 cm (31 in) seat pitch; other seating configurations range from 64 seats at 81 cm (32 in) to 72 seats at 76 cm (30 in); plus second cabin attendant's seat. Corporate version also available; typical seating for 44 passengers plus lounge area. Single baggage compartment at rear of cabin; two at front. Forward door, with a service door opposite on starboard side. Service door on each side at rear, that on port side replaced by a passenger door when cargo door is fitted at front. Two additional emergency exits (one each side); both rear doors also serve as emergency exits. Increased-capacity air conditioning system.

AVIONICS: *(ATR 72-500): Comms:* Rockwell Collins VHF transceiver and Mode 5 transponder; EADS Socata ELT; Loral CVR.

Radar: Honeywell weather radar.

Flight: Rockwell Collins VOR/ILS/MKR, ADF and DME; Honeywell EFIS (EADI and EHSI) and air data computer.

Instrumentation: Honeywell ASI, VSI and altimeter; Aeronetics RMI; Smiths standby compass; Thales standby horizon, altimeter and ASI.

DIMENSIONS, EXTERNAL: As ATR 42 except:
- Wing span .... 27.05 m (88 ft 9 in)
- Wing chord at tip .... 1.59 m (5 ft 2½ in)
- Wing aspect ratio 12.0
- Length overall .... 27.17 m (89 ft 1¾ in)
- Height overall .... 7.65 m (25 ft 1¼ in)
- Wheelbase .... 10.77 m (35 ft 4 in)
- Propeller ground clearance:
  - at OWE .... 1.21 m (3 ft 11½ in)
  - at max ramp weight .... 1.12 m (3 ft 8 in)
- Passenger door (rear, port): Height .... 1.73 m (5 ft 8 in)
  - Width .... 0.635 m (2 ft 1 in)
  - Height to sill .... 1.375 m (4 ft 6¼ in)
- Cargo door (fwd, port): Height .... 1.575 m (5 ft 2 in)
  - Width .... 1.295 m (4 ft 3 in)
  - Height to sill .... 1.12 m (3 ft 8 in)

DIMENSIONS, INTERNAL:
- Cabin: Length (excl flight deck, incl toilet and baggage compartments) .... 19.21 m (63 ft 0¼ in)
  - Cross-section .... as for ATR 42
  - Floor area .... 41.7 m^2 (449 sq ft)
  - Volume .... 76.0 m^3 (2,684 cu ft)
- Baggage/cargo compartment volume:
  - front (66 passengers) .... 5.8 m^3 (205 cu ft)
  - rear .... 4.8 m^3 (169.5 cu ft)
  - total .... 10.6 m^3 (374.5 cu ft)
- Containerised cargo volume:
  - 5 pallets .... 42.6 m^3 (1,504 cu ft)
  - 7 LD3 .... 31.4 m^3 (1,109 cu ft)

Air Tahiti ATR 72 ***NEW***/1044764

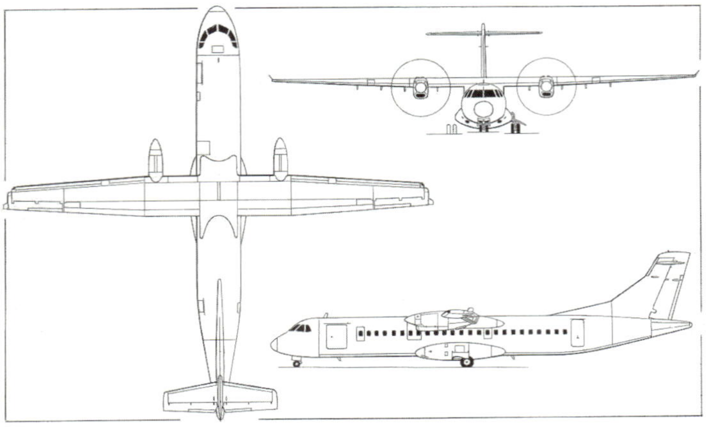

ATR 72–200 (two Pratt & Whitney Canada PW124B turboprops) *(Jane's/Dennis Punnett)* 0507153

Bulk cargo volume:
with 5 pallets ... 13.3 m^3 (470 cu ft)
with 7 LD3 ... 11.7 m^3 (413 cu ft)
Gross usable cargo volume ... 75.5 m^3 (2,666 cu ft)

AREAS: As ATR 42 except:
Wings, gross ... 61.0 m^2 (656.6 sq ft)
Ailerons (total) ... 3.75 m^2 (40.36 sq ft)
Flaps (total) ... 12.28 m^2 (132.18 sq ft)
Spoilers (total) ... 1.34 m^2 (14.42 sq ft)

WEIGHTS AND LOADINGS:
Operating weight empty ... 12,950 kg (28,550 lb)
Max fuel weight ... 5,000 kg (11,023 lb)
Max payload: standard ... 7,050 kg (15,543 lb)
optional ... 7,350 kg (16,204 lb)
Max T-O weight: standard ... 22,000 kg (48,501 lb)
optional ... 22,500 kg (49,604 lb)
Max ramp weight: standard ... 22,170 kg (48,876 lb)
optional ... 22,670 kg (49,979 lb)
Max landing weight: standard ... 21,850 kg (48,171 lb)
optional ... 22,350 kg (49,273 lb)
Max zero-fuel weight: standard ... 20,000 kg (44,092 lb)
optional ... 20,500 kg (45,195 lb)
Max wing loading: standard ... 360.7 kg/m^2 (73.87 lb/sq ft)
optional ... 368.9 kg/m^2 (75.55 lb/sq ft)
Max power loading: standard ... 5.36 kg/kW (8.81 lb/shp)
optional ... 5.49 kg/kW (9.01 lb/shp)

PERFORMANCE:
Max cruising speed at FL160, at 97% MTOW:
standard ... 276 kt (511 km/h; 318 mph)
optional ... 275 kt (509 km/h; 316 mph)
Econ cruising speed at FL230, at 95% MTOW ... 248 kt (459 km/h; 285 mph)
Service ceiling ... 7,620 m (25,000 ft)
Service ceiling, OEI, ISA + 10°C, 97% MTOW ... 4,330 m (14,200 ft)
T-O balanced field length:
at S/L, ISA: basic ... 1,223 m (4,015 ft)
optional ... 1,290 m (4,235 ft)
at FL30, ISA +10°C, at T-O weight, 68 passengers, both ... 1,300 m (4,265 ft)
Landing field length at S/L, at MLW: basic ... 1,048 m (3,438 ft)
optional ... 1,067 m (3,500 ft)
Runway ACN for flexible runway, category B ... 13
Still air range, reserves for 87 n mile (161 km; 100 mile) diversion and 45 min:
max payload: basic ... 715 n miles (1,324 km; 822 miles)
optional ... 890 n miles (1,648 km; 1,024 miles)
max fuel and zero payload ... 1,956 n miles (3,622 km; 2,251 miles)

OPERATIONAL NOISE LEVELS:
Flyover ... 79.0 EPNdB
Sideline ... 83.2 EPNdB
Approach ... 92.2 EPNdB

UPDATED

BELL/AGUSTA

BELL/AGUSTA AEROSPACE COMPANY (BAAC)

PO Box 901073, Fort Worth, Texas 76101, USA
Tel: (+1 817) 278 96 00
Fax: (+1 817) 278 97 26
Web: http://www.bellagusta.com
MANAGING DIRECTOR: Louis P Bartolatta Jr
EXECUTIVE MANAGING DIRECTORS:
Don Barbour (BA609 Programme)
Antonio Giovannini (AB139 Programme)

PARTICIPATING COMPANIES
Bell Helicopter Textron: see under USA
Agusta: see under Italy

Bell and Agusta announced on 8 September 1998 that they had agreed to establish a joint venture to manage development of two new aircraft: the BA609 tiltrotor, previously a Bell and Boeing programme, and the AB139, a new helicopter announced on the same day. Following approval of both boards, a definitive agreement was signed on 6 November 1998. Bell is the majority shareholder and will undertake final assembly for AB139s delivered to North America. Agusta, which has built Bell helicopters under licence since 1952, is investing and participating in development of the BA609, manufacturing some components and assembling those sold in Europe and certain other parts of the world. Additionally, AgustaWestland is responsible for the AB139's development and certification, with participation by Bell. A military version was revealed in July 2000. Flight testing of the AB139 began in February 2001, followed by the BA609 in March 2003.

UPDATED

BELL/AGUSTA AB139

TYPE: Multirole medium utility helicopter.

PROGRAMME: Announced at Farnborough Air Show, 8 September 1998, as joint venture between Agusta and Bell; to complement, rather than replace, Bell 412. Full-scale mockup unveiled at Paris Air Show 12 June 1999. AgustaWestland responsible for development, certification to JAR/FAR 29 and transition to production, with participation by Bell on a 75:25 per cent work-share basis; final assembly by AgustaWestland at Vergiate, and by Bell (possibly at Mirabel, Canada) for American and Pacific Rim customers. No designated "prototype"; first preproduction aircraft (01, later I-ACOI) undertook maiden flight on 3 February 2001 followed by second aircraft (02, later I-ATWO) on 4 June 2001 and third (03, later I-EPIC) on 22 October 2001; Assembly of first production aircraft began in late November 2001; this (I-ANEW) demonstrated at Farnborough in July 2002. Three preproduction aircraft and one tie-down helicopter (TDH) undertook flight test programme

First production Bell/Agusta AB139 after addition of upturned tailplane tips ***NEW***/0589281

Bell/Agusta AB139 interior *NEW*/0589280

leading to Italian certification on 20 June 2003, following 1,600 hour ground and flight test programme (including those flown by first preproduction aircraft which was lost in crash on 22 April 2002) and 750 hours completed on the TDH. FAA certification achieved 20 December 2004. Full-scale mockup of AB139 Military unveiled at Farnborough International 2000 in July 2000; development may follow certification of civilian version, but not being actively promoted, some of role having transferred to A 149.

Risk-sharing collaborators include GKN Westland (tail rotor drive train), Honeywell (avionics), Kawasaki (transmission input module), Liebherr Germany (landing gear and air conditioning system), Pratt & Whitney Canada (power plant) and PZL Świdnik (airframe components).

CURRENT VERSIONS: **AB139:** Commercial law enforcement and SAR version, *as described*.

AB139 Military: Proposed multirole military helicopter with provision for armoured crew seats, electronic warfare protection, IR suppressors, two internal pintle-mounted machine guns and easily removable stub-wing weapons supports for gun pods, rocket launchers and AAMs.

CUSTOMERS: More than 80 ordered by 40 customers by December 2004. Launch customer Bristow Helicopters of UK announced order for two on 26 September 2000 for delivery in 2003. Hawker Pacific ordered four on 12 February 2001 for corporate, utility and offshore operations in the Arabian Gulf. Recent customers include the government of Namibia, which ordered two for VIP and multimission duties, of which the first was delivered on 16 September 2004; the Aga Khan Development Network (AKDN), which has ordered four for operations in South and Central Asia in connection with construction of three University of Central Asia campuses; and Evergreen International, which ordered two in June 2003. Selected for US Coast Guard 'Deepwater' programme vertical take-off/landing recovery and surveillance (VRS) requirement, with first delivery expected in 2012. Selection by Irish Department of Defence announced 3 December 2004 for Irish Air Corps utility helicopter requirement, expected to lead to firm order for four. Anticipated market for 900 over 20 years, some 55 per cent for military use; 34 per cent of sales projected in Europe, 23 per cent in Middle East, 18 per cent in Far East, 13 per cent in South America and 12 per cent in North America. Target of 20 deliveries by end of 2004.

COSTS: Commercial version US$7 million (2002). Estimated direct operating cost based on North American operation, US$757.30 per hour (2003).

DESIGN FEATURES: Design goals include high manoeuvrability and agility, low pilot workload, night/all-weather operation, low acoustic and infra-red emissions and mission flexibility for commercial and military operators. Intended for offshore support, medevac, corporate/VIP transport, SAR and military operations. Able to operate at maximum T-O weight from Class A helipads at 945 m (3,100 ft) at ISA + 20°C. Five-blade, fully articulated, ballistic tolerant main rotor and four-blade tail rotor. Some transmission and rotor elements based on AgustaWestland A129 Mangusta. Upturned tips added to horizontal stabilisers of production aircraft.

FLYING CONTROLS: Four-axis, digital AFCS.

LANDING GEAR: Heavy-duty, retractable tricycle type with twin wheels on nose unit; single wheels on main units, which retract into side sponsons.

POWER PLANT: Two Pratt & Whitney Canada PT6C-67C turboshafts, with FADEC, each rated at 1,252 kW (1,679 shp) for T-O and 1,142 kW (1,531 shp) maximum continuous; OEI ratings 1,286 kW (1,725 shp) for two minutes and 1,252 kW (1,679 shp) maximum continuous. Fuel tanks behind main cabin. Main transmission can run for up to 30 minutes without oil.

ACCOMMODATION: Up to 15 passengers on crashworthy seats in three rows of five, two forward facing, one rearward facing, in unobstructed cabin with flat floor; flight-accessible baggage compartment at rear of cabin (base of tailboom). Alternatively, six stretchers and four attendants in medevac configuration. Plug-type sliding door on each side of cabin, with separate crew doors.

SYSTEMS: Systems duplicated and separated. Main and tail rotor ice protection optional. Smiths Aerospace HUMS; Eaton Corporation Smart Zapper chip detection system for main transmission, intermediate and tail rotor gearboxes.

AVIONICS: Honeywell Primus Epic as core system. Provision for up to four 203 × 254 mm (8 × 10 in) high-definition colour active matrix liquid crystal displays for MFD, PFD and

Bell/Agusta AB139 instrument panel *NEW*/0589279

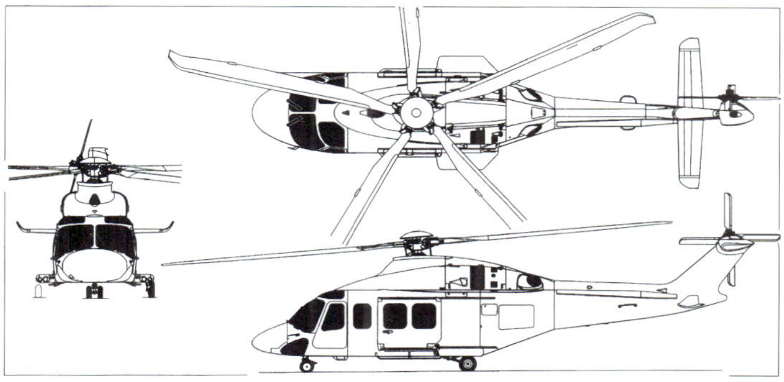

General arrangement of the Bell/Agusta AB139 *(Paul Jackson)* *NEW*/0589207

FLIR/video functions, and four-axis modular digital autopilot with flight director for hands-off operation and SAR modes.

DIMENSIONS, EXTERNAL:

Main rotor diameter	13.80 m (45 ft 3¼ in)
Tail rotor diameter	2.65 m (8 ft 8¼ in)
Length overall, rotors turning	16.65 m (54 ft 7½ in)
Fuselage: Length	13.53 m (44 ft 4¾ in)
Max width: across cabin	2.26 m (7 ft 5 in)
across sponsons	3.20 m (10 ft 6 in)
Height, rotors turning	4.95 m (16 ft 3 in)
Tail rotor ground clearance	2.30 m (7 ft 6½ in)
Cabin doors (two, each): Height	1.35 m (4 ft 5 in)
Width	1.68 m (5 ft 6 in)

DIMENSIONS, INTERNAL:

Cabin: Length	2.70 m (8 ft 10¼ in)
Max width	2.00 m (6 ft 6¾ in)
Max height	1.42 m (4 ft 8 in)
Volume	8.0 m^3 (283 cu ft)
Baggage compartment volume	3.4 m^3 (120 cu ft)

AREAS:

Main rotor disc	149.57 m^2 (1,610.0 sq ft)
Tail rotor disc	5.52 m^2(59.37 sq ft)

WEIGHTS AND LOADINGS:

Useful load	2,500 kg (5,512 lb)
Max external load	2,700 kg (5,952 lb)
Max T-O weight	6,000 kg (13,227 lb)
Max disc loading	40.11 kg/m^2 (8.22 lb/sq ft)

PERFORMANCE (estimated):

Never-exceed speed (VNE)	167 kt (309 km/h; 192 mph)
Max cruising speed	157 kt (291 km/h; 181 mph)
Max rate of climb at S/L	610 m (2,000 ft)/min
Service ceiling, OEI	2,926 m (9,600 ft)
Hovering ceiling OGE	3,660 m (12,000 ft)
Max range, no reserves	400 n miles (740 km; 460 miles)
Endurance	3 h 54 min

UPDATED

BELL/AGUSTA BA609

TYPE: Light utility tiltrotor.

DEVELOPMENT MILESTONES

Announced	Feb 96
Design submitted	May 97
First prototype material cut	Aug 97
Mockup unveiled	Jun 97
Ground-running began	6 Dec 02
First flight	7 Mar 03

PROGRAMME: Having earlier joined in partnership to develop the military V-22 tiltrotor, Bell Boeing revealed in February 1996 that studies were in progress for a nine-passenger civil tiltrotor aircraft in the 6,350 kg (14,000 lb) weight class, with the preliminary designation D-600. Subsequently, on 18 November 1996, the two companies announced that a joint venture was being established to design, develop, certify and market a six- to nine-passenger civil tiltrotor as the Bell Boeing 609.

Boeing withdrew as a partner on 1 March 1998 and Bell subsequently teamed with Agusta to develop, produce and market the tiltrotor as the BA609; this arrangement was

Bell/Agusta AB139 medium utility helicopter in prototype form *NEW*/0573344

Bell/Agusta BA609 during early test flying in 2003 ***NEW***/0589259

formally announced at the Farnborough Air Show in September 1998. Agusta is investing and participating in BA609 development and will be responsible for assembly of BA609s sold in Europe and elsewhere.

Preliminary design review completed May 1997. Manufacture of parts for prototypes began at Philadelphia, August 1997. Full-size mockup first exhibited at Paris Air Show, June 1997.

Ground-running trials began on 6 December 2002. First flight of the prototype (N609TR) took place (in the vertical mode only) on 7 March 2003, rescheduled from late 2002, following decision in April 2002 to slow development and certification as consequence of delays in Bell Boeing V-22 Osprey programme, the latter being leader in technology development. Prototype flew some 14 hours in first two months, with rotors angled up to 15° forward before being disassembled for examination; return to flying scheduled for 29 October 2004.

Four prototypes are being produced for 36-month flight test programme leading to certification in 2007 under FAR Pt 25 (fixed-wing aircraft) and Pt 29 (helicopters), plus Pt 21.17(b) Special Conditions for unique components. To be capable of single-pilot IFR operation. Prototype used primarily for expansion of flight envelope, while second, third and fourth airframes are dedicated to systems certification, avionics and icing approval, and FAA function and reliability, respectively. Second prototype scheduled to fly at AgustaWestland plant in the fourth quarter of 2004. Certification in FAA Powered Lift category, including two-crew IFR and flight into known icing approvals, expected in late 2007.

CURRENT VERSIONS: **BA609:** Initial version *as described.*

HV-609: Multimission version proposed by Bell and Lockheed Martin to satisfy US Coast Guard 'Deepwater' re-equipment programme as potential replacement for Dassault HU-25 Guardian, Eurocopter HH-65 Dolphin and Sikorsky HH-60J Jayhawk. Missions could include drug interdiction and SAR, with 30 to 50 HV-609s possibly being acquired.

UV-609: Utility version conceived by Bell for US Army combat role, including casualty evacuation, command and control, logistic support and light utility/troop transport tasks. Manufacturer also promoting UV-609 to US Marine Corps as potential training system for V-22 Osprey.

619: Projected 19-seat version aimed at commuter market; not launched by March 2003.

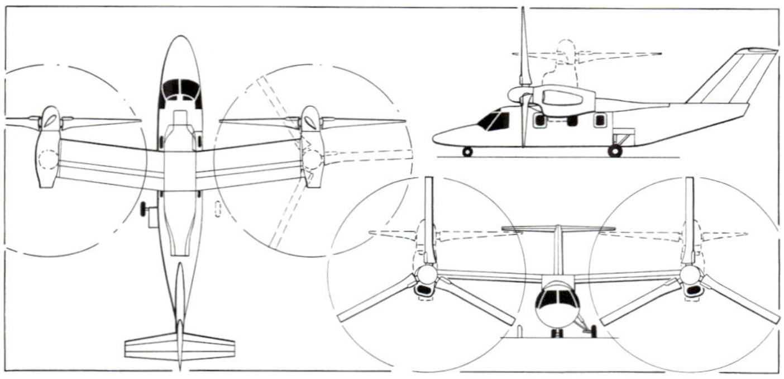

Bell/Agusta BA609 tiltrotor aircraft ***(James Goulding)*** 0044896

Bell/Agusta BA609 maiden flight ***NEW***/0573337

620: Proposed 22-seat version conceived by Bell Boeing team; no recent news received and concept may have lapsed.

CUSTOMERS: Order book opened 2 February 1997 at Heli Expo, with first order placed soon after by unspecified customer. Total of 70 ordered by 40 customers in 18 countries by March 2003; purchasers identified to date include Helitech Pty of Australia; Lider of Brazil; Canadian Helicopter Corporation and Northern Mountain Helicopters Inc of Canada; Petroleum Tiltrotors International of Dubai; Aero-Dienst GmbH of Germany; Mitsui of Japan; United Industries of South Korea; Helikopter Services of Norway; Lloyd's Investments of Poland; Bristow of the UK; Massachusetts Mutual Life Insurance, Austin Jet and Petroleum Helicopters of the USA; and Air Center Helicopters Inc of the US Virgin Islands. Briefing given to US Coast Guard, late 1997, followed by demonstration by XV-15 tiltrotor concept demonstrator aboard Coast Guard cutter *Mohawk* off Key West, Florida, in May 1999.

COSTS: US$8 million to US$10 million, depending on configuration.

DESIGN FEATURES: Combines the most favourable aspects of helicopter and aeroplane performance in passenger-carrying role. T tail configuration instead of endplate fin layout used by earlier tiltrotor designs (XV-15 and V-22 Osprey); this raises horizontal tailplane above rotor wake to minimise fore and aft pitching moment at transition phase of flight. Size of wing determined by requirement for it to hold all fuel for CG and safety considerations. Composites cross-shafts keep both proprotors turning in event of engine failure. Manual screwjack facility exists whereby the proprotors can be tilted into helicopter mode if cross-shafts fail. Designed using three-dimensional CATIA digital computer design system. Airframe has design life of 20,000 flight hours.

Refer also to Bell Boeing V-22 description for explanation of tiltrotor concept and for its extension to four-proprotor configuration.

FLYING CONTROLS: BAE Systems triplex digital fly-by-wire flight control system, with Dowty Aerospace actuators. T tail with conventional elevators; no rudder. Two-segment trailing-edge flaperons.

STRUCTURE: Aluminium fuselage structure with composites skinning; composites wing. Production fuselages, including cockpit, cabin and systems installation, will be built by risk-sharing partner Fuji Heavy Industries of Japan, which may also establish a third production line if substantial orders are won from the Japanese government; cabin doors and fuselage tailcone supplied by Kawasaki; wing and nacelles by Bell at Fort Worth, with final assembly line at new Bell tiltrotor manufacturing facility in Amarillo, Texas and at Agusta facility in Italy. Fuselage in three major sections; nose, centre and tail, with fuselage skin incorporating graphite stringers with Japanese Toray composites material. Same used for wing and nacelles, with upper and lower wing surfaces produced as single pieces.

LANDING GEAR: Retractable tricycle type, with twin nosewheels and single wheel on each main unit. Messier-Dowty overseeing design, development and manufacture of integrated landing gear system, including legs, wheels, tyres, brakes, brake control and landing gear control systems.

POWER PLANT: Two 1,447 kW (1,940 shp) Pratt & Whitney Canada PT6C-67A turboshaft engines, installed in tilting nacelles at wingtips, each driving a three-blade proprotor. Nacelle interface units by AMETEK Aerospace Systems. Rockwell Collins EICAS; modified oil system, with dual pumps, to generate sufficient oil pressure when operating in vertical mode; several planetary gears also removed to achieve direct drive 30,000 rpm output. Nacelle transition achieved in 20 seconds. Fuel in integral wing tanks; usable capacity 1,401 litres (370 US gallons; 308 Imp gallons). Provision for auxiliary fuel tanks.

ACCOMMODATION: Crew of two, side by side on flight deck, with dual controls. Maximum of nine passengers in standard aircraft. Crew and passenger door on starboard side, forward of wing. Accommodation pressurised and air conditioned; pressurisation differential 0.38 bar (5.50 lb/sq in). Transparencies by Sully Produits Speciaux of France.

SYSTEMS: Equipped for flight into known icing. Lucas Aerospace DC electrical power systems. Intertechnique brushless electric pumps and motor-operated shut-off valves.

AVIONICS: Rockwell Collins Pro Line 21 package as standard.

Comms: Dual VHF radios; Mode S transponder. Cockpit voice recorder included as standard.

Radar: Optional Rockwell Collins WXR-800 solid-state weather radar.

Flight: Dual VOR/ILS, DME and ADF, with integrated control of sensors by dual Rockwell Collins RTU-4200 radio tuning units. Rockwell Collins ALT-4000 radar altimeter optional. GPS included as standard, along with TCAS and FDR.

Instrumentation: Three 250 × 200 mm (10 × 8 in) active matrix colour LCD adaptive flight displays, including two primary flight displays and one multifunction display. Standby instrument system by Goodrich Aerospace.

Following data are provisional.

DIMENSIONS, EXTERNAL:

Proprotor diameter (each) .... 7.9 m (26 ft)
Width, rotors turning .... 18.3 m (60 ft)

Length overall ... 13.4 m (44 ft)
Max diameter of fuselage ... 1.75 m (5 ft 9 in)
Height to top of fin ... 4.6 m (15 ft)
Wheel track ... 3.0 m (10 ft)
Wheelbase ... 5.8 m (19 ft)
Distance between proprotor centres ... 10.0 m (33 ft)
Passenger door: Width ... 0.76 m (2 ft 6 in)

DIMENSIONS, INTERNAL

Cabin: Length ... 4.09 m (13 ft 5 in)
Max width ... 1.47 m (4 ft 10 in)
Max height ... 1.42 m (4 ft 8 in)
Baggage hold volume ... 1.4 m^3 (50 cu ft)

AREAS:

Rotor discs, each ... 49 m^2 (530 sq ft)

WEIGHTS AND LOADINGS:

Weight empty ... 4,765 kg (10,500 lb)
Max T-O weight ... 7,620 kg (16,800 lb)
Max disc loading ... 77.4 kg/m^2 (15.85 lb/sq ft)

PERFORMANCE:

Max cruising speed ... 275 kt (509 km/h; 316 mph)
Max rate of climb at S/L ... 457 m (1,500 ft)/min
Service ceiling ... 7,620 m (25,000 ft)
Service ceiling, OEI ... 4,875 m (16,000 ft)
Hovering ceiling, OGE ... 1,525 m (5,000 ft)
Range: normal fuel capacity with 2,500 kg (5,500 lb) payload at 250 kt (463 km/h; 288 mph) ... 750 n miles (1,389 km; 863 miles)
with auxiliary fuel tanks ... 1,000 n miles (1,852 km; 1,151 miles)
Endurance, normal fuel ... 3 h

UPDATED

BOEING/BAE SYSTEMS

BOEING/BAE SYSTEMS

PARTICIPATING COMPANIES

Boeing (Integrated Defense Systems): see under USA
BAE Systems: see under UK

Original McDonnell Douglas Corporation (MDC) and Hawker Siddeley companies initially associated in 1969 through US procurement of Harrier; relationship further developed with US Navy selection of BAE Systems Hawk; both types built at St Louis, although production of the Harrier II ended in 1997, leaving only remanufacture and upgrade work on the AV-8B. Joint work undertaken on advanced combat aircraft (later known as JSF), with additional participation of Northrop Grumman, until next-stage contracts were awarded to competing designs in November 1996. Boeing took over MDC in August 1997. Production of T-45 Goshawk continues, albeit at less than minimum economic rate.

UPDATED

BOEING/BAE SYSTEMS T-45 GOSHAWK

TYPE: Advanced jet trainer.

DEVELOPMENT MILESTONES

Official go-ahead	18 Nov 1981
First order	26 Jan 1988
First prototype material cut	Feb 1986
First flight	16 Apr 1988
First delivery	23 Jan 1992
Entered service	27 Jun 1992

PROGRAMME: Carrier-capable version of BAE Systems Hawk selected 18 November 1981 (from five other candidates) as winner of US Navy VTXTS (now T45TS) competition for undergraduate jet pilot trainer to replace T-2C Buckeye and TA-4J Skyhawk; original plan was for initial 54 'dry' (land-based) T-45Bs followed by 253 carrier-capable 'wet' T-45As; B model eliminated in FY84 in favour of 300 (and then 302) T-45As; FSD phase began October 1984; construction of two prototypes by Douglas began February 1986; funding approved 16 May 1986 for first three production lots. Lot 1 production contract (12 aircraft) awarded 26 January 1988; FSD prototypes made first flights 16 April (162787) and November 1988 (162788); original planned date for first deliveries (October 1989) delayed by further airframe and power plant changes requested by US Navy.

Announced 19 December 1989 that entire T45TS programme to be transferred to McDonnell Aircraft Co at St Louis; modified FSD prototypes made first flights September and October 1990; two Douglas production aircraft (163599 and '600) delivered to NATC Patuxent River, Maryland, on 10 October and 15 November 1990; first carrier landing (162787 on USS *John F Kennedy*) 4 December 1991; first McAir production aircraft (163601) flew at St Louis 16 December 1991 and handed over to USN 23 January 1992. Production one per month in 1995 following successful passing of US DoD Milestone III review on 17 January 1995 authorising full-rate production.

Digital/'glass cockpit' developed as 'Cockpit 21'; first to this standard (37th aircraft, 163635) made first flight 19 March 1994; planned production line introduction at 73rd aircraft, to be delivered October 1996, was delayed; flight trials of prototype aboard USS *John C Stennis* conducted April to May 1996 and successfully completed with 30 May approval for 'Cockpit 21' installation from 84th aircraft (165080), which first flew on 21 October 1997 and formally rolled-out on 31 October. Earlier aircraft to be retrofitted with 'Cockpit 21' from FY03, following a validation conversion of T-45A 163651 in 1998.

Beginning in 2000, all T-45s fitted with modified engine air intakes during routine servicing at NAS Kingsville; new intake has the top 'lip' extended forward, giving a raked profile, improving engine airflow at high angles of attack and reducing number of surges and compressor stalls.

CURRENT VERSIONS: **T-45A:** Baseline version, with analogue instruments.

T-45C: Digital ('Cockpit 21') avionics version with new computer, digital databus, data transfer cartridge and Litton GINA (GPS/INS assembly); first aircraft 165080; otherwise as T-45A; sole external difference is GPS antenna on spine of T-45C, immediately behind canopy.

CUSTOMERS: US Navy: two FSD prototypes and 203 production aircraft currently envisaged. Final delivery of FY00 was 127th production aircraft; 139th delivered to US Navy in October 2001; 150th on 8 August 2002, with total of 156 delivered by end of 2002. Long-term requirement is for 234 aircraft to sustain training up to 2035. Complete T45TS programme also involves 19 flight simulators (built by Hughes Training Inc); 48 computer-aided instructional devices, one training integration system mainframe, six electronic classrooms, 155 terminals, plus academic materials and contractor-operated logistic support. USN T45TS requirement stipulated 42 per cent size reduction in (TA-4 and T-2) training fleet, 25 per cent fewer flight hours, and 46 per cent fewer personnel.

Four training squadrons (VT) to equip: VT-21 and VT-22 of Training Wing 2 at Kingsville, Texas, in 1992–96; VT-9 and VT-23 (which redesignated VT-7 on 1 October 1999) of TW-1 at Meridian, Mississippi, 1997–2003. TW-2 assigned early aircraft; TW-1 receiving 90 'Cockpit 21' Goshawks (first official handover to VT-23 15 December 1997, aircraft having been received on 10 December). TW-2 will convert to the T-45C as its aircraft are retrofitted. Small number of T-45s wear 'Marines' titles to reflect the proportion (about 13 per cent) of USMC student pilots being trained on the type.

VT-21 operational 27 June 1992 for instructor training; four-aircraft operational evaluation begun by VT-21 on 18 October 1993 for one month first phase; student training begun 4 January 1994; first student flight 11 February 1994; first solo 23 March 1994; deployed to Miramar for deck landing course on aircraft carrier, September 1994; course graduated 5 October 1994. Second phase of operational evaluation (advanced tactics/weapons and carrier qualification) on USS *Dwight D Eisenhower* ended with clearance for fleet introduction being recommended on 5 July 1994. VT-23 (now VT-7) of TW-1 began student training on 8 July 1998; VT-9 began re-equipping in 2003. T-45 syllabus is 119 sorties (156 hours) plus 100 simulator sessions (95.4 hours).

In April 2000, 1,000th student graduated from T45TS course; fleet passed 200,000 hours in March 1999. By May 2001, T-45s had flown 342,000 hours, made 22,000 deck landings and trained 1,328 pilots; first aircraft exceeded 5,000 hours in April 2001. Fleet passed 450,000 hours and 28,500 carrier landings in February 2003, at which time total of 1,800 students had graduated.

COSTS: Original US$316 million for 12 aircraft in FY96 procurement augmented by US$6,830,421 for addition of 'Cockpit 21' modifications. Official unit cost

Boeing/BAE Systems T-45C Goshawk of VT-9/TW-1 climbing away after take-off *(Paul Jackson)* ***NEW***/1047213

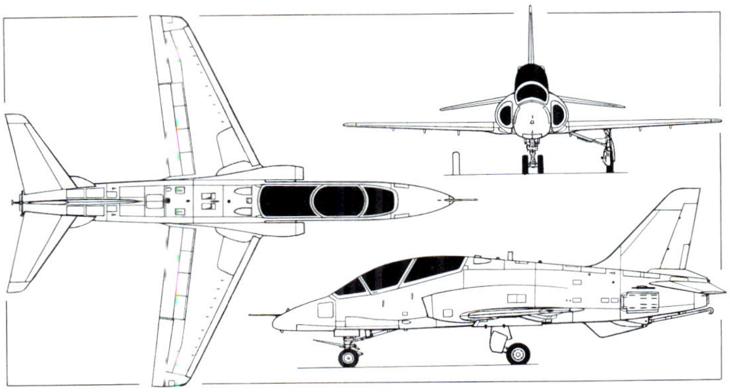

Boeing/BAE Systems T-45C Goshawk tandem-seat basic and advanced trainer with revised intake shape *(Jane's/Dennis Punnett)* 0099603

US NAVY PROCUREMENT

FY	Lot	Qty	First aircraft
88	1	12	163599
89	2	24	163611
92	3	12	163635
93	4	12	163647
94	5	12	165057
95	6	12	165069[1]
96	7	12	165081
97	8	12	165457
98	9	15	165469
99	10	15	165484
00	11	15	165598
01	12	14	165613
02	13	6	165627
03	14	8	
04	15	14	
05	16	8	
Total		**203**	

[1] 'Cockpit 21' installed from 12th of this batch (165080).

US$17.2 million (1999); FY02 purchase of six aircraft valued at US$180.6 million, with US$218.2 million appropriation in FY03 for eight aircraft.

DESIGN FEATURES: Carrier-capable adaptation of existing trainer design for cost and risk reduction. Generally as for two-seat BAE Systems Hawk Srs 60, but redesigned (including deeper and longer forward fuselage) and strengthened to accommodate new landing gear and withstand carrier operation (incidentally increasing fatigue life to 14,400 flying hours, although 28,800 hours achieved on static testing airframe); twin airbrakes of composites material; fin height increased by 15.2 cm (6 in) and single ventral fin added; rudder modified; tailplane span increased by 10.2 cm (4 in); wingtips squared off; nose tow launch bar added; underfuselage arrester hook, deployable 20° to each side of longitudinal axis. No provision for gun or outboard underwing hardpoints.

FLYING CONTROLS: Differences from two-seat BAE Systems Hawk include electrically actuated/hydraulically operated full-span wing leading-edge slats (operation limited to landing configuration); aileron/rudder interconnect; two fuselage-side airbrakes instead of one under fuselage and associated autotrim system for horizontal stabiliser when brakes deployed; BAE Systems yaw damper computer and addition of 'smurf' (side-mounted unit root fin), a small curved surface forward of each tailplane leading-edge root, to eliminate pitch-down during low-speed manoeuvres; Dowty actuators for slats and airbrakes.

STRUCTURE: BAE Systems (principal subcontractor) builds wings, centre and rear fuselage, fin, tailplane, windscreen, canopy and flying controls. Intended fatigue life of 14,000 hours. Batch of 54 composites tailplanes being built by Boeing for retrofit to extant T-45s.

LANDING GEAR: Wide-track hydraulically retractable tricycle type, stressed for vertical velocities of 7.28 m (23.9 ft)/s. Single wheel and long-stroke oleo (increased from 33 cm; 13 in of standard Hawk to 63.5 cm; 25 in) on each main unit; twin-wheel steerable nose unit with 40.6 cm (16 in) stroke. Articulated main gear, by AP Precision Hydraulics, is of levered suspension (trailing arm) type with a folding side-stay. Cleveland Pneumatic nose gear, with Sterer digital dual-gain steering system (high gain for carrier deck operations). Nose gear has catapult launch bar and holdback devices. Main units retract inward into wing, forward of front spar; nose unit retracts forward. All wheel doors sequenced to close after gear lowering; inboard mainwheel doors bulged to accommodate larger trailing arm and tyres. Gear emergency lowering by free-fall. Goodrich wheels, tyres and brakes. Mainwheel tyres size 24×7.7–10 (20 ply) tubeless; nosewheels have size 19×5.25–10 (12 ply) tubeless tyres. Tyre pressure (all units) 22.40 bar (325 lb/sq in) for carrier operation; reduced for land operation. Hydraulic multidisc mainwheel brakes with Dunlop adaptive anti-skid system.

POWER PLANT: One 26.00 kN (5,845 lb st) nominal rating Rolls-Royce Turbomeca F405-RR-401 (navalised Adour Mk 871) non-afterburning turbofan; installed rating 24.59 kN (5,527 lb st). Air intakes and engine starting as described for BAE Systems Hawk. Fuel system similar to BAE Systems Hawk, but with revision for carrier operation. Total internal capacity of 1,635 litres (432 US gallons; 360 Imp gallons). Provision for carrying one 591 litre (156 US gallon; 130 Imp gallon) drop tank on each underwing pylon.

ACCOMMODATION: Similar to BAE Systems Hawk, except that ejection seats are of Martin-Baker Mk 14 NACES (Navy aircrew common ejection seat) zero/zero rocket-assisted type.

SYSTEMS: Air conditioning and pressurisation systems, using engine bleed air. Duplicated hydraulic systems, each 207 bar (3,000 lb/sq in), for actuation of control jacks, slats, flaps, airbrakes, landing gear, arrester hook and anti-skid wheel brakes. No. 1 system has flow rate of 36.4 litres (9.6 US gallons; 8.0 Imp gallons)/min, No. 2 system a rate of 22.7 litres (6.0 US gallons; 5.0 Imp gallons)/min. Reservoirs nitrogen pressurised at 2.75 to 5.50 bar (40 to 80 lb/sq in). Hydraulic accumulator for emergency operation of wheel brakes. Pop-up Dowty Aerospace ram air turbine in upper rear fuselage provides emergency hydraulic power for flying controls in event of engine or No. 2 pump failure. No pneumatic system. DC electrical power from single brushless generator, with two static inverters to provide AC power and two batteries for standby power. Onboard oxygen generating system (OBOGS).

AVIONICS: Avionics and cockpit displays optimised for carrier-compatible operations.

Comms: Rockwell Collins AN/ARN-182 UHF/VHF, Honeywell APX-100 IFF.

Flight: AN/ARN-144 VOR/ILS by Rockwell Collins, Honeywell AN/APN-194 radio altimeter, Sierra AN/ARN-136A Tacan, BAE Systems yaw damper computer. T-45C version with digital avionics (aircraft No. 84 onwards) has revised navigation package comprising mission data input device (MDID); Litton LN-100G ring laser gyro and Rockwell Collins five-channel GPS linked by 12-state Kalman filter.

Instrumentation: US Navy AN/USN-2 standard attitude and heading reference system (SAHRS), Smiths Industries Mini-HUD (front cockpit), Racal Acoustics avionics management system, and Teledyne caution/warning system. Digital avionics from 84th production aircraft onwards: two 127 × 127 mm (5 × 5 in) Elbit monochrome multifunction screens in both cockpits, MIL-STD-1553B databus and Smiths HUD.

Mission: Electrodynamics airborne data recorder.

ARMAMENT: No built-in armament, but weapons delivery capability is incorporated. Single pylon under each wing for carriage of practice multiple bomb rack, rocket pods or auxiliary fuel tank. Provision for carrying single stores pod on fuselage centreline. CAI Industries weapon-aiming sight in rear cockpit.

DIMENSIONS, EXTERNAL:	
Wing span	9.39 m (30 ft 9¾ in)
Wing chord: at root	2.87 m (9 ft 5 in)
at tip	0.89 m (2 ft 11 in)
Wing aspect ratio	5.0
Length, overall, incl nose probe	11.98 m (39 ft 4 in)
Height overall	4.11 m (13 ft 6 in)
Tailplane span	4.59 m (15 ft 0¾ in)
Wheel track (c/l of shock-struts)	3.90 m (12 ft 9½ in)
Wheelbase	4.31 m (14 ft 1¾ in)
AREAS:	
Wings, gross	17.66 m^2 (190.1 sq ft)
Ailerons (total)	1.05 m^2 (11.30 sq ft)
Trailing-edge flaps (total)	2.50 m^2 (26.91 sq ft)
Airbrakes (total)	0.88 m^2 (9.47 sq ft)
Fin	2.61 m^2 (28.10 sq ft)
Rudder, incl tab	0.58 m^2 (6.24 sq ft)
Tailplane	4.43 m^2 (47.64 sq ft)
WEIGHTS AND LOADINGS:	
Weight empty	4,261 kg (9,394 lb)
Internal fuel: early production	1,312 kg (2,893 lb)
enhanced capacity	1,433 kg (3,159 lb)
Normal T-O weight	5,783 kg (12,750 lb)
Max T-O weight	6,123 kg (13,500 lb)
Max wing loading	346.7 kg/m^2 (71.02 lb/sq ft)
Max installed power loading	249 kg/kN (2.44 lb/lb st)
PERFORMANCE:	
Design limit diving speed at 1,000 m (3,280 ft)	575 kt (1,065 km/h; 662 mph)
Max true Mach No. in dive	1.04
Max level speed at 2,440 m (8,000 ft)	543 kt (1,006 km/h; 625 mph)
Max level Mach No. at 9,150 m (30,000 ft)	0.84
Carrier launch speed	121 kt (224 km/h; 139 mph)
Approach speed (typical)	125 kt (232 km/h; 144 mph)
Max rate of climb at S/L	2,440 m (8,000 ft)/min
Time to 9,150 m (30,000 ft), clean	7 min 40 s
Service ceiling	12,950 m (42,500 ft)
T-O to 15 m (50 ft)	1,100 m (3,610 ft)
Landing from 15 m (50 ft)	1,009 m (3,310 ft)
Ferry range	700 n miles (1,296 km; 805 miles)
g limits	+7.33/−3

UPDATED

EADS

EUROPEAN AERONAUTIC, DEFENCE AND SPACE COMPANY NV

Le Carré, Beechavenue 130-132, NL-1119 PR Schiphol-Rijk, Netherlands
Tel: (+31 20) 655 48 00
Fax: (+31 20) 655 48 01
Web: http://www.eads.net

PARTICIPATING COMPANIES

Aerospatiale Matra: EADS, France
CASA: EADS CASA, Spain
DaimlerChrysler Aerospace (DASA): EADS Deutschland, Germany

CHAIRMEN:
Arnaud Lagardière
Manfred Bischoff

CHIEF EXECUTIVES:
Phillipe Camus
Rainer Hertrich

CORPORATE COMMUNICATIONS:
D-81663 München
Tel: (+49 89) 60 70
e-mail: press@eads.net
SENIOR VICE-PRESIDENT, COMMUNICATIONS: Christian Poppe

Decision to merge Aerospatiale Matra of France and DASA of Germany was announced on 14 October 1999; previously mooted amalgamation of Spain's CASA with DASA was reconfirmed on 2 December 1999, increasing size of prospective company to 90 sites. Officially formed 10 July 2000, immediately becoming world's third-largest aerospace company. Asset distribution comprises 30.80 per cent stock flotation; 30.29 per cent directly, plus 2.75 per cent indirectly DaimlerChrysler; 5.53 per cent SEPI (Spanish state holding company); 30.29 per cent SOGEADE (Lagardière and French government jointly); and 0.34 per cent directly by French government.

EADS has five major operating divisions:

Airbus (International)

Military Transport Aircraft (Spain)

Aeronautics Comprising ATR, Eurocopter, EADS Socata, EADS EFW and EADS Sogerma Services

Defence & Security Systems Comprising EADS Military Aircraft (Germany), EADS Services, Missiles, EADS Systems & Defence Electronics and EADS Telecom

Space Company controls Socata and Eurocopter (100 per cent); also holds 80 per cent of Airbus, 56 per cent of Airbus Military Company and 51 per cent of EADS PZL in Poland; and has significant shares in Dassault (45.8 per cent), Eurofighter (43 per cent), ATR (50 per cent) and missile and space programmes. Aircraft produced by EADS are described in their appropriate national sections with the exception of the Mako, which is first to bear the EADS name.

Corporate headquarters established in Netherlands and EADS is subject to Dutch company law. Turnover in 2003 totalled €30,133 million; orders €61,150 million; backlog at 1 January 2004 €179,280 million. Workforce 109,135 total; by division is approximately 38 per cent Airbus, 26 per cent Aeronautics, 20 per cent Systems and Defence Electronics, 11 per cent Space and 4 per cent Military Transport Aircraft; remaining 1 per cent are HQ staff. By 2012, EADS work distribution planned to be 60 per cent commercial, 30 per cent military and 10 per cent space.

EADS Russia founded 25 March 2004 as joint programme (subject to Russian law) principally concerned with research and production for Airbus.

UPDATED

EADS MAKO

English name: Shark

TYPE: Light fighter/advanced jet trainer.

PROGRAMME: Joint Aermacchi/DASA AT-2000 programme launched in April 1989, Aermacchi withdrew in 1994; DASA maintained the programme in hope of a joint development venture with South Korea. Following the latter's expressed preference in 1996 for a partnership with Lockheed Martin to satisfy its KTX-2 requirement, DASA began seeking an alternative collaborator in addition to Denel in South Africa, which was to provide avionics. Hyundai of South Korea expressed interest in the venture in early 1997, intending to build non-composites components of the centre and rear fuselage and tail. Design revealed in October 1996, after completion of wind tunnel tests and when radar cross-section trials in hand. Feasibility study completed by DASA in late 1997. Named Mako in 1998, by which time was foreseen as a 'modular design concept' with different versions for air-to-air, air-to-ground and training missions.

Predefinition phase began in early 1998. Full definition began by mid-1999, following a decision by the DASA board; full-scale mockup shown at 1999 Paris Air Show; cockpit demonstrator (conventional and VR simulator) at Paris 2001. Potential suppliers represented on demonstrator comprised BAE Systems North America (throttle and stick), Goodrich (front ejection seat), Martin-Baker (rear ejection seat) and CDC (displays and control

Model of Mako-AT advanced trainer ***NEW***/0528669

system). MoUs signed at Dubai in February 2001 and at Paris in June 2001 with SNECMA (M88-2 engines), BAE Systems (FCS and control systems), BGT/Diehl (computer and weapons systems and training aids), APPH Precision Hydraulics (landing gear and hydraulics) and Fairey Hydraulics (flight control actuation). Original target partners (South Africa and South Korea) have not joined the programme and there is no unilateral home-market (Luftwaffe) interest.

Launch of 18-month definition phase, dependent upon joint development agreement with United Arab Emirates, had been expected at Dubai Air Show in November 2001, but was indefinately delayed; however, MoUs signed at that time with Autoflug (fuel systems); Flight Visions (displays and controls); Rockwell Collins (navigation and communications equipment, displays and controls). MoUs are for down-selection only, and do not confer subcontractor status. Financing based on risk and revenue sharing with major suppliers, national partners and industrial partners. MoU of 9 December 2002 requires GE Aircraft Engines to define and integrate variant of F414-GE-400 turbofan. In addition, Mako HEAT is candidate for AEJPT advanced European jet pilot training programme being considered by 12 air forces, including Luftwaffe. Separate interest in Mako LCA from Greece, UAE and Brazil. First of two prototypes to fly five years after launch decision; production start in following year; deliveries nine years after launch. A launch decision had not been made by mid-2004, although 1,200 hours of wind tunnel trials (19 separate campaigns) had been accumulated by that time.

CURRENT VERSIONS: **Mako-AT:** Advanced trainer. Original baseline advanced two-seat trainer, available with or without radar and internal gun. Capable of in-flight weapons system simulation, with generation of synthetic threats and targets. Also suitable as companion trainer in squadrons operating advanced fighters.

Mako-LCA: Light combat aircraft. Multirole fighter with **F1** single-seat (air-to-air) and **F2** two-seat (air-to-ground) versions. Modular systems concept allows aircraft to be tailored to customer's requirements. EADS perceives a world market for 1,600 light combat aircraft between 2009 and 2030.

Mako HEAT: High-energy advanced trainer. Launched at ILA, Berlin, May 2004 and directed at AEJPT 'Eurotrainer' air arms, with collaborative opportunities for European industry (secondary power system, fuel system ECS, hydraulics, electrics and landing gear), although powered by derated GE F414 engine. High specific excess power allows part of OCU fighter course (basic fighter manoeuvres, air combat tactics and manoeuvres, air-to-ground tactics and experience of reheat) to be 'downloaded' to trainer syllabus, with resultant cost savings, even for 4/5th generation combat aircraft. Thrust/weight ratio 0.92, or 1.08 at half-fuel.

CUSTOMERS: Predicted 400 sales by 2030, of which 40 per cent would be trainer version. Entered (1999) in Brazilian F-X BR fighter replacement competition (to replace Mirage III and F-5E). MoUs signed with UAE November 1999 and March 2001, but these do not involve payment. Candidate for 'Eurotraining' (AEJPT) initiative. Initial wave of international orders anticipated in 2008–2012, with second period of high interest foreseen in 2018.

COSTS: Estimated development cost US$1,300 million; expenditure €80 million by mid-2004, all provided by EADS and predecessors. Unit cost US$15 million to US$22 million, representing 15 per cent increase over competitors, although Mako HEAT will recoup this by 25 to 30 per cent reduction in fighter OCU costs per pilot due to syllabus changes, giving break-even after first six to eight pilots. Life cycle operating cost of Mako HEAT per hour is less than one-fifth of operational fighter.

DESIGN FEATURES: Modular design for ease of meeting different roles; simple structure permits low production cost. Conventional configuration described as 'transonic layout with supersonic performance'. Advanced trainer version is capable of combat missions, employing LO technology and optional thrust vectoring. LO features include chined forward section, wing/forebody blending and 'caret' engine air intakes. Structural life 16,000 hours or 30 years.

Wing leading-edge sweep 45°.

FLYING CONTROLS: Reprogrammable quadruplex digital FBW FCS with 'carefree handling'. Large, single-section flaperons; all-moving tailplane (tailerons); inset rudder; and full-span wing leading-edge slats. Four door-type airbrakes adjacent to jetpipe.

STRUCTURE: Centre-fuselage of aluminium; wing-, taileron- and fin skins of carbon fibre.

LANDING GEAR: Retractable tricycle type; single wheel on each leg.

POWER PLANT: *Mako AT, LCA:* One Eurojet EJ200 turbofan; installed derated to 75.0 kN (16,860 lb st) with afterburning; optionally with thrust vectoring (provision for which is made in the FCS and airframe structure); or with full afterburning (approximately 90 kN; 20,250 lb st), according to customer's requirement.

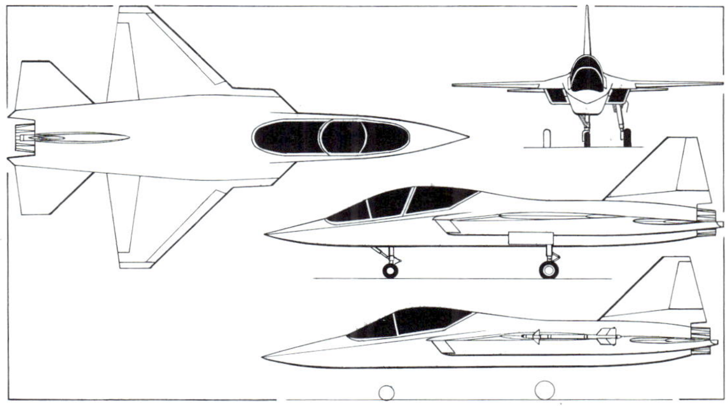

Provisional general arrangement of the Mako-AT advanced trainer, with additional side view (bottom) of single-seat Mako-LCA attack aircraft *(James Goulding)* 0137725

Mako HEAT: One General Electric F414M derated to 75 kN (16,860 lb st) with afterburning.

SNECMA M88-2 or -3 under consideration as alternative power plants for all versions. FCS incorporates provision for vectoring nozzle.

ACCOMMODATION: One or two pilots, according to version; layout and instrumentation generally as in Eurofighter Typhoon. Twin-seater has stepped cockpits, with 15° view over nose and 40° each side; seats inclined rearwards at 18°. Integral windscreen/front cockpit canopy slides forward; rear canopy slides aft.

SYSTEMS: OBOGS. APU. Optional OBIGGS.

AVIONICS: *Radar:* Optional 'AN/APG-67 class' multimode radar; other contenders include FIAR Grifo, Thales Avionics RD-400 and BAE Systems Bluehawk.

Instrumentation: Three MFDs in each cockpit, reflecting modern combat aircraft (such as Eurofighter) environment. Modular concept to allow easy upgrade or tailoring to individual customer requirements. Expected to include provision for FLIR, HMS and similar equipment.

Mission: Embedded in-flight simulation for combat training.

ARMAMENT: Optional Mauser BK 27 mm internal gun and seven low RCS external stores hardpoints: one at each wingtip, two under each wing and one below centreline. Centreline and inboard wing pylons stressed to 1,350 kg (2,976 lb); outboard to 675 kg (1,488 lb); wingtip for AAM only.

All data are provisional.

DIMENSIONS, EXTERNAL:

Wing span: plain tips	8.25 m (27 ft 0¾ in)
over missile rails	8.6 m (28 ft)
Wing aspect ratio	2.6
Length overall	13.75 m (45 ft)
Height overall	4.5 m (15 ft)
Wheel track	2.4 m (8 ft)
Wheelbase	4.8 m (16 ft)

AREAS:

Wings, gross	26.70 m² (287.4 sq ft)

WEIGHTS AND LOADINGS:

Weight empty: LCA: F1	6,020 kg (13,272 lb)
F2	6,370 kg (14,043 kg)
AT, HEAT	5,800 kg (12,787 lb)
Max weapon load	4,500 kg (9,920 lb)
Max internal fuel weight: LCA F1	3,300 kg (7,275 lb)
LCA F2, AT	3,000 kg (6,614 lb)
HEAT	2,500 kg (5,512 lb)
Mission T-O weight: HEAT	8,300 kg (18,298 lb)
Max T-O weight	13,000 kg (28,660 lb)
Max wing loading	486.9 kg/m² (99.72 lb/sq ft)
Max power loading (full afterburning)	149 kg/kN (1.46 lb/lb st)

PERFORMANCE (estimated):

Max level speed	M1.5
Service ceiling	15,240 m (50,000 ft)
T-O run	450 m (1,480 ft)
Landing run	750 m (2,460 ft)
Radius of action:	
interception, 54 min CAP plus dogfight	
plus reserves	100 n miles (185 km; 115 miles)
ASM launch (two), hi-hi-hi	648 n miles (1,200 km; 745 miles)
Ferry range: EJ200	2,000 n miles (3,704 km; 2,301 miles)
F414M	2,100 n miles (3,889 km; 2,416 miles)
g limits	+8/−3

UPDATED

EUROCOPTER

EUROCOPTER SAS (AN EADS COMPANY)

Aéroport International Marseille-Provence, F-13725 Marignane Cedex, France
Tel: (+33 4) 42 85 85 85
Fax: (+33 4) 42 85 85 00
Web: http://www.eurocopter.com
PRESIDENT AND CEO: Fabrice Brégier
DIRECTOR OF COMMUNICATIONS: Xavier Poupardin
CHIEF, PRESS & INFORMATION: Jean-Louis Espes

Eurocopter SA formed 16 January 1992 by merger of Aerospatiale and MBB (DASA) helicopter divisions. Share capital then held on two levels: Eurocopter Holding owned 60 per cent by Aerospatiale and 40 per cent by DASA; capital of Eurocopter SA held 75 per cent by Eurocopter Holding and 25 per cent by Aerospatiale. However, on 30 May 1997, the separate elements of Eurocopter were merged into a single management structure. Former Eurocopter France, Eurocopter International and Eurocopter Participations became a single entity (Eurocopter France, trading as Eurocopter), with Eurocopter Deutschland (in Germany) as a wholly owned subsidiary. Following formation of EADS, Eurocopter reconstituted on 18 September 2000 as simplified stock company (SAS) within EADS Aeronautics group. EADS CASA acquired 5 per cent share in December 2003, in return for relinquishing its share of Eurocopter España.

Eurocopter products cover 75 to 80 per cent of the range of helicopters in terms of size and capacity, but company intends to expand this to 95 per cent, although this aim set back by forced withdrawal from Euromil, particularly with co-operative development of Russian Mil Mi-38. Company claimed 57 per cent of civil helicopter market in 2001 and has averaged 15 per cent of military market in recent years. Employees total some 10,000. Orders received in 2003 for 293 helicopters, comprising 38 Colibris, 81 single-engine and 17 twin-engine Ecureuil/Fennecs, 27 EC 130s, 55 EC 135s, six BK 117s, 27 Dauphin/Panthers, four EC 145s, 10 EC 155s, eight Super Puma/Cougars (including EC 225/725s) and 20 NH 90s. Deliveries

Eurocopter studies include this transport helicopter based around a Sikorsky-type dynamic system and FBW controls from the NH90 programme ***NEW***/1047210

for the same period totalled 297. In 2004 orders were 332: 39 Colibris, 146 Ecureuil/Fennec/EC 130s (variants not specified), 67 EC 135s, 13 BK 117/EC 145s, 32 Dauphin/Panther/EC 155s, 15 Super Puma/Cougar/EC 225/EC 725s and 20 NH90s. With 35 sales of pre-owned helicopters, company gained 70 per cent exports in 2004; 51 per cent of sales in civil/parapublic sector.

Eurocopter subsidiaries employ some 1,250 persons and deliver about 150 aircraft per year; managed by Eurocopter Subsidiaries and Participations division (ownership percentage in parentheses): American Eurocopter (100), Eurocopter Canada (100), Hélibras (Brazil; which see; 76.5), Eurocopter España (Spain; see below; 60), Eurocopter Romania (see below; 51) and McAlpine (UK; 10) are all first-tier subsidiaries responsible for distribution, service and industrial work, including assembly and component manufacture (Canada and Hélibras), customisation (American, Canada, Hélibras and McAlpine) and repair and overhaul. Second-tier subsidiaries (distribution and service) comprise Eurocopter South East Asia (Singapore; 65), Eurocopter Mexico (100), Eurocopter Chile (100), Eurocopter South Africa (100), Australian Aerospace (known until 1 January 2003 as Eurocopter International Pacific; Australia; 100), and Euroheli (Japan; 10). Service subsidiaries are Eurocopter Philippines (70), Eurocopter Service Japan (51) and EuroAircraft Services (Malaysia; 34).

On 7 August 2003, American Eurocopter broke ground for a new, US$12 million plant at Columbus, Mississippi, for final assembly of up to 30 AStar/Ecureuils per year, plus manufacture of some AS 350, AS 355 and EC 120 components. Australian facility at Brisbane began assembly of 18 Tigers in April 2003 and assembly of EC 120s in June 2003.

Eurocopter
Address as above
Company headquarters and main production centre at Marignane (Marseille); works at La Courneuve (Paris). In 2002, these employed 5,057 and 771 personnel, respectively.

Eurocopter Deutschland
Industriestrasse 4, Postfach 1353, D-86609 Donauwörth, Germany
Tel: (+49 906) 710
Fax: (+49 906) 71 45 75
CHAIRMAN: Friedrich Dörhöfer
MEDIA RELATIONS EXECUTIVE: Christina Gotzhein

Industrial concern in charge of production and support of products originating in German part of Eurocopter. Wholly owned subsidiary with plants at Donauwörth (2,674 personnel in 2002) and Ottobrunn (634).

Current programmes are EC 135, BK 117C-1, EC 145 and Tiger; NH90 (industrialisation); also maintenance/upgrade work for CH-53G, Lynx and Sea King. Research and development programmes concentrate on noise abatement, fly-by-light technologies, all-weather capability of helicopters and optimisation of in-flight comfort and safety. Eurocopter Deutschland closely co-operates with other industrial partners in the aerospace industry as well as the German and French aerospace institutions DLR and ONERA, respectively.

Eurocopter Deutschland, the Federal Ministry of Defence and German Aerospace Research Establishment are jointly providing DM40 million to 50 million funding for development of **Helicopter Simulator for Technology, Operations and Research (HeSTOR)**, based on EC 135, to replace BO 105-based Advanced Technology Testing Helicopter System (ATTHeS) which was lost in an accident on 19 May 1995.

Eurocopter España SA
EADS CASA, Avenida de Aragón 404, E-28022 Madrid
PUBLIC RELATIONS EXECUTIVE: Francisco Salido

Established 28 September 2000, replacing former subsidiary Helicopteros España (HESA). Works at Quatro Vientos responsible for engineering, production and customer support, on par with French and German divisions. Owned 60 per cent Eurocopter and 40 per cent EADS CASA until 18 December 2003 announcement of CASA's buy-out in return for 5 per cent share of Eurocopter.

Eurocopter Romania SA
SC IAR SA, R-2200 Brasov
Agreement to form Eurocopter Romania signed with IAR (which see) on 22 December 2000. Joint venture (51 per cent Eurocopter-owned) markets Eurocopter products in Romania; performs subcontract work on support of Puma and Alouette for Eurocopter; and maintains, supports and upgrades Romanian and export Pumas. Authorised maintenance plant for EC 135 since November 2003.

Additionally, Eurocopter has entered into two alliances with manufacturers in Asia in order to develop specific products. These are regarded by the company as wholly Eurocopter aircraft.

Eurocopter/CATIC/ ST Aero
PARTICIPATING COMPANIES:
Eurocopter: International
CATIC: China
Singapore Technologies Aerospace: Singapore

This partnership formed to develop the EC 120 Colibri, production of which is now fully under way. Shares are Eurocopter 61 per cent (and programme leader), CATIC/HAMC 24 per cent and ST Aero 15 per cent.

Eurocopter/Kawasaki
PARTICIPATING COMPANIES:
Eurocopter: International
Kawasaki: Japan

Companies formed a partnership on 25 February 1977 to develop the BK 117 multipurpose helicopter. Low-rate production continues in both Germany and Japan. An upgraded version, the EC 145, was certified in December 2000.

UPDATED

EUROCOPTER 665 TIGER/TIGRE

TYPE: Attack helicopter.

PROGRAMME: France and Germany agreed in 1984 to develop a common combat helicopter; Eurocopter Tiger GmbH formed 18 September 1985 to manage development and manufacture for French and German armies; was not a full member of Eurocopter because it was working on a single government contract; executive authority for programme is DFHB (Deutsch Französisches Hubschrauberbüro) in Koblenz; procurement agency is German government BWB (Bundesamt für Wehrtechnik und Beschaffung).

Original 1984 MoU amended 13 November 1987; FSD approved 8 December 1987; main development contract awarded 30 November 1989, when name Tiger (Germany)/Tigre (France) adopted; five development aircraft built, including three unarmed aerodynamic prototypes, used also for core avionics testing (PT1, 2 and 3), one (PT4) in HAP (initially called Gerfaut) configuration and one (PT5) as UHT prototype; PT1 rolled out 4 February 1991; first flight 27 April 1991; fifth prototype flew on 21 February 1996, at which time the first four aircraft had accumulated 1,090 flying hours; further details below; total of 2,869 hours flown by five prototypes up to June 2001. Germany confirmed purchase of full 212 required, 1994, having considered cut to 138, but later reconsidered.

Industrialisation phase brought forward by two years to strengthen export prospects and Franco-German MoU signed 30 June 1995; timetable then was first deliveries in 1999 to France (approximately 10) and for export, but France announced spending moratorium in November 1995, postponing authorisation of further funding commitments until signature of a FFr2.5 billion (DM733.6 million) production investment contract on 20 June 1997. Deliveries then expected in 2001, but further delayed to July 2003 (for HAP; 2011 for HAC) by May 1996 defence plan, which envisaged procurement of only 25 Tigres in 2000–2002 budgets.

On 20 May 1998, France and Germany signed a commitment to order an initial joint batch of 160 Tigers. However, planned late 1998 placing of contracts was delayed by requirement of new German government to conduct a defence review; options included delaying ISD; or reducing numbers; or even cancelling UHT and procuring French HAP version. Production contract was finally signed on 18 June 1999 for the full 160 aircraft; first deliveries in 2002. HAP deliveries to include two in 2003, eight in 2004 and 10 per year in 2005–10; first production aircraft (9826) flew in Germany, 2 August 2002. Production of the first batch of 320 engines (plus 12 spares) began during 2000, and will continue through 2011. June 1999 contract also formalised German contract change from PAH2 to UHT and French change from HAP to HAP-F (*Finalisé*).

Spain became third member of Tiger programme in 2003 and integrated in airframe, engine, avionics and weapons development and manufacturing programmes.

Joint team at Marignane is flight testing basic helicopter, updating avionics during trials, and testing HAP variant; similar team at Ottobrunn is qualifying basic avionics, Euromep mission equipment package, and integrating weapons system. Rotor downwash problems resulted in trial forward positioning of horizontal stabiliser; by mid-1994 definitive solution adopted of reversion to original position, but halving area. By January 1998, the design had been frozen, and the development programme was more than 90 per cent complete. Development of HAD version involves Spanish participation from 2003 onwards.

First export order for Tiger confirmed 14 August 2001, when Australia announced selection to meet AIR 87 requirement. MoU on co-operation and exchange of information, to lead to production and delivery, signed by Eurocopter and Australian Defence Science and Technology Organisation on 12 July 2002.

CURRENT VERSIONS (GENERAL): Original partners require three versions in two basic configurations with about 80 per cent commonality: **U-Tiger** is basis of the UHT and HAC, both with mast-mounted sight and Trigat missiles; **HCP** (*Hélicoptère de Combat Polyvalent*) is basis of the HAP (roof sight and turreted gun). Other variants proposed to meet export requirements.

Tigre HAP: *Hélicoptère d'Appui et de Protection*; name Gerfaut dropped late 1993; escort and fire support version for French Army; armed with 30 mm Giat AM-30781 automatic cannon in undernose turret, with 150 to 450 rounds of ammunition; four Mistral air-to-air missiles and two pods each with twenty-two 68 mm unguided TDA rockets delivering armour-piercing darts, mounted on stub-wings, or 12-round rocket pod instead of each pair of Mistrals, making total of 68 rockets; roof-mounted sight, with TV, FLIR, laser range-finder and direct view optics sensors; image intensifiers integrated in helmets; and extended self-defence system. HAP configuration was approved by late 1998, permitting type qualification in December 2002. Deliveries from Marignane in 2003; final (37th) aircraft due in 2008.

UHT: *Unterstützungshubschrauber Tiger* (previously designated UHU); German Army multirole 'utility' or multirole anti-tank and fire-support helicopter for delivery from 2002; replaces dedicated anti-tank PAH-2 Tiger; type qualification due in December 2002; final assembly at Donauwörth. Underwing pylons for HOT 3 or (from 2006) Trigat missiles, Stinger self-defence missiles, unguided rockets, gun pod and extended self-defence system; mast-mounted TV/FLIR/laser ranger sight for gunner; nose-mounted FLIR for piloting. A mid-life upgrade for the UHT may integrate the Mauser 30 mm gun in a chin turret which traverses ±140° in azimuth and from +20 to −45° in elevation.

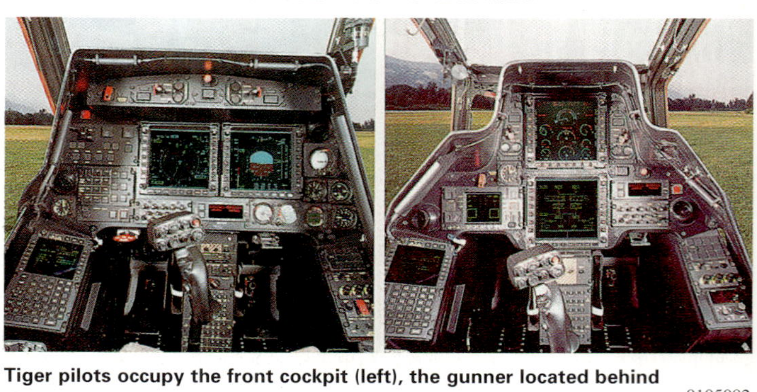

Tiger pilots occupy the front cockpit (left), the gunner located behind
0105082

Tigre HAC: *Hélicoptère Anti-Char*; anti-tank variant originally required by, but not yet ordered for French Army; final assembly at Donauwörth. Type qualification due in third quarter of 2011; same weapon options (except Mistral AAM in place of Stinger), mast-mounted sight and pilot FLIR system as UHT. A mid-life upgrade to the HCP could see addition of a mast-mounted automatic air surveillance and warning system, in the form of DAV pulse Doppler radar, together with HUMS and an IR jammer.

Export: A basic export version combining features of French and German versions; offered (unsuccessfully) to UK and Netherlands.

ARH Tiger (HCP): Hybrid Tiger variant to meet Australian Army Air 87 Requirement. Based on French HAP, with undernose Giat 30-781 30 mm cannon, roof-mounted sight and provision for underwing 70 mm rocket pods, but with added anti-tank capability. Australia also requires integration of the AGM-114 Hellfire ATM. Maximum mission weight of 6,100 to 6,300 kg (13,448 to 13,889 lb). A$1,300 million (US$674 million) contract for 22 signed 21 December 2001. First four from European production; fuselage sections of first aircraft joined 11 February 2003; first flight February 2004.

HCP Tiger (HCP): Export version based on French Army HAP with the same undernose gun turret and roof-mounted sight, HOT 3 and Trigat missiles; Hellfire optional. Strix roof sight (direct view optics optional) and additional laser designator plus video signal interfacing for Trigat operation. Either Mistral or Stinger air-to-air missiles. A mid-life upgrade to the HCP could see addition of a DAV mast-mounted air surveillance radar (pulse Doppler type) or a mast-mounted MMW radar for automatic ground and air surveillance. No gun pod option.

Tigre HAD: *Hélicoptère d'Appui-Destruction*. Multirole version for anti-tank, combat support, reconnaissance and escort roles; development cost estimated as €152 million (2002). Purchased by Spain. Uprated engines, increased MTOW to 6 tonne class, roof sight for Trigat ATMs. Additional weapons include 30 mm cannon, Mistral AAMs and 70 mm unguided rockets. HAD adopted by France in 2003, with final 43 of order for 80 Tigres to be delivered from 2009 onwards.

Tiger T800: LHTEC T800 or CTS800 engines proposed as an engine option for the Tiger. Turkey is sales prospect, as Army has reservations about growth potential of MTR390.

CURRENT VERSIONS (SPECIFIC): **PT1/F-ZWWW:** Aerodynamic prototype; basic avionics; first flight 27 April 1991. Successively fitted with aerodynamic mockups of mast-mounted and roof-mounted sights, nose-mounted gun and weapon containers. Relegated to ground fatigue testing and static display in early 1996 on completion of flight programme. Flown 502 hours.

PT2/F-ZWWY: HAP aerodynamic configuration; full core avionics; rolled out 9 November 1992; first flight 22 April 1993. Used for radar cross-section and detectability tests. Retrofit with HAP systems completed in November 1996; redesignated PT2R. Mistral launch trials at Landes ranges 14/15 December 1998; technical assessment by French Army at Valence between 17 May and 3 June 1999; rocket qualification, June 1999. Used for HAP version qualification (redesignated PT2R2) at Landes test centre between 4 April and 12 May 2000. Redesignated PT2X in 2001 to serve as multimission demonstrator, adding LFK/SAGEM sighting system for HOT 3 anti-tank missiles in addition to original Mistral missiles and rockets. Deck landing trials, May 2002, aboard FS *Siroco*, an amphibious landing ship.

PT3/9823: Full core avionics (including navigation and autopilot); first flight (as F-ZWWT) 19 November 1993. Retrofit with UHT systems began in February 1997; redesignated PT3R; Euromep C (see Avionics) from late 1997. HOT launches with mast sight at extreme range in night and smoke conditions, June 1999; hot weather trials at Bateen AB in Abu Dhabi September 1999. Moved back to France for HAC development.

PT4/F-ZWWU: HAP aerodynamic configuration and avionics (including roof sight, HUD and Topowl helmet sight; first Tiger with live weapons system); first flight 15 December 1994. Sighting system trials early 1995; Giat cannon trials (15 ground-based tests) completed at Toulon, April 1995; full testing began at CEV Cazaux, 21 September 1995 and, by late November, had demonstrated airborne cannon firing and launch of Mistral AAM (without seeker); by 1 January 1997 had fired eight Mistrals, 3,000 cannon rounds and 50 rockets; 1997 trials included two more Mistrals, rockets and tests of gun controls. Painted in three-tone disruptive camouflage. Winter trials in Sweden, early 1997 with skid/skis landing gear. Flown 296 hours to 1 December 1997; crashed during night low-level evaluation by Australian Army 17 February 1998.

PT5/9825: Full UHT avionics; first flight 21 February 1996. Undertook German Army weapon trials (Stinger, HOT 2 and 12.7 mm podded gun) in 1997 including the firing of six HOT 2s using Euromep Osiris mast-mounted sight. Retrofitted as PT5R with production-standard weapon system; first flight 8 October 1999. Weapon trials continued into 2004, including HOT, Trigat, 70 mm rocket and 12.7 mm machine gun.

First production Tiger in German UHT configuration *(Eurocopter/Patrick Penna)* ***NEW***/0096993

First production Tiger HAP S01/F-ZKDB *(Paul Jackson)* ***NEW***/0573388

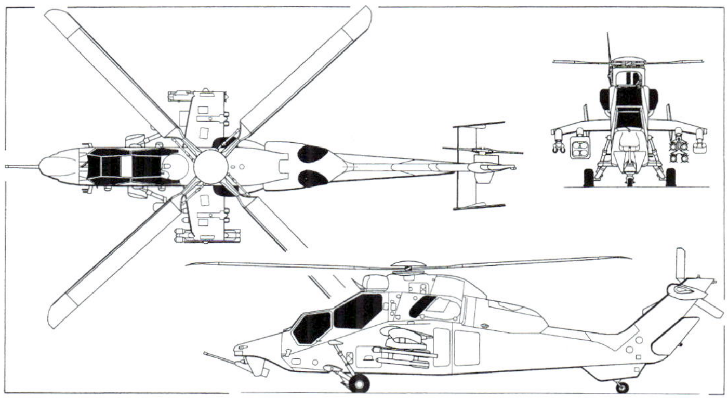

Eurocopter Tigre HCP combat escort and fire support helicopter *(Paul Jackson)* 0507609

PT6 : and PT7: Static test airframes for fatigue and crash-resistance trials.

PS1/F-ZVLJ: Pre-series HAP built at Marignane on production tooling; laid down in third quarter of 1998; first flight 21 December 2000. Tasks include validation of production methods and planned production configuration including 6.4 tonne MTOW extension.

UHT S01/9826: First true production aircraft; planned to fly on 1 March 2002, but not rolled out until 22 March; first flight 2 August 2002; used for six-month techeval/opeval trials, replacing PT5R. Pilot's visual system trials.

UHT S02: Electromagnetic compatibility trials at WTD 81, Greding; weapons qualification.

UHT S03: To WTD 61, Manching, for instructor training (with S04).

HAP S01: First production French Tiger, first flight (F-ZKDB) 26 March 2003; delivery to French Army in July 2003.

CUSTOMERS: Firm orders by early 2004 totalled 206: France and Germany 80 each; Australia 22 and Spain 24.

Original requirement was for 427 (France 75 HAP and 140 HAC, Germany 212 PAH-2); UHU (later UHT) version substituted for PAH-2s in 1993; French order amended by 1994 to 115 HAP and 100 HAC but may be reduced to overall total of 180; in mid-2001 French Army expressed preference for multirole Hélicoptère d'Appui-Destruction (HAD) version in place of two subvariants now on order. Germany committed to 212, of which 112 to be funded between 2001 and 2009; initial commitment of 20 May 1998 confirmed 80 each by France (70 HAP, 10 HAC) and Germany (80 UHT), as agreed by Franco-German Security Council on 9 December 1996. By 2002, official German sources suggesting full requirement only 110 and Eurocopter resigned to total of 240 between two launch partners. Confirmed in 2003 that France's initial order for 80 Tigres now to be 37 HAPs and 43 HADs (no HACs). First operational squardon (April 2006) to be formed within 5 Combat Helicopter Regiment (5 RHC) at Pau.

Germany's Tigers are required to equip four 48-aircraft regiments, each supporting an Army division, first being 1 Staffel of 36 Regiment at Fritzlar, where deliveries due to have begun in October 2004 with 16th UHT.

Joint training school at le Luc opened 1 July 2003 as the Ecole Franco-Allemand, or EFA with a Thomson Training and Simulation/STN Atlas aircrew training system, including six-axis motion simulators and wide-angle visual systems. Training course lasts 28 weeks for a crew chief, or 19 weeks for a pilot. Fleet of 14 German and 14 French Tigers will be assigned by 2006, with eleven simulators. In October 2000 there were reports that the German MoD was considering reducing its Tiger buy to 100 helicopters.

Australia selected Tiger for AIR 87 requirement on 14 August 2001; contract signed 21 December 2001 for 22 helicopters, of which final 18 assembled by Australian Aerospace at Brisbane, from where deliveries due between July 2005 and April 2008. First French-built components arrived at Brisbane 31 March 2003 and assembly work began on 10 April. Four complete French aircraft deliverable in pairs, December 2004 and early 2005, initial Marignane-built helicopter having flown on 20 February 2004 (temporarily as F-Z V L H). Exports of 200 Tigers thought possible.

Spanish government approval for €1,353 million purchase of 24 Tigers given 5 September 2003, price including logistic support and initial weaponry. Programme involves Spanish participation in Tiger through Eurocopter España; Spain designated sole source of all tailbooms and will assemble chosen HAD version; other Spanish firms to participate in overall Tiger programme, including engine (ITP), avionics (INDRA) and Trigat missile (LFK, INDRA, SENER and IZAR). Initial six HAP Tigers delivered in 2004–05, but earmarked for later conversion to HADs. Development of HAD configuration and integration of Spanish companies in Tiger programme to take four years before production begins of 18 HADs in Spain for 2007–11 delivery.

COSTS: Tiger current development cost, shared equally by France and Germany, reported DM2.2 billion. Production tooling cost FFr2.6 billion (US$500 million) (1996). Unit cost (1996) for UH estimated as US$11 million, including launchers and all government-furnished equipment. Initial batch of 160 assigned FFr21.5 billion, of which FFr13 billion for 80 German helicopters and FFr8.5 billion for 80 French (1998). Australian programme unit cost US$30.6 million (2001).

DESIGN FEATURES: Robust, tandem-seat design with pylon-mounted armament, representing current combat helicopter style. FEL (fibre elastomer) main rotor designed for simplicity, manoeuvrability and damage tolerance; has infinite life except for inspection of elastomeric elements at more than 2,500-hour intervals; hub consists of titanium centrepiece (including duct for mast-mounted sight) with composites starplates bolted above and below; flap and lead/lag motions of blades allowed by elastic bending of neck region and pitch change by elastic part of elastomeric bearings; lead/lag damping by solid-state viscoelastic damper struts faired into trailing-edge of each blade root; equivalent flapping hinge offset of 10.5 per cent gives high control power; SARIB passive vibration damping system between transmission and airframe; three-blade Spheriflex tail rotor has composites blades with fork roots; built-in ram air engine exhaust suppressors.

FLYING CONTROLS: Fully powered hydraulic flying controls by SAMM/Liebherr; Labinal/Electrométal servo trim; horizontal tail mounted beneath tail rotor; autopilot is part of basic avionics system (see under Avionics heading).

STRUCTURE: 80 per cent CFRP, block and sandwich and Kevlar sandwich; 6 per cent titanium and 11 per cent aluminium; airframe structure protected against lightning and EMP by embedded copper/bronze grid and copper bonding foil; stub-wings of aluminium spars with CFRP ribs and skins; titanium engine deck may be replaced by GFRP; airframe tolerates crash impacts at 10.5 m/s (34.4 ft/s) and meets MIL-STD-1290 crashworthiness standards; titanium main rotor hub centrepiece and tail rotor Spheriflex integral hub/mast; blade spars filament-wound; GFRP, CFRP skins and subsidiary spars and foam filling.

French plants building transmission, tail rotor, centre-fuselage (including engine installation), aerodynamics, fuel and electrical systems, weight control, maintainability, reliability and survivability; Eurocopter Deutschland responsible for main rotor, flight control and hydraulic systems, front and rear fuselage (including cockpits), prototype assembly, flight characteristics and performance, stress and vibration testing and simulation.

LANDING GEAR: Tailwheel type, non-retractable, with single wheel on each unit. Designed to absorb impacts of up to 6 m (20 ft)/s. Main gear by Messier-Bugatti, tail gear by Liebherr Aerotechnik.

POWER PLANT: Two MTU/Rolls-Royce/Turbomeca MTR 390 modular turboshaft engines mounted side by side above centre-fuselage, divided by armour plate 'keel' (engine first flown in Panther testbed 14 February 1991); power ratings are maximum T-O 958 kW (1,285 shp), super emergency 1,160 kW (1,556 shp), maximum continuous 873 kW (1,171 shp). MTR 390 Enhanced, with additional 14 per cent power, to be installed in 38th and subsequent French Tigres and all Spanish aircraft. LHTEC has proposed the T800-801 as a potential alternative power plant for export variants of the Tiger.

Self-sealing crashworthy fuel tanks, with explosion suppression and with non-return valves which minimise leakage in a crash; total capacity 1,360 litres (359 US gallons; 299 Imp gallons). Provision for two external tanks, one on each inboard pylon, each of approximately 350 litres (92 US gallons; 77 Imp gallons) capacity. Gearbox has specified 30 minutes' dry running capability (demonstrated 65 minutes, November 1994).

ACCOMMODATION: Crew of two in tandem, with pilot in front and weapons system operator at rear; full dual controls; both crew members can perform all tasks and weapon operation except that anti-tank missile firing only available to gunner. Armoured, impact-absorbing seats; stepped cockpits, with flat-plate windscreens and slightly curved non-glint transparencies.

SYSTEMS: Redundant hydraulic, electrical and fuel systems. Primary power generation by two 20 kVA alternators; DC power generation by two 300 A 28 V transformer/rectifiers and two 23 Ah Ni/Cd batteries.

Dual redundant AFCS provides four-axis command and stability augmentation. Basic AFCS modes: attitude hold, heading hold. Higher AFCS modes: Heading/acquire/hold, barometric altitude capture/hold, altitude acquire, airspeed hold, vertical speed acquire/hold, nav coupling, radar height hold, Doppler hover hold, line of sight acquisition/hold. Other AFCS functions: gun recoil force compensation, axis decoupling and tactical mode (follow-up trim on override of break-out forces).

AVIONICS: *Basic or core avionics* common to all three versions include bus/display system, com radio (French and German systems vary), autonomous nav system and radio/Doppler navaids, Thales TSC 2000 IFF Mk 12, NH 90-based ECM suite (including laser warning) and AFCS, all connected to and controlled through redundant MIL-STD-1553B data highway.

Flight: Navigation system, by Thales, Teldix and EADS, is fully redundant; system contains two Thales PIXYZ three-axis ring laser gyro units, two air data computers, two magnetic sensors, one Teldix/BAE Canada CMA 2012 Doppler radar, a radio altimeter and GPS providing data to Dornier EuroGrid digital map system; these sensors also provide signals for flight control, information display and guidance; integrated duplex AFCS by Thales and Nord Micro; AFCS computers produced by Thales, VDO-Luft and Litef.

Instrumentation: Colour liquid crystal flight displays showing symbology and imagery (two per cockpit for flight and weapon/systems information) by Thales and VDO-Luft; each crewman has central control/display unit for inputting all radio, electronic systems and navigation selections; digital map display system by Dornier and VDO-Luft (incorporating NH 90's Eurogrid map generation system); engine and systems data are fed into the databus for in-flight indication and subsequent maintenance analysis. BAE Systems Knighthelm fully integrated day and night helmet ordered for German Tigers; French Tigres have similar Thales Topowl helmet-mounted sights, with integrated night vision (image intensifiers), FLIR, video and synthetic raster symbology.

Mission: Euromep (European mission equipment package) includes SATEL Condor 2 pilot vision subsystem (PVS), air-to-air subsystem (Stinger or Mistral), mast-mounted sight and missile subsystem and Euromep management system all connected to separate MIL-STD-1553B data highway. Euromep Standard B avionics first flew February 1995 (PT5); Standard C testing began in late 1997 (PT3R). PVS has 40 × 30° instantaneous field of view (with ±110 × 35° total field of view) thermal imaging sensor steered by helmet position detector giving both crewmen day/night/bad weather vision, flight symbology and air-to-air aiming in helmet-mounted display; mast-mounted sight, gunner sight electronics and gunner's head-in target acquisition display and ATGW 3 subsystem connected by separate data highway; HOT 3 missile system also available. Thales armament control panel and fire-control computer.

HAP combat support mission equipment package includes SFIM STRIX gyrostabilised roof-mounted sight (with IRCCD IR channel) above rear cockpit; includes direct view optics with folding sight tube, television and IR channels and laser ranger/designator.

Self-defence: EADS C-model EW suite (as in NH90) is one element in HAP's fully integrated avionics suite. This has an EADS laser warning receiver, EADS missile launch warning device *(Lenkflugkörpersysteme)* and Thales EW processor and radar warning receiver. Also chaff and flare dispensers (with up to 144 cartridges, sequenced by a Saphir M system). UH is similar, but with option of fitting IR jammer.

ARMAMENT: See also accompanying table. Tiger has four outboard weapon stations for typical four HOT, four Trigat, two Mistral or Stinger launchers or pods of 12 or 22 rockets; optional 12.7 mm gun pod with 250 rounds or auxiliary fuel tank. HCP (HAP) options all include one 30 mm Giat AM-30781 automatic cannon with up to 450 rounds (traversing from +33 to −30° in elevation and through ±90° in azimuth). Hellfire ATM on Australian aircraft.

DIMENSIONS, EXTERNAL (UT: UHT and HAC, HCP: HAP, where different):

Main rotor diameter ... 13.00 m (42 ft 7¾ in)
Tail rotor diameter ... 2.70 m (8 ft 10¼ in)
Length overall, rotors turning ... 15.80 m (51 ft 10 in)
Length of fuselage: UT ... 14.08 m (46 ft 2¼ in)
  HCP: incl cannon ... 15.00 m (49 ft 2½ in)
    excl cannon ... 14.77 m (48 ft 5½ in)
Height: to top of rotor head ... 3.84 m (12 ft 7¼ in)
  to top of tail rotor disc ... 4.32 m (14 ft 2 in)
  to top of mast sight (UT only) ... 5.20 m (17 ft 0¾ in)
Width over weapon pylons ... 4.53 m (14 ft 10¼ in)
Wheel track ... 2.38 m (7 ft 9¾ in)
Wheelbase ... 7.65 m (25 ft 1 in)

AREAS:
Main rotor disc ... 132.70 m^2 (1,428.7 sq ft)
Tail rotor disc ... 5.72 m^2 (61.63 sq ft)

WEIGHTS AND LOADINGS:
Weight empty (HCP) ... 4,200 kg (9,259 lb)
Fuel weight: internal ... 1,080 kg (2,381 lb)
  external (two tanks) ... 555 kg (1,224 lb)
Mission T-O weight ... 5,300–6,100 kg (11,685–13,448 lb)
Max T-O weight: current ... 6,100 kg (13,448 lb)
planned ... 6,400 kg (14,109 lb)
Main rotor disc loading (max mission T-O weight) ... 46.0 kg/m^2 (9.41 lb/sq ft)

PERFORMANCE (UT and HCP, as above):
Never-exceed speed (VNE):
  UHT, HAC ... 171 kt (298 km/h; 185 mph)
  HCP, HAP ... 175 kt (322 km/h; 200 mph)
Max level speed: UT ... 140 kt (259 km/h; 161 mph)
  HCP ... 150 kt (278 km/h; 173 mph)
Cruising speed ... 124 kt (230 km/h; 143 mph)
Max rate of climb at S/L: UHT ... 642 m (2,106 ft)/min
  HAP ... 690 m (2,263 ft)/min
Vertical rate of climb at S/L: UHT ... 312 m (1,023 ft)/min
  HAP ... 384 m (1,259 ft)/min
Hovering ceiling OGE: UHT ... 3,200 m (10,500 ft)
  HAP ... 3,500 m (11,480 ft)
Range: on internal fuel ... 432 n miles (800 km; 497 miles)
  with ferry tanks ... 691 n miles (1,280 km; 795 miles)
Endurance: operational mission ... 2 h 50 min
  max internal fuel ... 3 h 25 min

UPDATED

EUROCOPTER SUPER PUMA MK I AND COUGAR MK I

Spanish military designations: HD.21 and HT.21
Swedish Air Force designation: Hkp 10

TYPE: Multirole medium helicopter.

PROGRAMME: First flight of AS 332 Super Puma (F-WZJA) 13 September 1978; six prototypes; deliveries began mid-1981; present version powered by Turbomeca Makila 1A1 introduced 1986; military versions renamed Cougar in 1990; first AS 332L (stretched fuselage) certified to French IFR Cat. II 7 July 1983 and delivered to Lufttransport of Norway; certified for flight into known icing 29 June 1983; FAA Cat. II certification with SFIM CDV 85 P4 four-axis AFCS and FAR Pt 25 Appendix C known icing clearance. Mk I continues in production as long as orders (mainly making good attrition) continue.

Some AS 332/532s built under licence by IPTN in Indonesia and TAI in Turkey; some assembled by CASA in Spain; and some equipped by F+W (now SF) in Switzerland. Super Puma/Cougar worldwide fleet had accumulated 2,134,000 hours by January 2001.

CURRENT VERSIONS: Designation suffixes: U: military unarmed utility, A: armed, S: anti-ship/submarine, C: military court (short) fuselage, L: long fuselage, military or civil.

AS 332L1 Super Puma: Standard Mk I civil version with long fuselage and airline interior for 20 passengers; UK CAA IFR certification 21 April 1992.

AS 532 Cougar 100: Simplified 9 tonne tactical transport; fixed landing gear reduces cruising speed by 5 kt (9 km/h; 6 mph); price reduced by 15 to 20 per cent. External tanks cannot be fitted. Two versions are proposed: **AS 532UB** Short fuselage version and **AS 532UE** stretched fuselage version (first flight due 2000, but not confirmed); only AS 532UE currently marketed. Brazilian Army announced as first customer (for eight) on 16 June 2001; deliveries were due to take place during January-June 2002.

AS 532UC Cougar: Military Mk I short fuselage unarmed utility; seats up to 21 troops and two crew; cabin floor reinforced for 1,500 kg/m^2 (307 lb/sq ft). Crashworthy self-sealing fuel tanks standard. Equipment includes IR suppressors, RWR, chaff dispensers, flare launcher and armoured protection for crew and troops.

AS 532UL Cougar: Military Mk I unarmed long fuselage version; cabin lengthened by 0.76 m (2 ft 6 in); extra fuel and two large windows in forward cabin plug; carries up to 25 troops and two crew. Crashworthy self-sealing fuel tanks standard. Equipment as per AS 532UC; sliding forward cabin doors can be fitted to allow carriage of forward-firing 7.62 mm machine guns.

AS 532AC Cougar: Armed AS 532UC (Mk I).

AS 532AL Cougar: Armed AS 532UL (Mk I).

TIGER WEAPON SYSTEM OPTIONS

Version	Primary weapon systems	Optional weapon systems
HAP	Turret-mounted 30 mm cannon Strix roof sight Helmet sight and clear sight (×2)	Two pods, 22 each 70 mm rockets plus two pods, 12 each submunitions rockets; or four Mistral AAMs; or four Mistral AAMs plus two pods, 22 each 70 mm rockets
HAC	Osiris roof sight Helmet sight/display (×2)	Eight HOT ATMs plus four Mistral AAMs; or eight Trigat ATMs plus four Mistral AAMs; or four Trigat ATMs plus four HOT ATMs plus four Mistral AAMs
UHT	Osiris roof sight Helmet sight/display (×2)	Four Stinger AAMs plus two 12.7 mm gun pods; or eight HOT or Trigat (or HOT/Trigat mix) ATMs; or two pods, 19 each 70 mm rockets
ARH	Turret-mounted 30 mm cannon Roof sight and laser designator/helmet sight Mistral AAM capability	Two Hellfire ATMs plus 26 off 70 mm rockets; or four Hellfire ATMs plus 33 off 70 mm rockets
HAD	Turret-mounted 30 mm cannon Roof sight Helmet sight/display (×2)	Four Mistral AAMs; or four pods of 68 mm or 70 mm rockets; or four Trigat ATMs

Eurocopter AS 332L1 Super Puma of Toho Air Service, Japan *(H Seo/Eurocopter)* ***NEW***/0546929

AS 532SC Cougar: Naval short Mk I version with ASW/ASV equipment; folding tail rotor pylon and main rotor blades; deck harpoon landing aid. Crew of two plus up to two operators.

AS 532UL Horizon: Battlefield surveillance radar helicopter; flight trials of small Orphée radar under an AS 330B Puma started 1986; French Army wanted 20 AS 532 Cougar Mk II fitted with radar, dedicated ECM and datalink (Orchidée programme). First flight of full-scale AS 532/Orchidée June 1990; programme cancelled for budgetary reasons August 1990; prototype without datalink flew 24 missions in Gulf War February 1991 in Operation Horus.

Horizon programme (*Hélicoptère d'Observation Radar et d'Investigation sur Zone*) with more effective operational concept and reduced development costs replaced Orchidée; development contract awarded to Eurocopter October 1992 for two (later increased to four) AS 532UL Horizons with same radar capabilities and ECM as Orchidée, but in AS 532UL Cougar Mk I, with standard ECM and longer mission endurance. Antenna rotates at either 2° 24' or 8°/s; radar range 108 n miles (200 km; 124 miles) with helicopter cruising at 97 kt (180 km/h; 112 mph) at 4,000 m (13,120 ft); normal operational range is 81 n miles (150 km; 93 miles); target speed resolution is 2 m (6.6 ft)/s. Military features include jet efflux deflector/diluters, anti-icing systems, NVG-compatible cockpit lighting, weather radar, encrypted communications, INS/GPS and active/passive countermeasures. First flight Horizon with full radar 8 December 1992; first delivery 24 June 1996; second (plus a ground station) in December 1996; third delivered in 1997 and fourth (plus second ground station) in 1998.

AS 332L2 Super Puma Mk II: Stretched version with Spheriflex rotor heads, new cockpit, enlarged rotors and more powerful Makila 1A2 engines. *Described separately.*

AS 532U2/A2 Cougar Mk II: Stretched version with Spheriflex rotor heads, new cockpit, enlarged rotors and more powerful Makila 1A2 engines. *Described separately.*

CUSTOMERS: Total orders for AS 332/532 and EC 225/725 series stood at 652, from 85 customers in 47 countries, by December 2002. Sales in 2003 totalled eight. Some 110 AS 332L1/L2 Super Pumas in civil offshore oil industry support operations; French military orders include six for French Air Force (three for nuclear test facilities in Pacific and three for government VIP flying); French Army (ALAT) has supplemented SA 330 Pumas with AS 532 Cougars; 22 delivered to *Force d'Action Rapide* between late 1988 and end 1991.

Export customers include Abu Dhabi (eight including two VIP of which five to be upgraded with Exocet, sonar and torpedoes under 1995 contract), Brazil (25 including six AS 532SCs, one AS 332M1 and eight ordered January 2001), Cameroon (one), Chile (army, two; navy, four AS 532SCs), China (six), Ecuador (eight including six army), Finland (three for border police), Gabon (two, Presidential Guard), Germany (three, border police), Greece (Ministry of Merchant Marine, four ordered August 1998 and first two delivered 21 December 1999; remaining pair March 2000), Greek Air Force (four plus two options in 2001 for delivery between 2001 and 2003); Indonesia (built under licence: see IPTN entry); Japan (three, army/VIP), Jordan (eight), South Korea (three army/VIP, one AS 332L1), Mexico (two VIP), Nepal (two, Royal Flight), Nigeria (two), Oman (two, Royal Flight), Panama (one VIP), Saudi Arabia (12 AS 532SCs), Singapore (36), Slovenian Army (two Uls delivered 2003), Spain (10 SAR HD.21s, two VIP HT.21s, 18 army tactical transport HT.21s plus 15 AS 532ULs ordered February 1996 and delivered from 1998), Sweden (12 SAR), Switzerland (15 AS 332Ms; plus 12 AS 532ULs ordered December 1998 for delivery from 2000; two to be built by Eurocopter and remainder subcontracted to RUAG Aerospace, which see), Togo (one), Turkey (30 AS 532ULs and 20 AS 532ALs in SAR configuration, including 28 assembled by TAI; first Turkish-built AS 532AL handed over 1 May 2000; deliveries to be completed by February 2003; first Turkish-built AS 532UL handed over 31 May 2000), Venezuela (six), Zaïre (one VIP). Competing for Taiwanese UH-1 and Greek UH-1/AB 205 replacement programmes. Bristow Helicopters acquired 31 examples of 19-passenger AS 332L for offshore oil support. AZAL of Azerbaijan ordered 22 AS 332L1s in 2003.

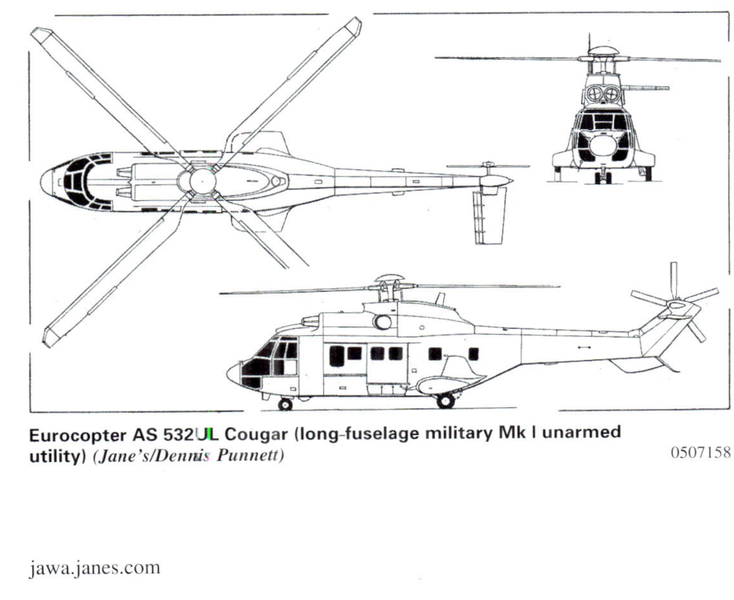

Eurocopter AS 532UL Cougar (long-fuselage military Mk I unarmed utility) *(Jane's/Dennis Punnett)* 0507158

COSTS: AS 332L1 US$14.25 million, IFR; direct operating cost US$1,542 per hour (both 1997).

DESIGN FEATURES: Four-blade fully articulated main rotor turning clockwise seen from above; five-blade tail rotor on starboard side of tailboom; engines mounted above cabin have rear drive into main transmission at 23,840 rpm; main rotor turns at 265 rpm, tail rotor at 1,278 rpm; various lateral sponsons available housing partly retracted main landing gear and combinations of additional fuel or pop-out floats; optional air conditioning system housed in casing on port forward flank of cabin; all civil versions certified for IFR category A and B to FAR Pt 29.

FLYING CONTROLS: Dual fully powered hydraulic controls with full-time autostabilisation and yaw damping; machine remains flyable with autostabiliser switched off; cyclic trimming by stick friction adjustment; inverted slot on tailplane to maintain attitude-holding effect at low climb speeds; large ventral fin; saucer fairing on rotor head to smooth wake of hub; four-axis SFIM 155 autopilot standard.

STRUCTURE: Conventional light alloy airframe with some titanium; crashworthy fuel system, impact-absorbing landing gear and other features; main rotor blades of GFRP with CFRP stiffening and Moltoprene filler; elastomeric drag dampers.

LANDING GEAR: Retractable tricycle high energy absorbing design by Messier-Bugatti; all units retract rearward hydraulically, mainwheels into sponsons on sides of fuselage; dual-chamber oleo-pneumatic shock-absorbers; twin-wheel self-centring nose unit, tyre size 7.00–6 (8 ply) tubeless, pressure 7.00 bar (102 lb/sq in); single wheel on each main unit with tyre size 615×225–10 (12 ply) tubeless or 640×230–10, pressure 9.00 bar (130 lb/sq in); hydraulic differential disc brakes, controlled by foot pedals; lever-operated parking brake; emergency pop-out flotation units can be mounted on main landing gear fairings and forward fuselage.

POWER PLANT: Two Turbomeca Makila 1A1 turboshafts, each with maximum contingency rating of 1,400 kW (1,877 shp) and take-off rating of 1,357 kW (1,820 shp). Maximum continuous rating of 1,184 kW (1,588 shp). Air intakes protected by a grille against ingestion of ice, snow and foreign objects; Centrisep multipurpose intake optional for flight into sandy areas. Transmission rated at 2,238 kW (3,000 shp) for T-O.

AS 532UC/AC have five flexible fuel tanks under cabin floor, with total usable capacity of 1,497 litres (395 US gallons; 329 Imp gallons); AS 532SC has total basic capacity of 2,141 litres (565 US gallons; 471 Imp gallons); AS 332L1/532UE/532UL/532AL have a basic fuel system of six flexible tanks with total capacity of 2,020 litres (533 US gallons; 444 Imp gallons) in the 332 and 2,003 litres (529 US gallons; 440 Imp gallons) in the 532; provision for additional 1,900 litres (502 US gallons; 418 Imp gallons) in four auxiliary ferry tanks installed in cabin; two external auxiliary tanks with total capacity of 650 litres (172 US gallons; 143 Imp gallons) standard on AS 532SC, optional on other versions; AS 532UB Cougar 100 has capacity of 1,533 litres (405 US gallons; 337 Imp gallons). For long-range missions (mainly offshore) in AS 332L1, a special internal auxiliary tank can be fitted in cargo sling well, in addition to the two external tanks, to raise total usable fuel capacity to 2,994 litres (791 US gallons; 658 Imp gallons). Refuelling point on starboard side of cabin; fuel system designed to avoid leakage following a crash; self-sealing tanks standard on military versions, optional on other versions; other options include a fuel dumping system and pressure refuelling.

ACCOMMODATION: One pilot (VFR) or two pilots side by side (IFR) on flight deck, with jump seat for third crew member or paratroop dispatcher; provision for composite light alloy/Kevlar armour for crew protection on military models; door on each side of flight deck and internal doorway connecting flight deck to cabin; dual controls, co-pilot instrumentation and crashworthy flight deck and cabin floors; maximum accommodation for 21 troops in AS 532UC, 20 passengers in AS 332L1, 24 in AS 332L2, and 25 troops in AS 532UL, 20 troops in AS 532UE; interiors available for VIP, air ambulance with six stretchers and 11 seated casualties/attendants; strengthened floor for cargo carrying, with lashing points; jettisonable sliding door on each side of main cabin or port side door with built-in steps and starboard side double door in VIP configuration; removable panel on underside of fuselage, at rear of main cabin, for longer loads; removable door with integral steps for access to baggage racks optional; hatch in floor below main rotor contains hook for slung loads up to 4,500 kg (9,920 lb) on internally mounted cargo sling; cabin and flight deck heated, ventilated and soundproofed; demisting, de-icing, washers and wipers for pilots' windscreens.

SYSTEMS: Two independent hydraulic systems, supplied by self-regulating pumps driven by main gearbox. Each system supplies one set of powered flying controls; left-hand system also supplies autopilot, landing gear, rotor brake and wheel brakes; hydraulically actuated systems can be operated on ground from main gearbox (when a special disconnect system is installed to permit running of port engine with rotors stationary), or by external power through ground power receptacle. Emergency landing gear lowering by standby pump.

Three-phase 200 V AC electrical power supplied by two 20 kVA 400 Hz alternators, driven by port side intermediate shaft from main gearbox and available on ground under same conditions as hydraulic ancillary systems; two 28.5 V DC transformer-rectifiers; main battery used for self-starting and emergency power in flight.

AVIONICS: *Comms:* Optional equipment includes VHF, UHF, tactical HF and HF/SSB radio and intercom.

Radar: Offshore models have nose-mounted radar; search and rescue version has nose-mounted RDR 1400 or chin-mounted 1500 search radar (as on Swedish Hkp 10s); naval ASW and ASV versions can have nose-mounted Thales Ocean Master radar, linked to a tactical table in the cabin.

Flight: ADF, VOR/ILS, radio altimeter, GPS, VLF Omega, Decca navigator and flight log, Doppler, SFIM 155 autopilot, with provision for coupling to self-contained navigation and landing systems; search and rescue version has Doppler, and Thales Nadir or Decca self-contained navigation system (Nadir Mk 2 in French Army and Greek versions), including navigation computer with SAR patterns, polar indicator, roller map display, hover indicator, route mileage indicator and groundspeed and drift indicator; SFIM CDV 155 autopilot coupler contains automatic nav track including search patterns, transitions and hover; multifunction video display shows radar and route images, SAR patterns and hover indication; Swedish Hkp 10s have Thales RAMS flight management system including BAE Systems AHRS, Decca Doppler and GPS. Optional Honeywell Mk XXII EGPWS.

Instrumentation: Full IFR instrumentation optional.

Mission: Naval ASW versions have Thomson Marconi HS 312 acoustics suite; upgraded FLASH version currently available.

EQUIPMENT: Optional fixed rescue hoist (capacity 275 kg; 606 lb) starboard side and electrical back-up hoist; equipment for naval missions can include sonar, MAD and sonobuoys.

ARMAMENT (OPTIONAL): Typical alternatives for army/air force missions are two 20 mm guns or two 7.62 mm machine guns or two 68 mm rocket launchers. Armament for naval missions includes two AM 39 Exocet missiles or two lightweight torpedoes.

DIMENSIONS, EXTERNAL:

Main rotor diameter .... 15.60 m (51 ft 2¼ in)

Tail rotor diameter .... 3.05 m (10 ft 0 in)

Main rotor blade chord ..... 0.60 m (1 ft 11½ in)
Length: overall, rotors turning ..... 18.70 m (61 ft 4¼ in)
fuselage, incl tail rotor:
AS 532UC/AC/SC/UB ..... 15.53 m (50 ft 11½ in)
AS 332L1/532AL/UL/UE ..... 16.29 m (53 ft 5½ in)
tail and rotors folded (AS 532SC) ..... 12.95 m (42 ft 5¾ in)
Width, blades folded:
AS 532UB/UC/AC/AL/UL/332L1 ..... 3.79 m (12 ft 5¼ in)
AS 532SC ..... 3.86 m (13 ft 3 in)
Width over sponsons:
AS 332L1/AS 532AL/UL/UC/AC ..... 3.38 m (11 ft 1 in)
AS 532SC ..... 3.56 m (11 ft 8¼ in)
Fuselage max width ..... 2.00 m (6 ft 6 in)
Height: overall ..... 4.92 m (16 ft 1¾ in)
blades and tail pylon folded:
AS 532UE/UC/AC/SC ..... 4.80 m (15 ft 9 in)
to top of rotor head ..... 4.60 m (15 ft 1 in)
Tailplane half-span ..... 2.11 m (6 ft 11 in)
Wheel track ..... 3.00 m (9 ft 10 in)
Wheelbase: AS 532UC/AC/SC ..... 4.49 m (14 ft 8¾ in)
AS 332L1/532AL/UL/UE ..... 5.25 m (17 ft 2¾ in)
Passenger cabin doors, each: Height ..... 1.35 m (4 ft 5 in)
Width ..... 1.30 m (4 ft 3¼ in)
Floor hatch, rear of cabin: Length ..... 0.98 m (3 ft 2¾ in)
Width ..... 0.70 m (2 ft 3½ in)

DIMENSIONS, INTERNAL:
Cabin:
Length: AS 532UC/AC/SC ..... 6.05 m (19 ft 10½ in)
AS 332L1/532AL/UL/UE ..... 6.81 m (22 ft 4 in)
Max width ..... 1.80 m (5 ft 10¾ in)
Max height ..... 1.55 m (5 ft 1 in)
Floor area: AS 532UC/AC/SC ..... 7.80 m^2 (84.0 sq ft)
AS 332L1/532AL/UL/UE ..... 9.18 m^2 (98.8 sq ft)
Usable volume:
AS 532UC/AC/SC ..... 11.4 m^3 (403 cu ft)
AS 332L1/532AL/UL/UE ..... 13.3 m^3 (470 cu ft)

AREAS:
Main rotor disc ..... 191.13 m^2 (2,057.3 sq ft)
Tail rotor disc ..... 7.31 m^2 (78.64 sq ft)

WEIGHTS AND LOADINGS: Weight empty (standard aircraft):
AS 532UC/AC ..... 4,330 kg (9,546 lb)
AS 532SC ..... 4,500 kg (9,920 lb)
AS 332L1/532AL/UL ..... 4,460 kg (9,832 lb)
AS 332UE ..... 4,485 kg (9,888 lb)
Max external payload ..... 4,500 kg (9,920 lb)
Max T-O weight:
internal load:
AS 532UC/AC/AL/UB/UL/SC/UB/UE ..... 9,000 kg (19,841 lb)
AS 332L1 ..... 8,600 kg (18,960 lb)
external slung load ..... 9,350 kg (20,615 lb)
Max disc loading, external load ..... 48.9 kg/m^2 (10.02 lb/sq ft)
Transmission loading at max T-O weight and power:
internal load:
AS 532UC/AC/AL/UB/UL/SC/UE ..... 4.03 kg/kW (6.61 lb/shp)
AS 332L1 ..... 3.85 kg/kW (6.32 lb/shp)
external load ..... 4.18 kg/kW (6.87 lb/shp)

PERFORMANCE:
Never-exceed speed (VNE) ..... 150 kt (278 km/h; 172 mph)
Cruising speed at S/L:
AS 532UC/AC/AL/UL/SC ..... 139 kt (257 km/h; 160 mph)
AS 532UB/UE ..... 134 kt (249 km/h; 155 mph)
AS 332L1 ..... 141 kt (261 km/h; 162 mph)
Max rate of climb at S/L:
AS 532UC/AC/AL/UL ..... 420 m (1,378 ft)/min
AS 532UB/UE ..... 432 m (1,417 ft)/min
AS 532SC ..... 372 m (1,220 ft)/min
AS 332L1 ..... 486 m (1,594 ft)/min
Service ceiling: AS 332L1 ..... 4,600 m (15,080 ft)
AS 532UC/AC/SC/AL/UL/UB/UE ..... 4,100 m (13,450 ft)
Hovering ceiling IGE:
AS 532AL/UL/AC/UC/SC/UB/UE ..... 2,800 m (9,180 ft)
AS 332L1 ..... 3,250 m (10,660 ft)
Hovering ceiling OGE:
AS 532AL/UL/AC/UC/SC/UB/UE ..... 1,650 m (5,420 ft)
AS 332L1 ..... 2,300 m (7,540 ft)
Range at S/L, standard tanks, no reserves:
AS 332L1 ..... 449 n miles (831 km; 516 miles)
AS 532UC/AC ..... 334 n miles (618 km; 384 miles)
AS532UB ..... 309 n miles (573 km; 356 miles)
AS 532SC ..... 492 n miles (911 km; 566 miles)
AS 532AL/UL ..... 455 n miles (842 km; 523 miles)
AS 332L1 ..... 470 n miles (870 km; 540 miles)
AS 532UE ..... 416 n miles (770 km; 478 miles)
Range with auxiliary internal tank:
AS 532UE ..... 486 n miles (900 km; 559 miles)
AS 532UB ..... 380 n miles (703 km; 437 miles)
Range at S/L with external (two 338 litre) and auxiliary (one 320 litre) tanks, no reserves:
AS 332L1 ..... 659 n miles (1,220 km; 758 miles)
AS 532UC/AC ..... 528 n miles (977 km; 607 miles)
AS 532AL/UL ..... 637 n miles (1,179 km; 733 miles)

UPDATED

EUROCOPTER SUPER PUMA MK II AND COUGAR MK II

Spanish Army designations: HU.21L

TYPE: Multirole medium helicopter.

PROGRAMME: Derived from Super Puma/Cougar Mk I (which see); first flight of development vehicle 6 February 1987; French certification 2 April 1992; UK BCAR 29 certification 16 November 1992; first delivery August 1993. Earlier AS 332L1 and 532U/A Cougar Mk I remain on production line. Further development is under consideration.

CURRENT VERSIONS: **AS 332L2 Super Puma Mk II:** Current production civil transport; innovations include Spheriflex rotor heads, super-emergency engine rating, EFIS flight deck and built-in health and usage monitoring system, duplex four-axis AFCS, and hydraulically powered standby electrics for 2 hours' operation after complete main generator failure. First delivery 28 August 1993.

AS 332L2 Super Puma Mk II VIP: Eight- to 15-passenger arrangement with attendant and fully equipped galley and lavatory; two four-seat lounges; fine materials and fittings throughout; range with eight passengers and attendant, 635 n miles (1,176 km; 730 miles). Used in 28 countries.

AS 332L2 Super Puma Mk II 'Jigsaw': 'Project Jigsaw' launched by Bristow Helicopters in 2001, under contract to BP, to modify Super Puma G-JSAR for offshore SAR. Modifications include 'glass cockpit' and four-axis coupled autopilot for IFR/day/night approaches, search patterns, transitions and hover; additional radios (HF, VHF/FM Marine Band and UHF); Chelton Homer and Becker Homer; NightSun searchlight (30 million cd); Sky Shout loudspeaker; Iridium satellite telephone; A5-1 chin turret with FLIR; rescue hoist; and extensive medical equipment. Completed aircraft exhibited at Helitech, Duxford, UK, 23 to 25 September 2003.

AS 532U2 Cougar Mk II: Unarmed military tactical transport; longest member of Cougar family; carries 29 troops and two-man crew.

AS 532A2 Cougar Mk II: Armed version. First order (for four) by French Air Force for combat SAR, 1995; known as Cougar RESCO (REcherche et Sauvetage en COmbat); prototype (F-ZVMC) undertook maker's trials in 1998; was delivered to CEAM military test unit on 9 September 1999 for one year of evaluation; second aircraft, to determine operational fit, was due to be delivered in 2000. Potential equipment includes RWR, missile approach warner, chaff, flares, rescue hoist, two outboard 20 mm cannon, two door-mounted 12.7 mm machine guns, encrypted Personnel Locator System,

First production EC 225, delivered to Algeria on 22 December 2004 ***NEW***/1047796

Eurocopter AS 332L2 Super Puma *(Patrick Penna)* ***NEW***/0576175

Manufactured image of Eurocopter EC 225 ***NEW***/0576170

Eurocopter AS 332L2 Super Puma of the Moroccan Gendarmerie *(Patrick Penna)* ***NEW***/0576177

third-generation NVG-compatible lighting, air-to-air refuelling capability. Nadir Mk 2 navigation computer, SFIM PA-165 four-axis digital autopilot, VOR, Tacan, DME, GPS, searchlight and FLIR. Alternative maximum T-O weight of 11,200 kg (24,692 lb) permits an 800 n mile (1,482 km; 920 mile), plus 30 min hover and 30 min reserve, unrefuelled radius of action to rescue two persons.

EC 225 Super Puma Mk II+: Growth version, announced June 1998, as competitor to Sikorsky S-92 in offshore support role; has military counterpart in **EC 725 Cougar Mk II+**. Cabin volume increased by 25 per cent to 20.0 m³ (706 cu ft) through increases of 35 cm (13¾ in) in height, 70 cm (2 ft 3½ in) in length and 25 cm (9¾ in) in width. Reinforced gearbox and 1,800 kW (2,414 shp) Makila 1A4 engines with FADEC produce 14 per cent power increase over earlier versions. Five-blade main rotor blades have composites spars and parabolic tips, diameter unchanged. MTOW increased to 10,400 kg (21,560 lb) for civil and 11,000 kg (24,251 lb) for military version (11,200 kg; 24,691 lb with external load). New integrated flight and display system (IFDS) using four 152 × 203 mm (6 × 8 in) multifunction LCDs plus four-axis autopilot. FAR/JAR-29 certification expected late 2003 and first deliveries to French Air Force, June 2004; targets not achieved, and first delivery effected 22 December 2004 when Algerian Government received initial production EC 225 for its VIP transport unit. First prototype (No. 2338, F-ZVLR, converted from standard Mk II) made maiden flight 27 November 2000, at which time Eurocopter revealed an order for four from the French Air Force, including one earlier Mk 2 raised to this standard. JAR 29 certification expected by end 2003. Launch customer for EC 225 is CHC Helikopter, which ordered one (plus one AS332L2) in early 2001. French Air Force took delivery of initial aircraft 15 January 2001 for trials. First pre-production EC 725 (with Makila 1A4 engines) flew in September 2002. First fully equipped production example due for delivery June 2004 to EH1/67 'Pyrénées' at Cazaux, followed by remaining three in the second half of 2004. Further order for 10 EC 725s for French Air Force Special Operations Command announced in late November 2002, with deliveries to be complete by end of 2006. Bristow Helicopters ordered two EC 225s, plus two options, on 12 January 2004.

CUSTOMERS: By December 2004, total of 682 of Super Puma/Cougar family had been ordered, including 22 EC 225/725s.

Production combined with Super Puma Mk I. Helikopter Service ordered four AS 332L2 Mk II in May 1992 and a further four in June 1997, latter delivered from 27 May 1998 onwards; RNethAF ordered 17 AS 532U2s October 1993 and deliveries began 1 April 1996 (with 400th AS 332/532) and last was received in October 1997; first of follow-on order for two for Bond Helicopters delivered 27 August 1998. Eventual 14 AS 532A2s required for combat SAR by French Air Force, first of four ordered delivered 9 September 1999; remainder to be delivered in three batches by 2015; these orders converted to EC 725 version late 2000. Saudi Arabia ordered 12 AS 532A2s in July 1996 under 'Al Fahd' contract for combat SAR, deliveries beginning in 1998; in-flight refuelling tests successfully completed during second half of 2000. VIP orders include Samsung (one AS 332L2), the German government (three AS 532U2s delivered from 4 November 1997 and operated by the Luftwaffe), and two AS 532L2s for Turkmenistan delivered in VVIP configuration in 2002. Three ordered for SAR by Hong Kong Government Flying Service for delivery mid-2001, the first being handed over 18 June 2001. Bond Offshore Helicopters Ltd of Aberdeen, Scotland, ordered six AS 332L2s on 10 December 2002, of which the first was delivered 19 May 2004.

COSTS: Bond Offshore Helicopters contract for six AS 332L2s valued at US$80 million (2002).

DESIGN FEATURES: Generally as for Mk I, but gross weight increased by 150 kg (331 lb) in 1993; composites plug in rear cabin area accommodates one extra row of seats and moves tail rotor rearwards; Spheriflex main and tail rotor heads with elastomeric spherical bearings and Kevlar retention bands; main rotor with longer blades having parabolic tips; enlarged composites sponsons can contain fuel, liferafts, air conditioning and pop-out floats.

FLYING CONTROLS: Fully powered hydraulically actuated with SFIM 165 four-axis digital AFCS and coupler.

STRUCTURE: Rear fuselage 55 cm (1 ft 9½ in) plug made of composites giving extra row of seats; large windows at rear of cabin; increased use of composites compared with Mk I; fuselage frames strengthened to retain same degree of crashworthiness.

LANDING GEAR: As Super Puma Mk I.

POWER PLANT: Two Turbomeca Makila 1A2 rear drive free turbines; maximum emergency power each (OEI 30 seconds) 1,573 kW (2,109 shp), 12 per cent higher than AS 332 Mk I; intermediate emergency power (OEI 2 minutes) 1,467 kW (1,967 shp); take-off power 1,376 kW (1,845 shp); maximum continuous power 1,236 kW (1,657 shp); protective intake grilles standard; optional Centrisep multipurpose intakes for dusty conditions. Transmission ratings: twin-engined maximum 2,410 kW (3,229 shp), maximum from single engine 1,666 kW (2,232 shp), maximum transient 20 seconds 2,651 kW (3,552 shp), maximum continuous 1,555 kW (2,084 shp); transmission has extended run-dry capability.

Standard fuel tankage under cabin floor 2,020 litres (535 US gallons; 444 Imp gallons); auxiliary tankage includes 324 litres (85.6 US gallons; 71.3 Imp gallons) in cargo hook well; sponson tanks each holding 325 litres (85.9 US gallons; 71.5 Imp gallons); 600 litres (159 US gallons; 132 Imp gallons) in cabin fuel tank; one to five internal ferry tanks each holding 475 litres (126 US gallons; 104.5 Imp gallons). Tankage (self-sealing tanks standard from 1997) AS 532A2 1,896 litres (500 US gallons; 417 Imp gallons); hook well tank 320 litres (84.5 US gallons; 70.4 Imp gallons); 325 litres (85.9 US gallons; 71.5 Imp gallons) in each sponson tank; crashproofing is standard in AS 532 and optional in 332; pressure refuelling optional. Military versions can be fitted with probe for in-flight refuelling; under consideration for EC 725.

ACCOMMODATION: Single pilot in DGAC Category B civil operation; single pilot plus licensed crewman in Category A VFR; two pilots in IFR; military operation one pilot in VFR, two

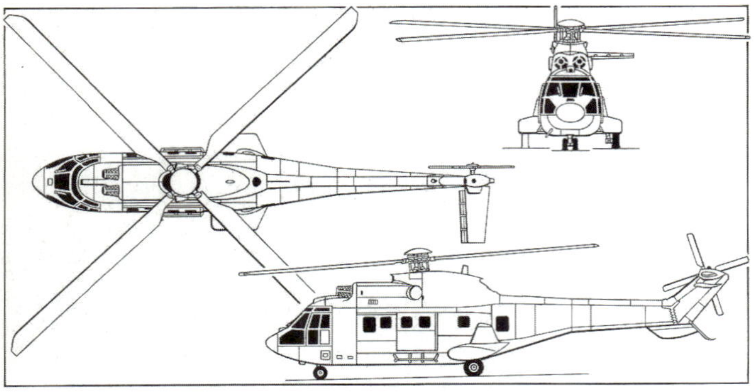

Eurocopter AS 332L2 Super Puma Mk II with Spheriflex main and tail rotor heads *(Jane's/Mike Keep)* 0507159

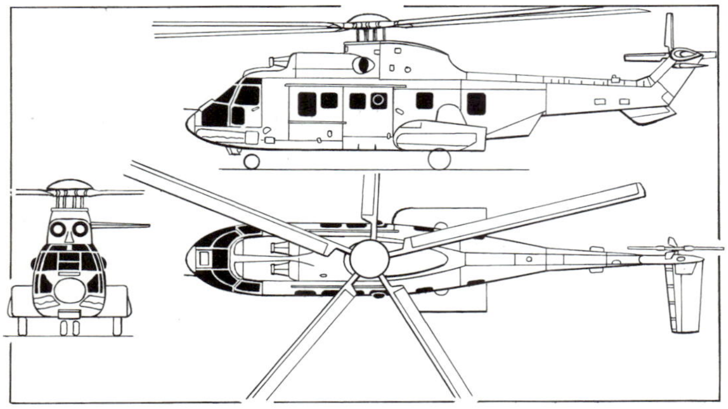

Eurocopter EC 225 Super Puma Mk II+ with five-blade main rotor ***NEW***/0589206

in IFR; civil transport 24 passengers and attendant in airline interior for 220 n miles (408 km; 253 miles) or 19 passengers at 81 cm (32 in) seat pitch for 350 n miles (648 km; 403 miles); military capacity, squad chief plus 28 troops; SAR version four crew plus 15 survivors over radius of 285 n miles (527 km; 328 miles). VIP version for eight to 15 passengers plus attendant; ambulance version holds 12 stretchers and four seated casualties plus attendant.

AVIONICS: *Instrumentation:* Civil and military cockpits have Sextant Avionique four-tube EFIS integrated flight and display system (IFDS) with four 15 × 15 cm (6 × 6 in) screens; subsystems include smart multimode display (SMDS), automatic flight control system (AFCS) and flight data system (FDS); military cockpit compatible with night vision goggles; civil and military IFR system offered, plus military communications radio; SAR system includes radar and FLIR coupled to display system as well as navigation computer with Doppler and GPS sensors; can also be fitted with inertial nav sensor for combat SAR configuration at customer's request; SAR system can include automatic search pattern, transition and hover hold.

Flight: Optional Bendix/King Mark XXII EGPWS.

Radar: RNethAF AS 532U2s have been retrofitted with Honeywell weather radar.

EQUIPMENT: Pop-out floats, rescue winch (272 kg; 600 lb), external sling. Military equipment (U2/A2) includes IR suppressors, flare/chaff dispensers, RWR, MAWS and armoured protection for crew and troops.

ARMAMENT: Armament choice as for AS 532A Cougar Mk I except for Exocet. Special sliding doors on forward cabin windows allow forward firing 7.62 mm machine guns to be fitted.

DIMENSIONS, EXTERNAL:
Main rotor diameter 16.20 m (53 ft 1½ in)
Tail rotor diameter 3.15 m (10 ft 4 in)
Length overall: rotors turning 19.50 m (63 ft 11 in)
main rotor folded 16.79 m (55 ft 0½ in)
Width: fuselage 2.00 m (6 ft 6 in)
over sponsons 3.38 m (11 ft 1 in)
overall, main rotor folded 3.86 m (12 ft 8 in)
Height: overall, tail rotor turning 4.97 m (16 ft 4 in)
to top of rotor head 4.60 m (15 ft 1 in)
Tailplane span 2.12 m (6 ft 11½ in)
Wheel track 3.00 m (9 ft 10 in)
Wheelbase 5.25 m (17 ft 2¾ in)
Sliding cabin doors, each: Height 1.35 m (4 ft 5 in)
Width 1.30 m (4 ft 3¼ in)
Floor hatch: Length 0.98 m (3 ft 2¾ in)
Width 0.70 m (2 ft 3½ in)

DIMENSIONS, INTERNAL:
Cabin: Max length 7.87 m (25 ft 10 in)
Floor length 6.15 m (20 ft 2¼ in)
Max width 1.80 m (5 ft 10¾ in)
Max height 1.45 m (4 ft 9 in)
Volume 15.0 m^3 (530 cu ft)

AREAS:
Main rotor disc 206.12 m^2 (2,218.7 sq ft)
Tail rotor disc 7.79 m^2 (83.88 sq ft)

WEIGHTS AND LOADINGS:
Manufacturer's weight empty: L2 4,686 kg (10,331 lb)
U2 4,945 kg (10,902 lb)
A2 5,012 kg (11,050 lb)
Useful load: L2 4,595 kg (10,130 lb)
U2 4,805 kg (10,593 lb)
A2 4,738 kg (10,445 lb)
EC 225 5,380 kg (11,861 lb)
EC 725 5,670 kg (12,500 lb)
Standard fuel weight: L2 1,596 kg (3,519 lb)
U2, crashworthy tanks 1,548 kg (3,412 lb)
Max normal T-O weight: L2 9,300 kg (20,502 lb)
U2 9,750 kg (21,495 lb)
EC 225 10,400 kg (22,928 lb)
EC 725 11,000 kg (24,250 lb)
A2 Combat SAR 11,200 kg (24,692 lb)
Max slung load: A2/U2, EC 725 5,000 kg (11,023 lb)
L2 4,500 kg (9,920 lb)
Max flight weight with slung load:
U2 10,500 kg (23,148 lb)
EC 225, EC 725 11,200 kg (24,692 lb)
Max disc loading, normal T-O weight:
L2 45.1 kg/m^2 (9.24 lb/sq ft)
U2 47.3 kg/m^2 (9.69 lb/sq ft)
U2, alternate 54.3 kg/m^2 (11.13 lb/sq ft)
Transmission loading at max T-O weight and power:
L2 3.85 kg/kW (6.34 lb/shp)
U2 with slung load 4.35 kg/kW (7.16 lb/shp)
U2 alternate 4.65 kg/kW (7.64 lb/shp)

PERFORMANCE:
Never-exceed speed (VNE):
L2/U2/EC 725 170 kt (315 km/h; 195 mph)
Fast cruising speed: L2 150 kt (277 km/h; 172 mph)
U2 147 kt (273 km/h; 170 mph)
EC 225 156 kt (289 km/h; 180 mph)
EC 725 160 kt (296 km/h; 184 mph)
Econ cruising speed: L2 136 kt (252 km/h; 157 mph)
U2 132 kt (244 km/h; 152 mph)
Rate of climb at 70 kt (130 km/h; 81 mph) at S/L:
L2 441 m (1,447 ft)/min
U2 384 m (1,260 ft)/min
Service ceiling (45.7 m; 150 ft/min climb), ISA:
L2 5,180 m (17,000 ft)
U2 4,100 m (13,450 ft)
Hovering ceiling IGE, normal T-O weight, ISA, T-O power: L2 3,120 m (10,240 ft)
U2 2,690 m (8,820 ft)
Hovering ceiling OGE, normal T-O weight, ISA, T-O power: L2 2,250 m (7,380 ft)
U2 1,900 m (6,240 ft)
Hovering ceiling OGE, slung load weight, ISA, T-O power: L2/U2 960 m (3,150 ft)
Range, no reserves, standard fuel, econ cruise:
L2 447 n miles (828 km; 514 miles)
U2 427 n miles (791 km; 491 miles)
Range, no reserves, max fuel, econ cruise:
L2 656 n miles (1,215 km; 755 miles)
U2 636 n miles (1,178 km; 732 miles)
EC 225, EC 725 783 n miles (1,450 km; 901 miles)
Ferry range:
EC 725 1,000 n miles (1,852 km; 1,150 miles)
Endurance, standard fuel, at 70 kt (130 km/h; 81 mph):
L2 4 h 54 min
U2 4 h 20 min
EC 725 6 h 30 min

UPDATED

EUROCOPTER AS 350 ECUREUIL/ASTAR AND AS 550 FENNEC

Brazilian Air Force designations: CH-50 and TH-5 Esquilo
Brazilian Army designation: HA-1 Esquilo
Brazilian Navy designation: UH-12 Esquilo
UK forces designation: Squirrel HT. Mk 1 and HT. Mk 2

TYPE: Light utility helicopter.

PROGRAMME: First flight (F-WVKH) powered by Lycoming LTS101 turboshaft 27 June 1974; first flight second prototype (F-WVKI) powered by Turbomeca Arriel 1A February 1975; first production version was AS 350B powered by 478 kW (641 shp) Arriel 1B and certified 27 October 1977; LTS101 powered AStar sold only in USA. AS 350BA Ecureuil was powered by 478 kW (641 shp) Turbomeca Arriel 1B and fitted with large main rotor blades of AS 350B2 (see below); maximum T-O weight increased by 150 kg (331 lb); AS 350B upgraded to AS 350BA in field; replaced AS 350B during 1992; French VFR certification 1991; UK and US certifications 1992, Japanese 1993. Production of AS 350BA ended in 1998.

Delivery of the 2,000th AS 350/550 (F-WWPZ, later ZJ270 of UK Defence Helicopter Flying School) effected in July 1997. In February 2001, delivery effected of 3,000th of Ecureuil family (including AS 355/555), which also was first EC 130 upgraded version. Total 10,986,000 Ecureuil flying hours by January 2001.

Eurocopter AS 350B3 *(Gérard Deulin)* ***NEW***/0576326

Built at Marignane, France, and under licence by Helibrás in Brazil (which see). See also the CHAIC Z-11 entry in the Chinese section.

CURRENT VERSIONS: **AS 350B2 Ecureuil:** Uprated engine and transmission; wide-chord new section main and tail rotor blades originally developed for AS 355 twin; certified 26 April 1989; known as **SuperStar** in North America; supplied to UK Ministry of Defence as **AS 350BB**, designated **Squirrel HT. Mk 1** with normal instrumentation and **Squirrel HT. Mk 2** with provision for pilot's NVGs.

AS 350B1, 350BA and 350B versions can be upgraded to 350B2.

AS 350B3 Ecureuil: Current variant; 632 kW (847 shp) Arriel 2B and wide-chord tail rotor; optimised for high-altitude operation; uprated transmission and digital engine control. First flight (F-WWPB) 4 March 1997; French VFR certification December 1997 after 150 hour test programme first delivery, to Osterman Helicopter, January 1998; Australian and New Zealand certification achieved late 1998. By 1 June 2001, 180 AS 350B3s were in service with 120 customers, in 25 countries; 100th AS 350B3 delivered 14 December 1999 to Japanese customer and 400th in 2003.

AS 350D AStar: Version of 350B for North American market; improved LTS101 engine, with 6 per cent more power, offered early 1999.

AS 350 Firefighter: Conair system able to pick up water load in 30 seconds while hovering over water; demonstrated 1986; Isolair system certified September 1995.

AS 550C3 Fennec: Light anti-tank version armed with four TOW missiles; has same power plant as AS 550A3 and 2 hour 45 minute endurance.

AS 550A3 Fennec: Light attack version of AS 350B2 with 3-hour endurance; powered by 632 kW (847 shp) Arriel 2B; standard features include taller landing gear, sliding doors; NVG-compatible cockpit; reinforced airframe; provision for armoured seats and engine cowlings; can be armed with single 20 mm cannon or two rocket launchers.

AS 550U3 Fennec: Military utility version with Arriel 2B power plant; capable of performing observation, commando and transport duties.

EC 130: *Described separately.*

CUSTOMERS: Total 3,028 AS 350s, AS 355s and EC 130s ordered by at least 70 countries by January 2001, of which 2,675 then in service; 1,948 single-engined Ecureuils in service at May 2000.

Military customers include Singapore armed forces (10 AS 550B2 and 10 AS 550C2); Australia (RAAF 18 for training; RAN six utility); Danish Army (12 AS 550C2 with ESCO HeliTOW system ordered 1987, delivered 1990); French Army ALAT (originally expressed need for up to 100 to replace Alouette IIs); French CEV (flight test centre), and FBS Ltd, which ordered 38 AS 350BBs, with deliveries beginning November 1996 and service entry in April 1997. Of these, 26 are Squirrel HT. Mk 1s with Defence Helicopter Flying School at Shawbury and 12 are NVG-compatible HT. Mk 2s with 670 Squadron, Army Air Corps, at Middle Wallop; by June 2001 the fleet had amassed 100,000 hours. Other orders include 11 for Brazilian Navy for liaison and patrol work; 30 for Brazilian Air Force for training; 36 for Brazilian Army (16 for liaison and 20 for training and firefighting); six AS 350B3 for Chilean Army; eight for China (see also CHAIC Z-11) and 40 AS 350B2 options for US Customs Service (augmenting five B2s delivered from 21 October 1998 to supplement 19 earlier AStars); six AS 550C3; 13 AS 350B3 ordered by United Arab Emirates in 2000; and four AS 350B3 for the South African Police Air Wing, of which last two delivered in January 2002. Orders in 1997–2003 were 102, 73, 95, 137, 103, 132 and 81. Deliveries in 2003 were 122.

COSTS: AS 350B3 US$1.4 million (2002).

DESIGN FEATURES: Conventional pod-and-boom design with power plant above cabin; fin, underfin and twin horizontal stabilising surfaces. Starflex bearingless glass fibre main rotor head; all versions now have lifting section composites main rotor blades. Main rotor turns at 394 rpm, tail rotor at 2,086 rpm.

FLYING CONTROLS: Single fully powered controls with accumulators to delay manual reversion following hydraulic failure until airspeed can be reduced; cyclic trim by adjustable stick friction; inverted aerofoil tailplane to adjust pitch attitude in climb, cruise and descent; saucer fairing on rotor head to smooth wake; swept fins above and below tail. Variety of autostabilisers and autopilots available (see Avionics).

STRUCTURE: Main rotor head and much of airframe of glass fibre and aramids; main rotor blades automatically manufactured in composites; self-sealing composites fuel tank in military versions.

LANDING GEAR: Steel tube skid type. Taller version standard on military aircraft. Emergency flotation gear optional.

POWER PLANT: AS 350B2 powered by one 546 kW (732 shp) Turbomeca Arriel 1D1 with transmission rated at 440 kW (590 shp) for T-O; AS 350B3/AS 550C3 powered by one 632 kW (847 shp) Turbomeca Arriel 2B turbine engine controlled by FADEC, with transmission rated at 500 kW (671 shp) for T-O; AS 350D powered by AlliedSignal LTS101 sold only in USA. Plastics fuel tank (self-sealing on AS 550) with capacity of 540 litres (143 US gallons; 119 Imp gallons). Optional 475 litre (125.5 US gallon; 104.5 Imp gallon) auxiliary tank in cabin.

Soloy offers conversion using Rolls-Royce 250-C30M turboshaft; see *Jane's Aircraft Upgrades*.

ACCOMMODATION: Two individual bucket seats at front of cabin and four-place rear bench standard, optional two-place front bench seat; optional ambulance layout; large forward-hinged door on each side of versions for civil use; optional sliding door at rear of cabin on port side; (sliding doors standard on military version); baggage compartment aft of cabin, with full-width upward-hinged door on starboard side; top of baggage compartment reinforced to provide work platform on each side.

SYSTEMS: Hydraulic system includes four single-body servo units, operating at 40 bar (570 lb/sq in) pressure, and accumulators to delay reversion to manual control; electrical system includes a 4.5 kW engine-driven starter/generator, a 24 V 16 Ah Ni/Cd battery and a ground power receptacle; cabin air conditioning system optional.

AVIONICS: *Comms:* VHF/AM radios, HF/SSB transponder and ICS.

Flight: VOR/ILS, ADF, marker beacon receiver and DME. SFIM PA 85T31 (IFR), GPS; full-colour LCD VEMD (vehicle and engine multifunction display) on AS 350B3 and AS 550C3.

EQUIPMENT: Options include 907 kg (2,000 lb) capacity cargo sling (1,160 kg; 2,557 lb for AS 350B2/550, 1,400 kg; 3,086 lb on AS 350B3), 135 kg (297 lb) capacity electric hoist (optional 204 kg; 450 lb on AS 350B3), a TV camera for aerial filming, SX-16 searchlight, Wescam and FLIR observation systems, and a 735 litre (194 US gallon; 161 Imp gallon) Simplex agricultural spraytank and boom system.

ARMAMENT: (AS 550C3): Provision for wide range of weapons, including 20 mm Giat M621 gun, FN Herstal TMP twin 7.62 mm and 12.7 mm machine gun pods, Thomson Brandt 68.12 launchers for twelve 68 mm rockets, Forges de Zeebrugge launchers for seven 2.75 in rockets, and ESCO HeliTOW anti-tank missile system.

DIMENSIONS, EXTERNAL:
- Main rotor diameter ... 10.69 m (35 ft 0¾ in)
- Main rotor blade chord ... 0.35 m (13¾ in)
- Tail rotor diameter ... 1.86 m (6 ft 1¼ in)
- Tail rotor blade chord: AS 350D ... 0.185 m (7¼ in)
  - AS 350B2/B3/550C3 ... 0.205 m (8 in)
- Length: overall, rotors turning ... 12.94 m (42 ft 5½ in)
  - fuselage ... 10.93 m (35 ft 10½ in)
- Width: fuselage (max) ... 1.80 m (5 ft 10¾ in)
  - overall, blades folded (horizontal stabiliser span) ... 2.53 m (8 ft 3¾ in)
- Height overall: AS 350/B2/B3/D ... 3.14 m (10 ft 3½ in)
  - AS 550C3 ... 3.34 m (10 ft 11½ in)
- Tailplane span ... 2.53 m (8 ft 3½ in)
- Skid track: AS 350/B2/B3/D ... 2.17 m (7 ft 1½ in)
  - AS 550C3 ... 2.28 m (7 ft 5¾ in)
- Cabin doors (civil versions, standard, each):
  - Height ... 1.15 m (3 ft 9¼ in)
  - Width ... 1.10 m (3 ft 7¼ in)

DIMENSIONS, INTERNAL:
- Cabin: Length ... 2.42 m (7 ft 11¼ in)
  - Width at rear ... 1.65 m (5 ft 5 in)
  - Height ... 1.35 m (4 ft 5 in)
  - Floor area ... 2.60 m^2 (28.0 sq ft)
- Baggage compartment volume ... 1.0 m^3 (35 cu ft)

AREAS:
- Main rotor disc ... 89.75 m^2 (966.1 sq ft)
- Tail rotor disc ... 2.72 m^2 (29.25 sq ft)

WEIGHTS AND LOADINGS:
- Weight empty: AS 350B2 ... 1,171 kg (2,582 lb)
  - AS 350B3 ... 1,175 kg (2,590 lb)
  - AS 350D ... 1,123 kg (2,475 lb)
  - AS 550C3 ... 1,220 kg (2,689 lb)
- Max T-O weight:
  - AS 350B2/B3/550C3 ... 2,250 kg (4,960 lb)
  - AS 350D ... 1,950 kg (4,300 lb)
- Max weight with slung load:
  - AS 350B2 ... 2,500 kg (5,511 lb)
  - AS 350B3/AS 550C3 ... 2,800 kg (6,173 lb)
  - AS 350D ... 2,100 kg (4,630 lb)
- Max disc loading:
  - AS 350D with slung load ... 23.4 kg/m^2 (4.79 lb/sq ft)
  - AS 350B2/550 clean ... 25.1 kg/m^2 (5.13 lb/sq ft)
  - AS 350B2 with slung load ... 27.9 kg/m^2 (5.71 lb/sq ft)
  - AS 350B3/550C3 with slung load ... 30.6 kg/m^2 (6.28 lb/sq ft)
- Transmission loading at max T-O weight and power:
  - AS 350B2 clean ... 5.11 kg/kW (8.40 lb/shp)
  - AS 350B2 with slung load ... 5.68 kg/kW (9.34 lb/shp)
  - AS 350B3/550C3 clean ... 4.50 kg/kW (7.39 lb/shp)
  - AS 350B3/550C3 with slung load ... 5.50 kg/kW (9.04 lb/shp)

PERFORMANCE (AS 350D at normal T-O weight, AS 350B2/B3/550 at 2,200 kg; 4,850 lb):
- Never-exceed speed (VNE) at S/L:
  - AS 350/B2/B3/550 ... 155 kt (287 km/h; 178 mph)
  - AS 350D ... 147 kt (272 km/h; 169 mph)
- Max cruising speed at S/L:
  - AS 350B2 ... 134 kt (248 km/h; 154 mph)
  - AS 350B3 ... 140 kt (259 km/h; 161 mph)
  - AS 350D ... 124 kt (230 km/h; 143 mph)
  - AS 550C3 ... 133 kt (246 km/h; 153 mph)
- Max rate of climb at S/L:
  - AS 350B2 ... 534 m (1,752 ft)/min
  - AS 350B3/550C3 ... 618 m (2,028 ft)/min
- Service ceiling: AS 350B2 ... 4,800 m (15,750 ft)
  - AS 350B3/550C3 ... 5,280 m (17,323 ft)
- Hovering ceiling:
  - IGE: AS 350B2 ... 3,200 m (10,500 ft)
    - AS 350B3/550C3 ... 4,260 m (13,976 ft)
  - OGE: AS 350B2 ... 2,550 m (8,360 ft)
    - AS 350B3/550C3 ... 3,630 m (11,909 ft)
- Range with max fuel at recommended cruising speed, no reserves:
  - AS 350B2 ... 360 n miles (666 km; 414 miles)
  - AS 350B3 ... 357 n miles (661 km; 410 miles)
  - AS 550C3 ... 350 n miles (648 km; 402 miles)

UPDATED

EUROCOPTER EC 130B

TYPE: Light utility helicopter.

PROGRAMME: Derived from AS 350 Ecureuil; PT-1 prototype first flew (unannounced) June 1999 and registered F-WQEV in May 2000; second prototype (F-WQES) assisted in 200 hour test programme, including high-altitude testing in Albuquerque, USA, September 2000; JAA certification 14 December 2000; FAA certification 21 December 2000; revealed only in February 2001, when initial production aircraft (also 3,000th of Ecureuil family) handed over to Blue Hawaiian Helicopters at HeliExpo. Four preproduction helicopters delivered to launch customers in first half of 2001.

CURRENT VERSIONS: **EC 130B4:** Initial production version.

CUSTOMERS: Announced orders include Blue Hawaiian (10 in 2001), Rocky Mountain Helicopters (10 in 2001), Mont Blanc Hélicoptères (two) and five to UK customers; deliveries started in June 2001 and totalled 10 in 2001, 30 in 2002 and 27 orders in 2003.

COSTS: US$1.6 million (2002).

DESIGN FEATURES: As for AS 355B3, but considerably modified external appearance, with windscreen side panels, doors and landing gear from EC 120, plus Fenestron based on EC 135, although symmetrical, carried by redesigned tailboom. Dual hydraulic system from AS 355N. Cabin width increased by 25 cm (10 in) to give additional 23 per cent internal space, accommodating seven or eight seats and enlarged (10 per cent) baggage compartment; optional air conditioning. Retains AS 355B3 engine and transmission, but automatic system (in conjunction with dual-channel FADEC, plus third, digital, back-up channel) matches rotor speed to flight conditions for noise reduction; 84.3 EPNdB flyover rating is 7 dB below ICAO limit and 0.5 dB under special Grand Canyon National Park restrictions achieved by use of parabolic rotor tips.

LANDING GEAR: Emergency flotation system available. Tailskid. Handling wheels.

POWER PLANT: One 543 kW (728 shp) Turbomeca Arriel 2B1 turboshaft with transmission rated at 632 kW (847 shp) for T-O, with FADEC. Fuel capacity 540 litres (143 US gallons; 119 Imp gallons) in single tank.

Eurocopter EC 130B4 *(H Fabre)* ***NEW***/0576249

ACCOMMODATION: Pilot and up to seven passengers in transport role; or pilot and one or two stretcher patients plus two medical attendants in casualty evacuation configuration. Dual controls.

SYSTEMS: 4.5 kW 28 V DC starter-generator. 15 Ah battery; 28 V DC cabin outlet.

AVIONICS: *Comms:* Shadin 8800 T altitude encoder; Bendix/King KT 76C Mode C transponder and NAT AA95 ICS.

Flight: Two Bendix/King KX 165A nav/com/glideslope; Garmin GNS 430 GPS; Artex 100 HM ELT; AIM 1110 HSI; Bendix/King KI 525 HSI; full-colour VEMD (Vehicle and Engine Multifunction Display).

DIMENSIONS, EXTERNAL:
Main rotor diameter ... 10.69 m (35 ft 0¾ in)
Length: overall, rotors turning ... 12.64 m (41 ft 5½ in)
fuselage ... 10.68 m (35 ft 0½ in)
Fuselage max width ... 2.03 m (6 ft 8 in)
Height: to top of fin ... 3.61 m (11 ft 10¼ in)
to top of rotor head ... 3.34 m (10 ft 11½ in)
Tailplane span ... 2.73 m (8 ft 11½ in)
Skid track ... 2.31 m (7 ft 7 in)
Cabin door: Height ... 1.18 m (3 ft 10½ in)
Max width: front ... 0.84 m (2 ft 9 in)
rear ... 0.92 m (3 ft 0¼ in)

DIMENSIONS, INTERNAL:
Cabin: Length ... 2.19 m (7 ft 2¼ in)
Max width ... 1.865 m (6 ft 1½ in)
Max height ... 1.28 m (4 ft 2½ in)
Volume ... 3.7 m^3 (131 cu ft)
Baggage hold volume:
front, port ... 0.29 m^3 (10.2 cu ft)
front, starboard ... 0.25 m^3 (8.8 cu ft)
rear ... 0.57 m^3 (20.1 cu ft)

AREAS:
Main rotor disc ... 89.75 m^2 (966.1 sq ft)

WEIGHTS AND LOADINGS:
Weight empty ... 1,360 kg (3,000 lb)
Max T-O weight: internal load ... 2,400 kg (5,291 lb)
external load ... 2,800 kg (6,173 lb)
Max disc loading:
internal load ... 26.74 kg/m^2 (5.48 lb/sq ft)
external load ... 31.20 kg/m^2 (6.39 lb/sq ft)
Max power loading:
internal load ... 4.42 kg/kW (7.27 lb/shp)
external load ... 5.16 kg/kW (8.48 lb/shp)

PERFORMANCE (two occupants):
Never exceed speed (VNE) ... 155 kt (287 km/h; 178 mph)
Max operating speed ... 127 kt (235 km/h; 146 mph)
Max rate of climb at S/L ... 698 m (2,290 ft)/min
Service ceiling ... 7,010 m (23,000 ft)
Hovering ceiling: IGE ... 5,875 m (19,280 ft)
OGE ... 5,320 m (17,460 ft)

OPERATIONAL NOISE LEVELS (ICAO Annex 16 Ch 8):
Average ... 86.8 EPNdB
Overflight ... 84.3 EPNdB

UPDATED

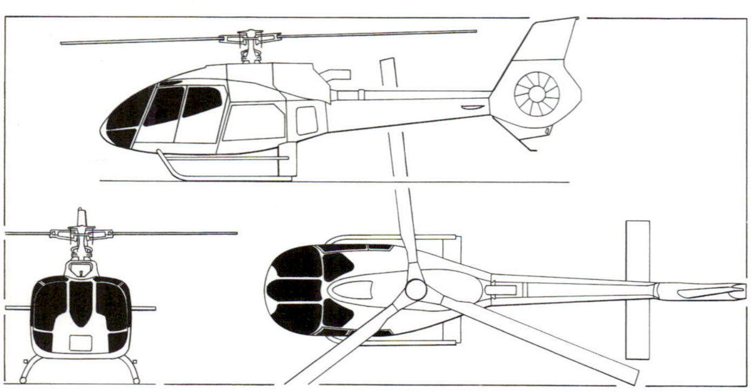

Eurocopter EC 130B (Turbomeca Arriel turboshaft) *(Paul Jackson)* 0062408

EUROCOPTER AS 355 ECUREUIL 2/TWINSTAR AND AS 555 FENNEC

Brazilian Air Force designations: CH-55 and VH-55 Esquilo
Brazilian Navy designation: UH-12B Esquilo
Royal Air Force name: Twin Squirrel

TYPE: Light utility helicopter.

PROGRAMME: Twin-engined version of AS 350. First flight of first prototype (F-WZLA) 28 September 1979; details of early production AS 355E/F versions in 1984–85 *Jane's*; AS 355F superseded by AS 355F1 in January 1984; AS 355F2 certified 10 December 1985; AS 355N powered by two Turbomeca TM 319 Arrius 1A certified in France in 1989, UK and USA in 1992; deliveries of this version began early 1992.

CURRENT VERSIONS: Current production AS 355 Ecureuil 2 and AS 555 Fennec are powered by two Turbomeca Arrius1A turboshafts.

AS 355N Ecureuil 2: Current civil production version, adaptable for passengers, cargo, police, ambulance, slung loads, carrying harbour pilots, working on high-tension cables and other missions; Category A OEI performance; known as **TwinStar** in USA.

Details refer to AS 355N, except where stated.

AS 555UN Fennec: Military utility version and French Army ALAT IFR pilot trainer.

AS 555AN Fennec: Armed version; later machines adapted for centreline-mounted 20 mm gun and pylon-mounted rockets.

Eurocopter AS 355N in police role *(Gérard Deulin)* ***NEW***/0576191

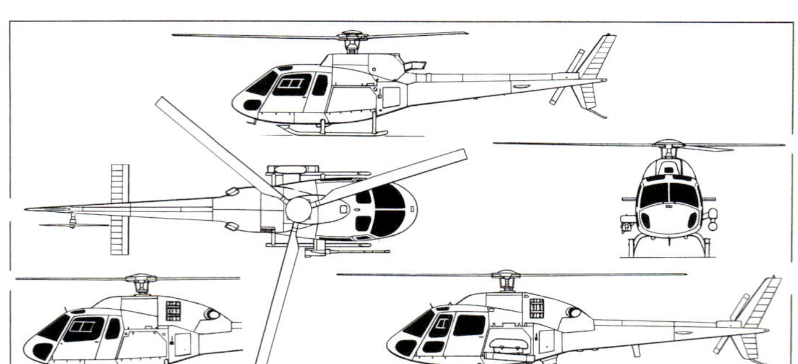

Eurocopter AS 555AN Fennec twin-turbine armed helicopter with side view (top) of single-engined AS 350B3 Ecureuil and scrap view (bottom) of AS 355N civil twin *(Jane's/Dennis Punnett)* 0085583

AS 555MN Fennec: Naval unarmed version. Can carry a 360° chin-mounted radar.

AS 555SN Fennec: Armed naval version for ASW and over-the-horizon targeting, operating from ships of 600 tonnes upwards; armament includes one lightweight homing torpedo or the cannon and rockets of land-based versions; avionics include Honeywell RDR-1500B 360° chin-mounted radar, Thales Nadir 10 navigation system, Thales RDN 85 Doppler and SFIM 85 T31 three-axis autopilot.

CUSTOMERS: Total 592 twin-engined Ecureuils in service at May 2000. French Air Force ordered 52 Fennecs, first eight as AS 355F1s powered by Rolls-Royce 250; flown by 67e Escadre d'Hélicoptères at Villacoublay and other units for communications and, with side-mounted Giat M61 20 mm gun pod, by ETOM 68 in Guyana; delivery of remaining 44 (AS 555AN powered by Arrius) began 19 January 1990; from 24th onwards, provision for centrally mounted 20 mm cannon and T-100 sight, plus Mistral missiles; French Army ordered 18 AS 555UNs for IFR training (delivery from February 1992); Brazilian Air Force has 13 AS 555s: 11 with armament, designated CH-55, and two VIP transports, designated VH-55; Brazilian Navy acquired 11 UH-12Bs; Brazilian AS 555s assembled by Helibrás in Brazil (which see); four more navies ordered eight AS 555s in 1992, two in 1993, four in 1994 and one in 1995. Two AS 355F1s, leased from OSS of Hayes, UK, supplied to the RAF's No. 32 (The Royal) Squadron at Northolt on 1 April 1996 for VIP transport; third subsequently added. Total of six AS 3555Ns ordered by Malaysian Navy on 13 October 2001 to replace Westland Wasps with deliveries at end 2003. Orders in 1997–2003 totalled nine, 20, 10, 13, 17, four and 17. Total of seven delivered in 2003.

DESIGN FEATURES: Starflex main rotor; two engine shafts drive into combiner gearbox containing freewheels, main rotor turns at 394 rpm and tail rotor at 2,086 rpm; otherwise substantially as AS 350.

FLYING CONTROLS: Powered, full dual flying controls without manual reversion; trim by adjustable stick friction; inverted aerofoil tailplane; swept fins above and below tailboom.

STRUCTURE: Light metal tailboom and central fuselage structure; thermoformed plastics for cabin structure.

LANDING GEAR: As for AS 350B2/550.

POWER PLANT: Two Turbomeca Arrius 1A turboshafts, each rated at 357 kW (479 shp) for take-off and 302 kW (406 shp) maximum continuous; 388 kW (520 shp) 30 minutes OEI emergency; 407 kW (547 shp) 2.5 minutes OEI emergency; full authority digital engine control (FADEC) allows automatic sequenced starting of both engines, automatic top temperature and torque limiting and preselection of lower limits for practice OEI operation. Transmission rating 511 kW (686 shp) for T-O. Two integral fuel tanks, with total usable capacity of 730 litres (193 US gallons; 160 Imp gallons), in body structure. Optional 475 litre (126 US gallon; 105 Imp gallon) auxiliary tank in cabin.

ACCOMMODATION: As for single-engined Ecureuil and Fennec.

SYSTEMS: As for AS 350B2/550, except two hydraulic pumps, reservoirs and tandem powered flying control units and electric generators.

AVIONICS: Options include a second VHF/AM and radio altimeter; provision for IFR system, SFIM 85 T31 three-axis autopilot and CDV 85 T3 nav coupler.

EQUIPMENT: Casualty installations, TV, FLIR, searchlight and 204 kg (500 lb) capacity winch available. Cargo hook optional; capacity 1,134 kg (2,500 lb). See also Current Version descriptions.

ARMAMENT (AS 555): Optional alternative weapons include TDA or Forges de Zeebrugge rocket packs, Matra or FN Herstal machine gun pods or a Giat M621 20 mm gun. Naval version carries one homing torpedo in ASW role, or SAR winch.

DIMENSIONS, EXTERNAL: As for AS 350B2/550C3, except:

Width: fuselage	1.87 m (6 ft 1½ in)
overall, blades folded	3.05 m (10 ft 0 in)
Tailplane span	3.05 m (10 ft 0 in)

DIMENSIONS, INTERNAL: *As for AS 350B2/550C3*

WEIGHTS AND LOADINGS:

Weight empty	1,436 kg (3,166 lb)
Max sling load	1,134 kg (2,500 lb)
Max T-O weight: internal load	2,600 kg (5,732 lb)
max sling load	2,600 kg (5,732 lb)
Max disc loading: internal load	29.0 kg/m² (5.93 lb/sq ft)
external load	29.0 kg/m² (5.93 lb/sq ft)
Transmission loading at max T-O weight and power:	
internal load	5.09 kg/kW (8.36 lb/shp)
external load	5.09 kg/kW (8.36 lb/shp)

PERFORMANCE (ISA):

Never-exceed speed (VNE)	150 kt (278 km/h; 172 mph)
Max cruising speed at S/L:	
AS 355N	120 kt (222 km/h; 138 mph)
AS 355UN/AN/MN/SN	118 kt (219 km/h; 136 mph)
Econ cruising speed	117 kt (217 km/h; 135 mph)
Max rate of climb at S/L	384 m (1,260 ft)/min
Rate of climb at S/L, OEI	156 m (512 ft)/min
Service ceiling	3,800 m (12,460 ft)
Service ceiling, OEI	1,450 m (4,760 ft)
Hovering ceiling: IGE	2,000 m (6,560 ft)
OGE	750 m (2,460 ft)
Radius, SAR, two survivors	70 n miles (129 km; 80.5 miles)
Range with max fuel at S/L, no reserves:	
AS 355N	385 n miles (713 km; 443 miles)
AS 355UN/AN	386 n miles (714 km; 444 miles)
AS 355MN/SN	350 n miles (648 km; 402 miles)
Endurance, no reserves:	
one torpedo, or cannon or rocket pods	2 h 20 min
cannon plus rockets	1 h 50 min

UPDATED

EUROCOPTER AS 365N DAUPHIN 2 AND AS 565 PANTHER

US Coast Guard designation: HH-65A Dolphin
Israel Defence Force name: Dolpheen

TYPE: Light utility helicopter.

PROGRAMME: Certified for VFR in France November 1989; 500th Dauphin (all versions) delivered in November 1991 was 83rd AS 365N2. Dauphin family had flown 2.86 million flying hours by June 2001.

CURRENT VERSIONS: **AS 365N2 Dauphin 2:** Recent baseline version; now produced in parallel with AS 365N3.

Details below refer to N2 version.

AS 365N3 Dauphin: Improved hot-and-high version; technology for N3 and N4 versions developed by experimental DGV (Dauphin *Grande Vitesse*) F-WDFK; DGAC certification in October 1997 and first delivery to Noordzee Helikopter Vlaanderen of Belgium in early 1998. Turbomeca Arriel 2C engines offer maximum T-O weight up to 48°C (118°F), rather than N2's 36°C (97°F). FADEC as standard. Orders had reached 40 by mid-2000. Upgrade of AS 365N1 and 365N2 is offered by Eurocopter.

AS 365N4 Dauphin: Redesignated EC 155; see following entry.

AS 366G1 (HH-65A Dolphin): US Coast Guard version (99 built, plus two ex-trials aircraft bought for evaluation by Israel as Dolpheen); trial installation of 895 kW (1,200 shp) LHTEC T800 turboshaft ordered February 1990, but abandoned November 1991 when LTS101 performance promised to improve; re-engining with French Arriels also declined. For description see 1991–92 and earlier *Jane's*.

AS 565UB Panther: Army version; last described in 2001–02 edition.

AS 565SB Panther: Naval version; last described in 2001–02 edition.

Ambulance/EMS: Flight crew of two plus two or four stretchers along cabin sides loaded through rear doors; up to four seats can replace stretchers on one side; doctor sits in middle with equipment; rear doors open through 180° instead of sliding.

Offshore support: Pilot plus up to 12 passengers, autopilot and navigation systems, pop-out floats.

DTV5: *(Démonstrateur Technologie Véhicule):* Demonstrator of new Dauphin variant; new rotor of 13.50 m (44 ft 3½ in) diameter; Tigre transmission; and two 809 kW (1,085 shp) Turbomeca TM 333-2E1 turboshafts. Conversion of Dauphin N4 prototype, subsequently renamed Dolphin VHS (very high speed).

Dolphin FWB: Testbed for fly-by-wire technology used in NH 90.

CUSTOMERS: Total 684 AS 365/366/565 ordered for civil and military use by over 187 customers from 53 countries by January 2001; 599 then in service; totals include 50 produced as Harbin Z-9/9A in China and 101 in HH-65A Dolphin US Coast Guard versions; Chinese production by HAMC (which see) continuing. Dauphin and Panther orders in

Taiwan Police Eurocopter AS 365N2 Dauphin *(Gérard Deulin)* ***NEW***/0576327

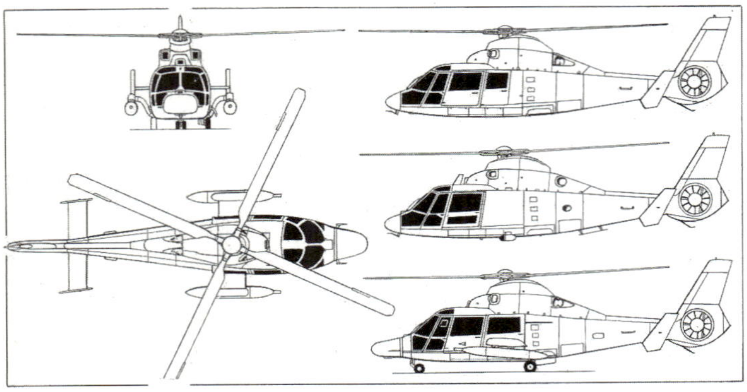

Military Panthers: Eurocopter AS 565AA with side views (centre) of AS 366 HH-65A Dolphin and (top) AS 365N2 Dauphin 2 *(Jane's/Dennis Punnett)* 0507161

1998–2003 totalled 28, 18, 18, 10, seven and 27. Eurocopter has entered offset agreement with LOT of Slovak Republic to allow modification of AS 365s for East European customers; first sale announced 1998. Recent customers include the Mexican Navy, which ordered two Panthers, plus eight options, in 2003, for delivery in 2005, and Greece (six AS 365N3s).

DESIGN FEATURES: Progressively tapered rear fuselage with stabilising surfaces comprising horizontal tail with endplate fins, plus central fin; Fenestron anti-torque rotor; engines above cabin.

AS 365N2 and N3 have Starflex main rotor hub with four blades; 11-blade Fenestron; main rotor blades have quick disconnect pins for manual folding; ONERA OA 212 (thickness/chord ratio 12 per cent) at root to OA 207 (thickness/chord ratio 7 per cent) at tip; adjustment tab near tip; leading-edge of tip swept at 45°; main rotor rpm 350; Fenestron rpm 3,665; rotor brake standard. AS 365N3 can be retrospectively fitted with EC 155's five-blade Spheriflex main rotor and 10-blade Fenestron rotor with unevenly spaced blades to reduce vibration.

FLYING CONTROLS: Hydraulic dual fully powered; cyclic trim by adjustable stick friction damper; fixed tailplane, with endplate fins offset 10° to port; IFR systems available.

STRUCTURE: Mainly light alloy; machined frames fore and aft of transmission support; main rotor blades have CFRP spar and skins with Nomex honeycomb filling; Fenestron duct and fin of CFRP and Nomex/Rohacell sandwich; nose and power plant fairings of GFRP/Nomex sandwich; centre and rear fuselage assemblies, flight deck floor, roof, walls and bottom skin of fuel tank bays of light alloy/Nomex sandwich.

LANDING GEAR: Hydraulically retractable tricycle type; twin-wheel self-centring nose unit retracts rearward; single wheel on each main unit; main legs retract into troughs in fuselage without cover doors; all three units have oleo-pneumatic shock-absorber; mainwheel tyres 15×6.00–6 or 380×150–6 (6 ply) tubeless, pressure 8.60 bar (125 lb/sq in); nosewheel tyres size 13×5.00–4, pressure 5.50 bar (80 lb/sq in); hydraulic disc brakes.

POWER PLANT: *AS 365N2:* Two Turbomeca Arriel 1C2 turboshafts, each rated at 551 kW (739 shp) for T-O and 471 kW (631 shp) maximum continuous, mounted side by side aft of transmission with stainless steel firewall between them. Transmission rating 965 kW (1,294 shp) for T-O and maximum continuous.

AS 365N3: Two FADEC-equipped Turbomeca Arriel 2C turboshafts, each rated at 635 kW (851 shp) for T-O, 597 kW (800 shp) maximum continuous and 729 kW (977 shp) for 30 seconds.

Standard fuel in four tanks under cabin floor and fifth tank in bottom of centre-fuselage; total capacity 1,135 litres (300 US gallons; 249.5 Imp gallons); provision for auxiliary tank in baggage compartment, with capacity of 180 litres (47.5 US gallons; 39.5 Imp gallons); or ferry tank in place of rear seats in cabin, capacity 475 litres (125.5 US gallons; 104.5 Imp gallons); refuelling point above landing gear door on port side. Oil capacity 14 litres (3.7 US gallons; 3.1 Imp gallons).

ACCOMMODATION: Standard accommodation for pilot and co-pilot or passenger in front, and two rows of four seats to rear; high-density seating for one pilot and 12 passengers; VIP configurations for four to six persons in addition to pilot; three forward-opening doors on each side; freight hold aft of cabin rear bulkhead, with door on starboard side; cabin heated and ventilated.

SYSTEMS: Air conditioning system optional. Duplicated hydraulic system, pressure 60 bar (870 lb/sq in). Electrical system includes two 4.8 kW starter/generators, one 24 V 27 Ah battery and two 250 VA 115 V 400 Hz inverters.

AVIONICS: *Comms:* Optional transponder.

Radar: Optional weather radar.

Flight: SFIM 155 duplex autopilot; optional VOR, ILS, ADF, DME, GPS and SFIM CDV 85 autopilot coupler. Optional Bendix/King Mark XXII EGPWS.

Instrumentation: Two-pilot IFR instrument panel; optional EFIS.

EQUIPMENT: Includes 1,600 kg (3,525 lb) capacity cargo hook, and 275 kg (606 lb) capacity hoist with 90 m (295 ft) cable.

DIMENSIONS, EXTERNAL (N2 and N3):
Main rotor diameter: 11.94 m (39 ft 2 in)
Fenestron diameter 1.10 m (3 ft 7¼ in)
Main rotor blade chord: basic 0.385 m (1 ft 3¼ in)
outboard of tab 0.405 m (1 ft 4 in)
Length: overall, rotor turning 13.73 m (45 ft 0½ in)
fuselage 11.63 m (38 ft 2 in)
Width, rotor blades folded 3.25 m (10 ft 8 in)
Height: to top of rotor head 3.47 m (11 ft 4½ in)
overall (tip of fin) 4.06 m (13 ft 4 in)
Wheel track 1.90 m (6 ft 2¾ in)
Wheelbase 3.64 m (11 ft 11¼ in)
Main cabin door (fwd, each side):
Height 1.16 m (3 ft 9½ in)
Width 1.14 m (3 ft 9 in)
Main cabin door (rear, each side):
Height 1.16 m (3 ft 9½ in)
Width 0.87 m (2 ft 10¼ in)
Baggage compartment door (stbd):
Height 0.51 m (1 ft 8 in)
Width 0.73 m (2 ft 4¾ in)

DIMENSIONS, INTERNAL:
Cabin: Length 2.30 m (7 ft 6½ in)
Max width 1.92 m (6 ft 3½ in)
Max height 1.16 m (3 ft 9¾ in)
Floor area 4.20 m² (45.2 sq ft)
Volume 5.0 m³ (176 cu ft)
Baggage compartment volume 1.6 m³ (56 cu ft)

AREAS:
Main rotor disc 111.97 m² (1,205.2 sq ft)
Fenestron disc 0.95 m² (10.23 sq ft)

WEIGHTS AND LOADINGS:
Weight empty: N2 2,281 kg (5,028 lb)
N3 2,302 kg (5,075 lb)
Max T-O weight: N2 4,250 kg (9,369 lb)
N3 4,300 kg (9,480 lb)
Max disc loading: N2 37.9 kg/m² (7.78 lb/sq ft)
N3 38.4 kg/m² (7.87 lb/sq ft)
Transmission loading at max T-O weight and power:
N2 4.40 kg/kW (7.24 lb/shp)

PERFORMANCE:
Never-exceed speed (VNE):
N2 and N3 155 kt (287 km/h; 178 mph)
Max cruising speed:
N2 and N3 150 kt (278 km/h; 173 mph)
Econ cruising speed: N2 137 kt (254 km/h; 158 mph)
N3 148 kt (274 km/h; 170 mph)
Max rate of climb at S/L: N2 408 m (1,339 ft)/min
Rate of climb at S/L, OEI: N2 78 m (256 ft)/min
N3 183 m (602 ft)/min
Service ceiling: N2 3,700 m (12,140 ft)
N3 4,700 m (15,420 ft)
Service ceiling, OEI: N2 1,200 m (3,940 ft)
N3 1,870 m (6,140 ft)
Hovering ceiling:
IGE: N2 2,000 m (6,560 ft)
N3 2,620 m (8,600 ft)
OGE: N2 1,200 m (3,940 ft)
N3 1,150 m (3,780 ft)
Max range with standard fuel:
N2 464 n miles (859 km; 534 miles)
N3 440 n miles (815 km; 506 miles)
Endurance with standard fuel: N2 4 h 18 min
N3 4 h 6 min

UPDATED

EUROCOPTER EC 155B

TYPE: Light utility helicopter.

PROGRAMME: Programme launched September 1996, as further development of Dauphin 2. Announced at 1997 Paris Air Show, when known as AS 365N4 Dauphin 2. First flight 17 June 1997 as conversion of DGV testbed F-WDFK (see AS 365N Dauphin 2 entry); 1,000 flying hours achieved by February 1998. New, EC 155 designation revealed at HAI convention, February 1998. First production EC 155 (F-WWOZ) flew 11 March 1998. JAR certification received 9 December 1998, FAR certification was due late 1999. Certification for single pilot in IFR issued 25 January 2000.

CURRENT VERSIONS: **EC 155B:** Baseline version; replaced in 2002; MTOW 4,800 kg (10,582 lb) and Arriel 2C1 engines.

EC 155B1: Upgraded version from 2002; *as described.* Features include new engine

Eurocopter EC 155B1 police helicopter *(Patrick Penna)* ***NEW***/0576186

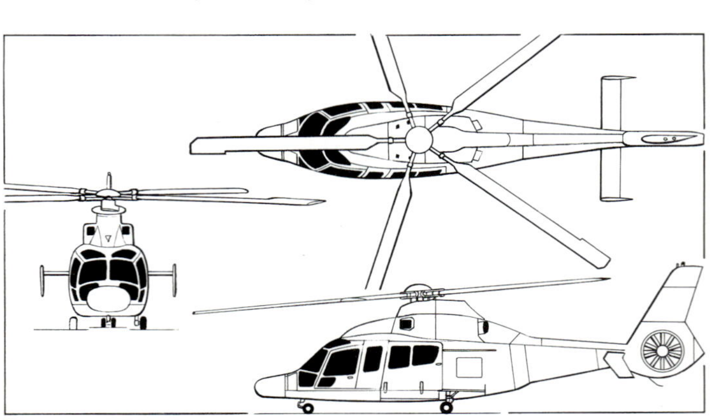

Eurocopter EC 155B *(James Goulding)* 0130860

cowlings, new hydraulic cooling system, chip detectors, cargo fire protection, jettisonable cockpit doors; fixed cockpit footsteps and Thales AHV 16 radar altimeter display. Available with standard Corporate, Offshore and Parapublic equipment packages.

EC 155B HTT: *Hélicoptère Tous Temps* (all-weather helicopter). Technology demonstrator (F-WQEZ), first flown 15 October 2002, for evaluation of helicopter navigation, ground collision avoidance and autopilot systems, aimed at relaxation of rules for rotary-wing IFR operations. Equipment includes a GPS/DGPS-based position-finding system which communicates with a ground station at Eurocopter's Marignane facility; four-axis autopilot, permitting steep approaches in zero visibility; and a mission computer with database-stored terrain maps which manages three-dimensional flight plans, coupled with ground collision avoidance system, to ensure safe terrain clearance during en route flight in zero visibility. Mission information is displayed on two large cockpit displays with three-dimensional mapping. Eurocopter expects that the technologies being evaluated in the HTT will be incorporated in its range of helicopters from 2005.

CUSTOMERS: Sales in 2000–03 totalled seven, 13, five and 10. Total of nine delivered in 2003. First customer, German Border Guard (Bundesgrenzschutz; BGS), ordered 13 for delivery between 16 March 1999 and 2000 and further two in February 2002 for delivery in 2003. German Interior Ministry ordered two for Baden-Württemberg regional government in February 2000 for delivery in March 2001; Hong Kong Government Flying Service ordered five of which last delivered 17 December 2002; Shell Nigeria ordered six in 2000; first aircraft delivered on 25 September 2001. Swedish Helicopter Service ordered three in June 2001 for delivery between October and December 2002. Sales in 2001 including three for COHC (China), two for SFC (Vietnam) and three for SHS (Sweden). Further 13 purchased by BGS in February 2002. AZAL of Azerbaijan ordered four in 2003. Royal Thai Police ordered two on 6 February 2004, for delivery in VIP configuration by January 2005.

COSTS: Royal Thai Police order for two valued at US$25 million (2004).

DESIGN FEATURES: Compared to earlier Dauphin models, has bulged sliding cabin doors, redesigned cabin windows and 40 per cent larger cabin area. Five-blade Spheriflex main rotor and 10-blade Fenestron rotor with unevenly spaced blades to reduce noise.

FLYING CONTROLS: As AS 365N.

STRUCTURE: As AS 365N.

LANDING GEAR: As AS 365N.

POWER PLANT: Two FADEC-equipped Turbomeca Arriel 2C2 turboshafts, each rated at 697 kW (935 shp) for T-O, 645 kW (865 hp) max continuous power and 780 kW (1,046 shp) for 30 seconds. Standard fuel in six tanks, total capacity 1,257 litres (332 US gallons; 277 Imp gallons); provision for auxiliary tank in baggage compartment, with capacity for 180 litres (47.6 US gallons; 39.6 Imp gallons); or ferry tank in place of rear seats in cabin, capacity 460 litres (121.5 US gallons; 101 Imp gallons); refuelling point above landing gear door on port side. Oil capacity 14 litres (3.7 US gallons; 3.1 Imp gallons).

ACCOMMODATION: Standard accommodation for pilot and co-pilot or passenger in front, and three rows of four seats to rear; high-density seating for one pilot and up to 14 passengers; VIP configurations for between four and eight persons in addition to pilot; 12 cabin seats in offshore version; up to six stretchers in casevac role; one crew door and one large sliding door on each side; freight hold aft of cabin rear bulkhead, with door on both sides. Option of hinged cabin door on VIP versions.

SYSTEMS: Electrical system includes two starter/generators, each 160 A, 28 V DC and 43 Ah Ni/Cd battery. Duplicated hydraulic system. Optional 10 kVA alternator.

AVIONICS: *Comms: Corporate:* Twin Collins 422A VHF/AM; Collins TDR 94D transponder; NAT AA20-431 passenger address system; Team BA 1920 passenger interphone. Kannad 406AF ELT. Optional Maritime NPX 138, Collins HF 9100 HF/SSB and Collins TDR 94D transponder; Racal CPT 609 ELT. *Offshore:* Twin Collins 422A; TRD 94D; NPX 138 standard; AA20-431; BA 1920; Kannad 406. Optional second TDR 94D and CPT 609. *Parapublic:* Twin Collins 422A; TDR 94D; AA20-431; BA 1920; Kannad 406. Optional NPX 138, HF 9100, CPT 609.

Radar: Corporate: Telephonics RDR 2000 with VRU. *Offshore:* Telephonics RDR 1400C standard; RDR 2000 optional replacement. *Parapublic:* RDR 1400C.

Flight: Corporate: Universal UNS-1D flight management system. Twin Collins VIR 432 VOR/ILS; Collins DME 442; Collins ADF 462. Optional Honeywell CAS 66A TCAS, Euroavionics Euronav 3 moving map, Racal V 694 voice alerting system. *Offshore:* UNS-1D; Twin Collins VIR 432; DME 442; ADF 462. Optional Trimble TNL 2101 (replacing UNS-1D), Bendix CAS 66A TCAS, Euronav 3, V 694. *Parapublic:* UNS-1D; Twin VIR 432; DME 442; ADF 462. Optional CAS 66A, Euronav 3, V 694.

Instrumentation: As AS 365N, but optional eighth MFD available (152 × 203 mm; 6 × 8 in) for mission equipment display.

Mission: German Border Guard aircraft to have Hellas obstacle warning system.

EQUIPMENT: Standard role packages, included in equipped weight, comprise:

Corporate: Air conditioning, hinged cabin doors, fin and belly strobe lights, retractable passenger steps both sides, improved soundproofing and carpets.

Offshore: Windscreen washer, high-intensity lights, emergency flotation gear, strobe lights, liferaft stowage and improved soundproofing.

Parapublic: Air Equipment SAR hoist (272 kg; 600 lb, 90 m; 295 ft), Spectrolab SX-16 searchlight to port, strobe lights, loudhailer and siren, rappelling rope stays, improved soundproofing and 12 cabin seats.

DIMENSIONS, EXTERNAL:

Main rotor diameter	12.60 m (41 ft 4 in)
Fenestron diameter	1.10 m (3 ft 7¼ in)
Main rotor blade chord: basic	0.385 m (1 ft 3¼ in)
outboard of tab	0.405 m (1 ft 4 in)
Length: overall, rotor turning	14.30 m (46 ft 11 in)
fuselage	12.71 m (41 ft 8½ in)
Tailplane span; and width, rotor blades folded	3.48 m (11 ft 5 in)
Height: to top of rotor head	3.64 m (11 ft 11¼ in)
overall (tip of fin)	4.35 m (14 ft 3¼ in)
Wheel track	1.90 m (6 ft 2¾ in)
Wheelbase	3.91 m (12 ft 10 in)
Cockpit door (each side): Height	1.34 m (4 ft 4¾ in)
Width	1.19 m (3 ft 10¾ in)
Baggage compartment door (stbd):	
Height	0.69 m (2 ft 3¼ in)
Width	0.73 m (2 ft 4¾ in)

DIMENSIONS, INTERNAL:

Cabin: Length	2.55 m (8 ft 4½ in)
Max width	2.10 m (6 ft 10¾ in)
Max height	1.34 m (4 ft 4¾ in)
Floor area	5.09 m^2 (54.8 sq ft)
Volume	6.7 m^3 (237 cu ft)
Baggage compartment: Floor area	2.95 m^2 (31.8 sq ft)
Volume	2.5 m^3 (88 cu ft)

AREAS:

Main rotor disc	124.69 m^2 (1,342.1 sq ft)
Fenestron disc	0.95 m^2 (10.23 sq ft)

WEIGHTS AND LOADINGS:

Weight: empty	2,615 kg (5,765 lb)
equipped: Corporate	3,153 kg (6,951 lb)
Offshore	3,027 kg (6,673 lb)
Parapublic	2,876 kg (6,340 lb)
Max underslung load	1,600 kg (3,527 lb)
Max T-O weight	4,850 kg (10,692 lb)
Max disc loading	38.9 kg/m^2 (7.97 lb/sq ft)

PERFORMANCE:

Never-exceed speed (VNE)	175 kt (324 km/h; 201 mph)
Max cruising speed at S/L	144 kt (267 km/h; 166 mph)
Econ cruising speed at FL60	146 kt (270 km/h; 168 mph)
Max rate of climb at S/L	354 m (1,161 ft)/min
Service ceiling	4,570 m (15,000 ft)
Hovering ceiling: IGE	2,290 m (7,520 ft)
OGE at 4,400 kg (9,700 lb)	2,365 m (7,760 ft)
Max range with standard fuel, no reserves:	
standard tanks	424 n miles (785 km; 487 miles)
one auxiliary tank	488 n miles (903 km; 561 miles)
Endurance: standard tanks	4 h 0 min
one auxiliary tank	4 h 35 min

UPDATED

EUROCOPTER EC 135 AND EC 635

TYPE: Light utility helicopter.

PROGRAMME: First flight on 15 October 1988 of technology prototype, D-HBOX, previously known as BO 108, powered by two Rolls-Royce 250-C20R turboshafts with conventional tail rotor; new all-composites bearingless tail rotor tested during 1990; Eurocopter announced in January 1991 that BO 108 was to succeed BO 105; first flight of second prototype (D-HBEC) powered by two Turbomeca TM 319-1B Arrius, 5 June 1991; production main and tail rotors flight tested during 1992 in preparation for certification programme; design revised late 1992 to increase maximum seating to seven; advanced Fenestron adopted.

Two preproduction prototypes D-HECX and D-HECY made first flights respectively 15 February and 16 April 1994 powered by Turbomeca Arrius 2B and P&WC PW206B intended as production alternatives; third preproduction prototype (D-HECZ) made first flight 28 November 1994, powered by Arrius 2B, and subsequently made type's US debut at HeliExpo '95 in Las Vegas in January 1995; total flight time of first three preproduction EC 135s was nearly 1,600 hours by end 1996, by which time all three preproduction prototypes had been retired. VFR certification to JAR 27 16 June 1996 and to FAR Pt 27 with Category A provisions and for both engine options, 31 July 1996; IFR certification awarded jointly by DGAC (France) and LBA (Germany) on 9 December 1998, while that for CAA attained late 2000; LBA (JAA) single pilot IFR certification achieved 2 December 1999; certified in 17 countries by June 2000; first two production aircraft delivered to Deutsche Rettungsflugwacht on 31 July 1996.

In November 2003 an agreement was signed with Eurocopter Romania for assembly of EC 135's at its Brasov, Romania facility.

CURRENT VERSIONS: **EC 135P1:** Pratt & Whitney engine version (PW206B). First two (D-HQQQ and D-HYYY; c/ns 0005 and 0006) delivered on 31 July 1996 to Deutsche Rettungsflugwacht.

EC 135P2: Introduced August 2001. PW206B2 engine with improved contingency ratings.

EC 135T1: Turbomeca engine version. First (0010/N4037A) delivered to USA in November 1996. Early helicopters had 435 kW (583 shp) Arrius 2B engines, later replaced by 500 kW (670 shp) Arrius 2B1. Uprated Arrius 2B1A engine certified April 2001.

EC 135T2: Deliveries from September 2002. Arrius 2B2 engines.

EC 135 ACT/FHS: Active control technology and flying helicopter simulator; German fly-by-light trials programme; first flight of modified production EC 135 (D-HECV) from Eurocopter's Ottobrunn facility 28 January 2002.

EC 135 APH: Advanced Police Helicopter. Unified mission fit offered by McAlpine Helicopters of UK, 1997; allows simple outfitting with sensors and equipment, according to tasking, using underfuselage pod; typical equipment, including loudspeakers, searchlights, microwave downlink and multisensor turret, can be fitted externally; TV and video equipment carried internally. Pod reduces maximum speed by 5 kt (9 km/h; 6 mph).

EC 635: Military version; mockup was conversion of first preproduction EC 135 (D-HECX). Offered (unsuccessfully) to South Africa and unveiled at Aerospace Africa Air Show on 28 April 1998. First customer was Portuguese Army, which ordered nine EC 635T1s on 22 October 1999 for delivery from June 2001. However, these retrospectively cancelled on 14 August 2002, following delay in post-delivery modifications; six sold to Jordan and delivered from July 2003 onwards; Jordan has further seven options.

CUSTOMERS: Announced customers include Boeing Executive Flight Operations (two in mid-2002) plus one option, Bond, UK (15), Czech Police Aviation Department (eight, for delivery between 2003 and 2008), Deutsche Rettungsflugwacht (German Air Rescue; two); EuropAvia, Switzerland; MFL, France; Helicap of Paris, France (14 EMS versions); Basque Police, Spain; Bavarian Police, Germany (9 with PW206 engines); Sultan of Pahang,

Eurocopter EC 135 underlaying a medical evacuation ***(Wolfgang Obrusnik)*** ***NEW***/0576202

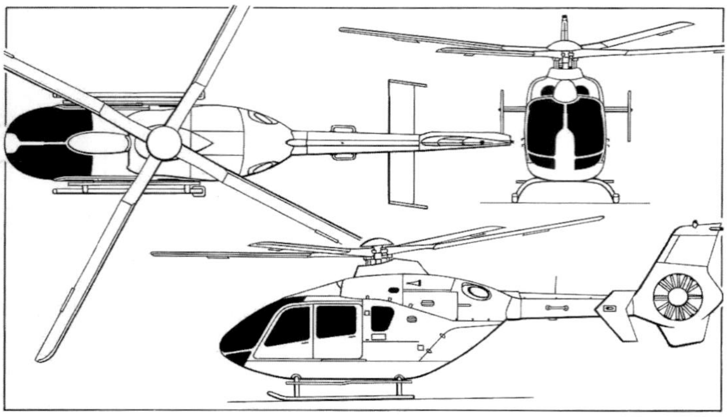

Eurocopter EC 135 seven-seat, twin-engined helicopter *(Jane's/Mike Keep)*
0507504

Malaysia; OAMTC Austrian Automobile Club (five, delivered from 10 August 1997 followed by order for 11 on 27 September 2000 for delivery between December 2000 and early 2002, and three ordered in June 2003); DAO, Argentina; Laidlaw, UK; Hood, UK; Manichi Press, Japan; Elifriulia, Italy; Proeteus, France; Air Service 51, France; Center for Emergency Medicine of Pittsburgh, Pennsylvania (one); Norsk Luftambulanse (six); Petroleum Helicopters Inc (three); Temsco of Ketchikan, Alaska (one); TexAir of Houston, Texas (one); ADAC, Germany (12 for EMS); Osterman; Swedish Police (seven, first aircraft handed over 10 August 2001) Haier, China (one in VIP configuration); Helicap, France (two, in EMS configuration) of which first delivered 12 September 2002); Helicsa, Spain (six, in EMS configuration); and Romanian Ministry of the Interior (five, of which the first three were delivered in December 2003).

Law enforcement users include Mecklenburg-West Pomerania, Germany (two), Rheinland-Palatinate, Germany (one, delivered 10 July 2002, plus two options), Saxony Police (one), Abu Dhabi Police (one), Chile Police (one), Greek Police (two), Travis County Police Department, USA (one), plus UK police forces (seven). German Army ordered 15 at cost of DM95 million in August 1997 for delivery in 2000 to replace Alouette II in training role of which the first was delivered on 13 September 2000. German Border Guard (Bundesgrenzshutz; BGS) ordered nine (subsequently increased to 11) in December 1997; first three delivered 21 September 2000, with eight in service by end of 2001.

Eurocopter expects to win 700 sales out of a world market for 1,350 during the period 1998 to 2007. More than 300 on order by 92 customers in 25 countries by January 2003. 100th delivered (to Bavarian Police) on 16 June 1999, and 250th to Spanish operator CoyotAir on 25 October 2002, and 300th to UK distributor McAlphine Helicopter Ltd, for North Midlands Police, in September 2003, some two-thirds with Turbomeca engines. Orders in 1999–2003 have comprised 30, 40, 38, 48 and 55. Total of 55 delivered in 2003.

COSTS: Operating cost 25 per cent lower than BO 105; development programme funded by Eurocopter Deutschland and Eurocopter Canada, suppliers, and German Ministries of Economics and Research and Technology; flyaway cost US$2.39 million (1996); programme unit cost DM6.67 million (US$4.4 million), Bavarian police (1996). Czech contract for eight valued at €36 million (2003).

DESIGN FEATURES: Designed to FAR Pt 27 including Category A and European JAR 27; pod-and-boom configuration, with Fenestron; forward flight stability by two horizontal and four vertical (fin, underfin and two endplates) surfaces; four-blade FVW bearingless main rotor, single-piece rotor head/mast; rotor rpm are variable; composites blades mounted on controlled flexibility composites arms giving flap, lag and pitch-change freedom; control demands transmitted from rods to root of blade by rigid CFRP pitch cuffs; main rotor blades have DM-H3 and –4 aerofoils with non-linear twist and tapered transonic tips; main rotor axis tilted forward 5°.

Airframe drag 30 per cent lower than BO 105 by clean and compact external shape; cabin height retained by shallow two-stage transmission; vibration reduced by ARIS mounting between transmission and fuselage; all dynamically loaded components to have 3,500 hours MTBR or be maintained on-condition. Fenestron has 10 asymmetrically spaced blades.

Second BO 108 prototype had EFIS-based IFR system; fuselage stretched 15 cm (5.9 in) and interior cabin width extended by 10 cm (3.9 in); main rotor diameter extended to 10.20 m (33 ft 5½ in); for EC 135, tail rotor replaced in 1992 by New Generation Fenestron with 11 fixed flow-straightening vanes in fan efflux designed to avoid momentum losses and improve fan figure of merit; vanes are swept relative to radius and fan has different number of blades to avoid shocks and reduce noise; fan blade tip speed is only 185 m (607 ft)/s; maximum T-O weight increased to 2,720 kg (5,997 lb).

FLYING CONTROLS: Conventional hydraulic fully powered controls with integrated electrical SAS servos; objective is single-pilot IFR with cost-effective stability augmentation. Electric cyclic trim system.

STRUCTURE: Airframe mainly Kevlar/CFRP sandwich composites, except aluminium alloy sidewalls, pod lower module and cabin floor, tailboom and around cargo area; some titanium in engine bay; composites tailplane.

LANDING GEAR: Skid type; ground handling wheels can be fitted.

POWER PLANT: Choice of turboshaft engines. Turbomeca-engined aircraft have two Arrius 2B2s, each giving 452 kW (606 shp) at T-O, 426 kW (571 shp) maximum continuous, 528 kW (708 shp) OEI continuous, 580 kW (777 shp) for 2 minutes with OEI and 609 kW (816 shp) for 30 seconds. Alternative power plant is two Pratt & Whitney Canada PW206B2s, each giving 463 kW (621 shp) at T-O, 419 kW (562 shp) maximum continuous, 528 kW (708 shp) OEI continuous, 580 kW (777 shp) for 2 minutes with OEI and 609 kW (816 shp) for 30 seconds. Both types of engine have FADEC. Transmission rating 616 kW (826 shp) maximum T-O, 567 kW (760 shp) maximum continuous, 353 kW (473 shp) OEI continuous, 513 kW (687 shp) for 2 minutes with OEI and 526 kW (705 shp) for 30 seconds.

Fuel capacity of first 249 aircraft 673 litres (178 US gallons; 148 Imp gallons) of which 663 litres (175 US gallons; 146 Imp gallons) are usable. Capacity 700 litres (185 US gallons; 154 Imp gallons) from No. 250. Additional long-range tank optional, usable capacity 198.5 litres (52.4 US gallons; 43.7 Imp gallons). Optional self-sealing fuel tanks. Oil capacity 8 litres (2.0 US gallons; 1.75 Imp gallons).

ACCOMMODATION: Seven persons, including one or two pilots, in standard version, or six persons in VIP version; optional max capacity of eight. Four-point harnesses for front seats; three-point harnesses for remaining seats. Forward-hinged doors for two front occupants; sliding doors for five persons in cabin. Rear of pod has clamshell doors for bulky items/cargo; flights permissible with clamshell doors removed; optional window in each rear door. Unobstructed cabin interior. EMS variant can accommodate one pilot with two stretcher cases and two seated medical staff/attendants; alternative layouts for one, one, three, or two, one, three, or two, two, two.

SYSTEMS: Redundant 28 V DC electrical supply systems to JAR/FAR 27 standards; two 160 A 28 V starter/generators and 24 V 17 Ah Ni/Cd batteries in Arrius 2B variant, two 200 A 28 V starter/generators and 24 V 25 Ah Ni/Cd batteries in PW206B variant. Fully redundant dual hydraulic systems. NATO standard external power connector.

AVIONICS: Bendix/King Gold Crown and Thales equipment. Total of 10 standard avionics and equipment packages available.

Radar: Provision for integrated weather radar.

Flight: Air data computer; SFIM automatic flight control system (AFCS); GPS. Honeywell combined solid-state flight data and cockpit voice recorder tested on EC 135 late 1998. BGS aircraft have EADS-Dornier Hellas obstacle warning system. Optional Bendix/King Mark XXII EGPWS available for EC 135 APH.

Instrumentation: Liquid crystal dual-screen (Thales SMD45) vehicle and engine management displays with AN equipment.

Mission: Optional FLIR and NVGs for military, police and ambulance roles. BGS EC 135s are first to be equipped with Hellas laser obstacle warning system.

EQUIPMENT: Options include cargo hook, external loudspeakers and searchlights, rescue winch (230 kg; 507 lb) with 50 m (164 ft) cable, emergency floats, sand filter, wire-strike protection system and light armour protection.

ARMAMENT: By 2003, approved to carry FN Herstal HMP 400 12.7 mm machine gun, Giat NC621 20 mm gun or 12-round 70 mm rocket launcher on pylons.

DIMENSIONS, EXTERNAL:

Main rotor diameter	10.20 m (33 ft 5½ in)
Fenestron diameter	1.00 m (3 ft 3¼ in)
Length: overall, rotor turning	12.16 m (39 ft 10¾ in)
fuselage: incl boom	10.20 m (33 ft 6½ in)
excl boom	5.87 m (19 ft 3 in)
Height: overall	3.51 m (11 ft 6 in)
to top of rotor head	3.35 m (11 ft 0 in)
Width:	
without rotor blades (tailplane span)	2.65 m (8 ft 8¼ in)
fuselage (max)	1.56 m (5 ft 1½ in)
Skid length	3.20 m (10 ft 6 in)
Skid track	2.00 m (6 ft 6¾ in)
Ground clearance: fuselage	0.40 m (1 ft 3¾ in)
tailboom underfin	0.66 m (2 ft 2 in)

DIMENSIONS, INTERNAL:

Cabin and cockpit:	
Length: normal	3.06 m (10 ft 0½ in)
with EMS floor extension	4.11 m (13 ft 5¾ in)
Max width	1.50 m (4 ft 11 in)
Max height	1.26 m (4 ft 1½ in)
Floor area	4.3 m^2 (46 sq ft)
Volume: EC 135	4.8 m^3 (170 cu ft)
EC 635	4.6 m^3 (162 cu ft)
Baggage compartment: Length	1.05 m (3 ft 5¼ in)
Max width	1.23 m (4 ft 0½ in)
Max height	0.70 m (2 ft 3½ in)
Floor area	1.20 m^2 (12.9 sq ft)
Volume	1.1 m^3 (39 cu ft)

AREAS:

Main rotor disc	81.71 m^2 (879.5 sq ft)
Fenestron disc	2.84 m^2 (30.52 sq ft)

WEIGHTS AND LOADINGS:

Weight empty: EC 135	1,490 kg (3,284 lb)
EC 635	1,530 kg (3,373 lb)
Max fuel: standard (No. 250 on)	560 kg (1,235 lb)
with long-range tank	720 kg (1,587 lb)
Max external load	1,360 kg (2,998 lb)
Max T-O weight: normal	2,835 kg (6,250 lb)
with external load	2,900 kg (6,393 lb)
Max disc loading: normal	34.7 kg/m^2 (7.11 lb/sq ft)
with external load	35.5 kg/m^2 (7.27 lb/sq ft)
Transmission loading at max T-O weight and power:	
Normal T-O	4.61 kg/kW (7.57 lb/shp)
T-O with external load	4.71 kg/kW (7.73 lb/shp)

PERFORMANCE (A: Arrius 2B2, B: PW206B2, at normal MTOW):

Never-exceed speed (VNE):	
A, B	140 kt (259 km/h; 161 mph)
Max cruising speed: A, B	138 kt (256 km/h; 159 mph)
Econ cruising speed: A	129 kt (239 km/h; 148 mph)
B	124 kt (230 km/h; 143 mph)
Max rate of climb at S/L: A, B	457 m (1,500 ft)/min
Rate of climb at S/L, OEI: A, B	65 m (215 ft)/min
Max certified altitude: A, B	3,050 m (10,000 ft)
Service ceiling, OEI: A	3,050 m (10,000 ft)
B	2,925 m (9,600 ft)
Max certified hovering ceiling, IGE:	
A, B	3,050 m (10,000 ft)
Hovering ceiling, OGE: A, B	2,195 m (7,200 ft)
Range at S/L, standard fuel (700 litres):	
A	335 n miles (620 km; 385 miles)
B	349 n miles (646 km; 401 miles)
Ferry range with long-range tank:	
A	431 n miles (798 km; 460 miles)
B	450 n miles (833 km; 517 miles)
Endurance at S/L, standard fuel (700 litres):	
A	3 h 24 min
B	3 h 47 min

UPDATED

EUROCOPTER EC 120 B COLIBRI

English name: Hummingbird
Spanish Air Force designation: HE.25
Chinese designation: HC120

TYPE: Light utility helicopter.

PROGRAMME: Definition phase of original P120L launched 15 February 1990; subsequently redesigned with 500 kg (1,102 lb) lower gross weight and new engine and rotor; development contract signed October 1992; redesignated EC 120, January 1993; design definition completed mid-1993; assembly of first of two prototypes began at Eurocopter France at Marignane in early 1995; first flight (F-WWPA) 9 June 1995; second prototype (F-WWPD) flown 17 July 1996; certification to JAR 27 was achieved on 16 June 1997 (following Arrius 2F engine approval by DGAC on 22 January 1997); FAR Pt 27 certification on 28 January 1998; operations in cold weather certified late 1998; first production EC 120 (c/n 1005) flew 5 December 1997 before delivery to Japanese distributor Nosaki on 23 January 1998 for eventual use (as JA120B) by Nosaki Sangyo, Osaka.

Eurocopter has 61 per cent share and responsible for instrument panel, landing gear, seats, rotor system, transmission, final assembly, flight test and certification; CATIC (China National Aero-Technology Import & Export Corporation in the form of Hafei Aviation Industry Company; 24 per cent) for cabin structure and doors, engine cowlings, pod central and intermediate structure and fuel system; and Singapore Technologies Aerospace (15 per cent) for tailboom, fin, horizontal stabiliser, Fenestron, general doors and instrument pedestal. Joint design team working in Eurocopter France at Marignane. Global support system, with 13 product centres, being initiated. Australian assembly centre established as part of offsets for 2001 purchase of Tiger attack helicopters by Australian Army, with first two Australian-assembled EC 120 BSS completed in 2003. Agreement signed with AVIC II of China in November 2003 for establishment of production line at Harbin with target output of 20 per year. By August 2002, total fleet time stood at 139,000 hours.

CUSTOMERS: More than 300 ordered by 211 customers in 28 countries by late 2002. Recent customers include the Spanish Air Force, which ordered 15 in 1999 for basic training; deliveries began 26 July 2000 and were completed in July 2001; and Air Logistics, which took delivery of six in 2001 for offshore operations in the Gulf of Mexico. Three ordered for Indonesian Navy and 12 for Indonesian Air Force; deliveries began June 2001. 100th aircraft handed over 19 April 2000 to a German customer; 200th to a Swedish customer on 29 March 2001; 300th to Australian customer 8 November 2002. Orders in 1999–2003 were 36, 61, 81, 38 and 38. Total of 41 delivered in 2003. Target production is for 900 by 2008; total market estimated at 1,600 to 2,000 during production life.

COSTS: US$840,000 (1999). Spanish contract for 15 valued at €15 million (1999), including training and spares.

DESIGN FEATURES: Pod-and-boom layout; horizontal stabilisers, fin and underfin for directional stability; anti-torque Fenestron. Three-blade main rotor on Spheriflex hub rotating clockwise (nominal rotor speed 406 rpm) integrated with main shaft and transmission; two-stage reduction gear; rotor brake; eight asymmetrically positioned bladed New Generation Fenestron (nominal tail rotor speed 4,567 rpm). External noise almost 7 dB below ICAO limits. Single level cabin floor. Engine mounted to left of main rotor mast to improve balance by counteracting main rotor downwash.

FLYING CONTROLS: Control forces on collective and cyclic reduced by three electrically actuated hydraulic servos operating at 37 bar (537 lb/sq in). Tab on each main rotor blade trailing-edge at three-quarter span.

STRUCTURE: Main rotor blades have carbon fibre rib with foam filler and glass fibre skin, plus stainless steel attachment bushes and leading-edge. Titanium alloy Spheriflex head and shaft made as single composites assembly; metal centre-fuselage; light alloy skid landing gear; crashworthy seats and fuel system. Light alloy tail rotor shaft.

LANDING GEAR: 'Moustache' configuration of fixed skids, having sweptback forward supports and full-width boarding step. Optional 1.50 m (4 ft 11 in) skis for snow operations.

POWER PLANT: One Turbomeca TM 319 Arrius 2F engine selected for first 300 EC 120s; rating 376 kW (504 shp) for T-O, 335 kW (449 shp) max continuous. Transmission rating 330 kW (442 shp). Fuel capacity 416 litres (109.9 US gallons; 91.5 Imp gallons) in two tanks (one located beneath cabin floor and one above baggage compartment); usable fuel 406 litres (107 US gallons; 89.3 Imp gallons).

ACCOMMODATION: Pilot and four passengers (or pilot, patient and paramedic in HEMS configuration); two front seats and one rear bench seat. Large door, front, starboard; hinged front and rear sliding door, port. Compartment below engine on same level as cabin floor, accessible from cabin, by external side door and rear door. Seating conforms to new FAR Pt 27 requirements for 30 *g* vertical and 18 *g* horizontal deceleration. Dual controls optional. Accommodation ventilated and heated; air conditioning optional.

SYSTEMS: VEMD (vehicle and engine multifunction display), fitted as standard, is a fully duplex three-module processing system using two glass screens to monitor performance and maintenance requirements. Electrical system includes 4.5 kW 28 V DC starter/generator and Ni/Cd battery.

AVIONICS: To customer's requirements. Six standard packages available for passenger transport (two), law enforcement, training (two) and utility.

Comms: Bendix/King: KX 165 nav/com/glideslope and secondary KY 196A VHF, Jolet JE2 NG ELT, Bendix/King KT 76A transponder and Team SIB 120 interphone in typical passenger configuration.

Flight: Trimble TNL 2101 GPS with moving map in typical passenger configuration.

Instrumentation: Standard fit of VEMD, ASI, NR/NF dual indicator and altimeter.

EQUIPMENT: Optional Aerazur emergency flotation bags, and cable cutters. Optional pilot-controllable swivelling landing light. Optional windscreen wipers.

DIMENSIONS, EXTERNAL:
Main rotor diameter ... 10.00 m (32 ft 9¾ in)
Main rotor blade chord ... 0.26 m (10¼ in)
Fenestron diameter ... 0.75 m (2 ft 5½ in)
Fenestron blade chord ... 0.06 m (2¼ in)
Length overall, rotors turning ... 11.52 m (37 ft 9½ in)
Fuselage: Length ... 9.60 m (31 ft 6 in)
Max width of cabin ... 1.50 m (4 ft 11 in)

Eurocopter EC 120 B Colibri *(Gérard Deulin)* ***NEW***/0576317

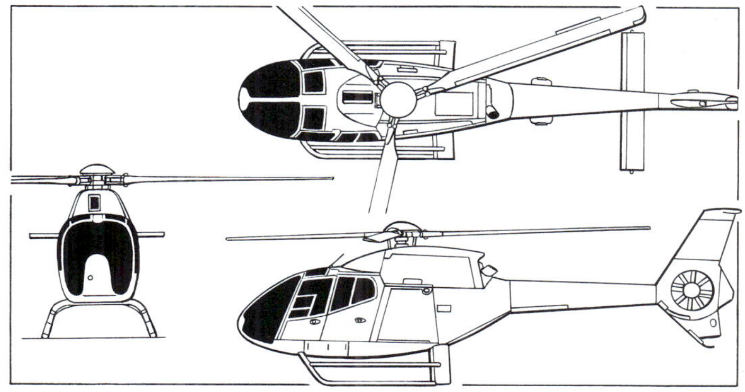

Eurocopter EC 120 B Colibri single-engined light helicopter *(James Goulding)* 0099566

Max depth, excl skids:
to engine cowling ... 1.67 m (5 ft 5¾ in)
to rotor head ... 2.52 m (8 ft 3¼ in)
Tailplane span ... 2.60 m (8 ft 6¼ in)
Height: overall ... 3.40 m (11 ft 1¾ in)
to top of rotor head ... 3.08 m (10 ft 1¼ in)
to top of engine cowling ... 2.23 m (7 ft 3¾ in)
Skid track ... 2.07 m (6 ft 9½ in)
Height of skids ... 0.56 m (1 ft 10 in)
Height of boarding step ... 0.50 m (1 ft 7¾ in)
Large door, starboard: Width ... 1.15 m (3 ft 9¼ in)
Height ... 1.16 m (3 ft 9¾ in)
Front door, port: Width ... 0.81 m (2 ft 8 in)
Height ... 1.16 m (3 ft 9¾ in)
Sliding door, port: Width ... 0.90 m (2 ft 11½ in)
Height ... 1.16 m (3 ft 9¾ in)

DIMENSIONS, INTERNAL:
Cabin: Length ... 2.30 m (7 ft 6½ in)
Max width ... 1.35 m (4 ft 5¼ in)
Max height ... 1.25 m (4 ft 1¼ in)
Floor area ... 3.0 m^2 (32.29 sq ft)
Volume ... 2.14 m^2 (75.6 sq ft)
Baggage compartment: Length ... 1.31 m (4 ft 3½ in)
Max width ... 0.72 m (2 ft 4¼ in)
Max height ... 1.02 m (3 ft 4¼ in)
Volume ... 0.8 m^3 (28.25 cu ft)

AREAS:
Main rotor disc ... 78.54 m^2 (845.4 sq ft)
Fenestron disc ... 0.44 m^2 (4.73 sq ft)

WEIGHTS AND LOADINGS:
Weight empty: standard ... 960 kg (2,116 lb)
typical passenger fit ... 1,007 kg (2,220 lb)
typical law enforcement fit ... 985 kg (2,172 lb)
advanced training fit ... 1,021 kg (2,251 lb)
Max sling load ... 700 kg (1,543 lb)
Max useful load ... 755 kg (1,664 lb)
Max T-O weight: internal load ... 1,715 kg (3,781 lb)
external load ... 1,800 kg (3,968 lb)
Max disc loading:
internal load ... 21.8 kg/m^2 (4.47 lb/sq ft)
external load ... 22.9 kg/m^2 (4.69 lb/sq ft)
Transmission loading at max T-O weight and power:
internal load ... 5.20 kg/kW (8.54 lb/shp)
external load ... 5.45 kg/kW (8.96 lb/shp)

PERFORMANCE (at 1,715 kg; 3,781 lb):
Never-exceed speed (VNE) ... 150 kt (278 km/h; 173 mph)
Cruising speed: max ... 122 kt (226 km/h; 140 mph)
econ ... 103 kt (191 km/h; 119 mph)
Max rate of climb at S/L ... 366 m (1,200 ft)/min
Service ceiling ... 5,180 m (17,000 ft)
Hovering ceiling: IGE ... 2,820 m (9,260 ft)
OGE ... 2,320 m (7,620 ft)
Range, no reserves ... 393 n miles (727 km; 452 miles)
Endurance, no reserves ... 4 h 32 min

UPDATED

EUROCOPTER BK 117

TYPE: Light utility helicopter.

PROGRAMME: Developed jointly by partners; four prototypes, first flight 13 June 1979; one preproduction aircraft, first flight 6 March 1981; first flights of production aircraft 24 December 1981 (JQ1001 in Japan) and 23 April 1982 (in Germany); certified in Germany and Japan 9 and 17 December 1982 respectively, followed by US FAA 29 March 1983 (FAR Pt 29, Categories A and B, including Amendments 29-1 to 29-16); deliveries began early 1983. See 1991–92 and previous editions for earlier A series and B-1. BK 117 fleet reached 1,274,000 flying hours by January 2001, at which time 416 delivered and 379 remaining in service.

In October 2001, Eurocopter announced plans for transfer of production from Donauwörth to Trento, Italy, where seven to be produced annually by Avionline (jointly owned by Aerosud Elicotteri and Helicopters Italia); marketing and delivery remain unchanged.

CURRENT VERSIONS: **BK 117 B-2:** Production completed in 1998.

BK 117 C-1: Turbomeca Arriel 1E2 engines first flown in converted BK 117A/B (F-WMBB) 6 April 1990; German LBA certification 17 January 1992; US FAA certification 7 December 1992; first delivery (to USA) December 1992. French DGAC certification 15 July 1993. Performance similar to that of BK 117 B-2; better hot-and-high performance.

Eurocopter BK 117 C-1 *(Wolfgang Obrusnik)* ***NEW***/0576310

Production BK 117 C-1 additionally features higher transmission and engine OEI ratings, new tail rotor blades and variable rotor speed to improve tail rotor thrust and reduce external noise, and torque matching system to reduce pilot workload.

Prototype C-1 (D-HECD) built 1992; exports began, to USA, December 1992. German LBA certification achieved 28 April 1994, followed by US FAA certification 29 September 1994, Italian RAI certification 24 November 1994, and UK CAA certification 28 July 1995.

Japanese (Kawasaki-built) version, which includes both re-engine and above improvements, was approved 8 June 1995 by Japanese CAB.

Details below apply to BK 117 C-1 version.

BK-117 C-2: Described separately as EC 145.

NBK-117: Designation of aircraft built by IPTN (now Dirgantara) under November 1982 agreement with MBB. Three only; no longer produced.

BK 117 P5: Advanced technology demonstrator in Japan; fly-by-wire control system.

EC-Futura BK117: Technology demonstrator, first flown July 1999 at Ottobrunn, used as part of All Weather Rescue Helicopter programme.

EC 145: See separate entry.

CUSTOMERS: Total of 303 (including prototypes) built by MBB/Eurocopter up to early 2003, comprising 71 BK 117 A-1s, 50 A-3s, 18 A-4s, 104 B-1s, nine B-2s and 51 C-1s; Kawasaki total was 132 by early 2003, mostly B-1/B-2s, but including some 20 C-1s; none further known by early 2004. Total production 435 by early 2004. Orders 1997–2003 were seven, five, three, seven, two, three and six. Customers include TsENTROSPAS of Russia (two, delivered 1998 and December 2000), Greek Fire Brigade (two, delivered June 2001), Telmex of Mexico (2001), First Security Bank of USA (October 2002), Virginia State Police (2002) and Italian EMS operators Aeroveneta, (2000) and Alidaunia (2001).

COSTS: BK 117 B-2 US$2.815 million (1996); BK 117 C-1 US$3.1 million (1996).

DESIGN FEATURES: Pod-and-boom configuration, latter mounted high for cabin access via rear freight doors; tail rotor mounted at fintip; large auxiliary fins at tip of each horizontal stabiliser; engines above cabin.

System Bölkow four-blade main rotor head, almost identical to that of BO 105; main rotor blades similar to but larger than those of BO 105, with NACA 23012/23010 (modified) section; optional two-blade folding. Two-blade semi-rigid tail rotor with MBB-S102E performance/noise optimised blade section; rotor rpm 383 (main), 2,169 (tail).

FLYING CONTROLS: Equipped as standard for single-pilot VFR operation; dual controls and dual VFR instrumentation optional; rotor brake and yaw CSAS standard on German-built versions, optional on Kawasaki aircraft; options include IFR instrumentation, two-axis (pitch/roll) CSAS and Honeywell SPZ-7100 dual digital AFCS. Mast moment indicator discourages excessive cyclic control inputs.

STRUCTURE: Main rotor has one-piece titanium hub with pitch-change bearings; fail-safe GFRP blades with stainless steel anti-erosion strip. Tail rotor, mounted on port side of central fin, has GFRP blades of high-impact resistance. Main fuselage pod and tailboom are aluminium alloy with single-curvature sheets and (on fuselage) bonded aluminium sandwich panels; secondary fuselage components are compound curvature shells of Kevlar sandwich. Engine deck, to which tailboom is integrally attached, forms cargo compartment roof and is of titanium adjacent engine bays. Detachable tailcone carries main fin/tail rotor support, and horizontal stabiliser with offset endplate fins.

Eurocopter responsible for rotor systems, tailboom, tail unit, skid landing gear, hydraulic system, engine firewall and cowlings, powered controls and systems integration; Kawasaki for fuselage, transmission, fuel and electrical systems, and standard equipment. Components single-sourced and exchanged for separate assembly lines at Donauwörth and Gifu; some components and accessories interchangeable with those of BO 105 (which see), from which hydraulic-powered control system is also adapted.

LANDING GEAR: Non-retractable tubular skid type, of aluminium construction. Skids are detachable from cross-tubes. Ground handling wheels standard. Emergency flotation gear, settling protectors and snow skids optional.

POWER PLANT: BK 117 B-2 has two Honeywell LTS 101-750B-1 turboshafts, each rated at 410 kW (550 shp) for take-off and maximum continuous power, 441 kW (592 shp) for 30 minutes with OEI. BK 117 C-1 has two Turbomeca Arriel 1E2 turboshafts each rated at 550 kW (738 shp) for take-off, 516 kW (692 shp) maximum continuous and 574 kW (770 shp) for 2½ minutes OEI.

Kawasaki KB 03 main transmission rated at 736 kW (986 shp) for twin-engine T-O, 632 kW (848 shp) maximum continuous; for single-engine operation, 550 kW (738 shp) allowed for 2½ minutes, 404 kW (542 shp) for maximum continuous (see also BK 117 C-1 improvements).

Fuel in four flexible bladder tanks (forward and aft main tanks, with two supply tanks between), in compartments under cabin floor. Two independent fuel feed systems for engines and common main fuel tank. Total standard fuel capacity 697 litres (184 US gallons; 153 Imp gallons), of which 685 litres (181 US gallons; 151 Imp gallons) are usable; single or twin internal auxiliary fuel tanks, each of 200 litres (52.8 US gallons; 44.0 Imp gallons) capacity, and two external auxiliary fuel tanks, each of 150 litres (39.6 US gallons; 33.0 Imp gallons), optional.

ACCOMMODATION: Pilot and up to six (executive version), seven (Eurocopter standard version) or nine passengers (Kawasaki version). High-density layouts available for up to 10 passengers in addition to pilot. Level floor throughout cockpit, cabin and cargo compartment. Jettisonable forward-hinged door on each side of cockpit, pilot's door having an openable window. Jettisonable rearward-sliding passenger door on each side of cabin, lockable in open position. Fixed steps on each side. Two hinged, clamshell doors at rear of cabin, providing straight-in access to cargo compartment. Rear cabin window on each side. Aircraft can be equipped for offshore support, medical evacuation (one or two stretchers side by side and up to six attendants), firefighting, search and rescue, law enforcement, cargo transport or other operations.

SYSTEMS: Ram air and electrical ventilation system. Fully redundant tandem hydraulic boost system (one operating and one standby), pressure 103.5 bar (1,500 lb/sq in), for flight controls. System flow rate 8.1 litres (2.14 US gallons; 1.78 Imp gallons)/min. Bootstrap/oil reservoir, pressure 1.70 bar (25 lb/sq in). Main DC electrical power from two 150 A 28 V (200 A 28 V for C-1) starter/generators (one on each engine) and a 24 V 25 Ah Ni/Cd battery. AC power provided by inverter; second AC inverter optional; emergency busbar provides direct battery power to essential services, external DC power receptacle.

AVIONICS: *Comms:* VHF-AM/FM, UHF and HF radios to customer's requirements.

Flight: Long-range navaids optional.

Radar: Multirole radar optional.

Instrumentation: Basic instrumentation for single-pilot VFR operation includes airspeed indicator with two electrically heated pitot tubes and static ports, two hydraulic pressure indicators, encoding altimeter, instantaneous vertical speed indicator, 4 in artificial horizon, 3 in standby artificial horizon, HSI, (4 in and 3 in artificial horizons and HSI optional on Kawasaki-built aircraft), gyro magnetic heading system, magnetic compass, ambient air temperature thermometer and clock.

EQUIPMENT: Standard basic equipment includes rotor brake, annunciator panel, two master caution lights, rotor rpm/engine fail warning control unit, fuel quantity indicator and low-level sensor, outside air temperature indicator, engine and transmission oil pressure and temperature indicators, two exhaust temperature indicators, dual torque indicator, triple tachometer, two N1 tachometers, full internal and external lighting, ground handling wheels, pilot's windscreen wiper, floor covering, interior panelling and sound insulation, ashtrays, map/document case, tiedown rings in cabin and cargo compartment, engine compartment fire warning indicator, engine fire extinguishing system, portable fire extinguisher, first aid kit, and single-colour exterior paint scheme.

Optional equipment includes long-range fuel tanks, high-density seating arrangement, bleed air heating system, crashworthy seats, emergency flotation gear, settling protectors, snow skids, main rotor blade folding kit, dual pilot operation kit, stretcher installation, external cargo hook for 1,500 kg (3,307 lb) maximum load, rescue hoist (90 m; 295 ft cable, maximum load 270 kg; 595 lb, variable winch speed), Spectrolab SX 16 remotely controlled searchlight, external loudspeaker, sand filter and kits for rescue, law enforcement, firefighting and VIP transport.

Active anti-vibration system certified in Japan in mid-1997; available as option or for retrofit.

DIMENSIONS, EXTERNAL:
Main rotor diameter .... 11.00 m (36 ft 1 in)
Tail rotor diameter .... 1.96 m (6 ft 5 in)
Tail rotor blade chord .... 0.22 m (8¾ in)
Length: overall, both rotors turning .... 13.01 m (42 ft 8¼ in)
fuselage, tail rotor blades vertical .... 9.93 m (32 ft 7 in)
Fuselage: Max width .... 1.60 m (5 ft 3 in)
Height: overall, both rotors turning .... 3.85 m (12 ft 7½ in)
to top of main rotor head .... 3.36 m (11 ft 0¼ in)
Tailplane span (over endplate fins) .... 2.71 m (8 ft 10¾ in)
Ground clearance: tail rotor .... 1.89 m (6 ft 2½ in)
fuselage .... 0.35 m (1 ft 1¾ in)
Skid track .... 2.50 m (8 ft 2½ in)

DIMENSIONS, INTERNAL:
Combined cabin and cargo compartment:
Max length .... 3.02 m (9 ft 11 in)
Width: max .... 1.59 m (5 ft 2½ in)
min .... 1.27 m (4 ft 2 in)
Height: max .... 1.28 m (4 ft 2½ in)
min .... 0.99 m (3 ft 3 in)
Useful floor area .... 3.70 m^2 (39.8 sq ft)
Volume .... 5.0 m^3 (177 cu ft)
Cabin only: Length .... 2.56 m (8 ft 4¾ in)

AREAS:
Main rotor blades (each) .... 1.76 m^2 (18.94 sq ft)
Tail rotor blades (each): B-2 .... 0.098 m^2 (1.05 sq ft)
C-1 .... 0.108 m^2 (1.16 sq ft)
Main rotor disc .... 95.03 m^2 (1,022.9 sq ft)
Tail rotor disc .... 3.00 m^2 (32.24 sq ft)

WEIGHTS AND LOADINGS:
Basic weight empty: B-2 .... 1,745 kg (3,846 lb)
C-1 .... 1,764 kg (3,890 lb)
Fuel: standard usable .... 558 kg (1,230 lb)
incl first auxiliary tank .... 718 kg (1,583 lb)
Max T-O weight: internal payload .... 3,350 kg (7,385 lb)
external payload .... 3,500 kg (7,716 lb)

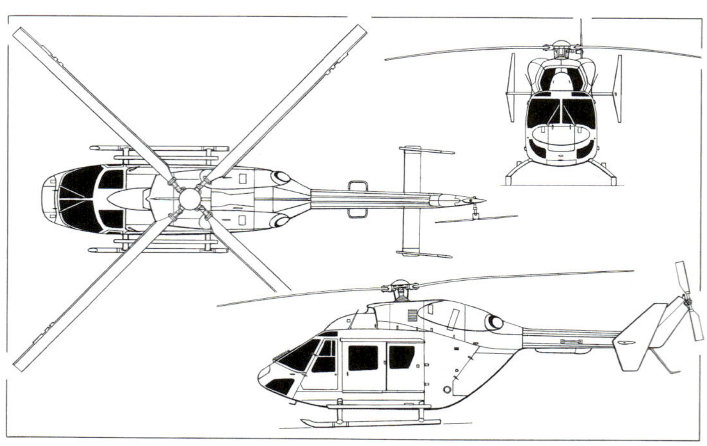
Eurocopter BK 117 B-2 twin-turboshaft multipurpose helicopter *(Jane's/Dennis Punnett)* 0507163

Max disc loading:
- internal payload: 35.25 kg/m² (7.22 lb/sq ft)
- external payload: 36.8 kg/m² (7.54 lb/sq ft)

Transmission loading at max T-O weight and power:
- internal payload: 4.55 kg/kW (7.48 lb/shp)
- external payload: 4.76 kg/kW (7.83 lb/shp)

PERFORMANCE (main values for BK 117 C-1; A: at gross weight of 3,000 kg; 6,614 lb, B: at 3,200 kg; 7,055 lb, C: at 3,350 kg; 7,385 lb):

Never-exceed speed (VNE) at S/L:
- A: 150 kt (277 km/h; 172 mph)
- B, C: 140 kt (259 km/h; 161 mph)

Max cruising speed: A: 135 kt (250 km/h; 155 mph)
- B: 134 kt (248 km/h; 154 mph)
- C: 133 kt (246 km/h; 153 mph)

Econ cruising speed: A: 127 kt (235 km/h; 146 mph)
- B: 126 kt (233 km/h; 145 mph)
- C: 125 kt (231 km/h; 144 mph)

Max rate of climb at S/L: A: 655 m (2,150 ft)/min
- B: 587 m (1,925 ft)/min
- C: 538 m (1,765 ft)/min

Max certified altitude: A, B, C: 4,575 m (15,000 ft)

Service ceiling: A, B: 5,480 m (18,000 ft)
- C: 5,090 m (16,700 ft)

Hovering ceiling IGE (zero wind):
- A: 3,690 m (12,100 ft)
- B: 3,050 m (10,000 ft)
- C: 2,530 m (8,300 ft)

Hovering ceiling IGE (20 kt; 37 km/h; 23 mph) crosswind: A: 3,200 m (10,500 ft)
- B: 2,530 m (8,300 ft)
- C: 1,920 m (6,300 ft)

Hovering ceiling OGE: A: 3,520 m (11,550 ft)
- B: 3,000 m (9,840 ft)
- C: 1,480 m (4,860 ft)

Range: C: 292 n miles (540 km; 336 miles)

Endurance: B, C: 2 h 50 min

UPDATED

EUROCOPTER EC 145

Japanese designation: BK 117C-2

TYPE: Light utility helicopter.

PROGRAMME: Incorporation of EC 135 technology into BK 117 (see separate entry for Eurocopter BK 117) began in 1997; named EC 145 in late 1999, but retains engineering designation **BK 117C-2** and is marketed in Japan as such. First flight of German aircraft (D-HMBK) (unannounced) 12 June 1999; first flight of Japanese prototype 15 March 2000; third prototype (D-HMBL) joined programme 14 April 2000; fourth (D-HMBM) on 27 October 2000. Kawasaki builds tail section; Eurocopter responsible for forward section. Certification by LBA received 12 December 2000; commercial launch at Paris Air Show, June 2001; FAA certification awarded 14 February 2002, coincident with type's formal 'introduction' at HeliExpo, Orlando, Florida.

French Gendarmerie Eurocopter EC 145 ***(Eurocopter/Wolfgang Obrusnik)*** ***NEW***/0567773

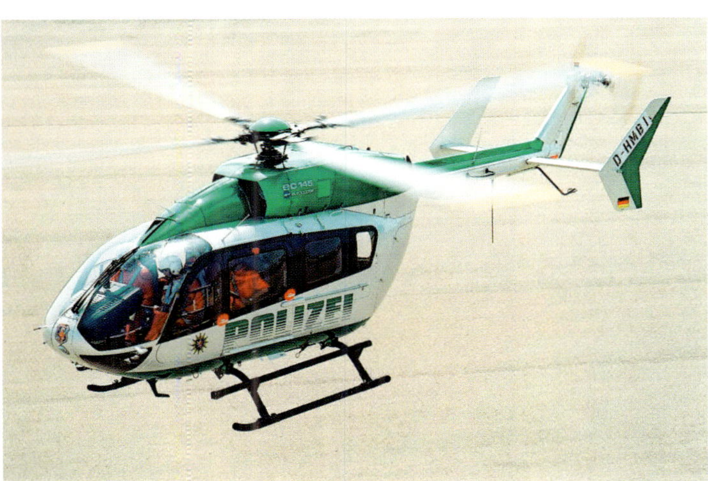

Eurocopter EC 145 of Hesse Police, Germany ***NEW***/0567772

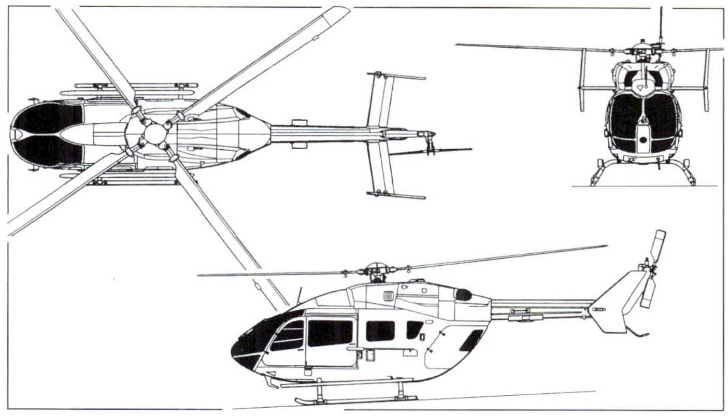

Eurocopter EC 145/Kawasaki BK 117C-2 ***(Paul Jackson)*** 0526898

CUSTOMERS: Launch customer was French Sécurité Civile, which ordered 32 in December 1997 for delivery between 2001 and 2006 to replace Alouette III; two preproduction examples delivered May 2001 for familiarisation; first production example (F-ZBPA) formally handed over at Nimes-Garons 24 April 2002; initial batch of 15 received by mid-2003. Second customer is French Gendarmerie, with firm order for eight placed in 1999; deliveries began 2002 and initial batch of five received by mid-2003. ADAC (German Automobile Club) ordered two in June 2001 to become civilian launch customer; delivery in 2002. Eight ordered in 2001, including four for Rega HEMS in Switzerland, delivered from 21 November 2002. Two for Hesse Police (Germany) delivered from April 2002.

COSTS: Sécurité Civile contract valued at US$170 million. Flyaway cost reported as US$4.9 million (2000).

DESIGN FEATURES: Fuselage redesigned forward of engines; new nose, based on EC 135, provides improved visibility. Main rotor blades are same diameter as BK 117, but have EC 135 profile. Redesigned tail rotor. Optimised for SAR and emergency roles.

STRUCTURE: As BK 117.

LANDING GEAR: As BK 117.

POWER PLANT: Two Turbomeca Arriel 1E2 turboshafts, each rated at 550 kW (738 shp) for take-off, 516 kW (692 shp) maximum continuous and 574 kW (770 shp) for 2½ minutes' OEI. Main transmission rated at 776 kW (1,040 shp) for twin-engine T-O, 632 kW (848 shp) maximum continuous; for single-engine operation 551 kW (739 shp) allowed for 2½ minutes, 404 kW (542 shp) for maximum continuous. Standard fuel contained in main tank, usable capacity 741.5 litres (195.7 US gallons; 163.1 Imp gallons) and left and right supply tanks, usable capacities 59 litres (15.6 US gallons; 13.0 Imp gallons) and 67 litres (17.7 US gallons; 14.7 Imp gallons respectively, for total capacity of 867.5 litres (229.0 US gallons; 190.8 Imp gallons). Optional long-range tanks increase usable capacity to 1,086 litres (287 US gallons; 239 Imp gallons).

ACCOMMODATION: Compared with BK 117, cabin is more spacious through removal of centre post and door supports.

SYSTEMS: As BK 117.

AVIONICS: Standard packages available for single/dual pilot IFR (*as described*), single pilot VFR and single pilot IFR.

Comms: Twin Becker ACU 5100 control panels and VCS 40A VHF-AM, IC 3100 intercom, MST 67A Mode S transponder, Artex C-406-2 ELT. Optional second IC 3100 and C406-2.

Radar: Optional Telephonics RDR 1400C weather radar and SMD 45 display.

Flight: Bendix/King KRA 405B radar altimeter; DMS 44A DME, twin VNS 41A VOR/ILS. Optional second KRA 405B, Chelton DFS 43A direction finder, Freeflight 2101 GPS and dual Canadian Marconi CMA 3000 management system.

Instrumentation: Four SMD 45 LCD displays (two primary, two navigation).

EQUIPMENT: As BK 117.

DIMENSIONS, EXTERNAL: As BK 117C-1, except:
- Main rotor diameter: 11.0 m (36 ft 1 in)
- Tail rotor diameter: 1.96 m (6 ft 5¼ in)
- Length: overall, both rotors turning: 13.03 m (42 ft 9 in)
  - fuselage, tail rotor blades vertical: 10.19 m (33 ft 5¼ in)
- Fuselage max width: 1.84 m (6 ft 0½ in)
- Height: overall, both rotors turning: 3.96 m (13 ft 0 in)
  - to top of main rotor head: 3.45 m (11 ft 3¾ in)
- Tailplane span (over endplate fins): 3.12 m (10 ft 2¾ in)
- Ground clearance: tail rotor: 2.00 m (6 ft 6¾ in)
  - fuselage: 0.45 m (1 ft 5¾ in)
- Skid track: 2.40 m (7 ft 10½ in)

DIMENSIONS, INTERNAL:
- Cabin: Length: excl cockpit: 3.42 m (11 ft 2½ in)
  - incl cockpit: 4.55 m (14 ft 11¼ in)
  - Max width: 1.40 m (4 ft 7 in)
  - Max height: 1.22 m (4 ft 0 in)
  - Floor area: excl cockpit: 4.72 m² (50.8 sq ft)
    - total: 5.43 m³ (58.5 sq ft)
  - Volume (total): 6.8 m³ (240 cu ft)

AREAS:
- Main rotor disc: 95.03 m² (1,022.9 sq ft)
- Tail rotor disc: 3.02 m² (32.48 sq ft)

WEIGHTS AND LOADINGS:
- Weight empty: 1,792 kg (3,951 lb)
- Max usable fuel: standard: 694 kg (1,530 lb)
  - optional: 869 kg (1,916 lb)
- Max T-O weight, internal or external payload: 3,585 kg (7,903 lb)
- Max underslung load: 1,500 kg (3,307 lb)
- Max disc loading: 37.7 kg/m² (7.73 lb/sq ft)
- Transmission loading at max T-O weight and power: 4.88 kg/kW (8.02 lb/shp)

PERFORMANCE (A: at gross weight of 2,200 kg; 4,850 lb, B: at 2,700 kg; 5,952 lb, C: at 3,000 kg; 6,614 lb; D: at 3,585 kg; 7,903 lb):

Never-exceed speed (VNE):
- A, B, C: 150 kt (278 km/h; 173 mph)
- D: 145 kt (268 km/h; 166 mph)

Max cruising speed: A: 138 kt (256 km/h; 159 mph)
- B: 137 kt (254 km/h; 158 mph)
- C: 136 kt (252 km/h; 156 mph)

D ... 133 kt (246 km/h; 153 mph)
Normal cruising speed: A ... 127 kt (235 km/h; 146 mph)
B ... 129 kt (239 km/h; 148 mph)
C ... 130 kt (241 km/h; 150 mph)
D ... 131 kt (243 km/h; 151 mph)
Max rate of climb: A ... 978 m (3,210 ft)/min
B ... 780 m (2,560 ft)/min
C ... 674 m (2,210 ft)/min
D ... 488 m (1,660 ft)/min
Service ceiling: A, B, C ... 5,485 m (18,000 ft)
D ... 5,240 m (17,200 ft)
Hovering ceiling IGE: A, B ... 5,485 m (18,000 ft)
C ... 4,695 m (15,400 ft)
D ... 2,925 m (9,600 ft)
Hovering ceiling OGE: A ... 5,485 m (18,000 ft)
B ... 5,120 m (16,800 ft)
C ... 4,345 m (14,250 ft)
D ... 370 m (1,215 ft)
Range at normal cruising speed, no reserves:
standard fuel:
B ... 380 n miles (705 km; 438 miles)
C ... 378 n miles (700 km; 435 miles)
D ... 367 n miles (680 km; 422 miles)
optional fuel:
C ... 472 n miles (875 km; 543 miles)
D ... 461 n miles (855 km; 531 miles)
Endurance at 65 kt (120 km/h; 75 mph), no reserves:
standard fuel:
B ... 3 h 55 min
C ... 3 h 50 min
D ... 3 h 35 min
optional fuel:
C ... 4 h 50 min
D ... 4 h 30 min

UPDATED

EUROFIGHTER

EUROFIGHTER JAGDFLUGZEUG GMBH

Am Söldermoos 17, D-85399 Hallbergmoos, München, Germany
Tel: (+49 811) 80 15 55
Fax: (+49 811) 80 15 57
e-mail: eurofighter.pr@ibm.net
Web: http://www.eurofighter.com
CHAIRMAN: Pablo de Bergia
MANAGING DIRECTOR: Bob Haslam
CEO: Filippo Bagnato
DIRECTOR, MARKETING SALES SUPPORT: Andrew Lewis
VICE-PRESIDENT, COMMUNICATIONS: Ian Bustin

Eurofighter GmbH formed to manage EFA (European Fighter Aircraft) programme June 1986, followed shortly after by Eurojet Turbo GmbH to manage engine programme. Eurofighter GmbH is owned by Alenia (Italy), BAE Systems (UK) and EADS (formerly CASA; Spain and DASA; Germany); development workshares are 21, 33 and 46 per cent, respectively. Eurojet Turbo participants are Fiat Aviazione (Italy), ITP (Spain), MTU-München (Germany) and Rolls-Royce (UK). Radar is provided by the Euroradar consortium of BAE Systems (UK), FIAR (Italy), EADS (Germany) and ENOSA (Spain). NETMA (NATO Eurofighter and Tornado Management Agency) supervises the programme on behalf of the customer air forces.

All four participating countries agreed on 22 December 1997 to proceed with production and signed the appropriate authorisation on 30 January 1998. The aircraft was formally named Typhoon on 2 September 1998 although, initially, that title was only used for marketing outside Europe. However, on 23 July 2002, Typhoon name was formally adopted by RAF.

On 4 November 1999, (then) four partners announced impending formation of Eurofighter International (EFI) as dedicated sales organisation with target of securing half of available market for 800 combat aircraft over following 30 years. Relations with NETMA remain unchanged. First export commitment issued by Greece on 7 March 2000.

Former subsidiary marketing organisation, Eurofighter International, was disbanded in 2002. Deliveries to customer air forces began following formal type acceptance on 30 June 2003.

UPDATED

German Eurofighter GT003, one of five transferred to JG 73's base at Laage on 30 April 2004 ***NEW***/0589268

EUROFIGHTER TYPHOON

Spanish Air Force designations: C.16 and CE.16
Royal Air Force designations: Typhoon T. Mk 1, T. Mk 1A and F. Mk 2
TYPE: Multirole fighter.

DEVELOPMENT MILESTONES

Requirement issued	Dec 83
Official go-ahead	23 Nov 88
First order (multinational)	22 Dec 97
First prototype material cut	May 98
Rolled out (unofficially)	mid-92
First flight	27 Mar 94
Public debut	11 Jun 95
First flight, pre-production	5 Apr 02
Production go-ahead	30 Jan 98
Certification	30 Jun 03
First flight, production	13 Feb 03
First delivery (Luftwaffe)	17 Feb 03
Entered service (multinational)	30 Jun 03

PROGRAMME: Began with outline staff target for common combat aircraft issued December 1983 by air chiefs of staff of France, Germany, Italy, Spain and UK; initial feasibility study launched July 1984; France withdrew July 1985, shareholdings then being readjusted to 33 per cent each to UK and Germany, 21 per cent Italy and 13 per cent Spain; project definition phase completed September 1986; definitive ESR-D (European Staff Requirement — Development) issued September 1987, giving military requirements in greater detail; definition refinement and risk reduction stage completed December 1987; main engine and weapons system development contracts signed 23 November 1988.

After various political delays were overcome, first (instrumented) production aircraft flew (in Italy) on 5 April 2002; initial 'delivery' was 'presentation' of first full production aircraft GT001/9831 to Luftwaffe at manufacturer's German plant, the aircraft not flying until 13 February 2003, following which, on 17 February, it was delivered to a Luftwaffe technical school for ground crew familiarisation. Also by 17 February, a full production (two-seat) aircraft had flown in each of the four participating countries. On 30 June 2003 Type Acceptance certificate was issued and aircraft was formally handed over to participating air arms at national ceremonies.

In December 2002, unofficial reports claimed UK was attempting to bring forward the ground attack enhancements scheduled for Tranche 3 into the second international production batch, allowing the RAF to cut back or eliminate its final batch of 88 aircraft. Statements throughout 2003 strengthened this impression and agreement on Tranche 2 had not been reached by scheduled date of end-2003. Eventually signed 17 December 2004.

Four partner air forces initially agreed to name their aircraft Eurofighter (or EF2000), using Typhoon name for export marketing. However, on 23 July 2002, RAF also adopted Typhoon during a ceremony at Farnborough Air Show.

Typhoon is participating in several fighter competitions, as detailed in the adjacent table.

No formal roll-out; DA1 and DA2 remained unflown for some 18 months after completion for exhaustive cross-checking of flight control system (FCS). First flight eventually achieved on 27 March 1994, but FCS development resulted in later aircraft flying out of sequence, DA6 being fourth to fly (31 August 1996), by which time earlier aircraft had completed 241 sorties. DA5 flew Eurofighter's 500th sortie (almost 450 hours total airborne time) on 21 October 1997; the 750th sortie (over 630 hours) was flown in June 1998 and the 1,000th in May 1999. Total 1,135 sorties and 931 hours by fourth quarter of 1999; 1,000th hour flown February 2000. Some 90 per cent of flight envelope had been explored by April 1999.

Defensive aids subsystem development contract awarded to Euro-DASS 13 March 1992, but Germany and Spain initially declined to participate; production contract allocated June 2001. 'A' version of ECR 90 radar first flew in nose of modified BAe One-Eleven testbed (ZE433) at Bedford, 8 January 1993; 'C' version is first ECR 90 packaged to fit Eurofighter; flown in One-Eleven from July 1996. First development standard radar delivered to DASA in June 1996; flight testing (in DA5) began 24 February 1997. Radar named Captor in 2000; first production unit delivered in February 2001. Electronically scanned version to fly in One-Eleven by 2003, for possible service in 2010.

Major airframe fatigue test (AFT) fuselage at Ottobrunn achieved 6,000 hours in May 1995 and target of 18,000 hours (equivalent to 6,000 hours of service use) on 4 September 1998.

Production version of EJ200 engine certified 8 March 2001. Thrust vectoring nozzles for EJ200 had undergone 78 hours of bench testing by mid-2000. Study launched by ITP of Spain in 1994, using private funding. Trial system demonstrated 23° 30' vector angle and 110°/s slew rate; advantages of installation for regular use would include 3 per cent

Eurofighter Typhoon T. Mk 1 of No. 29 Squadron, RAF ***NEW***/1047800

Second Spanish two-seat Eurofighter CE.16-02 in the markings of Ala 11 ***NEW***/0589267

First Typhoon T. Mk 1 in RAF hands, BT003/ZJ802 of No. 17 Squadron *(Paul Jackson)* ***NEW***/0573384

reduction in cruise drag, 7 per cent improvement in sustained turn rate, 7 per cent increase in installed thrust, 25 per cent reduction of take-off run and 3 per cent reduction of mission fuel burn.

Three production Tranches subdivided into eight Blocks, each of increased capability: Tranche 1 (148 aircraft) begins with PSP 1 (Block 1) software standard, suitable only for basic air defence training; PSP 2 (Block 2) configuration (2003–04) fully air defence capable (AMRAAM, ASRAAM, AIM-9) through addition of direct voice input (DVI), digital datalink (MIDS) and MLA, plus limited ground attack capability and interim defensive aids subsystem (DASS); PSP 3 (Block 5), representing 2005 production, gives 'swing role' capability with unguided air-to-surface weaponry (augmented by Paveway II and ALARM modifications for only UK; GBU-16 for Italy), enhanced situational awareness and improved survivability through incorporation of PIRATE passive tracking system, helmet-mounted display, towed radar decoy and full sensor fusion. Following five instrumented production aircraft, PSP 1 was to have been achieved in June 2002 on delivery of second two-seater for RAF (production number BT002), which is seventh series production (SPA7). PSP 2 to have followed in December 2003 with SPA44 (fourth RAF single-seat; BS004); and PSP 3 in April 2005 with SPA115 (20th German single-seat; GS020). See adjacent table

Tranche 2 aircraft, totalling 236, begin with Block 8 (Batch 4; 2008 production), having new hardware standard, including updated mission computer; Block 10 (Batch 5; 2008-09 production) will be to EOC 1 (enhanced operational capability) standard (adding Paveway II and III, IRIS-T, AIM-120C-5, plus full-standard DASS and other interoperability features

TRANCHE 1 BREAKDOWN

Batch	Software	Hardware	Block	Qty
IPA				2 S; 3 T
1	PSP1	CA1A	1	31 T
2	PSP2	CA1	2	78 S; 13 T
2	PSP3	CA1	5	16 S; 5 T
Total				**148**

Notes: Batch 1 (including 13 RAF) to be refurbished to Batch 2 standard after initial training use. Batch 2 includes 39 RAF.

IPA3/9803, the German development two-seat Eurofighter ***NEW***/0573380

GS002/9839/3007, first German single-seat Eurofighter for operational service, conducting its maiden flight on 22 October 2004 ***NEW***/0590159

including AIM-9P for Austria) and EOC 2 (Block 15; Batch 6; 2010-11) will add Meteor AAM, Storm Shadow, Brimstone, Taurus and GPS/IN guided weapons and reconnaissance capability; increased MTOW of 24,500 kg (54,013 lb) under consideration.

Tranche 3, comprising 236 aircraft, envisaged as Block 20 and 25, built 2011 to 2014 as Batches 7 and 8, has yet to be defined. By late 2003, UK requirements for earlier incorporation of Tranche 3 capabilities had cast programme into doubt.

In April 2002, industrial partners completed a 12-month study into further development of Eurofighter to maintain operational effectiveness beyond 2040. Options addressed (some previously mooted unofficially) included electronically scanned radar, two-dimensional thrust vectoring, extended range and signature reduction. These to be applicable to Tranche 2 and 3 aircraft. Germany and UK seeking to add long-range attack capability in Tranche 3; BAE Systems announced completion in May 2002 of wind tunnel trials of conformal fuel tanks offering 25 per cent extension of range and mockup shown at Farnborough in July 2002.

CURRENT VERSIONS (GENERAL): **Single-seater:** Standard version. Known to RAF as **Typhoon F. Mk 2** and to Spanish Air force as **C.16**.

Two-seater: Combat-capable conversion trainer. Slightly reduced internal fuel capacity. Known to RAF as **Typhoon T. Mk 1** (first 12 flying aircraft: IPA1/ZJ699 and ZJ800 to ZJ810), **T. Mk 1A** (ZJ811 and subsequent) and to Spanish Air Force as **CE.16**. Also known as EF2000B.

CURRENT VERSIONS (SPECIFIC): **DA1/9829:** (DASA-built at Ottobrunn; airframe No. 01; Luftwaffe serial number 9829.) By road to Manching 11 May 1992; first flight, 27 March 1994, using Phase 0 software; planned transfer to Warton for handling and envelope expansion trials (wearing UK serial number ZH586) cancelled; remained at Manching.

Transferred to Getafe, Spain, 8 April 2003 and reserialled XC.16-02. First Eurofighter flight with IRIS-T (dummy) AAMs 27 August 2003.

DA2/ZH588: (BAE at Warton; airframe No. 02.) First engine run 30 August 1992, first flight 6 April 1994; assigned to envelope expansion, 'carefree' handling and load trials; nine sorties to June 1994, when stood-down for FCS upgrade; reflown 17 May 1995 with Phase 2 version of flight software and made Eurofighter's world public debut (Paris, 11 June 1995, static) and UK debut (Fairford 22 July 1995, flying).

'Carefree handling' trials completed by mid-2002; DASS decoy trials.

DA3/MMX602: (Alenia at Turin/Caselle; airframe No. 04.) First with EJ200 power plants (series -01A) for engine trials (originally scheduled from March 1993 but postponed) and gun/weapon release trials. Initial flight 4 June 1995 on Phase 1 software (20° AoA and +6 *g* limits).

DA4/ZH590: (BAE; airframe No. 03.) First two-seat and first with full avionics (including ECR 90) for radar development, and 'carefree' handling trials. Rolled out 4 May 1994; first flight 14 March 1997; 14 sorties up to end of first phase of trials on 8 September 1997.

DASS ESM trials 2002.

DA5/9830: (DASA.) Construction begun 2 November 1992; maiden flight 24 February 1997; first with ECR 90 radar; autopilot and weapons trials. DS-X radar software standard upgraded to DS-C1 in June 1997, at which time first EJ200-03A engines fitted.

DA6/XCE.16-01: (EADS CASA at Seville.) Second two-seat; performance (including 'carefree' handling), environmental systems, MIDS integration and helmet integration trials.

First flight 31 August 1996; high-temperature trials at Morón, Spain, from 20 July 1998; trials of LCSS (see Systems) cooled aircrew clothing, June 1999; environmental trials (with DA1), including ground tests at Boscombe Down, completed May 2000. Direct voice input trials begun 2001. Refuelling trials with Spanish KC-130H Hercules mid-2002. Suffered double flame-out of 03A-standard engines on 21 November 2002 and abandoned by crew. Total 324 hours in over 340 sorties. Spanish testing resumed in April 2003 upon transfer of DA1 from Germany.

DA7: (Alenia.) Nav/com, performance and weapons integration trials. First flight 27 January 1997. First firing of AIM-9L Sidewinder 15 December 1997; first AIM-120 AMRAAM jettison two days later; first (1,000 litre) external wing tank jettison 17 June 1988.

Further eight ground testing part-airframes and five instrumented production aircraft.

IPA1/ZJ699: (UK.) Two-seat; first production Eurofighter Typhoon. Production numbers PT001/BTT1. Final assembly began 8 September 2000; delivery was due in late 2001; however, first flight postponed to 15 April 2002; first supersonic flight by a production aircraft, 26 April 2002; short lay-up after six sorties for addition of refuelling probe, flight test instrumentation and paint. Defensive aids trials. Planned 414 sorties to FOC.

IPA2/MMX614: (Italy.) Two-seat; PT002; air-to-surface weapon and sensor fusion trials. Final assembly began 3 November 2000. First flight 5 April 2002. Planned 301 sorties to FOC.

IPA3/9803: (Germany.) Two-seat; PT003; air-to-surface weapon integration. Rolled out 2 March 2002. First flight 8 April 2002; 100th flight 17 February 2004. Planned 408 sorties to FOC.

IPA4/C.16-20: (Spain.) Single-seat; PS001; AAM and gun trials; environmental system trials. Planned 170 sorties to FOC. First components delivered March 2003; first flight

Italian instrumented production aircraft IPA2/MMX614 ***NEW***/0573381

27 February 2004 as initial production single-seat Eurofighter. Transferred to Vidsel, Sweden, for cold weather trials 9 December 2004.

IPA5/ZJ700: (UK.) Single-seat; PS002; air-to-air and -surface weapons trials. Planned 223 sorties to FOC. First flight 7 June 2004.

Early production aircraft:

GT001/9831 : (to become 3001) (Germany.) Rolled out 14 November 2002; 'presented' to Luftwaffe 9 December 2002; first flight 13 February 2003; delivered to Air Force Technical School No. 1 at Kaufbeuren for ground training 17 February 2003.

IT001/CSX55092: (Italy.) First flight 14 February 2003. Initially assigned to EMC clearance.

BT001/ZJ800: (UK.) First flight 14 February 2003. To QinetiQ at Boscombe Down for EMC clearance.

ST001/CE.16-01: (Spain.) First flight 17 February 2003.

GS001: (Germany). First single-seat. Rolled out 14 July 2004.

GS002/9839 : (to become 3007) (Germany). Second single seat and first non-instrumental aircraft to fly, on 22 October 2004.

IS001/MMX7235: (Italy). First single-seat.

BS001/ZJ910: (UK). First single-seat.

SS001: (Spain). First single-seat.

CUSTOMERS: Originally declared requirements for 765 (UK and Germany 250 each; Italy 165 and Spain 100). In January 1994, Spain announced firm requirement for 87; Germany revised needs to 180 (including at least 40 fighter-bombers post-2012) under January 1996 work-share agreement. Final total is 620 aircraft, including 1,382 engines; plus options on 90 and 183, respectively (see accompanying table). MoUs 6 and 7 (production and logistical support) of 3 December 1997 and Production Investment contract of 10 December followed on 22 December 1997 by political agreement of all four nations to fund production phase.

Production contract (Supplement 1) for all 620 aircraft (plus 90 options) signed 30 January 1998; Supplement 2 agreement of 18 September 1998 authorised first 148 of these on fixed-price terms, together with 363 engines. UK parliament told in early 2002 that all four participating countries were reviewing aircraft delivery schedule.

Deliveries to air forces to have begun in June 2002 with RAF and Italy; this delayed one year to 30 June 2003, at which time 28 aircraft (eight German, five Italian, four Spanish and 11 UK) were earmarked for delivery before end of 2003, this not being achieved.

IPA4/C.16-20, initial single-seat aircraft from series production, making its first flight on 27 February 2004 ***NEW***/0576659

Italian Eurofighter IT002 arriving at Grosseto on 16 March 2004 for service with 4° Stormo ***NEW***/0589295

Eurofighter IPAS/ZJ700 taking off for its maiden flight at Warton, UK, on 7 June 2004 ***NEW***/0589296

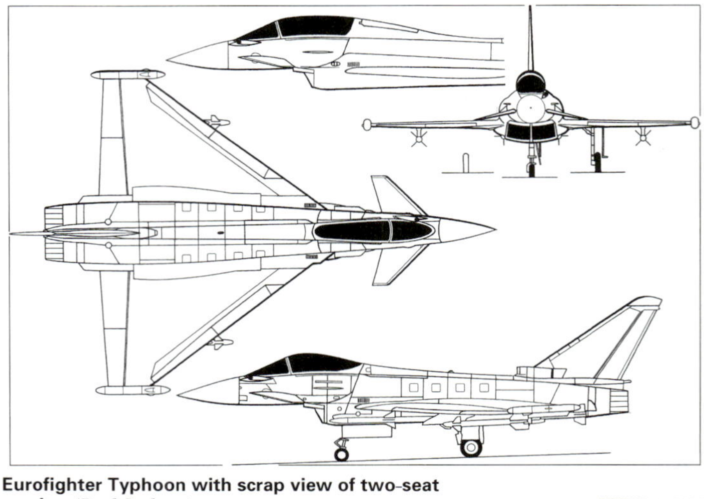

Eurofighter Typhoon with scrap view of two-seat version *(Paul Jackson)* ***NEW***/0573238

German Eurofighter 2000B GT003/9833 taking off on its maiden flight on 30 June 2003, the day of the type's service acceptance ***NEW***/0573383

First Spanish Eurofighter ST001/CE.16-01 wearing the fin badge of Ala 11 ***NEW***/0573382

RAF acceptance of ET001/ZJ800 rescheduled to December 2002, but not effected until 30 June 2003. First RAF unit, under Wg Cdr David Chan, is Operational Evaluation Unit (OEU, also known as No. 17 (Reserve) Squadron) at BAE Warton, where first 16 pilots converted using four aircraft in 1,300 flying hours; formation date of September 2002 delayed to 1 July 2003; first public showing of No. 17 Squadron aircraft (ZJ802 'AB') at Royal International Air Tattoo, Fairford, 18–20 July 2003 (had first flown only on 16th). RAF aircraft issued with Release to Service on 13 May 2004, allowing operational test and OCU syllabus validation flying to begin. In mid-2005 Coningsby received OEU (previously scheduled for January 2004) and formed No. 29 (Reserve) Squadron as an OCU, these with four and five aircraft, respectively (those of No. 29 being delivered from April 2004 onwards, before official formation). Coningsby to host two operational squadrons: No. 3 in 2006 and No. 11 in 2007. Deliveries of Tranche 2 aircraft to No. 6 Squadron at Leuchars begin 2008, two more units to follow at that base, while establishment of two at Leeming remains uncertain. First flight of Typhoon in No. 29 Squadron markings undertaken at Warton on 20 May 2004. First of seven combat squadrons were to have been air defence tasked; remainder to comprise three air defence, two multirole and one offensive support. However, by 2003, RAF was seeking all seven in multirole task by 2011, with Tranche 1 aircraft relegated to training and trials.

RAF announced in May 2000 that no ammunition would be procured for Eurofighter's cannon, despite installation of Mauser in Tranche 1 aircraft. Decision had been reversed by 2002.

German acceptances of squadron aircraft began 4 August 2003. First wing is JG 73 at Laage (formal conversion begun 1 October 2003 under auspices of manufacturer at Manching); initial instructor training completed 2 April 2004; unit moved with five aircraft to Laage on 30 April 2004; followed in 2005, by JG 74 at Neuberg, both in air defence role; third wing is attack unit JG 31 at Nörvenich (from 2007); fourth is air defence JG 71 at Wittmund (from 2009); fifth is JG 33 at Büchel (2012–2015) in attack role; final German delivery in 2014, but further two squadrons (single- or two-seat to be decided) may be obtained for air-to-ground operations.

Italian re-equipment begins with 4° Stormo at Grosseto (9° Gruppo and 20° Gruppo), other bases being Cameri (53° Stormo) and Trapani (37° Stormo). Initial acceptance (IT001) for technical training at Cameri 20 February 2004. First squadron aircraft (IT002/MM55093) accepted by 4° Stormo on 16 March 2004.

Deliveries to Spanish Air Force began 5 September 2003 upon hand-over of first aircraft, CE.16-01, although initial pilot training continued with manufacturer at Getafe until April 2004; formal acceptance ceremony on 9 October 2003. First squadron, from 2004, is Escuadron 113 of Ala 11 at Morón de la Frontera, to where initial three aircraft were delivered on 27 May 2004.

Other details in accompanying table.

Export orders also being solicited; various programmes, requirements and requests for proposals detailed separately. Eurofighter predicts export of 400 Typhoons between 2005 and 2030, worth in excess of £35 billion (2000).

Contract for Eurofighter ground support system signed 24 May 2000. UK has four training rigs, numbered ZJ695 to ZJ698. Aircrew Synthetic Training Aids programme contract, worth €1.07 billion, signed 27 April 2001, including 18 full mission simulators and nine cockpit trainers, deliverable from March 2004 to all four participants.

COSTS: Total development cost estimated as US$21 billion by 1998. Combined development investment estimated (1996) as DM18 billion (US$12.25 billion); non-recurring production investment estimated (1996) as DM12 billion (US$8.15 billion). UK expenditure up to 31 March 2001 on design, development and first batch of 55 aircraft was £5,444 million.

Flyaway price of DM75 million to DM85 million (US$51 million to US$58 million) and system price of DM150 million to DM170 million (US$102 million to US$116 million) at 1996 levels. UK NAO estimated (1996) £15.4 billion for 250 RAF aircraft, including £9.5 billion for production (unit cost £38 million) and this reaffirmed (US$58 million/£37 million) in mid-1998. UK programme cost officially estimated at £16.1 billion (mid-1999); of this £5,444 million had been committed by 31 March 2001.

DESIGN FEATURES: Agile fighter; subsonic instability exceeds 35 per cent (as achieved by Grumman X-29 research aircraft). Collaborative design by BAE, DASA, Alenia and EADS CASA, incorporating some design and technology (including low detectability) from BAe EAP programme. No official requirement for thrust vectoring (TV), supercruise or high order of 'stealthiness'; however, TV nozzle for Eurofighter is under private development; supercruise was 'inadvertently' demonstrated at high altitude in 1997; and RAF has confirmed that aircraft meets low-observables specification.

Low-wing, low-aspect ratio tailless delta with 53° leading-edge sweepback; underfuselage box with side-by-side engine air intakes, each with fixed upper wedge/ramp and vari-cowl (variable position lower cowl lip) with Dowty actuators.

Wing section NACA66 (mod); taper ratio 2.15; dihedral 1°. Foreplane section RAE 104.

Intended service life, 6,000 hours or 25 years. Integrated structural health-and-usage monitoring system (first in any combat aircraft) calculates structural fatigue at 20 positions on the airframe 16 times per second during flight. Maintainability features include 10 mmh/fh and single engine change by four engineers in 45 minutes. Operational turn-round by six ground crew in 25 minutes. Germany, and possibly UK, will contract-out maintenance on 'power by the hour' arrangement with private industry.

FLYING CONTROLS: Two-segment automatic slats on wing leading-edges, inboard and outboard flaperons on trailing-edges; all-moving foreplanes below windscreen; rudder; hydraulically actuated airbrake aft of canopy, forming part of dorsal spine; Liebherr primary flight control actuators. Full-authority quadruplex ACT (active control technology) digital fly-by-wire flight control system (team leader was DASA; Bodenseewerk and ENOSA flight control computer) combines with mission adaptive configuring and aircraft's instability in pitch to provide required 'carefree' handling, gust alleviation and high sustained manoeuvrability throughout flight envelope; pitch control provided by symmetric operation of foreplanes and wing flaperons; roll control primarily by differential operation of flaperons; yaw control by rudder. Cross-feeds between controls to optimise performance and handling qualities provide artificial longitudinal stability; yaw control via rudder; no manual reversion.

STRUCTURE: Fuselage, wings (including inboard flaperons), wing-fuselage fairings, fin and rudder mainly of CFC (carbon fibre composites) except for foreplanes, outboard flaperons and exhaust nozzle fairings (titanium); nose radome and fintip (GFRP); leading-edge slats, wingtip pods, fin base, fin leading-edge, rudder trailing-edge, cockpit side strake and canopy-to-airbrake fairing (aluminium-lithium alloy); and canopy surround (aluminium). CFC constitutes 70 per cent of surface area, with metal 15 per cent, GFRP 12 per cent and other materials 3 per cent.

Individual wing composed of 438 kg (966 lb) carbon fibre, 174 kg (384 lb) aluminium, 179 kg (395 lb) titanium, 33 kg (73 lb) fasteners, 17 kg (37 lb) miscellaneous items and 24 kg (53 lb) tip pod, while installations and equipment add further 120 kg (265 lb) and 154 kg (340 lb), respectively.

Manufacture includes such advanced techniques as superplastic forming and diffusion bonding; EADS CASA-led joint structures team. On development aircraft (only) UK responsible for front fuselage, foreplanes, starboard leading-edge slats and flaperons; Germany the centre-fuselage and fin, Italy the port wing, port leading-edge and flaperons, and stages 2 and 3 of rear fuselage; Spain the rear fuselage stage 1; and Spain and UK the starboard wing; no duplication of tooling.

Work-share on production aircraft involves UK for front fuselage, canards, windscreen, canopy, dorsal fairing inboard flaperons, fin and rear fuselage stage 1; Germany for centre-fuselage; Italy for port wing, outboard flaperons and rear fuselage stages 2 and 3; and Spain for starboard wing and leading-edge slats. Assembly at Caselle (Italy), Getafe (Spain), Manching (Germany) and Warton (UK).

LANDING GEAR: Dowty Aerospace retractable tricycle type with SICAMB mainwheels and Magnaghi/OMA nose gear; Dunlop Aviation wheels, brakes and braking system; Ultra Electronics landing gear computer. Single-wheel main units retract inward into fuselage; nosewheel unit forward. Nosewheel steering is subfunction of DFCS. Tyre sizes 28×9.5R15 main; 18×7.75R6 nose. Elektro Metall braking parachute at base of fin.

POWER PLANT: Two Eurojet EJ200 advanced technology turbofans (each of approximately 60 kN; 13,490 lb st dry and 90 kN; 20,250 lb nominal thrust with afterburning), mounted side by side in rear fuselage with ventral intakes. EJ200-01A initially; -1C for early flight tests; -03A first flown June 1997 (DA5); -03B followed in late 1998; and 03Z in December 1999. Staged EJ200 improvements available (but not funded) to 103 kN (23,155 lb st) (designated EJ 230) and then 117 kN (26,300 lb st). MTU FADEC. Lucas Aerospace fuel management system; Aeroquip GmbH fuel ducts; Autoflug sensors; Teldix flowmeters; VDO computer and gauges; Smiths fuel measurement system. First two development aircraft originally powered by two Turbo-Union RB199-122 (Mk 104E) afterburning turbofans (each more than 71.2 kN; 16,000 lb st). Both were retrofitted with EJ200s in 1998.

EUROFIGHTER TYPHOON REQUIREMENTS BY PARTNER NATIONS

	Delivery	Germany		Italy		Spain		UK		Totals		
Tranche		**S**	**T**	**S**	**T**	**S**	**T**	**S**	**T**	**S**	**T**	**Comb**
1 Batch 11	2003–04	0	9	0	7	1	5	1	13	2	34	36[1,2]
1 Batch 12	2005–06	28	7	19	3	11	3	36	5	94	18	112
2	2006–10	58	10	43	3	27	6	83	6	211	25	236[3]
3	2010–14	61	7	44	2	33	1	75	13	213	23	236[4]
Subtotals		**147**	**33**	**106**	**15**	**72**	**15**	**195**	**37**	**520**	**100**	**620**
		180		**121**		**87**		**232**				
Options		nil		9		16		65				90
Totals		**180**		**130**		**103**		**297**				**710**

Notes: S is single-seat T is two-seat
German aircraft numbered 9803 (IPA3), then 3001 onwards for both single- and two-seat versions; temporary flight test and evaluation serial numbers from 9831 onwards. RAF Tranche 1 are ZJ700 and ZJ910 to ZJ945 for single-seat, and ZJ699 and ZJ800 to ZJ815 for 17 of two-seat aircraft. Austrian aircraft (all single-seat) coded 7L-WA to 7L-WR

[1] Including five instrumented trials aircraft, one (17th from international production) for static testing, designated TR001
[2] Plus 363 engines and 247 radars
[3] Plus 519 engines and 256 radars
[4] Plus 500 engines and 256 radars

EUROFIGHTER TYPHOON EXPORT ORDERS AND PROSPECTS

Nation	Requirement	Timeframe	Status
Requests for proposals (RFP):			
Austria	18		Tenders for 24 fighters invited 10 October 2001. Proposals received from Eurofighter, F-16 and Gripen 23 January 2002 Selection announced 2 July 2002, but suspended 9 September 2002 following fall of government Competition reopened 28 February 2003; parliamentary decision 11 June 2003; presidential approval given 21 August 2003 and contract became offective 22 August for 18 single-seat Tranche 2 aircraft to be delivered in 2007 (four) and 2008 (14)
Greece	60 plus 30 option		Selection confirmed April 1999 EF Office Athens opened July 1999. Intentions confirmed March 2000. Order delayed by financial constraints, April 2001 and competition re-opened in 2004
South Korea	40	Deliveries 2004–2007	EF Office Seoul opened April 1998 RFP received June 1999 Proposal submitted 8 September 1999 Flight evaluation expected early 2000 Final RFP submitted 28 June 2000 Flight Evaluation Team December 2000
Norway	48–60	Deliveries 2010	EF Office Oslo opened March 1999 RFP received 15 February 1999 Proposal submitted 1 June 1999 Offsets proposal submitted 1 December 1999 Competition suspended 2000, but industrial participation agreement signed 28 January 2003 and several Norweigan companies received subcontracts
Requests for information (RFI):			
Australia	60	Decision after 2000	Discussions under way in 1999 F-35 JSF selected
Czech Republic	Possible 36	Long-term requirement	RFI received June 1999 RFI response September 1999 Gripen selected
Netherlands	60–70	Decision 2005 Deliveries 2010	RFI received June 1999 RFI response October 1999
Poland	60	Long-term requirement	RFI received June 1999 RFI response July 1999 F-16 selected
Saudi Arabia	possible 50–70	Decision after 2000	Early discussions under way in 1999
Singapore	20		RFI response submitted late 1999 Evaluation late 2001 Shortlisted 10 October 2003 (with Rafale and F-15)

Internal fuel capacity classified, but believed to total approximately 5,700 litres (1,506 US gallons; 1,254 Imp gallons) in two fuselage tanks and two integral wing tanks. Two-seat trainer lacks forward transfer tank, but partly offsets loss of capacity with auxiliary tank in the enlarged spine. Pressure refuelling point below fuselage, immediately behind air intake. Provision for in-flight refuelling and up to three suspended, external fuel tanks: two 1,000 litre (264 US gallon; 220 Imp gallon) or 2,000 litre (528 US gallon; 440 Imp gallon) underwing, plus one 1,000 litre (264 US gallon; 220 Imp gallon) centreline tank. Only the smaller tanks are rated for supersonic flight.

In early 1998, UK was reported to be designing upper fuselage conformal tanks to increase combat radius to 1,500 n miles (2,778 km; 1,726 miles); work subcontracted to GKN Engage in Australia. Wind tunnel trials of conformal tanks completed May 2002, size 1,500 litres (396 US gallons; 330 Imp gallons) each, providing for additional total of 2,400 kg (5,291 lb) of fuel, for 25 per cent range increase. Conformal tanks can be installed or removed in 75 minutes. Tranche 2 aircraft have structural and piping modifications to accept conformal tanks.

ACCOMMODATION: Pilot(s) on Martin-Baker Mk 16A zero/zero ejection seat(s). Single-piece Aerospace Composite windscreen and single-piece, rear-hinged canopy on both versions. Optional liquid-cooled vest for pilot. Helmet-mounted display. Anti-*g* trousers augmented by pressure breathing system. Two equipment/baggage stowage bays, each 0.01 m^3 (0.35 cu ft), above air intakes, port and starboard.

SYSTEMS: Responsibility for systems delegated to Eurofighter GmbH; participants are UK for electrics, Spain for environmental control and Germany for hydraulics.

Flight control system combines stability and control augmentation; good handling characteristics; high agility; and carefree handling, while incorporating safety features such as low-speed automatic recovery, emergency *g* override, *g* onset limitation, disorientation recovery capability and automatic reversion. System is controlled by four computers and features primary and secondary actuation. FCS features automatic revision through various back-up modes and is integrated with other systems through avionics (STANAG 3910) and utilities control (STANAG 3838) databusses. FCS software updated to Phase 2 in late 1995; Phase 2A 'carefree' handling throughout the subsonic flight envelope, including 25° AoA and over +6 *g* (cleared by DA2 in January 1998); Phase 2B allows carriage of heavy stores and comprises 2B1 (flown April 1999) for 28° AoA and +7.25 *g*, and 2B2 for 30–35° and 9 *g* (first flown 7 July 2000). Phase 3 is IOC standard in 2001 (9 *g* envelope), DA2, DA4 and DA6 receiving Phase 3R1 in that year; Phase 4 covers air-to-surface weapons; and full combat capability Phase 5. FCS incorporates auto-recovery mode ('panic button') for immediate return to straight-and-level flight in emergency. Ada language, apart from time-critical subroutines in Assembler.

Electrical system has Lucas Aerospace as leading supplier and is designed to minimise risk of total power loss by using high level of redundancy and by system positioning. Engine start, systems test, avionics activation and alignment, need no external power. Elements include ZF accessories drive gearbox, two AC and two DC engine-driven Hamilton Sundstrand generators, Honeywell APU in forward port wingroot, Varta battery in corresponding starboard position, two Ferranti Technologies transformer/rectifier units and power converter from same company.

Environmental control system (ECS), with Normalair Garrett as leading supplier, provides conditioned and partly conditioned air to cockpit canopy seal, anti- and de-misting, pilot's anti-*g* clothing, radar, FLIR, avionics and general equipment. Main oxygen provided by molecular sieve generation system (MSOGS). Precooler (at base of fin), heat exchanger, cold air unit and MSOG all located in aircraft's spine. Liquid cooling subsystem (LCSS) connects aircrew vest to ECS.

Hydraulic system (Magnaghi as leading supplier) comprises two independent circuits supplying power to flight control system (Dowty Boulton Paul actuators), landing gear (including nosewheel steering and brakes), port and starboard utilities, gun, canopy, airbrake and refuelling probe.

Utilities control system is integrated within overall system architecture and provides for continuous monitoring and fault detection, comprising front computer, fuel computer, secondary power system computer, landing gear computer and maintenance data panel. Integrated monitoring and recording system constantly checks status of all other systems, airframe and engine, to provide rapid, onboard fault diagnosis; functions include crash-survivable memory, bulk storage device, video-voice recording, mission data loading and portable data store, maintenance data panel, portable maintenance data store and air-to-ground relay of data.

AVIONICS: BAE Systems has overall team leadership for avionics development and integration. All avionics, flight control and utilities control systems integrated through STANAG 3910 databus highways with appropriate redundancy levels, using fibre optics and microprocessors. Some functions activated by direct voice input, with 100 word vocabulary.

Comms: Rohde & Schwarz or Elmer VHF/UHF communications, both secure and non-secure. BAE video and voice recorder. EADS and Cossor IFF; BAE antennas.

Radar: Euroradar ECR 90 Captor coherent, multimode, pulse-Doppler radar.

Flight: Smiths mission data loader and radar altimeter. Litton laser INS/GPS; Marconi SpA microwave landing system, plus ILS and differential global navigation satellite system; BAE Systems TERPROM and ground proximity warning; Tacan; provision for terrain reference navigation; Elmer SpA crash survival memory unit.

Instrumentation: Special attention given to reducing pilot workload. New cockpit techniques simplify safe and effective operation to limits of flight envelope while monitoring and managing aircraft and its operational systems, and detecting/identifying/attacking desired targets while remaining safe from enemy defences. This achieved through high level of system integration and automation, including HOTAS controls; BAE wide-angle (30° azimuth; 25° elevation) HUD able to display, in addition to other symbology, FLIR pictures from PIRATE sensor mounted externally to port side of cockpit; helmet-mounted sight (HMS), with helmet tracking system and Smiths Aerospace direct voice input (DVI), with 200-word vocabulary, for appropriate functions; and three Smiths Industries multifunction head-down colour CRT displays (MHDD) and Smiths glareshield standby displays. EADS, EDS Defence and CANAVA digital map generator; Teldix cockpit interface unit.

Mission: Alenia and Computing Devices nav/attack computers. Eurofirst (FIAR consortium) PIRATE (Passive Infra-Red Airborne Tracking Equipment) port side of windscreen. Secure MIDS datalink. (UK requirement for MIDS reconfirmed in November 2003.) RAF aircraft to have optional reconnaissance capability with a long-range electro-optical pod, for which SR(A) 1368 was issued in 1995.

Self-defence: Advanced integrated defensive aids subsystem (DASS), contracted to Euro-DASS consortium, led by BAE Systems and including Indra (Spain) and Elettronica (Italy); includes RWR and active jamming pod at each wingtip, laser warning receiver (each side, adjacent to windscreen), missile approach warning (wing leading-edges, inboard, and at rear base of fin), Elettronica Aster/GAMESA/CelsiusTech expendables (flares in flap actuator fairings; chaff in aileron actuator fairings) and towed radar decoys. (Germany initially did not join Euro-DASS, but is now to adopt standard equipment, with contract awarded October 2001). Two BAE Systems radar decoys in the starboard wingtip pod, each on a 100 m (320 ft) fibre optic cable. Italy considering Cross Eye ECM system as alternative to towed decoy. Spain also declined to join Euro-DASS, but is now participating; UK and Spain are only nations to have LWR. Initial contract for 103 systems placed June 2001; to be augmented by German requirement and addition of EADS to consortium.

EQUIPMENT: Hella lighting; Logic anti-collision beacons.

ARMAMENT: Total of 13 external stores stations: five (including one wet) under fuselage and four (including one wet) under each wing. Internally mounted 27 mm Mauser gun on starboard side with 150 rounds.

EUROFIGHTER TYPHOON WEAPONS INTEGRATION
(as envisaged before Tranche 2 agreement)

Type	Mission	Manufacturer	Features	Max load	Tranche 1 IOC/FOC	Tranche 2 EOC 1	Tranche 2 EOC 2
27 mm cannon	Short range defence	Mauser	Internally mounted. 150 rounds linkless system	1	X		
Meteor	Beyond Visual Range Air-to-Air Missile (BVRAAM)	EADS, MBDA, Saab	Ramjet propulsion Active radar seeker	6			X
AIM-120A/B AMRAAM	Advanced Medium Range AAM	Raytheon	Active radar seeker	6	X		
AIM-120C 5 AMRAAM	Advanced Medium Range AAM	Raytheon	Active radar seeker	6			X
ERAAM	Extended Range AAM	Raytheon	Dual pulse rocket motor	6	superceded by METEOR		
FMRAAM	Future Medium Range AAM	Raytheon	Liquid fuel ramjet propulsion	6	superceded by METEOR		
AIM-9L/I AIM-9L/I-1	Short range AAM	Raytheon, Ford Aerospace, Bodenseewerkgerätetechnik (BGT)		6	X		
ASRAAM	Advanced Short Range AAM	MBDA	Focal-plane array seeker	6	X		
IRIS-T	Advanced Short Range AAM	BGT	Infrared imaging seeker Thrust vector control	4 (option for 6)	X	X	
ALARM	Anti-radiation Missile	MBDA	Loiter capability	(6) currently are only 4 planned in Tranche 2		?	X
Penguin	Anti-ship Missile	Kongsberg Defence & Aerospace	Mk 3 version range 50 km	4			
Harpoon	Anti-ship Missile	Boeing	Low level flight, active radar seeker	4			
Brimstone	Anti-tank Missile	MBDA	Millimetric wave seeker	(18) currently are only 6 planned in Tranche 2			X
Taurus	Cruise Missile	EADS/Celsius/Bofors Missiles	Terrain-following, range 150–350 km	2			X
Storm Shadow	Cruise Missile	MBDA	Terprom terrain-following, range 350+ km	2			X
Paveway II (UK), GBU-10/16	Laser Guided bomb	Portsmouth Aviation Raytheon	Mk 83 (GBU 16)/Mk 84 (GBU 10) Mk 13/20 (PWII UK) bomb with laser guidance	(4) up to 5 (2000 lbs bombs) 7 (1000 lbs bombs) are planned depending on LDP position	X PW II initial capability X GBU 10/16	X PW II full capability	
Enhanced Paveway II (UK)	GPS and Laser Guided bomb			7		X	
GBU 32 JDAM	GPS Guided bomb	Boeing	Mk 83 bomb	5		X	
Lightning	Laser Designator Pod	Zeiss		1		partially	full
Paveway III (UK), GBU 24	Laser Guided bomb	Raytheon	Mk 84/BLU 109 bombs with advanced laser guidance	(3) up to 5 are planned depending on LDP position			X
BL755	Cluster bomb	Hunting Engineering	147 bomblets	6			
Bombs 500–2000 lb	Unguided bombs			(12) up to 5 (2000 lbs) 7 (1000 lbs) are possible			
Rockets CRV-7	Unguided rockets	Bristol Aerospace, Canada	19 rockets per pod	4			

Notes: Introduction indicated by 'X'

Planned weapons (with maximum load in parentheses) include:

Air-to-air: Meteor (six), AIM-120 AMRAAM (six), ERAAM (six), FMRAAM (six), AIM-9L Sidewinder (six), ASRAAM (six) and IRIS-T (six).

Air-to-air surface. ALARM (six), Penguin (four), Harpoon (four), Brimstone (18), Taurus, Storm Shadow, GBU-10 (four), GBU-16 (four), Paveway III (three), CRV-7 (four pods of 19 rockets each), 500 lb bombs (12) and reduced quantity of larger weapons up to 2,000 lb Mk 84 (four).

Typical weapon combinations for specific roles could include the following:

Air superiority: Six BVRAAMs, two SRAAMs, internal cannon, plus two 2,000 litre and one 1,000 litre external fuel tanks.

Interdictor/Strike: Two cruise missiles; four BVRAAMs, internal cannon plus two 2,000 litre and one 1 000 litre tanks.

Defence suppression: Six ARMs, four BVRAAMs, two SRAAMs, internal cannon and centreline 1,000 litre tank.

Multirole: Two Paveway LGBs, two ARMs, three BVRAAMs, two SRAAMs, internal cannon, laser designation pod and two 2,000 litre and one 1,000 litre tanks.

Close air support: Eighteen Brimstones, four BVRAAMs, two SRAAMs, internal cannon and one 1,000 litre centreline tank.

Maritime attack: Six ASMs, four BVRAAMs, two SRAAMs, internal cannon and one 1,000 litre centreline tank.

Phased introduction of weapons (as envisaged before final determination of Tranche 2) is shown in an accompanying table.

DIMENSIONS, EXTERNAL:
Wing span over ECM pods .... 11.28 m (37 ft 0 in)
Wing aspect ratio 2.5
Length: Fuselage .... 14.89 m (48 ft 10¼ in)
overall .... 15.96 m (52 ft 4¼ in)
Height overall .... 5.28 m (17 ft 4 in)
Wheelbase .... 4.07 m (13 ft 4¼ in)

AREAS:
Wings, gross .... 51.2 m^2 (551.1 sq ft)
Foreplanes (total) .... 2.40 m^2 (25.83 sq ft)

WEIGHTS AND LOADINGS (approx):
Weight empty .... 11,150 kg (24,582 lb)
Internal fuel load .... 4,500 kg (9,920 lb)
External stores load (weapons and/or fuel):
normal .... 6,500 kg (14,330 lb)
overload .... 7,500 kg (16,535 lb)
Max T-O weight: interceptor .... 16,000 kg (35,274 lb)
attack .... 21,000 kg (46,297 lb)
overload .... 23,500 kg (51,809 lb)
Max wing loading: interceptor .... 312.5 kg/m^2 (64.01 lb/sq ft)
attack .... 410.2 kg/m^2 (84.01 lb/sq ft)
overload .... 459.0 kg/m^2 (94.01 lb/sq ft)
Max power loading: interceptor .... 89 kg/kN (0.87 lb/lb st)
attack .... 117 kg/kN (1.14 lb/lb st)
overload .... 130 kg/kN (1.28 lb/lb st)

PERFORMANCE (approx):
Acceleration at low level, 200 kt to M1.0 .... 30 s
Max level speed M2.0
Time to 10,670 m (35,000 ft)/M1.5 .... up to 2 min 30 s
Service ceiling .... 16,765 m (55,000 ft)
Runway requirement .... 700 m (2,300 ft)
T-O run, air combat mission .... 300 m (985 ft)
Combat radius, single-seat:
ground attack, lo-lo-lo .... 325 n miles (601 km; 374 miles)
ground attack, hi-lo-hi with three LGBs, designator pod and seven AAMs .... 750 n miles (1,389 km; 863 miles)
air defence with 3 hour CAP .... 100 n miles (185 km; 115 miles)
air defence with 10 minute loiter .... 750 n miles (1,389 km; 863 miles)
g limits with full internal fuel and four AIM-120s .... +9/−3

UPDATED

EUROMIL

EUROMIL

PARTICIPATING COMPANIES
Mil: see under Russian Federation
Kazan: Russian Federation
GENERAL DIRECTOR: Vladimir Yablokov

In July 2003, Eurocopter withdrew from Euromil as a consequence of a Russian law making it impossible for the European company to hold management or intellectual rights. The Eurocopter share was taken over in equal parts by Mil and Kazan, making Euromil a wholly Russian company.

UPDATED

EUROMIL MI-38

TYPE: Medium transport helicopter.
PROGRAMME: Fuselage of demonstrator (plus two static test airframes) were substantially complete at Kazan plant by mid-1997, but engine and transmission suppliers had not then been selected. Programme then delayed by Russian financial collapse. P&WC now supplying 2,461 kW (3,300 shp) PW127 turboshafts for demonstrator and production-standard PW127T/S variant for Western exports; Klimov TVA-3000 for CIS operators, subject to completion of development. Revised timetable called for Mi-38 demonstrator (PT-1) first flight in 2001; four prototypes to follow by 2003; and series production from 2006, following FAR/JAR 29 certification; this further delayed, maiden flight target having moved to 2002, according to report in September 2001. By October 2002, initial sortie target was January 2003, yet in that month a further delay to mid-2003 was announced, citing burden of extensive ground testing required.

Model of the projected Mi-38 medium transport helicopter *(Paul Jackson)* 0105078

Mil (which see for Mi-38 description) responsible for development, including general design, drawings, component testing and flight testing; Eurocopter leading in flight deck, avionics system and passenger accommodation and responsible for preparation for export from Russia; Kazan manufactures fuselage and rotor blades and undertakes final assembly for domestic market. Major Russian subcontractors include Krasny Oktyabr for transmission and Stupino for rotor head.

EUROTILT

EUROPEAN TILTROTOR PROGRAMME

PARTICIPATING COMPANIES
AgustaWestland: Italy/UK
Eurocopter: International

TYPE: Multimission tiltrotor.
PROGRAMME: Beginning in 1987, the European Commission sponsored feasibility and definition studies for a tiltrotor transport, then known as EuroFAR. It was hoped by Eurocopter that this ground-work could be incorporated in a forthcoming venture, announced in October 1998, for which other European partners were sought. Eurocopter envisaged a 19/20 seat aircraft with an MTOW of 25,000 kg (55,116 lb) and a service-entry date of around 2015. In early 1999 it undertook a study, sponsored by the French defence ministry, into the military potential of tiltrotors.

Eurocopter's concept of a European tiltrotor has been rejected 0093643

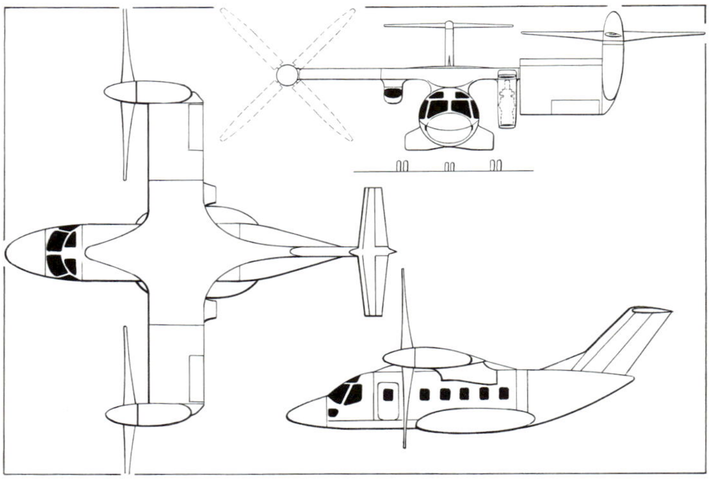

Agusta Erica; current studies are based on the tilting outboard wing (independent of proprotor angle) envisaged by this design *(James Goulding)* 0081740

By 1999, 33 companies from nine European countries had joined with Eurocopter to apply for EC funding, initially to build a test rig for a proposed 10,000 kg (22,046 lb), 12/19-seat tiltrotor, provisionally known as Eurotilt. Envisaged shares in the project, which could have led to a prototype flying in 2004–05 and entry into service in 2008, included France 36 per cent, Spain 25.7 per cent, Germany 22.3 per cent and Italy 11.6 per cent. Work-shares included Fiat Avio for transmission, CASA for wings and Rolls-Royce Turbomeca for RTM 332 engines. Performance parameters were 300 kt (556 km/h; 345 mph) cruising speed over ranges between 200 and 800 n miles (370 and 1,481 km; 230 and 920 miles) and a service ceiling of 7,620 m (25,000 ft). Development cost (1999) estimated as US$1.05 billion, of which US$87.5 million for ground test rig.

Meanwhile, in July 1999, Agusta announced the 10-tonne, 20-seat Erica Tiltrotor and was seeking €90 million of funding from EU's 5th Framework research programme to build a ground test vehicle (GTV). Testing was envisaged five years after GTV go-ahead; prototype first flight in further two years. Backing received from 16 companies, including GKN Westland (UK), ZF Luftfahrttechnik (Germany), IAI (Israel), Aermacchi (Italy), NLR (Netherlands), Gamesa (Spain) and Saab (Sweden).

Erica was second-generation tiltrotor, incorporating improvements over BA609 and V-22. Engines mounted inboard, underwing, driving connecting shafts to proprotors at wingtips. Outboard half of wing was to tilt with proprotors, obviating blocking effect of fixed wing and increasing vertical lift by 12 per cent. This improvement would permit smaller proprotor diameter, increasing cruising speed and allowing almost conventional rolling landings, with outboard wing tilted by only 5 to 7° to prevent ground contact.

In October 1999, European Commission rejected separate funding of competing European tiltrotors and urged merger of Erica with Eurotilt.

Common research project, known as 2Gether (second-generation European tilting highly efficient rotorcraft), submitted to European Commission on 31 March 2000 by Eurocopter, Westland and Agusta (since merged as AgustaWestland). Timetable envisaged initial €95 million study between 2000 and 2004, followed by demonstrator first flight in 2005 (additional €250 million) and production programme (€1,000 million) leading to series manufacture beginning in 2010. Launch version would have had 10-tonne MTOW and carried 20 passengers.

European Commission rejected 2Gether as too costly and insufficiently innovative. New development proposal submitted in March 2001; cost €40 million to €60 million, with industry providing half; features tilting outer wing, as proposed by Agusta; early research devoted to overcoming additional technical difficulties of this configuration; flight demonstration could begin in 2007–08. Study team also to include Mecaer and Teleavio of Italy; Gamesa and Sener of Spain; UK's FHL; ZF of Germany; ONERA, DLR, NLR and CIRA research institutes; Israel Aircraft Industries; and Pratt & Whitney Canada. These joined by Flight Science and Technology Laboratory at Liverpool University, UK, which inaugurated in June 2001 a flight simulator to be used in developing a control system for Eurocopter's Eurotilt submission.

Various research efforts merged into RHILP (Rotorcraft Handling qualities, Interaction and Load Prediction) project, combining Eurocopter (in France) and Eurocopter Deutschland, plus research organisations CIRA (Italy), DLR (Germany), NLR (Netherlands), ONERA (France) and University of Liverpool (UK). Evaluation and risk-minimisation project began March 2000 and so far embraces four work packages: WP1 for handling qualities definition, employing moving base simulator at Liverpool (first four with Bell XV-15 parameters, but progressing to Eurotilt in August 2002); WP2 analysing specific flight phenomena using 1:7 scale model at Eurocopter, Marignane; WP3 for loads alleviation; and WP4 use of Marignane's SPHERE simulator for piloted simulation tests. Completion due in 2003, although RHILP is only one of several Critical Technology Projects envisaged up to 2005 and its results will be incorporated in HOST flight mechanics software configured to represent Eurotilt.

By 2003, AgustaWestland Erica had evolved with smaller, but highly twisted four-blade proprotors of 7.4 m (24 ft) diameter, enabling the aircraft to take-off and land in aeroplane mode. This rotor design is optimised for speeds up to 350 kt (648 km/h; 403 mph), compared with 275 kt (509 km/h; 316 mph) for current tiltrotors. Performance and handling will be further enhanced by allowing the outboard wing panels to be tilted independently of the engine/proprotor pods at their tips, with the related benefits of a wider transition corridor between helicopter and aeroplane and improved autorotation characteristics. Pods will remain firmly coupled to each other, however. Capacity will be 19 passengers, or 22 plus 2.2 tonnes of luggage in STOL mode only.

UPDATED

IBIS

IBIS AEROSPACE LTD

PARTICIPATING COMPANIES
Aero Vodochody: see under Czech Republic
AIDC: see under Taiwan
EUROPEAN SALES OFFICE:
250 70 Odolena Voda, Czech Republic
Tel: (+420 2) 86 03 31 19
Fax: (+420 2) 83 97 00 55
US SALES OFFICE:
804 Water Street, Kerrville, Texas 78028
Tel: (+1 830) 257 82 00
Fax: (+1 830) 257 82 01
e-mail: ibisaerospace@ktc.com
Web: http://www.ibisaerospace.com
VICE-PRESIDENT OF MARKETING: Jeffrey V Conrad

FAR EAST SALES OFFICE:
1141-4, Lane 68, Fu-Hsing North Road, Taichung, Taiwan
Tel: (+886 4) 22 56 28 61
Fax: (+886 4) 22 56 23 25

Ibis Aerospace Ltd is a 50-50 joint venture company created by an agreement of 15 March 1997 under which Aero Vodochody and AIDC are co-developing and producing the Ae 270 Spirit single-turboprop utility aircraft. By early 2003, three prototypes were flying. Final assembly is only in the Czech Republic, at least initially.

UPDATED

IBIS AE 270 SPIRIT

TYPE: Light utility turboprop.

PROGRAMME: Announced by Aero Vodochody of Czech Republic in early 1990, originally as L-270; configuration modified 1991; originally planned in two versions (Ae 270 U and Ae 270 MP), but revised late 1993 and name Ibis introduced; design frozen 1995. Chief designer Jan Mikula. Three flying prototypes (Nos. 1, 3 and 5), plus one each for static and fatigue testing (Nos. 2 and 4, respectively); to be certified under FAR Pt 23 (Normal category) and be suitable for FAR Pt 135 single-pilot IFR. Wings for first prototype (Ae 270 P) received from Taiwan August 1999. Roll-out delayed by late arrival of some components, but took place on 10 December 1999; first flight (OK-EMA) 25 July 2000 and to VZLU test centre 25 October 2000. Third prototype (second flying, OK-SAR) rolled out 2 November 2001; first flight 23 December 2001 (not announced until 11 January 2002) followed by fifth prototype (third flying, first production conforming, OK-LIB) on 24 January 2003, although 'official' first flight was not until 25 February 2003. Sixth prototype (fourth flying, OK-INA) first flown 22 September 2003, and made the type's US debut at the NBAA Convention in Orlando 6 October 2003. Total of more than 1,000 hours logged in some 885 sorties by May 2004.

Certification of Ae 270 P originally expected mid-2002, followed by Ae 270 HP in fourth quarter; first deliveries (Ae 270 P) anticipated in October 2002, with Ae 270 HP deliveries commencing in early 2003. This not achieved, and at NBAA Convention at Orlando, Florida, in September 2002, Ibis announced suspension of Ae 270 P in order to concentrate resources on Ae 270 HP, with intention of EASA certification in late August 2004, followed by FAR Pt 23 (single-pilot) approval in September or October 2004. First delivery expected in late 2004. Total of 15 deliveries scheduled in 2005, 25 in 2006, and 35 in 2007.

CURRENT VERSIONS: **Ae 270 P:** Basic pressurised version, with 634 kW (850 shp) flat rated P&WC PT6A-42A engine, Honeywell avionics and retractable landing gear.

Ae 270 HP: High-performance version, announced 7 October 2000 at NBAA Convention in New Orleans; PT6A-66A engine offering improved speed, climb and altitude performance and greater fuel efficiency. Aircraft 05 is prototype for this version, which will be the first to be certified. *As described.*

Ae 270 W: Non-pressurised model (previously called Ae 270 U), with fixed landing gear, Walter M 601F engine and Czech avionics. Plans for this version in abeyance in 2001.

Third flying Ae 270 (OK-LIB) ***NEW***/0546943

Fourth flying Ibis Ae 270 Spirit (OK-INA) *(Paul Jackson)* ***NEW***/0576305

Ae 270 FP : and FW: Wheeled-float versions of P and W. Not currently promoted.

CUSTOMERS: Initial orders announced at Paris Air Show, June 2001, involved 51 aircraft for six regional distributors (four in USA, one each in Australia and South Africa). Firm orders for 76, plus six options, by May 2004.

COSTS: US$2.195 million (2004).

DESIGN FEATURES: Sized between Socata TBM 700 and Pilatus PC-12. High-aspect ratio low wing, circular cabin windows, sweptback vertical tail. Fence added to upper surface of wings, approx 1.12 m (3 ft 8 in) outboard, by 2003.

Medium-speed aerofoil section (thickness/chord ratio 17 per cent at root, 12 per cent at tip); leading-edge sweepback 7° 42'; dihedral 6°; incidence 3°; twist 3°.

FLYING CONTROLS: Conventional and manual. Elevators trimmed mechanically, ailerons and rudder electromechanically; trim tab in rudder. Wide-span single-slotted Fowler flaps (70 per cent of trailing-edge; positions 0, 20 and 40°) actuated hydraulically. Autopilot optional.

STRUCTURE: All-metal semi-monocoque, with fail-safe structural elements in fuselage and two-spar wings. Fuselage of conventionally formed sheet metal bulkheads, stringers and skin; wings (built by AIDC) have formed sheet metal ribs, stringers and chemically milled skin. Production in Czech Republic (including engine compartment by Moravan, cabin floor panels by LZ and rear fuselage, fin and rudder by EV-AT) except for wings and landing gear, produced in Taiwan by AIDC; final assembly by Aero Vodochody. Other suppliers include: Chelton-Systems (EFIS display), De Crane Aircraft Seating (crew and cabin seats); Gumotex Breclav (de-icer boots), Honeywell (air conditioning and pressurisation components), Intertechnique (fuel gauges), Jihlavan (hydraulic system), Mikrotechna Modrany (instrumentation), Moravan (wheels and brakes), PPG Industries (windows) SimCom (pilot and technical training), Southstar Aircraft Interiors (interior completion), and UNIS (automated electronic monitoring systems).

LANDING GEAR: Retractable tricycle type by Technometra Radotin, with Moravan wheels and brakes; steerable nosewheel (60° by brakes, 15° by rudder pedals). Bulged nosewheel doors inward retraction for mainwheels, rearward for nosewheel. Oleo-pneumatic shock-absorbers in all units; hydraulic disc brakes on mainwheels. Dunlop tubeless tyres: size 6.50–10,

Ae 270 combined passenger (open) and freight door *(Paul Jackson)* ***NEW***/0576304

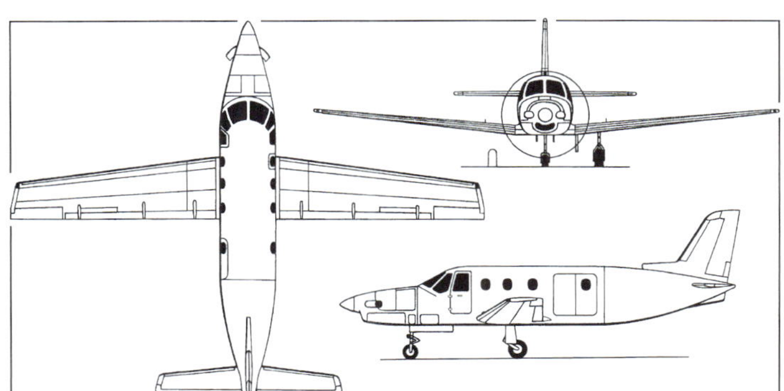
Ibis Aerospace Ae 270 W *(Jane's/Mike Keep)* ***NEW***/0580555

Ae 270 P Spirit ***NEW***/0546942

pressure 6.90 bar (100 lb/sq in) on mainwheels, size 6.00–6, pressure 3.80 bar (55 lb/sq in) on nosewheel. Minimum ground turning radius (based on nosewheel) 4.00 m (13 ft 1½ in).

POWER PLANT: One 634 kW (850 shp) (flat rated) Pratt & Whitney Canada PT6A-66A turboprop, driving a Hartzell HC-D4N-3/D9511FAKX four-blade, constant-speed, feathering and reversing propeller.

Integral fuel tank in each wing centre-section, combined usable capacity 1,170 litres (309 US gallons; 257 Imp gallons). Gravity refuelling point in top of each wing. Oil capacity 5.7 litres (1.5 US gallons; 1.25 Imp gallons) in Ae 270 P.

ACCOMMODATION: Flight crew of two standard, but to be certified for single-pilot operation. Main cabin suitable for up to eight passengers or 1,200 kg (2,645 lb) of cargo, or combinations of both. Six/seven-seat business or four-seat club layouts permit inclusion of lavatory. Medevac configuration provides accommodation for two stretcher cases and two medical attendants. Forward-opening crew door at front on port side; upward-opening passenger/cargo door on port side aft of wing with optional airstair door inset; downward-opening door extension, immediately to rear to provide freight loading aperture; overwing emergency exit on starboard side. Baggage door on starboard side at rear. Cockpit and cabin air conditioning, pressurisation and windscreen heating standard.

SYSTEMS: Electrical power provided by 28 V 250 A DC engine-driven starter/generator and 24 V 42 Ah lead-acid battery; 28 V DC external power connector. Standby generator optional. Hydraulic system, pressure 150 bar (2,175 lb/sq in), for actuation of flaps, mainwheel brakes and landing gear extension/retraction; flow rate 11 litres (2.9 US gallons; 2.4 Imp gallons)/min. Landing gear and mainwheel brakes also controllable by separate emergency hand-operated valves and parking brake.

Honeywell air conditioning and pressurisation system maintains differential of 0.30 bar (4.4 lb/sq in) up to altitude 7,500 m (24,600 ft). Gumotex pneumatic (engine bleed air) de-icing of wing and tailplane leading-edges; electric de-icing of windscreens, propeller blades; hot air de-icing of engine air intake; stall warning sensor and pitot tube standard on both models. Eros emergency oxygen system for crew and passengers.

AVIONICS: Standard flight, navigation and engine instrumentation (VFR or IFR) to comply with FAR Pt 23.

Radar: Weather radar optional.

Flight: IFR package standard for business versions, including VOR/ILS, ADF, Trimble GPS, S-Tec S-55 autopilot, and Ryan 9900BX TCAS.

Instrumentation: Chelton Systems FlightLogic synthetic vision EFIS optional.

DIMENSIONS, EXTERNAL:

Wing span	13.82 m (45 ft 4 in)
Wing chord: at root	1.88 m (6 ft 2 in)
at tip	1.04 m (3 ft 5 in)
Wing aspect ratio	9.1
Length overall	12.23 m (40 ft 1½ in)
Fuselage: Length	12.19 m (40 ft 0 in)
Max width	1.60 m (5 ft 3 in)
Max depth	1.75 m (5 ft 9 in)
Height overall	4.78 m (15 ft 8¼ in)
Elevator span	5.40 m (17 ft 8½ in)
Wheel track	2.83 m (9 ft 3½ in)
Wheelbase	3.53 m (11 ft 7 in)
Propeller diameter: Hartzell	2.44 m (8 ft 0 in)
McCauley	2.39 m (7 ft 10 in)
Propeller ground clearance:	
Hartzell	0.41 m (1 ft 4¼ in)
McCauley	0.44 m (1 ft 5¼ in)
Passenger/cargo door (port, rear):	
Height	1.30 m (4 ft 3¼ in)
Width	1.25 m (4 ft 1¼ in)
Height to sill	1.30 m (4 ft 3¼ in)
Crew door (port, fwd): Height	1.20 m (3 ft 11¼ in)
Width	0.70 m (2 ft 3½ in)
Height to sill	1.30 m (4 ft 3¼ in)
Emergency exit (stbd, overwing):	
Height	0.71 m (2 ft 4 in)
Width	0.50 m (1 ft 7¾ in)

DIMENSIONS, INTERNAL:

Cabin: Length: excl flight deck	4.98 m (16 ft 4 in)
incl flight deck	5.50 m (18 ft 0½ in)
Max width	1.45 m (4 ft 9 in)
Max height	1.36 m (4 ft 5½ in)
Volume	7.5 m^3 (265 cu ft)
Baggage compartment volume	0.73 m^3 (25.8 cu ft)

AREAS:

Wings, gross	21.00 m^2 (226.0 sq ft)
Ailerons (total)	1.00 m^2 (10.76 sq ft)
Trailing-edge flaps (total)	4.22 m^2 (45.42 sq ft)
Spoilers (total)	0.325 m^2 (3.50 sq ft)
Fin, incl dorsal fin	1.96 m^2 (21.10 sq ft)
Rudder, incl tab	1.11 m^2 (11.95 sq ft)
Tailplane	2.975 m^2 (32.02 sq ft)
Elevators (total)	1.885 m^2 (20.29 sq ft)

WEIGHTS AND LOADINGS (Ae 270 HP):

Weight empty, equipped	2,300 kg (5,071 lb)
Max fuel weight	1,034 kg (2,280 lb)
Max T-O weight	3,700 kg (8,157 lb)[1]
Max landing weight	3,700 kg (8,157 lb)
Max zero-fuel weight	3,500 kg (7,716 lb)
Max wing loading	176.2 kg/m^2 (36.09 lb/sq ft)
Max power loading	5.84 kg/kW (9.60 lb/shp)

[1] *To be increased retrospectively to 3,992 kg (8,800 lb)after initial certification*

PERFORMANCE (Ae 270 HP):

Max cruising speed: at S/L	220 kt (407 km/h; 253 mph)
at FL 200	270 kt (500 km/h; 311 mph)
Stalling speed, engine idling:	
flaps up	85 kt (158 km/h; 98 mph)
flaps down	66 kt (123 km/h; 76 mph)
Max rate of climb at S/L	521 m (1,710 ft)/min
Max certified altitude	9,140 m (30,000 ft)
T-O run	270 m (890 ft)
T-O to 15 m (50 ft)	550 m (1,805 ft)
Landing from 15 m (50 ft)	500 m (1,640 ft)
Landing run	167 m (1,545 ft)
Range with max fuel at FL 300:	
30 min reserves	1,610 n miles (2,981 km; 1,852 miles)
NBAA IFR reserves	1,420 n miles (2,629 km; 1,634 miles)

UPDATED

ILYUSHIN/IRKUT/HAL

ILYUSHIN/IRKUT/HAL

PARTICIPATING COMPANIES
Ilyushin: Russian Federation
Irkut: Russian Federation
HAL: India

Agreement announced between these three companies in June 2001 for joint design, development and co-production of a multirole transport aircraft, now known as IRTA-21. It will be funded 50 per cent by HAL and 50 per cent by the two Russian companies, with total investment in the region of US$310 million. Production lines in both countries are planned. Under agreement of January 2002, both HAL and Russian partners will transfer conceptual draft for series production and technology information during the design, development and productionisation planning stages. HAL will create infrastructure for design, development and manufacturing programme and commercial production in India; similar facilities in Russia will be established by Ilyushin and Irkut.

UPDATED

ILYUSHIN/IRKUT/HAL IRTA-21

TYPE: Medium transport/multirole.

PROGRAMME: This Indian requirement for a multirole transport aircraft (MTA) became the subject of a joint venture between HAL and Ilyushin (Russian designation Il-214) following the opening of negotiations in March 1999; joint agreement signed later in the same year; MoU in late 2000 followed in June 2001 by joint design, development and co-production agreement involving these two companies plus Irkut. Aircraft redesignated in 2003 as IRTA-21 (Indo-Russian Transport Aircraft, 21st century).

Configuration now differs considerably from the original design of 1996. Programme is included in Russian government's aeronautical development plan for 2001–15. Full-scale development launched in 2003, leading to flight testing during 2008–09 and production start in 2010 for deliveries in 2011.

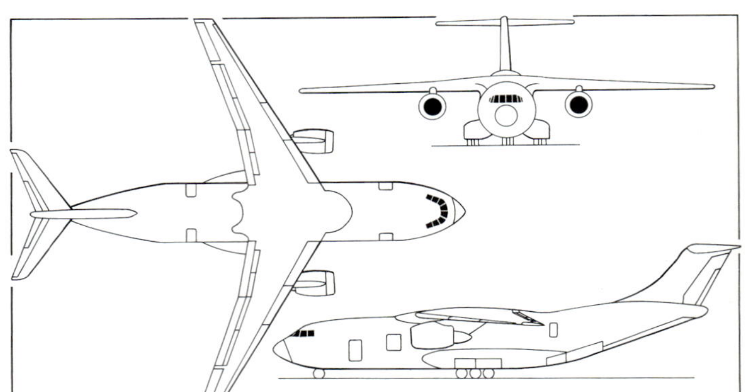

Russia and India's new multirole transport aircraft, the IRTA-21 ***(Paul Jackson)*** 0062411

CURRENT VERSIONS: **Airliner:** With cabin windows and seating for 94 to 104.

Tactical transport: Military version with rear-loading ramp; provision for 82 troops or 75 paratroops.

Commercial freighter: With rear-loading ramp.

CUSTOMERS: Required by Indian and Russian air forces; manufacture by HAL and Irkut. Initial Indian Air Force commitment for 45; estimated Russian requirement for 100.

COSTS: Development estimated as US$320 million for basic version. Unit cost US$25 million to US$29 million. Target price is 20 per cent less than for comparable aircraft.

DESIGN FEATURES: Sweptback high-wing monoplane with all-swept T tail; twin, underslung podded engines; upswept rear fuselage with rear-loading ramp. Intended to replace ageing An-12, An-26 and An-32.

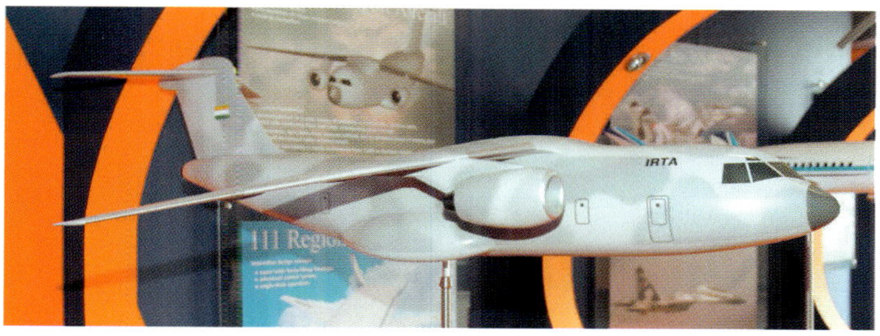

IRTA-21 model displayed at Paris in June 2003, wearing its revised designation ***(Paul Jackson)*** ***NEW***/0580809

FLYING CONTROLS: Digital fly-by-wire.

POWER PLANT: Two turbofans, each of some 108 to 118 kN (24,000 to 26,500 lb st). Options include CFM56, Aviadvigatel PS-12 (under development), Rolls-Royce BR715, PW6000 or IAE V2500.

AVIONICS: Integrated suite includes flight and navigation complex, flight deck information and control, com radios and sighting system.

Following details are provisional:

DIMENSIONS, EXTERNAL:
- Wing span ... 32.25 m (105 ft 9¾ in)
- Length overall ... 35.20 m (115 ft 5¾ in)
- Height overall ... 11.00 m (36 ft 1 in)

DIMENSIONS, INTERNAL:
- Cargo hold: Length ... 13.85 m (45 ft 5¼ in)
  - Max width at floor ... 3.45 m (11 ft 3¾ in)
  - Max height ... 3.40 m (11 ft 1¾ in)

WEIGHTS AND LOADINGS (design):
- Max payload ... 20,000 kg (44,090 lb)
- Max fuel weight ... 23,000 kg (50,706 lb)
- Max T-O weight ... 55,000 kg (121,250 lb)
- Max landing weight ... 49,000 kg (108,026 lb)

PERFORMANCE (estimated):
- Max cruising speed ... 470 kt (870 km/h; 541 mph)
- Normal cruising speed ... 432 kt (800 km/h; 497 mph)
- Airdropping speed ... 135 kt (250 km/h; 155 mph)
- Max operating altitude ... 12,000 m (39,370 ft)
- T-O to 15 m (50 ft) ... 1,450 m (4,760 ft)
- Landing from 15 m (50 ft) ... 1,350 m (4,430 ft)
- Range: airliner with max payload ... 1,889 n miles (3,500 km; 2,174 miles)
  - tactical transport with 4,500 kg (9,920 lb) payload and auxiliary fuel ... 3,239 n miles (6,000 km; 3,728 miles)

UPDATED

NH INDUSTRIES

NH INDUSTRIES SARL

Le Quatuor, Batiment C, 42 route de Galice, F-13090 Aix-en-Provence, France
Tel: (+33 4) 42 95 97 00
Fax: (+33 4) 42 95 97 48/49
e-mail: nhi90@nhindustries.com
Web: http://www.nhindustries.com
GENERAL MANAGER: Gérard Maitrepierre

PARTICIPATING COMPANIES:
- **AgustaWestland:** International
- **EADS Eurocopter:** International
- **EADS Eurocopter Deutschland:** Germany
- **Stork Fokker AESP:** Netherlands

NHIndustries established August 1992 to manage design and development of NH90 and is also responsible for production, logistic support, marketing and sales; in October 1995 workshares were slightly amended to Agusta 28.2 per cent, Eurocopter Deutschland 23.7 per cent, Eurocopter France 41.6 per cent, Fokker Aircraft 6.5 per cent. NFT (Norway) joined in 1994 as risk-sharing partner of Eurocopter France. Further amendment occurred on 21 June 2001, when Portugal joined programme and allocated 1.2 per cent share to OGMA. Netherlands activities transferred to Fokker Aerostructures BV following bankruptcy of the former Fokker Aircraft. Industrial ownership (and production share) is now EADS 61.7 per cent, Finmeccanica (Agusta) 31.6 per cent, Stork Fokker 5.5 per cent and Portuguese industry (as Eurocopter and Agusta subcontractor) 1.2 per cent.

Joint agency NAHEMA (NATO Helicopter Management Agency) formed February 1992 by the four governments, within NATO framework, to manage programme; now additionally represents Portuguese government; NAHEMA located alongside NH Industries in Aix-en-Provence.

UPDATED

First Finnish NH90 flew at Marignane on 15 September 2004 ***NEW***/1047703

Fourth NH90, 9890, in TTH guise with rear ramp ***NEW***/0113511

NH INDUSTRIES NH90

Swedish military designation: Hkp 14

TYPE: Multirole medium helicopter.

DEVELOPMENT MILESTONES

Requirement issued	83–84
Design began	Dec 86
Official go-ahead	26 Apr 90
First order (multinational)	30 Jun 00
Rolled out	29 Sep 95
First flight	18 Dec 95
Official first flight	15 Feb 96
Public debut	May 96
First flight, production	4 May 04
First delivery (planned)	04

PROGRAMME: Initial studies by NIAG (NATO Industrial Advisory Group) SG14 in 1983–84; September 1985 MoU between defence ministers of France, Germany, Italy, Netherlands and UK (USA and Canada having already withdrawn) preceded 14 month feasibility/predefinition study for new naval/army NH90 (NATO helicopter for the 1990s); initial design phase approved December 1986; second MoU in September 1987 led to predefinition phase and completion of weapons system definition in 1988; UK withdrew from programme April 1987; German workshare reduced early 1990, Italian participation renegotiated later 1990; French and German launch decision 26 April 1990; two ministerial MoUs in December 1990 and June 1991 cover design and development responsibilities; intercompany agreement signed March 1992; design and development contract signed 1 September 1992; five prototypes and GTV (ground test vehicle) completed, with three prototypes produced in France, TTH in Germany and NFH in Italy; assembly of first prototype began at Marignane, October 1993; first flight 18 December 1995. All five prototypes flying by December 1999.

Development suspended May 1994; restarted July 1994, but all new-development items subject to rigorous cost examination; preference for off-the-shelf components. Programme doubts dispelled by inclusion of NH90 in French defence plan adopted in June 1996, which requires first NFH delivery to Aéronavale in 2004 and first TTH to ALAT in 2011, although delivery of the last ALAT aircraft has been delayed to 2025. Bid for production investment and Lot 1 manufacture submitted 5 October 1997; four nations signed political authorisation for first production batch of 154 (amended later to 151) helicopters on 19 May 1998; pricing proposals submitted by NHI to NAHEMA on 15 July 1998; industrial contract was due in mid-1999 to permit first deliveries in 2003 for Netherlands Navy, but was delayed and latter date rescheduled to 2007.

In September 1999, Germany revealed a cut of 24 in its total order and now plans to receive first TTH in April 2004 and NFH in 2007, instead of withdrawing current Westland Lynx in 2003. Italian deliveries (both versions) to begin in April 2004, continuing until 2017–18. NH90 built on parallel production lines in France, Germany and Italy, with peak rate of 40 helicopters per year anticipated. Aircraft for Finland, Norway and Sweden will be completed on fourth assembly line at Patria's Jamsa site. Netherlands confirmed in November 1998 that its helicopters would be built in Italy; German NFHs built in France. Further, in 1998, a basic NH90 variant was proposed to which 'Kits on Option' could be added to produce specific variants at lower cost than purpose-build.

Rescheduled production contract signature, at Berlin Air Show, 8 June 2000, replaced by MoU committing to 243, plus 55 options and minimum buy of 366 from stated requirements for 595. Production investment and production contract finally signed in Paris on 30 June 2000 on these terms and quoting cost for 298 as €6.6 billion, with industrial participants providing 25 per cent of production investment. Portuguese contract signed at Paris Air Show on 21 June 2001 added fifth member to consortium and 10 aircraft to previous figures, for new totals of 253 firm, 376 minimum purchase and 605 stated requirement. This further increased by 52 in September 2001 on selection of NH90 for Nordic Standard Helicopter Project, programme offsets including production by Saab of 200 forward fuselages. Most recent order, placed on 29 August 2003, covered 20 helicopters for Greece.

On 6 October 1995, NAHEMA contracted NHI to undertake additional work and national customisation, involving integration of T700 engine to meet Italian requirement; design of rear ramp installation for TTH version; and reinforcement for carriage of stores up to 700 kg (1,543 lb). Associated national requirements include study of command post version;

Initial production NH90, a TTH for German Army, making its first flight on 4 May 2004 ***NEW***/0589294

cannon pod; sand filter; radiometer; second VHF/FM radio for TTH; and sonobuoy data relay, Tacan and rear ramp for NFH. Total value of €58.23 million slightly amends national workshares. Italy formally confirmed T700 engine selection in June 2000; Germany and Netherlands simultaneously confirmed RTM322, which earlier selected by France.

First five aircraft had flown more than 2,500 hours by end of November 2003. Flight envelope expanded by early 2002 to 190 kt (352 km/h; 219 mph), 6,096 m (20,000 ft) altitude and T-O weight of 10,600 kg (23,369 lb). The aircraft has also demonstrated fast rolling landings at 50 kt (93 km/h; 58 mph) and landings on slopes of up to 12°. Flight test activity in 2003 mainly dedicated to mission system qualification. Initial production aircraft (TTH for German Army, TGEA01, temporarily serialled 9891) first flew at Donauwörth 4 May 2004 and made public debut at ILA, Berlin, 11 May.

A crash safety test utilising an NH90 centre fuselage section was successfully undertaken at Ottobrunn, Germany, on 24 October 2002, shortly after the first two serial production fuselages were mated at Donauwörth and Vergiate; a similar event took place at Marignane by the end of 2002, thus activating three of the four planned assembly lines.

CURRENT VERSIONS (GENERAL): **NFH : (NATO Frigate Helicopter)** Naval version, primarily for autonomous ASW and ASVW; additional applications include OTH targeting, vertrep, SAR, transport and anti-air warfare support; designed for all-weather/severe ship motion environment; fully integrated mission system for crew of three (optionally four); ECM, anti-radar and IR protection systems. Italy considering a unique sensor fit from FIAR/Honeywell consortium. Agusta, with its EH101 and S-61 experience, is responsible for NFH development and integration.

First flight of NH90 PT3 with FBW system, December 2003 ***NEW***/0563252

TTH : (Tactical Transport Helicopter) Land-based army/air force version, primarily for tactical transport, airmobile operations and SAR with up to 20 armed troops or 2.5 tonnes of supplies; additional applications include tactical support, special EW, airborne command post, VIP transport and training; defensive weapons suite; rear-loading ramp/door, loading winch and rolling strips to be provided for French, Italian and German armies to accommodate light armoured anti-tank missile vehicle; 4-tonne external underslung load hook; high manoeuvrability and survivability for NOE operation near front line. Eurocopter Deutschland is responsible for development of the TTH.

C-SAR: Combat search and rescue version, yet to be fully defined, but probably equipped with refuelling probe, 'bolt-on' armour on lower fuselage, two external fuel tanks, flotation gear, secure datalink, integrated countermeasures, FLIR and four FIM-92 Stinger self-defence AAMs.

'Kits on Option': To accommodate differing budgetary and operational requirements, a baseline configuration has been agreed, to which specific packages could be added. French shipborne utility version combining troop seats and ramp could replace Super Frelon. This version could also form basis of dedicated VIP variant, and even of a civil variant. Eurocopter France is responsible for development of the basic helicopter.

Netherlands NFH: All 20 equipped with common core avionics, FCS, 360° radar, FLIR, ESM, ECM and sonics, but only 14 with full mission system; rest with provision for, but without, LRUs. Proposal to equip six for night special operations/insertion duties. To serve aboard eight M-type frigates, four LCF frigates, two LPDs and two AORs.

MH90: Version proposed to satisfy Canadian requirement for up to 28 maritime helicopters to replace CH-124 Sea King; NHIndustries has teamed with Thales and Lockheed Martin Canada to bid, with latter acting as prime contractor.

CURRENT VERSIONS (SPECIFIC): **PT1/F-ZWTH:** Prototype; common basic configuration; assembled at Marignane; rolled out 29 September 1995; first flight 18 December 1995 with RTM 322 engines; formal debut (official first flight) 15 February 1996; public debut at Berlin, May 1996; tactical (green) colours; with 160 hours logged, PT1 flew to Agusta's Cascina Costa plant on 29 July 1997 for installation of Alfa Romeo Avia/GE T700-T6E engines before an 18-month test programme funded by Italy; first flew as such on 13 March 1998; had amassed 205 hours with these engines by August 2000, including sea trials with French 'La Fayette' class frigate *Courbet*, making 62 deck landings in two days (22/23 July 1998). Flown with Italian serial MMX612. Total 365 hours by 31 August 2000, when scheduled flight activities were completed. Employed as static demonstration airframe.

PT2/F-ZWTI: Common basic configuration; rolled out November 1996 at Marignane; first flight 19 March 1997; naval (light grey) colours; first with fly-by-wire controls. Initially flew for 5 hours with mechanical back-up flight controls, then first flew with

NH90 PT2, used for naval development and demonstration flying ***(Paul Jackson)*** ***NEW***/0572248

NH90 landing trials aboard *Etna* ***NEW***/0572243

analogue FBW on 2 July 1997; digital FBW installed early 1998 (retaining mechanical backup) and first flown on 15 May. PT2 also first with automatic electric blade folding intended for NFH, and due to receive IR exhaust suppressors. External load tests in fourth quarter of 2002. Total 513 hours (including 235 in FBW mode) by 21 November 2003.

PT3/F-ZWTJ: Common basic configuration; first with core avionics system with AFCS, navigation, communications, IFF and full 'glass' cockpit; retains mechanical control back-up; first flight at Marignane 27 November 1998. Core avionics trials with production-standard LCDs; replaced PT2 in extending flight envelope. De-icing system tests began 13 April 2001, lasting 20 hours. Total 970 hours (including 300 FBW) by 21 November 2003. Became first helicopter in world to fly with full quadruplex FBW control system, December 2003; now representative of production standard NH90.

PT4/9890: TTH configuration; first flight 31 May 1999, at Ottobrunn. First NH90 to be fitted with C-model prototype of DaimlerChrysler Aerospace electronic warfare suite as one element in a fully integrated avionics suite, with LCD cockpit displays, FLIR, helmet sight, tactical control system, secure communications, and radar (with digital mapping). Cleared operation of rear loading ramp and doors open in flight up to 140 kt (259 km/h; 161 mph). Night flights with FLIR and rescue hoist demonstration. Cold weather trials at Iqaluit, Canada, January-February 2003. Total 387 hours (including 180 FBW) by 21 November 2003.

PT5/MMX613: NFH mission system; first flight at Cascina Costa 22 December 1999. Performed flotation system test. Accomplished initial NH90 deck operating campaign aboard replenishment tanker *Etna*, 2–8 May 2001, with 77 landings and 42 decklock engagements, all in FBW mode. Third series of deck trials, May 2003 (first with deck hook). Total 260 hours (including 181 FBW) by 21 November 2003.

GTV: Ground test vehicle at Cascina Costa. Instrumented to monitor 300 parameters. First run 28 September 1995; 539 hours accumulated by June 2000.

Radome of NH90 NFH *(Gérard Deulin)* ***NEW***/0572241

NH90 flight deck 0113512

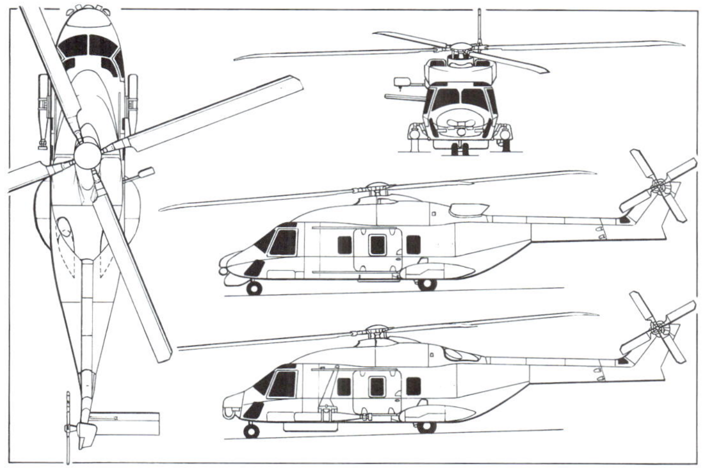

NHIndustries NH90 NFH, with additional side view (upper) of TTH *(Jane's/Mike Keep)* 0121430

CUSTOMERS: See table. Requirements originally estimated as 726 (France 220, Germany 272, Italy 214 and Netherlands 20); reduced to 647 in July 1996, then to 642 in 1998 and 595 in late 1999. These comprise Germany 181 TTH and 38 NFH (= 219); France 133 TTH and 27 NFH (= 160); Italy 150 TTH and 46 NFH (= 196); and Netherlands 20 NFH. Additionally, over 600 export orders are anticipated; NHI is bidding to meet Singapore requirement for 12 NFHs, while also competing for orders from Canada, Chile, Malaysia and Saudi Arabia. Selected for Nordic Standard Helicopter Project: Finland (**NH90-TTT**) and Norway announced decision on 13 September 2001, with Sweden following on 18 September. Greece ordered 20 NH90 (with 14 on option) on 29 August 2003. Oman ordered 20 TTHs in July 2004 for delivery by 2008. Australia added 12, designated **MRH90**, on 31 August 2004.

COSTS: Development cost €1,376.15 million (FFr 9.6 billion) (January 1988 values) agreed September 1992, provided by governments and industry: €421.83 million French government plus €161.24 million Eurocopter France; €256.41 million Germany plus €74.59 million Eurocopter Deutschland; €89.44 million Netherlands plus €2.64 million Fokker; and €370 million by Italian government. Additional development, agreed October 1995, adds €58.23 million. By January 1997, an estimated 75 per cent of R&D costs had been disbursed. Projected flyaway prices (1995) are FFr90 million for TTH and FFr145 million for NFH, but further 15 per cent cut demanded by French military procurement office in February 1996. Initial batch of French NFHs have 2000 unit cost of FFr200 million, including spares and tax; NHI quotes FFr115 million, pre-tax. Netherlands paying NLG1.7 billion (US$510 million) giving a unit price of US$25.5 million per aircraft. Cost of 10 TTH for Portugal quoted as €200 million.

DESIGN FEATURES: Compact external dimensions with large cabin (larger and slightly higher than Sikorsky Black Hawk). Low vulnerability/detectability, with aerodynamic and low-observables design, reduced maintenance requirements, and day/night operability within temperature range of −40 to +50°C. Specification requires fewer than 250 failures per 1,000 flight hours; 97.5 per cent mission reliability rate; 87 per cent availability; MTBF better than 4 hours; and under 2.5 mmh/fh (excluding engines). Service life of 10,000 hours over 30 years.

Titanium Spheriflex main rotor hub with elastomeric spherical thrust bearings; four blades with advanced aerofoils and curved tips; main rotor turns anticlockwise at 256.6 rpm with 219 m (719 ft)/s tip speed; bearingless tail rotor with four blades of similar design and construction; tail rotor with cross-beam hub turns at 1,235.4 rpm with 207 m (679 ft)/s tip

NH90 ORDERS AND OPTIONS

Customer	TTH	NFH	Total	First flight and location
Australia				
Air Force	12		**12**	
Finland				
Air Force	20		**20**	15 Sep 04 M/H
France				
Navy		27	**27**	M
Germany				
Air Force	30(24)		**30(24)**	D
Army	50(30)		**50(30)**	4 May 04 D
Greece				
Army	20(14)		**20(14)**	
Italy				
Air Force	— (1)		**— (1)**	V
Navy	10	46	**56**	V
Army	60		**60**	15 Sep 04 V
Netherlands				
Navy		20	**20**	V
Norway				
Air Force		14(10)	**14(10)**	V
Oman	20			
Portugal				
Army	10		**10**	
Sweden				
Helicopter Wing	13 (7)	5	**18 (7)**	H
Totals	**245(76)**	**112(10)**		
	357 (86)			

Notes: Options shown in parentheses. Second contract will be for 68 TTH for France, completing guaranteed purchase of 366 by original four partners. German Air Force total includes 23 for combat SAR.
Locations are D = Donauwörth, Germany; H = Halli, Finland (except first Finnish); M = Marignane, France; V = Vergiate, Italy

speed; automatic electronic folding of main rotor blades and tail pylon in NFH; manual blade folding (automatic optional) in TTH. Large unobstructed internal volume.

FLYING CONTROLS: Quadruplex fly-by-wire controls eliminate cross-coupling between control axes. NH90 is the world's first FBW transport helicopter. Flight control system developed by Eurocopter France; optimised for NOE flight and minimal vulnerability to small arms damage.

STRUCTURE: All-composites fuselage; fail-safe design of structure, rotating parts and systems for high safety levels. First set of main and tail rotor blades completed for ground testing May 1994; blades have multibox structure and glass and carbon fibre skin; leading-edge box and roved parts of blade prefabricated to reduce time in main mould; resin used reduces material ageing in damp and hot regions; blades designed and manufactured using CATIA CAD/CAM.

Eurocopter France builds rotor, rear gearbox, auxiliary gearbox, fairings and produces wiring, electrical system and air conditioning. Eurocopter Deutschland builds front and centre fuselage and fuel system. Agusta responsible for rear fuselage, main gearbox, hydraulic system and rear ramp; Stork Aerospace for tailboom (first completed unit delivered mid-2003), sliding doors, landing gear and fairings and intermediate gearbox. Assembly lines in France, Germany (TTH only), Italy and Finland (Patria) for the Nordic helicopters. Portuguese industries supply Agusta and Eurocopter with cabin steps, some electrical wiring and engine supports. Saab awarded contract in February 2002 for supply of 200 NH90 forward fuselage sections.

LANDING GEAR: Retractable crashworthy tricycle gear by DAF Special Products with twin-wheel nose unit (tyre size 6.00–6; 10 ply) and single-wheel main units (tyre size 615×225–10; 12 ply); all units retract rearwards; four-balloon emergency flotation gear effective up to Sea State 3. NFH has harpoon-type deck haul-down gear.

POWER PLANT: Two engines; required power from each engine during normal operation at S/L is 1,790 kW (2,400 shp) for 30 minutes, 1,663 kW (2,230 shp) maximum continuous, OEI maximum contingency (2½ minutes) 1,955 kW (2,622 shp), emergency (30 seconds) 2,158 kW (2,895 shp) and continuous (1 h) 1,802 kW (2,417 shp). Engine RFP issued April 1992. Decided on 25 January 1994 that NFH and TTH will be developed and qualified with 1,566 kW (2,100 shp) RTM 322-01/9 made by European consortium of Turbomeca, Rolls-Royce, MTU and Piaggio; during development, the 1,521 kW (2,040 shp) GE T700-T6E1 is being qualified by Alfa Romeo and GE to meet Italian military requirements. In June 2001, US$1 billion contract signed by Eurocopter for supply of production RTM 322-01/9 engines, including initial tranche of 254 power plants for installation in aircraft for France, Germany and Netherlands. Engine control by FADEC.

Transmission rating 2,925 kW (3,922 shp) with both engines, 2,170 kW (2,910 shp) for 30 seconds OEI. Modular main gearbox features integrated lubrication system with innovative integral monitoring and diagnostic system and, like the two accessory gearboxes, can run dry for 30 minutes.

Fuel system has eight Uniroyal crash-resistant self-sealing cells and AFG management system. Total internal capacity approx 2,500 litres (660 US gallons; 550 Imp gallons). Provision for pressure refuelling and hover refuelling. Provision for two external tanks (modified from Alpha Jet) each of 310 litres (82 US gallons; 68 Imp gallons) plus four ferry tanks in cabin.

ACCOMMODATION: Minimum crew one pilot VFR and IFR; NFH crew, pilot, co-pilot/TACCO on flight deck and one system operator in cabin; optionally, two pilots on flight deck and one TACCO, one SENSO in cabin; TTH, two pilots or one pilot and one crewman on flight deck and 14 to 20 equipped troops or one 2 tonne tactical vehicle in cabin. Sogerma crew seats and Autoflug composites passenger seats withstand 22 *g* or 11 m (35 ft)/s vertical descent crash loading. Swedish helicopters have 'high cabin', allowing more room for treatment of patients during SAR missions.

SYSTEMS: Full redundancy for all vital systems; hydraulic system has two Vickers mechanically driven and one electrohydraulic and one electric pumps; redundant flight control hydraulic system has two main separated and independent circuits; independent utilities system provides hydraulic power for landing gear and ancillary operations; electrically driven pump as back-up in event of one circuit failing; other circuit permanently supplied in parallel by a utility system pump. Electrical system by Thales has two Saft batteries and three 40 kVA Auxilec AC generators (two driven by main gearbox, one by remote accessory gearbox) feeding DC busses through transformer-rectifiers; 70 kW Microturbo Saphyr 100 APU for electrical engine starting, environmental control on ground and emergency use in flight; Kollmorgen-Artus emergency DC generator (driven by remote accessory gearbox); fire detection and suppression in engine bays, APU and cabin. Microtecnica air conditioning system (with two identical independent units) operating in three zones: cockpit, cabin (both vapour cycle) and main avionics bays (filtered external air). Anti-icing of main blades (standard on TTH, optional on NFH) by heated leading-edge mat (heating carbon coating and not metal parts). Anti-icing system also on tail rotor blades and horizontal stabiliser, and on air intakes. Ice protection system effective in maximum conditions defined by DEF-STAN 000–970. Dunlop Aerospace Ice Protection and Composites selected in mid-2002 to design, develop and produce anti-iced engine air intakes for NH90, with serial production starting in 2003.

AVIONICS: Core avionic system based on dual MIL-STD-1553B data highways (one for the core avionics system, one for the mission system) allows integration of aircraft and mission equipment through several computers. Eurocopter France responsible for basic avionics; Eurocopter Deutschland for TTH avionics; Agusta-Westland for NFH mission system.

Comms: Integrated management of communications and identification systems. Two V/UHF; V/UHF DF/homer; HF SSB and Thales TSC 2000 IFF. Elmer intercom.

Radar: NFH has Thales ENR 360° track-while-scan surveillance radar with target recognition capability; TTH has Honeywell weather radar. Telephonics APS-134B(V)3 Ocean Eye multimode surveillance radar chosen for Swedish helicopters.

Flight: Twin INS with embedded GPS and Doppler ground speed sensor; twin air data computers; provision for radio altimeter and MLS. SFIM/Alenia flight control computer and Thales/BAE Systems microwave landing system.

Instrumentation: NVG compatible. Five 200 mm (8 in) square Thales LCDs in cockpit of NFH; up to three more in cabin; four (fifth optional) in TTH. TTH has digital map. Central warning system, two remote frequency indicators and conventional analogue back-up instruments. Provision for Thales Topowl binocular helmet-mounted display and sight.

Mission: NFH: Alenia or Thales FLASH dipping sonar, high focal-length tactical FLIR, IFF interrogator, Link 11 datalink including ship-to-air umbilical, ESM and telebrief; electronic warfare subsystem. Dedicated video/audio network distributes tactical information to all crew members. TTH: FLIR. Tactical mission system for Swedish NH90 being developed by Saab under terms of agreement signed with NH Industries on 1 February 2002.

Self-defence: Provision for MAWS, RWR, LWR, Matra chaff/flares and IR jammer. TTH has self-protection suite with passive threat warning system known as EWS (electronic warfare suite) with EADS laser warning receiver, LFK AN/AAR-60 MILDS missile launch warning system and Thales EW processor and radar warning receiver. Active elements of self-protection suite include chaff/flare dispensers and IR jammer. Norway selected ITT Avionics AN/ALQ-211 SIRFC in first quarter 2002. Swedish NH90 self-protection EW suite developed jointly by Saab Avionics of Sweden and Avitronics of South Africa. Includes radar, missile and laser warning capabilities and countermeasures dispensing equipment.

EQUIPMENT: Rosemount windscreen washers/wipers; Siren 5 tonne cargo hook and rescue hoist. NFH has Magnaghi 5 ton deck restraint harpoon and automatic main rotor blade and tail folding.

ARMAMENT: TTH can have area suppression and self-defence armament; NFH to carry air-to-surface missiles weighing up to 400 kg (882 lb) and anti-submarine torpedoes (including Murène 90); air-to-air missiles optional. Italian NFH will carry Marte Mk 2/S.

DIMENSIONS, EXTERNAL:

Main rotor diameter	16.30 m (53 ft 5½ in)
Main rotor blade chord	0.65 m (2 ft 1½ in)
Tail rotor diameter	3.20 m (10 ft 6 in)
Tail rotor blade chord	0.32 m (1 ft 0½ in)
Length: overall, rotors turning	19.565 m (64 ft 2¼ in)
fuselage (NFH)	16.09 m (52 ft 8¼ in)
folded (NFH)	13.50 m (44 ft 3½ in)
Height: folded (NFH)	4.16 m (13 ft 7¾ in)
overall, tail rotor turning	5.31 m (17 ft 5 in)
Width: max (incl tailplane):	
NFH clean	4.515 m (14 ft 9¾ in)
NFH with torpedoes	4.625 m (15 ft 2 in)
over mainwheel fairings	3.63 m (11 ft 11 in)
over torpedoes (NFH)	3.85 m (12 ft 7½ in)
fuselage max	2.60 m (8 ft 6¼ in)
folded (NFH)	3.63 m (11 ft 11 in)
Tailplane half-span (stbd)	2.79 m (9 ft 1¾ in)
Wheel track	3.20 m (10 ft 6 in)
Wheelbase	6.08 m (19 ft 11½ in)
Rear-loading ramp: Length	1.58 m (5 ft 2¼ in)
Width	1.78 m (5 ft 10 in)
Cabin door: Width	1.60 m (5 ft 3 in)
Height	1.50 m (4 ft 11 in)

DIMENSIONS, INTERNAL:

Cabin: Length: excl rear ramp	4.80 m (15 ft 9 in)
incl rear ramp	6.17 m (20 ft 3 in)
Min usable width	2.00 m (6 ft 6¾ in)
Min usable height: standard	1.58 m (5 ft 2¼ in)
'high cabin'	1.82 m (5 ft 11¾ in)
Volume	18.0 m^3 (636 cu ft)

AREAS:

Main rotor disc	208.67 m^2 (2,246.1 sq ft)
Tail rotor disc	8.04 m^2 (86.57 sq ft)

WEIGHTS AND LOADINGS:

Weight empty: basic: NFH	6,287 kg (13,860 lb)
TTH	5,945 kg (13,106 lb)
equipped: TTH	6,200 kg (13,669 lb)
Standard fuel: NFH (usable)	2,036 kg (4,489 lb)
Max payload: TTH	2,500 kg (5,511 lb)
T-O weight, TTH, NFH:	
typical mission	10,000 kg (22,046 lb)
max	10,600 kg (23,369 lb)
Max disc loading	50.8 kg/m^2 (10.40 lb/sq ft)
Transmission loading at max T-O weight and power	3.62 kg/kW (5.96 lb/shp)

PERFORMANCE (TTH at 1,000 m; 3,280 ft ISA +15°C, NFH at S/L ISA +10°C, except where stated):

Max cruising speed:	
NFH	157 kt (291 km/h; 181 mph)
TTH	160 kt (298 km/h; 185 mph)
Econ cruising speed: NFH	132 kt (244 km/h; 152 mph)
TTH	140 kt (259 km/h; 161 mph)
Max rate of climb at S/L: NFH	528 m (1,732 ft)/min
TTH	522 m (1,713 ft)/min
Rate of climb at S/L, OEI: NFH	61 m (200 ft)/min
TTH	45 m (148 ft)/min
Hovering ceiling: IGE: NFH	3,140 m (10,300 ft)
TTH	2,960 m (9,720 ft)
OGE: NFH	2,515 m (8,260 ft)
TTH	2,355 m (7,720 ft)
Mission loiter: NFH, 50 n miles (93 km; 58 miles) from base, 20 min reserves	3 h 20 min
Range: max operational:	
NFH	490 n miles (907 km; 563 miles)
TTH	430 n miles (796 km; 494 miles)
ferry: NFH	655 n miles (1,213 km; 753 miles)
TTH	650 n miles (1,203 km; 748 miles)
Radius of action:	
TTH with 2,000 kg; 4,409 lb payload, 30 min reserves	160 n miles (296 km; 184 miles)
Max endurance: NFH	4 h 45 min
TTH	4 h 35 min

UPDATED

Iran

AII

AVIATION INDUSTRIES OF IRAN

Km 6, Karaj Old Road (PO Box 14195-111), next to Kan River, Tehran
Tel: (+98 21) 681 26 07 and 681 26 08
Fax: (+98 21) 680 92 68
e-mail: info@aii-co.com
Web: http:/www.aii-co.com
MANAGING DIRECTOR: Dr S H Pourtakdoust
COMMERCIAL DIRECTOR: A Sardari

AII was formed in April 1993. It is affiliated to the Industrial Development and Renovation Organisation (IDRO) of the Iranian Ministry of Industries and Mining to produce, repair and maintain various types of aircraft. Following the AVA-101 glider, it flew its first powered aircraft, the AVA-202 two-seat trainer, in mid-1997.

In 1998 the company began construction of its own 75 ha (185.3 acre) airport facility at Azadi, some 65 n miles (120 km; 75 miles) from Tehran; this was completed in 2000, and will be used to support future programmes. AII plans to establish an aeroclub at this location for pilot training and customer demonstrations, and it may eventually become a base for future private owner/operators of the company's products.

Under the designation AVA-303, AII is undertaking licensed assembly of the Polish PZL M18B Dromader agricultural aircraft to supplement those already purchased directly from PZL. By the end of 2002 the first set of Polish-supplied components had been received and assembled. The programme is expected to yield 12 to 15 aircraft for the Iranian Ministry of Agriculture, built at a new production line to be located at Azadi. AII plans some design changes, including a modified cockpit canopy.

Other programmes stated to be current in April 2004 included a new 19-seat transport (AVA-606, also to be a joint venture); a small helicopter (AVA-505) similar in appearance to the HESA Shahed 278; a new six-seat aircraft (AVA-404); and studies on a joint-venture project for a 150-passenger transport.

UPDATED

AII AVA-202

TYPE: Primary prop trainer/sportplane.

PROGRAMME: Intended for domestic market to avoid dependence on foreign imports; prototype (originally known as IR-02) made first flight on 3 June 1997; second had been built by 3 January 1999 when one was lost in a fatal accident while testing an emergency parachute. Series production launched in 2000. Made its public debut (fourth aircraft shown) at the Iran International Air Show in October/ November 2002, at which time certification still awaited. Iranian certification to JAR-VLA (Normal category) standard was expected by April 2004.

CUSTOMERS: Iranian aero clubs.

DESIGN FEATURES: Similar configuration to US Van's RV-6A, but with greater wing span; meets JAR 22 and JAR-VLA standards. Intended for pilot training, general aviation, surveillance and recreational flying.

NASA aerofoil sections: 63_2-215 (wing), 63_2-015 (horizontal tail) and 63-010 (vertical tail).

FLYING CONTROLS: Conventional and manual. Trim tab in port elevator. Electrically actuated single-slotted flaps.

STRUCTURE: Mainly aluminium alloy; steel alloys in some areas. Single-spar wings. Glass fibre engine cowling.

LANDING GEAR: Non-retractable tricycle type. Self-sprung, fuselage-mounted cantilever legs; castoring nosewheel with shimmy damper. Hydraulic mainwheel disc brakes. Streamlined composites fairings over all three wheels.

POWER PLANT: One Textron Lycoming O-320-B2B flat-four engine (119 kW; 160 hp at 2,700 rpm), driving a Sensenich M 74 DM two-blade, fixed-pitch metal propeller. Optional 112 kW (150 hp) Lycoming O-320-E2D. Integral fuel tank in each wing, combined capacity 120 litres (31.7 US gallons; 26.4 Imp gallons).

ACCOMMODATION: Two seats side by side. Dual controls. Wraparound windscreen; one-piece rearward-sliding canopy. Cabin heated and ventilated. Rear baggage compartment with internal access.

AVA-202 trainer/sportplane in manufacturer's livery ***NEW***/0576854

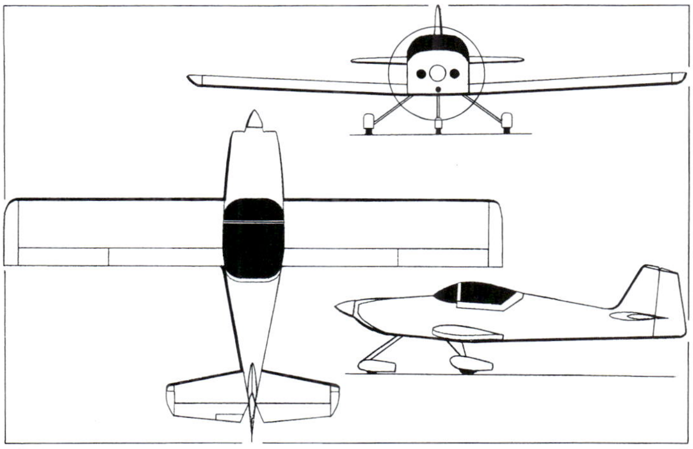

Aviation Industries of Iran AVA-202 trainer *(Paul Jackson)* 0062613

SYSTEMS: 12 V electrical system.

AVIONICS: To customer requirements; can include Bendix/King KX 155 VHF nav/com radio, ELT, and ADF/VOR/GPS.

DIMENSIONS, EXTERNAL:
- Wing span ... 8.74 m (28 ft 8 in)
- Wing chord, constant ... 1.24 m (4 ft 0¾ in)
- Wing aspect ratio ... 7.0
- Length overall ... 6.02 m (19 ft 9 in)
- Height overall ... 2.04 m (6 ft 8¼ in)
- Propeller diameter ... 1.88 m (6 ft 2 in)

AREAS:
- Wings, gross ... 10.87 m² (117.0 sq ft)
- Vertical tail surfaces (total) ... 1.04 m² (11.19 sq ft)
- Horizontal tail surfaces (total) ... 2.40 m² (25.83 sq ft)

WEIGHTS AND LOADINGS:
- Weight empty ... 500 kg (1,102 lb)
- Max T-O and landing weight ... 750 kg (1,653 lb)
- Max wing loading ... 69.0 kg/m² (14.13 lb/sq ft)
- Max power loading ... 6.29 kg/kW (10.33 lb/hp)

PERFORMANCE (estimated):
- Max level speed at S/L ... 140 kt (259 km/h; 161 mph)
- Max cruising speed, 75% power ... 135 kt (250 km/h; 155 mph)
- Stalling speed, flaps down ... 45 kt (84 km/h; 52 mph)
- Max rate of climb at S/L ... 457 m (1,500 ft)/min
- Service ceiling ... 6,400 m (21,000 ft)
- T-O run ... 250 m (820 ft)
- Range with max fuel ... 540 n miles (1,000 km; 621 miles)

UPDATED

AII AVA-303

TYPE: Agricultural sprayer.

PROGRAMME: Assembly and eventual licensed manufacture of PZL M-18B Dromader (which see), to supplement aircraft previously imported from Poland; first set of Polish-supplied components had been assembled by end of 2002. New production line to be located at Azadi airport facility.

CUSTOMERS: Iranian Ministry of Agriculture. Programme may yield 12 to 15 new aircraft, depending upon requirement.

DESIGN FEATURES: AII plans some design changes, including a modified cockpit canopy.

VERIFIED

AII AVA-404

In design stage 1996 as twin-turboprop light transport, but not proceeded with. Designation now re-allocated to a new six-seat design, though no details revealed by early 2004.

UPDATED

AII AVA-606

The 19-seat transport project announced by AII in 2002 now has the designation AVA-606. No details had been revealed by early 2004.

NEW ENTRY

AIO

AVIATION INDUSTRIES ORGANISATION OF THE ISLAMIC REPUBLIC OF IRAN

107 Gharani Avenue, Tehran
Tel: (+98 21) 881 04 81 to 881 04 87
Fax: (+98 21) 882 50 46
e-mail: aviation-id@isiran-net.com

AIO, a subdivision of the Iranian Ministry of Defence, is the policy maker, co-ordinator and planner for the manufacturing, overhaul, repair and support of aircraft by Iran's aviation industry. Created to counter problems arising from the Gulf wars and subsequent economic sanctions, its primary concern has been in supplying spares for aircraft in service and obtaining the technologies required for manufacturing various types of aircraft for the Iranian armed forces. However, it has now assumed a much greater role in aircraft development and production, including civil programmes such as the Ir.an-140.

According to Jane's sources in December 2001, the Iranian government was to consider a bill calling for the establishment of an Aerospace Organisation of the Defence and Armed Forces Logistics Ministry, into which all aerospace manufacturing and research units, as well as Iranian MoD factories, would be merged. No later information regarding this proposal had emerged by mid-2004.

Earlier proposals for a co-production programme involving the Ilyushin Il-114 have apparently lapsed, but talks between Iranian officials and the Tupolev JSC have continued concerning possible Iranian participation (including local assembly) in the Tu-204 and manufacture of up to 100 Tu-334s over a 15-year period.

SUBSIDIARIES:

Iran Aircraft Manufacturing Industries (IAMI)

HEAD OFFICE: 5th floor, 107 Gharani avenue (PO Box 14155-5568), Tehran
Tel: (+98 21) 882 73 75 and 882 83 55
Fax: (+98 21) 882 83 55
WORKS: PO Box 83145-311, Shahin Shahr, Esfahan
Tel: (+98 311) 221 42 18
Fax: (+98 311) 221 42 19
e-mail: info@hesaco.com
MANAGING DIRECTOR: Mohammed Eslami

Otherwise known as Hevapeimasazi, or **HESA**, this state-owned company is controlled jointly by the Iranian Ministry of Defence and Ministry of Industries and Mining, and is the main aircraft production centre of the AIO, with divisions responsible for trainers, helicopters, the Ir.an-140 transport and UAVs. Major programmes include the HT-80, a proposed replacement for (but based on) the Pilatus PC-7 Turbo-Trainer.

Esfahan factory construction initiated 1975 by Bell Helicopter Textron for production of Bell 214, but suspended 1978 after Islamic revolution when only 11 per cent completed. Reactivated 1983; began piston engine manufacture 1988 and Northrop F-5B overhaul 1989. Helicopter (Bell AH-1J and other) and hovercraft overhaul started 1993; first production piston engine completed in 1996.

Following a February 1996 agreement, a factory was set up, with Ukrainian government assistance, for licensed assembly of the Antonov An-140 twin-turboprop transport, now known by the Iranian designation Ir.an-140 Faraz. A similar agreement with MAPO organisation, for manufacture of 60 Motor-Sich TV7-117VMA turboprop engines, was announced at the 1997 Paris Air Show and valued at US$180 million. These engines were said to be for installation on 'Ukrainian- and Russian-built' airliners.

Qods Aviation Industries (QAI)

PO Box 13445-873, 4 Km Karaj Road, Tehran
Tel: (+98 21) 450 33 06
Fax: (+98 21) 450 33 07
e-mail: mohajer@isiran.com

Established 1984 to design, develop and manufacture surveillance/reconnaissance and attack RPV/UAVs and aerial targets (see *Jane's Unmanned Aerial Vehicles and Targets*). Subsidiary of IAMI/HESA. Also manufactures parachutes for various applications. Factory area 40,000 m^2 (430,550 sq ft); 1998 workforce (latest known figure) approximately 400.

Iran Aircraft Industries (IACI)

PO Box 14155-1449, Tehran
Tel: (+98 21) 603 56 06
Fax: (+98 21) 600 00 75
e-mail: karhava@isiran.com
MANAGING DIRECTOR: Nasser Marzebani
AERO-ENGINE PROGRAMMES MANAGER: Eng Amiri

Established 1970; first Iranian aviation company to obtain ISO 9002 certification. Known locally as **SAHA**, factory has overhauled, maintained and repaired approximately 4,000 aircraft and aero-engines. Like IAMI/HESA, it is controlled by the Ministry of Defence. It occupies a 45,000 ha (111,195 acre) site, and had a 1998 workforce of more than 1,000. It is designated to produce the Tu-204, and the Tu-334 and its D-436 engines, if these programmes come to fruition. Earlier aircraft projects included the Fajr (Dawn) piston-engined trainer described briefly in earlier editions of *Jane's All the World's Aircraft*.

Turbine Engine Manufacturing (TEM)

Manufactures engine components at a 100,000 m^2 (1.08 million sq ft) site with shop floor area of 33,000 m^2 (355,200 sq ft). Also has well-equipped research laboratories and active design team. May be producer of Tolloue 4 and 5 small jet engines, unveiled in late 1999 as power plants for UAVs and aerial targets.

Iran Helicopter Support and Renewal Company (IHSRC)

PO Box 13185-1688, Mehrabad Airport Road, Meraj Avenue, Tehran
Tel: (+98 21) 600 34 14
Fax: (+98 21) 601 58 62
e-mail: panha@isiran.com
MANAGING DIRECTOR: Nasser Akhavan
DEPUTY MANAGING DIRECTOR: Abdollah Kheirollahi

Established 1969 with assistance of Agusta (Italy) and Bell Helicopter (USA); overhaul and maintenance of Agusta-Bell 205, 212, 214A/B/C; Bell 206/206L JetRanger/LongRanger and AH-1J/JT Cobra series; Boeing CH-47C Chinook; and Sikorsky SH-3D Sea King and RH-53D Sea Stallion. In 1999, it was stated that experience gained with extensive rebuild and refurbishment of AH-1J SeaCobras had given IHSRC (known locally as **PANHA**) the potential for full manufacture, if so required. Current upgrade programmes (see *Jane's Aircraft Upgrades*) include those for Bell 205 (Shabaviz) and AH-1J (Project 2091).

UPDATED

IACI (SAHA) SHAFAGH

English name: The light before dawn
Russian name: Rassvet

TYPE: Advanced jet trainer/light attack jet.

PROGRAMME: Existence revealed at Aero 2000 air show in Tehran; name also transliterated as Safak or Shafak. Programme controlled by Tehran's Aviation University Complex (part of the Malek Ashtar University of Technology). Design, to IRGC specifications, based on earlier (twin-engined) proposals from Russian OKB headed by Fatidin Mukhamedov, with characteristically unorthodox configuration. Wind tunnel testing of one-seventh scale model in Tehran followed by completion of full-size mockup. Integration of landing gear, hydraulic system and avionics under way in late 2002. Displayed in model form at Iran International Air Show in late 2002, at which time prototype roll-out targeted for 2008. Further details could emerge at November 2004 air show.

CUSTOMERS: Potentially, Russian Air Forces (to replace Aero L-39 C) and Iranian Air Force.

DESIGN FEATURES: Close-coupled configuration. Circular wing centre-section with pronounced LERX and tapered outer panels; twin outward-canted fins and rudders and (probably all-moving) tailplane halves.

POWER PLANT: One RD-33 turbofan (General Electric J79 turbojet in initial design studies).

ACCOMMODATION: Crew of two in tandem; Russian (Zvezda K-36D ?) ejection seats.

AVIONICS: Row of three MFDs; HUD; HOTAS controls on centrally mounted column.

ARMAMENT: Seven external stores stations (one on centreline and three beneath each wing).

DIMENSIONS, EXTERNAL (approx):
Wing span .... 12 m (39.4 ft)
Length overall .... 14 m (45.9 ft)

WEIGHTS AND LOADINGS (approx):
Weight empty .... 5,000 kg (11,023 lb)

PERFORMANCE (estimated):
Max level speed M1.3

UPDATED

Underside view of Shafagh model showing external weapon stations *(Robert Hewson)* 0531277

View emphasising the unique circular centre-section of the Shafagh's wing *(Robert Hewson)* 0531276

IAMI (HESA) AZARAKHSH

English name: Lightning

TYPE: Attack fighter.

PROGRAMME: Development said to have begun in 1986; programme name *Oaj* (Zenith); stated to have passed acceptance tests by April 1997 and completed flight testing in June. Designers' names given as Morteza Sanai and Morteza Satari. Announced by Islamic Republic News Agency (IRNA) in late September 1997, quoting Army Chief of Staff Maj Gen Ali Shahbazi, that the aircraft had taken part, earlier that month, in domestic military exercises, during the course of which it had dropped two 113 kg (250 lb) napalm bombs.

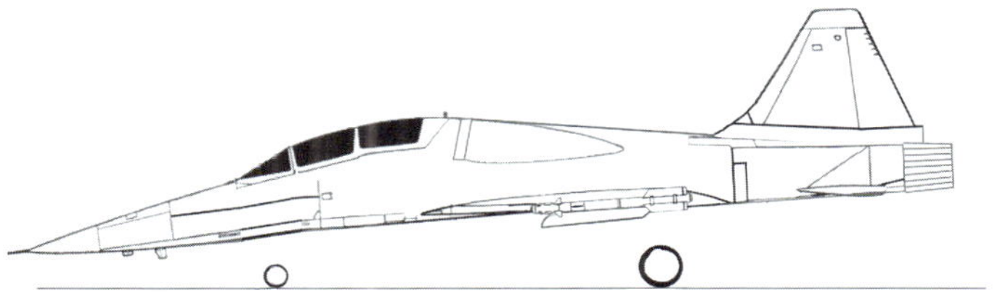

Possible appearance of Azarakhsh, based on some reports *(Jane's Defence Weekly)* 0067681

Announced at that time that production was to begin 'soon', but this now known to be incorrect.

In June 1999, Air Force Gen Habibollah Baghai told Iranian journalists that series production had begun, but stated by Lt Gen Iraj Ossareh (IRIAF Deputy Commander) in May 2000 that four aircraft completed at that time, with a fifth under construction. Current plans then to build up to 30 over next three years. Taxi trials reported completed by May 2001, with first flight believed imminent; at that time, fifth prototype said to be completed and sixth under construction. An Iranian news report of 15 August 2001, stated that production had been deferred pending resolution of 'serious structural problems'. At the same time, a parallel fighter programme known as *Owaz* was also reported. Aircraft, and any news of progress, were absent from Iranian air show in November 2002; but reported in late 2003 that AIO had announced plans during that year to begin low-rate series production of up to 30.

Relationship of Azarakhsh to Iran's other programmes involving F-5 modifications - Saeghe (which see) and Project Simorgh (conversion of F-5As to two-seat F-5Bs) - remains unclear.

Following details highly provisional and unconfirmed:

DESIGN FEATURES: In early 2000, Azarakhsh was reported to be a 10 to 15 per cent scaled-up, reverse-engineered Northrop F-5F, re-engined with two (sic) Klimov RD-33 turbofans, having shoulder-mounted air intakes. (However, earlier reports had claimed it to be based on the McDonnell Douglas F-4 Phantom.)

STRUCTURE: Extensive use of composites reported.

ACCOMMODATION: Two seats in tandem.

AVIONICS: *Radar:* Reported to be locally modified Phazotron N019ME Topaz with BVR capability and enhanced air-to-ground performance.

Self-defence: Said to have chaff/flare dispenser(s) in fairing beneath rear fuselage.

WEIGHTS AND LOADINGS (approx):

Weight empty .... 8,000 kg (17,637 lb)
Max external stores load .... 4,000–4,400 kg (8,818–9,700 lb)

PERFORMANCE (estimated):

Max level speed .... M1.6 class

UPDATED

IAMI (HESA) HT-80

TYPE: Basic turboprop trainer.

PROGRAMME: In addition to current programme to rebuild Pilatus PC-7s under Iranian designation S-68, in 2002–03 HESA was in conceptual ('pre-preliminary') design stage with HT-80 as eventual successor to the Swiss trainer in IRIAF service. Roll-out of first example tentatively forecast for 2005 or 2006. HESA may seek international partners (China one potential option).

DESIGN FEATURES: Generally similar to PC-7.

POWER PLANT: Current (2002) studies based on Walter M 601T (flat rated to 559 kW; 750 shp) or equivalent Motor-Sich turboprop, although P&WC PT6A said to be preferred. Four-blade propeller. Capability for 30 seconds of inverted flight.

ACCOMMODATION: Crew of two in tandem on stepped, SKS-94 lightweight ejection seats.

AVIONICS: Conventional (analogue) instrumentation.

Following data estimated and provisional:

DIMENSIONS, EXTERNAL:

Wing span .... 10.40 m (34 ft 1½ in)
Length overall .... 9.78 m (32 ft 1 in)
Height overall .... 3.30 m (10 ft 10 in)
Tailplane span .... 3.60 m (11 ft 9¾ in)
Wheel track .... 2.60 m (8 ft 6¼ in)

HESA model of the HT-80 Iranian development of the PC-7, targeted for roll-out in about 2005 *(Robert Hewson)* 0530184

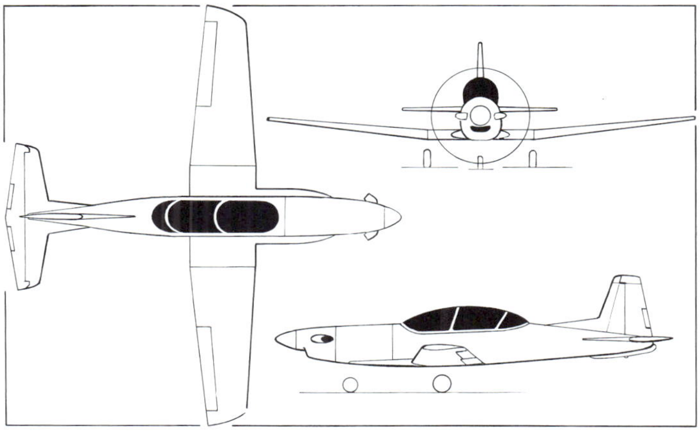

Provisional general arrangement of the IAMI (HESA) HT-80 turboprop trainer *(James Goulding)* ***NEW***/0552825

Wheelbase .... 2.32 m (7 ft 7¼ in)

WEIGHTS AND LOADINGS (A: Aerobatic, U: Utility category):

Max T-O weight: A .... 1,900–2,068 kg (4,188–4,559 lb)
U .... 2,700–2,883 kg (5,952–6,356 lb)
Max landing weight: A .... as MTOW
U .... 2,565–2,733 kg (5,655–6,025 lb)

PERFORMANCE (estimated; A and U as above):

Max operating speed:
A .... 240–324 kt (444–600 km/h; 276–372 mph)
U .... 270 kt (500 km/h; 310 mph)
Max cruising speed at S/L:
A, U .... 208–211 kt (385–391 km/h; 239–243 mph)
Stalling speed: A .... 71 kt (132 km/h; 82 mph)
U .... 83 kt (154 km/h; 96 mph)
Max rate of climb at S/L:
A .... 702–798 m (2,303–2,618 ft)/min
U .... 393 m (1,290 ft)/min
Max operating altitude: A, U .... 7,620 m (25,000 ft)
T-O to 15 m (50 ft) at S/L: A .... 320–425 m (1,050–1,395 ft)
U .... 1,180 m (3,870 ft)
Landing from 15 m (50 ft) at S/L:
A .... 480–550 m (1,575–1,805 ft)
U .... 800 m (2,625 ft)
g limits: A .... +7.0/–3.5
U .... +4.5/–2.5

UPDATED

IAMI (HESA) IR.AN-140 FARAZ

English name: Elevation

TYPE: Twin-turboprop airliner.

PROGRAMME: Licence-built Antonov An-140. Selected by Iran in 1995, against competition from 12 other contenders; initial agreement leading to Iranian manufacture signed February 1996; first stage is assembly from imported KhGAPP CKD kits at rate of up to 12 per year (first two kits received by early 2000); long-term objective is gradual progress to local parts manufacture; Iranian engineers installed at Antonov from 1997. Contract announced late 1998 for initial quantity of 80 aircraft by second quarter 2009, for Iranian domestic (civil and military) use only, but possibility of additional 25 production for export to CIS states has been suggested. In May 1999, Iran Electronics and Leninets (St Petersburg) signed contract for Iran 140 avionics development. Assembly of first aircraft (c/n 90-01) began at Esfahan March 1999 and completed by September 2000; maiden flight achieved on 7 February 2001. Second aircraft (of initial batch of eight) rolled out 26 December 2002, maiden flight in March 2003. Iranian certification awarded 14 April 2003; entered service with Kish Air on Tehran to Esfahan route shortly afterwards.

Total of 14 originally scheduled for completion from CKD kits by end of 2004, but due to funding delays only three completed at that time, with a fourth due by March 2005. Revised plan calls for additional four by March 2006 and next eight by March 2007, with annual output rate of 12 to be achieved in 2008. Offered for passenger (including VIP) civil transport, aeromedical (ambulance or flying hospital) use, cargo/passenger or parachutist carrying, photogrammetry, geosurvey, environmental research, fisheries protection and pollution control; military versions could include staff or personnel transport, coastal and maritime patrol, radar patrol/missile guidance, elint and crew training.

CURRENT VERSIONS: **Ir.An-140:** First three Iranian-assembled aircraft. *Following description applies to this version.*

Ir.An-140-100: Iranian production switching to this version from fourth aircraft onwards. KhGAPP scheduled to deliver first four CKD kits in 2004.

Stretched version: Projected version with 3.80 m (12 ft 5½ in) longer fuselage seating up to 68 passengers; 2,237 kW (3,000 shp) engines, 22,400 kg (49,383 lb) MTOW and 917 n mile (1,700 km; 1,056 mile) range.

Ir.An-140MPA: Maritime patrol version; under collaborative design development with three other companies in 2002–03 (undisclosed, but thought to be KhGAPP, Antonov and Thales). Major changes (assumed to be based on Ir.An-140-100 airframe) would be: bubble window each side of front fuselage; undernose FLIR turret; ventral search radar; underwing stations for weapons or auxiliary fuel tanks; MAD tailboom; and ESM suite. Internally, cabin configured for up to four systems operator positions. Roles could be extended to ASuV and/or ASW missions.

Ir.An-140TC: Tactical transport; under similar collaborative design development to MP version; rear fuselage redesigned to incorporate rear-loading ramp. Internal capacity for (typically) three 106 mm AA guns, three light vehicles or five LD3 containers; or up to 36 paratroops and a jumpmaster.

CUSTOMERS: Launch customers initially reported as Iran Air and Iran Asseman Airlines, of which latter signed order for 20 in October 1999, but Iran Air later declined to operate the type. Stated in October 2000 that orders had been received for 127. Kish Air eventually became first recipient, with delivery of first aircraft 2003. As at February 2005, two of the three aircraft then completed were in use by Safiran Airlines, which has ordered four. Saha

First Ir.An-140 assembled by IAMI (HESA), c/n 90-01 *(KhGAPP)* ***NEW***/1047209

Airlines is also a customer. Estimated Iranian market for 80. Likely also to be procured by Iranian Air Force and Navy (potential roles tactical transport, maritime patrol and EW).

COSTS: Reportedly US$9.5 million unit price (2003); set-up of production line US$273 million.

POWER PLANT: Two AI-30 Series 1 (Motor-Sich-built Klimov/ZMKB Progress TV3-117VMA-SBM1) turboprops, each flat-rated at 1,864 kW (2,500 shp) and driving an AV-140 six-blade propeller. Fuel capacity 5,460 litres (1,442 US gallons; 1,201 Imp gallons).

AVIONICS: Supplied initially by Avio Pribor; Thales (Sextant) suite under negotiation in 2002–03.

Note that some of following data from HESA, including dimensions, vary from those provided by Antonov for An-140:

DIMENSIONS, EXTERNAL:
- Wing span ... 25.505 m (83 ft 8¼ in)
- Length: overall ... 22.605 m (74 ft 2 in)
  - fuselage ... 22.46 m (73 ft 8¼ in)
- Fuselage max diameter ... 2.82 m (9 ft 3 in)
- Height overall ... 8.035 m (26 ft 4¼ in)
- Elevator span ... 9.385 m (30 ft 9½ in)
- Wheel track (c/l of shock struts) ... 3.18 m (10 ft 5¼ in)
- Wheelbase ... 8.07 m (26 ft 5¾ in)
- Distance between propeller centres ... 8.20 m (26 ft 10¾ in)
- Propeller diameter ... 3.72 m (12 ft 2½ in)
- Passenger door (rear, port): Height ... 1.68 m (5 ft 6¼ in)
  - Width ... 0.915 m (3 ft 0 in)
- Service door (rear, stbd): Height ... 1.29 m (4 ft 2¾ in)
  - Width ... 0.62 m (2 ft 0½ in)
- Cargo door (fwd, stbd): Height ... 1.29 m (4 ft 2¾ in)
  - Width ... 0.985 m (3 ft 2¾ in)
- Emergency exit (fwd, port): Height ... 0.51 m (1 ft 8 in)
  - Width ... 0.50 m (1 ft 7¾ in)

DIMENSIONS, INTERNAL:
- Passenger cabin:
  - Length:
    - excl galley/lavatory and baggage area ... 11.60 m (38 ft 0¾ in)
    - incl galley/lavatory and baggage area ... 14.51 m (47 ft 7¼ in)
  - Max width ... 2.60 m (8 ft 6¼ in)
    - Width at floor ... 2.28 m (7 ft 5¾ in)
  - Aisle width ... 0.46 m (1 ft 6 in)
  - Max height ... 1.90 m (6 ft 2¾ in)
- Underfloor cargo hold: Length ... 3.98 m (13 ft 0¾ in)
  - Width ... 1.45 m (4 ft 9 in)
  - Max depth ... 0.545 m (1 ft 9½ in)
- Baggage/cargo compartment volume:
  - rear ... 6.1 m^3 (215 cu ft)
  - underfloor ... 3.0 m^3 (106 cu ft)
- Overhead bins volume (total) ... 2.3 m^3 (81 cu ft)

WEIGHTS AND LOADINGS:
- Weight empty ... 11,800 kg (26,015 lb)
- Max fuel weight ... 4,370 kg (9,634 lb)
- Max payload ... 6,000 kg (13,228 lb)
- Baggage/cargo compartment capacity:
  - rear ... 1,040 kg (2,293 lb)
  - underfloor ... 560 kg (1,235 lb)
- Max T-O weight ... 21,000 kg (46,297 lb)
- Max landing weight ... 19,100 kg (42,108 lb)
- Max zero-fuel weight ... 17,800 kg (39,242 lb)
- Max power loading ... 5.63 kg/kW (9.26 lb/shp)

PERFORMANCE:
- Max cruising speed at FL130 ... 295 kt (546 km/h; 339 mph)
- Max rate of climb at S/L ... 420 m (1,378 ft)/min
- T-O distance, 10° flap ... 1,350 m (4,430 ft)
- Landing distance, 40° flap ... 1,350 m (4,430 ft)
- Range at optimum cruising speed, 45 min reserves:
  - with 52 passengers ... 1,112 n miles (2,060 km; 1,280 miles)
  - with 46 passengers ... 1,430 n miles (2,650 km; 1,646 miles)
  - with 34 passengers ... 1,987 n miles (3,680 km; 2,286 miles)
- Ferry range ... 2,105 n miles (3,900 km; 2,423 miles)

OPTIONAL NOISE LEVELS (estimated):
- T-O, flyover ... 85.0 EPNdB
- Sideline ... 82.5 EPNdB
- Approach ... 89.5 EPNdB

UPDATED

IAMI (HESA) PARASTU

English name: Swallow

TYPE: Basic prop trainer.

PROGRAMME: Revealed at Mehrabad Air Show in February 1996, when six or seven examples were reported to have been completed; reverse-engineered Raytheon (Beech) F33A Bonanza, with modifications that include performance-improving winglets. Programme initiated September 1987; first flight 5 April 1988. According to an IRNA announcement in September 1997, series production was then due to begin in the near future. However, according to a senior IRIAF spokesman in May 2000, the 11th aircraft was then in manufacture, with full-scale production imminent. Fourteen or more were reported to be in service by April 2001. Low-rate production possibly continuing in 2003–04.

UPDATED

IAMI (HESA) SANJAGHAK

English name: Dragonfly

TYPE: Side-by-side ultralight.

PROGRAMME: Exhibited November 2002 at Iranian Air Show on Kish Island, when described as "eight years old" Based on imported kits of Australian Moyes Dragonfly from German distributor, though with local variations.

DESIGN FEATURES: Braced high-wing monoplane; pod and boom fuselage; pusher engine mounted aft of wing.

FLYING CONTROLS: Manual. Junkers-type ailerons (approximately three-quarters of each half-span); these and elevators actuated by pushrods, rudder by cables.

STRUCTURE: Main frame of metal tube, with fabric-covered wings and tail surfaces. V-strut and wire-braced wings; wire-braced tailplane.

LANDING GEAR: Tricycle type; fixed.

POWER PLANT: Typically, single 38.8 kW (52 hp) Rotax 462 or 47.8 kW (64.1 hp) Rotax 582 flat-four engine; two-blade pusher propeller. Fuel tank between seats.

DIMENSIONS, EXTERNAL:
- Wing span ... 10.15 m (33 ft 3½ in)
- Wing chord, constant ... 1.60 m (5 ft 3 in)
- Length overall ... 5.70 m (18 ft 8½ in)
- Height overall ... 1.95 m (6 ft 4¾ in)

HESA Sanjaghak two-seat ultralight ***(Robert Hewson)*** ***NEW***/0552927

AREAS:
Wings, gross .... 15.20 m² (163.6 sq ft)
WEIGHTS AND LOADINGS:
Weight empty .... 150 kg (331 lb)
Max T-O weight .... 465 kg (1,025 lb)
PERFORMANCE:
Max level speed .... 54 kt (100 km/h; 62 mph)
Max rate of climb at S/L .... 180 m (591 ft)/min
T-O and landing run .... 80 m (265 ft)
Range .... approx 243 n miles (450 km; 279 miles)
g limits .... +6/–4

UPDATED

IAMI (HESA) SHAHED 278

TYPE: Light utility helicopter.
PROGRAMME: Three prototypes, first of which made maiden flight late 1997 or early 1998; most-flown prototype had accumulated 200 hours by March 2001. Public debut at Iran International air show in October/November 2002 as HESA 278, having previously been reported as Panha 287.
DESIGN FEATURES: Four-seater; intended for dual military/civil use, though designed primarily to meet military requirements. Designed entirely in Iran, according to manufacturer, although some sources report use of locally made airframe and dynamic components of Bell 206 JetRanger.
POWER PLANT: Single turboshaft. Prototypes have evaluated Rolls-Royce 250 and equivalents from Turbomeca and a Russian supplier, but final selection apparently not made by late 2002. Fuel capacity 287 litres (75.8 US gallons; 63.1 Imp gallons).
WEIGHTS AND LOADINGS
Weight empty .... 682 kg (1,504 lb)
Normal T-O weight .... 1,451 kg (3,198 lb)
PERFORMANCE:
Max level speed: at S/L .... 129 kt (240 km/h; 149 mph)
at FL50 .... 124 kt (230 km/h; 142 mph)
Max rate of climb at S/L .... 762 m (2,500 ft)/min
Service ceiling .... 6,400 m (21,000 ft)

HESA Shahed 278 light utility helicopter *(Robert Hewson)* 0530166

Hovering ceiling: IGE .... 6,218 m (20,400 ft)
OGE .... 4,865 m (15,960 ft)
Range .... 183 n miles (340 km; 211 miles)

UPDATED

IHSRC (PANHA) SANJAQAK

English name: Dragonfly
TYPE: Two-seat helicopter.
PROGRAMME: Announced January 1997 and reported in mid-1997 to be nearing flight test at a state research centre in Esfahan; max T-O weight 450 kg (992 lb). According to a Jane's source in March 2001, the Sanjaqak bears a close resemblance to the Bell 47 and is now in service, possibly for pilot training. It will be noted that IAMI (HESA) produces a Dragonfly ultralight aeroplane which employs slightly different English transliteration.

UPDATED

DORNA

H F DORNA COMPANY

20 6th Sarvestan Street (PO Box 16315-345), Pasdaran Avenue, Tehran 16619
Tel: (+98 21) 285 48 27
Fax: (+98 21) 284 18 31
e-mail: info@hfdorna.com
Web: http://www.hfdorna.com
MANAGING DIRECTOR: Yaghoob Antesary
CHIEF DESIGNER: Farid Najmabadi

Company established March 1989 to specialise in aircraft design and development, and in composites materials technology. It has 3,000 m² (32,300 sq ft) of factory space and had a workforce of 79 in July 2001. Current product is two-seat Blue Bird. Dorna planned to open a private pilot flying training school, equipped with Blue Bird, during 2004.

UPDATED

DORNA D139-PT1 BLUE BIRD

TYPE: Two-seat lightplane.
PROGRAMME: Launched 1994; prototype manufacture started mid-1994; first flight was made on 27 July 1998; prototype had accumulated 32 hours' flying in 45 flights by end of November that year. Certification to JAR 21 standards by the Iranian CAA was awarded on 2 December 2001, and initial production is under way.
CUSTOMERS: First six production aircraft sold to Iranian civil aeroclubs for pilot training; manufacture of these was in process in mid-2001. Stated orders for 10 and options for 25 more as at February 2004.
COSTS: Programme cost by November 1998 estimated at US$1.5 million excluding R&D; basic aircraft quoted at US$108,000 in December 2001.
DESIGN FEATURES: Conventional low-wing, fixed-gear lightplane, designed for low-cost, easy-to-fly/maintain operation.
Wings have constant chord, NACA 63-215 aerofoil section, 2° 30' dihedral from roots, 2° incidence and –2° twist.
FLYING CONTROLS: Conventional and manual. Ailerons and elevators actuated by pushrods; rudder by cables; elevators have electric trim. Plain flaps.

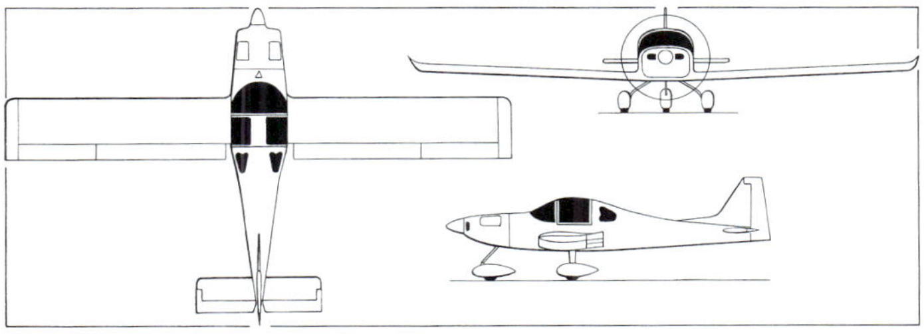

Dorna Blue Bird two-seat light aircraft *(James Goulding)* 0062614

STRUCTURE: All-composites. Two-spar wing, single-spar horizontal tail, monocoque fuselage.
LANDING GEAR: Non-retractable tricycle type, with Goodyear 5.00–5 tyres (6 ply on mainwheels, 4 ply on nosewheel). Fuselage-mounted cantilever self-sprung steel leg on each unit. Cleveland mainwheel brakes. Composites speed fairings on all three wheels.
POWER PLANT: One 84.6 kW (113.4 hp) Rotax 914 F3 flat-four engine driving an MT Propeller MTV-21-A/175-05 two-blade variable-pitch propeller at 5,800 rpm; three-blade MTV-6-A/172-08 variable-pitch propeller (as on prototype) optional. Fuel tank in each wing, combined capacity 128.7 litres (34.0 US gallons; 28.3 Imp gallons); gravity filling point on each tank. Oil capacity 3.0 litres (0.8 US gallon; 0.7 Imp gallon).
ACCOMMODATION: Two seats side by side, with baggage compartment aft of seats. Narrow hardtop roof to cabin; access via upward-opening window each side, each with smaller non-opening window to rear.
SYSTEMS: Electrical system: 12 V 20 A integral AC generator, 12 V 40 A external alternator and 35 Ah battery.
AVIONICS: *Comms:* Radio.
Instrumentation: Conventional VFR.
DIMENSIONS, EXTERNAL:
Wing span .... 9.45 m (31 ft 0 in)
Wing chord, constant .... 1.17 m (3 ft 10 in)
Wing aspect ratio 8.1

Rotax-engined Dorna Blue Bird ***NEW***/0546944

Length overall: 6.17 m (20 ft 3 in)
Fuselage max width: 1.12 m (3 ft 8 in)
Height overall: 2.17 m (7 ft 1½ in)
Wheel track: 1.94 m (6 ft 4½ in)
Wheelbase: 1.51 m (4 ft 11½ in)
Propeller diameter: 1.72 m (5 ft 7¾ in)
Propeller ground clearance: 0.18 m (7 in)

DIMENSIONS, INTERNAL:
Cockpit: Length: 1.09 m (3 ft 7 in)
Max width: 1.02 m (3 ft 4¼ in)

AREAS:
Wings, gross: 11.06 m^2 (119.0 sq ft)
Ailerons (total): 0.95 m^2 (10.23 sq ft)
Trailing-edge flaps (total): 1.54 m^2 (16.58 sq ft)
Fin: 0.42 m^2 (4.52 sq ft)
Rudder: 0.37 m^2 (3.98 sq ft)
Tailplane: 0.84 m^2 (9.04 sq ft)
Elevators (total): 0.515 m^2 (5.54 sq ft)

WEIGHTS AND LOADINGS:
Weight empty, equipped: 490 kg (1,080 lb)
Max fuel weight: 90 kg (198 lb)
Max T-O and landing weight: 750 kg (1,653 lb)
Max wing loading: 67.8 kg/m^2 (13.89 lb/sq ft)
Max power loading: 8.87 kg/kW (14.58 lb/hp)

PERFORMANCE (estimated):
Never-exceed speed (VNE): 173 kt (321 km/h; 200 mph)
Max level speed at FL120: 122 kt (225 km/h; 140 mph)
Max cruising speed at FL120: 113 kt (209 km/h; 130 mph)
Econ cruising speed at FL120: 87 kt (161 km/h; 100 mph)
Stalling speed at S/L, flaps down: 46 kt (84 km/h; 52 mph)
Max rate of climb at S/L: 229 m (750 ft)/min
Service ceiling: 4,265 m (14,000 ft)
T-O run: 275 m (900 ft)
Landing run: 244 m (800 ft)
Range with max fuel: 695 n miles (1,287 km; 800 miles)

UPDATED

FACI

FAJR AVIATION AND COMPOSITES INDUSTRY

Km 5, Karaj Road, PO Box 13445-885, Tehran
Tel: (+98 21) 465 94 61
Fax: (+98 21) 465 94 60
e-mail: faci@fajr-ind.com
Web: http://www.fajr-ind.com
GENERAL DIRECTOR: Mohammad Bani Najarian

FACI established 1991, with objective of developing all-composites aircraft in Iran. Plant has 30,000 m^2 (323,000 sq ft) of covered space at Mehrabad International Airport; total workforce is 150. Its first design, the F-3, entered production in 2001 and made its public debut in October 2002. Production continuing in 2004.

UPDATED

FACI FAJR F-3

English name: Dawn

TYPE: Four-seat lightplane.

PROGRAMME: Design began 1992; construction of prototype started 1993; first flight 26 January 1995; further airframe for static testing; JAR 23-standard domestic type certificate awarded September 2000; construction of first production aircraft began 2001; first deliveries due June 2002. Development plans so far announced include introduction of retractable landing gear and some weight reduction measures to improve cruise performance; flight tests of modified aircraft were scheduled for June 2002, but not reported by early 2004.

CUSTOMERS: Domestic orders for 20 by January 2003. Five reportedly completed by that time, of which two then delivered. Production had reached at least 10 by April 2004. Ordered by Iranian Border Patrol, which is reported to have requirement for 10.

DESIGN FEATURES: Intended for pilot training and utility roles. Designed for low cost and easy maintenance.

Horstmann-Quast HQ-42E advanced laminar aerofoil section, with thickness/chord ratio of 16 per cent and 2° 17' sweepback. Dihedral 6° 30' from roots; twist −2° 30'; incidence 2° 18'. Raked wingtips.

FLYING CONTROLS: Conventional and manual. Tab on each aileron. Electrically actuated trim tab on rudder and each elevator. Fixed-incidence tailplane. Electrically actuated Fowler flaps.

STRUCTURE: All-composites; monocoque fuselage; high-stress parts of glass fibre airframe are reinforced with carbon fibre.

LANDING GEAR: Tricycle-type; retractable. Goodyear tyres: size 15×6.00–6 (main) and 5.00×5 (6 ply) (nose); pressures 4.41 bar (64 lb/sq in) and 3.38 bar (49 lb/sq in) respectively. Nose gear is steerable (±18°) through a mechanical linkage to the rudder pedals. Parker Hannifin brakes are direct-operation, hydraulically actuated disc type. Minimum ground turning radius 2.75 m (9 ft 0¼ in).

POWER PLANT: One Textron Lycoming AEIO-540-L1B5 flat-six engine, flat rated at 201 kW (270 hp), driving a three-blade Hoffmann HOV 123K-V-K200-AH constant-speed propeller. Fuel capacity 212 litres (56.0 US gallons; 46.6 Imp gallons) standard; 292 litres (77.1 US gallons; 64.2 Imp gallons) optional. Oil capacity 15.1 litres (4.0 US gallons; 3.3 Imp gallons).

ACCOMMODATION: Four persons, in pairs. Dual controls. Two upward-hinged window/doors of glass fibre. Cabin heated and ventilated.

SYSTEMS: 24 to 28 V DC electrical system powered by Gill 70 A 10 Ah alternator. Oxygen system.

AVIONICS: To customer's requirements. Options include Bendix/King VHF com, ADF and GPS.

DIMENSIONS, EXTERNAL:
Wing span: 10.50 m (34 ft 5½ in)
Wing chord: at root: 1.76 m (5 ft 9¼ in)
at tip: 0.94 m (3 ft 1 in)
Wing aspect ratio: 7.9
Length overall: 8.07 m (26 ft 5¾ in)
Height overall: 3.05 m (10 ft 0 in)
Tailplane span: 3.87 m (12 ft 8¼ in)
Wheel track: 2.08 m (6 ft 10 in)
Wheelbase: 1.78 m (5 ft 10 in)
Propeller diameter: 2.10 m (6 ft 10¾ in)
Propeller ground clearance: 0.175 m (7 in)

DIMENSIONS, INTERNAL:
Cabin: Length: 2.18 m (7 ft 1¾ in)
Max width: 1.25 m (4 ft 1¼ in)
Max height: 1.53 m (5 ft 0¼ in)

AREAS:
Wings, gross: 14.02 m^2 (150.9 sq ft)

The FACI Fajr F-3 four-seat low-wing lightplane ***NEW***/0593248

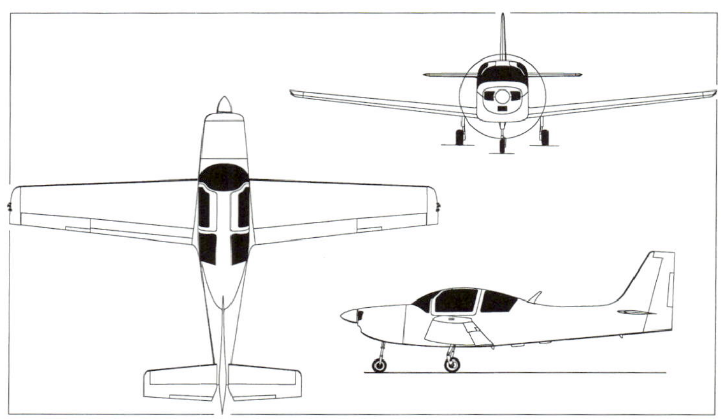

General arrangement of the FACI F-3 0110841

Ailerons (total, incl tabs): 0.46 m^2 (4.95 sq ft)
Flaps (total): 0.95 m^2 (10.23 sq ft)
Fin: 2.33 m^2 (25.08 sq ft)
Rudder (incl tab): 0.91 m^2 (9.80 sq ft)
Tailplane: 2.95 m^2 (31.75 sq ft)
Elevators (total, incl tabs): 1.16 m^2 (12.49 sq ft)

WEIGHTS AND LOADINGS:
Weight empty: 1,100 kg (2,425 lb)
Max fuel weight: standard: 141 kg (311 lb)
optional: 204 kg (450 lb)
Max T-O and landing weight: 1,580 kg (3,483 lb)
Max wing loading: 112.7 kg/m^2 (23.08 lb/sq ft)
Max power loading: 7.85 kg/kW (12.90 lb/hp)

PERFORMANCE:
Never-exceed speed (VNE): 200 kt (370 km/h; 230 mph)
Max level speed at FL40: 160 kt (296 km/h; 184 mph)
Cruising speed at 65% power at FL90: 144 kt (267 km/h; 166 mph)
Stalling speed, power off:
flaps up: 67 kt (125 km/h; 78 mph)
flaps down: 56 kt (104 km/h; 65 mph)
Service ceiling: 5,480 m (18,000 ft)
Max rate of climb at S/L: 314 m (1,030 ft)/min
T-O run: 310 m (1,020 ft)
T-O to 15 m (50 ft): 370 m (1,215 ft)
Landing from 15 m (50 ft): 720 m (2,365 ft)
Landing run: 360 m (1,180 ft)
Range, 65% power at FL100, standard fuel: 610 n miles (1,129 km; 702 miles)
Endurance: 6 h 15 min
g limits: +4.4/−1.76

UPDATED

IRGC

IRAN REVOLUTIONARY GUARD CORPS

Development of the Shahed 274 helicopter is reported to have been undertaken by the Pasdaran, or Revolutionary Guards. The organisation may also be responsible for a new transport helicopter, though no details of this had emerged by mid-2004.

UPDATED

IRGC SHAHED 274

TYPE: Light utility helicopter.

PROGRAMME: Sponsored (as X-5) by the Institute of Industrial Research and Development of the IRGC; reportedly due to have made its first flight in 1997. First aircraft (71-832) handed over to IRGC 16 September 1999. International public debut, in Tehran, 30 December 2000. At least two more (74-001 and '002) in service by end of 2001. Further public appearance in air show on Kish Island, October/November 2002. Current status still uncertain in mid-2004.

CUSTOMERS: Total of 20 (other sources suggest 30) were reportedly planned to be built by end of 2004.

DESIGN FEATURES: Two-blade main and tail rotors; fully enclosed cabin and tailboom; upper and lower vertical fins. Intended applications include observation, rescue and light cargo-carrying.

LANDING GEAR: Twin-skid type.

POWER PLANT: One 313 kW (420 shp) Rolls-Royce 250-C20B turboshaft.

ACCOMMODATION: Seats for five persons including pilot. Forward-opening crew door and passenger door each side; baggage door aft of latter on port side.

Following data provisional and unconfirmed.

DIMENSIONS, EXTERNAL:
- Main rotor diameter ..... 10 m (33 ft)
- Fuselage length ..... 9 m (30 ft)
- Height overall ..... 3 m (10 ft)

WEIGHTS AND LOADINGS:
- Weight empty ..... 1,000 kg (2,200 lb)
- Max T-O weight ..... 1,500 kg (3,300 lb)

First photograph released of the Shahed 274 five-seat helicopter ***(Aviation Industries Journal of Iran)*** 0095152

PERFORMANCE:
- Max level speed ..... 97 kt (180 km/h; 112 mph)
- Ceiling ..... 5,200 m (17,060 ft)
- Max range ..... 324 n miles (600 km; 372 miles)

UPDATED

IRGC TRANSPORT HELICOPTER

TYPE: Multirole medium helicopter.

PROGRAMME: According to a government announcement, a new Iranian-built transport helicopter made its first 'official' flight on 11 August 2001. No name or designation, or other details, were given, the announcement adding only that the aircraft was intended for airlifting military personnel and for civilian relief operations. This event may be connected with reports from industry sources at about the same time that Iran was considering either the remanufacture or new production of the Bell 214, large numbers of which were supplied to Iran before 1979. Plans for in-country licensed manufacture of the Bell 214 were thwarted following that year's revolution.

VERIFIED

IRIAF

ISLAMIC REPUBLIC OF IRAN AIR FORCE (YA HOSSEIN PROJECT/OWJ INDUSTRIAL COMPLEX)

PO Box 13885-193, Mehrabad Airport, Tehran
Tel: (+98 21) 452 08 10
Fax: (+98 21) 452 08 53

The Ya Hossein, also known as the Owj Industrial Complex, is an agency of the Iranian MoD and IRIAF, established in the early 1990s. Its activities have included design and development of various aircraft components and subsystems in both metal and composites, plus assorted ground equipment. Its most significant current project is the Tazarve jet trainer, which made its first flight in 2001 and its public debut at the end of 2002.

UPDATED

IRIAF SAEGHE

English name: Lightning

TYPE: Attack fighter.

PROGRAMME: Reportedly under development by IRIAF's MATSA technology and electronics centre at Mehrabad. Modified Northrop F-5E, rebuilt in twin-finned configuration and probably incorporating experience gained in Project Simorgh work on building new fuselage sections to convert in-country F-5As into two-seat F-5Bs. Photo of model shown on Iranian Students' News Agency website in 2003. First Saeghe airframe reportedly about 70 per cent complete by November 2002 and expected to be completed within next six months. Relationship to IAMI (HESA) Azarakhsh not yet clear.

UPDATED

IRIAF (YA HOSSEIN) TAZARVE

English name: Eagle

TYPE: Basic jet trainer.

PROGRAMME: First design studies early 1990s, leading to first prototype (named *Dorna*: crane), flown in or about 1995; substantially redesigned second prototype (*Tondar*: thunder) flown

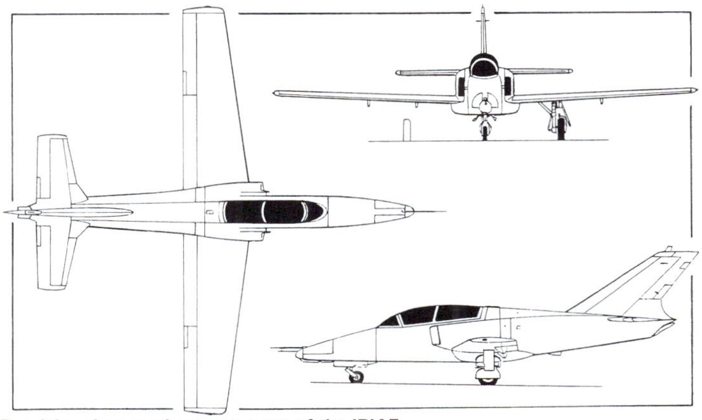

Provisional general arrangement of the IRIAF Tazarve ***(Michael Badrocke)*** ***NEW***/0567875

1998, leading to further wing and fuselage refinements; Tazarve is third (and reportedly final) configuration, embodying further wing, fuselage and systems improvements; manufacture began 1998 and first flight said to have taken place in mid-2001; public debut October/November 2002 at airshow on Kish Island, in full IRIAF insignia and wearing unexplained code 778-2 on vertical fin. Designation JT2-2 has been suggested for this aircraft. Fourth prototype under construction at that time and then due to fly during 2003.

CUSTOMERS: Under development for IRIAF, which has ordered five pre-series aircraft for delivery by 2005 and plans acquisition of 25 production examples thereafter.

DESIGN FEATURES: Intended to provide basic, advanced and lead-in fighter training, with secondary capability for light attack. Low/mid-mounted wings, lateral air intakes, conventional sweptback tail unit. Designed for 6,000 hour operational life.

Iran's Tazarve all-new jet trainer prototype ***(Robert Hewson)*** ***NEW***/0567232

FLYING CONTROLS: Hydraulically boosted ailerons, elevators and rudder; hydraulic flaps with push/pull rod controls.

STRUCTURE: Mainly GFRP/nickel sandwich and carbon fibre. Wings, fuselage and tail sections built as single pieces and bolted together; wing and tail structures resin-bonded. All load-bearing elements of main structure are carbon fibre; parts such as fairings, panels and doors are of GFRP. Nickel layer is sandwiched within composites for protection against lightning strikes and static electricity. MIL-SPEC design criteria applied throughout.

LANDING GEAR: Retractable tricycle type, nosewheel rearward, mainwheels inward into engine intake trunks.

POWER PLANT: One 12.68 kN (2,850 lb st) General Electric J85-GE-17 non-afterburning turbojet in prototypes; alternative such as Honeywell TFE731-2A turbofan being sought for production aircraft. Fuel capacity 704 kg (1,552 lb) in wing tanks and 300 litres (79.3 US gallons; 66.0 Imp gallons) in fuselage tanks.

ACCOMMODATION: Crew of two in tandem on Martin-Baker Mk 15 ejection seats; rear seat elevated. Wraparound windscreen; one-piece framed canopy, with built-in MDC, opening sideways to starboard. Accommodation fully pressurised and air conditioned.

AVIONICS: *Comms:* Locally produced UHF/VHF radio.

Instrumentation: Includes VOR, ILS, DME, ADF and Tacan, sourced from inventory stores for aircraft such as F-4 and PC-7.

DIMENSIONS, EXTERNAL:

Wing span	8.04 m (26 ft 4½ in)
Length overall	10.70 m (35 ft 1¼ in)
Height overall	3.63 m (11 ft 11 in)

WEIGHTS AND LOADINGS:

Weight empty	2,550 kg (5,622 lb)
Max T-O weight	4,000 kg (8,818 lb)

PERFORMANCE:

Max level speed at 5,485 m (18,000 ft)	350 kt (648 km/h; 402 mph)
Stalling speed	85 kt (158 km/h; 98 mph)
Service ceiling	11,582 m (38,000 ft)
T-O run	600 m (1,970 ft)
Landing run	650 m (2,135 ft)
Max range	405 n miles (750 km; 466 miles)
g limits	+8.0/−3.8

UPDATED

Israel

AD & D

AERO-DESIGN & DEVELOPMENT LTD

PO Box 565, Gad Finestein Road, Rehovot Industrial Park, IL-76102 Rehovot
Tel: (+972 8) 931 55 33
Fax: (+972 8) 931 55 32
e-mail: ad_d@netvision.net.il
Web: http://www.ad-d.co.il
CEO: Haggai Salomons
DIRECTORS:
Itzhak Dvir
Israel Gottesman
Danny Israeli

In addition to the Hummingbird VTOL lifting vehicle, AD & D developed a UAV version known as the Hornet, which began tethered flight testing in January 1999. Development of the CityHawk is now being continued by Urban Aeronautics Ltd (UrbanAero). AD & D became a subsidiary of Elbit Systems in 2003; activities include development of that company's Dagger and Skylark UAVs. The Hummingbird programme had been placed on hold by February 2004 after completing a few hours of flight testing.

UPDATED

IAI

ISRAEL AIRCRAFT INDUSTRIES LTD

Ben-Gurion International Airport, IL-70100 Tel-Aviv
Tel: (+972 3) 935 31 11
Fax: (+972 3) 935 85 16
Web: http://www.iai.co.il
CHAIRMAN: Ori Orr
PRESIDENT AND CEO: Moshe Keret
EXECUTIVE VICE-PRESIDENT AND COO: Ovadia Harari
CORPORATE VICE-PRESIDENT: Menham Shmul (General Manager, Military Aircraft Group)
DEPUTY CORPORATE VICE-PRESIDENT: Dr David Harari (Director, Research and Development)
DEPUTY CORPORATE VICE-PRESIDENT, MARKETING: Shimon Eckhaus
DIRECTOR OF CORPORATE INDUSTRIAL SERVICES: Menham Tadmor
DEPUTY CORPORATE VICE-PRESIDENT, COMMUNICATIONS: Doron Suslik

Tel: (+972 3) 935 85 09
Fax: (+972 3) 935 85 12
e-mail: hpaz@hdq.iai.co.il

Founded 1953 as Bedek Aviation; ceased being unit of Ministry of Defence and became government-owned corporation 1967; name changed to Israel Aircraft Industries 1 April 1967; number of divisions reduced from five to four February 1988 (Aircraft, Electronics, Technologies and Bedek Aviation). Further restructuring 1994 and now comprises four independent operating Groups: Commercial Aircraft, Military Aircraft, Bedek Aviation and Electronics. Covered floor space 680,000 m^2 (7.32 million sq ft); total workforce 14,500 in August 2001, of whom one-third are engineers or academics. IAI has ISO 9001 certification, and its products and technologies are fully backed by integrated logistic support, training and a complete after-sales service.

Approved as repair station and maintenance organisation by Israel Civil Aviation Administration, US Federal Aviation Administration, UK Civil Aviation Authority and Israel Air Force. Products include aircraft of own design; in-house produced airframe systems and avionics; service, upgrading and retrofit packages for civil and military aircraft and helicopters; other activities include space technology, missile and ordnance development, and seaborne and ground equipment.

Sold Galaxy Aerospace to General Dynamics on 1 May 2001, this becoming part of Gulfstream.

IAI comprises four principal divisions:

Bedek Aviation Group: Provides total aircraft, engine and component repair, overhaul, retrofit, modification, conversion and support services to airlines and leasing companies. On 21 February 2002, IAI and Hindustan Aeronautics Limited of India signed a co-operation agreement for HAL to participate in Bedek's Boeing 737 freighter conversion programme.

Military Aircraft Group: Develops and manufactures unmanned air vehicles; integrates and upgrades military fixed-wing aircraft and helicopters; designs and manufactures helicopter structures, electrical harnesses and crash-attenuating aircraft seats.

Electronics Group: Specialises in electronics, electro-optical and inertial systems, and components for military and commercial applications.

Commercial Aircraft Group: *Described separately.*

VERIFIED

COMMERCIAL AIRCRAFT GROUP

Ben-Gurion International Airport, IL-70100 Tel-Aviv
Tel: (+972 3) 935 33 37 and 935 71 49
Fax: (+972 3) 935 42 26
GENERAL MANAGER: M Boness

Established 1994. Design, development, certification and manufacture of civil aircraft; hydraulic, electromechanical and pneumatic components; dynamic systems; and composites material structures (all backed by full computer-aided design, engineering and manufacturing processes).

Gulfstream 100 (formerly IAI Astra) and Gulfstream 200 (formerly IAI Galaxy) are built in Israel and delivered to US for outfitting and sale. Gulfstream 150 under development.

VERIFIED

IAI C-5WA AIRTRUCK

TYPE: Twin-turboprop freighter.

PROGRAMME: IAI submission chosen in principle by Federal Express of USA against competition from Saab (Sweden) and Ayres (USA) in mid-1998 as replacement for Fokker Friendship parcel transports. FedEx purchase would be 100, but IAI is seeking further 100 sales to reduce unit price to FedEx-imposed ceiling of US$10 million. Risk-sharing partners also needed to ensure programme launch. Certification and first customer deliveries scheduled for four years after launch, with initial production rate of four per month. In 2004, the programme remained dormant.

UPDATED

URBANAERO

URBAN AERONAUTICS LTD

PO Box 190, Airport City, Ben-Gurion Airport, IL-70151 Tel-Aviv
Tel: (+972 3) 683 57 98
Fax: (+972 3) 518 76 45
e-mail: info@urbanaero.com
Web: http://www.urbanaero.com
PRESIDENT: Dr Rafi Yoeli
VICE-PRESIDENT, OPERATIONS: Omri Knoller
MANAGER, PUBLIC RELATIONS AND MARKETING: Janina Frankel-Yoeli

Company formed in 2000 (previously known as Romeo Yankee Ltd), specifically to develop urban aerial vehicles. Currently has CityHawk technology demonstrator under test; has also begun to develop X-Hawk (civil) and TurboHawk (military) variants, as well as unmanned derivatives.

UPDATED

URBANAERO CITYHAWK

TYPE: Lifting vehicle.

PROGRAMME: Announced April 2000; CityHawk prototype/demonstrator completed and undergoing tethered testing since August 2002. Engines had been run at 80 per cent power by September 2002, achieving lift-off of rear duct. During further trials between January and September 2003, performed 10 hover flights at low level to test patented vane control system.

CURRENT VERSIONS: **CityHawk:** POC demonstrator; *as described.*

X-Hawk: Civil derivative; *described separately.*

DESIGN FEATURES: VTOL research vehicle: two-seat extrapolation of AD & D Hummingbird concept. T-O/landing area compatible with typical urban parking spaces and garages; significantly quieter than a helicopter due to ducted fan design. Foreseen applications of this technology include personal urban transportation, police and traffic patrol, ambulance and urban evacuation, air taxi, and agricultural/environmental monitoring.

FLYING CONTROLS: In-house-developed fly-by-wire, controlled manually from left-hand seat. Differential thrust between fans for pitch and forward speed control; combined thrust of both fans controls vertical speed (rise/descent). Roll and yaw control by cascades of vanes on upper and lower surfaces of each fan.

STRUCTURE: Mainly (90 per cent) Kevlar/graphite composites shell.

LANDING GEAR: Two large, low-pressure tyres at front; two spring-mounted wheels at rear.

POWER PLANT: Two fans, each driven by four 23.1 kW (31 hp) Zanzottera Mz34i single-cylinder piston engines. Engine redundancy allows vehicle to fly down to a landing even if each fan loses one engine. Fuel capacity 60 litres (15.9 US gallons; 13.2 Imp gallons).

ACCOMMODATION: Side-by-side cockpits for two persons under separate canopies.

AVIONICS: Full instrument suite with telemetry system.

DIMENSIONS, EXTERNAL:
Length ... 4.70 m (15 ft 5in)
Max width ... 2.20 m (7 ft 2½ in)
Max height ... 1.80 m (5 ft 10¾ in)

WEIGHTS AND LOADINGS:
Max T-O weight ... 570 kg (1,256 lb)

PERFORMANCE (estimated):
Max level speed ... 80-90 kt (148-166 km/h; 92-103 mph)
Max operating altitude ... 2,440 m (8,000 ft)
T-O/landing footprint ... 2.5 × 5.5 m (8.2 × 18 ft)
Endurance ... almost 1 h

UPDATED

URBANAERO X-HAWK

TYPE: Lifting vehicle.

PROGRAMME: Commercial derivative of CityHawk.

CURRENT VERSIONS: **X-Hawk:** Planned, FAA-certified commercial aerial utility vehicle with modular payload bay and twin turboshaft engines; *as described.* Funding being sought for prototype construction.

DESIGN FEATURES: VTOL rescue vehicle. Two ducted fans and two pusher fans, all connected to pair of turboshaft engines mounted in rear nacelles. Cockpit on port side, payload bay on starboard side, between main lifting ducts. Potential civil applications include medical access and evacuation, powerline maintenance, bridge inspection and ship-to-shore taxi operation.

POWER PLANT: Twin turboshaft engines, total power 969 kW (1,300 shp).

ACCOMMODATION: Four persons including pilot. Inboard-hinged, upward-opening canopy each side.

Following data are provisional:

DIMENSIONS, EXTERNAL:
Length ... 5.70 m (18 ft 8½ in)
Max width ... 3.00 m (9 ft 10 in)
Max height ... 2.20 m (7 ft 2½ in)

DIMENSIONS, INTERNAL:
Payload bay floor area ... 3.50 m² (37.7 sq ft)

WEIGHTS AND LOADINGS:
Weight empty ... 1,157 kg (2,550 lb)
Max fuel weight ... 431 kg (950 lb)
Payload with max fuel ... 340 kg (750 lb)[1]
Max T-O weight ... 1,927 kg (4,250 lb)[1]

[1] *On single engine*

PERFORMANCE (estimated):
Max level speed ... 80-100 kt (148-185 km/h; 92-115 mph)
Max altitude ... 3,660 m (12,000 ft)
Mission radius for 30 min loiter ... 100 n miles (185 km; 115 miles)
Endurance ... 1 h 30 min

UPDATED

CityHawk during 2003 hover tests ***NEW***/0576828

UrbanAero CityHawk prototype ***NEW***/0593128

Four-seat X-Hawk in mockup form ***NEW***/0576829

Additional angle of X-Hawk mockup ***NEW***/0576830

Italy

AERONAUTICA MACCHI

AERONAUTICA MACCHI SPA
(A FINMECCANICA COMPANY)
Via Ing Paolo Foresio 1, I-21040 Venegono Superiore (VA)
Tel: (+39 0331) 86 59 12
Fax: (+39 0331) 86 59 10

Original Macchi company, founded by Eugenio Macchi on 1 May 1913 in Varese, produced famous line of high-speed flying-boats and seaplanes. Aeronautica Macchi became holding company for Aermacchi SpA (see following entry); SICAMB (Società Italiana Costruzioni Aeronautiche Martin-Baker; aerostructures and equipment manufacturing, including licensed production of Martin-Baker ejection seats) at Latina; and Logic (aircraft equipment and controls). A 27.4 per cent holding in Aeronautica Macchi was acquired by Aeritalia, now Alenia (and part of Finmeccanica), in 1983, the balance being held by the Foresio family. In January 1997, Aermacchi acquired the SIAI-Marchetti company from Alenia-Finmeccanica and on 19 December 2002, Finmeccanica obtained control of Aeronautica Macchi by increasing its 27.4 per cent shareholding to 94.6 per cent through acquisition of Foresio's 67.2 per cent share holding, this becoming effective on 1 July 2003. Holdings in SICAMB and Logic have been disposed of.

Also obtained with the SIAI-Marchetti agreement was Finmeccanica's nacelle manufacturing activities for Airbus airliners and Dassault Falcon business jets, previously the responsibility of Alenia at Turin. Related activities formerly undertaken by SIAI included Airbus A300 and 310 inspection doors and tailcones; overhaul and repair of various types of aircraft (notably C-130 Hercules, DHC-5 Buffalo and Cessna Citation II); participation in national or multinational programmes, producing parts for Eurofighter Typhoon (carbon fibre structures, titanium engine cowlings, ECM pods and weapon pylons), Alenia G222/C-27J (wing and tailerons), Panavia Tornado (wing furniture) and AMX. Macchi makes engine nacelles for Airbus A 318, 319, 320, 321 and 330, Dassault Falcon 900 and Falcon 2000; 1,000th Airbus nacelle delivered in December 2000. Contracted to build nacelles (in collaboration with Hurel Dubois) for Embraer 170/190 family. Will produce (with P&W Canada) complete nacelle and thrust reverser for Dassault Falcon 7X. Was risk-sharing partner in Fairchild Dornier (now AvCraft) 328JET, having supplied 100th fuselage (including earlier turboprop version) in 1998. Designed engine nacelles for Trent 900 and GP7200 versions of Airbus A380; first component delivered 6 May 2003. Space activities include elements of Ariane 5.

Workforce in 2003 totalled over 2,200, including 1,800 in Aermacchi. Activities in civil field accounted for over 50 per cent of turnover by 2000.

UPDATED

AERMACCHI SPA (SUBSIDIARY OF AERONAUTICA MACCHI SPA)

Via Ing Paolo Foresio 1, PO Box 101, I-21040 Venegono Superiore (VA)
Tel: (+39 0331) 81 31 11
Fax: (+39 0331) 82 75 95
e-mail: info@aermacchi.it
Web: http://www.aermacchi.it
CHAIRMAN: Dott Ing Giorgio Brazzelli
DEPUTY CHAIRMAN: Giorgio Zappa
MANAGING DIRECTOR: Giovanni Bertolone

Aermacchi is aircraft manufacturing company of Aeronautica Macchi group; plants at Venegono airfield occupy total area of 274,000 m^2 (2,949,300 sq ft), including 113,000 m^2 (1,216,300 sq ft) covered space; flight test centre has covered space of 5,100 m^2 (54,900 sq ft) in total area of 28,000 m^2 (301,400 sq ft). Subsidiary company is Logic avionics.

Aermacchi was one of two Italian assembly centres for AMX (an international venture also involving Brazil's Embraer), having produced 36, plus nine AMX-Ts; also has important roles in Eurofighter Typhoon and Yakovlev Yak-130 programmes, offering Westernised version of latter as M-346. In January 1996, Aermacchi obtained production rights to the former Valmet L-90TP Redigo turboprop trainer, which was redesignated M-290TP Redigo. On 1 January 1997, Aermacchi acquired all training activities and programmes of SIAI-Marchetti, including the SF.260 and S.211 (but not SF.600 Canguro, which obtained by VulcanAir) and support of earlier S.205s and S.208s remaining in service. Production was then transferred from Sesto Calende to the Aermacchi plant at Venegono. Continues to promote S.211, including S.211AD 'glass cockpit' upgrade based on MB-339FD, although no recent sales achieved. Allied with Dassault, EADS and Saab in undertaking feasibility study (contract signed 2 December 2002) for 12-nation Eurotrainer military training scheme). Since 1960, 2,000 Aermacchi trainers have been bought by 40 countries.

Aermacchi continues to promote the S.211, including a digital avionics upgrade *(Paul Jackson)* ***NEW***/0573408

AERMACCHI SF-260

Uruguayan Air Force designation: T-260

TYPE: Basic prop trainer/attack lightplane.

PROGRAMME: Originated as F.250 designed by Stelio Frati and made by Aviamilano; SF.250 prototype (I-RAIE) flew 15 July 1964, combining wings of F.8L Falco and fuselage of F.15 Picchio with modified canopy. One only, followed by two SF.260 prototypes (I-ALLA and HB-ELB) before project passed to SIAI-Marchetti.

Initial civil variants were SF.260 and SF.260B (Waco TS-250-3 Meteor on US market), followed by increased wing span SF.260C; SF.260D was later civil variant; SF.260M military trainer, sold widely in several subvariants; current versions developed (unsuccessfully) for USAF's Enhanced Flight Screener requirement; still in low-volume production as a military trainer following transfer of design rights to Aermacchi in 1997 as SF-260 (with hyphen). By 1999, SF-260s had flown more than 1,700,000 hours.

CURRENT VERSIONS: **SF-260E:** Direct-injection engine, strengthened airframe and 100 kg (220 lb) higher aerobatic weight than superseded SF.260D. Wing aerodynamically refined to preserve original stalling speed, despite increased MTOW; automatic fuel management. Canopy height and width increased. Certified by RAI on 21 January 1992 and by FAA on 17 August 1994.

Main description applies to SF-260E and Warrior.

SF-260F: As SF-260E, but carburetted engine. Promotion discontinued.

Aermacchi SF-260E development aircraft *(Paul Jackson)* ***NEW***/0573406

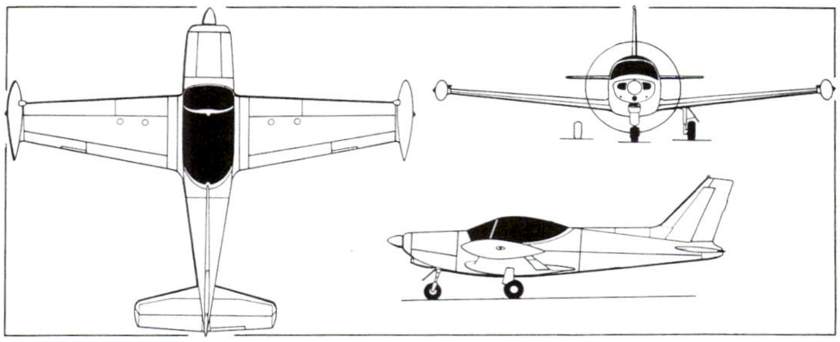

Aermacchi SF-260E/F basic trainer *(Paul Jackson)* 0011076

SF-260E Warrior: Two underwing pylons, for up to total 300 kg (661 lb) of external stores, and cockpit stores selection panel. Able to undertake a wide variety of roles, including low-level strike; forward air control; forward air support; armed reconnaissance; and liaison. Export versions are SF-260EU.

In November 2003, 30 SF-260EAs were ordered by the Italian Air Force to replace SF-260AMs of 71° Stormo at Latina in the primary training role; deliveries contracted from early 2005, agreement also including buy-back of 21 SF-260AMs.

SF-260TP: Turboprop-powered version. No recent production known.

CUSTOMERS: Total of some 880 produced (including turboprop version) for customers in 27 countries. Civil operators include Sabena, Royal Air Maroc and Alitalia. Six SF-260Fs ordered by Air Force of Zimbabwe, early 1997, of which last delivered in October 1998; these were first of type built at Venegono. In June 1998, 12 SF-260Es bought by Venezuelan Air Force, and delivered to 14 Grupo at Maracay from late 2000. Delivery of 13 SF-260Es to Uruguayan Air Force began in September 1999, following first flight on 23 July, and completed in December 2000. Mexico placed order in early 1998, initially for 13, but subsequently 30, of which first six delivered in early 2000. Five Warrior variants of SF-260EU to Mauritania in 2000.

DESIGN FEATURES: Side-by-side, piston-engined aerobatic trainer. Low-mounted wings with forward-swept trailing-edges and tip tanks; sweptback fin.

Wing section NACA 64_1-212 (modified) at root, 64_1-210 (modified) at tip; dihedral 6.5° 20' from roots.

FLYING CONTROLS: Conventional and manual. Frise ailerons each with servo tab; trim tabs for all three axes; four-way trim button on each stick top; electrically operated slotted flaps; controls operated by dual cables.

STRUCTURE: All-metal stressed skin structure; wing skin formed of butt-jointed panels flush riveted; single main spar and auxiliary rear spar; press-formed ribs; wings bolted together on centreline and attached to fuselage by six bolts. Fatigue life 15,000 hours.

LANDING GEAR: Electrically retractable tricycle type, with manual emergency actuation. Inward-retracting main gear, of trailing arm type, and rearward-retracting nose unit, each embodying Magnaghi oleo-pneumatic shock-absorber (Type 2/22028 in main units). Nose unit is of leg and fork type, with coaxial shock-absorber and torque strut. Cleveland P/N 3080A mainwheels, with size 6.00–8 tube and tyre (6 ply rating), pressure 2.45 bar (36 lb/sq in). Cleveland P/N 40–77A nosewheel, with size 5.00–5 tube and tyre (8 ply rating), pressure 1.96 bar (28 lb/sq in). Cleveland P/N 3000–500 independent hydraulic single-disc brake on each mainwheel; parking brake. Nosewheel steering (±20°) operated directly by rudder pedals.

POWER PLANT: One 194 kW (260 hp) Textron Lycoming AEIO-540-D4A5 flat-six driving a Hartzell HC-C2YK-1BF/8477-8R two-blade, constant-speed, metal propeller.

Fuel in two light alloy tanks in wings each holding 49.5 litres (13.1 US gallons; 10.9 Imp gallons); and two permanent wingtip tanks each holding 72 litres (19.0 US gallons; 15.85 Imp gallons). Total internal fuel capacity 243 litres (64.2 US gallons; 53.3 Imp gallons), of which 235 litres (62.1 US gallons; 51.7 Imp gallons) are usable. Individual refuelling point on top of each tank. In addition, SF-260W may be fitted with two 80 litre (21.1 US gallon; 17.5 Imp gallon) auxiliary tanks on underwing pylons. Oil capacity (all models) 11.4 litres (3.0 US gallons; 2.5 Imp gallons).

ACCOMMODATION: Two pilots side by side in adjustable seats, with full blind-flying panel on right and reduced panel and radio on left. Third seat in rear. All three seats equipped with lap belts and shoulder harnesses. Baggage compartment aft of rear seat. Upper portion of sliding canopy tinted. Emergency canopy release handle for front seat occupant. Steel tube windscreen frame for protection in the event of an overturn. Air conditioning, oxygen and enlarged canopy optional.

SYSTEMS: Foot-powered hydraulics for mainwheel brakes only. No pneumatic system. 24 V DC electrical system of single-conductor negative earth type, including 70 A Prestolite engine-mounted alternator/rectifier and 24 V 24 Ah battery, for engine starting, flap and landing gear actuation, fuel booster pumps, electronics and lighting. Sealed battery compartment in rear of fuselage on port side. Connection of an external power source automatically disconnects the battery. Heating system for carburettor air intake. Cabin heating, and windscreen de-icing and demisting, by heat exchanger using engine exhaust muffler. Additional manually controlled warm air outlets for general cabin heating. Optional R134a gas air conditioning system and autopilot both available from 2000.

AVIONICS: *Comms:* Bendix/King or Rockwell Collins airways radio with dual com and nav; in SF-260E, avionics mounted aft of cabin and battery moved forward.

AERMACCHI SF-260 MILITARY CUSTOMERS

Operator	Variant	Qty	First Aircraft
Belgium	MB	36	ST-01
	D	9	ST-40
Bolivia	CB	6	FAB-180
Brunei	W	2	129
Burkina Faso		4[2]	
Burundi	C	3	9U-RUA
Congo Republic	MC	21	AT-101
Dubai	TP	6	401
	W	1	H
Ethiopia	TP	11	156?
Haiti	TP	6	1271
Ireland	WE	11	222
Italy	260	3	MM91004
	AM	45	MM54418
	EA	30	MM
Libya	W	240	
Mauritania	EU	4[1 and 5]	5T-MAK
Mexico	E	30[1]	—
Myanmar	M	20	2001
Philippines	MP	30	601
	WP	16	631
	TP	18	701
Singapore	W	12	151
	MS	16	120
Somalia	C	6	6O-SBC
	W	10	
Sri Lanka	TP	9[3]	CT121
Thailand	MT	19	15–01
Tunisia	CT	6	W41501
	WT	12	W41401
Turkey	D	41	773
Uganda		3[4]	AF504
Uruguay	EU	13[1]	610
Venezuela	EU	12[1]	0027
Zambia	MZ	9	AF-507
Zimbabwe	C	17	R9900
	W	14	R3260
	F	6[1]	
Subtotal		**757**	
various civil		125	
Total		**882**	

[1] Built by Aermacchi
[2] Plus others second-hand
[3] Plus three second-hand
[4] Plus two or three from Libya
[5] Warrior

Notes: Nicaragua received some from Libya; Singapore donated 19 to Indonesia in 2002.

EQUIPMENT: Military equipment to customer's requirements. External stores can include one or two reconnaissance pods with two 70 mm automatic cameras, or two emergency supply containers or two external tanks. Landing light in nose, below spinner.

ARMAMENT: Warrior has two underwing hardpoints, able to carry external stores on NATO 356 mm (14 in) racks up to a maximum of 300 kg (661 lb) when flown as a single-seater. Typical alternative loads can include gun pods, each with 7.62 mm FN machine guns and 500 rounds; or 0.50 Browning machine gun; two Thiokol LUU-2/B parachute flares; bombs up to 150 kg (331 lb); Mk 76 practice bombs; two cartridge throwers for 70 mm multipurpose cartridges, F 725 flares or F 130 smoke cartridges.

DIMENSIONS, EXTERNAL:
- Wing span over tip tanks .... 8.35 m (27 ft 4¾ in)
- Wing chord: at root .... 1.60 m (5 ft 3 in)
  - mean aerodynamic .... 1.325 m (4 ft 4¼ in)
  - at tip .... 0.785 m (2 ft 7 in)
- Wing aspect ratio (excl tip tanks) 6.3
- Wing taper ratio 2.2
- Length overall .... 7.10 m (23 ft 3½ in)
- Fuselage: Max width .... 1.10 m (3 ft 7¼ in)
  - Max depth .... 1.04 m (3 ft 5 in)
- Height overall .... 2.41 m (7 ft 11 in)
- Elevator span .... 3.01 m (9 ft 10½ in)
- Wheel track .... 2.275 m (7 ft 5½ in)
- Wheelbase .... 1.66 m (5 ft 5¼ in)
- Propeller diameter .... 1.93 m (6 ft 4 in)
- Propeller ground clearance .... 0.32 m (1 ft 0½ in)

DIMENSIONS, INTERNAL:
- Cabin: Length .... 1.66 m (5 ft 5¼ in)
  - Max width .... 1.00 m (3 ft 3¼ in)
  - Height: from seat cushion:
    - to canopy .... 0.98 m (3 ft 2½ in)
    - to enlarged canopy .... 1.07 m (3 ft 6 in)
  - Volume .... 1.5 m³ (53 cu ft)
- Baggage compartment volume .... 0.18 m³ (6.4 cu ft)

AREAS:
- Wings, gross .... 10.10 m² (108.7 sq ft)
- Ailerons (total, incl tabs) .... 0.76 m² (8.18 sq ft)
- Trailing-edge flaps (total) .... 1.18 m² (12.70 sq ft)
- Fin .... 0.76 m² (8.18 sq ft)
- Dorsal fin .... 0.16 m² (1.72 sq ft)
- Rudder, incl tab .... 0.60 m² (6.46 sq ft)
- Tailplane .... 1.46 m² (15.70 sq ft)
- Elevator, incl tab .... 0.96 m² (10.30 sq ft)

WEIGHTS AND LOADINGS:
- Manufacturer's basic weight empty .... 779 kg (1,717 lb)
- Baggage capacity .... 20 kg (44 lb)
- Max fuel weight: wing internal and tip tanks (all versions) .... 169 kg (372.5 lb)
  - underwing tanks .... 115 kg (254 lb)
- Max T-O weight: E .... 1,200 kg (2,645 lb)
  - Warrior .... 1,350 kg (2,976 lb)
- Max wing loading: E .... 118.8 kg/m² (24.33 lb/sq ft)
  - Warrior .... 128.7 kg/m² (26.36 lb/sq ft)
- Max power loading: E, M .... 6.18 kg/kW (10.17 lb/hp)
  - Warrior .... 6.70 kg/kW (11.02 lb/hp)

PERFORMANCE (at 1,100 kg; 2,425 lb):
- Never-exceed speed (VNE) .... 236 kt (437 km/h; 271 mph)
- Max level speed at S/L .... 182 kt (337 km/h; 209 mph)
- Cruising speed at 75% power at 1,830 m (6,000 ft) .... 175 kt (324 km/h; 201 mph)
- Stalling speed, flaps down .... 59 kt (110 km/h; 68 mph)
- Max rate of climb at S/L .... 548 m (1,800 ft)/min
- Service ceiling .... 6,100 m (20,000 ft)
- T-O run .... 275 m (905 ft)
- Landing run .... 270 m (885 ft)
- Range with max internal fuel .... 710 n miles (1,314 km; 817 miles)
- Max range, internal and external fuel .... 1,090 n miles (2,018 km; 1,254 miles)
- g limits .... +6/–3

UPDATED

AERMACCHI M-311

TYPE: Basic jet trainer/light attack jet.

PROGRAMME: Announced at Farnborough International, July 2004, as substantially upgraded version of S.211, incorporating features of S.211A upgrade.

S.211 designed by SIAI-Marchetti and revealed in June 1977; first flight (I-SITF) 10 April 1981. Sold to Haiti (four; not used and to US civilian owners), Philippines (24, including 14 assembled locally) and Singapore (30, comprising six kits and 24 locally-built). Last delivery in 1991; promotion continued, including unsuccessful evaluation for USAF/USN JPATS programme. Acquired by Aermacchi in January 1997, which proposed modifications for existing machines, as detailed in *Jane's Aircraft Upgrades*.

DESIGN FEATURES: Compared with S.211, has strengthened airframe for increased manoeuvre envelope and longer fatigue life, plus new systems and maintenance philosophy. HOTAS controls; HUD, MFDs and embedded simulation capability.

Shoulder-wing NASA GAW-1 aerofoil; drooped wingtips; sweepback 15° 30' at quarter-chord; thickness/chord ratio 15 per cent at root, 13 per cent at tip; anhedral 2° from root; twist –3° 17'.

FLYING CONTROLS: Variable incidence tailplane; servo tab and horn balanced elevator; boosted ailerons powered by single hydraulic actuator in fuselage with electrically actuated trim bias in aileron linkage; trimmable rudder; electrically actuated Fowler flaps; hydraulically actuated airbrake under centre-fuselage.

STRUCTURE: Two-spar, one-piece metal wing forming integral tank in torsion box and bolted to fuselage; upper and lower skins formed in single sheets; 60 per cent of external surfaces of fuselage in composites. Intake trunking made of GFRP moulding reinforced with helically wound CFRP tapes and metal-sprayed on inner surface.

LANDING GEAR: Hydraulically retractable tricycle type, of Messier/Magnaghi design. Oleo-pneumatic shock-absorber in each unit. All units retract forward into fuselage (main units turning through 90° to lie flat in undersides of engine air intake trunks). Nosewheel steerable ±18°; powered steering system may be offered. Mainwheels size 6.50-8; nosewheel size 5.00-5 with water-deflecting chined tyre. Designed for sink rate of 4 m (13 ft)/s. Wheel brakes actuated hydraulically, independently of main hydraulic system. Provision for emergency free-fall extension.

POWER PLANT: One 14.19 kN (3,190 lb st) Pratt & Whitney Canada JT15D-5C turbofan with automatic ignition, chip detectors and compressor bleed valve to eliminate surge; noise and emissions meet FAR Pt 36 and ICAO rules. Lateral intake each side of fuselage, with splitter plate. Fuel in integral wing tanks and fuselage bladder tank; total capacity approx 910 litres (240 US gallons; 200 Imp gallons). Single gravity refuelling point in top surface of starboard wing. Electric fuel pump for engine starting and emergency use. Fuel and oil systems permit inverted flight. Provision for two 270 litre (71.3 US gallon; 59.4 Imp gallon) drop tanks on inboard underwing stores points. Oil capacity 10 kg (22 lb).

ACCOMMODATION: Two pilots in tandem, stepped up 28 cm (11 in) in rear, on Martin-Baker Mk 10 lightweight zero/zero ejection seats. Blast screen between seats. Pressurised and air conditioned cockpit under one-piece framed canopy opening sideways to starboard.

Model of projected Aermacchi M-311 ***(Paul Jackson)*** ***NEW***/0589241

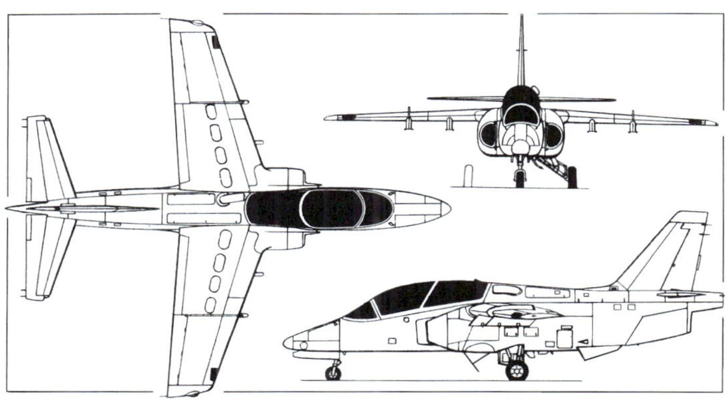

Aermacchi M-311 basic jet trainer ***(Jane's/Dennis Punnett)*** ***NEW***/0589205

SYSTEMS: ECS for cockpit pressurisation and air conditioning, using engine bleed air for heating. Maximum pressure differential 0.24 bar (3.5 lb/sq in). Hydraulic system pressure 207 bars (3,000 lb/sq in), for actuation of airbrake, landing gear, Freon compressor and aileron boost, and independent actuation of wheel brakes. Primary electrical system is 28 V DC, using engine-driven starter/generator; Ni/Cd battery; two static inverters supply AC power for instruments and avionics. External power receptacle in port side of lower fuselage aft of wing. Demand type main oxygen system, at 124 bars (1,800 lb/sq in) pressure, sufficient to supply both occupants for 5 hours, plus bottles for emergency oxygen supply.

AVIONICS: *Comms:* Dual V/UHF transceivers, intercom, and IFF or ATC transponder, standard.

Instrumentation: HUD and three-screen EFIS.

ARMAMENT: Four underwing hardpoints, stressed for loads of up to 350 kg (772 lb) inboard and 250 kg (551 lb) outboard, plus centreline pylon stresed for 90 kg (198 lb); maximum external load 1,090 kg (2,403 lb); (inboard only) two bombs or napalm containers of up to 300 kg; or four 74 mm cartridge throwers; or parachuting containers; S.211A can have special container or 12.7 mm gun pod on fuselage centreline pylon.

DIMENSIONS, EXTERNAL:
- Wing span ... 8.51 m (27 ft 11 in)
- Wing chord: at root ... 2.15 m (7 ft 0¾ in)
  - mean aerodynamic ... 1.65 m (5 ft 5 in)
  - at tip ... 1.00 m (3 ft 3¼ in)
- Wing aspect ratio ... 5.7
- Length overall ... 9.53 m (31 ft 3¾ in)
- Height overall ... 3.73 m (12 ft 2¾ in)
- Tailplane span ... 3.96 m (13 ft 0 in)
- Wheel track ... 2.29 m (7 ft 6 in)
- Wheelbase ... 4.02 m (13 ft 2¼ in)

AREAS:
- Wings, gross ... 12.60 m^2 (135.6 sq ft)
- Airbrake ... 0.42 m^2 (4.52 sq ft)
- Vertical tail surfaces (total) ... 2.01 m^2 (21.64 sq ft)
- Horizontal tail surfaces (total) ... 3.38 m^2 (36.38 sq ft)

WEIGHTS AND LOADINGS:
- Weight empty ... 2,200 kg (4,850 lb)
- Max fuel weight: internal ... 696 kg (1,534 lb)
  - external ... 390 kg (860 lb)
- Max T-O weight: trainer, clean ... 3,100 kg (6,834 lb)
  - armed version ... 4,000 kg (8,818 lb)
- Max wing loading:
  - trainer, clean ... 246.0 kg/m^2 (50.39 lb/sq ft)
  - armed version ... 317.5 kg/m^2 (65.03 lb/sq ft)
- Max power loading:
  - trainer, clean ... 218 kg/kN (2.14 lb/lb st)
  - armed version ... 282 kg/kN (2.76 lb/lb st)

PERFORMANCE:
- Never-exceed speed (VNE) ... M0.80 (400 kt; 740 km/h; 460 mph EAS)
- Stalling speed, flaps down ... 84 kt (156 km/h; 97 mph)
- Max rate of climb at S/L ... 1,478 m (4,850 ft)/min
- Service ceiling ... 12,200 m (40,000 ft)
- T-O run ... 470 m (1,545 ft)
- Landing run ... 570 m (1,870 ft)
- Max range on internal fuel, reserves 10% ... 750 miles (1,389 km; 863 miles)
- Ferry range with max internal and external fuel, 10% reserves ... 970 n miles (1,796 km; 1,116 miles)
- *g* limits: clean ... +7/–3
  - with external stores ... +5/–2.5

UPDATED

AERMACCHI MB-339

TYPE: Basic jet trainer/light attack jet.

DEVELOPMENT MILESTONES

Programme launched	11 Feb 75
First flight	12 Aug 76
First flight, production	20 Jul 78
First delivery (IAF)	8 Aug 79

PROGRAMME: MB-339A selected by Italian Air Force on 11 February 1975 in preference to MB-338; one static test airframe and two prototypes: first flights (MM588) 12 August 1976 and (MM589) 20 May 1977; first flight of production MB-339A (MM54438) 20 July 1978; initial two deliveries (MM54439 and 54440) on 8 August 1979; first batch of 51 included three used as radio calibration aircraft at Pratica di Mare and 14 PANs for Frecce Tricolori aerobatic team.

MB-339C developed as an uprated MB-339 with a Viper 680 power plant; experimental installation in I-MABX, first flown on 9 June 1983; initial MB-339C first flight (I-AMDA)

Aermacchi MB-339A of 61° Stormo, IAF, in special colours *(Paul Jackson)* ***NEW***/0573407

Most recent Italian Air Force version, the MB-339CD2 *(Paul Jackson)* ***NEW***/0558334

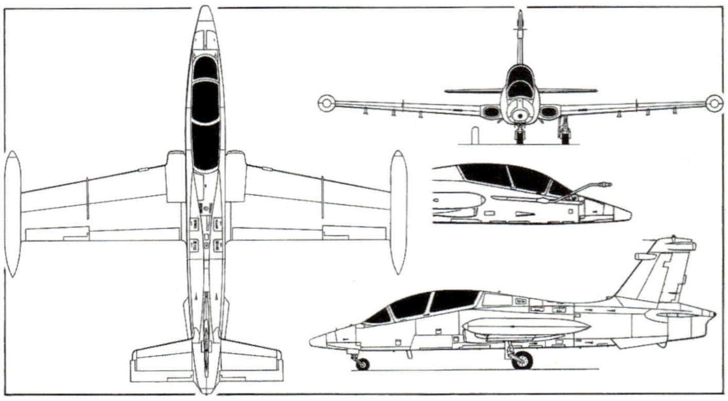

Aermacchi MB-339CD advanced trainer and attack aircraft, with additional starboard side view of optional refuelling probe *(Jane's/Dennis Punnett)* 0507612

17 December 1985; production aircraft (I-TRON) 8 November 1989. The 200th MB-339 (a 339CE for Eritrea) was delivered on 14 April 1997. MB-339s had flown some 500,000 hours by late 2002.

CURRENT VERSIONS: **MB-339A:** Initial military basic/advanced trainer. Final deliveries (small attrition replacement batch for Italy) in 1995. Main operator is 61° Stormo Scuola Volo Basico Iniziale su Aviogetti at Lecce-Galatina.

Lit110 billion MLU initiated in mid-1999 includes structural modifications on some 70 aircraft to extend service lives from 10,000 to 15,000 hours (from 20 to 30 years). Features of MLU include Litton Italia LISA-FG GPS/INS; new integrated air data transducer; provision for AN/ARC-150(V) Have Quick secure radio; Elmer crash recorder; ELT; instrument panel modifications; radio switches on control column; formation flying lights; higher-capacity batteries connected in parallel with starter/generator; provision for target towing; anti-skid brakes; nosewheel splash-guard; repositioned IFF antenna; airborne stress gauge; and improved access, by means of additional removable panels, to wing/fuselage joint for inspection and engine maintenance. Depot maintenance interval will increase to 1,500 hours after MLU. Prototype (CSX54453) first flown 20 December 1999; 'production' will take between eight and 10 years for whole fleet; first pair (MM54497 and MM54505) redelivered in 2002. As part of trials for expanded role of Italian MB-339s, AIM-9L Sidewinder AAM launch trials began mid-2003.

MB-339PAN : (*Pattuglia Aerobatica Nazionale: : national aerobatic team***)**Special version of MB-339A for Italian Air Force; smoke generator added; wingtip tanks deleted. Official debut 27 April 1982. Total 25 built as, or converted to, PAN. Included in MLU.

MB-339AM: Special anti-ship version armed with AOSM Marte Mk 2A missile; avionics, equivalent to MB-339C, include new inertial navigator, Doppler radar, navigation and attack computers, head-up display and multifunction display. Prototype converted from MB-339A; qualification completed January 1995. No known orders.

MB-339B: Powered by 19.57 kN (4,400 lb st) Viper Mk 680-43; larger tip tanks. Prototype (demonstrator I-GROW) modified in 1993 with two LCD EFIS displays and air-to-air refuelling (AAR).

MB-339RM : (*Radiomisure: : radio calibration***)**Three produced for Italian Air Force, 1981; withdrawn and transferred to training as MB-339As.

T-Bird II: One demonstrator modified for US JPATS competition; delivered to Lockheed factory at Marietta, Georgia, on 20 May 1992. Power plant, one Rolls-Royce Viper 680-582 of 17.79 kN (4,000 lb st) with noise reduction kit. Rejected in favour of Pilatus PC-9 (Raytheon T-6A Texan II), June 1995. Continues in use as chase aircraft.

MB-339CB: For Royal New Zealand Air Force; Viper 680 engine. Served with 14 Squadron at Ohakea in training and light attack roles until 13 December 2001, when unit disbanded and aircraft offered for sale being subsequently bought by Malaysia.

MB-339CD : (C Digital) Developed for Italian Air Force advanced/fighter lead-in training against order for 15. Rolled out 12 April 1996; first flight (conversion of MB-339A MM54544; renumbered MMX606) 24 April 1996. Rolls-Royce Viper Mk 632-43 engine; removable in-flight refuelling probe (first phase of tanker trials completed March 1996 by development MB-339 I-GROW); new avionic architecture based on a single central mission computer, MIL-STD-1553B digital databus, one ring laser gyro platform with embedded GPS, EFIS cockpits with HUD in each, three identical liquid crystal colour MFDs and HOTAS controls. Other details generally as MB-339CB, but (compared with MB-339A) nose lengthened; elevators modified to give handling characteristics closer to the fast jets to which students will progress; some structural parts and systems modified; and pilots' escape system updated. Achieved full operational capability with IAF in October 1998 after 148 sorties by prototype.

Contract signed in August 2001 for further 15 MB-339CDs, designated **MB-339CD2**; prototype was conversion of MMX606, first flown in new guise on 6 November 2001; new features of Lot 2 include 8.33 kHz radios, onboard simulation of electronic warfare (RWR, MAWS, chaff/flare), IFF Mode S, crash data recorder, ELT and underwater acoustic beacon, multifunctional information distribution system, full NVG capability, digital map system and autonomous air combat manoeuvring instrumentation. All new features being retrofitted to first batch, and contract includes full-mission simulator installed at Lecce-Galatina air base in December 2002. Initial aircraft of Lot 2 (MM55077) first flew

MB-339 PRODUCTION

Customer	Variant	Qty	First Aircraft	First Delivered
Argentine Navy	AA	10[2]	0761	1980
Eritrean Air Force	CE	6	409[3]	Apr 1997
Ghana Air Force	A	2	G800	1987
		2	G802	1994
Italian Air Force	A[1]	80	MM54438	19 Feb 1979
		20	MM54533	1986
		4	MM55052	1994
		2	MM55058	1995
	CD	15	MM55062	18 Dec 1996
		15	MM55077	Sep 2002
Malaysian Air Force	AM	13	M34-01	1983
New Zealand Air Force	CB	18	NZ6460	13 Mar 1991
Nigerian Air Force	AN	12	301	Jun 1984
Peruvian Air Force	AP	16	452	Nov 1981
UAE (Dubai) Air Force	A	2	431	24 Mar 1984
		5	433	1987
Subtotal		**222**		
Prototypes		2	MM588	1976
Demonstrator	A	1	I-NEUN	1980
	B	1	I-GROW	1984
	C	1	I-AMDA	1985
	K	1	I-BITE	1980
Total		**228**		

[1] Also 339PAN and 339RM
[2] One converted to JPATS demonstrator N339L; others lost or withdrawn from service; including at least two believed transferred to UAE
[3] Eritrean aircraft serial numbered in reverse order

24 May 2002 and delivered September 2002. Clearance issued in July 2002 for carriage of autonomous air combat manoeuvring instrumentation (AACMI) pod on port outer wing pylon. Sidewinder AAM carriage qualified 2003.

In June 2003 the MB-339CD prototype made the first flight of an onboard embedded simulation system funded by five European air forces. Designated RTP11.12, this allows pilots to train against simulated adversaries, in a virtual scenario including a threat environment, while flying a real aircraft — otherwise WaSiF: weapon simulation in flight.

Detailed description applies to the above version.

MB-339CE: Variant of MB-339CD; Viper 680 engine. Six ordered by Eritrea on 7 November 1995; deliveries began April 1997.

MB-339FD : (Full Digital) Export equivalent of CD, with Viper 680 engine. Offered to Australia in 1994; unsuccessful in competition against the BAE Systems' Hawk. Also offered to Brazil and South Africa as Xavante/Impala replacement. Purchase by Venezuela of an initial eight announced 7 July 1998, but decision subsequently reversed and competition reopened: no further announcement by early 2003.

CUSTOMERS: See table. Initial fifteen CDs ordered by Italian Air Force; delivery of two aircraft on 18 December 1996 followed by further 10 in 1997 and three in 1998 (last in November); first four to Reparto Sperimentale Volo at Pratica di Mare for trials before issue to 61° Stormo; some assigned to base flights at Gioia del Colle and Novara-Cameri for continuation training, including in-flight refuelling. Further 15 supplied in 2002–2003, comprising seven by December 2002 (initially on CB1 software configuration) and balance by December 2003.

DESIGN FEATURES: Conventional subsonic jet trainer with tandem, stepped cockpits and low, moderately sweptback wing. Designed to MIL-A-8866A for 15,000 hours service life. Intended maintenance requirement of 3.5 mmh/fh; scheduled maintenance at 150 and 300 hour intervals; inspection at 1,800 hours.

Wing section NACA 64A-114 (mod) at centreline, 64A-212 (mod) at tip; quarter-chord sweepback 8° 29'.

FLYING CONTROLS: Conventional. Power-assisted ailerons with artificial feel, plus servo tabs to assist emergency manual reversion; elevators and rudder actuated manually by pushrods; electrically controlled servo tab for control assistance and trimming on elevator; hydraulically actuated single-slotted flaps; two ventral strakes under tail; electrohydraulically actuated airbrake panel under forward fuselage; wing fence ahead of aileron inboard edge; both pilots have HUD; rear pilot elevated sufficiently to be able to aim guns and air-to-surface weapons and fly visual approaches (see nav/attack system under Avionics heading).

STRUCTURE: All-metal; stressed skin wings with main and auxiliary spars and spanwise stringers; bolted to fuselage; tip tanks permanently attached; rear fuselage detachable by four bolts for engine access.

LANDING GEAR: Hydraulically retractable tricycle type with oleo-pneumatic shock-absorbers, suitable for operation from semi-prepared runways. Hydraulically steerable nosewheel retracts forward; main units retract outward into wings. Low-pressure mainwheel tubeless tyres size 545×175-10 (14 ply); nosewheel tubeless tyre size 380×150-4 (6 ply). Emergency extension system. Hydraulic disc brakes with anti-skid system. Minimum ground turning radius 8.63 m (28 ft 3¾ in).

POWER PLANT: One Rolls-Royce Viper Mk 632-43 turbojet, rated at 17.79 kN (4,000 lb st). Fuel in two-cell rubber fuselage tank, capacity 781 litres (206 US gallons; 172 Imp gallons), and two wingtip tanks with combined capacity of 1,000 litres (264 US gallons; 220 Imp gallons). Total internal usable capacity 1,781 litres (470.5 US gallons; 392 Imp gallons). Self-sealing tanks optional. Single-point pressure refuelling point in port side of fuselage, below wing trailing-edge. Gravity refuelling points on top of fuselage and each tip tank. Provision for two drop tanks, each of 330 litres (87.1 US gallons; 72.6 Imp gallons) usable capacity, on centre underwing stations. Optional refuelling probe on starboard side of cockpits.

ACCOMMODATION: Crew of two in tandem, on Martin-Baker Mk 10LK zero/zero ejection seats in pressurised cockpit. Independent or rear-seat command ejection. Design in accordance with MIL-STD-203F. Rear seat elevated 32.5 cm (1 ft 1 in). Rearview mirror for each occupant. Two-piece moulded transparent canopy, opening sideways to starboard.

SYSTEMS: Pressurisation system maximum differential 0.24 bar (3.5 lb/sq in); cockpit designed for 40,000 pressurisation cycles. Bootstrap-type air conditioning system, also providing air for windscreen and canopy demisting. Hydraulic system, pressure 172.5 bar (2,500 lb/sq in), for actuation of flaps, aileron servos, airbrake, landing gear, wheel brakes and nosewheel steering. Back-up system for wheel brakes and emergency extension of landing gear. Main electrical DC power from one 28 V 9 kW engine-driven starter/generator and one 28 V 6 kW secondary generator; power distribution via five 28 V busses. Two 25 V 24 Ah Ni/Cd batteries for engine starting. Fixed-frequency 115/26 V AC power from two 600 VA single-phase static inverters; provision for additional inverter for three-phase AC. External power receptacle. Low-pressure demand oxygen system, operating at 28 bar (400 lb/sq in). Anti-icing system for engine air intakes.

AVIONICS: Representative fit includes following:

Comms: Honeywell AN/APX-100 IFF; Elmer SRT-651/N VHF/UHF transceiver.

Flight: Honeywell H-746-G GPS/INS, Rockwell Collins ADF-462 and VIR-432 VOR/ILS/MKR; Thales AN-490 tacan.

Instrumentation: Three MFDs; Thales LCD each cockpit; Alenia/Honeywell radalt.

Mission: Kaiser Sabre head-up display; Logic stores management system; Sekay video camera.

Self-defence: Optional RWR, chaff/flare dispenser and MAWS.

ARMAMENT: Up to 1,814 kg (4,000 lb) of external stores on six underwing hardpoints. Four inner hardpoints each stressed for up to 454 kg (1,000 lb) load, and two outer hardpoints each for up to 340 kg (750 lb) load. RNZAF aircraft fitted for AIM-9 Sidewinder and AGM-65 Maverick. Provision on two inner stations for installation of two Macchi gun pods, each containing either a 30 mm DEFA 553 cannon with 120 rounds or a 12.7 mm AN/M-3 machine gun with 350 rounds.

Other typical loads can include two Magic or AIM-9 Sidewinder air-to-air missiles on two outer stations; six general purpose or cluster bombs of appropriate weights; six AN/SUU-11A/A 7.62 mm Minigun pods, each with 1,500 rounds; six Matra 155 launchers, each for eighteen 68 mm rockets; six AN/LAU-68/A or AN/LAU-32G launchers, each for seven 2.75 in rockets; six Aerea AL-25-50 or AL-18-50 launchers, each with twenty-five or eighteen 50 mm rockets respectively; six Aerea AL-18-80 launchers, each with twelve 81 mm rockets; four AN/LAU-10/A launchers, each with four 5 in Zuni rockets; four TDA 100-4 launchers, each with four 100 mm TDA rockets; six Bristol Aerospace LAU-5002 launchers for CRV-7 high-velocity rockets; six Aerea BRD bomb/rocket dispensers; six Aermacchi 11B29-003 bomb/flare dispensers; six TDA 14-3-M2 adaptors, each with six BAP 100 anti-runway bombs or BAT 120 tactical support bombs. Marte 2A anti-ship missile completed MB-339 qualification trials in February 1995.

Following data are for MB-339CD.

DIMENSIONS, EXTERNAL:

Wing span over tip tanks .... 11.22 m (36 ft 9¾ in)
Wing aspect ratio 6.5
Length overall .... 11.24 m (36 ft 10½ in)
Height overall .... 3.945 m (12 ft 11¼ in)
Elevator span .... 4.16 m (13 ft 7¾ in)
Wheel track .... 2.48 m (8 ft 1¾ in)
Wheelbase .... 4.37 m (14 ft 4 in)

AREAS:

Wings, gross .... 19.30 m^2 (207.7 sq ft)
Ailerons (total) .... 1.33 m^2 (14.32 sq ft)
Trailing-edge flaps (total) .... 2.55 m^2 (27.45 sq ft)
Airbrake .... 0.52 m^2 (5.60 sq ft)
Fin .... 2.21 m^2 (23.78 sq ft)
Rudder, incl tab .... 0.68 m^2 (7.32 sq ft)
Tailplane .... 3.38 m^2 (36.38 sq ft)
Elevators (total incl tabs) .... 0.98 m^2 (10.55 sq ft)

WEIGHTS AND LOADINGS:

Weight empty .... 3,335 kg (7,352 lb)
Max weapon load .... 1,814 kg (4,000 lb)
Max fuel weight: internal .... 1,430 kg (3,153 lb)
external .... 530 kg (1,168 lb)
T-O weight, clean .... 4,950 kg (10,913 lb)
Max T-O weight with external stores .... 6,350 kg (14,000 lb)
Max wing loading .... 329.0 kg/m^2 (67.39 lb/sq ft)
Max power loading .... 357 kg/kN (3.50 lb/lb st)

PERFORMANCE (at trainer clean T-O weight, except where indicated):

Never-exceed speed (VNE) .... M0.82 (500 kt; 926 km/h; 575 mph)
Max level speed at S/L:
clean .... 470 kt (870 km/h; 541 mph)
with four 500 lb bombs .... 440 kt (815 km/h; 506 mph)
Max level speed at 9,140 m (30,000 ft) .... M0.77 (441 kt; 815 km/h; 508 mph)
Max speed for landing gear extension .... 175 kt (324 km/h; 202 mph)
T-O speed .... 100 kt (185 km/h; 115 mph)
Approach speed over 15 m (50 ft) obstacle .... 102 kt (189 km/h; 117 mph)
Stalling speed, 20% fuel .... 86 kt (160 km/h; 99 mph)
Max rate of climb at S/L .... 1,631 m (5,350 ft)/min
Service ceiling .... 13,715 m (45,000 ft)
T-O run .... 620 m (2,035 ft)
Landing run at S/L, 20% fuel .... 480 m (1,575 ft)

Radius of action, 10% reserves:
  with four 500 lb bombs:
    lo-lo-lo .... 156 n miles (289 km; 179 miles)
    hi-lo-hi .... 278 n miles (515 km; 829 miles)
  with two rocket launchers two gun pods and two gun pods, 5 min combat, hi-lo-hi ..... 280 n miles (519 km; 322 miles)
Ferry range, clean .... 1,122 n miles (2,078 km; 1,291 miles)
Max endurance with drop tanks .... 3 h 50 min
g limit .... +8.0/–4.0

UPDATED

AERMACCHI M-346

TYPE: Advanced jet trainer/light attack jet.

DEVELOPMENT MILESTONES

Design began	Jan 00
Announced	25 Jul 00
Rolled out (unofficially)	6 Jun 03
Rolled out	7 Jun 03
First flight	15 Jul 04

PROGRAMME: Announced 25 July 2000. Comprehensively redesigned and westernised version of Yakovlev Yak-130, on which Aermacchi had begun autonomous full-scale development in January 2000, conforming to requirements of Eurotrainer group. Three prototypes as follow-on to 300 trial flights with prototype Yak-130; mockup first shown at Paris, June 2001; maiden flight of P01/CMX615 scheduled for November 2003, following 7 June 2003 roll-out, but not achieved until 15 July 2004. P02 scheduled to follow in March 2005 and P03 (to production standard) in late 2006, these contributing to 700-hour test programme. A01 initial production aircraft to fly in late 2007 (preceded by static test and fatigue test airframes to production standard in 2005 and 2006, respectively).

Aermacchi shares design rights and has production and modification rights for the Yak-130. Development of the aircraft was transferred to Italy in mid-1998 to prevent furtherdelays resulting from the economic situation in Russia. Some funding provided by Italian Ministry of Industry. Italian military flight test centre (RSV) involved in certification testing, beginning with prototype's fourth sortie. Dott Massimo Luchesini is Aermacchi's AEM/Yak-130 programme director.

At Farnborough International, immediately following maiden flight, Aermacchi formally launched quest for industrial partners to share manufacture of any production aircraft.

CURRENT VERSIONS (SPECIFIC): **P01:** 001/CSX615. Handling and performance trials, including envelope expansion, high AoA and FCS proving.

P02: As P01, plus avionics testing, including onboard simulation, and flight refuelling. Fuselage removed from assembly jig on 11 December 2003.

P03: Avionics (including onboard simulation), definitive main landing gear, radar integration, electronic warfare and external stores trials.

CUSTOMERS: None. Estimated 600 sales in total market for 1,650 aircraft of this type over 25 years (2004).

DESIGN FEATURES: Extends flight envelope of existing jet trainers, notably turn and climb performance and usable AoA (40°). Related considerations are transonic aerodynamic configuration, twin-engine reliability, FBW control system, fighter-type cockpit and avionics, high reliability and low operational and maintenance costs. Able to 'download' some 30-40 per cent of training hours previously flown on combat aircraft at OCU.

Cockpit based on latest combat aircraft, such as Eurofighter Typhoon. 'Duck's beak' chine deleted from forward fuselage to simplify shape and facilitate optional addition of radar. Autonomous operation made possible by self-starting, OBOGS, BITE, APU and use of standard support equipment. Design to MIL-STD-1350A and MIL-A-8660A; maintenance to MIL-STD-470B; technical publications in accordance with applicable MIL-STD. Service life 16,000 hours. Wing dihedral 1°; tailplane anhedral 4°.

FLYING CONTROLS: FBW. Ailerons, elevators and rudder on primary control circuit, protected against double failure. Leading-edge flaps, double-slotted, constant chord trailing-edge flaps and spine airbrake on secondary control circuit, although first-mentioned are duplex. Smiths Aerospace primary actuators. Microtecnica actuators for leading- and trailing-edge flaps are rotary and ballscrew type, respectively. Airbrake failure mode is 'closed'.

STRUCTURE: Aluminium alloy skins, wings and fin; reinforced carbon fibre cockpit section and rear fuselage; metallic bonded composites ailerons, flaps, airbrake, rudder and elevons; Kevlar fin fillet, fin root fairings and tailcone; glass fibre nosecone and fintip. Low parts-count: is 2 t heavier than MB-339, but has 40 per cent fewer parts.

Aermacchi M-346 cockpit mockup ***NEW***/0546965

Aermacchi M-346 CMX615, the prototype ***NEW***/0589384

Wing comprises three main spars and two auxiliary spars with integral fuel tank. Main spars pick up one to three single-three machined mainframes.

LANDING GEAR: Nosewheel leg by Liebherr; tyre 18×5.5. Forward-retracting mainwheel legs from AMX fighter-bomber on first two prototypes; new units by Liebherr thereafter. Hydraulic actuation; two emergency accumulators.

POWER PLANT: Two 27.9 kN (6,280 lb st) Honeywell F124-GA-200 twin-shaft turbofans with dual-channel FADEC. Total fuel capacity approximately 2,500 litres (660 US gallons; 550 Imp gallons), comprising 394 litres (104 US gallons; 86.7 Imp gallons) in two-section tanks in each wing; main 1,493 litre (394 US gallon; 329 Imp gallon) centre fuselage tank; and 219 litre (57.9 US gallon; 48.2 Imp gallon) collector tank. Single-point pressure refuelling; can be refuelled with engines running. Two inboard wing pylons 'wet', each with provision for 580 litre (153 US gallon; 128 Imp gallon) tank. Secondo Mona fuel system. Optional refuelling probe.

ACCOMMODATION: Two Martin-Baker Mk 16D zero/zero ejection seats. HOTAS controls.

SYSTEMS: Teleavio/BAE Systems quadruplex, full authority reprogrammable FCS provides stability augmentation throughout flight envelope. Autopilot is part of FCS. Honeywell Normelair-Garrett environmental control system, providing 100 per cent conditioned air to cockpit, using engine or APU power; 'bootstrap' air cooler, precooler and dual heat exchangers, plus bleed leak detector; nominal cockpit pressure differential 0.25 bar (3.6 lb/sq in); Air Liquide OBOGS. Electrical system, by ASE and Logic, includes 20 kVA generator on each engine (100 per cent redundancy), two 9 kW transformer-rectifiers, one 5 kW 28 V DC emergency generator (Microturbo Rubis APU) and two Ni/Cd batteries. APU cleared for air-starts up to 6,100 m (20,000 ft); provides environmental control at 7 kg/s (15 lb/s) at 4.4 bar (64 lb/sq in) and 5 kW electrical power.

Two dual-redundant digital databusses conforming to MIL-STD-1553B: Avionics 1 for sensors and displays; Avionics 2 for radar, SMS, FLIR, AECM, RWR and HMD. Additional Armament databus to MIL-STD-1760 for weapons stations and SMS. Avionics 1 and 2 control is by mission computer symbol generator (MCSG) with failure back-up from

Prototype Aermacchi M-346 during its maiden flight on 15 July 2004 ***NEW***/0589385

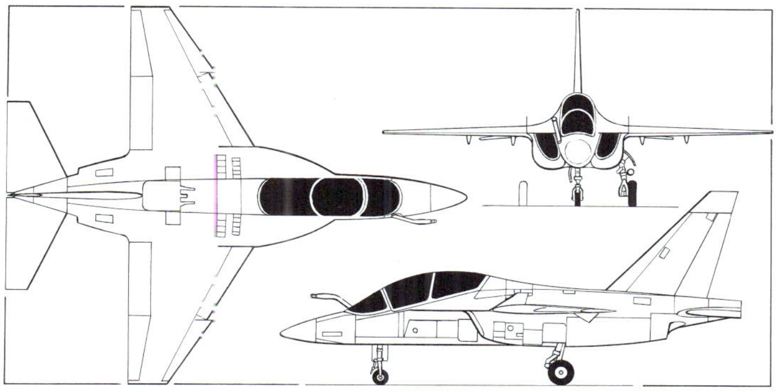

General arrangement of Aermacchi M-346 *(James Goulding)* ***NEW***/0552826

Prototype Aermacchi M-346 at its unofficial roll-out on 6 June 2003, the day prior to the formal presentation ***NEW***/0552831

miscellaneous computer (MISCO). Armament bus control by SMS acting as remote terminal of Avionics 2 bus. Air-to-air and air-to-ground attack computations by operational flight program (OFP) software in MCSG, latter also generating HUD symbology. Area navigation and flight director functions performed by embedded GPS inertial unit (EGI) and flight control computers, respectively.

Cockpit video recording provided by colour HUD TV camera, recording outside world with HUD symbology superimposed; associated VTR captures colour and monochrome video signals and is remotely controlled by a dedicated panel. MISCO also responsible for avionics system interface, aural warning generator for voice alert system (VAS), HUMS and gear down/locked confirmation tone. Crash-survivable memory unit (CSMU) interfaced via RS-422 to MISCO and audio system to record data for accident analysis. Two independent hydraulic systems by Microtecnica, both operating at 207 bar (3,000 lb/sq in): No.1 for flight controls, No. 2 for utilities and flight control back-up. Smiths/Microtecnica/OMA control surface actuators.

AVIONICS: Mission core system by Galileo Avionica.

Comms: Two Marconi VHF/UHF transceivers; Marconi IFF transponder. DbSystem intercom.

Radar: Provision in design.

Flight: Honeywell laser gyro INS with embedded GPS; Thales Tacan and Collins VOR/ILS/marker beacon receiver.

Instrumentation: Galileo Avionica raster/stroke HUD; three Galileo Avionica 127 × 127 mm (5 × 5 in) colour LCDs per cockpit; provision for HMD and NVGs. Up-front control panels for management of navigation and weapon delivery, communication, identification and radio navigation, plus optional FLIR.

Mission: Data transfer unit (DTU) for loading of mission plan. Centreline attachment for Vinten Vicon 19 reconnaissance pod, Litening FLIR, or ATLIS targeting pod. Cockpit video recorder. Radar and ECM simulation system integrated for advanced training.

Self-defence: Provision for RWR, chaff/flares and underwing ECM pod.

ARMAMENT: Nine hardpoints: one on centreline, rated at 600 kg (1,323 lb); four under each wing, rated at 1,050 kg (2,315 lb) (inboard), 550 kg (1,213 lb), 300 kg (661 lb) and 150 kg (331 lb) (wingtip). Optional cannon pod on centreline. Typical weapon loads include six Mk 82 (500 lb), four Mk 83 (1,000 lb) or two Mk 84 (2,000 lb) bombs; four GBU-12 (500 lb), GBU-16 (1,000 lb) or Opher Mk 82 guided bombs; six Rockeye or four BL 755 CBUs; six Durandal anti-runway dispensers; four BRD-4 or LAU-7/LAU-5002/LAU-32 rocket pods; four AIM-9 Sidewinder AAMs; four AGM-65 Maverick ASMs; or four Brimstone anti-armour missiles.

DIMENSIONS, EXTERNAL:
Wing span .... 9.72 m (31 ft 10¾ in)
Wing chord: at root (fuselage c/l) .... 3.87 m (12 ft 8¼ in)
at tip .... 0.97 m (3 ft 2¼ in)
mean aerodynamic .... 2.71 m (8 ft 10¾ in)
Wing aspect ratio 4.0
Length overall .... 11.49 m (37 ft 8¼ in)
Height, unladen .... 4.985 m (16 ft 4¼ in)
Tailplane span .... 4.83 m (15 ft 10¼ in)
Wheelbase .... 4.465 m (14 ft 7¾ in)
Wheel track .... 2.725 m (8 ft 11¼ in)

AREAS:
Wings, gross .... 23.52 m^2 (253.2 sq ft)

WEIGHTS AND LOADINGS:
Weight empty .... 4,610 kg (10,163 lb)
Max weapon load .... 3,000 kg (6,614 lb)
Max internal fuel weight .... 2,000 kg (4,409 lb)
Max T-O weight: trainer .... 6,700 kg (14,771 lb)
attack .... 9,500 kg (20,943 lb)
Max wing loading: trainer .... 284.9 kg/m^2 (58.34 lb/sq ft)
attack .... 403.9 kg/m^2 (82.73 lb/sq ft)
Max power loading: trainer .... 120 kg/kN (1.18 lb/lb st)
attack .... 171 kg/kN (1.68 lb/lb st)

PERFORMANCE:
Limiting speed .... 572 kt (1,059 km/h; 658 mph) EAS/M1.2
Max level speed at 1,525 m (5,000 ft) .... 590 kt (1,093 km/h; 679 mph)
Stalling speed, landing with 20% fuel, flaps down .... 90 kt (167 km/h; 104 mph)
Max rate of climb at S/L .... 6,100 m (20,000 ft)/min
Service ceiling .... 13,715 m (45,000 ft)
T-O run .... 290 m (955 ft)
Landing run .... 550 m (1,805 ft)
Radius of action:
air combat training with ACMI pod and one SRAAM, 20 min manoeuvring .... 100 n miles (185 km; 115 miles)
combat air patrol, two SRAAM, gun pod, plus two external fuel tanks, 2 h loiter .... 165 n miles (306 km; 190 miles)
close air support, hi-lo-lo-hi, two rocket pods, two ASMs, two Sidewinders, gun pod .. 270 n miles (500 km; 311 miles)
interdiction, hi-lo-lo-hi, two Mk 83 bombs, external fuel tanks .... 450 n miles (833 km; 518 miles)
Range, 10% reserves:
with max internal fuel .... 1,100 n miles (2,037 km; 1,265 miles)
with three external tanks .... 1,500 n miles (2,778 km; 1,726 miles)
g limits .... +8/−3

UPDATED

ALENIA

ALENIA AEROSPAZIO (A FINMECCANICA COMPANY)

Via Giulio Vincenzo Bona 85, I-00156 Roma
Tel: (+39 06) 41 72 31
Fax: (+39 06) 411 44 39
Web (1): http://www.finmeccanica.com
Web (2): http://www.alenia-aeronautica.it
PRESIDENT: Giorgio Zappa
SENIOR VICE-PRESIDENT, HEAD OF STRATEGY AND MILITARY PROGRAMMES: Carmelo Cosentino
SENIOR VICE-PRESIDENT, OPERATIONS: Gianni Cantini
HEAD OF AERONAUTICS DIVISION: Filippo Bagnato
SENIOR VICE-PRESIDENT, TECHNICAL AND PRODUCTION FACILITIES, AERONAUTICS DIVISION: Giuseppe Ragni
MARKETING COMMUNICATIONS OFFICER: Riccardo Rovere

Alenia Aerospazio, a Finmeccanica company, is the leading Italian aerospace designer and manufacturer. Employees total 12,000, engaged on projects and programmes for civil and military applications. Alenia Aerospazio is organised into two Divisions: Aeronautics and Space (Alenia Spazio SpA). On 14 April 2000, it signed a joint venture agreement with EADS to form the European Military Aircraft Company (EMAC), but early establishment was not achieved and intermittent discussions continued into 2002 before being abandoned. In January 2003, Alenia declined to take up offered 5 per cent stake in Airbus, but instead signed wide-ranging collaborative agreement with Boeing, with specific areas to be decided by committee formed in 2003.

Aeronautics Division (Alenia Aeronautica) dedicated to full range of activities, from design and production to modification and product support for both military and civil aircraft; most entail collaboration with other aerospace companies; in 2002, division had 7,173 employees and conducted almost half of its activities in the military sphere.

Division's activities fall into categories of Military Aircraft, Regional Aircraft and Aerostructures. It also operates Officine Aeronavali Venezia, specialising in maintenance, overhaul and modification of commercial and military aircraft.

In military sector, company designs and produces directly, or through international collaborations, combat and transport aircraft such as Tornado, Eurofighter Typhoon, C-27J, A400M and ATR 42 MP Surveyor; was also responsible for updating of the F-104S/ASA-M and TF-104G-M in service with Aeronautica Militare Italiana and has assembled AV-8B Harrier II Plus for the Italian Marina Militare.

In regional aircraft sector, activities include production of the ATR turboprop family developed with Aerospatiale Matra.

In the aerostructures sector, Alenia co-operates with other major aeronautical companies manufacturing structural parts for commercial aircraft such as B767, B777, B717, A321, A300/310, A330/A340, Falcon 900EX and Falcon 2000. Is 4 per cent risk-sharing partner in Airbus A380 programme, responsible for a fuselage section.

In the modification and maintenance field, Alenia Aerospazio has a full capability of design, production, installation and tests of complex parts and systems.

UPDATED

ALENIA/LOCKHEED MARTIN C-27J SPARTAN

TYPE: Twin-turboprop transport.

PROGRAMME: Conceived in 1960s as jet-powered, V/STOL transport to NATO requirement (NBMR-4), but this version not developed. Original Fiat G222 designed by Giuseppe Gabrielli; two prototypes (lacking the pressurisation standard on later aircraft) flew on 18 July 1970 (MM582) and 22 July 1971 (MM583), both at Turin; MM582 handed over to Italian Air Force for operational evaluation on 21 December 1971. First production G222 (MM62101) flew 23 December 1975; deliveries began 21 November 1976 with single aircraft to Dubai, and to Italy on 21 April 1978. Tenth production aircraft was first to be built at Naples; 27th (March 1979) was 22nd and last built at Turin. Main users in Italy are 2° and 98° Gruppi of the 46° Brigata Aerea at Pisa. Civil category R (equivalent to FAR Pt 25) certification granted to G222SAA by Italian airworthiness authority on 1 April 1997. Several subvariants for specific roles.

Improved version of G222 conceived during 1995 negotiations between Lockheed Martin and Alenia on potential offsets for proposed Italian purchase of C-130J Hercules; initially designated G222J, by reason of having C-130J flight deck features and improved (T64G) versions of the G222's engines with new four-blade propellers. Formally announced as a

Short landing demonstration by an Alenia C-27J Spartan *(Paul Jackson)* ***NEW***/0573405

DEVELOPMENT MILESTONES

G.222	
First flight	18 Jul 70
First flight, pre-production	23 Dec 75
First delivery (Dubai AF)	21 Nov 76
Subsequent versions:	
C-27J	
Design began	Sep 96
Programme launched	17 Jun 97
Announced	Feb 96
Rolled out	14 Jun 99
First flight	24 Sep 99
Certification	20 Jun 01
First flight, production	12 May 00

joint project in February 1996, when commonality with the C-130J was further increased by adoption of the Allison (now Rolls-Royce) AE 2100 as power plant, allied to six-blade propellers. Accordingly redesignated C-27J, to reflect the C-27A version of G222 delivered to the US Air Force. Feasibility phase February to September 1996; definition phase September 1996 to May 1997.

Development and certification costs being shared equally between Alenia Aerospazio and Lockheed Martin; latter responsible for propulsion systems, avionics, worldwide marketing and product support; Alenia for production, flight test and certification; promotion by Lockheed Martin Alenia Tactical Transport Systems. By 2003, Lockheed Martin reportedly seeking reduced participation through voluntary relegation to subcontractor.

Programme formally launched on 17 June 1997; 'propulsion test' prototype, a converted G222 demonstrator, rolled out at Turin/Caselle on 14 June 1999 and first flew (c/n 4043; I-CERX) 24 September 1999; initial testing completed second quarter of 2000; modified for certification trials and resumed flying on propulsion system, performance and handling evaluation.

Second C-27J and initial new-build aircraft, 4115/I-FBAX, first flew 12 May 2000; first with advanced flight deck and full avionics, new APU and new landing gear; achieved 54 hours/23 sorties in initial two months; available for sale after completion of civil certification trials. Third prototype, 4033/MMCSX62127, converted from Italian Air Force G222TCM, first flew 8 September 2000 returned to IAF after DGAA civil certification, which achieved on 20 June 2001. Military type certificate awarded 20 December 2001, following 445 sortie, 793 hour programme. First two batches, totalling 10, under construction (long lead items) by 1999. Final assembly remains in Italy only.

CURRENT VERSIONS: **Transport:** *As described.*

AEW: Airborne early warning version proposed in 1998, employing Ericsson Erieye system, as fitted to Saab S 100B Argus.

Firefighter: Proposed with 2,200 kg (4,850 lb) mission system and 6,800 kg (14,991 lb) of retardant liquid.

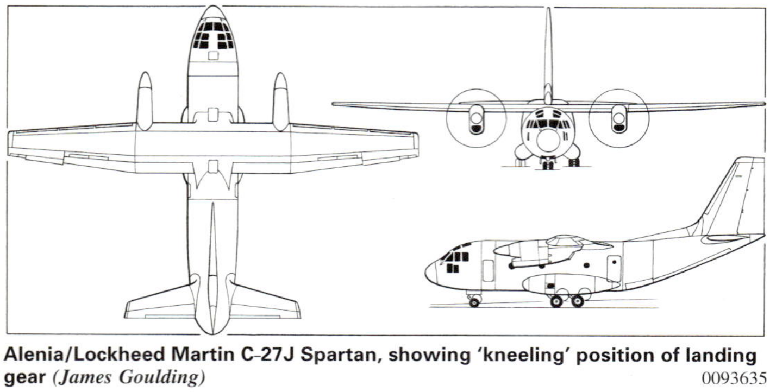

Alenia/Lockheed Martin C-27J Spartan, showing 'kneeling' position of landing gear *(James Goulding)* 0093635

Aerial Sprayer: Proposed with 2,200 kg (4,850 lb) mission system; capable of covering 50 hectares (124 acres) at 150 litres/hectare (16.0 US gallons; 13.0 Imp gallons/acre).

Maritime: Patrol version under consideration by Taiwan in 2003 as alternative to Lockheed Martin P-3 Orion anti-submarine aircraft.

CUSTOMERS: See table. Italian Air Force announced proposed launch order for 12 on 11 November 1999, although this not signed until 27 June 2002 and for initial quantity of five; deliveries planned to begin in mid-2005 and be completed in late 2006. Greek Air Force selected C-27J on 1 March 2002, proposing 12 plus three options, and signed contract for 12 on 17 January 2003; deliveries to begin in mid-2004 and end 12 months later. Up to 500 sales anticipated over a 20 year period, mostly to existing Hercules operators. Lockheed Martin co-markets the aircraft. Sales targets include Argentina, Brazil (12) and Taiwan (18 to 22); also promoted for US Army's 40-aircraft Aerial Common Sensor programme and to US Army National Guard (44) as replacement for Shorts Sherpas.

COSTS: €24.75 million (Greek programme unit cost, 2003).

DESIGN FEATURES: Conventional tactical transport configuration of high wing, pannier-mounted main landing gear and upswept rear fuselage with integral loading ramp.

Intended to complement the Lockheed Martin Hercules. Upgraded G222 with new, two-crew flight deck and increased performance. Propulsion system, cargo loading system and many of the avionics and flight controls are adapted from the C-130J Hercules. Compared with the G222, the C-27J is intended to provide increases of 35 per cent in range, 30 per cent cruise ceiling, 15 per cent high-speed cruise and over 200 per cent payload/range/speed; maintainability and reliability are scheduled to increase by 100 per cent for the engine, 275 per cent for propeller, 150 per cent for other systems and 30 per cent for avionics, resulting in a saving of 30 per cent in operating costs (including 5 per cent off fuel).

Wing has max thickness/chord ratio of 15 per cent. Dihedral 2° 30' on outer panels.

FLYING CONTROLS: Conventional; manually actuated ailerons and elevators; powered rudder. Ailerons each have inset servo tab. Two-section hydraulically actuated spoilers ahead of each outboard flap segment, used also as lift dumpers on landing. Double-slotted flaps extend over 60 per cent of trailing-edge. Spoilers and flaps fully powered by tandem hydraulic actuators. Rudder fully powered by tandem hydraulic actuators. Two tabs in each elevator; no rudder tabs.

STRUCTURE: Wing of aluminium alloy three-spar fail-safe box structure, built in three portions. One-piece constant chord centre-section fits into recess in top of fuselage and is secured by bolts at six main points. Outer panels tapered on leading- and trailing-edges. Upper surface skins are of 7075-T6 alloy, lower surface skins of 2024-T3 alloy. All control surfaces have bonded metal skins with metal honeycomb core.

Pressurised fail-safe fuselage of aluminium alloy stressed skin construction and circular cross-section. Easily removable stiffened floor panels. Cantilever safe-life tail surfaces of aluminium alloy, with sweptback three-spar fin and slightly swept two-spar variable incidence tailplane.

Subcontractors include Aermacchi (outer wings), Piaggio (wing centre-section), Agusta (tail unit), Magnaghi (landing gear) and Aeronavali Venezia (airframe components).

LANDING GEAR: Hydraulically retractable tricycle type, suitable for use from prepared runways, semi-prepared strips or grass fields. Main gear built by APPH; nose gear by Magnaghi. Steerable twin-wheel nose unit retracts forward. Main units, each consisting of two single wheels in tandem, retract into fairings on sides of fuselage. Oleo-pneumatic shock-absorbers. Gear can be lowered by gravity in emergency, the nose unit being aided by aerodynamic action and the main units by the shock-absorbers, which remain compressed in the retracted position. Oleo pressure in shock-absorbers is adjustable to permit variation in height and attitude of cabin floor from ground. Low-pressure tubeless tyres on all units, size 39×13 (14/16 ply) on mainwheels, 29×11.00–12 or 29×11.00–10 (10 ply) on nosewheels. Tyre pressures 4.41 bar (64 lb/sq in) on main units, 3.92 bar (57 lb/sq in) on nose unit. Hydraulic multidisc brakes.

POWER PLANT: Two Rolls-Royce AE 2100D2 turboprops, each rated at 3,460 kW (4,640 shp), driving Dowty R391 six-blade composites propellers. Fuel in integral tanks; two in outer wings, combined capacity 6,800 litres (1,796 US gallons; 1,495 Imp gallons); two centre-section tanks, combined capacity 5,200 litres (1,374 US gallons; 1,143 Imp gallons); crossfeed provision to either engine. Total overall fuel capacity 12,000 litres (3,170 US gallons; 2,638 Imp gallons).

ACCOMMODATION: Two-pilot crew on flight deck with third seat; provision for loadmaster or jumpmaster when required. Crew door, port, forward; Type III emergency door, starboard, forward; paratroop door each side, immediately rear of sponsons; rear loading ramp;

G222 AND C-27 CUSTOMERS

Customer	Version	Qty	First aircraft	Delivered
Argentine Army	G222	3	AE260	29 Mar 1977
Congo Air Force	G222	3		on order[4]
Dubai Air Force	G222	1	321	21 Nov 1976
Greek Air Force	C-27J	12[8]		2004
Italy: prototypes	G222	2	MM582	21 Dec 1971
Air Force	G222TCM	40[3]	MM62101	Apr 1978
	C-27J	5[7]		2005
Air Force	G222RM	4	MM62139	Jan 1983
Air Force	G222VS	2	MM62107	1978
SNPC	G222PROCIV	5[6]	MM62145	1987
Libyan Air Force	G222T	20	221	1981
Nigerian Air Force	G222	5	950	Sep 1984
Somali Air Force	G222	2	AM-94	1980
Venezuelan: Army	G222	1[1]	EV-8327	1983
Air Force	G222	6	1258	1984
Thai Air Force	G222	6	60307	2 May 1995
US Air Force	C-27A	10[5]	90–0170	17 Apr 1991
Alenia (demonstrator)	G222/C-27J	1	I-CERX	1983
	C-27J	1	I-FBAX	2000
Total		**129**[2]		

[1] Transferred to air force
[2] Prototypes and 22 early aircraft built at Turin; remainder at Naples
[3] Including 10 with provision for rapid conversion to 222SAA; two to Tunisian Air Force 19 May 2001; C-27J purchase involves Alenia buy-back of 39 G222s for possible resale
[4] Built, but awaiting completion of contractual details
[5] Transferred to US government, with civil registrations, 1999
[6] All reverted to transports in 2001
[7] Plus seven options
[8] Plus three options

emergency hatches (three) in roof, above flight deck and level with leading- and trailing edges.

Standard troop transport version has 34 foldaway sidewall seats and 12 stowable seats for 46 fully equipped troops (62 in high density). Paratroop transport can carry between 34 (normal) and 46 (maximum) fully equipped paratroops, and is fitted with 32 sidewall seats, plus eight stowable seats, door jump platforms and static lines. Cargo transport version can accept standard pallets of up to 2.24 m (88 in) wide, and can carry up to 9,000 kg (19,841 lb) of freight. Hydraulically operated rear-loading ramp and upward-opening door in underside of upswept rear fuselage, which can be opened in flight for airdrop operations. Five pallets of up to 1,000 kg (2,205 lb) each can be airdropped from rear opening, or single pallet of up to 5,000 kg (11,023 lb). Paratroop jumps can be made either from this opening or from rear side doors. Medical evacuation accommodation for 36 stretchers and six attendants. Entire accommodation pressurised.

SYSTEMS: Pressurisation system maintains a cabin differential of 0.41 bar (5.97 lb/sq in), giving a 1,200 m (3,940 ft) environment at altitudes up to 6,000 m (19,680 ft). Air conditioning system uses engine bleed air during flight; on ground, it is fed by compressor bleed air from APU to provide cabin heating to a minimum of 18°C. Honeywell 113 kW (152 hp) APU, installed in starboard main landing gear fairing, provides power for engine starting, hydraulic pump and alternator actuation, air conditioning on ground, and all hydraulic and electrical systems necessary for loading and unloading on ground.

Two independent hydraulic systems, each of 207 bar (3,000 lb/sq in) pressure. No. 1 system actuates flaps, spoilers, rudder, wheel brakes and (in emergency only) landing gear extension; No. 2 system actuates flaps, spoilers, rudder, wheel brakes, nosewheel steering, landing gear extension and retraction, rear ramp/door and windscreen wipers. Auxiliary hydraulic system, fed by APU-powered pump, can take over from No. 2 system in flight, if both main systems fail, to operate essential services. In addition, a standby hand pump is provided for emergency use to lower the landing gear and, on the ground, to operate the ramp/door and parking brakes.

Three 45 kVA alternators, one driven by each engine through constant-speed drive units and one by the APU, provide 115/200 V three-phase AC electrical power at 400 Hz. 28 V DC power is supplied from the main AC busses via two transformer-rectifiers, with 24 V 34 Ah Ni/Cd battery and static inverter for standby and emergency power. External AC power socket. Engine intakes anti-iced by electrical/hot air system. Pneumatically inflated de-icing boots on outer wing leading-edges, and fin and tailplane leading-edges, using engine bleed air. Liquid oxygen system for crew and passengers (with cabin wall outlets); this system can be replaced by a gaseous oxygen system if required. Emergency oxygen system available for all occupants in the event of a pressurisation failure.

AVIONICS: *Radar:* Northrop Grumman AN/APN-241.

Instrumentation: Five-screen EFIS based on C-130J flight deck. Radar integrated with moving map display. NVG-compatible.

DIMENSIONS, EXTERNAL:
Wing span ..... 28.70 m (94 ft 2 in)
Wing aspect ratio 10.0
Length overall ..... 22.70 m (74 ft 5½ in)
Height overall: unladen ..... 10.57 m (34 ft 8¼ in)
  fully laden ..... 9.70 m (31 ft 10 in)
Fuselage: Max diameter ..... 3.55 m (11 ft 7¾ in)
Tailplane span ..... 12.40 m (40 ft 8¼ in)
Wheel track ..... 3.67 m (12 ft 0½ in)
Wheelbase (to c/l of main units):
  unladen ..... 6.23 m (20 ft 5¼ in)
  fully laden ..... 6.40 m (21 ft 0 in)
Propeller diameter ..... 4.11 m (13 ft 6 in)
Distance between propeller centres ..... 9.50 m (31 ft 2 in)
Propeller/fuselage clearance ..... 1.04 m (3 ft 5 in)
Rear-loading ramp/door: Width ..... 2.45 m (8 ft 0½ in)
  Height ..... 2.25 m (7 ft 4½ in)
Crew door: Width ..... 0.70 m (2 ft 3½ in)
  Height ..... 1.52 m (4 ft 11¾ in)
Emergency door: Width ..... 0.53 m (1 ft 8¾ in)
  Height ..... 1.01 m (3 ft 3¾ in)
Paratroop doors: Width ..... 0.91 m (2 ft 11¾ in)
  Height ..... 1.92 m (6 ft 3½ in)
Emergency hatch:
  Flight deck: Width ..... 0.70 m (2 ft 3½ in)
    Length ..... 0.50 m (1 ft 7¾ in)
  Cabin (both): Width ..... 0.63 m (2 ft 0¾ in)
    Length ..... 0.90 m (2 ft 11½ in)

DIMENSIONS, INTERNAL:
Main cabin: Length ..... 8.58 m (28 ft 1¾ in)
  Width ..... 2.45 m (8 ft 0½ in)
  Height ..... 2.25 m (7 ft 4½ in)
  Floor area: excl ramp ..... 21.0 m² (226 sq ft)
    incl ramp ..... 25.7 m² (276 sq ft)
  Volume ..... 58.0 m³ (2,048 cu ft)

AREAS:
Wings, gross ..... 82.00 m² (882.6 sq ft)
Ailerons (total) ..... 3.65 m² (39.29 sq ft)
Trailing-edge flaps (total) ..... 18.40 m² (198.06 sq ft)
Spoilers (total) ..... 1.65 m² (17.76 sq ft)
Fin (incl dorsal fin) ..... 12.19 m² (131.21 sq ft)
Rudder ..... 7.02 m² (75.56 sq ft)
Tailplane ..... 19.09 m² (205.48 sq ft)
Elevators (total) ..... 4.61 m² (49.62 sq ft)

WEIGHTS AND LOADINGS:
Operating weight empty ..... 17,000 kg (37,479 lb)
Max payload: at 2.5 *g* ..... 9,000 kg (19,840 lb)
  at 2.25 *g* ..... 10,225 kg (22,542 lb)
  airdrop ..... 5,000 kg (11,023 lb)
Max fuel load ..... 9,400 kg (20,725 lb)
Max T-O weight ..... 31,800 kg (70,106 lb)
Max landing weight ..... 30,000 kg (66,138 lb)
Max cargo floor loading ..... 1,500 kg/m² (307.2 lb/sq ft)
Max wing loading ..... 387.8 kg/m² (79.43 lb/sq ft)
Max power loading ..... 4.60 kg/kW (7.55 lb/shp)

PERFORMANCE:
Max level speed ..... 325 kt (602 km/h; 374 mph)
Time to 4,570 m (15,000 ft) ..... 7 min
Initial cruising altitude ..... 8,380 m (27,500 ft)
Service ceiling ..... 9,145 m (30,000 ft)
Service ceiling, OEI ..... 3,355 m (11,000 ft)
T-O run ..... 410 m (1,345 ft)
T-O to 15 m (50 ft) ..... 640 m (2,100 ft)
Landing from 15 m (50 ft) ..... 690 m (2,265 ft)
Landing run ..... 390 m (1,280 ft)
Radius of action:
  with 46 paratroops ..... 1,100 n miles (2,037 km; 1,265 miles)
  with 5,000 kg (11,023 lb) airdrop load ..... 1,215 n miles (2,250 km; 1,398 miles)
Range:
  with max payload ..... 1,160 n miles (2,148 km; 1,334 miles)
  with 6,000 kg (13,228 lb) payload ..... 2,500 n miles (4,630 km; 2,877 miles)
  ferry ..... 3,200 n miles (5,926 km; 3,682 miles)
g limit ..... +3.0

UPDATED

ALISPORT

ALISPORT SRL

Via Confalonieri 22, I-23894 Cremella
Tel: (+39 039) 921 21 20
Fax: (+39 039) 921 21 30
e-mail: info@alisport.com
Web: http://www.alisport.com

Alisport markets the DEA-designed Yuma STOL lightplane.

NEW ENTRY

ALISPORT YUMA

TYPE: Side-by-side ultralight kitbuilt.
PROGRAMME: Development began 1994.
CURRENT VERSIONS: **Yuma:** *As described.*
Guérin G1: Derivative marketed by Espace Liberté of France.
DESIGN FEATURES: Objectives included better performance than existing STOL light aircraft, but without sacrifice of cruising speed; effective and harmonised controls; foldable wings.
High wings with V-struts; sweptback fin. Delivered as 85 per cent complete quick-build kit, less engine, propeller and instruments; quoted build time 200 hours.
FLYING CONTROLS: Conventional and manual. Large, cable-actuated control surfaces for effectiveness of response at slow speeds; 'spade'-type suspended aileron balance tabs. Flaps. Handley Page slats on wing leading-edge.
STRUCTURE: Generally of heat-treated and corrosion-proofed aluminium alloy; skins bonded before riveting. Welded tubular steel frame cockpit structure; steel alloy wing struts. Control surfaces covered in Dacron fabric. Wing leading-edge slats of carbon fibre. Composites tips to wing and empennage. Lexan windscreen; polycarbonate side windows.
LANDING GEAR: Tricycle type; fixed. Fuselage-mounted spring cantilever mainwheel legs with optional speed fairings. Steerable, telescoping nosewheel leg. Hydraulic disc brakes. Optional amphibious floats.
POWER PLANT: Choice of 59.6 kW (79.9 hp) Rotax 912 UL, 73.5 kW (98.6 hp) Rotax 912 ULS or 84.5 kW (113.3 hp) turbocharged Rotax 914 UL. Fuel in wing tanks and header tank, total capacity 87 litres (23.0 US gallons; 19.1 Imp gallons). GT two-blade wooden propeller.
EQUIPMENT: Optional glider towing hook and ventral baggage pod.
DIMENSIONS, EXTERNAL:
Wing span .... 9.75 m (31 ft 11¾ in)
Length overall .... 6.45 m (21 ft 2 in)
Height overall .... 2.45 m (8 ft 0½ in)
DIMENSIONS, INTERNAL:
Cabin max width .... 1.18 m (3 ft 10½ in)
AREAS:
Wings, gross .... 13.44 m^2 (144.7 sq ft)
WEIGHTS AND LOADINGS:
Weight empty .... 282 kg (622 lb)
Max T-O weight .... 450 kg (992 lb)
Max wing loading .... 33.5 kg/m^2 (6.86 lb/sq ft)
PERFORMANCE (Rotax 912 ULS engine):
Never-exceed speed (VNE) .... 100 kt (185 km/h; 115 mph)
Cruising speed at 75% power .... 86 kt (160 km/h; 99 mph)
Stalling speed, power off:
flaps up .... 30 kt (55 km/h; 35 mph)
flaps down .... 27 kt (50 km/h; 32 mph)
Max rate of climb at S/L .... 480 m (1,575 ft)/min
T-O run .... 30 m (100 ft)
Landing run .... 55 m (180 ft)
Endurance .... 4 h 18 m

UPDATED

Alisport Yuma ultralight kitplane operated in South Africa *(Patrick Allen)* ***NEW***/1110440

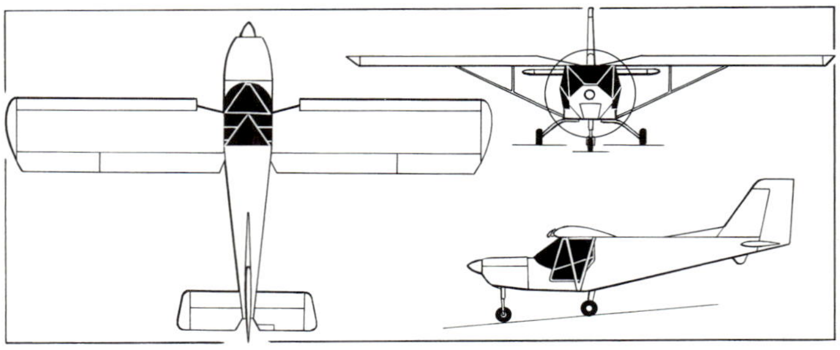
Alisport Yuma side-by-side ultralight *(James Goulding)* ***NEW***/0587527

ALPI

ALPI AVIATION SRL

Via Brigata Osoppo 180, I-33070 Vigonovo di Fontanafredda
Tel: (+39 0434) 37 04 96
Fax: (+39 0434) 25 39 38
e-mail: info@alpiaviation.com
Web: http://www.alpiaviation.com
CHIEF DESIGNER: Corrado Rusalen

FACTORY:
Via dei Templari 24, I-33080 San Quirino, Pordenone
US DISTRIBUTOR:
Orlando Sanford Aircraft Sales
1920 E Airport Boulevard, Sanford, Florida 32773
Tel: (+1 800) 276 77 61 and (+1 407) 322 36 62
Fax: (+1 407) 322 47 72
Web: http://www.airplane4sale.com

The Alpi Pioneer 300S is a derivation of the Asso V, designed by Vidor Guiseppe and formerly marketed by Rusalen & Rusalen. German marketing is by Flugtechnik at Damme. Alpi also produces the wooden Pioneer 200 and has a 5,000 m^2 (53,820 sq ft) factory at Pordenone; wooden components are sourced from Croatia.

UPDATED

ALPI PIONEER 200

Alpi Pioneer 200 registered in US *(Paul Jackson)* ***NEW***/0576834

TYPE: Side-by-side ultralight kitbuilt.
PROGRAMME: Draws on experience gained from Pioneer 300.
CUSTOMERS: At least 40 produced by early 2004. Exports include New Zealand and USA.
COSTS: Kit US$29,600; kit with builder's assistance US$43,600 (2004).
DESIGN FEATURES: Derived from Pioneer 300, but with more upright fin, untapered wing and fixed landing gear. Quoted build time 1,000 hours.
FLYING CONTROLS: Conventional and manual. Ailerons externally mass-balanced. Flight-adjustable tab in port elevator. Electrically operated flaps.
STRUCTURE: Fuselage of spruce with plywood skin overlaid with Dacron, except for fabric-covered elevators and rudder. Wing spar of spruce with plywood skin to leading-edge; remainder, including ailerons and flaps, fabric covered. Single-piece, forward-hinged canopy; glass fibre engine cowling.
LANDING GEAR: Tricycle type; fixed. Fuselage-mounted, tube metal spring cantilever mainwheel legs. Composites wheel fairings optional. Rubber-in-compression nosewheel shock absorber.
POWER PLANT: One 59.6 kW (79.9 hp) Rotax 912 UL flat-four driving two-blade fixed-pitch propeller or optionally 59.7 kW (80 hp) Jabiru 2200 and two- or three-blade variable pitch propellers. Fuel capacity 54 litres (14.3 US gallons; 11.9 Imp gallons), of which 50 litres (13.2 US gallons; 11.0 Imp gallons) are usable, in single centrally positioned fuel tank.
DIMENSIONS, EXTERNAL:
Wing span .... 7.55 m (24 ft 9¼ in)
Length overall .... 6.15 m (20 ft 2 in)
DIMENSIONS, INTERNAL:
Cockpit max width .... 1.05 m (3 ft 5¼ in)
AREAS:
Wings, gross .... 10.22 m^2 (110.0 sq ft)
WEIGHTS AND LOADINGS:
Weight empty .... 260 kg (573 lb)
Max T-O weight: Europe .... 450 kg (992 lb)
USA .... 499 kg (1,100 lb)
PERFORMANCE:
Never-exceed speed (VNE) .... 129 kt (240 km/h; 149 mph)
Max operating speed .... 111 kt (205 km/h; 127 mph)
Normal cruising speed at 75% power .... 100 kt (185 km/h; 115 mph)
Stalling speed, flaps down .... 34 kt (62 km/h; 39 mph)
Max rate of climb at S/L .... 457 m (1,500 ft)/min
T-O and landing run .... 100 m (330 ft)
Range with max fuel .... 351 n miles (650 km; 403 miles)
g limits .... +4/–2

UPDATED

ALPI PIONEER 300

TYPE: Side-by-side ultralight kitbuilt.
PROGRAMME: Derivation of Aerei Asso Vs. First flight February 1999 (Italian ultralight category; Mid-West engine); public debut 2 April 1999 and exhibited at Aero '99, Friedrichshafen, later in the same month, at which time total of 11 hours flown.
CURRENT VERSIONS: **300L:** Long-span version for European market; no longer produced by Alpi.
300S: Short-span version, predominantly for US market, *as described.*
300JS: Short span, Jabiru engine.
330: Inverted flight fuel system for aerobatic use. First known aircraft (I-7264) shown at Carpi Rally, 4 September 2004.
CUSTOMERS: At least 100 kits sold by early 2004; operated in France, Germany, Italy, Spain and UK, where PFA approval received 2004; first four 300Ss to US in 2001.
COSTS: Kit US$34,900; kit with builder's assistance $69,600 (2004).
DESIGN FEATURES: Streamlined, retractable gear, low-wing monoplane, adaptable for Ultralight or Experimental category operation. Tapered wings and horizontal tail surfaces;

First Alpi Pioneer 300 on UK civil register *(Paul Jackson)* ***NEW***/1047416

Alpi Pioneer 300JS short-wing, Jabiru-engined version *(Paul Jackson)* ***NEW***/0576849

sweptback fin; upturned wingtips; NACA 2315 aerofoil. Asso Vs has biconvex asymmetric wing section. Rearward-sliding canopy; fixed windscreen. Dual controls. Baggage shelf behind seats holds 27 kg (60 lb). Maximum roll rate 120°/s. Quoted build time 700 hours for standard kit; 350 hours for fast-build kit.

FLYING CONTROLS: Conventional and manual. Actuation by steel cables. Horn-balanced rudder. Electrically actuated flaps deflect 10, 20 and 30°. Trim tabs on rudder and port elevator.

STRUCTURE: Wooden airframe, including wing with single box spar; fuselage top and sides covered in carbon fibre shell, with Dacron covered undersides; latter replaced by removable carbon fibre shell on UK version. Integral fin with wooden spar and composites elsewhere; ailerons and flaps of wood with Dacron covering; tailplane wholly of composites;

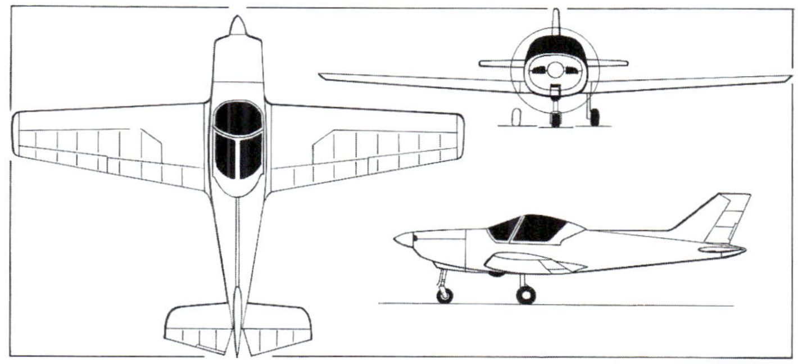

Alpi Pioneer 300 *(James Goulding)* 0137728

composites engine cowling and wingtips; Dacron-covered elevators and rudder; plywood-covered ailerons, flaps and tailplane.

LANDING GEAR: Retractable, tricycle type; steerable nosewheel with helical spring suspension. Mainwheels have levered oleo-pneumatic suspension (replacing rubber-in-compression in early aircraft). Mainwheels retract outwards; nosewheel rearwards; electric actuation with manual override. No doors, apart from fairing fixed ahead of nosewheel leg. Ingegno wheels with disc brakes; mainwheels size 4.00–6, nosewheel 4.00–4.

POWER PLANT: One 73.5 kW (98.6 hp) Rotax 912 ULS four-stroke, driving a two-blade GT-2/173/155 fixed-pitch propeller. Three-blade and variable-pitch propellers optional. Alternative Jabiru 3300, Rotax 914 (84.5 kW; 113.3 hp) and Mid-West AE110 engines. Fuel tank in each wing, combined capacity 80 litres (21.1 US gallons; 17.6 Imp gallons); 35 litre (9.2 US gallon; 7.7 Imp gallon) fuselage tank optional; 54 litre (14.3 US gallon; 11.9 Imp gallon) tank standard in US.

AVIONICS: To customer's choice.

DIMENSIONS, EXTERNAL:
Wing span ... 8.10 m (26 ft 7 in)
Length overall ... 6.25 m (20 ft 6 in)
Height overall ... 2.00 m (6 ft 6¾ in)

DIMENSIONS, INTERNAL:
Cockpit max width ... 1.05 m (3 ft 5¼ in)

AREAS:
Wings, gross ... 12.90 m^2 (138.9 sq ft)

WEIGHTS AND LOADINGS:
Weight empty ... 285 kg (628 lb)
Max T-O weight: ultralight ... 450 kg (992 lb)
experimental ... 544 kg (1,200 lb)

PERFORMANCE (Rotax 914 engine):
Never-exceed speed (VNE) and max level speed ... 145 kt (270 km/h; 167 mph)
Cruising speed at 75% power ... 135 kt (250 km/h; 155 mph)
Stalling speed ... 33 kt (60 km/h; 38 mph)
Max rate of climb at S/L ... 457 m (1,500 ft)/min
T-O and landing run ... 85 m (280 ft)
Range ... 702 n miles (1,300 km; 807 miles)
g limits ... +4/–2

UPDATED

DEA

DEA SRL

Renamed Alisport.

UPDATED

DFH

DF HELICOPTERS SRL

Via Priarona 82, I-15010 Cremolino-Ovada (AL)
Tel: (+39 143) 83 57 68
Fax: (+39 143) 865 55
e-mail: contact@df-helicopters.com
Web: http://www.df-helicopters.com
MANAGING DIRECTOR: Albert Frommer
PROJECT MANAGER: Guido Polidoro
QUALITY MANAGER: Walter Myrzik
MARKETING MANAGER: Milan Vuckovic

Original Dragon Fly company founded in 1993 by twin brothers Angelo and Alfredo Castiglioni specifically to produce the Dragon Fly 333 light helicopter, rights to which were taken over from general engineering company CRAE Elettromeccanica SpA. The company was taken over by German owners in 2001 and moved to a new 1,800 m2 (19,375 sq ft) factory at Ovada. The Dragon Fly 333 is being further developed into the Dragon Fly 334 GP.

UPDATED

DFH DRAGON 334

TYPE: Two-seat helicopter.

PROGRAMME: Developed originally by CRAE; design studies and manufacture of single-seat prototype 1985–88; ground and flight testing of this aircraft, 1989–90; two-seat Model 333AC prototype built and tested, 1991–93. Total of two single-seat and three two-seat prototypes, followed by four pre-series aircraft; production, as Dragon Fly 333, transferred to new factory from October 1994. Developed and tested by manufacturer to standards approaching FAR Pt 27; initial Italian certification is in ultralight class, but domestic VLR (very light rotorcraft) certification obtained 16 June 1996. Relaunched at Aero '03, Friedrichshafen, April 2003, under new ownership and new designation.

Prototype DFH Dragon 334 GP *(Paul Jackson)* ***NEW***/0580741

In 2001, a Dragon Fly was flown at the Cielo di Volo air show, powered by a 112 kW (150 hp) APU of undisclosed type.

CURRENT VERSIONS: **Dragon Fly 333:** Original version; discontinued. Powered by one Dragon Fly/Hirth F30A26AK four-cylinder two-stroke, rated at 82 kW (110 hp) for T-O, 70.8 kW (95 hp) maximum continuous.

Dragon Fly 333AC: Certified version to Italian VLR rules. Discontinued.

Héliot: RPV version developed in association with French companies Etudes et Développement Techniques (EDT) and CAC Systèmes; launched in June 1996 when prototype displayed at Eurosatory 1996 show at Le Bourget, Paris. No further production known. Discontinued.

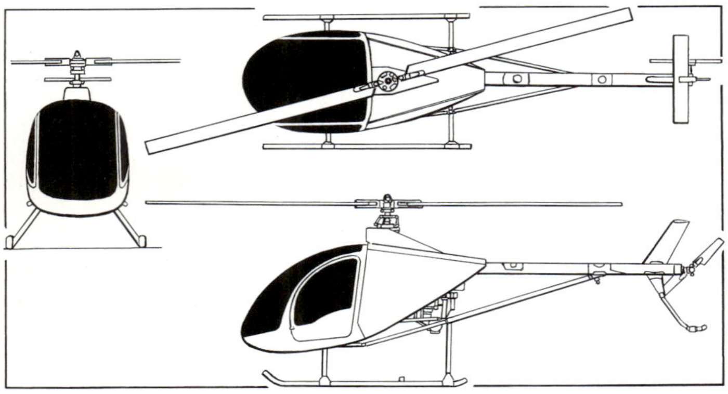

DFH Dragon 334 *(James Goulding)* ***NEW***/0589011

DFH X 334: Proof-of-concept vehicle for 334GP, with new engine, rotor controls and drive system. Flying by 2002.

Dragon 334 GP: Based on Dragon Fly 333 with commercial applications including traffic and border observation, aerial photography, powerline patrolling and crop-spraying. New rotor controls, drive system and tapered tail rotor blades, plus single-piece front/upper glazing. Prototype unflown by May 2003, when exhibited at Aero '03. *As described.*

CUSTOMERS: First delivery of Dragon Fly in May 1994 to Chinese Civil Protection Volunteers. Total of 80 ordered, of which some 70 delivered, by late 1998 to customers in Abu Dhabi, Australia, Belgium, Czech Republic, France, Germany, Italy (37), New Zealand, Portugal and Turkey. First delivery of Italian certified version was in 1997; initial operator was Venice Aero Club. Dragon 334 deliveries had not begun by May 2004, although the company reported that 15 had been built. 2004 production set at 50 units.

DESIGN FEATURES: Two-blade, semi-rigid main rotor and two-blade tail rotor; all blades of NACA 0012 aerofoil section; main rotor nominal speed 520 rpm. Can be road-towed on trailer with main blades folded. Optionally available in kit form.

STRUCTURE: Cabin is welded titanium frame with composites outer shell; aluminium alloy tailboom, rotor blades and landing skids. Full corrosion protection.

LANDING GEAR: Conventional twin-skid type. Emergency floats and skis under development.

POWER PLANT: One 84.6 kW (113.4 hp) Rotax 914 flat-six. Transmission driven through centrifugal clutch and three polymer V-belts. Fuel capacity 64 litres (16.9 US gallons; 14.1 Imp gallons) of which 57 litres (15.0 US gallons; 12.5 Imp gallons) are usable.

ACCOMMODATION: Side-by-side seats for two persons. Dual controls standard. Small baggage compartment below seats.

SYSTEMS: 12 V electrical system with engine-driven 75Ah generator and 12 V 19 Ah battery.

AVIONICS: *Comms:* Provision for transceiver and intercom.

Flight: Optional electric lateral trim.

Instrumentation: Standard VFR.

EQUIPMENT: Medevac and firefighting kits and external load hook under development.

DIMENSIONS, EXTERNAL:
Main rotor diameter ... 6.70 m (21 ft 11¾ in)
Tail rotor diameter ... 1.12 m (3 ft 8 in)
Length: overall, rotors turning ... 7.89 m (25 ft 10½ in)
fuselage ... 5.56 m (18 ft 3 in)
Height to top of rotor head ... 2.36 m (7 ft 9 in)
Skid track ... 1.55 m (5 ft 1 in)

DIMENSIONS, INTERNAL:
Cabin: Length ... 1.11 m (3 ft 7¾ in)
Max width ... 1.15 m (3 ft 9¼ in)
Max height ... 1.13 m (3 ft 8½ in)

AREAS:
Main rotor disc ... 35.21 m^2 (379.0 sq ft)
Tail rotor disc ... 0.985 m^2 (10.60 sq ft)

WEIGHTS AND LOADINGS:
Weight empty, standard ... 260 kg (573 lb)
T-O weight: normal ... 450 kg (992 lb)
max ... 550 kg (1,212 lb)

PERFORMANCE (estimated):
Max level speed at S/L ... 80 kt (148 km/h; 92 mph)
Cruising speed at S/L ... 70 kt (130 km/h; 81 mph)
Max rate of climb at S/L ... 381 m (1,250 ft)/min
Service ceiling ... 3,050 m (10,000 ft)
Hovering ceiling: IGE ... 2,050 m (6,720 ft)
OGE ... 1,450 m (4,760 ft)
Range with 20 min reserves ... 162 n miles (300 km; 186 miles)
Endurance ... 3 h

UPDATED

EURO ALA

EURO ALA

Fraz 5 Reparata, Zona Artigianale, I-64010 Civitella del Tronto (TE)
Tel: (+39 0861) 91 06 06
Fax: (+39 0861) 91 08 95
e-mail: info@euroala.it
Web: http://www.euroala.it
PRESIDENT: Alfredo Di Cesare

US SALES:
Euro ALA Advanced Light Aircraft Inc, 13 Crosley Lane, Suite 5, Sebring, Florida 33870
Tel: (+1 786) 417 89 60
e-mail: info@jetfoxusa.com
Web: http://www.jetfoxusa.com

The Advanced Light Aircraft (ALA) company was established in 1985 and changed name to Euroala in 1995. In 2003, it revealed that it was working on an improved Jet Fox, due for completion at the end of the year.

At the end of 2002, the company employed 15 workers at its 1,000 m^2 (10,764 sq ft) factory.

UPDATED

EUROALA JET FOX 97

TYPE: Side-by-side ultralight.

PROGRAMME: Earlier Jet Fox 91 first flown in 1991; 140 sold by 1993. Considerably improved Jet Fox 97 first flown 10 March 1997 and exhibited at Aero '97 at Friedrichshafen the following month. Jet Fox series certified in Germany 1993; France and Belgium in 1995 and Israel and Spain in 2001.

In January 2003 the company revealed that it was working on a new Jet Fox with wider cabin, new door opening system and aluminium wings with Fowler flaps, giving a cruise speed approaching 108 kt (200 km/h; 124 mph). Estimated first flight date given as end of 2003.

CUSTOMERS: Total of 135 built by the end of 2002, including remanufactured Jet Fox 91s. Ten sold in USA in 2002.

COSTS: US$43,100 with Rotax 912UL; US$43,900 with Rotax 912UL-S (both 2003).

DESIGN FEATURES: Considerably refined version of earlier design, but bears structural resemblance to Fox-C22; fully faired fuselage and engine; flaps. Wings rapidly detachable for transport.

FLYING CONTROLS: Conventional and manual. Flight-adjustable elevator trim.

STRUCTURE: Alloy tube and steel fuselage frame with composites shell; Dacron-covered aluminium wings and tail surfaces; glass fibre and carbon fibre for landing gear.

LANDING GEAR: Tricycle type; fixed. Fuselage-mounted spring cantilever mainwheel legs. Mainwheels 4.00-6. Cable-operated disc brakes. Steerable nosewheel, size 4.00-4. Speed fairings optional.

POWER PLANT: One 47.8 kW (64.1 hp) Rotax 582 UL two-stroke or 59.6 kW (79.9 hp) Rotax 912 UL four-stroke engine, driving a two-blade GT wooden propeller; 73.5 kW (98.6 hp) Rotax 912 UL-S also available. Single fuel tank behind seats, capacity 59 litres (15.6 US gallons; 13.0 Imp gallons).

EQUIPMENT: Optional ballistic parachute.

DIMENSIONS, EXTERNAL:
Wing span ... 9.78 m (32 ft 1 in)
Length overall ... 5.78 m (18 ft 11½ in)
Height overall ... 2.80 m (9 ft 2¼ in)

Euroala Jet Fox 97 ***NEW***/0538448

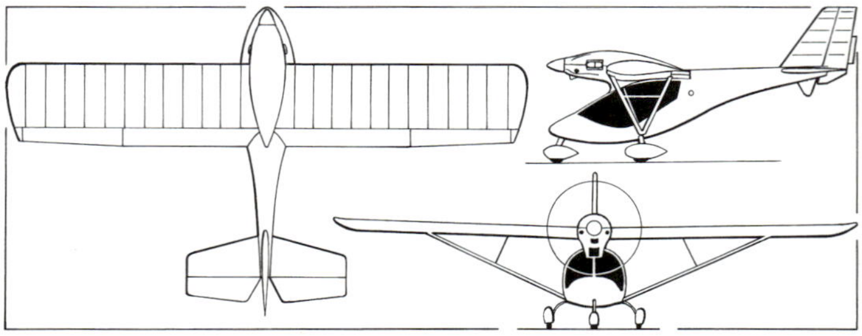

General arrangement of the Euroala Jet Fox 97 *(James Goulding)* 0054609

AREAS:
Wings, gross ... 14.62 m^2 (157.4 sq ft)

WEIGHTS AND LOADINGS (Rotax 912):
Weight empty, incl parachute ... 290 kg (639 lb)
Max T-O weight ... 450 kg (992 lb)

PERFORMANCE (Rotax 912):
Max level speed ... 94 kt (175 km/h; 109 mph)
Cruising speed ... 81 kt (150 km/h; 93 mph)
Stalling speed ... 33 kt (60 km/h; 38 mph)
Max rate of climb at S/L ... 360 m (1,181 ft)/min
T-O run ... 100 m (330 ft)
Landing run ... 120 m (395 ft)
Endurance ... 4 h 30 min
g limits ... +6/−3.5

UPDATED

FLYITALIA

FLYITALIA SRL

Via Gorizia 1, I-96100 Siracusa
Tel: (+39 0335) 709 65 12
e-mail: info@flyitalia.it
Web: http://www.flyitalia.it
COMMERCIAL DIRECTOR: Franco Buratto

Flyitalia is local distributor for two types of ultralight built in the Czech Republic: Aeropro Eurofox and Aerospool WT9 Dynamic.

In 2004, it planned to launch a third product known as the **MD3 Rider**, of which no details have yet been received, with the exception of a photograph.

NEW ENTRY

Flyitalia MD3 Rider two-seat ultralight ***NEW***/0572294

FLY LINE

FLY LINE SRL

Via del Commercio 16, I-33050 Torviscova (UD)
Tel/Fax: (+39 0431) 92 90 31
e-mail: info@flylineaviation.com
Web: http://www.flylineaviation.com and www.flylinesrl.it
DIRECTOR: Giuseppe Milito

Sig Milito designed aircraft for Fly Synthesis (Storch, Wallaby and Texan) before forming Fly Line to produce an aircraft similar in design to the last-mentioned. The prototype has since converted to ethanol power in collaboration with Baylor Institute for Air Science, USA.

UPDATED

FLY LINE M-10 REVENGE

TYPE: Side-by-side ultralight.

PROGRAMME: First flown (I-6924) mid-2002. Demonstrated at Paris Air Show, 15–22 June 2003.

CUSTOMERS: Fifth aircraft shipped to USA in July 2003 as demonstrator. Prototype converted to ethanol power and redesignated **Sniffer**; based at Waco, Texas, with Baylor Institute for Air Science as air pollution sampling platform.

COSTS: €41,000 flyaway, plus tax (2003).

DESIGN FEATURES: Intended to provide light sport aircraft performance at ultralight purchasing and operating costs. Low, constant-chord wing with upturned tips. Tapered rear fuselage and sweptback fin with low-mounted, constant-chord tailplane.

FLYING CONTROLS: Conventional and manual. Single-piece elevator with flight-adjustable tab; horn-balanced rudder. Plain flaps.

STRUCTURE: Composites throughout. Glass fibre, carbon fibre and Kevlar sandwich construction.

LANDING GEAR: Tricycle type; fixed. Fuselage-mounted spring cantilever mainwheel legs with speed fairings. Steerable, trailing link nosewheel, size 4.10/3.50-5. Hydraulic mainwheel brakes.

POWER PLANT: One 73.5 kW (98.6 hp) Rotax 912 ULS flat-four driving a three-blade, ground-adjustable pitch propeller. Fuel tanks in wings, total capacity 35 litres (9.2 US gallons; 7.7 Imp gallons).

EQUIPMENT: Provision for underwing hardpoints.

DIMENSIONS, EXTERNAL:
Wing span .... 9.03 m (29 ft 7½ in)
Length overall .... 6.93 m (22 ft 8¾ in)
Height overall .... 2.36 m (7 ft 9 in)

DIMENSIONS, INTERNAL:
Cockpit max width .... 1.17 m (3 ft 10 in)

AREAS:
Wings, gross .... 12.0 m^2 (129.2 sq ft)

Fly Line M-10 Revenge prototype *(Paul Jackson)* ***NEW***/0580837

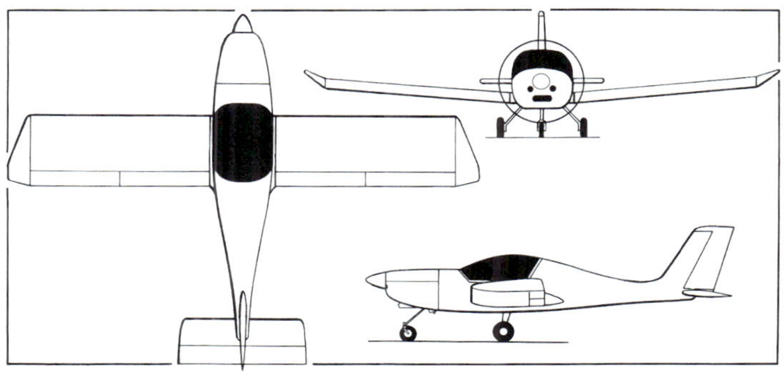

Fly Line Revenge two-seat ultralight *(James Goulding)* ***NEW***/0583689

WEIGHTS AND LOADINGS:
Weight empty .... 280 kg (617 lb)
Max T-O weight .... 450 kg (992 lb)

PERFORMANCE:
Never-exceed speed (VNE) .... 140 kt (260 km/h; 161 mph)
Cruising speed at 75% power .... 113 kt (210 km/h; 130 mph)
Stalling speed, power off, flaps down .... 32 kt (58 km/h; 36 mph)
g limits .... +4/–2

UPDATED

FLY SYNTHESIS

FLY SYNTHESIS SRL

Via Gorizia 63, I-33050 Gonars/Udine
Tel: (+39 0432) 99 24 82 and 99 35 57
Fax: (+39 0432) 93 12 30
e-mail: info@flysynthesis.com
Web: http://www.flysynthesis.com
OWNER AND DIRECTOR: Sonia Felice
PRODUCTION DIRECTOR: Rodolfo Ciotti
TECHNICAL, RESEARCH AND DEVELOPMENT DIRECTOR: Corrado Pinzana

The Storch ultralight produced by this company also provides the basis of the tailwheel/T tail Pipistrel Sinus motor glider and ultralight, built in Slovenia. US marketing undertaken by American Ghiles Aircraft.

At the end of 2002, FlySynthesis employed 18 in its 1,100 m^2 (11,850 sq ft) factory.

UPDATED

FLY SYNTHESIS STORCH

US Name: Lafayette Stork

TYPE: Side-by-side ultralight/kitbuilt.

PROGRAMME: First flown in 1991 and initially marketed as Rodaro Storch. Certified in Austria, Belgium, France, Germany, Greece, Hungary, Israel, Italy, Malaysia, Portugal, Slovenia and the USA.

CURRENT VERSIONS: **CL De Luxe**: Standard version, *as described*.

S: As CL De Luxe, but with shorter overall length (5.95 m; 19 ft 6¼ in) and wing span (9.32 m; 30 ft 7 in) and Rotax 912 UL or Jabiru 2200 engine. Flaps. Not available in kit form.

HS De Luxe: As S De Luxe, but with shorter overall length (5.75 m; 18 ft 10¾ in) and (8.71 m; 28 ft 7 in) wing span.

CUSTOMERS: 340 built by end of 2002.

COSTS: Ready-to-fly: CL De Luxe, HS De Luxe: (Rotax 582) €34,000, (Jabiru) €37,700; S: €43,700; kit: CL De Luxe, HS De Luxe (Rotax 582) €19,100, Jabiru €19,200 (all 2003).

DESIGN FEATURES: Laminar flow aerofoil section. Strut-braced wings fold alongside fuselage for storage. Quoted build time 180 hours.

Fly Synthesis Storch on beaching floats *(Paul Jackson)* ***NEW***/0583374

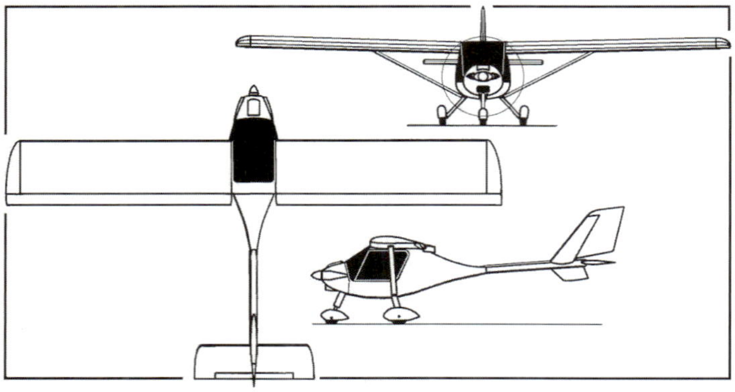

General arrangement of the Fly Synthesis Storch CL De Luxe with Rotax 582 engine *(Paul Jackson)* ***NEW***/0546946

FLYING CONTROLS: Manual. Full-span, Junkers-type ailerons and all-moving tailplane with anti-balance tab.

STRUCTURE: Composites main fuselage shell with welded steel tube frame in cabin area and tubular alloy tailboom; light alloy constant chord wing has 6082 light alloy spar and composites skin; all-composites wing optional.

LANDING GEAR: Tricycle type; fixed. Fuselage-mounted spring metal tube mainwheel legs with wheel fairings. Mainwheels 4.00–6, nosewheel 4.00-4. Drum brakes on CL De Luxe, disc brakes on S and HS De Luxe. Oil-damped, steerable nosewheel with maximum deflection ±40°.

POWER PLANT: CL De Luxe and HS De Luxe have one 47.8 kW (64.1 hp) Rotax 582 UL or 59.7 kW (80 hp) Jabiru 2200; S has one 59.6 kW (79.9 hp) Rotax 912 UL or 59.7 kW (80 hp) Jabiru 2200, each driving a two- or three-blade propeller. Standard fuel capacity 60 litres (15.9 US gallons; 13.2 Imp gallons), of which 56 litres (14.8 US gallons; 12.3 Imp gallons) are usable.

ACCOMMODATION: Cabin doors open upwards.

EQUIPMENT: BRS ballistic recovery parachute optional.

Data below refer to CL De Luxe with Jabiru 2200 engine.

DIMENSIONS, EXTERNAL:
Wing span .... 10.14 m (33 ft 3¼ in)
Length overall .... 6.25 m (20 ft 6 in)
Height overall .... 2.45 m (8ft 0½ in)

AREAS:
Wings, gross .... 13.02 m² (140.15 sq ft)

WEIGHTS AND LOADINGS:
Weight empty .... 258 kg (569 lb)
Max T-O weight .... 450 kg (992 lb)

PERFORMANCE:
Never-exceed speed (VNE) .... 97 kt (180 km/h; 112 mph)
Max level speed .... 86 kt (159 km/h; 99 mph)
Cruising speed at 75% power .... 84 kt (155 km/h; 96 mph)
Stalling speed .... 31 kt (57 km/h; 35 mph)
Max rate of climb at S/L .... 270 m (886 ft)/min
T-O run .... 105 m (345 ft)
Landing run .... 120 m (394 ft)
Range .... 356 n miles (659 km; 410 miles)
Endurance .... 3 h 18 min
g limits .... +4/–2

UPDATED

FLY SYNTHESIS TEXAN

TYPE: Side-by-side ultralight.

PROGRAMME: First flown 1999. Available factory-built only.

CURRENT VERSIONS: **Texan Top Class**: Standard version, *as described*.

Texan RG: Launched at Aero'03 Friedrichshafen where shown statically; retractable landing gear; mainwheels retract outward and nosewheel rearward. Dimensions as standard Texan, but increased vertical area at empennage provided by fin fillet and ventral strake. Empty weight 285 kg (628 lb); stalling speed 34 kt (62 km/h; 39 mph); never-exceed speed 151 kt (280 km/h; 174 mph).

CUSTOMERS: 25 built by January 2003, including five to Israel. Certified in Belgium, Israel, Italy and Portugal. In first quarter of 2003 a retractable landing gear version was being developed with a provisional never-exceed speed (VNE) of 151 kt (280 km/h; 174 mph); first flown May 2003 and production started thereafter.

COSTS: Kit €33,700; flyaway €51,700 (2003).

DESIGN FEATURES: Low-wing monoplane. Constant-chord wing of laminar flow aerofoil section; sweptback fin and rudder; rearward-sliding cockpit canopy with fixed windscreen. Quoted build time 280 hours.

FLYING CONTROLS: Manual. Ground-adjustable trim tab on starboard aileron; all-moving

Fly Synthesis Texan ***NEW***/0546949

Texan RG main landing gear retracted *(Paul Jackson)* ***NEW***/0583366

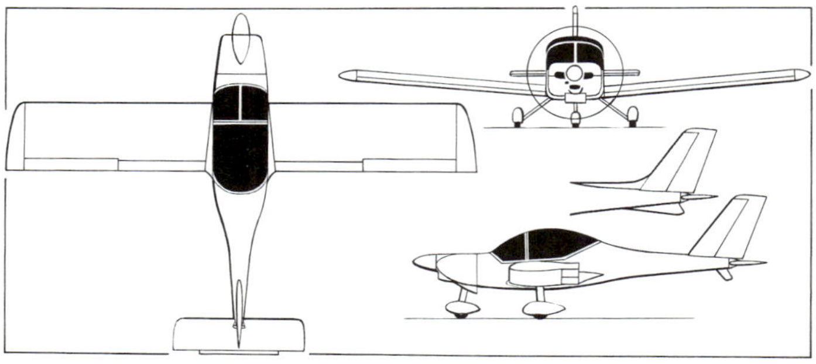

Fly Synthesis Texan two-seat ultralight, with scrap view of Texan RG empennage *(James Goulding)* ***NEW***/0583341

tailplane with anti-balance tab and mass balance weights; horn-balanced rudder. Four-position slotted flaps, maximum droop 45°.

STRUCTURE: Carbon fibre fuselage with Nomex sandwiched carbon fibre wings, tailplane and control surfaces.

LANDING GEAR: Tricycle type; fixed (except RG version). Fuselage-mounted spring cantilever mainwheel legs with wheel fairings. Mainwheels 4.00–6, nosewheel 4.00-4; streamlined fairings on all three wheels. Hydraulic disc brakes on mainwheels.

POWER PLANT: One 73.5 kW (98.6 hp) Rotax 912 ULS driving a two-blade GT-2/173/155 wooden or three-blade Pipistrel PBM300 BAM3D 165 composites propeller. Fuel in wingroot tanks, total capacity 72 litres (19.0 US gallons; 15.8 Imp gallons), of which 68 litres (18.0 US gallons; 15.0 Imp gallons) usable.

EQUIPMENT: BRS ballistic recovery parachute system optional.

DIMENSIONS, EXTERNAL:
Wing span .... 8.60 m (28 ft 2½ in)
Length overall .... 6.99 m (22 ft 11¼ in)
Height overall .... 2.61 m (8 ft 6¾ in)

AREA:
Wings, gross .... 12.00 m² (129.2 sq ft)

WEIGHTS AND LOADINGS:
Weight empty .... 275 kg (606 lb)
Max T-O weight .... 450 kg (992 lb)

PERFORMANCE (Rotax 912 ULS):
Never-exceed speed (VNE) .... 126 kt (235 km/h; 146 mph)
Cruising speed at 75% power .... 108 kt (200 km/h; 124 mph)
Stalling speed .... 34 kt (62 km/h; 39 mph)
Max rate of climb at S/L .... 300 m (984 ft)/min
T-O run .... 155 m (508 ft)
Landing run .... 140 m (460 ft)

UPDATED

FLY SYNTHESIS WALLABY

TYPE: Single-seat ultralight kitbuilt.
PROGRAMME: First flight took place in 1999. Certified in Belgium and Italy.
CUSTOMERS: Six built by January 2003.
COSTS: Kit €10,400, ready-to-fly €19,100 (2003).
DESIGN FEATURES: Strut-braced high-wing monoplane. Pod-and-boom fuselage. Wings fold for storage/transport. Open cockpit with windscreen; optional 'winter canopy'. Sweptback fin. Quoted build time 150 hours.
FLYING CONTROLS: Manual. All-moving tailplane with anti-balance tab and mass balance weights.
STRUCTURE: Utilises Storch wing and tail unit, but in composites and with reprofiled ventral fin. Glass fibre fuselage pod.
LANDING GEAR: Tricycle type; fixed. Fuselage-mounted spring metal tube cantilever mainwheel legs. Mainwheel size 14×4.
POWER PLANT: One 37.0 kW (49.6 hp) Rotax 503 UL driving a two-blade GT-2/170/90 propeller. Fuel capacity 60 litres (15.9 US gallons; 13.2 Imp gallons) in wingroot tanks.
ACCOMMODATION: Open, but 'winter canopy' available, as extra, to enclose occupants.
DIMENSIONS, EXTERNAL:
Wing span .... 9.42 m (30 ft 10¾ in)
Length overall .... 5.73 m (18 ft 9½ in)
Height overall .... 2.41 m (7 ft 10¾ in)
AREAS:
Wings, gross .... 12.70 m² (136.7 sq ft)
WEIGHTS AND LOADINGS:
Weight empty .... 208 kg (459 lb)
Max T-O weight .... 425 kg (937 lb)
PERFORMANCE:
Never-exceed speed (VNE) .... 71 kt (132 km/h; 82 mph)
Cruising speed at 75% power .... 51 kt (95 km/h; 59 mph)
Stalling speed .... 27 kt (50 km/h; 32 mph)
Max rate of climb at S/L .... 150 m (492 ft)/min
Landing run .... 90 m (295 ft)
g limits .... +4/–2

UPDATED

Fly Synthesis Wallaby *NEW*/0583661

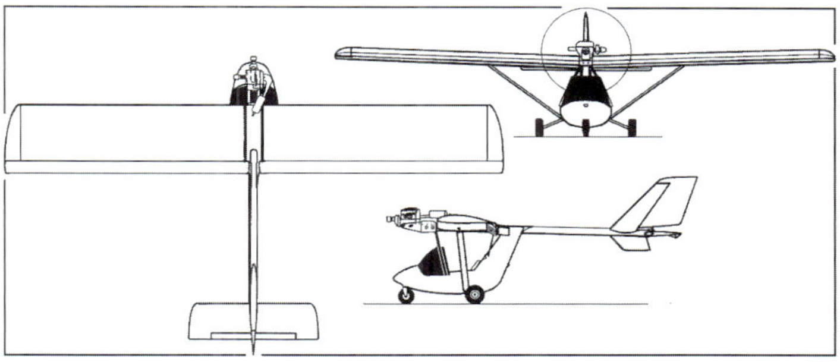

General arrangement of Fly Synthesis Wallaby *(Paul Jackson)* *NEW*/0546957

FRATI

STELIO FRATI

Via Noe 1, I-20100 Milano
Tel/Fax: (+39 02) 204 78 31
CHIEF DESIGNER: Dott Ing Stelio Frati

In 1998, Dott Ing Frati left General Avia to resume the activities of a freelance designer - a career successfully practised since 1950. His past designs include the F.8 Falco, F.15 Picchio, F.250 (now manufactured by Aermacchi as the SF-260), SF.600A Canguro (for SIAI-Marchetti, and now VulcanAir) and F.20 Pegaso. The F.8 Falco is marketed as a kit aircraft by Sequoia Aircraft Corporation (which see in US section).

Frati has designed a new four-seat, all-composites tourer with a 194 kW (260 hp) engine described under the III entry in this section; current projects are the F.1000 light jet and the F.2500 twin-turboprop.

UPDATED

FRATI F.1000

TYPE: Two-seat jet sportplane.
PROGRAMME: Concept, under development in early 2003.
DESIGN FEATURES: Family resemblance to previous Frati-designed jets Procaer F.400 Cobra and Promavia Jet Squalus. Unswept wings and tailplane; swept fin; small wing fences at mid-span; upturned wingtips. Dorsal intake protects engine from ingestion of foreign objects.
FLYING CONTROLS: Conventional and manual; trim tabs on elevators, ailerons and rudder.
LANDING GEAR: Retractable tricycle type.
POWER PLANT: One turbofan in 3.34 kN to 4.45 kN (750 to 1,000 lb st) class.
ACCOMMODATION: Two, side-by-side under bubble canopy with fixed windscreen.
DIMENSIONS, EXTERNAL:
Wing span .... 8.50 m (27 ft 10¾ in)
Wing aspect ratio 6.6
Length overall .... 8.65 m (28 ft 4½ in)
AREA:
Wings, gross .... 10.95 m² (117.9 sq ft)
WEIGHTS AND LOADINGS:
Weight empty .... 900 kg (1,984 lb)
Max T-O weight .... 1,450 kg (3,197 lb)

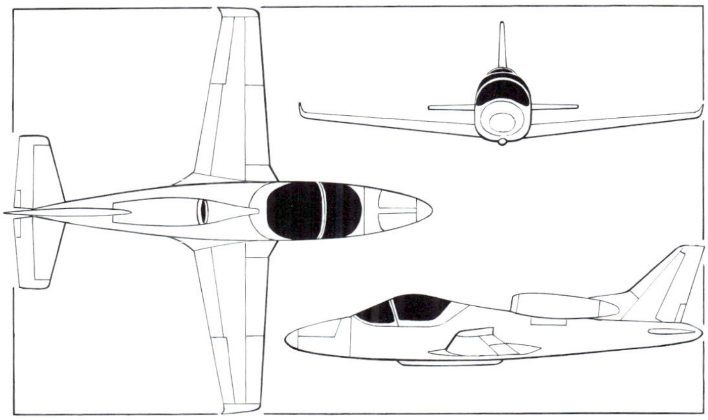

Frati F.1000 two-seat light jet *(James Goulding)* *NEW*/0552990

Mockup of Frati F.1000 *NEW*/0552062

Max wing loading .... 132.4 kg/m² (27.12 lb/sq ft)
PERFORMANCE (estimated):
Max level speed 310 kt (575 km/h; 357 mph)
Max cruising speed: 302 kt (560 km/h; 348 mph)
Max rate of climb at S/L 720 m (2,362 ft)/min
T-O run 350 m (1,150 ft)
Landing run 300 m (985 ft)
Endurance: 4 h

UPDATED

FRATI F.2500

TYPE: Twin-turboprop transport.
PROGRAMME: Under development in early 2003; derived from unbuilt F.3000 twin-turbofan commuter airliner.
DESIGN FEATURES: High wing; swept fin/rudder with dorsal fillet. Cabin section is square with constant frame size throughout length. Intended to meet FAR Pt 25 certification criteria. Design goals include STOL characteristics combined with adequate cruising speed and ease of manufacture in countries with limited aeronautical manufacturing experience.
Wing section NACA 63.
LANDING GEAR: Retractable tricycle type. Single wheel on each main unit; main units retract into small sponsons.
POWER PLANT: Two 1,026 kW (1,376 shp) Pratt & Whitney Canada PT6A-65 turboprops. Fuel contained in integral tanks in wings outboard of engines capacity 2,500 litres (660.4 US gallons; 550 Imp gallons).
ACCOMMODATION: Crew of two and 19 passengers in three-abreast 2+1 configuration, with overhead baggage bins; lavatory standard. Crew door at front of cabin, passenger door at rear, both on port side. Seven cabin windows on each side. In cargo configuration cabin can accommodate up to nine different types of container. Rear loading ramp with integral electrically driven winch to facilitate loading/unloading of cargo. Cockpit and cabin, or just cockpit, could be pressurised.
All data are provisional.
DIMENSIONS, EXTERNAL:
Wing span .... 21.00 m (68 ft 10¾ in)
Wing aspect ratio 9.7

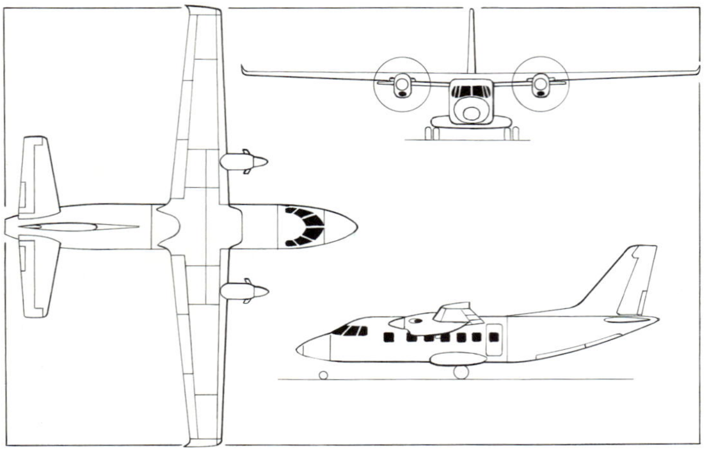

General arrangement of Frati F.2500 *(James Goulding)* ***NEW***/0552998

Length overall ... 16.80 m (55 ft 1½ in)
Height overall ... 6.40 m (21 ft 0 in)
Wheel track ... 3.91 m (12 ft 10¼ in)

DIMENSIONS, INTERNAL:
Cabin: Max height ... 1.80 m (5 ft 11 in)
Max width ... 1.93 m (6 ft 4 in)
Volume ... 24.07 m^3 (850.0 cu ft)

AREAS:
Wings, gross ... 45.50 m^2 (489.8 sq ft)

WEIGHTS AND LOADINGS (ESTIMATED):
Weight empty, equipped ... 4,700 kg (10,362 lb)
Payload ... 3,200 kg (7,055 lb)
Max fuel ... 2,000 kg (4,409 lb)
Max T-O weight ... 8,600 kg (18,959 lb)
Max wing loading ... 189.0 kg/m^2 (38.71 lb/sq ft)
Max power loading ... 4.19 kg/kW (6.89 lb/shp)

PERFORMANCE (estimated):
Max cruising speed at FL100 ... 250 kt (463 km/h; 288 mph)
Stalling speed, flaps down ... 70 kt (130 km/h; 81 mph)
Max rate of climb at S/L ... 762 m (2,500 ft)/min
Max rate of climb, OEI ... 244 m (800 ft)/min
Service ceiling ... 8,500 m (27,880 ft)
Service ceiling, OEI ... 4,000 m (13,120 ft)
T-O run ... 430 m (1,410 ft)
T-O to 15 m (50 ft) ... 700 m (2,300 ft)
Landing from 15 m (50 ft) ... 650 m (2,135 ft)
Landing run with reverse thrust ... 300 m (985 ft)
Range with 19 passengers, 45 min reserves ... 971 n miles (1,800 km; 1,118 miles)
Endurance: ... 5 h

UPDATED

HELISPORT

CH-7 HELISPORT SRL
Strada Traforo del Pino 102, I-10132 Torino
Tel: (+39 011) 899 95 65, 899 96 18 and 899 67 30
Fax: (+39 011) 899 96 18, 898 11 44 and 899 55 50
e-mail: kompress@tin.it
Web: http://www.ch-7helicopter.com

Previously known as EliSport, the company markets the Kompress, a development of the CH-6 open frame helicopter produced in 1987 by the Argentine designer Augusto Cicaré. US agent is Lancair (which see).

UPDATED

HELISPORT CH-7 KOMPRESS

TYPE: Two-seat ultralight helicopter kitbuilt.

PROGRAMME: CH-6 project acquired in 1989 by HeliSport; added new cockpit by sports car designer Marcello Gandini; renamed CH-7 Angel, two-seat version became Kompress.

CURRENT VERSIONS: **CH-7 Kompress:** Standard version, *as described.*

CH-7 Mariner: Float-equipped version. Multi-tube inflatable floats add 15 kg (33 lb) to empty weight.

CH-7 Angel: Single-seat version, powered by either 47.8 kW (64.1 hp) Rotax 582 UL (as Angel 582) or 59.6 kW (79.9 hp) Rotax 912 UL (as Angel 912). Max T-O weight of both versions is 300 kg (661 lb). No longer available.

CUSTOMERS: Some 140 flying by 2002, including 120 single-seat Angels built up to 1997. Owners in Australia, Germany, Italy, South Africa, Taiwan, USA and elsewhere.

COSTS: Basic kit, less engine and tax €70,000 (2004).

DESIGN FEATURES: Small 'penny-farthing' helicopter with two-blade, semi-rigid, teetering main and two-blade rotors; pod-and-boom fuselage; sweptback upper and lower fins, latter with tailskid. Main rotor transmission by single 10 cm (4 in) five-groove Goodyear belt. Main rotor has 10,000-hour life, on condition. Quoted build time 200 hours; quick-build kit 85 hours.

Rotor blade aerofoil sections NACA 8-H-12 (main) and NACA 63-014 (tail); 1.5 kg (3.3 lb) main rotor tip weights; max 520 rpm main, 2,808 rpm tail rotor.

FLYING CONTROLS: Conventional and manual, by pushrods and bell cranks.

STRUCTURE: Welded 4130 steel tube cabin frame, with nitrogen filling; glass fibre cockpit shell and Plexiglas or Lexan canopy; 2024T3 anodised aluminium alloy tailboom and landing skids. Rotor blades of composites (main) and aluminium alloy (tail).

LANDING GEAR: Fixed; twin-skid type.

POWER PLANT: One 84.6 kW (113.4 hp) Rotax 914 piston engine. Fuel capacity 40 litres (10.6 US gallons; 8.8 Imp gallons) standard in all versions; additional 19 litre (5.0 US gallon; 4.2 Imp gallon) tank optional.

EQUIPMENT: Optional hook beneath fuselage; max load 100 kg (220 lb).

DIMENSIONS, EXTERNAL:
Main rotor diameter ... 6.27 m (20 ft 6¾ in)
Tail rotor diameter ... 1.03 m (3 ft 4½ in)
Length: fuselage ... 5.31 m (17 ft 5 in)
overall, rotors turning ... 7.41 m (24 ft 4 in)

HeliSport CH-7 Kompress *(Paul Jackson)* ***NEW***/0583467

Fuselage max width ... 0.82 m (2 ft 8¼ in)
Height to top of rotor head ... 2.31 m (7 ft 7 in)
Skid track ... 1.50 m (4 ft 11 in)

DIMENSIONS, INTERNAL:
Cabin max width ... 0.76 m (2 ft 6 in)

AREAS:
Main rotor disc ... 29.90 m^2 (321.8 sq ft)
Tail rotor disc ... 0.83 m^2 (8.93 sq ft)

WEIGHTS AND LOADINGS:
Weight empty: Europe ... 245 kg (540 lb)
USA ... 281 kg (620 lb)
Max T-O weight: Europe ... 450 kg (992 lb)
USA ... 499 kg (1,100 lb)

PERFORMANCE:
Never-exceed speed (VNE) ... 112 kt (209 km/h; 129 mph)
Max cruising speed: ... 86 kt (160 km/h; 99 mph)
Max rate of climb at S/L: Europe ... 450 m (1,476 ft)/min
USA ... 347 m (1,140 ft)/min
Hovering ceiling: IGE ... 3,500 m (11,480 ft)
OGE ... 2,500 m (8,200 ft)
Service ceiling ... 5,000 m (16,400 ft)
Range: with normal fuel ... 298 n miles (552 km; 343 miles)
with optional fuel ... 388 n miles (719 km; 477 miles)
Endurance: ... 3 h 40 min

UPDATED

ICP

ICP SRL
Via Torino 12, Piovà Massaia (At)
Tel: (+39 0141) 99 65 03
Fax: (+39 0141) 99 65 06
e-mail: info@icp.it
Web: http://www.icp.it
PRESIDENT: Tancredi Razzano
TECHNICAL MANAGER: Stefano Agostinetto
QUALITY MANAGER: Renato Pescamona

The company was founded in 1991 and markets aircraft strongly resembling Zenair designs from the USA. It is also active in the automotive industry. Factory floor area 3,000 m^2 (32,300 sq ft); workforce 40 in January 2003. Production of the Amigo (Zenith Zodiac) has ended.

UPDATED

ICP MXP-740 SAVANNAH

TYPE: Side-by-side ultralight.

PROGRAMME: Strongly resembles Zenith CH 701, except for conventional rudder and elevators; see Ekoflug MXP-740 (Slovak Republic) for description. Version with

ICP Savannah Advanced wing retrofit *NEW*/0576239

ICP Bingo (Rotax 503 engine) *NEW*/0552049

amphibious floats is designated **Savannah Hydro**; with Rotax 503 engine is **Bingo 503**; with Simoni Victor 2 engine (first flown 19 October 2001) is **Super Bingo**; with HKS 700E is **Bingo 4T** (2003). ICP offers a retrofit tapered wing (span 8.00 m; 26 ft 3 in, area 9.40 m^2; 101.2 sq ft, root chord 1.50 m; 4 ft 11 in, tip chord 0.75 m; 2 ft 5½ in) to produce the **Savannah Advanced**. Savannah is certified for operation as an ultralight in Belgium, France, Germany, Israel, Italy, Luxembourg, Netherlands, Norway and Slovak Republic.

COSTS: €30,919 flyaway (2001).

UPDATED

ICP SA2 VIMANA

In 2003, drawings were released of a new lightplane under development. The aircraft is generally similar to the Savannah, but less angular in appearance.

NEW ENTRY

ICP SA2 Vimana *NEW*/0576331

ICP F28

ICP has commissioned the design of a four-seat light aircraft from Stelio Frati. No details of timescale have been announced.

The machine is of typical Frati appearance, with low, tapered wings, centre-hinged cabin doors and retractable tricycle landing gear.

Leading particulars include span of 8.80 m (28 ft 10½ in); length 6.80 m (22 ft 3¾ in); max level speed 172 kt (320 km/h; 199 mph); cruising speed 135 kt (250 km/h; 155 mph).

NEW ENTRY

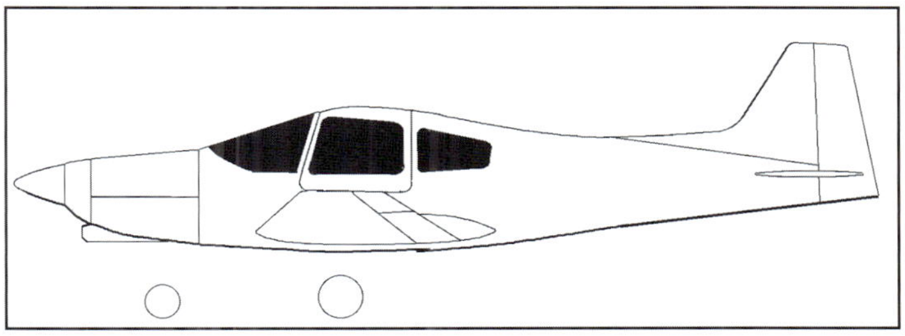

Profile of ICP F28 *(Paul Jackson)* *NEW*/0576218

III

INIZIATIVE INDUSTRIALI ITALIANE SPA

Via le Gorizia 6, I-00198 Roma
Tel: (+39 06) 854 63 41
Fax: (+39 06) 85 30 14 61
e-mail: treidirgen@skyarrow.com
Web: http://www.skyarrow.com
DIRECTOR: L Bartolotti Lauri

PRODUCTION PLANT:
Via Leonardo da Vinci 19/23, I-00016
Monterotondo Scalo, Roma
Tel: (+39 06) 90 03 55 45
Fax: (+39 06) 90 08 03 28
e-mail: sales@skyarrow.com

US AGENT:
Pacific Aerosystem Inc
5760 Chesapeake Court, San Diego, California 92123, USA
Tel: (+1 619) 571 14 41
Fax: (+1 619) 571 08 03
e-mail: pacaero@earthlink.net
PRESIDENT: Edward V Dempsey Jr

Company formed April 1947 in Trieste, as Meteor SpA. Main R&D and production facility then in Monfalcone, where two- and four-seat aircraft produced for training and sport flying, plus target aircraft and electronic systems for military market. Present company created in 1985; relocated to new facility in Monterotondo 1989; now produces pleasure boats and general aviation aircraft and is developing the Rondine unmanned surveillance system which is based on the Sky Arrow. Plans to market a range of low-wing lightplanes have been delayed.

UPDATED

III F-200 AND F-300

TYPE: Two-seat lightplane; four-seat lightplane.

PROGRAMME: Design (by Dott Ing Stelio Frati) started in 2000; initially envisaged as family of four, including D-130 primary trainer and a Sport version of F-300; manufacture of mockup, and wind tunnel testing, under way mid-2001; production scheduled to begin at end of 2002 but now postponed until after 2004, when kit versions will be offered. Certification of all variants to JAR/FAR Pt 23 IFR requirements will follow sometime thereafter.

CURRENT VERSIONS: **F-200**: Aerobatic trainer.

F-300: Tourer and business aircraft.

DESIGN FEATURES: Typical Stelio Frati design with low wing of tapered planform, and swept fin with long dorsal fillet.

FLYING CONTROLS: Conventional and manual; cable actuated. Horn-balanced elevators and rudder; trim tabs on each aileron and elevator; tab on rudder.

STRUCTURE: All composites; mostly carbon fibre, Kevlar and epoxy resin.

LANDING GEAR: Retractable tricycle type; main units retract inwards, nosewheel rearwards.

POWER PLANT: *F-200:* One 149 kW (200 hp) Textron Lycoming IO-360 flat-four driving a three-blade constant-speed propeller.

F-300: One 224 kW (300 hp) Textron Lycoming IO-540 flat-six with FADEC and three-blade constant-speed propeller.

ACCOMMODATION: Two persons, side by side in F-200; four in F-300, which has conventional cabin accessed via doors. Control sticks standard. Baggage compartment at rear of cabin on F-300, with external baggage door on port side aft of wing.

DIMENSIONS, EXTERNAL:
Wing span .... 9.60 m (31 ft 6 in)
Wing aspect ratio 7.0
Length overall .... 7.80 m (25 ft 7 in)
Height overall .... 2.88 m (9 ft 5½ in)

AREAS:
Wings, gross .... 13.15 m^2 (141.6 sq ft)

WEIGHTS AND LOADINGS:
Weight empty: F-200 .... 726 kg (1,600 lb)
F-300 .... 1,000 kg (2,205 lb)
Max T-O weight: F-200 .... 1,098 kg (2,420 lb)
F-300 .... 1,550 kg (3,417 lb)
Max wing loading: F-200 .... 83.44 kg/m^2 (17.09 lb/sq ft
F-300 .... 117.87 kg/m^2 (24.14 lb/sq ft)
Max power loading: F-200 .... 7.36 kg/kW (12.10 lb/hp)
F-300 .... 6.93 kg/kW (11.39 lb/hp)

PERFORMANCE:
Never-exceed speed (VNE):
F-200, F-300 .... 230 kt (426 km/h; 265 mph)
Max level speed at S/L:
F-200 .... 180 kt (333 km/h; 207 mph)
F-300 .... 205 kt (380 km/h; 236 mph)

Cruising speed at 75% power: at S/L:
F-200 .... 160 kt (296 km/h; 184 mph)
F-300 .... 180 kt (333 km/h; 207 mph)
at 4,575 m (15,000 ft):
F-300 .... 174 kt (322 km/h; 200 mph)
Stalling speed, flaps down, power off:
F-200 .... 50 kt (93 km/h; 57 mph)
F-300 .... 59 kt (109 km/h; 68 mph)
Max rate of climb at S/L:
F-200 .... 366 m (1,200 ft)/min
F-300 .... 482 m (1,580 ft)/min
Service ceiling: F-200 .... 4,575 m (15,000 ft)
F-300 .... 6,400 m (21,000 ft)
T-O run: F-200 .... 305 m (1,000 ft)
F-300 .... 280 m (920 ft)
Landing run: F-200 .... 274 m (900 ft)
F-300 .... 326 m (1,070 ft)
Range, 75% power, 30 min reserves:
F-200 .... 640 n miles (1,185 km; 736 miles)
F-300 .... 1,190 n miles
.... (2,204 km; 1,369 miles)
Endurance, 75% power, 30 min reserves:
F-200 .... 2 h 0 min
F-300 .... 6 h 30 min
g limits: F-200 .... +8.3/–4
F-300 .... +6.5/–3.2

UPDATED

III SKY ARROW

TYPE: Two-seat lightplane; tandem-seat ultralight.

PROGRAMME: First flight March 1993; certified by FAA to FAR Pt 23 and by ENAC (Italy) to JAR-VLA. primary aircraft sport category.

CURRENT VERSIONS: Designations refer to maximum T-O weights in kilograms.

450TS and 450TGS: For markets where MTOW for ultralights is 450 kg (including Czech Republic, Germany, Greece, Netherlands and Norway). TGS is German version. 450T and 450TG no longer promoted.

480T and 480TS: For markets (including Australia, Canada and New Zealand) where MTOW for ultralights is 480 kg.

500TF: For markets (including France) where MTOW for ultralights is 500 kg.

650TCS: JAR-VLA and FAR Pt 23 certified for day or night VFR. MTOW 650 kg. First production 650TC (I-TREI) built 1996; 13 TCs produced by mid-2001. None further by early 2003. TCS currently offered.

650TCNS: JAR-VLA and FAR Pt 23 certified for day or night VFR operation. FAA certificate awarded 3 March 2003. MTOW 650 kg. Total 28 built by late 2003, including first eight as TCN.

Description below applies to these versions, except where noted .

650 ERA: Environmental Research Aircraft equipped with real-time datalink connection to fixed or mobile ground stations for environmental monitoring. Joint testing by III and the Italian National Research Council's Biometeorology Institute scheduled for 2004 using 650 ERA I-LUND.

650 TCNS ERA: Environmental Research Aircraft variant equipped with surface surveillance sensors provided by the US National Oceanic Atmospheric Administration. Features fully retractable nosewheel and folding main landing gear.

650 TCNS RAWAS: Remotely Assisted Working Aerial System. Mission equipment can include gyrostabilised and traversable video cameras, and FLIR, IR linescan, vertical multispectral sensor, crop growth monitoring sensors, magnetometer or vertical panoramic still camera; operates as airborne relay station for RAWAS surveillance system, receiving and transmitting data at range up to 52 n miles (96 km; 60 miles) from mobile ground station. Fully retractable nosewheel and folding main landing gear.

RFG: 'Retractable Fixed Gear'. One aircraft (I-RAWH), demonstrated 2001; proof-of-concept for ERA and RAWAS versions, providing clear field of view for ventral sensors. Nosewheel retracts conventionally; regular Sky Arrow mainwheel legs splay outwards and upwards. Configuration being delivered in early 2003, when third version was under study.

CUSTOMERS: Over 200 built by 2003.

DESIGN FEATURES: Designed with objectives of good view from cockpit, lightness, ease of maintenance, ease of disassembly for transport and storage. Capable of being converted to an amphibian, with additional endplate fins on tailplane. High-mounted engine behind wing; pusher propeller; low-mounted pod fuselage below propeller carries T tail.

Wing aerofoil Gottingen 398 modified to enhance low-speed handling; constant wing chord; dihedral 1° 30'; twist 1°; strut-braced wings easily detachable at centreline joint.

FLYING CONTROLS: Conventional and manual. Ailerons and elevator are actuated by aluminium rods, rudder by cables. Electrically actuated half-span flaps and tabs; aileron deflections +20/–14°, elevator +22/–14°, rudder ±23°; flap deflections 10, 20 and 30°; centrally mounted elevator tab for trimming. Full dual controls (sidestick and rudder) and two throttles.

STRUCTURE: Fuselage, wings, tail unit and landing gear manufactured from carbon fibre sandwich with Kevlar for strengthening in some areas, such as the cabin. Fuselage constructed of right and left halves bonded together, containing six minor and two major

III Sky Arrow 450T *(Paul Jackson)* ***NEW***/0576653

III Sky Arrow 650TCNS 'Retractable Fixed Gear' prototype *(Paul Jackson)* ***NEW***/0576654

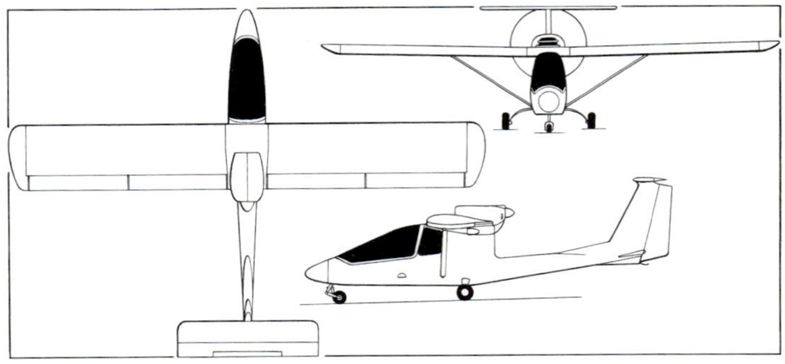

III Sky Arrow 650 very light aircraft *(James Goulding)* 0011089

bulkheads. Wings built round two C-section spars. Sky Arrow 450 and 480 series have aluminium wings which can be replaced by carbon fibre units. Horizontal tail surfaces of aluminium.

LANDING GEAR: Tricycle type; fixed. Fuselage-mounted spring; multilayer carbon- and glass fibre cantilever mainwheel legs with optional wheel fairings and hand-operated hydraulic disc brakes. Steel/carbon fibre castoring nose leg with rubber and spring shock-absorber. Twin amphibious floats optional.

POWER PLANT: *500TF:* One 59.6 kW (79.9 hp) Rotax 912 UL flat-four four-stroke, driving a two-blade fixed-pitch propeller via 1:2.27 reduction gear.

450TS, 450TGS and 480TS: One 73.5 kW (98.6 hp) Rotax 912 ULS driving a two-blade fixed-pitch propeller via 1:2.43 reduction gear; three-blade propeller standard on 450TGS.

650TCS : and 650TCNS: One 73.5 kW (98.6 hp) Rotax 912 S2 driving a Hoffmann HO17GHM-174 177CLD wood/composites, two-blade, fixed-pitch propeller.

Fuel capacity (all) 68 litres (18.0 US gallons; 15.0 Imp gallons), of which 67 litres (17.8 US gallons; 14.8 Imp gallons) are usable. Tank behind rear seat.

ACCOMMODATION: Two seats in tandem; sideways-opening tinted canopy; baggage compartments below and behind rear seat.

SYSTEMS: Dual hydraulic brakes. Electrical system includes 12 V 14 Ah battery; alternator.

AVIONICS: *Comms:* Single Bendix/King KX-125 nav com, dual optional; KT-76A transponder; altitude encoder; audio panel with voice-activated intercom; two Sigtronics headsets; ELT.

Flight: Optional Bendix/King KLX-1325A com/GPS with KI 204 indicator or Garmin GNS-430 nav/com/GPS with GI 106A indicator.

Instrumentation: ASI; altimeter; rate of climb indicator; attitude indicator; directional gyro; turn-and-bank indicator; magnetic compass; pitch trim indicator; tachometer; fuel quantity, fuel pressure, oil temperature, oil pressure, water temperature and vacuum gauges; ammeter/voltmeter; and flight hours meter, all standard. Manifold pressure indicator kit optional.

EQUIPMENT: Optional handicapped pilot kit comprises sidestick controller on port cockpit console for rudder and throttle inputs, enabling the aircraft to be flown by pilots unable to use their legs; baggage compartment accommodates folding wheelchair. Fitted trailer for road transportation, with hydraulic loading system, optional. Banner/glider towing hook, incorporating LCD micro video camera for monitoring towing operation, optional.

DIMENSIONS, EXTERNAL:
Wing span: all .... 9.70 m (31 ft 9¾ in)
Wing chord, constant .... 1.40 m (4 ft 7 in)
Wing aspect ratio 7.0
Length overall: all .... 7.60 m (24 ft 11¼ in)
Height overall: all .... 2.60 m (8 ft 6¼ in)
Tailplane span: all .... 2.80 m (9 ft 2¼ in)
Tailplane chord: all .... 0.75 m (2 ft 5½ in)
Wheel track .... 1.75 m (5 ft 9 in)
Propeller diameter (650) .... 1.75 m (5 ft 8¾ in)

AREAS:
Wings, gross: all .... 13.50 m² (145.3 sq ft)

WEIGHTS AND LOADINGS:
Weight empty: 450TGS .... 287 kg (633 lb)
450TS .... 286 kg (631 lb)
480TS .... 298 kg (657 lb)
500TF .... 335 kg (739 lb)
650TCS, 650TCNS .... 400 kg (882 lb)
Max T-O weight: 450TS, 450TGS .... 450 kg (992 lb)
480TS .... 480 kg (1,058 lb)
500TF .... 500 kg (1,102 lb)
650TCS, 650TCNS .... 650 kg (1,433 lb)
Max wing loading:
450TS, 450TGS .... 33.3 kg/m² (6.83 lb/sq ft)
480TS .... 35.6 kg/m² (7.28 lb/sq ft)
500TF .... 37.0 kg/m² (7.59 lb/sq ft)
650TCS, 650TCNS .... 48.2 kg/m² (9.86 lb/sq ft)
Max power loading:
450TS, 450TGS .... 6.12 kg/kW (10.06 lb/hp)
500TF .... 8.39 kg/kW (13.78 lb/hp)

650TCS, 650TCNS .... 8.84 kg/kW (15.53 lb/hp)

PERFORMANCE:

Never-exceed speed (VNE):

450TS, 450TGS, 480T, 480TS, 500TF .... 121 kt (224 km/h; 139 mph)
650TCS, 650TCNS .... 132 kt (244 km/h; 151 mph)

Max level speed at S/L:

450TS, 450TGS, 480TS, 500TF .... 108 kt (200 km/h; 124 mph)
650TCS, 650TCNS .... 107 kt (198 km/h; 123 mph)

Cruising speed at 75% power at S/L:

500TF .... 90 kt (167 km/h; 104 mph)
450TS, 450TGS, 480TS .... 92 kt (170 km/h; 106 mph)
650TCS, 650TCNS .... 97 kt (180 km/h; 112 mph)

Stalling speed, power off, flaps down:

450TS, 450TGS .... 33 kt (62 km/h; 38 mph)
480TS .... 34 kt (63 km/h; 40 mph)
500TF .... 35 kt (65 km/h; 41 mph)
650TCS, 650TCNS .... 40 kt (75 km/h; 47 mph)

Service ceiling: all .... 4,100 m (13,460 ft)

Max rate of climb at S/L:

450TS, 450TGS .... 384 m (1,260 ft)/min
480TS .... 372 m (1,220 ft)/min
500TF .... 274 m (900 ft)/min
650TCS, 650TCNS .... 258 m (846 ft)/min

T-O run: 450TS, 450TGS .... 100 m (330 ft)
480TS .... 115 m (380 ft)
500TF .... 145 m (475 ft)
650TCS, 650TCNS .... 177 m (585 ft)

T-O to 15 m (50 ft): 450TS, 450TGS .... 160 m (525 ft)
480TS .... 180 m (595 ft)
500TF .... 230 m (755 ft)
650TCS, 650TCNS .... 380 m (1,250 ft)

Landing run: 450TS, 450TGS .... 75 m (250 ft)
480TS .... 86 m (285 ft)
500TF .... 100 m (330 ft)
650TCS, 650TCNS .... 135 m (445 ft)

Landing from 15 m (50 ft): 450TG, 450TGS .... 170 m (560 ft)
480TS .... 188 m (620 ft)
500TF .... 200 m (445 ft)
650TCS, 650TCNS .... 215 m (705 ft)

Max range, 75% power, no reserves:

450TS, 450TGS, 480TS .... 340 n miles (630 km; 391 miles)
500TF .... 380 n miles (703 km; 437 miles)
650TCS, 650TCNS .... 329 n miles (610 km; 379 miles)

Endurance, 75% power, no reserves:

450TS, 450TGS, 480TS .... 3 h 40 min
500TF .... 4 h 10 min
650TCS, 650TCNS .... 3 h 25 min

g limits: all .... +3.8/−1.9

UPDATED

MAGNI

MAGNI GYRO

Via Puccini 10, I-21010 Besnate (VA)
Tel: (+39 0331) 27 48 16
Fax: (+39 0331) 27 48 17
e-mail: magnigyro@logic.it
Web: http://www.magnigyro.com
FOUNDER: Vittorio Magni

Founded as VPM in 1976 by Vittorio Magni, this company developed the single-seat MT5 and two-seat MT7 gyrocopters. It was renamed Magni Gyro in 1996 and currently produces a range of ultralight autogyros, including the open-cockpit single-seat M 18 Spartan and two-seat M 14 Scout and M 16 Tandem Trainer. In total, around 300 examples were active at the beginning of 2004.

UPDATED

MAGNI M 16 TANDEM TRAINER

TYPE: Two-seat autogyro ultralight.

PROGRAMME: Current production Magni aircraft.

CURRENT VERSIONS: **Tandem Trainer:** *As described.*

M16-2000T: One aircraft (c/n 022074) supplied by manufacturer in September 2003 and modified to United Kingdom BCAR Section T standards for 'Global Eagle' world circumnavigation expedition . However, due to certification difficulties, the attempt began on 26 April 2004 using a standard M 16, temporarily halting in India because of unfavourable weather and subsequently limiting objective to first UK-to Australia autogyro flight.

COSTS: Kit €75,895; ready-to-fly €79,065 both with Rotax 914 (2004).

DESIGN FEATURES: Open, tandem seating. Pod and boom fuselage with sweptback tailplane and triple fins; rudder on central fin only.

STRUCTURE: Chrome-alloy 4130 steel; TIG welded; glass fibre pod.

LANDING GEAR: Tricycle type; fixed. Fuselage-mounted, spring cantilever mainwheel legs with wheel fairings. Unfaired nosewheel.

POWER PLANT: One turbocharged 84.6 kW (113.4 hp) Rotax 914 UL flat-four with electric starter, driving a ground-adjustable pitch, three-blade, carbon fibre pusher propeller. Fuel in integral cockpit seat/tank of epoxy/glass fibre, capacity 50 litres (13.2 US gallons; 11.0 Imp gallons).

EQUIPMENT: Optional 150 litre (39.6 US gallon; 33.0 Imp gallon) agricultural spray gear.

DIMENSIONS, EXTERNAL:

Rotor diameter .... 8.23 m (27 ft 0 in)
Length overall .... 4.65 m (15 ft 3 in)
Height overall .... 2.60 m (8 ft 6¼ in)
Width overall .... 1.80 m (5 ft 10¾ in)
Propeller diameter .... 1.70 m (5 ft 7 in)

WEIGHTS AND LOADINGS:

Weight empty .... 270 kg (595 lb)
Max T-O weight .... 450 kg (992 lb)

PERFORMANCE:

Max level speed .... 100 kt (185 km/h; 115 mph)
Cruising speed .... 78 kt (145 km/h; 90 mph)
Max rate of climb at S/L .... 300 m (984 ft)/min
Service ceiling .... 3,500 m (11,480 ft)
T-O run .... 50 m (165 ft)
Landing run .... up to 30 m (100 ft)
Range, no reserves .... 260 n miles (481 km; 299 miles)
Endurance .... 3 h

UPDATED

MAGNI M 20 TALON

TYPE: Single-seat autogyro ultralight.

PROGRAMME: Prototypes flying; production had not begun by mid-2003, although promotion was continuing.

DESIGN FEATURES: Pod and boom fuselage with sweptback tailplane and triple fins; rudder on central fin only. M 18 Spartan can be upgraded to M 20 standard.

STRUCTURE: Welded 4130 chromoly main structure with glass fibre fuselage and wheel fairings. Two-blade main rotor of composites construction with mechanical prerotator.

LANDING GEAR: Tricycle type; fixed. As M 16.

POWER PLANT: One 59.6 kW (79.9 hp) Rotax 912 UL flat-four driving a ground-adjustable pitch, three-blade, carbon fibre pusher propeller. Fuel in integral cockpit seat/tank of epoxy/glass fibre, capacity 45 litres (11.9 US gallons; 9.9 Imp gallons).

DIMENSIONS, EXTERNAL:

Rotor diameter .... 7.62 m (25 ft 0 in)
Length overall .... 3.92 m (12 ft 10¼ in)
Height overall .... 2.37 m (7 ft 9¼ in)
Width overall .... 1.73 m (5 ft 8 in)
Propeller diameter .... 1.62 m (5 ft 3¾ in)

WEIGHTS AND LOADINGS:

Weight empty .... 205 kg (452 lb)
Max T-O weight .... 300 kg (661 lb)

PERFORMANCE:

Max level speed .... 96 kt (177 km/h; 110 mph)
Cruising speed .... 78 kt (145 km/h; 90 mph)

Magni M 16 Tandem Trainer *(Paul Jackson)* ***NEW***/0583480

Magni M 20 Talon single-seat ultralight autogyro 0054612

Max rate of climb at S/L .... 300 m (984 ft)/min
Service ceiling .... 3,500 m (11,480 ft)
T-O run .... 45 m (150 ft)
Landing run .... up to 30 m (100 ft)
Range, no reserves .... 156 n miles (289 km; 179 miles)
Endurance .... 3 h

UPDATED

MAGNI M 21

TYPE: Two-seat autogyro ultralight.

PROGRAMME: Development of M 19; prototype flying by mid-2002. No further details revealed by May 2004.

DESIGN FEATURES: Replacement for M 19, but featuring new configuration in form of twin tailbooms to improve stability at high speed and larger cabin.

UPDATED

PIAGGIO

PIAGGIO AERO INDUSTRIES SPA

Via Cibrario 4, I-16154 Genova Sestri, Genova
Tel: (+39 010) 648 11
Fax: (+39 010) 648 13 09
e-mail: commerciale@piaggioaero.it
Web: http://www.piaggioaero.com
CEO: José Di Mase
CHAIRMAN: Piero Ferrari
VICE-PRESIDENT, SALES: Massimo Isidori
GENERAL MANAGER: Giorgio Giorgi
DEPUTY DIRECTOR, MARKETING AND COMMERCIAL: Giuliano Felten

WORKS:
Genova Sestri and Finale Ligure (SV)

BRANCH OFFICE:
Via A Gramsci 34, I-00197 Roma

SUBSIDIARIES:
Piaggio Aero France SAS
Le Centaure-Nice Leader, 66 Route de Grenoble, F-06200 Nice, France
Tel: (+33 4) 89 88 77 77
Fax: (+33 4) 89 88 77 55
DIRECTOR: Alfonso Ruggiero

Piaggio America Inc
2 Exchange Street, Greenville, South Carolina 29605, USA
Tel: (+1 864) 277 39 79
Fax: (+1 864) 277 43 78
CEO AND PRESIDENT: Tom Appleton

Piaggio began aircraft production at Genoa Sestri in 1916 and later extended to Finale Ligure; Rinaldo Piaggio SpA formed 29 February 1964; covered floor area 50,000 m^2(538, 200 sq ft) at Sestri and 70,000 m^2(753,475 sq ft) at Finale Ligure. Workforce 775 and 525, respectively, in 2004, plus 35 at Naples research centre. Early in 1993, Rinaldo Piaggio restructured with Piaggio family retaining 19 per cent of stock; Alenia acquired 31 per cent holding in Piaggio in 1988 (adjusted to 24.5 per cent in 1992, but raised to 30.9 per cent in early 1993 following further restructuring and reorganisation). Industrie Aeronautiche and Meccaniche Rinaldo Piaggio SpA placed in administration in November 1994; following takeover in November 1998 by Turkish state holding company Tushav, which acquired 51 per cent of stock, was renamed Piaggio Aero Industries SpA, with balance of stock held by Italian investment group Aero Trust via Royal Bank of Canada (44 per cent), CSC of Italy (3 per cent) and the Buitoni family (2 per cent). However, Aero Trust share increased to 60 per cent in June 1999, returning company to Italian control. Sviluppo Italia, the Italian government industrial investment agency, acquired a 20 per cent share in Piaggio in Septemeber 2003, with an investment of €20 million. Turnover in 2002 was €138 million.

Two new subsidiaries were established in 2000: Piaggio Aero France SAS at Nice, France, which is a branch of Piaggio's Finale Ligure engineering department; and Piaggio America Inc, based in Greenville, South Carolina, but scheduled to move to Palm Beach International Airport, Florida, in first quarter 2004. In April 2003, Piaggio purchased Foxair, a corporate and air ambulance operator with a fleet of six aircraft (mainly Avantis), which is the North American support organisation for the P.180 Avanti.

Piaggio also licence-produces engines, including the RRT1 RTM 322 and mobile shelters; manufactures subassemblies of Alenia (Lockheed Martin) C-27J and Dassault Falcon 2000 and overhauls and supports Viper, T53 and T55 power plants.

UPDATED

PIAGGIO P.166

Production of the P.166 ended in 1994, although proposed new manufacture and projected upgrades caused its description to be retained in *Jane's All the World's Aircraft* until 2003-4.

In 2004, however, two further aircraft were built for the Italian Coast Guard and delivered to Catania-Fontanarossa as crew trainers, serial numbers CSX25183 and 25184. P.166 production thereby increased to 150, as under.

UPDATED

Variant	Qty	Remarks
P.166	27	Prototype and civil production
P.166AL-1	5	Italian civil and DL-3 prototype
P.166ML-1	37	Italian military
P.166M-APM	14	Italian Air Force
P.166B	4	Second civil variant
P.166CL-2	2	
P.166DL-2	2	Iraq
P.166S	20	South African Air Force
P.166DL-3	10	Including four to Somalia
P.166DL-3/APH	6	Italian Air Force
P.166DL-3/SEM1	12	Italian Coast Guard
P.166DL-3/SEM2	10	Italian Customs
P.166	1	untraced

PIAGGIO P.180 AVANTI

English name: Forward

TYPE: Business twin-turboprop.

PROGRAMME: Design studies began 1979; launched 1982; Gates Learjet became partner in 1983, but withdrew for economic reasons on 13 January 1986; all existing Learjet P.180 tooling and first three forward fuselages transferred to Piaggio; first flights of two prototypes I-PJAV 23 September 1986 and I-PJAR 14 May 1987; two static test fuselages; first Italian certification 7 March 1990; first flight full production P.180 (I-RAIH/N180BP), 30 May 1990; Italian and US certification 2 October 1990; first customer delivery (N180BP to Robert Pond) 30 September 1990; French certification March 1993. P.180 is certified to Italian RAI 223 and FAR Pt 23 including single pilot, night and day, VFR/IFR and flight into known icing. RVSM approval announced 24 July 2002. FAA Domestic RVSM and Canadian RVSM approvals granted in 2003.

On 5 October 2003 an Avanti established a new Class C.I.E world speed record while en route to the NBAA Convention, completing the 854 n miles (1,582 km; 983 mile flight from Chicago-Midway Airport to Orlando Executive Airport in 2 hours, 18 minutes, 12 seconds at an average speed of 374 kt (693 km/h; 430 mph).

CURRENT VERSIONS: Increased gross weight giving higher payload/range decided 1991 and early aircraft retrofitted with minor modifications to allow new weights; weights increased again in 1992. Modified version relaunched in 1997 incorporates changes including fin, rudder and foreplane of aluminium alloy construction and increased fuel capacity. Economy and deluxe interior versions in development in mid-1999.

CUSTOMERS: Total 30 (including prototypes) built by early 1995. 31st followed in early 1999; eight delivered in 2000, 12 in 2001, 16 in 2002 and 20 in 2003; target for 2004 was 22, of which two were delivered in the first three months. Italian Air Force ordered six for communications; first delivery MM62159 14 May 1993; further five (three for Italian Army; two for SNPC civil protection service) ordered 1994, of which first, MM62167, handed over on 28 July 1997 and delivered 29 August 1997, and second and third delivered March-April 1998, with two SNPC aircraft following in late 1998 and mid-1999. Additional six aircraft delivered for military use by mid-2004, including three for Italian Navy and at least one in navigation aids calibration role for IAF.

First new aircraft delivery since company reorganisation took place on 2 February 2000 to Aviation Services for Business Aircraft Inc, Portland, Oregon (N680JP). Recent customers include Italian car manufacturer Ferrari, which took delivery of the first of two aircraft (I-FXRB) on 18 May 2000 (these operated by Foxair); the Greek Ministry of Health, which ordered two in EMS configuration and, although built, these were not delivered; Locafit (Italy); CIT Group Equipment Financing (Arizona, USA); US charter broker Skyline Aviation, which ordered eight in 2002; Avia Aviation of Canada which has ordered six for delivery from early 2003; Pan Européenne Air Service of France, which ordered one for delivery in May 2003, plus one option; Euro SkyLink UK, which ordered two plus three options of which first (G-OESL) delivered at the European Business Aviation Convention and Exhibition at Geneva 26 May 2004; and Blue Panorama Airlines of Italy, which ordered two for delivery in April and July 2004; and Avantair, of Fairfield, New Jersey, which has ordered 29, of which the first will be delivered in 2005, followed by 11 in 2006 and the remainder in 2007. Total of 34 orders and eight options received in 2003.

COSTS: US$5.350 million (2004). Avantair contract for 29 valued at US$200 million (April 2004).

DESIGN FEATURES: Intended to provide jet-type speeds with turboprop economy. Three-surface control with foreplane and T tail to allow unobstructed cabin with maximum headroom to be placed forward of mid-mounted wing carry-through structure; pusher turboprops aft of cabin and wing reduce cabin noise and propeller vortices on wing, assisting in achievement

Piaggio P.180 Avanti business twin-turboprop ***(Paul Jackson)*** ***NEW***/0576610

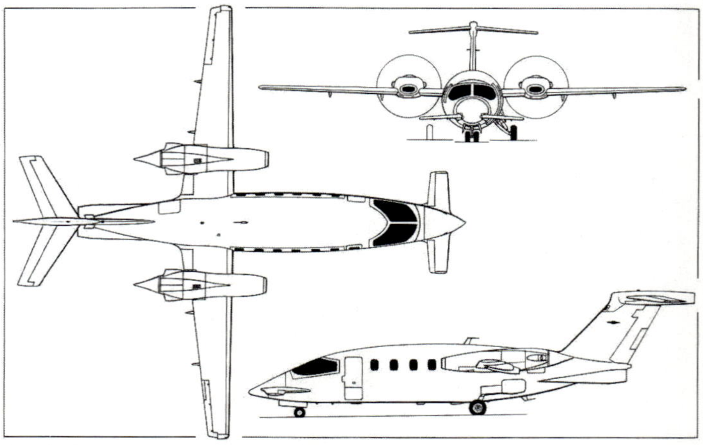

Piaggio P.180 Avanti corporate transport ***(Jane's/Dennis Punnett)*** 0507420

Piaggio Avanti of the Italian Air Force *(Paul Jackson)* **NEW**/0576641

of 50 per cent laminar flow; mid-wing avoids root bulges of low-set wings and spar does not pass through cabin; lift from foreplane allows horizontal tail to act as lifting surface and thereby reduce required wing area by 34 per cent.

Laminar flow wing section Piaggio PE 1491 G (mod) at root, PE 1332 G at tip; thickness/chord ratio 13 per cent at tip, 14.5 per cent at root; dihedral 2°; sweepback 1° 11' 24″; taper ratio 0.34; foreplane aerofoil Piaggio PE 1300 GN4 unswept; 5° anhedral on foreplane and tailplane; latter sweptback 29° 48' at 25 per cent chord. Tailplane swept 40° at 25 per cent chord.

FLYING CONTROLS: Manual. Aerodynamically and mass-balanced elevators; horn- and mass-balanced rudder. Variable incidence swept tailplane for trim; electrically actuated trim tab in starboard aileron and in rudder; two strakes under tail; electrically actuated outboard Fowler and inboard single-slotted flaps on wing synchronised with single-slotted flaps in foreplanes; three flap positions; dual control circuits; foreplane stalls before wing, providing pitch-down moment. All control surfaces have sealed gaps and are aerodynamically balanced. No dampers or stick-pusher.

STRUCTURE: Airframe 90 per cent aluminium alloy and 10 per cent composites. Fuselage precision stretch-formed in large seamless sections and inner structure matched to precise outer contour in innovative 'outside-in' construction technique; CFRP in high-stress areas; 'Working skin' wing main box, with machined spars and panels, plus integral stiffeners; two main spars; third spar runs from nacelle to fuselage centreline; aluminium leading-edges and aluminium and composites trailing-edges, both connected to main box. Similar construction of forward wing, which connected to lower fuselage at four points and has aluminium alloy leading-edges with electric de-icing blanket and full-depth honeycomb aluminium flaps. Tailplane has two-spar sandwich construction of graphite fabric on Nomex honeycomb core; elevators are single-spar and full-depth aluminium honeycomb with aluminium skins. Fin attached to tailcone bulkheads by four vertical aluminium machined spars; chemically milled aluminium sheet skins. Two-spar rudder, with aluminium alloy skin and carbon fibre and foam core.

Composites parts manufactured by Moreggio (nacelles) and Salver (horizontal stabiliser), wings and tail section and final assembly by Piaggio in Genoa. Interior completion of aircraft for the North American market is undertaken by Stevens Aviation of Greenville, South Carolina.

LANDING GEAR: Dowty Aerospace hydraulically retractable tricycle type, with single-wheel main units and twin-wheel nose unit. Main units retract rearward into sides of fuselage; nose unit retracts forward. Emergency hand pump. Dowty hydraulic shock-absorbers. Electro-hydraulic nosewheel steering available through ±20° on take-off and ±50° for taxying. Tyre sizes 5.50-10 (12-ply, main) and 5.00-5 (8-ply, nose). Goodrich hydraulic, multidisc carbon brakes.

POWER PLANT: Two 1,107 kW (1,485 shp) Pratt & Whitney Canada PT6A-66 turboprops, flat rated at 634 kW (850 shp), each mounted above wing in all-composites nacelle and driving a Hartzell five-blade constant-speed fully feathering reversible-pitch pusher propeller; propellers handed to counterrotate. Fuel in two fuselage tanks and two wing tanks; total fuel capacity 1,597 litres (422 US gallons; 351 Imp gallons), of which 1,583 litres (418 US gallons; 348 Imp gallons) are usable. Tanks divided into left and right groups (plus primary and secondary collector tanks) which are independent, except during optional pressure refuelling via single point on starboard centre-fuselage. Gravity refuelling point in upper part of fuselage.

ACCOMMODATION: Crew of one or two on flight deck; certified for single-pilot operation. Seating in main cabin for up to nine passengers, with galley, fully enclosed lavatory and coat storage area; choice of nine-passenger high-density or five-seat VIP cabins. Club passenger seats are armchair type, which can be reclined, tracked and swivelled, and locked at any angle. Foldaway tables can be extended between facing club seats. Two-piece wraparound electrically heated windscreen. Rectangular cabin windows, including one emergency exit at front on starboard side. Indirect lighting behind each window ring, plus individual overhead lights. Airstair door at front on port side is horizontally split, upper part forward-opening. Baggage compartment aft of rear pressure bulkhead, with upward-opening door immediately aft of wing on port side. Entire cabin area pressurised and air conditioned. New interior by Sergio Pininfarina under development in mid-1999; total of 15 interior options available.

SYSTEMS: Hamilton Sundstrand R70-3WG three-wheel air-cycle bleed air ECS, with maximum pressure differential of 0.62 bar (9.0 lb/sq in) maintaining sea level cabin altitude to 7,315 m (24,000 ft) and a 1,950 m (6,400 ft) cabin altitude at 12,500 m (41,000 ft). Single hydraulic system driven by electric motor, with hand pump for emergency back-up, for landing gear (up to 207 bar; 3,000 lb/sq in), brakes and steering (up to 83 bar; 1,200 lb/sq in). Electrical system powered by two 400 A 28 V starter/Lear Siegler generators and 25 V 38 Ah Ni/Cd battery, with triple-redundant essential bus. Two 250 VA static inverters for AC. External power receptacle above port mainwheel well. Scott 1.13 m³ (40 cu ft) oxygen system. Hot air anti-icing of main wing outer and inner leading-edges; electric anti-icing for foreplane and windscreen; rubber boot for engine intakes, with dynamic particle separator; propeller blades de-iced by engine exhaust.

AVIONICS: *Comms:* Dual Rockwell Collins VHF-22C transceivers; dual Rockwell Collins TDR-90 transponders.

Radar: Rockwell Collins WXR-840 weather radar standard, TWR-850 turbulence-detecting weather radar optional.

Flight: ADF-462, DME-42, ALT-55B radar altimeter, dual MCS-65 compasses, dual VIR-32 VOR/LOCGS/MCR, and dual RMI-36 (all Rockwell Collins); Goodrich Stormscope WX-1000E; dual Aeronetics RMI-3337 radio compasses. Rockwell Collins ADS-85 air data system and APS-65 autopilot with yaw damper. From c/n 1059 RVSM-compatible ADC-85A standard.

RVSM standard equipment replaces co-pilot's altimeter with ADDU including altitude encoding and preselect functions, and emergency power bus allowing operation of ADDU for minimum of 30 minutes in case of total electrical failure. ILS Cat. II kit optional, providing upgraded version of EFIS, autopilot and radio altimeter.

Instrumentation: Rockwell Collins EFIS-85B system, with two EFD-85 dual colour CRT MFDs for captain and central MFD-85B radar display; EHSI-74 colour display for co-pilot.

Mission: Optional Global/Wulfsberg Flitefone VI in-flight telephone.

DIMENSIONS, EXTERNAL:

Wing span	14.035 m (46 ft 0½ in)
Foreplane span	3.355 m (11 ft 0 in)
Wing chord: at root	1.82 m (5 ft 11¾ in)
at tip	0.62 m (2 ft 0½ in)
Foreplane chord: at root	0.79 m (2 ft 7 in)
at tip	0.55 m (1 ft 9¾ in)
Wing aspect ratio	12.3
Foreplane aspect ratio	5.1
Length overall	14.41 m (47 ft 3½ in)
Fuselage: Length	12.53 m (41 ft 1¼ in)
Max width	1.95 m (6 ft 4¾ in)
Height overall	3.98 m (13 ft 0¾ in)
Tailplane span	4.26 m (13 ft 11¾ in)
Wheel track	2.84 m (9 ft 4 in)
Wheelbase	5.79 m (19 ft 0 in)
Propeller diameter	2.16 m (7 ft 1 in)
Propeller ground clearance	0.795 m (2 ft 7¼ in)
Distance between propeller centres	4.125 m (13 ft 6½ in)
Passenger door (fwd, port): Height	1.345 m (4 ft 5 in)
Width	0.61 m (2 ft 0 in)
Height to sill	0.58 m (1 ft 10¾ in)
Baggage door (rear, port): Height	0.60 m (1 ft 11¾ in)
Width	0.70 m (2 ft 3½ in)
Height to sill	1.38 m (4 ft 6½ in)
Emergency exit (stbd): Height	0.67 m (2 ft 2¼ in)
Width	0.48 m (1 ft 7 in)

DIMENSIONS, INTERNAL:

Passenger cabin (excl flight deck):	
Length	4.55 m (14 ft 11¼ in)
Max width	1.85 m (6 ft 0¾ in)
Max height	1.75 m (5 ft 9 in)
Volume	10.6 m³ (375 cu ft)
Flight deck: Length	1.45 m (4 ft 9 in)
Volume	2.3 m³ (80 cu ft)
Baggage compartment: Floor length	1.70 m (5 ft 7 in)
Max length	1.95 m (6 ft 4¾ in)
Volume	1.25 m³ (44 cu ft)

AREAS:

Wings, gross	16.00 m² (172.2 sq ft)
Ailerons (total, incl tab)	0.66 m² (7.10 sq ft)
Trailing-edge flaps (total)	1.60 m² (17.23 sq ft)
Foreplane	2.19 m² (23.57 sq ft)
Foreplane flaps (total)	0.58 m² (6.30 sq ft)
Fin	4.73 m² (50.91 sq ft)
Rudder, incl tab	1.05 m² (11.30 sq ft)
Tailplane	3.83 m² (41.23 sq ft)
Elevators (total, incl tabs)	1.24 m² (13.35 sq ft)

WEIGHTS AND LOADINGS:

Weight empty, equipped	3,402 kg (7,500 lb)
Operating weight empty, one pilot	3,479 kg (7,670 lb)
Max usable fuel weight	1,271 kg (2,802 lb)
Max payload	907 kg (2,000 lb)
Payload with max fuel	589 kg (1,299 lb)

Max T-O weight 5,239 kg (11,550 lb)
Max ramp weight 5,262 kg (11,600 lb)
Max landing weight 4,965 kg (10,945 lb)
Max zero-fuel weight 4,309 kg (9,500 lb)
Max wing loading 327.4 kg/m² (67.07 lb/sq ft)
Max power loading 4.13 kg/kW (6.79 lb/shp)

PERFORMANCE:
Max operating Mach No (MMO) 0.70
Max operating speed (VMO) 260 kt (482 km/h; 299 mph) IAS
Max level speed at FL280 395 kt (732 km/h; 455 mph)
Manoeuvring speed 199 kt (368 km/h; 229 mph)
Max cruising speed with four passengers at mid-cruise weight:
at FL280 391 kt (724 km/h; 450 mph)
at FL350 368 kt (682 km/h; 423 mph)
at FL390 341 kt (632 km/h; 393 mph)
Stalling speed at max landing weight:
flaps up 109 kt (202 km/h; 125 mph)
flaps down 93 kt (172 km/h; 107 mph)
Max rate of climb at S/L 899 m (2,950 ft)/min
Rate of climb at S/L, OEI 230 m (755 ft)/min
Max certified altitude 12,500 m (41,000 ft)
Service ceiling 11,885 m (39,000 ft)
Service ceiling, OEI 7,590 m (24,900 ft)
T-O to 15 m (50 ft) ISA, S/L at max T-O weight 869 m (2,850 ft)
Landing from 15 m (50 ft) at max landing weight, no propeller reversal 872 m (2,860 ft)
Range:
at max cruise power, VFR reserves:
at FL280 1,195 n miles (2,213 km; 1,375 miles)
at FL350 1,482 n miles (2,744 km; 1,705 miles)
at FL390 1,721 n miles (3,187 km; 1,980 miles)
at max range power:
VFR reserves at FL300 1,540 n miles (2,852 km; 1,772 miles)
at FL350 1,683 n miles (3,117 km; 1,936 miles)
at FL390 1,793 n miles (3,320 km; 2,063 miles)
Max range, IFR reserves:
at FL300 1,304 n miles (2,415 km; 1,500 miles)
at FL350 1,420 n miles (2,629 km; 1,634 miles)
at FL390 1,509 n miles (2,794 km; 1,736 miles)

OPERATIONAL NOISE LEVELS (FAR Pt 36):
Appendix F 76.0 dB(A)
Appendix G 81.8 dB(A)

UPDATED

SAI

SOCIETA AERONAUTICA ITALIANA (ITALIAN AERONAUTICAL SOCIETY)

Via Fillipo Bottazzi 38, I-80126 Napoli
Tel: (+39 08) 23 49 45 80
Fax: (+39 08) 18 82 07 82
e-mail: info@saiaeronautica.com
Web: http://www.saiaeronautica.com

FOUNDERS:
Vincenzo Giordano
Gaetano Iervolino
Gastone Iervolino

INFORMATION OFFICER: Federica Tevvolino

SAI was established in 1997 to design, market and support ultralight aircraft; conduct aerodynamic research; and manage static and flight certification test programmes. Works at Vitulazio (CE) are being augmented by a new 10,000 m2 (107,640 sq ft) plant at San Salvatore Telesino (BN). Factory floor area 1,520 m² (16,350 sq ft); workforce six.

UPDATED

SAI G97 SPOTTER

TYPE: Multisensor surveillance lightplane; side-by-side ultralight kitbuilt.

PROGRAMME: Design began June 1997; construction of prototype G97 ultralight started November 1997; first flight June 1998, powered by 47.8 kW (64.1 hp) Rotax 582; first G97V lightplane (I-OPIS) had public debut at Aero '01, Friedrichshafen, in April 2001 and first flew 15 May 2001. Construction of first production G97 began September 2000; initial flight March 2001. Certification of G97V was due late 2001.

CURRENT VERSIONS: **G97V Spotter:** Certified light aircraft to Italian ENAC V.EL (elementary aircraft) standard. Differs from G97 prototype in having stretched forward fuselage.

G97 Spotter: As G97V, but in Ultralight category; assembled from kit.

CUSTOMERS: Six aircraft of all versions (including prototypes) flying by September 2001. Total 30 delivered to Italy, South America, South Africa, Spain and USA by mid-2003.

COSTS: G97 kit €32,000; ready-to-fly €56,100 with Rotax 912UL (2004).

DESIGN FEATURES: Advanced ultralight able to transition to certified aircraft category; intended for police/civil protection agency use, with helicopter-like view for crew assured by pusher propeller, side-by-side seating and absence of wing external bracing. Also suitable for private flying, training and light transport. Prime design objectives included sacrifice of speed to ensure utility; configuration overcomes related problems of ground cooling for engine, aerodynamic balance with large CG range, comfort in turbulent air and cabin noise. Computer modelling used in design, particularly with pod/boom interface and aerofoil tested in Department of Aircraft Design, Naples University.

Pod-and-boom fuselage; cantilever, constant chord wings, detachable for storage. Sweptback fin; tailplane below propeller slipstream.

Purpose-designed Giordano VG1-13H high-lift aerofoil; no sweep; no twist; thickness-chord ratio 13.1 per cent; dihedral 1°; incidence 2° 30'.

SAI G97 Spotter ultralight at Friedrichshafen in April 2003 *(Paul Jackson)* ***NEW***/0552051

SAI G97 Spotter ultralight kitbuilt *(Paul Jackson)* ***NEW***/0583535

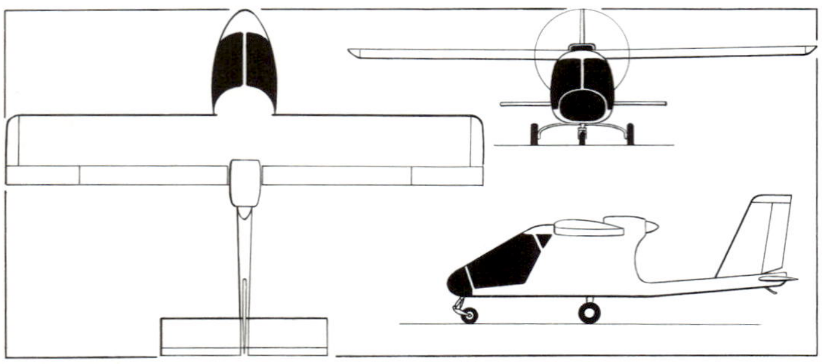

SAI G97 Spotter *(James Goulding)* 0121431

FLYING CONTROLS: Manual. Frise differential ailerons, rudder and all-moving tailplane with anti-balance tab and electrically actuated trim. Electrically actuated slotted flaps on 71 per cent of wing trailing-edge.

STRUCTURE: Mostly aluminium alloy structure and skin, except rudder and part of tailplane Dacron covered; wing/fuselage fairing, wingtips and tailcone of glass fibre.

LANDING GEAR: Tricycle type; fixed. Cantilever-sprung mainwheel legs of single carbon fibre/glass fibre piece, attached to fuselage by four bolts. Nosewheel with rubber shock-absorber. Hydraulic disc brakes on mainwheels operated by lever in cockpit. Steerable nosewheel. Main tyres size 6.00-5; nose 11×4.00-5.

POWER PLANT: One 59.6 kW (79.9 hp) Rotax 912 UL flat-four driving a Tonini two-blade, wooden propeller of 1.44 m (4 ft 8½ in) pitch. Two wing leading-edge fuel tanks in early aircraft, combined capacity 70 litres (18.5 US gallons; 15.4 Imp gallons), of which 66 litres (17.4 US gallons; 14.5 Imp gallons) are usable. By 2003, fuel capacity increased to 90 litres (23.8 US gallons; 19.8 Imp gallons). Option of 73.5 kW (98.6 hp) Rotax 912S or 74.0 kW (99 hp) Hirth 3701ES, and three-blade Helix propeller in later versions.

ACCOMMODATION: Two persons, side by side, with four-point harnesses. Upward-opening door each side. Dual controls, including central throttle; ventilation. Baggage compartment behind seats.

SYSTEMS: Electrical system has 12 V battery and 100 W alternator.

AVIONICS: To customer's requirements.

DIMENSIONS, EXTERNAL:
Wing span 8.25 m (27 ft 0¾ in)
Wing chord, constant 1.25 m (4 ft 1¼ in)
Wing aspect ratio 6.7
Length overall 6.24 m (20 ft 5½ in)
Height overall 2.30 m (7 ft 6½ in)
Tailplane span 2.90 m (9 ft 6¼ in)
Wheel track 2.14 m (7 ft 0¼ in)
Wheelbase 1.60 m (5 ft 3 in)
Propeller diameter: normal 1.66 m (5 ft 5¼ in)
max 1.72 m (5 ft 7¾ in)

DIMENSIONS, INTERNAL:
Cabin: Length 1.16 m (3 ft 9¾ in)
Max width 1.08 m (3 ft 6½ in)
Baggage volume 0.40 m³ (14.1 cu ft)

AREAS:
Wings, gross 10.30 m² (110.87 sq ft)
Tailplane 1.97 m² (21.20 sq ft)

WEIGHTS AND LOADINGS:
Weight empty: G97 283 kg (624 lb)
G97V 298 kg (657 lb)
Max T-O weight: G97 450 kg (992 lb)[1]
G97 alternative 530 kg (1,168 lb)[1]
G97V 530 kg (1,168 lb)
Max wing loading: G97 43.8 kg/m² (8.95 lb/sq ft)
G97 alternative, G97V 51.5 kg/m² (10.54 lb/sq ft)

Max power loading: G97 .... 7.56 kg/kW (12.4 lb/hp)
G97 alternative, G97V .... 8.90 kg/kW (14.62 lb/hp)

[1] *According to local regulations*

PERFORMANCE (Rotax 912 UL):
Never-exceed speed (VNE) .... 118 kt (220 km/h; 136 mph)
Max level speed at S/L .... 108 kt (200 km/h; 124 mph)
Cruising speed at 75% power .... 94 kt (175 km/h; 108 mph)
Econ cruising speed .... 86 kt (160 km/h; 99 mph)
Stalling speed, power off:
flaps up .... 38 kt (69 km/h; 43 mph)
flaps down .... 32 kt (58 km/h; 36 mph)
Max rate of climb at S/L .... 280 m (919 ft)/min
Service ceiling .... 3,505 m (11,500 ft)
T-O and landing run .... 85 m (280 ft)
Range with enlarged tanks .... 432 n miles (800 km; 497 miles)
g limits .... +4/−2

UPDATED

SG

SG AVIATION

Via Enzo Ferrari, I-04016 Sabaudia, Latina
Tel: (+39 0773) 51 52 15
Fax: (+39 0773) 51 14 07
e-mail: sg@storm-sg.it
Web: http://www.storm-sg.it
PRESIDENT: Giovanni Sassedo

NORTH AMERICAN AGENT:
SG Aviation America
26 McEwan Drive, Unit #6, Bolton, Ontario
L7E 1E6, Canada
Tel: (+1 905) 951 05 48
Fax: (+1 905) 951 06 49
e-mail: info@sgaviation.com
Web: http://www.sgaviation.com

In addition to the Storm series of light aircraft (based on an original Jean Pottier design); and Sea Storm, SG Aviation manufactures spare parts and subassemblies for the Italian Air Force in its 1,800 m^2 (19,375 sq ft) factory. The workforce numbered 70 in February 2003.

In April 2002, SG launched the high-wing Rally. By 2004, more than 1,000 complete aircraft and kits had been produced.

UPDATED

SG RALLY

TYPE: Side-by-side ultralight.

PROGRAMME: Aerodynamic studies began 1991; prototype flew in 1993, but development was suspended in favour of Sea Storm and Storm 300, both of which employed the Rally's wing profile. Work resumed in 2000; marketing launched in April 2002 to meet new, more liberal European sport pilot and US Experimental category legislation.

CURRENT VERSIONS: **Rally 105 S:** Experimental category or Ultralight version. *As described.*

Rally 105 UL: Ultralight version only. Empty weight 265 kg (584 lb); MTOW 435 kg (959 lb); reduced fuel and power.

CUSTOMERS: More than 20 sold by February 2004.

COSTS: Kit €25,100; ready-to-fly €54,000 (2004).

DESIGN FEATURES: High-mounted, strut-braced, constant-chord wings. Low tailplane and sweptback fin mounted on steeply waisted rear fuselage.

FLYING CONTROLS: Conventional and manual. Horn-balanced elevators and rudder. Flight-adjustable tab in starboard elevator. Slotted flaps.

STRUCTURE: All-composites fuselage, with aluminium wing. Wing section NACA 63-215 mod.

LANDING GEAR: Tricycle type; fixed. Fuselage-mounted spring cantilever mainwheel legs with wheel fairings; 5.00-5 wheels on all legs.

POWER PLANT: One 59.6 kW (79.9 hp) Rotax 912 UL-2 flat-four in Rally UL; 73.5 kW (98.6 hp) Rotax 912 ULS-2 in Rally S. Other suitable engines up to 93.2 kW (125 hp). Standard fuel capacity 80 litres (21.1 US gallons; 17.6 Imp gallons); can be increased to 130 litres (34.3 US gallons; 28.6 Imp gallons).

ACCOMMODATION: Gull-wing door on each side of cabin.

DIMENSIONS, EXTERNAL:
Wing span .... 9.18 m (30 ft 1½ in)
Length overall .... 7.08 m (23 ft 2¾ in)
Height overall .... 2.14 m (7 ft 0¼ in)

DIMENSIONS, INTERNAL:
Cabin max width .... 1.16 m (3 ft 9¾ in)

AREAS:
Wings, gross .... 12.48 m^2 (155.9 sq ft)

WEIGHTS AND LOADINGS:
Weight empty .... 290 kg (639 lb)[1]
Max T-O weight .... 450 kg (992 lb)[1]

[1] *if flown as ultralight*

PERFORMANCE:
Max level speed .... 130 kt (240 km/h; 149 mph)
Cruising speed at 75% power
.... 116 kt (215 km/h; 134 mph)
Stalling speed: flaps up .... 36 kt (65 km/h; 41 mph)
flaps down .... 30 kt (55 km/h; 35 mph)
Max rate of climb at S/L .... 390 m (1,280 ft)/min
T-O run .... 70 m (230 ft)
Landing run .... 110 m (360 ft)
Max range .... 809 n miles (1,500 km; 932 miles)
g limits .... +4.5/−2.2

UPDATED

SG Rally 105 S on floats *(Paul Jackson)* ***NEW***/0583534

SG STORM

TYPE: Side-by-side lightplane/kitbuilt.

PROGRAMME: Design started 11 February 1989 (Storm 280 SI); construction of prototype started 24 November 1990; first flight March 1991; certified to Aero Club Italia ULM category. Production aircraft manufacture began 29 March 1993, first flight July 1993; first delivery August 1993.

CURRENT VERSIONS: Total of six versions currently available, both above and below ultralight weight limit: **Storm 280 G,** has tricycle or tailwheel landing gear; **Storm 300 Special**, introduced in 1997, has tricycle gear. The **Storm Century** development of the Storm 300 Special was introduced in 2000 and is available in tricycle and tailwheel versions. Other modifications include a lengthened engine cowling, increased fuel capacity using wing tanks, and a baggage compartment behind rear cabin. Additionally, the 119 kW (160 hp) **Storm 400 Special** (400SP) is a stretched version of the 300 with a stronger spar which has been developed for the US market; see Power Plant paragraph for details of engines. The Storm 280 E, 280 SI, 300, 300 Focus, 320 E, 400 Focus and 400TI are no longer available.

Storm 500: Introduced in 2001 in two versions (basic CLK and retractable-gear SLK); is full-four-seat version, available only in tricycle form.

Storm RG: First flown 2001; development of 400 with conventional tailplane and modified fin with dorsal fillet. First deliveries in 2002. Prototype being tested with winglets; not standard.

Cyclone: Military version of Storm RG; hardpoint under each wing. Cockpit glazing comprises fixed windscreen; rearward sliding canopy; and fixed rear windows. No known production.

Sea Storm: Unrelated design; see separate entry.

CUSTOMERS: By early 2003, over 650 Storms ordered, of which more than 250 were flying. Around 80 per cent are tricycle landing gear versions.

COSTS: Kits: Storm 280G €19,450; Storm 300 Special €18,900; Storm Century €21,000; Storm 400 Special €22,500; Storm 500 €24,400, Storm RG €30,000; Storm Century '04 €23,200. Ready to fly: Storm 280G €61,200; Storm 300 Special €58,700; Storm Century €63,000; Storm RG €82,000; Storm Century '04 €68,700 (all 2004).

DESIGN FEATURES: Extensive use of modern materials and technology. All-aluminium fuselage and wings on 280 SI and 300 Special. Quoted build time for kits 600 hours.

Wing sections NACA 4415 (Storm 280 SI); GA 30U-6135 Mod (Storm 300 Special). Thickness/chord ratio 15 per cent; 3° dihedral; 3° incidence.

FLYING CONTROLS: Manual. All-moving tailplane with anti-balance tab on most versions; no rudder tab. Storm RG has conventional horizontal tail. Balanced Frise ailerons on 70 per cent of span. Electric trim controls. Flap settings: 280, 10/20/32°; 12/25/38° on 300 and 400 series; all models have Fowler flaps.

LANDING GEAR: Main legs are cantilever sprung. Tricycle versions have tubular steel sprung nosewheel leg steerable ±40°. Tailwheel leg of Storm 300 is metal sprung with a solid tyre. Hydraulic disc brakes on mainwheels. Speed fairings can be fitted. Matco tyres; mainwheels 15×6.00, pressure 2.00 bar (29 lb/sq in); tailwheel (mainwheel in Storm 300 Special) 15 cm (6 in), pressure 1.80 bar (26 lb/sq in). Storm RG has 5.00-5 wheels on all legs; main wheels retract inwards, nosewheel rearwards. No doors.

POWER PLANT: *Storm 280 G:* One 59.6 kW (79.9 hp) Rotax 912 UL or 84.6 kW (113.4 hp) Rotax 914 flat-four four-stroke driving either a two-blade GT166-42 fixed-pitch propeller or a variable-pitch IVO-Prop or AR-Plast propeller; or one 59.7 kW (80 hp) Moto Guzzi driving a two-blade GM fixed-pitch propeller. Fuel capacity for 280 and 300 series 85 litres (22.5 US gallons; 18.7 Imp gallons). Oil capacity 3.5 litres (0.92 US gallon; 0.77 Imp gallon).

Storm 300 Special and Century: One 59.6 kW (79.9 hp) Rotax 912 UL turbocharged four-cylinder four-stroke engine driving a two-blade Rospeller constant-speed metal propeller. Mid-West AE 110 Hawk 78.3 kW (105 hp) rotary engine optional. Fuel capacity of Storm 300 and 300 Special as Storm 280 G; Storm Century 100 litres (26.4 US gallons; 22.0 Imp gallons).

Storm 400 Special: One 119 kW (160 hp) Textron Lycoming O-320 flat-four; engines in the range 82.0 to 119 kW (110 to 160 hp) can be fitted. Fuel capacity 140 litres (37.0 US gallons; 30.8 Imp gallons); optional tanks increase this to 200 litres (52.8 US gallons; 44.0 Imp gallons).

Storm 500 CLK and 500 SLK: One 157 kW (210 hp) Textron Lycoming O-360. Fuel capacity 180 litres (47.6 US gallons; 39.6 Imp gallons); engines in the range 104 to 134 kW (140 to 180 hp) can be fitted.

SG Storm 400 *(Paul Jackson)* ***NEW***/0583363

SG Storm RG instrument panel *(Paul Jackson)* ***NEW***/0583391

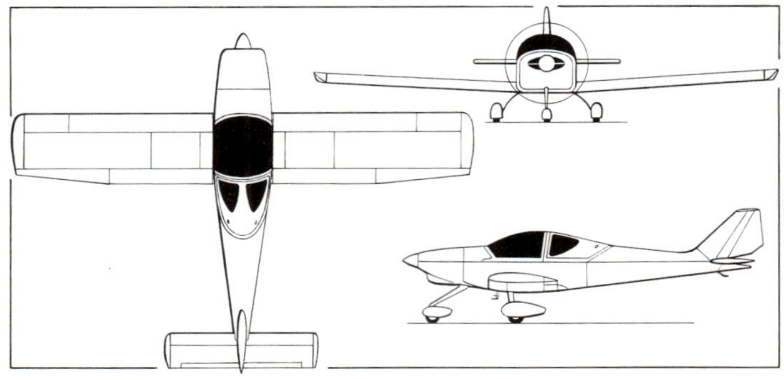

SG Storm, nosewheel version *(James Goulding)* 0011094

SG Storm 500 CLK under construction in the USA *(Paul Jackson)* ***NEW***/0583365

Storm RG: One 73.5 kW (98.6 hp) Rotax 912 ULS driving a Duc three-blade propeller or one 84.6 kW (113.4 hp) Rotax 914. Fuel capacity as Storm Century.

ACCOMMODATION: Two persons, side by side, with dual controls; single-piece canopy hinged upwards and forwards. Storm 400 TI can accommodate one large or two small additional passengers. Large baggage area behind cockpit.

SYSTEMS: 12 V battery.

AVIONICS: *Instrumentation:* Option of full IFR fit.

DIMENSIONS, EXTERNAL (A: 280 G, B: 300 Special, C: 400 Special, D: Century, E: 500 CLK; F: 500 SLK; G: RG with Rotax 912 ULS):

Wing span: A, B, D, G ..... 7.80 m (25 ft 7 in)
C ..... 7.98 m (26 ft 2¼ in)
E, F ..... 8.60 m (28 ft 2½ in)
Wing chord, constant: B ..... 1.30 m (4 ft 3 in)
Length overall: A, B ..... 6.70 m (21 ft 11¾ in)
C ..... 6.82 m (22 ft 4½ in)
D, G ..... 6.80 m (22 ft 3¾ in)
E, F ..... 7.22 m (23 ft 8¼ in)
Height overall: A, C ..... 2.15 m (7 ft 0¾ in)
B ..... 2.12 m (6 ft 11½ in)
G ..... 2.04 m (6 ft 8¼ in)

DIMENSIONS, INTERNAL:

Cabin max width: A, B, C, D, E, F, G ..... 1.10 m (3 ft 7¼ in)

AREAS:

Wings, gross: A, B, D, G ..... 9.98 m^2 (107.4 sq ft)
C ..... 10.45 m^2 (112.5 sq ft)
E, F ..... 11.35 m^2 (122.2 sq ft)

WEIGHTS AND LOADINGS:

Weight empty: A ..... 290 kg (639 lb)
B, G (ultralight) ..... 293 kg (646 lb)
C ..... 480 kg (1,058 lb)
D ..... 320 kg (705 lb)
E ..... 510 kg (1,124 lb)
F ..... 530 kg (1,168 lb)
G (experimental) ..... 350 kg (771 lb)
Max T-O weight: A, B, G (ultralight) ..... 450 kg (992 lb)
C ..... 850 kg (1,874 lb)
D ..... 560 kg (1,234 lb)
E ..... 975 kg (2,149 lb)
F ..... 995 kg (2,193 lb)
G (experimental) ..... 580 kg (1,278 lb)
Max wing loading: A ..... 47.0 kg/m^2 (9.63 lb/sq ft)
B ..... 45.1 kg/m^2 (9.24 lb/sq ft)
C ..... 81.3 kg/m^2 (16.66 lb/sq ft)
D ..... 56.1 kg/m^2 (11.49 lb/sq ft)
E ..... 85.9 kg/m^2 (17.59 lb/sq ft)
F ..... 87.7 kg/m^2 (17.96 lb/sq ft)
G (ultralight) ..... 35.1 kg/m^2 (7.18 lb/sq ft)
G (experimental) ..... 58.1 kg/m^2 (11.90 lb/sq ft)
Max power loading: A ..... 8.72 kg/kW (14.34 lb/hp)
B ..... 5.32 kg/kW (8.74 lb/hp)
C ..... 7.14 kg/kW (11.74 lb/hp)
D ..... 6.62 kg/kW (10.88 lb/hp)
E ..... 6.21 kg/kW (10.20 lb/hp)
F ..... 6.34 kg/kW (10.41 lb/hp)
G (ultralight) ..... 4.76 kg/kW (7.82 lb/hp)
G (experimental) ..... 7.89 kg/kW (12.97 lb/hp)

PERFORMANCE:

Never-exceed speed (VNE):
A ..... 162 kt (300 km/h; 186 mph)
B, C, D, G ..... 167 kt (310 km/h; 192 mph)
E, F ..... 172 kt (320 km/h; 198 mph)
Max operating speed (VMO):
A ..... 127 kt (235 km/h; 146 mph)
B ..... 152 kt (282 km/h; 175 mph)
D ..... 155 kt (287 km/h; 178 mph)
C, E, G ..... 157 kt (290 km/h; 180 mph)
F ..... 165 kt (305 km/h; 190 mph)
Cruising speed at 75% power:
A ..... 111 kt (205 km/h; 127 mph)
B, G ..... 143 kt (265 km/h; 165 mph)
C ..... 148 kt (274 km/h; 170 mph)
D ..... 150 kt (278 km/h; 173 mph)
E ..... 150 kt (277 km/h; 172 mph)
F ..... 158 kt (292 km/h; 181 mph)
Stalling speed, power off, flaps down:
A ..... 31 kt (57 km/h; 36 mph)
B ..... 29 kt (52 km/h; 33 mph)
C ..... 38 kt (70 km/h; 44 mph)
D, G ..... 30 kt (55 km/h; 35 mph)
E, F ..... 43 kt (78 km/h; 49 mph)
Max rate of climb at S/L: A ..... 330 m (1,083 ft)/min
B, D ..... 420 m (1,378 ft)/min
C, E, F ..... 450 m (1,476 ft)/min
G ..... 390 m (1,280 ft)/min
Service ceiling: A ..... 4,000 m (13,120 ft)
B, D, E ..... 6,096 m (20,000 ft)
G ..... 5,485 m (18,000 ft)
T-O run: A, D ..... 90 m (295 ft)
B ..... 120 m (395 ft)
C, E, F ..... 220 m (720 ft)
G ..... 110 m (360 ft)
Landing run: A ..... 150 m (495 ft)
B ..... 180 m (590 ft)
C ..... 200 m (660 ft)
D ..... 120 m (305 ft)
E, F ..... 250 m (820 ft)
G ..... 140 m (460 ft)
Range with max fuel:
A ..... 627 n miles (1,162 km; 722 miles)
B ..... 702 n miles (1,300 km; 807 miles)
C ..... 594 n miles (1,100 km; 683 miles)
D ..... 864 n miles (1,600 km; 994 miles)
E ..... 621 n miles (1,150 km; 714 miles)
F ..... 648 n miles (1,200 km; 745 miles)
G ..... 779 n miles (1,444 km; 897 miles)
g limits: A ..... +5.7/–3
B ..... +6/–3
C, E ..... +3.8/–2.0
D ..... +4.2/–2
F ..... +3.7/–1.9
G ..... +4.5/–2.0

UPDATED

SG SEA STORM

TYPE: Two-seat amphibian kitbuilt; four-seat amphibian kitbuilt. First flown 1999.

PROGRAMME: Derived from earlier ultralight version, with 59.7 kW (80 hp) engine and metal strut-braced wings, of which some 20 flying in Italy by 2002. First all-composites Sea Storm under construction in 2000/01 by SG's North American agent, Eric Dixon and first flew (N221ST) October 2001; later returned to Italy. Ultralight Sea Storm with carbon wing and bracing struts under construction in 2003.

CURRENT VERSIONS: **Z2**: Two-seat; baseline version.

Z4: Four-seat; increased wing span and power.

CUSTOMERS: Three sold in USA by April 2002; one Z2 to Australia in 2003. 26 sold (including earlier ultralight) by 2003.

COSTS: Kit UL €34,600; Z2 €34,600. Z4 €36,925 (2004).

DESIGN FEATURES: High, constant-chord wing, moderately sweptback tailplane and highly sweptback fin; pusher propeller immediately aft of wing trailing-edge; mid-positioned stub-wing with stabilising floats. Quoted kit build time 700 hours.

Wing dihedral 0° 5'; incidence 2° 30'.

FLYING CONTROLS: Conventional and manual. Fowler flaps. Electric trim.

STRUCTURE: Composites fuselage and aluminium wing with main and rear spars.

LANDING GEAR: Optional retractable wheel landing gear attached immediately outboard of wingroots; 6.00-6 wheels.

POWER PLANT: One pusher piston/rotary engine.

Z2: Option of 73.5 kW (98.6 hp) Rotax 912 ULS or 84.6 kW (113.4 hp) Rotax 914 flat-four or 70.8 kW (95 hp) Mid-West AE 100 rotary driving an adjustable-pitch propeller. Textron Lycoming engines in the 82.0 to 119 kW (110 to 160 hp) range are also suggested.

SG Sea Storm amphibian nearing completion in April 2001 *(Jane's/Susan Bushell)* 0110584

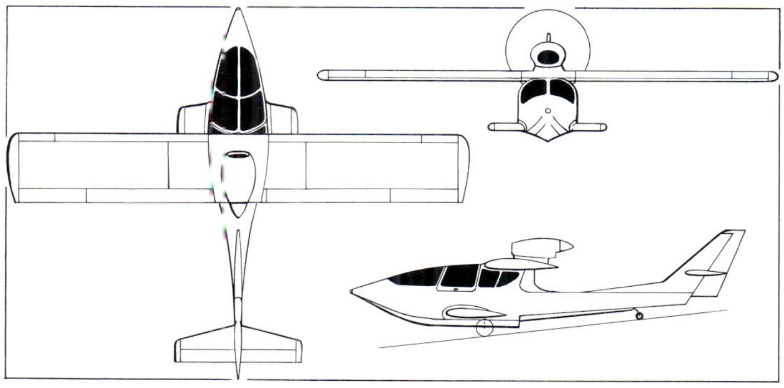

SG Sea Storm *(James Goulding)* 0093636

Standard fuel capacity 140 litres (37.0 US gallons; 30.8 Imp gallons), with optional extra 79 litre (20.9 US gallon; 17.4 Imp gallon) tanks available.

Z4: One 157 kW (210 hp) Textron Lycoming O-360; engines recommended in the range 149 to 224 kW (200 to 300 hp). Fuel capacity 200 litres (52.8 US gallons; 44.0 Imp gallons) as standard; optional 61 litre (16.1 US gallon; 13.4 Imp gallon) tanks available.

Idrovario hydraulic, flight-adjustable pitch, two- or three-blade propeller.

ACCOMMODATION: Two (side by side) or four persons in enclosed cabin with centreline-hinged door each side.

All data following (including dimensions) are approximate.

DIMENSIONS, EXTERNAL (A: Z2; B: Z4):
Wing span: A .... 9.60 m (31 ft 6 in)
B: .... 10.10 m (33 ft 1¾ in)
Length overall: A, B .... 7.44 m (24 ft 5 in)
Height overall: A, B .... 1.83 m (6 ft 0 in)
DIMENSIONS, INTERNAL:
Cabin max width: A, B .... 1.05 m (3 ft 5¼ in)
AREAS:
Wings, gross: A .... 12.96 m² (139.5 sq ft)
B .... 13.65 m² (146.9 sq ft)
WEIGHTS AND LOADINGS (A, B, as above):
Weight empty: A .... 510 kg (1,124 lb)
B .... 690 kg (1,521 lb)
Baggage capacity: A .... 30 kg (66 lb)
B .... 55 kg (121 lb)
Max T-O weight: A .... 810 kg (1,785 lb)
B .... 1,200 kg (2,645 lb)
Max wing loading: A .... 62.5 kg/m² (12.80 lb/sq ft)
B .... 87.9 kg/m² (18.01 lb/sq ft)
Max power loading:
A (Rotax 912) .... 11.02 kg/kW (18.11 lb/hp)
B (157 kW; 210 hp) .... 7.64 kg/kW (12.56 lb/hp)
PERFORMANCE (A, B, as above):
Never-exceed speed (VNE):
A .... 135 kt (250 km/h; 155 mph)
B .... 162 kt (300 km/h; 186 mph)
Max level speed: A .... 121 kt (225 km/h; 140 mph)
B .... 143 kt (265 km/h; 165 mph)
Cruising speed at 75% power at FL80:
A .... 111 kt (205 km/h; 127 mph)
B .... 125 kt (232 km/h; 144 mph)
Stalling speed:
flaps up: A .... 35 kt (64 km/h; 40 mph)
B .... 46 kt (85 km/h; 53 mph)
flaps down: A .... 33 kt (61 km/h; 38 mph)
B .... 40 kt (74 km/h; 46 mph)
Max rate of climb at S/L: A .... 240 m (787 ft)/min
B .... 330 m (1,082 ft)/min
Service ceiling: A .... 3,960 m (13,000 ft)
B .... 4,570 m (15,000 ft)
T-O run:
on land: A .... 180 m (591 ft)
B .... 220 m (722 ft)
on water: A .... 195 m (640 ft)
B .... 350 m (1,150 ft)
Landing run:
on land: A .... 200 m (656 ft)
B .... 250 m (820 ft)
on water: A .... 146 m (480 ft)
B .... 210 m (690 ft)
Range, no reserves:
A .... 917 n miles (1,700 km; 1,056 miles)
B .... 540 n miles (1,000 km; 621 miles)
g limits: A .... +4.2/−2.0
B .... +3.2/−1.8

UPDATED

TECNAM

COSTRUZIONI AERONAUTICHE TECNAM SRL

1a Traversa Via G Pascoli, I-80026 Casoria (Naples)
Tel: (+39 081) 758 32 10 and 758 87 51
Fax: (+39 081) 758 45 28
e-mail: tecnamca@tin.it
Web: http://www.tecnam.com
MANAGING DIRECTOR: Prof Luigi Pascale Langer
MARKETING AND SALES DIRECTOR: Paolo Pascale Langer

Company founded 1986, after Pascale brothers were released from the original Pascale company, Partenavia (see now VulcanAir), which had been placed under control of Alenia in 1981. Tecnam manufactures tailplane and other components of ATR 42/72 and parts of A 109, C-27J, SF-260, P.68, plus fuselage panels for Boeing. Workshops qualified to NATO AQAP4 and to Italian Aeronautical Register, JAR 21/F. Manufactures P92 Echo ultralight aircraft and P92-J very light aircraft and in 1997 launched production of the P96 and P96-J; P92 SeaSky amphibian added to range in 1998, P92-S in 1999, 2000 RG in 2000 and Sierra in 2003. Pacific Aerosystem of San Diego appointed US agents in 2001.

In April 2001, Tecnam announced that it was planning to add a four-seater to its range. By the end of 2003, had produced more than 1,000 kits and complete aircraft.

The company operates from an 11,000 m² (118,400 sq ft) facility at Napoli-Capodichino.

UPDATED

TECNAM P92 ECHO

TYPE: Side-by-side lightplane/kitbuilt.

PROGRAMME: First flight 14 March 1993. Prototype P92-J (I- TECN) first flown second quarter 1995 and received type certificate 10 November 1995. Production rate 10 per month by end 2001.

CURRENT VERSIONS: **P92:** Standard ultralight version (Italian ULM rules).

P92-J: Detail differences in airframe and equipment to comply with JAR-VLA airworthiness requirements. First delivery (002), to US distributor, second quarter 1996; 12 built by 1998; manufacture suspended until 2001, when replaced by upgraded P92-JS.

P92-JS: Upgraded version of P-92J with higher-powered Rotax 912 ULS and slightly higher max T-O weight plus increased fuel capacity.

P92-S and P92-ES Echo Super: Added to range in 1999; redesigned rear cabin with additional window; redesigned wing profile and wingtips, revised engine cowling and windscreen; improvements in performance. Jabiru-engined ultralight version is **P92-EM**. *As described.*

P92 Echo Short-Wing: Version for German market with lower empty weight of 275 kg (606 lb), max T-O weight 550 kg (1,212 lb). Bulged cockpit doors, but only one rear quarterlight each side.

Tecnam P92-ES Echo Super flown as an Australian ultralight *(Paul Jackson)* ***NEW***/0583390

Jabiru-engined Tecnam P92-EM kitplane *(Paul Jackson)* ***NEW***/0583386

P92 Echo 2000 RG: Retractable landing gear version introduced in 2000. Similar to P92-S, but with shorter (8.70 m; 28 ft 6½ in) wing span and reprofiled flaps and fintips. Fuselage lower surfaces redesigned (rounded and deeper) to accommodate retracted gear; bulged doors provide extra 4 cm (1½ in) of shoulder room.

P92 SeaSky: Amphibious floatplane version first flown 1 July 1997; strake below rear fuselage. Full Lotus floats with optional wheels for land operations. Rotax 912 ULS engine and increased empty and maximum T-O weights.

Echo 80: Either P92 or P92-S available with standard Rotax 912 UL engine.

Echo 2000 RG retractable landing gear version ***NEW***/0137829

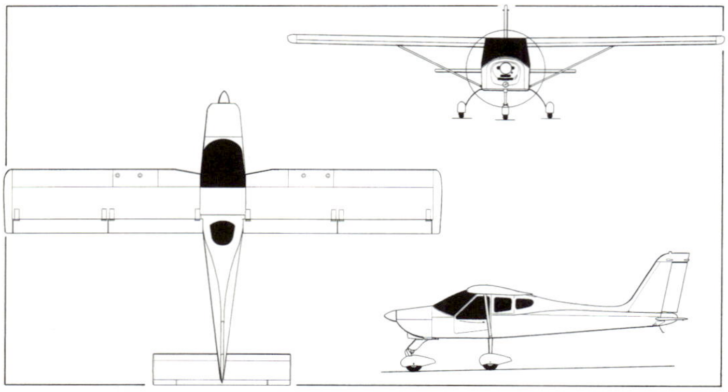

Tecnam P92-S Echo very light aircraft 0110602

Echo 100S: Either P92 or P92-S available with optional Rotax 912 ULS engine. Latter variant designated P92 Echo S 100S.

CUSTOMERS: Some 776 of all versions delivered by July 2003; six to Cambodian Air Force (first two September 1994); overseas orders currently account for 51 per cent of production, including many to Israel. P92-J and -JS deliveries totalled 34 by end of 2003. P92 2000 RG deliveries totalled 43 by end of 2003.

COSTS: P92 2000 RG kit US$32,100 less engine, propeller, fuel tanks and electrical wiring (2001); P92 Echo Short-Wing €50,924 with Jabiru 2200 (2003).

DESIGN FEATURES: Objectives were lightness, simplicity and accessibility for inspection and servicing. Braced single-spar high wing; aerofoil chosen for good performance at low Reynolds number; untapered with 1° 30' dihedral; underside of fuselage mostly flat for ease of kit assembly; large one-piece windscreen, inward-tapered inboard wing leading-edges, large windows in doors; rear view window (s).

FLYING CONTROLS: Manual. Frise ailerons; all-moving tailplane with anti-balance tab; electrically actuated flaps cover half trailing-edge.

STRUCTURE: Fuselage mainly metal except GFRP lower shell of engine cowling and wing leading-edge; ailerons and part of tailplane fabric covered; GFRP rudder tip; tailplane halves can be unpinned and removed quickly for transport; engine cowling removable by undoing four quick latches to reveal whole power plant.

LANDING GEAR: Tricycle type; fixed. Fuselage-mounted spring steel cantilever mainwheel legs with wheel fairings; hand-powered hydraulic disc brakes operated together from single lever in cockpit; nosewheel, with compressed rubber shock absorber, steered from rudder pedals; designed for grass field operation. Wheels 5.00-5 on P92-S and P92-JS; 400–6 on SeaSky. P92 2000 RG nosewheel is 11×4.00, has oleo-pneumatic shock-absorber and retracts rearwards; 14×4 mainwheels retract inward. Both tube legs are mounted on a single cantilever arm.

POWER PLANT: One 59.6 kW (79.9 hp) Rotax 912 UL (P92) or 912A (P92-J) or 73.5 kW (98.6 hp) Rotax 912 ULS flat-four engine; integrated reduction gear driving two-blade wooden propeller: GT-166/146 for Rotax 912 UL; GT-172/164 for Rotax 912 ULS landplane; GT-182/118 for SeaSky.

Limbach 2000 (59.7 kW; 80 hp) and Rotax 582 (47.8 kW; 64.1 hp) two-stroke offered as options for P92; P92 prototype had the latter; 59.7 kW (80 hp) Jabiru 2200 flat-four available from 2000.

P92 Echo Short-Wing has 59.7 kW (80 hp) Jabiru 2200 as standard, with options to fit 59.6 kW (79.9 hp) Rotax 912 UL or 73.5 kW (98.6 hp) Rotax 912 ULS.

Fuel capacity 70 litres (18.5 US gallons; 15.4 Imp gallons) P92 Echo short-wing has 59.7 kW (80 hp) Jabiru 2200 as standard, with options to fit 59.6 kW (79.9 hp) Rotax 912 UL or 73.5 kW (98.6) Rotax 912 ULS in two wing tanks, of which 60 litres (15.9 US gallons; 13.2 Imp gallons) are usable.

P92-JS fuel capacity 90 litres (23.8 US gallons; 19.18 Imp gallons).

ACCOMMODATION: Side-by-side seats with three-point harness; baggage space behind seats. Full dual controls and two throttles. Bulged doors on current versions give increased cabin width.

SYSTEMS: 14 V 55 A battery; 100 W alternator. Optional BRS-5 parachute in P92 2000 RG.

AVIONICS: Customer specified; optional IFR nav/com in P92-J.

DIMENSIONS, EXTERNAL (all versions, except where indicated):
- Wing span: P92 2000 RG, P92-JS ... 8.70 m (28 ft 6½ in)
  - P92, P92-S ... 9.30 m (30 ft 6 in)
  - P92-J, SeaSky ... 9.60 m (31 ft 6 in)
- Wing chord, constant portion ... 1.40 m (4 ft 7 in)
- Length overall: except P92 2000 RG ... 6.30 m (20 ft 8 in)
  - P92 2000 RG ... 6.40 m (21 ft 0 in)
- Height overall: except SeaSky ... 2.49 m (8 ft 2 in)
  - SeaSky ... 2.97 m (9 ft 9 in)
- Tailplane span ... 2.90 m (9 ft 6 in)
- Wheel track:
  - except SeaSky, P92 2000 RG, P92-JS ... 1.78 m (5 ft 10 in)
  - P92 2000 RG ... 1.73 m (5 ft 8 in)
  - P92-JS ... 1.80 m (5 ft 11 in)
- Float track: SeaSky ... 2.79 m (9 ft 2 in)
- Propeller diameter: 59.7 kW engine ... 1.66 m (5 ft 5¼ in)
  - 73.5 kW engine ... 1.72 m (5 ft 7¾ in)

DIMENSIONS, INTERNAL:
- Cabin max width ... 1.14 m (3 ft 9 in)

AREAS:
- Wings, gross:
  - except SeaSky, P92 2000 RG ... 13.20 m^2 (142.1 sq ft)
  - P92 2000 RG ... 11.98 m^2 (129.0 sq ft)
  - SeaSky ... 13.40 m^2 (144.2 sq ft)
- Tailplane ... 1.97 m^2 (21.20 sq ft)

WEIGHTS AND LOADINGS:
- Basic weight empty:
  - P92, P92-S, P92 2000 RG ... 281 kg (619 lb)
  - P92-J ... 305 kg (672 lb)
  - P92-JS ... 310 kg (683 lb)
  - SeaSky ... 330 kg (728 lb)
- Max T-O weight:
  - ultralight: P92, P92-S, P92 2000 RG ... 450 kg (992 lb)
    - SeaSky ... 500 kg (1,102 lb)
  - certified: P-92, P92-S, P92-JS ... 550 kg (1,212 lb)
  - P92-J ... 520 kg (1,146 lb)
  - SeaSky ... 600 kg (1,322 lb)

PERFORMANCE, POWERED (P92 in Echo 80 form, P92-S in Echo 100 form):
- Never-exceed speed (VNE):
  - except SeaSky and P92-JS ... 135 kt (250 km/h; 155 mph)
  - P92-JS ... 145 kt (270 km/h; 167 mph)
  - SeaSky ... 108 kt (200 km/h; 124 mph)
- Max level speed at S/L:
  - P92, P92-JS ... 119 kt (220 km/h; 137 mph)
  - P92-S ... 124 kt (230 km/h; 143 mph)
  - P92-J ... 110 kt (204 km/h; 127 mph)
  - P92 2000RG ... 135 kt (250 km/h; 155 mph)
  - SeaSky ... 92 kt (170 km/h; 106 mph)
- Cruising speed at 75% power at 2,200 propeller rpm:
  - P92 2000 RG ... 124 kt (230 km/h; 143 mph)
  - P92, P92-J, P92-JS ... 100 kt (185 km/h; 115 mph)
  - P92-S ... 105 kt (195 km/h; 121 mph)
  - SeaSky ... 84 kt (156 km/h; 97 mph)
- Stalling speed:
  - flaps up: P92 ... 36 kt (68 km/h; 42 mph)
    - P92-J ... 40 kt (74 km/h; 46 mph)
  - flaps down:
    - P92, P92-S, P92 2000 RG ... 33 kt (61 km/h; 38 mph)
    - P92-J, SeaSky ... 35 kt (64 km/h; 40 mph)
    - P92-JS ... 36 kt (67 km/h; 42 mph)
- Max rate of climb at S/L: P92 ... 330 m (1,082 ft)/min
  - P92-J ... 262 m (860 ft)/min
  - P92-JS ... 360 m (1,180 ft)/min
  - P92-S, P92 2000 RG ... 384 m (1,260 ft)/min
  - SeaSky ... 244 m (800 ft)/min
- Service ceiling: P92, P92-JS ... 4,265 m (14,000 ft)
  - P92 2000 RG, P92-S ... 4,510 m (14,800 ft)
  - P92-J ... 4,115 m (13,500 ft)
  - SeaSky ... 3,505 m (11,500 ft)
- T-O run:
  - on land: P92 ... 110 m (360 ft)
    - P92-J ... 122 m (400 ft)
    - P92-JS ... 143 m (470 ft)
    - P92-S ... 100 m (330 ft)
    - P92 2000 RG ... 140 m (460 ft)
    - SeaSky ... 122 m (400 ft)
  - on water: SeaSky ... 350 m (1,150 ft)
- Landing run:
  - on land:
    - except P92-J, P92 2000 RG, SeaSky ... 100 m (328 ft)
    - P92 2000 RG ... 110 m (360 ft)
    - P92-J ... 92 m (300 ft)
    - SeaSky ... 122 m (400 ft)
  - on water (SeaSky) ... 350 m (1,150 ft)
- Max range, no reserves:
  - except SeaSky and P92-JS ... 400 n miles (740 km; 460 miles)
  - P92-JS ... 450 n miles (833 km; 517 miles)
  - SeaSky ... 300 n miles (555 km; 345 miles)
- Endurance: P92 ... 4 h 30 min
  - P92-J ... 5 h
- *g* limits: P92-J, P-92-S ... +6/–3
  - P92-JS ... +3.8/–1.5
  - SeaSky ... +5.3/–2

PERFORMANCE, UNPOWERED:
- Best glide ratio: P92, P92-J ... 13

UPDATED

TECNAM P96 GOLF

TYPE: Side-by-side ultralight.

PROGRAMME: Prototype construction began July 1996; first flight March 1997. Uprated Golf 100 added in 1999.

CURRENT VERSIONS: **P96 Golf 80:** Ultralight version.

P96 Golf 100: Higher-powered ultralight version.

CUSTOMERS: The 200th Golf registered in Canada (C-ITEC) in February 2002; by late 2003 this figure had risen to 256.

DESIGN FEATURES: Emphasis on reduction of aerodynamic drag without compromising cost. Fuselage cross-sectional area allows easy access for taller occupants by reducing the height of structural carry-through elements. Wing lower surface and fuselage bottom are flush, obviating interference drag and complex fairings. Wing-to-fuselage point optimised by strakes at wingroots. Many parts common to P92 Echo.

FLYING CONTROLS: Manual. Frise ailerons; all-moving tailplane with anti-balance tab; electrically actuated flaps cover half of trailing-edge. Ground-adjustable tab on port aileron and rudder.

STRUCTURE: Fuselage includes a composites spine that follows through to the tailfin and keeps wetted area to a minimum while allowing for good pressure recovery aft of canopy.

Tecnam P96 Golf 100 *(Paul Jackson)* ***NEW***/0583382

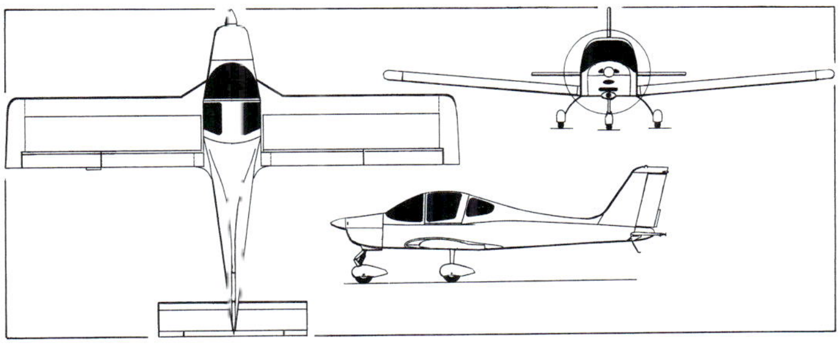

Tecnam P96 Golf two-seat low-wing ultralight aircraft (Rotax 912 piston engine) *(Paul Jackson)* 0062982

All-moving horizontal tail reduces drag to a minimum and provides optimum longitudinal stability for stick-free operation. Wing has 5° dihedral. Fabric-covered flaps, ailerons and tailplane; metal rudder.

LANDING GEAR: Tricycle type; fixed. Fuselage-mounted, spring cantilever mainwheel legs with wheel fairings; hydraulic disc brakes operated from single hand lever in cockpit. Levered suspension nosewheel leg with compressed rubber shock absorber; nosewheel steered by rudder pedals. Mainwheels 5.00-5, nosewheel 4.00-6. Speed fairings.

POWER PLANT: *Golf 80:* One 59.6 kW (79.9 hp) Rotax 912 UL flat-four engine with integrated reduction gear (1:2.27), driving a GT-2/166/146 single-piece two-blade wooden propeller.

Golf 100: 73.5 kW (98.6 hp) Rotax 912 ULS driving a GT-172/164 propeller.

Fuel capacity (both) 70 litres (18.5 US gallons; 15.4 Imp gallons).

ACCOMMODATION: Two seats side by side; baggage space behind seats. Fixed windscreen; canopy slides backwards on guiderails and can be opened with engine on and during flight. Full dual controls and twin throttles.

SYSTEMS: 100 W 12 V alternator and battery.

AVIONICS: To customer's requirements; provision for nav/com radio.

EQUIPMENT: Junkers ballistic parachute.

DIMENSIONS, EXTERNAL:
- Wing span ... 8.41 m (27 ft 7 in)
- Length overall ... 6.40 m (21 ft 0 in)
- Height overall ... 2.29 m (7 ft 6 in)
- Tailplane span ... 2.90 m (9 ft 6 in)
- Wheel track ... 1.78 m (5 ft 10 in)
- Wheelbase ... 1.60 m (5 ft 3 in)
- Propeller diameter ... 1.65 m (5 ft 5 in)

DIMENSIONS, INTERNAL:
- Cabin max width ... 1.13 m (3 ft 8½ in)

AREAS:
- Wings, gross ... 12.20 m² (131.3 sq ft)

WEIGHTS AND LOADINGS:
- Weight empty: Golf 80 ... 281 kg (619 lb)
- Max T-O weight: both ... 450 kg (992 lb)

PERFORMANCE, POWERED:
- Never-exceed speed (VNE) ... 140 kt (259 km/h; 161 mph)
- Max level speed: Golf 80 ... 121 kt (224 km/h; 139 mph)
  - Golf 100 ... 130 kt (241 km/h; 150 mph)
- Cruising speed at 75% power:
  - Golf 80 ... 105 kt (195 km/h; 121 mph)
  - Golf 100 ... 116 kt (215 km/h; 133 mph)
- Stalling speed, power off, flaps down:
  - both ... 33 kt (62 km/h; 38 mph)
- Max rate of climb at S/L: Golf 80 ... 271 m (890 ft)/min
  - Golf 100 ... 360 m (1,180 ft)/min
- Service ceiling: Golf 80 ... 4,000 m (13,120 ft)
  - Golf 100 ... 4,500 m (14,760 ft)
- T-O run: Golf 80 ... 110 m (361 ft)
  - Golf 100 ... 100 m (328 ft)
- Landing run: both ... 100 m (328 ft)
- Max range: Golf 80 ... 400 n miles (740 km; 460 miles)
- Endurance: Golf 80 ... 4 h 30 min
- Ultimate *g* limits: both ... +6/−3

PERFORMANCE, POWERED:
- Best glide ratio: Golf 80 ... 12

UPDATED

TECNAM P2002 SIERRA

TYPE: Side-by-side ultralight.

PROGRAMME: First publicly displayed (I-TEJF) at Aero '03, Friedrichshafen, April 2003. Undergoing approval by PFA in UK, 2004.

CURRENT VERSIONS: **P2002:** Standard version.

P2002-JF: For German market; *as described.*

CUSTOMERS: Registration of 32nd aircraft effected (in New Zealand) in May 2004.

COSTS: Kit £20,316, less engine (2004).

DESIGN FEATURES: Considerably upgraded Tecnam P96 Golf to meet JAR-VLA standards.

FLYING CONTROLS: Manual. All-moving tailplane. Slotted flaps.

STRUCTURE: Redesigned fuselage with reprofiled cabin area. New all-metal tapered wing has 5° dihedral and upturned tips.

LANDING GEAR: As P96 Golf.

POWER PLANT: One 73.5 kW (98.6 hp) Rotax 912 ULS-2 with 1:2.42 reduction gear driving a Hoffmann HO-17-GH, GT Propellers GT-2/173/VRR-FW101SRTC or similar two-blade, fixed-pitch propeller. Fuel capacity 100 litres (26.4 US gallons; 22.0 Imp gallons) in two wing tanks.

ACCOMMODATION: Two seats side by side. Canopy slides rearwards and can be opened with engine on and during flight.

DIMENSIONS, EXTERNAL:
- Wing span ... 8.60 m (28 ft 2½ in)
- Length overall ... 6.61 m (21 ft 8¼ in)
- Height overall ... 2.43 m (7 ft 11½ in)
- Tailplane span ... 2.90 m (9 ft 6¼ in)
- Wheel track ... 1.85 m (6ft 0¾ in)

DIMENSIONS, INTERNAL:
- Cabin max width ... 1.10 m (3 ft 7¼ in)

AREAS:
- Wings, gross ... 11.50 m² (123.8 sq ft)

WEIGHTS AND LOADINGS:
- Weight empty ... 320 kg (705 lb)
- Max T-O weight ... 580 kg (1,279 lb)

PERFORMANCE:
- Never-exceed speed (VNE) ... 156 kt (290 km/h; 180 mph)
- Max operating speed ... 130 kt (241 km/h; 150 mph)
- Cruising speed at 75% power ... 116 kt (215 km/h; 134 mph)
- Stalling speed, flaps down ... 39 kt (72 km/h; 45 mph)
- Max rate of climb at S/L ... 318 m (1,043 ft)/min
- Service ceiling ... 4,265 m (14,000 ft)
- T-O run ... 180 m (590 ft)
- Landing run ... 125 m (410 ft)
- *g* limits ... +6/−3

UPDATED

Tecnam P2002 Sierra nearing completion in the UK *(Paul Jackson)* ***NEW***/0589977

VULCANAIR

VULCANAIR SPA

Capodochino East Plant Via G Pascoli 7, I-80026 Casoria (NA)
Tel: (+39 081) 591 81 11
Fax: (+39 081) 591 81 72
e-mail: infomarketing@vulcanair.com
Web: http://www.vulcanair.com
PRESIDENT: Carlo de Feo
MARKETING MANAGER: Gianni de Stefano
DIRECTOR OF SALES: Remo De Feo

US MARKETING OFFICE:
VulcanAir Inc
1101 30th Street, NW, Suite 50, Washington, DC 20007, USA
Tel: (+1 202) 625 43 47
Fax: (+1 202) 625 43 53
e-mail: sales@vulcanair-usa.com
Web: http://www.vulcanair.usa.com

Samanta aircraft service company purchased by Carlo de Feo in 1997; Samanta renamed as VulcanAir in April 1998 when it purchased bankrupt Partenavia company. VulcanAir returned P.68 series to production in 1999 and continues to supply spares for Partenavia aircraft. Acquired rights to SF.600 Canguro from SIAI-Marchetti on latter's incorporation into Aermacchi in 1997. In June 1999, VulcanAir announced development of the VA 300 twin-diesel-engined light transport derivative of the P.68 and single-engined VF 600W Mission, the latter a new aircraft with some commonality with the SF.600A and P.68 series.

Workforce 160 in 2004; 60,000 m² (645,834 sq ft) production facility at Naples-Capodochino Airport.

UPDATED

VULCANAIR P.68, OBSERVER AND P.68 DIESEL

TYPE: Light utility twin-prop transport; multisensor surveillance twin-prop.

PROGRAMME: Prototype (I-TWIN) first flew 25 May 1970. Production of P.68 Victor and P.68B by Partenavia in Italy started 1978, followed by P.68C in 1980. Partenavia manufacture ended in 1994, but assets subsequently acquired by VulcanAir; 10 assembled in India up to 1999 by TAAL. Improvements announced in 1999 include Garmin avionics.

VulcanAir P.68C company demonstrator *(Paul Jackson)* ***NEW***/0576557

Engine runs of the first P.68 Diesel ***NEW***/1034622

VulcanAir Observer version of P.68 ***NEW***/1034621

CURRENT VERSIONS: **P.68C:** Basic version, as described.

P.68C TC: As P.68C, but with turbocharged engines for better hot-and-high performance.

P.68 Diesel: As P.68C, but with Snecma-Renault diesel engines. Engine runs under way by February 2004.

P.68 Observer 2: For use by government and specialised services for patrol, surveillance and search; largely transparent nose section with lowered, compact instrument panel; 63 × 46 cm (2 ft 1 in × 1 ft 6 in) underfuselage hatch can carry variety of electro-optical sensors; slightly different equipment from other versions.

P.68 TC Observer TC: Turbocharged version of Observer 2 for hot/high performance.

CUSTOMERS: Total 401 P.68s of all versions built in Italy by end 1998; these comprised 12 P.68s, 196 P.68Bs, 91 P.68Cs, 47 P.68C TCs, 39 Observers, seven Observer 2s, three Observer TCs and six untraced. VulcanAir began production batch of 20 (mainly Observer 2s) in 1998, these all complete by end of 2003. Eight Observer 2s sold to Italian police; first delivery (PS-B07) 11 November 1999. Further four (including three Observer 2s) registered in USA 2000–02, excluding demonstrator; others to Germany (two), Spain (three) and Italy (two) in 2003. Total 24 Observer 2s built by late 2003.

COSTS: US$475,000 (P.68C); US$485,000 (P.68 Observer 2); US$505,000 (P.68C-TC); US$515,000 (P.68 TC Observer), all IFR equipped (2001).

DESIGN FEATURES: High wing with NACA 63-3515 aerofoil section and Hoerner tips; dihedral 1°, incidence 1° 3'.

FLYING CONTROLS: Manual. Pushrod and cable actuated, with all-moving tailplane and anti-balance tab; trim tab in rudder; electrically operated single-slotted flaps.

STRUCTURE: Light alloy stressed skin fuselage with frames and longerons; stressed skin two-spar torsion box wing; metal stressed skin tailplane and fin; fuselage/wing fairings mainly GFRP.

LANDING GEAR: Tricycle type; fixed. Fuselage-mounted spring steel cantilever mainwheel legs; oleo suspension for nosewheel, steered from rudder pedals. Mainwheels Cleveland 40–142 with Goodyear 7.00-6 (8 ply) tyres; nosewheel Cleveland 40-77B with Goodyear 5.00-5 (6 ply) tyre 6.00-6 (8 ply) tyre optional; Cleveland Type 30–61 foot actuated heavy duty hydraulic disc brakes; streamlined wheel fairings optional. P.68 Observer 2 has larger mainwheel tyres as standard. Minimum ground turning radius 5.70 m (18 ft 8 in).

POWER PLANT: *P.68C:* Two 149 kW (200 hp) Textron Lycoming IO-360-A1B6 flat-four engines, each driving a Hartzell HC-C2YK-2CUF two-blade constant-speed fully feathering propeller.

P.68C TC: Two 157 kW (210 hp) Textron Lycoming TIO-360-C1A6D; same propellers as P.68C.

P.68 Diesel: Two 171.5 kW (230 hp) Snecma-Renault SR305-230 turbocharged flat-four diesel engines.

Fuel capacity 696 litres (184 US gallons; 153 Imp gallons) in integral wing tank, of which 670 litres (177 US gallons; 147 Imp gallons) usable; overwing gravity refuelling. Oil capacity 7.5 litres (2.0 US gallons; 1.7 Imp gallons) for each engine.

ACCOMMODATION: One or two pilots and five or six passengers; cabin has two forward-facing seats in middle and three-seat rear bench; club seating optional; baggage door at rear and pilot door at front on starboard side; passenger door to port in centre cabin; escape hatch on starboard side; baggage compartment accessible from inside cabin. P.68 Observer 2 has no front starboard door for pilots. Optional aerophotogrammetric floor hatch, 0.63 × 0.46 m (2 ft 0¾ in × 1 ft 6 in) on P.68C.

SYSTEMS: Two 28 V 130 Ah alternators and one 24 V 24 Ah battery; Goodrich pneumatic de-icing boots and eletro-thermal propeller de-icing optional; air conditioning optional.

AVIONICS: Garmin and Bendix/King avionics package standard.

Comms: GMA 340 audio panel with marker beacon receiver and intercom; GTX 320A transponder; control wheel PTT switch; micophone with headset; speakers with microphone jacks; avionics master switch; ELT 910 automatic ELT, all standard. NAT AMS44TL audio panel optional.

Radar: Bendix/King RDR 2000 vertical profile weather radar optional.

Flight: Dual Garmin GNS 430 com/nav/GS/GPS; Bendix/King KFC 150 two-axis autopilot with CWS instant disconnect; KEA 130 encoding altimeter; dual KI 209A VOR/LOC/GS/GPS indicators; KI 256 ADI; KI 525A HSI; KCS 55 compass, all standard. Options include GNS 530 (exchange for GNS 430); Goodrich WX500 Stormscope; Goodrich Skywatch; Bendix/King KRA 10A radar altimeter; KR 87 digital ADF; KN-62A DME; KI 229 RMI, all standard.

Instrumentation: ASI; VSI; electric turn-and-bank indicator; Vision Microsystems LCD graphic/digital monitoring system; magnetic compass; DVR 300I digital voice recorder/clock; gyro pressure gauge; annunciator panel; engine hours meter, all standard. Optional duplicated co-pilot's instruments comprise sensitive altimeter, ASI, vacuum turn-and-bank indicator, VSI and electric or vacuum attitude gyro indicator.

Mission: Observer 2 can carry FLIR, ATAL video surveillance pod with data downlink and SLAR. Aerial cameras, thermal imager and video camera can be operated through floor hatch.

EQUIPMENT: Standard equipment includes heated pitot and stall warning systems, 24 V 2 A auxiliary power socket, pilot and co-pilot storm windows, polished wing-mounted taxi and landing lights, wingtip recognition and strobe lights, ice inspection light on port wing, tail recognition and strobe lights, tiedown rings and control surface locks. Leather seats and tinted windows optional.

DIMENSIONS, EXTERNAL (C: P.68C, D: P.68 Diesel, TC: P.68C-TC, O: P.68 Observer 2, TCO: P.68 Observer TC):

Wing span	12.00 m (39 ft 4½ in)
Wing chord, constant	1.55 m (5 ft 1 in)
Wing aspect ratio	7.7
Length overall: C, D, TC	9.55 m (31 ft 4 in)
O, TCO	9.43 m (30 ft 11¼ in)
Height overall	3.40 m (11 ft 1¾ in)
Tailplane span	3.90 m (12 ft 9½ in)
Wheel track	2.40 m (7 ft 10½ in)
Wheelbase	3.65 m (11 ft 11¾ in)
Propeller diameter: all versions	1.83 m (6 ft 0 in)
Propeller ground clearance	0.77 m (2 ft 6¼ in)
Passenger doors (C, TC, each side):	
Port, fwd: max width	0.54 m (1 ft 9¼ in)
Port, centre: max width	0.82 m (2 ft 8¼ in)

Observer version of VulcanAir P.68 *(Paul Jackson)* ***NEW***/0558336

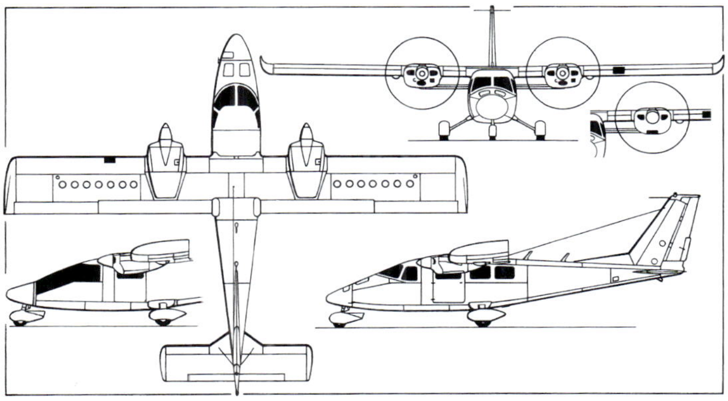

VulcanAir P.68C, with scrap view of Observer 2 nose and P.68 Diesel engine cowling *(Jane's/Mike Keep)* ***NEW***/0580553

Baggage door, stbd, aft:
  Max height ... 0.83 m (2 ft 8¾ in)
  Max width ... 0.80 m (2 ft 7½ in)

DIMENSIONS, INTERNAL:
 Cabin: Length ... 4.05 m (13 ft 3½ in)
  Max width ... 1.16 m (3 ft 9½ in)
  Max height ... 1.20 m (3 ft 11¼ in)
 Baggage compartment volume ... 0.56 m³ (19.8 cu ft)

AREAS:
 Wings, gross ... 18.60 m² (200.2 sq ft)
 Ailerons (total) ... 1.79 m² (19.27 sq ft)
 Trailing-edge flaps (total) ... 2.37 m² (25.51 sq ft)
 Fin ... 1.59 m² (17.11 sq ft)
 Rudder, incl tab ... 0.44 m² (4.74 sq ft)
 Tailplane, incl tab ... 4.41 m² (47.47 sq ft)

WEIGHTS AND LOADINGS:
 Weight empty equipped: C, O ... 1,320 kg (2,910 lb)
  TC, D, TCO ... 1,350 kg (2,976 lb)
 Baggage capacity ... 181 kg (400 lb)
 Max T-O weight ... 2,084 kg (4,594 lb)
 Max ramp weight ... 2,100 kg (4,630 lb)
 Max landing weight ... 1,980 kg (4,365 lb)
 Max zero-fuel weight ... 1,890 kg (4,167 lb)
 Max wing loading ... 112.0 kg/m² (22.94 lb/sq ft)
 Max power loading ... 6.99 kg/kW (11.49 lb/hp)

PERFORMANCE:
 Never-exceed speed (VNE):
  C, TC ... 193 kt (358 km/h; 222 mph)
  O ... 194 kt (359 km/h; 223 mph)
 Max level speed:
  D at S/L ... 159 kt (295 km/h; 183 mph)
  C, O, TC at S/L ... 173 kt (320 km/h; 199 mph)
  TCO at S/L ... 174 kt (322 km/h; 200 mph)
 Max cruising speed at 75% power:
  D at FL75 ... 154 kt (285 km/h; 177 mph)
  C, O at FL75 ... 165 kt (306 km/h; 190 mph)
  TC at FL120 ... 170 kt (315 km/h; 196 mph)
  TCO ... 165 kt (306 km/h; 190 mph)
 Stalling speed, power off:
  flaps up ... 68 kt (126 km/h; 79 mph)
  flaps down: O ... 57 kt (106 km/h; 66 mph)
  TC ... 61 kt (113 km/h; 71 mph)
  C, TCO ... 62 kt (115 km/h; 72 mph)
 Max rate of climb at S/L: C, O ... 378 m (1,240 ft)/min
  TC ... 465 m (1,525 ft)/min
  TCO ... 427 m (1,400 ft)/min
  D ... 530 m (1,740 ft)/min
 Max rate of climb, OEI: C, O ... 64 m (210 ft)/min
  TCO ... 73 m (240 ft)/min
  TC ... 88 m (290 ft)/min
  D ... 134 m (440 ft)/min
 Service ceiling: C ... 6,020 m (19,750 ft)
  O ... 6,000 m (19,685 ft)
  D, TC, TCO ... 6,100 m (20,000 ft)
 Service ceiling, OEI: C, O ... 1,753 m (5,750 ft)
  TC ... 4,420 m (14,500 ft)
  TCO ... 3,505 m (11,500 ft)
  D ... 3,660 m (12,000 ft)
 T-O run: C, O ... 240 m (790 ft)
  TC, TCO ... 230 m (755 ft)
 T-O to 15 m (50 ft): C, O ... 400 m (1,312 ft)
  TC ... 384 m (1,260 ft)
  TCO ... 415 m (1,362 ft)
 Landing from 15 m (50 ft): C, O ... 600 m (1,970 ft)
  TC ... 500 m (1,640 ft)
  TCO ... 485 m (1,591 ft)
 Landing run: all ... 200 m (656 ft)
 Range with max payload:
  TC ... 300 n miles (556 km; 345 miles)
  O with auxiliary fuel ... 590 n miles (1,093 km; 679 miles)
 Range with max fuel, best economy power:
  D at FL70 ... 1,513 n miles (2,802 km; 2,003 miles)
  C, O at FL70 ... 1,601 n miles (2,965 km; 1,842 miles)
  TCO at FL100 ... 1,376 n miles (2,550 km; 1,584 miles)
  TC at FL100 ... 1,104 n miles (2,046 km; 1,271 miles)
 Endurance with max fuel:, C, TC, O ... 11 h
  D ... 10 h
 g limits: C, O ... +3.74/−1.50
  TC, TCO ... +3.80/−1.52

UPDATED

VULCANAIR VA 300

TYPE: Light utility twin-prop transport.

PROGRAMME: Announced at Paris, June 1999, as a concept based on the VulcanAir AP.68TP Viator. Diesel engine test-flown on a tri-motor-configured P.68C in March 2001 and Phase 1 of the certification process has been completed; flying with the diesel engines running was due to start in late 2001, but the programme has been delayed pending delivery of engines and certification target date put back to 2005.

UPDATED

Retouched image, showing projected VulcanAir VA 300 0085593

VULCANAIR VF 600W MISSION

TYPE: Light utility turboprop.

PROGRAMME: Launched (as WF 600w) 1999; construction of first prototype began May 2001, at which time first flight planned for following September, but not achieved until 30 January 2003 (following roll-out on 31 May 2002 and first 'hop' on 4 December 2002); public debut

VulcanAir VF 600w Mission light utility turboprop *(Paul Jackson)* ***NEW***/0576506

(I-VAVF) at Paris Air Show 14 June 2003. Total of 88 flight hours recorded by June 2004, when the aircraft was grounded for modifications including a forward crew door and an aft passenger door on the starboard side, and an enlarged cargo door on the port side. Second prototype and one static test airframe then under construction, with maiden flight of second prototype schduled for October 2004, and EASA certification expected in March 2005, followed by FAA approval about two months later.

CUSTOMERS: Aimed at cargo/feeder market, law enforcement/security agencies, medevac surveillance, paratrooping and military logistics support. Order book scheduled to open in February 2005.

COSTS: Unit cost approximately US$1.2 million (2004). Estimated direct operating cost, based on utilisation of 1,000 hours per year, US$354.32 (2004).

DESIGN FEATURES: Single-engined derivative of Canguro. Nose-mounted engine; longer-span wings with new (modified NACA 63A3.515) aerofoil section; modified landing gear.

Following details summarise main differences from SF.600A.

LANDING GEAR: Tricycle type; fixed. Fuselage-mounted spring cantilever mainwheel legs; single mainwheels, with size 8.50–10 tyres, pressure 3.79 bar (55 lb/sq in); nosewheel tyre 6.50–8, pressure 5.17 bar (75 lb/sq in). Cleveland mainwheel disc brakes. Seaplane version being studied.

POWER PLANT: One 580 kW (778 shp) Walter M 601F-11 turboprop, driving an Avia V 510 five-blade constant-speed feathering and reversing metal propeller. Optional Pratt & Whitney Canada PT6A turboprop installation. Alternative power plants may be offered according to market requirements, Fuel capacity (two wing tanks plus reservoir) 1,338 litres (353 US gallons; 294 Imp gallons), of which 1,300 litres (343 US gallons; 286 Imp gallons) are usable.

ACCOMMODATION: Pilot and up to 10 passengers, or up to 15 passengers where local certification criteria permit. In cargo configuration the cabin can accommodate up to three standard Euro pallets. Quick-change cargo loading system with roller tracks, ball plates and latches optional. Pilot's door on port side; cargo/passenger sliding door on port side.

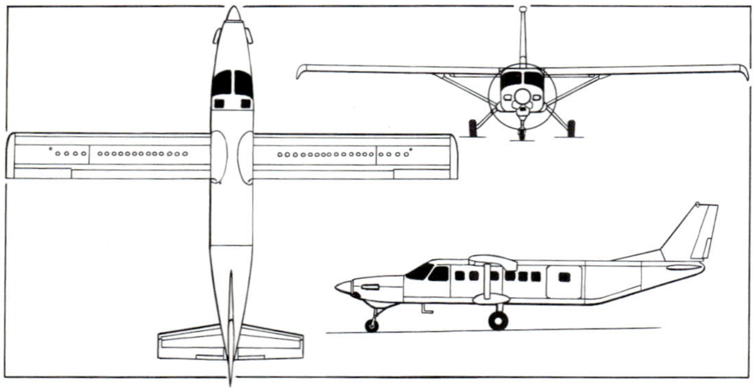

VulcanAir VF 600w Mission general arrangement *(James Goulding)* ***NEW***/0587526

VulcanAir VF 600w ***NEW***/1034625

AVIONICS: Bendix/King avionics suite standard, three-tube EFIS optional.

Comms: Bendix/King COM/NAV/GS/GPS, audio panel with marker beacon receiver and intercom, transponder, ELT 910.

Radar: INI 82A colour radar indicator.

Flight: KI 256 ADI, KI 229 RMI, KI 525A HSI, KN 62A DME, KR 87 ADF, KC 192 flight computer, KEA 130 encoding altimeter, VOR/LOC/GS indicator.

Instrumentation: Moritz engine instruments and annunciator panel, ASI, electric artificial horizon, sensitive altimeter, gyro pressure gauge with source indicator, rate of climb indicator, turn-and-bank indicator, flap position indicator, digital voice record clock.

DIMENSIONS, EXTERNAL:

Wing span	15.30 m (50 ft 2½ in)
Length overall	13.12 m (43 ft 0½ in)
Height overall	4.55 m (14 ft 11¼ in)
Tailplane span	5.00 m (16 ft 4¾ in)
Wheel track	3.50 m (11 ft 5¾ in)
Passenger door: Height	1.10 m (3 ft 7¼ in)
Width	0.60 m (1 ft 11½ in)
Cargo door: Height	1.40 m (4 ft 7 in)
Width	1.10 m (3 ft 7¼ in)
Service door: Height	1.40 m (4 ft 7 in)
Width	0.80 m (2 ft 7½ in)

DIMENSIONS, INTERNAL:

Cabin:	
Length, excl flight deck	5.08 m (16 ft 8 in)
Max width	1.23 m (4 ft 0¼ in)
Max height	1.17 m (3 ft 10 in)
Volume	7.3 m³ (258 cu ft)
Baggage compartment:	
Length	0.41 m (1 ft 4 in)
Max width	1.23 m (4 ft 0¼ in)
Max height	1.17 m (3 ft 10 in)
Volume	0.59 m³ (20.84 cu ft)

AREAS:

Wings, gross	24.80 m² (266.9 sq ft)

WEIGHTS AND LOADINGS:

Weight empty: passenger	2,000 kg (4,409 lb)
freighter	1,950 kg (4,299 lb)
Max payload	1,600 kg (3,527 lb)
Max fuel weight	1,040 kg (2,293 lb)
Max T-O weight	3,925 kg (8,653 lb)
Max ramp weight	3,946 kg (8,699 lb)
Max landing weight	3,720 kg (8,201 lb)
Max zero-fuel weight	3,600 kg (7,937 lb)
Max wing loading	158.27 kg/m² (32.42 lb/sq ft)
Max power loading	6.77 kg/kW (11.12 lb/shp)

PERFORMANCE:

Max level speed	190 kt (352 km/h; 219 mph)
Max cruising speed:	
at FL100	183 kt (339 km/h; 211 mph)
at FL200	166 kt (307 km/h; 191 mph)
Stalling speed, flaps down	61 kt (113 km/h; 71 mph)
Max rate of climb at S/L	274 m (900 ft)/min
Service ceiling	6,095 m (20,000 ft)
T-O run	433 m (1,420 ft)
FAA balanced field T-O length	616 m (2,020 ft)
FAA balanced field landing length	519 m (1,700 ft)
Landing run	244 m (800 ft)
Max range at FL200	1,190 n miles (2,204 km; 1,369 miles)
Endurance at FL200	6 h 48 min
g limits	+3.5/–1.4

UPDATED

Japan

FUJI

FUJI JUKOGYO KABUSHIKI KAISHA (FUJI HEAVY INDUSTRIES LTD)

Subaru Building, 7-2 1-chome, Nishi-Shinjuku, Shinjuku-ku, Tokyo 160-8316
Tel: (+81 3) 33 47 25 25
Fax: (+81 3) 33 47 25 88
e-mail: okak@sb.hq.subaru-fhi.co.jp
Web: http://www.fhi.co.jp
PRESIDENT: Takeshi Tanaka

UTSUNOMIYA MANUFACTURING DIVISION:
1-1-11 Yonan, Utsunomiya, Tochigi 320-8564
Tel: (+81 28) 684 70 55
Fax: (+81 28) 684 70 71
e-mail: nakagawan@ho.subaru-fhi.co.jp
GENERAL MANAGER: Youichi Sugimura
DEPUTY GENERAL MANAGER: Tsunenori Hoshi

AEROSPACE DIVISION:
SENIOR VICE-PRESIDENT AND GENERAL MANAGER: Hiroyuki Nakatsubo
VICE-PRESIDENT AND DEPUTY GENERAL MANAGER: Kisaburo Wani
DEPARTMENT GENERAL MANAGERS:
Kenichiro Usuki (Commercial Marketing and Sales)
Norihusa Matsuo (Defence Marketing and Sales)
Shunji Notake (Defence Market Development)
Takehiko Sarukawa (Administration)

Established 15 July 1953 as successor to Nakajima. Utsunomiya Manufacturing Division occupies 47.7 ha (117.9 acre) site, including 153,000 m² (1,646,880 sq ft) floor area; workforce 2,745 in April 2000.

Fuji producing Bell UH-1J (see *Jane's Aircraft Upgrades*); delivered last of 89 AH-1S attack helicopters (serial No. 73492) to JGSDF on 14 December 2000; now producing new T-7 trainer and will be prime contractor for JGSDF AH-64D; manufactured wings, tailplanes and canopies for Kawasaki T-4; produces tail units and canopies for Kawasaki OH-1; wings and tail units for Mitsubishi F-2; outer wings, nacelles and tail unit for ShinMaywa US-1A.

Commercial aircraft components produced are elevators for Boeing 737; spoilers, inboard and outboard ailerons for Boeing 747; outboard flaps for Boeing 757; wing/body fairings and main landing gear doors for Boeing 767 and 777, plus front centre wing box for 777; and complete wing sets for the Hawker Horizon business jet (which see). Company was selected in May 2000 as risk-sharing subcontractor to produce composites fuselages for Bell/Agusta BA609 tiltrotor, with deliveries to start in FY03; named in June 2002 as vertical tail source for Airbus A380; agreement in October 2003 with Eclipse Aviation to manufacture entire wing assembly for Eclipse 500. Other products include BQM-34AJKai (modified Firebee) and J/AQM-1 target drones and RPH-2 and FFOS unmanned helicopters (see *Jane's Unmanned Aerial Vehicles and Targets*).

Fuji has participated in such projects as design of the HOPE-X space shuttle and development of an NAL aero-spaceplane. Research continues towards an SST/HST (supersonic/hypersonic transport), including a thermal protection system, heat-resistant structures and composites materials.

UPDATED

FUJI T-7

TYPE: Basic turboprop trainer.

PROGRAMME: Modification (Fuji designation KM-2D) of JASDF piston-engined T-3, which it is intended to replace on an approximately one-for-one basis. Selected by JDA (in preference to Pilatus PC-7 Mk IIM) in third quarter 1998 to meet T-7 requirement; two prototypes included in FY99 budget request, but in late 1998 procurement deferred for one year. JDA reopened competition in August 1999, with bids in third quarter 2000; T-3Kai selected as basis for T-7, September 2000; first 11 T-7s approved in FY01 budget. A T-3Kai prototype was test flown and awarded JCAB type certificate. First two production aircraft (26-5901 and 26-5902; first flight July 2002) delivered to Air Development & Test Wing at Gifu in September 2002 and to training unit 12 Hikotai Kyolkudan at Hofu in April 2003; nine delivered by end of 2003, including seven in that year. At least 20 in service by end of 2004. Second training unit will be at Shizuhama.

CUSTOMERS: JASDF requirement for 47 T-7s over a 10-year period; 11 funded in FY01 defence budget; further 10 in FY02, nine in FY03 and 11 in FY04. To equip 11th and 12th Flying Training Wings.

Following details refer to T-3Kai.

DESIGN FEATURES: Modifications to engine cowling, wings and tail unit compared with T-3.

Fuji T-7 second production aircraft ***NEW***/1047797

Wing section NACA 23016.5 at root, NACA 23012 at tip; dihedral 6° from roots; sweepback 0° at quarter-chord.

FLYING CONTROLS: Conventional and manual. Plain ailerons, each with balance tab (port tab controllable for trim); controllable tab in each elevator; anti-balance tab in rudder. Single-slotted flaps.

STRUCTURE: Two-spar all-metal wing; all-metal semi-monocoque fuselage.

LANDING GEAR: Electrically retractable tricycle type.

POWER PLANT: One 336 kW (450 shp) Rolls-Royce 250-B17F turboprop; three-blade propeller. Two bladder-type fuel tanks in each wing, combined capacity 375 litres (99.0 US gallons; 82.4 Imp gallons).

ACCOMMODATION: Crew of two in tandem. Dual controls standard.

SYSTEMS: Include vapour cycle air conditioning system.

AVIONICS: Include VHF, UHF, ATC transponder, ICS and Tacan.

DIMENSIONS, EXTERNAL:
Wing span .... 10.04 m (32 ft 11¼ in)
Wing aspect ratio 6.1
Length overall .... 8.59 m (28 ft 2¼ in)
Height overall .... 2.96 m (9 ft 8½ in)
Tailplane span .... 3.71 m (12 ft 2 in)
Propeller diameter .... 2.12 m (6 ft 11½ in)

AREAS:
Wings, gross .... 16.50 m² (177.6 sq ft)
Fin .... 1.28 m² (13.78 sq ft)
Rudder, incl tab .... 0.66 m² (7.10 sq ft)
Tailplane .... 2.07 m² (22.28 sq ft)
Elevators (total, incl tabs) .... 1.39 m² (14.96 sq ft)

WEIGHTS AND LOADINGS:
Max T-O weight .... 1,585 kg (3,494 lb)
Max wing loading .... 96.1 kg/m² (19.67 lb/sq ft)
Max power loading .... 4.73 kg/kW (7.76 lb/shp)

PERFORMANCE:
Econ cruising speed at 915 m (3,000 ft) .... 161 kt (298 km/h; 185 mph)
Stalling speed, flaps and gear down .... 56 kt (104 km/h; 65 mph)
T-O to 15 m (50 ft) at S/L .... 608 m (1,995 ft)
Landing from 15 m (50 ft) at S/L .... 566 m (1,860 ft)

UPDATED

FUJI (BOEING) AH-64D

TYPE: Attack helicopter.

PROGRAMME: To replace existing JGSDF fleet of 80-plus AH-1Ss; original AH-X requirement said to be for up to 100, and government approval to acquire first 10 of selected design announced on 15 December 2000; these to be delivered by March 2006. RFP, issued 27 March 2001, resulted in submissions from Fuji/Boeing (AH-64D) and Mitsubishi/Bell (AH-1Z Viper), selection of former being announced on 27 August 2001. Selection of initially more expensive AH-64D said to be justified on grounds of better 20-year life-cycle costs, enabling smaller total number of aircraft to be procured; requirement accordingly now expected to be fulfilled by between 50 and 60 aircraft, of which only a proportion (possibly no more than one-third) to have Longbow radar.

Initial batch (possibly all of first 10) to be US-built, with Fuji as prime contractor for remainder; funding for first two included in FY02 budget allocations, followed by two more in each of FYs 03 and 04.

POWER PLANT: General Electric T700-GE-701C turboshafts confirmed.

SYSTEMS: Kanto (Smiths licence) electrical power generating system.

AVIONICS: To be decided.

ARMAMENT: To be decided.

UPDATED

JADC

JAPAN AIRCRAFT DEVELOPMENT CORPORATION

Daiichi Building, 1-2-3 Toranomon, Minato-ku, Tokyo 105-0001
Tel: (+81 3) 35 03 32 25
Fax: (+81 3) 35 04 03 68
Web: http://www.jadc.cr.jp
CHAIRMAN: Masamoto Tazaki
VICE-CHAIRMAN: Teiichi Nishikawa
SENIOR MANAGING DIRECTOR: Toshinori Nishi
MANAGING DIRECTORS:
Tsuneo Maede
Yoichi Sugimura

Known as CTDC (Civil Transport Development Corporation) from 30 March 1973 until 27 December 1982, JADC is a non-profit consortium established by airframe manufacturers Mitsubishi, Kawasaki and Fuji to promote commercial aircraft business with the support of the Japanese government; members now also include Nippi and ShinMaywa. Airframe production share (15 per cent of the Boeing 767 and 21 per cent of the 777) is managed by JADC's sister organisation, Commercial Aircraft Company (CAC). By end of 2002, CAC members had supplied 904 shipsets of parts for the 767 and 433 for 777.

Current research programmes include high-speed transport studies, advanced systems, innovative structures, next-generation avionics, advanced composites design and manufacturing and market research.

UPDATED

JADC YSX

TYPE: Regional jet airliner.

PROGRAMME: Feasibility studies intended for market slot between current regional jets and larger, single-aisle jets. JADC responsible for establishing configuration and structural parameters. Ministry of Economy, Trade and Industry set up working group (Aircraft Development Promotion Council) to consider possibility of linking YSX with Kawasaki C-X programme. No recent information.

UPDATED

KAC

KANEMATSU AEROSPACE CORPORATION

1-1-2 Shibaura, Minato-ku, Tokyo 105-8005
Tel: (+81 3) 55 40 87 05
Fax: (+81 3) 55 40 65 25
e-mail: pr@kanematsu.co.jp
Web: http://www.kanematsu.co.jp
PRESIDENT AND CEO: Tadashi Kurachi
SALES DIVISION MANAGER: Akiro Ohdoi

The Aerospace Department of Kanematsu's IT Division (workforce 785 at 31 March 2000) is prime contractor for outfitting Hawker 800s to JASDF specifications. Fuji is systems integrator and responsible for U-125/U-125A maintenance. Kanematsu is also Japanese distributor for Embraer ERJ-145.

UPDATED

KANEMATSU (HAWKER) 800

JASDF designations: U-125 and U-125A

TYPE: Multirole twin-jet.

PROGRAMME: Hawker 800 (which see) selected under JASDF H-X programme to replace Mitsubishi MU-2J and MU-2E in navaid calibration and SAR roles respectively; first U-125 delivered to JASDF 18 December 1992 and first U-125A (52-3003) delivered on 11 December 1994; three U-125As delivered by 31 January 1995 in preparation for formal handover. First six U-125As assembled in UK; seventh aircraft (from Wichita production line) delivered to Kanematsu November 1997; 21st arrived in Japan in January 2003 for fitting out.; 22nd in December 2003.

Hawker U-125A, displaying belly radar *(Mitsuhiro Kadota)* ***NEW***/0576559

CURRENT VERSIONS: **U-125:** For navaid flight check role, replacing MU-2J; operated by Flight Check Group at Iruma. Three ordered; all delivered from UK production.

U-125A: Search and rescue version, replacing MU-2E; 360° search radar, FLIR, airdroppable marker flares, liferaft and rescue equipment. Operated by detachments of the Air Rescue Wing at Chitose, Hyakuri, Komatsu and Naha, initially going to its Training Squadron at Komaki.

CUSTOMERS: JASDF (three U-125 and 24 U-125A ordered by FY04; options for total of 27 U-125As.

COSTS: ¥3.85 billion (2000).

DESIGN FEATURES: U-125A has deep observation 'patio' window each side of fuselage immediately ahead of wing, and dinghy/rescue pack dropping system via pressure door built into lower fuselage which is exposed for operation when landing gear is deployed.

AVIONICS (U-125A): *Radar:* Toshiba-built Raytheon 360° search radar.

Mission: Mitsubishi Electric IR imager in retractable underfuselage turret.

EQUIPMENT (U-125A): Flare and marker buoy dispenser; liferaft.

U-125/A PROCUREMENT AND DELIVERY (JASDF)
(at 1 January 2004)

Ordered			Delivered		
FY	**U-125**	**U-125A**	**CY**	**First Acft**	**Qty**
90	1		92	29-3041	1
91	1		93	39-3042	1
92	1		94	49-3043	1
92		3	95	52-3001	3
93		1	96	62-3004	1
94		1	97	72-3005	2
95		2	98	82-3007	3
96		3	99	92-3010	3
97		4	00	02-3013	3
98		3	01	12-3016	3
99		2	02	22-3019	2
00		2	03	32-3021	2
01		1	04	42-3022	1
02					
03		1			
04		1			
Totals	**3**	**24**			**26**

UPDATED

KAWASAKI

KAWASAKI JUKOGYO KABUSHIKI KAISHA (KAWASAKI HEAVY INDUSTRIES LTD)

Kobe Crystal Tower, 1-3 Higashi-Kawasaki-cho 1-chome, Chuo-ku, Kobe, Hyogo 650-8680
Tel: (+81 78) 371 95 30
Fax: (+81 78) 371 95 68
Web: http://www.khi.co.jp
AEROSPACE GROUP/GAS TURBINE & MACHINERY GROUP:
World Trade Center Building, 4-1 Hamamatsu-cho 2-chome, Minato-ku, Tokyo 105-6116
Tel: (+81 3) 34 35 21 11
Fax: (+81 3) 34 36 30 37
Web: http://www.khi.co.jp/aero/airplane.html

Its European name having no particular significance in Japan, the EC 145 is marketed locally with the original BK 117 designation ***NEW***/1047208

PRESIDENT: Chikashi Motoyama
MANAGING DIRECTOR AND PRESIDENT OF AEROSPACE COMPANY: Takashi Sugoh
MANAGING DIRECTOR AND PRESIDENT OF GAS TURBINE & MACHINERY GROUP: Takashi Yoshino
INFORMATION: Masayuki Hirata
WORKS: Gifu, Nagoya 1 and 2 (Aerospace Group); Akashi and Seishin (Gas Turbine & Machinery Group)

Company originated 1923 at Gifu when opened as Kakamigahara subplant for aircraft division of Kawasaki Dockyards. Present Gifu facility is on 712,000 m^2 (7.66 million sq ft) site with 291,000 m^2 (3.13 million sq ft) covered area. Nagoya 1 Works, opened December 1992, has site and covered areas of 71,000 and 10,000 m^2 (764,250 and 107,650 sq ft) respectively; corresponding figures for Nagoya 2 (opened 1979) are 18,000 and 6,000 m^2 (193,750 and 64,600 sq ft).

Kawasaki Aircraft Company has built many US aircraft under licence since 1955; amalgamated with Kawasaki Dockyard Company and Kawasaki Rolling Stock Manufacturing Company to form Kawasaki Heavy Industries Ltd 1 April 1969; Aerospace Group employs some 3,200 people; Gas Turbine & Machinery Group has some 800 aerospace-related employees. Kawasaki has had 25 per cent holding in Nippi since 1970; increased to 100 per cent with effect from 1 April 2003.

Kawasaki prime contractor on recently completed T-4 programme and for OH-1 observation helicopter; now also leading development of C-X/P-X transport/maritime patroller programme; co-developer and co-producer, with Eurocopter, of BK 117/EC 145 helicopter; those EC 145s produced in Japan continue to be marketed as the BK 117C-2. Is prime contractor for Japanese licensed production of CH-47 Chinooks for JGSDF and JASDF; was prime contractor for P-3C variants for JMSDF (now completed); also built MD Helicopters MD 500 under licence.

Subcontract work includes centre fuselage for Mitsubishi F-2; currently producing forward and centre-fuselage barrel sections and wing ribs for Boeing 767 and 777, plus rear centre wing box and rear pressure bulkhead for 777. Delivery of blended winglets for retrofit of early-model Boeing 737s (−300/400/500) began in October 2002. Acquisition of Nippi brings in additional component work for Boeing 747 (fuselage frames) and 777 (wing in-spar ribs and nosewheel doors); Mitsubishi F-2 (wing pylons); and ShinMaywa US-1AKai (main landing gear fairing). Kawasaki also responsible for design and sole-source manufacture of transmission for MD Helicopters Explorer; in 1999 signed agreement with Aerostructures of USA to manufacture cabin doors and tailcones for Bell/Agusta BA609. Risk-sharing partner on Embraer 170/175 and 190/195, for which wing components of former produced by Kawasaki from April 2001. Kawasaki Aeronáutica do Brasil established April 2002 and new Neiva-built facility 300 km (186 miles) north-west of São José dos Campos inaugurated 24 April 2003; first set of wings for 190/195 shipped to Embraer 10 October 2003. Nominated as prime contractor for maintenance and support of JASDF E-2C Hawkeyes, E-767 AWACS and C-130 Hercules. Selected on 5 June 2003 as prime contractor for 14 EH101 helicopters to be ordered by Japan (first aircraft ordered in FY03 and second in FY04; 11 to be JMSDF airborne mine countermeasures replacements for Sikorsky Sea Dragon, three as transports for use in Antarctica); licensed production and support agreement signed 22 September 2003. First four aircraft to be UK-built, Kawasaki then building Nos. 5 to 14; first delivery expected in 2007.

Kawasaki also extensively involved in satellites and launch vehicles; is member of International Aero Engines consortium and produces T53 and T55 engines under licence (see *Jane's Aero-Engines*); overhauls engines; and builds hangars, docks, passenger bridges and other airport equipment.

UPDATED

KAWASAKI C-X AND P-X

TYPE: Medium transport/multirole (C-X); maritime reconnaissance four-jet (P-X).

PROGRAMME: Initial development funding of ¥5.3 billion in FY01; RFP issued in May 2001. Foreign proposals for C-X included Airbus A310, Boeing C-17 and Lockheed Martin C-130J, but Kawasaki selected by JDA on 27 November 2001 to lead development of an indigenous design to meet both C-X and P-X requirements, with optimum degree of structural commonality; Fuji and Mitsubishi will be involved in programme, which could also lead to a 100- to 150-seat civil transport version for production post-2010. Two-year design phase begun in 2003, with prototype/preproduction development following from 2005; first flights targeted for 2006 (P-X) and 2007 (C-X); engineering and operational testing until 2009 for P-X, 2010 for C-X; service entry 2011 and 2012 respectively.

CURRENT VERSIONS: **C-X:** To fulfil long-standing JASDF requirement to replace current Kawasaki C-1A fleet with a transport of increased range and payload capacity. Desired features include a twin-turbofan power plant and a payload approximately double that of the Lockheed Martin C-130; earlier need for secondary aerial tanker capability no longer applicable. According to JDA's Technical Research and Development Institute (TRDI) in late 2000, design parameters include high-mounted wing, rear-loading ramp/door, digital AFCS and 'glass cockpit' avionics; outer wing, flight deck and tail unit to be common with those of JMSDF's P-X. Engine RFP issued by TRDI 16 December 2002; General Electric CF6-80C2 selected eight months later.

P-X: Outer wing, flight deck and tail unit as for C-X, but different fuselage cross-section, and wing to be low-mounted, carrying four (lower-powered) engines instead of two; AFCS to be fly-by-light as protection against electromagnetic interference (EMI).

CUSTOMERS: At least 44 C-Xs required by JASDF, to replace C-1As and C-130Hs; JMSDF needs about 80 P-Xs to replace Kawasaki (Lockheed Martin) P-3C fleet.

COSTS: ¥80 billion allocated to programme in FY03; further ¥86 billion (US$800 million) earmarked for FY04.

Following details are provisional:

POWER PLANT: Two 222 to 267 kN (50,000 to 60,000 lb st) turbofans in C-X; P-X would have four 49 to 67 kN (11,000 to 15,000 lb st) class turbofans. TRDI and IHI developing the XF7-10 53.4 kN (12,000 lb st) class engine for P-X; GE CF6-80C2L1F selection for C-X announced 12 August 2003, in preference to PW4000 and Trent 500; Goodrich nacelles.

AVIONICS: In process of selection; awaiting outcome in 2004 of joint MMA study with US Navy. CAE AN/ASQ-508(V) MAD chosen for P-X in February 2004.

DIMENSIONS, EXTERNAL:
Wing span .... 40 m (131.2 ft)
Length overall .... 40 m (131.2 ft)
Height overall .... 13 m (42.7 ft)

WEIGHTS AND LOADINGS:
Max payload: C-X .... 26,000 kg (57,320 lb)
Max T-O weight:
P-X .... 50,000-70,000 kg (110,230-154,325 lb)

PERFORMANCE (design):
Cruising speed:
C-X at FL426 .... 480 kt (889 km/h; 552 mph)
P-X at FL300 .... 450 kt (833 km/h; 518 mph)
Range: C-X with max payload .... 3,500 n miles (6,482 km; 4,027 miles)
P-X .... 4,300 n miles (7,963 km; 4,948 miles)

UPDATED

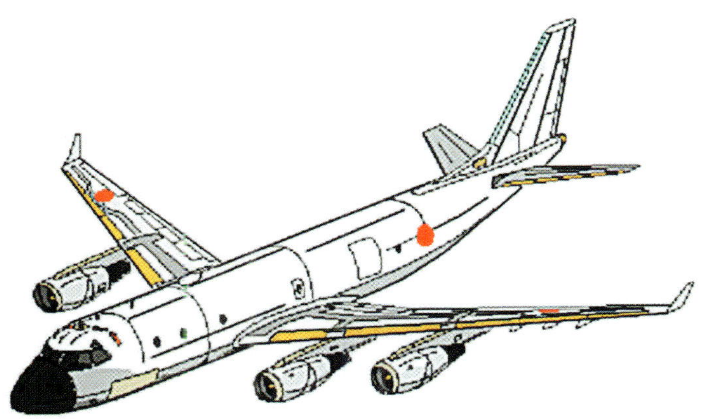

Artist's impression of P-X maritime patroller *(JMSDF)* ***NEW***/0577686

KAWASAKI T-4

PROGRAMME: Kawasaki named prime contractor 4 September 1981 by Japan Defence Agency; T-4 based on Kawasaki KA-851 design, by engineering team led by Kohki Isozaki; basic design studies completed October 1982; funding approved in FY83 and FY84 for four flying prototypes; prototype construction began April 1984; first flight of first XT-4 (56-5601) 29 July 1985; all four delivered between December 1985 and July 1986, preceded by static and fatigue test aircraft.

Production began FY86; first flight of production T-4, 28 June 1988; deliveries started 20 September 1988 to begin replacement of Lockheed T-33A and Fuji T-1A/B. Production completed March 2003; final aircraft (36-5812) rolled out 6 March 2003 and delivered to JASDF following day. Fuji and Mitsubishi each had 30 per cent share in programme.

UPDATED

KAWASAKI OH-1

TYPE: Armed observation helicopter.

PROGRAMME: Developed to replace OH-6Ds of JGSDF. Japan Defence Agency (JDA) awarded ¥2.7 billion (US$22.5 million) in FY92 to cover basic design phase of helicopter then provisionally designated OH-X; RFPs issued by JDA's Technical Research & Development Institute (TRDI) 17 April 1992; Kawasaki selected as prime contractor (60 per cent of programme) 18 September 1992, with Fuji and Mitsubishi (20 per cent each) as partners; Observation Helicopter Engineering Team (OHCET), formed by these three companies, began preliminary design phase 1 October 1992. Mockup made public 2 September 1994 under Japanese name *Kogata Kansoku* (new small observation [helicopter]).

Programme included six prototypes (four flying, two for ground test); first aircraft (32001) rolled out at Gifu on 15 March 1996 and made first flight 6 August 1996, followed by second prototype on 12 November; OH-1 designation assigned late 1996; first two XOH-1s handed over to JDA on 26 May and 6 June 1997; third flown on 9 January 1997, at which time earlier aircraft had accumulated some 30 and 20 hours, respectively; handed over 24 June 1997; fourth flown on 12 February 1997 and handed over 29 August 1997. Prototypes renumbered by 1999 from 32001-04 to 32601-04.

First three production OH-1s funded FY97 and ordered 1998; first prototype flown with more fuel-efficient TS1-10QT (replacing XTS1-10) engines, 30 March 1998. By early 1999, four prototypes had flown 450 hours and were due to complete further 450 hours by end of 1999, including operational evaluation at Akeno JGSDF base. First production OH-1 (32605) flown July 1999 and handed over to JGSDF at Gifu 24 January 2000.

Name 'Ninja' reportedly given in 2002, but not officially confirmed.

CURRENT VERSIONS: **OH-1:** Basic initial production version; *as described*.

Growth versions: Under study with more powerful engines (possibly LHTEC T800 or R-R/Turbomeca/MTU MTR 390) and uprated gearbox. **OH-1Kai** is possible candidate for AH-X requirement with tentative designation AH-2, armour-plated forward and centre fuselage, upgraded engines and transmission and additional weapons carriage.

CUSTOMERS: See table. Total of 24, including prototypes, ordered by FY04; at least 16 delivered by late 2003. Japan Ground Self-Defence Force requirement for 150 to 200.

Hiko Jikkentai (Flight Test Squadron) formed at Akeno with first four production aircraft on 27 March 2001; other deliveries by late 2002 included small numbers to Kasumigaura Bunko, Utsunomiya Bunko and Kyoiku Shien Hikotai (at Akeno), all of which are departments of the Army's Koku Gako (Aviation School). 1st Taisensha Herikoputatai (Anti-Tank Helicopter Squadron) at Tokashi began receiving OH-1s to replace OH-6Ds in 2003.

COSTS: Funding for development, prototypes and flight testing ¥2.5 billion in FY92, ¥10.2 billion in FY93, ¥50.1 billion in FY94 and ¥23.3 billion in FY95. Unit costs of first four production lots ¥1.924 billion (FY97), ¥2.018 billion (FY98), ¥2.229 billion (FY99) and ¥2.075 billion (FY00).

Kawasaki OH-1 armed observation helicopter *(Mitsuhiro Kadota)* ***NEW***/0576561

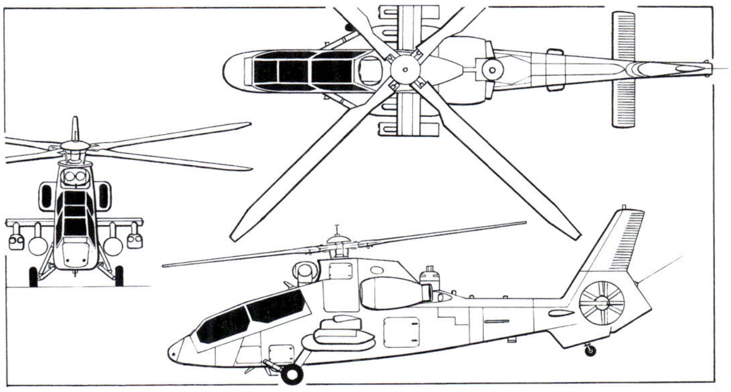

Kawasaki OH-1 (two Mitsubishi TS1 turboshafts) *(James Goulding)* 0093614

KAWASAKI OH-1 PROCUREMENT AND DELIVERY (JGSDF)

PROCUREMENT			DELIVERY			
FY	Qty	Cum total	FY	Qty	First aircraft	Cum total
93	2[1]	2				
95	2[1]	4				
97	3	7	97	4	32601	4
98	2	9				
99	3	12	99	3	32605	7
00	4	16				
01	2	18				
02	2	20				
03	2	22				
04	2	24				

[1] XOH-1 prototypes

DESIGN FEATURES: Kawasaki hingeless, bearingless and 20 mm ballistic-tolerant four-blade elastomeric main rotor and transmission system; Fenestron-type tail rotor with eight unevenly angled 'scissor' blades (35 and 55°); stub-wings for stores carriage. Active vibration damping system.

FLYING CONTROLS: Integrated AFCS and stability control augmentation system (SCAS).

STRUCTURE: Rotor blades and hub manufactured from GFRP composites; centre-fuselage and engines by Mitsubishi, tail unit/canopy/stub-wings/cowling by Fuji, rest by Kawasaki. Some 37 per cent of airframe (by weight) in GFRP/CFRP.

LANDING GEAR: Non-retractable tailwheel type. Provision for wheel/skis on main units.

POWER PLANT: Twin 659 kW (884 shp) FADEC-equipped Mitsubishi TS1-M-10 turboshafts (XTS1-10 originally in prototypes). Possibility of off-the-shelf alternative engines not ruled out. Transmission has 30-minute run-dry capability. Stub-wings can each carry a 235 litre (62.0 US gallon; 52.0 Imp gallon) auxiliary fuel tank.

ACCOMMODATION: Crew of two on tandem armoured seats (pilot in front). Flat-plate cockpit transparencies, upward-opening on starboard side for crew access.

AVIONICS: *Flight:* AFCS with stability augmentation and holding functions; dual HOCAS controls.

Instrumentation: Two Yokogawa Electric large, flat-panel, liquid crystal colour MFDs in each cockpit, linked to a MIL-STD-1553B databus; Shimadzu HUD in front cockpit.

Mission: Kawasaki electrically operated roof-mounted turret combining Fujitsu thermal imager, NEC real-time colour TV camera and NEC laser range-finder/designator; field of regard 110° in azimuth, 40° in elevation.

Self-defence: Spine-mounted IR jammer based on BAE Systems AN/ALQ-144.

ARMAMENT: Four Toshiba Type 91 (modified) lightweight, short-range, IR-guided AAMs on pylons under stub-wings for self-defence.

Following data all provisional:

DIMENSIONS, EXTERNAL:
- Main rotor diameter ... 11.6 m (38 ft 0¾ in)
- Main rotor blade chord, constant portion ... 0.38 m (1 ft 3 in)
- Wing span ... 3.0 m (9 ft 10 in)
- Fuselage: Length ... 12.0 m (39 ft 4½ in)
  - Max width ... 1.0 m (3 ft 3¼ in)
- Height: to top of rotor head ... 3.4 m (11 ft 1¾ in)
  - over tailfin ... 3.8 m (12 ft 5½ in)
- Tailplane span ... 3.0 m (9 ft 10 in)

AREAS:
- Main rotor disc ... 105.68 m^2 (1,137.5 sq ft)

WEIGHTS AND LOADINGS:
- Weight empty ... 2,450 kg (5,401 lb)
- Max weapons load ... 132 kg (291 lb)
- T-O weight: design ... 3,550 kg (7,826 lb)
  - max ... 4,000 kg (8,818 lb)
- Max disc loading ... 37.9 kg/m^2 (7.75 lb/sq ft)

PERFORMANCE:
- Max level speed ... 150 kt (277 km/h; 172 mph)
- Combat radius ... 108 n miles (200 km; 124 miles)
- Range ... 297 n miles (550 km; 342 miles)
- *g* limits ... +3.5/−1

UPDATED

KAWASAKI (BOEING) CH-47

JASDF/JGSDF designations: CH-47J and CH-47JA

TYPE: Medium lift helicopter.

PROGRAMME: FY84 defence budget approved purchase of three Boeing CH-47s, two for JGSDF and one for JASDF; first two built in USA and delivered second quarter 1986; Nos. 3 to 7 delivered as CKD kits for assembly in Japan; Kawasaki granted manufacturing licence for Japanese services' Chinooks; first CH-47Js delivered late 1986.

CURRENT VERSIONS: **CH-47J:** Generally similar to US CH-47D. Total 34 (from all sources) delivered to JGSDF by 1996, when CH-47JA introduced; procurement continued for JASDF.

Ground SDF CH-47JA built by Kawasaki *(Mitsuhiro Kadota)* ***NEW***/0576538

CH-47JA: Designation introduced to describe CH-47Js delivered to army aviation from 1996 with nose-mounted weather radar; first (52951) was 35th JGSDF Chinook, 50th Japanese Chinook; total 15 delivered by late 2002. More recent JASDF Chinooks reported as CH-47JA variant.

CUSTOMERS: See table. Total 71 (including US-built) delivered by late 2002. JGSDF had requirement (now exceeded) for 52, of which 53 in service by late 2002. Operated by 1st and 2nd Squadrons of 1st Helicopter Brigade at Kisarazu (CH-47J), 1st Composite Brigade on Okinawa (CH-47JA) and Aviation School at Akeno.

JASDF 18 CH-47J delivered by mid-2002. Operated by Air Rescue Wing flights at Iruma, Kasuga, Misawa and Naha (Okinawa).

COSTS: Unit cost ¥4.5 billion for CH-47J, ¥5.1 billion for CH-47JA (2000).

AVIONICS: *Flight:* GPS in JGSDF aircraft from 1993; Mitsubishi Precision INS.

KAWASAKI CH-47J/JA PROCUREMENT

FY	JASDF	First aircraft	JGSDF	First aircraft	Cum total
84	1	67-4471	2	52901	3
85	1	77-4472	3	52903	7
86	3	87-4473	4	52906	14
87	2	97-4476	4	52910	20
88	3	07-4478	5	52914	28
89	2	27-4481	5	52919	35
90	2	37-4483	5	52924	42
91	1	47-4485	3	52929	46
92			3	52932	49
93			2[2]	52951	51
94			2[2]	52953	53
95	1	87-4486	2[2]	52955	56
96			2[2]	52957	58
97			2[2]	52959	60
98			1[2]	52961	61
99	2	07-4487	2[2]	52962	65
00	1[2]	37-4489	2[2]	52964	68
01	1[2]		1[2]		70
02	2[2]		2[2]		74
03	4[2]		1[2]		79
04			2[2]		81
Totals	**26**		**55**		**81**

[1] First seven built in USA (two) or assembled from kits
[2] CH-47JA

UPDATED

MITSUBISHI

MITSUBISHI JUKOGYO KABUSHIKI KAISHA (MITSUBISHI HEAVY INDUSTRIES LTD)

Aerospace Administration Department, 5-1 Marunouchi 2-chome, Chiyoda-ku, Tokyo 100-8315
Tel: (+81 3) 32 12 31 11
Fax: (+81 3) 32 12 98 65
e-mail: QQ2100@hq.mhi.co.jp
Web: http://www.mhi.co.jp

NAGOYA AEROSPACE SYSTEMS WORKS:
10 Oye-cho, Minato-ku, Nagoya 455-8515
Tel: (+81 52) 611 21 21

PRESIDENT: Takashi Nishioka
EXECUTIVE VICE-PRESIDENT: Naochika Namba
GENERAL MANAGER, DEFENCE AIRCRAFT AND AERO-ENGINE DEPARTMENT: Kazuyuki Kato

Main Nagoya facility (previously known as Komaki North) was divided in 1989 into Aerospace Systems Works and Guidance & Propulsion Systems Works; former has 346,000 m^2 (3,724,300 sq ft) of floor space on a 67.3 ha (166.3 acre) site and had a January 2003 workforce of 3,902; latter occupies a 140.4 ha (346.9 acre) site with 84,530 m^2 (909,900 sq ft) of covered space and has approximately 1,700 employees. Oye plant manufactures aircraft fuselage components, spacecraft parts and other aero-related equipment; Tobishima plant responsible for aircraft fuselage subassembly, plus assembly and check-out of space systems; fuselage subassembly, final assembly, outfitting, flight test and repair undertaken by Komaki South plant.

Mitsubishi developed the MU-2 utility aircraft, Mu-300 business jet and F-2 support fighter, and is currently prime contractor for the F-2 and (Sikorsky) SH/UH-60 helicopter series; MH2000 helicopter restored to development flight test status in 2002. Studies under way in 2003–04 to evaluate potential successors to UH-60J for production start within 2005–2010 time frame. Previously, company was prime contractor for the T-2 supersonic trainer, F-4EJ Phantom II, F-1 close support aircraft and F-15J/DJ for the JASDF, with Fuji, Kawasaki, Nippi and ShinMaywa as principal subcontractors.

Preliminary design of 30-seat Mitsubishi airliner, shown in February 2004 ***NEW***/0576539

Mitsubishi F-2A attack fighter *(Mitsuhiro Kadota)* ***NEW***/0576540

In mid-2003, Mitsubishi received a contract from the Japanese Ministry of Economy, Trade and Industry (METI), expiring in March 2008, to design, build and flight test a technology demonstration prototype for a 30-passenger class regional jet.

Subcontract work includes participation in programmes for the Airbus A319/320 (shroud boxes since 2000), A330/340 (aft cargo doors from 2002), A380 (lower cargo doors: first delivered 2 October 2003); Boeing 737 (flaps); Boeing 747 (wing centre-section and flaps); Boeing 767/777 (fuselage panels and doors); Bombardier Global Express and Challenger 300 (wings); and components for the PW4000 turbofan engine. Negotiations to join Boeing 7E7 Dreamliner programme under way in 2003–04.

UPDATED

Prototype two-seat Mitsubishi F-2B *(Mitsuhiro Kadota)* ***NEW***/0576541

MITSUBISHI F-2

TYPE: Attack fighter.

PROGRAMME: Indigenous design, plus proposals based on F-15, F-16, F/A-18 and Tornado ADV, were originally considered; modified F-16C selected as Japan's FS-X replacement for Mitsubishi F-1 on 21 October 1987; Mitsubishi appointed prime contractor November 1988; initial contracts awarded for airframe design March 1989 and prototype active phased-array radar February 1990; General Electric F110-GE-129 Improved Performance Engine selected 21 December 1990. Programme delayed by questions of development sharing with General Dynamics (now Lockheed Martin Aeronautics Company (LMAC)) and technology transfer to Japan, but agreed at Japan 60 per cent and USA 40 per cent cost sharing (confirmed July 1996); first subcontracts to GD let February 1990 for design and development of rear fuselage, wing, leading-edge flaps, avionics and computer-based test equipment. Active phased-array radar, EW (ECM/ESM), mission computer and inertial reference system being developed using Japanese domestic technology.

Japan totally responsible for programme, including all funding; Japanese airframe subcontractors include Kawasaki and Fuji (see Structure). Programme involves four flying prototypes (two single-seat XF-2A first, then two tandem two-seat XF-2B) and two for static and fatigue test; construction began early 1994, final assembly mid-1994; first prototype (single-seat 63-0001) rolled out 12 January 1995; first flight 7 October 1995, followed by second prototype (63-0002) on 13 December 1995; third prototype (63-0003) flew on 18 April 1996 and fourth (63-0004; first in blue and grey colour scheme) on 24 May 1996. Japanese Cabinet approved 130-aircraft programme and allocated F-2 designation on 15 December 1995. Static test airframe (No. '991') under loading trials by 1995.

First XF-2A handed over to JDA on 22 March 1996, followed by '02 on 26 April 1997 and the two XF-2Bs on 9 August and 20 September 1997; prototypes re-serialled 63-8501/02/03/04 on 1 December 1997. Congressional approval of US participation in September 1996 followed in October by US$75 million Mitsubishi initial production contract to Lockheed Martin. First Lockheed Martin rear fuselage accepted by Mitsubishi in November 1998; first licence-built engines delivered by IHI in early 1999. Discovery in May 1998 of wing cracks, and evidence, when flying with two ASM-2 anti-ship missiles, of severe flutter, necessitated modifications to wingtips and pylon attachments, resulting in nine-month slippage of March 1999 scheduled completion date. Flight and static/fatigue testing rescheduled for completion in December 1999 and deliveries for FY99 (three) and FY00 (eight), but further wing crack problems revealed in mid-1999, causing completion of development testing to be extended to 30 June 2000 and first deliveries rescheduled to third quarter 2000. First production F-2A made maiden flight 12 October 1999 and delivered to JASDF on 25 September 2000. Total of 28 production F-2As and F-2Bs delivered by 31 March 2002; 36 by 31 March 2003.

Development of a dedicated air superiority version is under consideration, to replace the F-4EJKai.

CURRENT VERSIONS: **F-2A** Single-seat support fighter.

Description applies to F-2A except where indicated.

F-2B: Combat-capable two-seater.

CUSTOMERS: See table. Total of 69 production aircraft ordered by March 2003. JASDF (sole user) originally stated requirement for approximately 72 F-2As to replace F-1s in three support fighter squadrons, plus approximately 50 F-2Bs for OCU and possibly to replace T-2/2A. Current plans are to acquire 130 single/two-seaters (83 + 47, reduced from earlier figure of 141 by cancellation of 11 two-seaters).

First recipient originally due to begin conversion in FY99 and to be completely equipped by FY00; however, first delivery (03-8503) not made until 25 September 2000 (formal acceptance 3 October). Six F-2As and eight F-2Bs delivered by year-end: one F-2A temporarily to No. 1 Technical School at Hamamatsu and seven to Rinji F-2 Hikotai (temporary F-2 Squadron) at Misawa, which became 3 Squadron of 3 Wing at same base on 27 March 2001. Remainder of first 45 aircraft will be allocated to Matsushima OCU (19 Bs). In 2006, 6 Squadron of 8 Wing at Tsuiki will convert from F-1; followed in 2007 by 8 Squadron/3 Wing at Misawa (which, pending F-2s, has converted from F-1 to F-4EJKai Phantom).

COSTS: Total JDA expenditure (1988 to 1995) US$3.27 billion, including ¥75.7 billion (US$575 million) in FY92 to include first prototype and radar development; further ¥96.5 billion (US$804 million) in FY94 provided for three more flying prototypes and two for ground test. First two Mitsubishi contracts to GD totalled US$280.5 million; follow-on contract to Lockheed (5 February 1993) valued at US$74.2 million. FY96 batch of 11 cost US$1,081.58 million (US$98.32 million per unit); FY97 batch of eight cost US$797.47 million (US$99.68 million per unit); FY00 aircraft cost ¥11.8 billion each.

DESIGN FEATURES: Configuration based on Lockheed Martin F-16. New co-cured composites wing of Japanese design, with greater span, root chord and 25 per cent more area than that of F-16; tapered trailing-edge; slightly longer radome and forward fuselage to house new radar and other mission avionics; longer mid-fuselage and shorter jetpipes; increased-span tailplane; addition of brake-chute; adoption of increased performance engine.

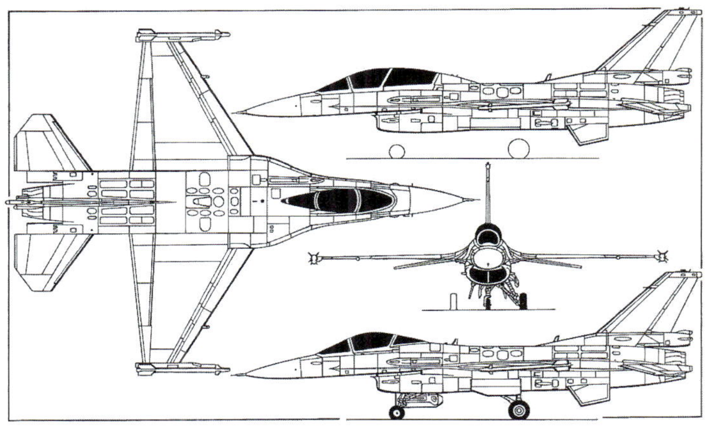

Mitsubishi F-2A, with additional side view (top) of two-seat F-2B *(Jane's/Mike Keep)* 0507619

Wing leading-edge sweepback 33° 12'; incidence 0°. Tailplane anhedral 8°.

FLYING CONTROLS: Full-span leading-edge flaps and trailing-edge flaperons; all-moving tailplane; rudder; twin, fixed ventral fins, canted outward 15°. Initially planned vertical canards deleted; CCV functions achieved instead by triple-redundant digital fly-by-wire system, developed jointly by Japan Aviation Electronics and Honeywell. Based on earlier Mitsubishi work with T-2 CCV testbed. Available modes include control augmentation, relaxed static stability, manoeuvre load control, decoupled yaw and manoeuvre enhancement. Single analogue back-up flight control system for roll and yaw control.

STRUCTURE: Composites structure wing (except leading-edge flaps), horizontal tail, fin (except leading-edge and base), rudder and landing gear doors; other structures also use advanced materials and structure technology, including Mitsubishi Rayon radar-absorbent material on nose, wing leading-edges and engine intakes; titanium in rear fuselage and tail unit.

Mitsubishi builds forward fuselage and wings; other Japanese airframe companies involved include Fuji (upper wing skins, wing fairings, radome, flaperons, engine air intakes and tail unit) and Kawasaki (fuselage mid-section, mainwheel doors and engine access doors). Lockheed Martin providing rear fuselage, 80 per cent of all port-side wing boxes, all leading-edge flaps, avionics systems, stores management systems and some test equipment; first LM production rear fuselage accepted 10 November 1998; first wing box delivered March 1999.

LANDING GEAR: Retractable tricycle type, with single wheel on each unit; mainwheels retract inward, nosewheel rearward. Mainwheel tyres size 27.75×8.75R145, pressure 22.06 bar (320 lb/sq in); nosewheel tyre size 18×5.7-8, pressure 20.69 bar (300 lb/sq in). Brake-chute in fairing at base of rudder.

POWER PLANT: One General Electric F110-GE-129 turbofan (131.2 kN; 29,500 lb st with afterburning), licence-built by IHI for production aircraft. Maximum internal fuel capacity 4,637 litres (1,225 US gallons; 1,020 Imp gallons), of which 4,588 litres (1,212 US gallons; 1,009 Imp gallons) are usable; reduced to 3,948 litres (1,043 US gallons; 868.5 Imp gallons), of which 3,903 litres (1,031 US gallons; 858.5 Imp gallons) usable, in F-2B. Maximum external fuel capacity (both) 5,678 litres (1,500 US gallons; 1,249 Imp gallons) (one 1,135.5 litre; 300 US gallon; 249.8 Imp gallon and two 2,271.25 litre; 600 US gallon; 499.6 Imp gallon tanks). No provision for in-flight refuelling.

SYSTEMS: Include onboard oxygen generation system (OBOGS).

AVIONICS: *Comms:* Magnavox AN/ARC-164 UHF transceiver; NEC V/UHF transceiver; Hazeltine AIFF; Kokusai Electric HF radio.

Radar: Mitsubishi Electric active phased-array radar.

Flight: Japan Aviation Electronics/Honeywell digital AFCS and laser IRS.

Instrumentation: Yokogawa 127 × 127 mm (5 × 5 in) colour LCD multifunction display and two 102 × 102 mm (4 × 4 in) liquid crystal colour MFDs; Shimadzu holographic wide-angle HUD.

Mission: Mitsubishi Electric mission computer. Sanders MPS III mission planning system.

Self-defence: Mitsubishi Electric integrated EW system.

MITSUBISHI F-2 PROCUREMENT AND DELIVERY (JASDF)

PROCUREMENT						DELIVERY						
FY	Lot	Qty		Cum total		CY	Qty		First aircraft		Cum total	
		F-2A	F-2B	F-2A	F-2B		F-2A	F-2B	F-2A	F-2B	F-2A	F-2B
92		1[1]	0	1								
93												
94		1[1]	2[1]	2	2							
95												
96	1	7	4	9	6	96	2	2	63-8501	63-8101	2	2
97	2	8	0	17	6	97						
98	3	2	7	19	13	98						
99	4	2	6	21	19	99						
00	5	0	9	21	28	00	6	4	03-8503	03-8103	8	6
01	6	8	4	29	32	01	13	nil	13-8508		21	6
02	7	6	2	35	34	02	nil	9		23-8107	21	15
03	8	6		41		03			33-8522	33-8116		
04	9	5		46								

[1] Prototypes; originally numbered 63-0001/02 (XF-2A) and 63-8501/02 (XF-2B)

ARMAMENT: One internal M61A1 Vulcan 20 mm multibarrel gun in port wingroot, plus 13 external stores stations: Sta 6 on centreline; Sta 1 (port) and 11 at each wingtip; and five under each wing (Sta 2, 3, 4L, 4, 5, 6, 8, 8R, 9 and 10); Flight Refuelling common rail launchers, built and installed by Nippi, configured initially for AIM-7F/M Sparrow medium-range air-to-air missiles (Sta 2 and 10); other armament expected to include AIM-9L or Mitsubishi AAM-3 air-to-air missiles (Sta 1, 2, 10 and 11) and ASM-1 and ASM-2 anti-shipping missiles (Sta 3, 4, 8 and 9); 500 lb bombs (Sta 4L and 8R); 340 kg bombs (Sta 4 and 8); CBU-87/B cluster bombs (Sta 4 and 8); and JLAU-3/A or RL-4 rocket launchers (Sta 4 and 8). Centreline and inboard underwing stations (5, 6 and 7) wet for carriage of drop tanks.

DIMENSIONS, EXTERNAL:
Wing span: over missile rails .... 11.125 m (36 ft 6 in)
excl missile rails .... 10.80 m (35 ft 5¼ in)
Wing chord: at root .... 5.27 m (17 ft 3½ in)
at tip .... 1.185 m (3 ft 10¾ in)
Wing aspect ratio 3.3
Length overall .... 15.52 m (50 ft 11 in)
Height overall .... 4.96 m (16 ft 3¼ in)
Tailplane span .... 6.045 m (19 ft 10 in)
Wheel track .... 2.36 m (7 ft 9 in)
Wheelbase .... 4.05 m (13 ft 3½ in)

AREAS:
Wings, gross .... 34.84 m^2 (375.0 sq ft)
Leading-edge flaps (total) .... 4.70 m^2 (50.59 sq ft)
Flaperons (total) .... 3.96 m^2 (42.63 sq ft)
Fin, incl dorsal fin .... 4.00 m^2 (43.06 sq ft)
Rudder .... 1.08 m^2 (11.63 sq ft)
Ventral fins (total) .... 1.50 m^2 (16.15 sq ft)
Horizontal tail surfaces (total) .... 7.05 m^2 (75.89 sq ft)

WEIGHTS AND LOADINGS:
Weight empty: F-2A .... 9,527 kg (21,003 lb)
F-2B .... 9,633 kg (21,237 lb)
T-O weight:
clean: F-2A .... 13,459 kg (29,672 lb)
F-2B .... 13,230 kg (29,167 lb)
with two short-range AAMs:
F-2A .... 13,713 kg (30,232 lb)
F-2B .... 13,412 kg (29,568 lb)
with four short-range and four medium-range AAMs:
F-2A .... 15,711 kg (34,637 lb)
F-2B .... 15,392 kg (33,934 lb)
with two short-range AAMs, six 500 lb bombs and two drop tanks: F-2A .... 19,888 kg (43,845 lb)
F-2B .... 19,569 kg (43,142 lb)
with two short-range AAMs, four anti-ship missiles and two drop tanks: F-2A .... 20,517 kg (45,232 lb)
F-2B .... 20,198 kg (44,529 lb)
Design max T-O weight with external stores:
F-2A, F-2B .... 22,100 kg (48,722 lb)
Design max landing weight:
F-2A, F-2B .... 18,300 kg (40,345 lb)
Design max wing loading .... 634.3 kg/m^2 (129.92 lb/sq ft)
Design max power loading .... 168 kg/kN (1.65 lb/lb st)

PERFORMANCE (F-2A):
Design max level speed: at high altitude .... approx M2.0
at low altitude .... approx M1.1
Typical combat radius:
F-2A .... more than 450 n miles (833 km; 518 miles)

UPDATED

MITSUBISHI (SIKORSKY) SH-60

JMSDF designations: SH-60J and SH-60K

TYPE: Naval combat helicopter.

PROGRAMME: Detail design of S-70B-3 version of Sikorsky SH-60B Seahawk (see US section), to meet JMSDF requirements, started August 1983; Japanese avionics and equipment integrated by Technical Research and Development Institute (TRDI) of JDA. First flight of first of two XSH-60J prototypes (8201), based on imported airframes, 31 August 1987; evaluation by 51st Air Development Squadron of JMSDF at Atsugi completed early 1991; first production SH-60J (8203) flown 10 May 1991, delivered 26 August.

CURRENT VERSIONS: **SH-60J:** Initial Japanese production version; *as described.*

Mitsubishi (Sikorsky) SH-60J of MSDF *(Mitsuhiro Kadota)* ***NEW***/0576545

SH-60K: Following trials begun by TRDI in 1995 on an all-composites rotor system with redesigned planforms and tips, offering 5 per cent increase in hovering efficiency and 30 per cent vibration reduction, FY97 budget requested funding for first prototype (then known as SH-60Kai), upgraded with new main rotor hub and blades, increased cabin cross-section (33 cm; 13 in longer and 15 cm; 6 in more headroom), active sonar, a tactical data processing and display system, and a laser-based shipboard automatic landing system. Prototype (8401) eventually rolled out 8 August 2001 and made its first hovering flight on 9 August.

CUSTOMERS: See table. JMSDF requirement for about 100 SH-60Js; 103 ordered, of which more than 80 in service, by 1 April 2002; half of force land-based. Three new aircraft approved in FY01 to offset delays in SH-60K programme; seven SH-60Ks approved in FY02. SH-60Js operated by Nos. 101, 121 and 123 Squadrons (at Tateyama); Nos. 122 and 124 (at Ohmura); No. 211 (training unit at Kanoya); squadron at Komatsushima undergoing conversion in 2001.

COSTS: Unit cost ¥5.07 billion (2000).

POWER PLANT: T700-401C engines manufactured by IHI.

ACCOMMODATION: Crew of three plus nine systems operators/observers.

AVIONICS (SH-60J): *Radar:* Japanese HPS-105 search radar.

Flight: Japanese automatic flight management system and ring laser gyro AHRS.

Instrumentation: Japanese controls and displays subsystem, datalink and tactical data processor.

Mission: Japanese HQS-104 sonar; Raytheon AN/ASQ-81D(V)4 MAD; Japanese HRR-1 sonobuoy receiver.

Self-defence: Japanese HLR-108C ESM. Some aircraft equipped with chaff/flare dispensers.

AVIONICS (SH-60K): Include Kanto (Smiths licence) enhanced flight data recorder.

EQUIPMENT: RAST, sonobuoys, rescue hoist and cargo sling.

ARMAMENT (SH-60J): Some being equipped with 7.62 mm machine guns for self-protection against terrorist attacks.

ARMAMENT (SH-60K): Anti-shipping missiles and gun.

Second SH-60K prototype *(Mitsuhiro Kadota)* ***NEW***/0576543

MITSUBISHI SH-60 PROCUREMENT (JMSDF)

SH-60J				SH60K		
FY	**Qty**	**First aircraft**	**Cum total**	**Qty**	**First aircraft**	**Cum total**
82	1[1]	8201	1			
83	1[1]	8202	2			
88	12	8203	14			
89	12	8215	26			
90	11	8227	37			
91	5	8238	42			
92	7	8243	49			
93	4	8250	53			
94	5	8254	58			
95	5	8259	64			
96	5	8265	70			
97	7	8271	77	1[2]	8401	1
98	7	8278	84	1[2]	8402	2
99	9	8285	93			
00	7	8294	100			
01	3	8301	103			
02				7	8403	9
03				7		16
04				7		23
Total	**103**		**103**	**23**		**23**

[1] Sikorsky-built XSH-60Js
[2] SH-60K prototypes

UPDATEDUPDATED

MITSUBISHI (SIKORSKY) UH-60

JASDF/JMSDF designation: UH-60J
JGSDF designation: UH-60JA

TYPE: Medium transport helicopter.

PROGRAMME: Detail design to Japanese requirements started April 1988; one US-built S-70A-12 imported, followed by two CKD kits for licence assembly (first flight 20 December 1989 at Sikorsky; delivered to JASDF 28 February 1991; second kit aircraft first flew February 1990, delivered to JASDF 29 March 1991). Remainder being built in Japan.

CURRENT VERSIONS: **UH-60J:** Standard search and rescue version for JASDF and JMSDF; can fly one hour search at 250 n miles (463 km; 288 miles) from base.

UH-60JA: JGSDF version; navigation/weather radar and FLIR night and adverse weather vision system. First (43101) completed late 1997; initial deliveries (four) January 1998.

Ground SDF Mitsubishi (Sikorsky) UH-60JA *(Mitsuhiro Kadota)* ***NEW***/0576547

JMSDF UH-60J rescue helicopter *(Mitsuhiro Kadota)* ***NEW***/0576548

CUSTOMERS: See table. JASDF has total requirement for 46 UH-60Js, JMSDF for 19; total of 26 in service with JASDF and 18 with JMSDF by 1 March 2003. JASDF aircraft operated by detached flights of Air Rescue Wing (HQ: Iruma); JMSDF UH-60Js to Atsugi Rescue Squadron in 1992; subsequently Kanoya, Tokushima, Shimofusa and Hachinohe.

JGSDF plans to procure up to 80 UH-60JAs in a US$2.67 billion programme announced early 1995; procurement began that year; JGSDF will deploy 50 with five district helicopter units alongside UH-1Js, with balance assigned to VIP transport and training; six delivered by 1 April 1999, this remaining as inventory total on 1 April 2000, despite more then awaiting acceptance. Total of 20 delivered by 1 March 2003. First two units are Nos. 1 and 12 Squadrons.

COSTS: Unit costs in FY00 were ¥3.37 billion for UH-60JA and ¥3.6 billion for air force UH-60J.

POWER PLANT: T700-401C turboshafts manufactured by IHI. External long-range fuel tanks.

ACCOMMODATION: Crew of four (JMSDF) or five (JASDF) plus up to 11 other persons. Bubble windows for pilot and on each side at front of main cabin.

AVIONICS: *Radar:* Nose-mounted Japanese search/weather radar.
Flight: Sikorsky self-contained navigation system.
Mission: Turret-mounted FLIR beneath nose.

EQUIPMENT: Rescue hoist, ETS and cargo sling.

MITSUBISHI UH-60 PROCUREMENT

FY	UH-60J (JASDF)	First aircraft	UH-60J (JMSDF)	First aircraft	UH-60JA (JGSDF)	First aircraft	Cum total
88	3[1]	18-4551					3
89	2	28-4554	3	8961			8
90	2	28-4556					10
91	4	38-4558	3	8964			17
92	2	58-4562	2	8967			21
93	1	68-4564	2	8969			24
94	2	68-4565	1	8971			27
95	2	78-4567	1	8972	2	43101	32
96	1	98-4569	2	8973	4	43103	39
97	3	08-4570	2	8975	4	43107	48
98	2	08-4573	2	8977	5	43111	57
99	2	18-4575			3	43116	62
00	2	28-4577			3	43119	67
01	2	48-4579	1	8979	2	43122	72
02	1				2		75
03	2				1		78
04	2				1		81
Totals	**35**		**19**		**27**		**81**

[1] One Sikorsky-built, two Mitsubishi-assembled from CKD kits

UPDATED

MITSUBISHI MH2000

TYPE: Light utility helicopter.

PROGRAMME: Launched in second half of 1995; applications include passenger and business transport, news gathering, law enforcement, search and rescue and emergency medical services. Four development aircraft (two flying, two for ground test). First flight 29 July 1996 (JQ 6003); second prototype (JQ 6004, later JA001M) flew late 1996. JCAB limited certification was awarded June 1997 and full VFR certification on 24 September 1999. First and second prototypes had flown approximately 800 hours (500 + 300) by April 1998. Initial production rate then planned for three per year; first production MH2000A (JQ 6005) was handed over to customer (Excel Air Service of Japan) on 1 October 1999. Four of first five (three customer aircraft and one demonstrator) recalled August 2000 when flaws discovered in metal engine covers; sixth MH2000 on production line at that time. Loss of one prototype due to tail rotor blade separation led to suspension of type certificate and redesign of tail rotor. Aircraft recertified with new rotor and resumed flight testing in 2002; Excel Air Service anticipated limited resumption of operations in 2003.

CURRENT VERSIONS: **MH2000:** Prototypes.

MH2000A: Production.

CUSTOMERS: Launch customer is Excel Air Service of Tokyo (one). Two others sold by October 1999, to National Aerospace Laboratory (delivered March 1999 as JA21ME) and a Japanese private customer. Anticipated (mainly Japanese) market for commuter, civil defence, media and cargo use. By July 2000, only three production aircraft had been registered in Japan. In 2003, Mitsubishi anticipating sales of 10 per year. Fifth production aircraft registered 21 April 2003.

DESIGN FEATURES: Single main rotor configuration (four blades with tapered tips); Fenestron-type tail rotor. Main gearbox and other drive mechanisms in overhead fairing are located aft of passenger cabin to minimise internal noise and vibration. Mid-mounted tailplane with angular endplate fins.

STRUCTURE: All-composites main and tail rotor blades.

LANDING GEAR: Conventional twin-skid type.

POWER PLANT: Two 653 kW (876 shp) Mitsubishi MG5-110 turboshafts, with digital electronic control permitting one-touch changeovers between high-speed and low-noise modes. Crash-resistant fuel tanks aft of passenger cabin, maximum usable capacity 1,132 litres (299 US gallons; 249 Imp gallons).

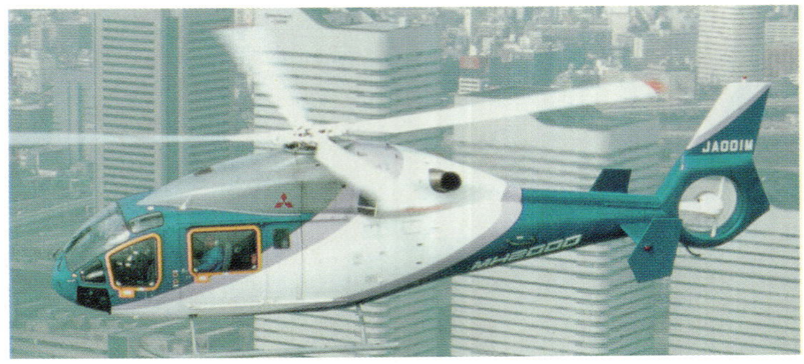

Second prototype Mitsubishi MH2000 twin-turboshaft commercial helicopter 0051108

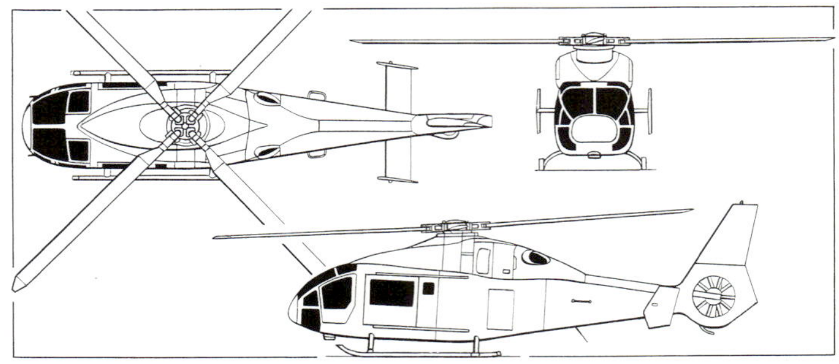

Mitsubishi MH2000A five/eight-passenger helicopter *(James Goulding)* 0062984

MH2000A instrument panel ***NEW***/0576551

MH2000A cabin interior ***NEW***/0576550

ACCOMMODATION: Flight crew of two. Main cabin has forward-facing, impact-absorbing seats for eight (standard) or six persons; five seats in club layout, with console, in VIP configuration; or two-person SAR team plus stretcher in search and rescue version. Crew door each side at front; large sliding door to main cabin on each side. Baggage compartment aft of fuel tanks, with external access.

AVIONICS: Avionics bay in rear fuselage aft of baggage compartment. Equipment to customer's choice; options to include GPS-based collision avoidance and AFCS.

DIMENSIONS, EXTERNAL:
- Main rotor diameter .... 12.20 m (40 ft 0¼ in)
- Length: overall, rotors turning .... 14.00 m (45 ft 11¼ in)
  - fuselage .... 12.20 m (40 ft 0¼ in)
- Height over vertical tail .... 3.80 m (12 ft 5½ in)
- Tail unit span .... 3.10 m (10 ft 2 in)
- Skid track .... 2.70 m (8 ft 10¼ in)
- Passenger sliding doors (each):
  - Width .... 1.20 m (3 ft 11¼ in)

DIMENSIONS, INTERNAL:
- Baggage compartment volume .... 2.24 m^3 (79 cu ft)

AREAS:
- Main rotor disc .... 116.90 m^2 (1,258.3 sq ft)

WEIGHTS AND LOADINGS:
- Weight empty .... 2,500 kg (5,512 lb)
- Max T-O weight .... 4,500 kg (9,920 lb)
- Max disc loading .... 38.5 kg/m^2 (7.88 lb/sq ft)

PERFORMANCE:
- Max level speed .... 151 kt (280 km/h; 174 mph)
- Cruising speed .... 135 kt (250 km/h; 155 mph)
- Hovering ceiling IGE .... 2,700 m (8,860 ft)
- Max range at cruising speed above, standard fuel,
  - no reserves .... 421 n miles (780 km; 484 miles)
- Max endurance, conditions as above .... 4 h

UPDATED

SHINMAYWA

SHINMAYWA KOGYO KABUSHIKI KAISHA (SHINMAYWA INDUSTRIES LTD)

1-1 Shinmeiwa-cho, Takarazuka, Hyogo 665-8550
Tel: (+81 798) 56 50 00
Fax: (+81 798) 56 50 01
e-mail: head@sa.shinmaywa.co.jp
Web: http://www.shinmaywa.co.jp
CHAIRMAN: Shiko Saikawa
PRESIDENT: Jushi Ide

AIRCRAFT DIVISION
1-1 Ohgi 1-chome, Higashinada-ku, Kobe 658-0027
Tel: (+81 78) 412 91 51
Fax: (+81 78) 431 36 95
EXECUTIVE OFFICER AND GENERAL MANAGER: Seiichi Yasumoto
WORKS: Konan and Tokushima

SALES OFFICES:
Business Development and Contract Head Office, Department No. 1:
3-2-43 Shitte, Tsurumi-ku, Yokohama-shi, Kanagawa 230-0003
Tel: (+81 45) 575 79 00
Fax: (+81 45) 575 79 01
DIRECTOR: Motoaki Nishimoto

Department No 2 (International Sales):
1-1 Ohgi 1-chome, Higashinada-ku, Kobe 658-0027
Tel: (+81 78) 412 91 51
Fax: (+81 78) 435 20 22
GENERAL MANAGER: Kazuya Makino

Former Kawanishi Aircraft Company (established 1928); became Shin Meiwa in 1949 and renamed ShinMaywa Industries in June 1992; major overhaul centre for Japanese and US military and commercial aircraft. Principal activities are production and/or upgrading of US-1A for JMSDF, and overhaul work on amphibians; upgraded US-1AKai made maiden flight in December 2003.

Manufactures external drop tanks for Mitsubishi F-2 and Kawasaki T-4; tailplanes for Mitsubishi-built SH-60J; internal cargo handling system for Kawasaki-built CH-47JA; fixed trailing-edges for Boeing 767, under subcontract to Vought; design and manufacture of wing/body fairings for Boeing 777. Support centre for U-36A (Learjet 36A); maintenance and technical support for JASDF U-4 (Gulfstream IV) since 1996.

Continues to study and look for partners to develop Amphibious Air Transport System, which is 30/50-passenger airliner powered by two wing-mounted turbofans with upper surface blowing; range would vary from 500 n miles (926 km; 575 miles) with full payload to 1,200 n miles (2,222 km; 1,381 miles) with full fuel; take-off distance 1,000 m (3,280 ft) on water and 800 m (2,624 ft) on soft ground; cruising speed between 300 and 360 kt (556 and 667 km/h; 345 and 414 mph).

UPDATED

SHINMAYWA US-1A

TYPE: Four-turboprop amphibian.

PROGRAMME: First flown 16 October 1974; first delivery (as US-1) 5 March 1975 (see 1985–86 *Jane's*); all now have T64-IHI-10J engines as US-1As.

CURRENT VERSIONS: **US-1A:** SAR amphibian, developed from PS-1 ASW flying-boat; manufacturer's designation **SS-2A.**

Data apply to US-1A except where indicated.

US-1AKai: Upgrade initiated in 1996: Rolls-Royce AE 2100J turboprops, six-blade Dowty propellers, modified wing with integral fuel tanks and composites floats, pressurised upper hull, Kawasaki FBW controls, 'glass cockpit', Thales Ocean Master radar, Honeywell RE-220 APU and LHTEC CTS800-based BLC compressor. Expected performance enhancements include maximum speed of 300 kt (555 km/h; 345 mph), service ceiling greater than 6,100 m (20,000 ft), maximum range of 2,500 n miles (4,630 km; 2,877 miles), and a reduction in take-off distance on water to 460 m (1,510 ft).

Parts manufacture for first aircraft began in April 2000; static and fatigue test airframes completed; former began testing in 2002, latter started in 2003; due to conclude in March 2005 and early 2006 respectively. First prototype (9901) rolled out 23 April 2003; first flight achieved on 18 December 2003; handed over to JMSDF's 51st Air Wing at Iwakuni 24 March 2004, for 27-month test programme. Second aircraft expected to fly in 2004. Service entry planned 2007, replacing US-1As, which will be retired.

US-2: Production designation of US-1AKai.

CUSTOMERS: Total of 20 US-1/1As ordered, 19 of which delivered by end of 2003. Orders since 1988 have been for attrition replacements and, more recently, due to phase-out of older aircraft. No. 71 SAR Squadron of the JMSDF maintains fleet structure of seven aircraft at Iwakuni and Atsugi bases.

JMSDF said to have requirement for 14 US-2s to replace existing US-1A fleet; JDA expected to request initial batch of seven subject to successful completion of flight test programme. First unit will be 71st Air Wing.

COSTS: US-1AKai development costs estimated at ¥66 billion (US$630 million) in 1996.

DESIGN FEATURES: Large turboprop-powered amphibian, suitable for maritime patrol and sea rescue missions. Boundary layer control system and extensive flaps for propeller slipstream deflection for very low landing and take-off speeds; low-speed control and stability enhanced by blowing rudder, flaps and elevators, and by use of automatic flight control system (see Flying Controls). Fuselage high length/beam ratio; V-shaped single-step planing bottom, with curved spray suppression strakes along sides of nose and spray suppressor slots in lower fuselage sides aft of inboard propeller line; double-deck interior. Large dorsal fin.

FLYING CONTROLS: Digital flight control system controlling elevators, rudder and outboard flaps. Hydraulically powered ailerons, elevators (with tabs) and rudder, all with feel trim. High-lift devices include outboard leading-edge slats over 17 per cent of wing span and large outer and inner blown trailing-edge flaps deflecting 60° and 80° respectively; outboard flaps can be linked with ailerons; inboard flaps, elevators and rudder blown by Mitsubishi BLC system. Two spoilers in front of outer flaps on each wing. Inverted slats on tailplane leading-edge.

STRUCTURE: All-metal; two-spar wing box. Wings built by Fuji, fuselage by ShinMaywa and floats by Showa (Mitsubishi, Kawasaki and ShinMaywa respectively for US-1AKai).

LANDING GEAR: Flying-boat hull, plus hydraulically retractable Sumitomo tricycle landing gear with twin wheels on all units. Steerable nose unit. Oleo-pneumatic shock-absorbers. Main units, which retract rearward into fairings on hull sides, have 40×14–22 tyres, pressure 7.79 bar (113 lb/sq in). Nosewheel tyres size 25×6.75–18, pressure 20.69 bar (300 lb/sq in). Three-rotor hydraulic disc brakes. No anti-skid units. Minimum ground turning radius 18.80 m (61 ft 8¼ in) towed, 21.20 m (69 ft 6¾ in) self-powered.

POWER PLANT: *US-1A:* Four 2,535 kW (3,400 shp) Ishikawajima-built General Electric T64-IHI-10J turboprops, each driving a Sumitomo-built Hamilton Sundstrand 63E60-27 three-blade constant-speed reversible-pitch propeller.

US-1AKai: Four 3,356 kW; (4,500 shp) Rolls-Royce AE 2100J turboprops, with FADEC and six-blade Dowty R414 propellers.

Fuel in five wing tanks, with total usable capacity of 11,640 litres (3,075 US gallons; 2,560.5 Imp gallons) and two fuselage tanks (10,849 litres; 2,866 US gallons; 2,386.5 Imp gallons); total usable capacity 22,489 litres (5,941 US gallons; 4,947 Imp gallons). Pressure refuelling point on port side, near bow hatch. Oil capacity 152 litres (40.2 US gallons; 33.4 Imp gallons). Aircraft can be refuelled on open sea, either from surface vessel or from another US-1A with detachable at-sea refuelling equipment.

ACCOMMODATION: Crew of three on flight deck (pilot, co-pilot and flight engineer), plus navigator, radio/radar operator and medical attendant's seats in main cabin. Latter can accommodate up to 20 seated survivors or 12 stretchers, one auxiliary seat and two observers' seats. Sliding rescue door on port side of fuselage, aft of wing.

SYSTEMS: Cabin air conditioning system. Two independent hydraulic systems, each 207 bar (3,000 lb/sq in). No. 1 system actuates ailerons, outboard flaps, spoilers, elevators, rudder and control surface feel; No. 2 system actuates ailerons, inboard and outboard flaps, wing leading-edge slats, elevators, rudder, landing gear extension/retraction and lock/unlock, nosewheel steering, mainwheel brakes and windscreen wipers. Emergency system, also of 207 bar (3,000 lb/sq in), driven by 24 V DC motor, for actuation of inboard flaps, landing gear extension/retraction and lock/unlock, and mainwheel brakes.

Honeywell 85-131J APU (RE-220 in US-1AKai) provides power for starting main engines and shaft power for 40 kVA emergency AC generator. Mitsubishi boundary layer control (BLC) system includes a C-2 compressor, driven by a 1,119 kW (1,500 shp) Ishikawajima-built General Electric T58-IHI-10-M2 gas turbine, housed in upper centre portion of fuselage, which delivers compressed air at 14 kg (30.9 lb)/s and pressure of 1.86 bar (27 lb/sq in) for ducting to inner and outer flaps, rudder and elevators (US-1AKai has LHTEC CTS800-4K turbine).

Electrical system includes 115/200 V three-phase 400 Hz constant-frequency AC and three transformer-rectifiers to provide 28 V DC. Two 40 kVA AC generators, driven by Nos. 2 and 3 main engines. Emergency 40 kVA AC generator driven by APU. DC emergency power from two 24 V 34 Ah Ni/Cd batteries.

De-icing of wing and tailplane leading-edges. Oxygen system for all crew and stretcher stations. Fire detection and extinguishing systems standard.

AVIONICS: *Comms:* HRC-121 HF, HRC-113 VHF and HRC-115-1 U/VHF radios; HIC-3B interphone; AN/APX-68-NB IFF transponder; RRC-22 emergency transmitter.

Radar: Thales OM-100 Ocean Master search radar.

Flight: HRN-101 ADF; OA-8697A/ARD UHF/DF; AN/ARN-118(V)2 Tacan; HRN-115-5 GPS nav system; HRN-107B-1 VOR/ILS receiver; AN/APN-171 (N2) radio altimeter; HPN-101B wave height meter; HSN-4 INS; N-TR-45 TAS transmitter; HRA-6-3 nav display; and N-ID-66/HRN BDHI.

EQUIPMENT: Marker launcher, 10 marine markers, six green markers, two droppable message cylinders, 10 float lights, pyrotechnic pistol, parachute flares, two flare storage boxes, binoculars, two rescue equipment kits, two droppable liferaft containers, rescue equipment launcher, lifeline pistol, lifeline, three lifebuoys, loudspeaker, hoist unit, rescue platform, lifeboat with outboard motor, camera, and 12 stretchers. Sea anchor in nose compartment. Stretchers can be replaced by troop seats.

DIMENSIONS, EXTERNAL:
- Wing span: 33.15 m (108 ft 9 in)
- Wing chord: at root: 5.00 m (16 ft 4¾ in)
  - at tip: 2.39 m (7 ft 10 in)
- Wing aspect ratio: 8.1
- Length overall: 33.46 m (109 ft 9¼ in)
- Height overall: 9.95 m (32 ft 7¾ in)
- Tailplane span: 12.36 m (40 ft 8½ in)
- Wheel track: 3.56 m (11 ft 8¼ in)
- Wheelbase: 8.33 m (27 ft 4 in)
- Propeller diameter: 4.42 m (14 ft 6 in)
- Rescue hatch (port side, rear fuselage):
  - Height: 1.58 m (5 ft 2¼ in)
  - Width: 1.46 m (4 ft 9½ in)

AREAS:
- Wings, gross: 135.82 m^2 (1,462.0 sq ft)
- Ailerons (total): 6.40 m^2 (68.90 sq ft)
- Inner flaps (total): 9.40 m^2 (101.18 sq ft)
- Outer flaps (total): 14.20 m^2 (152.85 sq ft)
- Leading-edge slats (total): 2.64 m^2 (28.42 sq ft)
- Spoilers (total): 2.10 m^2 (22.60 sq ft)
- Fin: 17.56 m^2 (189.0 sq ft)
- Dorsal fin: 6.32 m^2 (68.03 sq ft)
- Rudder: 7.01 m^2 (75.50 sq ft)
- Tailplane: 23.04 m^2 (248.00 sq ft)
- Elevators, incl tabs: 8.78 m^2 (94.50 sq ft)

WEIGHTS AND LOADINGS (search and rescue):
- Manufacturer's weight empty: 23,300 kg (51,367 lb)
- Weight empty, equipped: 25,630 kg (56,505 lb)
- Usable fuel: JP-4: 17,518 kg (38,620 lb)
  - JP-5: 18,397 kg (40,560 lb)
- Max oversea operating weight: 36,000 kg (79,365 lb)
- Max T-O weight: from water: 43,000 kg (94,800 lb)
  - from land: 45,000 kg (99,200 lb)
- Max wing loading: 331.3 kg/m^2 (67.85 lb/sq ft)
- Max power loading: 4.24 kg/kW (6.97 lb/shp)

Prototype ShinMaywa US-1AKai making its first flight on 18 December 2003 ***NEW***/0569771

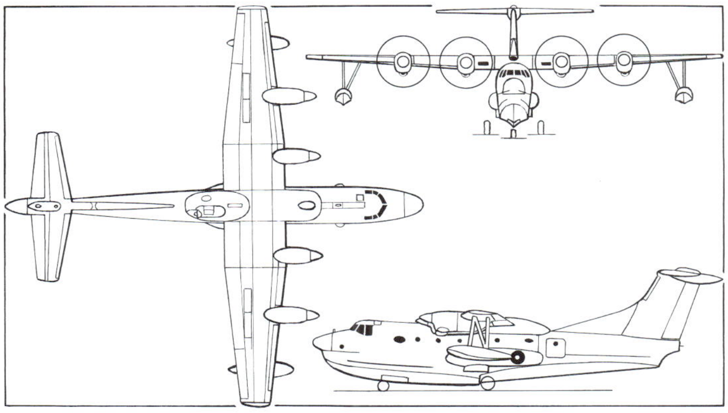

ShinMaywa US-2 (US-1AKai) SAR amphibian *(James Goulding)* ***NEW***/0583293

US-1/1A PROCUREMENT (JMSDF)

FY	Qty	Cum Total
72	1	1
73	2	3
77	1	4
78	2	6
79	1	7
80	1	8
82	1	9
83	1	10
84	1	11
88	1	12
92	1	13
93	1	14
95	1	15
96	1	16
97	1	17
99	1	18
01		18
02	1	19
04	1	20

PERFORMANCE (search and rescue, land T-O. A: at 36,000 kg; 79,365 lb weight, B: at 43,000 kg; 94,800 lb, C: at max T-O weight):

Max level speed: C ..... 276 kt (511 km/h; 318 mph)
Max level speed at FL100:
A ..... 282 kt (522 km/h; 325 mph)
Cruising speed at FL100:
C ..... 230 kt (426 km/h; 265 mph)
Max rate of climb at S/L: A ..... 713 m (2,340 ft)/min
C ..... 488 m (1,600 ft)/min
Service ceiling: A ..... 8,655 m (28,400 ft)
C ..... 7,195 m (23,600 ft)
T-O to 15 m (50 ft) on land, 30° flap, BLC on (ISA):
C ..... 655 m (2,150 ft)
T-O distance on water, 40° flap, BLC on (ISA):
B ..... 555 m (1,820 ft)
Landing from 15 m (50 ft) on land, AUW of 36,000 kg (79,365 lb), 40° flap, BLC on, with reverse pitch:
A ..... 810 m (2,655 ft)
Landing distance on water, AUW of 36,000 kg (79,365 lb), 60° flap, BLC on:
A ..... 220 m (720 ft)
Runway LCN requirement: B 42
Max range at 230 kt (426 km/h; 265 mph)
at FL100 ..... 2,060 n miles (3,815 km; 2,370 miles)

UPDATED

YANAGISAWA

GEN YANAGISAWA ENGINEERING SYSTEM CO LTD

5652-83 Sasaga, Matsumoto-shi, Nagano-ken 399-0033
Tel: (+81 263) 26 12 12
Fax: (+81 263) 26 13 13
e-mail: aviation@engineeringsystem.co.jp
Web: http://www.engineeringsystem.co.jp
PRESIDENT AND DESIGNER: Gennai Yanagisawa

US AGENT:
Ace Craft USA
Platte City, Montana
Web: http://www.acecraftusa.com

Mr Yanagisawa, whose nickname is 'Gen', developed a small, but comparatively high-powered, horizontally opposed two-cylinder engine while working for the Zenoa company. By 1985, this had been tested in a hang-glider, para-glider and model aircraft, but failed to find a market. Tests of a coaxial rotor RPV in 1987 were unsuccessful because of control problems, while a manned twin-engined design in 1992 failed to produce sufficient lift. By 1994, however, this had evolved into the BDH-1 (Boy's Dream Helicopter 1), exhibited at that year's Japan Aerospace Show. A three-engined BDH-2 followed in 1996 and was further modified as the BDH-3 which, although successful, was unable to sustain height with one engine inoperative. The BDH-4 was shown at AirVenture, Oshkosh, USA, in 1997 and continues to be refined. Manufacture is by ESCO, a general engineering company headed by Mr Yanagisawa, which has produced over 400 types of equipment since being founded in 1971. The assistance of a US agent was enlisted to collaborate on H-4 trials, such as ballistic parachutes and high-speed cruising, which are prohibited by Japanese air law. There were no fresh announcements made in 2003 and the aircraft did not appear at any major aviation trade shows that year.

UPDATED

Indoor trials of the Yanagisawa H-4 ***NEW***/0528746

YANAGISAWA GEN H-4

TYPE: Exoskelitor flying vehicle.

PROGRAMME: Four-engined development of BDH-3; basic version shown at AirVenture, Oshkosh, 1997; yaw control by small, electrically powered, auxiliary propellers; by 1998, yaw accomplished by differentiation of rotor rpm. Current configuration first flew in 1999. Marketed in kit form.

CURRENT VERSIONS: **H-4:** *As described.*

H-4 UAV: Radio-controlled version, first flown 22 March 2002.

COSTS: Kit US$30,900 (2004).

DESIGN FEATURES: 'Strap-on' personal helicopter with contrarotating twin blades of tapered planform and fixed pitch (thus no autorotation capability — replaced by ballistic parachute, if desired). Seat, with fuel tank at rear, supported by triangular tube framework landing gear and footrest. Gimballed platform above occupant's head houses four small motors, central transmission and rotor assembly with gears. Combined hand-grip control column and instrument panel attached to front of platform. Rotor speed 800 to 900 rpm. Quoted build time 40 hours.

FLYING CONTROLS: Two-axis control by tilting of engine/rotor platform; yaw control effected by thumb switch on instrument panel, operating electric motor which differentiates speed of rotors by 1 per cent.

STRUCTURE: Framework of 5 cm (2 in) aluminium tube; glass fibre seat and backpack; rotors of carbon fibre and Kevlar, with foam core.

LANDING GEAR: Four small 'wheels' each comprise multiple rollers in circular arrangement.

POWER PLANT: Four 124 cc (7.56 cu in) Gen 125 two-stroke, horizontally opposed, two-cylinder piston engines with independent ignition, positioned symmetrically on pivoting platform, each developing 7.5 kW (10 hp) at 8,500 rpm. Fuel, of 30:1 petrol and oil premix, in seatback tank, capacity up to maximum of 19 litres (5.0 US gallons; 4.2 Imp gallons).

DIMENSIONS, EXTERNAL:
Rotor diameter (each) ..... 4.00 m (13 ft 1½ in)

AREAS:
Rotor disc (each) ..... 12.57 m^2 (135.3 sq ft)

WEIGHTS AND LOADINGS:
Weight empty ..... 70 kg (154 lb)
Max T-O weight ..... 220 kg (485 lb)

PERFORMANCE:
Never-exceed speed (VNE) ..... 108 kt (200 km/h; 124 mph)
Cruising speed ..... 43 kt (80 km/h; 50 mph)
Service ceiling ..... 3,000 m (9,840 ft)
Endurance ..... up to 1 h 0 min

UPDATED

Jordan

KADDB

KING ABDULLAH II DESIGN AND DEVELOPMENT BUREAU

PO Box 928125, Amman 11190
Tel: (+962 6) 488 91 99
Fax: (+962 6) 488 96 99
e-mail (1): info@kaddb.com
e-mail (2): info@seabirdaviationjordan.com
Web (1): http://www.kaddb.com
Web (2): http://www.seabirdaviationjordan.com
DIRECTOR GENERAL: Maj Gen Khalid Jamokha
DIRECTOR, INTERNATIONAL AFFAIRS: Alec Mackenzie
COMMERCIAL OPERATIONS MANAGER, SEABIRD AVIATION JORDAN: Omar R Massarweh

Bureau established in August 1999 as basis of developing indigenous research and manufacturing industries to meet local and regional needs in defence and civil fields. Operates as a trading fund, with finance from defence budget and income from sales. Personnel total 200 (2003) at Amman and Commercial Operations Group at King Abdullah I Air Base, Marka.

Joint venture, known as Seabird Aviation Jordan and 51 per cent Australian-owned, formed in 2003 to assemble Seeker lightplane. KADDB has acquired intellectual property and world rights to manufacture and sales, leaving Seabird (Australia) responsible for further research and development, although components for Jordanian assembly continue to be supplied by Australia in the medium term. Initial operations at Marka Airport, but transferred to Queen Alia IAP in 2004. First deliveries - albeit from Australian production - in 2004.

Also in 2004, Seabird Jordan was negotiating to acquire Middle East assembly and support rights for Jabiru J400 four-seat ultralight/lightplane.

UPDATED

KADDB/SEABIRD JORDAN SB7L-360 SEEKER

TYPE: Multisensor surveillance lightplane.

PROGRAMME: Design (by Bill Whitney) began January 1985, construction January 1988; first flight of first (SB5N) prototype (VH-SBI, c/n 001, with Norton rotary engine) 1 October 1989, and of SB5E second prototype (VH-SBU, c/n 003, with Emdair engine) 11 January 1991 (c/n 002 was structural test airframe); fourth aircraft (c/n 004, VH-ZIG) was SB7L Seeker prototype, making first flight 6 June 1991; Seabird SB7L-235 certified by CASA 12 March 1993; definitive SB7L-360A (VH-OPT; first flight early 1993) has Textron Lycoming O-360 engine for higher performance and received Australian CAA type approval on 24 January 1994 and certification to FAR Pt 23 on 24 March 1999. Operational testing of Seeker included use as an airborne surveillance platform in October 1995 trials by the Australian Army. Components for an initial production batch of five aircraft completed by December 1997. Project then frozen and franchised producer sought, apparently being achieved in July 2000, on signature of agreement with Evektor-Aerotechnik. However, nothing further had been reported of this venture by mid-2003, although negotiations were said to be continuing.

Fifth aircraft (VH-SBO c/n 006) registered June 2003, being first to conform to production certification. On 17 August 2003, KADDB and Seabird revealed joint venture for assembly of Seekers at Marka. First aircraft, JY-SEE (the former VH-ZIG), shipped from Australia in September 2003 and flew in Jordan on 2 October; four to follow before end of that year, but no confirmation received that they had done so. Second arrival in Jordan, March 2004, was former VH-OPT. Output in 2004 set at 20 (none of which appeared to have been built), increasing to 60 in 2006, when new production premises to open at Queen Alia IAP.

CURRENT VERSIONS: **SB7L:** *As described.*

First aircraft of the reconstituted Iraq Air Force, Seabird Seeker YI-101 ***NEW***/1047385

SB9 Stormer: Projected light attack version with Rolls-Royce 250 turbine engine; tandem seats for two crew; chin turret below nose-mounted sensor package; pylon for gun pod, bombs or rocket launchers beneath each wing; maximum 250 kg (551 lb) on each pylon. MTOW 2,000 kg (4,409 lb), including 725 kg (1,598 lb) of sensors and weapons; range 550 n miles (1,018 km; 632 miles).

Sentinel: Proposed surveillance version of Seeker.

CUSTOMERS: Initial Jordanian order placed mid-2003 for two (2004 and 2005 deliveries) for Aqaba Special Economic Zone Authority in maritime surveillance role. Royal Jordanian Air Academy ordered two on 8 October 2003 and holds option on further pair for 2005 delivery. On 10 June 2004, Iraq Air Force purchased two Seekers from KADDB; requirement exists for fleet of 16 in surveillance role. Initial two aircraft, registered YI-101 and YI-102, both from previous Australian production, handed over 29 July 2004 and delivered from Jordan to Basrah, Iraq, by USAF C-130 Hercules on 3 August 2004, as equipment of 1 Air Reconnaissance Squadron.

Australian-built aircraft previously used by Country Energy (electricity distribution) and (from mid-2003) Oberon Air also in power- and pipeline-checking role.

COSTS: US$202,500 (2003).

DESIGN FEATURES: Braced high-wing monoplane with pod and boom fuselage, extensively glazed cabin and slightly sweptback vertical tail. Ventral and auxiliary fins added early 1993. Primary applications are observation/reconnaissance, offering helicopter-like view from cockpit and good low-speed handling and loiter capabilities; training and agricultural use also foreseen.

Wing section NACA 63_2-215 (modified); wedge tips; twist 3°, incidence 4° at root, 1° at tip; dihedral 2° 30' from roots.

FLYING CONTROLS: Conventional and manual. Rod-actuated slotted ailerons, one-piece horn-balanced elevator (with trim tab) and horn-balanced rudder; three-position (0, 20, 40°) slotted flaps on wing trailing-edge; fixed incidence tailplane. Inboard stall strips; vortex generators on wing upper surface near leading-edge and lower surface forward of ailerons; elevator bias trim.

Second Seabird SB7L Seeker in Jordan, JY-SEA ***NEW***/0563876

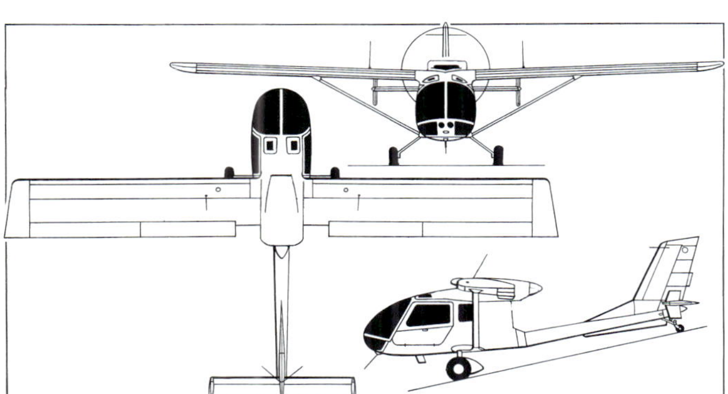

Seabird SB7L Seeker *(Jane's/Mike Keep)* 0093657

SEABIRD SEEKER PRODUCTION

c/n	Version	Registration
89001	SB5	VH-SBI. Withdrawn. Cancelled 6 Aug 04
002	SB5	static test airframe
90003	SB5	VH-SBU. Withdrawn. Cancelled 6 Aug 04
91004	SB7L	VH-ZIG, JY-SEE
92005	SB7L	VH-OPT, JY-SEA, YI-101
907006	SB7L	VH-SBO, YI-102

Notes: 007 to 011 in stock as components since 1996

STRUCTURE: Front fuselage mainly of 4130 chromoly steel tube with Kevlar non-load-bearing skin; aluminium alloy tubular tailboom; wings and tail unit conventional aluminium alloy stressed skin structures, former with single bracing strut and jury strut each side.

LANDING GEAR: Tailwheel type; fixed. Fuselage-mounted spring steel cantilever mainwheel legs with cutaway wheel fairings. Cleveland 8.00–6 mainwheels and fairings on cantilever spring steel legs; fully castoring Scott 8 in tailwheel with oil/nitrogen oleo strut and 210×65 McCreary tyre. Mainwheel tyre pressure 1.38 bar (20 lb/sq in); Cleveland disc brakes. Float gear to be developed.

POWER PLANT: One Textron Lycoming O-360-B2C flat-four engine, derated to 125 kW (168 hp) from 134 kW (180 hp) with lower compression ratio of 7.5:1 for Mogas fuel. Low propeller tip speed for minimal noise; Bishton BB177 two-blade fixed-pitch wood/composites pusher propeller. Fuel in two 92 litre (24.3 US gallon; 20.2 Imp gallon) integral wing tanks, each with overwing gravity filling point. Total 172 litres (45.4 US gallons; 37.8 Imp gallons) usable. Provision for auxiliary fuel tanks on underwing hardpoints. Oil capacity 7.5 litres (2.0 US gallons; 1.6 Imp gallons).

ACCOMMODATION: Side-by-side seats, adjustable fore and aft, for pilot and co-pilot or observer/passenger in extensively glazed cabin. Dual controls and four-point inertia reel seatbelts standard. Split (upward/downward-hinged) door each side, both removable. Ram air cabin ventilation. Space for 43 kg (95 lb) of baggage aft of seats.

SYSTEMS: 28 V electrical system, with 70 A alternator and Gill 246 battery, for engine start, instruments and lighting.

AVIONICS: *Comms:* Bendix/King KY 96A-61 VHF, KT 76A-13 transponder, AK 350 encoder and Seabird intercom standard. Optional KMA 24 audio panel and SL 70 transponder.

Flight: Optional Bendix/King KX 155 nav/com/glidescope, KI 208 VOR/LOC and KLN 90B GPS.

Mission: Potential surveillance equipment for Jordanian-built aircraft provided by Oberon Air Pty and FLIR Systems, including Ultra 7500 sensors and OFRIS geo-referencing system. Possible Broadband datalink.

EQUIPMENT: Hardpoint for 60 kg (132 lb) of external stores beneath each wing. Quick-change photo/survey modules, stretcher or 100 litre (26.4 US gallon; 22.0 Imp gallon) spraytank optional in place of right-hand seat.

DIMENSIONS, EXTERNAL:

Wing span ... 11.07 m (36 ft 4 in)
Wing chord, constant ... 1.22 m (4 ft 0 in)
Wing aspect ratio ... 9.4
Length overall ... 7.01 m (23 ft 0 in)
Fuselage max width ... 1.14 m (3 ft 9 in)
Height: to fintip ... 2.03 m (6 ft 8 in)
overall, propeller vertical ... 2.49 m (8 ft 2 in)
Tailplane span ... 2.90 m (9 ft 6 in)
Wheel track ... 2.03 m (6 ft 8 in)
Wheelbase ... 4.75 m (15 ft 7 in)
Propeller diameter ... 1.77 m (5 ft 9¾ in)
Cabin doors (two, each): Height ... 0.83 m (2 ft 8¾ in)
Width ... 0.98 m (3 ft 2½ in)
Height to sill ... 0.98 m (3 ft 2½ in)

DIMENSIONS, INTERNAL:

Cabin, incl baggage space:
Length ... 2.21 m (7 ft 3 in)
Max width ... 1.12 m (3 ft 8 in)
Max height ... 1.09 m (3 ft 7 in)
Baggage compartment volume ... 0.42 m^3 (15.0 cu ft)

AREAS:
Wings, gross .... 13.05 m² (140.5 sq ft)
WEIGHTS AND LOADINGS:
Basic weight empty .... 604 kg (1,332 lb)
Max fuel .... 115 kg (254 lb)
Baggage capacity .... 45 kg (99 lb)
Max T-O and landing weight .... 897 kg (1,977 lb)
Max wing loading .... 68.7 kg/m² (14.07 lb/sq ft)
Max power loading .... 7.16 kg/kW (11.77 lb/hp)
PERFORMANCE:
Never-exceed speed (VNE) .... 129 kt (238 km/h; 148 mph) CAS
Cruising speed at 75% power .... 112 kt (207 km/h; 129 mph)
Minimum patrol speed .... 65 kt (120 km/h; 75 mph) CAS
Stalling speed, flaps down .... 50 kt (93 km/h; 58 mph) IAS
Rate of climb: at S/L .... 288 m (944 ft)/min
at FL60 .... 173 m (567 ft)/min
Service ceiling .... 4,648 m (15,250 ft)
T-O run .... 265 m (870 ft)
Landing run .... 199 m (654 ft)
Range with reserves:
at min patrol speed .... 476 n miles (881 km; 547 miles)
at 65% power .... 470 n miles (870 km; 540 miles)
Endurance, with reserves: at min patrol speed .... 7 h 15 min
at 65% power .... 4 h 30 min

UPDATED

JAI

JORDAN AEROSPACE INDUSTRIES

Shmeisani, Sh Nasser bin Jameel Street, Building No 23 FL 1, PO Box 815570, Amman 11180
Tel: (+962 6) 556 05 11
Fax: (+962 6) 556 05 14
e-mail: info@jai.jo
Web: http://www.jai.jo
PRESIDENT AND CEO: Muayad al Samaraee
MARKETING DIRECTOR: Tamara Marsh

MANUFACTURING PLANT:
Queen Alia International Airport, Private Free Zone, Amman

Founded 2001 to provide aviation-related services to Middle East, Africa and Asia. By late 2003, had formed alliances with Remos and Germany, Lilienthal of Ukraine and SportCopter and Zenair of USA for local assembly of various light aircraft, as detailed. Awarded Production Certification No. 1 by Jordanian Civil Aviation Authority on 4 December 2003, initially for assembly of CH2000, first of which flew in Jordan on 14 December 2003. Had sold 25 aircraft (types not disclosed) by end of 2003.

The 6,200 m² (66,650 sq ft) Queen Alia plant is staffed by engineers trained and certified in Canada.

Associated ventures include flight training, research & development, aerial work (pipe/power-line monitoring, agricultural spraying), support of law enforcement, aerial tourism and spare parts supply.

Aircraft offered by JAI have been given local names, as summarised:

Barq S Copter: Single-seat autogyro. Version of Sport Copter Lightning. US manufacture by Sport Copter Inc (http://www.sportcopter.com). Rotax 503 UL engine; 37.0 kW; 49.6 hp. Weight 114 kg (252 lb) empty; 272 lb (600 lb) max T-O; max level speed 55 kt (101 km/h; 63 mph); max rate of climb 305 m (1,000 ft)/min; range 90 n miles (161 km; 100 miles).

Sport Copter 2 (Vortex 2S Copter) ***NEW***/0576337

JAI SAMA-CH2000 JY-MA8, first aircraft commercially built and flown in Jordan ***NEW***/0576332

Sport Copter Vortex (JAI Vortex 2 Copter) ***NEW***/0576335

Gulf Bird X-34: Three-seat, ultralight development of Lilienthal X-32.

Hawk-I CH801: Version of Zenith Super STOL CH 801 metal, four-seat lightplane. Production was scheduled to have begun in mid-2004.

RaLi: Version of Remos G-3 Mirage composites, two-seat ultralight.

Sama-CH2000: Version of AMD Zenith CH2T (CH 2000 Alarus) metal two-seat lightplane and primary trainer. Four ordered by Middle East Flight Academy at Amman for delivery from November 2003; demonstrator, US-built CH2000 N604AM, displayed at Dubai Air Show, 7 to 12 December 2003. First Jordanian-built, JY-MA8, flew 14 December 2003. Options include Sandel 3308 EFIS for instrument training.

RumBird X-32: Tandem-seat ultralight. Version of Lilienthal X-32 Bekas.

Sea Copter: Floatplane version of Vortex S Copter.

Vortex S Copter: Single-seat autogyro. Version of Sport Copter Vortex. Rotax 582 UL two-stroke, 48.0 kW (64.4 hp) but Subaru EA81 offered by JAI. Rotor diameter 7.62 m (25 ft 0 in); max T-O weight 345 kg (760 lb); max level speed 87 kt (161 km/h; 100 mph); max rate of climb 365 m (1,200 ft)/min; range 147 n miles (273 km; 170 miles). Optional enclosed cockpit with, or without, doors.

Vortex 2S Copter: Two-seat autogyro. Version of Sport Copter 2. Subaru EA81 flat-four engine, 82.0 kW (110 hp) or turbocharged 119 kW (160 hp). Weight 295 kg (650 lb) empty; 589 kg (1,300 lb) max T-O; max level speed 113 kt (209 km/h; 130 mph); max rate of climb 265 m (1,200 ft/min; range 269 n miles (498 km; 310 miles).

UPDATED

Kazakhstan

YAK ALACON

YAK ALACON JSC

ul Abay 76, Almaty 480057
Tel: (+7 3272) 55 74 00
Fax: (+7 3272) 63 12 07
e-mail: alacon@mailto.kz

This company was formed in conjunction with the Yakovlev design bureau of Russia to complete Russian certification of, market and manufacture the Yak-58. US marketing is supported by LMP Inc of Atlanta, Georgia, USA, the intention being to achieve FAA certification and assemble some aircraft in that country. Yak-58 production had previously been intended by Tbliaviamsheni (formerly TAM) at Tbilisi, Georgia, CIS, but the project was suspended in 1997.

VERIFIED

YAK ALACON YAK-58

TYPE: Six-seat utility transport.

PROGRAMME: Shown in model form at Moscow Air Show '90; full-scale mockup exhibited February 1991; prototype (RA-01003) initially reported to have flown at Tbilisi on 26 December 1993, but date later amended, without explanation, to 17 April 1994; it was lost in accident 27 May 1994; second prototype flew 10 October 1994; two more completed for trials at Zhukovsky test centre, plus two static test airframes, by January 1995; OKB reported receipt of 250 letters of intent to purchase; production by TASA at Tbilisi had begun by 1997 with initial batch of 20 for customers in Georgia, Kazakhstan and Uzbekistan. It is doubtful if many were completed before manufacture was suspended in 1997. One aircraft (4L-02010), fitted with Thomson-CSF radio surveillance equipment and operator's console, was exhibited at Paris Air Show in 1997.

Alacon plans to achieve certification of the Yak-58 by 2005, concentrating on M-14 engined version initially.

DESIGN FEATURES: Constant chord unswept wing with dihedral from roots and cambered tips; fuselage pod mounted above wing, with annular duct at rear to house air-cooled pusher engine; short twin booms carry sweptback and slightly toed-in tailfins and bridging horizontal tail surface. Small foreplane (67.5 cm; 2 ft 2½ in long) fitted to each side of cabin by 1995.

FLYING CONTROLS: Conventional and manual. Twin rudders and electrically actuated trailing-edge flaps.

LANDING GEAR: Pneumatically actuated retractable tricycle type; single wheel and low-pressure tyre on each trailing-link unit, for operation from unprepared strips; mainwheels retract inward, nosewheel forward; mainwheel tyre size 500×50, nosewheel tyre size 400×150.

POWER PLANT: One 265 kW (355 hp) VOKBM M-14PR nine-cylinder air-cooled radial engine enclosed in annular duct; three-blade variable-pitch pusher propeller. Location reduces noise in cabin.

Alternative engines are unspecified Teledyne Continental piston engine rated at 298 kW (400 hp) and unspecified RKBM TVD turboprop of 441 kW (592 shp).

ACCOMMODATION: Pilot and five passengers in pairs in enclosed cabin; large sliding door on starboard side, facilitating freight loading when passenger seats removed, or despatch of parachutists. Forward-hinged door on port side. Planned uses include business, taxi and ambulance transport; surveillance of forests, high-tension cables, oilfields and fisheries; mail and freight operation.

SYSTEMS: No hydraulic system. Two independent pneumatic systems for landing gear actuation and wheel brakes.

AVIONICS: Standard CIS equipment.

DIMENSIONS, EXTERNAL:
Wing span .... 12.70 m (41 ft 8 in)
Wing aspect ratio 8.1
Length overall .... 8.55 m (28 ft 0½ in)
Height overall .... 3.16 m (10 ft 4½ in)

AREAS:
Wings, gross .... 20.00 m^2 (215.3 sq ft)

WEIGHTS AND LOADINGS (power plants as specified):
Max payload: M-14, Continental .... 450 kg (992 lb)
TVD .... 500 kg (1,102 lb)
Max T-O weight: M-14 .... 2,100 kg (4,629 lb)
Continental .... 2,215 kg (4,883 lb)
TVD .... 2,250 kg (4,960 lb)
Max wing loading: M-14 .... 105.0 kg/m^2 (21.51 lb/sq ft)
Continental .... 110.8 kg/m^2 (22.68 lb/sq ft)
TVD .... 112.5 kg/m^2 (23.04 lb/sq ft)
Max power loading: M-14 .... 7.94 kg/kW (13.04 lb/hp)

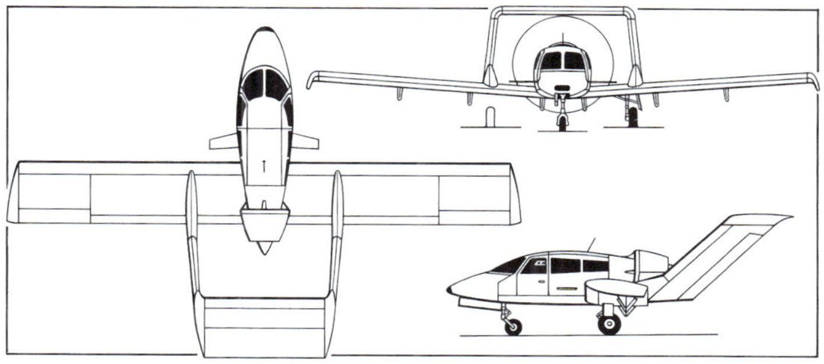

Yakovlev Yak-58 business aircraft (VOKBM M-14PR nine-cylinder piston engine) *(Mike Keep/Jane's)* ***NEW***/0015151

Yakovlev Yak-58 six-seat utility aircraft *(Paul Jackson)* 0103521

Continental .... 7.43 kg/kW (12.21 lb/hp)
TVD .... 5.10 kg/kW (8.38 lb/hp)

PERFORMANCE (estimated; power plants as specified):
Max level speed: M-14 .... 162 kt (300 km/h; 186 mph)
Continental .... 146 kt (270 km/h; 168 mph)
TVD .... 216 kt (400 km/h; 249 mph)
Normal cruising speed:
M-14 .... 154 kt (285 km/h; 177 mph)
Continental .... 118 kt (220 km/h; 137 mph)
TVD .... 189 kt (350 km/h; 217 mph)
Landing speed .... 67 kt (125 km/h; 78 mph)
Service ceiling: M-14, Continental .... 4,000 m (13,120 ft)
TVD .... 7,500 m (24,600 ft)
T-O run: M-14 .... 610 m (2,005 ft)
Continental .... 580 m (1,905 ft)
TVD .... 500 m (1,640 ft)
Landing run .... 600 m (1,970 ft)
Range:
M-14, Continental .... 540 n miles (1,000 km; 621 miles)
TVD .... 809 n miles (1,500 km; 932 miles)

UPDATED

Korea, South

KAI

KOREA AEROSPACE INDUSTRIES LTD

HEAD OFFICE: Seosomun-dong, Jung-gu, Seoul 100737
Tel: (+82 2) 20 01 31 14
Fax: (+82 2) 20 01 37 77
Web: http://www.koreaaero.com
PRESIDENT AND CEO: Kil, Hyoung-Bo
EXECUTIVE VICE-PRESIDENT: Ahn, Taek-Soon

CHANGWON PLANT: Seongju-dong, Changwon, Gyeongsangnam-do 641120
Tel: (+82 55) 280 61 14
Fax: (+82 55) 285 23 83

SACHEON PLANT 1: Yucheon-ri, Sanam-myeon, Sacheon, Gyeongsangnam-do 664942
Tel: (+82 55) 851 10 00
Fax: (+82 55) 851 10 05

SACHEON PLANT 2: Yongdang-ri, Sacheon-eup, Sacheon, Gyeongsangnam-do 664802
Tel: (+82 55) 851 29 99
Fax: (+82 55) 854 86 78

Korea Aerospace Industries was established on 1 October 1999 by consolidating the aerospace business of Daewoo Heavy Industries, Samsung Aerospace and Hyundai Space & Aircraft to create a single, unified aircraft manufacturer. In 2000, the Korean government designated KAI as the sole specialised aerospace defence industry in Korea, with exclusive and preferential rights in Korean aerospace defence programmes. Daewoo, Hyundai and Samsung each have a 28.1 per cent holding, with remaining 15.7 per cent held by creditors. KAI is headquartered in Seoul and operates three plants and three research facilities in the Sacheon and Changwon areas; total workforce was more than 3,200 in January 2003.

Business activities span both military and commercial sectors. Defence industry participation includes KT-1 basic trainer, T-50/A-50 advanced trainer, Korean Fighter Programme Phase 2, UAVs, Korean Multipurpose Helicopter (KMH) and Maritime Patrol Aircraft (MPA) programmes. In the commercial aircraft sector, KAI produces helicopters and exports airframe components.

Present and recent subcontract work includes A319/A320 wing top panels and A380 bottom panels for Airbus; wing rib assemblies for Boeing 747-400; stringers for Boeing 737/747/757; wing trailing-edges for Boeing 757/767; horizontal tail surfaces and trailing-edges for Bombardier de Havilland Dash 8; long-term contracts with Boeing, Airbus, Vought and Bombardier to produce and supply major aerostructures. MoU with Boeing in February 2003 for exclusive manufacture (2004 to 2010) of fuselages for AH-64 Apache attack helicopter; potential contract value US$350 million.

Signature of MoU between Daewoo and Korean Air Lines (KAL, which see) announced late August 2003, under which KAL would purchase former's stake in KAI for between 102 and 130 billion won (US$86 million to US$110 million) within next few months. However, MoU expired at year-end and negotiations still in stalemate in February 2004 with new discussions then scheduled for second quarter of that year. Negotiations with other potential investors (notably Hyundai, Samsung and EADS) continuing in early 2004.

UPDATED

KAI KT-1 WOONG-BEE

English name: Great Flight

TYPE: Basic turboprop trainer.

PROGRAMME: Started February 1988; built under KTX (Korean Trainer Experimental) programme to design by Daewoo and ADD (Agency for Defence Development); construction of first prototype began June 1991; nine prototypes (01-05 flying and 001-004 for static and fatigue test), of which 01, with 410 kW (550 shp) PT6A-25A engine, rolled out November 1991 and made first flight 12 December that year; 02, identical to 01, made first flight 5 February 1993; 03 (with PT6A-62) made first flight on 10 August 1995; named

KAI KT-1 turboprop trainer in RoKAF service ***NEW***/0576530

Woong-Bee in November 1995; 04, which first flew on 10 May 1996, further modified with nose shortened and horizontal tail surfaces remounted lower and farther aft; 03 also modified to this standard. Fifth (preproduction) prototype 05 flew for the first time on 16 March 1998; development and operational testing completed 18 September 1998 after 1,474 flying hours in 1,184 sorties. Series production began 1999; first aircraft handed over to RoKAF at Sachon AB on 7 November 2000 for use by 217 Tactical Training Squadron.

CURRENT VERSIONS: **KT-1:** Standard trainer,*as described*.

KT-1B: Export designation for Indonesian Air Force; some different avionics to customer's requirements and no Tacan or automatic rudder trim.

KT-1C: Proposed export version, with enhanced avionics suite integrated by CMC Electronics of Canada, based on that company's 25° field of view SparrowHawk HUD, HOTAS controls and a pair of CMC FV-4000 mission computers. 'All-glass' cockpit also features CMC up-front control panel and (in both cockpits) three 127 × 178 mm (5 × 7 in) MFDs. A cockpit demonstrator was displayed at the Seoul Air Show in late 2003. Other features include ability to carry a FLIR sensor and laser range-finder on the centreline stores station; weapon delivery system (simulated and live); fifth (underfuselage) hardpoint; a vapour-cycle ECS; and OBOGS.

KO-1: Forward air control (Korean Observer) and counter-insurgency version for RoKAF. Development started in 2000 by modifying aircraft 05 to FAC configuration asXKO-1 with weapon management system, HUD, armament management LCD MFD and four underwing hardpoints; first flight in this configuration 1 November 2001. Development delayed, but RoKAF given provisional approval in late 2003 to acquire 20; deliveries rescheduled to begin in late 2005 or early 2006.

CUSTOMERS: Total of 105 production aircraft approved by Republic of Korea Air Force (RoKAF). Contract 9 August 1999 for 85 KT-1 trainers, of which nearly all delivered by end of 2003, to replace Cessna T-37s. Follow-on option for 20 KO-1s for forward air control duties. Indonesian Air Force announced order for seven KT-1Bs on 26 February 2001, first six of which were delivered between 19 June and 26 July 2003; first aircraft (LD-0101) released to Indonesian Air Force 25 April 2003 in advance of official handover; IAF has option on further 13; to replace Beechcraft T-34Cs of No. 102 Squadron at Halim Perdanakusuma AB, Jakarta. Venezuelan order for up to 30 KO-1s in prospect in 2004.

COSTS: Reported unit costs of US$5 million for RoKAF, US$3.995 million for Indonesian Air Force (2003).

DESIGN FEATURES: Meets requirements of FAR/JAR 23 (Aerobatic Category) and MIL-F-8785C trainer category Class IV. Able to cover entire primary and basic training syllabus and forward air control missions.

Unswept low wing with NACA 63-218 (root) and 63-212 (tip) aerofoil sections; tandem cockpits; conventional unswept vertical and horizontal tail surfaces; four weapon hardpoints (four to six in KO-1); retractable tricycle landing gear.

FLYING CONTROLS: Primary control surfaces actuated mechanically, with electrically operated trim tab in starboard elevator and port aileron; automatic rudder trim (not on KT-1B) compensates for propeller-induced yaw to provide jet-like handling; rudder and one-piece elevator are horn-balanced. Airbrake under centre-fuselage and split flaps on wing trailing-edge are hydraulically operated.

LANDING GEAR: Hydraulically retractable tricycle type, with single wheel and oleo-pneumatic shock-absorber on each unit. Mainwheels retract inward; nosewheel is steerable ±18° and retracts rearward. Parker Hannifin mainwheels with 18×5.5 tyres (10 ply), pressure 9.65 bar (140 lb/sq in). Dunlop nosewheel with 5.00–5 tyre (14 ply), pressure 6.90 bar (100 lb/sq in). Parker Hannifin hydraulic mainwheel brakes.

POWER PLANT: One 708 kW (950 shp) (flat rated) Pratt & Whitney Canada PT6A-62 turboprop, driving a Hartzell HC-E4N-2/E9512CB-1 four-blade constant-speed fully feathering propeller. Single-lever combined control for engine fuel and propeller pitch.

Fuel in one 275.5 litre (72.8 US gallon; 60.6 Imp gallon) integral tank in each wing, giving total internal fuel capacity of 551 litres (146 US gallons; 121 Imp gallons). Gravity fuelling point in each wing upper surface. Provision for 189 litre (50.0 US gallon; 41.6 Imp gallon) auxiliary fuel tank on each inboard underwing station. Oil capacity 5.7 litres (1.5 US gallons; 1.25 Imp gallons). Fuel system permits up to 30 seconds of inverted flight.

ACCOMMODATION: Instructor and pupil in tandem cockpits on Martin-Baker Mk 16LF zero/zero ejection seats; rear seat elevated. One-piece canopy with MDC opens sideways to starboard.

SYSTEMS: Compressed bleed air air conditioning system for cockpit ventilation, heating, cooling and demisting; heating and cooling by wheel bootstrap air cycle system driven by engine compressor. Self-pressurised main and emergency hydraulic systems, operating pressure 207 bar (3,000 lb/sq in), flow rate 17.8 litres (4.7 US gallons; 3.9 Imp gallons)/min at 7,650 rpm. Pneumatic back-up system, also 207 bar (3,000 lb/sq in), for landing gear, wheel bay doors and flaps. Electrical power (28 V DC) available from engine-driven starter/generator and built-in battery, or from external power source. Fixed-geometry inertial separator for engine intake anti-icing. Diluter demand oxygen system provides 1,367 litres (48.2 cu ft) gaseous capacity for crew; selected and controlled individually from regulator/control panel in each cockpit.

KAI KT-1B Indonesian Air Force variant ***NEW***/0576526

Model of KAI KT-1C export version, showing additional centreline hardpoint ***(Paul Jackson)*** ***NEW***/0576524

Prototype XKO-1 forward air control version of KT-1 ***NEW***/0593867

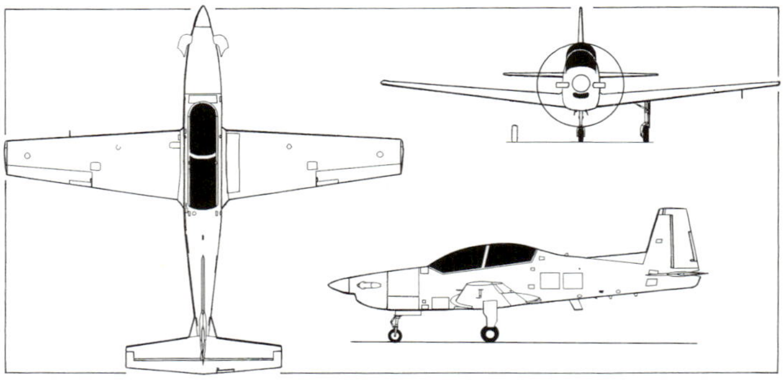

KAI KT-1 basic trainer (P&WC PT6A-62A turboprop engine) ***(Paul Jackson)*** 0062986

AVIONICS (KT-1): *Comms:* UHF/VHF radio, IFF, interphone and communication control system; optional ELT.

Flight: AHRS and Tacan; optional VOR/ILS.

Instrumentation: Smiths Aerospace integrated standby instrument system (ISIS) computes and displays attitude, altitude, airspeed, heading, Mach number and ILS within a single instrument using LCD technology.

AVIONICS (KT-1B): *Comms:* Commercial instead of MIL-STD UHF/VHF; ATC transponder replaces IFF; Tacan and flight load data recorder deleted.

AVIONICS (XKO-1): *Comms:* Integrated UHF/VHF radios, IFF, interphone and control system; optional ELT.

Flight: GPS/INS and Tacan; optional VOR/ILS.

Instrumentation: As KT-1, plus Elop HUD and Astronautics MFDs. Options include mission computer, AVTR and NVG-compatible cockpit lighting system.

ARMAMENT: Two stores stations under each wing. Four MBM Technology modular advanced lightweight training system (MALTS) practice bomb carriers on KT-1; XKO-1 equipped with underwing pylons for two 0.50 in machine gun pods (inboard); two LAU-3A or LAU-131A seven-round rocket launchers (outboard); two air-to-air training missiles (outboard); four Mk 81, Mk 82 or practice bombs (inboard/outboard); or two drop tanks (outboard). Provision for fuselage centreline station for laser range-finder, FLIR or chaff/flare dispenser pod.

DIMENSIONS, EXTERNAL:

Wing span	10.59 m (34 ft 9 in)
Wing chord: at root	2.06 m (6 ft 9 in)
at tip	0.97 m (3 ft 2 in)
Wing aspect ratio	7.0
Length overall	10.26 m (33 ft 8 in)
Height overall: KT-1	3.68 m (12 ft 1 in)
XKO-1	3.78 m (12 ft 5 in)
Tailplane span	4.17 m (13 ft 8 in)
Wheel track	3.54 m (11 ft 7 in)
Wheelbase	2.57 m (8 ft 5 in)
Propeller diameter	2.44 m (7 ft 11 in)
Propeller ground clearance	0.38 m (1 ft 3 in)

DIMENSIONS, INTERNAL:

Cockpits: Length (total)	3.20 m (10 ft 6 in)
Max width	0.99 m (3 ft 3 in)
Max height	1.52 m (5 ft 0 in)

AREAS:

Wings, gross	16.01 m^2 (172.3 sq ft)
Ailerons (total, incl tabs)	1.11 m^2 (11.95 sq ft)
Trailing-edge flaps (total)	2.22 m^2 (23.90 sq ft)
Fin, incl dorsal fin	1.35 m^2 (14.57 sq ft)
Rudder, incl tab	0.88 m^2 (9.43 sq ft)
Tailplane	2.32 m^2 (25.02 sq ft)
Elevators (total, incl tab)	1.19 m^2 (12.78 sq ft)

WEIGHTS AND LOADINGS:

Weight empty	1,910 kg (4,210 lb)
Max fuel weight	408 kg (900 lb)
Max T-O and landing weight:	
Aerobatic	2,540 kg (5,600 lb)
Utility	3,205 kg (7,065 lb)
with external stores	3,311 kg (7,300 lb)
Max wing loading:	
Aerobatic	158.7 kg/m^2 (32.50 lb/sq ft)
Utility	200.2 kg/m^2 (41.00 lb/sq ft)
with external stores	206.8 kg/m^2 (42.37 lb/sq ft)
Max power loading:	
Aerobatic	3.59 kg/kW (5.89 lb/shp)
Utility	4.53 kg/kW (7.44 lb/shp)
with external stores	4.68 kg/kW (7.68 lb/shp)

PERFORMANCE:

Never-exceed speed ... 350 kt (648 km/h; 402 mph)
Max level speed at 4,570 m (15,000 ft):
KT-1 ... 280 kt (518 km/h; 322 mph)
XKO-1 with external stores ... 244 kt (513 km/h; 280 mph)
Max operating speed ... 310 kt (574 km/h; 357 mph) IAS
Stalling speed, flaps and gear up ... 71 kt (132 km/h; 82 mph) IAS
Max rate of climb at S/L: clean ... 1,067 m (3,500 ft)/min
Service ceiling: clean ... 11,580 m (38,000 ft)
with external stores ... 9,140 m (30,000 ft)
T-O run ... 244 m (800 ft)
T-O to 15 m (50 ft) ... 494 m (1,620 ft)
Landing from 15 m (50 ft) ... 727 m (2,385 ft)
Landing run ... 397 m (1,300 ft)
Range at 7,620 m (25,000 ft), max internal fuel and 30 min reserves ... 900 n miles (1,666 km; 1,035 miles)
Ferry range at 6,100 m (20,000 ft) max internal and external fuel, 30 min reserves ... 1,118 n miles (2,070 km; 1,286 miles)
Endurance at 6,100 m (20,000 ft), max internal fuel and 30 min reserves ... up to 5 h
g limits: Aerobatic, clean ... +7/–3.5
with external stores ... +4.5/–2.3

UPDATED

KAI T-50 GOLDEN EAGLE

RoKAF designation (LIFT version): A-50

TYPE: Advanced jet trainer/light attack jet.

PROGRAMME: Begun by Samsung Aerospace (SSA) in 1992 under designation KTX-2 (Korean Trainer, Experimental); initial design assistance by Lockheed Martin as offset in F-16 Korean Fighter Programme; early design featured shoulder-mounted wings and twin tail unit; revised later to present configuration; basic configuration established mid-1995; full-scale development originally planned to begin in 1997, subject to finding risk-sharing partner; government go-ahead given on 3 July 1997; Samsung/Lockheed Martin agreement September 1997 to continue joint development until 2005; Lockheed Martin Aeronautics at Fort Worth responsible for wings, flight control system and avionics; development phase funded 70 per cent by South Korean government, 17 per cent by Samsung/KAI and 13 per cent by Lockheed Martin.

FSD contract, signed 24 October 1997, called for two static/fatigue test airframes and four flying prototypes (two T-50 and two A-50). Work split 55 per cent in USA, 44 per cent in South Korea and 1 per cent elsewhere. Preliminary design review (PDR) completed 12 to 16 July 1999; wind tunnel testing completed (4,800 hours) and aerodynamic design frozen November 1999. Critical design review (CDR) passed in August 2000. KTX-2 redesignated T-50 and A-50 in early 2000. T-50 International (TFI) established September 2000 (following July MoU) by KAI and LM Aero to market aircraft outside South Korea.

First prototype entered final assembly January 2001 and rolled out 31 October 2001; first flights 20 August (001) and 8 November 2002 (002). Static testing began 2 January 2002 and mostly completed by September 2003; fatigue tests began 22 July 2002 and completed first 'lifetime' of approx 8,330 hours in April 2003; second lifetime due for completion in June 2004. First supersonic flight (M1.05 at 12,200 m; 40,000 ft) 18 February 2003 on 60th sortie; 100th test flight 28 April 2003, on which date M1.2 achieved.

Developmental flight testing (105 flights) and initial integrated logistics assessment completed by first two prototypes by mid-2003; RoKAF initial operational assessment with these two aircraft completed in 26 sorties between 28 July and 14 August 2003. Follow-on operational assessment (approximately 64 flights) to continue until 2005. Third and fourth (lead-in fighter trainer) prototypes first flown on 29 August and 4 September 2003; ground-firing tests of gun started late October 2003; flight test of APG-67 radar began December 2003 (programme of approx 80 flights, ending in mid-2004). Other tasks scheduled during 2004 included engine air starts (begun March); high and maximum AoA;

Underview reveals T-50's appropriately avian paint scheme ***NEW***/0576862

stores carriage and weapon release; maximum speed and Mach number; maximum *g* loading; icing and other climatic trials; and continued avionics testing. By end 2003, prototypes had collectively flown 235 hours in 241 sorties; 300-hour landmark passed by February 2004.

Order for first 25 production T-50s placed 19 December 2003. Deliveries to RoKAF planned to begin in October 2005. Remaining 69 expected to be ordered in 2006 and delivered from 2008. In late 2003, T-50U variant (see below) offered to UAE; and Israeli MoD agreed to evaluate T-50 LIFT as potential TA-4 Skyhawk replacement.

Preliminary studies completed for possible 'F-50' fighter derivative, to enter development stage in about 2008 and targeted as Northrop F-5 replacement.

CURRENT VERSIONS (PLANNED): **T-50:** Advanced trainer; no internal gun or radar.

T-50 LIFT: Lead-in fighter trainer and light combat version with radar and internal gun. RoKAF designation **A-50**.

T-50U: Variant of T-50 LIFT proposed to United Arab Emirates in late 2003. Cockpits modified to include third, larger, MFD for commonality with UAE Air Force F-16s; HUD symbology similar to UAEAF Mirage 2000-9s; Thales TBA-6030 datalink; Terma EW suite controller; and customised mission simulation modes.

CUSTOMERS: Initial RoKAF requirement for 94 (50 T-50s and 44 A-50s), with options for up to 100 more, including further A-50s. Initially to replace RoKAF T-38s and F-5Bs; aimed also at F-5 replacement market. Israel and UAE invited to test-fly T-50 during 2004. Exports (from late 2006) estimated potentially at 600 to 800.

COSTS: Development programme cost estimated at US$2,000 million (1995); but reassessed in 1996 as US$1,500 million, and only US$1,200 million by early 1997. Initial October 1997 FSD contract valued at approximately US$1,270 million. Development phase re-estimated at US$1.8 billion to US$2.1 billion in mid-2000. Unit cost US$18 million to US$20 million for T-50, US$20 million to US$22 million for A-50 (2000).

DESIGN FEATURES: Mid-mounted, variable camber wings, swept back on leading-edges only; leading-edge root extensions (LERX); all-moving tailplane; sweptback fin leading-edge. Single turbofan engine, with twin side-mounted intakes. KAI developing fuselage and tail unit. Avionics include HUD and colour MFDs. Designed for service life of more than 10,000 hours.

FLYING CONTROLS: Digital fly-by-wire control of flaperons, leading-edge flaps, tailplane and rudder. Parker and Moog hydraulic actuators. Leading-edge manoeuvring flaps; split airbrake at rear of fuselage, between exhaust nozzle and tailplane halves.

STRUCTURE: Conventional aluminium alloy except for GFRP nose radome, control surfaces and some fairings. Lockheed Martin, avionics integration and wings for prototypes; KAI, fuselage, tail unit, production wings and final assembly at Sachon.

LANDING GEAR: Messier-Dowty hydraulically retractable tricycle type, with single wheel and oleo-pneumatic shock-absorber on each unit. Mainwheels retract into engine intake trunks, steerable nosewheel forward. Runway emergency arrester hook.

POWER PLANT: One General Electric F404-GE-102 turbofan (78.7 kN; 17,700 lb st with afterburning), equipped with FADEC. Seven internal fuel tanks (five in fuselage plus one in each wing); combined capacity 2,707 litres (715 US gallons; 595 Imp gallons). Provision for up to three 568 litre (150 US gallon; 125 Imp gallon) drop fuel tanks on centreline and/or inboard underwing stations.

ACCOMMODATION: Crew of two in tandem; stepped cockpits; Martin-Baker ejection seats. Rear-hinged, upward-opening single canopy; one-piece wraparound windscreen.

SYSTEMS: Dual-redundant, independent hydraulic systems (A and B), each at 214 bar (3,100 lb/sq in) pressure and with flow rate of 187 litres (49.4 US gallons; 41.1 Imp gallons)/min; System A has 207 bar (3,000 lb/sq in) emergency back-up system with 100 litres (26.4 US gallons; 22.0 Imp gallons)/min flow rate. Electrical system (30 to 40 kVA primary, 7 kVA for emergencies) powered by three 28 V, 100 A DC converters and one 500 W battery converter. Honeywell emergency power system (EPS) operates System A emergency back-up and 1 kVA emergency generator throughout flight envelope. Hamilton Sundstrand APU provides self-contained engine starting capability on ground and up to 6,100 m (20,000 ft). Digitally controlled Hamilton Sundstrand environmental control system for avionics cooling, capacity 5.6 kW. Litton self-generating oxygen life support system.

AVIONICS: *Comms:* Raytheon ARC-232 UHF/VHF radios; BAE Systems APX-118 IFF.

Radar: Lockheed Martin AN/APG-67(V)4 in A-50.

Flight: Digital fly-by-wire flight controls with HOTAS; Rockwell Collins VIR-130A VOR/ILS; nav/attack system for fighter lead-in training; Honeywell H-764G embedded GPS ring laser gyro INS; Tacan; HG-9550 radar altimeter.

Instrumentation: BAE Systems HUD; two Honeywell 127 × 127 mm (5 × 5 in) colour MFDs; Honeywell instrumentation displays (eight 76 mm; 3 in displays, including HSI, attitude indicator, electronic altimeter and Mach speed indicator).

Self-defence: A-50 provision for EW pods, chaff/flare dispenser and RWR.

Fourth (and second LIFT) prototype T-50/A-50 advanced trainer and light attack jet, which flew on 4 September 2003 ***NEW***/0562926

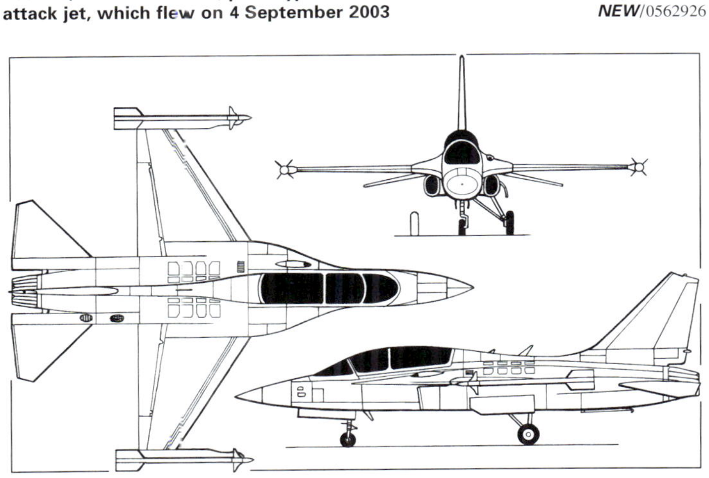

KAI T-50/A-50 Golden Eagle *(James Goulding)* 0589209

ARMAMENT: A-50 has internal General Dynamics 20 mm three-barrel Gatling-type cannon with 205 rounds (port LERX). Both versions have seven external stations (one on centreline, two under each wing and AAM rail at each wingtip) for AAMs, ASMs, rocket pods, bombs, munition dispensers, practice bombs or equipment, and training targets. Includes AIM-9 Sidewinder and TGM-65 Maverick training missiles, CBU-58 and Mk 20 cluster munition, ACMI (Air Combat Manoeuvring Instrumentation) pod, BDU-33 practice bombs and SUU-20 dispenser equipment, AGTS and TIX-3 training targets and Mk 82/83/84 series 500/1,000/2,000 lb bombs.

DIMENSIONS, EXTERNAL:
- Wing span: incl wingtip launcher .... 9.45 m (31 ft 0 in)
  - excl wingtip launcher .... 9.11 m (29 ft 10¾ in)
- Length overall .... 13.14 m (43 ft 1¼ in)
- Height overall .... 4.94 m (16 ft 2½ in)

WEIGHTS AND LOADINGS:
- Weight empty .... 6,350 kg (14,000 lb)
- Max internal fuel weight .... 2,220 kg (4,895 lb)
- Max external stores .... 4,309 kg (9,500 lb)
- Max T-O weight: clean .... 9,339 kg (20,589 lb)
  - with external stores .... 13,471 kg (29,700 lb)
- Max power loading: clean .... 119 kg/kN (1.16 lb/lb st)
  - with external stores .... 171 kg/kN (1.68 lb/lb st)

PERFORMANCE (design):
- Max level speed M1.5
- Stalling speed .... 105 kt (195 km/h; 121 mph)
- Max rate of climb at S/L .... 12,070 m (39,600 ft)/min
- Operational ceiling .... 12,200 m (40,000 ft)
- Service ceiling .... 14,780 m (48,500 ft)
- T-O run .... 345 m (1,130 ft)
- Landing run .... 707 m (2,320 ft)
- Max range .... 1,400 n miles (2,592 km; 1,611 miles)
- *g* limits .... +8/−3

UPDATED

KAI (LOCKHEED MARTIN) F-16C/D FIGHTING FALCON

TYPE: Multirole fighter.

PROGRAMME: Korean Fighter Programme (KFP) to co-produce 108 of 120 F-16s (80 Block 52D F-16Cs and 40 F-16Ds) announced in Seoul 28 March 1991; 120th aircraft delivered on 19 April 2001. Decision to produce a further 20 (Block 52 standard) announced 13 May 1999; approved July 2000 as KFP 2; deliveries were to have started in July 2003 and were due for completion in 2004.

COSTS: US$663 million (estimated) for additional 20 (1999).

POWER PLANT: One Pratt & Whitney F100-PW-229 turbofan.

AVIONICS: *Mission:* Lockheed Martin LANTIRN pods.

ARMAMENT: Includes AGM-84 Harpoon, AGM-88 HARM and AIM-120 AMRAAM missiles.

UPDATED

KAI KMH

TYPE: Multirole medium helicopter (requirement).

PROGRAMME: Initiated as part of Republic of Korea Army (RoKA) Force Improvement Plan, with objective of replacing 500MDs, UH-1Hs and AH-1Ss with advanced type usable in combat, light attack, command and control, liaison and passenger-carrying roles. Confirmed by Korean government as KMH (Korea Multirole Helicopter) acquisition programme in September 2001; Agency for Defence Development (ADD) confirmed as programme manager in 2002. Official go-ahead for KMH programme confirmed September 2003, but government ordered a re-evaluation of programme two months later, and budget constraints could result in down-sizing of requirement. Utility version to be developed first; subsequently planned attack variant, embodying maximum commonality, appeared in danger of cancellation by end of 2003. Earlier, KAI had approached AgustaWestland, Bell, Boeing, Eurocopter and Sikorsky as potential partners in KMH programme.

KMH officially became a joint international venture in March 2004 when RFP at last issued to seven manufacturers: AgustaWestland, Bell, Boeing, Eurocopter, Kaman, Sikorsky, and a Rosoboronexport consortium, requiring a 5,896 to 6,803 kg (13,000 to 15,000 lb) utility version for fielding in 2010, followed by an attack version two years later, both developed in partnership with KAI. Proposals required by 15 June 2004; ADD to issue own RFP for mission equipment later, as was KARI for power plant, rotor system and avionics. Final choice of partner then anticipated to be made in November 2004.

KAI KMH attack version model, shown at Singapore in February 2004 *(Paul Jackson)* ***NEW***/0576904

UPDATED

KAI (BELL) SB 427

First of only two Sacheon-built SB 427s, following its repurchase by Bell Canada *(Paul Jackson)* ***NEW***/0576533

TYPE: Light utility helicopter.

PROGRAMME: KAI's Sacheon plant is licensed local assembly centre for Bell 427s sold to customers in Southeast Asia; also supplies fuselages and cabins for Bell production line. Canadian and US FAA type certification obtained in November 1999 and January 2000 respectively; CAAC (China) type certificate April 2000; South Korean certificate for IFR operation in May 2000. Licence agreement provides for local manufacture of complete aircraft, but reports in late 2003 indicated that only two then completed by KAI, both repurchased by Bell and registered in Canada during October 2003 (c/n 58001 and '002). New KAI/Bell MoU in November 2003 provided for co-development (possibly to be joined later by Mitsui of Japan) of IFR variant with roomier cabin and improved cold-climate performance.

CURRENT VERSIONS: **SB 427:** Korean designation of standard Canadian-built Bell 427.

CUSTOMERS: First (Bell-built) 427 delivered to Broad Air Conditioning Company (China) in early 2000; second aircraft then due for delivery to Korea National Police in first quarter of 2001. Sales to China totalled 12 by December 2000, including five for Fei Ma Airlines, deliveries of which began in June 2001.

UPDATED

KAL

KOREAN AIR LINES CO LTD (AEROSPACE DIVISION)

KA Building, 1370 Gonghang-dong, Gangseo-gu, Seoul
Tel: (+82 2) 656 39 20
Fax: (+82 2) 656 39 17
e-mail: selbepr@koreanair.co.kr
Web: http://www.koreanair.co.kr
MANAGING VICE-PRESIDENT: Sang-Mook Suh
GENERAL MANAGER, PLANNING DEPARTMENT: Yong-An Yoon

KIMHAE PLANT:
103 Daejeo 2-dong, Gangseo-gu, Pusan 618142
Tel: (+82 51) 970 56 20
Fax: (+82 51) 970 50 62
MANAGING VICE-PRESIDENT: Sang-Mook Suh
GENERAL MANAGER: Yong-An Yoon

Korean Institute of Aeronautical Technology (KIAT)
461 Jonmin-dong, Yusung-ku, Taejon 305390
Tel: (+82 42) 868 61 14
Fax: (+82 42) 868 61 28
MANAGING VICE-PRESIDENT: Si-Yoong Yoo

Aerospace Division of Korean Air Lines established 1976 to manufacture and develop aircraft; occupies 64.75 ha (160 acre) site at Kimhae, including floor area of 250,000 m^2 (2.69 million sq ft); 2001 workforce about 2,150. Has overhauled RoKAF aircraft since 1978; programmed depot maintenance of US military aircraft in Pacific area began 1979, including structural repair of F-4s, systems modifications for F-16s, MSIP upgrading of F-15s and overhaul of C-130s. Began assembly in 1981 of first domestically manufactured fighter (Northrop F-5E/F), completing deliveries to RoKAF 1986. Since 1988, KAL has delivered wing components for Boeing 747/777, New-Generation 737 and MD-11 (last-named programme completed July1999); and fuselage components for MD-80/90 and Airbus A330/A340. Deliveries of nosesections for Boeing 717 started 28 March 1997, but suspended in 2001. KAL manufactured UH-60P helicopters between 1991 and 1999, under licence from Sikorsky.

Korean Institute of Aeronautical Technology (KIAT), established as division of KAL in 1978, has grown to become a major Korean aerospace industry R&D centre. It is to add a Helicopter Development Research Centre (HDRC) as a result of October 2001 agreement with Sikorsky to collaborate on KAL's submission in KMH (Korea Multirole Helicopter) programme (which see). Korean MoD's Agency for Defence Development (ADD) selected in 2002 to manage this programme. Official go-ahead announced September 2003 as 15 trillion won (US$13 billion) programme, despite continuing concerns over its economic viability; about 2 trillion won (US$1.7 billion) of this to be spent on R&D. Further concerns over cost voiced by government in November 2003, and re-evaluation of programme ordered. Aim is to complete development of utility version by 2010 and attack version (if still viable) by 2012. Draft programme plan was to have been completed by end of 2003 and selection of foreign partner by mid-2004, but programme status remained uncertain in early 2004.

As part of Korean industry development programme from 1988, KAL designed and developed Chang-Gong 91 light aircraft; co-developed (with McDonnell Douglas) the MD 520MK military helicopter derived from the MD 500; is a major member of KFP Korean Fighter Programme (wings and rear fuselages for F-16C/D); and domestically co-developed with Daewoo the KAI KT-1 basic trainer for the RoKAF. Following prolonged discussions in 2002–03 regarding possible integration of KAL and Korean Aerospace Industries (KAI, which see), it was announced in late August 2003 that MoU had been signed for KAL to acquire entire 28.1 per cent Daewoo stake in KAI. However, MoU expired at year-end without agreement on Daewoo purchase price; further discussions planned in second quarter of 2004.

UPDATED

KAL KMH

TYPE: Multirole medium helicopter (requirement).

PROGRAMME: Initiated as KAL's entrant in Korea Helicopter (KMH) competition for armed scout and attack roles, but M now indicating Multirole for appeal also to civil operators; early design assistance by Sikorsky; mockup displayed and first details released at Seoul Air Show in October 1996. Korean Army requirements subsequently underwent gradual change, leading to presentation to government by KAL of new conceptual design at 2001 Seoul Air Show. Basic concept now is to develop two versions (utility and attack) with common major subsystems such as engines, rotor system and drive train. KAL signed MoU with Sikorsky Aircraft on 16 October 2001, under which the two companies will participate in technical and strategic discussions to co-operate in product development, and in marketing initiatives for multimission helicopters to meet Korean Army requirements. KMH programme received official go-ahead September 2003, but re-evaluated by government two months later.

Following March 2004 issue of government RFP to seven potential partners (AgustaWestland, Bell, Boeing, Eurocopter, Kaman, Sikorsky and a Rosoboronexport consortium), KMH now officially a joint venture rather than entirely indigenous programme. Partner bids due mid-2004, with selection expected by year-end. KAI (which see) to be Korean prime contractor. See earlier editions for details of KAL design.

UPDATED

KARI

KOREA AEROSPACE RESEARCH INSTITUTE

Aircraft Division, PO Box 113, Yu-Sung, 305600 Taejon
Tel: (+42 860) 23 52
Fax: (+42 860) 20 06
e-mail: hanch@viva.kari.re.kr
Web: www.kari.re.kr
INFORMATION: Jong-Won Lee

The Aircraft Division of KARI is undertaking a study for a stratospheric airship and is engaged in the development of two four-seat light aircraft, as described in the accompanying entries.

UPDATED

KARI (VELOCITY) FIREFLY

TYPE: Four-seat kitbuilt.

PROGRAMME: Prototype made maiden flight 2002 and acting as US demonstrator 2003; second prototype employed in Korea at that time as avionics and equipment testbed. Shin Young Heavy Industries Company appointed by KARI in May 2003 to market kits in USA; choice of kit manufacturer still the subject of negotiations in late 2003. Prototype of revised design targeted to fly in early 2005.

DESIGN FEATURES: Originally used fuselage, foreplanes and landing gear of US Velocity XL; wings, fins and rudders developed by KARI. However, reported in November 2003 that KARI had modified design to include retractable landing gear, a slight increase in fuselage length, and a machined (instead of hand-moulded) wing built from non-foam GFRP. Said to be more stable at slow speeds than Velocity.

VERIFIED

KARI BORA

TYPE: Four-seat lightplane.

PROGRAMME: Four-year initial development phase began 21 August 1999 and ended 12 July 2003. Co-development by KARI and ultralight manufacturer Sung Jun Motoravia; latter seeking a US manufacturing partner in 2003. First flight targeted for second quarter 2004; flight testing for FAR Pt 23 certification will take place in USA.

DESIGN FEATURES: Wholly indigenous design. Fuselage similar to that of Firefly; forward-swept wings, with W leading-edge and turned-down tips; double-T tail unit.

FLYING CONTROLS: Conventional and manual; flaps fitted.

STRUCTURE: Composites construction.

POWER PLANT: One 186 kW (250 hp) piston engine.

ACCOMMODATION: Pilot and three passengers, in tandem pairs.

DIMENSIONS, EXTERNAL:

Wing span	10.67 m (35 ft 0 in)
Length overall	8.14 m (26 ft 8½ in)
Height overall	2.29 m (7 ft 6 in)

WEIGHTS AND LOADINGS (estimated):

Weight empty	726 kg (1,600 lb)
Max T-O weight	1,224 kg (2,700 lb)

PERFORMANCE (estimated):

Max level speed at S/L	200 kt (370 km/h; 230 mph)
Cruising speed at FL80	185 kt (343 km/h; 213 mph)
Stalling speed	56 kt (104 km/h; 64 mph)
Max rate of climb at S/L	335 m (1,100 ft)/min
Cruising altitude	2,440 m (8,000 ft)
T-O to 15 m (50 ft)	366 m (1,200 ft)
Landing from 15 m (50 ft)	336 m (1,100 ft)
Range	1,000 n miles (1,852 km; 1,151 miles)

UPDATED

KARI MLESA

TYPE: Helium non-rigid airship.

PROGRAMME: MLESA programme (Multipurpose Long Endurance Stratospheric Airship) launched in 2001 with investment estimated at some US$42 million over the period 2001 to 2007 (US$33.5 million from the South Korean government and US$8.5 million from the private sector). The project is supported by the Ministry of Commerce, Industry and Energy (MACIE). Completion of a quarter-scale (50 m; 164 ft long) unmanned version, to have a 3-hour endurance at 5,000 m (16,400 ft), was reported in November 2003. The electrically powered non-rigid is controlled automatically for station-keeping; the full-size (200 m; 656 ft) stratospheric version is intended to begin testing in the second half of 2007.

The subscale prototype has an electric motor, powered by a gas-turbine-driven generator, and has an MTOW of 2,800 kg (6,172 lb). In the 200 m Phase 2 airship, the electric motor's power will be derived from solar batteries and fuel cells. Operating at 20,000 m (65,620 ft), it is intended for use in communications relay, scientific research, observation and other applications.

NEW ENTRY

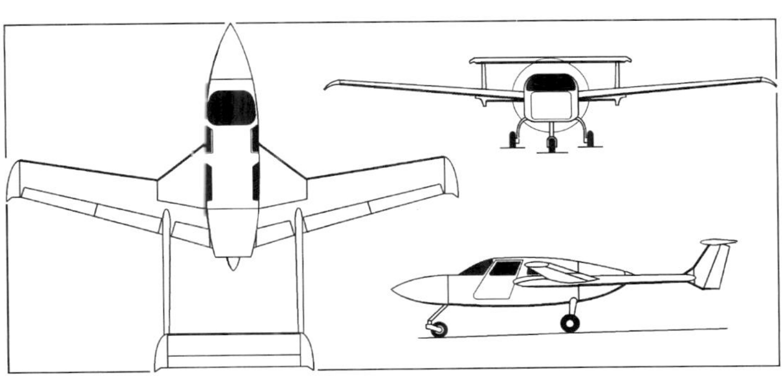

KARI Bora four-seat lightplane *(Paul Jackson)* ***NEW***/0567234

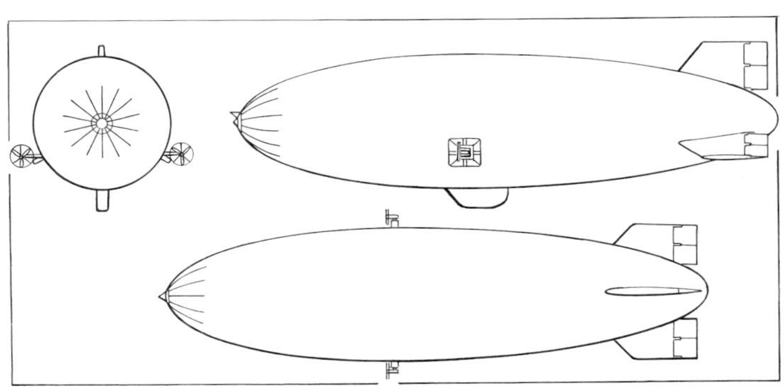

General appearance of the KARI subscale airship *(James Goulding)* ***NEW***/0567102

Lithuania

HELISOTA

HELISOTA LTD

J Bakanausko str 29E, LT-3018 Kaunas
Tel: (+370 37) 42 16 ?7
Fax: (+370 37) 42 04 20
e-mail: helisota@helisota.lt
Web: http://www.helisota.lt
GENERAL MANAGER: Josif Legenzof
DIRECTOR, INTERNATIONAL RELATIONS & MARKETING: Ruslanas Panovas
MANAGER, SPECIAL PROJECTS: Linas Elijosius

Helisota's principal business is maintenance, repair and overhaul of helicopters and some types of general aviation and light transport aeroplanes. However, in 2004 it announced plans to return to production the indigenous LAK-X lightplane.

NEW

HELISOTA LAK-X

TYPE: Side-by-side kitbuilt.

PROGRAMME: Design started (by Aeroplastika: see 1995-96 *Jane's All the World's Aircraft*) June 1988: construction of prototype LAK-XE (LY-XMH) begun by Avia Baltika inJanuary 1990; first flight 2 August 1992 with 50.7 kW (68 hp) Limbach L 1700 engine; prototype displayed at Moscow Air Show 1993; later uprated with 65 kW (87 hp) Limbach L 2400; horn balance added to rudder; three (one XE, two XA) completed by January 1996; first customer delivery (second XA prototype) May 1995; Avia Baltika pursued ground and flight test programme for JAR-VLA certification, and intended to produce kits and/or complete aircraft. Programme lapsed in late 1990s, but received in 2004.

CURRENT VERSIONS: **LAK-XA: (American)** Continental IO-240 engine, although second LAK-X fitted with PZL-Franklin 4A-235B3 engine to US order.

LAK-XE : (European) With choice of three engines, each of less than 74.6 kW (100 hp).

CUSTOMERS: Six reportedly built by Avia Baltika, of which two to USA (currently N8023J and N17EJ).

DESIGN FEATURES: Objectives were short field capability, high L/D range, good fuel economy and cabin comfort. Mid-mounted high-aspect ratio wing; laminar flow NACA 64 series aerofoil, thickness 15 per cent at root, 12 per cent at tip, with 6° dihedral, 4° incidence and 2° twist; fuselage of oval cross-section. Wings detachable for storage.

FLYING CONTROLS: Conventional and manual. Elevators and ailerons actuated by rods/levers, rudder by cables; electrically actuated flaps, maximum deflection 3° up at high cruising speed, 40° down for landing: aileron/flap mixer deflects ailerons 10° down with maximum flap deflection; electrically actuated elevator trim.

STRUCTURE: Wet layup epoxy/glass fibre/foam core composites basic airframe: fuselage moulded in left/right halves with six composites bulkheads and vertical fin webs and ribs; wing moulded in upper/lower shells with root rib. I-beam spar and rear shearweb; tailplane similar.

LANDING GEAR: Tricycle type; fixed. Fuselage-mounted, cantilever self-sprung legs of high-temperature prepreg epoxy/unidirectional glass fibre composites construction; single wheel, with fairing, on each unit; size 115 mm on mainwheels. 100 mm on nosewheel; tyre size 5.00-5 on mainwheels, 300 × 125 mm on steerable nosewheel (±180°); pressures 3.43 bars (50 lb/sq in) on mainwheels. 2.94 bars (43 lb/sq in) on nosewheel; hydraulic disc brakes.

POWER PLANT: **LAK-XA:** One 93.2 kW (125 hp) Teledyne Continental IO-240-B engine driving a Warp Drive ground-adjustable pitch propeller. Fuel capacity 85 litres (22.4 US gallons; 18.7 Imp gallons.

LAK-XE: One 69.4 kW (93 hp) Limbach L 2400 engine driving a Mühlbauer MTV-1 two-blade constant-speed (electric) composites propeller. Fuel in two integral wing tanks each with filter port; combined capacity 110 litres (29 US gallons; 24.2 Imp gallons) of which 100 litres (26.4 US gallons; 22 Imp gallons) are usable.

Avia Baltika-built LAK-X ***NEW***/0011109

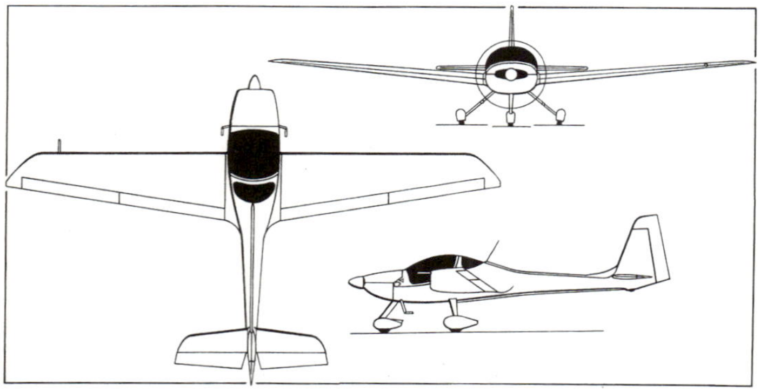

LAK-X two-seat lightplane *(Jane's/Mike Keep)* ***NEW***/0589204

ACCOMMODATION: Two seats side by side under single-piece forward-hinged bubble canopy; baggage compartment behind seats, capacity 30 kg (66 lb); accommodation is heated and ventilated.

SYSTEMS: Electrical system powered by 12 V engine-driven alternator: 12 V 45 Ah battery.

Data below apply to both versions except where noted:

DIMENSIONS, EXTERNAL:

Wing span	10.68 m (35 ft 0½ in)
Wing chord: at root	1.50 m (4 ft 11 in)
at tip	0.80 m (2 ft 7½ in)
Wing aspect ratio	11.3
Length overall: XA	7.00 m (22 ft 11½ in)
XE	6.95 m (22 ft 9½ in)
Height overall	2.20 m (7 ft 2½ in)
Tailplane span	3.30 m (10 ft 10 in)
Wheel track	2.10 m (6 ft 10¾ in)
Wheelbase	1.55 m (5 ft 1 in)

DIMENSIONS, INTERNAL:

Cabin: Length	1.40 m (4 ft 7 in)
Max width (incl wingroot armrests)	1.10 m (3 ft 7¼ in)
Max height	1.10 m (3 ft 7¼ in)
Baggage compartment volume	approx 0.50 m ^{3} (17.7 cu ft)

AREAS:

Wings, gross	12.05 m^2 (129.71 sq ft)
Ailerons (total)	1.10 m^2 (11.84 sq ft)
Trailing-edge flaps (total)	1.04 m^2 (11.19 sq ft)
Fin	1.48 m^2 (15.93 sq ft)
Rudder	0.71 m^2 (7.64 sq ft)
Tailplane	2.40 m^2 (25.83 sq ft)
Elevators (total)	0.86 m^2 (9.26 sq ft)

WEIGHTS AND LOADINGS:

Weight empty, equipped: XA	460 kg (1,014 lb)
XE	400 kg (882 lb)
Max T-O weight: XA	750 kg (1,653 lb)
XE	650 kg (1,433 lb)
Max wing loading: XA	62.2 kg/m^2 (12.75 lb/sq ft)
XE	53.9 kg/m^2 (11.05 lb/sq ft)
Max power loading: XA	8.05 kg/kW (13.22 lb/hp)
XE	9.38 kg/kW (15.41 lb/hp)

PERFORMANCE:

Max level speed at S/L:	
XA	135 kt (250 km/h; 155 mph)
XE	108 kt (200 km/h; 124 mph)
Max cruising speed: XA	118 kt (220 km/h; 136 mph)
XE	97 kt (180 km/h; 111 mph)
Econ cruising speed: XA	108 kt (200 km/h; 124 mph)
XE	86 kt (160 km/h; 99 mph)
Stalling speed:	
flaps up: XA	51 kt (95 km/h; 59 mph)
XE	46 kt (85 km/h; 53 mph)
flaps down: XA	44 kt (82 km/h; 51 mph)
XE	40 kt (75 km/h; 47 mph)
Max rate of climb at S/L: XA	300 m (984 ft)/min
XE	240 m (787 ft)/min
Service ceiling: XA, XE	2,000 m (6,560 ft)
T-O run: XA, XE	150 m (495 ft)
Landing run: XA	150 m (495 ft)
XE	100 m (330 ft)
Range with max payload and max fuel:	
XA, XE	459 n miles (850 km; 528 miles)
g limits: XA	+4/–2

NEW

Malaysia

CTRM

COMPOSITES TECHNOLOGY RESEARCH MALAYSIA SDN BHD

T-02, Third Floor, 2310 Century Square, Jalan Usahawan, 63000 Cyberjaya, Selangor Darul Ehsan
Tel: (+60 3) 83 13 51 00
Fax: (+60 3) 83 13 51 11
e-mail: asmuni@ctrm.com.my
Web: http://www.ctrm.com.my
WORKS: Batu Berendam, Malacca
CEO: Col (Retd) Rosdi Mahmud
COO: Francis Hiew Nyuk Fah
CORPORATE COMMUNICATIONS MANAGER: Asmuni Mat Saleh

Created in 1991, CTRM became an investor in Eagle Aircraft of Australia, via its subsidiary CTRM Aviation, formerly known as Eagle Aircraft (M) Sdn Bhd; component production for the two-seat Eagle 150B began in 1997, subsequently progressing to full local manufacture. This is undertaken in a 20,903 m^2 (225,000 sq ft) facility at Malacca by CTRM's wholly owned subsidiary CTRM Aero Composites (CTRM AC), formerly known as Aero-Composites Technologies Sdn Bhd. All Eagle production was transferred to Malaysia following closure of the Australian factory at the end of February 2002. CTRM is an 18 per cent investor in The Lancair Company, USA, and CTRM AC is currently also manufacturing composites components for the Lancair Columbia 300. It also is under contract to BAE Systems to produce composites wing leading- and trailing-edge panels for the Airbus A300 and A318/319/320/321.

CTRM's associate company Excelnet Sdn Bhd offers engineering solutions ranging from conceptual design to digital mockups and strength/stress analysis. It has recently been involved, with CTRM and BAE Systems, in developing the optionally piloted Eagle 150 Airborne Reconnaissance Vehicle (ARV), details of which are given in *Jane's Unmanned Aerial Vehicles and Targets*. Other current work includes wing life extension (spar modification) for the Scottish Aviation Bulldog, for which it has worldwide type certificate maintenance. In late 2003, CTRM and Eurocopter signed an agreement under which the former will overhaul and modify up to 80 former German Army BO 105 helicopters for EMS and various military roles over the ensuing five years. Initial batch comprises 30 aircraft.

UPDATED

EAM

EAGLE AIRCRAFT (MALAYSIA) SDN BHD (SUBSIDIARY OF CTRM)

POSTAL ADDRESS: Locked Bag 1028, Pejabat Pos Besar, 75150 Melaka
Tel: (+60 6) 317 03 48
Fax: (+60 6) 317 70 23
e-mail: razak.zain@ctrmeagle.com
Web: http://www.ctrm.com.my
CONTACT: Abd Razak Mohamed Zin

WORKS: Composites City, Batu Berendam Airport, 75350 Melaka
Tel: (+60 6) 317 10 07
Fax: (+60 6) 317 70 23

Formerly known as Eagle Aircraft (Malaysia) Sdn Bhd, CTRM Aviation was established in the mid-1980s to develop and manufacture a new light aircraft, also named Eagle, for the general aviation market. In 1993 it became a wholly owned subsidiary of CTRM, and overall owner (28 May 1993) of Eagle Aircraft Pty Ltd of Australia, where the aircraft was initially produced and/or assembled. Responsibility for the Eagle reverted to Malaysia following February 2002 closure of the Australian factory. Malaysian factory, completed March 1996, has floor area of 24,080 m^2 (259,200 sq ft).

UPDATED

EAM EAGLE 150

TYPE: Two-seat lightplane.

PROGRAMME: Launched 1981 with objective of producing first all-composites light aircraft in Australia; single-seat POC (proof-of-concept) aircraft now displayed at Power House Museum, Sydney; construction of two-seat preproduction prototype Eagle X started fourth quarter 1987; first flight (VH-XEG) second quarter 1988, with 58 kW (78 hp) Aeropower engine, replaced later by 74.5 kW (100 hp) Continental O-200; 200 hour test programme, meeting all original design criteria, completed by October 1988; plans for 1989 production start aborted; component manufacture began March 1991; production prototype (PPT1/VH-XEP) made first flight 6 November 1992; weight-restricted certification by Civil Aviation Safety Authority of Australia (CASA) 21 September 1993; European JAR-VLA certificate not awarded at that time and aircraft consequently redesigned with changes including increased span and chord on foreplane and mainplane flaps; redesigned and repositioned wing cuffs, with leading-edge extension outboard; vortex generators and redesigned horizontal stabiliser aerofoil section.

Series production in Australia, as Eagle X-TS, launched August 1993; first flight (VH-AHH) 23 October 1993 and first customer delivery December 1993 (VH-FPO to Department of Conservation and Land Management in Western Australia). Initial production in Australia only, but using some components manufactured by Eagle Aircraft (Malaysia) of which Eagle Aircraft Pty Ltd was a wholly owned subsidiary. First aircraft completed from Malaysian components (VH-PMI, c/n MA 00001) made maiden flight 27 March 2001. In December 2001 it was announced that production was moving to Malaysia and operations in Australia were being run down; this was achieved in late February 2002 and type certificate transferred to CTRM on 30 May 2002.

First Series 150 (converted production prototype VH-XEP) rolled out August 1997; certified on 13 November 1997 by CASA to JAR-VLA standards at MTOW of 640 kg (1,411 lb); this certification also valid in Malaysia. FAA certification achieved 11 February 1999, followed by New Zealand approval in mid-1999; also approved in Thailand; JAA certification expected in 2002. First night VFR instrumented aircraft certified and delivered in December 2000.

CURRENT VERSIONS: **Series 100:** Retrospective designation of former Eagle X-TS: production completed. Ten Series 100/Eagle X-TS built, including two for Malaysia (c/n 0003 and 0005) and one for John Roncz in USA (c/n 0010): five conversions (c/n 0002/3/5/7/8) to Series 150.

Series 150: Current production version, designated **150A** with IO-240-A engine and **150B** with IO-240-B. Engines have same maximum power of 93.2 kW (125 hp), but B version has increased mid-range power of almost 29 per cent to 69.4 kW (93 hp) and is 3.7 per cent quieter than A model. Available in Basic, Training and Executive variants, as below. US promotion devoted to 150B version only.

Basic: Standard Series 150 version; *as described*.

Training: As Basic, with addition of tachometer, turn co-ordinator, directional gyro, digital clock, cockpit internal lighting, transponder, nav/com/VOR, navigation and landing lights and Eagle external pinstriping and decals.

Executive: As Training, plus GPS moving map display, mainwheel spats, leather upholstery, cabin heater, baggage bins, sun visor and chart pocket.

Eagle ARV: Optionally piloted surveillance version (Aerial Reconnaissance Vehicle), co-developed by CTRM with BAE Systems Controls of California, which supplies complete modification kit including flight management system, sensor payload, datalink and ground control station. First three-aircraft system delivered to Royal Malaysian Air Force in late 2001. Further details in *Jane's Unmanned Aerial Vehicles and Targets*.

CUSTOMERS: Series 150 operators in Australia, Malaysia, New Zealand, USA and elsewhere. By January 2003, production (including Eagle X) totalled 46 in Australia (including two prototypes) and three in Malaysia. Two delivered to USA in 2002. No reports of orders or deliveries in 2003.

COSTS: Standard Eagle 150B US$136,950; Executive US$144,444 (2004). Total maintenance cost US$9.98 per hour over 2,000 hours' utilisation (2003).

DESIGN FEATURES: Intended primarily for *ab initio* training, recreational flying and surveillance. 'Tri-surface' configuration, with high-mounted mainplane, large low-mounted foreplane, and tailplane. Stall strips on foreplane ensure that it stalls before the mainplane.

Mainplane of tailored Roncz aerofoil section, thickness/chord ratio 16 per cent; no sweep; no dihedral.

FLYING CONTROLS: Conventional and manual. Slotted ailerons on mainplane, elevators on tailplane, and rudder, all with normal manual/mechanical actuation. Aileron travel +25 to −20°; elevator +25 to −24°; rudder ±23°. Pushrods on ailerons and elevators, cables on rudder. Electric pitch trim (tab in starboard elevator), manual roll trim; rudder tab. Electrically actuated single-slotted flaps on foreplane (full span; travel 0 to 35°) and mainplane (part span, range of travel +3° up to 38° down). Leading-edge stall strips, vortex generators and fences on mainplane, stall strips on foreplane. Flap and airflow control systems designed to achieve low stalling speed with relatively high wing loading, to provide good ride quality in turbulence.

STRUCTURE: Except for metal engine mounts and flight control rods, Eagle 150 built entirely of composites, with 90 per cent of structure bonded/assembled by EAM. Wings, fuselage and all control surfaces are Nomex honeycomb or high-density foams, sandwiched between multiple layers of carbon fibre; Kevlar reinforcement around wing leading-edges and shoulder of mainplane; cockpit is impact-resistant capsule of multilayered Kevlar and carbon fibre; carbon fibre spars. Entire structure uses Eagle-designed vinylester resins for strength/longevity/impact resistance and to minimise 'wet environment' problems inherent in standard epoxies.

LANDING GEAR: Tricycle type; fixed. Fuselage-mounted spring glass fibre/epoxy cantilever mainwheel legs and oleo nose leg. Choice of Cleveland or McCreary 5.00-5 wheel on each unit, with optional speed fairings. Cleveland hydraulic single-disc brakes on mainwheels. Castoring nosewheel, size 11×4.00, with Chen Shin tyre. Minimum ground turning radius 5.21 m (17 ft 1 in). Twin landing lights in nosewheel fairing.

POWER PLANT: One 93.2 kW (125 hp) Teledyne Continental IO-240-A or IO-240-B7B flat-four engine (see Current Versions above); respectively driving a McCauley 1A135BRM7054, 1A135BRM7057 or 1A135CRM7057 two-blade fixed-pitch metal propeller. Trials with MTV-7-D/175-112 three-blade propeller under way in early 2001.

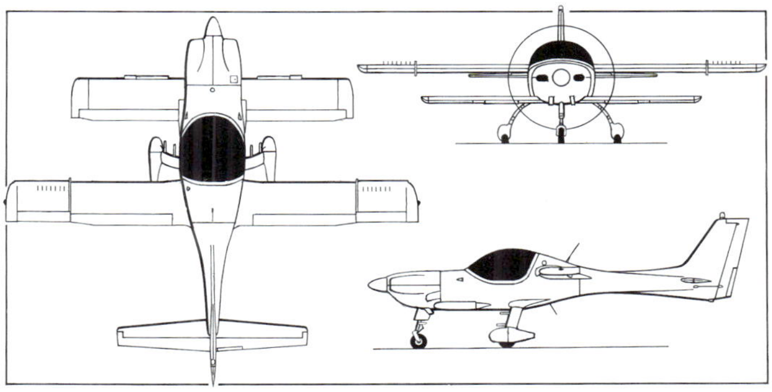

Eagle 150 two-seat light aircraft *(Jane's/Mike Keep)* 0092121

Eagle Series 150 operated in US *(Paul Jackson)* ***NEW***/0576502

Fuel capacity 102 litres (26.9 US gallons: 22.4 Imp gallons), of which 97 litres (25.6 US gallons: 21.3 Imp gallons) are usable. Oil capacity 5.7 litres (1.5 US gallons; 1.3 Imp gallons).

ACCOMMODATION: Two seats side by side with Y-shape control column operable from either seat. Adjustable rudder pedals. Bubble canopy hinged at front and opens upward. Four-point safety harnesses. Hat shelf and two baggage compartments.

SYSTEMS: Hydraulic system for manual brake actuation only; 12 V DC electrical system with 60 A alternator.

AVIONICS: Standard avionics suite as described below. Optional equipment as detailed under Current Versions. Night VFR package also available.

Comms: VHF com, intercom and ELT standard.

Instrumentation: ASI, altimeter, tachometer, oil pressure/temperature gauge, fuel pressure gauge, fuel quantity gauge, voltmeter/ammeter, OAT gauge, magnetic compass, pitch trim indicator, engine hours meter and flight hours meter.

DIMENSIONS, EXTERNAL:	
Wing span	7.16 m (23 ft 6 in)
Wing chord, constant	0.74 m (2 ft 5 in)
Wing aspect ratio	9.9
Foreplane span	4.88 m (16 ft 0 in)
Foreplane chord, constant	0.74 m (2 ft 5 in)
Length overall	6.45 m (21 ft 2 in)
Height overall	2.31 m (7ft 7 in)
Tailplane span	3.30 m (10 ft 10 in)
Wheel track	1.93 m (6 ft 4 in)
Wheelbase	1.40 m (4 ft 7 in)
Propeller diameter	1.78 m (5 ft 10 in)
DIMENSIONS, INTERNAL:	
Cabin: Length	1.37 m (4 ft 6 in)
Max width	1.07 m (3 ft 6 in)
Max height	0.86 m (2 ft 10 in)
AREAS:	
Wings, gross	5.20 m^2 (56.0 sq ft)
Foreplanes, gross	3.62 m^2 (39.0 sq ft)
Wing flaps (total)	0.90 m^2 (9.68 sq ft)
Foreplane flaps (total)	0.90 m^2 (9.68 sq ft)
Tailplane	1.49 m^2(16.00 sq ft)
WEIGHTS AND LOADINGS:	
Weight empty	429 kg (946 lb)
Baggage capacity: shelf	9 kg (20 lb)
compartments (total)	36 kg (80 lb)
Max T-O and landing weight	650 kg (1,433 lb)
Max wing/foreplane loading	73.7 kg/m^2 (15.09 lb/sq ft)
Max power loading	6.97 kg/kW (11.45 lb/hp)
PERFORMANCE (A: standard 150, B: 150 ARV):	
Never-exceed speed (VNE):	
A	165 kt (305 km/h; 189 mph)
Max level speed at S/L: A	130 kt (241 km/h; 150 mph)
B	133 kt (246 km/h; 153 mph)
Cruising speed at 75% power at FL20	
A	125 kt (232 km/h; 144 mph)
B	120 kt (222 km/h; 138 mph)
Loiter speed: B	65-80 kt (120-148 km/h; 75-92 mph)
Stalling speed at S/L:	
flaps up: A	52 kt (97 km/h; 60 mph)
B	56 kt (104 km/h; 64 mph)
flaps down: A	43 kt (80 km/h; 50 mph)
B	45 kt (84 km/h; 52 mph)

Max rate of climb at S/L: A, B .... 322 m (1,055 ft)/min
Service ceiling: A .... 4,575 m (15,000 ft)
B .... 4,880 m (16,000 ft)
T-O to 15 m (50 ft): A, B .... 349 m (1,145 ft)
Landing from 15 m (50 ft): A, B .... 365 m (1,200 ft)
Mission radius: B .... 135 n miles (250 km; 155 miles)
Range with max fuel: A .... 520 n miles (963 km; 598 miles)
Endurance at 60% power: A .... 5 h
Max endurance: B .... 10 h
g limits: JAR-VLA .... +3.8/−1.9
ultimate .... +8.55/−4.27

UPDATED

TAG

1 Alom Way, Sama Jaya Free Industrial Zone, Muara Tabuan, 93450 Kuching
Tel: (+60 82) 36 53 33
Fax: (+60 82) 36 73 33
e-mail: info@airships.cc
Web: http://www.airships.cc
PRESIDENT: Philip Yiin

TAG was formed to concentrate more than a decade of lighter than air technology and experience into a single R & D centre intended to become a regional and global focal point for future airship development. It is supported by its sister company, the Alom Group, which specialises in composites materials, structural applications and customised solutions. It has also acquired (for €13,500) the former CargoLifter 'Joey' prototype which, in 2004, was being modified to serve as a publicity vehicle. Manufacture is undertaken by Airship Systems Sdn Bhd. Other products include inflatable hangars and aerostats for various applications.

UPDATED

TAG AG SERIES

TYPE: Helium semi-rigid airship.
PROGRAMME: Range of new-generation airships under development, as described below.
CURRENT VERSIONS: **AG22U:** Remotely piloted, 22 m (72.2 ft) long unmanned airship. In flight test by April 2004 and expected to enter service by mid-year.
AG33mu: Manned technology demonstrator, 33 m (108.3 ft) long, with pilot-controlled mooring system and other innovative capabilities. Under construction in 2004.
AG90: 90 m (295.3 ft) long semi-rigid, able to carry up to 50 passengers. Design under way in early 2004; to be built later that year for 'a regional government'.
AG90DB: Hybrid vehicle project for ultra-large payloads and with partial aerodynamic lift.
AG100L: Lenticular design for logging industry and other aerial crane roles.
AG180: Projected heavy-lift semi-rigid, 180 m (590.6 ft) long and having 50 tonne payload capacity.
AG300: Projected stratospheric airship with ultra-long endurance.
DESIGN FEATURES: Conventional-shape envelope (except AG100L), and twin tailfins with elevators. Twin rear-mounted propulsion engines, located within tailfins. Third engine, mounted within bow, provides multidirectional thrust vectoring for directional control and terminal guidance for mooring system.
AVIONICS: Fly by light, developed by TAG.

UPDATED

Netherlands

DAC

DUTCH AEROPLANE COMPANY VOF

Pieter Zeemanweg 116, NL-3316 GZ Dordrecht
WORKS: Laan von Ypenburg 46, NL-2497 GB Den Haag
Tel: (+31 6) 54 94 15 23
Web: http://www.dac-ranger.nl
DIRECTORS:
Evert-Jan Cornet (Technical)
Jan Laurier (Information)
Alex Swagemakers (Propulsion)

DAC was established in 1999 to produce what is now known as the RangeR.

UPDATED

Prototype DAC RangeR on display at Aero '03, Friedrichshafen, in April 2003 ***(Paul Jackson)*** ***NEW***/0552687

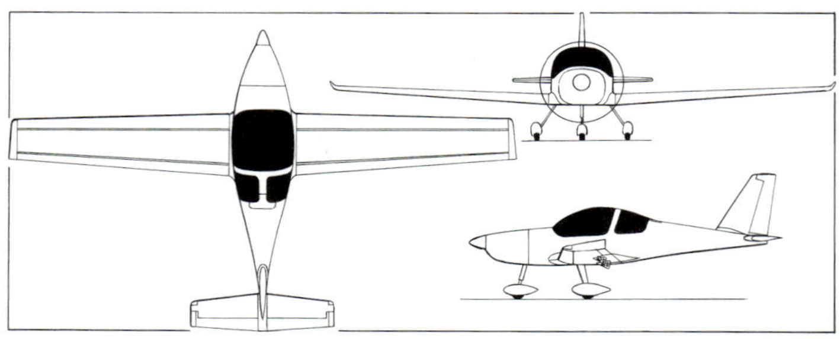

DAC RangeR two-seat composites lightplane ***(James Goulding)*** ***NEW***/0554616

DAC RANGER

TYPE: Side-by-side kitbuilt.
PROGRAMME: Based on Opel-powered Dieselis PL2 (F-PTDI), built in France by Paul Lucas and Serge Pennec, first flown September 1998, and shown at PFA International Rally, Cranfield, UK, in July 1999, where selected as basis of DAC venture. Martin Hollmann of US company ADI contracted by DAC to transform wood/composites Dieselis into RangeR, with changes to structure, fuselage size and wing section. Alex Swagemakers of TTA redesigned Opel 1.7 litre turbo-diesel engine for aeronautical use. Incomplete RangeR prototype (PH-DAC) exhibited at Aero '03, Friedrichshafen, April 2003; first flight then due in USA later in 2003 but this still had not taken place by late 2004.
COSTS: Marketed as fast-build kit, with customer's build-assistance option. Estimated kit cost €40,000 (2003), including engine and propeller, but excluding avionics.
DESIGN FEATURES: Objectives of RangeR redesign included JAR-VLA conformity, low construction cost, Diesel power, long range and easily disconnected wings for road transport and storage. Low, moderately tapered wing and tailplane; sweptback fin; upturned wingtips; fuselage tapers sharply forward and rear of cockpit.
FLYING CONTROLS: Conventional and manual. Horn-balanced rudder and single-piece elevator. Flaps; maximum deflection 30°.
STRUCTURE: Composites throughout. Fuselage in left and right halves bonded together, with fin ribs, wing spar bulkheads and firewall pre-installed. Wing main spars glued to bottom skin; forward ribs pre-positioned. Metal engine mount.
LANDING GEAR: Tricycle type; fixed. Cantilever main legs; speed fairings on all wheels.
POWER PLANT: One Opel Corsair Y17DT Diesel engine, with turbocharger and intercooler, rated at 67.1 kW (90 hp) at 4,000 rpm, driving a GT GT-2/174/VQS two-blade, wood/composites, fixed-pitch propeller via a 1:1.67 reduction gear. Fuel capacity 100 litres (26.4 US gallons; 22.0 Imp gallons).
ACCOMMODATION: Two persons, side by side beneath single-piece canopy/windscreen; two rear windows.
AVIONICS: To customer's specification.
DIMENSIONS, EXTERNAL:
Wing span .... 9.75 m (31 ft 11¾ in)
Wing aspect ratio 9.3
Length overall .... 6.00 m (19 ft 8¼ in)
Propeller diameter .... 1.74 m (5 ft 8½ in)
DIMENSIONS, INTERNAL:
Cabin max width .... 1.15 m (3 ft 9¼ in)
AREAS:
Wings, gross .... 10.20 m² (109.8 sq ft)
WEIGHTS AND LOADINGS (estimated):
Weight empty .... 500 kg (1,102 lb)
Max T-O weight .... 750 kg (1,653 lb)
Max wing loading .... 73.5 kg/m² (15.06 lb/sq ft)
Max power loading .... 11.18 kg/kW (18.36 lb/hp)
PERFORMANCE (estimated):
Max level speed at FL120 .... 150 kt (278 km/h; 173 mph)
Cruising speed at FL80 .... 132 kt (244 km/h; 152 mph)
Stalling speed: flaps up .... 49 kt (89 km/h; 56 mph)
flaps down .... 44 kt (81 km/h; 51 mph)
T-O run .... 153 m (500 ft)
Landing run .... 134 m (440 ft)
Range .... 1,241 n miles (2,300 km; 1,429 miles)

UPDATED

SSVOBB

STICHTING STUDENTEN VLIEGTUIGONTWIKKELING -BOUW EN -BEHEER

(Foundation for Student Aircraft Development, Manufacturing and Operation)
Kluyverweg 1, NL-2629 HS Delft
Tel: (+31 15) 278 10 57
Fax: (+31 15) 278 12 43
e-mail: ssvobb@lr.tudelft.nl
Web: http://www.delftaerospace.com/ssvobb
CHAIRMAN: Michiel Goossens
SECRETARY: Stijn Yperman
TREASURER: Flip Sondervan

IMPULS PROJECT:
PRODUCTION CO-ORDINATOR: Dieter Decoster
ENGINEERING CO-ORDINATOR: Joep Breuer

SSVOBB was formed in 1990 by The Society of Aerospace Students 'Leonardo da Vinci' of the Delft University of Technology, Faculty of Aerospace Engineering. Its main objective is to co-ordinate the design, manufacture and maintenance of aircraft by students. A replica of the 1937 Lambach HL II aerobatic biplane was constructed between 1989 and 1995, and work is proceeding on an original lightplane design, the Impuls. The projects are financed by funds and sponsorship.

UPDATED

SSVOBB IMPULS

TYPE: Two-seat lightplane.

PROGRAMME: Preliminary design started early 1994. Registration PH-VXM reserved March 1997. Main fuselage moulds completed early 1999. Main wings due to be completed by mid-2001, but this has slipped to 2005 due to redesign work; by early 2004 wing covering samples were being produced. The landing gear was completed by Grove Aircraft, California. In mid-2004 a new wind tunnel model was under construction. Certification to JAR 23 being undertaken under supervision of Prof dr ir Theo van Holten (flight performance); Prof dr ir J A Mulder (stability/control); and Prof dr ir MJL van Tooren (systems and structures)..

DESIGN FEATURES: Twin-boom layout with pod-type fuselage and high wing of parallel chord; swept fins.

FLYING CONTROLS: Conventional and manual. Actuation by pushrods (ailerons and elevator), with cable and pushrod interconnected twin rudders. Servo-actuated electric pitch trim and electrically operated single-slotted flaps.

STRUCTURE: All-composites wing with GFRP sandwich skin and spars; GFRP tail unit. Some GFRP parts will be produced by vacuum assisted resin injection process. Fuselage steel tube frame with glass fibre shell will be used to finalise design of cockpit and control systems.

Wing section NACA $63_3$415 with RA163CW3 flaps; twin tailbooms joined by NACA 0012 section horizontal stabiliser.

LANDING GEAR: Non-retractable tricycle type; aluminium cantilever sprung mainwheels. Non-steerable nosewheel; mainwheels have 45 cm (17¾ in) tyres pressurised to 3.14 bar (45.5 lb/sq in). Hydraulically operated brakes on mainwheels.

POWER PLANT: One 74.6 kW (100 hp) Mid-West AE110R rotary engine driving a three-blade fixed-pitch propeller. Fuel capacity 65 litres (17.2 US gallons; 14.3 Imp gallons) in single tank located between cockpit and engine. Oil capacity 3.0 litres (0.8 US gallon; 0.7 Imp gallon).

ACCOMMODATION: Side-by-side seating for pilot and passenger under one-piece, Perspex canopy. Baggage space behind seats.

DIMENSIONS, EXTERNAL:
- Wing span ... 7.75 m (25 ft 5 in)
- Wing chord, constant ... 1.03 m (3 ft 4½ in)
- Wing aspect ratio ... 7.5
- Length overall ... 6.37 m (20 ft 10¾ in)
- Fuselage: Length ... 3.32 m (10 ft 10¾ in)
  - Max width ... 1.20 m (3 ft 11¼ in)
- Height overall ... 2.15 m (7 ft 0¾ in)
- Tailplane span ... 2.10 m (6 ft 10¾ in)
- Wheel track ... 2.20 m (7 ft 2½ in)
- Wheelbase ... 1.75 m (5 ft 9 in)
- Propeller diameter ... 1.50 m (4 ft 11 in)

AREAS:
- Wings, gross ... 8.00 m² (86.1 sq ft)
- Ailerons (total) ... 0.66 m² (7.10 sq ft)
- Flaps (total) ... 1.06 m² (11.41 sq ft)
- Fins (total) ... 2.00 m² (21.53 sq ft)

Model of the SSVOBB Impuls two-seat light aircraft undergoing wind tunnel tests 0097639

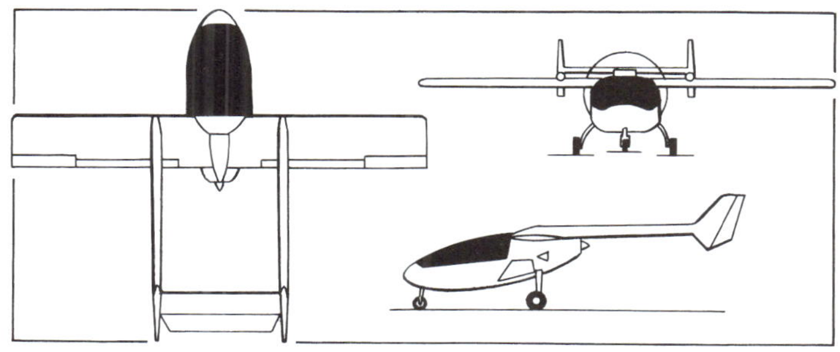

SSVOBB Impuls (MWAE rotary engine) *(Paul Jackson)* 0051122

- Rudders (total) ... 0.24 m² (2.58 sq ft)
- Tailplane ... 1.51 m² (16.25 sq ft)
- Elevators (total) ... 0.62 m² (6.67 sq ft)

WEIGHTS AND LOADINGS (estimated):
- Weight empty ... 375 kg (827 lb)
- Baggage capacity ... 30 kg (66 lb)
- Max T-O weight ... 575 kg (1,267 lb)
- Max wing loading ... 71.9 kg/m² (14.72 lb/sq ft)
- Max power loading ... 7.71 kg/kW (12.67 lb/hp)

PERFORMANCE (estimated):
- Max level speed at S/L ... 124 kt (230 km/h; 142 mph)
- Max cruising speed ... 113 kt (210 km/h; 130 mph)
- Stalling speed ... 52 kt (96 km/h; 60 mph)
- Max rate of climb at S/L ... 246 m (807 ft)/min
- T-O to 15 m (50 ft) ... 340 m (1,115 ft)
- Landing from 15 m (50 ft) ... 430 m (1,410 ft)
- Range with max fuel ... more than 378 n miles (700 km; 435 miles)
- Endurance ... 5 h
- *g* limits ... +4.4/–2.2

UPDATED

New Zealand

ALPHA

ALPHA AVIATION LTD

Hamilton Airport, Hamilton
CHAIRMAN: Richard Sealy
DIRECTORS: Richard Izard, David Stewart
CEO: Murray Dreyer

Alpha established December 2003 to build Robin R 2120 Alpha primary trainer in 14,000 m² (150,700 sq ft) new plant, employing up to 50 workers, at Hamilton. On 25 June 2004 announced imminent conclusion to purchase of rights, jigs and tools from Apex Aviation in France, anticipating first deliveries in 2005 and production build-up to 100 per year. Management includes Richard Izard of Robin dealership, Izard Pacific Aviation. Latter is shareholder, with HA Investments, Hammersmith Holdings and Hansen Aviation Services.

NEW ENTRY

ROBIN R 2120 ALPHA

TYPE: Primary prop trainer/sportplane.

PROGRAMME: Derived from original Robin HR 200/120B, which had first flown in 1971 and was built through much of 1970s; returned to production in 1993 with minor modifications as HR 200/120B. Final example of this version built in late 2000, by which time Robin 2120 and 2160 were in production.

R 2120 announced in early 2001. Combines fuselage, vertical tail surfaces, fuel capacity and power plant of Robin HR 200/120B with wings and horizontal tail surfaces of R 2160. Prototype (F-WZZX) completed mid-2000; demonstrator (F-WZZY) shown at Aero '01, Friedrichshafen, April 2001. First production aircraft (EC-HMG) to Spain by early 2001.

At Aero '03, in Friedrichshafen, Robin exhibited a fuselage mockup of an HR200 variant with Thielert Centurion 1.7 diesel engine driving a three-blade MTV-6-A/187-129 propeller; no details of planned production were given.

Design rights being acquired by Alpha Aviation of New Zealand, with intention of restarting production in 2005.

CURRENT VERSIONS: **R2120U:** Baseline (utility) version.

Alpha 120T: Marketing designation.

CUSTOMERS: Combined 331 HR 200/R 2120/R 2160s built by mid-2002 when production temporarily suspended; at least 12 of these were R 2120s. No production known in 2003. Deliveries in 2001 included six to New Zealand dealer Izard Pacific Aviation. Others to UK.

DESIGN FEATURES: Generally as Robin R 2120/Alpha 120T, but with extended rudder chord, underfin and other detail changes.

Wing section NACA 23015; dihedral 6° 20'; incidence 3° at root.

FLYING CONTROLS: Manual. Conventional fully balanced slotted ailerons and horn-balanced rudder without tabs. All-moving tailplane with anti-balance tabs each side. Electrically actuated slotted flaps.

STRUCTURE: Aluminium alloy wing spars and skinning; semi-monocoque fuselage. Anti-corrosion treatment standard.

Robin Alpha 120T (R 2120U) *(Paul Jackson)* ***NEW***/0580294

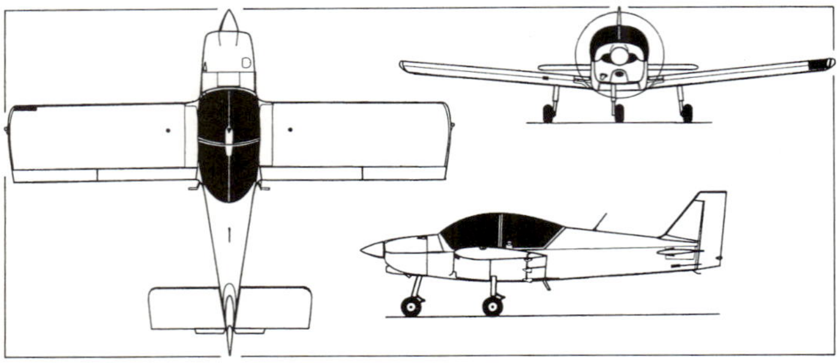

Robin R 2120 Alpha 0110870

LANDING GEAR: Non-retractable tricycle type with fairing, oleo-pneumatic shock-absorber and tyre (300×150) on each leg. Cleveland disc brakes on mainwheels. Nosewheel steering through rudder bar, self-centering in flight.

POWER PLANT: One 88.0 kW (118 hp) Textron Lycoming O-235-L2A flat-four engine, driving a Sensenich 74DM6S5-2-64 two-blade, fixed-pitch aluminium propeller. Provisional design studies completed for installation of 134 to 149 kW (180 to 200 hp) engine. Christen 603 inverted oil system and inverted fuel system standard. Fuselage fuel tank, capacity 120 litres (31.7 US gallons; 26.4 Imp gallons). If operated in Utility category, optional fuel tank of 160 litres (42.3 US gallons; 35.2 Imp gallons).

ACCOMMODATION: Two seats side by side on fore-and-aft adjustable, six-position bucket seats; five-point harnesses standard. Dual controls (sticks) standard.

SYSTEMS: 12 V 24 Ah electrical system.

DIMENSIONS, EXTERNAL:
Wing span .... 8.33 m (27 ft 4 in)
Wing aspect ratio 5.6
Wing chord, constant .... 1.555 m (5 ft 1¼ in)
Length overall .... 6.64 m (21 ft 9½ in)
Tailplane span .... 3.04 m (9 ft 11¾ in)
Tailplane chord .... 0.725 m (2 ft 4½ in)
Wheelbase .... 1.43 m (4 ft 8¼ in)
Propeller diameter .... 1.83 m (6 ft 0 in)

DIMENSIONS, INTERNAL:
Cabin max width .... 1.07 m (3 ft 6 in)

AREAS:
Wings, gross .... 12.50 m^2 (134.5 sq ft)
Ailerons, total .... 1.06 m^2 (11.41 sq ft)
Trailing-edge flaps, total .... 1.34 m^2 (14.42 sq ft)

WEIGHTS AND LOADINGS:
Weight empty .... 544 kg (1,199 lb)
Max T-O weight .... 800 kg (1,764 lb)
Max wing loading .... 64.0 kg/m^2 (13.11 lb/sq ft)
Max power loading .... 9.09 kg/kW (14.94 lb/hp)

PERFORMANCE:
Cruising speed at 75% power at 2,591 m (8,500 ft): .... 104 kt (193 km/h; 120 mph)
Max rate of climb at S/L .... 192 m (630 ft)/min
T-O to 15 m (50 ft) .... 490 m (1,610 ft)
Landing from 15 m (50 ft) .... 480 m (1,575 ft)
Range with max fuel .... 510 n miles (944 km; 586 miles)
g limits .... +6/−3.5

UPDATED

ROBIN R 2160

TYPE: Two-seat lightplane.

PROGRAMME: Certified in France in mid-1978 and in USA (FAR Pt 23 Aerobatic and Utility category) 15 November 1982. Some aircraft assembled in Canada (1983–85). Production then ceased, but restarted in France January 1994. Production expected to transfer to New Zealand in 2005.

CURRENT VERSIONS: **R 2160:** Baseline version, *as described below*.

R 2160i: Certified 8 July 1998, with 119 kW (160 hp) Textron Lycoming AEIO-320-D2B fuel-injected flat-four. First two to CATC training college in Thailand, 1998; at least six produced by 2004.

Alpha 160A: Aerobatic version.

Alpha 160Ai: Aerobatic version of R 2160i.

Alpha 160T: Basic trainer; introduced 2002.

CUSTOMERS: Total of over 133 sold in Europe, Australia, New Zealand, Canada and USA by late 2002, including 97 in first series. No known production in 2003, but one built for New Zealand in 2004.

DESIGN FEATURES: Generally as Robin R 2120/Alpha 120T, but with extended rudder chord, underfin and other detail changes.

Wing section NACA 23015; dihedral 6° 20'; incidence 3° at root.

FLYING CONTROLS: Manual. Conventional fully balanced slotted ailerons and horn-balanced rudder without tabs. All-moving tailplane with anti-balance tabs each side. Electrically actuated slotted flaps.

STRUCTURE: Aluminium alloy wing spars and skinning; semi-monocoque fuselage. Anti-corrosion treatment standard.

Robin R 2160 in Belgian ownership *(Paul Jackson)* ***NEW***/0572267

Robin HR 200/120B two-seat trainer, with additional scrap views of HR 200/160 forward fuselage and R 2160 empennage *(Paul Jackson)* 0137208

LANDING GEAR: Non-retractable tricycle type with fairing, oleo-pneumatic shock-absorber and tyre (300×150) on each leg. Cleveland disc brakes on mainwheels. Nosewheel steering through rudder bar, self-centering in flight.

POWER PLANT: One 119 kW (160 hp) Textron Lycoming O-320-D2A flat-four engine, driving a Sensenich 74DM6S5-2-64 two-blade, fixed-pitch aluminium propeller. Provisional design studies completed for installation of 134 to 149 kW (180 to 200 hp) engine. Christen 603 inverted oil system and inverted fuel system standard. Fuselage fuel tank, capacity 120 litres (31.7 US gallons; 26.4 Imp gallons). If operated in Utility category, optional fuel tank of 160 litres (42.3 US gallons; 35.2 Imp gallons).

ACCOMMODATION: Two seats side by side on fore-and-aft adjustable, six-position bucket seats; five-point harnesses standard. Dual controls (sticks) standard.

SYSTEMS: 12 V 24 Ah electrical system.

AVIONICS: To customer's choice, typically Bendix/King KX 155 nav/com, KI 208 VOR/LOC indicator, KT 76C transponder with ACK 30 altitude encoder, Garmin 150 XL GPS, and SPA 400 intercom system, microphone and handset jack plugs, and PTT switch on each control stick.

Instrumentation: Standard equipment includes ASI, sensitive altimeter, rate of climb indicator, artificial horizon, directional gyro, electric turn co-ordinator, magnetic compass, tachometer, eight-light annunciator/warning panel, OAT gauge, CHT gauge, EGT gauge, carburettor air temperature gauge (160A/T), oil pressure and temperature gauges, fuel quantity gauge, electronic fuel pressure management computer (160Ai), fuel pressure gauge (160Ai), flight time recorder, and *g* meter.

EQUIPMENT: Pitot heat, external power socket, wing-mounted taxi/landing lights, navigation lights, anti-collision strobe light, instrument panel lighting, and overall polyurethane paint with two-colour scheme, all standard.

DIMENSIONS, EXTERNAL:
Wing span .... 8.71 m (28 ft 7 in)
Wing chord, constant .... 1.56 m (5 ft 1¼ in)
Wing aspect ratio 5.8
Length overall .... 7.10 m (23 ft 3½ in)
Height overall .... 2.13 m (7 ft 0 in)
Tailplane span .... 3.03 m (9 ft 11¼ in)
Wheel track .... 2.91 m (9 ft 6½ in)
Wheelbase .... 1.44 m (4 ft 8½ in)
Propeller diameter .... 1.88 m (6 ft 2 in)

DIMENSIONS, INTERNAL:
Cabin: Max width .... 1.07 m (3 ft 6 in)
Height (seat cushion to canopy) .... 0.92 m (3 ft 0¼ in)

AREAS:
Wings, gross .... 13.01 m^2 (140.0 sq ft)

WEIGHTS AND LOADINGS:
Weight empty .... 555 kg (1,224 lb)
Max T-O weight: Aerobatic .... 800 kg (1,764 lb)
Utility .... 900 kg (1,984 lb)
Max baggage weight .... 40 kg (88 lb)
Max wing loading: Aerobatic .... 61.5 kg/m^2 (12.59 lb/sq ft)
Utility .... 69.2 kg/m^2 (14.17 lb/sq ft)
Max power loading: Aerobatic .... 6.71 kg/kW (11.02 lb/hp)
Utility .... 7.55 kg/kW (12.40 lb/hp)

PERFORMANCE (O-320 engine at Aerobatic max T-O weight):
Never-exceed speed (VNE) .... 180 kt (333 km/h; 207 mph)
Max level speed .... 138 kt (256 km/h; 159 mph)
Cruising speed:
at 75% power .... 131 kt (241 km/h; 151 mph)
at 65% power .... 120 kt (222 km/h; 138 mph)
Stalling speed: flaps up .... 52 kt (96 km/h; 60 mph)
flaps down .... 46 kt (86 km/h; 53 mph)
Max rate of climb at S/L .... 314 m (1,030 ft)/min
Service ceiling (30.5 m; 100 ft/min rate of climb) .... 4,575 m (15,000 ft)
T-O and landing run .... 230 m (755 ft)
T-O to 15 m (50 ft) .... 574 m (1,883 ft)
Landing from 15 m (50 ft) .... 440 m (1,444 ft)
Range at 65% power:
standard fuel .... 430 n miles (796 km; 494 miles)
optional fuel .... 558 n miles (1,033 km; 642 miles)
g limits .... +6/−3

UPDATED

PAC

PACIFIC AEROSPACE CORPORATION LIMITED

Private Bag HN 3027, Hamilton Airport, Hamilton
Tel:(+64 7) 843 61 44
Fax:(+64 7) 843 61 34
e-mail: pacific@aerospace.co.nz
Web: http://www.aerospace.co.nz
CEO: Paul Hebberd
MANAGING DIRECTOR: Brian Hare

Pacific Aerospace Corporation formed 1982 following acquisition of assets and undertakings of New Zealand Aerospace Industries; became wholly owned subsidiary of AeroSpace Technologies of Australia (75.1 per cent) and Lockheed Martin, USA (24.9 per cent). In October 1995, Aeromotive, a privately owned New Zealand company, purchased ASTA's 75.1 per cent share, and became sole shareholder in 2001. PAC maintains production and support facilities for its own aircraft, the CT4 Airtrainer series, Fletcher FU24 series, Cresco 08–600 and 08–750, and PAC 750XL; by March 2004 some 567 aircraft had been produced. Manufacturing facility also produces items for the Airbus A330/340. Workforce of 140 employed in the company's 24,155 m^2 (260,000 sq ft) facility.

UPDATED

PAC AIRTRAINER CT/4E

Royal Thai Air Force designation: BF 16

TYPE: Primary prop trainer/sportplane.

PROGRAMME: New Zealand redesign of Australian Victa Aircruiser; first flight 23 February 1972; deliveries began October 1973; 94 (plus two prototypes) built before production ended 1977 (75 CT/4A and 19 special military variants); production line reopened 1991 to build 12 civil CT/4Bs for Ansett Flying College and six for Thailand; last completed August 1992. First flight of CT/4C turboprop prototype (ZK-FXM, converted RAAF CT/4B) 21 January 1991.

CURRENT VERSIONS: **T350:** Turboprop version (previously known as CT/4C) with 313 kW (420 shp) Rolls-Royce 250-B17D (throttle limited to 261 kW; 350 shp) in lengthened nose. Certification was expected 1998 in anticipation of Southeast Asian order for 30, but this failed to materialise. Some interest still being shown in this version.

CT/4E: Developed version of CT/4B with more powerful engine; wing mounted slightly farther forward than on CT/4B; first flight (ZK-EUN, converted from RAAF CT/4A) 14 December 1991; NZ certification (FAR Pt 23 Amendment 36) 8 May 1992.

Detailed description applies to CT/4E.

CUSTOMERS: PAC previously built 75 CT/4As and 38 CT/4Bs; further production of CT/4E began in 1997, of which 13 supplied on 20 year lease to the Royal New Zealand Air Force as replacements for CT/4Bs; contract signed 18 August 1998; first (NZ1985) delivered to Pilot Training Squadron at Ohakea in same month; last (NZ1997) on 8 June 1999; by late 2000 the fleet had accumulated 10,000 flying hours. Also used by Red Checkers aerobatic team. Thailand has a requirement for 24 CT/4Es, initial order being for 12, deliveries of which began in July 1999 and were completed in June 2000 (final Thai aircraft first flew 11 May 2000). Further four Thai aircraft followed in August 2001. Total 32 (including demonstrator) built to late 2002; none further by early 2004, although further eight for Thailand then under construction. Two EFIS-equipped CT/4Es delivered to Singapore Youth Flying Club in June 2002.

DESIGN FEATURES: Metal translation of 1953-vintage wooden tourer. Conventional, low-wing lightplane. Tapered wings and constant-chord tailplane. Detachable wingtips to allow fitting of optional wingtip fuel tanks.

Wing section NACA 23012 (modified) at root, NACA 4412 (modified) at top; dihedral 6° 45' at chord line; incidence 3° at root; twist 3°; root chord increased by forward sweep of inboard leading-edges; small fence on outer leading-edges.

FLYING CONTROLS: Conventional and manual. Aerodynamically and mass-balanced bottom-hinged ailerons and statically balanced one-piece elevator actuated by pushrods; rudder by rod and cable linkage; ground-adjustable tab on rudder, electric trim control for elevator and rudder, electrically actuated single-slotted flaps.

STRUCTURE: Light alloy stressed skin except for Kevlar/GFRP wingtips and engine cowling; ailerons and flaps have fluted skins.

LANDING GEAR: Tricycle type; fixed. Fuselage-mounted spring steel cantilever mainwheel legs; steerable (±25°) nosewheel carried on telescopic strut and oleo shock-absorber. Main units have Dunlop Australia wheels and tubeless tyres size 15×6.00, pressure 1.65 bar (24 lb/sq in); nosewheel has tubeless tyre size 14×4.5, pressure 1.24 bar (18 lb/sq in). Single-disc toe-operated hydraulic brakes, with hand-operated parking lock. Minimum ground turning radius 2.90 m (9 ft 6 in). Landing gear designed to shear before any excess impact loading is transmitted to wing, to minimise structural damage in event of crash landing.

POWER PLANT: One 224 kW (300 hp) Textron Lycoming AEIO-540-L1B5 flat-six engine with inverted oil system three-blade Hartzell HC-C3YF-4BF/FC7663-2R or MT-Propeller MTV-9 series constant-speed metal propeller. Standard fuel capacity of 204.4 litres (54.0 US gallons; 45 0 Imp gallons) of which 200 litres (52.8 US gallons; 44.0 Imp gallons) are usable; wingtip tanks, each of 77.6 litres (20.5 US gallons; 17.1 Imp gallons) capacity, available optionally. Gravity fuelling point in each wing. Oil capacity 11.4 litres (3.0 US gallons; 2.5 Imp gallons).

ACCOMMODATION: Two seats side by side under hinged, fully transparent jettisonable Perspex canopy. Space to rear for optional third seat or 77 kg (170 lb) of baggage or equipment. Dual controls standard.

SYSTEMS: 28 V DC electrical system; ground power receptacle. Air conditioning optional.

AVIONICS: Garmin GNS 430 GPS/nav optional.

DIMENSIONS, EXTERNAL:
- Wing span: 7.92 m (26 ft 0 in)
- Wing chord: at root: 2.18 m (7 ft 2 in)
  - at tip: 0.94 m (3 ft 1 in)
- Wing aspect ratio: 5.2
- Length overall: 7.16 m (23 ft 5¾ in)
- Height overall: 2.59 m (8 ft 6 in)
- Fuselage: Max width: 1.12 m (3 ft 8 in)
  - Max depth: 1.40 m (4 ft 7¼ in)
- Tailplane span: 3.61 m (11 ft 10 in)
- Wheel track: 2.97 m (9 ft 9 in)
- Wheelbase: 1.66 m (5 ft 5½ in)

DIMENSIONS, INTERNAL:
- Cabin: Length: 2.74 m (9 ft 0 in)
  - Max width: 1.08 m (3 ft 6½ in)
  - Max height: 1.34 m (4 ft 5 in)

AREAS:
- Wings, gross: 11.98 m^2 (129.0 sq ft)
- Ailerons (total): 1.07 m^2 (11.56 sq ft)
- Flaps (total): 2.10 m^2 (22.60 sq ft)
- Fin: 0.78 m^2 (8.44 sq ft)
- Rudder, incl tab: 0.58 m^2 (6.26 sq ft)
- Tailplane: 1.51 m^2 (16.24 sq ft)
- Elevator: 1.56 m^2 (16.74 sq ft)

WEIGHTS AND LOADINGS:
- Weight empty, equipped: 807 kg (1,780 lb)
- Max fuel weight: 149 kg (328 lb)
- Max T-O weight: 1,179 kg (2,600 lb)
- Max wing loading: 98.38 kg/m^2 (20.15 lb/sq ft)
- Max power loading: 5.26 kg/kW (8.67 lb/hp)

PERFORMANCE:
- Never-exceed speed (VNE): 209 kt (387 km/h; 240 mph)
- Max level speed at S/L: 163 kt (302 km/h; 188 mph)
- Cruising speed at 2,590 m (8,500 ft) at 75% power: 152 kt (282 km/h; 175 mph)
- Stalling speed at S/L: flaps up: 60 kt (112 km/h; 69 mph)
  - flaps down: 44 kt (82 km/h; 51 mph)
- Max rate of climb at S/L, ISA: 558 m (1,830 ft)/min
- Service ceiling: 5,550 m (18,200 ft)
- Time to service ceiling: 13 min
- T-O run at S/L, ISA: 187 m (612 ft)
- Landing from 15 m (50 ft): 244 m (800 ft)
- Landing run: 169 m (553 ft)
- Range at S/L with max fuel at 75% power, ISA, no reserves: 520 n miles (963 km; 599 miles)

UPDATED

PAC CRESCO

TYPE: Agricultural sprayer.

PROGRAMME: Design began 1977; first flight of prototype (ZK-LTP) 28 February 1979; first flight of production aircraft early 1980; entered service January 1982 as Cresco 08-600; certified in Standard and Restricted categories 9 April 1984. First flight of PT6A-34AG version (ZK-TMN, c/n 010) 18 November 1992.

CURRENT VERSIONS: **Cresco 08-600:** Initial version, with 447 kW (599 shp) (flat rated) Honeywell LTP 101-600A-1A or -700A (derated to same power) turboprop. First nine aircraft (including prototype) only. No longer available.

Cresco 750: With higher-powered 559 kW (750 shp) Pratt & Whitney Canada PT6A-34AG engine; launched 1992 (first delivery, c/n 010, 23 December) as 08-600-34AG. However, aircraft are registered as 08-600s.

PAC 750XL: *Described separately.*

CUSTOMERS: Total 39 Crescos built by July 2003. One delivered to Kevron Pty Ltd in Australia during 1998 is used for geophysical exploration. Cresco operators located in New Zealand, Australia, Malaysia and Bangladesh.

COSTS: Approximately US$850,000 (2004).

DESIGN FEATURES: Turboprop development of earlier FU24 with approximately 60 per cent commonality of parts. Primary configuration as agricultural aircraft with appropriate hopper base; can also be used for firefighting. Some employed in sport parachuting, geophysical survey, freight and passenger services.

Wing section NACA 4415 (constant); 8° dihedral on outer wing panels; incidence 2°.

FLYING CONTROLS: Conventional and manual. Horn-balanced ailerons, horn-/mass-balanced elevator and rudder; electric elevator trim with manual override; all-moving tailplane with full-span anti-servo tab; ground-adjustable tab on rudder; single-slotted flaps; electrically operated through to 30°; row of vortex generators forward of each aileron.

STRUCTURE: Conventional light alloy; two-spar wing; cockpit area stressed for 25 *g* impact.

LANDING GEAR: Tricycle type; fixed. Wing-mounted cantilever main legs with integral oleo-pneumatic shock absorption. Cleveland wheels and hydraulic disc brakes on main units. Steerable nosewheel. Goodyear tyres, size 8.50–10 (10 ply) on mainwheels, pressure 2.62 bar (38 lb/sq in), nosewheel 8.5-6 (6 ply), pressure 2.07 bar (30 lb/sq in).

PAC CT/4E Airtrainer of Royal Thai Air Force 0137423

Pink PAC Cresco 750 VH-MOS used for parachutist dropping by Mile High P/L in Australia ***(Paul Jackson)*** ***NEW***/0552037

POWER PLANT: See Current Versions. Hartzell HC-B3TN-3D/T10282NS+4 three-blade constant-speed, fully feathering, reversible-pitch metal propeller. Four integral fuel tanks in wing centre-section: two coupled tanks totalling 283 litres (74.7 US gallons; 62.3 Imp gallons), of which 218 litres (57.5 US gallons; 48.0 Imp gallons) are usable and two coupled tanks totalling 281 litres (74.2 US gallons; 61.8 Imp gallons) all usable. Total usable fuel 499 litres (132 US gallons; 110 Imp gallons). Ferry configuration capacity 2,400 litres (634 US gallons; 528 Imp gallons) using hopper as auxiliary fuel tank. Four refuelling points in upper surface of each wing. Oil capacity 5.7 litres (1.5 US gallons; 1.25 Imp gallons).

ACCOMMODATION: Two-seat cockpit with sliding canopy; rear compartment, with port freight door, holds six passengers, equivalent freight or chemical hopper.

SYSTEMS: 24 V DC electric system. Air conditioning optional on 750XL.

EQUIPMENT: 1,893 litre (500 US gallon; 416 Imp gallon) liquid or 1,860 kg (4,100 lb) dry hopper on Cresco versions.

DIMENSIONS, EXTERNAL:
Wing span .... 12.81 m (42 ft 0 in)
Wing chord, constant .... 2.13 m (7 ft 0 in)
Wing aspect ratio 6.0
Length overall .... 11.07 m (36 ft 4 in)
Fuselage max width .... 1.22 m (4 ft 0 in)
Height overall .... 3.84 m (12 ft 7 in)
Tailplane span .... 4.93 m (16 ft 2 in)
Wheelbase .... 2.77 m (9 ft 1¼ in)
Wheel track .... 3.76 m (12 ft 4 in)
Propeller diameter .... 2.69 m (8 ft 10 in)
Propeller ground clearance .... 0.305 m (1 ft 0 in)
Cargo door (port): Height .... 0.94 m (3 ft 1 in)
Width .... 0.94 m (3 ft 1 in)
Height to sill .... 0.91 m (3 ft 0 in)

DIMENSIONS, INTERNAL:
Passenger/cargo compartment (aft of hopper):
Length .... 3.12 m (10 ft 3 in)
Max width .... 1.09 m (3 ft 7 in)
Max height .... 1.24 m (4 ft 1 in)
Floor area .... 2.79 m² (30.0 sq ft)
Volume: passenger version .... 3.4 m³ (120 cu ft)
cargo version .... 3.8 m³ (134 cu ft)
Hopper volume .... 1.86 m³ (66 cu ft)

AREAS:
Wings, gross .... 27.31 m² (294.0 sq ft)
Ailerons (total) .... 1.97 m² (21.18 sq ft)
Trailing-edge flaps (total) .... 2.92 m² (31.42 sq ft)
Fin .... 1.53 m² (16.47 sq ft)
Rudder .... 0.63 m² (6.77 sq ft)
Tailplane (excluding elevators) .... 3.13 m² (33.64 sq ft)
Elevator, incl tab .... 2.59 m² (27.92 sq ft)

WEIGHTS AND LOADINGS (Cresco 750):
Weight empty, equipped .... 1,338 kg (2,950 lb)
Max fuel .... 435 kg (960 lb)
Typical payload: Normal .... 1,111 kg (2,450 lb)
Agricultural (Restricted) .... 2,155 kg (4,750 lb)
Max disposable load (fuel + hopper):
Normal .... 1,524 kg (3,360 lb)
Agricultural (Restricted) .... 2,429 kg (5,356 lb)
Max T-O weight:
Normal .... 2,925 kg (6,450 lb)
Agricultural (Restricted) .... 3,742 kg (8,250 lb)
Max landing weight .... 2,925 kg (6,450 lb)
Wing loading: at Normal max T-O weight .... 107.1 kg/m² (21.94 lb/sq ft)
at Agricultural (Restricted) max T-O weight .... 137.1 kg/m² (28.08 lb/sq ft)
Power loading: at Normal max T-O weight .... 5.23 kg/kW (8.60 lb/shp)
at Agricultural (Restricted) max T-O weight .... 6.70 kg/kW (11.01 lb/shp)

PERFORMANCE (Cresco 750):
Never-exceed speed (VNE) .... 173 kt (320 km/h; 199 mph)
Max level speed at S/L:
750 .... 152 kt (282 km/h; 175 mph)
Max cruising speed at 75% power at 305 m (1,000 ft):
750 .... 140 kt (259 km/h; 161 mph)
Stalling speed, flaps down, power off:
750 .... 57 kt (106 km/h; 66 mph)
Max rate of climb at S/L: 750 .... 475 m (1,560 ft)/min
Max certified altitude .... 3,050 m (10,000 ft)
Absolute ceiling .... 8,230 m (27,000 ft)
T-O run: at MTOW .... 227 m (745 ft)
at 1,406 kg (3,100 lb) AUW .... 45 m (150 ft)
T-O to 15 m (50 ft) .... 325 m (1,065 ft)
Landing from 15 m (50 ft) .... 500 m (1,640 ft)
Landing run at 1,406 kg (3,100 lb) AUW with propeller reversal .... 86 m (285 ft)
Range at 75% power with standard fuel, no reserves:
normal fuel 750 .... 305 n miles (676 km; 420 miles)
additional fuel in hopper .... 2,037 n miles (3,774 km; 2,345 miles)
Endurance with standard fuel:
at 75% power .... 2 h 40 min
at 60% power .... 3 h 0 min

UPDATED

PAC 750XL

TYPE: Light utility turboprop.

PROGRAMME: Wide-body version of Cresco (described separately) with fuselage height overall raised 15 cm (6 in); design started January 2000 and prototype (ZK-XLA) first flown 5 September 2001. Optimised for parachute jumping. Other fields of use include carriage of freight, passengers and geophysical survey equipment and special operations.

Preliminary tests showed need for leading-edge root glove and increase of maximum flap angle from 30° to 40°. Certification by NZ CAA was achieved on 23 July 2003; FAA certification on 10 March 2004; EASA certification underway in mid-2004.

CUSTOMERS: Eighteen firm orders and 260 options held by early 2004. First production aircraft delivered to Great Lakes Skydiving Centre of Tanpo, New Zealand. Six delivered by July 2004, including one each to Switzerland and UK.

Transparent tambour parachute door, step and exterior hand-hold of parachutist-optimised PAC 750XL *(Paul Jackson)* ***NEW***/0589429

PAC 750XL carrying sport parachutists ***NEW***/0589476

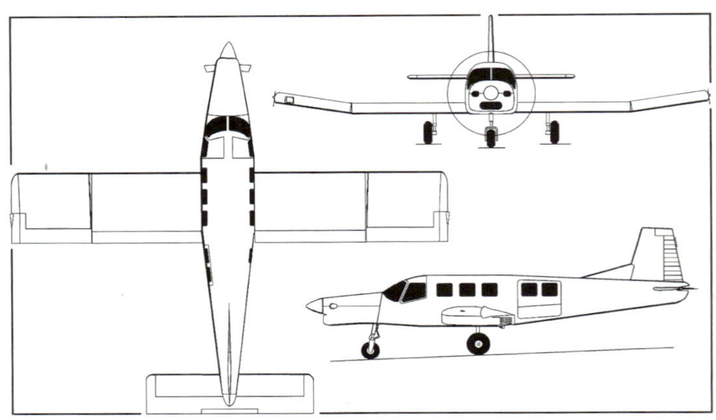

Pacific Aerospace Corporation 750XL general arrangement *(Paul Jackson)* ***NEW***/0552997

COSTS: Approximately US$1,000,000 (2004).

DESIGN FEATURES: Based on Cresco, but optimised for utility market, especially sport parachuting. Design objectives included good visibility from cockpit, ease of handling and high crosswind limit (15 kt).

Wing section NACA 4415 (constant); no dihedral on inner wing; 8° dihedral on outer wing panels; incidence 2°.

FLYING CONTROLS: Conventional and manual. Mass-balanced ailerons (travel +25/–10°), horn-/mass-balanced elevator (+30/8½°) and rudder; electric elevator trim with manual override; electric aileron trim (port); manual rudder trim; single-slotted flaps, deflection 21° for T-O, 40° for landing. Control surface actuation by cables; movements are ailerons +23°/–9° 30', elevators +10° 30'/–27° 30', rudder 20° port/25° starboard.

STRUCTURE: Conventional light alloy; single-spar wing; engine cowling of Kevlar and glass fibre; composites wingroot glove. Cockpit area stressed for 25 *g* impact.

LANDING GEAR: Tricycle type; fixed. Wing-mounted cantilever main legs with oleo-pneumatic shock absorption. Cleveland wheels and hydraulic disc brakes on main units. Goodyear tyres, size 8.50–10 (10 ply) on mainwheels, pressure 2.76 Bar (40 lb/sq in). Nosewheel 8.5-6 (6 ply), pressure 2.07 Bar (30 lb/sq in), steerable 20° port and 25° starboard. Minimum turning radius 12.27 m (40 ft 3 in).

POWER PLANT: One 559 kW (750 shp) Pratt & Whitney Canada PT6A-34 turboprop. Hartzell HC-B3TN-3D/T10282NS+4 three-blade constant-speed, fully feathering, reversible-pitch metal propeller. Four integral fuel tanks in wing centre-section, total capacity 861 litres (227 US gallons; 189 Imp gallons), of which 841 litres (222 US gallons; 185 Imp gallons) usable. Individual gross capacities: 284 litres (75.0 US gallons; 62.5 Imp gallons) front left; 293 litres (77.3 US gallons; 64.5 Imp gallons) front right; 142 litres (37.5 US gallons; 31.2 Imp gallons) rear left and same in rear right. Oil capacity 8.7 litres (2.3 US gallons; 1.9 Imp gallons). Four refuelling points in upper surface of each wing.

ACCOMMODATION: Cabin carries nine seated passengers at 86 cm (34 in) pitch or 18 floor-sitting parachutists (plus single pilot) and has 16 floor-mounted tiedown points and seat rails; cabin door, on port side behind wing, either two (fore- and aft-hinged) freight doors or tambour door composed of multiple, transparent, hinged elements, which slides upwards on tracks and can be closed by remote control by pilot; two upward-hinged doors on each side of cockpit for crew entry. Dual controls, but starboard control column removable. Max floor loading 730 kg/m^2 (150 lb/sq ft). Medevac version can carry two stretchers and three seated passengers, plus two flight crew.

SYSTEMS: 24 V DC electric system. Air conditioning optional.

AVIONICS: *Comms:* Garmin GMA 340 audio panel, GNC 250 com/nav and GTX 327 transponder.

Flight: Garmin GNS 430 GPS.

DIMENSIONS, EXTERNAL

Wing span	12.81 m (42 ft 0 in)
Wing chord, constant	2.13 m (7 ft 0 in)
Wing aspect ratio	6.0
Length overall	11.84 m (38 ft 10 in)
Fuselage:	
Max width	1.52 m (5 ft 0 in)
Max height	1.57 m (5 ft 2 in)
Height overall	4.04 m (13 ft 3 in)
Tailplane span	4.95 m (16 ft 3 in)
Wheelbase	3.17 m (10 ft 5 in)
Wheel track	3.68 m (12 ft 1 in)
Propeller diameter	2.69 m (8 ft 10 in)
Propeller ground clearance	0.41 m (1 ft 4 in)
Cabin door (port): Height	1.19 m (3 ft 11 in)*
Width	1.27 m (4 ft 2 in)
Height to sill	1.17 m (3 ft 10 in)

* *Slightly reduced by alternative door for parachuting*

DIMENSIONS, INTERNAL:

Cabin: Length	4.01 m (13 ft 2 in)
Max width	1.37 m (4 ft 6 in)
Max height	1.42 m (4 ft 8 in)

AREAS:

Wings, gross	28.34 m^2 (305.0 sq ft)
Ailerons (total)	2.04 m^2 (21.94 sq ft)
Trailing-edge flaps (total)	2.95 m^2 (31.74 sq ft)
Fin	1.80 m^2 (19.40 sq ft)
Rudder	1.08 m^2 (11.70 sq ft)
Tailplane (excluding elevators)	3.13 m^2 (33.64 sq ft)
Elevator, incl tab	2.59 m^2 (27.92 sq ft)

WEIGHTS AND LOADINGS:

Weight empty, equipped	1,406 kg (3,100 lb)
Max fuel weight	686 kg (1,512 lb)
Typical payload	1,996 kg (4,400 lb)
Max T-O weight	3,402 kg (7,500 lb)
Max landing weight	3,231 kg (7,125 lb)
Max wing loading	120.1 kg/m^2 (24.59 lb/sq ft)
Power loading at Normal max T-O weight:	6.09 kg/kW (10.00 lb/shp)

PERFORMANCE:

Never-exceed speed (VNE):	
cabin door closed	170 kt (314 km/h; 195 mph)
cabin door open	130 kt (240 km/h; 149 mph)
Max level speed at S/L	169 kt (313 km/h; 194 mph)
Stalling speed, flaps down, power off:	
at 3,402 kg (7,500 lb)	58 kt (108 km/h; 67 mph)
at 1,814 kg (4,000 lb)	45 kt (84 km/h; 52 mph)
Max rate of climb at S/L	547 m (1,795 ft)/min
Time to FL120	12 min
Time to S/L from FL120	3 min
Absolute ceiling	6,095 m (20,000 ft)
T-O run	259 m (850 ft)
T-O to 15 m (50 ft)	331 m (1,085 ft)
Landing from 15 m (50 ft)	426 m (1,400 ft)
Landing run	100 m (330 ft)
Range at 75% power with standard fuel, 45 min reserves	582 n miles (1,077 km; 669 miles)
Endurance with standard fuel, no reserves at 75% power	5 h 45 mins

UPDATED

Pakistan

PAC

PAKISTAN AERONAUTICAL COMPLEX

Kamra, District Attock

WORKS: F-6 Rebuild Factory; Mirage Rebuild Factory; **Kamra Avionics and Radar Factory**; Aircraft Manufacturing Factory (all at Kamra)

Tel: (+92 51) 927 01 11 to 927 01 17

Fax: (+92 51) 927 01 00

e-mail: info@pac.org.pk

Web: http://www.pac.org.pk

MANAGEMENT BOARD:

Air Vice-Marshal Aurangzeb Khan (Chairman)

Air Cdre Pervez Sadiq (Commercial)

Air Cdre Javed Iqbal Khan (Technical)

Badshah Gul (Finance)

MANAGING DIRECTORS:

Air Cdre Zakir Hussain Khan (AMF)

Air Cdre Khalid M Rajput (F-6RF)

Air Cdre Javed Khan (MRF)

Air Cdre Khalid Iqbal (KARF)

SALES AND MARKETING DIRECTOR: Arshad Majeed Khan

CHIEF TEST PILOT: Wg Cdr Ahmed Faiq

Pakistan Aeronautical Complex (PAC) is the nucleus of the aeronautical industry in Pakistan. Its professional standards and reliability enable domestic customers and foreign agencies alike to uncertake joint production with PAC in various fields. PAC Kamra comprises four factories: the Mirage Rebuild Factory (MRF), F-6 Rebuild Factory (F6-RF), Aircraft Manufacturing Factory (AMF) and Kamra Avionics and Radar Factory (KARF). All four are ISO 9000 certified.

Aircraft Manufacturing Factory (AMF): In operation since mid-1981, the AMF spearheads the aircraft manufacturing industry in Pakistan. It is the only aircraft manufacturing concern in Pakistan and is at present engaged in the manufacture of the light, robust, basic trainer-cum-surveillance aircraft Mushshak; the Super Mushshak — a variant with a more powerful engine, cockpit air conditioning, electric instruments and several other improvements; the Baaz and Ababeel low-speed target drones for anti-aircraft gunnery training; and development and manufacture (20 per cent), with NAMC of China, of the K-8 (Karakoram) jet trainer aircraft, already flying with the Pakistan Air Force. Manufacture of parts for Chengdu FC-1/JF-17 (eight of first 16 production aircraft) due to begin in 2005. First Kamra-assembled K-8 and JF-17 due for completion in 2006. In addition, the AMF has excellent facilities to rebuild and repair the Mushshak at factory level, along with after-sales spares support to existing Mushshak operators.

AMF also has the skills and capabilities of conventional/CNC machining, sheet metal forming, hand layup, contact suction hot press moulding, stretch moulding, gravity moulding, reaction injection moulding, mould/die and pattern making in woodwork, surface treatment/painting, heat treatment, tig welding (AC and DC), electric arc welding, gas welding, bronzing, soldering, tube banding, material testing, spectroanalysis, tensile testing, precision measurement chemical analysis, CNS calibration, co-ordinate measurement, mould/dies and fixture designing in tooling facility, CNC copy milling and EDM (spark erosion) milling.

A four-seat trainer for flying club use is said to be under development by the AMF, though no details of this programme have yet emerged.

Kamra Avionics and Radar Factory (KARF): KARF, the fourth factory of PAC, came into operation in 1987. It is an electronics centre with proven capabilities for rebuilding of radars, control and reporting centres (CRC), generators, and manufacture and repair of avionics systems. Currently, it is producing ESM equipment and airborne radar systems. All production activities are carried out in large, spaciously built shops equipped with state-of-the-art machinery. Apart from the production of BMJK/8602 RWR and Fiar Grifo 7 radar systems and the overhauling of pulse-Doppler radars, control and reporting centres and generators, KARF offers services in many fields such as surface mount technology, cable repair and manufacturing inspection/testing of microwave components, testing of gears and synchro bridge, environmental testing, testing hydraulic components, rebuilding of three-phase air conditioning systems and general engineering and painting facilities.

UPDATED

PAC (AMF) MUSHSHAK AND SUPER MUSHSHAK

English name: Proficient

TYPE: Three-seat lightplane.

PROGRAMME: Fifteen Mushshaks supplied complete from Sweden; 10 then assembled from SKD kits and 82 from CKD kits at Risalpur between 1975 and 1981; completely indigenous production followed, with 180 delivered by December 1996; 113 repaired at Kamra including 100 from 1980 to 1996; engines, instruments, electrical equipment and radios imported; complete airframe manufactured locally. Shahbaz (prototype 86–5147; first flight July 1987; English name Falcon) received US FAR Pt 23 certification 1989, but only four completed. Super Mushshak introduced 1997, entered production November 2000 and received domestic certification in 2002. Both versions still in production in 2004.

CURRENT VERSIONS: **Mushshak:** Standard production version; used for training, communications and observation; more than half a million hours flown by late 1999. AMF then upgrading first 16 (of 36) to Super Mushshak for PAF; 22 delivered by January 2003.

Super Mushshak: Further increase in power by adopting IO-540 flat-six engine. Prototype 95-5385X made first flight 15 August 1996. Revealed at Dubai Air Show in November 1997, at which time five completed. Pakistan Army has reportedly ordered 150, for delivery after completion of PAF order.

Description applies to above version, except where indicated.

CUSTOMERS: See table. Total of 315, comprising 15 imported aircraft plus 280 Mushshaks, 16 Super Mushshaks and four Shahbaz acquired/produced by May 2001 (first upgraded Super Mushshak redelivered to Pakistan Air Force 16 May). Mushshak deliveries to Iran in 1988–91; six supplied to Syria and three to Oman in 1994. Last-named became first export customer for Super Mushshak in December 2001; these delivered in mid-2002. Iranian Air Force reported to be considering upgrading its existing fleet to Supers. Sale of 20 Super Mushshaks to Saudi Arabia was reported in July 2004. Jordan and Yemen may also be potential export customers.

Instrument panel of the Super Mushshak 0084283

PAC Super Mushshak prototype *(Paul Jackson)* 0126891

MUSHSHAK PRODUCTION: (at January 2003)

Customer:	
At Risalpur:	
Pakistan Air Force/Army	92[1]
At Kamra:	
Pakistan Air Force	36[2]
Pakistan Army	117
Pakistan (Air Force and/or Army)	13
Iran	26[5]
Oman	8[4]
Syria	6
Pakistan Aeronautical Complex	4
Total	**302**[3]

[1] Saab kits (10 SKD, 82 CKD)
[2] First aircraft 83–5116 (numbers 5200–5299 not allocated); equip Nos. 1 and 2 Primary Flying Training Squadrons and three Headquarters Flights; being upgraded to Super in 2002–03
[3] Plus complete aircraft from Saab to PAF
[4] Includes five Super Mushshak; remaining three being upgraded to Super in late 2002/early 2003
[5] Includes one Super Mushshak donated to Air Force Academy in 2002

DESIGN FEATURES: Based on Swedish Saab Safari/Supporter, but without latter's armament option. Optimised for military support (originally including light attack) from semi-prepared airstrips; wing position aids ground observation.

Wing thickness/chord ratio 10 per cent; dihedral 1° 30'; incidence 2° 48'; sweepforward 5° from roots.

FLYING CONTROLS: Manual. Mass-balanced ailerons with servo tab in starboard unit; rudder with trim tab; all-moving mass-balanced tailplane with large anti-balance tab. Electrically actuated plain sealed flaps; leading-edge slots.

STRUCTURE: All-metal, except for GFRP tailcone, engine cowling panels, wing strut/landing gear attachment fairings and fintip.

LANDING GEAR: Tricycle type; fixed. Fuselage-mounted spring composites cantilever main legs; Cessna oleo-pneumatic nosewheel shock-strut. Goodyear wheels and Flight Custom tyres, size 6.00-6 (6 ply) on mainwheels and 5.00-5 (6 ply) on steerable (±30°) nosewheel. Tyre pressure (all units) 2.07 bar (30 lb/sq in). Parker Hannifin hydraulic disc brakes on main units.

POWER PLANT: *Mushshak:* One 149 kW (200 hp) Textron Lycoming IO-360-A1B6 flat-four engine, driving a Hartzell HC-C2YK-4F/FC7666A-2 two-blade constant-speed metal propeller.

Super Mushshak: One 194 kW (260 hp) Textron Lycoming IO-540-V4A5 flat-six; Hartzell HC-C2YK-1BF two-blade constant-speed propeller.

Fuel (both versions) in two integral wing fuel tanks, total capacity 178 litres (47.0 US gallons; 39.1 Imp gallons), of which 159 litres (42.0 US gallons; 35.0 Imp gallons) are usable. Oil capacity (both) 7.5 litres (2.0 US gallons; 1.6 Imp gallons). From 10 to 20 seconds inverted flight (limited by oil system) permitted.

ACCOMMODATION: Side-by-side adjustable seats, with provision for back-type or seat-type parachutes, for two persons beneath fully transparent upward-hinged canopy. Dual controls standard. Space aft of seats for 100 kg (220 lb) of baggage (with external access on port side) or, optionally, a rearward-facing third seat. Upward-hinged door, with window, beneath wing on port side. Cabin heated and ventilated in Mushshak, air conditioned in Super Mushshak.

SYSTEMS: Hydraulic reservoir for mainwheel brakes. 28 V DC electrical system (70 A alternator and 70 Ah battery).

AVIONICS: *Comms:* KT 76A/78A ATC transponder; KA 134 audio panel.

Flight: KX 125 nav/com; KLX 135 GPS/com; KR 87 ADF.

Instrumentation: Provision for full blind-flying instrumentation.

EQUIPMENT: Provision for six underwing attachment points, inner two stressed to carry up to 150 kg (330 lb) each and outer four up to 100 kg (220 lb) each. Options include ULV crop-spraying kit, target towing kit, or underwing supply/relief containers.

DIMENSIONS, EXTERNAL:
Wing span ... 8.85 m (29 ft 0½ in)
Wing chord: at root ... 1.21 m (3 ft 11¾ in)
outer panels, constant ... 1.36 m (4 ft 5½ in)
Wing aspect ratio ... 6.6
Length overall: Mushshak ... 7.00 m (22 ft 11½ in)
Super Mushshak ... 7.15 m (23 ft 5½ in)
Height overall ... 2.60 m (8 ft 6½ in)
Tailplane span ... 2.80 m (9 ft 2¼ in)
Wheel track ... 2.20 m (7 ft 2½ in)
Wheelbase: Mushshak ... 1.61 m (5 ft 3½ in)
Super Mushshak ... 1.58 m (5 ft 2¼ in)
Propeller diameter ... 1.93 m (6 ft 4 in)
Propeller ground clearance ... 7.62 cm (3 in)
Cabin door (port): Height ... 0.78 m (2 ft 6¾ in)
Width ... 0.52 m (1 ft 8½ in)

DIMENSIONS, INTERNAL:
Cabin: Max width ... 1.10 m (3 ft 7¼ in)
Max height (from seat cushion) ... 1.00 m (3 ft 3¼ in)

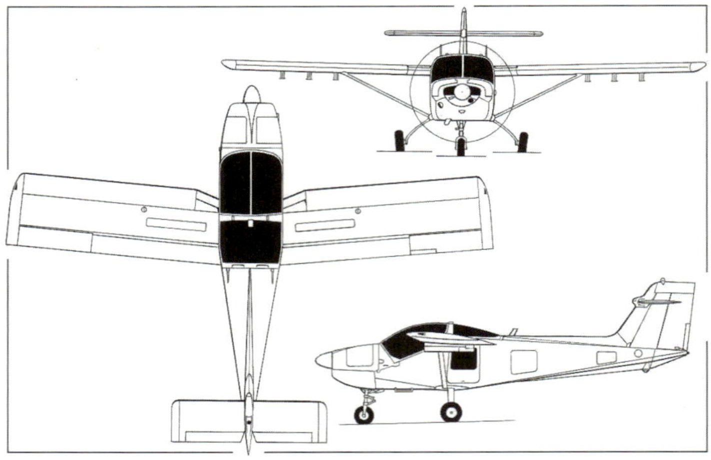

Pakistan Aeronautical Complex Mushshak two/three-seat light aircraft *(Jane's/Dennis Punnett)* 0507180

AREAS:
Wings, gross ... 11.90 m^2 (128.1 sq ft)
Ailerons (total) ... 0.98 m^2 (10.55 sq ft)
Flaps (total) ... 1.55 m^2 (16.68 sq ft)
Fin ... 0.77 m^2 (8.29 sq ft)
Rudder, incl tab ... 0.73 m^2 (7.86 sq ft)
Tailplane, incl tab ... 2.10 m^2 (22.60 sq ft)

WEIGHTS AND LOADINGS (Mushshak and Super Mushshak. A: Aerobatic, U: Utility, N: Normal category):
Weight empty, equipped: Mushshak ... 646 kg (1,424 lb)
Super Mushshak ... 760 kg (1,676 lb)
Max external stores load (both) ... 300 kg (661 lb)
Max T-O weight: A (both) ... 900 kg (1,984 lb)
U (Mushshak) ... 1,000 kg (2,205 lb)
U (Super Mushshak) ... 1,030 kg (2,270 lb)
N (Mushshak) ... 1,200 kg (2,645 lb)
N (Super Mushshak) ... 1,250 kg (2,755 lb)
Max wing loading: A (both) ... 75.6 kg/m^2 (15.48 lb/sq ft)
U (Mushshak) ... 84.0 kg/m^2 (17.20 lb/sq ft)
U (Super Mushshak) ... 86.6 kg/m^2 (17.73 lb/sq ft)
N (Mushshak) ... 100.8 kg/m^2 (20.65 lb/sq ft)
N (Super Mushshak) ... 105.0 kg/m^2 (21.51 lb/sq ft)
Max power loading (Mushshak):
A ... 6.04 kg/kW (9.92 lb/hp)
U ... 6.71 kg/kW (11.02 lb/hp)
N ... 8.05 kg/kW (13.23 lb/hp)
Max power loading (Super Mushshak):
A ... 4.64 kg/kW (7.63 lb/hp)
U ... 5.31 kg/kW (8.73 lb/hp)
N ... 6.45 kg/kW (10.60 lb/hp)

PERFORMANCE (Utility category):
Never-exceed speed (VNE):
both ... 196 kt (363 km/h; 226 mph)
Max level speed at S/L:
Mushshak ... 130 kt (240 km/h; 149 mph)
Super Mushshak ... 145 kt (268 km/h; 166 mph)
Cruising speed at S/L at 75% power:
Mushshak ... 115 kt (213 km/h; 132 mph)
Super Mushshak ... 130 kt (240 km/h; 149 mph)
Stalling speed, power off:
Mushshak, flaps up ... 60 kt (111 km/h; 69 mph)
Super Mushshak, flaps up ... 57 kt (105 km/h; 66 mph)
Mushshak, flaps down ... 54 kt (100 km/h; 63 mph)
Super Mushshak, flaps down
... 52 kt (96 km/h; 60 mph)
Max rate of climb at S/L:
Mushshak ... 372 m (1,220 ft)/min
Super Mushshak ... 518 m (1,700 ft)/min
Time to 1,830 m (6,000 ft): Mushshak ... 7 min 30 s
Time to 3,050 m (10,000 ft): Super Mushshak ... 10 min
Service ceiling: Mushshak ... 4,800 m (15,740 ft)
Super Mushshak ... 6,705 m (22,000 ft)
T-O run: Mushshak ... 150 m (495 ft)
Super Mushshak ... 183 m (600 ft)
T-O to 15 m (50 ft): Mushshak ... 252 m (825 ft)
Super Mushshak ... 275 m (900 ft)
Landing from 15 m (50 ft): Mushshak ... 314 m (1,030 ft)
Super Mushshak ... 296 m (970 ft)
Landing run: Mushshak ... 140 m (460 ft)
Super Mushshak ... 153 m (500 ft)
Range with max fuel:
Mushshak ... 580 n miles (1,074 km; 667 miles)
Super Mushshak ... 440 n miles (814 km; 506 miles)
Max endurance at 65% power at S/L, 10% reserves:
Mushshak ... 5 h 10 min
Super Mushshak ... 4 h 15 min
g limits (both): A ... +6/−3
U ... +5.4/−2.7
N ... +4.8/−2.4

UPDATED

Philippines

PADC

PHILIPPINE AEROSPACE DEVELOPMENT CORPORATION

PO Box 7395, Domestic Airport Post Office, Lock Box 1300, Domestic Road, Pasay City, Metro Manila
Tel: (+63 2) 832 37 57 and 23 81
Fax: (+63 2) 832 25 68
PRESIDENT: Reynato R Jose
SENIOR VICE-PRESIDENT: Josefina Laquindanum

PADC established 1973 as government arm for development of Philippine aviation industry; is now an attached agency of Department of Transportation and Communication (DOTC) and has technical workforce of about 200. Main activities are aircraft manufacturing and assembly; maintenance engineering, aircraft and spare parts sales; service centres for Britten-Norman Islander.

Past programmes have included licensed assembly of 67 Islanders (including 22 for Philippine Air Force), 24 Agusta S.211 jet trainers, 16 (of 18) SF-260TP turboprop trainers and 44 BO 105s. Six Lancair ESs and two Lancair IVs were being assembled from PAI kits in 1999 (latest information known) for delivery to the Philippine National Police (PNP).

PADC has maintenance/repair/overhaul centre for Rolls-Royce 250 series turbine engines and for Textron Lycoming and Teledyne Continental piston engines of up to 298 kW (400 hp). Its Maintenance and Engineering Department undertakes FMS work on 250-C30 engines for Sikorsky helicopters and 250-B17 turboprops and propellers for PAF Nomads.

PADC's 30/70 per cent joint venture with Eurocopter, known as Eurocopter Philippines, offers sales and after-sales service and assembly of Eurocopter helicopters from CKD kits. Two Ecureuils have been supplied to the PNP and one to the Department of Environment and Natural Resources (DENR).

PADC's latest known venture is a partnership with National Airmotive Corporation (NAC) of the USA to establish a centre in the Philippines to support the aviation needs of Southeast Asia (and especially the needs of the Philippine Air Force) regarding overhaul of Rolls-Royce T56/501 and 250 series engines, QEC overhaul and repair, propeller refurbishment and C-130 Hercules airframe components.

PAI

PACIFIC AERONAUTICAL INC

25 First Avenue, Mactan Export Processing Zone, Lapu-Lapu City 6015
Tel: (+63 32) 334 02 83/84/86
Fax: (+63 32) 334 02 85
e-mail: pacaero@skyinet.net
PRESIDENT: Robert C Fair

VICE-PRESIDENTS:
Wilfredo Dela Cruz (Production)
Augusto V Dayrit (Operations)

PAI continues to manufacture kits for Lancair aircraft (which see), and during 1999 (latest details known) shipped, on average, eight Lancair IV and two Lancair ES kits per month. Planned monthly rate for 2000 was three Lancair IV, two Lancair ES and six fast-build Legacy 2000 kits. The first batch of Legacy 2000 kits was shipped to the USA at the end of June 2000.

Poland

3XTRIM

ZAKŁADY LOTNICZE 3XTRIM SP. Z O.O. (3XTRIM AIRCRAFT FACTORY)

ulica Regera 109, PL-43 382, Bielsko-Biała
Tel: (+48 33) 818 92 51
Fax: (+48 33) 818 91 21
e-mail: biuro@3xtrim.pl
Web: http://www.3xtrim.pl
PRESIDENT: Dariusz Pietraszkiewicz
CHIEF DESIGNER: Jerzy Mastek

WORKS:
Bielsko-Biała and Szczecin; Warsaw factory planned

Company was formed in 1996 as WNKL A Kurbiel (former President, who died in July 2000), changing to its present title in November 1999, but has its roots in the SZD-PZL-Bielsko which for some 40 years has been the major producer of Polish high-performance sailplanes and gliders. Factories at Bielsko-Biała and Szczecin have floor areas of 700 and 1,500 m^2 (7,535 and 16,145 sq ft) respectively; combined workforce 20 in January 2003. Its main current products are the two-seat lightplanes and ultralights described in this entry, but glider production is under way and other (non-aerospace) products in composites are also manufactured.

UPDATED

3Xtrim 550 Trener ***NEW***/0576907

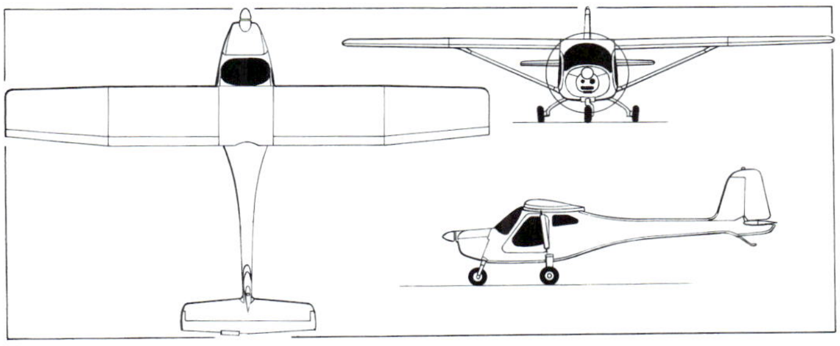

3Xtrim 550-VLA two-seat lightplane *(James Goulding)* 0093619

3XTRIM 450 ULTRA, 495 ULTRA PLUS AND 550 TRENER

TYPE: Side-by-side ultralight; two-seat lightplane.

PROGRAMME: Development of EOL-2 Racek 2, of which some 20 built. Design started 1996, prototype 450-UL making maiden flight May 1998; construction of first 550 Trener began 1997 (prototype SP-PUP made first flight 5 February 2000 and 150 hours flown by May 2001); static/fatigue testing October 1999 to January 2000; JAR-VLA tests completed 16 May 2000. 450 Ultra has Czech ultralight certificate and Polish ULA certification; VLA type certification for 550 Trener received 2003; French and German certification under way.

CURRENT VERSIONS: **450 Ultra:** Ultralight version. Reduction of MTOW to required 450 kg (992 lb) achieved by wider use of CFRP instead of GFRP and omission of some equipment and furnishing (for example, no door stays). Rotax 912 UL engine standard, but can be offered with 912 ULS for higher performance.

495 Ultra Plus: Designed to meet changes in ultralight classification; complies with JAR-VLA and satisfies new LIMAC AULA requirements. Identical to other versions except for Rotax 912A engine, wider use of CFRP and omission of some equipment and furnishing. Introduced for 2004.

550 Trener: For aero club and flying school (PPL) training; also available with IFR instrumentation as border patrol and reconnaissance aircraft.

3Xtrim 450 Ultra side-by-side ultralight ***NEW***/0538376

XXtrim: Modified version marketed in France by Aerotrophy. Reduced wing span and area (9.50 m; 31 ft 2 in and 11.90 m^2; 128.1 sq ft) and additional option of Jabiru engine. Performance generally comparable with that of Polish versions.

SF 45 SA Spirit: Version marketed in Germany by USA (which see).

CUSTOMERS: Total 30 450 Ultras and five 550 Treners ordered by December 2002, of which 26 delivered, starting 15 May 1998. Customer countries include France. At least 45 registered by mid-2004.

COSTS: 450 Ultra €48,800, 550 Trener €61,000, both including VAT (2002).

DESIGN FEATURES: Typical strut-braced high-wing cabin monoplane.

Wing section CAHI (TsAGI) R III (modified); dihedral 1° 30'; tailplane incidence −2° 30'.

FLYING CONTROLS: Conventional and manual (pushrod Frise-type ailerons and elevator, cable-operated rudder). Rudder and one-piece elevator horn-balanced; 450 Ultra has small, inset, electrically actuated trim tab in port half of elevator; tab in 550 Trener is twice as large and protrudes beyond trailing-edge. Three-position slotted flaps (0, 15 and 30°).

STRUCTURE: All-composites fuselage (glass and carbon fibre/epoxy cellular construction with plastics foam filling) with integral fin; GFRP and foam sandwich single-spar wing.

LANDING GEAR: Tricycle type; fixed. GFRP, self-sprung fuselage-mounted cantilever mainwheel legs and castoring (±15°), rubber-sprung nosewheel. Mainwheels are size 5.00-5, with 350×135 tyres on 550 Trener; size 4.00–6 on 450 Ultra. Tyre pressure (both) 2.50 bar (36 lb/sq in); hydraulic disc brakes. Floatplane version(s) planned.

POWER PLANT: *550 Trener:* One 73.5 kW (98.6 hp) Rotax 912 ULS flat-four engine, driving an MT-Propeller MT170R135-A2 two-blade fixed-pitch, KA-1/3P in-flight hydraulically controllable three-blade, or SR-2000 electrically controllable three-blade propeller.

450 Ultra: One 59.6 kW (79.9 hp) Rotax 912 UL and two-blade 3Xtrim 170 propeller with ground-adjustable pitch standard; optional controllable pitch propellers as for Trener. Rotax 912 ULS (see above) optional.

495 Ultra Plus: One 59.6 kW (79.9 hp) Rotax 912A; propeller options as for other versions.

Fuel capacity (all) 70 litres (18.5 US gallons; 15.4 Imp gallons) standard (65 litres; 17.2 US gallons; 14.3 Imp gallons usable), in single tank behind pilot's seat; optionally, two 43 litre (11.4 US gallon; 9.5 Imp gallon) tanks in same location. Oil capacity 2.5 litres (0.66 US gallon; 0.55 Imp gallon).

ACCOMMODATION: Two seats side by side; dual controls standard. Fully enclosed, heated and ventilated cabin with upward-opening door on each side. Baggage space behind seats.

SYSTEMS: Hydraulic system for brakes only; 12 V 18 Ah battery for electrical power; 250 W alternator optional.

AVIONICS: Bendix/King KY 97A and KT 76A or ICOM A-200 transponder; VFR instrumentation and stall warning system standard.

EQUIPMENT (EXCEPT ULTRA PLUS): Ballistic recovery parachute.

DIMENSIONS, EXTERNAL (A: 450 Ultra, B: 495 Ultra Plus, C: 550 Trener):
Wing span: A ... 9.60 m (31 ft 6 in)
B ... 9.80 m (32 ft 1¾ in)
C ... 10.03 m (32 ft 11 in)
Wing chord (all): at root ... 1.30 m (4 ft 3¼ in)
at tip ... 1.00 m (3 ft 3¼ in)
Wing aspect ratio: A ... 7.7
B ... 7.8
C ... 8.0
Length overall (all) ... 6.87 m (22 ft 6½ in)
Fuselage max width (all) ... 1.18 m (3 ft 10½ in)
Height overall (all) ... 2.405 m (7 ft 10¾ in)
Tailplane span (all) ... 2.75 m (9 ft 0¼ in)
Wheel track (all) ... 1.895 m (6 ft 2½ in)
Wheelbase (all) ... 1.48 m (4 ft 10¼ in)
Propeller diameter: MT, SR-2000 ... 1.70 m (5 ft 7 in)
KA-1 ... 1.60 m (5 ft 3 in)
Min propeller ground clearance ... 0.195 m (7¾ in)

AREAS:
Wings, gross: A ... 12.04 m^2 (129.6 sq ft)
B ... 12.31 m^2 (132.5 sq ft)
C ... 12.57 m^2 (135.3 sq ft)
Ailerons (total) ... 1.00 m^2 (10.76 sq ft)
Trailing-edge flaps (total): A ... 1.38 m^2 (14.85 sq ft)
C ... 1.50 m^2 (16.15 sq ft)
Fin ... 0.44 m^2 (4.74 sq ft)
Rudder ... 0.38 m^2 (4.09 sq ft)
Tailplane ... 1.10 m^2 (11.84 sq ft)
Elevator, incl tab ... 0.64 m^2 (6.89 sq ft)

WEIGHTS AND LOADINGS (A, B and C as above):
Weight empty: A ... 265 kg (584 lb)
B ... 295 kg (650 lb)
C ... 325 kg (717 lb)
Max fuel weight (all): standard ... 49 kg (108 lb)
max ... 61 kg (134 lb)
Max T-O weight: A ... 450 kg (992 lb)
B ... 495 kg (1,091 lb)
C ... 550 kg (1,212 lb)
Max wing loading: A ... 37.38 kg/m^2 (7.66 lb/sq ft)
B ... 40.21 kg/m^2 (8.24 lb/sq ft)
C ... 43.75 kg/m^2 (8.96 lb/sq ft)
Max power loading: A ... 7.55 kg/kW (12.41 lb/hp)
B ... 8.31 kg/kW (13.65 lb/hp)
C ... 7.48 kg/kW (12.29 lb/hp)

PERFORMANCE (A, B and C as above, Rotax 912 ULS engine):
Never-exceed speed (VNE):
A, B, C ... 116 kt (216 km/h; 134 mph)
Max cruising speed:
A, B ... 103 kt (190 km/h; 118 mph)
C ... 100 kt (185 km/h; 115 mph)
Cruising speed at 75% power:
A, B ... 94 kt (175 km/h; 109 mph)
C ... 91 kt (170 km/h; 106 mph)
Stalling speed, 30° flap: A ... 34 kt (63 km/h; 40 mph)
B ... 35 kt (65 km/h; 41 mph)
C ... 38 kt (70 km/h; 44 mph)
Max rate of climb at S/L: A, B ... 315 m (1,033 ft)/min
C ... 285 m (935 ft)/min
Service ceiling ... 4,000 m (13,120 ft)
T-O run ... 50 m (165 ft)
T-O to, and landing from, 15 m (50 ft):
A, B ... 225 m (740 ft)
C ... 250 m (820 ft)
Range with max standard fuel:
A, B, C ... 405 n miles (750 km; 466 miles)[1]
g limits: A, B, C ... +4.0/−2.0

OPERATIONAL NOISE LEVEL (ICAO Chapter 10, Appendix 16):
Average ... 67 dBA

[1] *C 675 n miles (1,250 km; 776 miles) with larger tanks*

UPDATED

AERO

AERO SP. Z O. O.

Wał Miedzeszyński 844, PL-03-942 Warszawa
Tel: (+48 22) 616 20 87
Fax: (+48 22) 617 85 28
e-mail: aero@post.pl
Web: http://www.aero.com.pl
PRESIDENT AND CHIEF DESIGNER: Ing Tomasz Antoniewski

UK DISTRIBUTOR:
S2T Aviation, 125 High Road, North Weald, Essex CM16 6EA
Tel: (+44 8700) 11 28 73
Fax: (+44 8700) 11 28 74
e-mail: info@s2taviation.com
Web: http://www.s2taviation.com

Company formed in 1994. Three P.220 Panda kits were ordered from Aerotechnik of Czech Republic that year, being given Polish designation suffix AT-2 (AT-1 having been a single-seat homebuilt design started in 1987). These all differed dimensionally and were variously fitted with Walter Mikron or 55.9 kW (75 hp) Limbach engines (first flight 12 December 1995 by c/n 001 SP-PUL, later SP-FUL). Third aircraft subsequently developed with more powerful Limbach and other modifications for production in Poland as AT-3; now available also with similarly rated Rotax power plant.

UPDATED

New instrument panel and ventilated cabin of the AT-3 R100 ***NEW***/0576928

AERO AT-3

TYPE: Side-by-side lightplane/kitbuilt.

PROGRAMME: Completed initially (c/n 003, SP-PUH) as P.220 S-AT2, but subsequently modified to AT-3 with 67.1 kW (90 hp) Limbach L-2400EB3AC engine, revised dimensions and higher weights; given new c/n of 001 and first flown in new configuration 18 January 1998. Received JAR-VLA certification 14 May 1999 and re-registered SP-GUH. Five production aircraft completed and five more under construction by May 2001. German certification then being sought; US planned later. Evaluated by Mooney in January 2004 as potential addition to its range in new light sport aircraft category, but dealership rights (in US and remainder of the world, except UK and Poland) were then acquired by AveoTech, whose Polish-built AveoSport was launched at Sun 'n' Fun in April 2004. However, this dealership had been suspended by June 2004, with new distributer possibly to be appointed.

CURRENT VERSIONS: **AT-3 L100:** With 74.6 kW (100 hp) Limbach L-2400DF1 flat-four engine and MT-160L-120-2C propeller. Four built by 2003, at which time fourth was under conversion to R100 version; none further expected.

AT-3 R100: With 73.5 kW (98.6 hp) Rotax 912 ULS flat-four and MT-165R-152-2M or GT-2/173/VRR-FW101SRTC propeller. Otherwise as AT-3 L100. Standard version from 2003. Marketed in US as AveoXTC.

AT-3SK: UK kit version.

AERO AT-3 PRODUCTION (AT MID-2004)

c/n	Registration	Type	Remarks
001	SP-TPC	L100	To R100 by 2004
002	SP-TPD	L100	
003	SP-TPE	L100	
004	SP-TPF	L100	To R100 in 2003; to
	N55XT	AveoSport	USA April 2004
005	SP-TPG	R100	US demonstrator 2004
006	G-UKAT	AT-3SK	Kitbuilt
007	VH-	R100	Australian owner
008	SP-EAR	R100	UK demonstrator
009		AT-2SK	Kitbuilt
010	SP-KOS	R100	

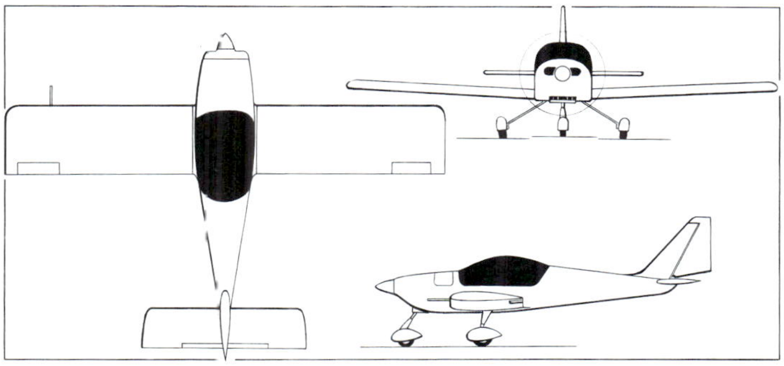

Aero (Antoniewski) AT-3 two-seat lightplane *(James Goulding)* 0121432

AveoSport: US version, which, with increased MTOW AveoX, has large winglets and is distributed by AveoTech of Denton Texas. Described separately.

CUSTOMERS: Initial customer Telekomunikacja Polska, for support of Polish aero clubs; first two in service by May 2001. Refer to table.

COSTS: Flyaway €79,000 (US$69,000), kit from €26,000 (US$22,000), excluding VAT (2003).

DESIGN FEATURES: Low, constant-chord wing and tailplane; sweptback fin. Large rudder cutaway for elevator travel.

Wing section NACA 4415.

Changes from AT-2/P.220 S include approximately 1 m² (10.8 sq ft) more wing area, 4 cm (1.6 in) longer cockpit, enlarged fin and rudder, and strengthened landing gear.

FLYING CONTROLS: Manual; piano-type top-hinged ailerons, horn-balanced rudder and all-moving elevator with central anti-balance tab; two-position split flaps (maximum deflection 40°).

STRUCTURE: Primarily metal, with some skins and fairings of carbon or glass fibre.

LANDING GEAR: Tricycle type; fixed. Cantilever, fuselage-mounted self-sprung main legs with 380×150/15×6.00 wheels and Cleveland hydraulic disc brakes; levered suspension nose leg with 5.00-4 wheel and rubber-in-compression shock absorption. Wheel speed fairings optional.

POWER PLANT: One Limbach or Rotax flat-four engine (see under Current Versions) with two-blade, fixed-pitch propeller. Fuel capacity (both) 70 litres (18.5 US gallons; 15.4 Imp gallons).

ACCOMMODATION: Two seats side by side; dual controls standard. Hatrack/baggage shelf behind seats, capacity 30 kg (66 lb). Upward-opening one-piece canopy, hinged at front.

AVIONICS: Bendix/King KY 97A transceiver and VFR instrumentation standard; IFR, including KY 197A, KR 87 ADF, KT 76A transponder and directional gyro, optional.

DIMENSIONS, EXTERNAL:
- Wing span ... 7.55 m (24 ft 9¼ in)
- Wing chord, constant ... 1.27 m (4 ft 2 in)
- Wing aspect ratio ... 6.1
- Length overall ... 6.25 m (20 ft 6 in)
- Height overall ... 2.23 m (7 ft 3¾ in)
- Elevator span ... 2.79 m (9 ft 1¾ in)
- Wheel track ... 2.26 m (7 ft 5 in)
- Wheelbase ... 1.36 m (4 ft 5½ in)
- Propeller diameter ... 1.65 m (5 ft 5 in)

AREAS:
- Wings, gross ... 9.30 m² (100.1 sq ft)

WEIGHTS AND LOADINGS:
- Weight empty: L100 ... 372 kg (820 lb)
- R100 ... 350 kg (772 lb)
- Max T-O weight: both ... 582 kg (1,283 lb)

Rotax-engined Aero AT-3 R100, now the standard version *(Aero/Frank Herzog)* ***NEW***/0576927

- Max wing loading: both ... 62.6 kg/m² (12.82 lb/sq ft)
- Max power loading: L100 ... 7.81 kg/kW (12.83 lb/hp)
- R100 ... 7.92 kg/kW (13.01 lb/hp)

PERFORMANCE:
- Never-exceed speed (VNE):
  - both ... 120 kt (223 km/h; 138 mph) IAS
- Max level speed: L100 ... 110 kt (205 km/h; 127 mph) IAS
  - R100 ... 116 kt (215 km/h; 133 mph) IAS
- Max cruising speed:
  - L100 ... 100 kt (185 km/h; 115 mph) IAS
  - R100 ... 108 kt (200 km/h; 124 mph) IAS
- Max manoeuvring speed (VA):
  - both ... 106 kt (197 km/h; 122 mph) IAS
- Stalling speed: both ... 45 kt (83 km/h; 52 mph) IAS
- Max rate of climb at S/L: L100 ... 210 m (689 ft)/min
  - R100 ... 222 m (728 ft)/min
- Service ceiling: both ... 3,800 m (12,460 ft)
- T-O run: L100 ... 170 m (560 ft)
  - R100 ... 160 m (525 ft)
- T-O to 15 m (50 ft):
  - L100 ... 465 m (1,525 ft)
  - R100 ... 440 m (1,445 ft)
- Landing from 15 m (50 ft):
  - both ... 445 m (1,460 ft)
- Landing run: L100 ... 160 m (525 ft)
  - R100 ... 150 m (495 ft)
- Max range: both ... 488 n miles (904 km; 561 miles)
- *g* limits: both ... +3.8/−1.5

UPDATED

AGROLOT

FUNDACJA AGROLOT (AGROLOT FOUNDATION)

aleja Krakowska 110/114, PL-00-971 Warszawa
Tel: (+48 22) 612 10 00
e-mail: agrolot@op.pl
CHAIRMAN: Andrzej Słociński

This foundation (established 1990) is responsible for the PZL-126P Mrówka programme previously managed by PZL Warszawa-Okecie. The prototype made its maiden flight in October 2000, some three years later than originally planned but full flight test programme not started until late 2003.

UPDATED

AGROLOT PZL-126P MRÓWKA 2001

English name: Ant

TYPE: Agricultural sprayer.

PROGRAMME: PZL-126 prototype (SP-PMA) built by PZL Warszawa-Okecie and made first flight 20 April 1990; this aircraft was used in 1994 to test a modified 44.7 kW (60 hp) PZL-F2A-120-C1 two-cylinder engine driving a two-blade fixed-pitch propeller. It was then withdrawn for rebuilding as the prototype PZL-126P planned operational version, which was also designed and developed at the PZL Warszawa-Okecie factory. Project is funded by Agrolot Foundation chief designer is Andrzej Słociński. First flight of the PZL-126P (SP-PMB) was first envisaged for 1997, but aircraft not completed until 1998, marked 'Mrówka 2001'; first flight was eventually made on 20 October 2000, but full flight test programme not started until late 2003. Agrolot hopes to complete it in 2005.

CURRENT VERSIONS: **PZL-126P 2001:** Under conversion since 1995. Increased wing span and area, more powerful engine, larger fuel tanks and increased payload compared with PZL-126 prototype. Intended as economical carrier of airborne systems for forestry protection (for example, combating pests, identifying diseased vegetation, detecting/controlling forest fires, spreading non-toxic small-volume selective-action agents for plant protection such as Trichrogramma parasitic wasp eggs laid in eggs of host insect). Other potential applications include environmental monitoring, glider towing and pipeline patrol; the possibility of an unmanned version with NBC detection equipment has also been reported.

Description applies to PZL-126P 2001.

DESIGN FEATURES: Single-engined low-wing monoplane with rectangular tail surfaces; meets requirements of FAR Pt 23. Semi-monocoque fuselage with equipment bay in lower portion aft of cockpit, accessible through hatch; upswept rear fuselage facilitates access to this bay. Quick-fastening lock beneath each wingtip for attachment of integral spraypod between main wing and tip. Aircraft can be dismantled quickly for towing on own landing gear by light all-terrain vehicle.

Wing aerofoil section NASA GA(W)-1; dihedral 3°.

FLYING CONTROLS: Manual. Three-section area-increasing flaps and flaperons on each wing, actuated by pushrods and interconnected by single central system located under cockpitfloor; elevator actuated by pushrods, rudder by cables from adjustable rudder pedals; trim tab on elevator, ground-adjustable tab on rudder.

STRUCTURE: All-metal except for GFRP/epoxy engine cowling, laminated wingtips and fairings. Single-spar wings and fixed tail surfaces; wings mounted on fuselage mandrels and offset spars connected with single pin; front part of fuselage has two 'tusks' for engine and nosewheel mounting.

LANDING GEAR: Non-retractable tricycle type. Self-springing cantilever mainwheel legs ofduralumin; castoring nosewheel with oleo-pneumatic shock-absorber. Wheel and tyre size 350×135 on all three units. Hydraulic differential mainwheel disc brakes.

POWER PLANT: One 74.6 kW (100 hp) Teledyne Continental O-200A flat-four engine, driving a McCauley 1A100MCM6950 two-blade metal propeller. Integral fuel tank in each wing, combined capacity 180 litres (47.6 US gallons; 39.6 Imp gallons).

ACCOMMODATION: Single-seat cockpit, fitted with airbag enabling in-flight seat adjustment. One-piece organic glass moulded canopy, opening sideways to starboard.

SYSTEMS: Hydraulic system for mainwheel brakes only; 24 V electrical system; hand pump to inflate seat cushion.

AVIONICS: *Comms:* Bendix/King KX 155 com/nav, KY 96A VHF transceiver and KT 76A transponder.

Flight: GPS 95XL global positioning; stall warning system.

Instrumentation: Prototype has IFR capability, but basic version fitted with VFR instrumentation. Instrument panel, with cover, can be lifted up for access to instruments and front part of cockpit.

EQUIPMENT: Spraypod can be fitted to outer end of each main wing panel, inboard of detachable tip. Pod is an integral unit with all necessary attachments for spraying, attached

Prototype PZL-126P Mrówka ***NEW***/1036586

by quick-fastening locks, electric multiplug socket supplying power to atomisers, and electric valve controlling outflow of liquid from tank. Optional third spray unit can be mounted under fuselage. Up to 35 ha (86.5 acres) can be sprayed in one flight with 70 litres (18.5 US gallons; 15.4 Imp gallons) of pesticide delivered at rate of 2 litres (0.53 US gallon; 0.44 Imp gallon)/ha.

Special biological spreader developed for dosing and spreading eggs of Trichrogramma wasp; eggs are carried in capsules in reel of tape covered with thin paper, spreader holding four such reels. Spreader is powered by electric motor and activated or stopped by push-button on throttle lever; 3 kg (6.6 lb) load of capsules is sufficient, at typical dispersal rate of four capsules every 50 m (164 ft) in rows 50 m apart, to seed an area of 800 ha (1,977 acres). Rotational mounting of spreader allows for full deflection and easy replacement of egg reels.

Aircraft can also carry miniaturised equipment such as photographic, video, thermal or other systems, coupled to satellite navigation system, monitoring and recording pictures or other signals from space.

DIMENSIONS, EXTERNAL:
Wing span: excl pods ... 7.66 m (25 ft 1½ in)
incl pods ... 8.46 m (27 ft 9 in)
Wing chord, constant:
excl extended flaps/flaperons ... 0.75 m (2 ft 5½ in)
incl extended flaps/flaperons ... 0.92 m (3 ft 0¼ in)
Wing aspect ratio ... 8.5

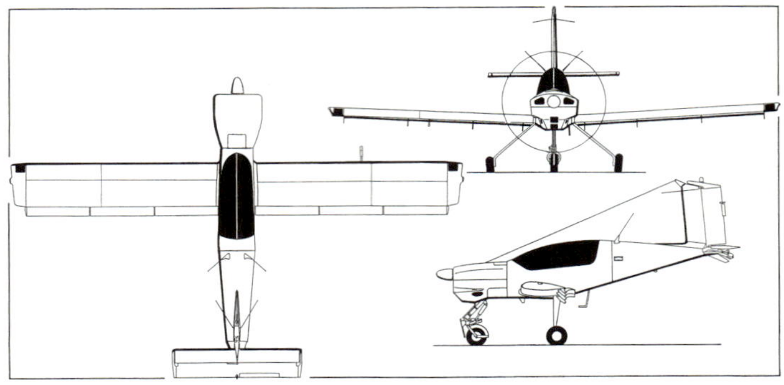

PZL-126P Mrówka light agricultural aircraft *(Paul Jackson)* 0011923

Length overall ... 5.25 m (17 ft 2¾ in)
Height overall ... 2.80 m (9 ft 2¼ in)
Tailplane span ... 2.20 m (7 ft 2½ in)
Wheel track ... 2.15 m (7 ft 0¾ in)
Wheelbase ... 1.35 m (4 ft 5¼ in)
Propeller diameter ... 1.75 m (5 ft 9 in)
AREAS:
Wings, gross: excl pods ... 6.87 m^2 (73.9 sq ft)
incl pods ... 7.27 m^2 (78.3 sq ft)
Flaperons (total) ... 0.48 m^2 (5.17 sq ft)
Trailing-edge flaps (total) ... 0.97 m^2 (10.44 sq ft)
Vertical tail surfaces (total) ... 0.73 m^2 (7.86 sq ft)
Horizontal tail surfaces (total) ... 1.08 m^2 (11.63 sq ft)
WEIGHTS AND LOADINGS:
Max T-O and landing weight ... 575 kg (1,267 lb)
Max wing loading:
without pods ... 83.7 kg/m^2 (17.14 lb/sq ft)
with pods ... 79.1 kg/m^2 (16.20 lb/sq ft)
Max power loading ... 7.72 kg/kW (12.68 lb/hp)
PERFORMANCE (estimated):
Max level speed ... 105 kt (195 km/h; 121 mph)
Operating speed ... 75 kt (140 km/h; 87 mph)
Max rate of climb at S/L ... more than 360 m (1,181 ft)/min

UPDATED

EADS PZL

EADS PZL WARSZAWA-OKĘCIE SA

aleja Krakowska 110/114, PL-00-971 Warszawa
Tel: (+48 22) 577 22 02
Fax: (+48 22) 577 22 03
e-mail: escons10@attglobal.net
Web: http://www.pzl-okecie.com.pl
PRESIDENT AND CEO: Carlos Navarro
VICE-PRESIDENT AND OPERATIONS DIRECTOR: Władisław Skorski
CHIEF MARKETING AND SALES EXECUTIVE: Zbigniew Wasiucionek, MSc(Ec)

Okęcie factory founded in 1928 and has subsequently built 71 different aircraft types, 48 of which went into series production. Currently occupies 279,373 m^2 (3,007,140 sq ft) site with 88,691 m^2 (954,660 sq ft) covered area. Responsible for light aircraft development and production, and for design and manufacture of associated agricultural equipment for its own aircraft and those built at other Polish factories; known as PZL Warszawa-Okęcie; became public stock company, entirely owned by Ministry of Industry and Trade, on 2 January 1995; by 1 January 2000 had produced 3,777 aircraft since 1945, including 33 in 1995, eight in 1996; 11 in 1997, 20 in 1998 and eight in 1999; workforce in 2002 was over 800. ISO 9001 status was awarded on 3 December 1999.

On 16 October 2001, as part of the contract to supply eight C-295M transports to the Polish Air Force, EADS CASA (which see) acquired a 61 per cent holding in EADS PZL, remaining 39 per cent being held by Polish Treasury and company employees. Facilities then refurnished and modernised for new activities. These will include increasing importance as an aerostructures centre and as a supplier of crop-spraying/firefighting aircraft and services. Components for the CASA C-295 and CN-235 transports are already being produced by Okęcie (first C-295 shipset delivered 22 May 2003), and extensive modernisation of the PZL-130TC-1 Orlik is to be undertaken for the Polish Air Force. Manufacture of small components for Airbus range starting in 2004. Discussions were under way in mid-2003 of a proposal to combine the activities of EADS PZL with those of EADS Socata.

Company continues to seek EC funding to develop the **PZL-240 Pelikan** biplane water bomber project, last reported in the late 1990s.

UPDATED

Wilga 2000 imported into the UK *(Paul Jackson)* ***NEW***/0547741

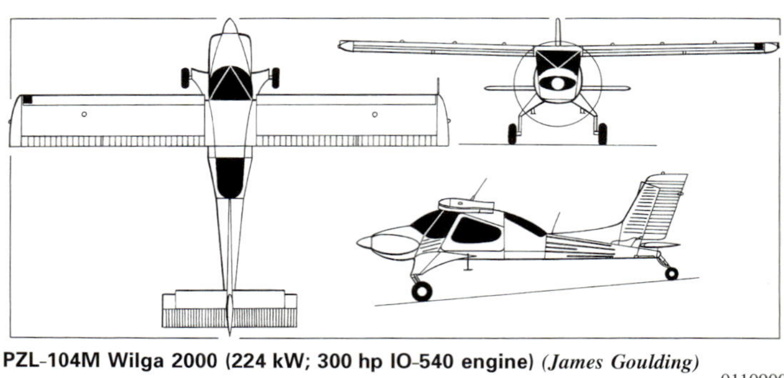

PZL-104M Wilga 2000 (224 kW; 300 hp IO-540 engine) *(James Goulding)* 0110900

EADS PZL PZL-104M WILGA 2000

English name: Oriole

TYPE: Four-seat lightplane.

PROGRAMME: First flight of prototype Wilga 1, 24 April 1962. Total of 962 of all earlier versions (except Indonesian licence-built Lipnur Gelatik) built by 1996, some of which stored and later supplied from stock, including one Wilga 80 in 1999.

Wilga 2000, improved version of Wilga 35, developed to appeal to Western markets, principally through use of Western engine and avionics. Wing and nose modifications tested on Wilga 35A SP-CSG. First flight of Wilga 2000 prototype (SP-PHG, later -WHG) 20 August 1996; first production (SP-AHV) in mid-1997; FAR Pts 23/36 certification achieved 25 February 2002. SP-PHG won 1st World Air Games, 1997.

CURRENT VERSIONS: **PZL-104M Wilga 2000:** Standard landplane version; *as described.*

PZL-104MN Wilga 2000: Designation of 2001 and subsequent production. FAR Pt 23 certification 25 February 2002. From 2003, production aircraft have 1,500 kg (3,306 lb) MTOW and approximately 925 kg (2,039 lb) empty weight, reflecting improvements of 100 kg (220 lb) and 50 kg (110 lb), respectively.

PZL-104MW Wilga 2000 Hydro: Twin-float version; first flight 11 September 1999. Maximum T-O weight 1,500 kg (3,306 lb); performance generally similar to landplane except T-O run from water 180 m (590 ft) and S/L climb rate 210 m (689 ft)/min. North American demonstrator under conversion in 2003 by Sealand Aviation of Vancouver, Canada; public debut in Anchorage, Alaska, May 2004.

CUSTOMERS: Ten reported orders for Wilga 2000 by late 1996, including four for Polish National Aeroclub and three for export. First two deliveries in 1997 to Polish Aeroclub and private owner; five built in 1998 and delivered under 1997 contract to Polish Border Guard, first one being handed over 1 June 1998; one built in 1999, delivered to UK as demonstrator in May 2000. None built in 2000. At least five built in 2001, increasing production to 14 (including prototype); no 2002 production; one built in 2003 and exported to Canada; total

15. Chinese interest reported in 2003; US law enforcement market being targeted early 2004 for Wilga 2000 equipped with IAI Taman POP-200 EO/IR sensor in belly-mounted turret.

DESIGN FEATURES: Suitable for wide variety of military, general aviation and flying club duties; STOL performance bestowed by flaps and slats allied to heavy-duty landing gear. High-mounted cantilever wings; braced tail unit; tall landing gear legs. In addition to Western equipment, Wilga 2000 features increased fuel, strengthened wing with integral fuel tanks and shorter mainwheel legs enclosed in fairings.

Wing section NACA 2415; dihedral 1°; incidence 8° 32'.

FLYING CONTROLS: Conventional and manual. Aerodynamically and mass-balanced slotted ailerons; tab on starboard aileron; aerodynamically, horn- and mass-balanced one-piece elevator and rudder; trim tab in centre of elevator; manually operated slotted flaps; fixed slat on wing leading-edge along full span. Aileron movement 26° up/16° down; elevator 38° up/18° down; elevator tab 20° up and down; rudder 26° left/right; flaps 0° up/21° down/44° down.

STRUCTURE: All-metal, with beaded skins; single-spar wings, with leading-edge torsion box; fuselage in two portions forward incorporating main wing spar carry-through structure; rear section is tailcone; cabin floor of metal sandwich, with paper honeycomb core, covered with foam rubber; aluminium tailplane bracing strut.

LANDING GEAR: Tailwheel type; fixed. Semi-cantilever main legs, of rocker type, have faired-in side Vs attached to lower fuselage and oleo-pneumatic shock-absorbers. Low-pressure tyres. Hydraulic brakes. Steerable tailwheel, carried on rocker frame with oleo-pneumatic shock-absorber. Metal ski landing gear optional. Airtech Canada CAP 3000 twin-float gear on Wilga 2000 Hydro.

POWER PLANT: One 224 kW (300 hp) Textron Lycoming IO-540-K1B5, -K1D5 or -K1J5D flat-six engine, driving a Hartzell HC-C3YR-1RF/F8468A-6R three-blade constant-speed propeller. Optional propeller EADS PZL SB 104M03037. Fuel in integral wing tanks, total capacity 400 litres (105 US gallons; 88.0 Imp gallons), of which 380 litres (100.4 US gallons; 83.6 Imp gallons) are usable. Oil capacity 11.4 litres (3.0 US gallons; 2.5 Imp gallons).

ACCOMMODATION: Passenger version accommodates pilot and three passengers, in pairs, with adjustable front seats. Baggage compartment aft of seats, capacity 30 kg (66 lb). Rear seats can be replaced by additional fuel tank for longer-range operation. Upward-opening door on each side of cabin, jettisonable in emergency.

AVIONICS: *Comms:* Bendix/King KX 155 main nav/com/glideslope and KY 96A standby VHF transceiver; KT 76A transponder; KMA 24 audio control/MKR; intercom.

Flight: Bendix/King KLN 89B GPS receiver and KR 87 ADF.

Instrumentation: ASI, VSI; ADF, VOR and twin indicators; artificial horizon; directional gyro; magnetic compass; altimeter; stall warning device; signalling and warning panel; clock; rpm tachometer; standard fuel/pressure/temperature gauges and indicators.

Mission: Polish Border Guard aircraft equipped with FLIR.

EQUIPMENT: Heavy-duty glider-towing hook, release handle and towing mirrors available as option.

DIMENSIONS, EXTERNAL:
Wing span ... 11.12 m (36 ft 5¾ in)
Wing chord, constant ... 1.40 m (4 ft 7 in)
Wing aspect ratio ... 8.2
Length overall: 2000 ... 8.10 m (26 ft 6¾ in)
2000 Hydro ... 8.52 m (27 ft 11½ in)
Height overall: 2000 ... 2.96 m (9 ft 8½ in)
2000 Hydro: on water ... approx 3.00 m (9 ft 10 in)
on land ... 3.57 m (11 ft 8½ in)
Tailplane span ... 3.70 m (12 ft 1¾ in)
Wheel track ... 2.75 m (9 ft 0¼ in)
Wheelbase ... 6.70 m (21 ft 11¾ in)
Distance between float c/l:
2000 Hydro ... 2.48 m (8 ft 1¾ in)
Propeller diameter ... 2.03 m (6 ft 8 in)
Passenger doors (each): Height ... 0.90 m (2 ft 11½ in)
Width ... 1.50 m (4 ft 11 in)

DIMENSIONS, INTERNAL:
Cabin: Length ... 2.20 m (7 ft 2½ in)
Max width ... 1.20 m (3 ft 10 in)
Max height ... 1.50 m (4 ft 11 in)
Floor area ... 2.20 m² (23.8 sq ft)
Volume ... 2.40 m³ (85 cu ft)
Baggage compartment ... 0.50 m³ (17.5 cu ft)

AREAS:
Wings, gross ... 15.00 m² (161.5 sq ft)
Tailplane ... 3.16 m² (34.01 sq ft)
Elevator, incl tab ... 1.92 m² (20.67 sq ft)

WEIGHTS AND LOADINGS:
Weight empty ... approx 925 kg (2,039 lb)
Baggage capacity ... 30 kg (66 lb)
Max T-O and landing weight: option ... 1,500 kg (3,306 lb)
FAA certification ... 1,400 kg (3,086 lb)
Max zero-fuel weight ... 1,360 kg (2,998 lb)
Max wing loading, option ... 100.0 kg/m² (20.48 lb/sq ft)
Max power loading, option ... 6.71 kg/kW (11.02 lb/hp)

PERFORMANCE (landplane):
Never-exceed speed (VNE) ... 131 kt (243 km/h; 151 mph)
Max level speed ... 113 kt (211 km/h; 131 mph)
Cruising speed at 75% power ... 103 kt (190 km/h; 118 mph)
Stalling speed: flaps up ... 58 kt (106 km/h; 66 mph)
flaps down ... 49 kt (89 km/h; 56 mph)
Max rate of climb at S/L ... 293 m (961 ft)/min
Service ceiling ... 4,118 m (13,510 ft)
T-O run ... 272 m (895 ft)
Landing run ... 150 m (495 ft)
Max range ... 809 n miles (1,500 km; 932 miles)

UPDATED

EADS PZL PZL-106BT TURBO-KRUK

English name: Turbo-Raven

TYPE: Agricultural sprayer.

PROGRAMME: Original piston-engined PZL-106 designed early 1972; first flight (SP-PAS), with IO-720 engine, 17 April 1973; last piston-engined aircraft built in 1990. First flight of turboprop-powered prototype (SP-PAA) 18 September 1985; FAR Pt 23 certification early 1994; features include taller fin and increased payload. A Turbo-Kruk was flown experimentally in 1997 with a 559 kW (750 shp) P&WC PT6A-34AG turboprop, modified cockpit and minor aerodynamic improvements.

CURRENT VERSIONS: **PZL-106BT-601:** Standard production version with Walter M 601D engine.

PZL-106BTU-34: Export version with PT6A-34AG turboprop; first flight (SP-PBW, 262nd Kruk built) 19 August 1998. Redesigned engine cowling, improved electrical system and new instrument panel. Aimed initially at Latin American market.

CUSTOMERS: Final aircraft, HC-CDU, delivered to Ecuador in November 2002. Export deliveries include Argentina (21), Egypt, Iran, South Africa (five) and Sudan. Overall total of 276 (all versions) built, comprising eight prototypes, 143 PZL-106As (of which two converted to -106B) between 1975 and 1981, 79 PZL-106BR/BSs and 46 PZL-106BTs.

UPDATED

EADS PZL PZL-110 KOLIBER 160

English name: Hummingbird

TYPE: Four-seat lightplane.

PROGRAMME: Licence-built and uprated version of Socata Rallye 100 ST; original (Morane-Saulnier) Rallye flew 10 June 1959; main production in France until 1983.

First PZL-110, modified to receive 86.5 kW (116 hp) PZL-F (Franklin) engine, first flown 18 April 1978; total 40 built of earlier Series I, II and III.

First flight of Koliber 150 prototype (SP-PHA) 27 September 1988; Polish type certificate for 150 awarded January 1989, FAA certificate for 150A February 1995. Koliber 160A introduced in 1998; production deliveries from 1999. FAR Pt 23 certification 13 February 2001.

CURRENT VERSIONS: **Koliber 160A:** Introduced 1998 and is now the only production version. First flight (SP-WGF) 15 April 1998; second prototype (SP-WGG) followed 25 May 1998. Two stock 150As converted to 160A prototypes and delivered to UK as demonstrators in 1998. First two production aircraft delivered to PZL International at North Weald, UK, in September and October 1999 for onward sale.

Following description applies to Koliber 160A.

Koliber 235A (PZL-111): More powerful version. Prototype maiden flight 14 September 1995; no production by mid-2004, but continues to be promoted.

CUSTOMERS: Kolibers sold in Canada, Denmark, Germany, Netherlands, Sweden, UK and USA. Total of 79 Koliber 110/150/150As built by 1995, of which five held in stock during 1997; no production in 1996–98. Eight production 160As delivered by late 2001, including one development aircraft, three supplied to UK distributor and two to USA; no known production in 2002 or 2003.

UPDATED

EKOLOT

EKOLOT

ulica Puaka 18, PL-38 400 Krosno
Tel/Fax: (+48 13) 436 38 97
e-mail: ekolot@ebd.pl
Web: http://www.ekolot.ebd.pl

ASSOCIATE:
SkyTec Aircraft, ulica Obornicka 4, PL-51 113 Wrocław
Tel: (+48 71) 352 51 18
e-mail: info@skytec-aircraft.pl
Web: http://www.skytec-aircraft.pl

MANAGING DIRECTOR: Radosław Sagan

Ekolot's current product range includes the JK-02 Fafik sailplane and the JK-04 RS Albatros and JK-05 Beetle ultralights.

UPDATED

EKOLOT JK-04 RS ALBATROS

English name: Albatross

TYPE: Side-by-side ultralight.

PROGRAMME: Under development by 2003, in collaboration with SkyTec Aircraft of Wrocław. Based upon, and retains most features of, JK-05 Beetle; main difference is low-wing configuration.

DESIGN FEATURES: Low-wing monoplane with tricycle landing gear, modified NN 1817 wing section, full-span flaperons (–6/+15/+28° settings) and Galaxy parachute recovery system. Wings detachable for transport and storage. Baggage space aft of seats.

STRUCTURE: As for JK-05 Beetle.

POWER PLANT: One 74.6 kW (100 hp) piston engine.

DIMENSIONS, INTERNAL:
Cabin max width ... 1.20 m (3 ft 11¼ in)

WEIGHTS AND LOADINGS:
Design max T-O weight ... 500 kg (1,102 lb)

PERFORMANCE (design):
Max level speed ... 113–124 kt (210–230 km/h; 130–143 mph)
Max cruising speed, 75% power ... 97–113 kt (180–210 km/h; 112–130 mph)
Max rate of climb at S/L ... 420–540 m (1,378–1,772 ft)/min
T-O run (grass) ... 80–120 m (265–395 ft)
T-O to 15 m (50 ft) (grass) ... 150–250 m (490–820 ft)

UPDATED

EKOLOT JK-05 BEETLE

Ekolot JK-05 Pathmaker wearing its Canadian marketing name *(Paul Jackson)* ***NEW***/0576842

Canadian marketing name: Pathmaker
French and Polish marketing name: Junior

TYPE: Side-by-side ultralight.

PROGRAMME: Design revealed publicly March 1997; composites airframe of JK-03 Junior prototype completed mid-1998; this aircraft made maiden flight 9 November 1999, powered by Suzuki motorcar engine. Public debut mid-2000 at air show in Blois, France, resulting in French orders for four production JK-03s. Improved version then developed as JK-05, making first flight 30 August 2001, powered by Rotax 582 UL engine. Verner 133 M and Rotax 912 UL introduced as alternative power plants in May and October 2002 respectively.

CURRENT VERSIONS: **JK-03 Junior:** Prototype and first four production aircraft.

JK-05 Beetle/Junior: Current production version; *as described*.

Pathmaker: Marketed in Canada by Skypaths Inc (856, 5th Concession Road West, Waterdown, Ontario L0R 2H2; *tel:* (+1 905) 659 05 55; *fax:* (+1 905) 659 00 05). Rotax 912 ULS engine and 65 litres (17.2 US gallons; 14.3 Imp gallons) fuel; max T-O weight 560 kg (1,234 lb); max range 410 n miles (760 km; 472 miles).

CUSTOMERS: Total 24 of Beetle/Junior series built by April 2003, including exports to Canada and France.

COSTS: €27,500 plus tax (2003).

DESIGN FEATURES: Braced high-wing monoplane (single I strut each side) with sweptback vertical tail; constant chord, non-swept wings (with leading-edge cutout at root to improve cockpit view) and horizontal tail. Small underfin.

Laminar flow wing section (modified NN 1817).

FLYING CONTROLS: Manual (pushrods and cables). Electrically actuated full-span slotted flaperons (6° up, 15 and 28° down); one-piece elevator, with electrically actuated central trim tab.

STRUCTURE: Stainless steel tube main frame; remainder of GFRP/CFRP-reinforced resin composites. Foam and carbon fibre wing main spar. Fin integral with fuselage. Can be assembled/disassembled for storage and transportation by two people in about 10 minutes.

LANDING GEAR: Tricycle type; fixed. Cantilever, fuselage-mounted, self-sprung composites main gear legs; steel nosewheel strut with rubber-in-compression shock-absorbers. All three wheels size 350×150 and fitted with speed fairings. Mainwheels have Bowden cable brakes; nosewheel steerable by rudder pedals.

POWER PLANT: One 47.8 kW (64.1 hp) Rotax 582 ULS or 58.2 kW (78 hp) Verner 133 M flat twin, or 59.6 kW (79.9 hp) Rotax 912 ULS flat-four; Aero Sail three-blade propeller.

Two fuel tanks aft of cockpit, combined capacity 60 litres (15.9 US gallons; 13.2 Imp gallons); gravity filler cap aft of wing on port side.

ACCOMMODATION: Window/door each side, hinged on centreline and opening upward; additional transparencies aft of these and in cabin roof. Baggage space aft of seats.

AVIONICS: Standard VFR instrumentation and flap position indicator. Options include intercom and ICOM IC-A110 radio.

EQUIPMENT: Galaxy recovery parachute in fuselage top-decking.

DIMENSIONS, EXTERNAL:
Wing span ... 10.00 m (32 ft 9¾ in)
Length overall ... 5.95 m (19 ft 6¼ in)
Height overall ... 2.30 m (7 ft 6½ in)

DIMENSIONS, INTERNAL:
Cabin max width ... 1.20 m (3 ft 11¼ in)

AREAS:
Wings, gross ... 9.72 m² (104.6 sq ft)

WEIGHTS AND LOADINGS (A: Rotax 582, B: Rotax 912, C: Verner 133 M):
Weight empty: A ... 265 kg (584 lb)
B ... 280 kg (617 lb)
C ... 290 kg (639 lb)
Max T-O weight: except Pathmaker ... 480 kg (1,058 lb)
Max wing loading: except Pathmaker ... 49.4 kg/m² (10.11 lb/sq ft)
Max power loading: A ... 10.04 kg/kW (16.50 lb/hp)
B ... 8.05 kg/kW (13.23 lb/hp)
C ... 8.25 kg/kW (13.55 lb/hp)

PERFORMANCE, POWERED (A, B and C as above):
Never-exceed speed (VNE):
all ... 113 kt (210 km/h; 130 mph)
Max level speed: A ... 97 kt (180 km/h; 112 mph)
B ... 113 kt (210 km/h; 130 mph)
C ... 103 kt (190 km/h; 118 mph)
Max cruising speed, 75% power:
A ... 84 kt (155 km/h; 96 mph)
B ... 97 kt (180 km/h; 112 mph)
C ... 89 kt (165 km/h; 103 mph)
Stalling speed, flaperons down:
all ... 31 kt (56 km/h; 35 mph)
Max rate of climb at S/L: A ... 240 m (787 ft)/min
B ... 480 m (1,575 ft)/min
C ... 300 m (984 ft)/min
T-O run: A, C ... 120 m (395 ft)
B ... 80 m (265 ft)
T-O to 15 m (50 ft): A, C ... 220 m (720 ft)
B ... 150 m (495 ft)
Landing run: all ... 100 m (330 ft)
Max range: A ... 243 n miles (450 km; 279 miles)
B ... 378 n miles (700 km; 435 miles)
C ... 270 n miles (500 km; 310 miles)
g limits ... +5/−3

PERFORMANCE, UNPOWERED:
Best glide ratio: power off ... 13–14
engine idling ... 9–10

UPDATED

PZL

POLSKIE ZAKŁADY LOTNICZE SP. Z.O.O (POLISH AVIATION FACTORY LLC)

ulica Wojska Polskiego 3, PL-39-300 Mielec
Tel: (+48 17) 788 64 56 and 788 79 21
Fax: (+48 17) 788 72 26 and 788 78 29
e-mail: pzl@ptc.pl
Web: http://www.pzlmielec.pl
PRESIDENT: Andrzej Szortyka
MANAGEMENT BOARD:
Marcin Suwala (Administrative Director)
Zigniew Dzialowski (Marketing & Development Director)
Kazimierz Strugala (Operations Director)
MARKETING AND SALES DIRECTOR: Marian Stachowicz

As the result of a restructuring process carried out by the Polish government, the former WSK-PZL Mielec and its subsidiary PZL Mielec Aircraft Company Ltd were subjected to liquidation proceedings in 1998. To replace the parent company, a new business entity under the above title came into operation on 23 October 1998. With its more than 60 years of history, a late 2002 workforce of 1,700, and the production facilities, technology, design and testing capabilities inherited from its predecessor, Polish Aviation Factory LLC remains the largest and best-equipped aircraft manufacturing plant in Poland. It is licensed to operate within the Europark Mielec special economic zone. In July 1999, its production facilities were certified by Poland's General Inspectorate of Civil Aviation under JAR 21 Subsection G, followed in September 1999 by design organisation approval under JAR 21 Subsection A. Quality management system (QMS) certification has been awarded by BAE Systems and Boeing. The factory also has ISO 9001 certification.

PZL's current indigenous aircraft programmes comprise the M18 Dromader, M26 Iskierka, M28B Bryza and M28 Skytruck. Four An-2s delivered to the Vietnamese People's Air Force in January 2002 following brief reopening of the Polish production line to complete aircraft abandoned a decade earlier; service support for some 12,000 previously delivered An-2s continues. Subcontract work continues for BAE Systems (Hawk), Boeing (757), Cessna (various), GKN Westland (for Boeing 737), Saab (Gripen) and Pratt & Whitney Canada (PT6).

UPDATED

PZL M18 DROMADER

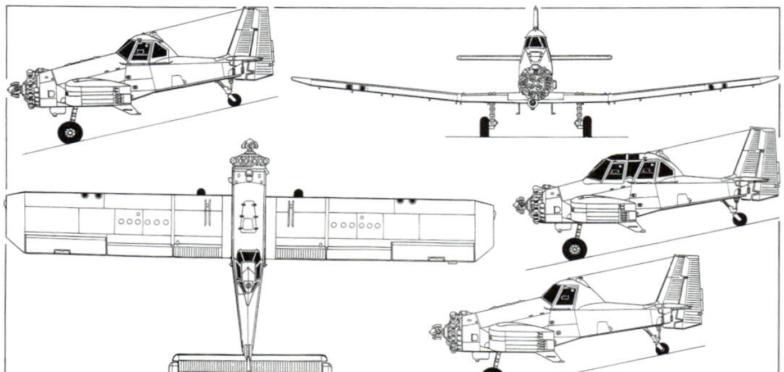

PZL M18B Dromader, with additional side elevations (right) of M18BS training version (upper) and original M18 (lower) 0062994

English name: Dromedary

TYPE: Agricultural sprayer.

PROGRAMME: Designed to meet requirements of FAR Pt 23; prototype (SP-PBW) first flew 27 August 1976; further two prototypes; M18 awarded Polish type certificate 27 September 1978; eight preproduction aircraft used for operational trials; initial US certification on 23 January 1981; M18A followed on 8 September 1987 and M18B on 19 December 1995. Later certified in Argentina, Australia, Brazil, Canada, China, former Czechoslovakia, France, former East Germany, Italy, Lithuania, New Zealand, Spain, USA and former Yugoslavia.

Initial M18 single-seat version produced from 1979 to 1984; followed by M18A two-seater produced 1984 to 1996) and M18AS training version. These superseded from mid-1996 by improved M18B and BS, which remain current standard variants.

CURRENT VERSIONS: **M18B:** Standard production version from mid-1996.

CUSTOMERS: Total of 731 built (all versions) by end of 2002, of which approximately 90 per cent for export; none further reported in 2003 or early 2004; sold to operators in Argentina, Australia, Brazil, Bulgaria, Canada, Chile, China, Cuba, former Czechoslovakia and East Germany, Greece (30 for firefighting), Hungary, Iran, Italy, Morocco, Nicaragua, Poland, Portugal, Spain, Swaziland, Trinidad, Turkey, USA (more than 200), Venezuela and former Yugoslavia.

UPDATED

PZL M26 ISKIERKA

English name: Little Spark
Export marketing name: Air Wolf

TYPE: Primary prop trainer/sportplane.

PROGRAMME: Two versions developed; first flight of first prototype (SP-PIA) with PZL-F engine 15 July 1986; first flight of Textron Lycoming-engined prototype (SP-PIB) 24 June 1987. Second prototype used as US demonstrator until destroyed on 17 August 1995. Polish certification obtained 26 October 1991; production of M-26 01 launched second quarter 1994 with initial 20 aircraft order from US distributor Melex; deliveries to USA began July 1996 with first production aircraft (N2601M). US FAR Pt 23 certification received 16 April 1998; also has Australian certification (both for Lycoming version). No recent orders, but remains available.

CURRENT VERSIONS: **M26 00:** PZL-F engine and Polish avionics. No production examples.

M26 01: Textron Lycoming AEIO-540-L1B5 engine and Bendix/King avionics. Most US aircraft are registered with type designation 'M2601'.

CUSTOMERS: Initial order for M26 01 by Melex USA Inc (20), of which two (fourth and fifth aircraft) delivered to Venezuelan National Guard Training Centre at Porlamar, Isla Margarita, in 1998. Short-listed by Israel Defence Force as Piper Super Cub replacement, but not selected. Sixth production aircraft to USA early 1999. Seventh production aircraft completed in August 2000 and retained in Poland; deliveries up to end of 2002 comprised four (plus second prototype and refurbished first prototype) to USA; two to Venezuela; current demonstrator; two due for delivery in early 2003 to Aeroclub of Poland, which has option for unspecified number of further deliveries.

UPDATED

PZL M28B BRYZA

English name: Breeze
NATO reporting name: Cash

TYPE: Twin-turboprop transport.

PROGRAMME: Developed by Antonov in former USSR as An-28; robust utility transport for service on Aeroflot's shortest routes, particularly those operated by An-2s into places relatively inaccessible to other fixed-wing aircraft; official Soviet flight testing completed 1972; first preproduction An-28 (SSSR-19723) originally retained same engines as prototype, but re-registered SSSR-19753 April 1975 when flown with current engines; production assigned to PZL Mielec 1978; temporary Soviet NLGS-2 type certificate awarded 4 October 1978 to second Soviet-built preproduction aircraft.

Polish manufacture of An-28 started with initial batch of 15; first flight (SSSR-28800) 22 July 1984; version received full Soviet type certificate 7 February 1986. Polish production now for domestic naval and military use only.

CURRENT VERSIONS: **M28B Bryza 1RM:** Originally designated An-28RM Bryza 1RM (*Ratownictwa Morskiego:* maritime reconnaissance). Prototype (SP-PDC) first flown June 1992; to Polish Navy June 1993 for evaluation; later upgraded and redelivered (as 0810) March 1999. First production example (serial number 1022) delivered 25 October 1994 and since upgraded; two further examples (1008 and 1017) ordered January 1998 and delivered March 1999 to No. 3 Naval Air Arm Division at Siemirowice. These were supplied, and earlier aircraft upgraded, with Bendix/King Silver Crown avionics. Further four aircraft (1006 and 1114 to 1116) delivered in 2000–01. Upgrade of all eight has continued with introduction of Hartzell five-blade propellers (mid-2000); and semi-retractable landing gear (first flown on 0810 on 29 January 2003) and modified nose housing new (RDR-2000) weather radar. In this configuration, aircraft is redesignated **1RMbis** (see below). Further plans include provision for additional fuel tankage.

M28B Bryza 1RMbis: Designation of anti-submarine version, to which all eight current 1RMs eventually to be upgraded. Additional fuel tankage and semi-retractable landing gear. FLIR turret in redesigned nose. First flight in this configuration (by 0810), 29 January 2003.

M28B Bryza 1TD: (*Towaru Desantowego:* cargo airdrop): Transport/paradrop version, designated for airdrop of paratroops or cargo containers, with folding seats, static lines and roller floor as appropriate. In medevac configuration, attachment provisions for six stretchers, oxygen mask container, resuscitator and medicare containers. Clamshell rear doors replaced by single door sliding forward under fuselage, similar to that of An-26. Previously known as An-28B1T, then An-28TD.

Prototype (SP-PDE) completed as B1 September 1991; to TD configuration May 1994 and to Air Force (as 0723) for evaluation May 1995; returned to PZL 1996. First production TD (1003) to Polish Air Force October 1994. Further 10 ordered for Polish Air Force 16 May 2001; deliveries began 26 March 2002 with 0203 and 0204; followed by 0205 and 0206 in May 2002, 0207 and 0208 in May 2003.

Polish Navy also operates two TDs (1117 and 1118), plus one aircraft (1007) originally converted from an An-28 as Bryza 2RF elint aircraft but delivered unequipped on 22 November 1996 and operated only in transport role.

M28M Bryza E: Two Polish Navy aircraft (0404 and 0405, originally delivered as An-28s in October 1988) equipped in 2000–01 with Swedish Space Corporation (Ericsson) MSS 5000 maritime surveillance system for ecological monitoring; 0405 recommissioned 28 November 2000; both redelivered to Polish Navy 18 January 2002; based at Gdynia. In these aircraft, single rear door is replaced by twin doors opening hydraulically inward, and additional side door installed in forward fuselage.

CUSTOMERS: Bryza deliveries to Polish Navy totalled 13 by May 2002 comprising eight 1RMs, two Es, two TDs and one 2RF; these equip the 3rd Division at Siemirowice. By end of 2003, Polish Air Force had seven Bryza 1TDs, with a further four to follow; these mainly with 13th Transport Regiment at Krakow, but two operated by 36 Squadron at Warsaw-Okecie in VIP transport role.

Two Skytrucks equivalent to Bryza 1RMbis ordered by Vietnam in October 2003; Indonesia reportedly negotiating at that time for eight Bryza 1RMbis and four Skytrucks.

New nose shape of the Bryza 1RMbis housing a FLIR *(R J Malachowski)* ***NEW***/0567714

Prototype Bryza 1RMbis, showing retracted gear, underbelly radome and undernose FLIR installation *(Piotr Butowski)* ***NEW***/0543003

POWER PLANT: Two WSK-Rzeszów 716 kW (960 shp) TWD-10B/PZL-10S turboprops, each driving a AW-24AN three-blade or Hartzell HC-B5MP-3D five-blade propeller; first aircraft with latter was 0810, redelivered to Polish Navy on 27 August 2000. PT6A-65B engines available optionally in two 670 litre (177 US gallon; 147.4 Imp gallon) outer wing tanks and one 620 litre (164 US gallon; 136.4 Imp gallon) centre-section tank, giving total capacity of 1,960 litres (518 US gallons; 431 Imp gallons); programme to install auxiliary tanks on landing gear sponsons under development in 2002.

ACCOMMODATION (BRYZA 1RM AND 1RMBIS): Pilot, co-pilot, engineer and three systems operators.

AVIONICS (BRYZA 1RM): *Comms:* LS-10C^3 subsystem; Bendix/King KT 76A transponder; NATO-compatible IFF.

Radar: PIT (Przemysłowy Instytut Telekomunikacji) ARS-400 360° search and surveillance radar (86 n mile; 160 km; 99 mile range), in underfuselage radome; Honeywell RDR-2000 weather radar in nose.

Flight: Include Bendix/King KNS 81S navigation system and KLN 90B GPS receiver.

Mission: Chelton DF 707-1 emergency locator system. SSC MSS 5000.

AVIONICS (BRYZA 1RMBIS): *Comms:* SC-10D2 IFF transponder.

Radar: PIT ARS-800 synthetic aperture search radar in underfuselage radome; Honeywell RDR-2000 weather radar in nose.

Flight: Bendix/King KFC 325 AFCS, KLN 900 GPS navigation system and KNR 634A nav receiver; LCR 92 laser gyro platform; Chelton DF system.

Polish Navy Bryza 1RM (marked as 1R) in standard grey camouflage *(Paul Jackson)* ***NEW***/0580745

Mission: SRM-800 maritime surveillance system; MAG-10 magnetometer; HYD-10 submarine detection (sonobuoy) system; LS10M automatic data transmission system; Star SAFIRE II FLIR turret in redesigned nose.

Self-defence: ESM-10 ESM system.

EQUIPMENT (BRYZA 1RM): ACR/RLB-14 radio marker buoys; two 100 kg (220 lb) SAB-100NM illumination flares; five six-seat Mewa-6 dinghies, including two for crew.

DIMENSIONS, EXTERNAL (AS M28 SKYTRUCK EXCEPT):

Propeller diameter: AW-24AN	2.80 m (9 ft 2¼ in)
Hartzell	2.82 m (9 ft 3 in)
Cargo door: Height	2.60 m (8 ft 6¼ in)
Width	1.20 m (3 ft 11¼ in)

DIMENSIONS, INTERNAL:

Cabin: Length (excl flight deck)	5.26 m (17 ft 3 in)
Max width	1.74 m (5 ft 8½ in)
Max height	1.72 m (5 ft 7¾ in)

WEIGHTS AND LOADINGS:

Basic weight empty	4,050 kg (8,929 lb)
Weight empty, equipped	4,350 kg (9,590 lb)
Max fuel weight	1,766 kg (3,893 lb)
Max payload	2,000 kg (4,409 lb)
Max T-O and landing weight	7,500 kg (16,534 lb)
Max power loading	5.24 kg/kW (8.61 lb/shp)

PERFORMANCE:

Never-exceed speed (VNE)	189 kt (350 km/h; 217 mph)
Max operating speed (VMO)	181 kt (335 km/h; 208 mph)
Unstick speed	73 kt (135 km/h; 84 mph)
Touchdown speed	76 kt (140 km/h; 87 mph)
Stalling speed, flaps up	67 kt (123 km/h; 77 mph)
Max rate of climb at S/L	540 m (1,772 ft)/min
Time to 3,000 m (9,840 ft)	6 min 43 s
Service ceiling	6,000 m (19,680 ft)
Service ceiling, OEI	3,500 m (11,480 ft)
T-O run	440 m (1,445 ft)
T-O to 10.7 m (35 ft)	570 m (1,870 ft)
Landing from 15 m (50 ft)	750 m (2,460 ft)
Landing run	330 m (1,085 ft)
Max range, 45 min reserves	809 n miles (1,500 km; 932 miles)
Max endurance at 3,000 m (9,840 ft)	6 h 30 min

UPDATED

PZL M28 05 SKYTRUCK

TYPE: Twin-turboprop transport.

PROGRAMME: Westernised development of the Antonov An-28, with Polish power plant and avionics replaced by P&WC PT6A-65B turboprops, Hartzell five-blade propellers and Bendix/King nav/com, weather radar and other equipment. Prototype conversion (SP-PDF) begun early 1991; first flight July 1993; Polish temporary type certificate to FAR Pt 23 Amendment 34 granted March 1994; permanent certificate March 1996, permitting MTOW increase to 7,000 kg (15,432 lb) for versions 01 and 02; subsequently certified to FAR Pt 23 Amendment 42 for increased gross weight version 05 (7,500 kg; 16,534 lb), a prototype of which was completed in 1996. Full FAA certification of M28 05 received April 2004; PZL expected to produce 10 Skytrucks during that year.

CURRENT VERSIONS: **M28 05 Skytruck:** *As described.*

M28B Bryza: Amended designation of current production military derivatives of An-28; described separately.

M28 03/04 Skytruck Plus: Provisional designation for projected stretched version.

CUSTOMERS: First customer delivery (HK-4066X to Latina de Aviación of Colombia) made in January 1996; firm orders totalled 12 at that time, including six for Venezuelan National Guard (deliveries to which began in December 1996); equip Air Detachment 5 at Caracas (four), AD 3 at Maracaibo (one) and AD 9 at Puerto Agoucho (one). One delivered to Aerogryf in Poland in mid-1996; and one to USA (subsequently to Honduras) in 1997; one to Venezuelan civil operator in 2000, but lost in July 2001. Venezuelan National Guard order subsequently (1997) increased to 12, second-batch deliveries beginning mid-1999 and completed (with single M28 02 Executive) in 2000. Twelve M28 05 ordered by Venezuelan Army; first deliveries April 2000; completed in 2001.

One (RAN-41) delivered to Royal Nepal Army on 7 November 2002 for 11th Brigade at Kathmandu; second ordered October 2003, with further two in prospect.

In late October 2003, Vietnamese People's Air Force ordered first two of 'at least eight to 10' Skytrucks required for maritime surveillance and border control missions; to be delivered before end of 2004, initially in transport configuration but then to be equipped with MSS-400 surveillance system (ARS-400 radar and CCS-400 command and control system) as fitted to Polish Navy Bryza 1R.

Indonesian government announced intention to buy 11 Skytrucks for Navy on 4 May 2004, this following a late 2003 contract for four for Indonesian Police. Former batch to have same MSS-400 maritime surveillance system as those for Vietnam. Indonesian Army also said to be interested.

DESIGN FEATURES: Inherited An-28's ability to operate from short, austere airfields; added Western engines and avionics to improve export sales prospects. Braced wings; will not stall because of action of automatic slats. Short stub-wing extends from lower fuselage to carry main landing gear and support wing bracing struts, curving forward and downward at front to serve as mudguards; underside of rear fuselage upswept, incorporating clamshell doors for passenger/cargo loading; twin fins and rudders, mounted on inverted-aerofoil, no-dihedral fixed incidence tailplane.

Wing section CAHI (TsAGI) R-II-14; thickness/chord ratio 14 per cent; constant chord, non-swept, no-dihedral centre-section, with 4° incidence; tapered outer panels with 2° dihedral, negative incidence and 2° sweepback at quarter-chord.

FLYING CONTROLS: Manual. Single-slotted mass and aerodynamically balanced ailerons (port aileron has trim tab), designed to droop with large, hydraulically actuated, two-segment double-slotted flaps; electrically actuated trim tabs in elevator have manual back-up; twin rudders each with electrically actuated trim tab; automatic leading-edge slats over full span of wing outer panels; slab-type spoiler forward of each aileron and each outer flap segment at 75 per cent chord; fixed slat under full span of tailplane leading-edge. If an engine fails, patented upper surface spoiler forward of aileron on opposite wing is opened automatically, resulting in wing bearing dead engine dropping only 12° in 5 seconds instead of 30° without spoiler; patented fixed tailplane slat improves handling during high angle of attack climb-out; under icing conditions, if normal anti-icing system fails, ice collects on slat rather than tailplane, to retain controllability.

STRUCTURE: Mostly metal; duralumin ailerons with fabric covering; duralumin slats with CFRP skins; CFRP spoilers and trim tabs. Air intakes lined with epoxy laminate. Two-spar wing with integral fuel tanks between spars.

LANDING GEAR: Non-retractable tricycle type, with single wheel and PZL oleo-pneumatic shock-absorber on each unit. Main units, mounted on small stub-wings, have wide tread balloon tyres, size 720×320, pressure 5.50 bar (80 lb/sq in). Steerable (±50°) and self-centring nosewheel, with size 6.5–10 Goodyear tyre, pressure 6.00 bar (87 lb/sq in) or 595×185–280 Stomil (Poland) tyre, pressure 3.50 bar (51 lb/sq in). Multidisc, anti-lock hydraulic brakes on main units, and inertial anti-skid units. Reinforced landing gear and brakes optional. Minimum ground turning radius 16.00 m (52 ft 6 in). Ski and float gear under development.

POWER PLANT: Two 820 kW (1,100 shp) Pratt & Whitney Canada PT6A-65B turboprops, each driving a Hartzell HC-B5MP-3D10876ASK five-blade, fully feathering propeller with reverse pitch. Two main fuel tanks in outer wings, each 660 litres (174 US gallons; 145 Imp gallons); two auxiliaries in wing centre-section, each 290 litres (76.6 US gallons; 63.8 Imp gallons); two auxiliary tanks in outer wings, each 189 litres (49.9 US gallons; 41.6 Imp gallons); total wing fuel thus 2,278 litres (602 US gallons; 501 Imp gallons). Auxiliary tanks gravity-filled; main tanks by pressure or gravity. Additional fuel can be carried in tank in main cabin. Oil capacity 16 litres (4.2 US gallons; 3.5 Imp gallons) per engine.

ACCOMMODATION: Pilot and co-pilot on flight deck, which has bulged side windows and electric anti-icing for windscreens, and is separated from main cabin by bulkhead with connecting door. Dual controls standard. Jettisonable emergency door on each side. Standard cabin layout of passenger version has seats for 19 people, with seven single seats on port side and six double seats on starboard side of aisle, at 72 cm (28 in) pitch. All seats easily foldable or removable for carriage of cargo. Aisle width 35.4 cm (13.9 in). Five passenger windows each side of cabin. Seats fold back against walls when aircraft is operated as a freighter or in mixed passenger/cargo role, seat attachments providing cargo tiedown points. Hoist of 500 kg (1,102 lb) capacity able to deposit cargo in forward part of cabin. Alternative interior configurations for 17 parachutists; or six stretchers, eight seated casualties and two medical attendants.

Entire cabin heated, ventilated, soundproofed and, optionally, air conditioned. Inward/upward-opening, hydraulically operated doors for passenger and cargo loading. Emergency exit on each side, plus a third aft of flight deck. Optional ventral pannier, increasing baggage/cargo capacity by 300 kg (661 lb).

SYSTEMS: No pressurisation or pneumatic systems. Hydraulic system powered by two engine-driven pumps, pressure 147 bar (2,132 lb/sq in), for flap, spoiler and (where fitted) cargo ramp actuation, mainwheel brakes and nosewheel steering, with emergency back-up system for spoiler extension and mainwheel braking. Optional air conditioning system,

First Skytruck of a pair ordered in 2003 by the air force of Vietnam *(Grzegorz Holdanowicz)* ***NEW***/0563155

W-3 SOKÓŁ PRODUCTION
(to 1 January 2002)

Version	Lot 1 1986–89	Lot 2 1989–93	Lot 3 1990–93	Lot 4 1992–98	Lot 5 1994–95	Lot 6 1996–98	Lot 7 1997–98	Lot 8 1999-	Total
W-3	10[1]	22[2]	17	3[3]					52
W-3A			1	4		12		2	19
W-3PPD							1		1
W-3AM				2		5	6		13
W-3RM			1	6					7
W-3W		1		5	17				23
W-3WA					3	2	10	8	23
W-3WARM							2	1	3
SRR-10						1			1
Unidentified							1		1
Static test		2[4]	1						3
Totals	**10**	**25**	**20**	**20**	**20**	**20**	**20**	**11**	**146**

[1] One converted as S-1RR Procjon prototype
[2] One to W-3U prototype; reverted to W-3 before delivery
[3] Two upgraded to W-3RL by WZL 1
[4] Israel Aircraft Industries and Instytut Lotnictwa

electrically powered with ground power supply back-up, reduces ambient temperatures of up to 40°C (104°F) by 10 to 12°C (18 to 22°F).

Primary electrical system is three-phase AC, with two 28 V DC 12 kW engine-driven starter/generators; two 200/115 V, 400 Hz, 1,500 VA inverters as secondary supply for windscreen heating. Two 26 Ah Ni/Cd batteries.

Electric anti-icing of flight deck windscreens, propellers, spinners and pitot heads, sufficient for short exposure to icing conditions. Oxygen system standard for crew, optional for passengers. No APU.

AVIONICS: Standard Gold Crown suite by Bendix/King.

Comms: Two KTR 908 VHF radios; two KMA 24 H70 audio selector intercoms; MST 67A transponder and encoding altimeter. Optional KHF 950 HF radio.

Radar: Optional RDR-2000 digital weather radar.

Flight: Navigation system based on two KNR 634 VOR/LOC/GS/MKR receivers for VOR/DME navigation with two EFS 50 EHSI; two KDF 806 ADF; LCR 92S12 laser gyro; two EFS 50 EADI. Optional KLN 900 GPS and FDR, KFC 325 AFCS and KCP 220 computer.

DIMENSIONS, EXTERNAL
Wing span .... 22.06 m (72 ft 4½ in)
Wing chord: at root .... 2.20 m (7 ft 2½ in)
mean aerodynamic .... 1.89 m (6 ft 2½ in)
at tip .... 1.10 m (3 ft 7¼ in)
Wing aspect ratio 12.3
Length overall .... 13.10 m (42 ft 11¾ in)
Fuselage: Length .... 12.68 m (41 ft 7¼ in)
Max width .... 1.90 m (6 ft 2¾ in)
Max depth .... 2.14 m (7 ft 0¼ in)
Height overall .... 4.90 m (16 ft 1 in)
Tailplane span .... 5.14 m (16 ft 10¼ in)
Wheel track .... 3.405 m (11 ft 2 in)
Wheelbase .... 4.40 m (14 ft 5¼ in)
Propeller diameter .... 2.82 m (9 ft 3 in)
Propeller ground clearance .... 1.06 m (3 ft 5¼ in)
Distance between propeller centres .... 5.20 m (17 ft 0¾ in)
Rear hydraulic doors:
Length .... 2.40 m (7 ft 10½ in)
Total width: at top .... 0.90 m (2 ft 11½ in)
at sill .... 1.20 m (3 ft 11¼ in)
Emergency exits (each):
Height .... 0.93 m (3 ft 0½ in)
Width .... 0.62 m (2 ft 0½ in)

DIMENSIONS, INTERNAL:
Cabin (excl flight deck): Length .... 5.26 m (17 ft 3 in)
Max width .... 1.74 m (5 ft 8½ in)
Max height .... 1.72 m (5 ft 7¾ in)
Floor area .... approx 7.5 m^2 (80.7 sq ft)
Volume .... approx 14.0 m^3 (494 cu ft)
Pannier (optional): Height .... 0.38 m (1 ft 3 in)
Width .... 1.53 m (5 ft 0¼ in)
Volume .... 1.35 m^3 (47.7 cu ft)

AREAS:
Wings, gross .... 39.72 m^2 (427.5 sq ft)
Ailerons (total) .... 4.33 m^2 (46.61 sq ft)
Trailing-edge flaps (total) .... 7.99 m^2 (86.00 sq ft)
Spoilers (total) .... 1.67 m^2 (17.98 sq ft)
Fins (total) .... 10.00 m^2 (107.64 sq ft)
Rudders (total, incl tabs) .... 4.00 m^2 (43.6 sq ft)
Tailplane .... 8.85 m^2 (95.26 sq ft)
Elevators (total, incl tabs) .... 2.56 m^2 (27.56 sq ft)

WEIGHTS AND LOADINGS:
Weight empty, equipped .... 4,100 kg (9,039 lb)
Max fuel load .... 1,766 kg (3,893 lb)
Max payload .... 2,000 kg (4,409 lb)
Fuel with max payload .... 1,085 kg (2,392 lb)
Payload with max fuel .... 1,319 kg (2,908 lb)
Max T-O and landing weight .... 7,500 kg (16,534 lb)
Max zero-fuel weight .... 6,600 kg (14,550 lb)
Max wing loading .... 188.8 kg/m^2 (38.67 lb/sq ft)
Max power loading .... 4.57 kg/kW (7.52 lb/shp)

PERFORMANCE (at 7,000 kg; 15,432 lb MTOW):
Max operating speed (VMO) .... 191 kt (355 km/h; 220 mph)
Econ cruising speed at FL984 .... 146 kt (270 km/h; 168 mph)
Max manoeuvring speed (VA) .... 132 kt (244 km/h; 152 mph)
Min control speed (VMCA) .... 83 kt (153 km/h; 96 mph)
Unstick speed .... 73 kt (135 km/h; 84 mph)
Approach speed .... 70 kt (130 km/h; 81 mph)
Stalling speed, flaps up .... 67 kt (123 km/h; 77 mph)
Max rate of climb at S/L .... 660 m (2,165 ft)/min
Time to FL984: air bleed off .... 6 min
air bleed on .... 9 min
Service ceiling .... 7,620 m (25,000 ft)
Service ceiling, OEI .... 3,750 m (12,300 ft)
T-O run .... 255 m (835 ft)
T-O to 10.7 m (35 ft) .... 325 m (1,065 ft)
Landing from 15 m (50 ft) .... 560 m (1,835 ft)
Landing run .... 260 m (855 ft)
Range at FL984, 45 min reserves:
with max payload .... 432 n miles (800 km; 497 miles)
with max fuel .... 809 n miles (1,500 km; 932 miles)
self-ferry, with additional tanks in cabin .... 1,673 n miles (3,100 km; 1,926 miles)
Endurance .... 6 h 12 min
g limits .... +3/−1

UPDATED

PZL-ŚWIDNIK SA

WYTWÓRNIA SPRZĘTU KOMUNIKACYJNEGO (TRANSPORT EQUIPMENT CORPORATION) PZL-ŚWIDNIK SA

aleja Lotników Polskich 1, PL-21-045 Świdnik
Tel: (+48 81) 751 20 61 and 751 35 05
Fax: (+48 81) 468 09 19 and 468 09 18
e-mail: hem@pzl.swidnik.pl
Web: http://www.pzl.swidnik.pl

PRESIDENT: Mieczysław Majewski
MANAGING DIRECTOR Jan Miroński
COMMERCIAL DIRECTOR: Ryszard Cukierman
DIRECTOR OF RESEARCH AND DEVELOPMENT: Jan Pysznicek
HEAD OF MARKETING: Andrzej Dyzma
HEAD OF SALES: Marzena Siwek
MARKETING MANAGER: Andrzej Stachyra

Świdnik factory recently completed 50 years of aircraft manufacture, having been established 1 January 1951; engaged initially in manufacturing wings and control surfaces for LiM-1 (MiG-15) jet fighter; began licensed production of Soviet Mi-1 helicopter in 1954 (first flight March 1956), building some 1,594 as SM-1s, followed by 86 Świdnik-developed SM-2s; design office formed at factory to work on variants/developments of SM-1 and original projects such as SM-4 Łątka. More than 7,300 helicopters of all types built to date, including four pre-series and 5 418 production Mi-2s from 1965; in addition, further six Mi-2 static test airframes produced. Eight more incomplete Mi-2s remained in storage in mid-2000; last Mi-2 delivered to United Arab Emirates Police in March 2002. Two PZL Kanias were delivered to the Polish Border Guard during 2003.

Świdnik works named after famous pre-war PZL designer Zygmunt Pulawski in September 1957; became a joint stock company on 4 January 1991, with Polish Treasury the major shareholder (now 74.39 per cent); workforce of 2,422 in December 2000. Buyer(s) for 36.7 per cent share in company still being sought, following failure to agree terms with AgustaWestland. Production of own designs concentrates on W-3 Sokół variants and SW-4; investors being sought in 2003–04 to launch production of fixed-wing I-23 lightplane designed by Instytut Lotnictwa. ISO 9001 status was achieved in 1994, and on 15 June 2000 Świdnik was granted AQAP-110 certification, having fully met NATO quality assurance requirements for design, development and production.

Świdnik manufactures PW-5 Smyk (from 1994) and PW-6 composites sailplanes (both recently sold to Egypt) and (also since 1994) central wing box for ATR 72; manufacturing fuselages for AgustaWestland A 109 Power and A 119 Koala between 1996 and 2002, fuselage components for Bell/Agusta AB139, cockpit modules for Dassault Mirage 2000–5, aircraft components for Ratier-Figeac, and baggage doors for Airbus A300/A320/A340. Co-operation agreement with Cessna, and start of Bell 412 tailboom production, announced in 2002. February 2004 MoU with Dirgantara to study feasibility of latter becoming components supplier for W-3 Sokół . In all, PZL-Świdnik is involved in some 20 international aerospace manufacturing programmes, and co-operates with 11 world aerospace companies, including Boeing, Eurocopter, Latécoère and Stork Fokker in addition to those already mentioned.

Subsidiaries include Heliseco, which operates over 100 helicopters on agricultural and similar tasks at home and in Egypt, Greece, Oman, Portugal, Spain and elsewhere.

UPDATED

W-3 SOKÓŁ CUSTOMERS
(at 1 January 2002)

Country	Customer	Qty	Version	First aircraft	First delivery
Czech Republic	Air Force	11	W-3A	0709	Sep 96
Germany	Saxony Police	2	W-3A	D-HSNA	Mar 94
Korea, South	Choong Nam Fire Dept	1	W-3A	119	Dec 99
	CitiAir	4	W-3AM	HL-9220	Dec 95
	Daegu Fire Dept	1	W-3AM	?	Nov 01
	Daewoo	1	W-3A	SP-FSU	not delivered
		1	W-3AM	HL-9257	Oct 96
	Heli Korea	4	W-3AM	HL-9259	Jul 97
Myanmar	Air Force	13	W-3	6501	Nov 90
Nigeria	Okada Air	1	W-3	5N-UYI	Dec 91
Poland	Air Force	7	W-3	0415	Jul 93
		5	W-3W	0516	Apr 94
		5	W-3WA	0618	?
	Army	17	W-3W	0601	Sep 94
		18	W-3WA	0806	97
	Heliseco	2	W-3	SP-SUB	Feb 97
		1	W-3AM	SP-SYI	Jul 96
	Lublin Medical Aviation Centre	1	W-3	SP-SXU	Feb 94
	Ministry of Defence	1	W-3/S-1RR	0203	96
		1	W-3RR/SRR-10	0720	98
		1	W-3PPD	0816	01
	Ministry of Interior	2	W-3	0315	Apr 90
		1	W-3A	0420	Jun 94
		1	W-3RM	0510	Apr 93
	Navy	2	W-3	0209	Jul 89
		5	W-3RM	0505	Jul 92
		3	W-3WARM	0813	Aug 98
	Polish Telecom	1	W-3	SP-SUI	Apr 92
	PZL Świdnik	1[2]	W-3	SP-SUA	Jan 89[1]
		1	W-3W prot	SP-SUW	Mar 93[1]
		1	W-3RM prot	0411/SP-SYG	Sep 95[1]
		1[3]	W-3A	SP-PSL	May 98[1]
	Zakopane Mountain Rescue	1	W-3	SP-SXT	Feb 93
		1	W-3A	SP-SXZ	Jan 96
Russia	Aeroflot	20[4]	W-3	SSSR-04101	Aug 88
	Krasnodarski Kray	1	W-3A?	RA-10801	Jul 03
UAE	Ministry of Internal Affairs	2	W-3A	?	02
Vietnam	SFC Vietnam	1	W-3AM	VN-417	Jan 98
Total		**143**			

[1] Date of registration
[2] Firefighting demonstrator
[3] Sextant AFDS avionics testbed
[4] 18 transferred to Heliseco from January 1996

PZL W-3 SOKÓŁ

English name: Falcon

TYPE: Multirole medium helicopter.

PROGRAMME: Developed under 1970 joint Polish/Soviet agreement; preliminary design by engineers from both countries at Mil OKB in Moscow 1972, where preliminary mockup completed 1975. Detail design and definitive mockup by Świdnik 1976; static/fatigue ground test airframe of 1978 followed by five flying prototypes, first of which (SP-PSA) made first flight 16 November 1979, and used in subsequent tiedown tests; remaining prototypes embodied changes resulting from tests; Soviet participation ended 1980. Manufacturer's flight trials resumed 6 May 1982 with second prototype (SP-PSB); third, fourth and fifth prototypes all made first flights in 1984, on 24 July (SP-PSC), 4 June (SP-PSD) and 26 November (SP-PSE) respectively; certification trials carried out in wide range of operating conditions, including heavy icing and extreme temperatures of −60 and +50°C. Provisional Polish certification 26 September 1988, followed by full certification in Poland 10 April 1990, and to Russian NLGW-1 and −2 regulations 17 December 1992; production started in 1985 and 20 early aircraft supplied to Aeroflot from 10 August 1988, but subsequently returned to Świdnik, most being reallocated to Heliseco for firefighting in Spain. Manufacture ended in 2001, although promotion continued for several years thereafter.

CUSTOMERS: See tables. Total of 143 (excluding five flying prototypes and four static test airframes) completed by early 2001, of which 131 then in service. Three reportedly purchased in early 2002 by Ukraine Border Guard. However, no new production reported since early 2001, although one sold from stock to Krasnodar regional government and handed over on 31 July 2003. ***UPDATED***

PZL SW-4

TYPE: Light utility helicopter.

PROGRAMME: Development began 1985; full-scale mockup completed 1987; major redesign undertaken 1989–90, using Allison (now Rolls-Royce) 250 engine in more streamlined fuselage with modified tail unit. Prototype (c/n 600102), rolled out December 1994, is non-flying testbed for ground and equipment tests; '101 is static test airframe, '103 and '104 are flying prototypes. First flight made by 600103 (red overall; later registered SP-PSW) on 26 October 1996 ('official' flight three days later).

Trials in 1997 demonstrated requirement for a new rotor head design, enlarged horizontal stabiliser and more robust hydraulic system. Following 70 hours of test-flying, SP-PSW was grounded in late 1997 for installation of SAMM-designed hydraulic flight control system, with which it was then due to return to flying in 1998. Second flying prototype (yellow overall), with improved skids, exhibited statically at Paris Air Show, June 1997; registered as SP-PSZ in October 1998, but not flown until early 2001. Some 640 hours (total) flown by July 2002, when certification programme almost complete; domestic certificate to JAR 27 awarded 14 November 2002. First five production aircraft started in July 2002. Second prototype shown at Paris in June 2001; first at Berlin in May 2002.

According to a mid-2003 report, Świdnik considering re-engining SW-4 with Pratt & Whitney Canada PW200 turboshafts, assembled locally by PZL-Rzeszów.

CURRENT VERSIONS: Will be marketed in two versions:

Economy: With one 336 kW (450 shp) Rolls-Royce 250-C20R engine and 1,600 kg (3,527 lb) normal T-O weight.

Second flying prototype of the SW-4 four/five-seat light helicopter (PZL-Świdnik/Krystyna Majkowska) ***NEW***/0538345

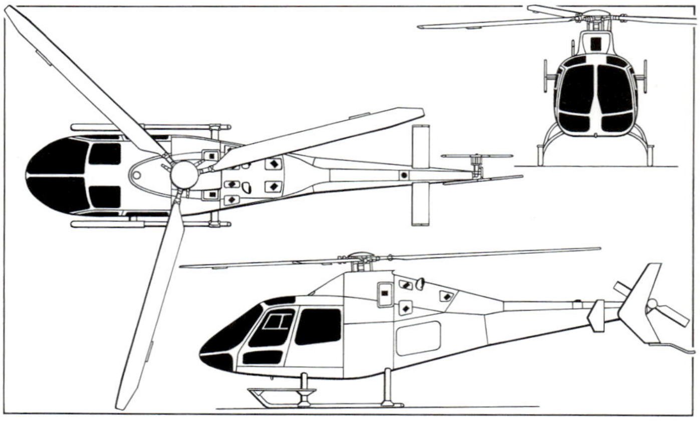

PZL-Świdnik SW-4 (Rolls-Royce 250-C20R/2 turboshaft) (Jane's/Mike Keep) 0100525

High Performance: With one 459 kW (615 shp) Pratt & Whitney Canada PW200/9 engine and 1,800 kg (3,968 lb) normal T-O weight.

Provision has been made for later installation of two engines to fulfil ICAO requirements for OEI operation (including hover).

CUSTOMERS: Polish Air Force requirement confirmed in mid-2002 for purchase of 47 by 2010, of which seven planned to enter service in 2005 and 14 in 2006; for use in training role. Two of first five production aircraft scheduled for Polish Air Force Academy; next three are commercial customers.

COSTS: US$750,000 (2003).

Cockpit of PZL Świdnik SW-4 ***NEW***/0051163

Following description applies to first flying prototype:

DESIGN FEATURES: Intended applications include passenger and cargo transport, medevac, border patrol, armed scout and training use.

Three-blade main rotor; arrowhead tailfin on port side, with two-blade tail rotor to starboard; narrow tailplane with small endplate fins; skid landing gear.

STRUCTURE: Main structural components of aluminium alloys and glass fibre epoxy. Cabin of foam-filled carbon and glass fibre panels in epoxy matrix; centre-fuselage is stressed box of honeycomb sandwich panels, to which fuel tank and main gearbox are attached. Glass fibre/epoxy main and tail rotor blades.

LANDING GEAR: Skid type; able to accommodate heavy landing sink rate of 3.1 m (10 ft)/s by elastic deflection of cross-tubes.

POWER PLANT: One 336 kW (450 shp) Rolls-Royce 250-C20R/2(SP) turboshaft. Transmission rating 336 kW (450 shp) for T-O, 283 kW (380 shp) maximum continuous; 30 minute run-dry capability. Standard fuel capacity 500 litres (132 US gallons; 110 Imp gallons) in tank below main gearbox.

ACCOMMODATION: One or two pilots plus three or four passengers; or one stretcher patient and a medical attendant. Dual controls can be added for training role. One front-hinged and one rearward-sliding door on each side of cabin. Baggage compartment, capacity 150 kg (331 lb), to rear of cabin, with external access.

AVIONICS: *Instrumentation*: Bendix/King or Garmin VFR standard, IFR optional.

EQUIPMENT: In law enforcement role, can be equipped with thermal imaging and TV cameras, searchlight, loudspeaker, special intercom and radio, and external hoist. For military roles, can be fitted with external support bars for up to 440 lb (970 lb) of specialist equipment.

DIMENSIONS, EXTERNAL:
- Main rotor diameter .... 9.00 m (29 ft 6¼ in)
- Tail rotor diameter .... 1.50 m (4 ft 11 in)
- Distance between rotor centres .... 5.25 m (17 ft 2¾ in)
- Length:
  - overall, both rotors turning .... 10.55 m (34 ft 7½ in)
  - fuselage: incl tailskid .... 9.075 m (29 ft 9¼ in)
    - excl tailskid .... 8.25 m (27 ft 0¾ in)
- Max width: cabin .... 1.515 m (4 ft 11¾ in)
  - over tail unit endplates .... 1.83 m (6 ft 0 in)
  - over skids .... 2.28 m (7 ft 5¾ in)
- Height to top of rotor head .... 2.97 m (9 ft 9 in)
- Skid track .... 2.20 m (7 ft 2½ in)
- Fuselage ground clearance .... 0.55 m (1 ft 9¾ in)

DIMENSIONS, INTERNAL:
- Cabin: Length .... 2.14 m (7 ft 0¼ in)
  - Max width .... 1.41 m (4 ft 7½ in)
  - Max height .... 1.27 m (4 ft 2 in)
  - Volume .... 3.85 m^3 (136 cu ft)
- Baggage compartment volume .... 0.85 m^3 (30.0 cu ft)

AREAS:
- Main rotor disc .... 63.62 m^2 (684.8 sq ft)
- Tail rotor disc .... 1.77 m^2 (19.05 sq ft)

WEIGHTS AND LOADINGS:
- Weight empty .... 1,050 kg (2,315 lb)
- Max payload: internal .... 550 kg (1,213 lb)[1]
  - on external sling .... 700 kg (1,543 lb)
- Max T-O weight: with sling load .... 1,800 kg (3,968 lb)
  - with internal payload .... 1,600 kg (3,527 lb)
- Max disc loading .... 26.7 kg/m^2 (5.47 lb/sq ft)
- Transmission loading at max T-O weight and power .... 5.07 kg/kW (8.33 lb/shp)

[1] *400 kg (882 lb) in main cabin, 150 kg (331 lb) in baggage compartment*

PERFORMANCE (A: at 1,600 kg; 3,527 lb AUW, ISA, B: at 1,800 kg; 3,968 lb):
- Never-exceed speed (VNE):
  - A, B .... 140 kt (260 km/h; 161 mph)
- Max cruising speed:
  - A, B .... 117 kt (218 km/h; 135 mph)
- Econ cruising speed:
  - A .... 102 kt (188 km/h; 117 mph)
  - B .... 103 kt (190 km/h; 118 mph)
- Max rate of climb at S/L: A .... 628 m (2,060 ft)/min
  - B .... 494 m (1,620 ft)/min
- Service ceiling: A .... 5,200 m (17,060 ft)
  - B .... 4,200 m (13,780 ft)
- Hovering ceiling IGE
  - A .... 3,300 m (10,825 ft)
  - B .... 2,150 m (7,050 ft)
- Hovering ceiling OGE:
  - A .... 2,200 m (7,220 ft)
  - B .... 910 m (2,980 ft)
- Range with max standard fuel, no reserves:
  - A .... 426 n miles (790 km; 490 miles)
  - B .... 409 n miles (758 km; 471 miles)
- Max endurance: A .... 5 h 35 min
  - B .... 5 h 12 min

UPDATED

PZL I-23 MANAGER

TYPE: Four-seat lightplane.

PROGRAMME: Launched as Instytut Lotnictwa development programme June 1992; funded by grant from State Committee for Scientific Research; construction started January 1994 of one static test (001) and two flying prototypes; first flight, by 002/SP-PIL, 20 February 1999; original name Ibis changed to Manager 1999; Polish type certificate BB-215 awarded 3 October 2001; second prototype had not been produced by early 2002, at which time first aircraft was re-registered SP-GIL; planned certification to Normal category FAR Pt 23 Amendment 42. PZL-Świdnik seeking production launch investment in 2003–04.

CURRENT VERSIONS (PLANNED): **I-23N:** Standard Normal category, retractable gear IFR version with dual controls.

Description applies to this version except where indicated.

I-23D: Normal category IFR executive or patrol version of I-23N with single controls and uprated, fuel-injection engine.

I-23S: Utility category basic trainer. Lower-powered engine, dual controls, VFR instrumentation, fixed landing gear and reinforced structure.

CUSTOMERS: Two ordered by Polish Ministry of Defence.

DESIGN FEATURES: Intended as safe, low-cost, low fuel consumption private owner aircraft. Conventional low-wing, tricycle-gear design with fully enclosed cabin.

Wing has ILL-217 (root) and ILL-213 (tip) aerofoil sections, 2° 30' leading-edge sweepback, 4° dihedral, 2° 30' incidence and –1° twist. Tailplane NACA 015 profile; fin NACA 0018.

Design permits use of any Textron Lycoming flat-four engine from 119 to 149 kW (160 to 200 hp), option for fixed or retractable landing gear, and equipment to customer's specifications.

FLYING CONTROLS: Conventional and manual. Ailerons deflect 22° up/17° down, elevators 30° up/24° down, rudder 21° left/right. Trim tab in port elevator. Flap settings 0/20/30°.

STRUCTURE: Mainly GFRP/CFRP/epoxy, with sandwich skins; CFRP for main spar and in region of landing gear attachments. Wing ribs and landing gear support plate of duralumin; steel fittings. Skin panels attached by adhesives, reinforced by bolts and rivets. First Polish aircraft with prepreg composites.

LANDING GEAR: Electrohydraulically retractable tricycle type on I-23N and D, with oleo-pneumatic shock-absorbers; steerable (±25°) nosewheel; mainwheels retract inward, nosewheel rearward, latter remaining half exposed when retracted; Parker wheels and brakes; Goodrich 5.00–5 tyres (main and nose), pressure 3.80 bar (55 lb/sq in) on mainwheels and 3.10 bar (45 lb/sq in) on nosewheel. Non-retractable gear on I-23S.

POWER PLANT: *I-23N:* One Textron Lycoming O-360-A1A flat-four (134 kW; 180 hp at 2,700 rpm), driving an MT-built Hartzell HC-C2YR-1BF/F766A-4 two-blade constant-speed propeller.

I-23D: 147 kW (200 hp) Lycoming IO-360-A1A.

I-23S: 118 kW (160 hp) Lycoming O-320-D2A and Sensenich fixed-pitch propeller.

Integral fuel tank in each wing, combined capacity 190 litres (50.2 US gallons; 41.8 Imp gallons) in I-23N/I-23D, 110 litres (29.1 US gallons; 24.2 Imp gallons) in I-23S. Gravity fuelling points on each wing. Oil capacity (all versions) 7.6 litres (2.0 US gallons; 1.7 Imp gallons).

ACCOMMODATION: Four persons (pilot and three passengers) on two rows of seats. Upward-opening door each side. Cabin heated and ventilated.

SYSTEMS: Electrohydraulic system for landing gear actuation; hydraulic system for mainwheel brakes only. Electrical power from 14 V 70 A alternator and 12 V 35 Ah battery. Heated pitot head.

AVIONICS: *Comms:* Bendix/King com transceiver and ATC transponder standard.

Flight: Bendix/King nav transceiver, ADF, GPS and directional gyro standard; optional full IFR system with KNS 81 integrated nav/RNav, DME and KCS 55A HSI.

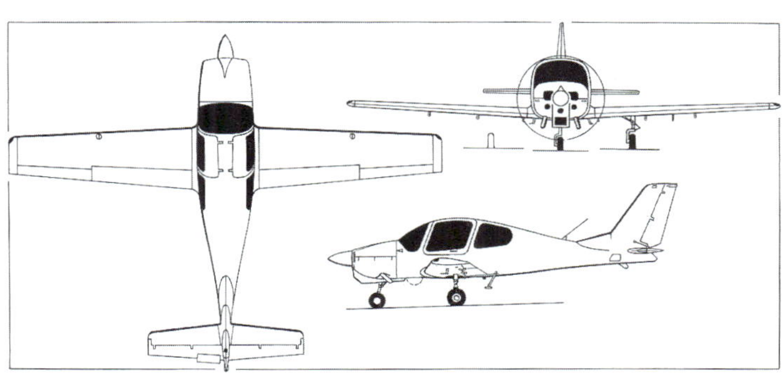

PZL I-23 Manager four-seat light aircraft *(Jane's/Mike Keep)* 0137337

Prototype I-23 Manager *(Paul Jackson)* ***NEW***/0552721

DIMENSIONS, EXTERNAL:
- Wing span .... 8.95 m (29 ft 4¼ in)
- Wing chord: at root .... 1.31 m (4 ft 3½ in)
  - mean .... 1.14 m (3 ft 8¾ in)
  - at tip .... 0.84 m (2 ft 9 in)
- Wing aspect ratio 8.0
- Length: overall .... 7.10 m (23 ft 3½ in)
  - fuselage .... 5.70 m (18 ft 8½ in)
- Height overall .... 2.49 m (8 ft 2 in)
- Tailplane span .... 3.20 m (10 ft 6 in)
- Wheel track .... 2.89 m (9 ft 5¾ in)
- Wheelbase .... 1.65 m (5 ft 5 in)
- Propeller diameter .... 1.83 m (6 ft 0 in)
- Propeller ground clearance .... 0.20 m (7¾ in)

DIMENSIONS, INTERNAL:
- Cabin: Length .... 2.35 m (7 ft 8½ in)
  - Max width .... 1.13 m (3 ft 8½ in)
  - Max height .... 1.20 m (3 ft 11¼ in)
- Baggage compartment:
  - Length .... 0.80 m (2 ft 7½ in)
  - Max width .... 0.68 m (2 ft 2¾ in)
  - Max height .... 0.35 m (1 ft 1¾ in)

AREAS:
- Wings, gross .... 10.00 m^2 (107.6 sq ft)
- Ailerons (total) .... 0.71 m^2 (7.64 sq ft)
- Trailing-edge flaps (total) .... 1.48 m^2 (15.93 sq ft)
- Fin .... 1.33 m^2 (14.32 sq ft)
- Rudder .... 0.95 m^2 (10.23 sq ft)
- Tailplane .... 2.08 m^2 (22.39 sq ft)
- Elevators (total) .... 0.83 m^2 (8.93 sq ft)

WEIGHTS AND LOADINGS (A: I-23N, B: I-23D, C: I-23S):
- Weight empty: A, B .... 690 kg (1,521 lb)
  - C .... 650 kg (1,433 lb)
- Fuel weight: A, B .... 130 kg (287 lb)
  - C .... 80 kg (176 lb)
- Max payload: A .... 330 kg (728 lb)
  - B .... 320 kg (705 lb)
  - C .... 300 kg (661 lb)
- Max T-O weight: A, B .... 1,150 kg (2,535 lb)
  - C .... 1,050 kg (2,315 lb)
- Max wing loading: A, B .... 115.0 kg/m^2 (23.55 lb/sq ft)
  - C .... 105.0 kg/m^2 (21.51 lb/sq ft)
- Max power loading: A .... 8.57 kg/kW (14.08 lb/hp)
  - B .... 7.71 kg/kW (12.68 lb/hp)
  - C .... 8.81 kg/kW (14.47 lb/hp)

PERFORMANCE (A, B and C as above):
- Never-exceed speed (VNE):
  - A .... 162 kt (300 km/h; 186 mph)
  - B .... 170 kt (315 km/h; 195 mph)
  - C .... 153 kt (285 km/h; 177 mph)
- Max cruising speed at S/L:
  - A .... 154 kt (285 km/h; 177 mph)
  - B .... 162 kt (300 km/h; 186 mph)
  - C .... 140 kt (260 km/h; 162 mph)
- Max manoeuvring speed: A, B, C .... 121 kt (225 km/h; 140 mph)
- Stalling speed: flaps up:
  - A, B .... 68 kt (125 km/h; 78 mph)
  - flaps down: A, B .... 62 kt (113 km/h; 71 mph)
- Max rate of climb at S/L: A .... 270 m (886 ft)/min
  - B .... 330 m (1,083 ft)/min
  - C .... 264 m (866 ft)/min
- T-O to 15 m (50 ft): A, B .... 480 m (1,575 ft)
- Landing from 15 m (50 ft): A, B .... 300 m (985 ft)
- T-O run: A .... 250 m (820 ft)
  - B .... 220 m (725 ft)
  - C .... 255 m (840 ft)
- T-O to 15 m (50 ft): A, B .... 480 m (1,575 ft)
- Landing from 15 m (50 ft): A, B .... 300 m (985 ft)
  - C .... 280 m (920 ft)
- Range with max fuel, 45 min reserves:
  - A .... 755 n miles (1,400 km; 869 miles)
  - B .... 782 n miles (1,450 km; 901 miles)
  - C .... 432 n miles (800 km; 497 miles)
- *g* limit .... +3.8

UPDATED

SSH

SERWIS SAMOLOTOW HISTORYCZNYCH-JANUSZ KARASIEWICZ (JANUSZ KARASIEWICZ HISTORIC AIRCRAFT SERVICE)

PL-43 385 Jasienica 829c
Tel: (+48 33) 815 34 91
Fax: (+48 33) 815 34 92
e-mail: jungmann@pro.onet.pl
MANAGING DIRECTOR: Janusz Karasiewicz
CUSTOMER SERVICE MANAGER: Jolanta Karasiewicz

GERMAN DISTRIBUTOR:
Bücker Flugzeugbau GmbH, Langwiederstrasse 1,
D-85757 Karlsfeld-Gröbenried
Tel: (+49 8131) 789 26
Fax: (+49 8131) 803 14
e-mail: rangsdorf@bueckerflug.de
Web: http://www.bueckerflug.de

SSH is certified by the Polish Aviation Authority as an approved production organisation under JAR 21, Subpart G. In addition to producing the Bü 131, it was building a prototype of a related aircraft, the Bü-133P Jungmeister, in late 2002. This was to have had the original Siemens Sh-14A radial engine, but has instead adopted the Czech LOM M 337 AK inline.

UPDATED

SSH (BÜCKER) T-131 JUNGMANN

TYPE: Aerobatic two-seat biplane.

PROGRAMME: Original German design first flew 27 April 1934; civil certification 3 October 1938; name refers to naval cadet; in addition to some 3,000 by parent firm (ending March 1941), licensed manufacture undertaken in Japan (339 K9W floatplanes and 1,037 Ki-86As), Spain (555 CASA 1.131s), Czechoslovakia (10 Tatra 131s and 300 Bu 131s, plus 260 Aero C.104s after 1945) and Switzerland (130 by Dornier). Last of some 5,700 were those produced in Spain, 1960.

SSH T-131PA Bücker Jungmann replica *NEW*/1034575

Partially completed T-131PA kitplane, showing fuselage structure *NEW*/1034581

SSH T-131 launched in March 1990, based on engineering drawings of Czech Bu 131D version (T and P in designations indicate Tatra and Poland). First of four pre-series aircraft flew 8 July 1994; first production aircraft exported to Austria in kit form December 1994 and made maiden flight 7 June 1997.

CURRENT VERSIONS: **T-131P:** Early version, with 78.3 kW (105 hp) Walter Minor 4-III four-cylinder inverted in-line engine and Hoffmann HO-40AHM-180 two-blade fixed-pitch metal propeller. Production complete.

T-131PA: More powerful version, with 103 kW (138 hp) LOM M 332AK four-cylinder, in-line engine. Prototype (SP-FPY) first flew 9 August 1995. Current baseline version; Polish certification under way in 2001 (latest information received), as step towards main objective of full standard certification in Germany.

Description applies to T-131PA.

CUSTOMERS: Three (including prototype) T-131Ps built 1994–95, of which two exported to Sweden. More than 20 T-131PAs registered by early 2002, including five in Austria; at least 10 to German owners, but retaining Polish registrations, three in Poland and one each to Brazil, Netherlands and Switzerland.

COSTS: Development €70,000 to €80,000 (2004).

DESIGN FEATURES: Single-bay biplane with interchangeable upper and lower wings, braced by pair of I-struts each side. This version based on Bü 131D, with ball bearings in control pivots and tailwheel, although Hirth 504A-2 engine replaced.

Wings have Göttingen aerofoil section (thickness/chord ratio 10.5 per cent) and 11° leading-edge sweepback; dihedral 3° 30' (upper), 1° 30' (lower); incidence −1° 30'; twist −1° 30'.

FLYING CONTROLS: Conventional and manual. Mass-balanced ailerons on all four wings; non-balanced rudder; trim tab in each elevator. No flaps.

STRUCTURE: Welded 4130 chromoly steel tube fuselage and tail unit; wooden wing spars (two) and ribs. Mostly Ceconite fabric covering except for metal engine cowling and around cockpit area. Fuel tank of glass fibre and carbon fibre.

LANDING GEAR: Tailwheel type; fixed. Two forward-raked, faired-in side Vs, each with telescoping shock absorber in forward element, hinged to lower longerons, plus half-axles attached to compression frame; sprung 200×50 tailwheel; mainwheels 420×150. Tyre pressures: main 1.50 bar (53 lb/sq in), tail 2.50 bar (88 lb/sq in). Hydraulic brakes.

POWER PLANT: One 104 kW (140 hp) LOM M 332AK four-cylinder in-line engine, driving an MT32.110/12 two-blade fixed-pitch wooden propeller. Fuselage fuel tank, capacity 85 litres (22.5 US gallons; 18.7 Imp gallons), of which 80 litres (21.1 US gallons; 17.6 Imp gallons) are usable. Oil capacity 8.5 litres (2.25 US gallons; 1.9 Imp gallons).

ACCOMMODATION: Two persons in tandem, open cockpits.

SYSTEMS: 24 V DC electrical system (generator and two batteries).

AVIONICS: Basic VFR standard; other avionics to customer's requirements.

DIMENSIONS, EXTERNAL:

Wing span	7.40 m (24 ft 3¼ in)
Wing chord (each, constant)	1.00 m (3 ft 3¼ in)
Wing aspect ratio	4.1
Length overall	6.62 m (21 ft 8½ in)
Height overall	2.25 m (7 ft 4½ in)
Tailplane span	2.50 m (8 ft 2½ in)
Wheel track	1.75 m (5 ft 9 in)
Wheelbase	4.10 m (13 ft 5½ in)
Propeller diameter	1.88 m (6 ft 2 in)
Propeller ground clearance	0.63 m (2 ft 0¾ in)

AREAS:

Wings, gross	13.50 m² (145.3 sq ft)
Ailerons (four, total)	2.00 m² (21.53 sq ft)
Fin	0.50 m² (5.38 sq ft)
Rudder	0.80 m² (8.61 sq ft)
Tailplane	1.25 m² (13.45 sq ft)
Elevators (total, incl tabs)	1.15 m² (12.38 sq ft)

WEIGHTS AND LOADINGS (approx):

Weight empty	470 kg (1,036 lb)
Fuel weight	63 kg (139 lb)
Max T-O and landing weight: Normal	720 kg (1,587 lb)
Utility	670 kg (1,477 lb)
Aerobatic	590 kg (1,300 lb)
Max wing loading	53.3 kg/m² (10.92 lb/sq ft)
Max power loading	6.90 kg/kW (11.33 lb/hp)

PERFORMANCE:

Never-exceed speed (VNE)	151 kt (280 km/h; 174 mph)
Max level speed	102 kt (190 km/h; 118 mph)
Max cruising speed at S/L	97 kt (180 km/h; 112 mph)
Econ cruising speed at S/L	86 kt (160 km/h; 99 mph)
Stalling speed, engine idling	46 kt (85 km/h; 53 mph)
Max rate of climb at S/L	210 m (689 ft)/min
Service ceiling	4,000 m (13,120 ft)
T-O run	180 m (590 ft)
T-O to 15 m (50 ft)	400 m (1,315 ft)

Landing from 15 m (50 ft) .... 300 m (985 ft)
Landing run .... 150 m (495 ft)
Max range, 30 min reserves .... 283 n miles (525 km; 326 miles)

UPDATED

SSH (BÜCKER) BÜ-133P JUNGMEISTER

English name: Young Champion

TYPE: Aerobatic single-seat biplane.

PROGRAMME: Entered service in 1936; also licence-built in Spain and Switzerland. Small number of replicas built in late 1960s by Bitz at Augsburg and Hirth at Nabern. SSH project initiated in late 1990s; prototype (SP-PUV) planned to fly in second quarter 2004. Perpetuates original German Bü 133C of 1935. (Correct German designation is unhyphenated.)

DESIGN FEATURES: Generally similar to those of Jungmann.

FLYING CONTROLS: As Jungmann.

STRUCTURE: Fabric-covered steel tube fuselage and tail unit, except for wooden turtledeck; two-spar wooden wings with composites leading-edges and metal interplane struts.

LANDING GEAR: Tailwheel type; fixed. Mainwheel tyres size 6.00–6.

POWER PLANT: One 160 kW (215 hp) LOM M-337AK six-cylinder inline engine.

ACCOMMODATION: Pilot only, in open cockpit.

DIMENSIONS, EXTERNAL:
Wing span .... 6.60 m (21 ft 7¾ in)
Length overall .... 6.02 m (19 ft 9 in)
Height overall .... 2.20 m (7 ft 2½ in)

AREAS:
Wings, gross .... 12.00 m² (129.2 sq ft)

WEIGHTS AND LOADINGS:
Weight empty .... 450 kg (992 lb)
Max T-O weight .... 600 kg (1,322 lb)
Max wing loading .... 50.0 kg/m² (10.24 lb/sq ft)
Max power loading .... 3.74 kg/kW (6.15 lb/hp)

PERFORMANCE (estimated):
Max level speed .... 145 kt (270 km/h; 167 mph)
Time to 1,000 m (3,280 ft) .... 1 min
Service ceiling .... 7,000 m (22,965 ft)
Range .... 324 n miles (600 km; 372 miles)

UPDATED

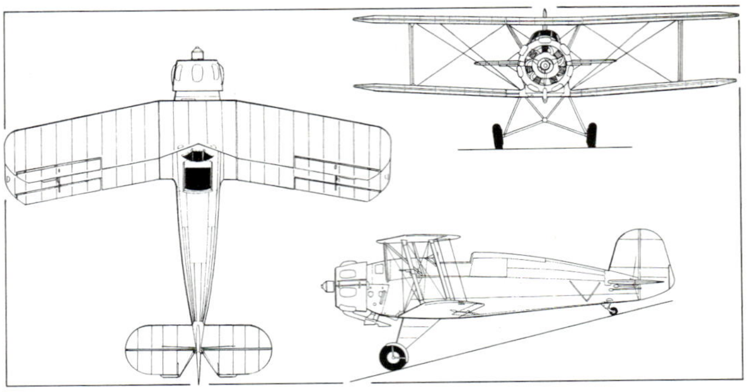

SSH (Bücker) Bü-133P Jungmeister 0110902

First SSH-built Jungmeister before covering *(Paul Jackson)* 0110937

ZRIPSL

ZAKŁAD REMONTÓW I PRODUKCJI SPRZĘTU LOTNICZEGO (AVIATION EQUIPMENT REPAIR AND PRODUCTION WORKS)

ulica Strażacka 60, PL-43 382 Bielsko-Biała
Tel/Fax:(+48 33) 815 01 10
e-mail: e.marganski@pro.onet.pl
Web: http://www.marganski.com.pl
MANAGING DIRECTOR: Edward Marganski

Edward Marganski has been connected with the design of several Polish light aircraft. His latest designs are the EM-10 Bielik, envisaged as a modern successor to the TS-11 Iskra jet trainer, and EM-11 Orka twin-engined utility aircraft.

Company established 1986 as first aeronautical private enterprise in Poland, initially inspecting and repairing wooden gliders and later extending capabilities to overhaul and repair of motor gliders and all-composites sailplanes. Co-operated with Krakow Aviation Museum in reconstructing four historical aircraft, followed by Grunau Baby reconstruction for a Swedish customer. Designed, built and flew Swift S-1 and MDM-1 Fox sailplanes under contract to those companies during 1990s, eventually producing nearly 30 of each type and modifying latter as motor glider. Bielik programme started 1999 and Orka two years later.

Non-aviation work has included composites components for motorcar bodies and, for Euros GmbH of Germany, 44 m (144.4 ft) diameter rotor blades for wind energy converters (WECs). In 2002, for customer in Belarus, designed and built components for a 1 MW WEC station said to be world's first to be based on the Magnus (rotating cylinder) effect. The factory has a floor area of 650 m² (7,000 sq ft) and a workforce (December 2002) of 27.

UPDATED

ZRIPSLEM-10 BIELIK AND FENIX

TYPE: Basic jet trainer.

PROGRAMME: Started as private venture under project name Iskra II, but renamed Bielik (a native Polish eagle) after press competition before public debut third quarter 2001 at exhibitions in Katowice and Radom. First flight (prototype SP-YEM) was then anticipated before end of 2001, but delayed by engine installation problems. Transported to Mielec for pre-flight preparation and tests in November 2002; first flight deferred until end of Polish winter; achieved 4 June 2003.

ZRiPSL EM-10 Bielik tandem-seat jet *(Paul Jackson)* ***NEW***/0580829

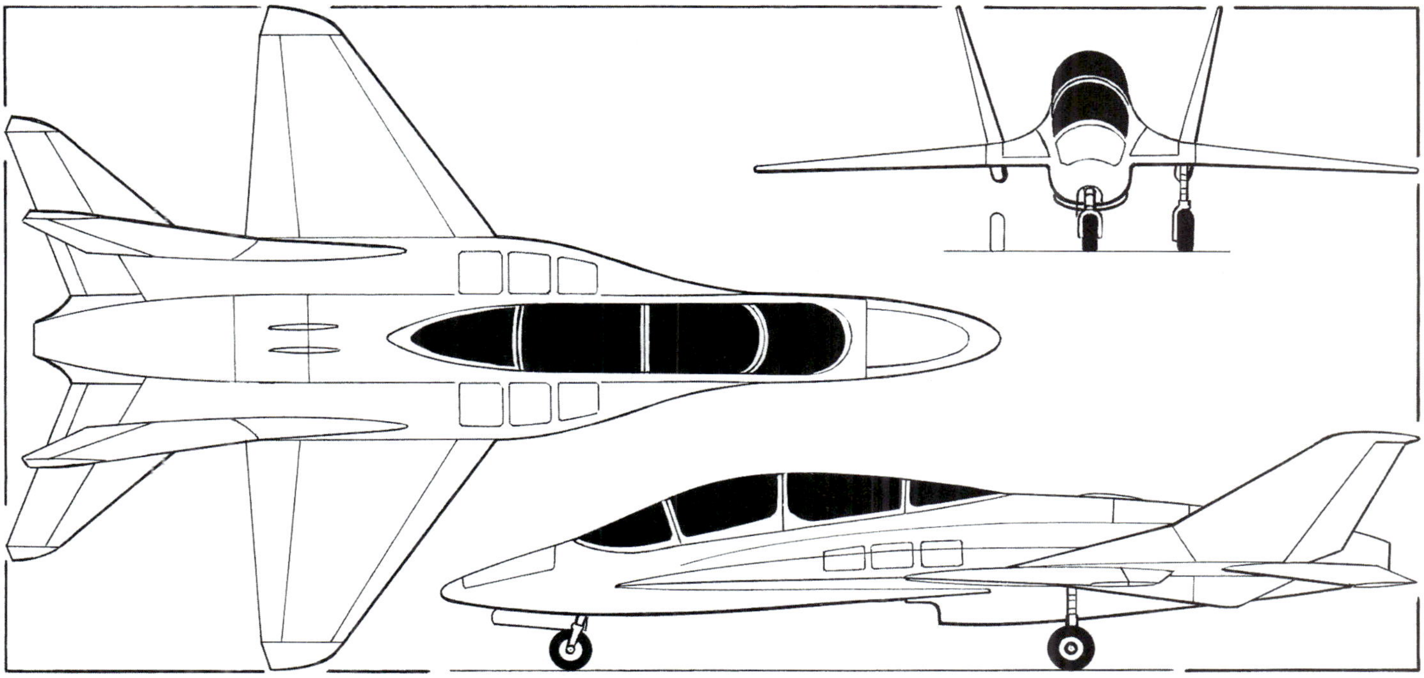

EM-10 Bielik twin-tailed, tandem-seat jet trainer *(James Goulding)* 0143736

Bielik cockpit detail *(Paul Jackson)* ***NEW***/0552867

CURRENT VERSIONS: **Bielik:** Military basic and advanced training version, aimed at Polish and foreign air forces as low weight and cost pilot trainer.

Fenix : (English name Phoenix) Civil general aviation version, including aerobatics; intended for export.

DESIGN FEATURES: Low T-O weight achieved by rejection of onboard armament in favour of electronic combat application simulation system. Mid-wing, blended wing/body design with forward-extending curved strakes; twin outward-canted fins and rudders. Designed for high-*g* and high-AoA manoeuvres.

Supercritical wing and tailplane aerofoil section, thickness/chord ratio 6 per cent; wing anhedral 3° 30'; tailplane anhedral 10°.

FLYING CONTROLS: Manual, with hydraulically boosted all-moving slab tailplane, with electrically actuated tabs, for pitch and roll control; twin rudders deflect outwards only. Full-span leading- and trailing-edge flaps.

STRUCTURE: All-composites (carbon fibre and epoxy resin) stressed-skin sandwich over most of airframe. Single-spar wing.

LANDING GEAR: Hydraulically retractable tricycle type, with single wheel on each unit. Wheel sizes: main 49.5 cm (19½ in), nose 15.2 cm (6 in). Mainwheels retract rearward, nosewheel forward. Mainwheel disc brakes.

POWER PLANT: One 12.75 to 17.65 kN (2,866 to 3,968 lb st) class turbojet in production aircraft; prototype has 13.52 kN (3,040 lb st) General Electric CJ610-6. Thrust vector control planned for production aircraft. Two fuel tanks in tandem in centre-fuselage, combined capacity 1,100 litres (291 US gallons; 242 Imp gallons).

ACCOMMODATION: Crew of two in tandem, on zero/zero ejection seats (zero height/150 kt; 278 km/h; 173 mph in prototype), in pressurised cockpit.

SYSTEMS: Hydraulic system, pressure 140 bar (2,030 lb/sq in), for landing gear, flap, airbrake and wheel brake actuation and slab tailplane control booster. Pneumatic system for emergency operation of trailing-edge flaps, landing gear extension and mainwheel brakes. 28 V DC electrical system. Crew oxygen system. Cockpit pressurisation system.

AVIONICS: Bielik basic trainer and Fenix to have standard nav/com avionics. Bielik advanced trainer version planned to be equipped with combat application electronic simulations of armament operation; real or virtual, radar- and IR-based observation and sighting systems; evaluation of sighting and weapon delivery techniques; and hostile air defence facilities.

DIMENSIONS, EXTERNAL:

Wing span	6.60 m (21 ft 7¾ in)
Wing aspect ratio	3.7
Length overall	9.00 m (29 ft 6¼ in)
Height overall	2.50 m (8 ft 2½ in)

AREAS:

Wings, gross	11.90 m² (128.1 sq ft)

WEIGHTS AND LOADINGS:

Weight empty	1,700 kg (3,748 lb)
Fuel weight	850 kg (1,874 lb)
Max T-O weight	2,500 kg (5,511 lb)
Max wing loading	210.1 kg/m² (43.03 lb/sq ft)
Max power loading (prototype)	185 kg/kN (1.81 lb/lb st)

PERFORMANCE (estimated):

Max Mach No.	0.9
Max level speed	594 kt (1,100 km/h; 684 mph)
Stalling speed	90 kt (165 km/h; 103 mph)
Max rate of climb at S/L	2,700 m (8,858 ft)/min
Ferry range with auxiliary tanks	1,349 n miles (2,500 km; 1,553 miles)

UPDATED

ZRIPSL EM-11 ORKA

English name: Whale

TYPE: Four-seat utility twin.

PROGRAMME: Started October 2001; revealed at Friedrichshafen air show, April 2003; prototype completed at that time; maiden flight achieved (SP-YEN) 8 August 2003; planned certification to JAR 23, Normal category. Seen as progenitor of family of 10 variants which would eventually include six-seat, pressurised, retractable-gear and amphibious versions, some of them with Diesel or turboprop power plants.

CURRENT VERSIONS (PLANNED): **Rotax 912 versions:** Four-seat executive/tourer; executive/cargo; rescue/ambulance; FLIR-carrying reconnaissance.

Diesel versions: Executive tourer; and E-O reconnaissance with radar and data downlink. Twin diesels of 101 to 149 kW (135 to 200 hp).

Turboprop versions: Executive with pressurised cabin and retractable gear; four-seat executive amphibian; stretched, six-seat executive with additional door; and amphibious version of last-named. Twin turboprops of 261 to 298 kW (350 to 400 shp).

DESIGN FEATURES: Multirole capabilities and near-STOL runway requirements. High-mounted, tapered wings with twin pusher engines; tadpole-shaped fuselage with sweptback fin and rudder; tapered, low-set horizontal tail surfaces. Underfuselage conformal planing hull attachment on amphibious versions, plus fixed floats underwing.

FLYING CONTROLS: Conventional and manual. Flaps and ailerons each constructed in left and right halves; horn-balanced elevator with flight-adjustable tab; horn-balanced rudder with adjustable tab.

STRUCTURE: All-composites.

LANDING GEAR: Tricycle type; fixed. Mainwheels, on fuselage-mounted, self-sprung cantilever legs, size 6.00-6; trailing-link nosewheel, size 5.00-5.

POWER PLANT: Two 73.5 kW (98.6 hp), wing-mounted Rotax 912 ULS flat-four engines, each driving a Woodcomp three-blade constant-speed pusher propeller. Two fuel tanks in each wing, combined capacity 240 litres (63.4 US gallons; 52.8 Imp gallons).

Diesel (101–149 kW; 135–200 hp) and turboprop (261–298 kW; 350–400 shp) engines envisaged for later versions.

ACCOMMODATION: Four seats in tandem pairs. Gull-wing windscreens/doors, hinged on centreline and opening upward. Baggage space aft of seats, expandable by folding or removal of three passenger seats.

AVIONICS: VFR or IFR, according to version.

DIMENSIONS, EXTERNAL:

Wing span	13.50 m (44 ft 3½ in)
Wing aspect ratio	10.9
Length overall	8.65 m (28 ft 4½ in)
Fuselage: Max width	1.30 m (4 ft 3¼ in)
Max depth	1.25 m (4 ft 1¼ in)
Height overall	2.62 m (8 ft 7¼ in)
Tailplane span	3.70 m (12 ft 1¾ in)
Wheel track	2.42 m (7 ft 11¼ in)
Wheelbase	3.38 m (11 ft 1 in)
Distance between propeller centres	2.97 m (9 ft 9 in)

AREAS:

Wings, gross	16.65 m² (179.2 sq ft)

WEIGHTS AND LOADINGS:

Weight empty	925 kg (2,039 lb)
Max payload	400 kg (882 lb)
Max T-O weight	1,375 kg (3,031 lb)
Max wing loading	82.6 kg/m² (16.92 lb/sq ft)
Max power loading	9.35 kg/kW (15.37 lb/hp)

PERFORMANCE (estimated):

Max level speed	173 kt (320 km/h; 199 mph)
Cruising speed	124 kt (230 km/h; 143 mph)
Min control speed (VMC)	59 kt (110 km/h; 68 mph)
Stalling speed, flaps down	49–52 kt (90–95 km/h; 56–59 mph)
Max rate of climb at S/L	240 m (787 ft)/min
Range:	
4 persons and 50% fuel	405 n miles (750 km; 466 miles)
2 persons and 100% fuel	809 n miles (1,500 km; 932 miles)

UPDATED

Instrument panel of ZRiPSL EM-11 Orka *(Paul Jackson)* ***NEW***/0580838

EM-11 Orka prototype at Friedrichshafen in April 2003 *(Paul Jackson)* ***NEW***/0552880

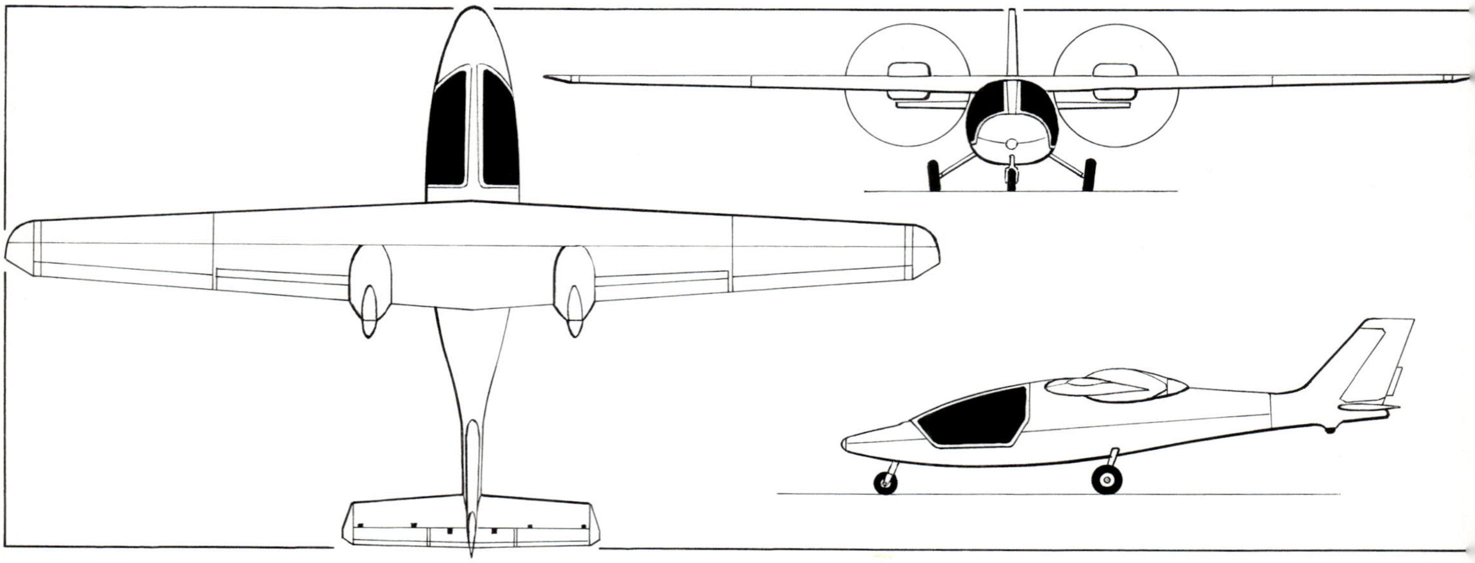

ZRiPSL Orka twin pusher-engined utility aircraft *(James Goulding)* ***NEW***/0558325

Romania

AEROSTAR

SC AEROSTAR SA
9 Condorilor Street, R-600302 Bacău
Tel: (+40 234) 57 50 70
Fax: (+40 234) 57 20 23 and 57 22 59
e-mail: aerostar@aerostar.ro
Web: http://www.aerostar.ro
PRESIDENT AND GENERAL DIRECTOR: Dipl Eng Grigore Filip
DIRECTORS:
Dipl Eng Ovidiu Buhai (Systems Division)
Dipl Eng Dorin Panfil (Industrial Products Division)
Dipl Eng Theodor Placinta (Technological Division)
Dipl Eng Serban Iosipescu (Aeronautical Products Division)
PUBLIC RELATIONS AND MARKETING: Doina-Gabriela Matanie

Factory celebrated 50th anniversary in 2003, having been founded 17 April 1953 as URA (later IRAv, then IAv Bacau), originally as repair centre for Romanian Air Force Yak-17/23, MiG-15/17/19/21 and Il-28/H-5 front-line military aircraft and Aero L-29/L-39 jet trainers; built first Romanian prototype of IAR-93 between 1972 and 1974. Five other specialised work sections: (1) landing gears, hydraulic and pneumatic equipment; (2) special production; (3) light aircraft; (4) engines and reduction gears; (5) avionics. Site area 45 ha (111.2 acres), including 25 ha (61.8 acres) of covered workshop space. Workforce 2,100 in April 2004.

Romanian State Ownership Fund (SOF) originally owned 69.99 per cent of Aerostar. In preparation for full privatisation, Aerostar reorganised 1997 into three semi-autonomous divisions: systems, commercial, and technology support, reporting to a strategic fourth division. Creation of a joint venture (Aerothom Electronics) with Thomson-CSF (now Thales) Communications announced June 1999; Aerostar owns 60 per cent of this company, formed to supply IFF equipment to Romanian MoD. Aerostar became fully privatised company 11 February 2000 as part of Aerostar-PAS-IAROM, a consortium of Aerostar management and employees (PAS) with IAROM, acquiring Aerostar stock previously held by SOF. The current shareholdings are IAROM 71 per cent; SIF 'Moldova' (financial investment company) 11 per cent; others 18 per cent. In 2003, SOF renounced the 'golden share' entitling it to veto matters affecting company's defence production capability, following harmonisation with EU regulations. Consortium agreed to take over all Aerostar's obligations and debts; not to dissolve company for 10 years; and to keep its core business intact for at least five years. Company turnover in 2003 was Lei 1,037 billion, of which exports accounted for more than 51 per cent.

Aerostar's main product for many years has been Iak-52 trainer, joined in 2001 by Aerostar 01 Festival. Manufactures airframes for German ULBI Wild Thing ultralight and components for Flug Werk replicas of Focke-Wulf Fw 190. North American P-51 Mustang replica available 2003 from Fighter Factory LLC, 13545 Sycamore Avenue, San Martin, California 95046, USA (*tel:* +1 408 683 25 31; *fax:* +1 408 683 25 33). Aerostar also responsible for Lancer and Sniper Romanian Air Force upgrade programmes for MiG-21s (completed April 2003) and MiG-29s.

Landing gears and/or hydraulic/pneumatic equipment have been produced for IAR-316B (Alouette III), IAR-330 (Puma), IAR-93, IAR-99, Romaero (BAC) One-Eleven and Iak-52. From 2003, producing sheet metal for Stork Aerospace contribution to manufacture of components for Airbus and Cessna. Engines include M-14P for Iak-52, M-14V26 for Ka-26 and RU-19A-300 APU for An-24; reduction gears include R-26. Avionics factory produces radio altimeters, radio compasses, marker beacon receivers, IFF and other radio/radar items.

UPDATED

Aerostar Iak-52TW in inverted flight ***NEW***/1034641

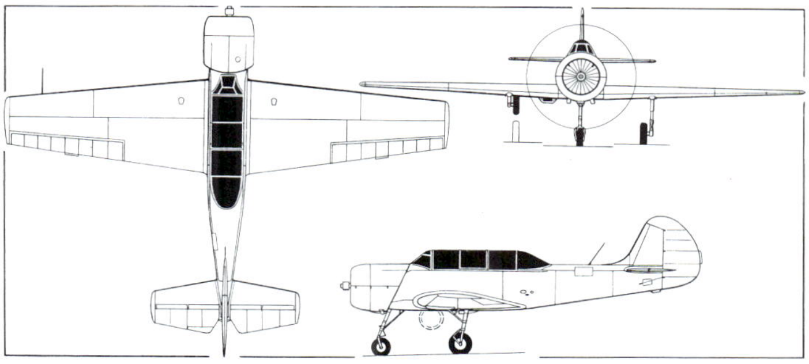

Aerostar (Yakovlev) Iak-52 tandem two-seat primary trainer *(Jane's/Dennis Punnett)* 0011952

AEROSTAR (YAKOVLEV) IAK-52

TYPE: Primary prop trainer/sportplane.

PROGRAMME: Design of Yak-52 in USSR began 1975; series production assigned to Comecon programme (Council for Mutual Economic Assistance), following Romanian-USSR intergovernmental agreement of 1974; designation is transliteration of 'Yak' in Romanian alphabet, but civil aircraft are registered as Yaks.

Construction began 1977 and first Romanian prototype (c/n 780102) made first flight May 1978; series production began 1979; first deliveries (to DOSAAF, USSR) 1975; 1,000th aircraft delivered in 1986 and 1,500th in 1990. After cessation of deliveries to USSR in 1991, production had almost come to a standstill, but interest reawakened by improved Iak-52W (first flight April 1999, debut at Paris Air Show, June 1999). Iak-52W began development 1998 and first flew (12201) March 1999; currently offered as alternative to standard version. Aerostar planned five-year programme of 30 per year; deliveries started August 1999. Tailwheel (TW) version announced early 2000; first flew ('26', c/n 0112226, later N52SD) 2 July 2001; launched at EAA AirVenture, Oshkosh, 24 to 30 July 2001; now only production version.

CURRENT VERSIONS: **Iak-52:** Standard model.

Iak-52W: Upgraded ('Westernised') version; introduced 1999. Main changes are increased fuel tankage, three-blade propeller, Western avionics, and metal-skinned control surfaces. Final delivery 2002; production completed.

Iak-52TW: New-build tailwheel version of Iak-52W, revealed early 2000. Production aircraft available from July 2001 (four sold by end of first day of promotion at Oshkosh); US marketing by GeSoCo Industries of Swanton, Vermont, USA. Features include extended (rounded) wingtips, uprated engine with underslung oil radiator and modified front shutters; each cockpit lengthened by 10 cm (4 in).

Detailed description applies to this version.

Yak-52M: Refurbishment project for Russian paramilitary flying training fleet of ROSTO (formerly DOSAAF). Prototype shown at Moscow, August 2003, but maiden flight ('01' at Ivanovo) not achieved until 16 April 2004. Modified wingtips and SKS-94MYa ejection system; new domed cockpit canopy. Wing span 9.50 m (31 ft 2 in); length 8.025 m (26 ft 4 in), height 2.875 m (9 ft 5¼ in), M-14Kh engine and MT-Propeller as standard, max T-O weight 1,423 kg (3,137 lb), aerobatic weight 1,315 kg (2,899 lb), fuel capacity 230 litres (60.8 US gallons; 50.6 Imp gallons) and range 486 n miles (900 km; 559 miles).

Yak-52TD: 'Tail-dragger' conversion of Iak-52, by Termikas of Lithuania, employing second-hand airframe, but costing one-third less than new Iak-52TW. Modifications include two integral wing tanks, combined capacity 170 litres (44.9 US gallons; 37.4 Imp gallons), in addition to standard tankage; propeller upgrade from two to three blades, as for

Prototype Yak-52TD conversion of original Iak-52 *(Paul Jackson)* ***NEW***/0580779

Iak-52TW; and landing gear doors. Prototype LY-TTD first flew 23 April 2004 as rebuild of 1988 Iak-52; shown at North Weald Aerofair, UK, May 2004; marketing by Yak UK Ltd.

CUSTOMERS: Estimated 1,815 of Iak-52, in 121 production batches, built by 1998, mainly for USSR and Romanian Air Force but latterly also for West European, Australian and American civil markets; Hungarian Air Force received 12 in early months of 1994, six of which upgraded by Aerostar in 2003 with some features (three-blade propellers and external lights) of Iak-52W. Production averaged 10 per year between 1992 and 1996 (almost all to USA or Hungary), augmenting large number of military surplus sales. Twelve delivered to Vietnam in 1997. No known deliveries since 1998.

Iak-52W/TW production began with 122nd batch (seven aircraft); total 12 Iak-52Ws (including prototype) built 1999-late 2000; subsequent production of 10 in batch 123, five in 124 and six in 125, bringing total to 28 by early 2004. Deliveries have included at least nine to USA and one to Australia. Iak-52TW deliveries began in mid-2001; some 20 built by early 2003; at least 16 to USA and one to Australia.

COSTS: Standard Iak-52TW US$150,000 (2003).

DESIGN FEATURES: Tandem-cockpit variant of Yak-50, with unchanged span and length, but with mainwheels fully retractable. Simple, robust, low-wing design with long landing gear legs to accommodate large propeller. Moderately tapered wings and tailplane; curved fin; 'glasshouse' canopy. Current production aircraft lack wingtip fairings. Strengthened wingtips for optional fitment of tip-tanks. Increased wing span compared with earlier versions.

Wing section Clark YH: thickness/chord ratio 14.5 per cent at root, 9 per cent at tip. Dihedral 2° from root; incidence 2°. Tailplane 0° dihedral, 1° 30' incidence.

FLYING CONTROLS: Conventional and manual. Actuation of mass-balanced slotted ailerons by pushrods; mass-balanced elevators by pushrods/cables; and horn-balanced rudder by cables; manually operated trim tab in port elevator; ground-adjustable tab on rudder and each aileron. Pneumatically actuated trailing-edge split flaps.

STRUCTURE: All-metal, including primary control surfaces. Single-spar wings; modified spar, post-1986 (and some retrofits), permits higher load factors. Wingtips strengthened to carry optional tip-tanks. Airframe life 5,000 hours.

LANDING GEAR: Semi-retractable tricycle type, with single wheel and oleo-pneumatic shock-absorber on each unit. Mainwheels (size 6.00-6) fully retractable, on Aerostar pneumatically actuated legs, retracting inward; flush doors. Cleveland wheels and Western tyres; Scott steerable tailwheel, with 10×3.50–4 tyre. Pneumatic actuation, nosewheel retracting rearward, main units forward. All three wheels remain fully exposed to airflow, against undersurface of fuselage and wings respectively, to offer greater safety in event of wheels-up emergency landing. Hydraulic mainwheel disc brakes, operated differentially from rudder pedals. Non-retractable plastics-coated duralumin skis, with shock-struts, can be fitted in place of wheels for winter operations.

POWER PLANT: One 298 kW (400 hp) Aerostar-built VOKBM (Bakanov) M-14P-XDK nine-cylinder air-cooled radial, driving a three-blade constant-speed MT-Propeller MTV-9-B-C/CL250-27 propeller. Two aluminium alloy fuel tanks, in wingroots forward of spar; collector tank in fuselage supplies engine during inverted flight.

Total internal fuel capacity 280 litres (74.0 US gallons; 61.6 Imp gallons). Oil capacity 10 litres (2.6 US gallons; 2.2 Imp gallons). Provision for wingtip tanks.

ACCOMMODATION: Tandem seats for pupil (at front) and instructor under long 'glasshouse' canopy, with separate rearward-sliding hood over each seat. Hooker Harness seat belts for both occupants. Dual controls standard. Seats and rudder pedals adjustable. Heating and ventilation standard. Externally accessible baggage and battery compartments on port side, aft of rear cockpit, plus permanently mounted retractable ladder for cockpit access.

SYSTEMS: Independent main and emergency pneumatic systems, pressure 50 bar (725 lb/sq in), for flap and landing gear actuation, engine starting and brake control. Electrical system (27 V DC) supplied by 3 kW engine-driven generator and two 27 V 23 Ah ASAM batteries; two static inverters in fuselage for 36 V AC power at 400 Hz. Oxygen system available optionally.

AVIONICS: *Comms:* Western instruments including Garmin IC-AQ200 com radio and GTX transponder, AmeriKing ELT and altitude encoder, and NAT AA80-20 intercom.

Flight: ARK-15M automatic radio compass, eight-channel ADF and GMK-1A gyrocompass.

Other avionics available at customer's option.

EQUIPMENT: Strobe and navigation lights, recessed landing and taxying lights; roll bar.

DIMENSIONS, EXTERNAL:	
Wing span	9.90 m (32 ft 5¾ in)
Wing chord: at root	2.10 m (6 ft 10¾ in)
mean aerodynamic	1.64 m (5 ft 4½ in)
at tip	1.08 m (3 ft 6½ in)
Wing aspect ratio	6.4
Length overall	7.98 m (26 ft 2¼ in)
Height overall, propeller turning	2.70 m (8 ft 10¼ in)
Wheel track	2.70 m (8 ft 10¼ in)
Propeller diameter:	2.40 m (7 ft 10½ in)
DIMENSIONS, INTERNAL:	
Cockpit: Max width	0.74 m (2 ft 5¼ in)
Max height	1.12 m (3 ft 8 in)
AREAS:	
Wings, gross	15.42 m² (166.0 sq ft)
Ailerons (total)	1.98 m² (21.31 sq ft)
Flaps (total)	1.03 m² (11.09 sq ft)
Fin	0.61 m² (6.57 sq ft)
Rudder	0.87 m² (9.36 sq ft)
Tailplane	1.32 m² (14.21 sq ft)
Elevators (total)	1.53 m² (16.47 sq ft)
WEIGHTS AND LOADINGS:	
Weight empty, equipped	980 kg (2,160 lb)
Max T-O weight	1,320 kg (2,910 lb)
Max wing loading	85.6 kg/m² (17.53 lb/sq ft)
Max power loading	4.43 kg/kW (7.27 lb/hp)
PERFORMANCE:	
Never-exceed speed (VNE)	226 kt (420 km/h; 261 mph)
Max level speed at S/L	162 kt (300 km/h; 186 mph)
Stalling speed, flaps down	46 kt (85 km/h; 53 mph)
Max rate of climb at S/L	360 m (1,181 ft)/min
Service ceiling	4,000 m (13,120 ft)
T-O run	140 m (460 ft)
Landing run	270 m (885 ft)
Range with max fuel at 500 m (1,640 ft) at cruising speed, 20 min reserves	540 n miles (1,000 km; 621 miles)
g limits	+7/−5

UPDATED

AEROSTAR FESTIVAL

TYPE: Side-by-side ultralight/sportplane.

PROGRAMME: Illustrated (as project) in 1999 company brochure, but no details then given. Made debut (as Aerostar 01) at Paris Air Show, June 2001, following first flight (YR-6138) on 31 May. Certification programme flown between 5 November 2001 and 9 February 2002, resulting in award of domestic certificate April 2002. Aircraft named Festival later that year. Improved design (see Design Features below) introduced 2003, first example (YR-6107) making maiden flight in September; also now certified by Romanian Airclub. Aircraft qualifies as Advanced Ultralight in Canada and as Light Sport Aircraft (LSA) in US.

CUSTOMERS: Initial orders received at time of debut (four for Romanian Airclub); planned eventual worldwide sales and output of 40 to 50 per year. First two production aircraft registered in USA (N1062N and N1062M) in July 2001.

COSTS: US$55,000 as LSA in US; US$40,000 as advanced ultralight in Canada; US$45,000 in Australia (all 2004).

DESIGN FEATURES: Designed to Canadian standard DS10141E; derivative of Air Light Wild Thing (see German section), originally utilising wings, nose, rear fuselage and tail unit from that aircraft. Principal differences are enlarged cabin, nosewheel landing gear, and relocation of wing in low-mounted position, without strut bracing, but still foldable alongside fuselage for transportation and storage. Subsequently improved with much-revised cockpit layout, upward-opening single-piece canopy and optional three-blade propeller.

Wing aerofoil section NACA 4415.

FLYING CONTROLS: Conventional and manual. Flaps and ailerons pushrod-actuated, elevators and horn-balanced rudder by cables; mechanically or electrically actuated elevator trim tab. Dual controls standard.

STRUCTURE: Aluminium alloy, except for fabric-covered flaps and ailerons.

LANDING GEAR: Tricycle type, fixed. Fuselage-mounted spring cantilever mainwheel legs, Matco hydraulic disc brakes and tyres; steerable, trailing-link nosewheel with rubber shock absorption.

POWER PLANT: One 59.7 kW (80 hp Jabiru 2200 engine standard, driving a two- or three-blade adjustable-pitch wooden propeller. Options include 59.6 kW (79.9 hp) Rotax 912 UL; 73.5 kW (98.6 hp) Rotax 912 ULS; 63.4 kW (85 hp) Limbach L 2000; or 89.5 kW (120 hp) Jabiru 3300. Fuel tank capacity (two wing tanks) 80 litres (21.1 US gallons; 17.6 Imp gallons). Oil capacity 2.5 litres (0.7 US gallon; 0.55 Imp gallon).

ACCOMMODATION: One-piece upward-opening canopy. Baggage compartment aft of seats.

AVIONICS: Basic VFR instrumentation (ASI, altimeter, variometer, turn-and-slip indicator and compass) standard; Bendix/King KY 92 radio optional.

DIMENSIONS, EXTERNAL:	
Wing span	9.17 m (30 ft 1 in)
Wing mean aerodynamic chord	1.525 m (5 ft 0 in)
Length overall	6.74 m (22 ft 1¼ in)
Height overall	2.47 m (8 ft 1¼ in)
Propeller diameter	1.50 m (4 ft 11 in)
AREAS:	
Wings, gross	13.97 m² (150.4 sq ft)
WEIGHTS AND LOADINGS (A: JABIRU 2200/ROTAX 912 UL, B: ROTAX 912 ULS):	
Weight empty: A	351 kg (774 lb)
B	354 kg (780 lb)
Max T-O weight: A, B	535 kg (1,179 lb)
PERFORMANCE (A and B as above):	
Never-exceed speed (VNE):	
A, B	116 kt (215 km/h; 133 mph)
Max level speed: A	99 kt (185 km/h; 115 mph)
B	105 kt (195 km/h; 121 mph)
Cruising speed: A	92 kt (170 km/h; 106 mph)
B	94 kt (175 km/h; 109 mph)
Stalling speed, flaps down: A	36 kt (66 km/h; 41 mph)
B	39 kt (71 km/h; 45 mph)
Max rate of climb at S/L: A	240 m (787 ft)/min

Second prototype Aerostar Festival side-by-side ultralight/sportplane ***NEW***/0593347

Instrument panel of Aerostar Festival ***NEW***/1034639

B ... 330 m (1,083 ft)/min
Service ceiling: A, B ... 4,000 m (13,120 ft)
T-O run: A ... 85 m (280 ft)
B ... 80 m (265 ft)
Landing run: A, B ... 90 m (295 ft)
Range with max fuel, 20 min reserves:
A ... 432 n miles (800 km; 497 miles)
B ... 405 n miles (750 km; 466 miles)
g limits: A, B ... +4/–2 (max)
... +6/–4 (ultimate)

UPDATED

AVIOANE

AVIOANE CRAIOVA SA

10 Aviatorilor Street, Judetul Dolj, R-207280 Ghercești
Tel: (+40 251) 40 29 01
Fax: (+40 251) 40 20 40
e-mail: avioane@acv.ro
Web: http://www.acv.ro
MANAGING DIRECTOR: Pantelimon Vilceanu
OPERATIONAL DIRECTOR: Mircea Popa
MARKETING MANAGER: Marius Cosma

Founded 1 February 1972 as Intreprinderea de Avioane Craiova, beginning aircraft manufacture with joint Romanian/Yugoslav IAR-93/Orao attack aircraft; changed to present name 17 December 2001; range of products and services for military and civil aviation includes equipment and parts manufacture for Fokker, GKN Westland and SABCA. Aircraft and equipment design and manufacture, repair and overhaul, life cycle management andintegrated logistics support. Granted ISO 9001 status 1998. Site area 1.70 ha (4.20 acres), including 47,500 m^2 (511,275 sq ft) shop floor area. Workforce had dwindled from 1,300 in April 2002 to about 870 by mid-2004. At that time, privatisation of the company was expected to begin in the following September, after transfer of 40 per cent of its assets to the Dolj County Administration Authority to repay earlier debts.

UPDATED

AVIOANE IAR-99 ŞOIM

English name: Hawk

TYPE: Advanced jet trainer/light attack jet.

PROGRAMME: Design by Institutul de Aviatie at Bucharest started 1975; three prototypes, of which S-001 made first flight 21 December 1985; S-002 was static test airframe; S-003 second flying prototype. Initial production batch of 17 (serial numbers 701–717), deliveries of which began 1988; two of these (708 and 709) completed for proposed avionics upgrade programme with Jaffe Aircraft of USA in 1991, which not pursued; one (712) equipped with Collins avionics in 1992; similar venture with IAI resulted in upgraded IAR-109 Swift prototype (7003, converted S-003), flown in November 1993, but no production of this version ensued.

Order for 24 new aircraft with Elbit modernised avionics announced in September 1998 and received Romanian MoD contract in 2000, following first flight of prototype (serial number 718) on 22 May 1997. First upgraded aircraft (718; actually constructed 1994) served as development demonstrator; production deliveries began with 719, which was rolled out on 31 July 2002 and handed over 1 August. However, defence cuts have subsequently reduced new-build order to 12, although some earlier aircraft will also be upgraded to new standard.

CURRENT VERSIONS: **Standard IAR-99:** Initial production version (17) for Romanian Air Force; in service with 67th Fighter-Bomber Group at Craiova. Three to be upgraded to IAR-99C standard and redelivered by 2006, after completion of ninth new-build aircraft.

IAR-99C: Elbit (Israel) teamed with Avioane to produce this version as lead-in fighter trainer for Romanian Air Force and for export. US$21 million contract for 24 aircraft announced in September 1998 and contained options for further 16 to follow. Order subsequently reduced to 12, of which first two (719 and 720) flown in August and October 2002; 721 and 722 had followed by mid-2003. Work continuing in 2003–04 on second batch of five, for delivery by 2006.

Following description applies to this version.

First production example of currently upgraded IAR-99C Şoim ***NEW***/0593377

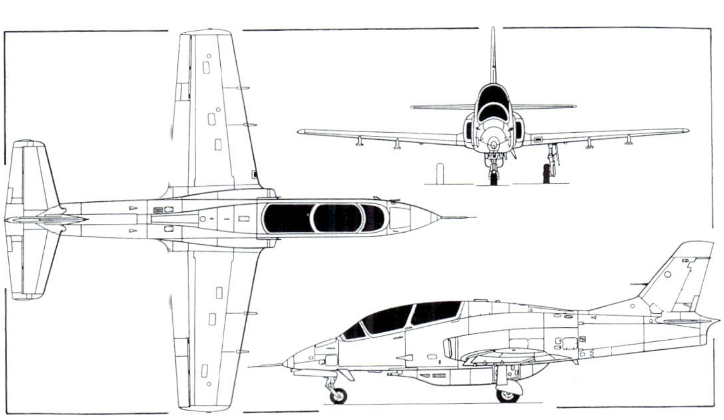

Antenna detail identifies the current IAR-99C Şoim *(Jane's/Dennis Punnett)* 0079327

CUSTOMERS: Total of 17 of original production version delivered to Romanian Air Force, of which about 13 believed still in service; deliveries of IAR-99C began in 2002. Lead-in fighter trainer (LIFT) for RomAF MiG-21 Lancer.

COSTS: IAR-99C approximately US$6 million (2002).

DESIGN FEATURES: Typical basic/advanced jet trainer with tandem, stepped cockpits and moderately tapered wings. Provision for armament increases versatility.

Wing section NACA 64$_1$A-214 (modified) at centreline; 64$_1$A-212 (modified) at tip; dihedral 3° from roots; quarter-chord sweepback 6° 35'; incidence 1° at root. Tail unit sweepback at quarter-chord 34° on fin, 9° 8' on tailplane.

Main feature of Elbit avionics suite is a data transfer system that processes navigational information received via datalink from other aircraft or from a ground station equipped with the same system. This information is presented on a simulated 'virtual radar' display in the front cockpit, so that the pilot can be trained in the use of radar without a need for the IAR-99 itself to be fitted with a radar. Current production IAR-99C can be identified by large blade antenna beneath nose on port side.

FLYING CONTROLS: Conventional and partly assisted. Statically balanced ailerons hydraulically actuated with manual reversion; mass-balanced elevators and rudder actuated mechanically by push/pull rods; servo tab in port aileron, trim tabs in rudder and each elevator, all operated electrically; ailerons deflect 15° up/15° down, elevators 20° up/10° down, rudder 25° to left and right. Hydraulically actuated single-slotted flaps, deflecting 20° for T-O and 40° for landing, retract gradually when airspeed reaches 162 kt (300 km/h; 186 mph); twin hydraulically actuated airbrakes under rear fuselage.

STRUCTURE: All-metal; aluminium honeycomb ailerons/elevators/rudder; semi-monocoque fuselage includes honeycomb panels for fuel tank compartments; machined wing skin panels form integral fuel tanks.

LANDING GEAR: Retractable tricycle type, with single wheel and oleo-pneumatic shock-absorber on each unit. Main-wheels retract inward, castoring nosewheel forward, all being fully enclosed by doors when retracted. Landing light in port wingroot leading-edge. Mainwheels fitted with tubeless tyres, size 552×164–10, pressure 7.5 bar (109 lb/sq in), and hydraulic disc brakes with anti-skid system. Nosewheel has tubeless tyre size 445×150–6, pressure 4.0 bar (58 lb/sq in).

POWER PLANT: One Turbomecanica Romanian-built Rolls-Royce Viper Mk 632-41M turbojet, rated at 17.79 kN (4,000 lb st). Fuel in two flexible bag tanks in centre-fuselage, capacity 900 litres (238 US gallons; 198 Imp gallons), and four integral tanks between wing spars, combined capacity 470 litres (124 US gallons; 103 Imp gallons). Total internal fuel capacity 1,370 litres (362 US gallons; 301 Imp gallons). Gravity refuelling point on top of fuselage. Provision for two drop tanks, each of 225 litres (59.5 US gallons; 49.5 Imp gallons) capacity, on inboard underwing stations. Maximum internal/external fuel capacity 1,820 litres (481 US gallons; 400 Imp gallons).

ACCOMMODATION: Crew of two in tandem, on Martin-Baker Mk 10L zero/zero ejection seats in pressurised and air conditioned cockpit. Rear (instructor's) seat elevated 35 cm (13.8 in). Dual controls standard. One-piece canopy with internal screen, opening sideways to starboard.

SYSTEMS: Engine compressor bleed air for pressurisation, air conditioning, anti-*g* suit and windscreen anti-icing system, and to pressurise fuel tanks. Hydraulic system, operating at pressure of 206 bar (2,990 lb/sq in), for actuation of landing gear and doors, flaps, airbrakes, ailerons and main-wheel brakes. Emergency hydraulic system for operation of landing gear doors, flaps and wheel brakes. Main electrical system, supplied by 9 kW 28 V DC starter/generator, with 28 V 36 Ah Ni/Cd battery, ensures operation of main systems, in case of emergency, and engine starting. Two static inverters (one 3 kVA and one 750 VA) supply two secondary AC networks: 115 V/400 Hz and 26 V/400 Hz. Oxygen system for two crew for 2 hours 30 minutes.

AVIONICS: Elbit Systems integrated suite, based on MIL-STD-1553B.

Comms: Dual VHF/UHF com; voice-activated intercom; IFF.

Flight: VOR/ILS; DME; ADF; Litton Italia INS with Trimble GPS nav.

Instrumentation: Modular multirole computer; HOTAS controls; head-up display with up-front control panel (HUD/UFCP), plus one LCD MFD and one CRT MFD, in front cockpit; two CRT MFDs (one as ASHM, or aft station HUD monitor) in rear (instructor's) cockpit; dual ADIs with HSIs; radar altimeter; Elta data transfer system (DTS) with pilot's virtual radar display; pilot's and instructor's display and sight helmets (DASH).

Self-defence: Elta RWR and ECM (jammer) pod; chaff/flare dispenser.

ARMAMENT: Five external stores stations. Centreline station can be occupied by ECM, laser designator or aerial reconnaissance pod, a drop fuel tank, or by a twin-barrel 23 mm GSh-23 gun pod with 200 rounds. All armament controlled by stores management system slaved to avionics mission computer. Four underwing hardpoints stressed for loads of 250 kg (551 lb) each. Typical underwing stores can include four 250 kg bombs; four triple carriers each for three 50 kg bombs (or two 100 kg and one 50 kg); laser- or IR-guided bombs (inboard pylons only); R-3, R-13, R-60 or Python 3 AAMs (outer pylons only); four L-16–57 launchers each containing sixteen 57 mm air-to-surface rockets; and auxiliary fuel tanks (see under Power Plant) on inboard pylons.

DIMENSIONS, EXTERNAL:
Wing span ... 9.85 m (32 ft 3¾ in)
Wing chord: at root ... 2.305 m (7 ft 6¾ in)
mean ... 1.965 m (6 ft 5¼ in)
at tip ... 1.30 m (4 ft 3¼ in)
Wing aspect ratio ... 5.2
Length overall ... 11.01 m (36 ft 1½ in)
Height overall ... 3.90 m (12 ft 9½ in)
Elevator span ... 4.12 m (13 ft 6¼ in)
Wheel track ... 2.685 m (8 ft 9¾ in)
Wheelbase ... 4.38 m (14 ft 4½ in)

AREAS:
Wings, gross ... 18.71 m^2 (201.4 sq ft)
Ailerons (total) ... 1.56 m^2 (16.79 sq ft)
Flaps (total) ... 2.54 m^2 (27.34 sq ft)

Fin, incl dorsal fin ..... 1.92 m² (20.67 sq ft)
Rudder ..... 0.63 m² (6.78 sq ft)
Tailplane ..... 3.125 m² (33.63 sq ft)
Elevators (total) ..... 1.25 m² (13.45 sq ft)
WEIGHTS AND LOADINGS:
Weight empty, equipped ..... 3,200 kg (7,055 lb)
Max fuel weight: internal ..... 1,100 kg (2,425 lb)
external ..... 350 kg (772 lb)
Max T-O weight: trainer ..... 4,400 kg (9,700 lb)
ground attack ..... 5,560 kg (12,258 lb)
Max wing loading: trainer ..... 235.2 kg/m² (48.17 lb/sq ft)
ground attack ..... 297.2 kg/m² (60.86 lb/sq ft)
Max power loading: trainer ..... 247.5 kg/kN (2.42 lb/lb st)
ground attack ..... 312.7 kg/kN (3.06 lb/lb st)
PERFORMANCE:
Max operating Mach number (MMO) 0.76
Max level speed at S/L:
trainer, clean ..... 467 kt (865 km/h; 537 mph)
Max speed for landing gear actuation ..... 135 kt (250 km/h; 155 mph)
Max rate of climb at S/L ..... 2,100 m (6,890 ft)/min
Service ceiling ..... 12,900 m (42,325 ft)
Min air turning radius ..... 330 m (1,083 ft)
T-O run: trainer ..... 450 m (1,477 ft)
ground attack ..... 960 m (3,150 ft)
T-O to 15 m (50 ft): trainer ..... 750 m (2,461 ft)
ground attack ..... 1,350 m (4,430 ft)
Landing from 15 m (50 ft): trainer ..... 740 m (2,428 ft)
ground attack ..... 870 m (2,855 ft)
Landing run: trainer ..... 550 m (1,805 ft)
ground attack ..... 600 m (1,969 ft)
Typical combat radius (one pilot, ventral gun, internal fuel only):
lo-lo-hi, four 16-round rocket pods, AUW 5,000 kg (11,023 lb) ..... 189 n miles (350 km; 217 miles)
hi-lo-hi, two 16-round rocket pods, two 50 kg and four 100 kg bombs, AUW 5,280 kg (11,640 lb) ..... 186 n miles (345 km; 214 miles)
hi-hi-hi, four 250 kg bombs, AUW 5,480 kg (12,081 lb) ..... 208 n miles (385 km; 239 miles)
Max range with internal fuel:
trainer ..... 593 n miles (1,100 km; 683 miles)
ground attack ..... 522 n miles (967 km; 601 miles)
Max endurance with internal fuel: trainer ..... 2 h 40 min
ground attack ..... 1 h 46 min
g limits ..... +7/–3.6

UPDATED

IAR

SC IAR SA

1 Aeroportului Street, R-507075 Brasov
Tel:(+40 268) 47 51 07
Fax: (+40 268) 47 69 81
e-mail: sales@iar.ro
Web: http://www.iar.ro
PRESIDENT AND CEO: Ion Georgescu
GENERAL DIRECTOR: Dipl Eng Neculai Banea
COMMERCIAL AND FINANCIAL DIRECTOR: Dipl Ec Ion Dumitrescu
MARKETING MANAGER: Stefan Paunescu

Factory, established 1925 and fully rebuilt 1968, continues work begun by IAR-Brasov; occupies 136 ha (336 acre) site, including 185,400 m² (1,995,625 sq ft) factory area, and had early 2003 workforce of about 1,300. Now the sole Romanian manufacturer of helicopters, it has produced more than 200 Alouette IIIs and 160 Pumas under French licence. Currently has full capacity and capabilities for design and development; manufacturing aircraft structural items; general assembling; integration of modern avionics and systems; maintenance, overhaul and repair works; modernisation programme; flight test and certification; and customer support. Certification to JAR 21, 145 and 147 was under way in 2003–04. Stated in mid-2004 that, as further stage in progress towards privatisation, company would be restructured as three separate companies, each having 64.8 per cent government ownership (Ministry of Economy and Trade) and the balance with private shareholders. One of the new companies, retaining the IAR name, would specialise in helicopter production and support, and would hold the present company's 51 per cent share of Eurocopter Romania.

IAR is a major supplier to the Romanian MoD. Main product is the IAR-330 L Puma, manufactured under Eurocopter licence; currently carrying out Puma SOCAT upgrading programme to meet Romanian MoD requirements for an advanced armed helicopter; other customers include UAE Air Force. Based on Eurocopter Romania SA joint venture created 21 June 2001, IAR provides worldwide Puma fleet logistic support, customisation work and upgrading such as Makila 1A1 re-engining; advanced avionics; new autopilot; and new optional equipment; local assembly of EC 135 began in 2003.

IAR also manufactures own-designed IAR-46S very light aircraft; has produced some 830 IS-28M2/GR motor gliders and IS-28/29/35 series gliders; spares for IAR-330 Puma and IAR-316 B Alouette III helicopters; aircraft components.

UPDATED

IAR IAR-46 S

TYPE: Two-seat lightplane.
PROGRAMME: Definition phase and marketing studies started early 1991; detail design began late 1991; development phase initiated mid-1992; first of two prototypes (YR-1037) shown at Paris Air Show June 1993 and made first flight November that year, followed by flight test evaluation, static test and ground testing for JAR-VLA compliance. Development of 02 second prototype began at end of 1995; this aircraft (YR-BVC) made first flight in April 1997; manufacturer's preliminary flight test programme completed August 1999; JAR-VLA certification (Utility category) by Romanian CAA began mid-1998 and received November 1999; YR-BVC registered to Airclub Romania in August 2000; FAA certified subsequently. Three IAR-46S production aircraft built and stored by late 1998; eventually registered April 2001. Limited production since then.
CURRENT VERSIONS: **IAR-46:** Prototypes 01 and 02; 59.6 kW (79.9 hp) Rotax 912 F3 engine; FAA certified.
IAR-46S: Entered production in 2000. First three (c/n 03 to 05) registered to Airclub of Romania in April 2001. Further five registered to same customer in 2002.

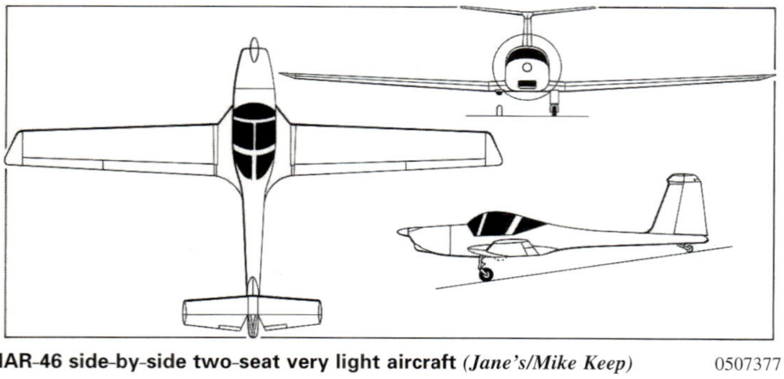

IAR-46 side-by-side two-seat very light aircraft *(Jane's/Mike Keep)* 0507377

Second prototype IAR-46 *(Paul Jackson)* 0062606

Detailed description applies to IAR-46S.
DESIGN FEATURES: Low-wing monoplane; GA(W)-1 aerofoil section) with high aspect ratio; raked tips; T tail. Dihedral 2° from centre-section; incidence 4° at root; no twist. Foldable horizontal tail surfaces.
FLYING CONTROLS: Conventional and manual. Actuation by pushrods and cables; trim tab in each elevator. Manually operated plain trailing-edge flaps.
STRUCTURE: All-metal except fabric covering on elevators and rudder and GFRP for non-stressed fairings. Wing has main and auxiliary spars. Fluted aluminium skins on ailerons and flaps.
LANDING GEAR: Retractable (manual/mechanical) single mainwheels with hydraulic shock-absorbers and hydraulic, toe-operated disc brakes; steerable, non-retractable tailwheel with rubber-in-compression shock-absorption. Rearward-retracting 5.00×5 Matco or Cleveland mainwheels, with 360×110 tyres, pressure 3.00 bar (43.5 lb/sq in). Tost 210×65 tailwheel and tyre, pressure 2.50 bar (36 lb/sq in).
POWER PLANT: One 73.5 kW (98.6 hp) Rotax 912 S3 (IAR-46S) flat-four engine, with 2.43:1 reduction gearing to a Hoffmann HO-V352F/170FQ+6 two-blade variable-pitch propeller. Fuel in single tank in fuselage, capacity 70 litres (18.5 US gallons; 15.4 Imp gallons).
ACCOMMODATION: Two adjustable backrest seats side by side; dual controls standard; adjustable rudder pedals. Fixed windscreen and rearward-sliding jettisonable canopy. Baggage compartment aft of seats. Cockpit heated and ventilated.
SYSTEMS: Electrical power supplied by 250 W 14 V DC alternator and 12 V 25 Ah battery. Hydraulic system for mainwheel brakes only, with parking brake valve.
AVIONICS: *Instrumentation:* Standard VFR instrumentation to JAR 22.1303 and JAR 22.1305. Options include horizon and directional gyros, turn co-ordinator and Becker nav/com and transponder avionics package.
EQUIPMENT: Anti-collision and position lights standard; ground power receptacle optional.
DIMENSIONS, EXTERNAL:
Wing span ..... 12.05 m (39 ft 6½ in)
Wing chord: at root ..... 1.40 m (4 ft 7 in)
at tip ..... 0.93 m (3 ft 0½ in)
Wing aspect ratio 9.4
Length overall ..... 7.85 m (25 ft 9 in)
Height overall ..... 2.15 m (7 ft 0½ in)
Tailplane span ..... 3.48 m (11 ft 5 in)
Wheel track ..... 1.59 m (5 ft 2½ in)
Wheelbase ..... 4.94 m (16 ft 2½ in)
Propeller diameter ..... 1.70 m (5 ft 7 in)
Propeller ground clearance ..... 0.295 m (11½ in)
DIMENSIONS, INTERNAL:
Cockpit: Length ..... 1.54 m (5 ft 0¾ in)
Max width ..... 1.04 m (3 ft 5 in)
Max height (from seat cushion) ..... 0.85 m (2 ft 9½ in)
AREAS:
Wings, gross ..... 13.87 m² (149.3 sq ft)
Ailerons (total) ..... 0.82 m² (8.83 sq ft)
Trailing-edge flaps (total) ..... 1.36 m² (14.64 sq ft)
Fin ..... 0.60 m² (6.46 sq ft)
Rudder ..... 0.89 m² (9.58 sq ft)
Tailplane ..... 1.64 m² (17.65 sq ft)
Elevators (total, incl tabs) ..... 1.10 m² (11.84 sq ft)
WEIGHTS AND LOADINGS:
Weight empty ..... 550 kg (1,213 lb)
Max fuel weight ..... 53 kg (117 lb)
Max T-O and landing weight ..... 750 kg (1,653 lb)
Max wing loading ..... 54.1 kg/m² (11.08 lb/sq ft)
Max power loading ..... 12.58 kg/kW (20.67 lb/hp)
PERFORMANCE:
Never-exceed speed (VNE) ..... 151 kt (280 km/h; 174 mph) IAS
Max level speed ..... 116 kt (215 km/h; 133 mph)
Max cruising speed ..... 103 kt (190 km/h; 118 mph) IAS

Econ cruising speed ... 86 kt (160 km/h; 99 mph) IAS
Max manoeuvring speed ... 93 kt (172 km/h; 107 mph)
Max speed for flap extension ... 76 kt (140 km/h; 87 mph)
Optimum climbing speed ... 60 kt (112 km/h; 70 mph)
Stalling speed: flaps up ... 50 kt (91 km/h; 57 mph) IAS
  flaps down ... 44 kt (81 km/h; 51 mph) IAS
Max rate of climb at S/L ... 180 m (591 ft)/min
Service ceiling ... 4,000 m (13,120 ft)
T-O run ... 185 m (610 ft)
T-O to 15 m (50 ft) ... 409 m (1,340 ft)
Landing from 15 m (50 ft) ... 171 m (560 ft)
Landing run ... 110 m (360 ft)
Range, with reserves ... 297 n miles (550 km; 341 miles)
g limits ... +4.4/–2.2

UPDATED

ROMAERO

SC ROMAERO SA

44 Bulevardul Ficusului, Sector 1 (PO Box 18), R-71544 Bucuresti 1
Tel: (+40 1) 232 37 35 and 232 60 60
Fax: (+40 1) 232 20 82
e-mail: management@romaero.rdsnet.ro
Web: http://www.romaero.ro
PRESIDENT AND MANAGING DIRECTOR: Francisc Toba
OPERATIONS DIRECTOR: Cristian Mihalache
DIRECTOR, SALES AND MARKETING: Misa Popic
PUBLIC RELATIONS: Carmen Gheorghiu

Romaero is an EASA- and FAA-approved airframe manufacturer and aircraft maintenance organisation with a reliable reputation as a partner of Western integrators. It was known initially for its production of the Britten-Norman/BNG Islander and Defender, which began in 1968. This was suspended in 1997 (more than 520 built), but was resumed in July 2002 with new contract for 24 aircraft during ensuing two years.

As a JAR/FAR 21-approved supplier, Romaero has manufactured and delivered items to Boeing (737/757 major subassemblies and components); Bombardier (415 water bomber cabin subassemblies and floats, Challenger 604 auxiliary fuel tanks and Learjet 45 tail units); BAE Systems (ATP and Avro RJ85/RJX detail parts and subassemblies); and IAI/Gulfstream (Astra SPX/G100 cabins and Galaxy/G200 rear fuselages). More recently, it entered into co-operation with BAE Systems to manufacture Nimrod subassemblies and parts and ATP large freight doors (LFD); and with Sabca to produce structural assemblies and subassemblies for the Airbus A380 programme. In 2004, installed LFDs in three ATP freighter conversions for Emerald Airways under BAE Systems subcontract.

In addition, Romaero's JAR/FAR 145-approved maintenance, repair and overhaul (MRO) status currently covers such work on various aircraft types, including the Islander, BAC One-Eleven, Boeing 727/737/757, Hawker 800 XP, Lockheed Martin C-130, McDonnell Douglas/Boeing DC-9 and MD-80 series, and Airbus A320.

Romaero's 364,347 m² (3,921,800 sq ft) site includes a 163,670 m² (1,761,725 sq ft) production facility and a further 147,096 m² (1,583,325 sq ft) of covered space. Workforce was approximately 1,300 in January 2003.

UPDATED

Russian Federation

AEROPRAKT

AEROPRAKT OOO (AEROPRACT JSC)

Ofis 76, ulitsa Novo-Sadovaya 44, 443110 Samara
Tel: (+7 8462) 92 68 01 and 34 36 32
e-mail: aeropract@samaramail.ru
Web: http://www.aeropract-samara.com
GENERAL DIRECTOR: Andrey G Pozdnyakov
CHIEF CONSTRUCTOR: Vladimir V Gaslov

Formed in 1974 to design and build gliders, small aircraft, flying-boats and amphibians, OKB Aeroprakt established a second centre at Kiev in 1986, only to be divided on dissolution of the USSR. Became LM Aeropract Samara, a joint Russo-Finnish venture, in 1991; KB Aeropract in 1993; Aeropract JSC in 1997 and Aeropract-Samara JSC in 2000.

The Russian Aeropract adopts a -ct ending in transliteration to differentiate it from the Aeroprakt now in Ukraine (which see). The two companies are no longer connected. Chief designer Igor Vakhrushev was killed in the crash of an Aeropract-23M in June 2000, after which production was suspended temporarily and the company split into two: Avantage which continued with the modified version of Aeropract-27, the 27M, shown at Moscow in August 2001 and Aeropract-Samara, which is responsible for the Aeropract-33, first flown at Moscow in August 2001.

UPDATED

Aeropract-27M Advantage two-seat lightplane *(Paul Jackson)* ***NEW***/0547744

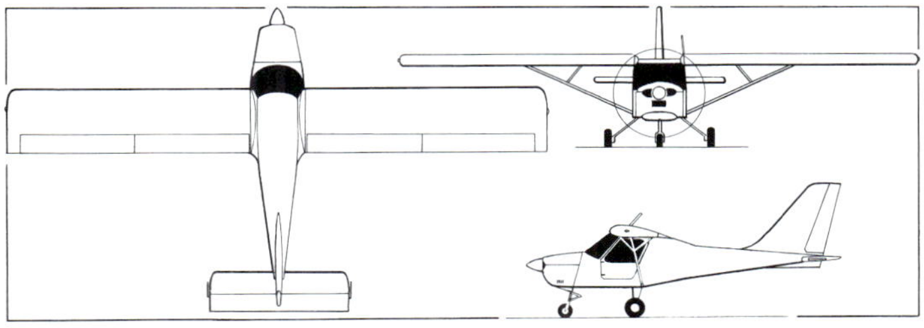

Aeropract-27M Advantage *(James Goulding)* 0121662

AEROPRACT-27

TYPE: Side-by-side lightplane/kitbuilt.

PROGRAMME: Prototype (FLARF-02647) first flew (in floatplane configuration) 22 June 1998; public debut at Gelendzhik 98 seaplane show, and in landplane version at MAKS, Moscow, August 1999; evaluated during 1999 and 2000 by AvtoZavod Flying Club in Togliatti. Seventh production airframe, under construction in mid-2001, is first with Rotax 914 engine.

CURRENT VERSIONS: **A-27:** Baseline version with Hirth engine.

A-27M Advantage: Version with 73.5 kW (98.6 hp) Rotax 912 ULS engine driving a Kiev ground-adjustable pitch propeller or 84.6 kW (113.4 hp) Rotax 914 ULS, driving a variable-pitch propeller. Several minor design changes, including more rounded forward lower fuselage surfaces, but description generally as for A-27, except where indicated.

COSTS: Kit US$14,000, excluding instruments (US$700), landing gear and Rotax 912 engine (US$14,000). Ex-works, complete, with wheels, US$35,500 (1999). Wheel landing gear US$1,000; skis US$250 to US$350; floats US$1,200 and US$1,650.

DESIGN FEATURES: Braced, high-wing monoplane with single main and two jury struts each side. Sweptback fin and rudder; unswept wings and horizontal tail. Meets JAR-VLA certification requirements.

FLYING CONTROLS: Manual. Ailerons, rudder and all-moving tailplane with anti-balance tab. Flaps. A-27M has fixed tailplane with elevators, and ailerons with external mass balances.

STRUCTURE: Tube metal fuselage with composites skin.

LANDING GEAR: Tricycle type; fixed. Cantilever, metal mainwheel legs; telescopic, leaf-sprung nosewheel leg. Mainwheel size 400×150; nosewheel size 300×125. Optional floats or skis.

POWER PLANT (A-27): One 70.8 kW (95 hp) Hirth F30 flat-four driving a four-blade, ground-adjustable pitch propeller. Fuel capacity: 80 litres (21.1 US gallons; 17.6 Imp gallons); (A-27M) 100 litres (26.4 US gallons; 22.0 Imp gallons).

ACCOMMODATION: Two persons, side by side; forward-hinged door each side. Luggage space behind seats.

AVIONICS: Standard VFR.

Data refer to both versions, except where indicated.

DIMENSIONS, EXTERNAL:
Wing span: A27 ... 10.60 m (34 ft 9¼ in)
  A-27M ... 10.00 m (32 ft 9¾ in)
Wing chord, constant ... 1.35 m (4 ft 5¼ in)
Length overall: A-27 ... 6.30 m (20 ft 8 in)
  A-27M ... 6.50 m (21 ft 4 in)
Height overall ... 2.22 m (7 ft 3½ in)

WEIGHTS AND LOADINGS:
Weight empty: A-27 ... 335 kg (739 lb)
  A-27M ... 325 kg (717 lb)
Max T-O weight: A-27 ... 600 kg (1,322 lb)
  A-27M ... 650 kg (1,433 lb)
Max power loading: A-27 ... 8.47 kg/kW (13.92 lb/hp)
  A-27M ... 8.84 kg/kW (14.53 lb/hp)

PERFORMANCE:
Max level speed: A-27 ... 95 kt (175 km/h; 109 mph)
  A-27M ... 97 kt (180 km/h; 112 mph)
Normal cruising speed: A-27 ... 70 kt (130 km/h; 81 mph)
  A-27M ... 70–81 kt (130–150 km/h; 81–93 mph)
T-O run ... 100 m (330 ft)
Range: A-27 ... 324 n miles (600 km; 372 miles)
  A-27M ... 540 n miles (1,000 km; 621 miles)

AEROPRACT-33

TYPE: Side-by-side lightplane/kitbuilt.

PROGRAMME: Prototype (FLARF-01033) first flown 14 November 2001.

CUSTOMERS: Total of six built by end 2003.

DESIGN FEATURES: Light tourer and sportplane with low-mounted, tapered wings, constant-chord tailplane and large, sweptback fin with small underfin. Meets JAR-VLA certification requirements.

FLYING CONTROLS: Conventional and manual. Horn-balanced elevator with central, flight-adjustable tab. Ailerons with external mass balances. Three-position (0, 20, 40°) flaps.

STRUCTURE: Generally of composites.

LANDING GEAR: Tricycle type; fixed. Cantilever mainwheel legs; leaf-sprung steerable

Prototype Aeropract-33 two-seat lightplane ***NEW***/0552056

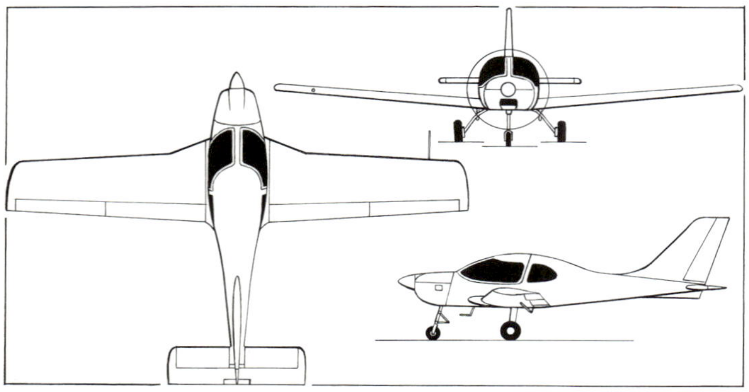

Aeropract-33 (Rotax 912 ULS engine) *(James Goulding)* ***NEW***/0554613

nosewheel, main wheels 4.00×150; nosewheel 300×125. Hydraulic brakes, optional floats and skis.

POWER PLANT: One 73.5 kW (98.6 hp) Rotax 912 ULS flat-four driving a three-blade flight adjustable Kievprop propeller; three aircraft have 84.6 kW (113.4 hp) Rotax 914 UL fitted. Fuel capacity 130 litres (34.3 US gallons; 28.6 Imp gallons).

ACCOMMODATION: Two persons, side by side, beneath two-piece, centreline-hinged canopy. Baggage stowage behind seats.

SYSTEMS: Electrical system with 12 V 25 Ah battery and PR-A12-24 transformer.

AVIONICS: *Comms:* Briz radio.

Instrumentation: Standard VFR. Optional GPS.

EQUIPMENT: MVEN KC-700 parachute recovery system.

DIMENSIONS, EXTERNAL:

Wing span ... 9.575 m (31 ft 5 in)
Wing chord: at root ... 1.745 m (5 ft 8¾ in)
at tip ... 1.00 m (3 ft 3¼ in)
Wing aspect ratio ... 7.3
Length overall ... 6.64 m (21 ft 9½ in)
Height overall ... 2.84 m (9 ft 4 in)
Tailplane span ... 2.84 m (9 ft 4 in)
Tailplane chord (constant) ... 0.80 m (2 ft 7½ in)
Wheel track ... 2.00 m (6 ft 6¾ in)
Wheelbase ... 1.65 m (5 ft 5 in)
Propeller diameter ... 1.85 m (6 ft 0¾ in)

DIMENSIONS, INTERNAL:

Cabin width ... 1.16 m (3 ft 9½ in)
Baggage volume ... 0.8 m³ (28 cu ft)

AREAS:

Wings, gross ... 12.56 m² (135.2 sq ft)
Ailerons (total) ... 1.28 m² (13.78 sq ft)
Flaps (total) ... 1.92 m² (20.67 sq ft)
Fin (incl dorsal fin) ... 2.01 m² (21.64 sq ft)
Rudder, incl tab ... 0.52 m² (5.60 sq ft)
Tailplane ... 2.24 m² (24.11 sq ft)
Elevator, incl tab ... 1.08 m² (11.63 sq ft)

WEIGHTS AND LOADINGS:

Weight empty ... 433 kg (955 lb)
Max payload ... 294 kg (648 lb)
Max T-O weight ... 750 kg (1,653 lb)
Max wing loading ... 59.7 kg/m² (12.23 lb/sq ft)
Max power loading: Rotax 912 ... 10.20 kg/kW (16.76 lb/hp)
Rotax 914 ... 8.87 kg/kW (14.58 lb/hp)

PERFORMANCE:

Never-exceed speed (VNE) ... 135 kt (250 km/h; 155 mph)
Max operating speed: Rotax 912 ... 116 kt (215 km/h; 134 mph)
Rotax 914 ... 119 kt (220 km/h; 137 mph)
Cruising speed at 75% power ... 103 kt (190 km/h; 118 mph)
Econ cruising speed ... 81 kt (150 km/h; 93 mph)
Stalling speed, power off, flaps down ... 44 kt (80 km/h; 50 mph)
Rate of climb at S/L: Rotax 912 ... 360 m (1,181 ft)/min
Rotax 914 ... 270 m (886 ft)/min
T-O run: Rotax 912 ... 150 m (492 ft)
Rotax 914 ... 100 m (330 ft)
Range ... 809 n miles (1,500 km; 932 miles)
g limits ... +4/–2

UPDATED

AEROPROGRESS/ROKS-AERO

AEROPROGRESS/ROKS-AERO

65A Volokolamskoye Shosse, 123424 Moskva
Tel: (+7 095) 145 80 44
Fax: (+7 095) 145 94 77
e-mail: aeroprogress@mtu-net.ru

PRESIDENT AND GENERAL DESIGNER: Evgeny P Grunin
DEPUTY GENERAL DESIGNER: Arnold I Andrianov
DEPUTY GENERAL DESIGNER, FOREIGN ECONOMIC RELATIONS: Alexander V Andreev
DESIGN BUREAU MANAGER: Sergei M Zhiganov

Known initially as ROS-Aeroprogress, this organisation was founded in 1990 to design and manufacture utility, commuter, amphibian, aerobatic, agricultural, firefighting, training and attack aircraft, WIG (wing-in-ground-effect) vehicles, replicas and other vehicles. ROKS-Aero Inc is the design bureau of Aeroprogress.

UPDATED

AEROPROGRESS/ROKS-AERO T-101 GRACH

English name: Rook

TYPE: Light utility turboprop.

PROGRAMME: Design started by Utility Aircraft Division in September 1991, as monoplane successor to Antonov An-2/3 biplane; construction of prototype started April 1992; first flight (FLARF-01466) 7 December 1994; no evidence of reported four additional prototypes, all of which are presumed to have been used for static testing.

Series manufacture by Moscow Aviation Production Organisation (MAPO) initiated January 1993; initial production aircraft shown unpainted at Lukhovitsy factory, August 1999; further seven then substantially complete. One T-101 delivered to Yakutia region in July 2000 on six-month lease, representing initial revenue-earning use; second supplied to undisclosed operator in early 2001 and third being prepared for delivery by late 2001. A Westernised version, developed via the Aeroprogress ROKS/Aero T-201, was unsuccessfully marketed as the Khrunichev T-201. In March 2002 Aeroprogress suggested it was looking for certification by 2004, but in June of that RSK 'MiG' admitted it was unable to provide the US$2 million needed to complete certification, even though several aircraft then partly complete on production line at LAPIK, Lukhovitsy, and Yakutia regional government interested in buying 10.

UPDATED

Aeroprogress/ROKS-Aero T-101 Grach prototype *(Paul Jackson)* ***NEW***/0547745

AEROSTATICA

AEROSTATIKA NAUCHNO-PROIZVODSTVENNAYA FIRMA (AEROSTATICA SCIENTIFIC-PRODUCTION COMPANY)

ulitsa Krylatskaya 31-1-315, 121614 Moskva
Tel: (+7 095) 158 48 18
Fax: (+7 095) 415 26 30
PRESIDENT: Dr Alexander N Kirilin

Aerostatica is the English transliteration of Aerostatika, and is used for marketing. The company's first two airships, the 01 and 02, were built in the late 1990s. More recently, it was developing an Aerostatica-200 twin-engined four-seater, intended to be followed by the larger Aerostatica-300, but neither reported to have flown by mid-2003.

AERO-VOLGA

AERO-VOLGA OOO NPO

ulitsa Kutyakova 6, Samara
Tel: (+7 8462) 32 62 63
e-mail: air@smrtic.ru
Web: http://www.aerovolga.ru

Two prototypes of this company's twin-engined L-6 amphibian have been flown: the original aircraft in June 2000 and a modified L-6M in September 2003. In 2003, Aero-Volga announced the similar LA-8 Flagman, which appears to be intended as the production version.

UPDATED

AERO-VOLGA LA-8 FLAGMAN

TYPE: Utility amphibian.

PROGRAMME: Project announced 2003 as development of L-6.

DESIGN FEATURES: High-wing amphibian. Wings strut-braced and with engines mounted on upper surface; stabilising floats at approximately three-quarter span. T-type tail. Downturned wingtips.

FLYING CONTROLS: Conventional and manual.

STRUCTURE: Composites, with monocoque hull.

LANDING GEAR: Nosewheel type; retractable. Mainwheels, size 476×168, nosewheel 400×150. Minimum ground turning radius 14.0 m (46 ft).

POWER PLANT: Two LOM V546 petrol engines, each 158 kW (212 hp). Optionally two 173 kW (232 hp) LOM M337B or 184 kW (247 hp) M337C six-cylinder piston engines. Fuel capacity 320 litres (84.5 US gallons; 70.4 Imp gallons); provision for two long-range tanks with combined capacity of 300 litres (79.3 US gallons; 66.0 Imp gallons).

ACCOMMODATION: Eight persons, including up to two pilots.

EQUIPMENT: Optional ballistic parachute.

DIMENSIONS, EXTERNAL:
Wing span .... 13.06 m (42 ft 10¼ in)
Length overall .... 10.80 m (35 ft 5¼ in)
Max width of fuselage .... 2.00 m (6 ft 6¾ in)
Height overall .... 3.40 m (11 ft 1¾ in)
Wheel track .... 1.18 m (3 ft 10½ in)
Wheelbase .... 3.48 m (11 ft 5 in)

Passenger cabin of Aero-Volga LA-2 Flagman ***NEW***/0567751

Aero-Volga Flagman instrument panel ***NEW***/0567750

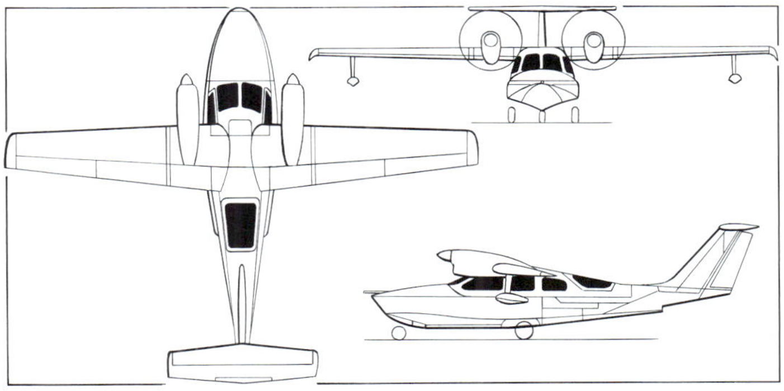

Aero-Volga LA-8 Flagman amphibian *(James Goulding)* ***NEW***/0572229

DIMENSIONS, INTERNAL:
Cabin: Length .... 3.45 m (11 ft 3¾ in)
Max width .... 1.64 m (5 ft 4½ in)
Max height .... 1.24 m (4 ft 0¾ in)
Volume .... 3.7 m³ (131 cu ft)

WEIGHTS AND LOADINGS:
Weight empty .... 1,400 kg (3,086 lb)
Max payload .... 670 kg (1,477 lb)
Max fuel weight .... 230 kg (507 lb)
Max T-O and landing weight .... 2,300 kg (5,070 lb)
Max power loading: .... 7.28 kg/kW (11.96 lb/hp)

PERFORMANCE:
Max level speed .... 140 kt (260 km/h; 164 mph)
Normal cruising speed .... 119 kt (220 km/h; 139 mph)
Stalling speed, power off, flaps up .... 65 kt (120 km/h; 76 mph)
Max rate of climb at S/L .... 336 m (1,102 ft)/min
T-O run: land .... 200 m (660 ft)
water .... 250 m (820 ft)
Range .... 647 n miles (1,200 km; 745 miles)
Endurance .... 5 h 40 min

UPDATED

ALBATROSS

TSENTR NAUCHNO-TEKHNICHESKOGO TVORCHESTVA ALBATROSS (ALBATROSS SCIENTIFIC-TECHNICAL WORKS)

ulitsa Tsentral'naya 22/126, 142092 Troitsk, Moskovskaya oblast
Tel/Fax: (+7 096) 601 66 76
GENERAL DIRECTOR: Valery Daryin

Original Aviacomplex company established in 1990 to build prototypes of the AS-2 two-seat lightplane. Renamed Albatros on 16 June 1991. Testing of the AS-2 was completed in collaboration with CAHI (TsAGI) and LII Flight Research Institute. Production by Aviacomplex and Myasishchev Engineering Bureau, with participation by other Russian companies. At 1999 Moscow Air Show (MAKS), a production AS-2 was shown under new designation of Albatros AS-3A and first details were revealed of a development, the Sigma-4, but no further details of either aircraft have been received from Albatross. However, the Sigma-4 was shown at the 2003 MAKS, credited to Elitar.

The AS-3A designation appears to have been re-applied to a considerably revised design, incorporating a pusher propeller, of which basic details have been received. In addition, Albatross produces plans for a variety of light engineering equipment as well as lathes for home use.

UPDATED

ALBATROSS AS-3A

TYPE: Tandem-seat ultralight.

PROGRAMME: Production aircraft built at Kazan.

CUSTOMERS: At least two flying with aero club at Podolsk, Moscow region.

COSTS: US$9,800 with engine; US$6,800 without engine (2004).

DESIGN FEATURES: High wing, open cockpit (with optional enclosure), pusher configuration with constant-chord, V-strut braced wings and strut-braced tailboom. Rear occupant's seat is raised.

Albatross AS-3A tandem-seat ultralight ***NEW***/0561694

FLYING CONTROLS: Conventional and manual. Flaps.
STRUCTURE: Duralumin tube and fabric; composites nacelle.
LANDING GEAR: Tricycle type; fixed.

POWER PLANT: One 44.7 kW (60 hp) piston engine driving a two-blade, pusher propeller.
EQUIPMENT: Ballistic parachute.
DIMENSIONS, EXTERNAL:

Wing span .... 9.80 m (32 ft 1¾ in)
Length overall .... 6.80 m (22 ft 3¾ in)

AREAS:

Wings, gross .... 14.85 m^2 (159.8 sq ft)

WEIGHTS AND LOADINGS:

Weight empty .... 240 kg (529 lb)
Max T-O weight .... 420 kg (925 lb)

PERFORMANCE:

Max level speed .... 65 kt (120 km/h; 75 mph)
Normal cruising speed .... 49 kt (90 km/h; 60 mph)
Max rate of climb at S/L .... 180 m (591 ft)/sec
T-O run .... 60 m (200 ft)
Range: with max fuel .... 540 n miles (1,000 km; 621 miles)
with two crew .... 162 n miles (300 km; 186 miles)

UPDATED

ARSENYEV

ARSENYEVSKOYE AVIATSIONNOYE PROIZVODSTVENNOYE PREDPRIYATIE IMENI N I SAZYKINA (ARSENYEV AVIATION PRODUCTION ENTERPRISE 'PROGRESS' NAMED FOR N I SAZYKIN)
prospekt Lenina 5, 692335 Arsenyev, Primovsky Kray
Tel: (+7 423) 612 48 97
Fax: (+7 423) 612 61 30
GENERAL DIRECTOR: Vladimir Pechyonkin

Originally designated GAZ 116, Arsenyev plant previously built the Mil Mi-24/25/35 series of combat helicopters, in parallel with Rostvertol, and was also manufacturer of the Yak-55 aerobatic lightplane; a modified version of the last-mentioned, the Technoavia SP-55, was returned to production in 1999.

Arsenyev is responsible currently for the Kamov Ka-50 (including, if ordered, the Ka-50-2 export version) and Mil Mi-34 helicopters, as well as Moskit missiles. It was reportedly assigned the new Kamov Ka-60, but that programme has been reallocated to RSK 'MiG'. Russian government shareholding is 51 per cent; personnel totalled some 5,000 in 2000.

A government decree of 31 December 1997 authorised the Arsenyev plant to offer the Ka-50 for export and supply spares and support for existing Mi-24/25/35 helicopters. Overseas delivery of P-15U, P-20, P-21 and P-22 cruise missiles is also covered.

Official plans announced on 12 May 2001 called for the Arsenyev plant to be incorporated in the proposed grouping to include RSK 'MiG', Tupolev and Kamov.

UPDATED

AVGUR

AVGUR VOZDUKHOPLAVATELNYI TSENTR (AUGUR AERONAUTICAL CENTRE)
ulitsa Stepana Shutova 4, 109380 Moskva
Tel/Fax: (+7 095) 359 10 01 and 359 10 65
e-mail: augur@pbo.ru
Web: http://www.augur.pbo.ru
WORKS: St Petersburg and Ulyanovsk
Tel/Fax (St Petersburg): (+7 812) 173 61 75
GENERAL DIRECTOR (ST PETERSBURG): Stanislav Vladimirovich Fedorov

Formed 29 October 1991 by the Moscow Aviation Institute and a group of private shareholders, this company is successor to AIG Kovanko, established in November 1986 by Mr Fedorov. Early achievements included first hot-air balloon flight over Moscow (AIGK Aist) in September 1989 and first Russian advertising aerostat (unmanned AP-60) in December 1989. Became joint stock company on 22 January 1993; St Petersburg division formed 4 November 1993, Ulyanovsk division 15 May 1996.

Current range of inflatables comprises BB, Au-2250 and VB hot-air balloons; Au-8 captive manned aerostat; Au-900 manned aerostat; Au-2, Au-5, Au-6 and Au-10 captive advertising aerostats; a range of small geostats; for RD-1.5, RD-2 and RD-2.5 remotely piloted small airships, see *Jane's Unmanned Aerial Vehicles and Targets*. Other activities include rental and leasing of airships and manufacture of ground-based inflatables of all kinds.

In 1997, Avgur became a founder member, with MAI, of the RosAeroSystems consortium.

VERIFIED

AVIA

NAUCHNO-PROIZVODSTVENNOE OBEDINENIE AVIA LTD (AVIA SCIENTIFIC-PRODUCTION ASSOCIATION JSC)
Studeni proezd 5, 129282 Moskva
Tel: (+7 095) 472 97 89 and 478 55 62
Fax: (+7 095) 479 16 09
e-mail: avia@msk.sitek.net
Web: http://www.npo-avia.ru
GENERAL DIRECTOR: Aleksandr Loshkarev
SALES DIRECTOR: Vadim Salimov

OKB AKKORD (Accord Design Bureau)
ulitsa Chaadaeva 1a, 603035 Nizhny Novgorod
Tel: (+7 8312) 46 71 41
e-mail: accord@nnov.cityline.ru
DIRECTOR: Yury Lakhtachev

EUROPEAN SALES AGENT:
Ground Support Equipment Srl
Viale del Vignola 44, I-00196 Roma, Italy
Tel: (+39 06) 322 28 77
Fax: (+39 06) 361 17 15
DIRECTOR: Roberto Salmoni

The Avia company was established on 20 February 1995 to design, manufacture and support light aircraft. Its associated design bureau (itself part of the Sokol plant) is responsible for the Accord range of light twin-engined aircraft. GSE of Italy was appointed as sales agent in 2000.

UPDATED

AVIA ACCORD-201

TYPE: Light utility twin-prop transport.
PROGRAMME: Original (lightweight) Accord design began May 1991; first flew 18 April 1994.

Production version is Accord-201; prototype flew 18 August 1997 at Nizhny Novgorod, initially with only VFR avionics and without floats. Demonstrations to prospective customers began December 1997; series of 12 flights from water, using floats, undertaken mid-1998. By early 1999, four more aircraft, including static test specimen, under construction and tooling in place for maximum possible production of 10 per month. Second prototype was due to fly in December 2001, but this has not taken place by May 2004, when prototype had accumulated 200 hours. Russian AP-23 certification anticipated October 1999, but delayed, eventually beginning on 20 April 2000. Experimental category certificate, granted mid-2000 with expiry date of 19 June 2001. Hartzell propellers received Russian certification in October 2000. Full certification expected mid-2002, using Becker avionics in place of original Bendix/King equipment. However, certification had not been completed by May 2004 but was then expected by year-end.

Avia Accord-201 as a landplane *(Sebastian Zacharias)* ***NEW***/0583381

Avia Accord-201 as a seaplane *(Yefim Gordon)* ***NEW***/0547732

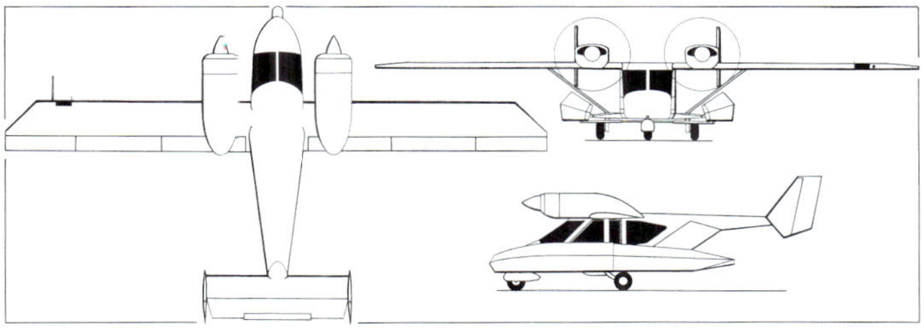

Avia Accord-201, showing position of retracted floats in nose and side views *(Paul Jackson)* 0079325

CURRENT VERSIONS: Offered for a variety of roles, including SAR (**201P**), aerial survey (**201AFS**), ecological monitoring (**201EM**), sigint (**201RC**) and jamming (**201REP**). Latest provisional version is the 201TDI, to be powered by two 172 kW (230 hp) SR-305 turbo diesels and due to fly in September 2004.

CUSTOMERS: First firm order received in May 1998 from Almazy Rossii-Sakha diamond mining company, which ordered two to be used for geomagnetic survey and exploration. Customers in Abu Dhabi and Liechtenstein by 2000; however, because of certification delays, former contract under renegotiation and latter cancelled by late 2001. No further sales reported, although in 2004 the company was forecasting production of 12–15 aircraft per year, at which time five were under construction.

COSTS: Standard aircraft US$298,000 on wheels; US$328,000 on floats; US$400,000–$500,000 for 201 TDI. Operating cost US$44 per hour (2000).

DESIGN FEATURES: High-wing monoplane with single bracing strut each side; unswept, high-lift wing section and blown flaps; pod and boom fuselage; cruciform tail surfaces with sweptback fin and rudder; mainwheels at tips of short stub-wings that support bracing struts. Constant-chord main wing panels, with engines on leading-edge; tapered outer panels.

Service life 15,000 hours/15 years. Wing section CAHI (TsAGI) P-II, with 12 per cent thickness/chord ratio at root, CAHI (TsAGI) P-III at tip; tip sweepback 45°; no dihedral, incidence 1°, no twist.

FLYING CONTROLS: Conventional and manual. Ailerons and elevators 100 per cent balanced; twin horn-balanced rudders; all cable actuated. Three-axis electric trim. Two-section slotted flaps each side.

STRUCTURE: All-metal; D16 aluminium alloy and high-strength stainless steel. Fuselage based on welded rectangular-section cabin frame of steel tubing; remainder of airframe conventional light alloy construction.

LANDING GEAR: Tricycle type; fixed. Mainwheel legs levered, with oleo-pneumatic shock absorbers, mounted on sponsons attached to lower fuselage at attachment of wing bracing struts; single wheel on each unit. Cleveland wheels and brakes; mainwheels type 40–142 with 8.00-6 tyres; nosewheel type 40–140 with 6.00-6 tyres; type 30–127 disc brakes. Minimum turning circle 3.2 m (10 ft 6 in). Maximum nosewheel steering angle ±60°. Optional floats (additional to, not replacing, fixed landing gear) retract upward electrically, outside mainwheels.

POWER PLANT: Two 157 kW (210 hp) Teledyne Continental IO-360-ES7B flat-six engines; Hartzell PHC-H3YF-2UF/FC7453 three-blade, constant-speed, fully feathering propellers. Fuel tank in each engine nacelle; total capacity 750 litres (200 US gallons; 165 Imp gallons). Total oil capacity 15.2 litres (4.0 US gallons; 3.3 Imp gallons).

In April 2004, two 169 kW (227 hp) SMA SR305 turbo-diesel engines were delivered for installation in one of the aircraft on the production line.

ACCOMMODATION: Pilot and up to six passengers in 2-3-2 seating, at 80 cm (31.5 in) pitch. Alternatively, pilot and three passengers and cargo, or all-cargo, or one stretcher patient, one or two attendants and medical equipment. Optional dual controls. Two large upward-hinged side doors. Baggage/cargo hold at rear of cabin, with door.

SYSTEMS: SAU-201 autopilot. Webasto Air Top 32 cabin heater. Electrical system includes two 27 V Teledyne N653344 DC generators and one 19 Ah Gill G-247 battery. Elipos-3 pneumatic de-icing system for wing, fin and tailplane leading-edges; Goodrich electric propeller de-icing. Janitrol cabin heater.

AVIONICS: Mainly Becker equipment:
Radar: Weather radar optional.

DIMENSIONS, EXTERNAL:
Wing span 13.75 m (45 ft 1¼ in)
Wing chord, constant 1.30 m (4 ft 3¼ in)
Wing aspect ratio 11.1
Length overall: landplane 8.12 m (26 ft 7¾ in)
amphibian 8.73 m (28 ft 7¾ in)
Height overall: landplane 2.94 m (9 ft 7¾ in)
amphibian 3.455 m (11 ft 4 in)
Span over tailfins 3.66 m (12 ft 0 in)
Wheel track 2.34 m (7 ft 8¼ in)
Float track 2.90 m (9 ft 6¼ in)
Wheelbase 2.845 m (9 ft 4 in)
Propeller diameter 1.93 m (6 ft 4 in)
Propeller ground clearance 1.25 m (4 ft 1¼ in)
Distance between propeller centres 2.80 m (9 ft 2¼ in)
Cabin doors (each): Height 1.10 m (3 ft 7¼ in)
Width: at top 0.75 m (2 ft 5½ in)
at bottom 1.50 m (4 ft 11 in)
Baggage/airdrop door: Height 0.90 m (2 ft 11½ in)
Width: at top 0.85 m (2 ft 9½ in)
at bottom 1.20 m (3 ft 11¼ in)

DIMENSIONS, INTERNAL:
Cabin: Length 3.50 m (11 ft 5¾ in)
Max width 1.31 m (4 ft 3½ in)
Max height 1.20 m (3 ft 11¼ in)
Floor area 3.50 m² (37.7 sq ft)
Volume 4.0 m³ (141 cu ft)
Baggage hold volume:
pilot and six passengers 0.50 m³ (17.7 cu ft)
pilot and four passengers 1.5 m³ (53 cu ft)
pilot and cargo 2.5 m³ (88 cu ft)

AREAS:
Wings, gross 17.00 m² (183.0 sq ft)
Ailerons (total) 1.17 m² (12.59 sq ft)
Trailing-edge flaps (total) 3.23 m² (34.77 sq ft)
Fins (total) 1.36 m² (14.64 sq ft)
Rudders (total) 1.36 m² (14.64 sq ft)
Tailplane 2.18 m² (23.47 sq ft)
Elevators (total) 1.78 m² (19.16 sq ft)

WEIGHTS AND LOADINGS:
Operating weight empty: landplane 1,350 kg (2,976 lb)
amphibian 1,530 kg (3,373 lb)
Max T-O weight 2,200 kg (4,850 lb)
Max wing loading 129.4 kg/m² (26.51 lb/sq ft)
Max power loading 7.03 kg/kW (11.55 lb/hp)

PERFORMANCE (ESTIMATED):
Never-exceed speed (VNE) 202 kt (375 km/h; 233 mph)
Max level speed at S/L:
landplane 161 kt (299 km/h; 186 mph)
amphibian 150 kt (277 km/h; 172 mph)
Max cruising speed at 75% power at 1,000 m (3,280 ft):
landplane 148 kt (275 km/h; 170 mph)
amphibian 138 kt (255 km/h; 158 mph)
Econ cruising speed at 3,000 m (9,840 ft):
landplane 119 kt (220 km/h; 137 mph)
amphibian 111 kt (205 km/h; 127 mph)
Stalling speed, flaps down, power off:
landplane 58 kt (106 km/h; 66 mph)
Max rate of climb at S/L:
landplane 445 m (1,460 ft)/min
amphibian 420 m (1,378 ft)/min
Service ceiling: landplane 6,100 m (20,020 ft)
amphibian 5,800 m (19,020 ft)
T-O run:
on land: landplane 240 m (790 ft)
amphibian 250 m (820 ft)
on water 360 m (1,185 ft)
Landing run on land:
landplane, amphibian 220 m (725 ft)
on water 190 m (625 ft)
Range with max fuel at 3,000 m (9,840 ft):
landplane 1,592 n miles (2,950 km; 1,833 miles)
amphibian 1,339 n miles (2,480 km; 1,541 miles)
g limits +3.8/–1.6

UPDATED

AVIACOR

AVIAKOR AVIATSIONNOYE ZAVOD OAO (AVIACOR AVIATION DEPOT JSC)

ulitsa Pskovskaya 32, 443052 Samara
Tel: (+7 8462) 27 03 87 and 27 04 44
Fax: (+7 8462) 27 06 91 and 27 04 77
e-mail: aviaprom@aviacor.ru
Web: http://www.aviacor.ru
GENERAL DIRECTOR: Gennady N Plotnikov
DIRECTOR, AN-140 PROGRAMME: Sergei V Kuznetsov

Founded in Voronezh during the 1930s and evacuated to Kuybishev (Samara) during Second World War, Aviacor (formerly GAZ 18) previously built the Tupolev Tu-154 three-turbofan transport. It has manufactured the Molniya-1 six-seat light aircraft and an air cushion vehicle derivative, and will produce the Tupolev Tu-354. It was also earmarked to build the Ukrainian Antonov An-70 and An-140 in Russia and to assist RSK 'MiG' with the Tu-334. The first An-140-100 (of six in initial batch) was rescheduled for construction in first quarter of 2002, but detailed discussions on a production timetable and contracts for delivery were not held with Antonov until 5 December 2001. First aircraft was eventually rolled out on 25 December 2003. Marketing is by International Aviation Project 140 (MAP-140), which formed 15 September 2003 as joint venture with Inter-AMI (subsidiary of KhGAPP at Kharkov).

It was announced in September 2001 that An-70 assembly had been reallocated to Polyot because of Aviacor's inability to provide start-up funding. Two new aircraft, known only as the Aist-2 and Aist-4, were reported to be under development, but nothing further has been heard. Subsidiary businesses are Aviacor-Service and Aviacor Repair. Company overhauls Tu-95MS strategic bombers and Tu-154 airliners (15 in 2000). Group members employ Westernised form of company name in place of direct transliteration, Aviakor.

The Kasatik light twin, developed by the co-located Feniks (Phoenix) Design Bureau during the 1990s, continues to benefit from intermittent promotion and, presumably, production. An example was shown at Moscow in August 2003.

Russian government shareholding is 25.5 per cent; controlling interest held by Siberian Aluminium Group since 1998. Undertakes subcontract work for Airbus and Boeing. Personnel in 2000 numbered some 4,000.

UPDATED

Feniks Kasatik three-seat light twin (Rotax 582 engines) *(Sebastian Zacharias)* ***NEW***/0567712

AVIASTAR

AVIASTAR, ULYANOVSKY AVIATSIONNYI PROMYSHLENNYI KOMPLEKS OAO (ULYANOVSK AVIATION INDUSTRIAL COMPLEX 'AVIASTAR' JSC)

prospekt Antonova 1, 432072 Ulyanovsk
Tel: (+7 8422) 20 25 75 and 29 10 22
Fax: (+7 8422) 20 35 06 and 29 23 65
e-mail: director@aviastar.link-ul.ru
Web: http:www.aviastar.link-ul.ru
CHAIRMAN AND GENERAL DESIGNER: Igor Shevchuk
DIRECTOR GENERAL: V V Mikhailov
MARKETING DIRECTOR: Aleksandr I Gladkov
EXECUTIVE DIRECTOR: Valery V Savotchenko

Founded in 1976, this 1.5 million m^2 (16.5 million sq ft) production facility, known as Plant 25 and named for D F Ustinov, began manufacture of the Antonov An-124 ultra-large airlifter in 1985, following with the Tupolev Tu-204 transport in 1987 and the related Tu-234 in 1996. It now takes the style Aviastar-SP.

In mid-1997, Aviastar agreed to merge with Tupolev design bureau and Aviacor. Tupolev merger effected 30 June 1999, forming Tupolev OAO with inputs from Russian government (51 per cent), Aviastar (43.6 per cent) and Tupolev Design Bureau (5.4 per cent). Aviastar now owned 74 per cent by Tupolev. Under Russian government plans announced on 12 May 2001, Aviastar will be incorporated in major grouping also including RSK 'MiG', Tupolev, Kamov, Aviacor and Sokol. Also in 1997, Aviastar Asia was formed in Taipei, Taiwan, as a joint venture with Asian interests to promote the Tu-204 in the Far East. Finance is provided by Aviastar Financial International, formed by Aviastar, Perm Motors and Moscow International Bank.

After five-year break, production of An-124 resumed with August 2000 delivery of one to Volga-Dniepr. State financing (Rb20 million) was received in 2000 for construction of a second assembly hall for An-124 production; however, by late 2001, Aviastar was urgently seeking new financial support to clear its debts. This reportedly achieved in early 2002 when Cato Avomatic (Egypt) and Leader Group (Russia) agreed to fund between 30 and 50 new Tu-204 airframes in 2003–05 initial step being batch of 25 agreed in December 2002. In October 2002, Ilyushin Finance announced intention of ordering 46 Tu-204s over six years; first three confirmed almost immediately. In September 2003 the plant agreed to accelerate production of five Tu-204s for China, which were required before the end of 2004, these being part of the Cato contract.

Tu-204 production rate remained at four per year in 2000, but entire plant operated at only 17 per cent capacity in 1999. Other activities include manufacture of bodies for Ikarus-280 bus and plans for major overhaul of LiAZ-677 buses.

UPDATED

BERIEV

BERIEVA AVIATSIONNYI KOMPANIYA (BERIEV AVIATION COMPANY)

ploshchad Aviatorov 1, 347923 Taganrog
Tel: (+7 86344) 499 01 and 498 39
Fax: (+7 86344) 414 54
e-mail: info@beriev.com
Web: http://www.beriev.com
GENERAL DIRECTOR: Viktor A Kobzev
HEAD OF MARKETING: Andrei Shishko
HEAD OF INFORMATION: Andrei I Salinkov

Original Beriev design bureau (OKB) founded in October 1934 by Georgy Mikhailovich Beriev (1902–79); except during part of the Second World War, 1942–45, it has been based at Taganrog, in northeast corner of Sea of Azov; since 1948 has been primary centre for Russian seaplane development. Beriev has manufactured more than 20 types of aircraft, most of which have entered series production. In 1990 was redesignated Taganrogsky Aviatsionnyi Nauchno-Tekhnichesky Kompleks imeni G M Berieva (TANTK: Taganrog Aviation Scientific-Technical Complex named for G M Beriev); became part of AVPK Sukhoi in 1996. As a consequence of this merger, the bureau's products are often manufactured at factories traditionally associated with Sukhoi, the Be-103 Bekas, for example, at Komsomolsk. On 1 January 1998 adopted style, 'Beriev Aviation Company' for international promotion, retaining TANTK in Russia. State shareholding is 38 per cent; personnel strength in 2001 was more than 2,000.

Beriev company now includes the experimental design bureau, experimental production facilities, a flight test complex, economic, financial and logistics support services, with test bases and proving grounds at the Black Sea and Sea of Azov. Its products are experimental prototypes of amphibious aircraft and wing-in-ground-effect (WIG) vehicles, together with test reports and technical documentation for their series production. It undertakes design and development of unconventional aircraft in response to requests for proposals from other companies, testing of aircraft and assemblies in maritime conditions, and training of aircrew and ground personnel for seaplane operation. In 2001, Beriev revealed it was developing a multirole amphibian designated **Be-112**.

UPDATED

BERIEV A-42

Similar to the Beriev A-40, this aircraft project was last described in the 1998–99 edition. Prototype 85 per cent complete at TANTK, Taganrog, by September 2000, at which time only Rb19.5 million assigned of Rb307 million required to complete development. By January 2002, Beriev was reporting that this prototype was an A-42P patrol and SAR variant; power plant quoted as two Progress D-27 propfans, each rated at 109.8 kN (24,690 lb st), which description applies to Beriev **A-45**. In this form, A-42P has MTOW of 96,000 kg (211,650 lb), combat load of 8,500 kg (18,739 lb) and range of 6,209 n miles (11,500 km; 7,145 miles). Main sensor suite to be Leninets Novella (Sea Dragon). Domestic needs are up to eight for military use and up to four for MChS civil protection agency; export interest (A-42PEh) from Canada, Chile, China, Finland, South Korea and Thailand. Announced in April 2002 that funding of A-42 to completion would begin in 2005, at earliest, although a September 2003 announcement predicted completion of the first aircraft in 2006. Intended avionics include Sea Snake search and attack radar and Aria-V flight and navigation system.

In mid-2002, Beriev was continuing promotion of an A-40Eh export version of the original aircraft as well as predicting a return of official interest to the A-40, which the Russian Navy abandoned in 1996, in favour of a maritime patrol version of Tupolev Tu-204 airliner.

Interest, but without full funding, continues to be maintained in Beriev's A-40 (illustrated) and A-42 flying-boats *(Yefim Gordon)* 0536625

UPDATED

BERIEV BE-103 BEKAS

English name: Snipe

TYPE: Six-seat amphibian.

PROGRAMME: Design started 1992; model exhibited and initial data released at Moscow Air Show '92; construction of preproduction batch of four began by KnAAPO at Komsomolsk-on-Amur 1994; prototype (RA-37019), with M-17F engines, displayed statically at Gelendzhik Hydro-aviation Show, on the Black Sea, 23 September 1996; first flew at Taganrog on 15 July 1997, but destroyed on its 28th sortie, during practice for Moscow Air Show on 18 August 1997. Second prototype (RA-03002) flew 17 November 1997; made type's first water take-off and landing on 24 April 1998 (aircraft's sixth sortie), but crashed on 29 April 1999 while returning from exhibition at Aero '99, Friedrichshafen; further two structure/static test airframes completed for use at Taganrog by early 1997.

Type certification rescheduled for second half of 1998 and first delivery for 1999, but these dates also not met; development being assisted by three preproduction aircraft, which completed by late 1997, although first of these (believed 03101) did not fly until 19 February 1999; 03103 shown at Gelendzhik in August 2000 when seen to have fixed wing leading-edge slats outboard of landing lights (as on 03102); by early 2001, 03103 had received nose radome. Total of five flying and two static test aircraft up to this time.

Be-103 development included, since 1997, in Rostov regional development programme, while funding provided by government's 1999–2000 Light Aircraft Development Programme; KnAAPO paid for VFR certification, which granted by Russian Aviation Register on 26 December 2001, after 161 hours of trials, including 59 water and 204 land take-offs. FAR/JAR 23 certification trials satisfactorily completed by April 2003 and confirmation expected to follow.

Marketed by **Takom Avia** consortium, formed by Beriev, KnAAPO and VO Mashinoexport of Moscow. In late 1999, consideration given to establishing final assembly centre in Malaysia, combining Russian airframes and US avionics and engines. Aerocorp International, appointed US dealer by 2002. Launch of US certification effort announced in September, but KnAAPO simultaneously revealed temporary suspension of production due to lack of orders. Named Bekas in February 2003. First production aircraft (03301) flew

US-registered Beriev Be-103 Bekas landing on Lake Parker, Lakeland, Florida *(Paul Jackson)* ***NEW***/0576920

First production Beriev Be-103 Bekas on Lake Winnibago, Oshkosh, before application of its US registration *(Paul Jackson)* ***NEW***/0576896

2 July 2003. US certification achieved 11 July 2003 (presented 31 July) and three aircraft air-freighted to Oshkosh by An-124 on 23 July for US distributor, Aerocorp, and immediate US debut at AirVenture '03, first becoming N13KL (03301).

CURRENT VERSIONS: **Be-103:** *As described.*

SA-20P: Described separately.

OSA: In September 2003, Beriev was reported to be working on design of a Be-103 variant with a single turboprop of between 298 and 373 kW (400 and 500 shp) under the provisional designation OSA (*Odnodvigatelnyi Samolet Amfibiya*: Single-Engined Amphibian), of which the SA-20P is regarded as a prototype.

CUSTOMERS: Projected sales of 350 to 450, including exports. Break-even at 250th. First reported order from Russian Border Guards for 20 (out of possible 200 required), originally for delivery from 1999; Forestry Service requires 30 to 40. First export order announced August 1999 in form of 10 for undisclosed customer; however, in January 2002 Beriev was still negotiating first firm orders, comprising possible six for forestry use and others for border patrol.

Initial production batch of 10; deliveries began July 2003; second batch of 20 due to have been started by late 2003.

COSTS: Basic price US$650,000 (2003), with IO-360 engines.

DESIGN FEATURES: Low-wing monoplane, with water-displacing wings of moderate sweep with large wingroot extensions; two-step boat hull; no stabilising floats; horizontal strake each side of nose plus vertical strake ahead of second hull step; wing centre-section increases lift during take-off and landing, partly compensating for absence of flaps. Engine pylon mounted on each side of rear fuselage, aft of wings; engines raised slightly on production version. Sweptback fin and rudder; tailplane mid-set on fin. Designed in compliance with AP-23 and FAR Pt 23 requirements; suitable for passenger and cargo transport, medical evacuation, patrol and ecological monitoring. Operable in wave heights up to 0.5 m (1½ ft) and on water as shallow as 1.25 m (4 ft). Service life 15,000 hours or 20 years.

Wing leading-edge sweep 22°; wing section NACA 2412M; dihedral 5° 3' on outer wings; incidence 1°. Fixed leading-edge slats. Fin leading-edge sweep 33°; tailplane 14° 40'.

FLYING CONTROLS: Manual. All-moving tailplane (with anti-balance tab) and ailerons actuated by pushrods; rudder by cables; two mass balances ahead of tailplane; mass balance (port side only) on rudder, which has flight-adjustable trim tab; electric trim; spring feel in tailplane control; no flaps. Control surface deflections: ailerons ±25°; tailplane +14/−6°; rudder ±27°.

STRUCTURE: All-metal semi-monocoque boat-type fuselage; all-metal single-spar wings. Extensive use of 1446 aluminium-lithium alloy.

LANDING GEAR: Pneumatically retractable tricycle type; single wheel on each unit; mainwheels retract forward into wing centre-section; nosewheel retracts forward; oleo-nitrogen shock-absorbers; brakes on mainwheels; nosewheel tyre size 400×150, mainwheel tyres 7.00-6; pressure 3.00 bar (43 lb/sq in) in all units; disc brakes; self-centring nosewheel steerable ±35°.

POWER PLANT: Two 157 kW (210 hp) Teledyne Continental IO-360-ES4 flat-four piston engines, each driving an MT-Propeller MTV-12-D-C-F-R-(M)/CFR183-17 three-blade, variable pitch (hydraulic), metal propeller. Later option will be two 221 kW (296 hp) VAZ-4263 piston engines and Avia AV-103 three-blade, variable-pitch propellers, with optional reversible pitch. Fuel tank in each wingroot, usable capacity 328 litres (86.6 US gallons; 72.1 Imp gallons), plus 6 litres (1.5 US gallons; 1.25 Imp gallons) unusable. Oil capacity 8 litres (2.0 US gallons; 1.7 Imp gallons) per engine.

ACCOMMODATION: Pilot and five passengers in pairs; dual controls optional; large upward-opening door on each side, hinged on centreline; folding seats facilitate entry/loading baggage/freight compartment aft of cabin. Optional ambulance configuration

Instrument panel of production Be-103 *(Paul Jackson)* ***NEW***/0576921

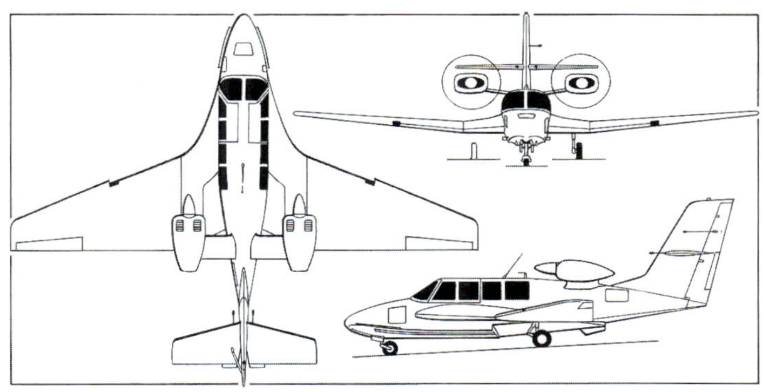

Beriev Be-103 Bekas twin-engined light multipurpose amphibian *(Jane's/Mike Keep)* 0132839

for pilot, one stretcher, three seated persons; all-cargo configuration for items up to 2.00 × 0.70 × 0.70 m (6 ft 6¾ in × 2 ft 3½ in × 2 ft 3½ in) in size; patrol; agricultural and ecological monitoring use.

SYSTEMS: Interior heated and ventilated. Pneumatic system, bottle capacity 6 litres (365 cu in) of air, pressure 49 bar (711 lb/sq in). Electrical system 27 V DC and three-phase 36 V 400 Hz AC, supplied by 3 kW DC generators, rectifiers and 25 Ah battery. Fire suppression system. Optional anti-icing.

AVIONICS: Integrated by Bendix/King. VFR standard; IFR optional. Version with Russian avionics may be offered in the future.

Comms: KY 196A VHF com radio, KMA 24 intercom, KX 165 nav/com/glideslope, KT 70 transponder and P-855 emergency locator beacon.

Radar: Optional RDR-2000 or RDR-1400 weather radar in extreme nose.

Flight: KN 63 DME, KR 87 ADF, KCS 55A compass, KRA 405 radar altimeter, KLN 89B GPS. Optional KFC 150 or KAP 140 autopilot.

Instrumentation: Conventional.

EQUIPMENT: Landing light in each wing leading-edge.

DIMENSIONS, EXTERNAL:

Wing span	12.72 m (41 ft 9 in)
Wing chord: at root	6.21 m (20 ft 4¾ in)
at tip	0.83 m (2 ft 9 in)
Wing aspect ratio	6.4
Length: overall	10.65 m (34 ft 11¼ in)
fuselage	9.96 m (32 ft 8 in)
Fuselage max width	1.35 m (4 ft 5¼ in)
Height overall	3.76 m (12 ft 4 in)
Tailplane span	3.90 m (12 ft 9½ in)
Wheel track	2.275 m (7 ft 5½ in)

Wheelbase ... 4.12 m (13 ft 6¼ in)
Propeller diameter ... 1.83 m (6 ft 0 in)
Distance between propeller centres ... 3.00 m (9 ft 10¼ in)
Passenger door (each): Height ... 0.90 m (2 ft 11½ in)
Width ... 1.30 m (4 ft 3¼ in)

DIMENSIONS, INTERNAL:
Cabin: Length ... 3.20 m (10 ft 6 in)
Max width ... 1.25 m (4 ft 1¼ in)
Max height ... 1.23 m (4 ft 0½ in)
Volume ... 4.6 m^3 (162 cu ft)
Baggage compartment: Max width ... 1.08 m (3 ft 6½ in)
Max height ... 0.88 m (2 ft 10½ in)
Volume ... 0.85 m^3 (30.0 cu ft)

AREAS:
Wings, gross ... 25.10 m^2 (270.2 sq ft)
Ailerons, total ... 0.80 m^2 (8.61 sq ft)
Fin ... 2.86 m^2 (30.78 sq ft)
Rudder ... 1.54 m^2 (16.58 sq ft)
Tailplane, total ... 3.68 m^2 (39.61 sq ft)

WEIGHTS AND LOADINGS:
Manufacturer's empty weight ... 1,680 kg (3,703 lb)
Weight empty, equipped ... 1,760 kg (3,880 lb)
Baggage capacity ... 50 kg (110 lb)
Max payload ... 332 kg (732 lb)
Max fuel weight ... 245 kg (540 lb)
Max T-O and landing weight ... 2,270 kg (5,004 lb)
Max wing loading ... 90.4 kg/m^2 (18.52 lb/sq ft)
Max power loading ... 7.25 kg/kW (11.91 lb/hp)

PERFORMANCE (US version):
Never-exceed speed (VNE) ... 130 kt (240 km/h; 149 mph)
Max cruising speed at FL98 ... 130 kt (240 km/h; 149 mph)
Econ cruising speed at FL98 ... 120 kt (222 km/h; 138 mph)
Stalling speed ... 59 kt (109 km/h; 68 mph)
Single-engine minimum control speed ... 62 kt (115 km/h; 71 mph)
Service ceiling ... 5,000 m (16,400 ft)
T-O run: on land ... 440 m (1,445 ft)
on water ... 620 m (2,035 ft)
Landing run: on land ... 240 m (790 ft)
on water ... 320 m (1,050 ft)
Range, with max fuel: at 500 m (1,640 ft) ... 637 n miles (1,180 km; 733 miles)
at FL98 ... 594 n miles (1,100 km; 683 miles)
Endurance, no reserves ... 6 h 30 min
g limits ... +3.45/–1.45

UPDATED

BERIEV BE-132

TYPE: Twin-turboprop light transport.

PROGRAMME: Original Be-30 first flown in prototype form on 3 March 1967; eight Be-30s built, but programme terminated when Aeroflot ordered Let L-410As from Czechoslovakia. Hard currency shortage revived programme 1993, with modestly upgraded version known as Be-32; one of original Be-30s (RA-67205) exhibited at 1993 Paris Air Show, as Be-32 demonstrator with original Russian TVD-10B engines; re-engined with PT6A-65B turboprops, RA-67205 first flew as Be-32K on 15 August 1995. Launch of Be-32K production continued to be discussed in 2001, although initial investment of Rb1,060 million required, plus Rb1,460 million in medium term; no announced orders.

Be-132 emerged from agreement between Motor-Sich of Ukraine, IAPO, Klimov and Beriev, signed at Paris Air Show in June 2001. Initial details released at MAKS, Moscow, August 2001. Completion of business plan and start of marketing were due for early 2002, but little evidence of progress 12 months later. Certification planned to AP-25.

Model of Beriev Be-132MK 0113983

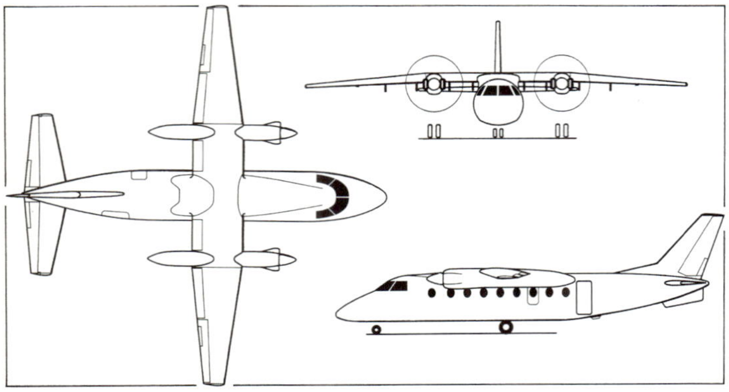

Beriev Be-132MK twin-turboprop multipurpose light transport *(Paul Jackson)* 0121080

CURRENT VERSIONS: **Be-132MK:** Initial version. Compared with Be-32, has more circular fuselage cross-section and increased length.

DESIGN FEATURES: General purpose light transport, optimised for operation from remote airfields. Conventional cantilever high-wing monoplane; three-section wings, with anhedral on outer panels; 42° sweptback vertical tail surfaces; engines at tip of centre-section each side. Ventral underfin.

FLYING CONTROLS: Conventional and manual. Trim tabs in both ailerons, both elevators and rudder. Double-slotted flaps in two sections each side, inboard and outboard of engines.

STRUCTURE: All-metal; semi-monocoque fuselage of rectangular section; spars and skin panels of wing torsion box are mechanically and chemically milled profile pressings; detachable bonded leading-edge; half of wings and most of tail unit covered with thin honeycomb panels stiffened with stringers; majority of fuselage made of adhesive-bonded panels; tips of wings and tail surfaces and wing/fuselage fillets of GFRP.

LANDING GEAR: Tricycle type; single wheel on each unit; nosewheel retracts forward, mainwheels rearward into engine nacelles; tyres size 720×320 on mainwheels, 500×150 on nosewheel; mainwheel brakes.

POWER PLANT: Two 1,103 kW (1,479 shp) Klimov VK-1500 turboprops; six-blade Aerosila AV-36 propellers; integral wing fuel tanks.

ACCOMMODATION: Basic seating for two crew and 26 passengers in two-plus-one configuration at 78 cm (30½ in) pitch. Carry-on baggage compartment on starboard side, aft of cabin seating, opposite forward-hinged door and airstairs; lavatory to rear.

DIMENSIONS, EXTERNAL:
Wing span ... 17.80 m (58 ft 4¾ in)
Length overall ... 17.925 m (58 ft 9¾ in)
Fuselage max width ... 1.70 m (5 ft 7 in)
Height overall ... 5.545 m (18 ft 2¼ in)
Tailplane span ... 7.40 m (24 ft 3¼ in)
Wheel track ... 6.00 m (19 ft 8¼ in)
Wheelbase ... 5.985 m (19 ft 7¾ in)
Propeller diameter ... 2.65 m (8 ft 8¼ in)
Cabin door: Height ... 1.30 m (4 ft 3 in)
Width ... 0.75 m (2 ft 5½ in)
Emergency exits (each): Height ... 0.92 m (3 ft 0¼ in)
Width ... 0.60 m (1 ft 11½ in)

WEIGHTS AND LOADINGS:
Max T-O weight ... 10,000 kg (22,046 lb)
Max power loading ... 4.54 kg/kW (7.45 lb/shp)

PERFORMANCE (estimated):
Max cruising speed ... 248 kt (460 km/h; 286 mph)
Econ cruising speed ... 232 kt (430 km/h; 267 mph)
Unstick speed ... 113 kt (210 km/h; 130 mph)
Landing speed ... 108 kt (200 km/h; 124 mph)
Service ceiling ... 6,000 m (19,680 ft)
T-O run ... 700 m (2,300 ft)
Landing run ... 450 m (1,480 ft)
Required runway length ... 1,500 m (4,925 ft)
Range:
with 26 passengers ... 712 n miles (1,320 km; 820 miles)
with max fuel ... 1,490 n miles (2,760 km; 1,715 miles)

BERIEV (BETAIR) BE-200 ALTAIR

TYPE: Twin-jet amphibian.

PROGRAMME: Initiated 1989 under design leadership of Alexander Yavkin. Russian government approval for purpose-designed water bomber granted 8 December 1990. Details announced, and model displayed, at 1991 Paris Air Show; full-scale mockup constructed 1991; development by Beta Air (Betair: **Be**riev **ta**ganrog **ir**kutsk) consortium, formed 27 November 1991 from Irkutsk Aircraft Production Association (53.5 per cent), Beriev OKB (20.5 per cent), ILTA Trade Finance SA of Geneva, Switzerland (5 per cent) and the Ukrainian bank Prominvest (15 per cent) plus unidentified partner(s) (6 per cent); included in Civil Aviation Development Programme and state forestry protection programme. Named Altair in February 2002.

Beriev responsible for development, design and documentation; systems bench-testing; static-, flight- and fatigue-testing of prototypes; certification and design support of serial production. Irkut duties comprise production preparation; manufacture of tooling; production of four prototypes and series aircraft; and spare parts manufacture. In May 2002, EADS signed joint marketing agreement with IAPO. Initial production and certification supported by Alenia of Italy under US$1.6 million programme financed by European Union. In March 2003, agreement reached on transfer of production to Beriev plant at Taganrog and in following June, EADS, Rolls-Royce Deutschland and Irkut study revealed potential market for 320 BR715-engined Altairs over 20 years, leading to decision to launch this variant.

First prototype (001) rolled out 11 September 1996, at Irkutsk; first flight scheduled 1997, but eventually achieved (from land) 24 September 1998; 'official' first flight 17 October 1998; aircraft transferred from Irkutsk to Taganrog (by then with 19 sorties and 26½ flying hours) on 27 April 1999 to begin certification trials, including water drops; exhibited at Paris (by then registered RA-21511) June 1999; first water landings and take-offs, 10 September 1999. Total 80 sorties, including 18 from water, by November 1999, when water operations ceased for winter. Experimental category certification achieved March 2000. Awarded limited category AP-25 certification for firefighting operations 10 August 2001, by which time had flown 337 sorties (100 from water) and achieved 202 flying hours. Second phase of testing begun October 2001 to certify patrol and passenger operations using BR715 engines, although most of that month occupied by a nine-country Far East sales tour. Prototype had accumulated 650 hours in nearly 600 sorties (216 from water) by May 2002. Demonstrated on Lake Sevran, Armenia, 15 August 2002; set six FAI climb-with-payload records in hydroplane class, 7 September 2002.

Second flying prototype (003) in firefighting configuration (Be-200ChS). Two test airframes, of which first ferried from Irkutsk to Taganrog by An-124 in March 1995, followed by second in 1997; these, respectively, static test (SI : *staticheskiy ispytaniya*) and fatigue test (RI : *resursnye ispytaniya*). Four production aircraft (0101 to 0104) in hand by late 1998; fifth was taking shape in early 2002. On original schedule, second prototype was due for delivery in late 1999 to Russian Emergencies Ministry. Repeated delays resulted in

RA-21515, first Beriev Be-200 Altair delivered to MChS *(Yefim Gordon)* **NEW**/0576918

first flight (003/RA-21512) on 27 August 2002, following which it was exhibited at Gelendzhik Gidroavia Salon, 4 to 8 September 2002.

First production aircraft (0101/RA-21515) rolled out 26 May and first flew 17 June 2003; Be-200 restricted type certificate issued by Interstate Aviation Committee on 20 June 2003; RA-21515 delivered to MChS 31 July 2003 and based at Zhukovsky, taking part in MAKS '03 air show in following month. Full certificate issued 29 December 2003. Second production aircraft began taxying trials on 26 February 2004.

Avionics development and scoop trials for water bomber version being undertaken by the Beriev Be-12P-200/Be-128 testbed since August 1996. TANTK also produced four Be-12Ps (*Pozharny:* firefighting) for concept and tactics development and two Be-12NKh (*Narodno-Khozyaistvennye:* National Economy) general purpose versions, both of which lost in 1994.

CURRENT VERSIONS: **Firefighting:** TsENTROSPAS/MChS designation **Be-200ChS:** Tanks under cabin floor of centre-fuselage, capacity 12 m^3 (423 cu ft) water; six tanks in cabin for 1.2 m^3 (42 cu ft) liquid chemicals; two retractable water scoops forward of step, two aft; 30 fully equipped smoke jumpers can be carried on seats along sidewalls of cabin, with jump-door at rear of cabin on starboard side; 12 tonnes of water scooped from seas in 14 seconds with waves up to 1.2 m (4 ft).

Fully fuelled, Be-200 can drop total 310,000 kg (683,420 lb) of water in successive flights when airfield to reservoir distance is 108 n miles (200 km; 125 miles) and reservoir to fire zone distance is 5.4 n miles (10 km; 6.2 miles); or 140,000 kg (308,640 lb) when distances are respectively 108 n miles (200 km; 125 miles) and 27 n miles (50 km; 31 miles). Tank emptying time 0.8 to 1.0 second; minimum dropping speed 119 kt (220 km/h; 137 mph). Tanks quickly removable when aircraft carries freight. Flight deck and cargo hold sealed against smoke ingress. ARIA-200M avionics features include water source/drop zone track memory, automatic glideslope and digital flight deck/ground fire crew communications. Also outfitted for SAR, with equipment including Orion 25S inflatable boat, naval radios, 600 W loudspeaker and searchlight, storage for 20 inflatable rafts and two four-seat motorboats, observer's station, 57 seats and attachment for 30 litters.

In 2004, Irkut was promoting a firefighting package of Be-200 assisted by fire reconnaissance Aeronautics Aerostar UAV supplied by Israeli manufacturer.

Passenger: Two flight crew, two cabin attendants, and 72 tourist class passengers four-abreast in pairs with centre aisle, at seat pitch of 75 cm (29.5 in).

Cargo: Payload 7,500 kg (16,534 lb) in unobstructed cabin 17.00 m (55 ft 9 in) long, 2.6 m (8 ft 6 in) wide and 1.9 m (6 ft 3 in) high.

Ambulance: Two flight crew, seven seated casualties/medical personnel, 30 stretchers in three tiers.

Patrol: Paramilitary version for possible use of Russian Frontier Guards revealed to be under development in early 2002.

Be-200PS: Search and rescue. Provisions included in Be-200ChS described above.

Be-200M: Redesignated Be-210 in 1998.

Be-200P: Projected anti-submarine version; combat radius 2,537 n miles (4,700 km; 2,920 miles); endurance 7 hours.

Be-210: Announced in 1998; development of Be-200 multirole amphibian for airline use in regions of minimal airport infrastructure. Seating for two pilots, two attendants and 72 passengers at 75 cm (29½ in) seat pitch. Other details may be assumed to be similar to Be-200, but airframe strengthened for extra fuel in wing centre-section (42,000 kg; 92,594 lb MTOW) and ferry range increased to 2,483 n miles (4,600 km; 2,858 miles), or 998 n miles (1,850 km; 1,149 miles) with full passenger load. In 2001, foreign investors, including Scorpion of Greece, showing interest in developing BR715-engined airliner version of Be-200; this may carry Be-210 designation.

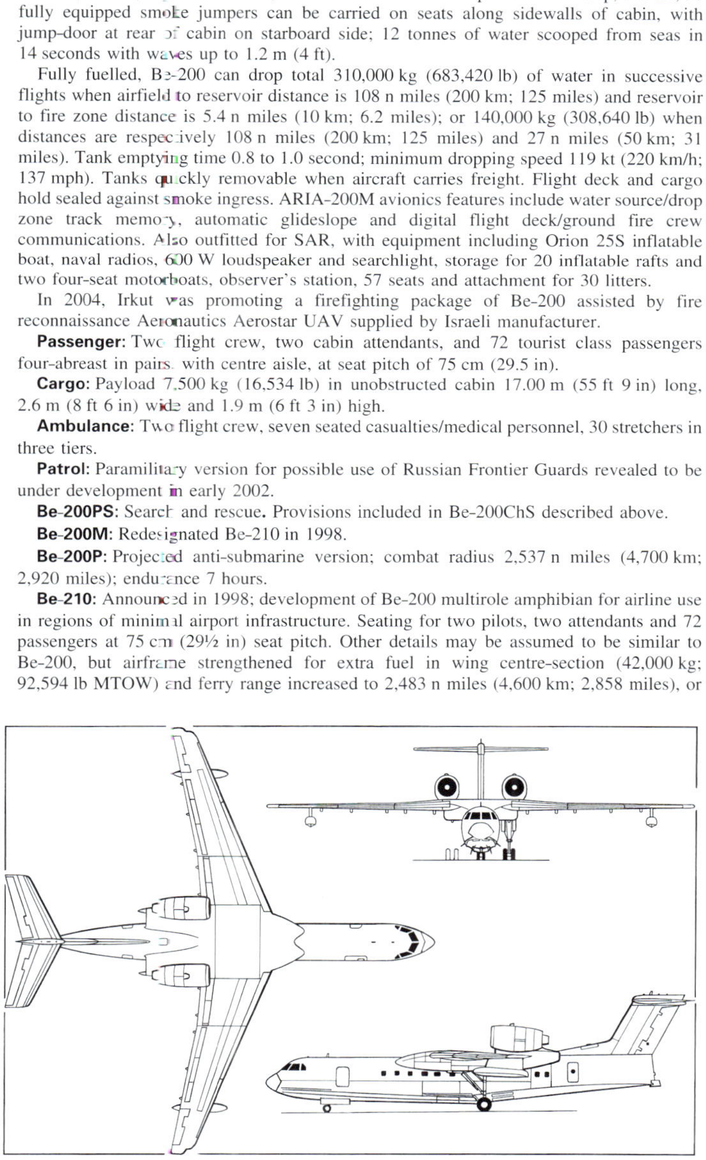

Beriev (Betair) Be-200 Altair civil utility amphibian *(Jane's/Mike Keep)* 0051205

CUSTOMERS: Seven ordered (of 20 required) by TsENTROSPAS/MChS (Russian Ministry of Emergency Situations) on 15 January 1997 for firefighting, with further orders to follow for SAR; decision in principle to acquire eighth, funded by cancellation of proposed two Antonov An74s. However, at time of maiden flight, only five orders were being claimed as firm. Will equip four SAR stations on Black Sea and Baltic and two in eastern Russia; first base was to have been Gelendzhik where two Altairs (0101 and 0102) scheduled to be in service by end of 2003. Beriev contracted in 2002 to establish training school at Taganrog for MChS pilots and technicians. However Zhukovsky (Ramenskoye) nominated as central base for MChS Altair fleet.

Russian state forest service/firefighting agency requires up to 54, of which near-term needs total 10 to 15; 50 required by Sakhalin regional administration; five by Irkutsk regional administration.

Potential market foreseen for more than 400 by 2010, of which over 60 per cent for export. South Korea evinced interest in 12 of maritime patrol version for police duties during 1998; China considering licensed production at Harbin; Italian interest in 15 reported mid-2000. Australia, France, Indonesia, Japan and Philippines expressing interest in 2001–02 and Be-200 demonstrated at Marseilles, France, 13 to 15 May 2002, and at Elefsis, Greece. Hawkins & Powers of USA negotiating for eight in early 2004.

COSTS: Basic price US$25 million (2000). Break-even at 46/48th aircraft. Development cost estimated as US$250 million, of which US$190 million expended by 2001.

DESIGN FEATURES: First Russian purpose-designed water-bomber. Derived from Beriev A-40, but underwing stabilising floats moved inboard from tips, twin-wheel main landing gear units, and no booster turbojets. Conforms to FAR Pt 25 criteria. Swept wings of moderate aspect ratio, with high-lift devices; single-step hull of high length/beam ratio, which reported to provide world's first variable-rise bottom, providing a considerable improvement in stability and controllability in the water, as well as a reduction in *g* loads when landing and taking off at sea; small wedge-shape boxes ('hydrodynamic compensators') aft of step aid 'unsticking' from water in wave heights up to 2.2 m (7 ft 2½ in); all-swept T tail; high-mounted engines protected from spray by strakes on each side of nose and by wings; large underwing pod each side of hull, faired into wingroot.

Wing leading-edge sweep 23° 13'; supercritical wing sections, thickness/chord ratio 16 per cent to 11.5 per cent.

FLYING CONTROLS: Conventional. Entire span of each wing trailing-edge occupied by aileron and two-section area-increasing single-slotted flaps; full-span leading-edge slats in three sections each side; five spoiler/lift dumper sections forward of flaps each side.

STRUCTURE: Hull made primarily of high-strength aluminium/lithium alloys; interior of composites; water tanks of ferric alloys of aluminium in firefighting version.

LANDING GEAR: Hydraulically retractable tricycle type. Twin-wheel main units, tyre size 950×300, pressure 9.80 to 10.30 bar (142 to 150 lb/sq in); twin nosewheels, tyre size 620×180, pressure 7.35 to 7.85 bar (106 to 114 lb/sq in). Mainwheels and nosewheels retract rearwards. Water rudder. Ground turning radius 17.4 m (57 ft 1 in). Nosewheel steering angle ±45°.

POWER PLANT: Two ZMKB Progress D-436TP turbofans, each 73.6 kN (16,550 lb st). Rolls-Royce BR715s and others under consideration as alternative engines. Fuel system by Pall (USA). Total oil capacity 22 litres (5.8 US gallons; 4.85 Imp gallons).

ACCOMMODATION: Two flight crew; up to 72 tourist class passengers at 75 cm (29.5 in) seat pitch and two attendants; or 10 to 32 first class and business class passengers at up to 102 cm (40 in) seat pitch, with provision for galley, lavatory and baggage stowage. Up to nine freight containers or, typically, six containers and 19 economy class seats. Cargo door, with integral passenger door, starboard side, forward, opens upwards; passenger door port, forward; emergency exits; forward and rear freight/baggage holds. Interior design by AIM Aviation (UK).

SYSTEMS: Barco Display Systems FMS system. All accommodation pressurised. Three hydraulic systems at 207 bar (3,000 lb/sq in); 150 litres (39.6 US gallons; 33.0 Imp gallons) of MGJ-5U fluid; flow rate 70 litres (18.5 US gallons; 15.4 Imp gallons)/min. Capacity of pneumatic system bottles 29 litres (1.02 cu ft). Three-phase 115/220 V 400 Hz AC electrical system; single-phase 115 V 400 Hz AC system; 27 V DC system; supplied by two engine-driven 60 kVA AC generators and three static inverters; three batteries. Gaseous oxygen bottle, pressure 147 bar (2,135 lb/sq in). Provision for de-icing tail unit, slats, engine air intakes and windscreen. TA-12 APU, operable up to 7,000 m (23,000 ft) in starboard wingroot. Propeller-driven emergency generator at base of fin.

AVIONICS: *Radar:* MN-85 weather radar in nose.

Flight: ARIA-200M digital flight and navigation system by American-Russian Integrated Avionics, a joint venture of AlliedSignal (now Honeywell) and Moscow Research Institute of Aircraft Equipment. INS standard.

Instrumentation: Honeywell EFIS displays, with six 152 × 203 mm (6 × 8 in) LCDs.

DIMENSIONS, EXTERNAL:

Wing span over winglets .... 32.78 m (107 ft 6½ in)
Wing chord: at root .... 5.58 m (18 ft 3½ in)

at tip .... 1.72 m (5 ft 7¾ in)
Wing aspect ratio 9.1
Length overall .... 31.43 m (103 ft 1½ in)
Fuselage: Length .... 29.18 m (95 ft 9 in)
Max diameter .... 2.86 m (9 ft 4½ in)
Height overall .... 8.90 m (29 ft 2½ in)
Tailplane span .... 10.115 m (33 ft 2¼ in)
Wheel track (c/l shock-absorbers) .... 4.30 m (14 ft 1¼ in)
Wheelbase .... 11.145 m (36 ft 6¾ in)
Width over stabilising floats .... 25.60 m (84 ft 0 in)
Passenger doors (each): Height .... 1.70 m (5 ft 7 in)
Width .... 0.90 m (2 ft 11½ in)
Cargo door: Height .... 1.76 m (5 ft 9¼ in)
Width .... 2.05 m (6 ft 8¾ in)
Emergency exits: Height .... 1.70 m (5 ft 7 in)
Width .... 0.90 m (2 ft 11½ in)

DIMENSIONS, INTERNAL:
Cabin, excl flight deck:
Length: passenger .... 17.00 m (55 ft 9 in)
cargo .... 18.70 m (61 ft 4¼ in)
Max width: passenger .... 2.40 m (7 ft 10½ in)
cargo .... 2.50 m (8 ft 2½ in)
Max height: passenger .... 1.80 m (5 ft 10¾ in)
cargo .... 1.89 m (6 ft 2½ in)
Floor area: passenger .... 39.0 m^2 (420 sq ft)
cargo .... 41.0 m^2 (441 sq ft)
Volume:
passenger:
forward baggage hold .... 8.8 m^3 (310 cu ft)
rear baggage hold .... 4.5 m^3 (159 cu ft)
main cabin and freight/baggage holds, total, cargo configuration .... 84.0 m^3 (2,966 cu ft)
cargo .... 80.8 m^3 (2,853 cu ft)

AREAS:
Wings, gross .... 117.44 m^2 (1,264.2 sq ft)
Ailerons (total) .... 3.56 m^2 (38.32 sq ft)
Flaps (total) .... 20.43 m^2 (219.91 sq ft)
Slats (total) .... 12.61 m^2 (135.74 sq ft)
Spoilers (total) .... 4.59 m^2 (49.41 sq ft)
Fin .... 12.60 m^2 (135.63 sq ft)
Rudder .... 4.60 m^2 (49.52 sq ft)
Tailplane .... 17.96 m^2 (193.33 sq ft)
Elevators (total) .... 6.96 m^2 (74.92 sq ft)

WEIGHTS AND LOADINGS:
Max payload .... 7,500 kg (16,534 lb)
Max fuel weight .... 12,260 kg (27,025 lb)
Max T-O and ramp weight:
cargo/passenger .... 42,000 kg (92,594 lb)
firefighting .... 37,200 kg (82,011 lb)
Max airborne weight (after water scooping) .... 43,000 kg (94,800 lb)
Max landing weight, land or water .... 35,000 kg (77,160 lb)
Max wing loading:
cargo/passenger .... 357.6 kg/m^2 (73.25 lb/sq ft)
firefighting .... 316.8 kg/m^2 (64.88 lb/sq ft)
max airborne weight .... 366.1 kg/m^2 (74.99 lb/sq ft)
Max power loading:
cargo/passenger .... 285 kg/kN (2.80 lb/lb st)
firefighting .... 253 kg/kN (2.48 lb/lb st)
max airborne weight .... 292 kg/kN (2.86 lb/lb st)

PERFORMANCE (A: passenger version; B: business; C: cargo/passenger; D: cargo; E: firefighting):
Max Mach No. in level flight: A, B, C, D 0.69
Never-exceed speed (VNE):
A, B, C, D .... 329 kt (610 km/h; 379 mph) EAS
Max level speed at 7,000 m (22,960 ft):
A, B, C, D .... 388 kt (720 km/h; 447 mph)
Max cruising speed:
A, C, D at 8,000 m (26,240 ft) .... 378 kt (700 km/h; 435 mph)
B at 10,000 m (32,800 ft) .... 378 kt (700 km/h; 435 mph)
E at 3,000 m (9,840 ft) .... 383 kt (710 km/h; 441 mph)
Econ cruising speed:
A, C, D at 8,000 m (26,240 ft) .... 297 kt (550 km/h; 342 mph)
B at 10,000 m (32,800 ft) .... 297 kt (550 km/h; 342 mph)
E at 3,000 m (9,840 ft) .... 324 kt (600 km/h; 373 mph)
Stalling speed: flaps up .... 116 kt (215 km/h; 134 mph)
flaps down .... 84 kt (155 km/h; 97 mph)
Max rate of climb at S/L .... 840 m (2,755 ft)/min
Rate of climb at S/L, OEI .... 168 m (550 ft)/min
Service ceiling: A, B, C, D .... 11,000 m (36,080 ft)
E .... 8,000 m (26,240 ft)
Service ceiling, OEI .... 5,500 m (18,040 ft)
T-O to 11 m (35 ft):
on land:
A, B, C, D .... 950 m (3,120 ft)
E .... 700 m (2,300 ft)
on water: A, B, C, D .... 1,400 m (4,595 ft)
E .... 1,000 m (3,280 ft)
Landing from 15 m (50 ft):
on land:
A, B, C, D .... 800 m (2,625 ft)
E .... 950 m (3,120 ft)
on water: A, B, C, D .... 950 m (3,120 ft)
E .... 1,300 m (4,265 ft)
Balanced runway requirement: all .... 1,800 m (5,905 ft)
Water scooping distance to 15 m (50 ft) .... 1,450 m (4,760 ft)
Range:
A with 72 passengers .... 998 n miles (1,850 km; 1,149 miles)
C, D with 6,500 kg (14,330 lb) freight, 1 h reserves .... 998 n miles (1,850 km; 1,149 miles)
E, ferry, with 1 h reserves .... 1,943 n miles (3,600 km; 2,236 miles)
all, with max fuel .... 2,078 n miles (3,850 km; 2,392 miles)

OPERATIONAL NOISE LEVELS:
T-O .... 96 EPNdB
Climb .... 90 EPNdB
Landing .... 98 EPNdB

UPDATED

BERIEV BE-310

TYPE: Regional airliner.

PROGRAMME: First details released at Farnborough International, July 2004. No known plans to manufacture a prototype.

DESIGN FEATURES: Landplane version of Be-210, itself a passenger conversion of the Be-200 amphibian; retains deep fuselage associated with boat hull.

POWER PLANT: Two D-436 or BR715 turbofans, as for Be-200/210.

ACCOMMODATION: Between 74 and 102 passengers, according to internal configuration.

DIMENSIONS, EXTERNAL:
Wing span .... 32.78 m (107 ft 6½ in)
Wing aspect ratio 9.1
Length overall .... 32.05 m (105 ft 1¾ in)
Height overall .... 8.90 m (29 ft 2½ in)
Wheel track .... 4.30 m (14 ft 1¼ in)
Wheelbase .... 11.145 m (36 ft 6¾ in)

DIMENSIONS, INTERNAL:
Cabin: Length .... 18.00 m (59 ft 0¾ in)
Max width at floor .... 3.10 m (10 ft 2 in)
Max height .... 2.00 m (6 ft 6¾ in)

AREAS:
Wings, gross .... 117.44 m^2 (1,264.2 sq ft)

WEIGHTS AND LOADINGS:
Max T-O weight .... 42,000 kg (92,594 lb)

PERFORMANCE:
Max level speed at FL262 .... 405 kt (650 km/h; 466 mph)
Econ cruising speed at FL262 .... 337 kt (625 km/h; 388 mph)
Max certified ceiling .... 10,000 m (32,800 ft)
T-O run .... 950 m (3,120 ft)
Landing run .... 1,030 m (3,380 ft)
Range with max payload .... 1,133 n miles (2,100 km; 1,304 miles)

UPDATED

Artist's impression of Beriev Be-310 regional airliner ***NEW***/0589252

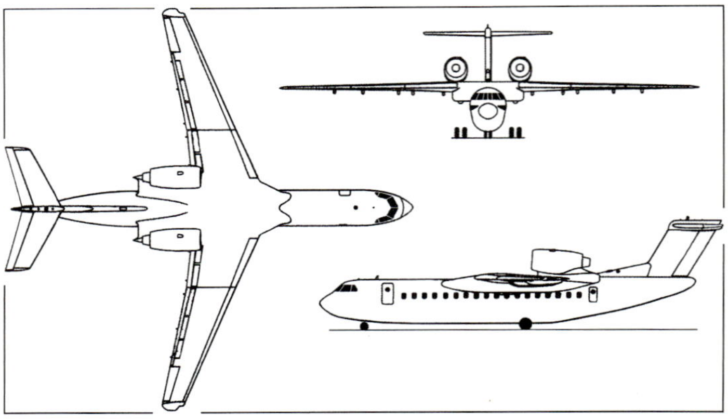

Beriev Be-310 twin-turbofan landplane ***NEW***/0589203

BERIEV (KNAAPO) SA-20P

TYPE: Six-seat amphibian.

PROGRAMME: Developed by KnAAPO at Komsomolsk as single-engined version of Beriev Be-103 Bekas for Russian Federation market. First aircraft under construction at KnAAPO by September 2001, when designation first announced. Prototype exhibited unflown at Gelendzhik Gidroavia Salon, September 2002. First flight 16 October 2002. Factory trials had been completed by April 2003; certification to Russian AP-23 and FAR Pt 23 due 2003, for first deliveries in 2004, although this timetable not met. Prototype transferred to KnAAPO Sports & Technical Club for experimental service.

Data generally as for Be-103, except that below.

COSTS: US$500,000 (2003).

DESIGN FEATURES: Single engine mounted in cabane above fuselage. Service life 12,000 hours or 15 years.

FLYING CONTROLS: Conventional horizontal tail surfaces (tailplane and separate elevator).

POWER PLANT: One 265 kW (355 hp) VOKBM M-14P nine-cylinder radial engine.

WEIGHTS AND LOADINGS:
Weight empty .... 1,600 kg (3,527 lb)
Useful load .... 634 kg (1,398 lb)
Max T-O weight .... 2,270 kg (5,004 lb)
Max wing loading .... 90.4 kg/m^2 (18.52 lb/sq ft)
Max power loading .... 8.58 kg/kW (14.10 lb/hp)

Beriev (KnAAPO) SA-20P single-engined amphibian ***NEW***/0576864

PERFORMANCE:
- Cruising speed .... 113 kt (210 km/h; 130 mph)
- Service ceiling .... 3,000 m (9,840 m)
- Range, 30 min reserves: with max fuel .... 556 n miles (1,030 km; 640 miles)
  - with max payload .... 277 n miles (513 km; 318 miles)
- T-O run: on land .... 260 m (855 ft)
  - on water .... 570 m (1,870 ft)

NEW ENTRY

CHERNOV

OPYTNYI KONSTRUKTORSKOYE BYURO CHERNOV B & M OOO (B & M CHERNOV EXPERIMENTAL DESIGN BUREAU JSC)

Tel: (+8 8462) 37 24 25
e-mail: chernovb@samaramail.ru
Web: http://www.samara.ru/~chernovb

The designs of Boris V Chernov have previously been associated with the No.1 Students' Design Bureau (SKB-1) at Samara. However, in August 1999, the Che-25 amphibian was shown at MAKS '99, credited to Chernov's own design office. The earlier Che-22 is now produced by Gidroplan, while the single-engine Che-23 has recently been completed. Other designs have been the Che-10 (1986), Che-15, Che-20 (1989) and Che-30 (1991), while Chernov was also associated with the Delta flex-wing and Albatros surface-skimmer. Work is currently in hand to design the Che-35 twin-engined flying-boat.

UPDATED

CHERNOV CHE-23

TYPE: Two-seat amphibian.

PROGRAMME: First flown 2003.

DESIGN FEATURES: Generally as Gidroplan Che-22.

FLYING CONTROLS: Manual.

STRUCTURE: Predominantly of metal.

LANDING GEAR: Optional amphibian configuration. Tailwheel/water rudder immediately to rear of second hull step.

POWER PLANT: One 47.8 kW (64.1 hp) Rotax 582 UL two-cylinder piston engine driving three-bladed propeller. Optionally, one 73.5 kW (98.6 hp) Rotax 912 ULS.

DIMENSIONS, EXTERNAL:
- Wing span .... 11.00 m (36 ft 1 in)
- Length overall .... 6.50 m (21 ft 4 in)
- Height overall .... 2.50 m (8 ft 2½ in)

WEIGHTS AND LOADINGS:
- Weight empty .... 295 kg (650 lb)
- Max T-O weight .... 570 kg (1,256 lb)

PERFORMANCE:
- Max level speed .... 81 kt (150 km/h; 93 mph)
- Normal cruising speed .... 65 kt (120 km/h; 75 mph)
- Stalling speed .... 30 kt (55 km/h; 35 mph)
- Max rate of climb at S/L .... 150 m (492 ft)/min

UPDATED

Che-30 prototype, a previous creation of Boris Chernov 0137984

Chernov Che-23 Rotax-powered amphibian ***NEW***/0583361

CHERNOV CHE-25

TYPE: Four-seat amphibian.

PROGRAMME: First flown 1995. Development of Che-22 Corvette, which is now produced by Gidroplan. Prototype (FLARF-17039), built by students' design bureau SKB-1 and first shown at Gelendzhik Hydro-aviation Show, Russia, September 1996. Che-25M shown at MAKS '99, Moscow, had redesigned empennage with raised tailplane and deletion of rudder's horn balance.

COSTS: US$44,000 (1997).

DESIGN FEATURES: Generally as Che-22. Operable on water in 0.8 m (30 in) swell.

FLYING CONTROLS: Manual, with full-span slotted flaperons. No tabs.

STRUCTURE: Single-spar wing of riveted duralumin; fuselage, tail unit and wingtip floats of vacuum-moulded GFRP sandwich. Wings braced by struts and wires; strut-braced tailplane. Airframe life 3,000 hours.

LANDING GEAR: Optional amphibian configuration; mainwheel legs, attached to fuselage sides, turn 90° forward for retraction; mainwheel tyre size 300×125; tailskid; cable brakes. Tailwheel/water rudder immediately to rear of second hull step.

POWER PLANT: Two 47.8 kW (64.1 hp) Rotax 582 UL-2V two-cylinder piston engines driving two-blade fixed-pitch propellers. Fuel capacity 130 litres (34.3 US gallons; 28.6 Imp gallons). Version, planned with Rotax 912 ULS of 73.5 kW (98.6 hp) and Sabaru EA-71.

ACCOMMODATION: Four persons; dual controls. Gull-wing door each side of cockpit.

DIMENSIONS, EXTERNAL:
- Wing span .... 12.60 m (41 ft 4 in)
- Wing chord, constant .... 1.40 m (4 ft 7¼ in)
- Wing aspect ratio 8.4
- Length overall .... 7.70 m (25 ft 3¼ in)
- Height overall (on wheels, fuselage horizontal) .... 2.50 m (8 ft 2½ in)

Che-25 with Rotax 912 powerplants ***NEW***/0583405

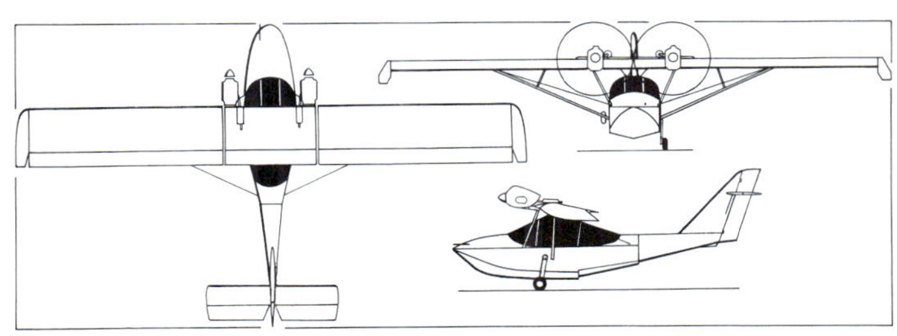

General arrangement of the Che-25 ***(Paul Jackson)*** 0084428

Che-25 four-seat amphibian ***(Yefim Gordon)*** ***NEW***/0552886

Tailplane span: 2.80 m (9 ft 2¼ in)
Wheel track (optional): 1.42 m (4 ft 8 in)
Propeller diameter: 1.84 m (6 ft 0½ in)
Distance between propeller centres: 1.84 m (6 ft 0½ in)

AREAS:
Wings, gross: 16.38 m^2 (176.3 sq ft)

WEIGHTS AND LOADINGS:
Weight empty: 590 kg (1,301 lb)
Max payload: 300 kg (661 lb)
Max T-O weight: 1,150 kg (2,535 lb)
Max wing loading: 70.2 kg/m^2 (14.38 lb/sq ft)
Max power loading: 12.03 kg/kW (19.77 lb/hp)

PERFORMANCE:
Max level speed: 97 kt (180 km/h; 112 mph)
Max cruising speed: 76 kt (140 km/h; 87 mph)
Unstick speed: 38 kt (70 km/h; 44 mph)
Max rate of climb at S/L: 300 m (984 ft)/min
Service ceiling: 4,000 m (13,120 ft)
T-O run: on land: 100 m (330 ft)
on water: 70 m (230 ft)
Range: 405 n miles (750 km; 466 miles)
g limits: +4/−2

UPDATED

Chernov Che-27 five-seat amphibian ***NEW***/0583356

CHERNOV CHE-27

TYPE: Five-seat amphibian.
PROGRAMME: First flown 2003.
DESIGN FEATURE: Practically indentical to Che-25M but with seating for five.
FLYING CONTROLS: Manual.
STRUCTURE: Predominantly of metal.
LANDING GEAR: Optional amphibian configuration. Tailwheel/water rudder immediately to rear of second hull step.
POWER PLANT: Two 47.8 kW (64.1 hp) Rotax 582 UL two-cylinder piston engines, each driving a three-blade propeller. Optionally, two 73.5 kW (98.6 hp) Rotax 912 ULS.

DIMENSIONS, EXTERNAL:
Wing span: 12.60 m (41 ft 4 in)
Length overall: 7.70 m (25 ft 3¼ in)
Height overall: 2.50 m (8 ft 2½ in)

WEIGHTS AND LOADINGS:
Weight empty: 590 kg (1,301 lb)
Max T-O weight: 1,150 kg (2,535 lb)

PERFORMANCE:
Max level speed: 97 kt (180 km/h; 112 mph)
Normal cruising speed: 76 kt (140 km/h; 87 mph)
Stalling speed: 38 kt (70 km/h; 44 mph)
Max rate of climb at S/L: 150 m (492 ft)/min
T-O run: on land: 100 m (330 ft)
on water: 70 m (230 ft)
Range: 405 n miles (750 km; 466 miles)
g limits: +4/−2

UPDATED

DUBNA (DMZ)

PROIZVODSTVENNO-TEKHNICHESKY KOMPLEKS DUBNENSKOGO MASHINOSTROITELNOGO ZAVOD AO (DUBNA MACHINE-BUILDING PLANT PRODUCTION TECHNICAL COMPLEX CO LTD)

ulitsa Zhukovskogo 2, 141980 Dubna, Moskovskaya oblast
Tel: (+7 09621) 515 21
Fax: (+7 09621) 232 99
e-mail: dmz@dubna.ru
Web: http://www.dmz.ru
GENERAL DIRECTOR: Aleksey Saushkin

Founded on 10 July 1939, the Dubna Plant is an enterprise of the General Department of Aircraft Industry of the State Committee on Defence Industries of the Russian Federation. In the 1960s, it launched airframe production of the MiG- 25 fighter and became a major producer of missiles for Eastern bloc and other countries. As part of its restructuring under the *Konversiya* policy, since reorganisation on current lines in 1994, it has produced components for the Sukhoi Su-29 two-seat aerobatic aircraft and developed the Dubna-1/2 series of light aircraft. Two Su-29 fuselages were built in 2000.

The single-seat Dubna-1 Shmel was simular to the current Osa. Dubna has an agreement to build light aircraft designed by MKB Raduga, whose precision-guided weapons are also manufactured by DMZ. By early 2004, there had been no further reports of the proposed Dubna-4 lightplane. The company had some 1,000 employees in early 2002 and is 60 per cent owned by the State. Other products include the AVP air cushion vehicle, satellite antennae, fabric treatment machines, cardboard box making machines and equipment for filling metal cans.

UPDATED

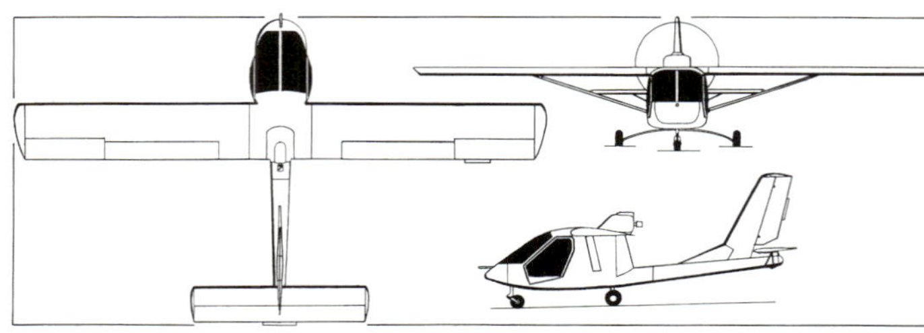

Dubna-2 Osa ***(Paul Jackson)*** 0092552

DUBNA-2 OSA

English name: Wasp
TYPE: Two-seat lightplane.
PROGRAMME: Designed by Taifun Experimental Design Bureau as DMZ-2; prototype (FLARF-01325) displayed at Moscow Air Show '95 and '97 with 59.6 kW (79.9 hp) Rotax 912 engine; promoted in China, 1996. By 1999, when definitive prototype FLARF-01268 shown at Moscow, detail changes included increased wheel track; repositioning of landing gear leg to below fuselage and of wing bracing strut to further forward; and Hirth engine, with additional fuel.

Production version of Dubna-2 Osa two-seat multipurpose light aircraft ***(Paul Jackson)*** 0092528

CUSTOMERS: Total 20 delivered by late 2001. Production of 12 planned for 2002, including four to United Arab Emirates and one each to Bulgaria and Greece; others to Saratov and Orenburg regional governments of Russia in agricultural spraying configuration. By early 2004, no confirmation that these deliveries had taken place, although it was reported in June 2004 that a follow-on batch of Dubna-2s was in the early stages of assembly.
COSTS: US$32,000 (2002).
DESIGN FEATURES: As Dubna-1, suitable for passenger, freight and mail transport, crop-spraying, forest surveillance, oil/gas pipeline and power line patrol, coastal fishery survey, aerial photography, *ab initio* pilot training and special missions.
FLYING CONTROLS: Conventional and manual. Flight-adjustable tab on elevator; ground-adjustable tabs on starboard aileron and rudder.
STRUCTURE: As Dubna-1.
LANDING GEAR: Tricycle type; fixed. Fuselage-mounted spring metal mainwheel legs; trailing-link nosewheel leg. Tyres size 400×150 mm on mainwheels, 310×135 mm on nosewheel. Floats and skis optional.
POWER PLANT: One 89.5 kW (120 hp) Hirth F30/4C flat-four driving two-blade pusher propeller. Fuel tank behind cabin, capacity 80 litres (21.1 US gallons; 17.6 Imp gallons).
ACCOMMODATION: Two seats side by side, with dual controls, in fully enclosed, extensively glazed cabin; forward-hinged window/door on each side; space for baggage/freight in compartment behind cabin.

DIMENSIONS, EXTERNAL:
Wing span: 9.92 m (32 ft 6½ in)
Wing aspect ratio: 9.0
Length overall: 5.97 m (19 ft 7in)
Width over doors: 1.25 m (4 ft 1¼ in)
Height overall: 2.91 m (9 ft 6½ in)
Tailplane span: 3.30 m (10 ft 10 in)
Wheel track: 2.40 m (7 ft 10½ in)
Wheelbase: 1.90 m (6 ft 2½ in)

AREAS:
Wings, gross: 10.92 m^2 (117.5 sq ft)

WEIGHTS AND LOADINGS:
Weight empty: 296 kg (653 lb)
Baggage capacity: 30 kg (66 lb)
Max T-O weight: 576 kg (1,270 lb)
Max wing loading: 52.7 kg/m^2 (10.81 lb/sq ft)
Max power loading: 9.60 kg/kW (15.88 lb/hp)

PERFORMANCE:
Max level speed: 89 kt (165 km/h; 102 mph)
Nominal cruising speed: 62 kt (115 km/h; 71 mph)
Landing speed: 41 kt (75 km/h; 47 mph)
Unstick speed: 46 kt (85 km/h; 53 mph)
Max rate of climb at S/L: 240 m (790 ft)/min

Service ceiling 4,000 m (13,120 ft)
T-O run 120 m (395 ft)
Max range 270 n miles (500 km; 310 miles)
g limits +4.0/–2.5

UPDATED

DUBNA-4

TYPE: Four-seat utility twin.

PROGRAMME: Designed by MKB Raduga; based on Dubna-2; completion of prototype planned for 2002, but no announcement known by early 2003. Optimised for use by Ministry of Emergency Situations and Forestry Service.

COSTS: Estimated US$55,000 to US$58,000 (2002).

ELITAR

VVV-AVIA

Volokolamskoye shosse 4, 125993 Moskva
Tel: (+7 095) 959 74 50
GENERAL DIRECTOR: Viktor A Tarhov

WORKS:
Samara

DESIGN BUREAU:
Ehlitar — 101 OOO, Korp 2, ulitsa B Cheremushkinskaya, 117448 Moskva
GENERAL DIRECTOR: Aleksey B Kondratev
MANAGER: Igor Bezrodnov

At the August 1999 MAKS air show, Ehlitar (which adopts the Western form, Elitar and was formed in 1997) displayed the newly flown IE-101 high wing lightplane and released initial details of the low-wing IE-201 Senator.

In addition to these light aircraft, Elitar was also developing at that time the E-301 and E-401 high-wing, twin-engined light transports, seating four to five and nine persons, respectively. A four-seat business 'triplane', the Premier, has been under design since 2000.

At Moscow in August 2003, Elitar unveiled the 202 and also displayed the previously known Sigma-4, formerly credited to the Albatross bureau. Company announcements have also mentioned the IE-203 floatplane project.

Members of the Elitar design and production team work for MiG, Molniya and Sukhoi design bureaux and have been responsible for the M-5 Oktyabr and M-9 Marafon ultralights.

Manufacturing of at least some Elitar designs was originally assigned to VVV-Avia, but by 2003 this company appeared to have formed a subsidiary at Samara, known as ZEVS-Avia, specifically to produce the Elitar-202. The earlier '101 and '201 seem no longer to be promoted.

UPDATED

Elitar-202 on show at MAKS '03 *(Sebastian Zacharias)* ***NEW***/0567691

Elitar Sigma-4 two-seat ultralight *(Robert Hewson)* ***NEW***/0567690

ELITAR-202

TYPE: Two-seat lightplane.

PROGRAMME: First shown at MAKS '03, Moscow, August 2003. Built by Samara ZEVS-Avia at Samara. Certification then expected in November 2003 but no evidence that this has occurred.

CUSTOMERS: Ten built and 12 more on order by August 2003; company foresees market for 600 aircraft in Russia and will produce up to 25 per month, depending on demand.

COSTS: US$60,000 (2004).

DESIGN FEATURES: Low-wing cabin monoplane of composites construction, optimised for high cruising speed and long range; constant-chord, cantilever mainplanes; sweptback fin with underfin.

LANDING GEAR: Tricycle type; fixed. Fuselage-mounted spring metal cantilever mainwheel legs with wheel fairings. Trailing-link nosewheel.

POWER PLANT: One 73.5 kW (98.6 hp) Rotax 912 ULS flat-four. Fuel capacity 110 litres (29.0 US gallons; 24.2 Imp gallons).

DIMENSIONS, EXTERNAL:
Wing span 8.30 m (27 ft 2¾ in)
Length overall 5.80 m (19 ft 0¼ in)

AREAS:
Wings area, gross 9.10 m^2 (98.0 sq ft)

WEIGHTS AND LOADINGS:
Weight empty 480 kg (1,058 lb)
Max T-O weight 750 kg (1,653 lb)
Max wing loading 82.4 kg/m^2 (16.88 lb/sq ft)
Max power loading 10.20 kg/kW (16.76 lb/hp)

PERFORMANCE:
Never-exceed speed (VNE) 146 kt (270 km/h; 168 mph)
Max level speed 113 kt (210 km/h; 130 mph)
Max cruising speed 108 kt (200 km/h; 124 mph)
Econ cruising speed 81 kt (150 km/h; 93 mph)
Stalling speed: flaps up 63 kt (115 km/h; 72 mph)
flaps down 57 kt (105 km/h; 66 mph)
Max rate of climb at S/L 210 m (689 ft)/min
T-O run 200 m (660 ft)
T-O to 15 m (50 ft) 620 m (2,035 ft)
Landing from 15 m (50 ft) 600 m (1,970 ft)
Landing run 180 m (590 ft)
Service ceiling 3,000 m (9,840 ft)
Range 351 n miles (650 km; 403 mph)
g limits +3.8/–1.9

UPDATED

ELITAR SIGMA-4

TYPE: Side-by-side ultralight.

PROGRAMME: Believed to have originated in Albatross design bureau. Marketed by Elitar from 2003.

CUSTOMERS: One aircraft noted with LII experimental department at Zhukovsky, Moscow region; another with Sebastopol Aero Club, Ukraine.

DESIGN FEATURES: Wing-and-boom configuration, with underslung crew pod braced to wings by single streamlined strut each side. Small underfin.

FLYING CONTROLS: Conventional and manual.

STRUCTURE: Generally of composites.

LANDING GEAR: Tricycle type; fixed. Fuselage-mounted spring cantilever mainwheel legs. Horizontal spring cantilever nosewheel leg with castoring wheel.

POWER PLANT: One 75.3 kW (98.6 hp) Rotax 912 ULS flat-four mounted above and ahead of the pod, driving a three-blade propeller. Fuel capacity 65 litres (17.2 US gallons; 14.3 Imp gallons).

DIMENSIONS, EXTERNAL:
Wing span 9.80 m (32 ft 1¾ in)

DIMENSIONS, INTERNAL:
Cabin max width 1.22 m (4 ft 0 in)

AREAS:
Wings, gross 11.00 m^2 (118.4 sq ft)

WEIGHTS AND LOADINGS:
Weight empty 320 kg (705 lb)
Max T-O weight 540 kg (1,190 lb)

PERFORMANCE:
Max level speed 113 kt (210 km/h; 130 mph)
Cruising speed 122 kt (170 km/h; 106 mph)
Stalling speed 50 kt (70 km/h; 44 mph)
Max rate of climb at S/L 360 m (1,181 ft)/min
T-O run 77 m (255 ft)
Range 378 n miles (700 km; 435 miles)

UPDATED

GIDROPLAN

GIDROPLAN OOO (HYDROPLANE LTD)

18 km, Moskovskoye Shosse, Litera A, A', Samara 443072
Tel: (+7 8462) 53 90 19
Fax: (+7 8462) 53 39 38
e-mail 1: hyd@mail.radiant.ru
e-mail 2: hydroplane@smrk.ru
Web: http://www.hydroplane.ru

EXECUTIVE DIRECTOR: Pavel Piotnitsa
CHIEF DESIGNER: Evgeny Ungerov

Gidroplan Ltd was formed in 1995 and has marketed amphibian lightplanes designed by the Chernov bureau.

This venture was placed on a firmer footing in late 2000 when Gidroplan invested the equivalent of US$500,000 in acquiring 3,500 m^2 (37,675 sq ft) of working area from the Samara Instrument Bearing Plant and installing production equipment. Manufacturing activities include the Tsikada (Cicada) twin-engined ultralight agricultural aircraft, some six of which were manufactured before production ceased (although it remains available to special order and was shown at Moscow in August 2003) and the Che-22 Corvette amphibian.

In recent years, designers associated with Gidroplan have produced flying prototypes of three lightplanes not promoted outside Russia. These are the **A-17** single-seat ultralight (260 kg; 573 lb MTOW, 22.4 kW; 30 hp engine, believed from SGAU); Chernov **Che-40** flying-boat (two Rotax 582 engines, briefly described in the 1999–2000 edition of *Jane's All the World's Aircraft* and now confirmed to have been built); and **Ul'tralayt** twin-tailboom ultralight (600 kg; 1,322 lb MTOW, Rotax 912 S) built in 1998 by E Yungerovym and S Seleznevym.

UPDATED

GIDROPLAN CHE-22 KORVET

English name: Corvette
US marketing name: Pelican

TYPE: Three-seat amphibian/kitbuilt.

PROGRAMME: Developed on basis of four earlier amateur-built, Vikhr-engined Che-20s, winners of many light aircraft competitions; design and construction of prototype started 1988, first flight 1989; construction of production aircraft started 1992, first flight 1993; more than 5,000 hours of test and training flights, and more than 120 amateur pilots trained on prototype and production aircraft in first five years. Promotion also undertaken by Refly of St Petersburg; launched as Refly Pelican at EAA AirVenture, Oshkosh, 2000, but no further US promotion known.

Upgraded version under development and certification during 2001 features cantilever wing, strengthened tail unit with increased rudder area, new control system, redesigned engine support and modified instrument panel; flight testing was completed at the end of 2001 and Russian certification was awarded on 26 December 2001, followed by JAR/FAR Pt 23 certification in early 2002.

CURRENT VERSIONS: **Corvette 503:** With two Rotax 503 UL two-cylinder two-stroke engines.

Corvette 582: With two Rotax 582 UL two-cylinder two-stroke engines.

Corvette 912: With one Rotax 912 UL four-cylinder four-stroke engine.

Corvette-L: Shown at Moscow in 2003. Miniature floats suspended from main wingtip floats.

Refly Pelican: Generally as for Corvette 582.

CUSTOMERS: First preproduction delivery to a forest control organisation 1990. One (Che-22R-2 c/n 006, RP-X1548 of OVOS WLL Co) sold in Philippines. Others to China, Spain and Vietnam. Total 60 built by February 2002, including 10 in 2000. First Pelican in US (N27NS, ex-RA-02777) registered in March 2001.

COSTS: Standard aircraft US$35,000 to US$55,000 (2002).

DESIGN FEATURES: Designed to JAR-VLA, FAR Pt 23 and AP-23 requirements, for maximum simplicity of design and construction; primarily for forestry applications, but suitable for patrol, inspection and ecological monitoring of hunting and fishing locations, border patrol, training and business operations. Quoted build time 1,500 hours, or 50 hours for fast build kit.

Braced parasol monoplane, with light central cabane and single main bracing strut and bracing wire each side; downturned wingtips serve as stabilising floats, but each has suspended miniature float to prevent wingtip contacting water at speed. Flying-boat hull with two steps; unswept strut-braced horizontal tail surfaces mounted on fin of sweptback vertical surfaces; engine(s) mounted on leading-edge of wing centre-section. US demonstrator has fin of reduced height. Unswept constant chord wing; CAHI (TsAGI) P-IIIA section, thickness/chord ratio 15 per cent; incidence 4° 30'; dihedral 1° 30'. Airframe life 3,000 flying hours.

FLYING CONTROLS: Conventional and manual, with pushrod actuation. Full-span slotted flaperons, elevators and horn-balanced rudder; fixed tab on port elevator.

STRUCTURE: Single-spar wing of riveted duralumin; fuselage, tail unit and integral wingtip floats of vacuum-moulded GFRP sandwich; flaperons fabric-covered.

LANDING GEAR: Optional amphibian configuration; spring cantilever mainwheel legs, attached to fuselage sides, turn 90° forward for retraction, tyre size 350×120; GFRP skis optional.

POWER PLANT: Two 37 kW (49.6 hp) Rotax 503 UL-2V or two 47.8 kW (64.1 hp) Rotax 582 UL or 59.6 kW (79.9 hp) Rotax 912 UL. Two-blade wooden or three-blade GFRP fixed-pitch propeller(s) of Russian manufacture. Two fuel tanks in wing centre-section, 100 litres (26.4 US gallons; 22.0 Imp gallons); gravity fuelling.

ACCOMMODATION: Pilot and two passengers under large glazed blister canopy, giving exceptional field of view. Detachable windscreen panels. Gull-wing canopy/door each side.

SYSTEMS: 12/24 V electrical system.

AVIONICS: *Comms:* VHF radios.

Flight: Garmin GPS.

EQUIPMENT: Optional BRS parachute.

DIMENSIONS, EXTERNAL:

Wing span	11.70 m (38 ft 4¾ in)
Wing chord, constant	1.40 m (4 ft 7¼ in)
Wing aspect ratio	8.3
Length overall	6.70 m (21 ft 11¾ in)
Height overall	2.20 m (7 ft 2¾ in)

Gidroplan Che-22 Korvet-L (Corvette-L), fitted with suspended hydroplane-type wingtip floats *(Sebastian Zacharias)* ***NEW***/0583379

Hydroplane Cicada (Gidroplan Tsikada) remains available to special order *(Sebastian Zacharias)* ***NEW***/0583466

Gidroplan Korvet displayed at Gelendzhik in August 2002 *(Yefim Gordon)* ***NEW***/0552879

Tailplane span	2.50 m (8 ft 2½ in)
Wheel track	1.60 m (5 ft 3 in)
Wheelbase	4.55 m (14 ft 11 in)
Propeller diameter	1.50 m (4 ft 11 in)
Canopy doors (each): Height	0.70 m (2 ft 3½ in)
Width	0.80 m (2 ft 7½ in)
Height to sill	0.63 m (2 ft 0¾ in)

DIMENSIONS, INTERNAL:

Cabin: Length	1.50 m (4 ft 11 in)
Max width	1.10 m (3 ft 7¼ in)
Max height	1.20 m (3 ft 11¼ in)

AREAS:

Wings, gross	16.40 m^2 (176.5 sq ft)
Flaperons (total)	3.40 m^2 (36.60 sq ft)
Fin	0.55 m^2 (5.92 sq ft)
Rudder	0.35 m^2 (3.77 sq ft)
Tailplane	1.68 m^2 (18.08 sq ft)
Elevators (total)	0.80 m^2 (8.61 sq ft)

WEIGHTS AND LOADINGS (A: Corvette 503, B: Corvette 582, C: Corvette 912):

Weight empty: A, B, C	360 to 480 kg (794 to 1,058 lb)
Max fuel: A, B, C	58 kg (128 lb)
Max T-O weight: A, B, C	675 kg (1,488 lb)
Max wing loading: A	21.95 kg/m^2 (4.50 lb/sq ft)
C	29.27 kg/m^2 (5.99 lb/sq ft)
Max power loading: A	4.86 kg/kW (7.99 lb/hp)
C	8.05 kg/kW (13.23 lb/hp)

PERFORMANCE:

Max level speed: A	81 kt (150 km/h; 93 mph)
B, C	86 kt (160 km/h; 99 mph)
Cruising speed: A	59 kt (110 km/h; 68 mph)
B, C	65 kt (120 km/h; 75 mph)
Max rate of climb at S/L: A	300 m (985 ft)/min
B	420 m (1,378 ft)/min
C	240 m (787 ft)/min
Service ceiling: A, B, C	3,000 m (9,840 ft)
T-O run: on land: A	90 m (295 ft)
B	70 m (230 ft)
C	100 m (328 ft)
on water: A	110 m (361 ft)
B	90 m (295 ft)
C	120 m (394 ft)
Range with standard fuel:	
A	232 n miles (430 km; 267 miles)
B	243 n miles (450 km; 279 miles)
C	324 n miles (600 km; 372 miles)
g limits	+3.8/−1.5
Permissible wave height: A, B, C	0.5 m (1 ft 7¾ in)

UPDATED

IAPO

IRKUTSKOYE AVIATSIONNOYE PROIZVODSTVENNOYE OBEDINENIE OAO (IRKUTSK AVIATION INDUSTRIAL ASSOCIATION JSC)

Renamed Irkut.

ILYUSHIN

MEZHDUNARODNYYI AVIATSIONNYI KOMPANIYA ILYUSHINA (ILYUSHIN INTERNATIONAL AVIATION COMPANY)

Incorporating aircraft design and production companies, Ilyushin MAK formed April 2000 by government decree. Initial members are Ilyushin Aviation Complex and VASO; following negotiations with Uzbekistan, will be expanded to include TAPO as second production plant; previous Russian and Uzbek government agreement of 1998 covered collaboration in developing and marketing the Il-76 and Il-114. These plans overtaken in May 2001 by wider national strategy of merging Ilyushin, Beriev, Mil, Sukhoi and Yakovlev.

AVIATSIONNYI KOMPLEKS IMENI S V ILYUSHINA OAO (Aviation Complex named for S V Ilyushin JSC)
Leningradsky prospekt 45G, 125190 Moskva
Tel: (+7 095) 157 35 73
Fax: (+7 095) 212 02 75
e-mail: ilyushin@online.ru
CHAIRMAN OF THE BOARD OF DIRECTORS AND GENERAL DESIGNER: Genrikh V Novozhilov
GENERAL DIRECTOR AND CEO: Victor V Livanov
HEAD OF INTERNATIONAL RELATIONS AND CHIEF DESIGNER: Igor Ya Katyrev
DEPUTY GENERAL DESIGNER, MARKETING, BUSINESS DEVELOPMENT AND FOREIGN ECONOMIC RELATIONS: Vladimir A Belyakov

Ilyushin OKB is named after Sergei Vladimirovich Ilyushin, who died 9 February 1977, aged 82. Bureau (OKB-240) was founded 17 April 1933; has been headed by Genrikh Novozhilov since 1970. About 60,000 aircraft of Ilyushin design have been built. Personnel in 2001 totalled about 3,500; by that time, activities were 90 per cent civil orientated.

Funding problems and bank collapses have affected Ilyushin and some of its suppliers, delaying the establishment of an Il-114 leasing arm and the acquisition of Western systems for the Il-96T, which was eventually abandoned. However, Bureau is jointly designing Il-214 twin-jet transport in collaboration with HAL of India (see International section). It is a partner, with Boeing and Sukhoi, in the RRJ regional jet programme announced in August 2001 and selected for production in 2003. Collaboration with Boeing began in 1997 and resulted in August 2001 opening of joint design training centre in Moscow to provide engineers, mainly from Ilyushin, for Boeing's design and technical research centres in Russian capital.

UPDATED

ILYUSHIN IL-76

Production of the Il-76 is, effectively, at an end, the Tashkent plant in Uzbekistan having a stock of some 40 uncompleted airframes with which to satisfy orders in the medium-term, including any for the stretched Il-76MF placed by the Russian Air Forces or its Il-76TF civil counterpart. Russian and Uzbekistani governments agreed joint marketing of Il-76 through Roseboronexport in March 2003.

Il-76MF prototype (first flown August 1995) underwent Russian military evaluation in first half of 2003. Tashkent's first-stage flight test report issued on 24 April 2003 prior to immediate transfer of aircraft to Akhtubinsk test centre, although trials there were curtailed, and not due to have resumed until July 2004. In late 2003, Ilyushin ordered a further two aircraft from Tashkent to be delivered to Russian Air Forces before end of 2005 for operational evaluation.

Associated requirement is transfer of production line from Uzbekistan to Russia (VASO plant at Voronezh), US$130 million cost of which may result in further delay. Some 70 MFs required by Russian Air Forces from 2006 onwards. Additional Chinese interest in 38 aircraft of which first batch of 20 provisionally agreed in September 2003.

Il-76 is also the basis of the Beriev A-50 AEW aircraft. Recent orders for similar aircraft are expected to involve conversion of existing Il-76/A-50s, one such being request by Israel Aircraft Industries in late 2002 for three to be converted with Phalcon AEW avionics for India.

Meanwhile, an initial batch of 10 to 14 existing Russian military Il-76MDs may be upgraded by VASO to Il-76MD-90 standard with PS-90A-76 turbofans replacing original D-30KPs, each upgrade costing US$14 million; two prototypes were being converted in late 2003/early 2004. Civil version is Il-76-90VD, indicating Volga-Dnepr cargo airline.

UPDATED

ILYUSHIN IL-78M

NATO reporting name: Midas

There have been no recent conversions of Il-76 to this airborne tanker configuration. However, India ordered six Il-78MKMs in February 2001 and these were converted from stocks of uncompleted Il-76s held at TAPO in Uzbekistan. First two were officially transferred to India in March 2003.

UPDATED

ILYUSHIN IL-96-300

TYPE: Wide-bodied airliner.

PROGRAMME: First of two flying prototypes (SSSR-96000) flew at Khodinka 28 September 1988 and exhibited at Paris in June 1989; second (SSSR-96001) first flew 28 November 1989; further airframe used for static and fatigue testing; all three built at GAZ 30, Khodinka; areas of commonality with Il-86 permitted planned test programme to be reduced to 750 flights totalling 1,200 hours; route proving trials by second production Il-96-300 (SSSR-96005) conducted late 1991; production at VAPO (now VASO), Voronezh; certification received 29 December 1992.

Following initially poor reliability, Aeroflot (ARIA) serviceability improved, as evidenced by TBO increase from 1,600 to 4,130 hours between 1997 and 1999. PS-90A engine in-flight shutdown had improved to one per 29,000 hours by 1999.

Westernised Il-96M/T versions developed during 1990s, but abandoned in mid-2001.

CURRENT VERSIONS: **Il-96-300:** *As described.*

First production Ilyushin Il-96-300 in service with Atlant Soyuz *(Yefim Gordon)* ***NEW***/0573403

Model of Ilyushin Il-96-400, showing Aviadvigatel NK-93 engines *(Sebastian Zacharias)* ***NEW***/0567711

Il-96-300V: Extended-range version on order by Vnukovo Airlines, 1999. Increased fuel for additional 1,200 miles (2,222 km; 1,380 miles).

Il-96PU: One aircraft (RA-96012) built for use of Russian president; VIP interior, additional communications facilities by OAO Relero of Omsk and medical centre. Second aircraft, designated Il-96PU(M), authorised by Russian government on 6 May 2000; this has different interior layout; first flown 20 April 2003 with registration RA-96016; delivered to Rossiya 14 May 2003 for internal fitments by UK company Diamonite Aircraft Furnishing; intended 30 August 2003 delivery date as replacement 'Air Force One'.

Il-96-300 Freighter: Design under way by early 2000; development partnership also includes VAPO and Russian banks. To carry 70,000 kg (154,325 lb) payload over 4,859 n miles (9,000 km; 5,592 miles); uprated PS-90 engines each 176.5 kN (39,683 lb st). Estimated unit cost US$30 million (2000).

Il-96-400T: Under development by mid-2000 (when designation Il-96-3XX briefly employed). Equivalent to Il-96M/T, but employing only Russian engines and avionics. Power plant envisaged as Aviadvigatel PS-90A modified with additional 14.67 kN (3,300 lb st) to allow full range at MTOW of 270,000 kg (595,250 lb), including 92,000 kg (202,825 lb) of cargo. Also shown in model form with 177 kN (39,683 lb st) Samara NK-93 turbofans. However, details published in August 2003 quote only standard 156.9 kN (35,275 lb st) PS-90As and 68,000 kg (149,915 lb) payload (386 to 436 passengers) and range of 3,077 n miles (5,700 km; 3,541 miles) with max payload or 5,237 n miles (9,700 km; 6,027 miles) with maximum passengers, taking off in 2,830 m (9,285 ft) and cruising at 469 kt (870 km/h; 540 mph) at FL430. Proposed to convert 10 uncompleted Il-96Ms to this standard for lease from Ilyushin Finance by Atlant Soyuz cargo airline; provisional agreement signed February 2001, envisaging 2003 service-entry; however, retrospective installation of cargo door could weaken airframe; Soyuz signed order for four —400Ts in August 2001, first being former Il-96T prototype, due in service early 2002; second — 400 to follow 24 months thereafter. Also in 2001, Aeroflot issued letter of intent for four— 400Ts to be delivered from 2006, and was considering passenger version. Volga-Dnepr reported in 2003 to have interest in further two.

Il-98: Envisaged in late 1990s as Il-96 version with two Western engines (PW4000, RR Trent or GE90). By 2004, was being depicted with Samara NK-93s.

CUSTOMERS: Aeroflot Russian International Airlines (six delivered by 1996); Domodedovo Airlines (four on order, of which third was delivered 16 April 1999); Russia State Transport Company (three; one delivered 1995; one in 2003; third completed, but not delivered); and Atlant Soyuz (received first production aircraft in 1999, following trials use). Total 12 production aircraft delivered: 10th in February 1996; 11th in April 1999; 12th in 2003.

Aeroflot signed lease agreement with Ilyushin Finance in November 1999 for further six -300s, plus sale, refurbishment and leaseback of one of original six; contract involves uncompleted aircraft on production line, permitting deliveries between 2001 and 2003. Target was not achieved but order remained in place, despite Aeroflot commitment to Il-96-400T. In March 2003, it was announced that VASO had received from Ilyushin

Projected Ilyushin Il-98, powered by two Samara NK-93 turbofans ***NEW***/0589286

Finance Company an initial payment of Rb180 million to fund restart of work on the eight Il-96s then on the assembly line, these comprising six for Aeroflot and two freighters for Atlant-Soyuz. Deliveries to begin 18 months later and proceed at two-monthly intervals. Aeroflot signed leasing agreement with Ilyushin Finance on 30 December 2003 for six to be delivered over a six-month period, beginning mid-2005. China Xinjiang Airlines requires three Il-96-300s and was continuing negotiations in early 2002; Kras Air of Russia revealed requirement in late 2001.

COSTS: Il-96-300 US$39 million (2003). Il-96-400T US$38 million, flyaway (2001).

DESIGN FEATURES: Superficial resemblance to Il-86, but new design, with different engines to overcome performance deficiencies of original Il-86; new structural materials and state-of-the-art technology intended to provide life of 60,000 hours and 12,000 landings; no lower deck passenger entry. Current development aiming at range of 6,475 n miles (12,000 km; 7,450 miles) with 300 passengers.

Conventional wide-bodied airliner, with low wing (and winglets) and four pod-mounted engines.

Supercritical wings, with 30° sweep at quarter-chord; sweepback at quarter-chord 37° 30' on tailplane, 45° on fin.

FLYING CONTROLS: Triplex fly-by-wire, with manual reversion; each wing trailing-edge occupied by, from root, double-slotted inboard flap, small inboard aileron, two-section single-slotted flaps, and outboard aileron used only as gust damper and to smooth out buffeting; seven-section full-span leading-edge slats on each wing; three airbrakes forward of each inboard trailing-edge flap; six spoilers forward of outer flaps; inboard pair supplement ailerons, others operate as airbrakes and supplementary ailerons; variable incidence tailplane; two-section rudder and elevators, without tabs.

STRUCTURE: Basically all-metal, including new high-purity aluminium alloy, with composites flaps, maindeck floors and underfloor holds of honeycomb and CFRP; inner wings three-spar, outer panels two-spar; each wing has seven machined skin panels, three top surface, four bottom, with integral stiffeners; circular-section semi-monocoque fuselage; leading- and trailing-edges of fin and tailplane of composites. Some components manufactured by PZL (Polish Aircraft Factory), Poland.

LANDING GEAR: Retractable four-unit type. Forward-retracting steerable twin-wheel nose unit; three four-wheel bogie main units. Two of latter retract inward into wingroot/fuselage fairings; third is mounted centrally under fuselage, to rear of others, and retracts forward after the bogie has itself pivoted upward 20°. Oleo-pneumatic shock-absorbers. Nosewheel tubeless tyres size 1,260×460; mainwheel tubeless tyres size 1,300×480. Tyre pressure (all) 11.65 bar (169 lb/sq in).

POWER PLANT: Four Aviadvigatel PS-90A turbofans, each 156.9 kN (35,275 lb st), on pylons forward of wing leading-edges. Thrust reversal standard. Integral fuel tanks in wings and fuselage centre-section, total capacity 148,260 litres (39,166 US gallons; 32,613 Imp gallons).

ACCOMMODATION: Pilot, co-pilot and flight engineer; two seats for supplementary crew or observer. Ten or 12 cabin staff. Basic all-tourist configuration has two cabins for 66 and 234 passengers respectively, nine-abreast at 87 cm (34.25 in) seat pitch, separated by buffet counter, video stowage and two lifts from galley on lower deck. Two aisles, each 55 cm (21.65 in) wide. Two lavatories and wardrobe at front; six more lavatories, a rack for cabin staff's belongings and seats for cabin staff at rear. Seats recline, and are provided with individual tables, ventilation, earphones and attendant call button. Indirect lighting is standard. 235-seat mixed class version has front cabin for 22 first class passengers, six-abreast in pairs, at 102 cm (40 in) seat pitch and with aisles 75.5 cm (29.7 in) wide; centre cabin with 40 business class seats, eight-abreast at 90 cm (35.4 in) seat pitch and with aisles 56.5 cm (22.25 in) wide; and rear cabin for 173 tourist class passengers, basically nine-abreast at 87 cm (34.25 in) seat pitch, with aisle width of 55 cm (21.65 in).

Future Aeroflot standard, defined late 1999, is 12 first class, 21 business class and 180 economy seats, plus complete interior renewal. Refurbishment cost (including some Western avionics) US$1.5 million per aircraft.

Passenger cabin is entered through three doors on port side of upper deck, at front and rear and forward of the wings. Opposite each door, on starboard side, is emergency exit door. Lower deck houses front cargo compartment for six ABK-1.5 (LD3) containers or igloo pallets, central compartment aft of wing for 10 ABK-1.5 containers or pallets, and tapering compartment for general cargo at rear. Three doors on starboard side provide separate access to each compartment. Galley and lifts are between front cargo compartment and wing, with separate door aft of front cargo compartment door.

SYSTEMS: Four independent hydraulic systems, using fireproof and explosion-proof fluid, at pressure of 207 bar (3,000 lb/ sq in). APU in tailcone for engine starting and air conditioning on ground.

AVIONICS: *Flight:* Triplex flight control and flight management systems, together with a head-up display, permit fully automatic en route control and operations in ICAO Cat. IIIa minima. Duplex engine and systems monitoring and failure warning systems feed in-flight information to both the flight engineer's station and monitors on the ground. Autothrottle is based on IAS, without angle of attack protection.

Instrumentation: Primary flight information is presented on dual twin-screen colour CRTs, fed by triplex INS, a satellite-based and Omega navigation system and other sensors. Another electronic system provides real-time automatic weight and CG situation data.

DIMENSIONS, EXTERNAL:
Wing span: excl winglets 57.66 m (189 ft 2 in)
over winglets 60.11 m (197 ft 2½ in)
Wing aspect ratio 9.5
Length overall 55.35 m (181 ft 7¼ in)
Fuselage: Length 51.15 m (167 ft 9¾ in)
Max diameter 6.08 m (19 ft 11½ in)
Height overall 17.55 m (57 ft 7 in)
Tailplane span 20.57 m (67 ft 6 in)
Wheel track 10.40 m (34 ft 1½ in)
Wheelbase 20.07 m (65 ft 10 in)
Passenger doors (three): Height 1.83 m (6 ft 0 in)
Width 1.07 m (3 ft 6 in)
Height to sill: Nos. 1 and 2 4.54 m (14 ft 10¾ in)
No. 3 4.80 m (15 ft 9 in)
Emergency exit doors (three):
Height 1.825 m (5 ft 11¾ in)
Width 1.07 m (3 ft 6 in)
Cargo compartment doors (front and centre):
Height 1.825 m (5 ft 11¾ in)
Width 1.78 m (5 ft 10 in)
Height to sill: front 2.34 m (7 ft 8¼ in)
centre 2.48 m (8 ft 1¾ in)
Cargo compartment door (rear):
Height 1.38 m (4 ft 6¼ in)
Width 0.97 m (3 ft 2¼ in)
Height to sill 2.74 m (9 ft 0 in)
Galley door: Height 1.20 m (3 ft 11¼ in)
Width 0.80 m (2 ft 7½ in)

DIMENSIONS, INTERNAL:
Cabins, excl flight deck: Height 2.60 m (8 ft 6¼ in)
Max width Approx 5.70 m (18 ft 8½ in)
Volume 350.0 m^3 (12,360 cu ft)
Cargo hold volume: front 37.1 m^3 (1,310 cu ft)
centre 63.8 m^3 (2,253 cu ft)
rear 15.0 m^3 (530 cu ft)

AREAS:
Wings, gross 391.60 m^2 (4,215.1 sq ft)
Vertical tail surfaces (total) 61.00 m^2 (656.60 sq ft)
Horizontal tail surfaces (total) 96.50 m^2 (1,038.75 sq ft)

WEIGHTS AND LOADINGS:
Basic operating weight 117,000 kg (257,950 lb)
Max payload 40,000 kg (88,185 lb)
Max fuel 114,900 kg (253,310 lb)
Max T-O weight 216,000 kg (476,200 lb)
Max landing weight 175,000 kg (385,800 lb)
Max zero-fuel weight 157,000 kg (346,125 lb)
Max wing loading 551.6 kg/m^2 (112.97 lb/sq ft)
Max power loading 344 kg/kN (3.37 lb/lb st)

PERFORMANCE (estimated):
Normal cruising speed at FL330-FL400 459–486 kt (850–900 km/h; 528–559 mph)
Approach speed 140–146 kt (260–270 km/h; 162–168 mph)
Balanced T-O runway length 2,600 m (8,530 ft)
Balanced landing runway length 1,980 m (6,500 ft)
Range, with UASA reserves: with max payload ... 4,050 n miles (7,500 km; 4,660 miles)
with 30,000 kg (66,140 lb) payload 4,860 n miles (9,000 km; 5,590 miles)
with 15,000 kg (33,070 lb) payload 5,940 n miles (11,000 km; 6,835 miles)

OPERATIONAL NOISE LEVELS: Il-96-300 is designed to conform with ICAO Chapter 3 Annex 16 noise requirements

UPDATED

ILYUSHIN IL-100

TYPE: Utility turboprop twin.

PROGRAMME: Announced February 2000, when in basic design stage. Draft design complete by October 2000. First flight due two years after go-ahead, which conditional upon 50 firm commitments. Manufacture of production aircraft at Lukhovitsy (LMZ). Estimated sales minimum of 200; potential 1,200 over 10 years. Following positive response from potential market, Ilyushin began search in late 2001 for launch customer to assist with costs of development. Nothing further had been heard of the aircraft by the end of 2003.

CURRENT VERSIONS: **Il-100:** Baseline version; 12 seats.

Il-100A: Stretched version; 18–20 seats.

COSTS: Development estimated as US$20 million to US$30 million. Unit cost US$1 million (2000).

DESIGN FEATURES: Replacement for Antonov An-2 utility biplane, but with 40 per cent reduction in direct operating cost. Accordingly, stated to be "in Cessna Caravan I class". Rugged, unpressurised and simple to maintain; STOL performance. High-wing, T-tail configuration with unswept rear fuselage, but only port side freight door. Airframe life 60,000 cycles. Choice of Russian (Kuznetsov NK-123) or Western (RR 250 or PT6) turboprop engines.

WEIGHTS AND LOADINGS:
Max payload 1,500 kg (3,307 lb)

PERFORMANCE (estimated):
Cruising speed 200 kt (370 km/h; 230 mph)
T-O run 400 m (1,315 ft)
Landing run 260 m (855 ft)

UPDATED

ILYUSHIN IL-103

TYPE: Four-seat lightplane.

PROGRAMME: Exhibited in model form at Moscow Aerospace '90; programme go-ahead 1990; first flight (RA-10321) 17 May 1994; second prototype (RA-10302) flew 30 January 1995 at LMZ (now LAPIK), Lukhovitsy, following August 1993 decision to establish a production line there; pre-series batch comprised three flyable and two static test aircraft, former having flown 414 sorties in two years after maiden flight; first production aircraft flew 30 January 1995.

Type certificate by Russian authorities under AP-23 received 15 February 1996; production approval certificates issued 7 March 1996 and 7 April 1998; airworthiness certificate granted 9 July 1997. Amended Russian certification granted 4 December 1997, following addition of GPS, radio compass and AHRS to avionics suite.

FAR Pt 23 certification achieved 9 December 1998 (first by a Russian-built aircraft) after 110-hour/208-sortie programme, but with 1,150 kg (2,535 lb) MTOW limit. Annex of 6 September 1999 allows operation from unpaved runways. FAR MTOW increased to 1,310 kg (2,888 lb) on 13 September 2002 (c/n 0501 onwards); retrofit available for earlier aircraft.

International marketing by MGA Ltd of Moscow (www.mga.ru). Initial export order, six for Peruvian Air Force, agreed June 1999 and finalised in April 2000; deliveries had not begun by August 2000, due to lack of payment, but all reported in service by late 2001.

CURRENT VERSIONS: **Il-103-01:** Baseline VFR version for Russian Federation market.

Il-103-10: Export version with fully upgraded avionics suitable for international airways navigation.

Ilyushin Il-103 wearing the name of General Designer Genrikh Novozhilov *(Paul Jackson)* **NEW**/0547746

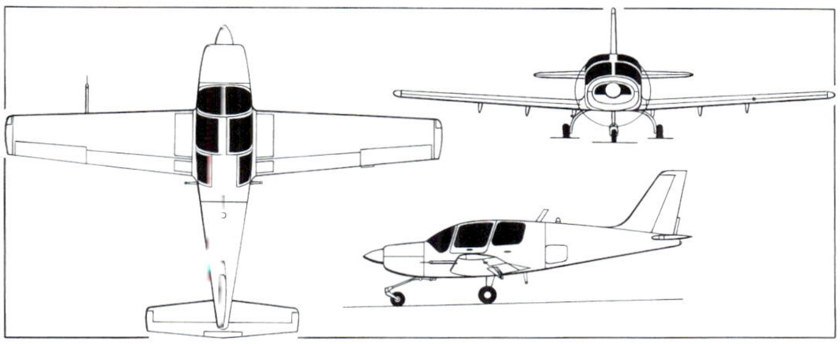

Ilyushin Il-103 two/five-seat multipurpose light aircraft *(James Goulding)* 0079319

Il-103-11: Export version with partly upgraded avionics suitable for local air navigation. Initial certification is for a max T-O weight of 1,285 kg (2,832 lb), as reflected in weight and performance data below. The Il-103 will later be certified at 1,310 kg (2,880 lb) and, ultimately, 1,460 kg (3,218 lb).

Il-103LL: Flying testbed (*Letny Laboratoriya*) fitted in mid-2003 with Russkaya Avionika LCD navigation display and imported inertial platform.

Il-103SKh: Crop-sprayer (*selskokhozyaistvenny:* agricultural) version; first flown 29 March 2000.

Il-103P: Surveillance version announced June 2002 in response to interest from Kazakhstan for Caspian Sea patrol. Undisclosed sensors have range of 40 km (64 miles) from cruising height of 3,000 m (9,840 ft). Potential for armament, including two machine guns.

CUSTOMERS: Initial operators (by 1999) included Il-Service at Myachkovo, Cherepovets Aviation Club, Avialesookhrana (forestry), Special Ecological Aviation Centre, Tatarstan National Flying Club, Civil Aviation Academy and St Petersburg Methodological Centre. Sales in 2000 comprised four to GP Bellesavia of Belarus in September for forest firefighting; one to Il-Service; and one for Vladimirsky Aerial Forest Protection. Despite prediction of Russian market for 1,500 and plans to produce 100 per year, sales stagnated in 2001; manufacture by LAPIK was reported, in mid-2000, temporarily to have ceased.

Avialine requires fleet of 12 for air taxi service from Moscow-Domodedovo. Federal Fire Service interested in 30. Uzbekistan government announced requirement for 100 in crop-spraying role, late 1999. Russian Federal Air Transport service reportedly ordered 30 in 1999 for issue to several flying schools, but these failed to materialise. Total 23 delivered by January 2002, including 10 exports (Belarus, four, and Peru, six). Further sales in 2002 included 23 to South Korean Air Force (signed 16 May 2002) and three to Laos. Korean deliveries then due to begin in September 2003, but Laotian funding was still awaited in late 2003.

COSTS: US$164,000 to 210,000 (2000).

DESIGN FEATURES: Conventional low-wing monoplane, with non-retractable landing gear, originally to meet DOSAAF requirement for 500 military/civil pilot trainers. Designed for daytime VFR flying, in non-icing conditions and ambient air temperature of −35 to +45°C; intended service life 12,000 hours/20,000 landings/15 years.

Wing sweep 0° at quarter-chord, slight dihedral from roots, twist 0°, thickness/chord ratio 16 per cent at root, 15 per cent at tip, taper ratio 1.9.

FLYING CONTROLS: Conventional and manual. Actuation by pushrods, except cable-actuated rudder; horn-balanced rudder and single-piece elevator; single-slotted trailing-edge flaps, 0° or 10° deflection; electrically actuated elevator trim tab. Deflection angles: ailerons +25/−20°, elevator +25/−20°, rudder ±25°. Ground-adjustable trim tabs on starboard aileron and rudder.

STRUCTURE: All-metal basically aluminium alloy, except for titanium firewall frame and wingroot attachments; bonded GFRP wingtips, elevator and rudder tips and elevator tab; and small amounts of magnesium alloys and Al-Li alloys. Single-spar wings, with front and rear false spars, integral fuel tanks and detachable leading-edge, mounted at sides of fuselage. Main spar is riveted beam with extruded caps; false spars stamped from sheet alloy; rolled sheet stringers. Semi-monocoque front/centre fuselage with inbuilt wing carry-through structure; separate rear fuselage; separate tailcone. Longitudinal fuselage members comprise two reinforced spars in cabin floor, two side-ribs of centre wing section and three beams for landing gear attachment. Two-spar, single-piece tailplane attached to fuselage by four bolts; two-spar fin. Detachable wings, fin and tailplane.

LANDING GEAR: Non-retractable tricycle type, with single wheel on each unit. Cantilever spring nose and mainwheel legs of titanium alloy; castoring nosewheel with shimmy damper; oleo-pneumatic nosewheel leg tested on RA-10300 in 1998. Mainwheel tyres size 400×150-115, pressure 3.95 bar (57 lb/sq in) on KT-214-1 wheels; nosewheel tyre size 310×135-99, pressure 3.43 bar (50 lb/sq in) on K-290 wheels. Optional tyres 6.00-6 and 5.00-5, respectively, on Parker 40-75B and 40-77B wheels. Multidisc hydraulic anti-lock brakes on mainwheels, pedal-actuated. Turning radius (outboard wheel) 4.70 m (15 ft 6 in). Floats and skis optional.

POWER PLANT: One 157 kW (210 hp) Teledyne Continental IO-360-ES2B flat-six engine; Hartzell BHC-C2YF-1BF/F8459A-8R two-blade, constant-speed (hydraulic), metal propeller. (Manual pitch control on three development aircraft; automatic pitch selection on production aircraft.) Alternative 194 kW (260 hp) Textron Lycoming or similarly rated Teledyne Continental engine under consideration. Fuel and oil systems suitable for inverted flight; two main fuel tanks in wingroots, total capacity 200 litres (52.8 US gallons; 44.0 Imp gallons); supply tank, capacity 3 litres (0.8 US gallon; 0.66 Imp gallon), in fuselage forward of wing front spar carry-through; gravity refuelling points in wingroots.

ACCOMMODATION: Two forward-folding seats side by side at front of cabin; bench seat for two adults or three children at rear; optional control wheel in place of standard stick; optional dual controls; optional front seats for parachutes; space for 220 kg (485 lb) freight with rear bench seat removed; two gull-wing window/doors, hinged on centreline, at front of canopy. Rear windows removable for ground emergency exit. Unrestricted access to baggage hold.

SYSTEMS: Cabin ventilated and heated and windscreen demisted electrically by fan heater. Electrical system 27 V DC, with 1,800 W 60 A generator and 25 Ah battery.

AVIONICS: *Comms:* Yurok VHF radio as standard; P-855A1 optional extra. Il-103-11 equipment includes UBD transponder. Il-103-10 has Bendix/King KX 165 nav/com/glideslope (replacing Russian radios) and KT 76A transponder.

Flight: BUR-4 flight data recorder.

Il-103-11 includes MKS-1 compass and Bendix/King KR 87 ADF and KLN 89B GPS. Il-103-10 additionally equipped with VOR/ILS, KN 63 DME, KMA 24 audio control/MKR, KCS 55A compass and encoding altimeter.

DIMENSIONS, EXTERNAL:

Wing span	10.56 m (34 ft 7¾ in)
Wing aspect ratio	7.6
Wing chord: at root	1.825 m (5 ft 11¾ in)
at tip	0.96 m (3 ft 1¾ in)
Length overall	8.00 m (26 ft 3 in)
Fuselage: Length	7.81 m (25 ft 7½ in)
Max height	1.42 m (4 ft 8 in)
Max width	1.40 m (4 ft 7 in)
Height overall	3.135 m (10 ft 3½ in)
Tailplane span	3.90 m (12 ft 9½ in)
Wheel track	2.405 m (7 ft 10¾ in)
Wheelbase	2.045 m (6 ft 8½ in)
Propeller diameter	1.93 m (6 ft 4 in)
Baggage door: Width	0.70 m (2 ft 3½ in)
Height	0.34 m (1 ft 1¼ in)

DIMENSIONS, INTERNAL:

Cabin: Length	2.65 m (8 ft 8¼ in)
Max height	1.27 m (4 ft 2 in)
Max width	1.30 m (4 ft 3 in)

AREAS:

Wings, gross	14.71 m^2 (158.4 sq ft)
Ailerons (total)	1.137 m^2 (12.24 sq ft)
Flaps (total)	2.204 m^2 (23.72 sq ft)
Fin	0.84 m^2 (9.04 sq ft)
Rudder	0.56 m^2 (6.03 sq ft)
Tailplane	1.48 m^2 (15.93 sq ft)
Elevator	1.56 m^2 (16.79 sq ft)

WEIGHTS AND LOADINGS:

Weight empty	900 kg (1,984 lb)
Baggage capacity	60 kg (132 lb)
Max payload	270 kg (595 lb)
Max fuel	150 kg (330 lb)
Max T-O weight	1,310 kg (2,888 lb)
Max wing loading	89.1 kg/m^2 (18.24 lb/sq ft)
Max power loading	8.37 kg/kW (13.75 lb/hp)

PERFORMANCE:

Never-exceed speed (VNE)	183 kt (340 km/h; 211 mph)
Max level speed	119 kt (220 km/h; 137 mph)
Cruising speed	97 kt (180 km/h; 112 mph)
Stalling speed: flaps up	64 kt (117 km/h; 73 mph)
10° flap	60 kt (111 km/h; 69 mph)
Max rate of climb at S/L	190 m (623 ft)/min
Max certified altitude	3,000 m (9,840 ft)
T-O run	380 m (1,250 ft)
Landing run	320 m (1,050 ft)
Max range at cruising speed, pilot and 270 kg (595 lb) payload, 30 min reserves	432 n miles (800 km; 497 miles)
g limits: Utility	+4.4/−1.8
Aerobatic	+6/−3

OPERATIONAL NOISE LEVELS: Designed to conform with GOST 23023-85 and ICAO Annex 16

UPDATED

ILYUSHIN IL-112

TYPE: Twin-turboprop transport.

PROGRAMME: Initiated 1994, with possibility of Russian government funding as alternative to Ukrainian Antonov An-38; design since enlarged and considerably refined; planned manufacture by KAPP at Kumertau and (Il-112V) VASO at Voronezh.

Revealed July 1999 that Il-112V selected by Russian defence ministry as potential replacement for Antonov An-24/26, Yak-40 and Let L 410; significant performance improvements to be achieved by redesign begun in same month. Presidential approval granted April 2000. First flight originally due in 2002, but revised timetable called only for final decision in that year between Il-112 and competitors. In April 2003, Il-112VT confirmed by Russian Air Forces selection committee as competition winner, overcoming

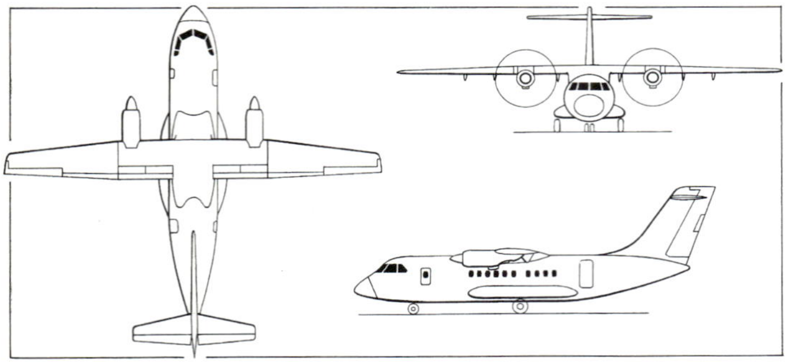

Ilyushin Il-112 twin-turboprop short-range passenger transport as originally envisaged *(James Goulding)* 0092139

Illustration of Ilyushin Il-112V in 2004 showed a deeper, shorter fuselage than depicted previously ***NEW***/0589285

opposition from Sukhoi Su-80, MiG-110, Myasishchev M-60 and Tupolev Tu-136. Four-year development programme launched following 17 November 2003 approval by Air Forces C-in-C of aircraft's technical specification. This called for preliminary design to be complete by early 2004, full scale development to begin in mid-2004 and first Il-112VT to fly in 2006 at VASO plant.

CURRENT VERSIONS: **Il-112:** Passenger transport, *as described in detail.*

Il-112 business transport: Basically as Il-112 airline version, but cabin divided into three sections, separated by curtains. Front cabin contains four pairs of facing seats, with tables between, and centre aisle, a wardrobe on the port side between the door and flight deck bulkhead, inward-opening door to flight deck, and large baggage compartment on starboard side with outside access.

Central compartment has four armchairs on port side, with tables between, three-seat sofa on starboard side to rear of cabinets; aisle width between armchairs and sofa 86 cm (34 in); wardrobe at rear on each side. Rear section contains main passenger door with airstairs on port side, two buffets, seat for attendant, and access to rear baggage compartment and lavatory.

Il-112T: Freighter with rear-loading ramp; lavatory and seat for loadmaster at front of unobstructed main hold; ramp forms rear wall of hold when stowed. Typical payloads include four 1AK-1.5 (LD3) or five 3AK-1 containers or five PA-1.5 pallets. Airstair door at front of hold on port side; service door opposite; emergency exit on each side in centre of hold.

Il-112VT *(voenny transportnny: military transport)*: Passenger/freighter for Russian armed forces. Rear ramp; provision for 35 passengers, or 18 stretchers or 34 paratroops. Patrol/surveillance version also to be produced. Detail design under way by early 2001; selected for production in April 2003.

CUSTOMERS: Requirement by Russian Air Forces could total 100 to 120; estimated potential for 200 civil sales.

DESIGN FEATURES: Based on Il-114 airframe. Conventional high-wing monoplane with T tail; constant-chord wing centre-section, tapered outer panels; sweptback fin and rudder; circular-section pressurised fuselage.

FLYING CONTROLS: Entire trailing-edge of each wing made up of horn-balanced aileron and two-section, area-increasing flaps; two-section spoilers forward of flap on each side of centre-section. Horn-balanced rudder and elevators; tab in rudder.

LANDING GEAR: Retractable tricycle type; twin-wheel nose unit; single wheel on each main unit, retracting into large fairing outside fuselage pressure cell.

POWER PLANT: Two 1,838 kW (2,466 shp) Klimov TV7-117SM turboprops in Il-112VT.

ACCOMMODATION: Flight crew of two; standard seating for 40 passengers, with 11 pairs of seats on port side and nine on starboard side of 45 cm (17.7 in) wide centre aisle. Maximum 60 in military version. Baggage/freight compartment with outside access on starboard side at front of cabin. Main airstair passenger door at rear of cabin on port side, with service door opposite; emergency exits at front of cabin on port side, in centre on starboard side. Access to rear baggage compartment and lavatory at rear of cabin; buffet and seat for attendant.

SYSTEMS: Aerosila TA-14 APU.

DIMENSIONS, EXTERNAL:
- Wing span: except Il-112V ... 22.55 m (73 ft 11¾ in)
  - Il-112V ... 23.45 m (76 ft 11¼ in)
- Length: except Il-112V: overall ... 21.78 m (71 ft 5½ in)
  - fuselage ... 20.00 m (65 ft 7½ in)
  - Il-112V ... 20.65 m (67 ft 9 in)
- Height overall: except Il-112V ... 7.63 m (25 ft 0¼ in)
  - Il-112V ... 7.90 m (25 ft 11 in)
- Tailplane span ... 7.14 m (23 ft 5 in)
- Wheel track ... 3.80 m (12 ft 5½ in)
- Passenger door: Height ... 1.70 m (5 ft 7 in)
  - Width ... 0.76 m (2 ft 6 in)
- Service door: Height ... 1.38 m (4 ft 6¼ in)
  - Width ... 0.72 m (2 ft 4¼ in)
- Forward baggage door: Height ... 1.30 m (4 ft 3 in)
  - Width ... 0.96 m (3 ft 1¾ in)
- Emergency exits (each): Height ... 0.91 m (2 ft 11¾ in)
  - Width ... 0.51 m (1 ft 8 in)

DIMENSIONS, INTERNAL:
- Hold: Max height ... 2.40 m (7 ft 10½ in)
  - Max width ... 2.40 m (7 ft 10½ in)

WEIGHTS AND LOADINGS (A: passenger transport, B: business transport, C: Il-112V):
- Weight empty, equipped: A, B, C ... 9,100 kg (20,062 lb)
- Max payload: A, B ... 4,000 kg (8,818 lb)
  - C ... 6,000 kg (13,227 lb)
- Max T-O weight: A ... 14,530 kg (32,033 lb)
  - B ... 13,490 kg (29,740 lb)
  - C ... 16,700 kg (36,817 lb)
- Max power loading: A ... 3.95 kg/kW (6.49 lb/shp)
  - B ... 4.67 kg/kW (6.03 lb/shp)
  - C ... 4.54 kg/kW (7.46 lb/shp)

PERFORMANCE (estimated):
- Nominal cruising speed at FL250
  - A, B ... 297–324 kt (550–600 km/h; 342–373 mph)
  - C ... 351 kt (650 km/h; 404 mph)
- T-O run: A ... 520 m (1,710 ft)
  - B ... 500 m (1,640 ft)
  - C ... 650 m (2,135 ft)
- T-O balanced field length: A ... 980 m (3,215 ft)
  - B ... 940 m (3,085 ft)
  - C ... 1,200 m (3,940 ft)
- Landing balanced field length: A, B ... 1,000 m (3,280 lb)
- Landing run: A, B ... 400 m (1,315 ft)
- Range: A ... 809 n miles (1,500 km; 932 miles)
  - B ... 1,943 n miles (3,600 km; 2,236 miles)
  - C: with 6,000 kg (13,227 lb) payload ... 540 n miles (1,000 km; 621 miles)
    - with 2,000 kg (4,409 lb) payload ... 3,239 n miles (6,000 km; 3,728 miles)

UPDATED

ILYUSHIN IL-114

TYPE: Twin-turboprop airliner.

DEVELOPMENT MILESTONES

Design submitted	1986
First flight	29 Mar 90
Certification	26 Apr 97
First flight, production	7 Aug 92

PROGRAMME: Design finalised 1986, as replacement for aircraft in An-24 class; was scheduled to enter service (with Aeroflot's Tashkent division) in 1992. Prototype (SSSR-54000) first flew at Khodinka 29 March 1990; second prototype (SSSR-54001) flew at Khodinka 24 December 1991, but lost in accident on 5 July 1993, resulting in withdrawal of government funding. Series production by TAPO at Tashkent, Uzbekistan, initial aircraft comprising Nos. 0101, 0103 and 0105 for flight development plus 0102 for static tests and 0104 for dynamic tests; first production aircraft flew 7 August 1992. Certification delayed by loss of second prototype and deferred certification of TV7 engines, but finally achieved on 26 April 1997. Negotiations reportedly under way in 1996 for production at Esfahan by Iran Aircraft Manufacturing Company; this venture lapsed upon selection of rival Antonov An-140. Recent development effort centred on -100 version.

In 1998, Russian and Uzbek governments agreed to promote Il-114; Ilyushin and Inkombank signed provisional agreement for production funding (although the bank's trading licence was revoked shortly afterwards, resulting in further delays); and Uzbekistan Airlines took delivery of production aircraft, making first commercial flight on 27 August 1998.

CURRENT VERSIONS: **Il-114:** *As described in detail.*

Former Uzbekistan Airlines Ilyushin Il-114 following transfer to Vyborg *(Yefim Gordon)* ***NEW***/0573404

Ilyushin Il-114 flight deck *(Yefim Gordon)* ***NEW***/0552041

Interior of Vyborg Il-114 *(Yefim Gordon)* ***NEW***/0552042

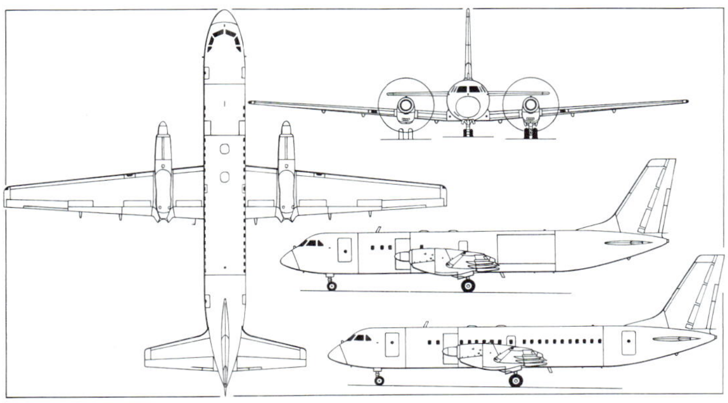

Ilyushin Il-114 short-range passenger transport, with additional side view (upper) of Il-114T freighter *(Jane's/Mike Keep)* 0011989

Il-114T: Cargo version developed for Uzbekistan Airlines; port freight door, size 3.25 × 1.715 m (10 ft 8 in × 5 ft 7½ in) in rear fuselage; removable roller floor; cargo attendants' cabin at forward end of freight hold accommodates up to two persons, is smokeproof and variable in size. MTOW and performance as for Il-114 passenger version, except where otherwise indicated. First production example (RA-91005) flew at Tashkent 14 September 1996. Second (UK-91004) converted by 1998, but lost in accident 5 December 1999. Prototype delivered to LII at Zhukovsky in early 2000 for certification trials; these were continued in Yakutia in early 2002.

Il-114-N200S: Rear loading ramp version; none yet built.

Il-114-100: With 2,051 kW (2,750 shp) P&WC PW127H turboprops, Hamilton Sundstrand 586F-7 six-blade propellers, Sextant avionics and new systems; primarily for export; English inscriptions and Imperial calibration; designated **Il-114PC** until late 1997; passenger and cargo (**Il-114-100T**) versions envisaged. Performance generally as Il-114, but with increased range and economy. Weight empty, equipped (including two crew) 16,100 kg (35,494 lb). Joint venture agreed by Ilyushin and P&WC on 16 June 1997; first flight at Tashkent (UK-91009) 26 January 1999; CIS Interstate Aviation Committee type certificate awarded 27 December 1999.

Il-114M: With TV7M-117 turboprops, increased maximum T-O weight and 7,000 kg (15,430 lb) payload.

Il-114MA: Version of Il-114M with P&WC engines to carry 74 passengers at 324–351 kt (600–650 km/h; 373–404 mph) on 1,079 n mile (2,000 km; 1,242 mile) stages.

Il-114P and Il-114MP: Maritime patrol versions. Described in 2001–02 and previous editions.

Il-114FK: Military reconnaissance/cartographic survey version. Described in 2001–02 and previous editions.

Il-114PR: Signals intelligence and electronic warfare version; announced October 2000.

Il-140: Tactical air control (Il-140) and maritime patrol/SAR (Il-114M) versions under study by October 2000. No further reports, but at International Maritime Defence Show at St Petersburg in June 2003, Radar-MMS showed a model of an Il-114 with Il-20M style of sigint/LOROP camera fairings, dorsal (Il-22 style) and ventral antenna pods and pylon-mounted SLAR pods.

CUSTOMERS: Uzbekistan Airlines purchased two pre-series aircraft in 1994; later modified with extra lavatory, new galley and accommodation reduced from 64 to 52; each flew 300 hours before engine overhaul time expiry and grounding; airline operated both Il-114Ts on service trials in 1998 and had total requirement for 10 Il-114/Ts, three of which on order in Il-114-100 configuration by April 2000 for delivery in March and April 2001, but failed to appear; first had been received by February 2003, with remaining two due in second half of that year, although further postponed, with first rescheduled for June 2004. Interest expressed by Bulgarian airlines; order for three Il-114-100s from Singaporean purchaser reported early 2000; delivery planned in 2001, although nothing further heard. Vyborg ordered three new aircraft in 2001 for operations from St Petersburg; deliveries planned for 2002 in 64-seat configuration, but first receipts, from August 2002, were two former Uzbekistan Airlines aircraft following their modification for flights in western European airspace. Vyborg services from St Petersburg began early May 2003 and "at least two" new aircraft were rescheduled for 2004 delivery. Three (excluding prototype) owned by Ilyushin Bureau.

Total of 15 flying or substantially complete January 1998, most awaiting orders; delivery totals officially quoted as 1992 two, 1993 one, 1994 one, 1997 one and 1998 one. No further new production by end of 2001, when seven remained on assembly line, including two Il-114Ts; situation similar in late 2003; by early 2002, components sets for first 40 production aircraft had been manufactured wholly or in part. Transaero opened negotiations in 2001 for between five and 10.

COSTS: US$10 million for Il-114; US$8 million for Il-114T (2000).

DESIGN FEATURES: Conventional low-wing monoplane; only fin and rudder swept; slight dihedral on wing centre-section, much increased on outer panels; operation from unpaved runways practical. Service life of 30,000 cycles/30,000 hours/30 years, with overhaul at 6,000 hour intervals.

FLYING CONTROLS: Manual actuation for all except elevator; each wing trailing-edge occupied entirely by aileron, with servo and trim tabs, and hydraulically actuated double-slotted trailing-edge flaps, inboard and outboard of engine nacelle; two airbrakes (inboard) and spoiler (outboard) forward of flaps; spoilers supplement ailerons differentially in event of engine failure during take-off. Elevator control is FBW with back-up cable actuation. Trim and servo tabs in rudder, trim tab in each elevator.

STRUCTURE: Approximately 10 per cent of airframe by weight made of composites; two-spar wings; removable leading-edge on outer panels; circular-section aluminium alloy semi-monocoque fuselage built as five subassemblies; metal tail unit (CFRP tailplane and fin boxes planned for later aircraft).

LANDING GEAR: Retractable tricycle type, with twin wheels on each unit. All retract forward hydraulically; emergency extension by gravity. Oleo-pneumatic shock-absorbers. Tyres size 620×80 on nosewheels, 880×305 on mainwheels. Nosewheels steerable ±55°. Disc brakes on mainwheels. All wheel doors remain closed except during retraction or extension of landing gear.

POWER PLANT: Two 1,839 kW (2,466 shp) Klimov TV7-117S turboprops (with potential to increase to 2,088 kW; 2,800 shp), each driving a low-noise six-blade Stupino SV-34 CFRP propeller. Integral fuel tanks in wings, capacity 8,780 litres (2,319 US gallons; 1,931 Imp gallons).

ACCOMMODATION: Flight crew of two, plus stewardess. Emergency exit window each side of flight deck. Four-abreast seats for 64 passengers in main cabin, at 76 cm (30 in) seat pitch, with central aisle 45 cm (17¾ in) wide. Provision for rearrangement of interior for increased seating, removal of seats for cargo-carrying, and lengthening of fuselage for 70 to 75 passengers. Two passenger doors on port side: airstair door at front of cabin, further door at rear, both opening outward. Galley, cloakroom and lavatory at rear; emergency escape slide by service door on starboard side. Type III emergency exit over each wing. Service doors at front and rear of cabin on starboard side. Baggage compartments forward of cabin on starboard side and to rear of cabin, plus overhead baggage racks. Optional carry-on baggage shelves in lobby by main door at front.

SYSTEMS: TsNPKO-114 autopilot began trials on Il-114-100 in 2001. Dual-redundant pressurisation and air conditioning system using bleed air from both engines; maximum differential 0.44 bar (6.4 lb/sq in). Two independent hydraulic systems, pressure 207 bar (3,000 lb/sq in), for landing gear actuation, wheel brakes, nosewheel steering, airbrakes and flaps. Three-phase 115/220 V 400 Hz AC electrical system powered by 40 kW alternator on each engine. Secondary 24 V DC system. Wing and tail unit leading-edges de-iced electrically by patented pulse wave system. Electrothermal anti-icing system for propeller blades and windscreen. Engine air intakes de-iced by hot air. APU in tailcone.

AVIONICS: Digital avionics for automatic or manual control by day or night, including automatic approach and landing in limiting weather conditions (ICAO Cat. I and II).

Instrumentation: Two colour CRTs for each pilot for flight and navigation information. Centrally mounted CRT for engine and systems data.

Flight: Barco computer for FMS.

DIMENSIONS, EXTERNAL:

Wing span	30.00 m (98 ft 5¼ in)
Wing aspect ratio	11.0
Length overall	26.875 m (88 ft 2 in)
Fuselage: Length	26.20 m (85 ft 11½ in)
Max diameter	2.86 m (9 ft 4½ in)
Height overall	9.185 m (30 ft 1½ in)
Tailplane span	11.10 m (36 ft 5 in)
Wheel track	8.40 m (27 ft 6½ in)
Wheelbase	9.13 m (29 ft 11½ in)
Propeller diameter	3.60 m (11 ft 9¾ in)
Propeller ground clearance	0.50 m (1 ft 7¾ in)
Propeller fuselage clearance	0.97 m (3 ft 2¼ in)
Passenger doors (each): Height	1.70 m (5 ft 7 in)
Width	0.90 m (2 ft 11¼ in)
Service door (front): Height	1.30 m (4 ft 3¼ in)
Width	0.96 m (3 ft 1¾ in)
Service door (rear): Height	1.38 m (4 ft 6¼ in)
Width	0.72 m (2 ft 4¼ in)
Emergency exit (each): Height	0.91 m (3 ft 0 in)
Width	0.51 m (1 ft 8 in)

DIMENSIONS, INTERNAL:

Length between pressure bulkheads	22.24 m (72 ft 11½ in)
Cabin: Length	18.93 m (62 ft 1¼ in)
Width: max	2.64 m (8 ft 8 in)
at floor	2.28 m (7 ft 5¾ in)
Max height	1.92 m (6 ft 3½ in)
Cargo cabin volume (Il-114T)	76.0 m³ (2,684 cu ft)

AREAS:

Wings, gross	81.90 m² (881.6 sq ft)

WEIGHTS AND LOADINGS:

Operating weight empty	15,000 kg (33,070 lb)
Max payload	6,500 kg (14,330 lb)
Max fuel	6,500 kg (14,330 lb)
Max T-O weight	23,500 kg (51,808 lb)
Max ramp weight	23,600 kg (52,029 lb)
Max wing loading	286.9 kg/m² (58.77 lb/sq ft)
Max power loading	6.39 kg/kW (10.50 lb/shp)

PERFORMANCE:

Max level speed	270 kt (500 km/h; 310 mph)
Cruising speed	254 kt (470 km/h; 292 mph)
Approach speed	100 kt (185 km/h; 115 mph)
Landing speed	87 kt (160 km/h; 100 mph)
Optimum cruising height: Il-114	7,600 m (24,940 ft)
Il-114T	6,000 m (19,680 ft)
T-O run: paved	1,360 m (4,465 ft)
Landing run: paved or unpaved	1,260 m (4,135 ft)
Range, with reserves:	
with 64 passengers	540 n miles (1,000 km; 621 miles)
with 1,500 kg (3,300 lb) payload	2,590 n miles (4,800 km; 2,980 miles)

UPDATED

INTERAVIA

INTERAVIA RADONEZH AO (INTERAVIA RADONEZH JSC)

Repikhovo, Sergiev-Posad, Moskovskaya oblast
Tel: (+7 254) 305 04
Fax: (+7 095) 258 40 50
GENERAL DIRECTOR: Sergey Esayan

Interavia has marketed the I-1L high-wing lightplane and I-3 aerobatic sportplane and designed what is now the Smolensk (Technoavia) Finist. Following a stagnation of sales, deliveries of the I-3 resumed in 2004. The company's 3,000 m^2 (32,300 sq ft) plant at Repikhovo, near Zagorsk, is a former medium-range ballistic missile factory. Majority shareholding in Interavia is held by Eagle Aerospace of Switzerland.

NEW ENTRY

Interavia I-3 aerobatic aircraft in tandem-seat form *(Paul Jackson)* ***NEW***/0054614

INTERAVIA I-3

TYPE: Aerobatic two-seat sportplane.

PROGRAMME: Designed by former Sukhoi personnel, also responsible for Su-26, Su-29 and Su-31 aerobatic types. Batch of 50 I-3s begun by Tushinsky Mashinostroitelnyi Zavod AO (Tushino Machine-Building Plant Ltd), Moscow, in early 1990s to order of Chechnya, but embargoed by Russian government. Five were completed as Technoavia SP-91s, but definitive SP-95 version failed to materialise; however, four SP-91s exported to USA, where one incorrectly registered as an SP-95.

Prototype I-3 exhibited at Moxcow Air Show in 1993. Exports to USA began in 1996, some being registered as E-3s. Production transferred from Tushino to Repikhovo.

CUSTOMERS: Total of at least 25 sold by 2003, of which four to European owners and at least eight to USA. Some 15 then remained incomplete, including two earmarked for export to Finland in 2004.

COSTS: US$85,000 to US$90,000 (2004).

DESIGN FEATURES: Conventional low-wing monoplane, capable of unlimited aerobatics: tapered multispar wings of symmetrical section, without dihedral or anhedral; unswept tail surfaces with pointed rudder tip; clear-view blister canopy over one or two seats. Convertible from two-seater to single-seater in 1 hour by removal of front cabin bay with control panel, installation of top fuselage panel and changing of canopy.

FLYING CONTROLS: Manually actuated conventional horn-balanced elevators and horn-balanced rudder; 80 per cent span flaperons. Ground-adjustable tab on rudder; two suspended balance tabs on each flaperon.

STRUCTURE: All-metal; semi-monocoque fuselage; fluted skin on fin, tailplane and all control surfaces.

LANDING GEAR: Tailwheel type, fixed; cantilever mainwheel legs. Mainwheel tyres size 400×150; tailwheel tyre size 200 × 80 mm.

POWER PLANT: One 265 kW (355 hp) VOKBM M-14P nine-cylinder air-cooled radial engine; three-blade wooden controllable-pitch propeller. Fuel tank in each wingroot; combined capacity 120 litres (31.7 US gallons; 26.4 Imp gallons).

ACCOMMODATION: One or two seats in tandem.

DIMENSIONS, EXTERNAL:
Wing span .... 8.10 m (26 ft 7 in)
Wing aspect ratio 5.7
Length overall .... 6.72 m (22 ft 0½ in)
Tailplane span .... 2.80 m (9 ft 2¼ in)
Propeller diameter .... 2.40 m (7 ft 10½ in)

AREAS:
Wings, gross .... 11.54 m^2 (124.2 sq ft)

WEIGHTS AND LOADINGS:
Weight empty .... 760 kg (1,675 lb)
Max T-O weight .... 1,063 kg (2,343 lb)
Max wing loading .... 92.1 kg/m^2 (18.86 sq ft)
Max power loading .... 4.02 kg/kW (6.60 lb/hp)

PERFORMANCE:
Never-exceed speed (VNE) .... 242 kt (450 km/h; 280 mph)
Max level speed .... 189 kt (350 km/h; 217 mph)
Max rate of climb at S/L .... 660 m (2,165 ft)/min
Service ceiling .... 5,000 m (16,400 ft)
T-O run: paved runway .... 200 m (656 ft)
unpaved runway .... 262 m (860 ft)
Landing run: paved runway .... 452 m (1,483 ft)
unpaved runway .... 477 m (1,565 ft)
Range with max fuel .... 377 n miles (70 km; 435 miles)
g limits: single seat .... +12/−10
two seats .... +10/−8

NEW ENTRY

IRKUT

NAUCHNO-PROIZVODSTVENNAYA KORPORATSIYA IRKUT OAO (IRKUT SCIENTIFIC-PRODUCTION CORPORATION JSC)

ulitsa Novatorov 3, 664020 Irkutsk
Tel: (+7 3952) 32 29 09
Fax: (+7 3952) 32 29 11
e-mail: market@irkut.ru
Web: http://www.irkut.ru
PRESIDENT: Alexey I Fedorov
GENERAL DIRECTOR: Vladimir V Kovalkov
CHIEF, PUBLIC RELATIONS: Elena Fedorova

Founded on 28 March 1932 and commissioned on 24 August 1934, as GAZ 125 (becoming GAZ 39 in 1941), Irkut has built some 6,500 aircraft of 16 types from Antonov, Ilyushin, MiG, Petlyakov, Sukhoi, Tupolev and Yakovlev bureaux and supplied them to 21 countries. In recent years, it has manufactured MiG-23UB trainers (1970–85); 165 kits for Indian-assembled MiG-27MLs; Su-27UB trainers (from 1986); and is currently responsible for producing the Su-30 fighter (since 1991) and Beriev Be-200 amphibian. Su-30 customers include China and India; offers Su-27UBM and Su-30KN upgrades to older aircraft. Series manufacture of the Yak-112 lightplane has been abandoned, although company has been allocated prospective manufacture of the Ilyushin/HAL Il-214 twin-jet transport. Also offers upgrades of Su-30, Mi-8, Mi-24, Aero L-39, Il-76 and Il-103.

Irkut shareholders are Brunswick UBS Warburg Nominees (25.72 per cent), Forpost Commercial Bank (20.60 per cent), FTK Company (20.35 per cent), APVK Sukhoi (14.70 per cent), Aerocom (10.18 per cent), other companies (4.9 per cent) and individuals (3.55 per cent). Subsidiary companies are Irkut Aviation Industrial Association (aircraft manufacturing plant), Beriev (39.57 per cent holding), Beta-Air (66.15 per cent), Russian Avionics (51 per cent), Irkut Aviastep (wholly owned), Itela (51 per cent), Techserviceavia (51 per cent), Gidroaviasalon (30 per cent) and Irkut private pension fund (wholly owned).

Is member of AVPK Sukhoi. Known from April 1989 until 2002 as IAPO (Irkutsk Aviation Industrial Association), having become a joint stock company in October 1992, but on 19 December 2002, shareholders approved a company change of name to Irkut NPK OAO, IAPO becoming a subsidiary. In 1997 was first Russian aviation enterprise to gain ISO 9002 status and in April 2001 became first Russian military production enterprise to issue short-term debt notes on financial market.

In 2000, IAPO branched out into design and manufacture of its own products in the form of the prototype A-002 autogyro. Expansion of civil programmes is strategic objective. The Irkut 111 design is for an ultra-wide-bodied regional airliner with twin fins and three-aisle configuration for only 106 passengers.

Under government plans announced in May 2001, IAPO is to join the aerospace group also including Sukhoi, Ilyushin, Mil, Yakovlev and their associated factories. In August 2003, Irkut and Yakovlev announced plans for a joint venture company to manage their eventual merger, agreed in October 2003 by means of a 75.5 per cent share of Yakovlev stock, valued at US$45 million. Employees in 2003 totalled 22,000 including 15,000 at IAPO factory; further 1,500 firms supply plant with materials. Irkut planned 20 per cent public stock flotation in 2004.

UPDATED

IRKUT A-002

TYPE: Three-seat autogyro.

PROGRAMME: Project announced October 2000; first product of Irkutsk Light Aircraft Design Bureau (OKB Legkoi Aviatsy OAO), although an experimental version was tested in 1998. Designer: Andrey Tatarnikov. Almost complete prototype shown at Moscow, August 2001. Maiden flight 21 April 2002. Production launched August 2002.

CUSTOMERS: First deliveries were due in late 2003 to Susuman Susumanzoloto OAO mining company. Sales contracts then held for 20 aircraft. However, in October 2003, Irkút delayed deliveries to 2004 and curtailed production to five, pending further development.

COSTS: US$120,000 (2003).

DESIGN FEATURES: Conforms to FAR Pt 27 and Russian AP-27. Streamlined, cabin autogyro which can be stored and serviced in a motorcar garage. Choice of engines, including those using motorcar fuel. Classic configuration, with pusher propeller; single, sweptback fin with large rudder; broad track landing gear. Engine-driven pre-rotation of rotor.

FLYING CONTROLS: Manual. All-moving tailplane mounted at one-third of fin height; horn- and mass-balanced rudder.

STRUCTURE: Generally of metal. Rudder and elevator of composites.

LANDING GEAR: Tricycle type; fixed. Cantilever-spring mainwheels; rubber-in-compression shock-absorber on nosewheel. Mainwheel tyres 400×150; nose 300×125. Hydraulic disc

Early test flight of the Irkut A-002 ***NEW***/0593355

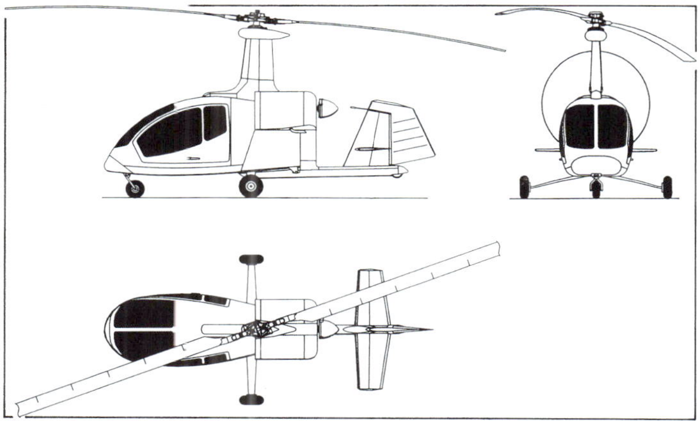

General arrangement of the Irkut A-002 autogyro 0121087

brakes on mainwheels. Safety tail skid on prototype; auxiliary tailwheel on production version.

POWER PLANT: Prototype has one 157 kW (210 hp) Teledyne Continental IO-320 flat-six driving a fixed-pitch, two-blade propeller. Alternative engines of 134 to 157 kW (180 to 210 hp) and three-blade propeller can be installed. Subaru motorcar engine under consideration in April 2003.

ACCOMMODATION: Two front seats, side by side, with dual controls; single rear seat. Upward-hinged door each side. Heating and ventilation provided.

DIMENSIONS, EXTERNAL:

Rotor diameter	9.80 m (32 ft 1¾ in)
Length of fuselage	4.98 m (16 ft 4 in)
Height, rotor removed	3.17 m (10 ft 4¾ in)
Wheel track	2.40 m (7 ft 10½ in)
Wheelbase	2.015 m (6 ft 7¼ in)
Propeller diameter	1.90 m (6 ft 2¾ in)

AREAS:

Rotor disc	75.43 m^2 (811.9 sq ft)

WEIGHTS AND LOADINGS:

Weight empty	500 kg (1,102 lb)
T-O weight: normal max	800 kg (1,763 lb)
overload	850 kg (1,873 lb)
Max rotor loading, normal	10.61 kg/m^2 (2.17 lb/sq ft)
Max power loading, normal	5.11 kg/kW (8.40 lb/hp)

PERFORMANCE (157 kW 210 hp) engine:

Max level speed	113 kt (210 km/h; 130 mph)
Min flying speed	22 kt (40 km/h; 25 mph)
Max rate of climb at S/L	150 m (492 ft)/min
Service ceiling	3,000 m (9,840 ft)
T-O run: jump start	zero
normal	40 m (135 ft)
Range with 200 kg (441 lb) payload	270 n miles (500 km; 310 miles)
Endurance	6 h

UPDATED

IRKUT (AVIASTEP) 111

TYPE: Regional jet airliner.

PROGRAMME: Developed by Aviastep bureau (General Director: Sergey Bogdanov) in Moscow and promoted by Irkut. Design began in 2000; first details released in 2002; model shown at Paris, June 2003. No further information by early 2004.

DESIGN FEATURES: Refer to the accompanying photograph. Ultra-wide body, with aerofoil characteristics, partly blended with conventional wings and podded engines. Lift/drag ratio of 17 at cruising speeds, compared with 13 to 15 of conventional airliner designs. Fuselage width demands V tail for control effectiveness.

FLYING CONTROLS: Two rudders on V tail; elevators in 'beaver tail' between finroots.

POWER PLANT: Two turbofans.

ACCOMMODATION: Typically, 100 passengers and 9,800 kg (21,605 lb) of freight. Maximum 106 passengers in single class.

DIMENSIONS, INTERNAL:

Cabin: Length	10.0 m (33 ft)
Max width	7.0 m (23 ft)

WEIGHTS AND LOADINGS:

Max T-O weight	50,000 kg (110,250 lb)

PERFORMANCE:

Normal cruising speed	M0.8
Approach speed	124 kt (230 km/h; 143 mph)
Max cruising altitude	12,000 m (39,370 ft)
Required runway length	1,800 m (5,900 ft)
Range: with max payload	1,725 n miles (3,200 km; 1,985 miles)
with 50 passengers and 4,700 kg freight	3,510 n miles (6,500 km; 4,040 miles)

UPDATED

Proposed Irkut (Aviastep) 111 *(Paul Jackson)* ***NEW***/0573461

KAMOV

KAMOV OAO (KAMOV JSC)

ulitsa 8 Marta 8a, Lyubertsy, 140007 Moskovskaya oblast
Tel: (+7 095) 700 32 04 and 171 37 43
Fax: (+7 095) 700 30 71 and 700 31 10
e-mail: kb@kamov.ru
Web: http://www.kamov.ru

GENERAL DESIGNER: Sergei V Mikheyev, PhD
DEPUTY GENERAL DESIGNER: Beniamin A Kasyanikov
CHIEF DESIGNERS:
Boris Gubarev (Ka-115)
Vyacheslav Krygin
Evgeny Pak
Aleksandr Piorzhnikov
Grigory Yakemenko (Ka-50)
Evgeny Sudarev

Formed in 1948 by Prof Dr Ing Nikolai Ilyich Kamov, this OKB originated in an experimental autogyro production plant operating in Moscow from 1940 to 1943. Kamov's initial brief was to design helicopters for naval use. Since the Ka-10 of 1949, Kamov helicopters have had coaxial contrarotating rotors; only the Ka-60/62, under development, have single main rotor, with anti-torque Fenestron. It was reported in September 2000 that a new transport helicopter, to compete against Western NH90 and S-92, is under development, but no further reports had been received by early 2004.

Since 1996, Kamov has been a member of the RSK 'MiG' consortium. It is 49 per cent state-owned, and its design bureau and experimental construction plant employed 2,500 people in 1998. Under plans announced in May 2001 by the Russian government, Kamov and MiG will be in the major aerospace grouping which also includes Tupolev, Aviastar, Aviacor and those factories producing Kamov helicopters, namely the Arsenyev factory (Ka-50) and the Kumertau factory (Ka-32 family). In 2002–03, Kamov was in the process of moving its flight test activities from the factory site at Lyubertsy to Chkalovsky, in the Shchelkovo area of Moscow Region, to escape the restrictions of urban development.

In September 2003, the Geralsy Group of Saudi Arabia announced plans for local manufacture of undisclosed Kamov helicopters. During the following month, Kamov agreed with Chinese industrial representatives on joint development of a fifth-generation combat helicopter, using Ka-50 and Ka-52 as a basis.

The affiliated Aero-Kamov Air Transportation Company (Tel/Fax: (+7 095) 700 31 60) operates a fleet of Ka-32s for diverse tasks, including firefighting, and has helicopters based in Russia, Canada and South Africa.

UPDATED

KAMOV KA-29 AND KA-31

NATO reporting name: Helix-B

TYPE: Medium transport helicopter; AEW helicopter.

PROGRAMME: Developed for AV-MF, following cancellation of proposed joint-service, tandem-rotor, multirole V-50 and its replacement by what became Ka-50, meeting Army requirement only. Ka-252TB prototype (also known as Izdelie D2B or Izdelie 502) first flew 28 July 1976, possibly with Ka-25 nose or original narrow Ka-27/Ka-32 nose. Production at Kumertau (KAPP) from 1984.

Entered service with Northern and Pacific Fleets 1985; photographed on board assault ship *Ivan Rogov* in Mediterranean 1987, thought to be Ka-27B and given NATO reporting name 'Helix-B'; identified as Ka-29 combat transport at Frunze (Khodinka) Air Show, Moscow, August 1989; Ka-31 radar picket version completed initial shipboard trials on aircraft carrier *Admiral of the Fleet Kuznetsov* (then *Tbilisi*) 1990.

CURRENT VERSIONS: **Ka-29TB : ('Helix-B')** Armed derivative for day/night, VFR and IFR, transport and close support of seaborne assault troops; in-the-field conversion from one role to the other. Non-retractable landing gear and 50 cm (20 in) wider armoured flight deck. Reportedly used by Experimental Combat Group in Chechen War in 1996. No recent production known.

Kamov Ka-31 with radar antenna deployed *(Yefim Gordon)* ***NEW***/0573402

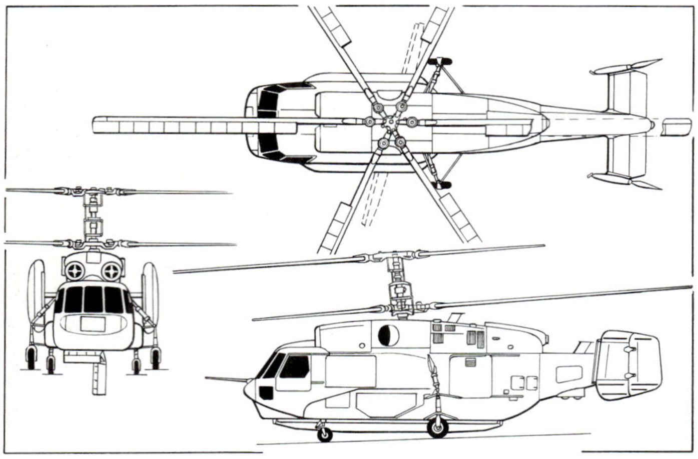

Kamov Ka-31, showing radar deployed in nose view *(James Goulding)*
0130861

Detailed description generally as for Ka-32 except as under.

Ka-31 (formerly Ka-29RLD: *radiolokatsyonnogo dozora*: radar picket helicopter): Development began 1980; first flown October 1987; two examples (031 and 032) tested on *Admiral of the Fleet Kuznetsov*; state testing completed in 1996; limited production launched (for Indian Navy) at Kumertau Aircraft Plant, Bashkirskaya, 1999. Indian aircraft have 12-channel Kronshtadt GPS with Abris digital moving map and a 152 × 203 mm (6 × 8 in) AMLCD screen.

Basic airframe of Ka-27 (which see in 2001–02 and earlier *Jane's*) with broader flight deck; E-801 or E-801M (export) Oko (eye) early warning radar system by Radio Engineering Institute, Nizhny Novgorod, includes large rotating radar antenna (area 6.0 m^2; 64.5 sq ft) that stows flat against underfuselage and deploys downward, turning through 90° into vertical plane before starting to rotate at 6 rpm; landing gear retracts upward to prevent interference, nosewheels into long fairings. Once system has been switched on, antenna extended and operation mode selected, data on air targets flying below helicopter's altitude, and on water surface situation, are acquired, evaluated and transmitted automatically to command centre, requiring only two crew (pilot and navigator, latter monitoring — but not operating — the system) in helicopter. Kronshtadt Kabris GPS navigation and display system. Loiter speed 54 to 65 kt (100 to 120 km/h; 62 to 75 mph) at up to 3,500 m (11,480 ft); loiter duration 2 h 30 min. Maximum surveillance radius 54 to 81 n miles (100 to 150 km; 62 to 93 miles) for fighter-size targets, 135 n miles (250 km; 155 miles) for surface vessels; up to 20 targets tracked simultaneously. Antenna can be retracted manually or explosively jettisoned in the event of a forced landing.

Two large panniers starboard side of cabin, fore and aft of main landing gear on helicopter numbered 032 (forward panniers only on 031); starboard airstair-type cabin door, aft of flight deck, divided horizontally into upward- and downward-opening sections, with box fairing in place of window; hatch window deleted above starboard rear pannier; new TA-8Ka APU positioned above rear of engine bay fairing, with slot-type air intake at front of housing, displacing usual ESM and IR jamming pods, gives radar and antenna an independent power supply. Tyre size 620×180 on mainwheels, 480×200 on nosewheels. Tailcone extended by fairing for flight recorder; no armour, gun door, stores pylons or outriggers.

Ka-33: Utility transport. Civilianised version of Ka-29TB shipborne assault transport. Designation revealed at Moscow Air Show in August 1997; no further details released and no known conversions.

CUSTOMERS: Total of 59 Ka-29s built for Russian Federation Naval Aviation (about 45) and Ukrainian Navy (about 12). Following 1996 evaluation, four Ka-31s ordered in August 1999 by Indian Navy for delivery in 2001 and basing aboard aircraft carriers and 'Krivak' class destroyers; further five ordered February 2001. Additional 12 may be required. First flight of Indian Ka-31 16 May 2001; by October 2001, first two Indian airframes delivered from KAPP to Kamov at Moscow for avionics installation; flight trials completed of first two Indian aircraft by September 2002, with hand-over following soon after; second pair delivered in April 2003, remainder being due by end of that year. In October 2002, Kamov reported a second export customer for Ka-31s in addition to Indian Navy.

COSTS: Indian Navy batch of four priced at Rs4 billion (US$92 million) (2000); second five cost US$108 million (2001).

POWER PLANT: Two Klimov TV3-117VMA turboshafts, each 1,633 kW (2,190 shp). Engines started by APU. Fuel tanks filled with reticulated polyurethane foam for fire suppression.

ACCOMMODATION: Wider flight deck than Ka-27 for two crew; three flat-plate windscreen glazings instead of two-piece curved transparency; 350 kg (772 lb) of armour around cockpit and engines; main cabin port-side door, aft of landing gear, divided horizontally into upward- and downward-opening sections, lower section forming step when open, to facilitate rapid exit of up to 16 assault troops; four stretcher patients, seven seated casualties and medical attendant in ambulance role; internal or slung cargo provisions.

AVIONICS: *Comms:* Two UHF and HF radios.

Radar: Primary radar in port side of nose.

Flight: INS; Doppler box under tailboom; IFF ('Slap Shot'). Indian helicopters have Khronstadt Kabris GPS.

Mission: Undernose Shturm-V missile guidance and LLTV pods; ESM 'flower pot' above rear of engine bay fairing.

Self-defence: L-166V IR jammer ('Hot Brick'); chaff/flare dispensers.

EQUIPMENT: Station-keeping light between ESM and jammer.

ARMAMENT: Four-barrel Gatling-type GShG-7.62 7.62 mm machine gun, with 1,800 rounds, flexibly mounted behind downward-articulated door on starboard side of nose; four pylons on outriggers, for two four-round packs of 9M114 Shturm (AT-6 'Spiral') ASMs and two UV-32-57 57 or B-8V20 80 mm rocket pods. Alternative loads include four rocket packs, two pods each containing a 23 mm gun and 250 rounds, or two ZAB-500 incendiary bombs. Internal weapons bay for torpedo or bombs. Provision for 30 mm Type 2A42 gun above port outrigger, with 250-round ammunition feed from cabin.

DIMENSIONS, EXTERNAL (A: Ka-29, B: Ka-31):
Rotor diameter (each) .... 15.90 m (52 ft 2 in)
Blade length, aerofoil section (each) .... 5.45 m (17 ft 10½ in)
Blade chord .... 0.48 m (1 ft 7 in)
Vertical separation of rotors .... 1.40 m (4 ft 7 in)
Length overall, excl noseprobe and rotors:
A .... 11.30 m (37 ft 1 in)
B .... 11.25 m (36 ft 11 in)
Height overall: A .... 5.40 m (17 ft 8½ in)
B .... 5.60 m (18 ft 4½ in)
Width: between centrelines of outboard pylons .... 5.65 m (18 ft 6½ in)
over tailfins and centred rudders .... 3.65 m (12 ft 0 in)
of flight deck .... 2.20 m (7 ft 2 in)
Mainwheel track .... 3.50 m (11 ft 6 in)
Nosewheel track: A .... 1.41 m (4 ft 7½ in)
B .... 2.41 m (7 ft 11 in)
Wheelbase: A .... 3.00 m (9 ft 10 in)
B .... 3.05 m (10 ft 0 in)

AREAS:
Rotor disc (each) .... 198.50 m^2 (2,136.6 sq ft)

WEIGHTS AND LOADINGS (A, B as above):
Weight empty: A .... 5,520 kg (12,170 lb)
Max load: A, internal .... 2,000 kg (4,409 lb)
A, external .... 4,000 kg (8,818 lb)
Max combat load: A .... 1,800 kg (3,968 lb)
Normal T-O weight: A .... 11,000 kg (24,250 lb)
B .... 12,500 kg (27,557 lb)
Max T-O weight: A, internal load .... 11,500 kg (25,353 lb)
B .... 12,500 kg (27,557 lb)
Max airborne weight:
A, external slung load .... 12,600 kg (27,775 lb)

PERFORMANCE (A, B as above):
Max level speed at S/L: A .... 151 kt (280 km/h; 174 mph)
B .... 135 kt (250 km/h; 155 mph)
Nominal cruising speed: A .... 130 kt (240 km/h; 149 mph)
B .... 119 kt (220 km/h; 137 mph)
Loitering speed: B .... 54 kt (100 km/h; 62 mph)
Max rate of climb at S/L: A .... 888 m (2,910 ft)/min
Service ceiling: A .... 4,300 m (14,100 ft)
B .... 3,500 m (11,480 ft)
Loitering altitude: B .... 3,500 m (11,480 ft)
Hovering ceiling OGE: A .... 3,000 m (9,840 ft)
Combat radius, with six to eight attack runs over target:
A .... 54 n miles (100 km; 62 miles)
Range:
A, max standard fuel .... 248 n miles (460 km; 285 miles)
B .... 324 n miles (600 km; 372 miles)
A, ferry .... 400 n miles (740 km; 460 miles)
Loitering endurance: B .... 2 h 30 min

UPDATED

KAMOV KA-32

NATO reporting name: Helix-C

TYPE: Multirole medium helicopter.

PROGRAMME: Development of Ka-27/32 began 1969; first flight of common prototype 1973; first Ka-32 (SSSR-04173) flew 8 October 1980; prototype of utility version shown at Paris Air Show June 1985; new military versions first exhibited at Moscow Air Show '95; Ka-32S and Ka-32T versions in production by KAPP; other conversions by Kamov at Lyubertsy. Klimov VK-3000 turboshaft was to be certified in 2001 as alternative power plant, but no installations have been reported.

CURRENT VERSIONS: Most recent production has been of Ka-21A1 version; earlier types included Ka-32T, Ka-32S, Ka-32K and 'A2, 'A3 and 'A7 derivatives of Ka-32A.

Ka-32A: Assemblies and systems of basic Ka-32 modified in 1990–93 to meet all requirements of Russian NLG-32-29 and NLG-32-33 and US FAR Pt 29/FAR Pt 33 airworthiness standards in categories A and B. First flight September 1990; Russian type certificates obtained for Ka-32A and its TV3-117VMA engines in June 1993. Production began 1996. Larger tyres. Optional pressure fuelling with reduced fuel capacity. Maximum accommodation for 13 passengers. Advanced avionics available, including Canadian Marconi dual CMA-900 flight management system, with EFIS, AFCS, CMA-2012 Doppler velocity sensor and CMA-3012 GPS sensor. Modification of helicopters to Ka-32A standard started by Kamov 1994.

Ka-32A1: Firefighting version of Ka-32A, first flown 12 January 1994. Equipped with Canadian or Russian variants of 'Bambi Bucket', capacity 5,000 litres (1,320 US gallons; 1,100 Imp gallons). Two operated by Moscow fire service, with doorway-mounted steerable water cannon and three types of rescue cage, able to lift two, 10 or 20 people from roofs of

Kamov Ka-32A1 emergency helicopter of Moscow fire service *(Yefim Gordon)*
NEW/0528670

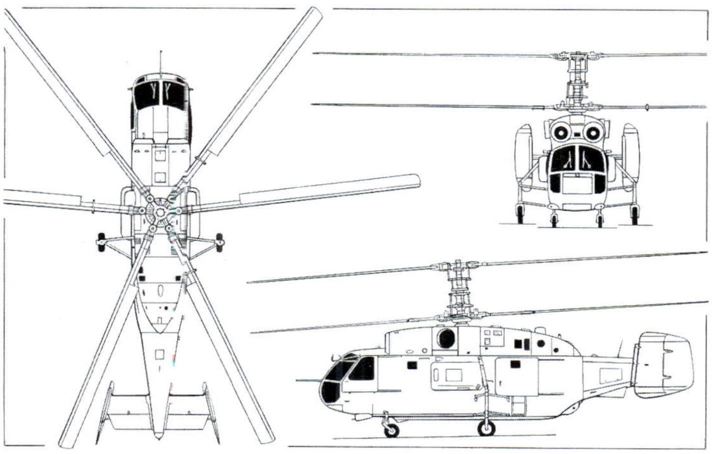

Kamov Ka-32T ('Helix-C') utility helicopter (two Klimov TV3-117V turboshafts) *(Jane's/Dennis Punnett)* 0507189

tall buildings. Other equipment includes searchlights and loudspeakers. Fire service aircraft and others flown by anti-riot police controlled by Aviatika Concern ISC, set up by Moscow city authorities and private investors to develop urban air transport system. Several on lease to South Korean forestry department have Simplex 10900-050 system, including 2,955 litre (780 US gallon; 650 Imp gallon) belly tank which can be refilled in 1½ minutes, plus 152 litre (40.2 US gallon; 33.4 Imp gallon) retardant tank. Ka-32 can also be fitted with 10900-055 system with two panniers totalling 5,000 litres (1,320 US gallons; 1,100 Imp gallons) of water and 250 litres (66.0 US gallons; 55.0 Imp gallons) of retardant.

One Moscow fire service helicopter (RA-31073) retrofitted with large, forward-facing nose boom for fire suppression in tall buildings; trials completed April 2001; shown statically at Moscow Salon, August 2001. System developed by Soyuz Federal Centre of Double Technologies at Dzerginsk; water supply of 2,800 litres (740 US gallons; 616 Imp gallons) carried in two underslung tanks or helicopter can be connected to fire vehicle on ground for unlimited supply; hose boom movable in vertical plane only.

Ka-32-10: Announced 25 May 2001; projected 24-seat civil version with enlarged cabin; internal payload 4,000 kg (8,818 lb). Target certification date 2004.

Ka-32A11BC: Built in accordance with requirements of Transport Canada. FAR Pt 29 certification gained 11 May 1998, but full clearance achieved 26 February 1999, after installation of dual actuators in flight control system; first Russian helicopter to gain Western certification. Two development aircraft delivered to VIH Logging in May 1997; flew 4,000 hours up to February 1999; also used for firefighting; further 15 on order by 1998.

Ka-32A12: Version approved by Aviation Register of Switzerland.

Ka-32M: Under development by Kamov, to increase lifting capability to 7,000 kg (15,432 lb); retrofit with 1,839 kW (2,466 hp) TV3-117VMA-SB3 engines. Probably replaced in planning by Ka-32-10.

CUSTOMERS: Aeroflot and its successors; operators in Bulgaria (32S), Canada (32A), Laos (air force; six Ka-32T), Papua New Guinea (32A), South Africa (32A), Switzerland (32A), Yemen (32S/T). Estimated 132 Ka-32s in civilian use, of which 50 were abroad in 1998. Between December 1993 and November 2000, 36 imported by LGI of South Korea for operation by Forestry Service (23 Ka-32Ts), National Maritime Police Agency (eight Ka-32Ss), Kyonggi Provincial Fire & Disaster HQ (two Ka-32Ts) and Kyongsang Buk-do Fire Defence Aviation Corps, National Parks and Ulsan Fire Defence HQ (one Ka-32T each); further 20 expected in settlement of Russian debt, of which 10 reportedly ordered in March 2001 and three delivered by end of 2001. Remaining seven were to follow in 2004, for total of 46.

Following lease of two (later three) Ka-32s, Cyprus government announced intention, August 2001, to purchase three. Algerian Air Force acquiring unknown number, three of which (two Ka-32T and one Ka-32S) noted at Sankt Peterburg in August 2002.

DESIGN FEATURES: Conceived as completely autonomous 'compact truck', to stow in much the same space as Ka-25 with rotors folded, despite greater power and capability, and to operate independently of ground support equipment; special attention paid to ease of handling with single pilot; overall dimensions minimised by use of coaxial rotors, requiring no tail rotor, and twin fins on short tailboom; upper rotor turns clockwise, lower rotor anti-clockwise; rotor mast tilted forward 3°; twin turbines and APU above cabin, leaving interior uncluttered; lower fuselage sealed for flotation.

FLYING CONTROLS: Dual hydraulically powered flight control systems, without manual reversion; spring stick trim; yaw control by differential collective pitch applied through rudder pedals; mix in collective system maintains constant total rotor thrust during turns, to reduce pilot workload when landing on pitching deck, and to simplify transition to hover and landing; twin rudders intended mainly to improve control in autorotation, but also effective in co-ordinating turns; flight can be maintained on one engine at maximum T-O weight.

STRUCTURE: Titanium and composites used extensively, with particular emphasis on corrosion resistance; fully articulated three-blade coaxial contrarotating rotors have all-composites blades with carbon fibre and glass fibre main spars, pockets (13 per blade) of Kevlar-type material, and filler similar to Nomex; blades have non-symmetrical aerofoil section; each has ground-adjustable tab; each lower blade carries adjustable vibration damper, comprising two dependent weights, on root section, with further vibration dampers in fuselage; tip light on each upper blade; blades fold manually outboard of all control mechanisms, to folded width within track of main landing gear; rotor hub is 50 per cent titanium/50 per cent steel; rotor brake standard; all-metal fuselage; composites tailcone; fixed incidence tailplane, elevators, fins and rudders have aluminium alloy structure, composites skins; fins toe inward approximately 25°; fixed leading-edge slat on each fin prevents airflow over fin stalling in crosswinds or at high yaw angles.

LANDING GEAR: Four-wheel type. Oleo-pneumatic shock-absorbers. Castoring nosewheels. Mainwheel tyres size 600×180 (Ka-32); 620×180, pressure 10.80 bar (156 lb/sq in) (Ka-32A). Nosewheel tyres size 400×150 (Ka-32); 480×200, pressure 5.90 bar (85 lb/sq in) (Ka-32A). Skis optional.

POWER PLANT: Two 1,633 kW (2,190 shp) Klimov TV3-117V (Ka-32) or TV3-117VMA (Ka-32A) turboshafts, with automatic synchronisation system, side by side above cabin, forward of rotor driveshaft. Main gearbox brake standard. Oil cooler fan aft of gearbox. Cowlings hinge downward as maintenance platforms. Fuel in tanks under cabin floor and inside container each side of centre-fuselage; capacity of main tanks 2,180 litres (576 US gallons; 480 Imp gallons); maximum capacity with two underfloor auxiliary tanks 3,450 litres (911 US gallons; 759 Imp gallons). Single-point pressure refuelling behind small forward-hinged door on port side, where bottom of tailboom meets rear of cabin.

ACCOMMODATION: Pilot and navigator side by side on air conditioned flight deck, in adjustable seats. Rearward-sliding jettisonable door with blister window each side. Seat behind navigator, on starboard side, for observer, loadmaster or rescue hoist operator. Alcohol windscreen anti-icing. Direct access to cabin from flight deck. Heated and ventilated main cabin of Ka-32 can accommodate freight or 16 passengers, on three folding seats at rear, six along port sidewall and seven along starboard sidewall (13 passengers in Ka-32A). Lifejackets under seats. Fittings to carry four stretchers. No provisions for lavatory or galley. Pyramid structure can be fitted on floor beneath rotor driveshaft to prevent swinging of external cargo sling loads. Rearward-sliding door aft of main landing gear on port side, with steps below. Emergency exit door opposite. Hatch to avionics compartment on port side of tailboom.

SYSTEMS: Three hydraulic systems: main system supplies servos, mainwheel brakes and hydraulic winch when fitted; standby system supplies only servos after main system failure; auxiliary system supplies brakes after main system failure and adjusts height of helicopter fuselage above ground; it can also be connected to main system for checking all functions on ground. Electrical system includes two independently operating AC generators and two batteries which cut in automatically or manually via inverters after AC generating system failure. After failure of either generator, the other is switched automatically to supply both circuits. Two rectifiers supply DC power. Electrothermal de-icing of entire profiled portion of each blade switches on automatically when helicopter enters icing conditions. Hot air engine intake anti-icing. APU in rear of engine bay fairing on starboard side, for engine starting and to power all essential hydraulic and electrical services on ground, eliminating need for GPU.

AVIONICS: *Flight:* Include electromechanical flight director controlled from autopilot panel, Doppler hover indicator, two HSI and air data computer. Fully coupled three-axis autopilot can provide automatic approach and hover at height of 25 m (82 ft) over landing area, on predetermined course, using Doppler. Radar altimeter. Doppler box under tailboom.

EQUIPMENT: Doors at rear of fuel tank bay provide access to small compartment for auxiliary fuel, or liferafts which eject during descent in emergency, by command from flight deck. Container each side of fuselage, under external fuel containers, for emergency flotation bags, deployed by water contact. Optional rescue hoist, capacity 300 kg (661 lb), between top of door opening and landing gear. Optional external load sling, with automatic release and integral load weighing and stabilisation systems. Firefighting version of Ka-32T demonstrated in 1996.

DIMENSIONS, EXTERNAL:

Rotor diameter (each)	15.90 m (52 ft 2 in)
Length overall: excl rotors	11.27 m (36 ft 11¾ in)
rotors folded	12.25 m (40 ft 2¼ in)
Width, rotors folded	4.00 m (13 ft 1½ in)
Height to top of rotor head	5.45 m (17 ft 10½ in)
Wheel track: mainwheels	3.515 m (11 ft 6½ in)
nosewheels	1.41 m (4 ft 7½ in)
Wheelbase	3.03 m (9 ft 11¼ in)
Cabin door: Height	approx 1.20 m (3 ft 11¼ in)
Width	approx 1.20 m (3 ft 11¼ in)

DIMENSIONS, INTERNAL:

Cabin: Length	4.52 m (14 ft 10 in)
Max width	1.30 m (4 ft 3 in)
Max height	1.24 m (4 ft 0¾ in)

AREAS:

Rotor disc (each)	198.50 m² (2,136.6 sq ft)

WEIGHTS AND LOADINGS:

Weight empty	6,610 kg (14,573 lb)
Max payload: internal	3,700 kg (8,157 lb)
external	5,000 kg (11,023 lb)
Normal T-O weight	11,000 kg (24,250 lb)
Max flight weight with slung load	12,700 kg (27,998 lb)

PERFORMANCE (Ka-32):

Max level speed	140 kt (260 km/h; 162 mph)
Max cruising speed	130 kt (240 km/h; 149 mph)
Service ceiling	6,000 m (19,680 ft)
Hovering ceiling OGE	3,500 m (11,480 ft)
Hovering ceiling OGE, OEI	1,705 m (5,600 ft)
Range with max standard fuel	432 n miles (800 km; 497 miles)
Max range with auxiliary fuel	612 n miles (1,135 km; 705 miles)
Endurance with max standard fuel	4 h 30 min
Max endurance with auxiliary fuel	6 h 25 min

UPDATED

KAMOV KA-50 CHERNAYA AKULA

English name: Black Shark
NATO reporting name: Hokum

TYPE: Attack helicopter.

PROGRAMME: Project launched in December 1977 as **V-80** (*Vertolyet* 80: Helicopter 80); first prototype (010) built by Kamov bureau and hovered at Lyubertsy 17 June 1982 and flew on

Kamov Ka-50 numbered '25', apparently recently built *(Yefim Gordon)* ***NEW***/0558800

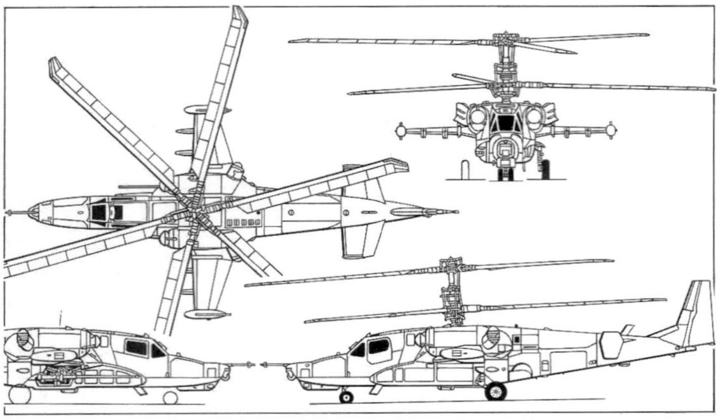

Ka-50 ('Hokum') single-seat combat helicopter with scrap view of gun installation on starboard side *(Jane's/Mike Keep)* 0507190

23 July, powered by TV3-117V engines; second prototype (011) flew 16 August 1983 with TV3-117VMA engines and mockup of Shkval tracking system, Merkury LLLTV, cannon and K-041 sighting system; both prototypes wore painted 'windows' to simulate fictitious rear cockpits. Initially reported in West in mid-1984, but first photograph did not appear (US Department of Defense's *Soviet Military Power*) until 1989.

First prototype lost in fatal accident on 3 April 1985; replaced by third prototype (012) with Mercury LLTV system for state comparative test programme against Mil Mi-28, which completed in August 1986. Two preproduction **V-80Sh-1** s (014 and 015) were first to be built at Arsenyev and introduced UV-26 chaff/flare dispensers; second had K-37-800 ejection system and mockup of LLLTV in articulated turret. Ordered into production in December 1987. Further three for continued development work comprised 018 (first flown at Arsenyev 22 May 1991), 020 'Werewolf' and 021 'Black Shark'. (Export marketing name was originally Werewolf, but changed to Black Shark by 1996.) State tests of Ka-50 began in mid-1991 and type was commissioned into Russian Army Aviation in August 1993 for trials at 4th Army Aviation Training Centre, Torzhok. In August 1994, the Ka-50 was included in the Russian Army inventory by Presidential decree, judged winner of the fly-off against Mi-28. The Mi-28 was nominally terminated on 5 October 1994 but the competition continued.

Further army evaluation followed when first two of four production Ka-50s were funded in 1994 and officially accepted on 28 August 1995; third and fourth received in 1996; these four numbered 20 to 23 (prompting pre-series 021 to be renumbered 024 to avoid confusion). Arsenyev production was to have increased to one per month during 1997, but this did not occur. The original Ka-50 (and rival Mi-28A) were overtaken by the issue of a revised requirement which emphasised night capability — favouring the two-seat Mi-28. The initial order for 15 Ka-50s was reportedly cancelled in September 1998, with procurement postponed until 2003. Three deployed to Mozdok during 1999 for use in Chechnya, but not used operationally. Two returned to theatre in December 2000; first firing of weapons against guerrilla forces was on 6 January 2001 (operating in conjunction with Mil Mi-24s); helicopters returned to Torzhok in March 2001. Unspecified modifications, found necessary as a consequence of operational deployment, had been incorporated by November 2002, according to a Kamov announcement.

Klimov VK-3000 turboshaft offered as alternative power plant.

CURRENT VERSIONS: **Ka-50** ('Hokum'): *As described.*

Ka-50N (*Nochnoy:* Nocturnal): Also reported as Ka-50Sh. Night-capable attack version; essentially a single-seat Ka-52. Programme began 1993; originally based on TpSPO-V and Merkury LLLTV systems, which tested on Ka-50 development aircraft. Ka-50N first reported April 1997 as conversion of prototype 018 with Thomson-CSF Victor FLIR turret above the nose and Arbalet (crossbow) mast-mounted radar, plus second TV screen in cockpit; FLIR integrated with Uralskyi Optiko-Mekhanicheskyi Zavod (UOMZ) Shamshit-50 (Laurel-50) electro-optic sighting system, incorporating French IR set. First flight variously reported as 4 March or 5 May 1997; programmed improvements included replacement of PA-4-3 paper moving map with digital equivalent; by August 1997, FLIR turret was repositioned below nose and Arbalet was removed; by mid-1998, had IT-23 CRT display replaced by TV-109, and HUD removed and replaced by Marconi helmet display. Proposed new cockpit shown in September 1998, having two Russkaya Avionika 203 × 152 mm (8 × 6 in) LCDs and central CRT for sensor imagery. Indigenous avionics intended for any local production orders; French systems as interim solution and standard for export. Republic of Korea Army evaluated both the Ka-50N and the baseline Ka-50. In 1999, preproduction aircraft 014 was exhibited with a UOMZ GOES sensor turret in place of Shkval.

Ka-50-2: Designation applies to three quite different aircraft. Basic Ka-50-2 is a variant of the Ka-50 single-seater, though the designation is also applied to two twin-seat aircraft; first of these was a version of the Ka-52 Alligator and, as such, was described in 2001–02 and earlier *Jane's*. All Ka-50-2s differ from the baseline Ka-52 in retaining attack and anti-tank role using 12 laser beam-riding AT-8 Vikhr ATGMs or 16 Rafael NT-D ATGMs; avionics to be supplied by Israel Aircraft Industries, Lahav Division; 024 used as demonstrator. The basic Ka-50-2 was proposed to China, Finland, India, South Korea, Malaysia, Myanmar, Poland, South Africa, Syria and Turkey.

Second variant of Ka-50-2 is another two-seater, intended to have conventional stepped tandem cockpits; is offered to armed forces which do not accept the single-seat or side-by-side two-seat layouts. A further subvariant of the tandem-seat Ka-50-2, the **Erdoğan** (Turkish for Born Fighter), was proposed to Turkey jointly by Kamov and Israel Aircraft Industries. This would have been fitted with longer-span wings and feature a NATO-compatible Giat 621 turret containing a single 20 mm cannon which would fold down below the belly of the helicopter in flight, for a 360° arc of fire; it would fold to starboard for landing, and could be fired directly forward, even when folded. TV3-117VMA-02 engines. Ten Turkish pilots flew Alligator '061' at Antalya, Turkey, in early 1999 as part of evaluation process; requirement was for 145. Named as second choice when Bell AH-1Z selected, and negotiations briefly reopened in mid-2002, following impasse in agreement with USA.

Ka-52: Two-seat version; described in 2001–02 and earlier *Jane's*.

CUSTOMERS: Four for Russian Army service trials, plus eight flying prototype and pre-series helicopters; all delivered. Further 10 ordered in 1997 budget and six in 1998, of which first three were due for delivery before end of 1998; initial helicopter eventually completed in June 1999, two more being due by mid-2000. By early 2004, it was still unclear if helicopters from the first batch of 10 had been delivered to Army Aviation. Two operational Ka-50s shown at Moscow Salon in August 2001 may have been repainted trials aircraft. One army helicopter lost in accident, 17 June 1998; attributed to rotor clash.

COSTS: Unit price of Ka-50N quoted as between US$12 million and US$15 million in mid-1999.

DESIGN FEATURES: World's first single-seat close support helicopter. Coaxial, contrarotating and widely separated semi-rigid three-blade rotors, with swept blade tip, attached to hub by steel plates; small fuselage cross-section, with nose sensors; flat-screen cockpit, heavily armour protected by combined steel/aluminium armour and spaced aluminium plates, with rearview mirror above windscreen; small sweptback tailfin, with inset rudder and large tab; high-set tailplane on rear fuselage, with endplate auxiliary fins; retractable landing gear; mid-set unswept wings, carrying ECM pods at tips; four underwing weapon pylons; engines above wingroots; high agility for fast, low-flying, close-range attack role; partially dismantled can be air-ferried in Il-76 freighter. Much of fuselage skin formed by large hinged door panels, providing access to interior equipment from ground level.

FLYING CONTROLS: Kamov coaxial design; generally as Ka-32.

STRUCTURE: Fuselage built around steel torsion box beam, of 1.0 m (3 ft 3¼ in) square section. Wing centre-section passes through beam. Cockpit mounted at front of beam, gearbox above and engines to sides. Carbon-based composites materials constitute 35 per cent by weight of structure, including rotors. Approximately 350 kg (770 lb) of armour protects pilot, engines, fuel system and ammunition bay; canopy and windscreen panels are 55 mm (2¼ in) thick bulletproof glass.

LANDING GEAR: Hydraulically retractable tricycle type; twin-wheel steerable nose unit and single mainwheels all semi-exposed when up; all wheels retract rearward; low-pressure tyres.

POWER PLANT: Two 1,633 kW (2,190 shp) Klimov TV3-117VMA turboshafts with VR-80 main reduction gearbox and two PVR-800 intermediate gearboxes, with air intake dust filters and exhaust heat suppressors. Later use of 1,838 kW (2,465 shp) TV3-117VMA-SB3 turboshafts intended. Two primary fuel tanks, filled with reticulated foam, inside fuselage box beam. Total internal capacity approximately 1,800 litres (485 US gallons; 404 Imp gallons). Front tank feeds port engine; rear feeds starboard and APU. Each tank protected by layers of natural rubber. Provision for four 500 litre (132 US gallon; 110 Imp gallon) underwing auxiliary fuel tanks. Transmission remains operable for 30 minutes after oil system failure.

ACCOMMODATION: Double-wall steel armoured cockpit, able to protect pilot from hits by 20 and 23 mm gunfire over ranges as close as 100 m (330 ft). Interior black-painted for use with NVGs. Specially designed Zvezda K-37-800 ejection system, ostensibly for safe ejection at any altitude (actually from 100 m; 330 ft); following explosive separation of rotor blades and opening of cockpit roof, pilot is extracted from cockpit by large rocket; alternatively, he can jettison doors and stores before rolling out of cockpit sideways. Associated equipment includes automatic radio beacon, activated during ejection, inflatable liferaft and NAZ-7M survival kit.

SYSTEMS: All systems configured for operational deployment away from base for up to 12 days without need for maintenance ground equipment; refuelling, avionics and weapon servicing performed from ground level. AI-9V APU for engine starting, and ground supply of hydraulic and electrical power, in top of centre-fuselage. Anti-icing system for engine air intakes, rotors, AoA and yaw sensors; de-icing of windscreen and canopy by liquid spray.

PrPNK Rubikon (L-041) piloting, navigation and sighting system based on five computers: four Orbita BLVM-20-751s for combat and navigation displays and target designation, plus one BCVM-80-30201 for WCS. Incorporates PNK-800 Radian navigation system, with C-061K pitch and heading data, IK-VSP-VI-2 speed and altitude and PA-4-3 automatic position plotting subsystems. Series 3 Tester U3 flight data recorder. Ekran BITE and warning system. KKO-VK-LP oxygen system with 2 litre (0.07 cu ft) supply for 90 minutes. Electrical supply from two 400 kW generators at 115 V 400 Hz three-phase AC; 500 W converter; rectifiers for 27 V DC supply.

AVIONICS: Integrated by NPO Elektro Avtomatika.

Comms: Two R800L1 and one R-868 UHF transceivers, SPU-9 intercom, P-503B headset recorder, Almaz-UP-48 voice warning system and HF com/nav; IFF ('Slap Shot').

Flight: INS; autopilot; Doppler box under tailboom; ARK-22 radio compass; A-036A radio altimeter.

Instrumentation: Conventional instruments; ILS-31 HUD; moving map display (Kronshtadt Abris on some aircraft); small IT-23MV CRT beneath HUD, with rubber hood, to display only FLIR and monochrome LLLTV imagery. Pilot has Obzor-800 helmet sight effective within ±60° azimuth and from −20 to +45° elevation; when pilot has target centred on HUD, he pushes button to lock sighting and four-channel digital autopilot into one unit. Displays compatible with OVN-1 Skosok NVGs.

Mission: To reduce pilot workload and introduce a degree of low observability, target location and designation are assigned to other aircraft; equipment behind windows in nose includes I-25IV Shkval-V daylight electro-optical search and auto-tracking system, laser marked target seeker and range-finder; FOV ±35° in azimuth +15 to −80° in elevation. FLIR turret to be added in nose for use with NVGs.

Self-defence: L150 Pastel RWR in tailcone, at rear of each wingtip EW pod and under nose; total of 512 chaff/flare cartridges (in four UV-26 dispensers) in each wingtip pod. L-140 Otklik laser detection system; L-136 Mak IR warning.

ARMAMENT: Four BD3-UV pylons on wings. Up to 80 S-8 80 mm air-to-surface rockets in four underwing B8V20A packs or 20 S-13 122 mm rockets in four B-13L pods; or up to 12 9A4172 Vikhr-M (AT-12) tube-launched laser-guided ASMs with range of 8 to 10 km (5 to 6.2 miles) capable of penetrating 900 mm of reactive armour; or mix of both; Vikhr launched from trainable UPP-800 mounts, which can be depressed to −12°; single-barrel 30 mm 2A42 gun on starboard side of fuselage, with up to 470 armour-piercing or high-explosive fragmentation rounds, can be depressed from +3° 30' to −37° in elevation and traversed from −2° 30' to +9° in azimuth hydraulically and is kept on target in azimuth by tracker which turns helicopter on its axis; two ammunition boxes in centre-fuselage. Front box contains 240 AP rounds, rear box 230 HE rounds. Selectable rapid (550 to 600 rds/min) or slow (350 rds/min) fire, with bursts of 10 or 20 rounds. Provision for alternative weapons, including UPK-23-250 23 mm gun pods, Igla or R-73 (AA-11 'Archer') AAMs, Kh-25MP (AS-12 'Kegler') ARMs, FAB-500 bombs or dispenser weapons.

DIMENSIONS, EXTERNAL:

Rotor diameter (each)	14.50 m (47 ft 7 in)
Length overall, rotors turning	16.00 m (52 ft 6 in)
Fuselage length, excl noseprobe	14.20 m (46 ft 7 in)
Wing span	7.34 m (24 ft 1 in)
Height overall	4.93 m (16 ft 2 in)
Tailplane span	3.16 m (10 ft 4½ in)
Wheel track: main	2.67 m (8 ft 9 in)
nose	0.34 m (1 ft 1½ in)
Wheelbase	4.19 m (13 ft 9 in)

AREAS:

Rotor disc (each)	165.13 m^2 (1,777.4 sq ft)

WEIGHTS AND LOADINGS:
Weight empty .... 7,800 kg (17,196 lb)
Max external stores .... 3,000 kg (6,610 lb)
Normal T-O weight: Ka-50 .... 9,800 kg (21,605 lb)
Erdoğan .... 9,800 kg (21,605 lb)
Max T-O weight: Ka-50 .... 10,800 kg (23,810 lb)
Erdoğan .... 11,300 kg (24,912 lb)
PERFORMANCE:
Max speed:
in shallow dive .... 210 kt (390 km/h; 242 mph)
in level flight .... 162 kt (300 km/h; 186 mph)
in sideways flight .... 43 kt (80 km/h; 49 mph)
in backward flight .... 48 kt (90 km/h; 55 mph)
Cruising speed .... 146 kt (270 km/h; 168 mph)
Vertical rate of climb at 2,500 m (8,200 ft) .... 600 m (1,970 ft)/min
Service ceiling .... 5,500 m (18,040 ft)
Hovering ceiling OGE .... 4,000 m (13,120 ft)
Range: combat .... 243 n miles (450 km; 279 miles)
with max internal fuel .... 280 n miles (520 km; 323 miles)
with 4 auxiliary tanks:
Ka-50 .... 594 n miles (1,100 km; 683 miles)
Erdoğan .... 626 n miles (1,160 km; 720 miles)
Endurance: standard fuel, 10 min reserves .... 1 h 40 min
with 2 auxiliary tanks .... 2 h 50 min
g limit .... +3.5

UPDATED

KAMOV KA-52 ALLIGATOR

Only one prototype of this two-seat combat helicopter has been built. Attempts to secure a customer continued in 2001–02 with the offer of a Ka-52K variant to South Korea. A second stage of State Tests was due to have begun in early 2003, up to which time, it was confirmed, the Arbalet radar had not been installed (except for display purposes). Radar trials were confirmed to have begun by November 2003. Irregular financing expected to cause trials to be extended over an indefinite period, but in a surprise announcement of December 2003, C-in-C of Russian Air Forces revealed plans for delivery of 12 Ka-52s (and some Mil Mi-28s) to begin in early 2004.

UPDATED

KAMOV KA-60 KASATKA

English name: Killer Whale

TYPE: Medium transport helicopter.

PROGRAMME: Original coaxial rotor, twin tail, single-engined V-60 won Soviet Army lightweight helicopter and Mi-8 replacement competition against heavier, twin-engined Mil Mi-36 in 1982; subsequently, design considerably modified to achieve greater speed through adoption of single main rotor and Fenestron-type of tail rotor with eleven blades. First flight originally due 1993, but programme slowed by funding shortages, and priority changed to promotion of civil variant (see following entry for Ka-62). Ka-60 officially revealed at Lyubertsy on 29 July 1997, when prototype close to completion; first flew (601) 10 December 1998; second sortie 21 December; first official flight 24 December; all were hovering flights; international debut at MAKS '99, Moscow, August 1999; first 'forward flight' 24 December 1999. Further testing was intermittent, due to irregular Ministry of Defence funding; production versions of RD-600V turboshaft installed mid-2002. At this time, it was stated only that the prototype had completed "several" flights, although State Trials were not due to begin until early 2003, which subsequently postponed to 2004. No. 601 had flown some 50 hours by October 2003.

Conflicting reports quote both Arsenyev and Ulan Ude as prospective production lines. However, LMZ (now LAPIK, part of RSK 'MiG') was reported in April 2000 to be preparing for production and in mid-2001 was building second prototype, which to be completed as a trainer in **Ka-60U** configuration. This entered final assembly in July 2002, although RD-600 engines were not due to have been received until late 2002. Displayed, apparently unflown, (marked as 602) at MAKS '03, Moscow, August 2003. Series production LAPIK was due to begin in 2003, later delayed to 2005. Meanwhile, third prototype to be completed in 2004 in assault transport configuration, all three to participate in almost 400 sorties of State Tests, lasting two years.

In August 2002, it was announced that power plant will be changed to Klimov VK-1500 to increase participation by RSK 'MiG' group.

A smaller variant of Ka-60 was reported in mid-2001 to have been offered to Russian Navy.

CURRENT VERSIONS: Pilot and aircrew training (Ka-60U), utility, shipborne over-the-horizon targeting (Ka-60K) reconnaissance (Ka-60R) anti-tank and anti-helicopter versions

Kamov Ka-60 instrument panel ***(Sebastian Zacharias)*** ***NEW***/0567710

Kamov Ka-60 prototype ***(Yefim Gordon)*** ***NEW***/0573401

Second prototype Kamov Ka-60 ***(Yefim Gordon)*** ***NEW***/0573400

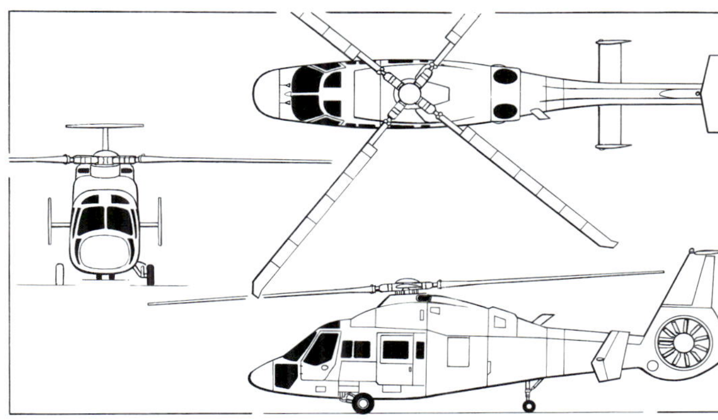

Kamov Ka-60 military multipurpose helicopter ***(James Goulding)*** 0062678

proposed; role of Ka-60R, with Shamshit target acquisition system, transferred to Kamov Ka-52.

CUSTOMERS: None. Long-term interest maintained by Russian Army; reportedly evaluated by Iran. Russian requirements estimated as over 100 Ka-60Us between 2001 and 2010.

COSTS: US$1.7 million, Ka-60U (2000).

DESIGN FEATURES: Generally as for Ka-62 (which see), but with IR- and radar-absorbent coatings, reduced rotor speed, low-IR exhausts.

POWER PLANT: Prototype has RKBM Rybinsk RD-600V turboshafts, as Ka-62. Production engine will be 1,103 kW (1,479 shp) Klimov VK-1500. RRTM RTM322 or GE CT7 available in export versions.

ACCOMMODATION: Up to 16 infantry troops; or six stretchers and three attendants. Pilot (starboard) and co-pilot/gunner (port) side by side. Provision for dual controls; control stick top common with Ka-50/52.

SYSTEMS: All Russian; Western equivalents optional for export.

AVIONICS: As above, including Pastel RWR and Otklik laser warning system.
Radar: Arbalet MMW antenna in nose.
Instrumentation: Three-screen EFIS.

EQUIPMENT: Cargo hook.

ARMAMENT: One-piece transverse boom through cabin, to rear of doors, optional to provide suspension for total of two B-8V-7 seven-round 80 mm rocket pods, two 7.62 mm or 12.7 mm gun pods, or similar armament.

DIMENSIONS, EXTERNAL: As Ka-62 except:
Fuselage length .... 13.465 m (44 ft 2 in)

WEIGHTS AND LOADINGS: Generally as Ka-62 except:
Max payload: external .... 2,750 kg (6,062 lb)

UPDATED

KAMOV KA-62

TYPE: Multirole medium helicopter.

PROGRAMME: Funded under Russian programme for development of civil aviation for 2000. Construction of prototype Ka-62 (then known as V-62) began early 1990, but apparently abandoned; one Ka-60 military version (which see) and two Ka-62s intended to undertake flight trials, although second of basic type completed as Ka-60U, delaying debut of civil version; will be certified under Russian AP and FAR Pt 29A/B standards. Republic of Buryatia officially requested Moscow for production rights for UUAP (which see) in February 2000. In April 2001, Turkish Ministry of Public Health was discussing contract for six Ka-62s, with total value of US$31.5 million. Russian government's 2002–10 aviation plan includes Rb62 million to develop Ka-62 and Rb51 million to launch production at UUAP and, possibly, RSK 'MiG' (LAPIK). Initial deployment will be in Khabarovsk Federal District of Russian Far East, following June 2002 agreement between Kamov and Far East Investment Company for manufacture of Ka-62 at unspecified plant(s) and marketing in Asia/Pacific region.

Model of Kamov Ka-62 civil multirole helicopter, confusingly wearing military camouflage *(Yefim Gordon)* ***NEW***/0528671

No prototype had been built by 2004, priority of Ka-62 indicated by official announcement in October 2003 that building of a prototype (which would be fourth Ka-60/62) only could "not be ruled out".

CURRENT VERSIONS: **Ka-62:** Basic model for domestic market.

Detailed description applies to the above.

Ka-62M: To be certified to Western standards, for sale outside Russian Federation and Associated States (CIS); two 1,212 kW (1,625 shp) General Electric T700/CT7-2D1 turboshafts; five-blade main rotor; avionics to be developed by Aviapribor. Production version is expected to be manufactured by KnAAPO at Komsomolsk.

Ka-64 Sky Horse: This development of the Ka-60/62 series is reported to be a joint venture with Agusta of Italy intended for export. Features include a conventional tail rotor, modified passenger cabin, Western avionics and option of General Electric CT7-2DL, LHTEC T800 or RRTI RTM 322 turboshaft engines. Production would be by UUAP at Ulan-Ude.

DESIGN FEATURES: Originated as military transport; all main systems and components duplicated, with main and secondaries routed on opposite sides of airframe; transmission resistant to 12.7 mm bullets; main blades to 23 mm shells; run-dry gearboxes. Advanced technology main rotors with sweptback tips. Production versions will have slower-turning five-blade rotor. Yaw control by 11-blade fan-in-fin. Reverse tricycle landing gear.

STRUCTURE: Composites account for 60 per cent, by weight, of structure, including blades of main rotor; fuselage sides, doors, floor and roof, tailboom, fin, vertical stabilisers, and fan blades of carbon-reinforced Kevlar.

LANDING GEAR: Retractable reverse tricycle type; single KT-217 mainwheels retract inward and upward into bottom of fuselage; twin rear wheels retract forward into tailboom; shock-absorber in each unit. Optional inflatable pontoons for emergency use on water.

POWER PLANT: Basic Ka-62 has two RKBM Rybinsk RD-600V turboshafts, each 956 kW (1,282 shp) max continuous; 1,140 kW (1,529 shp) emergency rating. However 1,103 kW (1,479 shp) Klimov VK-1500 may be offered in parallel with adoption of this engine for Ka-60. Fuel tanks under floor, capacity 1,450 litres (383 US gallons; 319 Imp gallons). General Electric T700/CT7-2D1 offered as alternative to RD-600V.

ACCOMMODATION: Crew of one or two, side by side; optional bulkhead divider between flight deck and cabin; up to 14 passengers in four rows; forward-hinged door each side of flight deck; large forward-sliding door and small rearward-hinged door each side of cabin; baggage hold to rear of cabin. VIP configuration to be available, with five to nine seats and refreshment bar.

SYSTEMS: Interior heated and air conditioned. Thermoelectric de-icing system optional. Ivchenko AI-9V APU originally proposed, but replacement Aerosila TA-14 under development.

AVIONICS: Optional Russian or Western.

EQUIPMENT: Stretchers, hoist above port cabin door, cargo tiedowns, and other items as necessary for variety of roles, including transport of slung freight; air ambulance/operating theatre; search and rescue; patrol of highways, forests, electric power lines, gas and oil pipelines; survey of ice areas; surveillance of territorial waters, economic areas and fisheries; mineral prospecting; and servicing of offshore gas and oil rigs.

DIMENSIONS, EXTERNAL:
Main rotor diameter ... 13.50 m (44 ft 3½ in)
Tail rotor diameter ... 1.40 m (4 ft 7 in)
Length overall, rotors turning ... 15.60 m (51 ft 2¼ in)
Fuselage length ... 13.465 m (44 ft 2 in)
Height: overall ... 4.60 m (15 ft 1 in)
to top of rotor head ... 3.80 m (12 ft 5½ in)
Width over endplate fins ... 3.00 m (9 ft 10 in)
Wheel track ... 2.50 m (8 ft 2½ in)
Wheelbase ... 5.445 m (17 ft 10¼ in)
Cabin doors (each) Height ... 1.30 m (4 ft 3¼ in)
Width ... 1.25 m (4 ft 1¼ in)

DIMENSIONS, INTERNAL:
Cabin, excl flight deck: Length ... 3.40 m (11 ft 2 in)
Width ... 1.78 m (5 ft 10 in)
Height ... 1.30 m (4 ft 3¼ in)

AREAS:
Main rotor disc ... 143.10 m^2 (1,540.3 sq ft)
Tail rotor disc ... 1.54 m^2 (16.57 sq ft)

WEIGHTS AND LOADINGS:
Max payload: internal: 62, 62M ... 2,000 kg (4,409 lb)
external: 62, 62M ... 2,500 kg (5,510 lb)
T-O weight: 62, normal ... 6,000 kg (13,228 lb)
62, internal load, max ... 6,250 kg (13,779 lb)
62 external load and 62M, internal load, max ... 6,500 kg (14,330 lb)
62M, external load, max ... 6,750 kg (14,880 lb)
Max disc loading: 62 ... 41.9 kg/m^2 (8.58 lb/sq ft)
62M ... 47.1 kg/m^2 (9.66 lb/sq ft)

PERFORMANCE (estimated, at max T-O weight with internal load):
Max level speed:
62, 62M ... 162 kt (300 km/h; 186 mph)
Cruising speed: 62, 62M ... 148 kt (275 km/h; 171 mph)
Max rate of climb at S/L: 62 ... 625 m (2,050 ft)/min
62M ... 690 m (2,263 ft)/min
Rate of climb at S/L, OEI: 62, 62M ... 123 m (403 ft)/min
Service ceiling: 62 ... 5,150 m (16,900 ft)
62M ... 5,500 m (18,040 ft)
Hovering ceiling: IGE: 62 ... 2,900 m (9,520 ft)
OGE: 62 ... 2,100 m (6,880 ft)
62M ... 2,500 m (8,200 ft)
Range with max standard fuel at 2,000 m (6,560 ft),
no reserves: 62 ... 332 n miles (615 km; 382 miles)
62M ... 353 n miles (655 km; 407 miles)
Range with auxiliary fuel at 2,000 m (6,560 ft),
no reserves: 62 ... 566 n miles (1,050 km; 652 miles)
62M ... 553 n miles (1,025 km; 637 miles)

UPDATED

KAMOV KA-226A SERGEI

TYPE: Light utility helicopter.

PROGRAMME: Turbine version of Ka-26, of which 816 built between 1968 and 1977. Announced at 1990 Helicopter Association International convention, Dallas, USA. Developed originally for Russian TsENTROSPAS disaster relief ministry, which is providing significant funding. First flight (RA-00199), at Lyubertsy, 3 September 1997; 'official' first flight on following day. Flew total of four sorties by 31 December 1997. Began AP-29 certification testing on 28 March 2001. State ground testing of Ka-226 second prototype completed at Strela's Orenburg plant on 6 March 2000. KAPP built two prototypes, of which first was rolled out at Kumertau on 29 May 1998; first production aircraft from KAPP was due to have flown in first quarter of 2002 but remained under construction in mid-2002. Planned certification in third quarter of 2002 was not achieved, being reportedly "seriously delayed". Provisional certification awarded by Interstate Aviation Committee (MAK) on 18 August 2003. Prototype destroyed in ground resonance incident, November 2002, but apparently replaced by demonstrator '199', displayed at Moscow, August 2003.

Named Sergei in 1999, honouring politician Sergei Shoigu, but programme also guided by Sukhoi General Designer, Sergei Mikheev. By mid-2000, Moscow regional government had provided Rb12 million in development funding and was beginning disbursements under second programme valued at Rb4 million. Initial deliveries due in first half of 2000, but not effected.

Prototypes built jointly by Kamov, Strela, KAPP and Ufa Motors, with final assembly by KAPP and Strela for production aircraft. Strela scheduled to have delivered five preproduction helicopters to Kamov at Lyubertsy by mid-2000; these for MChS Rossii, but not supplied until 2003, when one exhibited at MAKS '03. KAPP designated second production plant; first batch of five under construction by 2001. Motor-Sich of Zaporozhye, Ukraine, negotiated with Kamov in June 2000 to build Ka-226s powered by indigenous ZMKB AI-450 engine; agreement on AI-450 installation signed 15 August 2001. On 19 October 2001, however, Motor Sich announced it would source all Ka-226 components with Ukrainian industry, if decision to proceed were taken. Programme launch was reportedly imminent in late 2002. MoU on use of Turbomeca Arrius 2G signed in August 2001 with NPO Saturn and French manufacturer; collaboration agreement followed on 16 April 2002, and engines delivered in December 2003, with intention of certifying Arrius Ka-226 in late 2004 after trials of prototype. Batch of Rolls-Royce engines ordered by Kamov in July 2000.

CURRENT VERSIONS: **Ka-226A:** *As described.*

Ka-226AG: Gazprom specific configuration.

Ka-226-50: Designation first revealed in September as 'improved' version, but appears to apply to all production aircraft (at least from KAPP).

Ka-226U: Dual control trainer. First order placed early 2003 for one helicopter to fly in 2004, following manufacture at Kamov's own plant. Progress AI-450 turboshafts.

CUSTOMERS: Orders by January 2002 totalled 66: Gazprom 50 (of which 22 firm), Moscow City 10, TsENTROSPAS five and Bashkiriyan one. This had increased to 80 by August 2003.

Identified requirements include up to 20 for City of Moscow for patrol and medevac; some 250 for MChS Rossii/TsENTROSPAS disaster relief organisation; and up to 75 for Gazprom in gasfield support role. Firm order for 25 reportedly received from

Kamov Ka-226A '199' in August 2003, fitted with a square-backed cabin *(Yefim Gordon)* ***NEW***/0573399

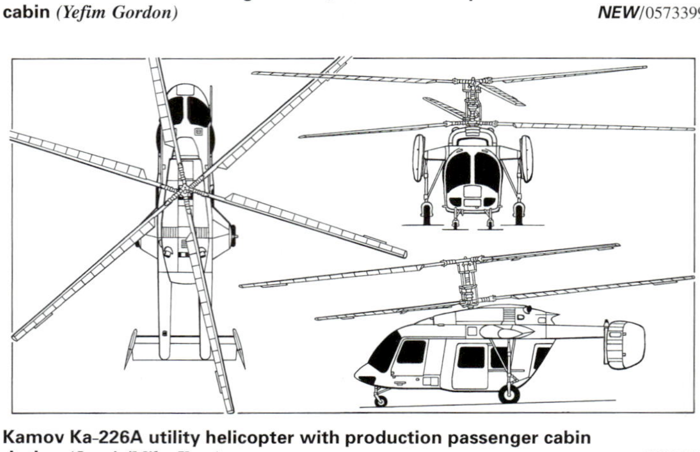

Kamov Ka-226A utility helicopter with production passenger cabin design *(Jane's/Mike Keep)* 0062676

Strela-built Kamov Ka-226 in air ambulance configuration for MChS *(Yefim Gordon)* ***NEW***/0558795

TsENTROSPAS by 1997, but quantity had reduced to 10 by 1999; by mid-2000 this quoted as five firm (to be first five production aircraft) and further 15 to be ordered by 2002; manufacture by Strela which had completed first two (including one in medevac configuration) by early 2002, but in October 2003 it was reported that all five had been "frozen". Bashkiriyan local government ordered one Ka-226-50 from KAPP in September 2001; this accepted 28 December 2001 (when still not cleared for flight) and due for trials at Zhukovsky before service entry. Funds for 22 of initial Gazprom order for 50 had been transferred by 2001, this initial batch, built by Strela, to have been received by 2005 (although formal signing of order for 50 was undertaken at Moscow Salon in August 2001). First two for Gazprom scheduled for delivery in 2004 in baseline configuration; remainder modified to specific customer requirements, designated Ka-226AG.

Moscow city government signed US$1.5 million order for 10 in December 2001, delivery over two years, but later announcement indicated that funds had not been earmarked; Moscow's helicopters built by KAPP. City allocated initial Rb33 million in 2002 and intended to receive three helicopters in 2003, four in 2004 and final three in 2005. Subsequently amended to first two receipts in 2004. Ka-226 was beaten by Kazan Ansat (which see) in competition to supply new training helicopter to Russian armed forces, announced September 2001, although small number of Ka-226s required by Russian Navy.

COSTS: US$1.5 million (2003). Batch of five for MChS worth Rb187 million (2003). Development cost Rb108 million (1999).

DESIGN FEATURES: Classic Kamov utility helicopter, featuring interchangeable mission pods. Refined development of Ka-26/126; new rotor system with hingeless hubs and glass fibre/carbon fibre blades; changes to shape of nose, twin tailfins and rudders, and passenger pod; passenger cabin has much larger windows and remains interchangeable with variety of payload modules including agricultural systems with hopper capacity of 1,000 litres (264 US gallons; 220 Imp gallons); Kamov BP-226 transmission; new rotor system, interchangeable with standard coaxial system, will become available later.

FLYING CONTROLS: Assisted by irreversible hydraulic actuators. Automatic rotor constant-speed control; conventional four-channel control (longitudinal, lateral, cyclic and differential pitch). Two endplate fins and rudders, toed inward 15°; fixed horizontal stabiliser.

STRUCTURE: Primarily of aluminium alloys, steel alloys and composites sandwich panels of GFRP with honeycomb filler. Rotor blade overhaul interval 2,000 hours; total life 6,000 hours, but to be extended by increments to 18,000 hours.

LANDING GEAR: Non-retractable four-wheel type. Main units at rear, carried by stub-wings. All units embody oleo-pneumatic shock-absorber. Mainwheel tyres size 595×185, pressure 2.50 bar + 0.50 (36.25 lb/sq in + 7.25); forward tyres size 300×125, pressure 3.50 bar + 0.50 (50.75 lb/sq in + 7.25). Forward units of castoring type, without brakes. Rear wheels have pneumatic brakes.

POWER PLANT: Two 335 kW (450 shp) Rolls-Royce 250-C20R/2 turboshafts, side by side aft of rotor mast, with individual driveshafts to rotor gearbox. Two 335 kW (450 shp) Rolls-Royce 250-C20B engines in prototypes. Transmission rating 626 kW (840 shp). Alternatively, two Progress (ZMKB) AI-450 turboshafts, each 331 kN (444 shp) or two Turbomeca Arrius 2G (500 kW; 670 shp) or Klimov VK-800 turboshafts (588 kW; 789 shp).

Standard fuel capacity 770 litres (203 US gallons; 169 Imp gallons), in tanks above and forward of payload module area. Provision for two external tanks, on sides of fuselage, total capacity 320 litres (84.5 US gallons; 70.4 Imp gallons).

ACCOMMODATION: Fully enclosed and lightly pressurised flight deck, with rearward-sliding door each side; normal operation by single pilot; second seat and dual controls optional. Cabin ventilated, and warmed and demisted by air from combustion heater, which also heats passenger cabin when fitted. Space aft of cabin, between main landing gear legs and under transmission, can accommodate variety of interchangeable payloads. Cargo/passenger pod has two bench seats, each accommodating three persons; one bench faces forward, the other, rear; baggage compartment behind rear wall. Seventh passenger beside pilot on flight deck. Provision for cargo sling. Ambulance pod accommodates two stretcher patients, two seated casualties and medical attendant. For agricultural work, chemical hopper (capacity 1,000 litres; 264 US gallons; 220 Imp gallons) and dust spreader or spraybar are fitted in this position, on aircraft's CG. (Flight deck pressurisation protects crew against chemical ingress.) Aircraft can also be operated with either an open platform for hauling freight or hook for slinging loads at end of a cable or in a cargo net.

SYSTEMS: Single hydraulic system, with manual override, for control actuators. Main electrical system 27 V 3 kW DC, with back-up 40 Ah battery; secondary system 36/115 V AC with two static inverters; 115/200 V AC system with 16 kVA generator (6 kVA to power agricultural equipment and rotor anti-icing). Electrothermal rotor blade de-icing; hot air engine air intake anti-icing; alcohol windscreen anti-icing; electrically heated pitot. Pneumatic system for mainwheel brakes, tyre inflation, agricultural equipment control, pressure 39 to 49 bar (570 to 710 lb/sq in). Oxygen system optional.

AVIONICS: Cockpit instrumentation and avionics to customer's choice, including Bendix/King equipment for IFR flight.

Comms: Optional Bendix/King KY 196A VHF radio; transponder.

Flight: Optional Bendix/King KN 53 ILS; KR 87 ADF; KLN 90B GPS; LCR 92 laser AHRS; and ADF.

EQUIPMENT: Specially equipped payload modules available for variety of roles, including ambulance and agricultural duties.

ARMAMENT: Optional provision for light weapons.

DIMENSIONS, EXTERNAL:
- Rotor diameter (each) ... 13.00 m (42 ft 7¾ in)
- Length of fuselage ... 8.10 m (26 ft 7 in)
- Width over stub-wings ... 3.22 m (10 ft 6¾ in)
- Height to top of rotor head ... 4.15 m (13 ft 7½ in)
- Wheel track: nosewheels ... 0.90 m (2 ft 11½ in)
  - mainwheels ... 2.56 m (8 ft 4¾ in)
- Wheelbase ... 3.48 m (11 ft 5 in)
- Passenger pod: Length ... 2.35 m (7 ft 8½ in)
  - Width ... 1.40 m (4 ft 7 in)
  - Height ... 1.54 m (5 ft 0¾ in)

DIMENSIONS, INTERNAL:
- Passenger pod: Length ... 2.04 m (6 ft 8¼ in)
  - Width ... 1.28 m (4 ft 2¼ in)
  - Height ... 1.40 m (4 ft 7 in)

AREAS:
- Rotor disc (each) ... 132.70 m^2 (1,428.4 sq ft)

WEIGHTS AND LOADINGS:
- Weight empty ... 1,952 kg (4,304 lb)
- Max underslung payload ... 1,300 kg (2,865 lb)
- Max internal fuel ... 600 kg (1,322 lb)
- Auxiliary fuel ... 256 kg (564 lb)
- T-O weight: normal ... 3,100 kg (6,835 lb)
  - max ... 3,400 kg (7,495 lb)
- Transmission loading at max T-O weight and power: normal ... 4.95 kg/kW (8.14 lb/shp)
  - max ... 5.43 kg/kW (8.92 lb/shp)

PERFORMANCE (estimated):
- Never-exceed speed (VNE) ... 115 kt (214 km/h; 133 mph)
- Max level speed ... 111 kt (205 km/h; 127 mph)
- Max cruising speed ... 105 kt (194 km/h; 120 mph)
- Max rate of climb at S/L ... 610 m (2,000 ft)/min
- Rate of climb, OEI ... 96 m (315 ft)/min
- Vertical rate of climb at S/L ... 168 m (550 ft)/min
- Service ceiling ... 6,200 m (20,340 ft)
- Hovering ceiling OGE ... 2,500 m (8,200 ft)
- Range: with max payload ... 16 n miles (30 km; 18 miles)
  - with max internal fuel ... 324 n miles (600 km; 372 miles)
  - with auxiliary fuel, no reserves ... 480 n miles (890 km; 552 miles)
- Max endurance with internal fuel, no reserves ... 4 h 42 min

UPDATED

KAPO

KAZANSKOYE AVIATSIONNOYE PROIZVODSTVENNOYE OBEDINENIE IMENI S P GORBUNOVA (KAZAN AIRCRAFT PRODUCTION ASSOCIATION NAMED FOR S P GORBUNOV)

ulitsa Dementiev 1, 420036 Kazan, Respublika Tatarstan
Tel: (+7 8432) 54 24 32
Fax: (+7 8432) 54 36 93
e-mail: kapo-imin@mail.ru
GENERAL DIRECTOR: Nail G Khairoullin

KAPO (formerly GAZ 22) established in Moscow on 14 May 1927 and moved to Kazan in November 1941; has built more than 18,000 aircraft of 34 types, including the Tu-4, Tu-16, Tu-22, Tu-104 and Il-62. It currently produces the Tupolev Tu-214 and will manufacture the Tu-330 freighter, if it is ordered, and is assigned the Tu-324 regional airliner. In 2000, KAPO belatedly delivered one Tu-160 strategic bomber, completion of which had been delayed by collapse of the former USSR; a second was to follow in 2002, but was delayed at least until 2003 and a further two are planned. In April 2002, Kazan received the first of 15 in-service Tu-160s to undergo upgrading; two Tu-22Ms also in stock.

Under aerospace industry restructuring plans announced in 2001, KAPO is to be joined with RSK 'MiG', Kamov, Tupolev and their associated plants (Sokol, Aviakor, Aviastar and KAPP). The plant also produces customer goods, including leather processing machinery. Nothing further has been heard of 2002 discussions on the possibility of aerostructures subcontracting for Airbus.

UPDATED

KAPP

KUMERTSKOYE AVIATSIONNOYE PROIZVODSTVENNOYE PREDPRIYATIE (KUMERTAU AVIATION PRODUCTION ENTERPRISE)

ulitsa Novozarinskaya 15A, 453350 Kumertau, Respublika Bashkortostan
Tel: (+7 34761) 422 53 and 423 00
Fax: (+7 34761) 439 13
GENERAL DIRECTOR: Boris S Malyshev
EXPORT DEPARTMENT: R M Rafikov

KAPP has manufactured helicopters, aeroplanes and related equipment at Kumertau since 1962. Major programmes have involved the Kamov Ka-26 and Ka-27/28/29/32 helicopter series, Myasishchev M-17/M-55, wings for the Tupolev Tu-154M transport, and unmanned air vehicles including the Tu-243 Reis-D reconnaissance system. KAPP is the only manufacturer of carbon fibre/glass fibre rotor blades in the Russian Federation and Associated States (CIS).

Current production includes main rotor blades and associated bushes for the Ka-226. KAPP built second and third prototype Ka-226s and is building the type for Moscow city government, which plans to receive the first two of 10 in 2004, and Bashkiriyan local government (one). Ka-32 remains in low-volume production; Ka-31 manufactured for India. KAPP has also been chosen to manufacture the Ilyushin Il-112 transport for the Russian Air Forces. Helicopter overhaul activities have included Indian Ka-28s.

The plant is to become part of the consortium led by RSK 'MiG' under aerospace restructuring plans announced by the Russian government in May 2001 and was due for privatisation in 2004. Employment in 2003 was 5,600.

UPDATED

KAZAN

KAZANSKY VERTOLETNYI ZAVOD OAO (KAZAN HELICOPTER PLANT JSC)

420085, ulitsa Tetsevskaya, Kazan
Tel: (+7 8432) 71 81 81
Fax: (+7 8432) 71 82 82
e-mail: market@kazanhelicopters.com
Web (1): http://www.kazanhelicopters.com
Web (2): http://www.euromil/.ru
GENERAL DIRECTOR: Aleksandr P Lavrentyev
MARKETING DIRECTOR: Valery A Pashko
CHIEF ENGINEER: Igor S Bougakov
CHIEF DESIGNER: Aleksei Stepanov
ANSAT PROGRAMME DIRECTOR: Valery B Kartashev

GAZ 387 founded in Leningrad in 1940 and re-established at Kazan on 15 August 1941 following evacuation. Built 11,344 Polikarpov Po-2 biplanes up to 1947, when production turned to combined harvesters. Since 1951, however, Kazan (KVZ) has marketed and built Mil helicopters comprising 30 Mi-1s, 3,257 Mi-4s, 4,066 Mi-8s, 2,620 Mi-8M/Mi-17s, 273 Mi-14s and (up to 2001) 1,094 Mi-8MTV/Mi-17-1Vs. Exports began in 1956 and now encompass over 80 countries and some 4,000 aircraft; the Mi-17 accounts for 1,200 exports to 30 countries.

Kazan became a joint stock company in 1994, then owned 38 per cent by the state, 62 per cent by its employees; now 29.9 per cent state-owned. At the 1995 Paris Air Show, Kazan exhibited a mockup of the first product of its own newly formed (1993) design office, the Ansat light multipurpose helicopter. Modifications and upgrades to Mi-17 have generated new business for Kazan. Russian certification for helicopter design awarded February 1997. A second light helicopter, the Aktai, was revealed at Moscow in August 1997 and was due to fly in 2004. In February 2003, Kazan offered to establish a production line for the Mi-52 if Mil is unable to secure funding. Kazan was also a co-founder of Euromil, with Mil Moscow Helicopter Plant, Eurocopter and Klimov, and is responsible for manufacturing parts for, and assembling, the Mi-38 prototypes the first of which flew in December 2003. However, in July 2003, Eurocopter had withdrawn from Euromil, allowing Kazan to acquire half its shareholding.

UPDATED

Slovakian military Mi-17V-5 in executive transport colour scheme *(Sebastian Zacharias)* ***NEW***/0576280

Kazan-built Mil Mi-172 for South Korea ***NEW***/0552693

KAZAN/MIL MI-17

TYPE: Medium transport helicopter.

Following entry refers specifically to the above variant; see Mil entry for other Mi-8M/Mi-17 versions.

PROGRAMME: Kazan's Mi-17 development intended to compete with latest Western helicopters (such as Sikorsky S-92, NHI NH 90 EHI EH 101 and Eurocopter AS 332) combining advanced avionics and systems with well-proven airframe and dynamic system. Prototype (RA-70937, converted from Mi-8MTV) displayed at 1995 Paris Air Show with features including widened forward, port, door; additional starboard door; and rear loading via short ramp and two clamshell doors. Further modified, with large, single-piece rear-loading ramp and other changes, for display at ILA '96, Germany, in search and rescue form, designated Mi-17MD Night. By 1997, '70937 had gained a spine-mounted IR jammer and flight deck armour and was stated by Kazan to have the dual designation Mi-17MD/Mi-8MTV-5. At the 1999 Moscow Air Show it had lost its registration and become an 'Mi-8MTV5-1' with the alternative designations (for export) Mi-17V-5. First flight in this guise was in January 1998; first production Mi-17V-5 flew August 1999.

Second example (RA-70877) shown at LIMA, Malaysia, November 1999, marked as Mi-17-1V, although with full Mi-17MD airframe modifications, was conversion of prototype Mi-17KF (Honeywell 'glass cockpit' avionics), destined for South Korean police force (eventually handed over, January 2000). Mi-17KF redesignated Mi-172.

Kazan demonstrator RA-70898, rebuilt in Mi-172 guise in 2000 (following accident) with partial 'glass cockpit' supplied by Elbit, GOES FLIR chin turret and additional external fuel tank (900 litres; 238 US gallons; 198 Imp gallons) each side, increasing range to 600 n miles (1,111 km; 690 miles); optimised for offshore oil support and maritime patrol; displayed at Farnborough, July 2000.

In 2002, Kazan began production of Mi-17V-5 with Klimov VK-2500 engines (replacing 1,471 kW; 1,973 shp TV3-117VMs) for increased high-altitude performance, including increase of 620 m (2,040 ft) in OGE hover and 200 m (660 ft) in service ceiling; can also take off with one engine inoperative; features Saphir APU to permit engine starting at 6,000 m (19,680 ft). Optional equipment includes NVGs, FLIR and searchlight, plus firefighting, SAR and medevac interiors. VK-2500 version certified, and deliveries began, mid-2003.

A further upgrade, designated Mi-8MTV-7 is under development.

CURRENT VERSIONS: **Mi-17V-5 (Mi-8MTV5-1):** *As described.*

Mi-17V-7: Proposed upgraded version with VK-2500 engines, uprated transmission, new composites main blades, new tail rotor, 'glass cockpit', increased service ceiling and extra 500 kg (1,102 lb) payload. Available from 2005.

CUSTOMERS: Four armed Mi-171V exported to Rwanda, 1999, with a fifth (Mi-172) aircraft in VIP configuration. South Korea (one Mi-172 for police force as Russian debt repayment); second was due to follow in 2003, also fitted with Honeywell avionics. Indian Ministry of Defence ordered 40 Mi-17-1Vs from Kazan in May 2000; deliveries completed 2001. Six Mi-17-1Vs delivered in Colombia in first half of 2002.

COSTS: Colombian order for six valued at US$36.1 million (2002).

DESIGN FEATURES: Generally similar to latest versions of Mi-17. Main differences include rear loading ramp and pointed 'Dolphin' nose with increased volume to allow space for radar. Ramp allows disembarkation of 36 equipped troops in less than 15 seconds; helicopter can fly with ramp open for supply dropping. Conventional pod and boom configuration; five-blade main rotor, inclined forward 4° 30' from vertical; interchangeable blades of basic NACA 230 section, twist 5°, thickness/chord ratio 13 per cent at root, 11.38 per cent at tip; spar failure warning system; drag and flapping hinges a few inches apart; flapping compensator and centrifugal blade droop stop; rotor brake standard; non-folding blades carried on machined spider; pendulum vibration damper; three-blade port tail rotor. Transmission comprises VR-14 two-stage planetary main reduction gearbox (engine input 15,000 rpm; oil capacity 49 litres; 12.9 US gallons; 10.8 Imp gallons), intermediate (8A-1515-000) and tail rotor (2Y6-1517-000) gearboxes, and drives off main gearbox for tail rotor, fan, AC generator, hydraulic pumps, compressor and tachometer generators; correct rotor speed maintained automatically by system that also synchronises output of the two engines; tail rotor pylon forms small vertical stabiliser; horizontal stabiliser near end of tailboom.

FLYING CONTROLS: Mechanical system, with irreversible hydraulic boosters; main rotor collective pitch control linked to throttles.

STRUCTURE: All-metal; semi-monocoque fuselage of riveted construction. Main rotor blades of aluminium alloy, each with extruded spar carrying root fitting, 21 honeycomb-filled trailing-edge pockets and blade tip, with titanium abrasion shielding on some of leading-edge; balance tab on each blade; each tail rotor blade made of spar andhoneycomb-filled trailing-edge. Composites blades and elastomeric hub components under development.

LANDING GEAR: Non-retractable tricycle type; self-centring twin-wheel nose unit, locked in flight; single wheel on each main unit; oleo-pneumatic (gas) shock-absorbers. Mainwheel tyres 865×280 type K2-110, pressure 5.4 bar (78 lb/sq in); nosewheel tyres 595×185 type KT-97/3, pressure 4.4 bar (64 lb/sq in). Pneumatic shoe-type brakes on mainwheels. Optional mainwheel fairings. Optional Aerozur four-bag inflatable emergency floats.

POWER PLANT: Two Klimov VK-2500 turboshafts; installed ratings of 1,986 kW (2,663 shp) in emergency; 1,765 kW (2,367 shp) for T-O. Engine cowling side panels form maintenance platforms when open, with access via hatch on flight deck.

Standard fuel capacity of 2,725 litres (720 US gallons, 599 Imp gallons) in two main external tanks and internal feed tank. Up to four ferry tanks in cabin, each 915 litres (242 US gallons; 201 Imp gallons). Engine oil capacity 15 litres (4.0 US gallons; 3.3 Imp gallons).

ACCOMMODATION: Two pilots side by side on flight deck, with provision for flight engineer's station. Windscreen de-icing standard. Up to 36 troops in cabin. Hydraulically actuated rear ramp.

Kazan Mi-172 fitted with Honeywell avionics ***NEW***/0552694

Mil Mi-172 VIP interior *(Valery Solomakhin/Kazan)* ***NEW***/0593937

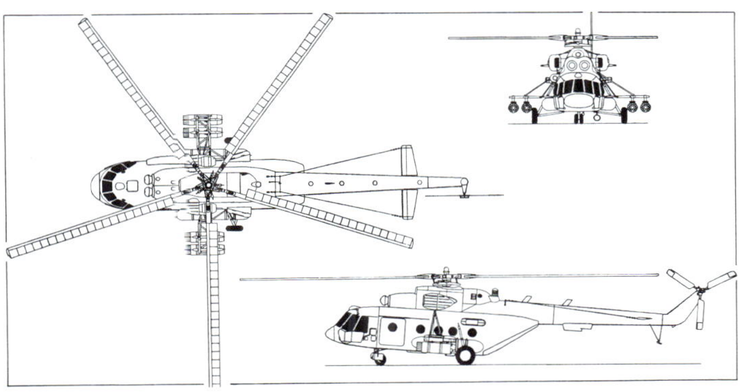

Kazan/Mil Mi-17V-5 0143323

Sliding, jettisonable passenger doors at front of cabin; electrically operated rescue hoist can be installed at this doorway.

SYSTEMS: AP-34B autopilot. Saphir 5K/G APU, mounted externally, starboard side; AC electrical supply from two 40 kW three-phase 115/220 V 400 Hz GT40P488 generators; two 25 V 20 Ah batteries. KO-50 kerosene heater in cabin. Main and auxiliary hydraulic systems, operating pressure between 44 and 64 bar (640 to 924 lb/sq in). Pneumatic system for braking; pressure 4 to 5 bar (58 to 73 lb/sq in); reservoir volume 10 litres (2.6 US gallons; 2.2 Imp gallons). Engine air intake de-icing standard. KKO oxygen system.

AVIONICS: *Comms:* Yadro-1A1 and KHF 950 VHF; P-863 VHF/UHF; ACR-500 VHF/UHF; Orlan-85ST UHF; LVD CO-70 or CO-96 transponder.

Radar: Type 8A-813, RDR 1400 or RDR 2000 weather radar optional.

Flight: Radio altimeter; radio compass; VOR/ILS; VIM-95; CD-75 VND-94 or Bendix/King KN 62 DME.

Self-defence (optional): ASO-2V chaff/flare dispensers under tailboom and L166V1AE IR jammer (NATO 'Hot Brick') at forward end of tailboom.

EQUIPMENT: Optional medical evacuation and SAR fitments. LPG-150M winch for cargo loading; optional SLG-300 300 kg (661 lb) external SAR winch.

ARMAMENT: Optional armament on two external pylons. Typical loads include UPK-23-250 23 mm cannon pods, 500 kg bombs and rocket pods.

DIMENSIONS, EXTERNAL:
- Main rotor diameter ... 21.295 m (69 ft 10½ in)
- Main rotor blade chord ... 0.52 m (1 ft 8½ in)
- Tail rotor diameter ... 3.91 m (12 ft 9 ⅞ in)
- Length: overall, rotors turning ... 25.31 m (83 ft 1¼ in)
  - fuselage ... 18.99 m (62 ft 3½ in)
- Width over weapon pylons ... 7.20 m (23 ft 7½ in)
- Tailplane span ... 3.705 m (12 ft 1¾ in)
- Height: overall, rotors turning ... 5.545 m (18 ft 2¼ in)
  - to top of rotor head ... 4.865 m (15 ft 11½ in)
- Wheel track: main ... 4.51 m (14 ft 9½ in)
  - nose ... 0.30 m (11¾ in)
- Wheelbase ... 4.28 m (14 ft 0½ in)
- Fwd passenger door (stbd): Height ... 1.405 m (4 ft 7¼ in)
  - Width ... 0.825 m (2 ft 8½ in)
- Fwd passenger door (port): Height ... 1.405 m (4 ft 7¼ in)
  - Width ... 1.215 m (3 ft 11¾ in)
- Rear cargo ramp: Height ... 1.56 m (5 ft 1½ in)
  - Width ... 2.30 m (7 ft 6½ in)

DIMENSIONS, INTERNAL:
- Cabin: Length ... 5.34 m (17 ft 6¼ in)
  - Width ... 2.34 m (7 ft 8¼ in)
  - Height ... 1.80 m (5 ft 10¾ in)

AREAS:
- Main rotor disc ... 356.16 m^2 (3,833.7 sq ft)
- Tail rotor disc ... 12.01 m^2 (129.28 sq ft)

WEIGHTS AND LOADINGS:
- Weight empty, equipped ... 7,468 kg (16,464 lb)
- Max payload:
  - internal ... 4,000 kg (8,820 lb)
  - external, on sling ... 4,500 kg (9,920 lb)
- Max T-O weight ... 13,000 kg (28,660 lb)
- Max disc loading ... 36.5 kg/m^2 (7.48 lb/sq ft)

PERFORMANCE (TV3-117VM engines):
- Never-exceed speed (VNE) ... 162 kt (300 km/h; 186 mph)
- Max level speed ... 135 kt (250 km/h; 155 mph)
- Max cruising speed ... 124 kt (230 km/h; 143 mph)
- Econ cruising speed ... 70 kt (130 km/h; 81 mph)
- Max forward rate of climb at S/L ... 624 m (2,047 ft)/min
- Service ceiling: TV3 ... 6,000 m (19,680 ft)
  - VK-2500 ... 6,200 m (20, 340 ft)
- Hovering ceiling, OGE: TV3 ... 3,980 m (13,055 ft)
- VK-2500 ... 4,600 m (15,100 ft)
- Range with 5% reserves:
  - with 4,000 kg (8,820 lb) payload ... 189 n miles (350 km; 217 miles)
  - ferry: normal fuel ... 386 n miles (715 km; 444 miles)
    - with max cabin fuel ... 890 n miles (1,650 km; 1,025 miles)

UPDATED

KAZAN ANSAT

English name: Light

TYPE: Light utility helicopter.

PROGRAMME: Design begun at Kazan in 1993; design subcontracts to Kazan State Technical University for structural strength and aerodynamic calculations; Aviacon Scientific and Production Centre for rotor; and Aeromekhanica for transmission. Name in Tatar language variously translated as *easy*, *simple*, *ethereal* and *light*.

Fuselage mockup exhibited at 1995 Paris Air Show; considerably revised engineering mockup (001) at Paris '97; by August 1998, now marked '01', this had accumulated 10 hours of ground running with engines and rotors; total 800 hours by February 2003.

First flight scheduled for late 1997, but initial designated flight trials aircraft (02) exhibited at Farnborough in September 1998, still unflown. First flight (02) was 12 minute hover on 17 August 1999; initial forward flight on 6 October 1999. Trials halted in November 1999, after 4 hours, due to gearbox problems; resumed in second quarter of 2000 with strengthened and redesigned main transmission, scarfed engine exhausts and new identity '902'. Total 120 hours up to February 2003.

Third (second flying) prototype (03) was to have joined the programme in late 1999, but not completed until August 2001. First flown 27 December 2001, this is to pre-production

Third (second flying) Ansat prototype wearing its most recent identity ***NEW***/0552692

First two flying Ansat prototypes *(Valery Solomakhin/Kazan)* ***NEW***/0593304

Mockup of Ansat-U training version *(Sebastian Zacharias)* ***NEW***/0576279

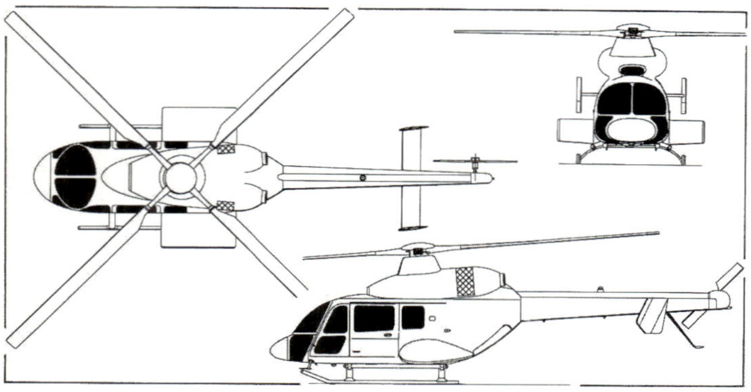

Kazan Ansat light multipurpose helicopter fitted with enlarged sponsons *(Paul Jackson)* 0507684

standard with small, detachable, pannier tanks, increased fin area, PW207 engines, additional side window and flatter windscreen combined with revised nose shape; will add 400 hours to trials programme. Certification flight testing began in October 2002, for completion before end of 2003; by early 2003 had been renumbered '904'.

On 14 September 2001, Ansat declared winner of competition to supply 100 training helicopters to Russian armed forces by 2015 (beating Mi-34 and Ka-226).

CURRENT VERSIONS: Can be optimised for transport (including underslung load), ambulance, SAR, training, patrol and other duties. Provision for attachment of sponsons or tanks to sides of cabin.

Ansat: *As described.*

Ansat-U: Optimised for training (*uchebni*). Aircraft No. 5 to this standard, including dual controls and wheel landing gear. Endplate fins deleted from horizontal tail. Expected to have TV-500A turboshafts.

Ansat-AG: Version for Gazprom pipeline inspection.

Ansat Nablyudatel: (Ansat Observation) Mockup of proposed scout helicopter based on Ansat was shown at Moscow Salon in August 2001. GOES 521 turret in nose with Svishch sensors integrated by Optooil ZAO. Max useful load 1,500 kg (3,307 lb); max T-O weight 3,500 kg (7,716 lb); max level speed 162 kt (300 km/h; 186 mph); range 383 n miles (710 km; 441 miles) on internal fuel, 691 n miles (1,280 km; 795 miles) with external tanks; service ceiling 5,700 m (18,700 ft); hovering ceiling 3,800 m (12,460 ft). Powered by two PW207K turboshafts, each rated at 522 kW (700 shp).

CUSTOMERS: Armed forces of Russia to purchase 100 by 2015. Russian Federal Border Service (Federaliya Pogranichiya Sluzhba) requirement for 100 notified in 1997. Under consideration by Gazprom in 2000 for pipeline inspection (100 required). Total 12 civil sales reported by late 2002, including one for export.

COSTS: US$2.0 million for utility version (2003). Kazan's development expenditure had reached Rb200 million by mid-2000.

DESIGN FEATURES: Optimised for aerodynamic efficiency, survivability in emergency landing (1 m; 3 ft/s power-off touchdown speed) and off-base routine maintenance. Traditional metal structure for simplicity and reliability. Meets FAR Pt 29 Category A and Russian AP-29 requirements. First Russian helicopter with shock-absorbing seats for passengers. Conventional configuration, with high-mounted tailboom carrying fixed horizontal stabiliser and twin fins; power plant above cabin. Hingeless main rotor hub with glass fibre torsion bar; four main blades; two-blade tail rotor. Two-stage, VR-23 main rotor reduction gear in magnesium case ahead of engines; ratio 16.4; rotation speed 365.4 rpm; blade tip speed 220 m (722 ft)/s. Transmission rating 769 kW (1,031 shp). Tail rotor speed 2,000 rpm via single stage conical geabox. Rotor brake. Manual blade folding.

Main rotor aerofoil section NACA 23012.

STRUCTURE: Aluminium alloy fuselage; sparing use of composites; layered glass fibre main rotor blades, window frames and nosecone.

LANDING GEAR: Twin skids with Kazan transverse shock-absorbers; tail bumper to protect anti-torque rotor. Wheels optional; tricycle configuration, with Yaroslav tyres and Gidroagregat (Balashikha) brakes. Optional emergency flotation system.

POWER PLANT: Two P&W Rus XRK206S turboshafts, each rated at 477 kW (640 shp) for T-O, 418 kW (560 shp) max continuous, in prototypes. Production version with PW207Ks, rated at 470 kW (630 shp) for T-O, 410 kW (550 shp) max continuous, 529 kW (710 shp) for 30 s, 470 kW (630 shp) continuous OEI and 491 kW (659 shp) 2 minutes OEI. FADEC standard. Fuel capacity 700 litres (185 US gallons; 154 Imp gallons) in either external panniers or underfloor. Optional internal ferry fuel.

Alternatively, two Salyut TV-500A turboshafts, each 478 kW (641 shp).

ACCOMMODATION: Up to 11 persons, including one or two pilots, on energy-absorbing seats; or two stretcher patients and three attendants; or internal or externally slung freight. Two forward-hinged doors each side of flight deck; two horizontally split doors each side of cabin, forward; baggage bay behind cabin, with rear-facing door. Baggage door also used for loading stretchers of medical variant. Accommodation ventilated and heated; optional air conditioning.

SYSTEMS: Avionika FBW controls comprise quadruplex electronic system and duplex hydraulic system. Automatic flight control is standard on all piloting functions and optional on navigation functions. Current FBW system to be replaced by KSU-A digital control system. Main transmission drives two alternators (each 200 V, 400 Hz), two generators (each 27 V), two fans and two hydraulic fuel pumps for separate systems. Electrical system 27 V, with battery; optional AC system, with second battery. Electric de-icing optional.

AVIONICS: Standard Russian avionics and instruments (apart from prototypes' BAE Systems North America engine parameters display, which to be replaced by Ulyanovsk BISK-A system); full Western avionics fit available as an option.

Radar: Provision in nosecone.

EQUIPMENT: Cabin tie-down points. Optional rescue hoist and sling.

ARMAMENT: Optional external rocket and machine gun pods.

DIMENSIONS, EXTERNAL:

Main rotor diameter	11.50 m (37 ft 8¾ in)
Main rotor blade chord	0.32 m (1 ft 0½ in)
Tail rotor diameter	2.10 m (6 ft 10½ in)
Length: overall, rotors turning	13.54 m (44 ft 5 in)
fuselage	11.06 m (36 ft 3½ in)
fuselage, tail rotor turning	11.54 m (37 ft 10¼ in)
pod	6.95 m (22 ft 9½ in)
Height to top of rotor head	3.40 m (11 ft 1¾ in)
Width: fuselage	1.80 m (5 ft 10¾ in)
over optional sponsons	3.61 m (11 ft 10¼ in)
Skid track	2.50 m (8 ft 2½ in)

DIMENSIONS, INTERNAL:

Cabin: Length	3.15 m (10 ft 4 in)
Max width	1.65 m (5 ft 5 in)
Max height	1.30 m (4 ft 3¼ in)
Volume	6.7 m^3 (237 cu ft)

AREAS:

Main rotor disc	103.87 m^2 (1,118.0 sq ft)
Tail rotor disc	3.46 m^2 (37.28 sq ft)

WEIGHTS AND LOADINGS:

Weight empty, equipped	1,900 kg (4,189 lb)
Max fuel weight	540 kg (1,190 lb)
Max payload: internal	1,000 kg (2,204 lb)
external	1,300 kg (2,866 lb)
T-O weight, internal or external load:	
normal	3,000 kg (6,613 lb)
max	3,300 kg (7,275 lb)
Max disc loading	31.8 kg/m^2 (6.51 lb/sq ft)
Transmission loading at max T-O weight and power	4.29 kg/kW (7.06 lb/shp)

PERFORMANCE:

Never-exceed speed (VNE)	153 kt (285 km/h; 177 mph)
Max level speed	148 kt (275 km/h; 171 mph)
Max cruising speed	135 kt (250 km/h; 155 mph)
Max rate of climb at S/L	960 m (3,150 ft)/min
Max rate of climb at S/L, OEI	240 m (787 ft)/min
Service ceiling	5,700 m (18,700 ft)
Service ceiling, OEI	2,600 m (8,540 ft)
Hovering ceiling, OGE	3,300 m (10,820 ft)
Range: max normal fuel	342 n miles (635 km; 394 miles)
ferry fuel	647 n miles (1,200 km; 745 miles)
Endurance	3 h 20 min

UPDATED

KAZAN AKTAI

English name: White Colt

TYPE: Three-seat helicopter.

PROGRAMME: Work began 1996. Project revealed at Moscow Air Show, 19 August 1997, with display of mockup. Second, engineering mockup built 1998. Kazan decided in mid-2001 to begin building a prototype to fly in 2003, but this target slipped to 2004.

CURRENT VERSIONS: Passenger, medical, evacuation and military configurations planned.

COSTS: "Less than US$950,000" (2003). Estimated (2000) market for over 1,000. Development cost (2000) US$10 million.

DESIGN FEATURES: Intended for light transport, patrol, medical evacuation and training. Certification will be to FAR Pt 27. Conventional pod and boom configuration with T tail.

Prototype Kazan Aktai at Moscow Salon in August 2003 *(Sebastian Zacharias)* ***NEW***/0576278

Instrument panel of Kazan Aktai *(Sebastian Zacharias)* ***NEW***/0576276

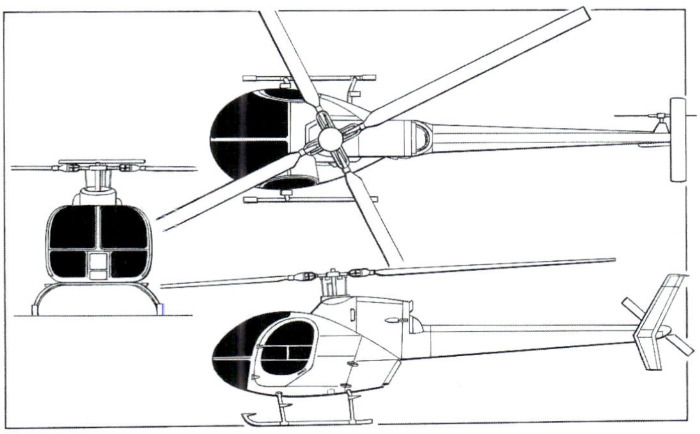

Kazan Aktai light helicopter *(James Goulding)* 0131345

Semi-articulated three-blade main rotor; tapered blades on production version; max (emergency) rotation speed 2,586 rpm; rotor brake; no provision for folding; Krasnyy Oktyabr VR-10 main gearbox, transmission rating 216 kW (290 hp); engine input 6,000 rpm. XP-10 gearbox for two-blade tail rotor.

STRUCTURE: Extensive use of composites throughout, including main rotor blades.

LANDING GEAR: Conventional, non-retractable skids on arched support tubes; spring shock absorbtion.

POWER PLANT: One 201 kW (270 hp) VAZ-4265 rotary motorcar engine mounted above the cabin and behind the gearbox and operating on 92/93 grade petrol. Manual rpm control. Fuel capacity 300 litres (79.2 US gallons; 66.0 Imp gallons). Refuelling point port side. Oil capacity 20 litres (5.3 US gallons; 4.4 Imp gallons).

ACCOMMODATION: Pilot and two passengers, all side by side, in individual seats; forward-hinged door each side. Twin clamshell doors at rear of pod, below tailboom, provide access to flat freight floor; one casualty stretcher can be loaded through rear doors when passenger seat is removed. Alternative access via door (including circular window) on starboard side. Optional, rear-facing seat in luggage bay. Optional heating and air conditioning.

SYSTEMS: Electrical system with accumulator and M-16 generator, max rating 16 kW. Anti-icing of main rotor blades.

AVIONICS: *Comms:* Yurok radio.

Flight: KG-1B navigation system; GPS.

EQUIPMENT: Inset toe-steps for maintenance access.

DIMENSIONS, EXTERNAL:

Main rotor diameter	10.00 m (32 ft 9¾ in)
Main rotor blade mean chord	0.24 m (9½ in)
Tail rotor diameter	1.48 m (4 ft 10¼ in)
Length overall, rotors turning	10.88 m (35 ft 8¼ in)
Fuselage: Length	8.35 m (27 ft 4¾ in)
Max width	1.70 m (5 ft 7 in)
Max height	1.48 m (4 ft 10¼ in)
Height overall	2.69 m (8 ft 10 in)
Tailplane span	1.44 m (4 ft 8¾ in)

Aktai engine and rear door details *(Sebastian Zacharias)* ***NEW***/0576277

Skid track	2.10 m (6 ft 10¾ in)
Passenger doors, both:	
Max height	1.00 m (3 ft 3¼ in)
Max width	0.80 m (2 ft 7½ in)
Baggage door, starboard:	
Max height	0.60 m (1 ft 11½ in)
Max width	0.70 m (2 ft 3½ in)

DIMENSIONS, INTERNAL:

Cabin: Length	1.20 m (3 ft 11¼ in)
Max width	1.70 m (5 ft 7 in)
Floor area	2.80 m^2(30.1 sq ft)
Volume	4.0 m^3 (141 cu ft)
Freight compartment: Length	1.20 m (3 ft 11¼ in)
Max width	1.00 m (3 ft 3¼ in)
Max height	1.48 m (4 ft 4 in)
Volume	2.8 m^3 (99 cu ft)

AREAS:

Main rotor blades, each	0.99 m^2 (10.66 sq ft)
Tail rotor blades, each	0.11 m^2 (1.18 sq ft)
Main rotor disc	78.54 m^2 (845.4 sq ft)
Tail rotor disc	1.72 m^2 (18.52 sq ft)
Fins, each	0.40 m^2(4.31 sq ft)
Tailplane	0.35 m^2 (3.77 sq ft)

WEIGHTS AND LOADINGS:

Weight empty	605 kg (1,334 lb)
Max payload	300 kg (661 lb)
Max T-O weight: normal	1,050 kg (2,314 lb)
max	1,150 kg (2,535 lb)
Max disc loading	14.64 kg/m^2 (3.00 lb/sq ft)
Max power loading	5.71 kg/kW (9.39 lb/hp)

PERFORMANCE (estimated):

Never-exceed speed (VNE)	135 kt (250 km/h; 155 mph)
Max level speed at S/L	97 kt (180 km/h; 112 mph)
Econ cruising speed	84 kt (155 km/h; 96 mph)
Max rate of climb at S/L	336 m (1,102 ft)/min
Service ceiling	4,000 m (13,120 ft)
Hovering ceiling: OGE	1,000 m (3,280 ft)
Max range	243 n miles (450 km; 279 miles)
Endurance	3 h

UPDATED

KAZAN ANSAT 3 MAKSIMUM

English name: Maximum

TYPE: Medium transport helicopter.

PROGRAMME: Initial details released, and cabin mockup shown, at Moscow Salon in August 2003. To be certified to AP-29 civil standards.

DESIGN FEATURES: Development of Ansat into medium helicopter category.

POWER PLANT: Two turboshaft engines mounted above cabin. Option of 956 kW (1,282 shp) RD-600, 1,103 kW (1,479 shp) VK-1500 or 736 kW (987 shp) VK-800.

ACCOMMODATION: Two flight crew and up to 17 passengers on troop seats in cabin.

DIMENSIONS, EXTERNAL:

Main rotor diameter	14.50 m (47 ft 6¾ in)
Tail rotor diameter	2.70 m (8 ft 10¼ in)
Fuselage max width	1.80 m (5 ft 10¾ in)
Height to top of fin	3.73 m (12 ft 2¾ in)
Tailplane span	2.85 m (9 ft 4¼ in)
Wheel track	2.83 m (9 ft 3½ in)
Wheelbase	3.55 m (11 ft 7¾ in)

DIMENSIONS, INTERNAL:

Cabin floor area	8.95 m^2 (96.3 sq ft)

AREAS:

Main rotor disc	165.10 m^2 (1,777.1 sq ft)
Tail rotor disc	5.72 m^2 (61.57 sq ft)

Interior of Ansat 3 Maksimum *(Sebastian Zacharias)* ***NEW***/0576220

WEIGHTS AND LOADINGS:
- Max payload .... 1,700 kg (3,748 lb)
- Max T-O weight .... 5,678 kg (12,517 lb)

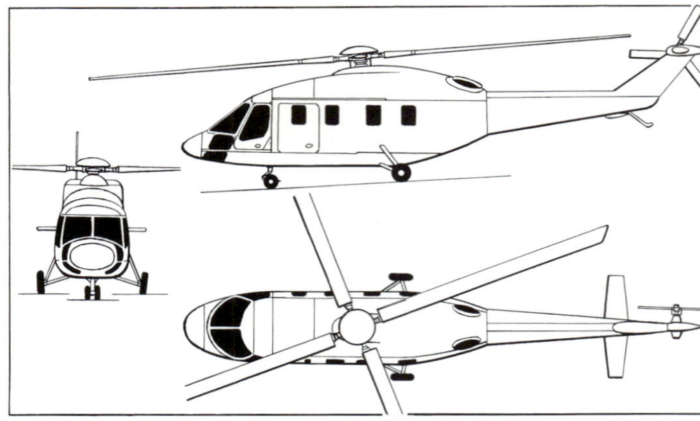

Ansat 3 Maksimum medium helicopter *(James Goulding)* ***NEW***/0576723

- Max disc loading .... 34.4 kg/m^2 (7.04 lb/sq ft)

PERFORMANCE:
- Max level speed .... 151 kt (280 km/h; 174 mph)
- Normal cruising speed .... 135 kt (250 km/h; 155 mph)
- Hovering ceiling: IGE .... 5,000 m (16,400 ft)
  - OGE .... 3,500 m (11,480 ft)
- Service ceiling .... more than 6,000 m (19,680 ft)
- Range with 30 min reserves .... more than 459 n miles (850 km; 528 miles)

NEW ENTRY

KHRUNICHEV

GOSUDARSTVENNYI KOSMICHESKII NAUCHNO-PROIZVODSTVENNYI TSENTR IMENI M V KHRUNICHEVA (STATE RESEARCH AND PRODUCTION SPACE CENTRE NAMED FOR M V KHRUNICHEV)

ulitsa Novozavodskaya 18, 121309 Moskva
Tel: (+7 095) 145 98 02
Fax: (+7 095) 145 92 03
e-mail: proton@online.ru
Web: http://www.khrunichev.com
GENERAL DIRECTOR: Anatoly I Kiselyov
DEPUTY GENERAL DIRECTORS:
Alexander V Lebedev
Anatoly A Kalinin
DIRECTOR, AVIATION PROGRAMME: Maksim Glazkov

Khrunichev Aviatekhnika
Tel: (+7 095) 145 93 33
Fax: (+7 095) 145 99 53
e-mail: proton@online.ru
Web: http://www.aircraft.avias.com
CHIEF DESIGNER: Evgeny P Grunin
DEPUTY CHIEF DESIGNER: Arnold I Andrianov
AVIATION DEPARTMENT MANAGER: Sergei M Zhiganov
RESEARCH AND DEVELOPMENT DIVISION MANAGER: Yuri I Polatsky
DESIGN DIVISION MANAGER: Mikhail M Vasyuk

The Aviation Department of this prominent rocket and space vehicle design and manufacturing centre was formed in August 1994 to develop and manufacture general purpose aircraft. It comprises a design bureau, experimental facilities and operational and maintenance services. The main task of the Aviation Department is to develop to the series production stage aircraft designed on the basis of foreign and domestic components, and then to market them worldwide.

The T-411 Aist (in production, notably for the export market) and the-T-501 are priority programmes, according to an announcement of April 2004. This also confirmed suspension of the T-440 Mercury turboprop twin and the relegation of various agricultural aircraft to the longer term. Several other designs have failed to progress beyond the mockup or prototype stage, including the T-21, T-415 and T-507.

During 2001, the company launched a joint venture with Sherpa Aircraft of the USA to market the T-411 in North America. It also builds Sherpa 200 and 300 airframes for distribution in the US, manufacture of the first '200 having begun in June 2003.

UPDATED

KHRUNICHEV T-411 AIST

English name: Stork

TYPE: Utility kitbuilt.

PROGRAMME: Programme started in 1994 when licence acquired from Aeroprogress/ROKS-Aero to produce the T-411 Aist-2 (see 1997–98 and previous editions); Aist-2 design started September 1992; construction of prototype began April 1993; first flight (marked only 'T-411') 10 November 1993 in standard form; 15 March 1994 with ski landing gear. Khrunichev prototype (RA-01585), with several detail changes, first exhibited at Moscow Air Show 1997. Second prototype reported nearing first flight in January 2002. Initially designated T-411 Wolverine, which name later transferred to T-421. North American promotion (in kit form) taken over by Sherpa Aircraft in 2001 and first two (unassembled) kits for US market were displayed at AirVenture, Oshkosh, in July 2003.

Manufacture intended by RSK 'MiG', whose Voronin plant contracted in March 2000 to build 15 for delivery in 2001, but delivery target slipped to first quarter 2002, with all 15 to be handed over during following six months. However, by June 2002, talks between Khrunichev and RSK 'MiG' still had not reached a satisfactory conclusion, although first aircraft then nearing completion.

In November 2003, Khrunichev reached agreement with Infrakos of Kazakhstan to produce T-411s at Baikonur, where the RSK-assembled aircraft will be completed. Certification was then expected by end 2004. At the same time an improved version with 2,500 kg (5,511 lb) max T-O weight and undisclosed increased fuel capacity was under consideration. By March 2003, deliveries were stated to be due for completion by end 2003, but this slipped and only two were delivered (in kit form) that year with the remaining 13 scheduled for delivery during 2004.

Second of two T-411s for US assembly *(Paul Jackson)* ***NEW***/0583380

Prototype Khrunichev Aist (VOKBM M-14X radial) *(Paul Jackson)* ***NEW***/0552719

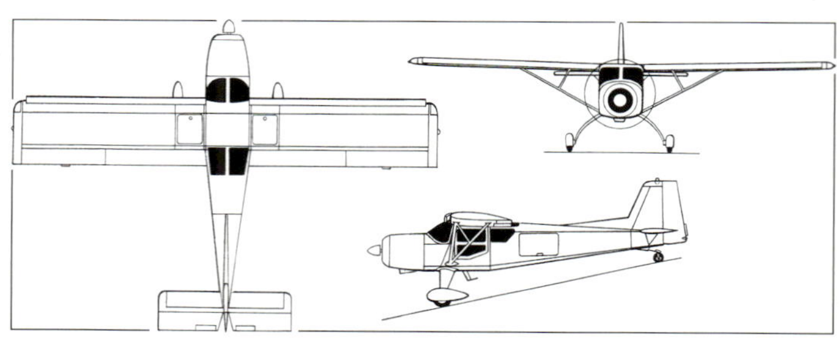

Khrunichev T-411 Aist four/five-seat light aircraft *(James Goulding)* 0079312

CURRENT VERSIONS: **T-411:** *As described.*

T-411A: Agricultural version.

T-411P: Floatplane version with twin floats, for operations in seas with wave height up to 0.35 m (1 ft 1¾ in); length overall 10.00 m (32 ft 9¾ in); maximum take-off weight 1,650 kg (3,637 lb); payload 440 kg (970 lb); cruising speed 113 kt (210 km/h; 130 mph); T-O run 315 m (1,033 ft); T-O to 15 m (50 ft) 475 m (1,558 ft); landing from 15 m (50 ft) 305 m (1,000 ft); landing run 195 m (640 ft); maximum range 607 n miles (1,125 km; 699 miles).

T-411F: As T-411F, but with amphibious floats.

T-411L: Ski-equipped version.

T-411LY: Textron Lycoming TIO-540 powered version, see Power Plant below.

T-411 Turbo: Walter M 601B turboprop. Incomplete prototype shown in Moscow, August 2001. An upgraded M 601E-powered example has also been mooted.

T-421: Described in 2001–02 *Jane's*.

CUSTOMERS: Orders received from Russian forestry, fisheries and national emergency ministries, Australia, Malaysia and USA from which approximately 70 orders received, including aircraft supplied in kit form.

COSTS: US$130,000 with M-14 engine; US$20,000 more with Lycoming engine (2002); this rose to "not less than US$180,000" by January 2004.

DESIGN FEATURES: Conventional strut-braced high-wing monoplane. Constant-chord unswept wings with fixed leading-edge slat, slotted ailerons and single-slotted, three-position (0, 20, 40°) trailing-edge flaps; V bracing struts. Rectangular-section fuselage. Unswept tail unit with dorsal fin; constant-chord horizontal tail surfaces. Compared with the Aeroprogress Aist-2, the Khrunichev version has slightly stretched airframe and more roomy cabin; different landing gear and power plant.

Wing section NACA 23011; dihedral 2° from root; incidence 3° 30'.

FLYING CONTROLS: Conventional and manual. Actuation by pushrods and cables; electrically controlled trim tab in each elevator; fixed full-span leading-edge slats; slotted flaps.

STRUCTURE: Fuselage structure of 4130 steel tube covered with Dacron synthetic fabric and (cabin and engine cowling) aluminium alloys. Two-spar, constant-chord wings; metal skin on leading-edge and over fuel tanks at root, Dacron covered between spars. Metal tail surfaces, Dacron covered.

LANDING GEAR: Tailwheel type; fixed. Fuselage-mounted spring steel cantilever mainwheel legs with wheel fairings. Self-centring castoring tailwheel. Mainwheel tyres 600×180; tailwheel 310×135. Brakes on mainwheels.

POWER PLANT: One 268 kW (360 hp) VOKBM M-14X or M-14Kh nine-cylinder radial with Mühlbauer MTV-9 three-blade variable-pitch propeller, or 261 kW (350 hp) Textron Lycoming TIO-540 flat-six with Hartzell three-blade variable-pitch propeller. Two main and two auxiliary fuel tanks in wingroots, total capacity 340 litres (89.8 US gallons; 74.8 Imp gallons); gravity fuelling. Oil capacity 22.0 litres (5.8 US gallons; 4.8 Imp gallons).

ACCOMMODATION: Pilot and three or four passengers, on side-by-side front seats and rear bench seat; starboard front seat and rear bench removable, or bench seat foldable, for carrying equivalent freight; convertible into ambulance for one stretcher patient, one seated casualty and medical attendant in addition to pilot. Dual controls standard. Baggage/cargo compartment aft of rear seats, with large upward-opening door on port side, used also for loading freight or stretcher. Forward-hinged jettisonable door each side of cabin. Ventilation and heating standard.

SYSTEMS: Pneumatic system for starting engine and operating mainwheel brakes; pressure 50 bar (725 lb/sq in). Electrical system includes GSR-3000M engine-driven 27 V DC generator and 28 Ah 12CAM-28 battery.

AVIONICS: Russian or Western (Honeywell) avionics available, including GPS. Optional autopilot.

DIMENSIONS, EXTERNAL:
- Wing span ... 13.02 m (42 ft 8½ in)
- Wing aspect ratio ... 7.0
- Length overall ... 9.45 m (31 ft 0 in)
- Height overall ... 2.60 m (8 ft 6¼ in)
- Wheel track ... 2.59 m (8 ft 6 in)
- Wheelbase ... 6.50 m (21 ft 4 in)
- Propeller diameter ... 2.03 m (6 ft 8 in)
- Baggage door: Max height ... 0.65 m (2 ft 1½ in)
  - Width ... 1.22 m (4 ft 0 in)

DIMENSIONS, INTERNAL:
- Cabin: Length ... 2.94 m (9 ft 7¾ in)
  - Max width ... 1.27 m (4 ft 2 in)
  - Max height ... 1.30 m (4 ft 3 in)

AREAS:
- Wings, gross ... 24.30 m² (261.6 sq ft)

WEIGHTS AND LOADINGS:
- Weight empty ... 544 kg (1,200 lb)
- Max payload ... 361 kg (795 lb)
- Max fuel weight ... 255 kg (562 lb)
- Max T-O weight ... 1,680 kg (3,703 lb)
- Max wing loading ... 69.1 kg/m² (14.16 lb/sq ft)
- Max power loading ... 6.27 kg/kW (10.30 lb/hp)

PERFORMANCE:
- Never-exceed speed (VNE) ... 151 kt (280 km/h; 174 mph)
- Cruising speed ... 110 kt (204 km/h; 127 mph)
- Stalling speed: flaps up ... 61 kt (112 km/h; 70 mph)
  - flaps down ... 51 kt (93 km/h; 58 mph)
- Max rate of climb at S/L ... 444 m (1,457 ft)/min
- Service ceiling ... 3,000 m (9,840 ft)
- T-O run ... 85 m (279 ft)
- Landing run ... 136 m (446 ft)
- T-O to 15 m (50 ft) ... 210 m (690 ft)
- Landing from 15 m (50 ft) ... 331 m (1,090 ft)
- Range with max fuel ... 648 n miles (1,200 km; 745 miles)

UPDATED

KHRUNICHEV T-419 STREKOZA

English name: Dragonfly

TYPE: Agricultural sprayer.

PROGRAMME: Mockup shown at Moscow Salon, August 2001. (Designation **T-411SKh** has also been used). Reported in May 2002 that prototype to fly within a few months, but no further progress apparent by mid-2004.

COSTS: US$100,000 (2002).

DESIGN FEATURES: Similar to company's earlier T-517 Farmer, using strut-braced high wing and underslung crew and payload pod. Intended for wide variety of uses, including passenger and cargo transportation, agricultural spraying, pipeline inspection and police and border guard patrols.

FLYING CONTROLS: Conventional and manual. Horn-balanced elevators and rudder. Flaps.

STRUCTURE: Principally of metal with external bracing.

LANDING GEAR: Tailwheel type; fixed.

POWER PLANT: One 268 kW (360 hp) VOKBM M-14X nine-cylinder radial engine. Fuel in two wing tanks.

ACCOMMODATION: Pilot and three passengers in cabin.

DIMENSIONS, EXTERNAL:
- Wing span ... 13.20 m (43 ft 3¾ in)
- Length overall ... 8.55 m (28 ft 0½ in)

WEIGHTS AND LOADINGS:
- Max payload ... 420 m (926 lb)
- Max fuel weight ... 280 m (617 lb)
- Max T-O weight ... 1,700 kg (3,747 lb)
- Max power loading ... 6.33 kg/kW (10.41 lb/hp)

PERFORMANCE:
- Normal cruising speed ... 97 kt (180 km/h; 112 mph)
- T-O run ... 90 m (295 ft)
- T-O to 15 m (50 ft) ... 250 m (820 ft)
- Landing from 15 m (50 ft) ... 280 m (919 ft)
- Landing run ... 136 m (446 ft)
- Range ... 594 n miles (1,100 km; 683 miles)

UPDATED

Khrunichev T-419 mockup *(Paul Jackson)* ***NEW***/0552718

KHRUNICHEV T-501

English name: Dragonfly

TYPE: Light utility turboprop.

PROGRAMME: In April 2004, Khrunichev announced that it had assigned priority to development of an extensively modified version of the Aeroprogress/ROKS-Aero T-101 Grach, which is now envisaged with two turboprop engines each of some 969 kW (1,300 hp); accommodation for up to 12 passengers; 2,700 n mile (5,000 km; 3,106 mile) range; and a price of US$1.2 million.

UPDATED

KNAAPO

GOSUDARSTVENNOYE UNITARNOYE PREDPRIYATIE KOMSOMOLSKOE-NA-AMURE AVIATSIONNOYE PROIZVODSTVENNOYE OBEDINENIE IMENI YU A GAGARINA (STATE UNITARY ENTERPRISE KOMSOMOLSK-ON-AMUR AIRCRAFT PRODUCTION ASSOCIATION NAMED FOR Y A GAGARIN)

ulitsa Sovetskaya 1, 681018 Komsomolsk-na-Amure
Tel: (+7 42172) 762 90
Fax: (+7 42172) 634 51 and 298 51
e-mail(1): knaapo@kmscom.ru
e-mail(2): knaapo@gin.ru
Web: http://www.knaapo.ru

GENERAL DIRECTOR: Viktor I Merkulov
TECHNICAL DIRECTOR: Aleksandr I Pekarsh
CHIEF, ADVANCED DEVELOPMENTS DEPARTMENT: Sergei Drobyshev

A major production centre for Sukhoi aircraft at the present time, KnAAPO has manufactured combat aircraft of the single-seat Su-27 series, including the naval Su-33 and Su-27KUB, and the Su35/35UB/37, and is also responsible for the S-80 STOL transport, Beriev Be-103 light amphibian and, when placed in production, Kamov Ka-62M helicopter. In September 2001 announced own SA-20P programme, being modified version of Be-103, with single engine.

Known as GAZ 126 when established in 1934, it has built more than 10,000 aircraft for 20 countries, beginning in May 1936. Test airfield is Dzemgi. Other products include small aluminium boats, audio-video equipment, televisions, polygraphy equipment and wooden furniture, and refurbishment of tramcars. Operates own transport airline with four An-12s, one An-26, two An-32s, two (leased to Atlant-Soyuz) Il-76s, two Tu-134s and three Mi-8s.

UPDATED

KULON-2

KULON-2
Zhukovsky, Moskovskaya oblast 129075
Tel: (+7 095) 287 14 71
Fax: (+7 095) 287 76 11
e-mail: Kulon@Mega.ru

At MAKS '03, Moscow, August 2003, this firm showed mockups of an agricultural sprayer and the Chirok light flying-boat. The latter, previously credited to the Pegas design bureau at Zhukovsky's LII flight institute, had first been shown in 1997, but has not progressed to flight.

Kulon-2's partners are Eysk-Aero, Legkaya Aviatsiya Taganroga and Aviaton (which previously built a prototype of the Merkury light twin).

VERIFIED

Kulon-2 SKhS agricultural aircraft *(Sebastian Zacharias)* ***NEW***/0567752

KULON-2 SKHS

TYPE: Agricultural sprayer.

PROGRAMME: Shown at Moscow, August 2003. First flight then scheduled for October 2003.

DESIGN FEATURES: *Sel'skokhuzyaistvennyi Samolet* (SKhS): agricultural aircraft. Envisaged as replacement for Antonov An-2s used in this role. To be certified to AP-23.

Typical sprayer configuration, with low-mounted, strut-braced wings and hopper ahead of raised cockpit. Sweptback fin.

FLYING CONTROLS: Conventional and manual. Horn-balanced rudder and elevators. Flaps.

LANDING GEAR: Tailwheel type; fixed. Mainwheels on side Vs with half-axles sprung inside centre of fuselage. Tailwheel on long cantilever tubular leg.

POWER PLANT: One 265 kW (355 hp) VOKBM M-14P radial engine driving a two-blade propeller.

DIMENSIONS, EXTERNAL:
Wing span .... 12.50 m (41 ft 0 in)
Length overall .... 8.50 m (27 ft 10½ in)

WEIGHTS AND LOADINGS:
Max chemical load .... 1,000 kg (2,205 lb)
Max T-O weight .... 2,200 kg (4,850 lb)

PERFORMANCE:
Normal operating speed .... 65–86 kt (120–160 km/h; 75–99 mph)
Max rate of climb at S/L .... 240 m (787 ft)/min

UPDATED

LVM

PROIZVODSTVENNOYE OBEDINENIE LEGKIE VERTOLETI MI OAO (MI-LIGHT HELICOPTERS PRODUCTION ASSOCIATION JSC)
a/ya 34, 103051 Moskva
Tel: (+7 095) 209 16 68
Fax: (+7 095) 200 25 34
GENERAL DIRECTOR: Grigory Bado

LVM, which adopted the export trading title of Mi-Light, was formed in April 1993 with responsibility for design, testing, series manufacture, sales, after-sales service, repair, spares provisioning, leasing and general support of the Mi-34 light helicopter. Activities also included the Mi-52 light rotorcraft. A further light design, the Mi-60, was shown in mockup form at the Moscow Salon in August 2001. Since then, however, there have been no indications received of further marketing activity.

Participating companies comprised Mil Moscow Helicopter Plant, NIAT - Moscow Aviation Technology Research Institute, Moscow Aviation Industry State Design and Research Institute, St Petersburg Aviation Industry State Design and Research Institute, Perm Motors, Sazykin Arsenyev Aviation Production Plant, Avitek, Aviastar (Ulyanovsk) and Aeromechanica Research and Production Association.

UPDATED

MIG

AVIATSIONNYI NAUCHNO-PROMYSHLENNYI KOMPLEKS MIG (ANPK MIG; MIG AVIATION SCIENTIFIC-INDUSTRIAL COMPLEX)
a/ya 125299, Leningradskoye shosse 6, Moskva
Tel: (+7 095) 155 22 49
Fax: (+7 095) 943 00 27
ADVISER TO GENERAL DIRECTOR: Rotislav A Belyakov
CHIEF DESIGNER: Lev Shengelaya
MIG-AT PROGRAMME DIRECTOR: Vasily Shtykalo
DIRECTOR OF STRATEGIC PLANNING, INFORMATION AND MARKETING: Alexander I Ageev
CHIEF, PUBLIC RELATIONS AND ADVERTISING DEPARTMENT: Svyatoslav Yu Ribas

This bureau originated in GAZ 1 (factory No. 1) founded in 1893. The experimental design bureau, founded by Artem I Mikoyan, dissociated from the factory in 1939 and later became officially the Mikoyan Aviation Scientific-Industrial Complex 'MiG' (the final letter indicating Mikhail I Gurevich). Currently known as A Mikoyan OKB Engineering Centre. It developed 250 projects and built 120 prototypes. More than 60,000 aircraft of MiG design have been manufactured, including 11,000 for export and MiG aircraft have at one time or another held 72 world records.

In 1995, the Moscow Aircraft Production Organisation and ANPK MiG were merged into the single state company MAPO 'MiG'. The original 'MiG' acronym continued to be used, though MiG lost its separate legal identity and was prevented from selling its own aircraft. On 4 July 1997 the legal status of MAPO 'MiG' was restored. Following a change of title, the bureau is now subordinate to RSK 'MiG', described separately.

Seeking to further its aerodynamic research, MiG proposed in late 2002 to return to airworthy condition the twice-flown 1.44 technology demonstrator, subject to interest and financial contribution by engine manufacturer, Saturn/Lyulka.

UPDATED

Second MiG-29K prototype in revised colour scheme *(Yefim Gordon)* ***NEW***/0546969

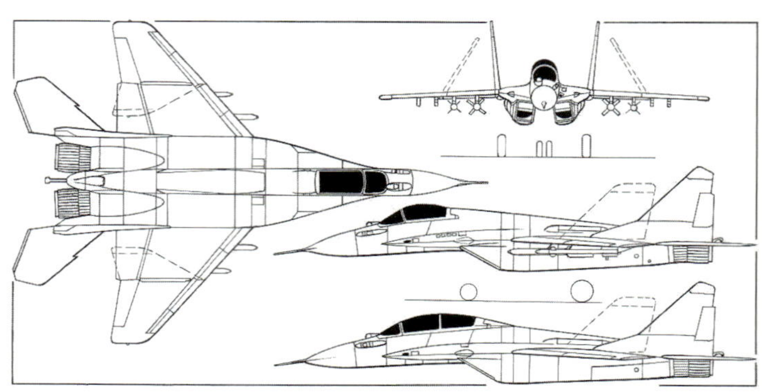

Navalised 'Fulcrums', the MiG-29K and side view (lower) of MiG-29KUB *(James Goulding)* 0131833

MIG-29

NATO reporting name: Fulcrum
Indian Air Force name: Baaz (Eagle)

TYPE: Multirole fighter.

PROGRAMME: Production of land-based MiG-29s ended in mid-1990s, leaving substantial stock of semi-complete airframes available to meet export orders. Activity currently centres on upgrades for existing aircraft.

Prospects for new production were long dependent upon Indian order for carrier-based MiG-29K. By 2001 — in advance of formal order — MiG was constructing two pre-series MiG-29Ks, one MiG-29KUB and static test airframe in single-seat configuration. Sokol plant (described separately) contracted in 2001 to build folding wing and nose section for production aircraft.

Eventually, on 20 January 2004, contract signed with India for 12 MiG-29Ks and four MiG-29KUB two-seat trainers, worth in excess of US$700 million, for delivery by 2007–08. Option exists for further 30 to be supplied by 2015. Production by LAPIK at Lukhovitsy.

Following data generally for MiG-29K. Earlier versions described in Jane's Aircraft Upgrades.

CURRENT VERSIONS: **MiG-29K** (Factory index 9.31; K for *korabelnyy;* ship-based): Maritime version, used for ski-jump take-off and deck landing trials on carrier *Admiral of the Fleet Kuznetsov* (formerly *Tbilisi*), beginning 1 November 1989; two new-build prototypes, using second-generation MiG-29M structure; completely redesigned, mainly steel, wing using modified aerofoil section and of increased area (increased span, reduced leading-edge sweep) with double slotted flaps, drooping flaperons and more powerful leading-edge flaps; new spar in front of original wing box; new strengthened centre-section without overwing louvres; upward-folding outer wing panels; RD-33K turbofans with 92.2 kN (20,723 lb st) contingency rating for ski-jump take-offs. Fuel capacity reduced to 5,670 litres (1,498 US gallons; 1,247 Imp gallons). First flown (16188 '311') 23 June 1988; second prototype was 27579 '312'.

State Acceptance Trials suspended due to funding difficulties, early 1992. Further development ended initially when not selected for deployment on *Admiral of the Fleet Kuznetsov,* but resumed at Zhukovsky in September 1996, in expectation of order from

India. First prototype currently grounded; second returned to flight status in support of MiG-29M programme. Proposed naval **MiG-29KU** (9.62) trainer derivative with two separate stepped cockpits remained unbuilt.

'MiG-29K-2002' (Factory index 9.41): The original MiG-29SMTK (factory index 9.17K; effectively a MiG-29SMT with the 9.31's folding wing and landing gear), previously offered to India along with the former helicopter carrier *Admiral Gorshkov*, is understood to have been replaced by a new, multirole, carrierborne variant based more closely on the MiG-29K/M, albeit without the expensive Al-Li alloys. Development of new MiG-29M variants (single- and two-seat) continued in 2003 with MiG-290VT (thrust-vectoring ground testbed) and MiG-29M2 (29MRCA) which first flew on 26 September 2001 after conversion to tandem configuration.

With a MIL-STD-1553B-type bus and open systems avionics architecture, the MiG-29K-2002 is compatible with a wide range of Russian and Western weapons, and may feature the colour displays and GPS-based navigation system of the MiG-29SMT. This variant, possibly designated MiG-29 MTK, is claimed to be able to perform 90 per cent of its missions with a 10 kt (18 km/h; 11 mph) wind-over-deck, even in tropical conditions using new autothrottle. Notable features of the new aircraft are its improved wing with enlarged trailing-edge flaps, digital FBW system, emergency fuel draining and much-reduced folded span of 5.80 m (19 ft 0¼ in), achieved by positioning the fold line much closer in to the wingroot, and by adding upward-folding tailplanes. The aircraft can also fold its radome (up and back), reducing overall length to 14.13 m (46 ft 4¼ in). Accordingly, 44,570-tonne *Admiral Gorshkov* , which India has also purchased, can carry a full air wing of 24 MiG-29Ks (plus six helicopters), or (according to some sources) as many as 30.

A projected **MiG-29K-2008** upgrade configuration could add Zhuk-MF phased-array radar, RD-133 turbofans, a computer upgrade and additional electro-optic, radar, IIR and reconnaissance pods. together with take-off performance improvements. In December 1999, it was reported that India had selected the MiG-29K for use aboard *Admiral Gorshkov*, with a quantity of Kh-35 anti-ship missiles and Kh-31P ARMs. Proposed local manufacture by HAL appears no longer to be viable.

MiG-29KUB (Factory index 9.47): Revised carrier-borne two-seat trainer design offered to India, based on MiG-29K-2002 with reduced-span, inboard wing fold and folding tailplanes. Assumed to feature original stepped tandem cockpits of MiG-29KU. Some reports suggest enlarged tailfins with integral fuel tanks.

DESIGN FEATURES: Emphasis from start on high manoeuvrability, to counter US F-15, F-16 and F-18, with target destruction at distances from 200 m (660 ft) to 32 n miles (60 km; 37 miles), and with effective air-to-surface capability. All-swept mid-wing configuration, with wide ogival wing leading-edge root extensions (LERX), 40 per cent of lift provided by lift-generating centre-fuselage, twin tailfins carried on booms outboard of widely spaced engines with wedge intakes. Gap between roof of each intake and skin of wingroot extension for boundary layer bleed.

Fire-control and mission computers link radar with laser range-finder and infra-red search/track sensor, in conjunction with helmet-mounted target designator. Difficult to get into stable flat spin. reluctant to enter normal spin, recovers as soon as controls released; wing leading-edge sweepback 73° 30' on LERX, 42° on outer panels; anhedral approximately 3°. Tailfins canted outward 6°; leading-edge sweep 47° 50' on fins, 50° on horizontal surfaces; anhedral 3° 30'. Fins canted outwards 6°. Engine replacement time 2 hours; preflight preparation 20 minutes. Design flying life 2,500 hours.

FLYING CONTROLS: Digital fly-by-wire.

STRUCTURE: Approximately 7 per cent of airframe, by weight, of composites; remainder metal; three-spar wings with three 'false spars' two ahead of, one behind, torsion box; 16 stringers and skins reinforced by stringers; trailing-edge wing flaps, ailerons and vertical tail surfaces of carbon fibre honeycomb; approximately 65 per cent of horizontal tail surfaces aluminium alloy, remainder carbon fibre; semi-monocoque all-metal fuselage built around 10 mainframes in three subassemblies, with forward (frames 1 to 3), centre (frames 4 to 7) and rear sections, the latter including the engine bays; fuselage sharply tapered and downswept aft of flat-sided cockpit area, with ogival dielectric nosecone mounting PVD-18 main pitot boom (PVD-7 auxiliary pitot mounted on side of nose); small vortex generator each side of nose helps to overcome early tendency to aileron reversal at angles of attack above 25°; tail surfaces carried on slim booms alongside engine nacelles.

LANDING GEAR: Hydromash retractable tricycle type, with oleo-pneumatic shock-absorbers. Mainwheels retract forward into wingroots, turning through 90° to lie flat above leg; nosewheels, on trailing-link oleo, retract rearward between engine air intakes. Hydraulic retraction and extension, with mechanical emergency release.

POWER PLANT: Two Klimov/Sarkisov RD-33 Series 3M 'marinised' turbofans, each 85.3 kN (19,180 lb st) with afterburning. Engines mounted 4° nose-up, and nacelles toe-in by 1° 30'. Engine ducts canted at approximately 9°, with wedge intakes, sweptback at approximately 35°, under wingroot leading-edge extensions. Refuelling probe and increased internal capacity.

ACCOMMODATION: Pressurised cockpit enclosed by frames 1 and 2. Pilot only, on 16° rearward-inclined K-36DM Series 2 zero/zero ejection seat, giving −14° view forward over the nose, and under hydraulically actuated, rearward-hinged transparent blister canopy in high-set cockpit. Sharply inclined one-piece curved windscreen of electrically de-iced triple glass. Three internal mirrors provide rearward view.

SYSTEMS: Two independent hydraulic systems powered by NP-103A variable-displacement pumps, driven by the engine accessory gearboxes, pressurised to 207 bar (3,000 lb/sq in), with 80 litres (21 US gallons; 17.5 Imp gallons) fluid. Main system powers one chamber of each control surface actuator, leading-edge and trailing-edge flaps, stick-pusher, artificial feel unit, landing gear extension/retraction, nosewheel steering and APU exhaust door; back-up system powers second chamber of each control surface actuator and stick-pusher, and can be powered by an emergency NS-58 pump.

Three separate pneumatic systems, with main system powering the wheel-brakes, canopy, fuel shut-offs and brake parachute actuator and jettison; and emergency system operating mainwheel brakes, and allowing emergency gear extension; final system pressurises hydraulic tanks and avionics bays.

Air conditioning uses bleed air cooled by heat exhangers and turbocooler. System also pressurises pilot's suit, demists screen and cools the gun bay.

Ozh-65 glycol-based liquid cooling system for radar.

AVIONICS: Expected to incorporate French, Russian and Indian components.

Radar: Phazotron-NIIR Zhuk-M (Zhuk-Meh in export form), able to track 20 targets simultaneously and engage four. N011M Bars offered as basis for alternative radar for MiG-29K.

Mission: Helmet-mounted sight.

ARMAMENT: R-73 (AA-11 'Archer'), R-27 (AA-10A 'Alamo'), and/or R-77 (AA-12 'Adder') AAMs; Kh-35 (AS-20 'Kayak'), Kh-31 (AS-17 'Krypton') and Kh-29 (AS-14 'Kedge') ASMs; KAB-250KR TV-guided bombs; FAB-500 and FAB-250 free-fall bombs; and rocket pods. One 30 mm Gryazev/Shipunov GSh-301 (TKB-687/9A4071K) single barrel gun in port wingroot leading-edge extension, with 150 AO-18 rounds.

DIMENSIONS, EXTERNAL:
Wing span ... 11.99 m (39 ft 4 in)
Wing aspect ratio ... 3.4
Length overall: incl noseprobe ... 17.37 m (56 ft 11¾ in)
folded ... 14.13 m (46 ft 4¼ in)
Height overall ... 5.18 m (17 ft 0 in)

AREAS:
Wings, gross ... 42.00 m² (452.1 sq ft)

WEIGHTS AND LOADINGS:
Max weapon load ... 4,500 kg (9,921 lb)
Max fuel load ... 5,240 kg (11,552 lb)
T-O weight: normal ... 18,550 kg (40,895 lb)
max ... 22,400 kg (49,383 lb)
Max wing loading ... 533.3 kg/m² (109.24 lb/sq ft)
Max power loading ... 131 kg/kN (1.29 lb/lb st)

PERFORMANCE:
Max level speed:
at height ... 1,296 kt (2,400 km/h; 1,491 mph)
at S/L ... 670 kt (1,240 km/h; 771 mph)
Max rate of climb at S/L ... 17,760 m (58,260 ft)/min
Service ceiling ... 17,500 m (57,420 ft)
Range:
combat: with max internal fuel ... 459 n miles (850 km; 528 miles)
with external tanks ... 702 n miles (1,300 km; 807 miles)
ferry ... 1,889 n miles (3,500 km; 2,174 miles)
g limits ... +8

UPDATED

MIG-AT

TYPE: Advanced jet trainer/light attack jet.

DEVELOPMENT MILESTONES

Design began	late '80s
Programme launched	25 Jun 90
Official go-ahead	May 92
Rolled out	18 May 95
First flight	16 Mar 96
Official first flight	21 Mar 96

PROGRAMME: Design started late 1980s; start of work formally authorised on 25 June 1990 by State Commission for Military-Industrial Affairs, in parallel with contracts to competing design bureaux; selected in May 1992 as one of two finalists (with Yak-130, following rejection of single-engined Sukhoi S-54 and Myasishchev M-200 contenders) as replacements for Aero L-29 Delfin and L-39 Albatros; originally planned (as MiG-ATTA, a designation also applied to abortive joint venture with Promavia) with T tail and R-35 (DV-2) engines, with wing by Daewoo, landing gear by Messier-Bugatti, avionics by Sextant, Thomson-CSF (now Thales) and SFENA. Under October 1992 and subsequent agreements, prototypes and pre-series aircraft are powered by Larzac 04R20 engines supplied by SNECMA of France; production aircraft for domestic market could have Larzacs licence-built by Chernyshev of Moscow, though there is a new Russian engine - the 16.7 kN (3,748 lb st) Aviadvigatel (Soyuz/CIAM) RD-1700 - which is stated to be 2.5 times cheaper than the Larzac and which may remedy early shortfalls in performance, ceiling and range. Suggested that MiG-AT has bureau Type number 08, perhaps with original T-tail configuration being 8.1 and later version as 8.2, prototype configuration being 8.21.

Prototype rolled out 18 May 1995; high-speed taxi trials at Zhukovsky August 1995; first 5 minute hop 16 March 1996; first flight 21 March 1996, piloted by Roman Taskaev; three aircraft to be used for flight test programme; two static test airframes; by 27 March 2001, first prototype had flown 496 sorties and second, 47. This increased by April 2002 to combined total of 770. Russian Air Forces began acceptance trials in December 2002 at 929th Test Centre, Akhtubinsk. Production of first series of 16 at RSK 'MiG' (Voronin) plant well advanced with seven in final assembly by December 1998. Subsequent progress

MiG-AT second prototype ***NEW***/0576272

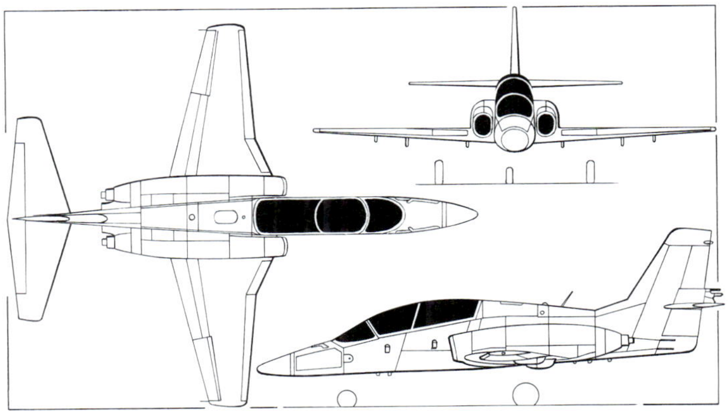

MiG-AT advanced trainer (*James Goulding*) 0130862

delayed by funding difficulties; total of six airframes (including static test) completed by late 2001, although only two then flown. Fourth flying aircraft handed over to MiG (for completion) by mid-2001. However, by end of 2003 only two aircraft flown, but third then expected to fly in early 2004, with fourth following six months later.

Following initial series of static trials on second (non-flying) airframe, operation at 5,600 kg (12,346 lb) cleared in 2001; subsequent increases planned to 6,500 kg (14,330 lb), then 7,800 kg (17,196 lb). By 2003, second prototype had received full-standard digital FBW system and cockpit displays, including 152 × 203 mm (6 × 8 in) LCD MFD.

Russian requirement for 200 to 250 trainers in this category. Request for funds for first 10 aircraft made by Russian Air Forces, late 1996, at which time 10 engines ordered from Turbomeca-SNECMA (following initial five supplied for two prototypes). Further 20 engines ordered late 1997 for FFr230 million (US$39 million), while Sextant contracted to supply 30 nav/attack systems. Competition against Yak-130 continued throughout 1990s; Russian Defence Ministry committee responsible for deciding between two contenders held first meeting on 17 February 2002, having been constituted six days earlier; Yak announced (formally, on 10 April 2002) as winner, but MiG encouraged to continue development of AT for possible export. However, in February 2003, Russian Air Forces C-in-C declared MiG-AT still in contention for production contract.

Actively promoted in India (evaluated in Russia by IAF pilots, December 2001) and South Africa; eliminated from Indian and South African competitions by BAE Hawk. Greek requirement for 40 trainers was targeted in 2001. Also assessed by Algeria and UAE. MiG-AT aircraft now being marketed as one element in AT System, incorporating simulators, computerised class rooms, part-task simulators, and video training packages, all using unified software. In August 2003, RSK 'MiG' and PZL signed enabling agreement for MiG-AT licensed production in Poland to replace TS-11 Iskra.

CURRENT VERSIONS (GENERAL): **MiG-AT:** Basic trainer; suitable for combat use of unguided weapons against ground and sea targets, and air-to-air missiles in conjunction with helmet-mounted target designation system. Pylons for three weapons/stores in trainer role, seven for combat training. Version for domestic use (ATR) has Russian-built engines and avionics; export version (ATF) has Sextant Topflight avionics (French).

Production version has longer nose than prototypes, but with pitots repositioned to slightly ahead of windscreen.

MiG-UTS (MiG-ATR): Planned Russian Air Forces' version, with Russian avionics and RD-1700 engines.

MiG-ATS: Combat trainer, with increased capabilities and underwing hardpoints able to launch guided missiles, using guidance equipment pod. Second flying prototype (823 '83') to this standard and has Russian avionics.

MiG-AS: Single-seat light tactical fighter version with built-in gun and radar for all-weather use of AAMs and ASMs; intended to compete in international market with BAE Hawk 200. Single-seater, aerodynamically similar to original MiG-AT, with rear cockpit covered by metal fairing of similar outline to original canopy.

MiG-AP: Patrol coastguard fishery protection law enforcement and SAR co-ordination aircraft; based on MiG-AS; nose-mounted search radar.

All versions can be modified for deck operation on aircraft carriers at customer's request as **MiG-ATK** (French avionics), **MiG-ATSK** (armed trainer), or **MiG-ASK** (single seater). Span with optional wing folding 6.50 m (21 ft 4 in). Provision for deck arrester hook and catapult launching.

MiG-AT can undertake the following training missions, all with take-off at 50 per cent fuel (plus 150 kg; 331 lb reserves):

Basic flying training: Between 30 and 35 flight manoeuvres, three roller landings and straight-in final landing at up to 27 n miles (50 km; 31 miles) from base, altitude up to 6,000 m (19,680 ft).

Ground attack: Four diving attacks, three advanced manoeuvre attacks and circuit/landing against visual target up to 43 n miles (80 km; 50 miles) from base, transiting at up to 2,500 m (8,200 ft).

Interception/combat manoeuvring: Loiter, directed interception, attack and close combat, day VMC/IMC at up to 140 n miles (260 km; 161 miles) from base, altitudes up to 8,000 m (26,240 ft).

CURRENT VERSIONS (SPECIFIC): **AT-1/821 '81':** Prototype; first flight 16 March 1996; first 'official' flight 21 March 1996. Has been modified since its initial flights, briefly gaining in overwing fence level with the leading-edge discontinuity. This since removed, but intakes have been redesigned and now project forward of the wing leading-edge; now have oval cross-section and blunter lips, rather than appearing to be sharp, inverted Us from the front. New intake configuration is understood to have first flown during May 1999; shown at MAKS '99, Moscow, August 1999. Rear cockpit now filled with test instrumentation. Tested with external fuel tanks. Conventional (non-FBW) controls. By 2001 had replaced pitot booms with shorter units not protruding beyond nosecone.

AT-3/823 '83': Second flying prototype shown (then still unflown) at MAKS '97, Moscow, August 1997; nominally representative of intended UTS in some respects, and of ATS armed trainer configuration in others. Fitted with centreline and underwing hardpoints; weapon carriage trials; Russian avionics and displays. Fitted with three-channel, digital FBW system in early 2000 (first Russian aircraft thus equipped); original 70 kg (154 lb) nose counterweight then removed, giving neutral stability. Original intake configuration with overwing fences was replaced by current design and aircraft resumed flying (with desert camouflage scheme) in June 2001, appearing at Paris later that month and first displaying wingtip missile rails.

AT-4/824 '84': Third flying prototype; under construction by 1999, but not flown up to early 2004. Features new Sextant avionics suite (full Topflight system, including compatibility with helmet-mounted displays); completion deferred pending availability of RD-1700 engines, although may fly initially with one RD-1700 and one Larzac.

COSTS: US$10 million to US$12 million (2002); development cost US$200 million.

DESIGN FEATURES: Intended to have manoeuvrability comparable with front-line combat aircraft, and service life of 15,000 flying hours or 30 years, with not fewer than 25,000 landings, maximum angle of attack of 25°, and sustained 5.4 *g* turn in 4 km (2.5 mile) radius at M0.7. Able to fly OEI at any stage, including take-off. Reconfigurable FBW FCS to simulate both the handling and limits of other front-line combat types. Onboard simulation of manoeuvring target, meteorological conditions and system failures via HUD, as well as specialised training for all operational modes of individual types of modern combat aircraft of Russian or foreign manufacture.

Conventional low-wing monoplane; wing-root leading-edges swept forward, with engine air intakes initially overwing to minimise risk of FOD; sweptback vertical tail surfaces; unswept tailplane; tailcone comprises two front-hinged door-type airbrakes.

FLYING CONTROLS: Avionika KSU-821 integrated hybrid (digital/analogue) multichannel fly-by-wire flight control system with multiple redundant air data computer. System incorporates autopilot, autothrottle and automatic stall-protection systems. Three-axis controls; three leading-edge slat sections and single-section slotted trailing-edge flaps on each wing. (Second prototype also flown with leading-edge flaps, which may be adopted). Split two-section rudder above and below tailplane.

STRUCTURE: One-piece, three-panel tapered wing of aluminium alloy, with some honeycomb skin. Three-section fuselage of aluminium alloy, but reportedly with 40 per cent of skin in CFRP/GFRP, including access panels on spine, three-piece engine covers, nose avionics bay cover and parts of wing/nacelle and nacelle/fuselage fairings. CFRP intake ducts and CFRP honeycomb wing trailing-edge control surfaces (elevators and flaps) and fin box, rudders, tailplane and elevators. Fin leading-edge of aluminium alloy and integral with rear fuselage. Titanium alloy rear fuselage forming 'open-box' with sidewalls acting as engine bay fire protection. Split airbrakes and ejection seat mounting tracks also of titanium with intake lips in stainless steel.

LANDING GEAR: Retractable tricycle type; nosewheel offset to starboard; single wheel on each trailing-link unit; wide-track main units retract inward, nosewheel forward; mainwheel tyre size 660×200, pressure 8.80 bar (128 lb/sq in); nosewheel tyre size 500×150, pressure 4.90 bar (71 lb/sq in); high-efficiency brakes; operation practicable from unpaved surfaces of bearing ratio 6 kg/cm^2 (85.3 lb/sq in).

POWER PLANT: Two Turbomeca-SNECMA Larzac 04-R20 turbofans, each 14.12 kN (3,175 lb st), mounted above wingroots. MRT-931 FADEC controls. Ratings can be reduced by 30 to 40 per cent for primary training. Export version offered with Larzac 04-Y3, having thrust increased to 16.7 kN (3,748 lb st) in joint development venture between Turbomeca and Klimov. Change to new Chernyshev-built Soyuz RD-1700 originally planned by 2002. One main fuselage fuel tank and one (three-section) tank in each inner wing. Total capacity 2,390 litres (632 US gallons, 527 Imp gallons). Single-point pressure refuelling point in lower rear fuselage adjacent to wing trailing-edge, with five gravity refuelling points (one above the fuselage tank and two above each wing). Tanks pressurised with engine bleed air to ensure high-altitude supply.

ACCOMMODATION: Two crew in tandem, on Zvezda K-93-LT zero/zero, 60 m (200 ft) inverted flight ejection seats operable at up to 485 kt (900 km/h; 559 mph) and M1.5 at altitudes up to 15,000 m (49,215 ft); minimum inverted flight ejection height 50 m (165 ft). Rear seat raised by 400 mm (15.75 in) to improve occupant's forward view. View over the nose −17° from cockpit, −7° from the rear. One-piece birdproof canopy, hinged on starboard side, with 'lock the canopy' warning. Provision for blinds for IFR flight training. Some sources suggest provision of 'built-in ladders', but no evidence for any more than retractable steps.

SYSTEMS: Dual hydraulic systems, driven by independent Messier-Bugatti PR70660-20 engine-driven pumps, actuate control surfaces, landing gear and wheel doors. General system supplied by starboard engine actuates one chamber of each electrohydraulic control actuator and hydraulic actuators of other systems; booster system supplied by port engine actuates other chamber of control actuators. Auxiliary booster system operated by NS-65 emergency hand pump. Electrical system supplies primary 27 V DC and 200/115 V 400 Hz three-phase secondary AC. Two 9 kW Auxilec 8044-31 DC starter/generators with Auxilec WDII-014 control and protection unit; two AC static inverters; 15CTSS-45B lead-zinc batteries.

AVIONICS: Integrated by Russian GoSNIIAS avionics research institute and Sextant Avionique, France. Multifunctional central computer, with all data integrated through MIL-STD-1553B equivalent databus.

Comms: ERA 2000 VHF/UHF transceiver, SG100 IFF, SPU-821 intercom, ALMAZ-UBS voice recorder.

Radar: Osa installation under development for optional fitment.

Flight: Automatic control system; UMT 33 air data unit; Totem 3000 laser-gyro INS; NSS 100S-1 GPS. NR 510A 101 VOR/ILS receiver/marker; NC-12B Tacan; EVS 915 video recorder; Tester U3A flight data recorder.

Instrumentation: Two MFD 55 multifunctional liquid-crystal colour CRT displays with buttons in each cockpit; helmet-mounted displays; front-cockpit wide-field HUD with input from colour video and TV camera; HSI/ADI.

ARMAMENT (MIG-AS): Nine (including wingtips) hardpoints for up to 2,000 kg (4,410 lb) of guided and unguided missiles, guns and bombs, including R-73E (AA-11 'Archer'), R-77 (AA-12 'Adder'), AIM-9L Sidewinder or Magic AAMs; Kh-29TD (AS-14 'Kedge') or Kh-31AE/PE (AS-17 'Krypton') ASMs; UB-16 (16 × 57 mm) or UB-8M (20 × 80 mm) rocket pods; cluster weapons; 100 to 500 kg bombs; UPK-23 twin-barrel 23 mm gun pods; eight Vikhr anti-tank missiles.

DIMENSIONS, EXTERNAL:

Wing span	10.16 m (33 ft 4 in)
Wing aspect ratio	5.7
Length overall	12.01 m (39 ft 4¾ in)
Height overall	4.42 m (14 ft 6 in)
Wheel track	3.80 m (12 ft 5¾ in)
Wheelbase	4.43 m (14 ft 6½ in)

AREAS:

Wings, gross	17.67 m^2 (190.2 sq ft)

WEIGHTS AND LOADINGS:

Fuel: nominal	850 kg (1,874 lb)
intermediate	1,280 kg (2,822 lb)
max	1,680 kg (3,704 lb)
Normal T-O weight: training	4,610 kg (10,163 lb)
Max T-O weight: training	5,210 kg (11,486 lb)
combat	7,800 kg (17,196 lb)
combat(alternate)	8,150 kg (17,967 lb)
Max wing loading:	
Normal T-O	260.9 kg/m^2 (53.43 lb/sq ft)
Max T-O	396.2 kg/m^2 (81.14 lb/sq ft)
Max power loading:	
Normal T-O	163 kg/kN (1.60 lb/lb st)
Max T-O	248 kg/kN (2.43 lb/lb st)

PERFORMANCE (estimated; A: basic training, B: advanced training):

Max Mach No.: A, B	0.85
Max level speed:	
A, B:	
at 2,500 m (8,200 ft)	540 kt (1,000 km/h; 621 mph)
at S/L	460 kt (850 km/h; 528 mph)
T-O speed: A	97 kt (180 km/h; 112 mph)
B	97–119 kt (180–220 km/h; 112–137 mph)
Landing speed:	
A, B	94–125 kt (175–232 km/h; 109–144 mph)
Max rate of climb:	
at S/L: A	1,680 m (5,510 ft)/min
B	4,140 m (13,580 ft)/min
at 5,000 m (16,400 ft): A	1,320 m (4,330 ft)/min
B	2,400 m (7,875 ft)/min
Service ceiling: A, B	14,000 m (45,940 ft)
T-O run: A	540 m (1,772 ft)
B	310 m (1,020 ft)
Landing run:	
on concrete: B	570 m (1,870 ft)
on unpaved runway: B	540 m (1,775 ft)

Range at M0.5 at 6,000 m (19,680 ft):
A, B ... 647 n miles (1,200 km; 745 miles)
Ferry range: A, B ... 1,400 n miles (2,600 km; 1,615 miles)
A, B (alternate) ... 1,079 n miles (2,000 km; 1,242 miles)
g limits: A, B ... +8.0/–2.0
Sustained *g* limit at M0.7 in turn:
at S/L: A ... +4.5
B ... +8
at 5,000 m (16,400 ft): A ... +3.4
B ... +5
Max side wind for T-O/landing ... 29 kt (54 km/h; 33 mph)

UPDATED

MIG-110

TYPE: Twin turboprop transport.

PROGRAMME: Design began February 1992; shown in model form at Moscow Air Show 1993; development received government approval January 1994 and proceeded at low priority until 1997, when completion of MiG-AT design released staff and facilities for more rapid progress; by late 1998, mockup assembled and one-third of construction documentation was complete. On 27 October 1998, ANL Handelsgesellschaft of Austria signed LoI for funding of MiG-110 design and prototype construction; ANL intends to assemble aircraft in Austria from Russian kits. Interest was being maintained in late 2001, involving Austrian purchase of MiG-29 fighters. RSK 'MiG' confirmed continuation of negotiations in December 2002, but no more recent announcements were known by early 2004.

Main production previously intended by Sokol at Nizhny Novgorod; however, by late 2002 Voronin production plant of RSK 'MiG' was claiming responsibility for series production and announced intention to begin fabrication of prototype during 2003. MiG delayed formation of prototype committee from late 2000 to early 2001. Promotion continued throughout 2001, but little progress reported towards construction of prototype; further mockup conference held in June 2002, when targets reset as completion of design documentation by end of 2002; prototype assembly begun late 2003 (not confirmed as having taken place); first flight 2005; and certification in 2007.

Contender in Russian Air Forces competition to replace Antonov An-24/An-26; Myasishchev M-60 (EhMZ), Su-80 and Tu-136 are rival designs.

CURRENT VERSIONS: **MiG-110:** Baseline version.

MiG-110A: Austrian-assembled version.

MiG-110M: Western avionics and engines.

MiG-110VT (*Vcyenno Transportnaya:* Military Transport): Proposed light tactical transport for Russian Air Forces.

MiG-110PR: Proposed SAR/reconnaissance variant for Russian Air Forces.

CUSTOMERS: Some 200 "ordered" by 20 Russian airlines up to February 2000; estimated world market for 1,900, including 1,200 in CIS.

COSTS: Programme cost estimated as US$100 million in late 2000; unit cost US$7.5 million (2002); break-even after 40 to 50 aircraft.

DESIGN FEATURES: Optimised for ease of rear access to cargo hold and comparatively wide track landing gear. Cabin can be equipped for passenger, combined cargo/passenger or cargo operations, day or night in all weathers.

High wing monoplane; anhedral centre-section between twin booms carrying engines at front and twin fins at rear; horizontal tail surfaces between fin tips; winglets for reduced drag; fuselage pod slung from centre-section, with upward-hinged rear fuselage section and loading ramp. Intended service life 40,000 hours.

FLYING CONTROLS: Conventional and manual. Twin rudders.

LANDING GEAR: Tricycle type, with twin wheels on each unit. Mainwheels retract rearwards into booms; nosewheel retracts forward. Optional floatplane version.

POWER PLANT: Two 2,059 kW (2,762 shp) Klimov TV7-117S Srs 2 turboprops; six-blade Stupino SV-34 propellers.

ACCOMMODATION: Two crew; seats for up to 48 passengers; or provisions for 15 passengers and 3,500 kg (7,715 lb) of freight, or 5,000 kg (11,023 lb) of freight, including containers (up to four ZAK-1) and vehicles loaded via a rear ramp after tailcone is hinged upwards. Accommodation pressurised. Main passenger door, with integral stairs, port side, front.

DIMENSIONS, EXTERNAL:
Wing span ... 25.00 m (82 ft 0¼ in)
Length overall ... 18.90 m (62 ft 1 in)
Height overall ... 5.385 m (17 ft 8 in)

DIMENSIONS, INTERNAL:
Cargo cabin: Length ... 7.40 m (24 ft 3¼ in)

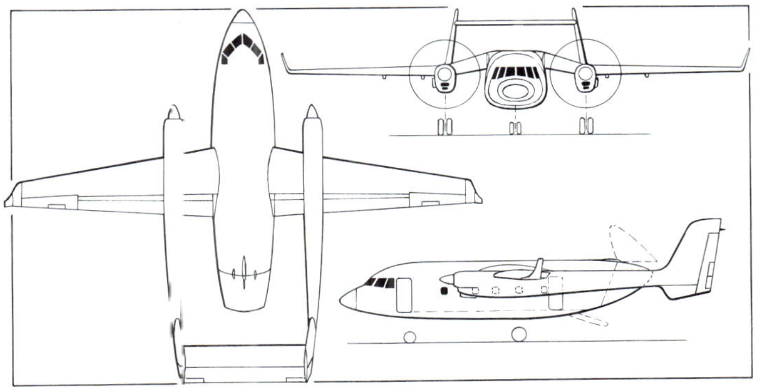

MiG-110 multipurpose transport *(James Goulding)* 0062665

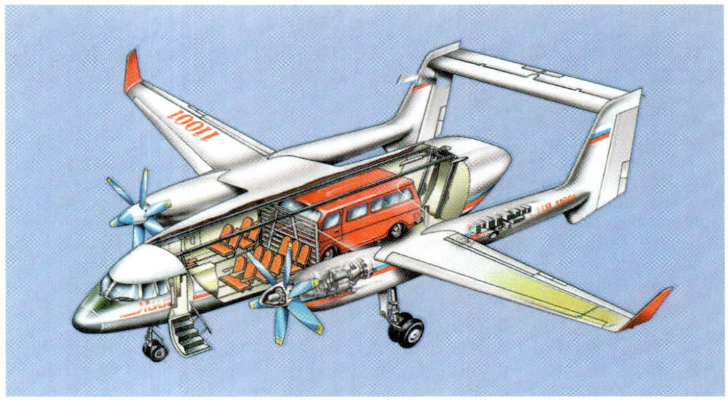

Partial cutaway of the MiG-110 in combi configuration shows several minor differences from the official general arrangement drawing 0121091

Max width ... 2.20 m (7 ft 2½ in)
Height: min ... 2.20 m (7 ft 2½ in)
max ... 2.76 m (9 ft 0¾ in)

WEIGHTS AND LOADINGS:
Max payload: paved runway ... 5,500 kg (12,125 lb)
unpaved runway ... 3,500 kg (7,716 lb)
Fuel weight: normal ... 2,000 kg (4,409 lb)
max ... 3,500 kg (7,716 lb)
Max T-O weight: paved runway ... 15,300 kg (33,730 lb)
unpaved runway ... 12,850 kg (28,329 lb)

PERFORMANCE (estimated):
Max level speed ... 297 kt (550 km/h 341 mph)
Max cruising speed ... 270 kt (500 km/h; 310 mph)
Cruising altitude ... 7,000 m (22,960 ft)
Balanced field length ... 1,000 m (3,280 ft)
Range, 30 min reserves: with 4,500 kg (9,921 lb) payload ... 1,042 n miles (1,930 km; 1,199 miles)
with full fuel and 2,660 kg (5,864 lb) payload ... 2,173 n miles (4,025 km; 2,501 miles)

UPDATED

MIG-201

TYPE: Light utility twin-prop transport.

PROGRAMME: First details announced in mid-2003. Construction of prototype then scheduled to begin in late 2004 or early 2005. Due to enter production at Lukhovitsy plant (LAPIK) in 2005–06.

CURRENT VERSIONS: **MiG-201:** *As described.*

MiG-201M: Alternative power plant of two 324 kW (435 hp) Teledyne Continental GTSIO-520F flat-six piston engines.

MiG-201MT1: Turboprop version with two 560 kW (751 shp) Walter M 601Es.

MiG-201MT2: Alternative turboprop version with two Pratt & Whitney Canada PT6A-42s, each rated at up to 634 kW (850 shp). Range 755 n miles (1,400 km; 869 miles) with max payload.

COSTS: US$650,000 with M-9 engines; US$1,400,000 with PT6A-42 engines.

DESIGN FEATURES: Replacement for Antonov An-2 utility biplane, but with twin engines. Rugged construction and short-field performance with maximum of 10 occupants or equivalent freight.

High, constant-chord wing; moderately sweptback fin with ventral fillet; low-mounted, constant-chord tailplane. Service life 40,000 hours.

STRUCTURE: Generally of metal.

LANDING GEAR: Tricycle type; retractable.

POWER PLANT: Two 327 kW (434 hp) VOKBM M-9FS nine-cylinder radial engines, each driving an MT-Propeller MTV-9 three-blade, variable-pitch propeller. Version powered by liquid natural gas may be offered.

ACCOMMODATION: Provision for nine passengers.

DIMENSIONS, EXTERNAL:
Freight door (port): Height ... 1.22 m (3 ft 11¼ in)
Width ... 1.30 m (4 ft 3¼ in)

DIMENSIONS, INTERNAL:
Cabin: Length ... 4.50 m (14 ft 9 in)
Max width ... 1.52 m (4 ft 11¾ in)
Max height ... 1.35 m (4 ft 5¼ in)

WEIGHTS AND LOADINGS:
Weight empty, equipped ... 2,850 kg (6,283 lb)
Max payload ... 840 kg (1,852 lb)
Max fuel weight ... 650 kg (1,433 lb)
Max T-O weight ... 3,900 kg (8,598 lb)

PERFORMANCE:
Max level speed ... 205 kt (380 km/h 236 mph)
Econ cruising speed ... 162 kt (300 km/h; 186 mph)
Service ceiling ... 6,000 m (19,680 ft)
T-O run ... 390 m (1,280 ft)
Range: with 840 kg (1,852 lb) payload ... 297 n miles (550 km; 341 miles)
with 550 kg (1,213 lb) payload ... 1,025 n miles (1,900 km; 1,180 miles)

UPDATED

MIL

MOSKOVSKY VERTOLETNY ZAVOD IMIENI M L MILYA OAO (MOSCOW HELICOPTER PLANT NAMED FOR M L MIL JSC)

Sokolnichyesky val, Stroenie 2, 107113 Moskva
Tel: (+7 095) 264 90 83
Fax: (+7 095) 264 55 71
e-mail: mvz@mi-helicopter.ru
Web: http://www.mi-helicopter.ru
GENERAL DIRECTOR: Yury M Andrianov
GENERAL MANAGER: Leonid Zapolsky
DESIGNER GENERAL (ACTING): Aleksei Samusenko
FIRST DEPUTIES:
S A Kolupayev
Aleksei G Samusenko

OKB founded 1947 by Mikhail Leontyevich Mil, who was involved with Soviet gyroplane and helicopter development from 1929 until his death on 31 January 1970, aged 60. His original Mi-1, first flown September 1948 and introduced into service 1951, was first series production helicopter built in former USSR. Twelve basic types of Mil helicopter have been built in 100 subvariants, representing total production of some 30,000, of which over 4,500 exported to 80 countries; 95 per cent of all helicopters in Russian Federation and Associated States (CIS) are Mil designs. Mil helicopters have set 96 world records. The Mil design bureau has its own experimental workshops in Moscow and flight test facilities at Chkalovsky airfield. The company established a joint venture, Euromil, with Eurocopter and others, for development and production of the Mil Mi-38, but Eurocopter withdrew in 2003. The Mi-38 flew at Kazan in December 2003; promotion and marketing is largely the responsibility of the Kazan company.

In 2001, Mil designers reported to be working on seven projects, comprising Mi-30, Mi-32, Mi-54 and Mi-58 transports; and Mi-40, Mi-42 and Mi-44 combat helicopters. Of these, only details of the Mi-40, Mi-54 and Mi-58 have been released. In mid-2003, Mil was anticipating an early launch of the Mi-54. However, main initiative of 2003 was launch of upgrade programme for Mi-2, designated Mi-2A and featuring turboshaft engine.

Mil design and production facilities eventually to be integrated into group comprising Mil Moscow, Kazan and Rostov (Rostvertol) plants, the whole to be included in the Russian aerospace grouping led by Sukhoi. In preparation for this, Mil and Rostvertol planning formation of joint company to be known as Mil Helicopters. The lighter types of Mil helicopter were marketed by the separate entity, LVM, but this appears no to have been active recently.

Mil forced into receivership in 1999; new management, imposed by Federal Financial Recovery Service, returned company to solvency and paid staff wage arrears; bankruptcy cancelled 23 February 2001 after effective return to State control when debts bought-out by Gosincor (Ministry of Property) and Inter-Regional Bank.

UPDATED

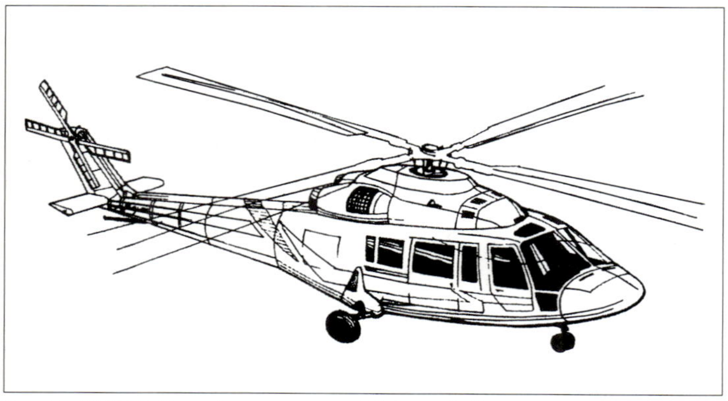

Proposed Mil Mi-54 twelve-passenger, 4.5 tonne helicopter ***NEW***/0576268

MIL MI-17 (MI-8M), MI-19, MI-171 AND MI-172

NATO reporting name: Hip

TYPE: Multirole medium helicopter.

PROGRAMME: Development of original Mi-8 began May 1960, to replace piston-engined Mi-4; first of two prototypes (V-8), with single AI-24V turboshaft and four-blade main rotor, flew 24 June 1961, given NATO reporting name 'Hip-A'; further prototype (V-8A; 'Hip-B'), with two production standard TV2-117 engines and four-blade main rotor, flew 2 August 1962; change to five-blade rotor for production. Total 7,300 Mi-8s and subsequent improved Mi-17s and Mi-171/172s marketed and delivered from Kazan since 1967, with more than 11,000 (about 3,700 Mi-8T and 7,300 Mi-17) from Ulan-Ude since 1970 for civil and military use, including 3,500 exported to 70 countries.

Prototype Mi-17, known initially as Mi-18, completed 1975 with basic Mi-8 airframe, and power plant and dynamic components of Mi-14. First flew 17 August 1975. Entered service with former Soviet forces in 1977 as Mi-8MT. First displayed at 1981 Paris Air Show as Mi-17 for civil use and export; exports began (to Cuba) 1983; Mi-8MTV followed in 1987; production of improved versions continues at Kazan and Ulan-Ude plants. Kazan has undertaken further development of the Mi-17 and these variants are described separately.

CURRENT VERSIONS: **Mi-17** ('Hip-H'): Basic designation of series, with suffix M for military versions. Mi-17P is basic civil version with 28 seats and rectangular cabin windows.

Detailed description applies to basic versions, except where indicated. Note that all helicopters of Mi-8/Mi-17 series in Russian military service are known as Mi-8s of various subtypes, regardless of engines fitted.

Mi-8MT ('Hip-H'): Designation of standard Mi-17s in RFAS military service. Twin or triple stores racks, but normal armament is 40 × 80 mm S-8 rockets in two BV-8-20A packs. Afghan experience led to adoption of nose armour, IR jammer, IR suppressors and provision for door-mounted PKT machine gun (rear starboard) and AGS-17 Plamya grenade launcher or NSV 12.7 mm Utyos heavy machine gun (forward port cabin door).

Mi-8MT EW variants: More than 30 EW versions of the Mi-8MT serve, or have served, with RFAS armed forces.

Latvian Air Force Mil Mi-8MTV-1 *(Paul Jackson)* ***NEW***/0576267

Night attack Mil Mi-8MTKO *(Yefim Gordon)* ***NEW***/0558802

Mi-8MTPB (or **Mi-17TPB, Mi-17P, Mi-17PP**) ('Hip-H EW'): ECM (radar and communications jammer) and comint helicopter, with three jamming systems in D/F band range over 30° sector and other frequencies over 120°.

Mi-8MTV ('Hip-H'): (V=*visotnyi*: high-altitude); 1,633 kW (2,190 shp) TV3-117VM turboshafts for improved 'hot-and-high' operation. Built at Ulan-Ude (as **Mi-8AMT**) from 1991; 100 built by 1999. Civil version built at Kazan is **Mi-8MTV-1** ; missile-armed, radar-equipped military version with six-hardpoint stub-wing is **Mi-8MTV-2** ; export equivalent is **Mi-17-1V** , with optional armament, nose radar, flotation gear and firefighting equipment. Indian Mi-17-1V has enlarged side doors and a rear ramp to facilitate airdrops, as developed at Kazan (which see). Firefighting conversion by Airod in Malaysia adds detachable aluminium boom to starboard side, stowed facing forwards, but trainable through 90°, with hose nozzle at tip. Internal tanks, fillable via own trailing hose and onboard pump in 1½ minutes, carry 2,300 litres (608 US gallons; 506 Imp gallons) of water, which may also be dumped from belly port in 8 seconds. Addition of suffix **-GA** indicates civil version, often with rectangular 'Salon'-type cabin windows. **Mi-8MTV-1S** is VVIP variant for Rossiya state transport company; two believed ordered in 2001. An Mi-8MTV upgrade project was under way in mid-2002 that may result in adoption of more powerful VK-2500 engines, composites main rotor blades and updated avionics as well as new armament to be developed by 2010; this is expected to be offered in new-build form towards the end of the present decade.

Mi-8MTKO (or **Mi-8N**); ('Hip-H'): (O = *ochki nochnogo videniya*: night vision goggles). Dedicated night attack conversion of Mi-8MT/MTV to meet urgent operational requirement; tested in Chechnya ('Caucasian campaign') with aircraft equipping BEG (Boyevaya Eksperimentalnaya Gruppa: Combat Experimental Group) based at Mozdok; at least five aircraft thus upgraded by mid-2000. (Unit additionally equipped with four or six upgraded Mi-24s, and also parented trial deployment of Ka-50. Originally intended to be equipped with modified Ka-29TB and Ka-50). Upgrade produced by Mil Moscow plant, with collaboration of Russkaya Avionika, NPO Geofizika and Uralsky Optik Mechanichesky Zavod team, and tested at Russian Air Forces flight test centre at Chkalovskaya from April 1999, thereafter undergoing evaluation at Torzhok with the 696th IIVP, equipping BEG from September 1999. By 2003, first deliveries had been made to unit in North Caucasus district, with further deliveries planned in 2004 (in conjunction with Mi-24PN 'Hinds') to equip one night-capable regiment by year's end.

Aircraft has NVG-compatible cockpit and external lights and may also feature NVG-covert, formation-keeping external lights. Crew equipped with Geofizika ONV-1 NVGs, and also use gyrostabilised GOES-321M undernose turret containing AGEMA 1000 or SAGEM-321 FLIR, laser range-finder and LLTV. Incorporates cockpit and navigation system upgrade, with KNI-8 navigation and display system, MFI-68 (or, possibly, US-supplied) colour LCD MFDs, A737-00 GPS, full-standard MTO will gain new air data system and Saratov Industrial Automatics OKB PKV-M digital autopilot. Similar export version offered to India (which has already used its existing Mi-17s in the gunship role in Kashmir, and hopes to upgrade these with an Mi-35-type weapon aiming system, improved avionics and a 'night cockpit') and others as **Mi-17N.**

Mi-17-1VA: Version of Mi-8MTV-1 produced for Ministry of Health of former USSR as flying hospital

Mi-8MTV-3/Mi-172A: As Mi-8MTV-2, also from Kazan, but with four-hardpoint stub-wing and other equipment changes and planned for certification to FAR Pt 29 standards; TV3-117VM Srs 2 engines, giving maximum cruising speed of 118 kt (218 km/h; 135 mph) and service ceiling of 6,000 m (19,680 ft); air conditioning and heating systems, main and tail rotor blade de-icing, canopy demisting and heating of engine air intakes standard; options include flotation gear, Doppler, weather radar, DME, GPS, VOR, ILS, transponder and VIP interiors for seven, nine and 11 passengers. Standard seating for up to 26 passengers.

Mi-172A export version first exhibited at 1994 Singapore Air Show; based on 'salon' (rectangular window) cabin. Seven ordered by Mesco, India, 1995; others to Vietnam.

Mi-8MTV-7: Upgraded development by Kazan; announced 2002.

Mi-172AG: Announced mid-2000; proposed version with TV3-117SB3 engines, improved composites main rotor blades and upgraded navigation avionics.

Mi-17PI: As Mi-17P but single D-band jamming system able to jam up to eight sources simultaneously over 30° sector.

Mi-17PG: As Mi-17P but with H/I-band system for jamming pulse/CW and CW interrupted equipment.

Mi-17AMT: Provisional designation for proposed Ulan-Ude-built variant offered to Malaysia to meet eventual 40-aircraft requirement, with initial order for six CSAR-configured aircraft reportedly already placed. May have rear loading ramp of

Mil Mi-171 in civilian guise *(Yefim Gordon)* ***NEW***/0558799

Ulan-Ude plant's Mi-8AMT(Sh) demonstrator *(Paul Jackson)* ***NEW***/0532079

Mi-8MTV-5. Mi-17 will replace 31 S-61 Nuri helicopters, and two already in use with Malaysian Fire Department. SME Aerospace appointed as partner responsible for assembly, flight test and local certification.

Mi-8AMT(Sh) 'Terminator': Current counterpart of Mi-8MTV series, built at Ulan-Ude. Crew positions protected by armour plating. Armament can include up to eight Igla-V AAMs or 9M114/9M114M Shturm-V ('Spiral') ASMs, with thimble radome on nose and chin-mounted electro-optics pod; alternative armament configurations include up to four pods of B8V20A type, each containing 20 S-8 unguided rockets; or a mix of guided and unguided weapons. Prototype (RA-25755) exhibited at Farnborough in September 1996. First production helicopter almost complete by late 1998, but delayed by funding shortage. State testing began 10 April 2000, and had been completed by January 2001: funding assistance provided by Interprofavia. Offered for export as **Mi-171(Sh)** .Night-capable variant under development by mid-2001.

Mi-171: Export version of Mi-8AMT; first displayed 1989 Paris Air Show. Mostly Western avionics, including Honeywell and Rockwell Collins radios. Built at Ulan-Ude since 1991.

In August 2001, Ulan-Ude showed upgraded Mi-171A at Moscow Salon, featuring rear cargo ramp (straight-sided, compared with Kazan version) and enlarged starboard passenger door. Klimov VK-2500 turboshaft also installed on latest examples of Mi-171.

Mi-17V-5 (Mi-8VTV-5): modification. Described separately. Kazan.

Mi-17KF Kittiwake: Flight deck upgrade offered by Kazan and Kelowna of Canada.

Mi-18: Originally was the designation of the prototype Mi-17.

Mi-19: Command relay platform; similar to earlier Mi-9.

Mi-17Z-2: Converted from 'Hip-H' in former Czechoslovakia for electronic warfare ECM role; local designation (Z-1 is similarly tasked Antonov An-26).

Long-range modification: AEFT (Auxiliary External Fuel Tanks) system by Aeroton adds a further 1,900 litres (502 US gallons; 418 Imp gallons) in two internal tanks, plus 2,850 litres (753 US gallons; 626 Imp gallons) in six tanks on the stores pylons of Mi-8MT, AMT, MTV-1, civil MTV and Mi-17 variants.

Upgrades: Kazan offers three stages of upgrade for existing Mi-17 variants. The first can include new composites main rotor blades with a 6,000-hour service life, NVG-compatible cockpit and improved avionics; second could include more powerful 1,864 kW (2,500 shp) engines, uprated transmission and Delta-H 'X-type' tail rotor, latter refinements presaging increase in MTOW to 14,000 kg (30,864 lb). Kazan has also teamed with BAE Systems and Kelowna Flightcraft Ltd to offer aircraft with new mission system, including automated flight control system and 'glass cockpit', plus FLIR, LLTV and other sensors; this upgrade package (or elements of it) can be installed in the Mi-172.

Kronshtadt is offering an avionics and cockpit upgrade, with LCD screens, digital moving map and embedded GPS.

Israel Aircraft Industries announced **Peak-17** in February 2001, combining features from Kamov Ka-50-2 Erdogan and Mi-35 export 'Hind' with 'glass cockpit', improved self-defence Rafael NT-D Dandy and NT-S Spike anti-armour missiles.

Elbit Systems of Israel in 2001 was promoting **'Andor'** Mi-8/17 upgrade for civil and military applications. Modular modernisation concept entails addition of various items of equipment, including electro-optical sensor turret, MFDs, HOCAS, NVGs, HUD, EW systems and choice of Eastern or Western weaponry.

CUSTOMERS: Many operational alongside Mi-8s in RFAS armed forces; also operated in Algeria, Angola, Armenia, Bangladesh, Bulgaria, Burkina Faso, Cambodia, China, Colombia, Costa Rica, Croatia, Cuba, Czech Republic, Ecuador, Egypt, Ethiopia, Hungary, India, Indonesia, Kenya, North Korea, Laos, Macedonia, Malaysia, Mexico, Myanmar, Nicaragua, Pakistan, Papua New Guinea, Peru, Poland, Romania, Sierra Leone, Slovak Republic, Sri Lanka, Syria, Turkey, Uganda, Ukraine, Venezuela, Vietnam, Yugoslavia. More than 810 exported by Aviaexport. Recent announced sales include six Mi-17s to India Border Guards Service in 2003, with first delivery in September, and 42 Mi-171s to Algeria, also in 2003. Malaysia placed order for 10 Mi-171s in 2003 and could eventually obtain as many as 40 for the air force.

COSTS: Kazan quoted US$3.7 million, flyaway, basically equipped, in March 2001. In mid-2002, Ulan-Ude quoting price as US$3.5 million to US$4 million, depending on configuration and equipment. Sale of 42 Mi-171s to Algeria in 2003 said to be worth US$180 million.

DESIGN FEATURES: Conventional pod and boom configuration; clamshell rear-loading freight doors on most versions, but both Kazan and Ulan-Ude plants offer rear-loading ramp version; five-blade main rotor, inclined forward 4° 30' from vertical; interchangeable blades of basic NACA 230 section, solidity 0.0777; spar failure warning system; drag and flapping

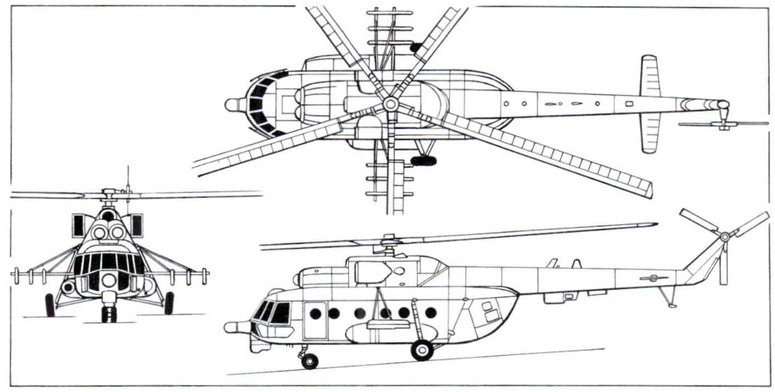

Mil Mi-8AMT(Sh) combat helicopter — the latest Mi-17 variant *(James Goulding)* 0100524

hinges a few inches apart; blades carried on machined spider; pendulum vibration damper; three-blade port tail rotor; transmission comprises VR-14 two-stage planetary main reduction gearbox, intermediate and tail rotor gearboxes, main rotor brake, and drives off main gearbox for tail rotor, fan, AC generator, hydraulic pumps and tachometer generators; correct rotor speed maintained automatically by system that also synchronises output of the two engines; tail rotor pylon forms small vertical stabiliser; horizontal stabiliser near end of tailboom. Civil versions have rectangular cabin windows.

FLYING CONTROLS: Mechanical system, with irreversible hydraulic boosters; main rotor collective pitch control linked to throttles.

STRUCTURE: All-metal; main rotor blades, mainly VD-76 aluminium alloy, each has extruded spar carrying root fitting, 21 honeycomb-filled trailing-edge pockets and blade tip, with titanium abrasion shielding on some of leading-edge; balance tab on each blade; each tail rotor blade made of spar and honeycomb-filled trailing-edge; semi-monocoque fuselage. Composites blades and elastomeric hub components under development.

LANDING GEAR: Non-retractable tricycle type; steerable twin-wheel nose unit, locked in flight; single wheel on each main unit; oleo-pneumatic (gas) shock-absorbers. Mainwheel tyres 865×280; nosewheel tyres 595×185. Pneumatic brakes on mainwheels; pneumatic system can also recharge tyres in the field, using air stored in main landing gear struts. Optional mainwheel fairings.

POWER PLANT: Two 1,397 kW (1,874 shp) Klimov TV3-117MT turboshafts in basic Mi-17; should one engine stop, output of the other increases automatically to contingency rating of 1,637 kW (2,195 shp), enabling flight to continue. Alternative high-altitude TV3-117VM engine has installed ratings of 1,545 kW (2,071 shp) in emergency; 1,397 kW (1,874 shp) for T-O; 1,250 kW (1,677 shp) normal; and 1,103 kW (1,479 shp) for cruising. Newest version of engine is TV3-117VMA-SB3, which believed to be available for new-build and upgraded aircraft. Deflectors on engine air intakes prevent ingestion of sand, dust and foreign objects; engine cowling side panels form maintenance platforms when open, with access via hatch on flight deck. VK-2500 engine installed on latest versions of Mi-171.

Standard fuel capacity of 2,615 litres (691 US gallons, 575 Imp gallons), comprising 2,170 litres (573 US gallons; 477 Imp gallons) in main external tanks and 445 litres (118 US gallons; 98.0 Imp gallons) in feed tank. Additional 1,830 litres (483 US gallons; 402 Imp gallons) optional auxiliary fuel in two tanks.

ACCOMMODATION: Two pilots side by side on flight deck, with provision for flight engineer's station. Military versions can be fitted with external flight deck armour. Windscreen de-icing standard.

Seats and bulkheads of basic version quickly removable for cargo carrying. Standard military versions have cargo tiedown rings in floor, winch of 150 kg (330 lb) capacity and pulley block system to facilitate loading of heavy freight, an external cargo sling system (capacity 3,000 kg; 6,614 lb), and 24 tip-up seats along sidewalls of cabin; six additional centreline seats optional. Military Mi-17-1V carries up to 30 troops, up to 20 wounded or 12 stretchers in ambulance role, or weapons on six outrigger pylons; civilian Mi-17 promoted as essentially a cargo-carrying helicopter, with secondary passenger transport role. Large windows on each side of flight deck slide rearward. Sliding, jettisonable main passenger door at front of cabin on port side; electrically operated rescue hoist (capacity 150 kg; 330 lb) can be installed at this doorway. Rear of cabin made up of clamshell freight-loading doors, which are smaller on commercial versions, with downward-hinged passenger airstair door centrally at rear. Hook-on ramps used for vehicle loading.

SYSTEMS: (Mi-17-1V/171): AI-9V APU for pneumatic engine starting; AC electrical supply from two 40 kW three-phase 115/220 V 400 Hz GT40/P-48 V generators. Engine air intake de-icing standard. Provision for oxygen system for crew and, in ambulance version, for patients. Freon fire extinguishing system in power plant bays and service fuel tank compartments, actuated automatically or manually. Two portable fire extinguishers in cabin.

AVIONICS: **(Mi-17-1V/171):** *Comms:* Baklan-20 and Yadro-1G1 com radio.

Radar: Type 8A-813 weather radar optional.

Flight: Type A-723 long-range nav.

Instrumentation: ARK-15M radio compass, ARK-UD radio compass, DISS-32-90 Doppler; AGK-77 and AGR-74V automatic horizons; BKK-18 attitude monitor; ZPU-24 course selector; A-037 radio altimeter.

Self-defence (optional): ASO-2V chaff/flare dispensers under tailboom and L-166V IR jammer (NATO 'Hot Brick') at forward end of tailboom.

EQUIPMENT: Options on military versions include external cockpit armour, triple-lobe engine nozzle IR suppressors and a VMR-2 fit for airdropping such stores as mines.

ARMAMENT: Various combinations of weaponry available, including 64 × 57 mm S-5 rockets in four UV-16-57 packs or 192 × 57 mm S-5 rockets in six UV-32-57 packs or four 9M17P Skorpion (AT-2 'Swatter') anti-armour missiles or six 9M14M Malyutka (AT-3 'Sagger') anti-armour missiles; can also operate with bombs and mines and able to use 23 mm GSh-23 gun packs, each with 250 rounds of ammunition. (AAMs and newer ASMs on Mi-8AMT(Sh).)

DIMENSIONS, EXTERNAL:

Main rotor diameter ... 21.29 m (69 ft 10½ in)
Tail rotor diameter ... 3.91 m (12 ft 9 ⅞ in)
Distance between rotor centres ... 12.66 m (41 ft 6 in)
Length: overall, rotors turning:
  Mi-17 ... 25.35 m (83 ft 2 in)
  Fuselage: Mi-17 ... 18.465 m (60 ft 7 in)
  Nose to tail rotor c/l: Mi-17 ... 18.425 m (60 ft 5½ in)
    Mi-172 ... 18.855 m (61 ft 10¼ in)
    Mi-17MD ... 18.99 m (62 ft 3¾ in)
  Fuselage, excl tail rotor ... 18.17 m (59 ft 7¼ in)
Width of fuselage, excl fuel tanks ... 2.50 m (8 ft 2½ in)

Tailplane span .... 3.705 m (12 ft 1¾ in)
Height: overall, rotors turning .... 5.545 m (18 ft 2¼ in)
to top of rotor head: Mi-8 .... 4.73 m (15 ft 6¼ in)
Mi-17 .... 4.745 m (15 ft 6¾ in)
Mi-17MD, Mi-171 .... 4.755 m (15 ft 7¼ in)
Mi-172 .... 4.865 m (15 ft 11½ in)
Wheel track: main .... 4.51 m (14 ft 9½ in)
Nose .... 0.30 m (11¾ in)
Wheelbase .... 4.28 m (14 ft 0½ in)
Fwd passenger door: Height .... 1.41 m (4 ft 7½ in)
Width .... 0.82 m (2 ft 8¼ in)
Rear passenger door: Height .... 1.70 m (5 ft 7 in)
Width .... 0.84 m (2 ft 9 in)
Rear cargo door: Height .... 1.82 m (5 ft 11½ in)
Width .... 2.34 m (7 ft 8¼ in)

DIMENSIONS, INTERNAL:
Passenger cabin: Length .... 6.36 m (20 ft 10¼ in)
Width .... 2.34 m (7 ft 8¼ in)
Height .... 1.80 m (5 ft 10¾ in)
Cargo hold (freighter, excl doors):
Length at floor .... 5.34 m (17 ft 6¼ in)
Width .... 2.29 m (7 ft 6¼ in)
Height .... 1.80 m (5 ft 10¾ in)
Volume .... 22.5 m^3 (795 cu ft)

AREAS:
Main rotor disc .... 356.16 m^2 (3,833.7 sq ft)
Tail rotor disc .... 12.01 m^2 (129.28 sq ft)

WEIGHTS AND LOADINGS:
Weight empty, equipped: Mi-17 .... 7,100 kg (15,653 lb)
Mi-171 .... 6,913 kg (15,241 lb)
Mi-17-1V .... 7,489 kg (16,510 lb)
Mi-17-1VA .... 7,586 kg (16,724 lb)
Mi-8AMT(Sh) .... 8,493 kg (18,724 lb)
Fuel weight, Mi-17-1V/171:
normal .... 2,027 kg (4,469 lb)
plus one auxiliary tank .... 2,737 kg (6,034 lb)
plus two auxiliary tanks .... 3,447 kg (7,600 lb)
Max payload:
internal: Mi-17, Mi-17-1V/171, Mi-8AMT(Sh) .... 4,000 kg (8,820 lb)
external, on sling:
Mi-17, Mi-17-1V .... 3,000 kg (6,614 lb)
Mi-171, Mi-8AMT(Sh) .... 4,000 kg (8,820 lb)
Mi-17-1V .... 5,000 kg (11,023 lb)
Normal T-O weight:
Mi-17, Mi-17-1V/171, Mi-8AMT(Sh) .... 11,100 kg (24,470 lb)
Max T-O weight:
Mi-17, Mi-8AMT(Sh) .... 13,000 kg (28,660 lb)
Mi-171 .... 12,000 kg (26,455 lb)
Mi-172 .... 11,878 kg (26,186 lb)
Max disc loading:
Mi-17, Mi-17-1V/171 .... 36.5 kg/m^2 (7.48 lb/sq ft)

PERFORMANCE:
Max level speed:
Mi-17/172, max AUW .... 135 kt (250 km/h; 155 mph)
Mi-17-1V/171, normal AUW .... 135 kt (250 km/h; 155 mph)
Mi-17-1V/171, max AUW .... 124 kt (230 km/h; 143 mph)
Max cruising speed:
Mi-17/172, max AUW .... 129 kt (240 km/h; 149 mph)
Mi-17-1V, normal AUW .... 135 kt (250 km/h; 155 mph)
Mi-171, normal AUW .... 124 kt (230 km/h; 143 mph)
Mi-171, max AUW .... 113 kt (210 km/h; 130 mph)
Service ceiling:
Mi-17M, normal AUW .... 5,600 m (18,380 ft)
Mi-17M, max AUW .... 4,400 m (14,440 ft)
Mi-17-1V, normal AUW .... 6,000 m (19,680 ft)
Mi-17-1V, max AUW .... 4,800 m (15,740 ft)
Mi-171, normal AUW .... 5,700 m (18,700 ft)
Mi-171, max AUW .... 4,500 m (14,760 ft)
Mi-172, max AUW .... 5,650 m (18,535 ft)
Hovering ceiling OGE:
Mi-17, max AUW .... 1,760 m (5,775 ft)
Mi-17M, normal AUW .... 3,900 m (12,795 ft)
Mi-17M, max AUW .... 1,500 m (4,920 ft)
Mi-17-1V/171, normal AUW .... 3,980 m (13,055 ft)
Mi-17-1V/171, max AUW .... 1,700 m (5,575 ft)
Mi-172, max AUW .... 3,300 m (10,825 ft)
Range with max standard fuel, 5% reserves:
Mi-17, normal AUW .... 267 n miles (495 km; 307 miles)
Mi-17, max AUW .... 251 n miles (465 km; 289 miles)
Mi-172, max AUW .... 267 n miles (495 km; 307 miles)
Range at 500 m (1,640 ft), max AUW, 30 min reserves:
Mi-17M/-1V, standard fuel .... 251 n miles (465 km; 289 miles)
Mi-171, standard fuel .... 329 n miles (610 km; 379 miles)
Mi-171, standard fuel plus one auxiliary tank .... 440 n miles (815 km; 506 miles)
Mi-17M, standard fuel plus two auxiliary tanks .... 545 n miles (1,010 km; 627 miles)
Mi-171, standard fuel plus two auxiliary tanks .... 575 n miles (1,065 km; 661 miles)
Mi-17M, standard fuel plus four auxiliary tanks .... 809 n miles (1,500 km; 932 miles)

UPDATED

MIL MI-26

NATO reporting name: Halo

TYPE: Heavy lift helicopter.

PROGRAMME: Development started early 1970s (initially as Mi-6M); aim was payload capability 1½ to 2 times greater than that of any previous production helicopter; first prototype flew 14 December 1977; first production aircraft rolled out October 1980; one of several prototype or preproduction Mi-26s (SSSR-06141) displayed at 1981 Paris Air Show; in-field evaluation, probably with military development squadron, began early 1982; fully operational 1983; export deliveries started (to India) June 1986; production continues at low rate, with manufacture and marketing by Rostvertol.

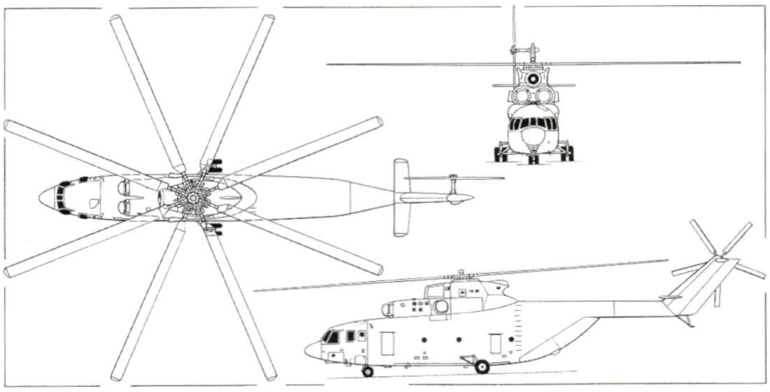

Mil Mi-26 multipurpose heavy-lift helicopter *(Jane's/Dennis Punnett)* 0507201

Mil Mi-26 in civilian operation *(Rostvertol)* ***NEW***/0576266

CURRENT VERSIONS: **Mi-26** (*Izdelie* 90): Basic military transport.

Detailed description applies to basic Mi-26, except where indicated.

Mi-26A: Modified military Mi-26, tested in 1985, with PNK-90 integrated flight/nav systems for automatic approach and descent to critical decision point, and other tasks. Not proceeded with.

Mi-26T: Basic civil transport (*Izdelie* 209), generally as military Mi-26. Production begun in 1985. Variants include **Geological Survey Mi-26** towing seismic gear, with tractive force of 10,000 kg (22,045 lb) or more, at 97 to 108 kt (180 to 200 km/h; 112 to 124 mph) at 55 to 100 m (180 to 330 ft) for up to 3 hours. The mockup of an Mi-26 two-crew flight deck was shown at the 1997 Moscow Air Show and was again displayed at Farnborough 2002, when Rostvertol said decision to install new avionics on helicopter dependent upon outcome of discussions undertaken at Farnborough; if go-ahead is given, new designation **Mi-26T2** will apply. New avionics suite will include PNK-26M flight-navigation system, incorporating five colour MFDs, two data input panels and a digital computer, plus GPS receiver and digital map and weather radar; increased automation will eliminate need for navigator/communications operator and flight engineer, although loadmaster will be retained. Military version will be adapted for night operations, using OVN-1 Skosok NVGs and GOES-321 gyrostabilised observation turret, containing a FLIR sensor and a laser range-finder. No designation has been announced for military versions.

Mi-26TS (*sertifitsyrovannyi*: certified): Mi-26T (*Izdelie* 219), but prepared for certification and marketed (in West as **Mi-26TC**) from 1996. One delivered to Samsung Aerospace Industries in South Korea on 13 September 1997; supplied with Twin Bambi Bucket fire-suppressant system and fulfils dual transport/firefighting roles. This version is subject of upgrade proposal involving installation of new avionics suite and other improvements that will reduce crew numbers from five to three and offer benefits in area of operational effectiveness; if implemented, is expected to result in improved helicopter becoming available in about 2006.

Mi-26MS: Medical evacuation version of Mi-26T, typically with intensive care section for four casualties and two medics, surgical section for one casualty and three medics, pre-operating section for two casualties and two medics, ambulance section for five stretcher patients, three seated casualties and two attendants; laboratory; and amenities section with lavatory, washing facilities, food storage and recreation unit. Civil version in use by MChS Rossii (Ministry of Emergency Situations). Alternative medical versions available, with modular box-laboratories or fully equipped medical centres that can be inserted into the hold for anything from ambulance to field hospital use. As field ambulance can accommodate up to 60 stretcher patients; or seven patients in intensive care, 32 patients on stretchers and seven attendants; or 47 patients and eight attendants in other configurations, which can include 12 bunks in four tiers forward, or patent Rostvertol box laboratory behind the first row of bunks, with 16 bunks behind.

The box includes an operating table, diagnostic equipment, anaesthetic and breathing equipment and other systems. Another configuration includes a larger theatre box by Heinkel Medizin Systeme and 12 stretchers behind, and the helicopter can be fitted with an X-ray laboratory or form the central element of a deployable air-portable field hospital.

Mi-26NEF-M: ASW version with search radar in undernose faired radome, extra cabin heat exchangers and towed MAD housing mounted on ramp.

Mi-26P: Transport for 63 passengers, basically four abreast in airline-type seating, with centre aisle; lavatory, galley and cloakroom aft of flight deck.

Mi-26PK: Flying crane (*kran*) derivative of Mi-26P with operator's gondola on fuselage side, next to cabin door on port side. First produced in 1997.

Mi-26PP: Reported ECM version. First noted 1986; current status unknown.

Mi-26S: Hastily developed version for disaster relief tasks following explosion at Chernobyl nuclear facility; equipped with deactivating liquid tank and underbelly spraying apparatus.

Mi-26TM: Flying crane, with gondola for pilot/sling supervisor under fuselage aft of nosewheels or under rear-loading ramp. First produced in 1992.

Mi-26TP: Firefighting (*pozharnyi*) version that appeared in 1994, with internal tanks able to dispense up to 15 000 litres (3,962 US gallons; 3,300 Imp gallons) fire retardant from one or two vents, or 17,260 litres (4,560 US gallons; 3,796 Imp gallons) of water from an underslung VSU-15 bucket, or from two linked EP-8000 containers. Can fill tanks on the ground using pumps with 3,000 litres (793 US gallons; 660 Imp gallons)/min throughput. One delivered to Moscow Fire Brigade on 19 August 1999.

Mi-26TZ: Tanker version that emerged in 1998, with 14,040 litres (3,710 US gallons; 3,088 Imp gallons) of T2, TS1 or R2 aviation fuel or DL, DZ or DA diesel oil fuel and 1,040 litres (275 US gallons; 228 Imp gallons) lubricants (in 52 jerry cans), dispensed through four 60 m (197 ft) hoses for aircraft, or 10 20 m (66 ft) hoses for ground vehicles. Conversion to/from Mi-26T takes 1 hour 25 minutes for each operation.

Mi-26M: Upgrade under development; all-GFRP main rotor blades of new aerodynamic configuration, new ZMKB Progress D-127 turboshafts (each 10,700 kW; 14,350 shp), and modified integrated flight/nav system with EFIS. Transmission rating unchanged, but full payload capability maintained under 'hot-and-high' conditions, OEI safety improved, hovering and service ceilings increased, and greater maximum payload (22,000 kg; 48,500 lb) for crane operations.

Two prototypes of a command support version of the Mi-26 are reported to have been built in 1988, with designation **Mi-27**. These feature new antennas along lower 'corner' of fuselage, blade and box-type and with long folded masts which are horizontal in flight, vertical when deployed on ground. Orders for production helicopters do not appear to have been placed.

CUSTOMERS: Nearly 300 said to have been built by 2001. Reportedly sold to about 20 countries; operators include Belarus (15), Cambodia (two), Congo Democratic Republic (one), India (10), Kazakhstan, North Korea (two), South Korea (one), Mexico (two second-hand) in 2000, Peru (three), Russian Army (35), Russian Ministry of Emergencies, Mil-Avia and Ukraine (20). Russian Army deliveries included four in 1994 (but none subsequently). One attrition replacement ordered in 2003 for Siberian division of MChS Rossii emergency relief service.

COSTS: US$8 million to US$10 million (Mi-26T) (2000).

DESIGN FEATURES: Largest ever production helicopter; empty weight comparable to that of Mi-6 and, as specified, is approximately 50 per cent of maximum T-O weight; weight saved by in-house design of main gearbox providing multiple torque paths, GFRP tail rotor blades, titanium main and tail rotor heads, main rotor blades of mixed metal and GFRP, use of aluminium-lithium alloys in airframe; conventional pod and boom configuration, but first successful use of eight-blade main rotor, of smaller diameter than Mi-6 rotor; payload and cargo hold size similar to those of Lockheed C-130 Hercules; auxiliary wings not required; rear-loading ramp/doors; main rotor rpm 132; main rotor spindle inclined forwards 4°.

FLYING CONTROLS: Hydraulically powered cyclic and collective pitch controls actuated by small parallel jacks, with redundant autopilot and stability augmentation system inputs. Fly-by-wire system flight tested 1994.

STRUCTURE: Eight-blade constant-chord main rotor; flapping and drag hinges, droop stops and hydraulic drag dampers; no elastomeric bearings or hinges; each blade has one-piece tubular steel spar and 26 GFRP aerofoil shape full-chord pockets, honeycomb filled, with ribs and stiffeners and non-removable titanium leading-edge abrasion strip; blades have moderate twist, taper in thickness toward tip, and are attached to small forged titanium head of unconventional design; each has ground-adjustable trailing-edge tab; five-blade constant-chord tail rotor, starboard side, has GFRP blades, forged titanium head; conventional transmission, with tail rotor shaft inside cabin roof; all-metal riveted semi-monocoque fuselage with clamshell rear doors; flattened tailboom undersurface; engine bay of titanium for fire protection; all-metal tail surfaces; swept vertical stabiliser/tail rotor support profiled to produce sideways lift; ground-adjustable variable incidence horizontal stabiliser.

LANDING GEAR: Non-retractable tricycle type; twin wheels on each unit; steerable nosewheels, tyre size 900×300; mainwheel tyres size 1,120×450. Retractable tailskid at end of tailboom to permit unrestricted approach to rear cargo doors. Length of main legs adjusted hydraulically to facilitate loading through rear doors and to permit landing on varying surfaces. Device on main gear indicates take-off weight to flight engineer at lift-off, on panel on shelf to rear of his seat.

POWER PLANT: Two 8,500 kW (11,399 shp) ZMKB Progress D-136 free-turbine turboshafts, side by side above cabin, forward of main rotor driveshaft. Air intakes fitted with particle separators to prevent foreign object ingestion, and have both electrical and bleed air anti-icing systems. Above and behind is central oil cooler intake. VR-26 fan-cooled main transmission, rated at 14,914 kW (20,000 shp), with air intake above rear of engine cowlings. System for synchronising output of engines and maintaining constant rotor rpm; if one engine fails, output of other is increased to maximum power automatically. Independent fuel system for each engine; fuel in eight underfloor rubber tanks, feeding into two header tanks above engines, which permit gravity feed for a period in emergencies; maximum standard internal fuel capacity 12,000 litres (3,170 US gallons; 2,640 Imp gallons); provision for four auxiliary tanks. Mi-26TS normal capacity is 13,020 litres (3,440 US gallons; 2,864 Imp gallons). Two large panels on each side of main rotor mast fairing, aft of engine exhaust outlet, hinge downward as work platforms.

ACCOMMODATION: Crew of five on flight deck: pilot (on port side) and co-pilot side by side, tip-up seat between pilots for flight technician, and seats for flight engineer (port) and navigator (starboard) to rear; upgrade proposal revealed in early 2001 involves installation of new avionics and will result in reduction of flight deck crew to three. Four-seat passenger compartment aft of flight deck. Loads in hold include two airborne infantry combat vehicles and a standard 20,000 kg (44,090 lb) ISO container; about 20 tip-up seats along each sidewall of hold; maximum military seating for 90 combat-equipped troops; alternative provisions for 60 stretcher patients and four/five attendants. Heated windscreen, with wipers; four large blistered side windows on flight deck; forward pair swing open slightly outward and rearward. Downward-hinged doors, with integral airstairs, at front of hold on port side, and each side of hold aft of main landing gear units. Hold loaded via downward-hinged lower door, with integral folding ramp, and two clamshell upper doors forming rear wall of hold when closed; doors opened and closed hydraulically, with back-up hand pump for emergency use. Two LG-1500 electric hoists on overhead rails, each with capacity of 2,500 kg (5,511 lb), enable loads to be transported along cabin; winch for hauling loads, capacity 500 kg (1,100 lb); roller conveyor in floor and load lashing points throughout hold. Flight deck fully air conditioned.

SYSTEMS: Two main and one emergency hydraulic systems, operating pressure 157 and 206 bar (2,276 and 2,987 lb/sq in). Electrical system three-phase 200/115 V 400 Hz; single-phase 115 V 400 Hz; three-phase, 36 V 400 Hz; single-phase 36 V 400 Hz; DC 27 V. TA-8V 119 kW (160 hp). APU under flight deck, with intake louvres (forming fuselage skin when closed) and exhaust on starboard side, for engine starting and to supply hydraulic, electrical and air conditioning systems on ground. Electrically heated leading-edge of main and tail rotor blades for anti-icing. Only flight deck pressurised.

AVIONICS: All items necessary for day and night operations in all weathers are standard.

Radar: Groza 7A813 weather radar in hinged (to starboard) nosecone.

Flight: Integrated PKV-26-1 flight/nav system and automatic flight control system, Doppler, map display, HSI, and automatic hover system. Optional GPS.

Self-defence: Military versions can have IR jammers and suppressors, IR decoy dispensers and colour-coded identification flare system.

EQUIPMENT: Hatch for load sling in bottom of fuselage, in line with main rotor shaft; sling cable attached to internal winching gear. Closed-circuit TV cameras to observe slung payloads. Specialised versions can utilise firefighting equipment.

ARMAMENT: None.

DIMENSIONS, EXTERNAL:

Main rotor diameter ... 32.00 m (105 ft 0 in)
Tail rotor diameter ... 7.61 m (24 ft 11½ in)
Length: overall, rotors turning ... 40.025 m (131 ft 3¾ in)
nose to turning tail rotor ... 35.91 m (117 ft 9¾ in)
fuselage, excl tail rotor ... 33.745 m (110 ft 8½ in)
Height: to top of rotor head ... 8.145 m (26 ft 8¾ in)
to top of fin ... 7.45 m (24 ft 5¼ in)
tail rotor turning ... 11.60 m (38 ft 0¾ in)
Width overall (outsides of mainwheels) ... 6.15 m (20 ft 2¼ in)
Tailplane span ... 6.02 m (19 ft 9 in)
Wheel track: c/l shock-absorbers ... 5.00 m (16 ft 4¾ in)
outer wheels ... 5.75 m (18 ft 10½ in)
Wheelbase ... 8.95 m (29 ft 4½ in)

DIMENSIONS, INTERNAL:

Freight hold:
Length: excl ramp ... 12.08 m (39 ft 7½ in)
ramp trailed ... 15.00 m (49 ft 2½ in)
Max width ... 3.25 m (10 ft 8 in)
Height ... 2.91–3.17 m (9 ft 6½ in–10 ft 4¾ in)
Floor area: excl ramp ... 39.3 m^2 (423 sq ft)
ramp trailed ... 49.2 m^2 (530 sq ft)
Volume: excl ramp ... 121.0 m^3 (4,273 cu ft)
ramp trailed ... 135.9 m^3 (4,799 cu ft)

AREAS:

Main rotor disc ... 804.25 m^2 (8,656.8 sq ft)
Tail rotor disc ... 45.48 m^2 (489.54 sq ft)

WEIGHTS AND LOADINGS:

Weight empty ... 28,200 kg (62,170 lb)
Max payload, internal or external ... 20,000 kg (44,090 lb)
Normal T-O weight:
except Mi-26TS ... 49,600 kg (109,350 lb)
Mi-26TS ... 49,650 kg (109,455 lb)
Max T-O weight ... 56,000 kg (123,450 lb)
Max disc loading: Mi-26TS ... 69.6 kg/m^2 (14.26 lb/sq ft)
Transmission loading at max T-O weight and power:
Mi-26TS ... 3.76 kg/kW (6.17 lb/shp)

PERFORMANCE (A: Mi-26, B: Mi-26M, C: Mi-26TS at normal T-O weight):

Max level speed: A ... 159 kt (295 km/h; 183 mph)
C ... 146 kt (270 km/h; 168 mph)
Normal cruising speed: A, C ... 137 kt (255 km/h; 158 mph)
Service ceiling: A ... 4,600 m (15,100 ft)
B ... 5,900 m (19,360 ft)
C ... 4,300 m (14,100 ft)
Hovering ceiling IGE:
A, ISA, with 5,100 kg (11,240 lb) payload ... 1,000 m (3,280 ft)
B, ISA + 15°C, with 12,300 kg (27,115 lb) payload ... 1,000 m (3,280 ft)
Hovering ceiling OGE, ISA: A ... 1,800 m (5,900 ft)
B ... 2,800 m (9,180 ft)
C ... 1,520 m (4,980 ft)
Range: A at 2,500 m (8,200 ft) ISA + 15°C, with 7,700 kg (16,975 lb) payload ... 270 n miles (500 km; 310 miles)
B at 2,500 m (8,200 ft) ISA + 15°C, with 13,700 kg (30,200 lb) payload ... 270 n miles (500 km; 310 miles)
A, S/L ISA, with max internal fuel at max T-O weight, 5% reserves ... 318 n miles (590 km; 366 miles)
A, S/L ISA, with four auxiliary tanks ... 1,028 n miles (1,905 km; 1,183 miles)

UPDATED

MIL MI-28

NATO reporting name: Havoc

TYPE: Attack helicopter.

DEVELOPMENT MILESTONES

Design began	1980
First flight	10 Nov 1982

PROGRAMME: Design started 1980 under Marat N Tishchenko; first of two flying Mi-28 prototypes (012) flew 10 November 1982; each prototype different: first and second (022) had upward-pointing exhaust diffusers and fixed undernose fairing for electro-optic equipment; first also had conventional three-blade tail rotor; second replaced this with the definitive 'Delta-H' configuration. The first Mi-28A (032) introduced the downward-pointing exhaust suppressors and flew in January 1988; second Mi-28A prototype (042) demonstrated at Moscow in 1992 and represented the intended production configuration. It had the definitive moving E-O sensor turret undernose, downward-pointing exhaust diffusers and wingtip electronics/chaff dispenser pods.

Following official assessment, Kamov Ka-50 was announced on 5 October 1994 as Russian Army's future light combat helicopter, in preference to Mi-28, and received order for small batch for State Tests. (Army Aviation was absorbed by Russian Air Forces on 1 January 2003.) However, Mi-28N programme was kept active by Rostvertol at Rostov-on-Don, which maintained readiness to launch production. By late 2003, Mi-28N's fortunes had revived and this confirmed in January 2004 by Air Forces C-in-C in announcement that it was "being considered" as next generation attack helicopter, partnered by small number of Ka-52 (two-seat Ka-50s) in intelligence-gathering role. Defence minister Sergey Ivanov declared in August 2004 that Mi-28 would be included in equipment of two mountain brigades to be formed in North Caucasus Military District. By mid-2004, official sources were referring to delivery of 50 Mi-28s by 2010 and long-term plans to acquire 500.

Mil Mi-28N prototype *(Rostvertol)* ***NEW***/0576258

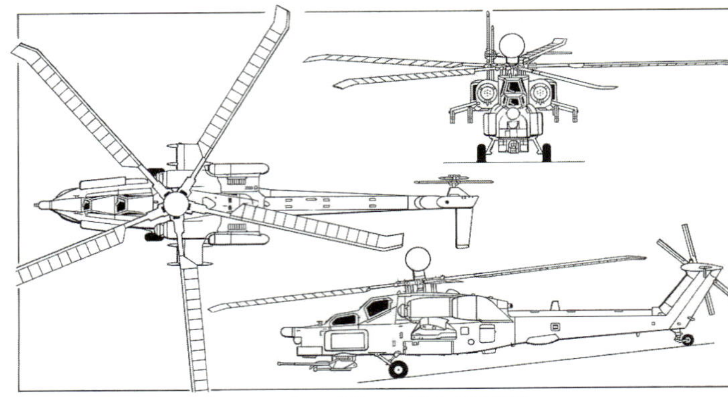

Mil Mi-28N combat helicopter (two Klimov TV3-117VMA turboshafts) *(James Goulding)* 0507528

CURRENT VERSIONS: **Mi-28:** First two prototypes with 1,434 kW (1,923 shp) TV3-117BM engines and VR-28 gearbox.

Mi-28A (Type 280): Basic version, *as described in detail*. Third and fourth aircraft built.

Mi-28N: (*Nochnoy*: Night): Unofficial names: Night Hunter and Night Pirate. Added night/all-weather operating capability. Russian Army funding announced January 1994; demonstrator (014) modified from first Mi-28 prototype (012); first hover 14 November 1995; formal roll-out 16 August 1996; first flight 30 April 1997. Problems reportedly experienced with main gear box caused long stand-down; by August 2003, prototype said to have accumulated only 20 flight hours.

Mast-mounted 360° scan millimetre wave Kinzhal V or Arbalet radar (pod soon enlarged in vertical plane); FLIR ball beneath missile-guidance nose radome and above new shuttered turret for optical/laser sensors, including Zenit low-light-level TV. EFIS cockpit. Armament of production version to include 9M114 Shturm (AT-6 'Spiral') or 9M120 Vikhr/Ataka (AT-12 'Swinger') ASMs and Igla (SA-16 'Gimlet') AAMs and R-73 AAMs. New composites rotor with sweptback blade tips added subsequently. Mi-28N introduced uprated VR-29 transmission and IKBO integrated flight/weapon aiming system, with automatic terrain-following and automatic target search, detection, identification and (in formations of Mi-28Ns) allocation; Ramenskoye Breo-28N mission control system.

Second Mi-28N prototype (Mi-28-IF) produced by Rostvertol with joint funding by Yugo-Zapadnyy Bank; rolled out at Rostov 9 March 2004; first flight ('02') 25 March 2004; press and VIP demonstration 31 March 2004; following factory tests, departed Rostov for State Tests in Moscow Region, 18 January 2005. To be joined by first production helicopter, which rolled out (c/n 0101) August 2004; due for completion of outfitting in first half of 2005. Official programme requires 600 sorties and will take between two and five years, according to funding priorities.

Mi-28NE: (*Nochnoy, Ehksport:* Night, Export): Version offered to, and rejected by, South Korea in 2000. Evaluated by Swedish Army in 2001 against Boeing AH-64 Apache and Eurocopter Tiger. Also offered to India and Malaysia, as well as Turkey, which still has a requirement for 50 attack helicopters, following collapse of attempt to purchase Bell AH-1Z King Cobra from USA. Is also candidate to fulfil China's requirement for an attack helicopter.

COSTS: Mi-28N development cost US$150 million (2000); unit cost approximately US$12 million (Rostvertol, 2004).

DESIGN FEATURES: Conventional gunship configuration, with two crew in stepped cockpits; original three-blade tail rotor superseded by low noise 'scissors' or 'Delta-H' type comprising two independent two-blade rotors set as narrow X (35°/145°) on same shaft with self-lubricating bearings; resulting flapping freedom relieves flight loads; agility enhanced by doubling hinge offset of main rotor blades compared with Mi-24; survivability emphasised; crew compartments protected by titanium and ceramic armour and armoured glass transparencies; single hit will not knock out both engines; new composites main rotor can withstand hit from round of up to 30 mm calibre; vital units and parts are redundant, widely separated and shielded by the less vital; multiple self-sealing fuel tanks in centre-fuselage enclosed in composites second skin, outside metal fuselage skin; no explosion, fire or fuel leakage results if tanks hit by bullet or shell fragment; energy absorbing seats and landing gear protect crew in crash landing at descent rate of 12 m (40 ft)/s; crew doors are rearward-hinged, to open quickly and remain open in emergency; parachutes are mandatory for Russian Federation and Associated States (CIS) military helicopter aircrew; if Mi-28 crew had to parachute, emergency system would jettison doors, blast away stub-wings, and inflate bladder beneath each door sill; as crew jumped, they would bounce off bladders and clear main landing gear; no provision for rotor separation; port-side door, aft of wing, provides access to avionics compartment large enough to permit combat rescue of two or three persons on ground, although it lacks windows, heating and ventilation.

Hand crank, inserted into end of each stub-wing, enables stores of up to 500 kg (1,100 lb) to be winched on to pylons without hoists or ground equipment; current 30 mm gun is identical with that of RFAS army ground vehicles and uses same ammunition; jamming averted by attaching twin ammunition boxes to sides of gun mounting, so that they turn, elevate and depress with gun; main rotor shaft has 5° forward tilt, providing tail rotor clearance; transmission capable of running without oil for 20 to 30 minutes; main rotor rpm 242; with main rotor blades and wings removed, helicopter is air-transportable in Il-76 freighter.

FLYING CONTROLS: Hydraulically powered mechanical type; horizontal stabiliser linked to collective; controls for pilot only.

STRUCTURE: Five-blade main rotor; blades have very cambered high-lift section and sweptback tip leading-edge; full-span upswept tab on trailing-edge of each blade; structure comprises numerically controlled, spirally wound glass fibre D-spar, blade pockets of Kevlar-like material with Nomex-like honeycomb core, and titanium erosion strip on leading-edge; each blade has single elastomeric root bearing, mechanical droop stop and hydraulic drag damper; four-blade GFRP tail rotor with elastomeric bearings for flapping; rotor brake lever on starboard side of cockpit; strong and simple machined titanium main rotor head with elastomeric bearings, requiring no lubrication; power output shafts from engines drive main gearbox from each side; tail rotor gearbox, at base of tail pylon, driven by aluminium alloy shaft inside composites duct on top of tailboom; sweptback mid-mounted wings have light alloy primary box structure, leading— and trailing-edges of composites; no wing movable surfaces; provision for countermeasures pod on each wingtip, housing chaff/flare dispensers and sensors, probably RWR; light alloy semi-monocoque fuselage, with titanium armouraround cockpits and vulnerable areas; composites access door aft of wing on port side;swept fin has light alloy primary box structure, composites leading— and trailing-edges; cooling air intake at base of fin leading-edge, exhaust at top of trailing-edge; two-position composites horizontal stabiliser.

LANDING GEAR: Non-retractable, tailwheel type; single wheel on each unit; mainwheel tyres size 720×320, pressure 5.40 bar (78 lb/sq in); castoring tailwheel with tyre size 480×200.

POWER PLANT: Two Klimov TV3-117VMA turboshafts, each 1,636 kW (2,194 shp), in pod above each wingroot; three jetpipes inside downward-deflected composites nozzle fairing on each side of third prototype shown in Paris 1989; upward deflecting type also tested. Deflectors for dust and foreign objects forward of air intakes, which are de-iced by engine bleed air. Internal fuel capacity 1,720 litres (454 US gallons; 378 Imp gallons). Provision for four external fuel tanks on underwing pylons.

ACCOMMODATION: Navigator/gunner in front cockpit; pilot behind, on elevated seat; transverse armoured bulkhead between; flat non-glint tinted transparencies of armoured glass; navigator/gunner's door on port side, pilot's door on starboard side.

SYSTEMS: Cockpits air conditioned and pressurised by engine bleed air. Duplicated hydraulic systems, pressure 152 bar (2,200 lb/sq in). 208 V AC electrical system supplied by two generators on accessory section of main gearbox, ensuring continued supply during autorotation. Low-airspeed system standard, giving speed and drift via main rotor blade-tip pitot tubes at −27 to +38 kt (−50 to +70 km/h; −31 to +43 mph) in forward flight, and ±38 kt (±70 km/h; ±43 mph) in sideways flight. Main and tail rotor blades electrically de-iced. Ivchenko AI-9V APU in rear of main pylon structure supplies compressed air for engine starting and to drive small turbine for preflight ground checks.

AVIONICS: *Comms:* UHF/VHF nav/com; small IFF fairing each side of nose and tail.

Instrumentation: Conventional IFR instrumentation, with autostabilisation, autohover, and hover/heading hold lock in attack mode; pilot has HUD and centrally mounted CRT for basic TV; aircraft designed for use with night vision goggles.

Mission: Radio for missile guidance in nose radome. Daylight optical weapons sight and laser range-finder in gyrostabilised and double-glazed nose turret above gun, with which it rotates through ±110°; wiper on outer glass protects inner optically flat panel.

Self-defence: Two fixed IR sensors on initial basic production Mi-28; IR suppressors, radar and laser warning receivers standard; optional countermeasures pod on each wingtip, housing chaff/flare dispensers and sensors, probably RWR. Mi-28N has integrated Vitebsk DASS with Pastel RWR, Mak IR warning system, Platan jammer and UV-26 flare dispensers.

EQUIPMENT: Two slots, one above the other on port side of tailboom, for colour-coded identification flares. Three pairs of rectangular formation-keeping lights in top of tailboom; further pair in top of main rotor pylon fairing.

ARMAMENT: One 2A42 30 mm turret-mounted gun (with 250 rounds in side-mounted boxes) in NPPU-28 mount at nose, able to rotate ±110°, elevate 13° and depress 40°; maximum rate of fire 900 rds/min air-to-air and air-to-ground. (New specially designed gun under development.) Two pylons under each stub-wing, each with capacity of 480 kg (1,058 lb), typically for total of sixteen 9M114 Shturm C (AT-6 'Spiral') radio-guided tube-launched anti-tank missiles and two UB-20 pods of eighty 80 mm S-8 or twenty 122 mm S-13 rockets or two UPK-23-250 gun pods. Alternative ATMs include Shipunov 9M120/9M121F Vikhr/Ataka and 9A-2200; up to eight 9M39 Igla-V AAMs in place of ATMs; in minelaying role can carry two KGMU-2 dispensers. Main 2A42 gun fired and guided weapons launched normally only from front cockpit; unguided rockets fired from both cockpits. (When fixed, gun can also be fired from rear cockpit.)

DIMENSIONS, EXTERNAL:

Main rotor diameter	17.20 m (56 ft 5 in)
Main rotor blade chord	0.67 m (2 ft 2½ in)
Tail rotor diameter	3.84 m (12 ft 7¼ in)
Tail rotor blade chord	0.24 m (9½ in)
Length overall, excl rotors, incl gun	17.01 m (55 ft 9¾ in)
Fuselage max width	1.85 m (6 ft 1 in)
Width over stub-wings	4.88 m (16 ft 0¼ in)
Height: overall	4.70 m (15 ft 5 in)
to top of rotor head	3.82 m (12 ft 6½ in)
Wheel track	2.29 m (7 ft 6¼ in)
Wheelbase	11.00 m (36 ft 1 in)

AREAS:

Main rotor disc	232.35 m² (2,501.0 sq ft)
Tail rotor disc	11.58 m² (124.65 sq ft)

WEIGHTS AND LOADINGS:

Weight empty, equipped: 28	7,900 kg (17,416 lb)
28A	8,095 kg (17,846 lb)
28N	8,590 kg (18,938 lb)
Fuel weight: standard internal	1,337 kg (2,947 lb)
with added tanks	1,782 kg (3,928 lb)
Normal T-O weight: 28	10,200 kg (22,487 lb)
28A	10,400 kg (22,928 lb)
28N	10,700 kg (23,589 lb)
Max T-O weight: 28	11,200 kg (24,691 lb)
28A, 28N	11,500 kg (25,353 lb)
Max disc loading: 28	48.2 kg/m² (9.87 lb/sq ft)
28A, 28N	49.5 kg/m² (10.14 lb/sq ft)

PERFORMANCE:

Max level speed: 28A	162 kt (300 km/h; 186 mph)
28N	172 kt (320 km/h; 199 mph)

Max cruising speed: 28A .... 143 kt (265 km/h; 164 mph)
28N .... 145 kt (270 km/h; 168 mph)
Max rate of climb at S/L .... 816 m (2,677 ft)/min
Service ceiling: 28A .... 5,800 m (19,020 ft)
28N (-SB3) .... 5,700 m (10,700 ft)
Hovering ceiling OGE: 28 .... 3,470 m (11,380 ft)
28A, 28N .... 3,600 m (11,820 ft)
28N (-SB3) .... 4,500 m (14,760 ft)
Radius of action, standard fuel, 10 min loiter at target, 5% reserves .... 108 n miles (200 km; 124 miles)
Range, max standard fuel, 10% reserves: all .... 234 n miles (435 km; 270 miles)
Ferry range, 5% reserves .... 593 n miles (1,100 km; 683 miles)
Endurance with max fuel .... 2 h
g limits .... +3/–0.5

UPDATED

MIL MI-34

NATO reporting name: Hermit

TYPE: Four-seat helicopter.

PROGRAMME: First flight 17 November 1986; two prototypes and structure test airframe completed by mid-1987, when exhibited for first time at Paris Air Show; first helicopter built in former USSR to perform normal loop and roll; series production began at Arsenyev plant in 1993; airframe manufactured by Carpathian Helicopter Production Association; marketed by Mi-Light Helicopters (LVM); Mi-34S/-34C completion was at Moscow plant of LVM, but subsequently reverted to Arsenyev; planned completion of 30 in 1994–95 hampered by lack of funding; five delivered in 1995, one in first half of 1996. State order for resumed manufacture 1996. Marketing initiative, early 1999, planned worldwide establishment of service centres in countries where more than 10 Mi-34s ordered. In mid-2003, production was stated to have temporarily ceased to allow transfer of manufacture to a fully constituted production line, the previous 22 helicopters having been manufactured as 'pre-production' aircraft. Production will resume with 276 kW (370 hp) power plant, development delays with which were due to have been overcome by early 2004.

Trials in 1999 by civil pilot training school at Omsk showed Mi-34 to be 2.8 times cheaper and more effective to operate than current fleet; school to acquire three Mi-34Cs and obtain up to 10 more in long term. Other schools expected to replace ageing Mi-2 with Mi-34.

In 2001, an upgraded variant was proposed with M-14 engine rated at 272 kW (365 hp), IFR instrumentation and auxiliary fuel tank. An agricultural variant was planned for debut in 2003.

CURRENT VERSIONS: **Mi-34S:** Basic version; marketed in Russia as Mi-34; certified by Interstate Aviation Committee Aviation Register (initially at 1,350 kg; 2,976 lb max T-O weight), with helicopter, engine and noise type certificates; meets FAR Pt 27 requirements. (Note that until 1999, all marketing literature for this version used the hybrid Roman/Cyrillic '**Mi-34C**' to indicate certified status.)

Description applies to Mi-34S (Mi-34C), except where indicated.

Mi-34L: Projected version with 261 kW (350 hp) Textron Lycoming TIO-540J piston engine.

Mi-34P (*patrulnyi:* patrol): Version of Mi-34S, equipped for police duties. Renewed interest in 2001 from Gazprom for pipeline patrol.

Mi-34A: Originally with 335 kW (450 shp) Rolls-Royce 250-C20R turboshaft; mockup, with luxury interior exhibited at Moscow Air Show'95. Promotion recommended in 1999, employing 376 kW (504 shp) Turbomeca TM 319 Arrius 2F turboshaft; MTOW 1,450 kg (3,196 lb); usable fuel increased to 340 litres (89.7 US gallons; 74.8 Imp gallons); accommodation for four passengers, plus pilot; 1.0 m³ (35 cu ft) baggage compartment. None yet built.

Mi-34M1/M2: Projected twin-turbine, six-passenger versions; MTOW 2,500 kg (5,511 lb).

Mi-34UT: Announced 2001. Dual control trainer. Unsuccessfully competed against Kazan Ansat in competition, in which decision announced September 2001, to provide 100 training helicopters for Russian armed forces.

Mi-234: VAZ rotary engine. Not built.

CUSTOMERS: Total of 22 delivered from Arsenyev by mid-2003, compared with estimated 425 called for in Russia's 1992–2000 civil aviation development plan. First three for Mayor's office, Moscow. Others used by Bashkir Airlines for oilfield support and Mi-Avia for patrol and training. One operated by Bosniac and Croat Federation Air Force. Three delivered to Nigerian Air Force in 2001, along with six Mi-35s; these were from first batch of six Mi-34s built at Arsenyev, after pause of several years; further five delivered to Nigeria by end of 2002. LVM (Mil's light helicopter marketing subsidiary) reportedly ordered 20 Mi-34s for construction at Arsenyev in 2001, but only delivery in that year apart from Nigerians was one to Sibneft. In June 2002, Russian sources reported foreign (assumed Nigerian) negotiations for "several dozen" Mi-34s although only known 2003 production commitment is follow-on batch of four for Nigeria and five for Omsk Civil Aviation Flying and Technical College by end of 2005.

Mil Mi-34P light utility helicopter *(Yefim Gordon)* ***NEW***/0532097

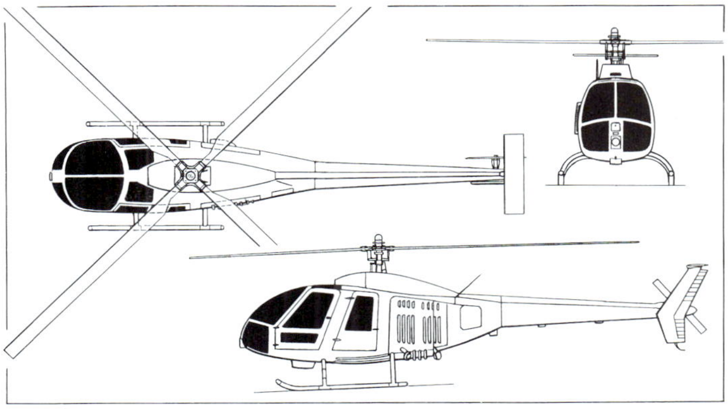

Mil Mi-34 two/four-seat training, competition and light transport helicopter *(Jane's/Dennis Punnett)* 0079310

COSTS: US$400,000, fully equipped (2003).

DESIGN FEATURES: Aerobatic helicopter; intended initially for training and international competition flying; conventional pod and boom configuration; piston engine of same basic type as that in widely used Yakovlev fixed-wing training aircraft and Kamov Ka-26 helicopters; suitable also for light utility, mail delivery, observation and liaison duties, and border patrol; later developments concentrate on light transport role.

Aerobatic capabilities include looping; backwards flight at 70 kt (130 km/h; 81 mph); and rotation about main rotor axis at 120°/s.

FLYING CONTROLS: Manual, with no hydraulic boost.

STRUCTURE: Semi-articulated four-blade main rotor with flapping and cyclic pitch hinges, but natural flexing in lead/lag plane; blades of GFRP with CFRP reinforcement, attached by flexible steel straps to head like that of Boeing MD 500; two-blade tail rotor of similar composites construction, on starboard side; riveted light alloy fuselage; sweptback tailfin with small unswept T tailplane.

LANDING GEAR: Conventional fixed skids on arched support tubes; small tailskid to protect tail rotor.

POWER PLANT: One 239 kW (320 hp) VOKBM M-14V-26V nine-cylinder radial air-cooled engine mounted sideways in centre-fuselage. Upgrade to 276 kW (370 hp) from 2004. Fuel capacity 176 litres (46.5 US gallons; 38.7 Imp gallons); system for inverted flight.

ACCOMMODATION: Normally one or two pilots, side by side, in enclosed cabin, with optional dual controls. Rear of cabin contains low bench seat, available for two passengers and offering flat floor for cargo carrying. Forward-hinged door on each side of flight deck and on each side of rear cabin.

SYSTEMS: Primary electric power provided by 27 V 3 kW engine-driven generator; secondary power supplies of 115 V AC, 400 Hz, single-phase and 36 V AC, 400 Hz, three-phase; 27 V 17 Ah battery.

AVIONICS: *Comms:* Briz VHF radio.

Flight: A-037 radio altimeter; ARK-22 radio compass.

Instrumentation: Magnetically slaved compass system incorporating radio magnetic indicator.

EQUIPMENT: Gyro horizon. Version used by Moscow Police has dual controls, two rear seats and loudspeaker under rear of pod.

DIMENSIONS, EXTERNAL:
Main rotor diameter .... 10.00 m (32 ft 9¾ in)
Main rotor blade chord .... 0.22 m (8¾ in)
Tail rotor diameter .... 1.48 m (4 ft 10¼ in)
Tail rotor blade chord .... 0.16 m (6¼ in)
Length overall, rotors turning .... 11.415 m (37 ft 5½ in)
Fuselage: Length .... 8.71 m (28 ft 7 in)
Max width .... 1.42 m (4 ft 8 in)
Height: overall .... 2.75 m (9 ft 0¼ in)
to fintip .... 2.45 m (8 ft 0½ in)
Skid track .... 2.175 m (7 ft 1½ in)
Fuselage ground clearance .... 0.36 m (1 ft 2¼ in)
Cockpit doors: Height .... 1.15 m (3 ft 9¼ in)
Max width .... 0.70 m (2 ft 3½ in)

AREAS:
Main rotor disc .... 78.70 m² (847.1 sq ft)
Tail rotor disc .... 1.72 m² (18.52 sq ft)

WEIGHTS AND LOADINGS:
Weight empty .... 950 kg (2,094 lb)
Fuel weight .... 128 kg (282 lb)
T-O weight: Aerobatic .... 1,100 kg (2,425 lb)
Normal .... 1,280 kg (2,822 lb)
Max .... 1,450 kg (3,196 lb)
Max disc loading .... 18.4 kg/m² (3.77 lb/sq ft)
Max power loading .... 6.08 kg/kW (9.99 lb/hp)

PERFORMANCE (Mi-34S/C at Normal TOW; Mi-34A at MTOW):
Max level speed: Mi-34S/C .... 113 kt (210 km/h; 130 mph)
Mi-34A .... 121 kt (225 km/h; 140 mph)
Max cruising speed:
Mi-34S/C .... 92 kt (170 km/h; 106 mph)
Mi-34A .... 113 kt (210 km/h; 130 mph)
Best climbing speed: Mi-34S/C .... 49 kt (90 km/h; 56 mph)
Service ceiling: Mi-34S/C .... 4,000 m (13,120 ft)
Mi-34A .... 5,000 m (16,400 ft)
Hovering ceiling, OGE: Mi-34S/C .... 900 m (2,960 ft)
Mi-34A .... 2,750 m (9,020 ft)
Range with max fuel at 500 m (1,640 ft), 5% reserves:
Mi-34S/C .... 192 n miles (356 km; 221 miles)
Mi-34A .... 297 n miles (550 km; 341 miles)
g limit .... +3

UPDATED

EUROMIL MIL MI-38

TYPE: Medium transport helicopter.

PROGRAMME: Design begun in 1983; model shown at 1989 Paris Air Show, when aircraft at mockup stage. Modifications in evidence by 1993 included fixed landing gear with wider track and reduced base. Under December 1992 agreement, Eurocopter was to integrate flight deck, avionics and passenger systems, and adapt Mi-38 for international market; Euromil joint stock company established September 1994 to advance collaboration, adding Kazan production plant (as main manufacturing and final assembly centre); funding for Euromil granted in October 1994 by European Bank for Reconstruction and Development. Sextant and Pratt & Whitney Canada added as risk-sharing parties for avionics and engine. Funding by Russian Ministry for Defence Industries 1996. In July 2003, however, Eurocopter withdrew from Euromil, which continues as an all-Russian company, but with Eurocopter continuing to supply avionics as a contractor. Mi-38 rotor blades began test flying on an Mi-17 in early 2001.

By 1997, Euromil was anticipating first flight in 1999 and start of production two years later, following FAR Pt 29 certification, but contracts for completion of demonstrator not signed until 18 August 1999, following unilateral decision of Euromil board in December 1998 to launch programme and fly demonstrator at Kazan in 2001. This subsequently slipped to 2002 and then to mid-2003, finally taking place on 22 December 2003, following roll-out on 20 November. Demonstrator (PT-1) is third airframe, following test articles at Mil Moscow and Kazan. Four further prototypes to follow, including two for static testing.

CURRENT VERSIONS: **Mi-38:** Pratt & Whitney engines.

Mi-38M: Proposed development version with 16,500 kg (36,376 lb) MTOW only when carrying underslung load.

Mi-382: Klimov engines.

Mi-382M: Proposed development version with 18,000 kg (39,683 lb) MTOW only when carrying underslung load of up to 8,000 kg (17,637 lb).

CUSTOMERS: Predicted sales of 200 in CIS, plus 100 exports.

COSTS: Estimated development cost (2001) US$500 million; unit cost US$12 million to US$16 million for export; US$11 million in Russian Federation (2002).

DESIGN FEATURES: Planned as replacement for Mi-8/17 series. Western engines optional. Conventional pod and boom configuration; power plant above cabin; six-blade main rotor with considerable non-linear twist and swept tips; main rotor has hydraulic drag dampers; single lubrication point, at driveshaft; rotor brake standard; Krasny Oktyabr transmission, comprising main, intermediate and tail gearboxes and tail rotor drive shaft; engine input 19,017 rpm; two independent two-blade tail rotors, set as narrow X on same shaft; sweptback fin/tail rotor mounting; small horizontal stabiliser; for day/night operation over temperature range −60 to +50°C.

FLYING CONTROLS: Fly-by-wire, with manual back-up.

STRUCTURE: Composites main and tail rotors by Kazan; low-profile titanium main rotor head, with elastomeric bearings, built by Stupino; fuselage, mainly composites, built by Kazan.

LANDING GEAR: Fixed tricycle type; single wheel on each main unit; oleo-pneumatic shock absorbers; twin, self-centring nosewheels; low-pressure tyres; optional pontoons for emergency use in overwater missions. Main tyres 950 mm diameter, pressure 5.88 bar (85 lb/sq in); nose tyres 600 mm diameter, pressure 4.90 bar (71 lb/sq in).

POWER PLANT: Those helicopters for CIS customers powered by two Klimov TVA-3000 (TV7-117 derivative) turboshafts, each rated at 1,838 kW (2,465 shp) for T-O; single-engine rating 2,610 kW (3,500 shp) and transmission rated for same power. Demonstrator to have two 2,461 kW (3,300 shp) P&WC PW127 turboshafts, which are also available, in PW127T/S form, as an option for Western customers. Power plant above cabin, to rear of main reduction gear; air intakes and filters in sides of cowling. Bag fuel tanks beneath floor of main cabin; provision for external auxiliary fuel tanks. Liquid petroleum gas fuel planned as alternative to aviation kerosene.

ACCOMMODATION: Crew of two on flight deck, separated from main cabin by compartment for majority of avionics; single-pilot operation possible for cargo missions. Lightweight seats for 30 passengers as alternative to unobstructed hold for 5,000 kg (11,020 lb) freight. Ambulance and air survey versions planned. Passenger door, forward, port; freight door, forward, starboard; cargo ramp at rear; hatch in cabin floor, under main rotor driveshaft, for tactical/emergency cargo airdrop and for cargo sling attachment; optional windows for survey cameras in place of hatch. Provision for hoist over port-side door, remotely controlled, hydraulically actuated rear cargo ramp, powered hoist on overhead rails in cabin, and roller conveyor system in cabin floor and ramp.

SYSTEMS: Air conditioning by compressor bleed air, or APU on ground, maintains temperature of not more than 25°C on flight deck in outside temperature of 40°C, and not less than 15°C on flight deck and in main cabin in outside temperature of −50°C. Three independent hydraulic systems; any one able to maintain control of helicopter in emergency. Electrical system has three independent generators, two at 12 kW DC and one 60 kW AC; optional fourth generator 40 or 60 kW AC; two batteries; electric rotor blade de-icing, main and tail. Independent fuel system for each engine, with automatic crossfeed; forward part of cowling houses VD-100 APU, hydraulic, air conditioning, electrical and other system components.

AVIONICS: Sextant equipment in export aircraft.

Radar: Weather/nav radar (range 54 n miles; 100 km; 62 miles).

Flight: Preset flight control system allows full autopilot, autohover and automatic landing. Avionics controlled by large central computer, linked also to automatic nav system with Doppler, ILS, satellite nav system, main radar, autostabilisation system and automatic radio compass.

First flight of Mil Mi-38, 22 December 2003 *(Valery Solomakhin)* ***NEW***/0569773

Prototype Mil Mi-38 *(Valery Solomakhin)* ***NEW***/0569774

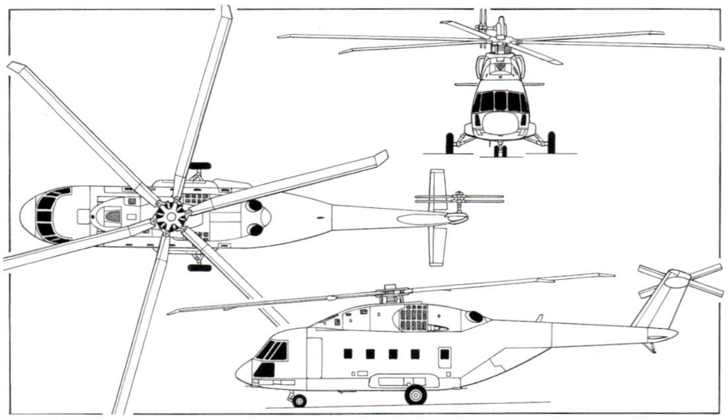

Mil Mi-38 medium transport helicopter *(Michael Badrocke)* ***NEW***/0576138

Instrumentation: Six colour CRTs for use in flight and by servicing personnel on ground. Equipment monitoring, failure warning and damage control system. Closed-circuit TV for monitoring cargo loading and slung loads. Options include low-cost electromechanical instrumentation based on that of Mi-8, sensors for weighing and CG positioning of cargo in cabin, and for checking weight of slung loads.

DIMENSIONS, EXTERNAL:
Main rotor diameter ... 21.10 m (69 ft 2¾ in)
Tail rotor diameter ... 3.84 m (12 ft 7¼ in)
Distance between rotor centres ... 12.755 m (41 ft 10¼ in)
Length: overall, rotors turning ... 25.20 m (82 ft 8¼ in)
fuselage ... 19.95 m (65 ft 5½ in)
Width, rotors folded ... 3.17 m (10 ft 4¾ in)
Height: to top of rotor head ... 5.20 m (17 ft 0¾ in)
to top of fin ... 5.56 m (18 ft 3 in)
Stabiliser span ... 4.20 m (13 ft 9½ in)
Wheel track ... 4.50 m (14 ft 9¼ in)
Wheelbase ... 5.17 m (16 ft 11½ in)
Forward freight door: Height ... 1.665 m (5 ft 5½ in)
Width ... 1.445 m (4 ft 9 in)
Forward passenger door: Height ... 1.66 m (5 ft 5¼ in)
Width ... 0.61 m (2 ft 0 in)
Floor hatch: Length ... 1.15 m (3 ft 9¼ in)
Width ... 0.75 m (2 ft 5½ in)

DIMENSIONS, INTERNAL:
Main cabin: Length ... 8.70 m (28 ft 6½ in)
Max width ... 2.30 m (7 ft 6½ in)
Width at floor ... 2.20 m (7 ft 2½ in)
Height: centre ... 1.80 m (5 ft 10¾ in)
rear ... 1.85 m (6 ft 1 in)
Volume ... 29.5 m^3 (1,042 cu ft)

AREAS:
Main rotor disc ... 349.67 m^2 (3,763.8 sq ft)
Tail rotor disc ... 11.58 m^2 (124.65 sq ft)

WEIGHTS AND LOADINGS (provisional):
Max payload: internal ... 5,000 kg (11,020 lb)
external ... 6,000 kg (13,228 lb)
Normal T-O weight ... 14,200 kg (31,305 lb)
Max T-O weight ... 15,600 kg (34,392 lb)
Max disc loading ... 44.6 kg/m^2 (9.14 lb/sq ft)
Transmission loading at max T-O weight and power ... 5.97 kg/kW (9.82 lb/shp)

PERFORMANCE (estimated, PW127 engines):
Max level speed ... 153 kt (285 km/h; 177 mph)
Max cruising speed ... 148 kt (275 km/h; 171 mph)
Service ceiling ... 5,500 m (18,040 ft)
Hovering ceiling OGE ... 2,500 m (8,200 ft)
Range, 30 min reserves:
With 5,000 kg (11,020 lb) payload ... 256 n miles (475 km; 295 miles)
with max internal fuel ... 442 n miles (820 km; 509 miles)
ferry, with auxiliary fuel ... 728 n miles (1,350 km; 838 miles)

UPDATED

MIL MI-52 SNEGIR

English name: Bullfinch

This four-seat light helicopter project was promoted during the 1990s, but nothing came of the venture. The company unveiled a mockup on 1 June 2000 and announced that it was then hoping to resuscitate the Snegir, with a view to flying a prototype in 2001, although this failed to occur. By mid-2001, Mil was still seeking US$15 million to build, test and certify two prototypes. In early 2003, Kazan helicopters plant offered to provide funds to launch production of Mi-52. However, a year later, no further announcements were known.

Mockup of Mil Mi-60 MAI exhibited by Rostvertol at Moscow Salon, August 2001 *(Paul Jackson)* 0121036

Single- (Mi-52-1) and twin-engined (Mi-52-2 with 199 kW; 266 hp VAZ-4265) versions are planned with commercial load capabilities of 350 kg (772 lb) and 400 kg (882 lb), respectively. A military trainer is also envisaged. Unit cost of baseline version estimated as US$300,000. Progress AI-450 turboshaft mentioned as alternative power source, early 2003.

UPDATED

MIL MI-60

TYPE: Three-seat helicopter.

PROGRAMME: Announced July 2000, when mockup under construction at Kazan helicopter factory; this exhibited at Moscow Salon in August 2001, promoted by Rostvertol, which considering establishing production line. Building of prototype was to have begun in 2001; development cost US$30 million, including US$2.58 million for scientific research and US$15.77 million for prototypes and preparation for series production; estimated unit cost US$140,000 to US$150,000. Programme included in Russian federal aviation plan for 2002–10, but no further announcements made by early 2004.

Development under leadership of M N Tishchenko of Moscow Aviation Institute, thus designated Mi-60 MAI: financed by Ministry of Education. Maintenance-free rotor hub; airframe TBO of 2,000 hours.

LANDING GEAR: Skid type.

POWER PLANT: One or two piston engine driving main gearbox via a belt; option of one 145 kW (195 hp) Textron Lycoming HIO-360-F1AD, one 177 kW (237 hp) VAZ-4261 or two 84.6 kW (113.4 hp) Rotax 914Fs.

DIMENSIONS, EXTERNAL:

Main rotor diameter	10.00 m (32 ft 9¾ in)
Main rotor blade chord	0.22 m (8¾ in)

WEIGHTS AND LOADINGS (estimated; L: Lycoming, R: Rotax, V: VAZ engines):

Fuel weight: L	79 kg (174 lb)
R	70 kg (154 lb)
V	71 kg (157 lb)
Max underslung load	90 kg (198 lb)
T-O weight: Normal: all	800 kg (1,764 lb)
Max: L	1,115 kg (2,458 lb)
R	1,260 kg (2,778 lb)
V	1,300 kg (2,866 lb)

PERFORMANCE (estimated; L, R, V as above):

Max level speed: L	108 kt (200 km/h; 124 mph)
R	113 kt (210 km/h; 130 mph)
V	121 kt (225 km/h; 140 mph)
Max cruising speed: L	94 kt (175 km/h; 109 mph)
R	103 kt (190 km/h; 118 mph)
V	100 kt (185 km/h; 115 mph)
Econ cruising speed: L, V	46 kt (85 km/h; 53 mph)
R	43 kt (80 km/h; 50 mph)
Max rate of climb at S/L:	
vertical: L	228 m (748 ft)/min
R	300 m (984 ft)/min
V	468 m (1,535 ft)/min
inclined: L	444 m (1,457 ft)/min
R	480 m (1,575 ft)/min
V	630 m (2,067 ft)/min
Hovering ceiling, IGE: L	1,800 m (5,900 ft)
R	2,000 m (6,560 ft)
V	3,200 m (10,500 ft)
Service ceiling: L	4,900 m (16,080 ft)
R	5,200 m (17,060 ft)
V	5,500 m (18,040 ft)
Range, all	216 n miles (400 km; 248 miles)
Endurance: L	3 h 40 min
R,V	4 h 20 min

UPDATED

MOLNIYA

NAUCHNO-PROIZVODSTVENNOYE OBEDINENIE MOLNIYA OAO (LIGHTNING SCIENTIFIC PRODUCTION ASSOCIATION JSC)

ulitsa Novoposelkovaya 6, 123459 Moskva
Tel: (+7 095) 492 92 35
Fax: (+7 095) 492 47 23
e-mail: molniya@dol ru
Web: http://www.buran.ru
MANAGING DIRECTOR: Aleksandr Bashilov
DEPUTY GENERAL DIRECTOR: Valery Verobzhev
CHIEF ENGINEER: Gennady G Kryuchkov
MARKETING DIRECTOR: Prof Mikhail Y Gofin
CHIEF OF MARKETING DEPARTMENT: Dr Vladimir I Fishelovich

To compensate for reduced funding for Buran space shuttle orbiter programme, in which it was much involved, Molniya has diversified its activities to include conveyor equipment (automatic car parking and wheelchair lifts) and aeroplane design. During the early 1990s, it projected a series of civil aircraft of vastly varying size (½ to 450 tonne payloads) all of 'triplane' (canard, wing and tailplane) configuration. The only one so far to fly has been the Molniya-1.

Development has been protracted, and full details last appeared in the 2001–02 *Jane's All the World's Aircraft*, followed by a brief update in 2002–03. The programme appeared to have stagnated by 2002, but in July 2003 an undisclosed investor committed to completing the aircraft's certification, then expected in March 2004. However, in October 2003, it was still being debated whether the VOKBM M-14 radial engine should be replaced by a Walter M 601 turboprop, and by May 2004; certification had still not been achieved. At this time the VOKBM M9F upgrade version of the M-14 engine was also under consideration. A total of 20 Molniya-1 airframes was built, comprising two prototypes, two for static test, one damaged beyond repair, one sold to an unknown customer (perhaps as an air cushion vehicle) and 14 remaining as uncompleted stock.

UPDATED

MVEN

MVEN OOO (MVEN LTD)

ul Dementyeva 2-V, 420036 Kazan
Tel/Fax: (+7 8432) 71 32 50
e-mail: mven@mven.kazan.ru
Web: http://www.mven.netfirms.com

Since establishment in 1990, MVEN has produced parachute ballistic recovery systems for various types of aircraft. In 1995 it was incorporated within the Scientific Research Institute of Aeronautical Technologies and began the manufacture of light aircraft fuselages and components under subcontract to Chernov and others. The company's first aircraft, the MVEN-1 agricultural sprayer was shown at the Gelendzhik Hydroaviation Salon in September 2002, three months before its maiden flight. MVEN is Russian dealer for Aeroprakt lightplanes built in Ukraine.

UPDATED

MVEN FERMER

English name: Farmer

TYPE: Agricultural sprayer.

PROGRAMME: Prototype built 2002. Length 6.58 m (21 ft 7 in). Prototype only. Shown at Gelendzhik, 4 to 8 September 2002. First flight 9 November 2002; official evaluation flight of 11 December gave clearance for continued company testing.

CURRENT VERSIONS: **MVEN-1:** Initial version; One 118 kW (158 hp) LOM M332B four-cylinder piston engine driving a three-blade propeller.

MVEN-2: Announced 2003, with target first flight date of May 2004 (apparently not achieved). Lycoming engine. Increased length and weights. Intended production version. *As described.*

DESIGN FEATURES: Classic low-wing sprayplane with chemical hopper ahead of raised cockpit giving good view over nose. Strut-braced, constant-chord wings and tailplane. Heavy-duty suspension and tyres.

Dihedral 5°; incidence 3°.

FLYING CONTROLS: Conventional and manual. No aerodynamic balances. Ailerons and flaps occupy full extent of wing trailing-edge. Flap deflections 0°, 20° and 40°.

STRUCTURE: Principally of glass fibre and carbon fibre.

LANDING GEAR: Tailwheel type; fixed. Centreline-hinged tube Vs attached to rubber-in-compression struts, upper ends of which anchored to fuselage sides. Hydraulic brakes on mainwheels. Mainwheels 595×185; tailwheel 250×80.

POWER PLANT: One 134 kW (180 hp) Textron Lycoming O-360 flat-four. Fuel capacity 70 litres (18.5 US gallons; 15.4 Imp gallons).

ACCOMMODATION: Two persons in tandem. Single, upward-hinged door, port side.

EQUIPMENT: Sprayer system with 13 atomisers on each of two underwing elements. MVEN ballistic parachute.

DIMENSIONS, EXTERNAL:

Wing span	10.40 m (34 ft 1½ in)
Wing chord, constant	1.25 m (4 ft 1¼ in)
Wing aspect ratio	8.3
Length overall	7.15 m (23 ft 5½ in)
Height overall	2.15 m (7 ft 0¾ in)
Tailplane span	3.76 m (12 ft 4 in)
Propeller diameter	1.93 m (6 ft 4 in)
Crew door: Width	0.70 m (2 ft 3½ in)

DIMENSIONS, INTERNAL:

Cockpit: Max width	1.22 m (4 ft 0 in)
Max height	0.74 m (2 ft 5 in)
Baggage space: Length	0.58 m (1 ft 10¾ in)
Max height	0.67 m (2 ft 2½ in)
Max width	0.57 m (1 ft 10½ in)

AREAS:

Wings, gross	13.00 m² (139.9 sq ft)
Ailerons (total)	1.65 m² (17.76 sq ft)
Flaps	2.78 m² (29.92 sq ft)
Fin	1.20 m² (12.92 sq ft)
Rudder	0.80 m² (8.61 sq ft)

WEIGHTS AND LOADINGS:

Weight empty	575 kg (1,268 lb)
Max chemical load	270 kg (595 lb)
Max T-O weight	980 kg (2,160 lb)
Max wing loading	75.4 kg/m² (15.44 lb/sq ft)
Max power loading	8.31 kg/kW (13.65 lb/hp)

MVEN-1 Fermer at its public debut, Gelendzhik, September 2002 *(Yefim Gordon)* ***NEW***/0552878

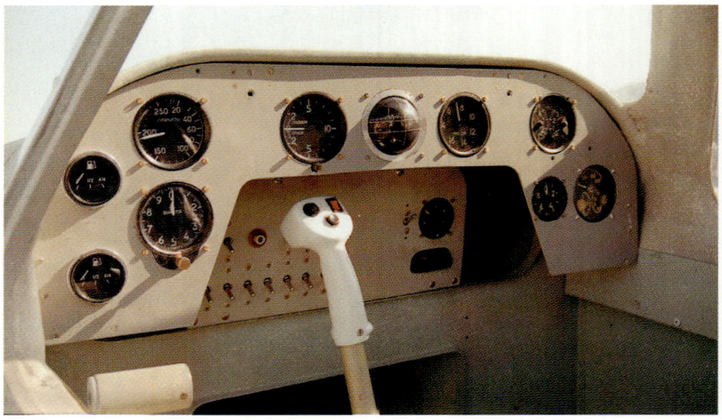

Cockpit of MVEN-1 *(Yefim Gordon)* 0527150

PERFORMANCE:

Max level speed	108 kt (200 km/h; 124 mph)
Normal operating speed	70–86 kt (130–160 km/h; 81–99 mph)
Max rate of climb at S/L	270 m (885 ft)/min
T-O run	180 m (590 ft)
Landing run	150 m (495 ft)
g limits	+6/–2

UPDATED

MYASISHCHEV

EKSPERIMENTALNYI MASHINOSTROITELNYI ZAVOD IMENI V M MYASISHCHEVA (EXPERIMENTAL ENGINEERING BUREAU NAMED FOR V M MYASISHCHEV)

140180 Zhukovsky-5, Moskovskaya oblast
Tel: (+7 095) 556 77 76 and 912 60 41
Fax: (+7 095) 556 55 83 and 728 41 30
e-mail: mdb@mail.sitek.ru
Web: http://www.corbina.ru/~kluka
GENERAL DESIGNER: Valery K Novikov
CHIEF DESIGNER, M-101: Evgeny Charsky
CHIEF DESIGNER, M-55: Leonid Sokolov
INFORMATION OFFICER: Stanislav G Smirnov
PUBLIC RELATIONS OFFICER: Antonina Frolova

The Myasishchev OKB was founded in 1952, and led by Prof Vladimir Mikhailovich Myasishchev until his death on 14 October 1978. In the 1970s and 1980s, the bureau was engaged in development of multipurpose subsonic high-altitude aircraft, but has more recently diversified into civil aviation. In 1981, the bureau was named after Prof Myasishchev.

Attempts to forge overseas alliances proved unsuccessful, but in April 2001 it was announced that the KASKOL group of companies, the Nizhny Novgorod Sokol Aviation Plant and Myasishchev had formed Novi Regionalnyi Samoletnyi (NoRS, New Regional Aeroplanes), which will produce the M-101T Gzhel business aircraft at Nizhny Novgorod.

In 2001, Myasishchev was reported to be preparing a response to an anticipated official requirement for a 130/170-seat short/medium-range airliner to be developed under the 2002–10 civil aviation plan and enter service in 2015. This would be based on the GP-60 series of aerodynamic studies and in April 2003 it announced that it was beginning an 18-month research programme in association with the Central Aerohydrodynamic Institute (TsAGI). Other aircraft in this family could include the 14-seat M-60 Svetozar, M-60 Perun (214-seat) and M-60 Kolovrat (35,000 kg; 77,161 lb payload). A 12-seat M-60-12 was stated in late 2003 to be under joint development with an unspecified Austrian company which was proposing to invest US$60 million in the venture.

UPDATED

MYASISHCHEV M-55

NATO Reporting name: Mystic

Russian defence ministry approved, in September 1999, plans to complete one of two unfinished Myasishchev M-55 high-altitude surveillance aircraft held at Smolensk production plant; funding assigned in 2001 for completion in 2002 to **M-55RM** standard, but in June 2002 it was reported that insufficient funds had been received to complete the aircraft in that year. The position was, apparently, unchanged at the start of 2004.

Also in 2001, India offered unspecified export variant designated **M-55RTR**, and Myasishchev announced plans for **M-55Kh** "space tourism" version with 3,000 kg (6,614 lb) S-XXI suborbital module carried on spine for launch at altitudes up to 19,000 m (62,340 ft). This also able to launch 300 kg (661 lb) satellites into 200 km (124 mile) Earth orbit by carrying 5.5 tonne booster at 10 per cent of ground-launch cost.

An unmanned surveillance version of M-55 has been proposed by Myasishchev. Two M-55s were flying in 2004, although the type has still to pass its State Tests, begun in 1998, but continually interrupted.

UPDATED

MYASISHCHEV M-101T GZHEL

TYPE: Light utility turboprop.

PROGRAMME: Derived from Nelli and M-70 projects; first shown in model form at Moscow Air Show '90; developed full-scale mockup exhibited at Moscow Air Show '92 with piston engine and at Moscow Air Show '93 with turboprop. Two static test airframes and three flying prototypes built at Nizhny Novgorod by Sokol; first (RA-15001) first flew 31 March 1995; this and second flying aircraft (RA-15003), shown at Moscow Air Show '95, lack ventral fin. Third prototype (RA-15004) to what then regarded as preproduction standard; first preproduction aircraft (RA-15101) displayed at Paris and Moscow Air Shows 1997. Initial batch of 15 at Sokol plant; 11 completed by September 2001; No. 6 (RA-15106) is first in full production configuration, shown at Moscow, August 2001; No. 7 for reliability tests; No. 8 previously intended as first for customer delivery. Trials aircraft crashed near Zhukovsky test aerodrome on 12 September 2001, following loss of longitudinal stability, putting programme back by six months. Certification trials were due to restart in May 2002, with completion scheduled by December 2002; certification to AP-23 eventually awarded 14 January 2003. In April 2002 it was reported that 12 aircraft had been built — 10 for flying trials and two for static testing. Cold weather testing at temperatures as low as –50°C (–58°F) was undertaken in Yakutia, Sakh Republic, in February 2004.

Russian certification to AP-23 in passenger category was expected late 1998 or early 1999, following delivery of two aircraft to State Civil Aviation Research Institute (Gos NII GA) on 23 January 1998 for short-range freight services from Moscow/Sheremetyevo by Fenix Air but delayed by lack of funding; meanwhile early production aircraft used on freight services by Sokol plant, completing over 500 missions by 2000, and one reported in use by Central African Airlines, based in Egypt. In November 2001, US$3 million was being sought to complete certification and further US$5 million to complete production tooling and retrospectively modify airframes already built. In April 2004, an agreement was announced with Mooney International for assembly of M-101Ts at Greenville for the US market.

Shortcomings in directional stability addressed by retrospective installation of additional two ventral fins and dorsal fillet on third prototype (RA-15004) in 1999, this also having extended wingtips. These additional tail surfaces standardised from RA-15106 onwards.

Pre-series Myasishchev M-101T Gzhel *(Paul Jackson)* ***NEW***/0561644

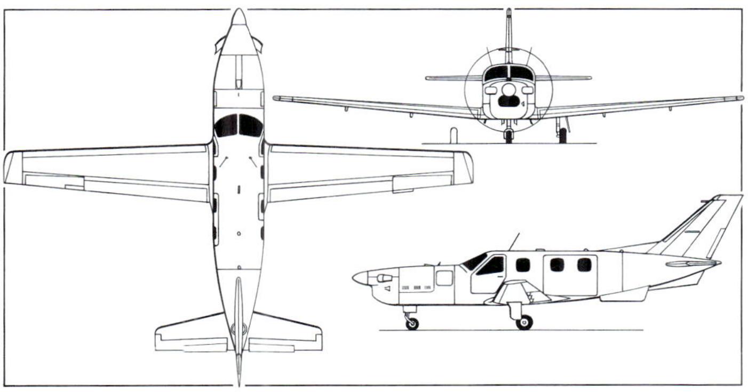

Myasishchev M-101T Gzhel six-seat light aircraft *(Jane's/Mike Keep)* 0121660

The aircraft is named for a type of fine porcelain made in the city of Gzhel. Some development funding was provided by Gzhel, other finance coming from Myasishchev, Sokol, Inkombank and a Czech bank. Marketing is by Gzhel-Avia.

CUSTOMERS: Undisclosed central African customer ordered one in early 2000; minimum estimated market for 300. Reported 70 to 80 orders by September. First Russian customer reported as Gazprom, which requires 10; company foresees production reaching 300 by end 2005, of which two-thirds would be exported. By April 2003, company reported 15 built, with 30 more under construction; delivery of first customer aircraft was due mid-2003, with eight scheduled to be handed over by end 2003 and a further 25 in 2004; by 2004, the company was stating that orders had been received from South Africa and Australia and production for the year would total seven (two for Russia and five abroad), followed by 18 in 2005. First confirmed receipt, in April 2004, by South African dealer, Rosavia Group of Johannesburg, although this was previously built (2001) aircraft RA-15106.

COSTS: US$1.6 million (2001). Hourly operating cost estimated at US$200 (2001). By mid-1999, Sokol plant had invested Rb102.4 million in M-101, and in 2000 was planning funding of US$2.5 million to complete certification.

DESIGN FEATURES: Intended for operation in remote areas; able to serve away from base aerodrome for up to 50 flying hours and fly from unpaved runways. Conventional all-metal low-wing monoplane with pressurised cabin; sweptback vertical tail surfaces and ventral fin. Designed in accordance with Russian AP-23 and US FAR Pt 23 airworthiness requirements. Service life of 15 years or 20,000 hours.

FLYING CONTROLS: Conventional and manual. Electrically operated trim tabs in port aileron, rudder and each elevator; ground-adjustable tab on starboard aileron. Double-slotted flaps operated hydraulically.

STRUCTURE: All-metal. Wing constructed around torsion box.

LANDING GEAR: Hydraulically retractable tricycle type; single wheel on each unit; nosewheel retracts rearward, mainwheels inward into wingroots and fuselage; mainwheels uncovered by doors when retracted; tyre size 500×150-9 on mainwheels, 400×150-5 on nosewheel; levered suspension legs; castoring nosewheel with shimmy damper; designed to use paved and unpaved runways.

POWER PLANT: One 559 kW (751 shp) Walter M 601 F32 turboprop, driving Avia Hamilton Sundstrand V-510 five-blade propeller. Power plant is protected against particle ingestion. Fuel in two wing tanks and two feeder tanks, with electrical pumps.

ACCOMMODATION: One or two pilots and four passengers, in pairs in pressurised cabin; max capacity, one pilot and seven passengers. Rear-hinged door to flight deck on port side; large door for passengers and freight loading aft of wing on port side; emergency exit on starboard side above wing; provision for rapid change to cargo/passenger, freight or ambulance configuration.

SYSTEMS: Hydraulic system for landing gear and flaps. Pressurisation system maintains cabin altitude of 2,400 m (7,880 ft). Pneumatic de-icing of wing, tailplane and fin leading-edges; electric anti-icing of propeller, pitot tubes and windscreen clear-view panel. Electrical system, 27 V DC, supplied by 250SG125Q1 starter/generator and 115 A/400 Hz single-phase secondary supply with Varta 24 V emergency battery.

AVIONICS: Russian or Western equipment for single-pilot VFR or two-pilot IFR.

Comms: Bendix/King KX 165 nav/com.

Flight: Bendix/King KLN 89B GPS, WX-900 Stormscope and autopilot all optional.

DIMENSIONS, EXTERNAL:

Wing span	13.00 m (42 ft 8 in)
Length overall	9.975 m (32 ft 8¾ in)
Height overall	3.45 m (11 ft 3¾ in)
Tailplane span	4.32 m (14 ft 2 in)
Wheel track	3.00 m (9 ft 10 in)
Wheelbase	2.885 m (9 ft 5½ in)
Propeller diameter	2.30 m (7 ft 6½ in)
Passenger/freight door: Height	1.15 m (3 ft 9¼ in)
Width	1.23 m (4 ft 0½ in)
Flight deck door: Max width	0.90 m (2 ft 11½ in)
Max height	0.975 m (3 ft 2½ in)
Emergency exit: Max width	0.485 m (1 ft 7 in)
Max height	0.665 m (2 ft 2¼ in)

DIMENSIONS, INTERNAL:

Cabin: Length	4.56 m (14 ft 11½ in)
Max width	1.32 m (4 ft 4 in)
Max height	1.26 m (4 ft 1½ in)
Volume	7.5 m^3 (265 cu ft)

AREAS:

Wings, gross	17.06 m^2 (183.6 sq ft)

WEIGHTS AND LOADINGS:

Weight empty	2,270 kg (5,004 lb)
Max payload	630 kg (1,389 lb)
Max fuel weight	450 kg (992 lb)
T-O weight: normal	2,900 kg (6,393 lb)
max	3,200 kg (7,054 lb)
Max wing loading	187.6 kg/m^2 (38.41 lb/sq ft)
Max power loading	5.18 kg/kW (8.51 lb/shp)

PERFORMANCE:

Max cruising speed	232 kt (430 km/h; 267 mph)
Stalling speed, flaps down	61 kt (113 km/h; 71 mph)
Cruising altitude	7,600 m (24,940 ft)
T-O run	380 m (1,250 ft)
Landing run	370 m (1,215 ft)
Range:	
with max payload	432 n miles (800 km; 497 miles)
with max fuel, 45 min reserves	755 n miles (1,400 km; 869 miles)

UPDATED

NAPO

GOSUDARSTVENNOYE UNITARNOYE PREDPRIYATIE NOVOSIBIRSKOYE AVIATSIONNOYE-PROIZVODST-VENNOYE OBEDINENIE IMENI V P CHKALOVA (STATE UNITARY ENTERPRISE NOVOSIBIRSK AIRCRAFT PRODUCTION ASSOCIATION NAMED FOR V P CHKALOV)

ulitsa Polzunova 15, 630051 Novosibirsk
Tel: (+7 3832) 79 80 95 and 77 37 06
Fax: (+7 3832) 77 10 26
e-mail napa@mail.cis.ru

GENERAL DIRECTOR: Nikolai I Bobritsky
MARKETING DIRECTOR: Valery Sadaev

Founded in 1936 as GAZ 153, NAPO has built about 36,000 aircraft of many types, including I-16, Yak-7, Yak-9, MiG-15, MiG-17, MiG-19, Yak-28, Su-9, Su-15 and, from 1972, Su-24 multirole combat aircraft. The plant is wholly state-owned, but 74.5 per cent of shares are to be transferred to Sukhoi Aviation Holding Company. NAPO's current products include the Antonov An-38 (available with local TVD-20 engines from 2002) and Sukhoi Su-32 (Su-27IB) as well as motor boats. Is promoting Su-24 modernisation programmes for export customers. Was designated to build Su-49 abandoned piston trainer and was nominated in March 2003 to produce the Sukhoi Russian Regional Jet, beginning in 2006.

UPDATED

ORION-AVIA

ORION-AVIA 000

Tyumen oblast.

A brief media report in November 2003 contained some details of an aircraft designed by the employees of Orion-Avia in the Tyumen district of the Russian Federation and at that time entering series manufacture.

NEW ENTRY

ORION-AVIA SK-12 ORION

TYPE: Four-seat amphibian.

PROGRAMME: First flown 2002. Completed flight tests at ROSTO air sport club by late 2003 and awarded flight certification. Series manufacture began late 2003.

LANDING GEAR: Electro-mechanically retractable type. Provision for skis.

POWER PLANT: Two piston engines.

ACCOMMODATION: Four persons, including pilot(s).

DIMENSIONS, EXTERNAL: None announced.

WEIGHTS AND LOADINGS: None announced.

PERFORMANCE:

Max level speed ... 108 kt (200 km/h; 124 mph)
Normal cruising speed ... 76 kt (140 km/h; 87 mph)
Unstick speed ... 38 kt (70 km/h; 44 mph)
T-O run, land ... 60 m (200 ft)
Range ... 540 n miles (1,000 km; 621 miles)

NEW ENTRY

OSKBES MAI

OTRASLEVOYE SPETSIALNOYE KONSTRUKTORSKOYE BYURO EKSPERIMENTALNOGO SAMOLOTSTROENIYA, MOSKOVSKOGO AVIATSIONNOGO INSTITUTA (SPECIAL DESIGN SECTOR BUREAU OF EXPERIMENTAL AIRCRAFT CONSTRUCTION, MOSCOW AVIATION INSTITUTE)

Volokolamskoye shosse 4, 125871 Moskva
MARKETING ADDRESS: As for RSK 'MiG'
Tel: (+7 095) 158 47 53
e-mail: oskbes@com2com.ru
Web: http://www.com2com.ru/oskbes
CHIEF DESIGNER: Nikolai P Goryunov

MAI founded 1930 and immediately began aircraft design and construction. Current design bureau formed by students in 1967 and was engaged in creation of piloted and unmanned light aircraft, which were developed according to requirements of various subsidiaries of the Ministries of Aviation Industry, Civil Aviation and Defence. By official decree of 11 August 1982, the Design Bureau was transformed into an experimental students' aircraft construction design bureau of MAI, although with a full-time engineering staff, to which students and lecturers were attached as required. By late 1980s, six experimental types had been built, comprising MAI-920 Kvant aerobatic lightplane, Photon research jet, PS-01 Komar and Elf-D UAVs and Elf and Junior lightplanes.

During late 1980s turned to aircraft production, first with Photon design association and later the Aviatika joint-stock company. As part of the Aviatika joint stock company, the MAI Design Bureau (OB) received the name of a 'Low-Sized Aircraft Experimental Design Bureau' (OKB MA). MAI Development Plant and MAPO (Moscow Aircraft Production Organisation) were used as an industrial base. Application to Avia Register and Air Industry Department of the Ministry of Industry of Russian Federation on 27 July 1992 resulted in award on 17 February 1993 of certificate N2R9 for the rights to develop civil aircraft.

Basis for the Aviatika aircraft manufacture at MAPO (now RSK 'MiG') plant is licence agreement of 10 April 1995. Following name change to Aviatika, new light aircraft certificate (NQ P-9A) issued on 20 April 1994. During 1994–1995, OSKBES MAI (OKB MA) developed a technical offer for a light six-seat business aircraft designated Aviatika-950. In early 1996, Aviatika JSC changed its business plan and dispensed with services of OSKBES (to the OSKBES (OKB MA)), although latter continued work with MAPO.

Aviatika transferred all aircraft design, engineering and intellectual rights to the Moscow State Aviation Institute (MAI) under Contract N9 13/39 of 19 February 1998 and an additional agreement, N2 1-13/39 dated 3 April 1998. Subsequently, OSKBES MAI transferred manufacturing license for Aviatika series aircraft to MAPO. Has now designed aircraft or subvariants; more than 280 of MAI-890 built.

UPDATED

MAI-205

TYPE: Single-seat autogyro ultralight.

PROGRAMME: Prototype completed 1994; designated Aviatika-MAI-890A and powered by a 47.8 kW (64.1 hp) Rotax 582 UL. Developed MAI-205 (marked only as '205') first flew 26 January 2001. Flight testing continuing in 2003 at Khodinka, Moscow.

COSTS: US$40,000 in agricultural version (2001). Development cost US$200,000.

DESIGN FEATURES: Conventional configuration, with fully enclosed cockpit, tail surfaces carried on tubular tailboom; two-blade autorotating rotor with optional pre-rotation for 'jump start'; rotor column folds for storage. Control movements emulate aeroplane, rather than autogyro, providing safer type conversion for former fixed-wing pilots. Ground-adjustable tab on rudder and each elevator.

Rotor blade aerofoil NACA 23012; blade angle 4° for running take-off and flight; 10° for 'jump start'.

MAI-205 single-seat autogyro ***NEW***/0593288

STRUCTURE: Fuselage of aluminium tube with glass fibre fairings. Strut and wire-braced, fabric-covered tailplane.

LANDING GEAR: Tricycle type; fixed. Twin cantilever mainwheel legs attached to main boom to rear of cabin. All wheels 300×125. Brakes on mainwheels. Steerable nosewheel.

POWER PLANT: One 84.6 kW (113.4 hp) Rotax 914 ULS piston engine mounted behind top of cabin pod; two-blade pusher propeller. Fuel tank under engine, capacity 24 litres (6.3 US gallons; 5.3 Imp gallons).

EQUIPMENT: Two 50 litre (13.2 US gallon; 11.0 Imp gallon) tanks. Optional Micronair AU7000 atomisers.

DIMENSIONS, EXTERNAL:

Rotor diameter ... 9.00 m (29 ft 6¼ in)
Height overall ... 2.87 m (9 ft 5 in)
Fuselage length ... 5.32 m 17 ft 5½ in)
Wheel track ... 1.52 m (4 ft 11¾ in)
Wheelbase ... 1.81 m (5 ft 11¼ in)

WEIGHTS AND LOADINGS:

Weight empty ... 370 kg (816 lb)
Max T-O weight ... 530 kg (1,168 lb)

PERFORMANCE:

Max level speed ... 62 kt (115 km/h; 71 mph)
Max rate of climb at S/L ... 162 m (531 ft)/min
Service ceiling ... 1,750 m (5,740 ft)
T-O run ... 30 m (100 ft)
Landing run ... 15 m (50 ft)
g limits ... +3.5/–0.5

UPDATED

MAI-217

TYPE: Two-seat lightplane.

PROGRAMME: Development under way at OSKBES by 2000. Promotion continuing in 2003, using heavily modified MAI-910.

CURRENT VERSIONS: **MAI-217:** *As described.*

MAI-217B (*Boyevoy:* combat) Proposed military version with cockpit armour and provision for up to 200 kg (441 lb) of light weapons.

DESIGN FEATURES: Designed to JAR-VLA criteria principally for agricultural applications. Side-by-side lightplane with strut-braced, mid-positioned, constant-chord wing. Uses some Aviatika-MAI-910 components. Also suitable for training and surveillance. Wings foldable for storage in 5 minutes. Convertible to tailwheel configuration by two persons in under 2 hours.

POWER PLANT: One 73.5 kW (98.6 hp) Rotax 912 ULS flat-four driving a two-blade, fixed-pitch propeller.

EQUIPMENT: Optional 200 litre (52.8 US gallon; 44.0 Imp gallon) chemical tank and spraybar.

ARMAMENT: Optional light bombs and rocket pods.

DIMENSIONS, EXTERNAL:

Wing span ... 11.32 m (37 ft 1¾ in)
Length overall ... 6.50 m (21 ft 4 in)
Height overall ... 1.90 m (6 ft 2¾ in)

AREAS:

Wings, gross ... 13.58 m² (146.2 sq ft)

WEIGHTS AND LOADINGS (estimated):

Weight empty ... 325 kg (717 lb)
Max T-O weight ... 700 kg (1,543 lb)

Artist's impression of MAI-217 ***NEW***/0593289

Max wing loading .... 51.5 kg/m² (10.56 lb/sq ft)
Max power loading .... 9.53 kg/kW (15.65 lb/hp)

PERFORMANCE (estimated):
Max level speed .... 108 kt (200 km/h; 124 mph)
Landing speed .... 39 kt (72 km/h; 45 mph)
T-O run .... 100 m (330 ft)
Landing run .... 140 m (460 ft)
Range .... 432 n miles (800 km; 497 miles)

UPDATED

AVIATIKA-MAI-223 KITYONOK

English name: Whale Calf

TYPE: Side-by-side-seat ultralight/kitbuilt.

PROGRAMME: Announced in 2003 with an intended first flight date during mid-year, followed by first deliveries before year end. By May 2003, three prototypes were under construction.

COSTS: Kit US$14,500, excluding engine (2003). Flyaway US$27,000 to US$30,000 with Rotax 582; US$34,000 to US$37,000 with Rotax 912 (2003).

DESIGN FEATURES: Intended to conform to JAR-VLA regulations. Constant-chord, V-strut braced parasol wing is forward-swept at approximately 6°, and can be folded for storage and transportation.

FLYING CONTROLS: Conventional and manual. Electrically operated trim tab and flaps; ground-adjustable rudder tab.

STRUCTURE: Fuselage, control surfaces and tail unit are predominantly of fabric-covered aluminium alloy. Two-spar wing has glass fibre skin and aluminium alloy ribs.

LANDING GEAR: Tailwheel type; fixed. Sweptback cantilever mainwheel leg attached to each side of fuselage at wing strut intersection; castoring tailwheel. Hydraulic disc brakes.

POWER PLANT: One 73.5 kW (98.6 hp) Rotax 912 ULS2 flat-four driving a ground-adjustable propeller. Alternatively, a 37.0 kW (49.6 hp) Rotax 503 UL or 47.8 kW (64.1 hp) Rotax 582 UL can be fitted.

SYSTEMS: 18 Ah Teledyne G-25 battery.

DIMENSIONS, EXTERNAL:
Wing span .... 8.40 m (27 ft 6¾ in)
Width, wing folded .... 3.20 m (10 ft 6 in)
Length overall .... 6.00 m (19 ft 8¼ in)
Height overall .... 2.10 m (6 ft 10¾ in)

AREAS:
Wings, gross .... 11.40 m² (122.7 sq ft)

WEIGHTS AND LOADINGS:
Weight empty .... 250 kg (551 lb)
Max fuel weight .... 40 kg (88 lb)
Max T-O weight .... 450 kg (992 lb)

PERFORMANCE (estimated, Rotax 912 ULS):
Max operating speed .... 103 kt (190 km/h; 118 mph)
Normal cruising speed .... 92 kt (170 km/h; 106 mph)
Stalling speed, flaps down .... 36 kt (65 km/h; 41 mph)
Max rate of climb at S/L .... 360 m (1,181 ft)
T-O run .... 100 m (330 ft)
Landing run .... 90 m (295 ft)
Range, with max fuel .... 432 n miles (800 km; 497 miles)

UPDATED

Artist's impression of Aviatika-MAI-223 ***NEW***/0552885

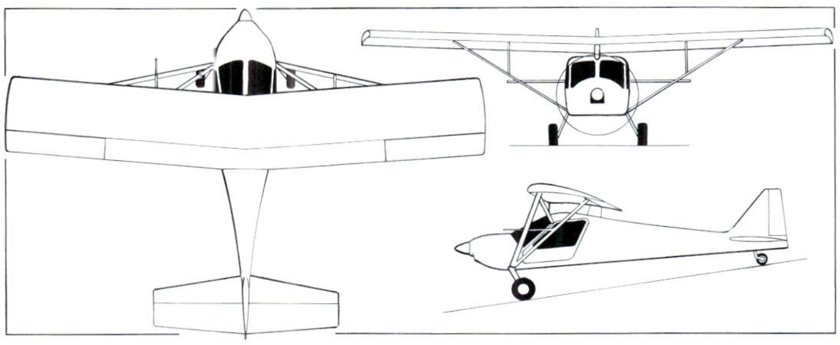

Aviatika-MAI-223 side-by-side ultralight ***NEW***/0558326

AVIATIKA-MAI-890

TYPE: Single-seat ultralight biplane; side-by-side ultralight biplane.

PROGRAMME: Open cockpit Yunior (Junior) first flew July 1987 as single-seater with 5.68 m (18 ft 7½ in) span and 280 kg (617 lb) MTOW. Developed as MAI-89, 1989, span 8.11 m (26 ft 7½ in); MTOW 340 kg (750 lb). Production version, MAI-890, flew 1990 and entered series manufacture in 1991; although this halted temporarily in mid-1990s, it was

Aviatika-MAI-890U two-seat ultralight ***NEW***/0593277

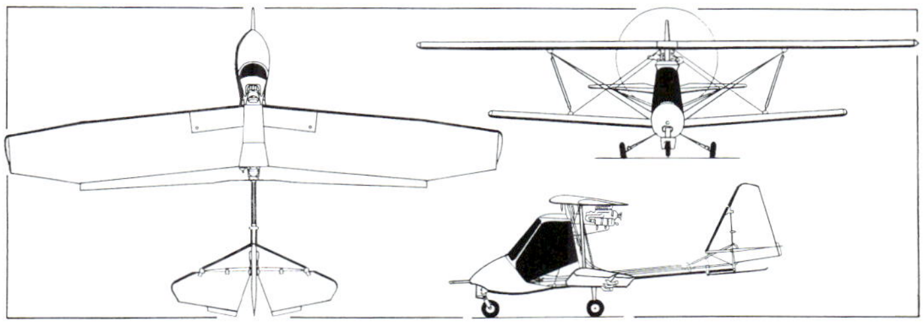

Aviatika-MAI-890 (Rotax piston engine) ***(Paul Jackson)*** 0051198

announced in 2002 that production would continue. Russian certification awarded in December 1999. Manufactured by RSK 'MiG'.

CURRENT VERSIONS: **Aviatika-MAI-890:** Single-seater.

Aviatika-MAI-890U: Side-by-side two-seater (*uchebny*: training). First flown August 1991 (47.8 kW; 64.1 hp Rotax 582 UL). Normally with Rotax 912 UL engine (first flight 16 September 1992). Empty weight 298 kg (657 lb); MTOW 540 kg (1,190 lb); length 5.49 m (18 ft 0¼ in). Floatplane version completed trials on Lake Shartash, Ekaterinburg, September 2000. Certification agreement reported to be nearly complete in June 2004; with estimated final certification date of end 2005.

Aviatika-MAI-890SKh: Single-seat crop-sprayer version (*selskokhozyaistvenny*: agricultural); Rotax 912 ULS engine; marketing name **Farmer**. Chemical tank under engine and spraybar aft of lower wings; weight of agricultural equipment 28 kg (62 lb); chemicals 60 kg (132 lb) (105 litres; 27.7 US gallons; 23.1 Imp gallons) maximum with Rotax 582 engine.

Certified version received Avia Register approval on 26 April 2002. Changes include conformal chemical tank, enhanced spraying equipment, ventilated cockpit with filter, stall warner, reinforced landing gear, improved fuel system and FLYdat engine instrument system.

Aviatika-MAI-890USKh: Two-seat crop-sprayer version based on MAI-890U. Tank volume 100 litres (26.4 US gallons; 22.0 Imp gallons).

Aviatika-MAI-890S: Rotax 912 ULS 73.5 kW (98.6 hp) flat-four. Prototype (ZU-BZO) converted in South Africa, 1999, by Aeroserve Pty Ltd. Regarded by OSKBES as standard 890.

CUSTOMERS: More than 280 delivered by early 2003, of which some 60 were agricultural variants; eight sold during last six months of 2001, including six equipped for crop spraying. Approximately 30 imported by a South African dealer in 1995; these known locally as **890CSH Mai** and **890U Mai**. Total market for agricultural variant estimated at 800 to 1,000.

COSTS: US$16,000 to US$34,000 (2002).

DESIGN FEATURES: Strut- and wire-braced biplane; unspinnable; conventional tail surfaces carried on tubular boom; engine mounted under trailing-edge of upper wing; slight sweepback on all wings; dihedral on lower wings only. Semi-aerobatic and capable of flying in rough air conditions. Designed to JAR-VLA and FAR Pt 23 standards. Two cockpit doors, one jettisonable; provision for backpack parachute(s).

FLYING CONTROLS: Conventional and manual. Full-span ailerons on lower wings; large-area rudder and elevators, each with ground-adjustable tab.

STRUCTURE: Aircraft grade aluminium and titanium alloys, alloy steels; fabric covering unaffected by sun's radiation or atmospheric precipitation.

LANDING GEAR: Tricycle type; fixed. Twin cantilever mainwheel legs attached to main boom to rear of cabin; single wheel on each unit. Brakes on mainwheels. Optional skis or floats. All wheels 300×125.

POWER PLANT: One 73.5 kW (98.6 hp) Rotax 912 ULS four-cylinder piston engine; VV-98E-11 two-blade pusher propeller. A 37.0 kW (49.6 hp) Rotax 503 UL-2V may also be fitted. Standard fuel 50 litres (13.2 US gallons; 11.0 Imp gallons); provision for 55 litre (14.5 US gallon; 12.1 Imp gallon) auxiliary tank on single-seat version.

EQUIPMENT: Provision for up to 120 kg (265 lb) payload on four attachments, under engine mountings (60 kg; 132 lb), or underbelly (100 kg; 220 lb), or at lower wingtips (each 45 kg; 99 lb), including agricultural dusting and spraygear for Aviatika-MAI-890USKh Farmer. Optional BRS ballistic parachute system.

Following details apply to certified 890SKh with Rotax 912 ULS engine.

DIMENSIONS, EXTERNAL:
Wing span, upper .... 8.11 m (26 ft 7½ in)
Length overall, incl pitot .... 5.32 m (17 ft 5½ in)
Height to tip of fin .... 2.25 m (7 ft 4½ in)

AREAS:
Wings, gross .... 14.29 m² (153.8 sq ft)

WEIGHTS AND LOADINGS:
Weight empty .... 331 kg (730 lb)
Max T-O weight .... 540 kg (1,190 lb)

PERFORMANCE:

Max level speed ... 70 kt (130 km/h; 81 mph)
Cruising speed ... 49 kt (90 km/h; 60 mph)
Landing speed ... 41 kt (75 km/h; 47 mph)
Max rate of climb at S/L ... 159 m (522 ft)/min
T-O run ... 90 m (295 ft)
Landing run ... 130 m (430 ft)
Endurance ... 2 h 53 min
g limits ... +5/−2.5

UPDATED

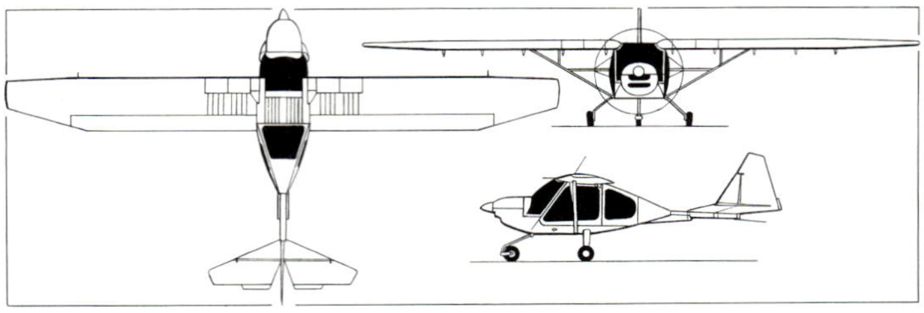

Aviatika-MAI-910 with original rear fuselage *(James Goulding)* 0011968

AVIATIKA-MAI-910 INTERFLY

TYPE: Two-seat lightplane.

PROGRAMME: Design started and construction of first prototype began 1993; first flight 22 June 1995 in MAI-910 configuration. Promotion began at MAKS '99, Moscow, August 1999; by mid-2003, flight testing was nearly complete.

Original version had boom-mounted empennage. Boom subsequently covered with removable fairings, for more conventional appearance, and ventral strake added.

DESIGN FEATURES: Designed to JAR-VLA standards. Braced high-wing monoplane; single bracing strut and jury strut each side; fully enclosed cabin pod; strut- and wire-braced conventional tail unit carried on interchangeable rear fuselage of conventional thickness or on short boom. Wings fold.

Wing section CAHI (TsAGI) Sh-1-14; 1° leading-edge sweepback; thickness/chord ratio 14 per cent; constant chord except for tapered tips; no dihedral; incidence 5°.

FLYING CONTROLS: Conventional and manual. Actuation by cables; long-span flaperon on each wing; ground-adjustable tab on rudder and each elevator.

STRUCTURE: Metal structure with fabric-covered wings and tail surfaces, metal-skinned fuselage. Wings have single tubular spar, diameter 102 mm (4 in), and stamped sheet metal ribs. Materials primarily aluminium alloy, with stainless steel and titanium alloy where appropriate.

LANDING GEAR: Tricycle type; fixed. Fuselage-mounted, spring cantilever mainwheel legs; trailing-link nosewheel. Mainwheel tyres 400×150 or 300×120; nosewheel tyre 300×120; tyre pressure (all) 2.53 bar (37 lb/sq in). Mechanical brakes with 300×120 tyres; hydraulic brakes with 400×150 tyres. Nosewheel steerable through ±50°. Tailwheel gear optional.

POWER PLANT: One 59.6 kW (79.9 hp) Rotax 912 UL flat-four; Russian two-blade fixed-pitch propeller. Two fuel tanks, each 25 litres (6.6 US gallons; 5.5 Imp gallons); one in each wingroot. Oil capacity 6 litres (1.6 US gallons; 1.3 Imp gallons).

ACCOMMODATION: Two seats side by side in fully enclosed cabin with wide field of view for observation, patrol, search and aerial photography. Passenger seat folds for carriage of stretcher or lengthy freight. Large rearward-sliding window/door each side. Space for baggage or freight behind seats. Cabin heated and ventilated.

Aviatika-MAI-910 light passenger/freight aircraft, showing new rear fuselage ***NEW***/0593291

SYSTEMS: 12 V AC electrical system; Teledyne battery.

AVIONICS: *Comms:* UHF radio.

Flight: GPS.

DIMENSIONS, EXTERNAL:

Wing span ... 10.73 m (35 ft 2½ in)
Wing chord: at root ... 1.10 m (3 ft 7¼ in)
at tip ... 0.70 m (2 ft 3½ in)
Wing aspect ratio ... 10.3
Length overall ... 5.26 m (17 ft 3 in)
Height overall ... 2.50 m (8 ft 2½ in)
Wheel track ... 2.20 m (7 ft 2¾ in)
Wheelbase ... 1.60 m (5 ft 3 in)
Propeller diameter ... 1.90 m (6 ft 2¾ in)
Propeller ground clearance ... 0.30 m (11¾ in)
Doors (each): Height ... 0.87 m (2 ft 10¼ in)
Max width ... 1.10 m (3 ft 7¼ in)
Baggage door: Length ... 0.75 m (2 ft 5½ in)
Width ... 0.70 m (2 ft 3½ in)

DIMENSIONS, INTERNAL:

Cabin: Max width ... 1.20 m (3 ft 11¼ in)
Max height ... 1.08 m (3 ft 6½ in)
Floor area ... 1.20 m² (12.9 sq ft)
Baggage/freight volume ... 0.70 m³ (24.7 cu ft)

AREAS:

Wings, gross ... 11.20 m² (120.6 sq ft)

WEIGHTS AND LOADINGS:

Weight empty ... 380 kg (838 lb)
Max fuel ... 40 kg (88 lb)
Max T-O weight ... 620 kg (1,366 lb)
Max wing loading ... 55.4 kg/m² (11.34 lb/sq ft)
Max power loading ... 10.40 kg/kW (17.09 lb/hp)

PERFORMANCE:

Never-exceed speed (VNE) ... 121 kt (225 km/h; 139 mph)
Max level speed at S/L ... 84 kt (155 km/h; 96 mph)
Max and econ cruising speed ... 75 kt (140 km/h; 87 mph)
Stalling speed ... 41 kt (76 km/h; 48 mph)
Max rate of climb at S/L ... 210 m (690 ft)/min
Service ceiling ... 3,000 m (9,840 ft)
T-O run ... 150 m (495 ft)
T-O to 15 m (50 ft) ... 300 m (985 ft)
Landing from 15 m (50 ft) ... 150 m (495 ft)
Range ... 216 n miles (400 km; 248 miles)
g limits ... +4/−2

UPDATED

POLYOT

PROIZVODSTVENNOYE OBEDINENIE POLET (POLYOT PRODUCTION ASSOCIATION)

ulitsa Bogdan Khmelnitsky, 226, 644021 Omsk
Tel: (+7 3812) 37 17 76
Fax: (+7 3812) 33 79 15
e-mail: polyot@omsknet.ru
Web: http://www.omsknet.ru/polyot
GENERAL DIRECTOR: Valentin Ziatsev

DESIGN BUREAU:
Tel: (+7 3812) 53 92 01
Fax: (+7 3812) 53 88 31
CHIEF DESIGNER: Viktor V Markelov

Established in 1941, Polyot is one of the largest aerospace enterprises in Russia, with facilities extending over more than 15 km² (5.8 sq miles) and up to 20,000 employees. During the Second World War it manufactured 3,405 Tupolev Tu-2 bombers and Yakovlev Yak-7 and Yak-9 fighters in 2½ years; post-war products included 758 Ilyushin Il-28 jet bombers and 58 Tupolev Tu-104 jet transports. It remains state-owned. The Antonov An-74 series of STOL transports is in low-rate production.

Polyot manufactured SS-4, SS-7 and SS-11 strategic missiles, warheads, and first-stage engines for Zenith and Energiya rocket launch vehicles. Other products include many spacecraft and a variety of consumer goods under *konversiya* programmes.

In June 1997, Polyot and Baranov Engine Manufacturing Association relaunched the Antonov An-3 programme (An-2 with 1,140 kW; 1,529 shp OMKB TVD-20 turboprop). Certification of factory to build An-3 received by May 2000, but funding unavailable for production line and only conversions of An-2s being considered. By mid-2003, An-3s were serving in seven companies and Polet expected to sell its annual allocation of 10, increasing this to 15 in 2004. Cost is US$800,000 to US$900,000. An-3 work was interrupted during the first half of 2001 when Polyot repaired the prototype Antonov An-70 after it had crashed following take-off from Omsk. Revealed in September 2001 that Polyot selected to assemble An-70s ordered for Russian Air Forces, although that requirement was later cast into doubt; first An-70 originally intended to fly in 2003. Required investment to begin An-70 production (Polyot to build cargo hold) quoted in late 2001 as Rb3 billion, of which Russian government and Omsk regional government each donating Rb1,050 million.

In May 2002, Polyot revealed its recent work on the Yula 'strap-on' personal helicopter, although little further has been heard of the proposal.

UPDATED

RIDA

PKB RIDA ANO

e-mail: reda@mail.ru
Web: http://www.reda.ru
CHAIRMAN: Aleksey Ivanter

Rida design bureau established in 1989 and designed the P10 Ptenets lightplane; produced a prototype of the Pony two-seat amphibian; and planned a four-seat derivative, the Prize.

More recently, two new programmes have been announced: the Aleks-251 and a larger, Proect 256 (Project 256) 11-seat transport powered by two VOKBM M-14 radial engines. The **Rida 256** was reported in March 2003 to be under development for the Australian Coast Guard, with first flight projected for late 2005 or early 2006. The project is said to be funded by Australian backers. Data include span 18.80 m (61 ft 8¼ in); MTOW 5,700 kg (12,566 lb); cruising speed 210 kt (390 km/h; 242 mph); range with 1,600 kg (3,527 lb) payload 270 n miles (500 km; 310 miles).

NEW ENTRY

Artists's impression of Rida Aleks-251 ***NEW***/0583657

RIDA ALEKS-251

TYPE: Six-seat amphibian.

PROGRAMME: Announced 2003, when first flight planned in second half of 2004. Reportedly developed for an undisclosed Russian Federation ministry, which has a requirement for up to 800.

DESIGN FEATURES: Parasol wing amphibian with single-step hull; cruciform empennage with sweptback fin. Constant-chord wing and tailplane; outrigger floats retract to form wingtips when not in use. Operable on water with wave height of 0.45 m (1 ft 6 in).

FLYING CONTROLS: Conventional and manual. Horn-balanced rudder. Flaps.

LANDING GEAR: Optional tricycle type.

POWER PLANT: Two LOM M332S four-cylinder engines, each 127 kW (170 hp), projecting forward from undersides of wing. Fuel tanks in wings; total capacity 800 litres (211 US gallons; 176 Imp gallons).

ACCOMMODATION: Pilot and five passengers; seven passengers at maximum capacity.

DIMENSIONS, EXTERNAL:
Wing span .... 13.20 m (43 ft 3¾ in)
Length overall .... 10.10 m (33 ft 1¾ in)
Wheel track (optional) .... 2.35 m (7 ft 8½ in)
Distance between propeller centres .... 3.00 m (9 ft 10 in)

DIMENSIONS, INTERNAL:
Cabin: Length .... 3.00 m (9 ft 10 in)
Max width .... 1.40 m (4 ft 7 in)
Max height .... 1.30 m (4 ft 3¼ in)

WEIGHTS AND LOADINGS:
Weight empty .... 1,250 kg (2,756 lb)
Max T-O weight .... 2,100 kg (4,629 lb)
Max power loading .... 8.29 kg/kW (13.61 lb/hp)

PERFORMANCE:
Max level speed .... 143 kt (265 km/h; 165 mph)
Normal cruising speed .... 121 kt (225 km/h; 140 mph)
Econ cruising speed .... 108 kt (200 km/h; 124 mph)
Unstick speed .... 59 kt (110 km/h; 68 mph)
Max rate of climb at S/L .... 360 m (1,181 ft)/min
T-O run .... 320 m (1,050 ft)
Landing run .... 240 m (790 ft)
Range: at max cruising speed .... 863 n miles (1,600 km; 994 miles)
at econ cruising speed .... 1,079 n miles (2,000 km; 1,242 miles)
ferry .... 2,699 n miles (5,000 km; 3,106 miles)
Endurance .... 25 h

NEW ENTRY

Artist's impression of Rida 256 amphibian ***NEW***/0583656

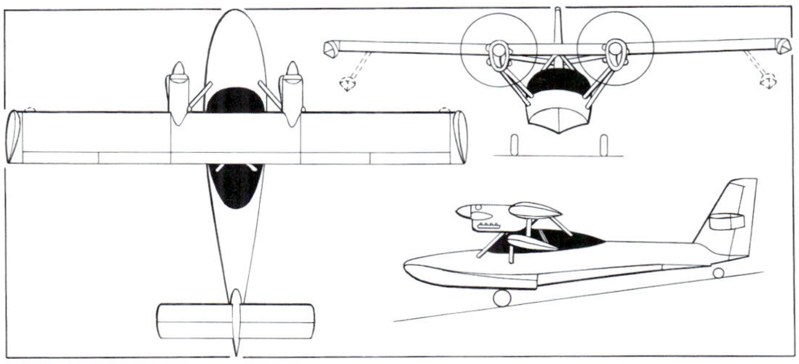

Rida Aleks-251 six-seat amphibian *(James Goulding)* ***NEW***/0583645

ROSAEROSYSTEMS

ROSAEROSYSTEMS S.R.A (DIVISION OF AUGUR AERONAUTICAL CENTRE)

ulitsa Stepana Shutova 4, 109380 Moskva
Tel/Fax:(+7 095) 359 10 01 and 359 10 65
e-mail: augur@pbo.ru
Web: http://www.rosaerosystems.pbo.ru
CHAIRMAN: Gennady Verba

RosAeroSystems was formed in October 1997 by Avgur, Moscow Aviation Institute and the Russian Aeronautical Society (RVO) as a consortium to combine the efforts of lighter than air manufacturing and operation. Foreign partners include Kubicek (Czech Republic), Voliris (France), and the US companies Worldwide Aeros and General Aeronautics Corporation. It offers a wide range of manned and unmanned airships and other aerostats under current design; any manufacture of these would be undertaken by Avgur, with whom it is co-located, or other Russian aviation companies. RosAeroSystems' workforce totalled more than 50 in 2001. RosAeroSystems International Ltd was formed as a US marketing organisation in 2002.

VERIFIED

ROSAEROSYSTEMS AU-11

TYPE: Helium non-rigid.

PROGRAMME: Construction (by Avgur) began January 1997. Planned to fly in September 1998, but delayed; eventually made first flight August 2001

Description as for Au-12 except as follows.

DESIGN FEATURES: Single-seat version of Au-12; otherwise generally differs only in dimensions, power plant, gondola and range/endurance performance.

POWER PLANT: One 20.9 kW (28 hp) VAZ-1111 motorcar engine and two-blade ducted propeller; 15° up/90° down thrust vectoring; 2.5 kW (3.4 hp) stern thruster engine and auxiliary fuel tank as in Au-12.

The Au-11, which made its maiden flight in August 2001 *(Paul Jackson)* 0126889

ACCOMMODATION: Pilot only. Upward-opening door on port side of gondola. Two to four ground crew.

DIMENSIONS, EXTERNAL:
Envelope: Length .... 27.50 m (90 ft 2¾ in)
Max diameter .... 6.875 m (22 ft 6¾ in)
Fineness ratio 4.0
Height overall .... 9.40 m (30 ft 10 in)

Au-11 gondola thrust-vectoring vanes *(Paul Jackson)* 0126887

Au-11 single-seat gondola *(Paul Jackson)* 0126888

DIMENSIONS, INTERNAL:
Envelope volume ... 669.0 m^3 (23,626 cu ft)
Ballonet volume (max) ... 123.0 m^3(4,344 cu ft)
AREAS:
Tail surface area (total) ... 26.50 m^2 (285.2 sq ft)
WEIGHTS AND LOADINGS:
Weight empty ... 465 kg (1,025 lb)
Max payload ... 160 kg (353 lb)
Max T-O weight ... 870 kg (1,918 lb)
PERFORMANCE (estimated):
Max level speed ... 43 kt (80 km/h; 49 mph)
Cruising speed ... 32 kt (60 km/h; 37 mph)
Pressure ceiling ... 2,000 m (6,560 ft)
Range at cruising speed:
standard fuel ... 162 n miles (300 km; 186 miles)
with auxiliary fuel tank ... 432 n miles (800 km; 497 miles)
Endurance: at max speed ... 2 h 30 min
at cruising speed ... 5 h

UPDATED

ROSAEROSYSTEMS AU-12

TYPE: Helium non-rigid.

PROGRAMME: Two-seat enlarged version of single-seat Au-11. Avgur prototype flown in 1998.

CURRENT VERSIONS: **Au-12:** *As described*; cruciform tailfins.

Au-12M: First displayed in 2003; X tailfin configuration. No other details known.

CUSTOMERS: Modified Au-12 envelope delivered to Voliris of France (which see) for that company's Voliris 900. Two Au-12s ordered in early 2003 by Russian Ministry of Transport and Communications for experimental traffic monitoring programme in Moscow; delivery in 2004.

DESIGN FEATURES: Intended for advertising, demonstration or patrol flights with purpose-designed onboard equipment, command uplink and information downlink. Space for two advertising displays on envelope sides. Conventional envelope shape; cruciform tailfins, each with elevator or rudder. Design is compliant with Russian Federation and US FAA airship design criteria.

FLYING CONTROLS: Engine thrust vectored in vertical plane by movable louvres; combined manual/electromechanical rudder and elevator control (cables); stern thruster in lower tailfin. Two internal air ballonets.

STRUCTURE: Envelope of modern composites fabric on Lavsan base, with polyurethane skin and titanium oxide UV protection. Tail surfaces have tubular frames with high-strength, stretch-wrapped fabric skins. Reinforced nose with mooring cone.

LANDING GEAR: Single non-retractable and self-centring wheel, on telescopic leg, under gondola. Provision for flotation bags.

POWER PLANT: One 64.9 kW (87 hp) Subaru 1.8 piston engine, with 30° up/105° down, five-vane cascade-type thrust vectoring, mounted on rear of gondola; 2.5 kW (3.4 hp) stern thruster electric motor and propeller mounted on port side of lower tailfin. Provision for 20 litre (5.3 US gallon; 4.4 Imp gallon) auxiliary fuel tank.

ACCOMMODATION: Pilot and one passenger in gondola. Upward-opening door on each side. Ground crew of two or three.

AVIONICS: Basic VFR or IFR instrumentation. Radio modem or satcom datalink for patrol/surveillance and other special missions.

EQUIPMENT: Depending on mission, can include such items as radiation and chemical monitoring devices; electromagnetic and heat flow gauges; TV camera with video downlink; NP-1 scientific package including photon spectrometer, radiometer, polarimeter and space navigation set; and onboard computer.

DIMENSIONS, EXTERNAL:
Envelope: Length ... 31.40 m (103 ft 0¼ in)
Max diameter ... 7.85 m (25 ft 9 in)
Fineness ratio ... 4.0
Height overall ... 11.30 m (37 ft 1 in)
Advertising panel (each side):
Length ... 20.00 m (65 ft 7½ in)
Depth ... 6.00 m (19 ft 8¼ in)
DIMENSIONS, INTERNAL:
Envelope volume ... 996.0 m^3 (35,173 cu ft)
Ballonet volume (two, total) ... 250.0 m^3 (8,829 cu ft)
AREAS:
Advertising area (each side) ... 120.0 m^2 (1,291.7 sq ft)

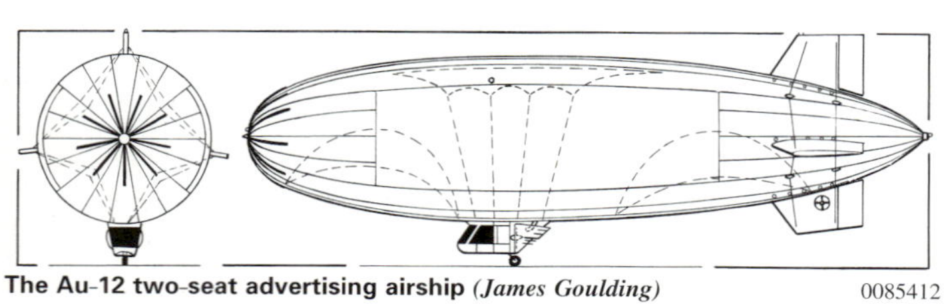

The Au-12 two-seat advertising airship *(James Goulding)* 0085412

RosAeroSystems Au-12M (new tailfin configuration) shown at MAKS '03, Moscow, August 2003 *(Sebastian Zacharias)* ***NEW***/0572210

WEIGHTS AND LOADINGS:
Weight empty ... 620 kg (1,367 lb)
Max payload ... pilot + 50 kg (110 lb)
PERFORMANCE:
Max level speed ... 43 kt (80 km/h; 50 mph)
Cruising speed ... 32 kt (60 km/h; 37 mph)
Cruising altitude ... 1,000 m (3,280 ft)
Pressure ceiling ... 2,000 m (6,560 ft)
Range at cruising speed above:
standard fuel ... 135 n miles (250 km; 155 miles)
with auxiliary fuel tank ... 324 n miles (600 km; 372 miles)
Endurance: at max speed ... 2 h
at cruising speed above ... 5 h

UPDATED

ROSAEROSYSTEMS DPD-5000

TYPE: Helium semi-rigid.

PROGRAMME: Details first made public August 1995 and in preliminary design by 1997. Still being offered to risk-sharing partners in 2004, but no evidence of any manufacture.

UPDATED

ROSAEROSYSTEMS DTS-N1

TYPE: Helium semi-rigid.

PROGRAMME: Revealed at Moscow Air Show August 2001 in association with Tsiolkovsky Airships company (named after celebrated Russian rocket and spaceflight engineer Konstantin Tsiolkovsky, hence alternative designation **DZ-N1** for Anglicised version, Ziolkovsky). Apparently had not progressed beyond design stage by mid-2004.

UPDATED

ROSAEROSYSTEMS MD-900

TYPE: Helium semi-rigid.

PROGRAMME: Details first made public August 1995; preliminary design by 1997 and still being refined in 2001; company seeking risk-sharing partners in 2003, but none announced by mid-2004.

UPDATED

ROSAEROSYSTEMS PD-360

TYPE: Helium semi-rigid.

PROGRAMME: Development of PD-220 design (1999–2000 *Jane's*), which it apparently supersedes (*patrulnyi dirizhabl*: patrol airship). Revealed 1998 as PD-300, later apparently redesignated PD-3000 but later quoted with original designation; re-sized and larger gondola shown in 2003, now called PD-360. Reported by Novosti news agency in November 2000 as "built and tested" at Avialine Scientific and Technical Design Bureau in Moscow, but photographic confirmation not yet seen (mid-2004).

DESIGN FEATURES: Designed for patrol, inspection, surveillance, people/payload transportation, search and rescue, photographic reconnaissance, agricultural roles, and other civil or paramilitary applications. Conventional shape envelope; X configuration tailfins, each with elevator or rudder. Compliant with Russian Federation and US FAA airship design criteria.

FLYING CONTROLS: Vectored thrust from cruise engines for fore-and-aft control. Duplex fly-by-wire control of rudders/elevators. Two internal air ballonets for pressure control; up to 150 litres (39.6 US gallons; 33.0 Imp gallons) water ballast for trim control.

STRUCTURE: Envelope of high-strength polyurethane-coated nylon with titaniaum dioxide UV protection; all-composites gondola.

LANDING GEAR: Single, non-retractable and self-centring mainwheel under rear of gondola; small nosewheel.

POWER PLANT: Two 127 kW (170 hp) LOM M332C four-cylinder inline cruise engines, each driving a two-blade propeller and with cascade-type thrust vectoring.

ACCOMMODATION: Pilot and co-pilot, with dual controls; detachable bulkhead, aft of which are passenger seats, cargo or mission equipment. Ground crew of four to six.

SYSTEMS: Include envelope anti-icing.

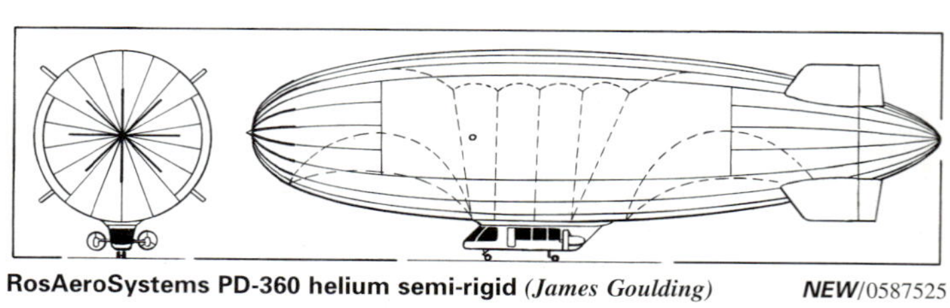

RosAeroSystems PD-360 helium semi-rigid *(James Goulding)* ***NEW***/0587525

AVIONICS: *Comms:* Dual nav/com radios; transponder.

Flight: Computer-aided flight controls; GPS navigation; ADF; marker beacon receiver. Autopilot optional.

Instrumentation: IFR standard.

EQUIPMENT: Could include video cameras, searchlight, rescue hoist, loudhailers and other appropriate mission equipment.

DIMENSIONS, EXTERNAL:
- Envelope: Length ..... 48.80 m (160 ft 1¼ in)
- Max diameter ..... 12.10 m (39 ft 8½ in)
- Fineness ratio 4.0
- Gondola: Length ..... 9.67 m (31 ft 8¾ in)
- Width at floor ..... 1.70 m (5 ft 7 in)
- Max depth ..... 2.25 m (7 ft 4½ in)

DIMENSIONS, INTERNAL:
- Envelope volume ..... 3,600 m^3 (127,130 cu ft)
- Ballonet volume (two, total) ..... 1,275 m^3 (45,025 cu ft)

WEIGHTS AND LOADINGS:
- Weight empty ..... 2,150 kg (4,740 lb)
- Max water ballast ..... 150 kg (331 lb)
- Max payload ..... 1,400 kg (3,086 lb)

PERFORMANCE (estimated):
- Max level speed ..... 59 kt (110 km/h; 68 mph)
- Max cruising speed ..... 49 kt (90 km/h; 56 mph)
- Cruising altitude ..... 1,500 m (4,920 ft)
- Pressure ceiling ..... 3,000 m (9,840 ft)
- Range: at 38 kt (70 km/h; 43 mph) cruising speed ..... 863 n miles (1,600 km; 994 miles)
- ferry ..... 1,619 n miles (3,000 km; 1,864 miles)
- Endurance: at max speed ..... 5 h
- at cruising speed above ..... 24 h

UPDATED

ROSTVERTOL

ROSTOVSKY VERTOLETNYI PROIZVODSTVENNYI KOMPLEKS OAO 'ROSTVERTOL' ('ROSTVERTOL' ROSTOV HELICOPTER PRODUCTION COMPLEX JSC)

ulitsa Novatorov 5, 344038 Rostov-na-Donu
Tel: (+7 8632) 31 74 93
Fax: (+7 8632) 45 05 35
e-mail: rostvertol@gin.ru
Web: http://www.rostvertolplc.ru

DIRECTOR GENERAL: Boris N Slyusar
CHIEF ENGINEER: Igor A Semyonov
DEPUTY DIRECTOR-GENERAL, MARKETING AND EXPORT SALES: Andrey B Shibitov

The company now known as Rostvertol was founded on 1 July 1939, and began by manufacturing wooden propellers. It progressed to aircraft production during the Second World War and to helicopter work in the mid-1950s. Past products at Rostov-on-Don included the UT-2M (1944), Po-2, Yak-14 glider, Il-10M, Il-40, Mi-1, Mi-6, Mi-10 and Mi-10K. Rostvertol is wholly privately owned.

Government decree of 19 February 1996 authorised Rostvertol to export spares and auxiliary equipment for Mil Mi-24 and Mi-25; and to export new Mi-26, Mi-28 and Mi-35 helicopters. Second prototype Mi-28N completed in early 2004, to be followed by further three for State Tests, then series manufacture beginning 2006. Handed over on 3 February 2004 first five 'Hinds' upgraded to night-capable Mi-24PN configuration for Russian Air Forces. Also involved in Mi-60 project.

Rostvertol also active in helicopter overhaul and upgrading. However, independent export sales authorisation rescinded 23 July 1999, and Rostvertol exports now made through official Russian agencies. Company has exported 650 aircraft.

In April 2003, established Aviakompania 'Rostvertol' ('Rostvertol' Aviation Company) to provide helicopter airlift, firefighting and maintenance services. An alliance with Mil is anticipated, under the new title of Mil Helicopters.

UPDATED

RSK 'MIG'/P A VORONIN PRODUCTION CENTRE

FEDERALNOYE GOSUDARSTVENNOYE UNITARNOYE PREDPRIYATIE, ROSSIYSKAYA SAMOLETOSTROITEL'NAYA KORPORATSIYA 'MIG' (RUSSIAN AIRCRAFT-BUILDING CORPORATION 'MIG' FEDERAL STATE UNITARY ENTERPRISE)

1-ñ Botkinsky proezd, Dom 7, 125040 Moskva
Postal address: a/ya No 1, 103045 Moskva
Tel: (+7 095) 155 22 39
Fax: (+7 095) 158 29 84
e-mail: mig@migavia.ru
Web: http://www.migavia.ru

GENERAL DIRECTOR: Valery A Toryanin
FIRST DEPUTY GENERAL DESIGNER, MIG BUREAU: Vladimir I Barkovsky
DEPUTY GENERAL DIRECTOR, MARKETING AND SALES: Vladimir P Vypryazhkin
DEPUTY DIRECTOR GENERAL, LUKHOVITSY PLANT: Yury A Tsupko

PUBLIC RELATIONS AND MEDIA CENTRE:
Tel: (+7 095) 207 04 76
Fax: (+7 095) 207 07 57
e-mail: migpress@mail.ru
DEPUTY DIRECTOR GENERAL, PUBLIC RELATIONS: Vyacheslav N Mileshko

Original Moscow Aircraft Production Organisation (MAPO) renamed by official decree of 8 December 1999 as RSK 'MiG' and its Botkinsky manufacturing plant dedicated to former director (1938–1982) Pavel A Voronin. Up to 1999, MAPO (formerly MMZ No. 30 Znamaya Truda [Banner of Labour]) had built 25,000 aircraft of 40 types, from the Nieuport IV of 1913 to the MiG-29. On 5 October 1999, it was nominated as a production centre for the Tupolev Tu-334 airliner in return for contributing 50 per cent of certification costs. By late 2003, lack of progress by the company in launching Tu-334 manufacture had prompted government consideration of alternative production centre.

Service activities include complete maintenance support and pilot training; is also involved in Chinese Chengdu FC-1 fighter programme, China.

Aircraft production is at the Voronin centre, which comprises original Botkinsky Street premises alongside the Mikoyan design complex in Moscow; at Kalyazin; and at LAPIK, Lukhovitsy. Personnel in 2003 totalled 14,500.

Members of RSK 'MiG' are:

GENERAL DIRECTORATE:
- A I Mikoyan OKB (R&D engineering centre), Moscow (descibed separately)
- P A Voronin Production Centre, comprising
- LAPIK, Lukhovitsy (described separately)
- Botkinsky proezd, Moscow
- Kalyazin Machine-Building Plant, Kalyazin
- A V Fedotov Flight Test Centre, Zhukovsky
- Aircraft Logistic Support Centre
- Marketing and Sales Centre
- Economic and Financial Centre
- Strategic Development and Planning Centre

ENGINE-BUILDING DIVISION:
- V V Chernyshev Machine-Building Enterprise, Moscow
- V Ya Klimov Plant, St Petersburg
- Krasny Oktyabr Machine-Building Enterprise, St Petersburg
- Soyuz Machine-Building Design Bureau, Tushino

INSTRUMENT-BUILDING DIVISION:
- Ryazan State Instrument Plant, Ryazan
- Pribor JSC, Kursk
- Elektroavtomatika Design Bureau, St Petersburg

HELICOPTER-BUILDING DIVISION (VERTOLETOSTROITELNYI DIVIZION):
- Kamov JSC, Lyubertsy

CIVIL AIRCRAFT-BUILDING DIVISION (DIVIZION GRAZHDANSKOGO SAMOLETOSTROENIYA):
- MiG-MGA, Moscow (MiG-Malaya Grazhdanskaya Aviatsiya)

PRODUCTS AND SERVICES DIVISION:
- MiG-Resurs JSC, Moscow
- MiG-Rost JSC, Moscow

GROUND SERVICING AND TEST EQUIPMENT DIVISION:
- Aviatest JSC, Rostov-on-Don

LAPIK — LUKHOVITSY AVIATSIONNYI PROIZVODSTVENNO-ISPTATELNNYI KOMPLEKS (LUKHOVITSY AVIATION PRODUCTION-TESTING COMPLEX):
140500 Lukhovitsy, Moskovskaya oblast
Tel: (+7 09663) 113 76
Fax: (+7 09663) 111 80
e-mail: lmz-avia@mtu-net.ru

Founded at Tretyakovo airfield in 1953, this state-owned subsidiary of RSK 'MiG' functioned as a flight test centre until 1968 (and continues to do so for the MiG-29). It then was assigned to Znamaya Truda production plant, replacing final assembly and test airfield at Khodinka, in the Moscow suburbs, and consequently acquired extensive equipment for manufacturing high-strength aluminium, titanium and composites components in support of RSK 'MiG' production, being renamed Lukhovitsy Mashinostroitelnyi Zavod (LMZ — Lukhovitsy Machine-Building Plant). On 7 March 1996 became first (and, currently, only) Russian factory with light aircraft production approval certificate. Assembly continues at low rate of the Aeroprogress T-101 Grach and Aviatika (MAI) lightplanes. Manufacture of the Ilyushin Il-103 and Interavia I-1L continues intermittently. Also participated in manufacture of parts for Sukhoi Su-29 and Su-31 sportplanes, being responsible for complete assembly from January 2001 onwards; four built in 2001.

Announced in June 2000 that Lukhovitsy will produce Ilyushin Il-100. By mid-2000, was preparing to build a prototype Kamov Ka-62. Naval MiG-29Ks for India will be built at LAPIK, as will MiG-110 medium transport, despite previous allocation to Sokol.

Other activities include flying school for commercial or private pilot training on Il-103; freight handling; and production of jetskis and suntanning beds.

In 2001, construction began at Lukhovitsy of a 12,600 m^2 (135,625 sq ft) Tu-334 assembly hall; this was commissioned in January 2003, launching move of almost all of Voronin plant out of central Moscow.

LMZ adopted current name of LAPIK by January 2002.

UPDATED

SAZ

SARATOVSKY AVIATSIONNY ZAVOD ZAO (SARATOV AVIATION PLANT JSC)

c/o ulitsa Pskovskaya 32, 443052 Samara
Tel: (+7 84 62) 92 66 55
Web: http://www.yak40.com/about.htm

Founded in 1929 as agricultural machinery manufacturer. First aircraft built at Saratov flew on 28 October 1938. During the Second World War, the Saratov plant (GAZ 292) manufactured

Yakovlev Yak-1 and Yak-3 piston-engined fighters, continuing post-war with Yak-11, La-15, Yak-23, Yak-25 and Yak-27 military jet aircraft and, from May 1952, Mi-2 helicopter. In the 1960s and 1970s, production centred on the Yak-40 three-turbofan transport, with Yak-36 and Yak-141 V/STOL prototypes and Yak-38 V/STOL fighters also manufactured in the 1970s and 1980s. Yak-42 entered series production in 1982, although there have been no recent completions. Yak-54 trainer/sporting aircraft production began in 1994, but in early 2001, SAZ was unsuccessfully, as it transpired, attempting to resuscitate an American order for 48 which had previously been terminated after only six deliveries. Of five Yak-54s to have been built in 2001, only one completed before venture was abandoned for lack of funds. Yak-56 primary trainer is also earmarked for production at SAZ, as are Yak-142, Yak-242 (since displaced in Yakovlev's plans by the MS-21 air liner) and Moukamedov A-209M light (3,900 kg; 8,598 lb), nine-seat turboprop twin.

Meanwhile, manufacturing is turning increasingly to agricultural tractors and road sweepers. SAZ is earmarked for amalgamation within Sukhoi holding group under 2001 plans for Russian aerospace industry rationalisation. Personnel in 2003 totalled over 8,000, of whom 2,800 were engineers. Company affairs are managed by Aviacor at Samara and various schemes to recapitalise the plant have been considered including, in September 2003, a fact-finding visit by US Naval Air Systems Command.

SAZ also worked on a completely new type of lifting-body jet aircraft known as the EKIP project. It was the first company in the Russian Federation and Associated States (CIS) to receive a production certificate for commercial aircraft in compliance with international standards.

UPDATED

SGAU

SAMARSKY GOSUDARSTVENNYI AEROKOSMITSESKY UNIVERSITET (SAMARA STATE AEROSPACE UNIVERSITY)

Moskovskoye shosse 34, 443086 Samara
Tel: (+7 8462) 35 18 26
Fax: (+7 8462) 35 18 36

This prominent centre of aerospace learning includes a design bureau, SKB-1, in which students can develop their skills. Although this has concentrated on the small-scale manufacture of lightplanes, it has also designed — under leadership of Albert Elatonsev, its director — an executive transport designated Aist 92 (Stork 92).

STUDENTSKOYE KONSTRUCTORSKOYE BURO 1(STUDENT DESIGN BUREAU 1):
Moskovskoye shosse 34, Korpus 3, 443086 Samara
Tel: (+7 8462) 35 72 81
e-mail: skbl@ssau.ru
Web: http://www.aero.ssau.ru/fla/skb1
DIRECTOR: Professor V M Shakhmistov

Bureau founded 1955; has built Sverchock-1 gyroplane (1971), A-13 motor glider (1974), Shmel lightplane (1977), Strekoza lightplane (1978), Bat glider (1978), A-10 glider (1982) and A-18 Rusich motor glider (1987). More recent designs, previously described in detail, have been Che-15 and Che-25 (Chernov and Gidroplan manufacture) S-202 amphibian, A-16 aerobatic monoplane (1997), S-302 twin-engined amphibian (1995), S-400 four-seat amphibian (1998), Favorit two-seat ultralight (1998) and Crechet two-seat lightplane (1999). Current projects are the Yastreb and a flex-wing trike named Brazhnik.

UPDATED

SKB-1 YASTREB

English name: Hawk
TYPE: Two-seat lightplane.
PROGRAMME: Prototype (RA-2362K) shown at MAKS '03, Moscow, August 2003.
CURRENT VERSIONS: **Sport:** Single-seat variant with 40 litre (10.6 US gallon; 8.8 Imp gallon) fuel capacity.
Trainer: Two-seat variant.
Seaplane: Two seats and twin floats.
Sprayer: Seating and fuel as for Sport variant.
DESIGN FEATURES: Low-wing configuration with mid-mounted, strut-braced tailplane; sweptback fin.
Wing section NASA GA(W)-1; thickness/chord ratio 17 per cent; dihedral 3°; twist 4°.
FLYING CONTROLS: Conventional and manual. Plain rudder and horn-balanced elevators. Flaps, maximum deflection 35°.
STRUCTURE: Generally of composites.
LANDING GEAR: Tailwheel type; fixed. Fuselage-mounted spring cantilever mainwheel legs.
POWER PLANT: One 73.5 kW (98.6 hp) Rotax 912 ULS flat-four driving a three-blade propeller via a 2.4286 reduction gear. Fuel capacity 100 litres (26.4 US gallons; 22.0 Imp gallons).
ACCOMMODATION: One or two (in tandem) persons, according to variant. Large canopy with separate windscreen and rear transparency.
EQUIPMENT: Sprayer version equipped with hopper for 230 litres (60.8 US gallons; 50.6 Imp gallons) of chemicals.
DIMENSIONS, EXTERNAL:
Wing span .... 10.795 m (35 ft 5 in)
Wing chord, constant .... 1.30 m (4 ft 3¼ in)
Wing aspect ratio 8.3
Length overall .... 7.25 m (23 ft 9½ in)
Height overall .... 1.89 m (6 ft 2½ in)
Tailplane span .... 3.30 m (10 ft 10 in)
Tailplane chord, constant .... 0.80 m (2 ft 7½ in)

Prototype SKB-1 Yastreb multirole lightplane at Moscow in August 2003 *(Sebastian Zacharias)* ***NEW***/0567744

Wheel track .... 1.96 m (6 ft 5¼ in)
Wheelbase .... 4.805 m (15 ft 9¼ in)
DIMENSIONS, INTERNAL:
Cabin: Max width .... 0.71 m (2 ft 4in)
Max height .... 1.23 m (4 ft 0½ in)
AREAS:
Wings, gross .... 14.00 m² (150.7 sq ft)
WEIGHTS AND LOADINGS (A: Sport, B: Trainer, C: Seaplane, D: Sprayer):
Weight empty: A .... 360 kg (794 lb)
B .... 390 kg (860 lb)
C .... 450 kg (992 lb)
D .... 430 kg (948 lb)
Max T-O weight: A .... 450 kg (992 lb)
B .... 600 kg (1,322 lb)
C .... 660 kg (1,455 lb)
D .... 750 kg (1,653 lb)
PERFORMANCE (A, B, C, D as above):
Never-exceed speed (VNE):
A .... 140 kt (260 km/h; 161 mph)
Max level speed: A .... 124 kt (230 km/h; 143 mph)
Max cruising speed: A .... 97 kt (180 km/h; 112 mph)
Unstick speed, 20° flap: A .... 30 kt (55 km/h; 34 mph)
B .... 32 kt (60 km/h; 37 mph)
C .... 35 kt (65 km/h; 40 mph)
D .... 38 kt (70 km/h; 43 mph)
Landing speed, 35° flap: A .... 35 kt (65 km/h; 40 mph)
B .... 38 kt (70 km/h; 43 mph)
C, D .... 40 kt (75 km/h; 47 mph)
Max rate of climb at S/L: A .... 420 m (1,378 ft)/min
B .... 300 m (984 ft)/min
C .... 240 m (787 ft)/min
D .... 210 m (689 ft)/min
T-O run: A .... 70 m (230 ft)
B .... 90 m (295 ft)
C .... 100 m (330 ft)
D .... 120 m (395 ft)
Range: A, D .... 216 n miles (400 km; 248 miles)
B, C .... 594 n miles (1,100 km; 683 miles)
g limits: A .... +5.2/–2.6
B .... +4.0/–2.0
C .... +3.8/–1.9
D .... +3.5/–1.7

UPDATED

SMOLENSK

SMOLENSKY AVIATSIONNYI ZAVOD OAO (SMOLENSK AIRCRAFT PLANT PLC)

ulitsa Frunze 74, 214006 Smolensk
Tel: (+7 0812) 61 85 31
Fax: (+7 0812) 61 95 50
e-mail: smaz@sci.smolensk.ru
Web: http://www.smaz.ru
GENERAL DIRECTOR: Aleksandr Miroshkin
CHIEF ENGINEER: Yury Litvinov
MARKETING MANAGER: Vladim Merzlyakov

Smolensk Aircraft Plant (SmAZ) was established in 1926 as an aviation repair facility. In 1934, Buro Osovikh Konstruktsii, OKB for experimental aircraft, relocated from Moscow to Smolensk GAZ 35, specialising in manufacture of prototype and experimental aircraft, including BOK series of stratospheric and ultra-long-range vehicles. Evacuated 1941-1943 and later overhauled Il-2 attack fighter. After Second World War, built Myasishchev M-17/M-55, wings for Yak-40 transports, and more than 500 Yak-18Ts from 1973. Built small number of early production Yak-42s alongside main series at Saratov (SAZ). Constructed Buran space shuttle. Government share was 25.5 per cent.

Production of Yak-18T was resumed 1993, five years after entire original fleet had been scrapped; more than 60 built up to 1996, when 15 placed in storage due to financial problems; production resumed in 1998; deliveries have included three to Ulyanovsk Aviation School in June 2001. Built three Yak-112 lightplanes before project transferred entirely to IAPO in Irkutsk. Other work included manufacture of RPVs and cruise missiles; wings for the Yak-42, upgrading of Yak-40s for business use (including additional fuel capacity), and development and production (from 1993) of Technoavia SM-92 Finist seven-seat STOL aircraft. By 2002, was responsible for marketing Technoavia products.

Has built wing and empennage for first three Sukhoi Su-38L agricultural aircraft and will be responsible for series production. Currently producing one further M-55, which has repeatedly failed to secure sufficient official annual funding to ensure completion. Also participated in Yak-130 programme, building prototype. Overhauls some 10 to 15 Yak-18Ts per year. Non-aerospace activities include manufacture of cloth processing machines and medical equipment.

In July 2004 was deleted from list of Strategic Enterprises of Russian State, permitting shares to be offered for private sale.

UPDATED

SMOLENSK SM-92 FINIST

TYPE: Light utility transport.

PROGRAMME: Design to FAR Pt 23 and JAR 23 standards, started July 1992; construction of first of two prototypes (RA-44482) began January 1993; first flight, as SM-92 Finist, made 28 December 1993; second Finist (RA-44484) completed round-the-world flight through Europe, Atlantic, Canada, Alaska and Siberia in August 1995 covering 16,200 n miles (30,000 km; 18,640 miles) in 160 flying hours; components production at SmAZ, Smolensk. Sixth aircraft (RA-44493) completed as **SM-92P** armed version (described in 2001–02 and previous editions). AP-23 certification received by late 1998 and Hungarian FAR Pt 23 certification in August 2000.

By 2003, RA-44484 was operating in Australia, equipped with wingtip pods.

In 1991, Moravan (Zlin) of Czech Republic (which see) bought plans of SM-92 for US$500,000 and launched local production as Z 400 Rhino. Russian manufacture appears to have ceased. According to Russian folklore, the Finist is a falcon with major powers.

CURRENT VERSIONS: **SM-92**: *As described.*

SM-92P: Border guard version. Described in 2001–02 and previous editions. Prototype, RA-44493, displayed at Moscow Air Show '95; second aircraft (RA-44494) complete by 1997, but converted to SM-92 Finist. No further manufacture known.

SMG-92 Turbo Finist: Described separately.

SM-92T Turbo Finist: Described separately.

Z 400 Rhino: Intended for production by Zlin in Czech Republic; one airframe sold to Moravan for this purpose, but project thwarted by unavailability of selected engine.

CUSTOMERS: By February 2004, 21 Finists and Turbo Finists had been built and five more were on order. Production of 16 for Russian Federation border guard (which eventually requires 300) was disrupted by problems at Smolensk aircraft factory; alternative manufacturer was sought, but nothing further reported. Regional government of Yakutia ordered 14 in October 1997, but none reported in service by 2003. First delivery (aircraft No. 3, RA-44485) to Mike Crymble in UK, 21 January 1995; RA-44487 to Sport-Para Centrum, Antwerp, Belgium, 6 July 1995.

COSTS: US$240,000 (2001).

DESIGN FEATURES: Rugged light transport; in approximate class of DHC-2 Beaver, but less powerful and lower in cost than most used Beavers. Sweptback fin and rudder; small dorsal fin; tailplane mounted on fin, with single bracing strut each side. Wide CG range.

Basic aircraft accommodates up to seven persons with baggage. Convertible under field conditions to transport 600 kg (1,323 lb) freight; two stretcher patients and attendant with medical equipment; to drop six trainee parachutists or four firefighters with parachutes and firefighting equipment. Can carry hopper for 600 kg (1,323 lb) agricultural chemicals in cabin, with spraybar, or cameras for forest surveillance, patrolling electric power lines, gas pipeline inspection and similar duties. Wingtip fuel pods optional; tanks have aerofoil-size central cutout and are fitted slightly inboard of wingtip.

Special P-301 M-15 wing section by Technoavia and CAHI (TsAGI). No sweep; thickness/chord ratio 15 per cent; dihedral 2°; incidence 3° at root, 1° at tip.

FLYING CONTROLS: Conventional and manual. Ailerons and elevators pushrod-actuated; rudder cable-actuated. Electrically operated, three-position (0, 20, 40°), two-section single-slotted flaps on each wing; horn-balanced rudder and elevators, with fluted skin; trim tab on starboard elevator; ground-adjustable tab on rudder; three-section tab along entire trailing-edge of each aileron, centre section being adjustable on ground.

STRUCTURE: All-aluminium, stressed-skin semi-monocoque construction. Simple, reliable structure, with no expensive or exotic materials; repair possible under field conditions. Airframe life 10,000 hours or 20,000 landings, with 'on condition' extension.

LANDING GEAR: Tailwheel type; fixed. Fuselage-mounted spring steel cantilever mainwheel legs with medium-pressure mainwheel tyres, size 600×180 on KT-317 wheels. Steerable, semi-castoring tailwheel with 255×110 tyre on tubular steel strut. Tyre pressure 2.45 bar (35.5 lb/sq in) on mainwheels; 2.95 bar (43 lb/sq in) on tailwheel. Wheel/skis and amphibious or plain floats optional. Pedal-operated disc brakes, with parking lock. Minimum turning radius 6.6 m (21 ft 8 in).

POWER PLANT: One VOKBM M-14Kh air-cooled nine-cylinder radial engine, rated at 265 kW (355 hp) for take-off and 213 kW (286 hp) maximum continuous, driving a Mühlbauer MTV-3-B-C/L250-21 three-blade variable-pitch propeller. Engine TBO 1,000 hours; total life 3,000 hours. Two fuel tanks in wing leading-edge, total capacity 392 litres (104 US gallons; 86.2 Imp gallons); usable capacity 380 litres (100.4 US gallons; 83.5 Imp gallons). Wingtip tanks optional, capacity 200 litres (52.8 US gallons; 44.0 Imp gallons) each. Oil capacity 30 litres (7.9 US gallons; 6.6 Imp gallons).

ACCOMMODATION: Pilot and six passengers, in pairs; quickly removable seats with folding armrests and back; dual controls standard for pilot training; adjustable rudder pedals. Small baggage container (or medical equipment stowage for ambulance version) on port side at rear of cabin. Forward-hinged, jettisonable door each side of flight deck; large rearward-sliding passenger/freight door on port side of cabin, openable in flight. Blister windows to flight deck and cabin. Steps on mainwheel legs and cable handholds for access to flight deck; removable tubular steel ladder beneath cabin door.

Convertible in field to transport 600 kg (1,323 lb) of freight; two stretcher patients and two attendants; six trainee parachutists or four smoke-jumpers with parachutes and firefighting equipment. Provision for carrying hopper for 600 kg (1,323 lb) of agricultural chemicals in cabin, or cameras for forest surveillance, patrolling electric power lines, gas pipeline inspection and similar duties. Clearance adequate for underbelly pannier. Cabin heated and ventilated.

SYSTEMS: Pneumatic system for engine starting, pressure 49 bar (710 lb/sq in). Electrical system provides 36/115 V AC power at 400 Hz and 28.5 V DC power, with 20NKBN-25 battery.

AVIONICS: *Comms:* Two Bendix/King KY 96A VHF transceivers, KA 134 audio control, KR 87A ADF and KT 76A transponder; Garmin GPS 150.

DIMENSIONS, EXTERNAL:
Wing span ... 14.60 m (47 ft 10¾ in)
Wing chord, constant ... 1.40 m (4 ft 7 in)
Wing aspect ratio ... 10.4
Length overall ... 9.30 m (30 ft 6¼ in)
Height: overall ... 3.08 m (10 ft 1¼ in)
fuselage horizontal ... 3.94 m (12 ft 11 in)
Tailplane span ... 5.53 m (18 ft 1¾ in)
Wheel track ... 2.95 m (9 ft 8¼ in)
Wheelbase ... 6.33 m (20 ft 9¼ in)
Propeller diameter ... 2.50 m (8 ft 2½ in)
Propeller ground clearance (tail up) ... 0.22 m (8¾ in)
Freight door: Height ... 1.12 m (3 ft 8 in)
Width ... 1.35 m (4 ft 5¼ in)

DIMENSIONS, INTERNAL:
Cabin: Length ... 3.40 m (11 ft 1¾ in)
Max width ... 1.27 m (4 ft 2 in)
Max height ... 1.38 m (4 ft 6¼ in)
Volume ... 5.2 m³ (184 cu ft)

AREAS:
Wings, gross ... 20.44 m² (220.0 sq ft)
Ailerons (total) ... 2.38 m² (25.62 sq ft)
Flaps (total) ... 3.28 m² (35.31 sq ft)
Fin ... 2.13 m² (22.93 sq ft)
Rudder ... 1.63 m² (17.55 sq ft)
Tailplane ... 3.73 m² (40.15 sq ft)
Elevators (total) ... 2.91 m² (31.32 sq ft)

WEIGHTS AND LOADINGS:
Operating weight empty ... 1,500 kg (3,307 lb)
Payload: max ... 600 kg (1,323 lb)
with full fuel ... 550 kg (1,212 lb)
Max T-O and landing weight ... 2,350 kg (5,180 lb)
Max wing loading ... 115.0 kg/m² (23.55 lb/sq ft)
Max power loading ... 8.87 kg/kW (14.59 lb/hp)

PERFORMANCE:
Never-exceed speed (VNE) ... 140 kt (260 km/h; 161 mph)
Max level speed ... 124 kt (230 km/h; 143 mph)
Max cruising speed ... 108 kt (200 km/h; 124 mph)
Econ cruising speed ... 92 kt (170 km/h; 106 mph)
Stalling speed, power off:
flaps up ... 63 kt (115 km/h; 72 mph)
T-O flap ... 57 kt (105 km/h; 66 mph)
full flap ... 54 kt (100 km/h; 62 mph)
Max rate of climb at S/L ... 300 m (985 ft)/min
Service ceiling ... 3,000 m (9,840 ft)
T-O and landing run on grass ... 250 m (820 ft)
Range, 40 min reserves:
with max payload:
at max level speed ... 332 n miles (615 km; 382 miles)
at econ cruising speed ... 594 n miles (1,100 km; 683 miles)
with max fuel and 550 kg (1,212 lb) payload ... 647 n miles (1,200 km; 745 miles)
Endurance with max fuel ... 7 h 30 min

UPDATED

Smolensk (Technoavia) SM-92 Finist *(Sebastian Zacharias)* ***NEW***/0589284

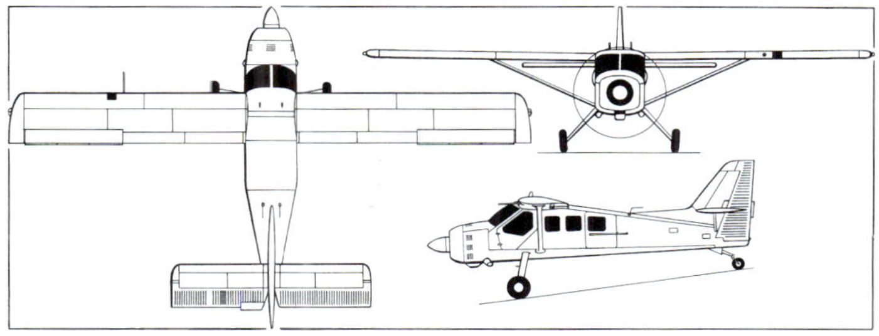

Smolensk (Technoavia) SM-92 Finist seven-seat STOL aircraft *(James Goulding)* 0062758

SMOLENSK (TECHNOAVIA) SM-92T TURBO FINIST

TYPE: Light utility turboprop.

PROGRAMME: Developed by Technoavia from SM-92 Finist as full turbo version. Prototype (conversion of RA-44482) made first flight 6 March 2002 piloted by Mikhail Moltchanuk. Publicly demonstrated at MAKS' 03; in production at Smolensk.

CUSTOMERS: Four completed by February 2004.

DESIGN FEATURES: Turboprop aircraft for multipurpose use, including paradropping, towing and ambulance. Optional wing hardpoints for cargo containers.

FLYING CONTROLS: As SM-92 Finist.

STRUCTURE: As SM-92 Finist.

LANDING GEAR: Tailwheel type; fixed. Optional skis; amphibian landing gear under development.

POWER PLANT: One 559 kW (750 shp) Walter M 601E turboshaft driving an Avia V-508E/99B three-blade, fully feathering reversible-pitch propeller. Two main wing fuel tanks, total capacity 800 litres (211 US gallons; 176 Imp gallons). Optional wingtip fuel tanks, each of 200 litres (52.8 US gallons; 44.0 Imp gallons).

ACCOMMODATION: Pilot and seven passengers or 10 parachutists. Cargo version also available.

SYSTEMS: Optional de-icing system.

AVIONICS: *Comms:* Radio, transponder and GPS system intended to be of US origin.
Radar: Optional weather radar.
Flight: Optional autopilot.

Data generally as for SM-92 Finist, except that below.

DIMENSIONS, EXTERNAL:
Wing span ... 9.925 m (32 ft 6¾ in)
Wheelbase ... 6.59 m (21 ft 7½ in)

WEIGHTS AND LOADINGS:
Weight empty ... 1,550 kg (3,417 lb)

Smolensk (Technoavia) SM-92T Turbo Finist and its progenitor, the SM-92 *(Sebastian Zacharias)* ***NEW***/0589239

Max T-O and landing weight .... 3,000 kg (6,613 lb)
Max power loading .... 5.37 kg/kW (8.82 lb/shp)

PERFORMANCE:
Max operating speed: .... 162 kt (300 km/h; 186 mph)
Normal cruising speed .... 140 kt (260 km/h; 162 mph)
Stalling speed, power off: flaps up .... 68 kt (125 km/h; 78 mph)
flaps down .... 60 kt (110 km/h; 69 mph)
Max rate of climb at S/L .... 457 m (1,500 ft)/min
Service ceiling .... 5,940 m (19,500 ft)
T-O run .... 250 m (820 ft)
T-O to 15 m (50 ft) .... 457 m (1,500 ft)
Landing run .... 280 m (920 ft)

UPDATED

SMOLENSK (TECHNOAVIA) SMG-92 TURBO FINIST

TYPE: Light utility turboprop.

PROGRAMME: Developed from SM-92 Finist by Aerotech Slovakia, with emphasis on sport parachutist dropping; commissioned and financed by G-92 Commerce Ltd, of Nagybanyai 61, H-1025 Budapest, Hungary (e-mail: budata@euroweb.hu). Work began in early 2000, with first flight (newly built aircraft, HA-YDF) 7 November 2000; Hungarian type certificate awarded 14 December 2000; delivery to Wingglider Ltd in UK, January 2001. Second conversion, HA-YDG, produced from RA-44485, exhibited at Aero '01, Friedrichshafen, April 2001, and stationed with Skydive Center, Bad Saulgau, Germany, from June 2001. By mid-2002 Finist prototype RA-44482 had also received a Walter M 601 engine (with different design of cowling) as well as wingtip pods. Total of four completed by end February 2003 (excluding conversion of prototype); none further reported.

Airframe and engine mounting produced by SmAZ at Smolensk; engine installation and cowling, paradrop step and handrails in vicinity of cabin door, new electrical system, instrument panel and engine controls by Aerotech Slovakia AS at Bratislava. Marketing by PB Aircraft, St Moritz, Switzerland (Web: http://www.turbofinist.com).

POWER PLANT: One 400 kW (536 shp) Walter M 601D-2 turboshaft, driving an Avia V-508D-2 three-blade, fully feathering, reversible-pitch propeller. Optional ferry tank.

ACCOMMODATION: Pilot and 10 parachutists; alternatively, pilot and up to seven passengers on lightweight seats.

Data generally as for SM-92 Finist, except that below.

DIMENSIONS, EXTERNAL:
Length overall .... 9.925 m (32 ft 6¾ in)
Wheelbase .... 6.59 m (21 ft 7½ in)

WEIGHTS AND LOADINGS:
Weight empty .... 1,450 kg (3,197 lb)
Max T-O weight: paradrop .... 2,700 kg (5,952 lb)
passenger .... 2,350 kg (5,180 lb)
Max landing weight .... 2,350 kg (5,180 lb)

PERFORMANCE:
Never-exceed speed (VNE) .... 164 kt (305 km/h; 189 mph)
Max operating speed (VMO) .... 143 kt (265 km/h; 165 mph)
Max level speed at 915 m (3,000 ft) .... 159 kt (295 km/h; 183 mph)
Cruising speed .... 130 kt (240 km/h; 149 mph)

Technoavia SMG-92 Turbo Finist conversion by Aerotech ***NEW***/0105658

Max rate of climb at S/L .... 457 m (1,500 ft)/min
Stalling speed, engine off:
flaps up .... 63 kt (115 km/h; 72 mph)
flaps down .... 54 kt (100 km/h; 63 mph)
Time to 4,000 m (13,120 ft) .... 11 min
Max operating altitude .... 5,940 m (19,500 ft)
Descent from 4,000 m (13,120 ft) .... 5 min 30 s
T-O run .... 250 m (820 ft)
T-O to 15 m (50 ft) .... 457 m (1,500 ft)
Landing run .... 280 m (920 ft)
Landing run with propeller reversal .... 145 m (475 ft)
Max range with optional fuel .... 324 n miles (600 km; 372 miles)

UPDATED

SMOLENSK (TECHNOAVIA) SM-2000

TYPE: Light utility turboprop.

PROGRAMME: Turboprop version of SM-2000P development of Yakovlev Yak-18T, combining aerodynamic refinements with stretched fuselage; intended to be main version. Prototype (RA-44445) first flown 17 July 2003 and made public debut at MAKS '03, Moscow, August 2003.

CUSTOMERS: One built and two on order by February 2004.

COSTS: US$600,000 including weather radar, IFR avionics and de-icing (2003).

DESIGN FEATURES: As Yak-18T and SM-2000P, but with longer fuselage, additional cabin door and wingtip pods. Airframe life 10,000 hours or 20 years.

FLYING CONTROLS: As SM-2000P, but incorporating electric trim on all three axes.

STRUCTURE: As SM-2000P but with extra cabin door, wing tip pods for weather radar and landing lights.

Prototype Smolensk (Technoavia) SM-2000 turboprop *(Sebastian Zacharias)* ***NEW***/0589420

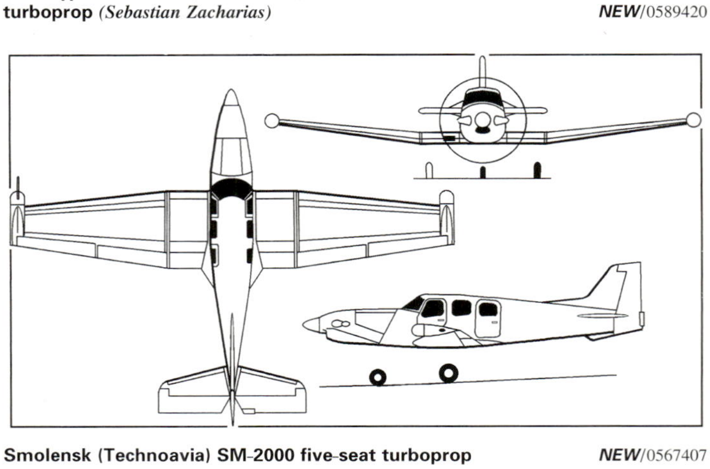

Smolensk (Technoavia) SM-2000 five-seat turboprop ***NEW***/0567407

LANDING GEAR: As SM-2000P. Pneumatic emergency landing gear extension and braking.
POWER PLANT: One 560 kW (751 shp) Walter M 601E turboprop driving an Avia V-508E/99B three-blade, fully feathering reversible-pitch propeller. Fuel capacity 845 litres (223 US gallons; 186 Imp gallons).
ACCOMMODATION: Five persons, including one or two pilot(s); rear three in semi-club arrangement with single forward-facing rear seat. Folding table. Crew door each side; passenger door port, rear.
SYSTEMS: BFGoodrich de-icing. Oxygen system.
AVIONICS: IFR standard fit.
Radar: Weather radar optional in starboard wingtip pod.

DIMENSIONS, EXTERNAL:
Wing span over tip pods ... 11.88 m (38 ft 11¾ in)
Length overall ... 9.435 m (30 ft 11½ in)
Max width of fuselage ... 1.28 m (4 ft 2½ in)
Height overall ... 3.41 m (11 ft 2¼ in)
Wheel track ... 3.12 m (10 ft 2¾ in)
Wheelbase ... 1.95 m (6 ft 4¾ in)
Propeller diameter ... 2.50 m (8 ft 2½ in)
Crew doors: Width ... 0.78 m (2 ft 6¾ in)
Passenger door: Width ... 0.66 m (2 ft 2 in)

DIMENSIONS, INTERNAL:
Cabin: Length between bulkheads ... 3.20 m (10 ft 6 in)
Max width ... 1.20 m (3 ft 11¼ in)
Max height ... 0.90 m (2 ft 11½ in)
Floor area ... 3.4 m² (37 sq ft)
Volume ... 3.5 m³ (124 cu ft)

WEIGHTS AND LOADINGS:
Weight empty ... 1,360 kg (2,998 lb)
Max T-O weight ... 2,100 kg (4,629 lb)
Max power loading ... 3.75 kg/kW (6.16 lb/shp)

PERFORMANCE:
Never-exceed speed (VNE) ... 259 kt (480 km/h; 298 mph)
Max operating speed (VMO) ... 208 kt (385 km/h; 239 mph)
Max cruising speed at FL98 ... 216 kt (400 km/h; 249 mph) TAS
Max rate of climb at S/L ... 660 m (2,165 ft)/min
T-O run ... 150 m (495 ft)
Landing run ... 350 m (1,150 ft)
Range with max fuel, 30 min reserves ... 809 n miles (1,500 km; 932 miles)

UPDATED

SMOLENSK (TECHNOAVIA) SM-2000P

TYPE: Six-seat utility transport.
PROGRAMME: Technoavia SM-94 developed in early 1990s as six-seat version of Yak-18T, incorporating Western avionics and aerodynamic refinements, including all-metal wings and empennage with all surfaces reprofiled to give squarer appearance, three-blade propeller, increased tankage and two-piece windscreen replacing multipane design. SM94-1 development aircraft (RA-44486) shown at MAKS, Moscow, in 1997, but progress subsequently slowed by financial considerations.
Prototype SM-2000P made first flight 21 March 2002 and displayed at MAKS '03. Designation SM-2000P differentiates it from the further developed SM-2000 turboprop, described separately.
CUSTOMERS: Two sold and two on order by February 2004.
DESIGN FEATURES: Upgraded version of Yak-18T.
Wing section Clark YH 14.5 at root and NACA 23012 at tip.
STRUCTURE: All-metal airframe with carbon fibre control surfaces.
LANDING GEAR: Retractable tricycle type, as fitted to Yak-18T, but with increased shock absorber pressure.
Data generally as for Yakovlev Yak-18T, except where indicated.
POWER PLANT: One 268 kW (360 hp) VOKBM M-14X aircooled nine-cylinder radial engine driving an MT-Propeller MTV-9-B-C/CL-260-27 three-blade constant-speed propeller.
Fuel in wing tanks, total 760 litres (201 US gallons; 167 Imp gallons).
ACCOMMODATION: Six persons, including pilot(s).

DIMENSIONS, EXTERNAL:
Wing span ... 11.76 m (38 ft 7 in)
Length overall ... 8.48 m (27 ft 9¾ in)
Height overall ... 3.35 m (11 ft 0 in)
Tailplane span ... 4.02 m (13 ft 2¼ in)
Wheel track ... 3.12 m (10 ft 2¾ in)
Propeller diameter ... 2.50 m (8 ft 2½ in)
Propeller ground clearance ... 0.16 m (6¼ in)
Wheel track ... as Yak-18T

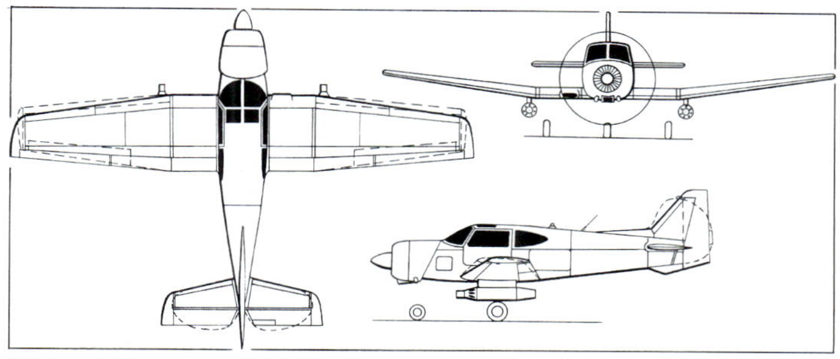
Smolensk (Technoavia) SM-2000P, with Yak-18T configuration shown by broken lines *(James Goulding)* 0015130

AREAS:
Wings, gross ... 19.45 m² (209.4 sq ft)
Ailerons (total) ... 1.40 m² (15.07 sq ft)
Flaps (total) ... 2.84 m² (30.57 sq ft)
Fin ... 2.35 m² (25.30 sq ft)
Rudder ... 1.15 m² (12.38 sq ft)
Tailplane ... 3.86 m² (41.55 sq ft)
Elevators (total) ... 1.74 m² (18.73 sq ft)

WEIGHTS AND LOADINGS:
Weight empty ... 1,225 kg (2,700 lb)
Max payload ... 500 kg (1,102 lb)
Max T-O weight: utility ... 2,100 kg (4,629 lb)
training ... 1,600 kg (3,527 lb)

PERFORMANCE:
Max level speed ... 156 kt (290 km/h; 180 mph)
Max cruising speed ... 140 kt (260 km/h; 162 mph)
Stalling speed: flaps up ... 65 kt (120 km/h; 75 mph)
flaps down ... 54 kt (100 km/h; 63 mph)
Service ceiling ... 4,000 m (13,120 ft)
T-O run ... 400 m (1,315 ft)
Landing run ... 350 m (1,150 ft)
Range with max fuel ... 1,349 n miles (2,500 km; 1,553 miles)
g limits ... +6/–3

UPDATED

SMOLENSK (TECHNOAVIA) SP-55M

TYPE: Aerobatic single-seat sportplane.
PROGRAMME: Initial batch of five aircraft under construction in 2000 at Progress plant at Arsenyev; Russian prototype unregistered; initial production aircraft, RA-44547, delivered in January 2001 to Richard Goode Aerobatics in UK.
COSTS: €115,000 (2003).
DESIGN FEATURES: Mid-wing configuration with symmetrical section of original Technoavia design; no dihedral, anhedral or incidence. Development of Yak-55M, last described in 1998–99 *Jane's All the World's Aircraft*. Modifications include redesigned rear fuselage turtledeck; new fin and tailplane fillets; reprofiled wingtips; redesigned flying surfaces with composites skin; steel engine firewall; new engine cowling; new instrument panel with US instruments and avionics; modified fuel and oil systems to FAR Pt 23 requirements; modified pneumatic system; pilot-controlled oil and carburettor heat doors; new entry steps and assist handle on fuselage and landing gear legs; inclined seat for increased *g* tolerance; new safety harness; baggage compartment and smoke system.
FLYING CONTROLS: Conventional and manual. Near-full-span ailerons have narrower chord than those of Yak-55M, without horn balances; elevators and rudder of greater area than Yak-55M. Trim tab on each elevator. Suspended balance tab below each aileron.
STRUCTURE: Generally similar to Yak-55M. Metal, single-spar wing with composites-covered ailerons and composites tips. Metal fuselage with steel firewall; composites-covered elevator and rudder.
LANDING GEAR: Tailwheel type; fixed. Fuselage-mounted spring steel cantilever mainwheel legs on all three units; KT-98 mainwheels with 400×150 tyres; lockable, free-castoring tailwheel.
POWER PLANT: One 265 kW (355 hp) VOKBM M-14P air-cooled, nine-cylinder radial engine driving an MT-Propeller MTV-9-B-C/CL260-27 three-blade, constant-speed (hydraulic)

Prototype Smolensk SM-2000P *(Yefim Gordon)* ***NEW***/0525971

Smolensk (Technoavia) SP-55M early production aircraft *(Sebastian Zacharias)* ***NEW***/0589396

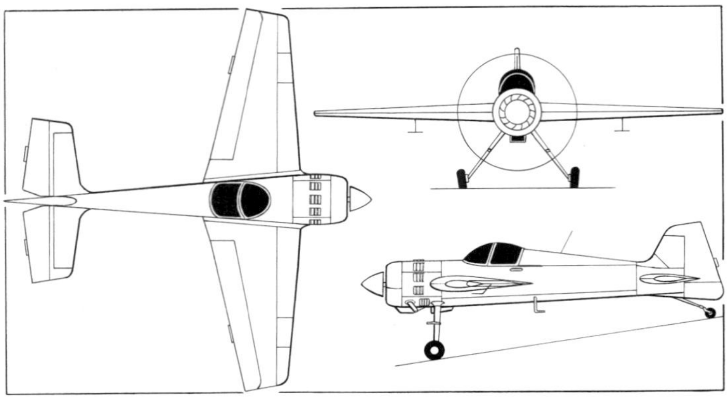

Smolensk (Technoavia) SP-55M single-seat sportplane *(James Goulding)* 0525057

metal propeller. Fuel in wing tanks, total capacity 120 litres (31.7 US gallons; 23.4 Imp gallons); inverted fuel and oil systems.

ACCOMMODATION: One person in inclined seat with recess for backpack parachute; fixed windscreen and one-piece sliding canopy.

SYSTEMS: Electrical system comprises lightweight 10 Ah generator and storage battery, both of US origin

AVIONICS: Avionics fit of US origin.

EQUIPMENT: Optional smoke generation system; or glider hook.

DIMENSIONS, EXTERNAL:
- Wing span ... 8.00 m (26 ft 3 in)
- Wing aspect ratio ... 5.3
- Length overall ... 7.48 m (24 ft 6½ in)
- Height overall ... 2.225 m (7 ft 3½ in)
- Tailplane span ... 3.51 m (11 ft 6¼ in)
- Wheel track ... 2.76 m (9 ft 0½ in)
- Propeller diameter ... 2.50 m (8 ft 2½ in)

AREAS:
- Wings, gross ... 12.17 m^2 (131.0 sq ft)
- Ailerons (total) ... 2.66 m^2 (28.63 sq ft)
- Fin ... 0.40 m^2 (4.31 sq ft)
- Rudder ... 1.02 m^2 (10.98 sq ft)
- Tailplane ... 1.40 m^2 (15.07 sq ft)
- Elevators (total) ... 1.79 m^2 (19.27 sq ft)

WEIGHTS AND LOADINGS:
- Weight empty ... 690 kg (1,521 lb)
- Max T-O weight: training ... 855 kg (1,885 lb)
  - ferry ... 955 kg (2,105 lb)
- Max wing loading: training ... 70.3 kg/m^2 (14.39 lb/sq ft)
  - ferry ... 78.5 kg/m^2 (16.07 lb/sq ft)
- Max power loading: training ... 3.23 kg/kW (5.31 lb/hp)
  - ferry ... 3.61 kg/kW (5.93 lb/hp)

PERFORMANCE:
- Never-exceed speed (VNE) ... 194 kt (360 km/h; 223 mph)
- Max level speed ... 173 kt (320 km/h; 199 mph)
- Stalling speed ... 57 kt (105 km/h; 66 mph)
- Max rate of climb at S/L ... 1,080 m (3,543 ft)/min
- Max roll rate ... 360°/s
- T-O run ... 170 m (558 ft)
- Range with max fuel at econ cruising speed at 1,000 m (3,280 ft), 7% fuel reserves ... 415 n miles (770 km; 478 miles)
- *g* limits ... +9/–6

UPDATED

SOKOL

SOKOL NIZHEGORODSKY AVIATSTROITELNYI ZAVOD AOOT (SOKOL NIZHNY NOVGOROD AIRCRAFT MANUFACTURING PLANT JSC)

ulitsa Chaadaeva 1, 603035 Nizhny Novgorod
Tel: (+7 8312) 29 32 16
Fax: (+7 8312) 29 89 20
e-mail: wings@kis.ru
Web: http://www.sokolplant.ru
DEPUTY CHAIRMAN: Vasily H Pankov
GENERAL DIRECTOR: Mikhail Shibaev
CHIEF ENGINEER: Valery Drobyshevski
HEAD OF MARKETING: Aleksandr E Zaitsev

Founded (as GAZ 21) on 1 February 1932, this production plant concentrated on fighter aircraft manufacture until the start of the *konversiya* programme led to retooling for civilian production; registered as a JSC on 22 September 1994. Joined Kaskol group of companies in 2000.

Built 17,691 fighters up to end of Second World War. From 1949, in collaboration with the Mikoyan OKB, it manufactured 2,148 MiG-15s (1949–51), 2,470 MiG-17s (1951–54), 1,125 MiG-19s (1955–58), 5,278 (or 5,752) MiG-21s (1957–85) and 1,186 MiG-25s (1966–85). Russian Ministry of Property holds 38 per cent of shares, which equivalent to 50.5 per cent of voting rights; Apone SA (16.99 per cent) Kaskol Group (15.67 per cent) and Vostokvneshtorg Group (7.00 per cent) are remaining major shareholders. Current or recent products include the MiG-29UB (from kits supplied from Moscow), MiG-31 and Yak-130 jet trainer; first preproduction aircraft completed in mid-2003. It is also responsible for supplying kits for upgrading 125 Indian Air Force MiG-21s to MiG-21-93 standard, with an initial order for 36. Agreement reached in July 2000 on upgrading half of Russian Air Forces' 280 MiG-31s.

Civilian programmes include manufacture of the Myasishchev M-101 Gzhel six-seat turboprop light aircraft, a prototype AeroRIC Dingo amphibious aircraft with air cushion landing gear and Avia Accord twin-engined light multipurpose aircraft; plans also exist for Myasishchev M-202 production. M-101 marketing is first task of New Regional Aircraft Company (NORS), formed 18 April 2001 by Sokol, Myasishchev and Kaskol financial group. AeroRIC Research-Industrial Company is co-located with Sokol; by 2001 its Dingo project was in suspension because of financial shortfall.

Plans to build MiG-110 twin-turboprop transport apparently not overtaken by announcement in early 2000 of its reallocation to RSK 'MiG' (LAPIK plant) and discussions on continued participation in programme under way with RSK MiG in mid-2000. Transfer of MiG-110 to LAPIK was reaffirmed in late 2002. Agreement of 2001 covers production of wings and noses for navalised MiG-29K for India, plus complete MiG-29KUB trainers. Production of the Eurospace F-15-F Excalibur four-seat touring aircraft was halted at an early stage when a contract for almost 100 was cancelled; some 20 airframes remained at Nizhny Novgorod in June 2000, when 10 sold to Turbine Design in USA.

Other products, in conjunction with AeroRIC, include Volga-2 WIGE craft, air cushion vehicles and Sokol 330 hydrofoil. Offers 'familiarisation' flights in MiG-29UB, MiG-21U and L-29 Delfin. Employment (including part-time) was 9,000 in 2001.

UPDATED

STRELA

STRELA PROIZVODSTVENNOYE OBEDINENIE (STRELA PRODUCTION ASSOCIATION)

ulitsa Shevchenko 26, 46005 Orenburg
Tel: (+7 3532) 35 72 09 and 35 61 74
Fax: (+7 3532) 35 54 60
GENERAL DIRECTOR: S I Grachev

The wholly state-owned Strela plant was established as GAZ 47 on 25 October 1941 and built the UT-2, Yak-6, ShChe-2 and Ts-25 glider between then and 1946. Returned to aviation manufacture in 1950 with Yak-14, Mi-1, La-17 up to 1964, by which time heavily involved in missile work. Currently manufactures Yakhont missile and Kamov Ka-226 helicopter. Assembled prototype Ka-226; plans for Ka-226 series manufacture included first three or four production helicopters in 2001, but production investment of Rb300 million was not announced until January 2002, first two Ka-226s being complete by late March 2002. Factory has orders for 27 Ka-226s including 22 for Gazprom, of which Nos. 3 to 13 were under construction in August 2003. It has produced replica Yak-3/Yak-9s and Mitsubishi A6M2s (Zeros) for export (20 Yaks and three Zeros built by December 2003, including two of latter for Hollywood film use and one in December 2003 for Japanese customer). A replica Yak-7 was under construction in 2003.

UPDATED

SUKHOI (AVPK SUKHOI)

GOSUDARSTVENNOYE UNITARNOYE PREDPRIYATIE AVIATSIONNYI VOYENNO-PROMYSHLENNYI KOMPLEKS SUKHOI (STATE UNITARY ENTERPRISE, AVIATION MILITARY-INDUSTRIAL COMPLEX SUKHOI)

a/ya 604, ulitsa Polikarpova 23B, 125284 Moskva
Tel: (+7 095) 941 76 87
Fax: (+7 095) 945 68 06
e-mail: apvk@sukhoi.org
Web: http://www.sukhoi.org
GENERAL DIRECTOR: Mikhail A Pogosyan
PRESS OFFICER: Yury Chervakov

AVPK (Aviatsionnyi Voyenno-Promyshlenny Kompleks) Sukhoi was created, under presidential decree, in August 1996, to bring together the Sukhoi and Beriev design bureaux, the production plants at Irkutsk (IAPO), Komsomolsk-on-Amur (KnAAPO), Novosibirsk (NAPO) and Taganrog (TANTK), and two banks, Oneximbank and Inkombank. The grouping was formally inaugurated on 30 December 1996. As originally constituted, the group comprises the wholly state-owned KnAAPO and NAPO; plus Sukhoi OKB (see following entry), IAPO and TANTK/Beriev, APVK Sukhoi having a controlling interest in three last-mentioned.

In October 2001, decree by Russian president created Sukhoi Aviation Holding Company, which became successor to AVPK Sukhoi, conducting initial board meeting on 29 October 2003. New structure combines State holdings in various associated design and production organisations (KnAAPO, NAPO, Irkut, TANTK and Sukhoi Design Bureau) into single entity.

Opytnyi Konstruktorskoye Buro Sukhogo OAO (Sukhoi Experimental Design Bureau JSC)

a/ya 604, ulitsa Polikarpova 23B, 125284 Moskva
Tel: (+7 095) 941 78 41
Fax: (+7 095) 945 66 06
CHAIRMAN: Yury Koptev
DEPUTY CHAIRMAN: Vladimir Chernov
GENERAL DESIGNER: Mikhail Petrovich Simonov
CHIEF DESIGNER: Vladimir Babak

OKB named for Pavel Osipovich Sukhoi, who headed it from 1939 until his death in September 1975. It remains one of two primary Russian centres for development of fighter and attack aircraft, and it is notable that 60 per cent of front-line Russian Air Forces' aircraft are of Sukhoi design. Bureau has widened its activities to include civilian aircraft, under *konversiya* programme, establishing separate division in 2000. This offsets the drastic reduction in combat aircraft production and competition in the upgrade and overhaul market.

Sukhoi Advanced Technologies
GENERAL DIRECTOR: Boris V Rakitin

This division is responsible for promotion and sales of Sukhoi light aircraft, such as the Su-29, Su-31, Su-38L and Su-49. Related activities include conversion of Su-26s and Su-29s with SKS-94 extraction systems.

Nauchno-Proizvodstvennoe Kontsern Shturmoviki Sukhogo (Sukhoi Stormovik Scientific-Production Concern)
GENERAL DIRECTOR: Vladimir Babak

Offers of upgrades of Su-25 'Frogfoot' to Russian and export customers.

Grazhdanskie Samolety Sukogo (Sukhoi Civil Aircraft)
ulitsa Polikarpova 23B Korpus 2, 125284 Moskva
Tel: (+7 095) 727 19 88
Fax: (+7 095) 946 05 40
e-mail: info@scac.ru
Web: http://www.scac.ru
GENERAL DIRECTOR: Andrey Il'in

Subdivision formed May 2000 as wholly owned element of Sukhoi Design Bureau; on 5 June 2000 signed agreement with Alliance Aircraft Corporation for joint development and production of StarLiner regional airliner. However, Sukhoi unilaterally cancelled agreement on 2 November 2000. In August 2001, Sukhoi revealed RRJ (Russian Regional Jet) programme in collaboration with Boeing. This selected in 2003 for production. Is also developing SSBJ supersonic business jet, a programme begun in 1987, at one time involving partnership with Gulfstream of US.

UPDATED

OPYTNYI KONSTRUKTORSKOYE BURO SUKHOGO AOOT (SUKHOI EXPERIMENTAL DESIGN BUREAU JSC) SUKHOI SU-27

NATO reporting name: Flanker
TYPE: Air-superiority fighter.
PROGRAMME: Development of long-range heavy fighter to meet PFI (*perspektivnyi frontovoy istrebitel:* advanced frontal fighter) requirement began 1969 under leadership of Pavel Sukhoi; augmented in high-low mix by Mikoyan LPFI (later MiG-29). T10 based on 'integral' layout of unbuilt T4MS. Construction of prototype began in 1974 under Mikhail Simonov's supervision, and it was flown 20 May 1977 by Vladimir Ilyushin. Development undertaken by nine flying **Su-27**'Flanker-As': four prototypes (T10-1 and T10-2) produced in OKB workshops; T10-3 and T10-4 built by Komsomolsk plant and assembled by OKB); five development (T10-5, -6, -9, -10 and -11) and one static test airframe at Komsomolsk. Individual prototype designations duplicated some competing unbuilt T10 configurations. Prototypes had curved wingtips, rearward-retracting nosewheel and tailfins mounted centrally above engine housings.

Some variation between prototypes; T10-1 originally flew with 'clean' wing; four fences then added, followed by anti-flutter weights on wingtips and fins; inboard fences subsequently removed. T10-3 (first flight 23 August 1979) and T10-4 (first flight 31 October 1979) used AL-31F engines, others had AL-21F-3s. Canted tailfins from T10-3. T10-5, -6, -9, -10, and -11 had eight pylons (two below each wing) and -9 and -11 had long-chord ogival radomes; these five aircraft collectively known as T10-5 subvariant. Original configuration revealed poor controllability with inadequate stability in roll and yaw, and with poor high AoA capability; two pilots lost their lives before major airframe redesign resulted in T-10S production configuration (known internally as T10 junior).

Construction of new model began 1979, with first flight of prototype (T10-7) 20 April 1981, followed by second prototype (T10-12); T10-8 was T10S-0 static test airframe; series production by KnAAPO at Komsomolsk began with T10-15, first flown 2 June 1982; entry into service 22 June 1985 with air defence regiment co-located with Komsomolsk factory airfield; official type acceptance 23 August 1990 (Council of Ministers decree); ground attack role observed in 1991. Following purchase of complete aircraft, China signed licensed production agreement on 6 December 1996, covering further 200. Production continues for export. In late 1999, Sukhoi quoted 567 of Su-27 family in service in eight countries.

Sukhoi Su-27SMK demonstrator *(via Yefim Gordon)* ***NEW***/1047023

CURRENT VERSIONS: **Su-27 ('Flanker-B'):** Single-seat land-based production version for air defence force (PVO); full-span leading-edge flaps, trailing-edge flaperons, 5 per cent increase in wing area, straight leading-edges, new aerofoil, square wingtips, carrying anti-flutter weights which doubled as AAM launchers; wider-spaced, uncanted tailfins outboard of engine housings; flatter canopy of reduced cross-section; extended tailcone instead of flat 'beaver-tail'; forward-retracting nosewheel; first flown (T10-7) 20 April 1981. Standard radar tracks 10 targets simultaneously, engages only one. There was early, but probably erroneous, speculation that PVO aircraft were designated Su-27P, with **Su-27S** designation applied to Frontal Aviation aircraft. All aircraft can carry Sorbtsya-S active ECM jammer pods on wingtips in place of wingtip launch rails. **Su-27S** designation little used, but differentiates production (Series) from prototype and preproduction. By 2003, Su-27S designation being used to identify those aircraft which were awaiting Su-27SM modifications. Four initial production aircraft (T10-15, -17, -18 and -22) used for State Acceptance Tests were part of small initial batch featuring horizontally cropped fin caps.

Detailed description applies to the above version, except where indicated.

'Su-27RV' ('Flanker-B'): Six replacement aircraft for Russian Knights aerobatic team feature GPS and Western-compatible communications equipment.

Su-27SK ('Flanker-B'): Export version of basic Su-27, using air-to-ground capabilities not exploited by Soviet/Russian Su-27s and using same weapons options and downgraded avionics. Armament, totalling up to 4,000 kg (8,818 lb), includes 250 kg and 500 kg bombs, B-8M1 packs of 20 × 80 mm rockets, B-13L packs of five 122 mm rockets, S-250FM 250 mm rockets, KMGU-2 cluster bombs, or podded SPPU-22 30 mm gun with optional downward-deflecting barrel for air-to-ground and air-to-air use. Dimensions, weights and performance generally similar to Su-27 but with reinforced landing gear giving increased (33,000 kg; 72,752 lb) MTOW. Chinese aircraft locally designated **J-11**, though this strictly applies only to locally manufactured examples. First Chinese-assembled aircraft flight tested in December 1998. Some later export aircraft may have upgraded radar and 6,200 kg (13,670 lb) weapon load, or even 8,000 kg (17,637 lb) according to some manufacturers' brochures.

Russian-managed upgrade of Su-27SK began 2004, including radar upgrade to N-1001VE standard with attendant ability to engage target with two Vympel RVV-AE AAMs simultaneously. Addition of indigeous Pl-12 AAM has not been attempted, neither has expected an attempt made to add multirole capability.

Su-27SM: Mid-life update configuration in parallel to Su-27UBM and Su-30KN; developed by Ramenskoye Instrument Design Bureau and Sukhoi. First conversion of Su-27S for Russian Air Forces flew at Komsomolsk on 27 December 2002; second followed by mid-2003. Upgrade adds MIL-STD 1553B digital databus (first in Russian military aircraft), two 152 × 203 mm (6 × 8 in) LCDs, central 101 × 127 mm (4 × 5 in) screen, A737 satellite navigation and new computer for N-001 radar, plus software for compatibility with AA-12 'Adder' AAMs and several ASMs/LGBs. No structural or engine improvements, but K-36D-3.5E ejection seat. First five Su-27SMs delivered to Lipetsk military test centre for operational training, 25 December 2003. Further seven delivered to new air regiment in Far East district on 23 December 2004 and 17 more due during 2005, of which 10 funded in that year. Second stage upgrade planned for 2005, adding L175 Khibiny EW system, 'quiet' radar and electro-optic targeting pod. Initial 20 first-stage upgrades to have been completed by end of 2003, when IOC was scheduled, although year-end target had fallen to "10 to 12" according to mid-2003 announcement.

Sukhoi Su-27 'Flanker-B' of Russian Knights demonstration team *(Yefim Gordon)* ***NEW***/1047025

Chinese Sukhoi Su-27UBK from the final batch ***NEW***/1047019

First Indonesian Su-27SK *(TNI AU)* ***NEW***/1043439

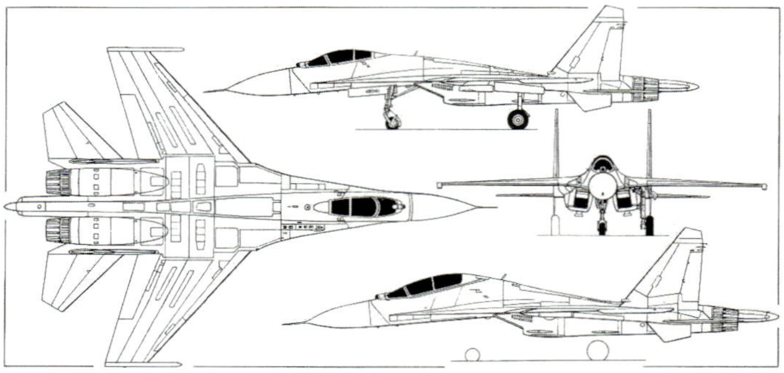

Sukhoi Su-27P, with added side elevation (bottom) of two-seat Su-27B ***(Jane's/Dennis Punnett)*** 0507210

Su-27SMK: Alternatively known as Su-27SK Upgrade; stated to raise aircraft to 4+ fighter generation. Available for previously exported (single-seat) Su-27SKs and with features developed for Russian Air Forces' Su-27SM. Demonstrator '305' first flew early 2003. Changes from baseline Su-27SK include upgraded fire control system for land, sea and air targets; LCD cockpit instrumentation; improved navigation integrated with Glonass and Navstar satellite systems; new elint system to facilitate ARM launching; new electro-optical sighting system with mode for semi-active laser homing air-to-surface missiles; expanded selection of missiles; in-flight refuelling; automatic mission data loading; and increased serviceability features.

Fire control update centred on SUV-UE air-to-air system (enhanced version of SUV-27, standard fit) with additional air-to-surface channel for active radar homing of anti-ship missiles, plus SUV-P air-to-surface system, additionally capable of displaying data to cockpit MFDs. SUV-UE has upgraded RLPK radar sighting system, and new OEPS electro-optic sighting and improved SILS-27M HUD. SUV-P based on standardised high-speed computers and integrated with SUO-30PK armament control system.

Data display based on two MFDs, multifunction LCD panel and modernised HUD. Some electromechanical indicators retained as backups.

Weapon options include previous R-27ER1, R-27ET1 MRAAMs and R-73E SRAAMs, now augmented by RVV-AE with active radar homing (total six); Kh-29T (TE) and Kh-29L TV/laser-homing ASMs (four), KAB-500Kr (four) or KAB-1500Kr (one) LGBs, Kh-31P passive ARMs (four) and Kh-31A active radar ASMs (four).

Su-27PD: Long-endurance test aircraft with retractable AAR probe, offset IRST and recontoured tail 'sting'; sometimes described as single-seat Su-30, originally Su-27P.

Su-30KI: The Su-30KI prototype (4002, wearing the basic Chinese dark grey colour scheme, but with a disruptive camouflage superimposed on the wings, tailplanes and outer surfaces of the tailfins) first flew on 28 June 1998, and was then sent to Chkalov Flight Test Centre at Akhtubinsk, where it was evaluated and used for launch trials of the RVV-AE (AA-12 'Adder') AAM, and for avionics and refuelling probe verification. The Su-30KI was described as a 'single-seat Su-30', and was tailored to meet an Indonesian requirement, before the Asian economic crisis halted that programme. Extent to which the aircraft incorporated the 'extended endurance' features of the Su-30 (apart from refuelling probe, GPS and Western VOR/DME navigation equipment) remains uncertain, however. Su-30KI also used as basis of Sukhoi's latest proposal to upgrade Russia's in-service Su-27s, replacing Su-27SM. Upgrade programme still confusingly referred to by Sukhoi as the Su-30KI, although Russian Air Forces understood to use neither the Su-30 designation nor the KI suffix.

Su-27UB ('Flanker-C'): Tandem two-seat trainer version of 'Flanker-B' with full combat capability (Sukhoi designation T10U); four prototypes (first of which for static testing) built at Komsomolsk; first flown 7 March 1985 (T10U-1). Series manufacture by Irkutsk Aircraft Production Association began 1986; first production UB (T10U-4) flew on 10 September 1986. Instructor in raised rear seat with 6° view forward over the nose; taller fin; length same as 'Flanker-B'; overall height 6.36 m (20 ft 10¼ in), maximum combat load 8,000 kg (17,637 lb). 1,500 kg increase in empty weight, no reduction in internal fuel capacity. Export version is **Su-27UBK**.

Su-27UBM: Russian Air Forces upgrade. First modified Su-27UB (1201 '20') delivered from Irkutsk plant to LII at Zhukovsky 6 March 2001; further seven conversions planned before end of same year; two completed by September 2001; parallel programme to Su-30KN upgrade. Equipment includes new computer, GPS, three 152 × 203 mm (6 × 8 in) MFDs, RVV-AE (AA-12) missiles and ground mapping mode for N001 radar. Later improvements may include radar-absorbent paint and refuelling probe.

'Aircraft 02-01' ('Flanker-C'): Second Su-27UB flying prototype (Komsomolsk 0201) converted as in-flight refuelling systems testbed with retractable probe and provision for centreline 'buddy' pod; first flown 6 June 1987. Later became Su-27PU.

Su-27M: Advanced development of Su-27. Still the official designation for the aircraft known to the OKB as the Su-35.

Su-27K: *Described separately.*

Su-27IB: *Described separately.*

Su-27PU: Two prototypes only; production as **Su-30**.

CUSTOMERS: Approximately 567 in service by December 1999. About 395 in service with Russian Air Forces (or 340 plus 10 with training units by mid-2000, according to some sources); these reports refer to active inventory; evidence suggests that main production batches for former USSR contained up to 600 Su-27s and 140 Su-27UBs.

China received 26 (24 Su-27SK single-seat, two Su-27UBK two-seat) in 1991–92, followed by 24 more (14 single-seat, 10 two-seat) in 1995–96; original agreements called for licensed manufacture of 200 as J-11, at planned rate of 50 a year, in new purpose-built factory (owned by Shenyang Aircraft Company, which see) from 1998 (when first two shipsets delivered from Komsomolsk; first Chinese-assembled Su-27 rolled out December 1998; only eight delivered by late 2000, following considerable quality control problems, leading to the procurement of 28 extra Russian-built Su-27UBKs, delivered 2000–02, first four on 14 December 2000, second four in December 2000; to Nanjing/Yuxikou air base, where 10 supplied in 2001 and balance of 10 following in 2002. China's licence agreement does not allow exports. The first 50 aircraft assembled from Russian-supplied kits of diminishing completeness and the Chinese aircraft will always include at least 30 per cent Komsomolsk content; first 10 kits stated to have been 100 per cent complete but, by 2000, China seeking more in this state to expedite deliveries. No licence has been granted for Chinese production of the AL-31 engine. China's first 50 Su-27s were initially based at Wuhan, with the 9th Fighter Regiment, 3rd Air Division though 24 subsequently moved to Liancheng, where 17 were damaged (three beyond repair) in an April 1997 hurricane. Two more were written off in accidents. According to Russian sources, 75 of 78 complete Su-27 K/UBKs had been delivered by end of 2002 and all 95 sets of components representing first J-11 batch by end of 2003. At that time, China had not ordered balance of 105 sets, nor shown signs of switching local manufacture to Su-30, as once expected.

Formation take-off by Su-27UB 'Flanker-Cs' of Russian Knights *(Yefim Gordon)* ***NEW***/1047024

Cockpit of Su-27SMK *(via Yefim Gordon)* ***NEW***/1047020

SU-27 CIS (USSR) PRODUCTION

Batch	Qty	Remarks
'Flanker-B' at Komsomolsk		
8	20	Full quantity unconfirmed
9	20	Including T10-30 (0906)
10	20	1003 with sidestick controller
11	20	Some to Ukraine
12	20	
13	20	Some to Ukraine
14	20	Some to Ukraine
15	20	Some to Ukraine
16	20	1602 to prototype Su-27M
17	20	Some to 582 IAP
18	20	
19	20	Some to 582 IAP and Ukraine
20	20	Some to 159 and 582 IAP
21	20	Some to 159 IAP and Ukraine
22	20	Some to 159 IAP
23	20	Some to 159 IAP
24	20	2404/T10-41; 2405 Su-27LMK; Some to 159 IAP
25	20	
26	20	Some to 159 and 582 IAP
27	20	Some to 159 and 582 IAP
28	20	
29	20	
30	20	Some to 61 IAP
31	20	Some to 237 GvTsPAT
32	20	
33	20	
34	20	Some to 61 IAP
35	20	
36	20	
37	20	3711 Su-27P; 3720 Su-27PD
'Flanker-C' at Irkutsk		
03	10	Some to Ukraine
04	10	
05	10	
06	10	
07	10	Some to 582 IAP
08	10	Some to 237 GvTsPAT
09	10	Some to 61 and 582 IAP
10	10	
11	10	Some to 582 IAP
12	10	Some to 582 IAP; 1201 first Su-27UBM
13	10	
14	10	
15	10	
16	10	Some to 237 GvTsPAT

Notes: Data from unofficial sources. Batches 9 and 10 may have contained fewer aircraft. IAP is Interceptor Air Regiment; GvTsPAT is Guards Centre for Display of Aviation Equipment

Vietnam received first batch of six (including one Su-27UB) in 1995; further six (two plus four) ordered, including two UBs delivered by airfreight on 1 December 1997 and second pair, representing balance of follow-on order, which destroyed in An-124 crash six days later. Latter replaced by single-seat aircraft, giving fleet of nine plus three; Su-30s also delivered. Upgrade expected, since Vietnam requires aircraft to use R-77 AAMs, Kh-29, Kh-31 and Kh-59 ASMs. Kazakhstan received 32 from Russia, including six in 1997, four in January 1999 and four during 2000 (with 12 then outstanding), as compensation for return of Tu-95s and for alleged environmental and ecological damage. Belarus has 22 or 23 (though these may have been sold-on, perhaps to Angola); Ukraine 66 to 70 and Uzbekistan 30; others reported y went to Armenia, Azerbaijan and Georgia; Syria has requested 14 (some reports suggest 17 in service others that only four were in service by mid-2000, two at Minkah AB and two at Damascus).

Ethiopia received the first of eight second-hand from Russia in late 1998. Eritrea obtained four Su-27Ss and two Su-27UBs from Ukraine in 2002. In late 1999, there were reports that Angola had taken delivery of eight Su-27s at Catumbela in August, with the balance of seven expected imminently. Pilots had reportedly trained in Belarus, the presumed source of the aircraft, though technical support came from Ukraine. Indonesia announced in March 2003 that two Su-27s and two Su-35s would be purchased immediately as prelude to acquisition of total of 48 'Flanker' versions over following four years. Order signed 24 April 2003 and two Su-27SKs (and two Su-30MKs) delivered 27 August 2003 to Iswahyudi. Further six SKs (and two Su-30MKs) to follow.

On 26 November 1995, the first of two Su-27s was delivered from Belarus to the USA inside an An-124 for an unknown purpose, believed to be support of military training exercises. The UK is reported by one source to have obtained a "Su-27P" from Ukraine in 2001.

DESIGN FEATURES: Developed to replace Yak-28P, Su-15 and Tu-28P/128 interceptors in APVO, for dual-role ground attack/air combat and to escort Su-24 deep-penetration strike missions; basic requirement was effective engagement of F-15 and F-16 and other future aircraft and cruise missiles; exceptional range on internal fuel made flight refuelling unnecessary until Su-24s received probes when more range was required in the escort role; external fuel tanks still not considered necessary; all-swept blended fuselage/mid-wing configuration, with long curved wing leading-edge root extensions, lift-generating fuselage, twin tailfins and widely spaced engines with wedge intakes; rear-hinged doors in intakes hinge up to prevent ingestion of foreign objects during take-off and landing; integrated fire-control system with pilot's helmet-mounted target designator; exceptional high-Alpha performance; basic wing sweepback 42° on leading-edge, 37° at quarter-chord; no dihedral or incidence.

FLYING CONTROLS: Four-channel analogue SDU-27 fly-by-wire, with no mechanical back-up; artificial feel; relaxed longitudinal stability; no ailerons; full-span leading-edge flaps and plain inboard flaperons controlled manually for take-off and landing, computer-controlled in flight; differential/collective tailerons operate in conjunction with flaperons and rudders for pitch and roll control; flight control system limits *g* loading to +9 and normally limits angle of attack to 30 to 35°; angle of attack limiter can be overruled manually for certain flight manoeuvres; large door-type airbrake in top of centre-fuselage.

STRUCTURE: All-metal, with extensive use of aluminium-lithium alloys and titanium but no composites; comparatively conventional three-spar wings; basically circular section semi-monocoque fuselage with load-bearing spine, sloping down sharply aft of canopy; cockpit high-set behind drooped nose; large ogival dielectric nosecone; long rectangular steel blast panel forward of gun on starboard side, above wingroot extension; two-spar fins and horizontal tail surfaces; uncanted vertical surfaces on narrow decks outboard of engine housings; fin extensions beneath decks form parallel, widely separated ventral fins.

LANDING GEAR: Hydraulically retractable tricycle type, made by Hydromash, with single wheel on each unit; KT-156D mainwheels turn 90° while retracting forward into wingroots;

SU-27 EXPORT PRODUCTION

Country	Type	Qty	First Aircraft	Remarks
China	Su-27SK	24	'05'	Batch 38 and 39; delivered from Komsomolsk 1991–92
	Su-27SK	14	'35'	Batch 39; 1995–96
	Su-27SK	95	'01'	All kits received by early 2004. Local assembly in progress
	Su-27UBK	2	'01'	Delivered 1991–92 from Irkutsk
	Su-27UBK	10	'03'	1995–96
	Su-27UBK	28	'51'	2000–02
Indonesia	Su-27SK	2	TS-2701	2003
		6		On order
Vietnam	Su-27SK	5	6001	
		4		Incl two replacements for Su-27UBKs
	Su-27UBK	1	8521	
		4		Two lost in airfreight delivery

Notes: China was to have assembled further 105 aircraft, possibly as Su-30s, but this now unlikely to proceed.

hydraulically steerable non-braking KN-27 nosewheel, with mudguard, also retracts forward; mainwheel tyres 1,300×350, pressure 12.25 to 15.70 bar (178 to 227 lb/sq in); nosewheel tyre 680×260, pressure 9.30 bar (135 lb/sq in); hydraulic carbon disc brakes with two-signal anti-skid system; electric brake cooling fan in each mainwheel hub; further brake in nosewheel; brake-chute housed in fuselage tailcone.

POWER PLANT: Two Saturn/Lyulka AL-31F turbofans, each 122.6 kN (27,557 lb st) with afterburning. Large spring-loaded auxiliary air intake louvres in bottom of each three-ramp engine duct near primary wedge intake; two rows of small vertical louvres in each sidewall of wedge, and others in top face; fine grille of titanium hinges up from bottom of each duct to shield engine from foreign object ingestion during take-off and landing.

By 2004, Salyut offering AL-31F upgrade for export in form of 130 kN (29,320 lb st) AL-31FM1, with 138 kN (31,085 lb st) AL-31FM2 in prospect. Simultaneously announced that AL-31F service life increased to 1,500 hours, with 1,000 hours guaranteed between overhauls.

Fuel in four integral tanks: three in fuselage and one split between each outer wing. Max internal fuel capacity approximately 11,775 litres (3,110 US gallons; 2,590 Imp gallons); normal operational fuel load 6,600 litres (1,744 US gallons; 1,452 Imp gallons). Higher figure represents internal auxiliary tank for missions in which manoeuvrability not important. No provision for external fuel tanks, except in versions where specifically indicated. Pressure or gravity fuelling. Flight refuelling capability optional; Su-27UB operated as buddy tanker during development of system.

ACCOMMODATION: Pilot only, on Zvezda K-36DM Series 2 zero/zero ejection seat, under large rear-hinged transparent blister canopy, with low sill; 14° view downward over the nose.

SYSTEMS: Automatically regulated cockpit air conditioning. Two independent, duplicated, hydraulic systems, pressure 275 bar (4,000 lb/sq in), for actuation of control surfaces, airbrake, landing gear, wheel brakes, air intake ramps and FOD screens. Electrohydraulic (flight control) parts of system quadruplicated. APU in top of rear fuselage for ground and emergency in-flight power. Pneumatic system pressure 210 bar (3,045 lb/sq in) for back-up landing gear extension avionics bay pressurisation and canopy operation and sealing. Electrical supply 27 V DC, 115 V and 200 V 400 Hz AC; two type NKBN-25 Ni/Cd batteries. AC supplied by two integral engine-driven generators with three-phase and single-phase converters. Gaseous oxygen for 4 flight hours.

AVIONICS: Systems integrated by NPO Elektroavtomatika.

Communications: R-800 UHF radio, R-864 HF, intercom and cockpit voice recorder. SO-69 ATC transponder; various IFF fits, according to production batch.

Flight: PNK-10 flight instrumentation and navigation suite encompasses the usual, traditional flight instruments (the IK-VSP altitude and speed data system), SAU-10 autopilot, Ts-050 computer, ARK-19 or ARK-20 ADF, a radio altimeter and A-317 Uron SHORAN. Aircraft capable of ICAO Cat. I autoland.

Radar: RLPK-27 radar sighting system with NIIP N001 Mech (Sword, NATO 'Slot Back') track-while-scan coherent pulse Doppler look-down/shoot-down radar (long-chord twist-cassegrain antenna diameter approximately 1.0 m; 3 ft 4 in) with search range of up to 54 n miles (100 km; 62 miles), tracking range 35 n miles (65 km; 40 miles), in forward hemisphere against MiG-21 size target (ability to track 10 targets and engage two simultaneously in current Su-27SK/UB upgrade configurations) and TsVM-80 computer.

Instrumentation: Integrated fire-control system enables radar, IRST and laser range-finder to be slaved to pilot's helmet-mounted target designator and displayed on wide-angle HUD; autopilot able to restore aircraft to right-side-up level flight from any attitude when 'panic button' depressed.

Mission: Duplex SUV-27 weapons targeting complex integrates radar and OEPS-27 electro-optic sighting system with Model 36Sh OLS-27 IR search/track (IRST) sensor, range 27 n miles (50 km; 31 miles), collimated laser range-finder, range 4.3 n miles (8 km; 5 miles), functioning through common optics in transparent housing forward of windscreen and NSTs-27 Schchel-3U helmet-mounted sight. Beryuza tactical/GCI datalink. Provision for reconnaissance pod on centreline pylon.

Self-defence: SPO-15LM Beryoza 360° radar warning antennas, outboard of each bottom air intake lip and at tail. Gardeniya active ECM jamming system. Three banks of APP-50 chaff/flare dispensers (total 96 cartridges) in bottom of long tailcone extension and top of tailsting. Tailcone widened to provide extra chaff/flare dispensers from batch 18 (mid-1987).

ARMAMENT: One 30 mm Gryazev/Shipunov 9A-4071K GSh-30-1 gun in starboard wingroot extension, with 150 rounds. Up to 10 AAMs in air combat role, on tandem pylons under fuselage between engine ducts, beneath each duct, under each centre-wing and outer-wing, and at each wingtip. Typically, two short-burn semi-active radar homing R-27R (NATO AA-10A 'Alamo-A') in tandem under fuselage; two short-burn IR homing R-27T (AA-10B 'Alamo-B') on centre-wing pylons; and long-burn semi-active radar homing R-27ER (AA-10C 'Alamo-C') or IR R-27ET (AA-10D 'Alamo-D') beneath each engine duct. The four outer pylons carry either R-73A (AA-11 'Archer') or R-60 (AA-8 'Aphid') close-range IR AAMs. R-33 (AA-9 'Amos') AAMs optional in place of AA-10s. Up to eight 500 kg bombs, sixteen 250 kg bombs or four launchers for S-8, S-13 or S-25 rockets.

DIMENSIONS, EXTERNAL (Su-27):

Wing span ... 14.70 m (48 ft 2¾ in)
Wing aspect ratio ... 3.5
Length overall, excl nose probe ... 21.94 m (71 ft 11½ in)
Height overall ... 5.93 m (19 ft 5½ in)
Fuselage: Max width ... 1.50 m (4 ft 11 in)
Tailplane span ... 9.88 m (32 ft 5 in)
Distance between fin tips ... 4.30 m (14 ft 1¼ in)
Wheel track ... 4.36 m (14 ft 3¾ in)
Wheelbase ... 5.88 m (19 ft 3½ in)

AREAS:

Wings, gross ... 62.04 m^2 (667.8 sq ft)
Wing leading-edge flaps (total) ... 4.60 m^2 (49.51 sq ft)
Flaperons (total) ... 4.90 m^2 (52.74 sq ft)
Fins (total) ... 11.90 m^2 (128.10 sq ft)
Rudders (total) ... 3.50 m^2 (37.67 sq ft)
Horizontal tail surfaces (total) ... 12.30 m^2 (132.40 sq ft)

WEIGHTS AND LOADINGS (A: Su-27, B: Su-27UB, C: Su-27SK, D: Su-27SMK):

Operating weight empty: A ... 16,380 kg (36,110 lb)
B ... 17,500 kg (38,580 lb)
Max fuel weight: A ... 9,400 kg (20,723 lb)
Normal T-O weight: A ... 23,000 kg (50,705 lb)
B ... 24,140 kg (53,220 lb)
D, with two R-73 and two R-27 missiles ... 23,700 kg (52,250 lb)
Max T-O weight: A, C, D ... 33,000 kg (72,750 lb)
B ... 33,500 kg (73,850 lb)
Max wing loading: A, C, D ... 531.9 kg/m^2 (108.94 lb/sq ft)
B ... 540.0 kg/m^2 (110.60 lb/sq ft)
Max power loading: A, C, D ... 135 kg/kN (1.32 lb/lb st)
B ... 124 kg/kN (1.22 lb/lb st)

PERFORMANCE:

Max level speed::
at height:
A, B ... M2.35 (1,350 kt; 2,500 km/h; 1,550 mph)
C ... M2.3 (1,319 kt; 2,443 km/h; 1,518 mph)
D ... M2.17 (1,241 kt; 2,300 km/h; 1,429 mph)
at S/L:
A, B ... M1.1 (725 kt; 1,345 km/h; 835 mph)
C, D ... M1.14 (756 kt; 1,400 km/h; 870 mph)
Stalling speed: A ... 108 kt (200 km/h; 125 mph)
Acceleration (D):
from 324 to 594 kt (600 to 1,100 km/h; 373 to 683 mph) ... 15 s
from 594 to 701 kt (1,100 to 1,300 km/h; 683 to 808 mph) ... 12 s
Rate of roll: A ... approx 270°/s
Service ceiling: A, D ... 18,000 m (59,060 ft)
B ... 17,500 m (57,420 ft)
C ... 18,500 m (60,700 ft)
T-O run: A ... 450 m (1,475 ft)
B ... 550 m (1,805 ft)
C, D ... 650 m (2,135 ft)
Landing run: A, D ... 620 m (2,035 ft)
B ... 700 m (2,300 ft)
Combat radius: A, B ... 810 n miles (1,500 km; 930 miles)
D ... 840 n miles (1,560 km; 970 miles)
Range with max fuel:
A, C ... 1,985 n miles (3,680 km; 2,285 miles)
B ... 1,620 n miles (3,000 km; 1,865 miles)
Range (D):
at S/L, internal fuel ... 745 n miles (1,380 km; 857 miles)
at S/L, with external tanks (dropped when empty) ... 894 n miles (1,656 km; 1,029 miles)
at height, internal fuel ... 2,046 n miles (3,790 km; 2,355 miles)
at height, with external tanks (dropped when empty) ... 2,370 n miles (4,390 km; 2,727 miles)
max, with flight refuelling ... 2,807 n miles (5,200 km; 3,231 miles)
g limit (operational): A, B, C, D ... +9

UPDATED

SUKHOI SU-34

NATO reporting name: Fullback
Export designation: Su-32
Design bureau designation: Su-27IB

TYPE: Attack fighter.

PROGRAMME: Side-by-side two-seat long-range fighter-bomber (*istrebitel bombardiroshchik*) Su-27 variant intended as tactical strike/attack replacement for Su-24 and Su-25; project designation T10V; redesignated Su-34 by Sukhoi ("to stress father-and-son relationship" to Su-24), but Russian Air Forces retained Su-27IB until late 2003, when Su-34 adopted, with Su-32 reserved for export. Su-32FN/Su-32MF assigned to proposed export versions but, in 2000, all Su-34s were redesignated Su-32.

Conceptual design ordered 21 January 1983, production authorised 19 June 1986. Designed under the direction of Rollan Martirosov. Prototype (T10V-1 '42') built in Sukhoi's own workshops and first flown 13 April 1990; first seen in Tass photograph showing this aircraft approaching (but not landing on) the carrier *Admiral of the Fleet*

Plan view of Sukhoi Su-32 *(Yefim Gordon)* ***NEW***/1047010

Sukhoi Su-32 T10V-7 *(Yefim Gordon)* ***NEW***/1047011

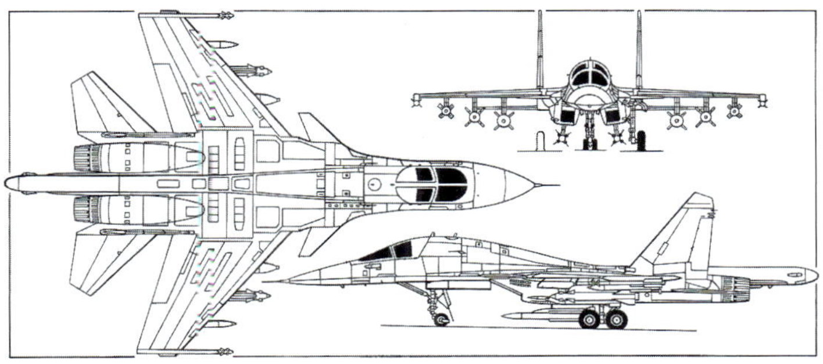

Three-view drawing of Sukhoi Su-27IB twin-turbofan theatre bomber *(Jane's/Mike Keep)* 0507380

Kuznetsov; described as deck landing trainer, but no wing folding or deck arrester hook; although with foreplanes and twin nosewheels like Su-27K. Designation Su-27KU quoted, though dedicated side-by-side carrier trainer then officially known as T10KM-2 or Su-27KM-2 for which the new side-by-side cockpit had been designed. Russian unofficial name 'Platypus'; exhibited to Russian Federation and Associated States (CIS) leaders at Machulishche airfield, Minsk, February 1992, with simulated attack armament on 10 external stores pylons (under each intake duct, on each wingtip, three under each wing); Kh-31A/P (AS-17 'Krypton') ASMs under ducts, R-73A (AA-11 'Archer') AAMs on wingtips; a 500 kg laser-guided bomb inboard, TV/laser-guided Kh-29 (AS-14 'Kedge') ASM on central pylon and RVV-AE (R-77; AA-12 'Adder') AAM outboard under each wing.

Production originally planned for Irkutsk, but eventually located at Novosibirsk. First series production aircraft flew 28 December 1994. Four were in assembly at Novosibirsk by early 1997. Twelve were once scheduled for delivery by 1998; intention was to replace all Su-24s by 2005; reconnaissance and electronic warfare versions reportedly under development. In September 2000, following increase in funding, Russian Air Forces C-in-C predicted first Su-32 deliveries in 2004, although by early 2002 this had slipped further to "after 2005". Statement of August 2003 indicated delivery between 2006 and 2010 of "a batch" of aircraft to Russian Air Forces, perhaps numbering "several dozen", although first receipts in 2005 or 2006, intended to be 10 aircraft of which half to go to bomber regiment in Chita region of Siberia (possibly 48th Division of 14th Air Force). In mid-2001, Sukhoi predicted completion of State testing of Su-32 by end of 2001, using five aircraft based at Akhtubinsk and Zhukovsky. Trials tempo intensified in 2002, work including tests of Sh141 mission avionics suite as replacement for earlier systems found to have been unsatisfactory. However, first phase of State Tests not completed until mid-2003, when second phase begun; this involved three aircraft at Akhtubinsk test centre and four at Zhukovsky.

Side-by-side cockpit to form basis of proposed Su-30-2 long-range interceptor and Su-33UB (Su-27KUB) carrier trainer (which see).

CURRENT VERSIONS (SPECIFIC): **T10V-1 '42'**: First flying prototype; detailed above. Converted from Su-27UB airframe by Sukhoi OKB workshops, with new nose built at Novosibirsk, and reportedly fitted with Su-33 main landing gear.

T10V-2 '43': Flown 18 December 1993; first aircraft to be built at Novosibirsk; introduced twin mainwheel bogies. Sometimes described as first production Su-34, in that it had Su-35-type four-hardpoint wing panels and larger internal fuel cells, reinforced wing centre-section, new main landing gear and fixed-geometry engine air intakes. However, Sukhoi now identifies this aircraft as Su-32 prototype, designed to meet Russian Air Forces requirement of 1998.

T10V-3: Static test air frame

T10V-4 '44': First flown late 1996. Reported at Leninets radar plant, Pushkino, early 1997; first with full avionics and weapons systems, except EW package. Exhibited at Paris Air Show, June 1997. Also became known as Su-32.

T10V-5 '45': First Su-27IB with full Leninets mission avionics fit. Sometimes described as first series-produced T10V. First flew 28 December 1994.

T10V-6 '46': First flown January 1998. Being upgraded in 2005 to full production standard, matching T10V-8.

T10V-7 '47': First flown 22 December 2000. Shown at Moscow in August 2003. Being upgraded in 2005 to full production standard, matching T10V-8.

T10V-8: "First flight" 20 December 2003, but may have flown shortly before. Full range of avionics and sensors.

CURRENT VERSIONS (GENERAL): Proposed tactical jammer to replace Yak-28PP and Su-24MP.

Su-32: Export version.

Su-34: Russian Air Forces version.

Details generally as for single-/tandem-seat Su-27, except those below.

DESIGN FEATURES: One third heavier empty weight, with 50 per cent increase in MTOW, 30 per cent increase in internal capacity, 10 per cent increase in mid-section. Completely new and wider front fuselage built as titanium armoured tub, 17 mm ($^{11}/_{16}$ in) thick; armour adds 1,480 kg (3,262 lb); new EFIS cockpit containing two seats side by side; side-by-side arrangement avoids some duplication of controls and instruments, while promoting better crew co-operation. New avionics suite integrated by Ramenskoye Instrument-making Design Bureau; wing extensions taken forward as chines to blend with dielectric nose housing nav/attack and terrain-following/avoidance radar; deep fairing behind wide humped canopy; small foreplanes; louvres on engine air intake ducts reconfigured; new landing gear; broader chord and thicker tailfins, containing fuel; no ventral fins; and a longer, larger diameter tailcone. This has been raised and now extends as a spine above the rear fuselage to blend into the rear of the cockpit fairing. It houses at its tip a rearward-facing radar to detect aircraft approaching from the rear.

LANDING GEAR: Retractable tricycle type; strengthened twin nosewheel unit with KN-27 wheels, tyre size 680×260, farther forward than on Su-27 and retracting rearward; main units have small tandem KT-206 wheels with tyres size 950×400, carried on links fore and aft of oleo. New down-lock fairings. Twin cruciform brake-chutes repositioned in spine to rear of spine/fairing juncture.

POWER PLANT: Two Saturn/Lyulka AL-31F turbofans; each 74.5 kN (16,755 lb st) dry and 122.6 kN (27,557 lb st) with afterburning. Later, two AL-31FM or AL-35F turbofans, each 125.5 to 137.3 kN (28,220 to 30,865 lb st) with afterburning. Production version to use 175 kN (39,240 lb st) AL-41F with FADEC and TVC, according to some sources. Additional fuel in tailfins and increased capacity No. 1 tank raising total to 12,100 kg (26,676 lb) plus provision for three external tanks totalling 7,200 kg (15,873 lb). Retractable flight refuelling probe beneath port windscreen.

ACCOMMODATION: Two crew side by side on modified K-36DM zero/zero ejection seats with built-in massage function. Access to cockpit via built-in extending ladder to door in nosewheel bay; area protected with 17 mm ($^2/_3$ in) thick titanium armour; lavatory and galley with air-stove inside deep fuselage section aft of cockpit.

AVIONICS: *Radar:* Leninets B004 multifunction phased-array radar with high resolution; rearward-facing radar in tailcone.

Instrumentation: Colour CRT, multifunction displays and helmet-mounted sight for pilot and navigator.

Mission: Built-in UOMZ EO IRST sighting system with TV and laser channels, optimised for air-to-ground use. Separate Geofizika podded thermal imaging system planned. New Argon main computer. Sorbtsya active ECM jamming pods under test on Su-27IB prototype 1995.

Self-defence: TsNIRTI electronic warfare system.

ARMAMENT: One 30 mm GSh-301 gun with 150 rounds. Twelve pylons for high-precision self-homing and guided ASMs, comprising Kh-59ME Ovod, Kh-31P/Kh-31A(P) (AS-17 'Krypton'), Kh-29T/TE/L (AS-14 'Kedge') and Kh-41/3M80 Moskit; and KAB-500 and KAB-1500 laser-guided bombs with ranges of 0 to 135 n miles (250 km; 155 miles); R-27 (AA-10 'Alamo'), R-73 (AA-11 'Archer') and RVV-AE (R-77; AA-12 'Adder') AAMs.

DIMENSIONS, EXTERNAL:

Wing span .... 14.70 m (48 ft 2¾ in)
Wing aspect ratio 3.5
Foreplane span .... 6.40 m (21 ft 0 in)
Length (without probe) .... 23.335 m (76 ft 6¾ in)
Height overall .... 6.50 m (21 ft 4 in)
Wheel track .... 4.40 m (14 ft 5¼ in)
Wheelbase .... 6.60 m (21 ft 7¾ in)

AREAS:

Wings, gross .... 62.00 m² (667.4 sq ft)

WEIGHTS AND LOADINGS (Su-32):

Max external stores .... 8,000 kg (17,637 lb)
T-O weight: normal .... 38,240 kg (84,304 lb)
max .... 44,350 kg (97,774 lb)

PERFORMANCE:

Max speed:
at height .... M1.8 (1,025 kt; 1,900 km/h; 1,180 mph)
at S/L .... M1.14 (756 kt; 1,400 km/h; 870 mph)
Service ceiling .... 15,000 m (49,220 ft)
Combat radius (internal fuel):
hi-hi-hi .... 594 n miles (1,100 km; 683 miles)
lo-lo-lo .... 324 n miles (600 km; 372 miles)
Range with max internal fuel .... 2,159 n miles (4,000 km; 2,485 miles)

UPDATED

SUKHOI SU-30 (SU-27PU)

NATO reporting name: Flanker-F Variant 1

TYPE: Air superiority fighter.

PROGRAMME: Design started 1986; proof-of-concept T10PU (T10U-2 'Aircraft 01-02'; see Su-27 entry) first flew 6 June 1987; construction of two prototypes (T10PU-5 '05' and T10PU-6 '06', converted from Su-27UBs T10U-5 and 0201/T10U-6) began at Irkutsk in 1987, as Su-27PU; first flown 31 December 1989; prototype flew 7,252 n miles (13,440 km; 8,351 miles) in 15 hours 42 minutes non-stop during round trips Moscow-Novaya Zemlia-Moscow and Moscow-Komsomolsk-Moscow. First pre-series Su-27PU flew at Irkutsk on 14 April 1992; initial two aircraft (27596 and 27597 c/ns 0101 and 0102), without military equipment, delivered to 'Test Pilots' aerobatic team at Zhukovsky flight test centre, possibly as Su-27PUDs. By 1999, 27597 was flight-testing MFI-68 152 × 203 mm (6 × 8 in) LCDs for Russian 'Flanker' upgrade programmes while 27596 had been used in support of the Su-30MK programme, renumbered '603'.

CURRENT VERSIONS: **Su-30:** (Sukhoi T10PU, *Izdelie* 10-4PU): Unofficial OKB designation for basic two-seat long-range interceptor for Russian Air Forces (to which it is still the Su-27PU); deliveries under way by 1996 to 54 Interceptor Air Regiment at Savostleyka advanced training base, though production very limited, and unit relies heavily on Su-27 and Su-27UB. Apparently, five Su-30s are in frontline use (Red 50, 51, 52, 53 and 54). Designed for mission of 10 hours or more with two in-flight refuellings; systems proved for extended duration sorties, including group missions with four Su-27s; only Su-30 would operate radar, enabling it to assign targets to Su-27s by radio datalink; can carry bombs and rockets but not guided air-to-surface weapons. Su-27UB training capability retained. Canards and thrust vectoring to be optional (see Su-30MKI). Export designation **Su-30K** (T10-4PK). Situation confused by tendency to describe all current Irkutsk-built two-seaters as Su-30s, even standard Su-27UB trainers and by emergence of upgrade using MiG-29SMT cockpit, using the same Su-30K designation. This Su-30K was reported to have been tested at Akhtubinsk by June 2000 but may be the same as the Su-30KM described below. Reported in early 2002 that 'Russian Knights' aerobatic team to convert to Su-30s from Su-27s.

Su-30KI: Single-seat configuration for Indonesia, subsequently offered more widely as an upgrade. Described in main Su-27 entry.

Su-30I: One foreplane-equipped Su-27PU prototype only; served as Su-30MKI prototype; believed offered as upgrade for Su-27 and as naval trainer, but neither taken up.

Su-30K-2: Variant of Su-33UB and described in that entry.

Su-30KN: An undelivered production Su-30 (0302 '302') first flew in early March 1999 as the testbed (also referred to as Su-30K and Su-30KM) for a Russian Air Forces upgrade of UB Su-27s and Su-30s. Prime purpose of 'Project 302' was to convert fighter into multirole attack aircraft by adding terrain-mapping and moving target indication to the N001 radar. This achieved by adding new bypass circuit (*obvodnoy kanal*, abbreviated to *Oko* — 'eye'). Cockpit initially unchanged, apart from MFI-55 127 × 127 mm (5 × 5 in) MFDs, an SUV-30K weapons control system comprising a new MVK computer added to existing SUV-27 system (permitting new types of AAM and ASM to be carried) and an A737 GPS; joint venture is undertaken by Sukhoi, Irkutsk (IAPO) and Russkaya Avionika and has high commonality with MiG-29SMT upgrade. Prototype tested at Air Forces Research Institute from mid-1999. State certification awarded 9 November 2001. Added weapons are Kh-29T short-range, or two Kh-59ME long-range TV-guided ASMs; up to six KAB-500KR bombs; four Kh-31P ARMs; and six Kh-31P/A anti-ship missiles. By late 2000, the Su-30KM designation had been replaced by Su-30KN and potential maritime capabilities were being stressed, including possible future compatibility with Kh-59, Yakhont and Alfa ASMs. Up to November 2001, upgrades comprised prototype '302' and three air force aircraft ('51' being the first), all used for trials and modernised at IAPO's expense; order signed by Russian Air Forces in October 2001 for eight upgrades in 2002, 10 in 2003 and 12 in 2004. However, by August 2003, only five upgraded aircraft had been redelivered. Su-27UBM and Su-27SM are parallel programmes. 12 Su-30Ks sought by Vietnam believed to be to this standard with Kh-29, Kh-31 and Kh-59 ASMs, KAB-500 PGMs and R-77 AAMs.

Su-30KNM: Subsequent modifications to 0302 involve larger (152 × 203 mm; 6 × 8 in), MFI-68 screens (three for pilot and four for WSO), Pero ('Feather') phased-array radar and equivalent of MIL-STD-1553B databus. Project began in 2002. Upgrade of air force aircraft could begin 2005.

Su-30M: (Sukhoi T-10PM): Multirole version; described separately.

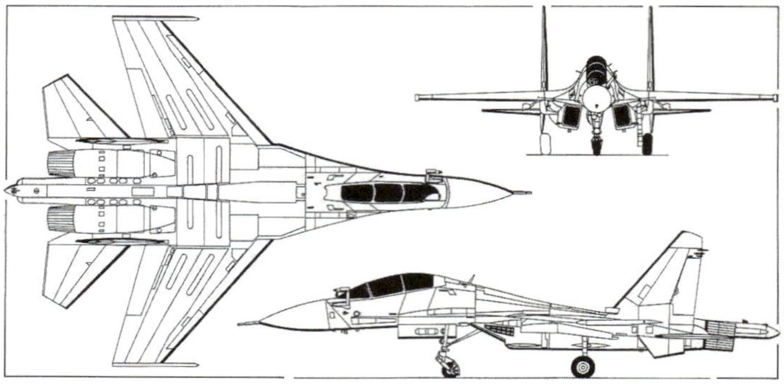

Sukhoi Su-30 two-seat long-range combat aircraft *(Paul Jackson)* 0507633

CUSTOMERS: See table. Indian aircraft to be upgraded to Su-30MKI in 2004, according to one report.

DESIGN FEATURES: Development of Su-27/27UB, with latter's tandem seating and new avionics, and without Su-35's advanced radar, foreplanes (in basic version), advanced control system and new power plant. Designed for effective engagement of fighters at long distances from base, and to destroy bombers and intercept cruise missiles. Integral configuration similar to Su-27UB, with unstable aerodynamic characteristics. Automatic control system standard.

FLYING CONTROLS: As Su-27UB.

STRUCTURE: As Su-27UB.

LANDING GEAR: As Su-27UB.

POWER PLANT: As Su-27UB, but flight refuelling probe and buddy refuelling capability standard.

ACCOMMODATION: Two crew in tandem in identical cockpits, on K-36DM zero/zero ejection seats, with rear seat raised.

SYSTEMS: As Su-27UB, except gaseous oxygen for 10 hours' flight.

AVIONICS: *Radar:* NIIP N001 Myech ('Slot Back') coherent pulse Doppler look-down/shoot-down radar offered, detection range up to 54 n miles (100 km; 62 miles), tracking range 35 n miles (65 km; 40 miles); ability to track 10 targets and engage two simultaneously; probably not available on current in-service aircraft.

Flight: New navigation system based on GPS, Loran and Omega.

SU-30 PROTOTYPES AND PRODUCTION

Batch	Qty	Identity	Remarks
Russia			
-	(1)	T10PU-5 '05'	Converted from T10U-5; became Su-30MKK
-	(1)	T10PU-6 '06'	Converted from T10U-2; became Su30MKI
1	1	'596'	Became '603' for Su-30MK programme
1	1	'597'	Testbed at Zhukovsky (LII)
2/3	5	'50' — '54'	54 IAP PVO, Savostleyka
3	1	'302'	Upgrade prototype (Su-30KN)
India			
PU	8	SB001-008	Delivered March-April 1997
K	10	SB009-018	Ordered September 1998. Delivered October 1999

Sukhoi Su-30 of 54 Regiment *(Yefim Gordon)* ***NEW***/1047009

Instrumentation: Integrated fire-control system enables radar, IRST and laser range-finder to be slaved to pilot's helmet-mounted target designator and displayed on wide-angle HUD.

Mission: Provision for fitting foreign-made airborne and weapon systems at customer's request.

Self-defence: SPO-15LM Beryoza 360° radar warning system; chaff/flare dispensers.

ARMAMENT: One 30 mm GSh-301 gun, with 150 rounds; 12 hardpoints for up to six R-27R1E and R-27T1E (AA-10 'Alamo') radar homing and IR long-range AAMs, and six R-73E (AA-11 'Archer') IR close-range AAMs; alternative RVV-AE (R-77; AA-12 'Adder') AAMs; unguided bombs or rockets as Su-27; reconnaissance or EW pods.

DIMENSIONS, EXTERNAL: As Su-27UB

WEIGHTS AND LOADINGS:

Fuel weight: normal	5,090 kg (11,222 lb)
max	9,400 kg (20,723 lb)
Max combat load	8,000 kg (17,635 lb)
Max external stores	8,000 kg (17,635 lb)
Normal T-O weight	24,550 kg (54,123 lb)
Max T-O weight	33,000 kg (72,752 lb)

PERFORMANCE:

Max level speed:	
at height	M2.35 (1,350 kt; 2,500 km/h; 1,550 mph)
at S/L	M1.14 (756 kt; 1,400 km/h; 870 mph)
Service ceiling	17,500 m (57,420 ft)
T-O run	550 m (1,805 ft)
Landing run	700 m (2,300 ft)
Combat range: with max internal fuel	1,619 n miles (3,000 km; 1,865 miles)
with one in-flight refuelling	2,805 n miles (5,200 km; 3,230 miles)
g limit	+8.5

UPDATED

SUKHOI SU-30M

NATO reporting name: Flanker

TYPE: Multirole fighter.

PROGRAMME: Design started 1991; demonstration and development work by Su-27UB 0806 '321' and 0403 '56'. 'Blue 56' first flew in Su-30M configuration on 14 April 1992, but it was not then canard-equipped and did not have full avionics package. Conversion of first true prototype ('06') began 1993, using 0201/T10PU-6; 0101 '603' (formerly first prototype Su-27PU-5) first demonstrated at Berlin Air Show 1994; thrust vectoring and canards under development as options 1997. Canards and AL-37PP engines first flown on '56' 1 July 1997, this regarded by Sukhoi as first flight of Su-30MK; second prototype (06) flew 23 April 1998. In production for India.

Su-30M designation initially associated with canard fitment. However, this now appears to identify multirole aircraft with upgraded airframe capable of 38,800 kg (85,539 lb) MTOW, irrespective of aerodynamic configuration. Chinese Su-30MKKs are described by Sukhoi as Su-30MK family members, despite having twin nosewheels and square-topped fins previously regarded as Su-35 features.

CURRENT VERSIONS: **Su-30M** (Flanker-F Variant 2): Basic version.

Su-30MK: Irkutsk-built. As Su-30M, for export.

Su-30MK: Designation re-used for advanced two-seater derived from Su-27SK by KnAAPO plant and combining two-seat Su-30 concept with avionics, canards and thrust vectoring of Su-37. Design, under A F Barkovsky, began 1994. Known internally (by KnAAPO) as Su-35UB (T10UBM) and Su-37UB. First and second true Su-30MKs, 01 (produced through the retrofit of canards to '56') and 06 (converted from T10PU-6 in Sukhoi OKB's own workshop) first flown on 1 July 1997 and 23 April 1998, respectively. Equipped with foreplanes and AL-37FP thrust-vectoring engines; demonstrated to Indian officials at Zhukovsky, 15 June 1998 as Su-30MK-1 and Su-30MK-6, respectively. Su-30MK-1 had the new twin-wheel nosegear, which may become a feature of the production MKI, but was lost while displaying at Paris Air Show on 12 June 1999. Additionally, 01 was exhibited at Aero India, December 1998. AL-37FP power plant, as specified for India, extends length by 40 cm (15¾ in) and incurs weight penalty of 110 kg (243 lb), engine life remaining unchanged at 5,000 hours (1,000 hours TBO). Nozzle movement is 15° up or down. Flight control system helps pilot to set power and thrust vector

First Indonesian Su-30 *(TNI AU)* ***NEW***/1043411

for each engine, according to required manoeuvre. Indian radar choice will be between improved version of Phazotron N010 (Zhuk-27) and NIIP N011M multimode, dual-frequency radar with electronically scanned antenna. Expected to be known as Su-30MKR if produced (at Irkutsk) for Russian Air Forces.

Su-30MKI (Flanker-H): Version for India in four configurations, initially referred to as Su-30MKI, MKII, MKIII and MKIV. Indian contract signed 30 November 1996. First eight delivered March 1997 to basic Su-30PU (Su-30K or even Su-37UB) standard, with AL-31F engines; eight for 1998 delivery were expected to have French Sextant avionics including VEH 3000 HUDs, high-resolution colour LCD MFDs, a new flight data recorder, a Totem ring laser gyro dual INS with embedded GPS, Israeli EW equipment, a new UOMZ OLS-30 electro-optic targeting system and rearward-facing radar in tailcone; 12 originally to have been delivered 1999 were to add canards, as on Su-37; final 12 (originally scheduled for delivery in 2000) were to have AL-37FP engines with single-axis thrust-vectoring nozzles inclined outwards 32° from the centreline for improved yaw control, especially in single-engined case. AL-37PP claimed to offer 3-D thrust vectoring, with nozzle actuation via the fuel, and not the hydraulic, system.

Completion and delivery of balance of 32 was repeatedly delayed; decision taken that all these would be completed to final standard before delivery. (In interim, India contracted on 18 December 1998 for 10 standard Su-30Ks which had been cancelled by Indonesia; all had been delivered by October 1999.) First prototype full-specification Su-30MKI flew at Irkutsk on 26 November 2000; Irkutsk built four prototypes, last of which handed over to Sukhoi design bureau on 11 August 2001.

First production aircraft flew 28 December 2001. Under renegotiated contract, IAF to receive six full-specification aircraft by 2002 and balance of 26 in batches beginning 18 to 24 months later; thereafter eight 1997-delivery aircraft will be raised to full standard. First two departed Irkutsk inside An-124 transport 22 June 2002. However, 'full-specification' still being received in three standards, first 10 being Stage I; these 10 formally accepted into service by 20 Squadron at Pune 27 September 2002. Stages II and III, (12 in 2003 followed by 10 in 2004), to introduce additional weapons and upgraded flight control system. Licensed production of 140 (since reduced to 120) Su-30MKIs by HAL was agreed in September 2000, followed by contract signature in Irkutsk on 28 December 2000. Sometimes referred to by KnAAPO as Su-35UB.

Su-30MKK ('Flanker-G'): Second K stands for *Kitaya*, or China. Two-seat, multirole version, with an N001VE radar (with expanded air-to-ground capabilities, including mapping); 'glass cockpit' with two 178 × 127 mm (7 × 5 in) MFI-9 colour LCD MFDs in front, plus single MFI-9 and 204 × 152 mm (8 × 6 in) MFI-10 in rear; ILS-31 HUD; A737 GPS; expanded EW capability; provision for various new TV- and EO-based targeting pods.

First Su-27PU (T10PU-5) was refurbished and rebuilt to serve as an Su-30MKK development aircraft, first flying in its new guise on 9 March 1999. First KnAAPO-built prototype ('501') flew 19 May 1999 (sometimes reported as 20 February 1999), with second ('502', in basic Chinese camouflage scheme but marked simply as Su-30MK, not MKK) following later in 1999. '501' and '502' representative of planned production configuration, with tall, flat-topped Su-35-type tailfins, retractable in-flight refuelling probes and (according to some sources) Su-35-type radomes. Further pair ('503' and '504') built by mid-2001.

China ordered 38 KnAAPO-built Su-30MKKs and placed supplementary order (initially quoted as 24, but later known to be 38) in June 2001. First batch of 10 Su-30MKKs left

Su-30MKK testbed '504', used to develop the MKK2 version *(Yefim Gordon)* ***NEW***/0558791

Chinese Air Force montage showing Sukhoi Su-30MKKs ***NEW***/1047003

Russia on delivery to China on 20 December 2000; nine followed in March/April 2001 and 10 more on 21 August. Later Chinese aircraft, known as Series III, will switch to Zhuk-MS radars. Russian weapon deliveries began in January 2001 with Kh-59ME (AS-18 'Kazoo'), Kh-29T (AS-14 'Kedge') and Kh-31P (AS-17 'Krypton') ASMs and KAB-500Kr guided bombs. Early Su-30MKKs variously reported with 'Three Swords' Air Regiment at Yuxikou, near Nanjing or at Wuhu, Anhui; those delivered in August 2001 were supplied to Cangzhou, Hebei. Chinese local assembly of Su-30MKK (possibly as J-13) was once expected, but now in doubt.

Su-30MKK2: First shown at MAKS, Moscow, August 2003; optimised for naval operations; development work undertaken by some of four retained Chinese Su-30MKKs ('501' to '504'). Improved precision attack capability and expanded command and control potential. Implied capability with Kupol (unconfirmed) M400 reconnaissance pod (carrying various options including SLAR, TV or LOROP); and UOMZ Sapsan-E electro-optic/laser targeting pod. Improved version of NIIP N001VE radar (simultaneous aerial engagements and ground mapping). Other weapons include KH-31A, Kh-29T/EE/L, KAB-500Kr/1500Kr, R-27, R-73 and RVV-AE. Deliveries began (six aircraft) February 2004; further 18 due before end of 2004.

Su-30MKK3: Further Chinese variant for naval use; deliveries due from 2005. MTOW 38,000 kg (83,775 lb); NIIR Zhuk-MSE radar; Kh-59ME and Kh-31A anti-ship missiles. Initial order for 24 aircraft.

Su-30MKM: Malaysian version; US$900 million contract initialled 19 May 2003. Manufacture by Irkut. Avitronics of South Africa contracted in April 2004 to supply 18 multisensor warning systems.

CUSTOMERS: Indian Air Force initial US$1.8 billion order for 40 signed 30 November 1996; deliveries from Irkutsk began to No. 24 'Hunting Hawks' Squadron at Pune in March 1997; declared operational 11 June 1997; 10 more ordered September 1998, increasing total to 50, including 18 unmodified Su-30Ks for training. First MKIs to 20 'Lightning' Squadron in May 2002. Option taken up on licensed production of 140 more by HAL, India; however statement of November 2002 noted IAF had reduced quantity to 120, but this later discounted. Final deliveries in December 2004.

On 29 August 1997, Indonesia signed for eight 'single-seat Su-30s' and four two-seat, but this was cancelled on 9 January 1998. Interest reportedly renewed in 2001, and in 2003 plan to acquire 48 of 'Flanker' family was announced, initial purchase being two each of Su-27 and Su-30. Malaysia ordered 18 on 19 May 2003 and Vietnam contracted for four (plus eight options) on 3 December 2003, the four Vietnamese delivered in December 2004.

COSTS: Indian aircraft quoted as US$20 million each (flyaway, 1998), with US$8 million extra per aircraft to integrate Indian-specific systems and avionics. Chinese order for 40 Su-30MKKs valued at US$1.5 billion (2002).

Description refers to two-seat Su-30MK which generally as for Su-30, except as follows:

DESIGN FEATURES: Improvement on combat capabilities of Su-30 by compatibility with high-precision guided air-to-surface weapons with standoff launch range up to 65 n miles (120 km; 75 miles), in addition to Su-30's ability to engage two airborne targets simultaneously.

POWER PLANT: Two Saturn/Lyulka AL-35F turbofans, each 123 kN (27,558 lb st).

AVIONICS: In addition to standard Su-30 systems, Su-30M has more accurate navigation system, a TV command guidance system, a guidance system for anti-radiation missiles, a larger monochrome TV display system in rear cockpit for ASM guidance, and ability to carry one or two pods, typically for laser designation or ARM guidance in association with Pastel RWR and APK-9 datalink. Western avionics, guidance pods and weapons can be fitted optionally. Sextant Avionique package for Indian aircraft includes VEH3000 or Elop HUD, Totem or Sigma 9SN/MF INS/GPS and liquid-crystal multifunction displays (six 127 × 127 mm; 5 × 5 in MFD 55 and one 152 × 152 mm; 6 × 6 in MFD 66 per aircraft).

ARMAMENT: One 30 mm GSh-301 gun, with 150 rounds; 12 external stations for 8,000 kg (17,635 lb) of stores, including FAB-250, FAB-500, OFAB-250-270, OFAB-100-120 and guided KAB-500KR and KAB-1500KR bombs; B-8M-1 (20 × 80 mm), B-13L (5 × 130 mm) and O-25 (single 266 mm) rocket packs; up to six R-27ER (AA-10C 'Alamo-C'), R-27ET (AA-10D 'Alamo-D') or RVV-AE (R-77; AA-12 'Adder') medium-range AAMs; or two R-27ETs and six R-73E (AA-11 'Archer') IR homing close-range AAMs; and a variety of air-to-surface weapons such as four ARMs, six guided bombs or short-range missiles with TV homing, six laser homing short-range missiles, or two long-range missiles with TV command guidance; these include Kh-29L/T (AS-14 'Kedge'), Kh-31A/P (AS-17 'Krypton') and Kh-59M; (AS-18 'Kazoo') with APK-9 pod.

DIMENSIONS, EXTERNAL: As Su-27 except:
Height overall .... 6.355 m (20 ft 10¼ in)

WEIGHTS AND LOADINGS (Su-30MK):
Weight empty .... 17,700 kg (39,022 lb)
Max external weapon load .... 8,000 kg (17,637 lb)
Max internal fuel weight .... 9,640 kg (21,253 lb)
T-O weight: normal
24,960 kg (55,027 lb) max .... 34,500 kg (76,059 lb)
overload .... 38,800 kg (85,539 lb)

PERFORMANCE (Su-30MK):
Max level speed:
at height .... (1,144 kt; 2,120 km/h; 1,317 mph)
at S/L .... M1.14 (729 kt; 1,350 km/h; 839 mph)
Max rate of climb at S/L .... 13,800 m (45,275 ft)/min
Service ceiling .... 17,300 m (56,760 ft)
T-O run, normal weight .... 550 m (1,805 ft)
Landing run with parachute .... 750 m (2,460 ft)
Combat range:
internal fuel .... 1,620 n miles (3,000 km; 1,865 miles)
with one in-flight refuelling .... 2,805 n miles (5,200 km; 3,230 miles)
g limit .... +9

UPDATED

SUKHOI SU-33UB (SU-27KUB)

TYPE: Multirole fighter.

PROGRAMME: Su-33 (Su-27K 'Flanker-D') single-seat naval fighter flew 17 August 1987; some 18 built in early 1990s for service aboard *Admiral of the Fleet Kuznetsov*; no further production.

Need for dedicated trainer for Su-33 became increasingly clear, and development of T10KM-2-based **Su-27KUB** (*korabelnyi uchebno boevoi:* as T10KU shipborne fighter trainer) began in late 1990s. Formally acknowledged 21 October 1998. Probably being developed as a private venture, with no firm commitment from Russian Navy. Likely to become **Su-33UB,** following redesignation of carrierborne single-seat Su-27K as Su-33.

Three prototypes, with noses built by KnAAPO, as T10KU-1, -2, and -3, incorporating lessons from the *Kuznetsov*'s 1996 Mediterranean cruise. Construction of T10KU-1 began in 1998 mating a new nose and new wings and tailplanes to an existing T10K prototype (T10K-4). Powered by AL-31F engines, the aircraft first flew on 29 April 1999 and made first arrested landing on dummy deck at NIUTK ('Nitka') test centre, Saki, 3 September 1999; first take-off from deck ramp followed on 6 September; first landing and take-off from carrier *Kuznetsov* on 6 October 1999. Initially unmarked, but later camouflaged and given side number '21'.

T10KU-2 and -3 reported to be under construction in 1999, perhaps using new-build airframes. By early 2004, first of these reported to have flown, with two further airframes undergoing static tests (probably not destined to fly) and two more on order. T10KU-1 test flown by Indian pilots, September 1999, but Su-27 judged too large for planned carriers. Further test series at Saki begun December 2000. In 2003, '21' was fitted with Fazotron-NIIP Zhuk-MSFEh radar, implying use as a testbed.

Any production is likely to be by KnAAPO; baseline variant due to be extrapolated to produce trainer, reconnaissance and AEW versions, the last-mentioned with a phased-array mounted on the spine, between the composites antenna tailfins. Increased thrust, thrust-vectoring AL-31FP, AL-31FM or AL-41F engines mooted for production version.

CURRENT VERSIONS: **Su-33UB:** *As described.*

Su-30K-2: Two-seat interceptor version based on Su-33 fuselage under construction at Komsomolsk, late 1999; due to fly late 2000, but no further reports received.

DESIGN FEATURES: Has navalised features of Su-33, including folding wing (with fold further outboard and with a larger fold angle). Some sources suggest that the Su-33UB's tailplanes do not fold, since they reach only as far as the new outboard wing fold. Double slotted flaps, unslotted, 'adaptive' leading-edge, arrester hook, datalink and carrier landing system. However, compared with Su-27IB, forward fuselage is slightly narrower, with seats closer together and has much less pronounced dorsal hump. New 'glass cockpit' with five colour

SU-30M PROTOTYPES AND PRODUCTION

Batch	Qty	Identity	Remarks
-	(1)	'56'; later '01'	Lost 12 June 1999
-	(1)	06	Previously T10PU-6
China			
MKK	4	501-504 Trials aircraft; apparently not part of contract	
MKK	38	'01'-38	Ordered 1999. Delivered 2000–01 from Komsomolsk
MKK	38	'39'-'76'	Ordered June 2001; delivered 2002–03
MKK2	28		Ordered January 2003; delivery from January 2004
India			
MKI	4	'02' — '05'	Development, built 2000–01
MKI	10	SB019-028	Delivered from Irkutsk 20 May 2002
MKI	12	SB029-040	Delivered from Irkutsk October to December 2003
MKI	10	SB041-048	Delivered from Irkutsk in December 2004
MKI	140		Production in India from 2007
Indonesia			
MK	2	TS-3001	Delivered August 2003
	2		On order from Komsomolsk
Malaysia			
MKM	18		Deliveries 2006-08 from Irkutsk
Vietnam			
MK	4		Deliveries December 2004 from Komsomolsk

Prototype Sukhoi Su-33UB in Russian Navy colours *(Yefim Gordon)* ***NEW***/0558786

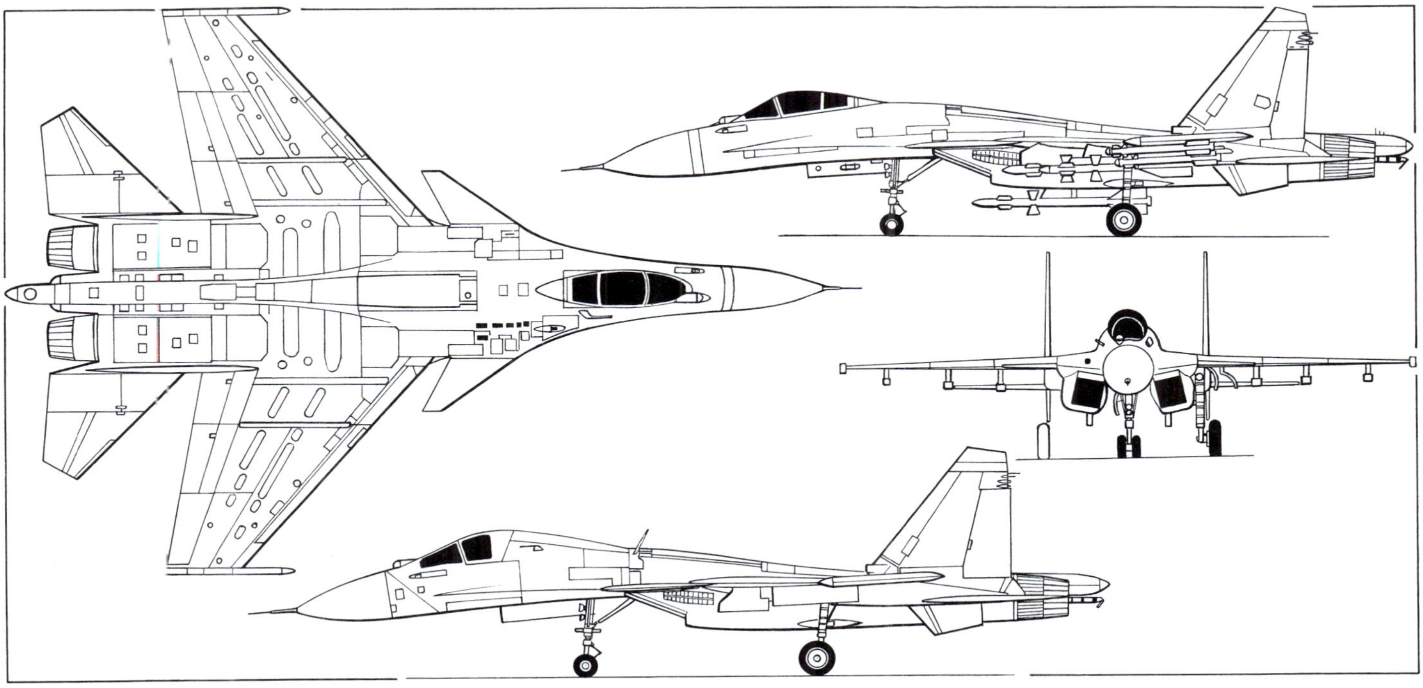

Sukhoi Su-33 single-seat carrierborne fighter, with additional side view (lower) of two-seat Su-33UB operational trainer *(James Goulding)* 0075937

LCD displays (one 53 cm; 21 in diagonally; rest 38 cm; 15 in) with provision for central or sidestick and with helmet-mounted sighting system. Aircraft has OBOGS and OBIGGS and so does not need oxygen or nitrogen bottles. N014 solid-state, phased-array radar, with enhanced air-to-ground and over-water capabilities, planned eventually, but prototypes have ballast or Zhuk-MS and production aircraft may initially use NIIR N610-27 (Zhuk 27); circular-section radomes replace flattened 'platypus' nose associated with Su-27IB. Some reports suggest that the Su-33UB's new wing is 12 per cent larger in span (16 m; 52½ ft) and area (70 m²; 750 sq ft), as are the canards and tailplanes. Production Su-27KUB will feature a higher seat, square-section, lengthened tail-sting, possibly mounting rear warning radar; tailcone folds (upwards) to reduce stowed length; prototypes have the standard Su-33 tailcone.

UPDATED

SUKHOI SU-35 AND SU-37 (SU-27M)

NATO reporting name: Flanker

TYPE: Multirole fighter.

PROGRAMME: Development of Su-27M authorised on 29 December 1993 by Council of Ministers. Experimental version of Su-27 with foreplanes (T10-24) flew May 1985; improved FBW system and refuelling probe tested by T10U-2. Five prototypes produced by conversion of production Su-27s, retaining single nosewheel and standard tailfins: T10M-1 '701' (ex-1602), then lacking radar and weapon control system (successively T10S-70, T10M, Su-27M, Su-35) flew 28 June 1988; T10M-2 flown 18 January 1989; T10M-5, T10M-6 '706' and T10M-7 '707'; used mainly by NII VVS at Akhtubinsk, flown by service pilots.

Production at KnAAPO, Komsomolsk, beginning with static test airframe T10M-4; first flight (T10M-3 '703') 1 April 1992; latter exhibited at 1992 Farnborough Air Show. Of further six ordered (T10M-8 to 13; '708' to '713'), final aircraft was cancelled. NIIP N011M Bars phased-array radar tested in '711' and '712'. Large-scale series production originally planned 1996–2005 as interim fighter pending availability of (also cancelled) Mikoyan MFI.

Three production aircraft (Blue 86, 87 and 88) delivered to Akhtubinsk from Komsomolsk in 1996 or 1997. In November 2002 it was announced that six Su-35s would be delivered to 237th Aviation Display Regiment at Kabinka, Moscow. No evidence that these arrived.

KnAAPO revealed in 2001 that production Su-35s and Su-35UBs will have thrust vectoring AL-31FP engines, thereby negating Su-37 designation.

CURRENT VERSIONS: **Su-35** (Flanker-E Variant 1): Baseline single-seater; *as described.*

Su-35UB: Two-seat (tandem) derivative of the Su-35 with same FCS, canard foreplanes, tall square-topped tailfins, 12-pylon wing, Zhuk main AI radar and N012 rearward-facing, tailcone-mounted radar. Developed as a demonstrator and trainer for the Su-35, the construction of the prototype (Blue 801) at Komsomolsk may have been prompted by the needs of Sukhoi's campaign to sell the Su-35 to South Korea. The aircraft first flew on 7 August 2000 and was reported to be undergoing trials at Akhtubinsk in October 2000. Has 123 kN (27,558 lb st) AL-31FP (AL-31F) thrust-vectoring engines.

Su-35UB has 38,800 kg (83,775 lb) MTOW; 8,000 kg (17,637 lb) combat load; 12,400 litres (3,276 US gallons; 2,728 Imp gallons) internal fuel load; 1,090 kt (2,020 km/h; 1,255 mph) max speed; 1,619 n mile (3,000 km; 1,864 mile) unrefuelled range; wing span 14.70 m (48 ft 2¾ in); length overall 21.94 m (71 ft 11½ in) (shorter tailboom); and height overall 6.355 m (20 ft 10¼ in).

Su-37 (Flanker-E Variant 2): Technology demonstrator for proposed production fighter resulting from vectored-thrust programme of the Su-27UB-PS, Su-27UBL, Su-27LMK No. 2405 and the penultimate prototype Su-35 ('711') which made its first flight, with nozzles fixed, on 2 April 1996 (or 12 April, according to some sources). This aircraft had previously been first of two N011M radar testbeds, flying in blue/grey camouflage. Designated Su-37 by Sukhoi bureau. By September 1996, when demonstrated at Farnborough Air Show (Western debut), in tropical camouflage, '711' had made 50 flights with hydraulically actuated nozzles able to move ±15° in pitching plane at rate of 30°/s under control of aircraft's flight control system. Probably toed-out 32° from the centreline, like Su-30 MKI, generating powerful yawing moment when actuated differentially. An emergency pneumatic system returns the nozzles to level flight setting in the event of system failure.

Any production Su-37 was expected to feature uprated AL-31FU engines (142.2 kN; 31,967 lb st). In 1999, Sukhoi was using designation Su-37MR for proposed production version, apparently reflecting a major avionics upgrade then under way. By 2001, '711' had been retrofitted with non-vectoring AL-31F engines, revised Ramenskoye instrumentation and a new flight control system which was stated to provide TV-type agility without recourse to vectoring; used in trials for Indian Su-30MKI Stage 3.

Second Su-37 reportedly converted from T10M-12 and first flew in mid-1998, initially powered by AL-31F; installation of thrust-vectoring AL-37FPs was expected in late 1998; status of this aircraft, if still extant, remains unclear. Development of AL-31F engine being

Sukhoi Su-35UB prototype *(Yefim Gordon)* **NEW**/0558792

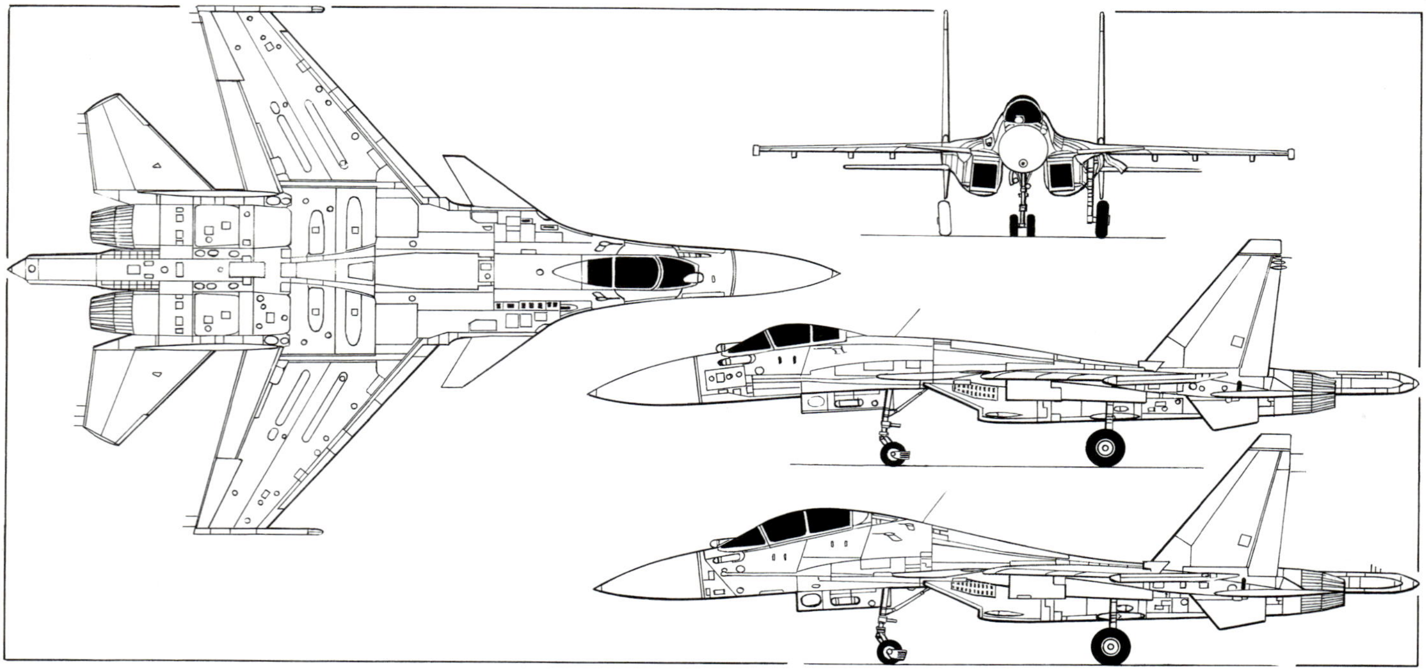

Sukhoi Su-35 single-seat counter-air and ground attack fighter, with additional side view (lower) of Su-35UB *(James Goulding)* 0525932

SU-35 PROTOTYPES AND PRODUCTION

Batch	Qty	Identity	Remarks
-	(1)	T10M-1 '701'	Previously Su-27 1602
-	(1)	T10M-2 '702'	Previously Su-27
-	(1)	T10M-5 '705'	Previously Su-27
-	(1)	T10M-6 '706'	Previously Su-27
-	(1)	T10M-7 '707'	Previously Su-27
10	1	T10M-4	Static test airframe
10	1	T10M-3 '703'	First Komsomolsk-built
11	1	T10M-8 '708'	
11	1	T10M-9 '709'	
11	1	T10M-10 '710'	
11	1	-	Static test airframe
11	1	T10M-11 '711'	Converted to Su-37; reverted to Su-35; crashed 19 December 2002
12	3	'86' *et seq*	Production
n/k	1	T10M-12 '712'	Radar testbed
n/k	1	'801'	Su-35UB

Notes: Former Su-27 aircraft may include 2209 and 2219

undertaken unilaterally by MMPP Salyut, version promising 10 to 15 per cent extra thrust having first flown on 24 January 2002 (on Su-27 testbed and lacking Klivit thrust vectoring nozzle).

Su-37 "Terminator": In November 2003, Sukhoi announced that it is developing an upgraded version of Su-37, with capabilities equivalent to a "4++" generation fighter, as a means to securing further export orders before the next-generation Su-47 is available in 2010-12. AL-31F engines, although AL-41F also quoted in relation to **Su-35BM**, which appears to be an alternative designation.

CUSTOMERS: Once scheduled for entry into Russian Air Forces service as Su-27M from 1995 onwards, for effective operation until 2015–2020; programme in suspension. Indonesia announced imminent purchase in March 2003 of two Su-35s and two Su-27s, to be followed over four years by further 44 'Flankers' of undisclosed type.

COSTS: Estimated US$35 million to US$40 million (2002).

DESIGN FEATURES: Advanced multirole development of Su-27 to counter latest versions of USAF F-15 Eagle and F-16 Fighting Falcon, with better dogfighting characteristics, higher AoA limits, lighter weight and new BVR armament; proposed to include 216 n mile (400 km; 248 mile) range AAM-L (one AAM-L contender was Novator KS- 172). Also planned to have greater autonomy from GCI control. Airframe, power plant, avionics and armament all upgraded; quadruplex digital fly-by-wire controls under development by Avionika, though prototypes retain analogue system; longitudinal static instability; 'tandem triplane' layout, with foreplanes; double-slotted flaperons; taller, square-tip twin tailfins with integral fuel tanks; reprofiled front fuselage for larger-diameter radar antenna; enlarged tailcone for rearward-facing radar; twin-wheel nose landing gear; axisymmetric thrust-vectoring nozzles under development for use on production aircraft (see Su-37).

STRUCTURE: Higher proportion of carbon fibre and aluminium-lithium alloy in fuselage; composites used for components such as leading-edge flaps, nosewheel door and radomes.

POWER PLANT: Production Su-27M planned to use two Saturn/Lyulka AL-35FP turbofans; each 123 kN (27,558 lb st) with afterburning; prototypes retain standard AL-31F. Increased internal tankage through use of welded aluminium-lithium tanks and new tanks in tailfins and finroots. Total fuel capacity 12,900 litres (3,408 US gallons; 2,838 Imp gallons). Retractable flight refuelling probe on port side of nose.

ACCOMMODATION: Pilot(s), on Zvezda K-36D-3.5E zero/zero ejection seat, this now angled back 30°.

AVIONICS: *Radar:* Originally planned to incorporate NIIP N011 Zhuk-27 multimode low-altitude terrain-following/avoidance radar, search range up to 54 n miles (100 km; 62 miles) against advancing target, 30 n miles (55 km; 34 miles) against retreating target; able to track 15 targets and engage four to six simultaneously. Phazotron Zhuk-Ph phased-array radar under development as alternatives or for retrofit; search range for fighter-size targets 75 n miles (140 km; 87 miles) with simultaneous tracking of 24 air targets and ripple-fire engagement of six to eight; N012 rearward-facing radar, range approximately 2 n miles (4 km; 2.5 miles), may enable firing of rearward-facing IR homing AAMs.

Flight: Fully automatic flight modes and armament control against ground, maritime and air targets, including automatic low-altitude flight and automatic target designation. RPKB nav system includes laser-gyro INS and Glonass GPS.

Instrumentation: Two-seat variant has side-by-side MFI-10-5 screens and HUD in front cockpit; vertically stacked MFI-10-5 screens in rear.

Mission: New-type IRST moved to starboard; small external TV pod; all combat flight phases computerised. Shown at Farnborough with GEC-Marconi TIALD (thermal imaging airborne laser designator) night/adverse visibility pod fitted for possible future use.

Self-defence: Entirely new integrated EW suite with ECM, including active jammer and wingtip Sorbtsa-S G- or J-band ECM/ESM pods; Pastel RWR; Mak IR-based MAWS.

ARMAMENT: One 30 mm GSh-30 gun in starboard wingroot extension, with 150 rounds. Mountings for up to 14 stores pylons, including R-27 (AA-10 'Alamo-A/B/C/D'), R-40 (AA-6 'Acrid'), R-60 (AA-8 'Aphid'), R-73E (AA-11 'Archer') and RVV-AE (R-77; AA-12 'Adder') AAMs, Kh-25ML (AS-10 'Karen'), Kh-25MP (AS-12 'Kegler'), Kh-29T (AS-14 'Kedge'), Kh-31P (AS-17 'Krypton') and Kh-59 (AS-18 'Kazoo') ASMs, S-25LD laser-guided rockets, S-25IRS IR-guided rockets, GBU-500L and GBU-1500L laser-guided bombs, GBU-500T and GBU-1500T TV-guided bombs, KMGU cluster weapons, KAB-500 bombs and rocket packs. Maximum weapon load 8,200 kg (18,077 lb).

DIMENSIONS, EXTERNAL:
- Wing span over ECM pods ... 15.16 m (49 ft 8¾ in)
- Wing aspect ratio ... 3.5
- Length overall ... 22.185 m (72 ft 9½ in)
- Height overall ... 6.36 m (20 ft 10¼ in)

AREAS:
- Wings, gross ... 62.04 m² (667.8 sq ft)

WEIGHTS AND LOADINGS:
- Weight empty ... 17,000 kg (37,479 lb)
- Max fuel weight ... 10,250 kg (22,597 lb)
- Max combat load ... 8,500 kg (18,739 lb)
- T-O weight: normal max ... 34,000 kg (74,957 lb)
- overload ... 38,800 kg (85,539 lb)
- Max wing loading (normal) ... 548.4 kg/m² (112.32 lb/sq ft)
- Max power loading (normal) ... 139 kg/kN (1.36 lb/lb st)

PERFORMANCE:
- Max level speed:
  - at height ... M2.35 (1,350 kt; 2,500 km/h; 1,555 mph)
  - at S/L ... M1.14 (756 kt; 1,400 km/h; 870 mph)
- Service ceiling ... 17,200 m (56,440 ft)
- Balanced runway length ... 1,200 m (3,940 ft)
- Range: with max internal fuel ... 1,835 n miles (3,400 km; 2,112 miles)
  - with flight refuelling ... 3,401 n miles (6,300 km; 3,914 miles)
- *g* limit ... +9

UPDATED

SUKHOI T-50 PAKFA

TYPE: Multirole fighter.

PROGRAMME: Successor programme to original quest for fifth-generation fighter, for which MiG 1.42 and Sukhoi S-37/Su-47 Berkut demonstrators were built. Development began in 1998 to meet Russian Air Forces TTZ (tactical technical assessment) of the same year. Sukhoi-led co-operative project launched May 2001; initially termed LFS (*legkiy frontovoy samolet*: Light Frontline Aircraft), although competition also known as LFI (*legkiy frontovoy istrebitel*: Light Frontline Fighter). Air Forces' project name is PAKFA (*perspektivnyi aviatsionnyi kompleks frontovoi aviatsyi*: Prospective Aviation Complex for Frontal Aviation).

Examining committee established 10 January 2002 to assess competing bids from RSK 'MiG' and AVPK Sukhoi, each of which nominated Yakovlev as associate. On 26 April 2002, Russian Federation Ministry of Industry, Science and Technology declared Sukhoi as lead developer of new aircraft, to be assisted by RSK 'MiG' and Yakovlev as subcontractors. In May 2002, development agreement signed by AVPK Sukhoi, Sukhoi OKB, State Research Institute of Aviation Systems (Gos-NIIAS), Central Aero- and Hydrodynamic Institute (TsAGI), Research Institute of Aero Engine Technology and Production (TsIAM), Central Research Institute of Material (VIAM), National Institute of Aviation Technologies (NIAT), Lyulka/Saturn engine bureau, Ramenskoye Instrument Design Bureau (RPKB), Aviapribor holding company, Aviakosmitcheskoye Oborudvanyye and Vympel and Strela weapon companies. Several manufacturing plants expected to join at later stage. Participation of KnAAPO production plant was agreed in June 2002.

Sukhoi Su-47 Berkut technology demonstrator *(Yefim Gordon)* ***NEW***/1047174

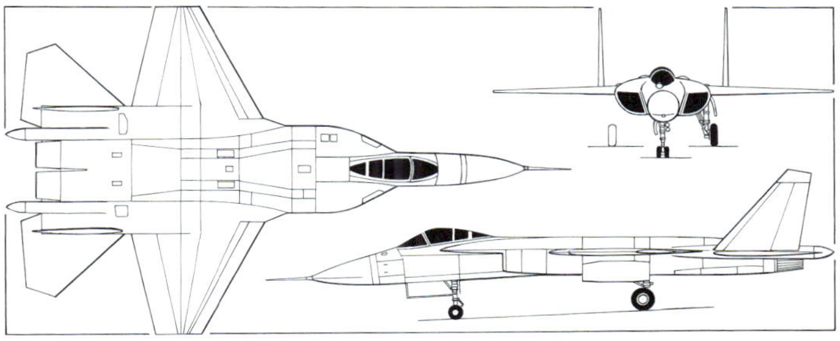

Provisional general arrangement of Sukhoi T-50 PAKFA *(James Goulding)* ***NEW***/0567236

Timetable for PAKFA is draft design by end of 2002; first flight in 2006; production from 2010. Official funding equivalent to US$1.5 billion promised for R & D, but this considered inadequate, representing some 20 per cent of needs, excluding production investment. Unit cost expected to be US$35 million to US$40 million (2002 prices), based on production of between 500 and 600 aircraft.

Parameters believed to include 20 tonne MTOW — between MiG-29 and Su-27 types it is due to replace — supersonic cruising speed, low observables, high manoeuvrability and short-field performance. Engine choice yet to be made, but NPO Saturn is offering a derivative of AL-41F1, which was delivered to the Gromov test centre at Zhukovsky in November 2003 to be installed in a flight test aircraft. Suggested that improved AL-31F will power initial PAKFAs, however.

In January 2003, India discussed terms of invitation by Russia to join PAKFA design team. Preliminary details and a probable configuration were unofficially circulating by mid-2003, showing an aircraft similar to the Su-47 Berkut, apart from more conventional wings.

DIMENSIONS, EXTERNAL:
- Wing span ... 15.5 m (51 ft)
- Length overall ... 23.0 m (75 ft)

WEIGHTS AND LOADINGS:
- T-O weight: normal ... 24,000 kg (53,000 lb)
- max ... 33,000 kg (72,750 lb)

PERFORMANCE:
- Max level speed ... 1,350 kt (2,500 km/h; 1,553 mph)
- Supercruise speed ... M1.6+
- Radius of action ... 647 n miles (1,200 km; 745 miles)
- Ferry range ... 2,159 n miles (4,000 km; 2,485 miles)

UPDATED

SUKHOI SU-29

TYPE: Aerobatic two-seat sportplane.

PROGRAMME: Announced at Moscow Air Show '90; design started 1990; construction of first of three prototypes and two static test airframes began 1991; prototype first flew 1991, first production aircraft May 1992; entered service July 1992. AP-23 type certificate awarded in 1994. Assessed by Russian Air Forces experimental centre at Akhtubinsk in 1996–97, but no orders placed.

Following structural failure of Su-29 wing in 1996, remedial manufacturing practices in place and recertified by 1998. Resumed civilian export production now featured M-14PF engine and new propeller, but M-9F engine introduced in 1999. Retrospective installation of SKS-94 pilot's emergency extraction system offered by Sukhoi from 2000.

Production transferred to LAPIK division of RSK 'MiG' (formerly LMZ) at Lukhovitsy from early 2001, earlier aircraft having been built by Sukhoi Advanced Technologies using components manufactured by LMZ, Dubna and others. Only one built in 2001.

CURRENT VERSIONS: **Su-29:** Basic two-seat training/aerobatic aircraft.

Description applies to baseline Su-29.

Su-29AR: Aircraft for Argentina have German propeller, Swiss canopies and US wheel assemblies and avionics, including GPS.

Argentine Air Force Sukhoi Su-29AR *(FAA)* ***NEW***/0576534

Sukhoi Su-29 *(Paul Jackson)* ***NEW***/0576535

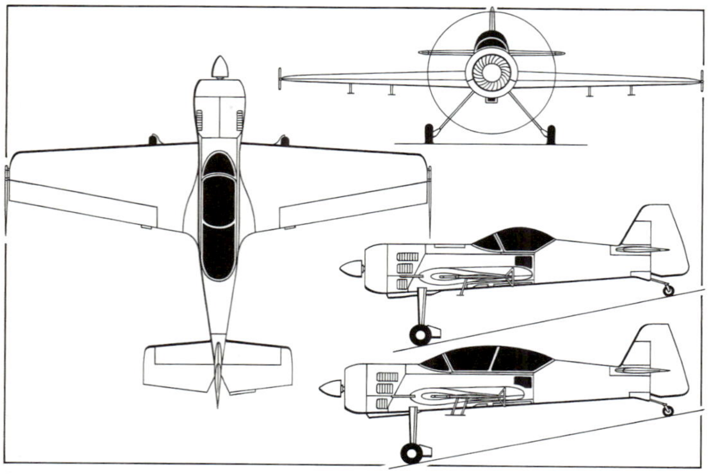

Sukhoi Su-29 with additional side view (upper) of Su-31T *(Jane's/Mike Keep)* 0015120

Su-29KS: Development vehicle for Zvezda SKS-94 lightweight crew extraction system. Weight empty, equipped 800 kg (1,764 lb). First exhibited at 1994 Farnborough Air Show. One only (RA-01485); carries designation 'Su-29KS'.

Su-29M: Production version from 1999 (initial series of 10 under construction); M-9F engine; weights as Su-29; no extraction system.

Su-49: Proposed military trainer.

CUSTOMERS: Total 52 basic Su-29s and one Su-29KS built and sold by 1997; many exported to Pompano Air Center, Florida, USA, for reassembly and delivery worldwide (including eight in 1992; 12 in 1993; seven in 1994 and four in 1995); others to Australia (three), Italy, South Africa (two) and UK. Relaunched production from 1999 (initial batch of 10). Eight ordered by Argentine Air Force, for training, March 1997; delivery between September 1997 and August 1998 to Escuadrilla Cruz del Sur (Southern Cross Squadron) based at Mendoza AFB. By mid-2001, Sukhoi had built 166 aerobatic light aircraft of the Su-26/29/31 series, of which all but 10 had been exported, including 80 to the USA. Sukhoi presented two to ruler of unspecified Gulf state, but these not used and eventually sold. Deliveries scheduled in 2002 to customers in the Czech Republic, Germany, Russia, South Korea and Switzerland were, apparently, not effected and there was no known production in 2003.

COSTS: US$220,000 to US$260,000 (2002).

DESIGN FEATURES: Typical aerobatic competition aircraft; mid-wing of specially developed symmetrical section, variable along span, slightly concave in region of ailerons to increase their effectiveness; leading-edge somewhat sharper than usual to improve responsiveness to control surface movement. Two-seat development of Su-26M single-seat aerobatic competition aircraft; wing span and overall length increased; improved aerodynamics and reduced stability margin for enhanced manoeuvrability. Service life 1,250 hours.

Wing leading-edge sweepback 3° 28', symmetrical section, thickness/chord ratio 16 per cent at root, 12 per cent at tip, dihedral 0°, incidence 0°.

FLYING CONTROLS: Conventional and manual. Elevators and rudder horn-balanced; elevator trim; two suspended triangular balance tabs under each aileron. Later Su-29s have Su-31-type wing with ailerons extending to wingtips. No flaps.

STRUCTURE: Composites comprise more than 60 per cent of airframe weight; one-piece wing, covered with honeycomb composites panels; foam-filled front box spar with CFRP booms and wound glass fibre webs; channel section rear spar of CFRP; titanium truss ribs; plain ailerons have CFRP box spar, GFRP skin and foam filling; fuselage has basic welded truss structure of VNS-2 high-strength stainless steel tubing; lower nose section of truss removable for wing detachment; quickly removable honeycomb composites skin panels; light alloy engine cowlings; integral fin and tailplane construction same as wings; rudder and elevator construction same as ailerons; titanium exhaust, battery box and firewall; forged magnesium control linkages. Aircraft assembled by Sukhoi from components produced by LMZ, Dubna (DMZ) and NPO Technologia (at Omsk).

LANDING GEAR: Tailwheel type; fixed. Fuselage-mounted spring titanium alloy cantilever mainwheel legs; mainwheels size 400×150, with hydraulic disc brakes; steerable tailwheel, on titanium spring, connected to rudder. Optional composites fairings for mainwheels.

POWER PLANT: One 265 kW (355 hp) VOKBM M-14PT or 294 kW (394 hp) M-14PF nine-cylinder radial engine; three-blade MT-Propeller MTV-3-8-S/L250-21 or MTV-9-260 propeller. Current production employs VOKBM M-9F (M-14 derivative) rated at 309 kW (414 hp). Steel tube engine mounting. Fuel tank in fuselage forward of front spar; capacity 63 litres (16.6 US gallons; 13.8 Imp gallons); tank in each wing leading-edge; capacity 106.5 litres (28.15 US gallons; 23.4 Imp gallons); total fuel capacity 276 litres (72.9 US gallons; 60.6 Imp gallons); gravity fuelling. Oil capacity 20 litres (5.3 US gallons; 4.4 Imp gallons). Fuel and oil systems adapted for inverted flight; pneumatic engine starting system.

ACCOMMODATION: Pilot only for aerobatic competition, two persons in tandem for training. Canopy normally opens sideways to starboard; but upward and rearward in emergency to jettison. Dual controls standard. Space for 5 kg (11 lb) baggage in rear fuselage.

SYSTEMS: Electrical system 24/28 V, with 3 kW generator, batteries and external supply socket.

AVIONICS: *Comms:* Briz VHF radio; optional Becker or Bendix/King com/nav and Garmin GPS.

EQUIPMENT: Optional provision for smoke generation.

DIMENSIONS, EXTERNAL:

Wing span	8.20 m (26 ft 10¾ in)
Wing chord: at root	1.985 m (6 ft 6¼ in)
at tip	1.04 m (3 ft 4¾ in)
Wing aspect ratio	5.5
Length overall	7.285 m (23 ft 10¾ in)
Height overall	2.885 m (9 ft 5¾ in)
Tailplane span	2.90 m (9 ft 6¼ in)
Wheel track	2.40 m (7 ft 10½ in)
Wheelbase	5.08 m (16 ft 8 in)
Propeller diameter	2.50 m (8 ft 2½ in)
Propeller ground clearance	0.425 m (1 ft 4¾ in)

DIMENSIONS, INTERNAL:

Cockpit: Length	2.60 m (8 ft 6¼ in)
Max width	0.82 m (2 ft 8¼ in)
Max height	1.05 m (3 ft 5¼ in)

AREAS:

Wings, gross	12.20 m² (131.4 sq ft)
Ailerons (total)	2.32 m² (24.97 sq ft)
Fin	0.28 m² (3.01 sq ft)
Rudder	0.90 m² (9.69 sq ft)
Tailplane	0.98 m² (10.55 sq ft)
Elevators (total)	1.56 m² (16.79 sq ft)

WEIGHTS AND LOADINGS (two persons):

Weight: empty	735 kg (1,620 lb)
empty, equipped	780 kg (1,720 lb)
Max fuel	207 kg (456 lb)
Max T-O weight: pilot only	860 kg (1,896 lb)
two persons	1,200 kg (2,645 lb)
Max wing loading	98.4 kg/m² (20.15 lb/sq ft)
Max power loading: M-14PT	4.53 kg/kW (7.45 lb/hp)
M-14PF	4.09 kg/kW (6.71 lb/hp)
M-9F	3.89 kg/kW (6.39 lb/hp)

PERFORMANCE (M-14PT engine):

Never-exceed speed (VNE)	242 kt (450 km/h; 279 mph)
Max level speed	175 kt (325 km/h; 202 mph)
Landing speed	65 kt (120 km/h; 75 mph)
Stalling speed	62 kt (115 km/h; 72 mph)
Max rate of climb at S/L	960 m (3,150 ft)/min
Service ceiling	4,000 m (13,120 ft)
Max rate of roll	360°/s
T-O run	120 m (395 ft)[1]
Landing run	380 m (1,250 ft)[1]
Range with max fuel	647 n miles (1,200 km; 745 miles)
g limits	+12/−10

[1] *at 914 kg (2,015 lb) AUW*

UPDATED

SUKHOI SU-31

TYPE: Aerobatic single-seat sportplane.

PROGRAMME: Design started 1991; prototype construction began 1992, flew June 1992 as Su-29T, demonstrated at 1992 Farnborough Air Show; followed by two more prototypes and two static test airframes; first production aircraft by Sukhoi Advanced Technologies (RA-01405) flown 1994. Su-31M2 introduced in 1999. Manufacture transferred to RSK 'MiG' (LAPIK at Lukhovitsy) in early 2001; two Su-31Ms, one Su-31 built in 2001; contract for 10 Su-29s/Su-31Ms for an undisclosed buyer pending in early 2002.

In XXI World Aerobatic Championships, June 2001, Su-31s gained first place and seven of next 14 places.

CURRENT VERSIONS: **Su-31:** Basic version; non-retractable landing gear. Alternatively known as **Su-31T** (*Turnirnyi:* Competition). See Su-29 drawings for side view.

Su-31M: As Su-31T but with Zvezda SKS-94 pilot's extraction system under modified canopy with deeper frame. Empty weight 760 kg (1,676 lb); normal T-O weight 880 kg (1,940 lb). Prototype RA-01486 converted from Su-31T.

Su-31M2: Improved version of Su-31M with larger wing, airframe weight reduced by approximately 80 kg (176 lb), ergonomically redesigned cockpit, and upgraded Zvezda SKS-94M pilot's extraction system. Prototype scheduled to fly in mid-2003.

Su-31X: Export version of Su-31T.

Su-31U: As Su-31T but retractable landing gear. None yet built.

Sukhoi Su-31M single-seat aerobatic competition aircraft *(Paul Jackson)* ***NEW***/0576508

Su-31ChM: Improved version of Su-31M2 under development in 2003 for World Aerobatic Championships competitions; features include airframe weight reduced by 25 kg (55 lb), and 313 kW (420 hp) VOKBM M-9F engine.

CUSTOMERS: Total 25 (including five Su-31Ms) built by late 1998 and exported to Australia, Brazil, Italy, Lithuania, South Africa, Spain, Switzerland, Ukraine, UK and USA. Su-31Ms operated in Italy, Russia and Switzerland. Production batch of five Su-31M2s, 1999–2000, of which at least two to USA; RSK 'MiG' to build initial batch of nine at its LAPIK division (three Su-29s and six Su-31s). Total of six Su-31s were scheduled for delivery in 2002, including two Su-31Ms for the Czech Republic and one Su-31M for a Swiss customer.

COSTS: US$220,000 to US$260,000 (2003).

DESIGN FEATURES: Basically single-seat version of Su-29 with uprated engine; new landing gear; 35° inclination of seat enables pilot to employ repeatedly a *g* load of +12/−10, giving advantages in controlling aircraft within limited flying area and to perform very complicated manoeuvres; improved field of view; two baggage compartments.

STRUCTURE: More than 70 per cent composites by weight; centre-fuselage is welded truss of high-strength stainless steel tube, with detachable skin panels of honeycomb-filled composite sandwich; rear fuselage is semi-monocoque of composites with honeycomb filler; one-piece two-spar wing with carbon fibre main spar, titanium ribs, covered with honeycomb sandwich skin; tail unit is all-composites; mainwheel legs titanium.

LANDING GEAR: As Su-29.

POWER PLANT: One 294 kW (394 hp) VOKBM M-14PF nine-cylinder radial engine in Su-31T/M; MTV-9 three-blade propeller. Basic fuel capacity 78 litres (20.6 US gallons; 17.2 Imp gallons), in fuselage tank; provision in Russian version for 210 litre (55.5 US gallon; 46.2 Imp gallon) centreline drop tank for ferrying; alternatively, in export version, two tanks in wings, total capacity 200 litres (52.8 US gallons; 44.0 Imp gallons).

ACCOMMODATION: Pilot only; windscreen and separate canopy, opening as Su-29, except single-piece windscreen/canopy and pilot extraction system in Su-31M and Su-31M2.

DIMENSIONS, EXTERNAL (Su-31T):
Wing span ... 7.80 m (25 ft 7 in)
Wing chord: at root ... 1.99 m (6 ft 6¼ in)
at tip ... 1.04 m (3 ft 4¾ in)
Wing aspect ratio ... 5.2
Length overall ... 6.83 m (22 ft 4¾ in)
Height overall ... 2.76 m (9 ft 0¾ in)
Tailplane span ... 2.90 m (9 ft 6¼ in)
Wheel track ... 2.40 m (7 ft 10½ in)
Wheelbase ... 4.90 m (16 ft 1 in)
Propeller diameter ... 2.50 m (8 ft 2½ in)
Propeller ground clearance ... 0.425 m (1 ft 4¾ in)

AREAS (Su-31T): As Su-29 except:
Wings, gross ... 11.83 m² (127.3 sq ft)

WEIGHTS AND LOADINGS (Su-31T, except where otherwise indicated):
Weight: empty ... 680 kg (1,499 lb)
empty, equipped: Su-31T ... 740 kg (1,631 lb)
Su-31M ... 750 kg (1,653 lb)
Max fuel: internal ... 53 kg (117 lb)
external ... 209 kg (461 lb)
T-O weight: normal ... 835 kg (1,841 lb)
max ... 968 kg (2,134 lb)
Max wing loading ... 82.0 kg/m² (16.80 lb/sq ft)
Max power loading: Su-31T/M ... 3.25 kg/kW (5.34 lb/hp)

PERFORMANCE (Su-31T):
Never-exceed speed (VNE) ... 243 kt (450 km/h; 280 mph)
Max level speed ... 178 kt (330 km/h; 205 mph)
Stalling speed ... 61 kt (113 km/h; 71 mph)
T-O speed ... 60 kt (110 km/h; 69 mph)
Landing speed ... 62 kt (115 km/h; 72 mph)
Max rate of climb at S/L ... 1,440 m (4,725 ft)/min
Service ceiling ... 4,000 m (13,120 ft)
Max rate of roll ... 400°/s
T-O run ... 110 m (360 ft)
Landing run ... 300 m (985 ft)
Range, internal fuel ... 156 n miles (290 km; 180 miles)
Ferry range ... 432 n miles (800 km; 497 miles)
g limits ... +12/−10

UPDATED

SUKHOI SU-38L

TYPE: Agricultural sprayer.

PROGRAMME: Design started, as Su-38, under Boris Rakitin, August 1993; originally based closely on Su-29 sportplane; construction of prototype began January 1994 but curtailed due to financial situation in Russia. Promotion resumed in 1998, under S-38L designation and with considerable change of detail design, including replacement of VOKBM M-14P radial engine. Announced June 2000 that order signed to build prototype, now known as Su-38L. Prototypes fabricated by SmAZ; target of five complete (including two static test) aircraft by second quarter of 2002; first two under assembly at Sukhoi OKB workshops by September 2000. First flight (01) 27 July 2001; public debut at Moscow Salon, 14–19 August 2001, when 01 shown with (static) and without (flying) winglets and then unflown 02 exhibited statically. Compared with earlier drawings and mockup, prototypes

Prototype Sukhoi Su-38L *(Yefim Gordon)* ***NEW***/0576509

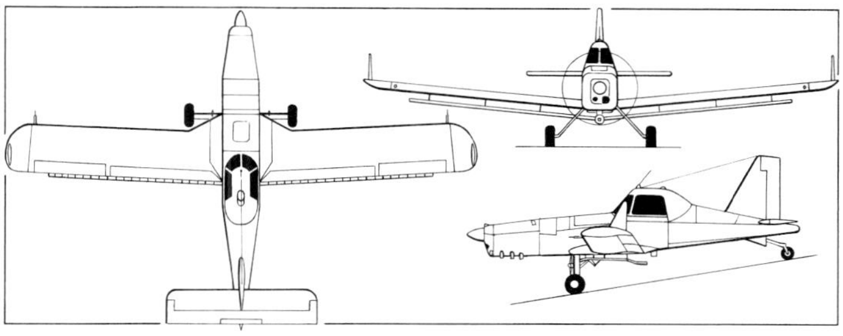

Sukhoi Su-38L agricultural aircraft *(James Goulding)* 0121659

have increased length due to additional fuselage plug ahead of wings. By June 2002 two (of three planned) prototypes had completed 60 missions of projected 100- to 150-sortie flight programme, intended to lead to initial certification in September or October 2003. Third prototype's maiden flight delayed from August 2002 to end 2002/early 2003. Initial production by SmAZ with target of 20 in 2003, rising to 40 in 2004, and potential for up to 100 aircraft per year. Second production line intended by Razdanmash AOZT of Armenia, subject to acquisition of US$23 million for flight test and tooling.

CUSTOMERS: Estimated market for 500 in Russia.

COSTS: Unit cost US$100,000 to US$150,000 (2002); direct operating cost US$300 per hour (2001).

DESIGN FEATURES: Conventional low-wing monoplane. Constant-chord wings swept slightly forward and with winglets and root nibs; sweptback fin, plus underfin doubling as tailwheel mount. Chemical hopper between engine and cockpit, capacity 500 litres (132 US gallons; 110 Imp gallons). Service life 3,000 hours/10 years.

FLYING CONTROLS: Conventional and manual. Horn-balanced single-piece elevator and rudder; trim tab in elevator; trailing-edge flaps.

LANDING GEAR: Tailwheel type; fixed. Mainwheel size 8.00-6, tailwheel size 310×135. Hydraulic brakes.

POWER PLANT: One 184 kW (247 hp) LOM M337S six-cylinder engine operating on a 1:2 mix of A-76 and A-92 mogas; LOM B-546 three-blade ground-adjustable propeller. Plans for uprated, 186 kW (250 hp) version in production aircraft. Fuel tank in each wing; total capacity 210 litres (55.5 US gallons; 46.2 Imp gallons).

ACCOMMODATION: Pilot only on energy-absorbing seat; baggage shelf behind seat. Cockpit is pressurised to prevent ingress of chemicals, and incoming air is filtered.

EQUIPMENT: Transland underwing spraybar. Cable cutter/deflector standard.

DIMENSIONS, EXTERNAL:
Wing span ... 11.53 m (37 ft 10 in)
Length overall ... 8.10 m (26 ft 6¾ in)
Height overall ... 2.655 m (8 ft 8½ in)

WEIGHTS AND LOADINGS:
Normal T-O and landing weight ... 1,200 kg (2,646 lb)

PERFORMANCE (estimated):
Operating speed ... 81–97 kt (150–180 km/h; 93–112 mph)
Stalling speed ... 41 kt (76 km/h; 48 mph)
Typical operating altitude ... 1 to 15 m (3 to 50 ft)
Ferry range:
with spraybars ... 486 n miles (900 km; 559 miles)
without spraybars ... 648 n miles (1,200 km; 745 miles)

UPDATED

SUKHOI SU-80

TYPE: Twin-turboprop transport.

PROGRAMME: First, largest and most advanced design by Sukhoi under *konversiya* programme of former Soviet industry; to replace L-410, An-28, Yak-40 and An-24; certification intended to FAR Pt 25, JAR 25 and AP-25. Work began 1989 on medical evacuation aircraft, then designated S-80, under order from Ministry of Public Health, by Sukhoi-Europe/Asia joint stock company, with founder members Sukhoi Design Bureau ASIC, KnAAPO (which see), Rybinsk Motor Engineering Design Bureau, Ramenskoye Instrument Engineering Design Bureau, and Instrument Engineering R&D Institute. Included in government's civil aviation plan, but minimal funding by public money. Model displayed 1989 Paris Air Show; funding ended with collapse of USSR; project revived 1992, with priority on more marketable passenger and passenger/cargo variants, with imported engines, propellers and avionics.

Manufacture of flying prototype and test airframe under way at KnAAPO's Komsomolsk-on-Amur plant by 1993; Rybinsk TVD-500 turboprop engines originally intended; agreement with General Electric to fit CT7-9 turboprops early 1995; design further refined and military versions reintroduced 1996; first flight scheduled second half 1996 but repeatedly postponed; prototype (RA-82911, in Su-80GP configuration) shown on production line on 15 May 1998 to representatives of Border Guards, MoD, and Magadan, Khabarovsk and Petropavlovsk-Kamchatski airlines; was handed over to Sukhoi for flight test preparation in January 2000 and relocated to Zhukovsky; first flight 4 September 2001; had flown 40 sorties by May 2002.

Second prototype was due to follow in early 2002, but unreported by early 2004; two static airframes (02 and 04) for structural testing, of which first was delivered to SIBNIA at Novosibirsk in early 1998 and second was to have been completed in third quarter of 2001. By early 2002, had been redesignated Su-80. 900-sortie certification flight test programme will lead to Russian AP-25 and FAR Pt 25 in 2004. First four production aircraft being manufactured at KnAAPO by early 2002; these have 1.40 m (4 ft 7 in) fuselage stretch, compared with prototype, increasing passenger capacity from originally 26 to 30. Discussions were under way during 2003 for possible licensed production in China, South Korea and the United Arab Emirates.

CURRENT VERSIONS: **Su-80GP:** Basic cargo/passenger (*gruzo/passazhirski*) version.

Detailed description applies specifically to Su-80GP; generally to all versions.

Su-80M: Medical evacuation (10 casualties) version.

Su-80P: Passenger (*passazhirski*) version.

Su-80PT: Patrol transport, embodying Leninets Strizh (martin) avionics suite, with undernose 360° search radar and rotating electro-optic (FLIR/LLLTV) sensor turret under centre-fuselage; one operator's console in cabin. Able to perform 6 to 9 hour patrol missions over sea and land frontiers up to 189 n miles (350 km; 217 miles) from base; personnel and cargo transportation; and airdropping up to 20 paratroops and/or cargo. Flight crew of two or three.

Prototype Sukhoi Su-80 twin-turboprop transport *(Paul Jackson)* ***NEW***/0576627

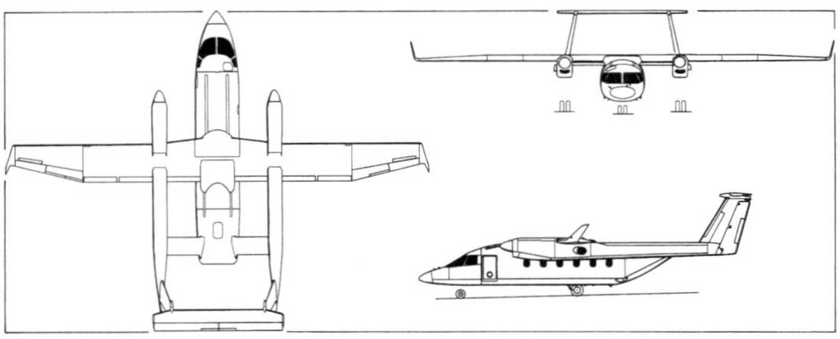

Production version of Sukhoi Su-80 0100419

Su-80TD: Troop, paratrooop (total 21), freight and medevac transport, generally similar to civil S-80GP, but with mechanised cargo-handling equipment.

CUSTOMERS: Market for 160 to 200 domestic sales estimated in 2001. Interest from 15 domestic customers in late 2002; export prospects include Argentina, China, Czech Republic, Indonesia, Malaysia, Thailand and Vietnam. Proposed for Russian Air Forces' future tactical military transport aircraft requirement and for Malaysian maritime patrol aircraft requirement. Total of 64 orders held by late 2002.

COSTS: Notionally funded by Russian Federation government but, by early 1997, KnAAPO had invested over Rb20 billion in development, compared with Rb1.5 billion received from official sources. Of US$20 million expended by mid-2000, two-thirds provided by KnAAPO; thereafter, Sukhoi providing funds for flight test programme, estimated as further US$20 million. Flyaway price US$5.5 million to US$6 million (2002 estimate).

DESIGN FEATURES: Utility freighter with unobstructed rear access for loading and wide track landing gear. Basically conventional high-wing, podded fuselage, twin-boom, rear-loading configuration, but with short tandem-wing surfaces between each tailboom and rear fuselage; unswept wings of high-aspect ratio with no dihedral or anhedral; large-span constant chord inner panels; small sweptback winglets on tapered outer panels; sweptback vertical tail surfaces, toed slightly inward, with bridging horizontal surfaces. Systems, accessories and components of Su-25, Su-27 and Su-35 embodied in Su-80. Automatic built-in systems testing.

FLYING CONTROLS: Conventional and manual. Actuation by rods and cables; flaps in three sections per wing with one section inboard of boom; split ailerons with two trim tabs on port side and one on starboard, electrically actuated; electrically actuated trim tab in each rudder.

STRUCTURE: Materials used in construction comprise 70 per cent aluminium alloy, 8 per cent composites, 6 per cent steel, 6 per cent titanium, 2.5 per cent stainless steel, 7 per cent other non-metals and 0.5 per cent other metals.

Fuselage reinforcement band in line with propellers.

LANDING GEAR: Retractable tricycle type; main units retract rearwards and are enclosed by four sequenced doors per side; nose unit retracts rearwards with two sequenced doors; twin wheels on each unit; main units retract into tailbooms; mainwheel tyre size 660×200-356; nosewheel tyres 500×170-254; nosewheels steerable ±38°.

POWER PLANT: Two 1,305 kW (1,750 shp) General Electric CT7-9B turboprops; production aircraft with engines built locally by GE-Saturn Aero Engines; Hamilton Sundstrand 14RF-35 feathering and reversible-pitch four-blade propellers. Provision for engine to function as APU with propeller stationary. Two fuel tanks, total capacity 3,656 litres (966 US gallons; 804 Imp gallons).

ACCOMMODATION: Two pilots and 30 passengers in four-abreast seating at 76 cm (30 in), or 3,300 kg (7,275 lb) of freight or mission equipment. All accommodation pressurised and air-conditioned. Baggage space, lavatory, wardrobe or galley at front, as specified by customer. Rear ramp space utilised for luggage racks. Available in 'Salon' configuration with nine, 12 or 16 passenger seats. Typical freighter has a row of seats behind flight deck and unobstructed main hold for a small vehicle or cargo. Door at centre of cabin on port side; hydraulically actuated; rear-loading ramp; emergency door on starboard side, opposite main door.

SYSTEMS: Electrical system developed by Lucas Aerospace and Auxilec. Anti-icing system by Goodrich. APU for autonomous operation at remote, unprepared sites.

AVIONICS: Integration by Elektroavtomatika.

Comms: Com/nav, identification and ATC equipment. VOR/DME/ILS for ICAO Cat. II operation of Russian manufacture.

Flight: Elektroavtomatika PNK-80 navigation system. AFCS and autopilot and satellite nav system of Russian manufacture.

Instrumentation: Russian-manufactured five-screen EFIS proposed for production aircraft.

EQUIPMENT: Options include equipment for air photography.

ARMAMENT (Su-80PT): Typically 23 mm GSh-23L gun pod pylon-mounted on starboard side of cabin; four underwing pylons for electro-optical ASM, 20-round rocket pack, cluster of eight Vikhr tube-launched missiles and R-60 (AA-8 'Aphid') self-defence AAM.

DIMENSIONS, EXTERNAL:
Wing span .... 23.17 m (76 ft 0¼ in)
Wing chord: at root .... 2.16 m (7 ft 1 in)
at tip .... 1.20 m (3 ft 11¼ in)
Wing aspect ratio 12.2
Length overall .... 18.26 m (59 ft 10¾ in)
Height overall .... 5.74 m (18 ft 10 in)
Tailplane span .... 7.00 m (22 ft 11½ in)
Wheel track .... 6.30 m (20 ft 8 in)
Wheelbase .... 6.34 m (20 ft 9½ in)
Propeller diameter .... 3.35 m (11 ft 0 in)
Propeller ground clearance .... 1.10 m (3 ft 7¼ in)
Distance between propeller centres .... 6.30 m (20 ft 8 in)
Passenger door/rear loading ramp:
Height .... 1.82 m (5 ft 11¾ in)
Width .... 1.89 m (6 ft 2½ in)
Crew door/Type II emergency exit:
Height .... 1.275 m (4 ft 2¼ in)
Width .... 0.76 m (2 ft 6 in)
Type II emergency exits (each):
Height .... 1.315 m (4 ft 3¾ in)
Width .... 0.51 m (1 ft 8 in)
Type III emergency exits (each):
Height .... 0.915 m (3 ft 0 in)
Width .... 0.51 m (1 ft 8 in)

DIMENSIONS, INTERNAL:
Cabin: Length, excl ramp .... 7.75 m (25 ft 5 in)
Max width .... 2.17 m (7 ft 1½ in)
Max height .... 1.82 m (5 ft 11¾ in)
Volume, incl ramp .... approx 30.65 m^3 (1,082 cu ft)

AREAS:
Wings, gross .... 44.36 m^2 (477.5 sq ft)
Ailerons (total) .... 3.72 m^2 (40.04 sq ft)
Trailing-edge flaps (total) .... 7.10 m^2 (76.42 sq ft)
Fins (total) .... 5.75 m^2 (61.89 sq ft)
Rudders (total) .... 2.55 m^2 (27.45 sq ft)
Horizontal tail surfaces .... 8.17 m^2 (87.94 sq ft)
Elevators (total) .... 3.22 m^2 (34.66 sq ft)

WEIGHTS AND LOADINGS (A: Su-80GP passenger; B: Su-80GP cargo):
Max payload: A .... 2,730 kg (6,019 lb)
B .... 3,300 kg (7,275 lb)
Max fuel: A, B .... 2,870 kg (6,327 lb)
Payload with max fuel: A .... 1,280 kg (2,822 lb)
B .... 1,630 kg (3,594 lb)
Max T-O weight: A, B .... 14,200 kg (31,306 lb)
Max landing weight: A, B .... 14,350 kg (31,636 lb)
Max wing loading: A, B .... 326.8 kg/m^2 (66.95 lb/sq ft)
Max power loading: A, B .... 5.56 kg/kW (9.13 lb/shp)

PERFORMANCE (estimated):
Max cruising speed .... 254 kt (470 km/h; 292 mph)
Normal cruising speed .... 232 kt (430 km/h; 267 mph)
Max certified altitude .... 6,000 m (19,685 ft)
Service ceiling .... 7,600 m (24,934 ft)
T-O run .... 930 m (3,051 ft)
T-O balanced field length .... 1,300 m (4,265 ft)
Landing run with propeller reversal .... 640 m (2,100 ft)
Range: with max payload:
A .... 702 n miles (1,300 km; 807 miles)
B .... 567 n miles (1,050 km; 652 miles)
Range with max fuel: A, B .... 1,673 n miles (3,100 km; 1,926 miles)
Ferry range: A .... 1,754 n miles (3,250 km; 2,019 miles)
B .... 1,776 n miles (3,290 km; 2,044 miles)

OPERATIONAL NOISE LEVELS: Designed to conform to FAR Pt 36 standards

UPDATED

SUKHOI RRJ

TYPE: Regional jet airliner.

PROGRAMME: On 13 April 2001, Rosaviakosmos and The Boeing Company agreed to joint development and marketing of the Russian Regional Jet (RRJ; transliterated into Cyrillic as RRZh, but also known by Sukhoi as *grazhdanskie samolet*: Civil Aircraft). At the Paris Air Show on 21 June 2001 it was announced that AVPK Sukhoi (Sukhoi Civil Aircraft) will lead the design and manufacture of the aircraft, with Ilyushin holding responsibility for certification, and Boeing handling marketing, sales, leasing and after-sales support. Three firms signed agreement 20 July 2001.

Provisional configuration and specification announced 13 August 2001, on eve of Moscow Salon. Initial phase of feasibility studies between June and December 2001; second phase completed in July 2002. Announced as winner of Russian Aviation & Space Agency (*Rosaviakosmos*) tender on behalf of Russian government for 200 regional jets on 12 March 2003; agreement on long-term co-operation between Sukhoi and Boeing signed next day. System suppliers conference 21–23 January 2004. Preliminary design review conducted with Boeing participation between 23 and 30 April 2004. Launch was due late 2004, on achievement of 50 orders; first flight is expected in 2006, with Russian AP-25 and FAR/JAR-25 certification and service entry in 2007. Irkut originally intended as sole production centre, but in February 2003 KnAAPO was nominated to build RRJ-95 version

Artist's impression of Sukhoi RRJ-75 ***NEW***/1047428

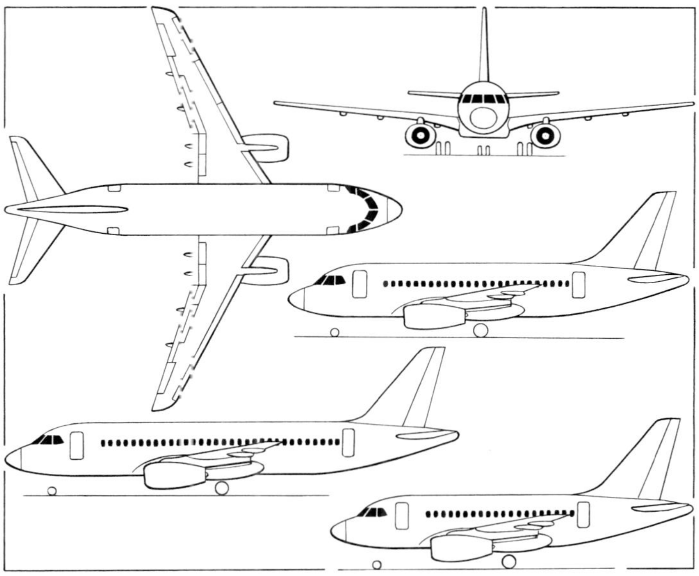

Sukhoi RRJ-75, with additional side views of RRJ-95 (centre) and RRJ-60 (bottom) *(James Goulding)* ***NEW***/0580543

as second production centre, with target combined initial output of 60 aircraft per year. RRJ-75 and -95 to be certified simultaneously, with -60 to follow. RRJ Airline Council established November 2003 with Western members to advise on specification.

CURRENT VERSIONS: **RRJ-60:** Formerly RRJ-55. Shrunk version; two fuselage sections, of 1.00 m (3 ft 3½ in) and 1.50 m (4 ft 11 in) respectively, removed fore and aft of wing.

RRJ-60LR: Long-range version.

RRJ-75: Baseline version.

RRJ-75LR: Long-range version.

RRJ-95: Stretched version; two plugs, one of 1.50 m (4 ft 11 in) and one of 2.00 m (6 ft 6¾ in) respectively, inserted fore and aft of wing.

RRJ-95LR: Long-range version.

RBJ: Proposed business jet version, under study in early 2004.

CUSTOMERS: Aeroflot has initial requirement for up to 30, having signed MoU in August 2001. Production expected to total 600 by 2020, and final total of 800, of which some 40% are expected to be ordered by CIS airlines, the remainder for export customers. Russian leasing company FLK ordered 30 in August 2003.

COSTS: Development costs estimated at US$670 million, plus US$350 million for engine (both 2003). Unit cost estimated at US$10 million to US$20 million (2003).

DESIGN FEATURES: Low-mounted, sweptback wings with podded engine on each. Sweptback horizontal and vertical tail surfaces. Airframe design optimised on RRJ-75 with goal of maximum commonality across the range of models.

STRUCTURE: Primarily aluminium alloy. Announced suppliers/supply partnerships include: B/E Aerospace (cabin interior); Aerospace/OAO Respirator (oxygen system); Air Cruiser (emergency rescue equipment); Curtiss-Wright (fire protection system); Goodrich (wheels and brakes); Hamilton Sundstrand (electrical system); Honeywell/MMPP Salyut (APU); Intertechnique/OAO Apris (fuel system); IPECO (pilots' seats); Messier Dowty/OAO Gidromash (landing gear); Liebherr/OAO Ehlektropribor (control system); Liebherr/OAO Teploobmennik (air-conditioning, pressurisation control and anti-icing systems); Parker Aerospace (hydraulic system); Sully (windows); and Thales/OAO Aviaproborholding (avionics).

LANDING GEAR: Retractable tricycle type. Twin wheels on each unit, with option of four-wheel bogey on main units.

POWER PLANT: Two Snecma/NPO Saturn SM146 turbofans, with FADEC, pod-mounted beneath wings, each rated at 62.4 kN (14,035 lb st) to 70.8 kN (15,920 lb st), according to version; 68.5 kN (15,395 lb st) to 77.5 kN (17,433 lb st) with automatic power reserve. 60 to 77.8 kN (14,000 to 16,000 lb st), according to version. Fuel capacity 13,135 litres (3,470 US gallons; 2,889 Imp gallons).

ACCOMMODATION: RRJ-60 seats up to 63 passengers; RRJ-75 seats up to 78 passengers; RRJ-90 seats up to 93 passengers, all in single-class accommodation at 81 cm (32 in) seat pitch, five abreast. Narrow- and wide-aisle configurations available. Alternative two-class configuration for baseline RRJ-75 has eight first class seats in forward section and 62 tourist class seats in main cabin. Baggage bins allow 0.08 m³ (2.74 cu ft) per passenger. Underfloor baggage compartments beneath front and rear of cabin.

AVIONICS: Cat. IIIa landing capability.

All data are provisional.

DIMENSIONS, EXTERNAL:

Wing span ... 27.80 m (90 ft 2¾ in)
Wing aspect ratio ... 9.8
Length overall: RRJ-60 ... 23.87 m (80 ft 4½ in)
 RRJ-75 ... 26.37 m (86 ft 11¼ in)
 RRJ-95 ... 29.87 m (98 ft 5 in)
Fuselage: Max width ... 3.46 m (11 ft 4¼ in)
 Max depth ... 3.62 m (11 ft 10½ in)
Height overall ... 10.28 m (32 ft 10½ in)
Tailplane span ... 10.03 m (32 ft 10¾ in)
Wheel track ... 5.75 m (18 ft 0½ in)
Wheelbase: RRJ-75 ... 9.34 m (32 ft 0¾ in)
Main passenger door (port, fwd):
 Height ... 1.83 m (6 ft 0 in)
 Width ... 0.86 m (2 ft 9¾ in)
Passenger door (port, rear): Height ... 1.83 m (6 ft 0 in)
 Width ... 0.76 m (2 ft 6 in)
Service doors (stbd, fwd and rear): Height ... 1.65 m (5 ft 5 in)
 Width ... 0.76 m (2 ft 6 in)
Baggage doors: Height ... 0.89 m (2 ft 11 in)
 Width ... 1.22 m (4 ft 0 in)

DIMENSIONS, INTERNAL:

Cabin:
 Length: RRJ-60 ... 14.65 m (48 ft 0¾ in)
  RRJ-75 ... 17.15 m (56 ft 3¼ in)
  RRJ-95 ... 20.65 m (67 ft 9 in)
 Width: max ... 3.26 m (10 ft 8¼ in)
  at floor ... 3.00 m (9 ft 10 in)
 Max height ... 2.12 m (6 ft 11½ in)
Overhead bins (per passenger) ... 0.06 m³ (2.12 cu ft)
Baggage hold:
 forward:
  Length: RRJ-60 ... 2.475 m
   RRJ-75 ... 3.475 m
   RRJ-95 ... 4.975 m
  Height: all ... 1.015 m
  Volume: RRJ-60 ... 4.97 m³
   RRJ-75 ... 6.97 m³
   RRJ-95 ... 9.99 m³
 aft:
  Length: RRJ-60 ... 3.47 m
   RRJ-75 ... 4.97 m
   RRJ-95 ... 6.97 m
  Height: all ... 1.015 m
  Volume: RRJ-60 ... 6.11 m²
   RRJ-75 ... 9.13 m²
   RRJ-95 ... 13.15 m²

AREAS:

Wings, gross ... 77.00 m² (828.8 sq ft)
Flaps (total) ... 13.90 m² (149.62 sq ft)
Fin ... 15.40 m² (165.76 cu ft)
Tailplane ... 19.60 m² (210.97 sq ft)

WEIGHTS AND LOADINGS:

Basic operating weight:
 RRJ-60, RRJ-60LR ... 21,900 kg (48,281 lb)
 RRJ-75 ... 22,910 kg (50,508 lb)
 RRJ-75LR ... 23,170 kg (51,081 lb)
 RRJ-95 ... 24,470 kg (53,947 lb)
 RRJ-95LR ... 24,715 kg (54,487 lb)
Max payload:
 RRJ-60, RRJ-60LR ... 7,100 kg (15,653 lb)
 RRJ-75, RRJ-75LR ... 9,140 kg (20,150 lb)
 RRJ-95, RRJ-95LR ... 11,870 kg (26,169 lb)
Max T-O weight: RRJ-60 ... 35,790 kg (78,517 lb)
 RRJ-60LR ... 38,950 kg (85,065 lb)
 RRJ-75 ... 38,820 kg (85,473 lb)
 RRJ-75LR ... 42,280 kg (93,200 lb)
 RRJ-95 ... 42,520 kg (93,707 lb)
 RRJ-95LR ... 45,880 kg (101,159 lb)
Max landing weight: RRJ-60, RRJ-60LR ... 31,620 kg (69,710 lb)
 RRJ-75, RRJ-75LR ... 34,960 kg (77,073 lb)
 RRJ-95, RRJ-95LR ... 39,385 kg (86,829 lb)
Max zero-fuel weight: RRJ-60, RRJ-60LR ... 29,000 kg (63,934 lb)
 RRJ-75 ... 32,050 kg (70,658 lb)
 RRJ-75LR ... 32,310 kg (71,231 lb)
 RRJ-95 ... 36,340 kg (80,116 lb)
 RRJ-95LR ... 36,585 kg (80,656 lb)
Max wing loading: RRJ-60 ... 462.5 kg/m² (94.73 lb/sq ft)
 RRJ-60LR ... 501.1 kg/m² (102.63 lb/sq ft)
 RRJ-75 ... 503.5 kg/m² (103.13 lb/sq ft)
 RRJ-75LR ... 549.0 kg/m² (112.45 lb/sq ft)
 RRJ-95 ... 552.0 kg/m² (113.06 lb/sq ft)
 RRJ-95LR ... 595.9 kg/m² (122.05 lb/sq ft)

PERFORMANCE:

Cruising speed: max ... M0.81
 normal ... M0.78
Approach speed at max landing weight:
 RRJ-60, RRJ-60LR ... 120 kt (222 km/h; 139 mph)
 RRJ-75, RRJ-75LR ... 127 kt (235 km/h; 146 mph)
 RRJ-95, RRJ-95LR ... 135 kt (250 km/h; 155 mph)
Initial cruise altitude ... 10,670 m (35,000 ft)
Max certified altitude ... 12,497 m (41,000 ft)
T-O field length: RRJ-60 ... 1,360 m (4,462 ft)
 RRJ-60LR ... 1,595 m (5,233 ft)
 RRJ-75 ... 1,612 m (5,289 ft)
 RRJ-75LR ... 1,645 m (5,397 ft)
 RRJ-95 ... 1,685 m (5,528 ft)
 RRJ-95LR ... 1,980 m (6,496 ft)
Landing field length:
 RRJ-60, RRJ-60LR ... 1,313 m (4,308 ft)
 RRJ-75, RRJ-75LR ... 1,404 m (4,606 ft)
 RRJ-95, RRJ-95LR ... 1,525 m (5,003 ft)
Range with max payload:
 RRJ-60 ... 1,861 n miles (3,546 km; 2,203 miles)
 RRJ-60LR ... 2,810 n miles (5,204 km; 3,233 miles)
 RRJ-75 ... 1,799 n miles (3,361 km; 2,088 miles)
 RRJ-75LR ... 2,767 n miles (5,124 km; 3,184 miles)
 RRJ-95 ... 1,771 n miles (3,279 km; 2,038 miles)
 RRJ-95LR ... 2,495 n miles (4,620 km; 2,871 miles)

UPDATED

TECHNOAVIA

NAUCHNO-KOMMERCHESKY FIRMA TECHNOAVIA (TECHNOAVIA SCIENTIFIC AND COMMERCIAL FIRM)
5 Magistralnaya Str, 14 build 1, 123007 Moscow
Tel/Fax: (+7 095) 748 11 84 and 11 87
e-mail: info@tehnoavia.ru
Web: http://www.tehnoavia.ru

Marketing of Technoavia aircraft has been taken over by Smolensk. Products currently offered are the SM-92 Finist family and SM-2000 development of the Yakovlev Yak-18.

UPDATED

TUPOLEV

TUPOLEV OAO (TUPOLEV JSC) AVIATSIONNY NAUCHNO-TEKHNISHESKY KOMPLEKS IMENI A N TUPOLEVA OAO (AVIATION SCIENTIFIC-TECHNICAL COMPLEX NAMED FOR A N TUPOLEV JSC)
REGISTERED ADDRESS: Adreulitsa Bakrushina 23, Korpus 1, 113054 Moskva
Tel: (+7 095) 238 34 13
Fax: (+7 095) 238 68 41
POSTAL ADDRESS: Naberezhnaya Akademika Tupoleva 17, 105005 Moskva
Tel: (+7 095) 267 25 33
Fax: (+7 095) 261 71 71
e-mail: tu@tupolev.ru
Web: http://www.tupolev.ru
PRESIDENT: Igor Shevchuk
HEAD OF ECONOMIC RELATIONS: Evgeny Efimov

Under decree of 30 June 1999, Tupolev PSC (Public Stock Company) formed 30 July 1999, combining Tupolev design bureau and Aviastar production plant at Ulyanovsk, with financial interest from Russian government amounting to 51 per cent, plus 43.6 per cent from Aviastar. PSC also holds intellectual rights to Tupolev designs previously owned by government. New entity was recertified by Interstate Aviation Committee Aviation Registrar in December 2000. Kazan plant will be incorporated later.

Other activities include sales and leasing of jet transports, aircraft operation and after-sales support. Tupolev Aviation Leasing Company supplies Tu-204, Antonov An-124 and Yakovlev Yak-40 transports.

Government plans to rationalise Russian aerospace industry, announced May 2001, call for incorporation of Tupolev (with Aviakor, Aviastar and Kazan plants) into RSK 'MiG' grouping, also including Kamov bureau and Sokol, KAPP and Arsenyev plants.

Tupolev Bureau was founded by Andrey N Tupolev (1888–1972) on 22 October 1922 and concentrated primarily on large military and civil aircraft until the early 1990s; Andrey A Tupolev (1925–2000) succeeded his father as General Designer; since 1922, Bureau has designed 300 aircraft, of which nearly 90 were built as prototypes and 35 placed in production; current effort is 80 per cent civil programmes.

It was suggested in 1998 that Tupolev had begun preliminary design of a new bomber to replace the Tu-160 and, probably, Tu-95/142; this would enter service some time after 2010 and may be based on earlier studies, such as the Tu-202 bomber of the 1980s or the more recent Tu-404 700/850-seat airliner. Continued work was confirmed in late 2002, although as a private study and "based on a new concept", which "is at the very earliest stage". In addition, Bureau remains heavily committed to the development of a civil supersonic transport aircraft and of cryogenic fuels for jet and turboprop transports.

Tupolev's head office, main design bureau and experimental facility are in Moscow; Tomilino branch and flight research centre at Zhukovsky; and design offices at Samara, Kazan and Voronezh.

UPDATED

TUPOLEV TU-160

NATO reporting name: Blackjack
Unofficial name: Belyi Lebed (White Swan)

TYPE: Strategic bomber.

PROGRAMME: Designed as Article 70 (or Aircraft K) under leadership of Valentin I Bliznuk; programme began 1967, but relaunched following issue of more modest specification in 1970; competitors were Myasishchev M-18 and Sukhoi T-4MS; derived from unbuilt Tu-135 bomber and Tu-144 derivatives; won design competition in 1972; incorporated some features from M-18. First of two prototypes (70-01) observed by intelligence source at Zhukovsky flight test centre 25 November 1981 (photographed from landing airliner); first flew 18 or 19 December 1981; first exceeded M1.0 February 1985; second (production-standard) aircraft first flew 7 October 1984. Third prototype (70-03) set world records for altitude, speed in a closed circuit and weight-to-altitude on 31 October 1989.

Tupolev Tu-160 'Blackjack' *(Yefim Gordon)* ***NEW***/0576742

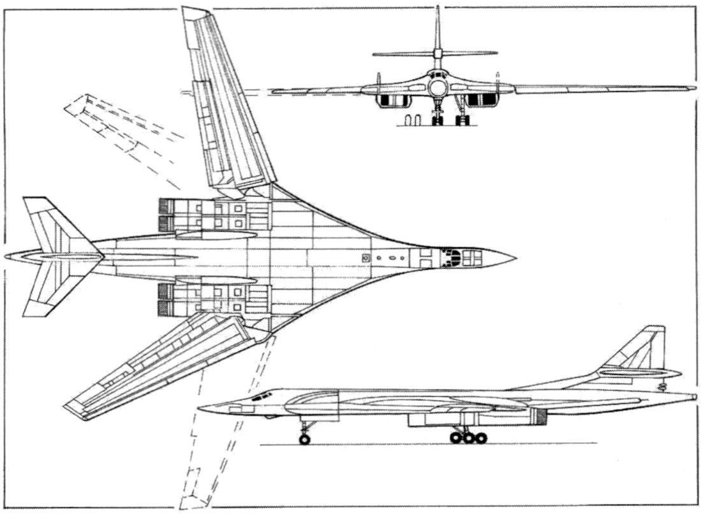

Tupolev Tu-160 strategic bomber *(James Goulding)* 0507635

Further nine closed-circuit records established 22 May 1990 by aircraft '70-304'; total 57 records including 44 world records, held by Tu-160s at various times.

Second production aircraft lost in pre-delivery take-off crash, March 1987; US Defense Secretary Frank Carlucci invited to inspect 12th aircraft built, at Kubinka airbase, near Moscow, 2 August 1988; deliveries to 184th Guards Heavy Bomber Aviation Regiment, Priluki airbase, Ukraine, began April 1987; equipment of 1096th HBAR at Engels from 16 February 1992, but only six received before production at Kazan airframe plant terminated 1992; of 100 aircraft due to be built, at least 32 (including prototypes) accounted for by mid-1990s; unconfirmed reports suggest total of 40 having flown, plus three uncompleted at Kazan.

In 2001, Russia was reported to be formulating a Tu-160 upgrade, to be undertaken at Kazan, which includes provision for a conventionally armed cruise missile. Upgrade of 15 aircraft was agreed late 2001 and formally announced at Kazan on 18 January 2002; work will extend service lives to 2030; first upgrade candidate was delivered to Kazan on 5 April 2002. In October 2002, Tupolev confirmed this to be a two-stage upgrade, the first part (2003–04) concentrating on the addition of new weapons and the second, at an unspecified date, involving new avionics for navigation, attack and electronic warfare, with high degree of commonality with parallel Tu-22M and Tu-142 (Tu-95MS) upgrades.

CURRENT VERSIONS: **Tu-160** ('Blackjack'): Strategic bomber.

Tu-160SK: Commercialised, demilitarised version, as carrier component of Burlak aviation space launch system; Burlak-Diana two-stage rocket, carrying payload, under fuselage on centreline mount. Announced at Singapore Air Show '94; proposed by Russian partners MKB-Raduga, OKB MEI and Tupolev, with German company OHB-System. At the 1995 Paris Air Show, Tu-160 0401 was exhibited statically (the type's Western debut) with a model Burlak rocket below the fuselage. Development of the system continued even after the German government withdrew funding in 1998, and may form the basis of the Ukrainian/PIC HAAL-2000 programme.

Tu-160M: Proposed stretched variant carrying two 2,700 n mile (5,000 km; 3,107 mile) range hypersonic Kh-90 missiles.

Tu-160P: Proposed very long-range escort fighter.

Tu-160PP: Proposed escort jammer.

Tu-160R: Proposed strategic reconnaissance platform.

CUSTOMERS: In January 2001, Russia had seven operational Tu-160s; a further eight undergoing refurbishment for delivery over following months; and one under construction.

Of original deliveries, Ukraine government seized 19 at Priluki on achieving independence, pre-empting planned transfer to Engels; purchase of these by Russia was subject to protracted negotiations, and aircraft deteriorated in storage; March 1996 agreement on transfer of 10 best airframes was not implemented; attempts to purchase eight failed in March 1998 and Russia then supposedly abandoned hopes of expanding Tu-160 fleet. However, October 1999 announcement revealed eight to be returned to Russia for refurbishment (together with three Tu-95MSs, for a total of US$285 million) and these were delivered between 5 November 1999 and 21 February 2000. Six others scrapped with US assistance, last being destroyed on 4 February 2001. August 2000 report mentioned further three Tu-160s which Ukraine could return to Russia in part-payment for natural gas deliveries; these later destoyed. Tenth and final remaining Ukrainian Tu-160 was flown to Poltava Kondratyuk-Shagray aerospace museum on 5 April 2000.

Russia maintained force of six (declared as ALCM carriers under START) at Engels (where 1096th HBAR was redesignated as part of the 121st Guards HBAR, within 22 Air Division, in 1994), plus flying testbed at Zhukovsky; at least four more were derelict at Zhukovsky by 1995. Despite this, 1998 US estimates of Russian combat forces reported total holding of nominally serviceable aircraft as 25, though this would not seem to have been possible. Eight more had been due to re-enter service from 2001 after purchase from Ukraine, while plan announced in June 1999 to complete one unfinished Tu-160 at Kazan. This aircraft ('07' *Aleksandr Molodchyi*) was delivered on 5 May 2000. Another, brand new Tu-160 (the first of three more incomplete, unfinished aircraft at Kazan) was due to be delivered in late 2002 (although this was not achieved, because of funding shortage, and postponed to 2004), and the remaining two will follow, increasing the fleet to 18. At one time there were plans to refurbish some (perhaps four) of the grounded aircraft at Zhukhovsky. Ex-Ukrainian aircraft failed to enter service when planned due to poor condition and need for extensive refurbishment; requirements were still being discussed in early 2002. Plans not adopted for Tu-160 to re-enter limited production, to meet a stated requirement for 25 operational 'Blackjacks' by 2003, allowing formation of a second regiment. Original six at Engels are also named: '01' *Mikhail Gromov* (crashed

18 September 2003; second Tu-160 loss); '02' *Vasily Retsetnikov*; '03' *Pavel Taran*; '04' *Ivan Yargin*; '05' *Il'ya Muromets* (1) and '06' *Il'ya Muromets* (2). The 10th Tu-160 (fleet number '16') was named *Aleksey Plokhov* on 17 April 2003 and another is to be named *Valery Chkalov*.

On 3 March 1999, the Russian Commonwealth Aerospace Technology Consortium (RCATC) was authorised by the Ukrainian government to sell three demilitarised Ukrainian Air Force Tu-160s, plus spares, to Platforms International Corporation of the USA, with which it has finalised a strategic partnership. The US$20 million deal includes a 20 per cent interest in Orbital Network Services Corporation, which plans to use the aircraft as reusable communications satellite launchers in its HAAL-2000 High-Altitude Air Launch programme. The aircraft would probably be modified to Tu-160SK standards and continue to be based at Priluki, maintained and flown by Ukrainian crews, but flown to customer countries for individual space launch missions.

DESIGN FEATURES: Intended for high-altitude standoff role carrying ALCMs and for defence suppression, using short-range attack missiles similar to US Air Force SRAMs, along path of bomber making low-altitude penetration to attack primary targets with free-fall nuclear bombs or missiles; this implies capability of subsonic cruise/supersonic dash at almost M2 at 18,300 m (60,000 ft) and transonic flight at low altitude. About 20 per cent longer than USAF B-1B, with greater unrefuelled combat radius and much higher maximum speed; low-mounted variable geometry wings, with very long and sharply swept fixed root panel; small diameter circular fuselage; horizontal tail surfaces mounted high on fin, upper portion of which is pivoted one-piece all-moving surface; large dorsal fin; engines mounted as widely separated pairs in underwing ducts, each with central horizontal V wedge intakes and jetpipes extending well beyond wing centre-section trailing-edge; manually selected outer wing sweepback 20, 35 and 65°; when wings fully swept, inboard portion of each trailing-edge flap hinges upward and extends above wing as large fence; unswept tailfin; sweptback horizontal surfaces, with conical fairing for brake-chute aft of intersection.

FLYING CONTROLS: Quadruplex fly-by-wire with mechanical reversion. Full-span leading-edge flaps, long-span double-slotted trailing-edge flap and inset drooping aileron on each wing; five-section spoilers forward of flaps; all-moving vertical and horizontal one-piece tail surfaces.

STRUCTURE: Slim and shallow fuselage blended with wing-roots and shaped for maximum hostile radar signal deflection; 20 per cent titanium, including leading-edges and wing centre-section spar box.

LANDING GEAR: Twin nosewheels retract rearward; main gear comprises two bogies, each with three pairs of wheels; retraction very like that on Tu-154 airliner; as each leg pivots rearward, bogie rotates through 90° around axis of centre pair of wheels, to lie parallel with retracted leg; gear retracts into thickest part of wing, between fuselage and inboard engine on each side; so track relatively small. Nosewheel tyres size 1080×400; mainwheel tyres size 1260×425.

POWER PLANT: Four purpose-designed Samara NK-321 turbofans, each 137.3 kN (30,865 lb st) dry, 245 kN (55,115 lb st) with afterburning. In-flight refuelling probe retracts into top of nose. Fuel in centre-section spar box and in outer wings.

ACCOMMODATION: Four crew members in pairs, on individual Zvezda K-36LM zero/zero ejection seats, in pressurised compartment; one window each side of flight deck can be moved inward and rearward for ventilation on ground; flying controls use fighter-type sticks rather than yokes or wheels; crew enter via extending ladder in nosewheel bay. Cooking facilities, folding bed and lavatory.

AVIONICS: Systems utilise around 100 digital processors and eight digital nav computers.

Radar: Obzor (NATO 'Clam Pipe') nav/attack radar in slightly upturned dielectric nosecone with separate Sopka radar providing terrain-following capability.

Flight: K-042K astro-inertial nav with map display.

Instrumentation: Analogue instruments. No HUD or CRTs.

Mission: OPB-15 strike sight fairing with flat glazed front panel, under forward fuselage, for video camera to provide visual assistance for weapon aiming.

Self-defence: Baikal self-protection system, with integrated RHAWS, chaff/flare dispensers in tailcone and active jamming.

ARMAMENT: No guns. Internal stowage for free-fall bombs, mines, short-range attack missiles or ALCMs; two tandem 12.80 m (42 ft) long weapon bays; MKU-6-5U rotary launcher for six (or up to 12) Kh-55MS (AS-15 'Kent') or RKV-500B (AS-15B 'Kent-B') ALCMs or 12 to 24 Kh-15P (AS-16 'Kickback') SRAMs in each bay. Aircraft upgraded from 2002 onwards have ability to launch Kh-555 missiles; later plans envisage carriage of up to 12 non-nuclear Kh-101 ALCMs, when available.

DIMENSIONS, EXTERNAL:

Wing span: fully spread (20°)	55.70 m (182 ft 9 in)
35° sweep	50.70 m (166 ft 4 in)
fully swept (65°)	35.60 m (116 ft 9¾ in)
Wing aspect ratio: fully spread	8.6
Length overall	54.10 m (177 ft 6 in)
Height overall	13.10 m (43 ft 0 in)
Tailplane span	13.25 m (43 ft 5¾ in)
Wheel track	5.40 m (17 ft 8½ in)
Wheelbase	17.88 m (58 ft 8 in)

DIMENSIONS, INTERNAL:

Weapons bay (each): Volume	43.0 m^3 (1,518 cu ft)

AREAS:

Wings, gross: fully swept	360.00 m^2 (3,875.0 sq ft)
fully spread	approx 400.00 m^2 (4,305.6 sq ft)
moving areas, fully swept (total)	approx 180.00 m^2 (1,937.5 sq ft)

WEIGHTS AND LOADINGS:

Weight empty	110,000 kg (242,505 lb)
Weight empty, equipped	117,000 kg (257,940 lb)
Max fuel	171,000 kg (376,990 lb)
Max weapon load	40,000 kg (88,185 lb)
Normal T-O weight	267,600 kg (589,950 lb)
Max T-O weight	275,000 kg (606,260 lb)
Max landing weight	155,000 kg (341,710 lb)
Max power loading	280 kg/kN (2.75 lb/lb st)

PERFORMANCE:

Max level speed at 12,200 m (40,000 ft)	M2.05 (1,200 kt; 2,220 km/h; 1,380 mph)
Cruising speed at 13,700 m (45,000 ft)	M0.9 (518 kt; 960 km/h; 596 mph)
Max rate of climb at S/L	4,200 m (13,780 ft)/min
Service ceiling	15,000 m (49,200 ft)
T-O run at max AUW	2,200 m (7,220 ft)
Landing run at max landing weight	1,600 m (5,250 ft)
Radius of action at M1.5	1,080 n miles (2,000 km; 1,240 miles)
Max unrefuelled range	6,640 n miles (12,300 km; 7,640 miles)
g limit	+2

UPDATED

TUPOLEV TU-44

TYPE: Light utility twin-prop transport.

PROGRAMME: Announced at Moscow, August 2003; joint venture by Tupolev (design) and SKB ATIC (programme management); production jigs by ATIC and Composite-Avia. Four flying prototypes, plus two static/fatigue test airframes then planned for completion in 2004, but none reported by mid-year; certification targeted for third quarter of 2005. Twin-float version also planned.

CUSTOMERS: First two aircraft reportedly ordered by unidentified Russian customer by January 2004.

COSTS: Development costs estimated at approximately US$3 million (2003).

DESIGN FEATURES: Robust passenger and cargo transport suitable for operation from semi-prepared airstrips. Adaptable for paramilitary support, with provision for light armament or weapons on two underwing pylons. High wing braced by single strut each side; taper on trailing-edges. Sweptback fin with fillet; tapered tailplane. Engines attached below inboard wing.

FLYING CONTROLS: Conventional and manual. Flight-adjustable trim tabs in rudder, each elevator and each aileron. Rudder and elevators horn-balanced. Three-section flaps each side, two inboard components divided by engine nacelle.

STRUCTURE: Metal fuselage main framework; up to 90 per cent of remainder in "modern composite materials" by SKB ATIC; final assembly by VASO at Voronezh.

LANDING GEAR: Tricycle type; fixed. Wide-track, fuselage-mounted, self-sprung cantilever mainwheel legs. Floats for future seaplane version have been wind-tunnel tested at CAHI (TsAGI).

POWER PLANT: Two LOM M337C four-cylinder inline engines, each 184 kW (247 hp) and each driving a two-blade propeller. Normal fuel capacity 350 litres (92.5 US gallons; 77.0 Imp gallons). Optional underwing auxiliary tanks for additional 120 litres (31.7 US gallons; 26.4 Imp gallons).

ACCOMMODATION: Pilot plus up to seven passengers. Cargo door port side, behind wing trailing-edge; emergency exits opposite, and above flight deck. Baggage compartment behind cabin.

DIMENSIONS, EXTERNAL:

Wing span	13.40 m (43 ft 11½ in)
Length overall	10.10 m (33 ft 1½ in)
Height overall	4.40 m (14 ft 5¼ in)
Wheel track	3.20 m (10 ft 6 in)

DIMENSIONS, INTERNAL:

Cabin: Length	4.40 m (14 ft 5¼ in)
Max width	1.56 m (5 ft 1½ in)
Max height	1.50 m (4 ft 11 in)
Baggage compartment: Length	1.80 m (5 ft 10¾ in)
Max width	1.20 m (3 ft 11¼ in)
Max height	1.10 m (3 ft 7¼ in)

WEIGHTS AND LOADINGS:

Max payload	600 kg (1,333 lb)
Max T-O and landing weight	2,100 kg (4,629 lb)

PERFORMANCE (estimated):

Max level speed	194 kt (360 km/h; 224 mph)
Max certified altitude	6,000 m (19,680 ft)
Range at 167 kt (310 km/h; 193 mph) at FL30:	
with five passengers	648 n miles (1,200 km; 745 miles)
ferry with normal fuel	1,187 n miles (2,200 km; 1,367 miles)

UPDATED

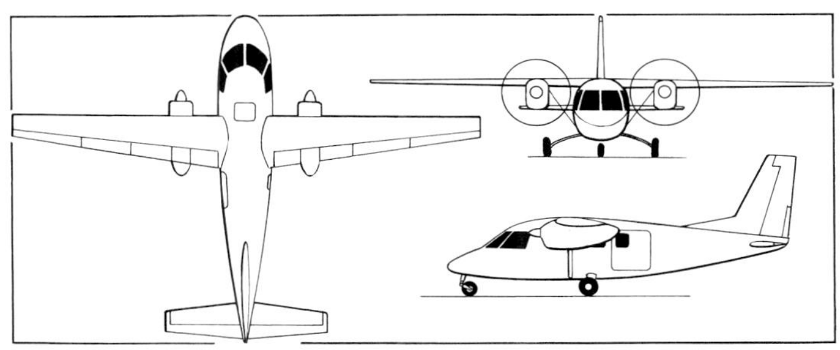

Tupolev Tu-44 utility twin *(James Goulding)* ***NEW***/0569765

TUPOLEV TU-156

TYPE: Technology demonstrator.

PROGRAMME: On 23 April 1994, Russian government allocated funding for conversion of three Tu-154s to Tu-156 standard, delivery of 12 NK-89 turbofans and six cryogenic fuel systems; an installation to supply LNG was to be established at Samara. Order reduced to one Tu-156; delivery originally scheduled for 1998, but had not taken place by mid-2004; federal funding of Rb7.5 million and Rb17 million was planned for allocation in 2001 and 2002 respectively to cover construction of a ground test rig for the cryogenic fuel system; total cost of development programme estimated at Rb200 million to 2005; to be funded by Rosaviakosmos (70 per cent), with balance shared equally between Tupolev and Kuznetsov. According to a March 2003 report, half the required funds had been allocated to the Tu-156 programme by that time, but the completion date had been further extended. By mid-2004, reports appeared to indicate that cryogenic propulsion development programme was focused on Tu-204K, though increased funding unlikely before 2006.

UPDATED

TUPOLEV TU-204 AND TU-214

TYPE: Twin-jet airliner.

PROGRAMME: Development, to replace Tu-154 and Il-62, announced 1983; preliminary details available 1985; programme finalised 1986; first prototype (SSSR-64001), with PS-90AT engines, flown 2 January 1989 by Tupolev chief test pilot A Talalakine; two more prototypes (RA-64003 and 004) followed, plus two (002 and 005) for structural and fatigue testing. Second version, with RB211-535E4-B engines, flew 14 August 1992 (fourth flying prototype, RA-64006) and was demonstrated at Farnborough Air Show in following month.

Prototype Tupolev Tu-204-300 (Tu-324) *(Yefim Gordon)* ***NEW***/0580750

Tupolev Tu-204C freighter *(Yefim Gordon)* ***NEW***/0580749

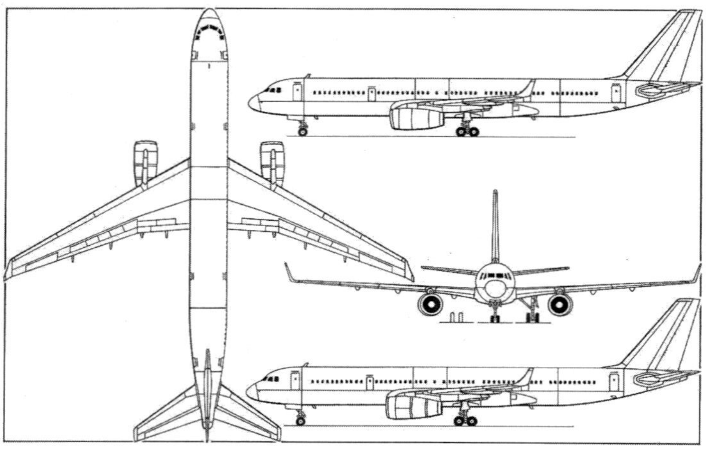

Tupolev Tu-204-200 medium-range transport (two Aviadvigatel PS-90A turbofans), with additional side view (top) of Tu-204-220 *(Jane's/Mike Keep)* 0507381

DEVELOPMENT MILESTONES

Tu-204	
Design began	1983
Programme launched	1986
First flight	2 Jan 89
Production go-ahead	1990
Certification	12 Jan 95
Entered service (Vnukovo Airlines)	23 Feb 96
Subsequent versions:	
Tu-204-300	
First flight	17 Aug 03
Tu-214	
First flight	21 Mar 96
Public debut	Sep 96
Certification	29 Dec 00
Entered service (Dal'avia)	23 May 01
Tu-234	
First flight	8 Jul 00

Production of basic Tu-204 series by Aviastar at Ulyanovsk began 1990, Tu-214 by KAPO at Kazan in 1994; Russian certification of basic Tu-204 received 12 January 1995; used initially only for freight services. First revenue-earning passenger flight, Moscow to Mineralnye Vody, operated by Vnukovo Airlines 23 February 1996. Restrictions on PS-90-engined Tu-204s were eventually removed in mid-1998. Deliveries outside CIS began on 2 November 1998 with handover of -120 and -120C freighter to Air Cairo on seven-year lease.

As at mid-2004, the variants certified to AP-25 and JAR 25 standards and in production were stated to be the Tu-204-100, Tu-204-120 and Tu-214; in production, with certification expected by year-end, was the Tu-204-300.

Avionics upgrade under discussion with NII in mid-2000; goal is six LCD, two-pilot flight deck; estimated cost US$5 million to US$8 million. First application on Tu-204-300.

CURRENT VERSIONS: **Tu-204:** Basic medium-haul airliner for up to 210 passengers or maximum payload of 21,000 kg (46,295 lb); 158.3 kN (35,580 lb st) Aviadvigatel PS-90A turbofans. Marketed 1989. RVSM approval granted 28 December 2000.

Tu-204C: Cargo version (27 t capacity) of basic Tu-204. (Western marketing designation is Tu-204C, but correct Cyrillic designation, used in Russian literature, is **Tu-204S**.) Certification received in August 2003.

Tu-204-100: Extended-range passenger version of basic Tu-204; additional fuel in wing centre-section; dimensions, payload and power plant unchanged; maximum T-O weight 103,000 kg (227,070 lb). Marketed 1993. First flight 9 August 2001. Aviastar production.

Tu-204-100C: Freighter, as Tu-204-100. Payload increased at expense of reduced maximum range. Marketed 1994. Bulk freight volume 180.0 m^3 (6,357 cu ft), not including fore and aft underfloor holds.

Tu-204-120: As Tu-204–100, but with 189.4 kN (42,580 lb st) Rolls-Royce RB211-535B75 turbofans; maximum T-O weight 103,000 kg (227,070 lb); Russian avionics; prototype (RA-64006) flew 14 August 1992. First production aircraft (RA-64027) flew 7 March 1997; gained limited Russian certification on 16 July 1997 and full certification 12 months later. JAA certification effort began May 2001. First five Sirocco 120/120Cs have Russian avionics; later aircraft are Phase II with Honeywell items, including VIA 2000 suite. Aviastar production.

Tu-204-120C: As -100C, but RB211 engines; two for Air Cairo (via Sirocco). First aircraft (RA-64028) flew 10 November 1997; domestic certification completed by August 1998. JAA (now EASA) certification effort began May 2001 and had completed initial testing by beginning of 2004; approval anticipated by end of that year. First delivery 2 November 1998. Aviastar production. High gross weight version (**Tu-204-120CE?**), with 30,000 kg (66,139 lb) payload, certified 2004.

Tu-204-200: Further increase in payload and T-O weight; small increase in fuel in wing centre-section and adjacent baggage hold; dimensions and power plant as Tu-204-100; strengthened landing gear; marketed 1994. First example of Series 200 was a Tu-214 freighter, first flown in March 1996; first true Tu-204-200 (RA-64036) was nearing completion at Ulyanovsk in late 1998; PS-90 engines. By mid-2000, **Tu-214** designation was also being used for this version.

Tu-204-200C: As Tu-204-200. Payload increased at expense of reduced maximum range. Marketed 1994. Also designated **Tu-214C**.

Tu-204-200C³ (Cargo, Converted, Containerised): Combi version, marketed as **Tu-214C³**.

Tu-204-220: As Tu-204-200, but 191.7 kN (43,100 lb st) Rolls-Royce RB211-535E4 or RB211-535F5 turbofans; Tupolev-funded variant with 185.5 kN (41,700 lb st) Pratt & Whitney PW2240 turbofans; promotion began 1998. Is Tu-214 variant, designated **Tu-224** by KAPO.

Tu-204-220C: Cargo version of Tu-204-220.

Tu-204-300: Short-fuselage (162-passenger), extended-range version of Tu-204-100. Prototyped by KAPO (static test only) as Tu-234 (see below). First flight test aircraft (RA-64026, built by Aviastar at Ulyanovsk) rolled out as Tu-234, minus engines, in mid-1996; fitted with PS-90s in early 2000 and reportedly first flew 8 July 2000 (ahead of prototype), although this can not be confirmed. Option for 20 signed by Transaero at MAKS air show, August 2003. Aviastar first production example (PS-90P engines) in final assembly May 2003; maiden flight (RA-64026) 17 August 2003 and public debut at MAKS 2003 following day. Certification targeted for 2004.

Tu-204K: Projected dual-fuel version. Previously known as Tu-204-600, then Tu-206. Inherits earlier experience gained in earlier Tu-136 and Tu-156 cryogenic propulsion programmes; 22,500 kg (49,604 lb) of liquefied natural gas (LNG) in conformal dorsal tanks, plus 5,500 kg (12,125 lb) Avgas in wings. Under development.

Tu-214: Longer-range, increased-weight version of Tu-204-100, produced by KAPO. Prototype (RA-64501) rolled out at Kazan 5 February 1996; first flew 21 March 1996; public debut at Farnborough Air Show, September 1996. Provisional certification awarded 1998. PW2037 being assessed as an alternative to PS-90A. By early 2000, Kazan stock included seven partly complete Tu-214s from initial production batch of 18. Flight testing of prototype suspended after some 300 sorties, but resumed in May 2000 with further series of 85 flights leading to 'second class' AP-25 certification in March 2000 and full approval on 29 December 2000. First of two aircraft for Dal'avia (s/n 64502) made its first flight on 10 April 2001, and was delivered on 23 May 2001, for service on routes between Moscow and Asia; second received 20 October 2001.

Tu-214F (previously **Tu-214C** or **Tu-214-200C**): Dedicated cargo version (30 t capacity); 3.405 × 2.18 m (11 ft 2 in × 7 ft 1¾ in) door. Alternative designation **Tu-214F** quoted in 2002. Atlant-Soyuz may become launch customer.

Tu-214D: Long-range, 150-passenger version, first mentioned mid-2002. Auxiliary fuel tanks in underfloor holds, potentially extending range to 4,967 n miles (9,200 km; 5,716 miles).

Tu-214VSSN: VVIP version (*vozdushnoe sudno specialnogo naznacheniya*) for Russian president. Fourth production Tu-214 (RA-64504) is in this configuration; first flight 22 June 2002.

Tu-216: Dual-fuel KAPO counterpart of Tu-204K.

Tu-224: KAPO equivalent of Tu-204-220. Remarks as for Tu-234 but has RB211-535E4 engines. Not yet built.

Tu-234: KAPO prototype for Tu-204-300. Announced 1994; trunk route version with fuselage shortened by 5.80 m (19 ft 0¼ in): 3.00 m (9 ft 10 in) forward and 2.80 m (9 ft 2¼ in) aft of wing; short-range and mid-range models for 166 passengers, long-range model for 99 to 162 passengers; 158.3 kN (35,580 lb st) PS-90A engines. Prototype (rebuild of original Tu-204 prototype RA-64001) rolled out on 24 August 1995 at Moscow Air Show, but used only for static testing.

CUSTOMERS: Total of approximately 37 (excluding six flying or static prototypes) built by beginning of 2004, including six 204C, ten 204-100, two 204-120, three 204-120C, one 204-200, one 204-300, six 214 and eight unidentified/not yet determined. Current operators at that time are given in the table.

There have been many more commitments and announced intentions which have failed to materialise. According to KAPO announcements in 2003, orders for the Tu-214 were received during the year from the Russian MoD (three), GTK Rossiya (three), Dal'avia (two), Omskavia (four), Aviast (four, including one cargo), and the Bulgarian Carriers Air Sofia and BH Air (total of four). Four Iranian airlines (Aria, Caspian, Iran Air Tours and Kish Air) placed an order for 20 Tu-204-100s in April 2004.

Many early Tu-204s have changed ownership since delivery. In addition, Tupolev has four aircraft, while two were used only for static testing. KMV (Kaminvodyavia) aircraft owned by Perm engine plant, which took two Tu-204s in lieu of cash payment for PS-90s; one Perm aircraft sold to Sibir in December 1999. Siberia Airlines also planning to acquire three Tu-214s from Kazan line.

COSTS: With PS-90 engines and Russian avionics US$27 million (2001); with RB211-535 engines US$38 million (1996).

DESIGN FEATURES: Conventional low/mid-wing configuration, with all surfaces swept back, and winglets; wing dihedral from roots; semi-monocoque oval section pressurised fuselage; torsion box of fin forms integral fuel tank, used for automatic trimming of CG in flight; design life 45,000 flights, 60,000 flight hours or 20 years.

TUPOLEV TU-204/214 PRODUCTION
(at early 2004)

Line No.	Registration	Operator or Fate
1	RA-64001	Storage at Zhukovsky
2	nil	Static test
3	RA-64003	Storage at Zhukovsky
4	RA-64004	Storage at Zhukovsky
5	nil	Scrapped at Novosibirsk (static test)
6	RA-64006	Storage at Ulyanovsk (Aviastar)
7	RA-64007	Storage at Vnukovo (Vnukovo)
8	RA-64008	Derelict at Vnukovo (Vnukovo)
9	RA-64009	Storage at Ulyanovsk (Vnukovo)
10	RA-64010	Storage at Ulyanovsk (Vnukovo)
11	RA-64011	Sibir
12	RA-64012	Storage at Ulyanovsk (Vnukovo)
13	RA-64013	Storage at Ulyanovsk (Vnukovo)
14	RA-64014	Kras Air
15	RA-64015	Kras Air
16	RA-64016	Kavminvodyavia (KMV)
17	RA-64017	Sibir
18	RA-64018	Kras Air
19	RA-64019	Kras Air
20	RA-64020	Kras Air
21	RA-64021	Air Rep
22	RA-64022	Kavminvodyavia (KMV)
23	SU-EAH	Air Cairo
24	LY-AGT	Aviastar
25	SU-EAI	Air Cairo
26	RA-64026	Tupolev
27	SU-EAF	Air Cairo
28	SU-EAG	Air Cairo
29	SU-EAJ	Air Cairo
30	nil	Storage at Ulyanovsk
31		Sirocco (assigned)
32	RA-64032	Air Rep
33	SU-EAK	Sirocco/Heavylift
34	nil	Incomplete
35	nil	Incomplete
36	RA-64036	Incomplete
501	RA-64501	Tupolev
502	RA-64502	Dal'avia
503	RA-64503	Dal'avia
504	RA-64504	Rossiya
505	RA-64505	Rossiya
506	RA-64506	Rossiya
507	RA-64507	Dal'avia

Wing section supercritical; sweepback 28°; thickness/chord ratio 14 per cent at root, 9 to 10 per cent at tip; negative twist.

FLYING CONTROLS: Triplex digital fly-by-wire, with triplex analogue back-up; conventional 'Y' control yokes selected after evaluation of alternative sidestick on Tu-154 testbed; inset aileron outboard of two-section double-slotted flap on each wing trailing-edge; two-section upper surface airbrake forward of each centre-section flap; five-section spoiler forward of each outer flap; four-section leading-edge slat over full span of each wing; conventional rudder and elevators; no tabs.

STRUCTURE: Approximately 18 per cent of airframe by weight of composites; three-piece two-spar wing, with metal structure, part composites skin; carbon fibre skin on spoilers, airbrakes and flaps; glass fibre wingroot fairings; all-metal fuselage, utilising aluminium-lithium and titanium; nose radome and some access panels of composites; extensive use of composites in tail unit, particularly for leading-edges of fixed surfaces and for rudder and elevators.

LANDING GEAR: Hydraulically retractable tricycle type; electrohydraulically steerable twin-wheel nose unit (±10° via rudder pedals; ±70° by electric steering control) retracts forward; four-wheel bogie main units retract inward into wing/fuselage fairings. Rubin KT-196 wheels on all units. Carbon disc brakes, electrically controlled. Tyre size 1,070×390 on mainwheels, 840×290 on nosewheels. Original Russian (NIIShP) or Goodyear radial tyres progressively replaced from 2000 by Michelin M12601-01 type.

POWER PLANT: Two turbofans (see notes on versions), underwing in composites cowlings. Tu-204-200 series carries fuel in six integral tanks in wings, in centre-section and adjacent baggage hold, and in tailfin, total capacity 40,730 litres (10,760 US gallons; 8,960 Imp gallons); torsion box of fin forms integral fuel tank for automatic trimming of CG in flight, as well as for optional standard use.

ACCOMMODATION: Can be operated by pilot and co-pilot, but Aeroflot specified requirement for a flight engineer; provision for fourth seat for instructor or observer. Three basic single-aisle passenger arrangements in Tu-204-200 series: (1) 184 seats, with 30 business class seats six-abreast at pitch of 96 cm (38 in) at front, and 154 tourist class seats six-abreast at pitch of 81 cm (32 in) at rear; (2) 200 six-abreast tourist class seats at pitch of 81 cm (32 in); (3) 212 seats six-abreast at tourist class pitch of 81 cm (32 in). Tu-234 has 162 seats in short-range version, 145 seats at same pitch in mid-range and 102 seats in long-range version, all at 81 cm (32 in) pitch. All configurations have buffet/galley and lavatory immediately aft of flight deck; 184-seat and 200-seat configurations have two more lavatories in tourist cabin and a further large buffet/galley and service compartment at rear of passenger accommodation; 212-seat configuration has two lavatories and galley at rear; overhead stowage for hand baggage. Other layouts optional, with increased galley and lavatory provisions.

Passenger doors at front and rear of cabin on port side; service doors opposite. Type I emergency exit doors fore and aft of wing each side; inflatable slide for emergency use at each of eight doors. Tu-204-200 has two underfloor baggage/freight holds for total of eight LD-3-46 international class containers, three in forward hold, five in rear hold (AK-07 containers in Tu-204 and 204-100). Fully automatic container loading system; manual back-up.

SYSTEMS: Triplex fly-by-wire digital control system, with triplex analogue back-up; Barco Display Systems selected in 2000 to supply FMS computers. Three independent hydraulic systems, pressure 207 bar (3,000 lb/sq in). Ailerons, elevators, rudder, spoilers and airbrakes operated by all three systems; flaps, leading-edge slats, brakes and nosewheel steering operated by two systems; landing gear retraction and extension by all three systems. Electrical power supplied by two 200/115 V 400 Hz AC generators and a 27 V DC standby system. Type TA-12-60 APU in tailcone.

AVIONICS: Avionics of Russian design or, optionally, integrated by Sextant Avionique or Honeywell, with standard nav/com equipment by Rockwell Collins.

Comms: VHF and HF radio, intercom.

Radar: Weather radar.

Flight: KSPNO-204 triplex automatic flight control and navigation system (being upgraded in 2002), VOR, DME, automatic approach and landing system, for operation to ICAO Cat. IIIa minima; Series 100/200 have RLG INS (Litton LTN-101 or Honeywell HG115OBD02), GPS (Litton LTN-2001) and TCAS II (Honeywell).

Instrumentation: EFIS equipment comprises two colour CRTs for flight and nav information for each pilot, plus two central CRTs for engine and systems data. All six to have been replaced by LCDs by end of 2002.

DIMENSIONS, EXTERNAL (A: 204, B: 204-100 (and 204C unless otherwise stated), C: 204-120 (and 204-120C unless otherwise stated), D: 204-200 and Tu-214, E: 204-220, F: 204-300 short-range, G: 204-300 mid-range, H: 204-234 long-range, J: 224):

Wing span: all ... 41.80 m (137 ft 1¾ in)
Wing aspect ratio: all ... 9.6
Length overall: A, B, C, D, E ... 46.16 m (151 ft 5¼ in)
F, G, H, J ... 40.20 m (131 ft 10¾ in)
Fuselage: Max width ... 3.80 m (12 ft 5½ in)
Max depth ... 4.10 m (13 ft 5½ in)
Height overall: all ... 13.90 m (45 ft 7¼ in)
Tailplane span: A, B, C, D, E ... 15.00 m (49 ft 2½ in)
Wheel track: A, B, C, D, E ... 7.82 m (25 ft 8 in)
Wheelbase: A, B, C, D, E ... 17.00 m (55 ft 9¼ in)
Passenger doors (each): Height: all ... 1.85 m (6 ft 0¾ in)
Width: all ... 0.84 m (2 ft 9 in)
Service doors (each): Height: all ... 1.60 m (5 ft 3 in)
Width: all ... 0.65 m (2 ft 1½ in)
Cargo door (appropriate versions):
Height ... 2.18 m (7 ft 1¾ in)
Width ... 3.405 m (11 ft 2 in)
Emergency exit doors (each):
Height: all ... 1.44 m (4 ft 8¾ in)
Width: all ... 0.61 m (2 ft 0 in)
Baggage holds:
Height to sill: all ... 2.71 m (8 ft 10¾ in)

DIMENSIONS, INTERNAL:
Cabin, excl flight deck:
Length: A, B, C, D, E ... 30.18 m (99 ft 0 in)
Max width: all ... 3.57 m (11 ft 8½ in)
Max height: all ... 2.28 m (7 ft 6 in)
Fwd cargo hold:
Height: all ... 1.16 m (3 ft 9¾ in)
Volume: A, B, C ... 11.0 m³ (388 cu ft)
Rear cargo hold:
Height: A, B, C, D, E ... 1.16 m (3 ft 9¾ in)
Volume: A, B, C ... 15.4 m³ (544 cu ft)

AREAS:
Wings, gross: all ... 184.20 m² (1,982.7 sq ft)

WEIGHTS AND LOADINGS:
Operational weight empty: A ... 58,300 kg (128,530 lb)
B ... 58,800 kg (129,630 lb)
C ... 59,300 kg (130,735 lb)
D ... 61,600 kg (135,805 lb)
E ... 59,000 kg (130,070 lb)
Tu-214 freighter ... 57,000 kg (125,665 lb)
Max payload: A, B, C ... 21,000 kg (46,296 lb)
204C ... 27,000 kg (59,524 lb)
204-120C ... 25,000 kg (55,116 lb)
D, E (weight limited), J ... 25,200 kg (55,555 lb)
Tu-214 freighter ... 30,000 kg (66,139 lb)
F, space limited (196 seats) ... 19,565 kg (43,132 lb)
F, G ... 18,000 kg (39,682 lb)
H ... 12,000 kg (26,455 lb)
Payload with max fuel: 204C ... 13,900 kg (30,644 lb)
204-120C ... 13,200 kg (29,101 lb)
Max baggage/freight:
fwd hold: E ... 3,625 kg (7,990 lb)
rear hold: E ... 5,075 kg (11,190 lb)
Max fuel: A ... 24,000 kg (52,910 lb)
B, C, D, E, J ... 32,700 kg (72,090 lb)
F ... 24,300 kg (53,572 lb)
G, H ... 36,000 kg (79,366 lb)
Max ramp weight: E ... 111,750 kg (246,360 lb)
Max T-O weight: A ... 94,600 kg (208,550 lb)
B, C ... 103,000 kg (227,075 lb)
D, E, J ... 110,750 kg (244,155 lb)
F ... 88,000 kg (194,005 lb)
G, H ... 103,000 kg (227,070 lb)
Max landing weight: B, C, F, G, H ... 88,000 kg (194,005 lb)
D ... 93,000 kg (205,030 lb)
Max zero-fuel weight: E ... 84,200 kg (185,625 lb)
Max wing loading: A ... 513.6 kg/m² (105.19 lb/sq ft)

B, C, G, H ... 559.2 kg/m² (114.53 lb/sq ft)
D, E, J ... 601.2 kg/m² (123.15 lb/sq ft)
F ... 477.7 kg/m² (97.85 lb/sq ft)
Max power loading: A ... 299 kg/kN (2.93 lb/lb st)
B ... 325 kg/kN (3.19 lb/lb st)
C ... 269 kg/kN (2.63 lb/lb st)
D ... 350 kg/kN (3.43 lb/lb st)
E ... 289 kg/kN (2.83 lb/lb st)
F ... 278 kg/kN (2.73 lb/lb st)
G, H ... 325 kg/kN (3.19 lb/lb st)

PERFORMANCE:
Nominal cruising speed at FL364-397:
A, B, C, D, E ... 437–459 kt (810–850 km/h; 503–528 mph)
F, G, H ... 448–459 kt (830–850 km/h; 515–528 mph)
Approach speed: F ... 119 kt (220 km/h; 137 mph)
G, H ... 122 kt (225 km/h; 140 mph)
Cruising height ... 12,100 m (39,700 ft)
T-O run: F ... 1,450 m (4,760 ft)
G, H ... 2,050 m (6,725 ft)
Required T-O runway at max T-O weight:
A ... 2,030 m (6,660 ft)
B (except 204C), D, E ... 2,045 m (6,710 ft)
204C ... 2,160 m (7,090 ft)
C ... 1,830 m (6,005 ft)
F, G, H ... 2,500 m (8,200 ft)
J ... 2,630 m (8,630 ft)
Required landing runway: F, G, H ... 2,000 m (6,565 ft)
Range:
with max payload:
A ... 1,312 n miles (2,430 km; 1,509 miles)
B ... 2,321 n miles (4,300 km; 2,671 miles)
Tu-204C ... 2,321 n miles (4,300 km; 2,671 miles)
C ... 2,213 n miles (4,100 km; 2,547 miles)
Tu-204-120C ... 1,808 n miles (3,350 km; 2,081 miles)
D ... 2,343 n miles (4,340 km; 2,696 miles)
Tu-214 passenger ... 3,374 n miles (6,250 km; 3,883 miles)
E ... 2,483 n miles (4,600 km; 2,858 miles)
H ... 3,455 n miles (6,400 km; 3,976 miles)
with design payload:
B ... 3,509 n miles (6,500 km; 4,038 miles)
F ... 1,403 n miles (2,600 km; 1,615 miles)
G ... 3,617 n miles (6,700 km; 4,163 miles)
H ... 4,990 n miles (9,250 km; 5,750 miles)
with max fuel:
Tu-204C ... 3,671 n miles (6,800 km; 4,225 miles)
Tu-204-120C ... 3,542 n miles (6,560 km; 4,076 miles)
D ... 3,887 n miles (7,200 km; 4,473 miles)
H ... 4,967 n miles (9,200 km; 5,716 miles)
J ... 3,617 n miles (6,700 km; 4,163 miles)

UPDATED

TUPOLEV TU-324 AND TU-414

TYPE: Regional jet airliner.

PROGRAMME: Announced at 1995 Paris Air Show; proposed manufacture by KAPO at Kazan, Tatarstan, from 2003. Specification revised 1996 and further refined in 1997 when governments of Russia and Tatarstan agreed on 20 August to 50:50 sharing of development costs, to which latter had contributed some US$70 million by late 2002. Tatarstan placed symbolic first order for a presidential transport (since superseded by Tu-214 VSSN, which see). Mockup completed by late 1997; technical documentation issued 2000. Both versions entered for 2002–03 national RRJ competition staged by Rosaviacosmos and eventually won by Sukhoi. Tupolev and KAPO planned to continue development provided non-federal funding can be found; reported in March 2003 that Tu-324 had by then been allocated some US$38 million in state funding. Further investment of US$70 million by Tatarstan government reported in July 2003, coupled with statement that prototype manufacture was about to begin. However, Tupolev spokesman confirmed in February 2004 that Tu-324 progress had been halted following loss of RRJ contract, and attention now focused on Tu-414 variant.

Model of Tupolev Tu-324 *(Paul Jackson)* ***NEW***/1047206

Model of Tupolev Tu-414 *(Paul Jackson)* 0526930

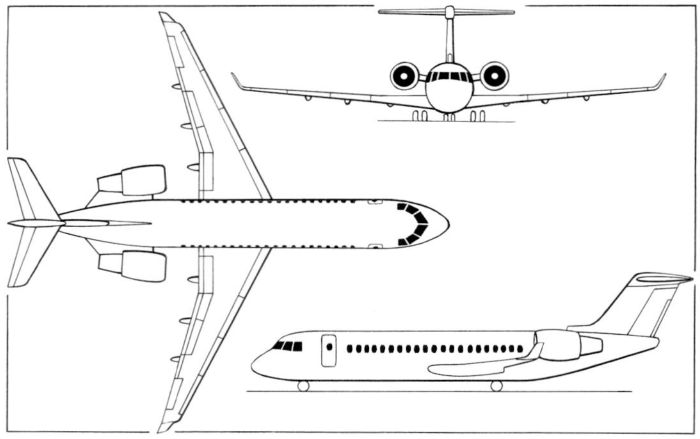

Tupolev Tu-324R twin-turbofan (AI-22) regional transport *(James Goulding)* 0079303

CURRENT VERSIONS: **Tu-324A** (*Administrativny:* executive): VIP transport, carrying up to 10 passengers.

Tu-324R: Regional airliner: 52 seats in single-class configuration: 10 + 34 or 36 in two-class. In abeyance.

Tu-414: Project for stretching to 72/75-seat airliner, possibly with Ivchenko Progress D-436T1 or Rolls-Royce Deutschland BR710-48 turbofans. Initial design completed December 2003; 'electronic mockup' planned for first half of 2004. Maximum commonality with Tu-324, but no dimensions, weights or performance data available by mid-2004.

CUSTOMERS: Estimated market for 400 to 450, including 150 Tu-324A.

COSTS: Estimated total programme cost US$330 million; Tu-324A US$10 million, export versions US$15 million to US$20 million (2001).

DESIGN FEATURES: Wings scaled down from Tu-204; illustrations show the aircraft with and without winglets. Is first Russian aircraft programme designed entirely with use of computers.

LANDING GEAR: Messier-Dowty main contractor.

POWER PLANT: Two General Electric CF34-3B1 turbofans, each 41.0 kN (9,220 lb st), in pods on sides of rear fuselage. Alternatives under consideration include Progress AI-22 and Soyuz TRDD R-126-300.

AVIONICS: Honeywell main supplier.

DIMENSIONS, EXTERNAL:
Wing span ... 23.20 m (76 ft 1½ in)
Length overall: Tu-324A ... 23.00 m (75 ft 5½ in)
Tu-324R ... 25.50 m (83 ft 8 in)
Height overall ... 7.10 m (23 ft 3½ in)

WEIGHTS AND LOADINGS:
Max payload: Tu-324A ... 1,800 kg (3,968 lb)
Tu-324R ... 5,500 kg (12,125 lb)
Max T-O weight: Tu-324A, Tu-324R ... 23,700 kg (52,249 lb)

PERFORMANCE (estimated, AI-22 engines):
Max cruising speed ... 448–459 kt (830–850 km/h; 516–528 mph)
Max certified altitude ... 11,600 m (38,060 ft)
Balanced runway length ... 1,800 m (5,910 ft)
Range:
Tu-324R: with 52 passengers ... 1,349 n miles (2,500 km; 1,553 miles)
with 46 passengers ... 1,619 n miles (3,000 km; 1,864 miles)
Tu-324A, 8 passengers ... 4,022 n miles (7,450 km; 4,629 miles)

UPDATED

TUPOLEV TU-330

TYPE: Twin-jet freighter.

PROGRAMME: Announced early 1993 as replacement for Antonov An-12; government support announced 23 April 1994; first 10 to be built at Kazan; first flight scheduled 1997, for service from 1998; neither target achieved. Production at both KAPO and Samara. Stated by commander of Russian Military Transport Aviation to be one of three basic types to be operated by his force in 21st century. Production drawings released by Tupolev to KAPO at Kazan in late 1997 and construction then begun. By early 2000 the venture was being offered as a potential partnership.

Tu-330 included in federal plans for civil aviation up to 2015, requiring investment of equivalent of US$340 million over seven years. Tupolev expenditure to early 2000 totalled Rb80 million. Kazan plant hoped for production launch in 2002, but funding still awaited

Model of proposed Tupolev Tu-330 with PS-90A engines *(Paul Jackson)* ***NEW***/1047205

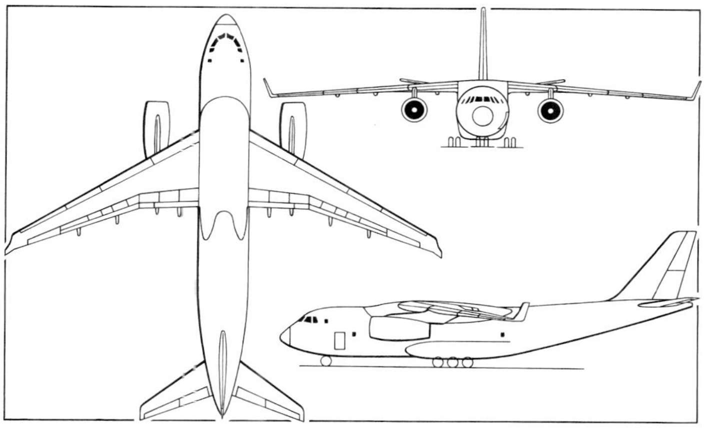

Tupolev Tu-330 twin-turbofan freight transport *(Jane's/Mike Keep)* 0559411

in mid-2004. However Russian Air Forces' reluctance to accept An-70 seen to improve Tu-330's chances of go-ahead, and named in 2003 as one of two main contenders (with Il-214) to replace An-12.

CURRENT VERSIONS: **Tu-330**: *As described.*

Tu-330K (previously **Tu-338**): Proposed cryogenic-fuelled version with Samara NK-94 engines and 22,600 kg (49,824 lb) of liquid natural gas in conformal dorsal tank. Technical proposals completed, but not yet funded (mid-2004).

COSTS: Unit price US$25 million to US$35 million (2000, estimate). Estimated cost of two flying and two ground test aircraft, plus certification, US$100 million (2003).

DESIGN FEATURES: See accompanying three-view drawing: said to have up to 75 per cent commonality with Tu-204. Wing design basically similar to Tu-204; engine pylons as Tu-204; rear-loading ramp. Potential secondary applications include refuelling tanker and maritime patroller.

FLYING CONTROLS: Each wing has four-section leading-edge flaps, three-section trailing-edge flaps, eight spoilers forward of flaps, and aileron. Variable incidence tailplane. Elevators and rudder each in two sections.

LANDING GEAR: Retractable tricycle type; twin-wheel, rearwards-retracting nose unit; each main unit has three pairs of wheels (one pair with brakes) in tandem. Able to operate from concrete or grass.

POWER PLANT: Two Aviadvigatel PS-90A turbofans. Rolls-Royce and Pratt & Whitney turbofans under consideration for export sales. Tu-330K is projected for LNG transportation, with three tanks in the cabin for 22,800 kg (50,265 lb) of LNG.

ACCOMMODATION: Flight crew of two.

AVIONICS: Include six colour MFDs.

DIMENSIONS, EXTERNAL:
Wing span over winglets .... 43.50 m (142 ft 8½ in)
Wing aspect ratio 10.6
Length overall .... 42.00 m (137 ft 9½ in)
Height overall .... 14.00 m (45 ft 1¼ in)
Rear-loading ramp: Length .... 4.00 m (13 ft 1½ in)
Width .... 4.00 m (13 ft 1½ in)

DIMENSIONS, INTERNAL:
Cargo hold:
Length: excl ramp .... 19.50 m (63 ft 11½ in)
incl ramp .... 23.50 m (77 ft 1¼ in)
Width at floor .... 4.00 m (13 ft 1½ in)
Height: min .... 3.55 m (11 ft 7¾ in)
max .... 4.00 m (13 ft 1½ in)
Volume .... 350.0 m^3 (12,360 cu ft)

AREAS:
Wings, gross .... 177.80 m^2 (1,913.8 sq ft)

WEIGHTS AND LOADINGS:
Payload: in hold .... 30,000 kg (66,139 lb)
on ramp .... 5,000 kg (11,023 lb)
max .... 35,000 kg (77,162 lb)
Max T-O weight .... 103,500 kg (228,175 lb)
Max wing loading .... 582.1 kg/m^2 (119.22 lb/sq ft)

PERFORMANCE (estimated):
Nominal cruising speed at FL360
.... 431–458 kt (800–850 km/h; 497–528 mph)
T-O and landing distance .... 2,200 m (7,220 ft)
Nominal range:
with 15,000 kg (33,069 lb) payload .... 3,779 n miles (7,000 km; 4,349 miles)
with 20,000 kg (44,090 lb) payload .... 3,020 n miles (5,600 km; 3,480 miles)
with 30,000 kg (66,139 lb) payload .... 1,620 n miles (3,000 km; 1,865 miles)

UPDATED

TUPOLEV TU-334 AND TU-336

TYPE: Twin-jet airliner.

DEVELOPMENT MILESTONES

Launched	1986
Rolled out	25 Aug 95
First flight	8 Feb 99

PROGRAMME: Launched 1986, with target first flight in 1991 and service entry in 1995; chief designer Igor Kaligin. First prototype Tu-334 (RA-94001) eventually rolled out at Zhukovsky during Moscow Air Show, 25 August 1995; then scheduled to fly 1996 with D-436T1 engines, but delayed by funding shortages. Refinancing by Ukrainian government in early 1997 resulted in new target date of May/June 1997 for first Tu-334s built by Aviant at Kiev; this also missed; second and third identical, but non-flying, prototypes under construction at Zhukovsky by 1996; second used for static tests at CAHI (TsAGI) 1996; unidentified test airframe delivered to Aviatest LNK at Riga from Kiev in mid-2000 to simulate 20 year fatigue life; further two prototypes and three production aircraft under

Prototype Tupolev Tu-334 *(Paul Jackson)* ***NEW***/0580780

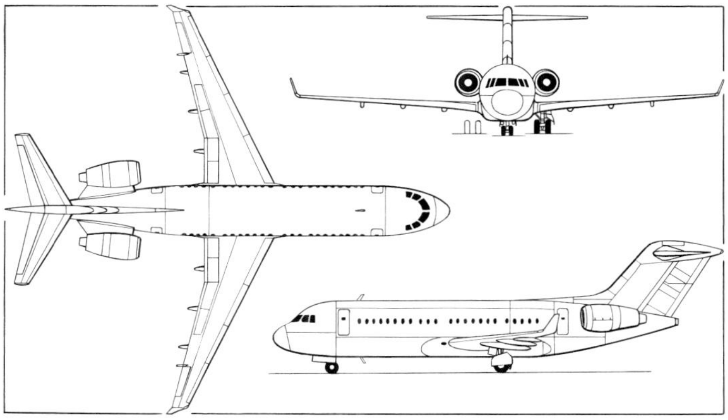

Tupolev Tu-334-100 twin-turbofan medium-range transport *(James Goulding)* 0084429

construction at Kiev by early 1999 (three for full assembly by Aviant; part of fuselage and wings of other two for delivery to RSK 'MiG' at Lukhovitsy). Subsequent input from Aviant of these components only, instead of originally planned second assembly line.

RA-94001 eventually flown 8 February 1999 and had achieved eight sorties by time of public debut at Moscow in August 1999. Plans for 200 flights in 2001 not achieved (114 only), and this total (of planned 900) not reached until late 2003. By mid-2003, second flying prototype (Aviant-built 005) said to be more than 90 per cent complete; was rolled out 2 August 2003; made maiden flight (RA-94005) on 21 November 2003 and delivered to Zhukovsky for flight trials 27 November. Completion of third development aircraft (003, first by RSK 'MiG'), anticipated by June 2003, but apparently negated by Rosaviakosmos announcement in January 2004 allocating Tu-334 production to four-company consortium of KAPO, Aviant, Aviastar and Tavia (Taganrog). Russian certification awarded in late 2003.

AP-25/FAR Pt 25 certification programme was due for completion in 2001, but victim of above-mentioned delays, created by erratic funding and delayed delivery of D-436 engines; certification now seen as unlikely before late 2004 at earliest. TANTK at Taganrog was to have been second source for -100, but Russian government decree of 5 October 1999 nominated RSK 'MiG' (which is to contribute half of certification costs, first payment being made in January 2002); first Taganrog fuselage transported to Voronin Production Centre, Moscow, arriving 7 April 2000. Completion and final assembly was to have been in second quarter of 2001, but this not achieved (see above); Taganrog horizontal tail delivered early 2002. Half of all assemblies and components for RSK 'MiG' production to be supplied by Aviant; other components supplied by Kazan and Ulyanovsk factories. Approval for RSK 'MiG' production awarded 31 October 2002.

On 16 June 1997 Rolls-Royce Deutschland agreed loan of BR710 engines to build prototype of Tu-334-120; first flight of -120 had been expected late 1998, but delayed by funding shortages. Revised plans for BR710s to go into -100 prototype, after short test programme with D-436s, also thwarted by aircraft's continued grounding. R-R support now directed at versions powered by BR715 engines.

During mid-1997, AIO of Iran showed interest in establishing a kit assembly line at Esfahan for 100 Tu-334s (of which 50 for local airlines) over 15 years, based on eventual fuselage detailed assembly in Iran, with complete wings from Ukraine and empennage from Russia. No recent news received concerning this venture.

Five versions being promoted in 2003–04 were Tu-334-100, -100D, -120D, -200 and -220.

CURRENT VERSIONS: **Tu-334-100**: Basic version, with D-436T1 engines, for 72 mixed class or 102 tourist class passengers. Prototypes are of this version.

Tu-334-100C: Cargo version of -100; up to six 2.24 × 3.17 m (88 × 125 in) pallets in underfloor compartments.

Tu-334-100D: Extended-range 102-seat version, announced at Moscow Air Show '95. Basically similar to Tu-334-100, but with increased wing span and uprated D-436T2 engines; increased fuel and maximum T-O weight. Not yet built.

Tu-334-120: As Tu-334-100, but two 88.9 kN (19,995 lb st) Rolls-Royce Deutschland BR715-56 turbofans. Third flying prototype to be in this configuration.

Tu-334-120D: Rolls-Royce Deutschland BR715-56 engines; increased weight and range. Announced 1998.

Tu-334-200: Basically similar to Tu-334-100D and previously known as **Tu-354**; D-436T2 engines; fuselage lengthened by 3.90 m (12 ft 9½ in) to accommodate up to 126 passengers at 81 cm (32 in) pitch; increased wing span; four-wheel bogies on main landing gear units. Construction of four by Aviacor in Samara reportedly begun by early 1997, but no recent news from this quarter and Aviacor no longer mentioned in connection with Tu-334 programme.

Tu-334-220: Further increased MTOW; BR715-56 engines. Announced 1998.

Tu-334K: Design study (2003) for cryogenically fuelled short-range version carrying 7,000 kg (15,432 lb) of liquid natural gas and 2,000 kg (4,409 lb) of Avgas; Rolls-Royce Deutschland BR 710C engines; range 1,079 n miles (2,000 km; 1,242 miles) with 102 passengers.

CUSTOMERS: Provisional orders in 1995 for approximately 40 aircraft, for Rossiya Airlines, Bashkiri Airlines, Tyumen Airlines and Tatarstan Airlines. MoU for 20 Tu-334-120s signed by Aeroflot in June 2000; other letters of intent from Atlant Soyuz (two) and Pulkovo (15

Tupolev Tu-334, cutaway drawing key

1 Glass fibre radome
2 Radome hinge, opens to starboard
3 Short-range radio navigation localiser antenna
4 Weather radar scanner
5 Scanner mounting/tracking mechanism
6 ILS glideslope antenna
7 Sloping front pressure bulkhead
8 Radar equipment modules
9 Rudder pedals
10 'Ram's-horn' control column, three-channel digital fly-by-wire flight control system
11 Instrument panel, six-screen EFIS displays
12 Instrument panel shroud
13 Windscreen wipers
14 Electrically heated windscreen panels
15 Overhead systems switch panel
16 Transponder antennae
17 Starboard side engineering panel
18 Cockpit doorway
19 Third crew member's seat for optional flight engineer
20 Co-pilot's seat
21 Centre console, incorporating engine throttle levers
22 Direct-vision opening side window panel
23 Captain's seat
24 Observer/instructor's folding seat
25 Side console panel
26 Incidence vane
27 Nose landing gear wheel bay
28 Pitot head
29 Fully steerable twin nosewheels, forward retracting
30 Nosewheel leg doors with taxying lights
31 Retractable landing light
32 Forward lavatory
33 Hand basin
34 Forward galley
35 Starboard forward service/ emergency door
36 CAS antenna
37 Forward cabin bulkhead
38 Cabin attendants' folding seats (two)
39 Forward entry door
40 Door-mounted escape slide
41 Underfloor avionics equipment racks
42 Wing inspection light
43 Forward underfloor freight hold
44 First class passenger seats (eight), optional layout for eight first-class and eighty-four tourist-class passengers
45 VHF antenna
46 SNS antennae
47 Curtained cabin divider, removable
48 Cabin window panels
49 Underfloor baggage containers
50 Six-abreast tourist-class seating, standard layout for 102 all-tourist class passengers
51 Forward freight hold door
52 Floor beam structure
53 Cabin wall trim panelling
54 Cabin insulation blankets
55 Satcom antenna
56 ADF antennae
57 Conditioned air risers to overhead ducts
58 Fuselage frame and stringer structure
59 Underfloor air distribution ducting
60 Wing front spar/fuselage attachment joint
61 Ventral environmental control system conditioning pack
62 Wing centre-section with integral fuel tank
63 Front spar attachment built-up fuselage main frame
64 Pressure floor above wing centre-section box
65 Upper anti-collision beacon
66 Starboard wing integral fuel tanks
67 Pressure refuelling connections
68 Four-segment leading-edge slat
69 Slat guide rails
70 Slat torque shaft and screw jacks
71 Wingtip vent tank
72 Starboard navigation (green) and strobe (white) lights
73 Winglet
74 Static dischargers
75 Aft navigation light (white)
76 Starboard aileron
77 Aileron hydraulic actuators
78 Flap outboard guide rail fairing
79 Starboard outer slotted flap segment, extended
80 Flap screw jacks, guide rails and carriages
81 Outboard spoiler panels (two) with roll control function
82 Flap operating torque shaft from central hydraulic drive motor
83 Inboard spoiler/lift dumper panels (two)
84 Starboard inboard slotted flap segment, extended
85 Dual TRSP antennae
86 Rear spar attachment built-up fuselage main frame
87 Composites floor panels with continuous seat rails
88 Rear cabin tourist class seating
89 Overhead baggage lockers
90 Rear freight hold door
91 Cabin ceiling trim/lighting panels
92 Passenger service units in overhead panels
93 Starboard side rear service door/ emergency exit
94 Galley/buffet unit
95 Rear lavatory compartment
96 Rear pressure bulkhead
97 Fuselage tailcone frame and stringer structure
98 Engine mounting and fin support attachment main frame
99 Fin front spar attachment joint
100 Composites fin-root fairing
101 Starboard engine nacelle
102 Hinged engine cowlings
103 Starboard thrust reverser, deployed
104 Translating cowling/reverser cover
105 Composites fin leading-edge panels
106 Two-spar fin torsion box structure
107 Fin ribs and stringers
108 Trimming tailplane hydraulically operated screw jacks (three)
109 Tailplane sealing plate
110 Tailplane pivot mounting
111 Elevator hinge controls, digital control with emergency mechanical backup
112 Starboard tailplane
113 Starboard elevator
114 Tailplane pivot mounting tail fairing
115 Port elevator, composites structure
116 Static dischargers
117 Two-spar torsion box tailplane structure
118 Tailplane rib and stringer structure
119 Composites tailplane leading-edge panels
120 Two-segment rudder
121 Rudder hydraulic actuators, digital control with emergency mechanical backup
122 Rudder, composites structure
123 APU exhaust
124 Auxiliary power unit (APU)
125 Fin rear spar attachment joint
126 Fin spar mounting bulkhead and APU bay firewall
127 Hydraulic equipment bay
128 Engine pylon tail fairing
129 Port core engine (hot stream) exhaust
130 Fan air (cold stream) exhaust
131 Fan air by-pass duct
132 Thrust reverser cascades
133 Reverser screw jack operating drive mechanism
134 Ivchenko Progress D-436T1 turbofan
135 Optional Rolls-Royce Deutschland BR715-56 turbofan
136 Engine accessory equipment gearbox
137 Hinged lower cowling panel
138 Main engine mounting
139 Intake lip bleed air de-icing
140 Acoustically lined intake duct
141 Crew wardrobe
142 Rear door with emergency escape slide
143 Rear cabin window panels
144 Wingroot composites trailing-edge fairing
145 Inboard flap track
146 Port inboard slotted flap segment
147 Inboard spoilers/lift dumpers
148 Spoiler hydraulic actuators
149 Main landing gear mounting auxiliary spar
150 Mainwheel bay
151 Wing rear spar/fuselage attachment joint
152 Hydraulic jack
153 Main leg pivot mounting
154 Inboard flap carriage and track
155 Flap, composites structure
156 Flap extended position
157 Port outboard spoilers with roll control function
158 Spoiler panel composites structure
159 Flap track fairings
160 Flap shroud panel
161 Aileron hydraulic actuators
162 Port aileron, composites structure
163 Aft navigation light (white)
164 Port winglet
165 Port navigation (red) and strobe (white) lights
166 Four-segment leading-edge slat
167 Two-spar torsion box wing structure
168 Wing rib structure
169 Wing bottom skin/stringer panel with access covers
170 Port wing integral fuel tank
171 Leading-edge slat guide rails
172 Twin mainwheels; four-wheel bogie on Tu-334-220
173 Trailing link axle beam
174 Mainwheel leg strut and shock-absorber
175 Leading-edge slat screw jack, torque shaft operated from central hydraulic drive motor
176 Immersed fuel pumps
177 Fuel venting channel box section stringer
178 Wingroot machined cruciform section skin joint fitting
179 Conditioning system ground connection
180 Landing light
181 Composites wingroot leading-edge fairing

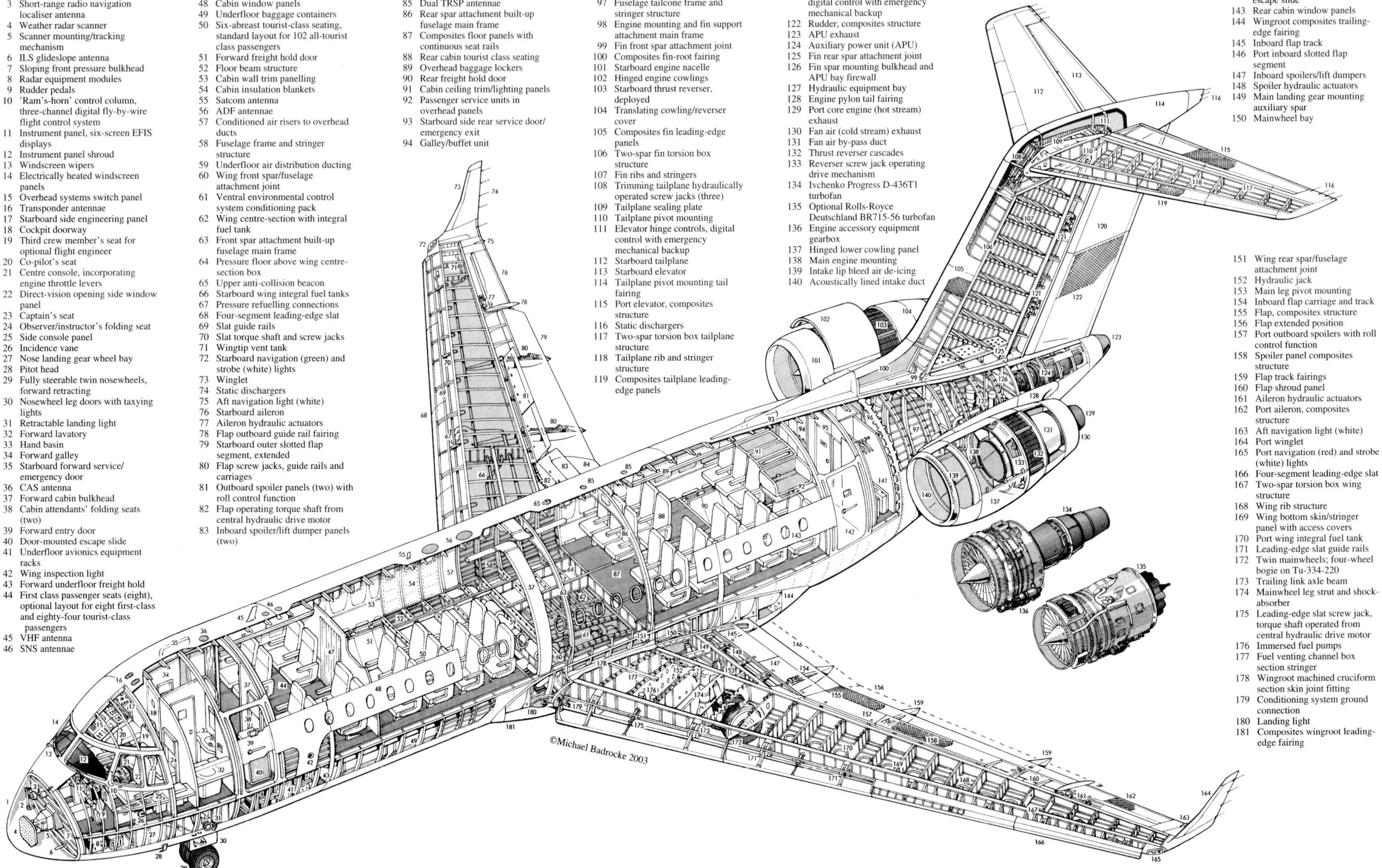

to 20) during first half of 2002. First firm orders (five Tu-334-100s each for 10-year lease by Aerofraht and Atlant Soyuz Airlines) announced by RSK 'MiG' at MAKS air show, August 2003; deliveries, to latter, to begin in 2005. CIS market estimated as about 350 by Rosaviakosmos. Tupolev website reported orders from 12 airlines as at November 2003.

COSTS: Tu-334-100 US$15 million to US$18 million; Tu-334-200 US$20 million (both 2003). Break-even at 48th aircraft, according to RSK 'MiG'.

DESIGN FEATURES: Replacement for Tu-134 and Yak-42, to meet requirements of Russian Federation and Associated States (CIS) airlines. Wings have much in common with those of Tu-204 and the fuselage is shortened version of Tu-204, with identical flight deck. Configuration is all-swept low/mid-wing, with rear-mounted engines and T tail; circular-section semi-monocoque fuselage; wings have dihedral from roots. Short-stroke landing gear for minimal door sill height, to facilitate baggage handling at airports with limited loading capabilities. Service life 60,000 hours.

Wings have supercritical section, 24° sweepback, with winglets.

FLYING CONTROLS: Main and standby fly-by-wire; emergency hydraulic and mechanical back-up, except for ailerons. Two-section, single-slotted, trailing-edge flaps, two-section airbrakes forward of inner flap and two-section spoilers forward of outer flap on each wing; four-section leading-edge slat over full span each wing; conventional ailerons, elevators and two-section rudder; no tabs. Variable-incidence tailplane.

STRUCTURE: Composites and other lightweight materials make up 20 per cent of structure by weight. Metal alloy wing, fin and horizontal tail, each constructed on two spars, have composites leading-edges and covered by milled panels. Elevators, ailerons, airbrakes, spoilers, flaps, rudder and floor panels of composites honeycomb. Fuselage of metal alloy with riveted skin. Wings by Aviant at Kiev (Ukraine); final assembly by RSK 'MiG' at Lukhovitsy.

LANDING GEAR: Retractable tricycle type; twin wheels on each unit of Tu-334-100, four-wheel bogies on main units of Tu-334-200; main units retract inward into wing/fuselage fairings; trailing-link mainwheel legs on Tu-334-100; nosewheels retract forwards, forward elements of two-section door each side remaining closed except during cycling. Some recent Tupolev illustrations have shown all versions of Tu-334 with four-wheel main bogies. Tyre size 1,070×390-480R on mainwheels; 680×260-355R on nosewheels. Stout grille behind mainwheels of prototype to prevent ice and slush entering turbofans.

POWER PLANT: *Tu-334-100:* Two Ivchenko Progress D-436T1 turbofans, each rated at 73.6 kN (16,535 lb st).

Tu-334-100D/-200: Two D-436T2 turbofans, each rated at 80.4 kN (18,078 lb st).

Tu-334-120/-120D/-220: Rolls-Royce Deutschland BR715-56 turbofans each rated at 88.9 kN (19,995 lb st). D-436T1 production by consortium of Motor-Sich at Zaporozhye (Ukraine), Salyut and UMPO.

Fuel in integral wing tanks, total capacity (Tu-334-100) 9,540 kg (21,032 lb) or (-100D/-120D/-200/-220) 15,700 kg (34,613 lb).

ACCOMMODATION (TU-334-100 AND -120): Crew of two (optionally three) on 'Tu-204 compliant' flight deck; provision for fourth seat for instructor or observer. Three basic single-aisle passenger arrangements: (1) 72 seats, with 12 seats four-abreast in first class cabin at front, at pitch of 102 cm (40 in) and with 73 cm (28.75 in) aisle, and 60 tourist class seats six-abreast at 81 cm (32 in) pitch with 49.5 cm (19.5 in) aisle; (2) 102 seats, all tourist, at 81 cm (32 in) pitch; and (3) 74 seats, with eight business class and 66 in economy class. All three configurations have buffet/galley, coat compartment and lavatory immediately behind flight deck, a further lavatory and service compartment at rear; 72-seater has additional galleys at front and rear; overhead stowage for hand baggage. Other arrangements at customer's option. Passenger doors at front and rear of cabin on port side; service doors opposite. Underfloor baggage/freight holds; doors on starboard side.

ACCOMMODATION (TU-334-200): Flight deck unchanged. Two basic single-aisle passenger arrangements: (1) 110 seats, with eight seats four-abreast in first class cabin at front, at pitch of 99 cm (39 in), and 102 tourist class seats six-abreast at 81 cm (32 in) pitch; (2) 126 seats, all tourist, at 81 cm (32 in) pitch. Facilities as Tu-334-100/120, plus emergency exit over wing each side.

SYSTEMS: APU in tailcone. Hydraulic system for actuation of flying controls. Air conditioning unit in wing centre-section.

AVIONICS: *Radar:* Nose-mounted weather radar.

Instrumentation: EFIS standard, with six CRTs. Satcom optional. Equipment for landings in ICAO Cat. IIIa conditions.

DIMENSIONS, EXTERNAL:
Wing span: 100, 120 ... 29.77 m (97 ft 8 in)
100D, 120D, 200, 220 ... 32.61 m (107 ft 0 in)
Wing aspect ratio: 100, 120 ... 10.2
100D, 120D, 200, 220 ... 10.6
Length overall: 100, 120, 120D ... 31.26 m (102 ft 6¾ in)
100D ... 31.76 m (104 ft 2½ in)
200, 220 ... 35.16 m (115 ft 4½ in)
Fuselage: Max width ... 3.80 m (12 ft 5½ in)
Max depth ... 4.10 m (13 ft 5½ in)
Height overall: 100 ... 9.38 m (30 ft 9¼ in)
100D ... 9.60 m (31 ft 6 in)
Tailplane span ... 11.12 m (36 ft 5¾ in)
Wheel track (c/l shock struts): all ... 4.80 m (15 ft 9 in)
Wheelbase: 100, 120 ... 11.75 m (38 ft 6¾ in)
Baggage door:
forward: Height ... 1.03 m (3 ft 4½ in)
Width ... 1.165 m (3 ft 9¾ in)
Height to sill ... 1.26 m (4 ft 1½ in)
rear: Height ... 0.95 m (3 ft 1½ in)
Height to sill ... 0.93 m (3 ft 0½ in)

DIMENSIONS, INTERNAL:
Cabin: Length: 100, 120 ... 17.84 m (58 ft 6¼ in)
Height:
above aisle: all ... 2.185 m (7 ft 2 in)
beneath hand baggage racks: all ... 1.70 m (5 ft 7 in)
Max width: all ... 3.57 m (11 ft 8½ in)
Volume: 100, 120 ... 118.0 m³ (4,167 cu ft)
Baggage holds:
Max length: 100, 120, forward ... 4.10 m (13 ft 5¼ in)
100, 120, rear ... 3.00 m (9 ft 10 in)
Max width: 100, 120 ... 2.50 m (8 ft 2½ in)
Max height: 100, 120 ... 1.20 m (3 ft 11¼ in)
Volume: 100, forward ... 11.7 m³ (413 cu ft)
Total volume: 100, 120 ... 16.2 m³ (572 cu ft)
100D, 120D ... 15.8 m³ (558 cu ft)
200, 220 ... 23.3 m³ (823 cu ft)

AREAS:
Wings, gross: 100, 120 ... 83.23 m² (895.8 sq ft)
100D, 120D, 200, 220 ... 100.0 m² (1,076.4 sq ft)

WEIGHTS AND LOADINGS:
Operating weight empty: 100 ... 28,950 kg (63,824 lb)
100D ... 31,920 kg (70,371 lb)
120D ... 32,320 kg (71,253 lb)
200 ... 33,975 kg (74,902 lb)
220 ... 34,325 kg (75,674 lb)
Design payload:
100, 100D, 120D ... 9,690 kg (21,363 lb)
200, 220 ... 11,970 kg (26,389 lb)
Max payload:
100, 100C, 100D, 120, 120D ... 9,690 kg (21,363 lb)
200, 220 ... 15,000 kg (33,069 lb)
Max fuel: 100 ... 9,540 kg (21,032 lb)
100D, 120D, 200, 220 ... 15,700 kg (34,613 lb)
Max T-O weight: 120, 120C ... 46,100 kg (101,630 lb)
100 ... 47,900 kg (105,600 lb)
120D ... 53,750 kg (118,498 lb)
100D ... 54,420 kg (119,975 lb)
200 ... 54,470 kg (120,085 lb)
220 ... 54,800 kg (120,815 lb)
Max landing weight: 100 ... 43,500 kg (95,900 lb)
100D ... 44,600 kg (98,326 lb)
Max wing loading:
120D ... 537.5 kg/m² (110.09 lb/sq ft)
120 ... 553.9 kg/m² (113.44 lb/sq ft)
100 ... 575.5 kg/m² (117.87 lb/sq ft)
200 ... 544.7 kg/m² (111.56 lb/sq ft)
100D, 120D ... 544.2 kg/m² (111.46 lb/sq ft)
220 ... 548.0 kg/m² (112.24 lb/sq ft)
Max power loading: 100 ... 325 kg/kN (3.19 lb/lb st)
120 ... 334 kg/kN (3.27 lb/lb st)
100D ... 338 kg/kN (3.31 lb/lb st)
120D ... 302 kg/kN (2.96 lb/lb st)
200 ... 339 kg/kN (3.32 lb/lb st)
220 ... 308 kg/kN (3.02 lb/lb st)

PERFORMANCE (estimated):
Nominal cruising speed at FL360: 100, 100D, 200 ... 431-442 kt (800-820 km/h; 497-510 mph)
Balanced runway length at 30°C:
100 ... 2,100 m (6,890 ft)
Range with max payload, 60 min reserves:
100 ... 1,101 n miles (2,040 km; 1,267 miles)
100D ... 1,630 n miles (3,020 km; 1,876 miles)
120D ... 1,544 n miles (2,860 km; 1,777 miles)
200 ... 529 n miles (980 km; 608 miles)
220 ... 448 n miles (830 km; 515 miles)
Range with design payload:
100, 100C ... 1,700 n miles (3,150 km; 1,957 miles)
100D, 120D ... 2,213 n miles (4,100 km; 2,547 miles)
120 ... 1,685 n miles (3,122 km; 1,939 miles)
200, 220 ... 1,187 n miles (2,200 km; 1,367 miles)

OPERATIONAL NOISE LEVELS (ICAO, estimated):
T-O ... 85.8 EPNdB
Approach ... 97.4 EPNdB
Sideline ... 92.0 EPNdB

UPDATED

TUPOLEV TU-444

TYPE: Supersonic business jet.

PROGRAMME: Design concept completed by December 2003; development planned to begin in first half of 2004, subject to obtaining start-up funding. Subscale supersonic passenger aircraft (SPS), to carry up to six people including pilot(s); first stage towards prospective international (Europe-wide), large-capacity (provisionally SPS-2) supersonic transport.

COSTS: Approximately US$1 billion estimated to develop Tu-444; unit cost (target) US$50 million).

NEW ENTRY

UNIKOMTRANSO

UNIKOMTRANSO AO (UNICOMTRANSO JSC)

ulitsa Sovetskoi Armii 16540, 443090 Samara
Tel: (+7 8462) 33 07 29
e-mail: unicomtranso@mail.ru
Web: http://www.gointernational.com/unicom
GENERAL DIRECTOR: Oleg N Voronkov

Established 1990 as Unikomtrans, a production co-operative operating an An-12 and An-74 on air freight operations. Became Unikomtranso AO on 6 April 1994 and made first issue of shares in 2000. Business includes aircraft upgrades, light aircraft design, construction of lightplane prototypes and flight- and technical training on light aircraft. Manufacture in series assigned to Progress Scientific & Technical Complex under agreement of 16 March 2000.

First aircraft, the Model 11 Frigat, was exhibited at Moscow Salon in August 1995; remains available. Don was first shown in August 2001. On 3 March 2003, Unikomtranso JSC and Express Aircraft Company LLC of USA signed five-year agreement covering Russian manufacture of Express 2000, first five kits being due to arrive from USA in mid-2003. Unkcomtranso to complete 90 aircraft as agricultural sprayers for use in Samara region.

VERIFIED

UNIKOMTRANSO DON

TYPE: Agricultural sprayer.

PROGRAMME: Designed by S F Tsarkov, whose previous products included Leader, Frigat and Flamingo lightplanes. Production at Progress plant, Samara. Two-seat prototype (c/n 1010101; registration pending) shown at Moscow, August 2001. Following flight tests during early 2002, modified as **Don-M** and entered production as such in June 2002.

CURRENT VERSIONS: **Single-seat:** With 300 litre (79.3 US gallon; 66.0 Imp gallon) chemical tank. Starboard side pilot's door.

Two-seat: With 200 litre (52.8 US gallon; 44.0 Imp gallon) chemical tank; dual controls. Additional port side door for rear occupant.

DESIGN FEATURES: Pod-and-boom configuration with pusher propeller and high, constant chord wing, having single bracing strut each side. Tailplane, of inverted aerofoil section, braced to fin by struts. Sweptback, narrow chord fin. Can be dismantled or reassembled by two persons in 1 hour. Airframe protected for 25 years of open-air operation.

FLYING CONTROLS: Conventional and manual. Pushrod and cable actuation. No aerodynamic balances. Ground-adjustable tabs on starboard aileron and port elevator. Plain flaps.

STRUCTURE: Metal throughout.

LANDING GEAR: Tricycle type; fixed. Fuselage-mounted spring cantilever main legs with hydraulic wheel brakes. Trailing link nosewheel mounted rigidly on forward-projecting horizontal arm. All tyres size 400×150. Optional floats or skis.

POWER PLANT: Production Don-M has one VAZ-2112-Avia adapted motorcar engine rated at 101 kW (136 hp). Prototype had 73.5 kW (98.6 hp) Rotax 912 ULS flat-four driving a three-blade Kiev ground-adjustable pitch propeller. Fuel capacity 50 litres (13.2 US gallons; 11.0 Imp gallons).

ACCOMMODATION: One or two seats (in tandem).

EQUIPMENT: Chemical tank behind seats(s). Spraybar beneath wings; coverage rate of 1 hectare (2.5 acres)/min.

DIMENSIONS, EXTERNAL:
Wing span ... 10.40 m (34 ft 1½ in)
Length overall ... 6.40 m (21 ft 0 in)
Height: to fintip ... 2.20 m (7 ft 2½ in)
propeller turning ... 2.80 m (9 ft 2¼ in)
Tailplane span ... 2.70 m (8 ft 10¼ in)
Wheel track ... 2.00 m (6 ft 6¾ in)
Wheelbase ... 2.15 m (7 ft 0¾ in)

WEIGHTS AND LOADINGS:
Weight empty ... 420 kg (926 lb)
Max T-O weight ... 750 kg (1,653 lb)
Max power loading (Rotax 912) ... 10.20 kg/kW (16.77 lb/hp)

PERFORMANCE:
Never-exceed speed (VNE) ... 90 kt (166 km/h; 103 mph)
Max level speed ... 80 kt (148 km/h; 92 mph)
Normal cruising speed ... 65 kt (120 km/h; 75 mph)
Min operating speed ... 43 kt (80 km/h; 50 mph)
Unstick speed ... 27 kt (50 km/h; 31 mph)
T-O run ... 250 m (820 ft)
Landing run ... 100 m (330 ft)
Range ... 135 n miles (250 km; 155 miles)
Endurance ... 2 h 0 min

UPDATED

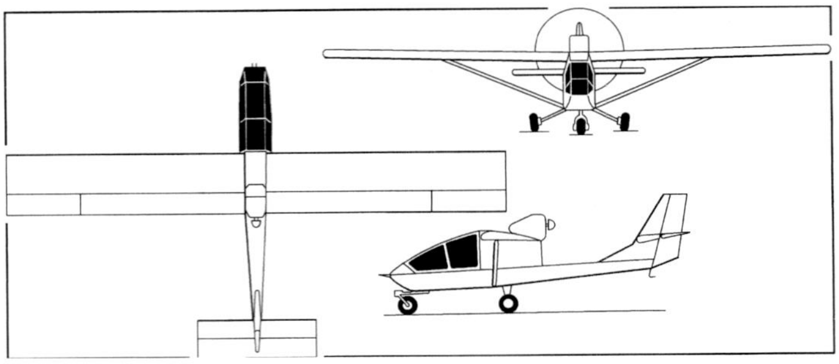

Unikomtranso Don agricultural sprayer *(Paul Jackson)* ***NEW***/0558327

Prototype Unikomtranso Don on display at Moscow in August 2001 *(Paul Jackson)* 0525059

UUAP

ULAN-UDENSKY AVIATSIONNYI ZAVOD OAO (ULAN-UDE AVIATION PLANT JSC)

ulitsa Khorinskaya 1, 670009 Ulan-Ude, Buryatia
Tel: (+7 3012) 25 74 75 and 25 35 55
Fax: (+7 3012) 25 21 47
e-mail: sales@uuaz.ru
Web: http://www.uuaz.ru
GENERAL DIRECTOR: Leonid Ya Belykh
TECHNICAL DIRECTOR: Sergey V Solomin
DEPUTY DIRECTOR-GENERAL, SALES AND MARKETING: Tsydyp Ts Galdanov
MANAGER, COMMERCIAL DEPARTMENT: Andrei Polyutin

Founded on 1 January 1939 as GAZ 99, the Ulan-Ude Aviation Plant formed on 26 October 2000 and was formally incorporated as such on 23 January 2001, having previously been known as Ulan-Ude Aviation Industrial Association. It manufactures a wide range of items, from helicopters and aeroplanes to spares and domestic equipment. Is only Russian Federation plant currently producing both fixed- and rotary-wing aircraft. Employees in 2000 totalled 5,056.

The factory has built I-16, Yak-3, Pe-2, La-5, La-7, La-9 and MiG-15 combat aircraft and, more recently, An-24 transports, Yak-25RV reconnaissance aircraft and MiG-27 fighter-bombers, together totalling 8,000 aeroplanes. Helicopter production began in 1956, progressing through Ka-15, Ka-18 and Ka-25 to the 1970 launch of Mi-8 manufacture; total of 3,700 Mi-8 series built.

Current products include modern developments of the Mil Mi-8/Mi-17 series of helicopters (including ten Mi-171s delivered to China in 1999 and 21 to Iran in 2000–01); and the Sukhoi Su-25UBK combat trainer and its related Su-39 attack aircraft. In 2001 revealed Mi-171 variant with rear loading ramp of own design, thus directly competing with Kazan's Mi-8MTV-5. Gained Far East order for five Mi-171Sh (Mi-8AMTSh) combat helicopters in 2001 and sold 10 Mi-171s to Malaysia. Four Mi-171s to Pakistan, June 2002, with 12 more following by end of that year. Supplied four Mi-171s to Slovak Police. Also in 2001, was building two Su-25TMs with own funds. Will build Kamov Ka-62 and Ka-64.

Also undertakes overhaul and upgrades, including five refurbished Mi-8AMTs supplied to Iranian Navy in 1999. Government shareholding is 38 per cent. Joint venture with SME Aviation of Malaysia, 1999, for promotion of Mi-8/Mi-17 and Su-25.

UPDATED

VASO

VORONEZHSKOYE AKTSIONERNOYE SAMOLETOSTROITELNOYE OBSHCHESTVO OAO (VORONEZH SHAREHOLDER AIRCRAFT-BUILDING SOCIETY JSC)

ulitsa Tsiolkovskoga 27, 394029 Voronezh
Tel: (+7 80732) 49 93 97
Fax: (+7 80732) 49 90 17
e-mail: vaso@email.ru
Web: http://www.vaso.ru
GENERAL DIRECTOR: Vyacheslav A Silikov
CHIEF DESIGNER: Mikhail N Shushpanov
HEAD OF MARKETING AND SALES: V A Tulupov

VASO is an integral part of the Ilyushin Aviation Complex, responsible for production of the Ilyushin Il-96 transport aircraft; after almost ceasing due to lack of funding, manufacture was resumed in early 2003 to meet an Aeroflot order for eight. In July 2004, VASO directors authorised a follow-on batch of five Il-96s to meet possible future orders. Production in 2005 intended for Aeroflot and Cubana, latter having ordered two VVIP aircraft for Fidel Castro. Further plans, as yet unfunded, but estimated to cost US$130 million, call for establishment of assembly line for Ilyushin Il-76, currently built in Uzbekistan. Transfer will produce an estimated 10,000 jobs.

VASO plans to form a joint holding with Ilyushin, initially by uniting the state shareholdings (30 per cent and 60 per cent, respectively) in each company; this will later be incorporated within AVPK Sukhoi as part of aviation industry restructuring plans. Russian government holding is 20 per cent. Future products could also include Antonov An-148 and Ilyushin Il-112V. Overhaul and modification activities have included conversion in 2003 of two Russian Air Forces Il-76MDs to -90 standard with PS-90A-76 turbofans.

Plant originated in GAZ 18, established 1932, building Tupolev TB-3 and DB-3. Original plant evacuated to Samara (then Kuybushev) in 1941, forming basis of current Aviacor. Re-established at Voronezh in 1947 as GAZ 84; built Ilyushin Il-10, Il-28, Tupolev Tu-16, Antonov An-12 and Tu-128 heavy interceptor and Tu-144 supersonic airliner and Ilyushin Il-86 wide-bodied airliner. Until recently known as VAPO. Plans to build Tupolev Tu-54 agricultural aircraft. Also produces boats, agricultural machinery, washing machines, food containers, motorcar silencers and other consumer items.

UPDATED

VITEK

KOMPANIYA VITEK (VITEK COMPANY)

ulitsa Generala Panfilova D20, Korpus 3, 125080 Moskva
Tel: (+7 095) 158 23 67
Fax: (+7 095) 158 65 14
e-mail: info@vitek-ltd.ru
Web: http://www.vitek-ltd.ru
DIRECTOR: Yury D Bazhanov
CHIEF ENGINEER: Eduard Babenko

Vitek is a building materials company which also operates a light aviation division under the title OAO Istrin Experimental Mechanical Plant. Promotion continues of the Ovod lightplane, although only two have been built; cost is US$25,000 (2002). In March 2002, the company announced that it had completed design of its Ezhik agricultural aircraft. A three-seat lightplane and four/six-seat amphibian are also being designed.

Company has 3,500 m^2 (37,675 sq ft) manufacturing facility in Moscow suburb of Istrinsky, capable of producing 30 lightplanes per year. It is Russian agent for the Canadian Safari kitbuilt helicopter.

UPDATED

VITEK EZHIK

English name: Hedgehog

TYPE: Agricultural sprayer.

PROGRAMME: Design begun in 1999 and completed by March 2002; construction of prototype then begun; public debut was then planned for August 2003 Moscow Salon but was rescheduled for the first half of 2004 and later revised to "Summer" 2004.

DESIGN FEATURES: Economic sprayer, with same payload as Sukhoi Su-38L, but at half purchase price. Raised cockpit with good forward/downward view. Constant-chord, low-mounted, cantilever wings; sweptback tail surfaces with mutual wire bracing; narrow, deep fuselage. Fuel tanks designed to collapse in event of impact.

FLYING CONTROLS: Conventional and manual. No horn balances.

LANDING GEAR: Tailwheel type; fixed. Two side Vs hinged to lower longerons, plus half-axles attached to compression frame.

POWER PLANT: Initially intended to utilise one 154 kW (207 hp) LOM M337A six-cylinder piston engine operating on A-76 motorcar fuel; later revised to 157 kW (210 hp) Walter Minor M-337A.

ACCOMMODATION: Pilot only, in enclosed cockpit behind 'unbreakable' glass.

EQUIPMENT: Chemical hopper; eight rotary atomisers beneath wings. Motorised pump and sprayer being sourced from UK; distribution atomisers from USA; emergency purge takes 3 seconds. Wire cutting system fitted in front of cockpit.

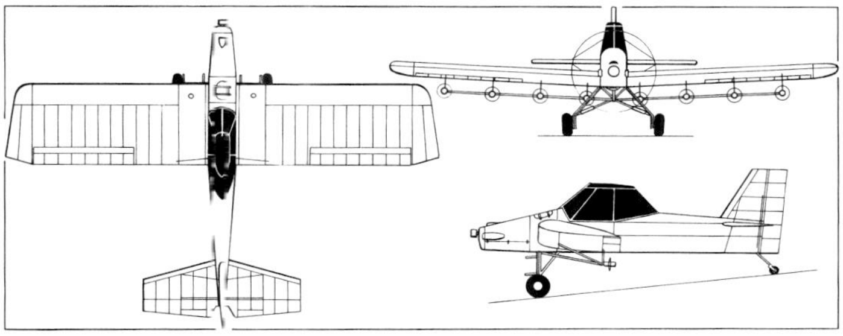

Vitek Ezhik agricultural sprayer *(James Goulding)* ***NEW***/0525933

DIMENSIONS, EXTERNAL:

Wing span	12.00 m (39 ft 4½ in)[1]
Wing chord, constant	2.20 m (7 ft 2½ in)
Length overall	8.325 m (27 ft 3¾ in)
Height overall	3.35 m (11 ft 0 in)
Tailplane span	4.30 m (14 ft 1¼ in)

[1] *Company literature also gives 11.00 m (36 ft 1 in)*

AREAS:

Wings, gross	24.25 m^2 (261.0 sq ft)

WEIGHTS AND LOADINGS:

Weight empty	520 kg (1,146 lb)
Payload: normal	300 kg (661 lb)
max	360 kg (794 lb)
Max T-O weight	960 kg (2,116 lb)
Max wing loading	39.6 kg/m^2 (8.11 lb/sq ft)
Max power loading	6.22 kg/kW (10.22 lb/hp)

PERFORMANCE:

Max level speed	97 kt (180 km/h; 112 mph)
Spraying speed	49–81 kt (90–150 km/h; 56–93 mph)
T-O run	90 m (295 ft)
Landing run	55 m (180 ft)

UPDATED

YAKOVLEV

OPYTNO-KONSTRUKTORSKOYE BYURO IMENI A S YAKOVLEVA OAO (EXPERIMENTAL DESIGN BUREAU JSC NAMED FOR A S YAKOVLEV)

Leningradsky prospekt 68, 125315 Moskva
Tel: (+7 095) 158 36 61
Fax: (+7 095) 787 28 44
e-mail: yakokb@cityline.ru
Web: http://www.yak.ru

CHAIRMAN AND PRESIDENT: Oleg F Demchenko

GENERAL DIRECTOR: Aleksandr G Efanov

FIRST DEPUTY GENERAL DIRECTOR: Nikolay N Dolzhenkov

DEPUTY GENERAL DIRECTORS:
Arkady I Gurtovoy (External Economic Relations and Marketing)
Roman P Taskaev (Flight Testing)

CHIEF DESIGNERS:
Konstantin F Popovich (Yak-130)
Aleksei G Rakhimbaev (Yak-40, Yak-42)
Dmitry K Drach (Yak-52M, Yak-54, Yak-58, Yak-152, Yak-112)
Andrey I Matveev (MC-21)
Yuri I Yankevich (UAVs)

DEPARTMENTAL CHIEFS:
Anatoly S Ivanov (Marketing and Sales)
Evgeny M Tarasov (External Relations)
Yuri V Zasypkin (Information/Press Service)

More than 200 aircraft designs and variants have been completed by Yakovlev since 1927, of which about 100, in 40 main types, have been manufactured in series. About 70,000 aircraft of Yakovlev design have been built, including 40,000 in Second World War and Chinese CJ-6s.

By mid-2000, Yakovlev working on design of short/medium-range, 130 to 170 seat transport conforming to Russian aviation technology plan up to 2015 and provisionally designated MC-21. Having been suspended in the 1990s, the **Yak-48**, was under renewed study by 2003 as the basis of a 30/50-seat replacement for the Yak-40 in standard- and stretched-fuselage versions, powered by two ZMKB AI-222 turbofans. Company plans for resumed manufacture of the Yak-42 centre on installation of Ivchenko Progress D-436 turbofans in Yak-42D variant.

Production of the Yak-58 lightplane is planned in Kazakhstan by the related Yak Alacon company. Yakovlev designers are working with HAIG of China on the projected JL-15 advanced trainer.

In October 2003, Irkut and Yakovlev signed letter of intent on merger, former to purchase 75.5 per cent share of Yakovlev for US$45 million; the purchase was finalised on 29 May 2004 at a price of US$70 million.

UPDATED

YAKOVLEV YAK-130

TYPE: Advanced jet trainer/light attack jet.

PROGRAMME: One of designs by five OKBs to meet official Russian requirement for 200 aircraft to replace Aero L-29 and L-39 Albatros for all aspects of flying training, basic to combat simulation, and combat; others included Sukhoi S-54, Myasishchev M-200 and MiG-AT. Known initially as Yak-UTS; work began in 1991; preliminary down-selection (with MiG-AT) in 1995; developed in partnership with Aermacchi of Italy with definition completed in December 1999 resulting in a common basic configuration, from which all current versions are derived; originally designated Yak/AEM-130 outside Russian Federation and Associated States (CIS) but renamed Aermacchi M-346 in July 2000.

Yak-130 announced as winner of Russian Air Forces' trainer competition on 10 April 2002 (decision having been reached on 16 March), in preference to MiG-AT. Yakovlev bureau funded 84 per cent of programme to this stage. Russian Air Forces' variant required for service from 2010 onwards. Aermacchi has rights to establish a second assembly line in Italy and contribute tail section and wing assemblies and FBW control system and avionics to both production lines, all components to be single-sourced. Yakovlev designed, and will manufacture, wing, tail surfaces, canopy, and digital FBW FCS and will integrate avionics and weapon systems on non-Russian Federation aircraft.

First preproduction Yak-130 on static display at Paris in June 2003 *(Paul Jackson)* ***NEW***/0558333

DEVELOPMENT MILESTONES

Design began	1991
Selected by Russian Air Forces	16 Mar 02
Official go-ahead	1995
First flight	25 Apr 96
Public debut	30 Nov 94
Production go-ahead	10 Apr 02

Yak-130D development aircraft shown to press 30 November 1994; first flight 25 April 1996 at Zhukovsky (later registered RA-43130). Flew 52 sorties (36 hours) in first year, including six without winglets, pending fitment of differently shaped units. To Italy for flight trials July/August 1997, during which (18 July) the 100th sortie was flown; low-speed, high-AoA programme (30 flights) completed in Italy by prototype 31 January 1999; AoAs of up to 41° (maximum), 35° in stabilised flight and 28° in landing configuration were flown successfully. Aircraft returned to Russia for maintenance, followed by evaluation at Akhtubinsk and Armavir test centres; again in Italy by 2000. Total 305 sorties and 254 hours flown by mid-2001; 350 sorties by mid-2002; painted in tactical camouflage for display at MAKS, Moscow, August 2001. Flown 485 sorties up to February 2003. By end of 2003, RA-43130 was still the only known flyable Yak-130.

Ten planned preseries aircraft, slightly smaller than prototype, were to comprise three for Zhukovsky test centre for evaluation originally planned to begin in 1998 and seven for air force evaluation planned for 1999 although development programme transferred to Italy, mid-1998, to escape financial instability in Russia, leading to far-reaching reappraisal. By September 1999 it was expected that funding for four preproduction prototypes would soon be obtained, together with three aircraft for evaluation; contract for initial four Yak-130-01s (including two — later reduced to one, the second airframe — for static tests) signed February 2001 between Yakovlev and Sokol production plant, all to be built in 2003–04; initial aircraft ('01') rolled out at Sokol plant on 30 May 2003; exhibited, unflown, at Paris in June 2003; first flight targeted for November 2003, with certification testing to be complete by end of 2004, but maiden flight not accomplished until 30 April 2004 (with AI-222-25 engines); static test airframe delivered by Sokol to Yakovlev Design Bureau in January 2004; series production from 2005. First prototype was to receive RD-2500 engines in third quarter of 2001, while 10 interim AI-222 engines were due to be delivered in 2002 from Motor-Sich for use in pre-series aircraft. However, only first two AI-222s had been received from Ukraine by early 2003 for static trials; first flyable engines then due in

October 2003. Nevertheless, in April 2003, Russian Air Forces predicting receipt of three Yak-130s by end of that year (not achieved) and balance of seven in 2004 to complete test and evaluation fleet. Announced 20 June 2003 that Russian President had approved manufacture of 200 Yak-130s.

Navalised version proposed for carrier training; tender now includes Penza (Russia)/CAE Electronics (Canada) simulators and Yak-152 for screening and primary training.

CURRENT VERSIONS: **Yak-130D:** Development prototype RA-43130.

Yak-130UBS: For Russian air force; indigenous equipment; service entry originally scheduled for 2000.

Yak/AEM-130: Hybrid export version; also known as AY-130; programme frozen. Items listed under Systems and Avionics refer mainly to this version. Viewed as low flyaway cost solution to Western trainer requirements by using Yak's good aerodynamic configuration and by addressing poor man-machine interface, low thrust-to-weight ratio and some other "problem areas".

Yak-133: Single-seat light attack version with additional fuel in lieu of rear cockpit (1,050 litres; 277 US gallons; 231 Imp gallons); radius of action up to 594 n miles (1,100 km; 684 miles) with 1,500 kg (3,307 lb) weapon load; 27.5 kN (6,173 lb st) AI-222-28 engines. Design study only.

Yak-133IB: Fighter-bomber (*Istrebitel-Bombardirovshchik*) with nose radar, possibly Leninets Kinzhal.

Yak-133R: Reconnaissance (*Razvedchik*) version.

Yak-133PP: Electronic warfare (*Postanovshchik Pomekh*) version. Nose radar and underwing jamming pods.

Aermacchi M-346: Under development in Italy. Marketing agreement assigns NATO to Aermacchi and CIS to Yakovlev; rivals will compete in remaining markets.

CUSTOMERS: Initial batch of 10 being delivered in 2003–04 for evaluation at Krasnodar Military Aviation Institute. Interest expressed by Slovak Republic, which received demonstration in July 1997; Slovakia announced on 5 August 1998 that debts owed by Russia will be repaid in part by delivery of 12 Yak-130s, although no firm contract yet placed. The Yak-130 remains a competitor in Slovakia's search for a subsonic multirole fighter with the AMX, L 159, Hawk and MiG-AT, but has not been officially selected. Its Slovak engines give it competitive advantage. Algerian interest was reported in early 2002; Indian prospects for licensed production raised in January 2003; Mexican interest also in 2003. Sales anticipated of up to 630, of potential market for 2,300, plus 200 or more for Russian Air Forces.

COSTS: US$12 million to US$15 million, depending on standard (2002). US$70 million reportedly already spent by Aermacchi and US$30 million by Yakovlev, with another US$100 million of R&D spending to follow.

DESIGN FEATURES: Designed from outset to conform to Western specifications for structure, flight controls and maintenance (MSG-3 and MIL-STD-470B). All-swept mid-wing monoplane, except for straight wing and tailplane trailing-edges; no dihedral or anhedral; two-piece full-span automatic leading-edge slats originally on prototype, but dogtooth leading-edge discontinuity tested on RA-43130 will be adopted on production version; all-moving low-mounted tailplane, with dogtooth leading-edge; forward-hinged, door-type airbrake on spine, ahead of fin; engines in ducts under wingroots, beneath LERX extending almost to windscreen; optional air intake screens obviate ingestion damage when operating from unpaved airstrips. Wing leading-edge sweepback 31°. Prototype's winglets reduced in size and moved inboard to wingroot as 'LEX fence'; may be deleted in production version. Strakes ahead of windscreen by 1999 replicate characteristics of planned rounded nose. Series production versions will be 1 tonne lighter than the prototype with a more slender, shallower fuselage, lacking chines.

Front cockpit of the first preproduction Yak-130 *(Paul Jackson)* ***NEW***/0558332

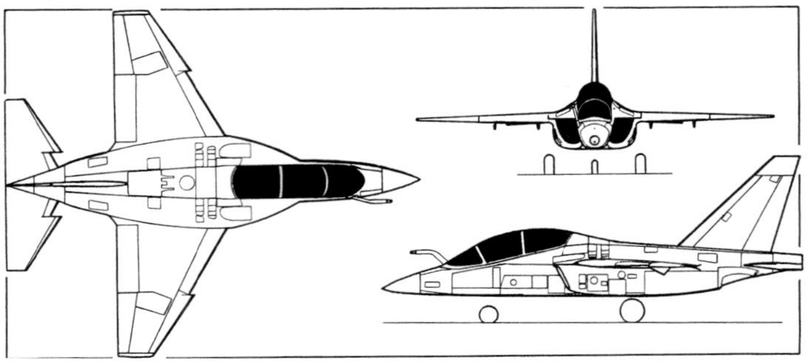

Production version of Yakovlev Yak-130 advanced trainer/combat aircraft *(Paul Jackson)* 0051417

Service life 10,000 hours; extension planned to 15,000 hours, or 30 years.

FLYING CONTROLS: Avionika full-authority, analogue, four-channel fly-by-wire system, with digital 'fourth channel' and 'carefree handling', slaved to laser-gyro platform, developed from that of Yak-141 V/STOL combat aircraft. Demonstrator has two selectable FCS models, one simple, one in MiG-29 class; will be replaced by digital system developed by BAE Italy and Lear Astronics. Prototype inherently stable; production aircraft intended to have 5 per cent longitudinal instability to reproduce handling characteristics of MiG-29/Su-27 series. Handling characteristics can be reprogrammed to simulate other types. Max AoA 42°.

STRUCTURE: Mainly of light alloys, but carbon fibre composites for most control surfaces. Structural design accords to MIL-STD-1530A and MIL-A-8660A.

Combat version has Kevlar armour around power plant, cockpits and avionics compartment aft of cockpits.

LANDING GEAR: Retractable tricycle type; single wheel on each unit; mainwheels retract into engine ducts; low-pressure tyres. Nosewheel steering to MIL-STD-8812 Class A. Anti-skid system.

POWER PLANT: Prototype fitted with two Povazské Strojàrne ZMK DV-2S (Klimov RD-35) turbofans each 21.58 kN (4,852 lb st). Autonomous engine starting; 20 second negative *g* fuel supply.

Production version of DV-2, designated ZMKB Progress AI-222-25, built in Russia by MMPP Salut and with FADEC, considered as interim power plant until availability of Soyuz RD-2500 and fitted in preproduction aircraft; both production engines rated at 24.5 kN (5,512 lb st).

Two fuel tanks in wings; one in centre-fuselage; combined capacity approximately 2,250 litres (594 US gallons; 495 Imp gallons), but normal training fuel restricted to approximately 1,060 litres (280 US gallons; 233 Imp gallons). Provision for two external fuel tanks, each of 560 litres (148 US gallons; 123 Imp gallons) beneath wings. Single-point pressure refuelling; gravity filling optional. Detachable aerial refuelling probe. Prototype has intake FOD doors of MiG-29 type which will be option for production version. Production aircraft may receive new 2,500 kg Tushino-Soyuz RD-2500 engines, based on existing RD-1700. Other engine choices include the Klimov RD-35, Lyulka/Saturn AL-55 and Zaparozhe's AI-222.

ACCOMMODATION: Two crew in tandem under blister canopy, on Zvezda K-36LT3.5Ka zero/zero ejection seats with built-in computer, side-force rocket engine and extreme attitude performance; export option of Martin-Baker Mk 16 standard on M-346; explosive cord canopy penetration; rear seat raised. Accommodation air conditioned and pressurised up to differential of 0.26 bar (3.8 lb/sq in). Front pilot has maximum view over nose of −16°; rear pilot −6°.

SYSTEMS: Flight control system includes four BAE Systems computers and provides artificial stability, flying qualities in accordance with MIL-STD-1797A, controllability up to at least 35° AoA, carefree handling (*g* limitation, stall/spin prevention, AoA limitation) and adaptability to various degrees of automation and autopilot modes; includes automatic reversionary modes in the event of damage or failure; and built-in test capability. Honeywell GTCP 36-150 APU, providing normal and emergency electric, hydraulic and pneumatic power. Zvezda OBOGS. Dual independent hydraulic systems, pressure 207 bar (3,000 lb/sq in); single failure leaves 75 per cent control surface movement; separate accumulators for emergency landing gear extension and brake. Dual Auxilec 20 kVA generators (one per engine); two Auxilec 6 kW transformer/rectifiers; one APU-driven 5 kW 28 V DC generator; system rating 200/115 V AC, 400 Hz. Two Ni/Cd batteries. Anti-icing system for windscreen and engine inlet guide vanes. Fire suppression system for engine and APU.

AVIONICS: Western avionics optional to replace standard Russian equipment. Primary suppliers Leninets (Russia) and GFSA/FIAR (Italy).

Comms: Dual V/UHF transceivers; intercom.

Radar: Optional. Kopyo and Osa under consideration.

Flight: Nav computer, laser gyro INS with embedded GPS, air data system, short-range radio nav Tacan and VOR/ILS, ADF and radio altimeter; IFF.

Instrumentation: HUD in front cockpit as part of collimated flight and sighting display in conjunction with pilot's helmet-mounted target designator; two Elektroavtomatika 130 mm (5.1 in) square colour multifunction liquid-crystal displays in each cockpit (three 152 × 203 mm; 6 × 8 in MFDs in export aircraft) with standby electromechanical flight/nav instruments. Optional helmet-mounted display. NVG compatibility and HOTAS controls.

Mission: Optional laser/TV weapon guidance system; Platan and Sapsan under consideration. Mission data entry system; optical weapon aiming computer and weapons control system; simulator of moving targets, guidance commands, weapons preparation and tactical situations; flight data and crew actions recorder; TV monitor of eyes and hands positions, with video recorder, in front cockpit.

Self-defence: Provision for chaff/flare dispenser, RWR and active electronic countermeasures.

ARMAMENT (OPTIONAL): Seven (optionally nine, including wingtip AAM stations) hardpoints for up to 3,000 kg (6,614 lb) of rockets, guns, missiles, guided and unguided bombs, including eight 250 kg bombs, four 500 kg laser-guided bombs, submunitions containers, Kh-25ML (AS-10 'Karen') or AGM-65 Maverick ASMs, R-73 (AA-11 'Archer'), AIM-9L Sidewinder or Magic 2 AAMs.

DIMENSIONS, EXTERNAL (production aircraft):

Wing span	9.725 m (31 ft 10¾ in)
Wing aspect ratio	4.0
Length overall	11.495 m (37 ft 8½ in)
Height overall	4.76 m (15 ft 7½ in)
Wheel track	2.53 m (8 ft 3½ in)
Wheelbase	3.90 m (12 ft 9½ in)

AREAS:

Wings, gross	23.52 m^2 (253.2 sq ft)

WEIGHTS AND LOADINGS (production aircraft):
- Weight empty ... 4,600 kg (10,141 lb)
- Max weapon load ... 3,000 kg (6,614 lb)
- Max internal fuel weight ... 1,750 kg (3,858 lb)
- T-O weight: trainer, typical ... 5,700 kg (12,566 lb)
  - attack, normal ... 8,090 kg (17,835 lb)
  - attack, max ... 9,000 kg (19,841 lb)
- Max wing loading: trainer ... 242.3 kg/m² (49.64 lb/sq ft)
  - attack/trainer ... 382.7 kg/m² (78.37 lb/sq ft)
- Max power loading: trainer ... 116 kg/kN (1.14 lb/lb st)
  - attack/trainer ... 184 kg/kN (1.80 lb/lb st)

PERFORMANCE (clean, estimated):
- Max level speed ... 572 kt (1,060 km/h; 659 mph)
- Unstick speed ... 108 kt (200 km/h; 124 mph)
- Landing speed ... 103 kt (190 km/h; 118 mph)
- Stalling speed, 20% fuel ... 89 kt (165 km/h; 103 mph) CAS
- Max rate of climb at S/L ... 4,500 m (14,763 ft)/min
- Service ceiling ... 12,500 m (41,020 ft)
- T-O run ... 340 m (1,115 ft)
- Landing run ... 550 m (1,805 ft)
- Radius of action, 5 min combat: interdiction, two Mk 83 1,000 lb bombs, two AAMs, gun pod, two external tanks, hi-lo-hi ... 440 n miles (815 km; 506 miles)
  - close air support, two Mk 82 500 lb and two Mk 83 1,000 lb bombs, two AAMs, gun pod, hi-lo-hi, 30 min loiter ... 120 n miles (222 km; 138 miles)
  - combat air patrol, two AAMs, gun pod, two external tanks, 2 h loiter at 11,600 m (38,060 ft) ... 240 n miles (444 km; 276 miles)
- Range with max internal fuel ... 1,079 n miles (2,000 km; 1,242 miles)
- Max rate of roll ... 200°/s
- Sustained *g* limit at M0.8 at 4,575 m (15,000 ft) with 50% internal fuel ... +4.8
- *g* limits ... +8/–3

UPDATED

YAKOVLEV YAK-3M AND YAK-9U-M

TYPE: Aerobatic single-seat sportplane.

PROGRAMME: Reproduction Second World War fighter. Prototype of original wooden-skinned Yak-3 flew 1943; all-metal version flew 1945; deliveries to Soviet air forces began July 1944, totalled 4,848 (of 36,737 Yakovlev single-engined Second World War fighters); described often as lightest weight and most agile monoplane of 1939–45 period; able to turn 360° in 18.5 seconds. All-metal version now available, with Western power plant and uprated instrumentation; new-build 'prototype' (0470101, later N854DP) displayed 1993 Paris Air Show; 11 built by Strela at Orenburg to meet orders from Gunnell Museum, USA; one later to New Zealand (then Australia in 1999) and one to South Africa.

Production turned in 1996 to Yak-9U of similar appearance. Original Yak-9 entered service in 1942 (before Yak-3) and 16,769 were built. Total of eight all-metal Yak-9U-M replicas exported to USA by mid-1999, of which five then flying; one subsequently to France; distribution by Shadetree Aviation of Carson City, Nevada (Tel: +1 775 841 10 68; Web: http://www.shadetreeusa.com). Ninth Yak-9 completed by Strela in late 2001 for export to US, being registered in June 2003.

In all, 21 reproductions built by 2002, according to Yakovlev information, apparently including one untraced Yak-7.

UK importer, Richard Goode Aerobatics, commissioned four further Yak-9s for delivery in 2002, three having already been sold by May 2002. None of these had appeared by early 2004, however.

Note that further Yak-3UAs have been produced by other sources through conversion of Yak-11s, mostly with Pratt & Whitney R-1830/R-2000 Twin Wasp radial engines.

COSTS: US$385,000 plus tax (2004).

Following data apply to Yak-9.

DESIGN FEATURES: Precise reproduction in metal of Second World War airframe, except for repositioned carburettor air intake above engine cowling to suit changed engine. Conventional cantilever low-wing configuration, with tapered, round-tipped wings. Mainwheel tyres size 600×180.

POWER PLANT: Reconditioned 925 kW (1,240 hp) Allison V-1710-39 12-cylinder V liquid-cooled piston engine; three-blade propeller. Fuel capacity 526 litres (139 US gallons; 116 Imp gallons).

DIMENSIONS, EXTERNAL:
- Wing span ... 9.20 m (30 ft 2¼ in)
- Wing aspect ratio ... 5.7
- Length overall ... 8.55 m (28 ft 0½ in)
- Height overall ... 2.39 m (7 ft 10 in)

AREAS:
- Wings, gross ... 14.85 m² (159.8 sq ft)

WEIGHTS AND LOADINGS:
- Weight empty ... 2,135 kg (4,707 lb)
- Max T-O weight ... 2,600 kg (5,732 lb)
- Stalling speed: clean ... 103 kt (191 km/h; 119 mph)
  - landing configuration ... 81 kt (150 km/h; 94 mph)

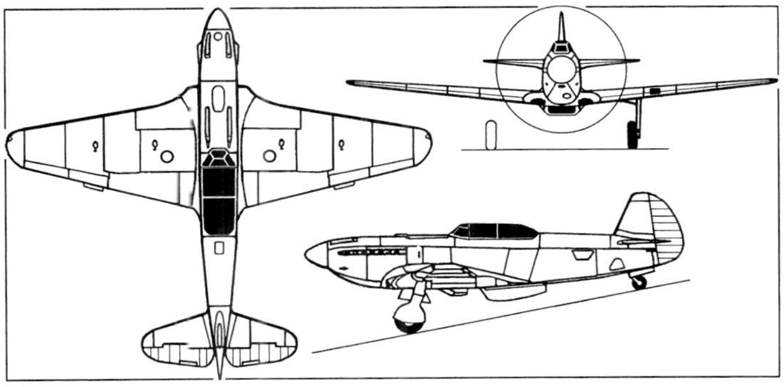

Yakovlev Yak-9U reproduction *(Paul Jackson)* ***NEW***/0580703

Reproduction Yak-9U operated in the USA *(Paul Jackson)* ***NEW***/0580744

- Max wing loading ... 175.1 kg/m² (35.86 lb/sq ft)
- Max power loading ... 2.81 kg/kW (4.62 lb/hp)

PERFORMANCE:
- Max level speed ... 377 kt (698 km/h; 434 mph)

UPDATED

YAKOVLEV YAK-18T

TYPE: Four-seat lightplane.

PROGRAMME: Prototype first flew mid-1967 as extensively redesigned cabin version of veteran Yak-18 trainer; over 700 built at Smolensk Aircraft Plant (SmAZ), including trainer, ambulance, communications and light freight versions; production resumed by Smolensk in 1993 to meet new contracts and one converted to Technoavia SM-94 prototype (see 2000–01 and previous editions). Second production run ended in 1996 due to bankruptcy of SmAZ. In 1998, work recommenced on 15 stored, part-built airframes, which being completed with two-piece windscreen, designed for Technoavia SM-94. New aircraft delivered as either SM-94 or distributed in Western Europe (eight reserved) by Richard Goode Aerobatics of
White Waltham, UK (+44 1963 36 27 46; fax +44 1963 36 37 51; e-mail richard.goode@russianaeros.com).Production continues, with some airframes destined to be completed as Smolensk (Technoavia) SM-2000 six-seat turboprops.

CUSTOMERS: First resumed (1993) deliveries to Luxembourg, Philippines, Russia, Switzerland, Turkey, UAE, UK and USA. Recent deliveries include three to Ulyanovsk Aviation School in June 2001. Production in 2001 reportedly 10 to 12 per year; total of 80 built between 1993 and mid-2001, increasing Yak-18T total to 613, not including some 6,638 Yak-18s and 1,796 Chinese CJ-6s.

DESIGN FEATURES: Adaptation of aerobatic trainer. Conventional cantilever low-wing monoplane; three-section two-spar wings; semi-monocoque fuselage of basically square section; strut- and wire-braced tail surfaces. Airframe life 5,000 hours or 20 years.

Wing section Clark YH, thickness/chord ratio 14.5 per cent at root, 9.3 per cent at tip; dihedral 7° 20′ on outer panels only.

FLYING CONTROLS: Conventional and manual. Plain flaps and elevator trim.

STRUCTURE: All-metal construction, except for fabric covering on parts of outer wing panels, ailerons and tail surfaces.

LANDING GEAR: Retractable tricycle type; single wheel on each unit; pneumatic retraction, K-141/T-141 mainwheels inward into centre-section, Type 44-1 nosewheel rearward; Hydromash oleo-nitrogen shock-absorbers; castoring nosewheel is non-steerable; Yaroslavl tyres size 500×150 on mainwheels, 400×150 on nosewheel; tyre pressure (all) 2.95 bar (43 lb/sq in); pneumatic plate-type brakes on mainwheels. Minimum ground turning radius 2.52 m (8 ft 3¼ in).

POWER PLANT: One 265 kW (355 hp) VOKBM M-14P air-cooled radial piston engine; optional 294 kW (394 hp) M-14PF in 1999 version; V-530TA-D35 two-blade variable-pitch metal propeller; wingroot fuel tanks, combined capacity 190 litres (50.2 US gallons; 41.8 Imp gallons); gravity fuelling points on wing upper surface. Optional usable fuel capacities of 321 litres (84.7 US gallons; 70.6 Imp gallons), 360 litres (95.3 US gallons; 79.4 Imp gallons) or 614 litres (162 US gallons; 135 Imp gallons) available in 1999 version. Oil capacity 20 litres (5.3 US gallons; 4.4 Imp gallons). Inverted fuel and oil systems optional.

Yakovlev Yak-18T *(Paul Jackson)* ***NEW***/0580777

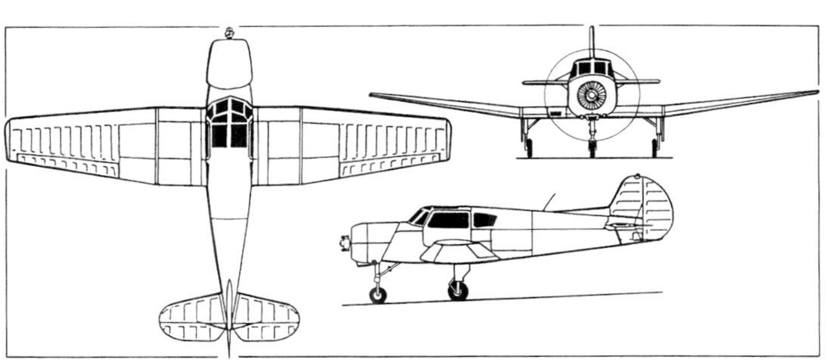

Yakovlev Yak-18T (VOKBM M-14P radial engine) *(Paul Jackson)* 0015144

ACCOMMODATION: Cabin of main production version seats four persons in pairs, with forward-hinged jettisonable door each side; rear bench seat removable for freight carrying; ambulance configuration accommodates pilot, stretcher patient and medical attendant; baggage compartment aft of rear seat. Option to increase seating to six has not been pursued. Cabin heated and ventilated.

SYSTEMS: Pneumatic system for landing gear and flaps, operating pressure 49 bar (711 lb/sq in); 36/115 V AC electrical system at 400 Hz, 28.5 V DC and 20NKBN-25 battery for red panel lighting, nav and landing lights, and fintip anti-collision beacon.

AVIONICS: *Comms:* Baklan-5 UHF radio, intercom.

Instrumentation: ARK-15M radio compass, radio altimeter and flight recorder standard. Customised avionics available in 1999 version.

EQUIPMENT: Optional smoke system for 1999 version.

DIMENSIONS, EXTERNAL:
Wing span ... 11.16 m (36 ft 7¼ in)
Wing chord: at root ... 2.00 m (6 ft 6¾ in)
at tip ... 1.06 m (3 ft 5¾ in)
Wing aspect ratio ... 6.6
Length overall ... 8.39 m (27 ft 6½ in)
Height overall ... 3.40 m (11 ft 1¾ in)
Tailplane span ... 3.54 m (11 ft 7½ in)
Wheel track ... 3.12 m (10 ft 2¾ in)
Wheelbase ... 1.955 m (6 ft 5 in)
Propeller diameter ... 2.40 m (7 ft 10½ in)
Propeller ground clearance ... 0.16 m (6¼ in)
Cabin doors (each): Height ... 0.88 m (2 ft 10¾ in)
Width ... 1.18 m (3 ft 10½ in)
Baggage door: Height ... 0.50 m (1 ft 7¾ in)
Width ... 1.00 m (3 ft 3¼ in)

DIMENSIONS, INTERNAL:
Cabin: Max width ... 1.28 m (4 ft 2¼ in)
Max height ... 1.25 m (4 ft 1¼ in)

AREAS:
Wings, gross ... 18.80 m^2 (202.4 sq ft)
Ailerons (total) ... 1.92 m^2 (20.66 sq ft)
Flaps (total) ... 1.60 m^2 (17.22 sq ft)
Fin ... 0.72 m^2 (7.73 sq ft)
Rudder ... 0.98 m^2 (10.57 sq ft)
Tailplane ... 1.95 m^2 (21.00 sq ft)
Elevators (total) ... 1.235 m^2 (13.30 sq ft)

WEIGHTS AND LOADINGS (A: instructor and one pupil, B: four persons):
Weight empty ... 1,217 kg (2,683 lb)
Max payload: A ... 306 kg (675 lb)
B ... 338 kg (745 lb)
Max fuel ... 140 kg (308 lb)
Max T-O weight: A ... 1,500 kg (3,307 lb)
B ... 1,650 kg (3,637 lb)
Max landing weight: B ... 1,650 kg (3,637 lb)
Max wing loading: A ... 79.8 kg/m^2 (16.34 lb/sq ft)
B ... 87.8 kg/m^2 (17.98 lb/sq ft)
Max power loading: A ... 5.66 kg/kW (9.32 lb/hp)
B ... 6.22 kg/kW (10.25 lb/hp)

PERFORMANCE:
Never-exceed speed (VNE) ... 162 kt (300 km/h; 186 mph)
Max level speed: A, B ... 159 kt (295 km/h; 183 mph)
Max cruising speed: B ... 135 kt (250 km/h; 155 mph)
Econ cruising speed ... 113 kt (210 km/h; 130 mph)
Landing speed ... 68 kt (125 km/h; 78 mph)
Max rate of climb at S/L: B ... 300 m (985 ft)/min
Service ceiling: A, B ... 5,520 m (18,120 ft)
T-O run: B ... 260–280 m (855–920 ft)
Landing run: B ... 345–365 m (1,135–1,200 ft)
Range with standard max fuel, with reserves:
A, B ... 313 n miles (580 km; 360 miles)
g limits: A ... +6.4/–3.2

UPDATED

YAKOVLEV YAK-54

TYPE: Aerobatic two-seat sportplane.

PROGRAMME: Yak-54 announced 1992; prototype at 1993 Paris Air Show; first flown 23 December 1993; third prototype at Paris in 1995. In production from 1995 at Saratov Aviation Plant (SAZ), with maximum capacity of more than 100 per year. However, production suspended in 1998; to resume after certification achieved, early 2002; initial batch of five funded (and to be marketed) by ZAO Gorki Yu-2 company, including two for Promety (Prometheus) aerobatic group (representing total 2002 production) and one for export to Australia; a Slovakian customer has also been mentioned. However, in February 2002, SAZ again suspended production, having built only one aircraft (in 2001), this then due to receive Russian certification at end of 2002 (confirmation not yet received), following improvements to aerodynamics aimed at enhancing basic handling qualities and aerobatic capability. Following resumption of manufacture, Saratov had delivered three Yak-54s to Central Aeroclub of Samara by December 2003, and planned to manufacture seven or eight in 2004, including one for the Russian Aerobatic Team, four to France and two to Germany.

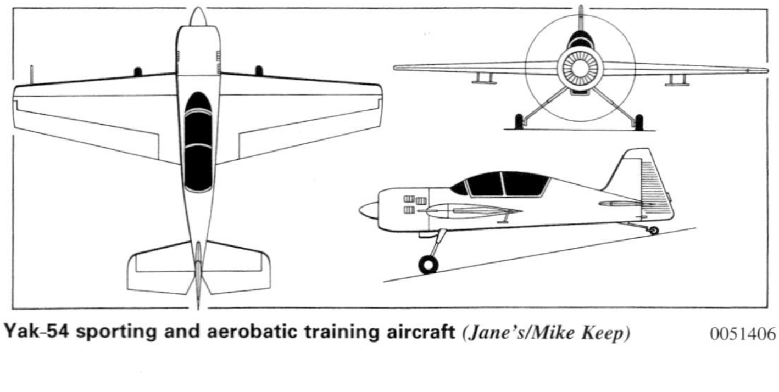

Yak-54 sporting and aerobatic training aircraft *(Jane's/Mike Keep)* 0051406

Aerobatic display by Yak-54 *(Paul Jackson)* ***NEW***/0580759

CURRENT VERSIONS: **Yak-54:** *As described.*

Yak-56: Primary trainer derivative; also designated **Yak-54M**; redesignated **Yak-152** in late 2000 (although Yak-54M designation was still being used in 2001); 294 kW (394 hp) M-14PF, retractable tricycle landing gear and SKS-94 crew ejection system. Prototype under construction in 1999. Wing span 8.80 m (28 ft 10½ in); length overall 7.30 m (23 ft 11½ in); height overall 2.80 m (9 ft 2¼ in); empty weight 950 kg (2,094 lb); fuel weight 200 kg (441 lb); maximum take-off weight 1,300 kg (2,866 lb); max level speed 202 kt (375 km/h; 233 mph) manoeuvring speed 243 kt (450 km/h; 280 mph); stalling speed 54 kt (100 km/h; 63 mph); maximum range 540 n miles (1,000 km; 621 miles); *g* limits +9/–7.

In October 2001, Yak-152 was unsuccessful in competition against Sukhoi Su-49 to provide new equipment for ROSTO (successor to DOSAAF) paramilitary training organisation. Despite this, prototype intended to fly in 2003 and type to enter production at Saratov (SAZ). Commitment reconfirmed by Yakovlev's chairman in April 2002, but engine reverted to 265 kW (355 hp) M-14Kh and MTOW 1,320 kg (2,910 lb). By mid-2004, first flight had not taken place.

Yak-57: Single-seat sportplane; under development by 1999.

CUSTOMERS: Total 48 Yak-54s ordered in January 1997 by Northwest Aerobatic Center, Ephrata, Washington, USA, and Dancing Bear air show team; 15 built up to 1998 suspension of production; five (of six exported) flying in USA, plus one (second-hand) delivered to Australia during 1999. In March 2002, Yak-54 fleet stated to be four in Russia, five in US, plus two destroyed in accidents. US marketing by Yakovlev Aircraft USA, Scotts Mills, Oregon; e-mail: info@yak54.com. Kaiser Flugzeugbau of Germany planned acquisition of evaluation Yak-54 in 2004 with a view to becoming a distributor.

COSTS: Yak-152 programme US$2 million to US$3 million (2002). Yak-54 US$150,000 to US$170,000 (2002); Yak-152 US$160,000 to US$170,000 (2002).

DESIGN FEATURES: Optimised for aerobatics; derived from Yak-55. Conventional mid-wing configuration; symmetrical section; no dihedral, anhedral or incidence; almost full-span ailerons, elevators and rudder all horn-balanced; each aileron has large suspended balance tab. Designed on basis of systems and units of Yak-55M.

FLYING CONTROLS: Conventional and manual. Ailerons occupy 90 per cent of wing trailing-edge and have both horn balance and suspended tab; horn-balanced tail surfaces.

STRUCTURE: All-metal; two-spar wings; semi-monocoque fuselage; conventional tail unit.

LANDING GEAR: Tailwheel type; fixed. Fuselage-mounted titanium spring cantilever mainwheel legs and small wheels, tyre size 400×150; tailwheel tyre size 200×80.

POWER PLANT: One 265 kW (355 hp) VOKBM M-14P nine-cylinder air-cooled radial engine; MTV-9 three-blade variable-pitch propeller.

ACCOMMODATION: Two seats in tandem under continuous transparent canopy, hinged to starboard.

DIMENSIONS, EXTERNAL:
Wing span ... 8.16 m (26 ft 9¼ in)
Wing aspect ratio ... 5.2
Length overall ... 6.91 m (22 ft 8 in)

AREAS:
Wings, gross ... 12.89 m^2 (138.75 sq ft)

WEIGHTS AND LOADINGS:
Max T-O weight: one pilot ... 850 kg (1,874 lb)
two occupants ... 990 kg (2,182 lb)
Max wing loading ... 76.8 kg/m^2 (15.73 lb/sq ft)
Max power loading ... 3.74 kg/kW (6.15 lb/hp)

PERFORMANCE:
Never-exceed speed (VNE) ... 243 kt (450 km/h; 280 mph)
Stalling speed ... 60 kt (110 km/h; 69 mph)
Rate of roll ... 345°/s
Max rate of climb at S/L ... 900 m (2,950 ft)/min
Ferry range ... 377 n miles (700 km; 435 miles)
g limits ... +9/–7

UPDATED

YAKOVLEV YAK-SKH

TYPE: Agricultural sprayer.

PROGRAMME:
Announced late 2000, when still at project stage; provisional SKh designation indicates *selskokhozyaistvenny:* agricultural; derived from Yak-54 aerobatic sportplane, retaining M-14 engine. Single-seat version to be standard; second seat optional. No further announcements known by mid-2004.

UPDATED

YAKOVLEV/ILYUSHIN MS-21

TYPE: Regional jet airliner.

PROGRAMME: Originated in Yak-242, which proposed as successor to Yak-42 and included in State Programme for Aviation, with maiden flight scheduled for 1998. Project lapsed, but revived in 2003 when re-launched as MS-21 (*Magistral'nyi Samolet 21 Veka*: airliner of the 21st Century), with design inputs from Ilyushin, having gained official support against Il-214 and Irkut 111 in competition decided on 21 July and formally announced 18 August 2003. Also envisaged as Tu-154M replacement.

Ilyushin expected to become lead partner on MS-21 and sign development contract with Rosaviakosmos; Yakovlev then to become subcontractor to Ilyushin. However, in

Cutaway model of proposed Yakovlev/Ilyushin MS-21 *(Paul Jackson)* ***NEW***/1047439

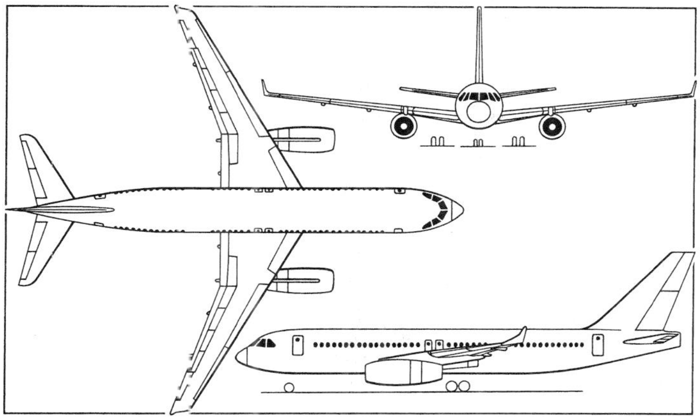

Yakovlev Yak-242, on which the MS-21 is based *(Jane's/Mike Keep)* ***NEW***/0572221

December 2004, Ilyushin confirmed Yakovlev as "lead developer". Irkut to join programme as risk-sharing partner. First flight two years from go-ahead and service entry three years thereafter. Production will be by Aviastar.

CURRENT VERSIONS: **MS-21-100:** With 132 seats in three classes (12+24+96).

MS-21-200: With 156 seats in two classes.

MS-21-300: With 174 seats in single class.

MS-21K: Freighter; 17 t payload.

MS-21KP: Combi; cargo compartment length 21.43 m (70 ft 3¾ in).

CUSTOMERS: Predicted market for 600 MS-21s over 20 years, including 400 for Russian operators.

COSTS: Development estimated as US$460 million (2003), including US$200 million from Russian government.

DESIGN FEATURES: Low-wing airliner with two podded engines underwing. Cruising speed in excess of M0.8; significant use of composites materials. To conform to equivalents of FAR Pt 25 for airframe; Pt 33 engines; Pt 34 emissions; and Pt 36/ICAO Chapter 4 noise. Intended 70 per cent engine, systems and avionics commonality with Ilyushin/Irkut/HAL Il-214. Service life 60,000 hours and 30,000 landings over 25 years.

Supercritical wing section designed in collaboration with TsAGI; leading-edge sweepback 25°; winglets.

FLYING CONTROLS: Digial fly-by-wire controls. Double-slotted flaps.

LANDING GEAR: Tricycle type; hydraulically retractable. Four-wheel main bogies; tyre size 950×300; pressure 7.85 bar (114 lb/sq in). Twin nosewheel, size 660×200, pressure 8.82 bar (128 lb/sq in).

POWER PLANT: Two 118 kN (26,450 lb st) Aviadvigatel PS-12 turbofans (derated PS-90A).

ACCOMMODATION: Two flight deck crew. Seating for 130 to 170 passengers. Combi and freighter versions with cargo door ahead of wing, port side; accommodates eight P1, P2 or P6 pallets.

All data are provisional.

DIMENSIONS, EXTERNAL:	
Wing span	36.25 m (118 ft 11¼ in)
Wing aspect ratio	11
Length overall	38.00 m (124 ft 8 in)
Max diameter of fuselage	4.10 m (13 ft 5½ in)
Freight door (optional, port, front):	
Width	2.90 m (9 ft 6¼ in)
Height	2.10 m (6 ft 10¾ in)
AREAS:	
Wings, gross	120.0 m² (1,291.7 sq ft)
WEIGHTS AND LOADINGS:	
Operating weight empty, equipped	38,400 kg (84,655 lb)
Max payload	18,000 kg (39,680 lb)
Max fuel weight	22,000 kg (48,500 lb)
Max T-O weight: 100	65,800 kg (145,065 lb)
200	71,100 kg (156,750 lb)
300	72,000 kg (158,735 lb)
PERFORMANCE:	
Cruising speed	459 kt (850 km/h; 528 mph)
Range: 100	2,537 n miles (4,700 km; 2,920 miles)
200	2,969 n miles (5,500 km; 3,417 miles)
300	2,429 n miles (4,500 km; 2,796 miles)

UPDATED

Serbia and Montenegro

UTVA

VAZDUHOPLOVNA INDUSTRIJA UTVA (SHELDRAKE AIRCRAFT INDUSTRY)

Jabuki put bb, YU-26000 Pančevo
Tel: (+381 13) 31 53 83
Fax: (+381 13) 31 98 59
e-mail: utvaai@ptt.yu; utva@verat.net
Web: http://www.utvaaviation.co.yu

GENERAL MANAGER: Dipl Ing Tomislav Bjelogrlic
PRODUCTION MANAGER: Dipl Ing Dragan Turkalj
AIRCRAFT DEVELOPMENT MANAGER: Dipl Ing Tonko Mihovilovic

Utva formed at Zemun, 5 June 1937; to Pančevo 1939. The 86,000 m² (925,700 sq ft) factory was heavily damaged during NATO air strikes on 24 March 1999, at which time development work on the G-4M Super Galeb, Lasta 2 primary/basic trainer and Utva-96 ceased. In late 2003 the company announced that it was restarting work on the Utva-96; in early 2004 it was announced that Utva would be separated from its parent company Lola prior to privatisation, scheduled for 7 October 2004. The company is also developing unmanned aerial vehicles (UAVs) and aerial targets.

UPDATED

UTVA-96

TYPE: Four-seat lightplane.

PROGRAMME: Resumption announced in 2003 as a development of earlier two-seat Utva-75 trainer used by air forces and flying clubs of the former Yugoslavia, at which time two prototypes were under construction.

DESIGN FEATURES: Compared to earlier Utva-75, has strengthened fuselage to meet FAR Pt 23 requirements and redesigned engine cowling.

FLYING CONTROLS: Conventional and manual. Fluted skin on ailerons, flaps, fin, rudder and elevators; rudder and elevators horn-balanced; ground-adjustable tab on rudder; Flettner trim tab on each aileron; Teleflex actuator for elevator control tab.

STRUCTURE: All-metal semi-monocoque.

LANDING GEAR: Tricycle type; fixed. Wing-mounted, cantilever main legs with trailing-link suspension and oleo-pneumatic shock-absorbers. Small tail bumper. Tyres, size 7.00-8 on mainwheels; size 6.00-6 on nosewheel; hydraulic brakes.

POWER PLANT: One 224 kW (300 hp) Textron Lycoming IO-540-L1A5D flat-six. Total fuel 153 litres (40.4 US gallons; 33.7 Imp gallons) in integral wing tanks.

ACCOMMODATION: Pilot and three passengers in two pairs of seats; single-piece front windscreen and larger side windows. Separate baggage compartment.

AVIONICS: Integrated system to meet IFR requirements.

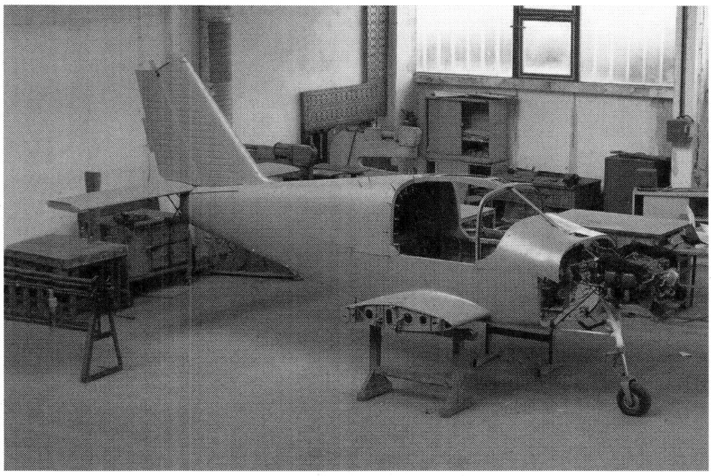

Prototype Utva-96 under construction ***NEW***/0567734

DIMENSIONS, EXTERNAL:	
Wing span	9.73 m (31 ft 11 in)
Wing chord	1.55 m (5 ft 1 in)
Wing aspect ratio	6.5
Length overall	7.11 m (23 ft 4 in)
Height overall	3.15 m (10 ft 4 in)
Tailplane span	3.80 m (12 ft 5½ in)
Wheel track	2.58 m (8 ft 5½ in)
Wheelbase	1.99 m (6 ft 6¼ in)
Propeller diameter	1.93 m (6 ft 4 in)
Propeller ground clearance	0.295 m (11¾ in)
AREAS:	
Wings, gross	14.63 m² (157.5 sq ft)
Ailerons (total)	1.38 m² (14.85 sq ft)
Flaps (total)	1.61 m² (17.33 sq ft)
Vertical tail surfaces (total)	1.78 m² (19.16 sq ft)
Horizontal tail surfaces (total)	3.34 m² (35.95 sq ft)

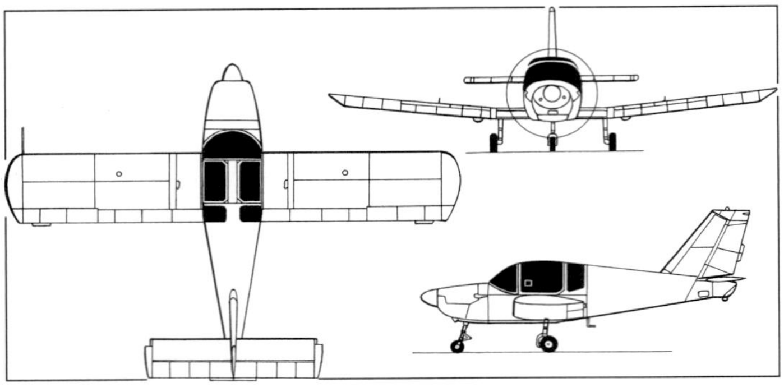

Utva-96 four-seat lightplane *(James Goulding)* ***NEW***/1047690

WEIGHTS AND LOADINGS (estimated):
Weight empty ... 710 kg (1,565 lb)
Baggage capacity ... 50 kg (110 lb)
Max T-O weight ... 1,210 kg (2,668 lb)
Max wing loading ... 82.7 kg/m² (16.94 lb/sq ft)
Max power loading ... 5.40 kg/kW (8.89 lb/hp)

PERFORMANCE (estimated):
Max operating speed ... 103 kt (190 km/h; 118 mph)
Max cruising speed ... 94 kt (175 km/h; 109 mph)
Stalling speed ... 54 kt (100 km/h; 62 mph)
Max rate of climb at S/L ... 240 m (787 ft)/min
Service ceiling ... 4,000 m (13,120 ft)
T-O run ... 200 m (660 ft)
T-O to 15 m (50 ft) ... 400 m (1,315 ft)
Landing from 15 m (50 ft) ... 340 m (1,115 ft)
Landing run ... 180 m (590 ft)
Range with max fuel ... 324 n miles (600 km; 373 miles)

UPDATED

Singapore

ST AERO

SINGAPORE TECHNOLOGIES AEROSPACE LTD (DIVISION OF SINGAPORE TECHNOLOGIES ENGINEERING LTD)

540 Airport Road, Paya Lebar, Singapore 539938
Tel: (+65) 62 87 11 11
Fax: (+65) 62 80 97 13 and 280 82 13
e-mail: comms.aero@stengg.com
Web: http://www.stengg.com
CHAIRMAN AND CEO: C B Lim
DEPUTY CHAIRMAN: S F Boon
PRESIDENT: K K Tay
DEPUTY PRESIDENT AND COO: Y S Ho
SENIOR VICE-PRESIDENTS:
W S Chiang (Marketing)
K P Chong (AMM)
S G Lim (EMS)
C J Sew (CERO)
S J Tan (Engineering and Development)
HEAD OF CORPORATE COMMUNICATIONS: Audrey Tan

Formed early 1982 as government-owned Singapore Aircraft Industries Ltd, controlled by Ministry of Defence Singapore Technology Holding Company Pte Ltd; renamed Singapore Aerospace April 1989; 15,000 m² (161,450 sq ft) new facility at Paya Lebar opened October 1983.

ST Aero, the aerospace arm of Singapore Technologies Engineering Ltd, has more than 5,000 employees worldwide. In the two decades up to 2004, it had redelivered more than 3,500 commercial aircraft, 6,000 engines and 600,000 components. Currently organised into Commercial Business Group, Military Business Group and Engineering and Development Centre. Provides comprehensive high-echelon commercial and military aircraft maintenance and engineering services, including structural modification and refurbishment, engine and aircraft component repair and overhaul, precision manufacturing, spares and material support.

ST Aero is a 15 per cent partner in EC 120 helicopter programme with Eurocopter and CATIC (see under International); is manufacturing 100 shipsets of nosewheel doors worth US$12 million for Boeing 777 (option for another 100). Partners Boeing in freighter conversions of 757 (17 delivered to DHL) and MD-11 (13 for UPS); seeking participation in Airbus A380. Recently upgraded F-5Es and F-5Fs of RSAF to F-5S and F-5T standard with new avionics and WDNS (IOC January 1998); now partnering IAI and Elbit of Israel for avionics upgrade (first flight June 2001) of 48 F-5E/Fs of Turkish Air Force; and involved with Lockheed Martin, BAE Systems, FIAR, Rafael, Thales and others in promoting 'Falcon ONE' upgrade for F-16A/B Fighting Falcon.

In 2001, ST Aero restructured into the following three operating segments, whose composition in 2004 was as follows:

Aircraft Maintenance and Modification (AMM)
ST Aviation Services Co Pte Ltd (SASCO)
ST Mobile Aerospace Engineering Inc* (MAE)
ST Aerospace Engineering Pte Ltd (STA Engineering)
Pacific Flight Services Pte Ltd (PFS)
San Antonio Aerospace LP (SAA)
Bournemouth Aviation Services Co Ltd (BASCO)
Aerospace Engineering Services Pty Ltd (AES)
* Subsidiary of Singapore Technologies Engineering Ltd

Component/Engine Repair and Overhaul (CERO)
ST Aerospace Engines Pte Ltd (STA Engines)
ST Aerospace Systems Pte Ltd (STA Systems)
Singapore Precision Repair and Overhaul Pte Ltd (S-PRO)
Turbine Overhaul Services Pte Ltd (TOS)
Turbine Coating Services (TCS)
Composite Technology International Pte Ltd (CTI)

Engineering and Materials Services (EMS)
ST Aerospace Supplies Pte Ltd (STA Supplies)
Airline Rotables Ltd (ARL)
ST Aerospace International Structures Pte Ltd (STAIS)
VisionTech Engineering Pte Ltd (VTE)
ST Aviation Resources Pte Ltd (STAR)
Singapore British Engineering Pte Ltd (SBE)
Asian Aerospace Pte Ltd (AAPL)
Eurocopter South East Asia Pte Ltd

UPDATED

Slovakia

AEROPRO

AEROPRO S R.O.

Kostolna 42, SK-949 01 Nitra
Tel/Fax: (+421 37) 652 63 55
e-mail: aeropro@stonline.sk
Web: http://www.aeropro.sk
GERMAN DISTRIBUTOR:
Ikarusflug Leichtflugzeuge GbR
Mennwanger Strasse 3, D-88682 Salem
Tel: (+49 7553) 17 70
Fax: (+49 7553) 605 38
Web: http://www.ikarusfug.de

Light aviation supplies company Aeropro has been manufacturing aircraft since 1992. In 1995 it took over production of the Eurofox ultralight formerly marketed by EV-AT of Czech Republic.

UPDATED

AEROPRO EUROFOX

TYPE: Side-by-side ultralight (as Fox).

PROGRAMME: Adapted in 1990 from SkyStar (Denney) Kitfox, primarily for sale in Germany by Ikarusflug; renamed Eurofox in 1995. Certification to German Bauvorschriften für Ultraleichtflugzeuge (BFU) and Slovakian ultralight requirements was achieved in March 1996. Certification of Eurofox Pro (now Eurofox Space); now also certified in Belgium, Czech Republic, France, Italy and Netherlands. The 1999 models introduced increased cabin width and option of Rotax 912 ULS with provision for glider towing.

CURRENT VERSIONS: **Eurofox Basic:** Tailwheel ultralight version; two-blade propeller. MTOW 450 kg (992 lb).

Eurofox Space: Introduced (as Eurofox Pro) 1999; JAR-VLA version with 480 kg (1,058 lb) MTOW; nosewheel. Improvements include increased cabin width, increased leg room (partly through raised instrument panel) and modified engine cowling.

Aeropro Eurofox Space seeking US buyers at AirVenture, Oshkosh *(Paul Jackson)* ***NEW***/0589261

Note that many aircraft are still registered with the original type name **Fox**.

CUSTOMERS: The 100th was built in 2001 for a Dutch owner; company reported 146 built by mid-2004, with examples flying in Belgium, Canada, Czech Republic, France, Germany, Italy, Morocco, Netherlands, Poland, Portugal, Slovakia and USA.

COSTS: Eurofox Space €19,290 (2003), including tax, basic.

DESIGN FEATURES: Extensive redesign of SkyStar Kitfox (which see). Main changes are lengthened fuselage to improve longitudinal stability, NACA 4412 modified wing section, Junkers combined flaps/ailerons, more spacious cabin dimensions and revised landing gear.

FLYING CONTROLS: Conventional rudder and elevator, manually actuated (cables and pushrods respectively); 80 per cent span, rod-actuated mass-balanced Junkers-type flaperons, maximum deflection 20°.

STRUCTURE: Folding wings have two alloy tubing spars with diagonal bracing, alloy full and half ribs, glass fibre flaps and leading-edges and Ceconite covering. Fuselage of welded chromoly steel tubing.

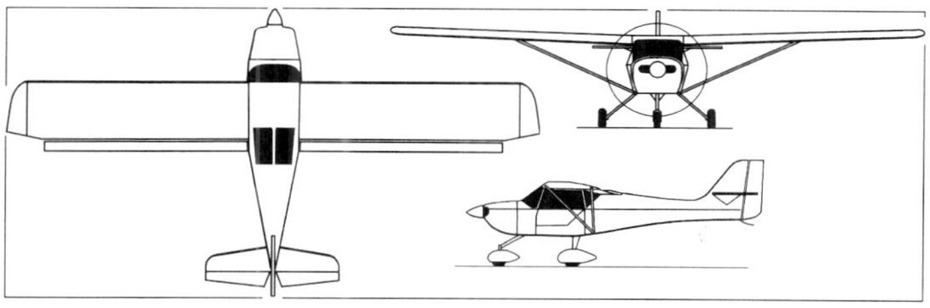

Aeropro Eurofox Space in its nosewheel version, with and without wheel fairings *(Paul Jackson)* ***NEW***/0589202

LANDING GEAR: Tailwheel type (fixed) on Eurofox Basic, with cantilever main legs, hydraulic disc brakes and 15×6.00-6 (optionally 14×4/4.00-6) mainwheel tyres; tailwheel tyre size 210×65 mm. Eurofox Space has tricycle gear with steerable 4.00-4 nosewheel with rubber-in-compression suspension.

POWER PLANT: One 59.6 kW (79.9 hp) Rotax 912 UL flat-four four-stroke driving Woodcomp SR200 three-blade fixed-pitch wooden propeller or Faturik Fiti Eco Competition three-blade composites propeller. Alternatively, one 73.5 kW (98.6 hp) Rotax 912 ULS. Fuel capacity 58 litres (15.3 US gallons; 12.8 Imp gallons), of which 56 litres (14.8 US gallons; 12.3 Imp gallons) are usable.

SYSTEMS: Electrical system with 12 V 16 Ah battery and 250 W alternator.

AVIONICS: Optional avionics by Becker and Honeywell, including GPS, to customer's choice.

EQUIPMENT: USH 520 or BRS5UL4 parachute recovery system optional.

DIMENSIONS, EXTERNAL (A: Eurofox, B: Eurofox Space):

Wing span: A, B .... 9.125 m (29 ft 11¼ in)
Wing chord, constant portion, incl flaperons .... 1.30 m (4 ft 3¼ in)
Length:
overall: A .... 6.00 m (19 ft 8¼ in)
B .... 5.73 m (18 ft 9½ in)
wings folded: A .... 6.10 m (20 ft 0¼ in)
B .... 6.20 m (20 ft 4 in)
Height overall:
tailwheel (top of wing) .... 1.72 m (5 ft 7¾ in)
nosewheel (over tail) .... 2.28 m (7 ft 5¾ in)
Width overall, wings folded .... 2.40 m (7 ft 10½ in)
Tailplane span .... 2.32 m (7 ft 7¼ in)
Wheel track .... 1.94 m (6 ft 4½ in)
Wheelbase: A .... 4.10 m (13 ft 5½ in)
B .... 1.23 m (4 ft 0½ in)
Propeller diameter: Woodcomp .... 1.68 m (5 ft 6¼ in)
Faturik .... 1.60 m (5 ft 3 in)

DIMENSIONS, INTERNAL:
Cabin max width: A .... 1.06 m (3 ft 5¾ in)
B .... 1.12 m (3 ft 8 in)

AREAS:
Wings, gross: A, B .... 11.40 m² (122.7 sq ft)

WEIGHTS AND LOADINGS:
Weight empty (with parachute recovery system):
A .... 279 kg (615 lb)
B .... 270 kg (595 lb)
Max T-O weight: A .... 450 kg (992 lb)
B .... 480 kg (1,058 lb)

PERFORMANCE (at 450 kg; 992 lb):
Never-exceed speed (VNE) .... 99 kt (185 km/h; 115 mph)
Cruising speed at 75% power .... 86 kt (160 km/h; 99 mph)
Stalling speed .... 35 kt (65 km/h; 41 mph)
Max rate of climb at S/L .... 300 m (984 ft)/min
T-O to 15 m (50 ft) .... 191 m (625 ft)
Range with max fuel .... 378 n miles (700 km; 435 miles)

UPDATED

AEROSPOOL

AEROSPOOL SPOL. S R.O.

Letiskova 10, SR-971 03 Prievidza
Tel: (+421 46) 518 32 00
Fax: (+421 46) 518 32 50
e-mail: aerospool@aerospool.sk
Web: http://www.aerospool.sk
GENERAL DIRECTOR: Jan Hrabovsky

GERMAN DISTRIBUTOR:
Ikarusflug Leichtflugzeuge GbR
Mennwanger Strasse 3, D-88682 Salem
Tel: (+49 7553) 17 70
Fax: (+49 7553) 605 38
Web: http://www.ikarusflug.de

Aerospool's current product is the WT-9 Dynamic ultralight, but it also repairs and manufactures sailplanes for Schempp-Hirth and produces components for lightplanes assembled in Europe and USA. Factory covered area 1,700 m² (18,300 sq ft); workforce more than 70 in 2003.

Skyshop Inc of Stuart, Florida, became US distributor for WT-9 in 2003.

UPDATED

AEROSPOOL WT-9 DYNAMIC

TYPE: Side-by-side ultralight/kitbuilt.

PROGRAMME: First flight 1999. Developed from Aerospool Impulz, designed by Tadeas Wala, winner of 1999 Slovak national prize for industrial design and subsequently marketed by Impulse Aircraft (which see). Public debut of Dynamic at Friedrichshafen, April 2001. Prototype (D-MXWT) in Speed configuration; c/n DY003 (D-MZKZ), in Club version, also exhibited.

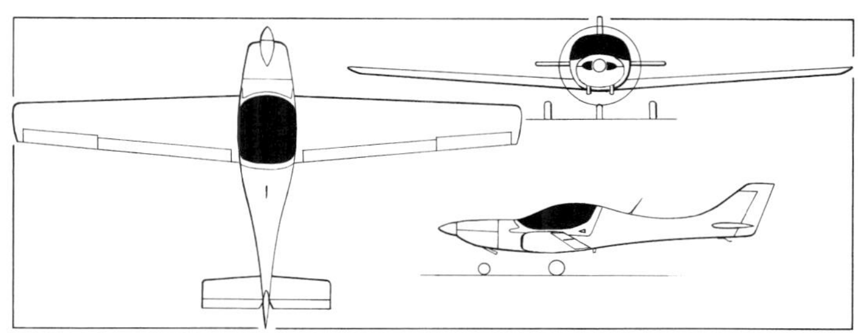

Aerospool WT-9 Dynamic two-seat ultralight *(James Goulding)* 0525761

CURRENT VERSIONS: **Club:** Fixed gear and Rotax 912 UL/A/F engine. Also available from 2002 as **Club S** with more powerful Rotax 912 ULS/S.

Tow: More powerful Rotax and aero-tow equipped.

Speed: Same engine as Tow and Club S, but retractable gear.

CUSTOMERS: Total of 50 ordered by January 2004.

COSTS: €74,900 (Club), €79,500 (Tow) and €87,600 (Speed), all including VAT (2004); avionics extra.

DESIGN FEATURES: Conventional low-wing monoplane; tapered wings; sweptback vertical tail surfaces.

US-registered Aerospool WT-9 Dynamic *(Paul Jackson)* ***NEW***/0572481

Wing aerofoil section MS(1)-0313.

FLYING CONTROLS: Conventional and manual. No tabs; horn-balanced rudder. Slotted flaps (settings 0/15/40°).

STRUCTURE: All-composites (aramid, glass and carbon fibre sandwich).

LANDING GEAR: Tricycle type, fixed in Club and Tow versions, electrohydraulically retractable (but no wheel doors) in Speed version. All three wheels size 14×4 on Club and Tow, which have self-sprung cantilever main legs. On Speed, levered suspension main legs retract inward, with 15×6.00–6 wheels and rearward-retracting 13×5.00–6 nosewheel. Hydraulic mainwheel brakes (all versions).

POWER PLANT: *Club:* One 59.6 kW (79.9 hp) Rotax 912 UL/A/F and Kremen SR 2000 three-blade, variable-pitch wooden propeller.

Club S, Tow and Speed: One 73.5 kW (98.6 hp) Rotax 912 ULS/S; propeller Kremen SR 2000 or Rospeller 2-40 three-blade carbon fibre.

Fuel capacity (all versions) 75 litres (19.8 US gallons; 16.5 Imp gallons); inboard integral tank in each wing.

ACCOMMODATION: Dual controls. Front-hinged, upward-opening one-piece canopy. Space for 10 kg (22 lb) baggage behind seats.

AVIONICS: Optional, to customer's choice.

EQUIPMENT: Ballistic recovery parachute optional.

DIMENSIONS, EXTERNAL:

Wing span ... 9.00 m (29 ft 6¼ in)
Length overall ... 6.37 m (20 ft 10¾ in)
Height overall ... 2.00 m (6 ft 6¾ in)
Propeller diameter ... 1.70 m (5 ft 7 in)

DIMENSIONS, INTERNAL:

Cabin: max width ... 1.15 m (3 ft 9¼ in)
baggage volume ... 0.09 m³ (3.2 cu ft)

AREAS:

Wings, gross ... 10.35 m² (111.4 sq ft)

WEIGHTS AND LOADINGS:

Weight empty: Club ... 264 kg (582 lb)
Tow ... 267 kg (589 lb)
Speed ... 274 kg (604 lb)
Max T-O weight: all ... 450 kg (992 lb)[1]

[1] *472.5 kg (1,041 lb) permitted in Germany*

PERFORMANCE:

Never-exceed speed (VNE): all ... 145 kt (270 km/h; 167 mph)
Normal operating limit speed (VNO) ... 135 kt (250 km/h; 155 mph)
Stalling speed: flaps up ... 42 kt (77 km/h; 48 mph)
flaps down ... 36 kt (65 km/h; 41 mph)
Max rate of climb at S/L: Club ... 384 m (1,260 ft)/min
Tow, Speed ... 311 m (1,020 ft)/min
T-O run: Club ... 76 m (250 ft)
Tow, Speed ... 71 m (235 ft)
T-O to 15 m (50 ft): Club ... 254 m (835 ft)
Tow, Speed ... 237 m (780 ft)
Landing from 15 m (50 ft): all ... 255 m (840 ft)
Landing run: all ... 149 m (490 ft)
Max range: all ... 647 n miles (1,200 km; 745 miles)
g limits: all ... +5/–3

UPDATED

EKOFLUG

EKOFLUG SPOL S. R. O.

Košice Airport, SR-070 20
Tel/Fax: (+421 95) 42 31 53
DIRECTOR: Ivo Vacek

Ekoflug builds the CH 701 ultralight which is distributed in Italy by ICP.

VERIFIED

EKOFLUG MXP-740

TYPE: Side-by-side ultralight.

PROGRAMME: Modified version of Zenith Zenair STOL CH 701, with increased wing span and Rotax 912 engine as standard.

CURRENT VERSIONS: **MXP-740:** Baseline version.

Savannah: Italian and UK marketing name; Experimental category. *As described.*

Savannah Turbo: With 89.4 kW (120 hp) engine modified by ICP.

Savannah Hydro: Floatplane; 500 kg (1,102 lb) MTOW.

Bingo: Italian marketing name; Ultralight category; formerly Savannah Light.

CUSTOMERS: Exported to Belgium, France, Germany, Israel, Netherlands, Norway, Tanzania, UK and elsewhere.

COSTS: UK kit prices ex-Italy: airframe £12,250; airframe and Jabiru 2200 engine £19,750; both including VAT (2004).

DESIGN FEATURES: As CH 701; wing section NACA 65018 (modified).

FLYING CONTROLS: Flaperon deflections: 20° take-off; 40° landing.

POWER PLANT: One 59.6 kW (79.9 hp) Rotax 912 UL flat-four engine; DUC two- or three-blade propeller. Alternatively, 59.7 kW (80 hp) Jabiru 2200. Fuel capacity 80 litres (21.1 US gallons; 17.6 Imp gallons).

Data follow for Jabiru engine.

DIMENSIONS, EXTERNAL:

Wing span ... 9.00 m (29 ft 6¼ in)

Ekoflug-built, ICP-sourced MXP-740 Savannah marketed in UK by Sandtoft Ultralights *(Paul Jackson)* ***NEW***/0589283

Length overall ... 6.10 m (20 ft 0¼ in)
Height overall ... 2.90 m (9 ft 6¼ in)

AREAS:

Wings, gross ... 12.96 m² (139.5 sq ft)

WEIGHTS AND LOADINGS:

Weight empty ... 260 kg (573 lb)
Max T-O weight ... 450 kg (992 lb)

PERFORMANCE:

Max level speed ... 78 kt (145 km/h; 90 mph)
Max cruising speed ... 65 kt (120 km/h; 75 mph)
Stalling speed: flaps down ... 28 kt (52 km/h; 32 mph)
T-O run ... 40 m (135 ft)
g limits ... +6/–4

UPDATED

Slovenia

AMS

AMS-FLIGHT D.O.O.

Kavčičeva 4, SI-1000 Ljubljana
Tel: (+386 4) 535 14 00
Fax: (+386 4) 535 14 05
e-mail: ams@ams-flight.si
Web: http://www.ams-flight.si
SALES MANAGER: Matjaz Slana
PRODUCTION MANAGER: Ales Cebavs

As well as producing parts for Glaser-Dirks sailplanes and distributing that company's DG-303 and DG-505 AMS-Flight has taken over production and marketing for the former Technoflug TFK-2 Carat. By 1 January 2004, AMS had produced 1,017 sailplanes and motorgliders, including six Carats. At this time the workforce totalled 50. Some 94 per cent of production is exported.

UPDATED

AMS CARAT A

TYPE: Motor glider.

PROGRAMME: First flown 16 December 1997; flight testing carried out during 1998. Initial production aircraft displayed at Aero '99, Friedrichshafen, in April 1999 under name of Technoflug TFK-2 Carat. Folding propeller certified 7 November 2000; certified by LBA 15 July 2001. Promotion and marketing transferred to AMS, which displayed fourth aircraft at Aero '03, Friedrichshafen, April 2003.

Production AMS Carat A *(Paul Jackson)* ***NEW***/0561688

CUSTOMERS: Six built by 1 January 2004, including exports to Germany and USA. Examples up to 21st had been registered by mid-2004.

COSTS: US$69,635 (2001) without avionics.

DESIGN FEATURES: Low wing of 21.3 aspect ratio; optional winglets; T tail. Uses wings and horizontal tail (but not water tank) of Schempp-Hirth Discus sailplane, with main spar passing under pilot's knees; wings and tailplane can be detached for storage and aircraft will fit into sailplane trailer. Optional winglets.

FLYING CONTROLS: Conventional and manual. Schempp-Hirth airbrakes on upper wing surfaces and 'turbulators' below.

STRUCTURE: Generally of glass fibre and carbon fibre composites. Wings have PVC foam cores.

LANDING GEAR: Tialwheel type; retractable. Mainwheels, size 4.00–4 in, retract inwards and forwards electrohydraulically with manual override; fixed steerable 210×65 mm tailwheel. Hydraulic disc brakes. Parking brake.

POWER PLANT: One 1,800 cc (109.8 cu in) Sauer E1S conversion of Volkswagen four-stroke, air-cooled motorcar engine, producing 40 kW (53.6 hp) at max continuous power, driving a 1.40 m (4 ft 7 in) Technoflug KS-F3-1A/140-1 fixed-pitch propeller, blades of which are held forward by gas damping springs located under the spinner when not turning. Other planned engines include 59.6 kW (79.9 hp) Rotax 912. Fuel capacity 45 litres (12.0 US gallons; 10.0 Imp gallons) in single fuel tank situated between cockpit and engine.

SYSTEMS: Electric starter and generator; 12 V 16 Ah battery. Electric fuel pump.

AVIONICS: Suggested package includes ATR720A radio, Honeywell transponder and Filser DX50FAI GPS.

DIMENSIONS, EXTERNAL:
- Wing span .... 15.00 m (49 ft 2½ in)
- Length overall .... 6.21 m (20 ft 4½ in)

DIMENSIONS, INTERNAL:
- Baggage compartment volume .... 0.1 m³ (3.53 cu ft)
- Baggage capacity .... 145 kg (320 lb)

AREAS:
- Wings, gross .... 10.58 m² (113.9 sq ft)

WEIGHTS AND LOADINGS:
- Weight empty .... 325 kg (717 lb)
- Max T-O weight .... 470 kg (1,036 lb)

PERFORMANCE, POWERED:
- Never-exceed speed (VNE) .... 135 kt (250 km/h; 155 mph)
- Normal cruising speed at 75% power .... 116 kt (214 km/h; 133 mph)
- Stalling speed .... 44 kt (80 km/h; 50 mph)
- Max rate of climb at S/L .... 210 m (689 ft)/min
- T-O run .... 226 m (741 ft)
- Landing run .... 137 m (450 ft)
- Range with max fuel, 30 min reserves .... 486 n miles (900 km; 559 miles)
- *g* limits .... +5.3/–2.65

PERFORMANCE, UNPOWERED:
- Stalling speed .... 44 kt (80 km/h; 50 mph)
- Best glide ratio 35
- Min rate of sink .... 0.75 m (2.46 ft)/s

UPDATED

PIPISTREL

PIPISTREL D. O. O.

ulica Štrancarjeva 11, SL-65270 Ajdovsciná
Tel: (+386 5) 366 38 73
Fax: (+386 5) 366 12 63
e-mail: pipistrel@siol.net
Web: http://www.pipistrel.si
GENERAL MANAGER: Ivo Boscarol

Pipistrel originated in mid-1980s when partners Ivo Boscarol and Bojan Sajovic began building flex-wing trikes. First major design was Ajdovsciná, of which 32 built between 1989 and December 1992, including exports to Italy (these first private Yugoslav exports of light aircraft). Also began producing ground-adjustable pitch propellers, of which 1,400 built by 2000. Pipistrel Plus flex-wing entered production in 1990; more than 200 now sold. Supplied flex-wings to Croatian Army in 1991. Spider flex-wing entered production in 1992; more than 150 produced.

Company formally established as Pipistrel d.o.o. on 18 November 1992 and supplied flex-wings to Italian Army's special forces in following year. Twister flex-wing introduced 1998. Moved to new premises at Ajdovsciná aerodrome in 2001; workforce 18. Production rate four aircraft per month in mid-2003. Construction of new assembly facility at Ajdovscina airport (more than 2,000 m²; 21,528 sq ft) started 5 September 2003; when finished, will permit increased output of eight to 10 per month.

In 2000, Pipistrel began design of **Arcus** ultralight, no details of which have yet been published. The company is also developing the Taurus self-launching ultralight sailplane which has a 'pop-up' Rotax 503 engine and made its maiden flight on 15 May 2004.

UPDATED

PIPISTREL SINUS

TYPE: Side-by-side ultralight motor glider/kitbuilt.

PROGRAMME: Design begun 1994; project leader Franco Orlando. Prototype (S5-NBP) built and certified in Slovenia; public debut at Aero '95, Friedrichshafen, April 1995, but not flown until 1996. Series production from 1999. Marketed in France by Zen ULM and elsewhere. Kit versions introduced 1 April 2004 (51 per cent kit to meet FAA requirements and quick-build QBK).

CURRENT VERSIONS: **Sinus 503:** 37.0 kW (49.6 hp) Rotax 503 UL-2V two-stroke and Pipistrel variable-pitch, folding propeller. Certified in Australia, Belgium, Colombia, Croatia, France, Hungary, Italy, Japan, Poland and Slovenia.

Sinus 912: 59.6 kW (79.9 hp) Rotax 912 UL-2 flat-four with Pipistrel variable-pitch, folding propeller. *As described.* Certified in same countries as 503.

CUSTOMERS: Sold in 1[illegible] European countries, including 54 to French customers by 2002. Others to Australia, Brazil, Colombia, Congo, Eritrea, Japan, Kenya, Mexico, New Caledonia, Reunion Islands, Santo Domingo, South Africa, Thailand, USA and Venezuela. Total sales (including Virus) at least 143 by August 2004.

COSTS: Sinus 503 €46,900; Sinus 912 €57,900 ready to fly. Kit versions: 912 51 per cent €49,500, QBK €53,400; 503 51 per cent €40,000, QBK €44,000. (All 2004).

DESIGN FEATURES: Meets JAR 22 specifications. Pod-and-boom fuselage with high cantilever wing (except on prototype, which was strut-braced), upturned wingtips, and strut-braced T tail. In-flight-adjustable pitch propeller for improved engine-off performance; can be feathered for gliding.

Aerofoil section Orlando-Venuti IMD 029-b.

Wing is employed by Albastar Apis single-seat sailplane.

Pipistrel Sinus flown in USA *(Paul Jackson)* ***NEW***/0589416

FLYING CONTROLS: Conventional elevator and rudder, manually operated. Flight-adjustable trim tab in starboard elevator. Flaperons on inboard 90 per cent of wing trailing-edge, deflections –5, +9, +18°. Schempp-Hirth-type spoilers in upper wing, one per side.

STRUCTURE: Principally of glass fibre, carbon fibre and Kevlar. Quoted build time 200 hours for QBK kits.

LANDING GEAR: Tailwheel type; fixed. Fuselage-mounted spring cantilever main legs with speed fairings. Disc brakes on mainwheels.

POWER PLANT: One Rotax piston engine, as under Current Versions. Pipistrel Vario in-flight-adjustable pitch two-blade composites propeller. Fuel in two tanks, combined capacity 60 litres (15.9 US gallons; 13.2 Imp gallons).

ACCOMMODATION: Two persons, side by side in enclosed cockpit, each with upward-opening door.

EQUIPMENT: BRS ballistic parachute systems optional.

Data for Sinus 912.

DIMENSIONS, EXTERNAL:
- Wing span .... 14.97 m (49 ft 1¼ in)
- Length overall .... 6.60 m (21 ft 7¾ in)
- Height overall .... 1.70 m (5 ft 7 in)

DIMENSIONS, INTERNAL:
- Cabin max width .... 1.10 m (3 ft 7¼ in)

AREAS:
- Wings, gross .... 12.26 m² (132.0 sq ft)

WEIGHTS AND LOADINGS:
- Weight empty .... 284 kg (626 lb)
- Max T-O weight .... 450 kg (992 lb)[1]

[1] *472.5 kg (1,041 lb) in France if fitted with ballistic parachute; 544 kg (1,200 lb) in Australia (AUF rules)*

PERFORMANCE, POWERED:
- Never-exceed speed (VNE) .... 121 kt (225 km/h; 139 mph)
- Max level speed .... 119 kt (220 km/h; 137 mph)
- Cruising speed .... 108 kt (200 km/h; 124 mph)
- Stalling speed: flaps up .... 36 kt (66 km/h; 41 mph)
- flaps down .... 34 kt (63 km/h; 40 mph)
- Max rate of climb at S/L .... 390 m (1,280 ft)/min
- Service ceiling .... 8,800 m (28,875 ft)
- T-O run .... 88 m (290 ft)
- Range .... 647 n miles (1,200 km; 745 miles)
- *g* limits .... +4/–2

PERFORMANCE, UNPOWERED:
- Best glide ratio at 51 kt (95 km/h; 59 mph), folded propeller 27
- Min rate of sink at 49 kt (90 km/h; 56 mph), folded propeller .... 1.03 m (3.4 ft)/s

UPDATED

Pipistrel Virus two-seat ultralight *(Paul Jackson)* ***NEW***/0589415

PIPISTREL VIRUS

TYPE: Side-by-side ultralight.

PROGRAMME: Variant of Pipistrel Sinus; development began 1997; first flown (S5-PBG) 1999; series production from 2000. Certified in France, Italy and Slovenia. Company demonstration team created in 2003 with three Virus 912s (S5-PCA, 'PBR and 'PCP).

CURRENT VERSIONS: **Virus 912:** Certified version; *as described.*

COSTS: €57,900 (2004).

DESIGN FEATURES: Tricycle landing gear and shorter span. Otherwise generally as Sinus.

FLYING CONTROLS: As Sinus.

LANDING GEAR: Tricycle type; fixed. Fuselage-mounted spring cantilever mainwheel legs with wheel fairings; steerable nosewheel.

POWER PLANT: Available only with Rotax 912 UL flat-four (59.6 kW; 79.9 hp) and Pipistrel Vario in-flight-adjustable, featherable two-blade propeller. Fuel as Sinus.

DIMENSIONS, EXTERNAL:
Wing span ... 12.36 m (40 ft 6½ in)
Length overall ... 6.40 m (21 ft 0 in)
Height overall ... 2.00 m (6 ft 6¾ in)

AREAS:
Wings, gross ... 11.00 m² (118.4 sq ft)

WEIGHTS AND LOADINGS:
Weight empty ... 285 kg (628 lb)
Max T-O weight ... 450 kg (992 lb)[1]

[1] *472.5 kg (1,041 lb) in France if fitted with ballistic parachute*

PERFORMANCE, POWERED:
Never-exceed speed (VNE) ... 135 kt (250 km/h; 155 mph)
Max level speed, fixed prop ... 130 kt (240 km/h; 149 mph)
Cruising speed, fixed prop ... 121 kt (225 km/h; 140 mph)
Stalling speed: flaps up ... 40 kt (74 km/h; 46 mph)
flaps down ... 35 kt (64 km/h; 40 mph)
Max rate of climb at S/L ... 432 m (1,417 ft)/min
Service ceiling ... 8,100 m (26,575 ft)
T-O run ... 96 m (315 ft)
Range ... 680 n miles (1,260 km; 782 miles)
g limits ... +4/–2

PERFORMANCE, UNPOWERED:
Best glide ratio at 59 kt (110 km/h; 68 mph) ... 24
Min rate of sink at 52 kt (96 km/h; 60 mph) ... 1.72 m (5.6 ft)/s

UPDATED

South Africa

DART

PO Box 223, Botha's Hill, Kwazulu Natal 3660
Tel: (+27 82) 564 77 65
Fax: (+27 31) 765 51 88
e-mail: info@dart-industries.com
Web: http://www.dart-industries.com
PRODUCTION DIRECTOR: Peter Dale
SALES DIRECTOR: Dave Dormer

Dart Industries purchased the Dragonfly programme from Slipstream International of the USA in March 2004.

NEW ENTRY

DART DRAGONFLY

TYPE: Side-by-side ultralight kitbuilt.

PROGRAMME: Prototype first flight 16 June 1980, designed by Bob Walters. Initially marketed by Viking as kit or as plans. Programme sold to Slipstream International, then transferred to Dart in March 2004.

CURRENT VERSIONS: **Dragonfly Mark I:** Original configuration, with non-retractable mainwheels at tips of foreplane. EAA Oshkosh Outstanding Design, 1980.

Dragonfly Mark II: In parallel production, for operation from unprepared strips and narrow taxiways. Main landing gear in short non-retractable cantilever units under wings, with individual hydraulic toe brakes, and increased foreplane and elevator areas; wheel track 2.44 m (8 ft).

Dragonfly Mark III: Flight tested in 1985; non-retractable, narrow-track tricycle landing gear with steerable nosewheel and cantilever mainwheel legs attached to fuselage. **Mark III Millenium** displayed at EAA AirVenture, Oshkosh, July 2001 (N204AS, then unflown) as

Viking/Slipstream/Dart Dragonfly Mark II built in Australia *(Paul Jackson)* ***NEW***/1047354

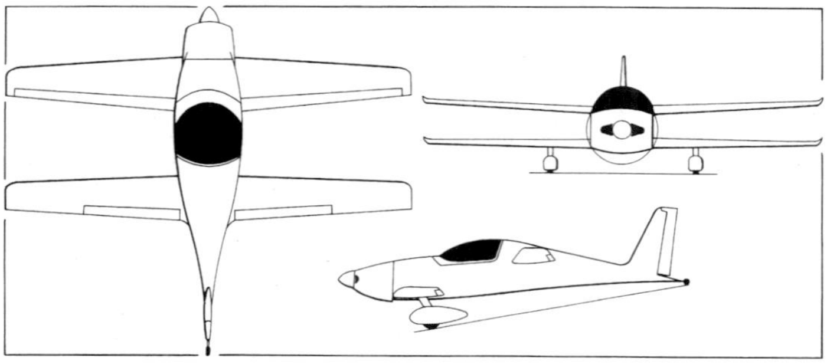

Dart Dragonfly Mark II *(Paul Jackson)* 0016522

Tricycle gear SlipStream Dragonfly Mk III Millenium *(Paul Jackson)* 0526988

conversion of earlier aircraft. Recommended engines: 59.7 kW (80 hp) Jabiru 2200 flat-four or 89 kW (120 hp) Jabiru 3300 flat-six.

CUSTOMERS: More than 100 kits delivered, together with 2,000 sets of plans; some 70 kitbuilt and more than 430 plans-built aircraft flying.

COSTS: Standard kit, including Jabiru 2200 engine, US$38,500 (2003).

DESIGN FEATURES: Unconventional shoulder-wing monoplane with equal-span foreplane; latter also mounting landing gear in one of two alternative locations.

Quoted build time 1,000 hours, or 500 hours for quick-build kit.

Wing section Eppler 1213. Foreplane GU25 section. Thickness/chord ratio 15 per cent; dihedral 3°; no incidence or sweepback. Foreplane t/c 17 per cent; anhedral 3° on Mk I or dihedral on Mk II; incidence –1°; no sweepback.

FLYING CONTROLS: Manual. Ailerons on inboard trailing-edge of wing; near full-span elevators on foreplane; horn-balanced rudder. Flettner tab on each foreplane.

STRUCTURE: Composites wing; foreplane and tail unit structures of styrene foam, glass fibre, carbon fibre and epoxy. Semi-monocoque fuselage, formed (not carved) from 12.5 mm (½ in) thick urethane foam, with strips of 18 mm (¾ in) foam bonded along edges to allow large-radius external corners. Fuselage covered with glass fibre inside and out.

LANDING GEAR: See Current Versions; employs choice of sprung steel or glass fibre landing gear. Brakes fitted.

POWER PLANT: Mks I and II have one 44.5 kW (60 hp) 1,835 cc modified Volkswagen motorcar engine; 1,600 cc engine, rated at 33.5 kW (45 hp), optional. Fuel capacity 56.8 litres (15.0 US gallons; 12.5 Imp gallons).

DIMENSIONS, EXTERNAL:
Wing span ... 6.71 m (22 ft 0 in)
Foreplane span ... 6.71 m (22 ft 0 in)
Length overall ... 5.79 m (19 ft 0 in)
Height overall: Mk I ... 1.22 m (4 ft 0 in)
Mk III ... 1.68 m (5 ft 6 in)

DIMENSIONS, INTERNAL:
Cabin: Max width ... 1.09 m (3 ft 7 in)
Height ... 0.79 (2 ft 7 in)

AREAS:
Wings, gross ... 4.24 m² (45.6 sq ft)
Foreplane, net ... 4.33 m² (46.6 sq ft)

WEIGHTS AND LOADINGS (Millenium, Jabiru 2200):
Weight empty ... 272 kg (600 lb)
Max T-O weight ... 567 kg (1,250 lb)

PERFORMANCE (Millenium, Jabiru 2200):
Never-exceed speed (VNE) ... 156 kt (290 km/h; 180 mph)
Cruising speed ... 113 kt (209 km/h; 130 mph)
Stalling speed ... 48 kt (89 km/h; 55 mph)
Max rate of climb at S/L ... 259 m (850 ft)/min
T-O to 50 ft (15 m) ... 366 m (1,200 ft)
Landing from 50 ft (15 m) ... 610 m (2,000 ft)
Range ... 434 n miles (804 km; 500 miles)
g limits ... +4.4/–2.0

UPDATED

DENEL

DENEL AVIATION (DIVISION OF DENEL (PTY) LTD)

PO Box 11, Kempton Park 1620, Gauteng
Tel: (+27 11) 927 34 16
Fax: (+27 11) 395 15 24
e-mail: info@aviation.denel.co.za
Web: http://www.denel.co.za
GROUP CHAIRMAN AND COO: Sandile D M Zungu
CEO: Victor G M Moche
EXECUTIVE DIRECTOR, AEROSPACE GROUP: Knox Msebenzi
DIVISIONAL GENERAL MANAGER: Trevor van Zyl
PROMOTIONS OFFICER, CORPORATE MARKETING: Claire Nelson-Esch

Predecessor Atlas Aircraft Corporation founded 1964 by Bonuskor as private company; delivered first Impala Mk 1 (Aermacchi MB-326) jet trainers to SAAF 1966; manufacturing, design and development facilities for airframes, engines, missiles and avionics; developed Cheetah from Mirage III, Rooivalk from Puma, Oryx (Super Puma) from Puma, V3B and V3C dogfight missiles, and many weapons installations. Incorporated into Armscor Group 1969; restructuring of Armscor on 1 April 1992 created Denel as self-sufficient commercial industrial group, in which Atlas Aviation became military aircraft manufacturing branch of Simera in Denel Aerospace Group. Renamed Denel Aviation in April 1996. Further reorganisation of 2004 placed Aerospace Group as one of three major Denel businesses (with Land Systems Group and Commercial & IT). Aerospace Group comprises:

Airframe Manufacturing: At Kempton Park (part of Denel Aviation); Gripen landing gear, rear fuselage and pylons Hawk tailplane, airbrake and flaps; Rooivalk complete aircraft; Boeing 737, 747 and 777 minor parts.

Denel Aerospace Systems: At Irene, Centurion. Ground-based air defence and UAVs.

Aerospace Engineering: At Kempton Park and Irene. Aerospace structural and avionics engineering, manufacture and testing.

Aircraft Logistics: At Kempton Park. Support, repair and refurbishment, including C-130 Hercules, Rooivalk and Cheetah.

OTB Test Range: At Bredasdorp. Weapons performance; satellite launch support.

Denel Optronics: Sighting and surveillance systems.

MoU in late 1998 provided for co-production of, and export licence for, Agusta A 119 Koala helicopter; July 2000 agreement with Agusta, covering airframe manufacture and systems installation for 25 of 30 A109s ordered by South African Air Force, was extended in 2001 to include airframes of 20 A 109Ms for Sweden. This followed in December 2001 by agreement with AgustaWestland to produce and market A 109LUH, A 109 Power and A 119 Koala for customers in specified countries in Southeast Asia, the Middle East, South America, Africa and elsewhere. Final assembly of first Italian-built A109LUH for South African Air Force (I-PLUH/4001) began in early 2003; first assembled locally by Denel (4006) made maiden flight 9 September 2004.. Beginning May 2001, Denel manufacturing, under a US$3.6 million contract, main landing gear portion of lower centre-fuselage for Swedish Air Force Lot 3 Gripen combat aircraft, plus additional sections for the 28 Gripens ordered by the South African Air Force. Initial SwAF order is for 40 units, with options for a further 60.

The South African government search for a strategic equity partner for a partial privatisation of Denel Aviation ended on 12 October 2000 with the announcement that BAE Systems had been selected. South African government approval for BAE Systems to acquire a 30 per cent holding in Denel Aviation for R375 million (US$36.8 million) was announced in May 2002, but not taken up. Alternative equity partner sought in 2003; in November that year, Denel announced intention to separate its aerostructures and component manufacturing operations as part of a move to restore profitability.

First Denel-built A 109LUH on its maiden flight, 9 September 2004 ***NEW***/1047693

UPDATEDUPDATED

DENEL AH-2A ROOIVALK

English name: Red Kestrel

TYPE: Attack helicopter.

PROGRAMME: Considerably modified development of French SA 330 Puma. XDM and ADM prototypes reconfigured during 1995 with original Topaz engines replaced by more powerful Makila 1K2s with digital engine control, making first flight in this form June 1995. Content for EDM prototype awarded in March 1994; more than 1,500 hours flown by XDM, ADM and EDM by December 1998. Airframe changes included modified IR exhaust suppression and intakes on XDM; fuselage construction unchanged on XDM and ADM.

Submitted (partnered by Marshall Aerospace) for British Army requirement in 1993–94, but subsequently eliminated. SAAF order for 12 signed in July 1996. First production Rooivalk (c/n 1001, SAAF serial number 670) left airframe jig 31 July 1997 and made first public flight 17 November 1998; handed over to SAAF on 17 November 1998 and joined No. 16 Squadron on 6 January 1999, formal acceptance being on 7 May 1999. Six delivered by June 2001 and all 12 by March 2004; trials at Test Flight Development Centre, Bredasdorp.

A commissioning order signed with the SAAF on 27 August 2003 provides for new functions in the communications system; addition of a missile approach warning system (MAWS); addition of the Mistral ATM; updates to HUMS and mission planning; and establishment of depot level repair capability. One development aircraft used for deck

First production AH-2A Rooivalk armed with eight Mokopa ATMs and four 68 mm rocket pods *(Jane's/Kenneth Munson)* ***NEW***/0589164

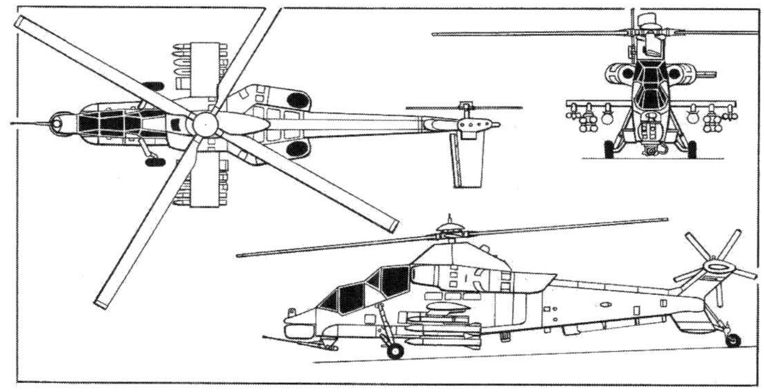

Denel Rooivalk combat support helicopter ***(James Goulding)*** 0507637

handling trials aboard support ship SAS *Drakensberg* in late 2002 and early 2003.

MoU with Airod of Malaysia provided for co-production in that country if ordered by Royal Malaysian Air Force.

CURRENT VERSIONS: **XDM** (Experimental Development Model): First prototype (originally with Topaz engines and designated XH-2); later refitted with 1,356 kW (1,819 shp) Makila 1A1 engines. Design began late 1984 to meet South African Air Force user requirement specification; public roll-out 15 January and first flight 11 February 1990; programme temporarily halted later in 1990, but restarted as mainly company-funded venture. Served as mechanical systems aircraft.

ADM (Advanced Demonstration Model): Second prototype, with fully integrated avionics and full weapons system; power plant as for XDM; first flight second quarter 1992 and later registered ZU-AHC. Weapons development aircraft, including testing of Denel Mokopa (Black Mamba) long-range (more than 10 km; 6.2 miles) laser-guided anti-tank missile. First round of weapons trials (F2 cannon, HOT 3 missiles and 68 mm rockets) completed 1997. First airborne firing of Mokopa 21 September 1999; full integration of this weapon completed late 2001.

EDM (Engineering Development Model): Third (pre-production) aircraft (672), with 1,376 kW (1,845 shp) Makila 1K2 engines from outset. First flight 17 November 1996. IR signature and weight of intakes and exhaust reduced; MTOW reduced; up to 16 Mokopa anti-tank missiles on stub-wings. Digital avionics include AFCS, Sextant ring laser gyro INS and GPS receiver, single monochrome control and display unit (CDU), two Thales (Sextant) colour MFDs, health and usage monitoring system (HUMS), Thales (Sextant) Topowl helmet-mounted sight displays (HMSD) and Kentron tracking system. SAGEM Nightowl nose-mounted targeting sight. First round of avionics qualification completed 1997; second phase started November 1997. Sea level performance trials completed late 2001.

AH-2A: SAAF production version; previously known as CSH-2 but redesignated in 1998. First four aircraft delivered to Block 1A standard; Nos. 5–10 Block 1B; Nos. 11 and 12 Block 1E. Block 1A is basic aircraft with functional mechanical and avionics systems; Block 1B added Topowl and Nightowl optical systems and night flying capability; Block 1E is definitive configuration, with fully operable helmet sight and weapon systems. Block 1A and 1B aircraft being updated to Block 1E standard; four so updated and two others in process by March 2003. First two Block 1E aircraft redelivered to No. 16 Squadron by September 2003 and fully operational with nine Block 1E aircraft by March 2004, remaining three then in retrofit for delivery later. Two Block 1E aircraft being used for further development purposes. Squadron carrying out OT&E on weapons system during 2004.

Description applies to EDM/production AH-2A except where indicated.

Maritime version: Projected version, shown as model at Farnborough Air Show, September 1998, though no commitment yet to build. Would have chin-mounted, 360° maritime search radar in place of undernose cannon; vision-enhancing E-O suite in nose; tailwheel moved forward 2 m (6.6 ft) to facilitate deck operations; flotation gear on forward sponsons and tailboom; manually operated blade folding; and enhanced ECM and communications systems. Shorter stub-wings would support four underwing Penguin or Exocet anti-ship missiles, plus tip-mounted AAMs for self-defence.

CUSTOMERS: South African Air Force, 12 ordered; possible longer-term requirement for up to 36. No. 16 Squadron reactivated 1 January 1999 at Bloemspruit AB as only squadron. Initial delivery 6 January 1999; deliveries originally scheduled at four per year until third quarter 2001 but protracted; total of 10 delivered by October 2002, with remaining two delivered by March 2004.

Intent to order eight reconfirmed by Malaysian Prime Minister in 1998, but not yet contracted (late-2004); possible requirement for up to 24. Algeria, Australia, Singapore and Turkey have also expressed interest (named **RedHawk** for unsuccessful Australian Army bid), and ADM demonstrated to Saudi Arabia in August 1997, but no export orders up to late 2004.

COSTS: According to September 2002 report, R&D costs were R126 million, one-off restructuring costs R124 million and production costs R179 million; further R305 million written off. Additional funding for R661 million (US$81.2 million) early 2003 to complete full commissioning to Block 1E standard for service acceptance.

DESIGN FEATURES: Based on reverse engineering of SA 330 Puma dynamics system to Super Puma equivalent standard; engines moved aft to clear stepped tandem cockpits; rear drive taken inboard and forward to modified transmission. Nose-mounted target acquisition turret and chin-mounted gun; non-swept stub-wings for weapon carriage. With Makila 1K2 engines, four-blade fully articulating main rotor rotates at 267 to 290 rpm; five-blade, starboard-mounted tail rotor rotates at 1,290 rpm. Rotor brake fitted. Fixed leading-edge slat on horizontal stabiliser.

Low radar/visual/IR/acoustic signatures; optimised for NOE operation; NVG-compatible 'glass cockpit'; primary missions anti-armour and close air support; integrated digital nav/attack system.

FLYING CONTROLS: Duplex digital AFCS with automatic height hold, hover capture and hover hold; HOCAS (hands on collective and stick) controls.

STRUCTURE: Crash-resistant primary structure, of aluminium alloy, is of I-beam and monocoque construction; rotor blades and fuselage secondary structures of composites sandwich. Integral access ladders in fuselage; CFRP engine cowls form work platforms.

LANDING GEAR: Forward-mounted Messier-Dowty main gear with two-stage high-absorption main legs; fully castoring tailwheel at base of lower fin. Mainwheel tyres size 615×225–10. Gear designed to withstand landing impact of up to 6 m (20 ft)/s.

POWER PLANT: Two licence-built Turbomeca Makila 1K2 turboshafts, with digital control, each rated at 1,376 kW (1,845 shp) for T-O, 1,420 kW (1,904 shp) continuous power and 1,573 kW (2,109 shp) for 30 seconds emergency operation. Rear drive turned to drive forward into rear of transmission. Infra-red heat suppressors on exhausts; particle separators on intakes. Main gearbox mounted on vibration isolation system using tuned beam to isolate fuselage from rotor vibrations (said to have one of lowest vibration levels in its class).

Transmission rated at 2,243 kW (3,008 shp) for T-O and 1,826 kW (2,449 shp) maximum continuous from both engines; single-engine transmission ratings are 1,660 kW (2,226 shp) maximum contingency and 1,491 kW (2,000 shp) maximum continuous.

Fuel tankage (self-sealing) in fuselage, under stub-wings (three 618 litre; 163.3 US gallon; 135.9 Imp gallon tanks for total of 1,854 litres; 489.8 US gallons; 407.8 Imp gallons). Pressure refuelling and defuelling. Provision for 750 litre (198 US gallon; 165 Imp gallon) drop tank on each inboard underwing station.

ACCOMMODATION: Pilot (rear) and co-pilot/weapons officer in stepped, tandem cockpits with Martin-Baker armoured, crashworthy seats and armour protection. Access to cockpits via upward-opening gull-wing starboard side window panels. All transparencies flat-plate or single-curvature. Dual flight controls.

SYSTEMS: Environmental control system for cockpit air conditioning. Two independent hydraulic systems, each at 175 bar (2,538 lb/sq in) pressure; at 170 bar (2,465 lb/sq in), flow rate is 12 litres (3.2 US gallons; 2.6 Imp gallons)/min in starboard system and 27 litres (7.1 US gallons; 5.9 Imp gallons)/min in port system.

Electrical power from two 20 kVA alternators providing 200 V three-phase and 115 V single-phase AC at 400 Hz, with two transformer-rectifiers and two 24 V 31 Ah batteries for 28 V DC power. Full or partial electric de-icing/demisting optional.

Crew oxygen system, and fire detection and extinguishing systems, standard.

AVIONICS: All-digital, interfaced to dual-redundant mission computers and MIL-STD-1553B databusses; stores management system conforms to MIL-STD-1760A. Avionics integrator is ATE (Advanced Technologies and Engineering), supported by 15 subcontractors.

Comms: Frequency-agile transceivers: dual V/UHF (30 to 400 MHz) for normal use and single HF (2 to 30 MHz) for NOE flights; intercom/audio system; IFF transponder.

Flight: Duplex four-axis digital AFCS, incorporating (in EDM) dual Thales (Sextant) Stratus three-axis strapdown ring laser gyro AHRS, Sextant NSS 100-1 eight-channel GPS receiver, Doppler radar velocity sensor, J-band radar altimeter, heading sensor unit (two magnetometers), air data unit and omnidirectional airspeed sensor, all interfaced to two redundant navigation computers. Autopilot has one-touch autohover and attitude hold. Nav/attack system programmed by preloaded cartridge; can hold up to five flight plans (100 waypoints), which can be edited in flight by either crew member.

Instrumentation: Integrated flight management system, with (in EDM) two 160 × 160 mm (6.3 × 6.3 in) Thales (Sextant) MFD 66 liquid crystal colour MFDs, two Sextant Topowl helmet-mounted sights/displays, and a position management system in each cockpit. Both cockpits have a back-up basic instrument panel for 'get home' capability in the event of computer or power failure. MFDs show flight control, navigation (including moving map on pilot's display), threat warning/EW, fire/weapons control and TDATS imagery, and can copy to each other. Cockpit instruments compatible with NVGs.

Mission: Nose-mounted, gyrostabilised turret contains target detection, acquisition and tracking system (TDATS) incorporating a three fields of view TV and FLIR with automatic guidance and tracking, an LLTV camera and a laser range-finder; associated equipment includes crew members' Thales (Sextant) helmet sights, missile command link and tracking goniometer. Helmet sights display both flight and weapon data, and can cue both the TDATS turret and the gun turret. Cumulus (South Africa) and Thales (Pilkington Optronics) PNVS (Pilot Night Vision Sensor), a two-axis turret with 40 × 30° field of view FLIR, offered as export alternative to Sextant helmet sights.

Self-defence: Optimised ECM can include radar and laser warning receivers, chaff/flare dispensers and RF, IR and laser jammers.

ARMAMENT: Long- or short-barrel version of 20 mm Armscor F2; cleared also for 30 mm weapon. F2 has up to 700 rounds of ammunition, firing rate of 740 rds/min, and slew rate of 90°/s. Gun linked to TDATS and helmet-mounted sight display (HMSD). Three underwing stores stations each side. Two or four M159 19-tube launchers for Forges de Zeebrugge FZ90 70 mm unguided rockets and/or four-round launchers for up to 16 Mokopa anti-tank missiles (semi-active laser version) on inner four underwing pylons; two or four Mistral IR air-to-air missiles on outboard pylons.

DIMENSIONS, EXTERNAL:
- Main rotor diameter ... 15.58 m (51 ft 1½ in)
- Tail rotor diameter ... 3.05 m (10 ft 0 in)
- Length: fuselage, excl gun, tail rotor turning ... 16.39 m (53 ft 9¼ in)
  - overall, rotors turning ... 18.73 m (61 ft 5½ in)
- Wing span (to c/l of AAM pylons) ... 6.355 m (20 ft 10¼ in)
- Fuselage:
  - Max width (excl engine fairings) ... 1.28 m (4 ft 2½ in)
  - Max depth (belly to top of rotor head) ... 4.00 m (13 ft 1½ in)
- Height: over tail rotor ... 4.445 m (14 ft 7 in)
  - to top of rotor head ... 4.59 m (15 ft 0¾ in)
  - overall ... 5.185 m (17 ft 0¼ in)
- Wheel track ... 2.78 m (9 ft 1½ in)
- Wheelbase ... 11.77 m (38 ft 7½ in)

AREAS:
- Main rotor disc ... 190.60 m^2 (2,052.1 sq ft)
- Tail rotor disc ... 7.27 m^2 (78.23 sq ft)

WEIGHTS AND LOADINGS:
- Weight empty ... 5,730 kg (12,632 lb)
- Max internal fuel weight ... 1,469 kg (3,238 lb)
- External weapons load with full internal fuel ... 1,563 kg (3,446 lb)
- Max external load (with 1,000 kg; 2,205 lb fuel) ... 2,032 kg (4,480 lb)
- Typical mission T-O weight ... 7,500 kg (16,535 lb)
- Max T-O weight ... 8,750 kg (19,290 lb)
- Max disc loading ... 45.9 kg/m^2 (9.40 lb/sq ft)
- Transmission loading at max T-O weight and power (Makila 1K2) ... 3.90 kg/kW (6.41 lb/shp)

PERFORMANCE (at 7,500 kg; 16,535 lb combat weight, except where indicated. A: ISA at S/L, B: ISA + 27°C at 1,525 m; 5,000 ft):
- Never-exceed speed (VNE):
  - A, B ... 167 kt (309 km/h; 192 mph)
- Max cruising speed: A ... 150 kt (278 km/h; 173 mph)
  - B ... 130 kt (241 km/h; 150 mph)
- Max sideways speed: A ... 50 kt (93 km/h; 58 mph)
- Max rate of climb at S/L: A ... 798 m (2,620 ft)/min
  - B ... 760 m (2,493 ft)/min
- Rate of climb at S/L, OEI: A ... 344 m (1,128 ft)/min
- Service ceiling: A ... 6,100 m (20,000 ft)
  - B ... 5,150 m (16,900 ft)
- Hovering ceiling IGE: A ... 5,850 m (19,200 ft)
  - B ... 3,110 m (10,200 ft)
- Hovering ceiling OGE: A ... 5,455 m (17,900 ft)
  - B ... 2,410 m (7,900 ft)
- Excess hover power margin OGE, S/L anti-tank mission ... +39%
- Range with max internal fuel, no reserves:
  - A ... 380 n miles (704 km; 437 miles)
  - B ... 507 n miles (940 km; 584 miles)
- Range at max T-O weight with external fuel:
  - A ... 680 n miles (1,260 km; 783 miles)
  - B ... 720 n miles (1,335 km; 829 miles)
- Endurance with max internal fuel, no reserves:
  - A ... 3 h 36 min
  - B ... 4 h 55 min
- Endurance at max T-O weight with external fuel:
  - A ... 6 h 52 min
  - B ... 7 h 22 min
- *g* limits ... +2.6/–0.5

UPDATED

LAMMER GEYER

LAMMER GEYER AVIATION

Wonderboom Airport, Pretoria
Tel: (+27 11) 468 41 85 or (+44 7792) 36 63 63
e-mail: pwareham@hotmail.com
Web: http://www.jupiteraircraft.co.za
DIRECTOR: Peter Wareham

The company's first aircraft flew on 1 December 2002 and is now available as a kit.

UPDATED

LAMMER GEYER JUPITER

TYPE: Four-seat kitbuilt.

PROGRAMME: Project begun in 1996. Prototype first flew (ZU-CNH) 1 December 2002. Marketing began June 2003, but was interrupted shortly thereafter when Mr Wareham and his aircraft returned to the UK.

COSTS: Kit estimated US$32,000, excluding engine, instruments and avionics (2003). Will be available in subkits.

DESIGN FEATURES: Intended to possess STOL performance and operate over long ranges from airfields with density altitudes of some 2,440 m (8,000 ft). Detail design, by former RAF flying instructor Peter Wareham, obviates features of other aircraft which "aviators find frustrating", including cabin width, view from rear seats (raised 15 cm; 6 in), all-round view (bulged canopy and downward-sloping engine cowling) and access to switches and controls. Limited aerobatic capability.

Jupiter prototype on an early test flight ***NEW***/0552876

Prominently curved fuselage in profile; constant-chord wing with upturned tips; sweptback fin with fillet and underfin.

Wing section NACA 64 series.

FLYING CONTROLS: Conventional and manual. Flaps.

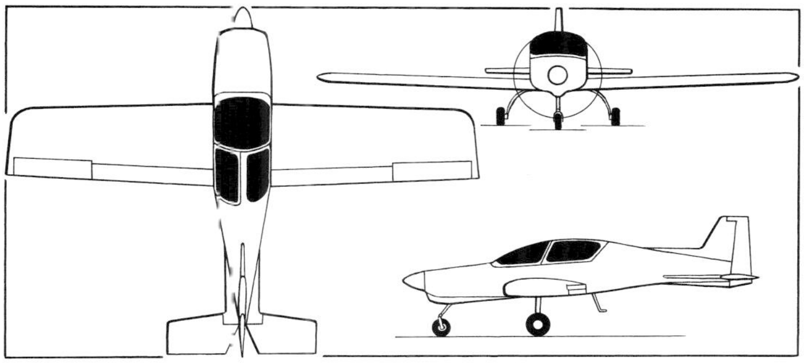

Lammer Geyer Jupiter (Teledyne Continental IO-360 flat-six) *(James Goulding)* ***NEW***/1047691

STRUCTURE: Composites throughout.

LANDING GEAR: Tricycle type; fixed. Fuselage-mounted spring glass fibre cantilever mainwheel legs with Cleveland brakes and 6.00-6 wheels. Nosewheel is glass fibre quarter ellipse cantilever sprung, with castoring 5.00-5 wheel.

POWER PLANT: One 157 kW (210 hp) Teledyne Continental IO-360 flat-six. Fuel capacity 326 litres (86.1 US gallons; 71.7 Imp gallons). Other engines under consideration included modified Chevrolet V-6 and V-8 automobile engines.

ACCOMMODATION: Four persons in two side-by-side pairs. Large door permits stretcher access.

AVIONICS: To customer/builder's requirements.

EQUIPMENT: Optional glider towing hook.

DIMENSIONS, EXTERNAL:
- Wing span ... 9.80 m (32 ft 1¾ in)
- Wing aspect ratio ... 6.2
- Length overall ... 7.25 m (23 ft 9½ in)
- Height overall ... 2.36 m (7 ft 9 in)

DIMENSIONS, INTERNAL:
- Cabin max width ... 1.17 m (3 ft 10 in)

AREAS:
- Wings, gross ... 15.50 m^2 (166.8 sq ft)
- Horizontal tail surfaces (total) ... 2.53 m^2 (27.23 sq ft)

WEIGHTS AND LOADINGS:
- Weight empty ... 560 kg (1,235 lb)
- Max T-O weight ... 1,025 kg (2,259 lb)
- Max wing loading ... 66.1 kg/m^2 (13.54 lb/sq ft)
- Max power loading ... 6.55 kg/kW (10.76 lb/hp)

PERFORMANCE (estimated):
- Max level speed ... 152 kt (281 km/h; 175 mph)
- Cruising speed at 75% power ... 137 kt (253 km/h; 157 mph)
- Stalling speed, flaps down ... 44 kt (80 km/h; 50 mph)
- Max rate of climb at S/L ... 600 m (1,970 ft)/min
- Service ceiling ... 8,230 m (27,000 ft)
- T-O run ... 190 m (625 ft)
- Range at econ cruising speed ... 1,080 n miles (2,000 km; 1,242 miles)
- *g* limits at 820 kg (1,800 lb) AUW ... +6/–3.5

UPDATED

RAVIN

RAVIN COMPOSITE AIRCRAFT MANUFACTURES

e-mail: jan@saravin.com
Web: http://www.saravin.com

DIRECTORS:
Jan Troskie
Leon Joubert
Hannes van Ark
Francois Jordaan

VERIFIED

RAVIN RAVIN 500

TYPE: Four-seat kitbuilt.

PROGRAMME: Transformation of Piper PA-24 Comanche (first flown 24 May 1956) into composites. Project originated during Jan Troskie's ownership of a Comanche 260C and following joint construction of Tri-R KIS composites kitplane with Leon Joubert, who later designed Ravin's avionics suite.

Prototype Ravin 500 composites re-creation of Piper Comanche ***NEW***/0552882

Prototype ZU-CTW first flown 15 September 2002; further aircraft (ZU-DCT) registered in 2003. Also available as complete aircraft.

CUSTOMERS: Total of three built by July 2004.

DESIGN FEATURES: Based on Piper Comanche, but some 7 per cent smaller and with redesign to enhance aerodynamics (wing profile), drag reduction (engine cowling), occupant's comfort and operating costs. Original Fowler flaps replaced by plain units; horizontal tail surfaces changed to conventional tailplane and elevator.

Can be equipped with standard Comanche engine, engine mount, propeller, landing gear, flight control systems, avionics and instruments or with more modern equivalents.

Airframe compatible with conversion to PA-30 Twin Comanche.

Kit version quoted build time 500 hours.

FLYING CONTROLS: Conventional and manual. Plain flaps.

STRUCTURE: Composites throughout.

LANDING GEAR: Tricycle type; retractable.

POWER PLANT: One 194 kW (260 hp) Textron Lycoming IO-540 flat-six. Fuel capacity 606 litres (160 US gallons; 133 Imp gallons) in wing tanks.

ACCOMMODATION: Four persons in tandem side-by-side pairs.

DIMENSIONS, EXTERNAL:
- Wing span ... 10.40 m (34 ft 1½ in)
- Wing aspect ratio ... 7.4
- Length overall ... 7.42 m (24 ft 4 in)

AREAS:
- Wings, gross ... 14.60 m^2 (157.2 sq ft)

WEIGHTS AND LOADINGS:
- Weight empty ... 850 kg (1,874 lb)
- Max T-O weight ... 1,620 kg (3,571 lb)
- Max wing loading ... 111.0 kg/m^2 (22.73 lb/sq ft)
- Max power loading ... 8.36 kg/kW (13.73 lb/hp)

PERFORMANCE:
- Max level speed ... 210 kt (389 km/h; 242 mph) IAS
- Cruising speed at 75% power:
  - S/L ... 185 kt (342 km/h; 213 mph) IAS
  - at FL65 ... 196 kt (363 km/h; 226 mph)
- Stalling speed: clean ... 64 kt (119 km/h; 74 mph) IAS
  - flaps and gear down ... 56 kt (104 km/h; 65 mph) IAS
- Range with 10% reserves ... 2,000 n miles (3,704 km; 2,301 miles)

UPDATED

SKYJUMPER

Box 911, Melville, Johannesburg
Tel: (+27 11) 444 52 10
e-mail: jeff@skyjumper.com
Web: http://www.skyjumper.co.za

Skyjumper produces a single-seat powered aerostat of the same name.

UPDATED

SKYJUMPER SKYJUMPER

TYPE: Helium non-rigid airship.

PROGRAMME: Under test for approximately two years; patent pending in mid-2004.

DESIGN FEATURES: Spherical envelope, from which is suspended, on a rotating attachment, an open capsule containing pilot seat, controls and pusher engine. Docking bay and mooring ropes also available.

Advertised for 'tethered rides', but can also be flown untethered. Can be customised for such potential applications as advertising, border patrol, sport flying, game watching, film platform, reconnaissance, scientific research and wildlife conservation.

FLYING CONTROLS: Rudder for directional control; ballast for altitude control.

STRUCTURE: Envelope of airship-standard composites fabric; pilot pod of carbon fibre.

LANDING GEAR: Airbag beneath pilot pod.

POWER PLANT: One 13.4 kW (18 hp) piston engine with electric starting, driving a three-blade composites pusher propeller.

Skyjumper pilot's gondola ***NEW***/0589251

Skyjumper personal non-rigid airship ***NEW***/0589230

DIMENSIONS, EXTERNAL:
Envelope diameter .... 8.70 m (28 ft 6½ in)
DIMENSIONS, INTERNAL:
Envelope volume .... 345.0 m^3 (12,184 cu ft)
WEIGHTS AND LOADINGS:
Max pilot weight .... 100 kg (220 lb)
PERFORMANCE:
Max level speed .... 16 kt (30 km/h; 18 mph)
Cruising speed .... 8 kt (15 km/h; 9 mph)

UPDATED

Spain

AEROCOPTER

AEROCOPTER SL

Pol Ind el Rubial 90, E-03400 Villena (Alicante)
Tel: (+34 679) 47 79 55
Fax:: (+34 965) 80 85 62
e-mail: info@aerocopter-europa.com
Web: http://www.aerocopter-europa.com
PRESIDENT: Pedro Esteban Poverda
MANAGER: Carlos Figuero Cano

AeroCopter showed the first production Futura autogyro at the Aero '03 exhibition at Friedrichshafen, Germany, in April 2003. The company employs a workforce of eight in its 800 m^2 (8,611 sq ft) factory.

UPDATED

AEROCOPTER FUTURA

TYPE: Two-seat autogyro ultralight.
PROGRAMME: Design began in 1998; two prototypes flown in 2002. Initial production aircraft (unregistered; black colour scheme) flew immediately after debut at Friedrichshafen, Germany, in April 2003. Second production aircraft, incorporating detail changes including cantilever spring main landing gear (unregistered; red colour scheme) exhibited at Blois, France, in September 2004.
COSTS: €59,000 (2004).
DESIGN FEATURES: Objectives were "spectacular and aggressive" appearance, combined with latest technology and high level of standard equipment. Twin-tailboom configuration with enclosed cabin; tailbooms and main landing gear mounted on anhedral aerofoil-shape fairings. Main rotor pre-rotation by Bowden cable. An 'F-16 style' air intake replaced upper air intake on initial production aircraft and fairing added to rotor mast.

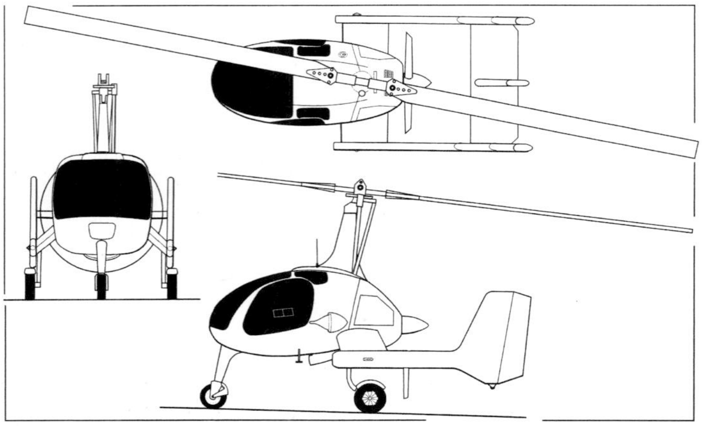

AeroCopter Futura two-seat autogyro in revised form ***(Paul Jackson)*** ***NEW***/0589201

Second production AeroCopter Futura on display at Blois, France, in September 2004 ***(Bob Sage)*** ***NEW***/1047700

FLYING CONTROLS: Twin rudders, plus all-moving central fin, actuated by Bowden cables.
STRUCTURE: Generally of glass fibre, with chrome-molybdenum tube cabin framework. Main rotors designed and manufactured by ASII in USA. Forward fuselage and glazing based on Rotorway Exec helicopter.
LANDING GEAR: Tricycle type; fixed. Mainwheels 400×100; nosewheel 4.00–6. Hydraulic disc brakes on mainwheels. Steerable faired-in nosewheel directly linked to rudder. Wheel guards on main wheels.
POWER PLANT: One 74.6 kW (100 hp) fuel-injected BMW R 1100 S flat-four driving a Kremen three-blade, flight-adjustable (electric), wooden pusher propeller via a reduction drive including a centrifugal clutch in initial production aircraft but replaced by 104 kW (140 hp) Subaru EA-81. Optionally, an 84.6 kW (113.4 hp) Rotax 914 turbo can be fitted. Propeller options include three-blade Kiev examples and three-blade in-flight-adjustable Kaspar.
DIMENSIONS, EXTERNAL:
Rotor diameter .... 8.53 m (28 ft 0 in)
Width overall .... 2.10 m (6 ft 10¾ in)
Length overall .... 4.00 m (13 ft 1½ in)
Height overall .... 3.10 m (10 ft 2 in)
AREAS:
Rotor disc .... 56.74 m^2 (610.7 sq ft)

WEIGHTS AND LOADINGS:
Weight empty ..... 275 kg (606 lb)
Max T-O weight ..... 550 kg (1,212 lb)
PERFORMANCE:
Max level speed ..... 103 kt (190 km/h; 118 mph)
Normal cruising speed ..... 86 kt (160 km/h; 99 mph)
T-O run ..... 100 m (330 ft)
Landing run ..... 5 m (20 ft)
Endurance ..... 4 h

UPDATED

CAG

CONSTRUCCIONES AERONAUTICAS DE GALICIA

Aeródromo de Vilaframil, E-27797 Ribadeo (Lugo)
Tel: (+34 982) 15 60 02
Fax: (+34 982) 15 60 09
e-mail: informacion@cag-sl.com
Web: http://www.cag-sl.com
PRESIDENT: Antonio Castelo Silvira

CAG's Toxo kitbuilt can be completed by amateur builders in the company's 1,600 m² (17,225 sq ft) factory. In July 2003, it entered into a joint venture with Mooney, whereby the US company will assemble the aircraft and assist with development of a four-seater version.

UPDATED

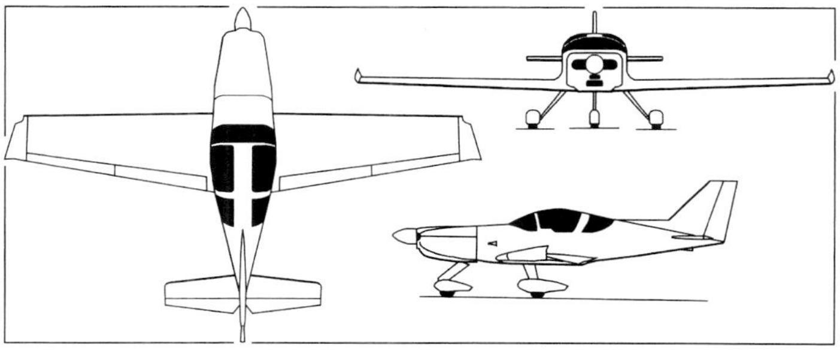

CAG Toxo II (Jabiru 2200 flat-four) *(Paul Jackson)* ***NEW***/0589200

CAG TOXO

English name: Gorse Bush

TYPE: Two-seat lightplane; side-by-side ultralight kitbuilt.

PROGRAMME: Designed by José Luis Miro. Original Toxo (EC-YYU) flew in 1999, followed by second prototype (EC-YYV); Toxo II (EC-ZFB) followed in September 2001.

CURRENT VERSIONS: **Toxo**: Ultralight; approved in France and Spain; similar registration under way in Austria, Germany and Netherlands in 2003. Typically 59.7 kW (80 hp) Jabiru 2200 flat-four engine and Duc fixed-pitch propeller. Fuel capacity 60 litres (15.9 US gallons; 13.2 Imp gallons).

Toxo II: VLA category. *As described.*

Mooney Speedster: US version, fuselage 15 cm (6 in) wider and 10 cm (4 in) taller than Spanish version; currently in abeyance.

Toxo IV: Proposed four-seat version.

CUSTOMERS: Total 22 delivered by early 2004 to customers in Brazil, France, Germany, Portugal and Spain.

COSTS: VLA €89,477 (2003) excluding tax, complete; VLA kit €54,458 (2003) including Jabiru 3300 engine, excluding tax.

DESIGN FEATURES: Experimental/VLA or Ultralight category; engines between 59.7 and 134 kW (80 and 180 hp). Wide speed range and high aerodynamic efficiency, including short field performance and good stability. Streamlined, tapered low-wing design with sweptback fin and upturned wingtips.

FLYING CONTROLS: Conventional and manual. Pushrod actuation. Fowler slotted flaps. Variable incidence (electrical) tailplane for trimming. (Horn-balanced elevators and alternative all-moving tailplane of prototypes replaced by mass-balanced elevators for production Toxo II).

STRUCTURE: Mainly of vacuum-bonded glass fibre and honeycomb; aluminium and steel fittings. Wing has T-beam composites spar and integral fuel tanks.

LANDING GEAR: Tricycle type; fixed. Fuselage-mounted spring cantilever mainwheel legs with wheel fairings. Steerable, trailing link nosewheel with wheel fairing. Mainwheels size 5.00. Brakes and parking brake.

POWER PLANT: To customer's requirements. Typically 89 kW (120 hp) Jabiru 3300 driving a Warp Drive Airmaster three-blade, constant-speed propeller. Rotax 914S installation (73.5 kW; 98.6 hp) under development. One with Rotax 912 delivered early 2004. Fuel capacity 150 litres (39.6 US gallons; 33.0 Imp gallons) in VLA version.

CAG Toxo II two-seat lightplane *(Paul Jackson)* ***NEW***/0589428

ACCOMMODATION: Two persons, side by side, each with centreline-hinged, upward-opening door; separate windscreen; baggage space to rear of seats; rear quarterlight each side. Dual controls. Cockpit heated and ventilated.

AVIONICS: *Comms:* Optional Becker AR-4201 radio, Zennheiser intercom and Becker ATC 4401-175 Mode S transponder.

Flight: Optional Bendix/King moving map display.

DIMENSIONS, EXTERNAL:
Wing span: Ultralight ..... 7.96 m (26 ft 1½ in)
VLA ..... 7.00 m (22 ft 11½ in)
Wing chord, mean ..... 1.00 m (3 ft 3¼ in)
Length overall ..... 5.40 m (17 ft 8½ in)
Height overall ..... 2.00 m (6 ft 6¾ in)
DIMENSIONS, INTERNAL:
Cabin max width ..... 1.05 m (3 ft 5¼ in)
AREAS:
Wings, gross: Ultralight ..... 8.70 m² (93.65 sq ft)
VLA ..... 7.00 m² (75.3 sq ft)
WEIGHTS AND LOADINGS:
Weight empty: Ultralight ..... 280 kg (617 lb)
VLA ..... 340 kg (750 lb)
Max T-O weight: Ultralight ..... 472.5 kg (1,041 lb)
VLA ..... 650 kg (1,433 lb)
Max wing loading: VLA ..... 92.9 kg/m² (19.02 lb/sq ft)
Max power loading: VLA ..... 7.27 kg/kW (11.94 lb/hp)
PERFORMANCE (Ultralight with Jabiru 2200 engine, VLA with Jabiru 3300 engine):
Never-exceed speed (VNE) ..... 216 kt (400 km/h; 248 mph)
Max level speed: VLA ..... 184 kt (340 km/h; 211 mph)
Cruising speed: max:
Ultralight ..... 130 kt (240 km/h; 149 mph)
VLA ..... 155 kt (287 km/h; 178 mph)
normal: Ultralight ..... 115 kt (213 km/h; 132 mph)
VLA ..... 140 kt (259 km/h; 161 mph)
Stalling speed: Ultralight ..... 35 kt (64 km/h; 40 mph)
VLA ..... 38 kt (70 km/h; 44 mph)
Max rate of climb at S/L (fixed pitch prop):
Ultralight ..... 305 m (1,000 ft)/min
VLA ..... 457 m (1,500 ft)/min
Service ceiling: VLA ..... 4,500 m (14,760 ft)
T-O to 15 m (50 ft): Ultralight ..... 190 m (625 ft)
VLA ..... 150 m (495 ft)
Landing from 15 m (50 ft): Ultralight, VLA ..... 250 m (820 ft)
g limits: Ultralight ..... +6/−4
VLA ..... +10/−8

UPDATED

COLYAER

CONSTRUCCIONES LIGERAS Y AERONÁUTICAS SL

Pombal-Carreter Adina, E-36979 Portonovo, Pontevedra
Tel: (+34 986) 72 78 53
Fax: (+34 986) 72 78 54
e-mail: info@colyaer.com
Web: http://www.colyaer.com
DIRECTORS:
Martin Uhia Lima
José Manuel Bujan

Company formed 11 November 1995 to develop and market Martin 3 lightplane as landplane and flying-boat. Manufacture is in 1,300 m² (14,000 sq ft) workshop at Portonovo.

VERIFIED

COLYAER MARTIN 3

TYPE: Side-by-side ultralight.

PROGRAMME: Developed from basic design of Canelas Ranger, produced in 1983 by Martin Uhia Lima and manufactured in small numbers. Prototype was flying-boat; second was landplane. Following formation of Colyaer company, third prototype built and design information computerised for production.

Third prototype Colyaer Martin 3 ***NEW***/0552875

CURRENT VERSIONS: **M3-S100:** Motor glider; greater wing area.

M3-S110: Shorter span; higher cruising speed.

Mascato: Amphibian; described separately.

DESIGN FEATURES: Intended to provide side-by-side seating for two, good outward visibility, spacious cockpit, 108 kt (200 km/h; 124 mph) cruising speed, Ultralight or JAR-VLA compatibility, range of 809 n miles (1,500 km; 932 miles), composites construction and competitive price.

High wing, without bracing struts, mounted on pod-and-boom fuselage, with T tail. Engine and pusher propeller mounted behind cockpit.

Wing section NACA 643-618.

FLYING CONTROLS: Conventional and manual. Electrically operated flaps, max deflections −7° for long-distance cruises; +50° for short landing.

STRUCTURE: Composites throughout. Carbon and glass fibre cantilever mainwheel legs; Kevlar, Nomex and epoxy resins.

LANDING GEAR: Tricycle type; fixed. Fuselage-mounted spring cantilever mainwheel legs with wheel fairings. Trailing-link nosewheel with wheel fairing.

POWER PLANT: Suitable engines between 45 and 75 kW (60 and 100 hp). Standard fuel capacity 40 litres (10.6 US gallons; 8.8 Imp gallons). Two optional tanks increase capacity to 90 litres (23.8 US gallons; 19.8 Imp gallons) or 130 litres (34.3 US gallons; 28.6 Imp gallons).

DIMENSIONS, EXTERNAL:
Wing span ... 10.40 m (34 ft 1½ in)
Length overall ... 5.85 m (19 ft 2¼ in)

DIMENSIONS, INTERNAL:
Cockpit max width ... 1.10 m (3 ft 7¼ in)

AREAS:
Wings, gross: S100 ... 12.00 m^2 (129.2 sq ft)
S110 ... 10.50 m^2 (113.0 sq ft)

WEIGHTS AND LOADINGS:
Weight empty ... 270 kg (595 lb)
Max T-O weight ... 450 kg (992 lb)

PERFORMANCE, POWERED:
Max level speed: S100 ... 97 kt (180 km/h; 112 mph)
S110 ... 108 kt (200 km/h; 124 mph)
Stalling speed: both ... 36 kt (65 km/h; 41 mph)
Max rate of climb at S/L:
S100 ... 300 m (984 ft)/min
S110 ... 240 m (787 ft)/min
T-O and landing run: S100 ... 120 m (395 ft)
S110 ... 140 m (460 ft)
Range with max optional fuel ... 809 n miles (1,500 km; 932 miles)
g limits ... +6/−4

PERFORMANCE, UNPOWERED:
Stalling speed ... 36 kt (65 km/h; 41 mph)
Best glide ratio: S100 ... 23
S110 ... 17
Min rate of sink: S100 ... 1.2 m (3.9 ft)/s
S110 ... 1.5 m (4.9 ft)/s

UPDATED

COLYAER MASCATO

TYPE: Two-seat amphibian ultralight kitbuilt.

PROGRAMME: Derivation of Martin 3. Prototype completed in early 2003.

CURRENT VERSIONS: **Mascato S100:** Sole version currently marketed.

Data generally as for Martin 3 S100, except where stated.

FLYING CONTROLS: Flap angle −5° for cruising.

LANDING GEAR: Single-step hull with outrigger floats at two-thirds' span. Max wave height 30 cm (1 ft).

DIMENSIONS, AREAS, WEIGHTS, PERFORMANCE POWERED: As Martin 3 S100, except
T-O and landing run ... 140 m (460 ft)

PERFORMANCE, UNPOWERED:
Best glide ratio ... 20
Min rate of sink ... 1.5 m (4.9 ft)/s

UPDATED

Colyaer Mascato ultralight amphibian ***NEW***/0552874

EADS CASA

CONSTRUCCIONES AERONAUTICAS SA (MILITARY TRANSPORT AIRCRAFT DIVISION OF EADS)

Avenida de Aragón 404, PO Box 193, E-28022 Madrid
Tel: (+34 91) 585 70 00
Fax: (+34 91) 585 76 66/7
e-mail: communications@casa.es
Web (1): http://www.casa.eads.net
Web (2): http://www.eads-nv.com

Works: Madrid (Barajas), Getafe, Toledo (Illescas), Seville (San Pablo and Tablada) and Cádiz (Puerto Real and Bahia)

PRESIDENT AND CEO: Francisco Fernández Sáinz

HEAD OF DIVISION:
Francisco Fernández Sáinz (Military Transport Aircraft)
Pablo de Bergia (Military Aircraft)
Pedro Méndez (Space)

DIRECTOR, COMMERCIAL COMMUNICATIONS: Miguel Sánchez

CASA, founded in 1923 to build French Breguet XIXs under licence for Spain's military aviation arm, was owned 99.2852 per cent by the Spanish state holding company Sociedad Estatal de Participaciones Industriales (SEPI), 0.7098 per cent by DASA (Germany) and 0.005 per cent by others. It was announced on 11 June 1999 that SEPI and DaimlerChrysler had signed an MoU to unite CASA and DASA in a new single company. This company was subsequently created as EADS by Aerospatiale Matra and DaimlerChrysler Aerospace and joined by CASA as a founder member on 2 December 1999. As a result, EADS CASA was restructured. In October 2001, the company acquired a 41 per cent holding in PZL Warszawa-Okęcie. A 5 per cent share in Eurocopter, and thence a 40 per cent stake in the newly formed Eurocopter España, was established in late 2003; this company responsible initially for final assembly of 24 Eurocopter Tigers for Spanish Army.

Military Transport Aircraft Division: This division is the Military Transport Aircraft Division of EADS and is responsible for the design, manufacture and marketing of light and medium transport aircraft such as the C-212, CN-235 and C-295. Some structural components for the CN-235 and C-295 are now manufactured by ENAER of Chile, under an eight-year co-operation agreement signed on 21 June 2001.

EADS CASA has been designing and manufacturing light and medium military transport aircraft for over 30 years for such roles as paratroop drops, aerial delivery, medevac, logistic transport or maritime patrol. These aircraft have seen service in more than 50 countries. Its former Airbus Division was fully responsible for design, development and manufacture of diverse structural parts integrated into all Airbus aircraft, having specialised in horizontal stabilisers, as well as in components manufactured in carbon fibre. The latter capability enables it to play an essential role within construction of the A380 megaliner, in which EADS participates with an 80 per cent stake. EADS CASA is the subcontractor responsible for designing and producing a tanker conversion kit for four Airbus A310MRTTs ordered by the German Air Force (first kit delivery was due in November 2002); is leading the A330MRTT air tanker consortium bidding for the UK Future Strategic Tanker Aircraft (FSTA) requirement; and in November 2001 launched (for availability in 2005) a development programme for an advanced air refuelling boom system (ARBS) with fly-by-wire control.

It also leads the future A400M heavy military transport aircraft programme, managed by Airbus Military Company in which EADS has a major stake. The A400M has been designed according to the shared requirements of seven air forces, and as a European reply to the replacement of current military transport aircraft such as the C-130 Hercules and C.160 Transall.

The division offers its FITS (Fully Integrated Tactical System) for diverse applications such as maritime patrol and surveillance, or anti-submarine/anti-surface warfare. First recipient of FITS was Spanish Air Force P-3B fleet (first of five upgraded aircraft rolled out early May 2003); selected also by Brazilian Air Force to retrofit nine P-3As.

A new factory was inaugurated at Bahia de Cadiz on 2 July 2003. Replacing the older facility at Puntales, it has a total area of 78,840 m^2 (848,625 sq ft), including 20,056 m^2 (215,880 sq ft) of covered accommodation.

Military Aircraft Division: Military Aircraft Division is a 43 per cent partner within the Eurofighter consortium, being fully responsible for engineering, structural tests and systems, as well as integrated logistic support, manufacture and prototype flight tests. It is currently in the development stage.

For series production Eurofighters, it manufactures starboard wings as well as port and starboard slats for every aircraft. Final assembly of all aircraft ordered by the Spanish Air Force is also carried out by EADS CASA.

EADS CASA has more than 40 years of experience working with military and commercial operators in areas of maintenance, overhaul and modernisation of aircraft and helicopters. Recently developed modernisation programmes for Spanish Air Force F-5, C-130 Hercules and Mirage F1 fleets are currently being offered to other countries. Modernisation programmes concerning F-18 and F-5 avionics are currently under way.

Space Division: This division develops structural, thermal and power distribution subsystems for launchers, satellites and orbital infrastructures. It also takes part in military satellite and mini-satellite programmes for the Spanish MoD and participates in the programmes for the development of new technologies, both in Spain and abroad.

UPDATED

CASA C-212 SERIES 400

US military designation (Series 200): C-41A

TYPE: Twin-turboprop transport.

PROGRAMME: Design studies for C-212 began 1964; design accepted by Spanish Air Force 1967; two prototypes (plus one static test) ordered 24 September 1968; first flights 26 March and 23 October 1971. Series 100 and 200 no longer produced, but four Series 200 in service with US Special Operations Forces belatedly allocated C-41A designation in 2002. Series 300 certified to FAR Pts 25, 121 and 135 in December 1987 but no longer offered; replaced by Series 400 from 1998. Former name Aviocar now discontinued.

CURRENT VERSIONS: **100:** Initial version; 158 built (CASA two prototypes plus 128, IPTN 28 from Spanish CKD kits).

200: Succeeded Srs 100 from 1980; 151 built by CASA. Final examples produced in Indonesia by Dirgantara.

300: Differed from Series 200 in having winglets to improve climb performance; enlarged nose for increased payload; and enhanced cruise performance; 52 built.

400: Initiated mid-1995, but not formally launched until June 1997 (Paris Air Show), following first flight (EC-212) on 4 April that year. EFIS and re-rated TPE331-12JR engines maintaining T-O power under hot/high conditions without APR; increased maximum payload; other improvements as detailed under Design Features below. Spanish type certificate 30 March 1998. Available in configurations described below.

400 Airliner: Standard seating for 26 passengers, or 24 if lavatory included.

Detailed description applies to 400 Airliner except where indicated.

400 Utility: Standard seating for 23 passengers, or 21 if lavatory included, with maximum capacity for 26.

400M: Military troop/cargo/general purpose transport.

Patrullero: Special missions (anti-submarine and maritime patrol) versions; described separately.

CASA C-212-400 PRODUCTION AND ORDERS
(at June 2004)

Country	Customer	Total
Australia	Skytraders	2[6]
Dominican Republic	Air Force	3[4]
Lesotho	Defence Force	1[5]
Paraguay	Air Force	1[7]
Spain	Min of Agriculture, Fisheries and Food	3[3]
Suriname	Air Force	2[2]
USA	EADS CASA (demonstrator)	2
	Phillips Alaska/BP	1
Venezuela	Navy	3[1]
Totals		**18**

[1] Delivered May 1998
[2] Includes one Patrullero; first delivered December 1998, second (Patrullero) in May 1999
[3] Patrullero. First delivered December 1998; second in March 2001; third in May 2003
[4] Third aircraft delivered 12 December 2000
[5] Delivered 18 September 2001
[6] Deliveries began April 2004; Antarctic operation
[7] Delivered 21 February 2004; ambulance conversion kit

Flight deck of the C-212 Series 400 0051420

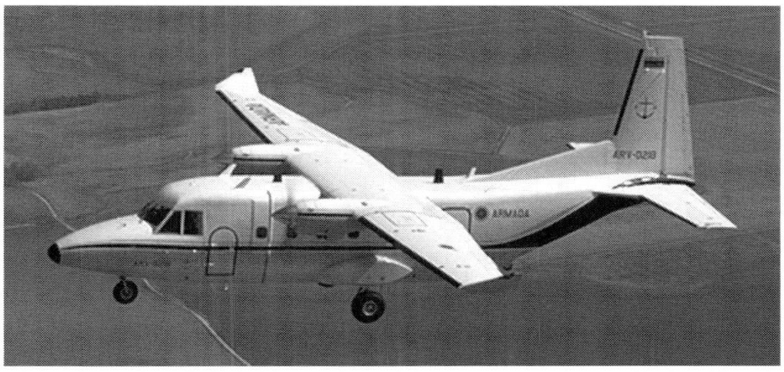
Venezuelan Navy CASA C-212-400M ***NEW***/0558590

CUSTOMERS: See table. Total 474 of all versions sold by June 2004; total includes up to 26 Series 200s still to be completed by Dirgantara of Indonesia. C-212s of all Series have been purchased or leased by over 40 military and 50 civil operators in 38 countries and had accumulated over 2.5 million hours by 1999.

DESIGN FEATURES: Typical high-wing, rear-loading military transport.

Wing section NACA 65$_3$-218; no dihedral; incidence 2° 30'; swept winglets canted upwards at 45°; meets FAR Pt 36 noise limits.

Series 400 features engine change (see Power Plant); avionics upgrade; underfloor avionics boxes relocated to former nose baggage compartment, facilitating ventral installation of a 360° scan radar and eliminating need for special-shape 'platypus' nose of earlier Patrullero versions, which allowed only 270° scan; relocated pressure refuelling point; main cabin refurbished; cargo winch option.

FLYING CONTROLS: Conventional and manual. Trim tab in port aileron; trim and geared tabs in rudder and each elevator; double-slotted flaps.

STRUCTURE: All-metal light alloy fail-safe structure; unpressurised; two-spar tailplane and fin. Wing centre-section, forward and rear passenger doors and dorsal fin manufactured by AISA.

LANDING GEAR: Non-retractable tricycle type, with single mainwheels and single steerable nosewheel. CASA oleo-pneumatic shock-absorbers. Goodyear wheels and tyres, main 11.00–12 (10 ply), nose 2–×7.7 (8 ply). Tyre pressure 3.86 bar (56 lb/sq in) on main units, 4.14 bar (60 lb/sq in) on nose unit. Goodyear hydraulic disc brakes on mainwheels. No brake cooling. Anti-skid system optional.

POWER PLANT: Two 690 kW (925 shp) Honeywell TPE331-12JR-701C engines. Dowty Aerospace R-334/4-82-F/13 four-blade, constant-speed, fully feathering, reversible-pitch propellers.

Fuel in four integral wing tanks, with total capacity of 2,040 litres (539 US gallons; 449 Imp gallons), of which 2,000 litres (528 US gallons; 440 Imp gallons) usable. Gravity refuelling point above each tank. Single pressure refuelling point in starboard mainwheel fairing. Additional fuel can be carried in one 1,000 litre or two 750 litre (264 or 198 US gallon; 220 or 165 Imp gallon) optional ferry tanks inside cabin, and/or two 500 litre (132 US gallon; 110 Imp gallon) auxiliary underwing tanks. Oil capacity 4.5 litres (1.2 US gallons; 1.0 Imp gallon) per engine.

ACCOMMODATION: Crew of two on flight deck; cabin attendant in civil version. Ergonomically redesigned flight deck; improved cabin soundproofing, sidewall design, lighting and toilet; 400M, foldaway seats along each side of cabin.

For troop transport role, main cabin can be fitted with 25 inward-facing seats along cabin walls, to accommodate 24 paratroops with instructor/jumpmaster; or 25 fully equipped troops. As ambulance, cabin is normally equipped to carry 12 stretcher patients and four medical attendants. As freighter, up to 2,950 kg (6,504 lb) of cargo can be carried in main cabin, including two LD1, LD727/DC-8 or three LD3 containers, or light vehicles. Cargo system, certificated to FAR Pt 25, includes roller loading/unloading system and 9 *g* barrier net. Photographic version equipped with two RC-20/30 vertical cameras and darkroom. Navigation training version has individual desks/consoles for instructor and five pupils, in two rows, with appropriate instrument installations.

Civil passenger transport version has standard seating for up to 26 in mainly three-abreast layout at 72 cm (28.5 in) pitch, with provision for quick change to all-cargo or mixed passenger/cargo interior. Lavatory, galley and 400 kg (882 lb) capacity baggage compartment standard, plus additional 150 kg (330 lb) in nose bay. VIP transport version can be furnished to customer's requirements.

Forward/outward-opening door on port side immediately aft of flight deck; forward/outward-opening passenger door on port side aft of wing; inward-opening emergency exit opposite each door on starboard side. Additional emergency exit in roof of forward main cabin. Two-section underfuselage loading ramp/door aft of main cabin can be opened in flight for discharge of paratroops or cargo, and can be fitted with optional external wheels for door protection during ground manoeuvring. Interior of rear-loading door can be used for additional baggage stowage in civil version. Entire accommodation heated and ventilated; air conditioning optional.

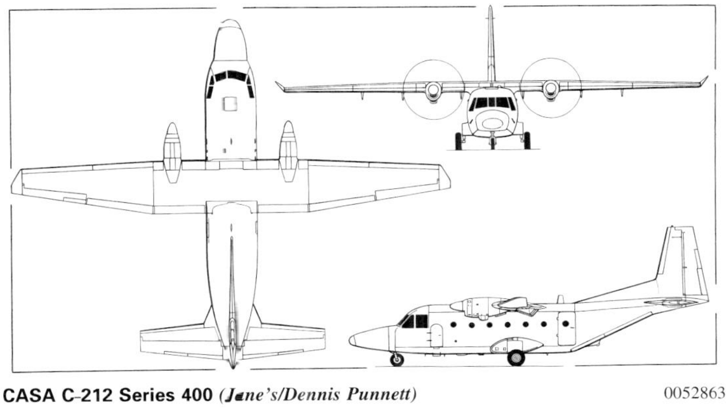
CASA C-212 Series 400 *(Jane's/Dennis Punnett)* 0052863

SYSTEMS: Freon cycle or (on special mission versions) engine bleed air air conditioning system optional. Hydraulic system, operating at service pressure of 138 bar (2,000 lb/sq in), provides power via electric pump to actuate mainwheel brakes, flaps, nosewheel steering and rear cargo ramp/door. Hand pump for standby hydraulic power in case of electrical failure or other emergency. Electrical system supplied by two 9 kW starter/generators, three batteries and three static converters. Pneumatic boot and engine bleed air de-icing of wing and tail unit leading-edges; electric de-icing of propellers and windscreens. Oxygen system for crew (including cabin attendant); two portable oxygen cylinders for passenger supply. Engine and cabin fire protection systems.

AVIONICS: *Comms:* Rockwell Collins VHF, ATC transponder, intercom and PA system standard; ELT; Rockwell Collins HF, UHF and second transponder, and Fairchild CVR, optional.

Radar: Honeywell weather radar standard. RDR-1500B search radar in ventral radome on Patrullero versions.

Flight: Rockwell Collins VOR/ILS, ADF, DME and radio altimeter, and Honeywell AFCS and directional gyro, standard; second Rockwell Collins ADF, Global Omega nav, Dorne & Margolin marker beacon receiver and Fairchild flight data recorder optional. Flight management system (FMS) incorporates VOR, ADF, DME and GPS nav receiver.

Instrumentation: EFIS with four CRTs; IEDS (integrated engine data system) with two colour LCDs. Blind-flying instrumentation standard.

EQUIPMENT: 1,000 kg (2,205 lb) capacity cargo winch optional.

ARMAMENT (MILITARY VERSIONS, OPTIONAL): Two machine gun pods or two rocket launchers, or one launcher and one gun pod, on hardpoints on fuselage sides (capacity 250 kg; 551 lb each).

DIMENSIONS, EXTERNAL:
Wing span ..... 20.27 m (66 ft 6 in)
Wing chord: at root ..... 2.50 m (8 ft 2½ in)
at tip ..... 1.25 m (4 ft 1¼ in)
Wing aspect ratio 10.0
Length overall ..... 16.15 m (52 ft 11¾ in)
Fuselage max width ..... 2.30 m (7 ft 6½ in)
Height overall ..... 6.60 m (21 ft 7¾ in)
Tailplane span ..... 8.40 m (27 ft 6¾ in)
Wheel track ..... 3.10 m (10 ft 2 in)
Wheelbase ..... 5.46 m (17 ft 11 in)
Propeller diameter ..... 2.79 m (9 ft 2 in)
Propeller ground clearance (min) ..... 1.44 m (4 ft 8¾ in)
Distance between propeller centres ..... 5.27 m (17 ft 3¾ in)

Passenger door (port, rear): Height .... 1.58 m (5 ft 2¼ in)
Width .... 0.70 m (2 ft 3½ in)
Crew and servicing door (port, fwd):
Max height .... 1.10 m (3 ft 7¼ in)
Width .... 0.58 m (1 ft 10¾ in)
Rear-loading door: Max length .... 3.66 m (12 ft 0 in)
Max width .... 1.68 m (5 ft 6¼ in)
Max height .... 1.80 m (5 ft 10¾ in)
Emergency exit (stbd, fwd): Height .... 1.10 m (3 ft 7¼ in)
Width .... 0.58 m (1 ft 10¾ in)
Emergency exit (stbd, rear): Height .... 0.95 m (3 ft 1½ in)
Width .... 0.56 m (1 ft 10 in)

DIMENSIONS, INTERNAL:
Cabin (excl flight deck and rear-loading door):
Length: passenger .... 7.27 m (23 ft 10¼ in)
cargo/military .... 6.55 m (21 ft 5¾ in)
Max width .... 2.10 m (6 ft 10¾ in)
Max height .... 1.80 m (5 ft 10¾ in)
Floor area: passenger .... 13.5 m^2 (145 sq ft)
cargo/military .... 12.2 m^2 (131 sq ft)
Volume: passenger .... 23.8 m^3 (840 cu ft)
cargo/military .... 22.0 m^3 (777 cu ft)
Cabin: volume incl flight deck and rear-loading door .... 30.4 m^3 (1,074 cu ft)
Baggage compartment volume .... 3.6 m^3 (127 cu ft)

AREAS:
Wings, gross .... 41.00 m^2 (441.3 sq ft)
Ailerons (total, incl tab) .... 1.875 m^2 (20.18 sq ft)
Trailing-edge flaps (total) .... 2.14 m^2 (23.03 sq ft)
Fin, incl dorsal fin .... 6.58 m^2 (70.83 sq ft)
Rudder, incl tab .... 2.16 m^2 (23.25 sq ft)
Tailplane .... 12.57 m^2 (135.30 sq ft)
Elevators (total, incl tabs) .... 1.78 m^2 (19.16 sq ft)

WEIGHTS AND LOADINGS:
Manufacturer's weight empty .... 3,780 kg (8,333 lb)
Weight empty, equipped (cargo) .... 4,550 kg (10,031 lb)
Max cargo payload .... 2,950 kg (6,504 lb)
Max fuel: standard .... 1,600 kg (3,527 lb)
with underwing auxiliary tanks .... 2,400 kg (5,291 lb)
Max T-O and landing weight .... 8,100 kg (17,857 lb)
Max ramp weight .... 8,150 kg (17,967 lb)
Max zero-fuel weight .... 7,500 kg (16,535 lb)
Max cabin floor loading .... 732 kg/m^2 (150 lb/sq ft)
Max wing loading .... 197.6 kg/m^2 (40.46 lb/sq ft)
Max power loading .... 6.04 kg/kW (9.92 lb/shp)

PERFORMANCE (Series 400M):
Max cruising speed at FL100 .... 195 kt (361 km/h; 224 mph)
Econ cruising speed at FL100 .... 163 kt (302 km/h; 188 mph)
Time to FL100 .... 8 min
Service ceiling .... 7,925 m (26,000 ft)
Service ceiling, OEI .... 3,275 m (10,740 ft)
T-O to 15 m (50 ft) .... 587 m (1,925 ft)
Landing from 15 m (50 ft) .... 527 m (1,730 ft)
Required runway length for STOL operation .... 402 m (1,320 ft)
Range:
with max payload .... 233 n miles (431 km; 268 miles)
with max standard fuel and 2,000 kg (4,409 lb) payload .... 800 n miles (1,481 km; 920 miles)
with max standard and auxiliary fuel and reduced payload .... 1,375 n miles (2,546 km; 1,582 miles)

UPDATED

CASA C-212 PATRULLERO

TYPE: Maritime surveillance twin-turboprop.

DESIGN FEATURES: Special missions versions of C-212 (various Series).

CURRENT VERSIONS: **MP:** Maritime patrol version; belly-mounted 360° search radar and FLIR/TV turret. Other features include operator's console, datalink via Inmarsat, external loudspeakers, searchlight, observation bubble windows, camera, and seating for 10 passengers. Initial C-212-400MP deliveries to Spanish Ministry of Agriculture, Fisheries and Food (one in December 1998 and a second on 29 March 2001) and Suriname Air Force (one in May 1999); third for Spanish MAFF delivered in May 2003.

Pollution control: Equipment as for MP, plus dedicated sensor for detection, measurement and control of harvest plagues, marine resources and pollution. This equipment has been fitted in Swedish Coast Guard and Portugese Air Force C-212s.

CUSTOMERS: See table with main C-212-400 entry. More than 56 Patrulleros of all Series sold by mid-2004.

POWER PLANT: As relevant C-212 Series. Two 500 litre (132 US gallon; 110 Imp gallon) underwing auxiliary fuel tanks.

ACCOMMODATION: Flight crew of two; mission sensor operator and observer in MP. Two observer stations at bubble windows in rear of cabin. Operator console with search radar, FLIR/TV, moving map and communications control. Radar repeater display and searchlight control added to flight deck. Galley, lavatory and rest area equipped with commuter seats or stretchers.

AVIONICS (MP): *Comms:* Single VHF, UHF/VHF and HF radios; intercom; IFF.

Radar: 360° scan underfuselage radar.

Flight: Redundant flight management system with dual GPS nav.

Instrumentation: Four-screen EFIS; IEDS.

Mission: FLIR/TV turret, Inmarsat datalink, electronic cartography and camera in MP. Additional to, or independent of, this equipment, Pollution control version can be fitted with a SLAR, IR/UV scanner or microwave radiometer.

EQUIPMENT: Survival kit and raft launcher in rear ramp; searchlight on fuselage hardpoint.

ARMAMENT: Hardpoint (capacity 500 kg; 1,102 lb) on each side of fuselage, on which can be carried machine gun pods or rocket launchers.

PERFORMANCE:
Endurance with underwing fuel tanks .... 8 h 30 min

UPDATED

CASA C-295

Spanish Air Force designation: T.21

TYPE: Twin-turboprop transport.

PROGRAMME: Stretched derivative of Airtech CN-235M (which see). Initiated November 1996, after market survey, as unilateral development of CASA/IPTN CN-235. Announced at Paris Air Show, June 1997. Prototype (EC-295 *Ciudad de Getafe*), a converted CN-235, flew for first time 28 November 1997 and had accumulated 801 hours in 379 sorties and 555 landings by 17 December 1999. First production C-295 (EC-296 *Ciudad de Sevilla*), designated S-1, made its maiden flight on 22 December 1998; had flown 516 hours in 232 sorties and 379 landings by 17 December 1999.

Certified by INTA (for military operations) on 30 November 1999; DGAC certification received 3 December 1999 and FAA (FAR Pt 25) certification on 17 December 1999; first customer aircraft (XT.21-02) demonstrated at Paris in June 2001.

CURRENT VERSIONS: **C-295M:** Military version; *as described*. Ordered by air forces of Algeria, Brazil, Jordan, Poland and Spain.

C-295MPA/ASW Persuader: Maritime patrol and anti-submarine version; equipped with CASA Fully Integrated Tactical System (FITS) mission suite. Ordered by air forces of Algeria and UAE (four each).

CUSTOMERS: See table. Spanish Air Force (launch customer) contract signed January 2000; deliveries began with handover of first (second production) aircraft, T.21-02 '35-40', to No. 35 Wing at Getafe; six delivered by October 2002 and scheduled to be completed in 2004. Four FITS-equipped C-295MPA/ASW ordered by UAE (Abu Dhabi) to meet its Shaheen

CASA C-295 refuelling from a KC-130H Hercules ***NEW***/0589297

1 requirement. Polish Air Force ordered eight (plus four on option) to replace An-26s of 13th Transport Squadron; delivery to end in 2005. Jordanian deliveries began September 2003. Brazilian 12 to replace obsolete C-115 Buffalos.

Selected by Swiss Air Force in December 2000 (two C-295M: SwFr 109 million funding approved 27 May 2004). Kenyan Air Force reportedly negotiating for three in late 2003.

C-295 ORDERS AND DELIVERIES (at June 2004)

Customer	Qty	Version	Announced	Delivery
Algeria	6	M[1]	03	04
	4	MPA/ASW	03	?
Brazil	12	M	4 Nov 02	04
Jordan	2	M	12 Feb 03	Sep 03
Poland	8	M	28 Aug 01	15 Aug 03
Spain	9	M	30 Apr 99	15 Nov 01
UAE	4	MPA/ASW	Mar 01	?

Notes
[1] Includes two VIP

COSTS: UAE order reportedly valued at US$140 million (2001). Eight for Poland cost US$212 million (2001); 12 for Brazil approximately US$270 million (2002); two for Jordan US$45 million (2003).

DESIGN FEATURES: Typical rear-loading, high-wing military transport. Conforms to FAR Pt 25 and MIL-STD-7700C. Compared with CN-235, has fuselage lengthened by six frames (three forward and three aft of wing); wing reinforced for higher operating weights and to support up to six optional underwing stores stations; additional 1,450 litres (383 US gallons; 319 Imp gallons) of fuel in wings; reinforced landing gear with twin nosewheels to facilitate operation from unprepared airstrips; modernised avionics identical to those of CN-235 Srs 300.

STRUCTURE: Provision for three hardpoints under each wing, capacities 800 kg (1,764 lb) inboard, 500 kg (1,102 lb) centre and 300 kg (661 lb) outboard.

LANDING GEAR: Main gear as CN-235 but reinforced for higher operating weights; twin nosewheels. Designed for sink rate of 3.05 m (10 ft)/s at normal MTOW and 2.74 m (9 ft)/s at overload MTOW; and operation from semi-prepared runways down to CBR-2 category.

POWER PLANT: Two Pratt & Whitney Canada PW127G turboprops, each with T-O rating of 1,972 kW (2,645 shp) normal, 2,177 kW (2,920 shp) with APR. Hamilton Sundstrand HS-568F-5 six-blade composites propellers with autofeathering and synchrophasing. Fuel capacity 7,650 litres (2,021 US gallons; 1,683 Imp gallons). Optional in-flight refuelling probe on starboard side of flight deck.

ACCOMMODATION: Crew of two; dual controls standard. Two or three longitudinal rows of foldaway seats in main cabin of C-295M for 71 troops; 48 paratroops (each 130 kg; 287 lb with full equipment); as a 12-stretcher intensive care unit or, in medevac configuration, 27 stretchers and four medical personnel. Lavatory, at front on port side, standard. In cargo role, can accommodate up to five 2.24 × 2.74 m (88 × 108 in) pallets, including one on rear ramp/door; or one 88 × 108 in and three 2.24 × 3.18 m (88 × 125 in) pallets; or three Land Rover-type light vehicles. Paratroop door each side at rear of cabin. Crew door aft of flight deck on starboard side, with emergency exit opposite on port side.

Persuader version accommodates, in addition to flight deck crew, two (MPA) or four (ASW) mission sensor operators, at identical, universal and configurable consoles; two observers at bubble windows in rear of cabin; radar repeater added to flight deck; galley, lavatory, and rest area equipped with commuter seats.

SYSTEMS: Pressure differential 0.38 bar (5.58 lb/sq in), giving cabin atmosphere equivalent to 2,354 m (7,723 ft) at altitude of 7,620 m (25,000 ft).

AVIONICS: All-digital, and compatible with NVGs. Thales (Sextant) Topdeck displays standard. Thales is overall systems integrator.

Comms: Two Thales (Sextant) U/VHF-AM/FM and one HF radio; Honeywell IFF transponder and solid-state FDR; ELT; air data system. Honeywell solid-state CVR. Alternative com/nav/ident functions optional.

Radar: Honeywell RDR-1400C colour weather radar with search and beacon modes and ground mapping vertical navigation capability.

Flight: Four-dimensional FMS; dual ADU 3008 air data units; dual AHRS; dual multimode receivers; dual GPS receivers; dual multifunction control and display units; Honeywell enhanced GPWS. Two central data processor/interface units and standard digital busses (ARINC 429 and MIL-STD-1553B). Options include Thales (Sextant) Totem 3000 laser gyro navigation system, enhanced TCAS, Cat. II ILS, MLS and satcom.

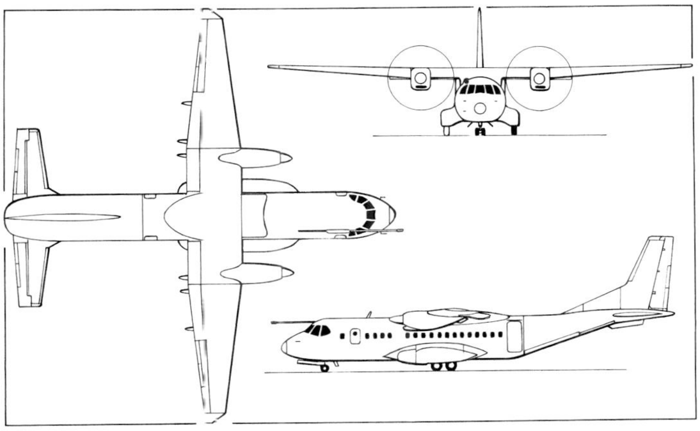

C-295M derivative of the CN-235 *(James Goulding)* 0110905

CASA C-295M of the Polish air Force's 13th Air Transport Squadron *(Grzegorz Holdanowicz)* ***NEW***/0560102

Instrumentation: Four 152 × 203 mm (6 × 8 in) Thales (Sextant) LCDs; one integrated electronic standby instrument; two IFC cabinets with autopilot module; radar altimeter. Video images from E-O sensors such as FLIR or cameras can be displayed on the four LCDs. Full NVG compatibility. Provision for two Thales (Sextant) HUDs.

Mission: CASA FITS suite for Persuader versions includes a central (and optionally redundant) tactical processor; four identical multifunctional consoles, interconnected via a local area network, each with its own RISC processor identical to the tactical processor; a 51 cm (20 in) high-resolution LCD; two touchscreens; and a keyboard and trackball; plus such peripherals as a pilot's presentation screen, recording and data loading units, and an armament and sonobuoy management panel.

Self-defence: Self-protection suite optional for C-295 ASW.

EQUIPMENT (PERSUADER): Sonobuoy or flare and marker launcher, survival kit and raft launcher in rear ramp; externally mounted searchlight.

ARMAMENT (PERSUADER): Three hardpoints under each wing (capacities 800, 500 and 300 kg; 1,764, 1,102 and 661 lb) for torpedoes, ASMs or depth charges.

DIMENSIONS, EXTERNAL:
Wing span ... 25.81 m (84 ft 8¼ in)
Length overall ... 24.46 m (80 ft 3 in)
Height overall ... 8.66 m (28 ft 5 in)
Tailplane span ... 10.60 m (34 ft 9¼ in)
Wheel track ... 3.98 m (13 ft 0¾ in)
Wheelbase ... 8.435 m (27 ft 8 in)
Propeller diameter ... 3.94 m (12 ft 11 in)
Distance between propeller centres ... 7.00 m (22 ft 11½ in)
Paratroop doors (two, each):
Height ... 1.75 m (5 ft 9 in)
Width ... 0.90 m (2 ft 11½ in)
Crew door (fwd, stbd): Height ... 1.27 m (4 ft 2 in)
Width ... 0.73 m (2 ft 4¾ in)
Emergency exit (fwd, port): Height ... 0.91 m (3 ft 0 in)
Width ... 0.50 m (1 ft 7¾ in)

DIMENSIONS, INTERNAL:
Cabin, excl flight deck:
Length: excl ramp/door ... 12.69 m (41 ft 7½ in)
incl ramp/door ... 15.73 m (51 ft 7¼ in)
Max width ... 2.70 m (8 ft 10¼ in)
Width at floor ... 2.36 m (7 ft 9 in)
Max height ... 1.90 m (6 ft 2¾ in)
Volume available for cargo ... 56.86 m³ (2,008 cu ft)

WEIGHTS AND LOADINGS (A: normal, B: overload):
Max payload: A ... 7,050 kg (15,543 lb)
B ... 9,250 kg (20,393 lb)
Max T-O weight: A ... 21,000 kg (46,297 lb)
B ... 23,200 kg (51,147 lb)
Max landing weight: A ... 20,700 kg (45,635 lb)
B ... 23,200 kg (51,147 lb)
Max zero-fuel weight: A ... 18,500 kg (40,784 lb)
B ... 20,700 kg (45,634 lb)
Max power loading: A ... 5.32 kg/kW (8.75 lb/shp)
B ... 5.89 kg/kW (9.67 lb/shp)

PERFORMANCE (C-295M at normal MTOW except where indicated):
Max cruising speed at optimum altitude ... 260 kt (482 km/h; 299 mph)
Time to optimum cruising altitude ... 12 min
Absolute ceiling ... 9,145 m (30,000 ft)
Service ceiling, OEI ... 4,125 m (13,540 ft)
T-O run at S/L: A, ISA ... 844 m (2,770 ft)
A, ISA + 20°C ... 934 m (3,065 ft)
T-O to 15 m (50 ft) at S/L:
A, ISA ... 1,025 m (3,365 ft)
A, ISA + 20°C ... 1,103 m (3,620 ft)
Landing from 15 m (50 ft) ... 729 m (2,390 ft)
Landing run ... 420 m (1,380 ft)
Range, ISA, reserves for 45 min hold at 460 m (1,500 ft):
with max payload:
A ... 840 n miles (1,555 km; 966 miles)
B ... 690 n miles (1,277 km; 794 miles)
with max fuel:
A with 2,940 kg (6,482 lb) payload ... 2,477 n miles (4,587 km; 2,850 miles)
B with 8,000 kg (17,637 lb) payload ... 1,160 n miles (2,148 km; 1,334 miles)
Ferry range ... 2,700 n miles (5,000 km; 3,107 miles)
g limits: A ... 2.53
B ... 2.25

UPDATED

VOL MEDITERRANI

VOL MEDITERRANI SL

Finca 'Les Umbertes', Apartat de Correus 83, E-08180 Moià, Barcelona
Tel: (+34 93) 830 12 52
Fax: (+34 93) 830 12 92
e-mail: volmediterrani@esqual.com
Web: http://www.esqual.com
DIRECTOR: Marta Boronat

Company headquarters are at the 'Les Umbertes' aerodrome, in the Barcelona region. Former co-director Francesc Velasco was killed in a flying accident on 20 October 2003.

UPDATED

VOL MEDITERRANI VM-1 ESQUAL

English name: Shark

TYPE: Side-by-side ultralight/kitbuilt.

PROGRAMME: Prototype EC-YZZ flown 2000. UK debut at PFA Rally, Cranfield, June 2002 (demonstrator EC-ZFF).

CURRENT VERSIONS: **VM-1:** Original version. Initial 73 aircraft, built up to mid-2003. Wing span 9.25 m (30 ft 4¼ in); length 5.86 m (19 ft 2¾ in).

From 74th aircraft onwards, has wider chord ailerons and smaller wingtip fairings for slightly reduced span. Prototype was EC-ZFX fitted with new wing, including fixed landing gear.

VM-1C: 'Carbon' version of VM-1, adding carbon fibre main spar, electric elevator trim and dual controls.

VM-1P: At least two VM-1s registered in Portugal.

Fixed gear: Nosewheel landing gear. Fuselage-mounted spring composites mainwheel legs and trailing-link nosewheel, all with wheel fairings. Mainwheels 3.50–6; nose 4.00–4. *As described.*

Retractable gear: Tailwheel type. Retractable mainwheels; fixed tailwheel with speed fairing. Prototype EC-ZFX had flown by mid-2002.

CUSTOMERS: More than 80 built and 100 sold by early 2004.

COSTS: Kit less engine and instruments: VM-1 €30,545, VM-1C €37,245; Rotax full kit: VM-1 €44,800, VM-1C €51,580; basic equipped, Rotax flyaway: VM-1 €51,964, VM-1C €57,964. All prices not including tax (2003).

DESIGN FEATURES: High-performance ultralight with low landing speed achieved by refinement of wing design. Low-mounted, tapered wing and mid-tailplane. Sweptback fin. No composites work required for assembly.

FLYING CONTROLS: Conventional and manual; ailerons and elevator pushrod actuated. Variable incidence (electric) tailplane for trim. Slotted flaps; maximum deflection 45°; flaps also deflect upwards to optimise cruising speed.

STRUCTURE: Glass fibre monocoque fuselage with integral fin; cockpit area reinforced. Steel tube centre-section. Wing main spar of carbon fibre or glass fibre (according to customer's choice; latter with 10 kg; 22 lb weight saving), with secondary trailing-edge spar and glass fibre sandwich skin.

LANDING GEAR: As described under Current Versions. Hydraulic brakes.

POWER PLANT: Option of 73.5 kW (98.6 hp) Rotax 912 ULS (standard) or 59.7 kW (80 hp)

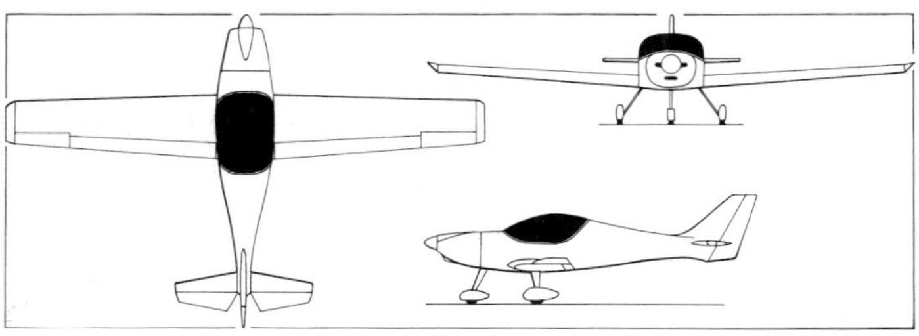

Vol Mediterrani VM-1 Esqual side-by-side kitbuilt *(Paul Jackson)* ***NEW***/0576480

Vol Mediterrani VM-1 Esqual development aircraft EC-ZFX fitted with current style wing and fixed landing gear ***NEW***/0576499

Jabiru 2200 flat-four or 89 kW (120 hp) Jabiru 3300 flat-six; last-mentioned adds 8.5 kg (19 lb) to empty weight. Warp Drive three-blade ground-adjustable pitch propeller standard. Airmaster propeller optional. Fuel in two wing tanks, combined capacity 75 litres (19.8 US gallons; 16.5 Imp gallons).

EQUIPMENT: Optional Galaxy ballistic parachute.

DIMENSIONS, EXTERNAL:

Wing span	9.25 m (30 ft 4¼ in)
Length overall	6.08 m (19 ft 11½ in)
Height overall	1.94 m (6 ft 4½ in)

DIMENSIONS, INTERNAL:

Cabin max width	1.10 m (3 ft 7¼ in)

AREAS:

Wings, gross	9.20 m² (99.0 sq ft)

WEIGHTS AND LOADINGS (Rotax 912 ULS and glass fibre wing spar):

Weight empty	275 kg (606 lb)
Max T-O weight	450 kg (992 lb)

PERFORMANCE (Rotax 912 ULS and three-blade propeller; J2200 and J3300, fixed-pitch, where different):

Max level speed: Rotax	146 kt (270 km/h; 168 mph)
J2200	124 kt (230 km/h; 143 mph)
J3300	148 kt (275 km/h; 171 mph)
Cruising speed at 75% power:	
Rotax	135 kt (250 km/h; 155 mph)
J2200	113 kt (210 km/h; 130 mph)
J3300	132 kt (245 km/h; 152 mph)
Stalling speed, flaps down	35 kt (64 km/h; 40 mph)
Max rate of climb at S/L: Rotax	549 m (1,800 ft)/min
J2200	305 m (1,000 ft)/min
J3300	457 m (1,500 ft)/min
T-O run: Rotax	95 m (315 ft)
J2200	130 m (430 ft)
J3300	105 m (345 ft)
Landing run	95 m (315 ft)
Range	567 n miles (1,050 km; 652 miles)
g limits	+9/–3 (ultimate)

UPDATED

Sweden

SAAB

SAAB AB

SE-581 88 Linköping
Tel: (+46 13) 18 00 00
Fax: (+46 13) 18 00 11
e-mail: infosaab@saab.se
Web: http://www.saab.se

PRESIDENT AND CEO: Åke Svensson
EXECUTIVE VICE-PRESIDENT AND CFO: Göran Sjöblom
SENIOR VICE-PRESIDENT, COMMUNICATIONS: Jan Nygren
PRESS DIRECTOR: Anders Stålhammar

Svenska Aeroplan AB founded at Trollhättan 1937 to make military aircraft; amalgamated 1939 with Aircraft Division of Svenska Järnvägsverkstäderna rolling stock factory at Linköping; renamed Saab Aktiebolag May 1965; merged with Scania-Vabis 1968 to combine automotive interests; Malmö Flygindustri acquired 1968. Bid for Celsius AB 16 November 1999 and acquisition completed 8 March 2000; enlarged company reorganised in 2000 into six main business areas and several independent operations:

Saab Aerospace: Combining military aircraft, future aerospace systems and commercial programmes. Described in further detail.

Saab Aviation Services: Commercial aircraft maintenance and engine and component maintenance. Comprises Celsius Aviation Services, Saab Aircraft and Saab Aircraft Leasing; holds type certificates for Saab 340 and Saab 2000 regional turboprop airliners, production of which ended in 1999.

Saab Systems and Electronics: Electronic warfare, simulation and training and radar control, including Combitech Systems and Ericsson Saab Avionics.

Saab Bofors Dynamics: Missiles, anti-armour and underwater systems.

Saab Technical Support and Services: Including aircraft maintenance.

Saab Ericsson Space: Digital, microwave and mechanical products.

It was announced on 30 April 1998 that Saab's owner, Investor AB, had approved the sale to British Aerospace plc (now BAE Systems) of a 35 per cent voting shareholding. Investor remained Saab's leading owner, with 36 per cent of the votes and 20 per cent of the capital. Purchase rights to the remaining 29 per cent of votes and 45 per cent of capital continued to be offered to Investor shareholders. Announced 7 December 2004 that from 1 January 2005, BAE share in Saab AB reduced to minimum of 20 per cent; at same time, Saab took lead in export marketing through previously equally shared Gripen International company.

More than 4,000 military and commercial aircraft delivered since 1940; has held dealership for MD Helicopters (formerly Schweizer/Hughes, McDonnell Douglas and Boeing) products in Scandinavia and Finland since 1962. In September 2000, Saab announced entry into design of UAVs, in conjunction with Swedish Defence Materiel Administration and other local industries.

SAAB AEROSPACE:

SE-581 88 Linköping
Tel: (+46 13) 18 00 00
Fax: (+46 13) 18 24 11
e-mail: coms@saab.se
Web: http://www.saabaerospace.com

GENERAL MANAGERS:
Lennart Sindahl (Aerosystems)
Pontus Kallén (Aerostructures)
Kjell Johnsson (Support)
COMMUNICATIONS DIRECTOR: Peter Larsson

GRIPEN INTERNATIONAL:
SE-581 88 Linköping
Tel: (+46 13) 18 40 00
Fax: (+46 13) 18 00 55
Web: http://www.gripen.com
MANAGING DIRECTOR: Johan Lehander
DEPUTY MANAGING DIRECTOR: Kjell Möller
MARKETING DIRECTOR: Bob Kemp
COMMUNICATIONS DIRECTOR: Owe Wagermark

Web: http://www.gripen.co.uk

On 14 July 2003, Saab's Aerospace division was restructured into two business units and a support organisation:

Saab Aerosystems: Operating areas include defence marekt, including Gripen, advanced air combat systems such as UAVs, and subsystems for other aircraft.

Saab Aerostructures: Responsible for fuselage and structural components manufacture for Gripen, Airbus, Boeing, helicopters and other aircraft.

Saab Support: Group services for business units at Linköping, co-ordinated with other sites.

Aerostructures work includes floor assemblies, pylons and landing gear doors for the Airbus A340-500/600 series and components for A320. Saab also partner in A380 and A400M programmes. Builds A380 wing panels (first mid- and outer leading-edge delivered 24 April 2003); crew doors and wingtip panels for Boeing 737; and wing components for Boeing 777. Developing new tactical system for NH90 helicopter under contract placed in early 2002. Saab group sales include 40 per cent exports and are 70 per cent defence-related.

Main activity of Aerosystems is development and production of Gripen multirole fighter.

Original Industrigruppen JAS AB (JAS Industry Group) formed 1981 to represent (then) Saab-Scania (65 per cent share), Ericsson (16 per cent), Volvo Flygmotor (15 per cent) and FFV Aerotech (4 per cent) in JAS 39 Gripen programme; continues to act as contractor for Försvarets Materielverk (Defence Materiel Administration, FMV) and co-ordinated JAS 39 Gripen programme within Sweden. Volvo Flygmotor became Volvo Aero in 1994 and FFV Aerotech share passed to Celsius in 1990 through merger with Bofors. Employees of Saab Aerospace totalled 4,120 in 2001. International marketing of JAS 39 Gripen is now responsibility of Saab and BAE Systems, acting jointly since November 1995; current members of JAS Industry Group are subcontractors.

Workshare on export orders would be Saab 55 per cent/BAE 45 per cent under agreement signed 12 June 1995. BAE involvement also includes manufacture of main landing gear, the first of which was delivered from the Brough factory in April 1996 and incorporated in 49th production Gripen. From May 1999, BAE also responsible for assembly of wing attachment unit and associated component join-up for two-seat aircraft (single-seaters to follow).

UK and Swedish export guarantee departments signed an agreement in August 1996 to finance any third-party sales. A contract of 16 November 1995 provides for Danubian Aircraft Company of Hungary to manufacture Gripen components at Tököl as part of a wider Swedish-Hungarian defence information exchange agreement signed 18 December 1995. First Hungarian parts (tailcone fittings) delivered April 1996. Further Swedish-Hungarian agreement of 22 January 1997 provided for comprehensive industrial offsets in the event of a Gripen purchase. First South African parts — Denel-built assembly set for stores pylon — handed over April 2000. Contract awarded to Denel in March 2001 for manufacture of section of centre fuselage for 40 Lot 3 Gripens, plus 28 for SAAF aircraft. Polish part manufacture (metal subassemblies) began July 2000 at PZL (Polish Aircraft Factory).

Gripen International formed 3 September 2001 to strengthen Saab-BAE Systems marketing ventures; is jointly owned and staffed, but registered in Sweden.

UPDATED

SAAB JAS 39 GRIPEN

English name: Griffin

TYPE: Multirole fighter.

PROGRAMME: Funded definition and development began June 1980; initial proposals submitted 3 June 1981; government approved programme 6 May 1982; initial FMV development contract 30 June 1982 for five prototypes and 30 production aircraft, with options for next 110; overall go-ahead confirmed second quarter 1983; first test runs of RM12 engine January 1985; Gripen HUD first flown in Viggen testbed February 1987; study for two-seat JAS 39B authorised July 1989.

First of five single-seat prototypes (39-1) rolled out 26 April 1987; made first flight 9 December 1988 but lost in landing accident after fly-by-wire problem 2 February 1989; only six sorties flown. Subsequent first flights 4 May 1990 (39-2), 20 December 1990 (39-4), 25 March 1991 (39-3) and 23 October 1991 (39-5); the 2,000th Gripen sortie was flown (by 39-4) on 22 December 1995; modified Viggen (37-51) retired at end of 1991 after assisting with avionics trials (nearly 250 flights); two single-seat fatigue test airframes (39-51 discarded 1993; 39-52 began 16,000 hour programme, August 1993 and achieved 8,000 hours in early 1996). Second production batch (Lot 2: 110 aircraft) approved 3 June 1992; first production Gripen (39101) made first flight 10 September 1992 and joined test programme in lieu of 39-1; flight test programme in 1995–96, included high-AoA (at least 28° achieved) and spin trials by 39-2 and trials of an APU (for Lot 2 production) by 39-4. All development work in the original (Lot 1) contract had been completed by late 1996; total programme was over 1 800 hours in 2,300 sorties by six aircraft. By 1996 had demonstrated M1.08 cruise without reheat. Follow-on trials with mockup aerial refuelling probe conducted by 39-4 on eight sorties between 2 and 17 November 1998 from RAF VC10 K. Mk 4. First flight of JAS 39C with refuelling probe extended June 2004. Captive flight of two KEPD 150 Taurus SOMs on inner wing pylons of 39145 of F7 conducted on 27 August 1998. Live Raytheon AIM-120 AMRAAM firings by 39-5 in April 1998.

Initial production aircraft for Swedish Air Force (39102) made first flight 4 March 1993 and was handed over to FMV 8 June 1993; flight control software modified following loss of 39102 in crash on 8 August 1993 and installed from December 1994; further software upgrade to new-generation P11 standard (introducing 11 filters to prevent pilot-induced oscillations) first flown 22 March 1995 in trials aircraft and installed in test Gripens from late 1995; in production aircraft (known then as R11) built after early 1996 (and retrofitted from 1997 as R11:9); modified control stick introduced with production aircraft 39108 (first flight 11 April 1995). R12:3 flight control software under trial by late 1999, bringing Gripen up to original design goals; installed during following year as R12:4. Next stage is R14, increasing MTOW by some 1,000 kg (2,205 lb).

PP12 display processor for Lot 2 colour displays first flown August 1995. Thrust-vectoring under consideration for Gripen; first stage is proposed participation in further trials programme of Rockwell/DASA X-31. JAS 39B prototype rolled out 29 September 1995; first flight 29 April 1996.

Initial 30 production JAS 39As delivered 1993–96 (five in 1994; six in 1995, comprising 39108 to 112 and 120; and last of balance — 39129 — on 13 December 1996); by 2002, these all upgraded to Lot 2 standard; deliveries of second lot of 110 began 19 December 1996, and completed in 2003; Swedish Parliament authorised third batch of 64 on 13 December 1996; contract formally placed on 26 June 1997; deliveries between 2003 and 2007. Saab contracted on 24 November 1999 to supply new EWS 39 defensive aids suite, designed by Ericsson Saab Avionics and with substantial CelsiusTech content. In same month, Gripen made first flight with FADEC, as destined for Lot 3 aircraft.

First unit was F7 Wing at Såtenäs; maintenance training begun May 1994 at Linköping; conversion scheduled to begin 1 October 1995 but postponed to 1996, with pilot training centre at Såtenäs officially opened 9 June 1996; Gripen IOC achieved September 1997, following three-week field exercise by 2/F7 Squadron. Last of F7's previous Viggens withdrawn in October 1998. By mid-2002 SwAF Gripens had flown 34,000 sorties and 25,000 hours. SwAF received 100th Gripen on 12 March 2001 and 113 by 31 December that year. Final JAS 39A (and first JAS 39C) delivered 6 September 2002.

Gripen being promoted in several fighter competitions, as described in Customers section. On 18 November 1998, it was announced that the aircraft had been selected for purchase by the South African Air Force and this was confirmed by contract signature on 3 December 1999. In connection with SAAF requirement, Gripen completed initial series of trials with helmet-mounted display (Thales Optronics Guardian) in February 2001.

First production JAS 39D Gripen conducting its maiden flight on 2 June 2004 ***(Janne Gustavsson)*** ***NEW***/1047046

Initial production JAS 39C armed with AMRAAM and KEPD 350 dispensers underwing ***(Jonas Tillgren)*** ***NEW***/1047040

First Saab JAS 39C Gripen in Czech Air Force colours making its maiden flight on 18 December 2004 ***NEW***/1043428

CURRENT VERSIONS: **JAS 39A:** Standard single-seater. Final aircraft delivered 6 September 2002.

Description applies to JAS 39A except where indicated.

JAS 39B: Two-seater with 0.655 m (2 ft 1¾ in) fuselage plug and lengthened cockpit canopy. Primary roles conversion and tactical training, but also combat-capable. Not used for instruction until January 2002, when first students from pilot training began conversion. Avionics essentially as for JAS 39A and both cockpits identical, except no HUD in rear; instead, HUD image from front seat can be presented on flight data display in rear cockpit. Boosted environmental control system (ventral air intake replaces twin scoops on fuselage sides); inflatable airbag protects rear occupant during pre-ejection canopy fracturing. Reduced internal fuel; no internal gun.

Prototype entered final assembly 1 September 1994; first flight (39800) 29 April 1996; first production two-seater (39801) completed final assembly on 29 February 1996 and flew on 22 November; production deliveries began with 39802 on 19 May 1998.

Fatigue test specimen (39-71) also built; began a simulated 16,000 hour programme in February 1996.

JAS 39C and D: New features warrant revised designations for JAS 39A and B; improvements include full-colour displays, FADEC, helmet-mounted sight, new (Modular Airborne Computer System) processor for PS-05/A radar, Saab Dynamics IR-OTIS IR search and tracking system, in-flight refuelling probe and enhanced EW systems. IR-OTIS tested on Saab Viggen in 1997.

Originally planned for introduction at start of Lot 3, but final 19 Lot 2 aircraft (plus one testbed, 39207/39-6, first flown 29 April 2002) completed to interim standards with refuelling probes and upgraded computers and designated JAS 39C. First production aircraft (JAS 39C 39208) flown 14 August 2002 and to FMV for service trials 6 September 2002. Initial JAS 39D production aircraft (39815) flew 2 June 2004.

An electronically scanned radar antenna is under development for a potential Gripen MLU in 2010. By 2000, plans in hand to deliver all 14 Lot 3 two-seat Gripens in **JAS 39D C²** configuration with entirely redesigned rear cockpit for command and control duties.

JAS 39 EBS HU: Export basic standard, Hungary. Ordered February 2003 in form of remanufactured JAS 39A/Bs.

Enhanced Gripen: Envisaged for 2010; first details mid-2001. Could include improved EW systems with laser warning, missile approach warning linked to towed decoys and EW function incorporated in electronically scanned radar. Range enhanced by standardisation on two-seat airframe, with fuel in place of rear cockpit, and/or conformal tanks; new engine, possibly EJ200, M88 or F414, under consideration. Navalised Gripen reportedly under consideration by India in 2002.

CUSTOMERS: Swedish Air Force requirement was originally 280, to equip 16 squadrons, second eight replacing JA and AJS Viggens; this reduced when third batch authorisation (December 1996) covered only four squadrons, thus amending requirement to 204 (including 28 JAS 39B two-seat versions); however, prototype JAS 39B added to first batch as conversion of aircraft under production, further amending contracts to 175 single-seat and 29 two-seat.

First 30 ordered with prototypes and full-scale development 30 June 1982; next 110 (of which first — 39131 — flew on 20 August 1996) include 14 JAS 39Bs and 20 JAS 39Cs; black radomes on 39109 to 39127; low-visibility markings from 39128, augmented by light grey (previously medium grey) radomes from 39131. Third lot of 64 (also including 14 two-seaters) for delivery between 2003 and 2007 are to JAS 39C/D standard.

Some Lot 3 improvements incorporated early in final 20 Lot 2 aircraft, (which redesignated JAS 39C), comprising refuelling probe and substantially improved computer (unofficially known as 'Mk 3') including D96 (MACS) processor to simplify possible later upgrade to colour displays.

Total 120 (including two-seat) delivered by July 2002. First operational squadron was 2/F7, September 1997, followed by 1/F7 12 months later. F10 at Angelholm followed in 1999–2000, conversion of pilots (by F7) having begun on 11 March 1999, followed by arrival of first two aircraft on 30 September 1999; first squadron (2/F10) operational September 2000, with second following in mid-2001. However, F10 disbanded by 31 December 2002 and transferred its equipment to F17 at Kallinge, latter being inaugurated as Gripen unit on 14 June 2002. F4 at Froson-Ostersund to convert in 2004, while F21 at Luleå received first two Gripens on 17 January 2002 (although its second squadron will not become operational until 2005). From 1 January 2004, 2/F17 ('172 Squadron') is Swedish Air Force Rapid Reaction Unit, for worldwide deployment at 30 days' notice, equipped with eight JAS 39As. Second squadron of F17 receiving JAS 39Cs, beginning 15 September 2004, these to assume SAFRRU role in 2006.

Saab JAS 39A Gripen *(Katsuhiko Tokunaga)* ***NEW***/1047037

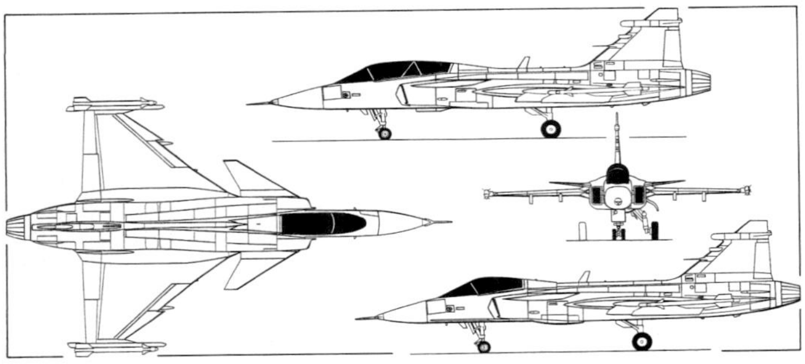

JAS 39A Gripen multirole combat aircraft for the Swedish Air Force, with additional side view (top) of two-seat JAS 39B *(Jane's/Dennis Punnett)* 0064907

Exports of some 250 anticipated over 20 years from 1996; by then, Saab and BAE engaged in 12 export campaigns; presentations made to Austria, Brazil, Czech Republic, Chile and Poland in 1996–97. Promotion in Philippines, Slovenia and South Africa, 1997. Details of competitions summarised below.

Australian Air 6000 requirement received Gripen response on 1 February 2002, but Lockheed Martin F-35 selected June 2002.

Austrian RFP received October 2001 for up to 30 new fighters; Gripen proposal delivered 22 January 2002, based on either 24 single- and four two-seat aircraft for early delivery; or 12 leased aircraft as lead-in to later delivery of 24 plus six new-build Gripens. Revised proposals, delivered to Austria on 30 April 2002, envisaged accelerated delivery between mid-2005 and mid-2007, but Eurofighter Typhoon eventually selected by parliament on 2 July 2002, although almost immediately suspended.

Brazilian RFP received August 2001 for potential 24 aircraft. Revised proposals delivered to Brazil on 3 May 2002. Programme suspended 2003.

Czech Republic RFP issued 27 December 2000; invitation to tender for new fighter issued 9 January 2001; Saab response delivered 31 May 2001 for 24 or 36 aircraft with 150 per cent offsets. Plan officially confirmed on 10 December 2001 for 24 Gripens valued at Kcs50 billion, including spares and training; first squadron to be operational in 2005 and second and final in 2008. Formal approval given 22 April 2002, contract signing due before end of year, but programme was abandoned in November 2002 and replaced by quest for 12 to 14 aircraft, with Gripen still among contenders. Gripen lease recommendation made by evaluation committe on 17 December 2003, and on 14 June 2004, Czech government agreed lease of 14 Gripens (12 JAS 39Cs and two JAS 39Ds) between 2005 and 2015. Initial flight of Czech Gripen (actually fourth JAS 39C of contract, 39237) conducted on 18 October 2004; deliveries scheduled from 1 May 2005. Contract valve Kcs 19.65 billion. Aircraft have refuelling probes, OBOGS, colour cockpit displays with English language, NATO communications and weapons capability.

Hungary received Gripen proposal on 30 November 2000, this amended on 9 February 2001 with lease proposal; selected Gripen for air defence role on 10 September 2001, initially with lease of 14 aircraft (including two tandem-seat) for 10 years at cost of Ft130 billion to Ft140 billion. MoU signed 23 November 2001 and agreement finalised 20 December 2001 requiring initial deliveries in fourth quarter of 2004 and last in June 2005. However, negotiations reopened on 3 September 2002, following election of new Hungarian government. Parliamentary authorisation granted 10 March 2003, formalising agreement signed with Swedish Defence Material Administration on 3 February 2003 for lease/purchase of 12 single-seat and two tandem-seat aircraft with increased (multirole) capabilities, but at penalty of 2006 start to deliveries. Aircraft are remanufactured Swedish JAS 39A/Bs, designated JAS 39 EBS HU; lease is 10 years, with purchase option. Modifications from Swedish standard include NATO standard communications, retractable refuelling probe, strengthened wings for LGB carriage and instrument calibration in Imperial units. Assembly of first aicraft began 1 October 2004. Operating unit based at Kecskemét.

Polish negotiations begun September 2001 in search for 60 new fighters. Final assembly of required 44 to 60 Polish Gripens would have been by PZL Aircraft Factory, under 1997 agreement, as covered by June 1999 proposal; however, revised offer of January 2001 restricted to five-year lease of 16 aircraft, including two trainers. Poland ordered Lockheed Martin F-16s in December 2002.

South Africa selected Gripen on 18 November 1998 for planned purchase of 28; order placed 15 September 1999 for nine two-seat Gripens, with option on further 19 of single-seat version; formal signature 3 December 1999, but first deliveries postponed at that time from 2002 to mid-2006. However, only one two-seat Gripen due in that year for trials, followed by remaining eight between November 2007 and September 2009. Single-seat Gripens due between November 2009 and late 2011.

COSTS: Planned cost of SEK25.7 billion in 1982, increased to SEK48.5 billion in 1991 by inflation; FMV has reported total cost increase of SEK9.3 billion for period 1982–2001; total budget SEK60.2 billion decided by Swedish Parliament 1993. SEK22.7 billion spent by 1 July 1993, including SEK14.5 billion to IG JAS. Up to SEK300 million approved late 1991 for JAS 39B development. Costs for 300-aircraft programme (subsequently reduced to 204) estimated in 1994 as SEK15 billion for development and SEK48 billion for production. South African purchase of 28 estimated programme cost R10.9 billion (US$1.9 billion) (1998).

DESIGN FEATURES: Intended to replace AJ/SH/SF/JA/AJS versions of Saab Viggen, in that order, and remaining J 35 Drakens; to operate from 800 m (2,625 ft) Swedish V90 road strips; simplified maintenance and quick turnround with ground crew comprising one technician and five conscripts.

By 2000, Gripen demonstrating 12 mmh/fh; 7.6 hours MTBF readiness rate; and 10 minute turnround in fighter configuration, 20 minutes in attack role. Targets are 10 mmh/fh and 9.0 hours MTBF.

Mid-mounted delta wing, with squared tips for missile rails, has three-section leading-edge, of which inboard and outboard sections sweptback at 55° and centre at 52°; foreplanes, independently movable, have leading-edge sweep of approximately 58°. Sweptback fin carries various antennas. Moderately high cockpit. Near-rectangular engine air intakes, each with splitter plate.

FLYING CONTROLS: BAE Systems (Lot 1) or Lockheed Martin SA11 (Lot 2) triplex fly-by-wire system with Moog electrically signalled servo valves on powered control units; Saab Combitech aircraft motion sensors and throttle actuator; mini-stick and HOTAS controls.

Leading-edge with dog-tooth and automatic flaps (one inboard/one outboard of dog-tooth, inner one outboard of canard) on Lucas Aerospace 'geared hinge' rotary actuators; two elevon surfaces at each trailing-edge; individual all-moving foreplanes, which also 'snowplough' for aerodynamic braking after landing; airbrake each side of rear fuselage.

STRUCTURE: Airframe is 60 per cent aluminium by weight, 6 per cent titanium and 5 per cent other metals; most of remainder is carbon fibre. First 3½ carbon fibre wing sets produced by BAE; all subsequent carbon fibre parts made by Saab, including wing boxes, foreplanes, fin and all major doors and hatches. BAE produces landing gear and, from 1999, responsible for assembly of centre fuselage for two-seat version, with single-seat to follow. Denel of South Africa contracted in 2000 to design and produce stores pylons for export Gripens. PZL-Mielec of Poland began delivering tailcones in late 2000. By 2002, late production JAS 39As had received additional 200 kg (441 lb) of strengthening to landing gear and wings for MTOW increase to 14,000 kg (30,864 lb) and service life increase from 3,000 to 4,000 hours.

LANDING GEAR: AP Precision Hydraulics retractable tricycle gear, single mainwheels retracting hydraulically forward into fuselage; steerable twin-wheel nose unit retracts rearward. Goodyear wheels. Carbon disc brakes and ABS anti-skid units. Nosewheel braking. Entire gear designed for high rate of sink. Mainwheel tyres 25.5×8.0–14 (16 ply); nosewheel 14×5.5–6 (8 ply).

POWER PLANT: One General Electric/Volvo Flygmotor RM12 (F404-GE-400) turbofan, rated initially at approximately 54 kN (12,140 lb st) dry and 80.5 kN (18,100 lb st) with afterburning. RM12UP version, in Lot 3 aircraft, incorporates FADEC, improved flame holder and redesigned turbine. Fuel in integral tanks in fuselage and wings. Intertechnique fuel management system; Dowty fuel health monitoring system for emergency (leak/battle damage) conservancy management. Optional FRL telescopic, retractable, hydraulically actuated in-flight refuelling probe mounted in port engine air intake; available on export aircraft and fitted to Swedish single-seat aircraft from 39207 (106th) onwards. External fuel carriage option of 1,136 litre (300 US gallon; 250 Imp gallon) EDTL (export drop tank, large).

ACCOMMODATION: Pilot only in JAS 39A/C, on Martin-Baker Mk 10L zero/zero ejection seat. Hinged canopy (opening sideways to port) and one-piece windscreen by Lucas Aerospace. Two seats in tandem in JAS 39B/D; command sequence in two-seat aircraft ejects rear occupant first, simultaneously inflating an airbag between the two cockpits to protect the rear pilot from Perspex splinters.

SYSTEMS: Hymatic environmental control system for cockpit air conditioning, pressurisation and avionics cooling. Hughes-Treitler heat exchanger. Two hydraulic systems, with Dowty equipment and Abex pumps. Hamilton Sundstrand main electrical power generating system (40 kVA constant speed, constant frequency at 400 Hz) comprises an integrated drive generator, generator control unit and current transformer assembly. Lucas Aerospace auxiliary and emergency power system, comprising gearbox-mounted turbine, hydraulic pump and 10 kVA AC generator, to provide auxiliary electric and hydraulic power in event of engine or main generator failure. In emergency role, the turbine is driven by engine bleed or APU air; if this is not available, the stored energy mode, using thermal batteries, is

SWAF JAS 39 CONTRACTS

Batch	Qty	Type	Serials	Remarks
Prototypes (5)	5	JAS 39	39-1 to 39-5	Plus 39-6 (below)
Lot 1 (30)	29	JAS 39A	39101 to 39129	39130 completed as JAS 39B
	1	JAS 39B	39800	
Lot 2 (110)	76	JAS 39A	39131 to 39206	
	14	JAS 39B	39801 to 39814	
	1	JAS 39C	39207/39-6	Testbed
	19	JAS 39C	39208 to 39226	Interim standard
Lot 3 (64)	50	JAS 39C	39227 to 39276	Including export diversions, of which 39224 to '245 to Czech Republic as 9234 to 9245
	14	JAS 39D	39815 to 39828	Including export diversions, of which 39819 and '820 to Czech Republic as 9819 and 9820
Total	**209**			

SPK 39 reconnaissance pod mockup fitted on centreline pylon of Saab Gripen *(FMV)* ***NEW***/0577277

selected automatically. APU and air turbine starter for engine starting, cooling air and standby electrical power. Original APU was Microturbo TGA15-090; changed to TGA15-328 from 40th JAS 39A and retrofitted to all in early second batch; Hamilton Sundstrand APU from 106th production single-seat aircraft (39207) and will be retrofitted. Optional OBOGS on export aircraft. Lot 3 Gripens have single Ericsson Saab Avionics GECU general electronic control unit, replacing previous three controllers for air, fuel and hydraulic systems.

AVIONICS: Service entry with E11 avionics software; upgraded via E12, E12.5 and E14 to E15 in 2001 and E15.1 (full AMRAAM and additional air-to-ground modes) in 2002.

Comms: CelsiusTech dual VHF/UHF transceivers and IFF in early aircraft; Rohde & Schwarz Series 6000 in third production lot and last 20 Lot 2 aircraft; option for installation in entire fleet; Series 6000 is element of tactical radio system (TARAS), providing secure communications via Data Link 39. Export aircraft to have Avitronics (Grintek) GUS 1000 audio management system.

Radar: Ericsson/BAE PS-05/A multimode pulse Doppler target search and acquisition (lookdown/shootdown) radar (weight 156 kg; 344 lb). For fighter missions, system provides fast target acquisition at long range; search and multitarget track-while-scan; quick scanning and lock-on at short ranges; and automatic fire control for missiles and cannon. In attack and reconnaissance roles, operating functions are search against sea and ground targets; mapping, with normal and high resolution; and navigation. Upgrade with electronically scanned antenna is intended in 2010.

Flight: Ericsson SDS 80 central computing system (D80 computer, Pascal/D80 high-order language and programming support environment; upgraded D80E computer flown mid-1994 and introduced from 39108; D96 computer from 92nd single-seat aircraft, 39193); three MIL-STD-1553B databusses, one of which links flight data, navigation, flight control, engine control and main systems; Honeywell laser INS and radar altimeter; Nordmicro air data computer. BAE three-axis strapdown gyromagnetic unit provides standby attitude and heading information. Navigation data fusion in prospect via NINS (new integrated navigation system), under development by 1998; this to be followed by NILS (new integrated landing system) for autonomous Cat. I landing capability.

Instrumentation: Ericsson EP-17 electronic display system, incorporating Kaiser (Hughes in Lot 1) wide-angle HUD and using advanced diffraction optics to combine symbology and video images; display processor (replacing PP1 and PP2 in Lot 1) makes colour imagery possible; this facility is not required on Swedish Lot 2 aircraft), but colour capability introduced from 106th single-seat, 39207 (which will be regarded as sixth 'prototype' 39-6); three Ericsson CRT HDDs, each 120 × 150 mm (4¾ × 6 in), but 152 × 203 mm (6 × 8 in) MFID 68 active matrix LCDs in Lot 3. Left-hand (flight data) HDD normally replaces all conventional flight instruments; central display shows computer-generated map of area surrounding aircraft with tactical information superimposed; right-hand CRT is a multisensor display showing information on targets acquired by radar, FLIR and weapon sensors. Minimum of conventional analogue instruments for back-up only. Thales Guardian helmet-mounted display (HMD) undertook compatibility trials at Linköping in early 2001. BAE Systems and Cumulus of South Africa developing HMD for delivery with SAAF Gripens in 2007.

Mission: Rafael/Zeiss Optronik Litening targeting and navigation pod (Swedish designation PWS 39) under development in 2000 for carriage under starboard air intake trunk, forward of wing leading-edge, providing heat picture of target on right-hand HDD; first test flight, February 2003; trials complete by late 2004. IR-OTIS (a combined TRST and FLIR) under development by Saab Avionics for installation ahead of windscreen, slightly offset to port. CelsiusTech datalink shares information with up to four aircraft simultaneously. Contract for SEK600 million awarded to Saab in late 2001, for 680 kg (1,500 lb) Modular Airborne Reconnaissance System 39 (SPK 39) pods to be delivered from 2004 onwards and become operational in 2005 with F17 (four pods), F7 (two) and F21 (two).

Vinten Vicon 70 Srs 72C modular reconnaissance pod available for export aircraft.

Self-defence: CelsiusTech RWR in early aircraft replaced by second-generation (Gen 2) version in 2000. CelsiusTech countermeasures, including chaff/flare and jamming. Saab EWS 39 electronic warfare suite ordered 1999 to replace existing system in Swedish aircraft, and is similar to export equipment; includes BOL 500 (BO2D) towed radar decoy under port wing, two pylon-mounted BOP 402 (BOP B) dispensers, laser warning system and missile approach warning.

ARMAMENT: Internally mounted 27 mm Mauser BK27 automatic cannon in port side of lower front fuselage (shoulder station 4P) and two wingtip-mounted Rb74 (AIM-9L) Sidewinder IR AAMs (stations 1P and 1S) standard. (No internal gun in JAS 39B.) Six other external hardpoints (2P, 2S, 3P and 3S beneath wings, one on centreline and 4S below starboard air intake trunk). Maximum loads 110 kg (243 lb) each on 1P and 1S; 600 kg (1,323 lb) each on 2P and 2S; 1,300 kg (2,866 lb) each on 3P and 3S; 250 kg (551 lb) on 4S; and 1,100 kg (2,425 lb) on 5 (centreline). Weapons include short- and medium-range air-to-air missiles such as Rb74, MICA or Rb99 (AIM-120) AMRAAM; air-to-surface missiles such as Rb75 (Maverick); anti-shipping missiles such as Saab RBS 15F; DWS 39 (BK 90) munitions dispenser; KEPD 150 and KEPD 350 SOMs; conventional or retarded bombs; air-to-surface rockets; or external fuel tanks.

Swedish MoD contract of October 2001 covers integration of LGBs, of which GBU-10, -12 and -16 (and associated Mk 82, 83 and 84 bombs) cleared by mid-2004. AGM-65G Maverick trials in 2004; AGM-65H and K to follow, along with AIM-9L/M and AIM-120B/C5. IRIS-T AAM being integrated for 2004 IOC with Swedish AF. New weapons under consideration for integration include Brimstone and Meteor. MUPSOW standoff weapon and V-3E A-Darter AAM for South Africa. Agreement of June 2001 adds Rafael weapons to potential armoury, initially Python 4 AAM and Spice guided bomb. Saab Aerosystems commissioned in November 2003 to integrate Meteor AAM.

DIMENSIONS, EXTERNAL:

Wing span, incl missile rails	8.40 m (27 ft 6¾ in)
Length, excl pitot tube: JAS 39A	14.10 m (46 ft 3 in)
JAS 39B	14.755 m (48 ft 5 in)
Height overall	4.50 m (14 ft 9 in)
Wheel track	2.40 m (7 ft 10½ in)
Wheelbase: JAS 39A	5.20 m (17 ft 0¾ in)
JAS 39B	5.90 m (19 ft 4¼ in)

WEIGHTS AND LOADINGS:

Operating weight empty: JAS 39A	6,622 kg (14,600 lb)
JAS 39C	approx 6,820 kg (15,036 lb)
JAS 39B	8,000 kg (17,637 lb)
Internal fuel weight	2,268 kg (5,000 lb)
T-O weight, clean	approx 8,500 kg (18,740 lb)
Max T-O weight with external stores:	
R12 software	12,500 kg (27,557 lb)
R14 software	14,000 kg (30,864 lb)

PERFORMANCE:

Min flying speed	less than 100 kt (185 km/h; 115 mph)
Max level speed	supersonic at all altitudes
T-O and landing strip length	approx 800 m (2,625 ft)
Combat radius	approx 432 n miles (800 km; 497 miles)
g limit	+9

UPDATED

Switzerland

ACEAIR

ACEAIR SA

Via Cantonale 35D, CH-6928 Manno
Tel: (+41 91) 605 55 46
Fax: (+41 91) 605 55 19
e-mail: info@aceair.ch
Web(1): http://www.aceair.ch
Web(2): http://www.aeriks.com
Web(3): http://www.aeris.ch
CHAIRMAN: Ugo Wyss
GENERAL MANAGER AND CEO: Antonio Latella
CHIEF ENGINEER AND DESIGNER: Ing Igor Medici
PRODUCT MANAGER: Ing Massimo Stoppa

Aceair's unorthodox Aeriks A-200 kitbuilt sportplane was unable to make its debut as scheduled at the EAA's AirVenture at Oshkosh in July 2001, but made a successful first flight in May 2002. Kit production began in May 2003.

UPDATED

ACEAIR AERIKS A-200

TYPE: Tandem-seat sportplane kitbuilt.

PROGRAMME: Concept phase began June 1997; design started 1998; officially launched April 1999 after trials with two radio-controlled scale models; engineering phase, dynamic (radio-controlled) model flight tests and wind tunnel test activity completed October 1999. Full-scale engineering mockup shown at EAA AirVenture, Oshkosh, July 2000; detail design started two months later; prototype construction began mid-2001; debut planned for EAA AirVenture at Oshkosh July 2001, but failed to occur. Rolled out 21 April 2002 and first flight eventually made (HB-YKS) on 29 May 2002. Original name of Aeris was changed to Aeriks in July 2002 due to pre-existing trade name protection in US. Public

Aceair Aeris/Aeriks prototype ***NEW***/0567026

debut at EAS rally, Lodrino, 16–18 August 2002. Production of kits began in May 2003 following Aero 2003 show in Friedrichshafen. Prototype had flown more than 60 hours in 56 flights by March 2004; Swiss Experimental category type approval still awaited at that time

CUSTOMERS: More than 10 kits sold by December 2003, including three to Swiss customers.

COSTS: US$59,000 (2003).

DESIGN FEATURES: Three lifting surfaces aerodynamic configuration; single pusher engine. Straight-tapered, low-swept (4° at 25 per cent chord) main wing amidships, with 2° dihedral from roots, features natural laminar flow sections based on NASA NLF(1)-0215F; main wing twist (2° 30' washout from root chord) and planform are selected for stable low-speed characteristics. Foreplane aerofoil Wortmann FX-75-141 (4° sweepback at quarter-chord), 6° anhedral. Lifting surfaces relative setting angles optimised for low drag at normal cruising speed. Standard symmetrical sections on T-tail configured, 30° swept horizontal surface. Separate fins above and below fuselage (lower portions serve as over-rotation bumper and propeller strike protector); rudder in upper vertical fin.

Cockpit instruments of Aceair Aeriks ***NEW***/0567025

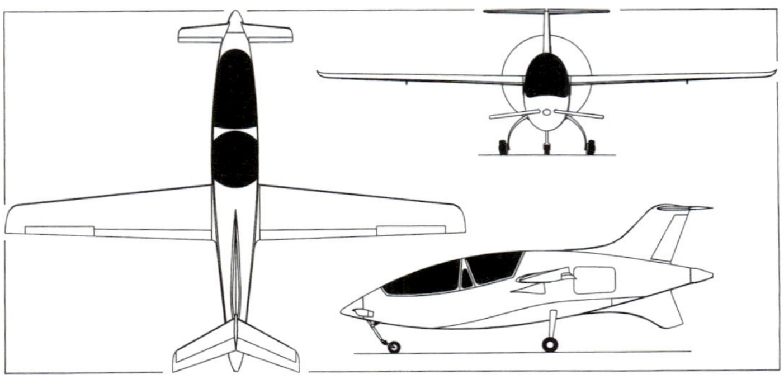

Aceair Aeriks (GIAE-110R engine) *(Paul Jackson)* 0533394

Quoted build time 1,000 to 1,500 hours, plus 200 to 400 hours for fully retractable landing gear option.

FLYING CONTROLS: Mechanical via push/pull rods for both ailerons and T-tail-mounted elevators; ⅛ in steel cables for rudder control. Foreplane flaps have electric trim. Electrically actuated single-slotted flaps on main wing, synchronised with plain flaps on foreplanes; all flaps can be partially deflected for T-O; landing flap setting is 40° on main wing, 20° on foreplane (relative to wing chord).

STRUCTURE: All-composites. Main wings detachable for storage and/or transportation; C-section carbon fibre main spars and rear auxiliary spars; 30 per cent chord single-slotted flap and 25 per cent chord aileron in each wing; glass fibre ribs. Monocoque fuselage. Primary structures built from wet laminated glass fabric or prepreg carbon fabric, laid-up on CNC milled female moulds; all shells of honeycomb core sandwich. Glass fibre flaps and control surfaces with carbon fibre reinforcement. Primary foreplanes and tail assembly of carbon fibre C-spars, composites ribs and glass fibre-based honeycomb core sandwich skins.

LANDING GEAR: Tricycle type. Electrically actuated nose gear retracts rearward into fuselage; non-retractable main gear units on cantilever self-sprung glass fibre struts with Cleveland heavy-duty disc brakes and wheels. Nosewheel steerable by differential braking. Hydraulic mainwheel brakes and parking brake.

POWER PLANT: One Diamond Engines GIAE-110R rotary engine (78.3 kW; 105 hp), driving a three-blade constant-speed MT-Propeller MTV-7-D/RD157-106 pusher propeller via an aluminium driveshaft. Maximum propeller speed 2,500 rpm. Engine is mounted far aft of passenger cabin and has two NACA-type scoop air inlets under main wingroot and ventral airscoop for main radiator cooling. Electrically actuated cowl flap integral with port-side NACA inlet. Fuel in single 110 litre (29.1 US gallon; 24.2 Imp gallon) integral fuselage tank aft of passenger seat.

ACCOMMODATION: Pilot and co-pilot or passenger in tandem in individual cockpits; dual sidestick controls. Separate one-piece flush-type canopies, hinged on starboard side. Baggage compartment, capacity 20 kg (44 lb), under passenger seat. Central carbon fibre roll bar for flip-over protection.

SYSTEMS: Single voltage (14 V DC) electrical system powered by two engine-driven generators to provide power for gear and flap extension/retraction; main bus powers flaps, trim, landing gear, main fuel pump and other flight-essential equipment.

AVIONICS: To customer's requirement. Can be equipped to full IFR standard.

EQUIPMENT: Anti-collision strobe light and navigation light in each wingtip; one white strobe light in tail tip.

DIMENSIONS, EXTERNAL:	
Wing span	8.00 m (26 ft 3 in)
Foreplane span	1.99 m (6 ft 6¼ in)
Wing aspect ratio	10.3
Foreplane aspect ratio	6.6
Length overall	6.50 m (21 ft 4 in)
Height over tailplane	2.465 m (8 ft 1 in)
Elevator span	2.20 m (7 ft 2½ in)
Wheel track	1.50 m (4 ft 11 in)
Wheelbase	2.72 m (8 ft 11 in)
Fuselage ground clearance at mainwheels	0.57 m (1 ft 10½ in)
Propeller diameter	1.57 m (5 ft 1¾ in)
Propeller ground clearance	0.565 m (1 ft 10¼ in)
DIMENSIONS, INTERNAL:	
Cockpits: Max combined length	2.75 m (9 ft 0¼ in)
Max width	0.74 m (2 ft 5¼ in)
Max height	1.00 m (3 ft 3¼ in)
AREAS:	
Wings, gross	6.20 m² (66.7 sq ft)
Foreplanes, gross	0.60 m² (6.46 sq ft)
Ailerons (total, incl tabs)	0.375 m² (4.04 sq ft)
Fins (total)	1.35 m² (14.53 sq ft)
Rudder	0.24 m² (2.58 sq ft)
Tailplane	1.06 m² (11.41 sq ft)
Elevators (total, incl tabs)	0.425 m² (4.57 sq ft)
WEIGHTS AND LOADINGS:	
Weight empty	400 kg (882 lb)
Fuel weight	82 kg (181 lb)
Payload with max fuel	200 kg (441 lb)
Max T-O weight	650 kg (1,433 lb)
Max wing/foreplane loading	95.59 kg/m² (19.58 lb/sq ft)
Max power loading	8.31 kg/kW (13.65 lb/hp)
PERFORMANCE (prototype, measured, March 2004):	
Never-exceed speed (VNE)	200 kt (370 km/h; 230 mph)
Max level speed at S/L	155 kt (287 km/h; 178 mph)
Manoeuvring speed (VA)	130 kt (241 km/h; 150 mph)
Max cruising speed at FL80	135 kt (250 km/h; 155 mph)
Stalling speed at S/L: flaps up	66 kt (122 km/h; 76 mph)
flaps down (two-seat)	60 kt (111 km/h; 69 mph)
flaps down (solo)	57 kt (104 km/h; 65 mph)
Max rate of climb: at S/L	296 m (970 ft)/min
at FL100	107 m (350 ft)/min
Service ceiling	3,660 m (12,000 ft)
T-O run at S/L	540 m (1,775 ft)
T-O to 15 m (50 ft) (JAR 23.59) at S/L	690 m (2,265 ft)
Landing from 15 m (50 ft) (JAR 23.75) at S/L	525 m (1,725 ft)
Landing run at S/L	285 m (935 ft)
Max range at FL80, no reserves	650 n miles (1,203 km; 748 miles)
Max endurance at FL80, no reserves	5 h
g limits	+4.4/−1.8

UPDATED

INTRACOM

INTRACOM AERO DESIGN GMBH

Chemin de Faguillon 1, CH-1223 Geneva, Switzerland
Tel: (+41 22) 786 27 57
Fax: (+41 22) 786 27 62
e-mail: intracom@bluewin.ch
Web: http://www.intracom.ch
PRESIDENT AND CEO: Nik Schmidt

Intracom Aero Design, formed in 1994, is the aircraft design and manufacturing arm of General Machinery SA, an international company with diverse interests in engineering, transport, travel, ecology and development of new technologies. It joined with the Khrunichev company of Russia to offer the GM-17 light transport in Western markets but, in November 2001, it formed a new alliance with Technoavia, and its associated SmAZ production plant, of the same country. The GM-17 had its public debut at MAKS '03, Moscow, in August 2003.

It is planned to produce a diverse family of closely related aircraft, including GM-11/12 turboprop trainer/light attack aircraft, GM-16 six-seat amphibian and GM-19 light utility turboprop. Priority is afforded to the GM-19, which will mark a break away from collaboration with Russia and is expected to be a wholly Western European venture.

VERIFIED

INTRACOM GM-17 VIPER

TYPE: Light utility turboprop.

PROGRAMME: Designed within Khrunichev bureau in Moscow by Evgeny P Grunin. Considerably modified Piper PA-31P Pressurised Navajo (PA-31T version also suitable for conversion.

Intracom GM-17 Viper initial preproduction aircraft *(Sebastian Zacharias)* ***NEW***/0580811

Intracom had previously proposed a collaborative project with Piper, to be known as PA-31TXL, but this was rejected by the US company.

First flight (GM-17-000, registered RA-01559; the former N38RG) 6 December 2000; 10 sorties and 6 hours flown by February 2001, when laid-up for static testing. Further two prototypes were to have followed in 2001, using Navajos G-OIEA and SE-IGB as basis; programme delayed by change of partner company to Technoavia (chief designer Vyatcheslav Kondratiev) in November 2001 and renaming of aircraft from Feniks (Phoenix) to Viper. Khrunichev prototype transferred to SmAZ at Smolensk for disassembly and rebuild; changes included replacement of winglets by tip tanks and revised nose contours; reflown January 2003 as GM-17-000-1; flew 60 hours before relegation to static testing in April 2003. Three PA-31P airframes (GM-17-001, 002 and 003) delivered to Technoavia in

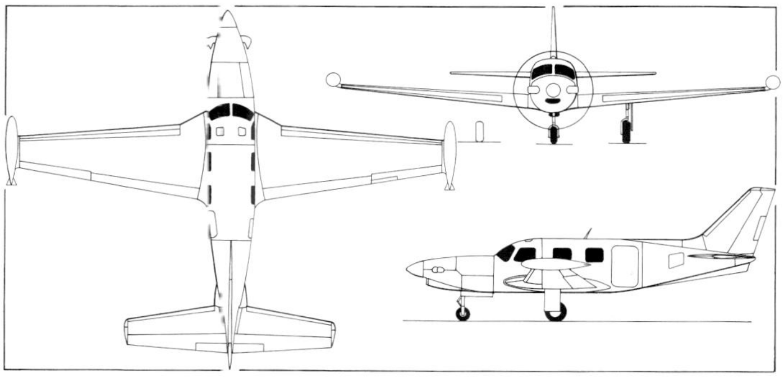

Intracom GM-17 Viper fitted with wing tanks *(Paul Jackson)* ***NEW***/0567290

November 2001 and entered modification at SmAZ; GM-17-001/RA-44471 flown March 2003 and accumulated 60 hours in first three months, Russian demonstrator and exhibition at Moscow in August 2003; GM-17-002 scheduled for August 2003 first flight and initial customer delivery in October 2003; GM-17-003 flown 2003 as European demonstrator, fitted with enlarged wingtip pods containing radar (starboard) and landing light. GM-17-004, final pre-series, for December 2003 first flight and February 2004 delivery.

Intended conversion of further 28 aircraft between 2004 and 2006, when new version of GM-17 will be available. All 32 initial aircraft to have Experimental category certification. However, in early 2004 Technoavia withdrew from the project because of a financial dispute and advised Rosaviakosmos to withdraw flight permits for both Viper prototypes, further claiming that the design is unable to meet FAR Pt 23/AP-23 certification standards because of age and lack of Piper support.

COSTS: US$860,000 (2003).

DESIGN FEATURES: Single-engined turboprop for passenger and light freight transport; suitable for executive ferry, medical evacuation, patrol, geological survey and training.

Low-mounted, tapered wing with wingtip tanks or winglets; mid-positioned, tapered tailplane and sweptback fin.

Wing section NACA 63_2415 at root; 63_1212 at tip; leading-edge sweepback 5°; thickness chord ratio 14 per cent at root, 12 per cent at tip; dihedral 5°; incidence 1° 30'; twist –2° 30'. Tailplane incidence 2°. wingroot nib sweepback 28°.

FLYING CONTROLS: Conventional and manual. Trim tab in starboard aileron, rudder and both elevators. Flap settings 20° for take-off; 40° for landing. Aileron deflections +24°/–14°; rudder ±35°; elevators +16°/–20°.

STRUCTURE: All retained airframe and landing gear components overhauled before reassembly. Revised structure includes fireproof steel bulkhead (Frame 3) behind engines; new engine cowlings forward of Frame 4; replacement wing panels in former engine positions; tailplane root fillets; new window frames; ventral fin; revised tailplane incidence (+2°, replacing –2°); composites fintip replaced fuel tanks and lines, oil system, electrical system (retaining some PA-31 parts), fire warning system and hydraulic system, de-icing system and oxygen system (reduced capacity).

LANDING GEAR: Tricycle type; hydraulically actuated; mainwheels retract inward; nosewheel rearward; single wheel on each unit. Nosewheel steerable ±28°; free castoring angle ±80°. Goodyear mainwheels and 6.50–10 (8-ply) tyres; Cleveland nosewheel with 6.00-6 (8-ply) tyres. Hydraulic brakes.

POWER PLANT: One 560 kW (757 shp) Walter M 601E turboprop, driving a five-blade V510 constant-speed, fully reversing propeller. (GM-17-000 and 001 have M 601D engine and V508E/99 three-blade propellers). Total fuel capacity 520 litres (137 US gallons; 114 Imp gallons) in wing tanks. Optional wingtip tanks of Piper design 125 litres (33.0 US gallons; 27.5 Imp gallons) each. Max fuel 960 litres (253 US gallons; 211 Imp gallons).

ACCOMMODATION: Two pilots and six passengers; forward four passengers in club arrangement; leather seats; lavatory and baggage area at rear. Externally accessed baggage compartment in nose. Single door, port, rear. Emergency exit, starboard. Cabin protected by up to 25 mm (1 in) of soundproofing material. Accommodation pressurised and air-conditioned. Emergency oxygen masks.

SYSTEMS: Hydraulic system, pressure 150 bar (2,175 lb/sq in); 27 V electrical system; emergency oxygen. De-icing system on all leading-edges; pneumatic on initial aircraft, but alcohol or electrical system for future installation.

AVIONICS: *Comms:* Garmin GNS 430 and GNS 530 VHF/NAV. Garmin GTX 327 transponder. C406-1 ELT.

Radar: Optional Bendix/King RDR 2000 weather radar in pod on starboard wing.

Flight: Bendix/King KR 87 radio compass, KI 228 ADF KRA 10, HSI 525, Sandel SN-3 colour map and S-Tec 55 autopilot.

EQUIPMENT: Four landing lights as standard; further optional light in wingtip tanks, if fitted.

DIMENSIONS, EXTERNAL:
Wing span over optional tanks ... 13.04 m (42 ft 9½ in)
Wing chord: at root ... 2.28 m (7 ft 5¾ in)
at tip ... 0.985 m (3 ft 2¾ in)
Wing aspect ratio ... 5.4
Length: overall ... 10.56 m (34 ft 7¾ in)
fuselage ... 9.64 m (31 ft 7½ in)
Fuselage: max width ... 1.44 m (4 ft 8¾ in)
max height ... 1.625 m (5 ft 4 in)
Height overall ... 4.075 m (13 ft 4½ in)
Tailplane span ... 6.045 m (19 ft 10 in)
Tailplane chord ... 1.155 m (3 ft 9½ in)
Wheel track ... 4.19 m (13 ft 9 in)
Wheelbase ... 2.63 m (8 ft 7½ in)
Propeller diameter ... 2.50 m (8 ft 2½ in)
Propeller ground clearance ... 0.30 m (11¾ in)
Passenger door: Height ... 1.32 m (4 ft 4 in)
Width ... 0.78 m (2 ft 6¾ in)
Emergency exit: Height ... 0.55 m (1 ft 9¾ in)
Width ... 0.41 m (1 ft 4¼ in)
Nose baggage door: Height ... 1.20 m (4 ft 8¾ in)
Width ... 0.70 m (2 ft 3½ in)

DIMENSIONS, INTERNAL:
Cabin (incl flight deck and baggage):
Length ... 4.40 m (14 ft 5¼ in)
Max width ... 1.32 m (4 ft 4 in)
Max height ... 1.32 m (4 ft 4 in)
Baggage hold volume: cabin ... 0.62 m^3 (22 cu ft)
nose ... 0.57 m^3 (20 cu ft)

AREAS:
Wings, gross ... 31.30 m^2 (336.9 sq ft)
Ailerons (total) ... 1.235 m^2 (13.29 sq ft)
Trailing-edge flaps (total) ... 3.135 m^2 (33.74 sq ft)
Winglets (total) ... 2.86 m^2 (30.78 sq ft)
Fin ... 2.60 m^2 (8 ft 6¼ in)
Rudder (incl tab) ... 0.865 m^2 (9.31 sq ft)
Tailplane ... 6.72 m^2 (72.33 sq ft)
Elevators (total) ... 2.45 m^2 (26.37 sq ft)

WEIGHTS AND LOADINGS:
Weight empty ... 1,756 kg (3,871 lb)
Max T-O and landing weight ... 3,200 kg (7,054 lb)
Max payload ... 1,000 kg (2,205 lb)
Max wing loading ... 102.2 kg/m^2 (20.94 lb/sq ft)
Max power loading ... 5.71 kg/kW (9.39 lb/shp)

PERFORMANCE (at max T-O weight except where indicated):
Max level speed ... 232 kt (430 km/h; 267 mph)
Normal cruising speed ... 221 kt (410 km/h; 254 mph)
Econ cruising speed ... 140 kt (260 km/h; 162 mph) IAS
Stalling speed, flaps down ... 67 kt (123 km/h; 77 mph)
Max rate of climb at S/L ... 426 m (1,398 ft)/min
Service ceiling ... 7,165 m (23,500 ft)
T-O run ... 620 m (2,035 ft)
Landing run with propeller reversal ... 340 m (750 ft)
Range with tip tanks, 40 min reserves:
high speed, low altitude ... 674 n miles (1,250 km; 776 miles)
at FL195, 205 kt
... 1,295 n miles (2,400 km; 1,491 miles)
ferry ... 1,598 n miles (2,960 km; 2,409 miles)
Endurance, absolute ... 7 h 50 min

UPDATED

INTRACOM GM-19

TYPE: Light utility turboprop.

PROGRAMME: Construction of prototype was to begin in 2004.

DESIGN FEATURES: Wings and empennage of Piper PA-31 married to new fuselage, but incorporating systems already employed in GM-17 Viper for simplicity, low risk and minimum cost. High-mounted, tapered wing; mid-mounted, tapered tailplane; and fin with large fillet.

FLYING CONTROLS: Conventional and manual. Trim tabs in elevator, starboard aileron and rudder.

STRUCTURE: Generally of metal.

POWER PLANT: One 560 kW (757 shp) Walter M 601E turboprop. Fuel capacity 1,220 litres (322 US gallons; 268 Imp gallons).

ACCOMMODATION: Crew door each side; main door to port, with emergency exit opposite.

SYSTEMS: Generally as for GM-17 Viper.

AVIONICS: Generally as for GM-17 Viper.

DIMENSIONS, EXTERNAL:
Wing span ... 13.92 m (45 ft 8 in)
Wing aspect ratio ... 7.8
Length overall ... 10.46 m (34 ft 3½ in)
Height overall ... 4.84 m (15 ft 10½ in)
Tailplane span ... 6.045 m (19 ft 10 in)
Wheel track ... 3.40 m (11 ft 1½ in)
Wheelbase ... 3.94 m (12 ft 11 in)
Propeller diameter ... 2.30 m (7 ft 6½ in)
Main door: Height ... 1.65 m (5 ft 5 in)
Width ... 1.20 m (3 ft 11¼ in)

DIMENSIONS, INTERNAL:
Cabin (incl cockpit): Length ... 5.20 m (17 ft 0¾ in)
Max width ... 1.30 m (4 ft 3¼ in)
Max height ... 1.65 m (5 ft 5 in)

AREAS:
Wings, gross ... 24.8 m^2 (266.9 sq ft)

WEIGHTS AND LOADINGS:
Weight empty ... 1,400 kg (3,086 lb)
Max payload ... 1,400 kg (3,086 lb)
Max fuel weight ... 976 kg (2,151 lb)
Max T-O weight ... 3,500 kg (7,716 lb)
Max wing loading ... 141.1 kg/m^2 (28.91 lb/sq ft)
Max power loading ... 6.20 kg/kW (10.19 lb/shp)

PERFORMANCE:
Max cruising speed ... 194 kt (360 km/h; 224 mph)

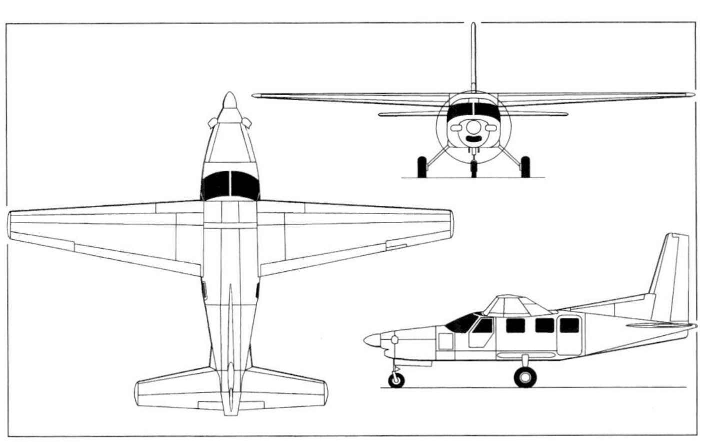

Provisional general arrangement of Intracom GM-19 *(Paul Jackson)* ***NEW***/0567296

Stalling speed, power off:
  flaps up .... 67 kt (123 km/h; 77 mph)
  flaps down .... 59 kt (109 km/h; 68 mph)
Max rate of climb at S/L .... 570 m (1,870 ft)/min
T-O run .... 330 m (1,085 ft)
Landing run with propeller reversal .... 230 m (755 ft)

Range:
  with max fuel .... 1,436 n miles (2,660 km; 1,652 miles)
  with max payload .... 944 n miles (1,750 km; 1,087 miles)
Endurance .... 8 h

UPDATED

LIGHTWING

LIGHT WING AG

Spichermattstrasse 14, CH-6370 Stans
Tel: (+41 41) 611 05 85
Fax:: (+41 41) 611 70 47
e-mail: info@lightwing.ch
Web: http://www.lightwing.ch
CEO: Marco Trüssel
TECHNICAL DIRECTOR: Alois Amstutz
DESIGNER: Hans Gygax

Having designed what is now marketed as the Ikarus C 42, Hans Gygax has produced a slightly larger, utility version. This is marketed by Light Wing AG, which was formed in September 2000 to seek launch funding for large-scale production in several applications, particularly glider towing. Manufacture is at Buochs aerodrome.

UPDATED

First production Lightwing AC4 on show at Aero '03 at Friedrichshafen in April 2003 *(Paul Jackson)* ***NEW***/0552057

Lightwing AC4 two-seat lightplane, baseline version ***NEW***/0587541

LIGHTWING AC4

TYPE: Two-seat lightplane.

PROGRAMME: Prototype first flew June 1998. First production aircraft shown, substantially complete, at Aero '03, Friedrichshafen, April 2003.

CURRENT VERSIONS: **AC4:** Baseline version; *as described*. Capabilities include touring and medical evacuation.

AC4 Schlepp: Glider tug; 93.2 kW (125 hp) turbocharged Rotax 914. Rate of climb, laden, 300 m (984 ft)/min.

AC4 SPC: US Sport Plane Class; otherwise as baseline AC4.

AC8: Light transport ("air van"); under development by 2003.

DESIGN FEATURES: Meets Swiss 'EcoLight' category requirements for 560 kg (1,234 lb) MTOW; certifiable to JAR requirements. High wing with V-strut bracing; tailplane braced to fin.

FLYING CONTROLS: Conventional and manual. Flaps. Flight-adjustable trim tab on port elevator.

STRUCTURE: Glass fibre fuselage built around full-length, large diameter, circular section metal boom. Metal tailplane, elevators, fin and rudder, with composites tips; metal wing, flaps and ailerons, with composites, upturned wingtips.

LANDING GEAR: Tricycle type; fixed. Fuselage-mounted spring cantilever mainwheel legs with wheel fairings; steerable nosewheel with wheel fairing.

POWER PLANT: One 73.5 kW (98.6 hp) Rotax 912 ULS flat-four driving a Warp Drive three-blade, carbon fibre, fixed-pitch propeller via 1:2.4 reduction gearing. Optional Neuform ground-adjustable propeller on first production aircraft. Fuel capacity 100 litres (26.4 US gallons; 22.0 Imp gallons).

ACCOMMODATION: Two persons, side-by-side; large stowage space behind seats. Broad, upward-hinged door each side.

EQUIPMENT: Optional retractable towing line in rear fuselage.

DIMENSIONS, EXTERNAL:
  Wing span .... 9.60 m (31 ft 6 in)
  Wing aspect ratio 7.3
  Length overall .... 6.95 m (22 ft 9½ in)
  Height overall .... 2.30 m (7 ft 6½ in)
DIMENSIONS, INTERNAL:
  Cabin: Length .... 1.90 m (6 ft 2¾ in)
    Max width .... 1.27 m (4 ft 2 in)

AREAS:
  Wings, gross .... 12.70 m² (136.7 sq ft)
WEIGHTS AND LOADINGS:
  Weight empty .... 300 kg (661 lb)
  Max T-O weight .... 560 kg (1,234 lb)
  Max wing loading .... 44.1 kg/m² (9.03 lb/sq ft)
  Max power loading .... 7.62 kg/kW (12.52 lb/hp)
PERFORMANCE:
  Never-exceed speed (VNE) .... 113 kt (210 km/h; 130 mph)
  Cruising speed at 75% power .... 100 kt (185 km/h; 115 mph)
  Stalling speed, power off, flaps down .... 34 kt (63 km/h; 40 mph)
  Max rate of climb at S/L .... 439 m (1,440 ft)/min
  T-O run .... 92 m (300 ft)
  Landing run .... 98 m (320 ft)
  Range with max fuel .... 810 n miles (1,500 km; 932 miles)
  g limts .... +4/–2

UPDATED

MSW

MSW AVIATION

Rigackerstrasse 24, CH-5610 Wohlen
Tel: (+41 56) 622 18 07
Fax: (+41 56) 611 00 55
e-mail: info@mswaviation.com
Web: http://www.mswaviation.com
PRESIDENT: Max Vogelsang

MSW Aviation was formed in 1991. Its latest product is the Votec 322 kitbuilt aerobatic two-seater. Construction assistance offered to builders at new 250 m² (2,691 sq ft) facility opened in October 2004 at Birrfeld airfield, near company headquarters. Similar facility available at Schwäbisch Hall in southern Germany.

UPDATED

MSW Votec 322 two-seat sportplane *(Paul Jackson)* ***NEW***/0589417

MSW VOTEC 322

TYPE: Tandem-seat sportplane kitbuilt.

PROGRAMME: Designed by Max Vogelsang; name indicates Vogelsang Technik, 320 hp, 2 seats. Inspired by, and derived from, DR 107 One Design single-seater, one of which acquired by MSW and flown as HB-YJM. Prototype Votec 322 (HB-YJY) made first flight 6 April 2001 and public debut at Friedrichshafen Air Show later that month. Certified by Swiss FOCA.

CUSTOMERS: Five sold and two flying by April 2003; third (HB-YLA) then complete. Customers in Austria, Germany and Switzerland.

COSTS: Kit €155,000 basic, excluding engine, propeller and instruments (2003).

DESIGN FEATURES: Low-wing monoplane with symmetrical aerofoil section and angular flying surfaces. Available only as a kit, but MSW provides factory-based builder's assistance programme.

FLYING CONTROLS: Conventional and manual. Near full-span ailerons; horn-balanced rudder and elevators; in-flight-adjustable trim tab in each elevator.

STRUCTURE: Steel tube fuselage main frame; wooden wings; carbon fibre fuselage skin and tail unit.

LANDING GEAR: Tailwheel type; fixed. Fuselage-mounted spring cantilever mainwheel legs with wheel fairings; ; steerable tailwheel. Mainwheel speed fairings.

POWER PLANT: One 246 kW (330 hp) Textron Lycoming AEIO-540 flat-six engine, driving an MT-Propeller MTV-9 three-blade constant-speed propeller. Fuel in integral 40 litre (10.6 US gallon; 8.8 Imp gallon) tank in each wingroot, plus 108 litre (28.5 US gallon; 23.8 Imp gallon) fuselage tank; total capacity thus 188 litres (49.7 US gallons; 41.4 Imp gallons).

ACCOMMODATION: Two persons in tandem under one-piece canopy which opens sideways to starboard.

AVIONICS: Described as 'glass cockpit' standard.

DIMENSIONS, EXTERNAL:
- Wing span .... 7.30 m (23 ft 11½ in)
- Length overall .... 6.50 m (21 ft 4 in)
- Height overall .... 2.50 m (8 ft 2½ in)

AREAS:
- Wing area .... 10.50 m² (113.0 sq ft)
- Ailerons (total) .... 1.50 m² (16.15 sq ft)

WEIGHTS AND LOADINGS:
- Weight empty .... 600 kg (1,323 lb)
- Max T-O weight .... 950 kg (2,094 lb)
- Max power loading .... 3.86 kg/kW (6.34 lb/hp)

PERFORMANCE:
- Never-exceed speed (VNE) .... 220 kt (407 km/h; 253 mph)
- Max cruising speed .... 174 kt (322 km/h; 200 mph)
- Stalling speed .... 55 kt (102 km/h; 64 mph)
- Max rate of climb at S/L .... 1,067 m (3,500 ft)/min
- Max rate of roll .... 420°/s
- T-O run .... 112 m (370 ft)
- Landing run .... 350 m (1,150 ft)
- Range .... 500 n miles (926 km; 575 miles)
- Endurance .... 3 h
- *g* limits .... ±10

UPDATED

PICCARD

20 Avenue de Florimont, CH-1006 Lausanne
Tel: (+41 21) 320 77 30
Fax: (+41 21) 320 77 00
e-mail (1): info@bertrandpiccard.com
e-mail (2): info@solar-impulse.com
Web: http://www.solar-impulse.com
PRESIDENT: Dr Bertrand Piccard

PROJECT SCIENTIFIC ADVISER:
Ecole Polytechnique Fédérale de Lausanne
Les Terraces 1-3, CH-1015 Lausanne
Tel: (+41 21) 693 22 22
Fax: (+41 21) 693 64 00
Web: http://www.epfl.ch

In 2003, Bertrand Piccard initiated the Solar Impulse project, the main objective of which is to make a round-the-world flight in an aircraft powered entirely by solar energy. An associate in the project is Brian Jones, Piccard's co-pilot in the 1999 round-the-world flight of the *Breitling Orbiter 3* balloon.

UPDATED

PICCARD SOLAR IMPULSE

TYPE: Technology demonstrator.

PROGRAMME: Announced 28 November 2003. First prototype to be designed and built 2004-05; first flight planned for early 2006; milestone of 36-hour flight, including complete night in the air, targeted for 2007. Second prototype to fly at end of that year; ultra-long-distance and duration flights planned from then on, leading to round-the-world attempt in 2009. Swiss Federal Institute for Technology (EFPL) will supply an engineer/pilot and act as scientific partner, following a feasibility study which began in March 2003.

In addition to achieving the flight itself, the venture aims to exploit and develop solar energy technology to make it possible for an aircraft to sustain very long-term flight without the use of any kind of embarked fossil fuel and without emitting any pollutants. It defines sustainable development as "the ability of present generations to provide for their own needs without jeopardising the ability of future generations to provide for theirs".

Computerised impression of Solar Impulse as envisaged in 2004 ***NEW***/0589242

DESIGN FEATURES: Published impressions indicate that the aircraft will have the general configuration of a high aspect ratio, T-tailed sailplane, with the entire wing upper surface covered in solar cells. Wing span is described as being longer than that of a Boeing 747.

UPDATED

PILATUS

PILATUS AIRCRAFT LTD

PO Box 992, CH-6371 Stans
Tel: (+41 41) 619 62 06
Fax: (+41 41) 610 92 30
e-mail: info@pilatus-aircraft.com
Web: http://www.pilatus-aircraft.com
CHAIRMAN: Peter Küpfer
PRESIDENT AND CEO: Oscar J Schwenk
VICE-PRESIDENTS:
Oskar Bründler (Controlling, Finance and Administration)
John Senior (R&D)
Jim Roche (Marketing and Sales)
Ignaz Gretener (General Aviation)
Marcus Kälin (Maintenance)
Max Zuberbühler (Logistics)
Hans Niederberger (Manufacturing)
Anton Waldispühl (Aircraft Assembly)

PC-12 MARKETING:
Pilatus Business Aircraft Ltd
Jefferson County Airport, 11755 Airport Way,
Broomfield, Colorado 80021, USA
Tel: (+1 303) 465 90 99
Fax: (+1 303) 465 91 90

Operators of Pilatus PC-6B2/H4s include French Army Aviation *(Paul Jackson)* ***NEW***/1047446

Web: http://www.pc12.com
PRESIDENT AND CEO: Thomas Bosshard

In December 1998, as part of a move to dispose of its aerospace activities, Oerlikon-Bührle (now renamed Unaxis) revealed that it was seeking a buyer for Pilatus. New owners were named on 1 March 2001 as a consortium comprising Jörg Burkart, IHAG Holding, Hoffman-LaRoche, Karl Nicklaus and Unaxis.

Current products include PC-6 Turbo Porter, PC-9M Turbo Trainer, PC-12 and PC-21. Headquarters and main plant at Stans, Switzerland. Wholly owned subsidiaries include Pilatus Business Aircraft Ltd in USA, Pilatus Australia Pty Ltd, in Australia and TSA Transairco and Altenrhein Aviation in Switzerland. There are also offices in UK, Malaysia and Abu Dhabi.

UPDATED

PILATUS PC-6 TURBO PORTER

US Army designations: UV-20A Chiricahua and AU-23A Peacemaker
Royal Thai Air Force designation: B.JH 2

TYPE: Light utility turboprop.

PROGRAMME: First flight piston-engined prototype 4 May 1959 (see contemporary *Jane's*); PC-6/A, A1, A2, B, B1, B2 and C2-H2 with various turboprops (see 1974-75 *Jane's All the World's Aircraft*); superseded by PC-6/B2-H4, introduced mid-1985. Production ended in 2000; c/n 939 departed to USA in October 2000.

Manufacture reinstated; c/n 940 first flew 12 May 2003 and exported to France for CERP de Maubeuge as F-GRUB.

Full description of PC-6/B2-H4 in 2001-02 All the World's Aircraft and in current Jane's Aircraft Upgrades; following is shortened version.

CURRENT VERSIONS: **PC-6/B2-H4:** Gross weight for FAR Pt 23 passenger carrying increased by 600 kg (1,323 lb) over immediately previous B2-H2, giving up to 570 kg (1,257 lb) greater payload for CAR 3 operations; changes include turned-up wingtips, enlarged dorsal fin, uprated mainwheel shock-absorbers, new tailwheel assembly and slight airframe reinforcement; H4 changes can be retrofitted to B2-H2.

CUSTOMERS: Total of 528 built by late 2004, including 45 of piston-engined versions and 92 of early turboprop versions under licence manufacture by Fairchild in USA; some 290 remain in service in more than 50 countries. Six built 2003-04 and delivered to CERP de Maubeuge, France; Jereb Airservice, Germany; Gardenia, Italy; Yasi Aviation, Indonesia (two); and CDP du Centre, Orleans, France. See tables.

DESIGN FEATURES: Rugged, light transport with STOL performance and easily accessible, versatile cabin. Braced, constant-chord, high wing with single strut each side; unswept fin with large dorsal fillet.

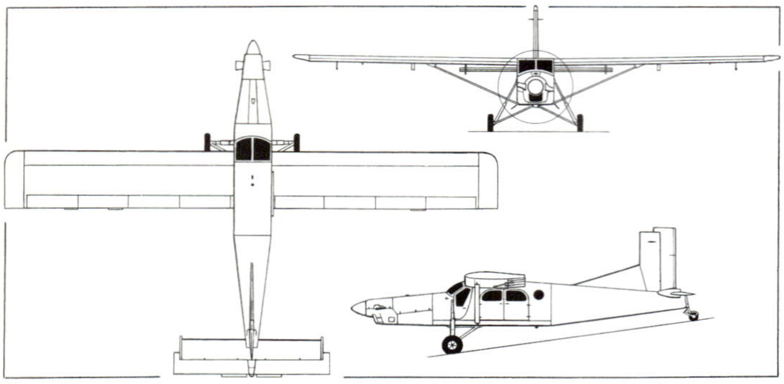

Pilatus PC-6/B2-H4 Turbo Porter with optional port-side double door *(Jane's/Dennis Punnett)* 0507227

PILTUS PC-6 ANNUAL PRODUCTION

		Swiss Prod	US Prod	Cum total
Year	**Qty**	**Cum**		
1959	4	4		4
1960	5	9		9
1961	22	31		31
1962	14	45		45
1963	14	59		59
1964	13	72		72
1965	23	95		95
1966	23	118	4	122
1967	35	153	22	179
1968	26	179	6	211
1969	27	206	8	246
1970	3	209	nil	249
1971	7	216	9	265
1972	5	221	15	285
1973	nil	221	5	290
1974	9	230	22	321
1975	15	245	1	337
1976	22	267		359
1977	17	284		376
1978	8	292		384
1979	5	297		389
1980	3	300		392
1981	5	305		397
1982	4	309		401
1983	9	318		410
1984	14	332		424
1985	2	334		426
1986	10	344		436
1987	5	349		441
1988	1	350		442
1989	7	357		449
1990	7	364		456
1991	4	368		460
1992	12	380		472
1993	12	392		484
1994	12	404		496
1995	1	405		497
1996	2	407		499
1997	3	410		502
1998	9	419		511
1999	6	425		517
2000	5	430		522
2003	1	431		523
2004	5	436		528
Totals		**430**	**92**	**528**

Wing section NACA 64-514 (constant) with span-increasing wingtips; dihedral 1°; incidence 2°.

LANDING GEAR: Non-retractable tailwheel type. Oleo shock-absorbers in all units. Steerable (±25°)/lockable tailwheel. Main tyres size 11.00–12 (8 ply), pressure 1.38 bar (20 lb/sq in), or 7.50–10 (8 ply). Tailwheel with size 5.00-4 (6 ply) tyre, pressure 3.24 bar (47 lb/sq in). Goodyear hydraulic disc brakes. Pilatus wheel/ski gear and (since July 1994) plain or amphibious floats optional.

POWER PLANT: One 507 kW (680 shp) Pratt & Whitney Canada PT6A-27 turboprop (flat rated at 410 kW; 550 shp at S/L), driving a Hartzell HC-B3TN-3D/T-10178 C or CH, or T10173 C or CH, three-blade constant-speed fully feathering reversible-pitch propeller with Beta mode control; four-blade propeller optional for operations in noise-sensitive areas. Standard fuel in integral wing tanks, usable capacity 644 litres (170 US gallons; 142 Imp gallons). Two underwing auxiliary tanks, each of 242.3 litres (64.0 US gallons; 53.3 Imp gallons), available optionally. For self-ferry, up to three 189 litre (50.0 US gallon; 41.6 Imp gallon) fuel tanks can be carried in main cabin.

ACCOMMODATION: Cabin has pilot's seat forward on port side, with one passenger seat alongside, and is normally fitted with six quickly removable seats, in pairs, to rear of these for additional passengers. Up to 11 persons, including pilot, can be carried in 2-3-3-3 high-density layout; or up to 10 parachutists; or two stretchers plus three attendants in ambulance configuration. Floor is level, flush with door sill, with seat rails. Forward-opening door beside each front seat. Large rearward-sliding door on each side of main cabin; fixed, external mounting step, starboard side. Optional double door, without central pillar, on port side. Hatch in floor 0.58 × 0.90 m (1 ft 10¾ in × 2 ft 11½ in), openable from inside cabin, for aerial camera or supply dropping. Starboard side door in rear fuselage permits stowage of six passenger seats or accommodation of freight items up to 5.0 m (16 ft 5 in) in length. Walls lined with lightweight soundproofing and heat insulation material. Adjustable heating and ventilation systems. Dual controls optional.

EQUIPMENT: Generally to customer's requirements, but can include gear for parajumping role; stretchers for ambulance role; aerial photography and survey gear. Agricultural version no longer produced.

DIMENSIONS, EXTERNAL:
- Wing span ..... 15.87 m (52 ft 0¾ in)
- Wing chord, constant ..... 1.90 m (6 ft 3 in)
- Wing aspect ratio 8.4
- Length overall ..... 10.95 m (35 ft 11 in)
- Height overall (to top of rudder), tail down ..... 3.20 m (10 ft 6 in)
- Wheel track ..... 3.00 m (9 ft 10 in)
- Wheelbase ..... 7.87 m (25 ft 10 in)
- Propeller diameter ..... 2.56 m (8 ft 5 in)

DIMENSIONS, INTERNAL:
- Cabin, from back of pilot's seat to rear wall:
  - Length ..... 2.30 m (7 ft 6½ in)
  - Max width ..... 1.16 m (3 ft 9½ in)
  - Max height (at front) ..... 1.28 m (4 ft 2½ in)
  - Height at rear wall ..... 1.18 m (3 ft 10½ in)
  - Floor area ..... 2.67 m^2 (28.6 sq ft)
  - Volume ..... 3.3 m^3 (107 cu ft)

AREAS:
- Wings, gross ..... 30.15 m^2 (324.5 sq ft)

WEIGHTS AND LOADINGS:
- Weight empty, equipped ..... 1,270 kg (2,800 lb)
- Max fuel weight: internal ..... 508 kg (1,120 lb)
  - underwing ..... 392 kg (864 lb)
- Max payload:
  - with reduced internal fuel ..... 1,130 kg (2,491 lb)
  - with max internal fuel ..... 1,062 kg (2,341 lb)
  - with max internal and underwing fuel ..... 571 kg (1,259 lb)
- Max T-O weight, Normal (CAR 3):
  - wheels (standard) ..... 2,800 kg (6,173 lb)
  - skis ..... 2,600 kg (5,732 lb)
- Max landing weight: wheels ..... 2,660 kg (5,864 lb)
  - skis ..... 2,600 kg (5,732 lb)
- Max cabin floor loading ..... 488 kg/m^2 (100 lb/sq ft)
- Max wing loading (Normal):
  - wheels ..... 92.9 kg/m^2 (19.03 lb/sq ft)
  - skis ..... 86.2 kg/m^2 (17.67 lb/sq ft)
- Max power loading (Normal):
  - wheels ..... 6.83 kg/kW (11.22 lb/shp)
  - skis ..... 6.34 kg/kW (10.42 lb/shp)

PERFORMANCE (Normal category):
- Never-exceed speed (VNE) ..... 151 kt (280 km/h; 174 mph) EAS
- Max cruising speed at S/L ..... 125 kt (232 km/h; 144 mph)
- Econ cruising speed at 3,050 m (10,000 ft) ..... 115 kt (213 km/h; 132 mph)
- Stalling speed, power off:
  - flaps up ..... 58 kt (108 km/h; 67 mph) EAS
  - flaps down ..... 52 kt (96 km/h; 60 mph) EAS
- Max rate of climb at S/L ..... 308 m (1,010 ft)/min
- Max operating altitude ..... 7,620 m (25,000 ft)
- Service ceiling ..... 6,250 m (20,500 ft)
- T-O run at S/L: normal and STOL ..... 197 m (650 ft)
- T-O to 15 m (50 ft) at S/L: normal ..... 475 m (1,560 ft)
  - STOL ..... 440 m (1,445 ft)
- Landing from 15 m (50 ft) at S/L at MLW, with propeller reversal ..... 315 m (1,035 ft)
- Landing run at S/L at MLW, with propeller reversal ..... 127 m (420 ft)
- Max range at 115 kt (213 km/h; 132 mph) at 3,050 m (10,000 ft), no reserves:
  - with max payload ..... 394 n miles (730 km; 453 miles)
  - with standard internal fuel ..... 500 n miles (926 km; 576 miles)
  - with standard internal and underwing fuel ..... 811 n miles (1,503 km; 933 miles)
- *g* limits ..... +3.58/−1.43

UPDATED

PILATUS PC-9M ADVANCED TURBO TRAINER

Slovene Military Aviation name: Hudournik (Swift)
Royal Thai Air Force designation: BF. 19

TYPE: Basic turboprop trainer.

PROGRAMME: Design began May 1982; aerodynamic elements tested on PC-7 1982–83; first flights by two preproduction PC-9s: HB-HPA on 7 May 1984 and HB-HPB on 20 July 1984; aerobatic certification 19 September 1985. PC-9M introduced early 1997.

CURRENT VERSIONS: **PC-9**: Standard production version until 1997.

PC-9/A: Australian version of PC-9.

PC-9B: German target towing version of PC-9, operated for German Air Force by Aircraft GmbH (previously Condor Flugdienst); increased fuel for a 3 hour 20 minute mission; two Meggitt RM-24 winches on inboard pylons with targets stowed aft of winch; TAS 06 acoustic scoring system.

PC-9 Mk II: Modified for US Air Force/Navy JPATS trainer programme; became Beech (T-6A Texan II).

PC-9M: Upgraded version introduced 1997 and now the standard production model. Enlarged dorsal fin fairing (as on PC-7 Mk II M) improves longitudinal stability and reduces

Pilatus PC-9B target facilities aircraft *(Paul Jackson)* ***NEW***/0576511

Slovene Pilatus PC-9M Hudournik *(Paul Jackson)* ***NEW***/0576510

stick forces; modified wingroot fairings improve low-speed characteristics; 'exciter strips' on wing leading-edges improve stall characteristics; new engine/propeller control system separates flight idle and ground idle modes, reducing acceleration time and improving handling.

Detailed description applies to PC-9M.

Hudournik (Swift): Dedicated weapons training version in service with Slovene Military Aviation, modified after delivery by Radom Aviation Systems of Israel. First example to this standard (L9-61) made first flight early May 1999. Mission systems installation includes Litton INS-100G inertial navigation with GPS; HOTAS controls; Flight Visions HUD with up-front control and rear cockpit video repeater; a central flight data recorder; chaff/flare dispenser and control panel; Honeywell EFIS and displays; new weapon selection and armament control unit and displays; MIL-STD-1553B dual databus; and RS-422/ARINC 429 datalink. Maximum external stores load increased to 1,250 kg (2,756 lb).

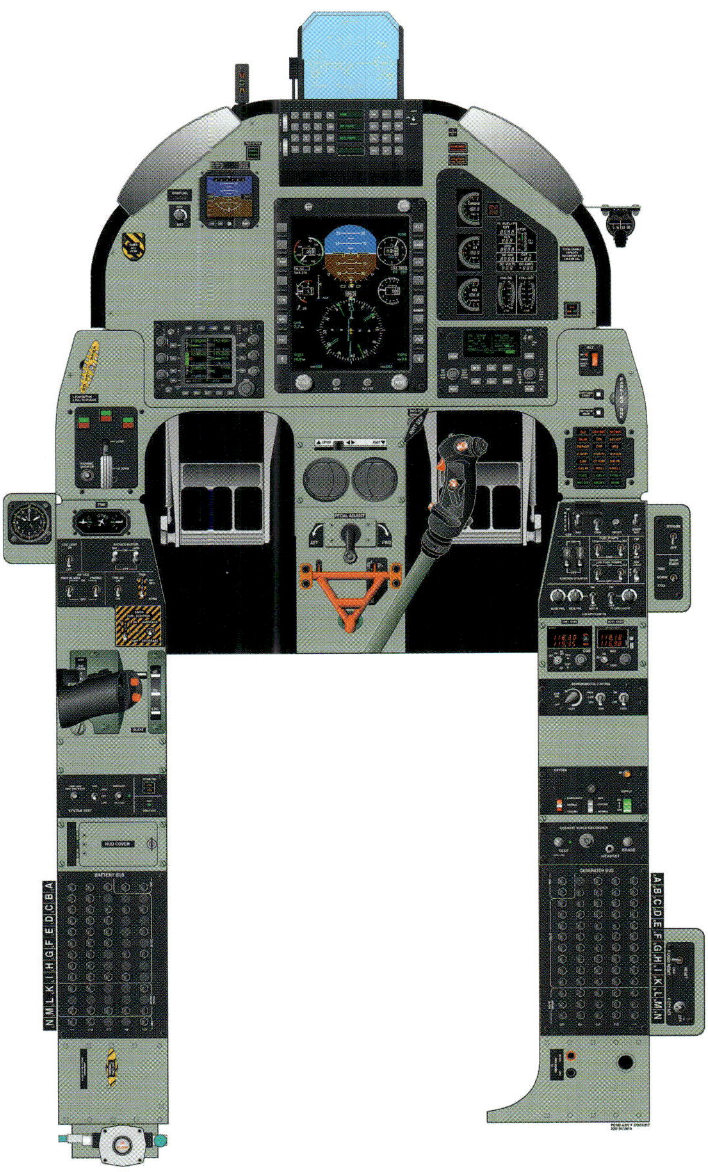

PC-9M front cockpit (IFR advanced avionics with HUD) ***NEW***/1111626

CUSTOMERS: Total of 273 ordered and 267 built by mid-2004. Saudi aircraft prepared for delivery by British Aerospace; first pair from second batch (Nos. 31 and 32) handed over in UK on 4 December 1995. Order for eight for Irish Air Corps announced 16 January 2003, for delivery and operational capability by mid-2004. First aircraft (260) flew 6 February 2004. Bulgarian Air Force ordered six, for delivery during 2004; first aircraft (663) flew on 20 September 2004.

DESIGN FEATURES: Typical turboprop trainer; stepped cockpits; aerobatic; performance and handling ease student's transition to jets. Meets FAR Pt 23 (Amendment 52) and special Swiss federal civil conditions for both Aerobatic and Utility categories; complies with selected parts of US military training specifications.

Conventional, low-wing monoplane; parallel chord wing centre-section; tapered outer panels; tailplane LERX.

Wing section PIL15M825 at root, IL12M850 at tip; quarter-chord sweepback 1°; dihedral 7° from centre-section; incidence 1° at root; twist −2°.

FLYING CONTROLS: Conventional and manual. Cable-operated elevator and rudder; ailerons operated by pushrods; mass-balanced, electrically actuated trim tab in each aileron and starboard half of elevator; electrically actuated trim/anti-balance tab in rudder controlled from rocker switch on power control lever. Trim aid device (TAD) automatically adjusts rudder trim tab setting in response to changes in engine torque or aircraft airspeed. This counteracts aircraft yaw induced by variations in effect of propeller slipstream, depending upon airspeed and engine power setting. TAD can be selected on or off by switch on trim control panel on left-hand side panel of front cockpit. Electrically operated split flaps, deflections 0°, 23° (take-off) and 50° (landing), hydraulically operated 'perforated' airbrake under centre-fuselage.

STRUCTURE: All-metal with some GFRP wing/fuselage fairings; one-piece wing with auxiliary spar, ribs and stringer-reinforced skin.

LANDING GEAR: Retractable tricycle type, with hydraulic actuation. Mainwheels retract inward to lie semi-recessed in wing centre-section covered by bulged doors; nosewheel retracts rearward; all units enclosed when retracted. Oleo-pneumatic shock-absorber in each leg. Hydraulically actuated nosewheel steering. Goodrich wheels and tubeless tyres, with

PILATUS PC-9 PRODUCTION (EXCLUDING BEECH T-6)

Customer	Version	Qty	First aircraft	First delivery
Military:				
Angola	PC-9	4	(c/n 115)	1987
Australia	PC-9/A	67[1]	A23-001	Oct 1987
Bulgaria	PC-9M	6	663	2004
Croatia	PC-9M	17[4]	054	1997
Cyprus	PC-9	2	901	1989
Iraq	PC-9	20	5801	1987
Ireland	PC-9M	8	260	2004
Myanmar	PC-9	10	3601	Apr 1986
Oman	PC-9M	12	426	1999
Saudi Arabia	PC-9	50	2201	Jan 1987
Slovenia	PC-9M	9[5]	L9-61	Nov 1998
Switzerland	PC-9	14[2]	C-401	1987
Thailand	PC-9	36	1/34	1991
US Army	PC-9	3[3]	91-071	1991
Subtotal		**258**		
Civil:				
Pilatus	PC-9	2	HB-HPA	
	PC-9 Mk II	1	HB-HPC	
	PC-9M	1	HB-HPJ	
Condor	PC-9B	10[6]	D-FAMT	Sep 1990
BAE	PC-9	1	ZG969	Jul 1989
Total		**273**		

[1] First 17 (after two from Pilatus) assembled by Hawker de Havilland from Swiss kits; final 48 locally built
[2] First two for evaluation; returned to Pilatus
[3] Transferred to Slovenia in 1995; now L9-51 et seq
[4] Plus three second-hand (051 et seq)
[5] Modified to Hudournik standard in Israel after delivery
[6] Plus ex-demonstrator, August 2000

Goodrich multipiston hydraulic disc brakes on mainwheels. Main tyres 20×4.4 (8 ply), nose tyre 16×4.4 (8 ply). Low-pressure tyres optional. Parking brake.

POWER PLANT: One 857 kW (1,150 shp) Pratt & Whitney Canada PT6A-62 turboprop, flat rated at 708 kW (950 shp), driving a Hartzell HC-D4N-ZA/09512A four-blade constant-speed fully feathering propeller. Single-lever engine control. Propeller is controlled by a constant-speed unit (CSU) and a variable-speed propeller governor system (VPGS). VPGS provides a linear engine power response resulting in balanced engine loads and reduced yaw moments. VPGS automatically varies the propeller speed between 1,750 and 2,000 rpm in response to engine torque. CSU adjusts the blade pitch to achieve a propeller speed as determined by the VPGS.

Fuel in two integral tanks in wing leading-edges, total capacity 540 litres (143 US gallons; 119 Imp gallons), of which 518 litres (137 US gallons; 114 Imp gallons) usable. Overwing refuelling point on each side. Fuel system includes 12 litre (3.2 US gallon; 2.6 Imp gallon) aerobatics tank in fuselage, forward of front cockpit, which permits up to 60 seconds of inverted flight. Provision for two 154 or 248 litre (40.7 or 65.5 US gallon; 33.9 or 54.5 Imp gallon) drop tanks on centre underwing attachment points. Total oil capacity 16 litres (4.2 US gallons; 3.5 Imp gallons).

ACCOMMODATION: Two Martin-Baker Mk 11A adjustable ejection seats, each with integrated personal survival pack and fighter-standard pilot equipment. Stepped tandem arrangement with rear seat elevated 15 cm (5.9 in). Seats operable, through canopy, at zero height and speeds down to 60 kt (112 km/h; 70 mph) EAS. Anti-*g* system. One-piece acrylic Perspex windscreen; one-piece framed canopy, incorporating roll-over bar, opens sideways to starboard. Dual controls standard. Cockpit heating, cooling, ventilation and canopy demisting standard. Space for 25 kg (55 lb) of baggage aft of seats, with external access.

SYSTEMS: Vapour cycle air conditioning system and engine bleed air for cockpit heating/ventilation and canopy demisting. Fairey Systems hydraulic system, pressure 207 bar (3,000 lb/sq in), for actuation of landing gear, mainwheel doors, nosewheel steering and airbrake; system maximum flow rate 18.8 litres (4.97 US gallons; 4.14 Imp gallons)/min. Bootstrap oil/oil reservoir, pressurised at 3.45 to 207 bar (50 to 3,000 lb/sq in). A supplementary high-pressure, one-shot nitrogen storage bottle provides power to lower the landing gear in an emergency. This bypasses the main hydraulic systems by feeding directly into shuttle valves in the down side of the main door and landing gear actuators.

Primary electrical system (28 V DC operational, 24 V nominal) powered by 30 V 300 A starter/generator and 24 V 40 Ah Ni/Cd battery. Ground power receptacle. Electric anti-icing of pitot tube, static ports and AoA transmitter standard; electric de-icing of propeller blades. Diluter demand oxygen system, capacity 1,350 litres (47.6 cu ft) selected and controlled individually from panel in each cockpit. OBOGS standard.

AVIONICS: Standard equipment can be augmented or replaced by alternative avionics to customer's requirements.

Comms: Single or dual VHF and/or UHF and/or radios. Radio management system controls audio services from com, nav and interphone systems. Advanced IFR option includes transponder and ELT.

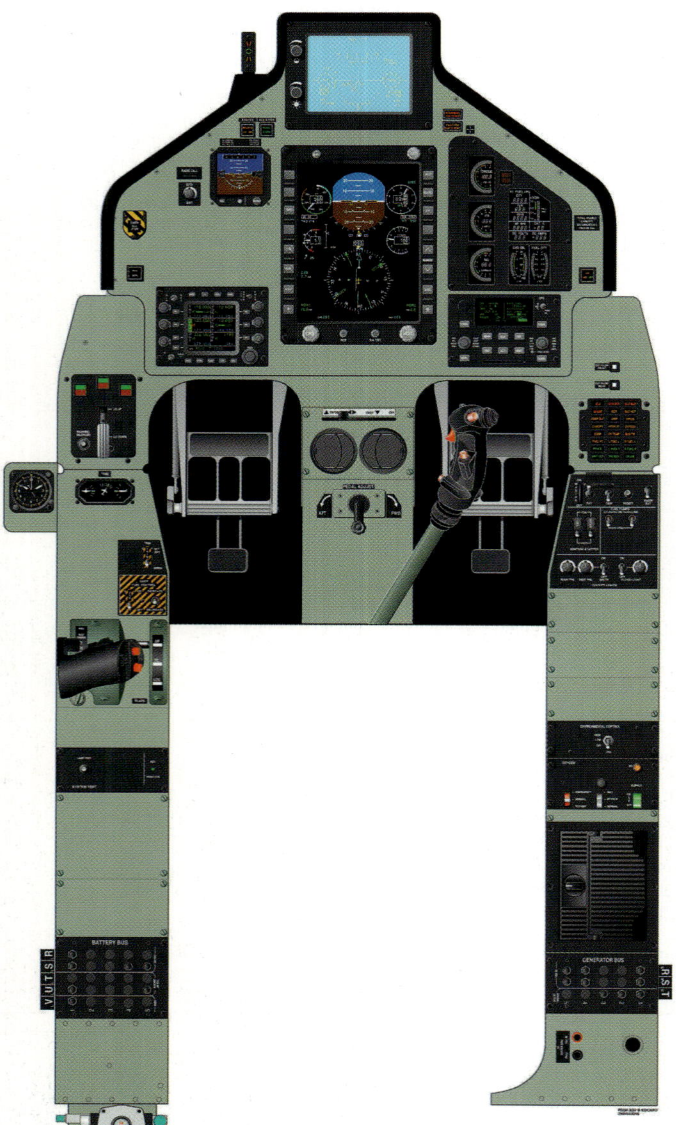

PC-9M rear cockpit ***NEW***/1111627

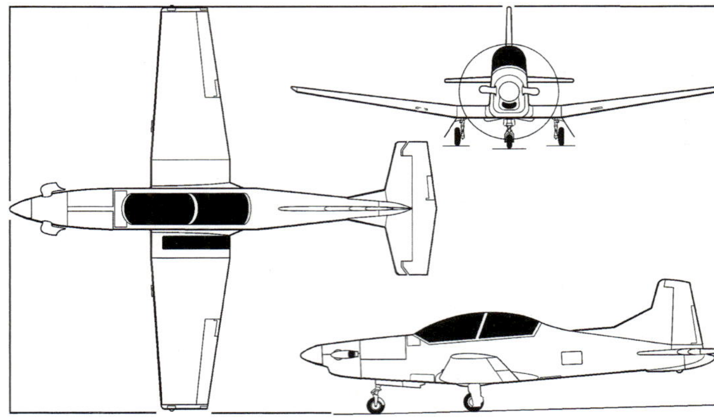

Pilatus PC-9M turboprop trainer ***NEW***/0567335

Flight: Advanced IFR option includes VHF Nav 1, VHF Nav 2, Tacan/DME, ADF, GPS, hybrid INS, air data computer and radar altimeter.

Instrumentation: Standard IFR installation includes Mach/airspeed indicators; VSIs; main (encoding) and standby altimeters; standby attitude indicator; standby magnetic compass; combined aileron/rudder/elevator trim indicators; accelerometer indicators; electric clocks; and engine indication system with LCD displays.

Advanced IFR configuration has CMC Electronics HUD with TEAC solid-state video recorder, Meggitt 152 × 203 mm (6 × 8 in) active matrix LCD flight display, rear cockpit repeater and video recorder.

Independent secondary flight display system in each cockpit presents essential flight and navigation data in case of primary equipment failure.

EQUIPMENT: Three hardpoints under each wing: inboard and centre stations each stressed for 250 kg (551 lb), outboard stations for 110 kg (242.5 lb) each. Retractable 250 W landing/taxying light in each main landing gear leg bay. Optional equipment can include smoke generators for aerial displays, target towing kit, electronic jammer for EW training, and blind-flying hood.

ARMAMENT: FN Herstal package for Hudournik/Swift comprises HMP-250 12.7 mm gun pods; LAU-7A seven-round unguided rocket launchers; Alkan Type 65 adaptor and dispensers for IBDU-33 practice bombs; and a laser range-finder pod.

DIMENSIONS, EXTERNAL:

Wing span	10.125 m (33 ft 2½ in)
Wing chord: mean aerodynamic	1.65 m (5 ft 5 in)
mean geometric	1.61 m (5 ft 3½ in)
Wing aspect ratio	6.3
Length overall	10.13 m (33 ft 3 in)
Fuselage max width	0.97 m (3 ft 2¼ in)
Height overall	3.26 m (10 ft 8⅓ in)
Tailplane span	3.665 m (12 ft 0¼ in)
Wheel track	2.54 m (8 ft 4 in)
Wheelbase	2.31 m (7 ft 7 in)
Propeller diameter	2.44 m (8 ft 0 in)
Propeller ground clearance	0.18 m (7 in)

DIMENSIONS, INTERNAL:

Baggage compartment volume	0.14 m^3 (4.9 cu ft)

AREAS:

Wings, gross	16.29 m^2 (175.3 sq ft)
Ailerons (total)	1.57 m^2 (16.90 sq ft)
Trailing-edge flaps (total)	1.77 m^2 (19.05 sq ft)
Airbrake	0.30 m^2 (3.23 sq ft)
Fin	0.86 m^2 (9.26 sq ft)
Rudder, incl tab	0.90 m^2 (9.69 sq ft)
Tailplane	1.80 m^2 (19.38 sq ft)
Elevator, incl tab	1.60 m^2 (17.22 sq ft)

WEIGHTS AND LOADINGS (A: Aerobatic, U: Utility category with underwing stores):

Basic weight empty, equipped	1,725 kg (3,803 lb)
Max underwing stores load	1,040 kg (2,293 lb)
Max T-O weight: A	2,350 kg (5,180 lb)
U	3,200 kg (7,054 lb)
Max ramp weight: A	2,360 kg (5,203 lb)
U	3,210 kg (7,076 lb)
Max landing weight: A	2,350 kg (5,180 lb)
U	3,100 kg (6,834 lb)
Max zero-fuel weight: A	2,000 kg (4,409 lb)
Max wing loading: A	144.3 kg/m^2 (29.55 lb/sq ft)
U	196.4 kg/m^2 (40.23 lb/sq ft)
Max power loading: A	3.32 kg/kW (5.45 lb/shp)
U	4.52 kg/kW (7.42 lb/shp)

PERFORMANCE (A and U as above, propeller speed 2,000 rpm):

Max operating Mach number (MMO): A, U	0.65
Max operating speed (VMO):	
A, U	320 kt; (593 km/h; 368 mph) EAS
Max cruising speed:	
A at S/L	271 kt (501 km/h; 311 mph)
A at 3,050 m (10,000 ft)	297 kt (550 km/h; 342 mph)
A at 7,620 m (25,000 ft)	300 kt (556 km/h; 345 mph)
Max manoeuvring speed:	
A	205 kt (380 km/h; 236 mph) EAS
U	200 kt (370 km/h; 230 mph) EAS
Max speed with flaps and/or landing gear down:	
A, U	150 kt (278 km/h; 172 mph) EAS
Stalling speed, engine idling:	
flaps and gear up:	
A	77 kt (143 km/h; 89 mph) EAS
U	90 kt (167 km/h; 104 mph) EAS
flaps and gear down:	
A	69 kt (128 km/h; 80 mph) EAS
U	80 kt (149 km/h; 93 mph) EAS
Max rate of climb at S/L: A	1,247 m (4,090 ft)/min
Time to 4,575 m (15,000 ft): A	4 min 5 s

Max operating altitude: A, U .... 7,620 m (25,000 ft)
Service ceiling: A, U .... 11,580 m (38,000 ft)
T-O run at S/L: A .... 243 m (795 ft)
T-O to 15 m (50 ft) at S/L: A .... 390 m (1,280 ft)
Landing from 15 m (50 ft) at S/L at MLW:
A .... 700 m (2,295 ft)
Landing run at S/L at MLW:
A (normal braking act on) .... 351 m (1,150 ft)
Max range at 210 kt (389 km/h; 242 mph) at 7,620 m (25,000 ft), 5% and 20 min fuel reserves:
A .... 830 n miles (1,537 km; 955 miles)
Max range, no reserves .... 1,065 n miles (1,970 km; 1,225 miles)
Max endurance at 110 kt (204 km/h; 127 mph) IAS, conditions as above: A .. 4 h 30 min
g limits: A .... +7/–3.5
U .... +4.5/–2.25

UPDATED

PILATUS PC-21

TYPE: Basic turboprop trainer.

PROGRAMME: Project launched November 1998; development began January 1999; provisional designation chosen to indicate 21st Century technology, but was retained for commercial use; first brief details revealed late 1999; modified PC-7 Mk II development aircraft (HB-HMS) flying by October 2000 with a sweptback fin, spoilers, power management system and five-blade propeller; later received PC-21 avionics and mission management system.

Prototype (HB-HZA) rolled out 1 May 2002; first flight 1 July 2002; public debut at Royal International Air Tattoo, Fairford, UK, 20–21 July 2002, at which time the aircraft had flown some 14 hours, including ferry time; more than 500 hours flown by October 2004; appeared at Farnborough Air Show from 22 July 2002. Second development aircraft (HB-HZB) flew 7 June 2004 and made first public appearance at RIAT, Fairford, UK, July 2004. Civil VFR certification achieved December 2004, followed by IFR in late 2005. Production to start in late 2005.

CUSTOMERS: Estimated market for 1,000 aircraft in class over 20-year period, of which Pilatus hopes to capture 50 per cent; Australia, Singapore, UAE and UK seen as primary marketing targets for launch orders. Marketing strategy includes complete Pilatus-operated training packages, employing PC-21s.

COSTS: Company-funded development cost estimated at SFr200 million; unit cost SFr10 million (2004).

DESIGN FEATURES: Optimised for advanced flying (pilot) and weapons training (pilot and WSO), but at turboprop cost; high wing loading. Avionics able to simulate aiming and release of multiple weapons (in excess of aircraft's actual carrying capability) without resort to training rounds, although addition of underwing stores pylons is possible. Jet-like response assisted by digital power management system which schedules full power at 200 kt (370 km/h; 230 mph) and above. Propeller and spinner angled 4° down and 4° starboard to offset full torque, while wing and fin slightly offset for similar reason; optional automatic yaw compensation. Design aims included superior aerodynamic performance; integrated and cost-effective training system; and (30-year) life-cycle support cost not exceeding current turboprop trainers.

Wing leading-edge sweepback 12° 42'; tailplane sweepback approximately 16°; dihedral from roots.

FLYING CONTROLS: Conventional. Manually actuated rudder and elevator; hydraulic ailerons. Door-type spoiler in each wing upper surface, ahead of outboard one-third of flaps, to enhance roll. Flight-adjustable trim tabs in rudder, each aileron and each half of tailplane. Flaps hydraulically actuated.

STRUCTURE: New design; three-spar wing; wing leading-edges of impact-absorbent material for birdstrike resistance.

LANDING GEAR: Retractable tricycle type, with single wheel on each unit. Retraction inwards (mains) and rearward (nose). Mainwheel tyres size 20×14.4R12, pressure 16.6 bar (240 lb/sq in); nosewheel tyre size 10×4.4R8, pressure 11.0 bar (160 lb/sq in).

POWER PLANT: One 1,193 kW (1,600 shp) Pratt & Whitney Canada PT6A-68B turboprop, driving a Hartzell E8991KX five-blade graphite/titanium propeller.

ACCOMMODATION: Two in tandem on Martin-Baker Mk 16L zero zero ejection seats in pressurised cockpit. Two-section acrylic canopy manufactured by Mecaplex of Switzerland, with thickened front for birdstrike resistance without resort to separate windscreen. Rear seat height increased in comparison with PC-9; front occupant has –11° forward view; rear occupant –4° 40'.

SYSTEMS: OBOGS, ECS and VCCS standard.

Prototype PC-21 over the Alps ***NEW***/0552741

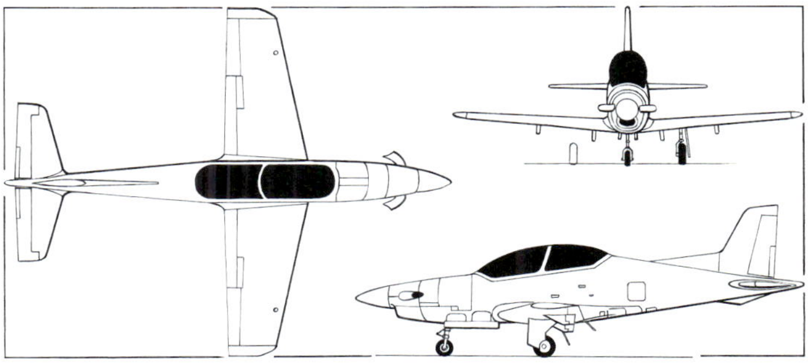

Pilatus PC-21 advanced turboprop trainer *(James Goulding)* 0536697

AVIONICS: *Instrumentation:* 'Glass cockpit' featuring three 152 × 203 mm (6 × 8 in) AMLCD PFDs and two 76 mm (3 in) Meggitt AMLCD secondary flight displays in each cockpit; Flight Visions SparrowHawk HUD with FVD-4000 HUD symbol generator standard in front cockpit; HUD repeater in rear cockpit; displays are NVG-compatible. Digital recording of all displays for post-flight debriefing.

Mission: Embedded simulation and mission planning/debriefing. Integrated ground-based training environment.

ARMAMENT: Provision for external stores on one centreline and four underwing hardpoints.

DIMENSIONS, EXTERNAL:
Wing span .... 8.765 m (28 ft 9 in)
Length overall .... 11.19 m (36 ft 8½ in)
Height overall .... 3.915 m (12 ft 10 in)
Fuselage: Max depth .... 2.015 m (6 ft 7¼ in)
Max width .... 1.00 m (3 ft 3¼ in)
Tailplane span .... 4.00 m (13 ft 1½ in)
Wheel track .... 2.735 m (8 ft 11¾ in)
Wheelbase .... 2.495 m (8 ft 2¼ in)

WEIGHT AND LOADINGS:
Weight empty .... 2,250 kg (4,960 lb)
T-O weight: Aerobatic (clean) .... 3,100 kg (6,834 lb)
with underwing tanks .... 3,600 kg (7,936 lb)
max .... 4,250 kg (9,370 lb)
Max external load .... 1,150 kg (2,535 lb)
Max power loading .... 3.56 kg/kW (5.86 lb/shp)

PERFORMANCE (estimated):
Max operating speed .... (370 kt; 685 km/h; 426 mph)
Design diving speed .... (420 kt; 778 km/h; 483 mph)
Max level speed at FL100 .... 340 kt (630 km/h; 391 mph)
Stalling speed, flaps and gear down .... 80 kt (148 km/h; 92 mph)
Max certified altitude .... 7,620 m (25,000 ft)
Service ceiling .... 11,582 m (38,000 ft)
g limit .... +8

UPDATED

First and second (nearest) PC-21 prototypes in formation ***NEW***/1111625

PILATUS PC-12

TYPE: Light utility turboprop.

PROGRAMME: Announced at NBAA Convention October 1989; first flight P.01 (HB-FOA) 31 May 1991; first flight of P.02 second prototype (HB-FOB) 28 May 1993; Swiss certification to FAR Pt 23 Amendment 42 (covering FAR Pt 135 commercial and Pt 91 general operations) received 30 March 1994, FAA type approval 15 July 1994, and FAR Pt 25 certification for flight into known icing conditions in 1995. Deliveries (N312BC to Carlston Leasing Corporation in USA) began September 1994.

Higher gross weight option (4.5 tonnes, hence **PC-12/45**) introduced in 1996 and gained FAA certification on 31 July 1996; became production standard by 1997. FAA FAR Pt 135 approval for commercial single-engined IFR operations announced in third quarter 1997; production increased from three to four per month in August 1997. First airline scheduled service operator (1997) was Kelner Airways of Canada. Total of 76 ordered and 51 delivered in 1998 (including 100th in April); 29 ordered in first half of 1999; 200th sale announced at NBAA Convention. Worldwide fleet hours passed 675,000 in October 2003. Now certified in 20 countries. 1,500th Pilatus single-engined turboprop, manufactured in mid-2001, was PC-12 N377PC.

CURRENT VERSIONS: **Standard:** Nine-passenger commuter or passenger/cargo combi.

Detailed description applies to standard PC-12/45 except where indicated.

Executive: Six to eight passenger seats.

PC-12M and Spectre: Multi-mission versions; *described separately*.

CUSTOMERS: Completion of 300th production aircraft effected in 2001; 400th in (N500ZP) in August 2003; and 500th in 2004. Deliveries in 2000 and 2001 totalled 70 each, 45 in 2002, and 55 in 2003. Some 59 per cent of first 400 aircraft are registered in USA; further 14 per cent in Canada (including nine with Royal Canadian Mounted Police). Sales total 18 per cent in air ambulance configuration; 8 per cent airline use; and 74 private/corporate operators. US Drug Enforcement Agency (DEA) received two in transport configuration in December 1999 and October 2000, and placed further orders for ICE surveillance version. Other customers include those in Argentina (Border Guard), Australia (Royal Flying Doctor Service with 17 aircraft, including three delivered in early 2002), Austria, Bangladesh, Belize, Bermuda, Brazil, Canada, Denmark, France, Germany, Greece, India, Japan, Kenya, Mexico, Netherlands, Norway, Poland, Russia, South Africa (Red Cross), Switzerland, USA and Zimbabwe. Recent customers include the Bulgarian Air Force, which ordered one PC-12M for delivery in 2004.

First single-engine turboprop fractional ownership scheme involves PC-12s of Alpha Flying's PlaneSense programme. Similar scheme, based on hourly use, introduced 2001 by Lions Air TimeJet of Zurich, Switzerland.

COSTS: Basic price US$2,712,760 (2002). Direct operating cost US$359.29 per hour (mid-2003).

DESIGN FEATURES: Low-wing monoplane with tapered wing and T tail; latter's mounting reduces trim changes with power; CG range 13 to 46 per cent of MAC. Modifications following early flight trials include introduction of winglets, increased wing span, paired elevators instead of single surface, sweptback tailplane/elevator tips, addition of tailplane/fin bullet fairing, and enlarged dorsal fin and ventral strakes.

Wing sections (modified NASA GA(W)-1 series), LS(1)-0417-MOD at root and LS(1)-0313 at tip.

FLYING CONTROLS: Conventional and manual. Actuation by pushrods and cables; servo tab in each aileron; electrically actuated Flettner tab in rudder; short-span mass-balanced ailerons; electrically actuated Fowler flaps cover 67 per cent of wing trailing-edge; electrically actuated (dual redundant) variable incidence tailplane.

STRUCTURE: Aluminium alloy primary skin and structure; composites for ventral strakes and dorsal fin (Kevlar/honeycomb sandwich), wingtips (glass fibre), fairings, engine cowling (glass fibre/honeycomb sandwich) and interior trim; titanium firewall; two-spar wing with integral fuel tankage; airframe proved for 20,000 hour life; complete internal and external corrosion protection.

All parts fabricated in Switzerland, but wings and fuselage assembled by OGMA in Portugal before return to Swiss production line. North American interiors installed in USA finishing centre; others outfitted at Stans.

LANDING GEAR: Pilatus hydraulically retractable tricycle type, with single wheel on each unit; nosewheel mechanically steerable ±12° by rudder pedals, or ±60° with differential braking. Emergency extension system; toe-operated brakes; parking brake. Suitable for operation from grass strips. Goodrich wheels and low-pressure tyres on all units: size 22×8.50-10 on main gear, 17.5×6.25-6 on nose unit; tyre pressure 4.14 bar (60 lb/sq in) all round. Propeller ground clearance maintained with nose leg compressed and nosewheel tyre flat. Trailing-link main gear retracts inward into wings, nose gear rearward under flight deck. Ground turning radius about wingtip 9.8 m (32 ft 2 in), about main gear 4.53 m (14 ft 10 in).

POWER PLANT: One 1,197 kW (1,605 shp) Pratt & Whitney Canada PT6A-67B turboprop, flat rated to 895 kW (1,200 shp) for T-O and 746 kW (1,000 shp) for climb and cruise. Hartzell HC-E4A-3D/E10477K constant-speed, fully feathering reversible-pitch four-blade aluminium propeller, turning at 1,700 rpm. Two integral fuel tanks in wings, total capacity 1,540 litres (407 US gallons; 339 Imp gallons), of which 1,522 litres (402 US gallons; 335 Imp gallons) are usable. Gravity fuelling point in top of each wing. Oil capacity 11 litres (2.9 US gallons; 2.4 Imp gallons).

Current PC-12 flight deck, with optional co-pilot's EFIS ***NEW***/1111620

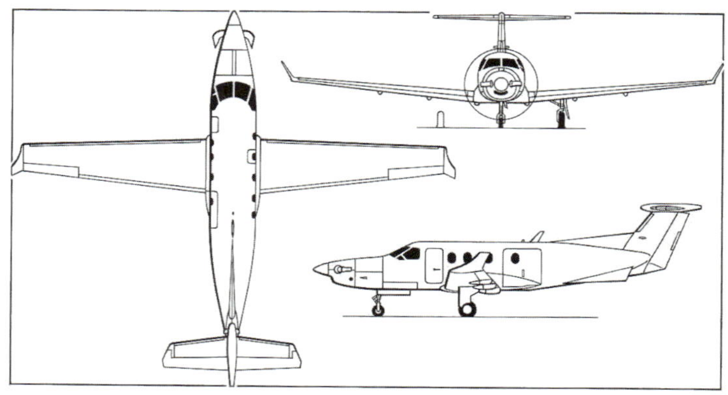

Pilatus PC-12/45 pressurised light utility and business transport ***(Jane's/Mike Keep)*** 0507439

ACCOMMODATION: Two-seat flight deck: approved for single pilot, with dual controls; second flight instrument panel optional. Limit of nine passengers under FAR Pt 23, or executive layout for six to eight, both with lavatory. Two or three stretcher patients, plus life support systems and medical attendants, in ambulance configuration. Downward-opening airstair crew/passenger door at front, upward-opening cargo door at rear, both on port side; Type III emergency exit above wing on starboard side.

SYSTEMS: Normalair Garrett engine bleed air ECS, maximum pressure differential 0.4 bar (5.8 lb/sq in), maintaining a 2,440 m (8,000 ft) cabin altitude at 7,620 m (25,000 ft) and a 3,050 m (10,000 ft) cabin altitude at 9,150 m (30,000 ft). Vickers Systems (Germany) hydraulic system, pressure 207 bar (3,000 lb/sq in), for landing gear actuation. Electrical power system (28 V DC) supplied by 300 A engine-driven starter/generator, 130 A back-up alternator and 24 V 40 Ah Ni/Cd or 24 V 42 Ah lead-acid battery. Second battery optional. Goodrich pneumatic boot de-icing of wing and tailplane leading-edges; Goodrich electric anti-icing of windscreen, propeller blades, pitot probes and stall warning sensor; exhaust air anti-icing of engine air intake. Oxygen system for crew and passengers.

AVIONICS: *Comms:* Bendix/King dual KX 165A VHF nav/com transceivers, KMA 24H audio control/voice activated intercom system, KT 70 Mode S transponder, KR 21 marker beacon receiver, and SEPRE-IESM kannad 121 ELT, all standard; Bendix/King KHF 950 HF radio optional.

Pilatus PC-12 in 'night sky' colour scheme ***NEW***/1111623

PC-12 (Spectre) ICE surveillance aircraft *(US Bureau of Immigration & Customs Enforcement)* ***NEW***/1047204

Radar: Honeywell RDR 2000 weather radar standard, mounted in wing pod.

Flight: Honeywell KFC 325 autopilot system, KN 63 DME, KR 87 ADF, KNI 582 RMI, KRA 405 radar altimeter, KEA 130A encoding altimeter, Bendix/King KLN 90B GPS and Litef LCR-92 AHRS standard; second LCR-92, WX 1000E Stormscope, TCAS 66A, Mk VI GPWS, KMD 850 MFD, IHAS 8000 traffic and terrain awareness system, Max-Viz EVS-1000 enhanced vision system, and co-pilot's EFIS, all optional.

Instrumentation: Bendix/King EHI 40/50 EFIS with 102 mm (4 in) display standard; 127 mm (5 in) EFIS on pilot and co-pilot panels, optional. Central Advisory and Warning System (CAWS) standard.

DIMENSIONS, EXTERNAL:
Wing span ... 16.23 m (53 ft 3 in)
Wing aspect ratio ... 10.2
Length overall ... 14.40 m (47 ft 3 in)
Height overall ... 4.27 m (14 ft 0 in)
Elevator span ... 5.21 m (17 ft 1 in)
Wheel track ... 4.52 m (14 ft 10 in)
Wheelbase ... 3.53 m (11 ft 7 in)
Propeller diameter ... 2.67 m (8 ft 9 in)
Propeller ground clearance ... 0.32 m (1 ft 0½ in)
Passenger door: Height ... 1.35 m (4 ft 5 in)
Width ... 0.635 m (2 ft 1 in)
Cargo door: Height ... 1.32 m (4 ft 4 in)
Width ... 1.35 m (4 ft 5 in)
Emergency exit: Height ... 0.66 m (2 ft 2 in)
Width ... 0.48 m (1 ft 7 in)

DIMENSIONS, INTERNAL:
Cabin: Length, excl flight deck ... 5.16 m (16 ft 11 in)
Max width ... 1.52 m (5 ft 0 in)
Width at floor ... 1.295 m (4 ft 3 in)
Max height ... 1.45 m (4 ft 9 in)
Volume ... 9.34 m³ (330 cu ft)
Baggage compartment volume ... 1.13 m³ (40 cu ft)

AREAS:
Wings, gross ... 25.81 m² (277.8 sq ft)

WEIGHTS AND LOADINGS
Weight empty: commuter/airliner ... 2,600 kg (5,732 lb)
Basic operating weight: single pilot ... 2,690 kg (5,930 lb)
executive ... 2,903 kg (6,400 lb)
Combi ... 2,536 kg (5,591 lb)
Max usable fuel ... 1,226 kg (2,703 lb)
Max payload: commuter/airliner ... 1,410 kg (3,108 lb)
Executive ... 1,243 kg (2,740 lb)
Combi ... 1,474 kg (3,250 lb)
Payload with max fuel: executive ... 388 kg (855 lb)
Max ramp weight ... 4,520 kg (9,965 lb)
Max T-O and landing weight ... 4,500 kg (9,920 lb)
Max zero-fuel weight ... 4,100 kg (9,040 lb)
Max wing loading ... 174.3 kg/m² (35.71 lb/sq ft)
Max power loading ... 5.03 kg/kW (8.27 lb/shp)

PERFORMANCE:
Max operating Mach number (MMO) ... 0.48
Max operating speed (VMO) ... 240 kt (444 km/h; 276 mph) CAS
Max cruising speed at 7,620 m (25,000 ft) ... 270 kt (500 km/h; 311 mph)
Stalling speed:
flaps and gear up ... 92 kt (171 km/h; 106 mph) IAS
flaps and gear down ... 64 kt (119 km/h; 74 mph) IAS
Max rate of climb at S/L ... 512 m (1,680 ft)/min
Max certified altitude ... 9,150 m (30,000 ft)
T-O run ... 450 m (1,480 ft)
T-O to 15 m (50 ft) ... 701 m (2,300 ft)
Landing from 15 m (50 ft) at MLW ... 558 m (1,830 ft)
Landing run at MLW ... 288 m (945 ft)
Max range at 9,150 m (30,000 ft), VFR reserves:
commuter/airliner ... 2,261 n miles (4,187 km; 2,602 miles)
executive ... 2,197 n miles (4,070 km; 2,529 miles)
combi ... 2,282 n miles (4,226 km; 2,626 miles)
g limits: flaps up ... +3.4/−1.36
flaps down ... +2.0

UPDATED

PILATUS PC-12M AND SPECTRE

TYPE: Multisensor surveillance turboprop.

PROGRAMME: Announced at Dubai Air Show in November 1995; then named Eagle, but name discontinued in 2002; renamed Spectre in 2003. Second PC-12, HB-FOB, converted as demonstrator (first flight October 1995). Second demonstrator HB-FOG (c/n 134) followed in September 1996 with airframe modifications and different equipment and third HB-FOL (c/n 166) in 1999. New Spectre version unveiled in 2003.

CURRENT VERSIONS: **PC-12M:** Multi-mission variant, formerly named PC-12 Eagle; features enhanced electrical power generator to integrate optional power consuming equipment according to customer requirements. Typical missions include navaid calibration, environmental survey, aerial photography/mapping and medevac, with quick-change facility permitting change of role to cargo, passenger/cargo, VIP transport or medevac in less than three hours. Specific modifications to demonstrator aircraft are as follows:

HB-FOB: Ventral pannier carrying a variety of electro-optical sensors. A sensor management system (SMS) controls and monitors sensors and displays data on two consoles (one electro-optical, one radar, for comint and elint) inside cabin; includes three COTS processors. Consoles have four displays: colour displays for navigation, situational awareness and ESM/IR/TV; and one 325 mm (14½ in) square LCD XGA high-resolution colour display for synthetic aperture radar (SAR). Sensor operator's map display is 'North up', featuring coastlines, rivers, roads and political boundaries.

Sensor options include: (1) WF-160DS turret with FLIR having 7.2 × 9.8° wide field of view and 2.16 × 2.95° narrow field, plus daylight camera with 20 to 280 mm zoom lens; (2) SAR installations of several types, either mechanical or electronically scanned (one or two dimensions, from 45 to 200 cm; 18 to 79 in long, giving up to 36 cm; 14¼ in resolution); and (3) RISTA (Reconnaissance, Infra-red Surveillance, Target Acquisition), derived from US Army's Airborne Minefield Detection System and combining FLIR (for battle damage assessment) and IR linescan (covering a 7 n mile; 13 km; 8 mile ground track) with real-time downlink to ground station.

HB-FOG: Sensors include WF-160DS IR/E-O system with tracker and Geotrack cueing; Raytheon Sea Vue SV 1021 radar with ISAR capability 1997 upgrade for SAR; 1998 upgrade for track-while-scan of 100 targets and MTI on SAR; communications digital datalink; 108 n mile (200 km; 124 mile) range video downlink to ground station; and sensor management system.

In late 2001, HB-FOG was delivered to Gruppe Rüstung of Swiss Defence Procurement Agency and based at Emmen for trials connected with new FLORAKO air defence ground environment. Pannier removed and fintip extension not fitted; standard PC-12 winglets. Transferred to Swiss Air Force in 2005.

HB-FOL: Modified in 1999 as airborne cellphone relay station, with large ventral pannier, ventral strakes and fintip extension.

PC-12 Spectre: Multirole variant; prototype (N146PC), modified and mission equipped by PILBAL of Denver, Colorado. Announced 27 July 2003 at Airborne Law Enforcement Association meeting at Wichita, Kansas, having flown earlier that month. First customer was US Bureau of Immigration and Customs Enforcement. Retains PC-12 capabilities, but additionally incorporates retractable turret beneath the non-pressurised part of rear fuselage for FLIR Systems Safire II, or Safire III or Wescam MX-15 sensors and operator's position in cabin with two display monitors, video recorder and audio communications panel with

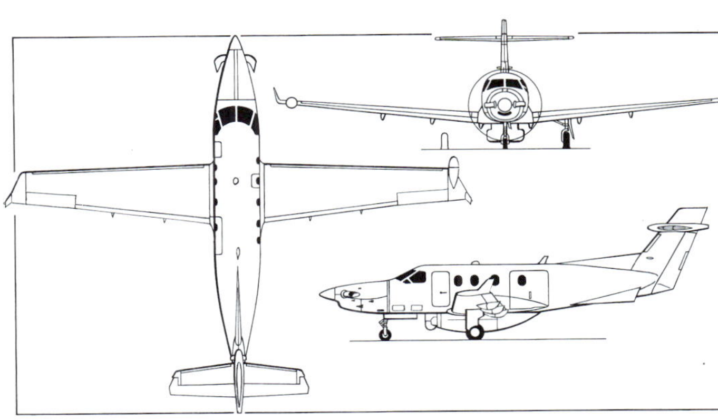

Pilatus PC-12 Eagle showing taller fin and ventral pannier fitted to second prototype *(James Goulding)* 0051495

Prototype Pilatus PC-12 Spectre, showing sensor turret deployed ***NEW***/0567336

broadband communications link. Spectre does not require aerodynamic modifications (enlarged fin or strakes) although N146PC retains large wingtips.

PC-12 M also capable of carrying such other sensors as 2 to 18 GHz (optionally 40 GHz) elint, including automatic jammer; forward- and backward-looking sensor management systems; a passive missile warning system, including automatic chaff/flare dispensers or active laser countermeasures; radar optimised for land- or sea surveillance; and a bidirectional secure datalink to either land, sea or air units.

CUSTOMERS: One delivered to Swiss Defence Procurement Agency. Pilatus in negotiations with several countries, unconfirmed reports citing Brunei, Malaysia and Thailand. US Bureau of Immigration & Customs Enforcement acquiring fleet of Spectres, known as **PC-12 ICE** (Immigration & Customs Enforcement); initial aircraft were N541PB in March 2004 and N497PC in June 2004; variant formally announced at ICE presentation on 14 May 2004, when PC-12 was stated to be replacement for ICE's Beech King Air fleet.

DESIGN FEATURES: FLIR turret at front of ventral pannier; winglets replaced by conventional wingtips on HB-FOB (weather radar in starboard tip pod); undertail strakes enlarged (to maximum depth of 53 cm; 21 in) to maintain stability. HB-FOG fitted 1997 with new 'tiplets' (wingtips with small vertical winglets), newly defined undertail strakes and additional fin area above the elevator, increasing overall height by 48 cm (19 in). HB-FOG and HB-FOL fitted with fintip extensions above tailplane which, with ventral strakes, is mandatory if large ventral pannier is installed. Can undertake various reconnaissance and surveillance missions, but retains ability to revert to passenger and other missions at short notice.

DIMENSIONS, EXTERNAL: As standard PC-12/45 except: PC-12M:
Wing span .... 16.115 m (52 ft 10½ in)
Height overall (with fintip extension) .... 4.74 m (15 ft 6½ in)

WEIGHTS AND LOADINGS: As standard PC-12/45 except:
Weight empty, Spectre .... 3,020 kg (6,659 lb)
Max payload .... 1,402 kg (3,091 lb)

PERFORMANCE:
Max cruising speed at FL200: PC-12 Eagle .... 230 kt (426 km/h; 265 mph)
PC-12M, Spectre .... 240 kt (444 km/h; 276 mph)
Stalling speed, flaps and gear down .... 65 kt (121 km/h; 76 mph) IAS
Max rate of climb at S/L .... 408 m (1,340 ft)/min
Max operating altitude .... 9,145 m (30,000 ft)
T-O run .... 450 m (1,475 ft)
Landing run .... 280 m (920 ft)
Max range at FL300 at 200 kt (370 km/h; 230 mph), 45 min reserves: PC-12 Eagle .... 1,633 n miles (3,080 km; 1,913 miles)
Time on station, two pilots, max fuel, loiter speed of 110 kt (203 km/h; 129 mph) with 30 min reserves:
at FL50 .... 7 h 35 min
at FL100 .... 8 h 32 min
at FL150 .... 9 h 24 min
at FL200 .... 10 h 15 min

UPDATED

RUAG

RUAG AEROSPACE (SUBSIDIARY OF RUAG HOLDING)

PO Box 301, CH-6032 Emmen
Tel: (+41 41) 268 41 11
Fax: (+41 41) 260 25 88
e-mail: marketing.aerospace@ruag.com
Web: http://www.ruag.com
CEO: Peter Schneuwly
MARKETING MANAGER: René Diehl

In 1996, the former Swiss Federal Aircraft Factory Emmen merged with the industrial division of the Swiss Air Force Logistic Command and parts of the Federal Ordnance Office. On 1 January 1999, it was converted from a federal ordnance establishment into a public diversified private company within the RUAG group and renamed Swiss Aircraft and Systems Enterprise Corporation. This title was changed to RUAG Aerospace in 2001, with sites at Emmen, Dübendorf, Stans, Zweisimmen, Interlaken, Lodrino (Zürich) Airport and Alpnach. In 2002, RUAG Aerospace had 1,710 in workforce and turnover of SwFr368 million; factory area totalled 92,000 m^2 (990,279 sq ft) and office area 25,000 m^2 (269,098 sq ft). Major shareholder is Swiss government.

Main activities are the maintenance, manufacture, modernisation and assembly under licence of military and civil aircraft and guided missiles. R&D activities involve four departments, of which Aerodynamics and Flight Mechanics department has four wind tunnels for speeds up to M4 to M5, and test cells for piston and jet engines with and without afterburners. Structural and Systems Engineering department deals with aircraft and space hardware. Electronics and Missile Systems department deals with all systems aspects of aircraft avionics and missiles. Fourth R&D department is for prototype fabrication, flight test, instrumentation and system environmental testing. The Production department can handle mechanical, sheet metal and composites and electronic, electrical, electromechanical and electro-optical subassemblies.

Recent programmes have included structural improvements to Swiss Air Force Tiger F-5E/Fs and the integration of new systems (smoke systems, target towing, missiles); manufacture and delivery of Rafale drop tanks to Dassault Aviation; assembly of 19 BAe Hawk Mk 66s; integration of AS 532UL Cougars for the Swiss Air Force and assembly of 32 single- and two-seat F-18 Hornets for the Swiss Air Force; it also established engineering support and manufactured leading-edges, rudders and its own low-drag pylons for the last-named.

Current work includes the serial production of rudders for the Boeing 717, wingtips for the Airbus A319/A320/A321, wing outer fixed trailing-edges for the Airbus A380, the tailplane for the Pilatus PC-12 and the manufacture and ground support of the Ranger UAV and its hydraulic launcher for Switzerland and Finland. During 2001 and 2002 it was assembling 10 of the follow-on order for 12 AS 532UL Cougars ordered for the Swiss Air Force in December 1998.

RUAG Aerospace is sole manufacturer of Dragon anti-tank missile; TOW missile programme began 1996; Stinger missile programme began 1989.

RUAG Aerospace is accredited to maintain GE J85 (Tiger), Adour (Hawk), GE F404 (F/A-18), Makila (Cougar) and PT6A (PC-6, PC-7, PC-9) engines in own workshops. JAR 145 accreditation allows the company to carry out maintenance work at all levels. This includes inspections, repairs, modifications, prototyping and fault elimination. Other specialities include non-destructive testing procedures such as ultrasonic, eddy current and X-ray as well as comprehensive service packages for systems re-engineering.

In collaboration with Contraves, RUAG Aerospace co-manufactures all payload fairings for Ariane IV and V and Atlas V space launchers on behalf of the ESA, ESTEC, CNES and private industry.

UPDATED

Taiwan

AIDC

AEROSPACE INDUSTRIAL DEVELOPMENT CORPORATION

111-6 Lane 68, Fu-Hsing North Road, Taichung 407
Tel: (+886 4) 22 84 28 25
Fax: (+886 4) 22 84 23 70
e-mail: aidc@ms.aidc.com.tw
Web: http://www.aidc.com.tw
CHAIRMAN: Tao-Yu Sun
VICE-PRESIDENT, BUSINESS: Chou Yeuan-Yuan
MANAGER, PUBLIC RELATIONS: Michael H L Chang

Established 1 March 1969; became subsidiary of Chung Shan Institute of Science and Technology (CSIST) under Ministry of National Defense on 1 January 1983 and known as Aero Industry Development Center until June 1996. New 35,700 m² (384,270 sq ft) assembly facility opened in 1989. Current facilities at Taichung, Kangshan and Sha Lu; total site area 109.3 ha (270 acres); workforce was 4,400 in mid-2001.

From 1 July 1996 was officially transferred under Ministry of Economic Affairs with title as above; completed production of IDF Ching-Kuo fighter in January 2000. AIDC is a participant (with 5 per cent share) in the Sikorsky S-92 Helibus (which see), responsible for the cockpit structure, and (from 1999) is sole-source supplier of crew and passenger doors for the S-76. Other subcontract work includes wing design and manufacture of the Ibis Aerospace Ae 270 (10 sets produced by August 2004, when production suspended); Boeing 717-200 (tail unit; first example delivered mid-December 1997); Dassault Falcon 900 and 2000 (rudders); Eurocopter/CATIC/ST Aero EC 120 Colibri (rear fuselage components); Alenia C-27J Spartan (tail unit); Learjet 45 (tail unit); and Bombardier Challenger 300 (rear fuselage and tail unit). Assembled 13 Bell OH-58Ds for Republic of China Army (first one delivered 16 November 1999; last in third quarter 2001). It also participates in such programmes as the Lockheed Martin F-16 combat aircraft, and the Honeywell TFE1042 and Rolls-Royce 572K gas-turbine engines. However, despite this range of work, AIDC revenue had reportedly shrunk from about US$750 million in 1999 to less than US$300 million in 2001, and additional subcontract work was sought in 2002. Bell offered AIDC a share in manufacture (tailbooms, elevators and engine cowlings) of AH-1Z and UH-1Y in September 2003, leading to agreement 27 February 2004 for initial low-rate production of 53 tailbooms and elevators, with engine cowlings contracted elsewhere. Under same agreement, AIDC to become upgrade centre for Taiwan's existing AH-1W fleet, and will be future certified maintenance centre for these aircraft. Future attack helicopter competition for Taiwan pits AH-1Z against AH-64D Apache Longbow; down-selection of winner expected in mid-2004; up to 75 required.

As a government-owned commercial entity, AIDC is dedicated to research and development for Taiwan's aerospace industry (military and commercial aircraft). Full privatisation, originally planned for fourth quarter of 1999, has been repeatedly postponed following a mid-2000 company restructuring based on three core activities: defence systems and technologies, aerostructures and engines. In 2002, Taiwan government set a 'must keep' privatisation deadline of December 2003, but by then had again deferred a decision by a further year or more. Further restructuring begun in July 2004 under new five-year business plan, leading to reduction in number of separate divisions from 58 to 27.

UPDATED

CSIST

PO Box 90008-11-15, Fu-Hsing North road, Taichung 40722
Tel: (+886 4) 23 74 65 77
Fax: (+886 4) 23 74 65 45
Web: http://www.csist.gov.tw
PROJECT MANAGER: Gene Ho

The CSIST was created on 1 July 1968 as an agency of the Taiwan Ministry of National Defence. The Aeronautical Systems Research Division, which became a part of CSIST in January 1983, occupies a 9.3 ha (23 acre) site and has a workforce of approximately 700.

UPDATED

CSIST GOLDEN EAGLE 600

TYPE: Two-seat helicopter.

PROGRAMME: Launched in 2001; funded by Ministry of Economic Affairs. Some assistance provided by American SportsCopter (ASII) of USA. Prototype first flew 3 August 2003 and made public debut at Taipei Aerospace Technology Exhibition later that month. Institute hoped to complete 20 hour initial flight test programme by end of that year. Domestic and possibly US certification to be obtained. Local manufacturer will be sought for production aircraft.

COSTS: Target cost US$170,000 (2003).

DESIGN FEATURES: Of broadly similar overall configuration to Robinson R22. Two-blade, teetering main rotor and two-blade tail rotor.

LANDING GEAR: Twin-skid type.

POWER PLANT: One 134 kW (180 hp) Textron Lycoming HO-360-C1A piston engine.

WEIGHTS AND LOADINGS:
Max T-O weight ..... 810 kg (1,785 lb)

PERFORMANCE (design):
Max level speed ..... 100 kt (185 km/h; 115 mph)
Cruising speed ..... 80 kt (148 km/h; 92 mph)
Max rate of climb at S/L ..... more than 300 m (984 ft)/min
Service ceiling ..... 4,260 m (14,000 ft)

UPDATED

YOSHINE

YOSHINE HELICOPTER CO LTD

PO Box 53-150, Taichung 40300
Tel: (+886) 939 92 87 19
Fax: (+886) 423 16 35 33
e-mail (1): clin@mail.com
e-mail (2): yoshine@ezycopter.com
Web (1): http://www.ezycopter.com
Web (2): http://www.yoshine.com.tw
PRESIDENT: Charles Lin

Yoshine, which is a sister company of Modus Verticraft, was established to develop a safer personal helicopter equipped with a coaxial rotor. An agreement was reached with LA Systems of the UK to set up EzyCopter UK Ltd as a dealer/distributor in mid-2004. Other distributors include Helitrade (Italy), Plasgold (Philippines) and Remax (USA).

UPDATED

YOSHINE EZYCOPTER

TYPE: Single-seat ultralight helicopter kitbuilt.

PROGRAMME: As of mid-2004, first of two AIDC-built prototypes was due to fly in fourth quarter of that year. By December 2004, Yoshine had not published a picture of a completed helicopter.

CUSTOMERS: Advance orders for 300 reported in 2004.

COSTS: Kit US$50,000 (2004).

DESIGN FEATURES: Few details released. Employs Yoshine YX-24 coaxial rotor head with twin two-blade semi-rigid rotors. No tail rotor.

FLYING CONTROLS: Standard collective and cyclic pitch controls.

POWER PLANT: Two 18.6 kW (25 hp) Hirth piston engines with correlated throttle and collective controls.

Coaxial rotor head of Yoshine EzyCopter ***NEW***/0589165

WEIGHTS AND LOADINGS (estimated):
Weight empty ..... 150 kg (331 lb)
Max T-O weight ..... 350 kg (771 lb)

UPDATED

Turkey

ALP

ALP HAVACILIK (ALP AVIATION)

Organize Sanayi Bölgesi, 8 Cadde, TR-26110 Eskişehir
Tel: (+90 222) 236 13 00 and 236 12 61
Fax: (+90 222) 236 12 85
e-mail: alphavacilik@alp.com.tr
Web: http://www.alp.com.tr
MANAGING DIRECTOR: Ahmet Aytekin

Formed 29 July 1999 as joint venture company by Sikorsky Aircraft Turkey Inc and The Alpata Group (50 per cent each) to manufacture high-technology, precision-machined aerospace and defence components and assemblies for global market in modern 6,500 m²

(70,000 sq ft) facility. Company has 16 CNC machine tools and 2004 workforce of more than 80. Is exporting helicopter components to Sikorsky USA under two contracts awarded on 15 November 1998 with total potential value of more than US$60 million. Also a supplier to TAI (which see) and TUSAŞ Engine Industries.

UPDATED

TAI

TURKISH AEROSPACE INDUSTRIES (TUSAS HAVACILIK VE UZAY SANAYII A S)
PO Box 18, Kavaklidere TR-06692, Ankara
Tel: (+90 312) 811 18 00
Fax: (+90 312) 811 14 25 and 811 14 08
e-mail: bcates@tai.com.tr
Web: http://www.tai.com.tr
GENERAL MANAGER: Kaya Ergenç
GENERAL CO-ORDINATOR: Lt Gen (Ret'd) Vural Avar
EXECUTIVE DIRECTORS:
Cemal Nadir Sayin (Production and Materials)
Aydin Oniz (Business Development and Management)
Dr Fatih Tezok (Design and Engineering)
MARKETING MANAGER: Yilmaz Güldoğan
PUBLIC RELATIONS CO-ORDINATOR: Nursel Köran
PUBLIC RELATIONS SPECIALIST: Bican Çelik Ates

Turkish Aerospace Industries (TAI), established on 15 May 1984, is a majority-owned Turkish company made up of Turkish (51 per cent) and US (49 per cent) partners. Shareholders are Turkish Aircraft Industries Inc (49 per cent), Turkish Armed Forces Strengthening Foundation (1.9 per cent), Turkish Aeronautical Association (0.1 per cent), Lockheed Martin of Turkey Inc (42 per cent) and General Electric International Inc (7 per cent). Workforce in mid-2003 was 1,700.

TAI's aircraft facilities, located in Ankara, occupy a 23 ha (56.8 acre) site, including more than 160,000 m^2 (1.72 million sq ft) of covered space. They are furnished with high-technology machinery and equipment that provide extensive capabilities ranging from parts manufacturing to aircraft assembly, flight test and delivery from aerodrome and works at Akinci. Quality control meets world standards including NATO AQAP-110, ISO-9001:2000, AS EN 9100 and AECMA-EASE.

With experience in aircraft and aerostructures manufacturing business, TAI is a supplier for Lockheed Martin (F-35); MD Helicopters (MD 902 fuselages); Boeing (737 wingtips and flight deck panels); Airbus (A319/320/321 fuselage panels); Eurocopter (Super Puma/Cougar, and rear doors and engine cowlings for EC 135); Sikorsky (S-70A tailbooms, tail rotor pylons and tailplanes, and tailplanes for S-76); EADS CASA; AgustaWestland; Sonaca; Northrop Grumman; and others.

TAI's experience includes co-production of 278 F-16 fighters, CN-235 light transports (59), SF-260D trainers (34) and AS 532 Cougar (28), as well as design and development of UAVs, target drones and the ZIU agricultural aircraft. As a full member of Airbus Military SAS, TAI is engaged (7.15 per cent workshare) in the development of the A400M, for which it will produce forward and centre fuselages, ailerons, spoilers, paratroop doors, hatches and tailcone. In 2002 it became the seventh international participant in the Lockheed Martin JSF programme.

TAI's core business also includes modernisation, modification and systems integration programmes and after-sales support. Major programmes of this kind are structural modification and EW retrofit of Turkish Air Force F-16s; modification of S-2E Tracker maritime patrol aircraft into firefighting aircraft (new avionics, Honeywell TPE331 turboprops and a 4,500 kg; 9,921 lb capacity water tank); CN-235 modification for Open Skies; CN-235 and Black Hawk modifications for Special Forces (Yarasa programme for Black Hawk); modification of CN-235 platforms for MPA/MSA missions for the Turkish Navy and Coast Guard (Meltem II programme); Cougar modification and modernisation; and 'glass cockpit' retrofit of S-70 helicopters. Contract work (US$49 million) to modify the four Boeing 737-700 AEW&C aircraft ordered by the Turkish Air Force (Peace Eagle programme) is taking place between July 2003 and June 2008. An agreement to produce centre fuselage sections for the AB139 helicopter was signed with AgustaWestland in April 2004.

UPDATED

TAI (AIRTECH) CN-235M

PROGRAMME: Under a contract with EADS CASA of Spain, TAI produced 50 CN-235 M-100 twin-turboprop transports for the Turkish Air Force between 1991 and 1998, manufacturing 92 per cent of the total airframe, including 20 per cent made of composites. Production continued with six as surveillance aircraft for the Turkish Navy and three for maritime patrol with the Turkish Coast Guard, ordered on 23 September 1998 and delivered from 23 December 2001 onwards. Airframe percentage in these nine aircraft was increased to 95. All had been delivered by June 2003. A further five fuselages for the Spanish assembly line were delivered to EADS CASA between 11 June 2002 and 5 February 2003. Latest (December 2003) contract is for three paratroop doors, delivered to EADS CASA in February and March 2004.

UPDATED

TAI (BELL) AH-1Z KING COBRA

TYPE: Attack helicopter.
PROGRAMME: RFP issued 30 May 1997. Turkish Army ATAK programme. Bids from Boeing (Apache Longbow); AgustaWestland (A 129 International); Eurocopter (Tiger); Bell AH-1Z (King Cobra); and Kamov/IAI (Ka-50-2 Erdoğan). AH-1Z selected as preferred option in July 2000, with reserve choice of Ka-50-2. Contract negotiations with Bell protracted due to US insistence on retention of US (instead of proposed Turkish indigenous) mission computer; supposedly resolved in second quarter 2002 by agreement for co-production of US equipment, with contract signature then expected during third quarter. However, in early September 2002, Turkish defence procurement organisation said to be re-evaluating Kamov Ka-50-2 in wake of further delays over access to AH-1Z source codes and performance guarantees. Talks resumed with Kamov in mid-2003, and later in year with Bell; latter still attempting to find an acceptable solution to problems in early months of 2004, but in May Turkish Defence Minister announced cancellation of attack helicopter programme.

UPDATED

TAI (EUROCOPTER) COUGAR MK I

PROGRAMME: Within the framework of the Eurotai consortium, TAI co-produced 28 of 30 Cougar Mk I multirole medium helicopters (10 for the Turkish Army and 20 for the Turkish Air Force), under a 1997 contract (Phenix II programme) valued at US$434 million. These comprise 24 AS 532ULs (Army six utility, four SAR; Air Force 14 SAR) and six AS 532ALs (Air Force, combat SAR). The first two helicopters were search and rescue aircraft, manufactured in France by Eurocopter and delivered in April and May 2000; the TAI programme covered fabrication, assembly, systems integration and flight test. First Turkish-assembled Cougar (an AS 532UL) was handed over to the Army on 31 May 2000; deliveries were completed on 27 June 2003. Equip Nos. 101, 125 and 210 Squadrons of Turkish Air Force. Army Cougars serve with 1st, 2nd and 3rd Aviation Regiments; Aegean Army Regiment; Cyprus Turkish Peace Forces Army Aviation Command; and Army Aviation School.

UPDATED

Ukraine

AEROKOPTER

AEROKOPTER OOO (AEROCOPTER LTD)
ulitsa Lishchevinkov 27, 36014 Poltava
Tel: (+38 0532) 66 98 58 and 66 87 60
Fax: (+38 0532) 66 07 10
e-mail: slcopter@kot.poltava.ua

Aerokopter was established on 14 December 1999 to develop light rotorcraft. It was briefly associated with what, on 3 May 2000, became the separate Aviaimpex design bureau, also active in Ukraine.

VERIFIED

AEROKOPTER ZA-6 SAN'KA

TYPE: Two-seat helicopter.
PROGRAMME: First flown in 1999. No further reports by early 2004.
CURRENT VERSIONS: **ZA-6**: As described.
AK-1: Development; described separately.
CUSTOMERS: Operators include Ukraine AeroClub.
DESIGN FEATURES: Conventional pod-and-boom helicopter with five-blade main rotor and two-blade tail rotor mounted on port side. High agility. Power plant, fuel tanks and gear box mounted externally, to rear of crew pod.
FLYING CONTROLS: Manual.
STRUCTURE: Metal and composites.
LANDING GEAR: Twin skids.
POWER PLANT: One 119 kW (160 hp) Subaru EJ22 flat-four. Fuel capacity 80 litres (21.1 US gallons; 17.6 Imp gallons).
ACCOMMODATION: Two persons, side by side.
DIMENSIONS, EXTERNAL:
Main rotor diameter .... 6.00 m (19 ft 8¼ in)

Aerokopter ZA-6 San'ka *(Aeroklub Ukrainy)* ***NEW***/0580863

Main rotor blade chord .... 0.15 m (6 in)
Tail rotor diameter .... 1.24 m (4 ft 0¾ in)
Tail rotor blade chord .... 0.115 m (4½ in)
AREAS:
Main rotor disc .... 28.27 m^2 (304.3 sq ft)
Tail rotor disc .... 1.21 m^2 (13.00 sq ft)
WEIGHTS AND LOADINGS:
Weight empty .... 400 kg (882 lb)
Max T-O weight .... 750 kg (1,653 lb)
Max disc loading .... 26.5 kg/m^2 (5.43 lb/sq ft)
Max power loading .... 6.29 kg/kW (10.33 lb/hp)
PERFORMANCE:
Max level speed .... 103 kt (190 km/h; 118 mph)
Max cruising speed .... 76 kt (140 km/h; 87 mph)
Econ cruising speed .... 49 kt (90 km/h; 56 mph)
Max rate of climb at S/L:
at optimal forward speed .... 642 m (2,106 ft)/min

657 rpm main rotor speed .... 444 m (1,457 ft)/min
vertical ascent .... 282 m (925 ft)/min
Service ceiling .... 2,100 m (6,880 ft)
Range with max fuel .... 270 n miles (500 km; 310 miles)

UPDATED

AEROKOPTER AK-1

TYPE: Two-seat helicopter

PROGRAMME: Two-stage development of ZA-6, initially involving one of the latter (registered GL-0478, shown at Manufacturing & Security Exhibition, Kiev, September 2002), designated AK-1-5 (five-blade main rotor). At the same event, the uncompleted AK-1-3 prototype was shown with the definitive three-blade rotor, starboard side tail rotor and twin vertical fins mounted on horizontal tail surfaces. Marketing agreements are in place with agents in Australia, France, Italy and Spain, but no further reports had been received by 2004.

UPDATED

Prototype AK-1-3, showing three-blade main rotor and starboard side tail rotor *(Yefim Gordon)* 0527149

Aerokopter AK-1-5 five-blade prototype *(Yefim Gordon)* 0527152

AEROPRAKT

AEROPRAKT OOO (AEROPRAKT JSC)

a/ya 112, 03148 Kiev
Tel: (+38 044) 457 92 93
Fax: (+38 044) 457 91 59
e-mail: air@prakt.kiev.ua
Web: http://www.aeroprakt.kiev.ua
MANAGER: Oleg V Litovchenko
CHIEF DESIGNER: Yuri V Yakovlev

US AGENT:: Spectrum Aircraft Corporation
PO Box 1381, Sebring, Florida 33872
Tel: (+1 941) 314 97 38
Fax: (+1 941) 314 02 85
e-mail: jhunter@stratc.net
Web: http://www.spectrumaircraft.com

Aeroprakt was formed in Kiev in 1986, after Yuri Yakovlev, who had designed the T-8 lightplane for the Aeroprakt organisation of Samara, Russia, was invited to work at Antonov ASTC. Yakovlev then produced the A-20 Chervonets and joined with fellow engineer Oleg Litovchenko in an initially part-time venture. Established in present, commercial form in 1991. Retaining the 'k for Kiev' in its name, the company now employs only even-numbered designations for its aircraft, the latest to fly being the A-28. Company has agencies in Australia, Belgium, Canada, Denmark, Germany, Iran, Italy, Latvia, New Zealand, Poland, Russia, South Africa, Spain, Turkey, UAE, UK and USA. Design and production personnel total 50; factory space 2,000 m^2 (21,525 sq ft), although some production subcontracted to Aviant and Antonov. More than 100 Aeroprakt aircraft have been constructed.

UPDATED

AEROPRAKT-20

US marketing name: Spectrum SA-20 Vista

TYPE: Tandem-seat ultralight/kitbuilt.

PROGRAMME: Design started September 1990; construction of prototype began November 1990; first flight 5 August 1991; construction of first production aircraft started September 1991, first flown 15 August 1993; prototype appeared abroad for first time in Hodkovice, Czech Republic, at European Microlight Championship, August 1993, gaining ninth place; fourth production A-20 won third place at World Microlight Championship, Poznan, Poland, August 1994; seventh production A-20 gained second place at EMC '95, Little Rissington, UK; initial production aircraft was placed first at EMC, Hungary, 2002.

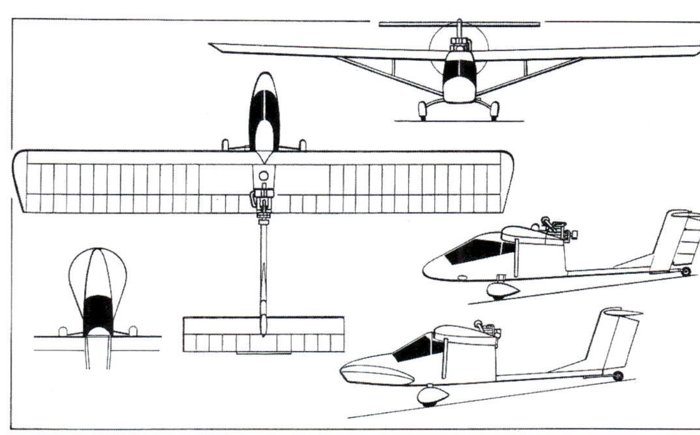

Aeroprakt-20, with additional side and scrap plan views of the radiometer-equipped A-20S *(James Goulding)* 0507648

CURRENT VERSIONS: **SA-20 Vista:** Baseline aircraft, *as described in detail.*

A-20R912 Sky Cruiser: Described separately.

A-20S: Variant under development by 1997, with large multichannel radiometer built into its nose for a wide range of ecological survey work. No known manufacture.

A-20SKh: Agricultural (*selbskokhozyaistvennyi:* agricultural) with 59.6 kW (79.9 hp) Rotax 912 UL, three-blade ground-adjustable pitch Ivoprop propeller, sweptback (1° 30') wing, two chemical tanks, each 90 litres (23.8 US gallons; 19.8 Imp gallons) scabbed to fuselage sides, and streamlined spraybar with four or six atomisers. Prototype construction began July 1997; first flight September 1997; no subsequent production reported. A fuller description has been published.

SA-20 Vista STOL: With full-span slotted flaperons and 47.8 kW (64.1 hp) Rotax 582 UL; cruising speed 76 kt (140 km/h; 87 mph).

SA-20 Vista SS: Rotax 582 and wing span of 10.02 m (33 ft 4 in); empty weight 222 kg (490 lb); cruising speed 87 kt (185 km/h; 100 mph); range 205 n miles (379 km; 236 miles) or 399 n miles (740 km; 460 miles) with auxiliary fuel.

A-30 Vista Speedster: Prototype under test at Sebring, Florida, by early 2002. Apparently not proceeded with.

CUSTOMERS: First customer delivery August 1993; exports to Czech Republic, Germany, Hungary, Jordan, Poland, Russia, Singapore and United Arab Emirates; last-mentioned acquired seven for Umm-al Qaiwain Flying Club. Total of 41 A-20s of all types produced by early 2004.

COSTS: Quick-build kit: Vista US$26,500; Vista STOL US$28,500; Vista SS US$33,000; all including engine, propeller and instruments (2004).

DESIGN FEATURES: High-wing, T-tail configuration, optimised for rapid disassembly for transport by trailer. Tail surfaces mounted on aluminium tube which detaches at joint with the nacelle.

Unswept wing; P-IIIa-15 wing section, thickness/chord ratio 15 per cent; chamfered tips; dihedral 1° 30'; incidence at roots 3° 30', twist 3°.

FLYING CONTROLS: Manual. Single-piece, rod-operated flaperons extend full length of wing trailing-edge; no tabs. Single-piece elevator with protruding, in-flight adjustable tab. Full-height rudder with no tab.

STRUCTURE: Riveted aluminium wing structure; leading-edge D box closed by I-section main spar; stamped wing ribs; bent sheet rear false spar; partially fabric covered. Control surfaces similar to wings. All-composites honeycomb sandwich fuselage pod.

LANDING GEAR: Tailwheel type; fixed. Fuselage-mounted spring cantilever mainwheel legs with wheel fairings. Steerable tailwheel.

POWER PLANT: One 37.0 kW (49.6 hp) Rotax 503 UL-2V two-stroke piston engine driving a KievProp three-blade, ground-adjustable pitch pusher propeller via 3:1 reduction gearbox. Rotax 462, 582 (47.8 kW; 64.1 hp), 618 and 912 engines optional. Fuelling point on starboard side of nacelle. Standard capacity 38 litres (10.0 US gallons; 8.4 Imp gallons); optional 90 litres (23.8 US gallons; 19.8 Imp gallons).

EQUIPMENT: Ballistic recovery parachute.

DIMENSIONS, EXTERNAL:
Wing span .... 11.34 m (37 ft 2½ in)
Length overall .... 6.72 m (22 ft 0½ in)
Height: cabin roof .... 1.74 m (5 ft 8½ in)
propeller turning .... 2.20 m (7 ft 2½ in)

AREAS:
Wings, gross .... 15.70 m^2 (169.0 sq ft)

WEIGHTS AND LOADINGS:
Weight empty .... 218 kg (481 lb)
Max T-O weight .... 455 kg (1,003 lb)

PERFORMANCE (Rotax 582 engine):
Max level speed .... 86 kt (160 km/h; 99 mph)
Stalling speed, flaperons down .... 26 kt (48 km/h; 30 mph)
Max rate of climb at S/L .... 300 m (984 ft)/min
T-O and landing run .... 80 m (265 ft)
Max range .... 216 n miles (400 km; 248 miles)
Endurance .... 4 h 30 min
g limits .... +4/–2

UPDATED

AEROPRAKT-20R912 SKY CRUISER

US marketing names: Vista Cruiser and Varlet

TYPE: Tandem-seat ultralight/kitbuilt.

PROGRAMME: See Aeropraakt-20; designation indicates increased engine power provided by flat-four Rotax 912. Alternatively designated **Aeropraakt-20M**. Construction of prototype Sky Cruiser began February 1997; first flight 11 May 1997. Gained second place in 1st World Air Games, Turkey, September 1997; second place in 1998 World Microlight Cup, Hungary; and third place, 2nd World Air Games, Spain.

CURRENT VERSIONS: **Vista Cruiser:** Baseline version. *As described.*

Super Cruiser: Increased span (11.34 m; 37 ft 2½ in) and MTOW 548 kg (1,210 lb); optional fuel increase to 220 litres (58.1 US gallons; 48.4 Imp gallons). Performance generally unchanged.

Varlet: Announced 2003; for US LSA category. Wing span 9.40 m (30 ft 10 in); wing area 13.00 m^2 (139.9 sq ft); weight empty 229 kg (505 lb); MTOW 453 kg (1,000 lb); cruising speed 111 kt (206 km/h; 128 mph); stalling speed, landing configuration 33 kt (62 km/h; 38 mph); max rate of climb 427 m (1,400 ft)/min; T-O run 85 m (280 ft); range 434 n miles (804 km; 500 miles). Alternative Experimental category version has 8.53 m (28 ft 0 in) wing span.

CUSTOMERS: Total 11 aircraft ordered and built by mid-2002.

COSTS: Quick-build kit (40 hours) including engine, propeller and instruments (2004): Vista Cruiser, US$42,250; Varlet US$47,800 (2004).

DESIGN FEATURES: Development of A-20 with increased power, shorter wing and tailplane spans, engine cowling, constant-speed propeller, fully balanced control surfaces, smaller wheels and more streamlined wing struts and landing gear. Sweepback 1° 30'.

Description of A-20 applies generally, except as below.

POWER PLANT: One 73.5 kW (98.6 hp) Rotax 912 ULS flat-four driving a three-blade Ivoprop constant-speed propeller. Standard fuel capacity 128 litres (33.8 US gallons; 28.8 Imp gallons).

DIMENSIONS, EXTERNAL:
Wing span .... 10.17 m (33 ft 4½ in)
Length overall .... 6.72 m (22 ft 0½ in)
Height: cabin roof .... 1.68 m (5 ft 6¼ in)
propeller turning .... 2.33 m (7 ft 7¾ in)

DIMENSIONS, INTERNAL:
Cockpit max width .... 0.64 m (2 ft 1¼ in)

AREAS:
Wings, gross .... 14.00 m^2 (150.7 sq ft)

WEIGHTS AND LOADINGS:
Weight empty .... 249 kg (550 lb)
Max T-O weight .... 455 kg (1,003 lb)

PERFORMANCE:
Max level speed .... 113 kt (210 km/h; 130 mph)
Cruising speed .... 100 kt (185 km/h; 115 mph)
Stalling speed, landing configuration .... 30 kt (55 km/h; 34 mph)
Max rate of climb at S/L .... 360 m (1,181 ft)/min
T-O run .... 61 m (200 ft)
Landing run .... 77 m (250 ft)
Range .... 608 n miles (1,126 km; 700 miles)
Endurance .... 12 h 0 min
g limits .... +4/–2

UPDATED

AEROPRAKT-22

French marketing name: Vision
Russian name: Sharik (Balloon)
UK marketing name: Foxbat
US marketing name: Valor

TYPE: Side-by-side ultralight/kitbuilt.

PROGRAMME: Design started February 1990, construction of prototype began September 1994; first flight 21 October 1996; certified to German BFU-95; production began in early 1999. Approved for kit assembly in UK.

CURRENT VERSIONS: **Aeropraakt-22 Sharik:** Original version, with 59.6 kW (70.9 hp) Rotax 912 UL engine and fuselage fuel tank.

Following versions have uprated engine.

Foxbat: UK version, marketed from 2000 by Small Light Aeroplane Company Ltd at Otherton, Staffordshire (www.foxbat.co.uk). Prototype, G-FBAT (16th airframe) first flown after kit assembly in UK 12 August 2000. BCARS certification in early 2002; several built under auspices of PFA.

Valor: US version.

Vision: Marketed in France by Aerotrophy of Montagu.

Description applies to Valor.

CUSTOMERS: More than 50 A-22s built by early 2002 and delivered to Australia, France, Georgia, Germany, Ireland, Italy, Latvia, Luxembourg, Mexico, Poland, Russia, UAE, UK and USA.

COSTS: UK kit cost £32,500 including Rotax 912 ULS engine and Newton propeller, builder's assistance, basic instruments and tax (2004). US kit cost US$42,500 with Rotax 912 UL engine.

DESIGN FEATURES: Constant-chord wings and horizontal tail surfaces. Wings swept forward 2° 30'; P-IIIa-15 wing section, thickness/chord ratio 15 per cent; chamfered tips; dihedral 1° 30'; incidence at root 4°; twist 2° 30'.

FLYING CONTROLS: Manual, by pushrods and cables. Full-span slotted flaperons with trim tab to starboard; single-piece elevator with tab; and sweptback rudder with ground-adjustable tab.

STRUCTURE: Riveted aluminium wing structure; leading-edge D box closed by I-section main spar; stamped wing ribs; bent sheet rear false spar; Ceconite covering on wings (except

Spectrum Aircraft SA-20 Vista Cruiser *(Paul Jackson)* ***NEW***/0580758

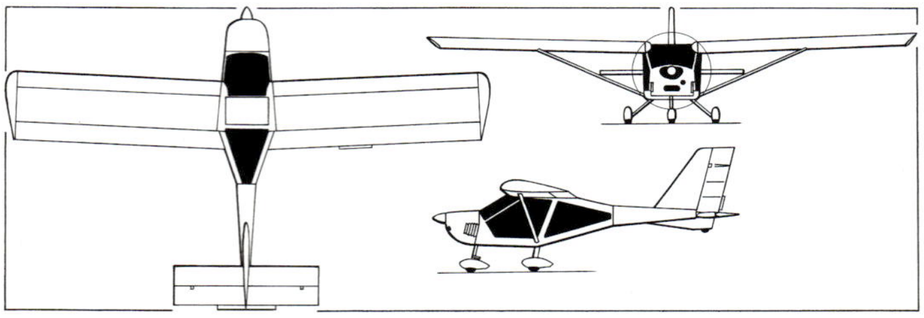

Aeropraakt-22 ultralight *(James Goulding)* 0054615

SA-20 Vista *(Paul Jackson)* ***NEW***/0525065

Foxbat version of Aeropraakt-22 for UK market *(Paul Jackson)* ***NEW***/0580757

metal leading-edge), rudder and elevator. Fin and tailplane similar to wings. Anodised aluminium fuselage with profiled sheet and fluted skin, stamped bulkheads, steel and aluminium tubing. Extensive glazing. Glass fibre engine cowling, wheel spats, wing fillets and fintip.

LANDING GEAR: Tricycle type; fixed. Fuselage-mounted spring steel mainwheel legs with wheel fairings. Hydraulic mainwheel brakes. Steerable nosewheel; small tailwheel protects ventral strake from nose-high landings.

POWER PLANT: One 73.5 kW (98.6 hp) Rotax 912 ULS or 59.6 kW (79.9 hp) Rotax 912 UL flat-four driving a KievProp three-blade ground-adjustable pitch propeller or (Foxbat) Newton two-blade. Fuel tank in each wingroot; combined capacity 90 litres (23.8 US gallons; 19.8 Imp gallons). Optional capacity 135 litres (35.6 US gallons; 29.7 Imp gallons).

ACCOMMODATION: Two persons, side by side. Large upward-hinged window/door on each side.

EQUIPMENT: Ballistic parachute.

Note that data vary slightly between German, UK and US variants.

DIMENSIONS, EXTERNAL:
- Wing span ... 10.10 m (33 ft 1¾ in)
- Length overall ... 6.30 m (20 ft 8 in)
- Height overall ... 2.40 m (7 ft 10½ in)
- Tailplane span ... 3.00 m (9 ft 10 in)
- Propeller diameter ... 1.68 m (5 ft 6 ¼ in)

DIMENSIONS, INTERNAL:
- Cabin: Length ... 1.60 m (5 ft 3 in)
  - Max width ... 1.30 m (4 ft 3¼ in)
  - Height ... 1.10 m (3 ft 7¼ in)

AREAS:
- Wings, gross ... 13.70 m² (147.5 sq ft)

WEIGHTS AND LOADINGS:
- Weight empty, equipped ... 260 kg (573 lb)
- Max T-O weight ... 450 kg (992 lb)

PERFORMANCE (Rotax 912 UL):
- Max level speed ... 92 kt (170 km/h; 106 mph)
- Cruising speed ... 86 kt (160 km/h; 99 mph)
- Stalling speed ... 30 kt (55 km/h; 35 mph)
- Max rate of climb at S/L ... 300 m (984 ft)/min
- T-O and landing run ... 90 m (295 ft)
- Range with max fuel ... 594 n miles (1,100 km; 683 miles)
- *g* limits ... +4/−2

UPDATED

AEROPRAKT-24

US marketing name: Viking

TYPE: Three-seat amphibian kitbuilt.

PROGRAMME: Design began September 2000; first flight 25 October 2001.

CURRENT VERSIONS: **Viking:** *As described*

Twin Viking: Twin-engined version; programme currently suspended. Two Rotax 582s, each 47.8 kW (64.1 hp); empty weight 431 kg (950 lb); MTOW 726 kg (1,600 lb); max level speed 87 kt (161 km/h; 100 mph); T-O run 138 m (450 ft); stalling speed 33 kt (62 km/h; 38 mph).

CUSTOMERS: Five sold by early 2004.

COSTS: US$65,500, complete, quick-build (2004). Basic kit US$39,000, plus tax (2004).

DESIGN FEATURES: Braced high-wing monoplane, capable of meeting JAR-VLA certification criteria, which not yet pursued. Several detail changes from version promoted in late 1990s, but not built, including weight in Experimental category; single bracing strut and jury strut each side; flying-boat hull with single step; sweptback vertical tail with unswept horizontal surfaces at tip.

Wing section P-IIIa-15; no dihedral; incidence 4° 30′ at mean aerodynamic chord; twist −3°.

FLYING CONTROLS: Conventional and manual.Mass-balanced ailerons and elevator; ailerons and flaps separate, and both slotted; flap deflections 10, 20 and 30°. Actuation by pushrods, except cable-operated rudder and elevator trim tab.

STRUCTURE: Generally as for A-20 and A-22. Hull and integral fin of glass fibre and epoxy resin sandwich with foam core, reinforced by internal bulkheads and beams forming three separate, watertight compartments. Engine and metal wing, with fabric covering, both attached to central, metal box-pylon supported by two stainless steel bracing struts within the cabin. Composites floats constructed integrally with pylons. Metal tailplane with fabric covering.

LANDING GEAR: Tailwheel type, with retractable, but externally stowed, Matco 6.00-6 mainwheels with disc brakes. Faired tailwheel doubles as water rudder. Cantilever spring steel mainwheel legs attached to hull bulkhead. Initial manual retraction replaced by electric motor from 2004. Stabilising float under each outer wing.

POWER PLANT: One 73.5 kW (98.6 hp) Rotax 912 ULS or 84.5 kW (113.3 hp) Rotax 914 flat four projecting from wing centre-section. Fuel capacity 87 litres (23.0 US gallons; 19.1 Imp gallons) in tank behind cabin. Optionally, to provide additional baggage capacity, fuselage tank replaced by two wing tanks, combined capacity 90 litres (23.8 US gallons; 19.8 Imp gallons).

ACCOMMODATION: Two seats forward; removable seat at rear of cabin. Sliding door each side openable in flight. Luggage stowage in forward watertight compartment; accessed via door in upper surface.

DIMENSIONS, EXTERNAL:
- Wing span ... 11.04 m (36 ft 2¾ in)
- Wing chord, constant ... 1.40 m (4 ft 7 in)
- Wing aspect ratio ... 7.9
- Length overall ... 7.28 m (23 ft 10½ in)
- Height: top of engine cowlings ... 2.425 m (7 ft 11½ in)
  - fintip ... 2.68 m (8 ft 9½ in)
- Tailplane span ... 3.60 m (11 ft 9¾ in)
- Tailplane chord, constant ... 0.80 m (2 ft 7½ in)
- Wheel track ... 2.165 m (7 ft 1¼ in)
- Propeller diameter ... 1.82 m (5 ft 11½ in)
- Cabin doors (each): Height ... 0.63 m (2 ft 0¾ in)
  - Width ... 1.00 m (3 ft 3¼ in)

DIMENSIONS, INTERNAL:
- Cabin: Length ... 2.20 m (7 ft 2½ in)
  - Max width ... 1.10 m (3 ft 7¼ in)
  - Max height ... 1.20 m (3 ft 11¼ in)
- Forward baggage hold: Length ... 0.90 m (2 ft 11½ in)
  - Max width ... 1.05 m (3 ft 5¼ in)
  - Max height ... 0.75 m (2 ft 5½ in)
  - Volume ... 0.15 m³ (5.3 cu ft)

AREAS:
- Wings, gross ... 15.50 m² (166.8 sq ft)
- Ailerons (total) ... 1.78 m² (19.16 sq ft)
- Trailing edge flaps (total) ... 2.44 m² (26.26 sq ft)
- Fin ... 1.05 m² (11.30 sq ft)
- Rudder ... 0.72 m² (7.75 sq ft)
- Tailplane ... 2.88 m² (31.00 sq ft)
- Elevator, incl tab ... 1.44 m² (15.50 sq ft)

WEIGHTS AND LOADINGS:
- Weight empty ... 460 kg (1,014 lb)
- Max T-O weight ... 750 kg (1,653 lb)
- Max wing loading ... 50.0 kg/m² (10.24 lb/sq ft)
- Max power loading:
  - Rotax 912 ... 10.20 kg/kW (16.76 lb/hp)
  - Rotax 914 ... 8.88 kg/kW (14.59 lb/hp)

PERFORMANCE:
- Never-exceed speed (VNE) ... 97 kt (180 km/h; 112 mph)
- Max level speed ... 78 kt (145 km/h; 90 mph)
- Cruising speed at 75% power ... 65 kt (120 km/h; 75 mph)
- Stalling speed: flaps up ... 46 kt (85 km/h; 52 mph)
  - flaps ... 38 kt (70 km/h; 43 mph)
- Max rate of climb at S/L ... 150 m (495 ft)/min
- Service ceiling ... 3,500 m (11,480 ft)
- T-O to 15 m (50 ft): land ... 150 m (492 ft)
  - water ... 200 m (660 ft)
- Landing from 15 m (50 ft): land ... 350 m (1,150 ft)
  - water ... 400 m (1,315 ft)
- Range with max fuel ... 378 n miles (700 km; 435 miles)
- Endurance ... 4 h 30 min
- *g* limits ... +4/−2

UPDATED

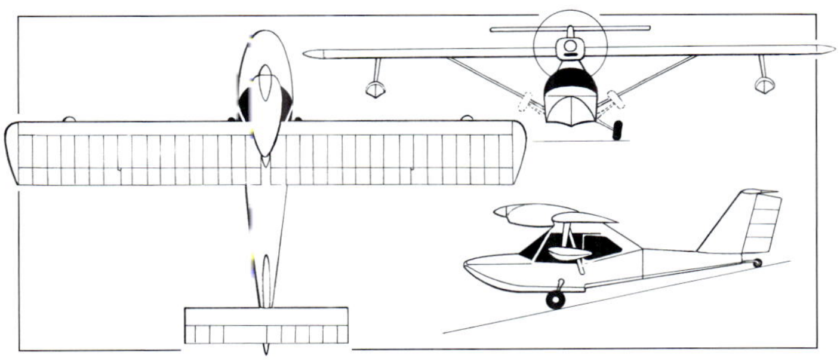

Aeroprakt-24 kitbuilt amphibian *(James Goulding)* 0558328

Aeroprakt-24 landing ***NEW***/0538384

AEROPRAKT-26

US marketing name: Twin Vista

TYPE: Tandem-seat ultralight twin/kitbuilt.

PROGRAMME: Development of A-20 with two engines; commissioned by Gulf Aviation Technologies; prototype construction began March 1996; first flight 18 November 1997. Initially marketed in US as Vulcan.

CURRENT VERSIONS: **SA-26 Twin Vista:** Lower-powered variant; Rotax 503 engines. Marketed in US by Spectrum Aircraft.

Twin Vista SS: With Rotax 582 engines.

Vulcan: Aeroprakt-36; *described separately.*

CUSTOMERS: More than 15 sold by late 2002 to customers in Guatemala, Panama, UAE and USA.

COSTS: Kit US$39,500; fast-build kit US$44,250; both including engines, propellers, basic digital instruments, paint and fabric (2004).

Description of A-20 applies generally, except as below.

DESIGN FEATURES: Introduces twin-engine safety margins to A-20 design; able to take off on one engine. Sweepback increased to 3°; tailfin height and rudder chord both increased.

POWER PLANT: Two 47.8 kW (64.1 hp) Rotax 582 UL two-cylinder two-stroke engines, each driving a three-blade, ground-adjustable pitch Ivoprop propeller. Alternatively, two 38.3 kW (51.6 hp) Rotax 462 or 34.0 kW (45.6 hp) Rotax 503 twin-piston engines. Fuel capacity 90 litres (23.8 US gallons; 19.8 Imp gallons) standard; 180 litres (47.5 US gallons; 39.6 Imp gallons) optional in Twin Vista SS.

SYSTEMS: Electrical system with 12 V DC, 14 Ah battery for electric starter.

DIMENSIONS, EXTERNAL:
- Wing span ... 11.34 m (37 ft 2½ in)
- Length: fuselage ... 5.75 m (18 ft 10½ in)
  - overall ... 6.72 m (22 ft 0½ in)
- Height overall ... 1.90 m (6 ft 2¾ in)
- Propeller diameter ... 1.70 m (5 ft 7 in)

DIMENSIONS, INTERNAL:
- Cabin: Length ... 2.20 m (7 ft 2½ in)
  - Max width ... 0.64 m (2 ft 1¼ in)

A-26 Twin Vista SS assembled by a US builder *(Paul Jackson)* ***NEW***/0552709

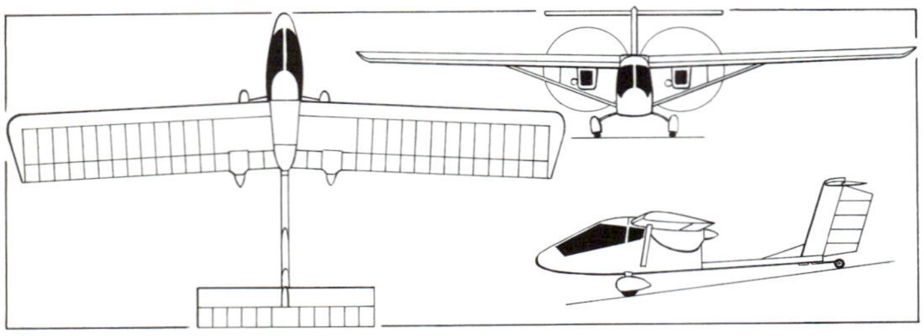

Aeroprakt-26 twin-engined ultralight *(James Goulding)* 0051502

Max height ... 1.20 m (3 ft 11¼ in)

AREAS:
Wings, gross ... 15.70 m² (169.0 sq ft)

WEIGHTS AND LOADINGS:
Weight empty: Twin Vista ... 300 kg (660 lb)
Twin Vista SS ... 318 kg (701 lb)
Max T-O weight ... 550 kg (1,212 lb)[1]

[1] *or local ultralight limit*

PERFORMANCE (Twin Vista SS):
Max level speed ... 97 kt (180 km/h; 112 mph)
Normal cruising speed ... 96 kt (177 km/h; 110 mph)
Stalling speed, landing configuration ... 30 kt (55 km/h; 34 mph)
Max rate of climb at S/L ... 549 m (1,800 ft)/min
T-O run ... 31 m (100 ft)
Landing run ... 107 m (350 ft)
Range with normal fuel ... 173 n miles (321 km; 200 miles)
Endurance ... 6 h 0 min
g limits ... +4/–2

UPDATED

AEROPRAKT-28

US marketing name: SA-28 Victor

TYPE: Four-seat kitbuilt twin.

PROGRAMME: Project began in December 1997; first flight October 1999; after almost 100 hours, prototype shipped to Dubai for one year of field tests. Initial eight aircraft were reported to be in production by 2000. Following Dubai trials, prototype refurbished at Kiev and registered to John Hunter of Sebring, Florida, in April 2001. US Marketing by Spectrum Aircraft.

CURRENT VERSIONS: **A-28:** Prototype. Two 59.6 kW (79.9 hp) Rotax 912 UL engines.

SA-28 Victor: US production version. Several changes, including altered fuselage shape. Optional nosewheel configuration, with and without retraction option. As described.

Prototype Aeroprakt-28 on the US civil register *(Paul Jackson)* 0525069

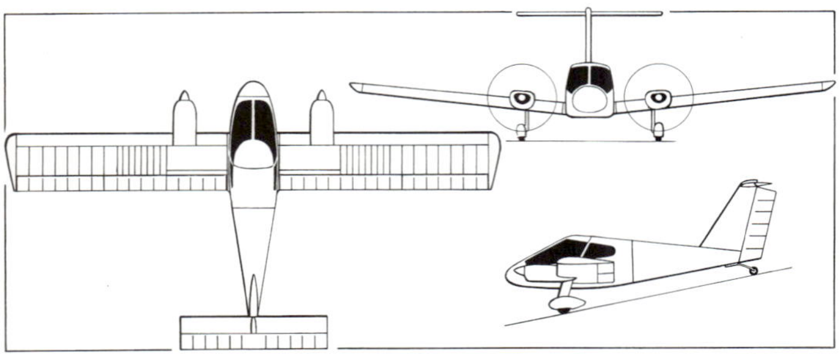

Aeroprakt-28 (two Rotax 912 piston engines) *(James Goulding)* 0051503

SA-28 Victor Special: Rotax 914 engines; each 84.5 kW (113.3 hp). Rate of climb 457 m (1,500 ft)/min; cruising speed 126 kt (233 km/h; 145 mph).

COSTS: US$114,500 including engines, propellers and instruments (2004).

DESIGN FEATURES: Provides twin-engined reliability for flight over hazardous terrain, with 300 kg (661 lb) payload and 10 hour endurance.

Low-wing, twin-engined configuration with T tail; sole four-seater in current Aeroprakt range.

Unswept, constant-chord wings of P-IIIa-15 section; thickness/chord ratio 15 per cent; washout 5°; chamfered wingtips. Dihedral 5°; incidence 5°; twist 3°. US production version has NACA 633-618 laminar flow aerofoil.

FLYING CONTROLS: Manual. Conventional ailerons, slotted flaps and rudder; single-piece, mass-balanced elevator. Flight-adjustable trim tabs on fin and rudder. Flap deflections 0, 14, 33 and 40°.

STRUCTURE: Composites cabin with riveted 2024-T3 aluminium monocoque rear fuselage; tailplane, of composites ribs and skin built on metal spar, is reinforced version of A-20 unit; aluminium rudder structure with Stitts Poly-Fiber covering. Aluminium wings with fabric-covered ailerons. Composites engine cowlings and wheel fairings. Aluminium 6061-T256 mainwheel legs.

LANDING GEAR: Tailwheel type; fixed. Wing-mounted cantilever main legs with Citröen motorcar shock-absorbers and wheel fairings. Mainwheels size 6.00-6; tailwheel 260×85.

POWER PLANT: Two 73.5 kW (98.6 hp) Rotax 912 ULS flat-four engines. Fuel in wings, total capacity 180 litres (47.6 US gallons; 39.6 Imp gallons). Aeroprakt three-blade ground-adjustable pitch propellers.

ACCOMMODATION: Up to four persons in side-by-side pairs. Two door/windscreens open forwards on centreline hinges.

DIMENSIONS, EXTERNAL:
Wing span ... 12.10 m (39 ft 8½ in)
Wing chord, constant ... 1.40 m (4 ft 7 in)
Wing aspect ratio ... 8.8
Length overall ... 7.045 m (23 ft 1¼ in)
Height overall ... 2.33 m (7 ft 7¾ in)
Tailplane span ... 1.17 m (3 ft 10 in)
Wheel track ... 3.60 m (11 ft 9¾ in)
Wheelbase ... 4.90 m (16 ft 1 in)
Propeller diameter ... 1.85 m (6 ft 0¾ in)

DIMENSIONS, INTERNAL:
Cabin: Length ... 2.10 m (6 ft 10¾ in)
Max width ... 1.16 m (3 ft 9½ in)
Max height ... 1.28 m (4 ft 2½ in)

AREAS:
Wings, gross ... 16.70 m² (179.8 sq ft)

WEIGHTS AND LOADINGS:
Weight empty ... 608 kg (1,340 lb)
Max T-O weight ... 1,000 kg (2,204 lb)
Max wing loading ... 59.9 kg/m² (12.26 lb/sq ft)
Max power loading ... 6.80 kg/kW (11.18 lb/hp)

PERFORMANCE:
Never-exceed speed (VNE) ... 156 kt (289 km/h; 180 mph)
Max cruising speed ... 109 kt (201 km/h; 125 mph)
Stalling speed, landing configuration ... 40 kt (73 km/h; 45 mph)
Max rate of climb at S/L ... 366 m (1,200 ft)/min
T-O run ... 107 m (350 ft)
Landing run ... 122 m (400 ft)
Range with max fuel ... 908 n miles (1,681 km; 1,045 miles)
Endurance ... 12 h 30 min
g limits ... +4/–2

UPDATED

AEROPRAKT-36

US marketing name: Vulcan

TYPE: Tandem-seat twin/kitbuilt.

PROGRAMME: Commissioned by Spectrum Aircraft of US in February 2002; design began September 2002; cockpit mockup approved December 2002; assembly of prototype began May 2003; first flight 2004. Kit-deliveries began immediately. Promoted at AirVenture, Oshkosh, July 2003. Development of Aeroprakt-26 Vulcan.

CUSTOMERS: Several deposits received from US customers by early 2004.

DESIGN FEATURES: Generally as for A-26, but with two close-centreline-thrust engines.

Wing section P-IIIa-15; dihedral 1° 30'; incidence 2° 30' at mean aerodynamic chord; twist 3°; sweptback 3°.

FLYING CONTROLS: Manual. Mass-balanced ailerons, rudder and all-moving tailplane with integral, electric anti-balance/trim tab. Flaps; deflections 10, 20, 30 and 40°. Actuation by pushrods, except cables for mid-portion of tailpane linkage.

STRUCTURE: Single-spar, strut-braced wing of aluminium alloy with steel fittings; composites wingtip and leading-edge, latter forming fuel tank; aluminium alloy bracing struts with composites end fairings. Forward and centre fuselage of glass fibre and foam sandwich monocoque, reinforced by bulkheads and beams, with frames and stringers, plus integral fin and rudder, both of aluminium. Single-spar, all-moving tailplane of aluminium, except composites mass-balance fairing. All control surfaces aluminium-skinned.

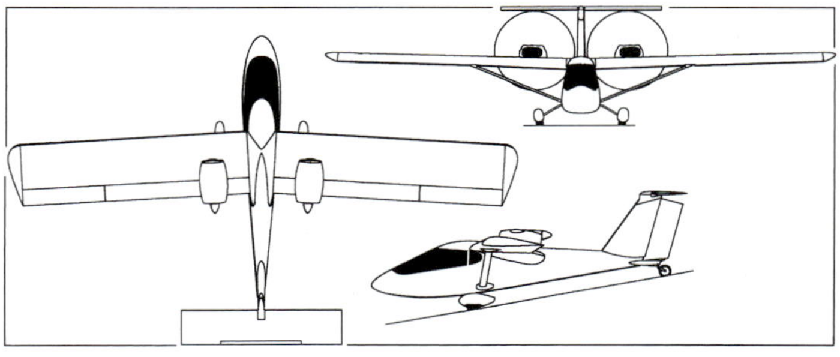

Aeroprakt-36 Vulcan *(Paul Jackson)* ***NEW***/0580702

LANDING GEAR: Tailwheel type; fixed. Fuselage-mounted, spring cantilever mainwheel legs with speed fairings and Matco 6.00-6 wheels with disc brakes. Steerable tailwheel; 3.00-4.

POWER PLANT: Two 73.5 kW (98.6 hp) Rotax 912 ULS flat-four piston engines driving two-blade, pusher propellers. Fuel tanks in inboard wing leading edges, combined capacity 120 litres (31.7 US gallons; 26.4 Imp gallons).

ACCOMMODATION: Two in tandem. Starboard-hinged canopy replaceable by stub windscreen for open-cockpit flying. Rear baggage compartment with 56 cm (1 ft 6 in) square external door, port.

DIMENSIONS, EXTERNAL:
Wing span ... 11.40 m (37 ft 4¾ in)
Wing chord, constant ... 1.40 m (4 ft 7 in)
Wing aspect ratio ... 8.3
Length overall ... 7.34 m (24 ft 1 in)
Height overall ... 1.98 m (6 ft 6 in)
Tailplane span ... 3.60 m (11 ft 9¾ in)
Propeller diameter ... 1.80 m (5 ft 10¾ in)

DIMENSIONS, INTERNAL:
Cockpit: Length ... 2.15 m (7 ft 0¾ in)
Max width ... 0.70 m (2 ft 3½ in)
Max height ... 1.10 m (3 ft 7¼ in)

AREAS:
Wings, gross ... 15.68 m² (168.8 sq ft)
Ailerons (total) ... 1.54 m² (16.58 sq ft)
Flaps (total) ... 2.72 m² (29.28 sq ft)
Fin ... 0.97 m² (10.44 sq ft)
Rudder ... 0.73 m² (7.86 sq ft)
Tailplane ... 2.88 m² (31.00 sq ft)

WEIGHTS AND LOADINGS:
Weight empty ... 450 kg (992 lb)
Max T-O weight ... 750 kg (1,653 lb)
Wing loading ... 47.8 kg/m² (9.79 lb/sq ft)
Power loading ... 5.10 kg/kW (8.38 lb/hp)

PERFORMANCE:
Never-exceed speed (VNE) ... 290 km/h
Max level speed ... 140 kt (260 km/h; 162 mph)
Cruising speed at 75% power ... 124 kt (230 km/h; 143 mph)
Econ cruising speed ... 97 kt (180 km/h; 112 mph)
Stalling speed: flaps up ... 46 kt (85 km/h; 53 mph)
flaps down ... 38 kt (70 km/h; 44 mph)
Max rate of climb at S/L ... 660 m (2,165 ft)/min
Rate of climb at S/L, OEI ... 240 m (787 ft)/min
Service ceiling ... 8,535 m (28,000 ft)
T-O run ... 100 m (330 ft)
T-O to 15 m (50 ft) ... 200 m (660 ft)
Landing from 15 m (50 ft) ... 300 m (985 ft)
Landing run ... 150 m (495 ft)
Range, 30 min reserves ... 513 n miles (950 km; 590 miles)
g limits ... +4.4/–1.76

UPDATED

ANTONOV

AVIATSIONNY NAUCHNO-TEKHNICHESKY KOMPLEKS IMENI O K ANTONOVA (AVIATION SCIENTIFIC-TECHNICAL COMPLEX NAMED FOR O K ANTONOV)

ulitsa Tupoleva 1, 03062 Kiev
Tel: (+38 044) 454 31 49
Fax: (+38 044) 442 41 44
e-mail: info@antonov.com
Web: http://www.antonov.com

GENERAL DESIGNER AND PRESIDENT: Pyotr Balabuyev
FIRST DUPTY GENERAL DESIGNER: Dmytro Kiva
DIRECTOR GENERAL: Volodymir Korol
EXECUTIVE DIRECTOR, ANTONOV AIRLINES: Konstantin Lushakov
HEAD OF MARKETING AND SALES: Alexander Kiva
PUBLIC RELATIONS OFFICER: Andrey Sovenko

Antonov OKB was founded in 1946 by Oleg Konstantinovich Antonov, who died 4 April 1984, aged 78. More than 22,000 aircraft of over 100 types and versions of Antonov design have been built; more than 1,500 have been exported, to 42 countries. Other production includes trolley-buses, trams and racing bicycles.

Antonov continues to be engaged in designing and building new aircraft prototypes as well as modifications of earlier designs; and the provision of operational and product support and engineering work on extending the service life of exisiting aircraft. It also provides service such as basic and conversion training for flight and maintenance crews; and international air charter transportation, particularly of outsize cargoes. On-site training of personnel is available. Antonov also participates in international co-operation in the field of aircraft and equipment design and manufacture as well as the development of land vehicles.

Current projects include the An-148 twin-turbofan airliner and An-70 military transport.

Antonov received Aviation Register approval on 30 December 1992 to develop civil aircraft. It also operates its own cargo airline with a fleet of one An-225, seven An-124s, one An-22, three An-12s, and one each An-26 and An-74. Company parents the Medium Transport Aircraft International Consortium created by Russia and Ukraine in February 1996 and formally established on 18 May 1999.

By November 2003, Antonov-designed aircraft held 477 international records, including 124 gained by the An-225 in a single sortie on 1 September 2001.

Production plants associated with ANTK Antonov include Aviastar, Aviakor, Aviant, Kharkov Aircraft Manufacturing Company, Polyot (Omsk) and NAPO.

UPDATED

ANTONOV AN-38

TYPE: Twin-turboprop transport.

PROGRAMME: Requirement for 25–30 seat development of An-28 emerged during 1989 sales tour of India. Development of all-new An-38 approved by Soviet Ministry of Aviation, late 1990. Details announced, and model displayed, at 1991 Paris Air Show; initial batch of six built, not by Antonov, but at production factory, NAPO, Novosibirsk, Russia: one prototype (01001; first flight 23 June 1994, with TPE331 engines), four trials aircraft and one (01002) for static testing at Kiev; certification to AP-25 granted 22 April 1997. In December 1995, Antonov and NAPO formed joint venture company, Siberian Antonov Aircraft, to produce, market and provide after-sales service for the An-38. Indian demonstration tour undertaken in July and August 1997, followed by appearance at Aero India in December 1998 and February 2001.

Production at NAPO was suspended between 2000 and 2003, but in March 2003 an agreement was signed by the plant, Moscow Leasing Company and Volgograd-SpetzAvia for resumption with a batch of five.

CURRENT VERSIONS: **An-38-100:** With Honeywell TPE331 engines. First and second (01003; exhibited Moscow 1997) flying aircraft to this standard. Trials of international navigation avionics completed March 2000.

An-38-110: Reduced avionics fit in comparison with –100.

An-38-120: Enhanced avionics fit in comparison with –100; equipment includes VOR/DME, Opal-B voice recorder and SPPZ-2000 ground proximity warner. NAPO-Aviation's aircraft to this standard.

An-38-200: With Omsk MKB 'Mars' TVD-20-03 engines. Third and fourth prototypes were planned to this standard when engine development complete; however, minor problems with Aerosila AV-36M propeller delayed programme, but maiden flight achieved (at NAPO) 11 December 2001. Equipment standard as for An-38-120, but with addition of TCAS-2000 traffic collision avoidance system. State Tests were completed on 28 November 2002.

Antonov An-38 RA-41910 on lease to Layang Layang Aerospace of Malaysia *(Paul Jackson)* ***NEW***/0583346

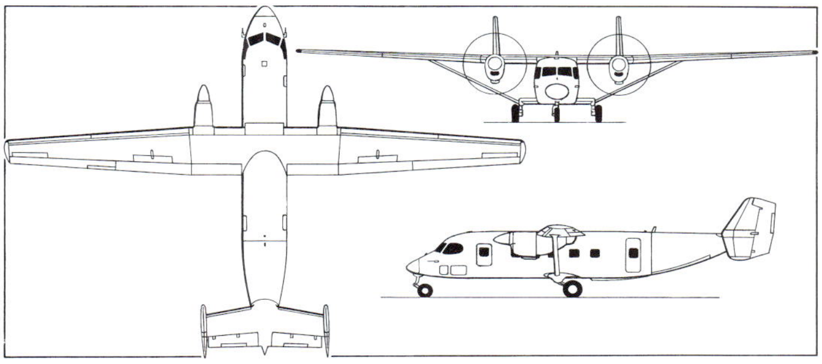

Antonov An-38-100 27-seat transport *(Jane's/Mike Keep)* 0051508

An-38D: Proposed military version for parachutist dropping, medical evacuation and general light transport.

An-38K: Convertible version of An-38-100; large upward-hinging side door at rear on port side; able to carry four LD-3 (KMP-500) or five LD-3K containers (= *konteinernyi*); cargo handling equipment removable for conversion to 30-passenger transport.

Versions with RKBM TVD-1500 engines said to be under construction in 2000, but none had emerged by mid-2002. All versions can be equipped for aerial photography (An-38F: *fotografiya*), survey (An-38GF *geofizichesky*), forest patrol (An-38D: *desantnyi*), VIP transport, ambulance (An-38S: *sanitarnyi*; six stretchers, nine seated, with attendant) and fishery/ice patrol duties (An-38LR: *ledovoi razvedki*). An assault transport, also designated An-38D, capable of carrying 22 paratroops, 26 troops or 3 tonnes of cargo, was revealed to be in the design stage in August 2001; –100 or –200 engine options will be available.

CUSTOMERS: Eight Srs 100s produced by mid-2000: two prototypes (one at Antonov; one at NAPO), one static test airframe and five with airlines (Vostok, three; Alrosa, two); by mid-2004, no further production had been undertaken. Two Vostok aircraft to Malaysia during 2001 for lease by Layang Layang Aerospace for tourist flights, cargo transport and aerial photography.

First three (subsequently increased to eight) An-38-100s ordered for Vostok Airlines and received by mid-1995 for one year of intensive trials before passenger certification. Second firm customer is Chukotavia (10; although initial batch is two); letters of intent from Petropavlovsk-Kamchatsky, Merninsky, Novosibirsky, Ulyanovsky and Nikolaevsk-na-Amur. In 1999–2000, second prototype was being operated by NAPO-Aviatrans, the airline of the NAPO aircraft factory; this also to Layang Layang by 2003. By July 2003, Layang Layang was discussing order for six An-38s subsequent to type's certification in Malaya. In 1998, Siberia Airlines was considering purchase of two. Alrosa-Avia of Zhukovsky has ordered five for diamond mining support, of which first in service by early 2000; second followed in July 2000. Indian Air Force interest expressed in initial six to 10; estimated market for 40 with Indian regional airlines (2001), of which 20 covered by letters of intent. Interest in -200 from Vietnam Airlines, which in 2001 signed lease for NAPO's own -120; this still in Vietnam in 2004. First customer for Srs 200 is expected to be Kemerovo Airlines, which required two for delivery in late 2003.

COSTS: An-38-100 basic price US$4 million (2000); An-38-200 US$3.5 million (2003).

DESIGN FEATURES: Developed from PZL Mielec (Antonov) An-28 to replace An-24s, Let L 410s and Yak-40s. New high-efficiency engines; lengthened passenger cabin; optional weather radar and automatic flight control system; improved sound and vibration insulation; reduced external noise; wheel or ski landing gear; rear cargo door and cargo handling system; able to operate from unpaved runways; operating temperatures from –45 to +45°C, including 'hot-and-high' conditions. Service life 30,000 hours. Maintenance (including overhaul) requirement 4 man-hours/flying hour.

FLYING CONTROLS: Conventional and manual. Single-slotted mass and aerodynamically balanced ailerons (port aileron has trim tab), designed to droop with large, hydraulically actuated, two-segment double-slotted flaps; electrically actuated trim tabs in each elevator have manual back-up; twin rudders each with electrically actuated trim tab; automatic leading-edge slats over full span of wing outboard of engines; slab-type spoiler forward of each aileron and each outer flap segment at 75 per cent chord.

LANDING GEAR: Tricycle type; fixed. Mainwheels 610×320-330; nosewheel 600×320-254.

POWER PLANT: Two Honeywell TPE331-14GR-801E turboprops, each 1,118 kW (1,500 shp), driving Hartzell HC-B5MA five-blade propellers rotating at 1,522 rpm; or two Omsk MKB 'Mars' TVD-20 turboprops, each 1,029 kW (1,380 shp), driving Aerosila AV-36M quiet, constant-speed reversible-pitch propellers rotating at 1,827 rpm.

ACCOMMODATION: Two crew side by side on flight deck; passenger cabin equipped normally with 26 seats, basically three-abreast, with centre aisle; 27 seats at 75 cm (29½ in) pitch optional; ambulance version for six stretchers, eight seated casualties and medical attendant, executive versions with eight to 10 seats and forest surveillance/paradrop version for 26 smoke-jumpers or trainee paratroops (reduced to 22 with full kit) available; seats and baggage compartment can be folded quickly against cabin wall to provide clear space for 2,500 kg (5,510 lb) of freight. Maximum practical cargo dimensions in combi variant are 1.40 m (4 ft 7 in) height and 1.05 m (3 ft 5¼ in) width with seats removed or 0.95 m (3 ft 1½ in) with seats stowed against walls. Door with airstairs on port side, with service door opposite; emergency exit each side. Optional cargo door under upswept rear fuselage slides forward under cabin for direct loading/unloading of freight.

AVIONICS: Russian or Western equipment; latter comprises Bendix/King Silver Crown range; former listed below.

Comms: SO-72 transponder.

Radar: A813Ts weather radar.

Flight: BSFK-1 navigation system; VEM-72PB-3A altimeter and A-037 radar altimeter; twin US-450K airspeed indicators; SAU-28 AFCS; M3 GPS; VMD-94 DME; SPPZ-200 GPWS; TCAS-2000 TCAS; and KURS-93M autoland.

EQUIPMENT: Hand-operated travelling overhead winch in cabin: capacity 500 kg (1,102 lb).

DIMENSIONS, EXTERNAL:

Wing span	22.065 m (72 ft 4¾ in)
Length overall	15.67 m (51 ft 5 in)
Height overall	5.05 m (16 ft 6¾ in)
Span over tailfins	5.14 m (16 ft 10½ in)
Wheel track	3.515 m (11 ft 6½ in)
Wheelbase	6.345 m (20 ft 9¾ in)
Distance between propeller centres	5.58 m (18 ft 3¾ in)
Propeller diameter: Hartzell	2.85 m (9 ft 4 in)
AV-36	2.65 m (8 ft 8¼ in)
Cargo ramp:	
Length	2.40 m (7 ft 10½ in)
Width: at floor	1.40 m (4 ft 7 in)
at rear	1.00 m (3 ft 3¼ in)
Passenger door: Width	0.70 m (2 ft 3½ in)
Height	1.55 m (5 ft 1 in)

DIMENSIONS, INTERNAL:

Cargo hold: Length, excl ramp	7.83 m (25 ft 8¼ in)
Width: max	1.74 m (5 ft 8½ in)
at floor	1.55 m (5 ft 1 in)
Max height	1.70 m (5 ft 7 in)
Volume, incl ramp	24.7 m^3 (872 cu ft)

WEIGHTS AND LOADINGS (A: An-38-100, B: An-38-200, K: An-38K):

Weight empty: A	5,735 kg (12,643 lb)
Max payload: A, B	2,500 kg (5,510 lb)
K	3,200 kg (7,055 lb)
Max fuel: A	2,210 kg (4,872 lb)
Max T-O weight: A	9,500 kg (20,943 lb)
B	9,930 kg (21,891 lb)

PERFORMANCE (estimated, with TPE331 engines):

Max level speed	219 kt (405 km/h; 252 mph)
Nominal cruising speed:	
A, B	205 kt (380 km/h; 236 mph)
Nominal max cruising altitude: A, B	4,200 m (13,780 ft)
T-O run: A	350 m (1,150 ft)
K	480 m (1,575 ft)
Landing run: A	270 m (885 ft)
K	440 m (1,445 ft)
Balanced field length: A	895 m (2,940 ft)
B	1,050 m (3,445 ft)
Range at FL100, no reserves:	
with max payload: A	513 n miles (950 km; 590 miles)
B	540 n miles (1,000 km; 621 miles)
with max fuel (1,650 kg; 3,638 lb payload):	
A	944 n miles (1,750 km; 1,087 miles)
B	961 n miles (1,780 km; 1,106 miles)

UPDATED

ANTONOV AN-70

TYPE: Medium transport/multirole.

PROGRAMME: Development began 1975 to replace some An-12s remaining in air force service from 2002–2003; initial all-metal, 5.0 m (16 ft 5 in) fuselage diameter design featured four D-236 turboprops and tail gunner's position, and was in competition with Yakovlev design (then known as Yak-44). Significant redesign began in 1986, upon appointment of Pyotr Balabuyev as designer, resulting in postponement of planned 1986 first flight.

Announced by *Izvestia* 20 December 1988; at 1991 Paris Air Show Antonov OKB reported prototype being assembled at Kiev; funding by Russian (80 per cent) and Ukrainian governments under agreement of 24 June 1993; preliminary details released and model displayed at Moscow Aero Engine and Industry Show April 1992.

Prototype rolled out at Antonov's Kiev plant on 20 January 1994; first flight (01-01) 16 December 1994 (was also delivery flight to Gostomel test airfield); this aircraft lost during fourth sortie following in-flight collision with chase An-72 on 10 February 1995. Second prototype (without nose-mounted instrumentation boom) produced by upgrading of static test airframe (for which replacement under construction in 2000); rolled out 24 December 1996; first flight (01-02/UR-NTK) 24 April 1997; international debut at Moscow Air Show, August 1997; handed over to Russian Air Forces' test centre at Akhtubinsk, August 1998. Had flown 380 of planned 780-sortie test programme by January

Antonov An-70 tactical transport *(Paul Jackson)* ***NEW***/0583347

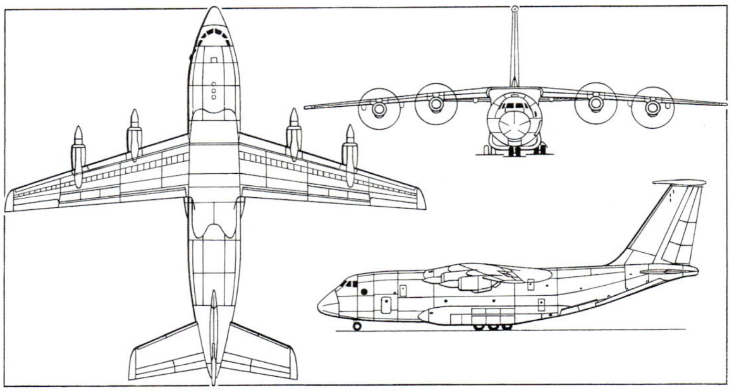

Antonov An-70 tactical transport *(James Goulding)* 0130863

2001; high AoA trials mid-2000; first stage of State testing completed October 2000, confirming safety in all flight regimes. Crash-landed and fuselage broken into two parts immediately after take-off from Omsk on 27 January 2001 following double engine failure. Airframe transferred to Polet aircraft plant and repaired at cost of US$3 million; reflown 5 June 2001; certification rescheduled to first quarter of 2002; appeared at MAKS '01, Moscow, August 2001. By June 2004, stated to require further 100 sorties to complete testing, which target set for end of 2005.

Bilateral agreement between Russia and Ukraine revised 18 May 1999, and underlined decision taken to order first 10 (and 50 engines) before planned production at Aviant plant, Kiev, from 1999 and Aviacor plant, Samara, from 2000; each plant building initial batch of five, which then intended to enter service by 2003. Russian production specification issued 4 December 1999; Russian government decree of 4 October 2000 covers purchase of 164 An-70s, but, at same time, Samara regional government announced that local manufacturers were withdrawing from An-70 project. In late 2003, power-plant manufacturing partner, Salyut of Moscow, proposed a revised compressor design for D-27 engine which would address Russian military concern over efficiency and reliability. Other sources reported simultaneously that these issues already solved in Ukraine. Ukrainian government resolution guaranteeing An-70 purchase passed 12 October 2000; total of 65 to be obtained by 2018. First five for Ukraine ordered from Aviant on 2 April 2001. However, in September 2001, Polyot was allocated Russian An-70 final assembly, augmented by Novosibirsk's NAPO and Voronezh's VASO. Aviacor no longer involved. Russian share of series production is 72 per cent; Ukraine 28 per cent. Assembly of first Polet-built An-70 formally began on 10 October 2002. Polet (Russian) and Aviant plants signed agreement in June 2004 for joint manufacture of An-70 in Ukraine using Russian-built components.

By September 2002, An-70 had flown 550 hours in 440 sorties towards anticipated late 2003 completion of State Tests, but unofficial reports were by that time citing alleged discovery of 382 deficiencies, of which 95 rated essential for pre-delivery rectification. These included engine reliability, the prototype having suffered 52 in-flight shutdowns during its 440 sorties. Russian Air Forces' disinclination to fund An-70 was being discussed in November 2002, and C-in-C voiced outright rejection of purchase at January 2003 press conference, claiming An-70 unsuitable for its intended task of replacing An-12, and pledging support for rival Tupolev Tu-330. Joint (Russo-Ukrainian) acceptance testing was frozen on 25 February 2003; resumed following signature of agreement by Cs-in-C of Russian and Ukrainian air forces, 16 June 2004. By April 2003 however, it was reported unofficially that Air Forces' support had been resumed, but with proviso of rewritten TTZ (tactical and technical specification) detailing changes to avionics and equipment and extending development period. Optimism short-lived, and despite support from some political quarters, An-70 funding was removed from 2004 budget at military request.

Faced with shortfall in Russian financial support, Ukrainian government considered unilateral financial support in 2003 national budget, allowing production launch, but decided against the proposal in November 2002, although continued flight testing did receive official funds. Trials at Feodosia test centre included February 2003 air-drop within 18 seconds of four pallets totalling 34.5 tonnes; operations planned from airstrips with bearing strength of between 6 and 9 kgf/cm^2 (85 to 128 lbf/sq in). Short field trials at Feodosia between 18 July and 4 August 2003.

Ukrainian parliament decision of February 2004 to complete An-70 development and production in two-stage programme between 2004 and 2022 assigns some Hr1 billion up to 2006 for development (Hr130 million; US$25 million), production investment (Hr450 million; US$85 million) and two production aircraft (Hr430 million; US$80 million). Initial expenditure in 2004 was Hr243 million (US$46 million). Phase 2, from 2007 to 2022, not yet costed. Report of May 2004 implausibly indicated that Ukrainian Air Force would receive first two production An-70s before end of that year.

Medium Transport Aircraft International Consortium (MTA) formed February 1996 to co-ordinate development, certification, sales and after-service, comprising four Ukrainian and six Russian companies: ANTK Antonov (designer and component manufacturer), ZMKB Progress (engine designer), Aviant (airframe manufacturer) and Motor Sich (engine manufacturer) from Ukraine; Aviapribor (flight control system manufacturer),

Elektroavtomatika (avionics), Leninets (airborne monitoring and diagnostic system) and Aerosila (propfan) from Russia. Refer to 'Structure'. Following inconclusive discussions with Russia over international promotion, Ukraine unilaterally announced Poland as partner for Western sales of An-70 in November 2003. Was also hoping to obtain Chinese order.

CURRENT VERSIONS: **An-70:** Military STOL transport; proposed production version. Stated to have double the payload of Lockheed Martin C-130J, with similar STOL and rough-field performance, yet only 40 kt (74 km/h; 46 mph) slower than the Boeing C-17A Globemaster III, which carries 15 per cent more payload.

Detailed description applies to baseline An-70.

An-70-100: Proposed military STOL transport, as An-70 but with two-crew cockpit.

An-77: Proposed military STOL transport for export customers; as An-70-100 but with cockpit for two or three crew. Runway length of 1,900 m (6,235 ft) required with 35,000 kg (77,161 lb) payload for 2,051 n mile (3,800 km; 2,361 mile) range.

An-70T: Commercial transport, generally as An-70, with improved runway capability but no requirement for STOL. Two or three crew. To carry 35,000 kg (77,161 lb) payload 2,051 n miles (3,800 km; 2,361 miles) from 1,900 m (6,235 ft) runway, or 20,000 kg (44,092 lb) for 2,915 n miles (5,400 km; 3,555 miles) from 1,300 m (4,265 ft) runway.

An-170: Heavy transport derivative carrying 45,000–50,000 kg (99,210–110,230 lb) of cargo.

CUSTOMERS: Requirements originally expressed by Russian Air Forces for up to 500 and by Ukrainian Air Force for 100. This modified by 1999 to 164 for Russia and 65 for Ukraine, with envisaged in-service date of 2002. German requirement potentially for 75 aircraft. By mid-1999 potential civil operators had signed documents of intent for some 100 further aircraft. Chinese interest reported in 2000. Czech Republic reportedly finalising a contract for three in 2002, delivery to be from Omsk at one per year, beginning 2005. Hungarian and Indian interest reported in late 2002, the latter considering licensed production.

COSTS: Military version US$60 million (2002), but estimated in 2003 that this could be reduced to "under US$40 million". Civil An-70T to cost 20 to 30 per cent less than military variant. Total of US$3,500 million reportedly spent on development by late 2002, of which Russian defence ministry then owed Antonov US$49 million; further US$86 million (of which US$61 million due from Russia) required to complete development and flight testing. Engines cost US$75 million up to 2002, and further US$30 million allegedly required to address recorded malfunctions during testing.

DESIGN FEATURES: First aircraft to fly powered only by propfans. Slightly larger than projected European FLA transport, much smaller than US Boeing C-17A; conventional high-wing configuration, with wings and tail surfaces slightly sweptback; supercritical wing section; anhedral from roots; loading ramp/doors under upswept rear fuselage with adjustable sill height and built-in cargo handling system; horizontal tail surfaces on rear fuselage; propfans mounted conventionally on wing leading-edge; propeller wash doubles wing lift during take-off and landing. Multiple-section control surfaces provide redundancy in event of battle damage or physical obstruction. Claimed features include independent operational capability at non-equipped airfields for 30 days. Design life 15,000 cycles and 45,000 flying hours in 25 years. Operable 3,500 hours per year, with eight to 10 man-hours of maintenance per flying hour. Cost-effective with only 200 flying hours per month.

FLYING CONTROLS: Prototypes have fly-by-wire system with four digital and six analogue channels; primary controls are quadruplex; back-up by unique fly-by-hydraulics system, in which pilot or autopilot inputs are relayed (via conventional 'mini-wheels') to actuators by commands in hydraulic control channels, unaffected by electromagnetic interference. Production aircraft will have a four-channel, all-digital primary FCS, rather than the hybrid digital/analogue system of the prototypes. Secondary FCS controls two independent flap systems, leading-edge slats and blown flaps. Three-section double-winged rudder. Double-slotted trailing-edge blown Fowler flaps in two sections on each wing; forward element maximum deflection 60°; intermediate settings 5, 10, 15, 20, 25, 30, 35, 40 and 50°. Three-section spoilers forward of each outer flap. Leading-edge has flaps inboard; slats centre and outboard. Two-section, double-hinged elevators forward section maximum deflections +28/–20°, rear section +50/–40°, to enhance low-speed authority; horizontal stabilisers are fitted with automatic leading-edge slats.

STRUCTURE: Approximately 28 per cent of airframe, by weight, made of Organit composites, including complete tail unit, ailerons and flaps. Fuselage stringer/skin joints are spot-welded and hot-bonded, manually.

Single source for all major components. Russian assembly by Polyot, which also to build cargo hold; NAPO producing the centre section; VASO, the wing. Aviant at Kiev to assemble Ukrainian aircraft and build flight deck, empennage and engine nacelles.

LANDING GEAR: Twin-wheel nose unit; each main unit has three pairs of wheels in tandem, retracting into large fairing on side of cabin; can operate from CBR6 unpaved surfaces. All tyres 1,120×450. Steel-steel brakes. Nosewheel turning angle ±55° for taxying; nosewheel turning radius 19.6 m (64¼ ft); wingtip turning radius 34.8 m (114¼ ft); required taxiway width for 180° turn 38.4 m (126 ft).

POWER PLANT: Four ZMKB Progress/Ivchenko D-27 propfans, each 10,290 kW (13,800 shp) built consortium including Motor-Sich and Russia's Salyut. Aerosyla Stupino SV-27 contrarotating propellers each with eight composites blades in front and six at rear. Reversible-pitch blades of scimitar form, with electric anti-icing.

ACCOMMODATION: Three flight crew (two pilots and flight engineer or 'tactical pilot') plus loadmaster; navigation station on captain's left is optionally operated by fourth member of flight crew; provision for converting cockpit for two-crew operation, with co-pilot operating flight engineer's station; seats in forward fuselage for two cargo attendants; freight loaded via rear ramp using four built-in, powered hoists (each of 3 tonne capacity) reaching out 6.6 m (22 ft) from aircraft. Hoists can be combined for heavier loads. Freight can be carried on PA-5.6 rigid pallets, PA-3, PA-4 and PA-6.8 flexible pallets, in UAK-2.5, UAK-5 and UAK-10 containers; unpackaged freight, wheeled and tracked vehicles, food and perishables can be carried; seats for 300 troops, or 206 stretchers, can be installed using optional, prefabricated (10 section) upper deck or optional, easily removable seven-section upper deck in cargo hold; vehicles, freight and paratroops can be airdropped; maximum single airdrop item weight 20,000 kg (44,092 lb); crew door at front of cabin on port side; two upper deck doors each side, front and rear; cargo hold pressurised and air conditioned.

SYSTEMS: Aircraft systems automated to simplify operation and decrease probability of crew errors. Electronpribor engine control system; Leninets monitoring and information system.

AVIONICS: Integrated by Avapribor, Leninets and Elektroavtomatika. Flight data, navigation and radio-navigation systems to ARINC 700 requirements; digital multiplex data interface equivalent to Western MIL-STD-1553B.

Comms: Integrated system by Gorkiski.

Flight: Ring laser INS; SKI-77 HUD; flight management system; designed for operation in adverse weather and for landing in ICAO Cat. II and IIIa conditions. BASK-70 onboard diagnostic system collects data from subsystems, registering and analysing about 8,000 parameters.

Instrumentation: Ten-screen EFIS by Elektroavtomatika comprises six main screens, each 200 × 200 mm (7¾ × 7¾ in), facing pilots and two each at navigation and flight engineer's stations, plus smaller secondary LCD screens and roof-mounted HUDs for pilot and co-pilot on production aircraft.

EQUIPMENT: Four electric hoists in hold.

DIMENSIONS, EXTERNAL:

Wing span	44.06 m (144 ft 6¾ in)
Length overall	40.73 m (133 ft 7½ in)
Fuselage diameter	5.60 m (18 ft 4½ in)
Height overall	16.38 m (53 ft 9 in)
Wheel track (bogie centres)	5.21 m (17 ft 1 in)
Wheelbase: front mainwheels	12.48 m (40 ft 11¼ in)
centre mainwheels	13.98 m (45 ft 10½ in)
rear mainwheels	15.48 m (50 ft 9½ in)
Propeller diameter	4.50 m (14 ft 9 in)
Propeller ground clearance (outer)	3.00 m (9 ft 10 in)
Rear-loading aperture: Height	4.10 m (13 ft 5½ in)
Width	4.00 m (13 ft 1½ in)

DIMENSIONS, INTERNAL:

Cargo hold:	
Floor length: excl ramp	18.60 m (61 ft 0¼ in)
incl loadable ramp	21.90 m (71 ft 10¼ in)
Max width	4.80 m (15 ft 9 in)
Max width at floor	4.00 m (13 ft 1½ in)
Height: max	4.40 m (14 ft 5¼ in)
min	4.10 m (13 ft 5½ in)
Floor area, incl ramp	89.0 m^2 (958 sq ft)
Volume: pressurised	400.0 m^3 (14,126 cu ft)
total	425.0 m^3 (15,008 cu ft)

WEIGHTS AND LOADINGS:

Weight empty	74,000 kg (163,140 lb)
Normal payload (incl 5,000 kg; 11,025 lb on ramp)	35,000 kg (77,161 lb)
Normal payload from unpaved runway	30,000 kg (66,138 lb)
Payload: max	47,000 kg (103,615 lb)
restricted runway option	35,000 kg (77,161 lb)
700 m runway option	20,000 kg (44,092 lb)
Max T-O weight	145,000 kg (319,675 lb)
Max power loading	3.52 kg/kW (5.79 lb/shp)

PERFORMANCE (estimated):

Cruising speed: long range	405 kt (750 km/h; 466 mph)
max short range	432 kt (800 km/h; 497 mph)
Min control speed (demonstrated)	53 kt (98 km/h; 61 mph)
Nominal cruising height	9,100-11,000 m (29,860–36,080 ft)
Runway length required:	
conventional T-O	1,550–1,800 m (5,085–5,905 ft)
short T-O	700 m (2,300 ft)
Range (runway length A: 1,800 m; 5,905 ft, B: 700 m; 2,300 ft)	
with 47 tonnes:	
A	1,619 n miles (3,000 km; 1,864 miles)
B	not an option
with 35 tonnes:	
A	2,699 n miles (5,000 km; 3,106 miles)
B	not an option
with 30 tonnes:	
A	3,239 n miles (6,000 km; 3,728 miles)
B	647 n miles (1,200 km; 745 miles)
with 20 tonnes:	
A	3,563 n miles (6,600 km; 4,101 miles)
B	1,619 n miles (3,000 km; 1,864 miles)
with max fuel: all options	4,319 n miles (8,000 km; 4,971 miles)

UPDATED

ANTONOV AN-72 AND AN-74

NATO reporting name: Coaler

TYPE: Twin-jet freighter.

PROGRAMME: Originated as military transport based on stillborn An-60 64–73 seat civil airliner designed to meet 1967 specification. Revised design, following issue of military requirement in 1968, included relocation of engines above wings. Prototype An-72, built at Kiev, flew (SSSR-19774) 31 August 1977, followed by five (one static test, one fatigue test, three flying) preproduction aircraft, 1979–81. Manufacture transferred to Kharkov, Ukraine, where first production An-72 flew 22 December 1985.

An-74 also produced at Kharkov from December 1986. An-74 announced February 1984, meeting requirement for operation in Arctic and Antarctic conditions. An-72P maritime patrol version demonstrated 1992. Production of An-74 also started by Polyot Industrial Association at Omsk, Russia, in 1993 (assisted by Progress at Arsenyev); first Polyot aircraft (RA-74050) was flown 25 December 1993. Development by Antonov of various new versions of An-74 started 1995.

CURRENT VERSIONS: **An-72A: ('Coaler-C')** STOL transport for military use. Compared with An-72 ('Coaler-A') prototypes, wing span increased by 6.00 m (19 ft 8¼ in), fuselage length by 1.50 m (4 ft 11 in), fuel load by 2,500 kg (5,512 lb). An-72 prototype 003/SSSR-19795

Antonov An-72/74 thrust reversers *(Paul Jackson)* ***NEW***/0589979

Antonov An-74TK-200 combi transport *(Paul Jackson)* ***NEW***/0589978

converted to An-74 and first flew (SSSR-780334) 29 September 1983. Additionally, on production version, tail unit and engine air intake de-icing improved over An-72 prototype; Buran radar in enlarged radome; advanced navigation aids, including inertial navigation system; provision for wheel/ski landing gear.

An-72P: Maritime patrol version; delivered to Ukraine during 1990s.

An-72-100: Civilianised An-72, certified 1997; upgraded avionics.

An-74: Designation refers, generally, to all civilian versions (as under), although initially intended for Polar research, ice floe reconnaissance and transport. Airframe identical with An-72 except for two blister windows at rear of flight deck and front of cabin on port side. Proving trials 1983–89. First preproduction An-74 (SSSR-58642) first flew at Kharkov on 26 June 1986; further five development aircraft built by 1989. Maiden flight of production An-74 (c/n 0706) December 1989; features include new APU, Buran-74 radar in larger radome and improved navigation aids. Type certificate awarded August 1991; initially operated by Yakutsk division of Aeroflot.

An-74-200: Freight version with D-36 Series 3A engines of unchanged rating; increased payload and maximum T-O weight. Able to carry four UAK-2.5 containers. Crew of four/five.

An-74T-200A ('Coaler-B'): Transport (*transportnyi*) version with longer hold; payload 10,000 kg (22,045 lb); roller conveyors in floor; crew of two. Range with maximum payload up to 728 n miles (1,350 km; 838 miles), with 3,500 kg (7,716 lb) payload 2,159 n miles (4,000 km; 2,485 miles). MTOW increased to 36,500 kg (80,468 lb).

An-74TK-200 ('Coaler-B'): Convertible transport/passenger aircraft (*transportnyi konvertiruyemnyi*) with twin seats for 52 passengers that fold against cabin walls, and with baggage racks, buffet/galley and lavatory. Certified by Interstate Aviation Committee. Alternative all-cargo or all-passenger or combi layouts. Typical combi options include 12 passengers plus 6,000 kg (13,228 lb) of freight and 20 plus 4,500 kg (9,921 lb). Built-in loading equipment. Crew of two. Range with 10,000 kg (22,045 lb) payload, 1 hour reserve, 459 n miles (850 km; 528 miles).

An-74TK-100/200 Salon: Business transport for 10 to 16 passengers, with increased cabin comfort. Equipment includes telephone, fax, video, bar, refrigerator, galley and separate rest area. Optional compartment for car at rear. Power plant and weights as basic An-74. Crew of four/five.

An-74T-100 ('Coaler-B'): As An-74T-200, with flight engineer and navigator stations (crew of four).

An-74TK-100: As An-74TK-200, with navigator and flight engineer stations (crew of four). Russian type certificate issued 4 August 1995.

An-74TK-100C: One VIP/medical evacuation aircraft (RA-74005) delivered to Gazprom on 26 February 2002. Air Ambulance Technology (Austrian) equipment. Designation remains '-100C' in both Cyrillic and Roman alphabets.

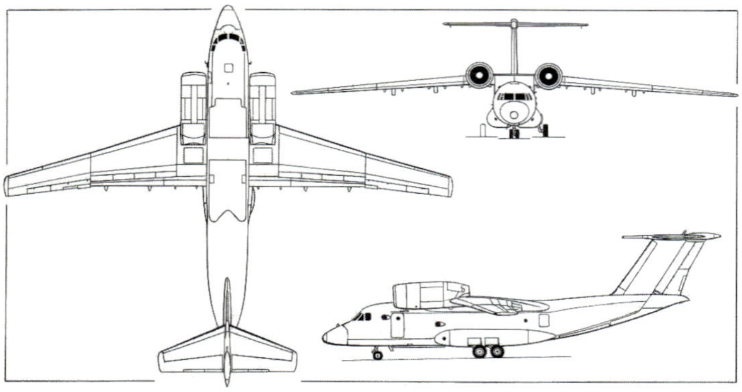

Antonov An-74 ('Coaler-B') STOL transport (two ZMKB Progress/Ivchenko D-36 turbofans) *(Jane's/Dennis Punnett)* 0507543

Antonov An-74TK-200 combi transport *(Paul Jackson)* ***NEW***/0583402

An-74TK-300: *Described separately.*

CUSTOMERS: More than 160 An-72/74s (including 96 military) built before An-74 production additionally established at Omsk; 20 in Russian Air Forces; four in Peruvian Air Force; 26 in Ukrainian Air Force; others in Estonia, Kazakhstan, Latvia, and Moldova. Order placed by Iran in 1997 for 12 An-74TKs for Presidential Guard; other recent deliveries to Laotian government, Ukraine Border Guard (one An-72P in 2002), MChS Rossii, Vitair and Gazpromavia. By end of 2003, operators included nine with 68 An-72s and eight with 62 An-74s. Two An-74TK-200s built at Kharkov in 1999–2000; about 180 built by 2003. Indian interest being pursued in 2000; Omsk received order for one for MChS Rossii and one of required three for Russian Federal Security Service, to have been delivered in 2003; by late 2003, Omsk had delivered total of only five aircraft since 1993 and had further two in various stages of completeness. Kharkov received orders from Egyptian and Libyan governments in July 2004 for combined total of five An-74TK-200s and five -300s.

COSTS: US$9 million (Omsk, 2000).

DESIGN FEATURES: Primary role as STOL replacement for turboprop An-26, with emphasis on freight carrying. High-wing, T-tail configuration, with upswept rear fuselage for freight access. Ejection of exhaust efflux over upper wing surface and down over large multislotted flaps gives considerable increase in lift; high-set engines avoid foreign object ingestion; special ramp/door as An-26; low-pressure tyres and multiwheel landing gear for operation from unprepared strips, ice or snow; sweptback fin and rudder.

Wing leading-edge sweepback 17°; anhedral approximately 10° on outer wings; normal T-O flap setting 25 to 30°, maximum deflection 60°.

FLYING CONTROLS: Conventional and manual. Two tabs in port aileron, one starboard; double-hinged rudder, with tab in lower portion of two-section aft panel; during normal flight only lower rear rudder segment is used; both rear segments used in low-speed flight; forward segment is actuated automatically to offset thrust asymmetry; horn-balanced and mechanically actuated, aerodynamically balanced elevators, each with two tabs; hydraulically actuated full-span wing leading-edge flaps outboard of nacelles; trailing-edge flaps double-slotted in exhaust efflux, triple-slotted between nacelles and outer wings; four-section spoilers forward of triple-slotted flaps; two outer sections on each side raised before landing, remainder opened automatically on touchdown by sensors actuated by weight on main landing gear; inverted leading-edge slat on tailplane linked to wing flaps.

STRUCTURE: All-metal; multispar wings mounted above fuselage; wing skin, spoilers and flaps mainly of aluminium, but flaps covered with titanium aft of engine nacelles; circular semi-monocoque fuselage, with rear ramp/door; tapered fairing forward of T tail fin/tailplane junction, blending into ogival rear fairing.

LANDING GEAR: Hydraulically retractable tricycle type, primarily of titanium. Rearward-retracting steerable twin-wheel nose unit. Each main unit comprises two trailing-arm legs in tandem, each with a single wheel, retracting inward through 90° so that wheels lie horizontally in bottom of fairings, outside fuselage pressure cell. Oleo-pneumatic shock-absorber in each unit. Low-pressure tyres, size 720×310 on nosewheels, 1,050×400 on mainwheels. Hydraulic disc brakes. Telescopic strut hinges downward, from rear of each side fairing, to support fuselage during direct loading of hold with ramp/door under fuselage.

POWER PLANT: Two ZMKB Progress/Ivchenko D-36 high-bypass ratio turbofans (Srs 2A in An-74; Srs 3A in An-74-200, T-100/200 and TK-100/200), each 63.74 kN (14,330 lb st). Integral fuel tanks between spars of outer wings. Thrust reversers standard.

ACCOMMODATION: An-72 has folding seats along sidewalls and removable central seats for 68 troops. Can carry 57 parachutists, and has provision for 24 stretcher patients. An-74's pilot and co-pilot/navigator side by side on flight deck, as in basic An-72, plus flight engineer, with provision for fourth person. Heated windscreen with two wipers. Bulged observation windows on port side for navigator and hydrologist. Flight deck and cabin pressurised and air conditioned. Main cabin designed primarily for freight, including four UAK-2.5 containers or four PAV-2.5 pallets each weighing 2,500 kg (5,511 lb). Provision for wardrobe and galley. Movable bulkhead between passenger and freight compartments, with provision for 1,500 kg (3,307 lb) of freight in rear compartment. Reinforced, movable bulkhead in combi versions protects passengers from shifting cargo in the event of sudden deceleration. Main crew and passenger door at front of cabin on port side. Emergency exit and servicing door at rear of cabin on starboard side.

Downward-hinged and forward-sliding rear ramp/door for loading trucks and tracked vehicles, and for direct loading of hold from trucks. It is openable in flight, enabling freight loads of up to 7,500 kg (16,535 lb), with a maximum of 2,500 kg (5,511 lb) per individual item, to be airdropped by parachute extraction system. In normal freight role, 1,000 kg (2,204 lb) of payload can be placed on ramp. Maximum size of containers up to 1.90 × 2.44 × 1.46 m (6 ft 3 in × 8 ft × 4 ft 9½ in), pallets up to 1.90 × 2.42 × 1.46 m (6 ft 3 in × 7 ft 11 in × 4 ft 9½ in). Main crew and passenger door at front of cabin on port side. Emergency exit and servicing door at rear of cabin on starboard side.

SYSTEMS: Air conditioning system with independent temperature control in flight deck and main cabin areas; used to refrigerate main cabin when perishable goods carried. Maximum cabin pressure differential 0.49 bar (7.1 lb/sq in). Hydraulic system for landing gear, flaps and ramp. Electrical system powers auxiliary systems, flight deck equipment, lighting and mobile hoist. Thermal de-icing system for leading-edges of wings and tail unit (including tailplane slat), engine air intakes and cockpit windows. Provision for TA-12 APU in starboard landing gear fairing. This can be used to heat cabin; under cold ambient conditions, servicing personnel can gain access to major electric, hydraulic and air conditioning components without stepping outside.

Antonov An-74TK-100, cutaway drawing key

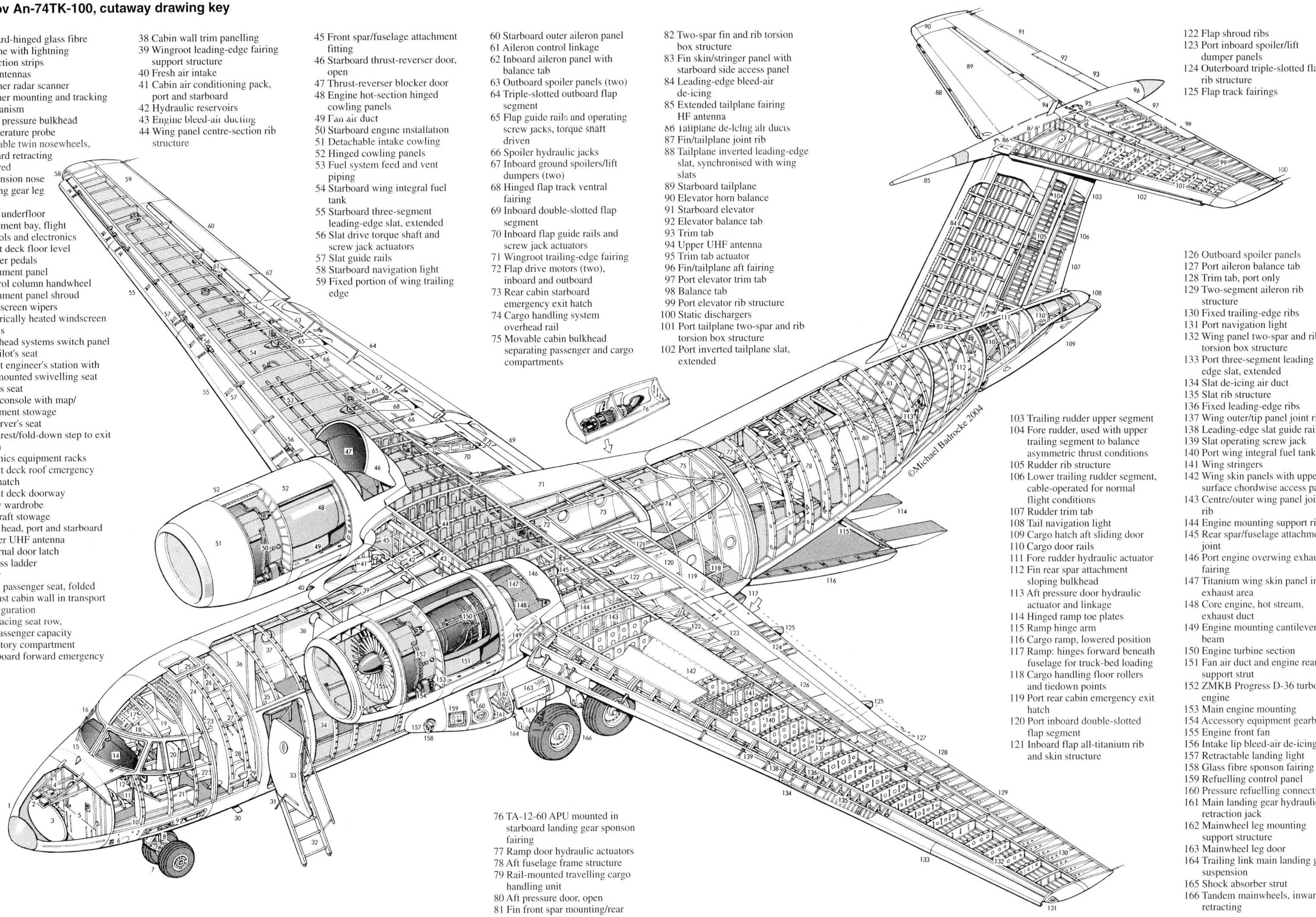

1 Upward-hinged glass fibre radome with lightning protection strips
2 ILS antennas
3 Weather radar scanner
4 Scanner mounting and tracking mechanism
5 Front pressure bulkhead
6 Temperature probe
7 Steerable twin nosewheels, forward retracting
8 Levered suspension nose landing gear leg strut
9 Nose underfloor equipment bay, flight controls and electronics
10 Flight deck floor level
11 Rudder pedals
12 Instrument panel
13 Control column handwheel
14 Instrument panel shroud
15 Windscreen wipers
16 Electrically heated windscreen panels
17 Overhead systems switch panel
18 Co-pilot's seat
19 Flight engineer's station with rail-mounted swivelling seat
20 Pilot's seat
21 Side console with map/ document stowage
22 Observer's seat
23 Headrest/fold-down step to exit hatch
24 Avionics equipment racks
25 Flight deck roof emergency exit hatch
26 Flight deck doorway
27 Crew wardrobe
28 Life raft stowage
29 Pitot head, port and starboard
30 Lower UHF antenna
31 External door latch
32 Access ladder
33 Door
34 Twin passenger seat, folded against cabin wall in transport configuration
35 Aft-facing seat row, 52-passenger capacity
36 Lavatory compartment
37 Starboard forward emergency exit
38 Cabin wall trim panelling
39 Wingroot leading-edge fairing support structure
40 Fresh air intake
41 Cabin air conditioning pack, port and starboard
42 Hydraulic reservoirs
43 Engine bleed-air ducting
44 Wing panel centre-section rib structure
45 Front spar/fuselage attachment fitting
46 Starboard thrust-reverser door, open
47 Thrust-reverser blocker door
48 Engine hot-section hinged cowling panels
49 Fan air duct
50 Starboard engine installation
51 Detachable intake cowling
52 Hinged cowling panels
53 Fuel system feed and vent piping
54 Starboard wing integral fuel tank
55 Starboard three-segment leading-edge slat, extended
56 Slat drive torque shaft and screw jack actuators
57 Slat guide rails
58 Starboard navigation light
59 Fixed portion of wing trailing edge
60 Starboard outer aileron panel
61 Aileron control linkage
62 Inboard aileron panel with balance tab
63 Outboard spoiler panels (two)
64 Triple-slotted outboard flap segment
65 Flap guide rails and operating screw jacks, torque shaft driven
66 Spoiler hydraulic jacks
67 Inboard ground spoilers/lift dumpers (two)
68 Hinged flap track ventral fairing
69 Inboard double-slotted flap segment
70 Inboard flap guide rails and screw jack actuators
71 Wingroot trailing-edge fairing
72 Flap drive motors (two), inboard and outboard
73 Rear cabin starboard emergency exit hatch
74 Cargo handling system overhead rail
75 Movable cabin bulkhead separating passenger and cargo compartments
76 TA-12-60 APU mounted in starboard landing gear sponson fairing
77 Ramp door hydraulic actuators
78 Aft fuselage frame structure
79 Rail-mounted travelling cargo handling unit
80 Aft pressure door, open
81 Fin front spar mounting/rear pressure bulkhead
82 Two-spar fin and rib torsion box structure
83 Fin skin/stringer panel with starboard side access panel
84 Leading-edge bleed-air de-icing
85 Extended tailplane fairing HF antenna
86 Tailplane de-icing air ducts
87 Fin/tailplane joint rib
88 Tailplane inverted leading-edge slat, synchronised with wing slats
89 Starboard tailplane
90 Elevator horn balance
91 Starboard elevator
92 Elevator balance tab
93 Trim tab
94 Upper UHF antenna
95 Trim tab actuator
96 Fin/tailplane aft fairing
97 Port elevator trim tab
98 Balance tab
99 Port elevator rib structure
100 Static dischargers
101 Port tailplane two-spar and rib torsion box structure
102 Port inverted tailplane slat, extended
103 Trailing rudder upper segment
104 Fore rudder, used with upper trailing segment to balance asymmetric thrust conditions
105 Rudder rib structure
106 Lower trailing rudder segment, cable-operated for normal flight conditions
107 Rudder trim tab
108 Tail navigation light
109 Cargo hatch aft sliding door
110 Cargo door rails
111 Fore rudder hydraulic actuator
112 Fin rear spar attachment sloping bulkhead
113 Aft pressure door hydraulic actuator and linkage
114 Hinged ramp toe plates
115 Ramp hinge arm
116 Cargo ramp, lowered position
117 Ramp: hinges forward beneath fuselage for truck-bed loading
118 Cargo handling floor rollers and tiedown points
119 Port rear cabin emergency exit hatch
120 Port inboard double-slotted flap segment
121 Inboard flap all-titanium rib and skin structure
122 Flap shroud ribs
123 Port inboard spoiler/lift dumper panels
124 Outerboard triple-slotted flap rib structure
125 Flap track fairings
126 Outboard spoiler panels
127 Port aileron balance tab
128 Trim tab, port only
129 Two-segment aileron rib structure
130 Fixed trailing-edge ribs
131 Port navigation light
132 Wing panel two-spar and rib torsion box structure
133 Port three-segment leading edge slat, extended
134 Slat de-icing air duct
135 Slat rib structure
136 Fixed leading-edge ribs
137 Wing outer/tip panel joint rib
138 Leading-edge slat guide rails
139 Slat operating screw jack
140 Port wing integral fuel tanks
141 Wing stringers
142 Wing skin panels with upper surface chordwise access panel
143 Centre/outer wing panel joint rib
144 Engine mounting support rib
145 Rear spar/fuselage attachment joint
146 Port engine overwing exhaust fairing
147 Titanium wing skin panel in exhaust area
148 Core engine, hot stream, exhaust duct
149 Engine mounting cantilevered beam
150 Engine turbine section
151 Fan air duct and engine rear support strut
152 ZMKB Progress D-36 turbofan engine
153 Main engine mounting
154 Accessory equipment gearbox
155 Engine front fan
156 Intake lip bleed-air de-icing
157 Retractable landing light
158 Glass fibre sponson fairing
159 Refuelling control panel
160 Pressure refuelling connection
161 Main landing gear hydraulic retraction jack
162 Mainwheel leg mounting support structure
163 Mainwheel leg door
164 Trailing link main landing gear suspension
165 Shock absorber strut
166 Tandem mainwheels, inward retracting

AVIONICS: *Comms:* HF com, VHF com/nav. 'Odd Rods' IFF standard.

Radar: Navigation/weather radar in nose.

Flight: ADF. Compatible with DME, Tacan, VOR, ILS and SP systems. Doppler-based automatic navigation system, linked to onboard computer, is preprogrammed before take-off on push-button panel to right of map display.

Instrumentation: Failure warning panels above windscreen display red lights for critical failures, yellow lights for non-critical failures, to minimise time spent on monitoring instruments and equipment.

An-74 has enhanced avionics, including GPS, GPWS and TCAS 2000 in latest production. Five-screen 'glass cockpit' with twin HUDs under development by 2003.

EQUIPMENT: Removable mobile winch, capacity 2,500 kg (5,511 lb), assists loading. Cargo straps and nets stowed in lockers on each side of hold when not in use. Provision for roller conveyors in floor.

DIMENSIONS, EXTERNAL:
Wing span ... 31.89 m (104 ft 7½ in)
Wing aspect ratio ... 10.3
Length overall ... 28.07 m (92 ft 1¼ in)
Fuselage: Max diameter ... 3.10 m (10 ft 2 in)
Height overall ... 8.65 m (28 ft 4½ in)
Wheel track ... 4.09 m (13 ft 5 in)
Wheelbase ... 8.68 m (28 ft 5¾ in)
Min loading clearance beneath rear fuselage ... 2.80 m (9 ft 2¼ in)
Distance between engine centrelines ... 4.15 m (13 ft 7½ in)
Crew/passenger door: Height ... 1.65 m (5 ft 5 in)
Width ... 0.90 m (2 ft 11¼ in)
Rear-loading door: Length ... 7.10 m (23 ft 3½ in)
Width ... 2.40 m (7 ft 10½ in)
Height to sill ... 1.54 m (5 ft 0¾ in)

DIMENSIONS, INTERNAL:
Cabin: Length, An-74: excl ramp ... 10.50 m (34 ft 5¼ in)
incl ramp ... 14.30 m (46 ft 11 in)
Width: at floor level ... 2.15 m (7 ft 0½ in)
max ... 2.50 m (8 ft 2½ in)
Height ... 2.20 m (7 ft 2½ in)
Floor area: An-74T ... 22.5 m^2 (242 sq ft)
Ramp area ... 8.2 m^2 (88.3 sq ft)
Volume: An-74T (total) ... 73.3 m^3 (2,589 cu ft)

AREAS:
Wings, gross ... 98.53 m^2 (1,060.6 sq ft)

WEIGHTS AND LOADINGS (A: An-72, C: An-74, D: An-74 Salon, E: An-74-200, F: An-74T-200 and An-74TK-200):
Weight empty: A ... 19,050 kg (42,000 lb)
F ... 21,820 kg (48,105 lb)
Max fuel: A, C, D, E, F ... 13,200 kg (29,100 lb)
Max payload: C, D ... 7,500 kg (16,535 lb)
A, E, F ... 10,000 kg (22,045 lb)
Max T-O weight:
from 1,800 m (5,905 ft) runway:
C ... 34,500 kg (76,060 lb)
from 600–800 m (1,970–2,630 ft) runway:
A ... 31,200 kg (68,784 lb)
C ... 34,800 kg (76,720 lb)
D, E, F ... 36,500 kg (80,468 lb)
Max landing weight: C, D, E, F ... 33,000 kg (72,750 lb)

PERFORMANCE (An-72. A: at T-O weight of 31,200 kg; 68,784 lb, B: at T-O weight of 27,500 kg; 60,625 lb on 1,000 m; 3,280 ft unprepared runway. C, D, E, F, An-74 series as above):
Max level speed at 10,000 m (32,810 ft):
A ... 380 kt (705 km/h; 438 mph)
Max level speed at 10,100 m (33,135 ft):
C, D, E, F ... 377 kt (700 km/h; 434 mph)
Cruising speed at 10,000 m (32,810 ft):
A, B ... 297–324 kt (550–600 km/h; 342–373 mph)
Approach speed: A ... 97 kt (180 km/h; 112 mph)
Service ceiling: A ... 10,700 m (35,100 ft)
B ... 11,800 m (38,720 ft)
F ... 10,100 m (33,140 ft)
Service ceiling, OEI: A ... 5,100 m (16,740 ft)
B ... 6,800 m (22,300 ft)
T-O run: A ... 930 m (3,055 ft)
B ... 620 m (2,035 ft)
T-O to 10.7 m (35 ft): A ... 1,170 m (3,840 ft)
B ... 830 m (2,725 ft)
Landing run: A ... 465 m (1,525 ft)
B ... 420 m (1,380 ft)
Max length of runway required:
C, D ... 1,200–1,800 m (3,940–5,905 ft)
E, F ... 1,400–2,150 m (4,595–7,055 ft)
Range, 45 min reserves:
A with max payload ... 430 n miles (800 km; 497 miles)
A with 7,500 kg (16,535 lb) payload ... 1,080 n miles (2,000 km; 1,240 miles)
A with max fuel ... 2,590 n miles (4,800 km; 2,980 miles)
B with 5,000 kg (11,020 lb) payload ... 430 n miles (800 km; 497 miles)
B with max fuel ... 1,760 n miles (3,250 km; 2,020 miles)
F with 10,000 kg (22,046 lb) payload ... 809 n miles (1,500 km; 932 miles)
F with 5,000 kg (11,023 lb) payload ... 1,943 n miles (3,600 km; 2,236 miles)
Range, 1 hour reserves:
C, F with 7,500 kg (16,535 lb) payload ... 944 n miles (1,750 km; 1,087 miles)
E with 7,500 kg (16,535 lb) payload ... 1,160 n miles (2,150 km; 1,336 miles)
C, E with 5,000 kg (11,020 lb) payload ... 1,511 n miles (2,800 km; 1,739 miles)
F with 5,000 kg (11,020 lb) payload ... 1,403 n miles (2,600 km; 1,615 miles)
C with max fuel and 800 kg (1,763 lb) payload ... 2,375 n miles (4,400 km; 2,734 miles)
D with max fuel and 16 passengers ... 2,429 n miles (4,500 km; 2,796 miles)
E with max fuel and 2,500 kg (5,511 lb) payload ... 2,294 n miles (4,250 km; 2,640 miles)
F with max fuel and 5,000 kg (11,020 lb) payload ... 2,321 n miles (4,300 km; 2,671 miles)
Endurance: F ... 6 h 50 min

UPDATED

ANTONOV AN-74-300

TYPE: Twin-jet freighter.

PROGRAMME: First variant, An-74TK-300, announced mid-1998. Prototype An-74TK-300 modified by KhGAPP from An-74 c/n 1910; work began December 1999; first official flight (UR-74300) 20 April 2001; international debut at Paris Salon, June 2001; certification trials completed July 2002 after 219 sorties; AP-25 certificate approved 9 September 2002.

CURRENT VERSIONS: An-74 derivative in which podded, underslung engines, incorporating thrust-reversers, replace the normal An-72/74 installation.

An-74TK-300: Baseline transport for 52 passengers. *As described.*

An-74MP-300: Proposal announced 2002. Maritime patrol version of An-74TK-300; also capable of carrying 22 paratroops, or 44 soldiers or 16 stretchers. Provision for GSh-23L gun, rockets and 100 kg bombs.

An-74TK-300D: Executive transport for 30 passengers.

CUSTOMERS: Initial order for two from Ukraina airline. MoU for 25 from Aeroflot, signed June 2002; Chinese interest in two aircraft reported late 2002.

COSTS: US$12 million to US$14 million (2003).

DESIGN FEATURES: Revised engine installation decreases T-O run, despite loss of Coanda effect; new power plants reduce maintenance requirements and fuel consumption (up to 29 per cent); new position simplifies access.

Cabin enlarged by repositioning of partitions; integral airstairs; upgraded air conditioning; improved instrumentation.

POWER PLANT: Two 63.7 kN (14,330 lb st) Progress D-36-4A turbofans.

AVIONICS: *Comms:* SO-72M transponder, Arlekin-DA HF radio, twin Orlan-85ST VHF, Lainer-MVL PA system, Opal-B CVR, twin Cospas ELT. Optional Arlekin-DA HF radio.

Radar: Buran-74T weather radar.

Flight: NVS-74TM Malva navigation system, SAU-74-03 autopilot, EGPWS, twin SH-3301 GPS, twin A-037 radar altimeters, twin KURS-93M VOR/ILS, twin SD-75 DME, Veer-M Shoran, twin ARK-22 ADF and TCAS 2000. Optional SAU-104/74 digital autopilot.

Instrumentation: 'Glass cockpit' FMS optional.

Generally as for An-74, except that below.

DIMENSIONS, EXTERNAL:
Tailplane span ... 10.00 m (32 ft 9¾ in)
Wheel track (tread centre) ... 4.07 m (13 ft 4¼ in)
Wheelbase (mean) ... 8.08 m (26 ft 6 in)
Distance between engine centrelines ... 8.90 m (29 ft 2½ in)

DIMENSIONS, INTERNAL (VIP version):
Cabin: Length ... 10.49 m (34 ft 5 in)
Width ... 2.20 m (7 ft 2½ in)
Height ... 2.50 m (8 ft 2½ in)
Volume ... 54.0 m^3 (1,907 cu ft)

AREAS:
Wings, gross ... 98.62 m^2 (1,061.5 sq ft)

WEIGHTS AND LOADINGS:
Max ... 10,000 kg (22,045 lb)
Max fuel weight ... 13,210 kg (29,123 lb)
Max T-O weight ... 36,500 kg (80,468 lb)
Max landing weight ... 33,000 kg (72,752 lb)

PERFORMANCE:
Max level speed ... 378 kt (700 km/h; 435 mph)
Max cruising speed ... 313 kt (580 km/h; 360 mph)
Service ceiling ... 10,100 m (33,140 ft)
Balanced field length ... 2,050 m (6,725 ft)
Range: with nine passengers and 2,300 kg (5,070 lb) cargo ... 1,781 n miles (3,300 km; 2050 miles)
with eight passengers ... 2,375 n miles (4,400 km; 2,734 miles)
ferry ... 2,807 n miles (5,200 km; 3,231 miles)

UPDATED

Prototype An-74-300 in Antonov house colours *(Paul Jackson)* ***NEW***/0554453

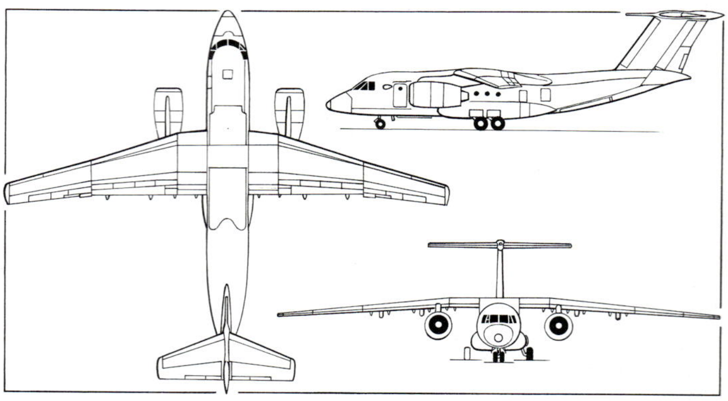

Antonov An-74T-300 STOL transport *(James Goulding)* 0079288

ANTONOV AN-124

NATO reporting name: Condor

TYPE: Outsize freighter.

PROGRAMME: Bureau design number 400; originally designated An-40. Prototype (SSSR-680125) first flew 26 December 1982; second aircraft static test airframe only. First production aircraft (SSSR-82002 *Ruslan*, named after giant hero of Russian folklore immortalised by Pushkin) exhibited 1985 Paris Air Show; lifted payload of 171,219 kg (377,473 lb) to 10,750 m (35,269 ft) on 26 July 1985, exceeding by 53 per cent C-5A Galaxy's record for payload lifted to 2,000 m and setting 20 more records. Entered service January 1986; set closed-circuit distance record 6 to 7 May 1987 by flying 10,880.625 n miles (20,150.921 km; 12,521.201 miles) in 25 hours 30 minutes.

Deliveries to VTA (Russian Air Forces transport arm), to replace An-22, began 1987; in September 1990, during Gulf crisis, an An-124 carried 451 Bangladeshi refugees from Amman to Dacca, after being fitted with chemical toilets, a 570 litre (150 US gallon; 125 Imp gallon) drinking water tank and foam rubber cabin lining in lieu of seats. Carried heaviest single commercial load transported by air: 135.2 tonnes in 1993; and heaviest commercial shipment moved in one flight; 146 tonnes in 1994.

Service life extension programme begun by Aviastar on first of 17 Antonov Airlines and Volga-Dnepr aircraft in 2000; includes new avionics; upgraded crew rest compartment; and cargo floor and loading equipment strengthening. First, RA-82078 of Volga-Dnepr, redelivered 14 March 2000. Continued production apparently was assured at Ulyanovsk (Aviastar plant) by Volga-Dnepr's requirement for expansion by one or two An-124s per year; company plans to register two future aircraft in UK, following JAR 25 certification. Volga-Dnepr dispute with Ulyanovsk local government resulted in temporary suspension of orders in October 2002, but in June 2003 this restarted following resolution of the dispute. In 2004, Volga-Dnepr predicted purchase of "several dozen" An-124s by 2017.

On 31 August 2004, Aviant and Aviastar plant of Russia signed agreement for further manufacture in Russia of up to 80 An-124s between 2006 and 2020, employing empennages built by Aviant; version will be An-124-100M-150.

CURRENT VERSIONS: **An-124:** Baseline transport.

Detailed description applies to above version.

An-124-100: Commercial transport; civil type certificate granted by AviaRegistr of Interstate Aviation Committee of CIS on 30 December 1992. Civil-operated An-124s are now to this standard. Maximum T-O weight restricted to 392,000 kg (864,200 lb) and maximum payload to 120,000 kg (264,550 lb).

An-124-100M: Upgrade developed by Antonov; includes main components of the An-124-100 programme: payload increased from 120 t to 150 t; take-off weight increased from 392 t to 402 t; flight range with 120 t cargo increased from 4,750 km to 5,300 km; service life increased up to 24,000 hours; integral equipment for loading and unloading single pieces of cargo weighing up to 40 t; fuselage structure reinforced to provide airlift of single pieces of cargo weighing up to 150 t; navigation system and radar updated; digital anti-skid braking system installed; reduced crew requirement from six to four members; and improved comfort level of crew rest cabin.

Zaporozhye Progress Engine Design Bureau (Ukraine) undertook development work to improve reliability, gas-dynamic stability and service life of the D-18T-3 engine. Its assigned service life is 12,000 hours at the first stage of operation, and 24,000 hours at the second stage; TBO is 6,000 hours. Corresponds to Part III of the ICAO noise regulations; Part IV is planned.

Prototype (RA-82079) completed at Ulyanovsk late 1995, but not flown until June 2000; delivered Volga-Dnepr 3 August 2000. Service life extension to 24,000 hours formally certified on 22 February 2001. Final details were agreed in November 2001 for upgrades of remainder of Volga-Dnepr, plus Antonov Airlines fleets. RA-82045 upgraded at Ulyanovsk and re-flown on 2 November 2003. New deliveries began with hand-over to Volga-Dnepr of RA-82081 on 25 May 2004 (rolled out 16 March, first flown 11 April 2004).

An-124-200: Proposed version with GE CF6-80C2 engines, each 263 kN (59,200 lb st).

An-124-100VS: Russian government approval given late 1998 to modify two An-124s to carry the Vozdushny Start booster, capable of placing a 1,630 kg (3,593 lb) satellite into 200 km (124 mile) orbit. Booster is extracted from An-124's rear by parachute before rocket ignition. Alternative (Anglicised) designation **An-124AL**. Aircraft include RA-82010. IOC expected by early 2003, but not achieved.

Aircraft transferred from Russian Air Forces to Ulyanovsk for modification, last arriving on 8 February 2001; first aircraft's public debut was at Ulyanovsk on 24 August 2001; second was due for completion before end of 2001; operated by Polet alongside further two ex-military transfers. Discussions under way in late 2003 for use of Biak Island, Indonesia, as Equatorial launch base.

An-124-300: Proposed stretched version under study in 2003. Fuselage lengthened by 5.90 m (19 ft 4¼ in), by plugs fore and aft of wing, to give 1,300 m^3 (45,900 cu ft) cargo volume; wing centre section increased in width to provide additional fuel volume and increase span to 79.90 m (262 ft 1½ in). Payload options to include 150,000 kg (330,700 lb) over 4,373 n miles (8,100 km; 5,033 miles) and 120,000 kg (263,550 lb) over 5,399 n miles (10,000 km; 6,213 miles). Engines may be of Western origin.

CUSTOMERS: Manufactured at Kiev by Aviant and (beginning with eighth production aircraft, SSSR-82005, in late 1985) by Aviastar at Ulyanovsk; 55 completely or substantially built by late 1995, comprising 19 at Kiev (including prototype, but not test airframe) and 36 at Ulyanovsk; by June 2000, only 18 and 34, respectively, had flown. Final Kiev aircraft (UR-CCX) flew 6 October 2003 and granted export licence in January 2004 for delivery to UAE Ministry of Defence; 36th Ulyanovsk An-124 flown 11 April 2004. Production at

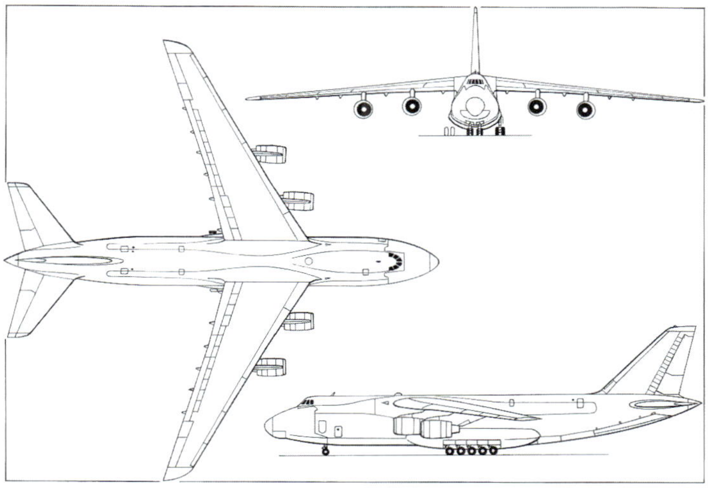

Antonov An-124 (four ZMKB Progress/Ivchenko D-18T turbofans) *(James Goulding)* 0507440

Kiev temporarily halted by 1991 after 16 aircraft and briefly resumed with two more in 1993–94. Ulyanovsk has continued throughout, producing 33 by 1995; in 2001, No. 35 was scheduled for delivery to Polet in 2002, but was not received until 18 June 2004. In early 2002, further two aircraft under construction at Aviastar for Volga-Dnepr, first of which (prospective RA-82081) formally ordered (company's 10th) on 24 July 2002 and delivered May 2004.

First international commercial operator was Antonov ASTC in collaboration with general sales agent Air Foyle of Luton, UK; others previously available to HeavyLift (UK) from Volga-Dnepr (Russia), for charter operations until partnership dissolved in 2000 and Air Foyle Heavylift joint venture launched in October 2001. Operators in 2003 comprised Antonov Airlines (seven), Polet (six), Volga-Dnepr (10), Russian Air Forces (11) and Libyan Cargo Airlines (two, delivered 2001 and 20 September 2002). Further 11 Russian Air Forces aircraft stored pending possible resale (of which one to Polet in 2003); four destroyed; one (prototype) in storage.

COSTS: US$38 million (2004).

DESIGN FEATURES: World's largest production aircraft; upward-hinged visor-type nose and rear fuselage ramp/door for simultaneous front and rear loading/unloading; titanium floor throughout constant-section main hold, which is lightly pressurised, with a fully pressurised cabin for passengers above; landing gear for operation from unprepared fields, hard packed snow and ice-covered swampland; steerable nosewheels and mainwheels permit turns on 45 m (148 ft) wide runway.

Service life initially 7,500 hours; contract for extension to 12,000 hours signed between Volga-Dnepr and Antonov ASTC, July 2000; aircraft delivered from 2000 have 24,000 hour airframe and engine life.

Supercritical wings, with anhedral; sweepback approximately 35° on inboard leading-edge, 32° outboard; all tail surfaces sweptback.

FLYING CONTROLS: All surfaces hydraulically actuated; manual control of automatic control systems, control surface actuators, control system manual linkage and trimming system. Two-section ailerons, three-section single-slotted Fowler flaps (two outer, one inner) and six-section full-span leading-edge flaps on each wing; small slot in outer part of two inner flap sections each side to optimise aerodynamics; 12 spoilers on each wing, forward of trailing-edge flaps (four lateral control spoilers outboard; four glissage spoilers; and four airbrakes inboard); no wing fences, vortex generators or tabs; hydraulic flutter dampers on ailerons; rudder and each elevator in two sections, without tabs but with hydraulic flutter dampers; fixed incidence tailplane; control runs (and other services) channelled along fuselage roof.

STRUCTURE: Basically conventional light alloy, but 5,500 kg (12,125 lb) of composites make up more than 1,500 m^2 (16,150 sq ft) of surface area, giving weight saving of more than 2,000 kg (4,410 lb); each wing has one-piece root-to-tip upper surface extruded skin panel, strip of carbon fibre skin panels on undersurface forward of control surfaces, and glass fibre tip; front and rear of each flap guide fairing of glass fibre, centre portion of carbon fibre; central frames of semi-monocoque fuselage each comprise four large forgings; fairings over intersection of fuselage double-bubble lobes in line with wing, from rear of flight deck to plane of fin leading-edge, primarily of glass fibre, with central, and lower underwing, portions of carbon fibre; other glass fibre components include tailplane tips, nosecone, tailcone and most bottom skin panels forming blister underfairing between main landing gear legs; carbon fibre components include strips of skin panels forward of each tail control surface, nose and main landing gear doors, some service doors, and clamshell doors aft of rear-loading ramp.

An-124 heavy transport operated by Antonov's own freight company *(Paul Jackson)* ***NEW***/0583401

LANDING GEAR: Hydraulically retractable nosewheel type, made by Hydromash, with 24 wheels. Two independent forward-retracting and steerable twin-wheel nose units, side by side. Each main gear comprises five independent inward-retracting twin-wheel units; front two units on each side steerable. Each mainwheel bogie enclosed by separate upper and lower doors when retracted. Nosewheel doors and lower mainwheel doors close when gear extended. All wheel doors of carbon fibre. Main gear bogies retracted individually for repair or wheel change. Mainwheel tyres size 1,270×510. Nosewheel tyres size 1,120×450.

Aircraft can 'kneel', by retracting nosewheels and settling on two extendable 'feet', giving floor of hold a 3.5° slope to assist loading and unloading. Process takes 3 minutes to lower aircraft and 6½ minutes to raise. Rear of cargo hold lowered by compressing main gear oleos. Carbon brakes normally toe-operated, via rudder pedals. For severe braking, pedals depressed by toes and heels. Turning radius (outboard wheels) 19.6 m (64 ft 4 in).

POWER PLANT: Four ZMKB Progress/Ivchenko D-18T turbofans, each 229 kN (51,590 lb st); thrust reversers standard. Optional hush kit certified to ICAO Chapter 3 in mid-1997. Engine cowlings of glass fibre; pylons have carbon fibre skin at rear end. All fuel in 10 integral tanks in wings, total capacity 348,740 litres (92,128 US gallons; 76,714 Imp gallons), not all of which is utilised in civil versions.

ACCOMMODATION: Crew and passenger accommodation on upper deck; freight and/or vehicles on lower deck. Flight crew of six, in pairs, on flight deck, with place for loadmaster in lobby area (10 to 12 cargo handlers and servicing staff carried on commercial flights). Pilot and co-pilot on fully adjustable seats, which rotate for improved access. Two flight engineers, on wall-facing seats on starboard side, have complete control of master fuel cocks, detailed systems instruments, and digital integrated data system with CRT monitor. Behind pilot are navigator and communications specialist, on wall-facing seats. Between flight deck and wing carry-through structure, on port side, are toilets, washing facilities, galley, equipment compartment, and two cabins for up to six relief crew, with table and facing bench seats convertible into bunks.

Aft of wing carry-through is passenger cabin for 88 persons. Hatches in upper deck provide access to wing and tail unit for maintenance when workstands not available. Flight deck and passenger cabin each accessible from cargo hold by hydraulically folding ladder, operated automatically with manual override. Rearward-sliding and jettisonable window each side of flight deck. Primary access to flight deck via airstair door, with ladder extension, forward of wing on port side. Smaller door forward of this and slightly higher. Door from main hold aft of wing on starboard side. Upper deck doors at rear of flight deck on starboard side and at rear of passenger cabin on each side. Emergency exit from upper deck aft of wing on each side.

Hydraulically operated visor-type upward-hinged nose takes 7 minutes to open fully, with simultaneous extension of folding nose loading ramp. When open, nose is steadied by reinforcing arms against wind gusts. No hydraulic, electrical or other system lines broken when nose is open. Radar wiring passes through tube in hinge. Hydraulically operated rear-loading doors take 3 minutes to open, with simultaneous extension of three-part folding ramp. This can be locked in intermediate position for direct loading from truck. Aft of ramp, centre panel of fuselage undersurface hinges upward; clamshell door to each side opens downward.

Completely unobstructed lower deck freight hold has titanium floor, attached 'mobilely' to lower fuselage structure to accommodate changes of temperature, with rollgangs and retractable attachments for cargo tiedowns. Load limits per tiedown fitting: 12,000 kgf (26,455 lbf) on main floor; 5,000 kgf (11,023 lbf) on rear ramp. Narrow catwalk along each sidewall facilitates access to, and mobility past, loaded freight. Payloads include largest CIS main battle tanks, complete missile systems, 12 standard ISO containers, oil well equipment and earth movers; HeavyLift/Volga-Dnepr aircraft previously transported Airbus wings in Europe. No personnel carried normally on lower deck in flight, because of low pressurisation, but can accommodate 360 troops (not including 88 on upper deck), plus two lavatories and oxygen bottles. Military aircraft equipped to airdrop up to 16 pallets, each of up to 4,500 kg (9,920 lb); or 268 paratroops in two passes. Medical evacuation capability is 288 stretchers and 28 attendants.

SYSTEMS: Automatic flight control system includes control loading for elevator and ailerons; stability augmentation for elevator and rudder; elevator trim and balance; elevator and rudder gear ratio system; and flight limit condition restriction for elevator. Entire interior of aircraft is pressurised and air conditioned. Maximum pressure differential 0.55 bar (7.8 lb/sq in) on upper deck, 0.25 bar (3.55 lb/sq in) on lower deck. Four independent hydraulic systems. Quadruple redundant fly-by-wire flight control system, with mechanical emergency fifth channel to hydraulic control servos. Special secondary bus electrical system. Landing lights under nose and at front of each main landing gear fairing. APU in rear of each landing gear fairing for engine starting, can be operated in the air or on the ground to open loading doors for airdrop from rear or normal ground loading/unloading, as well as for supplying electrical, hydraulic and air conditioning systems. Bleed air anti-icing of wing leading-edges. Electro-impulse de-icing of fin and tailplane leading-edges.

AVIONICS: *Radar:* Two dielectric areas of nose visor enclose forward-looking weather radar and downward-looking ground-mapping/nav radar.

Flight: Hemispherical dielectric fairing above centre-fuselage for satellite nav receiver; quadruple INS; Loran and Omega.

Instrumentation: Conventional flight deck equipment, including automatic flight control system panel at top of glareshield, weather radar screen and moving map display forward of throttle and thrust reverse levers on centre console. No electronic flight displays. Dual attitude indicator/flight director and HSIs, and vertical tape engine instruments.

EQUIPMENT: Two electric travelling cranes in roof of hold, each with two lifting points, offer total lifting capacity of 20,000 kg (44,092 lb). First trial of 30,000 kg (66,139 lb) system in December 1999; development then launched of 40,000 kg (88,185 lb) lifting system. Two winches each pull a 3,000 kg (6,614 lb) load. Small two-face mirror, of V form, enables pilots to adjust their seating position until their eyes are reflected in the appropriate mirror, which ensures optimum field of view from flight deck.

DIMENSIONS, EXTERNAL:
Wing span ... 73.30 m (240 ft 5¾ in)
Wing aspect ratio ... 8.6
Length overall ... 69.10 m (226 ft 8½ in)
Fuselage max width ... 7.28 m (23 ft 10½ in)
Height overall ... 21.08 m (69 ft 2 in)
Wheel track ... 8.00 m (26 ft 3 in)
Wheelbase (centre row mainwheels) ... 22.90 m (75 ft 1½ in)
Sill height:
front door: normal ... up to 2.79 m (9 ft 1¾ in)
kneeling ... 1.43 m (4 ft 8¼ in)
rear door, normal ... up to 2.85 m (9 ft 4¼ in)

DIMENSIONS, INTERNAL:
Cargo hold:
Length at floor:
excl ramps ... 36.48 m (119 ft 8¼ in)
incl rear ramp ... 41.54 m (136 ft 3½ in)
max ... 42.68 m (140 ft 0¼ in)
Width: at floor ... 6.40 m (21 ft 0 in)
max ... 6.63 m (21 ft 9 in)
at ceiling ... 4.26 m (13 ft 11¾ in)
Max height ... 4.40 m (14 ft 5¼ in)
Floor area, incl ramp ... 2,650 m^2 (2,852 sq ft)
Volume, incl ramp ... 1,160 m^3 (40,965 cu ft)
Passenger cabin: Width at floor ... 3.80 m (12 ft 5½ in)

AREAS:
Wings, gross ... 628.0 m^2 (6,760.0 sq ft)

WEIGHTS AND LOADINGS (A: basic An-124, B: An-124-100):
Operating weight empty: A ... 181,000 kg (399,025 lb)
Max payload (incl rear ramp): A ... 150,000 kg (330,700 lb)
B ... 120,000 kg (264,550 lb)
Max load on rear ramp: A and B ... 10,000 kg (22,046 lb)
Max fuel weight (An-124-210) ... 214,000 kg (471,790 lb)
Max T-O weight: A ... 402,000 kg (886,250 lb)
B ... 392,000 kg (864,200 lb)
Max ramp weight: B ... 398,000 kg (877,425 lb)
Max landing weight: B ... 330,000 kg (727,500 lb)
Max zero-fuel weight: A ... 325,000 kg (716,500 lb)
Max wing loading: A ... 640.1 kg/m^2 (131.11 lb/sq ft)
Max power loading: A ... 441 kg/kN (4.32 lb/lb st)

PERFORMANCE:
Max cruising speed: A ... 467 kt (865 km/h; 537 mph)
Normal cruising speed at FL328-394:
A ... 432–459 kt (800–850 km/h; 497–528 mph)
Airdropping speed range ... 127–216 kt (235–400 km/h; 146–249 mph)
Approach speed:
A ... 124–140 kt (230–260 km/h; 143–162 mph)
Max certified altitude ... 12,000 m (39,380 ft)
T-O balanced field length at max T-O weight:
A ... 3,000 m (9,840 ft)
T-O run: A ... 2,520 m (8,270 ft)
Landing run at max landing weight: A ... 900 m (2,955 ft)
Range: with max payload:
A ... 1,997 n miles (3,700 km; 2,299 miles)
B ... 2,591 n miles (4,800 km; 2,982 miles)
with 80,000 kg (176,375 lb) payload:
A ... 4,535 n miles (8,400 km; 5,219 miles)
B ... 4,211 n miles (7,800 km; 4,846 miles)
with 40,000 kg (88,184 lb) payload:
A ... 6,209 n miles (11,500 km; 7,145 miles)
B ... 5,885 n miles (10,900 km; 6,772 miles)
ferry, with max fuel:
A, B ... 7,559 n miles (14,000 km; 8,699 miles)

OPERATIONAL NOISE LEVELS:
Stated to meet ICAO requirements

UPDATED

ANTONOV AN-140

TYPE: Twin-turboprop airliner.

PROGRAMME: Announced at Paris in June 1993 as An-24 replacement; preliminary design finalised April 1994; two prototypes with TV3-117 engines and static test airframe constructed at Kiev; rolled out 6 June 1997; first flight (UR-NTO) at Kiev-Svyatoshino (landing at Gostomel flight test centre, as base for certification trials) 17 September 1997; second airframe (0102) for static tests; second flying prototype (UR-NTP) rolled out 11 December 1997 and flew 26 December. Third (0104; first production), due to fly at Kharkov, early 1999, and undertake electromagnetic compatibility and climatic tests but maiden sortie delayed until 11 October 1999; this aircraft (UR-PWO) flew 41 sorties towards certification.

Certification programme of 940 hours began August 1998 and completed in early 2000; included cold trials at Arkhangelsk (by UR-NTO between 29 March and 1 May 1999, and by UR-NTP in Yakutia for 10 days in January 2000) and hot-and-high trials in Uzbekistan

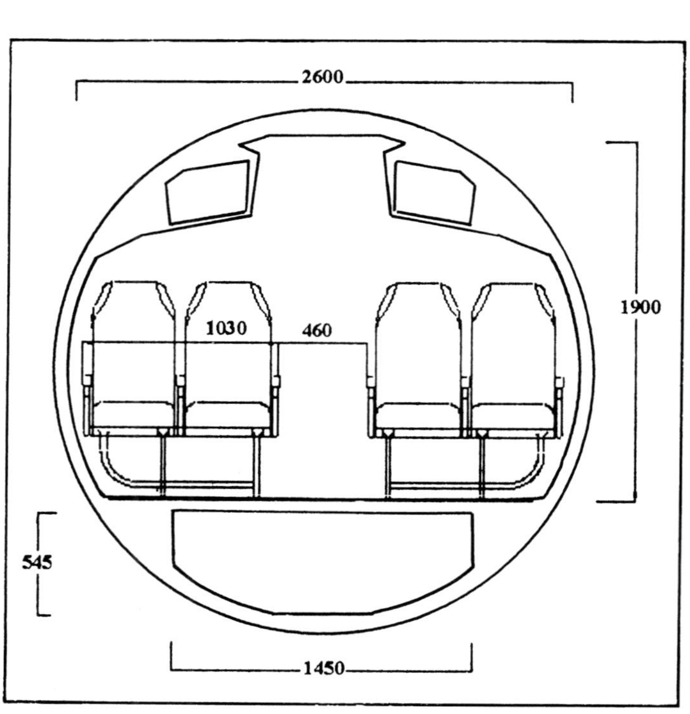

Cross-section of the An-140 cabin; dimensions in millimetres ***(Paul Jackson)***
0052938

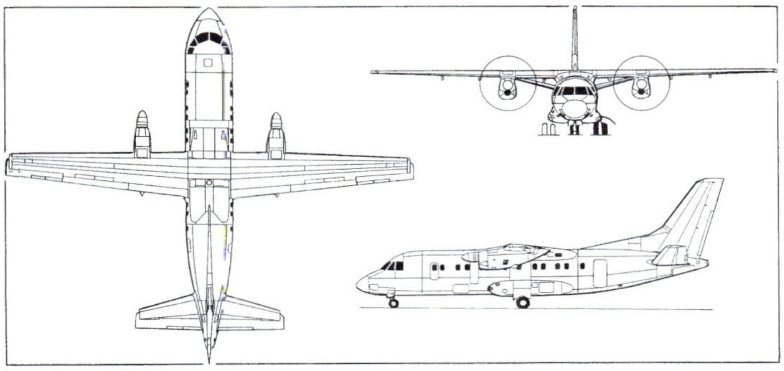

Antonov An-140-100 short-range transport *(James Goulding)* ***NEW***/0554426

and Kyrgystan (by UR-NTP, concluding 3 September 1999). Following first series of flight testing, tailplanes of prototypes modified to obviate propwash-induced vibration, gaining 15° dihedral and shortened elevator horn balance. An-140's 1,000th hour flown (by UR-NTP) on 12 January 2000. Trials concluded on 26 March 2000 after 1,286 hours in 1,138 sorties. Certification in Russia (AP-25) and Ukraine was achieved on 25 April 2000 coincident with that of TV3-117VMA-SBM1 engine and AV-140 propeller. Meets FAR Pt 25 airworthiness, Pt 34 emissions and Pt 36 noise requirements.

Series production began 1999 at KhGAPP, Kharkov (where wings for prototypes were built) and at Aviacor, Samara, Russian Federation.

Aviacor production intended to be 10 in 2000 (first in July) and to reach 30 per year by 2001, but reduced to combined total of eight in 2000–01; five under assembly at Aviacor by May 2000, but first (confusingly, RA-14001) not rolled out until 25 December 2003, being one of two ordered in 2003 by Samara Airlines.

KhGAPP deliveries began with UR-14001 to Odessa Airlines on 4 March 2002. This made type's Western debut at Farnborough in July 2002. Six production aircraft supplied by KhGAPP before change to An-140-100, first of which delivered on 26 April 2004.

Production share agreement of 1998 assigns empennage to Antonov; engine nacelles, wing and associated control surfaces to KhGAPP; landing gear to Aviagregat at Samara and Youzhmash at Dniepropetrovsk; and fuselage to Aviacor (incorporating Avio Interiors fittings from Italy).

Agreement signed December 1995 for assembly by HESA at rate of 12 a year from 1999, progressing to local parts manufacture, in new plant at Esfahan, Iran, with Ukrainian assistance; first two kits shipped to Iran by late 2000; initial aircraft first flown 7 February 2001. Iranian type certificate issued 14 April 2003.

CURRENT VERSIONS: **An-140:** Initial six production aircraft only; replaced in 2004 by -100.

An-140T: Proposed 6 tonne freighter with large door, port side, rear. Convertible **An-140TK** will be similar, but with 5 tonne payload and up to 50 passengers.

An-140VIP: Executive version; range 2,159 n miles (4,000 km; 2,485 miles).

An-140-100: Improved version; advanced engine nacelle design, increased tankage and wing span increased by 1.00 m (3 ft 3¼ in) giving 21,500 kg (47,399 lb) MTOW and further 162 n miles (300 km; 186 miles) of range. Standard version from 2004. (Series 100 originally proposed in 1997 as 68-seat version with 3.80 m; 12 ft 5½ in fuselage stretch.) *As described.*

An-142: Under development for 2001 first flight; forward-retracting rear loading ramp similar to An-26. No further reports.

Ir.An-140: Licence-built by HESA in Iran.

An-140 Military: Patrol, surveillance, photographic and similar variants proposed for military operators.

CUSTOMERS: Air Ukraine letter of intent for up to 40 by 2010 signed 17 September 1997 (first flight); initial four firm conversions made June 1999 (for manufacture at Kharkov), specifying deliveries from 2000; however, initial recipient from Ukrainian production was Ikar Airlines, with five on order from January 2000 contract, first being UR-PWO, delivered in 2001, following testing at Antonov; this soon disposed of, however, and no more supplied to Ikar. First order for 15 placed by republic of Sakha-Yakutia, Russian Federation, mid-1998, requiring 10 to be delivered to Sakha Airlines from Samara production in 2000–04. First Samara-built aircraft destined for Samara Airlines in first quarter of 2002, but not completed until 2004. Aeroflot letter of intent for 50 PW127-powered An-140As signed mid-1999 (specifying Samara production). Tyumenaviatrans reportedly holds option on 25 An-140s, but this had been relegated to expression of interest (also from Polar Airlines and Mirny) when An-140 toured Siberia in 2000. Odessa Airlines ordered five from Kharkov production, of which first (UR-14001, the former UR-PWO) entered revenue-earning service on 29 March 2002, having been handed over to Ukrtransleasing on 4 March; next delivery was fourth Kharkov, on 21 January 2003. Second and third Kharkov aircraft delivered to Aeromost (airline formed July 2002 by Kharkov plant and originally named Aeromist) on 3 and 22 August 2002. Third Aeromost delivered 15 September 2003. Third operator of Kharkov aircraft was Motor Sich, which received UR-14005 on 23 May 2003 and began services two days later. Seventh Kharkov machine was first An-140-100, delivered (UR-14007) to Il'ich-Avia on 26 April 2004, followed by UR-14008 to same airline on 21 August 2004. Volga-Dnepr signed letter of intent in October 2002 for five from Kharkov, due 2003–04 (but, apparently, not received), while Buryatia Airlines (two) and Sibaviatrans (three) placed orders in December 2002, and Sakha/Yakutia followed with four in January 2003. Five ordered by Air Libya in June 2003 and four by AZAL of Azerbaijan in June 2004. Launch order for Ir.an-140 is 20 for Iran Asseman Airlines; Iran Air also will be operator; Iranian licence initially for 80 aircraft, but interest reported, late 1998, in building further 160. Estimated sales of 645 by 2011, including 430 to Russia and 70 to Ukraine.

COSTS: US$7 million (2000).

DESIGN FEATURES: Light transport, designed to be capable of autonomous operation from airfields with unprepared runways at all altitudes and in all weathers, providing airline-standard comfort. International certification and various engine options to maximise sales prospects. Conventional high-wing monoplane; tapered wing with unswept leading-edge; sweptback fin and tailplane, latter with 15° dihedral; engines mounted underwing. Maintenance target of 6.5 mmh/fh. Service life 50,000 landings/50,000 hours/25 years.

FLYING CONTROLS: Conventional and manual. Control surfaces all horn balanced; ailerons with servo tab in each (plus trim tab starboard); elevator with two-section trim/servo tab each side; rudder with large single tab. Two-section flaps in each wing; auto spoiler ahead of each outboard flap section.

STRUCTURE: Largely of aluminium, with some titanium.

LANDING GEAR: Retractable tricycle type by Pivdennyi; twin Rubin wheels on each unit; nosewheels retract forward, mainwheels into fairings each side of lower fuselage. Mainwheels size 810×320-330; nosewheels 600×220-254. Rubin braking system. Able to operate from gravel or unpaved fields.

POWER PLANT: Two 1,839 kW (2,466 shp) Ivchenko TV3-117VMA-SBM1 (built at Zaporozhye, Ukraine, by Motor-Sich), driving AV-140 feathering and reversible propellers. FED fuel-management system; Star engine control system.

ACCOMMODATION: Flight crew of two, plus cabin attendant; basic seating for 52 passengers, four-abreast with centre aisle, at 78 cm (31 in) pitch, or 48 at 81 cm (32 in) pitch. Main passenger door with airstairs, at rear of cabin on port side, with service door (also emergency exit Type I) opposite; Type III emergency exit port side at front of cabin; cargo door (and Type I exit) starboard side, front. Coat stowage, galley and lavatory at rear of cabin. Baggage/freight compartment at rear of cabin, plus forward underfloor freight hold, with door on starboard side. Cargo door on starboard side, forward part of cabin floor reinforced, and detachable equipment provided, enabling 1,650 to 3,650 kg (3,638 to 8,046 lb) of bulk cargo and 36 to 20 passengers to be carried with forward rows of seats removed. Overhead baggage lockers. Accommodation ventilated and pressurised.

SYSTEMS: Motor-Sich AI-9-3B 16 kW APU in rear fuselage. FED hydraulic system; Nauka air conditioning. Kommunar anti-icing and air conditioning control systems. Auxilec generators; Eros oxygen system.

Avionika SAU 28-02 analogue flight control system. Digital ICVSP-140-01 ADS optional.

AVIONICS: *Comms:* Arlekin-D-A HF/ radio, twin Orlan-85ST VHF, AVSA-MVL PA system, Opal-B CVR, R-855A1 emergency VHF/UHF, SO-72M transponder, Satori ELT96 ELT. Muza-A entertainment system.

Radar: Buran A-140 weather radar

Flight: ARK-25 ADF, Veer-M Shoran, Kurs-93M VOR/ILS, RMI-3 RMI, SN-3301 GPS, VND-94 VOR, A037 D1 radio altimeter, TCAS-94 and SSPZ-2000 GPWS.

First Antonov An-140-100, wearing Il'ich-Avia colours *(Paul Jackson)* ***NEW***/0589976

DIMENSIONS, EXTERNAL:
Wing span: An-140 ... 24.505 m (80 ft 4¾ in)
An-140-100 ... 25.505 m (83 ft 8¼ in)
Length overall ... 22.605 m (74 ft 2 in)
Height overall ... 8.23 m (27 ft 0 in)
Wheel track (c/l shock-absorbers) ... 3.18 m (10 ft 5¼ in)
Wheelbase ... 8.125 m (26 ft 7¾ in)
Propeller diameter ... 3.72 m (12 ft 2½ in)
Distance between propeller centres ... 8.20 m (26 ft 10¾ in)
Main cabin door: Height ... 1.605 m (5 ft 3¼ in)
Width ... 0.91 m (2 ft 11¾ in)
Emergency exit: Height ... 1.19 m (3 ft 10¾ in)
Width ... 0.515 m (1 ft 8¼ in)
Service door: Height ... 1.29 m (4 ft 2¾ in)
Width ... 0.62 m (2 ft 0½ in)
Cargo door: Height ... 1.29 m (4 ft 2¾ in)
Width ... 0.99 m (3 ft 3 in)
Underfloor freight hold door:
Height ... 0.90 m (2 ft 11½ in)
Width ... 1.02 m (3 ft 4¼ in)

DIMENSIONS, INTERNAL:
Cabin: Length:
excl flight deck, galley/lavatory area ... 10.82 m (35 ft 6¼ in)
incl galley/lavatory and baggage ... 14.62 m (47 ft 11½ in)
Max width ... 2.60 m (8 ft 6¼ in)
Max height ... 1.90 m (6 ft 2¾ in)
Volume ... 65.5 m^3 (2,313 cu ft)
Aft baggage compartment: Volume ... 6.1 m^3 (215 cu ft)
Underfloor freight hold: Length ... 3.98 m (13 ft 0¾ in)
Max width ... 1.45 m (4 ft 9 in)
Max height ... 0.545 m (1 ft 9¾ in)
Volume ... 3.0 m^3 (106 cu ft)
Overhead baggage lockers:
Volume (total) ... 2.4 m^3 (85 cu ft)

WEIGHTS AND LOADINGS (An-140-100):
Operating weight empty ... 13,200 kg (29,101 lb)
Baggage capacity ... 1,840 kg (4,057 lb)
Max payload ... 6,000 kg (13,227 lb)
Max fuel weight ... 4,680 kg (10,318 lb)
Max T-O weight ... 21,500 kg (47,399 lb)
Max power loading ... 5.85 kg/kW (9.61 lb/shp)

PERFORMANCE (An-140-100):
Max cruising speed ... 290 kt (537 km/h; 334 mph)
Econ cruising speed ... 248 kt (460 km/h; 286 mph)

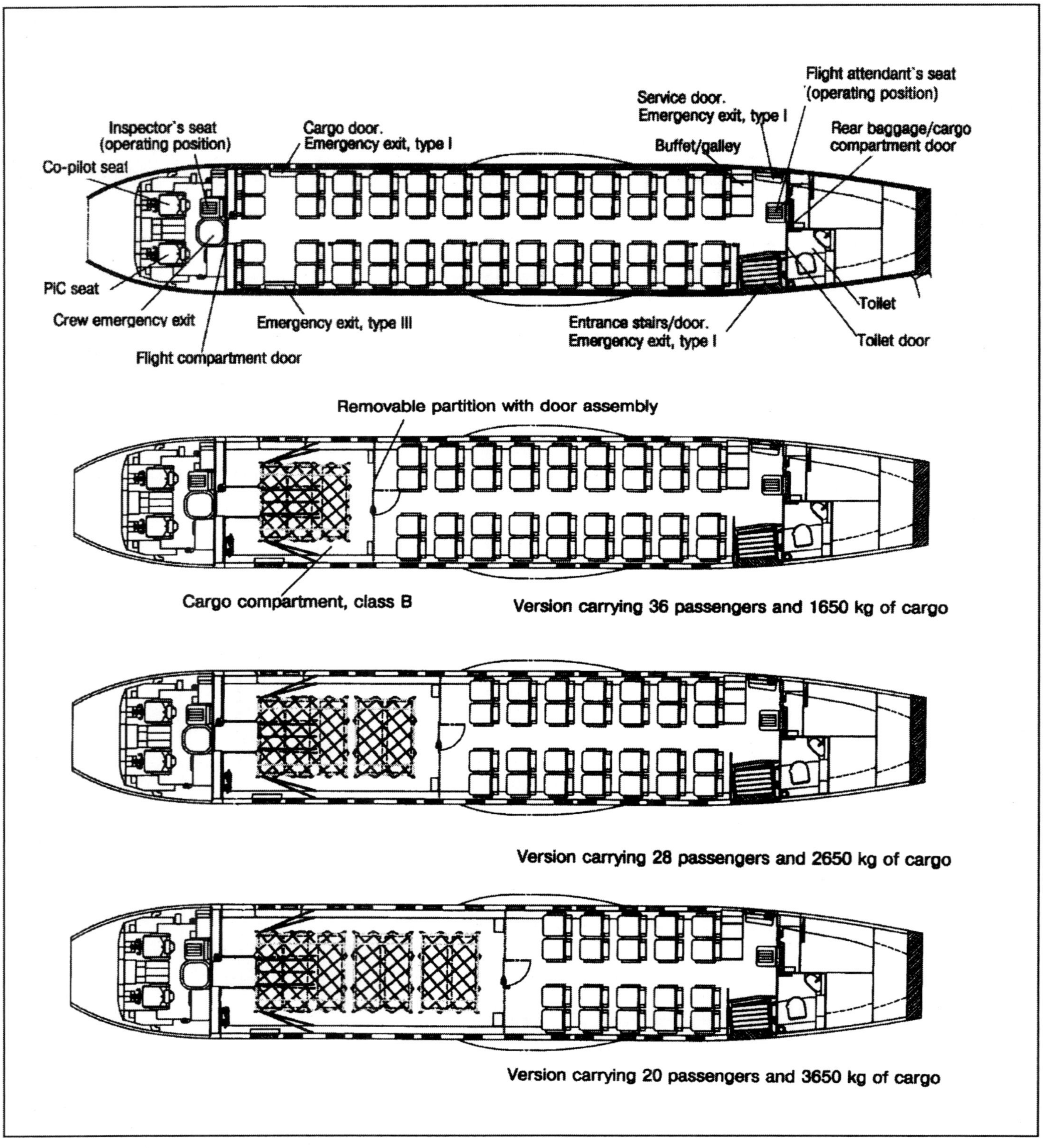

Interior options available for Antonov An-140 ***NEW***/0583343

Landing speed .... 109 kt (202 km/h; 125 mph)
Nominal cruising altitude FL200
Max certified altutude .... 7,620 m (25,000 ft)
T-O run .... 880 m (2,890 ft)
Landing run .... 530 m (1,740 ft)
Range, no reserves:
with 6,000 kg (13,227 lb) payload .... 928 n miles (1,720 km; 1,068 miles)
with 52 passengers .... 1,495 n miles (2,770 km; 1,721 miles)
with 42 passengers .... 1,835 n miles (3,400 km; 2,112 miles)
ferry .... 2,176 n miles (4,030 km; 2,504 miles)

UPDATED

ANTONOV AN-148

TYPE: Regional jet airliner

PROGRAMME: In September 2001, Antonov revealed that it was working on the design of an 80-seat, twin-jet (Progress D-36-5AF), 1,349 n mile (2,500 km; 1,553 mile) range airliner. Configuration is similar to the An-74T-300; stretched and shrunk versions are also planned.

The first of three (two flying; one static) Antonov-built prototypes was under construction at Kharkov by early 2002, with series manufacture planned to be by KhGAPP at Kharkov and VASO at Voronezh, Russia. Antonov and UUAP signed agreement for licensed production in February 2002. Ukrainian participants include Motor Sich and ZhEDB Progress. By September 2004, envisaged that Aviant would be sole source, employing subassemblies from KhGAPP (centre fuselage) and VASO (empennage).

Fuselage removed from jigs 28 February 2003; unofficially rolled out 25 September 2004; formal presentation 14 October 2004; certification due in 2006.

CURRENT VERSIONS: **An-148-100V:** Baseline version; 75 passengers at 81cm; 32 inch pitch (80 at 76 cm; 30 in) and range of 1,943 n miles (3,600 km; 2,236 miles).

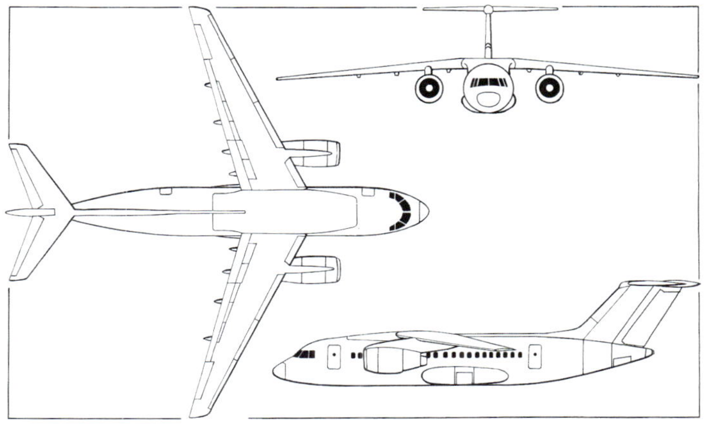

General arrangement of the Antonov An-148 *(James Goulding)* ***NEW***/0583344

Model of Antonov An-148 regional airliner *(Paul Jackson)* ***NEW***/0589975

An-148-100A: Short-haul version; 75 passengers; range 1,187 n miles (2,200 km; 1,367 miles).

An-148-100E: Long haul, 75-seat version; range 2,753 n miles (5,100 km; 3,169 miles).

An-148-100E1: Long-haul version; 55 passengers in three-class cabin over 3,779 n miles (7,000 km; 4,349 miles); additional fuel tanks in baggage hold.

An-148-100E2: Executive transport; between 10 and 30 passengers; range 4,697 n miles (8,700 km; 5,405 miles). Baggage hold fuel.

CUSTOMERS: Volga-Dnepr letter of intent for three signed October 2002, anticipating delivery in 2005–06. Aeroflot lease agreement for 30 An-74s will be transferred to An-148. Aerosvit letter of intent for 10, 2004.

COSTS: US$15 million to US$20 million (2002 estimate).

DESIGN FEATURES: Generally as for Antonov An-74, but FBW control system. Wing developed by KhGAPP. Intended parameters include 99.4 per cent dispatch reliability; 300 h/month utilisation; and 1.3 mmh/fh. AP/FAR/EASA Pt 25 certification; ICAO IV noise/enironmental impact.

POWER PLANT: Two progress (ZMKB) D436-148 turbofans, each of 65.6 kN (14,740 lb st). Provision for alternative CF34-8/10, PW800 and SM-146 engines.

ACCOMMODATION: Flight deck crew of two, plus jump-seat. Seating in cabin as per variant. Passenger door, front, port; service doors, forward, starboard and rear port. Front and rear

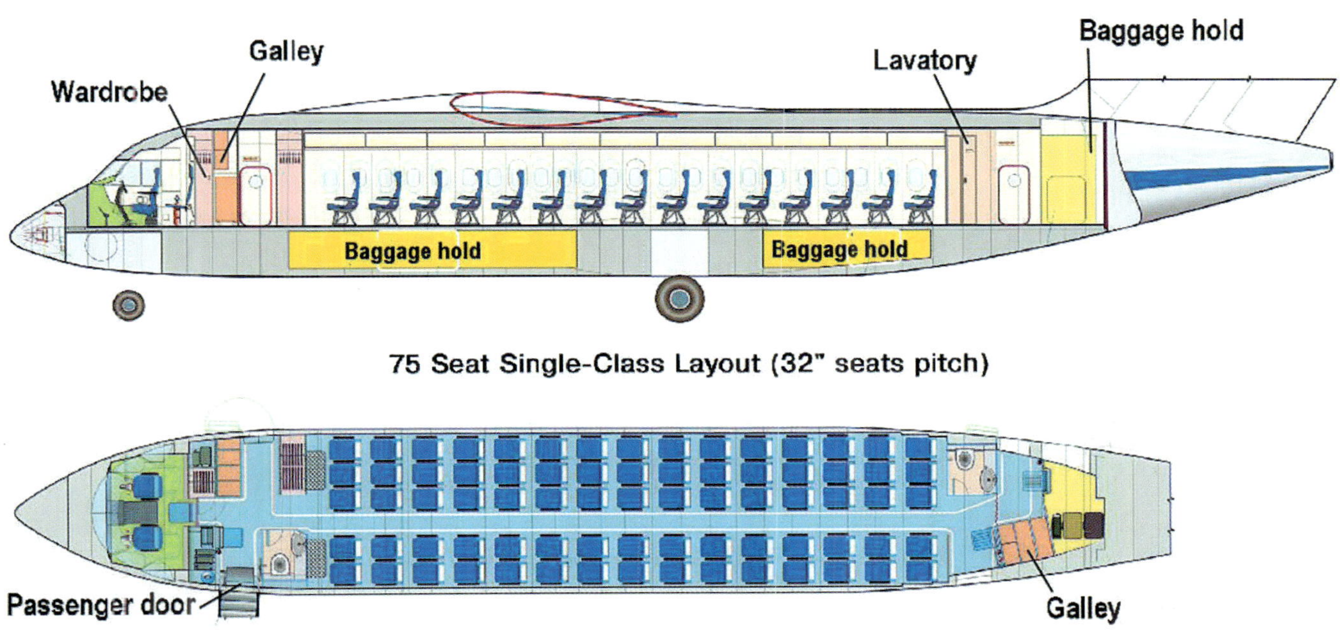

Antonov An-148 75-seat interior layout ***NEW***/0583533

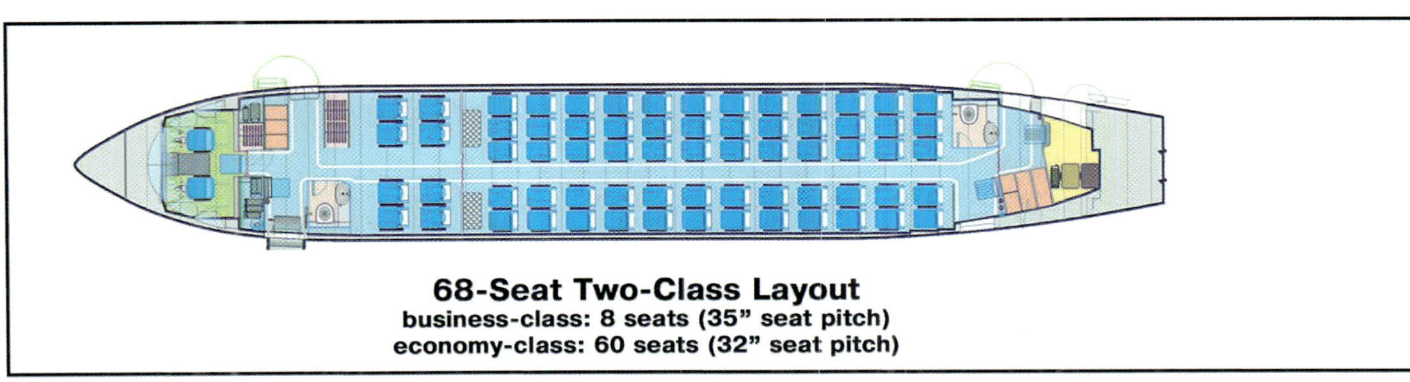

Mixed class seating option in Antonov An-148 ***NEW***/0583532

underfloor baggage holds, each with door, starboard. Cabin baggage compartment, rear, with door, starboard. Crew wardrobe, passenger wardrobe and lavatory forward; galley and lavatory rear, ahead of baggage compartment.
AVIONICS: Two-crew cockpit with five-screen display by Aviapribor.
DIMENSIONS, EXTERNAL:
Wing span .... 28.91 m (94 ft 10¼ in)
Length overall .... 29.13 m (95 ft 6¾ in)
Height overall .... 8.20 m (26 ft 10¾ in)
Baggage door, underfloor (forward, stbd):
Height .... 1.00 m (3 ft 3¼ in)
Width .... 1.50 m (4 ft 11 in)
Baggage door, underfloor (rear, stbd):
Height .... 1.00 m (3 ft 3¼ in)
Width .... 1.00 m (3 ft 3¼ in)
Baggage door, cabin (rear, stbd):
Height .... 1.00 m (3 ft 3¼ in)
Width .... 0.75 m (2 ft 5½ in)
DIMENSIONS, INTERNAL:
Baggage hold volume:
underfloor: forward .... 7.5 m^3 (265 cu ft)
rear .... 4.3 m^3 (152 cu ft)
cabin .... 2.8 m^3 (99 cu ft)
WEIGHTS AND LOADINGS:
Max payload .... 9,000 kg (19,842 lb)
Max T-O weight:
100A .... 36,800 kg (81,130 lb)
100V .... 39,500 kg (87,082 lb)
100E .... 42,600 kg (93,916 lb)
PERFORMANCE (An-148-100V, estimated):
Cruising speed: max .... 470 kt (870 km/h; 541 mph)
econ .... 443 kt (820 km/h; 510 mph)
Cruising altitude .... 12.500 m (41,000 ft)
Required runway length .... 1,800 m (5,905 ft)
Range: with max payload .... 1,906 n miles (3,530 km; 2,193 miles)
with max fuel .... 3,552 n miles (6,580 km; 4,088 miles)

UPDATED

AVIAIMPEX

AVIAIMPEX JSC

Kiev
CHIEF DESIGNER: Mikhail Yu Kuchin
YANHOL PROJECT CHIEF: Sergey Tyurin

Aviation division of TVT Corporation formed to introduce helicopter design and production to Ukraine. Initial liaison with Aerokopter Ltd of Poltava (formed 14 December 1999); however, on 3 May 2000 Aerokopter design bureau divided and both elements became autonomous. Aviaimpex also undertakes helicopter overhaul and maintenance.

First product was the three-seat Yanhol, completed in 2002, but about which nothing further has been reported.

Future plans include an eight-seat, turbine-powered helicopter.

UPDATED

AVIAIMPEX YANHOL

English name: Angel
TYPE: Three-seat helicopter.
PROGRAMME: Presented to the media on 31 August 2001, when prototype structurally complete. First flight planned for June 2002 (no confirmation received), followed by deliveries in 2003. By May 2002, Aviaimpex had revealed plans for gunship and UAV versions.
CUSTOMERS: Options placed on 400 (mostly in two-seat configuration) by early 2002, including Ukrainian Defence Ministry (reportedly 100 for pilot training), Internal Affairs Ministry (patrol), Georgian government and several city administrations, including Moscow Mayor's Office.
DESIGN FEATURES: Pod-and-boom configuration, with T tail and landing skids. Three-blade main and two-blade tail rotors.

Aviaimpex Yanhol light helicopter *(Yefim Gordon)* 0525944

COSTS: US$120,000 in Ukraine; US$150,000 export (2002). Operating cost US$70 per hour (2002).
POWER PLANT: Two Rotax 912 ULS flat-four piston engines, each 73.5 kW (98.6 hp).
WEIGHTS AND LOADINGS:
Max payload .... 350 kg (772 lb)
Max T-O weight .... 870 kg (1,918 lb)
PERFORMANCE:
Max cruising speed .... 89 kt (165 km/h; 103 mph)
Range with max fuel .... 183 n miles (340 km; 211 miles)

AVIANT

AVIANT, KIEVSKY GOSUDARSTVENNY AVIATSIONNY ZAVOD (AVIANT, KIEV STATE AVIATION PLANT)

prospekt Peremohy 100/1, 03062 Kiev
Tel: (+38 044) 454 52 01
Fax: (+38 044) 442 72 45
e-mail: aviant@carrier.kiev.ua
GENERAL DIRECTOR: Oleg Shevchenko
COMMERCIAL DIRECTOR: Oleksandr Grybanov
TRADE DEPARTMENT DIRECTOR: Valery Kuzenkov

The Kiev Aviation Plant has effectively terminated work on the Antonov An-32 twin-turboprop transport and An-124 heavy transport, the 19th and last Kiev-built An-124 having flown on 6 October 2003. However, it built two prototypes of the An-70 four-propfan transport and shares in production with Aviacor at Samara. The two firms will also manufacture the Tupolev Tu-334 twin-turbofan transport of which initial Aviant aircraft first flew 21 November 2003. Overhauls An-24/26, An-32 and An-124 aircraft. Offers An-3 turboprop conversion of Antonov An-2. Non-aviation activities include manufacture of trolley-buses, pressure chambers and customer goods. Recently contributed to manufacture of prototype Aviaimpex Yanhol helicopter prototype and is involved in manufacture of Aeros flex-wing trikes.

Aviant previously built 3,320 An-2s (1947–59), An-8s (1956–59), 1,028 An-24s (1959–78), 1,402 An-26s (1969–85), 123 An-30s (1973–79), 361 An-32s and 17 An-124s. An-32 remains available to order, using stock of uncompleted fuselages, nine of which were due for assembly in 2004 for Libya (two), Turkey and Sri Lanka.

UPDATED

Aviant's 19th and final An-124 making its initial flight ***NEW***/1033940

IKAR

FIRMA AEROKLUB 'IKAR' ('IKAR' AERO CLUB FIRM)

ulitsa Vyborgskaya 99, 03115 Kiev
Tel/Fax: (+38 044) 458 48 16
e-mail: karpes@aviaikar.kiev.ua
Web: http://users.i.com.ua/~a-sergey

'Ikar' Aero Club was founded in 1995 to develop, build and provide engineering and maintenance support for light aircraft. In addition to the Ai-10, it has built the Ai-9 Lis of similar configuration; and Aist pod-and-boom ultralight. Projects include the Zhuravlik powered hang-glider; four-seat Ai-20 Dedal; Ai-30 Chirok four-seat amphibian; Ai-40 Leleka agricultural sprayer; tandem-seat Ai-50 Skvorets; and Ai-60 (also known as Ai-240) four-seat light twin.

UPDATED

IKAR AI-10 IKAR

TYPE: Side-by-side ultralight.
PROGRAMME: First flown in 1999. Gained French ultralight approval in October 2002.
COSTS: US$35,000 to US$39,000, according to equipment (2001).
DESIGN FEATURES: High-mounted, constant-chord, strut-braced wing. Sweptback fin and mid-mounted tailplane.
FLYING CONTROLS: Conventional and manual. Flaps.
STRUCTURE: Metal airframe. Fabric-covered wing.
LANDING GEAR: Tailwheel type; fixed.
POWER PLANT: One Rotax flat-four driving a Petrenko P-912-170 three-blade, wooden propeller: 59.6 kW (79.9 hp) Rotax 912 UL; 73.5 kW (98.6 hp) 912 ULS; or 84.6 kW (113.4 hp) 914. Fuel capacity 88 litres (23.2 US gallons; 19.4 Imp gallons).
DIMENSIONS, EXTERNAL:
Wing span .... 9.06 m (29 ft 8¾ in)
Length overall .... 6.21 m (20 ft 4½ in)
Height overall .... 1.70 m (5 ft 7 in)
AREAS:
Wings, gross .... 11.74 m² (126.4 sq ft)
WEIGHTS AND LOADINGS:
Weight empty .... 284 kg (626 lb)
Max T-O weight: Rotax 912 ULS .... 510 kg (1,124 lb)
Rotax 912 UL .... 450 kg (992 lb)

Ikar Ai-10 side-by-side ultralight *NEW*/0583459

PERFORMANCE (Rotax 912 ULS):
Cruising speed .... 97 kt (180 km/h; 112 mph)
Stalling speed .... 27 kt (50 km/h; 32 mph)
Max rate of climb at S/L .... 360 m (1,181 ft)/min
T-O run .... 80 m (265 ft)
Landing run .... 60 m (200 ft)
Range with max fuel .... 540 n miles (1,000 km; 621 miles)
g limits .... +6/−3

UPDATED

KHGAPP

KHARKOVSKOYE GOSUDARSTVENNOYE AVIATSIONNOYE PROIZVODSTVENNOYE PREDPRIYATIE (KHARKOV STATE AVIATION PRODUCTION ENTERPRISE

ulitsa Sumskaya 134, 61023 Kharkov
Tel: (+38 057) 707 08 00
Fax: (+38 057) 707 08 39
e-mail: ksamc@ksamc.com
Web: http://www.ksamc.com
GENERAL DIRECTOR: Pavel O Naumenko
HEAD OF INFORMATION BUREAU: Vladim Araslanov

Established in 1926 as GAZ 135, Kharkov State Aircraft Manufacturing Company (KSAMC) has built military, passenger and freight aircraft, including the K-5, KhAI-1, K-7, Su-2, MiG-15UTI, Tu-104, Tu-124, Tu-134 and An-72/72P, plus Tu-141 UAV. Export of its products began in 1964. ISO 9002 was gained in 1998.

Current production is centred on the Antonov An-74 in various versions, including the An-74TK-300, with underslung engines, and the twin-turboprop An-140. Flew An-74T-200A subvariant on 24 December 2004, this being first of five for Egypt, plus two An-74TK-200S 'flying hospitals' for Syria, due to be delivered in 2005. Will also produce centre fuselages for Antonov An-148 production by Aviant.

Kharkov consortium also provides support services, including marketing, maintenance, flying training, technical training, spare parts provision, ground equipment, test equipment and turnkey maintenance facilities.

Antonov An-140 assembly at Kharkov *NEW*/1047202

UPDATED

LILIENTAL'

LILIENTAL' ZAO (LILIENTHAL JSC)

a/ya 10038, 61070 Kharkov
e-mail: lilienthal@xai.kharkov.ua
Web: http://ai.kharkov.com/~lilienthal
PRESIDENT: Anatoly Borodin
PRODUCTION DIRECTOR: Yury Komarovsky
SALES MANAGER: Alexey Voloshin

The company (which adopts the German spelling of aviation pioneer Otto Lilienthal for marketing in the West) and its predecessors have built 15 types of aircraft, from hang-gliders (first flight 4 September 1975) to twins. It is currently Ukraine's largest manufacturer of ultralights. Lilienthal was formed by Anatoly Borodin and Valentin Loskov out of the hang-glider club associated with the KhGAPP aircraft plant, transforming into an ultralight club in 1976 and establishing a design bureau within Kharkov factory in 1987. Lilienthal was established as a firm in 1991, manufacturing the Kh-37 (X-37) Chibis hang-glider and moving two years later to the Kh-32 (X-32) Bekas. By 2003, more than 100 of each had been built, production continuing at a rate of 10 per year. Manufacture is conducted in close association with related companies Otto Ltd and Gustav Ltd. Lilienthal is also a service centre for Rotax engines.

It will be noted that the Kh-designations of the Kharkov-designed aircraft of Lilienthal retain the Cyrillic 'X' for Western trading.

NEW ENTRY

Lilienthal X-32 Bekas tandem-seat ultralight *(Sebastian Zacharias)* *NEW*/0576741

LILIENTHAL X-32 BEKAS

English name: Snipe
TYPE: Tandem-seat ultralight.
PROGRAMME: First flown March 1993; certified in Ukraine in 1995. Available in AT (sportplane), UT (trainer), SKh (agricultural), H (floatplane) and N (skiplane) versions. Under a 2003 agreement with Jordan Aerospace Industries, is assembled and marketed in the Middle East as JAI RumBird X-32.
CUSTOMERS: Over 100 built by 2004.
COSTS: X-32AT US$15,510 for airframe, or US$22,300 with Rotax, US$33,800 with Rotax 912 and US$34,790 with Rotax 912S, plus tax (2004).
DESIGN FEATURES: Pod-and-boom pusher monophone, with constant-chord high wing braced by V struts and intermediate struts. T-tail with single bracing strut each side.
Wing section NACA 4412.

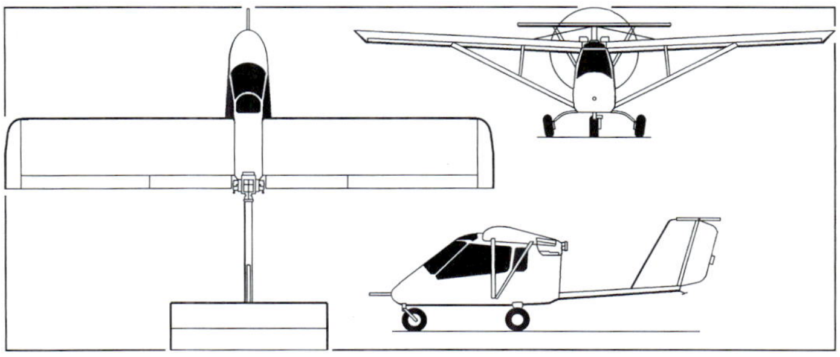

Lilienthal X-32 Bekas *(Paul Jackson)* *NEW*/0576137

FLYING CONTROLS: Conventional and manual. Slotted flaps.
LANDING GEAR: Tricycle type; fixed. Fuselage-mounted spring cantilever mainwheel legs. Trailing-link, castoring nosewheel. Mainwheel brakes.
POWER PLANT: Suitable for engines between 60 hp and 100 hp; typically 48.0 kW (64.4 hp) Rotax 582 UL, 59.6 kW (79.9 hp) Rotax 912 or 73.5 kW (98.6 hp) Rotax 912S. VPSh-2 Donchak (or similar) variable-pitch propeller. Normal fuel capacity 40 litres (10.6 US gallons; 8.8 Imp gallons) in fuselage tank; optional additional 50 litres (13.2 US gallons; 11.0 Imp gallons) in wing tanks of Rotax 912-engined versions; total 90 litres (23.8 US gallons; 19.8 Imp gallons) max optional capacity.

DIMENSIONS, EXTERNAL:
Wing span .... 9.00 m (29 ft 6¼ in)
Length overall .... 6.25 m (20 ft 6 in)
AREAS:
Wings, gross .... 12.33 m² (132.7 sq ft)
WEIGHTS AND LOADINGS:
Chemical load (X-32SKh) .... 120 kg (265 lb)
Max T-O weight: Rotax 582 .... 450 kg (992 lb)
Rotax 912 .... 550 kg (1,212 lb)
PERFORMANCE:
Max level speed: Rotax 582 .... 85 kt (158 km/h; 98 mph)
Rotax 912 .... 94 kt (175 km/h; 109 mph)
Cruising speed: Rotax 582 .... 67 kt (125 km/h; 78 mph)
Stalling speed, power off, flaps down:
Rotax 582 .... 31 kt (57 km/h; 36 mph)
Rotax 912 .... 37 kt (67 km/h; 42 mph)
Max rate of climb at S/L .... 240 m (787 ft)/min
T-O run .... 50 m (165 ft)
Landing run .... 30 m (100 ft)
Endurance: normal fuel .... 3 h 20 min
max fuel .... 7 h 30 min
g limits: Rotax 582 .... +4.4/–2.2
Rotax 912 .... +3.8/–1.9

NEW ENTRY

LILIENTHAL X-34 BEKAS

English name: Snipe
TYPE: Three-seat ultralight.
PROGRAMME: Development of X-32. Marketed and assembled in Middle East by Jordan Aerospace Industries as GulfBird X-34. Options include Rescuer medical evacuation version.

Lilienthal X-34 Bekas three-seat version ***NEW***/0576640

COSTS: Airframe US$19,990; with engine US$38,990 plus tax (2004).
DESIGN FEATURES: Generally as for X-32 Bekas, but enlarged cabin for two passengers behind pilot.
POWER PLANT: One 73.5 kW (98.6 hp) Rotax 912S flat-four. Fuel options as X-32.
DIMENSIONS, INTERNAL:
Cabin max width .... 1.00 m (3 ft 3¼ in)
AREAS:
Wings, gross .... 13.75 m² (148.0 sq ft)
WEIGHTS AND LOADINGS:
Weight empty .... 198 kg (437 lb)
Max T-O weight .... 555 kg (1,223 lb)
Max wing loading .... 40.4 kg/m² (8.27 lb/sq ft)
Max power loading .... 7.55 kg/kW (12.40 lb/hp)
PERFORMANCE:
Range .... 378 n miles (700 km; 435 miles)
g limits .... +3.8/–1.9

NEW ENTRY

MOTOR-SICH

OTKRITOYE AKTSIONERNOYE OBSHCHESTVO MOTOR-SICH (MOTOR-SICH JSC)

ulitsa 8 Marta 15, Zaporozh'e 69068
Tel: (+38 0612) 61 47 77
Fax: (+38 0612) 65 60 07
e-mail: motor@motorsich.com
Web: http://www.ukrainetrade.com/motorsich
GENERAL DIRECTOR: Vyacheslav Boguslaev

Long-established engine manufacturer Motor-Sich was confirmed in August 2000 as production centre for those Kamov Ka-226 utility helicopters required for use in Ukraine. These to be powered by indigenous ZMKB AI-450 engines and employ components sourced entirely from Ukrainian manufacturers.

UPDATED

United Kingdom

ATG

ADVANCED TECHNOLOGIES GROUP

No 2 Hangar, Cardington Field, Shortstown, Bedford MK42 0TJ
Tel: (+44 1234) 74 35 81
Fax: (+44 1234) 74 35 82
e-mail: airshiptech@compuserve.com
Web: http://www.atg-airships.com
PRESIDENT: Air Marshal Sir John Walker
EXECUTIVE CHAIRMAN AND TECHNICAL DIRECTOR: Roger Munk
MANAGING DIRECTOR: Peter Ward
SALES AND MARKETING DIRECTOR: Gordon Taylor
PUBLIC RELATIONS MANAGER: Joanna Amis

US SUBSIDIARY:
SkyCat Technologies Inc, 3201 Airpark Drive, Suite 101, Santa Maria, California 93455
Tel: (+1 805) 614 08 24
Fax: (+1 805) 614 09 24
e-mail: skycattech@bigfoot.com
Web: http://www.skycattech.com

ATG is the trading name, from June 2000, of Airship Technologies Services Ltd, which was created in February 1996 to revive the activities of the former Airship Industries company, responsible for the British-designed Skyship 500/600 series.

ATG's design and production team draws on experience in designing and operating more transport category, passenger certified airships than any other world operator. Current work is directed towards two totally new concepts in lighter-than-air design: SkyCat, a hybrid air vehicle, and StratSat, a stratospheric telecommunications platform. SkyCat has a fully amphibious design incorporating a sophisticated lifting body, ellipsoidal cross-section hull and a catamaran hover cushion landing system which equips it to land on a grass field, a swamp, sand, snow or open sea.

In March 2002, ATG announced the first flight of its AT-10 airship, intended for advertising work, as a camera platform, and for surveillance and pilot training applications. The AT-10, SkyCat 20 and StratSat will all be powered by ATG diesel engines ranging from 74.6 to 447 kW (100 to 600 hp). Of these, the 100 hp engine was ready for limited production in early 2004, the first series example completing its CAA acceptance test on 15 January. At that time the 600 hp engine had reached prototype completion and testing status. In late 2002, ATG announced receipt of a contract from Japanese engine maker IHI for a propulsion system for a low-altitude flight test airship, an initial stage in Japan's programme to develop an unmanned airship-based stratospheric telecommunications and surveillance system. This propulsion system had been delivered and tested by early 2004.

UPDATED

ATG AT-10

TYPE: Helium non-rigid airship.
PROGRAMME: First flight expected in early 2002; prototype registered G-OATG in November 2001; made first flight 28 March 2002. Phase 2 flight testing completed by July 2003, with some 111 hours in about 50 flights accumulated by end of August 2003. C of A (Airship Special Category) awarded 4 March 2004; work towards full public transport certification continuing at that time. More than 160 hours' flying had been completed by mid-May 2004.

Prototype ATG AT-10 four/five-passenger airship ***NEW***/0589387

CUSTOMERS: First sale announced in May 2001, to Chinese operator for advertising and police surveillance duties in Shanghai area. This airship reported as partially completed by September 2002.
COSTS: US$3 million (2001).
DESIGN FEATURES: Commercial airship with capability for advertising and camera platform use and for other applications such as pilot training. X configuration tailfins.
FLYING CONTROLS: Split-channel, optically signalled flight control system incorporating sidestick controllers, powered flight actuators and an Ada control system interface; provision for autopilot. Single ballonet centrally located over gondola for envelope pressure control.
STRUCTURE: Envelope of laminated translucent fabric with external catenary collar system supporting gondola. Provision for internal mercury illumination system.
LANDING GEAR: Non-retractable bicycle type, coil-sprung with oleo damping; twin wheels on each unit.
POWER PLANT: Two 74.6 kW (100 hp) (59.7 kW; 80 hp maximum continuous) ATG A-Tech 100 horizontally opposed, two-stroke, direct injection Diesel engines with vectorable-thrust ducted propellers. Supercharged induction system.
ACCOMMODATION: One or two pilot stations with sidestick controls; four or five passenger seats on multiple tracks; large flight deck and passenger cabin transparencies; large passenger door on port side. Rear of gondola contains lavatory, ballonet air system, pneumatics, electrical services and cabin heater.
SYSTEMS: Low-pressure pneumatic system to power flight control actuators and provide low susceptibility to lightning strike and electromagnetic interference. Electrical power (28 V DC) from 6 kVA alternator.
DIMENSIONS, EXTERNAL:
Length overall .... 41.40 m (135 ft 10 in)
Max diameter .... 10.70 m (35 ft 1¼ in)
Height overall .... 13.60 m (44 ft 7½ in)

DIMENSIONS, INTERNAL:
Envelope volume ... 2,500 m^3 (88,290 cu ft)
Gondola: Length ... 4.40 m (14 ft 5¼ in)
Max width ... 1.52 m (4 ft 11¾ in)
Width at floor ... 0.83 m (2 ft 8¾ in)
Min height ... 1.86 m (6 ft 1¼ in)
WEIGHTS AND LOADINGS:
Max payload ... 740 kg (1,631 lb)
PERFORMANCE (design):
Max level speed ... 60 kt (111 km/h; 69 mph)
Cruising speed ... 50 kt (93 km/h; 58 mph)
Pressure ceiling ... 2,745 m (9,000 ft)
Max range at 30 kt (55 km/h; 35 mph) ... 1,020 n miles (1,889 km; 1,173 miles)
Endurance at 30 kt (56 km/h; 35 mph) ... 34 h

UPDATED

ATG SKYCAT

TYPE: Hybrid non-rigid air vehicle.

PROGRAMME: Launched 29 June 2000, with three versions announced; name derived from 'sky catamaran'; numerals in designations indicate payload capacity in tonnes. Preceded by SkyKitten I, a 12.2 m (40 ft) long proof-of-concept prototype; first flight (retrospectively allocated, but did not carry, UK 'B Conditions' registration G-86-01 which eventually applied instead to one-tenth scale StratSat prototype) 28 June 2000; more than 30 short flights, from land and water, by early 2001; bow thruster added. SkyKitten II (13.5 m; 44.3 ft long, envelope volume 145 m^3; 5,120 cu ft) is one-seventh scale representation of SkyCat 20; 3.0 kW (4 hp) engines (being test-run in early 2004); retractable hover skirts (to be fitted after first flight); envelope shape identical to SkyCat 20. Shown statically at Paris, June 2001, registered G-86-02.

CURRENT VERSIONS: **SkyCat 20:** Initial operational version. Manufacture under way by mid-2000; not yet completed (mid-2004).

SkyCat 200: Second, and larger, operational version. Work under way on payload module. Industrial partner IAR Brasov (which see under Romania) was due to begin deliveries of major subassemblies in 2001, but not confirmed by mid-2004.

SkyCat 1000: Largest in current range; targeted to fly in late 2008. Subject in 2001 of US Army study contract to assess ultra-heavy-lift potential; completed study delivered to US Army in 2002.

CUSTOMERS: SkyCat 20 ordered by World Airships (UK).

COSTS: SkyCat 20 approximately US$25 million to US$28 million (2001).

DESIGN FEATURES: Brings together lighter than air and hovercraft technologies, enabling airship to land on virtually any flat land or water surface without need for landing infrastructure or ground crew. Larger versions seen as heavy-lift military or cargo-carrying vehicles (SkyCat 1000 has payload capacity of 10 Boeing 747s) and can be constructed in the open, eliminating need for costly hangars.

Image of SkyCat 20 in the livery of a possible future customer, the US Customs Service ***NEW***/0589395

SkyKitten II G-86-02 on display at Paris in June 2001 ***(Paul Jackson)*** 0126884

Elliptical cross-section created by twin hulls, allied to cambered longitudinal shape, provides up to 40 per cent of lift.

FLYING CONTROLS: Flight capability derived from combination of aerodynamic lift and helium buoyancy. All versions have dual-channel, optically signalled flight control system, plus pressure control by multiple ballonets fore and aft in each of the outer hulls.

STRUCTURE: Envelope of laminated fabric with internal catenary system supporting payload module. Internal diaphragms supporting envelope shape allow for limited compartmentalisation, further enhancing vehicle's fail-safe properties. Payload module located on centreline, with roll-on/roll-off freight capability in SkyCat 200 and 1000.

LANDING GEAR: Air-cushion landing system. Hover skirts on underside of outer hulls provide amphibious capability, as well as enhancing ground handling compared with conventional airships. Hover skirts are 'sucked in' for clean in-flight profile and enhanced all-round field of view. System shares use of ballonet fans with hull pressure system.

POWER PLANT: *SkyCat 20:* Four 447 kW (600 hp) (358 kW; 480 hp maximum continuous) ATG A-Tech 600 horizontally opposed, four-cylinder two-stroke, direct injection diesel engines, two mounted forward on hull and two on hull stern for cruise operation; all four in ducts with blown vanes to vector thrust for T-O, landing and ground handling. Supercharged induction system.

SkyCat 200: Four 5,966 kW (8,000 shp) (4,474 kW; 6,000 shp maximum continuous) turboshafts, in thrust-vectoring ducts positioned as for SkyCat 20.

SkyCat 1000: Six 11,185 kW (15,000 shp) (8,948 kW; 12,000 shp maximum continuous) turboshafts, two within each stern duct each driving a single propeller and one each side of hull forward.

ACCOMMODATION: Flight deck has side-by-side pilot stations, with sidestick controls and large transparencies for wide field of view; on larger models, also provides space (18.6 m^2; 200 sq ft in SkyCat 200, 83.6 m^2; 900 sq ft in 1000) for off-duty crew members.

SkyCat 20: Four passenger seats on individual tracks in payload compartment; large door each side with built-in steps; full-width door at rear for cargo use.

SkyCat 200 and 1000: Main deck provides clear load space for cargo/freight on a military-rated floor structure. Mezzanine decking can be provided to give multiple lower load area. Ramps (at rear on 200, fore and aft on 1000) provide roll-on/roll-off access to cargo area. Above door aperture is further accommodation space (37.2 m^2; 400 sq ft in 200, 195.1 m^2; 2,100 sq ft in 1000).

SYSTEMS: All versions have low-pressure pneumatic system as described for AT-10.

DIMENSIONS, EXTERNAL:
Length overall: 20 ... 81.30 m (266 ft 8¾ in)
200 ... 185.00 m (606 ft 11½ in)
1000 ... 307.00 m (1,007 ft 2½ in)
Max width: 20 ... 41.00 m (134 ft 6¼ in)
200 ... 77.00 m (252 ft 7½ in)
1000 ... 136.00 m (446 ft 2¼ in)
Height overall: 20 ... 23.90 m (78 ft 5 in)
200 ... 47.00 m (154 ft 2½ in)
1000 ... 77.00 m (252 ft 7½ in)
DIMENSIONS, INTERNAL:
Envelope volume: 20 ... 32,000 m^3 (1.13 million cu ft)
200 ... 475,500 m^3 (16.77 million cu ft)
1000 ... 2 million m^3 (70.63 million cu ft)
Payload deck:
Length: 20 ... 22.30 m (73 ft 2 in)
200 ... 49.40 m (162 ft 1 in)
1000 ... 80.80 m (265 ft 1 in)
Width: 20 ... 3.00 m (9 ft 10 in)
200 ... 7.50 m (24 ft 7¼ in)
1000 ... 12.20 m (40 ft 0¼ in)
Height: 20 ... 2.00 m (6 ft 6¾ in)
200 ... 5.00 m (16 ft 4¾ in)
1000 ... 8.00 m (26 ft 3 in)
WEIGHTS AND LOADINGS:
Max payload: 20 ... 20,000 kg (44,092 lb)
200 ... 200,000 kg (440,925 lb)
1000 ... 1,000,000 kg (2,204,625 lb)
PERFORMANCE (estimated):
Max level speed: 20 ... 80 kt (148 km/h; 92 mph)
200 ... 90 kt (166 km/h; 103 mph)
1000 ... 110 kt (203 km/h; 126 mph)
Cruising speed: 20 ... 70 kt (130 km/h; 81 mph)
20 ... 75 kt (139 km/h; 86 mph)
1000 ... 100 kt (185 km/h; 115 mph)
Pressure ceiling: 20, 200, 1000 ... 2,745 m (9,000 ft)1[1]
Range:
20 (max payload) ... 1,225 n miles (2,268 km; 1,409 miles)
20 (ferry) ... 4,000 n miles (7,408 km; 4,603 miles)
200 ... 3,225 n miles (5,972 km; 3,711 miles)
1000 ... 4,000 n miles (7,408 km; 4,603 miles)

[1] *5,485 m (18,000 ft) for SkyCat 20 high-altitude version*

UPDATED

BA

Hangar 43, Bentwaters, Rendlesham, Woodbridge, Suffolk IP12 2TW
Tel: (+44 1394) 42 02 C2
Fax: (+44 1394) 42 02 20
e-mail: info@bellaaviation.co.uk
Web: http://www.bellaaviation.co.uk
CEO: Mahesh Patel
CO-CEO: Captain Pant
GENERAL MANAGER: Anthony Preston
CHIEF DESIGNER ENGINEER: Dr Paul Welsh
DESIGN ENGINEER: Martino Felchero

Bella Aviation, which had been involved in manufacturing the CFM Shadow series for more than 20 years, announced in June 2004 that it had acquired rights to the product range of the former CFM Airborne (UK) Ltd and was to resume production immediately of the Shadow and Streak series. Bella then had seven personnel and a 670 m^2 (7,212 sq ft) assembly plant in a hangar at its 1,000 m^2 (10,764 sq ft) site on the former RAF Bentwaters airfield, and was working on new and improved additions to the Shadow/Streak family, total sales of which were nearly 400 at the time of the takeover. A 'next-generation Shadow' was in the design stage for a planned first flight in 2006.

Unfortunately, Bella announced on 20 August 2004 that it had suspended operations and laid off its staff.

UPDATED

BA SHADOW SERIES D AND E

TYPE: Tandem-seat ultralight/kitbuilt.

PROGRAMME: Shadow designed by David Cook and first flew as prototype in 1983; in production since 1984; type approval to BCAR CAP 482 Section S gained May 1985. Certification of Series E was continuing in 2003-04.

Unacceptable level of landing gear collapse incidents caused CAA to issue a warning about all Shadows and derivatives on 30 December 2003, due to concerns about airworthiness of landing gear/fuel tank arrangement, with intention to ground existing fleet on 31 March 2004 unless an acceptable modified gear could be fitted. Three separate groups developed improved landing gear arrangements during February and March, and on 30 April 2004 the PFA cleared the version developed by Danny Crosbie, which reinforces the original landing gear with twin carbon fibre arms attached outboard of retained drag bracing tubes, cross-braced by twin steel cables. By November 2004, 132 sets of this gear had been sold to existing Shadow/Streak owners. In addition, designer David Cook developed a 2024T3 aluminium cantilever gear with single cross-bracing wire which achieved PFA approval in late October and had some 30 retrofit orders at that time. BA produced a composites cantilever version which was to have been incorporated in new aircraft and be available for retrofit to existing Shadow variants.

CFM Shadow retrofitted with Crosbie landing gear ***(Paul Jackson)*** ***NEW***/1047694

Detail of Crosbie mainwheel legs ***(Paul Jackson)*** ***NEW***/1047695

Detail of Cook mainwheel legs ***(Brian Johnson)*** ***NEW***/1047696

Other changes proposed by BA included a wider cockpit with more headroom; fingertip trim control; and performance/payload improvements.

CURRENT VERSIONS: **Series D**: Rotax 447 or 503 engine. Dual controls standard. Suitable for cropspraying with 64 litre (16.8 US gallon; 14.0 Imp gallon) chemical tank and multisensor surveillance for photography (Hasselblad survey camera), video recording, linescan or thermal imagery, or closed-circuit TV microwave transmission to a command vehicle/station. Certified to BCAR CAP 482 Section S in 1998.

Series E: As Series D, but with electric starter, long-range fuel tanks and optional 59.6 kW (79.9 hp) Rotax 912 UL four-stroke engine. *As described.*

CUSTOMERS: Total of 16 Series D and three Series E delivered by June 2001, including one fitted with hand controls for a disabled pilot.

COSTS: Series D: Flyaway £26,705; kit, including Rotax 582, propeller and VFR instrumentation, £16,850 (2004).

DESIGN FEATURES: Design stated to have no 'defined stall' and to be unspinnable. Pod-and-boom layout; slightly tapered wings; three vertical stabilisers; sweptback tailplane with large rudder beneath; downturned wingtips. Options include agricultural spraygear and other specialised equipment. Quoted kit assembly time is 500 hours.

FLYING CONTROLS: Conventional and manual with cable actuation. Frise ailerons; full-span elevator with electrically actuated trim tab; fixed tab on rudder. Three-position flaps: 0, 15 and 30°.

STRUCTURE: Fuselage pod of Araldite-bonded Fibrelam. Wings comprise plywood formed leading-edge built on aluminium and plywood internal shearweb structure. Styrofoam formers provide both leading- and trailing-edge structures. Trailing surfaces covered with polyester cloth. Tailplane and rudder surfaces are polyester cloth-covered aluminium frames. Empennage connected to the monocoque by an aluminium boom.

LANDING GEAR: Tricycle type; fixed. Fuselage-mounted, spring moulded composites cantilever mainwheel legs; optional wheel fairings; castoring nosewheel; brakes. Options include wheel fairings and floats.

POWER PLANT: One 47.8 kW (64.1 hp) Rotax 582 UL two-stroke driving a three-blade Precision pusher propeller; or one 59.6 kW (79.9 hp) Rotax 912 UL four-stroke, driving a three-blade GSC propeller. Standard fuel capacity (582 UL) 62 litres (16.4 US gallons; 13.6 Imp gallons), (912 UL) 27 litres (7.1 US gallons; 5.9 Imp gallons). Options include a 30 litre (7.9 US gallon; 6.6 Imp gallon) auxiliary fuel tank.

EQUIPMENT: ASI, VSI, altimeter, tachometer, fuel and temperature gauges, slip ball indicator, electric elevator trim and front cockpit brakes standard. Alloy wheels, rear cockpit brakes, intercom, GPS and ballistic parachute recovery system optional. Modified cockpit for disabled pilots comprises combined throttle/roll hand control on left console, sidestick controller for pitch/rudder on right console, each with hand-operated brake lever.

Data below refer to Shadow Series E with Rotax 582 UL engine.

DIMENSIONS, EXTERNAL:	
Wing span	10.00 m (32 ft 9½ in)
Length overall	6.13 m (20 ft 1¼ in)
Height overall	1.75 m (5 ft 9 in)
Propeller diameter	1.32 m (4 ft 4 in)
AREAS:	
Wings, gross	15.25 m^2 (164.1 sq ft)
WEIGHTS AND LOADINGS:	
Weight empty	215 kg (474 lb)
Max T-O weight	425 kg (937 lb)
PERFORMANCE:	
Cruising speed at 75% power	86 kt (159 km/h; 99 mph)
Stalling speed, flaps down	34 kt (63 km/h; 39 mph)
Max rate of climb at S/L, solo	366 m (1,200 ft)/min
T-O to 15 m (50 ft)	130 m (425 ft)
Landing from 15 m (50 ft)	183 m (600 ft)
Range with standard fuel	370 n miles (685 km; 426 miles)

UPDATED

BA STREAK

TYPE: Tandem-seat ultralight/kitbuilt.

PROGRAMME: Construction of prototype started 1987 and this first flew June 1988; first flight of production aircraft (also known as Streak Shadow) October 1988. Also built under licence in South Africa, with Jabiru as engine option.

CURRENT VERSIONS: **Streak SA**: JAA Group A (Aeroplanes) certified version. *As described.*

Streak SLA: Microlight version. Differs from SA in having increased flap deflection and elevator gap seal to reduce minimum flying speed to comply with Microlight regulations.

CUSTOMERS: Total of 113 Streak SAs and seven Streak SLAs delivered by 11 June 2001. Recent customers include the Indian Air Force, which ordered 24 Rotax 912-engined SAs in September 1999 for delivery over a 15-month period for bird control at airfields.

COSTS: SA/SLA with Rotax 582 UL: Assembled £26,255. Kit £16,400. (All excluding VAT, 2004). Factory-assembled aircraft available for export only.

DESIGN FEATURES: Derivative version of Shadow (which see); has new wing design of reduced span, light airframe weight and more powerful engine; no 'defined stall' and is spin-resistant.

FLYING CONTROLS: See Structure. Electric trim.

STRUCTURE: Similar to Shadow but with foam/glass fibre wings (CFM aerofoil section), and control surfaces with aluminium alloy ribs and polyester fabric covering (ailerons, flaps, rudder and elevators).

POWER PLANT: Engine options and fuel capacities as for Shadow Series D/E.

DIMENSIONS, EXTERNAL:	
Wing span	8.53 m (28 ft 0 in)
Length overall	6.40 m (21 ft 0 in)
Height overall	1.75 m (5 ft 9 in)
AREAS:	
Wings, gross	13.01 m^2 (140.0 sq ft)
WEIGHTS AND LOADINGS:	
Weight empty	190 kg (419 lb)
Max T-O weight	408 kg (900 lb)
PERFORMANCE (Streak SA):	
Max level speed	94 kt (174 km/h; 108 mph)
Stalling speed	34 kt (63 km/h; 39 mph)
Max rate of climb at S/L	366 m (1,200 ft)/min
T-O to 15 m (50 ft)	130 m (426 ft)
Landing from 15 m (50 ft)	183 m (600 ft)
Range	385 n miles (713 km; 443 miles)

UPDATED

BA (CFM) Streak rflying with a unique and exempt cantilever main landing gear in 2004 *(Paul Jackson)* ***NEW***/0589311

BA STAR STREAK

TYPE: Tandem-seat ultralight/kitbuilt.

PROGRAMME: Development of Streak with reduced wing chord, wider cockpit and uprated engine. Prototype flew in 1992; first deliveries February 1994. Total of nine delivered by 11 June 2001.

CURRENT VERSIONS: **Star Streak:** Standard version

Star Streak 912: Higher-powered version with 59.6 kW (79.9 hp) Rotax 912UL engine.

As described.

COSTS: Star Streak: Assembled £28,495. Kit £18,195. (All excluding VAT, 2004). Factory-assembled aircraft available for export only.

POWER PLANT: Star Streak has one 47.8 kW (64.1 hp) Rotax 582 UL two-stroke; Star Streak 912 has one 59.6 kW (79.9 hp) Rotax 912 UL four-stroke. Fuel capacity (both) 27 litres (7.1 US gallons; 5.9 Imp gallons).

DIMENSIONS, EXTERNAL:

Wing span ... 8.53 m (28 ft 0 in)
Length overall ... 6.05 m (19 ft 10 in)
Height overall ... 1.75 m (5 ft 9 in)

AREAS:

Wings, gross ... 11.00 m² (118.4 sq ft)

WEIGHTS AND LOADINGS:

Weight empty ... 205 kg (452 lb)
Max T-O weight ... 408 kg (900 lb)

PERFORMANCE:

Cruising speed at 75% power ... 94 kt (174 km/h; 108 mph)
Stalling speed, flaps down ... 38 kt (71 km/h; 44 mph)
Max rate of climb at S/L ... 427 m (1,400 ft)/min
T-O to 15 m (50 ft) ... 122 m (400 ft)
Landing from 15 m (50 ft) ... 183 m (600 ft)
Range with max fuel ... 210 n miles (388 km; 241 miles)

UPDATED

CFM Star Streak with Crosbie main gear *(Paul Jackson)* ***NEW***/1047697

BAE

BAE SYSTEMS PLC

HEADQUARTERS: Warwick House, PO Box 87, Farnborough Aerospace Centre, Farnborough, Hampshire GU14 6YU
Tel: (+44 1252) 37 32 32
Fax: (+44 1252) 38 30 00
Web: http://www.baesystems.com

CHAIRMAN: Richard Olver

CEO: Mike Turner

CHIEF OPERATING OFFICERS:
Chris Geoghegan (Operational, Capital and Shared Services)
Ian King
Steve Mogford (Programmes)
Mark Ronald (North America)

MARKETING DIRECTOR: Mike Rouse

DIRECTOR, CORPORATE COMMUNICATIONS: Hugh Colver

BAE Systems was established on 30 November 1999, upon the merger of British Aerospace (BAe) and the Marconi Electronic Systems business, then part of GEC-Marconi.

Sales in 2003 totalled £12.6 billion; order book at 31 December 2003 was £46.0 billion (including joint ventures). Employees in 2003 totalled nearly 90,000 located in UK, USA, Sweden, Saudi Arabia, France, Italy, Germany, Australia and Canada.

Design and manufacturing of new aircraft is undertaken by Air Systems Group.

AIR SYSTEMS GROUP: Group formed January 2002 to manage BAE Systems' actvities in Hawk, Nimrod, Eurofighter Typhoon and Lockheed Martin F-35 JSF programmes. Personnel total 10,500. Nimrod structural upgrades are undertaken at the Woodford plant. New Business team involved with development of UAVs and RAF's Future Offensive Air System studies.

BROUGH: Brough, East Yorkshire HU15 1EQ
Tel: (+44 1482) 66 71 21
Fax: (+44 1482) 66 66 25

SAMLESBURY: Samlesbury, Balderstone, Lancashire BB2 7LF
Tel: (+44 1254) 81 23 71
Fax: (+44 1254) 7680 00

WARTON: Warton Aerodrome, Preston, Lancashire PR4 1AX
Tel: (+44 1772) 63 33 33
Fax: (+44 1772) 63 47 24

WOODFORD: Woodford Aerodrome, Cheshire SK7 1QR
Tel: (+44 161) 439 50 50
Fax: (+44 161) 955 30 08

UPDATED

BAE SYSTEMS HAWK 50, 60 AND 100 SERIES

RAF/FAA designations: Hawk T. Mks 1 and 1A
US Navy designation: T-45A and T-45C Goshawk
Canadian Forces designation: CT-155

TYPE: Advanced jet trainer/light attack jet.

PROGRAMME: HS P1182 Hawk first flew 21 August 1974; first-generation Hawk remains available and is marketed with advanced 100 Series and single-seat 200 Series (detailed separately) to meet customers' requirements; Hawk design leadership transferred from Kingston to Brough 1988, and final assembly and flight test from Dunsfold to Warton 1989. Brough subsequently assumed responsibility for final assembly, although first flights still at Warton.

Hawk 50 Series main exports made December 1980 to October 1985; largely supplanted by 60 Series; Hawk 100 enhanced ground attack export model announced mid-1982; first flight of 100 Series aerodynamic prototype (G-HAWK/ZA101 converted as Mk 100 demonstrator) 21 October 1987; trials of wingtip Sidewinder rails started at Warton in April 1990. Warton assembly line officially opened 24 October 1991.

CURRENT VERSIONS: **Hawk T. Mk 1:** Two-seater for RAF advanced flying and weapon training; 23.1 kN (5,200 lb st) Adour 151-01 (–02 in Red Arrows aircraft) non-afterburning turbofan; two dry underwing hardpoints; underbelly 30 mm gun pack; three-position flaps; simple weapon sight in some aircraft of No. 4 FTS.

Hawk T. Mk 1A: Contract January 1983 to wire 89 Hawks (including Red Arrows) for AIM-9L Sidewinder on each inboard wing pylon and optional activation of previously unused outer wing hardpoints; last conversion redelivered 30 May 1986.

RAF Hawk re-wing programme began 1989; initial 85 wings completed by BAe in 1993; delivery of second batch of 59 began November 1993 and completed in 1995. Rebuild programme by BAe for 80 RAF Hawks authorised 27 December 1998 to replace centre and rear fuselage by Mk 65 standard structures, extending service life to 2010; work undertaken at DARA St Athan and BAE Systems at Brough; first aircraft, XX348, redelivered 11 April 2000; last was XX242, delivered to Red Arrows on 27 August 2003. Aircraft comprised 11 Mk 1s, 62 Mk 1As and seven Mk 1Ws, retaining original designations. (Sole external indication of this upgrade is displacement of rear strobe light from centreline.) Avionics upgrade plans being formulated in Staff Requirement (Air) 449; primary aim is 'glass cockpit'. In 2003, UK MoD fleet included 132 Hawks.

Hawk T. Mk 1W: Following re-winging, 24 RAF Hawk T. Mk 1s gained the ability to carry stores on two underwing pylons, although not the centreline gun pod. The alternative designation **T. Mk 1FTS** is also used for this modification.

Hawk 50 Series: Initial export version with 23.1 kN (5,200 lb st) Adour 851 turbofan.

T-45 Goshawk: US Navy version; built by Boeing. *Described separately.*

Hawk 60 Series: Development of 50 Series with 25.4 kN (5,700 lb st) Adour 861 turbofan; leading-edge devices and four-position flaps to improve lift capability.

First of two follow-on BAE Systems Hawk Mk 115s built for Canada in 2004 *(Paul Jackson)* ***NEW***/1047017

Hawk 100 Series: Enhanced ground attack development of 60 Series, announced mid-1982, to exploit Hawk's stores carrying capability; two-seater, with perhaps pilot only on combat missions; 26.0 kN (5,845 lb st) Adour Mk 871 turbofan; new combat wing incorporating fixed leading-edge droop for increased lift and manoeuvrability from M0.3 to M0.7; full-width flap vanes; manually selected combat flaps; detail changes to wing dressing; structural provision for wingtip missile pylons; MIL-STD-1553B databus; advanced Smiths Industries HUDWAC and new air data sensor package with optional laser ranging and FLIR in extended nose; improved weapons management system allowing preselection in flight and display of weapon status; manual or automatic weapon release; passive radar warning; HOTAS controls; full-colour multipurpose CRT display in each cockpit; provision for ECM pod.

Demonstrator ZA101. Production prototype Mk 102D (ZJ100) flown 29 February 1992. Early orders from Abu Dhabi (placed 1989), Indonesia (signed June 1993), Malaysia (signed 10 December 1990) and Oman (signed 30 July 1990). FLIR, laser ranger and Sky Guardian RWR in Omani aircraft.

Selected by Australia in November 1996 as next-generation lead-in fighter trainer; order signed 24 June 1997 for 33 aircraft; Australian Hawk Mk 127s have new, advanced instrumentation resembling that of the F-18 Hornet and including an integrated Smiths system of three 127 mm (5 in) square colour screens in each cockpit, HUD, upgraded mission computer, engine life computer and stores management system. First Australian Hawk (A27-01) flew (in UK, temporarily as ZJ632) on 16 December 1999. First Australian-assembled aircraft flew (A27-010) on 12 May 2000; final Australian production delivery 8 August 2001; last UK-built aircraft delivered 5 October 2001.

BAE, Bombardier and partners T-6A Harvard II and Hawk Mk 115 as equipment of Canadian-based NATO Flying Training in Canada (NFTC) programme; contract agreed 17 November 1997 and formally issued on 12 May 1998 for 18, plus eight options, two of latter being taken up in 1999; further two stated to be required in early 2000, after Singapore joined programme; order total increased to 22 by early 2002. First aircraft, 155201, arrived in Canada 4 July 2000; officially received 6 July; instructor training began 12 July 2000. First flight of 22nd Hawk on 29 June 2004.

Hawk LIFT: Variant of Series 100 as lead-in fighter trainer (LIFT). Features include 'combat wing', Adour Mk 871 or Mk 951 engine, new three MFD 'glass cockpit' with NVG compatibility, MIL-STD-1553B digital databus, revised mission planning system/data transfer unit, INS/GPS, OBOGS, APU, HUMS, new HUD and HOTAS controls. Provision for FLIR and aerial refuelling. Initial customer is South Africa (Mk 120); first SAAF Hawk (and sole aircraft assembled in UK) first flew (ZJ970, later SAAF 250) 2 October 2003; delivered to South Africa inside Antonov An-22 17 October 2003.

Offered to RAF (**Mk 128**) to meet MFTS (military flying training system) requirement for some 31 to enter service in 2007 as lead-in to Eurofighter Typhoon, having open-architecture avionics and Adour Mk 951 engines. Maximum weapon load 3,000 kg (6,614 lb). On 30 July 2003, BAE announced formal RAF commitment to potential £800 million contract for 20 Mk 128s, and 24 options; design and development contract, worth £159 million, signed 22 December 2004, covering only two development aircraft to fly from mid-2005 onwards; production contract to follow by mid-2006, although 35 RAF serial numbers reserved in December 2004. Subject to satisfactory contract negotiations, which not completed 12 months later. Schedule was then for deliveries beginning 2008.

Hawk 200 Series: Single-seat multirole version. *Described separately.*

Hawk 100NDA: New Development Aircraft; based on Mk 127. Prototype, and testbed for Adour Mk 951 turbofan, first flew (ZJ951) at Warton 5 August 2002 (initially with interim standard engine). To Bredasdorp, South Africa, for further trials associated with SAAF Hawk Mk 120; first flew with Mk 951 engine in May 2003.

CUSTOMERS: Refer to table. BAE delivered 20 Hawks in 1996 and 22 in 1997; none in 1998; deliveries resumed in 1999 with 10 to Indonesia before supply of remaining six was suspended by UK government embargo of 11 September 1999; deliveries re-authorised in early 2000, but without certain US-sourced components. Deliveries in 2000 totalled 16 (eight Australian and eight Canadian), not including Australian assembly. In 2001, UK production line delivered 13 (three to Australia and 10 to Canada); 2002 production comprised two to Canada and one company trials aircraft; 2003 production totalled one for South Africa. Two for Canada in 2004.

Hawk LIFT selected for purchase by South Africa, as announced 18 November 1998; order for 12, plus 12 options, announced 15 September 1999 and signed 3 December 1999; these to be first Hawks with Adour Mk 951 engines. Bahrain announced selection of Hawk on 22 July 2002; contract signed 28 January 2003; version based on Mk 127, but with Adour Mk 951 engine; six aircraft, plus six options; part of training programme also including Slingsby Firefly piston lightplanes.

India requested quotation in September 1999 for 92 Hawks, later amended to 66, of which eight kits and 42 complete aircraft to be assembled by HAL at Bangalore; Anglo-Indian political reaffirmation of purchase plans on 12 December 2000; purchase agreement anounced 3 September 2003 for 24 UK-built Hawks and 42 assembled in India; designation Mk 132.

COSTS: South African purchase of 24 Hawk LIFTs estimated at R4.7 billion (1998). Canadian contract for 20 valued at £400 million (1999). Hawk centre/rear fuselage replacement for 80 RAF aircraft valued at £100+ million (2000). Eventual 44 RAF Mk 128s valued at £800 million (2003).

DESIGN FEATURES: Fully aerobatic two-seat advanced jet trainer, adaptable for ground attack and air defence; design capable of other optional roles, with wing improvements on developed Series to enhance combat efficiency. Low wing and mid-mounted, anhedral, sweptback tailplane; air intake on each side of fuselage, forward of wing leading-edge; single non-afterburning engine; elevated rear cockpit to enhance forward view; two strakes below rear fuselage; smurfs (refer Hawk 200 entry) on 100 Series; optional underwing hardpoints; wingtip AAM rails (100 Series).

Wing thickness/chord ratio 10.9 per cent at root, 9 per cent at tip; dihedral 2°; sweepback 26° on leading-edge, 21° 30' at quarter-chord.

FLYING CONTROLS: Conventional and assisted. Ailerons and one-piece all-moving tailplane actuated hydraulically by tandem actuators; rudder manually actuated, with electric trim tab. Hydraulically actuated double-slotted flaps, outboard 300 mm (12 in) of flap vanes normally deleted; small fence on each wing leading-edge; 100 and 200 Series use special 'combat wing' with full-width flap vanes (refer Hawk 200 entry); large airbrake under rear fuselage, aft of wings. Hydraulic yaw damper on 100 Series rudder.

STRUCTURE: Aluminium alloy; one-piece wing, with machined torsion box of two main spars, auxiliary spar, ribs and skins with integral stringers; most of box forms integral fuel tank; honeycomb-filled ailerons; composites wing fences; frames and stringers fuselage. Wing attached to fuselage by six bolts.

LANDING GEAR: Wide-track, hydraulically retractable tricycle type, with single wheel on each unit. AP Precision Hydraulics oleos and jacks. Main units retract inward into wing, ahead of front spar; castoring (optionally power-steered) nosewheel retracts forward. Dunlop mainwheels, brakes and tyres size 6.50–10 (14 ply) tubeless, pressure 9.86 bar (143 lb/sq in). (Hawk Srs 60 and 100 mainwheel pressure 17.23 bar; 250 lb/sq in at 9,100 kg; 20,061 lb

South African BAE Systems Hawk Mk 120, sole UK-built aircraft, '250' ***NEW***/1047067

Prototype BAE Systems Hawk T.Mk 1, pictured 30 years after its first flight, serving as a trials support aircraft with QinetiQ *(Paul Jackson)* ***NEW***/1047035

BAE SYSTEMS HAWK CUSTOMERS

Customer	Qty	Mark	First aircraft	Deliveries	Squadrons
Abu Dhabi	16	63	1001	Oct 1984 — May 1985	
	18	102	1051	Apr 1993 — Mar 1994	Khalif bin Zayed Air College
	4	63C	1017	Feb 1995 — Mar 1995	
Australia	33	127	A27-01	Apr 2000 — Oct 2001	76, 79
Bahrain	6	129			
Canada	22	115/CT-155	155201	Jul 2000-2004	419
Dubai	8	61	501	Mar 1983 — Sep 1983	III Shaheen
	1	61	509	Jun 1988	III Shaheen
Finland	50	51	HW301	Dec 1980 — Oct 1985	3/11, 3/21, 3/31,
	7	51A	HW351	Nov 1993 — Sep 1994	Koulutuslentolaivue, 2/Tukilentolaivue
India	24[1]	132			
	42[16]	132			
Indonesia	20	53	LL-5301	Sep 1980 — Mar 1984	103
	8	109	TT-1201	May 1996 — Mar 1997	1, 12
	16	209	TT-1205	Feb 1996 — Mar 1997	12
	16	209	TT-0217	Apr 1999 —	1
Kenya	12	52	1001	Apr 1980 — May 1982	(Laikipia AB)
Korea, South	20	67	67–496	Sep 1992 — Aug 1993	216
Kuwait	12	64	140	Nov 1985 — Sep 1986	12
Malaysia	10	108	M40-01	Jan 1994 — Sep 1995	3 FTC (15 Sqdn)
	18	208	M40-21	Aug 1994 — May 1995	6, 9, 3 FTC (15 Sqdn)
Oman	4	103	101	Dec 1993 — Jan 1994	6
	12	203	121	Dec 1994 — May 1995	6
Saudi Arabia	30	65	2110	Aug 1987 — Oct 1988	21, 37
	20	65A	7901	Mar 1997 — Dec 1997	79
South Africa	12	120/LIFT	250[1]	2005	85 CFS
Switzerland	20	66	U-1251	Nov 1989 — Nov 1991	Nil. Withdrawn from service 2002
UK	176	T. Mk 1	XX154	Nov 1976 — Feb 1982	4 FTS (19 & 208 Sqdns), Red Arrows, 100 Sqdn, FSATO, QinetiQ
	2[17]	128	ZK010	2008	—
USA	204[4]	T-45A/C	162787	Apr 1988 —	VT-7, VT-21, VT-22
Zimbabwe	8	60	600	Jul 1982 — Oct 1982	2
	5	60A	608	Jun 1992 — Sep 1992	2
Demonstrator	3	60/102D/102NDA	G-HAWK		—
	3	200/200/200RDA	ZG200		—
Total	**862**				

[1] Built at Brough, Hamble and Samlesbury, assembled at Warton or Brough and first flown at Warton; unless stated otherwise, remainder built at Kingston, assembled at Dunsfold (290) and Bitteswell (1)
[2] Laser nose
[3] 46 assembled by Valmet in Finland
[4] 19 assembled by F + W in Switzerland
[5] Production by Boeing in USA (see International section)
[6] Two first flown at Warton
[7] Wingtip Sidewinders
[8] 89 converted to T. Mk 1A
[9] 13 converted to 63A; two to 63B
[10] Removable refuelling probe
[11] Radar warning receiver
[12] First 12 assembled at Warton; remainder by BAe Australia at Williamtown, assisted by Hunter Aerospace, Hawker de Havilland and Qantas
[13] One first flown at Warton
[14] Five repurchased by BAE Systems in 1999
[15] Plus 12 options; all except first aircraft assembled in South Africa
[16] Assembled by HAL in India
[17] Two demonstrators. Minimum 20 and maximum 44 to be acquired, although 35 serial numbers initially assigned

T-O weight.) Nosewheel and tyre size 16×4.4 (8 ply) tubeless, pressure 8.27 bar (120 lb/sq in). Tail bumper fairing under rear fuselage. Anti-skid wheel brakes. Tail braking parachute, diameter 2.64 m (8 ft 8 in), on Mks 52/53 and all 60 and 100 Series aircraft.

POWER PLANT: One Rolls-Royce Turbomeca Adour non-afterburning turbofan, as described under Current Versions. Adour Mk 861A for Switzerland assembled locally by Sulzer Brothers. Adour Mk 951, available for new-build or retrofitted Hawks rated at 31.1 kN (7,000 lb st) and has doubled TBO of 4,000 hours. Engine starting by Microturbo integral gas-turbine starter. Fuel in one fuselage bag tank of 832 litres (220 US gallons; 183 Imp gallons) capacity and integral wing tank of 823 litres (217 US gallons; 181 Imp gallons); total fuel capacity 1,655 litres (437 US gallons; 364 Imp gallons). Pressure refuelling point near front of port engine air intake trunk; gravity point on top of fuselage. Provision for carrying one 455 or 591 litre (120 or 156 US gallon; 100 or 130 Imp gallon) drop tank on each inboard underwing pylon, according to Series.

ACCOMMODATION: Crew of two in tandem under one-piece, fully transparent, acrylic canopy, opening sideways to starboard. Fixed front windscreen able to withstand a 0.9 kg (2 lb) bird at 454 kt (841 km/h; 523 mph). Improved front windscreen fitted retrospectively to RAF Hawks, able to withstand a 1 kg (2.2 lb) bird at 528 kt (978 km/h; 607 mph); this installed on all current export aircraft. Separate internal screen in front of rear cockpit. Rear seat elevated. Martin-Baker Mk 10LH zero/zero rocket-assisted ejection seats, with MDC (miniature detonating cord) system to break canopy before seats eject. MDC can also be operated from outside the cockpit for ground rescue. Dual controls standard. Entire accommodation pressurised, heated and air conditioned.

BAE Systems Hawk 100 with wingtip Sidewinders and additional side view (top) of Hawk 60 Series *(Jane's/Dennis Punnett)* 0507441

SYSTEMS: BAE Systems cockpit air conditioning and pressurisation systems, using engine bleed air. Two hydraulic systems; flow rate: System 1, 36.4 litres (9.6 US gallons; 8.0 Imp gallons)/min; System 2, 22.7 litres (6.0 US gallons; 5.0 Imp gallons)/min. Systems pressure 207 bar (3,000 lb/sq in). System 1 for actuation of control jacks, flaps, airbrake, landing gear and anti-skid wheel brakes. Compressed nitrogen accumulators provide emergency power for flaps and landing gear at pressure of 2.75 to 5.50 bar (40 to 80 lb/sq in). System 2 dedicated to powering flying controls. Hydraulic accumulator for emergency operation of wheel brakes. Pop-up Hamilton Sundstrand ram air turbine in upper rear fuselage provides emergency hydraulic power for flying controls in event of engine or No. 2 pump failure. No pneumatic system.

DC electrical power from single 12 kW 30 V DC brushless generator, with two 3 kVA 115/26 V 400 Hz three-phase inverters to provide AC power and two batteries for up to 20 minutes of standby power. Gaseous oxygen system for crew; optional Hamilton Sundstrand (HS Marston) OBOGS first installed in Australian Hawk Mk 127s from 2000.

AVIONICS: *Comms:* Mk 1/Srs 50 includes Sylvania UHF and VHF; Cossor 2720 Mk 10A IFF in Finnish aircraft; Srs 60 has Rockwell Collins UHF and VHF, Magnavox UHF and Raytheon 2720 IFF; Srs 100 has Rockwell Collins AN/ARC-182 U/VHF, Magnavox AN/ARC-164 UHF and Raytheon 4720 IFF. Australian Mk 127 has twin Rockwell Collins AN/ARC-210 UHF/VHF and Raytheon 4720 IFF.

Flight: Mk 1/Srs 50 with Raytheon CAT 7000 Tacan, Cossor ILS having CILS.75/76 localiser/glideslope receiver and marker receiver; Rockwell Collins VOR/ILS and ADF, Rockwell Collins Tacan, Smiths-Newmark 6000-05 AHRS and Smiths radar altimeter in Srs 60; Srs 100 has BAE Systems INS300 inertial platform, Rockwell Collins AN/ARC-118 Tacan, Rockwell Collins VIR-31A VOR/ILS and Smiths 0103-KTX-1 radar altimeter, all integrated via dual redundant MIL-STD-1553B databus. Optional Skyforce Skymap II GPS-driven moving map in some RAF Hawks.

Instrumentation: Smiths-Newmark compass, BAE Systems gyros and inverter, two Honeywell RAI-4 4 in (100 mm) remote attitude indicators and magnetic detector system in Mk 1/Srs 50; Smiths 1500 Series HUDWAC in Srs 100; BAE Systems F.195 weapon sight in approximately 90 RAF aircraft; BAE ISIS 195 sight in Srs 50 and Srs 60, except Saab RGS2 in Finnish Mk 51.

Mission: BAE Systems camera and recorder in F.195-equipped RAF aircraft (requirement announced December 1997 to re-equip 41 with video recorders by late 1999); Vinten camera and recorder in Srs 50 and Srs 60. Srs 100 has Smiths 3000 Series colour MFD, GEC data transfer system and Vinten colour video recording system; plus BAE Systems Type 105H laser range-finder; optional FLIR.

Self-defence : (Series 100 only) Racal Prophet RWR in Mk 102s of Abu Dhabi; BAE Systems Sky Guardian in Indonesian, Malaysian and Omani aircraft, and retrofitted to Abu Dhabi's Mk 63s; optional chaff/flare dispenser at base of fin.

ARMAMENT: Underfuselage centreline-mounted 30 mm BAE Systems Aden Mk 4 cannon with 120 rounds (VKT 12.7 mm machine gun beneath Finnish aircraft), and two or four hardpoints under wing, according to Series. Provision for pylon in place of ventral gun pack. In RAF training roles, normal maximum external load is about 680 kg (1,500 lb), but the uprated Hawk 60 and 100 Series are cleared for an external load of 3,000 kg (6,614 lb), or 500 kg (1,102 lb) at 8 *g*. Typical weapon loadings on 60 Series include 30 mm or 12.7 mm centreline gun pod and four packs each containing eighteen 68 mm rockets; centreline reconnaissance pod and four packs each containing twelve 81 mm rockets; five 1,000 lb free-fall or retarded bombs; four launchers each containing four 100 mm rockets; nine 250 lb or 250 kg bombs; thirty-six 80 lb runway denial or tactical attack bombs; five 600 lb cluster bombs; four Sidewinder or two Magic air-to-air missiles; or four CBLS 100/200 carriers, each containing four practice bombs and four rockets. Vinten reconnaissance pod available for centre pylon. Similar options on 100 Series, plus wingtip air-to-air missiles. Mk 102s of Abu Dhabi/UAEAF can carry (but not designate for) two Alenia Marconi PGM-500 ASMs.

DIMENSIONS, EXTERNAL:
Wing span: Mk 1, Srs 50, Srs 60 ... 9.39 m (30 ft 9¾ in)
Srs 100 ... 9.08 m (29 ft 9½ in)
Srs 100 with Sidewinders ... 9.94 m (32 ft 7⅜ in)
Wing chord: at root ... 2.65 m (8 ft 8¼ in)
at tip ... 0.90 m (2 ft 11½ in)
Wing aspect ratio: Mk 1, Srs 50, Srs 60 ... 5.3
Srs 100 ... 4.9
Length: fuselage (incl jetpipe):
Mk 1, Srs 50, Mks 60-66 ... 10.775 m (35 ft 4¼ in)
Mk 67 ... 11.375 m (37 ft 3¼ in)
Srs 100 ... 11.40 m (37 ft 4¼ in)
fuselage and pitot:
Mk 1, Srs 50, Mks 60-66 ... 11.455 m (37 ft 7 in)
Mk 67, Srs 100 (non combat) ... 12.035 m (39 ft 5¾ in)
Srs 100 (incl chaff/flare dispenser) ... 12.095 m (39 ft 8 in)
overall:
Mk 1, Srs 50, Mks 60-66 ... 11.845 m (38 ft 10¼ in)
Mk 67, Srs 100 ... 12.425 m (40 ft 9¼ in)
Nose to pitot tip:
Mk 1, Srs 50, Mks 60-66 ... 0.68 m (2 ft 2¾ in)
Mk 67 ... 0.66 m (2 ft 2 in)
Srs 100 ... 0.635 m (2 ft 1 in)
Tailplane overhang ... 0.39 m (1 ft 3½ in)
Height overall ... 3.98 m (13 ft 0¾ in)
Tailplane span ... 4.39 m (14 ft 4¾ in)
Wheel track ... 3.47 m (11 ft 5 in)
Wheelbase ... 4.50 m (14 ft 9 in)

AREAS:
Wings, gross ... 16.69 m^2 (179.6 sq ft)
Ailerons (total):
Mk 1, 50 and 60 Series ... 1.05 m^2 (11.30 sq ft)
100 Series ... 0.97 m^2 (10.44 sq ft)
Trailing-edge flaps (total) ... 2.50 m^2 (26.91 sq ft)
Airbrake ... 0.53 m^2 (5.70 sq ft)
Fin: Mk 1, 50 and 60 Series ... 2.51 m^2 (27.02 sq ft)
100 Series ... 2.61 m^2 (28.10 sq ft)
Rudder, incl tab ... 0.58 m^2 (6.24 sq ft)
Tailplane ... 4.33 m^2 (46.61 sq ft)

WEIGHTS AND LOADINGS:
Weight empty: 60 Series ... 4,012 kg (8,845 lb)
100 Series ... 4,400 kg (9,700 lb)
Max weapon load (60, 100 Series) ... 3,000 kg (6,614 lb)
Max usable fuel weight:
internal ... 1,277 kg (2,815 lb)
external ... 930 kg (2,050 lb)
Max T-O weight: T. Mk 1 ... 5,700 kg (12,566 lb)
50 Series ... 7,350 kg (16,200 lb)
60, 100 Series ... 9,100 kg (20,061 lb)
Max landing weight: T. Mk 1 ... 4,649 kg (10,250 lb)
60 Series ... 7,650 kg (16,865 lb)
Max wing loading: T. Mk 1 ... 341.5 kg/m^2 (69.97 lb/sq ft)
50 Series ... 440.4 kg/m^2 (90.20 lb/sq ft)
60, 100 Series ... 545.4 kg/m^2 (111.70 lb/sq ft)
Max power loading: T. Mk 1 ... 246 kg/kN (2.42 lb/lb st)
50 Series ... 318 kg/kN (3.12 lb/lb st)
60 Series ... 359 kg/kN (3.52 lb/lb st)
100 Series ... 350 kg/kN (3.43 lb/lb st)

PERFORMANCE:
Never-exceed speed (VNE), clean:
at S/L ... M0.87 (575 kt; 1,065 km/h; 661 mph EAS)
at and above 5,180 m (17,000 ft) ... M1.2 (575 kt; 1,065 km/h; 661 mph EAS)
Max level speed at S/L:
50 Series ... 535 kt (990 km/h; 615 mph)
60 Series ... 545 kt (1,010 km/h; 627 mph)
100 Series ... 540 kt (1,001 km/h; 622 mph)
Max level flight Mach No. ... 0.88
Stalling speed, flaps down ... 96 kt (177 km/h; 110 mph)
Max rate of climb at S/L ... 3,600 m (11,800 ft)/min
Time to 9,150 m (30,000 ft), clean: 60 Series ... 6 min 54 s
100 Series ... 7 min 30 s
Service ceiling: 60 Series ... 14,020 m (46,000 ft)
100 Series ... 13,565 m (44,500 ft)
T-O run (clean): 60 Series ... 710 m (2,330 ft)
100 Series ... 640 m (2,100 ft)
Landing run (10% fuel load): 60 Series ... 550 m (1,800 ft)
100 Series ... 605 m (1,980 ft)
Combat radius, Aden gun pod and two Sidewinder AAMs:
with four 500 lb bombs and two 130 Imp gallon tanks ... 345 n miles (638 km; 397 miles)
with four 1,000 lb bombs ... 125 n miles (231 km; 143 miles)
Ferry range:
60 Series, with two 591 litre (156 US gallon; 130 Imp gallon) drop tanks ... 1,575 n miles (2,917 km; 1,812 miles)
100 Series, as above ... 1,360 n miles (2,519 km; 1,565 miles)
Endurance, 100 n miles (185 km; 115 miles) from base:
60 Series ... approx 2 h 42 min
100 Series ... approx 2 h 6 min
g limits: full internal fuel, all ... +8/−4
100 series, 60% fuel load:
1,360 kg (3,000 lb) external load ... +8/−4
2,721 kg (6,000 lb) external load ... +6/−3

UPDATED

BAE SYSTEMS NIMROD MRA. MK 4

TYPE: Maritime reconnaissance four-jet.

PROGRAMME: Original Hawker Siddeley HS 801 Nimrod first flew 23 May 1967 as adaptation of de Havilland D.H.106 Comet airliner. Total of two prototypes (converted Comet fuselages; now withdrawn), 46 Nimrod MR. Mk 1s and three Nimrod R. Mk 1 elint/sigint aircraft built for the RAF. Of these, 35 converted to MR. Mk 2 from 1979 onwards and 11 became Nimrod AEW. Mk 3s, but failed to enter service and were scrapped. MR. Mk 2 fleet of Nos. 42(R), 120, 201 and 206 Squadrons at Kinloss in 1996 totalled 28 (including three in long-term storage), plus one under conversion to R. Mk 1, four scrapped or used for ground instruction and two lost in accidents. Two original R. Mk 1s remain with No. 51 Squadron at Waddington.

Following RAF issue of Staff Requirement (Air) 420 for a Nimrod replacement, Dassault Atlantique 3, two versions of Lockheed Martin P-3 Orion (new production from manufacturer or refurbished P-3A/Bs from Loral) and upgraded Nimrod offered as potential solutions. UK official announcement on 25 July 1996 nominated last-mentioned proposal under commercial designation (since abandoned) Nimrod 2000. Although based on existing aircraft, the upgrade involves extensive (80 per cent) reconstruction of the airframe, plus incorporation of many new components, including engines, wings, landing gear and general systems, as well as new flight deck and detection systems. Contract MAR21a/100 awarded January 1997. Initial three development batch (DB) aircraft are conversions of stored Mk 2s, on which work began in early 1997; first delivery originally scheduled to RAF in 2001, followed by IOC in April 2003 on delivery of seventh aircraft; by early 1999, these dates had slipped 23 months due, in part, to excessive weight of new wing. New schedule called for first flight in June 2002; service deliveries in August 2004; IOC in March 2005; and final delivery in December 2008. Critical design review for air vehicle conducted late 1999 and finalised in early 2000, completing weapon system design. However, due to further slippage, prototype was given media debut on 16 August 2002 in expectation of maiden flight before year-end; delivery then stated to be late 2004.

By October 2002, first flight had been rescheduled for second half of 2003 due to discovery of "minor issues during airframe stress testing", although in-service date of March 2005 stated to be unchanged. BAE Systems and MoD announced change to Nimrod contract on 19 February 2003, this including halt to structural work on fourth and subsequent aircraft until risk reduction exercise completed on first three. This formalised by agreement signed 23 February 2004. Unofficial estimates of service-entry revised to 2009.

BAE Systems Nimrod MRA. Mk 4 radar housing ***NEW***/1047063

Maiden flight of first Nimrod MRA. Mk 4, 26 August 2004 ***NEW***/0577851

Ground testing and verification of engineering plans undertaken at Warton on Nimrod MR. Mk 1 prototype XV147. FR Aviation at Hurn, Bournemouth, selected for airframe relifing and upgrading as subcontractor to BAE, flying 'green' aircraft to Warton for mission avionics installation; first three (DB) aircraft (XV247/PA1, XV234/PA2 and XV242/PA3, for modification in that order) dismantled at RAF Kinloss and flown to Hurn by Antonov An-124 14 to 16 February 1997; redelivery to Warton originally due late 1999, January and February 2000. Fourth conversion subject, XV251/PA4, to Hurn under own power, 2 November 1998.

Programme revised in early November 1999, when FR Aviation contract cancelled and airframe work transferred to BAE at Woodford. PA1 completed at Hurn; PA2, PA3 and PA4 airfreighted to Woodford in An-124 in November/December 1999 for completion. Fifth subject, XV258/PA5, delivered directly ex-RAF to Woodford on 8 November 1999; further two (XZ284 and XV233) followed in 2000 and one (a former ground instructional airframe, XV253) in October 2001. Wing for initial aircraft delivered from Chadderton to Woodford in September 2000; second aircraft similarly fitted in December 2000.

Short Brothers and various BAE plants supply airframe components and subassemblies. Ground training systems supplied by Thomson TSL.

Boeing supplying and integrating TCCS (tactical command and sensor system); 15-month laboratory test began at Warton in November 1998.

Static test wing to be mounted on Nimrod AEW. Mk 3 fuselage for 1,000 hour trial, beginning 2004.

CURRENT VERSIONS (SPECIFIC): **PA1/ZJ516:** First flight 26 August 2004; landed at Warton. Aerodynamic, avionics and engine trials.

PA2/ZJ518: First flight 15 December 2004; landed at warton. Full mission system; avionics and mission systems tests; weapon carriage and release trials.

PA3/ZJ517: Mission systems testbed. Production standard, but additional flight test instrumentation fitted.

CUSTOMERS: Royal Air Force order was originally 21. In deference to comprehensiveness of rebuild, aircraft issued with new serial numbers ZJ514 to ZJ534. However, in March 2002 it was announced that the contract had been reduced to 18. Defence review of July 2004 reduced in-service fleet requirement to 12 aircraft, in addition to which unspecified number of reserves also required. Decision unlikely until signature of production agreement, due December 2005.

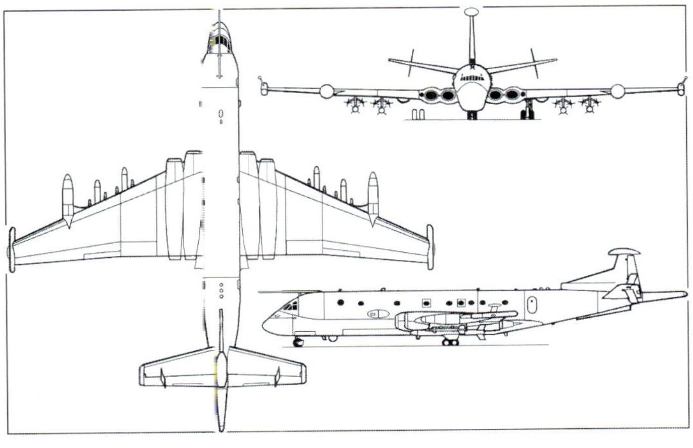

BAE Systems Nimrod MRA. Mk 4 maritime patrol aircraft *(Paul Jackson)* 0533724

COSTS: Ordered as £2 billion programme, including training system and initial logistic support, of which 75 per cent of work being placed with UK companies. Boeing TCCS mission system avionics contract valued at US$639 million (1996). Programme cost increased to £2.4 billion by March 1999, equivalent to 0.5 per cent above inflation. In mid-2000, BAE agreed to pay £46 million penalty because of 23-month programme slippage. BAE profits suffered £300 million charge in 2000 to offset expected Nimrod contract losses. Programme cost stated in 2002 to be £2.98 billion. Further £500 million charged to BAE accounts for 2002.

DESIGN FEATURES: World's first jet-powered, land-based maritime patrol and ASW aircraft on original 1969 service-entry; remains sole four-jet in this role. Optimised for anti-submarine warfare (ASW), anti-surface unit warfare (ASUW), search and rescue (SAR), maritime reconnaissance and provision of aid to civil authorities, including fisheries protection and operations against terrorism, drug smuggling and blockade running. Combines fast transit to operational area with low wing loading (and, hence, manoeuvrability) when on station. Can cruise on two or three engines to extend endurance. No special ground equipment required for off-base deployments.

Mid-mounted wing, with two engines in each root and auxiliary fuel tank in leading-edge at approximately two-thirds span; wingtip ESM pods; two hardpoints under each wing. Dihedral tailplane with auxiliary fins at one-third span to counter aerodynamic effects of electro-optical turret rotation; fin has large fillet and electrical equipment/decoy housing pod at tip. MAD tailboom. 'Double bubble' fuselage cross-section, with weapons bay below; latter restructured for Mk 4 and now in two, separated parts. Wing sweepback 20° at quarter-chord.

Nimrod Mk 4 upgrade extends operational lifetime by 25 years. Compared with original version, new power plant improves fuel consumption by over 20 per cent and thrust by 30 per cent. Larger wing and increased thrust restore performance despite 20 per cent increase in maximum take-off weight.

FLYING CONTROLS: Conventional control surfaces actuated hydromechanically. Plain flaps immediately outboard of engines.

STRUCTURE: All-metal, two-spar wing comprises centre-section, two stub-wings and two outer panels. Integrally machined stringers and machined spars. Fuselage is all-metal, semi-monocoque with pressurised upper component and unpressurised pannier including the weapons bay and nose radome. Pannier segments are free to move, so that structural loads are not transmitted to the pressure cell. Cantilever, all-metal tail surfaces surmounted by glass fibre fintip pod.

Nimrod MRA. Mk 4 introduces new wing centre box and wing inner panels of increased span designed by BAE Filton, built and assembled at Chadderton. Outer wings, to have been retained under BAE proposal, also new-build, as requested by RAF; designed by Filton, manufactured at Prestwick. Centre fuselage designed at Prestwick and built at Brough; lower (unpressurised) fuselage and front fuselage designed at Farnborough and built at Brough; rear fuselage designed by Dassault (France) and built at Brough; tailplane, elevators, rudder and weapons bay doors designed at Prestwick and Farnborough and manufactured at Prestwick and by FR Aviation. Flap actuation system designed by Dowty. Other new items include wingtip pods, fin fillet, fintip pod and (enlarged) finlets. Retained structures, principally the fuselage pressure hull and empennage, are receiving new protective treatment, although all pressure bulkheads and floors are new. Some 80 per cent of the Mk 4 will be newly built.

LANDING GEAR: Retractable tricycle type; all-new Dowty unit in Nimrod MRA. Mk 4. Four-wheel tandem bogie main units; twin nosewheels. Dunlop tyres, wheels and brakes.

POWER PLANT: Four 'marinised', non-afterburning Rolls-Royce BR710 Mk 101 (BR710B3-40) turbofans with FADEC and EICAS, each rated at 68.9 kN (15,500 lb st). Additional 15 to 17 per cent of internal fuel compared with Nimrod MR. Mk 2.

ACCOMMODATION: Two-person flight crew; eight-person tactical team (Tacco 1, Tacco 2, radar, communications, ESM, two acoustics and one sonobuoy loader/general assistant);

Prototype BAE Systems Nimrod MRA. Mk 4 prior to first flight ***NEW***/1047066

optional positions for two visual observers. Additional 13 seats for support personnel or replacement crew. Provision for further workstation. Crew areas, front to rear, are flight deck, lavatory, lateral observers' seats with bubble windows, tactical compartment, optional AEO compartment, galley and sonobuoy/general storage area, including crew door. Integral airstairs.

SYSTEMS: Smiths Aerospace flight management system (based on Boeing 737, but with vertical navigation capability). New Normalair-Garrett environmental control system (inlet at base of fin fillet) and hydraulic system. Hot air anti-icing system. Portable data storage system for fault analysis and rectification; compatible with Logistics Information Technology System. Honeywell APU in starboard wing trailing-edge. Lucas power-generation system; Normalair-Garrett (OBOGS) oxygen system. Ram-air turbine scabbed to fuselage side, ahead of port wingroot.

AVIONICS: Mission avionics integration by BAE Systems, significant proportion of new equipment being obtained from Airbus and Eurofighter Typhoon programmes. Nimrod MRA. Mk 4 programmed with 5.8 million lines of computer software code.

Comms: Digital communications subsystem by Telephonics: five V/UHF, two HF.

Radar: Thales Avionics Searchwater 2000MR maritime surveillance radar. Modes include pulse Doppler (air-to-air), inverse SAR, weather and swath and spot SAR.

Flight: Smiths Industries navigation and flight management system. Rockwell Collins TCAS II with Thales IFF Mode S transponder; Thales (Sextant) AFCS with autothrottle and vertical hold/select; EGPWS. Two Litton LN-100G laser INS with embedded GPS; Avionics Specialities ground proximity warning and collision-avoidance. Triplex air data system. Rockwell Collins ADF; BAE Systems ILS/VOR/MLS; Flight Data Co MMS, Smiths NAV/FMS, Thales radar altimeter and Rockwell Collins Tacan.

Instrumentation: Thales (Sextant) seven-screen full-colour LCD EFIS, plus EICAS. Based on that of Airbus A330/A340.

Mission: 'Fourth-generation' sensor suite, including Boeing's tactical command sensor subsystem and Smiths Industries' armament control system conforming to MIL-STD-1760. Mission system comprises four major subsystems linked by five dual redundant MIL-STD-1553B databuses: Communications, DASS, Armament Control and Tactical Command. Tactical display also capable of integrating weather radar, EO/FLIR imagery and defensive aids information. Seven identical, Boeing-supplied operator consoles integrated with Northrop Grumman (EOSDS) Nighthunter FLIR and LLLTV turret under forward fuselage; two consoles dedicated to acoustics, plus one other reconfigurable as such; EOSDS imagery is recorded and can be transmitted via satcom. Retains CAE ASQ-504(V) AIMS MAD and Computing Devices Canada/Ultra Electronics AQS-970 processor (derived from AQS-971 in Nimrod MR. Mk 2, but MAD now fully integrated with tactical system and with 64-buoy capability. Ultra Link 11, Rockwell Collins Link 16 (JTIDS), MBDA SHF/satcom, Frederick TTY, Raytheon UHF/satcom. Full range of sonobuoys dispensed from four Normalair-Garrett rotary launchers (each holding 10 Size A buoys) in the rear fuselage; two single-barrel launchers for high-level release when fuselage pressurised. Internal racks for 180 buoys (or 360 in emergency), including new SR(SA) 903 active sonobuoy. Active search sonobuoy system (ASSS) ordered 2002 in form of Ultra Electronics CAMBS VI. Elta EL/L-8300UK ESM provides all-round coverage, ±35° in the vertical plane, with rotating antenna beneath fuselage.

Self-defence: Integrated defensive aids subsystem (DASS) by BAE North America, including a Raytheon AN/ALE-50 (MoD) towed decoy, Vicon 78, Thales (Vinten) chaff/flare dispensers (eight chaff, four flare), BAE Systems AN/AAR-57 missile approach warning and BAE AN/ALR-56M RWR, last mentioned being separate from ESM system. Potential for later addition of DIRCM, laser warning receiver and integrated countermeasures system.

ARMAMENT: Smiths armament control system capable of accepting all current and known future maritime patrol weapons. Four underwing hardpoints, each capable of carrying two side-mounted, self-protection Sidewinder AAMs in addition to an anti-ship missile. Weapons bay, with two pairs of doors, is able to carry up to six lateral rows of ASW weapons, including up to nine Sting Ray torpedoes as well as bombs. Maximum load of Harpoon ASMs is four underwing and one in each of two weapon bays.

DIMENSIONS, EXTERNAL:
Wing span over tip pods ... 38.71 m (127 ft 0 in)
Length overall (excl refuelling probe) ... 38.63 m (126 ft 9 in)
Height overall ... 9.45 m (31 ft 0 in)
Tailplane span ... 14.51 m (47 ft 7¼ in)

DIMENSIONS, INTERNAL:
Cabin (incl flight deck): Length ... 26.82 m (88 ft 0 in)
Max width ... 2.95 m (9 ft 8 in)
Max height ... 2.08 m (6 ft 10 in)
Volume ... 124.1 m^3 (4,384 cu ft)

AREAS:
Wings, gross ... 235.80 m^2 (2,538.0 sq ft)
Ailerons (total) ... 5.63 m^2 (60.60 sq ft)
Tailplane ... 40.41 m^2 (435.00 sq ft)
Elevators, incl tabs ... 12.57 m^2 (135.30 sq ft)

WEIGHTS AND LOADINGS:
Weight empty ... 46,500 kg (102,515 lb)
Max weapon load ... more than 5,443 kg (12,000 lb)
Max fuel weight ... 50,122 kg (110,500 lb)
Max T-O weight ... 104,420 kg (230,205 lb)
Max zero-fuel weight ... 58,287 kg (128,500 lb)
Max wing loading ... 442.8 kg/m^2 (90.70 lb/sq ft)
Max power loading ... 379 kg/kN (3.71 lb/lb st)

PERFORMANCE:
Max operating Mach No. (MMO) ... 0.77
Service ceiling ... 12,800 m (42,000 ft)
Range with max internal fuel ... more than 6,000 n miles (11,112 km; 6,904 miles)
Endurance, unrefuelled ... more than 15 h
g limit ... +3

UPDATED

BNG

B-N GROUP LTD

Bembridge, Isle of Wight PO35 5PR
Tel: (+44 1983) 87 25 11
Fax: (+44 1983) 87 32 46
e-mail: enquiries@b-n.aero
Web: http://www.britten-norman.com
CHAIRMAN: Alawi Zawawi
DEPUTY CHAIRMAN: Maurice Hynett
CHIEF EXECUTIVE: William Hynett
COMMUNICATIONS: Kate Bucci

Pilatus Aircraft Ltd of Switzerland, which acquired Britten-Norman (Bembridge) Ltd 1979, sold the company on 21 July 1998 to investment group Litchfield Continental Ltd, at which time the original name Britten-Norman Ltd was re-adopted. Company sold almost immediately to Global Spill Management (environmental protection) which reformed as Biofarm (pharmaceutical manufacturer) in Romania. Agreement to purchase Romaero, long-term maker of Islander airframes, made January 1999, but not concluded. Proposed capital injection by UAE-based HSDP, early 2000, also failed to take place.

On 3 April 2000 Britten-Norman was placed in receivership; sale announced 26 April to Alawi Zawawi of Oman, and new company formed as B-N Group Ltd (BNG). Order backlog at that time stood at nearly £12 million. Employment stood at approximately 100 in early 2003.

On 28 January 2002, B-N Group signed a 15-year agreement with Bembridge Airfield Ltd under which BNG will manage and develop Bembridge Airfield, thus securing the site for the manufacture and development of the Islander and Defender range.

In August 2002, BNG signed a contract with Romaero for production of 24 Islander airframes over a two-year period, with the option for Defenders to replace some of the contracted Islanders. B-N Aircraft Ltd is design and manufacturing subsidiary; Fly BN is support subsidiary.

UPDATED

BNG BN2B ISLANDER

TYPE: Light utility twin-prop transport.

PROGRAMME: Prototype (G-ATCT) first flight 13 June 1965 with two 157 kW (210 hp) Rolls-Royce Continental IO-360-B engines and 13.72 m (45 ft) span wings; subsequently re-engined with Textron Lycoming O-540s and flown 17 December 1965; wing span also increased by 1.22 m (4 ft) to initial production standard; production prototype BN2 (G-ATWU) flown 20 August 1966; domestic C of A received 10 August 1967; FAA type certificate 19 December 1967; Romanian manufacture (see Romaero entry) began 1969. In July 2002, further 24 Islanders ordered by BNG from Romaero with option to include two three-engined Trislanders. Contract included completion of three Islanders in advanced state of manufacture when former B-N company entered receivership. First of these flew 2 September 2002 and delivered to UK on 5 September as BN2B-20.

CURRENT VERSIONS: **BN2 Islander:** Initial piston-engined production model (23 built: see earlier *Jane's*).

BN2A Islander: Piston version built from 1 June 1969 until 1989 (see *Jane's Aircraft Upgrades*). Production totalled 890.

BN2B Islander: Standard piston-engined version since 1979; higher maximum landing weight; improved interior design; available with two engine choices and optional wingtip fuel tanks as **BN2B-26** with O-540s and **BN2B-20** with IO-540s (BN2B-27 and -21 no longer available). Features include range of passenger seats and covers, more robust door locks, improved door seals and stainless steel sills, redesigned fresh air system to improve ventilation in hot and humid climates, smaller diameter propellers to reduce cabin noise, and redesigned flight deck and instrument panel. Total of 154 delivered by end December 1999.

B-N, in conjunction with Hartzell Propeller Inc, has been testing a three-blade scimitar-shaped propeller developed as part of the NASA-sponsored AGATE/GAP programme, using a BN-2B operated by the North East Police Air Support Unit. The aim of the test programme is to minimise the aircraft's noise footprint through enhanced propeller aerofoil efficiency and reduced tip speeds.

In early 2003 BNG was evaluating aero diesel engines, including the 261 kW (350 hp) SMA, for possible installation on future production Islanders, and for retrofit to existing Islanders and Trislanders.

Detailed description applies to BN2B version.

Series of modification kits available as standard or option for new production aircraft and can be fitted retrospectively to existing aircraft.

BN2B Defender: Military version. Design features as detailed under Turbine Islander and Defender entry. Customers include Indian Navy (six recently converted to turbine power), Belgian Army, Botswana Defence Forces and Jamaican Defence Force.

BN2C Islander: Proposed upgrade with 224 kW 300 hp piston engines. Hartzell scimitar-blade propellers, fuel tanks in wingtips, all-glass cockpit and enlarged cargo door. Under study in early 2004.

BN2T Turbine Islander: *Described separately.*

BN2T-4R Defender: *Described in 1997–98 and earlier editions.*

BN2T-4S Defender 4000: *Described separately.*

BN2A Mk III Trislander: *Described separately.*

CUSTOMERS: By December 2001, deliveries of Islanders and Defenders totalled 1,227. Recent customers have included the Botswana Defence Force (one attrition replacement), Hawker Pacific (Australia) and Tomen Aerospace (Japan). First to fly post-April 2000 purchase of company was G-BWNG exhibited at Farnborough Air Show, July 2000 having flown on 10 July 2000 and arrived at Bembridge on 19 July. First delivery under new ownership was a BN2B-26 delivered on 10 November 2000 to Aer Arann of Galway, Ireland, followed by another Islander delivered to Ryuku Air Commuter of Japan on 12 November 2000. Deliveries in 2001 included one BN2B-20 each to St Barth Commuter of St Barthelemy, Frisia Luftverkehr GmbH of Germany (four-blade propellers), Ryukyu Air Commuter and Kyokushin Air of Japan, and Calcasieu Parish Police Jury of the USA, last-named equipped with Micronair spraypods for mosquito control duties.

COSTS: Direct operating cost (2003): BN2B-26 £157.53 per hour; BN2B-20 £171.57 per hour (900+ hours annual utilisation; 0.90 mmh/fh).

DESIGN FEATURES: Robust STOL light transport, suitable for operation from semi-prepared airstrips. High, unswept constant-chord wings and tailplane; sweptback fin with small fillet; upswept rear fuselage. Three-blade propellers available on BN2B-20 and BN2B-26 giving quieter noise signature.

NACA 23012 wing section; no dihedral; incidence 2°; no sweepback.

FLYING CONTROLS: Conventional and manual. Actuation by pushrods and cables. Slotted ailerons, with starboard ground-adjustable tab; mass-balanced elevator; trim tabs in rudder and elevator; single-slotted flaps operated electrically; fixed incidence tailplane. Dual controls standard.

STRUCTURE: L72 aluminium-clad aluminium alloys; two-spar wing torsion box in one piece; flared-up wingtips; integral fuel tanks in wingtips optional; four-longeron fuselage of pressed frames and stringers; two-spar tail unit with pressed ribs.

LANDING GEAR: Non-retractable tricycle type, with twin wheels on each main unit and single steerable nosewheel. Cantilever main legs mounted aft of rear spar. All three legs fitted with oleo-pneumatic shock-absorbers. All five wheels and tyres size 16×7-7, supplied by Goodyear. Tyre pressure: main 2.41 bar (35 lb/sq in); nose 2.00 bar (29 lb/sq in). Foot-operated air-cooled Cleveland hydraulic brakes on main units. Parking brake. Wheel/ski gear available optionally. Minimum ground turning radius 9.45 m (31 ft 0 in).

POWER PLANT: Two Textron Lycoming flat-six engines, each driving a Hartzell HC-C2YK-2B or -2C two-blade constant-speed feathering metal propeller; optional three-blade Hartzell HC-C3YR-2UF/FC8468-8R. Propeller synchronisers optional. Standard power plant (BN2B-26) 194 kW (260 hp) O-540-E4C5, but 224 kW (300 hp) IO-540-K1B5 fitted at customer's option (BN2B-20).

Integral fuel tank between spars in each wing, outboard of engine. Total fuel capacity (standard) 518 litres (137 US gallons; 114 Imp gallons). Usable fuel 492 litres (130 US gallons; 108 Imp gallons). With optional fuel tanks in wingtips, total capacity is increased to 814 litres (215 US gallons; 179 Imp gallons). Additional pylon-mounted underwing auxiliary tanks, each of 227 litres (60.0 US gallons; 50.0 Imp gallons) capacity, available optionally. Refuelling point in upper surface of wing above each internal tank. Total oil capacity 22.75 litres (6.0 US gallons; 5.0 Imp gallons).

ACCOMMODATION: Up to 10 persons, including pilot, on side-by-side front seats and four bench seats. No aisle. Seat backs fold forward. Access to all seats via three forward-opening doors, forward of wing and at rear of cabin on port side and forward of wing on starboard side. Baggage compartment at rear of cabin, with port-side loading door in standard versions. Exit in emergency by removing door windows. Special executive layouts available.

Can be operated as freighter, carrying more than a ton of cargo; in this configuration passenger seats can be stored in rear baggage bay. In ambulance role, up to three stretchers and two attendants can be accommodated. Other layouts possible, including photographic and geophysical survey, parachutist transport or trainer (with accommodation for up to eight parachutists and dispatcher), firefighting, environmental protection and crop-spraying.

SYSTEMS: Janaero cabin heater standard. 45,000 BTU Janitrol combustion unit, with circulating fan, provides hot air for distribution at floor level outlets and at windscreen demisting slots. Fresh air boosted by propeller slipstream, is ducted to each seating position for on-ground ventilation. Electrical DC power, for instruments, lighting and radio, from two engine-driven 24 V 70 A self-rectifying alternators and a controller to main busbar and circuit breaker assembly. Emergency busbar is supplied by a 24 V 25 Ah heavy-duty lead-acid battery in the event of a twin alternator failure. Ground power receptacle provided. Optional electric de-icing of propellers and windscreen, and pneumatic de-icing of wing and tail unit leading-edges. Oxygen system available optionally for all versions.

BN2B-26 Islander used for surveillance and light transport within the Falkland Islands ***NEW*/0576513**

AVIONICS: *Comms:* Dual VHF nav/com with ILS; transponder; intercom, including second headset, and passenger address system, all standard. Optional ELT, HF/MF radios.

Radar: Optional weather radar in nose.

Flight: ADF, MKR, VOR, DME standard. Optional autopilot, EGPWS, TCAS and B-RNAV.

Instrumentation: IFR standard.

DIMENSIONS, EXTERNAL:

Wing span	14.94 m (49 ft 0 in)
Wing chord, constant	2.03 m (6 ft 8 in)
Wing aspect ratio	7.4
Length overall	10.86 m (35 ft 7¾ in)
Fuselage: Max width	1.21 m (3 ft 11½ in)
Max depth	1.46 m (4 ft 9¾ in)
Height overall	4.18 m (13 ft 8¾ in)
Tailplane span	4.67 m (15 ft 4 in)
Wheel track (c/l of shock-absorbers)	3.61 m (11 ft 10 in)
Wheelbase	3.99 m (13 ft 1¼ in)
Propeller diameter	1.98 m (6 ft 6 in)
Cabin door (front, port): Height	1.10 m (3 ft 7½ in)
Width: top	0.64 m (2 ft 1¼ in)
Height to sill	0.59 m (1 ft 11¼ in)
Cabin door (front, starboard):	
Height	1.10 m (3 ft 7½ in)
Max width	0.86 m (2 ft 10 in)
Height to sill	0.57 m (1 ft 10½ in)
Cabin door (rear, port): Height	1.09 m (3 ft 7 in)
Width: top	0.635 m (2 ft 1 in)
bottom	1.19 m (3 ft 11 in)
Height to sill	0.52 m (1 ft 8½ in)
Baggage door (rear, port): Height	0.69 m (2 ft 3 in)

DIMENSIONS, INTERNAL:

Passenger cabin, aft of pilot's seat:	
Length	3.05 m (10 ft 0 in)
Max width	1.09 m (3 ft 7 in)
Max height	1.27 m (4 ft 2 in)
Floor area	2.97 m^2 (32.0 sq ft)
Volume	3.7 m^3 (130 cu ft)
Baggage space aft of passenger cabin	1.4 m^3 (49 cu ft)
Freight capacity:	
aft of pilot's seat, incl rear cabin baggage space	4.7 m^3 (166 cu ft)
with four bench seats folded into rear cabin baggage space	3.7 m^3 (130 cu ft)

AREAS:

Wings, gross	30.19 m^2 (325.0 sq ft)
Ailerons (total)	2.38 m^2 (25.60 sq ft)
Flaps (total)	3.62 m^2 (39.00 sq ft)
Fin	3.41 m^2 (36.64 sq ft)
Rudder, incl tab	1.60 m^2 (17.20 sq ft)
Tailplane	6.78 m^2 (73.00 sq ft)
Elevator, incl tabs	3.08 m^2 (33.16 sq ft)

WEIGHTS AND LOADINGS (A: 194 kW; 260 hp engines, B: 224 kW; 300 hp engines):

Weight empty, equipped (without avionics):	
A	1,866 kg (4,114 lb)
B	1,925 kg (4,244 lb)
Max disposable load:	
with max standard fuel: A	1,128 kg (2,486 lb)
B	1,082 kg (2,386 lb)
with max optional fuel: A	1,069 kg (2,356 lb)
B	1,023 kg (2,256 lb)
Payload: with max standard fuel:	
A	774 kg (1,706 lb)
B	497 kg (1,096 lb)
with max optional fuel:	
A	715 kg (1,576 lb)
B	438 kg (966 lb)
Max fuel weight:	
standard: A, B	354 kg (780 lb)
with optional tanks in wingtips: A, B	585 kg (1,290 lb)
Max T-O and landing weight: A, B	2,993 kg (6,600 lb)
Max zero-fuel weight (BCAR):	
A, B	2,855 kg (6,300 lb)
Max floor loading, without cargo panels:	
A, B	586 kg/m^2 (120 lb/sq ft)
Max wing loading: A, B	99.2 kg/m^2 (20.31 lb/sq ft)
Max power loading: A	7.73 kg/kW (12.69 lb/hp)
B	6.70 kg/kW (11.00 lb/hp)

PERFORMANCE (A and B as above):

Never-exceed speed (VNE):	
A, B	183 kt (339 km/h; 211 mph) IAS
Max level speed at S/L:	
A	148 kt (274 km/h; 170 mph)
B	151 kt (280 km/h; 173 mph)
Max cruising speed at 75% power at FL70:	
A	139 kt (257 km/h; 160 mph)
B	142 kt (264 km/h; 164 mph)
Cruising speed at 67% power at FL90:	
A	134 kt (248 km/h; 154 mph)
B	137 kt (254 km/h; 158 mph)
Cruising speed at 59% power at FL120:	
A	130 kt (241 km/h; 150 mph)
B	132 kt (245 km/h; 152 mph)
Stalling speed:	
flaps up: A, B	50 kt (92 km/h; 57 mph) IAS
flaps down: A, B	40 kt (74 km/h; 46 mph) IAS
Max rate of climb at S/L: A	262 m (860 ft)/min
B	344 m (1,130 ft)/min
Rate of climb at S/L, OEI: A	44 m (145 ft)/min
B	60 m (198 ft)/min
Absolute ceiling: A	4,145 m (13,600 ft)
B	6,005 m (19,700 ft)
Service ceiling: A	3,445 m (11,300 ft)
B	5,240 m (17,200 ft)

Service ceiling, OEI: A 1,525 m (5,000 ft)
B 1,980 m (6,500 ft)
T-O run at S/L, zero wind, hard runway:
A 278 m (915 ft)
B 215 m (705 ft)
T-O run at 1,525 m (5,000 ft): A 396 m (1,300 ft)
B 372 m (1,225 ft)
T-O to 15 m (50 ft) at S/L, zero wind, hard runway:
A 371 m (1,220 ft)
B 355 m (1,165 ft)
T-O to 15 m (50 ft) at 1,525 m (5,000 ft):
A 528 m (1,735 ft)
B 496 m (1,630 ft)
Landing from 15 m (50 ft) at S/L, zero wind, hard runway: A, B 299 m (980 ft)
Landing from 15 m (50 ft) at 1,525 m (5,000 ft):
A, B 357 m (1,170 ft)
Landing run at 1,525 m (5,000 ft): A, B 171 m (560 ft)
Landing run at S/L, zero wind, hard runway:
A, B 140 m (460 ft)
Range, block-to-block, plus 10% plus 45 min reserves:
B, standard fuel: IFR 503 n miles (931 km; 578 miles)
VFR 639 n miles (1,183 km; 735 miles)
B, with optional tanks:
IFR 896 n miles (1,659 km; 1,031 miles)
VFR 1,075 n miles (1,990 km; 1,237 miles)

UPDATED

BNG BN2T TURBINE ISLANDER AND DEFENDER

UK Army Air Corps designation: Islander AL. Mk 1
Royal Air Force designation: Islander CC. Mk 2/2A

TYPE: Utility turboprop twin.

PROGRAMME: Turboprop Islander; first flight of prototype (G-BPBN) 2 August 1980 with two Allison (now Rolls-Royce) 250-B17Cs; British CAA certification received end of May 1981; first production aircraft delivered December 1981; FAR Pt 23 US type approval 15 July 1982; full icing clearance to FAR Pt 25 gained 23 July 1984.

CUSTOMERS: Total 65 aircraft delivered by December 1999, mainly of the Defender variant; by early 2004, further four aircraft under construction. Several intended BN2Ts completed as BN2Bs at Romanian factory before delivery to UK. Customers include the Rhine Army Parachute Association. For turbine (BN2T) version, customers include Moroccan Ministry of Fisheries, Pakistan Maritime Security Agency, Belgian Gendarmerie, Netherlands Police, Royal Air Force, UK Army Air Corps and Mauritius Coast Guard.

COSTS: Direct operating cost (2003) £158.05 per hour (900+ hours annual utilisation; 0.90 mmh/fh).

DESIGN FEATURES: As for piston-engined version; turboprops enable use of available low-cost jet fuel; low operating noise level; available for same range of applications as Islander. Defender has four underwing hardpoints for standard NATO pylons to attach fuel tanks, weapons and other stores; number of additional airframe options, including sliding door on rear port side which can be opened in flight, are also offered. Concept of Defender is to provide a low-cost airframe which can be fitted with best available sensors to meet operational needs of customers.

Description of BN2B Islander applies also to BN2T and Defender, except as follows:

POWER PLANT: Two 298 kW (400 shp) Rolls-Royce 250-B17C turboprops, flat rated at 238.5 kW (320 shp), and each driving a Hartzell three-blade constant-speed fully feathering metal propeller. Usable fuel 814 litres (215 US gallons; 179 Imp gallons). Pylon-mounted underwing tanks, each of 227 litres (60.0 US gallons; 50.0 Imp gallons) capacity, are available optionally for special purposes. Total oil capacity 5.7 litres (1.5 US gallons; 1.25 Imp gallons).

ACCOMMODATION: Generally as for BN2B. In ambulance role can accommodate, in addition to pilot, a single stretcher, one medical attendant and five seated occupants; or two stretchers, one attendant and three passengers; or three stretchers, two attendants and one passenger. Other possible layouts include photographic and geophysical survey; parachutist transport or trainer (with accommodation for up to eight parachutists and a dispatcher); and pest control or other agricultural spraying. Maritime Turbine Islander/Defender versions available for fishery protection, coast guard patrol, pollution survey, search and rescue, and similar applications. In-flight sliding parachute door optional.

AVIONICS: *Comms:* Optional maritime band and VHF transceivers.

Flight: VLF/Omega nav system; radar altimeter.

Self-defence (Military versions): Since at least 1994 Army Air Corps Islander AL. Mk 1s have had provision for AN/ALQ-144 IR jammer under fuselage. Lockheed Martin IRCM suite tested on AAC Islander in early 1998.

EQUIPMENT: According to mission, can include fixed tailboom or towed bird magnetometer, spectrometer, or electromagnetic detection/analysis equipment (geophysical survey); one or two cameras, navigation sights and appropriate avionics (photographic survey); 189 litre (50.0 US gallon; 41.5 Imp gallon) Micronair underwing spraypods complete with pump and rotary atomiser (pest control/agricultural spraying versions); dinghies, survival equipment and special crew accommodation (maritime versions). Optimal installations for British Army Islanders include photographic surveillance package of door-mounted Zeiss 610 camera, vertical Zeiss Trilens 80 mm and either F126 vertical, RMK vertical or two KS-153 vertical cameras; Vinten F143 panoramic is also available.

DIMENSIONS, EXTERNAL: As for BN2B, except:
Length overall: standard nose 10.86 m (35 ft 7¾ in)
weather radar nose 11.07 m (36 ft 3¾ in)
Propeller diameter 2.03 m (6 ft 8 in)

WEIGHTS AND LOADINGS:
Weight empty, equipped 1,832 kg (4,040 lb)
Payload with max fuel 689 kg (1,519 lb)
Max T-O weight 3,175 kg (7,000 lb)
Max landing weight 3,084 kg (6,800 lb)
Max zero-fuel weight 2,994 kg (6,600 lb)
Max wing loading 105.2 kg/m² (21.54 lb/sq ft)
Max power loading 5.33 kg/kW (8.75 lb/shp)

PERFORMANCE (standard Turbine Islander/Defender):
Max cruising speed:
at FL100 170 kt (315 km/h; 196 mph)
at S/L 154 kt (285 km/h; 177 mph)
Cruising speed at 72% power:
at FL100 150 kt (278 km/h; 173 mph)
at FL50 143 kt (265 km/h; 165 mph)
Stalling speed, power off:
flaps up 52 kt (97 km/h; 60 mph) IAS
flaps down 45 kt (84 km/h; 52 mph) IAS
Max rate of climb at S/L 320 m (1,050 ft)/min
Rate of climb at S/L, OEI 66 m (215 ft)/min
Service ceiling over 7,010 m (23,000 ft)
Absolute ceiling, OEI over 3,050 m (10,000 ft)
T-O run 255 m (840 ft)
T-O to 15 m (50 ft) 381 m (1,250 ft)
Landing from 15 m (50 ft) 339 m (1,110 ft)
Landing run 228 m (750 ft)
Range (IFR) with max fuel, reserves for 45 min hold plus 10%
590 n miles (1,093 km; 679 miles)
Range (VFR) with max fuel, no reserves 728 n miles (1,348 km; 838 miles)

UPDATED

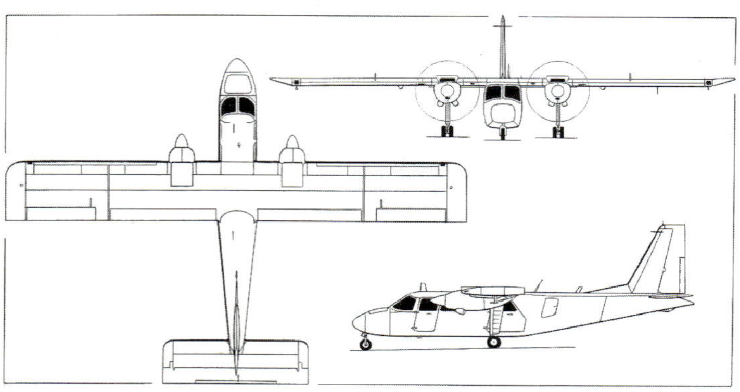

BNG BN2T Turbine Islander ***(Jane's/Dennis Punnett)*** 0507236

Army Air Corps BNG BN2T Islander AL. Mk 1 ***(Paul Jackson)*** ***NEW***/0576501

BNG BN2T-4S DEFENDER 4000

UK Army Air Corps designation: Defender

TYPE: Multisensor surveillance twin-turboprop.

PROGRAMME: Marketing experience with BN2T Defenders revealed that many military and government agencies expressed need for greater payload; announced 1994; prototype (G-SURV) made first flight 17 August 1994; public launch at Farnborough Air Show September 1994; CAA certification achieved 13 November 1995; CAA transport (passenger) certification 4 April 1996.

Britten-Norman and Orenda Recip Inc of Canada announced in September 1999 joint study to re-engine Defender with Orenda OE600 V-8 piston engines, resulting in improved performance and efficiency. Projected specification includes maximum T-O weight 4,241 kg (9,350 lb), maximum cruising speed 176 kt (326 km/h 202 mph) and maximum rate of climb 366 m (1,200 ft)/min.

CURRENT VERSIONS: **BN2T-4S Defender 4000:** Standard version, *as described.*

National Defender: Derivative version of Defender 4000 optimised for homeland defence role; features include increased sensor payload of 1,587 kg (3,499 lb).

CUSTOMERS: Seven aircraft registered to manufacturer in April 1996. Irish Ministry of Justice took delivery of one on 14 August 1997; aircraft is operated by the Irish Air Corps at

BNG Defender 4000 in service with Greater Manchester Police ***(GMPB)*** ***NEW***/0576495

BNG DEFENDER 4000 PRODUCTION

C/N	Reg/serial	Date	Operator
4005	G-SURV	17 Aug 94	(Pilatus) Britten-Norman
		19 Oct 98	Police Aviation Services
		02	Atlantic Air Transport
	ZG995	26 Mar 04	Army Air Corps
4006	G-SJCH	26 Feb 01	Hampshire Police
4007	(G-BWPM)		not yet completed (reserved as BN2T-4R)
4008	255	15 Aug 97	Garda Siochana (IAC)
4009	9M-TPS	23 Nov 97	Sabah Air
4010	ZG996	26 Mar 04	Army Air Corps
4011	G-GMPB	2 Jul 02	Greater Manchester Police
4012	ZG997	26 Mar 04	Army Air Corps
4013	(G-BWPW)		not yet completed
4014	(G-BWPX)		not yet completed

Notes: c/n 4001–04 are BN2T-4R; two conversions and two new-build

Instrument panel of production BNG Defender 4000 ***NEW***/0576496

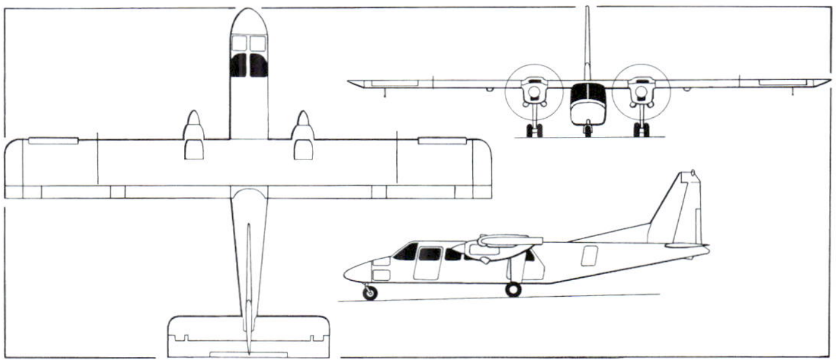
BNG BN2T-4S Defender 4000 ***(James Goulding)*** 0051535

Baldonnel, Dublin, on behalf of the Garda Siochana (National Police), and is equipped with a comprehensive suite of law enforcement communications equipment, thermal imager and observers' removable consoles. Sabah Air of Malaysia took delivery of one in November 1997; primarily equipped for aerial photographic work, with quick-change facility to 12-passenger transport. Police Aviation Services UK took delivery of a single (ex-demonstrator) example in November 1998, this later (2002) passed to Atlantic Air Transport, Coventry, for operation on behalf of the National Environmental Research council. Hampshire Police Authority took delivery of one (G-SJCH, named *Sir John Charles Hoddinott*) on 26 February 2001 for service with its Lee-on-Solent-based Air Support Unit, replacing a BN2B Islander. Greater Manchester Police ordered one (G-GMPB) on 15 January 2002, delivery of which was achieved on 2 July 2002. In March 2004, Army Air Corps purchased three (including prototype) for surveillance operations in Iraq.

COSTS: £2.5 million (2002). Direct operating cost (2003): £167.01 in 'cruise' operation or £147.52 in 'surveillance' role (900+ hours per year utilisation and 0.90 mmh/fh).

DESIGN FEATURES: As for Islander, but with enlarged wing, based on that of Trislander, although retaining original control surfaces; compared with BN2T, has tailplane and elevator of increased span; modified fin with larger fillet; fuselage stretched by means of 0.76 m (2 ft 6 in) plug forward of wing to seat up to 16; more powerful engines for greater sortie time and payload capacity; fuselage and landing gear strengthened; flight deck windows deepened for enhanced field of view; redesigned nose and tail unit; rear-sliding door with blister window.

POWER PLANT: Two Rolls-Royce 250-B17F turboprops, each flat rated at 298 kW (400 shp) and driving a three-blade propeller. Internal fuel capacity 1,131 litres (299 US gallons; 249 Imp gallons).

ACCOMMODATION: Flight crew of two on airline-type seats; sliding seat rails permit seat and equipment positioning anywhere within fuselage. Space for two or more consoles and operators in tandem along one side of cabin. Up to 16 troops/passengers in tactical transport role. Certified to carry pilot and 11 passengers, subject to local airworthiness requirements.

SYSTEMS: Electrical system includes two 200 A engine-driven generators. Full de-icing.

AVIONICS: *Comms:* Full range of open or secure voice com radios from UHF, VHF, HF and VHF-FM.

Radar: Modified nose can accommodate 68.5 cm (27 in), 360° rotating antenna for maritime, simple search or weather radar (BAE Systems Seaspray 2000 in prototype Defender 4000).

Flight: Fully integrated autopilot. GPS nav system, integrated with Omega or INS.

Mission: Sensors can include thermal imagers and/or hand-held or podded video or film cameras. Prototype on debut fitted with Agema FLIR under fuselage on starboard side; alternatives could include BAE Systems pod or FLIR 2000. Appropriate radars could include BAE Systems Seaspray, Racal Super Searcher, Thomson-CSF Detexis Ocean Master, Telephonics 143 or Litton 504(V)5.

ARMAMENT: Two hardpoints under each wing, inboard pair stressed for loads of up to 340 kg (750 lb) and outboard pair for up to 159 kg (350 lb) each. Typical weapons on inboard stations can include Sting Ray torpedo or Sea Skua anti-ship missile.

DIMENSIONS, EXTERNAL: As for BN2B, except:

Wing span .... 16.15 m (53 ft 0 in)
Wing aspect ratio 8.0
Length overall .... 12.20 m (40 ft 0½ in)
Height overall .... 4.36 m (14 ft 3½ in)
Tailplane span .... 5.31 m (17 ft 5 in)
Wheelbase .... 4.76 m (15 ft 7½ in)

AREAS:

Wings, gross .... 32.61 m^2 (351.0 sq ft)

WEIGHTS AND LOADINGS:

Weight empty, equipped .... 2,223 kg (4,900 lb)
Max usable internal fuel .... 908 kg (2,002 lb)
Payload with max fuel .... 724 kg (1,598 lb)
Max T-O and landing weight .... 3,855 kg (8,500 lb)
Max zero-fuel weight .... 3,765 kg (8,300 lb)
Max wing loading .... 109.6 kg/m^2 (22.45 lb/sq ft)
Max power loading .... 6.46 kg/kW (10.63 lb/shp)

First BNG Defender 4000 ***(Paul Jackson)*** ***NEW***/0554421

PERFORMANCE:
Max cruising speed: at S/L ... 154 kt (285 km/h; 177 mph)
at FL100 ... 176 kt (326 km/h; 202 mph)
Transit speed from base to patrol area ... 160 kt (296 km/h; 184 mph)
Cruising speed at 72% power:
at FL50 ... 149 kt (276 km/h; 171 mph)
at FL100 ... 160 kt (296 km/h; 184 mph)
Stalling speed, power off:
flaps up ... 53 kt (99 km/h; 61 mph) IAS
flaps down ... 47 kt (87 km/h; 54 mph) IAS
Max rate of climb at S/L ... 381 m (1,250 ft)/min
Rate of climb at S/L, OEI ... 61 m (200 ft)/min
Absolute ceiling ... 7,620 m (25,000 ft)
Absolute ceiling, OEI ... 3,660 m (12,000 ft)
T-O run ... 356 m (1,170 ft)
T-O to 15 m (50 ft) ... 565 m (1,855 ft)
Landing from 15 m (50 ft) ... 589 m (1,935 ft)
Landing run ... 308 m (1,015 ft)
Max range on internal fuel:
with IFR reserves ... 861 n miles (1,594 km; 990 miles)
with VFR reserves ... 1,006 n miles (1,863 km; 1,157 miles)
Max endurance on internal fuel ... 8 h 30 min

UPDATED

BNG BN2A MK III TRISLANDER

BNG announced resumption of Trislander production in February 2003. A demonstrator was under construction in mid-2003, based on a withdrawn airframe (6Y-JQE of Trans-Jamaican Airlines fitted with a new wing from a BN2T-4R); on completion, to undertake a worldwide demonstration tour to solicit sufficient firm orders to enable full-scale series production to be undertaken.

In early 2004 BNG was in discussion with potential partners in China for a possible joint venture to manufacture Trislanders there for the Chinese domestic market.

Integrity Aircraft of Australia is planning to engage the Romanian production plant to build a Trislander variant powered by a single Honeywell TPE331 turboprop in the tail position.

BNG also refurbishes Trislanders, including one resold to Rockhopper Aero in February 2004.

BNG Trislander from original production *(Mike McBride)* ***NEW***/0576476

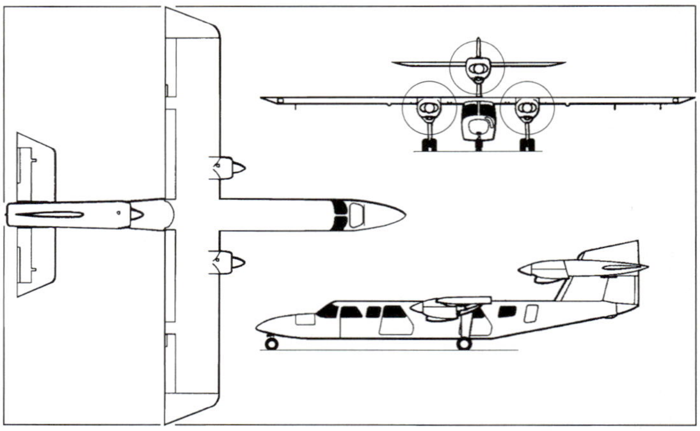

BNG Trislander general arrangement ***NEW***/0576477

UPDATED

CAMERON

CAMERON BALLOONS LTD

St John Street, Bedminster, Bristol BS3 4NH
Tel: (+44 117) 963 72 16
Fax: (+44 117) 966 11 68
e-mail: hcameron@cameronballoons.co.uk
Web (1): http://www.cameronballoons.co.uk
Web (2): http://www.thunderandcolt.co.uk
MANAGING DIRECTOR: Don Cameron
SALES DIRECTORS: Nick Purvis
Jim Howard
MARKETING DIRECTOR: Alan Noble
TECHNICAL DIRECTOR: Tom Sage
OPERATIONS AND PRODUCTION DIRECTORS: Lindsay Sadler
David McGibbon
PUBLIC RELATIONS: Hannah Cameron

Cameron Balloons, formed in 1970 and now part of the Cameron Holdings Group, holds CAA, FAA, French CNT and German Musterzulassungsschein type certificates for its hot-air balloons. It is the world's largest manufacturer of conventional and special-shape balloons and, by mid-2002 (latest information received), had produced nearly 5,500 from its main factory in Bristol, of which 375 or more (built or on order) were special shapes. A sister company in Dexter, Michigan, USA, had produced more than 1,200 by that time.

Details of Cameron's conventional-shape balloon range are currently available on the company website.

Cameron also designs and produces hot-air airships, being the first company to develop a craft of this type. Current production hot-air airships are predominantly the DP-80, DP-90, AS 105, the slightly more streamlined AS 105GD and the AS-120. Production of these had totalled 70 by June 2001, comprising three D-38 type, 25 DP type, 10 D-50 type, 20 D-96 type and 12 AS type, in addition to six earlier DG type helium airships. No further sales had been reported by mid-2004.

UPDATED

CFM

CFM AIRBORNE (UK) LTD

Unit 2D, Eastlands Industrial Estate, Leiston, Suffolk IP16 4LL
Tel: (+44 1728) 83 23 53 and 83 21 00
Fax: (+44 1728) 83 29 44
e-mail: CFMAirborne@yahoo.com

Cook Flying Machines (trading as CFM MetalFax Ltd) went into receivership during 1996; assets were purchased from the liquidator by a small group of investors and the company was relaunched as CFM Aircraft Ltd, to manufacture and market the Shadow series of microlights and Streak/Star Streak light aircraft in kitbuilt and production versions as described below. By June 2001, 397 CFM ultralights had been delivered to customers in 40 countries. In 1999, CFM secured what was then the world's largest ultralight order when the Indian Air Force contracted for 24 Streak Shadows, of which 18 had been delivered by June 2001.

Workforce was 7 in 2001, with annual production capacity of 12 to 18 assembled aircraft, and a similar number of kits, at the company's 465 m^2 (5,000 sq ft) factory at Leiston. Flight testing undertaken at Parham (Framlingham), to which all company operations were scheduled to relocate in the summer of 2001.

However, in mid-2002 CFM went into receivership. Assets acquired by Airborne Innovations of Texas, which had earlier purchased Laron Aviation Technologies, the former US distributor for CFM. CFM renamed CFM Airborne (UK) Ltd in 2003, but acquired in mid-2004 by Bella Aviation Ltd (BA, which see).

UPDATED

EUROPA

EUROPA AIRCRAFT 2004 LTD

Unit 2A, Dove Way, Kirby Mills Industrial Estate, Kirkbymoorside, North Yorkshire YO62 6NR
Tel:(+44 1751) 43 17 73
Fax:(+44 1751) 43 17 06
e-mail: John@europa-aircraft.com
CHAIRMAN: James S Tucker
TECHNICAL DIRECTOR: Andy Draper
SALES DIRECTOR: John Wheeler

Europa Aircraft company founder Ivan Shaw previously built three Rutan canards before designing Europa for European environment, including operating from grass strips. Don Dykins, formerly British Aerospace Chief Aerodynamicist, defined aerodynamics; Barry Mellers (Chief Designer of Slingsby Aviation Ltd) made structural calculations. In October 2001, company was bought from US parent Liberty Aerospace (which see) by its management, led by MD Keith Wilson, and renamed Europa Management (International). Europa is UK distributor for US-built Liberty XL-2.

Europa sales passed the 1,000 mark in late 2003, but on 6 July 2004 Europa Management was placed in receivership with debts of £700,000. It re-formed on 6 September 2004 as Europa Aircraft (2004).

UPDATED

EUROPA EUROPA

TYPE: Side-by-side lightplane-motor glider kitbuilt.

PROGRAMME: Design started January 1990; prototype (G-YURO) made first flight 12 September 1992. PFA certification achieved May 1993; produced mainly in kit form under PFA auspices, but two assembled by Europa Aviation in 1994 and further three followed by 1996; first customer-built aircraft (G-OPJK) flown 14 October 1995.

Reported in mid-2002 that Europa to modify XS version to comply with new FAA 'sportplane' category (MTOW 590 kg; 1,300 lb or less); may also offer factory-built aircraft, though had not done so by mid-2004.

Company announced sale of 1,000th kit on 24 November 2003; in 2004, introduced new ready-to-install cockpit module, said to save at least 200 hours of build time.

CURRENT VERSIONS: **Europa Classic:** Standard aircraft; has options of several types of engine, and monowheel, tailwheel or tricycle landing gear. Superseded by XS.

Production-standard Europa XS motor glider, with winglets *(Keith Wilson)* ***NEW***/0589192

Europa XS tri-gear celebrating the 1,000th sale *(Keith Wilson)* ***NEW***/0589310

Mono and tri-gear Europa XSs in formation *(Keith Wilson)* ***NEW***/0589166

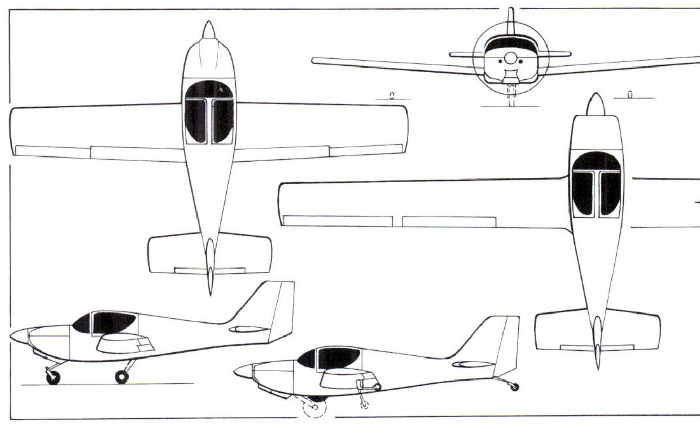

Standard (monowheel) version of Europa, with additional side elevation of tricycle option and partial plan view of Europa Motor Glider *(James Goulding)* 0015624

Europa XS: Improved version introduced June 1997. Features include higher aspect ratio premoulded wings; larger ailerons; enlarged cabin providing additional legroom and seat width; Rotax 912 ULS or turbocharged Rotax 914 UL engine in **Turbo XS**, with new cowlings giving improved cooling, lower drag, increased propeller clearance and better field of view from cockpit; non-steerable tailwheel permitting full rudder deflection during take-off and landing; and supplementary tank holding 35 litres (9.1 US gallons; 7.6 Imp gallons) of usable fuel. Maximum take-off weight increased by 32 kg (70 lb), making it possible to fit a child's seat in the baggage area; maximum speed 174 kt (322 km/h; 200 mph); maximum range more than 869 n miles (1,609 km; 1,000 miles). Prototype (G-EUXS) first flown 21 March 1997; public debut at EAA Convention at Oshkosh July 1997. Trial installation undertaken in 1999 on G-WWWG with 89 kW (120 hp) Wilksch Airmotive WAM 120 three-cylinder Diesel.

Europa motor glider Alternative long-span glider wings, interchangeable in quoted 5 minutes with standard wings, and featuring trailing-edge airbrakes, maximum extension 80°, inboard of the ailerons, but no flaps; public debut, installed on demonstrator G-ODTI, at PFA International Rally at Cranfield, July 1997; first flight in this configuration on 20 November 1998; production-standard wing, which made public debut at PFA International Rally, July 2000, has winglets and extended span of 14.40 m (47 ft 3 in).

CUSTOMERS: More than 1,000 kits sold by mid-2004 to customers in over 33 countries, of which more than 250 then flown.

COSTS: Airframe kits (XS monowheel) £18,295, (XS tri-gear) £18,795, (XS motor glider monowheel) £22,395, (XS motor glider tri-gear) £22,895; engine packages (Rotax 912 ULS) £11,650, (Rotax 914 UL) £16,250. Motor glider wing retrofit for XS, £10,350; cockpit module £1,875 extra. (All prices 2004).

DESIGN FEATURES: Objectives included low-cost, economical cruise at IAS up to 120 kt (222 km/h; 138 mph) over 500 n miles (926 km; 575 miles), grass field capability, and ability to rig and de-rig quickly for storage in a trailer similar to a shorter version of glider trailer. Rig/de-rig by six pip-pins, two for tail, two for each wing. Quoted kit building time without cockpit module option is 700 hours. Designed to JAR-VLA, stressed for maximum in-flight normal *g* of 4.3; proof factor of 2 used in design instead of the more usual 1.5, because of extensive use of composites materials in primary structure

Low/mid-wing layout; moderately tapered wings and tailplane; large, sweptback fin and short fuselage.

Aerofoil is laminar flow Dykins design with 12 per cent thickness/chord ratio and 2° 30' washout at tips. Modified Dykins section (15 per cent t/c ratio) on motor glider wings. Wing dihedral 3° (2° 30' on motor glider); no tail dihedral.

Speed Kit introduced in 1999 features outrigger mechanism and wheel fairings and flap hinge fairings for monowheel versions; and wheel, landing gear leg and flap hinge fairings for tricycle version resulting in 9 to 10 kt (17 to 18 km/h; 10 to 11 mph) speed increase for Europa XS.

FLYING CONTROLS: Manual. Conventional ailerons and rudder; all-moving tailplane for pitch control with tab geared for balance, under pilot control for trim. Ground-adjustable tabs in ailerons and rudder. Two control columns, one centrally mounted at each seat; two pairs of rudder pedals. Central console between seats has throttle, combined flap and landing gear levers. Pitch trim switch next to throttle. Slotted flaps with settings of 0° and 25°, electrically actuated on tricycle landing gear version.

STRUCTURE: General construction of GFRP.

LANDING GEAR: Monowheel version has large, single, semi-retractable mainwheel and steerable tailwheel; outriggers at about half-span mounted on nylon stalks which retract with flaps. Mainwheel uses standard 6 in hub, as in many light aircraft; and an 8.00–6 Tundra tyre. Tricycle landing gear version has fuselage-mounted, fixed, spring steel cantilever legs, castoring nosewheel, and toe brakes on port side rudder pedals; mainwheel tyre size 5.00-5, nosewheel 4.00–5. Tailwheel landing gear kit offered by Aero Developments of Kemble, UK.

POWER PLANT: One 59.6 kW (79.9 hp) Rotax 912 UL flat-four engine, directly driving a three-blade fixed-pitch Warp Drive propeller; propeller pitch is adjustable on ground to match operating environment (fine pitch for good take-off distances and rate of climb but higher noise and lower cruising speeds; reverse for coarse pitch); alternatively, one 73.5 kW (98.6 hp) Rotax 912 ULS or one 84.6 kW (113.4 hp) turbocharged Rotax 914 UL, driving a three-blade variable-pitch NSI propeller. Alternative Subaru/NSI EA81 installation in 73 kW (98 hp) and 88 kW (118 hp) form first flown (G-NDOL) 18 November 1995. MWAE rotary (73 kW; 98 hp), BMW 1100RS (67.1 kW; 90 hp) and 89 kW (120 hp) Wilksch WAM 120 engine installations under development, the first-mentioned installed in UK prototype G-YURO and UK- and US-built tricycle gear demonstrators (G-KITS and N496TG); BMW engine began ground running trials in the third quarter of 1996. Jabiru 3300 offered from 1998.

Normal fuel capacity 68.2 litres (18.0 US gallons; 15.0 Imp gallons); optional 104.6 litres (27.6 US gallons; 23.0 Imp gallons).

ACCOMMODATION: Enclosed cabin seating two side by side under individual upward-opening canopies, hinged on centreline. Baggage compartment at rear of cabin.

SYSTEMS: Hydraulics: mainwheel brake. Electrical: 12 V 30 Ah battery; alternator fit appropriate to engine.

AVIONICS: Customer choice.

DIMENSIONS, EXTERNAL:

Wing span: XS ... 8.28 m (27 ft 2 in)
motor glider ... 14.40 m (47 ft 3 in)
Width, wings removed ... 1.17 m (3 ft 10 in)
Wing aspect ratio: XS ... 7.2
motor glider ... 15.6
Length overall ... 5.84 m (19 ft 2 in)
Height overall: monowheel ... 1.32 m (4 ft 4 in)
tricycle ... 2.13 m (7 ft 0 in)
Tailplane span ... 2.44 m (8 ft 0 in)
Wheel track (tricycle) ... 1.83 m (6 ft 0 in)
Wheelbase (tricycle) ... 1.47 m (4 ft 9¾ in)
Propeller diameter ... 1.57 m (5 ft 2 in)
Propeller ground clearance: XS ... 0.39 m (1 ft 3½ in)
XS ... 0.39 m (1 ft 3½ in)

DIMENSIONS, INTERNAL:

Cabin:
Length XS ... 1.83 m (6 ft 0 in)
Max width ... 1.12 m (3 ft 8 in)
Max height ... 0.97 m (3 ft 2 in)
Baggage volume: XS ... 0.64 m^3 (22.5 cu ft)

AREAS:

Wings, gross: XS ... 9.48 m^2 (102.0 sq ft)
motor glider ... 13.29 m^2 (143.0 sq ft)
Ailerons (total): XS ... 0.62 m^2 (6.650 sq ft)
Trailing-edge flaps (total) ... 1.37 m^2 (14.70 sq ft)
Fin ... 1.03 m^2 (11.10 sq ft)
Rudder ... 0.48 m^2 (5.20 sq ft)

WEIGHTS AND LOADINGS (standard aircraft unless otherwise stated):

Weight empty, equipped:
XS monowheel ... 340 kg (750 lb)
XS tricycle ... 354 kg (780 lb)
motor glider ... 358 kg (790 lb)
Baggage capacity ... 36 kg (80 lb)

Max T-O and landing weight:
- XS and motor glider ... 621 kg (1,370 lb)[1]

Payload with full fuel ... 197 kg (435 lb)
Max wing loading: XS ... 65.6 kg/m² (13.43 lb/sq ft)
- motor glider ... 46.8 kg/m² (9.58 lb/sq ft)

Max power loading:
- Rotax 912 ULS ... 8.46 kg/kW (13.89 lb/hp)
- Rotax 914 UL ... 7.35 kg/kW (12.08 lb/hp)

[1] *657 kg (1,450 lb) permitted in USA only*

PERFORMANCE, POWERED (XS and motor glider with Rotax 912 ULS engine, unless otherwise stated):

Max cruising speed at 75% power:
- Rotax 912 ULS at FL80:
  - monowheel ... 140 kt (259 km/h; 161 mph)
  - tricycle ... 135 kt (250 km/h; 155 mph)
  - motor glider ... 130 kt (241 km/h; 150 mph)
- Rotax 914 UL at FL100:
  - monowheel ... 174 kt (322 km/h; 200 mph)
  - tricycle ... 166 kt (307 km/h; 191 mph)
- Rotax 914 UL at FL175 ... 151 kt (280 km/h; 174 mph)

Econ cruising speed:
- motor glider ... 100 kt (185 km/h; 115 mph)

Stalling speed, power off:
- flaps up ... 50 kt (92 km/h; 57 mph)
- flaps down ... 45 kt (83 km/h; 51 mph)

Max rate of climb at S/L:
- Rotax 912 ULS: XS ... 305 m (1,000 ft)/min
  - motor glider ... 335 m (1,100 ft)/min
- Rotax 914 UL: XS and motor glider ... 396 m (1,300 ft)/min

T-O run:
- Rotax 912 ULS:XS ... 180 m (590 ft)
  - motor glider ... 183 m (600 ft)
- Rotax 914 UL: XS and motor glider ... 153 m (500 ft)

Landing run: all ... 183 m (600 ft)

Range at econ cruising speed:
- standard fuel ... 732 n miles (1,355 km; 842 miles)
- optional fuel ... 1,093 n miles (2,024 km; 1,257 miles)

g limits (all): design ... +3.8/–1.9
- ultimate:XS ... +8.55/–4.27
  - motor glider ... +10.87/–4.27

PERFORMANCE, UNPOWERED (motor glider):
- Best glide ratio at 53 kt (98 km/h; 61 mph) IAS ... 27
- Min rate of sink at 47 kt (87 km/h; 52 mph) ... 1.02 m (3.33 ft)/s

UPDATED

FACL

FARNBOROUGH AIRCRAFT CORPORATION LTD

Building 304a, Cody Technology Park, Farnborough, Hampshire GU14 0LS
Tel: (+44 1753) 74 01 00
Fax: (+44 1753) 71 43 99
e-mail: info@farnborough-aircraft.com
Web: http://www.farnborough-aircraft.com
CHAIRMAN AND CEO: Andrew Taee
TECHNICAL DIRECTOR: Afandi Darlington
MARKETING DIRECTOR: Jonathan Sumner

Former Farnborough-Aircraft.com (FAC) applied for insolvency protection in first half of 2002; this rejected, but 51 per cent of shareholders voted on 8 July 2002 for restructuring under new company title FACL, as above. Workforce numbered 10 in early 2003. Reported in late 2004 that major investor Geoffrey Gallery had agreed to maintain financing project.

UPDATED

FARNBOROUGH F1

TYPE: Business turboprop.

PROGRAMME: Launched by Farnborough-Aircraft.com Ltd 14 October 1999, at which time initial wind tunnel testing had been completed. Further wind tunnel testing verified laminar wing design and performance of Fowler flap system. Assets of original company acquired by FACL mid-2002; investment to complete development apparently in place by late 2004, at which time more than 65,000 hours of development work completed and parts manufacture said to be under way, with maiden flight then predicted during first half of 2005.

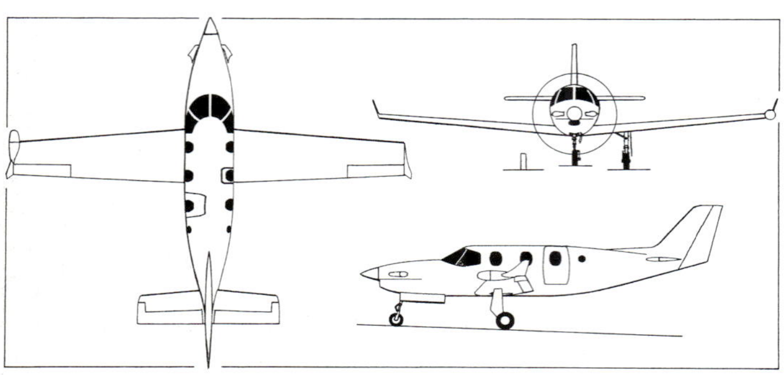

Farnborough F1 general arrangement *(Paul Jackson)* 0110907

Artist's impression of Farnborough F1 turboprop business aircraft ***NEW***/0538379

CUSTOMERS: Intended for air taxi operators providing on-demand point-to-point travel from small airfields. Initial market estimated to be 250 per year from 2007.

COSTS: Predicted unit cost, equipped US$2.2 million (2003).

DESIGN FEATURES: Designed with heavy reliance on computational fluid dynamics, backed up by wind tunnel tests. Design goals include ability to operate from small airstrips, carrying five passengers at up to 325 kt (602 km/h; 374 mph) over a range of 1,000 n miles (1,852 km; 1,150 miles) with NBAA IFR reserves. Sidestick controls; low-rpm propeller for low noise.

Wing aerofoil proprietary High-Lift, Laminar Flow sections: HLLF29155 at root, HLLF13130 at tip.

FLYING CONTROLS: Unpowered sidestick connected to conventional controls; 30 per cent chord Fowler flaps occupy approximately two-thirds span. Novel Fowler mechanism retracts almost entirely into wing.

STRUCTURE: Primarily carbon composites; fuselage of carbon fibre skins with foam core, moulded in two vertically split halves; two-spar carbon fibre wing with aluminium alloy landing gear pick-up ribs, flap tracks and aileron attachments.

LANDING GEAR: Retractable tricycle type; trailing-link main units retract inwards, nosewheel rearwards.

POWER PLANT: One 908 kW (1,218 shp) Pratt & Whitney PT6A-60A turboprop, flat rated at 634 kW (850 shp). Integral, gravity-fed fuel tank in each wing, combined capacity 965 litres (255 US gallons; 212 Imp gallons).

ACCOMMODATION: Pilot and five passengers in pressurised cabin, maximum pressure differential 0.55 bar (8.0 lb/sq in) maintaining sea level cabin altitude to 5,790 m (19,000 ft). Three cabin windows on each side; door on starboard side aft of wing; overwing emergency exit on starboard side. Optional lavatory at rear of cabin. Alternative configurations for commuter, medevac and cargo applications.

SYSTEMS: Hydraulic system for landing gear actuation. Electrical power (28 V DC) provided by 200 A starter/generator and 120 A alternator; 40 Ah main battery for engine start and systems power in event of engine failure. Wing leading-edges de-iced electrically. Environmental control system and supplemental oxygen system.

AVIONICS: 'Glass cockpit' compatible with existing and planned ATC procedures and GPS-based en-route navigation and airfield approach systems. Weather radar in wingtip pod.

DIMENSIONS, EXTERNAL:
- Wing span ... 12.39 m (40 ft 8 in)
- Wing aspect ratio ... 8.4
- Length overall ... 11.02 m (36 ft 2 in)
- Height overall ... 4.24 m (13 ft 11 in)

DIMENSIONS, INTERNAL:
- Cabin: Length ... 3.51 m (11 ft 6 in)
  - Max width ... 1.42 m (4 ft 8 in)
  - Max height ... 1.37 m (4 ft 6 in)

AREAS:
- Wings, gross ... 18.30 m² (197.0 sq ft)

WEIGHTS AND LOADINGS:
- Operating weight empty ... 1,786 kg (3,937 lb)
- Max fuel weight ... 758 kg (1,671 lb)
- Max payload ... 567 kg (1,250 lb)
- Payload with max fuel ... 157 kg (346 lb)
- Available fuel with max payload ... 348 kg (767 lb)
- Max T-O weight ... 2,700 kg (5,954 lb)
- Max ramp weight ... 2,714 kg (5,984 lb)
- Max landing weight ... 2,566 kg (5,656 lb)
- Max zero-fuel weight ... 2,353 kg (5,187 lb)
- Max wing loading ... 147.6 kg/m² (30.22 lb/sq ft)
- Max power loading ... 4.26 kg/kW (7.00 lb/shp)

PERFORMANCE (estimated):
- Max operating speed (V_{MO}) ... 285 kt (528 km/h; 328 mph)
- High cruising speed at FL300 ... 324 kt (600 km/h; 373 mph)
- Long-range cruising speed at FL300 ... 225 kt (417 km/h; 259 mph)
- Design manoeuvring speed (V_A) ... 150 kt (278 km/h; 173 mph)
- Stalling speed, flaps and landing gear down ... 59 kt (110 km/h; 68 mph)
- Time to FL250 ... 10 min
- Service ceiling ... 10,670 m (35,000 ft)
- T-O run at S/L ... 500 m (1,645 ft)
- Range with max fuel ... 1,478 n miles (2,737 km; 1,700 miles)
- Range with 363 kg (800 lb) payload, NBAA IFR reserves, 100 n mile (185 km; 115 mile) alternate ... 828 n miles (1,533 km; 952 miles)
- Ferry range ... 1,489 n miles (2,757 km; 1,713 miles)

UPDATED

FANWING

Marco Polo House, 3-5 Lansdowne Road, Croydon, Surrey CR0 2BX
Tel: (+44 7855) 37 40 05
e-mail: info@fanwing.com
Web: http://www.fanwing.com
DIRECTORS: Pat Peebles
Dikla Peebles

ITALIAN OFFICE:
Via Mandriola 10, 1-00143 Roma (RM)
Tel: (+39 06) 713 62 53

FanWing Ltd was formed on 30 October 1999 to develop manned and unmanned rotary-wing aircraft of a particularly unorthodox concept.

UPDATED

FANWING FW 8

TYPE: Technology demonstrator.

PROGRAMME: Began with small-scale, unmanned model designed in Italy and flown September 1998; several more scale versions subsequently, gradually increasing lifting efficiency and glide ratio; these led to August 2002 development contract from UK Department of Trade and Industry for commercial surveillance UAV version which began flight testing September 2003. Winglets added November 2003 and re-flown at 17.5 kg (38.6 lb) gross weight including 8 kg (17.6 lb) of payload. Plans for 2004 included doubling present weight to permit heavier payloads (including radar) and achieve vertical take-off; 'open-ended' enlargement had reached 21 kg (46.3 lb) gross with 12 kg (26.5 lb) payload by June 2004.

FW 8 ultralight is first planned application of FanWing principle to a piloted version. Future applications foreseen include agricultural sprayer, patrol, firefighting and other civil and military missions. Principle could also be applied to a medium-size passenger transport or freighter.

DESIGN FEATURES: Conventional wing is replaced by a combination of multiblade, backward-rotating cylinders, driven by a centrally mounted engine, immediately ahead of a narrow chord wing. Fan effect of the cylinders draws in air and accelerates it over the trailing-edge, generating lift even when the aircraft is stationary.

Lifting efficiency and performance increase with size, and wind tunnel testing has indicated that a full-sized FanWing could lift up to 2 tons weight with only 74.6 kW (100 hp) of power. Rotor rpm less than 1,000, thus producing low noise. Wing said to be unable to stall; lift declines gradually with forward speed until static lift limit is reached. Narrow chord wing is affected less by turbulence than a conventional wing, enhancing gust resistance. No sharp air flow separation from wing trailing-edge at any stage of the flight envelope. Other attributes include stable low/slow flying speeds and low noise levels.

FLYING CONTROLS: Three fan sections per wing. Leading-edge flaps provide principal aerodynamic control of both lift and thrust; by increasing the blanking effect of these on one wing and reducing them on the other, aircraft will turn without yawing.

Following provisional details apply to projected FW 8 ultralight.

POWER PLANT: Two 22.4 kW (30 hp) piston engines.

ACCOMMODATION: Pilot and two passengers.

DIMENSIONS, EXTERNAL:
Wing span .... 10.00 m (32 ft 9¾ in)

WEIGHTS AND LOADINGS:
Weight empty .... 350 kg (772 lb)

PERFORMANCE:
Max level speed .... 32 kt (60 km/h; 37 mph)
Ceiling .... 6,000 m (19,680 ft)
Endurance .... 10 h

UPDATED

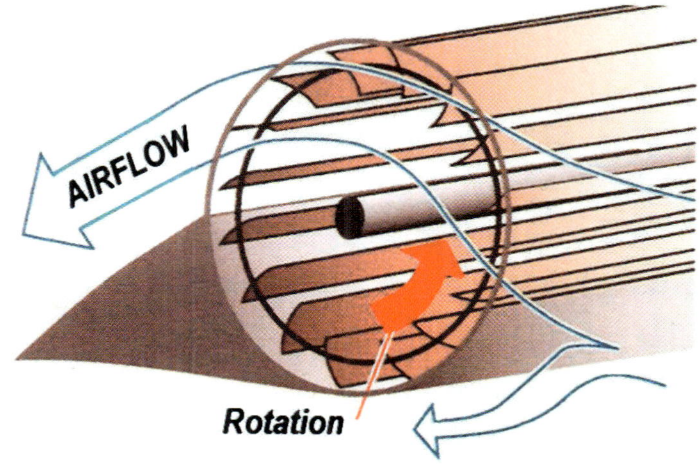

Principle of the FanWing ***NEW***/0589233

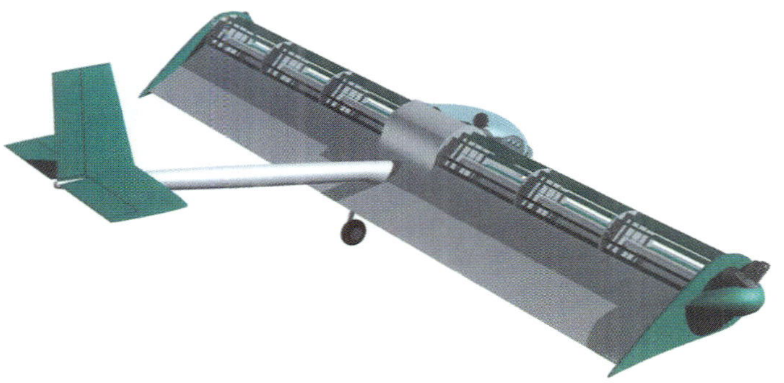
Projected FanWing FW 8 ultralight *(KMi - The Open University)* ***NEW***/0589236

Projected FanWing FW 8 ultralight *(KMi - The Open University)* ***NEW***/0589235

Radio-controlled FanWing test article *(Paul Jackson)* ***NEW***/0589232

QINETIQ

QINETIQ

A1 Cody Technology Park, Ively Road, Farnborough, Hampshire GU14 0LX
Tel: (+44 1252) 39 46 11
Fax: (+44 1252) 39 33 99
Web: http://www.qinetiq.com
CEO: Sir John Chisholm
MARKETING DIRECTOR: Brenda Jones
HEAD OF MEDIA RELATIONS: Joanna Sale

QinetiQ (pronounced 'kinetic') was created in July 2001 from the major portion of the former Defence Evaluation and Research Agency (DERA), itself derived from the earlier Royal Aircraft Establishment at Farnborough and elsewhere. One of Europe's largest science and technology organisations, its activities and facilities include indoor and outdoor ranges for air-, land- and sea-launched weapons trials; wind tunnels; underwater target ranges and marine testing facilities; automotive test tracks; and climatic testing laboratories. Previously structured as a public-private partnership (PPP), subsequently to be found a strategic investment partner as a prelude to flotation as a fully commercial company. Carlyle Group named as preferred bidder in September 2002. This became effective in December that year, with Carlyle acquiring a 33.8 per cent interest in QinetiQ and a further 3.7 per cent offered to company employees. The remaining 62.5 per cent was retained by the UK MoD, but is planned to be sold off during the ensuing three to five years.

QinetiQ is main sponsor for the huge QinetiQ 1 experimental balloon.

UPDATED

QINETIQ QINETIQ 1

TYPE: Helium balloon.

PROGRAMME: Announced in November 2001 as attempt on balloon world altitude record (first since present record of 34,668 m; 113,740 ft was established by two US Navy fliers on 4 May 1961) and for scientific study of cosmic rays and solar radiation in the stratosphere; planned to reach 40,235 m (132,000 ft). Launch window of July to September 2002 unable to be used due to unsuitable weather and high levels of solar activity; rescheduled for third quarter of 2003, following 72 hour countdown. During that period, improvements to envelope were made and tested (early 2003) in a 1:88 scale version. Attempted launch on 3 September 2003, from MV *Triton* off the Cornish coast, had to be aborted with balloon only half-inflated, when an envelope seam ruptured as it unwound from the storage drum.

Project Director Elson and co-pilot Prescot decided to use 2004 to investigate different launch methods and alternative launch sites to improve chances of future success. Ways to improve the balloon's envelope quality are also being studied, before making a second attempt in 2005.

STRUCTURE: Polyethylene envelope, reinforced with Kevlar straps. Carbon fibre crew platform.

ACCOMMODATION: Crew of two (Andy Elson and Colin Prescot, current holders of world balloon endurance record), on non-pressurised open flight deck platform.

EQUIPMENT: Cameras and real-time video downlink. Zvezda space suits, pressurised to 10,660 m (35,000 ft) equivalent altitude, and oxygen cylinders for crew.

Following data are approximate.

DIMENSIONS, EXTERNAL:
Height overall at launch .... 387 m (1,270 ft)

When launched, QinetiQ 1 will be the world's largest manned aircraft *(QinetiQ)* 0525367

DIMENSIONS, INTERNAL:
Envelope volume at 40,235 m (132,000 ft) .... 1.27 million m^3 (45 million cu ft)
WEIGHTS AND LOADINGS:
Weight empty, uninflated .... less than 3.05 tonnes (3 tons)
PERFORMANCE (estimated):
Max rate of ascent/descent .... 335 m (1,100 ft)/min
Time to 40,235 m (132,000 ft) .... 3–5 h

UPDATED

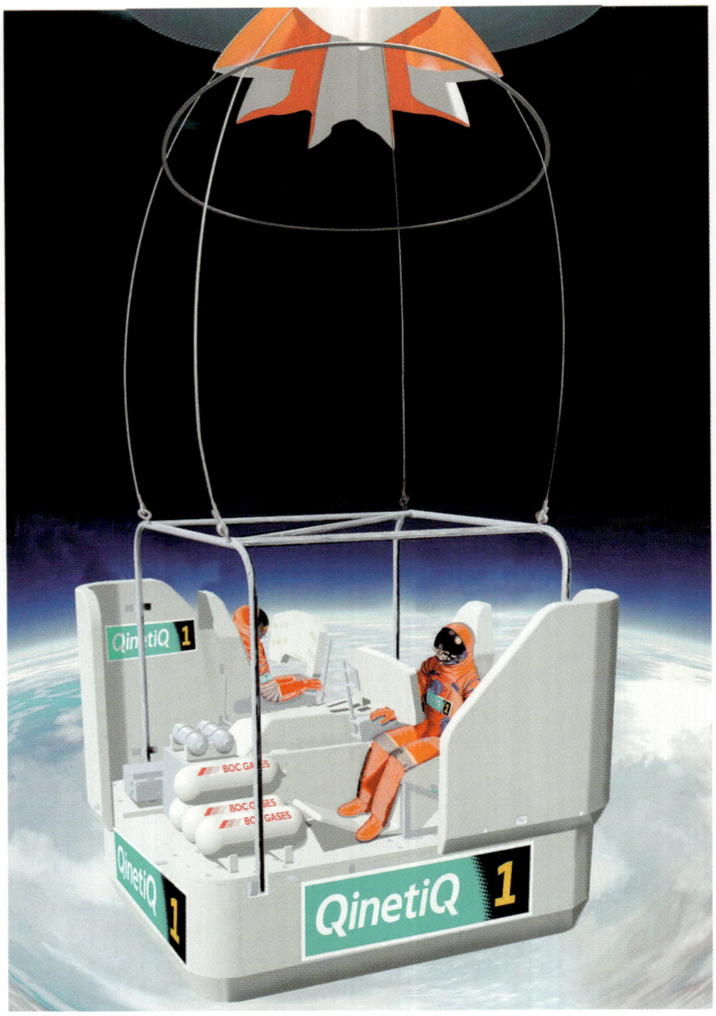

Artist's impression of QinetiQ 1 gondola *(QinetiQ)* 0525947

REALITY

REALITY AIRCRAFT LTD

Unit 1A, Amesbury Business Park, London Road, Amesbury, Wiltshire SP4 7LS
Tel: (+44 1980) 62 55 51
Fax: (+44 1980) 62 62 25
e-mail: info@realityaircraft.com
Web: http://www.realityaircraft.com
MANAGING DIRECTOR: Terry Francis
MARKETING DIRECTOR: Tom Lafferty
DIRECTOR: Kate Mather

Reality markets the Easy Raider, for which it has world rights, except in the USA, and the new (2003) Escapade. US partner, Just Aircraft LLC, manufacturers and markets the Easy Raider in North America and developed the Escapade. A related aircraft, the Just Summit, is marketed only in the USA.

UPDATED

REALITY EASY RAIDER

TYPE: Tandem-seat ultralight kitbuilt.

PROGRAMME: Development of Flying K (now Sky Raider LLC) Sky Raider II, incorporating over 100 modifications. Prototype G-SRII (Rotax 503 engine) flew 2001; conforms to BCAR Section S. Kit fabrication by Just Aircraft of USA, with which Reality holds joint design rights.

CURRENT VERSIONS: **Easy Raider 503:** One 37.0 kW (49.6 hp) Rotax 503 DCDI-2V two-stroke engine driving a Powerfin FL370T three-blade, ground-adjustable pitch, composites propeller. Initial version.

Easy Raider J2.2: One 59.7 kW (80 hp) Jabiru 2200 engine driving a two-blade Powerfin ground-adjustable pitch or two-blade Newton wooden propeller. Prototype G-OESY (c/n 0002) built late 2001.

Easy Raider R100: BMW R100 four-stroke motorcycle engine driving a Powerfin two-blade, ground-adjustable pitch propeller. Prototype G-SLIP (c/n 0004) under construction in 2002.

CUSTOMERS: Total 12 sold by mid-2004, of which 10 registered in UK. Continues in intermittent production.

COSTS: Kit £9,898 excluding engine; or £12,563 (Rotax), £14,500 approx (BMW) or £16,298 (Jabiru), including engine, cowlings and propeller, excluding tax (2003).

DESIGN FEATURES: High wing with two tubular main, and two secondary bracing struts to each wing. Tailplane braced with twin wires above and twin rods below. Large fin fillet, compared with Sky Raider, enables sideslipping with cockpit doors in place. Wings fold for storage. Quoted build time 350 hours.

FLYING CONTROLS: Conventional and manual. Frise ailerons, cable actuated. Flight-adjustable tab in port tailplane. Horn-balanced tail surfaces. Slotted flaps, deflections 0, 15, 30 and 40°.

STRUCTURE: Fabric-covered 4130 chromoloy steel tube fuselage with composites engine cowling. Wing built on large diameter metal tube leading-edge/spar with wooden ribs, fabric covering and composites downturned wingtips.

Jabiru-engined Reality Easy Raider *(Paul Jackson)* ***NEW***/0558568

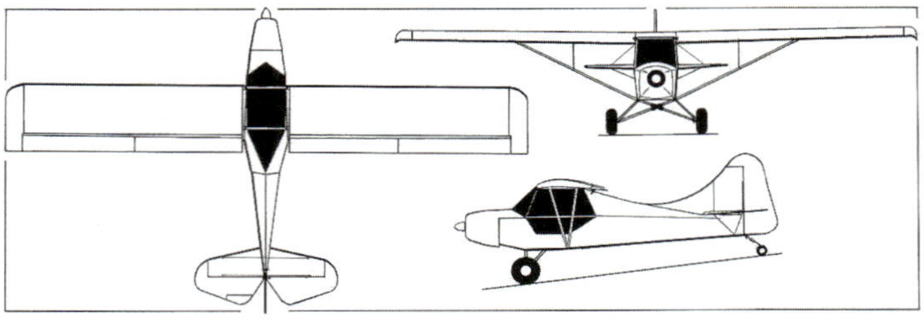

Reality Easy Raider (Rotax 503 engine) *(Paul Jackson)* 0137970

LANDING GEAR: Tailwheel type; fixed. Two faired-in side Vs hinged to lower longerons, plus half-axles, each with bungee shock-absorber, attached to compression frame. Mainwheels 8.00-6 (4 ply); Matco steerable tailwheel. Optional floats or amphibious floats.

POWER PLANT: As under Current Versions. Two fuel tanks in wings, total capacity 37.9 litres (10.0 US gallons; 8.3 Imp gallons) standard; 72 litres (19.0 US gallons; 15.8 Imp gallons) optional.

DIMENSIONS, EXTERNAL:
Wing span .... 8.53 m (28 ft 0 in)
Width, wings folded .... 2.49 m (8 ft 2 in)
Length: overall .... 5.18 m (17 ft 0 in)
wings folded .... 6.10 m (20 ft 0 in)
Height overall .... 1.75 m (5 ft 9 in)
DIMENSIONS, INTERNAL:
Cabin max width .... 0.67 m (2 ft 2½ in)
AREAS:
Wings, gross .... 9.75 m^2 (105.0 sq ft)

WEIGHTS AND LOADINGS (J: Jabiru 2200, R: Rotax 503):
Weight empty: J ... 225 kg (495 lb)
R ... 191 kg (420 lb)
Max T-O weight: J, R ... 450 kg (992 lb)
PERFORMANCE (J, R as above):
Max level speed: J ... 75 kt (139 km/h; 86 mph)
R ... 70 kt (129 km/h; 80 mph)
Normal cruising speed: J ... 60 kt (111 km/h; 69 mph)
R ... 55 kt (102 km/h; 63 mph)
Stalling speed, flaps down: J, R ... 34 kt (63 km/h; 40 mph)
Max rate of climb at S/L: J ... 213 m (700 ft)/min
R ... 122 m (400 ft)/min
T-O run: J ... 80 m (265 ft)
R ... 156 m (515 ft)
Landing run: J, R ... 139 m (460 ft)

UPDATED

REALITY (JUST PLANE) ESCAPADE

TYPE: Side-by-side ultralight kitbuilt.

PROGRAMME: US prototype, built by Just Aircraft LLC of Caldwell, Iadho, first flew (N116JA, Rotax engine) 22 February 2003 and made its public debut at Sun 'n' Fun, Lakeland, Florida, 2 to 8 April 2003; UK prototype (G-ESCA, Jabiru engine) followed on 2 June 2003, with debut at PFA International Rally, Kemble, 11 to 13 July 2003. G-ESCA awarded UK Permit to Fly on 18 September 2003. Grand Champion lightplane at Sun 'n' Fun 2004. Development is planned of a VLA version with 500 kg (1,102 lb) MTOW.

CURRENT VERSIONS: **Escapade:** *As described.*

Highlander: Introduced 2004; taller, square-cornered fin; and additional 91 cm (3 ft 0 in) wing span. 'Bush plane' featuring enlarged, balanced elevator, rudder and associated trim tab; extended, strengthened landing gear with tundra tyres and brakes; and gull-wing, flight-openable doors. First aircraft, N8056A, built from 13th Escapade kit.

CUSTOMERS: At least 45 kits sold by late 2004.

COSTS: Kite, excluding engine and instruments: tailwheel version £10,900; nosewheel version £11,500 (2004).

DESIGN FEATURES: Baseline requirements included wide, dual control cockpit, nose- or tailwheel options, high commonality with Easy Raider, wide C of G range, 70 kt cruise and 35 kt, stall and wide range of alternative engines.

Fuselage similar to Skystar Kitfox Lite[2], but with slight stretch and several detail improvements, married to wing of Reality Easy Raider. Wing braced by V struts; wire-braced empennage. Conforms to BCAR Section S.

FLYING CONTROLS: Conventional and manual. Cable actuation. Flight-adjustable tab in port elevator. Flaps; maximum deflection 40°. Optional electric elevator trim.

STRUCTURE: Chromoloy 4130 steel tube fuselage with fabric covering and composites engine cowling and aluminium turtledeck. Wing as for Easy Raider, with aluminium trailing edges and glass fibre leading edges to flaps and ailerons.

LANDING GEAR: Fixed; tricycle or tailwheel type. Two faired-in side Vs attached to lower longerons, plus half-axles, each with bungee cord shock-absorber, attached to compression frame. Mainwheels typically 6.00-6; solid, steerable Matco tailwheel or castoring nosewheel. Optional hydraulic disc brakes. Version with steerable nosewheel is planned.

POWER PLANT: Options include 37.0 kW (49.6 hp) Rotax 503 driving a three-blade Powerfin composites propeller; 73.5 kW (98.6 hp) Rotax 912 ULS; 59.6 kW (79.9 hp) Rotax 912 UL with two-blade Powerfin propeller; and 59.7 kW (80 hp) Jabiru 2200 with two-blade Powerfin or Newton wood propeller; and 89 kW (120 hp) Jabiru 3300. Usable fuel capacity 72 litres (19.0 US gallons; 15.8 Imp gallons).

Just Plane Highlander ***NEW***/1047189

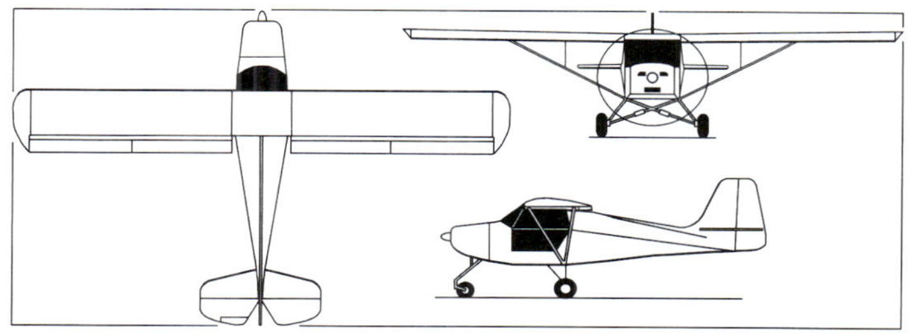

Reality (Just) Escapade side-by-side kitplane *(Paul Jackson)* ***NEW***/0587524

DIMENSIONS, EXTERNAL:
Wing span ... 8.69 m (28 ft 6 in)
Width, wings folded ... 2.42 m (7 ft 11½ in)
Length excl propeller ... 5.79 m (19 ft 0 in)
Height overall: tailwheel ... 2.18 m (7 ft 2 in)
nosewheel ... 2.18 m (7 ft 2 in)
DIMENSIONS, INTERNAL:
Cockpit max width ... 1.12 m (3 ft 8 in)
AREAS:
Wings, gross ... 10.03 m² (108.0 sq ft)
WEIGHTS AND LOADINGS:
Weight empty (Rotax 912) ... 240 kg (529 lb)
Max T-O weight: in UK ... 450 kg (992 lb)
in USA ... 558 kg (1,232 lb)
PERFORMANCE (Rotax 912):
Normal cruising speed ... 80 kt (148 km/h; 92 mph)
Stalling speed, power off, flaps down ... 34 kt (63 km/h; 40 mph)
Max rate of climb at S/L ... 305 m (1000 ft)/min
T-O run ... 90 m (295 ft)
Landing run ... 100 m (330 ft)
g limits ... +4/-2

UPDATED

US-built Just Plane Escapade *(Paul Jackson)* ***NEW***/1047191

SLINGSBY

SLINGSBY AVIATION LIMITED

Ings Lane, Kirkbymoorside, York YO62 6EZ
Tel: (+44 1751) 43 24 74
Fax: (+44 1751) 43 11 73
e-mail: sales@slingsby.co.uk
Web: http://www.slingsby.co.uk
MANAGING DIRECTOR: Jeff Bevan
CHIEF DESIGNER: David Goddard
SALES AND MARKETING MANAGER: Steven Boyd
BUSINESS DEVELOPMENT MANAGER: Peter Lovering

Formerly a subsidiary of ML Holdings, Slingsby was sold to Cobham plc in 1997 and is a component of Chelton Group within that holding. Company specialises in application of composites materials; was previously manufacturer of sailplanes but now concentrates on development and low-rate production of T67 Firefly series of aerobatic training aircraft, for which new orders received in 2001–02.

Other activities include design and manufacture of air cushion vehicles and airship gondolas, tailfins and propeller ducts in composites materials; and design, development and manufacture of high-performance composites structures for marine and aerospace industries. Supplier to Lockheed Martin of C-130J cockpit interior trim panels in sculptured composites. Other production includes components for Europa kitbuilt and propeller spinners for Bombardier Dash 8 Q400.

Work for UK MoD includes technical support for RAF Air Cadet gliders and Firefly aircraft.

VERIFIED

SLINGSBY T67 FIREFLY

US Air Force designation: T-3A Firefly

TYPE: Primary prop trainer/sportplane.

PROGRAMME: Current composites constructed Firefly developed from wooden Slingsby T67A (licence-built version of French Fournier RF6B; T67B gained CAA certification 18 September 1984; T67C was CAA certified 15 December 1987. Subsequent versions were T67M, T67M200 and T67M260.

Last-mentioned selected by USAF to meet Enhanced Flight Screener (EFS) requirement. Slingsby prime contractor for both acquisition contract and seven-year contractor logistic support (CLS) contract. Northrop Grumman subcontractor for final assembly at Hondo, Texas, and for operation of CLS activities at Hondo and USAF Academy at Colorado Springs. US Type Certificate (Aerobatic Category) awarded to T67M260 on 15 December 1993.

Low-rate manufacture only, since 1997, but expanded in 2001–02 to cover order for 16 T67M260s placed by Royal Jordanian Air Force and three for Bahrain.

CURRENT VERSIONS: **T67C:** Basic version, including earlier C1, C2 and C3 subvariants. One 119 kW (160 hp) Textron Lycoming O-320-D2A flat-four engine, driving a Sensenich M74DM6-O-64 two-blade, fixed-pitch, metal propeller. Fuel tank in each wing leading-edge, combined capacity 159 litres (42.0 US gallons; 35.0 Imp gallons). Nine T67Cs purchased by Netherlands government Civil Aviation Flying School for KLM and Royal Netherlands Navy pilot training; 12 T67Cs for Canadian Department of National Defence for military primary flying training; plus other T67C variants for UK schools.

T67M: Military variant. First flight of T67M Firefly 160 (G-BKAM) 5 December 1982; CAA certification 20 September 1983; designation changed to T67M Mk II as wing fuel tanks and two-piece canopy introduced. Powered by 119 kW (160 hp) Textron Lycoming AEIO-320-D1B flat-four engine, driving a Hoffmann HO-V72 two-blade constant-speed composites propeller.

Sold to Netherlands for military grading; Japan and UK for airline training; and Switzerland for aerobatic and general flying training. Used by RAF, RN and British Army for elementary flying training; Joint Elementary Flying Training School (JEFTS) formed at Topcliffe, July 1993, with first of eventual 18 (including some second-hand) T67M Mk IIs operated under contract by Hunting Aircraft Ltd (later Hunting Contract Services, and now Babcock HCS); transferred to Barkston Heath in April 1995. Now operating as Defence Elementary Flying Training School (DEFTS).

T67M200: Development of T67M; 149 kW (200 hp) Textron Lycoming AEIO-360-A1E engine; Hoffmann HO-V123 three-blade, variable-pitch, composites propeller; wing fuel capacity as for T67C. First flight 16 May 1985; CAA certification 13 October 1985. First customer Turkish Aviation Institute, Ankara (16 delivered from 1985); others ordered by Dutch operator King Air (three T67M200s, plus one T67M Mk II) as screening trainers for prospective RNethAF pilots; Royal Hong Kong Auxiliary Air Force (now Government Flying Service) (four); and Norwegian government's Flying Academy (six).

Bahrain Air Force Slingsby Firefly T67M260 ***NEW***/0563503

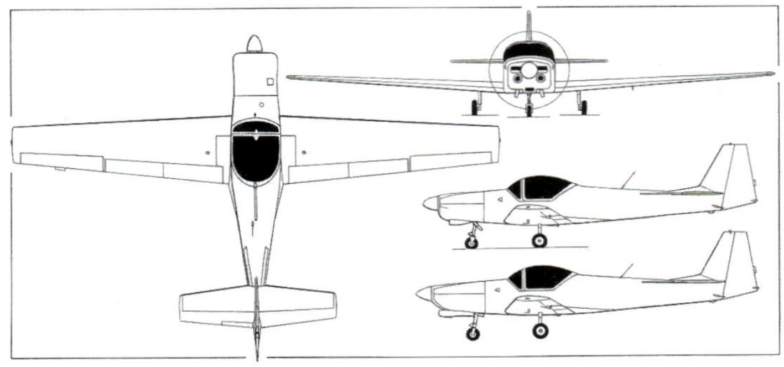

Slingsby T67M Mk II Firefly (Textron Lycoming AEIO-320-D1B engine) with additional side view (bottom) of T67M260 *(Jane's/Dennis Punnett)* 0507238

T67M260: Development of T67M; higher-powered Lycoming engine; electric elevator trim and electric flaps optional; cabin air conditioning system optional; higher maximum T-O and aerobatic weights, to allow 227 kg (500 lb) for two pilots and equipment plus full fuel load.

Prototype (G-BLUX) flown May 1991, evaluated at Wright-Patterson AFB; hot-and-high trials at USAF Academy in mid-1991. Preproduction aircraft (G-EFSM) flown September 1992. First flight of first USAF aircraft (92-0625/N7020D) 4 July 1993. T-3A type certificate awarded by CAA and FAA December 1993. Official USAF acceptance 25 February 1994; student pilot training began March 1994. Also acquired for UK military pilot training and by Bahrain and Jordan.

T-67 Mk 2: In design stage mid-2003. Shorter-span, swept wings with thinner aerofoil section, and EFIS instrumentation. Estimated maximum and cruising speeds of 180 kt (333 km/h; 207 mph) and 150 kt (278 km/h; 173 mph) respectively. Potential candidate for UK Military Flying Training System requirement.

CUSTOMERS: Total of 276 civil/military T67s of all subtypes (including 10 T67As) delivered to customers in 15 countries by December 2002 (one built in 1998, two in 1999, two in 2000, none in 2001 and 19 in 2002–03). Total production 281, including demonstrators and replacement for one USAF aircraft lost during testing, but excluding two T67B static test airframes.

US Air Force acquired 113 T-3As in three lots (38, 42 and 33) to replace Cessna T-41. USAF pilot conversion completed September 1993. Deliveries began January 1994 and were completed in November 1995. First operating unit (with 56 aircraft) was 1st Flight Screening Squadron of 12th Flying Training Wing based at Hondo, Texas; second and final operating unit (with 57 aircraft) is 557th FTS of USAF Academy at Colorado Springs; deliveries to this unit from August 1994; operational January 1995. All T-3As wore dual military/civilian identities. Remaining fleet of 110 grounded in July 1997 for fuel system modifications, and remained thus in early 2002, despite reassessment proving aircraft safe. In late 2000, USAF announced that it was contracting out future flight screening to civilian operators and in early 2002 was considering options for the T-3A fleet. The issue had not been resolved by mid-2004.

Hunting Contract Services took delivery of 25 T67M260s between June 1996 and March 1997 for extension to JEFTS tri-service training programme, followed by a further two in September 1999. JEFTS expanded to 18 M200s and 27 M260s, based at Barkston Heath (20), Middle Wallop, Cranwell (six) and Church Fenton (five), but 16 of M260 fleet sold in 2003 on conclusion of part of contract, RAF withdrawing and remainder forming DEFTS

Slingsby T67M260 Firefly of UK's DEFTS *(Paul Jackson)* ***NEW***/0589327

Slingsby T67M260 Firefly in the insignia of Babcock HCS *(Paul Jackson)* 0568391

T67 FIREFLY PRODUCTION
(at December 2003)

Type	First Aircraft	Qty
T67A	G-BIOW	10
T67B	G-BIUZ	14
T67C	G-BLRE	7
T67C-3	PH-SGA	21
T67M	G-BKAM	12
T67M MkII	PH-KIF	23
T67M200	G-BLUX	32
T67M260/T-3A	92–0625	114
T67M260	G-ZEIN	48
Total		**281**

at Barkston Heath and Middle Wallop, some wearing (on opposite sides of fin) badges of 674 Squadron AAC and 703 Squadron FAA.

One each delivered to Belize Defence Force and a UK civilian customer in 1996. Royal Jordanian Air Force ordered 16 T67M260s in September 2001; these delivered between 15 July and 20 December 2002 to replace Bulldogs with No. 4 Squadron at Al Mafraq AFB. Three T67M260s ordered (via BAE Systems) in March 2002 by Bahrain Amiri (now Royal Bahraini) Air Force; delivered in February 2003. Further Bahraini order in prospect.

DESIGN FEATURES: Conventional low-wing monoplane; design translated from wood to GFRP. Tapered wings and mid-mounted tailplane; sweptback fin.

Wing section NACA 23015 at root, 23013 at tip; dihedral 3° 30'; incidence 3°.

FLYING CONTROLS: Conventional and manual. Mass-balanced Frise ailerons, without tabs; mass-balanced elevators with manually operated port trim tab (electric trim optional); trailing-edge fixed hinge flaps; spin strakes forward of tailplane roots.

STRUCTURE: GFRP; single-spar wings with double skin (corrugated inner skin bonded to plain outer skin) and ribs in heavy load positions; frame and top-hat stringer fuselage; stainless steel firewall between cockpit and engine; fixed incidence tailplane of similar construction to wings (built-in VOR antenna); fin incorporates VHF antenna.

LANDING GEAR: Tricycle type; fixed. Wing-mounted cantilever main legs with integral oleo-pneumatic shock-absorber in each unit. Steerable nosewheel with integral shock absorber. Mainwheel tyres size 6.00–6, pressure 1.38 bar (20 lb/sq in). Nosewheel tyre size 5.00–5, pressure 2.55 bar (37 lb/sq in). Hydraulic disc brakes. Parking brake.

Data follow for T67M260/T-3A.

POWER PLANT: One 194 kW (260 hp) Textron Lycoming AEIO-540-D4A5 flat-six engine, driving a Hoffmann HO-V123K-KV/180DT three-blade constant-speed composites propeller. Fuel 159 litres (42.0 US gallons; 35.0 Imp gallons) in wing leading-edges; refuelling point in upper wing surface.

ACCOMMODATION: Two seats side by side, fixed windscreen and rearward-hinged upward-opening rear-canopy section. Optional raised canopy and lowered seats to accommodate crew wearing military-style helmets. Dual controls standard. Adjustable rudder pedals. Cockpit heated and ventilated. Baggage space aft of seats.

SYSTEMS: Hydraulic system for brakes only. Vacuum system for blind-flying instrumentation. Electrical power supplied by 28 V 70 A engine-driven alternator and 24 V 15 Ah battery.

AVIONICS: Optional avionics, available to customer requirements, include equipment by Bendix/King, up to full IFR standard. Also now available with Chelton Flight Systems general aviation EFIS.

Instrumentation: Standard avionics include artificial horizon and directional gyro, with vacuum system and vacuum gauge, electric turn and slip indicator, rate of climb indicator, recording tachometer, stall warning system, clock, outside air temperature gauge, accelerometer.

EQUIPMENT: Includes tiedown rings and towbar; cabin fire extinguisher, crash axe, heated pitot; instrument, landing, navigation and strobe lights. Optional equipment includes external power socket, and wingtip-mounted smoke system.

DIMENSIONS, EXTERNAL:
- Wing span ..... 10.59 m (34 ft 9 in)
- Wing chord: at root ..... 1.53 m (5 ft 0¼ in)
  - at tip ..... 0.83 m (2 ft 8¾ in)
- Wing aspect ratio 8.9
- Length overall ..... 7.57 m (24 ft 10 in)
- Height overall ..... 2.36 m (7 ft 9 in)
- Tailplane span ..... 3.40 m (11 ft 1¾ in)
- Wheel track ..... 2.44 m (8 ft 0 in)
- Wheelbase ..... 1.50 m (4 ft 11 in)
- Propeller diameter ..... 1.80 m (5 ft 10¾ in)

DIMENSIONS, INTERNAL:
- Cockpit: Length ..... 2.05 m (6 ft 8¾ in)
  - Max width ..... 1.08 m (3 ft 6½ in)
  - Max height ..... 1.08 m (3 ft 6½ in)

AREAS:
- Wings, gross ..... 12.63 m^2 (136.0 sq ft)
- Ailerons (total) ..... 1.24 m^2 (13.35 sq ft)
- Trailing-edge flaps (total) ..... 1.74 m^2 (18.73 sq ft)
- Fin ..... 0.80 m^2 (8.61 sq ft)
- Rudder ..... 0.82 m^2 (8.80 sq ft)
- Tailplane ..... 1.65 m^2 (17.76 sq ft)
- Elevators (incl tab) ..... 0.99 m^2 (10.66 sq ft)

WEIGHTS AND LOADINGS (with air conditioning installed):
- Weight empty ..... 794 kg (1,750 lb)
- Max fuel weight ..... 114 kg (252 lb)
- Max T-O and landing weight:
  - Utility/Aerobatic ..... 1,157 kg (2,550 lb)
- Max wing loading ..... 91.5 kg/m^2 (18.75 lb/sq ft)

Max power loading 5.97 kg/kW (9.81 lb/hp)
PERFORMANCE:
Never-exceed speed (VNE) 195 kt (361 km/h; 224 mph)
Max level speed at S/L 152 kt (281 km/h; 175 mph)
Max cruising speed at 75% power at FL85 140 kt (259 km/h; 161 mph)
Stalling speed, power off, flaps down 54 kt (100 km/h; 63 mph)
Max rate of climb at S/L 480 m (1,380 ft)/min
T-O run 334 m (1,095 ft)
T-O to 15 m (50 ft) 689 m (2,265 ft)
Landing from 15 m (50 ft) 984 m (3,228 ft)
Landing run 401 m (1,315 ft)
Range with max fuel at 65% power at FL80, allowances for T-O, climb and 30 min reserves 407 n miles (753 km; 468 miles)
Endurance at best econ setting at FL80, allowances as above 5 h 40 min
g limits +6/–3

UPDATED

WARRIOR

WARRIOR (AERO-MARINE) LTD

The Portway Centre, Old Sarum, Salisbury, Wiltshire SP4 6EB
Tel: (+44 1722) 42 18 60
Fax: (+44 1722) 42 18 61
e-mail: jlabouchere@centaurseaplane.com
Web: http://www.centaurseaplane.com
MANAGING DIRECTOR: James Labouchere

NORTH AMERICAN OFFICE(WARRIOR (AERO-MARINE INC):
23 Oceanwood Drive, Scarborough, Maine 04074, USA
Tel/Fax: (+1 207) 885 99 20
e-mail: dverrill@centaurseaplane.com
VICE-PRESIDENT AND DIRECTOR: David C Verrill

Company created in 1995 to develop and market Centaur six-seat amphibian; US subsidiary formed in January 2000 to manufacture and promote aircraft in North American market. Funding acquired by March 2002 from private investors (more than US$3 million), Maine Technology Institute (US$500,000) and UK Department of Trade and Industry (£450,000/US$640,000); further US$2 million then said to be required for marketing and US demonstration tour. First flight postponed 12 months from original estimate.

UPDATED

WARRIOR CENTAUR

TYPE: Six-seat amphibian.

PROGRAMME: First design studies 1992; Mk 1 hull design, and first flight by one-fifth scale model, in 1993; Mk 2 hull configuration developed and trialled 1995; dynamically refined (and still current) one-fifth model first flown 1996; cockpit mockup completed 1997; initial funding by Royal Aeronautical Society Handley Page award. Revealed at Farnborough Air Show, September 1998, when development and planning continuing. Funding for construction and testing of full-size prototype secured 1 September 2001. Prototype, manufactured by Warrior in association with Maine Composites Inc, began in December 2001 and was then expected to fly by end of 2003; however, by then postponed to fourth quarter 2004 (not achieved due to withdrawal of funding pledge). Design changes have included new contours on rear fuselage and outer wings, and deletion of wing bracing struts. Fuselage shipped to company's US facility in October 2003 for installation of systems and power plant. Specifications for military and coast guard search and rescue were proposed in March 2004. Certification to JAR/FAR 23 anticipated 24 months after first flight. Production aircraft will be built in the UK and USA.

CUSTOMERS: Five orders by March 2002. Production target is 68 in first full year; 172 in second.

COSTS: Expected to be between US$525,000 and US$575,000 at 2002 prices and rates. Seat-mile cost estimated at 35 per cent less than comparable amphibians.

By December 2003, the prototype Centaur had been moved to Sanford Airport, Maine, for installation of power plant and systems ***NEW***/0569573

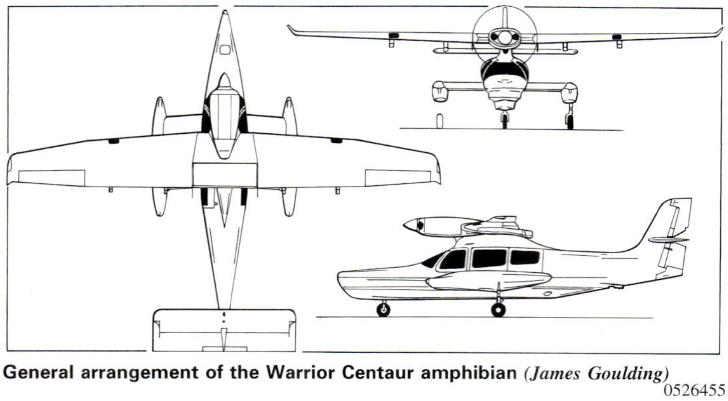

General arrangement of the Warrior Centaur amphibian *(James Goulding)* 0526455

Following description applies to full-size aircraft:

DESIGN FEATURES: Concepts proven and refined using CATIA software. Objectives of design included low hydrodynamic drag and structural weight, use of composites materials, folding wing, comfortable operation in coastal wave conditions (0.9 to 1.2 m; 3 to 4 ft wave height), competitive seat-mile costs and payload-range performance, safety, saline impermeability and access to all areas frequented by boats.

Slender boat hull without planing step improves rough-water ride and increases aerodynamic efficiency; parasol-mounted high-lift wing with centrally installed engine nacelle; stub-wing/sponsons with end-mounted floats; conventional tail surfaces. Main panels of wing fold back manually (optionally hydraulically) to within beamwidth of sponsons to permit docking in standard 12 m (40 ft) yacht berths.

FLYING CONTROLS: Conventional and manual. Electrohydraulically actuated, slotted trailing-edge flaps.

STRUCTURE: Composites construction (carbon and glass fibres). Top surfaces of stub-wings laminated with energy absorbent tread; door sills reinforced with stainless steel.

LANDING GEAR: Electrohydraulically retractable tricycle type in amphibian, nosewheel retracting rearward, mainwheels outward into sponson tip-floats. Landing gear can be omitted in 'pure' seaplane version, with corresponding increase in payload. Stern-mounted 7 kW (9 hp) water-jet for manoeuvring in marine environment.

POWER PLANT: One 224 kW (300 hp) Textron Lycoming IO-540-C or 261 kW (350 hp) IO-540-J2B flat-six engine; three-blade, constant-speed, variable-pitch propeller. Further plans include 410 kW (550 shp) PT6A-powered turboprop version. One fuel tank in each sponson: see under Weights and Loadings for details.

ACCOMMODATION: Pilot and one passenger in cockpit; five passenger seats in flying-boat cabin, four in amphibian (forward-facing or club layout). Baggage compartment behind seats. Three doors on port side; two on starboard.

SYSTEMS: Electrically driven hydraulic actuation of wing folding (optional), landing gear (optional) and flaps. Electrical systems powered by 28 V DC alternator and 24 V lead-acid batteries. Air conditioning optional.

AVIONICS: Honeywell APEX 1000 standard; marine HF radio, two-axis autopilot and bow-mounted weather radar.

DIMENSIONS, EXTERNAL:
Wing span 13.65 m (44 ft 9½ in)
Width over sponsons 4.54 m (14 ft 10¾ in)
Length overall 11.15 m (36 ft 7 in)
Height overall, on land 3.78 m (12 ft 4¾ in)
Wheel track 4.09 m (13 ft 5 in)

DIMENSIONS, INTERNAL:
Cabin, excl cockpit: Length 3.81 m (12 ft 6 in)
Max internal load length 4.27 m (14 ft 0 in)
Max width 1.37 m (4 ft 6 in)
Max height 1.22 m (4 ft 0 in)
Volume 3.74 m^3 (132 cu ft)
Baggage compartment volume 1.59 m^3 (56 cu ft)

WEIGHTS AND LOADINGS (A: 300 hp, B: 350 hp engine):
Weight empty: A 1,207 kg (2,661 lb)
B 1,274 kg (2,809 lb)
Baggage capacity 90 kg (200 lb)
Fuel weight: max 354 kg (780 lb)
with max payload 68 kg (150 lb)
Payload: max:
amphibian 641 kg (1,414 lb)
flying-boat 718 kg (1,583 lb)
with max fuel 264 kg (583 lb)
Max T-O weight from water: A 1,920 kg (4,232 lb)
B 2,100 kg (4,629 lb)

PERFORMANCE (estimated, A and B as above):
Cruising speed:
at max cruise power at FL50, 75% MTOW: A 127 kt (235 km/h; 146 mph)
B 149 kt (276 km/h; 171 mph)
at recommended cruise power at S/L: A 119 kt (220 km/h; 137 mph)
Stalling speed, power off:
flaps up: A 51 kt (95 km/h; 59 mph)
flaps down: A 46 kt (86 km/h; 53 mph)
B 48 kt (89 km/h; 56 mph)
Time to FL100: at max cruise power:
A 13 min 24 s
B 11 min
T-O run:
on water: A 173 m (569 ft)
T-O to 15 m (50 ft):
on land: A 286 m (940 ft)
B 289 m (950 ft)
on water: A 362 m (1,190 ft)
B 365 m (1,200 ft)
Landing from 15 m (50 ft) at 1,700 kg (3,748 lb):
on land: A 348 m (1,140 ft)
Range at max cruising speed, VFR reserves:
with max payload: A 150 n miles (278 km; 172 miles)
with 545 kg (1,202 lb) (amphibian) or 615 kg (1,356 lb) payload (seaplane):
A 552 n miles (1,022 km; 635 miles)
B 595 n miles (1,101 km; 684 miles)
with 340 kg (750 lb) payload:
A and B 1,200 n miles (2,222 km; 1,380 miles)

UPDATED

United States

AAI

AMERICAN AUTOGYRO INC

3000 South Palo Verde Road, Buckeye, Arizona 85326
Tel: (+1 623) 393 94 51
Fax: (+1 623) 393 97 02
e-mail: sales@americanautogyro.com
Web: http://www.americanautogyro.com
PRESIDENT: James P Mayfield III

The company was formed in December 2002, as a division of Groen Brothers Aviation, designer and builder of larger autogyros, but was merged back into Groen Brothers Aviation on 1 November 2004. AAI's initial objective has been to develop stability augmentation and other safety features for existing autogyros; in February 2003, it began design of its own Sparrowhawk autogyro.

UPDATED

AAI SPARROWHAWK

TYPE: Two-seat autogyro/kitbuilt.

PROGRAMME: Following incremental development on several testbeds, launched at AirVenture, Oshkosh, July 2003, when proof-of-concept vehicle N974J (flown February 2003) was displayed. Kit deliveries then scheduled to begin in late 2003, although stability augmentation kits for gyrocopters already flying were available from April 2003. Factory-complete aircraft also available; law enforcement/aerial patrol conversion (EJ22 engine) first flew in third quarter of 2003.

AAI-built production (N4132R) prototype first flown 25 February 2004. By April 2004, over 1,000 flight hours had been accumulated by the various prototypes; first customer-built kit first flown (in Russia) 23 September 2004.

CUSTOMERS: Total of 43 sales by April 2004. Dealers appointed in the Bahamas, Canada, Costa Rica, Ecuador, India, New Zealand, South Africa and South Korea by September 2004.

COSTS: Experimental category kit US$27,390; Light Sport Aircraft quick-assembly kit US$40,700 or US$46,200 flyaway; law enforcement version US$62,384, assembled and equipped (all 2004). Operating cost US$20 per hour (2003).

DESIGN FEATURES: Development aircraft based on RAF 2000 pod, but modified substantially with large all-moving rudder and fixed cruciform horizontal stabiliser with endplates. Rudder/stabiliser intersection in line with propeller boss for optimum dynamic stability, assisted by rudder self-centring springs.

SparrowHawk is unconnected with Canadian RAF company.

FLYING CONTROLS: Dual control sticks for roll and cyclic rotor control; rudder pedals also used for ground steering, in conjunction with differential braking.

STRUCTURE: Aluminium frame with glass fibre pod and honeycomb composites empennage; aluminium keel and mast; tubular steel landing gear; bonded metal rotor blades; carbon composites propeller.

LANDING GEAR: Tricycle type with steerable nosewheel and safety tailwheel; fixed. Cable-operated brakes. All wheels 6.00-6.

POWER PLANT: One 110 kW (148 hp) Subaru EJ22 flat-four driving a two-blade, fixed-pitch, carbon fibrepusher propeller and rotor prerotate drive; dual electronic ignition systems. Fuel capacity 95 litres (25.0 US gallons; 20.8 Imp gallons) of which 87 litres (23.0 US gallons; 19.2 Imp gallons) are usable. Oil capacity 3.3 litres (0.87 US gallons; 0.72 Imp gallons).

AVIONICS: Bendix/King or Micro-Air equipment to customer's choice.

INSTRUMENTATION: AMPtronics Skydat LCD instrument displays for all flight, rotor and engine functions.

DIMENSIONS, EXTERNAL:	
Rotor diameter	9.14 m (30 ft 0 in)
Rotor blade chord	0.22 m (8½ in)
Fuselage: Length	4.14 m (13 ft 7 in)
Max width (over landing gear)	1.83 m (6 ft 0 in)
Height: overall	2.97 m (9 ft 9 in)
to top of fuselage	1.90 m (6 ft 3 in)
Wheel track	1.68 m (5 ft 6 in)
Wheelbase	1.46 m (4 ft 9½ in)
Doors (each side): Width	0.91 m (3 ft 0 in)
Height to sill	0.84 m (2 ft 9 in)
DIMENSIONS, INTERNAL:	
Cabin: Length	1.55 m (5 ft 1 in)
Max height	1.35 m (4 ft 5 in)
Max width	1.13 m (3 ft 8½ in)

American Autogyro SparrowHawk N4132R, officially registered as an RAF 2000 in which form it was built in 1996 *(Paul Jackson)* ***NEW***/0589418

Instrument console of SparrowHawk N4132R *(Paul Jackson)* ***NEW***/0589447

AREAS:	
Rotor disc	65.64 m² (706.5 sq ft)
Vertical tail	1.07 m² (11.50 sq ft)
Horizontal tail	1.02 m² (11.00 sq ft)
WEIGHTS AND LOADINGS:	
Weight empty	386 kg (850 lb)
Max T-O weight	612 kg (1,350 lb)
PERFORMANCE:	
Never-exceed speed (VNE)	100 kt (185 km/h; 115 mph)
Max cruising speed	70 kt (130 km/h; 81 mph)
Max rate of climb at S/L	152 m (500 ft)/min)
Service ceiling	3,050 m (10,000 ft)
T-O run	92–153 m (300–500 ft)
T-O to 15 m (50 ft)	305 m (1,000 ft)
Landing from 15 m (50 ft)	92 m (300 ft)
Range at 75% power	245 n miles (453 km; 281 miles)

UPDATED

ABC

AMERICAN BLIMP CORPORATION

1900 North-East 25th Avenue, Suite 5, Hillsboro, Oregon 97124
Tel: (+1 503) 693 16 11
Fax: (+1 503) 681 09 06
Web: http://www.americanblimp.com
PRESIDENT AND CEO: James Thiele
EXECUTIVE VICE-PRESIDENT: Charles Ehrler
CHIEF ENGINEER: Rudy Bartel

AIRBORNE SURVEILLANCE GROUP:
302 Ritchie Highway, Severna Park, Maryland 21146-1910
Tel: (+1 410) 544 65 07 and 544 65 08
Fax: (+1 410) 544 65 09
e-mail: Jud4Blimps@aol.com
VICE-PRESIDENT, MARKETING: E Judson Brandreth Jr

OPERATING COMPANY:
The Lightship Group, 5728 Major Boulevard, Suite 314, Orlando, Florida 32819
Tel: (+1 407) 363 77 77
Fax: (+1 407) 363 09 62
e-mail: info@Lightships.com
Web: http://www.lightships.com

In 1995, ABC announced the design of a series of airships larger than the A-60, designated A-1-50 (marketing designation A-150). These are a refinement of the earlier A-120 conceptual design with a new technology envelope. The civil version incorporates the ABC internal illumination system for advertising and is designated a Lightship; the non-illuminating surveillance version is designated SPECTOR. Lightship envelopes remain translucent for internal illumination, whereas SPECTOR envelopes are opaque. Each is offered in three sizes to optimise costs with mission requirements. Other structural elements are identical.

The same envelope technology was introduced as an option for production on existing A-60+ Lightships from 1996 (first contract, for seven envelopes to be supplied by ILC Dover, announced 18 March 1996).

On 23 July 2002, ABC announced that it had acquired the Virgin Lightships interests in the former's affiliate The Lightship Group (TLG), following Virgin's decision to exit the commercial airship market. TLG is now a wholly owned ABC subsidiary, and operates 16 ABC airships around the world.

Company is currently developing airship designs for medium-altitude (2,000 to 3,000 m; 6,560 to 9,840 ft) operation with military and commercial payloads for surveillance and communications relay. It has also, on behalf of the Jet Propulsion Laboratory, been studying the use of a helium airship as a possible vehicle for unmanned exploration of Titan, the largest of Saturn's moons. Named Aerover, the projected airship would be approximately 10 m (33 ft) long, 2.44 m (8 ft) in diameter and weigh about 100 kg (220 lb).

In 1999, ABC increased its manufacturing area by 30 per cent with a 410 m² (4,400 sq ft) extension at Hillsboro, which now covers almost 2,325 m² (25,000 sq ft). At the same time, a

280 m² (3,000 sq ft) fabrication shop was being readied at the Tillamook flight test centre. In August 2003, ABC's combined fleet of 17 A-60+/A-1-50 airships achieved a cumulative total of 150,000 flight hours.

On 21 November 2000, ABC purchased the type certificate (originally issued on 17 November 1992) for the Colt GA 42 non-rigid helium airship. It is believed that three airships of this type were included in the transfer.

UPDATED

ABC LIGHTSHIP A-60+

TYPE: Helium non-rigid airship.

PROGRAMME: A-50 prototype N5132A registered 31 March 1988 and made first flight 9 April 1988; first flight of production A-60 June 1989; A-60 certified by FAA 18 May 1990 for day/night and VFR/IFR flight; first delivery (to Virgin Lightships Ltd) November 1990; A-60+ certified third quarter 1991 by FAA; APU upgrade certified by FAA, and FLIR Systems Inc gyrostabilised IR camera approved by FAA, in 1993; all A-60s converted to A-60+ in 1993. Certified also in Argentina, Australia, Brazil, Canada, Germany, Italy and South Africa; approved for operation in China, Japan and Turkey. Airship 05 reached 10,000 flight hours on 28 October 1999, and two others had achieved this total by February 2000. A Lightship Group two-man British crew in 016/N606LG set a new FAI Class BA absolute world speed record for gas airships on 19 January 2000, reaching 50.11 kt (92.80 km/h; 57.66 mph) over a 1 km course.

One A-60+ loaned by The Lightship Group was used in January 2000 for UK-based flight trials fitted with a DERA (now QinetiQ) Mineseeker ultra-wideband, ground-penetrating radar for detection of undersurface metal and plastics mines. Subsequently, after the addition of a Wescam 16DS-M dual-sensor (daylight TV and FLIR) payload, this airship conducted a six-week programme (4 October to 14 November 2000) from bases in Italy and Kosovo, surveying 30 sites in the latter locality for mines and other unexploded ordnance on behalf of the United Nations Mine Action Co-ordination Centre (UN MACC). As a result of the success of this operation, a Mineseeker Foundation was set up as a joint venture with DERA/QinetiQ which, in April 2001, announced plans to raise £10 million to purchase five similarly equipped airships.

CURRENT VERSIONS: **A-60:** Initial production version. All since converted to A-60+.

A-60+: Current version from 1991; larger envelope; payload increased by 181 kg (400 lb) compared with A-60; Lightsign message board option from 1992. Also offered in **SPECTOR 19** form (which see).

Description applies to A-60+.

CUSTOMERS: Total of 23 registered by October 2004: see table; 018 made its first flight 4 August 2000; 01 badly damaged by storm on 7 September 2001 and may not be repairable.

DESIGN FEATURES: Conventional-shape envelope; cruciform tail surfaces; gondola suspended by 12 catenary cables each attached to external patches, eliminating need for internal cables, gastight fittings and bellow sleeves. Can be inflated without a net by attaching ballasted gondola before adding helium; gondola can be removed from inflated Lightship without a net by ballasting catenary cables.

FLYING CONTROLS: Single internal ballonet; rudder or elevator on each of four tailfins; primary flight controls for left-hand front seat only.

STRUCTURE: Outer envelope skin of Dacron/Mylar with separate urethane film inner gastight bladder and single ballonet. All structural attachments to sewn outer bag, such as nose mooring, fin base and guy wires, car catenary and handling lines, made with webbing reinforcements sewn directly to envelope.

LANDING GEAR: Single twin-wheel unit beneath gondola; small tailwheel at base of lower vertical fin. Operating area diameter required: 229 m (750 ft)

POWER PLANT: Twin 59.7 kW (80 hp) Limbach L 2000 engines, pusher-mounted to enhance propulsive efficiency and reduce noise. Standard capacity of rear-mounted fuel tank is 280 litres (74.0 US gallons; 61.6 Imp gallons); can be refuelled in the field without mooring.

ACCOMMODATION: Two single seats and rear bench seat in gondola (pilot plus three adult passengers or two adults and two small children.) Ground crew of 13.

SYSTEMS: 28 V 70 A electrical system with two 28 V 35 A alternators. APU capable of delivering 2.5 kW at 110 V certified December 1993 and now standard. Used mainly to power internal illumination lights, but can also be used for other airborne equipment.

EQUIPMENT: Two 1 kW bulbs for internal illumination system. Certified for gyrostabilised TV or IR camera mount, complete with microwave downlink for live broadcast; can be operated simultaneously with APU.

DIMENSIONS, EXTERNAL:
Overall: Length ... 39.01 m (128 ft 0 in)
Width ... 10.97 m (36 ft 0 in)
Height ... 13.41 m (44 ft 0 in)
Envelope: Length ... 39.01 m (128 ft 0 in)
Max diameter ... 10.01 m (32 ft 10 in)
Fineness ratio ... 3.9

Lightship A-60+ advertising airship ***NEW***/0558597

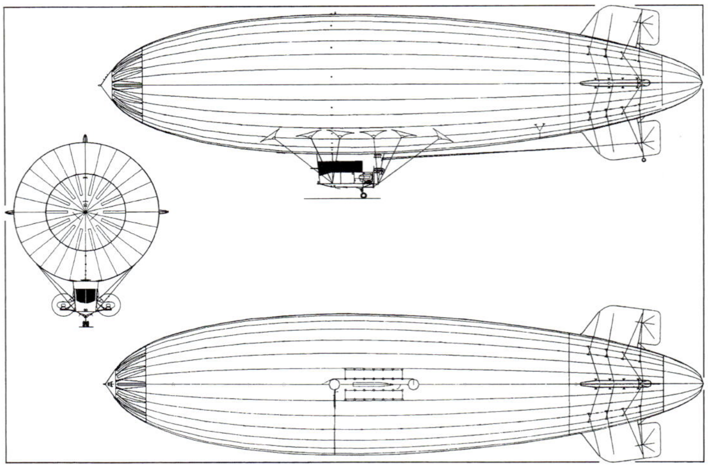

ABC Lightship A-60+ ***(Paul Jackson)*** ***NEW***/0568971

Advertising banner areas:
Each side: Length ... 27.43 m (90 ft 0 in)
Max depth ... 6.10 m (20 ft 0 in)
Underside, forward:
Length ... 6.10 m (20 ft 0 in)
Width ... 7.62 m (25 ft 0 in)
Gondola: Length ... 3.96 m (13 ft 0 in)
Width ... 1.52 m (5 ft 0 in)
Height (incl landing gear) ... 2.90 m (9 ft 6 in)
Propeller diameter ... 1.52 m (5 ft 0 in)

DIMENSIONS, INTERNAL:
Envelope volume ... 1,926 m³ (68,000 cu ft)
Gondola: Cabin length ... 2.68 m (8 ft 9½ in)
Cabin height ... 1.93 m (6 ft 4 in)

AREAS:
Tailfins (four, total) ... 42.74 m² (460.0 sq ft)

WEIGHTS AND LOADINGS:
Total weight empty ... 1,393 kg (3,070 lb)
Max buoyancy ... 1,814 kg (4,000 lb)
Max dynamic lift ... 113 kg (250 lb)
Max useful lift, ISA at 660 m (2,000 ft) ... 680 kg (1,500 lb)
Max gross weight ... 1,993 kg (4,394 lb)

PERFORMANCE:
Max level speed ... 47 kt (88 km/h; 55 mph)
Cruising speed ... 28 kt (51 km/h; 32 mph)
Max rate of ascent ... 488 m (1,600 ft)/min

A-60+ PRODUCTION AND LOCATION
(at October 2004)

c/n[1]	Identity	Owner	Location
01	N2012P *Snoopy 1*	Icarus Aircraft Inc	USA
03	N2017A/I-TIRE *Spirit of Europe* 1/G-TLEL	Lightship Europe	UK
04	N2022B/ZS-ODY/N460LG	Lightship Group	USA
05	N560VL *Skeeter*	Virgin Lightships	USA
06	N660VL *Van Heusen*	Lightship Group	USA
07	N760VL *Snoopy* 2/N760AB	Icarus Aircraft Inc	Out of service (USA)
08	N860AB	Lightship Group	Out of service (Brazil)
09	N960AB	Beijing Orient	Out of service (USA)
010	N3119W/VH-ZIC *Spirit of the South Pacific*/N610LG	Lighship Group	USA
011	N11ZP	Lightship Group	Out of service (USA)
012	N12ZP *Spirit of Europe 2*	Virgin Lightships	Japan
014	N614LG/PT-MKJ *Spirit of the Americas*/N614LG	Lightship America	USA
015	N605LG	Lightship Group	USA
016	N606LG/G-OLEL	Lightship Europe	UK
017	N607LG *Horizon*	Icarus Aircraft Inc	USA
018	N618LG	Lightship Group	USA
019	N619LG	ABC	USA
020	N620LG	ABC	USA
021–023	N621LG-N623LG[2]	ABC	USA

[1] 02 and 013 not built
[2] Reservations only; not completed by mid-2004

Service ceiling .... 2,225 m (7,300 ft)
Max rate of descent .... 396 m (1,300 ft)/min
Min T-O distance .... 112 m (366 ft)
Max range at 28 kt (51 km/h; 32 mph) .... 425 n miles (787 km; 489 miles)
Max endurance at 28 kt (51 km/h; 32 mph) .... 15 h

UPDATED

ABC LIGHTSHIP/SPECTOR SERIES

TYPE: Helium non-rigid airships.

PROGRAMME: Three envelope sizes offered for both civil and surveillance versions; SPECTOR name applies to surveillance versions. First to be built, a Lightship A-1-50 (N5132A, c/n 001), made its first flight on 8 January 1997. FAA type certificate issued 3 October 1997, at which time a total of 683 hours had been flown; 10,000-hour milestone passed in early 2004. Marketing designations are A-150 and A-170; designation on Type Certificate and registration documents of current version is A-1-50.

On 13/14 September 2004, an A-150 piloted by Carl Harbuck and Douglas McFadden established (subject to FAI confirmation) a new world class duration record of 24 hours 39 minutes 55 seconds, exceeding the previous record by more than 10 hours. Some 125 litres (33.0 US gallons; 27.5 Imp gallons) of fuel - enough for more than three additional hours' flying - remained at the end of the flight.

CURRENT VERSIONS: **Lightship A-60+/SPECTOR 19:** See A-60+ entry.

Lightship A-1-50/SPECTOR 42: Initial production version; *as described.*

Lightship A-170/SPECTOR 48. One completed and two others nearing completion by early November 2004. First example took part in late September and early October 2004, in a week-long joint ABC/ARINC demonstration on behalf of the Joint Land Attack Cruise Missile Defense Elevated Netted Sensor (JLENS) Project Office that included a 24-hour endurance flight over the Pentagon and support of a joint force protection mission. Equipment on board included Rapid Aerostat Initial Deployment System (RAIDS) surveillance cameras and sensors of the type used in support of troops in Afghanistan and Iraq. ARINC has also proposed an unmanned version for a similar role.

CUSTOMERS: Seven A-1-50s and one A-170 completed by October 2004; see table.

COSTS: SPECTOR 42 approximately US$3.5 million (2002).

DESIGN FEATURES: As described for A-60+; larger gondola; also seat tracks for quick reconfiguration of cabin water ballast trim system.

FLYING CONTROLS: Standard as for A-60+; dual controls and autopilot optional. Trimming with ballonet accomplished by transferring water to the nosecone tank from the main tank in the rear portion of the gondola. Certified for day/night, VFR/IFR, single-pilot operation.

STRUCTURE: Single-walled ILC Dover envelope with single (26 per cent of volume) ballonet; Tedlar outer film plus inner ripstop fabric/helium barrier film laminated material with heat-sealed seams. Envelopes translucent on Lightship versions, opaque for SPECTOR; those for SPECTOR series have carbon black in the material for ease of maintenance. Control surfaces and fins of fabric-covered aluminium; gondola of aluminium and fabric-covered steel tubing.

LANDING GEAR: Non-retractable tricycle type. Spring-damped main units each with single wheel and tyre; slightly raised nose unit with twin wheels/tyres; small tailwheel at base of lower vertical fin. Required operating area diameter: 305 m (1,000 ft).

POWER PLANT: *SPECTOR 42 and 48:* Two 134 kW (180 hp) Textron Lycoming IO-360-B1G6 flat-four engines, each driving an MTV-25-B-C five-blade constant-speed, reversible-pitch tractor propeller. Rear-mounted fuel tank, standard capacity 560 litres (148 US gallons; 123 Imp gallons). Airship can be refuelled in the field without mooring.

ACCOMMODATION: Single seats for pilot and co-pilot (optional) or passenger on flight deck; five single seats and a three-person bench seat in main cabin. Lavatory and galley optional. Door on port side, aft of pilot station; emergency exit opposite on starboard side. Ground crew of 10 to 12.

SYSTEMS: Electrical system is 28 V DC, with two 90 A alternators; additional 2.5 kW/110 V at 60 Hz available from removable APU.

AVIONICS: Primary suite from Bendix/King Silver Crown range, including:

Comms: KY 196 com, KT 76A transponder and KMA 24H audio panel.

Flight: KX 155 nav/com and Apollo 820 GPS receiver.

EQUIPMENT: Two 1 kW bulbs for internal illumination system. Gyrostabilised camera mount, as in A-60+. Gondola hardpoint provisions for external installation of a variety of sensor equipment. Standard seat tracks in main cabin facilitate sensor/electronics rack installations.

DIMENSIONS, EXTERNAL (A: A-1-50/SPECTOR 42, B: A-170/SPECTOR 48):
Overall: Length: A .... 50.29 m (165 ft 0 in)
B .... 53.64 m (176 ft 0 in)
Width: A, B .... 14.02 m (46 ft 0 in)
Height: A, B .... 16.76 m (55 ft 0 in)
Envelope: Length: A .... 49.38 m (162 ft 0 in)
B .... 52.73 m (173 ft 0 in)
Max diameter: A, B .... 13.11 m (43 ft 0 in)
Fineness ratio: A 3.8
B 4.0
Advertising banner areas:
Each side: Length .... 35.97 m (118 ft 0 in)
Max depth .... 8.53 m (28 ft 0 in)
Underside, forward:
Length .... 8.53 m (28 ft 0 in)
Width .... 9.75 m (32 ft 0 in)
Gondola: Length: A, B .... 8.08 m (26 ft 6 in)

ABC Lightship A-1-50 with 10-place gondola ***NEW***/1048026

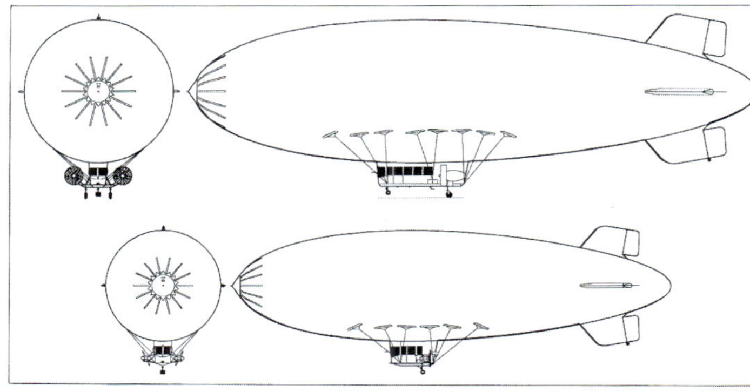

Comparative views, to scale, of the Lightship A-1-50 (top) and A-60+ ***(Paul Jackson)*** 0507571

Max width: A, B .... 6.25 m (20 ft 6 in)
Max height: A, B .... 3.35 m (11 ft 0 in)
Propeller diameter: A, B .... 1.68 m (5 ft 6 in)

DIMENSIONS, INTERNAL:
Envelope volume: A .... 4,248 m^3 (150,000 cu ft)
B .... 4,814 m^3 (170,000 cu ft)
Ballonet volume: A .... 1,104 m^3 (39,000 cu ft)
B .... 1,257 m^3 (44,387 cu ft)
Gondola:
Cabin overall length:
A, B .... 4.79 m (15 ft 8½ in)
Main cabin length: A, B .... 4.27 m (14 ft 0 in)
Average cabin width: A, B .... 1.52 m (5 ft 0 in)
Cabin height: A, B .... 1.93 m (6 ft 4 in)

WEIGHTS AND LOADINGS:
Total weight empty: A .... 2,860 kg (6,305 lb)
B .... 2,897 kg (6,387 lb)
Max payload: A .... 663 kg (1,461 lb)
B .... 993 kg (2,189 lb)
Max buoyancy: A .... 4,055 kg (8,939 lb)
B .... 4,595 kg (10,131 lb)
Max dynamic lift: A, B .... 340 kg (750 lb)
Max useful lift at 656 m (2,000 ft), ISA:
A .... 1,633 kg (3,600 lb)
B .... 2,253 kg (4,967 lb)
Max T-O weight: A .... 4,394 kg (9,689 lb)
B .... 4,981 kg (10,981 lb)

PERFORMANCE (B: estimated):
Max level speed: A .... 52 kt (96 km/h; 59 mph)
B .... 50 kt (92 km/h; 57 mph)
Cruising speed: A .... 40 kt (74 km/h; 46 mph)
Max single-engine speed: A .... 33 kt (61 km/h; 38 mph)
B .... 31 kt (57 km/h; 35 mph)
Max rate of ascent: A, B .... 488 m (1,600 ft)/min
Max rate of descent: A .... 488 m (1,600 ft)
B .... 427 m (1,400 ft)
Service ceiling: A, B .... 3,050 m (10,000 ft)
Max rate of descent: A .... 488 m (1,600 ft)/min
Max range at 35 kt (65 km/h; 40 mph):
A .... 534 n miles (990 km; 615 miles)
B .... 437 n miles (811 km; 503 miles)
Max endurance at 35 kt (65 km/h; 40 mph):
A, B .... approx 15 h

UPDATED

A-1-50 PRODUCTION AND LOCATION
(at October 2004)

c/n	Identity	Owner	Location
101	N151AB	Lightship Group	USA
102	N152LG/N151LG *Bud I*/PR-ANA	Lightship Group	Brazil
103	N153LG/TC-AVK	KOC Holdings	Out of service
104	N154ZP	(destroyed 22 May 2004)	
105	N155LG *Trumpasaurus*	Lightship Group	USA
106	N156LG	Virgin Lightships	USA
107	N157LG	ABC	USA

ADAM

ADAM AIRCRAFT INDUSTRIES

12876 East Jamison Circle, Englewood, Colorado 80112
Tel: (+1 303) 406 59 00 and (+1 866) 232 62 47
Fax: (+1 303) 406 59 50
e-mail: info@adamaircraft.com
Web: http://www.adamaircraft.com

CEO: George F 'Rick' Adam
COO: Cecil Miller
PRESIDENT: John C Knudsen
EXECUTIVE VICE-PRESIDENT: Chris Finnoff
VICE-PRESIDENTS:
Richard Boone (Manufacturing and Test Engineering)
John Hamilton (Marketing)
Matthew Kull (Operations)
Bill Mermelstein (Propulsion)
Dennis Olcott (Design Engineering)
Mike Smith (Finance)
Tom Wiesner (Flight Operations)
Joe Wilding (Advanced Development)

Adam Aircraft Industries formed in 1998 and is developing the A500 centreline-thrust twin-engine business aircraft and the A700 twin-engined business jet derivative, which it will manufacture in a new 2,975 m^2 (32,000 sq ft) facility at Denver Centennial Airport, Colorado. By mid-2000 the workforce numbered 12 but had risen to 72 by October 2001 and 132 by July 2002.

Having designed its own seats for the A500, Adam has begun an auxiliary business as a seat producer to the aerospace industry.

VERIFIED

Third prototype Adam A500 ***NEW***/0589399

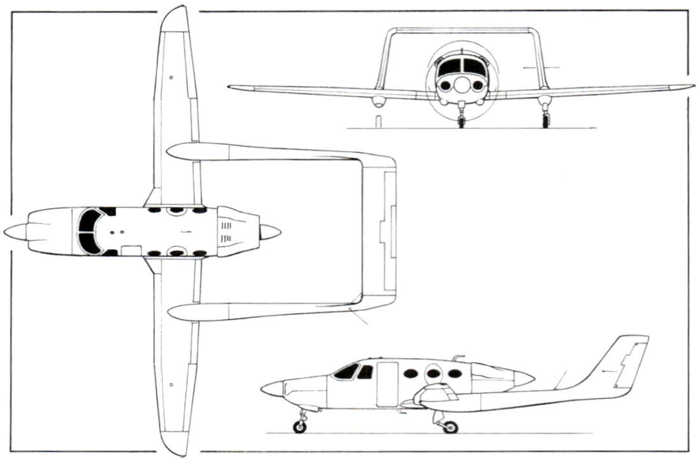

Adam A500 business twin-prop prior to addition of winglets *(Michael Badrocke)* 0536696

ADAM A500

TYPE: Business twin-prop.

PROGRAMME: Designed by Burt Rutan, with designation M-309 denoting his 309th aircraft design; development started September 1999; proof-of-concept aircraft (N309A), manufactured by Scaled Composites Inc, first flown at Mojave, California, 21 March 2000; formally rolled out 5 April 2000; total of 80 flight hours completed by the time of public debut at EAA AirVenture 2000 at Oshkosh in July 2000 and 155 hours by EAA AirVenture 2001; second aircraft, designated A500, and in proposed production configuration, first flew (N500AX) 11 July 2002 at Centennial Airport, Denver. First production fuselage, completed January 2002, used for promotional display; three conforming production aircraft participating in FAA FAR Pt 23 IFR certification, using FAA's Certification Process Improvement programme including flight into known icing; no carry-forward certification from M-309 trials, and this aircraft had been withdrawn from use by mid-2002. Second A500 prototype (N501AX, first with winglets,) flew 18 February 2003; third prototype (N502AX) flew 12 December 2003; FAA type inspection authorisation received, 30 September 2004; sixth airframe, first destined for customer, rolled out 10 October 2004. Certification expected in November or December 2004 in USA with delivery of two aircraft scheduled before the end of the year and 40 in 2005.

CURRENT VERSIONS: **M-309:** Original proof-of-concept model, described in 2002–03 *Jane's*.

A500: Production version with changes to wing span, length, flying controls and door positioning; *as described.*

A family of aircraft, ultimately stretched to seat 19 passengers, is planned.

CUSTOMERS: Order backlog representing 24 months' production held by June 2004.

COSTS: Unit cost US$1,065,000 (both 2004).

DESIGN FEATURES: Twin-boom configuration with swept fins and high-set tailplane; unswept low wings have dihedral on outboard panels and small winglets. Tractor/pusher power plant simplifies handling with single engine failure.

FLYING CONTROLS: Conventional and manual. Flaperons of M-309 replaced by separate ailerons and Fowler flaps. Trim tab in each aileron (flight adjustable). Full-width elevator with tab and external mass balance. Twin rudders have no tabs. No flaps on proof-of-concept aircraft; production aircraft will have separate flaps.

STRUCTURE: Single-cure graphite carbon composite with no secondary bonds or fasteners. Proof-of-concept aircraft has three-spar main wing spar, but production aircraft will have a single-piece spar, with wing mounted lower than on POC aircraft to minimise spar carry-through in cabin. Wing attached by four bolts to fuselage.

Adam A500 interior ***NEW***/0589402

LANDING GEAR: Retractable tricycle type, with single wheel on each unit; trailing link suspension on main units. All wheels retract rearwards; mainwheels 6.00–6, nosewheel 15×6.0–6. Hydraulic brakes.

POWER PLANT: Two 261 kW (350 hp) Teledyne Continental TSIO-550-E flat-six piston engines, with optional TCM/Aerosance FADEC, mounted in centreline thrust configuration and driving a three-blade Hartzell FC7663D-2R scimitar-bladed propeller at the front and a three-blade Hartzell FLC7663DF-2RX at rear. Fuel capacity 946 litres (250 US gallons; 208 Imp gallons) in outer wing tanks of which 871 litres (230 US gallons; 192 Imp gallons) are usable.

ACCOMMODATION: Pilot and five passengers in pressurised cabin on three pairs of leather seats. Door with integral airstair on port side, ahead of wing leading edge; emergency window on each side. Integrated ventral baggage pod optional.

SYSTEMS: Pressurisation system maintains a 2,440 m (8,000 ft) cabin environment at 7,620 m (25,000 ft). Vapour-cycle air conditioning system. Optional TKS liquid de-icing system for wing and tailplane leading-edges; windscreen and propeller have electrically heated de-icing.

AVIONICS: *Comms:* Dual Garmin GNS 430 GPS/NAV/COM/ILS, Garmin GMA 340 audio panel and Garmin GTX 327 transponder, all standard; GTX 330 transponder, GDL 49 WX Datalink transceiver, and Aircell AST 3500 satcom with wireless handset, optional.

Radar: Weather radar in wing-mounted pod optional.

Flight: S-Tec System FiftyFive X autopilot standard; L-3 Comm WX-500 Stormscope and Skywatch, and Ryan 99009 BX TCAD, optional.

Instrumentation: Avidyne FlightMax Entegra system comprising two 26 cm (10.4 in) diagonal, high-resolution, sunlight-readable screens, one each for MFD and PFD functions, each incorporating ADAHRS, EADI and EHSI functions; dual Vision Microsystems VM1000 EMS.

DIMENSIONS, EXTERNAL:
Wing span .... 13.41 m (44 ft 0 in)
Wing aspect ratio 10.9
Length overall .... 11.18 m (36 ft 8 in)
Height overall .... 2.90 m (9 ft 6 in)
Wheel track .... 3.57 m (11 ft 8½ in)
Wheelbase .... 3.08 m (10 ft 1¼ in)
Propeller diameter .... 1.93 m (6 ft 4 in)

DIMENSIONS, INTERNAL:
Cabin: Length .... 4.15 m (13 ft 7¼ in)
Max width .... 1.37 m (4 ft 6 in)
Max height .... 1.29 m (4 ft 3 in)
Volume .... 7.4 m^3 (262 cu ft)

AREAS:
Wings, gross .... 15.33 m^2 (165.0 sq ft)

WEIGHTS AND LOADINGS:
Weight empty .... 1,533 kg (3,380 lb)
Max fuel .... 499 kg (1,100 lb)
Max T-O weight .... 2,857 kg (6,300 lb)
Max wing loading .... 186.4 kg/m^2 (38.18 lb/sq ft)
Max power loading .... 5.48 kg/kW (9.00 lb/hp)

PERFORMANCE:
Max level speed at FL220 .... 250 kt (463 km/h; 288 mph)
Cruising speed at 75% power at FL220 .... 230 kt (426 km/h; 265 mph)
Econ cruising speed at 60% power at FL220 .... 200 kt (370 km/h; 230 mph)
Stalling speed .... 70 kt (130 km/h; 81 mph)
Max rate of climb at S/L .... 549 m (1,800 ft)/min
Rate of climb at S/L, OEI .... 122 m (400 ft)/min
Service ceiling .... 7,620 m (25,000 ft)
Service ceiling, OEI .... 4,570 m (15,000 ft)
Range with max fuel, NBAA IFR reserves:
max cruise power .... 1,050 n miles (1,944 km; 1,208 miles)
economy power .... 1,150 n miles (2,129 km; 1,323 miles)
max range, VFR reserves .... 1,470 n miles (2,722 km; 1,691 miles

UPDATED

Adam A500, cutaway drawing key

1 Hartzell three-blade 'scimitar' constant-speed propeller
2 Spinner
3 Propeller hub pitch change mechanism
4 Engine bay cooling air intakes
5 Two-piece carbon fibre composite (CFC) cowling panels
6 Continental TSIO-550 flat-six turbocharged engine with single-lever FADEC control
7 Exhaust-driven turbocharger
8 Pressurisation air duct
9 Nosewheel leg pivot mounting
10 Aft-retracting nosewheel
11 Torque scissor links
12 Exhaust pipe and fairing
13 Nose landing gear retraction link/breaker strut, hydraulically operated
14 Engine bay baffle plate
15 Twin magnetos
16 Accessory equipment compartment
17 Engine electrical unit
18 Hydraulic brake reservoir
19 Engine mounting subframe
20 Cabin conditioned air delivery duct
21 Forward pressure bulkhead/ engine bay firewall
22 Cabin air distribution duct and vents
23 Rudder pedals
24 Nosewheel housing
25 Engine mounting support structure
26 Windscreen panels
27 Instrument panel shroud
28 Standby compass
29 Instrument panel, Aviadyne Flightmax triple-screen LCD displays
30 Circuit breaker panel
31 Sidestick controller, fully duplicated controls
32 Pilot's sidestick controller armrest
33 Cabin door with integral airstairs
34 Door internal latch
35 Door support cables/handgrips
36 Entry doorway
37 Pilot's seat
38 Safety harness
39 Second pilot or forward passenger seat
40 GPS antennas
41 Stowage locker
42 Starboard wing integral fuel tanks
43 Toray CFC wing skin panelling
44 Fuel filler cap
45 Ventral pitot head
46 Starboard navigation light
47 Winglet
48 Aileron mass balance
49 Starboard aileron
50 Flap operating link, torque shaft driven
51 Outboard ventral flap hinge
52 Starboard outboard flap segment
53 Fuselage CFC structure, skin panels with honeycomb core stiffening
54 Starboard emergency escape window hatch
55 Cabin wall trim linings
56 Plug-in video display, optional entertainment system
57 Aft-facing passenger seats, optional four-seat club layout
58 Underfloor flap drive motor
59 Flap drive torque shaft
60 Wing centre-section carry through spar tunnels
61 Port emergency escape window hatch
62 Rear passenger seats
63 Hand baggage space behind seats
64 Rear engine mounting support structure
65 Cabin rear bulkhead
66 Cabin ceiling trim and lighting panel
67 Upper VHF communications antenna
68 Engine bay firewall
69 Aft engine air intakes
70 Aft Continental TSIO 550 turbocharged engine
71 24 V battery, one per engine bay
72 Aft engine exhaust pipe and fairing
73 CFC engine cowling panels
74 Generator
75 Propeller spinner
76 Aft Hartzell three-blade propeller
77 Starboard CFC tailboom
78 Tailplane control cables
79 Starboard rudder hinge control
80 Elevator cable quadrants
81 Starboard rudder
82 Tailfin; integral structure with tailboom
83 Control cable pulleys
84 Anti-collision beacon
85 Single-piece elevator
86 Elevator trim tab
87 Elevator mass balance
88 Elevator hinge control link
89 CFC horizontal tailplane with honeycomb stiffened skin panels
90 Detachable leading-edge
91 Rudder interconnecting cable
92 Rudder trim tab
93 Port rudder
94 Rudder mass balance
95 Fin CFC spar and rib structure
96 VOR localiser antenna
97 Port rudder hinge control link
98 Port tailboom CFC skin and frame structure
99 ELT antenna
100 Main landing gear hydraulic power unit
101 Mainwheel bay
102 Hydraulic retraction jack
103 Port outboard single-slotted flap segment
104 Flap rib structure
105 Aileron trim tab
106 Port aileron rib structure
107 Aileron mass balance
108 Detachable wingtip fairing
109 Port winglet
110 Port navigation light
111 Aileron hinge control
112 Detachable leading-edge panel
113 Port wing fuel filler
114 Wing bottom skin panel with access hatches
115 Two-spar and rib all-CFC wing torsion box structure
116 Port wing outboard fuel tank
117 Leading-edge aileron control rod
118 Port mainwheel, aft retracting
119 Trailing link mainwheel suspension
120 Shock absorber strut
121 Main landing gear leg pivot mounting
122 Wing panel dry bay
123 Inboard fuel tank bay
124 Port inboard single-slotted flap segment
125 Inboard flap operating link and hinge mounting
126 Fuel collector box and pump
127 Wingroot/tank bay end rib
128 Aileron control rod linkage and fuselage pressure seal
129 Extended chord leading-edge fairing

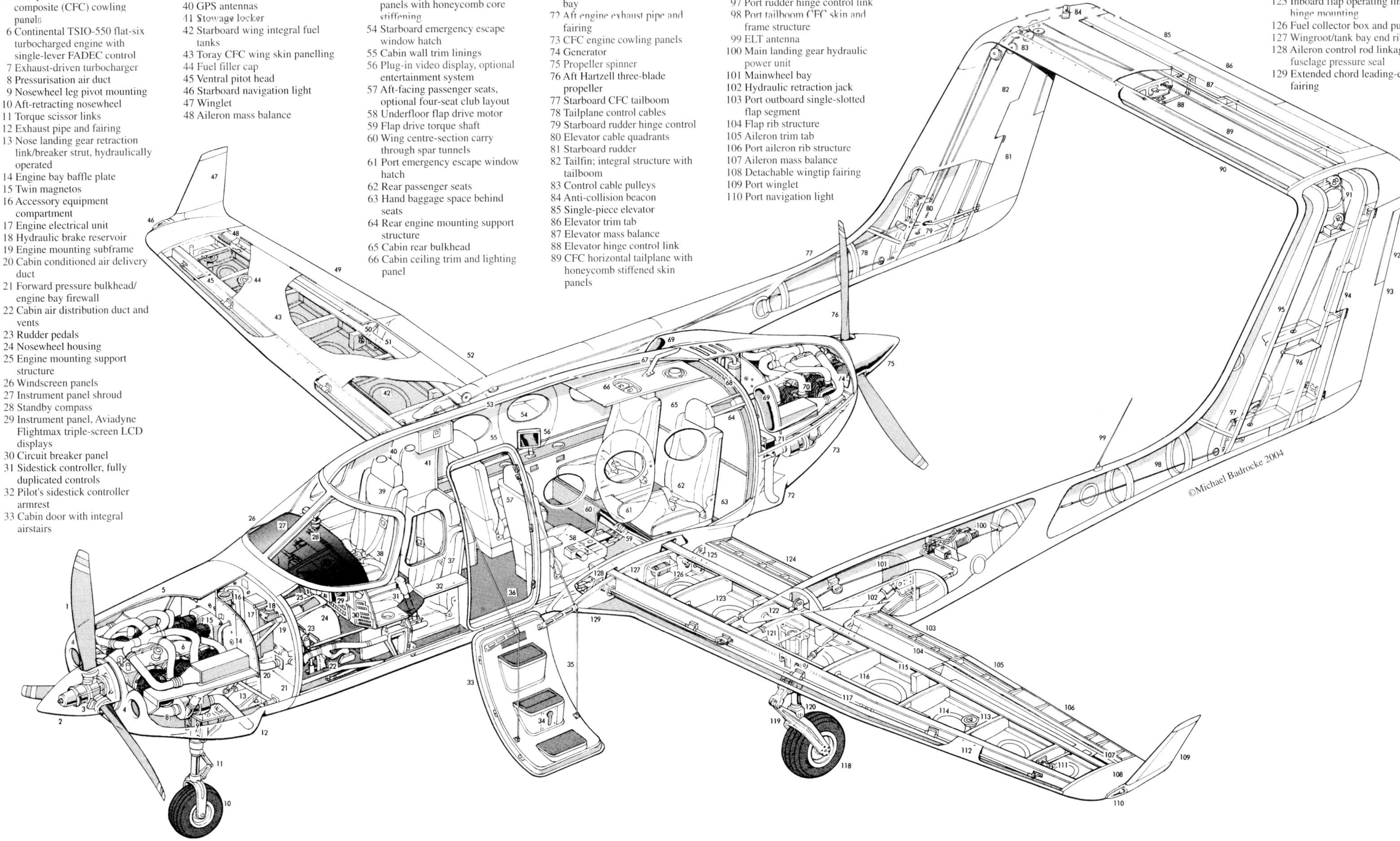

NEW/0581425

Adam A500, cutaway drawing key

ADAM A700

TYPE: Light business jet.

PROGRAMME: Announced on 21 October 2002; first flight of the proof of concept prototype (N700JJ, which simultaneously registered as - but not painted as - N700AJ) 27 July 2003, followed by public debut at EAA AirVenture, Oshkosh, 29 July 2003; FAA certification to FAR/JAR 25 and first customer deliveries in late-2005.

CUSTOMERS: Total of 205 ordered by October 2004, including 75 for launch customer air taxi operator Pogo, 30 for private buyers, and 100 for undisclosed fractional ownership operator.

COSTS: Introductory price US$1,995,000 (2003).

DESIGN FEATURES: Design draws on A500 (which see). Engines pylon-mounted at rear of fuselage.

FLYING CONTROLS: Similar to A500. Twin tabs in elevator and on one rudder. Two-section flaps (inboard and outboard of boom) on each wing.

STRUCTURE: Monocoque fuselage of carbon fibre composites with honeycomb stiffening. Two-spar wing with integral fuel tank.

LANDING GEAR: Generally as A500; mainwheels 6.50–8; nosewheel 6.00–6.

POWER PLANT: Two Williams International FJ33-4A turbofans, each rated at 5.34 kN (1,200 lb st). Fuel capacity (production) 1,249 litres (330 US gallons; 275 Imp gallons).

ACCOMMODATION: One or two pilots in cockpit, plus four passengers in club seating in cabin, separated by divider and refreshment centre. Integral airstair on port side ahead of wing leading edge, with emergency exits on both sides of fuselage at centre of cabin. Lavatory fitted in aft of cabin. Baggage compartment in nose, forward of pressure bulkhead, accessible externally. Second baggage compartment at rear of cabin. Accommodation is pressurised and air conditioned.

SYSTEMS: As A500, but with de-icing protection for wing, horizontal tail, windscreen and engine inlets with known icing certification. Dual 200 Ah starter-generators; dual bus electrical system with separate bus for emergency operation. Pressurisation system, maximum pressure differential 0.57 bars (8.33 lb/sq in), maintains a sea level cabin environment to 3,780 m (12,400 ft) and a 2,440 m (8,000 ft) cabin altitude to 12,500 m (41,000 ft).

AVIONICS: Avidyne Entegra II integrated avionics with three-tube EFIS comprising two PFDs and one MFD.

All data are provisional.

DIMENSIONS, EXTERNAL:
- Wing span ... 13.41 m (44 ft 0 in)
- Length overall ... 12.42 m (40 ft 9 in)
- Height overall ... 2.92 m (9 ft 7 in)

DIMENSIONS, INTERNAL:
- Cabin (between pressure bulkheads):
  - Length ... 4.88 m (16 ft 0 in)
  - Max width ... 1.37 m (4 ft 6 in)
  - Max height ... 1.31 m (4 ft 3½ in)
  - Volume ... 6.9 m³ (245 cu ft)
- Baggage hold, volume: Front ... 0.85 m³ (30.0 cu ft)

WEIGHTS AND LOADINGS:
- Basic operating weight ... 2,204 kg (4,860 lb
- Max fuel ... 1,003 kg (2,211 lb)
- Max T-O weight ... 3,447 kg (7,600 lb)
- Max landing weight ... 3,356 kg (7,400 lb)
- Max ramp weight ... 3,470 kg (7,650 lb)
- Max zero-fuel weight ... 3,175 kg (7,000 lb)
- Payload with max fuel ... 329 kg (725 lb)
- Fuel with max payload ... 295 kg (650 lb)

Adam A700 prototype on an early test flight ***NEW***/0589414

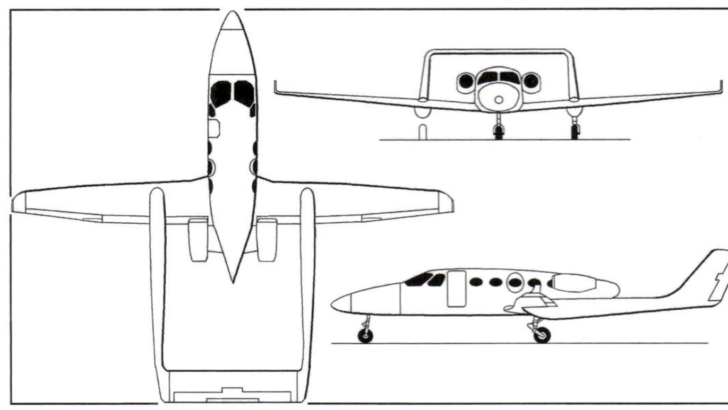

Adam A700 business jet ***(Paul Jackson)*** ***NEW***/0568988

- Max power loading ... 323 kg/kN (3.17 lb/lb st)

PERFORMANCE:
- Max cruising speed at FL350 ... 340 kt (630 km/h; 391 mph) TAS
- Service ceiling ... 12,500 m (41,000 ft)
- T-O run ... 899 m (2,950 ft)
- Service ceiling OEI ... 9,140 m (30,000 ft)
- Max rate of climb, OEI ... 165 m (540 ft)/min
- Range, 45 min reserves:
  - IFR ... 1,100 n miles (2,037 km; 1,265 miles)
  - VFR ... 1,450 n miles (2,685 km; 1,668 miles)

UPDATED

ADI

ADI

5 Harris Court, Building S, Monterey, California 93940
Tel: (+1 831) 649 62 12
Fax: (+1 831) 649 57 38
e-mail: aircraft@mbay.net
Web: http://www.aircraftdesigns.com
PRESIDENT: Martin Hollmann

This company produces kits for the Hollmann-designed Stallion; sells plans for the Hollmann HA-2M Sportster and Hollmann Bumble Bee gyrocopters (see 1992–93 *Jane's*, Private Aircraft section); and is a major contributor to the Lancair series of kitbuilt aircraft (which see), as well as others including Seawind, Pulsar Sport 150, Prowler, Thunder Mustang, Kitfox and Condor. Covered plant area is 560 m² (6,025 sq ft) and full-time employees numbered two in 2003.

UPDATED

ADI SUPER STALLION

TYPE: Six-seat kitbuilt.

PROGRAMME: Construction of prototype began in 1990; first flew (N408S) July 1994.

CURRENT VERSIONS: **Super Stallion**: *As described.*

Turbine Stallion: One aircraft fitted with Walter M601 turboprop by Turbine Design of Deland, Florida.

CUSTOMERS: Four production aircraft registered and 54 kits sold by July 2001 including one to New Zealand; seven flying by early 2002; two on US civil register by July 2004.

COSTS: US$98,225 (2004) excluding engine, instruments and avionics.

DESIGN FEATURES: High-performance tourer for amateur construction. Designed to FAR Pt 23. Extensive use of composites. Employs Lancair IV wings and landing gear. Wing easily removed for storage and transportation. Winglets introduced from 2000. Quoted build time 1,500 hours.

Laminar flow, cantilever wing with 2° washout at tip. Wing section NACA 64-212 at tip, Jacosky RXM5-217 at root. Optional winglets.

FLYING CONTROLS: Conventional and manual. Flight-adjustable tab in port elevator; horn-balanced rudder with ground-adjustable tab. Fowler flaps droop to 30° and are 2.84 m (9 ft 4 in) long. Aileron length 1.78 m (5 ft 10 in).

STRUCTURE: Prebuilt centre-fuselage area of welded 4130 steel tubing; remainder of airframe primarily of graphite/Nomex honeycomb core/epoxy.

LANDING GEAR: Tricycle type; retractable. Hydraulic power is provided by an electric motor. Cleveland 6.00–6 mainwheels and tyres; nosewheel 5.00×5. Hydraulic disc brakes.

Prototype ADI Super Stallion, following retrofit with winglets ***(Paul Jackson)*** 0110737

POWER PLANT: Choice of either 224 kW (300 hp) Teledyne Continental IO-550-G or 261 kW (350 hp) turbocharged Teledyne Continental TSIO-550-B six-cylinder engine driving a McCauley or Hartzell three-blade constant-speed propeller. Fuel capacity 681 litres (180 US gallons; 150 Imp gallons) in two wing tanks. Other engine options include 268 kW (360 hp) Textron Lycoming TIO-540-AE2A, 560 kW (751 shp) Walter M 601E and reconditioned 410 kW (550 shp) PT6A-20.

ACCOMMODATION: Two pilots and up to four passengers in three pairs of seats. Two rear pairs can be swivelled to provide club seating arrangement. Alternatively, centre row of seats can be quickly removed for carriage of cargo. Large (two-part, horizontally split) door on port side. Removable 1.88 m (6 ft 2 in) × 0.91 m (3 ft 0 in) panel on starboard side for loading. Baggage shelf.

AVIONICS: To customer's specification.

DIMENSIONS, EXTERNAL:
- Wing span without winglets ... 10.67 m (35 ft 0 in)
- Wing chord: at root ... 1.52 m (5 ft 0 in)
  - at tip ... 0.91 m (3 ft 0 in)
- Wing aspect ratio ... 8.8
- Length overall ... 7.62 m (25 ft 0 in)
- Max width of fuselage ... 1.27 m (4 ft 2 in)
- Height overall ... 2.90 m (9 ft 6 in)
- Tailplane span ... 4.17 m (13 ft 8 in)
- Tailplane chord: at root ... 0.91 m (3 ft 0 in)
  - at tip ... 0.56 m (1 ft 10 in)
- Wheelbase ... 2.34 m (7 ft 8 in)
- Wheel track ... 2.08 m (6 ft 10 in)

DIMENSIONS, INTERNAL:
Cabin: Length 3.30 m (10 ft 10 in)
Max width 1.24 m (4 ft 1 in)
AREAS:
Wings, gross 13.00 m² (140.0 sq ft)
Vertical tail surfaces (total) 1.78 m² (19.20 sq ft)
Tailplane 3.07 m² (33.00 sq ft)
WEIGHTS AND LOADINGS:
Weight empty 998 kg (2,200 lb)
Max payload 499 kg (1,100 lb)
Max T-O weight 1,724 kg (3,800 lb)
Max wing loading 132.5 kg/m² (27.14 lb/sq ft)
Max power loading: 300 hp 7.71 kg/kW (12.67 lb/hp)
350 hp 6.61 kg/kW (10.86 lb/hp)

PERFORMANCE:
Max operating speed (VMO):
300 hp 220 kt (407 km/h; 253 mph)
350 hp 266 kt (493 km/h; 306 mph)
Max cruising speed at 2,500 rpm:
300 hp 200 kt (370 km/h; 230 mph)
350 hp 256 kt (474 km/h; 295 mph)
Stalling speed, power off, flaps down 62 kt (115 km/h; 71 mph)
Max rate of climb at S/L: 300 hp 488 m (1,600 ft)/min
350 hp 792 m (2,600 ft)/min
Service ceiling 9,750 m (32,000 ft)
T-O run 549 m (1,800 ft)
Landing run 213 m (700 ft)
Landing from 15 m (50 ft) 366 m (1,200 ft)
Range at cruising speed, no reserves 2,346 n miles (4,345 km; 2,700 miles)

UPDATED

AEROCOMP

AEROCOMP

800 Kemp Street, Merritt Island, Florida 32952
Tel/Fax: (+1 321) 453 66 41 and 452 71 68
e-mail: info@AerocompInc.com
Web: http://www.AerocompInc.com
OWNERS:
Ron Lueck
Stephen Young

Formed in 1993 and previously known for its production of floats, Aerocomp markets the Comp Air 3 and 4 kitbuilts, together with six-, seven-, eight- and 10-seat versions, known respectively as the Comp Air 6, 7, 8 and 10. Aerocomp acquired production rights for the Merlin GT and E-Z Flyer from Merlin Aircraft Inc, and has passed the latter to Blue Yonder Aviation of Canada (which see).

The company had 22 employees in 2003 in its 1,580 m² (17,000 sq ft) factory. In February 2002, Aerocomp announced that it was working on a jet-powered kitbuilt, the CA-J and its turbine-power comp Air 12 derivative.

By July 2004, more than 500 Aerocomp kits had been sold.

UPDATED

AEROCOMP CA-J COMP AIR JET AND COMP AIR 12

TYPE: Business monojet kit built; business turboprop kitbuilt.

PROGRAMME: Announced 11 February 2002, and fuselage mockup shown at Sun 'n' Fun April 2002, when order book opened. Completion of first prototype was expected in the third quarter of 2003 with initial kit deliveries following during the first quarter of 2004; this subsequently slipped and Comp Air Jet prototype (N41861) made first taxi test runs on 30 June 2004 and 30-minute maiden flight on 20 July 2004, piloted by Ron Lueck.

CURRENT VERSIONS: **Comp Air Jet**: *As described.*

Comp Air 12: Turbine-engined version announced late 2002. Powered by a nose-mounted 746 kW (1,000 shp) Pratt & Whitney PT6.

COSTS: Kit cost US$399,000 including engine; estimated cost to finished standard US$600,000 (both 2004).

DESIGN FEATURES: Low-wing monoplane with wing set towards rear of passenger cabin. Wing leading-edge sweepback approx 2°; trailing-edge sweepforward approx 7°; tailplane similar. Tailplane set at base of fin, above fuselage. Engine inlets mounted at rear of fuselage behind cabin.

Modified NACA 65-0215 laminar flow airfoil.

FLYING CONTROLS: Conventional and manual.

STRUCTURE: Uses same structural technology as other aircraft in the Aerocomp product line, but company suggests builders use professional assistance in kit completion; quoted completion time six to eight months. Extensive use of carbon fibre hybrid sandwich throughout structure.

LANDING GEAR: Tricycle layout. Retractable.

POWER PLANT: One factory remanufactured ZMKB Progress AI-25 turbofan rated at 15.12 kN (3,400 lb st) for T-O and 11.12 kN (2,500 lb) during cruise. Design will also accommodate Pratt & Whitney JT12-8 or CJ610 engines. Fuel capacity 1,817 litres (480 US gallons; 400 Imp gallons).

ACCOMMODATION: Pilot and seven to nine passengers. Main door on port side, immediately behind cockpit.

SYSTEMS: Pressurisation maintains 3,050 m (10,000 ft) cabin altitude at 9,115 m (29,900 ft) service ceiling.

All data are provisional.

AVIONICS: *Comms:* UPS Apollo MX20 MFD; Garmin GTX330 Mode S transponder.

Flight: BFG WX500 Stormscope weather radar; WSI InFlight weather service system.

Instrumentation: Specially designed digital engine gauges from Electronics International.

DIMENSIONS, EXTERNAL:
Wing span 13.87 m (45 ft 6 in)
Length overall 11.43 m (37 ft 6 in)
Height overall 4.42 m (14 ft 6 in)

Artist's impression of Comp Air 12 ***NEW***/0589405

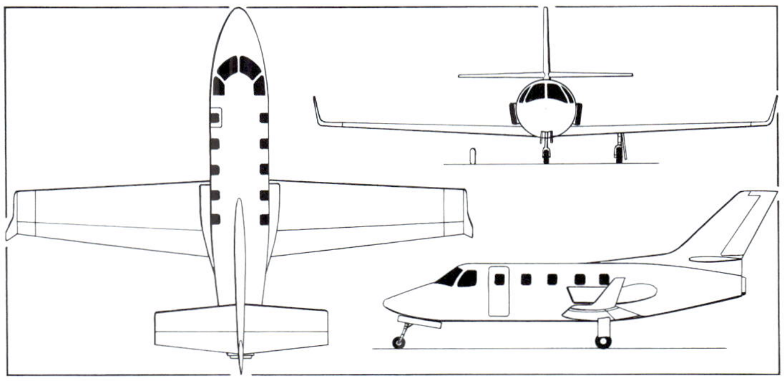

Aerocomp Comp Air Jet *(James Goulding)* ***NEW***/1047689

Aerocomp CA-J Comp Air Jet beginning its third sortie, 22 July 2004 ***NEW***/0589412

DIMENSIONS, INTERNAL:
Cabin max height 1.78 m (5 ft 10 in)
AREAS:
Wings, gross 25.46 m² (274.0 sq ft)
WEIGHTS AND LOADINGS:
Weight empty 2,268 kg (5,000 lb)
Max T-O weight 4,445 kg (9,800 lb)
Max wing loading 174.6 kg/m² (35.77 lb/sq ft)
Max power loading 294 kg/kN (2.88 lb/lb st)
PERFORMANCE (estimated):
Never-exceed speed (VNE) 369 kt (684 km/h; 425 mph)
Normal cruising speed 348 kt (644 km/h; 400 mph) TAS
Stalling speed 70 kt (130 km/h; 81 mph)
Max rate of climb at S/L 1,765 m (5,800 ft)/min
Service ceiling 9,115 m (29,900 ft)
T-O run 610 m (2,000 ft)
Landing run 855 m (2,800 ft)
Range 955 n miles (1,770 km; 1,100 miles)

UPDATED

AEROCOMP COMP AIR 3

TYPE: Three-seat kitbuilt.

PROGRAMME: Announced 1998; smaller version of Comp Air 4.

CUSTOMERS: At least one flying by early 2003.

COSTS: Kit US$27,595 (2003).

Data generally as for Comp Air 4, except as noted.

STRUCTURE: Wings removable for storage. Tapered wings, optional.

POWER PLANT: One 119 kW (160 hp) Textron Lycoming O-320 in prototype; engines in range 112 to 186 kW (150 to 250 hp) recommended. Fuel capacity: standard 170 litres (45.0 US gallons; 37.5 Imp gallons), optional 757 litres (200 US gallons; 167 Imp gallons).

ACCOMMODATION: Pilot and passenger side by side; third occupant, or cargo, in back of cockpit.

DIMENSIONS, EXTERNAL:
Wing span 10.52 m (34 ft 6 in)
Wing aspect ratio 6.6
Length overall 7.32 m (24 ft 0 in)
Max width of fuselage 0.91 m (3 ft 0 in)
Height overall 2.46 m (8 ft 1 in)
DIMENSIONS, INTERNAL:
Cabin max width 1.09 m (3 ft 7 in)

AREAS:
Wings, gross ... 16.35 m² (176.0 sq ft)
WEIGHTS AND LOADINGS:
Weight empty ... 590 kg (1,300 lb)
Max T-O weight ... 1,111 kg (2,450 lb)
Max wing loading ... 68.0 kg/m² (13.92 lb/sq ft)
PERFORMANCE (119 kW; 160 hp Textron Lycoming engine):
Max operating speed ... 152 kt (281 km/h; 175 mph)
Normal cruising speed ... 126 kt (233 km/h; 145 mph)
Stalling speed ... 40 kt (73 km/h; 45 mph)
Max rate of climb at S/L ... 335 m (1,100 ft)/min
Service ceiling ... 4,724 m (15,500 ft)
T-O run ... 107 m (350 ft)
Landing run ... 183 m (600 ft)
Range ... 725 n miles (1,342 km; 834 miles)

UPDATED

AEROCOMP COMP AIR 4

TYPE: Four-seat kitbuilt.
PROGRAMME: First flew 3 April 1995. Public debut, then named Comp Monster, at Sun 'n' Fun 1995, when powered by 82 kW (110 hp) Hirth F 30 four-cylinder two-stroke engine.
CURRENT VERSIONS: **150G**: Powered by 112 kW (150 hp) Textron Lycoming O-320.
180G: 134 kW (180 hp) Textron Lycoming O-360; *as described*.
180SF: Float-equipped version of 180G.
Trainer: 'Two plus two' seat trainer version.
CUSTOMERS: Total of 60 sold and 26 flying by end 2003.
COSTS: US$47,995 with 150 hp, US$46,995 with 180 hp Lycoming; US$26,995 basic kit without engine, propeller and instruments; Trainer kit US$29,995. Fast-build option US$1,995 (all 2003).
DESIGN FEATURES: Easy-to-build composites aircraft; quoted build time 350 to 400 hours. High, braced wing; unswept, and with straight or, optionally, tapered trailing-edge.
Various construction options, including tricycle landing gear, tapered wing, enlarged flaps and 1,451 kg (3,200 lb) MTOW.
Modified Clark Y aerofoil wings with turned-down tips.
FLYING CONTROLS: Conventional and manual. Electric trim. Flaps of two optional sizes.
STRUCTURE: Composites construction with carbon and Kevlar reinforcement. Single-strut braced wings; braced tailplane. Optional tapered wing. Fuselage width upgrades available: 1.17 m (3 ft 10 in) and 1.21 m (3 ft 11½ in).
LANDING GEAR: Fixed; with nosewheel, tailwheel, floats or amphibious gear, last-mentioned using Matco wheels and hydraulic brakes with 5.00–5 in tyres. Non-retracting 0.15 m (6 in) Matco tailwheel acts as water rudder. Standard version has fuselage-mounted spring steel cantilever mainwheel legs with Matco brakes and 6.00-6 wheels; optional tailwheel is 0.20 m (8 in) on cantilever leg.
POWER PLANT: One 134 kW (180 hp) Textron Lycoming O-360-A1A flat-four engine, driving a Sensenich 76 × 57 three-blade metal propeller. Options exist for engines ranging from 82 to 186 kW (110 to 250 hp); prototype employed 82 kW (110 hp) Hirth 95. Fuel capacity 197 litres (52.0 US gallons; 43.3 Imp gallons).
ACCOMMODATION: Pilot and three passengers in two side-by-side pairs. Glass-panelled door on each side of cabin for improved downwards visibility. Optional baggage pod below fuselage.
DIMENSIONS, EXTERNAL:
Wing span ... 10.67 m (35 ft 0 in)
Wing chord (constant version) ... 1.88 m (6 ft 2 in)
Length overall ... 7.92 m (26 ft 0 in)
Height overall ... 2.44 m (8 ft 0 in)
Tailplane span ... 3.05 m (10 ft 0 in)
Doors (each): Height ... 0.76 m (2 ft 6 in)
Width ... 0.91 m (3 ft 0 in)
DIMENSIONS, INTERNAL:
Cabin: Length ... 2.74 m (9 ft 0 in)
Max width ... 1.08 m (3 ft 6½ in)
Max height ... 1.30 m (4 ft 3 in)
AREAS:
Wings, gross ... 19.70 m² (212.0 sq ft)
WEIGHTS AND LOADINGS (G: 180G landplane, SF: 180SF floatplane):
Weight empty: G ... 630 kg (1,390 lb)
SF ... 753 kg (1,660 lb)
Baggage capacity ... 81.6 kg (180 lb)
Max T-O weight: both ... 1,292 kg (2,850 lb)
Max wing loading: both ... 65.6 kg/m² (13.44 lb/sq ft)
Max power loading: both ... 9.64 kg/kW (15.83 lb/hp)
PERFORMANCE:
Max operating speed:
G ... 152 kt (281 km/h; 175 mph)
SF ... 110 kt (204 km/h; 127 mph)
Normal cruising speed at 70% power:
G ... 135 kt (249 km/h; 155 mph)
SF ... 100 kt (185 km/h; 115 mph)
T-O speed ... 36 kt (68 km/h; 42 mph)
Stalling speed: G ... 34 kt (63 km/h; 39 mph)
SF ... 35 kt (65 km/h; 40 mph)
Max rate of climb at S/L: G ... 442 m (1,450 ft)/min
SF ... 366 m (1,200 ft)/min

Aerocomp Comp Air 4 four-seat kitbuilt *(Paul Jackson)* 0525786

Service ceiling ... 4,880 m (16,000 ft)
T-O run: G ... 122 m (400 ft)
SF ... 221 m (725 ft)
Landing run ... 168 m (550 ft)
Range ... 660 n miles (1,222 km; 759 miles)

UPDATED

AEROCOMP COMP AIR 6

TYPE: Six-seat kitbuilt.
PROGRAMME: Development of Comp Air 4; first flew January 1996.
CURRENT VERSIONS: **CA6G**: Normal landing gear, *as described*.
CA6SF: Float-equipped version.
CA6AF: Amphibian.
CA6TW: Tapered wing.
CA6TWHG: Tapered wing, high gross (1,451 kg; 3,200 lb MTOW).
CA6SHG: Super high gross (1,632 kg; 3,600 lb MTOW), wide-bodied version.
CUSTOMERS: At least 115 flying by early 2002.
COSTS: Kit US$34,995, without engine (2003).
DESIGN FEATURES: As Comp Air 4. Quoted build time 350 hours.
Data for CA6G generally as Comp Air 4, except those below.
STRUCTURE: Wings removable for storage. Optional wide-body fuselage, enlarged flap and tapered wing upgrades available.
LANDING GEAR: Fixed; with tricycle, tailwheel or float options. Mainwheels 6.00–6 in. Optional spring 7075-T6 aluminium alloy cantilever mainwheel legs in place of composites standard version.
POWER PLANT: Prototype had 164 kW (220 hp) Franklin engine; design compatible with engines in the 164 to 224 kW (220 to 300 hp) class. Automotive engine conversions are also possible. Fuel capacity 310 litres (82.0 US gallons; 68.3 Imp gallons).
ACCOMMODATION: Four adults plus full fuel and baggage; or four adults plus two children, with reduced fuel or baggage.
DIMENSIONS, EXTERNAL:
Wing span ... 10.52 m (34 ft 6 in)
Wing aspect ratio ... 5.6
Length overall ... 7.62 m (25 ft 0 in)
Height overall ... 2.44 m (8 ft 0 in)
DIMENSIONS, INTERNAL:
Cabin: Length ... 3.30 m (10 ft 10 in)
Max width ... 1.08 m (3 ft 6½ in)
Max height ... 1.30 m (4 ft 3 in)
AREAS:
Wings, gross ... 19.70 m² (212.0 sq ft)
WEIGHTS AND LOADINGS:
Weight empty ... 721 kg (1,590 lb)
Max T-O weight ... 1,293 kg (2,850 lb)
Max wing loading ... 65.6 kg/m² (13.44 lb/sq ft)
Max power loading ... 7.88 kg/kW (12.95 lb/hp)
PERFORMANCE (164 kW; 220 hp Franklin engine):
Max operating speed ... 152 kt (281 km/h; 175 mph)
Normal cruising speed ... 143 kt (266 km/h; 165 mph)
Stalling speed ... 42 kt (78 km/h; 48 mph)
Max rate of climb at S/L ... 518 m (1,700 ft)/min
Service ceiling ... 5,480 m (18,000 ft)
T-O run ... 107 m (350 ft)
Landing run ... 168 m (550 ft)
Range ... 695 n miles (1,287 km; 800 miles)

UPDATED

Aerocomp Comp Air 6G six-seat kitbuilt *(Paul Jackson)* ***NEW***/0589330

AEROCOMP COMP AIR 7

TYPE: Utility kitbuilt.
PROGRAMME: Development of Comp Air 6. First shown 1998.
CURRENT VERSIONS: **Comp Air 7P**: Piston-engine version.
Comp Air 7T: Turbine version.
Comp Air 7SL: Stretched version, introduced in 2001, with 30 cm (12 in) fuselage stretch and equal addition to wing chord; narrow window immediately to rear of front doors.
Comp Air 7SL T: Turbine version of 7SL.
CUSTOMERS: At least 80 flying or under construction by early 2002, including 25 turbine versions; by January 2004, 58 Comp Air 7s and 18 Comp Air 7Ts had flown.
COSTS: Comp Air 7 US$49,995; Comp Air 7T US$54,995; Comp Air 7SL US$64,995; all minus engine (2003).
Data generally as for Comp Air 6, except those below.
DESIGN FEATURES: Suited for bush operations. Quoted build time 700 hours.
FLYING CONTROLS: Horn-balanced tail surfaces. Ground-adjustable tab on each aileron and on rudder. Trim tab in starboard elevator.
LANDING GEAR: Tailwheel type; fixed. Fuselage-mounted spring composites cantilever mainwheel legs with optional wheel fairings. Mainwheels 8.00–6. Optional aluminium alloy landing gear.

Aerocomp Comp Air 7SLX *(Paul Jackson)* ***NEW***/0589470

POWER PLANT: *Comp Air 7:* One Textron Lycoming TIO-540 rated between 194 and 261 kW (260 and 350 hp), driving a two-blade metal propeller. Fuel capacity 333 litres (88.0 US gallons; 73.3 Imp gallons); optional extra tank increases capacity to 455 litres (120 US gallons; 100 Imp gallons).

Comp Air 7T and *7SL:* One 490 kW (657 shp) Walter M 601D turboprop driving an Avia three-blade, constant-speed, feathering propeller; optionally a five-blade Avia Hamilton propeller can be used. Fuel capacity 568 litres (150 US gallons; 125 Imp gallons) in 7T, 719 litres (190 US gallons; 158 Imp gallons) in 7SL.

ACCOMMODATION: Pilot and up to six passengers in enclosed cabin. Third (rear passengers') door, starboard side.

DIMENSIONS, EXTERNAL (Comp Air 7 and Comp Air 7T):
Wing span: CA 7P, CA 7T ... 10.67 m (35 ft 0 in)
CA 7SL ... 10.06 m (33 ft 0 in)
Wing aspect ratio: CA 7P, CA 7T ... 6.9
CA 7SL ... 5.7
Length overall: CA 7P ... 8.08 m (26 ft 6 in)
CA 7T ... 8.99 m (29 ft 6 in)
CA 7SL ... 9.30 m (30 ft 6 in)
Height overall: CA 7P, CA 7T ... 2.44 m (8 ft 0 in)
CA 7SL ... 2.74 m (9 ft 0 in)

DIMENSIONS, INTERNAL:
Cabin max width: CA 7P standard ... 1.08 m (3 ft 6½ in)
CA 7P optional; CA 7T and CA 7SL standard ... 1.17 m (3 ft 10 in)
CA 7P and CA 7SL optional ... 1.21 m (3 ft 11½ in)

AREAS:
Wings, gross: CA 7P, CA 7T ... 16.54 m² (178.0 sq ft)
CA 7SL ... 17.74 m² (191.0 sq ft)

WEIGHTS AND LOADINGS:
Weight empty: CA 7P ... 953 kg (2,100 lb)
CA 7T ... 1,157 kg (2,550 lb)
CA 7SL: standard ... 1,179 kg (2,600 lb)
high gross upgrade ... 1,315 kg (2,900 lb)
Max T-O weight: CA 7P ... 1,678 kg (3,700 lb)
CA 7T: standard ... 1,710 kg (3,770 lb)
high gross upgrade ... 2,177 kg (4,800 lb)
CA 7SL: standard ... 2,086 kg (4,600 lb)
high gross upgrade ... 2,358 kg (5,200 lb)
Max landing weight:
CA 7T high gross upgrade ... 2,087 kg (4,600 lb)
Max wing loading:
CA 7P ... 101.5 kg/m² (20.79 lb/sq ft)
CA 7T: standard ... 103.4 kg/m² (21.18 lb/sq ft)
high gross upgrade ... 131.7 kg/m² (26.97 lb/sq ft)
CA 7SL: standard ... 117.6 kg/m² (24.08 lb/sq ft)
high gross upgrade ... 132.9 kg/m² (27.23 lb/sq ft)
Max power loading:
CA 7T: standard ... 3.49 kg/kW (5.74 lb/shp)
high gross upgrade ... 4.45 kg/kW (7.31 lb/shp)
CA 7SL: standard ... 4.26 kg/kW (7.00 lb/shp)
high gross upgrade ... 4.82 kg/kW (7.91 lb/shp)

PERFORMANCE:
Never-exceed speed (VNE):
CA 7P ... 191 kt (354 km/h; 220 mph)
CA 7T ... 223 kt (413 km/h; 257 mph) IAS
CA 7SL ... 206 kt (381 km/h; 237 mph) IAS
Max operating speed:
CA 7P ... 178 kt (330 km/h; 205 mph)
Max cruising speed:
CA 7T ... 239 kt (443 km/h; 275 mph)
CA 7SL ... 217 kt (402 km/h; 250 mph)
Stalling speed: CA 7 ... 46 kt (86 km/h; 53 mph)
CA 7T ... 48 kt (89 km/h; 55 mph)
CA 7SL ... 53 kt (97 km/h; 60 mph)
Max rate of climb at S/L:
CA 7P ... 457 m (1,500 ft)/min
CA 7T ... 1,219 m (4,000 ft)/min
CA 7SL ... 914 m (3,000 ft)/min
Service ceiling: CA 7 ... 7,620 m (25,000 ft)
T-O run: CA 7P ... 145 m (475 ft)
CA 7T ... 122 m (400 ft)
CA 7SL ... 183 m (600 ft)
Landing run: CA 7 ... 244 m (800 ft)
Range:
CA 7P ... 1,100 n miles (2,037 km; 1,265 miles)
CA 7SL ... 900 n miles (1,666 km; 1,035 miles)
g limits: CA 7T ... +6/–4

UPDATED

AEROCOMP COMP AIR 8

TYPE: Utility kitbuilt.

PROGRAMME: Prototype first flew mid-June 1999 and publicly displayed at Oshkosh in July 1999.

CUSTOMERS: 20 flying or under construction by late 2003. All believed to be CA 8T turboprop versions.

COSTS: Kit US$74,995 (2003), excluding engine.

DESIGN FEATURES: Generally similar to Comp Air 7. Quoted build time 800 hours.

FLYING CONTROLS: Mass-balanced controls for high-speed handling.

STRUCTURE: Generally as Comp Air 7, but with strengthened carbon fibre fuselage and tail unit. Tapered wings standard. Fuselage width upgrades available.

LANDING GEAR: Tailwheel type; fixed. Fuselage-mounted spring composites cantilever mainwheel legs with optional wheel fairings. Optional aluminium alloy landing gear. Tricycle gear and floatplane versions also available.

POWER PLANT: One 490 kW (657 shp) Walter M 601D turboprop driving an Avia three-blade, feathering, constant-speed propeller. Fuel capacity 681 litres (180 US gallons; 150 Imp gallons).

ACCOMMODATION: Pilot and five adults in three pairs of bucket seats, plus rear bench seat for two children.

DIMENSIONS, EXTERNAL:
Wing span ... 10.97 m (36 ft 0 in)
Wing aspect ratio ... 5.5
Length overall ... 9.60 m (31 ft 6 in)
Height overall ... 2.46 m (8 ft 1 in)

DIMENSIONS, INTERNAL:
Cabin max width: standard ... 1.17 m (3 ft 10 in)
wide option ... 1.21 m (3 ft 11½ in)

AREAS:
Wings, gross ... 22.02 m² (237.0 sq ft)

WEIGHTS AND LOADINGS:
Weight empty: standard ... 1,315 kg (2,900 lb)
high gross option ... 1,406 kg (3,100 lb)
Max T-O weight: standard ... 2,177 kg (4,800 lb)
intermediate ... 2,359 kg (5,200 lb)
high gross option ... 2,540 kg (5,600 lb)
Max wing loading: standard ... 98.9 kg/m² (20.25 lb/sq ft)
intermediate ... 107.1 kg/m² (21.94 lb/sq ft)
high gross option ... 115.4 kg/m² (23.63 lb/sq ft)
Max power loading: standard ... 4.45 kg/kW (7.31 lb/shp)
intermediate ... 4.82 kg/kW (7.91 lb/shp)
high gross option ... 5.19 kg/kW (8.52 lb/shp)

PERFORMANCE:
Never-exceed speed (VNE) ... 206 kt (381 km/h; 237 mph) IAS
Max cruising speed at FL210 ... 217 kt (402 km/h; 250 mph) IAS
Normal cruising speed ... 182 kt (338 km/h; 210 mph)
Stalling speed, power off, flaps down ... 42 kt (78 km/h; 48 mph)
Max rate of climb at S/L ... 610 m (2,000 ft)/min
Service ceiling ... 7,620 m (25,000 ft)
T-O run ... 122 m (400 ft)
Landing run ... 183 m (600 ft)
Range ... 990 n miles (1,833 km; 1,139 miles)

UPDATED

Aerocomp Comp Air 8T six-plus-two kitbuilt *(Paul Jackson)* ***NEW***/0589376

AEROCOMP COMP AIR 10

TYPE: Utility kitbuilt.
PROGRAMME: Announced 1997; based on Comp Air 6.
CURRENT VERSIONS: **CA10:** Original version, no longer available.
CA10XL T: Turbine version. *As described.*
CUSTOMERS: Eleven flying by early 2004.
COSTS: US$84,995 turbopowered kit version (2003).
Data generally as for Comp Air 6, except that below.
DESIGN FEATURES: Twin outward-canted fins in addition to conventional horizontal tail surfaces. Quoted build time 800 hours. Optional HP (high performance) wing and two-stage increased MTOW options.
FLYING CONTROLS: Conventional and manual. Twin rudders operating in unison. Balance tab in starboard elevator.
STRUCTURE: Extensive use of composites throughout. Single-strut braced two-spar wings. Quoted build time 800 hours.
LANDING GEAR: Tricycle type; fixed. Fuselage-mounted spring 7075-T6 aluminium alloy cantilever mainwheel legs. Mainwheel size 8.00–6 in; nosewheel 6.00–6 in. Float and tailwheel configurations optional.
POWER PLANT: One 490 kW (657 shp) Walter M 601D driving an Avia three-blade, constant-speed propeller. Standard fuel capacity of 455 litres (120 US gallons; 100 Imp gallons); optionally 681 litres (180 US gallons; 150 Imp gallons). Oil capacity 5.7 litres (1.5 US gallons; 1.25 Imp gallons).
ACCOMMODATION: Pilot and up to nine passengers; passenger door, forward, each side; horizontally split, two-piece freight door, starboard, rear side; extra passenger door on port rear side. Optional luggage pannier.

DIMENSIONS, EXTERNAL:
Wing span .... 11.43 m (37 ft 6 in)
Length overall .... 9.45 m (31 ft 0 in)
Height overall .... 2.64 m (8 ft 8 in)
Wheel track .... 2.79 m (9 ft 2 in)

DIMENSIONS, INTERNAL:
Cabin: Length .... 3.96 m (13 ft 0 in)
Max width .... 1.52 m (5 ft 0 in)
Max height .... 1.35 m (4 ft 5 in)

AREAS:
Wings, gross .... 23.60 m^2 (254.0 sq ft)

WEIGHTS AND LOADINGS:
Weight empty:standard .... 1,247 kg (2,750 lb)
high gross .... 1,429 kg (3,150 lb)
Baggage capacity: standard .... 136 kg (300 lb)
Max T-O and landing weight:
standard .... 2,358 kg (5,200 lb)
intermediate .... 2,540 kg (5,600 lb)
high gross .... 2,721 kg (6,000 lb)
Max wing loading: standard .... 100.0 kg/m^2 (20.47 lb/sq ft)
intermediate .... 107.6 kg/m^2 (22.05 lb/sq ft)
high gross .... 115.3 kg/m^2 (23.62 lb/sq ft)
Max power loading: standard .... 4.82 kg/kW (7.91 lb/shp)
intermediate .... 5.19 kg/kW (8.52 lb/shp)
high gross .... 5.56 kg/kW (9.13 lb/shp)

PERFORMANCE:
Max level speed .... 186 kt (346 km/h; 215 mph)
Max cruising speed at 75% power .... 167 kt (309 km/h; 192 mph)
Normal cruising speed at 65% power .... 152 kt (282 km/h; 175 mph)
Stalling speed: flaps up .... 51 kt (95 km/h; 59 mph)
flaps down .... 46 kt (86 km/h; 53 mph)
Max rate of climb at S/L .... 762 m (2,500 ft)/min
Service ceiling .... 7,620 m (25,000 ft)
T-O run .... 152 m (500 ft)
Landing run .... 183 m (600 ft)
Range with max fuel at 65% power .... 795 n miles (1,472 km; 915 miles)
Endurance at 65% power, 60 min reserves .... 3 h 42 min

Comp Air CA10XL T utility kitbuilt *(Paul Jackson)* ***NEW***/0589436

UPDATED

AEROCOMP MERLIN GT

TYPE: Side-by-side kitbuilt.
PROGRAMME: Developed by Macair Aircraft in Canada; later Merlin Aircraft. Rotax 582-powered version first flown January 1985; 912-powered version in January 1993.
CURRENT VERSIONS: Available as floatplane, trainer, agricultural sprayer.
CUSTOMERS: At least 300 flying by mid-2002 (latest data received).
COSTS: Complete kits: Standard (with Rotax 582) US$27,995; Rotax 912-powered version US$35,995; Rotax 912S-powered version US$36,995 (2003).
DESIGN FEATURES: Strut-braced high wing; braced tailplane. No aerodynamic balances or tabs on empennage. Wings fold for storage. Quoted build time 400 hours.
FLYING CONTROLS: Manual. Full span Junkers flaperons; conventional rudder and elevator. Actuation by pushrods and cables.
STRUCTURE: Fuselage of welded 4130 chromoly steel tubing, with wooden floor and glass fibre engine cowling, otherwise fabric-covered. Metal wings with fabric covering.
LANDING GEAR: Tailwheel type; fixed. Two side Vs hinged to lower longerons, plus half-axles sprung under centreline with bungee cord; hydraulic brakes; steerable tailwheel. Full Lotus or Superfloats floats optional.
POWER PLANT: One 47.8 kW (64.1 hp) Rotax 582 engine, driving a two- or three-blade propeller; or a choice of 59.6 kW (79.9 hp) Rotax 912, 73.5 kW (98.6 hp) Rotax 912 ULS, 55.0 kW (73.8 hp) Rotax 618, 74.6 kW (100 hp) Canadian Automotive (CAM) 100, 59.7 kW (80 hp) Jabiru 2000 or 82.0 kW (110 hp) Formula Power Subaru EA81 engines. Fuel capacity 61 litres (16.0 US gallons; 13.3 Imp gallons).
SYSTEMS: 12 V electrical system, with battery.
ACCOMMODATION: Two persons, side by side with baggage stowage behind seats. Upward-opening door each side. Dual controls.

DIMENSIONS, EXTERNAL:
Wing span .... 9.75 m (32 ft 0 in)
Wing chord, constant .... 1.52 m (5 ft 0 in)
Wing aspect ratio 6.0
Length overall .... 6.10 m (20 ft 0 in)
Height overall .... 1.98 m (6 ft 6 in)
Tailplane span .... 2.15 m (7 ft 0½ in)
Wheel track .... 2.21 m (7 ft 3 in)
Wheelbase .... 6.10 m (20 ft 0 in)

DIMENSIONS, INTERNAL:
Cabin max width .... 1.04 m (3 ft 5 in)

AREAS:
Wings, gross .... 14.86 m^2 (160.0 sq ft)

WEIGHTS AND LOADINGS (Rotax 912):
Weight empty .... 279 kg (615 lb)
Max baggage capacity .... 45 kg (100 lb)
Max T-O weight: landplane .... 590 kg (1,300 lb)
floatplane .... 635 kg (1,400 lb)
Max wing loading: landplane .... 39.7 kg/m^2 (8.13 lb/sq ft)
floatplane .... 42.7 kg/m^2 (8.75 lb/sq ft)
Max power loading: landplane .... 9.89 kg/kW (16.25 lb/hp)
floatplane .... 10.65 kg/kW (17.5 lb/hp)

PERFORMANCE (Rotax 912):
Never-exceed speed (VNE) .... 104 kt (193 km/h; 120 mph)
Cruising speed .... 81 kt (150 km/h; 93 mph)
Stalling speed .... 34 kt (63 km/h; 38 mph)
Max rate of climb at S/L .... 455 m (1,500 ft)/min
T-O distance .... 30 m (100 ft)
Landing distance .... 46 m (150 ft)
Max range .... 350 n miles (648 km; 402 miles)

UPDATED

Aerocomp Merlin GT *(Paul Jackson)* 0089492

AEROCOPTER

AEROCOPTER INC

300 Brickstone Square, Andover, Massachusetts 01810-1435
Tel: (+1 978) 749 99 99
Fax: (+1 978) 749 88 88
Web: http://www.aerocopter.com
CHAIRMAN: Rouzbeh Yassini

VICE-PRESIDENTS:
George Syrovy (Engineering)
Siamak Yassini (Business Development)

Founded in August 2000 by Syrovy and Siamak Yassini, AeroCopter is developing the Humming, which is described as a "third generation" air transport with VTOL capability. Public debut was at the December 2001 NBAA Convention in New Orleans. The project remains active, but there are no known plans to construct a prototype.

UPDATED

AEROCOURIER

AEROCOURIER GROUP

10504 Southwest Indianola Road, Augusta, Kansas 67010
Tel: (+1 316) 733 52 38
e-mail: questions@aerocouriergroup.com
Web: http://www.aerocouriergroup.com

VICE-PRESIDENTS:
John Guernsey (Engineering/Certification)
Justin Ladner (Engineering/Operations)
Mike Hahn (Operations)
MEDIA RELATIONS: Cindy Bagaus

AeroCourier was founded in April 2001 to design and produce a light freight aircraft in the Cessna 208 Caravan class. Detail design, and also production of up to 1,500 shipsets of fuselages and wings, has been subcontracted to Dirgantara of Indonesia. Flight testing was to be at Augusta Municipal Airport, Wichita, Kansas; assembly plant for production aircraft in Wichita or Enid, Oklahoma. The project continues, but has not yet progressed to the construction of a prototype.

UPDATED

AEROLITES

AEROLITES INC

12104 David Road, Welsh, Louisiana 70591
Tel/Fax: (+1 337) 734 38 65
e-mail: aerolites@centurytel.net
Web: http://www.aerolites.com
PRESIDENT: Daniel J Roche

UPDATED

AEROLITES AEROMASTER AG

TYPE: Agricultural sprayer ultralight.

PROGRAMME: First flown 1993.

CUSTOMERS: Total of 28 flying by mid-2004, including at least one in South Africa.

COSTS: Kit US$24,995 (2004).

DESIGN FEATURES: Strut-braced aluminium wing has ladder construction with internal diagonal bracing tested to +6/−4 *g* ; aluminium ribs slot into pockets within fabric covering. Quoted build time 130 hours.

FLYING CONTROLS: Conventional and manual.

STRUCTURE: Fuselage of welded 4130 chromoly with fabric covering.

LANDING GEAR: Tailwheel type; fixed.

POWER PLANT: One 47.8 kW (64.1 hp) Rotax 582 two-cylinder engine driving Warp Drive carbon fibre three-blade propeller through C type 3:1 reduction gear. Fuel capacity 38 litres (10.0 US gallons; 8.3 Imp gallons).

EQUIPMENT: 12-nozzle SprayMiser CDA system fitted for agricultural operations; belly tank capacity 114 litres (30.0 US gallons; 25.0 Imp gallons). Optional ballistic parachute.

DIMENSIONS, EXTERNAL:
Wing span ... 8.74 m (28 ft 8 in)
Length overall ... 5.64 m (18 ft 6 in)
Height overall ... 2.11 m (6 ft 11 in)

AREAS:
Wings, gross ... 13.54 m² (145.7 sq ft)

WEIGHTS AND LOADINGS:
Weight empty ... 193 kg (425 lb)
Max T-O weight ... 453 kg (1,000 lb)

PERFORMANCE (with SprayMiser system fitted):
Never-exceed speed (VNE) ... 95 kt (177 km/h; 110 mph)
Normal cruising speed ... 65 kt (121 km/h; 75 mph)
Spray speed ... 56 kt (105 km/h; 65 mph)
Stalling speed ... 28 kt (52 km/h; 32 mph)
Max rate of climb at S/L ... 244 m (800 ft)/min
T-O run ... 153 m (500 ft)
Landing run ... 61 m (200 ft)
Range with max fuel ... 160 n miles (296 km; 184 miles)
Swath width ... 9–30 m (30–100 ft)

UPDATED

Aerolites Aeromaster agricultural ultralight *(Geoffrey P Jones)* 0092107

AEROLITES AEROSKIFF

TYPE: Two-seat amphibian ultralight kitbuilt.

CUSTOMERS: Seven flying by mid-2004.

COSTS: Kit US$24,795 (2004).

DESIGN FEATURES: Pusher layout. Wings detach for storage. Quoted build time 190 hours.

FLYING CONTROLS: Conventional and manual. High-lift flaps.

STRUCTURE: Glass fibre fuselage and fabric-covered, strut-braced wings.

LANDING GEAR: Tailwheel type; tailwheel also retracts for water landings. Drum brakes. Floats mounted on tubular outriggers at wing mid-point.

POWER PLANT: One 47.8 kW (64.1 hp) Rotax 582 water-cooled engine is standard; optionally, 55.0 kW (73.8 hp) Rotax 618. Fuel capacity 45 litres (12.0 US gallons; 10.0 Imp gallons).

SYSTEMS: Optional bilge pump and electric starter.

DIMENSIONS, EXTERNAL:
Wing span ... 9.04 m (29 ft 8 in)
Length overall ... 6.81 m (22 ft 4 in)
Height overall ... 2.13 m (7 ft 0 in)

AREAS:
Wings, gross ... 14.49 m² (156.0 sq ft)

WEIGHTS AND LOADINGS:
Weight empty ... 256 kg (565 lb)
Max T-O weight ... 510 kg (1,125 lb)

PERFORMANCE (Rotax 582):
Never-exceed speed (VNE) ... 82 kt (152 km/h; 95 mph)
Normal cruising speed ... 56 kt (105 km/h; 65 mph)
Stalling speed ... 33 kt (62 km/h; 38 mph)
Max rate of climb at S/L ... 183 m (600 ft)/min
T-O run: water ... 168 m (550 ft)
land ... 134 m (440 ft)
Landing run: land ... 92 m (300 ft)
Range ... 140 n miles (259 km; 161 miles)

UPDATED

Aerolites Aeroskiff ***NEW***/0561756

AEROLITES BEARCAT

TYPE: Single-seat ultralight kitbuilt.

CUSTOMERS: Total of 35 completed by early 2004.

COSTS: Kit US$14,695, including engine and propeller (2004).

DESIGN FEATURES: Replica Corben Baby Ace (1930s vintage). Parasol wing is strut-braced and similar to AeroMaster, but struts are joined to fuselage at separate points rather than meeting at single point. Quoted build time 125 hours.

FLYING CONTROLS: Conventional and manual.

STRUCTURE: Fabric-covered 4130 chromoly fuselage and aluminium extended-wing spar.

LANDING GEAR: Tailwheel type; fixed. Bungee shock absorption on main legs; optional hydraulic disc brakes.

POWER PLANT: One 31.0 kW (41.6 hp) Rotax 447 in-line two-cylinder piston engine, driving a two-blade propeller. Optionally, 37.0 kW (49.6 hp) Rotax 503 UL-2V. Fuel capacity 19 litres (5.0 US gallons; 4.2 Imp gallons).

DIMENSIONS, EXTERNAL:
Wing span ... 9.14 m (30 ft 0 in)
Length overall ... 5.33 m (17 ft 6 in)
Height overall ... 1.98 m (6 ft 6 in)

AREAS:
Wings, gross ... 13.94 m² (150.0 sq ft)

WEIGHTS AND LOADINGS:
Weight empty ... 136 kg (300 lb)
Max T-O weight ... 340 kg (750 lb)

PERFORMANCE:
Never-exceed speed (VNE) ... 95 kt (177 km/h; 110 mph)
Max operating speed ... 65 kt (121 km/h; 75 mph)
Normal cruising speed ... 59 kt (109 km/h; 68 mph)
Stalling speed ... 24 kt (44 km/h; 27 mph)
Max rate of climb at S/L ... 137 m (450 ft)/min
T-O run ... 84 m (275 ft)
Landing run ... 61 m (200 ft)
Range with max fuel ... 125 n miles (231 km; 143 miles)

UPDATED

AEROSTAR

AEROSTAR AIRCRAFT CORPORATION LLC

10555 Airport Drive, Coeur d'Alene Airport, Hayden Lake, Idaho 83835-9742
Tel: (+1 208) 762 03 38 or (+1 800) 442 42 42
Fax: (+1 208) 762 83 49
e-mail: info@aerostarjet.com
Web (1): http://www.aerostarjet.com
Web (2): http://www.aerostaraircraft.com
PRESIDENT: Steve Speer
VICE-PRESIDENT: James S Christy
SENIOR PROJECT ENGINEER: William V Leeds

Aerostar Aircraft Corporation was formed in 1991 by former Ted Smith employees and acquired the rights to the Aerostar series of pressurised piston-engined twins from Piper in the same year. It offers upgrades, spares support and second-hand sales, and is planning to extend its 2,975 m² (32,000 sq ft) factory to 10,400 m² (112,000 sq ft) to accommodate production of a jet version.

A new company, Aerostar Jet LLC, was planned for launch in 2003 for the development of the FJ-100; however this was contingent upon receipt of US$50 million to fund development, certification and production and no confirmation had been received by February 2004.

UPDATED

AEROSTAR FJ-100

TYPE: Business jet.

PROGRAMME: Jet derivative of former Piper/Ted Smith Aerostar 600/700 series, of which over 1,000 were built; will be certified as a new-build design, but first flight-test example will be modified from an existing piston-engined aircraft. First flight originally scheduled for August 2001, with certification and first deliveries in July 2002; but revised to maiden flight in second half of 2004 and deliveries in 2006.

CURRENT VERSIONS: Family planned, ranging from four to eight seats.

CUSTOMERS: Total 25 deposits received by April 2001.

COSTS: Development costs estimated at US$40 million. Unit price US$2 million (2003). Estimated direct operating cost US$339.14 per hour (2001).

DESIGN FEATURES: Improved, jet-powered version of original piston-engined Aerostar; FAR Pt 23 compliant.

FLYING CONTROLS: Conventional and manual. Single-slotted Fowler flaps; maximum deflection 45°. Electrically operated trim tabs on port aileron, rudder and both elevators.

STRUCTURE: Metal throughout; semi-monocoque fuselage with flush-riveted skins; two-spar wing mounted mid-fuselage with 2° dihedral; contains integral fuel tanks. Mid-mounted tailplane; glass fibre dorsal fillet.

LANDING GEAR: Retractable, hydraulically actuated tricycle type; oleo-pneumatic suspension. Electrically actuated nosewheel steering. Dual calliper disc brakes on main wheels.

POWER PLANT: Two 5.35 kN (1,200 lb st) Williams FJ33-1 turbofans on elastomeric engine mounts on pylon on each side of rear fuselage. Fuel in integral wing tanks, plus bladder-type fuselage tank between rear of cabin and baggage compartment; all three tanks crossfeed for balanced fuel load.

ACCOMMODATION: Pilot and up to seven passengers in pairs of seats; standard layout is four club seats in cabin. Door on port side of fuselage behind cockpit. Baggage compartment to rear of cabin; optional lavatory on starboard side at rear of cabin; optional fold-out table.

SYSTEMS: 28 V DC battery for engine starting and as emergency/back-up source; DC external power point on aft fuselage underside. Cabin pressurised to 0.52 bar (7.6 lb/sq in) at 3,050 m (10,000 ft) and cooled using Freon 134 vapour-cycle system. De-icing system to be fitted to engine inlets, surface leading edges and windscreen; exact type to be determined.

AVIONICS: Cockpit modelled on original Aerostar; instrument panel canted to enhance readability by solo pilot.

Comms: Dual VHF nav/com; Dual Mode S transponders; audio control panel.

Radar: Colour weather radar.

Flight: GPS; DME; three-axis autopilot.

DIMENSIONS, EXTERNAL:	
Wing span	11.18 m (36 ft 8 in)
Wing aspect ratio	7.6
Length overall	11.76 m (38 ft 7 in)
Height overall	4.42 m (14 ft 6 in)
Tailplane span	4.37 m (14 ft 4 in)
Wheel track	4.37 m (14 ft 4 in)
Wheelbase	3.10 m (10 ft 2 in)
DIMENSIONS, INTERNAL:	
Cabin: Length	4.23 m (13 ft 10½ in)
Max width	1.16 m (3 ft 9½ in)
Max height	1.21 m (3 ft 11½ in)
Internal baggage compartment volume	0.68 m³ (24.0 cu ft)
External baggage compartment volume	0.85 m³ (30.0 cu ft)
AREAS:	
Wings, gross	16.54 m² (178.0 sq ft)
WEIGHTS AND LOADINGS:	
Weight empty	1,905 kg (4,200 lb)
Max fuel weight	1,131 kg (2,494 lb)
Max T-O and landing weight	3,311 kg (7,300 lb)
Max ramp weight	3,334 kg (7,350 lb)
Max zero-fuel weight	2,676 kg (5,900 lb)
Max wing loading	200.2 kg/m² (41.01 lb/sq ft)
PERFORMANCE:	
Max operating speed: S/L to FL290	270 kt (500 km/h; 311 mph CAS)
above FL290	M0.71
High cruising speed	402 kt (745 km/h; 463 mph)
Long-range cruising speed	360 kt (667 km/h; 414 mph)
Manoeuvring speed	167 kt (309 km/h; 192 mph)
Stalling speed, landing configuration	79 kt (147 km/h; 91 mph)
Max rate of climb at S/L	945 m (3,100 ft)/min
Rate of climb at S/L, OEI	268 m (880 ft)/min
Service ceiling	12,495 m (41,000 ft)
T-O run	594 m (1,950 ft)
Landing run	497 m (1,630 ft)
Max range: IFR	1,700 n miles (3,148 km; 1,956 miles)
NBAA IFR with 100 n mile (185 km; 115 mile) alternate	1,500 n miles (2,778 km; 1,726 miles)
g limits	+3.6/–1.6

Artist's impression of FJ-100 (two Williams FJ33-1 turbofans) 0084552

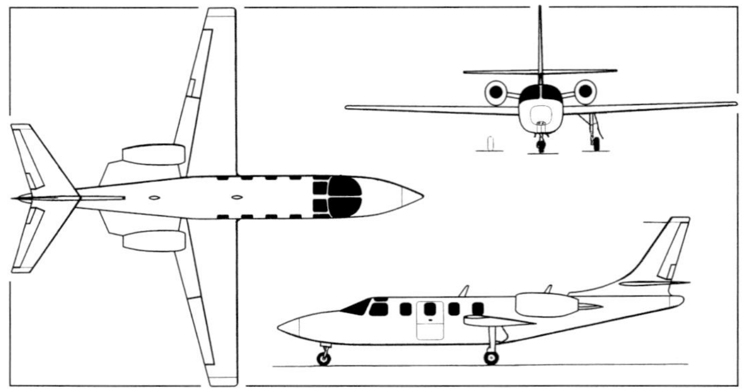

Aerostar FJ-100 general arrangement ***(Paul Jackson)*** 0084553

UPDATED

AEROS

WORLDWIDE AEROS CORPORATION

6005 Yolanda Avenue, Tarzana, California 91356
Tel: (+1 818) 344 39 99
Fax: (+1 818) 344 39 33
e-mail: info@aerosml.com
Web: http://www.aerosml.com
CEO: Igor Pasternak
EXECUTIVE VICE-PRESIDENT, SALES AND MARKETING: Frederick Edworthy
VICE-PRESIDENT, ENGINEERING: Andrei Rybkin
AEROSCRAFT DEVELOPMENT MANAGER: Stayne Hoff

Formed in Ukraine in 1988, Aeros transferred its activities to the USA and was incorporated in Delaware in 1992, relocating to California in October 1993. Aeros' technical capabilities are based on more than 20 years of research (including that initiated before 1988 by former Soviet design bureaux) into LTA technologies, especially those relating to cargo airships. Its current operation combines manufacturing and flight test facilities, R&D, management and marketing. Aeros Airship Company subsidiary formed September 1999 to lease and operate Aeros airships; operations began, with an Aeros 40B, in November 1999. It has also produced a wide range of manned and unmanned civil and military aerostats for commercial, advertising, surveillance and scientific applications.

The company's current airship product is the Aeros 40B. Development continues of the Aeroscraft, a partially buoyant lifting body; this programme encompasses the 80- to 200-passenger Aeros-ML (D1) and the Aeros-D4 and -D8 cargo versions with a carrying capability of up to 725,750 kg (1.6 million lb).

Stratospheric airship R&D is an important part of company business for the present and future. It is currently designing and manufacturing such a craft for the government of South Korea.

UPDATED

AEROS AEROS-ML

TYPE: Helium rigid airship.

PROGRAMME: Current incarnation of former D1 project, now envisaged as mass passenger transport with cabin comfort comparable with cruise liners. No launch date yet announced. Aeros announced in April 2004 that it expected to bid an Aeroscraft variant for the forthcoming DARPA Walrus programme, which studies the feasibility of an outsize strategic military airlift vehicle.

DESIGN FEATURES: Exterior illumination on hull sides and tailfins. Gondola/cabin integrated within hull structure; 'glass cockpit' controls; turboprop engines; air cushion landing gear system to eliminate need for ground handling crew.

FLYING CONTROLS: Digital flight control system; five tailfins, each with electrically actuated control surface; twin canards at nose. Pressurisation control by internal ballonets in lower hull (12), helium valves and water ballast. Ballonets separated from helium by air/helium barrier.

STRUCTURE: Hull of tubular trusses and bonded composites panels (approx 1,000); aluminium alloy and composites cabin; aluminium alloy semi-monocoque tailfins and canards.

LANDING GEAR: Air cushion system for assisted T-O and vacuum mooring.

POWER PLANT: Two 1,044 kW (1,400 shp) class, thrust-vectoring Rolls-Royce turboprops; plus two electric motors to power bow thrusters and tail control surfaces.

ACCOMMODATION: Passenger cabin inside hull, seating from 80 to 200 people. Polycarbonate and acrylic windscreens and windows in hull sides.

SYSTEMS: AC electrical system.

EQUIPMENT: Provision for hull-side and vertical fin illuminations. Navigation lights and anti-collision beacons.

DIMENSIONS, EXTERNAL (approx):	
Hull: Length overall	81.08 m (266 ft)
Max diameter	35.36 m (116 ft)
Propeller diameter (PT6A)	7.92 m (26 ft)
DIMENSIONS, INTERNAL (approx):	
Hull max volume	25,000 m³ (882,875 cu ft)

Computer graphic depicting the Aeros-ML as it might appear if selected for the DARPA/US Army Walrus programme ***NEW***/1047173

Ballonet volume (12, total) ... 5,000 m^3 (176,575 cu ft)
Cabin: Length ... 38.10 m (125 ft)
Max width ... 24.99 m (82 ft)
Max height ... 2.50 m (8.2 ft)

WEIGHTS AND LOADINGS:
Weight empty ... 20,340 kg (44,842 lb)
Useful load ... 12,712 kg (28,025 lb)

PERFORMANCE (estimated):
Max level speed ... 152 kt (281 km/h; 174 mph)
Max rate of ascent ... 610 m (2,000 ft)/min
Pressure ceiling ... 2,440 m (8,000 ft)
Max rate of ascent and descent ... 610 m (2,000 ft)/min
Max range ... 2,400 n miles (4,444 km; 2,761 miles)
Max endurance ... 24 h

UPDATED

AEROS 40B SKY DRAGON

TYPE: Helium non-rigid airship.

PROGRAMME: Prototype; first flight (N818AC) 11 September 1998; exported to China (Thakral Media Corporation) in November 1998; operating in Hong Kong 1999. Three more (one for Argos Medien AG of Germany and two more for China) under construction in 1999–2001, of which N819AC for China destroyed in accident on 9 January 2001; N820AC for Argos Medien registered in January 2001, delivered in February 2002 but reverted to US registry in August 2003. FAA certification 21 June 2000; STC for external payload mounting awarded January 2001. Certified by LBA (Germany) January 2002. Second of Chinese pair still not completed by early 2005.

CUSTOMERS: Highest reported constructor's number by early 2005, but only six examples identified.

DESIGN FEATURES: Göttingen 409 root rib profile; X configuration tail unit, each with ruddervator. One ellipsoid ballonet forward and one aft (together, 22 per cent of total volume). Internal illumination by three 400 W lamps: one above gondola and one in each ballonet.

FLYING CONTROLS: Fly-by-wire electronically controlled, pneumatically actuated ruddervators, with manual back-up.

STRUCTURE: Envelope made up of 1,020 heat-sealed panels of a transparent, nylon-based, 7 ply polyurethane coated fabric; tailfins aluminium alloy with fabric covering. Aluminium alloy semi-monocoque gondola, with polycarbonate sheet transparencies.

LANDING GEAR: Two mainwheels, with shock-absorbers.

POWER PLANT: Two 93 kW (125 hp) Teledyne Continental IO-240-B8 flat-four engines; MTV-7-D/LD170-12 variable- and reversible-pitch, three-blade wooden pusher propellers. Fuel capacity 287 litres (75.9 US gallons; 63.2 Imp gallons), of which 267 litres (70.5 US gallons; 58.7 Imp gallons) are usable.

Aeros 40B Sky Dragon twin-engined airship in flight ***NEW***/1047138

ACCOMMODATION: Pilot and four passengers (pilot and co-pilot only in prototype). Front two seats fully articulating and reclining; three bucket seats to rear. Full-depth windscreen; door, with upper and lower windows, in each side of gondola.

SYSTEMS: Pressure maintained by two air valves in each ballonet, each pneumatically actuated by an electrical blower and having mechanical return springs. Three emergency rip panels for fast deflation. Two ballast compartments in gondola. Electrical systems 28 V DC and 100 V AC. Nosecone mast mooring; ground handling crew of 12.

AVIONICS: Digital 'glass cockpit' package includes audio panel, voice annunciator, colour LEDs, GPS/com, second com, and transponder/encoder; readings displayed in both digital and analogue form. Provisions for full IFR package.

EQUIPMENT: Halon fire extinguisher and first aid kit. Navigation lights (bow, stern, port and starboard); anti-collision beacons (one each port, starboard and beneath gondola); two propeller lights; two red ceiling lights in gondola cabin. Three optional lamps (see Design Features) for internal envelope illumination.

DIMENSIONS, EXTERNAL:
Length overall ... 43.59 m (143 ft 0 in)
Envelope: Length ... 42.61 m (139 ft 9½ in)
Max diameter ... 10.60 m (34 ft 9½ in)
Fineness ratio ... 4.0
Height overall ... 13.35 m (43 ft 9½ in)
Banner (each side): Length ... 22.86 m (75 ft 0 in)
Height ... 7.92 m (26 ft 0 in)
Propeller diameter ... 1.70 m (5 ft 7 in)

DIMENSIONS, INTERNAL:
Envelope volume ... 2,508 m^3 (88,570 cu ft)
Ballonet volume (two, total) ... 549.2 m^3 (19,396 cu ft)
Gondola cabin: Length ... 1.96 m (6 ft 5 in)
Max width ... 1.57 m (5 ft 2 in)
Max height ... 1.73 m (5 ft 8 in)

AREAS:
Banner area (each side) ... 185.34 m^2 (1,995.0 sq ft)

WEIGHTS AND LOADINGS:
Weight empty ... 1,914 kg (4,220 lb)
Useful load ... 676 kg (1,490 lb)

PERFORMANCE:
Max level speed ... 44 kt (82 km/h; 51 mph) IAS
Max rate of climb ... 610 m (2,000 ft)/min
Max rate of descent ... 610 m (2,000 ft)/min
Max certified altitude ... 2,885 m (7,500 ft)
Pressure ceiling ... 3,000 m (9,840 ft)
Endurance: at 45% power ... 6 h 42 min
max ... 24 h

UPDATED

AG CAT

ALLIED AG CAT PRODUCTIONS

PO Box 482, 301 West Walnut Street, Walnut Ridge, Arkansas
Tel: (+1 870) 886 21 11
Fax: (+1 870) 886 97 86
e-mail: agcat@bscn.com
PRESIDENT: Frank D Kelley

Grumman G-164 Ag Cat radial-engined agricultural biplane first flew 22 May 1957 and certified 20 January 1959; produced under subcontract by Schweizer; rights passed to Gulfstream American upon its separation from Grumman, but planned transfer of manufacture to Savannah, Georgia, forestalled by Schweizer's outright purchase of programme in January 1981. Total of over 2,600 built, latterly only to specific order, before programme sold to Ag Cat Corporation of Malden, Missouri, in January 1996. Type Certificate again transferred on 22 February 2001 to current owner.

Allied Ag Cat supports the worldwide fleet of some 2,000 G-164 variants, some being G-164D, certified 26 March 1979 with Pratt & Whitney PT6A turboshaft, or similarly-powered Super B (certified 23 December 1985); services include remanufacture of existing airframes. From 2007, Allied Ag Cat plans to reintroduce new production aircraft (12 in first year; 24 thereafter), principally Ag Cat Super-B Turbine with 507 kW (680 shp) or 559 kW (750 shp) PT6A-15AG/-34AG engines and 1,514 litre (400 US gallons; 333 Imp gallon) chemical hopper.

NEW ENTRY

Early versions of Grumman/Schweizer Ag Cat can be remanufactured by Allied Ag Cat Productions *(Paul Jackson)* ***NEW***/1047429

AHA

ADVANCED HYBRID AIRCRAFT INC

PO Box 144, Eugene, Oregon 97440
Tel/Fax: (+1 604) 541 80 64
e-mail: bnb@ahausa.com
Web: http://www.ahausa.com
MANAGER AND DIRECTOR: Bruce N Blake

Formerly based in Isle of Man, UK; relocated to Australia in 1993; to USA in 1996; and to British Columbia, Canada, in 1998 (while retaining a presence in Oregon, USA). A facility at Boundary Bay Airport, Vancouver, was established in 2002. Venture capital for airship development, expected in early 1997, did not materialise, but reported that AHA signed letter of intent on 6 October 1998 with Xiao Bang Group (XBG) of China initiating 10 year joint venture to build and market new buoyant aircraft in China. Intention is to manufacture first Light Utility and a two-seat LV2 Hornet in USA, then to establish assembly plant in Zhuhai after training Chinese workforce. A formal agreement was signed in February 1998 with the Blimp Group of Buenos Aires, Argentina, to promote AHA buoyant aircraft. In March 2001, a similar agreement was signed with Denebola (DSA) SA de CV of Ensenada, Mexico.

In late 2002, AHA was contracted to design, build and test a prototype of the Parabounce Propbike man-powered airship. New investment to revive the projected Advanced Airship Corporation ANR still had not been located by mid-2004.

Latest design study is for an outsize heavy-lift hybrid named **Behemuth**, conceived to meet the 30 US ton payload requirements of the DARPA Walrus programme. This would achieve vertical lift by means of three dedicated lift fans, each driving a 18 m (60 ft) diameter rotor accommodated inside a large Bernoulli duct located within the full depth of the hull volume. By combining these with two propulsion engines mounted on wings with high-lift flaps, the Behemuth could operate with both vertical and horizontal thrust. After vertical take-off with combined hull and wing lift, the vertical thrusters could be powered down and the ducts closed off. Alternatively, using its wheeled landing gear, the craft would be capable of STOL operation (T-O angle 12°). Internally, there is a 61 × 6.1 m (200 × 20 ft) cargo bay each side of the vertical lift ducts, with a combined volume of about 4,248 m³ (150,000 cu ft), and twin roll-on/roll-off ramps at each end. To hold itself on the ground without the use of external anchors and ballast (a DARPA requirement), the Behemuth retracts its landing gear and settles onto two lengthwise air-inflated pontoons and operating the centre lift fan in reverse pitch at a moderate throttle setting.

UPDATED

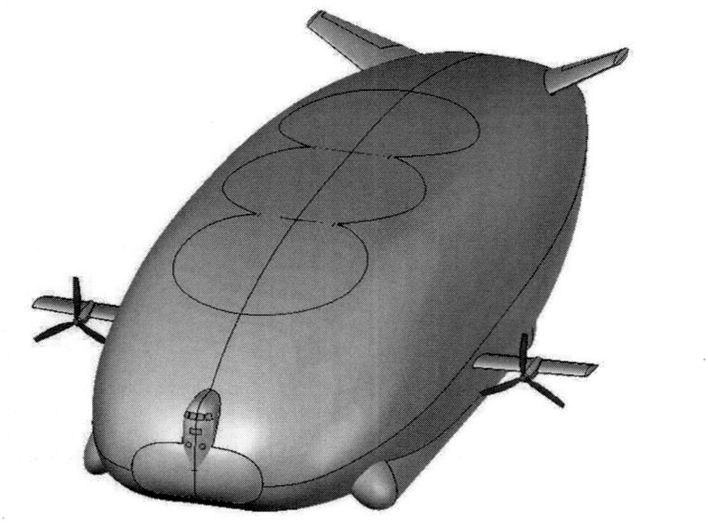

Computerised image of AHA's projected Behemuth heavy lifter ***NEW***/1114589

AHA HORNET

TYPE: Helium non-rigid hybrid airship.

PROGRAMME: Design of Hornet LV (leisure variant) initiated October 1991; funding for development and manufacture, which will be to FAA P-8110-2 airship design criteria standards, was still being sought in mid-2004.

CURRENT VERSIONS: **Hornet LV2:** Two-seat LV, to be offered also as single-seat LV kit. Meets FAA '49 per cent rule' to qualify for Experimental category operation. To be used for promotions at major world air shows and elsewhere.

Light Utility: More robust, four-seat version; described separately.

Hornet AW: Aerial work version, planned to follow if sales of LV2 are successful; intended specifically to meet needs of electronic news-gathering organisations. Expected to provide capabilities similar to those of current twin-engined helicopters but at about half the acquisition and operating cost.

COSTS: LV2, US$162,100 (uncertified) or US$289,400 (FAA certified); AW (basic configuration), US$1.46 million (2004).

DESIGN FEATURES: Two-seat gondola/cabin for LV2, with fixed tricycle landing gear; able to taxi like an ultralight, mast-moored by one person (mainly an automated operation), and remain on ground when vacated by pilot.

Designed to operate at heaviness of about 50 per cent, compared with conventional types with which operation at or near equilibrium is essential; one half of total lift provided aerostatically by buoyancy of the helium-filled envelope, the other half being generated aerodynamically by stub-wings attached to sides of envelope. Hybrid buoyant aircraft do not require ballast and have flexibility to carry greater or lesser load, thus providing greater productivity.

FLYING CONTROLS: Fly-by-wire; automatic envelope pressure control.

STRUCTURE: Major subcontractors expected to be ILC Dover (envelope) and Scaled Composites (gondola).

LANDING GEAR: Non-retractable tricycle type.

POWER PLANT: Two 67.1 kW (90 hp) MWE AE 100R rotary engines in LV2, mounted on stub-wings that can be tilted upward to vector thrust 30° downward for climb. Hornet AW would have four 74.6 kW (100 hp) AE 100Rs and 45° thrust vectoring.

ACCOMMODATION: Two seats in LV2, one in LV.

DIMENSIONS, EXTERNAL:
Envelope:
Length overall: LV2 ... 19.51 m (64 ft 0 in)
AW ... 30.48 m (100 ft 0 in)
Max diameter: LV2 ... 4.88 m (16 ft 0 in)

Scale model of the AHA Hornet LV2 0507470

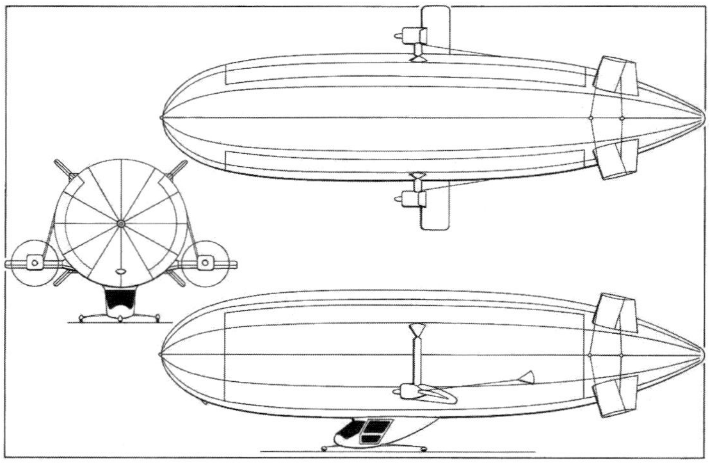

General arrangement of the AHA Hornet LV2 ***(Jane's/Mike Keep)*** 0507294

AW ... 6.10 m (20 ft 0 in)
Fineness ratio: LV2 ... 4.0
AW ... 5.0
Wing span: LV2 ... 9.14 m (30 ft 0 in)
AW ... 9.21 m (30 ft 2½ in)
Wing chord, constant: LV2 ... 1.52 m (5 ft 0 in)
AW ... 1.25 m (4 ft 1¼ in)

DIMENSIONS, INTERNAL:
Envelope volume: LV2 ... 250.0 m³ (8,830 cu ft)
AW ... 623.0 m³ (22,000 cu ft)
Max ballonet volume: LV2 ... 37.5 m³ (1,325 cu ft)
AW ... 124.6 m³ (4,400 cu ft)

WEIGHTS AND LOADINGS:
Disposable load: LV2 ... 209 kg (460 lb)
AW ... 427 kg (941 lb)
Buoyancy: LV2 ... 249 kg (550 lb)
AW ... 560 kg (1,235 lb)
Max heaviness: LV2 ... 249 kg (550 lb)
AW ... 720 kg (1,587 lb)
Max T-O weight: LV2 ... 499 kg (1,100 lb)
AW ... 1,280 kg (2,822 lb)

PERFORMANCE (estimated):
Max level speed: LV2 ... 61 kt (113 km/h; 70 mph)
AW ... 81 kt (150 km/h; 93 mph)
Pressure ceiling: LV2 ... 1,525 m (5,000 ft)
AW ... 2,135 m (7,000 ft)
Range: LV2, AW ... 200 n miles (370 km; 230 miles)

UPDATED

AHA LIGHT UTILITY

TYPE: Helium non-rigid hybrid airship.

PROGRAMME: Development of Hornet.

CURRENT VERSIONS: **Light Utility:** For similar applications to Hornet LV2, but more robust and intended to operate from rough fields and away from built-up areas.

Light Utility.XR Patroller: Long-range version, designed for coastal patrol and drug interdiction duties. Presented to Mexican Navy in June 2001.

Patroller Four: Enlarged version of XR; single PT6 turboprop engine; payload 4,534 kg (10,000 lb); nominal crew of eight to 10; four-day endurance (more with in-flight refuelling); optional retractable landing gear.

COSTS: Light Utility (four- to six-seat) US$972,000 (2004); XR prices on application.

DESIGN FEATURES: Generally as for Hornet.

FLYING CONTROLS: As for Hornet.

LANDING GEAR: Non-retractable tricycle type.

POWER PLANT: *Light Utility:* Two non-vectoring 74.6 kW (100 hp) MWE AE 100R rotary engines, mounted on stub-wings at rear of gondola.

Technology demonstration model of the AHA Light Utility airship 0016730

Light Utility. XR: Two 97 kW (130 hp) DAIR 100 two-cylinder Diesel engines; plus stern thruster.

ACCOMMODATION: Four seats, including pilot(s), standard; six or eight seats optional. Four crew plus bunks, galley and lavatory in XR.

DIMENSIONS, EXTERNAL (A Light Utility, B: Light Utility. XR, C: Patroller Four):
- Envelope:
  - Length overall: A ... 30.48 m (100 ft 0 in)
    - B ... 38.10 m (125 ft 0 in)
    - C ... 56.39 m (185 ft 0 in)
  - Max diameter: A ... 7.49 m (24 ft 7 in)
    - B ... 10.24 m (33 ft 7 in)
    - C ... 15.24 m (50 ft 0 in)
  - Fineness ratio: A ... 4.1
    - B, C ... 3.7
- Wing span: A ... 5.92 m (19 ft 5 in)
- Wing chord, constant: A ... 1.44 m (4 ft 8½ in)
- Height overall: B ... 13.89 m (45 ft 7 in)

DIMENSIONS, INTERNAL:
- Envelope volume: A ... 1,200.6 m³ (42,400 cu ft)
  - B ... 2,406.9 m³ (85,000 cu ft)
  - C ... 8,000 m³ (282,500 cu ft)
- Max ballonet volume: A ... 155.7 m³ (5,500 cu ft)
- Gondola (C): Length ... 12.19 m (40 ft 0 in)
- Max height ... 1.93 m (6 ft 4 in)
- Floor area ... 25.08 m² (270.0 sq ft)
- Volume ... 50.4 m³ (1,780 cu ft)

WEIGHTS AND LOADINGS (B):
- Disposable load ... 431 kg (950 lb)
- Buoyancy ... 1,080 kg (2,380 lb)
- Max heaviness ... 359 kg (791 lb)
- Max T-O weight ... 1,436 kg (3,166 lb)

PERFORMANCE (estimated A, B and C as above):
- Max level speed: A ... 51 kt (94 km/h; 58 mph)
  - B, all three engines ... 75 kt (139 km/h; 86 mph)
  - C, all three engines ... 79 kt (146 km/h; 91 mph)
- Max cruising speed, two engines, at 75% power:
  - B ... 52 kt (96 km/h; 60 mph)
  - C ... 64 kt (119 km/h; 74 mph)
- Econ cruising speed, one engine, at 75% power:
  - B ... 41 kt (76 km/h; 47 mph)
  - C ... 36 kt (67 km/h; 41 mph)
- Pressure ceiling: A, C ... 2,745 m (9,000 ft)
- Range: A ... 300 n miles (555 km; 345 miles)
  - B (two engines, 50 kt; 93 km/h; 58 mph cruise) ... 1,000 n miles (1,852 km; 1,150 miles)
  - B (one engine, 30 kt; 56 km/h; 35 mph cruise) . 3,000 n miles (5,556 km; 3,452 miles)
  - C (35 kt; 65 km/h; 40 mph IAS cruise) ... 4,200 n miles (7,778 km; 4,833 miles)
- Endurance, no reserves, conditions as above:
  - B (two engines) ... 24 h
  - B (one engine) ... 100 h
  - C (one engine) ... 120 h

UPDATED

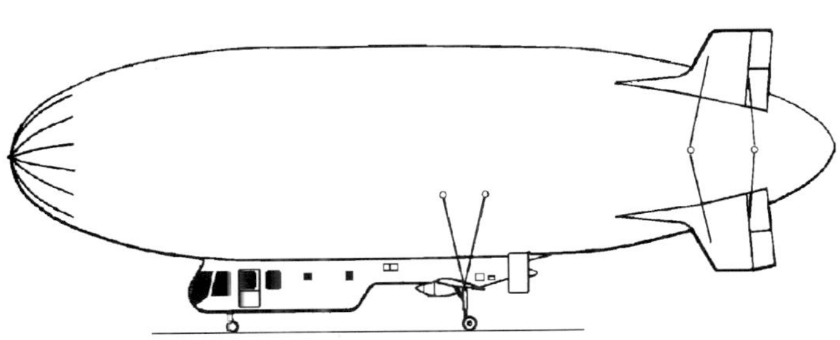

AHA Light Utility.XR Patroller 0525375

AHA PROMETHEUS GLOBAL CRUISER

TYPE: Hybrid rigid air vehicle.

PROGRAMME: Conceptual study in 2004. Named after Titan in Greek mythology who stole fire from the gods.

COSTS: US$300 million to develop prototype; approx US$500 million for certification (2004).

DESIGN FEATURES: Conceived as aerial equivalent of cruise liner. 'Shark-like' hull shape, with canards and triple (inverted Y) tail surfaces with elevators and split rudder.

FLYING CONTROLS: Vectored thrust propulsion. Canards and empennage computer-controlled to counteract gusting. Stabilisers alleviate heave, pitch and roll. As back-up, ascent/descent thrusters can be powered up to offset excessive climb rates.

STRUCTURE: Forward and aft hull sections are quasi-monocoque, double-curvature honeycomb sandwich shells of co-cured carbon/glass fibres and epoxy resin; thickness variable as required for local rigidity and resistance to buckling. Habitated volume of 2024-T3 and 7075-T6 aluminium alloys, with honeycomb panel walls and floors; ceilings and minor features of fabric.

POWER PLANT: Two 2,983 kW (4,000 shp) engines, each driving a ducted propulsion unit. Fuelled by liquid hydrogen gas (LHG).

ACCOMMODATION: Based on requirement for 500 tourist class passengers and 100 crew.

DIMENSIONS, EXTERNAL:
- Length overall ... 301.8 m (990 ft)
- Width overall ... 52.7 m (173 ft)
- Height overall ... 41.5 m (136 ft)
- Hull: Max width ... 45.3 m (148.5 ft)
  - Max height ... 30.2 m (99 ft)
- Propeller duct diameter ... 9.1 m (30 ft)

DIMENSIONS, INTERNAL:
- Volume: fwd hull ... 100,950 m³ (3,565,000 cu ft)
  - aft hull ... 136,006 m³ (4,803,000 cu ft)
  - habitated ... 55,643 m³ (1,965,000 cu ft)
  - total ... 292,600 m³ (10.333 million cu ft)

WEIGHTS AND LOADINGS:
- Airframe ... 77,632 kg (171,150 lb)
- Liquid H_2 ... 74,884 kg (165,090 lb)
- Payload (100 crew + 500 pax) ... 84,454 kg (186,190 lb)
  - Gross aerostatic lift:
    - 90% full ... 211,140 kg (465,485 lb)
    - 100% full ... 236,970 kg (522,430 lb)

PERFORMANCE (estimated):
- Max cruising speed ... 100 kt (185 km/h; 115 mph)
- Range, no reserves ... 23,900 n miles (44,260 km; 27,500 miles)
- Endurance ... 239 h/10 days

NEW ENTRY

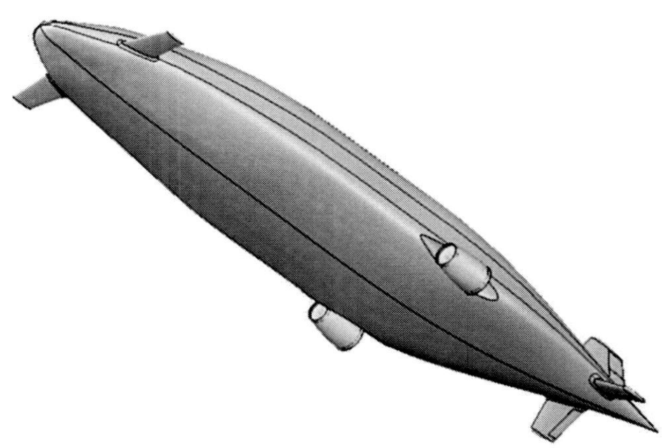

Computer graphic depicting the Prometheus Global Cruiser ***NEW***/1048630

AIR

AIRCRAFT INVESTOR RESOURCES LLC

2121 Redbird Drive, Las Vegas, Nevada 89134
Tel: (+1 702) 682 23 89
e-mail: jeff@epicaircraft.com
Web: http://www.epicaircraft.com

CHAIRMAN: Rick Schrameck
PRESIDENT, R&D AND MANUFACTURING: Bob Fair
INFORMATION SOURCE: Jeff Sanders

UPDATED

AIR EPIC LT

TYPE: Business turboprop.

PROGRAMME: Announced at AirVenture, Oshkosh, July 2003; first flight (N370JP) 17 July 2004. Initially available as experimental category kitbuilt with Walter or PT6A engine; certified version to follow, with FAA approval expected in the first quarter of 2006.

CURRENT VERSIONS: **Epic LT:** *As described.*

TAM-AIR Jet: Twin-turbofan versions produced in collaboration with TAM of Georgia. Described separately.

COSTS: Experimental category kit US$800,000 to US$850,000 with Walter engine; US$1.2 million to US$1.25 million with PT6A. Certified version US$2 million (PT6A).

DESIGN FEATURES: "Full people; full fuel; full speed" capability intended. Designed for owner/operator requiring luxury motorcar standard of comfort; interior styled by Nissan Design Centre. Optional winglets.

FLYING CONTROLS: Conventional and manual. Flaps.

STRUCTURE: Generally of carbon fibre.

POWER PLANT: One turboprop: 560 kW (751 shp) Walter M 601 or 895 kW (1,200 shp) Pratt & Whitney Canada PT6A-67 according to version. Fuel capacity 1,041 litres (275 US gallons; 229 Imp gallons).

ACCOMMODATION: Six persons, including pilot(s); rear occupants in 'club four' configuration. Cabin pressurised.

DIMENSIONS, EXTERNAL:
Wing span ..... 12.56 m (41 ft 2½ in)
Length overall ..... 11.00 m (36 ft 1¼ in)
Height overall ..... 3.93 m (12 ft 10¾ in)

AREAS:
Wings, gross ..... 17.63 m² (189.8 sq ft)

DIMENSIONS, INTERNAL:
Cabin: Length ..... 4.88 m (16 ft 0 in)

Prototype AIR Epic LT ***NEW***/0583651

Max width ..... 1.30 m (4 ft 3 in)
Max height ..... 1.37 m (4ft 6 in)

WEIGHTS AND LOADINGS:
Weight empty ..... 1,651 kg (3,640 lb)
Payload with max fuel ..... 499 kg (1,100 lb)
Max T-O weight ..... 2,871 kg (6,330 lb)

PERFORMANCE (estimated, PT6A engine):
Cruising speed: max ..... 350 kt (648 km/h; 403 mph) IAS
econ ..... 289 kt (535 km/h; 333 mph) IAS
Time to FL300 ..... 22 min
Max certified altitude ..... 9,450 m (31,000 ft)
Range, 45 min reserves:
at max cruising speed ..... 1,396 n miles (2,585 km; 1,606 miles)
at econ cruising speed ..... 1,608 n miles (2,978 km; 1,850 miles)

UPDATED

AIR Epic LT protype on an early test flight ***NEW***/1047795

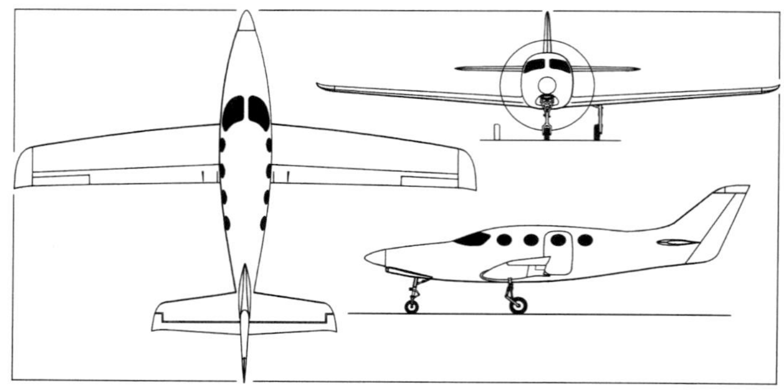

AIR Epic LT business turboprop ***(Paul Jackson)*** ***NEW***/0589193

AIRDALE FLYER

AIRDALE FLYER COMPANY

4317 Aviation Way #9, Caldwell, Idaho 83605
Tel: (+1 208) 459 62 54
e-mail: ukav8r@mindspring.com
Web: http://www.airdale.com
MARKETING MANAGER: Steve Winder

Airdale's current product is the Skygear Comet.

NEW ENTRY

AIRDALE COMET

TYPE: Side-by-side ultralight/kitbuilt.

PROGRAMME: Design started by Paul Kenneth in mid-1990s; made US debut at Sun 'n' Fun in 2001. Originally promoted as ready to fly aircraft in South Korea, but no sales from that source known; main US marketing awaiting announcement of new FAA Sportplane Licence category in 2004.

CURRENT VERSIONS: **Comet 21:** Baseline version, *as described.*

Comet UL: Lighter empty weight version, powered by Rotax 582 engine.

CUSTOMERS: Total of 35 sold by early 2002 to customers in Canada, Italy, Japan and USA.

DESIGN FEATURES: Braced high-wing monoplane, similar in appearance to Cessna 152. Available ready to fly or in kit form; quoted build time 300 hours.

FLYING CONTROLS: Conventional and manual. Frise ailerons; elevator electric trim. Fowler flaps.

STRUCTURE: Prewelded 4130 chromoly steel tube fuselage inside two CFRP half-shells; pop-riveted, single-spar, all-aluminium wing with drooped tips; all-CFRP tail unit. Single-strut bracing of wings and one-piece tailplane.

LANDING GEAR: Tricycle type; fixed. Rubber-sprung, steerable nosewheel; fuselage-mounted cantilever sprung steel main legs. Hydraulic brakes.

POWER PLANT: *Comet 21:* One 59.7 kW (80 hp) Jabiru 2200 engine, driving two-blade propeller.

Comet UL: One 47.8 kW (64.1 hp) Rotax 582 UL.

Usable fuel capacity (two tanks) 38 litres (10.0 US gallons; 8.3 Imp gallons) standard, 76 litres (20.0 US gallons; 16.6 Imp gallons) optional.

ACCOMMODATION: Pilot and passenger. Forward-opening window/door each side; one-piece Plexiglas windscreen.

AVIONICS: VFR or IFR instrumentation to customer's choice.

DIMENSIONS, EXTERNAL:
Wing span ..... 8.69 m (28 ft 6 in)
Length overall ..... 5.79 m (19 ft 0 in)
Height overall ..... 2.44 m (8 ft 0 in)

DIMENSIONS, INTERNAL:
Cabin max width ..... 1.12 m (3 ft 8 in)

AREAS:
Wings, gross ..... 10.59 m² (114.0 sq ft)

US Comet SP demonstrator at Sun 'n' Fun, 2001 ***(Jane's/Susan Bushell)***
0137948

WEIGHTS AND LOADINGS (A: Comet 21, B: Comet UL):
Weight empty: A ..... 236 kg (520 lb)
B ..... 218 kg (480 lb)
Max T-O weight: A ..... 499 kg (1,100 lb)
B ..... 449 kg (990 lb)
Max wing loading: A ..... 47.1 kg/m² (9.65 lb/sq ft)
B ..... 42.4 kg/m² (8.68 lb/sq ft)
Max power loading
A ..... 8.37 kg/kW (13.75 lb/hp)
B ..... 9.40 kg/kW (15.44 lb/hp)

PERFORMANCE (at respective MTOW, A and B as above):
Never-exceed speed (VNE):
A, B ..... 117 kt (217 km/h; 135 mph)
Max cruising speed at 215 m (700 ft) at 75% power:
A ..... 104 kt (193 km/h; 120 mph)
B ..... 96 kt (177 km/h; 110 mph)
Stalling speed:
flaps up: A, B ..... 40 kt (73 km/h; 45 mph) CAS
flaps down: A ..... 33 kt (62 km/h; 38 mph) CAS
B ..... 34 kt (63 km/h; 39 mph) CAS
Max rate of climb at S/L: A ..... 366 m (1,200 ft)/min
B ..... 283 m (930 ft)/min
Service ceiling: A ..... 3,960 m (13,000 ft)
T-O run: A ..... 76 m (250 ft)
B ..... 92 m (300 ft)
Landing run: A, B ..... 54 m (175 ft)

UPDATED

AIR COMMAND

AIR COMMAND INTERNATIONAL INC

PO Box 1177, Caddo Mills Municipal Airport Building B, Caddo Mills, Texas 75135
Tel: (+1 903) 527 33 35
Fax: (+1 903) 527 38 05
e-mail: aircmd@aircommand.com
Web: http://www.aircommand.com

Formed in May 1992, Air Command produces kits of autogyros.

In 2003, Air Command had a 465 m² (5,000 sq ft) factory and seven employees.

VERIFIED

AIR COMMAND COMMANDER ELITE

TYPE: Two-seat autogyro kitbuilt.

PROGRAMME: Construction of tandem-seat F-30 began in late 1999; made first flight 1 March 2000; public debut at Sun 'n' Fun, April 2000. A side-by-side version of the F-30 has been introduced.

CURRENT VERSIONS: **Elite F-30:** Basic version, with Hirth F-30 engine.

Elite F-30ES: As F-30, but side-by-side seating.

Elite Mazda: As F-30, but with more powerful engine; withdrawn 2002.

CUSTOMERS: Over 200 kits sold by end 2002.

COSTS: F-30 US$22,475; F-30ES US$21,792, both including engine (2004).

DESIGN FEATURES: Open cockpit autogyro with wide range of options to enable individual requirements to be met; many subassemblies preconstructed by factory. Quoted build time 100 hours.

STRUCTURE: Pre-drilled and anodised 6061-T6 tubular airframe; 5 × 5 cm (2 × 2 in) tubing used for mast. Optional glass fibre enclosure. F-30 has aluminium rudder and composites horizontal stabiliser and tailfins; Dragon Wings tip-weighted rotor blades.

LANDING GEAR: Fixed tricycle type with glass fibre speed fairings. Aluminium suspension system with Hager wheels and brakes on all wheels. Nosewheel is fully castoring with leg of 4130 steel tube and 6.00×6 Azusa tyres; 7.6 cm (3 in) tailwheel. Mainwheel hydraulic brakes.

POWER PLANT: One 82.0 kW (110 hp) Hirth F-30 driving a four-blade Warp Drive propeller. Fuel capacity 37.9 litres (10.0 US gallons; 8.3 Imp gallons) in two seat tanks; optional 68.1 litre (18.0 US gallon; 15.0 Imp gallon) seat tank.

ACCOMMODATION: Pilot and passenger in tandem on Elite F-30; side by side on Elite F-30ES.

SYSTEMS: Basic engine instrumentation is provided with each kit. Options include hydraulic pre-rotator and clutch reduction drive on F-30; and pre-rotator, C Box upgrade and electric start on F-30ES. Optional New Horizons Components gyro recovery system rocket-powered ballistic parachute.

DIMENSIONS, EXTERNAL:

Rotor diameter: F-30	8.53 m (28 ft 0 in)
F-30ES	8.84 m (29 ft 0 in)
Rotor blade chord: F-30	0.19 m (7½ in)
Fuselage length: F-30	4.09 m (13 ft 5 in)
F-30ES	3.25 m (10 ft 8 in)
Height to top of rotor head: F-30	2.74 m (9 ft 0 in)
F-30ES	2.13 m (7 ft 0 in)

Air Command Commander Elite *(Paul Jackson)* ***NEW***/0587515

Width: F-30	1.88 m (6 ft 2 in)
F-30ES	1.70 m (5 ft 7 in)

AREAS:

Rotor disc: F-30	57.2 m² (616.0 sq ft)
F-30ES	61.4 m² (661.0 sq ft)

WEIGHTS AND LOADINGS:

Weight empty: F-30	222 kg (490 lb)
F-30ES	141 kg (310 lb)
Max payload: both	191 kg (420 lb)
Max T-O weight: F-30	578 kg (1,275 lb)
F-30ES	412 kg (910 lb)
Max disc loading: F-30	10.1 kg/m² (2.07 lb/sq ft)
F-30ES	6.7 kg/m² (1.38 lb/sq ft)
Max power loading: F-30	7.05 kg/kW (11.59 lb/hp)
F-30ES	5.04 kg/kW (8.27 lb/hp)

PERFORMANCE:

Max operating speed: F-30	95 kt (177 km/h; 110 mph)
F-30ES	73 kt (135 km/h; 84 mph)
Cruising speed:	
F-30	65–70 kt (121–129 km/h; 75–80 mph)
F-30ES	56 kt (105 km/h; 65 mph)
Min flying speed: F-30	18 kt (33 km/h; 20 mph)
F-30ES	20 kt (36 km/h; 22 mph)
Max rate of climb at S/L: F-30	366 m (1,200 ft)/min
Service ceiling: both	3,050 m (10,000 ft)
T-O run: both	31–76 m (100–250 ft)
Landing run: both	0–6 m (0–20 ft)

UPDATED

AIR TRACTOR

AIR TRACTOR INC

PO Box 485, Municipal Airport, Olney, Texas 76374
Tel: (+1 940) 564 56 16
Fax: (+1 940) 564 23 48
e-mail: airmail@airtractor.com
Web: http://www.airtractor.com
PRESIDENT: Leland Snow
VICE-PRESIDENT, FINANCE: David Ickert
VICE-PRESIDENT, SALES: Kristin Edwards

Air Tractor agricultural aircraft based on 50-year experience of Leland Snow, who produced Snow S-2 series, which later became Rockwell S-2R; AT-300/301/302 (587 built) no longer in production; 2,000th aircraft delivered in 2002. Eight versions available, powered by various P&W PT6A and R-1340 engines. Company now has 130 employees and total of 13,660 m² (147,000 sq ft) of manufacturing space at three plants.

UPDATED

AIR TRACTOR AT-401 AIR TRACTOR

TYPE: Agricultural sprayer.

PROGRAMME: AT-401 developed 1986 from AT-301, with increased wing span and larger hopper. AT-401A version with Polish PZL-3S radial engine abandoned, with just one aircraft produced. AT-401B version replaced the 401 and has Hoerner wingtips and increased wing span.

CURRENT VERSIONS: **AT-401B:** Standard version; *as described*.

Increased T-O weight: Optional version with landing gear of AT-402A and 517 kg (1,140 lb) of additional disposable load.

CUSTOMERS: By December 2003, 198 AT-401s, one AT-401A and 71 AT-401Bs delivered to Argentina, Australia, Brazil, Canada, Colombia, Mexico, Spain and USA. No reported deliveries in 2002, but two 401Bs registered in 2003. Production of AT-400/400A totalled 72 and 14, respectively.

COSTS: Standard AT-401B US$277,500; with customer-supplied engine US$235,300 (both 2004).

DESIGN FEATURES: Purpose-designed sprayer with cantilever low wing and hopper at CG. Constant-chord, unswept wings and tailplane; latter braced; moderately sweptback fin. High-mounted cockpit with protective reinforcement.

Wing aerofoil NACA 4415; dihedral 3° 30'; incidence 2°.

FLYING CONTROLS: Conventional and manual. Balance tabs on ailerons, elevators and rudders; ailerons droop 10° when electrically operated Fowler flaps deflected to their maximum of 26°.

STRUCTURE: Two-spar wing structure of 2024-T3 light alloy, with alloy steel lower spar cap; bonded doubler inside wing leading-edge to resist impact damage; glass fibre wingroot fairings and skin overlaps sealed against chemical ingress; wing ribs and skins zinc chromated before assembly; flaps and ailerons of light alloy. Fuselage of 4130N steel tube, oven stress relieved and oiled internally, with skin panels of 2024-T3 light alloy attached by Camloc fasteners for quick removal; rear fuselage lightly pressurised to prevent chemical ingress; cantilever fin and strut-braced tailplane of light alloy, metal-skinned and sealed against chemical ingress.

LANDING GEAR: Tailwheel type; fixed. Fuselage-mounted, cantilever heavy-duty E-4340 spring steel main gear, thickness 28.6 mm (1.125 in); flat spring suspension for castoring and lockable tailwheel. Cleveland mainwheels with tyre size 8.50–10 (8 ply), pressure 2.83 bar (41 lb/sq in). Tailwheel tyre size 5.00–5. Cleveland four-piston brakes with heavy-duty discs. Optional AT-402A-type landing gear.

POWER PLANT: One remanufactured 447 kW (600 hp) Pratt & Whitney R-1340 air-cooled radial engine with speed ring cowling or optional RamAir Scoopkit, driving a Hamilton Sundstrand 12D40/6101A-12 two-blade propeller; optional propellers include a Pacific Propeller 22D40/AG200-2 Hydromatic two-blade constant-speed metal and Hydromatic 23D40 three-blade propeller. Air Tractor has designed and is producing new FAA-approved replacement crankshaft for R-1340; other new replacement parts available include main and thrust bearings, and blower (impeller) bearings. Over 300 replacement crankshafts delivered by early 1997. 477 litre (126 US gallon; 105 Imp gallon) fuel tanks. Oil capacity 30 litres (8.0 US gallons; 6.7 Imp gallons). Orenda OE600 V-8 liquid-cooled piston engine of 447 kW (600 hp) tested in December 1999; development towards certification continued, with aim of deliveries during 2001.

ACCOMMODATION: Single seat with nylon mesh cover in enclosed cabin which is sealed to prevent chemical ingress. Downward-hinged window/door on each side. 'Line of sight' instrument layout, with swing-down lower instrument panel for ease of access for instrument maintenance. Baggage compartment in bottom of fuselage, aft of cabin, with door on port side. Cabin ventilation by 0.10 m (4 in) diameter airscoop.

SYSTEMS: 24 V electrical system, supplied by 35 A engine-driven alternator; 60 A alternator optional.

AVIONICS: Optional avionics include Bendix/King KX 155 nav/com/glideslope and ACK E-01 emergency locator transmitter.

EQUIPMENT: Agricultural dispersal system comprises a 1,514 litre (400 US gallon; 333 Imp gallon) Derakane vinylester resin/glass fibre hopper mounted in forward fuselage with hopper window and instrument panel-mounted hopper quantity gauge; 0.97 m (3 ft 2 in) wide Transland gatebox; Transland 5 cm (2 in) bottom loading valve; Agrinautics 6.4 cm (2½ in) spraypump with Transland on/off valve and five-blade variable-pitch plastics fan, and 38-nozzle stainless steel spray system with streamlined booms; swath width 24.39 m (80 ft 0 in). Ground start receptacle and hopper rinse system are standard.

Optional equipment includes night flying package comprising strobe and navigation lights, night working lights, retractable 600 W landing light in port wingtip; and ferry fuel system. Alternative agricultural equipment includes Transland 22358 extra high-volume spreader, Transland 54401 NorCal Swathmaster, and 40 extra spray nozzles for high-volume spraying. Three-piece safety plate glass centre windshield and washer system optional, as is a new crew seat.

DIMENSIONS, EXTERNAL:

Wing span	15.57 m (51 ft 1¼ in)
Wing chord, constant	1.83 m (6 ft 0 in)

Radial-engined Air Tractor AT-401 agricultural sprayer ***NEW***/1048110

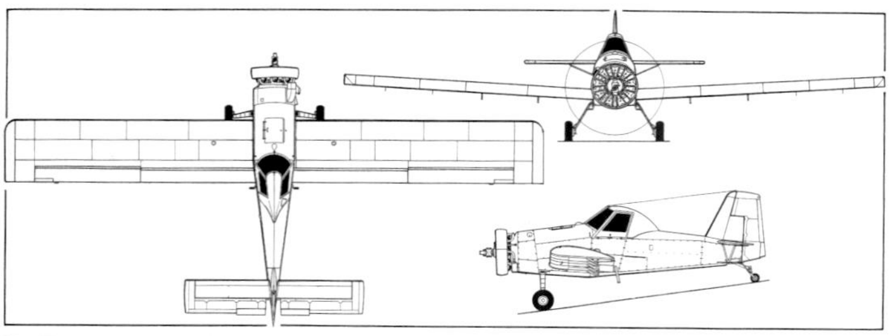

Air Tractor AT-401B (Pratt & Whitney R-1340 radial engine) *(Paul Jackson)* 0051564

Wing aspect ratio 8.5
Length overall .... 8.23 m (27 ft 0 in)
Height overall .... 2.59 m (8 ft 6 in)
Propeller diameter: standard .... 2.77 m (9 ft 1 in)
optional .... 2.59 m (8 ft 6 in)

AREAS:
Wings, gross .... 28.43 m^2 (306.0 sq ft)
Ailerons (total) .... 3.55 m^2 (38.20 sq ft)
Trailing-edge flaps (total) .... 3.75 m^2 (40.40 sq ft)
Fin .... 0.90 m^2 (9.70 sq ft)
Rudder .... 1.30 m^2 (14.00 sq ft)
Tailplane .... 2.42 m^2 (26.00 sq ft)
Elevators, incl tabs .... 2.36 m^2 (25.40 sq ft)

WEIGHTS AND LOADINGS (S: standard, IGW: increased gross weight):
Weight empty, spray equipped .... 1,925 kg (4,244 lb)
Max T-O weight: S .... 3,565 kg (7,860 lb)
IGW .... 4,082 kg (9,000 lb)
Max landing weight: S .... 2,721 kg (6,000 lb)
IGW .... 3,175 kg (7,000 lb)
Max wing loading: S .... 125.4 kg/m^2 (25.69 lb/sq ft)
IGW .... 143.6 kg/m^2 (29.41 lb/sq ft)
Max power loading: S .... 7.97 kg/kW (13.10 lb/hp)
IGW .... 9.13 kg/kW (15.00 lb/hp)

PERFORMANCE (at standard max T-O weight, except where indicated):
Max cruising speed at S/L, hopper empty .... 135 kt (251 km/h; 156 mph)
Cruising speed at FL40 .... 124 kt (230 km/h; 143 mph)
Typical working speed .... 104–122 kt (193–225 km/h; 120–140 mph)
Stalling speed at 2,721 kg (6,000 lb) AUW:
flaps up .... 64 kt (118 km/h; 73 mph)
flaps down .... 53 kt (99 km/h; 61 mph)
Stalling speed as usually landed .... 47 kt (87 km/h; 54 mph)
Max rate of climb at S/L:
at max landing weight .... 335 m (1,100 ft)/min
at max T-O weight .... 158 m (520 ft)/min
T-O run .... 402 m (1,320 ft)
Range at econ cruising speed at FL80, no reserves .... 547 n miles (1,014 km; 630 miles)

UPDATED

AIR TRACTOR AT-402 TURBO AIR TRACTOR

TYPE: Agricultural sprayer.

PROGRAMME: Follow-on from AT-400 of 1980, AT-402 first flight August 1988; certified November 1988; first delivery late 1988. Current 402A and 402B have Hoerner wingtips and increased span.

CURRENT VERSIONS: **AT-402A:** Introduced in mid-1997, supplementing AT-402B, powered by a 410 kW (550 shp) P&WC PT6A-11AG. Aimed at first-time turbine buyer and priced accordingly, complete with spray dispersal equipment, lights and air conditioning.

AT-402B: Combines fuselage, tail surfaces and landing gear of AT-400 with turboprop engine and wing of AT-401B.

Description refers to AT-402B, and is generally as for AT-401, except that below.

CUSTOMERS: Total of 68 AT-402s, 105 AT-402As and 33 AT-402Bs delivered by December 2003. Production continued in 2004.

COSTS: Standard AT-402A with PT6A-11AG US$468,500; standard AT-402B US$562,500; or US$265,500 with customer-supplied PT6A-15AG, -27, -28 or -34AG engine (2004).

DESIGN FEATURES: Broadly as AT-401. All versions have steel alloy lower wing spar caps for long fatigue life and reinforced leading-edges to prevent birdstrike damage. Size 29×11.0–10 high-flotation tyres and wheels as standard.

POWER PLANT: One 507 kW (680 shp) P&WC PT6A-15AG, -27 or -28 turboprop, either new or customer-furnished, driving a Hartzell HCB3TN-3D/T1028N+4 three-blade constant-speed reversible-pitch propeller. Standard fuel capacity 644 litres (170 US gallons; 142 Imp gallons); optional fuel tankage 818 litres (216 US gallons; 180 Imp gallons) or 886 litres (234 US gallons; 195 Imp gallons).

SYSTEMS: 250 A starter/generator and two 24 V 21 Ah batteries.

EQUIPMENT: Hopper and gatebox size as for AT-401; optional equipment includes Transland extra high-volume dispersal system; engine-driven air conditioning system.

DIMENSIONS, EXTERNAL:
Length overall .... 9.32 m (30 ft 7 in)
Height: propeller turning .... 3.40 m (11 ft 2 in)
to top of cockpit .... 2.90 m (9 ft 6 in)
Tailplane span .... 5.23 m (17 ft 2 in)
Wheel track .... 2.62 m (8 ft 7 in)

WEIGHTS AND LOADINGS:
Weight empty, spray equipped, both .... 1,783 kg (3,930 lb)
Certified gross weight (FAR Pt 23) .... 3,175 kg (7,000 lb)
Typical operating weight (CAM 8):
AT-402A .... 3,901 kg (8,600 lb)
AT-402B .... 4,159 kg (9,170 lb)
Max landing weight, both .... 3,175 kg (7,000 lb)
Max wing loading .... 137.2 kg/m^2 (28.10 lb/sq ft)
Max power loading .... 7.70 kg/kW (12.65 lb/shp)

PERFORMANCE:
Max level speed at S/L:
clean .... 174 kt (322 km/h; 200 mph)
with dispersal equipment .... 160 kt (298 km/h; 185 mph)
Cruising speed at 283 kW (380 shp) at FL80 .... 142 kt (264 km/h; 164 mph)
Typical working speed .... 113–126 kt (209–233 km/h; 130–145 mph)
Stalling speed at 2,721 kg (6,000 lb) AUW:
flaps up .... 64 kt (118 km/h; 73 mph)
flaps down .... 53 kt (99 km/h; 61 mph)
Stalling speed at 2,041 kg (4,500 lb) typical landing weight .... 46 kt (86 km/h; 53 mph)
Max rate of climb at S/L, dispersal equipment installed: AUW of 2,721 kg (6,000 lb) ..... 495 m (1,625 ft)/min
AUW of 3,565 kg (7,860 lb) .... 320 m (1,050 ft)/min
T-O run at 3,565 kg (7,860 lb) AUW .... 247 m (810 ft)
Landing run at 2,041 kg (4,500 lb) AUW .... 122 m (400 ft)
Range at econ cruising speed at FL80, no reserves .... 573 n miles (1,062 km; 660 miles)

UPDATED

Air Tractor AT-402A Turbo Air Tractor ***NEW***/1048111

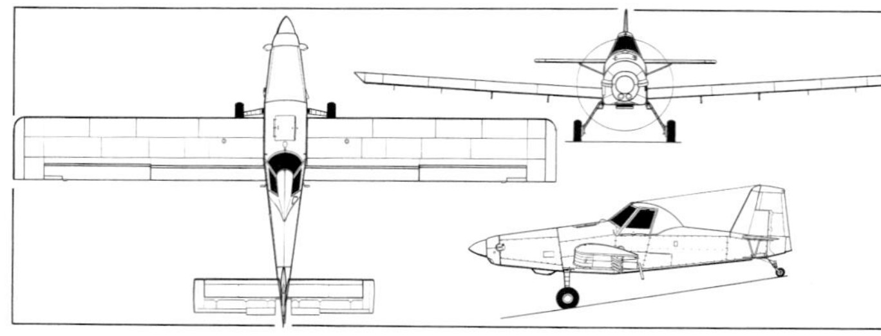

Air Tractor AT-402A/AT-402B Turbo Air Tractor *(Paul Jackson)* 0079012

AIR TRACTOR AT-502 TURBO

TYPE: Agricultural sprayer.

PROGRAMME: Developed as stretched AT-401. First flight of AT-502 April 1987; certified 23 June 1987.

CURRENT VERSIONS: **AT-502:** Original version.

Description refers to AT-502, and is generally as for AT-401, except that below.

AT-502A: Similar to AT-502B but with 820 kW (1,100 shp) PT6A-45R; slow turning (1,425 rpm) five-blade Hartzell (HCB5MP-3C/M10876AS) propeller; enlarged vertical tail surfaces. For operation in mountainous terrain or short strips. Prototype first flight February 1992; certified April 1992; 26 delivered by end 2003.

AT-502B: Stronger wing than AT-502, with 0.61 m (2 ft) longer span and Hoerner wingtips, which are stated to increase width of spray pattern by 0.91 m (3 ft); 303 delivered by end of 2003.

AT-503A: Dual control, tandem-seat trainer; PT6A-34AG engine. Prototype designated AT-503. In low-rate production.

CUSTOMERS: Total of 551 AT-500 series delivered by December 2003, including out-of-production AT-501 (nine), AT-502 (208), AT-503 (one), plus AT-503A (three). Production continued in 2004.

COSTS: Standard AT-502B US$614,500 with PT6A-15AG, or US$661,000 with-34AG; or

Air Tractor AT-502 at work ***NEW***/1048112

US$311,900 with customer-supplied PT6A-15AG, -27, -28, -34AG (2004).

DESIGN FEATURES: See AT-401. 1,892 litre (500 US gallon; 416 Imp gallon) hopper handles low-density nitrogen-based fertilisers such as urea; safety glass centre windscreen with wiper. Alloy steel lower spar cap and bonded doubler on inside of wing leading-edge for increased resistance to impact damage, and glass fibre wingroot fairings.

LANDING GEAR: Tailwheel type; fixed. Fuselage-mounted, heavy-duty E-4340 spring steel main gear, thickness 33.2 mm (1.31 in); flat spring for castoring and lockable tailwheel. Cleveland mainwheels, with high-flotation tyres size 29×11.0–10, pressure 3.45 bar (50 lb/sq in); tailwheel tyre size 5.00–5; Cleveland six-piston brakes with heavy-duty discs.

POWER PLANT: One 507 kW (680 shp) Pratt & Whitney Canada PT6A-15AG, PT6A-27 or PT6A-28, or 559 kW (750 shp) PT6A-34 or PT6A-34AG turboprop, driving a Hartzell HCB3TN-3D/T10282–4 three-blade metal propeller. Standard fuel capacity 644 litres (170 US gallons; 142 Imp gallons). Optional capacities 818 litres (216 US gallons; 180 Imp gallons) and 886 litres (234 US gallons; 195 Imp gallons). AT-502A has 818 litre (216 US gallon; 180 Imp gallon) tanks as standard.

ACCOMMODATION: One or two persons (see Current Versions). Has quickly detachable instrument panel and removable fuselage skin panels for ease of maintenance.

SYSTEMS: Three 24 V 42 Ah batteries and 250 A starter/generator.

AVIONICS: *Comms:* Optional avionics include Bendix/King KX 155 nav/com/glideslope and KR 87 ADF, KY 196 VHF radio, KT 76A transponder and ACK E-01 emergency locator transmitter.

EQUIPMENT: Agricultural dispersal system comprises a 1,892 litre (500 US gallon; 416 Imp gallon) hopper mounted in forward fuselage with hopper window and instrument panel-mounted hopper quantity gauge; 0.97 m (3 ft 2 in) wide Transland gatebox; Transland 6.4 cm (2½ in) bottom loading valve; Agrinautics 6.4 cm (2½ in) spraypump with Transland on/off valve and five-blade variable-pitch plastics fan and 38-nozzle stainless steel spray system with streamlined booms. Optional dispersal equipment includes 7.6 cm (3 in) spray system with 119 spray nozzles, and automatic flagman; spray swath 25.91 m (85 ft 0 in).

Standard equipment includes safety glass centre windscreen panel, ground start receptacle, three-colour polyurethane paint finish, strobe and navigation lights; windscreen washer and wiper; and twin nose-mounted landing/taxi lights. Optional equipment includes engine-driven air conditioning system, night flying package, comprising night working lights, retractable 600 W landing light in port wingtip; fuel flowmeter, fuel totaliser and ferry fuel system. Alternative agricultural equipment includes Transland 22356 extra high-volume spreader, Transland 54401 NorCal Swathmaster, 40 extra spray nozzles for high-volume spraying and eight- or 10-unit Micronair Mini Atomiser unit; hopper rinse tank is standard. New crew seat is also optional.

DIMENSIONS, EXTERNAL:

Wing span: AT-502A/502B	15.85 m (52 ft 0 in)
AT-502	15.24 m (50 ft 0 in)
Wing chord, constant	1.83 m (6 ft 0 in)
Wing aspect ratio	8.7
Length overall	10.11 m (33 ft 2 in)
Height: propeller turning	3.63 m (11 ft 11 in)
to top of cockpit	3.12 m (10 ft 3 in)
Tailplane span	5.23 m (17 ft 2 in)
Wheel track	3.12 m (10 ft 3 in)
Wheelbase	6.64 m (21 ft 9½ in)
Propeller diameter: AT-502B	2.69 m (8 ft 10 in)

AREAS:

Wings, gross: AT-502A/502B	28.99 m^2 (312.0 sq ft)
AT-502	27.87 m^2 (300.0 sq ft)
Ailerons (total)	3.53 m^2 (38.0 sq ft)
Trailing-edge flaps (total)	3.75 m^2 (40.4 sq ft)
Fin	0.90 m^2 (9.7 sq ft)
Rudder	1.30 m^2 (14.0 sq ft)
Tailplane	2.41 m^2 (26.0 sq ft)
Elevators (total, incl tab)	2.44 m^2 (26.3 sq ft)

WEIGHTS AND LOADINGS:

Weight empty, spray equipped	1,949 kg (4,297 lb)
Max T-O weight	4,399 kg (9,700 lb)
Max landing weight	3,629 kg (8,000 lb)
Max wing loading:	
AT-502A/502B	151.8 kg/m^2 (31.09 lb/sq ft)
Max power loading: 502A	5.37 kg/kW (8.82 lb/shp)
502B (PT6A-34AG)	7.87 kg/kW (12.93 lb/shp)
502B (PT6A-15AG)	8.68 kg/kW (14.26 lb/shp)

PERFORMANCE (AT-502B with spray equipment installed):

Never-exceed speed (VNE) and max level speed at S/L, hopper empty	156 kt (290 km/h; 180 mph)
Cruising speed at FL80, 283 kW (380 shp)	136 kt (253 km/h; 157 mph)
Typical working speed	104–126 kt (193–233 km/h; 120–145 mph)
Stalling speed at 3,629 kg (8,000 lb) AUW:	
flaps up	72 kt (132 km/h; 82 mph)
flaps down	59 kt (110 km/h; 68 mph)
Stalling speed at 1,978 kg (4,360 lb) typical landing weight	46 kt (86 km/h; 53 mph)
Max rate of climb at S/L, AUW of 3,629 kg (8,000 lb):	
with PT6A-15AG	311 m (1,020 ft)/min
with PT6A-34AG	360 m (1,180 ft)/min
Max rate of climb at S/L, AUW of 4,309 kg (9,500 lb):	
with PT6A-15AG	232 m (760 ft)/min
with PT6A-34AG	282 m (925 ft)/min
T-O run at AUW of 3,629 kg (8,000 lb):	
with PT6A-15AG	236 m (775 ft)
with PT6A-34AG	222 m (730 ft)
T-O run at AUW of 4,309 kg (9,500 lb):	
with PT6A-15AG	356 m (1,170 ft)
with PT6A-34AG	302 m (990 ft)
Range with max fuel	538 n miles (998 km; 620 miles)

UPDATED

AIR TRACTOR AT-602

TYPE: Agricultural sprayer.

PROGRAMME: Prototype first flew 1 December 1995. Certification completed 6 June 1996 and deliveries began the following month. Aircraft fits into Air Tractor range between AT-502B and AT-802A, being AT-502 with increased wing and tail spans, plus taller fin.

CUSTOMERS: Total of 121 sold by early December 2003. Production continued in 2004.

COSTS: Standard AT-602 with PT6A-60AG US$869,500; with customer-supplied engine US$383,500 (2004).

Description generally as for AT-401, except that following.

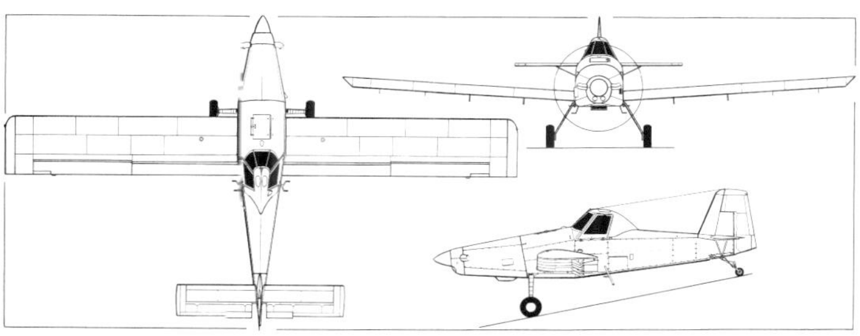

Air Tractor AT-502B agricultural aircraft *(Paul Jackson)* 0079013

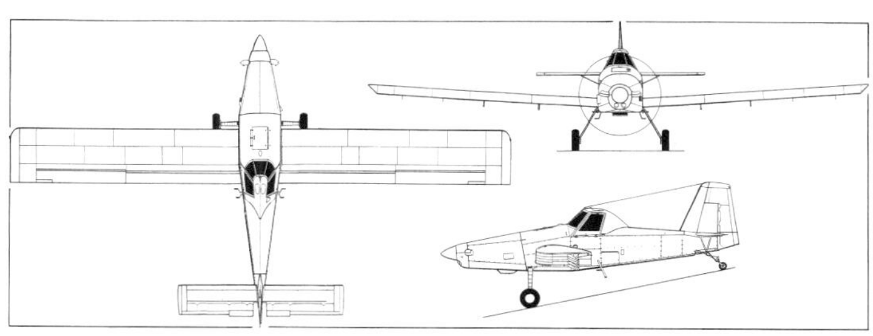

Air Tractor AT-602 (P&WC PT6A-60AG engine) *(Paul Jackson)* 0079015

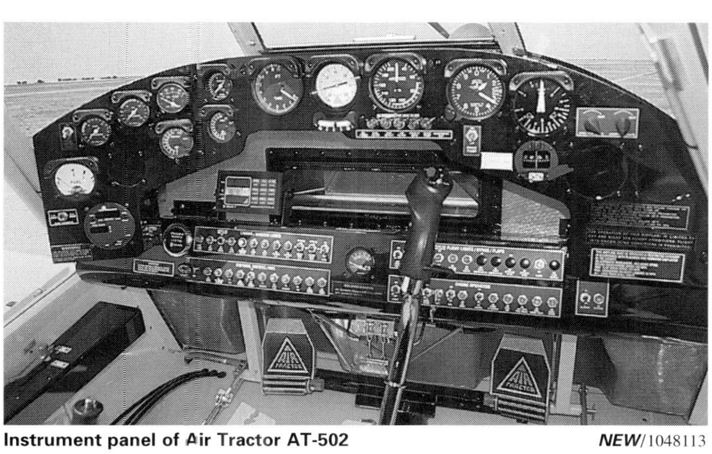

Instrument panel of Air Tractor AT-502 ***NEW***/1048113

Air Tractor AT-602 agricultural sprayer ***NEW***/1048114

POWER PLANT: Choice of 783 kW (1,050 shp) Pratt & Whitney Canada PT6A-60AG turboprop, or PT6A-65AG of 966 kW (1,295 shp), driving a Hartzell five-blade constant-speed reversing propeller. Fuel capacity 818 litres (216 US gallons; 180 Imp gallons); optional fuel capacity 1,105 litres (292 US gallons; 243 Imp gallons).

EQUIPMENT: Glass fibre hopper, capacity 2,385 litres (630 US gallons; 525 Imp gallons); Transland 0.96 m (3 ft 2 in) wide gatebox; five-blade ground-adjustable spraypump fan; pump shut-off valve. Optional crew seat.

DIMENSIONS, EXTERNAL:	
Wing span	17.07 m (56 ft 0 in)
Wing chord, constant	1.83 m (6 ft 0 in)
Wing aspect ratio	9.3
Length overall	10.41 m (34 ft 2 in)
Height: propeller turning	3.71 m (12 ft 2 in)
to top of fin	3.38 m (11 ft 1 in)
to top of cockpit	3.17 m (10 ft 5 in)
Tailplane span	5.66 m (18 ft 7 in)
Wheel track	3.07 m (10 ft 1 in)
Wheelbase	6.71 m (22 ft 0 in)
AREAS:	
Wings, gross	31.22 m² (336.0 sq ft)
WEIGHTS AND LOADINGS:	
Weight empty, equipped	2,540 kg (5,600 lb)
Max T-O weight	5,670 kg (12,500 lb)
Max landing weight	5,443 kg (12,000 lb)
Max wing loading	181.6 kg/m² (37.20 lb/sq ft)
Max power loading	7.25 kg/kW (11.90 lb/shp)
PERFORMANCE:	
Never-exceed speed (VNE)	189 kt (350 km/h; 217 mph)
Working speed	130 kt (241 km/h; 150 mph)
Stalling speed: flaps up	86 kt (160 km/h; 99 mph)
flaps down	75 kt (139 km/h; 87 mph)

UPDATED

AIR TRACTOR AT-802

TYPE: Agricultural sprayer.

PROGRAMME: Design started July 1989; optional configuration as firefighter; first flight of prototype (N802LS) 30 October 1990; second aircraft flew November 1991, with PT6A-45R and configured as agricultural model with spraybooms, pump and Transland gatebox. Production deliveries started second quarter 1993.

CURRENT VERSIONS: **AT-802**: Tandem two-seater powered by P&WC PT6A-45R; certified for gross weight of 6,804 kg (15,000 lb) 27 April 1993. PT6A-65AG and -67AG versions certified April 1993 at maximum T-O weight of 7,257 kg (16,000 lb).

AT-802A: Single-seat version; third production aircraft in this configuration; can be powered by refurbished PT6A-65AG or -67AG engine. First flight 6 July 1992; FAA certification gained 17 December 1992; certified for gross weight of 6,804 kg (15,000 lb) with PT6A-45R on 27 April 1993. PT6A-65AG versions certified March 1993 at maximum T-O weight of 7,257 kg (16,000 lb).

AT-802AF Fire Boss: Amphibious firefighting version; Wiplane amphibious floats with water-scooping capability.

AT-802F: Two-seat firefighting version; PT6A-67AG engine. FAA certification at 5,670 kg (12,500 lb) maximum T-O weight 17 December 1992; certified at 7,257 kg (16,000 lb) on 27 April 1993, giving useful load of 3,987 kg (8,790 lb).

Data apply to AT-802F version, except where indicated.

CUSTOMERS: Total 166 registered by early 2004. Deliveries include 20 to Australia, 28 to Europe and eight to Canada plus two to Saudi Arabia for oil slick eradication. Total of 56 engaged in firefighting duties in six countries during 2002. Recent customers include Compañia de Extincion General de Incendios SA (CEGISA) of Spain, which purchased the first two production AT-802AF Fire Bosses in late 2003, leasing one to Conair of Canada. Forest Protection Limited (FPL) of New Brunswick, Canada, which ordered six AT-802Fs, one purchased after lease during 2001, two for delivery in 2002 and three scheduled for delivery in 2003, to replace Grumman Avenger firebombers; and the US State Department's International Narcotics and Law Enforcement Division, which ordered eight AT-802s for delivery from April 2002, for drug eradication duties in Colombia in a joint programme with the Colombian National Police. These aircraft are modified with self-sealing fuel tanks, armoured engine compartments and cockpits, with bulletproof windscreens, engine fire-extinguishing systems; oxygen system, GPS and HF radios.

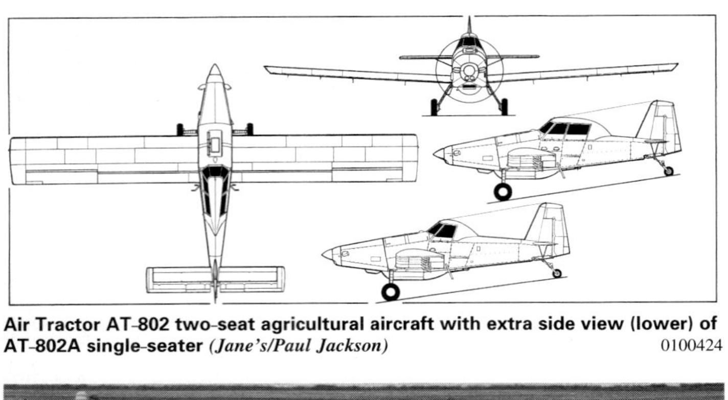

Air Tractor AT-802 two-seat agricultural aircraft with extra side view (lower) of AT-802A single-seater ***(Jane's/Paul Jackson)*** 0100424

Air Tractor AT-802A of Forest Protection Ltd, Canada ***NEW***/0576911

COSTS: Standard AT-802A US$996,500 with new PT6A-65AG; with customer-supplied PT6A-65AG or -67AG US$463,500. Standard AT-802AF US$1,245,000 with PT6A-67AG; or US$625,500 with customer-supplied PT6A-67AG. Two-seat version with dual controls US$32,500 extra (all 2003).

DESIGN FEATURES: Generally as for AT-401. Largest aircraft built by company to date; full dual controls for training; also designed for firefighting; programmable logic computer with cockpit control panel and digital display enables pilot to select coverage level and opens hydraulically operated 'bomb bay' drop doors to prescribed width, closing them when selected amount of retardant released. Drop doors adjust automatically for changing head pressure and aircraft acceleration to provide a constant flow rate and even ground coverage.

Wing aerofoil section NACA 4415; dihedral 3° 30'; incidence 2°.

FLYING CONTROLS: Manually operated ailerons, elevators and rudder with balance tabs; electrically operated Fowler trailing-edge flaps deflect to maximum 30°.

STRUCTURE: Two-spar wing of 2024-T3 light alloy with alloy steel upper and lower spar caps and bonded doubler on inside of leading-edge for impact damage resistance; ribs and skins zinc chromated before assembly; glass fibre wingroot fairing and skin overlaps sealed against chemical ingress; flaps and ailerons of light alloy. Fuselage of welded 4130N steel tube, oven stress relieved and oiled internally, with skin panels of 2024-T3 light alloy attached by Camloc fasteners for quick removal; rear fuselage lightly pressurised to prevent chemical ingress. Cantilever fin and strut-braced tailplane of light alloy, metal skinned and sealed against chemical ingress.

LANDING GEAR: Tailwheel type; fixed. Fuselage-mounted, cantilever heavy-duty E-4340 spring steel main legs, thickness 44.5 mm (1.75 in); flat spring suspension for castoring and lockable tailwheel. Cleveland mainwheels with tyre size 11.00-12 (10 ply), pressure 4.14 bar (60 lb/sq in). Tailwheel tyre size 6.00-6. Cleveland eight-piston brakes with heavy-duty discs.

POWER PLANT: One Pratt & Whitney Canada PT6A-65AG or -67AG turboprop, rated at 966 kW (1,295 shp) for -65AG and 1,007 kW (1,350 shp) for -67AG, both driving a Hartzell five-blade feathering and reversible-pitch constant-speed metal propeller. Fuel in two integral wing tanks, total usable capacity 961 litres (254 US gallons; 211 Imp gallons); optional tanks increase capacity to 1,438 litres (380 US gallons; 317 Imp gallons). Engine air is filtered through two large pleated paper industrial truck filters.

ACCOMMODATION: One or two seats in enclosed cabin, which is sealed to prevent chemical ingress and protected with overturn structure. Four downward-hinged doors, two on each side. Windscreen is safety-plate auto glass, with washer and wiper. Air conditioning system standard.

SYSTEMS: Hydraulic system, pressure 207 bar (3,000 lb/sq in).

AVIONICS: Choice of avionics suites by Bendix/King and Garmin.

Comms: Bendix/King KX 155A nav/com; KY 99A or KY 196A transceivers; KLX 135 GPS/com; KT 76A transponder; KMA 24H audio control panel; Garmin GNS 430 or GNS 530 nav/com/GPS; GTX 320 transponder; GMA 340 audio panel.

Flight: Bendix/King KT 87 ADF; KCS 55A compass; KN 64 DME.

EQUIPMENT: Two removable Derakane vinylester hoppers forward of cockpit and 227 litre (60 US gallon; 50 Imp gallon) gate tank in ventral bulge, for agricultural chemical, fire retardant or water; total capacity 3,028 litres (800 US gallons; 666 Imp gallons). Electronic spray monitoring system standard. Transland high-volume fertilising gate 0.96 × 0.25 m (3 ft 2 in × 10 in) optional.

DIMENSIONS, EXTERNAL:	
Wing span	17.68 m (58 ft 0 in)
Wing chord, constant	2.07 m (6 ft 9½ in)
Wing aspect ratio	8.4
Length overall	10.95 m (35 ft 11 in)
Height overall	3.89 m (12 ft 9 in)
Tailplane span	6.03 m (19 ft 9½ in)
Wheel track	3.10 m (10 ft 2 in)
Wheelbase	7.25 m (23 ft 9½ in)
Propeller diameter (-65AG)	2.92 m (9 ft 7 in)
AREAS:	
Wings, gross	37.25 m² (401.0 sq ft)
Ailerons (total)	4.61 m² (49.60 sq ft)
Trailing-edge flaps (total)	5.54 m² (59.60 sq ft)
Fin	1.24 m² (13.40 sq ft)
Rudder	1.57 m² (16.90 sq ft)
Tailplane	3.44 m² (37.00 sq ft)
Elevators (total, incl tab)	3.00 m² (32.30 sq ft)
WEIGHTS AND LOADINGS (AT-802A):	
Weight empty, equipped: sprayer	2,867 kg (6,320 lb)
firefighter	3,270 kg (7,210 lb)
Max T-O and landing weight	7,257 kg (16,000 lb)
Max wing loading	194.8 kg/m² (39.90 lb/sq ft)
Max power loading (-67AG)	7.21 kg/kW (11.85 lb/shp)
PERFORMANCE:	
Max level speed at S/L	182 kt (338 km/h; 210 mph)
Max cruising speed at FL80	192 kt (356 km/h; 221 mph)
Stalling speed, power off, flaps down, at max landing weight	79 kt (147 km/h; 91 mph)
Max rate of climb at S/L	259 m (850 ft)/min
Service ceiling	3,965 m (13,000 ft)
T-O run	610 m (2,000 ft)
Range with max fuel	696 n miles (1,289 km; 800 miles)

UPDATED

AIR TRACTOR S-22 SURVEYOR

TYPE: Light utility turboprop.

PROGRAMME: Design started 2000; first delivery then expected in mid-2004, but development slowed in 2002, and on hold in 2004.

CUSTOMERS: US Department of the Interior has ordered one, with potential requirement for up to 15 for aerial survey of water-fowl.

DESIGN FEATURES: High wing; intended for passenger/freight carrying and aerial applications duties in remote areas; design goals include exceptional cockpit visibility and high cruising speed.

LANDING GEAR: Amphibious landing gear standard.

POWER PLANT: One Pratt & Whitney Canada PT6A turboprop.

ACCOMMODATION: Standard accommodation for up to 10 passengers.

UPDATED

AMD

AIRCRAFT MANUFACTURING & DEVELOPMENT

415 Airport Road, Heart of Georgia Regional Airport, PO Box 639, Eastman, Georgia 31023
Tel: (+1 912) 374 27 59
Fax: (+1 912) 374 27 93
e-mail: info@newplane.com
Web: http://www.NewPlane.com
PRESIDENT: Mathieu Heintz
MANUFACTURING DIRECTOR: Shawn Fallin
SALES MANAGER: Lisa Lewis

AMD was established in 1999 to produce and market the Zenair Zenith CH 2000 and CH2T. Its purpose-built, 2,600 m² (28,000 sq ft) factory became operational in December 1999; the first US-produced CH 2000 was completed in January 2000.

On 1 March 2001, AMD first flew the CH 640 kitbuilt, which draws on the CH 2000, but is now marketed by Zenair.

UPDATED

AMD ZENITH CH 2000 ALARUS

TYPE: Two-seat lightplane.

PROGRAMME: Prototype C-FQCU first flown 26 June 1993; Canadian and US certification of prototypes 26 and 31 July 1994, respectively; first two production models (C-FRSK and C-FRSV) completed April 1994. First delivery September 1994; US type certification 25 July 1995 in Utility category and 29 January 2003 in Normal category. A test example fitted with 104.5 kW (140 hp) Lycoming O-320 was produced in 1995. Certified for IFR operations and spins on 2 October 1996. JAA certification received March 2000. Company revealed it would produce a lighter version during 2001, with 95 litre (25.0 US gallons; 20.8 Imp gallon) fuel capacity; this appears not to have occurred.

CURRENT VERSIONS: **CH 2000:** Standard production model, *as described*.

CH2T: Basic model intended for flight training market; avionics are not included.

CUSTOMERS: Total of 43 delivered by time production moved from Canada to US in June 1999. First US-built aircraft (N17KA, one of 11 CH2Ts for KeyFlite Academy at Utica, New York, and Nashua, New Hampshire) received C of A on 14 January 2000 and delivered following day. Additionally, Zenair built production aircraft 0063 to 0065. A total of 106 built by August 2004.

COSTS: CH 2000 US$129,000 (2004). Provisional cost of lightweight version US$84,900 (2001).

DESIGN FEATURES: Side-by-side two-seat and smaller-span derivative of Tri-Z CH 300. Designed to conform to FAR Pt 23 (JAR-VLA equivalent level of safety). Conventional, low-wing configuration, with mid-mounted tailplane and sweptback fin.

Wing section LS (1) 0417 (mod).

FLYING CONTROLS: Manual. All-moving tailplane deflects 12° up and 9° 30' down with anti-balance tabs; all-moving fin/rudder with horn balance deflects 21° 30' each way. Electrically actuated split flaps lower to 50°; ailerons deflect ±15°.

STRUCTURE: Semi-monocoque fuselage of aluminium alloy, with stressed skins. Wings and tailplane aluminium skinned.

LANDING GEAR: Non-retractable tricycle type with steerable (±14°) nosewheel. Shock-cord absorption on nosewheel. Cleveland 5.00-5 in wheels with hydraulic disc brakes on mainwheels. Optional wheel fairings.

POWER PLANT: One 86.5 kW (116 hp) Textron Lycoming O-235-N2C, driving metal Sensenich 72-CK-0-46 or 72-CK-0-48 two-blade propeller. Fuel capacity 106 litres (28.0 US gallons; 23.3 Imp gallons) standard of which 104 litres (27.5 US gallons; 22.9 Imp gallons) are usable. 129 litres (34.0 US gallons; 28.3 Imp gallons) optional. Oil capacity 5.7 litres (1.5 US gallons; 1.25 Imp gallons).

SYSTEMS: 12 V 60 A, heavy-duty battery.

ACCOMMODATION: Two persons, side by side, with dual controls; upward-hinged door each side; fixed windscreen; rear side windows. AmSafe inflatable Restraint system certified for use in April 2004. Baggage shelf behind seats.

AVIONICS: *Comms:* Based around Garmin GNS 430 COM/VOR/GPS, GI 106A CDI, GMA 340 audio panel and GTX 327 transponder, plus Bendix King KX155 NAV/COM. Options include PS 100II intercom.

DIMENSIONS, EXTERNAL:
- Wing span ... 8.79 m (28 ft 10 in)
- Wing aspect ratio ... 5.4
- Length overall ... 7.01 m (23 ft 0 in)
- Height overall ... 2.08 m (6 ft 10 in)
- Propeller diameter ... 1.83 m (6 ft 0 in)

DIMENSIONS, INTERNAL:
- Cabin max width ... 1.17 m (3 ft 10 in)

AREAS:
- Wings, gross ... 12.73 m² (137.0 sq ft)

WEIGHTS AND LOADINGS:
- Weight empty ... 533 kg (1,085 lb)
- Baggage weight ... 18 kg (40 lb)
- Max T-O weight: Utility ... 728 kg (1,606 lb)
  - Normal ... 767 kg (1,692 lb)
- Max wing loading: Utility ... 57.2 kg/m² (11.72 lb/sq ft)
  - Normal ... 60.3 kg/m² (12.35 lb/sq ft)
- Max power loading: Utility ... 8.43 kg/kW (13.84 lb/hp)
  - Normal ... 8.88 kg/kW (14.59 lb/hp)

PERFORMANCE:
- Never-exceed speed (VNE) ... 139 kt (257 km/h; 160 mph)
- Econ cruising speed ... 99 kt (183 km/h; 114 mph)
- Stalling speed ... 48 kt (89 km/h; 55 mph)
- Max rate of climb at S/L ... 229 m (750 ft)/min
- Service ceiling ... 3,660 m (12,000 ft)
- T-O to 15 m (50 ft) ... 500 m (1,640 ft)
- Landing run ... 183 m (600 ft)
- Landing from 15 m (50 ft) ... 555 m (1,820 ft)
- Range with max payload: standard fuel ... 471 n miles (872 km; 542 miles)
  - optional fuel ... 851 n miles (1,577 km; 980 miles)
- *g* limits, Utility category ... +4.4/−2.2

UPDATED

AMD Zenith CH 2000 Alarus *(Paul Jackson)* ***NEW***/1047015

AMERICAN CHAMPION

AMERICAN CHAMPION AIRCRAFT CORPORATION

PO Box 37, 32032 Washington Avenue, Highway D, Rochester, Wisconsin 53167
Tel: (+1 262) 534 63 15
Fax: (+1 262) 534 23 95
e-mail: aca-sales@mia.net
Web: http://www.americanchampionaircraft.com
PRESIDENT AND CEO: Jerry K Mehlhaff
GENERAL MANAGER: Dale Gauger
CHIEF ENGINEER: Jerry Mehlhaff Jr

ACAC offers new-build Citabria (now called Aurora, Adventure and Explorer), Super Decathlon and Scout, all designed by Aeronca. Model 7 (based on Second World War L-3 Grasshopper liaison/observation machine) certified 18 October 1945. Series later marketed by Champion Aircraft Corporation (1954); from 1970 by Bellanca Aircraft Corporation; and then, in 1982, Champion Aircraft Company. Assets transferred to Jerry Mehlhaff's American Champion Aircraft Corporation, which restarted production in 1990. By end of 2003, various manufacturers had built 14,654 of the Model 7 and Model 8 families, American Champion being responsible for 667. In addition, ACAC offers its new metal spar wing for retrofit on all models, including the original 7AC Champion. A new model, the 7ACA, was displayed at Oshkosh in July 2000; the company is investigating ways of making the aircraft financially viable for production.

UPDATED

AMERICAN CHAMPION PRODUCTION
(at 1 January 2004)

Version	Certification	Number Built	Recent Shipments[1]	ACA Total				
		To 1989	To 2003	2000	2001	2002	2003	
7AC	18 Oct 45	7,200	7,200					
7ACA	30 Apr 71	71	71					
7BCM	16 Sep 47	509	509					
7CCM	12 Jul 48	226	226					
7DC	12 Jul 48	184	184					
7EC	30 Nov 49	773	773					
7ECA	5 Aug 64	1,353	1,390	3	2	3	2	37
7GC	30 Oct 58	171	171					
7GCA/AA	17 Nov 59	396	478	23	8	12	9	82
7GCB/BA	28 Apr 60	195	195					
7GCBC	3 Dec 65	1,215	1,372[2]	22	21	13	12	157
7HC	6 Aug 59	39	39					
7JC	24 Feb 60	25	25					
7KC	10 Jan 61	4	4					
7KCAB	21 Dec 66	624	624					
8GCBC	30 Apr 74	360	451	23	6	11	8	91
8KCAB	16 Oct 70	642	942	25	19	14	32	300
Totals		**13,987**	**14,654**	**96**	**56**	**53**	**63**	**667**

[1] Shipment totals differ slightly from annual production totals
[2] Including 16 aircraft duplicating previous Bellanca c/ns

AMERICAN CHAMPION 7ACA CHAMP

TYPE: Two-seat lightplane.

PROGRAMME: Model 7AC Champion was initial production version (7,200 built with Continental A-65 engine) from 1945 onwards; Model 7ACA Champ certified 30 April 1971 with 44.7 kW (60 hp) Franklin engine, but only 71 built by Bellanca.

American Champion proposing relaunch with 59.7 kW (80 hp) Jabiru 2200 flat-four; new-build prototype (N82AC) displayed at Oshkosh in July 2000, shortly after first flight; will remain experimental while company explores demand. Target maximum take-off weight is 533 to 559 kg (1,220 to 1,232 lb) and maximum level speed 100 kt (185 km/h; 115 mph); cost estimated in region of US$60,000 (2000 prices), or US$45,000 for quick-build kit, if offered. At Sun 'n' Fun, in April 2002, the company confirmed that investigations were continuing, with emphasis then on propeller optimisation. Certification will be in FAA Light Sport Aircraft category.

UPDATED

Prototype for proposed relaunch of American Champion 7ACA Champ *(Paul Jackson)* 0525806

AMERICAN CHAMPION 7ECA CITABRIA AURORA

TYPE: Two-seat lightplane.

PROGRAMME: Original Aeronca Model 7 and successors built in 16 subvariants, with various letter suffixes. Currently the lowest powered of the American Champion range, the 7EC was certified on 30 November 1949 and 7ECA on 5 August 1964; 773 of former and 1,369 of latter had been built up to 1984; was reintroduced into range during 1995 as the cheapest aircraft. Current production aircraft has improved ventilation and heating; redesigned instrument panel and quick-jettison door; and dual controls and brakes.

CUSTOMERS: Total three delivered in 2002, two in 2003 and one in the first six months of 2004, increasing overall production to 1,391.

COSTS: Standard aircraft US$73,900 (2003).

DESIGN FEATURES: High-wing cabin monoplane; wings braced by V-struts and two secondary struts each side; wire-braced fin and tailplane; constant-chord wings; sweptback fin with (current versions) squared tip.

Wing section NACA 4412; dihedral 1°.

FLYING CONTROLS: Conventional and manual. Horn-balanced elevators and rudder. Flight-adjustable trim tab in port elevator. No flaps. Control surface movements: ailerons +27° 30'/−19°; elevators ±24°; rudder ±25°. Aileron suspended balance tabs (spades) optional.

STRUCTURE: Stainless steel exhaust system. Fuselage and empennage are welded chromoly steel tube with Dacron covering; two-spar wing has aluminium spars and ribs, Dacron covering, GFRP tips and steel tube V struts.

LANDING GEAR: Tailwheel type; fixed. Fuselage-mounted, cantilever spring steel mainwheel legs; mainwheels 6.00–6 (4 ply); tailwheel 5.00, steerable. Hydraulic, toe-operated disc and parking brakes. Main wheel fairings optional. Floatplane option includes auxiliary fins above and below tailplane.

American Champion 7ECA Citabria Aurora *(Paul Jackson)* ***NEW***/0576910

POWER PLANT: One 88 kW (118 hp) Textron Lycoming O-235-K2C flat-four piston engine driving a Sensenich 72 CKS8-52 fixed-pitch propeller. Fuel in two wing tanks, total capacity 136 litres (36.0 US gallons; 30.0 Imp gallons) of which 132 litres (35.0 US gallons; 29.1 Imp gallons) usable; overwing refuelling point for each tank. Oil capacity 5.7 litres (1.5 US gallons; 1.3 Imp gallons).

ACCOMMODATION: Pilot and passenger in tandem; five-point safety harness; quick jettison door on starboard side; pilot's port window hinged upwards and outwards. Space for baggage behind seats. Dual controls. Options include split door for photographic work, external bottom-hinged baggage door, greenhouse roof and tinted cabin windows, wide rear seat and rear seat heater.

SYSTEMS: Electric starter; 60 A alternator; 12 V gel-cell battery; voltage regulator with protector. Optional lighting system.

AVIONICS: To customer's choice.

Comms: Optional Communication Group includes Garmin GNC 250XL com/nav, GTX 327 transponder, broadband transmitting antenna, twin headset jacks, cabin speaker and noise cancelling microphone. Intercom system, front and rear PTT switches and ELT also optional.

Flight: Full vacuum gyro package and electric attitude indicator and directional gyro set optional.

Instrumentation: ASI, sensitive altimeter, magnetic compass, fuel gauge, oil temperature and pressure gauges and recording tachometer all standard. VSI, electric turn co-ordinator, OAT gauge, EGT gauge, CHT gauge, combination EGT/CHT eight-probe scanner and electric clock optional.

EQUIPMENT: Optional equipment includes Standard Operation Group comprising navigation lights, landing light, cabin light, dual wingtip strobe lights and bullet propeller spinner; deluxe Duraplush fabric/vinyl interior; landing gear leg steps; speed kit including wing strut fairings; CFP-2 corrosion protection; remote mounted oil filter; and oil quick drain, all optional.

DIMENSIONS, EXTERNAL:
- Wing span .... 10.21 m (33 ft 6 in)
- Wing aspect ratio 6.8
- Length overall .... 6.73 m (22 ft 1 in)
- Height overall .... 2.35 m (7 ft 8½ in)
- Propeller diameter .... 1.83 m (6 ft 0 in)

DIMENSIONS, INTERNAL:
- Cabin: Length .... 2.69 m (8 ft 10 in)
  - Max width .... 0.76 m (2 ft 6 in)
  - Max height .... 1.19 m (3 ft 11 in)

AREAS:
- Wings, gross .... 15.33 m² (165.0 sq ft)

WEIGHTS AND LOADINGS:
- Weight empty, equipped .... 508 kg (1,120 lb)
- Max baggage weight .... 45 kg (100 lb)
- Max T-O weight .... 793 kg (1,750 lb)
- Max wing loading .... 51.8 kg/m² (10.61 lb/sq ft)
- Max power loading .... 9.03 kg/kW (14.83 lb/hp)

PERFORMANCE:
- Never-exceed speed (VNE) .... 140 kt (260 km/h; 162 mph)
- Max level speed at S/L .... 104 kt (193 km/h; 120 mph)

Cruising speed at 75% power	100 kt (185 km/h; 115 mph)
Stalling speed	46 kt (84 km/h; 52 mph)
Max rate of climb at S/L	226 m (740 ft)/min
Service ceiling	3,500 m (11,500 ft)
T-O run	138 m (450 ft)
T-O to 15 m (50 ft)	272 m (890 ft)
Landing from 15 m (50 ft)	233 m (765 ft)
Landing run	122 m (400 ft)
Range with max fuel, allowance for start, taxi, S/L T-O, climb and descent, no reserves:	
at 75% power	556 n miles (1,030 km; 640 miles)
at 55% power	617 n miles (1,142 km; 710 miles)
g limits	+5/–2

UPDATED

AMERICAN CHAMPION 7GCAA CITABRIA ADVENTURE

TYPE: Two-seat lightplane.

PROGRAMME: American Champion 7GCAA Adventure relaunched 1997. Model 7GCA originally certified 17 November 1959 and 7GCAA 30 July 1965; these built up to 1980.

CUSTOMERS: First series totalled 396 built; American Champion delivered 12 in 2002, nine in 2003 and seven in the first six months of 2004, increasing overall production to 488.

COSTS: US$84,900 (2004)

DESIGN FEATURES: As for Aurora.

FLYING CONTROLS: As for Aurora except elevator movements +28/–24°.

STRUCTURE: As for Aurora.

LANDING GEAR: As for Aurora.

POWER PLANT: One 119 kW (160 hp) Textron Lycoming O-320-B2B flat-four engine driving a two-blade fixed-pitch Sensenich 74DM6S8-1-56 propeller. Fuel capacity as for Aurora, oil capacity 7.6 litres (2.0 US gallons; 1.67 Imp gallons).

ACCOMMODATION: As for Aurora.

SYSTEMS: As for Aurora.

AVIONICS: As for Aurora.

EQUIPMENT: As for Aurora.

DIMENSIONS, EXTERNAL: As for Aurora, except:

Propeller diameter	1.88 m (6 ft 2 in)

DIMENSIONS, INTERNAL: As for Aurora

AREAS: As for Aurora

WEIGHTS AND LOADINGS: As for Aurora, except:

Weight empty	544 kg (1,200 lb)
Max power loading	6.66 kg/kW (10.94 lb/hp)

PERFORMANCE:

Never-exceed speed (VNE)	140 kt (260 km/h; 162 mph)
Max level speed	121 kt (225 km/h; 140 mph)
Max cruising speed at 75% power	117 kt (217 km/h; 135 mph)
Stalling speed, power off	46 kt (84 km/h; 52 mph)
Max rate of climb at S/L	356 m (1,167 ft)/min
Service ceiling	4,575 m (15,000 ft)
T-O run	108 m (355 ft)
T-O to 15 m (50 ft)	186 m (610 ft)
Landing from 15 m (50 ft)	230 m (755 ft)
Landing run	122 m (400 ft)
Range with max fuel:	
at 75% power, no reserves	482 n miles (893 km; 555 miles)
at 55% power	595 n miles (1,102 km; 685 miles)
g limits	+5/–2

UPDATED

American Champion 7GCAA Citabria Adventure *(Paul Jackson)* ***NEW***/0576926

AMERICAN CHAMPION 7GCBC CITABRIA EXPLORER

TYPE: Two-seat lightplane.

PROGRAMME: Formerly Citabria 150S, certified 3 December 1965.

CURRENT VERSIONS: **Explorer:** *As described.*

High Country Explorer: Announced July 2004; powered by Superior Air Parts Inc Vantage flat-four, unleaded fuel piston engine, derated to 127 kW (170 hp) and delivering 108 kW (145 hp) at 2,285 m (7,500 ft), compared with 87 kW (117 hp) of standard 7GCBA power plant. Prototype flown mid-2004; demonstrated 549 m (1,800 ft)/min climb at S/L.

CUSTOMERS: Total of 13 delivered in 2002, 12 in 2003, and 11 in the first six months of 2004, increasing overall production total to 1,365.

COSTS: Standard aircraft US$98,900 (2004).

DESIGN FEATURES: As for Aurora. Specific 7GCBC features include optional wheel fairings. NACA 4412 aerofoil section. Dihedral 2°; incidence 1°.

FLYING CONTROLS: As for Aurora, except elevator movements +28/–24°. Flaps; maximum deflection 35°.

STRUCTURE: Aluminium alloy (front) and steel tube (aft) bracing struts.

LANDING GEAR: As for Aurora.

POWER PLANT: As for Adventure. Fuel and oil as for Aurora.

ACCOMMODATION: As for Aurora.

SYSTEMS: As for Aurora.

AVIONICS: As for Aurora. Optional IFR package comprises heated pitot head. Bendix/King KR 22 marker beacon receiver, panel lighting and gyro backlighting upgrade.

American Champion 7GCBC Citabria Explorer *(Paul Jackson)* ***NEW***/0554457

American Champion 7GCBC Explorer *(Paul Jackson)* ***NEW***/0576894

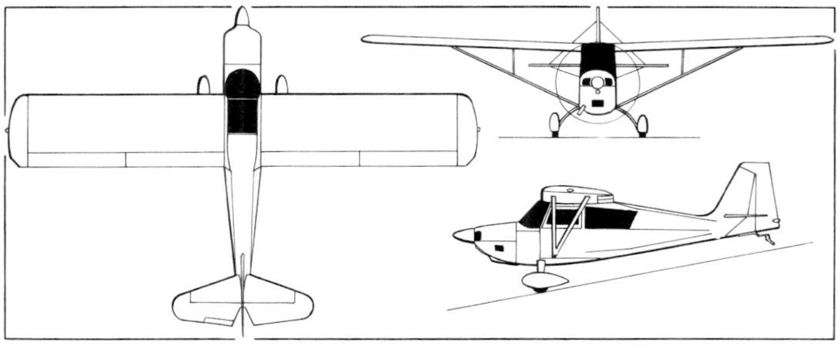

General arrangement of the American Champion 7GCBC Citabria Explorer *(James Goulding)* 0525808

EQUIPMENT: As for Aurora.

DIMENSIONS, EXTERNAL:

Wing span	10.49 m (34 ft 5 in)
Wing chord, constant	1.52 m (5 ft 0 in)
Wing aspect ratio	6.9
Length overall	6.74 m (22 ft 1 in)
Height overall	2.35 m (7 ft 8½ in)
Wheel track	1.93 m (6 ft 4 in)
Wheelbase	4.90 m (16 ft 1 in)
Propeller diameter	1.85 m (6 ft 1 in)

DIMENSIONS, INTERNAL: As for Aurora

AREAS:

Wings, gross	15.97 m^2 (171.9 sq ft)
Ailerons (total)	1.53 m^2 (16.50 sq ft)
Fin	0.65 m^2 (7.02 sq ft)
Rudder	0.63 m^2 (6.83 sq ft)
Tailplane	1.14 m^2 (12.25 sq ft)
Elevators (total, incl tab)	1.35 m^2 (14.58 sq ft)

WEIGHTS AND LOADINGS:

Weight empty, equipped	567 kg (1,250 lb)
Baggage capacity	45 kg (100 lb)
Max T-O and landing weight	816 kg (1,800 lb)
Max wing loading	51.1 kg/m^2 (10.47 lb/sq ft)
Max power loading	6.85 kg/kW (11.25 lb/hp)

PERFORMANCE:

Never-exceed speed (VNE)	140 kt (261 km/h; 162 mph)
Max level speed at S/L	115 kt (212 km/h; 132 mph)
Cruising speed:	
at 75% power, at optimum height	111 kt (206 km/h; 128 mph)
at 65% power	107 kt (198 km/h; 123 mph)
Stalling speed, flaps up	45 kt (82 km/h; 51 mph)
flaps down	40 kt (74 km/h; 46 mph)
Max rate of climb at S/L	344 m (1,130 ft)/min
Service ceiling	4,725 m (15,500 ft)
T-O run	126 m (412 ft)
T-O to 15 m (50 ft)	200 m (656 ft)
Landing from 15 m (50 ft)	226 m (740 ft)
Landing run	110 m (360 ft)
Range with max fuel, allowance for start, taxi, S/L T-O, climb and descent, no reserves:	
at 75% power	431 n miles (799 km; 496 miles)
at 55% power	521 n miles (966 km; 600 miles)
g limits	+5/–2

UPDATED

AMERICAN CHAMPION 8GCBC SCOUT

TYPE: Two-seat lightplane.

PROGRAMME: Model 8 series of Champion developed by Bellanca as heavy-duty version of Model 7; second subvariant was 8GCBC, which received type approval on 30 April 1974.

CURRENT VERSIONS: **Scout:** Fixed-pitch propeller.

Super Scout: Constant-speed propeller.

CUSTOMERS: Total of 11 delivered in 2002, eight in 2003, and nine in the first six months of 2004, increasing total production to 460. Customers include the Department of Conservation and Land Management of Western Australia, which operates eight Scouts on fire spotting duties, replacing Piper Super Cubs.

COSTS: Scout: US$116,900; Scout CS: US$118,900 (both 2004).

DESIGN FEATURES: Upgraded version of Bellanca/Champion Scout, with new lighter and stronger metal spar wing with 300 per cent less deflection than previous wooden wings; circuit breakers; modern avionics; revised interior; and high-gloss weather-resistant exterior finish. Can operate from short fields while towing glider or banner; low speed assists pipeline, border patrol, forestry and wildlife management roles.

Wing section NACA 4412; dihedral 1°; incidence 1°.

FLYING CONTROLS: As for Explorer. Control surface movements: ailerons +27° 30'/–19°; elevators +29/–26°; rudder ±25°. Four-position trailing-edge flaps droop 7, 16, 21 and 27°.

STRUCTURE: As for Explorer, but metal belly skin optional.

LANDING GEAR: As for Aurora; main tyres 8.50–6 (4/6 ply); tail 5.00–5 (4 ply). Optional EDO 89–2000 floats or skis.

POWER PLANT: One 134 kW (180 hp) Textron Lycoming O-360-C1G flat-four engine, driving either a McCauley 1A200HFA80 fixed-pitch (in Scout) or Hartzell HC-C2YR-1BF/F7666A constant-speed (Super Scout) propeller; three-blade 1.83 m (6 ft 0 in) Hartzell HC-C3YR-1RF/F7282 constant-speed propeller optional for Super Scout. Standard fuel capacity as for Aurora; optional additional tank increases usable total to 265 litres (70.0 US gallons; 58.3 Imp gallons). Oil capacity 7.5 litres (2.0 US gallons; 1.7 Imp gallons).

ACCOMMODATION: As for Aurora. Heated and ventilated.

SYSTEMS: As for Aurora.

AVIONICS: As for Aurora.

EQUIPMENT: External baggage door standard. Options include cropduster package, long-range fuel tank and glider tow assembly.

DIMENSIONS, EXTERNAL (A: fixed-pitch propeller, B: constant-speed propeller):

Wing span ... 11.02 m (36 ft 2 in)
Wing chord, constant ... 1.52 m (5 ft 0 in)
Wing aspect ratio ... 7.3
Length overall: A ... 6.93 m (22 ft 9 in)
B ... 7.01 m (23 ft 0 in)
Height overall: A ... 2.65 m (8 ft 8½ in)
B ... 2.96 m (9 ft 8½ in)
Tailplane span ... 3.10 m (10 ft 2¼ in)
Propeller diameter: A ... 1.93 m (6 ft 4 in)
B ... 2.03 m (6 ft 8 in)

DIMENSIONS, INTERNAL: As for Aurora

AREAS:

Wings, gross ... 16.70 m² (180.0 sq ft)
Ailerons (total) ... 1.57 m² (16.88 sq ft)
Flaps (total) ... 2.25 m² (24.26 sq ft)

WEIGHTS AND LOADINGS (A and B as above):

Weight empty: A ... 597 kg (1,315 lb)
B ... 635 kg (1,400 lb)
Baggage capacity ... 45 kg (100 lb)
Max T-O weight: A and B, Normal ... 975 kg (2,150 lb)
A, Restricted ... 1,179 kg (2,600 lb)
Max wing loading:
A and B, Normal ... 58.3 kg/m² (11.94 lb/sq ft)
A, Restricted ... 70.5 kg/m² (14.44 lb/sq ft)
Max power loading:
A and B, Normal ... 7.28 kg/kW (11.94 lb/hp)
A, Restricted ... 8.80 kg/kW (14.44 lb/hp)

PERFORMANCE (A and B as above):

Never-exceed speed (VNE) ... 140 kt (260 km/h; 162 mph)
Max level speed at S/L:
A ... 117 kt (217 km/h; 135 mph)
B ... 121 kt (225 km/h; 140 mph)
Cruising speed at 75% power:
A ... 106 kt (196 km/h; 122 mph)
B ... 113 kt (209 km/h; 130 mph)
Stalling speed: flaps up ... 47 kt (87 km/h; 54 mph)
flaps down: A ... 44 kt (81 km/h; 50 mph)
B ... 43 kt (79 km/h; 49 mph)
Max rate of climb at S/L: A ... 329 m (1,080 ft)/min
B ... 328 m (1,075 ft)/min
Service ceiling: A ... 4,420 m (14,500 ft)
B ... 5,180 m (17,000 ft)
T-O run: A ... 155 m (510 ft)
B ... 149 m (490 ft)
T-O to 15 m (50 ft) ... 312 m (1,025 ft)
Landing from 15 m (50 ft): A, B ... 376 m (1,235 ft)
Landing run: A, B ... 128 m (420 ft)
Range with standard tankage, no reserves:
A, at 75% power ... 333 n miles (618 km; 384 miles)
A, at 75% power, optional fuel ... 680 n miles (1,260 km; 783 miles)
A, at 55% power, optional fuel ... 779 n miles (1,444 km; 897 miles)
B, at 75% power ... 360 n miles (668 km; 415 miles)
B, at 75% power, optional fuel ... 732 n miles (1,355 km; 842 miles)
B, at 55% power, optional fuel ... 838 n miles (1,551 km; 964 miles)

UPDATED

2004-built American Champion 8GCBC Super Scout *(Paul Jackson)* ***NEW***/0561585

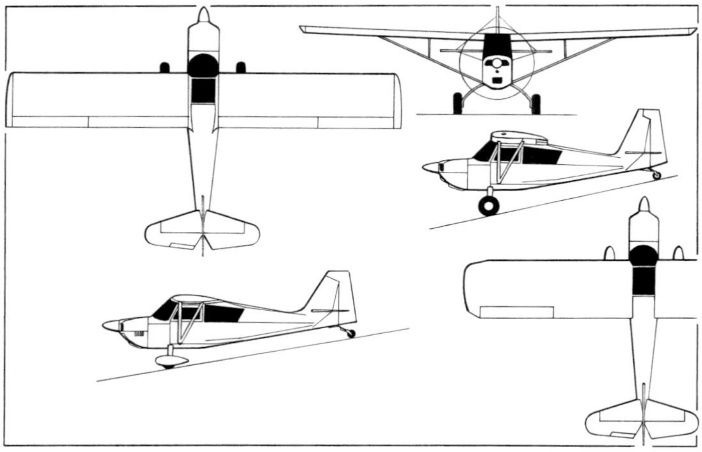

American Champion 8GCBC general arrangement, with scrap plan view and additional side view (lower) of 8KCAB *(James Goulding)* 0525811

2004-built American Champion 8GCBC Scout *(Paul Jackson)* ***NEW***/0576902

AMERICAN CHAMPION 8KCAB SUPER DECATHLON

TYPE: Two-seat lightplane.

PROGRAMME: Original Decathlon was powered by a 112 kW (150 hp) Lycoming IO-320 or AEIO-320: certified 16 October 1970 in Normal and Acrobatic categories. Super introduced by American Champion; first flight (N38AC) July 1990. The Fixed Pitch Decathlon (Sensenich propeller) is no longer marketed.

CUSTOMERS: Total of 14 delivered in 2002, 32 in 2003, and 22 in the first six months of 2004, increasing overall production to 964.

COSTS: Standard price US$113,900 (2003).

DESIGN FEATURES: Generally as for Aurora. Cleared for limited inverted flight; constant-speed propeller. Wing section NACA 1412 (modified); dihedral 1°; incidence 1° 30'.

FLYING CONTROLS: As for Aurora, but with balanced elevators. Control surface movements: ailerons ±30°; elevators ±20°; rudder ±30°.

STRUCTURE: As for Aurora. Optional metal wing spar.

LANDING GEAR: As for Aurora. Mainwheel tyres 6.00–6 or 8.00–6 (4 ply), tail tyre 5.00–5 (4 ply). Cleveland disc brakes; optional wheel fairings.

POWER PLANT: One 134 kW (180 hp) Textron Lycoming AEIO-360-H1B flat-four engine, driving a Hartzell HC-C2YR-4CF/FC7666A-2 two-blade constant-speed propeller. Fuel capacity 151 litres (40.0 US gallons; 33.3 Imp gallons), of which 148 litres (39.0 US

American Champion 8KCAB Super Decathlon from 2004 production *(Paul Jackson)* ***NEW***/0576877

gallons; 32.5 Imp gallons) are usable. Oil capacity 9.5 litres (2.5 US gallons; 2.1 Imp gallons).

ACCOMMODATION: As for Aurora, but competition Hooker harness optional.

SYSTEMS: As for Aurora, plus accelerometer and electric fuel boost pump.

AVIONICS: As for Aurora.

DIMENSIONS, EXTERNAL:

Wing span	9.75 m (32 ft 0 in)
Wing chord, constant	1.63 m (5 ft 4 in)
Wing aspect ratio	6.1
Length overall	6.98 m (22 ft 10¾ in)
Height overall	2.36 m (7 ft 9 in)
Tailplane span	3.10 m (10 ft 2¼ in)
Propeller diameter	1.88 m (6 ft 2 in)

DIMENSIONS, INTERNAL: As for Aurora

AREAS:

Wings, gross	15.71 m² (169.1 sq ft)

WEIGHTS AND LOADINGS:

Weight empty	608 kg (1,340 lb)
Baggage capacity	45 kg (100 lb)
Max T-O weight	816 kg (1,800 lb)
Max wing loading	52.0 kg/m² (10.64 lb/sq ft)
Max power loading	6.09 kg/kW (10.00 lb/hp)

PERFORMANCE:

Never-exceed speed	173 kt (321 km/h; 200 mph)
Max level speed at FL70	135 kt (249 km/h; 155 mph)
Cruising speed:	
at 75% power	128 kt (237 km/h; 147 mph)
at 55% power	111 kt (206 km/h; 128 mph)
Stalling speed	46 kt (86 km/h; 53 mph)
Max rate of climb at S/L	390 m (1,280 ft)/min
Service ceiling	4,815 m (15,800 ft)
T-O run	151 m (495 ft)
T-O to 15 m (50 ft)	276 m (904 ft)
Landing from 15 m (50 ft)	321 m (1,051 ft)
Landing run	130 m (425 ft)
Range, no reserves:	
at 75% power	509 n miles (944 km; 587 miles)
at 55% power	542 n miles (1,006 km; 625 miles)
g limits	+6/−5

UPDATED

AMERICAN HOMEBUILTS'

AMERICAN HOMEBUILTS' INC

10419 Vander Karr Road Hebron, Illinois 60034
Tel: (+1 815) 648 46 17
Fax: (+1 815) 648 16 07
e-mail: info@AmericanHomebuilts.com
Web: http://www.AmericanHomebuilts.com
PRESIDENT Steve Nusbaum

American Homebuilts' was formed in November 1996 to market the John Doe and its derivatives, designed by Mr Nusbaum.

UPDATED

AMERICAN HOMEBUILTS' VAQUERO JOHN DOE

TYPE: Tandem-seat kitbuilt.

PROGRAMME: Following experience gained with Ericson Stork ultralight built during late 1980s, design work on John Doe started August 1992 and construction of proof-of-concept aircraft (with Continental A-80–8 engine) began 1993, latter first flown (N10CY) May 1994; prototype (N20CY; designated John Doe Spirit and with Continental IO-240B engine) first flown April 1997 and later fitted with LOM M132A, with which first flown May 1998 as part of two-year test programme. First production aircraft (N30CY) first flown March 2002; kit deliveries began immediately.

CUSTOMERS: Three flying by mid-2004.

COSTS: Kit US$19,500 without coverings, instruments or power plant (2004); estimated cost to first flight US$35,000.

DESIGN FEATURES: Designed for operation from 'bush' airstrips; many repairs can be carried out 'off base'. Ribs replaceable without removing wing. Quoted build time 500 hours.

Wing aerofoil section NACA4415.

FLYING CONTROLS: Conventional and manual. Ailerons and flaps are interchangeable, as are elevators and rudders. Patented flap/aileron mechanism allows slotted Frise ailerons to droop progressively (8, 15 and 22°) as three-position flaps (15, 25 and 40°) are lowered. Slats and stall fences also fitted to reduce stall speeds further. Stabiliser incidence variable.

American Homebuilts' Vaquero John Doe *(Paul Jackson)* ***NEW***/0587516

STRUCTURE: Fabric-covered welded 4130 steel tube fuselage and two-spar 6061-T6 aluminium I-beam wing; nose area metal panelled. High wing, with Horton tips, supported by two streamline 6061-T6 V-struts.

LANDING GEAR: Tailwheel type; fixed. Sprungsteel suspension. Cleveland wheels and disc brakes. Mainwheel tyre size 8.50–6. Maule tailwheel. Welded attachment points for floats are standard.

POWER PLANT: One 89.5 kW (120 hp) LOM 132A driving a fixed-pitch IVO V-231 or LOM 72"-13" three-blade propeller; alternatively a 93 kW (125 hp) Teledyne Continental IO-240B flat-four or 74.6 kW (100 hp) Teledyne Continental O-200A can be fitted; design will accept other engines in the 48.5 to 119 kW (65 to 160 hp) range. Fuel capacity 98 litres (26.0 US gallons; 21.7 Imp gallons) in two wing tanks.

ACCOMMODATION: Pilot and passenger in tandem; can be flown solo from rear seat. Upward-hinged Lexan doors fitted on both sides of fuselage.

DIMENSIONS, EXTERNAL:

Wing span	9.32 m (30 ft 7 in)
Wing chord, constant	1.30 m (4 ft 3 in)
Wing aspect ratio	7.2
Length overall	6.63 m (21 ft 9 in)
Height overall	2.06 m (6 ft 9 in)
Wheel track	1.68 m (5 ft 6 in)
Wheelbase	5.11 m (16 ft 9 in)
Propeller diameter (IVO)	1.78 m (5 ft 10 in)
Propeller ground clearance, tail up	0.56 m (1 ft 10 in)

DIMENSIONS, INTERNAL:

Cabin: Length	2.18 m (7 ft 2 in)
Max width	0.69 m (2 ft 3 in)
Max height	1.19 m (3 ft 11 in)
Cargo compartment volume	0.51 m³ (18.0 cu ft)

AREAS:

Wings, gross	12.13 m² (130.55 sq ft)

WEIGHTS AND LOADINGS:

Weight empty	399 kg (880 lb)
Max T-O weight	680 kg (1,500 lb)
Max wing loading	56.1 kg/m² (11.49 lb/sq ft)
Max power loading	7.30 kg/kW (12.00 lb/hp)

PERFORMANCE (LOM engine):

Never-exceed speed (VNE)	139 kt (257 km/h; 160 mph)
Max operating speed	122 kt (225 km/h; 140 mph)
Max cruising speed	113 kt (209 km/h; 130 mph)
Normal cruising speed	100 kt (185 km/h; 115 mph)
Stalling speed, flaps down	27 kt (49 km/h; 30 mph)
Max rate of climb at S/L	290 m (950 ft)/min
Service ceiling	3,960 m (13,000 ft)
T-O run	37 m (120 ft)
T-O to 15 m (50 ft)	137 m (450 ft)
Landing from 15 m (50 ft)	61 m (200 ft)
Landing run	38 m (125 ft)
Range with max fuel	391 n miles (724 km; 450 miles)
g limits	+4.4/−3

UPDATED

AMV

AMV AIRCRAFT

18 Via Tranquila, Aliso Vieho, California 92656
Tel/Fax: (+1 949) 457 11 84
Web: http://www.AMVaircraft.com
OWNER: Attila Melkuti

Mr Melkuti plans to fly an aircraft he describes as the world's first practical flying machine to have a single ducted propeller and be capable of short or vertical take-off and landing.

VERIFIED

AMV 211

TYPE: Technology demonstrator.

PROGRAMME: Prototype completed in July 2003 shown statically at AirVenture, Oshkosh, later same month, following which preparations were to be made for flight. Registration N211VT reserved in 1999, but only activated in September 2003. After initial ground tests, engine coding was improved to prevent over-heating during hovering and modification were effected to the roll control system. Experimental category airworthiness inspection in May 2004; maiden flight to follow.

CURRENT VERSIONS: **Prototype:** As described.

Production: Slightly enlarged version with tandem seats turboshaft engine in 447 and 597 kW (600 to 800 hp) class.

Six-seat: Certified, six-seat business aircraft with two turboshafts, each 597 to 671 kW (800 to 900 shp).

DESIGN FEATURES: Duct for horizontally mounted, five-blade propeller below fuselage, with wings outboard. Twin fins on tailbooms, each with rudder. Beaver tail with full-width elevator, including anti-balance tab.

For vertical take-off, propeller plane is horizontal; after rising, craft tilts forward by undisclosed means, gradually gathering forward speed. When propeller plane reaches some 26° inclination, wing is horizontal, creating lift.

Wings NACA 65_2-215 at root; 64_1-112 at tip; no dihedral; incidence 1°; no twist.

FLYING CONTROLS: Full-span ailerons, elevator, and rudders balanced by weight (horn type). No flaps, but control louvers placed in the propeller slipstream for control during vertical flight and transition to horizontal flight. Rudders controlled by cables, elevator and ailerons

Prototype AMV 211 VTOL aircraft *(Paul Jackson)* ***NEW***/0561707

Details of the AMV 211's ducted propeller *(Paul Jackson)* ***NEW***/0561706

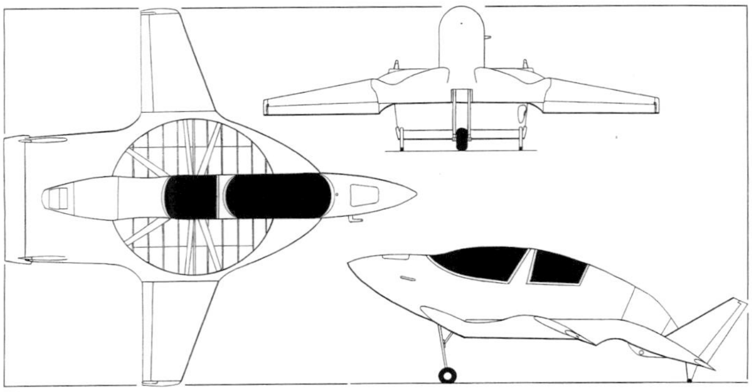

AMV 211 technology demonstrator *(Paul Jackson)* ***NEW***/0580858

by pushrods and bell cranks. Elevator equipped with electrically actuated trim tab. Control louvre, which controls pitch during hover, is mechanically connected to elevator roll controls for hover are mechanically connected to the ailerons. Torque (yaw) control during hover is achieved by two louvres moving in opposite directions these are located centrally in the propeller's slipstream, and are electrically actuated.

STRUCTURE: Load-bearing structure is welded, chromoly tubing. Fuselage wings, and other aerodynamic surfaces are moulded graphite epoxy composites. Propeller blades are graphite and Kevlar.

LANDING GEAR: Tricycle type; semi-retractable. Monowheel and twin, fixed tailwheels one beneath each fin. Monowheel hydraulically turned some 40 degrees rearwards for horizontal flight. Disk brakes on mainwheel.

POWER PLANT: One 336 kW (450 hp) Mazda turbocharged rotary engine. Fuel capacity 189 litres (50.0 US gallons; 41.6 Imp gallons) in four wing and fuselage tanks.

ACCOMMODATION: Two persons in tandem with individual canopies.

SYSTEMS: Electric system 12 V; main and standby batteries. Hydraulic handpump, operating at 103 bar (1,500 lb/sq in) for landing gear actuation and movement of louvres between VTOL and horizontal flight mode.

DIMENSIONS, EXTERNAL:

Wing span	6.10 m (20 ft 0 in)
Length overall	6.10 m (20 ft 0 in)
Height overall	2.13 m (7 ft 0 in)
Propeller diameter	2.39 m (7 ft 10 in)

WEIGHTS AND LOADINGS:

Weight empty	570 kg (1,256 lb)
Max T-O weight	907 kg (2,000 lb)
Max power loading	2.43 kg/kW (4.00 lb/hp)

PERFORMANCE (estimated):

Max level speed	243 kt (451 km/h; 280 mph)
Cruising speed at 60% power	217 kt (402 km/h; 250 mph)
Max rate of climb at S/L	914 m (3,000 ft)/min
Service ceiling	7,620 m (25,000 ft)
Range	1,000 n miles (1,852 km; 1,150 miles)

UPDATED

ANGEL

ANGEL AIRCRAFT CORPORATION

1410 Arizona Place SW, Orange City, Iowa 5141
Tel: (+1 712) 737 33 44
Fax: (+1 712) 737 33 99
e-mail: angelair@orangecitycomm.net
Web: http://www.angelaircraft.com
PRESIDENT: Carl A Mortenson

MARKETING:
Kansair Inc, 1533 20th Park Place, Emporia, Kansas 66801
Tel: (+1 620) 342 35 00
e-mail: kansair@osprey.net
PRESIDENT: Jerry Waddell

Angel 44 was developed by The King's Engineering Fellowship (TKEF) through donations, and designed by Carl Mortenson, with Ed Mortenson as chief engineer. TKEF holds Type Certificate. Production Certificate awarded to Angel Aircraft Corporation on 23 September 2003. Plant covered area 1,300 m² (14,000 sq ft).

UPDATED

ANGEL 44

TYPE: Light utility twin-prop transport.

PROGRAMME: Design started November 1972; prototype construction began January 1977; prototype (N44KE) built on production tooling; first flight 13 January 1984; structural testing began 1990; certification to FAR Pt 23 (Normal category) for day, night, VFR and IFR conditions gained 20 October 1992. Production began December 1993; second aircraft completed July 1994 but held in abeyance, pending acquisition of larger production premises. Second Angel eventually registered in August 2002 to Kansair Inc as N442KB, having been displayed at AirVenture, Oshkosh, in previous month. Third (N442KC) completed early 2004 and awarded airworthiness certificate on 24 February; company-owned.

DESIGN FEATURES: Designed partly, but not exclusively, for missionary aviation; other commercial uses include air taxi, air ambulance, air observation/patrol, fishery/pipeline/border inspection, tracking, mining/oil/rubber/forestry operations, ranching and firefighting control. Design goals include STOL capability, operation from soft and rough fields, easy repair in field, and easy loading of bulky cargo.

Vortex generators on wing outboard panels for enhanced lift; small wing endplates. Crashworthiness built and tested in key structures; seats dynamically tested to absorb 20 *g* vertically and 26 *g* horizontally; cabin structure includes areas of double-wall and tested for overturn impact survivability. Extra-large propeller spinners assist airflow around rear fuselage.

Wing section NACA 23018–23010 with modified leading-edge; sweepback 15° 36' at leading-edge, 11° at quarter-chord; dihedral 5° 24'; incidence 3° at root, –0° 22' at tip.

FLYING CONTROLS: Manual. Actuation by cables. Almost full-span, hydraulically actuated semi-Fowler flaps deflecting to maximum 37°. Lateral control initiated by small ailerons at wingtips known as trimmerons (first 3 to 5° of bank), further bank being induced by spoilers in form of 11 small plates immediately forward of flaps on each wing; elevators and rudder, both with horn balanced and trim tabs (one in each side of elevator).

STRUCTURE: Riveted aluminium alloy and welded 4130/4340 steel tube; wing has built-up capstrip spars and 19 die-formed ribs each side; broad-chord fin and rudder with large dorsal fin; CFRP spinners; GFRP/epoxy in some areas.

LANDING GEAR: Retractable tricycle type. Electrohydraulic retraction, mainwheels inward into wingroots (no wheel doors), nosewheel rearward. Emergency extension by gravity. Cleveland wheels; McCreary mainwheel tyres size 8.50–10 (pressure 2.41 bar; 35 lb/sq in), McCreary or Goodyear nosewheel tyre size 8.50–6 (pressure 1.03 bar; 15 lb/sq in). Hydraulic Cleveland disc brakes. Min ground turning radius 5.11 m (16 ft 9 in) due to fully-castoring nosewheel.

POWER PLANT: Two 224 kW (300 hp) Textron Lycoming IO-540-M1C5 flat-six engines, each driving a Hartzell HC-E3YR-2ATLF/FLC7468 three-blade, constant-speed (hydraulic), feathering pusher propeller. Two wing fuel tanks, gravity flow cross-fed, with total capacity 849 litres (224 US gallons; 187 Imp gallons). Refuelling point in top of each wing. Oil capacity 22.7 litres (6.0 US gallons; 5.0 Imp gallons).

ACCOMMODATION: Accommodation for two pilots and four passengers, rearmost facing sideways, others on three individual seats. Seats can be removed for carrying cargo, including four 208 litre (55 US gallon; 45.8 Imp gallon) drums. Four large windows and one smaller circular window on each side of cabin. Horizontally divided clamshell door on port side at front of cabin with folding step in lower element: emergency exit on starboard side. Heating and window air vents standard. Compartment for baggage at rear of fuselage, with door on port side.

SYSTEMS: Hydraulic system, with electric pump, for landing gear and flap actuation. Electrical system with two alternators includes 12 V DC battery in nose.

AVIONICS: IFR. Bendix/King and Garmin avionics, including twin com/nav transceivers, glideslope, ADF, transponder, ELT and GPS Weather radar, Loran C and HF com optional.

DIMENSIONS, EXTERNAL:

Wing span	12.16 m (39 ft 10¾ in)
Wing aspect ratio	7.06
Wing chord: at root	2.59 m (8 ft 6 in)
at tip	1.00 m (3 ft 3½ in)

Third Angel 44, completed 2004 ***NEW***/1047170

Length overall ... 10.21 m (33 ft 6 in)
Height overall ... 3.51 m (11 ft 6 in)
Tailplane span ... 5.30 m (17 ft 4½ in)
Wheel track ... 3.95 m (12 ft 11¾ in)
Wheelbase ... 4.67 m (15 ft 4 in)
Propeller diameter ... 1.93 m (6 ft 4 in)
Propeller ground clearance at T-O ... 0.23 m (9.01 in)
Cabin door (port, fwd): Height ... 0.91 m (3 ft 0 in)
Width ... 1.04 m (3 ft 5 in)
Height to sill ... 0.91 m (3 ft 0 in)
Baggage door: Height ... 0.41 m (1 ft 4 in)
Width ... 0.61 m (2 ft 0 in)
Height to sill ... 1.17 m (3 ft 10 in)
Emergency exit door (stbd, fwd):
Height ... 0.48 m (1 ft 7 in)
Width ... 1.04 m (3 ft 5 in)
Max height ... 1.32 m (4 ft 4 in)

DIMENSIONS, INTERNAL:
Cabin, incl flight deck:
Length ... 3.51 m (11 ft 6 in)
Max width ... 1.07 m (3 ft 6 in)
Max height ... 1.14 m (3 ft 9 in)
Volume ... 2.38 m^3 (84.0 cu ft)
Baggage compartment volume (aft) ... 0.28 m^3 (10 cu ft)

AREAS:
Wings, gross ... 20.94 m^2 (225.4 sq ft)
Ailerons (total) ... 0.39 m^2 (4.20 sq ft)
Trailing-edge flaps (total) ... 4.05 m^2 (43.60 sq ft)
Spoilers (total) ... 0.37 m^2 (4.00 sq ft)
Rudder, incl tab ... 1.41 m^2 (15.15 sq ft)
Tailplane (total) ... 6.23 m^2 (67.02 sq ft)
Elevators (total, incl tabs) ... 2.76 m^2 (29.67 sq ft)

WEIGHTS AND LOADINGS:
Weight empty ... 1,760 kg (3,880 lb)
Baggage capacity (rear) ... 91 kg (200 lb)
Max fuel weight ... 612 kg (1,350 lb)
Max T-O and landing weight ... 2,631 kg (5,800 lb)

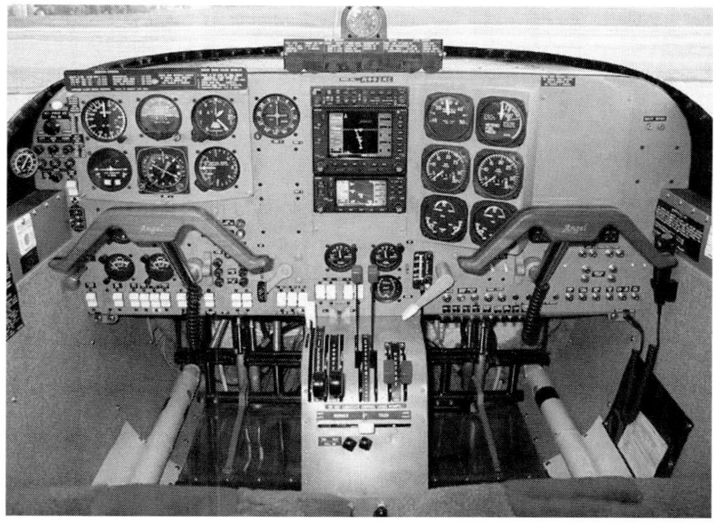

Instrument panel of Angel 44 No. 3 ***NEW***/1047169

Max wing loading ... 125.6 kg/m^2 (25.73 lb/sq ft)
Max power loading ... 5.87 kg/kW (9.67 lb/hp)

PERFORMANCE:
Never-exceed speed (VNE) ... 209 kt (387 km/h; 240 mph)
Max level speed at S/L ... 180 kt (333 km/h; 207 mph)
Cruising speed at 65% power at FL115 ... 169 kt (313 km/h; 195 mph)
Stalling speed, flaps and gear down:
power off ... 57 kt (106 km/h; 66 mph)
power on ... 51 kt (95 km/h; 59 mph)
Max rate of climb at S/L ... 410 m (1,345 ft)/min
Rate of climb, OEI ... 60 m (196 ft)/min
Service ceiling ... 6,265 m (20,560 ft
Service ceiling, OEI ... 1,580 m (5,180 ft)
T-O run ... 201 m (660 ft)
T-O to 15 m (50 ft) ... 428 m (1,405 ft)
Landing from 15 m (50 ft) ... 320 m (1,050 ft)
Landing run ... 173 m (570 ft)
Range with max fuel, VFR reserves:
65% power ... 1,248 n miles (2,311 km; 1,436 miles)
45% power ... 1,605 n miles (2,972 km; 1,847 miles)
35% power ... 1,720 n miles (3,185 km; 1,979 miles)
Endurance, VFR reserves: with eight occupants ... 3 h
with max fuel and three occupants:
65% power ... 7 h 54 min
45% power ... 11 h 18 min
35% power ... 13 h 6 min

UPDATED

ASII

AMERICAN SPORTSCOPTER INTERNATIONAL INC

Hangar 21A, PO Box 14608, Newport News, Virginia 23608
Tel: (+1 757) 872 87 78
Fax: (+1 757) 872 87 71
e-mail: asii@asiicopter.com
Web: http://www.ultrasport.rotor.com

FAR EAST OFFICE:
Light's American Sportscopter Inc
656 Jianshing Road, Bei Chiu, Taichung 404, Taiwan
Tel: (+886 4) 22 07 56 78
Fax: (+886 4) 22 08 77 88
e-mail: info@lasi.com.tw
Web: http://www.ultrasport.rotor.com

ASII was founded in 1990. North and South American markets are covered by US office.

UPDATED

ASII ULTRASPORT 254 AND 331

TYPE: Single-seat ultralight helicopter kitbuilt.

PROGRAMME: Prototype (254 version) first flight 24 July 1993 and publicly displayed at Oshkosh that year. Two prototypes built and tested. Prototype 331 (N331UV) first flew December 1993. LBA certification was expected by early 2002: version is only marketed in countries where ultralight helicopters are regulated to company's satisfaction. In late 2002, it was announced that Huzhou Taixing Aviation Technology of Huzhou, China was producing helicopters from kits, including examples of Ultrasports.

CURRENT VERSIONS: Designations reflect empty weight in pounds.

Ultrasport 254: Single-seat, ultralight to FAR Pt 103.

Ultrasport 331: Initial experimental category version to FAR Pt 21.191(g); single-seat 'growth' model. Meets FAA 51 per cent amateur-built kit rules.

Ultrasport 331H: with 48.5 kW (65 hp) Hirth H32 added to range in 2003 and replaces the Ultrasport 331.

CUSTOMERS: Total of 70 single-seaters sold by end 2002, of which more than 50 were flying.

COSTS: 254: US$34,900; 331 US$37,900 (2003).

DESIGN FEATURES: Design objective of 254 was basic weight not to exceed 115 kg (254 lb) in order to comply with FAR Pt 103. Two-blade composites construction main rotor with tip weights for momentum conservation in event of engine failure, 8° linear twist and infinite life; shielded two-blade tail rotor; tailplane with fins at tip; tail rotor drive carried in narrow streamlined tailboom. Centrifugal sprag clutch for starting engages rotors at 2,000 engine rpm and automatically disengages in the event of engine failure. Quoted build time 80 hours. Broad (2.44 m; 8 ft 0 in) skid track helps to prevent rollovers.

ASII Ultrasport 331 single-seat ultralight helicopter with doors removed *(Paul Jackson)* 0110714

FLYING CONTROLS: Conventional collective, cyclic and yaw pedals. Floor-mounted cyclic option available since 1998; early models had top-mounted stick.

STRUCTURE: Generally of epoxy resin, graphite fabric and Nomex honeycomb; aluminium tailboom.

LANDING GEAR: Aluminium skids stressed for landings at up to 2.5 *g*; floats (weight 18 kg; 40 lb) optional.

POWER PLANT: *254:* One 41.0 kW (55 hp) Hirth 2703 dual-carburettor two-stroke engine with pull starter and 12:1 planetary transmission. Normal fuel capacity 19 litres (5.0 US gallons; 4.1 Imp gallons).

331: One 48.5 kW (65 hp) Hirth 2706 dual-carburettor two-stroke engine with electric starter (also option for 254) or Hirth H32 of same power in 331H; 12:1 planetary transmission. Fuel capacity 38 litres (10.0 US gallons; 8.3 Imp gallons).

ACCOMMODATION: Single seat partially enclosed (254 version); enclosed (331 version).

SYSTEMS: Electrical: 12 V battery; 14 V alternator fit appropriate to engine.

AVIONICS: Customer choice.

DIMENSIONS, EXTERNAL:
Main rotor diameter 6.40 m (21 ft 0 in)
Main rotor blade chord 0.17 m (6¾ in)
Tail rotor diameter 0.76 m (2 ft 6 in)
Tail rotor blade chord 0.05 m (2 in)
Fuselage: Length, main rotor folded 5.84 m (19 ft 2 in)
Skid track 2.44 m (8 ft 0 in)
Width: 331H 1.98 m (6 ft 6 in)
Height overall 2.39 m (7 ft 10 in)
DIMENSIONS, INTERNAL:
Cabin: Length 1.32 m (4 ft 4 in)
Max width 0.76 m (2 ft 6 in)
Max height 1.50 m (4 ft 11 in)
AREAS:
Main rotor disc 32.2 m^2 (346.4 sq ft)
WEIGHTS AND LOADINGS:
Weight empty: 254 114 kg (252 lb)
331 150 kg (330 lb)
Max T-O and landing weight: 254 260 kg (575 lb)
331 294 kg (650 lb)
PERFORMANCE:
Max level speed: 254 55 kt (101 km/h; 63 mph)
331 90 kt (167 km/h; 104 mph)
Cruising speed: 254, 331 55 kt (101 km/h; 63 mph)
Max rate of climb at S/L 305 m (1,000 ft)/min
Service ceiling 3,660 m (12,000 ft)
Hovering ceiling: IGE 3,290 m (10,800 ft)
OGE 2,135 m (7,000 ft)
Range, normal fuel, 55 kt:
254 65 n miles (120 km; 75 miles)
331 130 n miles (241 km; 150 miles)

UPDATED

ASII ULTRASPORT 496

TYPE: Two-seat ultralight helicopter kitbuilt.

Details as for 254 and 331, except as follows:

PROGRAMME: Developed from the 331, the Ultrasport 496 (N496AS) first flew in July 1995. Deliveries began in April 1997.

CURRENT VERSIONS: **Ultrasport 496:** Original version; no longer available.

Ultrasport 496RT: Announced 2002. Uses 84.6 kW (113.4 hp) Rotax 914 turbocharged engine with electric start and quadruple carburettors.

Ultrasport 496H: Replaced original 496 in late 2002; *as described.*

Sportscopter 600: New Zealand assembly; initial four registered in July 1999.

Vigilante: Remotely piloted version developed by Science Applications International Corporation (SAIC). **Vigilante 496** prototype (N496UV) evaluated as optionally piloted vehicle (OPV) by US Navy during first and second quarters of 1998. Improved **Vigilante 502** is UAV only. Further details in *Jane's Unmanned Aerial Vehicles and Targets.*

CUSTOMERS: By end 2001, 135 had been delivered of which 35 were then flying. No further information received.

COSTS: 496 US$62,900; 496RT US$77,900 (2003).

ASII Ultrasport 496 *(Paul Jackson)* ***NEW***/0587517

DESIGN FEATURES: Generally as for 254/331. Quick-build kit quoted build time 80 hours; dual controls standard.

STRUCTURE: Infinite-life composites rotor blades and fuselage. Shaft-driven four-blade composites tail rotor. Vertically mounted direct-drive helical spur gears 11:1 two-stage main transmission.

LANDING GEAR: Aluminium skids; track as for 254/331. Floats optional.

POWER PLANT: Original model had one 85.8 kW (115 hp) Hirth F30 quad-carburettor engine with electric starter; 75.0 kW (101 hp) Hirth H30E version designated 496H demonstrated at Sun 'n' Fun 2002 and has now supplanted F30. Fuel capacity 67 litres (17.6 US gallons; 14.7 Imp gallons).

EQUIPMENT: Ballistic parachute under consideration.

DIMENSIONS, EXTERNAL:
Main rotor diameter 7.01 m (23 ft 0 in)
Main rotor blade chord 0.17 m (6¾ in)
Tail rotor diameter 0.76 m (2 ft 6 in)
Tail rotor blade chord 0.05 m (2 in)
Length, blades folded 6.02 m (19 ft 9 in)
Height 2.51 m (8 ft 3 in)
Width overall 2.21 m (7 ft 3 in)
DIMENSIONS, INTERNAL:
Cabin: Length 1.35 m (4 ft 5 in)
Max width 1.22 m (4 ft 0 in)
Max height 1.50 m (4 ft 11 in)
AREAS:
Main rotor disc 38.60 m^2 (415.5 sq ft)
WEIGHTS AND LOADINGS:
Weight empty: 496H 272 kg (600 lb)
496RT 299 kg (660 lb)
Max T-O and landing weight: 496H 512 kg (1,130 lb)
496RT 499 kg (1,100 lb)
PERFORMANCE (at max T-O weight, ISA):
Never-exceed speed (VNE) 90 kt (167 km/h; 104 mph)
Max cruising speed 61 kt (113 km/h; 70 mph)
Max rate of climb at S/L 305 m (1,000 ft)/min
Hovering ceiling: IGE 3,290 m (10,800 ft)
OGE 2,135 m (7,000 ft)
Range 130 n miles (240 km; 149 miles)
Endurance 3 h

UPDATED

ATG

AVIATION TECHNOLOGY GROUP, INC

8001 South InterPort Boulevard, Suite 310, Englewood, Colorado 80112-5951
Tel: (+1 303) 799 41 97
Fax: (+1 303) 799 48 13
e-mail: info@avtechgroup.com
Web: http://www.avtechgroup.com
PRESIDENT: George E Bye
EXECUTIVE VICE-CHAIRMAN: Horst Bergmann
EXECUTIVE VICE-PRESIDENT, OPERATIONS: Charlie Johnson
VICE-PRESIDENT, ENGINEERING: Dr Robert S Wolf
DIRECTOR, MARKETING: Bob Uhle
MANAGER, PROGRAMME: George Bullard
COMMUNICATIONS MANAGER: Sara Newton

Aviation Technology Group was formed in 1998 to develop the ATG Javelin two-seat jet; it was incorporated in June 2000. A 9,290 m^2 (100,000 sq ft) factory is being constructed at Double Eagle Airport, Albuquerque, New Mexico; workforce numbered 45 in May 2004 and was scheduled to rise to 70 by January 2005.

UPDATED

ATG-1 JAVELIN

TYPE: Two-seat jet sportplane.

PROGRAMME: Announced early 2001, at which time initial wind tunnel testing was being undertaken; this finished 28 February 2001. Construction of full-scale, non-flying mockup being undertaken during mid-2001. A second series of wind tunnel tests was scheduled for late 2001 and 2002, to be followed by a maiden flight during late 2005. One (N104TG) non-conforming and two conforming prototypes to fly 1,400-hour test programme. Certification to FAR Pt 23 Amdt 52 (Aerobatic) and first deliveries intended early 2007. Target production rate 75 aircraft per year.

At the NBAA Convention in Las Vegas on 18 October 2004 ATG and Israel Aircraft Industries announced a strategic co-operation agreement for the design development and manufacture of the AJT variant, with IAI investing equity and matching funds in ATG and final assembly of the AJT to take place in the USA and Israel as part of an integrated system for advanced and lead-in pilot training that will include flight simulators, ground-based systems, maintenance and logistics infrastructures. Initial target is Israel Defence Force/Air Force requirement for replacement of Fouga Magister (IAI Tzukit) and advanced trainer versions of A-4 Skyhawk.

All data are provisional.

Mock-up of ATG-1B Javelin two-seat jet sportplane and advanced jet trainer *(Paul Jackson)* ***NEW***/0561708

CURRENT VERSIONS: **ATG-1A:** Initial design. Mid-wing configuration and separate windscreen. Replaced by ATG-1B.

ATG-1B (known as Javelin): Mockup unveiled 12 July 2002 and shown at NBAA Convention, Orlando, Florida, in September 2002. Low wing with leading-edge 'dog-tooth' and two-piece canopy.

Javelin HDI (Homeland Defense Interceptor): Single-seat version armed with pylon-mounted AIM-9 Sidewinder or air-to-air Stinger missiles, was announced in early 2002. Quoted maximum speed of M1.6.

Javelin AJT (Advanced Jet Trainer): Two-seat version equipped with ejection seats and head-up displays to complement Javelin HDI.

UCAV: Unmanned combat version announced in July 2002; later dropped.

CUSTOMERS: Production forecasts range from 711 to 1,638 units (2002 estimate). Deposits (US$25,000 each) taken on 26 aircraft by July 2002. By October 2004, the company had five distributorships (of intended six in USA) and orders for more than 80 aircraft, including 12 received during the NBAA Convention in Las Vegas that month, representing planned production until the end of 2007.

COSTS: US$2.50 million (2004) including type rating course will rise to US$2.50 million once distributor network becomes active. Development costs estimated at US$120 million up to certification. HDI version provisionally costed at under US$10 million.

DESIGN FEATURES: Extensive use of aluminium provides optimum strength to weight ratio.

FLYING CONTROLS: Conventional hydraulic and manual reversionary mode. Flaps. Electrically trimmed slab-type stabiliser and twin 20° outward-canted vertical stabilisers.

ATG-1B Javelin advanced jet trainer and private sportplane ***NEW***/0590454

STRUCTURE: Fuselage of modular aluminium structure with swept cantilevered aluminium wing with leading edge root extensions. Quarter-chord sweep: wings 33° 10'; tailplane 37° 0'; fins 41° 0'. Sweptback tail; two aft body strakes.

LANDING GEAR: Hydraulically operated retractable tricyle type with oleo-pneumatic trailing-link struts. Anti-skid brakes.

POWER PLANT: Two Williams FJ33 turbofan engines, each 6.7 kN (1,500 lb st). Fuel capacity 1,060 litres (280 US gallons; 233 Imp gallons).

ACCOMMODATION: Pilot and passenger in tandem. Dual controls. Max cabin pressure differential 0.61 bar (8.8 lb/sq in). Maintains 2,438 m (8,000 ft) at 13,716 m (45,000 ft). Air conditioning standard. Production aircraft will have side hinged canopies with integral explosive cords.

SYSTEMS: Combination of de-icing/anti-icing systems is planned: electro-expulsive de-icing for wings, bleed air heat for engine inlets, electric de-icing for windscreen and electric anti-icing for probes.

AVIONICS: Full IFR capability with dual EFIS displays. Options include head-up display and airborne video camera systems.

Comms: VHF/UHF radios; Mode S transponder.

Radar: Colour weather radar.

Flight: E-TCAS, TAWS and GPS integrated FMS with waypoint map. Three axis autopilot.

Instrumentation: Avidyne FlightMax Entegra two-tube EFIS comprising one PFD and one MFD, each 264 mm (10.4 in), in each cockpit.

DIMENSIONS, EXTERNAL:
- Wing span .... 6.71 m (22 ft 0 in)
- Wing aspect ratio 4.0
- Length overall .... 10.82 m (35 ft 6 in)
- Height overall .... 3.20 m (10 ft 6 in)
- Tailplane span .... 4.27 m (14 ft 0 in)

AREAS:
- Wings, gross .... 11.15 m² (120.0 sq ft)
- Horizontal tail .... 3.25 m² (35.00 sq ft)
- Vertical tail (each) .... 3.96 m² (42.60 sq ft)

Instrument panel of ATG Javelin ***NEW***/0561754

WEIGHTS AND LOADINGS:
- Weight empty .... 1,588 kg (3,500 lb)
- Max fuel .... 850 kg (1,875 lb)
- Baggage capacity .... 91 kg (200 lb)
- Max T-O weight .... 2,699 kg (5,950 lb)
- Max wing loading .... 242.1 kg/m² (49.58 lb/sq ft)
- Max power loading .... 202 kg/kN (1.98 lb/lb st)

PERFORMANCE:
- Never exceed speed (VNE) M0.96
- Max operating speed .... 500 kt (926 km/h; 575 mph) EAS
- Max cruising speed .... 528 kt (978 km/h; 608 mph) TAS/M0.92
- Stalling speed, power off; flaps down .... 98 kt (181 km/h; 113 mph) IAS
- Max rate of climb at S/L .... 3,655 m (12,000 ft)/min
- Service ceiling .... 13,716 m (45,000 ft)
- Range with max fuel, 100 n mile NBAA reserve: at long-range cruising speed .... 1,250 n miles (2,315 km; 1,438 miles)
  - at M0.92 .... 830 n miles (1,537 km; 955 miles)
- Ferry range, 30 min VFR reserves .... 1,500 n miles (2,778 km; 1,726 miles)
- Endurance .... 3 h 22 min
- *g* limits .... +6/–3

UPDATED

AUC

AMERICAN UTILICRAFT CORPORATION

554 Briscoe Boulevard, Lawrenceville, Georgia 30045
Tel: (+1 678) 376 08 98
Fax: (+1 678) 376 90 93
e-mail: marketing@utilicraft.com
Web: http://www.utilicraft.com
FOUNDER, PRESIDENT AND CEO: John DuPont
VICE-PRESIDENT, GENERAL MANAGER: R Darby Boland
VICE-PRESIDENT, OPERATIONS: Thomas Dapogny

Company formed August 1990 to design and produce Freight Feeder cargo transport. Headquarters at Gwinnett County, Atlanta, Georgia; final assembly planned in San Juan Pueblo, New Mexico, in a purpose-built 7,430 m² (80,000 sq ft) facility funded in part by US$11 million investment TSAY Native American casino and construction company.

US$50 million of financing secured during June 2001 but economic slowdown in late 2001 affected progress with construction of prototype Freight Feeder; however on 25 June 2002 AUC signed an MoU with Averitt Express for lease of capacity of first 25 aircraft, and announced a more aggressive merger and acquisition strategy. In March 2003, AUC announced the formation of a new division in conjunction with Averitt, AUC X-press, offering air cargo services with a fleet of Cessna 404 and Beechcraft 1900 aircraft, which will be supplemented by the FF-1080 in due course; agreement since lapsed.

In July 2003, AUC signed a further MoU, this time with Alpine Air Express, for development of the FF-1080; again agreement has lapsed.

In November 2003, AUC signed an MoU with WSI Hong Kong to cover intended purchase of 300 FF-1080s, including exclusive Far East sales and support; in January 2004, WSI signed an initial firm purchase agreement for 36, deliveries to begin in 2006.

UPDATED

AUC FF-1080 FREIGHT FEEDER

TYPE: Twin-turboprop freighter.

PROGRAMME: Design started 1990; patents filed 1991; original capacity for four LD3 containers increased to six in 1997. Two flying prototypes planned; first flight due 18 months after programme financing in place, followed by FAR Pt 25 certification. Mock up completed end 1999 at which time pre-production prototype intended for completion early 2002 and first flight was due before end 2002, with certification in 2003. FAA Pt 25 certification; this was revised following market downturn at end 2001.

In November 2003, company signed MoU with WSI Hong Kong for acquisition of 300 aircraft; purchase agreement for first 36 signed January 2004. Initial deliveries due to begin 2006, dependent on completion of pre-production prototype in 2004 (which was not achieved).

Manufacture of first 36 aircraft planned to run concurrently with certification programme. By mid-2004, AUC was in negotiations with Metalcraft of Cedar City, Utah, which planned fuselage production, plus initial assembly at AUC plant, Lawrenceville, Georgia.

CURRENT VERSIONS: **FF-1080-100:** Short fuselage version; four LD3 containers; PW121 engines. Not intended for construction at present; weights and loadings in 1999–2000 *Jane's*.

FF-1080-200: Standard version; *as described*.

FF-1080-500: Enlarged version; specifications to be announced by end 2004.

CUSTOMERS: Company estimates market for 5,000 aircraft and aims to capture 10 per cent. Target market is Cessna 208 replacement/growth and F27 all-cargo market. Discussions in hand with several unidentified potential customers; first production batch to comprise 36 aircraft. WSI Hong Kong (see above) is new launch customer. In June 2004 an LOI was signed with Global Air Group of Brisbane, Australia, to cover purchase of 50-plus option on 50 FF-1080s.

COSTS: FF-1080-200 FAR Pt 25 certification estimated US$75 million (2004). Standard flyaway version preliminary price US$7.5 million (2004).

DESIGN FEATURES: High-mounted wings, tapered outboard of overwing engine nacelles; box-section fuselage with ventral pannier; upswept rear fuselage with rear-loading door; angular tail surfaces with large dorsal fin fairing. Onboard freight management system.

FLYING CONTROLS: Conventional and manual. Upper surface blowing (USB) via overwing engine mounting.

STRUCTURE: All-aluminium construction. Final assembly by AUC using components provided by management partners under subcontract. Interest in component manufacture shown by companies in USA. Fuselage subassemblies constructed by Metalcraft Technologies Inc of Utah. Wing subassemblies designed by Aerostructures Corporation. Other risk-sharing partners include Goodrich AAR Cargo Systems, MPC Products, HS Dynamic Controls, Michelin Aircraft Tire Corporation, Securaplane Technologies, Lord Mounts, Hi-Temp Insulation Inc and General Electrodynamics Corp.

LANDING GEAR: Non-retractable tricycle type; twin nosewheels and tandem pairs of mainwheels.

POWER PLANT: *FF-1080-200:* Two 2,051 kW (2,750 shp) Pratt & Whitney Canada PW127F turboprops, each driving a Hamilton Sundstrand 568F six-blade propeller. FADEC fitted as standard. Fuel capacity 6,853 litres (1,810 US gallons; 1,507 Imp gallons), optional fuel 10,601 litres (2,800 US gallons; 2,332 Imp gallons).

FF-1080-100: PW121 engines each rated at 1,567 kW (2,100 shp).

ACCOMMODATION: Flight crew of two on IPECO seats. Main cargo hold accommodates up to six LD3 containers. Cargo roller floor from AAR Cargo Systems. Large cargo double door on port side, forward of wing; rear-loading door; crew airstair door on port side of flight deck. Flight deck pressurised; main cargo bay unpressurised. Additional capacity in cargo pannier and nose compartment.

SYSTEMS: Securaplane smoke detection system; HS Dynamic Controls anti-icing and de-icing systems; electrical system integration by Thales Avionics Electrical Systems. Shaw Aero Devices fuel management system.

AVIONICS: Flight and engine displays and autopilot by Meggitt Avionics of UK; GPS by UPS Aviation Technologies.

DIMENSIONS, EXTERNAL:
- Wing span .... 27.10 m (88 ft 11 in)
- Length overall .... 23.80 m (78 ft 1 in)
- Height overall .... 9.04 m (29 ft 8 in)
- Side cargo door: Height .... 2.13 m (7 ft 0 in)
  - Width .... 2.59 m (8 ft 6 in)
- Rear cargo door: Height .... 2.13 m (7 ft 0 in)
  - Width .... 1.68 m (5 ft 6 in)

DIMENSIONS, INTERNAL:
- Cargo bay: Length .... 13.31 m (43 ft 8 in)
  - Max width .... 1.93 m (6 ft 4 in)
  - Max height .... 2.13 m (7 ft 0 in)

Impression of the mid-range Freight Feeder 200 0097154

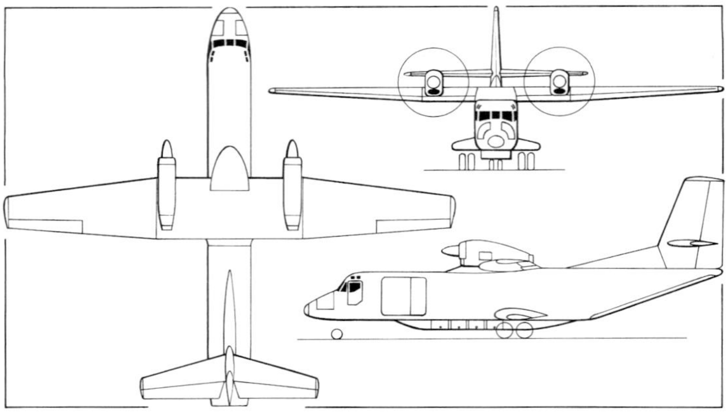

General appearance of the AUC FF-1080–200 Freight Feeder *(James Goulding)* 0015552

Volume: cargo bay .... 54.8 m^3 (1,936 cu ft)
pannier .... 5.61 m^3 (198 cu ft)
nose compartment .... 5.47 m^3 (193 cu ft)

WEIGHTS AND LOADINGS (estimated):
Weight empty: 200 .... 8,117 kg (17,894 lb)
500 .... 14,515 kg (32,000 lb)
Payload:
for 400 n mile (740 km; 460 mile) range:
200 .... 6,619 kg (14,592 lb)
500 .... 9,979 kg (22,000 lb)
for max range: 200 .... 4,019 kg (8,860 lb)
500 .... 7,348 kg (16,200 lb)
Max T-O weight: 200 .... 17,236 kg (38,000 lb)
Max landing weight: 200 .... 16,783 kg (37,000 lb)
Max zero-fuel weight: 200 .... 12,247 kg (27,000 lb)
Max power loading: 200 .... 4.21 kg/kW (6.91 lb/shp)

PERFORMANCE (estimated):
Cruising speed: 200 .... 250 kt (463 km/h; 287 mph)
500 .... 270 kt (500 km/h; 311 mph)
Stalling speed, power off, 200:
T-O flap .... 85 kt (158 km/h; 98 mph)
flaps down .... 73 kt (136 km/h; 84 mph)
Minimum single-engine control speed (VMC):
200 .... 82 kt (152 km/h; 95 mph)
Max rate of climb at S/L: 200 .... 671 m (2,200 ft)/min
Rate of climb at S/L, OEI: 200 .... 213 m (700 ft)/min
Service ceiling: 200 .... 7,620 m (25,000 ft)
Service ceiling, OEI: 200 .... 4,572 m (15,000 ft)
T-O run: 200 .... 558 m (1,830 ft)
T-O to 15 m (50 ft): 200 .... 856 m (2,810 ft)
Landing from 15 m (50 ft): 200 .... 634 m (2,080 ft)
Landing run: 200 .... 396 m (1,300 ft)
Range: with max fuel:
200 .... 1,500 n miles (2,778 km; 1,726 miles)
500 .... 2,000 n miles (3,704 km; 2,301 miles)
with max payload:
200 .... 500 n miles (926 km; 575 miles)

UPDATED

AVEOTECH

AVEOTECH INTERNATIONAL LTD

2040 Sky Lane, Denton, Texas 76207
Tel: (+1 940) 594 37 87
e-mail: info@aveoaeroplanes.com
Web: http://www.aveoaeroplanes.com
VICE-PRESIDENT, TECHNICAL SERVICES & MANUFACTURING: Don Back
MARKETING DIRECTOR: James King

AveoTech holds marketing rights to the Polish-built Aero AT-3 in all countries except the UK.

NEW ENTRY

AVEOTECH AVEOSPORT, X AND XTC

TYPE: Side-by-side ultralight/kitbuilt.

PROGRAMME: Version of Aero AT-3, manufactured in Poland, for US LSA category. Marketing launched at Sun 'n' Fun, Lakeland, Florida, April 2004, following delivery of US demonstrator (N55XT) in previous month.

CURRENT VERSIONS: **AveoSport:** LSA/Sport Pilot version. *As described.*

AveoX: Kit version. MTOW increased to 582 kg (1,283 lb), as for AT-3, but with upturned wingtips of AveoSport.

AveoXTC: AT-3 (span 7.55 m; 24 ft 9½ in) marketed in US under JAR-VLA bilateral certification agreement with Poland.

CUSTOMERS: Early sales of US versions included one to Venezuela.

COSTS: AveoSport US$95,000; AveoXTC US$69,500; AveoX kit US$25,000, or US$37,000 fast-build or US$49,000 with builder's assistance less engine and instruments (2004).

DESIGN FEATURES: Low-mounted, constant-chord wing with upturned tips; sweptback fin with significant (Pottier-type) rudder undercut for tailplane travel. Forward-hinged, single-piece canopy.

First US-registered AveoTech AveoSport at Sun 'n' Fun, April 2004 *(Paul Jackson)* ***NEW***/0576838

AveoSport instrument panel *(Paul Jackson)* ***NEW***/0576837

FLYING CONTROLS: Manual. All-moving tailplane with anti-balance tab and horn-balanced rudder. Split flaps; max deflection 40°.

STRUCTURE: Primarily metal, with some composites fairings, including wingtips and engine cowling.

LANDING GEAR: Tricycle type; fixed. Fuselage-mounted, cantilever sprung mainwheel legs and levered suspension nosewheel with rubber-in-compression shock absorber. Wheel speed fairings.

POWER PLANT: One 73.5 kW (98.6 hp) Rotax 912S flat-four driving an Aveo Sail three-blade propeller (GT-2 optional). Fuel tank ahead of cockpit, capacity 70 litres (18.5 US gallons; 15.4 Imp gallons).

DIMENSIONS, EXTERNAL:
Wing span .... 8.08 m (26 ft 6 in)
Length overall .... 6.25 m (20 ft 6 in)
Height overall .... 2.23 m (7 ft 3¾ in)

WEIGHTS AND LOADINGS:
Weight empty .... 318 kg (700 lb)
Max T-O weight .... 558 kg (1,232 lb)

PERFORMANCE:
Never-exceed (VNE) and max cruising speed .... 115 kt (213 km/h; 132 mph)
Cruising speed .... 108 kt (200 km/h; 124 mph) IAS
Stalling speed, flaps down .... 44 kt (82 km/h; 51 mph)
Max rate of climb at S/L .... 274 m (900 ft)/min
T-O and landing run .... 183 m (600 ft)
Range with max fuel .... 460 n miles (851 km; 529 miles)

NEW ENTRY

AVIABELLANCA

AVIABELLANCA AIRCRAFT CORPORATION

PMB 47, 2315 B Forest Drive, Annapolis, Maryland 21401
Tel: (+1 410) 266 55 18
Fax: (+1 410) 266 86 97
e-mail: avbellanca@comcast.net
Web: http://www.aviabellancainc.com
PRESIDENT: August T Bellanca

The original Bellanca Aircraft Corporation was established in 1927 by Giuseppe Mario Bellanca and produced many high-wing and low-wing cabin monoplanes such as the Cruisair and Cruisemaster both before and after the Second World War, as well as several one-off transocean aircraft. Bellanca left the original firm to set up Bellanca Development Company in 1954, which was renamed Bellanca Aircraft Engineering in 1963 and AviaBellanca in 1983. AviaBellanca's designer is August Bellanca, son of Giuseppe, who died in 1960. New, all-composites August Bellanca-designed Skyrocket II introduced 1975 and set five NAA/FAI closed circuit speed records; improved Skyrocket III made public debut at Oshkosh 1997. Some 3,800 Bellanca-designed aircraft are registered in the USA.

By August 2003, the company was engaged in raising a second round of finance to complete development and testing of the Skyrocket and as production working capital.

UPDATED

AviaBellanca Skyrocket II prototype *(Jim Koepnick/AviaBellanca)* 0062654

AVIAT

AVIAT AIRCRAFT INC

672 South Washington, PO Box 1240, Afton, Wyoming 83110
Tel: (+1 307) 885 31 51
Fax: (+1 307) 885 96 74
e-mail: aviat@aviataircraft.com
Web: http://www.aviataircraft.com
CHAIRMAN AND PRESIDENT: Stuart Horn
DIRECTOR OF SALES AND MARKETING: Bob James
NATIONAL SALES MANAGER: Mark James

The Pitts Aerobatics company was acquired by Christen Industries in 1983 along with manufacturing and marketing rights for the Pitts Special aerobatic aircraft. In turn, Aviat Aircraft Inc acquired Christen Industries in 1991 and production and type certificates for the Christen range. Aviat itself was bought by Stuart Horn in December 1995 and type certificates passed to Sky International Inc on 10 January 1996. The company has acquired manufacturing rights to the 1930s/1940s Globe Swift and Monocoupe 110, but a return to production has not been pursued. Aviat delivered a total of 83 aircraft in 1998, 85 in 1999, 91 in 2000, 57 in 2001, 55 in 2002 and 64 in 2003. Factory floor area 6,503 m^2 (70,000 sq ft). Workforce stood at 58 in 2002.

UPDATED

Aviat Husky A-1B, official lead aircraft of the USA's 2003 National Air Tour *(Paul Jackson)* ***NEW***/0561705

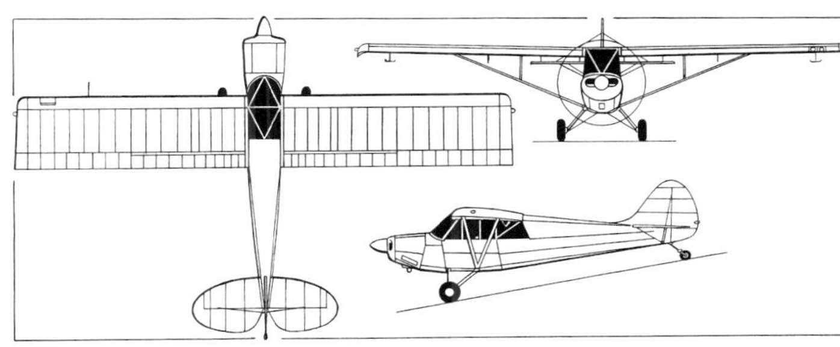

Aviat Husky A-1 (Textron Lycoming O-360 flat-four) *(James Goulding)* 0015590

AVIAT HUSKY A-1

TYPE: Two-seat lightplane.

PROGRAMME: First flight (N6070H) 1986; FAA certification of original A-1 version under FAR Pt 23 achieved by Christen Industries 1 May 1987, this having 816 kg (1,800 lb) MTOW. Type Certificate passed to Aviat Inc, then White International on 3 December 1992, then Sky International on 10 January 1996.

CURRENT VERSIONS: **A-1:** Original version, now superseded; 398 built.

A-1A: Prototype N15LF. Certified 28 January 1998, with MTOW increased by 41 kg (90 lb). Total 69 built, 1998–99 only.

A-1B: Introduced as mission-specific version initially intended for government agencies. Has higher maximum take-off weight of 907 kg (2,000 lb). Certified 28 January 1998. Became sole production version in 2001.

CUSTOMERS: Production for 2000 totalled 80 (four A-1A and 76 A-1B); for 2001, 50 (all A-1Bs); 2002, 34 (all A-1Bs); and 2003, 37 (all A-1Bs). Some 730 built by mid-2004, including 467 A-1/A-1As.

COSTS: A-1A US$127,500; A-1B US$139,700 (2003).

DESIGN FEATURES: Conventional high-wing monoplane; constant-chord wing with V-strut bracing and auxiliary struts; wire-braced empennage with elliptical surfaces.

Wing has modified Clark Y US 35B section; optional drooped Plane Booster wingtips and vortex generator kit.

FLYING CONTROLS: Conventional and manual. Symmetrical section ailerons with spade-type mass balance; trim tabs in elevators; slotted high-lift flaps. Fixed tailplane; trim by adjustable bungee. Control surface movements: ailerons ±20°; elevators +29/–15°; rudder ±25°; flaps 30°.

STRUCTURE: Tubular welded 4130 steel fuselage. Wing has two aluminium spars, metal ribs and metal leading-edge, Dacron covering overall. A-1B has 7075-T76 aluminium rear spar, replacing 6061-T6. Twin bracing struts each side of wings and wire- and strut-braced tail unit. Light alloy slotted flaps and ailerons, with Dacron covering. Fuselage and tail have chrome molybdenum steel tube frames, covered in Dacron except for metal skin to rear fuselage. Seven-coat corrosion protection.

LANDING GEAR: Tailwheel type; fixed. Two faired side Vs and half-axles hinged to lower longeron, sprung under centreline, with internal (under front seat) bungee cord shock-absorption. Cleveland mainwheels, tyres size 8.00-6 as standard; 6.00-6 or 8.50-6 tyres and 24×10-6 or 26×10.5-6 Tundra tyres optional. Cleveland mainwheel hydraulic disc brakes, toe-operated. Steerable leaf-spring tailwheel. Optional EDO 89-2000, Baumann BF-2100 and Wipline 2100 floats, Aero M1500, M1800, M2000 or M3000H wheel replacement skis, Aero wheel skis and Flairdyne hydraulic wheel-retracting skis.

Aviat Husky A-1B *(Paul Jackson)* ***NEW***/0576870

POWER PLANT: One 134 kW (180 hp) Textron Lycoming O-360-A1P flat-four engine, driving a Hartzell HC-C2YK-1BF two-blade constant-speed metal propeller. Fuel in two metal tanks, one in each wing, total capacity 208 litres (52.0 US gallons; 45.75 Imp gallons), of which 189 litres (50.0 US gallons; 41.6 Imp gallons) are usable. Fuel filler point in upper surface of each wing, near root.

ACCOMMODATION: Enclosed cabin seating two in tandem, on ergonomic lumbar and side support seats, with dual controls. Custom leather seats optional. Five-point safety harness. Downward-hinged door on starboard side, with upward-hinged window above. Optional skylight window in roof. A-1B has luggage door on starboard side as standard; optional on A-1A. Rear compartment for extra 13.6 kg (30 lb) baggage optional (structural provision standard) from 2000; external access starboard side, below rear window; certified retrofitting kits available.

SYSTEMS: 12 V electrical system includes lights, 22 Ah battery, and 60 A alternator.

AVIONICS: VFR standard; Loran C transponder, GPS, nav/com, intercom, and IFR instrumentation, including Garmin GNS530 GPS/com and VM 1000 panel, optional.

Instrumentation: Sensitive altimeter, ASI, magnetic compass, digital tachometer, oil temperature/pressure gauge, manifold pressure gauge and CHT gauge all standard.

EQUIPMENT: Rear seat heater and windscreen defroster standard on A-1B.

DIMENSIONS, EXTERNAL:

Wing span	10.82 m (35 ft 6 in)
Wing aspect ratio	6.9
Length overall	6.88 m (22 ft 7 in)
Height overall	2.01 m (6 ft 7 in)
Propeller diameter	1.93 m (6 ft 4 in)

DIMENSIONS, INTERNAL:

Baggage hold volume	0.28 m^3 (10 cu ft)

AREAS:

Wings, gross	17.00 m^2 (183.0 sq ft)
Ailerons (total)	1.43 m^2 (15.40 sq ft)
Trailing-edge flaps (total)	2.09 m^2 (22.50 sq ft)
Fin	0.43 m^2 (4.66 sq ft)
Rudder	0.62 m^2 (6.76 sq ft)
Tailplane	1.48 m^2 (15.90 sq ft)
Elevators, incl tabs	1.31 m^2 (14.10 sq ft)

WEIGHTS AND LOADINGS (A-1B where different):
Weight empty ... 540 kg (1,190 lb)
Baggage capacity ... 23 kg (50 lb)
Max T-O weight:
landplane: A-1A ... 857 kg (1,890 lb)
A-1B ... 907 kg (2,000 lb)
floatplane: A-1A ... 943 kg (2,079 lb)
A-1B ... 998 kg (2,200 lb)
Max wing loading:
landplane: A-1A ... 50.4 kg/m^2 (10.33 lb/sq ft)
A-1B ... 53.4 kg/m^2 (10.93 lb/sq ft)
floatplane: A-1A ... 55.5 kg/m^2 (11.36 lb/sq ft)
A-1B ... 58.7 kg/m^2 (12.02 lb/sq ft)
Max power loading:
landplane: A-1A ... 6.39 kg/kW (10.50 lb/hp)
A-1B ... 6.76 kg/kW (11.11 lb/hp)
floatplane: A-1A ... 7.03 kg/kW (11.55 lb/hp)
A-1B ... 7.44 kg/kW (12.22 lb/hp)

PERFORMANCE (landplane; A-1B where different):
Never-exceed speed (VNE) ... 132 kt (245 km/h; 152 mph)
Max level speed ... 126 kt (233 km/h; 145 mph)
Cruising speed: at 75% power at 1,220 m (4,000 ft) ... 122 kt (225 km/h; 140 mph)
at 55% power ... 113 kt (209 km/h; 130 mph)
Stalling speed, A-1B, flaps down:
power on ... 37 kt (69 km/h; 43 mph)
power off ... 46 kt (86 km/h; 53 mph)
Landing speed: A-1A ... 42 kt (77 km/h; 48 mph)
A-1B ... 51 kt (94 km/h; 58 mph)
Max rate of climb at S/L ... 457 m (1,500 ft)/min
Service ceiling ... 6,100 m (20,000 ft)
T-O run: flaps down ... 61 m (200 ft)
T-O to 15 m (50 ft), flaps down ... 122 m (400 ft)
Landing from 15 m (50 ft), flaps down ... 244 m (800 ft)
Landing run, full flaps ... 107 m (350 ft)
Range: A-1A at 75% power, no reserves ... 608 n miles (1,126 km; 700 miles)
A-1B at 55% power, no reserves ... 695 n miles (1,287 km; 800 miles)

UPDATED

AVIAT HUSKY PUP A-1B

TYPE: Two-seat lightplane.

PROGRAMME: Developed as simplified, cheaper version of Husky for daytime VFR only. Prototype (N180HY) first flown 7 March 2002; public debut at EAA Sun 'n' Fun at Lakeland, Florida, in April 2002. Certification achieved 18 August 2003, at which time a small batch of completed aircraft was being held at the factory for immediate issue to buyers.

CUSTOMERS: Three delivered in 2003.

COSTS: US$115,000 VFR (2003).

DESIGN FEATURES: As Husky except no flaps, and ailerons do not droop.

FLYING CONTROLS: Generally as Husky, but with flaps deleted and modified engine cowling.

STRUCTURE: As Husky.

LANDING GEAR: As for Husky A-1; mainwheel size 6.00–6 (4-ply), optionally 8.00–6 (4- or 6-ply). Cleveland wheels and brakes.

POWER PLANT: One 119.3 kW (160 hp) Textron Lycoming O-320-D2A flat-four, driving a Sensenich 74DM6S8-0-58 two-blade, fixed-pitch, metal propeller. Fuel as for Husky.

ACCOMMODATION: Two in tandem. Optional rear compartment, as Husky.

AVIONICS: VFR panel standard.
Comms: Becker com, GTX 320 transponder, PM 10000 II intercom.
Flight: Garmin GPS III.
Instrumentation: ASI, altimeter, compass, digital tachometer, oil pressure/temperature gauge, EGT/CHT gauge and ammeter all standard.

EQUIPMENT: Two-colour paint scheme and polished wheel hub caps standard.

DIMENSIONS, EXTERNAL (As for Husky except):
Propeller diameter ... 1.88 m (6 ft 2 in)

WEIGHTS AND LOADINGS:
Weight empty ... 517 kg (1,140 lb)
Max T-O weight: ... 907 kg (2,000 lb)
Max wing loading: ... 53.4 kg/m^2 (10.93 lb/sq ft)
Max power loading: ... 7.61 kg/kW (12.50 lb/hp)

PERFORMANCE:
Never-exceed speed (VNE) ... 133 kt (246 km/h; 153 mph)
Cruising speed ... 122 kt (225 km/h; 140 mph)
Stalling speed, flaps up ... 43 kt (79 km/h; 49 mph)
Service ceiling ... 5,180 m (17,000 ft)
Max rate of climb at S/L ... 241 m (790 ft)/min
T-O run ... 186 m (610 ft)
Landing run ... 181 m (595 ft)
Range, with reserves ... 869 n miles (1,609 km; 1,000 miles)
Endurance ... more than 7 h

UPDATED

Prototype Aviat Husky Pup *(Paul Jackson)* ***NEW***/0576848

AVIAT PITTS S-2C

TYPE: Aerobatic two-seat biplane.

PROGRAMME: Developed from Pitts S-2 series of two-seat biplanes using wing design from Curtis Pitts. Uses S-2B basic structure modified with differently-shaped wingtips, empennage tips, ailerons and lower fuselage.

Pitts S-1 first flew September 1944; available only from plans until S-1S certified 13 February 1973, followed by S-1T on 24 July 1975. Two-seat S-2 added to original Type Certificate on 8 March 1971; S-2A on 11 June 1971 (205 built); S-2S 29 May 1981; and S-2B 6 April 1983. S-2C certified in Normal and Aerobatic categories on 4 June 1998.

CUSTOMERS: Total 63 built by December 2003; four delivered in 2002 and three in 2003.

COSTS: Basic price US$194,295 (2003).

DESIGN FEATURES: Compact, positive-stagger, single-bay biplane; constant-chord wings, upper of which sweptback; wire-braced, elliptical tail surfaces. Optimised for aerobatic performance. Differs from S-2B in having symmetrical ailerons; squared fin-, tailplane- and wingtips; metal fuselage undersides removable for easier maintenance. Stretch-formed compound-curved engine cowling. Rate of roll is greater than 300°/s.

Wing sections NACA 6400 series on upper wing, 00 series on lower wings. Both wings have 1° 30' incidence and lower wing has 3° dihedral.

FLYING CONTROLS: Conventional and manual. Rudder and elevators aerodynamically balanced. Compared to S-2B, the S-2C only has a trim tab on starboard elevator. Control surface deflections; ailerons +28° 30'/−22°; elevator ±27°; rudder ±30°.

STRUCTURE: Fuselage 4130 steel tube with wooden stringers, aluminium top decking and side panels; remainder Dacron covered. Steel tube, Dacron-covered fixed tail surfaces; Dacron-covered control surfaces.

LANDING GEAR: Tailwheel type; fixed. Two faired mainwheel legs with half-axles sprung under centreline with internal bungee suspension. Mainwheels 5.00–5; tyre pressure 2.34 bar (34 lb/sq in).

POWER PLANT: One 194 kW (260 hp) Textron Lycoming AEIO-540-D4A5 flat-six driving a Hartzell HC-C3YR-1A/C7690C or E 'Claw' three-blade aerobatic constant-speed (hydraulic) composites propeller. Fuel capacity 110 litres (29.0 US gallons; 24.0 Imp gallons) of which 106 litres (28.0 US gallons; 23.3 Imp gallons) are usable; for aerobatic flight, the 19 litre (5.0 US gallon; 4.2 Imp gallon) wing tank is not used. Oil capacity 11.4 litres (3.0 US gallons; 2.5 Imp gallons).

ACCOMMODATION: Pilot and passenger in tandem under one-piece rearward-sliding canopy; single-piece windscreen. Optional replacement upper decking for single-seat operation, comprising front cockpit flush fairing and shortened canopy.

EQUIPMENT: Options include custom paint and exterior design, electronic gauges, custom parachutes, and canopy covers.

DIMENSIONS, EXTERNAL:
Wing span ... 6.10 m (20 ft 0 in)
Length overall ... 5.41 m (17 ft 9 in)
Height overall ... 1.96 m (6 ft 5 in)
Wheel track ... 1.54 m (5 ft 0¾ in)
Wheelbase ... 4.14 m (13 ft 7 in)
Propeller diameter ... 2.03 m (6 ft 8 in)

DIMENSIONS, INTERNAL:
Cabin: Length ... 2.11 m (6 ft 11 in)
Max width ... 0.71 m (2 ft 4 in)
Max height ... 1.19 m (3 ft 11 in)

AREAS:
Wings, gross ... 11.85 m^2 (127.5 sq ft)

WEIGHTS AND LOADINGS:
Weight empty ... 524 kg (1,155 lb)
Baggage capacity ... 9 kg (20 lb)
Max T-O and landing weight ... 771 kg (1,700 lb)
Max wing loading ... 65.1 kg/m^2 (13.33 lb/sq ft)
Max power loading ... 3.98 kg/kW (6.54 lb/hp)

PERFORMANCE:
Never-exceed speed (VNE) ... 185 kt (342 km/h; 212 mph)
Max level speed ... 169 kt (313 km/h; 194 mph)
Manoeuvring speed ... 134 kt (248 km/h; 154 mph) IAS
Cruising speed at 55% power ... 150 kt (278 km/h; 173 mph)
Stalling speed, power off, flaps up ... 56 kt (104 km/h; 65 mph)
Max rate of climb at S/L ... 884 m (2,900 ft)/min
Max roll rate ... more than 300°/s
T-O run ... 169 m (555 ft)
T-O to 15 m (50 ft) ... 262 m (860 ft)

Aviat Pitts S-2C (Textron Lycoming flat-six) 0525815

Aviat Pitts S-2C *(Paul Jackson)* ***NEW***/0576867

Landing from 15 m (50 ft) .... 366 m (1,200 ft)
Landing run .... 229 m (750 ft)
Max range at 55% power, 30 min reserves .... 300 n miles (555 km; 345 miles)
Endurance, 30 min reserves .... 1 h 36 min
g limits .... +6/–5

UPDATED

AVIAT (CHRISTEN) EAGLE II

TYPE: Tandem-seat biplane kitbuilt.

PROGRAMME: Originated as rival to Pitts S-1, designed by Frank Christensen. First flown in February 1977. Kit production began October 1977, initially by Christen Industries.

CUSTOMERS: More than 350 completed and flown by mid-2001, including 320 in the United States. Promotion and production continuing in 2003.

COSTS: Kit US$85,000 (2002) less engine and propeller.

DESIGN FEATURES: Generally as for S-2B (see 2001–02 edition); similar to S-2C. Available in 24 individual subassemblies; quoted build time 1,800 hours.

FLYING CONTROLS: Conventional and manual. Ailerons on both sets of wings.

STRUCTURE: Fabric-covered wings with wooden spars and ribs plus metal leading- and trailing-edges. Welded 4130 steel tube fuselage, fabric covered aft of cockpit; metal panelled before.

LANDING GEAR: Tailwheel type; fixed. Fuselage-mounted, spring steel cantilever mainwheel legs. Hydraulic brakes.

POWER PLANT: One 149 kW (200 hp) Textron Lycoming AEIO-360-A1D driving a Hartzell HC-C2YK-4/C7666A-2 constant-speed two-blade metal propeller. Fuel capacity 94.6 litres (25.0 US gallons; 20.8 Imp gallons), of which 90.8 litres (24.0 US gallons; 20.0 Imp gallons) are usable.

ACCOMMODATION: Pilot and passenger in tandem; flown solo from rear seat. Optional single-seat upper decking assembly comprises front cockpit flush fairing and shortened canopy.

SYSTEMS: Inverted oil system and manual fuel pump system as standard.

DIMENSIONS, EXTERNAL:
Wing span .... 6.07 m (19 ft 11 in)
Length overall .... 5.46 m (17 ft 11 in)
Height overall .... 1.98 m (6 ft 6 in)

Aviat Eagle II with standard double canopy *(Paul Jackson)* **NEW**/0576899

AREAS:
Wings, gross .... 11.61 m² (125.0 sq ft)

WEIGHTS AND LOADINGS:
Weight empty .... 465 kg (1,025 lb)
Max T-O and landing weight .... 715 kg (1,578 lb)
Max baggage weight .... 13.6 kg (30 lb)
Max wing loading .... 61.6 kg/m² (12.62 lb/sq ft)
Max power loading .... 4.80 kg/kW (7.89 lb/hp)

PERFORMANCE:
Never-exceed speed (VNE) .... 182 kt (338 kt; 210 mph)
Max operating speed .... 159 kt (296 km/h; 184 mph)
Cruising speed .... 143 kt (266 km/h; 165 mph)
Stalling speed .... 51 kt (90 km/h; 58 mph)
Max rate of climb at S/L .... 640 m (2,100 ft)/min
Service ceiling .... 5,180 m (17,000 ft)
Max roll rate .... 187°/s
T-O run .... 244 m (800 ft)
Landing from 15 m (50 ft) .... 480 m (1,575 ft)
Range with 30 min reserves .... 380 n miles (703 km; 437 miles)
g limits .... +9/–6

UPDATED

AVIATION DEVELOPMENT

AVIATION DEVELOPMENT INTERNATIONAL LTD

7172 SR 38 NE, Bloomingburg, Ohio 43106
Tel: (+1 740) 437 76 88
Fax: (+1 740) 335 89 96
e-mail: rjs5653@sbcglobal.net

JOINT OWNERS:
Rick Schneider
Steve Sollars

Aviation Development International was formed in 1997 to build and market the Alaskan Bushmaster kitbuilt. The company has a 1,486 m² (16,000 sq ft) facility at Highland County Airport, Hillsboro.

Another company product is the Loadmaster, an Aermacchi AL-60 re-engined with a VOKBM M-14P nine-cylinder radial of 294 kW (394 hp). It is planned to convert two further AL-60s with a 560 kW (751 shp) Walter M 601 turboprop.

UPDATED

AVIATION DEVELOPMENT ALASKAN BUSHMASTER

TYPE: Four-seat kitbuilt.

PROGRAMME: Developed during 1980s by Alaskan bush pilot Rick Schneider to replace his Piper Super Cub. Initial design based on Merlin Explorer. Original production ceased in 1990 but restarted in 1997.

CUSTOMERS: 12 original examples and one new-build version flying by April 2001, at which time 10 further aircraft on order. One registered to Southland Holdings of Wilmington, Delaware, in July 2002; three on US register in August 2004.

COSTS: Kit US$22,500, excluding engine, propeller, instruments and upholstery.

DESIGN FEATURES: Piper Tri-Pacer family development. No welding, machining or heavy forming required during construction. Quoted build time 600 hours.

FLYING CONTROLS: Conventional and manual. Stainless steel control cables. Horn-balanced elevators and rudder; flight-adjustable tabs in both elevators.

STRUCTURE: Fabric-covered fuselage and tail of welded 4130 steel tube; fabric-covered extruded aluminium wings, flaps and ailerons. Wing braced with two extruded aluminium lift struts; tailplane and fin braced by wires.

LANDING GEAR: Tailwheel type; fixed. Spring aluminium main legs and tailwheel. Bungee cord suspension. Mainwheels 8.50-6. Options include Tundra tyres, Schneider Wheel Skis, skis and floats.

POWER PLANT: Prototype had 149 kW (200 hp) Textron Lycoming O-320 flat-four engine driving a two-blade fixed-pitch propeller; recommended engine power between 112 and 224 kW (150 and 300 hp), including engines from Teledyne Continental and LOM. Standard fuel capacity 114 litres (30.0 US gallons; 25.0 Imp gallons) per tank; option of two, four or six wing tanks, taking maximum possible capacity to 681 litres (180 US gallons; 150 Imp gallons).

ACCOMMODATION: Pilot and three passengers in two pairs of seats. Dual controls. Door each side, plus freight door to port, rear.

Alaskan Bushmaster fitted with 119 kW (160 hp) LOM engine *(Paul Jackson)* **NEW**/1047166

DIMENSIONS, EXTERNAL:
Wing span, without tips .... 11.89 m (39 ft 0 in)
Wing aspect ratio 7.2
Length overall: O-320 engine .... 7.09 m (23 ft 3 in)
LOM 322 engine .... 7.49 m (24 ft 7 in)
Height overall .... 2.08 m (6 ft 10 in)
Tailplane span .... 3.48 m (11 ft 5 in)
Wheel track .... 2.29 m (7 ft 6 in)

DIMENSIONS, INTERNAL:
Cabin max width .... 1.04 m (3 ft 5 in)

AREAS:
Wings, gross .... 19.51 m² (210.0 sq ft)
Ailerons (total) .... 1.77 m² (19.00 sq ft)
Flaps (total) .... 2.15 m² (23.10 sq ft)
Horizontal tail surfaces (total) .... 3.58 m² (38.50 sq ft)

WEIGHTS AND LOADINGS:
Weight empty, typical .... 680 kg (1,500 lb)
Baggage capacity .... 91 kg (200 lb)
Max T-O weight .... 1,361 kg (3,000 lb)
Max wing loading .... 69.7 kg/m² (14.29 lb/sq ft)
Max power loading, 119 kW (160 hp) engine .... 11.41 kg/kW (18.75 lb/hp)

PERFORMANCE (119 kW; 160 hp engine):
Never-exceed speed (VNE) .... 130 kt (241 km/h; 150 mph)
Normal cruising speed .... 100 kt (185 km/h; 115 mph)
Stalling speed: flaps up .... 38 kt (70 km/h; 43 mph)
flaps down .... 31 kt (57 km/h; 35 mph)
Max rate of climb at S/L .... 305 m (1,000 ft)/min
T-O run .... 244 m (800 ft)
Landing run .... 91 m (300 ft)
Range with standard fuel .... 536 n miles (993 km; 617 miles)

UPDATED

AVID

AVID AIRCRAFT INC

Avid Aircraft was formed in 1983 by Dean Wilson to produce the Avid Flyer Mark IV, Catalina, Bandit and Magnum. In May 1999, the company was sold to Jim and Kim Tomash who restarted production in Ennis, Montana. In November 2003 it was reported that Avid had closed its factory.

Support of existing aircraft, and production of a new version of Flyer, was transferred to Airdale Flyer Company of Coldwell, Idaho.

UPDATED

AVID FLYER MARK IV

TYPE: Side-by-side ultralight kitbuilt.

PROGRAMME: First flown 1983; available as single kit, or six separate kits to spread cost of purchase. Strongly influenced design of Indaer-Peru Chuspi light aircraft (see under Peru in 1992–93 *Jane's*), and the SkyStar series of kitbuilt aircraft. Avid **Bandit** identical, except for 454 kg (1,000 lb) maximum T-O weight.

CUSTOMERS: More than 1,600 Avid Flyer kits delivered, with over 900 flying.

COSTS: Kit: High Gross STOL US$15,740, Aerobatic Speedwing US$14,740, both firewall back (2003).

DESIGN FEATURES: Strut-braced, high-wing ultralight of traditional configuration and construction. Two forms available (interchangeable), as original **High Gross STOL** with unique near full-span auxiliary aerofoil flaperons, and shorter span **Aerobatic Speedwing** using new wing section; cruising and stalling speeds raised with Aerobatic Speedwing fitted. Wings fold for storage. Quoted build time 650 hours.

FLYING CONTROLS: Manual. Junkers flaperons (see Design Features), elevators with adjustable trim tab, and rudder.

STRUCTURE: Aluminium wing spars and plywood ribs, covered with heat-shrunk Dacron. Welded steel tube fuselage, rudder, tailplane and elevators. Dacron covered except for fuselage nose which has premoulded GFRP cowlings. Fin integral with fuselage.

LANDING GEAR: Non-retractable tricycle or tailwheel landing gear, with Tundra tyres and brakes. Wheel rim size 12.7 cm (5 in). Steerable tailwheel. Mainwheels have bungee cord shock absorption. Optional Aqua 1500 floats, skis and wheel/skis. Wider landing gear available as option, introduced 2001.

POWER PLANT: One 47.8 kW (64.1 hp) Rotax 582 two-stroke engine, driving a two- or three-blade propeller; fixed-pitch wooden propeller for STOL, three-blade ground-adjustable for Speedwing. Fuel capacity 53 litres (14.0 US gallons; 11.7 Imp gallons) for High Gross STOL and 68 litres (18.0 US gallons; 15.0 Imp gallons) for Aerobatic Speedwing; option to double capacity of each. Similar capacity fuel tanks may be added in port wing.

DIMENSIONS, EXTERNAL (A: STOL, B: Aerobatic Speedwing):
Wing span: A ... 9.11 m (29 ft 10½ in)
B ... 7.30 m (23 ft 11½ in)
Length overall ... 5.46 m (17 ft 11 in)
Height overall: tailwheel ... 1.80 m (5 ft 11 in)
tricycle ... 2.11 m (6 ft 11 in)

DIMENSIONS, INTERNAL:
Cabin max width ... 1.00 m (3 ft 3½ in)

AREAS (A and B as above):
Wings, gross: A ... 11.38 m² (122.5 sq ft)
B ... 9.04 m² (97.3 sq ft)

WEIGHTS AND LOADINGS (A and B as above):
Weight empty: A ... 200–231 kg (440–510 lb)
B ... 231 kg (510 lb)
Baggage capacity ... 16 kg (35 lb)
Max T-O weight: A ... 522 kg (1,150 lb)
B ... 476 kg (1,050 lb)

Avid Aircraft Avid Flyer Speedwing in tailwheel configuration and with Rotax 582 engine, built by Mr G E Laucht *(Paul Jackson)* ***NEW***/0561699

PERFORMANCE (A and B as above):
Max level speed at 1,525 m (5,000 ft):
A, B ... 117 kt (217 km/h; 135 mph)
Max cruising speed: A ... 78 kt (145 km/h; 90 mph)
B ... 100 kt (185 km/h; 115 mph)
Stalling speed:
A, flaps down, engine idling ... 28 kt (52 km/h; 32 mph)
B ... 37 kt (68 km/h; 42 mph)
Max rate of climb at S/L: A ... 518 m (1,700 ft)/min
B ... 366 m (1,200 ft)/min
Service ceiling: A, B ... 3,810 m (12,500 ft)
T-O run: A ... 27 m (90 ft)
B ... 38 m (125 ft)
Landing run: A ... 46 m (150 ft)
B ... 152 m (500 ft)
Range, no reserves: A ... 295 n miles (547 km; 340 miles)
B ... 491 n miles (910 km; 566 miles)

AVID MAGNUM

TYPE: Side-by-side kitbuilt.

CUSTOMERS: About 200 under construction with around 100 completed.

COSTS: Kit US$26,990 without engine (2003).

DESIGN FEATURES: High-wing braced monoplane, similar to Avid Flyer, but in certified category, using a Textron Lycoming engine. Recent upgrades include larger rudder and new external baggage door. Quoted build time 750 hours.

FLYING CONTROLS: Similar to Avid Flyer.

STRUCTURE: Similar to Avid Flyer, with Poly Fiber covering.

LANDING GEAR: Available as tricycle or tailwheel, with Cleveland wheels and brakes; wheel fairings. Spring aluminium landing gear standard. Optional floats.

POWER PLANT: One Textron Lycoming engine in 80.5 to 134 kW (108 to 180 hp) range, including O-235, O-320 and O-360. Tricycle version also accepts 119 kW (160 hp) Subaru EJ22 engine. Fuel capacity 110 litres (29.0 US gallons; 24.1 Imp gallons) of which 106 litres (28.0 US gallons; 23.3 Imp gallons) are usable. Optional 45 litre (12.0 US gallon; 10.0 Imp gallon) wingtip tanks available.

ACCOMMODATION: Two seats side by side, with dual controls, plus optional jump seat for small adult or two children in 0.81 m³ (28.5 cu ft) baggage area.

DIMENSIONS, EXTERNAL:
Wing span ... 10.06 m (33 ft 0 in)
Wing chord, constant ... 1.37 m (4 ft 6 in)
Wing aspect ratio ... 8.0
Width, wings folded ... 2.59 m (8 ft 6 in)
Length overall ... 6.40 m (21 ft 0 in)
Height overall ... 1.86 m (6 ft 1¼ in)

DIMENSIONS, INTERNAL:
Cabin max width ... 1.12 m (3 ft 8 in)

AREAS:
Wings, gross ... 12.63 m² (136.0 sq ft)

WEIGHTS AND LOADINGS:
Weight empty ... 442 kg (975 lb)
Baggage capacity ... 68 kg (150 lb)
Max T-O weight ... 794 kg (1,750 lb)
Max wing loading ... 62.8 kg/m² (12.87 lb/sq ft)
Power loading (119 kW; 160 hp) ... 6.66 kg/kW (10.94 lb/hp)

PERFORMANCE (119 kW; 160 hp O-320 engine):
Never-exceed speed (VNE) ... 134 kt (249 km/h; 155 mph)
Cruising speed ... 113 kt (209 km/h; 130 mph)
Stalling speed ... 31 kt (58 km/h; 36 mph)
Max rate of climb at S/L ... 549 m (1,800 ft)/min
Service ceiling ... more than 6,100 m (20,000 ft)
T-O run ... approx 46 m (150 ft)
T-O to 15 m (50 ft) ... 69 m (225 ft)
Landing run ... 76 m (250 ft)
Range ... 450 n miles (833 km; 517 miles)

AVIPRO

AVIPRO AIRCRAFT LTD

3536 East Shangri-La Road, Phoenix, Arizona 85028
Tel: (+1 602) 971 37 68
Fax: (+1 602) 971 38 96
e-mail: info@bearhawkaircraft.com
Web: http://www.bearhawkaircraft.com
PRESIDENT: Budd Davisson

At AirVenture 2003, AviPro launched a kitbuilt version of the previously plans-only Bearhawk. The first set of components was shipped from the company's factory at Atlixco, Mexico, on 22 July 2002.

VERIFIED

Barrows Bearhawk built from plans only *(Paul Jackson)* ***NEW***/0561698

AVIPRO BEARHAWK

TYPE: Four-seat kitbuilt.

PROGRAMME: Prototype RB-4 (N6890R) first flew with Lycoming O-360 engine in December 1993, but not promoted until 1995; plans marketed by designer, Robert Barrows; second prototype (N33RB) flew 27 October 1999 with Lycoming O-540 engine and large cargo doors; by mid-2003, eight Bearhawks were flying (including two prototypes with total of 1,000 hours). First plans-built aircraft (N156RM) flown by Robert Marek on 4 May 2001; others completed with Ford V-6 and Rover V-8 motorcar engines. Two-seat Bearhawk 2 (N289R) prototype first flown by Mr Barrows in mid-2003, but decision on commerical launch had not then been taken.

Marketing of AviPro kits began in 2003, with first airframe displayed at AirVenture, Oshkosh, from 29 July 2003. FAA 51% rule certification received end November 2003; first production kit (N303AP) made first flight 4 December 2003.

CURRENT VERSIONS: **Bearhawk:** Base model, *as described.*

Bearhawk Patrol: Tandem-seat two-place version, available as plans only.

CUSTOMERS: 31 kits delivered by January 2004 and kit deliveries then set at two per month; in addition 600 sets of plans had been sold by August 2004 with 200 under construction.

COSTS: Kit US$27,700, excluding engine, propeller and instrument (2003).

DESIGN FEATURES: Four-seat utility design capable of carrying full load of occupants and fuel; cabin size exceeds Cessna 172. High, strut-braced wing and mutually braced empennage. Aerofoil section NACA 4412.

FLYING CONTROLS: Conventional and manual. Horn-balanced rudder and elevators. Large flaps. Flight-adjustable tabs in both elevators.

STRUCTURE: Welded steel tube fuselage with fabric covering. All aluminium wing with fabric-covered control surfaces attached to rear spar. Metal engine cowling from Pitts Special.

LANDING GEAR: Tricycle type; fixed. Montana Floats optional. Mainwheels 7.00–6. Steerable tailwheel, size 2.80/2.50–4. Rubber-in-compression suspension.

POWER PLANT: Recommended piston engines include 119 kW (160 hp) Textron Lycoming

O-320; 134 kW (180 hp) O-360; 194 kW (260 hp) O-540; and Teledyne Continental O-470 of 179 kW (240 hp). Three-blade propeller, typically Hartzell F7693F.

Normal fuel capacity 208 litres (55.0 US gallons; 45.8 Imp gallons in two wing tanks. Optional capacity 257 litres (68.0 US gallons; 56.6 Imp gallons).

ACCOMMODATION: Four persons in two side-by-side pairs, with baggage area to rear. Rear seat removable. Door each side, front, split horizontally; front- and rear-hinged double doors, starboard, provide access to rear seats and baggage.

DIMENSIONS, EXTERNAL:

Wing span	10.06 m (33 ft 0 in)
Wing aspect ratio	6.1
Length overall	7.16 m (23 ft 6 in)
Rear doors (starboard): Width	1.83 m (6 ft 0 in)

DIMENSIONS, INTERNAL:

Cabin: Length	2.95 m (9 ft 8 in)
Max width	1.07 m (3 ft 6 in)

AREAS:

Wings, gross	16.72 m (180.0 sq ft)

WEIGHTS AND LOADINGS:

Weight empty	522 to 612 kg (1,150 to 1,350 lb)
Max T-O weight: landplane	1,134 kg (2,500 lb)
floatplane	1,224 kg (2,700 lb)
Max wing loading: landplane	67.8 kg/m^2 (13.89 lb/sq ft)
floatplane	73.2 kg/m^2 (15.00 lb/sq ft)

PERFORMANCE:

Never-exceed speed (VNE)	152 kt (281 km/h; 175 mph)
Normal cruising speed at 60% power:	
260 hp engine	139 kt (257 km/h; 160 mph)
180 hp engine	122 kt (225 km/h; 140 mph)
Stalling speed, flaps down	35 kt (65 km/h; 40 mph)
Max rate of climb at S/L	518 m (1,700 ft)/min
T-O and landing run	107 m (350 ft)
Range at 50% power	782 n miles (1,448 km; 900 miles)

First AviPro Bearhawk built from a kit *(Paul Jackson)* ***NEW***/1047187

UPDATED

AVOCET

AVOCET AIRCRAFT LLC

Ten Bay Street, Westport, Connecticut 06880
Tel: (+1 203) 454 56 56
Web: http://www.avocetprojet.com

CHAIRMAN: Carey Robinson Wolchok
CEO: David Tait
COO: Mark Biagetti
VICE-PRESIDENT, BUSINESS DEVELOPMENT: Dr Winston Mahabir
VICE-PRESIDENT, INTEGRATED MARKETING AND SALES: Jeffrey R Myers
ISRAEL AIRCRAFT INDUSTRIES PROGRAMME MANAGER: Ofer Shifris

Avocet Aircraft LLC was formed in 2002 to develop a new-generation light jet. On 4 August 2003, Avocet and Israel Aircraft Industries signed an MoU leading to joint development and marketing of the Avocet ProJet, with the aim of IAI holding responsibility for design, development certification and manufacture and Avocet handling marketing, customer relations, training, delivery and product support. Formal go-ahead for IAI to join the project was intended mid-2004, following completion of a definition study. However, IAI declined to back the project, which has been suspended, pending alternative finance.

UPDATED

AVOCET PROJET

TYPE: Light business jet.

PROGRAMME: Original timetable called for first flight in fourth quarter 2005; FAA FAR Pt 23 certification for single pilot and known icing operation, in fourth quarter 2006, followed by first customer deliveries in January 2007. Programme suspended in mid-2004 when Israel Aircraft Industries withdrew its offer of support, reportedly because Avocet had not provided its own US$100 million investment and failed to secure a substantial number of firm orders.

Artist's impression of Avocet ProJet Light business jet ***NEW***/0561753

UPDATED

BARR

BARR AIRCRAFT

900 Airport Road, Montoursville, Pennsylvania 17754
Tel: (+1 570) 368 36 55
Fax: (+1 570) 368 20 95
e-mail: six@barraircraft.com
Web: http://www.barraircraft.com

MANAGING DIRECTOR: James L Barr

Barr Aircraft was formed in 1958 and holds supplemental type certificates for the Piper PA-12, PA-14, PA-20 and PA-22 series of aircraft and is presently developing the BarrSix, which first flew on 4 November 2003. The company employs four people in its 557 m^2 (6,000 sq ft) facility.

UPDATED

BARR BARRSIX

TYPE: Six-seat kitbuilt.

PROGRAMME: Prototype (N33W) was due to fly late 1999, but delayed by hangar construction; new completion target was late 2000 but this slipped into 2003; wing and tail mating undertaken in September 2001; final assembly took place from September 2002. Engine test runs began in March 2003, followed by taxi tests in July 2003; first flight expected in September 2003, but did not take place until 4 November 2003. By August 2004, it had accumulated 90 hours of flying time.

CUSTOMERS: Launch orders being sought once prototype is flying.

COSTS: Kit US$89,900; estimated cost of completed aircraft US$134,900 (2004); quick-build option adds US$69,900 (2004).

DESIGN FEATURES: Designed to meet FAR Pt 23 and 51 per cent kitplane rules; extensive use of composites results in lightweight design. High wing with single bracing strut each side. Quoted build time 2,500 hours.

Dihedral 1° 44'; incidence 1° 30' at wing root; −1° 30' at tip.

FLYING CONTROLS: Conventional and manual. Frise-type ailerons travel +21°/−14°30' NACA single-slotted three-position (10, 20, 40°) flaps; horn-balanced rudder (travel ±27°13') and elevators (travel +21°/−17°).

STRUCTURE: Primary structure is 7781 pre-preg 36–38 per cent resin with Nomex core and graphite rod packs. Fuselage has central keel fitted with load-bearing rods.

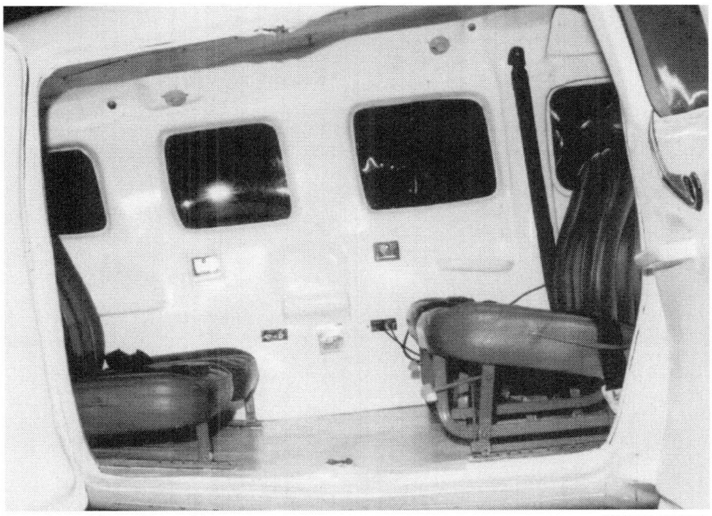

Interior of BarrSix cabin ***NEW***/0587523

LANDING GEAR: Tricycle type; fixed. Fuselage-mounted spring cantilever mainwheel legs; steerable nosewheel; speed fairings optional. Skis, floats and tailwheel layout optional.

POWER PLANT: One 298 kW (400 hp) Textron Lycoming IO-720-A1B flat-eight driving a Hartzell HC-E3YR-1RF/F8468A-6R constant-speed propeller. Fuel capacity 341 litres (90.0 US gallons; 74.9 Imp gallons), optionally increased to 526 litres (140 US gallons; 116 Imp gallons). Oil capacity 19.9 litres (5.25 US gallons; 4.4 Imp gallons).

ACCOMMODATION: Pilot and up to five passengers; club seating for four in rear cabin. Passenger/cargo double door on starboard side; optional door in port side for pilot. Dual controls.

AVIONICS: Onboard computer will provide aeronautical charts and e-mail services.

Prototype BarrSix at Lock Haven, former headquarters of Piper Aircraft ***NEW***/0587522

DIMENSIONS, EXTERNAL:
- Wing span: 10.92 m (35 ft 10 in)
- Wing chord: at root: 1.63 m (5 ft 4 in)
  - at tip: 1.07 m (3 ft 6 in)
- Wing aspect ratio: 7.4
- Length overall: 9.12 m (29 ft 11 in)
- Height overall: 2.18 m (7 ft 2 in)
- Tailplane span: 3.96 m (13 ft 0 in)
- Wheelbase: 1.76 m (5 ft 9¼ in)
- Propeller diameter: 2.11 m (6 ft 11 in)
- Propeller ground clearance: 0.25 m (9¾ in)
- Passenger door: Height: 1.02 m (3 ft 4 in)
  - Width: 0.94 m (3 ft 1 in)
- Cargo double doors: Height: 0.98 m (3 ft 2½ in)
  - Width: 1.12 m (3 ft 8 in)

DIMENSIONS, INTERNAL:
- Cabin: Length: 3.89 m (12 ft 9 in)
  - Max width: 1.35 m (4 ft 5 in)
  - Max height: 1.24 m (4 ft 1 in)

AREAS:
- Wings, gross: 16.17 m^2 (174.0 sq ft)
- Ailerons (total): 1.61 m^2 (17.32 sq ft)
- Trailing-edge flaps (total): 2.69 m^2 (29.00 sq ft)

WEIGHTS AND LOADINGS:
- Weight empty: one seat: 997 kg (2,197 lb)
  - fully IFR equipped: 1,134 kg (2,501 lb)

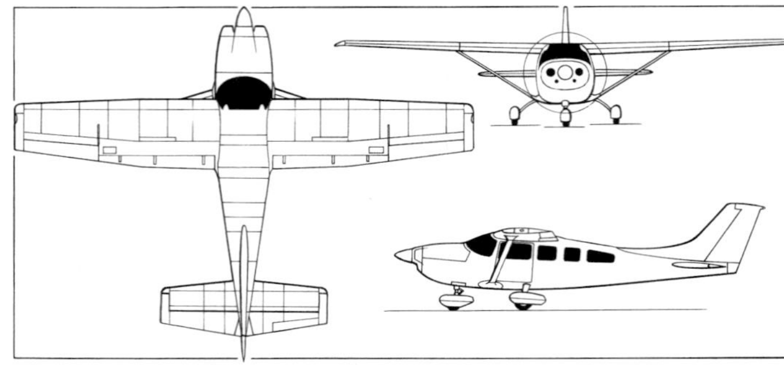

Barr BarrSix six-seat tourer with optional pilot's door *(James Goulding)* 0100430

- Baggage capacity: 91 kg (200 lb)
- Max T-O weight: 2,041 kg (4,500 lb)
- Max wing loading: 124.8 kg/m^2 (25.86 lb/sq ft)
- Max power loading: 6.85 kg/kW (11.25 lb/hp)

PERFORMANCE:
- Never-exceed speed (VNE): 238 kt (441 km/h; 274 mph)
- Max level speed: 206 kt (381 km/h; 237 mph)
- Max cruising speed: 196 kt (362 km/h; 225 mph)
- Normal cruising speed: 180 kt (333 km/h; 207 mph)
- Stalling speed, power off:
  - flaps up: 62 kt (115 km/h; 71 mph)
  - flaps down: 40 kt (73 km/h; 45 mph)
- Max rate of climb at S/L: 314 m (1,030 ft)/min
- Service ceiling: 5,940 m (19,500 ft)
- T-O run: 274 m (900 ft)
- T-O to 15 m (50 ft): 549 m (1,800 ft)
- Landing from 15 m (50 ft): 427 m (1,400 ft)
- Landing run: 229 m (750 ft)
- Range with max fuel: 1,251 n miles (2,317 km; 1,440 miles)
- Endurance: 6 h 24 min
- *g* limits: +3.86/−1.96

UPDATED

BEDE

BEDECORP LLC

6440 Norwalk Road, Unit F, Medina, Ohio 44253
Tel: (+1 330) 725 99 11
Fax: (+1 330) 239 27 39
e-mail: flybede@aol.com
Web: http://www.jimbede.com
PRESIDENT: James 'Jim' R Bede

Bedecorp specialises in design of high-performance aircraft for private ownership; civilian version of a previous product, BD-10, was transferred to Vortex Aircraft Company as the Phoenix. In June 2000, Bede announced it was working on the BD-17 Nugget, an all-metal low-wing single-seater and is actively marketing the BD-6 development of the BD-4; details of the BD-6 appeared in *Jane's* 1977–78 edition. Bedecorp is also offering plans for the BD-4.

In mid-2003, Bedecorp revealed that it is working on the BD-18 two-place derivative of the BD-17 Nugget.

UPDATED

Bede BD-17 Nugget single-seat sportplane *(Paul Jackson)* ***NEW***/1047165

BEDE BD-17 NUGGET AND BD-18 NUGGET

TYPE: Single-seat sportplane kitbuilt; side-by-side sportplane kitbuilt.

PROGRAMME: Announced June 2000; prototype (N624BD) completed by end of that year and first flew 11 February 2001. Certification programme under way. Public debut at Sun 'n' Fun, April 2001.

CURRENT VERSIONS: **BD-17:** Single-seat version, *as described.*

BD-18: Two-seat derivative, announed mid-2003. First flight was scheduled for June 2003 but slipped. Construction as BD-17.

COSTS: BD-17 US$19,990, BD-18 US$28,500, both less engine (2003).

CUSTOMERS: At least 17 BD-17 kits sold and four aircraft registered by August 2003; three of which were completed by that time.

DESIGN FEATURES: Intended to be extremely easy to build, with approximately 110 parts. Optional folding wings. Constant-chord wings. Fin fillet.

Wing section (BD-18) NACA 64416A (mod).

FLYING CONTROLS: Conventional and manual. No flaps on prototype BD-17; production kits have flaps. BD-18 maximum flap deflection 30°.

STRUCTURE: All-metal honeycomb sandwich. Honeycomb fuselage covered with 5 mm (0.20 in) aluminium sheet; control surfaces of urethane foam. Constant-chord wings have extruded tube spar.

LANDING GEAR: Fixed; choice of tricycle or tailwheel layout. Fuselage-mounted spring plywood mainwheel legs. Optional castoring nosewheel on forward-raked, spring steel tube cantilever leg. Speed fairings optional. Differential braking. Prototype BD-17 has 5.00-5 mainwheel tyres and 25 cm (10 in) nosewheel tyre.

POWER PLANT: *BD-17:* Power plants in the range 33.6 to 59.7 kW (45 to 80 hp) being considered; prototype has 44.7 kW (60 hp) HKS-700 two-cylinder four-stroke engine. Fuel capacity 76 litres (20.0 US gallons; 16.7 Imp gallons).

BD-18: Power plants in the range 59.7 to 89.5 kW (80 to 120 hp) recommended, though engines as large as 134 kW (180 hp) are being considered. Prototype has 59.7 kW (80 hp) Jabiru 2200. Fuel capacity 163 litres (43.0 US gallons; 35.8 Imp gallons).

ACCOMMODATION: Pilot only in BD-17, pilot and passenger side-by-side in BD-18, rearward-sliding canopy.

All data provisional.

DIMENSIONS, EXTERNAL:
- Wing span: BD-17: 6.55 m (21 ft 6 in)
  - BD-18: 7.77 m (25 ft 6 in)

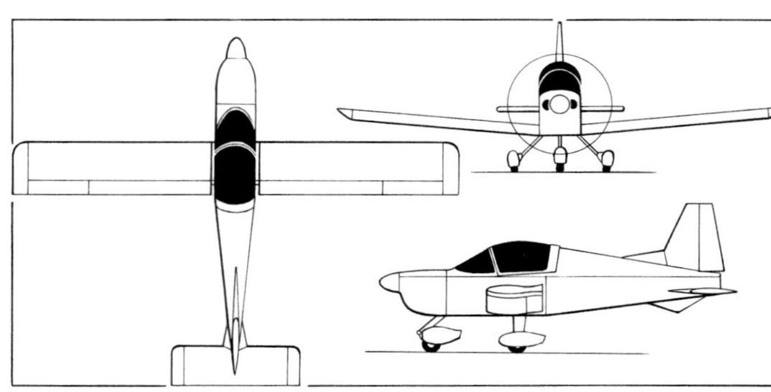

Bede BD-17 Nugget single-seat kitbuilt *(James Goulding)* 0121112

- Wing chord, constant: BD-17: 0.76 m (2 ft 6 in)
  - BD-18: 1.03 m (3ft 4½ in)
- Wing aspect ratio: BD-17: 8.6
  - BD-18: 7.7
- Length overall: BD-17: 5.47 m (17 ft 11½ in)
  - BD-18: 5.97 m (19 ft 7 in)
- Height overall: both: 2.36 m (7 ft 9 in)

DIMENSIONS, INTERNAL:
- Cockpit max width: BD-17: 0.89 m (2 ft 11 in)
  - BD-18: 1.19 m (3 ft 11 in)

AREAS:
- Wings, gross: BD-17: 4.97 m^2 (53.5 sq ft)
  - BD-18: 7.86 m^2 (84.6 sq ft)

WEIGHTS AND LOADINGS (BD-18 with 89.4 kW; 120 hp engine):
- Weight empty: BD-17: 254 kg (560 lb)
  - BD-18: 306 kg (675 lb)
- Max T-O weight: BD-17: 430 kg (950 lb)
  - BD-18: 567 kg (1,250 lb)

PERFORMANCE (estimated; BD-18 with 89.4 kW; 120 hp engine):
Never-exceed speed (VNE) ... 169 kt (313 km/h; 195 mph)
Cruising speed: BD-17 ... 129 kt (238 km/h; 148 mph)
BD-18 ... 137 kt (254 km/h; 158 mph)
Stalling speed, flaps up:
BD-17, BD-18 ... 55 kt (102 km/h; 63 mph)
flaps down: BD-17 ... 48 kt (89 km/h; 55 mph)
BD-18 ... 47 kt (87 km/h; 54 mph)
Max rate of climb at S/L: BD-17 ... 290 m (950 ft)/min
BD-18 ... 442 m (1,450 ft)/min
T-O run: BD-17 ... 213 m (700 ft)
BD-18 ... 168 m (550 ft)
Landing run: BD-17 ... 189 m (620 ft)
BD-18 ... 165 m (540 ft)
Range, max fuel and reserves:
BD-17 ... 825 n miles (1,528 km; 950 miles)
BD-18 ... 912 n miles (1,689 km; 1,050 miles)

UPDATED

BEECH

RAYTHEON AIRCRAFT COMPANY

At the September 2002 NBAA Convention, held in Orlando, Florida, Raytheon Aircraft (under which heading further details will be found) announced its intention to revert to separate marketing of its Beech and Hawker lines of private and executive aircraft, reducing the emphasis on its corporate name.

Beech Aircraft Corporation founded in 1932 by Walter and Olive Ann Beech; trading name **Beechcraft**; became wholly owned subsidiary of Raytheon on 8 February 1980 and incorporated within Raytheon Aircraft Company upon its foundation on 15 September 1994. In 2003, the former Beechjet 400 was transferred to the Hawker marque.

Prototype Beech T-6B N3000B displayed at Farnborough in July 2004 *(Paul Jackson)* ***NEW***/0589365

BEECH (3000) T-6 TEXAN II

Canadian Forces name: CT-156 Harvard II

TYPE: Basic turboprop trainer.

PROGRAMME: US version of Pilatus PC-9 Advanced Turbo Trainer (which see). For participation in USAF/USN Joint Primary Aircraft Training System (JPATS) competition, Beech and Pilatus reached agreement on joint approach in August 1990; Beech received two standard PC-9s from Pilatus, one of which (N26BA) converted as engineering development prototype and completed more than 260 hours' flight testing to reduce programme risk and develop engineering design baseline before return to Pilatus. It was followed by two Beech-built production prototypes (Beech designation PD373; later, Beech 3000), first flights December 1992 (N8284M/PT-2) and July 1993 (N209BA/PT-3); PT-2 used to complete flight test programme and evaluate systems performance; PT-3 incorporated several improvements and was principal aircraft for USAF/USN flight evaluation; one prototype and four production aircraft completed more than 1,400 hours of flight testing to achieve FAA certification on 30 July 1999. Promotional name for the purposes of competition was **Beech Mk II**.

Selection as JPATS winner announced 22 June 1995; total requirement for 782 aircraft (454 USAF, 328 US Navy) by 2017; all to be built in USA; contract valued at US$4.7 billion awarded February 1996; allocated designation T-6 Texan II (in advance of T-4 and T-5) to honour North American AT-6 Texan of Second World War era; Canadian name reflects British Commonwealth name for AT-6. By mid-2004, Beech data showed USAF with 413 firm commitments and further 196 anticipated; and USN 90 firm and 279 anticipated; increasing US contracts to 503 firm and 475 planned, or 978 in total, despite official figure remaining 782.

First metal cut February 1997; roll-out of manufacturing development aircraft (N23262/95-3000/PT-4) 29 June 1998; first flight 15 July 1998; first delivery (95-3003/PT-7) to Randolph AFB, June 1999, for technical evaluation; further two were to have followed in third quarter of 1999 for six month multiservice operational test and evaluation programme (by first two Lot-3 aircraft), but were delayed by engineering fault with PT6 engine. First handover was 95-3004 on 1 March 2000; first squadron of 12th FTW, the 559th FTS, began equipping on 23 May 2000.

Initial production rate 12 aircraft per year, rising to maximum of 59 by 2004; Milestone III authorisation to proceed with full-rate production granted by USAF on 3 December 2001. Early aircraft to 12th FTW at Randolph AFB for instructor training; IOC with USAF at Moody AFB, Georgia, in June 2001, with student pilot training beginning on 10 October; all USAF units to be fully equipped by 2011. First two aircraft for US Navy formally accepted at Wichita factory 31 August 2002; delivered to Pensacola, Florida, on 1 November 2002 and assigned to Training Wing Six; following initial training of instructors and maintenance personnel, formal student tuition began on 5 August 2003 with VT-4 and VT-10 squadrons. Delivery of 200th T-6A took place at beginning of September 2003, by which time USAF had received 108 and US Navy 21, balance comprising 45 for Greece and 26 for NATO. FlightSafety selected on 21 April 1997 to provide associated ground training system at cost over 24 years of US$500 million.

By 1 July 2004, 266 aircraft in service (155 USAF, 40 USN, 26 Canadian and 45 Greek) had flown 211,900 hours with 98.5 per cent mission reliability, demonstrating 2.54 mmh/fh and sortie turnround (Canada) of 5 to 8 minutes. Deliveries to US armed forces by December 2005 totalled 172 Air Force and 43 Navy.

USAF T-6A TEXAN II PROCUREMENT AT 1 JANUARY 2005

FY	Lot	Qty	First Aircraft	Order	Delivery
95	1	1[1]	95-3000	Jun 95	1999
95	2	3[2]	95-3001	Jul 96	1999–2000
96	3	6	95-3004	Sep 96	2000
97	4	15	96-3010	Apr 97	2000
98	5	22	98-3025	Feb 98	2000–01
99	6	22	99-3547	May 99	2001–02
00	7	29	00-3569	Jun 00	2002–03
01	8	35	01-3598	Apr 01	2003–04
02	9	40	02-3633	Jan 02	2005
03	10	35	03-3673	Dec 02	2006
04	11	47	04-3708?	Dec 03	2007
05	12	50	05-3755?	Dec 04	2008
06	13	57[2]	06-3804?		2009
Total		**362**			

Notes: Serial number prefixes of early aircraft do not necessarily indicate year of order, as is usual in USAF. Early aircraft numbered 95-3000 to 3009, 96-3010 to 3013 and 97-3014 onwards

[1] Engineering and manufacturing development aircraft
[2] 95-3001 subsequently transferred to US Navy as 165958, but remained with Raytheon Beech

US NAVY T-6A TEXAN II PROCUREMENT

FY	Lot	Qty	First Aircraft
—	—	(1)	165958[1]
00	1	11	165959
01	2	24	165970
02	3	7	165994
03	4	4	166001
Total		**46**[2]	

[1] Formerly USAF 95-3001; bailed to Raytheon Beech
[2] Total excludes 165958

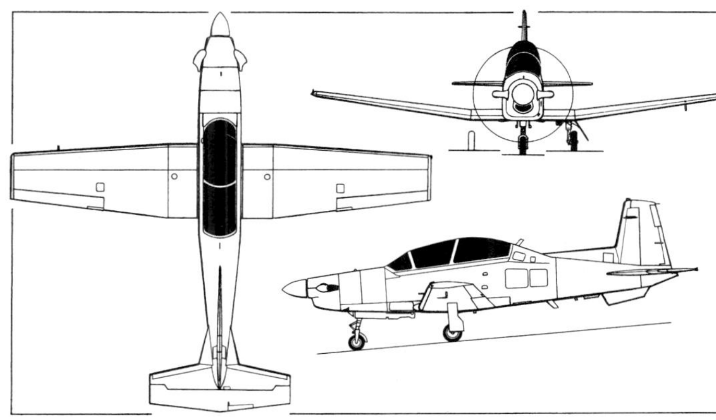

Beech T-6A Texan II primary trainer *(Paul Jackson)* ***NEW***/1047789

CURRENT VERSIONS: **T-6A Texan II**: USAF/USN version, *as described*.

T-6A-1/CT-156 Harvard II: Version for NATO Flying Training in Canada (NFTC) programme; generally similar to Texan II, but with blind-flying hood, dual VOR and ADF, and back-up VHF comm in place of UHF.

T-6B: Trainer and light attack (**AT-6B**) derivative of T-6A announced at Farnborough International in July 2002. Commonality with T-6A is 85 per cent. Incorporates revised cockpit management systems, with avionics suite provided by Flight Visions; latter includes FV-4000 modular mission display processor (MMDP), SparrowHawk HUD, stores management system and associated multifunction displays (MFDs) and six underwing pylons. Weaponry including guns, bombs and rockets is available, permitting use in light attack/counter-insurgency role. Prototype N3000B first flown 2004 and exhibited at Farnborough International, July 2004.

Beech/Pilatus PC-9 Mk II: Designation of Greek aircraft as initially announced by Raytheon, although generally known as T-6; first 20 are in USAF/US Navy configuration, remaining 25 in Greek-specified New Trainer Aircraft (NTA) configuration, with weapon sighting system plus provision for carriage of weaponry and auxiliary fuel tanks on three stores stations beneath each wing. Weapons trials, including tests with gun and rocket armament, conducted at Eglin AFB, Florida, in first half of 2002. NTA variant is not to FAR Pt 23.

CUSTOMERS: USAF initial operator is 12th Flying Training Wing at Randolph AFB, Texas; second is 479th Flying Training Group at Moody AFB, Georgia (49 aircraft during 2001–02); subsequent deliveries to 47th FTW, Laughlin AFB, Texas (96 aircraft during 2002–04, of which first four delivered 15 November 2002); 71st FTW, Vance AFB, Oklahoma (91 aircraft during 2005–06), 14th FTW, Columbus AFB, Mississippi (89 aircraft during 2006–08) and 80th FTW, Sheppard AFB, Texas (69 aircraft during 2008–09). Initial Navy aircraft included in USAF Lot 7; first deliveries to US Navy on 1 November 2002, with training beginning on 5 August 2003; operating units will include TW-4 at Corpus Christi, Texas, TW-5 at Whiting Field, Florida, and TW-6 at Pensacola, Florida. US Navy aircraft allocated serial numbers 165958 to 166285.

Chilean Air Force signed a letter of intent in late 1996 for future purchase of 16 to 25 Beech Mk IIs, but has still to place firm order.

Bombardier Services of Canada ordered 24 T-6A-1s in December 1997 for its NFTC programme for delivery to No. 2 FTS at Moose Jaw between April and December 2000; first delivery (156103) 29 February 2000; NFTC inaugurated 6 July 2000; all 24 aircraft handed over by end 2000. Further two aircraft, ordered in mid-2002, were delivered in 2003. On 9 October 1998, Greek Air Force announced selection of T-6 and ordered 45, delivered

Beech T-6A Texan II of 12th Flying Training Wing formating at Sun 'n' Fun 2004 with an earlier generation of Texan, the AT-6 or US Navy/Marines SNJ *(Paul Jackson)* ***NEW***/0583088

T-6A TEXAN II DELIVERIES

CY	USAF	USN	NATO/Canada[1]	Greece	Total
1999	1				1
2000	19		24	6	49
2001	37			10	47
2002	40	7[2]		13	60
2003	28	22	2	16	68
Total	**125**	**29**	**26**	**45**	**225**

[1] Known locally as CT-156 Harvard II
[2] Includes one ex-USAF, also counted earlier

between 2000 and 2003; option held on further five has still to be converted to firm order. First delivery (001) 17 July 2000 to 361 Squadron at Kalamata; training of student pilots began September 2001.

Raytheon holds marketing rights to PC-9 Mk II in all countries except Switzerland.

COSTS: Programme US$7 billion, of which US$362 million contracted by March 1998 for first 47 aircraft, comprising one manufacturing development aircraft and 46 production aircraft in five lots. Flyaway cost US$5.41 million over 740 aircraft (1998). Contracts for first 168 aircraft and associated ground training systems amount to US$852.6 million. Lot 10 batch of 35 aircraft cost US$169.9 million, including associated training devices and manuals. Contract for two additional NFTC aircraft plus support valued at approximately US$11.6 million.

DESIGN FEATURES: Certified to FAR Pt 23 in Aerobatic category, including 15s inverted flying and 5s intentional zero *g*. Prohibited manoeuvres are slow roll, stall turn (hammerhead), vertical roll, sustained vertical nose-down and knife-edge.

Approximately 90 per cent redesign of PC-9. Initial improvements in T-6 prototypes comprised new aft fuselage for improved flying qualities; new canopy shape for pressurisation and strengthening for birdstrike resistance; new engine cowling; single-point refuelling; zero/zero ejection seats; cockpit redesign to accommodate 95 per cent of body sizes; laser-initiated canopy fracture system; improved seat mountings; increased engine power, plus digital engine control to replicate jet response; continuous inertial particle separator for engine intake; HFC air conditioning; addition of large aft-fuselage avionics bay; and digital avionics with active matrix LCDs.

Following JPATS selection, further changes involved OBOGS; maintenance-free hydraulic accumulator; redesigned hydraulics system, wheels and brakes for simpler maintenance; improved corrosion protection; 18,720 hour fatigue life in JPATS mission profiles; 4 m (13 ft)/s landing sink rate provision; removable vertical stabiliser for improved maintenance; on-condition hot section inspections; and 4,500 hour engine TBO. NATO T-6As in Canada additionally provided with cold weather start capability.

FLYING CONTROLS: Conventional and manual. Selectable trim aid device (TAD) by Aeromach Labs automatically trims rudder in conjunction with throttle position to minimise effect of torque. Flight-adjustable rudder and elevator tabs; aileron trim by biased centring spring. Stall strip on starboard wing. Control deflections include rudder ±24°; elevators 18° up/16° down; ailerons 20° up/11° down; and flaps 23° for take-off, 50° for landing. Airbrake maximum deflection of 67° 30'.

STRUCTURE: Generally all-metal; durability and damage-tolerant (DADT) structure includes wing and tailplane leading-edges reinforced for bird resistance; 18,000 hour service life. Avionics bay behind rear cockpit; forward-hinged door each side. Airframe built entirely in USA.

LANDING GEAR: Goodrich wheels and brakes. Capable of withstanding sink rates up to 4 m (13 ft)/s. Mainwheels 20×4.4 (14 ply (tubeless); nosewheel 16×4.4 (8 ply) tubeless. Otherwise generally as for PC-9.

POWER PLANT: One 1,274 kW (1,708 shp) Pratt & Whitney Canada PT6A-68 turboprop flat rated at 820 kW (1,100 shp) max continuous (104 per cent), driving a Hartzell HC-E4A-2/E9612 four-blade, fully-feathering propeller at a constant 2,000 rpm. Raytheon/P&WC power management unit provides jet-type linear throttle response. T-6A has integral 348 litre (92.0 US gallon; 76.6 Imp gallon) tank in each wing leading-edge; tanks freely interconnected; total fuel capacity 697 litres (184 US gallons; 153 Imp gallons), of which 681 litres (180 US gallons; 150 Imp gallons) are usable. AT-6B fuel capacity 852 litres (225 US gallons; 187 Imp gallons). AT-6B has provision for two underwing auxiliary fuel tanks, each 265 litres (70.0 US gallons; 58.3 Imp gallons). Pressure refuelling/defuelling point in lower fuselage, adjacent to port wingroot; gravity refuelling points at three quarters' span in each wing upper surface. Oil capacity 17 litres (4.5 US gallons; 3.75 Imp gallons).

ACCOMMODATION: Two pilots in stepped cockpits on Martin-Baker Mk 16LA zero/zero ejection seats. Limited HOTAS controls in T-6A; full multifunction HOTAS in AT-6B. Three-piece, Pilkington Aerospace acrylic starboard-hinged canopy; integral windscreen 19 mm (¾ in) thick for resistance to strike by 1.8 kg (4 lb) bird at 270 kt (500 km/h; 311 mph); other panels 9 mm (⅓ in) thick. ATK canopy fracturing initiation unit and Teledyne McCormick fracturing system. Accommodation pressurised and air conditioned; warm air canopy defrosting/demisting. Baggage compartment in rear fuselage, behind avionics bay; upward-hinged door port side.

SYSTEMS: Generally as PC-9. Dowty Aerospace hydraulics; Enviro Systems air conditioning unit; Litton OBOGS. Cockpit pressurisation system has 0.24 bar (3.5 lb/sq in) max differential. Battery 28 V. Goodrich emergency power system.

AVIONICS: Honeywell primary contractor.

Comms: VHF/UHF transceiver, plus back-up VHF with separate antenna; intercom and interphone. MST 67A Mode S transponder. Back-up UHF in Canadian aircraft. T-6A's radio management unit replaced in AT-6B by up-front control panel below HUD.

Flight: Dual VNS-41 navigation systems (including VN-411B receivers) for VOR, localiser, glideslope and marker beacon functions through common antenna. DFS-43A ADF (DF-431 receiver and AT-434 loop/sense antenna); KLN 90 GPS (combined INS-GPS in AT-6B); and Rockwell Collins 252-channel DM-441B DME. Litef fibre optic gyro for AHRS. Goodrich collision warning system. Flight data recorder. Extended ADF navigation in Canadian aircraft. Integrated radar altimeter in AT-6B.

Instrumentation: Two independent EFIS 50 127 mm (5 in) square active matrix LCDs, plus control panel, in each cockpit (or one, plus 127 × 102 mm; 5 × 4 in MFD in AT-6B); attitude director indicator portion of EFIS provides primary attitude display, turn rate, mode selection annunciation, localiser/glideslope deviation as appropriate; horizontal situation indicator for primary heading display, primary navigation display, course select indication, navigation source annunciation, DME, localiser and glideslope deviation, remote map presentation and selected heading reference marker. Additional three 76 mm (3 in) square MFDs per cockpit provide engine and auxiliary instrument information, including fuel state and pressurisation (engine MFD being replaced in AT-6B by 152 × 203 mm; 6 × 8 in MFD). Standby instrumentation comprises airspeed, attitude, turn-and-slip and magnetic compass (electromechanical in T-6A, digital in AT-6B). Korry Electronics warning panels. Flight Visions SparrowHawk HUD and second 152 × 203 mm (6 × 8 in) active matrix, liquid crystal MFD (with moving map provision) in T-6B version. AT-6B is compatible with Gen 4 NVGs. Provision for three-screen EFIS version, with HUD and integrated display panel for advanced training.

Mission: Dual-mission computers, FN Herstal weapons control system and Avimo gunsight in AT-6B. Provision for laser range-finder.

Beech T-6B instrument panel, showing Flight Visions avionics ***NEW***/1047798

ARMAMENT: AT-6B version has three hardpoints under each wing. Typical loads include (a) two HMP 0.5 in machine gun pods; (b) two external fuel tanks; (c) six BDU-33 50 lb practice bombs; (d) two BDU-33s, two HMPs and two LAU-68 seven-round rocket pods; and (e) two Mk 82 500 lb bombs.

DIMENSIONS, EXTERNAL: As for PC-9M

AREAS:

Wings, gross	16.29 m² (175.3 sq ft)

WEIGHTS AND LOADINGS:

Weight empty	2,136 kg (4,709 lb)
Baggage capacity	36 kg (80 lb)
Max fuel weight	499 kg (1,100 lb)
Max underwing stores (AT-6B)	1,415 kg (3,120 lb)
Max T-O and landing weight	2,948 kg (6,500 lb)
Max ramp weight	2,971 kg (6,550 lb)
Max zero-fuel weight	2,495 kg (5,500 lb)
Max wing loading	181.0 kg/m² (37.08 lb/sq ft)
Max power loading	3.60 kg/kW (5.91 lb/shp)

PERFORMANCE:

Never-exceed speed (VNE)	350 kt (648 km/h; 402 mph)
Max level speed: at altitude	310 kt (575 km/h; 357 mph)
at low level	270 kt (500 km/h; 311 mph) IAS
Max operating Mach No.	0.67
Manoeuvring speed	236 kt (437 km/h; 272 mph)
Max cruising speed at 2,285 m (7,500 ft)	230 kt (426 km/h; 265 mph)
Approach speed	100 kt (185 km/h; 115 mph) IAS
Stalling speed, power off:	
flaps up	82 kt (152 km/h; 95 mph)
flaps down	74 kt (137 km/h; 86 mph)
Max rate of climb at S/L	1,372 m (4,500 ft)/min
Max certified altitude	9,450 m (31,000 ft)
Max cruising altitude, T-6	10,670 m (35,000 ft)
T-O run	437 m (1,435 ft)
T-O to 15 m (50 ft)	654 m (2,145 ft)
Landing from 15 m (50 ft)	1,030 m (3,380 ft)
Landing run	739 m (2,425 ft)
Training runway requirement	1,219 m (4,000 ft)
Range at altitude	850 n miles (1,574 km; 978 miles)
Endurance at max cruising speed	3 h
g limits	+7/−3.5

UPDATED

BEECH BONANZA A36

TYPE: Six-seat utility transport.

PROGRAMME: Descended from Model 35 of 1945 (10,403 built in V tail configuration), but now has little more than the name in common; further 1,911 late Model 33s built as Bonanzas, earlier version being Debonair. Model 36 certified 1 May 1968; developed from Bonanza E33A; A36 certified 24 October 1969 and introduced 1970; current subvariant announced 3 October 1983, succeeding model powered by 212.5 kW (285 hp) Continental IO-520-BB; certified in FAA Utility category.

CURRENT VERSIONS: **Bonanza A36:** Standard version; as described. Features introduced for 2001 model year include dual Garmin GNS 430 nav/comm/GPS.

Description applies to A36, except where otherwise stated.

Bonanza B36TC: Described separately.

CUSTOMERS: Total of some 3,568 Bonanza 36/A36s delivered by June 2004, including 89 in 1995, 83 in 1996, 85 in 1997, 71 in 1998, 77 in 1999, 85 in 2000, 63 in 2001, 51 in 2002, 55 in 2003 and 22 in the first six months of 2004.

COST: US$626,600 typically equipped (2004).

DESIGN FEATURES: Conventional low-wing cabin monoplane. Tapered wings with leading-edge gloves; tapered tailplane and sweptback fin.

Beech A36 Bonanza *(Paul Jackson)* ***NEW***/0576882

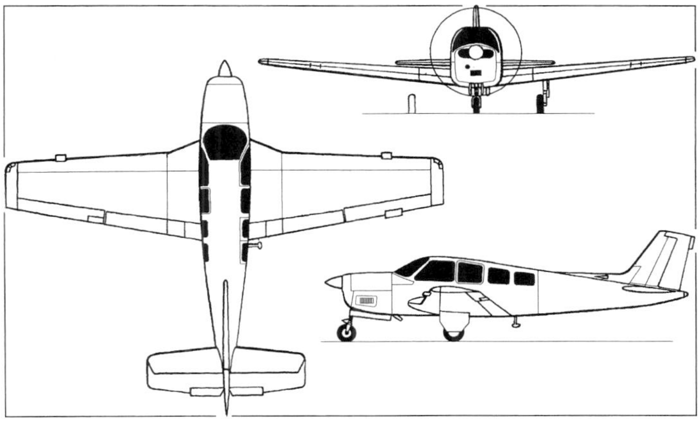

Beech Bonanza A36 four/six-seat utility aircraft *(James Goulding)* ***NEW***/0568990

Beech modified NACA 23016.5 wing section at root, modified 23012 at tip; dihedral 6°; incidence 4° at root and 1° at tip.

FLYING CONTROLS: Conventional and manual. Electrically actuated trim tab in each elevator; ground-adjustable tabs in ailerons and rudder; single-slotted, three-position flaps.

STRUCTURE: Light alloy, with two-spar wing torsion box and stressed-skin tail surfaces. Conventional construction.

LANDING GEAR: Electrically retractable tricycle type, with steerable nosewheel. Mainwheels retract inward into wings, nosewheel rearward. Beech oleo-pneumatic shock-absorbers in all units. Cleveland mainwheels, size 6.00–6, and tyres, size 7.00–6 (6 ply), pressure 2.28 to 2.76 bar (33 to 40 lb/sq in). Cleveland nosewheel and tyre, size 5.00–5 (4 ply), pressure 2.76 bar (40 lb/sq in). Cleveland ring disc hydraulic brakes. Parking brake. Magic Hand landing gear system optional.

POWER PLANT: One 224 kW (300 hp) Teledyne Continental IO-550-B Raytheon Special Edition flat-six engine, driving a Hartzell three-blade constant-speed metal propeller. The engine is equipped with an altitude-compensating fuel pump which automatically makes the fuel/air mixture leaner and richer during climb and descent respectively. Usable fuel capacity 280 litres (74.0 US gallons; 61.6 Imp gallons).

ACCOMMODATION: Enclosed cabin seating four to six persons on individual seats. Pilot's seat is vertically adjustable. Dual controls standard. Two rear removable seats and two folding seats permit rapid conversion to utility configuration. Optional club seating with rear-facing third and fourth seats, executive writing desk, refreshment cabinet, headrests for third and fourth seats, reading lights and fresh air outlets for fifth and sixth seats. Double doors of bonded aluminium honeycomb construction on starboard side facilitate loading of cargo. As an air ambulance, one stretcher can be accommodated with ample room for a medical attendant and/or other passengers. Extra windows provide improved view for passengers.

SYSTEMS: Optional 12,000 BTU refrigeration-type air conditioning system comprises evaporator located beneath pilot's seat, condenser on lower fuselage, and engine-mounted compressor. Air outlets on centre console, with two-speed blower. Electrical system supplied by 28 V 100 A alternator, 24 V 15.5 Ah battery; optional standby generator. Hydraulic system for brakes only. Standby vacuum pump standard. Pneumatic system for instrument gyros optional. Oxygen system and electric propeller de-icing optional.

AVIONICS: *Comms:* Dual Garmin GNS 430 nav/comms with GI 106 indicators or one each GNS 430 and GNS 530; Bendix/King KT 76C transponder; PS Engineering PMA7000M-S audio control system.

Flight: Bendix/King encoding altimeter; KFC 225 AFCS; KI 256 flight command indicator/gyro horizon; KCS 55A compass; Goodrich WX-500 Stormscope; Shadin ADC 200+ fuel and air data computer.

EQUIPMENT: Standard equipment includes LCD digital chronometer, EGT and OAT gauges, rate of climb indicator, turn co-ordinator, 3 in horizon and directional gyros, four fore- and aft-adjustable and reclining seats, armrests, headrests, single diagonal strap shoulder harness with inertia reel for all occupants, pilot's storm window, ultraviolet-proof windscreen and windows, sun visors, large cargo door, emergency locator transmitter, stall warning device, alternate static source, heated pitot, rotating beacon, three-light strobe system, carpeted floor, super soundproofing, control wheel map lights, entrance door courtesy light, internally lit instruments, coat hooks, glove compartment, in-flight storage pockets, approach plate holder, utility shelf, cabin dome light, reading lights, instrument post lights, control wheel map light, electroluminescent subpanel lighting, landing light, taxying light, full-flow oil filter, three-colour polyurethane exterior paint, external power socket, static wicks and towbar.

Optional equipment includes propeller de-icing, dual controls, leather seats, co-pilot's wheel brakes, air conditioning, fifth passenger seat, fresh air vent blower and ground com switch.

DIMENSIONS, EXTERNAL:

Wing span	10.21 m (33 ft 6 in)
Wing chord: at root	2.13 m (7 ft 0 in)
at tip	1.07 m (3 ft 6 in)
Wing aspect ratio	6.2
Length overall	8.38 m (27 ft 6 in)
Height overall	2.62 m (8 ft 7 in)
Tailplane span	3.71 m (12 ft 2 in)

Wheel track	2.92 m (9 ft 7 in)
Wheelbase	2.39 m (7 ft 10¼ in)
Propeller diameter: A36	2.03 m (6 ft 8 in)
Forward cockpit door: Height	0.91 m (3 ft 0 in)
Width	0.94 m (3 ft 1 in)
Baggage compartment door: Height	0.61 m (2 ft 0 in)
Width	0.99 m (3 ft 3 in)
Rear passenger/cargo door: Height	0.89 m (2 ft 11 in)
Width	1.14 m (3 ft 9 in)

DIMENSIONS, INTERNAL:

Cabin (aft of firewall, incl extended baggage compartment): Length	3.84 m (12 ft 7 in)
Max width	1.07 m (3 ft 6 in)
Max height	1.27 m (4 ft 2 in)
Volume	3.88 m³ (137 cu ft)

AREAS:

Wings, gross	16.80 m² (181.0 sq ft)
Ailerons (total)	1.06 m² (11.40 sq ft)
Trailing-edge flaps (total)	1.98 m² (21.30 sq ft)
Fin	0.93 m² (10.00 sq ft)
Rudder, incl tab	0.52 m² (5.60 sq ft)
Tailplane	1.75 m² (18.82 sq ft)
Elevators, incl tabs	1.67 m² (18.00 sq ft)

WEIGHTS AND LOADINGS:

Weight empty, standard:	1,148 kg (2,530 lb)
Basic operating weight (single pilot)	1,225 kg (2,700 lb)
Baggage capacity	181 kg (400 lb)
Max T-O weight	1,655 kg (3,650 lb)
Max ramp weight	1,661 kg (3,663 lb)
Max zero-fuel weight	1,592 kg (3,509 lb)
Payload with max fuel	313 kg (689 lb)
Max wing loading	98.5 kg/m² (20.17 lb/sq ft)
Max power loading	7.40 kg/kW (12.17 lb/hp)

PERFORMANCE:

Max level speed (min weight)	184 kt (340 km/h; 212 mph)
Max cruising speed (mid-cruise weight):	
at FL60	176 kt (326 km/h; 202 mph)
at FL80	174 kt (322 km/h; 200 mph)
at FL100	171 kt (317 km/h; 197 mph)
Normal cruising speed:	
at FL60	165 kt (306 km/h; 190 mph)
at FL80	167 kt (309 km/h; 192 mph)
at FL100	163 kt (302 km/h; 188 mph)
Long-range cruising speed:	
at FL60	134 kt (248 km/h; 154 mph)
at FL80	140 kt (259 km/h; 161 mph)
at FL100	143 kt (265 km/h; 165 mph)
Approach speed	80 kt (148 km/h; 92 mph)
Stalling speed, power off:	
flaps up	68 kt (126 km/h; 78 mph) IAS
30° flap	59 kt (109 km/h; 68 mph) IAS
Max rate of climb at S/L	375 m (1,230 ft)/min
Service ceiling	5,640 m (18,500 ft)
T-O field length	583 m (1,913 ft)
Landing from 15 m (50 ft)	442 m (1,450 ft)
Landing run	290 m (950 ft)
Range, pilot and two passengers, VFR reserves:	
max cruising speed : at FL60	671 n miles (1,242 km; 772 miles)
at FL80	713 n miles (1,320 km; 820 miles)
at FL100	751 n miles (1,390 km; 864 miles)
normal cruising speed: at FL60	736 n miles (1,363 km; 847 miles)
at FL80	746 n miles (1,381 km; 858 miles)
at FL100	775 n miles (1,435 km; 891 miles)
long-range cruising speed: at FL60	919 n miles (1,702 km; 1,057 miles)
at FL80	923 n miles (1,709 km; 1,062 miles)
at FL100	916 n miles (1,696 km; 1,054 miles)

UPDATED

BEECH TURBO BONANZA B36TC

TYPE: Six-seat utility transport.

PROGRAMME: Derived from A36 (see separate entry for Beech Bonanza A36). Certified as A36TC 7 December 1978; 271 A36TCs delivered; improved B36TC introduced 1982 following type certification on 15 January 1982.

CURRENT VERSIONS: **Bonanza B36TC:** final version. Discontinued.

CUSTOMERS: Total of 695 A/B36TCs registered by November 2002, including 14 per year in 1993–97, 22 in 1998, 20 in 1999, 18 in 2000, 26 in 2001 and five in 2002; none further in 2003 or 2004.

UPDATED

BEECH BARON 58

TYPE: Six-seat utility twin.

PROGRAMME: Prototype Beech 55 Baron flew 29 February 1960; total 3,728 built, not including 94 Model 56 Turbo Barons. Model 58 derived from E55 with 25 cm (10 in) fuselage stretch, plus internal rearrangement to lengthen cabin by 76 cm (30 in); certified in FAA Normal category 19 November 1969; marketing began in following month. Raytheon delivered 2,000th standard Model 58 (N558TH) in July 2001. Beech 58P Pressurised Baron and 58TC turbocharged version no longer produced.

CURRENT VERSIONS: **Baron 58:** Standard version, *as described*. Features introduced for 2001 model year include dual Garmin GNS 530 nav/comm/GPS.

Baron 58 Jaguar Special Edition: Discontinued.

CUSTOMERS: Operators include several commercial aviation flying schools. Total 2,818 Baron 58s delivered by June 2004, including 29 in 1995, 44 in 1996, 35 in 1997, 42 in 1998, 47 in 1999, 50 in 2000, 47 in 2001, 27 each in 2002 and 2003, and 11 in the first six months of 2004. (Incorporates 497 of 58P and 151 of 58TC versions.)

COST: US$1,099,600 typically equipped (2004).

DESIGN FEATURES: Conventional low-wing monoplane, ultimately derived from Beech Bonanza. Tapered wing with leading-edge gloves inboard of engines; sweptback fin.

Beech 58 Baron twin-prop *(Paul Jackson)* ***NEW***/0576888

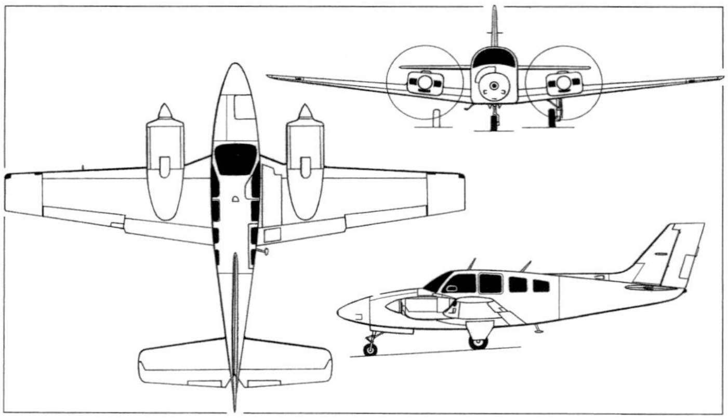

Beech Baron 58 (two Teledyne Continental IO-550-C piston engines) *(Jane's/Dennis Punnett)* ***NEW***/0568989

Wing section NACA 23015.5 at root, 23010.5 at tip; dihedral 6°; incidence 4° at root and 0° at tip.

FLYING CONTROLS: Conventional and manual. Manually operated trim tabs in elevators, rudder and port aileron; electrically operated single-slotted flaps. Control surface movements: ailerons ±20°; elevator ±30/–15°; rudder ±25°; flap settings 15° for approach, 28° maximum.

STRUCTURE: Light alloy with two-spar wing box; elevators have smooth magnesium alloy skins.

LANDING GEAR: Electrically retractable tricycle type. Main units retract inward into wings, nosewheel aft. Beech oleo-pneumatic shock-absorbers in all units. Steerable nosewheel with shimmy damper. Cleveland wheels, with mainwheel tyres size 6.50–8 (8 ply) tubed, pressure 3.59 to 3.96 bar (52 to 56 lb/sq in) or 19.5×6.75–8 (10 ply) tubeless. Nosewheel tyre size 5.00–5 (6 ply) tubed, pressure 3.79 to 4.14 bar (55 to 60 lb/sq in) or 15×6.0-6 (4 ply) tubed. Cleveland ring disc hydraulic brakes. Heavy-duty brakes optional. Parking brake.

POWER PLANT: Two 224 kW (300 hp) Teledyne Continental IO-550-C Raytheon Special Edition flat-six engines, each driving a Hartzell three-blade constant-speed fully feathering metal propeller. The standard fuel system has a usable capacity of 628 litres (166 US gallons; 138 Imp gallons). Optional 'wet wingtip' installation also available, increasing usable capacity to 734 litres (194 US gallons; 161.5 Imp gallons).

ACCOMMODATION: Standard version has four individual seats in pairs in enclosed, soundproofed, heated and ventilated cabin, with door on starboard side for pilot(s). Single diagonal strap shoulder harness with inertia reel standard on all seats. Pilot's vertically adjusting seat is standard. Co-pilot's vertically adjusting seat, folding fifth and sixth seats, or club seating comprising folding fifth and sixth seats and aft-facing third and fourth seats, are optional. Adjustable rudder pedals (retractable on starboard side). Executive desk available as option with club seating. Baggage compartment in nose. Double passenger/cargo doors on starboard side of cabin (split vertically: forward-hinged at front, aft portion rear-hinged) provide access to space for baggage or cargo behind the third and fourth seats. Pilot's storm window. Openable windows adjacent the third and fourth seats are used for ground ventilation and as emergency exits. Windscreen defrosting standard.

SYSTEMS: Cabin heated by Janitrol 50,000 BTU heater, which serves also for windscreen defrosting. Oxygen system of 1,389 litres (49 cu ft) or 1,814 litres (64 cu ft) capacity optional. Electrical system includes two 28 V 100 A engine-driven alternators with alternator failure lights and two 12 V 25 Ah batteries. Two 100 A alternators optional. Hydraulic system for brakes only. Pneumatic pressure system for air-driven instruments, and optional wing and tail unit de-icing system. Oxygen system, cabin air conditioning and windscreen electric anti-icing optional.

AVIONICS: As for Beech A36 Bonanza, plus radar.

Radar: Honeywell RDR 2000 VP colour weather radar.

EQUIPMENT: Standard equipment includes ultraviolet-proof windscreen and cabin windows, super soundproofing, heated pitot head, instrument panel floodlights, navigation and position lights, steerable taxying light, dual landing lights, heated fuel vents, heated fuel and stall warning vanes and external power socket.

Options include alternate static source, internally illuminated instruments, strobe lights, electric windscreen anti-icing, wing ice detection light, static wicks, leather seat trim, cabin club seating, cockpit relief tube, cabin fire extinguisher and ventilation blower.

DIMENSIONS, EXTERNAL:

Wing span	11.53 m (37 ft 10 in)
Wing chord: at root	2.13 m (7 ft 0 in)
at tip	0.90 m (2 ft 11½ in)
Wing aspect ratio	7.2
Length overall	9.09 m (29 ft 10 in)
Height overall	2.97 m (9 ft 9 in)
Tailplane span	4.85 m (15 ft 11 in)
Wheel track	2.92 m (9 ft 7 in)
Wheelbase	2.72 m (8 ft 11 in)
Propeller diameter	1.96 m (6 ft 5 in)
Rear passenger/cargo doors:	
Max height	0.89 m (2 ft 11 in)
Width	1.14 m (3 ft 9 in)

Baggage door (fwd): Height ... 0.56 m (1 ft 10 in)
Width ... 0.64 m (2 ft 1 in)

DIMENSIONS, INTERNAL:
Cabin (incl rear baggage area):
Length ... 3.84 m (12 ft 7 in)
Max width ... 1.07 m (3 ft 6 in)
Max height ... 1.27 m (4 ft 2 in)
Floor area ... 3.72 m^2 (40.0 sq ft)
Volume ... 3.9 m^3 (137 cu ft)
Baggage compartment: nose ... 0.51 m^3 (18.0 cu ft)

AREAS:
Wings, gross ... 18.51 m^2 (199.2 sq ft)
Ailerons (total) ... 1.06 m^2 (11.40 sq ft)
Trailing-edge flaps (total) ... 1.98 m^2 (21.30 sq ft)
Fin ... 1.46 m^2 (15.67 sq ft)
Rudder, incl tab ... 0.81 m^2 (8.75 sq ft)
Tailplane ... 4.95 m^2 (53.30 sq ft)
Elevators, incl tabs ... 1.84 m^2 (19.80 sq ft)

WEIGHTS AND LOADINGS:
Weight empty ... 1,764 kg (3,890 lb)
Basic operating weight, typical ... 1,896 kg (4,180 lb)
Baggage capacity: cabin ... 181 kg (400 lb)
nose ... 136 kg (300 lb)
Max T-O weight ... 2,495 kg (5,500 lb)
Max ramp weight ... 2,506 kg (5,524 lb)
Max zero-fuel weight ... 2,365 kg (5,215 lb)
Payload with max fuel ... 213 kg (470 lb)
Fuel with max payload ... 140 kg (309 lb)
Max landing weight ... 2,449 kg (5,400 lb)
Max wing loading ... 134.8 kg/m^2 (27.61 lb/sq ft)
Max power loading ... 5.58 kg/kW (9.17 lb/hp)

PERFORMANCE (cruising speeds at average cruise weight):
Max level speed at S/L ... 208 kt (386 km/h; 239 mph)
High cruising speed, 2,200 rpm:
at FL60 ... 202 kt (374 km/h; 223 mph)
at FL80 ... 200 kt (370 km/h; 230 mph)
at FL100 ... 198 kt (367 km/h; 228 mph)
Normal cruising speed, 2,300 rpm:
at FL60 ... 190 kt (352 km/h; 219 mph)
at FL80 ... 192 kt (356 km/h; 221 mph)
at FL100 ... 189 kt (350 km/h; 217 mph)
Long-range cruising speed, 2,500 rpm:
at FL60 ... 159 kt (294 km/h; 183 mph)
at FL80 ... 164 kt (304 km/h; 189 mph)
at FL100 ... 167 kt (309 km/h; 192 mph)
Approach speed ... 95 kt (176 km/h; 109 mph)
Stalling speed, power off
flaps up ... 84 kt (156 km/h; 97 mph) IAS
flaps down ... 75 kt (139 km/h; 86 mph) IAS
Max rate of climb at S/L ... 518 m (1,700 ft)/min
Rate of climb at S/L, OEI ... 119 m (390 ft)/min
Time to FL100 ... 10 min
Service ceiling ... 6,305 m (20,680 ft)
Service ceiling, OEI ... 2,220 m (7,280 ft)
T-O field length ... 701 m (2,300 ft)
Landing from 15 m (50 ft) ... 747 m (2,450 ft)
Landing run ... 396 m (1,300 ft)
Range, pilot and four passengers, VFR reserves
... 802 n miles (1,485 km; 922 miles)
Range, pilot with two passengers, VFR reserves:
high cruising speed: at FL60 ... 885 n miles (1,639 km; 1,018 miles)
at FL80 ... 936 n miles (1,733 km; 1,077 miles)
at FL100 ... 981 n miles (1,816 km; 1,128 miles)
normal cruising speed: at FL60 ... 962 n miles (1,781 km; 1,107 miles)
at FL80 ... 979 n miles (1,813 km; 1,126 miles)
at FL100 ... 1,027 n miles (1,902 km; 1,181 miles)
long-range cruising speed: at FL60 ... 1,294 n miles (2,396 km; 1,489 miles)
at FL80 ... 1,276 n miles (2,363 km; 1,468 miles)
at FL100 ... 1,287 n miles (2,383 km; 1,481 miles)
Ferry range ... 1,379 n miles (2,553 km; 1,587 miles)

UPDATED

BEECH KING AIR C90B

TYPE: Business twin-turboprop.

PROGRAMME: King Air 90 first flew 21 November 1963. C90B announced at NBAA Convention, October 1991; superseded King Air 90, A90, B90, C90, C90-1, C90A; introduced four-blade McCauley propellers, special interior soundproofing, updated and redesigned interior, updated cockpit features and interior noise and vibration levels substantially reduced. A model delivered on 24 June 1996 to Jeld-Wen of Klamath Falls, Oregon, was the 5,000th King Air of all versions to be produced.

By mid-2004 some 6,000 King Airs of all models had been produced, and were operating in 105 countries; total fleet time stood at more than 40 million hours.

CURRENT VERSIONS: **C90B:** Standard version from 1991, but engineering designation remains 90A.

Detailed description applies to C90B.

King Air 200 and 350: Described separately.

CUSTOMERS: Total 1,706 commercial and 226 military King Air 90/A90/B90/C90/C90-1/C90A/C90B/C90SE delivered by June 2004, including 40 C90Bs and (now discontinued) C90SEs in 1995, 38 C90Bs in 1996, 41 C90Bs in 1997, 37 C90Bs in 1998, 41 C90Bs in 1999, 46 C90Bs in 2000, 41 C90Bs in 2001, 21 C90Bs in 2002, 18 C90Bs in 2003 and seven C90Bs in the first six months of 2004.

Beech King Air C90 in current house colours *(Paul Jackson)* ***NEW***/0576909

Beech King Air C90 business twin-turboprop *(Paul Jackson)* ***NEW***/0568417

COSTS: C90B US\$2.750 million, typically equipped (2004).

DESIGN FEATURES: Conventional low-wing, twin-engined monoplane. Wing section NACA 23014.1 (modified) at root, 23016.22 (modified) at outer end of centre-section, 23012 at tip; dihedral 7°; incidence 4° 48' at root, 0° at tip; tailplane 7° dihedral.

FLYING CONTROLS: Conventional and manual. Trim tabs on port aileron, in both elevators and rudders; single-slotted aluminium flaps.

STRUCTURE: Generally light alloy; magnesium ailerons. Internal corrosion-proofing.

LANDING GEAR: Hydraulically retractable tricycle type. Nosewheel retracts rearward, mainwheels forward into engine nacelles. Mainwheels protrude slightly beneath nacelles when retracted, for safety in a wheels-up emergency landing. Fully castoring steerable nosewheel with shimmy damper. Beech oleo-pneumatic shock-absorbers. Goodrich mainwheels with tyres size 8.50-10 (8 ply) tubeless, pressure 3.79 bar (55 lb/sq in). Goodrich nosewheel with tyre size 6.50–10 (6 ply) tubeless, pressure 3.59 bar (52 lb/sq in). Goodrich heat-sink and air-cooled multidisc hydraulic brakes. Parking brake. Minimum ground turning radius 10.82 m (35 ft 6 in).

POWER PLANT: Two 410 kW (550 shp) Pratt & Whitney Canada PT6A-21 turboprops, each driving a Hartzell four-blade constant-speed fully feathering propeller. Propeller auto ignition system, environmental fuel drain collection system, magnetic chip detector, automatic propeller feathering and propeller synchrophaser standard. Fuel in two tanks in engine nacelles, each with usable capacity of 231 litres (61.0 US gallons; 50.8 Imp gallons), and auxiliary bladder tanks in outer wings, each with capacity of 496 litres (131 US gallons; 109 Imp gallons). Total usable fuel capacity 1,454 litres (384 US gallons; 320 Imp gallons). Refuelling points in top of each engine nacelle and in wing leading-edge outboard of each nacelle. Oil capacity 13.2 litres (3.5 US gallons; 2.9 Imp gallons) per engine.

ACCOMMODATION: Two four-way adjustable seats side by side in cockpit with dual controls standard. Normally, four reclining seats in main cabin, in two pairs facing each other fore and aft. Standard furnishings include cabin forward partition, with fore and aft partition and coat rack, hinged nose baggage compartment door, seat belts and inertia reel shoulder harnesses for all seats. Standard accommodation for six passengers, four in club arrangement, one on sideways-facing seat adjacent door, and one on lavatory/seat in baggage area. Baggage racks at rear of cabin on starboard side, with lavatory on port side. Door on port side aft of wing, with built-in airstairs. Emergency exit on starboard side of cabin. Entire accommodation pressurised, heated and air conditioned. Electrically heated windscreen, windscreen defroster and windscreen wipers standard.

SYSTEMS: Pressurisation by dual engine bleed air system with pressure differential of 0.34 bar (5.0 lb/sq in), maintaining a sea level cabin environment to 3,373 m (11,065 ft). Cabin heated by 45,000 BTU dual engine bleed air system and auxiliary electrical heating system. Hydraulic system for landing gear actuation. Electrical system includes two 28 V 300 A starter/generators, 24 V 34 Ah air-cooled Ni/Cd battery with failure detector. Oxygen system, 623 litres (22 cu ft) or 1,814 litres (64 cu ft) capacity, optional. Vacuum system for flight instruments. Automatic pneumatic de-icing of wing/fin/tailplane leading-edges standard. Engine and propeller anti-icing systems standard. Engine fire detection and extinguishing system optional.

AVIONICS: Standard Rockwell Collins Pro Line II package with two-tube EFIS-84 in sectional instrument panel.

Comms: Dual VHF-22A transceivers with CTL-22 controls; dual TDR-64 transponders; dual DB Systems Model 415 audio systems; dual Lockheed Martin Fairchild CVR A1005; dual Flite-Tronics PC-250 inverters; edgelight radio panel; ELT; avionics master switch; ground clearance switch on com 1; control wheel push-to-talk switches; dual hand microphones and cockpit speakers; dual Telex headsets.

Radar: WXR-270 colour weather radar.

Flight: Dual VIR-32 VOR/LOC/GLS/MKR receivers with CTL-32 controls; ADF-60A with CTL-62 control; DME-42 with IND-42 indicator; RMI-30; dual MCS-65 compass systems. Goodrich WX-1000+ Stormscope optional. APS-65H autopilot/flight director with EADI-84 and EHSI.

Instrumentation: EFIS-84; ALT-50A radio altimeter; pilot's encoding altimeter; dual 2 in electric turn and bank indicators; co-pilot's 75 mm (3 in) horizon indicator; co-pilot's HSI. Dual blind-flying instrumentation with dual instantaneous VSIs; standby magnetic compass; digital OAT gauge; LCD digital chronometer clock; vacuum gauge; de-icing pressure gauge; cabin rate of climb indicator; cabin altitude and differential pressure indicator; flight hour recorder; automatic solid-state warning and annunciator panel. Primary and secondary instrument lighting systems.

EQUIPMENT: Standard equipment includes fresh air outlets; oxygen outlets with overhead-mounted diluter demand masks with microphones; removable low-profile lavatory with shoulder harness and lap belt; two cabin tables; cabin fire extinguisher; wing ice lights; dual landing lights; nosewheel taxying light; flush position lights; dual rotating beacons; rheostat-controlled white cockpit lighting; wingtip recognition lights; wingtip and tail strobe lights; and vertical tail illumination lights.

Optional equipment includes electric flushing lavatory; engine fire detection system; oxygen bottle; cabinet with three drawers and stereo tape deck storage; and pilot-to-cabin paging with four stereo speakers.

DIMENSIONS, EXTERNAL:
Wing span ... 15.32 m (50 ft 3 in)
Wing chord: at root ... 2.15 m (7 ft 0½ in)
at tip ... 1.07 m (3 ft 6 in)
Wing aspect ratio ... 8.6
Length overall ... 10.82 m (35 ft 6 in)
Height overall ... 4.34 m (14 ft 3 in)
Tailplane span ... 5.26 m (17 ft 3 in)
Wheel track ... 3.89 m (12 ft 9 in)
Wheelbase ... 3.73 m (12 ft 3 in)
Propeller diameter ... 2.29 m (7 ft 6 in)
Propeller ground clearance ... 0.34 m (1 ft 1½ in)
Passenger door: Height ... 1.30 m (4 ft 3½ in)
Width ... 0.69 m (2 ft 3 in)
Height to sill ... 1.22 m (4 ft 0 in)

DIMENSIONS, INTERNAL:
Pressurised area: Length ... 5.43 m (17 ft 10 in)
Cabin: Length ... 3.835 m (12 ft 7 in)
Max width ... 1.37 m (4 ft 6 in)
Max height ... 1.45 m (4 ft 9 in)
Floor area ... 6.50 m^2 (70.0 sq ft)
Volume: cabin ... 6.4 m^3 (227 cu ft)
including flight deck ... 8.98 m^3 (317 cu ft)
Baggage compartment volume, rear ... 1.4 m^3 (48 cu ft)

AREAS:
Wings, gross ... 27.31 m^2 (293.9 sq ft)
Ailerons (total) ... 1.29 m^2 (13.90 sq ft)
Trailing-edge flaps (total) ... 2.72 m^2 (29.30 sq ft)
Fin ... 2.20 m^2 (23.67 sq ft)
Rudder, incl tab ... 1.30 m^2 (14.00 sq ft)
Tailplane ... 4.39 m^2 (47.25 sq ft)
Elevators, incl tabs (total) ... 1.66 m^2 (17.87 sq ft)

WEIGHTS AND LOADINGS:
Weight empty ... 3,086 kg (6,803 lb)
Basic operating weight, typical ... 3,193 kg (7,040 lb)
Max fuel weight ... 1,167 kg (2,573 lb)
Max T-O weight ... 4,581 kg (10,100 lb)
Max ramp weight ... 4,608 kg (10,160 lb)
Max zero-fuel weight ... 4,178 kg (9,212 lb)
Max landing weight ... 4,354 kg (9,600 lb)
Fuel with max payload ... 430 kg (948 lb)
Max wing loading ... 167.8 kg/m^2 (34.36 lb/sq ft)
Max power loading ... 5.59 kg/kW (9.18 lb/shp)

PERFORMANCE:
Max level speed ... 249 kt (461 km; 286 mph)
High cruising speed
at FL160 ... 246 kt (456 km/h; 283 mph)
Long-range cruising speed at FL260 ... 196 kt (363 km/h; 226 mph)
Approach speed ... 101 kt (187 km/h; 116 mph)
Stalling speed, power off:
wheels and flaps up ... 88 kt (163 km/h; 101 mph) IAS
flaps down ... 78 kt (144 km/h; 90 mph) IAS
Max rate of climb at S/L ... 610 m (2,003 ft)/min
Rate of climb at S/L, OEI ... 151 m (494 ft)/min
Max certified altitude ... 9,150 m (30,000 ft)
Service ceiling ... 8,809 m (28,900 ft)
Service ceiling, OEI ... 4,511 m (14,800 ft)
Time to FL250 ... 22 min
T-O run ... 620 m (2,035 ft)
T-O field length ... 826 m (2,710 ft)
Accelerate/stop distance ... 1,113 m (3,650 ft)
Landing field length at max landing weight, with propeller reversal ... 698 m (2,290 ft)
Range, with NBAA IFR reserves, 100 n mile (185 km; 115 mile) alternate:
with max payload ... 191 n miles (353 km; 219 miles)
with max fuel ... 1,207 n miles (2,235 km; 1,389 miles)
with four passengers ... 1,040 n miles (1,926 km; 1,196 miles)
ferry range ... 1,264 n miles (2,340 km; 1,454 miles)

UPDATED

BEECH KING AIR B200

Israel Defence Force name: Tsofit (Thrush)
Swedish Air Force designation: Tp 101

TYPE: Utility turboprop twin.

PROGRAMME: Design of Super King Air 200 began October 1970; 'Super' prefix deleted from all 200, 300 and 350 series King Airs in 1996; first flight (c/n BB1) 27 October 1972; certified to FAR Pt 23 plus icing requirements of FAR Pt 25, 14 December 1973; design of B200 (prototype c/n BB343) began March 1980; production started May 1980; FAA certification 13 February 1981; on sale March 1981.

CURRENT VERSIONS: **King Air B200:** Baseline version.

Detailed description applies to B200.

King Air B200C: As B200 but with 1.32 × 1.32 m (4 ft 4 in × 4 ft 4 in) cargo door. Two, identified as military C-12R/AP, ordered by US Army on behalf of Greece, January 2000; special missions fit includes cameras for geophysical survey, removable for VIP transport; delivered mid-2002.

King Air B200T: Standard provision for removable tip tanks, adding total 401 litres (106 US gallons; 88.25 Imp gallons), making total 2,460 litres (650 US gallons; 541 Imp gallons). Span without tip tanks 16.92 m (55 ft 6 in). Total 38 built up to 1993. Four ordered by US Army Missile Command in October 1998 for delivery by June 2000; however, these appear to be part of batch of five for Israel Defence Force, also completed in June 2000 and increasing production total to 43. Further three built in 2002 for undisclosed customer; total 46, all converted on production line from B200s.

King Air B200CT: Combines tip tanks and cargo door as standard. Four built by 1983 for Chile (one) and Peru (Navy, three). All converted on production line from King Air 200. Israel Defence Force/Air Force ordered five, of which first was delivered on 4 September 2002, with deliveries completed in June 2003. All B200T/CTs converted on production line from King Air 200s.

Maritime patrol B200T: Production discontinued.

Beech 200 HISAR: Radar surveillance platform. Launched 1997; based on Beech 200T airframe; demonstrator (N4277E) undertook 16-country tour in 1997–98. First international sale to Traffic 2000 of Germany, November 1997, for North Sea environmental monitoring. Ventral radome for Hughes Integrated Synthetic Aperture Radar (HISAR) suitable for border surveillance, remote sensing, pollution monitoring, EEZ patrol and agricultural monitoring. Equipment operator's console in cabin.

C/RC/UC-12: Military versions.

King Air 300LW: Production discontinued..

King Air 350: *Described separately.*

CUSTOMERS: Deliveries by 30 June 2004 totalled 1,911 for commercial and private orders, plus 397 military versions (described in previous *Jane's All the World's Aircraft*) to US armed forces and foreign customers. Total of 23 King Air B200s delivered in 1994, 28 in 1995, 33 in 1996, 45 in 1997, 45 in 1998, 55 in 1999, 59 in 2000, 46 in 2001, 26 in 2002, 38 in 2003 and 12 in the first six months of 2004.

French Institut Géographique National has three B200Ts fitted with twin Wild RC-10 Superaviogon camera installations and Doppler navigation; maximum endurance 10 hours 20 minutes; high flotation landing gear; special French certification for maximum T-O weight 6,350 kg (14,000 lb) and maximum landing weight 6,123 kg (13,500 lb). Egyptian government acquired one King Air 200 in 1978 for water, uranium and other natural resources exploration over Sinai and Egyptian deserts as follow-up to satellite surveys; fitted with remote sensing gear, specialised avionics and special cameras. Navaid checking versions supplied to Taiwan government (one) and Malaysian government (two). Other special missions aircraft delivered to Taiwan Ministry of Interior May 1979; Royal Hong Kong Auxiliary Air Force (later Government Air Service) (two) 1986 and 1987; four King Air 200s operated by Swedish Air Force since 1988 as **Tp 101**.

King Air B200 selected by UK Serco Group plc for Royal Air Force Cranwell Multi-Activity Contract (MAC) and Multi-Engine Pilot Training Interim Solution (replacing BAe Jetstreams); first two (of seven) handed over on 12 December 2003, with remainder following by March 2004.

COSTS: B200 US$4.970 million (2004).

DESIGN FEATURES: Generally as for earlier versions. Wing aerofoil NACA 23018 to 23016.5 over inner wing, 23012 at tip; dihedral 6°; incidence 3° 48' at root, −1° 7' at tip; swept vertical and horizontal tail.

FLYING CONTROLS: Conventional and manual. Trim tabs in port aileron and both elevators; anti-servo tab in rudder; single-slotted trailing-edge flaps; fixed tailplane.

STRUCTURE: Two-spar light alloy wing; safe-life semi-monocoque fuselage.

Beech King Air B200 *(Paul Jackson)* ***NEW***/0576891

LANDING GEAR: Hydraulically retractable tricycle type, with twin wheels on each main unit. Single wheel on steerable nose unit, with shimmy damper. Main units retract forward, nosewheel rearward. Beech oleo-pneumatic shock-absorbers. Goodrich mainwheels and tyres size 18×5.5 (10 ply) tubeless, pressure 7.25 bar (105 lb/sq in). Oversize and/or 10 ply mainwheel tyres optional. Goodrich nosewheel size 6.50×10 (8 ply) tubeless, with tyre size 22×6.75-10, pressure 3.93 bar (57 lb/ sq in). Goodrich hydraulic multiple-disc brakes. Parking brake.

POWER PLANT: Two 634 kW (850 shp) Pratt & Whitney Canada PT6A-42 turboprops, each driving a Hartzell four-blade, constant-speed, reversible-pitch, metal propeller with autofeathering and synchrophasing. Bladder fuel cells in each wing, with main system capacity of 1,461 litres (386 US gallons; 321.5 Imp gallons) and auxiliary system capacity of 598 litres (158 US gallons; 131.5 Imp gallons). Total usable fuel capacity 2,059 litres (544 US gallons; 453 Imp gallons). Two refuelling points in upper surface of each wing. Wingtip tanks optional, providing an additional 401 litres (106 US gallons; 88.3 Imp gallons) and raising maximum usable capacity to 2,460 litres (650 US gallons; 541 Imp gallons). Oil capacity 29.5 litres (7.8 US gallons; 6.5 Imp gallons).

ACCOMMODATION: Pilot only, or crew of two side by side, on flight deck, with full dual controls and instruments as standard. Seven cabin seats standard, each equipped with seat belts and inertia reel shoulder harness; optional seats in baggage compartment raise passenger capacity to nine. Partition with sliding door between cabin and flight deck, and partition at rear of cabin. Door at rear of cabin on port side, with integral airstair. Large cargo door optional. Inward-opening emergency exit on starboard side over wing. Lavatory and stowage for baggage in rear fuselage. Maintenance access door in rear fuselage; radio compartment access doors in nose. Cabin is air conditioned and pressurised, with electric heat panels to warm cabin before engine starting.

SYSTEMS: Cabin pressurisation by engine bleed air, with a maximum differential of 0.44 bar (6.5 lb/sq in), maintaining a sea level cabin environment to 4,661 m (15,293 ft). Cabin air conditioner of 32,000 BTU capacity. Auxiliary electric cabin heating. Oxygen system for flight deck, and 623 litre (22 cu ft) oxygen system for cabin, with automatic drop-down face masks; 2,182 litre (77 cu ft) system optional. Dual vacuum system for instruments. Hydraulic system for landing gear retraction and extension, pressurised to 171 to 191 bar (2,475 to 2,775 lb/sq in). Separate hydraulic system for brakes. Electrical system has two 300 A 28 V starter/generators and a 24 V 34 Ah air-cooled Ni/Cd battery with failure detector. AC power provided by dual 300 VA inverters. Engine fire detection system standard; engine fire extinguishing system optional. Pneumatic de-icing of wings and tailplane standard. Anti-icing of engine air intakes by hot air from engine exhaust, electrothermal anti-icing for propellers.

AVIONICS: Standard Rockwell Collins Pro Line II package. Pro Line 21 available as option from 2003.

Comms: Cockpit-to-cabin paging standard. Bendix/King KHF 950 transceiver, Fairchild A-100A cockpit voice recorder standard, Wulfsberg Flitefone optional.

Radar: Rockwell Collins WXR-270 colour weather radar standard, WXR-840 or WXR-850 turbulence detecting radar optional.

Flight: Collins APS-65H autopilot/flight director, Universal UNS-1D and UNS-1K navigation management systems, with GPS optional.

Instrumentation: Rockwell Collins EFIS-84/B-14, pilot's ALT-80A encoding altimeter; dual maximum allowable airspeed indicators, and flight director standard. Options include Rockwell Collins three-tube EFIS-85B with MFD.

EQUIPMENT: Standard/optional equipment generally as for King Air C90B except fluorescent cabin lighting, one-place couch with storage drawers, flushing chemical toilet, cabin electric heating, cockpit/cabin partition with sliding doors, and airstair door with hydraulic snubber and courtesy light, standard. FAR Pt 135 operational configuration includes cockpit fire extinguisher and 2.2 m³ (77 cu ft) oxygen bottle with cockpit oxygen pressure indicator as standard. A range of optional cabin seating and cabinetry configurations is available, including quick-removable fold-up seats.

DIMENSIONS, EXTERNAL:
- Wing span: 16.61 m (54 ft 6 in)
- Wing chord: at root: 2.18 m (7 ft 1¾ in)
  - at tip: 0.90 m (2 ft 11½ in)
- Wing aspect ratio: 9.8
- Length overall: 13.36 m (43 ft 10 in)
- Height overall: 4.52 m (14 ft 10 in)
- Tailplane span: 5.61 m (18 ft 5 in)
- Wheel track: 5.23 m (17 ft 2 in)
- Wheelbase: 4.56 m (14 ft 11½ in)
- Propeller diameter: 2.39 m (7 ft 10 in)
- Propeller ground clearance: 0.43 m (1 ft 4¾ in)
- Distance between propeller centres: 5.23 m (17 ft 2 in)
- Passenger door: Height: 1.31 m (4 ft 3½ in)
  - Width: 0.68 m (2 ft 2¾ in)
  - Height to sill: 1.17 m (3 ft 10 in)
- Cargo door (optional): Height: 1.32 m (4 ft 4 in)
  - Width: 1.24 m (4 ft 1 in)
- Nose avionics service doors (port and stbd):
  - Max height: 0.57 m (1 ft 10½ in)
  - Width: 0.63 m (2 ft 1 in)
  - Height to sill: 1.37 m (4 ft 6 in)
- Emergency exit (stbd): Height: 0.66 m (2 ft 2 in)
  - Width: 0.50 m (1 ft 7¾ in)

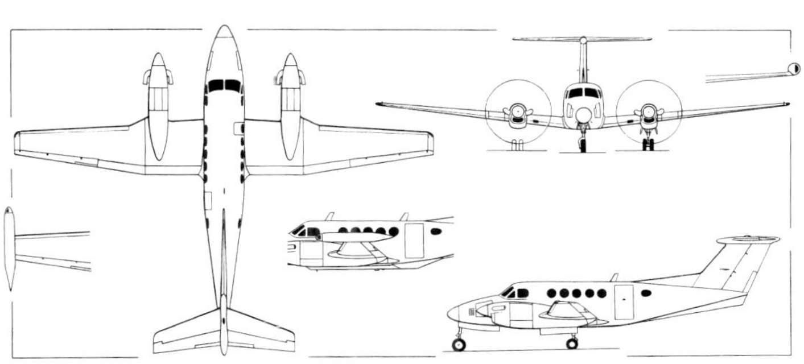

Beech King Air B200 twin-turboprop transport, with scrap views of wingtip tanks and centre-fuselage of photo survey aircraft for IGN *(Jane's/Dennis Punnett)* ***NEW***/0559361

DIMENSIONS, INTERNAL:
- Cabin:
  - Length: between pressure bulkheads: 6.71 m (22 ft 0 in)
    - aft of cockpit divider: 5.08 m (16 ft 8 in)
  - Max width: 1.37 m (4 ft 6 in)
  - Max height: 1.45 m (4 ft 9 in)
  - Floor area: 7.80 m² (84 sq ft)
  - Volume, including flight deck: 10.99 m³ (388 cu ft)
- Baggage hold, rear of cabin:
  - Volume: 1.57 m³ (55.3 cu ft)

AREAS:
- Wings, gross: 28.15 m² (303.0 sq ft)
- Ailerons (total): 1.67 m² (18.00 sq ft)
- Trailing-edge flaps (total): 4.17 m² (44.90 sq ft)
- Fin: 3.46 m² (37.20 sq ft)
- Rudder, incl tab: 1.40 m² (15.10 sq ft)
- Tailplane: 4.52 m² (48.70 sq ft)
- Elevators, incl tabs (total): 1.79 m² (19.30 sq ft)

WEIGHTS AND LOADINGS:
- Weight empty: 3,757 kg (8,283 lb)
- Basic operating weight, typical (one pilot): 3,869 kg (8,530 lb)
- Baggage capacity: 249 kg (550 lb)
- Max fuel: 1,653 kg (3,645 lb)
- Fuel with max payload: 721 kg (1,590 lb)
- Max T-O and landing weight: 5,670 kg (12,500 lb)
- Max ramp weight: 5,710 kg (12,590 lb)
- Max zero-fuel weight: 4,990 kg (11,000 lb)
- Max wing loading: 201.4 kg/m² (41.25 lb/sq ft)
- Max power loading: 4.48 kg/kW (7.35 lb/shp)

PERFORMANCE:
- Never-exceed speed (VNE): 259 kt (480 km/h; 298 mph) IAS
- Max operating Mach No.: 0.52
- High cruising speed at FL220: 292 kt (541 km/h; 336 mph)
- Long-range cruising speed at FL290: 221 kt (409 km/h; 254 mph)
- Stalling speed: flaps up: 99 kt (183 km/h; 114 mph) IAS
  - flaps down: 75 kt (139 km/h; 86 mph) IAS
- Approach speed: 103 kt (191 km/h; 119 mph)
- Max rate of climb at S/L: 750 m (2,460 ft)/min
- Rate of climb at S/L, OEI: 226 m (740 ft)/min
- Service ceiling: 10,670 m (35,000 ft)
- Service ceiling, OEI: 6,675 m (21,900 ft)
- T-O field length: 786 m (2,580 ft)
- Accelerate-stop distance: 1,040 m (3,411 ft)
- Landing distance: 867 m (2,845 ft)
- Range with NBAA IFR reserves, 100 n mile (185 km; 115 mile) alternate:
  - with max payload: 407 n miles (753 km; 468 miles)
  - with max fuel: 1,755 n miles (3,250 km; 2,019 miles)
  - with four passengers: 1,483 n miles (2,746 km; 1,706 miles)
- Ferry range: 1,818 n miles (3,366 km; 2,092 miles)

UPDATED

BEECH KING AIR 350

US Army designation: C-12S
JGSDF designation: LR-2

TYPE: Utility turboprop twin.

PROGRAMME: Replaced King Air 300 (1991-92 *Jane's*); first flight (N120SK) September 1988; introduced at NBAA Convention 1989; certified to FAR Pt 23 (commuter category); first delivery 6 March 1990; Russian certification to AP 23 in November 1995; FAA approval for operation from unprepared runways granted during 1997.

CURRENT VERSIONS: **King Air 350:** Baseline version.

Detailed description applies to King Air 350.

King Air 350C: Has 132 × 132 cm (52 × 52 in) freight door with built-in airstair passenger door.

King Air 350 Special Mission: Version available from last half of 2004 for homeland security, aerial photography and airways and ground-based navaid checking. Max take-off and landing weight 7,484 kg (16,500 lb), maximum ramp weight 7,530 kg (16,600 lb), max zero-fuel weight 5,897 kg (13,000 lb); provision for HISAR and SeaVue reconnaissance systems; wing fuel lockers; mission endurance seven hours.

Australian Army received King Air 350 VH-HPJ in 1998 as follow-on (to Douglas C-47 Dakota) test platform for Ingara SAR/MTI radar in ventral pannier; installation by Hawker Pacific.

RC-350 Guardian: Elint version, converted from 350 prototype 1991 by Beech Aircraft Corporation; mission avionics include Raytheon AN/ALQ-142 ESM, Watkins-Johnson 9195C communications interceptor, Honeywell laser INS, GPS receiver and Cubic secure digital datalink; can loiter on station at 10,670 m (35,000 ft) for more than 6 hours; can locate/monitor radar emitters in 20 MHz to 18 GHz range, and intercept communications within 20 to 1,400 MHz bandwidths. Wingtip pods house AN/ALQ-142 antennas; underfuselage bulge contains antenna for comint system.

LR-2: Japan Ground Self-Defence Force funded two in FY97 for liaison and reconnaissance (undisclosed sensor in ventral radome). Total requirement for 20, of which third funded in 1999, fourth in FY00, fifth in FY01 and sixth in FY02. First delivery (23051) 22 January 1999; initial operator is the HQ Flight of 1st Helicopter Brigade at Kisarazu.

C-12S: US Army version with quick-change cargo capability and seating for up to 15 passengers. By late 2000, no aircraft of this type had been identified in service.

CUSTOMERS: Total 394 King Air 350s and 350Cs delivered by June 2004; first 350C delivery in 1990 to Rossing Uranium, Namibia; 15 King Air 350 deliveries 1995, 27 in 1996, 30 in 1997, 42 in 1998, 45 in 1999, 46 in 2000, 32 in 2001, 24 in 2002, 24 in 2003 and eight in the first six months of 2004. Recent customers include the Royal Australian Air Force, which signed a lease contract on 20 November 2002 for seven King Air 350s for delivery from June 2003, to be operated by 32 Squadron for the RAAF School of Navigation at East Sale, replacing H.S. 748s and King Air 200s.

COSTS: 350 US$5.802 million (2004).

DESIGN FEATURES: Compared with King Air 300, fuselage stretched 0.86 m (2 ft 10 in) by plugs 0.37 m (1 ft 2½ in) forward of main spar and 0.49 m (1 ft 7½ in) aft; wing span increased by 0.46 m (1 ft 6 in) with NASA winglets 0.61 m (2 ft 0 in) high; two additional cabin windows each side. Can depart with full payload and full tanks. Raisbeck dual aft body strakes (DABS), standard on production aircraft from c/n FL-312 (N3165M) in first quarter 2001, reduce drag, improve handling and stability and relax or eliminate restrictions on operations with inoperative yaw damper.

FLYING CONTROLS: Automatic cable tensioner in aileron circuit and larger elevator bobweight; larger rudder anti-servo tab; ailerons and rudder cleaned up.

STRUCTURE: As for B200.

LANDING GEAR: As for B200.

POWER PLANT: Two 783 kW (1,050 shp) Pratt & Whitney Canada PT6A-60A turboprops, each driving a Hartzell four-blade, constant-speed, fully feathering, reversible-pitch, metal propeller. Bladder cells and integral tanks in each wing, with usable capacity of 1,438 litres (380 US gallons; 316.5 Imp gallons); auxiliary tanks inboard of engine nacelles, capacity 601 litres (159 US gallons; 132.5 Imp gallons). Total fuel capacity 2,040 litres (539 US gallons; 449 Imp gallons). No provision for wingtip tanks. Oil capacity 30.2 litres (8.0 US gallons; 6.7 Imp gallons).

ACCOMMODATION: Double club seating for eight passengers; optionally two more seats in rear of cabin and one passenger on side-facing lavatory seat making maximum 11 passengers; certified for maximum 17 occupants including crew. Ultra Electronics UltraQuiet active noise control system installed as standard from 1998.

SYSTEMS: As for B200, except for automatic bleed air-type heating and 22,000 BTU cooling system with high-capacity ventilation system; 2,182 litre (77 cu ft) oxygen system standard; hydraulic landing gear retraction and extension; two 300 A 28 V starter/generators with triple bus electrical distribution system. Ultra Electronics Ltd UltraQuiet active noise control system introduced as standard from 1998, comprising 12 loudspeakers, 24 microphones and a high-speed digital processor which cancel propeller noise and reduce in-flight cabin sound level to less than 80 dB(A).

AVIONICS: Rockwell Collins Pro Line II as core system, with three- or five-tube EFIS, TWR-850 Doppler weather radar and Universal UNS-1K or UNS-1D FMS both with 12-channel GPS. Pro Line 21 option available from 2003.

EQUIPMENT: Generally as for B200.

DIMENSIONS, EXTERNAL: As for B200 except:
Wing span over winglets ... 17.65 m (57 ft 11 in)
Wing aspect ratio ... 10.8
Length overall ... 14.22 m (46 ft 8 in)
Height overall ... 4.37 m (14 ft 4 in)
Propeller diameter ... 2.67 m (8 ft 9 in)
Propeller ground clearance ... 0.29 m (11½ in)
Emergency exit (each side of cabin, above wing):
Height ... 0.66 m (2 ft 2 in)
Width ... 0.50 m (1 ft 7¾ in)

DIMENSIONS, INTERNAL:
Cabin (excl cockpit): Length ... 5.94 m (19 ft 6 in)
Max width ... 1.37 m (4 ft 6 in)
Height ... 1.45 m (4 ft 9 in)
Volume ... 10.05 m^3 (355 cu ft)
Baggage volume: internal ... 1.6 m^3 (55 cu ft)
optional nacelle lockers ... 0.45 m^3 (16 cu ft)

AREAS:
Wings, gross ... 28.80 m^2 (310.0 sq ft)

WEIGHTS AND LOADINGS:
Weight empty ... 4,230 kg (9,326 lb)
Basic operating weight, typical (one pilot) ... 4,341 kg (9,570 lb)
Max fuel weight ... 1,638 kg (3,611 lb)
Max T-O and landing weight ... 6,804 kg (15,000 lb)
Max ramp weight ... 6,849 kg (15,100 lb)
Max zero-fuel weight ... 5,670 kg (12,500 lb)
Fuel with max payload ... 1,179 kg (2,600 lb)
Max wing loading ... 236.2 kg/m^2 (48.39 lb/sq ft)
Max power loading ... 4.35 kg/kW (7.14 lb/shp)

PERFORMANCE:
Max level speed ... 315 kt (584 km/h; 363 mph)
High cruising speed at FL240 ... 312 kt (578 km/h; 359 mph)
Long-range cruising speed at FL330 ... 237 kt (439 km/h; 273 mph)
Stalling speed at max landing weight, flaps and wheels down . 81 kt (150 km/h; 94 mph)
Approach speed ... 109 kt (202 km/h; 125 mph)
Max rate of climb at S/L ... 832 m (2,731 ft)/min
Rate of climb at S/L, OEI ... 168 m (552 ft)/min

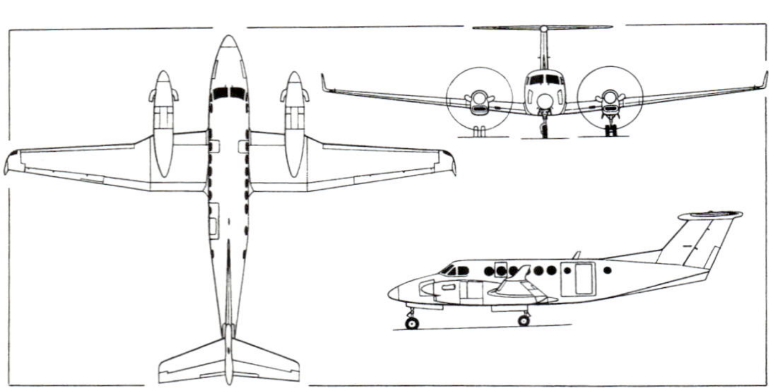

Beech King Air 350 *(Jane's/Dennis Punnett)* 0085295

Beech King Air 350 *(Paul Jackson)* ***NEW***/0576893

Time to FL250 ... 15 min
Service ceiling ... 10,670 m (35,000 ft)
Service ceiling, OEI ... 6,555 m (21,500 ft)
T-O balanced field length ... 1,006 m (3,300 ft)
Landing distance ... 821 m (2,692 ft)
Range, NBAA IFR reserves, 100 n mile (185 km; 115 mile) alternate:
with max payload ... 947 n miles (1,753 km; 1,089 miles)
with max fuel ... 1,598 n miles (2,959 km; 1,839 miles)
with four passengers ... 1,725 n miles (3,194 km; 1,985 miles)
Ferry range ... 1,814 n miles (3,359 km; 2,087 miles)

UPDATED

BEECH 390 PREMIER I

TYPE: Light business jet.

PROGRAMME: Design started early 1994 as PD374 (later PD390) and approved early 1995; originated in former Beech design offices, but was first aircraft to carry only the Raytheon name; brief details of 'new light business jet' revealed June 1995; launched at National Business Aircraft Association Convention in Las Vegas 26 September 1995 with full-scale fuselage/cabin mockup; wind-tunnel tests of one-eighth model conducted early 1996 at Boeing, Boeing V/STOL, NASA-Lewis and Wichita State University facilities; to compete with Cessna CitationJet.

First forward fuselage completed in February 1997 and mated to aft fuselage in April 1998; roll-out (N390RA, c/n RB-1) 19 August 1998; first flight 22 December 1998. Second aircraft (N704T) first flown 4 June 1999, followed by third (N390TC), first with complete interior, on 17 September 1999; public debut (N390TC) at National Business Aviation Association Convention at Atlanta, Georgia, October 1999; more than 720 flight test hours accumulated by 23 December 1999, at which time eight production aircraft were in final assembly; static testing of wing to 150 per cent of design load completed on 17 December 1999; four aircraft undertook 1,400 hour flight test programme culminating in FAA FAR Pt 23 certification on 23 March 2001, followed by German certification on 3 September 2001; FAA RVSM approval granted in May 2003; certified in Bermuda, Denmark, Mexico, Israel, South Africa and Switzerland in 2002, and in China in November 2003. Deliveries began with three aircraft in third quarter of 2001: RB-4, -6 and -7 to Tyrose Investments, Raytheon and Town & Country Food Markets, respectively. 100th Premier I production fuselage completed 24 July 2003. Target production rate 60 per year from 2003. Total flight time stood at more than 23,000 flight hours by May 2004.

CUSTOMERS: More than 300 orders received by October 2001 from customers in 27 countries, of which some 100 were from outside the USA and 51 from Europe, representing a backlog until 2005. Total of 100 delivered by April 2004, including 18 in 2001, 29 in both 2002 and 2003 and 12 in the first six months of 2004. Customers include Raytheon Travel Air, the fractional ownership subsidiary of Raytheon Aircraft, which has ordered 71 for delivery beginning 2001; the Jordan Grand Prix racing team, which has ordered one; and Aviation Leasing Group (ALG Transportation Inc) of London, which ordered three in August 2000, two of which will be used by the Civil Aviation Training Centre (CATC) in Thailand for training student pilots for Thai Airways International and other Pacific Rim carriers.

COSTS: US$5.250 million (2004).

DESIGN FEATURES: Conventional small business jet, developed with assistance of CATIA programmes. Rear-mounted engines, T tail and wing mounted below fuselage for additional cabin space. Sweepback 20° at 25 per cent chord; 2° 30' dihedral; tailplane sweepback 25° at 25 per cent chord.

FLYING CONTROLS: Conventional and manual. Activation via pushrods and cables. Pitch trim via electrically actuated, variable incidence tailplane and mechanically driven geared tab on each elevator; electrically actuated trim tab on each aileron; electrically actuated rudder trim tab. Electrically signalled, hydraulically powered, three-segment spoilers on upper surface of each wing augment aileron roll control; outboard and middle panels provide roll, airbrake and post-landing lift-dump functions; inboard panel provides lift-dump function only. Control surface maximum deflections: rudder 25°; rudder trim ±20°; elevators 20° up, 9° 36' down; elevator trim 3° 3.6' up, 12° 36' down; tailplane incidence 1° 24' leading-edge up, 7° leading-edge down; ailerons 15° 30' up, 12° 30' down; aileron trim ±20°; roll-sontrol spoilers, outboard panels 9°, mid panels 8° 3'; speedbrakes 23°; lift dumpers, inboard panels 60°, mid and outboard panels 45°. 75 per cent span, four-segment, electrically controlled Fowler flaps, deflections 0, 10, 20 and 30°. Rudder boost, for asymmetric thrust and yaw damper, standard.

STRUCTURE: Fuselage of graphite/epoxy laminate and honeycomb composites, formed by Cincinnati Milacron Viper automatic fibre-placement machines over aluminium mandrel, placing fibres at speeds up to 46 m (150 ft) per minute, enabling entire fuselage to be completed in one week; elimination of all internal frames, and skin thickness of 20 mm (0.78 in), increase cabin volume by 13 per cent and afford a weight saving of some 20 per cent over conventional alloy construction. Wing of aluminium alloy with six-spar wing box, manufactured using high-speed equipment capable of machining more than 93 m^2 (1,000 sq ft) of material per minute, and automatic riveting machines; with exception of three small bays along trailing-edge, entire wing is used for fuel storage. Ailerons and flaps of graphite/epoxy composites; fin has aluminium alloy spars and ribs with graphite/epoxy honeycomb skin; tailplane has one-piece, composites forward-and-rear spar with alloy centre rib, composites mid– and tip ribs and Nomex composites skin.

LANDING GEAR: Hydraulically actuated, retractable tricycle type with free-fall emergency extension system; single wheel on each unit. Mainwheel size 22×8.2 (12 ply); nosewheel 18×4.4 (6 ply). Mainwheels retract inwards; nosewheel forwards. Steerable nosewheel, maximum pedal-commanded deflection ±35°, increasing to ±45° with differential braking and asymmetric thrust. Hydraulic disc brakes with electric anti-skid system.

POWER PLANT: Two pod-mounted Williams FJ44-2A turbofans, each rated at 10.23 kN (2,300 lb st). Fuel contained in integral wing tanks, each of four sections, plus inboard collector tank; total capacity 2,092.6 litres (552.8 US gallons; 460.3 Imp gallons), of which 2,073.6 litres (547.8 US gallons; 456.1 Imp gallons are usable (for c/n RB-75 and later, gravity-filled), with gravity filling point on each wing. Single-point pressure refuelling/defuelling optional, with slightly reduced maximum and usable capacities.

ACCOMMODATION: Crew of one or two, with dual controls standard; six passengers in cabin, comprising four in standard club seating arrangement with tracking, swivelling and reclining capability and stowable writing tables and two on fixed forward-facing seats to rear; lavatory at rear, doubling as flight-accessible baggage compartment, maximum capacity 64 kg (140 lb). Refreshment/hang-up baggage cabinet on forward starboard side of cabin. Airstair door on port side to rear of flight deck; single plug-type emergency exit on starboard side. Three cabin windows on each side. Accommodation is air conditioned and pressurised. Externally accessible, unpressurised main baggage compartment to rear of cabin, with upward-opening door on port side, can accommodate large items such as skis; heating optional; forward baggage compartment in nose on port side with swing-up door.

SYSTEMS: Pressurisation system, maximum differential 0.58 bar (8.4 lb/sq in), maintains sea level cabin environment to 6,523 m (21,400 ft) and 2,440 m (8,000 ft) cabin altitude to 12,500 m (41,000 ft). Vapour cycle, ozone-safe R134a air conditioning system. Hydraulic

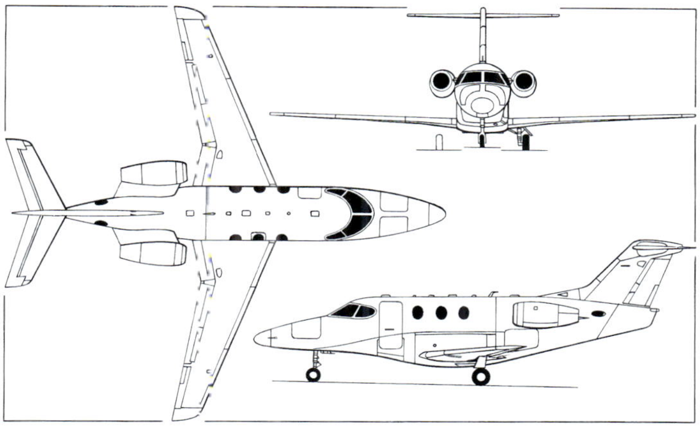

Beech Premier I (two Williams FJ44 turbofans) *(James Goulding)* 0126960

Beech Premier I light business jet ***NEW***/0568421

system, maximum pressure 207 bar (3,000 lb/sq in), for landing gear, brakes, anti-skid and spoilers. Electrical system comprises two 28 V 300 A engine-driven starter/generators, 24 V 40 Ah lead/acid main battery, 24 V 5 Ah standby battery and 28 V external power receptacle; system is configured so that load-shedding is primarily automatic in the event of failure of any or all main electrical power sources. Oxygen system, capacity 1,134 litres (40 cu ft) standard, 2,182 litres (77 cu ft) optional, with diluter-demand masks for crew and continuous flow masks for passengers. Engine bleed-air anti-icing for wing leading-edges and nacelle inlets; electromagnetic expulsion de-icing (EMED) for tailplane leading-edges, automatically activated by dual nose-mounted, heated, ice detectors; electrically heated windscreens (with silicone coating for rain dispersal), pitot tubes and AoA probes.

AVIONICS: Rockwell Collins Pro Line 21 EFIS avionics suite as core system.

Comms: Dual Rockwell Collins VHF-422A transceivers, TDR-94 Mode S transponders and DB Model 438 audio systems; single CTL-23 nav/com tuning unit; four-speaker cabin paging unit.

Radar: Rockwell Collins WXR-800 colour weather radar.

Flight: Dual Rockwell Collins AHC-3000 AHRS, ADC-3000 air data computers, CDU-3000 control/display units and VIR-432 nav receivers; IAPS-3000 lightweight, integrated avionics processing system; FGC-3000 flight guidance system, FMS-3000 flight management system with database, ADF-462, DME-442, GPS-4000, ALT-4000 radio altimeter and MDC-3000 maintenance diagnostic computer.

Instrumentation: Rockwell Collins AFD-3010 integrated EFIS comprising two 254 × 203 mm (10 × 8 in) active matrix LCD adaptive flight displays providing PFD and MFD functions, with CRT HSI and back-up electromechanical rate/sensor/attitude instrument and ASI on right side; second PFD optional, but not mandatory for RVSM compliance.

DIMENSIONS, EXTERNAL:
Wing span ... 13.56 m (44 ft 6 in)
Wing aspect ratio ... 8.0
Length overall ... 14.02 m (46 ft 0 in)
Height overall ... 4.67 m (15 ft 4 in)
Tailplane span ... 4.90 m (16 ft 1 in)
Wheel track ... 2.79 m (9 ft 2 in)
Wheelbase ... 5.36 m (17 ft 7 in)
Crew/passenger door: Height ... 1.27 m (4 ft 2 in)
Width ... 0.64 m (2 ft 1½ in)

DIMENSIONS, INTERNAL:
Cabin:
Length: between pressure bulkheads ... 5.69 m (18 ft 8 in)
excl flight deck ... 4.11 m (13 ft 6 in)
excl flight deck and lavatory ... 3.40 m (11 ft 2 in)
Width: max ... 1.68 m (5 ft 6 in)
at floor ... 1.22 m (4 ft 0 in)
Max height ... 1.65 m (5 ft 5 in)
Volume excl flight deck and lavatory ... 8.9 m³ (315 cu ft)
Baggage volume:
external, nose ... 0.28 m³ (10.0 cu ft)
external, tailcone ... 1.25 m³ (44 cu ft)
internal, forward ... 0.08 m³ (3.0 cu ft)
internal, (lavatory) ... 0.56 m³ (19.9 cu ft)

AREAS:
Wings, gross ... 22.95 m² (247.0 sq ft)
Horizontal tail surfaces (total) ... 4.65 m² (50.00 sq ft)
Vertical tail surfaces (total) ... 4.78 m² (51.5 sq ft)

WEIGHTS AND LOADINGS:
Empty weight ... 3,683 kg (8,120 lb)
Basic operating weight ... 3,774 kg (8,320 lb)
Equipped basic operating weight, typical ... 3,812 kg (8,405 lb)
Baggage capacity:
external, nose ... 68 kg (150 lb)
external, tailcone ... 181 kg (400 lb)
internal, forward ... 27 kg (60 lb)
internal, (lavatory) ... 63.5 kg (140 lb)
Max fuel weight (usable) ... 1,665 kg (3,670 lb)
Max-T-O weight ... 5,670 kg (12,500 lb)
Max ramp weight ... 5,710 kg (12,590 lb)
Max landing weight ... 5,262 kg (11,600 lb)
Max zero-fuel weight ... 4,536 kg (10,000 lb)
Fuel with max payload ... 1,175 kg (2,590 lb)
Max wing loading ... 247.1 kg/m² (50.61 lb/sq ft)
Max power loading ... 277 kg/kN (2.72 lb/lb st)

PERFORMANCE:
Max operating speed:
S/L to FL270 ... 320 kt (593 km/h; 368 mph)
above FL270 ... M0.80
High cruising speed at FL330 ... 451 kt (835 km/h; 519 mph)
Long-range cruising speed at FL410 ... 369 kt (683 km/h; 425 mph)
Approach speed ... 119 kt (220 km/h; 137 mph)
Max operating altitude ... 12,500 m (41,000 ft)
Service ceiling, OEI ... 8,534 m (28,000 ft)
Time to FL370 ... 17 min
T-O field length ... 1,156 m (3,793 ft)
Landing distance ... 966 m (3,170 ft)
Range, NBAA IFR reserves, 100 n mile (185 km; 115 mile) alternate:
with max payload ... 826 n miles (1,529 km; 950 miles)
with max fuel ... 1,460 n miles (2,703 km; 1,680 miles)
with four passengers ... 1,290 n miles (2,389 km; 1,484 miles)
Ferry range ... 1,502 n miles (2,781 km; 1,728 miles)
g limits ... +3.2/−1.28

UPDATED

BELL

BELL HELICOPTER TEXTRON INC (SUBSIDIARY OF TEXTRON INC)

PO Box 482, Fort Worth, Texas 76101
Tel: (+1 817) 280 20 11
Fax: (+1 817) 280 23 21
Web: http://www.bellhelicopter.textron.com
CHAIRMAN EMERITUS: John R Murphey
CHIEF EXECUTIVE OFFICER: Michael Redenbaugh
PRESIDENT AND COO: Glenn E Hess
VICE-PRESIDENT, RESEARCH AND ENGINEERING: M D Sills
DIRECTOR, PUBLIC AFFAIRS AND ADVERTISING: Carl L Harris

During 1970–81, Bell Helicopter Textron was unincorporated division of Textron Inc; became wholly owned subsidiary of Textron Inc from 3 January 1982. Bell Helicopter Textron Canada (see Canadian section) formed at Montréal/Mirabel under contract with Canadian government October 1983; transfer to Mirabel, completed January 1987, of Bell 206B JetRanger and 206L LongRanger production. Production of Bell 212/412 transferred mid-1988 and early 1989 respectively; Bell 230, 430, 427 and 407 programmes also undertaken in Canada, although some of these now terminated. In mid-2003, Bell Helicopter had workforce of more than 7,000 and was valued at US$1.6 billion.

Bell and Boeing collaborate in design and manufacture of V-22 Osprey tiltrotor aircraft, as described in the following entry. New 41,800 m² (450,000 sq ft) factory at Amarillo International Airport, Texas, completed in 1999 as a Tiltrotor Assembly Centre (TAC) for the V-22 and the commercial BA609; latter tiltrotor, previously also a joint venture with Boeing, became a solely Bell programme on 1 March 1998 and is now (with the AB139) the core of a joint venture effort undertaken in conjunction with Agusta of Italy; an agreement establishing the Bell/Agusta Aerospace Company was signed in November 1998. Amarillo factory will also be final assembly centre for AH-1Z and UH-1Y helicopters to be remanufactured for US Marine Corps. Existing assembly hall being extended by 10,525 m² (113,300 sq ft) to accommodate AH-1Z and UH-1Y, with new 6,700 m² (72,000 sq ft) flight hangar also to be completed by February 2006.

Bell helicopters built in USA detailed here. Those currently built in Canada listed under Canada; other models built under licence by Dirgantara in Indonesia and Agusta in Italy; KAI is co-producing Bell 427 as SB 427 in Republic of Korea; Bell Helicopter Asia (Pte) Ltd is wholly owned Singapore-based company for marketing and support in Southeast Asia.

In late 2003, Bell announced intention to establish new organisation to be known as XWORX; this will concentrate on advanced development projects, such as production of a full-scale prototype of the HV-911 Eagle Eye UAV as well as a variety of tiltrotor concepts.

UPDATED

BELL 449 SUPERCOBRA AND KING COBRA

US Navy/Marine Corps designations: AH-1W and AH-1Z

TYPE: Attack helicopter.

PROGRAMME: Prototype Bell 209, derived from single-engined UH-1, first flew as tandem-seat combat aircraft on 7 September 1965. Built for US armed forces and export and under licence in Japan, as described in previous editions of *Jane's*. Universally known as HueyCobra.

First twin-engined Cobra was AH-1J SeaCobra, delivered from mid-1970; AH-1T Improved SeaCobra followed from 1977. All surviving US Marine Corps AH-1J SeaCobras withdrawn and 44 (including one ground-based trainer) AH-1T Improved SeaCobras converted to AH-1W to augment new production.

CURRENT VERSIONS: **AH-1W SuperCobra:** Bell flew AH-1T powered by two GE

Bell SuperCobra combat helicopter ***NEW***/0576493

DEVELOPMENT MILESTONES

Official go-ahead	Nov 1996
First order (development)	Nov 1996
Rolled out	20 Nov 2000
First flight	7 Dec 2000
Production go-ahead	23 Oct 2003
First delivery (trials)	Mar 2001

T700-GE-700; first flight of improved AH-1T+, including GE T700-GE-401 engines, 16 November 1983. USMC received 169 new-build examples as well as two maintenance trainers; 10 supplied to Turkey and 63 to Taiwan. Missions of AH-1W include anti-armour, escort, multiple-weapon fire support, armed reconnaissance, search and target acquisition.

AH-1W Upgrades: Following abandonment of the proposed Integrated Weapon System (IWS) project in July 1995 and the Marine Observation and Attack Aircraft programme which was intended to provide a replacement for both the AH-1W SuperCobra and the UH-1N Iroquois, the US Marine Corps has opted for a two-stage upgrade of the AH-1W, allowing it to be retained in the active inventory until about 2030. Phase 1 concerned installation of a Night Targeting System (NTS), under which USMC AH-1Ws fitted with the Israeli Tamam laser NTS for dual TOW/Hellfire day, night and adverse weather capability.

Conversion of a prototype (162533) was authorised in December 1991, with an initial batch of 25 sets being built by Tamam for delivery from January 1993; joint production with Kollsman was approved in May 1994. A total of 250 sets was required by the USMC, with further sets produced for Turkey and Taiwan. Deliveries of modified aircraft to operational units of the USMC began in June 1994.

A further improvement programme, involving installation of an Embedded Global Positioning System/Inertial Navigation System (EGI), was subsequently undertaken. Two prototype conversions (162532 and 163936) were delivered to test units for trials in November 1995 and March 1996, with EGI installed on new-build aircraft from Lot 9 onwards, as well as older AH-1Ws as a retrofit programme.

Phase 2 entails installation of the Bell 680 four-blade rotor, offering a 70 per cent reduction in vibration; formerly designated **AH-1W (4BW)**, but now known as **AH-1Z**. Initial trials of the four-blade rotor system were undertaken with AH-1W 161022; bench testing of the new drive system began in second quarter of 1999 and was completed in first quarter of 2000. Bell also demonstrated 30-minute run-dry capability of new intermediate and tail rotor gearboxes in March 2000. AH-1Z has new four-blade, all-composites, hingeless/bearingless rotor system; four-blade composites tail rotor; new transmission rated at 1,957 kW (2,625 shp) and new wing assemblies able to carry twice the number of anti-armour missiles, as well as more fuel and additionally permitting concurrent carriage of two air-to-air self-defence missiles. Horizontal tail surfaces increased in size; these initially had end plates, but they were subsequently deleted and will not be fitted to production helicopters.

Lockheed Martin selected to develop and manufacture AN/AAQ-30 Hawkeye advanced target sighting system (TSS), with work on US$8 million, 54 month, engineering development and integration programme beginning in July 1998. TSS features imaging technology by Wescam of Canada and Lockheed Martin's Sniper third-generation FLIR, as well as colour TV camera, laser ranger, spot-tracker and designator.

Also featured on the AH-1Z are 'glass cockpits'; Northrop Grumman (formerly Litton Industries) has been selected as prime contractor for this aspect of the upgrade. Digital transfer of information on tactical situation, weaponry and flight data will enable crew interchangeability and allow AH-1Z to be flown from either front or rear seat. Major subcontractors include Rockwell Collins, which will supply active matrix liquid crystal displays (AMLCDs); Smiths Industries (fire-control system); Meggitt Avionics (standby air data and inertial sensing devices); and BAE Systems (air data computers). Other elements of the upgrade include new stores management system, onboard systems monitoring, mission data loader, HOTCC (hands on throttle, collective and cyclic) controls, airborne target handover system and a new EW suite.

A US$310 million cost-plus-fixed-fee contract was awarded to Bell in November 1996, for design, development, fabrication, installation, test and delivery of three engineering development AH-1W SuperCobra Upgrade Aircraft. Assembly of first AH-1Z begun at Hurst, Texas, in April 1999, by which time 85 per cent of drawings had been released, with design work due for completion by end of 1999. Initial AH-1Z (166477, c/n 59001, previously 162549) completed final assembly in second quarter of 2000 and moved to Bell Flight Research Center at Arlington, Texas, for installation of instrumentation and functional testing that included restrained ground running which was completed in October 2000. Formal roll-out at Arlington on 20 November 2000, with first flight on 7 December. Second development aircraft (166478, c/n 59002, previously 163933) was due to fly in 2001, but handling quality problems that emerged early in flight test programme necessitated redesign of horizontal stabiliser assembly and caused delay; this eventually flew for first time on 4 October 2002, having been forestalled by third development aircraft (166479, c/n 59003, previously 162532), which made its maiden flight on 26 August 2002. By mid-November 2002, all three aircraft had accumulated 390 flight hours, demonstrating 160 kt (296 km/h; 184 mph) cruise and 220 kt (407 km/h; 253 mph) maximum speed. Programme includes flight test and evaluation at Patuxent River, Maryland, to where first AH-1Z was airlifted by C-5 Galaxy on 31 March 2001. Initial operational assessment began 12 February 2003 and was completed on 27 March, involving total of just under 12 flight hours and using third AH-1Z development aircraft; in May, first AH-1Z undertook series of high altitude tests (with first UH-1Y) at Alamosa, Colorado. Full operational evaluation expected to begin August 2004. Weapons testing will take place at Yuma Proving Ground, Arizona with other trials at China Lake, California. Testing of full-scale AH-1Z structural test article at Arlington began in April 2000; on 22 November 2002, significant milestone passed with completion of 20,000 hour fatigue life demonstration, but further fatigue and static loads evaluation to follow, with airframe also earmarked for survivability assessment on completion of test duty. Finalisation of the cockpit upgrade design occurred in FY99, with first order for remanufacture due to be placed in FY04. IOC scheduled for 2007, with peak production rate requiring 24 AH-1Ws to be upgraded annually. Initial deliveries will be to training unit HMT-303 at Camp Pendleton, California.

AH-1RO Dracula: Derivative of AH-1W for Romania, which intended to purchase initial batch of 96. Project abandoned by Bell in fourth quarter of 1999.

AH-1Z King Cobra: Version for Turkey, which planned to acquire 145 attack helicopters at cost of US$4 billion; type was formally selected in July 2000, but failure to reach agreement on cost and technology transfer details finally resulted in Turkey abandoning plans to go ahead with deal, decision to this effect being reached during meeting of Defence Industries Undersecretariat on 14 May 2004.

ARH-1Z: Designation allocated to version unsuccessfully proposed for Australian Army Project Air 87 armed reconnaissance helicopter.

NAH-1Z: Designation allocated to two of three engineering development aircraft, indicating permanent assignment to test duty.

MH-1W: In April 1998, Bell revealed a reconnaissance, armed escort and fire support 'multimission' version of the SuperCobra under this designation. Evolved in response to a perceived need for armed helicopters to undertake anti-drug operations, marketing efforts principally aimed at Latin American countries, with presentations to Argentina, Brazil, Chile, Colombia and Venezuela. Configuration includes a nose-mounted sighting system, with a FLIR Star Safire FLIR sensor, laser range-finder, video recorder and automatic target tracker. Proposed weaponry includes a 20 mm cannon as well as 0.50 in gun pods and up to four 70 mm rocket pods, but excludes anti-armour missiles and air-to-air missiles.

CUSTOMERS: US Marine Corps (see under Current Versions); total of 179 new-build AH-1W, including 10 diverted to Turkey.

Deliveries to USMC began on 27 March 1986 to Camp Pendleton, California, for HMLA-169, −267, −367 and −369, plus HMT-303 for training; further aircraft issued to USMC Reserve, beginning with HMA-775 (now HMLA-775) at Camp Pendleton from June 1992. Procurement augmented by remanufacture of 43 AH-1Ts to AH-1W for HMLA-167 and −269 at New River, North Carolina; last completed in 1993. Last new-build aircraft for USMC delivered in fourth quarter of 1998.

Turkish Land Forces received five AH-1Ws in 1990 and five in 1993; all diverted from USMC contracts. Taiwan signed letter of offer and acceptance February 1992, for 42 over five years (plus one ground trainer); deliveries began 1993 for training with USMC; first aircraft 501 (ex-164913); in service with 1st Attack Helicopter Squadron at Lung Tan and 2nd AHS at Shinsur. Further 21 AH-1Ws subject of Taiwanese re-order announced in mid-1997. Taiwan has existing requirement for up to 75 attack helicopters and is expected to announce decision in late 2004 or early 2005, following evaluation of AH-1Z and Boeing AH-64D Apache.

Remanufacture programme for USMC calls for 180 AH-1Ws to be produced, with procurement of first Low-Rate Initial Production (LRIP) batch of three helicopters in FY04, with delivery scheduled for 2006; another small LRIP batch will follow in FY05, with full-rate production starting in FY06, assuming successful conclusion of operational evaluation.

COSTS: Total value of 1997 follow-on order for 21 AH-1Ws by Taiwan US$479 million. In late 2002, total cost of remanufacturing 180 AH-1Zs and 100 UH-1Ys estimated to be US$6.2 billion.

DESIGN FEATURES: Essentially first-generation attack helicopter, with slightly stepped tandem seating and stub-wings for armament. Two-blade main rotor, similar to that of Bell 214, with strengthened rotor head incorporating Lord Kinematics Lastoflex elastomeric and Teflon-faced bearings. Blade aerofoil Wortmann FX-083 (modified); normal 311 rpm. Tail rotor also similar to that of Bell 214 with greater diameter and blade chord; normal 1,460 rpm. Rotor brake standard. Stub-wings have NACA 0030 section at root; NACA 0024 at tip; incidence 14°; sweepback 14.7°. AH-1Z incorporates new four-blade rotor system and transmission.

STRUCTURE: Main rotor blades have aluminium spar and aluminium-faced honeycomb aft of spar; tail rotor has aluminium honeycomb with stainless steel skin and leading-edge. Airframe conventional all-metal semi-monocoque.

LANDING GEAR: Non-retractable tubular skid type on AH-1W; AH-1Z has new, lighter design with rectangular cross tubes. Ground handling wheels optional.

POWER PLANT: Two General Electric T700-GE-401 turbo-shafts, each rated at 1,285 kW (1,723 shp). Transmission rating 1,515 kW (2,032 shp) for take-off; 1,286 kW (1,725 shp) continuous; AH-1Z transmission flat rated at 1,957 kW (2,625 shp). Fuel (JP5) for AH-1W contained in two interconnected self-sealing rubber fuel cells in fuselage, with protection from damage by 0.50 in ballistic ammunition, total usable capacity 1,128 litres (298 US gallons; 248 Imp gallons); AH-1Z has two internal cells; forward fuselage one contains 705 litres (186.3 US gallons; 155.1 Imp gallons), while aft fuselage cell contains 468 litres (123.7 US gallons; 103.0 Imp gallons); also has larger stub-wings, containing additional fuel and increasing total internal capacity to 1,560 litres (412 US gallons; 343 Imp gallons). Gravity refuelling point in forward fuselage, pressure refuelling point in rear fuselage. Provision for carriage on underwing stores stations of two or four external fuel tanks each of 291 litres (77.0 US gallons; 64.1 Imp gallons) capacity; or two 379 litre (100 US gallon; 83.3 Imp gallon) tanks; or two 100 and two 77.0 US gallon tanks; large tanks on outboard pylons only. AH-1Z can also operate with 871 litre (230 US gallon; 191.5 Imp gallon) external fuel tanks, allowing maximum external carriage of 2,498 litres (660 US gallons; 550 Imp gallons) at present, with potential for increase to 3,482 litres (920 US gallons; 766 Imp gallons). Oil capacity 19 litres (5.0 US gallons; 4.2 Imp gallons).

ACCOMMODATION: Crew of two in tandem, with co-pilot/ gunner in front seat and pilot at rear in AH-1W; crew stations are interchangeable in AH-1Z. Cockpit is heated, ventilated and air conditioned. Dual controls; lighting compatible with night vision goggles, and armour protection standard. Forward crew door on port side and rear crew door on starboard side, both upward-opening. Inflatable body and head restraint system by Simula of Phoenix, Arizona, nearing end of development in mid-1995; retrofit provisions installed in 1996 production, with system incorporated in 1997 production.

SYSTEMS: Three independent hydraulic systems, pressure 207 bar (3,000 lb/sq in), for flight controls and other services. Electrical system comprises two 28 V 400 A DC generators, two 24 V 34.5 Ah batteries and three inverters: main 115 V AC, 1 kVA, single-phase at 400 Hz, standby 115 V AC, 750 VA, three-phase at 400 Hz and a dedicated 115 V AC 365 VA single-phase for AIM-9 missile system. AiResearch environmental control unit.

AVIONICS: *Comms:* Two AN/ARC-210(V) radios, KY-58 TSEC secure voice set; AN/APX-100(V) IFF. AH-1Z to have AN/ARC-210(V) plus Embedded Communication Security (COMSEC) and embedded Mode 4 and Mode 5 IFF.

Radar: None at present installed, but possibility exists that pod-mounted version of Longbow radar will eventually be fitted as so-called Cobra Radar System. Mockup of possible installation displayed at Asian Aerospace show in Singapore, February 2002, this depicting pod-mounted radar sited above wing assembly. Longbow radar and associated AGM-114L RF Hellfire missile being offered to Taiwan, as part of an upgrade package for existing AH-1W fleet.

Flight: AN/ASN-75 compass set, AN/ARN-89B ADF, AN/ARN-118 Tacan on AH-1W and AN/ARN-153(V-4) Tacan on AH-1Z, AN/APN-154(V) radar beacon set and Teledyne AN/APN-217 Doppler-based navigation system being replaced by Embedded Global Positioning System/Inertial Navigation System (EGI), which adds Honeywell CN-1689(V) EGI and Rockwell Collins AN/ARN-153(V) Tacan. Installation began with Lot 9 production in October 1996 and also involved retrofit of earlier aircraft. AN/APN-194 radar altimeter on AH-1W, with AH-1Z having embedded radar altimeter in EGI. Rockwell Collins CDU-800 control/display unit and dual Rockwell Collins ICU-800 processors.

Instrumentation: Kaiser HUD compatible with PNVS-5 and Elbit ANVIS-7 night vision goggles. Helmet-mounted sighting/aiming device selected June 2002 in form of Thales TopOwl ; first helmet delivered to Northrop Grumman for testing in November 2002. Two 203 × 152 mm (8 × 6 in) Rockwell Collins flat-panel colour AMLCD multifunction displays installed on AH-1Z, which also has single 107 × 107 mm (4.2 × 4.2 in) dual-function display.

Mission: Tamam/Kollsman Night Targeting System (NTSF-65) comprising FLIR, laser range-finder/designator, TV camera, day/night video tracker and full in-flight boresighting; being retrofitted (within M-65 sighting system for TOW missiles) from 1993 (see Design Features); first redelivery in June 1994; alternative Boeing Electronic Systems NightHawk system also offered for export SuperCobras, following 1992–93 flight testing. Lockheed Martin AN/AAQ-30 Hawkeye multisensor electro-optical/IR targeting system to be installed in AH-1Z; this features large-aperture mid-wave FLIR, colour TV camera, laser range-finder/designator, laser spot tracker and on-gimbal inertial measurement system.

Self-defence: AN/APR-39(V) pulse radar signal detecting set, AN/APR-44(V) CW radar warning system, and AN/ALQ-144(V) IR countermeasures set. Dual AN/ALE-39 chaff system with one MX-7721 dispenser mounted on top of each stub-wing. Improved countermeasures suite in AH-1Z replaces AN/APR-39 and AN/APR-44 by AN/APR-39B(V)2 radar warning receiver. AN/AVR-2A laser detecting set (LDS) embedded in AN/AAR-47(V)2 missile warning set on AH-1Z, which also fitted with AN/ALE-47 chaff/flare dispensing subsystem. AH-1Z to have ITT AN/ALQ-211 SIRFC as optional alternative for export aircraft.

ARMAMENT: Electrically operated General Electric undernose A/A49E-7(V4) turret housing an M197 three-barrel 20 mm gun. A 750-round ammunition container is located in the fuselage directly aft of the turret; firing rate is approximately 650 rds/min; a 16-round burst limiter is incorporated in the firing switch. Either crew member can fire the gun, which can be slaved to a helmet-mounted sight/aiming device. Gun can be tracked 110° to each side, 18° upward, and 50° downward, but barrel length of 1.52 m (5 ft 0 in) makes it imperative that the M197 is centralised before wing stores are fired. Six wing stores stations with underwing attachments for up to four LAU-61A (19-tube), LAU-68A, LAU-68A/A, LAU-68B/A or LAU-69A (seven-tube) 2.75 in Hydra 70 rocket launcher pods; two CBU-55B fuel-air explosive weapons; four SUU-44/A flare dispensers; two M118 grenade dispensers; Mk 45 parachute flares; or two GPU-2A or SUU-11A/A Minigun pods.

Provision for carrying totals of up to eight TOW missiles, eight AGM-114 Hellfire missiles, two AIM-9L Sidewinder or AGM-122A Sidearm missiles, on outboard underwing stores stations. Canadian Marconi TOW/Hellfire control system enables AH-1W to fire both TOW and Hellfire missiles on same mission.

Following data applicable to AH-1W except where stated.

DIMENSIONS, EXTERNAL:

Main rotor diameter	14.63 m (48 ft 0 in)
Main rotor blade chord	0.84 m (2 ft 9 in)
Tail rotor diameter	2.97 m (9 ft 9 in)
Tail rotor blade chord	0.305 m (1 ft 0 in)
Distance between rotor centres	8.89 m (29 ft 2 in)
Wing span	3.28 m (10 ft 9 in)
Wing aspect ratio	3.7
Length: overall, rotors turning	17.68 m (58 ft 0 in)
fuselage	13.87 m (45 ft 6 in)
Width overall	3.28 m (10 ft 9 in)
Height: to top of rotor head	4.11 m (13 ft 6 in)
overall	4.44 m (14 ft 7 in)
Ground clearance, main rotor turning	2.74 m (9 ft 0 in)
Elevator span	2.11 m (6 ft 11 in)
Width over skids	2.24 m (7 ft 4 in)

AREAS:

Main rotor blades (each)	6.13 m^2 (66.0 sq ft)
Tail rotor blades (each)	0.45 m^2 (4.835 sq ft)
Main rotor disc	168.11 m^2 (1,809.6 sq ft)
Tail rotor disc	6.94 m^2 (74.70 sq ft)
Vertical fin	2.01 m^2 (21.70 sq ft)
Horizontal tail surfaces	1.41 m^2 (15.20 sq ft)

WEIGHTS AND LOADINGS:

Weight empty: AH-1W	4,953 kg (10,920 lb)
AH-1Z	5,579 kg (12,300 lb)
Mission fuel load (usable): AH-1W	946 kg (2,086 lb)
AH-1Z	2,615 kg (5,764 lb)
Max useful load (fuel and disposable ordnance):	
AH-1W	1,736 kg (3,828 lb)
AH-1Z	2,858 kg (6,300 lb)
Max T-O and landing weight:	
AH-1W	6,690 kg (14,750 lb)
AH-1Z	8,391 kg (18,500 lb)
Max disc loading: AH-1W	39.8 kg/m^2 (8.15 lb/sq ft)
AH-1Z	49.9 kg/m^2 (10.22 lb/sq ft)
Transmission loading at max T-O weight and power:	
AH-1W	4.42 kg/kW (7.26 lb/shp)
AH-1Z	4.29 kg/kW (7.05 lb/shp)

Bell AH-1Z third development aircraft ***NEW***/0576494

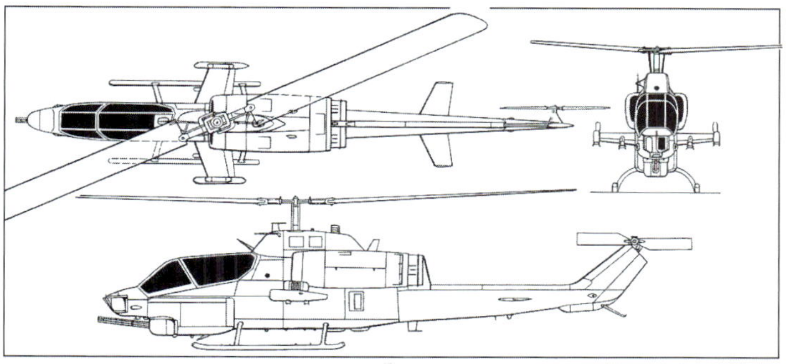

Bell AH-1W SuperCobra ***(Jane's/Dennis Punnett)*** 0507445

PERFORMANCE:

Never-exceed speed (VNE):	
AH-1W	190 kt (352 km/h; 219 mph)
AH-1Z	222 kt (411 km/h; 255 mph)
Max level speed at S/L	152 kt (282 km/h; 175 mph)
Max cruising speed: AH-1W	150 kt (278 km/h; 173 mph)
AH-1Z	160 kt (296 km/h; 184 mph)
Rate of climb at S/L, OEI: AH-1W	244 m (800 ft)/min
AH-1Z	327 m (1,074 ft)/min
Service ceiling	more than 4,270 m (14,000 ft)
Service ceiling, OEI	more than 3,660 m (12,000 ft)
Hovering ceiling: IGE	4,495 m (14,740 ft)
OGE	915 m (3,000 ft)
Mission radius, with 1,134 kg (2,500 lb) payload: AH-1Z	125 n miles (232 km; 144 miles)
Range at S/L with standard fuel, 20 minute reserves:	
AH-1W	280 n miles (518 km; 322 miles)
AH-1Z	370 n miles (685 km; 426 miles)
Max endurance with standard fuel: AH-1W	2 h 48 min
AH-1Z	3 h 30 min
g limits: AH-1W	+2.5/–0.5
AH-1Z	+2.8/–0.5

UPDATED

BELL QUAD TILTROTOR

TYPE: Multimission tiltrotor.

PROGRAMME: Bell announced in early 1999 that it was studying a proposed Quad TiltRotor (QTR) to meet Future Transport Rotorcraft (FTR) requirements. As projected, the aircraft would feature a fuselage approximately the size of that of a Lockheed Martin C-130-30, mated to two sets of wings, engines and tiltrotors from the Bell Boeing V-22 Osprey, the rear units mounted on stub wings to extend span and ensure adequate fuselage clearance. Rear tiltrotors could fold in cruising flight, with their engines providing supplemental thrust. The Quad TiltRotor would be able to accommodate up to 90 passengers, or an AH-64, AH-1Z, RAH-66, UH-1Y or UH-60 helicopter, or three HMMWVs, or up to eight 463L pallets.

Provisional specifications include VTOL maximum T-O weight 45,360 kg (100,000 lb); STOL maximum T-O weight 63,505 kg (140,000 lb); and a payload up to 18,144 kg (40,000 lb). The Quad TiltRotor was in the conceptual design phase in mid-2000, with water tunnel testing of a 1/48th scale model to visualise complex airflow patterns around the tandem wings and four tiltrotors completed.

In mid-2000 DARPA awarded Bell a US$400,000 phase 1 contract as part of a three-phase, US$6 million cost- sharing programme to study the feasibility of the QTR concept. Phase 1 involved a detailed technology study. In phase 2, a 1/14th scale hovering model of the QTR was test-flown. Phase 3 comprised wind tunnel tests to determine loads and aerodynamic performance of the full-size aircraft. The study is being undertaken by Bell's XWORX facility in Arlington, Texas.

The first of two demonstrators could be flown by 2006, with production deliveries starting in 2010. Potential customers include the US Marine Corps (to replace Sikorsky CH-53E helicopters and KC-130 Hercules) and USAF (MH-53J combat SAR/special forces helicopters).

UPDATED

Model of proposed Bell Quad TiltRotor *(Paul Jackson)* 0525825

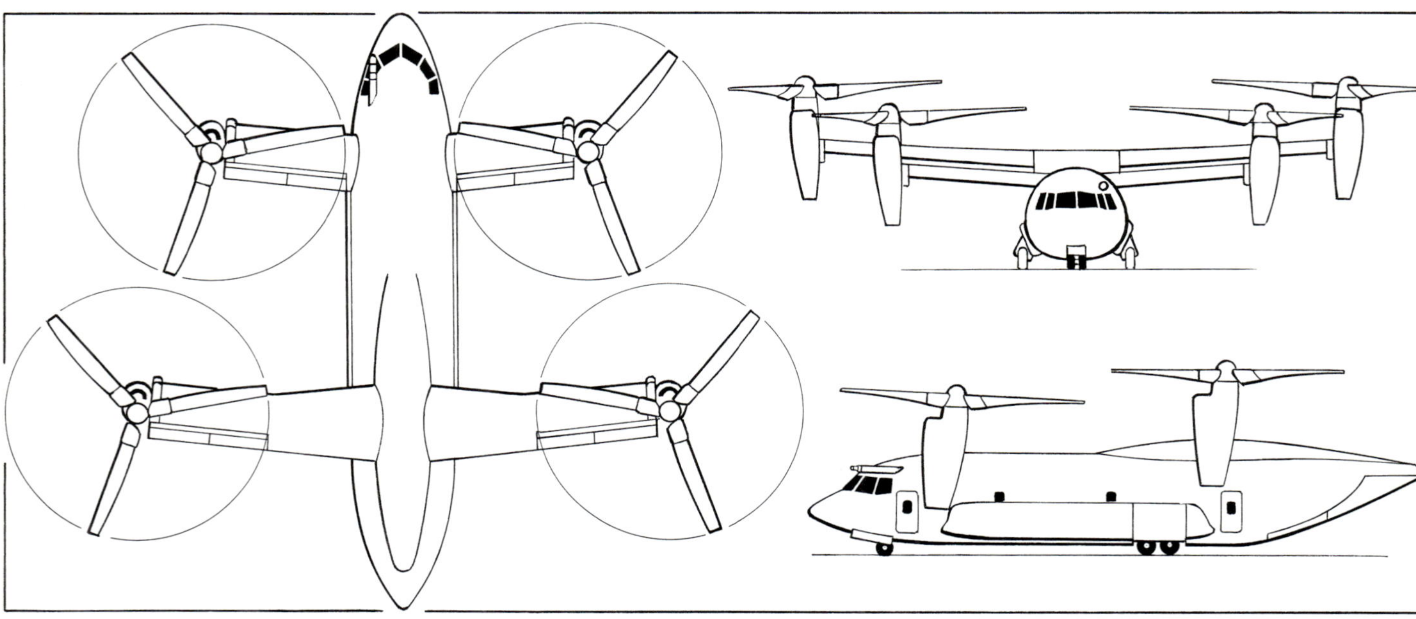

Bell Quad TiltRotor general arrangement *(James Goulding)* 0121269

BELL BOEING

BELL HELICOPTER TEXTRON AND THE BOEING COMPANY

Bell Boeing V-22 Joint Program Office, PO Box 70, Patuxent River, Maryland 20670-0070
Tel: (+1 301) 757 66 34
Web (1): http://www.boeing.com/rotorcraft/military/v22
Web (2): http://www.bellhelicopter.textron.com/products/tiltRotor/v22

BELL BOEING PROGRAMME DIRECTOR: Michael J Tkach
BELL PROGRAMME GENERAL MANAGER: Dennis Jarvi
V-22 PROGRAMME MANAGERS:
Col Craig Olson, USAF
Col Dick Wohls, USAF (Deputy for CV-22)
COMMUNICATIONS MANAGERS:
Gidge Dady (US Navy Public Affairs)
Carl Harris (Bell Helicopter Textron)
Douglas C Kinneard (Boeing)

UPDATED

BELL BOEING V-22 OSPREY

US Air Force designation: CV-22
US Navy designation: HV-22
US Marine Corps designation: MV-22
Manufacturer's model: 901

TYPE: Multimission tiltrotor.

PROGRAMME: Based on Bell/NASA XV-15 tiltrotor; initiated as US Department of Defense Joint Services Advanced Vertical Lift Aircraft (JVX), run by US Army, FY82; programme

Bell Boeing V-22 No. 10 dropping parachutists ***NEW***/0573398

DEVELOPMENT MILESTONES

Design began	26 Apr 1983
First order (prototypes)	2 May 1986
First flight	19 Mar 1989
First flight, pre-production	Feb 1997
First flight, production	30 Apr 1999
First delivery (Marine Corps)	14 May 1999

transferred to US Navy January 1983; 24 month US Navy preliminary design contract 26 April 1983; aircraft named V-22 Osprey January 1985; seven year full-scale development (FSD) began 2 May 1986 with order for six prototypes (Nos. 1, 3 and 6 by Bell at Arlington, Texas; Nos. 2, 4 and 5 by Boeing at Wilmington, Delaware) plus static test airframes.

Prototype (163911) first flew 19 March 1989; joined by four further aircraft by June 1991; sixth not flown; all now retired (last flight 27 March 1997 by 163913).

Osprey passed critical design review 13 December 1994; simultaneous defence review authorised V-22 production for both Marines and special forces, but latter version subsequently delayed, with decision to proceed with EMD phase not reached until January 1997. In meantime, contract for five low-rate initial production (LRIP) aircraft awarded June 1996.

All four EMD Ospreys flown in 1997–98. EMD Ospreys — see Current Versions (specific) — have significant changes from earlier aircraft, including substantial reduction in empty weight to approximately 14,800 kg (32,628 lb); aluminium cockpit cage, replacing titanium, but with smaller windows to preserve structural strength; upgraded flight controls; enhanced engine and drive system; improved tail unit construction (built by Aerostructures) including fibre placement aft fuselage; redesigned rotor system; absence of fin tuning weights; improved wing constructional techniques; redesigned wiring; and pyrotechnic escape hatches.

Manufacture of first LRIP MV-22B (165433) began 7 May 1997; splicing of three major fuselage sections took place in Philadelphia on 25 February 1998, with completed fuselage airlifted by C-17 to Arlington on 8 September 1998 for final assembly. Second LRIP aircraft also completed at Arlington, whereupon assembly transferred to new factory at Amarillo, Texas.

First flight of first LRIP MV-22B on 30 April 1999, followed by official roll-out and handover to US Marine Corps on 14 May, with delivery to New River later in May; subsequently to Patuxent River for flight testing. Next major milestone was operational evaluation; seven-month opeval began 2 November 1999 with first two LRIP MV-22Bs, which joined by two more by January 2000. These four aircraft accumulated 805 flight hours in 522 sorties by end of opeval in July 2000.

Opeval included trials in USS *Essex* with four MV-22Bs in first quarter 2000 and one aircraft temporarily sent to Kirtland AFB, New Mexico, in March 2000 for trials with USAF 58th Special Operations Wing, before programme halted on 8 April 2000 when type grounded following fatal crash of fourth LRIP MV-22B (165436) at Avra Valley Airport, Arizona. Subsequent investigation established most likely cause as 'power settling', a condition in which it becomes difficult to stop descent because of recirculating air from rotor downwash. Clearance to return to flight given on 25 May 2000, with opeval resuming on 5 June. Final phase included trials at China Lake, California, and New River, North Carolina, before being concluded in late July 2000.

Further brief grounding order imposed on 11 aircraft (four EMD and seven LRIP machines) on 25 August 2000 in wake of problem encountered with retaining assembly for one of the interconnect driveshaft couplings; this came loose in flight, necessitating precautionary landing and inspection before flight operations resumed on 1 September 2000.

Subsequent events included final assessment by Multiservice Operational Test Team that MV-22 was operationally effective and suitable, with announcement on 8 November 2000 recommending full fleet introduction. Thereafter, decision on whether to authorise full-rate production was expected in December, but was indefinitely put on hold following loss of eighth LRIP MV-22B (165440) near New River on 11 December and imposition of grounding order. Cause of accident reported to be hydraulic system failure in first instance, compounded by error in software inputs to flight control system.

At time of grounding on 11 December 2000, V-22 total flight time was 3,884 hours, during which aircraft had demonstrated speed of 342 kt (633 km/h; 394 mph), 7,620 m (25,000 ft) height, 27,442 kg (60,500 lb) MTOW and 3.9 *g* load factor; more than 1,000 hours added between late May 2002 and December 2003, after resumption of flight testing.

V-22 remained grounded throughout 2001, although production continued at low rate while various studies and reports prepared. In December 2001, it was decided to continue with the Osprey programme, subject to successful completion of further testing during 2002–04; in meantime, Bell Boeing directed to store total of 19 V-22s, comprising already completed examples and some to come from assembly line during 2002. On conclusion of test programme, these stored aircraft to be upgraded and fitted with latest safety features.

Flight testing resumed on 29 May 2002 at Patuxent River, with eventual total of seven (two EMD and five LRIP) aircraft to participate in 18-month programme that will evaluate

Bell Boeing V-22 Osprey deck landing ***NEW***/0573395

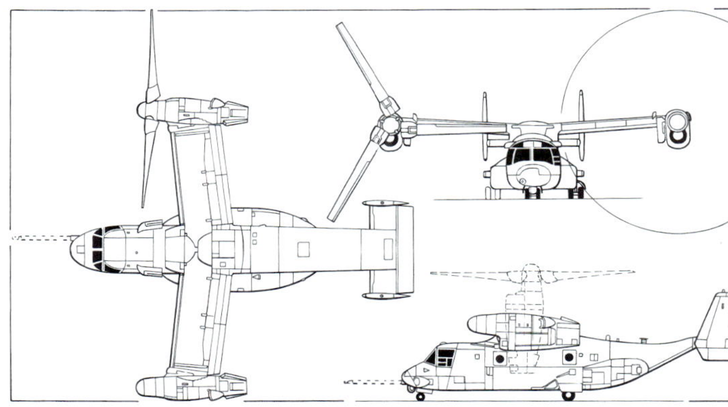

Bell Boeing MV-22B Osprey multimission tiltrotor with retractable refuelling probe ***(James Goulding)*** ***NEW***/0532096

improved hydraulic and electrical systems, plus upgraded software; test objectives also to include exploration of aspects including vortex ring state boundaries, formation flying and low-speed hover and landing conditions. Test programme will involve 1,800 flight hours, with icing, cargo handling and radar warning systems also to be studied. Flight activity initially used only the fourth EMD aircraft, but this joined by 21st Osprey (165443) which flew for first time at Amarillo, Texas, on 7 September 2002 and was flown to Patuxent River on 12 October. One week later, the number of flyable aircraft at Patuxent River increased to three, with the first post-grounding sortie by the second EMD machine. Subsequent developments include assignment of additional aircraft to test effort during 2003; these include 22nd and 24th Ospreys, with latter departing from Patuxent River on 7 November at start of six-month period of cold weather trials to be conducted from Canadian Forces Base Shearwater, Nova Scotia. At same time, aircraft 10 and 22 embarked on USS *Bataan* for shipboard testing (including assessment of effects of rotor downwash), following on from earlier deck-handling trials on USS *Iwo Jima* by aircraft 10 in mid-January 2003.

Other notable tests undertaken in 2003 included three-week visit by aircraft 21 to Fort Bragg, North Carolina, for series of airdrop trials; continuing expansion of flight envelope, with emphasis on high rates of descent; and exploration of vortex ring state characteristics. Despite a brief hiatus in March, when the test fleet was grounded for 10 days to replace suspect hydraulic lines, a Defence Acquisition Board programme review on 10 May reported favourably, although it did not authorise an increase in production rate. Following resumption of flight testing in May 2002, Osprey fleet passed 500th flight hour on 30 May 2003, this total rising to more than 1,000 hours by December 2003.

At Edwards AFB, the first CV-22 EMD aircraft returned to flight duty on 11 September 2002; second CV-22 EMD aircraft placed in Benefield Anechoic Chamber at Edwards for three-month assessment of electronic warfare antennae configuration that was completed on 5 October 2002; it returned to flight status on 14 July 2003. These will be joined in 2005 by two new-build test aircraft being purchased with FY02 funding; and by another LRIP aircraft that is to be modified to CV-22 configuration at cost of US$61 million.

US Marine Corps not now expecting to make decision to move to full-rate production before 2006–07, but will continue purchasing Ospreys at rate of 11 per year (including some CV-22Bs); MV-22B version will be to progressively improved standard, starting with Block A in 2003, incorporating re-routed and separated hydraulic lines and electrical wiring, as well as a redesigned nacelle assembly. The redesigned nacelle will initially be installed on aircraft 34 as part of the Block A upgrade, which affects the 25th to 49th Ospreys; with effect from the 50th aircraft, Block A enhancements will be incorporated during production. Block B standard MV-22 will be available from 2005 and will possess enhanced maintenance access, while the proposed Block C version will feature operational improvements, including telescopic in-flight refuelling probe, revised environmental control system and other changes intended to improve payload and performance capabilities.

Block A enhancements will be retrofitted to LRIP aircraft that have already been completed; these include eight at New River plus others at Amarillo production facility. This work is expected to begin in 2003, with all surviving LRIP aircraft having been upgraded to Block A standard by 2007.

CURRENT VERSIONS (SPECIFIC): **No.7/164939:** Assembly began 15 February 1995; fuselage sections mated at Boeing's Philadelphia plant on 1 August 1995; followed two months later by wing and nacelle mating at Fort Worth. Fuselage to Fort Worth for fitting of wings and engines. Ground vibration testing completed in late July 1996: tests of hydraulic lines concluded soon after, with integrated functional testing commencing in August. First flight in helicopter mode 5 February 1997 (at Fort Worth); first transition to conventional flight 6 March 1997; aircraft to V-22 Integrated Test Team at Patuxent River on 15 March 1997; ensuing objectives (structural, load and vibration tests) completed by mid-1998. Prepared at Arlington for trials of CV-22 version's auxiliary fuel tanks and terrain-following/terrain-avoidance radar system; modification work began 1 July 1999; flight status regained on 28 February 2000. Grounded throughout 2001, but resumed test

Bell Boeing V-22 No. 9 during flare dropping trials at Edwards AFB in 2004 *(Kevin Kidd, USAF)* ***NEW***/1047801

flying at Edwards on 11 September 2002. During 2003, trials work included evaluation of multimode radar, notable highlights occuring on 4 April, when it achieved low altitude target for aeroplane flight mode, and also demonstrated safe operation of terrain-following radar mode.

No. 8/164940: First flight 15 August 1997; to Patuxent River on 13 September 1997 and allocated to propulsion and systems testing and envelope expansion, including high-altitude trials and external loads demonstrations in 1998. During latter, it set new unofficial world record in August 1998 by carrying 4,536 kg (10,000 lb) external load at 220 kt (407 km/h; 253 mph). Grounded throughout 2001, but returned to flight at Patuxent River on 19 October 2002.

No. 9/164941: First flight 17 July 1997; to Patuxent River 30 October 1997; allocated to validation of FLIR, navigation and other mission systems, as well as government technical evaluation. Had achieved total of 318 flight hours in 140 sorties, when arrived at Arlington on 7 June 1999 to be remanufactured as prototype for CV-22 special missions version, at cost saving of US$50 million; work completed in mid-2000, with aircraft joining 418th Flight Test Squadron at Edwards AFB, California, on 18 September 2000 for 18-month trials programme. Grounded throughout 2001; used for electronic warfare testing in Benefield Anechoic Facility at Edwards during third quarter of 2002 and eventually resumed flight test duty on 14 July 2003, thereafter being used to evaluate mission-related equipment, including defensive countermeasures avionics and multimode radar. Chaff/flare trials at Edwards AFB, September-October 2004.

No. 10/164942: Wing and fuselage mating accomplished at Fort Worth in September 1996. To Patuxent River 15 February 1998, following first flight on 15 January; underwent modification in mid-1998, before participating (with No. 9) in operational evaluation during September and October 1998; performed sea trials in USS *Saipan* in January-February and USS *Tortuga* and *Saipan* in August-September 1999. First V-22 to resume flying after 17-month grounding, 29 May 2002.

CURRENT VERSIONS (GENERAL): **MV-22B:** Basic US Marine Corps transport; original requirement for 552 (now 360), to replace CH-46 Sea Knight and CH-53 Sea Stallion. First delivery (to Patuxent River via New River) in late May 1999 with three more handed over to test unit HMX-1 by January 2000. First unit was planned to be VMMT-204 at New River, North Carolina, which officially redesignated (from HMT-204) on 10 June 1999; received four opeval aircraft from HMX-1 at end of July 2000 to function as fleet replacement squadron and also train USAF pilots; training operations interrupted by subsequent grounding order. Task of conducting exhaustive operational evaluation of MV-22B during 2004 assigned to VMX-22, which activated at New River on 28 August 2003. The first deployable squadron was expected to be VMM-264, also at New River, but this may have changed as consequence of much-delayed fleet introduction. USMC plans called for 18 regular and four Reserve MV-22 squadrons (each with 12 aircraft), plus training unit already mentioned. IOC now set for 2005.

HV-22B: US Navy combat search and rescue (CSAR), special warfare and fleet logistics model. Requirement for 48 (originally 50). In April 2004, it was revealed that HV-22 designation to be dropped, with US Navy instead to receive suitably modified version of MV-22.

CV-22B: US Air Force long-range special missions aircraft to replace MH-53M helicopter and augment MC-130 (Hercules) with Air Force Special Operations Command (AFSOC). Original requirement for 80 reduced to 55 then 50. Current procurement plan calls for purchases to continue to FY17, but AFSOC urging accelerated programme to conclude by FY12. CV-22B should carry 12 troops or 1,306 kg (2,880 lb) internal cargo over 520 n mile (964 km; 599 mile) radius at 250 kt (463 km/h; 288 mph), with ability to hover OGE at 1,220 m (4,000 ft) at 35°C (95°F). EMD go-ahead authorised in January 1997, with award of US$490 million contract; critical design review completed in mid-December 1998.

Key feature of CV-22B will be ITT Avionics AN/ALQ-211 SIRFC (suite of integrated RF countermeasures); US$20.7 million LRIP contract awarded by Bell Boeing in January 2001 for four systems plus spares and engineering support. First SIRFC was due for delivery in 2002, with full-rate production decision dependent on successful outcome of flight testing.

V-22 also under consideration for USAF combat search and rescue (CSAR) mission as potential replacement for Sikorsky HH-60G.

CUSTOMERS: Production of 458 Ospreys planned, comprising 360 MV-22Bs for Marine Corps, 50 CV-22Bs for Air Force and 48 MV-22s for Navy; see table for details of initial procurement. Bell Boeing has proposed V-22 as suitable for UK's joint RAF/Navy Support Amphibious Battlefield Rotorcraft requirement, for which over 40 will be needed. Suitably modified V-22 could also satisfy Royal Navy's Future Organic Airborne Early Warning (FOAEW) requirement.

COSTS: Estimated cost (1991) to complete full-scale development, US$2,750 million. Unit cost said to be US$32.3 million (November 1997). Bell Boeing received US$1.5 billion contract in August 2002 to produce 20 tiltrotors, comprising nine MV-22s in each of FY01 and FY02 for delivery by September 2004, plus two production representative CV-22 test vehicles to be completed by October 2005. FY04 request includes US$817 million for nine MV-22s and two CV-22s. Separate US$490 million contract modification awarded by USN in December 1996 for EMD of CV-22 special operations version; acquisition costs of 50 CV-22s estimated at US$3.72 billion based on FY01-08 procurement plan. Flyaway cost of MV-22B reported to be US$68.4 million (September 2002); this stated to have risen to US$74 million in late 2003, despite start of implementation of cost-cutting efforts intended to reduce unit cost to US$58 million.

DESIGN FEATURES: Unconventional design, with proprotors and engines mounted at tips of wings. Fuselage optimised for transport, featuring upswept rear, with loading ramp and twin fins of moderate sweepback. High-mounted, constant-chord wings with slight forward sweep; unswept tailplane; prominent landing gear sponsons.

OSPREY PROCUREMENT

Year	Lot	MV-22B	CV-22B	Total
FY97	1	5		5
FY98	2	7		7
FY99	3	7		7
FY00	4	11		11
FY01	5	9		9
FY02	6	9	2	11
FY03	7	11		11
FY04	8	9	2	11
FY05	9	8	3	11
Total		**76**	**7**	**83**

Notes: FY02 CV-22B are dedicated test aircraft

Bell Boeing V-22 Osprey ***NEW***/0573397

During vertical take-off, wing begins to produce lift and ailerons, elevators and rudders become effective at between 40 and 80 kt (74 and 148 km/h; 46 and 92 mph). At this point, rotary-wing controls are gradually phased out by the flight control system. At approximately 100 to 120 kt (185 to 222 km/h; 115 to 138 mph), wing is fully effective and cyclic pitch control of proprotors is locked out.

In conversion from aeroplane flight to hover, fuselage and wing are free to remain in level attitude, eliminating tendency for wing to stall as speed decreases. Rotor lift fully compensates for decrease in wing lift. Because of great variability between aircraft and nacelle attitude, conversion corridor (range of permissible airspeeds for each angle of nacelle tilt) is very wide (about 100 kt; 185 km/h; 115 mph).

Engines are connected by shaft through wing. Under dual engine operations, shaft transmits very little power, but if one engine is lost, half remaining power is transferred to opposite proprotor. In event of double engine failure, can maintain proprotor rpm while descending without power. Pilot has options of making wingborne or rotorborne descent.

Engines, transmission and proprotors tilt through 97° 30' between forward flight and steepest approach gradient or tail-down hover; cross-shaft keeps both proprotors turning after engine loss. Three-blade, contrarotating proprotors have special high-twist tapered format blades with elastomeric bearings and powered folding mechanisms; separate swashplates produce respectively yaw and fore-and-aft translation in hover and sideways flight in level attitude; tip speed 202 m (662 ft)/s.

Wing-fold sequence from helicopter mode involves power-folding of blades parallel to wing leading-edge, tilting engine nacelles down to horizontal and rotating entire wing/engine/proprotor group clockwise on stainless steel carousel to lie over fuselage; entire procedure for stowage takes about 90 seconds and MV-22B occupies same amount of deck space as Sikorsky CH-53E.

FLYING CONTROLS: Three-lane fly-by-wire (Moog actuators) with automatic stabilisation, full autopilot and formation-flying modes. Conventional aeroplane stick, rudder pedals and throttle automatically function as in helicopter cyclic stick, yaw pedals and collective control. Automatic control of configuration change during transition and of transfer of control from aerodynamic surfaces to rotor-blade pitch changing; flaperons and ailerons droop during hover to reduce download on wing. Pilot also controls nacelle angle. Failure of one nacelle actuator automatically results in both nacelles reverting to helicopter mode for a vertical or run-on emergency landing.

Rotors have separate cyclic control swashplates for sideways flight and fore-and-aft control (symmetrical for forward and rearward flight and differential for yaw) in hover. Lateral attitude controlled in hover by differential rotor thrust. Integrated electronic cockpit with six electronic display screens; helicopter-style control columns rather than aileron wheels, but left-handed power levers move forward for full power in opposite sense to helicopter collective lever. Using nacelle control provides additional method of manoeuvring, completely independent of fuselage attitude or cyclic pitch. Minimum time to accomplish a full conversion from hover to aeroplane flight mode is 12 seconds.

STRUCTURE: Approximately 43 per cent of airframe is composites; main composites are Hercules IM-6 graphite/epoxy in wing and AS4 in fuselage and tail; proprotor blades of graphite/glass fibre; nacelle cowlings and pylon supports are of GFRP. Cabin has composites floor panels and aluminium window frames. Crew seats of boron carbide/polyethylene laminate.

Different design approach adopted for EMD and subsequent production aircraft, whereby integrated product teams (IPTs) were tasked with using best available material to reduce cost and weight and simultaneously improve quality. In consequence, fuselage is now a hybrid structure, with mainly aluminium frames and composites skins. Benefits of IPT process are a 1,200 kg (2,645 lb) reduction in weight and a reduction in cost of over 22 per cent, with part count falling by 36 per cent and fastener count by 34 per cent.

High-strength wing, torsion box made up from one-piece upper and lower skins with moulded ribs and bonded stringers; two-segment graphite single-slotted flaperons with titanium fittings; three-segment detachable leading-edge of aluminium alloy with Nomex honeycomb core. Wing locking and unlocking with Lucas Aerospace actuators; fuselage sponsons contain landing gear, air conditioning unit and fuel; tail unit of Hercules AS4 graphite/epoxy, built by Bell from first EMD aircraft onwards. Bell also contributes wing, nacelles, proprotor, ramp, overwing fairing and transmission systems and integrates engines. Boeing responsible for fuselage, landing gear, electric and hydraulic systems and integrates avionics.

LANDING GEAR: Dowty hydraulically actuated retractable tricycle type, with twin wheels and oleo-pneumatic shock-absorbers on each unit, Menasco Canada steerable nose unit. Main gear accommodated in sponsons on lower sides of centre-fuselage. Dowty Toronto two-stage shock-absorption in main gear is designed for landing impacts of up to 3.66 m (12 ft)/s normal, 4.48 m (14.7 ft)/s maximum, and has been drop tested to 7.32 m (24 ft)/s. Nosewheels retract rearward, mainwheels forward (was rearward on FSD aircraft only). Manual and nitrogen pressurised standby systems for emergency extension. Parker Bertea wheels and multidisc hydraulic carbon brakes. Main tyres 8.50-10 (12 ply) tubeless; nose tyres 18×5.7-8 (14 ply) tubeless.

POWER PLANT: Two Rolls-Royce T406-AD-400 (501-M80C) turboshafts, each with T-O and intermediate rating of 4,586 kW (6,150 shp) and maximum continuous rating of 4,392 kW (5,890 shp), installed in Bell-built hydraulically actuated tilting nacelles at wingtips and

Bell Boeing V-22 Osprey ***NEW***/0573396

Bell Boeing V-22 Osprey ***NEW***/0573394

driving a three-blade proprotor incorporating substantial proportion of graphite/epoxy and glass fibre materials. Nacelle configuration subject to redesign in 2002, with new unit due to first be installed on aircraft number 34 for test purposes; will subsequently be fitted to new production with effect from 50th Osprey.

Engine output transferred to each rotor via proprotor gearbox which also drives tilt-axis gearbox, each of which absorbs 315 kW (423 shp) for electric and hydraulic pressure generation. Tilt-axis gearboxes connected by segmented shaft, driving (in fuselage) single mid-wing gearbox and APU. Transmission T-O rating 3,408 kW (4,570 shp) in MV-22; 3,706 kW (4,970 shp) in USN and USAF versions. OEI rating 4,415 kW (5,920 shp); rotor rpm remains constant. With total engine failure, aircraft generators can maintain essential systems on 67 per cent of rotor rpm. Transmission has 30 minute run dry capability.

Each nacelle has a Honeywell IR emission suppressor at rear. Air particle separator and Lucas inlet/spinner ice protection system for each engine. Lucas Aerospace FADEC for each engine, with analogue electronic back-up control. Pratt & Whitney originally named as second production source for engines, starting with production Lot 5.

Fuel capacity varies according to role; basic internal usable fuel (JP-5) in four crash-resistant, self-sealing (nitrogen pressurised) ILC Dover flexible fuel tanks: one 1,809 litre (478 US gallon; 398 Imp gallon) forward cell in each sponson and a 333 litre (88.0 US gallon; 73.3 Imp gallon) feed tank in each outer wing; basic fuel (MV-22) 4,285 litres (1,132 US gallons; 943 Imp gallons). Additional fuel for long range contained in 1,196 litre (316 US gallon; 263 Imp gallon) cell in rear of starboard sponson (MV-22 option 5,481 litres; 1,448 US gallons; 1,206 Imp gallons) and four 280 litre (74.0 US gallon; 61.6 Imp gallon) auxiliary cells in each wing leading-edge (CV-22 baseline 7,722 litres; 2,040 US gallons; 1,699 Imp gallons). Self deployment aided by up to two 3,028 litre (800 US gallon; 666 Imp gallon) auxiliary tanks in cabin. (Not all versions have all tanks.) Pressure refuelling point in starboard sponson leading-edge; gravity point in upper surface of each wing. Simmonds fuel management system. In-flight refuelling probe in lower starboard side of forward fuselage. Fixed 4.88 m (16 ft) probe installed initially, but will be replaced by retractable 5.49 m (18 ft) probe on both CV-22B and MV-22B.

ACCOMMODATION: Normal crew complement of pilot (in starboard seat), co-pilot and crew chief in USMC variant. USAF CV-22B will have third seat for flight engineer. Flight crew accommodated on Simula Inc crashworthy armoured seats capable of withstanding strikes from 0.30 in armour-piercing ammunition, 30 *g* forward and 14.5 *g* vertical decelerations. Flight deck has overhead and knee-level side transparencies in addition to large windscreen and main side windows, plus an overhead rearview mirror.

Main cabin can accommodate up to 24 combat-equipped troops, on IAI — Golan Industries inward-facing crash-worthy foldaway seats, plus two gunners; up to 12 litters plus medical attendants; or a 9,070 kg (20,000 lb) cargo load with energy absorbing tiedowns. Cargo handling provisions include a 907 kg (2,000 lb) capacity cargo winch and pulley system and removable roller rails. Main cabin door at front on starboard side, top portion of which opens upward and inward, lower portion (with built-in steps) downward and outward. Full-width rear-loading ramp/door in underside of rear fuselage, operated by Parker Bertea hydraulic actuators. Emergency exit windows on port side; escape hatch in fuselage roof aft of wing.

SYSTEMS: Environmental control system, utilising engine bleed air; control unit in rear of port main landing gear sponson. For NBC protection, ECS provides 0.062 bar (0.9 lb/sq in) flight deck pressure differential, and 0.048 bar (0.7 lb/sq in) in cabin. Three hydraulic systems (two independent main systems and one standby), all at operating pressure of 345 bar (5,000 lb/sq in), with Parker Bertea reservoirs.

Electrical power initially supplied by two Leland 40 kVA constant frequency AC generators, although these being replaced by Smiths Aerospace unit from 2004 onwards; V-22 also has two 50/80 kVA variable frequency DC generators (one driven by APU), rectifiers, and a 24 Ah lead-acid battery. Latter provides 20 minutes' emergency flight power.

GE Aerospace triple redundant digital fly-by-wire flight control system, incorporating triple primary FCS (PFCS) and triple automatic FCS (AFCS) processors, and triple flight control computers (FCC) each linked to a MIL-STD-1553B databus; two PFCSs and one AFCS are fail-operational. FBW system signals hydraulic actuation of flaperons, elevator and rudders, controls aircraft transition between helicopter and aeroplane modes, and can be programmed for automatic management of airspeed, nacelle tilting and angle of attack.

Bell Boeing V-22 Osprey carrying artillery piece ***NEW***/0573424

Bell Boeing V-22 Osprey ***NEW***/0573425

FCCs provide interfaces for swashplate, conversion actuator, flaperon, elevator, rudder and engine nacelle primary actuators, flight deck central drive, force feel, and nosewheel steering. Dual 1750A processors for PFCS and single 1750A for AFCS incorporated in each FCC. Non-redundant standby analogue computer (in development aircraft only) provides control of aircraft, including FADEC and pylon actuation, in the event of FBW system failure.

Hamilton Sundstrand Turbomach T-62T-46-2 224 kW (300 shp) APU, in rear portion of wing centre-section, provides power for mid-wing gearbox which, in turn, drives two electrical generators and an air compressor. Anti-icing of windscreens and engine air intakes; de-icing of proprotors and spinners. Clifton Precision combined oxygen (OBOGS) and nitrogen (OBIGGS) generating systems for cabin and fuel tank pressurisation respectively. Systron Donner pneumatic fire protection systems for engines, APU and wing dry bays.

AVIONICS: *Comms:* VHF/AM-FM, HF/SSB and (USAF only) UHF secure voice com; IFF. CV-22B will have four DCS2000 radios, combining AN/ARC-210 transceiver and KY-58 communications security encoder. Crash position indicator also to be installed on CV-22B.

Radar: USAF and USN only. Raytheon AN/APQ-174D terrain-following, terrain-avoidance multifunction radar in offset (to port) nose thimble for USN; advanced version, designated AN/APQ-186, with lower-altitude TF capability, in USAF aircraft.

Flight: AN/ARN-153(V) Tacan, AN/ARN-147 VOR/ILS, AHRS, AN/APN-194(V) radar altimeter, OA-8697/ARC UHF/VHF automatic direction-finder, lightweight inertial navigation system (LWINS), miniature airborne GPS receiver (MAGR) and digital map displays; Jet Inc ADI-350W standby attitude indicator; L-3 Communications data acquisition and storage system. Two Control Data AN/AYK-14 mission computers, with Boeing/IBM software. Low probability of intercept/detection radar altimeter may be added to CV-22B version at later date.

Instrumentation: Elbit/BAE Systems North America AN/AVS-7 pilots' night vision and integrated display system. Four full-colour. multifunction displays (MFDs) in cockpit provide pilots with primary flight symbology to control and navigate aircraft, plus video imagery such as FLIR and digital map data; these originally relied on CRTs, but replaced by EFW Inc flat-panel active matrix liquid crystal displays (AMLCDs) on US Marine Corps MV-22Bs starting in FY99, followed by USN HV-22s. AMLCDs to be installed on USAF CV-22B from outset. Lighting compatible with night vision goggles. Additional AMLCD control display unit and engine indication and crew alerting system (CDU/EICAS) in centre of console.

Mission: Raytheon AN/AAQ-27 FLIR incorporating laser range-finder in undernose fairing, with wide and narrow fields of view. Additional equipment expected to satisfy unique mission requirements of special operations forces and Navy combat search and rescue.

Self-defence: Honeywell AN/AAR-47 missile warning system; AN/APR-39A radar warning system; IR warning system; BAE Systems North America AN/ALE-47 countermeasures dispenser system (CMDS). CV-22B for USAF to utilise ITT Avionics AN/ALQ-211 suite of integrated RF countermeasures (SIRFC), incorporating radar warning receiver, ESM radar location and jammers and will also have AN/AVR-2A laser detector

Bell Boeing V-22 Osprey and Boeing Hornet ***NEW***/0573426

system and a second, forward-firing, AN/ALE-47 dispenser system. Dedicated electronic warfare display (DEWD) unit also under development for CV-22B by Meggitt Avionics.

EQUIPMENT: Provision for internally stowed Goodrich rescue hoist over forward (starboard) cabin door on CV-22B. MV-22B initially to have hoist in cabin, for deployment through side door, but now expected to position hoist near the rear ramp. USMC and USAF Ospreys have three fast rope/rope ladders to facilitate insertion/extraction operations.

ARMAMENT: Provision stipulated in December 1994 for nose cannon; budget constraints resulted in this being deleted, but consideration again given in late 1998 to installing nose-mounted turreted defensive gun on MV-22B and this was evaluated in 2000, with General Dynamics Armament Systems 12.7 mm weapon based on GAU-19/A selected in September 2000. However, plans to install integrated gun again shelved in late 2002 on cost grounds, with decision taken to rely instead on manually operated 0.5 in machine guns.

DIMENSIONS, EXTERNAL:	
Rotor diameter, each	11.61 m (38 ft 1 in)
Rotor blade chord: at root	0.90 m (2 ft 11½ in)
at tip	0.56 m (1 ft 10 in)
Wing span: excl nacelles	14.02 m (46 ft 0 in)
incl nacelles	15.52 m (50 ft 11 in)
Width: rotors turning	25.40 m (83 ft 4 in)
stowed	5.61 m (18 ft 5 in)
Wing chord, constant	2.54 m (8 ft 4 in)
Wing aspect ratio, excl nacelles	5.5
Distance between proprotor centres	14.20 m (46 ft 7 in)
Length: fuselage, excl probe	17.47 m (57 ft 4 in)
overall, wings stowed/blades folded	19.08 m (62 ft 7 in)
Height: over tailfins	5.38 m (17 ft 8 in)
wings stowed/blades folded	5.56 m (18 ft 3 in)
overall, nacelles vertical	6.63 m (21 ft 9 in)
Tail span, over fins	5.61 m (18 ft 5 in)
Wheel track (c/l of outer mainwheels)	4.62 m (15 ft 2 in)
Wheelbase	7.62 m (25 ft 0 in)
Nacelle ground clearance, nacelles vertical	1.32 m (4 ft 4 in)
Proprotor ground clearance, nacelles vertical	6.35 m (20 ft 10 in)
Dorsal escape hatch: Length	1.02 m (3 ft 4 in)
Width	0.74 m (2 ft 5 in)
DIMENSIONS, INTERNAL:	
Cabin: Length	7.37 m (24 ft 2 in)
Max width	1.80 m (5 ft 11 in)
Max height	1.83 m (6 ft 0 in)
Usable volume	24.3 m^3 (858 cu ft)
AREAS:	
Rotor discs, each	105.36 m^2 (1,134.1 sq ft)
Rotor blades: each	4.05 m^2 (43.58 sq ft)
total	24.30 m^2 (261.5 sq ft)
Wing, total incl flaperons and fuselage centre-section	35.49 m^2 (382.0 sq ft)
Flaperons, total	8.25 m^2 (88.80 sq ft)
Fins, total	21.63 m^2 (232.80 sq ft)
Rudders, total	3.27 m^2 (35.20 sq ft)
Tailplane	8.22 m^2 (88.50 sq ft)
Elevators, total	4.79 m^2 (51.54 sq ft)
WEIGHTS AND LOADINGS:	
Weight empty	15,032 kg (33,140 lb)
Max fuel weight: MV-22 baseline	3,493 kg (7,700 lb)
MV-22 option	4,468 kg (9,850 lb)
CV-22 baseline	6,282 kg (13,850 lb)
CV-22 with cabin tanks	11,970 kg (26,390 lb)
Max internal payload (cargo)	9,072 kg (20,000 lb)
Cargo hook capacity: single	4,536 kg (10,000 lb)
two hooks (combined weight)	6,804 kg (15,000 lb)
Rescue hoist capacity	272 kg (600 lb)
Normal mission T-O weight: VTO	21,545 kg (47,500 lb)
STO	24,947 kg (55,000 lb)
Max STO weight for self-ferry	27,442 kg (60,500 lb)
Max floor loading (cargo)	1,464 kg/m^2 (300 lb/sq ft)
Max disc loading: VTO	111.3 kg/m^2 (22.79 lb/sq ft)
STO	130.2 kg/m^2 (26.67 lb/sq ft)
Transmission loading at max T-O weight and power	4.03 kg/kW (6.62 lb/shp)

PERFORMANCE:

Max level speed at S/L .... 275 kt (509 km/h; 316 mph)
Max cruising speed: at S/L, helicopter mode .... 100 kt (185 km/h; 115 mph)
Max forward speed with max slung load .... 214 kt (396 km/h; 246 mph)
Max rate of climb at S/L: vertical .... 332 m (1,090 ft)/min
inclined .... 707 m (2,320 ft)/min
Service ceiling .... 7,925 m (26,000 ft)
Service ceiling, OEI .... 3,441 m (11,300 ft)
Hovering ceiling OGE .... 4,331 m (14,200 ft)
T-O run at normal mission STO weight .... less than 152 m (500 ft)
Range:
amphibious assault .... 515 n miles (953 km; 592 miles)
VTO with 4,536 kg (10,000 lb) payload .... 350+ n miles (648+ km; 403+ miles)
VTO with 2,721 kg (6,000 lb) payload .... 700+ n miles (1,296+ km; 806+ miles)
STO with 4,536 kg (10,000 lb) payload .... 950+ n miles (1,759+ km; 1,093+ miles)
STO at 27,442 kg (60,500 lb) self-ferry gross weight, no payload .... 2,100 n miles (3,892 km; 2,418 miles)
g limits .... +4/–1

UPDATED

BLENNTEC

BLENNTEC (FORMERLY UPSHIP CORPORATION)

5198 Highway 84, Elba, Alabama 36323
Tel/Fax: (+1 334) 897 61 32
e-mail: airship@alaweb.com
Web: http://www.airship.us
PRESIDENT AND TECHNICAL DIRECTOR: Jesse Blenn

Company renamed BlennTEC in 2003, although UPship retained as airship trademark type name. First to be built will be a two-seat, 32 m (105 ft) example, planned to be marketed under new FAA Light Sport Aircraft standards for approximately US$75,000. Will have all main features of eight-seat, 50 m UPship described below, including bow thruster, semi-rigid construction and ground handling systems; will also be used for pilot training and to test cargo handling systems for Brazilian cargo market. This will allow further development and verification of 50- and 90-tonne payload cargo designs currently in progress (lengths 180 and 216 m; 590.6 and 708.7 ft respectively.

Additional efforts in 2003 included co-operative agreement in Brazil, and design work on a six-seat, 45 m (147.6 ft) UPship planned for construction by a Malaysian company for Far East market. Concurrently, design work is proceeding on an improved internal combustion engine, for which patents were pending in mid-2004. Initial product will be a 72 cc, 3.7 kW (5 hp) industrial version, expected sales revenue from which will be used to help finance airship development.

UPDATED

BLENNTEC UPSHIP 50 METRE

TYPE: Helium semi-rigid.

PROGRAMME: UPship design started 1989, original 750-001 later being enlarged as two-seat, 32 m proof-of-concept vehicle; further enlarged in 1999 with lower-drag envelope profile and three seats (or two occupants and heavier commercial or scientific payload) for planned US Type Certification. Now focusing again on initial improved 32 m (105 ft) version (UPship 001) to market under FAA Light Sport Aircraft rules from 2004. First version planned later for full (Argentinian) Type Certificate will be 50 m eight-seater.

CURRENT VERSIONS: **50 m UPship**: *Following description applies to this version except where indicated.*

CUSTOMERS: To be built and operated as aerial platforms for customer's specific applications; FAA-certified 50 m version targeted at advertising, tourism, surveillance and scientific markets; other sizes planned to meet sport aviation, cargo and passenger markets.

COSTS (ESTIMATED): 32 m US$75,000; 50 m between US$600,000 and US$1 million. Commercial operating costs (50 m) less than US$100,000 per month and expected leasing cost US$175,000 per month. (All 2003.)

DESIGN FEATURES: Study of strengths and weaknesses of past and present airships, focusing on aerodynamic and structural efficiency, including features not seen or proposed since 1930s. Listed features include better distribution of loads; lower internal pressures; better maintenance access; multiple helium compartments; and hardpoints for ground handling. Basic design scaleable to larger sizes than present airships, with almost all the advantages of rigid construction yet less costly or complex. Mechanical landing system permits greatly reduced ground handling crew (four for advertising, six for passenger operations).

Envelope shaped for minimum air resistance; propulsion engines mounted in inverted-V tailfins; nose-mounted thruster engine for all-speed control.

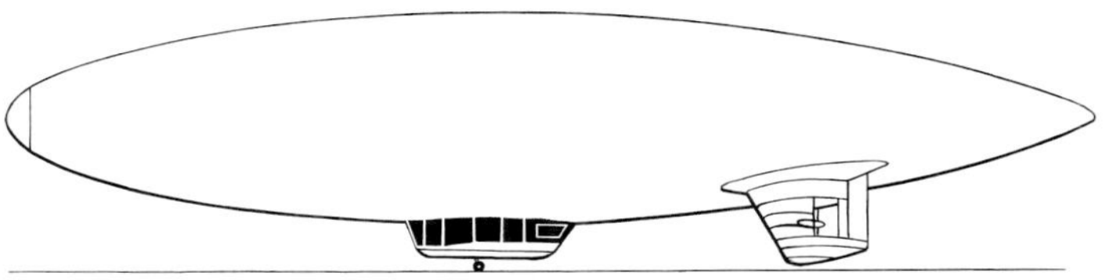

Side elevation of UPship 001 32 m small demonstrator *(James Goulding)* ***NEW***/1043410

STRUCTURE: Framework of aluminium and steel tubing, with some carbon composites; envelope of proprietary rip-stop construction, with internal divisions and five helium cells. US-patented tailfins deflect under excessive ground or air loads.

POWER PLANT: Three 59.7 kW (80 hp) Jabiru piston engines, with electric starting and fixed-pitch propellers; one (pusher) in each tailfin and one operating bow thruster. Fuel capacity 270 litres (71.3 US gallons; 59.4 Imp gallons).

ACCOMMODATION: Pilot and seven passengers in ventral gondola, with rear entry stair; quiet and vibration-free due to distance from engines. Eletrically heated cabin, with space provision for lavatory and/or rescue or other equipment.

DIMENSIONS, EXTERNAL:
Envelope: Length .... 50.0 m (164 ft 0½ in)
Max diameter .... 10.0 m (32 ft 9¾ in)

DIMENSIONS, INTERNAL:
Envelope volume .... 2,400 m³ (84,755 cu ft)
Helium volume .... 2,100 m³ (74,160 cu ft)

WEIGHTS AND LOADINGS (approx):
Weight empty .... 1,260 kg (2,778 lb)
Useful lift .... 840 kg (1,852 lb)
Thruster lift .... 200 kg (441 lb)
Total lift .... 2,100 kg (4,630 lb)

PERFORMANCE (estimated):
Max level speed .... 59 kt (110 km/h; 68 mph)
Cruising speed: at 75% power .... 54 kt (100 km/h; 62 mph)
at 50% power .... 47 kt (87 km/h; 54 mph)
at 25% power .... 37 kt (69 km/h; 43 mph)
Range: at 75% power .... 326 n miles (605 km; 375 miles)
at 50% power .... 399 n miles (740 km; 459 miles)
at 25% power .... 615 n miles (1,139 km; 707 miles)
Endurance, 20% fuel reserves:
at 75% power .... 5 h 30 min
at 50% power .... 8 h 15 min
at 25% power .... 16 h 30 min

UPDATED

BOEING

THE BOEING COMPANY

100 North Riverside Plaza, Chicago, Illinois 60606
Tel: (+1 312) 544 20 00
Web: http://www.boeing.com
PRESIDENT AND CEO: Harry C Stonecipher
NON-EXECUTIVE CHAIRMAN: Lewis E Platt
SENIOR VICE-PRESIDENT AND CHIEF TECHNOLOGY OFFICER: James M Jamieson
SENIOR VICE-PRESIDENT AND CHIEF FINANCIAL OFFICER: James A Bell
SENIOR VICE-PRESIDENT, COMMUNICATIONS: Tod R Hullin

Company founded July 1916. Currently organised into major business segments as detailed below, plus several smaller elements. Revenues for 2002 were US$54.1 billion.

Boeing and McDonnell Douglas merger under the Boeing name was completed on 4 August 1997, elevating Boeing to status of largest aerospace company in the world. Headquarters remained in Seattle, Washington, until September 2001, when moved to Chicago, Illinois. Expansion of global interests occurred in 1997, with Boeing entering into alliances with AVIC and Taikoo Aircraft Engineering Co of China; with CSA Czech Airlines to establish joint venture company and acquire stake in Aero Vodochody (although this venture was under dissolution in 2004); and with Instytut Lotnictwa of Poland for potential future collaboration on advanced technologies In January 1999, Boeing reached agreement with Netherlands-owned company MD Helicopters on previously announced disposal of light civil helicopter range. In January 2003, Boeing had 166,800 employees. Total of 358 aircraft delivered in 2003.

Following further reorganisation in latter half of 2002, operating components of The Boeing Company comprise:

INTEGRATED DEFENSE SYSTEMS Follows this entry
BOEING COMMERCIAL AIRPLANES Follows Integrated Defense Systems
PHANTOM WORKS Follows Boeing Commercial Airplanes

UPDATED

INTEGRATED DEFENSE SYSTEMS

PO Box 516, St Louis, Missouri 63166-0516
Tel: (+1 314) 232 02 32
Fax: (+1 314) 234 82 96
PRESIDENT AND CEO: James F Albaugh
VICE-PRESIDENT AND COO: David O Swain
MEDIA CONTACTS:
DIRECTOR, EXTERNAL COMMUNICATIONS: Walt Rice, *Tel:* (+1 314) 234 21 49
MEDIA RELATIONS, MANAGER: Jim Keller, *Tel:* (+1 314) 233 02 05
MANAGER MEDIA/INTERNATIONAL PROGRAMMES:
Mary Ann Brett, *Tel:* (+1 314) 234 71 11

Created in latter half of 2002 through merger of Military Aircraft and Missile Systems with Space and Communications, Integrated Defense Systems has overall responsibility for military business activities. Management of aircraft and helicopter programmes as detailed below reflects the military customer base, but several subordinate organisations exist to oversee other key areas.

Other elements include **Aerospace Support** (service and logistics support, plus modifications).

In early 2004, business was valued at US$27 billion and had over 78,000 employees.

Major business units of Integrated Defense Systems are as follows:

NAVAL SYSTEMS:
SENIOR VICE-PRESIDENT:
John A Lockard
(St Louis, Missouri)

AIR FORCE SYSTEMS:
SENIOR VICE-PRESIDENT:
George K Muellner (Long Beach, California)

ARMY SYSTEMS:
SENIOR VICE-PRESIDENT:
Roger A Krone
(Delaware County, Pennsylvania)

NASA SYSTEMS:
SENIOR VICE-PRESIDENT:
Mike Mott
(Houston, Texas)

MISSILE DEFENSE SYSTEMS:
SENIOR VICE-PRESIDENT: James W Evatt (Washington, DC)

HOMELAND SECURITY:
SENIOR VICE-PRESIDENT:
Rick Stephens
(Seal Beach, California)

SPACE AND INTELLIGENCE SYSTEMS:
SENIOR VICE-PRESIDENT:
Roger F Roberts
(Seal Beach, California)

AEROSPACE SUPPORT:
SENIOR VICE-PRESIDENT:
Pat Finneran
(St Louis, Missouri)

UPDATED

BOEING F-15E EAGLE

Israel Defence Force name: Ra'am (Thunder)
TYPE: Multirole fighter.

DEVELOPMENT MILESTONES

Official go-ahead	24 Feb 1984
First flight, production	11 Dec 1986
First delivery	12 Apr 1988
Entered service	28 Dec 1988

PROGRAMME: F-15 initially employed as air superiority fighter, having first flown on 27 July 1972. Demonstration of industry-funded Strike Eagle prototype (71-0291) modified from F-15B, including accurate blind weapons delivery, completed at Edwards AFB and Eglin AFB in 1982; product improvements for the F-15E were tested on four Eagles between November 1982 and April 1983, including first take-off at 34,019 kg (75,000 lb), 3,175 kg (7,000 lb) more than F-15C with conformal tanks; new weight included conformal tanks, three other external tanks and eight 500 lb Mk 82 bombs; 16 different stores configurations tested, including 2,000 lb Mk 84 bombs, and BDU-38 and CBU-58 weapons delivered visually and by radar.

Full programme go-ahead announced 24 February 1984; first flight of first production F-15E (86-0183) 11 December 1986; first delivery to Luke AFB, Arizona, 12 April 1988; first delivery 29 December 1988 to 4th Wing at Seymour Johnson AFB, North Carolina.

In mid-2001, USAF and Boeing examining various options for mid-life upgrade of F-15E known as Block 6; among the possibilities are adoption of active electronically scanned array (AESA) radar, update of radar warning receiver equipment, re-engining with version of General Electric F110 turbofan, installation of new wing structure incorporating carbon fibre skin that would provide increased take-off weight and accommodate two extra hardpoints.

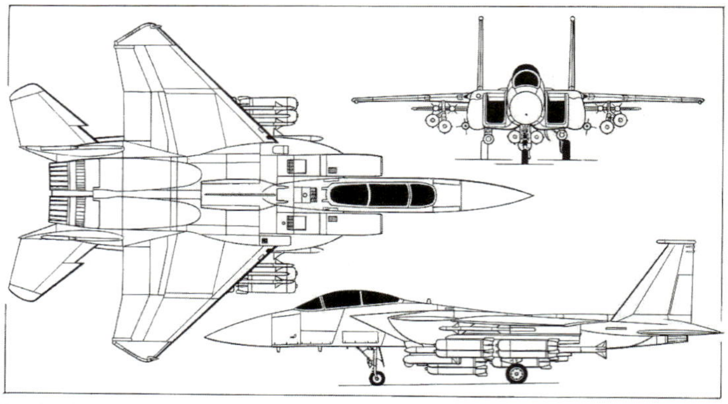

Boeing F-15E Eagle equipped for high ordnance payload air-to-ground mission *(Jane's/Dennis Punnett)* 0507280

CURRENT VERSIONS: **F-15E:** Basic version, *as detailed*.

F-15F: Proposed single-seat version, optimised for air combat; not built.

F-15H: Proposed export version, lacking specialised air-to-ground capability; supplanted by F-15S. Designation subsequently allocated to version for Greece, which eventually selected F-16 Fighting Falcon.

F-15I Ra'am: Israeli export version of F-15E; selected November 1993; confirmed 27 January 1994; 21 ordered 12 May 1994; option on four more converted to firm order in November 1995. First flight of initial (unpainted) aircraft on 12 September 1997; this was subject of formal roll-out and handover ceremony at St Louis on 6 November 1997, with first pair of aircraft to be delivered leaving St Louis on 16 January 1998 and arriving in Israel three days later; total of 16 delivered during 1998, with final nine aircraft following in 1999. Operating unit No. 69 Squadron. Tactical electronic warfare system deleted; replaced by Israeli-built SPS-2100 integrated system including active jamming, radar and missile warning, and dispenser subsystems.

Otherwise identical to USAF F-15E, with F100-PW-229 engines, LANTIRN pods, full capability AN/APG-70 radar, Kaiser holographic HUD, Litton ring laser INS and VHSIC central computer. Associated equipment includes four Sanders mission planning subsystems and one Sanders common mapping production system (CMPS) to assist ground planning, briefing and debriefing activities at total cost of US$6.2 million. F-15I includes significant number of co-produced components, such as airframe and wing subassemblies, heat exchangers and weapons pylons.

F-15K: Version conceived to satisfy Republic of Korea's F-X fighter requirement; preliminary bids submitted in June 2000. Selection was expected in second half of 2001, but decision delayed, with formal announcement being made on 19 April 2002, when Korea revealed intent to purchase 40 aircraft at total cost of US$4.2 billion. F-15K incorporates General Electric F110-GE-129 engine, AN/APG-63(V)1 radar, Raytheon AN/AWW-13 advanced data link pod for missile guidance and mid-course updating of SLAM-ER weapon, expanded weapons capability (including JDAM, SLAM-ER and AIM-9X) and improved environmental control system, plus Lockheed Martin TIGER Eyes sensor suite

Boeing F-15E Eagle of Lakenheath-based 48th FW *(Paul Jackson)* ***NEW***/0572315

including AN/AAS-42 infra-red search and track system, Boeing Joint Helmet-Mounted Cueing System and revised cockpit with seven 152 × 152 mm (6 × 6 in) Kaiser Electronics AMLCDs. Final assembly of initial aircraft to begin in May 2004, with first flight expected in March 2005. First pair of aircraft will be delivered in November 2005, followed by 10 in 2006, 16 in 2007 and 12 in 2008. Follow-on order for up to 80 more aircraft remains a possibility, since original requirement was for 120.

F-15L: Proposed less costly version offered to Israel by Boeing in unsuccessful attempt to secure order for new combat aircraft; F-16I eventually selected.

F-15S: Saudi Arabian export version of F-15E, lacking some air-to-air and air-to-ground capabilities; Saudi Arabian request for 72 aircraft approved by US government in December 1992; initially designated **F-15XP**; first funds assigned by US government on 23 December 1992; contract signature by Saudi government May 1993; planned delivery rate halved, early 1994, to one per month, beginning 1995. First F-15S flown 19 June 1995; official roll-out and handover 12 September 1995. First two examples delivered to Saudi Arabia in November 1995 were second and third built; total of 49 received by end of 1998 and 70 by end of 1999, with final two following in late July 2000.

Saudi versions comprise 24 optimised for air-to-air missions and 48 optimised for air-to-ground; AN/APG-70 radar. Despite planned restrictions, aircraft delivered with full F-15E capability, plus conformal fuel tanks and tangential stores attachments. Armament includes AGM-65D/G Maverick, AIM-7 Sparrow, AIM-9M and AIM-9S Sidewinder missiles, CBU-87 submunitions dispenser and GBU-10/12 bombs. Saudi programme includes about 154 Pratt & Whitney F100-PW-229 engines.

F-15SE: Proposed single-seat version of F-15E for USAF, for which pricing data supplied to Department of Defense as part of Quadrennial Defense Review. Not proceeded with.

F-15T: Version offered to Singapore, which requires up to 24 new fighters, with initial tranche of eight aircraft to be delivered in 2008–09. Original six contenders reduced to shortlist of three on 10 October 2003, with F-15T in competition with Dassault Rafale and Eurofighter Typhoon. Further evaluation planned, with final decision to be taken in late 2004 or early 2005. Is expected to feature multimode version of Raytheon AN/APG-63(V)3 AESA radar and more modern countermeasures suite as well as improved sensor systems and navigation equipment.

F-15U: Version conceived to satisfy United Arab Emirates requirement for long-range interdictor aircraft, in which it was competing against Lockheed Martin F-16 (selected), Dassault Rafale, Eurofighter Typhoon and Sukhoi Su-30MK. **F-15U Plus** proposal anticipated extended range, with additional 2,570 kg (5,665 lb) of fuel in thicker clipped-delta, 50° leading-edge sweep wing; more stores stations and internally situated IR navigation and targeting sensor suite in lieu of LANTIRN. Typical ordnance loads would have comprised nine 2,000 lb Mk 84 bombs or seven laser-guided GBU-24s.

F-15MANX: Stealthy, tail-less proposal based on technology developed for ACTIVE thrust-vectoring programme. Concept appears dormant.

CUSTOMERS: USAF funding for originally planned 392 reduced to 200; however, further nine funded in FY91 and FY92, comprising three Gulf War loss replacements and six with proceeds of sale to Saudi Arabia of 24 F-15C/Ds. Additional six in FY96 budget, despite not being requested by USAF Further six aircraft in FY97 and five in FY98, raising total USAF buy to 226, with another five in FY00 and five in FY01. Saudi Arabia 72 (F-15S); Israel 25 (F-15I); Republic of Korea 40 (F-15K).

Initial USAF combat-capable unit, 4th Wing, declared operational October 1989, currently with 333, 334, 335 and 336 Fighter Squadrons; and others with 57th Wing USAF Weapons School at Nellis AFB, Nevada; 90th FS of 3rd Wing at Elmendorf, Alaska, received first F-15E on 29 May 1991; 391st FS, sole F-15E squadron in multitype 366th Wing at Mountain Home AFB, Idaho, received first aircraft (reallocated from early production) 6 November 1991; 492nd and 494th FS of 48th FW at Lakenheath, UK, received first aircraft on 21 February 1992. Deliveries of Lot 7 began to 48th FW on 13 April 1994 and 209th USAF F-15E (92-0366) to 57th Wing on 11 July 1994. First aircraft from Lot 9 flown on 1 April 1999 and delivered to 57th Wing on 7 June; remaining five F-15Es from this batch all assigned to 48th Fighter Wing by November 1999; subsequent deliveries to 48th FW comprised all Lot 10 and Lot 11 aircraft, with final pair arriving on 26 August 2000. First Lot 12 aircraft (00-3000) delivered 21 June 2002; subsequently assigned to 48th FW; four more Lot 12 aircraft all to 48th FW by third quarter of 2003, along with first Lot 13 machine (01-2000). USAF had accepted 233 F-15Es by end of 2003, with final three to be delivered in 2004, at rate of one per quarter.

Initial contract placed with McDonnell Douglas by US government on 18 December 1992 for 72 Saudi aircraft; project name, Peace Sun IX.

COSTS: US$35 million, flyaway; US$2,000 million for 21 F-15Is (1993), Israel; US$288.5 million appropriation for five F-15Es in FY00. Total cost of South Korean purchase expected to be US$4.2 billion.

DESIGN FEATURES: Conceived as uncompromised air superiority fighter, but further developed for strike/attack mission, with provision for increased weight of ordnance. Typical 1970s fighter configuration, with twin fins positioned to receive vortex flow off wing and maintain directional stability at high angles of attack. Straight two-dimensional external compression engine air inlet each side of fuselage. High-mounted cockpit for all-round vision

Mission includes approach and attack at night and in all weathers; main systems of F-15E include high-resolution, synthetic aperture Raytheon AN/APG-70 radar, wide field of view FLIR, Lockheed Martin LANTIRN navigation (AN/AAQ-13) and targeting (AN/AAQ-14) pods beneath starboard and port air intakes respectively; air-to-air capability with AIM-7 Sparrow, AIM-9 Sidewinder and AIM-120 AMRAAM retained; rear cockpit has four multipurpose CRT displays for radar, weapon selection, and monitoring enemy tracking systems; front cockpit modifications include redesigned up-front controls, wide field of view HUD, colour CRT multifunction displays for navigation, weapon delivery, moving map, precision radar mapping and terrain-following. Engines have digital electronic control, engine trimming and monitoring; fuel tanks are foam-filled; more powerful generators; better environmental control.

NACA 64A aerofoil section with conical camber on leading-edge; sweepback 38° 42' at quarter-chord; thickness/chord ratio 6.6 per cent at root, 3 per cent at tip; anhedral 1°; incidence 0°.

USAF F-15E PROCUREMENT

FY	Batch	Qty	First aircraft
86	Lot 1	8	86-0183
87	Lot 2	42	87-0169
88	Lot 3	42	88-1667
89	Lot 4	36	89-0471
90	Lot 5	36	90-0227
91	Lot 6	36	91-0300
91	Lot 7	6	91-0600
92	Lot 8	3	92-0364
96	Lot 9	6	96-0200
97	Lot 10	6	97-0217
98	Lot 11	5	98-0131
00	Lot 12	5	00-3000
01	Lot 13	5	01-2000
Total		**236**	

Boeing F-15I Ra'am of 69 Squadron, Israel Defence Force *(Paul Jackson)* ***NEW***/0546853

FLYING CONTROLS: Powered. Plain ailerons and all-moving tailplane with dog-tooth extensions, both powered by National Water Lift hydraulic actuators; rudders have Ronson Hydraulic Units actuators; no spoilers or trim tabs; Moog boost and pitch compensator for control column; plain flaps; upward-opening airbrake panel in upper fuselage between fins and cockpit. Digital triple-redundant BAE Systems flight control system capable of automatic coupled terrain-following.

STRUCTURE: Wing based on torque box with integrally machined skins and ribs of light alloy and titanium; aluminium honeycomb wingtips, flaps and ailerons; airbrake panel of titanium, aluminium honeycomb and graphite/epoxy composites skin. F-15E version includes 60 per cent of earlier F-15 structure redesigned to allow 9 *g* and 16,000 hours fatigue life; superplastic forming/diffusion bonding used for upper rear fuselage, rear fuselage keel, main landing gear doors, and some fuselage fairings, plus engine bay structure.

LANDING GEAR: Hydraulically retractable tricycle type, with single wheel on each unit. All units retract forward. Cleveland nose and main units, each incorporating an oleo-pneumatic shock-absorber. Honeywell wheels and Michelin AIR X radial tyres on all units. Nosewheel tyre size 22×7.75–9 or 22×7.75R9 (26 ply) tubeless; mainwheel tyres size 36×11–18 or 36×11R18 (30 ply) tubeless; tyre pressure 21.03 bar (305 lb/sq in) on all units. Honeywell five-rotor carbon disc brakes.

POWER PLANT: Initially, two Pratt & Whitney F100-PW-220 turbofans, each rated for take-off at 104.3 kN (23,450 lb st), installed, with afterburning. USAF aircraft 135 onwards (90-0233), built from August 1991, have 129.4 kN (29,100 lb st) Pratt & Whitney F100-PW-229s, which also ordered for Saudi F-15S and Israeli F-15I. F-15K for South Korea fitted with General Electric F110-GE-129 turbofans. Air inlet controllers by Hamilton Sundstrand. Air inlet actuators by National Water Lift.

Internal fuel in foam-filled structural wing tanks and six Goodyear fuselage tanks, total capacity 7,643 litres (2,019 US gallons; 1,681 Imp gallons). Simmonds fuel gauge system. Optional conformal fuel tanks (CFT) attached to side of engine air intakes, beneath wing, each containing 2,737 litres (723 US gallons; 602 Imp gallons). Provision for up to three additional 2,309 litre (610 US gallon; 508 Imp gallon) external fuel tanks. Maximum total internal and external fuel capacity 20,044 litres (5,295 US gallons; 4,409 Imp gallons).

ACCOMMODATION: Two crew, pilot and weapon systems officer, in tandem on Boeing ACES II zero/zero ejection seats. Single-piece, upward-hinged, bird-resistant canopy; in-service modification undertaken by USAF replaced original windscreen with laminated polycarbonate material.

SYSTEMS: Lucas Aerospace generating system for electrical power, with Hamilton Sundstrand 60/75/90 kVA constant-speed drive units. Litton molecular sieve oxygen generating system (MSOGS) introduced in 1991 to replace liquid oxygen system. Honeywell air conditioning system. Three independent hydraulic systems (each 207 bar; 3,000 lb/sq in) powered by Abex engine-driven pumps; modular hydraulic packages by Hydraulic Research and Manufacturing Company. AlliedSignal APU for engine starting, and for provision of limited electrical or hydraulic power on the ground independently of main engines.

AVIONICS: *Comms:* Raytheon AN/ARC-164 UHF transceiver and UHF auxiliary transceiver with cryptographic capability; Honeywell AN/APX-101 IFF transponder; BAE Systems AN/APX-76 IFF interrogator with Litton reply evaluator. Some aircraft have AN/ARC-190 HF radio for very long-range communications. F-15K will be fitted with BAE Systems AN/APX-113 combined interrogator transponder (CIT) unit and Link 16 MIDS (Multifunctional Information Distribution System).

Radar: Raytheon AN/APG-70 I-band pulse Doppler radar provides air-to-air capability equal to F-15C, plus high-resolution synthetic aperture mode for air-to-ground. Republic of Korea F-15K to have AN/APG-63(V)1 radar. Raytheon studying possible application of AESA radar for export customers and retrofit to USAF aircraft as part of upgrade.

Flight: Triple redundant BAE Systems digital flight control system with automatic terrain-following standard. IBM CP-1075C very high-speed integrated circuit (VHSIC) central computer introduced in 1992 (replacing CP-1075). Honeywell AN/ASK-6 air data computer, Honeywell AN/ASN-108 AHRS, Honeywell CN-1655A/ASN ring laser gyro INS providing basic navigation data and serving as primary attitude reference system, Rockwell Collins AN/ARN-118 Tacan, Rockwell Collins HSI presenting aircraft navigation information on a symbolic pictorial display, Rockwell Collins AN/ARN-112 ILS receiver, Rockwell Collins ADF receiver, Dorne & Margolin glideslope localiser antenna and Teledyne Avionics angle of attack sensors. Rockwell Collins Miniaturised Airborne GPS Receiver installed from 1995. Latest aircraft to join USAF (FY96 and subsequent production) feature new embedded GPS/INS.

In July 1998, Boeing began flight test of commercial computing technology in F-15E as part of Phantom Works Bold Stroke project, whereby commercial technology is being used to provide non-proprietary computer systems for installation on military aircraft. F-15E installation embodied advanced display core processor (ADCP) using PowerPC hardware to replace existing central computer and multipurpose display processor; initial trial demonstrated same capabilities as current military hardware.

Instrumentation: FLIR imagery displayed on Kaiser IR-2394/A wide field of view HUD; Honeywell vertical situation display set using CRT to present radar, electro-optical identification and attitude director indicator formats to pilot under all light conditions; moving map display by Honeywell RP-341/A remote map reader. Honeywell digital map system intended to replace remote map reader from 1996. F-15I has Elbit helmet-mounted sight and other cockpit displays. Final 10 aircraft for USAF have Kaiser 125 × 125 mm (5 × 5 in) flat panel colour display (FPCD) instead of current CRT; this unit may also be

US F-15 PRODUCTION

	F-15A	F-15B	F-15C	F-15D	F-15DJ	F-15E	F-15I	F-15J	F-15K	F-15S	Total
Israel	19	2	18	13			25				77
Japan					12			10[1]			22
Saudi Arabia			55	19						72	146
South Korea									40[4]		40
USA	365[2]	59[3]	409	61		236					1,130
Total	**384**	**61**	**482**	**93**	**12**	**236**	**25**	**10**	**40**	**72**	**1,415**

Notes: Mitsubishi of Japan also produced 155 F-15Js and 36 F-15DJs to complete total JASDF procurement of 165 and 48, respectively

[1] Including eight kits to Mitsubishi
[2] Including 10 YF-15As
[3] Including two YF-15Bs, of which one converted to F-15E prototype
[4] On order for delivery during 2005–08.

retrofitted to earlier F-15Es. F-15K will feature four 152 × 152 mm (6 × 6 in) Multipurpose Displays (MPDs) and three 127 × 127 mm (5 × 5 in) FPCDs; all produced by Kaiser Electronics and all incorporating AMLCD technology.

Mission: Lockheed Martin LANTIRN externally mounted sensor package comprising AN/AAQ-13 navigation pod and AN/AAQ-14 targeting pod. Lockheed Martin Sniper XR advanced targeting pod is scheduled to be installed on USAF F-15E, but will not be available for some time; as interim solution, following hurried integration and evaluation in latter half of 2002, Rafael/Northrop Grumman Litening Extended Range (ER) system has been installed on some USAF aircraft; batch of 24 pod systems ordered in early 2003 at cost of US$32.6 million, with deliveries completed in April 2003. Lockheed Martin TIGER Eyes sensor suite to be installed on aircraft for Republic of Korea.

Self-defence: Northrop Grumman Enhanced AN/ALQ-135(V) internal countermeasures set provides automatic jamming of enemy radar signals; BAE Systems AN/ALR-56C RWR, Raytheon AN/ALQ-128 EW warning set, (initially) BAE Systems AN/ALE-45 chaff dispenser. Unique SPS-2100 EW system developed by Elisra and Rokar for Israeli F-15I. Sanders and ITT Avionics AN/ALQ-214 Integrated Defensive Electronic CounterMeasures (IDECM) under development and will be installed on F-15E; IDECM package incorporates Sanders AN/ALE-55 fibre optic towed decoy (FOTD). Saab BOL chaff dispensers to be adopted universally, with first contract awarded to BAE Systems in fourth quarter of 2001. F-15K will have BAE Systems AN/ALR-56C(V)1 RWR and AN/ALE-47 countermeasures dispenser system, plus Northrop Grumman AN/ALQ-135 internally mounted jamming system.

ARMAMENT: 20 mm M61A1 six-barrel gun in starboard wing-root, with 512 rounds. General Electric lead computing gyro. Provision on underwing (one per wing) and centreline pylons for air-to-air and air-to-ground weapons and external fuel tanks. Wing pylons use standard rail and launchers for AIM-9 Sidewinder (Israeli F-15I also compatible with Rafael Python 4) and AIM-120 AMRAAM air-to-air missiles; AIM-7 Sparrow and AIM-120 AMRAAM can be carried on ejection launchers on the fuselage or on tangential stores carriers on CFTs. Maximum aircraft load (with or without CFTs) is four each AIM-7 and AIM-9, or up to eight AIM-120. Single or triple rail launchers for AGM-65 Maverick air-to-ground missiles can be fitted to wing stations only.

Tangential carriage on CFTs provides for up to six bomb racks on each tank, with provision for multiple ejector racks on wing and centreline stations. Edo BRU-46/A and BRU-47/A adaptors throughout, plus two LAU-106A/As each side of lower fuselage. F-15E can carry a wide variety and quantity of guided and unguided air-to-ground weapons, including Mk 20 Rockeye (26), Mk 82 (26), Mk 84 (seven), BSU-49 (26), BSU-50 (seven), GBU-10 (seven), GBU-12 (15), GBU-15 (two), GBU-24 (five), CBU-52 (25), CBU-58 (25), CBU-71 (25), CBU-87 (25) or CBU-89 (25) bombs; JDAM, JSOW and WCMD weapons; SUU-20 training weapons (three); A/A-37 U-33 tow target (one); and B57 and B61 series nuclear weapons (five). Is only USAF strike aircraft able to carry GBU-28. An AN/AXQ-14 datalink pod is used in conjunction with the GBU-15; LANTIRN pod illumination is used to designate targets for laser-guided bombs; AGM-130 powered standoff bomb integrated in 1993; AGM-88 HARM capability added in 1996. AN/AWG-27 armament control system; Programmable Armaments Control Set adopted by final 10 new-build USAF aircraft and likely to be retrofitted to most earlier aircraft.

Pneumatic weapon ejection system under development in early 2000; makes use of compressed air for weapons separation from aircraft.

DIMENSIONS, EXTERNAL:
- Wing span ... 13.05 m (42 ft 9¾ in)
- Wing aspect ratio ... 3.0
- Length overall ... 19.43 m (63 ft 9 in)
- Height overall ... 5.63 m (18 ft 5½ in)
- Tailplane span ... 8.61 m (28 ft 3 in)
- Wheel track ... 2.75 m (9 ft 0¼ in)
- Wheelbase ... 5.42 m (17 ft 9½ in)

AREAS:
- Wings, gross ... 56.49 m^2 (608.0 sq ft)
- Ailerons (total) ... 2.46 m^2 (26.48 sq ft)
- Flaps (total) ... 3.33 m^2 (35.84 sq ft)
- Fins (total) ... 9.78 m^2 (105.28 sq ft)
- Rudders (total) ... 1.85 m^2 (19.94 sq ft)
- Tailplanes (total) ... 10.34 m^2 (111.36 sq ft)

WEIGHTS AND LOADINGS (F100-PW-220 engines):
- Operating weight empty (no fuel, ammunition, pylons or external stores) ... 14,515 kg (32,000 lb)
- Max weapon load ... 11,113 kg (24,500 lb)
- Max fuel weight: internal (JP4) ... 5,952 kg (13,123 lb)
  - CFTs (two, total) ... 4,265 kg (9,402 lb)
  - external tanks (three, total) ... 5,396 kg (11,895 lb)
  - max internal and external ... 15,613 kg (34,420 lb)
- Max T-O weight ... 36,741 kg (81,000 lb)
- Max landing weight:
  - unrestricted ... 20,094 kg (44,300 lb)
  - at reduced sink rates ... 36,741 kg (81,000 lb)
- Max zero-fuel weight ... 28,440 kg (62,700 lb)
- Max wing loading ... 650.5 kg/m^2 (133.22 lb/sq ft)
- Max power loading ... 176 kg/kN (1.73 lb/lb st)

PERFORMANCE:
- Max level speed at height ... M2.5
- Max combat radius ... 685 n miles (1,270 km; 790 miles)
- Max range ... 2,400 n miles (4,445 km; 2,762 miles)

UPDATED

BOEING F/A-18 SUPER HORNET

US Navy designations: F/A-18E, F/A-18F and EA-18G

TYPE: Multirole fighter.

DEVELOPMENT MILESTONES

Requirement issued	1991
First order (EMD contract)	June 1992
Rolled out	18 Sep 1995
First flight	29 Nov 1995
Production go-ahead	26 Mar 1997
First flight, production	6 Nov 1998
First delivery	18 Dec 1998
Entered service (VFA-122)	17 Nov 1999

PROGRAMME: Proposed 1991 as replacement for cancelled GD/MDC A-12 and follow-on for early F/A-18As and other USN/MC tactical aircraft as they phase out; based on earlier versions of Hornet; development funding approved by Congress for FY92; US$4.88 billion engineering and manufacturing development contract awarded June 1992, covering seven flight test aircraft (five Es; two Fs) and three ground test articles, plus associated 7½-year test programme; US$754 million award in 1992 to GE for F414 engine development.

Critical design review (CDR) undertaken 13 to 17 June 1994 at St Louis by team of independent government evaluators; successfully negotiated, with F/A-18E/F satisfying or surpassing all timescale, cost, technical, reliability and maintainability requirements.

Principal subcontractor Northrop Grumman launched assembly of first aircraft on 24 May 1994 with start of work on centre/aft fuselage section at Hawthorne, California. First forward fuselage section followed suit on new assembly line at St Louis, Missouri, 23 September 1994; completed 12 January 1995, except for wiring; mating with centre/aft section from Hawthorne effected 8 May 1995. Roll-out of prototype (E1/165164) on 18 September 1995; first flight 29 November 1995.

Prototype delivered to Naval Air Warfare Center at Patuxent River, Maryland, on 14 February 1996 for start of three year flight test programme involving seven aircraft; initial flight from Patuxent River made by E1 on 4 March 1996. Second F/A-18E prototype (E2/165165) made first flight on 26 December 1995 and first F/A-18F (F1/165166) on 1 April 1996. Supersonic speed exceeded for first time on 12 April 1996 when E1 achieved M1.1. Carrier suitability trials began mid-1996 and F1 completed three successful catapult launches at Patuxent River on 6 August that year.

By 1 February 1997, when the last development aircraft was delivered to Patuxent River, test fleet had flown 390 sorties, for a total of 631 flight hours. Subsequent milestones were passed on 13 May 1997 (1,000th flight hour); 9 September 1997 (1,500th flight hour); 24 September 1997 (1,000th sortie); 11 December 1997 (2,000th flight hour and 1,300th sortie); 1 July 1998 (3,000th hour) and 12 January 1999 (4,000th hour). First missile launch (an AIM-9 Sidewinder) accomplished by aircraft F2 on 5 April 1997; other weapons expended by mid-May 1997 included AIM-7 Sparrow, AIM-120 AMRAAM, SLAM/SLAM-ER, AGM-84 Harpoon, Mk 82 and Mk 83 bombs plus Rockeye CBUs. Successful deployment of AN/ALE-50 towed radar decoy also accomplished by mid-May 1997.

Flight test programme briefly halted in October 1997 for engine inspection after discovery of cracks in stator vanes. More intractable problem concerned wing-drop, with uncommanded departures from controlled flight evident as early as the seventh sortie in March 1996. Boeing and US Navy considered three solutions to eradicate this in January 1998, including fitting stall strips on upper surface; adding a chord-wise fence just inboard of the hinge fairing; and switching to a porous hinge fairing with slots that allow air to flow in both directions through wing fold fairing. Last-mentioned option selected in early 1998; subsequent flight testing confirmed efficacy of solution.

EMD phase of test programme completed end April 1999, by which time seven aircraft had accumulated 4,673 flight hours in 3,172 sorties; in the process, over 15,000 test points completed and 29 weapons configurations cleared for flight. By mid-2000, follow-on testing had increased totals to more than 5,500 flight hours in 3,800 flights.

Static testing began in August 1995 with airframe ST50; shock loading assessment from February 1996 with DT50; fatigue testing from 30 June 1997 with FT50, which completed first lifetime (6,000 hours) one month ahead of schedule, on 27 August 1998. ST50 transferred to Lakehurst, New Jersey, for series of six emergency barricade engagements; first successfully completed on 3 September 1997, but ST50 damaged during third test on 23 September when restraint cable failed. ST50 subsequently repaired and returned to test

Two-seat F/A-18F of training squadron VFA-122 *(Paul Jackson)* ***NEW***/0576488

CURRENT AND PLANNED US NAVY SUPER HORNET SQUADRONS

Squadron number	Variant	FY of conversion	Previous equipment	Remarks
VFA-2	F/A-18F	03	F-14D	
VFA-11	F/A-18E	07	F-14B	
VFA-14	F/A-18E	01	F-14A	
VFA-22	F/A-18E	03	F/A-18C	
VFA-31	F/A-18F	07	F-14D	
VFA-32	F/A-18F	05	F-14B	
VFA-41	F/A-18F	01	F-14A	
VFA-81	F/A-18E	05	F/A-18C	
VFA-86	F/A-18F	06	F/A-18C	
VFA-102	F/A-18F	02	F-14B	
VFA-103	F/A-18F	05	F-14B	
VFA-105	F/A-18E	07	F/A-18C	
VFA-115	F/A-18F	00	F/A-18C	First squadron to deploy
VFA-122	F/A-18E/F	00	—	Training unit
VFA-137	F/A-18E	03	F/A-18C	
VFA-143	F/A-18F	07	F-14B	
VFA-146	F/A-18E	08	F/A-18C	
VFA-154	F/A-18F	04	F-14A	
VFA-211	F/A-18F	06	F-14A	
VFA-213	F/A-18F	06	F-14D	

duty, for live-fire testing at China Lake, California; trials included firing large armour-piercing incendiary projectile through inlet duct into aft fuel tank, with resultant hole showing little evidence of leakage.

Approval for low-rate initial production (LRIP) of 62 aircraft in three lots given on 26 March 1997; first lot composed of eight F/A-18Es and four F/A-18Fs. Assembly of first LRIP F/A-18E (165533) started at Northrop Grumman in May 1997 and at Boeing in September 1997; final assembly began at St Louis on 19 June 1998, with mating of centre/aft and forward fuselage sections; first flight 6 November 1998, six weeks ahead of schedule; this aircraft officially accepted by US Navy on 18 December 1998 and flown to Patuxent River to join flight test programme before attachment to VX-9 squadron at China Lake, California, for operational evaluation. Latter programme began 27 May 1999 and involved total of 1,233 flight hours in 866 sorties by seven LRIP aircraft (three F/A-18Es and four F/A-18Fs) in six-month period, including testing of all mission capabilities in varying climates as well as operations at sea aboard USS *John C Stennis* and participation in 'Red Flag' exercise at Nellis AFB, Nevada. Results of operational evaluation announced 15 February 2000, with report stating aircraft to be "operationally effective and operationally suitable" and recommending fleet introduction with the US Navy.

First USN squadron was VFA-122, at Lemoore, California, as specialist FRS (Fleet Replacement Squadron), training pilots and ground crew; received first seven aircraft (of eventual 34) 17 November 1999. First operational squadron was VFA-115, which began transition from F/A-18C to F/A-18E in third quarter 2000; made first deployment in USS *Abraham Lincoln* from 24 July 2002 and experienced combat debut on 6 November 2002 when VFA-115 aircraft dropped four GPS-guided JDAMs on SAM sites in southern Iraq. See table for further details of transition programme.

CURRENT VERSIONS: **F/A-18E:** Single-seat. Initial aircraft (all LRIP examples plus those purchased in FY00 and FY01) to so-called Block 1 standard, with AN/APG-73 and introducing other items of equipment as they become available; subsequent aircraft (FY02 and later) to so-called Block 2 standard, incorporating revised forward fuselage as part of ECP 6038, with fibre optic data network and advanced crew station, including 205 × 255 mm (8 × 10 in) display, for NFO among other improvements. Will also eventually feature AN/APG-79 AESA, evaluation of which began at China Lake, California, in July 2003, clearing way for first deliveries to fleet units by 2005–06. Raytheon AN/ASQ-228 ATFLIR pod also a feature of Block 2 enhancements and completed operational evaluation in third quarter of 2003, with IOC following immediately; VFA-102 was first operational squadron.

F/A-18F: Two-seat. Also produced in Block 1 and Block 2 versions as detailed above.

EA-18G 'Growler': Programme launched as private venture, two-seat electronic warfare dedicated derivative of F/A-18F. Initially known by manufacturer as the **F/A-18 C^{2}W**; announcement of development made 7 August 1995. Boeing is prime contractor, with Northrop Grumman as principal subcontractor with special responsibility for integration of electronic warfare suite; minimal structural changes; wideband receiver pods will replace wingtip AIM-9 Sidewinder AAMs; other pods and antennae to be carried on weapon pylons; satcom receiver positioned behind cockpit.

US Navy requirement calls for new EW aircraft to replace Grumman EA-6B Prowler from 2009 onwards; existing EA-6B equipment such as AN/ALQ-99 jamming system pods to migrate to **EA-18G**, with US Navy seeking to acquire about 90 aircraft by 2012. Second multiyear procurement batch (FY05 to FY09) at present expected to include 128 two-seaters, of which about 56 should emerge as EA-18G EW aircraft, with first four LRIP examples to be funded in FY05. IOC target date in FY09.

Initial flight demonstration of first EMD F/A-18F, configured with three AN/ALQ-99 jamming pods and two fuel tanks, accomplished in November 2001; further trials in 2002 expanded performance envelope and provided data on noise and vibration characteristics. These initial evaluations confirmed suitability of Super Hornet for EW mission and paved way for award, in early 2004, of US$979 million contract to cover five-year SDD (system development and demonstration) programme. Majority of SDD testing to be undertaken by two F/A-18Fs that will be modified to pre-production EA-18G configuration.

Multimission capability will be retained by EA-18G, which will feature Improved Capability 3 (ICAP 3) EW suite, currently under test for installation on EA-6B Prowler and comprising AN/ALQ-99 jamming pods and AN/ALQ-218 digital receiver system. So-called Block 1 EA-18G will carry EW avionics in gun bay, with associated AN/ALQ-218 receivers in wingtip pods and up to five AN/ALQ-99F jamming pods; armament will include AGM-88 HARM for attack and AIM-120C AMRAAM for self-defence. Enhanced Block 2/3 versions expected to use New Technology (NT) jamming pod, for which it is hoped to secure development funds in 2006. Other Block 2/3 enhancements could include additional precision weapons, IDECM, new communications jammer and Joint Tactical Radio System.

CUSTOMERS: US Navy. Seven prototypes; approval for 12 LRIP aircraft in FY97 (Lot 1). See table for details of three LRIP batches and two multiyear procurement (MYP) batches; contract for first MYP batch signed on 15 June 2000, with second MYP contract awarded on 29 December 2003. All 62 LRIP aircraft delivered by end of third quarter of 2001, with first full-rate production Super Hornet (an F/A-18F) being handed over in late September; 100th Super Hornet (an F/A-18F) delivered 14 June 2002; 170 aircraft had been built by end of 2003. Broad requirement originally identified for over 1,000 by FY15, but Quadrennial Defense Review reduced this to minimum of 548 and maximum of 785; defence guidance planning proposal in first quarter of 2002 advocated further reduction to 460. Latest procurement plans (early 2004) anticipate purchase of 540 F/A-18E/Fs plus at least 90 EA-18Gs. Boeing seeking export sales, following approval by US government; Malaysia is prime candidate to be first export customer, with proposed sale involving total of 12–18 F/A-18Fs. Super Hornet will also be offered to Australia, which seeking new multirole fighter under Project Air 6000.

COSTS: Development estimated US$4.8 billion (1992); US$1,089 million in FY92 budget; US$943 million in FY93; approximately US$1,500 million in FY94 and US$1,348 million requested for FY95. US$2,600 million in provisional FY97 budget for first 12 aircraft; with US$2,100 million requested for FY98 procurement of 20 aircraft. US GAO estimated flyaway unit cost as being of the order of US$43.6 million in 1996, based on original planned procurement of 1,000. Most recent figures, for second MYP batch, put total cost at US$8.6 billion for 210 aircraft. Unit cost quoted as US$48 million in mid-2000, when Boeing pursuing measures to drive cost down to around US$40 million by 2005. Estimated cost of EA-18G approximately US$66 million.

DESIGN FEATURES: Generally as for first-generation Hornet. Stretched versions of F/A-18C/D; gross landing weight increased by 4,536 kg (10,000 lb); 0.86 m (2 ft 10 in) fuselage plug; wings photometrically increased in size to provide 9.29 m^2 (100.0 sq ft) extra area and 1.31 m (4 ft 3½ in) span increase; control surfaces disproportionately enlarged and dogtooth added to leading-edge for increased aileron authority. Wings 2.5 cm (1 in) deeper at root; larger horizontal tail surfaces; LEX size substantially increased in early 1993 (from total 5.8 m^2; 62.4 sq ft to 7.0 m^2; 75.3 sq ft, compared with 5.2 m^2; 56.0 sq ft on F/A-18C/D), ensuring full manoeuvre capability at beyond 40° AoA; also incorporates spoilers on upper surface of LEX as speedbrake and to increase nose-down control authority. Nevertheless, has 42 per cent fewer parts than immediate predecessor.

Additional 1,637 kg (3,600 lb) of internal and 1,406 kg (3,100 lb) of external fuel; 40 per cent extra range; further two (making 11) weapon hardpoints (stations 2 and 10, inboard of wingtips, for AAMs and ASMs of up to 520 kg; 1,146 lb); 'bring-back' weapons load increased to 4,082 kg (9,000 lb); additional survivability measures; air intakes redesigned and slewed to increase mass flow to more powerful F414-GE-400 engines and also changed to 'caret' shape to reduce radar signature. Incorporates several other 'affordable' stealth features to reduce radar cross-section, including saw-toothed doors and panels, realigned joints and edges and angled antennas.

FLYING CONTROLS: Full digital fly-by-wire controls using ailerons and tailerons for lateral control, plus flaps in flaperon form at low airspeeds; leading- and trailing-edge flaps scheduled automatically for high manoeuvrability, fast cruise and slow approach speed; horizontal stabilisers automatically assume neutral position if one is damaged, with pitch control then being passed to other surfaces; both rudders turned in at take-off and landing to provide extra nose-up trim effort; fly-by-wire returns towards 1 *g* flight if pilot releases

Boeing F/A-18F Super Hornet *(Paul Jackson)* ***NEW***/0576490

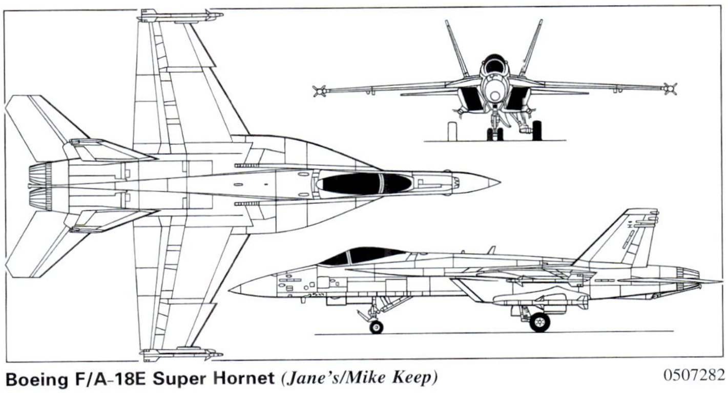

Boeing F/A-18E Super Hornet *(Jane's/Mike Keep)* 0507282

SUPER HORNET PROCUREMENT

FY	Batch	Block	F/A-18E	F/A-18F	Total
97	LRIP 1	52	8	4	12
98	LRIP 2	53	8	12	20
99	LRIP 3	54	14	16	30
00	MYP1	55	15	21	36
01	MYP1	56	14	25	39
02	MYP1	57	29	19	48
03	MYP1	58	12	33	45
Subtotal			**100**	**130**	**230**
04	MYP1	59		42	
05	MYP2	60			42
06	MYP2	61			42
07	MYP2	62			42
08	MYP2	63			42
09	MYP2	64			42
Total					**482**

Notes: Breakdown of aircraft funded from FY04 not available; second MYP contract includes option to increase number by up to six per year.

controls; lateral and then directional control progressively washed out as angle of attack reaches extreme values; height, heading and airspeed holds provided in fly-by-wire system; aircraft can land automatically using carrier-based guidance system. Bertea hydraulic actuators for trailing-edge flaps; Hydraulic Research actuators for ailerons; National Water Lift actuators for tailerons.

STRUCTURE: Multispar wing mainly of light alloy, with graphite/epoxy inter-spar skin panels and trailing-edge flaps; tail surfaces mainly graphite/epoxy skins over aluminium honeycomb core; graphite/epoxy fuselage panels and doors; titanium engine firewall. Northrop Grumman responsible for rear and centre fuselages and delivered 100th shipset to St Louis at end of January 2002; assembly and test at St Louis factory; CASA produces horizontal tail surfaces, flaps, leading-edge extensions, speedbrakes, rudders and rear side panels for all F/A-18s.

LANDING GEAR: Messier-Dowty retractable tricycle type, with twin-wheel nose and single-wheel main units. Nose unit retracts forward, mainwheels rearward, turning 90° to stow horizontally inside the lower surface of the engine air ducts. Bendix wheels and brakes. Nosewheel tyres size 22×6.6–10 (20 ply) tubeless, pressure 24.13 bar (350 lb/sq in) for carrier operations, 10.34 bar (150 lb/sq in) for land operations. Mainwheel tyres size 30×11.5–14.5 (24/26 ply) tubeless, pressure 24.13 bar (350 lb/sq in) for carrier operations, 13.79 bar (200 lb/sq in) for land operations. Ozone nosewheel steering unit. Nose unit towbar for catapult launch. Arrester hook, for carrier landings, under rear fuselage.

POWER PLANT: Two General Electric F414-GE-400 turbofans, each rated at approximately 97.9 kN (22,000 lb st) with afterburning. Internal fuel capacity (JP-5 fuel) 8,063 litres (2,130 US gallons; 1,774 Imp gallons). Wing tanks protected from combat damage by low-density foam. Provision for five 1,250 litre (330 US gallon; 275 Imp gallon) or Lincoln Composites 1,817 litre (480 US gallon; 400 Imp gallon) external tanks, giving maximum fuel capacity of 17,148 litres (4,530 US gallons; 3,772 Imp gallons) and ability to operate in air refuelling role using hose drum unit on centreline station. Normal operational fit anticipated as three external tanks of either size.

ACCOMMODATION: Pilot only in F/A-18E, on Martin-Baker SJU-5/6 NACES zero/zero ejection seat, in pressurised, heated and air conditioned cockpit. Upward-opening canopy, with separate windscreen, on all versions. Two pilots or pilot and Naval Flight Officer in F/A-18F. Pilot and Electronic Countermeasures Officer in EA-18G.

SYSTEMS: High commonality with F/A-18C/D, incorporating two separate and independent hydraulic systems, each at 207 bar (3,000 lb/sq in), but with more powerful actuators to accommodate enlarged control surfaces with increased deflections. Leland Electrosystems power generating system; provides 60 per cent more electrical power than in F/A-18C. Hamilton Sundstrand air conditioning; Vickers hydraulic pumps; oxygen system; fire detection and extinguishing systems. Honeywell GTC36-200 APU.

AVIONICS: Over 90 per cent commonality with F/A-18C, but differences include following:

Comms: Rockwell Collins AN/ARC-210 secure UHF/VHF radio. Multifunction information distribution system (MIDS) datalink, Rockwell Collins digital communication system and Hazeltine combined interrogator transponder to be installed as part of upgrade package; IOC of MIDS-configured aircraft occurred in late 2003.

Radar: Raytheon AN/APG-73 multimode, digital air-to-air and air-to-ground radar as standard. Raytheon AN/APG-79 active electronically scanned array (AESA) X-band radar under development for use on Super Hornet from 2006; first radar unit formally rolled out by Raytheon on 21 November 2002. Current plans anticipate procurement of 413 AESA radars, comprising 277 for new-build aircraft and 136 for retrofit to existing F/A-18E/Fs. Delivery of AN/APG-79-equipped aircraft expected to begin by 2005–06, following start of low-rate initial production in FY03, when eight radars will be acquired for trials and operational evaluation in 2006; second LRIP batch, funded in FY04, comprises 12 radars.

Flight: Litton embedded GPS; Smiths/Harris tactical aircraft moving map capability (TAMMAC); DRS Technologies deployable flight incident recorder set (DFIRS).

Instrumentation: Cockpit has 76 × 127 mm (3 × 5 in) touch-panel LCD upfront display and 159 mm (6¼ in) square colour LCD tactical situation display; also two 127 mm (5 in) square monochrome displays and will have monochrome programmable LCD in place of F/A-18C engine/fuel display; Kaiser AN/AVQ-28 HUD. Planar Advance/dpiX awarded joint development contract for Eagle-6 multifunction AMLCD in first half of 1998. Aft cockpit to be 'missionised' for strike operations, including 205 × 255 mm (8 × 10 in) AMLCD and two hand controllers for use by NFO.

Mission: Raytheon AN/ASQ-228 advanced targeting forward-looking infra-red (ATFLIR) targeting and navigation system subject of US$900 million EMD contract in March 1998. Low-rate production authorised in March 2001 with first units delivered 21 May 2002; operational evaluation completed in third quarter of 2003, with full rate

Single-seat F/A-18E (nearest) and two-seat F/A-18F Super Hornets of trials squadron VX-31, US Navy *(Jamie Hunter)* ***NEW***/1047787

production go-ahead in late 2003, when Raytheon awarded US$298 million contract for initial 88 pods. Total purchase then expected to be about 530 pods, down from original planned 574. Shared Reconnaissance Pod (SHARP) system, incorporating Recon/Optical CA-279 digital reconnaissance sensor system, under development. Initial deployment by VFA-14 and VFA-41 in USS *Nimitz* in early 2003, albeit with some key elements, such as high-resolution electro-optical/infra-red sensor and datalink, omitted. DRS Technologies WRR-818 cockpit video recording system.

Advanced mission computers based on commercial hardware and software and colour flat panel LCD displays are in development for incorporation in FY05, with DY4 Systems awarded US$1 million order to upgrade existing system which relies on Control Data Corporation AN/AYK-14 digital computers; upgrade will involve replacement of AN/AYK-14 by DMV-179 single board computers and PMC-642 fibre channel network interface module. Vision Systems International joint helmet-mounted cueing system (JHMCS) installed to target the AIM-9X high off-boresight missile; testing of JHMCS on Super Hornet began at China Lake, California, in 2001, with delivery of production articles beginning in 2002.

Self-defence: Management by BAE Systems (formerly Sanders) AN/ALQ-214(V)2 integrated defensive electronic countermeasures suite (IDECM); interfaces with Raytheon AN/ALR-67(V)3 radar warning receiver, BAE Systems (formerly Sanders) AN/ALE-55 fibre optic towed radar decoy (with triple dispenser stowed between jetpipes); and four BAE Systems AN/ALE-47 chaff/flare dispensers. However, delay and cost growth of IDECM means that first three operational squadrons have aircraft fitted with less sophisticated Raytheon AN/ALE-50 towed decoy in conjunction with AN/ALQ-165 ASPJ. AN/ALQ-214 RF countermeasures system should achieve IOC in July 2004, with AN/ALE-55 FOTD to begin operational evaluation in December 2005.

ARMAMENT: Full range of USN offensive and defensive ordnance can be employed, with new weapons added as they become available. Integration of Joint Air-to-Surface Stand-Off Missile (JASSM) now under way and should be completed by December 2007. At least 29 weapons combinations cleared before service entry. M61A2 20 mm cannon with 400 rounds; will be compatible with forthcoming AIM-9X Advanced Sidewinder missile.

DIMENSIONS, EXTERNAL (approx):
Wing span over missiles 13.62 m (44 ft 8½ in)
Wing aspect ratio 4.0
Width, wings folded 9.94 m (32 ft 7¼ in)
Length overall 18.38 m (60 ft 3½ in)
Height overall 4.88 m (16 ft 0 in)

AREAS:
Wings, gross 46.45 m^2 (500.0 sq ft)

WEIGHTS AND LOADINGS:
Operating weight empty: F/A-18E 14,552 kg (32,082 lb)
F/A-18F 14,875 kg (32,794 lb)
EA-18G 14,806 kg (32,642 lb)
Max fuel weight (JP-5): internal 6,354 kg (14,008 lb)
external 7,381 kg (16,272 lb)
Max external stores load: T-O 8,028 kg (17,700 lb)
landing 4,082 kg (9,000 lb)
T-O weight, attack mission 29,937 kg (66,000 lb)
Max wing loading 644.5 kg/m^2 (132.00 lb/sq ft)
Max power loading 153 kg/kN (1.50 lb/lb st)

PERFORMANCE (estimated):
Max level speed at altitude more than M1.8
Approach speed 125 kt (232 km/h; 144 mph)
Combat ceiling 15,240 m (50,000 ft)
Min wind over deck:
launching 30 kt (56 km/h; 34.5 mph)
recovery 15 kt (28 km/h; 17.5 mph)
Combat radius:
interdiction with two SLAM-ER, two AMRAAMs, two Sidewinders and three 1,817 litre (480 US gallon; 400 Imp gallon) external tanks, hi-hi-hi (including flight of SLAM-ER) 945 n miles (1,750 km; 1,087 miles)
fighter escort with four AMRAAMs, two Sidewinders and three 1,817 litre (480 US gallon; 400 Imp gallon) external tanks 795 n miles (1,472 km; 914 miles)
Combat endurance:
maritime air superiority, six AAMs, three 1,818 litre (480 US gallon; 400 Imp gallon) external tanks, 150 n miles (278 km; 173 miles) from aircraft carrier 2 h 15 min
g limit +7.5

UPDATED

BOEING C-17A GLOBEMASTER III

TYPE: Strategic transport.

DEVELOPMENT MILESTONES

Programme launched	28 Aug 1981
First order (development)	31 Dec 1985
First prototype material cut	2 Nov 1987
First flight	15 Sep 1991
Production go-ahead	20 Jan 1988
First flight, production	18 May 1992
First delivery	14 Jun 1993
Entered service (IOC)	17 Jan 1995

PROGRAMME: US Air Force selected McDonnell Douglas to develop C-X cargo aircraft 28 August 1981; full-scale development called off January 1982 and replaced on 26 July 1982 by slow-paced preliminary development order; development and three prototypes (one flying) ordered 31 December 1985; fabrication of prototype C-17A (T1/87-0025) began 2 November 1987; first production C-17A contract 20 January 1988; assembly started at Long Beach 24 August 1988; assembly of prototype completed 21 December 1990. Programme transferred from Douglas Aircraft Company to McDonnell Douglas Aerospace in 1992; to McDonnell Douglas Military Transport Aircraft in 1996 and to Boeing following merger of August 1997.

First flight (T1/87-0025) 15 September 1991 — also delivery to Edwards AFB; first flight of initial production aircraft (P1/88-0265) 18 May 1992; first delivery to operational unit 14 June 1993. First overseas service flight (P11/92-3291) to Mildenhall, UK, 25 May 1994.

C-17 was named Globemaster III on 5 February 1993; peak production rate 15 per year (but could be increased to 18); assembly in 102,200 m^2 (1.1 million sq ft) facility at Long Beach, California.

Boeing C-17A Globemaster III of the RAF *(Paul Jackson)* ***NEW***/0576865

Development flight testing completed 15 December 1994, by which time 16 production aircraft delivered and 22nd record set (further 13 world records set by 71st production aircraft during testing at Edwards in November 2001). Initial AMC Squadron (17th AS) received its 12th C-17A 22 December 1994; achieved IOC 17 January 1995 with acceptance of 13th (nominal spare) aircraft. Second squadron (14th AS) received its first aircraft 17 February 1995. Air Education and Training Command's 97th AMW at Altus AFB, Oklahoma, subsequently took delivery of eight aircraft between March 1996 and November 1997, all of which had previously been assigned to the 437th AW. Deliveries to 7th AS of 62nd AW at McChord AFB, Washington, began at end of July 1999. First C-17 for Air National Guard (02-1112) handed over at Long Beach on 17 December 2003 and allocated to 172nd AW at Jackson, Mississippi. C-17 fleet passed 500,000 flying hours in 2003, excluding more than 1,000 hours accumulated by the test-dedicated prototype aircraft.

CURRENT VERSIONS: **C-17A:** Basic standard.

Detailed description applies to C-17A.

C-17B: Not assigned. Used unofficially (along with C-17ER) to identify aircraft with extended-range fuel tanks (Lot 12 and upwards).

KC-17: Private venture tanker/transport project, offered unsuccessfully as replacement for USAF KC-135R/T Stratotanker.

BC-17X: Projected civil cargo version, known until 2000 as MD-17. No recent information, but appears dormant.

CUSTOMERS: US Air Force; original requirement 210; cut to 120 by 1991; capped at 40 in January 1994 for two year probationary period, during which contractor to achieve performance, cost and delivery targets; decision taken on 3 November 1995 to acquire the balance of 80 C-17s. Multiyear procurement contract signed 31 May 1996 for production through 2004; first aircraft from this contract (P41/97-0041) delivered 10 August 1998; last due to have been handed over on 30 November 2004.

Second multiyear procurement contract valued at US$9.7 billion for additional 60 aircraft announced on 15 August 2002, raising total USAF buy to 180, with negotiations under way between Boeing and the USAF for further 42 C-17s that could be purchased with FY08-10 funds, sustaining production at rate of 14 per year.

Basing plans were revealed soon after the decision to procure 120 in all, when the USAF announced that the 437th AW at Charleston AFB, South Carolina, and the 62nd AW at McChord AFB, Washington, will each receive 48 Globemaster IIIs, with these respectively being supported by Air Force Reserve Command personnel of the 315th AW and 446th AW. A further eight C-17As are used by the 97th Air Mobility Wing at Altus AFB, Oklahoma, for training duties. Finally, eight are assigned to the Air National Guard's 172nd AW at Jackson, Mississippi, first of which (P112/21112) handed over 17 December 2003 as first for an ANG unit. McGuire AFB, New Jersey and March Air Reserve Base, California, are expected to receive aircraft from the 60 that were ordered in 2002, but other locations earmarked to gain C-17s are Elmendorf AFB, Alaska, and Hickam AFB, Hawaii, both of which will eventually have eight each.

The first C-17A to be handed over to Air Mobility Command was the final aircraft from Lot 2 (P6/89-1192), which made its first flight on 8 May 1993 and was ferried to Charleston AFB for the 437th AW on 14 June 1993. Subsequent deliveries to Air Mobility Command comprise four Lot 3 aircraft between 26 August 1993 and 8 February 1994; four Lot 4 aircraft between 8 April and 20 August 1994; six Lot 5 aircraft between 28 September 1994 and 19 June 1995; six Lot 6 aircraft between 31 July 1995 and May 1996; six Lot 7 aircraft between August 1996 and June 1997; eight Lot 8 aircraft between August 1997 and May 1998; eight Lot 9 aircraft between August 1998 and February 1999; nine Lot 10 aircraft between April and December 1999, including the 50th C-17A, which was accepted by the USAF on 21 May 1999; 13 Lot 11 aircraft between March and December 2000; 19 Lot 12 aircraft (including four for RAF) between February 2001 and April 2002; 12 Lot 13 aircraft between May 2002 and early 2003; and 15 Lot 14 aircraft between early 2003 and 17 December 2003. By end of 2002, more than 90 C-17As had been received by the USAF; 100th C-17 (including three prototypes and four for UK) delivered on 8 November 2002, this being 93rd production aircraft for USAF (01-0193).

In early 1997, UK Ministry of Defence reportedly considering acquisition of a small number of C-17s for service with RAF in strategic airlift role; no formal requirement then existed, but cost data were received from the USA. Invitation to submit bids for supply on lease basis of up to four 'C-17 equivalent' Short Term Strategic Airlift aircraft issued September 1998, but subsequent criticism by other potential bidders that requirement was weighted in favour of C-17 prompted UK MoD to waive some requirements. Submissions made in early 1999, but competition then suspended in August 1999, when none of competing bids proved acceptable. Restarted, late 1999; announced that C-17 selected on 16 May 2000, with total of four aircraft to be acquired for No. 99 Squadron. These taken from Lot 12 production; assembly of first officially begun 28 June 2000, although lease agreement with Chase Manhattan Bank not signed until 12 January 2001. Lease duration set at seven years, with options for two one-year extensions. Each aircraft valued at £197 million. First aircraft (UK1/ZZ171/00-0201) accepted by RAF on 17 May 2001 at Long Beach; ferried to UK via Charleston AFB and arrived at Brize Norton 23 May. Final aircraft (UK4/ZZ174/00-0204) accepted 24 August 2001 and at Brize Norton by end of August; first operational sortie accomplished in early June; fleet in-service date officially 30 September 2001, at which time UK MoD requested information about availability of further two aircraft and Boeing allocated three further airframes (with provisional US civil registrations) in anticipation of early decision which failed to materialise because of a shortage of funds. However, in mid-2003, UK Ministry of Defence reported to be discussing lease of additional two to four C-17s with Boeing.

Canada has requirement for new strategic transport and is believed to be examining several options, including lease or outright purchase of C-17; Australia has also begun studies of requirements for strategic airlift aircraft. Version of the C-17 to be offered for Japanese C-X requirement.

COSTS: Multiyear procurement contract for 80 aircraft signed in May 1996 worth US$14.2 billion; separate US$1.6 billion contract followed in December 1996 for 320 F117 engines. Unit price proposed in November 1995 was US$190 million; recent figures (early 2000) refer to unit cost of US$199 million. Second multiyear procurement contract for 60 aircraft valued at US$9.7 billion.

C-17 PROCUREMENT

FY	Lot	Qty	Cum	Line Numbers	First aircraft	Delivery
—	Proto	1	(1)	T1	87-0025	15 Sep 1991
88	1	2	2	P1-P2	88-0265	18 May 1992
89	2	4	6	P3-P6	89-1189	7 Sep 1992
90	3	4	10	P7-P10	90-0532	26 Aug 1993
91	—	—	—	—	—	—
92	4	4	14	P11-P14	92–3291	8 Apr 1994
93	5	6	20	P15-P20	93–0599	29 Sep 1994
94	6	6	26	P21-P26	94–0065	31 Jul 1995
95	7	6	32	P27-P32	95–0102	3 Jul 1996
96	8	8	40	P33-P40	96–0001	28 Aug 1997
97	9	8	48	P41-P48	97–0041	10 Aug 1998
98	10	9	57	P49-P57	98–0049	22 April 1999
99	11	13	70	P58-P70	99–0058	17 March 2000
00	12	15	85	P71-P85	00–0171	8 Feb 2001
	12	4 (RAF)	89	UK1-UK4	00–0201/ZZ171	17 May 2001
01	13	12	101	P86-P97	01–0186	May 2002
02	14	15	116	P98-P112	02–1098	Early 2003
03	15[1]	15	131	P113-P127	03-3113	Early 2004
04	16	11	142	P128-P138	—	—
05	17	14	156	P139-P152	—	—
06	18	14	170	P153-P166	—	—
07	19	14	184	P167-P180	—	—
Total		**184 + 1**				

[1] Includes first seven aircraft from follow-on batch of 60 C-17s.

DESIGN FEATURES: Typical T tail, upswept rear fuselage, high-wing, podded-engine transport, deriving much detail, including externally blown flap system, from McDonnell Douglas YC-15 medium STOL transport prototypes, which involves extended flaps being in exhaust flow from engines during take-off and landing. Combines load-carrying capacity of Lockheed C-5 with STOL performance of Lockheed Martin C-130; required to operate routinely from 915 m (3,000 ft) long and 27.45 m (90 ft) wide runways, complete 180° three-point turn in 25 m (82 ft) and back-taxi up 1 in 50 gradient when fully loaded using thrust reversers. Structure designed to survive battle damage and protect crew; rear-loading ramp.

Supercritical wing with 25° sweepback; 2.72 m (8.9 ft) high NASA winglets, angled at 15° from vertical and with 30° sweep.

FLYING CONTROLS: First military transport with all-digital FBW control system and two-crew cockpit with central stick controllers; outboard ailerons and four spoilers per wing; four elevator sections; two-surface rudder split into upper and lower segments; full-span leading-edge slats; two-slot, fixed-vane, simple hinged flaps over about two-thirds of trailing-edge; small strakes under tail. Quadruple-redundant BAE Systems digital fly-by-wire flight control system, with mechanical back-up; redesign of FCS initiated in 2002, with BAE Systems selected in October to upgrade with CsLEOS real-time operating system (RTOS) as part of package of improvements to autopilot flight control system. This will be installed on new-build C-17s beginning with 121st aircraft, due for delivery in mid-2004.

STRUCTURE: Major subassemblies produced in factory at Macon, Georgia; some 50 subcontractors, of which 21 for airframe; subcontractors include Vought (composites ailerons, rudder, elevators, vertical and horizontal stabilisers, engine nacelles and thrust reversers), McCoak Metals (wing skins), Contour Aerospace (wing spars and stringers), Kaman Aerospace (wing ribs and bulkheads), Hitco Technologies (tailcone), Heath Tecna (wing-to-fuselage fillet), Aerostructures Hamble (composites flap hinge fairings and trailing-edge panels) and Northwest Composites (main landing gear pod panels). Raytheon was original supplier of composites winglets and landing gear doors, but replaced by Marion Composites with effect from 41st aircraft; Marion Composites also fabricates nose and tail radomes for last 80 aircraft of USAF order. C-17A structure is 69.3 per cent aluminium; 12.3 per cent steel alloys; 10.3 per cent titanium and 8.1 per cent composites. Wings of P1-P10 underwent local strengthening as consequence of load test of 1 October 1992; first rework to P8/90-0533 at McDonnell Douglas, Tulsa, between 3 January and 9 April 1994.

Proposal, early 1994, to design all-composites horizontal tail surfaces for weight-saving resulted in McDonnell Douglas (later Boeing) securing a US$40.7 million contract to build new unit; revised tail manufactured by Northrop Grumman (now Vought) using AS-4 carbon fibre for spars and skins and machined 7075 aluminium for ribs, resulting in assembly said to be 50 per cent cheaper and 20 per cent lighter than existing tail. Incorporates 2,000 fewer components and 42,000 fewer fasteners. Following testing on prototype, completed April 1999, it was incorporated on production aircraft beginning with P51/98-0051.

Landing gear pod entirely redesigned in 1995, drastically reducing number of parts and fasteners and simplifying process of attachment to aircraft; introduced on first Lot 8 aircraft (P33/96-0001) and then expected to save US$88 million over remainder of production programme for USAF. New, lighter and less costly engine nacelle, saving 113 kg (250 lb) per unit, flight tested at Edwards AFB between December 1997 and February 1998; first production aircraft with new nacelles (P41/97-0041) delivered to USAF on 10 August 1998.

LANDING GEAR: Hydraulically retractable tricycle type, with free-fall emergency extension; designed for sink rate of 3.81 m (12 ft 6 in)/s and suitable for operation from paved runways or unpaved strips. Mainwheel units, each consisting of two legs in tandem with three wheels on each leg, rotate 90° to retract into fairings on lower fuselage sides; tyre size 50×21.0–20 (30 ply) tubeless; pressure 9.52 bar (138 lb/sq in). Menasco (Goodrich from 41st aircraft) twin-wheel nose leg retracts forwards; tyre size 40×16–14 (26 ply) tubeless; pressure 10.69 bar (155 lb/sq in). Honeywell wheels and carbon brakes initially fitted; with effect from aircraft P90/01-0190, Messier-Bugatti wheels, tyres and brakes fitted as standard. Permitting C-17 to be certified to 278,959 kg (615,000 lb) MTOW, these are to be retrofitted to entire fleet. Minimum ground turning radius at outside mainwheels 17.37 m (57 ft 0 in); minimum taxiway width for three-point turn 27.43 m (90 ft 0 in); wingtip/tailplane clearance 74.24 m (237 ft 0 in).

POWER PLANT: Four Pratt & Whitney F117-PW-100 (PW2040) turbofans, with maximum flat rating of 179.9 kN (40,440 lb st), pylon-mounted in individual underwing pods and each fitted with a directed-flow thrust reverser deployable both in flight and on the ground. With effect from the 20th production aircraft, an improved version of the F117 was adopted, this embodying single-crystal turbine blade technology, a supercharged compressor and enhanced thermal barrier coatings; benefits include a 20 per cent reduction in cost of maintenance, increased reliability and slightly better sfc. Pratt & Whitney, Boeing and USAF seeking in late 2003 to improve engine reliability in wake of increase in unplanned removals and concerns over durability that emerged during Afghan and Iraq operations. Provision for in-flight refuelling. Two outboard wing fuel tanks of 21,210 litres (5,603 US gallons; 4,665 Imp gallons) each; two inboard wing fuel tanks of 30,056 litres (7,940 US gallons; 6,611 Imp gallons) each; total capacity 102,532 litres (27,086 US gallons; 22,554 Imp gallons). BAE Systems fuel pumps. With effect from the 71st (first Lot 12) aircraft, an extended-range fuel tank containment system (ERFCS) was adopted; this converts wing dry bay into additional fuel tank containing approximately 36,339 litres (9,600 US gallons; 7,994 Imp gallons) of fuel, thus increasing range with 18,144 kg (40,000 lb) payload by 900 n miles (1,667 km; 1,036 miles). ERFCS also installed on four UK aircraft.

ACCOMMODATION: Normal flight crew of pilot and co-pilot side by side and two observer positions on flight deck, plus loadmaster station at forward end of main floor; access to flight deck via downward-opening airstair door on port side of lower forward fuselage. Bunks for crew immediately aft of flight deck area; crew comfort station at forward end of cargo hold.

Main cargo hold able to accommodate US Army wheeled and tracked vehicles up to M1 main battle tank, including 5 ton expandable vans in two rows, or up to three AH-64 Apache attack helicopters, with loading via hydraulically actuated rear-loading ramp which forms underside of rear fuselage when retracted. Aircraft fitted with 27 stowable tip-up seats along each sidewall and another 48 seats carried on board which can be erected along the centreline; optionally up to 36 litters for medical evacuation mission or up to 90 passengers on 10-passenger pallets in addition to 54 sidewall seats. Air delivery system capability for nine 463L pallets plus two on ramp in single row; or logistics handling system for 18 463L pallets in double row. Airdrop capability includes single platforms of up to 27,215 kg (60,000 lb), multiple platforms of up to 54,430 kg (120,000 lb), container delivery system (CDS) for up to 40 CDS bundles or 40 Tri-wall Aerial Delivery System (TRIADS) containers, or up to 102 paratroops; aircraft originally configured for single-row airdrop, but dual-row capability operationally certified in 1998 and is standard feature of 51st and subsequent C-17s, as well as being retrofitted to earlier aircraft. Equipped for low-altitude parachute extraction system (LAPES) drops.

Cargo handling system includes rails for airdrops and rails/rollers for normal cargo handling, plus Goodrich winch. Each row of rails/rollers can be converted quickly by a single loadmaster from one configuration to the other. Total of 295 cargo tiedown rings, each stressed for 11,340 kg (25,000 lb), all over cargo floor forming grid averaging 74 cm (29 in) square. Three quick-erecting litter stanchions, each supporting three litters, permanently carried. Main access to cargo hold is via rear-loading ramp, which is itself stressed for 18,145 kg (40,000 lb) of cargo in flight. Underfuselage door aft of ramp moves upward inside fuselage to facilitate loading and unloading. Paratroop door at rear on each side; four overhead FEDS (flotation equipment deployment system) escape hatches, three of which are equipped with a liferaft.

SYSTEMS: Include Honeywell computer-controlled integrated environmental control system and cabin pressure control system; 2,440 m (8,000 ft) equivalent cabin pressure up to 11,280 m (37,000 ft); quad-redundant flight control and four independent 276 bar (4,000 lb/sq in) hydraulic systems; independent fuel feed systems; electrical system; Honeywell GTCP331-250G APU (at front of starboard landing gear pod), provides auxiliary power for environmental control system, engine starting, and on-ground electronics requirements; onboard inert gas generating system (OBIGGS) for the explosion protection system, pressurised by engine bleed air at 4.14 bar (60 lb/sq in) to produce NEA (nitrogen-enriched air) and governed by a Parker-Gull system controller; fire suppression system; smoke detection systems. Separate OBIGGS for ERFCS-configured aircraft. Next-generation OBIGGS being designed and developed at present and will be installed on all aircraft. All phases of cargo operation and configuration change capable of being handled by one loadmaster.

Electrical system includes single 90 kVA generator per engine and an APU, providing 115/200 V, three-phase, 400 Hz AC power; four 200 A transformer-rectifiers providing 28 V DC; single-phase, 1,000 VA inverter for ground refuelling and emergency AC; and two 40 Ah Ni/Cd batteries for APU starting and emergency DC. Aeromedical equipment provided with 60 Hz power.

Boeing C-17A Globemaster III of the RAF *(Paul Jackson)* ***NEW***/0576866

AVIONICS: *Comms:* Telephonics Corporation IRMS open architecture integrated radio management system; Rockwell Collins AN/ARC-210 secure radio (from P50/98-0050, replacing Raytheon AN/ARC-187 plus Honeywell KY 58 encoder); satcom; VHF-AM/FM; HF; secure voice and jam-resistant UHF/VHF-FM intercom; IFF/SIF; en route army/marine UHF LOS and satcom hook-up; cockpit voice recorder; crash position indicator. Automatic communications processor and multiband radio in new-build aircraft starting with P50/98-0050 and retrofitted to earlier C-17As. In December 1998, Rockwell Collins selected by Boeing to provide SAT-2000 Aero-1 satcom systems and CMU-900 communication management units for future global air traffic management (GATM) environment; initial order for support and preproduction hardware, with options to install equipment on all 120 aircraft. Production line GATM introduction at P71/00-0171, involving TCAS, Mode S IFF, ADS-A (automatic dependent surveillance) and Inmarsat Aero I satcom, plus liquid crystal displays replacing all head-down CRTs. GATM augmented from P97 by HF datalink and RNP 4 (required navigation performance) measures. Under development in 1999 were two SOLL II (special operations low-level) roll-on/roll-off communications suites for use on specific missions; and carry-on radio suite for use on special forces flights, beginning in 2002; SOLL II configured C-17 made combat debut in Iraq war of 2003.

Radar: Honeywell AN/APS-133(V) weather/mapping radar initially installed. New Honeywell RDR-4000M radar to be fitted from outset to all aircraft of follow-on order and will be retrospectively installed on older C-17s as their existing systems become inoperative.

Flight: Three Delco Electronics mission computers with MDC software and electronic control system on P1 to P40; beginning with P41/97-0041, two Lockheed Martin core integrated processors (CIPs) replaced Delco Electronics equipment and retrofitted on earlier aircraft by September 1998; additional memory added from P50/98-0050 onwards. Hamilton Sundstrand aircraft and propulsion data management computer; Honeywell dual air data computers; Litton warning and caution system; master warning system provides aural and voice alerts plus visual alerts on glareshields; General Dynamics automatic test equipment, and support equipment data acquisition and control system; GPS and four Honeywell inertial reference units, although new, embedded GPS INS introduced from P50/98-0050; VOR/DME Tacan; ILS/marker beacon; UHF-DF; ground proximity warning system; radar altimeter; flight data recorder; flight plan entry manually or via laptop computer. BAE Systems terrain awareness warning system (TAWS) to be installed in conjunction with Lockheed Martin Control Systems video integrated processor (VIP), currently under development, from P86 onwards; TAWS underwent initial flight testing from Edwards AFB in 2001–02 and will complete further trials in first half of 2002.

Instrumentation: Advanced digital avionics and four Honeywell full-colour cathode ray tube (CRT) displays; new Honeywell 152 × 152 mm (6 × 6 in) AMLCD colour multifunction displays (MFDs) in 71st and subsequent production aircraft; two BAE Systems full flight regime foldable head-up displays.

Mission: Litton integrated mission and communications keyboards (MCKs) and displays (MCDs); Sierra Technologies AN/APN-243A(V) station-keeping equipment (SKE 2000) fitted as standard commencing with P33/96-0001, retrofitted to earlier aircraft and upgraded from P86 (first Lot 13) onwards to increase capacity from 17 to 100 aircraft. Loadmaster's laptop computer from P41/97-0041 onwards; this replaced from P58/99-0058 by Dolch aircrew data device.

Self-defence: Development of defensive electronic systems completed in 1994. ATK, Lockheed Martin or Honeywell AN/AAR-47 missile approach warning system and associated BAE Systems AN/ALE-47 automatic flare dispenser installed on five aircraft; USAF has adopted this equipment across the fleet. Northrop Grumman laser-based large aircraft infrared countermeasures system (LAIRCM) currently being installed on production C-17s and will be retrofitted to older aircraft. Initial installation of single turret on batch 12 aircraft from third quarter 2003, with definitive LAIRCM (including three laser transmitters) to follow at later date.

EQUIPMENT: Removable crew armour can be fitted around flight deck and loadmaster's area.

DIMENSIONS, EXTERNAL:
Wing span: wings only ... 50.29 m (165 ft 0 in)
  at winglet tips ... 51.74 m (169 ft 9 in)
Wing aspect ratio ... 7.2
Length: overall ... 53.04 m (174 ft 0 in)
  fuselage ... 48.49 m (159 ft 1¼ in)
Height: to flight deck roof ... 7.34 m (24 ft 1 in)
  to winglet tips ... 6.93 m (22 ft 9 in)
  overall ... 16.79 m (55 ft 1 in)
Fuselage diameter ... 6.85 m (22 ft 6 in)
Tailplane span ... 19.81 m (65 ft 0 in)
Wheel track ... 10.26 m (33 ft 8 in)
Wheelbase: to front mainwheel ... 17.60 m (57 ft 8¾ in)
  to rear mainwheel ... 20.05 m (65 ft 9½ in)
Ground clearance under engine pods:
  inboard ... 2.71 m (8 ft 10½ in)
  outboard ... 2.35 m (7 ft 8½ in)
Distance between fuselage centreline and engine centreline: inboard .. 7.44 m (24 ft 5 in)
  outboard ... 13.94 m (45 ft 9 in)
Height of winglets ... 2.72 m (8 ft 11 in)

DIMENSIONS, INTERNAL:
Cargo compartment:
  Length, incl 6.05 m (19 ft 10 in) rear-loading ramp ... 26.82 m (88 ft 0 in)
  Loadable width ... 5.49 m (18 ft 0 in)
  Max height: under wing ... 3.96 m (13 ft 0 in)
    aft of wing ... 4.50 m (14 ft 9 in)
Height to sill ... 1.63 m (5 ft 4 in)
  Volume ... 591.8 m³ (20,900 cu ft)

AREAS:
Wings, gross ... 353.03 m² (3,800.0 sq ft)
Winglets (total) ... 3.33 m² (35.85 sq ft)
Ailerons (total) ... 11.83 m² (127.34 sq ft)
Tailplane ... 78.50 m² (845.00 sq ft)

WEIGHTS AND LOADINGS:
Operating weight empty:
  non-ERFCS aircraft ... 125,645 kg (277,000 lb)
  ERFCS aircraft ... 127,685 kg (281,500 lb)
Max payload (2.5 *g* load factor):
  non-ERFCS aircraft ... 76,655 kg (169,000 lb)
  ERFCS aircraft ... 75,250 kg (165,900 lb)
Max weight on rear-loading ramp ... 18,143 kg (40,000 lb)
Max usable fuel weight:
  non-ERFCS aircraft ... 82,125 kg (181,054 lb)
  ERFCS aircraft ... 110,990 kg (244,688 lb)
Max T-O weight: ERFCS aircraft
... 278,959 kg (615,000 lb)
Max wing loading: ERFCS aircraft ... 790.2 kg/m² (161.84 lb/sq ft)
Max power loading: ERFCS aircraft ... 388 kg/kN (3.80 lb/lb st)

PERFORMANCE:
Normal cruising speed at 8,535 m (28,000 ft) ... M0.74–0.77
Max cruising speed at low altitude ... 350 kt (648 km/h; 403 mph) CAS
Airdrop speed: at S/L ... 115–250 kt (213–463 km/h; 132–288 mph) CAS
  at 7,620 m (25,000 ft) ... 130–250 kt (241–463 km/h; 150–288 mph) CAS
Approach speed with max payload ... 115 kt (213 km/h; 132 mph) CAS
Service ceiling ... 13,715 m (45,000 ft)
Runway LCN (paved surface) ... better than 49
T-O field length at MTOW ... 2,360 m (7,740 ft)
Landing field length with 72,575 kg (160,000 lb) payload, using thrust reversal ... 915 m (3,000 ft)

Range with payloads indicated, with no in-flight refuelling
  18,144 kg (40,000 lb) payload:
  non-ERFCS aircraft ... 4,400 n miles (8,148 km; 5,063 miles)
  ERFCS aircraft ... 5,300 n miles (9,815 km; 6,099 miles)
  72,575 kg (160,000 lb), T-O in 2,286 m (7,500 ft), land in 915 m (3,000 ft), load factor of 2.25 *g*
  both: ... 2,400 n miles (4,444 km; 2,761 miles)
  self-ferry (zero payload), T-O in 1,128 m (3,700 ft), land in 701 m (2,300 ft), load factor of 2.5 *g*:
  non-ERFCS aircraft ... 4,700 n miles (8,704 km; 5,408 miles)
  ERFCS aircraft ... 6,250 n miles (11,575 km; 7,192 miles)

UPDATED

BOEING 737 AEW&C

TYPE: Airborne early warning and control system.

PROGRAMME: Adaptation of Boeing Business Jet (BBJ, which combines 737-700 fuselage with strengthened wing and landing gear of 737-800). Additional features include extra fuel

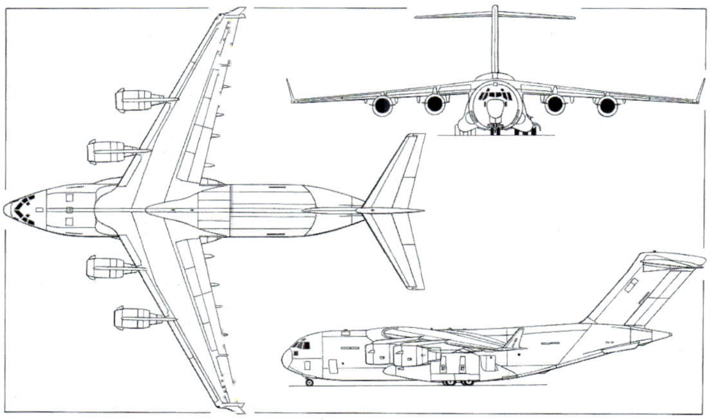

Boeing C-17A Globemaster III long-range heavy cargo transport *(Jane's/Dennis Punnett)* 0507286

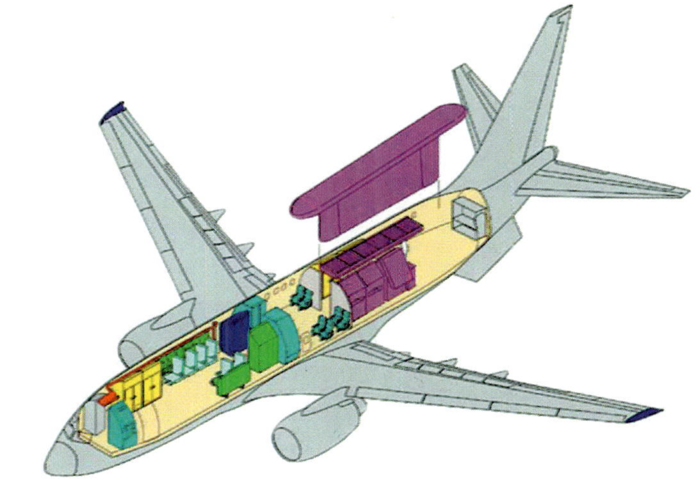

Interior diagram of Boeing 737 AEW&C *(Northrop Grumman)* ***NEW***/1047225

tanks in former baggage hold. Proposed for Australian Project Air 5077 Wedgetail by Boeing, Northrop Grumman Electronic Sensor Systems Division (ESSD) and BAE Australia. Competed against proposals from Lockheed Martin and Raytheon. Secured initial design activity contract worth about US$6 million in December 1997, paving way for submission of full tenders in early 1999. Selection of Boeing submission officially announced on 21 July 1999. RAAF initially planned to acquire seven aircraft for US$1.32 billion; cost concerns resulted in reduction to six plus one option in May 2000, but when contract signed on 20 December, this further reduced to four and three options, at reported cost of US$1.65 billion. Delivery of first two aircraft scheduled for November 2006, with operating unit to be No. 2 Squadron. Second pair to be handed over at end of 2007. Main base at Williamtown, New South Wales, with two aircraft permanently deployed to forward operating location at Tindal, Northern Territories; IOC to be achieved at end of 2008. Decision on exercising option for two additional aircraft was due 30 June 2003, but then extended to 30 June 2004; on 11 May 2004, it was announced that this would be converted to firm order, valued at US$224.25 million, for delivery by 2008. Preliminary design review of radar and IFF systems successfully completed in September 2001; further reviews of navigation system, mission computing hardware and other airborne mission systems undertaken by January 2002, with ground-based elements not scheduled to undergo PDR until 2004.

ESSD L-band multirole electronically scanned array (MESA) radar mounted above rear fuselage ('top hat' configuration) providing 360° coverage from stationary antenna 10.7 m (35 ft) long and 2.35 m (7 ft 9 in) high that weighs 2,853 kg (6,290 lb). Operating modes will include acute long-range or broad short-range scanning and track-while-scan; maximum detection range said to exceed 216 n miles (400 km; 249 miles). In late 1998, demonstrator BBJ temporarily fitted with full-scale replica of MESA radar, six operator consoles and equipment cabinets for inspection by Australian defence department officials; mockup also featured in-flight refuelling probe above cockpit and EW/ECM sensors. Definitive aircraft will, however, feature 10 operator consoles initially, with potential to add further two. Aerodynamic effects of MESA offset by two large strakes below rear fuselage. Electronic warfare self-protection (EWSP) system will include Northrop Grumman AN/AAQ-24(V) directed infra-red countermeasures system, plus chaff and flares; Elta providing advanced ESM/elint systems; these will be controlled by the ALR-2001 computer. Other mission equipment to include Link 11, JTIDS, Mode S IFF and satcom; flight deck tactical displays, three HF and eight VHF/UHF radios. Patrol endurance of 9 hours at 300 n miles (555 km; 345 miles) from base can be extended by airborne refuelling, with modification to include installation of flying boom receptacle and a removable probe. Maximum T-O weight 77,565 kg (171,000 lb); service ceiling 12,500 m (41,000 ft).

First Australian aircraft rolled out at Seattle on 31 October 2002; subsequently, flown to Georgetown, Delaware, on 4 January 2003 for installation of an auxiliary fuel system and tanks. Latter procedure took about three months to complete, whereupon aircraft returned to Boeing for structural modifications associated with installation of radar and mission systems. First MESA radar rolled out at beginning of November 2002 and installed on test range by Northrop Grumman; unit subsequently to Boeing for installation on 737, which took place on 22 October 2003. First flight of first aircraft (A30-001) accomplished on 20 May 2004; subsequently used for 27-week aerodynamic test phase, with second aircraft earmarked for test and evaluation of radar from May 2005. However, in late 2004 it was established that 'top hat' antenna would need additional height above the fuselage for improved radar coverage.

Republic of Korea interested in AEW-configured 737 as less costly solution to E-X requirement in lieu of Boeing 767, which considered too expensive; programme relaunched on 4 February 2004, after being suspended in 2002. Boeing proposal in competition with rival offerings from IAI/Elta (combination of Ilyushin Il-76 and Gulfstream G-550 platforms), L-3 Communications (Airbus A321) and Thales (Airbus A320). Total of four aircraft required by Korea, which is expected to take delivery of first two aircraft in 2009, with second pair following in 2011. Bids were due for submission in June/July 2004, with selection in November, followed by contract award in 2005.

After studying proposals for AEW aircraft involving Airbus A310/Phalcon and Boeing 737/MESA combinations, in early December 2000 Turkey announced selection of latter and revealed intent to obtain total of six aircraft (with option on two more) at cost of US$1.5 billion under 'Peace Eagle' programme; these will include some indigenous equipment. Contract signature was expected in early 2001, but negotiations continued throughout remainder of 2001 and 2002; contract finally signed in May 2003, at which time the number of aircraft to be purchased had been reduced to four (with two on option). First aircraft to be supplied by Boeing; remaining three will be fitted out by Turkish Aerospace Industries. Initial delivery target date is July 2007, with final aircraft going to TAI in June 2008.

Boeing originally forecasting potential sales of up to 50 737 AEW&C aircraft, with other possible customers including Italy, Singapore and Spain; by the beginning of 2002, this had fallen to 30 over next 10 years.

UPDATED

BOEING 737 MMA

TYPE: Maritime surveillance twinjet.

PROGRAMME: Design study for 737 MMA (Multi-mission Maritime Aircraft) revealed 18 April 2000 as potential replacement for Lockheed Martin P-3C Orion and EP-3E Aries II. Based on C-40A (see Boeing 737 entry under Boeing Commercial Airplanes), combining 737-900 wing and 737-800 fuselage, latter with internal weapons bay beneath forward section and former with hardpoints for air-to-surface missiles. Full range of maritime patrol equipment envisaged, including enlarged nose accommodating Raytheon AN/APS-137 search radar, up to seven operators' consoles and rotary sonobuoy launcher, plus additional fuel in aft baggage hold.

Maximum T-O weight 77,565 kg (171,000 lb); maximum transit speed 490 kt (907 km/h; 564 mph); mission radius 2,000 n miles (3,704 km; 2,301 miles). US Navy awarded US$493,000 concept exploration study contract to Boeing in July 2000; this followed in September 2002 by US$7 million contract for Phase 1 of component advanced development (CAD), during which air vehicle performance validated and mission system parameters were developed and analysed. Subsequently, in February 2003, a further contract, worth US$20.5 million, was awarded for Phase 2 of CAD, including additional performance analysis and continuing development of the associated mission system. Similar contracts to Lockheed Martin, which proposed an Orion derivative for the MMA.

On completion of Phase 2, Navy released request for proposals in June 2003; final bids were due for submission by 29 December 2003, with US Navy planning to award contract for the system development and demonstration (SDD) programme on 9 June 2004; in the event, announcement that Boeing proposal selected delayed until 14 June, at which time contract valued at US$3.89 billion awarded for SDD. SDD phase will entail manufacture of five airframes; one for airworthiness test, two for testing of mission systems and single examples for static and fatigue test purposes. SDD will continue until final quarter of FY08, with IOC expected in 2013; key programme milestones include decision to begin low-rate initial production which is due to be taken by 2008. Full-rate production not expected to begin until 2012 and will involve from 12 to 18 aircraft per year; total of 108 MMAs planned to be procured by US Navy.

Boeing also offering a version of the 737 for the joint German-Italian MPA-R programme, which could result in a contract for a total of 24 aircraft. Selection of winner was set for 2003, with contract award in 2004 and Boeing is seeking 'harmonisation', so that it can offer a single proposal to satisfy both the MMA and MPA-R requirements. Future of joint German-Italian programme appears uncertain, following German offer to buy surplus P-3C Orions from Netherlands in latter half of 2003.

UPDATED

BOEING 767 MILITARY VERSIONS

TYPE: Tanker-transport; airborne ground surveillance system.

PROGRAMME: Variants of Boeing 767 twin-turbofan airliner (which see under Boeing Commercial Airplanes for technical description).

CURRENT VERSIONS: **767 AWACS:** Production complete.

KC-767 Global Tanker Transport Aircraft (GTTA): Tanker-transport version announced by Boeing, February 1995, in anticipation of Japanese order; Boeing in discussions with Kawasaki by mid-1996, concerning co-operative venture to offer tanker version of the 767 to Japan ASDF. Kawasaki involvement then expected to take form of post-production work (including fitment of boom refuelling gear, extra tanks and associated plumbing) on 767-300 derivative. JASDF intended to request funding from FY99, but economic downturn and reductions in defence budget caused delay. Purchase intention reaffirmed 14 December 2001, by Japanese government, which announced plan to buy four aircraft. On 4 April 2003, Boeing announced signature of contract for first of planned four tanker-transports; this due for delivery in "Spring 2007", with remainder following at rate of one aircraft per year.

Version of 767 with three hose drum units offered unsuccessfully to UK Royal Air Force which had FSTA requirement for a new tanker aircraft to replace VC-10 and Tristar Airbus A330 derivative selected in late 2003.

Tanker-transport proposals most recently based on 767-200ER (see Boeing Commercial Airplanes entry); fuel dispensed through Boeing 'flying boom' and two Smiths Aerospace underwing pods; boom remotely controlled from cabin, assisted by CCTV and/or three-dimensional helmet-mounted display; proximity trials at Patuxent River NAS in June 2000 showed 767 to be stable refuelling platform. Fuel capacity 91,380 litres (24,140 US gallons; 20,101 Imp gallons), in standard (wing) tanks, plus 21,198 litres (5,600 US gallons; 4,663 Imp gallons) in supplementary underfloor tanks; total 112,578 litres (29,740 US gallons; 24,764 Imp gallons). As freighter (side cargo door and reinforced floor), can carry up to 18 463L pallets on main deck or 216 passengers, with additional cargo capacity on lower deck dependent upon auxiliary tank configuration.

First Boeing 767 tankers will be flown by Italian Air Force, which revealed intention in July 2001 to purchase four new-build aircraft as replacements for Boeing 707 tankers. Signature of final contract took place in early December 2002; delivery will be accomplished during 2005–08 and total cost, including option on two additional aircraft, is about US$618 million. Based on the 767-200ER commercial transport, it will be powered by CF6-80C2 turbofans and will feature a Boeing air-refuelling boom, a RARO II remote air-refuelling operator station, wing pods containing hose and drogue refuelling apparatus and a centreline hose and drogue system. It will also have a refuelling receptacle fitted as standard. First aircraft rolled out 3 July 2003 and subsequently flown to Wichita, Kansas, on 16 July for installation of refuelling equipment and operator's station. Conversion of final three aircraft to be undertaken in Italy.

Two different configurations were offered to Australia, to satisfy Project Air 5402 requirement. EADS also in contention with Airbus A330, with RAAF selecting Airbus proposal on 15 April 2004.

Model of Boeing 737 MMA *(Paul Jackson)* ***NEW***/0576903

Projected Boeing 767 tanker-transport *(Paul Jackson)* ***NEW***/0576886

A derivative of the Boeing 767 is also to begin process of replacing the USAF KC-135 Stratotanker. In 2001, proposal to lease 100 aircraft emerged, with this envisaging a 10-year period, after which the USAF could negotiate outright purchase; this won more support after September 2001 terrorist attack on USA, and USAF formally notified Congress in April 2002 of intention to begin negotiations immediately. On 23 May 2003, the US Department of Defense announced approval of lease arrangement and revealed more details of US$16 billion programme; lease said to be for "six years starting in 2006", at unit cost of US$131 million, plus US$7 million in lease-unique costs per aircraft; same announcement noted that agreement included provision to buy the aircraft outright for about US$4 billion when lease terminates in "2017". USAF then anticipated receiving first aircraft in FY06; longer-term goal is to acquire up to 500 new tankers, allowing eventual retirement of veteran KC-135. Deployment plans revealed in mid-June 2003 indicated that the bases at Fairchild, Washington; Grand Forks, North Dakota; and MacDill, Florida, would each have 32 aircraft assigned by 2011.

Lease deal subsequently became mined in controversy that led to dismissal of Boeing's chief financial officer Michael Sears and resignation of chief executive Philip Condit in November 2003 over concerns about unethical conduct; in combination with Congressional opposition, this resulted in revised deal that was made public on 6 November 2003. Under this, USAF would lease first 20 aircraft from 2006 and purchase further 80, to be delivered in 2008–14; total cost of this programme expected to be US$27.6 billion up to and including FY17, although delivery time scale would be extended to 2014, with only 43 aircraft on USAF strength by FY09 instead of originally planned 60. On 26 May 2004, US Defense Secretary Donald Rumsfield announced that no decision would be taken on lease/purchase of KC-767 until fresh studies completed in November 2004.

Boeing formed 767 Tanker Programs office in March 2001 and subsequently selected its Wichita, Kansas facility as the centre for tanker modification work.

Multimission Command and Control Aircraft (MC2A): Now designated Northrop Grumman E-10.

COSTS: Estimated US$216 million for Japanese 767 tanker (2001). Boeing price varies from US$150 million to US$225 million, according to quantity procured.

AVIONICS: Rockwell Collins selected by Boeing to provide avionics and support for KC-767. This will include communications and navigation equipment, as well as modification of commercial 767-400 large format display system (LFDS) and 767-200 electronic flight instrument system (EFIS) and engine indication and crew alert system (EICAS) to incorporate military mission requirements.

Comms: ACSS MASS traffic surveillance/rendezvous/station-keeping system and Mode S/IFF transponder. Rockwell Collins Link-16 and SATCOM terminal. UHF and VHF radio. Military TCAS. Boom interphone for secure communication with receiver aircraft.

Flight: Includes modified AN/ARN-153 Tacan.

Instrumentation: KC-767 will have Kaiser Electro-Optics HUD for remote aerial refuelling operator.

Data follow for Tanker-Transport.

WEIGHTS AND LOADINGS (estimated):

Operational weight, empty	90,720 kg (200,000 lb)
Max T-O weight	179,170 kg (395,000 lb)
Max ramp weight	179,625 kg (396,000 lb)
Max zero-fuel weight	117,934 kg (260,000 lb)

UPDATED

BOEING AL-1A

TYPE: Missile defence system.

PROGRAMME: Prototype YAL-1A (USAF serial number 00-0001) ordered on 12 November 1996; purchase of 747-400F airframe from Boeing confirmed on 30 January 1998. Formal authority to proceed received on 26 June 1998, after TRW successfully demonstrated laser firing and missile tracking. First metal cut on YAL-1A on 10 August 1999; rolled out at Everett on 12 December 1999; first flight 6 January 2000; delivered to Boeing Wichita on 21 January 2000 for outfitting with strengthened floor and modifications to take nose-mounted laser turret; subsequent critical design review completed in late April.

On completion of 1.6 million man-hours of work at Wichita, aircraft was originally expected to fly in February 2002, then be ferried to Edwards AFB, California, for installation of six-module COIL laser from May 2002, followed by trials against various missiles, with PDRR phase culminating with demonstration against a ballistic missile fired from Vandenberg AFB, California, in September 2003; however, by early 2002, a combination of changed defence priorities and technical issues had combined to cause delay in programme and the ballistic missile shoot-down demonstration had been postponed to 2004. By start of 2004, programme had experienced further slippage, with shoot-down demonstration now unlikely before early 2005.

Maiden flight following modification eventually achieved on 18 July 2002. YAL-1A completed short evaluation of aircraft performance and system operation from Wichita before being painted and prepared for delivery to Edwards AFB, where laser to be installed. Initial evaluation revealed problems with buffet and lateral acceleration forces acting on pylon for active ranging system pod located above flight deck; this was sufficiently severe to necessitate removal of pylon and pod after third flight and forced Boeing to redesign this unit. YAL-1A ferried to Edwards on 19 December 2002 for final preparations in advance of airborne laser firing tests; before that, laser unit to be tested on ground in special laboratory facility that contains an 'ironbird' 747 fuselage. Initial batch of chemicals for laser delivered to Edwards AFB on 4 December 2003, when 16,656 litres (4,400 US gallons; 3,664 Imp gallons) of hydrogen peroxide arrived; at that time, it was hoped to accomplish first laser 'light' in second quarter of 2004.

Design work on second aircraft had begun by beginning of 2004, but firm order then still awaited by Boeing; USAF intent to acquire as many as seven AL-1As now seems unlikely to be satisfied and programme may ultimately be limited to just two aircraft. IOC was planned for late 2007, but even limited contingency capability had been indefinitely postponed by February 2004.

Prototype YAL-1A arriving at Edwards AFB on 19 December 2002 ***(USAF)*** ***NEW***/0547142

CUSTOMERS: US Air Force (seven required).

COSTS: Programme cost (1997, estimated) US$5 billion for one concept prototype and one EMD (both eventually to be fully upgraded) and five production aircraft. Initial programme definition and risk reduction (PDRR) contract of November 1996 valued at US$1.1 billion. US$598 million requested for FY03, including US$85 million for long-lead items for EMD aircraft.

DESIGN FEATURES: Based on airframe of Boeing 747-400F (which see under Boeing Commercial Airplanes). Equipped with TRW multihundred kW chemical oxygen iodine laser (COIL) and Lockheed Martin beam control/fire-control system, for target acquisition, plus aiming and firing of laser. Active ranging system housed in pod. Intended to destroy theatre ballistic missiles during their boost phase, but with additional capability against low-flying cruise missiles; other potential applications began emerging in 1998, these including protection of high-value aircraft such as AWACS and Joint STARS by destroying SAMs and AAMs; imaging and reconnaissance with aid of optical telescope; SEAD, by combining intelligence data to engage hostile missile sites and radar control systems; and command and control through search/detection of infra-red signatures to aid cueing of other weapons. Laser able to fire 20 to 30 times per mission; titanium to be used in certain areas to protect against heat damage to undersides caused by laser exhaust gases. Mission crew expected to comprise four specialists at individual consoles at rear of forward compartment, specifically the mission crew commander, airborne surveillance officer, weapon system operator and special equipment operator; standard flight deck crew of two could be augmented for long missions.

UPDATED

BOEING 114 AND 414 CHINOOK

US Army designations: CH-47 and MH-47 Chinook
Royal Air Force designations: Chinook HC. Mk 2, HC. Mk 2A and HC. Mk 3
South Korean Air Force designation: HH-47D
Spanish Army designation: HT.17
JASDF and JGSDF designations: CH-47J and CH-47JA

TYPE: Medium-lift helicopter.

DEVELOPMENT MILESTONES

CH-47D	
Programme launched	1976
First flight	11 May 1979
First order	Oct 1980
First flight, production	26 Feb 1982
First delivery	31 Mar 1982
Entered service (US Army IOC)	28 Feb 1984

PROGRAMME: Design of all-weather medium transport helicopter for US Army began 1956; first flight of YCH-47A 21 September 1961. initial production concerned CH-47A and CH-47B. Performance increased in CH-47C by uprated transmissions and 2,796 kW (3,750 shp) T55-L-11A; internal fuel capacity increased; first flight 14 October 1967; delivered to US Army from 1968.

CURRENT VERSIONS: **CH-47D:** US Army contract to modify one each of CH-47A, B and C to prototype Ds placed 1976; first flight 11 May 1979; first production contract October 1980; first flight 26 February 1982; first delivery 31 March 1982; initial operational capability (IOC) achieved 28 February 1984 with 101st Airborne Division: first multiyear production contract for 240 CH-47Ds awarded 8 April 1985; second multiyear production contract for 144 CH-47Ds (including 25 MH-47Es) awarded 13 January 1989, bringing total CH-47D (and MH-47E) ordered to 472; two Gulf War attrition replacements authorised August 1992 (these new-build); seven ex-Australian rebuilds funded June 1993 for delivery January to November 1995. Additional new-build CH-47D ordered for US Army in June 1999 was delivered to Fort Hood, Texas, on 26 June 2002.

CH-47D update included strip down to bare airframe, repair and refurbish, fit Honeywell T55-L-712 turboshafts, uprated transmissions with integral lubrication and cooling, composite rotor blades, new flight deck compatible with night vision goggles (NVG), new redundant electrical system, modular hydraulic system, advanced automatic flight control system, improved avionics and survivability equipment, Solar T62-T-2B APU operating hydraulic and electrical systems through accessory gear drive, single-point pressure refuelling, and triple external cargo hooks. Principal external change is large, rectangular air intake in leading-edge of rear sail. Composites account for 10 to 15 per cent of structure. About 300 suppliers involved.

At maximum gross weight of 22,680 kg (50,000 lb), CH-47D has more than double useful load of CH-47A. Sample loads include M198 towed 155 mm howitzer, 32 rounds of ammunition and 11-man crew, making internal/external load of 9,980 kg (22,000 lb); D5 caterpillar bulldozer weighing 11,225 kg (24,750 lb) on centre cargo hook; US Army Milvan supply containers carried at up to 130 kt (256 km/h; 159 mph); up to seven 1,893 litre (500 US gallon, 416 Imp gallon), 1,587 kg (3,500 lb) rubber fuel blivets carried on three hooks.

MH-47D Special Operations Aircraft: Element of 160th Special Operations Aviation Regiment (at Hunter AAF, Georgia) equipped with 11 CH-47Ds modified to **MH-47D SOA** standard with refuelling probes (first refuelling July 1988), thermal imagers, Honeywell RDR-1300 weather radar, Rockwell Collins 'glass cockpits', improved communications and navigation equipment, and two pintle-mounted 7.62 mm six-barrel miniguns. Navigator/commander's station also fitted. It is intended to update all MH-47Ds to MH-47G configuration, incorporating technology improvements developed for the CH-47F.

GCH-47D: Limited number of Chinooks grounded for maintenance training at Fort Eustis, Virginia.

HH-47D: South Korean Air Force's 235 Squadron received six aircraft with this local designation for SAR.

JCH-47D: At least two aircraft (84-24159 and 90-0180) modified for temporary test tasks.

MH-47E: Special Forces variant with US Army Special Operations Command (USASOC); planned procurement 51, all originally to be deducted from CH-47D conversions, but only 25 initially received, plus prototype. Prototype development contract 2 December 1987; long-lead items for next 11 helicopters authorised 14 July 1989; firm order for 11, plus option on next 14, awarded 30 June 1991 Lot 2 (14 helicopters) confirmed 23 June 1992. Prototype (88-0267) flew 1 June 1990; delivered 10 May 1991; initial production aircraft (90-0414) flown 1992; first 11 (of 24 intended) originally due to be delivered from November 1992 to 2 Battalion of 160th Special Operations Aviation

US ARMY CH-47D PROCUREMENT

FY	Qty	First aircraft	Remarks
81	9	81-23381	
82	19	82-23762	
83	24	83-24102	
84	36[1]	84-24152	
85	48	85-24322	MYP1
86	48	86-1635	MYP1
87	48	87-0069	MYP1
88	48[2]	88-0062	MYP1
89	48	89-0130	MYP2
90	48[3]	90-0180	MYP2
91	48[4]	91-0230	MYP2
92	48[5]	92-0280	MYP2
	2[6]	92-0367	
93	7[7]	93-0928	
98	1[8]	98-2000	
Total	**482**		

Notes: Excludes three prototypes, of which two were issued with new serial numbers on remanufacture to definitive CH-47D standard
MYP1 First multiyear procurement contract, 8 April 1985
MYP2 Second MYP, 13 January 1989

[1] One crashed on test
[2] One to YMH-47E
[3] This batch includes one aircraft subsequently further modified to MH-47E configuration under separate contract
[4] Total includes six which were further modified to MH-47E under separate contract
[5] This batch includes two of original three prototypes (remaining prototype used as maintenance trainer at Fort Eustis, Virginia) and 11 from Italian production; total includes 18 which were further modified to MH-47E under separate contract
[6] New-build Gulf War attrition replacements, authorised 2 August 1992
[7] Former Australian Air Force CH-47Cs; contract 2 June 1993
[8] New-build aircraft; delivered 26 June 2002

US ARMY MH-47E PROCUREMENT

FY	Qty	First aircraft
90	1	90-0414
91	6	91-0496
92	4	92-0400
	14	92-0464
Total	**25**	

Notes: Excludes prototype (88-0267).

Regiment at Fort Campbell, Kentucky. Following mission software problems, deliveries began January 1994 with 91-0498 to Fort Campbell; last of original 26 (including prototype) received at Fort Campbell April 1995.

Mission profile 5½ hour covert deep penetration over 300 n mile (560 km; 345 mile) radius in adverse weather, day or night, all terrain with 90 per cent success probability. Equipment includes IBM-Honeywell integrated avionics with four-screen NVG-compatible EFIS; dual MIL-STD-1553B digital databusses; AN/ASN-145 AHRS; jamming-resistant radios; Rockwell Collins CP1516-ASQ automatic target handoff system; inertial AN/ASN-137 Doppler, Rockwell Collins AN/ASN-149(V)2 GPS receiver and terrain-referenced positioning navigation systems; Rockwell Collins ADF-149; laser (Raytheon AN/AVR-2), radar (Lockheed Martin AN/APR-39A) and missile (Honeywell AN/AAR-47) warning systems; ITT AN/ALQ-136(V) pulse jammer and Northrop Grumman AN/ALQ-162 CW jammer; BAE Systems M-130 chaff/flare dispensers; Raytheon AN/APQ-174A radar with modes for terrain-following down to 30 m (100 ft), terrain-avoidance, air-to-ground ranging and ground-mapping; Raytheon AN/AAQ-16 FLIR in chin turret; digital moving map display; Elbit ANVIS-7 night vision goggles; uprated T55-L-714 turboshafts with FADEC; increased fuel capacity; additional troop seating (44 maximum); OBOGS; rotor brake; 272 kg (600 lb) rescue hoist with 61 m (200 ft) usable cable; two six-barrel miniguns in forward port and starboard positions (also on MH-47D); provisions for Stinger AAMs using FLIR for sighting. This system largely common with equivalent Sikorsky MH-60K (which see). In November 2002, Smiths Aerospace received contract for more than US$3 million to demonstrate maintenance, analysis, safety and training (MAST) programme, incorporating integrated HUMS; contract for two systems to be installed on MH-47 in second quarter of 2003, with further systems subject to option.

MH-47E has nose of Commercial Chinook to allow for weather radar, if needed; forward landing gear moved 1.02 m (3 ft 4 in) forward to allow for all-composites external fuel pods (also from Commercial Chinook) that double fuel capacity; Brooks & Perkins internal cargo handling system. Plans are under way to update 23 surviving MH-47Es to MH-47G configuration to improve performance and extend service life. See also Chinook HC. Mk 3, below.

Chinook HC. Mk 2/2A: RAF version; Mk 1 designation CH47-352; all survivors of original 41 HC. Mk 1s upgraded to HC. Mk 1B; UK MoD authorised Boeing to update 33 (later reduced to 32) Mk 1Bs to Mk 2, equivalent to CH-47D, October 1989; changes include new automatic flight control system, updated modular hydraulics, T55-L-712F power plants with FADEC, stronger transmission, improved Solar 71 kW (95 shp) T62-T-2B APU, airframe reinforcements, low IR paint scheme, fuel system and standardisation of defensive aids package (IR jammers, chaff/flare dispensers, missile approach warning and machine gun mountings). Smiths Industries Health and Usage Monitoring System (HUMS) installed on fleet-wide basis at cost of about £100 million during 1997 to 2000.

Conversion continued from 1991 to July 1995. Chinook HC. Mk 1B ZA718 began flight testing Chandler Evans/Hawker Siddeley dual-channel FADEC system for Mk 2 in October 1989. Same helicopter to Boeing, March 1991; rolled out as first Mk 2 19 January 1993; arrived RAF Odiham 20 May 1993; C(A) clearance November 1993. Initial deliveries pooled at Odiham by Nos 7 and 27 (Reserve) Squadrons — latter for training; first delivery to No 18 Squadron at Laarbruch, Germany, 1 February 1994; to No 78 Squadron, Falkland Islands, February 1994. Final Mk 1 withdrawn from service, May 1994, at which time 11 Mk 2s received. Further three new-build Mk 2s ordered 1993 (as Lot 2), for delivery from late 1995; decision to order further 14 Mks 2A/3 for delivery 1997–2000 at cost of US$365 million announced March 1995; these comprise Lots 3 and 4, respectively; Mk 2A has dynamically tuned fuselage. Latest purchase raised total RAF procurement to 58. First Lot 3 HC. Mk 2A (ZH891) handed over in USA 6 December 1997; shipped to UK and arrived Boscombe Down for clearance trials 18 December 1997. Rest of Lot 3 delivered by end 1998. Retrofits announced in 1996 comprise Smiths HUMS and Thales Avionics RA 800 secure communications control system, with latter to be replaced by RA800M unit during 2003–05 in conjunction with Bowman radios. Thales also prime contractor for Chinook Night Enhancement Package (NEP) which was declared operational on 26 August 2002; this intended to improve capability at night by integrating optical and navigation sensors; improving NVG imagery and providing secure communications suite, with entire package incorporating provision of MIL-STD-1553 databus. Some equipment purchased for HC.Mk 3 version also used, including BAE Systems FLIR; mounted below nose section, sensor imagery is presented on two screens (at pilot and co-pilot positions). Thales AMS

Royal Air Force Boeing Chinook HC. Mk 2 *(Paul Jackson)* ***NEW***/0576497

2000 control display navigation unit also subject to upgrade and now incorporates combined GPS and INS, with extra display screen in main cabin that allows third member of crew to assist with navigation. Total of 27 (of 40) RAF Chinook HC. Mk 2/2As will be configured with NEP cockpit capability, although not all will be fully equipped. Current squadron dispositions are Nos. 7, 18 and 27 at Odiham; and No. 78 in Falkland Islands.

Chinook HC. Mk 3: Eight of 14 additional RAF Chinooks, announced March 1995, assigned to Special Forces; build standard as CH-47SD, but with MH-47E's large fuel panniers, weather radar and refuelling probes. First flight of initial aircraft (ZH897, as N2045G) in mid-October 1998. Following flight test, all transferred to temporary store at Shreveport, Louisiana, for six to eight months, pending fitment of avionics (including Sky Guardian RWR) which form subject of separate contract; delivery was expected to begin in February 2000, but UK refused to accept initial aircraft, citing software problems as cause. First example eventually accepted in December 2000; initial deliveries to UK were ZH898 and ZH899 to Bristol by sea, arriving 15 July 2001, with remainder following by December 2001 (apart from one retained by Boeing for trials); by mid-2002, two flying on Military Aircraft Release trials, one with RAF at Odiham for ground training and four in storage; RAF service entry with Joint Helicopter Command further delayed by UK Audit Office report of 7 April 2004, citing £260 million programme as "one of the most incompetent procurements of all time". Mk 3s restricted to heights above 152 m (500 ft) and flight only in cloudless weather because of lack of fully digital cockpit. Fleet requires £127 million upgrade before service-entry, then estimated in 2007. Possibility of US Army 'buy-back' deal to increase its own fleet of special operations Chinooks emerged in 2002; under this, UK could expect to receive enhanced MH-47G equivalent at a later date. No recent news received of this idea.

HT.17 Chinook: Spanish Army version.

Boeing 234: Commercial version, currently out of production.

Boeing 414: Export military version. Superseded by CH-47D International Chinook and CH-47SD versions (see below).

CH-47D International Chinook: Boeing 414-100 first sold to Japan; Japan Defence Agency ordered two for JGSDF and one for JASDF in 1984; first flight (N7425H) January 1986 and, with second machine, delivered to Kawasaki April 1986 for fitting out; co-production arrangement (see under Japan: Kawasaki **CH-47J**). International Chinook was available in four versions with combinations of standard or long-range (MH-47E-type) fuel tanks and T55-L-712 SSB or T55-L-714.

CH-47SD 'Super D': Latest export variant, first flown on 25 August 1999; embodies some improvements first installed on the MH-47E for US special operations forces, including increased max T-O weight. Honeywell T55-GA-714A turboshaft with Coltec Industries FADEC chosen as standard power plant; single-point pressure refuelling and jettison capability on both sides of aircraft, with fuel contained in two ballistic and crash-resistant tanks; total usable capacity is 7,828 litres (2,068 US gallons; 1,722 Imp gallons); CH-47SD also has Smiths digital fuel quantity gauging system in place of Ragen analogue system.

Simplified structure offers benefits in maintainability and reliability. Specific changes include machined frames instead of standard frames.

CH-47SD also incorporates modernised NVG-compatible cockpit with avionics control management system (ACMS), utilising proven military and commercial off-the-shelf equipment on single console to reduce pilot workload and with provisions for growth. ACMS organised into three separate functional groups, specifically air vehicle group (including engines, drive system, fuel, electrics, flight controls and warning indicators); mission group (including communications, navigation and survivability equipment such as INS/GPS, weather radar and digital map display); and pilotage group (including EFIS). Radar nose (but not necessarily radar) fitted as standard.

Avionics suite is comparable to that of baseline CH-47D, but features two embedded INS/GPS units as well as AN/ARN-147 VOR/ILS and AN/ARN-149 ADF, plus space and power provisions for Tacan. AN/ARC-210 frequency-hopping radios to be installed on CH-47SD for Taiwan.

Initial order from Singapore received shortly before March 1998 announcement of formal SD launch, with first acceptance (in US) accomplished in early June 2001. Second customer is Taiwan which, in January 2000, announced purchase of nine CH-47SDs for use by Army; delivery accomplished by October 2002.

CH-47F Improved Cargo Helicopter (ICH): Boeing programme for improved Chinook configuration for US Army involving the development and production of a new version that will remain operational and cost-effective until 2033, when last examples expected to be replaced by a new cargo helicopter to be developed from about 2015. Programme will coincide with the existing Chinook beginning to reach planned life cycle limits in 2002. Benefits envisaged include greater airframe and systems reliability arising from lower levels of vibration; reduced operating and support costs; reduced pilot workload; and more efficient cargo handling. Weights unchanged, although US Army aircraft, like CH-47D, will have peacetime MTOW of 22,680 kg (50,000 lb). Key goal of upgrade is ability to airlift 7,257 kg (16,000 lb) load over 50 n miles (92 km; 56 miles) under 35°C (95°F) and 1,220 m (4,000 ft) altitude daytime conditions.

Studies indicated that further remanufacture, rather than new production, offered best degree of affordability, with a key factor being fleet-wide fitment of the T55-GA-714A, an improved variant of the T55-L-714 engine, as installed in the MH-47E and the International Chinook. The T55-GA-714A results from a Honeywell (at Greer, South Carolina) kit upgrade programme begun December 1997 and affecting approximately 1,150 existing T55-L-712 engines in the Army inventory; upgrading brings power ratings up to those of the T55-L-714 and adds marinisation features.

Increased versatility also derived from replacement of the existing cargo handling system by integral flip-over roller panels in the cargo deck.

Cockpit modernisation by Rockwell Collins will include adding two dual-redundant MIL-STD-1553B databusses, four 127 mm (5 in) square MFD 255 EFIS LCDs, two 152 × 203 mm (6 × 8 in) MFD-268E1 situation awareness displays, two CDU 900 controller/processors, DR-200 data loader, moving digital map display, digital communications and electronic instrument displays. This allows for updated communications and navigation, enabling the Chinook to meet US Army Force XXI Battlefield requirements. Rockwell Collins selected by Boeing in April 1998 to provide integrated avionics suite, which will feature open architecture to facilitate future growth and updating; CH-47F will be first US Army helicopter to incorporate improved data modem (IDM), allowing it to link with digital battlefield and automatically send and receive information concerning targeting and aircraft position.

Under terms of Army development contract awarded in late 1997, two Chinooks (83-24107 and 83-24115) have been refurbished for the engineering and manufacturing development (EMD) phase, for which contract worth US$76 million was awarded to Boeing in May 1998. Subsequently, total of about 300 expected to be upgraded to CH-47F standard at maximum rate of 26 aircraft per year. Some new-build examples also to be received by US Army, starting with seven in FY04. Both EMD aircraft arrived for conversion on 5 January 1999; first (98-0011/M8001; ex-83-24107) officially rolled out on 11 July 2001, having flown for the first time on 25 June; second (98-0012/M8002; ex-83-24115) completed upgrade in September 2001, with EMD phase planned to end in December 2002; second CH-47F was delivered to the US Army on 7 May 2002 and flown to Fort Campbell, Kentucky, for further test flights. Low-rate initial production (LRIP) begun in FY03, with seven aircraft, six of which are to MH-47G standard, with most FY04 funding also directed towards MH-47G; only in FY05 does CH-47F get bigger share of appropriation, with total of 14 to be acquired alongside 12 MH-47Gs. Possibility exists that Army may now choose to convert almost all surviving CH-47Ds to CH-47F standard.

MH-47G: Upgraded special operations version embodying improvements developed for the CH-47F. Intention is to modernise existing MH-47D and MH-47E helicopters. Upgrades include a 'glass cockpit', with five colour MFDs, and MH-47G expected to have enhanced self-protection equipment in the form of a radio frequency jammer and a directed-infra-red countermeasures system. Raytheon ZSQ-1(V) electro-optical sensor system also to be installed; this will provide improved navigation, surveillance and targeting capability as well as enhanced mission performance and situational awareness by day and night. Some 'new-build' MH-47Gs also to be acquired, following decision to more than double the number of special operations-configured Chinooks to about 70 in all. First aircraft (00-02160) made initial flight on 15 March 2004 and was delivered to US Army on 7 May; this one of three being modernised to replace losses; further six MH-47Gs included in the first LRIP CH-47F batch, for delivery between October 2004 and March 2005, with 16 more in second batch and another 12 in third.

CH-47X: Proposed follow-on programme to CH-47F in response to diversion of funds from Future Transport Rotorcraft (FTR). MTOW would be in region of 31 tonnes (68,000 lb), requiring new engine in 4.5 MW (6,000 shp) class, as well as new dynamic components, including four-blade rotor system.

CUSTOMERS: US Army received five YCH-47A, 349 CH-47A, 108 CH-47B, 270 CH-47C and three CH-47D as new-build aircraft from US production plus 11 CH-47C from Italian production; further 479 CH-47D/MH-47E obtained from conversion programme involving CH-47A/B/C models as detailed in table elsewhere. CH-47F version also to be acquired in new-build form, with contract for initial batch of seven awarded to Boeing in December 2003. CH-47D, MH-47D and MH-47E inventory was just over 450 in mid-2002.

Exports include five CH-47Cs to Argentina (three Air Force; two Army); Australia 12 CH-47Cs with crashworthy fuel system (four refurbished as CH-47D and redelivered in 1995 to C Squadron of Army's 5 Aviation Regiment at Townsville; seven sold to US Army and converted to CH-47D for Hawaii National Guard; contract for further two signed 19 June 1998, with delivery to Townsville for further modification on 10 March 2000; these accepted by Army in February 2001); Canada nine CH-47Cs designated **CH-147**, delivered from September 1974, but withdrawn from use and seven sold to Netherlands in July 1993; Egypt finalised contract for four CH-47Ds on 20 August 1998, with delivery in mid-1999; Greece ordered seven CH-47Ds in October 1998, deliveries being effected between April and October 2001; Italy (two CH-47Cs, followed by 40 kits for local assembly and delivery to Iran and Italy); Japan (two plus four kits to initiate licensed production, with at least 72 more co-produced by Kawasaki); Netherlands, six ordered in December 1993 for delivery in 1998 (first, D-101, on 12 May 1998), plus seven ex-Canadian CH-147s upgraded to CH-47D (including nose radar) by Boeing and redelivered from August 1995; Spanish Army 19, designated **HT.17**, delivered to 5th Helicopter Transport Battalion (Bheltra-V) at Colmenar Viejo, Madrid (last six have Honeywell RDR-1400 weather radar), total of 17 upgraded by Boeing to CH-47D between August 1990 and November 2002; South Korea, total 30 International Chinooks, comprising 24 for Army (deliveries from June 1988 to

Spanish Air Force CH-47D Chinook with radar nose *(Paul Jackson)* ***NEW***/0576498

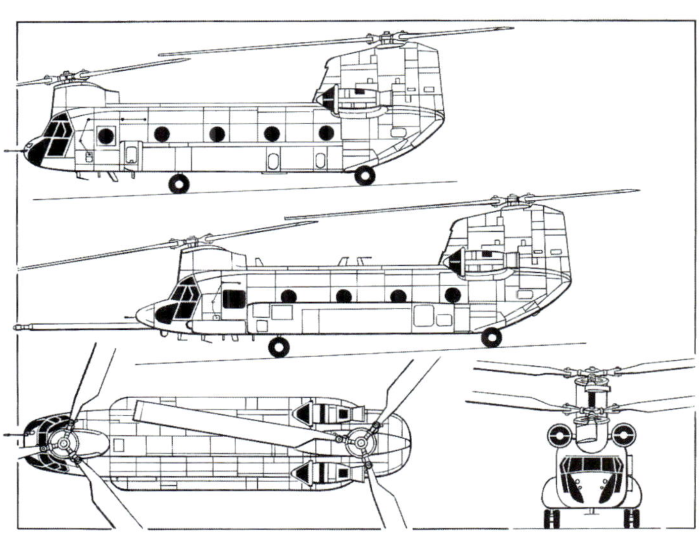

Boeing CH-47D military transport helicopter with additional side view (lower) of MH-47E special forces' variant *(James Goulding)* 0507658

CHINOOK PRODUCTION

Customer	YCH-47A	CH-47A	CH-47B	CH-47C	Model 234	CH-47D	CH-47F	Agusta	Kawasaki	Total
Argentina				5						**5**
Australia				12		2				**14**
Canada				9						**9**
Egypt						4		15		**19**
Greece						7		10		**17**
Iran								30		**30**
Iran (kits)				38						**38**
Italy								38		**38**
Italy (kits)				2						**2**
Japan				2					72	**74**
Japan (kits)				4						**4**
Libya								20		**20**
Morocco								9		**9**
Netherlands						6				**6**
Nigeria				5						**5**
Singapore						16				**16**
South Korea						30				**30**
Spain				19						**19**
Taiwan				3		9				**12**
Thailand						6				**6**
UK				41		17				**58**
USA	5	349	108	270		3	7	11		**753**
Civil					10					**10**
Not known				1						**1**
Total	**5**	**349**	**108**	**411**	**10**	**100**	**7**	**133**	**72**	**1,195**

Notes: CH-47Cs for Taiwan were built to notional civilian Model 234MLR standard
CH-47C figures include some helicopters produced as Model 414 International Chinook
Table excludes two airframes completed by Boeing for mock-up purposes
Kits for Iran and Italy were assembled in Italy by Meridionali/Agusta; kits for Japan assembled in Japan by Kawasaki
Nigerian CH-47Cs not delivered; one retained by Boeing as maintenance trainer; fate of others uncertain

CHINOOK REMANUFACTURE BY BOEING

	CH-47D	MH-47E	CH-47F	MH-47G	Total
Australia	4				**4**
Egypt	12				**12**
Greece	9				**9**
Netherlands	7				**7**
Spain	17				**17**
UK	32				**32**
USA	454	25	3	13	**495**
Total	**535**	**25**	**3**	**13**	**576**

Notes: US Army CH-47D total excludes three prototypes, but includes one that was further converted as MH-47E prototype
Egyptian and Greek aircraft originally built in Italy
CH-47F and MH-47G remanufacture now under way

November 1998); and six for Air Force (ordered early 1990; delivered November 1991 to March 1992); Singapore, six ordered April 1994 for delivery in 1996, (training operations conducted in 1996–97 at Grand Prairie, Texas, before transfer to Singapore base) followed by 10 CH-47SDs ordered in 1998 and delivered to 127 Squadron from June 2001 to January 2002; Taiwan (three, notionally civil Boeing 234MLR used by Army until June 2002, when transferred to National Fire Administration; finalised contract for nine CH-47SDs in January 2000; deliveries between November 2001 and October 2002); Thailand, six for Army (three ordered August 1988, three early 1990, deliveries February 1991 to March 1992; all are International Chinook); UK (58, see entries for Chinook HC. Mk 2/2A/3); 16 civilian, of which few left operating in original oilfield support role; two for trials; 45 in kits, comprising 40 for assembly in Italy and five for Japan.

Agusta sold licence-built CH-47Cs to Egypt (15, of which 12 upgraded to CH-47D standard by Boeing and redelivered in 2002–04; remaining three may follow); Greece (10, of which nine converted to CH-47D by Boeing between March 1992 and 1995; first redelivery in November 1993), Iran (30), plus 38 more from kits supplied by Boeing, Italy (38, comprising 22 CH-47C, 10 CH-47C Plus and six civilian Model 234s for Civil Protection Agency; latter operated by Italian Army), Libya (20), Morocco (nine) and Pennsylvania US Army National Guard (11). Kawasaki (which see) has firm orders for at least 72, with production continuing.

Total Chinook orders, including civil and five CH-47Cs that were not delivered to Nigeria, 1,195.

COSTS: Unit cost of CH-47SD quoted as US$30-32 million (2002). US Army estimates cost of CH-47F programme, including R&D, long-lead production and upgrade of 300 Chinooks, as approximately US$3.3 billion (1999). Procurement of seven new-build CH-47F valued at US$151.5 million (December 2003). Contract for 16 MH-47G conversions worth US$278.6 million awarded to Boeing in December 2003.

DESIGN FEATURES: Two three-blade intermeshing contrarotating tandem rotors; front rotor turns anti-clockwise, viewed from above; rotor transmissions driven by connecting shafts from engine gearbox, which is driven by rear-mounted engines; normal rotor speed 225 rpm. Classic rotor heads with flapping and drag hinges; manually foldable blades, using Boeing VR7 and VR8 aerofoils with cambered leading-edges; blades can survive hits from 23 mm HEI and API rounds; rotor brake optional.

Development of new, low-maintenance elastomeric rotor hub begun at end August 1999, following signature of contract with US Army; development, testing and installation to be completed over four-year period, with new assembly to be installed on all US Army regular and reserve force Chinooks. Also under consideration for CH-47F and is incorporated in CH-47SD. Will have longer fatigue life (4,500 hours), 75 per cent fewer parts and offer 70 per cent reduction in requirement for special maintenance tools; will also retain same rotor flight dynamics and be fully interchangeable with existing hub.

Constant cross-section cabin with side door at front; rear-loading ramp that can be opened in flight; underfloor section sealed to give flotation after water landing; access to flight deck from cabin; main cargo hook mounting covered by removable floor panel so that load can be observed in flight.

FLYING CONTROLS: Differential fore and aft cyclic for pitch attitude control; differential lateral cyclic pitch (from rudder pedals) for directional control; automatic control to keep fuselage aligned with line of flight. Dual hydraulic rotor pitch-change actuators: secondary hydraulic actuators in control linkage behind flight deck for autopilot/autostabiliser input; autopilot provides stabilisation, attitude hold and outer-loop holds.

STRUCTURE: Blades based on D-shaped glass fibre spar, fairing assembly of Nomex honeycomb core and crossply glass fibre skin.

LANDING GEAR: Non-retractable quadricycle type, with twin wheels on each front unit and single wheels on each rear unit. Oleo-pneumatic shock-absorbers in all units. Rear units fully castoring; power steering on starboard rear unit. All wheels are size 24×7.7, with tyres size 8.50–10, pressure 6.07 bar (88 lb/sq in). Single-disc hydraulic brakes on all six wheels. Provision for fitting detachable wheel/skis.

POWER PLANT: Two Honeywell T55-L-712 turboshafts, pod-mounted on sides of the rear pylon, each with a standard power rating of 2,237 kW (3,000 shp) and maximum rating of 2,796 kW (3,750 shp). Honeywell T55-L-712 SSB engine has standard power rating of 2,339 kW (3,137 shp) and maximum of 3,217 kW (4,314 shp). Standard in MH-47E and International Chinook are two Honeywell T55-L-714 turboshafts, each with a standard power rating of 3,108 kW (4,168 shp) continuous and emergency rating of 3,629 kW (4,867 shp). CH-47SD (and Netherlands CH-47Ds) have T55-GA-714A turboshafts, with maximum continuous rating of 3,039 kW (4,075 shp). FADEC installed on late production CH-47Ds and CH-47SD; CH-47F to have Goodrich FADEC system. From January 1991, more than 100 CH-47Ds fitted with engine air particle separator (also available for RAF variant). Transmission capacity (CH-47D/SD and MH-47E) 5,617 kW (7,533 shp) on two engines and 3,430 kW (4,600 shp) OEI; rotor rpm 225.

Self-sealing pressure refuelled crashworthy fuel tanks in external fairings on sides of fuselage. Total usable fuel capacity 3,914 litres (1,034 US gallons; 861 Imp gallons) in CH-47D. Provision for up to three additional long-range tanks in cargo area, each of 3,028 litres (800 US gallons; 666 Imp gallons); maximum fuel capacity (fixed and auxiliary) 12,998 litres (3,434 US gallons; 2,859 Imp gallons). Oil capacity 14 litres (3.7 US gallons; 3.1 Imp gallons).

Normal fuel (MH-47E/G, International Chinook and CH-47SD) 7,828 litres (2,068 US gallons; 1,722 Imp gallons) but MH-47E/G can operate with three long-range tanks in cargo area, each containing 3,028 litres (800 US gallons; 666 Imp gallons), raising total fuel capacity to 16,913 litres (4,468 US gallons; 3,720 Imp gallons). MH-47s have 8.97 m (29 ft

5 in) refuelling probe on starboard side of forward fuselage, which extends 5.41 m (17 ft 9 in) forward of the nose. Refuelling probe also an option for International Chinook.

ACCOMMODATION: Two pilots on flight deck, with dual controls. Lighting compatible with pilots' NVGs (Nite-Op in RAF variant). Jump seat for crew chief or combat commander. Jettisonable door on each side of flight deck. Depending on seating arrangement, 33 to 55 troops can be accommodated in main cabin, or 24 litters plus two attendants, or (see under Current Versions) vehicles and freight. Rear-loading ramp can be left completely or partially open, to permit transport of extra-long cargo and in-flight parachute or free-drop delivery of cargo and equipment.

Main cabin door, at front on starboard side, comprises upper hinged section which can be opened in flight, and lower section with integral steps. Upper section has a panel with window that is jettisonable. Triple external Breeze-Eastern C-160 cargo hook system on US Army aircraft, with centre hook rated to carry maximum load of 11,793 kg (26,000 lb) and the forward and rear hooks 7,711 kg (17,000 lb) each, or 10,433 kg (23,000 lb) in unison, while International Chinook and CH-47SD ratings are 12,701 kg (28,000 lb) for centre hook and 9,072 kg (20,000 lb) each for forward and rear hooks, or 11,340 kg (25,000 lb) in unison. Options are available for a power-down ramp and water dam to permit ramp operation on water, internal ferry fuel tanks, external rescue hoist, and windscreen washers.

SYSTEMS: Hydraulic system contains a utility system, a No. 1 flight control system and a No. 2 flight control system. Each includes a separate, variable delivery pump and a reservoir cooler module. Utility system contains a pressure control module and a return control module. Both flight control systems contain a power control module, incorporating pressure and return in one module for each system. Each flight control system can be driven by the utility system for ground checkout through a power transfer unit without intermixing of hydraulic fluids. All hydraulic systems have a pressurised reservoir to prevent pump cavitation and are serviced by a common filter.

The CH-47D has a significant reduction in the number of hydraulic lines and fittings; approximately 200 tubes and hoses eliminated, thereby obviating approximately 700 leak points. Majority of all lines now swaged together for permanent joining; Rosan fluid adaptors for hardpoint connections. All systems capable of being monitored, both in flight and on the ground, for servicing prechecks and fault isolation. Subsystems designed for pressurisation on demand; power transfer units now used for flight control ground system checkout. Electrical system includes two 40 kVA oil-cooled alternators driven by transmission drive system. Solar T62-T-2B APU drives a 20 kVA generator and hydraulic motor pump, providing electrical and hydraulic power for main engine start and system operation on the ground.

AVIONICS: Baseline CH-47D. Specific MH-47E and MH-47G avionics listed under those headings. Netherlands aircraft have Honeywell advanced cockpit management system (ACMS) and EFIS, latter adopted by Boeing as baseline for subsequent sales.

Comms: Honeywell AN/ARC-199 HF com radio, Rockwell Collins AN/ARC-186 VHF/FM-AM, Magnavox AN/ARC-164 UHF/AM com; C-6533 intercom (Netherlands aircraft have Telephonics Corp STARCOM); Honeywell AN/APX-100 IFF.

Flight: Honeywell AN/APN-209 radar altimeter; AN/ARN-89B ADF; BAE Systems AN/ASN-128 Doppler; AN/ARN-123 VOR/glideslope/marker beacon receiver; and AN/ASN-43 gyromagnetic compass. AFCS maintains helicopter stability, eliminating the need for constant small correction inputs by the pilot to maintain desired attitude. The AFCS is a redundant system using two identical control units and two sets of stabilisation actuators. RAF Chinooks have Racal RNS252 Super TANS INS including GPS.

Instrumentation: Flight instruments are standard for IFR, and include an AN/AQU-6A horizontal situation indicator. CH-47F incorporates Integrated Engine Crew Advisory System (IECAS).

Mission: Chelton 19-400 satellite communications antenna on some RAF helicopters.

Self-defence: Provisions for Lockheed Martin AN/APR-39A RWR, Sanders AN/ALQ-156 missile warning equipment and BAE Systems M-130 chaff/flare dispensers. RAF Chinooks have BAE Systems M-206/M-1 chaff/flare dispensers, BAE Systems ARI 18228 Sky Guardian RWR and (from 1990) Lockheed Martin AN/ALQ-157 IR jammers and Honeywell AN/AAR-47 missile approach warning equipment; Elisra SPS-65V-3 integrated airborne self-protection system chosen for installation on HC. Mk 3 version. Northrop Grumman AN/AAR-54(V) passive missile approach warning system adopted by Royal Netherlands Air Force, with installation accomplished during 2001–04.

EQUIPMENT: Hydraulically powered winch for rescue and cargo handling, rearview mirror, plus integral work stands and step for maintenance.

ARMAMENT: Provision for two machine guns or miniguns in crew door (starboard) and forward hold window (port).

DIMENSIONS, EXTERNAL:
Rotor diameter (each) ..... 18.29 m (60 ft 0 in)
Rotor blade chord (each) ..... 0.81 m (2 ft 8 in)
Distance between rotor centres:
CH-47SD ..... 11.85 m (38 ft 10¾ in)
Length:
overall, rotors turning:
MH-47E and CH-47SD ..... 30.14 m (98 ft 10¾ in)
fuselage: MH-47E and CH-47SD ..... 15.87 m (52 ft 1 in)
MH-47D/E, incl probe ..... 21.08 m (69 ft 2 in)
Width, rotors removed: MH-47E ..... 4.78 m (15 ft 8 in)
Height to top of rear rotor head:
CH-47SD ..... 5.70 m (18 ft 8½ in)
MH-47E ..... 5.59 m (18 ft 4 in)
Ground clearance, rotors turning:
rear approach ..... 5.77 m (18 ft 11 in)
Ground clearance, static:
rear approach:
MH-47E and CH-47SD ..... 4.90 m (16 ft 0¾ in)
forward approach: CH-47SD ..... 2.29 m (7 ft 6 in)
Wheelbase: MH-47E and CH-47SD ..... 7.87 m (25 ft 10 in)
Passenger door (fwd, stbd): Height ..... 1.68 m (5 ft 6 in)
Width ..... 0.91 m (3 ft 0 in)
Height to sill ..... 1.09 m (3 ft 7 in)
Rear-loading ramp entrance: Height ..... 1.98 m (6 ft 6 in)
Width ..... 2.31 m (7 ft 7 in)
Height to sill ..... 0.79 m (2 ft 7 in)

DIMENSIONS, INTERNAL (MH-47E and CH-47SD):
Cabin, excl flight deck: Length ..... 9.19 m (30 ft 2 in)
Width: mean ..... 2.29 m (7 ft 6 in)
at floor ..... 2.51 m (8 ft 3 in)
Height ..... 1.98 m (6 ft 6 in)
Floor area ..... 21.0 m^2 (226 sq ft)
Usable volume ..... 41.7 m^3 (1,474 cu ft)

AREAS:
Rotor blades (each) ..... 7.43 m^2 (80.0 sq ft)
Rotor discs (total):
MH-47E and CH-47SD ..... 525.3 m^2 (5,655 sq ft)

WEIGHTS AND LOADINGS:
Weight empty: MH-47E ..... 12,210 kg (26,918 lb)
CH-47SD ..... 11,550 kg (25,463 lb)
Useful load: MH-47E ..... 12,284 kg (27,082 lb)
CH-47SD ..... 12,944 kg (28,537 lb)
Max underslung load: CH-47SD ..... 12,700 kg (28,000 lb)
Max fuel weight:
MH-47E, CH-47SD ..... 6,815 kg (15,025 lb)
Max T-O weight:
MH-47E, CH-47SD ..... 24,494 kg (54,000 lb)
Transmission loading at max T-O weight and power:
MH-47E, CH-47SD ..... 4.36 kg/kW (7.17 lb/shp)

PERFORMANCE (at 22,680 kg; 50,000 lb):
Max level speed: MH-47E ..... 154 kt (285 km/h; 177 mph)
CH-47SD ..... 155 kt (287 km/h; 178 mph)
Max cruising speed at S/L:
MH-47E, CH-47SD ..... 140 kt (259 km/h; 161 mph)
Max rate of climb: MH-47E ..... 561 m (1,840 ft)/min
CH-47SD ..... 563 m (1,846 ft)/min
Service ceiling: MH-47E ..... 3,090 m (10,140 ft)
CH-47SD ..... 3,385 m (11,100 ft)
Hovering ceiling:
IGE: MH-47E ..... 2,985 m (9,800 ft)
CH-47SD ..... 2,835 m (9,300 ft)
IGE, ISA + 20°C (68°F): MH-47E ..... 2,410 m (7,900 ft)
CH-47SD ..... 2,180 m (7,160 ft)
OGE: MH-47E, CH-47SD ..... 1,675 m (5,500 ft)
OGE, ISA + 20°C (68°F):
MH-47E, CH-47SD ..... 1,005 m (3,300 ft)
Radius of action, MH-47E, deploy special forces team (1,814 kg; 4,000 lb) at 1,220 m (4,000 ft), 35°C (95°F) ambient temperature ..... 505 n miles (935 km; 581 miles)
Range:
MH-47E, self-deployment at 24,494 kg (54,000 lb) T-O weight ..... 1,260 n miles (2,333 km; 1,449 miles)
CH-47SD with 12,558 kg (27,686 lb) payload ..... 651 n miles (1,207 km; 750 miles)

UPDATED

BOEING AH-64 APACHE

US Army designations: AH-64A and AH-64D
Israel Defence Force name: AH-64A Pethen (Cobra); AH-64D Sharaf (Serpent)

TYPE: Attack helicopter.

DEVELOPMENT MILESTONES

Official go-ahead	Dec 1976
First flight	30 Sep 1975
First delivery (US Army)	26 Jan 1984

PROGRAMME: Original Hughes Model 77 entered for US Army advanced attack helicopter (AAH) competition; first flights of two development prototype YAH-64s 30 September and 22 November 1975; selected by US Army December 1976; named Apache late 1981.

Deliveries started 26 January 1984; 900th delivered in October 1995, at which time US Army had ordered 821 AH-64As (excluding prototypes) with export contracts totalling 104; latter total increased to 258 (116 AH-64A; 142 AH-64D) by end of 2002. Last of 821 AH-64As delivered to US Army on 30 April 1996; aircraft concerned was production vehicle 915, and manufacture of AH-64A variant terminated with completion of 937th example (for Egypt), in November 1996. Delivery of 1,000th Apache (including AH-64D rebuilds) effected on 30 March 1999.

Boeing awarded four year, US$15.9 million, contract on 27 May 1998 to design, manufacture and flight test new centre fuselage section incorporating advanced composites materials; Phantom Works responsible for design leadership, with support from Boeing facilities at Long Beach, Philadelphia and St Louis. If successful, new section from rear of aft cockpit to just behind engines will be simpler to manufacture as well as lighter and more durable than existing all-metal structure.

In separate development, on 5 October 1998, a modified AH-64D Apache Longbow prototype made initial flight with Rotorcraft Pilot's Associate (RPA) advanced cockpit management system. Also developed by Phantom Works over 60 month period, under terms of US$80 million advanced technology demonstration contract, this featured updated controls and displays, including Boeing-developed four-axis, full authority, advanced

Boeing AH-64D Apache Longbow *(Paul Jackson)* ***NEW***/1044805

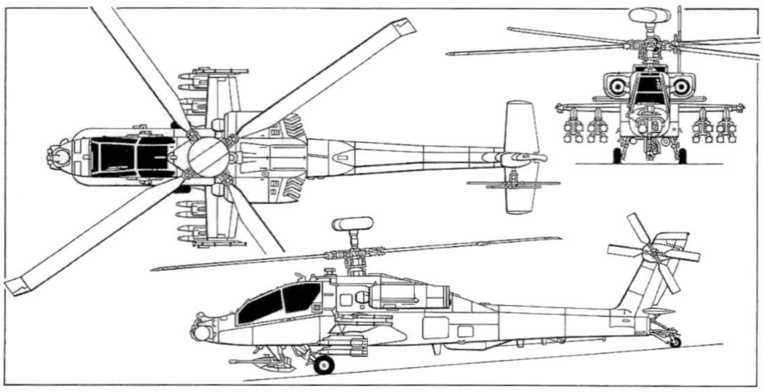

Boeing AH-64D Apache Longbow tandem-seat advanced attack helicopter *(Mike Keep)* 0507285

digital flight control system. Flight and mission data presented to pilots on three large multipurpose colour displays; RPA also features advanced data fusion and an advanced pilotage system, as well as the ability to recognise and respond to verbal commands. Company flight trials continued throughout remainder of 1998, with test aircraft then visiting US Army's Yuma Proving Grounds for demonstration flights in January-February 1999.

In 2001, Boeing proposed series of upgrades designed to keep AH-64D in operational force until 2030; these included adoption of new split-face power plant gearbox, new main rotor assembly using composite blades and titanium hub, composites fuselage structure elements for enhanced payload, and open architecture avionics with gradual introduction of Boeing RPA features such as digital navigation and communications equipment. Objective of upgrade package to restore original performance, reduce purchase price by about 30 per cent and cut operating costs by 50 per cent, but US Army not enthusiastic and instead chose to direct available funding towards near-term reliability and sustainability concerns, such as adoption of Lockheed Martin Arrowhead targeting and navigation sensor system. Some of these features seem likely to be adopted for future Block III upgrade.

Other enhancements, including digital avionics incorporating commercial off-the-shelf (COTS) technology, tested on preproduction AH-64D Apache Longbow first flown 12 July 2001; Boeing introducing some of these during production of the second multiyear batch of helicopters (see **Block II**).

CURRENT VERSIONS: **AH-64A:** Produced for US Army and export; 937 built. First 603 had two 1,265 kW (1,696 shp) T700-GE-701 engines. Total of 501 US Army examples being upgraded to AH-64D. Retrofit from 1993 with SINCGARS secure radios and GPS; first installed in Apaches of 5-501 Aviation Regiment at Camp Eagle, South Korea.

GAH-64A: Designation applies to at least 17 AH-64As that have been grounded and assigned as technical instruction training aids.

JAH-64A: Designation applied to seven AH-64As that have been used for special test duties; at least one has reverted to standard AH-64A.

AH-64B: Cancelled 1992. Was planned near-term upgrade of 254 AH-64As with improvements derived from operating experience in 1991 Gulf War, including GPS SINCGARS radios, target handover capability, better navigation, and improved reliability.

AH-64C: Previous designation for upgrade of AH-64As to near AH-64D standard, apart from omission of Longbow radar and retention of -701 engines; provisions for optional fitment of both. Army requested draft proposal, August 1991; funding for two prototype conversions awarded in September 1992. Designation abandoned late 1993.

AH-64D Apache Longbow: Current improvement programme involving 501 remanufactured helicopters for US Army and new-build examples for export. Based on Lockheed Martin and Northrop Grumman joint venture development of mast-mounted AN/APG-78 Longbow millimetre-wave radar and Hellfire missile with RF seeker; Northrop Grumman has lead on Longbow with Lockheed Martin taking principal role for Hellfire. Programme also includes more powerful engines, larger generators, MIL-STD-1553B databus allied to dual 1750A processors, and a vapour cycle cooling system for avionics; early user tests completed April 1990. First 284 US Army aircraft to **Block I** standard; balance as Block II.

Detailed description applies to AH-64D.

Full-scale development programme, lasting 51 months, authorised by Defense Acquisition Board August 1990, but airframe work extended in December 1990 to 70 months to coincide with missile development; supporting modifications being incorporated progressively; first flight of AH-64A (82-23356) with dummy Longbow radome 11 March 1991; first (89-0192) of six AH-64D prototypes flown 15 April 1992; second (89-0228) flew 13 November 1992; fitted with radar in mid-1993 and flown 20 August 1993; No.3 (90-0324) flown 30 June 1993; No.4 (90-0423) on 4 October 1993; No.5 (85-25477; formerly AH-64C No.1) 19 January 1994 (first Apache with new Hamilton Sundstrand lightweight flight management computer); No.6 (85-25408) flown 4 March 1994; last two mentioned lack radar. Following termination of AH-64C in late 1993, original plan was to convert 748 AH-64As to AH-64D, although only 227 (original AH-64D total) to carry Longbow radar; subsequent review, revealed at start of 1999, proposed reducing total number of conversions to 530 and increasing purchase of Longbow radars to 500. To date, US Army has funded total of 501 conversions.

Under original programme, AH-64D to equip 26 battalions; three companies, each with eight helicopters, per battalion. Longbow can track flying targets and see through rain, fog and smoke that defeat FLIR and TV. RF Hellfire, delivered to US Army from November 1996, can operate at shorter ranges; it can lock on before launch or launch on co-ordinates and lock on in flight; Longbow scans through 360° for aerial targets or scans over 270° in 90° sectors for ground targets; mast-mounted rotating antenna weighs 113 kg (250 lb). Longbow radar transmitter subject of redesign in late 1997 to overcome poor performance of some electrical components in low temperatures; eliminate lengthy and costly manual integration necessary to achieve required output; and avoid shortages of critical components which suppliers reported in 1995 that they would no longer provide. New transmitter meets or surpasses original specification and is fully interchangeable with original unit.

Further modifications include 'manprint' cockpit with large displays, air-to-air missiles, digital autostabiliser, integrated GPS/Doppler/INS/air data/laser/radar altimeter navigation system, digital communications, faster target handoff system, and enhanced fault detection with data transfer and recording. Cockpit displays initially monochrome, but these replaced by colour displays with effect from 27th production conversion (97-5027); first flight of AH-64D with colour displays on 12 September 1997. AH-64D No.1 made first Hellfire launch on 21 May 1993; first RF Hellfire launch 4 June 1994 (first combat use of RF Hellfire in opening phase of ground conflict during Gulf War of 2003); first demonstration of digital air-to-ground data communications with Symetrics Industries improved data modem, 8 December 1993.

Advanced acquisition phase contract for remanufacture programme, worth US$279.6 million and covering 18 helicopters (later increased to 24), awarded on 14 December 1995; predated by arrival of first two AH-64As (85-25387 and 85-25394) at Mesa on 27 November 1995 for stripping to basic fuselage in readiness for start of remanufacture in early 1996. First fuselage moved to final assembly area on 15 August 1996; first flight (85-25387 with new identity 96-5001) 17 March 1997; formal roll-out at Mesa on 21 March.

Multiyear contract, worth US$1.87 billion, covering 232 AH-64Ds (retrospectively including advanced acquisition aircraft) over five year period signed 16 August 1996; further 269 conversions to be acquired in second multiyear contract covering FY01 to FY05, which signed 29 September 2000, at total cost of US$2.3 billion.

First of 217 so-called **Block II** aircraft (02-5285) delivered to US Army on 25 February 2003; incorporates advanced avionics, digital enhancements and communications upgrades, permitting secure transfer of digitised battlefield information to air and ground forces. Future **Block III** upgrade expected to build on existing Block II architecture, but also likely to include other improvements such as image fusion of sensors, as well as ability to control UAVs and utilise newer, longer-range weapons; upgrade programme will affect at least 230 survivors of initial 284 Block I Apaches, with work likely to begin in 2008 and continue until 2012. New composites main rotor to be adopted as part of Block III enhancements, which estimated to cost about US$1.3 billion. A more durable and uprated 2,530 kW (3,400 shp) gearbox may also be installed from 2010.

AH-64D deliveries to US Army began 31 March 1997 and total of 18 handed over by end of 1997. Delivery of 24th and last Lot 1 aircraft accomplished 4 March 1998, with US Army having accepted total of 55 by end of 1998, 102 by end of 1999 and 159 by end of 2000; 100th remanufactured AH-64D delivered to US Army on 9 December 1999; 200th followed on 23 August 2001; 300th in second quarter of 2003; 350th in April 2004. Final example of initial multiyear contract and first helicopter from second multiyear contract both delivered to US Army on 3 April 2002; 500th AH-64D (383rd for US Army) delivered 26 August 2004, with production for US Army scheduled to end with acceptance of 501st example in July 2006.

Initial AH-64D battalion (1-227 AvRgt) at Fort Hood, Texas fully equipped by end July 1998 and attained combat-ready status on 19 November 1998, after eight month training programme at company and battalion level which included four live fire exercises and more than 2,500 flight hours; see table for details of actual and planned re-equipment programme.

AH-64DN: Designation assigned to 30 helicopters delivered to Royal Netherlands Air Force between May 1998 and April 2002.

WAH-64D: British Army version with Longbow radar and two Rolls-Royce/Turbomeca RTM 322 turboshafts; built by Westland in the UK.

CUSTOMERS: US Army 827 AH-64A (including six prototypes), of which last delivered 30 April 1996; see Programme and Current Versions for details. Confirmed export orders and firm commitments totalled 273 in late 2004. Israel ordered 18 in March 1990; first two delivered 12 September 1990 to 113 Squadron; further 24 ex-US Army AH-64As supplied for 113 and 127 Squadrons at Ramon; deliveries to 190 Squadron at Ramat David, 1995. After rejecting new-build AH-64D Apache Longbow on cost grounds, Israel considered upgrade of existing AH-64A to AH-64D; however, upgrade proposal later abandoned in favour of acquisition of new-build AH-64D; letter of offer and acceptance signed February 2001, covering nine helicopters (of which one was to be a remanufactured AH-64A). Contract for eight new-build AH-64Ds awarded on 29 May 2002, at which time additional three remanufactured helicopters included in overall deal. Subsequently, in mid-2004, it appeared Israel intends to upgrade all existing AH64As to AH-64D configuration.

Deliveries began in April 1993 to Saudi Arabia (12) and in 1994 to Egypt (24). Further orders from Greece (20) and United Arab Emirates (20), both in December 1991; six handed over to UAE on 30 October 1993; 14 followed in 1994, with 10 more received later; following formal request in mid-2002, UAE continuing to consider possibility of upgrading existing AH64As to AH-64D standard, incorporating Lockheed Martin Arrowhead piloting and weapon aiming system as well as enhanced AN/ALQ-211 EW suite. Total cost of upgrading 30 UAE helicopters would be approximately US$1.5 billion, but purchase of new-build examples not ruled out. Greek deliveries from February 1995 for training in USA; initial six Apaches ferried to New Orleans, Louisiana, for shipment to Greece on 9 June 1995. Greece subsequently revealed intent in 2002 to purchase 24 AH-64Ds, although eventual deal, finalised on 2 September 2003, covered only 12 helicopters, for

US ARMY AH-64D APACHE DEPLOYMENT (AT JULY 2004)

Unit	Base	IOC date
Regular Army Units:		
1-227 AVN	Robert Gray AAF, Tx	19 Nov 1998
2-101 AVN	Campbell AAF, Ky	Oct 1999
1-3 AVN	Hunter AAF, Ga	Mar 2001
1-2 AVN	Camp Page AAF, South Korea	Aug 2001
1-101 AVN	Campbell AAF, Ky	Dec 2001
6/6 CAV	Illesheim, Germany	May 2002
1-4 AVN	Hood AAF, Tx	Oct 2002
3/6 CAV	Desidero AAF, South Korea	Mar 2003
1-229 AVN[1]	Simmons AAF, NC	Jan 2004
2/6 CAV	Illesheim, Germany	Jun 2004

[1] 229 AVN inactivated May 2004; AH-64 battalion reassigned to 3rd Infantry Division and will eventually move to Hunter AAF, Ga

APACHE REMANUFACTURE PROGRAMME

Lot	Year	Qty	First aircraft
—	Prototypes	6	89-0192
—	Preproduction	2	
1	FY96	24	96-5001
2	FY97	24	97-5025
3	FY98	44	98-5049
4	FY99	66	99-5093
5	FY00	74	00-5159
6	FY01	52	01-5233
7	FY02	60	02-5285
8	FY03	74	03-5345
9	FY04	64	04-5419
10	FY05	19	05-5483
	Total	**509**[1]	

[1] Total excludes aircraft to be remanufactured for overseas operators, including Israel (4) and Egypt (35)

APACHE PROCUREMENT

	Qty	Remarks
US Army AH-64A		
FY73	3	Prototypes
FY79	3	Preproduction
FY82	11	
FY83	48	
FY84	112	
FY85	138	
FY86	116	
FY87	101	
FY88	77	
FY89	54	
FY90	132	
FY91	12	
FY92	10	
FY94	10	
Subtotal	**827**[1]	

delivery from first quarter of 2007. Another 12 AH-64As for Egypt approved early 1995, with Egypt revealing intent in July 2000 to upgrade 35 AH-64A to AH-64D. Contract signed in November 2001, with deliveries beginning in July 2003, although these do not have Longbow radar; by early 2004, eight Egyptian aircraft had been remanufactured.

Version of AH-64D Apache Longbow offered to British Army by consortium of McDonnell Douglas, Westland, Lockheed Martin, Northrop Grumman and Shorts; UK announced order for 67 on 13 July 1995; all except initial eight assembled by Westland (AgustaWestland) and described elsewhere. Netherlands signed contract on 24 May 1995 for 30 AH-64DN Apaches for Nos. 301 and 302 Squadrons at Gilze-Rijen; radar not required at outset, but formal request made in November 1999 and this will now be installed; US Army provided 12 AH-64As for use by No. 301 Squadron on lease basis for period 1996–99; all delivered 13 November 1996 and since returned. First AH-64DNs accepted by Netherlands at Mesa on 15 May 1998 and 21 had been delivered by late August 2001, with 30th and last handed over on 30 April 2002; No. 302 Squadron first unit to be equipped, with initial deliveries to Gilze-Rijen in mid-May 2000; No. 301 Squadron became operational in 2003, although Netherlands government in late 2003 announced intention to reduce size of fleet to about 24 helicopters. Singapore selected AH-64D on 14 June 1999; total of eight initially to be acquired, with AGM-114K Hellfire 2 laser-guided missile, but lacking Longbow radar; option on further 12 taken up on 23 August 2001, at which time it was revealed that all 20 will have Longbow radar; first example accepted at Mesa on 17 May 2002. Kuwait was authorised in late 1997 to receive 16 AH-64Ds under FMS programme; letter of acceptance due for signature in last quarter of 1998, but deal stagnated because of concern over inclusion of Longbow radar. Letter of offer and acceptance finally signed at beginning of September 2002, with package including eight Longbow radars. On 27 August 2001, Japan announced selection of AH-64D to satisfy its AH-X combat helicopter requirement; an initial batch of 10 helicopters is being acquired with FY02 funds, with the total buy likely to number 60, some of which will be fitted with Longbow radar. Future customers could include South Korea, which initially needs 18 combat helicopters, but could eventually purchase 36. Taiwan has indicated desire to obtain at least 30 new combat helicopters and apparently favours the AH-64D Apache Longbow; a decision was expected in 2004, but has been deferred by at least a year, with contract award not now likely until at least 2006. Following collapse of plan to acquire AH-1Z King Cobra, Turkey expected to invite new bids to supply 80 combat helicopters; AH-64D a candidate and likely to be in competition with A129 Mangusta and Eurocopter Tiger.

COSTS: Programme cost (807 aircraft at 1991 values) US$1,169 million. Egyptian Lot 2 request costed at US$318 million for 12 Apaches plus four spare Hellfire launchers, thirty-four 70 mm (2.75 in) rocket launchers, six spare T700 engines, one spare TADS/PNVS system and miscellaneous spares. Second multiyear procurement contract valued at US$2.3 billion, including training devices, spares, logistics and support services.

DESIGN FEATURES: Modern, tandem-seat, armoured and damage-resistant combat helicopter; is required to continue flying for 30 minutes after being hit by 12.7 mm bullets coming from anywhere in the lower hemisphere plus 20°; also survives 23 mm hits in many parts; target acquisition and designation sight (TADS) and pilot night vision sensor (PNVS) sensors mounted in nose; low-airspeed sensor above main rotor hub; avionics in lateral containers; chin-mounted Chain Gun fed from ammunition bay in centre-fuselage; four weapon pylons on stub-wings (six when air-to-air capability is installed); engines widely separated, with integral particle separators and built-in exhaust cooling fittings; four-blade main rotor with lifting aerofoil blade section and swept tips; blades can be folded or easily removed; tail rotor consists of two teetering two-blade units crossed at 55° to reduce noise; airframe meets full crash-survival specifications. Six AH-64s will fit in C-5 and three in C-17A.

Main transmission, by Litton Precision Gear Division, can operate for 1 hour without oil; tail rotor drive, by Aircraft Gear Corporation, has grease-lubricated gearboxes with Bendix driveshafts and couplings; gearboxes and shafts can operate for 1 hour after ballistic damage; main rotor shaft runs within airframe-mounted sleeve, relieving transmission of flight loads and allowing removal of transmission without disturbing rotor.

FLYING CONTROLS: Fully powered controls with stabilisation and automatic flight control system; automatic hover hold; tailplane incidence automatically adjusted by Hamilton Sundstrand control to streamline with downwash during hover and to hold best fuselage attitude during climb, cruise, descent and transition.

STRUCTURE: Main rotor blades (by Tool Research and Engineering Corporation, Composite Structures Divisions) tolerant to 23 mm cannon shells, have five U-sections forming spars and skins bonded with structural glass fibre tubes, laminated stainless steel skin and composites rear section; blades attached to hub by stack of laminated steel straps with elastomeric bearings. New rotor blades fabricated entirely from composites materials flown on Apache for first time on 6 November 2003. Northrop Grumman produced all fuselages, wings, tail, engine cowlings, canopies and avionics containers until 2003, when production of fuselage transferred to Korea Aerospace Industries; first completed unit accepted by Boeing in South Korea on 19 April 2004, thereafter being shipped to the final assembly facility in Mesa.

LANDING GEAR: Menasco trailing arm type, with single mainwheels and fully castoring, self-centring and lockable tailwheel. Mainwheel tyres size 8.50–10 (10 ply) tubeless, tailwheel tyre size 5.00–4 (14 ply) tubeless. Hydraulic brakes on main units. Main gear is non-retractable, but legs fold rearward to reduce overall height for storage and transportation. Energy-absorbing main and tail gears are designed for normal descent rates of up to 3.05 m (10 ft)/s and heavy landings at up to 12.8 m (42 ft)/s. Take-offs and landings can be made at structural design gross weight on terrain slopes of up to 12° (head-on) and 10° (side-on).

POWER PLANT: Two General Electric T700-GE-701C turboshafts, each rated at 1,409 kW (1,890 shp) for 10 minutes, 1,342 kW (1,800 shp) for 30 minutes, 1,238 kW (1,660 shp) maximum continuous and 1,447 kW (1,940 shp) 2½ minutes OEI. Engines mounted one on each side of fuselage, above wings, with key components armour-protected. Upper cowlings let down to serve as maintenance platforms. General Electric and US Army to undertake joint remanufacture programme of existing engines into T700-GE-701D standard from mid-2003; intended to alleviate poor levels of reliability, this will also increase power slightly and may be installed on more than 700 Apaches.

AH-64 modernisation may ultimately lead to installation of new engine on Block III helicopters. In addition, Boeing proposing adoption of new main rotor system allied to new drive system, offering improved performance, greater reliability and simplified maintenance.

Two crash-resistant fuel cells in fuselage, combined capacity 1,421 litres (375 US gallons; 312 Imp gallons). Modifications ordered September 1993 for carriage of four 871 litre (230 US gallon; 192 Imp gallon) Brunswick Corporation external tanks on 437 Apaches. Total internal and external fuel 4,910 litres (1,295 US gallons; 1,078 Imp gallons). New crashworthy, ballistically self-sealing internal auxiliary fuel tank entered evaluation phase in fourth quarter of 1997; tank holds 492 litres (130 US gallons; 108 Imp gallons) and is interchangeable with ammunition storage magazine, enabling all four weapons pylons to carry ordnance on long-range missions. Testing of preproduction system undertaken in 1998 in addition to formal test and qualification programme by US Army; total of 48 units ordered from Robertson Aviation in 1999; smaller 379 litre (100 US gallon; 83.3 Imp gallon) internal tank utilised by AH-64D, this permitting carriage of up to 300 rounds of 30 mm ammunition for Chain Gun. 'Black Hole' IR suppression system protects aircraft

Exports		First aircraft and date	
Israel (Lot 1)	18[2]	801	Sep 1990
Israel (Lot 2)	12[7]		2005
Saudi Arabia	12[3]	90-0291	Apr 1993
Egypt (Lot 1)	24[3]	3701	Feb 1994
Egypt (Lot 2)	12[3]	3725	1996
Greece (Lot 1)	20[3]	E-1001	Feb 1995
Greece (Lot 2)	12[6]		2007
UAE (Lot 1)	20[3]	050	Oct 1993
UAE (Lot 2)	10[3]	070	1996
Netherlands	30[4]	Q-01/98-0101	May 1998
UK	67[5]	ZJ166	Sep 1998
Singapore	20[6]		May 2002
Kuwait	16[6]		
Subtotal	**273**		
Total	**1,100**		

[1] 501 being remanufactured as AH-64D
[2] AH-64A; further 24 obtained from US Army
[3] AH-64A
[4] Non-radar AH-64D; 12 US Army AH-64A operated during 1996–99 on lease pending delivery of AH-64DN
[5] WAH-64D
[6] AH-64D
[7] AH-64D. Includes four remanufactured helicopters

APACHE PRODUCTION

Lot	FY	Qty US	Qty Export	First aircraft
—	73	3[1]		73-22247
—	79	3[2]		79-23257
1	82	11		82-23355
2	83	48		83-23787
3	84	112		84-24200
4	85	138		85-25351
5	86	116		86-8940
6	87	101		87-0407
7	88	67		88-0197
		10		88-0275
8	89	54	18	89-0192
9	90	60	6	90-0280
10	90	72	36	90-0415
11	91/92	22	34	91-0112
12	94	10	10	94-0328
13	95		12	95-0108
—	98 et seq		157[3]	98-0101
Subtotals	**827**		**273**	
Total			**1,100**	

[1] Prototypes; 73-22247 was static test airframe
[2] Preproduction
[3] AH-64D/WAH-64D; all other aircraft are AH-64A version. Total includes four remanufactured helicopters for Israel, which also to receive eight new-build AH-64Ds

from heat-seeking missiles: this eliminates an engine bay cooling fan, by operating from engine exhaust gas through ejector nozzles to lower the gas plume and metal temperatures.

ACCOMMODATION: Crew of two in tandem: co-pilot/gunner (CPG) in front, pilot behind on 48 cm (19 in) elevated seat. Crew seats, by Simula Inc, are of lightweight Kevlar. Northrop Grumman canopy, with PPG transparencies and transparent acrylic blast barrier between cockpits, is designed to provide optimum field of view. Crew stations are protected by Ceradyne Inc lightweight boron armour shields in cockpit floor and sides, and between cockpits, offering protection against 12.7 mm armour-piercing rounds. Sierracin electric heating of windscreen. Seats and structure designed to give crew a 95 per cent chance of surviving ground impacts of up to 12.8 m (42 ft)/s.

SYSTEMS: Honeywell totally integrated pneumatic system includes a shaft-driven compressor, air turbine starters, pneumatic valves, temperature control unit and environmental control unit. Fairchild Controls improved environmental control system comprises a distributed vapour-cycle cooling and heating unit, with two redundant systems incorporating dual-speed compressors, digital databus controllers and multiple heat exchangers, fans and control valves. Parker dual-hydraulic systems, operating at 207 bar (3,000 lb/sq in), with actuators ballistically tolerant to 12.7 mm direct hits. Redundant flight control system for both rotors. In the event of a flying control system failure, the system activates Honeywell secondary fly-by-wire control. Honeywell electrical power system, with two 45 kVA fully redundant engine-driven AC generators, two 300 A transformer-rectifiers, and URDC standby DC battery. Honeywell GTP 36-155(BH) 93 kW (125 shp) APU for engine starting and maintenance checking. DASA (TST) electric blade de-icing. Smiths Aerospace integrated electrical power management system (IEPMS) installed on AH-64D is currently being upgraded to incorporate multichannel remote interface unit (RIU) that will replace core electronics and wiring associated with conventional electrical control systems; improved IEPMS available from 2001 is being installed on approximately 300 Apaches for US Army. JGSDF Apaches will have Smiths Aerospace advanced electrical power management system (AEPMS), this to be manufactured under licence by Kanto Aircraft Instruments of Fujisawa, Japan.

AVIONICS: *Comms:* AN/ARC-164 UHF, AN/ARC-222 SINCGARS secure UHF/VHF; AN/ARC-220 UHF to be retrofitted; KY-28/58/TSEC crypto secure voice, C-8157 secure voice control; AN/APX-100 IFF unit with KIT-1A secure encoding; C-10414 Tempest intercom.

Radar: Lockheed Martin/Northrop Grumman AN/APG-78 Longbow mast-mounted 360° radar installed on most AH-64Ds, presenting up to 256 targets on tactical situation display;detects air targets in air-to-ground mode; air-to-air mode for flying targets only.

Flight: BAE North America AN/ASN-157 lightweight Doppler navigation system, Litton LR-80 (AN/ASN-143) strapdown AHRS, AN/ARN-89B ADF, GPS, Honeywell digital automatic stabilisation equipment (DASE), Astronautics Corporation HSI, Pacer Systems omnidirectional, low-airspeed air data system, remote magnetic indicator, BITE fault detection and location. Doppler system, with AHRS, permits nap-of-the-earth navigation and provides data for storing target locations. BAE Systems air data system, comprising two omnidirectional airspeed and direction sensors (AADSs) mounted on engine cowlings and a high integration air data computer (HIADC) installed in avionics bay.

Instrumentation: Honeywell all-raster symbology generator processes TV data from IR and other sensors, superimposes symbology, and distributes the combination to CRT and helmet-mounted displays; Honeywell AN/APN-209 radar altimeter video display unit. 'Manprint' (manpower integration) instrumentation including Litton Canada upfront display and two Honeywell 152 × 152 mm (6 × 6 in) monochrome CRT displays in each cockpit of early aircraft; from 27th AH-64D, Honeywell flat-panel, colour, active matrix LCD multipurpose displays (MPDs) installed in both cockpits, as well as in aircraft for UK and Netherlands. AH-64A's 1,200 cockpit switches reduced to approximately 200 on AH-64D.

Mission: Lockheed Martin target acquisition and designation sight and AN/AAQ-11 pilot's night vision sensor (TADS/PNVS) comprises two independently functioning, fully integrated systems mounted on nose.

TADS consists of a rotating turret (±120° in azimuth, +30/−60° in elevation) housing sensor subsystems, optical relay tube (being replaced under 1998 contract by Planar Advance/dpiX flat panel display) in the CPG's cockpit, three electronic units in the avionics bay, and cockpit-mounted controls and displays; used principally for target search, detection and laser designation, with CPG as primary operator (can also provide back-up night vision to pilot in event of PNVS failure). Once acquired by TADS, targets can be tracked manually or automatically for autonomous attack with gun, rockets or Hellfire missiles. TADS daylight sensor consists of TV camera with narrow (0° 50') and wide angle (4° 0') fields of view; direct view optics (4° narrow and 18° wide angle); laser spot tracker; and International Laser Systems laser range-finder/designator. New switchable eyesafe laser range-finder designator (SELRD) currently being developed by Kollsman Inc under US$2.8 million contract awarded at start of 2000; to be installed in existing TADS turret on AH-64A and AH-64D and will have 80 per cent commonality with Kiowa Warrior SELRD. Night sensor, in starboard half of turret, incorporates FLIR sight with narrow, medium and wide angle (3° 6', 10° 6' and 50°) fields of view.

PNVS consists of FLIR sensor (30 × 40° field of view) in rotating turret (±90° in azimuth, +20/−45° in elevation) mounted above TADS; electronic unit in the avionics bay; and pilot's display and controls; provides pilot with thermal imaging for nap-of-the-earth flight to, from and within battle area at night or in adverse daytime weather, at altitudes low enough to avoid detection.

Second-generation FLIR sensor to be installed on AH-64D; Lockheed Martin Arrowhead and Raytheon FIREsight systems conceived to satisfy this requirement. Arrowhead selected in late October 2000 to progress to EMD phase. Production is underway, with new sensor to be installed on late production AH-64Ds and retrofitted to existing helicopters, allowing fielding of first US Army unit by mid-2005. Arrowhead also available to overseas operators.

PNVS imagery displayed on monocle in front of one of pilot's eyes; flight information including airspeed, altitude and heading is superimposed on this imagery to simplify piloting. Monocle is part of Honeywell integrated helmet and display sighting system (HADSS) worn by both crew members. Symetrics Industries improved data modem for transmission of target data (and eventually real-time imagery) between helicopters, tactical jet, Joint STARS airborne command posts, HQs and ground units at 16,000 bits/s, plus radio frequency interferometer beneath radome for identification of hostile transmitters.

Self-defence: Aircraft survivability equipment (ASE) consists of Litton AN/APR-39 passive RWR, Sanders AN/ALQ-144 IR jammer, Raytheon AN/AVR-2 laser warning receiver, ITT AN/ALQ-136 radar jammer and chaff dispensers and Lockheed Martin AN/APR-48A radar frequency interferometer. Sanders AN/ALQ-212 Advanced Threat Infra-Red Countermeasures (ATIRCM) system and ITT AN/ALQ-211 suite of integrated RF countermeasures (SIRFC) system currently under development. ATIRCM combines next-generation directable IRCM system with Sanders AN/AAR-57 Common Missile Warning System (CMWS); contractor tests of SIRFC undertaken on Apache Longbow in latter half of 1999, followed by operational test and evaluation from early 2000. Elisra began flight test of passive airborne warning system in second half of 1999 and Israel plans to install this on its AH-64 fleet. IDF-AF AH-64D to use Elisra-supplied EW equipment, including radar, laser and IR missile warning systems, ECM and chaff and flare dispensers; system will provide audio warning and include visual threat indication via 100 × 100 mm (4 × 4 in) display. Laser warning system expected to be LWS-20V-2 unit. Upgraded Egyptian AH-64D fitted with Northrop Grumman AN/ALQ-162(V)6 'Shadowbox' high-band RF countermeasures system. Greek and Kuwaiti AH-64Ds to be fitted with BAE Systems HIDAS (Helicopter Integrated Defensive Aids System). Netherlands Apaches given upgraded self-protection systems in early 2004 in readiness for deployment to Afghanistan; package comprised Terma AMASE (Apache Modular Aircraft Survivability Equipment), incorporating AN/ALQ-213(V) integrated EU controller, colour threat displays and stub-wing, pod-mounted Northrop Grumman AN/AAR-54 missile warning system and expendables.

EQUIPMENT: Avpro of UK cleared Exint transport pod for use with AH-64 at start of 2000, but certification to carry personnel still required. In special forces insertion role, Apache can carry maximum of four pods, each able to accommodate 226 kg (500 lb) payload.

ARMAMENT: Boeing M230 Chain Gun 30 mm automatic cannon, located between the mainwheel legs in an underfuselage mounting with Smiths Industries electronic controls. Normal rate of fire is 625 rds/min of HE or HEDP (high-explosive dual-purpose) ammunition, which is interoperable with NATO Aden/DEFA 30 mm ammunition. Maximum ammunition load is 1,200 rounds. 'Sideloader' system installed in starboard forward avionics bay; cut normal loading time of 30 minutes by up to half and reduced number of personnel required from three to one. Gun mounting designed to collapse into fuselage between pilots in the event of a crash landing.

New electric turret under development by Boeing, which received two year, US$5 million contract in first half of 1999; objective is to achieve accuracy of 0.5 mrads compared with current 3.0 mrads. Gun, mount and feed system to be retained in conjunction with redesigned mechanical system featuring electric rather than hydraulic drive as well as digital control; result should be at least 10 per cent lighter and require one instead of two electrical boxes. HR Textron responsible for controls, with Boeing providing the rest.

Four underwing hardpoints, with Aircraft Hydro-Forming pylons and ejector units, on which can be carried up to 16 AGM-114 Hellfire anti-tank missiles or up to 76 2.75 in FFAR (folding fin aerial rockets) in their launchers or a combination of Hellfires and FFAR. New Joint Common Missile (JCM) to be developed as replacement for Hellfire from 2009; weapon will possess tri-mode seeker, combining semi-active laser capability with imaging infra-red and millimetre-wave radar. JCM expected to have 8.6 n mile (16 km; 10 mile) range when launched from helicopter. Compatibility with AAM such as Raytheon Stinger or Thales Starstreak under consideration, but appears unlikely in near-term future; provision of an AAM will require additional hardpoints if exisiting weapons capacity not to be compromised. Hellfire remote electronics by Rockwell Collins; Honeywell aerial rocket control system; multiplex (MUX) system units by Honeywell. Co-pilot/gunner (CPG) has primary responsibility for firing gun and missiles, but pilot can override his controls to fire gun or launch missiles.

DIMENSIONS, EXTERNAL:

Main rotor diameter	14.63 m (48 ft 0 in)
Main rotor blade chord	0.53 m (1 ft 9 in)
Tail rotor diameter	2.79 m (9 ft 2 in)
Length overall: tail rotor turning	15.54 m (51 ft 0 in)
both rotors turning	17.76 m (58 ft 3¼ in)
Wing span: clean	5.23 m (17 ft 2 in)
over empty weapon racks	5.82 m (19 ft 1 in)
Height: over tailfin	3.55 m (11 ft 7½ in)
over tail rotor	4.30 m (14 ft 1¼ in)
to top of rotor head	3.84 m (12 ft 7 in)
overall (top of air data sensor)	4.66 m (15 ft 3½ in)
overall, Longbow radar	4.95 m (16 ft 3 in)
Main rotor ground clearance (turning)	3.59 m (11 ft 9¼ in)
Distance between c/l of pylons:	
inboard pair	3.20 m (10 ft 6 in)
outboard pair	4.72 m (15 ft 6 in)
Tailplane span	3.40 m (11 ft 2 in)
Wheel track	2.03 m (6 ft 8 in)
Wheelbase	10.59 m (34 ft 9 in)

AREAS:

Main rotor disc	168.11 m^2 (1,809.5 sq ft)
Tail rotor disc	6.13 m^2 (66.0 sq ft)

WEIGHTS AND LOADINGS:

Weight empty:	
without Longbow	approx 5,165 kg (11,387 lb)
with Longbow	5,352 kg (11,800 lb)
Max fuel weight: internal	1,108 kg (2,442 lb)
external (four Brunswick tanks)	2,712 kg (5,980 lb)
Primary mission gross weight	7,480 kg (16,491 lb)
Design mission gross weight	8,006 kg (17,650 lb)
Max T-O weight: -701 engines	9,525 kg (21,000 lb)
-701C engines, ferry mission, full fuel	10,432 kg (23,000 lb)
Max disc loading	62.1 kg/m^2 (12.71 lb/sq ft)

PERFORMANCE (A: with -701 engines, without Longbow at 6,552 kg; 14,445 lb AUW, L: Apache Longbow at 7,530 kg; 16,601 lb with -701C engines):

Never-exceed speed (VNE)	197 kt (365 km/h; 227 mph)
Max level and max cruising speed:	
A	158 kt (293 km/h; 182 mph)
L	143 kt (265 km/h; 165 mph)
Max rate of climb at S/L: L	736 m (2,415 ft)/min
Max vertical rate of climb at S/L:	
A	762 m (2,500 ft)/min
L	450 m (1,475 ft)/min
Service ceiling: A	6,400 m (21,000 ft)
L	5,915 m (19,400 ft)
Service ceiling, OEI: A	3,290 m (10,800 ft)
Hovering ceiling:	
IGE: A	4,570 m (15,000 ft)
L	4,170 m (13,690 ft)
OGE: A	3,505 m (11,500 ft)
L	2,890 m (9,480 ft)
Max range, internal fuel: 30 min reserves:	
A	260 n miles (482 km; 300 miles)
L	220 n miles (407 km; 253 miles)
no reserves: L	257 n miles (476 km; 295 miles)
Ferry range, max internal and external fuel, still air, 45 min reserves	1,024 n miles (1,899 km; 1,180 miles)
Endurance at 1,220 m (4,000 ft) at 35°C	1 h 50 min
Max endurance, L: internal fuel	2 h 44 min
internal and external fuel	8 h 0 min

g limits at low altitude and airspeeds up to 164 kt (304 km/h; 189 mph) .... +3.5/–0.5

WEIGHTS FOR TYPICAL MISSION PERFORMANCE: (all without Longbow; A: anti-armour at 1,220 m/4,000 ft and 35°C, four Hellfire and 320 rounds of 30 mm ammunition; B: as A, but with 1,200 rounds; C: as A, but with six Hellfire and 540 rounds; D: anti-armour at 610 m/2,000 ft and 21°C, 16 Hellfire and 1,200 rounds; E: air cover at 1,220 m/4,000 ft and 35°C, four Hellfire and 1,200 rounds; F: as E but at 610 m/2,000 ft and 21°C, four Hellfire, 19 rockets, 1,200 rounds; G: escort at 1,220 m/4,000 ft and 35°C, 19 rockets and 1,200 rounds; H: escort at 610 m/2,000 ft and 21°C, 38 rockets and 1,200 rounds):

Mission fuel: A .... 727 kg (1,602 lb)
G .... 741 kg (1,633 lb)
E .... 745 kg (1,643 lb)
C .... 902 kg (1,989 lb)
B .... 1,029 kg (2,269 lb)
D .... 1,063 kg (2,344 lb)
H .... 1,077 kg (2,374 lb)
F .... 1,086 kg (2,394 lb)

Mission gross weight: A .... 6,552 kg (14,445 lb)
E .... 6,874 kg (15,154 lb)
G .... 6,932 kg (15,282 lb)
B, C .... 7,158 kg (15,780 lb)
D .... 7,728 kg (17,038 lb)
F .... 7,813 kg (17,225 lb)
H .... 7,867 kg (17,343 lb)

TYPICAL MISSION PERFORMANCE (A-H as above):

Cruising speed at intermediate rated power:
C .... 147 kt (272 km/h; 169 mph)
D .... 148 kt (274 km/h; 170 mph)
F .... 150 kt (278 km/h; 173 mph)
B .... 151 kt (280 km/h; 174 mph)
E, H .... 153 kt (283 km/h; 176 mph)
A .... 154 kt (285 km/h; 177 mph)
G .... 155 kt (287 km/h; 178 mph)

Max vertical rate of climb at intermediate rated power:
B, C .... 137 m (450 ft)/min
H .... 238 m (780 ft)/min
F, G .... 262 m (860 ft)/min
E .... 293 m (960 ft)/min
D .... 301 m (990 ft)/min
A .... 448 m (1,470 ft)/min

Mission endurance (no reserves): A, E, G .... 1 h 50 min
C .... 1 h 47 min
D, F, H .... 2 h 30 min
B .... 2 h 40 min

UPDATED

BOEING

BOEING COMMERCIAL AIRPLANES

1901 Oakesdale Avenue Southwest, Renton, Washington 98055
Tel: (+1 206) 746 11 11
Fax: (+1 206) 746 29 49
Web: http://www.boeing.com

PRESIDENT: Alan R Mulally
EXECUTIVE VICE-PRESIDENTS:
AIRPLANE PROGRAMS: James M Jamieson
COMMERCIAL AVIATION SERVICES: Michael Bair
VICE-PRESIDENT, COMMUNICATIONS: Tom Downey

Single-aisle aircraft are 717, 737 and 757; twin-aisle are 747, 767 and 777.

BOEING AIRLINER BACKLOG
(at 1 January 2004)

717-200	36	
Total 717		**36**
737-600	19	
737-700	428	
737-700BBJ	8	
737-800	358	
737-800BBJ	1	
737-900	17	
Total 737		**831**
747-400	14	
747-400ERF	6	
747-400F	17	
Total 747		**37**
757-200	8	
757-300	5	
Total 757		**13**
767-200ER	4	
767-300ER	21	
Total 767		**25**
777-200	9	
777-200ER	74	
777-200LR	5	
777-300	9	
777-300ER	71	
Total 777		**168**
Total backlog		**1,110**

BOEING AIRLINER FIVE-YEAR DELIVERIES

Year	717	737	747	757	767	777	MD-11	MD-80	MD-90	Total
2000	32	281	25	45	44	55	4		3	**489**
2001	49	299	31	45	40	61	2			**527**
2002	20	223	27	29	35	47				**381**
2003	12	173	19	14	24	39				**281**
2004	12	202	15	11	9	36				**285**

BOEING COMMERCIAL AIRPLANE GROUP ORDERS AND DELIVERIES
(at 1 January 2004)

	Orders Total	Deliveries Total	Orders in 2004	Deliveries in 2004	Backlog
707/720	1,010	1,010	0	0	
Total 707/720	**1,010**	**1,010**	**0**	**0**	**0**
717-200	169	137	8	12	32
Total 717	**169**	**137**	**8**	**12**	**32**
727	1,831	1,831	0	0	
Total 727	**1,831**	**1,831**	**0**	**0**	**0**
737-100	30	30	0	0	
737-200	1,114	1,114	0	0	
737-300	1,113	1,113	0	0	
737-400	486	486	0	0	
737-500	389	389	0	0	
737-600	73	56	1	3	17
737-700/C-40A	974	615	43	112	359
737-700BBJ	77	71	1	3	6
737-800	1,207	824	103	78	383
737-800BBJ	12	10	1	0	2

	Orders Total	Deliveries Total	Orders in 2004	Deliveries in 2004	Backlog
737-900	55	46	–2	6	9
Total 737	**5,530**	**4,754**	**147**	**202**	**776**
747-100	250	250	0	0	
747-200	393	393	0	0	
747-300	81	81	0	0	
747-400	449	440	–2	3	9
747-400D	19	19	0	0	
747-400ER	6	6	0	0	
747-400ERF	15	12	2	5	3
747-400F	111	91	10	7	20
747-400M	61	61	0	0	
Total 747	**1,385**	**1,353**	**10**	**15**	**32**
757-200	913	911	0	6	2
757-200M	1	1	0	0	
757-200PF	80	80	0	0	
757-300	55	55	0	5	
Total 757	**1,049**	**1,047**	**0**	**11**	**2**
767-200	128	128	0	0	
767-200ER	120	114	2	0	6
767-300	104	104	0	0	
767-300ER	515	502	1	9	13
767-300F	45	40	5	0	5
767-400ER	38	37	1	0	1
Total 767	**950**	**925**	**9**	**9**	**25**
777-200	91	84	0	2	7
777-200ER	414	350	12	22	64
777-200LR	5	0	0	0	5
777-300	64	55	2	2	9
777-300ER	99	10	28	10	89
Total 777	**673**	**499**	**42**	**36**	**174**
7E7-3	30	0	30		30
7E7-8	26	0	26		26
Total 7E7	**56**	**0**	**56**		**56**
Grand Totals	**12,653**	**11,556**	**272**	**285**	**1,097**

Notes: Table does not include one each of 727-100, 747-100, 757-200, 767-200 and 777-200 owned by Boeing. Annual order total is net (after deduction of cancellations)

UPDATED

BOEING 737

US Navy designation: C-40A Clipper

TYPE: Twin-jet airliner.

DEVELOPMENT MILESTONES

737-100	
First order	15 Feb 65
Rolled out	17 Jan 67
First flight	9 Apr 67
Certification	15 Dec 67
First Delivery	28 Dec 67
Entered service (Lufthansa)	10 Feb 68
Subsequent versions	
737-600	
First order	15 Mar 95
Rolled out	8 Dec 97
First flight	22 Jan 98
Certification	18 Aug 98
First delivery	18 Sep 98
Entered service (SAS)	25 Oct 98
737-700	
Go-ahead	17 Nov 93
First order	17 Nov 93
Rolled out	8 Dec 96
First flight	9 Feb 97
Certification	7 Nov 97
First delivery	17 Dec 97
Entered service (Southwest airlines)	18 Jan 98
737-800	
First order	5 Sep 94
Rolled out	30 Jun 97
First flight	31 Jul 97
Certification	13 Mar 98
First delivery	22 Apr 98
Entered service (Hapag-Lloyd)	24 Apr 98
737-900	
First order	10 Nov 97
Rolled out	23 Jul 00
First flight	3 Aug 00
Certification	17 Apr 01
First delivery	16 May 01
Entered service (Alaska Airlines)	27 May 01

PROGRAMME: Original Boeing 737 first flew 9 April 1967; -100 and -200 with Pratt & Whitney JT8D engines; -300 entered service with CFM56 engines, November 1984, followed by -400 and -500; 3,132nd and last of this generation delivered February 2000. These versions, including military T-43 and Surveiller, described in *Jane's Aircraft Upgrades*.

Current 'Next-Generation' of family (initially 737X) originated in 1991, when Boeing asked more than 30 airlines to help define improved series; company board authorised offer for sale June 1993; Southwest Airlines ordered 63 737-700s (32 converted from options for 737-300s) plus 63 new options (all, and more, taken up) 18 November 1993; roll-out (737-700) 8 December 1996; first flight (N737X) 9 February 1997; certification 7 November 1997, followed by first deliveries. CFM56-7B power plant first flew on Boeing 747 testbed on 16 January 1996. Flight test programme involved 10 aircraft: four 737-700s, three 737-800s and three 737-600s. FAA approval for 180-minute ETOPS granted in September 1999. By January 2000, B737s of all subtypes had flown over 100 million hours. 1,000th NG 737 (N418WN) first flew 1 November 2001 and delivered to Southwest Airlines on 13 November.

Further enhancements, incorporated in 737-900 demonstrator, completed in March 2002; features include quiet climb system, GPS landing and synthetic vision system.

CURRENT VERSIONS: **737-600:** Smallest of current 737 family. Known as 737-500X until officially launched 15 March 1995; 110 two-class passengers; final assembly of prototype began at Renton 29 August 1997; roll-out December 1997; first flight 22 January 1998; FAA certification 18 August 1998; first delivery (SE-DNM to launch customer SAS) 18 September 1998.

737-700: First to be ordered and manufactured; mid-size version of family, equivalent to previous ('Classic') 737-300, seating 126 passengers in two-class layout. First aircraft (N737X) rolled out 2 (officially 8) December 1996; first flight 9 February 1997, followed by second aircraft 27 February that year; second aircraft attained maximum certified altitude of 12,500 m (41,000 ft) for the first time on 19 March 1997; FAA certification 7 November 1997, with first delivery (fourth built, N700GS to Southwest Airlines) on 17 December 1997.

737-700IGW: Formerly 737-700X; increased gross weight version, based on Boeing Business Jet airframe, including forward cargo door, port. Available as **737-700C** (Convertible) and **737-700C/QC** (Convertible/Quick change); latter interchangeable between 140 single-class (81 cm; 32 in pitch) passengers and eight 2.24 × 3.18 m (88 × 125 in) pallets within one hour, compared to six hours of standard -700C. Max T-O weight 77,560 kg (171,000 lb); max landing weight 60,780 kg (134,000 lb); max zero-fuel weight 57,155 kg (126,000 lb); operating weight empty: -700C passenger 40,000 kg (88,190 lb), -700C cargo 38,365 kg (84, 580 lb), -700C.QC passenger 41,840 kg (92,240 lb), -700C/QC cargo 38,905 kg (85,770 lb). Fuel capacity 26,025 litres (6,875 US gallons; 5,725 Imp gallons); design range: -700C passenger 3,210 n miles (5,944 km; 3,694 miles), -700C/QC passenger 3,115 n miles (5,769 km; 3,584 miles), cargo (both versions) 2,880 n miles (5,333 km; 3,314 miles). Cargo volume: lower hold 27.4 m^3 (966 cu ft), main deck (cargo conversion only) 99.7 m^3 (3,520 cu ft).

Initial two ordered 29 August 1997 as **C-40A Clipper** to meet US Navy requirement for C-9B Skytrain II replacement. First C-40A flew 17 April 2000; first delivery to US Naval Reserve Fleet Logistics Support Squadron VR-59 at NAS/JRB Fort Worth, Texas 21 April 2001; three additional C-40As operated by VR-59, and two by VR-58 at NAS Jacksonville, Florida; sixth delivered 28 October 2002 and seventh in 2004.

Commercial launch customer for the 737-700C/QC was Saudi Aramco of Saudi Arabia, which took delivery of two in 2001.

First civil equivalent, designated **737-700C**, was N743A, first flown 18 September 2001 and delivered to ARAMCO on 31 October.

737-800: Known as 737-400X Stretch until launched 5 September 1994; seats 162

Boeing 737-700 of Southwest Airlines *(Paul Jackson)* ***NEW***/1044979

two-class passengers; roll-out (N737BX) 30 June 1997; first flight 31 July 1997; certified by FAA 13 March and JAA 9 April 1998; first delivery (D-AHFC to launch customer Hapag-Lloyd) 22 April 1998. Hapag-Lloyd became first commercial courier to operate 737-800 fitted with optional winglets in May 2001. Ryanair was recipient of 800th —800 on 1 September 2004.

737-800ERX:: Under study in early 2004. Would employ -900X's strengthened wing, plus stronger landing gear. MTOW 83,460 kg (184,000 lb).

737-900: Formerly 737-900X; launched 10 November 1997 with an order for 10, plus 10 options, from Alaska Airlines; largest 737 variant to date, with (compared to -800) stretch by means of 1.57 m (5 ft 2 in) forward plug and 1.07 m (3 ft 6 in) aft plug and strengthened fuselage; seating for 177 two-class passengers; deliveries from early 2001. One prototype/certification aircraft only; rolled out 23 July 2000; first flight (N737X) 3 August 2000; FAA certification achieved 17 April 2001 following 156 hours of ground testing and a two-aircraft flight test programme totalling 649 flight hours in 296 sorties; first delivery (Alaska Airlines) 16 May 2001 and entered service 27 May.

737-900X: Increased capacity, long-range version under study in 2001 and offered from 2002 to compete with Airbus A321 in European charter market; heavier gauge wing skins; additional Type II emergency exit aft of the wing, port side, to increase certification-limited maximum capacity to 204 or 210 passengers; increase in max take-off weight to 83,551 kg (184,200 lb); max payload range 2,710 n miles (5,018 km to 3,118 miles); CFM56-7B27/B1F engines and winglets.

Boeing Business Jet : (BBJ) Corporate versions, *described separately*.

Special missions: Military versions, *described separately*.

CUSTOMERS: As per table.

BOEING 737 NEXT-GENERATION ORDER BOOK
(at 1 January 2005)

Customer	Variant	First Order	First Delivery	600	700	800	900
Aeromexico	737-700	27 Dec 02	7 Oct 03		19		
Air Algérie	737-600	16 Jul 98	29 Apr 02	5			
	737-800	16 Jul 98	31 Jul 00			10	
Air Berlin	737-800	22 Dec 94	7 May 98			26	
Air China	737-700	31 Jul 01	17 Jun 03		11		
	737-800	30 Oct 97	17 Aug 99			11	
Air Europa	737-800	31 Oct 03	4 Oct 04			9	
Air Pacific	737-700	5 Sep 96	22 Sep 98		1		
	737-800	5 Sep 96	24 May 99			2	
Air Senegal	737-700	27 May 04	—		1		
AirTran	737-700	3 Jul 03	—		30		
Alaska Airlines	737-700	10 Nov 97	28 Jul 99		17		
	737-800	10 Nov 97	—			2	
	737-900	10 Nov 97	15 May 01				12
All Nippon	737-700	27 Jun 03	—		45		
American Airlines	737-800	21 Nov 96	7 Feb 99			124	
American Trans Air	737-800	30 Jun 00	29 Jun 01			20	
Ansett Worldwide	737-700	2 Sep 96	22 Mar 99		4		
Aramco Services Co	737-700	15 Dec 99	15 Sep 00		5		
Austrian	737-800	5 Mar 03	—			4	
Bavaria	737-700	30 Jan 95	27 Apr 98		6		
Bouillion Aviation	737-700	29 Jul 98	24 Jan 01		14		
	737-800	29 Jul 98	26 Apr 01			18	
Braathens	737-700	3 Feb 97	25 Aug 98		10		
Britannia Airways	737-800	18 Nov 98	6 Apr 00			4	
China Airlines	737-800	22 Dec 95	26 Oct 98			13	
China Eastern	737-700	2 Oct 01	10 Sep 02		10		
China Southern	737-700	2 Oct 01	16 Jan 04		8		
	737-800	2 Oct 01	22 Aug 02			12	
China Southwest	737-800	1 Jun 98	29 Oct 99			3	
China Yunnan	737-700	30 Oct 97	8 Dec 98		4		
CIT Leasing	737-700	14 Jun 99	11 Apr 01		12		
	737-800	14 Jun 99	26 Apr 01			19	
Continental	737-700	25 Jul 96	30 Mar 98		51		
	737-800	25 Jul 96	23 Jun 98			121	
	737-900	18 Mar 98	29 May 01				15
Copa Airlines	737-700	19 Jan 99	21 Jan 00		14		
	737-800	28 Jun 02	6 Oct 03			3	
Delta Air Lines	737-800	10 Jun 97	22 Oct 98			132	
Eastwind Airlines	737-700	20 Oct 97	13 May 98		2		
EasyJet	737-700	28 Jul 98	13 Oct 00		32		
El Al	737-700	1 Apr 98	11 Aug 99		2		
	737-800	1 Apr 98	24 Feb 99			3	

Customer	Variant	First Order	First Delivery	600	700	800	900
Ethiopian Airlines	737-700	28 Nov 02	24 Nov 03		3		
Garuda	737-700	15 Dec 99	—		18		
GATX Financial Corp	737-800	31 Jul 96	14 Dec 98			20	
GATX Jet Partners	737-800	31 May 00	29 Jun 00			18	
GECAS	737-600	22 Jan 96	24 Sep 99	7			
	737-700	22 Jan 96	30 Oct 98		92		
	737-800	22 Jan 96	29 Jul 98			65	
Germania	737-700	28 Mar 95	10 Mar 98		12		
GOL Airlines	737-800	17 May 04	—			17	
Hainan Airlines	737-800	30 Oct 97	23 Aug 99			19	
Hapag-Lloyd	737-800	18 Nov 94	22 Apr 98			37	
ILFC	737-600	2 Sep 96	17 Oct 03	15			
	737-700	25 Jul 95	25 Mar 98		107		
	737-800	25 Jul 95	10 Jun 98			92	
Itochu AirLease	737-800	11 Jul 96	21 Jan 99			10	
Jet Airways	737-700	14 Jun 99	17 May 01		4		
	737-800	11 Dec 96	24 Dec 98			11	
	737-900	14 Jun 99	6 May 03				2
Kenya Airways	737-700	7 Mar 00	11 Dec 02		2		
KLM	737-800	14 Mar 97	25 Feb 99			13	
	737-900	7 Sep 98	29 Jun 01				5
Korean Air	737-800	9 Jun 98	15 Aug 00			6	
	737-900	9 Jun 98	21 Nov 01				16
Lauda Air	737-600	31 Dec 98	4 May 00	2			
	737-700	8 Jan 99	19 Apr 01		2		
	737-800	8 Jan 99	28 Jul 98			4	
Linas Aereas Azteca	737-700	13 Feb 01	29 Jun 01		2		
LOT	737-800	9 Oct 96	—			2	
Luxair	737-700	28 Feb 03	9 Feb 04		3		
Maersk	737-700	16 Jun 94	2 Mar 98		12		
Midway Airlines	737-700	14 Jun 99	29 Sep 00		7		
Oman Air	737-700	30 Apr 01	25 Jun 02		1		
	737-800	30 Apr 01	1 Jul 03			2	
Pegasus Airlines	737-800	4 Sep 97	25 Mar 99			1	
Pembroke Capital Ltd	737-600	14 Mar 95	27 Apr 99	1			
	737-700	6 Mar 97	15 Dec 98		8		
Qantas	737-800	9 Nov 01	14 Jan 02			33	
Royal Air Maroc	737-700	30 Aug 96	16 Apr 99		5		
	737-800	30 Aug 96	10 Jul 98			14	
Ryanair	737-800	9 Mar 98	19 Mar 99			155	
SAS	737-600	14 Mar 95	18 Sep 98	28			
	737-700	14 Mar 95	3 Nov 99		6		
	737-800	14 Mar 95	3 May 00			23	
Shandong Airlines	737-700	12 Nov 03	—		3		
	737-800	12 Nov 03	—			4	
Shanghai Airlines	737-700	30 Oct 97	25 May 00		4		
	737-800	2 Oct 01	14 Aug 02			1	
Shenzhen Airlines	737-700	30 Oct 97	30 Sep 98		6		
	737-900	12 Nov 03	—				5
South African Airways	737-800	31 May 00	26 Jul 02			5	
Southwest	737-700	17 Nov 93	17 Dec 97		276		
Sumitomo Corp	737-800	11 Jul 96	3 May 99			2	
Sunrock	737-700	17 Jun 97	—		2		
	737-800	17 Jun 97	20 Apr 00			3	
Taiwan Air Force	737-800	30 Nov 98	8 Dec 99			1	
TAROM	737-700	17 Jun 99	27 Mar 01		4		
Tombo Aviation	737-700	31 Dec 96	2 Dec 98		7		
	737-800	31 Dec 96	30 Nov 99			8	
Transavia	737-700	15 Nov 01	18 Feb 03		4		
	737-800	6 Nov 95	16 Jun 98			13	
TunisAir	737-600	28 Oct 97	25 May 99	7			
Turkish Airlines	737-800	7 Oct 97	30 Oct 98			41	
US Navy	737-700	3 Sep 97	29 Sep 00		8		
	737-800	30 Jun 04	—			5	
Virgin Blue	737-800	31 Dec 02	13 Aug 03			17	
Westjet	737-600	21 May 04	—	3			
	737-700	23 Aug 00	18 Oct 02		34		
	737-800	23 Aug 00	—			5	
Wuhan Airlines	737-800	29 Jun 98	19 Mar 01			2	
Xiamen Airlines	737-700	30 Oct 97	21 Aug 98		11		
Various unidentified	737-700BBJ	22 Jan 96	23 Nov 98		77		
	737-800BBJ	19 Nov 97	28 Feb 01			12	
	737-600	29 Dec 04	—	5			
	737-700	14 Aug 03	3 Dec 04		26		
	737-800	18 Mar 02	31 Mar 04			22	
Subtotals				**73**	**1,044**	**1,219**	**55**
Total						**2,398**	

Notes: For deliveries see Group Orders and Deliveries table

COSTS: List prices (2004; all in millions): 737-600 US$44.0 to US$52.0; 737-700 US$50.5 to US$59.0; 737-800 US$61.5 to US$69.5; 737-900 US$64.5 to 74.0.

DESIGN FEATURES: Conventional, medium-size airliner with podded engines and sweptback wing and tail surfaces. Dihedral 6° at root; sweepback 25° at quarter-chord. Greater range and speed than previous 737s, with less noise and fewer emissions; wing area increased by some 25 per cent by means of 0.43 m (1 ft 5 in) increase in wing chord and about 4.83 m (15 ft 10 in) increase in wing span; new high-lift systems; larger tail surfaces; increased tankage gives US transcontinental range; new aircraft can use same runways, taxiways, ramps and gates as preceding variants; new variant of CFM56 turbofan derated from nominal thrust to suit smaller versions of the family. Noise on ground reduced by approximately 12 dB by new diffuser duct and cooling vent silencer on APU, new ECS fan and duct and new electrical/electronics cooling fan.

FLYING CONTROLS: Conventional and powered. All surfaces actuated by two independent hydraulic systems with manual reversion for ailerons and elevator; elevator servo tabs unlock on manual reversion; rudder has standby hydraulic actuator and system. Three outboard-powered overwing spoiler panels on each wing assist lateral control and also act as airbrakes. Variable incidence tailplane has two electric motors and manual standby.

Leading-edge Krueger flaps inboard and four sections of slats outboard of engines; two airbrake/lift dumper panels on each wing, inboard and outboard of engines; continuous-span, double-slotted trailing-edge flaps inboard and outboard of engines.

FAA Cat. II landing minima system standard using SP-300 dual digital integrated flight director/autopilot: Cat. IIIa capability optional.

STRUCTURE: Aluminium alloy dual-path fail-safe two-spar wing structure with corrosion-resistant 7055-T77 upper skin. Aluminium alloy two-spar tailplane. Graphite composites ailerons, elevators and rudder. Aluminium honeycomb spoiler/airbrake panels and trailing-edges of slats and flaps. Fuselage structure fail-safe aluminium. Elevators, rudder and ailerons contain graphite/Kevlar; other, unstressed, components in GFRP and CFRP include nosecone, wing/fuselage fairing, fin fillet, fintip and flap actuator fairings. Rears of engine nacelles are of graphite/Kevlar/glass fibre.

LANDING GEAR: Hydraulically retractable tricycle type, with Boeing oleo-pneumatic shock-absorbers; inward-retracting main units have no doors, wheels forming wheel well seal; nose unit retracts forward; free-fall emergency extension. Twin nosewheels have tyres size 27×7.75. Main units have heavy-duty twin wheels, H40×14.5–19 heavy-duty tyres, and Honeywell or Goodrich heavy-duty wheel brakes as standard. Mainwheel tyre pressure 13.45 to 14.00 bar (195 to 203 lb/sq in). Nosewheel tyre pressure 11.45 to 11.85 bar (166 to 172 lb/sq in).

POWER PLANT: *737-600:* Two CFM International CFM56-7B18 turbofans, each rated at 86.7 kN (19,500 lb st) standard, or two CFM56-7B22s, each rated at 101 kN (22,700 lb st) in high gross weight version.

737-700: Two CFM56-7B20s, each rated at 91.6 kN (20,600 lb st) standard, or two CFM56-7B24s, each rated at 101 kN (22,700 lb st) in high gross weight version.

737-800: Two CFM56-7B24s, each rated at 107.6 kN (24,200 lb st) standard, or two CFM56-7B27s, each rated at 121.4 kN (27,300 lb st) in high gross weight version.

737-900: Two CFM56-7B26s, each rated at 117 kN (26,300 lb st) standard, or two CFM56-7B27s, each rated at 121.4 kN (27,300 lb st) in high gross weight version.

Fuel capacity (all) 26 025 litres (6,875 US gallons; 5,725 Imp gallons).

ACCOMMODATION: *All:* Crew of two side by side on flight deck. One plug-type door at each corner of cabin, with passenger doors on port side and service doors on starboard side. Airstair for forward cabin door optional. Overwing emergency exit on each side. One or two galleys and one modular vacuum lavatory forward and one or two galleys and lavatories aft; all lavatories interconnected to single waste collection tank at port side rear. Lightweight interior, of crushed core materials, has movable class divider, overnight seating-pitch flexibility and modular passenger service unit (PSU) including fold-down video screen in underside of baggage bin. Centreline stowage bins optional for emergency equipment and crew baggage.

Two underfloor baggage holds, forward and aft of wing. Rear hold has provision for post-delivery installation of telescopic baggage conveyor system (when additional fuel tanks not fitted). One baggage door in starboard side of each hold.

737-600: Alternative cabin layouts seat from 110 to 132 passengers. Typical arrangements offer eight first class seats four-abreast at 91 cm (36 in) pitch and 102 tourist class seats six-abreast at 81 cm (32 in) pitch in mixed class; and 132 all-tourist class at 76 cm (30 in) pitch. Total overhead baggage capacity of 6.1 m^3 (216 cu ft), equivalent to 0.045 m^3 (1.6 cu ft) per passenger.

737-700: Alternative cabin layouts seat from 126 to 149 passengers. Typical arrangements offer eight first class seats four-abreast at 91 cm (36 in) pitch and 118 tourist class seats six-abreast at 81 cm (32 in) pitch in mixed class; and 149 all-tourist class at 76 cm (30 in) pitch. Total overhead baggage capacity of 7.0 m^3 (248 cu ft), equivalent to 0.05 m^3 (1.8 cu ft) per passenger. *C-40A* options comprise 121 passengers, all-cargo (eight pallets) and combinations of 70 passengers and three pallets.

737-800: Alternative cabin layouts seat from 162 to 189 passengers. Typical arrangements offer 12 first class seats four-abreast at 91 cm (36 in) pitch and 150 tourist class seats six-abreast at 81 cm (32 in) pitch in mixed class; and 189 all-tourist class at 76 cm (30 in) pitch. Total overhead baggage capacity of 9.3 m^3 (328 cu ft), equivalent to 0.05 m^3 (1.7 cu ft) per passenger.

737-900: Alternative cabin layouts seat from 177 to 189 passengers. Typical arrangements offer 12 first class seats four-abreast at 91 cm (36 in) pitch and 165 tourist class seats six-abreast at 81 cm (32 in) pitch in mixed class; and 189 all-tourist class at 81 cm (32 in) pitch.

SYSTEMS: Honeywell 131-9(B) APU with air start capability to maximum certified altitude and 90 kVA electrical load capability to 11,278 m (37,000 ft). Three-wheel air cycle environmental control system with optional ozone converter and digital cabin pressure controls.

AVIONICS: *Flight:* Satellite navigation standard. Optional satcom and dual FMS (single standard) integrated with GPS.

Instrumentation: Honeywell Air Transport Systems common display system (CDS) with five-screen flat-panel liquid crystal display (LCD) technology and programmable software, enables operators to emulate previous 737 electronic flight instrument system (EFIS) and 747-400/777 primary flight display-navigation display (PFD-ND) flight deck formats. Optional HUD.

DIMENSIONS, EXTERNAL:
Wing span (all versions): standard .... 34.31 m (112 ft 7 in)
with winglets .... 35.79 m (117 ft 5 in)
Wing chord: at root .... 5.71 m (18 ft 9 in)
at tip .... 1.25 m (4 ft 1¼ in)
Wing aspect ratio, standard 9.4
Length: overall: 600 .... 31.24 m (102 ft 6 in)
700 .... 33.63 m (110 ft 4 in)
800 .... 39.47 m (129 ft 6 in)
900 .... 42.11 m (138 ft 2 in)
fuselage: 700 .... 32.18 m (105 ft 7 in)
800 .... 38.02 m (124 ft 9 in)
900 .... 40.67 m (133 ft 5 in)
Height overall: 600, 700 .... 12.57 m (41 ft 3 in)
800, 900 .... 12.55 m (41 ft 2 in)
Tailplane span: all .... 14.35 m (47 ft 1 in)
Wheel track (c/l shock-struts): all .... 5.71 m (18 ft 9 in)
Wheelbase: 700 .... 12.60 m (41 ft 4 in)
800 .... 15.60 m (57 ft 2 in)
900 .... 17.17 m (56 ft 4 in)
Distance between engine centrelines .... 9.65 m (31 ft 8 in)
Main passenger door (port, fwd), all:
Height .... 1.83 m (6 ft 0 in)
Width .... 0.86 m (2 ft 10 in)
Height to sill: at OWE .... 2.74 m (9 ft 0 in)
at MTOW .... 2.59 m (8 ft 6 in)
Passenger door (port, rear):
Height: all .... 1.83 m (6 ft 0 in)
Width: all .... 0.76 m (2 ft 6 in)
Height to sill: 600, 700: at OWE .... 3.10 m (10 ft 2 in)
at MTOW .... 2.95 m (9 ft 8 in)
800, 900: at OWE .... 3.12 m (10 ft 2 in)
at MTOW .... 2.97 m (9 ft 9 in)
Emergency exits (overwing, port and stbd, each), all:
Height .... 0.96 m (3 ft 2 in)
Width .... 0.51 m (1 ft 8 in)
Service door (stbd, fwd), all:
Height .... 1.65 m (5 ft 5 in)
Width .... 0.76 m (2 ft 6 in)
Height to sill: at OWE .... 2.74 m (9 ft 0 in)
at MTOW .... 2.59 m (8 ft 6 in)
Service door (stbd, rear):
Height: all .... 1.65 m (5 ft 5 in)
Width: all .... 0.76 m (2 ft 6 in)
Height to sill: 600, 700: at OWE .... 3.10 m (10 ft 2 in)
at MTOW .... 2.97 m (9 ft 9 in)
800, 900: at OWE .... 3.12 m (10 ft 3 in)
at MTOW .... 2.97 m (9 ft 9 in)
Baggage hold door (stbd, fwd), all:
Height: door 1.30 m (4 ft 3 in) clear access .... 0.89 m (2 ft 11 in)
Width .... 1.22 m (4 ft 0 in)
Height to sill .... 1.45 m (4 ft 9 in)
Baggage hold door (stbd, rear), all:
Height: door 1.22 m (4 ft 0 in) clear access .... 0.84 m (2 ft 9 in)
Width .... 1.22 m (4 ft 0 in)
Height to sill: 700 .... 1.78 m (5 ft 10 in)
800, 900 .... 1.80 m (5 ft 11 in)

DIMENSIONS, INTERNAL:
Cabin, aft of flight deck to rear pressure bulkhead:
Length: 600 .... 21.79 m (71 ft 6 in)
700 .... 24.18 m (79 ft 4 in)

Boeing 737-600 serving Scandinavian Airlines System *(Paul Jackson)* ***NEW***/1047767

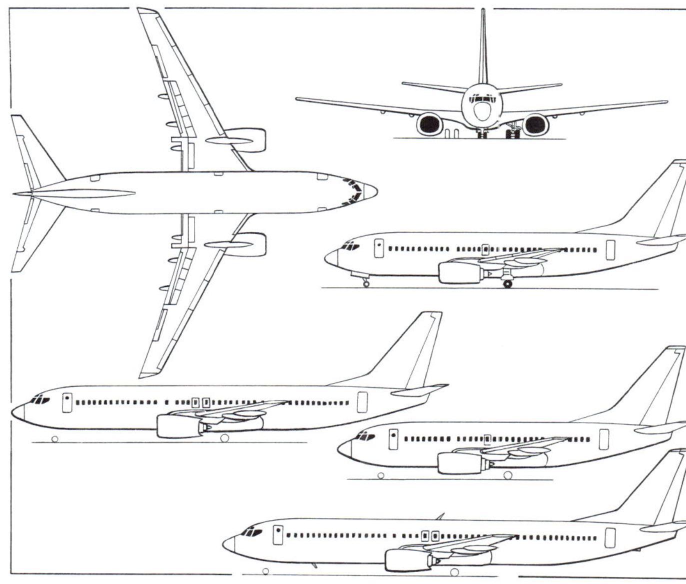

Boeing 737-700, with additional side views of stretched -800, shorter -600 and further stretched -900 (bottom) *(James Goulding)* 0089520

Boeing 737-800 of Air Berlin *(Paul Jackson)* ***NEW***/1047769

Boeing 737-900 in Korean Air's 'Boeing 7E7' colour scheme *(Paul Jackson)* ***NEW***/1047768

800 ... 30.02 m (98 ft 6 in)
Max height: all ... 2.13 m (7 ft 0 in)
Floor area: 600 ... 67.3 m^2 (725 sq ft)
700 ... 75.1 m^2 (808 sq ft)
800 ... 94.0 m^2 (1,012 sq ft)
Baggage hold:
Length: 700: front ... 4.67 m (15 ft 4 in)
rear ... 8.03 m (26 ft 4 in)
800: front ... 7.67 m (25 ft 2 in)
rear ... 10.87 m (35 ft 8 in)
900: front ... 9.25 m (30 ft 4 in)
rear ... 11.94 m (39 ft 2 in)
Max width, all: at roof ... 3.15 m (10 ft 4 in)
at floor ... 1.22 m (4 ft 0 in)
Min height, all ... 1.13 m (3 ft 8½ in)
Volume:
600: front ... 7.0 m^3 (248 cu ft)
rear ... 13.4 m^3 (472 cu ft)
700: front ... 11.5 m^3 (406 cu ft)
rear ... 16.9 m^3 (596 cu ft)
800: front ... 19.6 m^3 (692 cu ft)
rear ... 25.5 m^3 (899 cu ft)
900: front ... 23.8 m^3 (840 cu ft)
rear ... 28.7 m^3 (1,012 cu ft)

AREAS:
Wings, gross ... 125.00 m^2 (1,345.5 sq ft)
Vertical tail surfaces (total) ... 26.40 m^2 (284.2 sq ft)
Horizontal tail surfaces (total) ... 32.80 m^2 (353.1 sq ft)

WEIGHTS AND LOADINGS (600: A: CFM56-7B18 engines, B: CFM56-7B22s; 700: A: CFM56-7B20s, B: CFM56-7B24s; 800: A: CFM56-7B24s, B: CFM56-7B27s; 900: A: CFM56-7B26s, B: CFM56-7B27s):
Operating weight empty:
600, 110 passengers: A, B ... 37,104 kg (81,800 lb)
700, 126 passengers: A, B ... 38,147 kg (84,100 lb)
800, 162 passengers: A, B ... 41,145 kg (90,710 lb)
900, 177 passengers: A, B ... 42,493 kg (93,680 lb)
Max T-O weight:
600: A ... 56,245 kg (124,000 lb)
B ... 65,090 kg (143,500 lb)
700: A ... 60,330 kg (133,000 lb)
B ... 70,080 kg (154,500 lb)
800: A ... 70,535 kg (155,500 lb)
B ... 79,015 kg (174,200 lb)
900: A ... 74,840 kg (164,000 lb)
B ... 79,015 kg (174,200 lb)
Max ramp weight:
600: A ... 56,470 kg (124,500 lb)
B ... 65,315 kg (144,000 lb)
700: A ... 60,555 kg (133,500 lb)
B ... 70,305 kg (155,000 lb)
800: A ... 70,760 kg (156,000 lb)
B ... 79,245 kg (174,700 lb)
900: A ... 74,615 kg (164,500 lb)
B ... 79,245 kg (174,700 lb)
Max landing weight:
600 A, B ... 54,655 kg (120,500 lb)
700: A ... 58,060 kg (128,000 lb)
B ... 58,605 kg (129,200 lb)
800: A ... 65,315 kg (144,000 lb)
B ... 66,360 kg (146,300 lb)
900: A ... 66,360 kg (146,300 lb)
B ... 66,810 kg (147,300 lb)
Max zero-fuel weight:
600: A ... 51,480 kg (113,500 lb)
B ... 51,710 kg (114,000 lb)
700: A ... 54,655 kg (120,500 lb)
B ... 55,200 kg (121,700 lb)
800: A ... 61,690 kg (136,000 lb)
B ... 62,730 kg (138,300 lb)
900: A ... 62,730 kg (138,300 lb)
B ... 63,640 kg (140,300 lb)
Max wing loading:
600: A ... 450.0 kg/m^2 (92.16 lb/sq ft)
B ... 520.7 kg/m^2 (106.65 lb/sq ft)
700: A ... 482.6 kg/m^2 (98.85 lb/sq ft)
B ... 560.6 kg/m^2 (114.83 lb/sq ft)
800: A ... 564.3 kg/m^2 (115.57 lb/sq ft)
B ... 632.1 kg/m^2 (129.47 lb/sq ft)
900: A ... 595.1 kg/m^2 (121.89 lb/sq ft)
B ... 632.1 kg/m^2 (129.47 lb/sq ft)
Max power loading:
600: A ... 324 kg/kN (3.18 lb/lb st)
B ... 322 kg/kN (3.16 lb/lb st)
700: A ... 329 kg/kN (3.23 lb/lb st)
B ... 347 kg/kN (3.40 lb/lb st)
800: A ... 327 kg/kN (3.21 lb/lb st)
B ... 325 kg/kN (3.19 lb/lb st)
900: A ... 318 kg/kN (3.12 lb/lb st)
B ... 325 kg/kN (3.19 lb/lb st)

PERFORMANCE (A, B as above):
Max operating Mach No. (MMO): all ... 0.82
Cruising speed: all ... M0.785
Approach speed:
600: A, B ... 125 kt (232 km/h; 144 mph)
700: A ... 129 kt (239 km/h; 148 mph)
B ... 130 kt (241 km/h; 150 mph)
800: A ... 141 kt (261 km/h; 162 mph)
B ... 142 kt (263 km/h; 163 mph)
900: A, B ... 141 kt (261 km/h; 162 mph)
Max certified altitude: all ... 12,500 m (41,000 ft)
Initial cruising altitude, ISA +10°C:
600: A ... 12,500 m (41,000 ft)
B ... 12,220 m (40,100 ft)
700: A ... 12,500 m (41,000 ft)
B ... 11,700 m (38,400 ft)
800: A ... 11,675 m (38,300 ft)
B ... 10,955 m (35,940 ft)
900: A ... 11,215 m (36,800 ft)
B ... 10,820 m (35,500 ft)
T-O field length, S/L, 30°C:
600: A ... 1,616 m (5,300 ft)
B ... 1,796 m (5,890 ft)
700: A ... 1,744 m (5,720 ft)
B ... 1,677 m (5,500 ft)
800: A ... 2,100 m (6,890 ft)
B ... 2,308 m (7,570 ft)
900: A ... 2,591 m (8,500 ft)
B ... 2,439 m (8,000 ft)
Landing field length at max landing weight:
600: A, B ... 1,342 m (4,400 ft)[1]
700: A ... 1,418 m (4,650 ft)
B ... 1,433 m (4,700 ft)[1]
800: A ... 1,646 m (5,400 ft)
B ... 1,646 m (5,400 ft)
900: A, B ... 1,662 m (5,450 ft)
Design range:
600 with 110 passengers:
A ... 1,340 n miles (2,481 km; 1,542 miles)
B ... 3,050 n miles (5,648 km; 3,509 miles)
700 with 126 passengers:
A ... 1,540 n miles (2,852 km; 1,772 miles)
B[2] ... 3,260 n miles (6,037 km; 3,751 miles)
800 with 162 passengers:
A ... 1,990 n miles (3,685 km; 2,290 miles)
B[2] ... 2,940 n miles (5,444 km; 3,383 miles)
900 with 177 passengers:
A ... 2,060 n miles (3,815 km; 2,370 miles)
B[2] ... 2,745 n miles (5,083 km; 3,158 miles)

[1] *Category D Honeywell brakes*
[2] *with optional blended winglets*

UPDATED

BOEING 747-400

TYPE: Wide-bodied airliner.

PROGRAMME : World's first wide-body jet airliner; Pan American order for 25 announced 13 April 1966; official programme launch 25 July 1966; first flight 9 February 1969; FAA certification 30 December 1969; first delivery (to Pan Am) 12 December 1969; first route service New York-London flown 21 January 1970. 747-400 announced October 1985. In May 1990, Boeing decided to market only the -400; last -200 (a -200F Freighter for Nippon Cargo Air Lines) delivered 19 November 1991.

For all variants before 747-400, see *Jane's Aircraft Upgrades*. Production of earlier variants totalled 724 (205 -100, 45 SP, 393 -200 and 81 -300). Nineteen Pan American 747s modified as passenger/cargo **C-19A**s by Boeing Military Airplanes for Civil Reserve Air Fleet (see 1990–91 edition). Boeing board approved launch of 747-400IGW in December 1997.

Boeing 747's 35th anniversary of maiden flight occurred 9 February 2004, at which time 1,341 had been delivered, accumulating 35 billion n miles (65 billion km) and carrying 3.6 billion passengers since service entry in 1970.

Series 400 announced October 1985 as 747 development with extended capacity and range; design go-ahead July 1985; first order 22 October 1985; roll-out 26 January 1988; first flight 29 April 1988; certified with P&W PW4056 on 10 January 1989; first delivery 26 January 1989; entered service with Northwest Airlines 9 February 1989; certified with GE CF6-80C2B1F on 8 May 1989; R-R RB211-524G on 8 June 1989; R-R RB211-524H on 11 May 1990. Since May 1990, -400 is the only 747 marketed. 1,200th 747 delivered to British Airways on 17 February 1999.

CURRENT VERSIONS: **747-400:** Basic passenger version; standard and three optional gross weights (see below).

Detailed description applies to -400, except where indicated.

Boeing 747-400 of Air India *(Paul Jackson)* ***NEW***/1047763

DEVELOPMENT MILESTONES

747-100	
First order	13 Apr 66
Rolled out	30 Sep 68
First flight	9 Feb 69
Certification	30 Dec 69
First delivery	12 Dec 69
Entered service (Pan Am)	21 Jan 70
Subsequent versions	
747-400	
First order	22 Oct 85
Rolled out	26 Jan 88
First flight	29 Apr 88
Certification	10 Jan 89
First delivery	26 Jan 89
Entered service (Northwest Airlines)	9 Feb 89
747-400M	
First order	9 Apr 86
Rolled out	23 Mar 89
First flight	30 Jun 89
Certification	1 Oct 89
First delivery	1 Sep 89
Entered service (KLM)	Sep 89
747-400 Domestic	
First order	18 Dec 88
Rolled out	18 Feb 91
First flight	18 Mar 91
Certification	10 Oct 91
First delivery	10 Oct 91
Entered service (Japan Air Lines	Oct 91
747-400ER	
First delivery	19 Dec 00
Rolled out	17 Jun 02
First flight	31 Jul 02
Certification	29 Oct 02
First delivery	31 Oct 02
Entered service	7 Nov 02

747-400M Combi: Passenger/freight version; initial order 9 April 1986; rolled out 23 March 1989; first flight 30 June 1989; certified 1 October 1989; first delivery 1 September 1989 to KLM. Maximum 266 three-class passengers with freight, 413 without; port-side rear freight door; main deck limit is seven pallets at 27,215 kg (60,000 lb); underfloor and fuel capacities as for passenger 747; 49 delivered by 31 December 1996. For all gross weights, maximum landing weight 285,763 kg (630,000 lb) and maximum zero-fuel weight 256,280 kg (565,000 lb). All three engine options available.

747-400F: All-freight version. *Described separately.*

747-400 Domestic: Special high-density two-class 568-passenger version; first order 18 December 1988; rolled out 18 February 1991; first flown 18 March 1991; certified 10 October 1991 and delivered same day to Japan Air Lines (first of six) and later to All Nippon (six) and Japan Air System (one). Maximum T-O weight 272,155 kg (600,000 lb) but can be certified to 394,625 kg (870,000 lb). Structurally reinforced; no winglets; lower engine thrust; five more upper deck windows; revised avionics software and cabin pressure schedule; brake cooling fans; five pallets, 14 LD-1 containers and bulk cargo under floor; GE or P&W engines.

747-400 Performance Improvement Package (PIP): Announced April 1993, and first stage implemented in July 1993. Included gross weight increase of 2,268 kg (5,000 lb). Second stage, implemented in December 1993, included longer-chord dorsal fin made of CFRP, and wing spoilers held down more tightly to reduce profile drag and leakage. These improvements were immediately applied to production aircraft and are retrofittable; PIP flight tested in leased United Airlines 747-400 May 1993.

747-400ER: Offered (as 747-400IGW) from December 1997 in response to Qantas requirement, for which the carrier placed an order for six, with first delivery then scheduled for October 2002. One or two additional fuel tanks in hold. Range 7,500 n miles (13,890 km; 8,630 miles) with one additional tank; 7,700 n miles (14,260 km; 8,861 miles) with two. Structural strengthening around centrebody, wing/fuselage joint, flaps and landing gear.

Prototype N747ER (1,308th B747) rolled out 10 June 2002 (official ceremony on 17th); maiden flight 31 July 2002; became VH-OEE of Qantas.

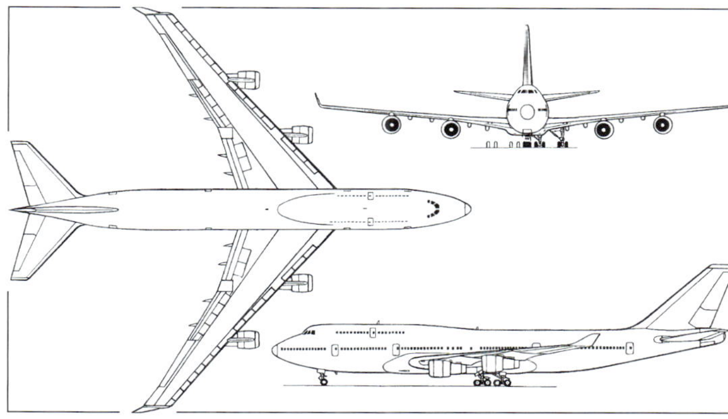

Boeing 747-400 advanced long-range airliner (General Electric CF6-80C2 engines) *(Jane's/Dennis Punnett)* 0526900

747-400 ERF: See following entry.

747-400X QLR: Study, since superseded, for developed, Quiet Longer Range version, initially designated 747-400X. Based on 747-400 airframe, but with revised (B777-style) flight deck, crew rest/passenger sleeping area in upper aft fuselage, increased provision for carry-on baggage in cabin, 747-400F's thicker gauge outboard wing with B767-400ER-style raked wingtips (span 68.66 m; 225 ft 3 in) instead of winglets, MD-11-type trailing-edge wedges (of which flight tests began in October 1998), strengthened fuselage sections and landing gear, and modifications to cargo and fuel systems to permit installation of additional fuselage tank forward of centre wing tank, with second additional tank of same capacity optional. Max ramp weight 418,665 kg (923,000 lb); max T-O weight 417,760 kg (921,000 lb); operating weight empty 186,425 kg (411,000 lb), max structural payload 65,315 kg (144,000 lb); fuel load 248,714 litres (65,705 US gallons; 54,710 Imp gallons), giving max range, with 416 passengers in three classes, of 7,980 n miles (14,779 km; 9,183 miles); alternative 396,900 kg (875,000 lb) MTOW and 7,500 n mile (13,890 km; 8,630 mile) range to comply with QC2 noise regulations. Alternative seating up to 524 (including 42 first class). Cruising Mach No is 0.86. By late 2002, QLR had generated little airline interest and Boeing had developed the proposal with greater range and payload, provisionally designating it **747-800X**.

Initial engine planned to be GE CF6-80C2B9F of 282 kN (63,300 lb st). New 'chevron' engine nacelles with serrated rear edges on core and fan nozzles promote mixing of bypass and core flows, and mixing of bypass flow and ambient air; combined with acoustic engine liners, these make QLR some 6 dB quieter (20 per cent on T-O; 40 per cent on approach), enabling it to meet QC2 noise standards.

QLR announced at Asian Aerospace, Singapore, 26 February 2002 (when 'chevron' design revealed). Studies replaced by 747 Advanced.

747-400XF QLR: Study for cargo version of Quiet Longer Range, with simplified and lightened handling system. Launched (then as Longer-Range 747-400 Freighter) April 2001 with order from ILFC. MTOW 417,760 kg (921,000 lb); range 5,150 n miles (9,537 km; 5,926 miles) with 112,810 kg (248,700 lb) payload. Cargo volumes (upper, lower and bulk) as for 747-400F. No extra fuel; capacity as for basic 747-400, GE-engined variant (203,325 litres; 53,765 US gallons; 44,769 Imp gallons). Typical cruising speed M0.845 on GE CF6-80C2B9F engines.

747X and 747X Stretch: Boeing cancelled development of its proposed 747X and 747X Stretch in March 2001.

747-800X: Under consideration by late 2002. Evolved from 747-400X QLR, with 1.98 m (6 ft 6 in) forward fuselage stretch; some 3,785 litres (1,000 US gallons; 833 Imp gallons) of additional fuel in tailplane tanks; between 20 and 40 more seats; and range increased to 8,000 n miles (14,816 km; 9,206 miles). Candidate engines in 276 to 285 kN (62,000 to 64,000 lb st) class. Developed into 747 Advanced.

747 Advanced: Revealed mid-2003 as further development of 747X QLR (and 747XF QLR) theme, drawing on Boeing 7E7 Dreamliner technology, including new engines. Service entry possible in 2009. Raked wingtips with overall span of 68.66 m (225 ft 3 in); trailing-edge wedge and fuselage plugs carried forward from previous studies, although last-mentioned to comprise 2.03 m (6 ft 8 in) ahead of wing and 1.52 m (5 ft 0 in) behind. Length 74.22 m (243 ft 6 in), height overall 19.38 m (63 ft 7 in). Weight empty 198,660 kg (437, 975 lb), max T-O 421,840 kg (930,000 lb), zero-fuel 269,430 kg (594,000 lb), max structural payload 70,770 kg (156,020 lb). Fuel capacity 225,705 litres (59,625 US gallons; 49, 648 Imp gallons); max operating speed M0

.86 with 448 passengers (24/85/339) over 8,090 n miles (14,982 km; 9,309 miles).

CUSTOMERS: As per table. Launch customer Northwest Orient Airlines ordered 10 -400s with PW4000s and 420-passenger interior October 1985; first delivery 26 January 1989.

BOEING 747-400 ORDER BOOK
(at 1 January 2005)

Customer	Variant	First order	First delivery	Engine	400	400D	400ER	400ERF	400F	400M
Air Canada	400M	20 Jan 89	4 Jun 91	PW4056						3
Air China	400	31 May 90	20 Mar 92	PW4056	6					
	400M	16 May 86	13 Oct 89	PW4056						8
Air China Cargo	400F	27 Jul 04		PW4056					2	
Air France	400	16 Dec 87	28 Feb 91	CF6-80	7					
	400ERF	24 Apr 01	31 Oct 02	CF6-80				2		
	400M	16 Dec 87	17 Sep 91	CF6-80						5
Air India	400	14 Aug 91	4 Aug 93	PW4056	6					
Air Namibia	400M	21 Apr 99	21 Oct 99	CF6-80						1
Air New Zealand	400	30 Jul 84	14 Dec 89	RB211-524	3					
	400	1 Mar 91	31 Oct 98	CF6-80	1					
All Nippon Airways	400	21 Oct 86	28 Aug 90	CF6-80	12					
	400D	21 Jan 86	13 Jan 92	CF6-80		11				
Amiri Flight	400	30 Nov 99	30 Nov 99	CF6-80	1					
Asiana Airlines	400	12 Jun 89	24 Jun 93	CF6-80	2					
	400F	3 Sep 90	4 Nov 94	CF6-80					6	
	400M	12 Jun 89	1 Nov 91	CF6-80						6
Atlas Air	400F	9 Jun 97	29 Jul 98	CF6-80					16	
British Airways	400	15 Aug 86	30 Jun 89	RB211-524	57					
Canadian Airlines Intnl	400	28 Jul 88	11 Dec 90	CF6-80	4					
Cargolux Airlines	400F	6 Dec 90	17 Nov 93	CF6-80					5	
	400F	13 Jun 95	8 Dec 98	RB211-524					9	
Cathay Pacific Airways	400	3 Jun 86	8 Jun 89	RB211-524	17					
	400F	28 Feb 90	1 Jun 94	RB211-524					6	
China Airlines	400	21 Jul 87	8 Feb 90	PW4056	13					
	400	28 Nov 02	none	CF6-80	4					
	400F	11 Aug 99	6 Jul 00	PW4056					21	
China Southern	400F	13 Feb 01	19 Jun 02	PW4062					2	
El Al	400	11 Dec 90	27 Apr 94	PW4056	4					
EVA Air	400	6 Oct 89	2 Nov 92	CF6-80	7					
	400F	6 May 99	20 Jul 00	CF6-80					3	
	400M	6 Oct 89	16 Sep 93	CF6-80						8
Garuda Indonesia	400	15 Nov 90	14 Jan 94	CF6-80	2					
GE Capital	400	22 Dec 95	18 Jun 99	CF6-80	1					
	400F	15 Dec 99	16 Oct 00	CF6-80					5	
ILFC	400	16 May 88	31 May 91	CF6-80	12					
	400	16 May 88	25 Sep 91	RB211-524	2					
	400F	30 Jan 90	14 Apr 99	CF6-80					1	
	400ERF	17 Apr 01	17 Oct 02	CF6-80				3		
Japan Airlines	400	21 Sep 87	25 Jan 90	CF6-80	37					
	400D	30 Jun 88	10 Oct 91	CF6-80		8				
	400F	7 Oct 02	none	CF6-80					2	
Japan Air SDF	400	23 Dec 87	17 Sep 91	CF6-80	2					
KLM	400	9 Apr 86	18 May 89	CF6-80	5					
	400ERF	1 Mar 99	31 Mar 03	CF6-80				3		
	400M	9 Apr 86	1 Sep 89	CF6-80						17
Korean Air Lines	400	29 Aug 86	13 Jun 89	PW4056	27					
	400F	11 Jun 90	6 Sep 96	PW4056					10	
	400ERF	31 Dec 01	13 Jun 03	PW4056				7		
	400M	14 Apr 88	27 Jun 90	PW4056						1
Kuwait Airways	400M	12 Apr 92	29 Nov 94	CF6-80						1
Lufthansa	400	21 May 86	23 May 89	CF6-80	24					
	400M	21 May 86	19 Sep 89	CF6-80						7
Malaysia Airlines	400	19 Oct 88	27 Sep 90	CF6-80	2					
	400	12 Jan 89	27 Aug 92	PW4056	17					
	400F	8 Jan 96	none	PW4056					2	
	400M	30 Oct 87	6 Oct 89	CF6-80						2
Mandarin Airlines	400	15 Sep 94	14 Jun 95	PW4056	1					
Nippon Cargo Airlines	400F	30 Jan 04		CF6-80					3	
Northwest Airlines	400	22 Oct 85	26 Jan 89	PW4056	16					
Omani Royal Flight	400	31 Jul 00	14 Dec 01	CF6-80	2					
Philippine Airlines	400	29 Oct 92	19 Nov 93	CF6-80	7					
	400M	18 Jan 96	29 Mar 96	CF6-80						1
Qantas Airways	400	2 Mar 87	11 Aug 89	RB211-524	21					
	400ER	19 Dec 00	31 Oct 02	CF6-80			6			
Saudia	400	18 Jun 95	24 Dec 97	CF6-80	5					
Singapore Airlines	400	27 Mar 86	18 Mar 89	PW4056	42					
	400F	16 Jan 90	5 Aug 94	PW4056					17	
South African Airways	400	6 May 89	19 Jan 91	RB211-524	6					
	400	30 Dec 98	30 Dec 98	CF6-80	2					

Customer	Variant	First order	First delivery	Engine	400	400D	400ER	400ERF	400F	400M
Thai Airways Intnl	400	16 Jun 87	21 Feb 90	CF6-80	18					
United Airlines	400	7 Nov 85	30 Jun 89	PW4056	44					
US Air Force										
(AL-1A)	400F	30 Jan 98	21 Jan 00	CF6-80					1	
UTA	400	3 Jul 86	22 Sep 90	CF6-80	1					
	400M	3 Jul 86	26 Jul 91	CF6-80						1
Virgin Atlantic										
Airways	400	20 Dec 96	17 Jun 97	CF6-80	9					
Subtotals					**449**	**19**	**6**	**15**	**111**	**61**
Total							**661**			

Notes: For deliveries see Group Orders and Deliveries table

COSTS: US$198 million to US$227 million (2004); Combi version US$202 million to US$228 million (2002).

DESIGN FEATURES: Wide-bodied extrapolation of Boeing intercontinental jet configuration of low wing and four podded engines, optimised for greater passenger numbers and increased efficiency. Twin-deck forward fuselage; four mainwheel bogies for weight distribution.

According to engine type, fuel burn per seat over 3,000 n mile (5,556 km; 3,452 mile) sector varies between 135.8 kg (299.3 lb) and 138.5 kg (305.4 lb).

Sweepback at quarter-chord 37° 30'; thickness/chord ratio 13.44 per cent inboard, 7.8 per cent at mid-span, 8 per cent outboard; dihedral at rest 7°; incidence 2°; winglets, canted 22° outward and swept 60°, increase range by 3 per cent; upper deck extended rearward by 7.11 m (23 ft 4 in).

FLYING CONTROLS: Conventional and powered.

Elevators: Four elevator sections mechanically linked with breakable shear devices; each elevator has dual hydraulic-powered control units; control feel and three individual autopilot input servos mounted on central elevator quadrant; all surfaces have position transmitters; feel computer-operated by pitot pressure and tailplane angle.

Rudder: Upper rudder surface operated by three hydraulic actuators served by two hydraulic systems, lower surface by two actuators fed by remaining two hydraulic systems; no balance weights; each rudder has separate yaw damper module; left and right digital air data computers provide signals for controlling rudder ratio changer on each rudder surface according to air data and tailplane angle; combined feel actuator, rudder centring and trim actuator in rear servo area; mechanical cable linkage between rudder pedals and aft actuator area; rudder trim control switches on centre console. Maximum rudder deflection ±30°.

Tailplane: Tailplane angle set by hydraulic motor-driven shaft and ball screw with primary and secondary hydraulic brakes; flight control unit and air data computer signals sent to tailplane through dual stabiliser, trim and rudder ratio modules, which automatically apply Mach trim, and by dual-stabiliser control modules; tailplane trim limits computed according to flap positions.

Lateral control: Pilot and co-pilot aileron linkage can be physically separated if necessary; all four ailerons operate at low speeds; outboard ailerons are locked out at cruising speed; the inboard spoiler panel on each wing used on ground only; remainder have variable ratio response and spoiler mixer units; there are trim, centring and feel units.

Leading-edge and trailing-edge devices: Three-section Krueger flaps inboard of engines; variable camber slats between (five-section) and outboard (six-section) of engines lie flat when retracted and adopt camber curvature when extended. Two flap assemblies on each wing, one inboard of engines and the other between engines; three sections, fore flap, mid-flap and aft flap, move rearwards as single flat panel up to 5° deflection; thereafter, three sections separate progressively to form three slots, and camber angles relative to each other increase progressively.

Automatic flight control system: Combines autopilot, flight director and automatic tailplane trim and sends commands through triple independent flight control computers; system automates all flight phases except take-off; dual digital air data computers; pilots' primary flight and navigation displays are large-size cathode-ray tubes; two engine indicating and crew alerting screens, one on main panels, one on console; three multifunction control and display panels control flight management system, navigation and communications; flight control computers (autopilot) and inertial reference units are triplicated; new features include full-time autothrottle and dual-thrust management system included in flight management computer; integrated radio control panels and automatic start and shutdown of APU.

STRUCTURE: Wing and tail surfaces are aluminium alloy dual-path fail-safe structures; advanced aluminium alloys in wing torsion box save 2,721 kg (6,000 lb); advanced aluminium honeycomb spoiler panels; CFRP winglets and main deck floor panels; advanced graphite/phenolic and Kevlar/graphite in cabin fittings and engine nacelles; frame/stringer/stressed skin fuselage with some bonding. Improved corrosion protection and further coverage with compound introduced from 1993.

LANDING GEAR: Twin-wheel nose unit retracts forward; main gear consists of four four-wheel bogies; two, mounted side by side under fuselage at wing trailing-edge, retract forward; two, mounted under wings, retract inward; nosewheel steerable up to 70° left or right from tillers; full rudder pedal travel gives up to 7° for use at high speed; two centre main legs steer up to 13° when nosewheels are steered more than 20° and speed is less than 20 kt (37 km/h; 23 mph); carbon disc brakes on all mainwheels, with individually controlled digital anti-skid units; one of three brake pressure supplies automatically selected; main and nose tyres H49×19.0–20 or -22 (32 ply). Minimum ground turning radius, with body gear steering, is 48.46 m (159 ft 0 in) at wingtip and 27.73 m (91 ft 0 in) at nosewheels.

POWER PLANT: Four turbofans for baseline 747-400, these comprise: 252 kN (56,750 lb st) Pratt & Whitney PW4056; 258 kN (57,900 lb st) General Electric CF6-80C2B1F or 276 kN (62,100 lb st) CF6-80C2B5F; or 258 kN (58,000 lb st) Rolls-Royce RB211-524G or 270 kN (60,600 lb st) RB211-524H. Further optional engines (subject to certification) are 267 kN (60,000 lb st) PW4060, 276 kN (62,000 lb st) PW4062 and 274 kN (61,500 lb st) CF6-80C2B1F1. For 747-400ER, initial engine choices are CF6-80C2B5F, PW4062 and RB211-524H8T.

Fuel in four main tanks in wings can feed to any engine; in addition there are a centre-wing tank and reserve tanks in outer wing; optional tailplane tank; vent and surge tanks in outer wings and starboard tailplane; jettison pumps in inner main tanks; APU fed from port inner tank; automatic refuelling through two receptacles under each wing leading-edge between engines; automatic condensate scavenging and flame arresters in vent outlets.

Basic fuel capacity 204,355 litres (53,985 US gallons; 44,952 Imp gallons) with P&W and R-R engines; 203,523 litres (53,765 US gallons; 44,769 Imp gallons) with GE engines. At alternative higher T-O weights above 394,625 kg (870,000 lb) use of 12,492 litre (3,300 US gallon; 2,748 Imp gallon) tailplane/centre section tank is mandatory; fuel capacity including tailplane tank is therefore 216,846 litres (57,285 US gallons; 47,700 Imp gallons) with P&W and R-R engines and 216,013 litres (57,065 US gallons; 47,516 Imp gallons) with GE engines.

Usable fuel in 747-400ER comprises 239,389 litres (63,240 US gallons; 52,658 Imp gallons) with GE engines; 240,222 litres (63,460 US gallons; 52,841 Imp gallons) with P&W and R-R engines. Volumes include two tanks in cargo hold each of 11,583 litres (3,060 US gallons; 2,548 Imp gallons).

Nitrogen-based flammability reduction system (FRS) trials with 747-400 tanks undertaken in 2003 with Honeywell/Parker Aerospace OBNGS; available from late 2006 or early 2007.

ACCOMMODATION: Two-crew flight deck, with seats for two observers; two-bunk crew rest cabin accessible from flight deck. Optional (but currently available on 90 per cent of B747-400 fleet) overhead cabin crew rest compartments above rear of main deck cabin (four bunks, four seats; eight bunks, two seats; two bunks, two seats, five sleeper seats). Typical 416-seat, three-class, long-range configuration accommodates 40 business class on upper deck; 23 first class in front cabin, 38 business class in middle cabin and 315 economy class in rear cabin on main deck. Maximum upper deck capacity 69 economy class. First class seating six abreast with two 86 cm (34 in) aisles, each twin-seat unit 1.45 m (4 ft 9 in) wide. Business passengers four abreast with 72 cm (28½ in) aisle and 1.37 m (4 ft 6 in) wide seat pairs on upper deck or two-three-two on lower deck with two 63 cm (24¾ in) aisles and 2.08 m (6 ft 10 in) triple seat. Economy seating three-four-three, with 49.5 cm (19½ in) aisles, two 1.51 m (4 ft 11½ in) triple seats and 2.07 m (6 ft 9½ in) quad seat. Five passenger doors on each side; upper deck emergency door each side. 747-400ER accommodates, typically, 500 in two-class arrangement (42 first, 458 economy or 416 three-class (as above).

Centre overhead stowage bins 0.16 m^3 (5.7 cu ft) volume per 1.02 m (40 in) long bin; outboard bins 0.45 m^3 (15.9 cu ft) volume per 1.52 m (60 in) long bin; 0.083 m^3 (2.95 cu ft) bin volume per passenger (three-class). Two modular upper deck lavatories, 14 on main deck, relocatable (six upper deck optional locations; 33 on lower deck) and vacuum-drained into four waste tanks with combined volume of 1,136 litres (300 US gallons; 250 Imp gallons). Single-point drainage. Basic galley configuration, one on upper deck, seven centreline and two sidewall on main deck; lavatories and galleys can be quickly relocated if required fittings are installed; advanced integrated audio/video/announcement system.

Underfloor freight: forward compartment, five 2.44 m (96 in) × 3.18 m (125 in) pallets (totalling 58.8 m^3, 2,075 cu ft) or 16 LD-1 containers (totalling 78.4 m^3; 2,768 cu ft); aft compartment, 14 LD-1 containers (totalling 68.6 m^3; 2,422 cu ft) or four pallets (totalling 47.0 m^3; 1,660 cu ft); and bulk storage behind aft compartment 23.6 m^3 (835 cu ft). Max capacity of 747-400 (16 LD-1s forward, 14 LD-1s aft and bulk storage in extreme rear) 170.6 m^3 (6,025 cu ft). Door to each of three areas, all starboard side. Optional cargo door, port rear, on Combi version. 747-400ER lower deck accommodates 129 m^3 (4,550 cu ft) within LD-1 containers, plus 22.3 m^3 (789 cu ft) of bulk cargo, when two fuel tanks fitted. Capacity of 747-400X QLR reduced by up to six containers when both additional fuel tanks fitted; capacity 158.5 m^3 (5,599 cu ft) with basic fuel; 137.0 m^3 (4,837 cu ft) with max fuel.

SYSTEMS: Each engine drives a hydraulic pump feeding an independent system; services are connected to supplies in such a way that loss of one supply cannot disable one system; two hydraulic systems also have air-driven pumps to maintain pressure and two have electric pumps; one electric pump can be run to provide braking when the aircraft is being towed on the ground; all four hydraulic reservoirs can be filled from a single location in the port main landing gear bay.

Hot air bled from the low-pressure and high-pressure compressors of all four engines is precooled by fan exit air and fed via a manifold to the cabin pressurisation and air conditioning system and to provide de-icing of wing leading-edge and engine nose cowling and to pressurise hydraulic tanks. Three conditioning packs in wing/fuselage fairing provide cabin air; five cabin zones, each with digital temperature control.

Each engine drives an integrated drive generator supplying 90 kVA power to respective AC busses; three generators are a dispatch item, but one will supply essential loads; APU drives two further generators; automatic start-up, load transfers and load shedding reduce crew workload; power systems may be isolated from each other for triple-channel Cat. III autoland.

Completely self-contained 1,081 kW (1,450 shp) P&WC PW901A APU, mounted clear of all flight-critical structure and flight controls in the extreme tail, drives two 90 kVA generators that can supply electrical power for whole aircraft; also supplies compressed air to operate pneumatic components; can run at up to 6,100 m (20,000 ft) and supply compressed air below 4,575 m (15,000 ft). Capabilities include maintenance of 24°C (75°F) ground cabin temperature in 38°C (100°F) ambient conditions.

Forward underfloor cargo compartment heated to 5°C by hot air exhausted from flight deck cooling equipment and avionics in main equipment centre, boosted as necessary by two electrical heaters; rear underfloor hold heated to minimum 5°C or 18°C (selected by crew) by engine bleed.

Overheat detection and automatic extinguishing provided in all lavatories; APU automatically shut down and fire extinguisher bottles initiated on detection of fire; each engine has three dual fire detectors in series and a fourth detector for overheating. Underfloor freight compartments and upper deck hold of Combi have smoke detectors and extinguisher systems; wheel wells have overheat detectors.

AVIONICS: Boeing launched development of new Flight Management Computer software in January 1993 to match existing aircraft to international Future Air Navigation System (FANS-1) during 1995. Standard avionics fit as follows:

Comms: Dual VHF and HF transceivers with Selcal; dual transponders; flight intercom with air-to-ground facility, connectable also to satcom system; cabin entertainment and passenger address and service units.

Radar: Colour weather radar transmitting in I- and G-bands.

Flight: Dual VOR; triple ILS receivers with single marker beacon receiver; dual ADF; dual DME; all nav radios automatically tuned by flight management computer system (FMCS). Automatic flight control system (AFCS) integrates autopilot, flight director and automatic stabiliser trim functions; dual digital air data computers with dual selectable pressure sensors, angle of attack sensors and total air temperature probes; FMCS allows crew to preselect flight plan using standard air traffic control language; FMCS incorporates database, updated every 28 days, which includes data on waypoints, airports, standard instrument departures (SIDs), standard terminal arrival routes (STARs), airline routes and information on specific geographic areas; triple ring laser gyro inertial reference units provide navigation input on EFIS, flight management displays or radio magnetic indicators; other systems include ground proximity warning, triple low-range radio altimeters and TCAS.

Central maintenance computer monitors over 75 electrical and electromechanical systems, performs tests and centralises maintenance data; failures are indicated in EICAS displays and stored for future reference for in-flight use or line or hangar maintenance. Satcom datalink allows ground crews to interrogate system for additional information while aircraft in flight.

Instrumentation: Electronic flight instrument system (EFIS) comprising six (left/right inboard/outboard and central upper/lower) 20.3 × 20.3 cm (8 × 8 in) integrated display units (IDU), two each for primary flight display (PFD), navigation display (ND) and engine indicating and crew alerting (EICAS) functions; all IDUs receive data from all three EFIS/EICAS interface units (EIU), updated via software data loader; PFD and EICAS primary formats automatically switch to inboard and lower IDUs respectively, with facility for manual selection of formats on different IDUs as required. B747-400 has 181 switches, 171 lights and 13 gauges, compared with 284, 555 and 132 of earlier variants; total 365 is below average 450 for typical two-crew jet transport.

DIMENSIONS, EXTERNAL:
Wing span: normal ... 64.44 m (211 ft 5 in)
with winglets ... 64.92 m (213 ft 0 in)
Wing span, fully fuelled ... 64.92 m (213 ft 0 in)
Wing chord: at root ... 14.63 m (48 ft 0 in)
at tip ... 4.06 m (13 ft 4 in)
Winglet height ... 0.89 m (2 ft 11 in)
Wing aspect ratio ... 7.7
Length: overall ... 70.67 m (231 ft 10¼ in)
fuselage ... 68.63 m (225 ft 2 in)
Max width of fuselage ... 6.50 m (21 ft 4 in)
Height overall:
at OWE: -400 ... 19.51 m (64 ft 0 in)
-400ER ... 19.58 m (64 ft 3 in)
at MTOW: -400 ... 18.77 m (61 ft 7 in)
-400ER ... 19.05 m (62 ft 6 in)
Tailplane span ... 22.17 m (72 ft 9 in)
Wheel track (c/l shock-struts):
outer pair ... 11.00 m (36 ft 1 in)
inner pair ... 3.84 m (12 ft 7 in)
Wheelbase: mean ... 25.60 m (84 ft 0 in)
to forward main bogie ... 24.07 m (78 ft 11½ in)
to rear main bogie ... 27.14 m (89 ft 0½ in)
Distance between engine centrelines:
outboard ... 41.66 m (136 ft 8 in)
inboard ... 23.37 m (76 ft 8 in)
Passenger doors (10, each):
Height: door, clear access ... 1.93 m (6 ft 4 in)
Width: door ... 1.19 m (3 ft 11 in)
clear access ... 1.07 m (3 ft 6 in)
Height to sill:
at OWE:
front: -400 ... 5.16 m (16 ft 11 in)
-400ER ... 5.21 m (17 ft 1 in)
rear: -400 ... 5.31 m (17 ft 5 in)
-400ER ... 5.38 m (17 ft 8 in)
at MTOW:
front: -400 ... 4.72 m (15 ft 6 in)
-400ER ... 4.75 m (15 ft 7 in)
rear: 400 ... 4.80 m (15 ft 9 in)
-400ER ... 4.98 m (16 ft 4 in)
Upper deck emergency door (two):
Height: door ... 2.01 m (6 ft 7¼ in)
clear access ... 1.83 m (6 ft 0 in)
Width ... 1.07 m (3 ft 6 in)
Height to sill:
at OWE: -400 ... 7.90 m (25 ft 11 in)
-400ER ... 7.95 m (26 ft 1 in)
at MTOW: -400 ... 7.52 m (24 ft 8 in)
-400ER ... 7.54 m (24 ft 9 in)
Baggage door (front hold): Height ... 1.68 m (5 ft 6 in)
Width ... 2.64 m (8 ft 8 in)
Height to sill: at OWE ... 3.10 m (10 ft 2 in)
at MTOW ... 2.69 m (8 ft 10 in)
Baggage door (rear hold): Height ... 1.68 m (5 ft 6 in)
Width ... 2.64 m (8 ft 8 in)
Height to sill: at OWE ... 3.17 m (10 ft 5 in)
at MTOW ... 2.82 m (9 ft 3 in)
Bulk loading door:
Height: max (front) ... 1.42 m (4 ft 8 in)
min (rear) ... 1.24 m (4 ft 1 in)
Width ... 1.12 m (3 ft 8 in)
Mean height to sill: at OWE ... 3.40 m (11 ft 2 in)
at MTOW ... 3.00 m (9 ft 10 in)
Combi cargo door (port):
Height (clear access) ... 3.05 m (10 ft 0 in)
Width ... 3.40 m (11 ft 2 in)
Height to sill: at OWE ... 5.26 m (17 ft 3 in)
at MTOW ... 4.87 m (16 ft 0 in)

Notes: *747-400ER baggage door heights to sill: add 50 to 75 mm (2 to 3 in)*

DIMENSIONS, INTERNAL:
Cabin (main): Max height ... 2.41 m (7 ft 11 in)
Passenger cabin volume ... 885.9 m³ (31,285 cu ft)

AREAS:
Wings, gross ... 541.16 m² (5,825.0 sq ft)
Ailerons (total) ... 20.90 m² (225.00 sq ft)
Trailing-edge flaps (total) ... 78.69 m² (847.00 sq ft)
Leading-edge flaps (total) ... 43.85 m² (472.00 sq ft)
Inboard spoilers (total) ... 12.78 m² (137.60 sq ft)
Outboard spoilers (total) ... 15.46 m² (166.40 sq ft)
Fin ... 77.11 m² (830.00 sq ft)
Rudder ... 21.37 m² (230.00 sq ft)
Tailplane ... 136.57 m² (1,470.00 sq ft)
Elevators (total, incl tabs) ... 30.38 m² (327.00 sq ft)

WEIGHTS AND LOADINGS (747-400: GB: CF6-80C2B1F, GM: CF6-80C2B5F, PB: PW4056, PM: PW4062, RB: RB211-524G, RM: 211-524M engines; all 416 passengers, five cargo pallets and 14 containers, 747-400ER: ER GM CF6-80C2B5F, ER PM: PW4062, ER RM: RB211-524H-8T with passengers/cargo as above):
Operating weight empty: GB ... 180,485 kg (397,900 lb)
GM ... 180,895 kg (398,800 lb)
PB ... 180,845 kg (398,700 lb)
PM ... 181,255 kg (399,600 lb)
RB ... 181,435 kg (400,000 lb)
RM ... 181,845 kg (400,900 lb)
ER GM/PM/RM ... 184,565 kg (406,900 lb)
Baggage/freight capacity, all:
forward compartment ... 26,490 kg (58,400 lb)
aft compartment ... 22,938 kg (50,570 lb)
bulk compartment ... 6,749 kg (14,880 lb)
Max structural payload:
ER GM, PM, RM ... 67,175 kg (148,100 lb)
Max fuel weight: GB ... 162,580 kg (358,425 lb)
GM ... 172,560 kg (380,425 lb)
PB, RB ... 163,250 kg (359,900 lb)
PM, RM ... 173,225 kg (381,900 lb)
ER, GM ... 192,190 kg (423,700 lb)
ER, PM, ER, RM ... 192,855 kg (425,175 lb)
Max T-O weight:
GB, PB, RB ... 362,875 kg (800,000 lb)
GM, PM, RM ... 396,895 kg (875,000 lb)
ER GM, PM, RM ... 412,770 kg (910,000 lb)
Max ramp weight:
GB, PB, RB ... 364,235 kg (803,000 lb)
GM, PM, RM ... 398,255 kg (878,000 lb)
ER, GM, PM, RM ... 414,130 kg (913,000 lb)
Max landing weight:
GB, PB, RB ... 260,360 kg (574,000 lb)
GM, PM, RM ... 295,745 kg (652,000 lb)
ER, GM, PM, RM ... 295,745 kg (652,000 lb)
Max zero-fuel weight:
GB, PB, RB ... 242,670 kg (535,000 lb)
GM, PM, RM ... 251,745 kg (555,000 lb)
ER, GM, PM, RM ... 251,745 kg (555,000 lb)
Max wing loading:
GB, PB, RB ... 670.5 kg/m² (137.34 lb/sq ft)
GM, PM, RM ... 733.4 kg/m² (150.21 lb/sq ft)
Max power loading:
GB, RB ... 352 kg/kN (3.45 lb/lb st)
GM, PB ... 359 kg/kN (3.52 lb/lb st)
PM ... 360 kg/kN (3.53 lb/lb st)
RM ... 368 kg/kN (3.61 lb/lb st)

PERFORMANCE (as above; landing at MLW):
Cruising Mach No. ... 0.85
Approach speed:
GB, PB, RB ... 146 kt (270 km/h; 168 mph)
GM, PM, RM ... 157 kt (291 km/h; 181 mph)
Initial cruising altitude:
GB, PB, RB ... 10,575 m (34,700 ft)
GM, PM, RM ... 10,000 m (32,800 ft)
T-O field length, 30°C (86°F): GB ... 2,820 m (9,250 ft)
GM ... 3,033 m (9,950 ft)
PB ... 2,820 m (9,250 ft)
PM ... 2,990 m (9,800 ft)
RB ... 2,850 m (9,350 ft)
RM ... 3,215 m (10,550 ft)
Landing field length:
GB, PB, RB ... 1,905 m (6,250 ft)
GM, PM, RM ... 2,180 m (7,150 ft)
Design range:
GB ... 6,185 n miles (11,454 km; 7,117 miles)
GM ... 7,260 n miles (13,445 km; 8,354 miles)[1]
PB ... 6,195 n miles (11,473 km; 7,129 miles)
PM ... 7,325 n miles (13,565 km; 8,429 miles)[1]
RB ... 6,040 n miles (11,186 km; 6,950 miles)
RM ... 7,170 n miles (13,278 km; 8,251 miles)[1]

[1] *Fuel volume limited*

UPDATED

BOEING 747-400F

USAF designation: AL-1A

TYPE: Four-jet freighter.

PROGRAMME: Initial order 13 September 1989; rolled out 8 March 1993; first flight (N6005C) 4 May 1993; FAA certification 22 October 1993; JAR certification followed; first delivery (Cargolux) 17 November 1993. During 2000, Boeing began design of a freighter conversion of passenger 747-400s as an eventual partner to the first 747-300F conversion, begun in that year. However, this would differ from new-production -400F in several respects.

CURRENT VERSIONS: **747-400F:** *As described.*

747-400ERF: Freighter version of 747-400ER; first order placed 17 April 2001; rolled out 5 September 2002; first flight (N5017Q/F-GIUA) 30 September 2002; certified 16 October 2002; deliveries began 17 October (Air France). Deliveries to Air France (two),

Boeing 747-400F Freighter in Cathay Pacific service *(Paul Jackson)* ***NEW***/1047748

ILFC (three) and KLM (three, beginning 31 March 2003). Others for undisclosed customer(s).

Differences from 747-400ER include 302,090 kg (666,000 lb) max landing weight; 277,145 kg (611,000 lb) zero-fuel weight; 164,380 kg (362,400 lb) empty operating weight; 112,765 kg (248,600 lb) structural max payload; 530 m³ (18,720 cu ft) containerised volume on main deck; 159 m³ (5,600 cu ft) lower deck containerised volume and 14.7 m³ (520 cu ft) bulk cargo volume. Usable fuel is 203,523 litres (53,765 US gallons; 44,769 Imp gallons) for GE-engined version and 204,355 litres (53,985 US gallons; 44,952 Imp gallons) for P&W and R-R-engined versions.

747-400XF: Variant of projected 747-400X QLR, now replaced by 747F Advanced.

747F Advanced: Counterpart of 747 Advanced, announced mid-2003, for possible service entry in 2009. Shorter upper deck (as -400F) and other Advanced features, but forward plug of 3.56 m (11 ft 8 in), giving length overall of 75.74 m (248 ft 6 in). Weight empty 182,800 kg (403,000 lb), max T-O 332,940 kg (734,000 lb), zero-fuel 314,790 kg (694,000 lb), max structural payload 132,000 kg (291,000 lb). Fuel capacity 204,366 litres (53,988 US gallons; 44,954 Imp gallons).

747-400SF: 'Special Freighter' conversion of former passenger aircraft, retaining enlarged upper floor as crew rest area, but most windows deleted. Freight door, port, rear; integrated flight deck displays. Payload capability 113,400 kg (250,000 lb); typically up to 30 pallets 2.44 × 3.20 m (8 ft × 10 ft 6 in) over 4,100 n miles (7,593 km; 4,718 miles). Launched January 2004, on receipt of initial order from Cathay Pacific for six, plus six options.

AL-1A: Anti-missile defence aircraft. *Described separately.*

CUSTOMERS: See table with main 747-400 entry. First 747-400F delivered to Cargolux 17 November 1993; total of 34 delivered by October 1999. Production of 16 envisaged in 2000. Recent customers include China Airlines, which ordered 13 on 11 August 1999, for delivery between 2000 and 2007 (first on 6 July 2000); EVA Air, three for delivery from 2000 (first on 20 July); and Cathay Pacific Airways, which ordered two in October 1999 (first delivered on 12 September 2000), subsequently increasing its order to five.

COSTS: List price US$202 million to US$228 million (2004).

DESIGN FEATURES: 747-200F fuselage (short upper deck) with additional changes combined with stronger and larger 747-400 wing; strengthened floor of short upper deck, as offered for -200F, also integrated into 747-400F; further developed freight handling system; total cargo volume 777.8 m³ (27,467 cu ft), of which 604.5 m³ (21,347 cu ft) on main deck, 158.6 m³ (5,600 cu ft) in lower hold and 48.3 m³ (520 cu ft) available for bulk cargo. Compared to -200F, empty weight saving of 2,000 kg (4,409 lb) has raised maximum revenue freight load to about 113,000 kg (249,125 lb), at which range is 4,400 n miles (8,149 km; 5,063 miles); fuel consumption more than 15 per cent lower than 747-200F. Same gross weights as passenger 747-400; maximum landing weight at optional T-O weight, 302,090 kg (666,000 lb); maximum zero-fuel weight, 276,690 kg (610,000 lb), can be increased on condition T-O weight is decreased.

ACCOMMODATION: Two-pilot crew, as 747-400. Upward-opening nose cargo door and optional port-side rear cargo door; underfloor cargo doors fore and aft of wing and bulk cargo door aft of rear underfloor door; two crew doors to port. Capacity for 30 pallets on main deck and 32 LD-1 containers plus bulk cargo under floor.

UPDATED

BOEING 757-200

US Air Force designations: C-32A and C-32B

TYPE: Twin-jet airliner.

PROGRAMME: Announced early 1978; has 707/727/737 fuselage cross-section and two large turbofans; Eastern Air Lines and British Airways ordered 21 firm and 24 optioned and 19 + 18 respectively 13 August 1978; first flight (N757A) 19 February 1982 powered by 166.4 kN (37,400 lb st) Rolls-Royce RB535Cs and designated 757-200; first Boeing airliner launched with foreign engine.

FAA certification 21 December 1982; CAA certification 14 January 1983; revenue services began 1 January 1983 (EAL) and 9 February 1983 (BA). First flight of 757 powered by P&W PW2037s, 14 March 1984; certified October 1984 and delivered to Delta; first 757 with RB535E4s delivered to EAL 10 October 1984; first extended-range model delivered to Royal Brunei Airlines May 1986; 757 with RB535E4 engines approved FAA ETOPS December 1986 (extended to 180 minutes July 1990); 757 with PW2037/2040 ETOPS approved April 1990 (180 minutes for PW2037 April 1992); Boeing windshear guidance and detection system approved by FAA January 1987. Certified for operation in the Russian Federation and Associated States (CIS) September 1993. On 14 February 2002, 1,000th Boeing 757 was delivered, being 148th for American Airlines.

Boeing announced on 16 October 2002 that manufacture of the 757 was to be terminated in late 2004. Final aircraft 1.050th overall (including prototypes) officially rolled out at Renton on 28 October 2004; delivery to Shangai Airlines then scheduled for April 2005, registered B-2876.

CUSTOMERS: As per table.

UPDATED

Swan song. Departure of a Song (Delta Air Lines subsidiary) Boeing 757-200 appropriately illustrates the type's departure from production in 2004 *(Paul Jackson)* ***NEW***/1047782

BOEING 757 ORDER BOOK
(at 1 January 2005)

Customer	First order	First Delivery	Variant	Engine	Qty
Air 2000/First Choice	12 Dec 86	27 Apr 97	757-200	RB211	5
Air Europe	2 Jul 82	30 Mar 83	757-200	RB211	15
Air Holland	25 Jun 87	9 Mar 88	757-200	RB211	3
Airtours International/My Travel	17 May 96	20 Mar 97	757-200	RB211	1
America West	6 Feb 87	10 Dec 87	757-200	RB211	4
American Airlines	25 May 88	17 Jul 89	757-200	RB211	126
American Trans Air	7 Sep 94	26 Sep 95	757-200	RB211	13
	30 Jun 00	4 Aug 01	757-300	RB211	12
Ansett Worldwide	7 Oct 87	15 Feb 89	757-200	RB211	18
	16 Nov 88	2 Dec 92	757-200	PW2037	3
	7 Oct 87	26 Jul 89	757-200PF	RB211	4
Arkia Israeli Airlines	31 Jul 98	31 Jan 00	757-300	RB211	2
Azerbaijan Airlines	30 Dec 97	27 Sep 00	757-200	RB211	2
Britannia Airways	20 Dec 91	10 Apr 92	757-200	RB211	11
British Airways	31 Aug 78	25 Jan 83	757-200	RB211	50
China Southern	1 Mar 88	22 Nov 88	757-200	RB211	19
China Southwest	31 May 90	5 Aug 92	757-200	RB211	12
China Xinjiang Airlines	30 Oct 97	30 Apr 98	757-200	RB211	6
Condor Flugdienst	29 Sep 88	19 Mar 90	757-200	PW2040	17
	9 Dec 96	10 Mar 99	757-300	RB211	13
Continental Airlines	8 Oct 90	12 May 94	757-200	RB211	41
	2 Jan 01	20 Dec 01	757-300	RB211	9
Delta Air Lines	12 Nov 80	5 Nov 84	757-200	PW2037	110
	14 Nov 89	9 Jan 92	757-200	PW2040	6
Eastern Air Lines	31 Aug 78	22 Dec 82	757-200	RB211	25
El Al	1 Oct 86	25 Nov 87	757-200	RB211	7
Ethiopian Airlines	9 Jun 89	25 Feb 91	757-200	PW2040	4
	9 Jun 89	24 Aug 90	757-200PF	PW2040	1
Far Eastern Air Transport	8 Nov 94	8 Nov 94	757-200	PW2037	7
GATX Capital Corp	31 Jul 96	25 Apr 97	757-200	RB211	3
GE Capital	13 Jun 00	5 Feb 01	757-200	RB211	4
GPA Group Ltd	22 Apr 88	30 Aug 91	757-200	RB211	12
Iberia Airlines	25 May 90	7 Jun 93	757-200	RB211	24
Icelandair	19 Oct 88	4 Apr 90	757-200	RB211	7
	16 Jun 97	18 Mar 02	757-300	RB211	1
ILFC	18 Jun 86	1 Apr 87	757-200	RB211	44
	25 Apr 88	25 Feb 92	757-200	PW2037	13
	25 Apr 88	28 May 91	757-200	PW2040	25
JMC Airline/Thomas Cook	3 May 00	24 Apr 01	757-300	RB211	2
Kawasaki Leasing	25 Apr 88	13 May 92	757-200	RB211	2
LTE	25 Apr 88	15 Apr 92	757-200	RB211	1
LTU	25 Aug 83	25 May 84	757-200	RB211	12
Mexican Government	16 Nov 87	16 Nov 87	757-200	RB211	1
Mid East Jet	27 Jan 97	27 Jan 97	757-200	PW2040	1
Monarch Airlines	19 Feb 81	21 Mar 83	757-200	RB211	7
Northwest Airlines	29 Nov 83	28 Feb 85	757-200	PW2037	56
	16 Jan 01	20 Jul 02	757-300	PW2037/40	16
Republic Airlines	1 Oct 85	6 Dec 85	757-200	RB211	6
R-R Aircraft Mgmt	1 Aug 98	21 May 99	757-200	RB211	4
Royal Air Maroc	5 Feb 86	15 Jul 86	757-200	PW2037	2
Royal Brunei Airlines	30 May 85	6 May 86	757-200	RB211	3
Royal Nepal Airlines	17 Feb 86	9 Sep 87	757-200	RB211	1
	17 Feb 86	15 Sep 88	757-200M	RB211	1
Shanghai Airlines	14 Dec 88	4 Aug 89	757-200	PW2037	13
Singapore Airlines	31 May 83	12 Nov 84	757-200	PW2037	4
Starflite Corp	15 Dec 99	17 Dec 99	757-200	RB211	1
Sterling European	13 Jan 89	2 Aug 90	757-200	RB211	3
Sunrock Aircraft	25 Apr 88	19 Jun 91	757-200	RB211	2
Transavia Airlines	23 May 89	22 Feb 93	757-200	RB211	3
Trans World Airlines	5 Feb 96	10 Feb 97	757-200	PW2037	14
Turkmenistan Airlines	20 Oct 95	29 Aug 96	757-200	RB211	3
United Airlines	26 May 88	24 Aug 89	757-200	PW2037	47
	26 May 88	17 Aug 90	757-200	PW2040	51
UPS	6 Nov 90	2 Jul 94	757-200PF	RB211	40
	31 Dec 85	17 Sep 87	757-200PF	PW2040	35
US Airways	25 Apr 89	26 Feb 93	757-200	RB211	23
USAF	8 Aug 96	29 May 98	757-200	PW2040	4
Uzbekistan Airways	20 Oct 95	19 Oct 96	757-200	PW2037	3
Xiamen Airlines	27 Oct 89	12 Aug 92	757-200	RB211	9
Total					**1,049**

Notes: For deliveries see Group Orders and Deliveries table

BOEING 757-300

TYPE: Twin-jet airliner.

PROGRAMME: Stretched 757-200. Launched 2 September 1996; major assembly began 9 August 1997; prototype (N757X; 804th B757; earmarked for Condor Flugdienst as D-ABOA) rolled out at Renton on 19 May 1998 (officially 31 May); first flight (transmitted live on the Internet) 2 August 1998. 912-hour, 356-sortie flight test programme undertaken by three aircraft. N757X/NU701, employed on initial airworthiness and basic controllability tests; D-ABOB/NU721, first flown 4 September 1998, for ground effect, autoland and fuel consumption tests; and D-ABOC/NU722, first flown 2 October 1998, initially for HERF testing, before installation of standard interior for a four-day period of simulated airline operations with launch customer Condor Flugdienst. FAA certification with 180-minute ETOPS approval 27 January 1999, followed by first delivery (D-ABOE) to Condor on 10 March 1999 and first revenue-earning service 19 March 1999. Achieved 8.7 hours daily utilisation and 99.64 per cent reliability rate in first 12 months of operation. All early aircraft with Rolls-Royce engines; first with PW2040 power (N753JM) was flown 20 February 2002, before delivery to Northwest Airlines.

CUSTOMERS: See table accompanying Boeing 757-200 entry. Launch customer Condor Flugdienst ordered 12 firm, with 12 options, on 2 September 1996; Icelandair ordered two, with Rolls-Royce engines, in June 1997, for delivery in second quarter 2001 and second quarter 2002. Boeing 757 production terminated in 2004.

UPDATED

BOEING 767

TYPE: Wide-bodied airliner.

PROGRAMME: Launched on receipt of United Air Lines order for 30 on 14 July 1978; construction of basic 220-passenger 767-200 began 6 July 1979; first flight (N767BA) 26 September 1981 with P&W JT9D turbofans; first flight fifth aircraft with GE CF6-80A 19 February 1982; 767 with JT9D-7R4D certified 30 July 1982; with CF6-80A 30 September 1982.

First delivery with JT9D (United Air Lines) 19 August 1982 (initial service 8 September); first delivery with CF6 (Delta) 25 October 1982. ETOPS approval for 767-200 with JT9D-7R4 or CF6-80A or -80A2 granted January 1987; ETOPS approval for 767-200 and -300 with PW4000 obtained April 1990; 180-minute ETOPS approval with PW4000 engines obtained August 1993. Joint 757/767 crew rating approved 22 July 1983. Boeing windshear detection and guidance system FAA approved for 767-200 and -300 February 1987.

Boeing 767-400 first flew 9 October 1999; FAA certification and 180-minute ETOPS granted 20 July 2000; JAA certification 24 July 2000; FAA common type rating with 767-200/300 and 757-200/300 issued 21 August 2000.

CURRENT VERSIONS: **767-200:** Basic model; no longer available; final delivery 24 February 1994. Medium-range variant (MTOW 136,080 kg; 300,000 lb) has reduced fuel; higher gross weight variant (142,880 kg; 315,000 lb) certified June 1983.

DEVELOPMENT MILESTONES

767-200	
First order	14 Jul 78
Rolled out	4 Aug 81
First flight	26 Sep 81
Certification	30 Jul 82
First delivery	19 Aug 82
Entered service (United Air Lines)	8 Sep 82
Subsequent versions	
767-200ER	
First order (Ethiopian Airlines)	16 Dec 82
Rolled out	14 Feb 84
First flight	6 Mar 84
Certification	Mar 84
First delivery	26 mar 84
Entered service (El Al)	27 Mar 84
767-300	
First order (Japan Airlines)	29 Sep 83
Rolled out	14 Jan 86
First flight	30 Jan 86
Certification	22 Sep 86
First delivery	25 Sep 86
Entered service (Japan Airlines)	20 Oct 86
767-300ER	
Official go-ahead	21 Jul 84
First order (American Airlines)	3 Mar 87
Rolled out	3 Nov 86
First flight	9 Dec 86
Certification	20 Jan 88
First delivery	19 Feb 88
Entered service (American Airlines)	3 Mar 88
767-400ER	
First order (Delta Air Lines)	20 Mar 97
Rolled out	26 Aug 99
First flight	9 Oct 99
Certification	20 Jul 00
First delivery	11 Aug 00
Entered service (Continental Airlines)	14 Sep 00

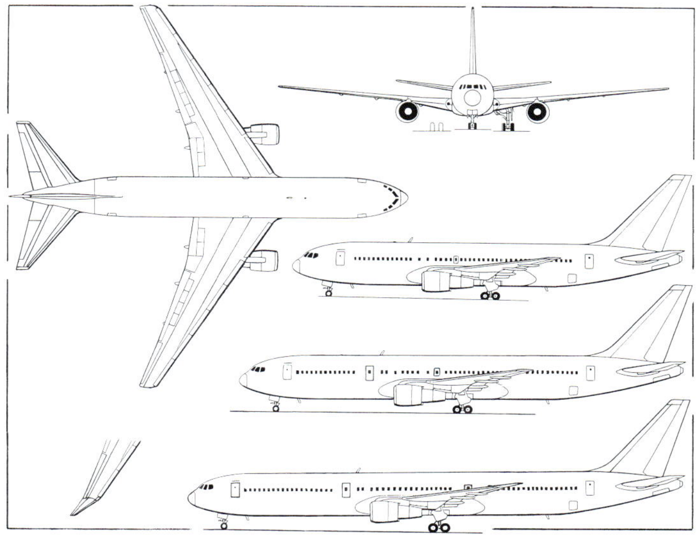

Boeing 767-200 wide-bodied airliner, with additional side view of stretched -300 and further stretched -400, including last mentioned's revised wingtips *(Jane's/Dennis Punnett)* 0085775

767-200ER: Extended-range version; announced January 1983; first flight 6 March 1984; basic -200ER with centre-section tankage and gross weight increased; first delivered to El Al 26 March 1984.

767-300: Stretched 269-passenger version, with 3.07 m (10 ft 1 in) plug forward of wing and 3.35 m (11 ft) plug aft, and same gross weight as 767-200; strengthened landing gear and thicker metal in parts of fuselage and underwing skin; same flight deck and systems as other 767s; same engine options as 767-200ER; first ordered (by Japan Airlines) 29 September 1983. First flight with JT9D-7R4D engines 30 January 1986; certified with JT9D-7R4D and CF6-80A2 22 September 1986. First delivery (Japan Airlines) 25 September 1986. British Airways ordered 11 in August 1987, later increased to total 25, with Rolls-Royce RB211-524H engines; delivered from 8 February 1990. No longer available.

767-300ER: Extended-range, higher gross weight version; development began January 1985; optional gross weights 172,365 kg (380,000 lb) and, from 1992, 186,880 kg (412,000 lb); further increased centre-section tankage. Engine choice CF6-80C2, PW4000, RB211-524H; structural reinforcement; certified late 1987. Launch customer American Airlines (15), delivered from 19 February 1988. New interior introduced late 2000; based on Boeing 777; first recipient Lauda Air.

767-300ERX: Further range extension, under study from 1998 to 2002, but not proceeded with; addition of tailplane fuel tank, capacity 7,571 litres (2,000 US gallons; 1,665 Imp gallons) would have increased range to 6,695 n miles (12,400 km; 7,705 miles).

767-300X: Proposed early 2002. Main feature would be rapidly changeable 767-400-type wingtip, allowing airlines to customise aircraft to individual routes, according to distance, reverting to standard tips for long range.

767-300 General Market Freighter: Described separately.

767-400ER: Stretched version. Features include strengthened wing with thicker ribs, spars and skin for increased MTOW; updated flight deck based on Boeing 777; fuselage lengthened 6.43 m (21 ft [illegible] in) by means of plugs forward (3.36 m; 11 ft 0¼ in) and aft (3.07 m; 10 ft 0¾ in) of centre-section; stringerless window belt with elliptical 777-type cabin windows; wing span increased by 4.42 m (14 ft 6 in) with highly sweptback (9° 50') wingtips of composites construction which reduce take-off distance, increase climb rate and improve fuel consumption; redesigned interior; with accommodation for up to 304 passengers; new landing gear with 46 cm (18 in) longer main legs, Boeing 777 brakes and 127 cm (50 in) tyres and revised, hydraulically actuated tail skid. Engine choice of CF6-80C2B7F1 or CF6-80C2B8, with PW4000 series as option; fuel capacity as currently offered on 767-300ER, enabling 767-400ER to operate most existing -300ER routes.

Offered from January 1997; launch customer Delta Air Lines announced intention to order 21 on 20 March 1997; confirmed 28 April 1997; Continental ordered 26 on 10 October 1997. Assembly of first aircraft began at Everett on 9 February 1999; roll-out 26 August 1999; first flight (N76400 No. 1) 9 October 1999. Four aircraft took part in test programme, comprising 1,150 flight hours and 1,200 ground testing hours: prototype used primarily to test and certify basic handling qualities; N76401 served as aerodynamics and avionics certification article; N87402, with full cabin interior used for systems development and certification; N47403 (first flown June 2000) for cabin entertainment and related evaluation. World tour (by N76400 No. 2) in July-August 2000. First delivery (N828MH), to launch customer Delta Air Lines, 11 August 2000; deliveries to Continental Airlines began 30 August 2000 with refurbished second prototype (N76401/N66051). Also on 30 August 2000, Delta received first 767-400ER with Rockwell Collins Large Format Display System, comprising six 203 × 203 mm (8 × 8 in) LCDs.

767-400ER Shrink: Under study in 1999 as alternative to 767-300ERX. Not proceeded with.

Longer Range 767-400ER: Extended-range version; launched (as 767-400ERX) 13 September 2000, but discontinued by 2002.

767 AWACS: Military version. *Described separately.*

KC-767A: Air refuelling tanker; USAF desicion announced 23 May 2004 to proceed with lease of 100. *Described separately.*

Northrop Grumman E-10: Ground surveillance version. *Described separately.*

767 AST: Boeing contracted by US Army on 11 October 1994 to supply company-owned 767 as Airborne Sensor Testbed for long wavelength infra-red surveillance system; operating contract was extended for 12 months in September 1998 at cost of US$4.1 million.

767 SF: Special Freighter conversion of 767-200 airliner; available from 2000; payload 39,010 kg (86,000 lb); freight door as 767-300F; strengthened floor, main landing gear and forward fuselage.

CUSTOMERS: As per table. Original prototype became 767 Airborne Surveillance Testbed (formerly AOA) for US Army. One reconfigured by E-Systems as medevac aircraft for Civil Reserve Air Fleet.

COSTS: US$108.5 million to US$120.0 million 767-200ER; US$123.5 million to US$136.5 million 767-300ER; US$135.0 million to US$148.5 million 767-400ER (all 2004).

Boeing 767-300ER of Australian Airlines, a Qantas subsidiary *(Paul Jackson)* ***NEW***/1047778

BOEING 767 ORDER BOOK
(at 1 January 2005)

Customer	Variant	Engine	First order	First delivery	Qty
Aeromaritime	200ER	PW	17 Jan 89	26 Jul 90	2
	300ER	PW	17 Jan 89	22 Aug 91	1
Air Algérie	300ER	GE	1 May 89	28 Jun 90	3
Air Canada	200	PW	11 Jul 79	30 Oct 82	10
	200ER	PW	11 Jul 79	18 Oct 84	9
	300ER	PW	31 Aug 89	10 Aug 93	6
Air China	200ER	PW	23 May 85	9 Oct 85	6
	300	PW	31 May 90	20 May 92	4
Air France	300ER	PW	7 Feb 92	14 May 93	3
Air Mauritius	200ER	GE	19 Jan 87	5 Apr 88	2
Air New Zealand	200ER	GE	30 Jul 84	3 Sep 85	3
	300ER	GE	1 Mar 91	11 Aug 93	5
Airtours International/My Travel	300ER	GE	17 Aug 93	16 Mar 94	3
Air Zimbabwe	200ER	PW	20 Jul 88	28 Nov 89	2
All Nippon Airways	200	GE	1 Oct 79	25 Apr 83	25
	300	GE	26 Dec 85	30 Jun 87	34
	300ER	GE	26 Dec 85	26 Jun 89	24
	300F		7 Aug 01	28 Aug 02	3
American Airlines	200	GE	15 Nov 78	4 Nov 82	13
	200ER	GE	15 Nov 78	18 Nov 85	17
	300ER	GE	3 Mar 87	19 Feb 88	58
Amiri Flight	300ER	GE	30 Nov 99	15 Dec 99	1
Ansett Australia	200	GE	17 Mar 80	7 Jun 83	5
Ansett Worldwide	200ER	PW	8 Sep 88	1 May 90	3
	200ER	GE	28 Sep 93	28 Sep 93	1
	300ER	PW	16 Nov 88	13 Dec 91	21
	300ER	GE	24 Jun 94	24 Jun 94	8
Asiana Airlines	300	GE	23 Dec 88	27 Sep 90	9
	300ER	GE	3 Sep 90	7 Nov 91	2
	300F	GE	3 Sep 90	23 Aug 96	1
Avianca	200ER	PW	24 Dec 80	26 Feb 90	2
BCC Leasing	300ER	PW	25 Sep 01	24 Oct 02	3
Braathens	200	PW	28 Apr 80	23 Mar 94	2
Britannia Airways	200	GE	31 Mar 80	6 Feb 84	11
	300ER	GE	23 Nov 94	15 May 96	7
British Airways	300ER	RR	14 Aug 87	8 Feb 90	28
Canadian Airlines Int'l	300ER	GE	7 Apr 87	15 Apr 88	14
China Airlines	200	PW	20 Mar 80	20 Dec 82	2
China Yunnan	300ER	RR	10 Jan 95	26 Jul 96	3
Condor Flugdienst	300ER	PW	16 Nov 88	13 Jul 91	11
Continental Airlines	200ER	GE	23 Nov 98	9 Nov 00	10
	400ER	GE	10 Oct 97	30 Aug 00	16
Delta Airlines	200	GE	15 Nov 78	25 Oct 82	15
	300	GE	15 Nov 78	7 Nov 86	24
	300	PW	20 Dec 90	17 Jun 93	4
	300ER	PW	22 Sep 88	9 Jun 90	31
	300ER	GE	10 Jun 97	18 Jun 98	22
	400ER	GE	10 Jun 97	11 Aug 00	21
Egyptair	200ER	PW	12 Jan 84	20 Jul 84	3
	300ER	PW	25 Oct 88	15 Aug 89	2
El Al Israel Airlines	200	PW	18 Mar 81	12 Jul 83	2
	200ER	PW	18 Mar 81	26 Mar 84	2
Ethiopian Airlines	200ER	PW	16 Dec 82	23 May 84	3
	300ER	PW	28 Nov 02	24 Nov 03	3
Eva Airways	200	GE	10 Jun 93	13 Jan 94	4
	300ER	GE	6 Oct 89	30 May 91	4
Flightlease	300ER	PW	15 Dec 99	4 Apr 00	4
GATX Capital Corp	200ER	GE	24 May 89	25 Jan 91	1
GECAS	300ER	GE	30 Aug 95	30 Aug 95	29
	300F	GE	15 Dec 99	28 Nov 01	1
GPA Ltd	200ER	PW	18 Apr 89	14 Jan 92	1
	300ER	GE	18 Apr 89	3 May 93	2
	300ER	PW	18 Apr 89	28 Feb 01	11
Gulf Air	300ER	GE	27 Apr 88	14 Jun 88	20
Hainan Airlines	300ER	PW	21 Aug 02	31 Oct 02	3
ILFC	200	GE	27 Aug 86	25 Aug 87	1
	200ER	GE	18 Feb 86	29 May 86	4
	300ER	GE	16 May 88	14 Jun 91	36
	300ER	PW	16 May 88	21 Mar 91	16
Itochu Air Lease Corp	300ER	GE	15 Apr 91	14 Aug 92	3
Itochu Corp	200ER	GE	4 Nov 93	1 Dec 94	4
Japan Airlines	200	PW	29 Sep 83	22 Jul 85	3
	300	PW	29 Sep 83	25 Sep 86	13
	300	GE	10 Jun 93	1 Aug 94	9
	300ER	GE	27 Nov 00	19 May 02	15
Kazakstan Airlines	200ER	GE	29 Dec 00	8 Feb 02	1
Kuwait Airlines	200ER	PW	18 Sep 84	20 Mar 86	3
LAM Mozambique	200ER	PW	1 Aug 90	24 Aug 93	1
LAN Chile	300ER	GE	28 May 97	29 Apr 98	3
	300F	GE	17 Nov 97	23 Sep 98	5
Lauda Air	300ER	PW	21 Apr 97	29 Apr 88	7
LOT Polish Airlines	200ER	GE	4 Nov 88	21 Apr 89	2
	300ER	GE	4 Nov 88	21 Aug 90	3
LTU	300ER	PW	21 Jan 88	2 Feb 89	5
Malev Hungarian Airlines	200ER	PW	21 Feb 91	30 Apr 93	2
Martinair Holland	300ER	PW	11 Mar 88	21 Sep 89	6
Mid East Jet	200ER	GE	4 Oct 96	4 Oct 96	1
Pacific Western	200	PW	21 Dec 78	4 Mar 83	2
Piedmont	200ER	GE	25 Jul 86	21 May 87	6
Qantas Airways	200ER	PW	7 Sep 83	3 Jul 85	7
	300ER	GE	24 Apr 87	30 Aug 88	22
Royal Brunei Airways	300ER	GE	8 Mar 96	8 Mar 96	2
SAS	200ER	PW	18 Jan 88	11 May 90	2
	300ER	PW	18 Jan 88	29 Mar 89	14
Shanghai Airlines	300	PW	14 May 93	22 Jul 94	4
Singapore Lease	300ER	GE	8 Aug 95	4 Aug 95	3

Customer	Variant	Engine	First order	First delivery	Qty
TACA Int'l Airlines	200	GE	22 May 86	22 May 86	1
	300ER	GE	27 Aug 91	26 Aug 91	2
Transbrasil	200	GE	30 Jul 81	23 Jun 83	3
TWA	200	PW	5 Dec 79	22 Nov 82	10
Turkmenistan Airlines	300ER	GE	11 Aug 03	21 Oct 04	1
unidentified	200ER	GE	15 Apr 02	27 Mar 03	8
	300ER	GE	12 Nov 04	none	3
	300F	GE	6 May 04	none	3
	400ER	GE	28 Sep 04	none	1
United Airlines	200	PW	14 Jul 78	19 Aug 82	19
	300ER	PW	19 May 89	18 Apr 91	37
United Parcel Service	300F	GE	15 Jan 93	12 Oct 95	32
US Airways	200ER	GE	4 Apr 89	22 May 90	6
Uzbekistan	300ER	PW	20 Oct 89	27 Nov 96	4
Varig	200ER	GE	18 Mar 86	2 Jul 87	6
	300ER	GE	8 Sep 98	21 Dec 89	4
Total					**950**

Notes: For deliveries see Group Orders and Deliveries table

DESIGN FEATURES: Low-wing wide-bodied airliner with twin, podded turbofans underwing. Boeing aerofoils; quarter-chord sweepback 31° 30'; thickness/chord ratio 15.1 per cent at root, 10.3 per cent at tip; dihedral 6°; incidence 4° 15'.

FLYING CONTROLS: Conventional and hydraulically powered. Inboard, all-speed (between inner and outer flaps) and outboard low-speed ailerons supplemented by flight spoilers (four-section outboard; two-section inboard) also acting as airbrakes and lift dumpers; single-slotted, linkage-supported outboard trailing-edge flaps, double-slotted inboard; track-mounted leading-edge slats; variable incidence tailplane driven by hydraulic screwjack; two-piece elevators each side; no trim tabs; roll and yaw trim through spring feel system; triple digital flight control computers and EFIS; Boeing windshear detection and guidance system optional. Control surface deflections: outboard ailerons +30/−15°, inboard ailerons ±20°, inboard flaps 61° (first element 36°), outboard flaps 36°, spoilers +60°, elevators +28/−20°, rudder ±26°; tailplane incidence +2/−12°.

STRUCTURE: Fail-safe structure. Conventional aluminium structure augmented by graphite ailerons, spoilers, elevators, rudder and floor panels; advanced aluminium alloy keel beam chords and wing skins; composites engine cowlings, wing/fuselage fairing and rear wing panels; CFRP landing gear doors; and aramid flaps and engine pylon fairings.

Subcontractors include Boeing Military Aircraft (wing fixed leading-edges); Northrop Grumman (wing centre-section and adjacent lower fuselage section; fuselage bulkheads); Vought Aircraft (horizontal tail); Canadair (rear fuselage); Alenia (wing control surfaces, flaps and leading-edge slats, wingtips, elevators, fin and rudder, nose radome); Fuji (wing/body fairings and main landing gear doors); Kawasaki (forward and centre fuselage; exit hatches; wing in-spar ribs); Mitsubishi (rear fuselage body panels and rear fuselage doors).

LANDING GEAR: Hydraulically retractable tricycle type; Menasco twin-wheel nose unit retracts forward; Cleveland Pneumatic main gear, with two four-wheel bogies, retracts inward; oleo-pneumatic shock-absorbers; Honeywell wheels and brakes; mainwheel tyres of current production versions H46×18.0-20 (26/28 ply for -200/300; 32 ply for -200ER/300ER); nosewheel tyres size H37×14.0-15 (22/24 ply) for all; steel disc brakes on all mainwheels; electronically controlled anti-skid units. Nosewheel steerable ±16°; ±65° for ground manoeuvring. Min taxiway width for U-turn 50.3 m (165 ft).

POWER PLANT: Two high-bypass turbofans in pods, pylon-mounted on the wing leading-edges.

General Electric options: 225 kN (50,600 lb st) CF6-80C2B2F, 251 kN (56,500 lb st) CF6-80C2B4F, 268 kN (60,200 lb st) CF6-80C2B6F, 276 kN (62,100 lb st) CF6-80C2B7F or F1 and 282 kN (63,500 lb st) CF6-80C2B8F.

Pratt & Whitney options: 233 kN (52,300 lb st) PW4052, 254 kN (57,100 lb st) PW4056, 268 kN (60,200 lb st) PW4060 and 282 kN (63,300 lb st) PW4062. P&W JT9D-7R4D of 213.5 kN (48,000 lb st) no longer offered.

Rolls-Royce options: 251 kN (56,400 lb st) RB211-524G4-T and 265 kN (59,500 lb st) RB211-524H2-T.

Fuel in one integral tank in each wing, and in centre tank, with total capacity of 63,216 litres (16,700 US gallons; 13,905 Imp gallons) in 200/300; 767-200ER and -300ER has additional 27,558 litres (7,280 US gallons; 6,062 Imp gallons) in second centre-section tank, raising total capacity to 90,774 litres (23,980 US gallons; 19,967 Imp gallons). In -400ER, tankage increased to total of 91,377 litres (24,140 US gallons; 20,101 Imp gallons). Refuelling point in port outer wing.

ACCOMMODATION: Operating crew of two on flight deck; observer's seat and optional second observer's seat. Basic accommodation in -200 models for 224 passengers, made up of 18 first class passengers forward in six-abreast seating at 96.5 cm (38 in) pitch, and 206 tourist class in seven-abreast seating at 81 cm (32 in) pitch. Window or aisle seats comprise 86 per cent of total. Type A inward-opening plug doors provided at both front and rear of cabin on each side of fuselage, with options of Type A, I or III emergency exits at various mid-cabin locations on each side. Total of five lavatories installed, two centrally in main cabin, two aft in main cabin, and one forward in first class section. Galleys situated at forward and aft ends of cabin. Alternative single-class layouts provide for 255 tourist passengers seven-abreast (two-three-two) at 81 cm (32 in) pitch (one overwing exit each side) and maximum (requiring two additional overwing emergency exits) 290, mainly eight-abreast (two-four-two), at 76 cm (30 in) pitch. Three-class layout for 181 passengers: 15 first class (two-one-two) at 152 cm (60 in) pitch; 40 business class (two-two-two) at 91 cm (36 in); and 126 tourist class (two-three-two) at 81 cm (32 in).

Basic accommodation in -300 models for 269 passengers, made up of 24 first class passengers forward in six-abreast seating at 96.5 cm (38 in) pitch, 245 tourist class in seven-abreast at 78.7 cm (31 in) pitch, six lavatories and five galleys. Alternatives include 286 in two-three-two seating at 81 cm (32 in) pitch and 218 in three-class layout comprising 18 first, 46 business and 154 tourist class passengers arranged as in -200. Maximum seating capacity in -300 models is 350 passengers at 71 cm (28 in), six lavatories and four galleys; capacities from 291 upwards require standard -300 door configuration (each side) of Type A front and rear and two Type Is overwing to be replaced by two Type As, plus third Type A ahead of wing and Type I adjacent to trailing-edge.

Accommodation in -400ER for 243 passengers in three-class configuration (16/36/189); 296 in two-class (24 first; 272 economy); or 409 in all-economy.

Underfloor cargo holds (forward and rear, combined) of -200 versions can accommodate, typically, up to 22 LD2 or 11 LD1 containers; 767-300 underfloor cargo holds can accommodate 30 LD2 or 15 LD1 containers. Starboard side forward and rear cargo doors of equal size on 767-200 and 767-300, but larger forward door standard on 767-200ER and 767-300ER and optional on 767-200 and 767-300. Bulk cargo door at rear on port side. Overhead stowage for carry-on baggage is 0.08 m³ (3.0 cu ft) per passenger. Cabin air conditioned, cargo holds heated.

SYSTEMS: Honeywell dual air cycle air conditioning system. Pressure differential 0.59 bar (8.6 lb/sq in). Electrical supply from two engine-driven 90 kVA three-phase 400 Hz constant frequency AC generators (120 kVA in -400ER), 115/200 V output. 90 kVA generator mounted on APU for ground operation or for emergency use in flight. Three hydraulic systems at 207 bar (3,000 lb/sq in), for flight control and utility functions, supplied from engine-driven pumps and a Honeywell bleed air-powered hydraulic pump or from APU. Maximum generating capacity of port and starboard systems is 163 litres (43 US gallons; 35.8 Imp gallons)/min; centre system 185.5 litres (49.0 US gallons; 40.8 Imp gallons)/min, at 196.5 bar (2,850 lb/sq in). Reservoirs pressurised by engine bleed air via pressure regulation module. Reservoir relief valve pressure nominally 4.48 bar (65 lb/sq in). Additional hydraulic motor-driven generator, to provide essential functions for extended-range operations, standard on 767-200ER and 767-300ER and optional on 767-200 and 767-300. Nitrogen chlorate oxygen generators in passenger cabin, plus gaseous oxygen for flight crew. APU in tailcone to provide ground and in-flight electrical power and pressurisation (Honeywell 331-400 of 120 kVA in 767-400ER). Anti-icing for outboard wing leading-edges (none on tail surfaces), engine air inlets, air data sensors and windscreen.

AVIONICS: *Radar:* Honeywell RDR-4A colour weather radar in aircraft for All-Nippon, Britannia and Transbrasil.

Flight: Standard ARINC 700 series equipment, including Honeywell VOR/ILS/marker beacon receivers, ADF, DME, RMI-743 radio magnetic indicator and radio altimeter. Honeywell IRS, FMCS and DADC, as described in Boeing 757 entry; dual digital flight management systems, and triple flight control computers, including FCS-700 flight control system; certified for Cat. IIIb landings; options include Boeing's windshear protection and guidance system.

Instrumentation: Honeywell EFIS-700 electronic flight instrument system.

DIMENSIONS, EXTERNAL:

Wing span: except 400ER ... 47.57 m (156 ft 1 in)
400ER ... 51.92 m (170 ft 4 in)
Wing chord: at root ... 8.57 m (28 ft 1¼ in)
at tip ... 2.29 m (7 ft 6 in)
Wing aspect ratio: except 400ER ... 8.0
400ER ... 9.3
Length:
overall: 200/200ER ... 48.51 m (159 ft 2 in)
300/300ER ... 54.94 m (180 ft 3 in)
400ER ... 61.37 m (201 ft 4 in)
fuselage: 200/200ER ... 47.24 m (155 ft 0 in)
300/300ER ... 53.67 m (176 ft 1 in)
400ER ... 60.08 m (197 ft 1½ in)
Fuselage max width ... 5.03 m (16 ft 6 in)
Height overall: 200ER/300 ER ... 15.85 m (52 ft 0 in)
400 ER: at OWE ... 17.02 m (55 ft 10 in)
at MTOW ... 16.69 m (54 ft 9 in)
Tailplane span ... 18.62 m (61 ft 1 in)
Wheel track (c/l shock absorbers) ... 9.30 m (30 ft 6 in)
Wheelbase: 200/200ER ... 19.69 m (64 ft 7 in)
300/300ER ... 22.76 m (74 ft 8 in)
Turning radius, 400ER: at nosewheel ... 29.3 m (96 ft)
at mainwheel (outer rims) ... 17.7 m (58 ft)
at wingtip ... 39.3 m (129 ft)
Distance between engine centres ... 15.85 m (52 ft 0 in)
Ground clearance under engine nacelles at MTOW ... 1.19 m (3 ft 11 in)
Passenger doors, 200ER/300ER (two, fwd and rear, port):
Height ... 1.88 m (6 ft 2 in)
Width ... 1.07 m (3 ft 6 in)
Passenger doors, 400ER (three, port):
Height, Width ... as 200ER/300ER
Height to sill: No. 1: at OWE ... 4.39 m (14 ft 5 in)
at MTOW ... 4.14 m (13 ft 7 in)
No 2: at OWE ... 4.60 m (15 ft 1 in)
at MTOW ... 4.42 m (14 ft 6 in)
No. 3: at OWE ... 5.18 m (17 ft 0 in)
at MTOW ... 4.90 m (16 ft 1 in)
Service door (two/three starboard):
Height ... 1.83 m (6 ft 0 in)
Width ... 1.07 m (3 ft 6 in)
Emergency exits (two, each): Height ... 0.97 m (3 ft 2 in)
Width ... 0.51 m (1 ft 8 in)
Cargo door (fwd, stbd):
Height: door ... 1.75 m (5 ft 9 in)
Clear access ... 1.70 m (5 ft 7 in)
Width: door ... 3.45 m (11 ft 7½ in)
Clear access ... 3.40 m (11 ft 2 in)
Height to sill, 400ER: at OWE ... 2.62 m (8 ft 7 in)
at MTOW ... 2.39 m (7 ft 10 in)
Cargo door (rear, stbd)
Height: door ... 1.75 m (5 ft 9 in)

Clear access 1.70 m (5 ft 7 in)
Width: door 1.92 m (6 ft 3½ in)
Clear access 1.78 m (5 ft 10 in)
Height to sill, 400 ER: at OWE 3.20 m (10 ft 6 in)
at MTOW 2.95 m (9 ft 8 in)
Bulk cargo door (port, rear):
Height: door 1.22 m (4 ft 0 in)
Clear access 1.10 m (3 ft 7½ in)
Width: door 0.96 m (3 ft 2 in)
Clear access 0.96 m (3 ft 2 in)
Height to sill, 400ER: at OWE 3.33 m (10 ft 11 in)
at MTOW 3.07 m (10 ft 1 in)

DIMENSIONS, INTERNAL:
Cabin, excl flight deck:
Length: 200/200ER 33.93 m (111 ft 4 in)
300/300ER 40.36 m (132 ft 5 in)
Max width 4.72 m (15 ft 6 in)
Height: max 2.87 m (9 ft 5 in)
min under baggage lockers 1.90 m (6 ft 3 in)
Floor area: 200/200ER 155.5 m^2 (1,674 sq ft)
300/300ER 184.7 m^2 (1,988 sq ft)
400ER 213.8 m^2 (2,301 sq ft)
Volume: 200/200ER 428.2 m^3 (15,121 cu ft)
300/300ER 483.9 m^3 (17,088 cu ft)
Volume, flight deck 13.5 m^3 (478 cu ft)
Cargo holds (containerised):
Length, 400ER: fwd 16.26 m (53 ft 4 in)
rear 14.73 m (48 ft 4 in)
Height above roller floor 1.68 m (5 ft 6 in)
Width at floor 3.66 m (12 ft 0 in)
Volume:
200/200ER: fwd 40.8 m^3 (1,440 cu ft)
rear 34.0 m^3 (1,200 cu ft)
300/300ER: fwd 54.4 m^3 (1,920 cu ft)
rear 47.6 m^3 (1,680 cu ft)
400ER: fwd 68.0 m^3 (2,400 cu ft)
rear 61.2 m^3 (2,160 cu ft)
Bulk cargo hold volume:
200ER/300ER 12.2 m^3 (430 cu ft)
400ER 9.8 m^3 (345 cu ft)
Combined baggage hold/bulk cargo hold volume:
200/200ER 87.0 m^3 (3,070 cu ft)
300/300ER 114.1 m^3 (4,030 cu ft)
Total cargo hold volume:
200/200ER 111.3 m^3 (3,930 cu ft)
300/300ER 147.0 m^3 (5,190 cu ft)

AREAS:
Wings, gross: except 400ER 283.3 m^2 (3,050.0 sq ft)
400ER 290.70 m^2 (3,129.0 sq ft)
Ailerons (total) 11.58 m^2 (124.60 sq ft)
Trailing-edge flaps (total) 36.88 m^2 (397.00 sq ft)
Leading-edge slats (total) 28.30 m^2 (304.60 sq ft)
Spoilers (total) 15.83 m^2 (170.40 sq ft)
Fin 30.19 m^2 (325.00 sq ft)
Rudder 15.95 m^2 (171.70 sq ft)
Tailplane 59.88 m^2 (644.50 sq ft)
Elevators (total) 17.81 m^2 (191.70 sq ft)

WEIGHTS AND LOADINGS (parameters as under):
767-200ER with 181 (three class) or 224 passengers:
2GB CF6-80C2B2F engines at basic TOW
2GM CF6-80C2B7F engines at max TOW
2PB PW4052 engines at basic TOW
2PM PW4062 engines at max TOW
767-300ER with 218 (three class) or 269 passengers:
3GB CF6-80C2B4F engines at basic TOW
3GM CF6-80C2B7F engines at max TOW
3PB PW4056 engines at basic TOW
3PM PW4062 engines at max TOW
3RB RB211-524G4 engines at basic TOW
3RM RB211-524H engines at max TOW
767-400ER with 269 (two class) passengers:
4GB CF6-80C2B7F1 engines at basic TOW
4GM CF6-80C2B8 engines at max TOW
4PB PW4056 engines at basic TOW
4PM PW4062 engines at max TOW
(identifiers such as '2GB' are not Boeing designations)
Operating weight empty:
2GB (181), 2GM (181) 84,960 kg (187,300 lb)
2GB (224), 2GM (224) 84,280 kg (185,800 lb)
2PB (181), 2PM (181) 85,005 kg (187,400 lb)
2PB (224), 2PM (224) 84,325 kg (185,900 lb)
3GB (218), 3GM (218) 90,810 kg (200,200 lb)
3GB (269), 3GM (269) 90,130 kg (198,700 lb)
3PB (218), 3PM (218) 90,855 kg (200,300 lb)
3PB (269), 3PM (269) 90,175 kg (198,800 lb)
3RB (218), 3RM (218) 91,625 kg (202,000 lb)
3RB (269), 3RM (269) 90,945 kg (200,500 lb)
4GB, 4GM 103,145 kg (227,400 lb)
4PB, 4PM 103,870 kg (229,000 lb)
Baggage capacity, underfloor:
200ER:
fwd: standard 9,798 kg (21,600 lb)
alternate 15,309 kg (33,750 lb)
rear: standard 8,165 kg (18,000 lb)
alternate 12,247 kg (27,000 lb)
300ER:
fwd: standard 13,063 kg (28,800 lb)
alternate 20,412 kg (45,000 lb)
rear: standard 11,431 kg (25,200 lb)
alternate 17,574 kg (38,745 lb)
Bulk hold capacity (all versions) 2,926 kg (6,450 lb)
Max fuel weight: 200, 300 51,130 kg (112,725 lb)
200ER, 300ER 73,635 kg (162,340 lb)
400ER 73,365 kg (161,740 lb
Max T-O weight: 2GB, 2PB 156,490 kg (345,000 lb)
2GM, 2PM 179,170 kg (395,000 lb)
3GB, 3PB, 3RB 172,365 kg (380,000 lb)
3GM, 3PM, 3RM 186,880 kg (412,000 lb)
4GB 181,435 kg (400,000 lb)
4GM, 4PM 204,115 kg (450,000 lb)
Max ramp weight: 2GB, 2PB 156,945 kg (346,000 lb)
2GM, 2PM 179,625 kg (396,000 lb)
3GB, 3PB, 3RB 172,820 kg (381,000 lb)
3GM, 3PM, 3RM 187,335 kg (413,000 lb)
4GB 181,935 kg (401,100 lb)
4GM, 4PM 204,570 kg (451,000 lb)
Max landing weight: 2GB, 2PB 126,100 kg (278,000 lb)
2GM, 2PM 136,080 kg (300,000 lb)
3GB, 3GM, 3PB, 3PM, 3RB, 3RM 145,150 kg (320,000 lb)
4GB, 4GM, 4PB, 4PM 158,760 kg (350,000 lb)
Max zero-fuel weight:
2GB, 2PB 114,755 kg (253,000 lb)
2GM, 2PM 117,935 kg (260,000 lb)
3GB, 3GM, 3PB, 3PM, 3RB, 3RM 133,810 kg (295,000 lb)
4GB, 4GM, 4PM 149,685 kg (330,000 lb)
Max wing loading:
2GB, 2PB 552.3 kg/m^2 (113.11 lb/sq ft)
2GM, 2PM 632.3 kg/m^2 (129.51 lb/sq ft)
3GB, 3PB, 3RB 608.3 kg/m^2 (124.59 lb/sq ft)
3GM, 3PM, 3RM 659.5 kg/m^2 (135.08 lb/sq ft)
4GB 624.2 kg/m^2 (127.84 lb/sq ft)
4GM, 4PM 702.2 kg/m^2 (143.82 lb/sq ft)
Max power loading: 2GB 348 kg/kN (3.41 lb/lb st)
2GM 324 kg/kN (3.18 lb/lb st)
2PB 336 kg/kN (3.30 lb/lb st)
2PM 318 kg/kN (3.12 lb/lb st)
3GB 343 kg/kN (3.36 lb/lb st)
3GM 338 kg/kN (3.32 lb/lb st)
3PB 339 kg/kN (3.33 lb/lb st)
3PM 332 kg/kN (3.25 lb/lb st)
3RB 344 kg/kN (3.37 lb/lb st)
3RM 353 kg/kN (3.46 lb/lb st)
4GB 328 kg/kN (3.22 lb/lb st)
4GM 361 kg/kN (3.54 lb/lb st)
4PM 362 kg/kN (3.55 lb/lb st)

PERFORMANCE:
Normal cruising speed, all versions M0.80
Approach speed at MLW:
2GB, 2PB 137 kt (254 km/h; 158 mph)
2GM, 2PM 142 kt (263 km/h; 163 mph)
3GB, 3GM, 3PB, 3PM 145 kt (269 km/h; 167 mph)
3RB, 3RM 148 kt (275 km/h; 171 mph)
Initial cruising altitude at max T-O weight:
2GB 11,550 m (37,900 ft)
2GM, 2PM 10,670 m (35,000 ft)
2PB 11,310 m (37,100 ft)
3GB 10,700 m (35,100 ft)
3GM, 3PM 10,180 m (33,400 ft)
3PB 10,730 m (35,200 ft)
3RB 10,730 m (35,200 ft)
3RM 10,210 m (33,500 ft)
Service ceiling, OEI: 2GB 5,090 m (16,700 ft)
2GM 4,205 m (13,800 ft)
2PB 5,395 m (17,700 ft)
2PM 4,845 m (15,900 ft)
3GB 4,510 m (14,800 ft)
3GM 3,930 m (12,900 ft)
3PB 4,785 m (15,700 ft)
3PM 4,085 m (13,400 ft)
3RB 3,900 m (12,800 ft)
3RM 3,505 m (11,500 ft)
T-O field length S/L, 86°F: 2GB 2,301 m (7,550 ft)
2GM, 3PB 2,485 m (8,150 ft)
2PB 2,180 m (7,150 ft)
2PM 2,439 m (8,000 ft)
3GB 2,530 m (8,300 ft)
3GM 2,713 m (8,900 ft)
3PM 2,652 m (8,700 ft)
3RB 2,500 m (8,200 ft)
3RM 2,896 m (9,500 ft)
Landing field length at MLW: 2GB 1,524 m (5,000 ft)
2GM 1,555 m (5,100 ft)
2PB 1,509 m (4,950 ft)
2PM 1,540 m (5,050 ft)
3GB, 3GM, 3PB, 3PM 1,677 m (5,500 ft)
3RB, 3RM 1,707 m (5,600 ft)
3RM 2,896 m (9,500 ft)
Design range:
2GB (181) 5,125 n miles (9,491 km; 5,897 miles)
2GB (224) 4,830 n miles (8,945 km; 5,558 miles)
2GM (181) 6,655 n miles (12,325 km; 7,658 miles)
2GM (224) 6,545 n miles (12,121 km; 7,531 miles)
2PB (181) 5,030 n miles (9,315 km; 5,788 miles)
2PB (224) 4,740 n miles (8,778 km; 5,454 miles)
2PM (181) 6,555 n miles (12,139 km; 7,543 miles)
2PM (224) 6,450 n miles (11,945 km; 7,422 miles)
3GB (218) 5,230 n miles (9,686 km; 6,018 miles)
3GB (269) 4,890 n miles (9,056 km; 5,627 miles)
3GM (218) 6,150 n miles (11,389 km; 7,077 miles)
3GM (269) 5,990 n miles (11,093 km; 6,893 miles)

3PB (218) .... 5,155 n miles (9,547 km; 5,932 miles)
3PB (269) .... 4,820 n miles (8,926 km; 5,546 miles)
3PM (218) .... 6,075 n miles (11,250 km; 6,991 miles)
3PM (269) .... 5,915 n miles (10,954 km; 6,806 miles)
3RB (218) .... 4,880 n miles (9,037 km; 5,615 miles)
3RB (269) .... 4,555 n miles (8,435 km; 5,241 miles)
3RM (218) .... 5,850 n miles (10,834 km; 6,732 miles)
3RM (269) .... 5,620 n miles (10,408 km; 6,467 miles)
4GB .... 4,375 n miles (8,102 km; 5,034 miles)
4GM .... 5,645 n miles (10,454 km; 6,496 miles)

UPDATED

BOEING 767-300 GENERAL MARKET FREIGHTER

TYPE: Twin-jet freighter.

PROGRAMME: First 767 specialised package freighter launched 15 January 1993 by United Parcel Service order; mockup completed early 1994; rolled out 8 May 1995; first flight (N301UP) 20 June 1995; certified 12 October 1995. The 767-300F for general operation was ordered by Asiana in November 1993 and differs from UPS version in having mechanical freight handling on main and lower decks, air conditioning for animals and perishables on main and forward lower decks and more elaborate crew facilities.

CUSTOMERS: UPS ordered 30 parcel freighters 15 January 1993; first delivery 12 October 1995; entered service 16 October 1995; last delivery on 9 September 1999; two of further 30 options taken up and delivered in October and November 2001. Asiana ordered two 767-300Fs in November 1993; contract reduced to one, which was delivered 23 August 1996. Five to LAN Chile between 23 September 1998 and 9 October 2001. One ordered by GECAS; one by undisclosed customer. Total 40.

COSTS: US$131.5 million to US$143.5 million (2004).

DESIGN FEATURES: Modifications include reinforced landing gear and internal wing structure; main deck floor strengthened to take 24 containers, each 2.24 × 3.18 m (88 × 125 in); no passenger windows; 2.67 × 3.40 m (8 ft 9 in × 11 ft 1¾ in) freight door forward to port; pilot-type rating and extensive component commonality with 757 Freighter.

POWER PLANT: Two General Electric CF6-80C2B6F/B7F, Pratt & Whitney PW4060/4062 or Rolls-Royce RB211-524H turbofans; 267 kN (60,000 lb st) each.

DIMENSIONS, INTERNAL:
Cabin: Length .... 39.80 m (130 ft 7 in)
Main deck container capacity .... 339.5 m^3 (11,990 cu ft)
Lower deck cargo capacity .... 92.9 m^3 (3,282 cu ft)

WEIGHTS AND LOADINGS:
Max T-O weight: standard .... 185,065 kg (408,000 lb)
optional .... 186,880 kg (412,000 lb)
Max ramp weight .... 187,335 kg (413,000 lb)
Max zero-fuel weight .... 140,160 kg (309,000 lb)
Max landing weight .... 147,871 kg (326,000 lb)
Max wing loading .... 659.5 kg/m^2 (135.08 lb/sq ft)
Max power loading .... 350 kg/kW (3.43 lb/lb st)

PERFORMANCE:
Range: with 40,823 kg (90,000 lb) payload .... 4,000 n miles (7,408 km; 4,603 miles)
with 50,800 kg (112,000 lb) payload .... 3,000 n miles (5,556 km; 3,452 miles)

UPDATED

BOEING 777

TYPE: Wide-bodied airliner.

DEVELOPMENT MILESTONES

777-200	
Programme launched	29 Oct 90
First order (United Airlines)	15 Oct 90
Rolled out	9 Apr 94
First flight	12 Jun 94
Certification	19 Apr 95
First delivery	15 May 95
Entered service (United Airlines)	7 Jun 95
Subsequent versions	
777-200ER	
Programme launched	29 Oct 90
First order	14 Jun 91
Rolled out	21 Aug 96
First flight	7 Oct 96
Certification	17 Jan 97
First delivery	6 Feb 97
Entered service (British Airways)	9 Feb 97
777-200LR	
Programme launched	29 Feb 00
First order (EVA Airways)	27 Jun 00
777-300	
Programme launched	26 Jun 95
First order	14 Jun 95
Rolled out	8 Sep 97
First flight	16 Oct 97
Certification	4 May 98
First delivery	22 May 98
Entered service (Cathay Pacific)	27 May 98
777-300ER	
Programme launched	29 Feb 00
First order (Japan Airlines)	31 Mar 00
Rolled out	14 Nov 02
First flight	24 Feb 03
Certification	16 Mar 04
First delivery (Air France)	29 Apr 04

United Airlines Boeing 777-200 *(Paul Jackson)* ***NEW***/1047777

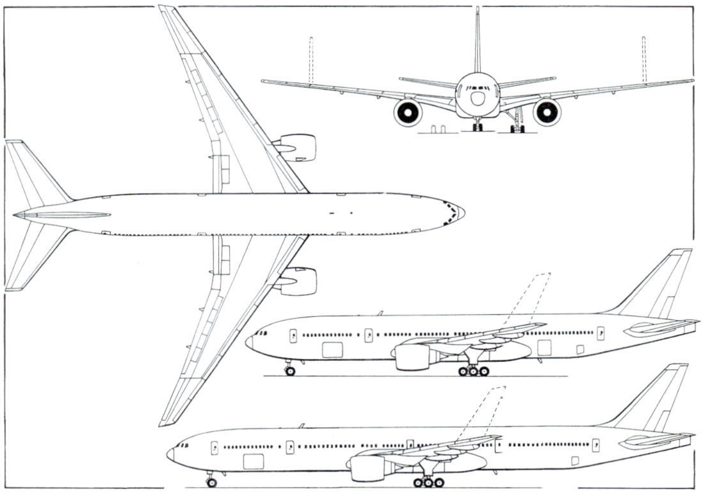

Boeing 777-200 twin-turbofan high-capacity airliner *(Jane's/Mike Keep)* 0015648

PROGRAMME: Formerly known as 767-X, initial variant now 777-200; Boeing board authorised firm offer for sale 8 December 1989; launch order by United Airlines 15 October 1990; Boeing launched production programme of initial market and Increased Gross Weight 777s and formed 777 Division 29 October 1990; configuration frozen March 1991; Boeing signed final agreement with Mitsubishi, Kawasaki and Fuji, making them risk-sharing programme partners for about 20 per cent of the 777 structure, on 21 May 1991; roll-out occurred 9 April 1994.

First flight (line no. WA001/N7771 with PW4084 engines), 12 June 1994, WA002 15 July, WA003 2 August, WA004 28 October, WA005 11 November; WA006/G-ZZZA (first with GE90, for British Airways) 2 February 1995, same day as FAA granted engine approval; PW-engined aircraft accumulated some 3,235 hours in 2,340 cycles in test programme, leading to joint FAA/JAA certification on 19 April 1995; FAA awarded 180-minute ETOPS approval 30 May 1995; first delivery, to United Airlines, on 15 May 1995; service entry with United on 7 June 1995 with inaugural revenue flight (by N777NA) from London to Washington, DC. Second GE90 aircraft for British Airways (WA010/G-ZZZB) joined WA006 in 1,750 hour, 1,260 cycle test programme; certification with GE engine 9 November 1995, with deliveries to BA commencing two days later.

Flight testing of Rolls-Royce Trent 800 on Boeing 747-100 testbed began late March 1995, with first flight of Trent-powered 777 (Boeing test aircraft) 26 May; certification and first delivery (to Thai International) on 3 April 1996 (formal hand-over 31 March), with ETOPS clearance following in October. The US National Aeronautic Association awarded Boeing and the 777 its 1995 Robert J. Collier Trophy for aeronautical achievement. On 2 April 1997, a Trent-powered 777 landed at Boeing Field having captured the eastbound world circumnavigation record in 41 hours 59 minutes with a single stop at Kuala Lumpur, Malaysia, its arrival at the latter also having secured the non-stop great circle distance record of 10,266.9 n miles (19,014.3 km; 11,814.9 miles).

Two new derivatives launched 29 February 2000 as 777-200LR and 777-300ER, both as extended-range versions of current production types. Anticipated market of 500 for these derivatives, 45 per cent of which with Asian operators.

CURRENT VERSIONS: **777-100X**: Under study in 2000 to meet Singapore Airlines requirements for shortened, 250-seat version. Not proceeded with.

777-200: Initial production version. Maximum T-O weight 247,210 kg (545,000 lb); up to 440 passengers at maximum density.

777-200ER: Formerly -200-IGW (Increased Gross Weight). Maximum T-O weight initially 286,895 kg (632,500 lb); increased to 293,930 kg (648 000 lb) in March 1998 and 297,555 kg (656,000 lb) in January 1999; additional 53,828 litres (14,220 US gallons; 11,840 Imp gallons) of fuel in centre-section tank; same passenger capacity as basic aircraft; configuration frozen January 1994, including strengthened wing, fuselage, empennage, landing gear and engine pylons. First flew (G-VIIA, one of two for British Airways) 7 October 1996 (GE engines); FAA/JAA ETOPS certification 5 February 1997; delivery 6 February. First Trent-powered -200ER (A6-EMI, for Emirates) flew 21 November 1996. January 1999 MTOW increase resulted from fitting 777-300 main landing gear and restricting CG travel to 4 per cent, for increased taxying weight. First flight with GE90-94B engines 12 June 2000; FAA certification 14 November 2000, followed immediately by first delivery (to Air France).

First 777 for private use is a -200ER (N777AS) first flown 3 November 1998 and delivered to Raytheon Systems at Waco, Texas, for outfitting; customer delivery to Middle East Jet due late 2000.

777-200LR: Ultra-long-range version (previously 777-200X) powered by two General Electric GE90-110B engines, each rated at 489 kN (110,000 lb/st); launched 29 February 2000. Maximum T-O weight 347,815 kg (766,800 lb); fuel capacity 202,292 litres (53,440 US gallons; 44,498 Imp gallons), with addition of tanks in rear cargo hold; range 9,100 n miles (16,853 km; 10,472 miles) with 301 passengers. Raked wingtips extend total span by 3.86 m (12 ft 8 in). Firm launch order from EVA Airways, 27 June 2000, for three. On 1 October 2001, Boeing announced suspension of development programme for up to 18 months this being resumed in March 2003; rolled out in early 2005; deliveries to Pakistan International Airlines scheduled to begin in January 2006.

777-200LRXF: Studies revealed in early 2004 for freighter carrying up to 110 tonnes.

777-250ERX: Under consideration by late 2002; length 68.6 m (225 ft); range 7,500 n miles (13,890 km; 8,630 miles) with up to 330 passengers.

BOEING 777 ORDER BOOK
(at 1 January 2005)

Customer	Variant	Engine	First Order	First Delivery	Qty
Air China	777-200	PW4000	24 Mar 97	26 Oct 98	10
Air France	777-200	GE90	20 Nov 96	27 Mar 98	18
	777-300ER	GE90	13 Nov 00	1 Jun 04	10
Air New Zealand	777-200ER	Trent 800	25 Aug 04	none	4
Alitalia	777-200ER	GE90	9 Nov 00	23 Aug 02	6
All Nippon Airways	777-200	PW4000	19 Dec 90	4 Oct 95	20
	777-200ER	PW4000	19 Dec 90	6 Oct 99	4
	777-300	PW4000	12 Sep 95	30 Jun 98	10
	777-300ER	GE90	18 Jul 00	28 Oct 04	6
American Airlines	777-200ER	Trent 800	21 Nov 96	21 Jan 99	54
Asiana Airlines	777-200ER	PW4000	20 Dec 96	24 May 02	7
British Airways	777-200	GE90	21 Aug 91	11 Nov 95	5
	777-200ER	GE90	21 Aug 91	6 Feb 97	24
	777-200ER	Trent 800	1 Aug 98	7 Jun 00	16
Cathay Pacific Airways	777-200	Trent 800	6 May 92	9 May 96	5
	777-300	Trent 800	6 May 92	22 May 98	12
China Southern Airlines	777-200	GE90	17 Dec 92	28 Dec 95	4
	777-200ER	GE90	17 Dec 92	28 Feb 97	2
Continental Airlines	777-200ER	GE90	12 May 93	28 Sep 98	16
Delta Air Lines	777-200ER	Trent 800	13 Nov 97	23 Mar 99	13
Egyptair	777-200ER	PW4090	23 Aug 95	23 May 97	5
El Al Israel Airlines	777-200ER	Trent 800	14 Dec 99	29 Jan 01	4
Emirates	777-200	Trent 800	4 Jun 92	5 Jun 96	3
	777-200ER	Trent 800	4 Jun 92	11 Apr 97	6
	777-300ER	GE90	19 Jul 04	none	4
Etihad Airways	777-300ER	GE90	30 Nov 04	none	5
EVA Airways	777-200LR	GE90	27 Jun 00	none	3
	777-300ER	GE90	27 Jun 00	none	12
Garuda Indonesian	777-200ER	GE90	25 Jun 96	none	6
GECAS	777-200ER	GE90	26 Sep 00	10 Oct 03	4
	777-300ER	GE90	26 Sep 00	none	14
ILFC	777-200ER	GE90	15 Dec 92	8 Jan 98	31
	777-200ER	PW4000	15 Dec 92	16 Jul 98	6
	777-200ER	Trent 800	2 Sep 97	none	4
	777-300	Trent 800	6 Mar 96	26 Sep 00	8
	777-300ER	GE90	28 Nov 00	29 Apr 04	18
Japan Air System	777-200	PW4000	29 Jun 93	3 Dec 86	7
Japan Airlines	777-200	PW4000	24 Jan 92	15 Feb 96	7
	777-200ER	GE90	27 Nov 00	11 Jul 02	11
	777-300	PW4000	22 Dec 95	28 Jul 98	8
	777-300ER	GE90	31 Mar 00	15 Jun 04	8
Kenya Airways	777-200ER	n/k	22 Mar 02	21 May 04	3
KLM	777-200ER	n/k	26 Aug 02	23 Oct 03	6
Korean Air Lines	777-200ER	PW4000	16 Dec 93	21 Mar 97	18
	777-300	PW4000	16 Dec 93	12 Aug 99	4
Kuwait Airways	777-200	GE90	10 Jul 96	30 Mar 98	2
Lauda Air	777-200ER	GE90	13 Dec 91	24 Sep 97	3
Malaysia Airlines	777-200ER	Trent 800	8 Jan 96	23 Apr 97	15
Mid East Jet	777-200ER	GE90	1 Oct 97	24 Nov 98	1
Pakistan International Airlines	777-200ER	GE90	14 Nov 02	29 Jan 04	3
	777-200LR	GE90	14 Nov 02	none	2
	777-300ER	GE90	14 Nov 02	none	3
Saudia	777-200ER	GE90	18 Jun 95	26 Dec 97	23
Saudi Oger	777-200ER	GE90	30 Nov 98	22 Oct 99	1
Singapore Aircraft Leasing	777-200ER	Trent 800	22 Dec 95	20 Mar 01	2
	777-300	Trent 800	22 Dec 95	12 Nov 99	4
Singapore Airlines	777-200ER	Trent 800	22 Dec 95	5 May 97	47
	777-300	Trent 800	22 Dec 95	10 Dec 98	12
	777-300ER	GE90	23 Dec 04	none	18
Thai Airways International	777-200	Trent 800	20 Jun 91	31 Mar 96	8
	777-300	Trent 800	22 Dec 95	23 Dec 98	6
unidentified	777-200ER	Trent 800	23 Dec 04	none	6
	777-300ER	GE90	29 Jun 04	none	1
United Airlines	777-200	PW4000	15 Oct 90	15 May 95	22
	777-200ER	PW4000	15 Oct 90	7 Mar 97	39
Vietnam Airlines	777-200ER	PW4000	31 Jan 02	20 Aug 03	4
Total					**673**

Notes: For delivery totals see Group Orders and Deliveries table

777-300: Initially known as 777 Stretch; revealed at Paris Air Show, 14 June 1995; launched by Boeing board 26 June 1995 when 36 commitments held; configuration frozen October 1995; major assembly started 7 April 1997; roll-out 8 September 1997; first flight (Rolls-Royce engines) 16 October 1997 (N5014K); first with P&W engines (HL-7534 of Korean Air) 4 February 1998; FAA certification achieved 4 May 1998 with 180-minute ETOPS approval; first delivery (B-HNH, 136th 777) to Cathay Pacific 22 May 1998.

Compared with first-generation 747s, 777-300 carries same number of passengers, but at two-thirds of fuel cost and with 40 per cent less maintenance. Features strengthened airframe, inboard wing and landing gear, ground-manoeuvring cameras on horizontal tail surfaces and wing/fuselage fairing; tailskid; Type A emergency door over each wing; and fuselage stretched by 19 frames, 10.13 m (33 ft 3 in) longer (5.33 m; 17 ft 6 in ahead of wing and 4.80 m; 15 ft 9 in aft of wing) than that of 777-200 to increase passenger capacity to 550 in single-class high-density configuration.

777-300ER: Extended range version of 777-300 (previously 777-300X); launched 29 February 2000; launch customer Japan Airlines ordered eight on 31 March 2000 for delivery from October 2003, but first recipient now ILFC/Air France on 29 April 2004 (beginning F-GSQA). Prototype assembly started 20 June 2002; rolled out 6 November 2002; official roll out 14 November 2002; first flight (WD501/N5017V) 24 February 2003; second prototype (WD502/N5016R) flew 6 April 2003; WD501 set twin-jet MTOW record of 351,350 kg (774,600 lb) at Edwards AFB, 19 May 2003 and subsequently exhibited at Paris Air Show, 15 to 22 June 2003. First customer aircraft (466th 777) rolled out 7 December 2003; first flew 10 January 2004; delivered Air France (F-GSQA) 29 April 2004, although owned by ILFC; first owned by Air France (F-GSQC) delivered 1 June 2004. Initial delivery to Japan Airlines was second prototype (429th 777) on 15 June 2004 (as JA731J); first prototype followed.

Both -200LR and -300ER have strengthened horizontal and vertical tail surfaces, strengthened wings with new tips which increase span to 64.80 m (212 ft 7 in) and strengthened landing gear, including semi-levered main gear and extendable (up to 25 cm; 9¾ in) nosewheel leg on -300ER. Modified landing gear permits steeper rotation, and thus operations from shorter runways; anti-tail strike system automatically moves elevator when runway contact is imminent. Early -300ERs temporarily limited to 344,865 kg (760,300 lb) MTOW.

777-300ER Stretch: Under study in 1999 as possible replacement for 747-400 with Asian carrier on routes to Europe; fuselage stretch of about 7.9 m (26 ft) to accommodate an additional 60 passengers, but airframe otherwise unchanged; range as for 777-300ER. Also referred to as **777-400X**. Not proceeded with.

CUSTOMERS: As per table.

COSTS: Estimated development cost US$4 billion (1990). US$165.0 million to US$182.5 million 777-200; US$173.5 million to US$195.5 million 777-200ER; US$202.0 million to US$225.5 million 777-200LR; US$191.5 million to US$218.0 million 777-300; US$218.0 million to US$245.5 million 777-300ER (all 2004). 777-300 estimated to burn 33 per cent less fuel and have 40 per cent lower maintenance cost than 747-100/200, resulting in direct operating cost savings of around 30 to 35 per cent.

DESIGN FEATURES: Objectives of design included replacement of McDonnell Douglas DC-10 Srs 10 and Lockheed TriStar in regional market, as well as DC-10 Srs 30 and Boeing 747SP in intercontinental service; also is replacement for early 747s. All features required for 180-minute ETOPS incorporated and tested in basic aircraft design. Cylindrical fuselage wider than 767 to allow twin aisle seating for from six- to 10-abreast; lavatories and overhead baggage bins designed to allow rapid change of layout.

Boeing 777-200ER of Kuwait Airways *(Paul Jackson)* ***NEW***/1047776

Boeing 777-200ER of Singapore Airlines *(Paul Jackson)* ***NEW***/1047775

Wing, of 31° 30' sweepback at quarter-chord, incorporates new technology to allow minimum M0.83 cruise in combination with high thickness for economical structure and large internal volume, long span for improved take-off and payload/range and large area for high cruise altitude and low approach speed; no winglets. Design included provision for outer 6.48 m (21 ft 3 in) of each wing to be folded to vertical to reduce gate width requirement at airports; this option not proceeded with.

FLYING CONTROLS: Fly-by-wire. Hydraulically actuated, with trim tab in rudder. Six-segment slats in each wing leading-edge. Single-slotted flaps mid-wing, double-slotted flaps inboard; flaperon between inboard and outboard flaps. Five-segment spoilers ahead of single-slotted flaps; two-segment spoilers ahead of double-slotted flaps.

Boeing's first airliner fly-by-wire system; fully powered control surface actuators (31 by Teijin Seiki America) electrically signalled from full FBW system; this signals slats, flaps, spoilers and control feel unit as well as inboard flaperons, outboard ailerons, elevators and rudder; system provides flight envelope protection as well as stabilisation and autopilot inputs, but the normal control columns and rudder pedals in cockpit are back-driven by the system to give the pilots direct appreciation of the activity of the automatic system.

In normal mode, flight guidance commands are generated by Rockwell Collins triple redundant digital autopilot/flight directors and the control laws and envelope protection commands are shaped by the Marconi Avionics triple digital primary flight computers; each of the three primary flight computers contains three 32-bit microprocessors (a Motorola 68040, an Intel 80486 and an AMD 29050), all three programmed in Ada to perform all FBW functions; with power supply and ARINC 629 modules, each microprocessor module constitutes a lane and the three lanes constitute a channel; each lane is compared with the others in its channel; the system not only has high fault tolerance, but allows deferred maintenance, by which failures can be carried over until the next scheduled maintenance.

Commands to the powered control units are produced by three BAE Systems and Teijin Seiki actuator control electronics units, which have a fourth analogue channel directly signalled from the sticks and pedals in the cockpit; normal operating mode is for the aircraft to be flown through autopilots, primary flight computers and actuator control electronics, which simultaneously back-drive the sticks and pedals in the cockpit; first degraded (secondary) mode is used if inertial units and standby attitude sensors all become disabled and the pilots take manual control through the digital primary flight computers; second degraded (direct) mode bypasses the main FBW system with the direct analogue link between cockpit and actuator control electronics; ultimate standby is mechanical control of tailplane incidence for the pitch axis and two wing spoiler panels for lateral control. Some powered control units produced by Parker-Bertea and Moog; tailplane trim module and hydraulic brake by Raytheon Systems.

Pitch axis control law is C* U, effectively tending to make the aircraft hold an airspeed and to respond in pitch attitude to a departure from that airspeed; trim changes due to configuration changes are suppressed; the system returns the bank angle to 35° if that angle is exceeded by the pilots and the controls then released; the system prevents exceeding the limiting airspeed and stalling; asymmetric thrust is automatically countered; the variable feel system adjusts control forces to warn of approach to flight envelope limits in manual flight; the FBW system is linked to the ARINC 629 dual triplex digital databusses (see also aircraft information management system under Avionics heading).

STRUCTURE: Composites of carbon and toughened resin used in skins of tailplane and fin torsion boxes and cabin floor beams; CFRP used for rudder, elevators, ailerons, flaps, engine nacelles and landing gear doors; hybrid composites in wingroot fairing; GFRP in fixed-wing leading-edge, tailplane and fin fore and aft panels, wing aft panels, engine pylon fairings and radome. Toughened materials have high damage resistance and allow simple low-temperature bolted repairs. Metal structure includes thick skins without need for tear straps; no bonding; single-piece fuselage frames; fuselage skin of advanced 2000-series aluminium alloy; wing, empennage and engine nacelle leading edges of 2000-series; wing top skin and stringers made in advanced 7055 aluminium alloy with greater compression strength; tailcone in standard 7000-series; 10 per cent of structure weight is composites.

Fully digital product definition with all parts created by Dassault/IBM CATIA CAD/CAM and communicated to manufacturing and publications; structure and systems integration, tube and cable run design completed before design release; 238 design/build teams have ensured that design, fabrication and test have proceeded concurrently for structure and systems. Whole aircraft defined in computer system; no mockup built.

Centre and rear fuselage barrel sections, tailcone, doors, wingroot fairing and landing gear doors made in Japan. Wing and tail leading-edges and moving wing parts, landing gear, floor beams, nose landing gear doors, wingtips, dorsal fin and nose radome made by Northrop Grumman, Kaman, Alenia (Italy), Embraer (Brazil), Short Brothers (UK), Singapore Aerospace Manufacturing, HDH and ASTA (Australia), Korean Air and other subcontractors. Boeing manufactures flight deck and forward cabin, basic wing and tail structures and engine nacelles; assembles and tests completed aircraft.

LANDING GEAR: Retractable tricycle type (Menasco/Messier-Bugatti joint design for main gear); two main legs carrying six-wheel bogies with steering rear axles automatically engaged by nose gear steering angle; max nosewheel angle ±70°; six-wheel bogies avoid need for third leg in fuselage and simplify braking system; twin-wheel steerable nose gear; mainwheel tyres H49×19.0–22 or 50×20.0R22 (32 ply); nosewheel tyres 42×17.0R18 (26 ply). Honeywell Carbenix 4000 mainwheel brakes arranged so that initial toe-pedal pressure used during taxying applies brakes to alternate sets of three wheels to save brake wear; full toe-pedal pressure applies all six brakes together. Minimum taxiway width for U-turn: 200 srs 48.2 m (158 ft); 300 srs 56.7 m (186 ft).

POWER PLANT: Two turbofans.

777-200: 343 kN (77,000 lb st) General Electric GE90-77B, 331 kN (74,500 lb st) Pratt & Whitney PW4074 or 343 kN (77,200 lb st) PW4077, or 327 kN (73,400 lb st) Rolls-Royce Trent 875 or 338 kN (76,000 lb st) Trent 877.

777-200ER: 377 kN (84,700 lb st) GE90-85B, 400 kN (90,000 lb st) GE90-90B, 417 kN (93,700 lb st) GE90-94B, 376 kN (84,600 lb st) PW4084, 401 kN (90,200 lb st) PW4090, 436 kN (98,000 lb st) PW4098, 372 kN (83,600 lb st) Trent 884, 400 kN (90,000 lb st) Trent 892 or 415 kN (93,400 lb st) Trent 895.

777-200LR: GE90-110B engines, each 489 kN (110,000 lb st), or GE90-115B1s of 512 kN (115,000 lb st).

777-300: PW4090, PW4098 or Trent 884, 892, as above.

777-300ER: GE90-115B1 engines, each 512 kN (115,000 lb st). All fuel of baseline version contained in integral tanks in wing torsion box, with reserve tank, surge tank and fuel vent and jettison pipes all inboard of wing fold; pressure refuelling points in lower wing leading edge, immediately outboard of engine pylons. Combined capacity of main, centre and reserve tanks in 777-200 is 117,348 litres (31,000 US gallons; 25,813 Imp gallons); 777-200ER and 777-300 fuel capacity increased by centre-section tank of 53,829 litres (14,220 US gallons; 11,841 Imp gallons) to maximum of 171,176 litres (45,220 US gallons; 37,653 Imp gallons).

Standard fuel of 777-200LR is 181,283 litres (47,890 US gallons; 39,877 Imp gallons), with provision for up to three 7,003 litre (1,850 US gallon; 1,540 Imp gallon) tanks in rear cargo compartment, increasing capacity to 204,185 litres (53,940 US gallons; 44,914 Imp gallons).

Fuel capacity of 777-300LR in all weight options is 181,283 litres (47,890 US gallons; 39,877 Imp gallons).

ACCOMMODATION: Two-pilot crew; cabin cross-section, which is between that of 747 and 767, chosen to allow widest selection of twin-aisle class and seating layouts ranging from six- to 10-abreast; galleys and lavatories can be located at a selection of fixed points in front and rear cabins or freely positioned within large footprints in which they can be moved in 2.5 cm (1 in) increments and attached to prepositioned mounting, plumbing and electric fittings; overhead bins open downward and provide each passenger with 0.08 m³ (3 cu ft) volume; bins can be removed without disturbing ceiling panels, ducts or support structure; advanced cabin management system simplifies cabin management and includes digital sound system of hi-fi quality.

Typical configurations for 200/200ER include 305 passengers in three-class layout: 24 first class (two-two-two-abreast), 54 business class (two-three-two) and 227 economy class (three-three-three); 367 in two classes: 14 business (two-three-two) and 353 economy (three-three-three); 375 in two classes: 30 first class (two-two-two at 97 cm; 38 in pitch) and 345 economy (two-five-two at 79 cm; 31 in or 81 cm; 32 in pitch); 400 in two classes: 30 first class, as immediately previously and 370 economy (three-four-three, pitches as previously); and 440 passengers in single class (three-four-three).

For 777-200LR, typically 279 in two classes (42 first, 237 economy) or 301 in three classes (16 first, 58 business, 227 economy).

For 777-300, options include 368 in three classes: 30 first (two-two-two at 152 cm; 60 in), 84 business (two-three-two at 97 cm; 38 in) and 254 economy (two-five-two at 79 cm; 31 in or 81 cm; 32 in); 386 in first (30), business (77) and economy (279) classes, as previously, except last-mentioned at three-four-three, with only four seats at minimum pitch; 451 passengers in two classes: 40 first, as above, and 411 economy (two-five-two, at usual pitches); 479, in first (44) and economy (435) classes, the former with 97 cm; 38 in pitch and latter three-four-three at usual pitches; and ultimate 550 single-class passengers, mostly two-four-two. Underfloor cargo compartments have mechanical handling system

Boeing 777-200ER of American Airlines *(Paul Jackson)* ***NEW***/1047771

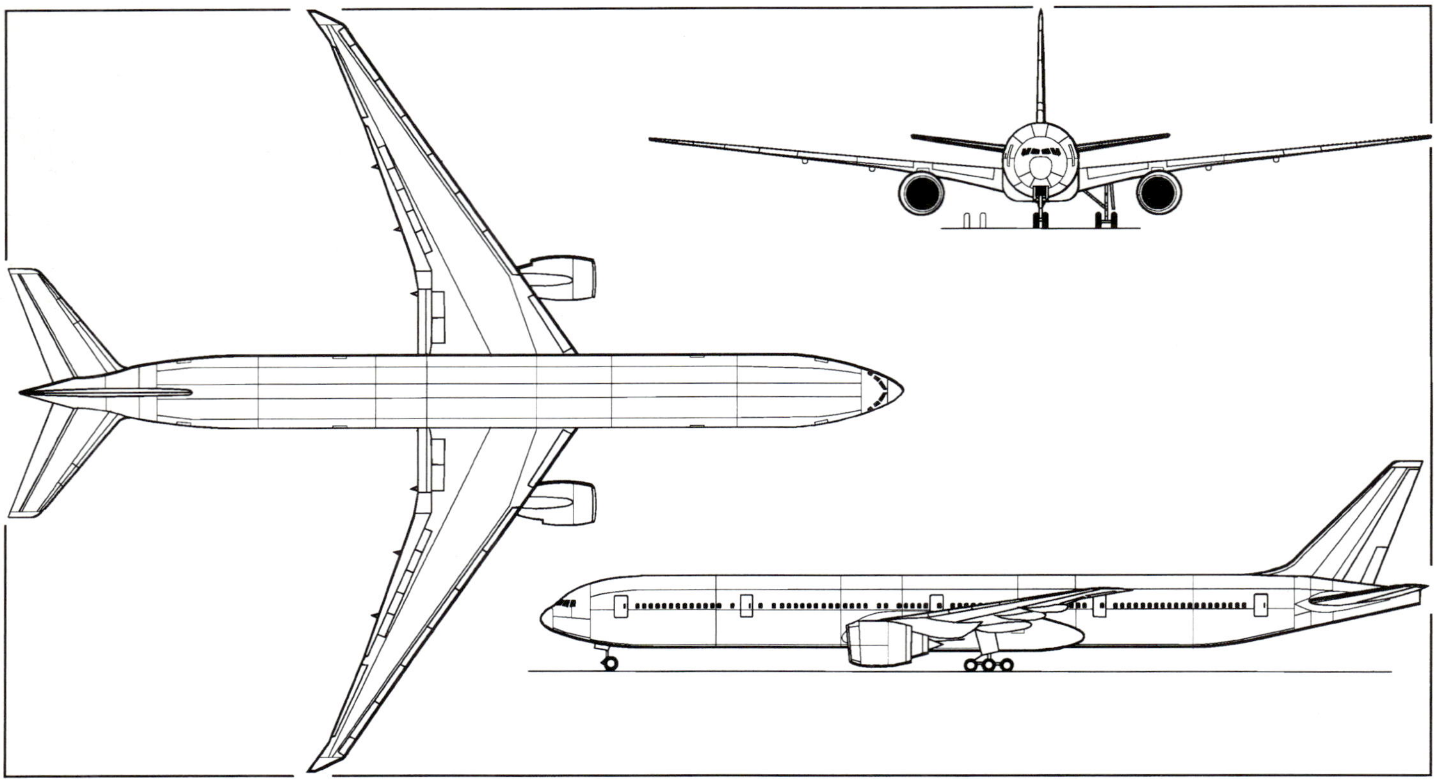

Boeing 777-300ER long-range airliner *(Paul Jackson)* **NEW**/1047709

and can accommodate all LD formats and 88 in or 96 in width pallets; up to 32 LD-3 containers plus 17.0 m^3 (600 cu ft) bulk cargo can be loaded in 777-200, or 44 LD-3s and some bulk cargo in 777-300.

Options for 777-300ER include 339 in two classes (56 first, 283 economy) and 370 in three classes (12/42/316).

Flight crew rest module on flight deck, port, contains two bunks. Optional underfloor crew rest modules available with six, eight or 10 bunks, and stowage space and require only electrical connection and hatch in passenger cabin floor, level with wing trailing-edge. Roof-mounted rest modules optional for ER and LR versions, occupying unused space between rows of baggage bins; first delivery May 2003.

Series 200 has four passenger doors each side; Series 300 has additional pair over wings.

SYSTEMS: Honeywell air drive unit, using bleed air from engines, APU or ground supply, drives central hydraulic system; cabin air supply and pressure control by Honeywell; Hamilton Sundstrand variable-speed, constant frequency AC electrical power generating system, with two 120 kVA integrated drive generators, one APU-driven generator and Honeywell ram air turbine system. Honeywell GTCP331-500 APU; Hamilton Sundstrand air conditioning; Smiths Industries ultrasonic fuel quantity gauging system and electrical load management system; optional wingtip folding by Raytheon Montek Division and Frisby Airborne Hydraulics.

AVIONICS: *Radar:* Honeywell weather radar standard.

Flight: Main navigation system is Honeywell air data and inertial reference (ADIRS) containing the Hexad skewed axis arrangement of six ring laser gyros; standby system is the secondary attitude and air data reference unit (SAARU) containing interferometric fibre optic gyros (using light transmitted in two directions along fibre optic paths), which produces a secondary flight director attitude display, airspeed and altimeter; both are linked to the ARINC 629 digital databus (777 being first aircraft thus equipped); Honeywell TCAS; Honeywell/BAE Systems Canada global navigation satellite sensor with 12-channel receiver; Honeywell/Racal multichannel satcom system optional. New, smaller flight computers installed from October 2003 onwards.

Dual Honeywell aircraft information management system (AIMS) contains the processing equipment required to collect, format and distribute onboard avionic information, including the flight management system (FMS), engine thrust control, digital communications management, operation of flight deck displays and monitoring of aircraft condition; both pilots and ground engineers can assess the condition of all onboard avionics systems.

Instrumentation: Six-screen EFIS using Honeywell 203 mm (18 in) ARINC D-size colour liquid crystal flat panel displays (two primary flight displays, two navigation displays and two engine/EICAS displays); three multipurpose control and colour display units on centre console provide interface with integrated aircraft information management system, which handles flight management, thrust control and communications control as well as all systems information.

Mission: Flight crew's 'electronic flight bag' (EFB) software installed from October 2003 (KLM Royal Dutch Airlines first) as option.

EQUIPMENT: Boeing 777-300 has Ground Maneuver Camera System with TV cameras in leading-edges of both horizontal stabilisers and underside of fuselage.

DIMENSIONS, EXTERNAL:

Wing span: 200, 200ER	60.93 m (199 ft 11 in)
200LR, 300ER	64.80 m (212 ft 7 in)
Wing aspect ratio	8.7
Length: fuselage: 200	62.94 m (206 ft 6 in)
300	73.08 m (239 ft 9 in)
overall: 200	63.73 m (209 ft 1 in)
300, 300ER	73.86 m (242 ft 4 in)
Fuselage: Length	62.74 m (205 ft 10 in)
Max diameter	6.20 m (20 ft 4 in)
Height: overall: 200LR:	
at OWE	18.75 m (61 ft 6 in)
at MTOW	18.49 m (60 ft 8 in)
300LR: at OWE	18.85 m (61 ft 10 in)
at MTOW	18.24 m (59 ft 10 in)
300	18.49 m (60 ft 8 in)
300ER	18.57 m (60 ft 11 in)
Tailplane span	21.52 m (70 ft 7½ in)
Wheel track (c/l shock absorbers)	10.97 m (36 ft 0 in)
Wheelbase: 200 srs	25.88 m (84 ft 11 in)
300 srs	31.22 m (102 ft 5¼ in)
Turning radius: at nosewheel: 200 srs	28.0 m (92 ft)
300 srs	33.8 m (111 ft)
at mainwheel (outer rims): 200 srs	15.5 m (51 ft)
300 srs	17.7 m (58 ft)
at wingtip: 200 srs	43.3 m (142 ft)
300 srs	45.1 m (148 ft)
Distance between engine centres	19.23 m (63 ft 1 in)
Ground clearance under engine nacelles at MTOW	5.13 m (16 ft 10 in)

Prototype Boeing 777-300ER displayed at Paris in June 2003 *(Paul Jackson)* **NEW**/0561723

Passenger doors (four/five each side, according to variant):
Height ... 1.88 m (6 ft 2 in)
Width ... 1.07 m (3 ft 6 in)
Height to sill: 200LR
No. 1 : at OWE ... 5.05 m (16 ft 7 in)
at MTOW ... 4.70 m (15 ft 5 in)
No. 2: at OWE ... 5.13 m (16 ft 10 m)
at MTOW ... 4.85 m (15 ft 11 in)
No. 3: at OWE ... 5.31 m (17 ft 5 in)
at MTOW ... 5.13 m (16 ft 10 in)
No. 4: at OWE ... 5.51 m (18 ft 1 in)
at MTOW ... 5.31 m (17 ft 5 in)
300 ER
No. 1: at OWE ... 5.13 m (16 ft 10 in)
at MTOW ... 4.80 m (15 ft 9 in)
No. 2: at OWE ... 5.21 m (17 ft 1 in)
at MTOW ... 4.93 m (16 ft 2 in)
Overwing ... inaccessible
No. 4: at OWE ... 5.31 m (17 ft 5 in)
at MTOW ... 5.11 m (16 ft 9 in)
No. 5: at OWE ... 5.66 m (18 ft 7 in)
at MTOW ... 5.18 m (17 ft 0 in)
Forward cargo door, stbd: Height ... 1.70 m (5 ft 7 in)
Width ... 2.69 m (8 ft 10 in)
Height to sill: 200LR
at OWE ... 3.10 m (10 ft 2 in)
at MTOW ... 2.79 m (9 ft 2 in)
300ER: at OWE ... 3.20 m (10 ft 6 in)
at MTOW ... 2.87 m (9 ft 5 in)
Rear cargo door, stbd:
Height ... 1.70 m (5 ft 7 in)
Width: standard ... 1.78 m (5 ft 10 in)
optional ... 2.69 m (8 ft 10 in)
Height to sill: 200LR
at OWE ... 3.40 m (11 ft 2 in)
at MTOW ... 3.23 m (10 ft 7 in)
300ER: at OWE ... 3.58 m (11 ft 9 in)
at MTOW ... 3.20 m (10 ft 6 in)
Bulk cargo door, stbd: Height ... 0.91 m (3 ft 0 in)
Width ... 1.14 m (3 ft 9 in)
Height to sill: 200LR
at OWE ... 3.61 m (11 ft 10 in)
at MTOW ... 3.40 m (11 ft 2 in)
300ER: at OWE ... 3.76 m (12 ft 4 in)
at MTOW ... 3.33 m (10 ft 11 in)

DIMENSIONS, INTERNAL:
Cabin: Length ... 49.10 m (161 ft 1 in)
Max width ... 5.87 m (19 ft 3 in)
Floor area: 200, 200ER ... 279.1 m^2 (3,004 sq ft)
Underfloor baggage hold:
Forward: Length: 200 ... 14.96 m (49 ft 1 in)
300 ... 20.29 m (66 ft 7 in)
Volume: 200 ... 80.5 m^3 (2,844 cu ft)
300 ... 107.4 m^3 (3,792 cu ft)
Aft: Length: 200 ... 11.35 m (37 ft 3 in)
300 ... 16.15 m (53 ft 0 in)
Volume: 200 ... 62.6 m^3 (2,212 cu ft)
200LR with max fuel ... 35.8 m^3 (1,264 cu ft)
300 ... 89.5 m^3 (3,160 cu ft)
Bulk: Length: 200, 300 ... 4.47 m (14 ft 8 in)
Volume: 200, 300 ... 17.0 m^3 (600 cu ft)
Height (all) ... 1.70 m (5 ft 7 in)
Total volume: 200 ... 160.2 m^3 (5,656 cu ft)
200LR with max fuel ... 133.3 m^3 (4,708 cu ft)
300 ... 213.8 m^3 (7,552 cu ft)

AREAS:
Wings, projected: short span ... 427.8 m^2 (4,605.0 sq ft)
extended span ... not disclosed
Ailerons (total) ... 7.11 m^2 (76.50 sq ft)
Trailing-edge flaps (total) ... 67.13 m^2 (722.60 sq ft)
Slats (total) ... 36.84 m^2 (396.50 sq ft)
Inboard spoilers (total) ... 8.67 m^2 (93.30 sq ft)
Outboard spoilers (total) ... 14.34 m^2 (154.40 sq ft)
Flaperons ... 6.69 m^2 (72.00 sq ft)
Horizontal tail surfaces, projected ... 101.26 m^2 (1,090.0 sq ft)
Vertical tail surfaces, projected ... 53.23 m^2 (573.00 sq ft)
Elevators, incl tabs (total) ... 25.48 m^2 (274.30 sq ft)
Rudder, incl tab ... 18.16 m^2 (195.50 sq ft)

WEIGHTS AND LOADINGS (parameters as under)::
777-200 with 305 passengers (24/54/227):
2GB ... GE90-77B engines at basic TOW
2GM ... GE90-77B engines at max TOW
2PB ... PW4074 engines at basic TOW
2PM ... PW4077 engines at max TOW
2RB ... Trent 875 engines at basic TOW
2RM ... Trent 877 engines at max TOW
777-200ER with 301 passengers (16/58/227):
2ERGB ... GE90-85B engines at basic TOW
2ERGM ... GE90-94 engines at max TOW
2ERPB ... PW4084 engines at basic TOW
2ERPM ... PW4090 engines at max TOW
2ERRB ... Trent 884 engines at basic TOW
2ERRM ... Trent 895 engines at max TOW
777-200LR:
2LRGB ... GE90-110B engines at max TOW
2LRGM ... GE90-115B engines at max TOW
777-300 with 368 passengers (30/84/254):
3PB ... PW4090 engines at basic TOW
3PM ... PW4098 engines at max TOW
3RB ... Trent 892 engines at basic TOW
3RM ... Trent 892 engines at max TOW
777-300ER with 370 passengers (12/42/316):
3ERGB ... GE90-115B1 engines at basic TOW
3ERGM ... GE90-115B1 engines at max TOW
(identifiers such as '2GB' are *not* Boeing designations)
Operating weight empty: 2GB ... 140,660 kg (310,100 lb)
2GM ... 140,795 kg (310,400 lb)
2PB ... 138,890 kg (306,200 lb)
2PM ... 139,025 kg (306,500 lb)
2RB ... 141,205 kg (311,300 lb)
2RM ... 141,385 kg (311,700 lb)
2ERGB ... 144,830 kg (319,300 lb)
2ERGM ... 145,015 kg (319,700 lb)
2ERPB ... 143,065 kg (315,400 lb)
2ERPM ... 143,835 kg (317,100 lb)
2ERRB ... 141,205 kg (311,300 lb)
2ERRM ... 141,385 kg (311,700 lb)
200LR ... 145,150 kg (320,000 lb)
3PB ... 158,030 kg (348,400 lb)
3PM ... 158,485 kg (349,400 lb)
3RB, 3RM ... 155,540 kg (342,900 lb)
300ER ... 167,830 kg (370,000 lb)
Max fuel weight: 200 ... 94,210 kg (207,700 lb)
200ER/300 ... 135,845 kg (299,490 lb)
200LR: basic ... 145,540 kg (320,860 lb)
max ... 162,410 kg (358,050 lb)
300ER ... 145,540 kg (320,860 lb)
Max T-O weight:
2GB, 2PB, 2RB ... 229,575 kg (506,000 lb)
2GM, 2PM, 2RM ... 247,205 kg (545,000 lb)
2ERGB, 2ERPB, 2ERRB, 3PB, 3RB ... 263,080 kg (580,000 lb)
2ERGM, 2ERPM, 2ERRM ... 297,555 kg (656,000 lb)
200LR ... 347,815 kg (766,800 lb)
3PM, 3RM ... 299,370 kg (660,000 lb)
3ERGB ... 345,050 kg (760,700 lb)
3ERGM ... 351,535 kg (775,000 lb)
Max ramp weight allowance:
200, 2ERB, 200LR, 300, 300ER ... 907 kg (2,000 lb)
2ERM ... 454 kg (1,000 lb)
Max landing weight: 200 ... 201,845 kg (445,000 lb)
200ER ... 213,190 kg (470,000 lb)
200LR ... 223,170 kg (492,000 lb)
300 ... 237,680 kg (524,000 lb)
300ER ... 251,290 kg (554,000 lb)
Max zero-fuel weight: 200 ... 190,510 kg (420,000 lb)
200ER ... 199,580 kg (440,000 lb)
200LR ... 209,105 kg (461,000 lb)
300 ... 224,525 kg (495,000 lb)
300ER ... 237,680 kg (524,000 lb)
Max wing loading:
2GB, 3PB, 2RB ... 536.5 kg/m^2 (109.88 lb/sq ft)
2GM, 3PM, 2RM ... 577.8 kg/m^2 (118.35 lb/sq ft)
2ERGB, 2ERPB, 2ERRB, 3PB, 3RB ... 614.9 kg/m^2 (125.95 lb/sq ft)
2ERGM, 2ERPM, 2ERRM ... 695.5 kg/m^2 (142.45 lb/sq ft)
3PM, 3RM ... 699.8 kg/m^2 (143.32 lb/sq ft)
Max power loading: 2GB ... 335 kg/kN (3.29 lb/lb st)
2GM ... 361 kg/kN (3.54 lb/lb st)
2PB ... 346 kg/kN (3.40 lb/lb st)
2PM ... 360 kg/kN (3.53 lb/lb st)
2RB ... 351 kg/kN (3.45 lb/lb st)
2RM ... 366 kg/kN (3.59 lb/lb st)
2ERGB ... 349 kg/kN (3.42 lb/lb st)
2ERGM ... 357 kg/kN (3.50 lb/lb st)
2ERPB ... 350 kg/kN (3.43 lb/lb st)
2ERPM ... 371 kg/kN (3.64 lb/lb st)
2ERRB ... 354 kg/kN (3.47 lb/lb st)
2ERRM ... 358 kg/kN (3.51 lb/lb st)
2LRGB ... 355 kg/kN (3.49 lb/lb st)
2LRGM ... 340 kg/kN (3.33 lb/lb st)
3PB, 3RB ... 328 kg/kN (3.22 lb/lb st)
3PM ... 343 kg/kN (3.37 lb/lb st)
3RM ... 374 kg/kN (3.67 lb/lb st)
3ERGB ... 337 kg/kN (3.31 lb/lb st)
3ERGM ... 344 kg/kN (3.37 lb/lb st)

PERFORMANCE:
Cruising speed: all ... M0.84
Approach speed: 200 ... 136 kt (252 km/h; 157 mph)
200ER ... 138 kt (256 km/h; 159 mph)
300 ... 149 kt (276 km/h; 171 mph)
300ER ... 150 kt (278 km/h; 173 mph)
Initial cruising altitude (ISA + 10°C):
2GB ... 12,010 m (39,400 ft)
2GM, 2PB ... 11,550 m (37,900 ft)
2PM, 2ERGB ... 11,155 m (36,600 ft)
2RB ... 11,645 m (38,200 ft)
2RM ... 11,370 m (37,300 ft)
2ERGM ... 10,575 m (34,700 ft)
2ERPB ... 10,820 m (35,500 ft)
2ERPM ... 10,270 m (33,700 ft)
2ERRB ... 11,005 m (36,100 ft)
2ERRM ... 10,455 m (34,300 ft)
3PB ... 10,975 m (36,000 ft)
3PM ... 10,425 m (34,200 ft)
3RB ... 11,245 m (36,900 ft)
3RM ... 10,395 m (34,100 ft)
3ERGM ... 9,845 m (32,300 ft)
Service ceiling, OEI (ISA +10°C):
2GB ... 5,515 m (18,100 ft)
2GM ... 4,724 m (15,500 ft)
2PB ... 4,940 m (16,200 ft)
2PM ... 4,877 m (16,000 ft)
2RB ... 4,816 m (15,800 ft)

2RM ... 5,365 m (17,600 ft)
2ERGB ... 3,995 m (13,100 ft)
2ERGM ... 3,719 m (12,200 ft)
2ERPB ... 4,359 m (14,300 ft)
2ERPM ... 3,660 m (12,000 ft)
2ERRB ... 4,755 m (15,600 ft)
2ERRM ... 3,749 m (12,300 ft)
3PB ... 4,572 m (15,000 ft)
3PM ... 3,719 m (12,200 ft)
3RB ... 4,816 m (15,800 ft)
3RM ... 3,505 m (11,500 ft)

T-O field length (30° C): 2GB ... 2,073 m (6,800 ft)
2GM ... 2,530 m (8,300 ft)
2PB, 2RB ... 2,164 m (7,100 ft)
2PM, 2RM ... 2,576 m (8,450 ft)
2ERGB ... 2,515 m (8,250 ft)
2ERGM ... 3,033 m (9,950 ft)
2ERPB ... 2,591 m (8,500 ft)
2ERPM ... 3,582 m (11,750 ft)
2ERRB ... 2,545 m (8,350 ft)
2ERRM ... 3,140 m (10,300 ft)
3PB ... 2,759 m (9,050 ft)
3PM ... 3,292 m (10,800 ft)
3RB ... 2,667 m (8,750 ft)
3RM ... 3,734 m (12,250 ft)
3ERGM ... 3,200 m (10,500 ft)

Landing field length: 2GB, 2GM ... 1,570 m (5,150 ft)
2PB, 2PM, 2RB, 2RM ... 1,555 m (5,100 ft)
2ERGB, 2ERGM ... 1,616 m (5,300 ft)
2ERPB, 2ERPM, 2ERRB, 2ERRM ... 1,601 m (5,250 ft)
300 ... 1,844 m (6,050 ft)

Design range:
2GB ... 3,985 n miles (7,380 km; 4,585 miles)
2GM ... 5,145 n miles (9,528 km; 5,920 miles)
2PB ... 3,955 n miles (7,324 km; 4,551 miles)
2PM ... 5,070 n miles (9,389 km; 5,834 miles)
2RB ... 4,100 n miles (7,593 km; 4,718 miles)
2RM ... 5,210 n miles (9,648 km; 5,995 miles)
2ERGB ... 5,810 n miles (10,760 km; 6,686 miles)
2ERGM ... 7,730 n miles (14,315 km; 8,895 miles)
2ERPB ... 5,695 n miles (10,547 km; 6,553 miles)
2ERPM ... 7,410 n miles (13,723 km; 8,527 miles)
2ERRB ... 5,840 n miles (10,815 km; 6,720 miles)
2ERRM ... 7,665 n miles (14,195 km; 8,820 miles)
200LR ... 9,150 n miles (16,945 km; 10,529 miles)
3PB ... 3,880 n miles (7,185 km; 4,465 miles)
3PM ... 5,710 n miles (10,574 km; 6,570 miles)
3RB ... 4,050 n miles (7,500 km; 4,660 miles)
3RM ... 5,955 n miles (11,028 km; 6,852 miles)
3ERGB ... 6,240 n miles (11,556 km; 7,180 miles)
3ERGM ... 7,880 n miles (14,593 km; 9,068 miles)

UPDATED

BOEING 7E7 DREAMLINER

TYPE: Wide-bodied airliner.

DEVELOPMENT MILESTONES

Programme launched	26 Apr 04
Marketing began	15 Dec 03
First order (All Nippon Airways	26 Apr 04

PROGRAMME: Originated in Super Efficient Airliner studies undertaken in 2001–02 in parallel with (then) higher-profile Sonic Cruiser; assumed prominence when latter abandoned in December 2002. Working designation of 7E7 to indicate 'Efficient'; named Dreamliner on 15 June 2003, following public Internet vote. Programme headquarters at Everett. Development effort will involve 15,000 hours of wind tunnel testing.

Airline commitments sought from early 2004 onwards and launch in latter part of that year; configuration freeze in second quarter of 2005; first flight 2007; service entry 2008.

On 15 December 2003, Boeing's Board of Directors approved the start of marketing of the 7E7 in the expectation of receiving sufficient proposals by airlines to warrant a formal launch to the programme during 2004. World market for airliners in the 7E7 class was then estimated as between 2,000 and 3,000 over the following 20 years. Launch customer All Nippon Airways announced order for 50 on 26 April 2004; 7E7 programme formally launched same day.

CURRENT VERSIONS: **7E7-8**: Baseline version; length 56.0 m (184 ft); span 58.8 m (193 ft); height 16.5 m (54 ft); 217 passengers in three-class layout. Range up to 8,500 n miles (15,742 km; 9,781 miles). MTOW 216,500 kg (477,300 lb). Service entry 2008.

7E7-3: Short range. Length and height as 7E7, but span reduced to 51.5 m (169 ft). Total 289 passengers in two classes. Range 3,500 n miles (6,480 km; 3,040 miles). MTOW 163,500 kg (360,450 lb), with variant-specific weight-saving measures employed. Service entry 2008.

7E7-9: Stretched; length 62.0 m (203 ft); span and height as 7E7-8. Up to 257 passengers in three-class layout. Range up to 8,300 n miles (15,370 km; 9,550 miles). MTOW 222,400 kg (490,300 lb). Service entry 2010.

All versions have fuselage max diameter of 5.74 m (18 ft 10 in) and cruising speed of M0.85.

CUSTOMERS: Boeing studies show 20-year market for between 2,000 and 3,000 mid-size airliners of 7E7 class. By mid-2004, Boeing had orders and deposits for nearly 200, representing almost first three years of production. ANA expected to divide order as 30 7E7-3s and 20 7E7-8s.

COSTS: Indicated in 2004 to be "close to" Boeing 767-300ER's price (2002) of US$120 million. ANA launch order for 50 valved at US$6 billion (2004), assuming list price.

DESIGN FEATURES: Intended to replace Boeing 767 and reduce seat/mile costs while providing increased versatility by enabling direct operations into smaller airports, thereby obviating the inconvenience of passenger transfers at hubs.

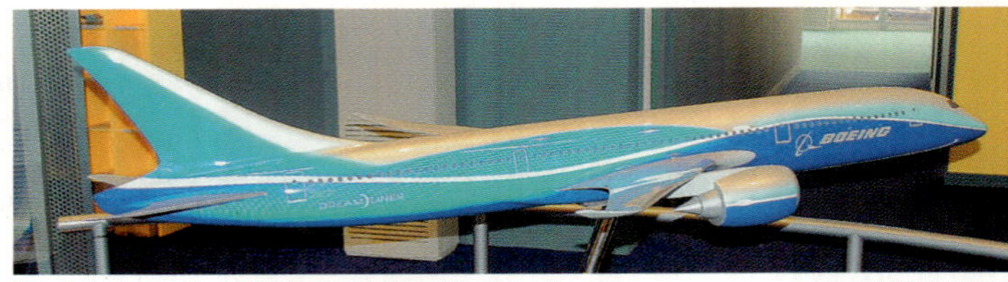

Model of Boeing 7E7 Dreamliner *(Paul Jackson)* ***NEW***/1044960

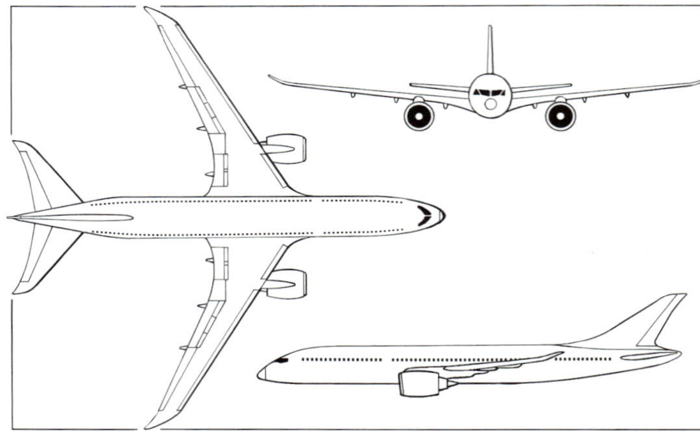

Non-specific general arrangement of Boeing 7E7 Dreamliner *(James Goulding)* ***NEW***/0568987

BOEING 7E7 ORDER BOOK
(at 1 January 2005)

Customer	Variant	Engine	First Order	First Delivery	Qty
Air New Zealand	7E7-8	RR	25 Aug 04	none	2
All Nippon Airways	7E7-3	RR	26 Jul 04	none	30
	7E7-8	RR	26 Jul 04	none	20
Blue Panorama	7E7-8	TBA	29 Dec 04	none	4
Total					**56**

Efficiency gains of 15 to 20 per cent, compared to Boeing 767, to be achieved by 17 per cent reduction in fuel burn, aerodynamic improvements and airframe weight reduction. Cruising Mach No 0.85.

Low wing configuration with two pylon-mounted turbofans underwing. Sweptback wing with increased sweep at tips, latter partly upturned. Pronounced sweep on fin leading- and trailing-edges.

FLYING CONTROLS: Outboard and inboard ailerons; two-section elevator; single-piece rudder. Three-section spoilers/airbrakes ahead of outboard flaps; two section spoilers/airbrakes ahead of inboard flaps.

STRUCTURE: Up to 65 per cent of airframe built by external suppliers. Some 35 per cent produced in Japan, with Mitsubishi contributing composites wingbox; Kawasaki the intermediate forward fuselage, main landing gear well and wing fixed trailing edge; Fuji responsible for centre wing box and integration of wheelwells. Vought and Alenia team producing 26 per cent, principally centre and rear fuselage sections and tailplane; Boeing producing almost 40 per cent at Australian, Canadian and US sites (forward fuselage, engine pylons and flight deck at Wichita, wing/fuselage fairings at Winnipeg, wing moving leading- and trailing-edges at Tulsa and Hawker de Havilland, Australia, and fin at Frederickson). Composites for "majority" of fuselage and wing, predominantly carbon fibre/epoxy but with titanium-graphite (TiGr) laminate in wings. Toray Industries prepreg carbon fibre/epoxy resin composite used for major sections of wing and fuselage. Possible use of new aluminium alloys for smaller structural pieces. Airframe 50 per cent composites by weight; 20 per cent aluminium; 15 per cent titanium; and 10 per cent steel.

BOEING 7E7 MAJOR SUPPLIERS

Alenia Aeronautica	Centre and aft fuselage; tailplane
Boeing (Everett)	Design and final assembly
Boeing (Frederickson)	Fin
Boeing (Tulsa)	Wing leading-edge
Boeing (Wichita)	Nose and forward fuselage
Boeing (Winnipeg)	Wing-to-body fairing
Boeing (HdH, Australia)	Wing moving trailing-edges
Dassault Systems	CATIA software
FR-HITemp	Fuel pumps and valves
Fuji	Centre wing box
General Electric	Engine option
Goodrich	Engine nacelles; fuel controls
Hamilton Sundstrand	Environmental controls; electrical power; power distribution; APU
Honeywell	Navigation systems; health management
Kawasaki	Intermediate forward fuselage; mainwheel wells; wing fixed trailing-edges
Mitsubishi	Wing box
Rockwell Collins	Communications, surveillance and display systems
Rolls-Royce	Engine option
Smiths Aerospace	Common core system
Toray Industries (Japan)	CFRP
Toray Industries (USA)	CFRP
Vought Aircraft	(as Alenia Aeronautica)

LANDING GEAR: Tricycle type; retractable. Messier-Dowty design and manufacture. Choice of two wheel/brake manufacturers to be offered. All-electric brakes.

POWER PLANT: Two turbofans, each in 235 to 311 kN (52,800 to 69,000 lb st) class, with bypass ratio between 9 and 12, pressure ratio 50:1 and fan diameter up to 2.92 m (9 ft 7 in). General Electric, Gen X and Rolls-Royce Trent 1000 options. Goodrich fuel management system, nacelles and thrust reverser. Kidde fire protection.

ACCOMMODATION: Eight-abreast (two-four-two) seating. Cabin max width 5.74 m (18 ft 10 in). Underfloor hold able to carry LD3 containers two abreast; five pallets and five LD3s in 7E7-3 and 7E7-8; six plus eight in 7E7-9. Cabin equivalent altitude 1,830 m (6,000 ft) with 20 to 30 per cent humidity. Pilot's controls by Kaiser Electroprecision.

SYSTEMS: Real-time structural monitoring by continuous collection of data from embedded sensors. Variable-frequency electrical system; 350 bar (5,000 lb/sq in) hydraulic system. Hamilton Sundstrand APU, environmental control system, power generating and starting systems.

AVIONICS: Rockwell Collins displays and communications/surveillance systems. Thales integrated flight displays. Honeywell crew information system/management system (CIS/MS). Avionics based around Smiths Aerospace common core system (CCS) comprising two dual-redundant common computing resources (CCR) cabinets containing fault-tolerant computing modules, each with ARINC 653 partitioned software operating environment. ARINC 664 deterministic ethernet/advanced communications network connects CCR to several remote data concentrators and avionics and utilities systems.

UPDATED

DOUGLAS PRODUCTS DIVISION (DIVISION OF BOEING COMMERCIAL AIRPLANE GROUP)

The Douglas Aircraft Company became Douglas Products Division when Boeing absorbed McDonnell Douglas in August 1997. Final MD-80 series, an MD-83, was delivered on 21 December 1999; MD-90 line closed at end of 2000; last MD-11 delivered 22 February 2001. Former MD-95 now marketed as Boeing 717. Having openly questioned the future of the 717, Boeing reconfirmed its commitment to continued production on 13 December 2001, albeit with lower production rate and revised delivery dates.

Interest in the BC-17X commercial freighter version of the C-17 Globemaster III appeared to be undergoing a revival in mid-2004.

UPDATED

BOEING 717

TYPE: Twin-jet airliner.

DEVELOPMENT MILESTONES

Announced	Jun 91
Prototype construction started	(late) 94
Launch order (ValuJet)	19 Oct 95
Rolled out	10 Jun 98
First flight	2 Sep 98
Certification	1 Sep 99
First delivery (AirTran)	23 Sep 99
Entered service	12 Oct 99

PROGRAMME: Announced at Paris Air Show 1991 as MD-95; potential airline customers briefed and manufacturing partners announced in Berlin, November 1994; modification of former Eastern Airlines DC-9-30 into development prototype began late 1994; design acquired by Boeing in August 1997 and renamed Boeing 717 in January 1998, re-using model number originally allocated to KC-135A tanker vesion of Boeing 707. Three flight test aircraft (T1, T2 and T3); first nose section (built by MDC) delivered to Huntington Beach on 11 December 1996 for T1, which began final assembly in May 1997 and was rolled out on 10 June 1998. BR 715 engine certified 1 September 1998.

First flight of prototype (N717XA; T1) 2 September 1998; second prototype (N717XB; T2) first flew 26 October 1998; third (N717XC; T3) on 16 December 1998, at which time first two had flown 361 hours in 193 sorties. First production aircraft (N717XD; P1) rolled out 23 January 1999; JAA/FAA certification awarded 1 September 1999 at end of five aircraft, 2,000 hour, 1,900 sortie programme; first aircraft (N942AT; third production) delivered to AirTran on 23 September 1999, after demonstration flights; entered service on Atlanta-Washington, DC route on 12 October 1999. Updated FMS certified 20 October 2000, adding GPS, fuel prediction, vertical guidance and automatic calculation of V-speeds. CIS certification awarded 20 February 2001. Delivery of 100th effected on 18 June 2002 as 37th for AirTran.

Production rate, set at five per month in 2001, had dropped to less than one per month in 2004. Boeing reported to be considering introducing a winglets option in an effort to boost flagging sales, which seen as due at least in part to competition from company's own 737-600.

CURRENT VERSIONS: **Boeing 717-200**: Initial (and so far only) production version, *to which following description applies*. Available in basic (**BGW**) and high (**HGW**) gross weight versions.

Boeing 717-300X: Proposed stretched version, formerly MD-95-50; studies suggest typical two-class seating for 130 passengers, with overall length increased to 42.16 m (138 ft 4 in) by addition of nine frames (five forward and four aft of wing); higher MTO and space-limited payloads weights; additional service door aft of wing; and 93.4 kN (21,000 lb st) BR 715 C1-30 engines. AirTran has expressed interest in converting some -200 options to this model. Under consideration late 2003 by Star Group (Air Canada, Austrian Airlines, Lufthausa and SAS); interest also reported from Delta, Iberia and Northwest Airlines. If launched, service entry is targeted for second half of 2006.

BOEING 717-200 ANNUAL ORDERS AND DELIVERIES

CY	Ordered	Delivered
1995	42	-
1996	-	-
1997	-	-
1998	41	-
1999	12	12
2000	21	32
2001	5	49
2002	32	20
2003	8	12
2004[1]	6	7
Total	**167**	**132**

[1] To 1 July

Boeing 717 Business Express: Corporate version of 717-200, unveiled at EBACE Convention in Geneva, Switzerland, on 7 May 2003. Configurable for 40 to 80 passengers in first and/or business class interior (typically, 60 passengers at 132 cm; 52 in seat pitch). Max range in HGW configuration, with auxiliary fuel and 60 passengers, 3,140 n miles (5,815 km; 3,613 miles). Complements BBJ family.

CUSTOMERS: See tables. Manufacturer foresees a market for over 3,000 aircraft in this class over 20 years. Launched 19 October 1995 with order for 50, plus 50 options, from ValuJet Airlines (later renamed AirTran). By 1 July 2004, orders stood at 167, of which 132 had been delivered. Six delivered during first half of 2004, during which period AirTran ordered a further six for delivery in 2005.

COSTS: AirTran order for 50 aircraft worth US$1 billion. List price US$38 million to US$42 million per aircraft (2004). Business Express US$27 million, basic; US$31 million in HGW configuration with five auxiliary fuel tanks (2003).

DESIGN FEATURES: All major elements of airframe based on DC-9/MD-80 series; systems and avionics are blend of low cost and advanced technology. Conventional low-wing, rear-engine, T-tail configuration.

Wing from DC-9-34, but with 1° 34' additional incidence; sweep 24° 30' at quarter-chord; thickness/chord ratio 11.6.

FLYING CONTROLS: Conventional and partly assisted. Elevator and ailerons are manually actuated via cables; rudder powered hydraulically with manual reversion; fly-by-wire trimming of rudder, two-section spoilers and elevator; three-section, double-slotted flaps; full-span, two-position, five-section leading-edge slats.

STRUCTURE: All-metal, two-spar wing with riveted spanwise stringers; glass fibre trailing-edges on wings, ailerons, flaps, elevators and rudder. Variations from MD-80/MD-90 include thicker skins on tail surfaces; MD-87 fuselage lengthened forward of wing by three frames 1.45 m (4 ft 9 in) and fin tip by 250 mm (10 in); wing/fuselage fillet extended forwards by three frames, using composites structure. Composites also for fuselage tailcone, fintip, elevator and aileron tabs, radome and wing trailing-edge panels; otherwise of 2024T3 aluminium alloy.

Partners are: Alenia (fuselage sections), Korean Air Lines Aerospace Division (nose structure and main passenger door/entry area), KAI (Hyundai) (wings, in conjunction with Boeing Toronto Ltd, which built initial sets of wings for flight test aircraft and early production units), BAE Systems (wing join and underwing barrel) Rolls-Royce Deutschland (power plant), Goodrich (engine nacelles), ShinMaywa Industries Ltd (horizontal tail surfaces and engine pylons), Fischer Advanced Composite Components GmbH (cabin furnishings), Andalucia Aerospacial (slats, landing gear components, aft pressure bulkhead), Israel Aircraft Industries SHL Servo Systems (landing gear), Honeywell (environmental control system, wheels and brakes, (flight guidance and avionics systems), Parker-Hannifin Corp (hydraulic and control systems), AIDC (empennage), Labinal (electric assemblies), Hamilton Sundstrand (electrical power generating system), and (in partnership with Hamilton Sundstrand) Auxiliary Power International Corporation (APU). Final assembly in Long Beach.

LANDING GEAR: Hydraulically retractable tricycle with steerable nosewheels; twin wheels on all legs. All-steel brakes; anti-skid units. Main tyres H41×15.0–19 (24 ply); nose tyres 26×6.6 (12 ply).

POWER PLANT: Two Rolls-Royce Deutschland BR 715 A1-30 turbofans, each 82.3 kN (18,500 lb st) at T-O at 30°C ambient, for 717-200 and Business Express BGW; BR 715 C1-30 of 93.4 kN (21,000 lb st) for 717-200 and Business Express HGW; optional 89.9 kN (20,000 lb st) available; switching between three thrust ratings does not involve engine hardware changes. Integrated drive generators system. Goodrich single-pivot door-type reversers for ground use only. Engine pylon based on MD-80 but thinner and without powered flap, although with extra frame for additional strength.

Standard (BGW) fuel capacity 13,904 litres (3,673 US gallons; 3,058 Imp gallons) in three tanks in wingroots and fuselage centre section. HGW has additional 1,628 litre (430 US gallon; 358 Imp gallon) tank in forward baggage hold and 1,022 litre (270 US gallon; 225 Imp gallon) tank in rear baggage hold for total volume of 16,667 litres (4,403 US gallons; 3,666 Imp gallons). Fuel recirculation system prevents wing upper surface ice accumulation and aids cooling.

ACCOMMODATION: Crew of two on advanced flight deck optimised for reduced parts count and high reliability. Cabin cross-section as for MD-80. Typical two-class seating for 106 passengers: eight first-class in four-abreast seating with 0.91 m (36 in) pitch and 98 standard

BOEING 717-200 ORDER BOOK

(at 1 January 2005)

Customer	First Order	First Delivery	Qty
Aerolineas de Baleares	16 May 00	22 Jun 00	3
AirTran	1 Dec 95	23 Sep 99	63
Bavaria International	4 May 98	29 Dec 99	5
Hawaiian Airlines	1 Mar 00	28 Feb 01	13
Jetstar	29 Dec 00	29 Dec 00	3
Midwest Airlines	15 Apr 02	28 Feb 03	25
Pembroke Capital	31 Dec 98	17 Aug 00	26
Turkmenistan Airlines	25 Jul 00	31 Jul 01	7
TWA	31 Dec 98	18 Feb 00	24
Total			**169**

Boeing 717-231 of launch customer, AirTran *(Paul Jackson)* ***NEW***/1047749

class with 0.81 m (32 in) pitch in five-abreast arrangement in new modern cabin, alternatively 117 single-class in 0.79/0.81 m (31/32 in) pitch. Seating designed by Avio Interiors and complies with 16 *g* impact regulations. Cabin designed with inputs from 500 airline executives, flight attendants and passengers; interior, manufactured by Fischer Advanced Composite Components of Austria, features wider and deeper overhead baggage bins, full-grip handrail throughout length of cabin. Two vacuum-operated lavatory units at rear of cabin, plus optional further unit at front for first-class. Galley units positioned at front of cabin. Underfloor baggage and cargo hold; latter reduced in HGW version by auxiliary fuel tanks. Main passenger door, fwd, port, with optional airstair stowage below; service door opposite; aft door on centreline with rearward-facing ramp stairs; two Type III emergency exits overwing on each side. Front and rear underfloor baggage hold doors, starboard.

SYSTEMS: Honeywell dual air cycle air conditioning and pressurisation system with digital cabin air controllers, utilising engine bleed air, maximum differential 0.54 bar (7.77 lb/sq in). Three-wheel air cycle machine and modified water separator in rear fuselage. Two separate 207 bar (3,000 lb/sq in) hydraulic systems for operation of spoilers, flaps, slats, rudder, landing gear, nosewheel steering, brakes, thrust reversers and ventral stairway. Maximum flow rate 30.3 litres (8.0 US gallons; 6.7 Imp gallons)/min. Airless bootstrap-type reservoirs, output pressure 2.07 bar (30 lb/sq in). Pneumatic system, for air conditioning/pressurisation, engine starting and ice protection, utilises engine bleed air and/or APU. Electrical system includes two 35/40 kVA integrated drive generators, plus 60 kVA APU generator. Oxygen system of diluter demand type for crew on flight deck; continuous flow chemical canister type with automatic mask presentation for passengers. Anti-icing of wing, engine inlets and tailplane by engine bleed air. Electric windscreen de-icing. Thermal anti-icing of leading-edges. APIC APS 2100 APU.

AVIONICS: Honeywell Versatile Integrated Avionics VIA 2000 computer as core avionics management system; full Cat. IIIa capability standard; Cat. IIIb capability available optionally, with addition of radar altimeter, ILS receiver and inertial reference unit, SFE ARINC 700 avionics.

Flight: Honeywell flight management system (FMS), inertial reference system (IRS), digital flight guidance system (DFGS), digital air data computer and windshear detection system. To be certified for Cat. IIIb automatic landings. Flight computer system upgraded and certified 20 October 2000; changes include improved GPS.

Instrumentation: Six-tube EFIS with 203 × 203 mm (8 × 8 in) LCD screens providing navigation, flight management and systems data. Flight deck features include MD-11 AFCS with glareshield-mounted controls enabling crew to fly the aircraft automatically with only push-button and thumbwheel inputs. Simplified integrated flightcrew warning and alerting panel (IFWAP) overhead control panel with four LCDs replacing 13 gauges, meters and switch panels. Central fault display system for reduced maintenance time.

DIMENSIONS, EXTERNAL:
Wing span ... 28.45 m (93 ft 4 in)
Wing chord: at root ... 5.44 m (17 ft 10 in)
  at tip ... 1.12 m (3 ft 8 in)
Wing aspect ratio ... 8.7
Length: overall ... 37.80 m (124 ft 0 in)
  fuselage ... 34.34 m (112 ft 8 in)
Fuselage max width ... 3.34 m (10 ft 11½ in)
Height overall: at OWE ... 8.86 m (29 ft 1 in)
  at MTOW ... 8.76 m (28 ft 9 in)
Tailplane span ... 11.23 m (36 ft 10 in)
Wheel track ... 4.88 m (16 ft 0 in)
Wheelbase ... 17.60 m (57 ft 8¾ in)
Distance between engine centrelines ... 6.60 m (21 ft 8 in)
Passenger door (fwd, port):
  Height ... 1.83 m (6 ft 0 in)
  Width ... 0.86 m (2 ft 10 in)
  Height to sill: at OWE ... 2.46 m (8 ft 1 in)
    at MTOW ... 2.21 m (7 ft 3 in)

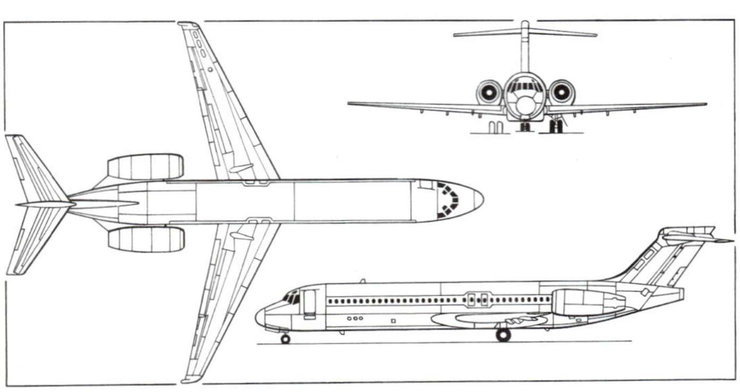

Boeing 717-200 airliner (two Rolls-Royce Deutschland BR 715 turbofans) *(James Goulding)* 0126948

Aft door (centreline):
  Height ... 1.83 m (6 ft 0 in)
  Width ... 0.70 m (2 ft 3¾ in)
  Height to sill: at OWE ... 2.01 m (6 ft 7 in)
    at MTOW ... 1.83 m (6 ft 0 in)
Service door (fwd, stbd):
  Height ... 1.22 m (4 ft 0 in)
  Width ... 0.69 m (2 ft 3 in)
  Height to sill: at OWE ... 2.46 m (8 ft 1 in)
    at MTOW ... 2.21 m (7 ft 3 in)
Emergency exits (above wings, two per side):
  Height ... 0.91 m (3 ft 0 in)
  Width ... 0.51 m (1 ft 8 in)
  Height to sill: at OWE ... 2.87 m (9 ft 5 in)
    at MTOW ... 2.77 m (9 ft 1 in)
Baggage door:
  forward:
    Height ... 1.27 m (4 ft 2 in)
    Width ... 1.34 m (4 ft 4¾ in)
    Height to sill: at OWE ... 1.30 m (4 ft 3 in)
    at MTOW ... 1.09 m (3 ft 7 in)
  rear:
    Height ... 1.27 m (4 ft 2 in)
    Width ... 0.91 m (3 ft 0 in)
    Height to sill: at OWE ... 1.35 m (4 ft 5 in)
    at MTOW ... 1.17 m (3 ft 10 in)

DIMENSIONS, INTERNAL (BGW: standard, HGW: with extended-range tanks):
Cabin: Length ... 23.44 m (76 ft (11 in)
  Max width ... 3.33 m (10 ft 11 in)
  Width at floor ... 3.12 m (10 ft 3 in)
  Height at aisle ... 2.03 m (6 ft 8 in)
Underfloor baggage/freight hold:
  BGW: front:
    Length ... 11.04 m (36 ft 2½ in)
    Height ... 0.99 m (3 ft 3 in)
    Volume ... 18.3 m³ (646 cu ft)
    rear: Length ... 5.21 m (17 ft 1 in)
    Height ... 0.99 m (3 ft 3 in)
    Volume ... 8.2 m³(289 cu ft)
  HGW:
    front: Volume ... 14.9 m³ (527 cu ft)
    rear: Volume ... 5.7 m³ (203 cu ft)
Business Express total volume:
    BGW ... 34.1 m³ (1,204 cu ft)
    HGW ... 19.9 m³ (703 cu ft)

AREAS:
Wings, gross ... 92.97 m² (1,000.7 sq ft)

WEIGHTS AND LOADINGS (BGW: standard, HGW: with extended-range tanks):
Operating weight empty: BGW ... 30,618 kg (67,500 lb)
  HGW ... 31,071 kg (68,500 lb)
Basic operating weight, Business Express:
  BGW ... 31,071 kg (68,500 lb)
  HGW ... 32,146 kg (70,870 lb)
Max structural payload: BGW ... 12,020 kg (26,500 lb)
  HGW ... 14,515 kg (32,000 lb)
    Business Express: BGW ... 11,567 kg (25,500 lb)
    Business Express: HGW ... 13,440 kg (29,630 lb)
Max fuel weight: BGW ... 11,162 kg (24,609 lb)
  HGW ... 13,381 kg (29,500 lb)
Fuel for max payload:
  Business Express BGW ... 5,738 kg (12,650 lb)
  Business Express HGW ... 9,435 kg (20,800 lb)
Max T-O weight: BGW ... 49,895 kg (110,000 lb)
  HGW ... 54,885 kg (121,000 lb)
Max ramp weight: BGW ... 50,349 kg (111,000 lb)
  HGW ... 55,340 kg (122,000 lb)
Max landing weight: BGW ... 45,359 kg (100,000 lb)
  HGW ... 49,895 kg (110,000 lb)
Max zero-fuel weight: BGW ... 42,638 kg (94,000 lb)
  HGW ... 45,586 kg (100,500 lb)
Max wing loading: BGW ... 536.7 kg/m² (109.92 lb/sq ft)
  HGW ... 590.4 kg/m² (120.92 lb/sq ft)
Max power loading: BGW ... 303 kg/kN (2.97 lb/lb st)
  HGW ... 294 kg/kN (2.88 lb/lb st)

PERFORMANCE:
Max limiting Mach No. (MMO) ... 0.82
Max operating speed (VMO) at FL 260 ... 340 kt (629 km/h; 391 mph)

Normal cruising speed:
BGW, HGW ... 438 kt (811 km/h; 504 mph) (M0.77)
Approach speed: BGW ... 132 kt (244 km/h; 152 mph)
HGW ... 139 kt (257 km/h; 160 mph)
Initial cruise altitude, ISA +10°C:
BGW ... 10,424 m (34,200 ft)
HGW ... 9,815 m (32,200 ft)
Max certified altitude ... 11,280 m (37,000 ft)
Service ceiling, OEI: BGW ... 5,365 m (17,600 ft)
HGW ... 4,360 m (14,300 ft)
T-O field length at S/L, ISA +10°C:
BGW ... 1,677 m (5,500 ft)
HGW ... 1,753 m (5,750 ft)
Landing field length at MLW: BGW ... 1,418 m (4,650 ft)
HGW ... 1,524 m (5,000 ft)
Max range at FL342, domestic reserves, 106 passengers and baggage:
BGW ... 1,415 n miles (2,620 km; 1,628 miles)
HGW ... 2,055 n miles (3,805 km; 2,364 miles)
Range, Business Express:
30 passengers:
BGW ... 2,180 n miles (4,037 km; 2,508 miles)
HGW ... 3,330 n miles (6,167 km; 3,832 miles)
60 passengers:
BGW ... 2,060 n miles (3,815 km; 2,370 miles)
HGW ... 3,140 n miles (5,815 km; 3,613 miles)
80 passengers:
BGW ... 1,980 n miles (3,667 km; 2,278 miles)
HGW ... 2,770 n miles (5,130 km; 3,187 miles)

OPERATIONAL NOISE LEVELS (ICAO Annex 16 Ch 3):
717-200: T-O with cutback ... 81.4 dB
Approach ... 91.4 dB
Sideline ... 89.2 dB
Business Express:
T-O: BGW ... 80.4 EPNdB
HGW ... 82.1 EPNdB
Approach: BGW ... 91.6 EPNdB
HGW ... 91.5 EPNdB
Sideline: BGW ... 91.4 EPNdB
HGW ... 91.6 EPNdB

UPDATED

PHANTOM WORKS

PO Box 2515, Seal Beach, California 90740
Tel: (+1 562) 797 20 20
PRESIDENT: George Muellner
MANAGER, MEDIA RELATIONS: David Phillips

Tasked with improving Boeing's competitive position through use of innovative technologies, improved processes and creation of new products, the Phantom Works was originally established by McDonnell Douglas at St Louis. Technologies and expertise have expanded in support of company-wide activities, including commercial transport aircraft; this division had workforce of 4,700 in January 2003, with personnel at every major Boeing facility.

Phantom Works activities are spread over six sites: Huntsville (operations analysis, advanced military aircraft and missiles, manufacturing technology & prototyping and advanced support concepts), Mesa (advanced rotorcraft), Philadelphia (advanced rotorcraft), Seattle (advanced commercial aircraft and information & engineering technology), Southern California (advanced tankers and transports; space and communications) and St Louis (computing technology).

In addition to the programmes described in detail, Phantom Works is also involved with Naval Unmanned Combat Air Vehicle (UCAV-N), Solar Orbit Transfer Vehicle, X-31A Vector, X-37 Reusable Spaceplane, X-40 Space Maneuver Vehicle, X-43 Hyper-X and X-45 UCAV.

UPDATED

BOEING BLENDED WING BODY LARGE COMMERCIAL TRANSPORT

TYPE: New-concept airliner.

PROGRAMME: In 1996, shortly before absorption by Boeing, McDonnell Douglas released details of the BWB-1-1 (blended wing body) study giving the most comprehensive analysis yet of the flying wing's advantages for very large transports. At least half-a-dozen derivatives under consideration at the beginning of 2002, ranging in passenger capacity from 180 to 570. An earlier incarnation, capable of carrying 800 passengers over a 7,000 n mile (12,964 km; 8,055 mile) range, gave benefits that included a 15.2 per cent reduction in T-O weight and 12.3 per cent less empty weight; 27.5 per cent reduction in fuel burn; 27.0 per cent reduction in installed thrust; and 20.6 per cent better lift:drag. These achieved by increasing wing area by 27.9 per cent (reducing loading by 33.6 per cent) and having a 19.2 per cent greater span. Tests of a radio-controlled 5 m (17 ft) span scale BWB-1-1 were undertaken by Stanford University, California, at El Mirage dry lake in July 1997 as part of a US$2.3 million programme to evaluate flight control laws for a flying wing. Working in tandem with Cranfield Aerospace, Boeing expects to test a 17 per cent scale low-speed vehicle in the UK and the USA. Expected to fly for the first time in early 2004, this will have a span of about 9.15 m (30 ft), three small engines and a slightly redesigned planform incorporating a cranked-arrow wing leading-edge. In addition to flight testing, it will probably also undergo wind tunnel testing at NASA Langley, Virginia, or Ames, California.

One potential application is a BWB tanker for multipoint aerial refuelling. Equipped with three 'smart' booms, two hose/drogue refuelling points and automated refuelling capabilities, BWB tanker would be able to accommodate simultaneous air-to-air refuelling of multiple conventional aircraft or UAVs. Fuel carried in wing tanks; maximum payload space would be available for up to 23 conventional pallets and 40 troops.

As a C^2 ISR (command, control, intelligence, surveillance, reconnaissance) platform, the BWB would provide increased loiter time, large interior space suitable for battle management control rooms, and ample exterior locations for conformal phased-array antennas for broadband communications with no increase in radar signature. These capabilities make the BWB additionally suitable as a long-range standoff weapons platform.

Other possible military derivatives could include a version to satisfy the USAF's emerging C-X future cargo aircraft requirement, also known as the Air Mobility Command-X (AMC-X) project.

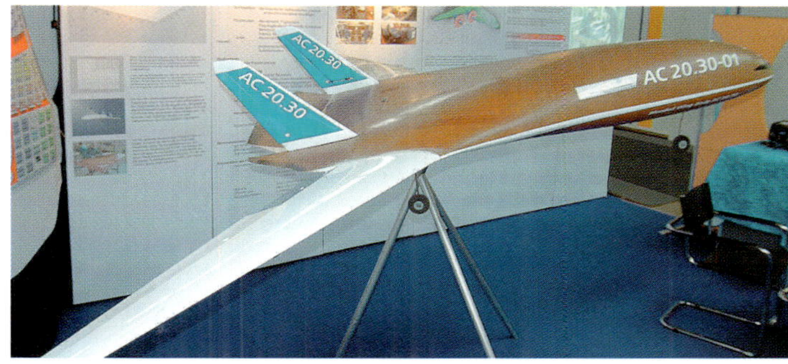

Research into the BWB configuration has been independently undertaken in Germany by Hamburg University of Applied Science, whose AC 20.30 model (1:30 scale) first flew on 16 December 2003. Studies predict service entry of a 900-seat BWB in 2030 *(Paul Jackson)* ***NEW***/0580878

Several possible BWB derivatives have emerged as candidates for production, including the **BWB-250**, with accommodation for about 260 passengers, and the **BWB-450** with room for about 480 passengers. *Estimated data below refer to the 800-seat Boeing BWB-1-1 proposal powered by three 275 kN (61,900 lb st) turbofans.*

DIMENSIONS, EXTERNAL:
Wing span: excl winglets ... 85 m (280 ft)
incl winglets ... 88 m (289 ft)
Wing aspect ratio ... 5.1
Length overall ... 49 m (161 ft)
Height overall ... 15 m (50 ft)

AREAS:
Wings: trap ... 728 m² (7,840 sq ft)
gross ... 1,423 m² (15,325 sq ft)

WEIGHTS AND LOADINGS:
Weight empty ... 167,750 kg (369,800 lb)
Weight empty, equipped ... 186,900 kg (412,000 lb)
Max payload ... 104,800 kg (231,000 lb)
Max fuel weight ... 122,500 kg (270,000 lb)
Max T-O weight ... 373,300 kg (823,000 lb)
Max zero-fuel weight ... 291,650 kg (643,000 lb)
Fuel burn over 7,000 n miles with 800 passengers ... 96,820 kg (213,450 lb)
Max wing loading ... 513 kg/m² (105 lb/sq ft)
Max power loading ... 452 kg/kN (4.4 lb/lb st)

PERFORMANCE:
Normal cruising speed ... M0.85
Max approach speed ... 150 kt (278 km/h; 173 mph) EAS
Initial cruising altitude ... 10,665 m (35,000 ft)
T-O field length ... 3,353 m (11,000 ft)
Range with 800 passengers ... 7,000 n miles (12,964 km; 8,055 miles)

UPDATED

BOEING X-50A DRAGONFLY CANARD ROTOR/WING

TYPE: New-concept rotorcraft.

PROGRAMME: The X-50A Canard Rotor/Wing (CR/W) began as a McDonnell Douglas Helicopters project for a VTOL reconnaissance and surveillance UAV in 1992; concept based on NASA-funded studies into high-speed rotorcraft and earlier (Hughes) rotor/wing studies. Design undertaken by Phantom Works personnel at Mesa, Arizona, under a March 1998 DARPA contract calling for the manufacture and testing of two prototypes. US Marine Corps is reported to have expressed interest in a manned version armed with missiles and other weaponry for use in the escort role.

As currently envisaged, it will possess VTOL attributes of the helicopter and the ability to fly like a conventional fixed-wing aircraft, with the addition of foreplanes being a key factor in bestowing this potential. Power will be provided by a single turbofan engine. In helicopter mode, exhaust gases will pass through channels in the two-blade rotor before exiting through vents in the blade tips. In aeroplane mode, as speed increases, exhaust efflux progressively transfers from driving the rotor to a nozzle at the rear, thus providing forward thrust.

With no tail rotor, directional control in helicopter mode will rely on what Boeing calls 'reaction drive', in which exhaust gases are diverted to nozzles on both sides of the tail; for aeroplane mode, conventional control surfaces will be provided, although consideration is being given to employing reaction jets for directional control in both modes of flight.

Following vertical take-off as a helicopter, the CR/W will accelerate to about 122 kt (225 km/h; 140 mph) before making the transition to aeroplane. This will be accomplished with the assistance of flap surfaces on the foreplanes and the aft-mounted wings. These will increase the amount of lift generated by the wings and reduce rotor loadings, whereupon the

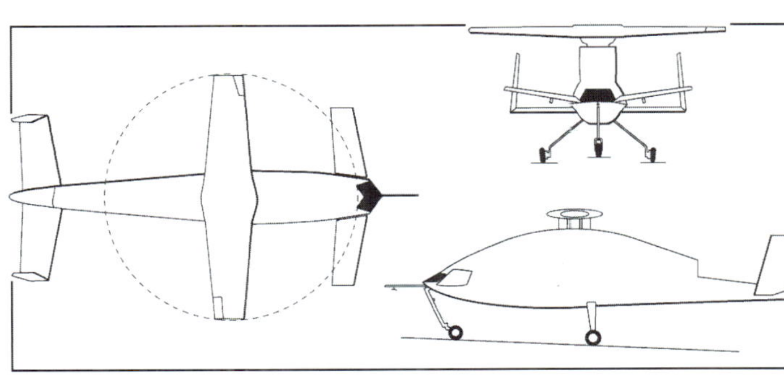

General arrangement of the Boeing X-50A Dragonfly demonstrator *(Paul Jackson)* ***NEW***/0546966

rotor will be slowed to a stop before being locked in position across the fuselage to act as an extra lifting surface. The flaps will then be retracted, with the three lifting surfaces sharing the load. For landing, the process will be reversed, enabling the CR/W to operate safely from confined areas, such as a carrier deck or helicopter platform. In conventional aeroplane mode, maximum speed is predicted to exceed 375 kt (695 km/h; 432 mph).

Initial technology demonstrator is X-50A subscale, unmanned aerial vehicle (UAV); first prototype was to have flown in 2001, but postponed; roll-out May 2002; maiden flight delayed, following discovery of high cyclic loads during ground testing that necessitated repairs to rotor bearings as well as adoption of a modified swashplate design and use of steel instead of aluminium for greater stiffness; eventually accomplished on 4 December 2003 at Yuma Proving Ground, Arizona. First flight lasted 80 seconds and involved vertical take-off, brief hover and vertical landing. Further test flights scheduled for January 2004 delayed by range conflict at Army proving ground, but were expected to resume in February. Test vehicle not to exceed 150 kt (278 km/h; 172 mph). X-50A will take-off vertically and operate as helicopter up to 60 kt (111 km/h; 69 mph), when canard and stabiliser begin to provide lift; transition to conventional wing-borne flight will occur at 120 kt (222 km/h; 138 mph). X-50A prototype is powered by a 3.11 kN (700 lb st) Williams F112 turbofan. Any subsequent production CR/W is likely to use advanced composites materials and be built in both manned and unmanned versions; Boeing studies include a 1,100 kg (2,425 lb) maritime UAV and a 10,000 to 11,000 kg (22,046 to 24,250 lb) manned aircraft that could function as an armed escort fighter in support of the MV-22B Osprey.

DIMENSIONS, EXTERNAL:
- Wing span/rotor diameter ... 3.66 m (12 ft 0 in)
- Foreplane span ... 2.71 m (8 ft 10¾ in)
- Length overall ... 5.39 m (17 ft 8 in)
- Height overall ... 1.98 m (6 ft 6 in)
- Tail unit span ... 2.47 m (8 ft 1¼ in)

WEIGHTS AND LOADINGS:
- Weight empty ... 574 kg (1,265 lb)
- Fuel weight ... 66 kg (146 lb)
- Max payload ... 91 kg (200 lb)
- Max T-O weight ... 645 kg (1,423 lb)

PERFORMANCE (estimated):
- Max level speed ... 150 kt (278 km/h; 172 mph)

UPDATED

BOEING ADVANCED TACTICAL TRANSPORT

TYPE: Medium transport/multirole.

PROGRAMME: Design study revealed by McDonnell Douglas (now Boeing) in September 1996 for a potential replacement for the C-130 Hercules. Original study envisaged aircraft with four 8,950 kW (12,000 shp) turboprops fixed to wing, which tilted 15° for take-off and 45° for landing; potential in-service date of 2020. Those studies provided basis for No-Tail Advanced Theater Transport (NOTAIL ATT) Tilt-wing Super Short Takeoff and Landing (SSTOL) or 'Super Frog' as project originally known. Advanced Medium Transport (AMT) name adopted by mid-2002, although had evolved further, into Advanced Tactical Transport (ATT), by early 2003.

Concept revealed in September 1998, with 39 m (128 ft) span aircraft designed to carry up to four times the payload of the C-130J Hercules and use runways as short as 183 m (600 ft). Baseline requirement is delivery of 27,215 kg (60,000 lb) load on to 229 m (750 ft) of rough airstrip, 1,220 m (4,000 ft) AMSL in 35°C (95°F) ambient temperature. Signature of co-operative research and development agreement (CRADA) between Boeing's Phantom Works and USAF Research Laboratory at Wright-Patterson AFB, Ohio, in December 1998 paved way for development and demonstration of 'enabling technologies' that could lead to production-configured aircraft; subsequently, in late 2000, Boeing also concluded development agreement with DARPA.

Fuselage interior width is 6.4 m (21 ft) and proposed ATT will be able to accommodate various loads up to a maximum of about 36,285 kg (80,000 lb), including one AH-64 Apache helicopter; or up to 10 cargo pallets; or 40 fully equipped soldiers. Wing features four widely separated podded turboprop engines, each driving an eight-blade propeller. Wing originally pivoted about lateral axis near rear spar, with tilt used to increase lift during take-off and landing. Directional control of tail-less design uses similar system to B-2 bomber, which has split flaps. By early 2000, however, design had progressed to forward-swept (7 to 9°) wing, pivoted at leading-edge and tilting downwards at trailing-edge, with further refinements by second quarter of 2001 adding small horizontal tail surfaces and adopting C-17-style fuselage.

Initial wind tunnel testing of subscale model completed by end of 1998, followed by further trials with a 7 per cent subscale model during 1999; latter included tethered and untethered tests which began at Gray Butte, California, on 5 June. Further wind tunnel testing of latest configuration completed in January 2001, with Boeing planning to conduct manned simulations and powered wind tunnel tests with larger models in 2001–03. In first quarter of 2003, Boeing revealed that it was considering resurrecting a McDonnell Douglas YC-15 prototype and modifying it to serve as a tilt-wing technology demonstrator. The most significant alterations would involve installation of a forward-cranked wing that could rotate the trailing-edge downwards by up to 20° and the switch from turbofan to turboprop engines. If it goes ahead, the demonstrator's primary objectives would be concerned with proving the viability of the tilt-wing super STOL transport aircraft concept and validating use of cyclic controls on a fixed-wing aircraft.

Boeing believes that production aircraft could be operational in 2015, with ATT a potential candidate for USAF MC-X special operations aircraft requirement.

UPDATED

SHARED SERVICES GROUP

2810 160th Avenue SE, Bellevue, Washington 98008
Tel: (+1 425) 865 51 66
Fax: (+1 425) 865 29 58
PRESIDENT: Richard Stephens
PUBLIC RELATIONS DIRECTOR: Karen H Burt

Shared Services Group provides information management services, personnel management and training, travel services and computing resources to Boeing operating divisions and government customers on a worldwide basis. It has over 18,000 personnel worldwide.

UPDATED

BOEING BUSINESS JETS

PO Box 3707, Seattle, Washington 98124-2207
Tel: (+1 206) 655 98 00
Fax: (+1 206) 655 97 00
e-mail: business.jets@boeing.com
PRESIDENT: Steven Hill
COMMUNICATIONS DIRECTOR: Sandra Augers
MARKETING DIRECTOR: Charles Colburn

PARTICIPATING COMPANIES:
Boeing Company: see this entry
General Electric Company: see *Jane's Aero Engines*

In July 1996 The Boeing Company and The General Electric Company announced formation of a joint venture, Boeing Business Jets, to develop and market corporate versions of the Next-Generation 737, deliveries of which began in 1998.

Company headquarters is at the corporate air services facility of Boeing Field's Flight Center.

UPDATED

BOEING BUSINESS JET (BBJ)

US Air Force designations: C-40B and C-40C

TYPE: Large business jet.

PROGRAMME: Launched July 1996. Aircraft are assembled at Boeing Commercial Airplane Group's Renton facility and supplied to Boeing Business Jets, which hands them over in 'green' condition to customers and delivers them to DeCrane Aircraft at Georgetown, Delaware, for installation of long-range fuel tanks before the aircraft are delivered to the customer's chosen completion centre for interior outfitting and painting; designated completion centres are DeCrane Aircraft and Associated Air Center in Dallas, Texas; Greenpoint Technologies in Seattle, Washington; Arkansas; Lufthansa Technik in Hamburg, Germany; and Jet Aviation in Basle, Switzerland, but other completion centres may be used at the discretion of the customer. After completion, Boeing ferries the aircraft to the customer's base and carries out crew training.

First BBJ (101st N-G 737, N737BZ) rolled out 26 July 1998; first flight 4 September 1998; FAA and JAA certification achieved 29 October 1998; supplementary type certificate for long-range tanks awarded 20 May 1999, following demonstration non-stop flight of 6,252.5 n miles (11,580 km; 7,200.4 miles) in 13 hours 57 minutes 42 seconds. First delivery 23 November 1998.

US Air Force ordered first of up to seven C-40Bs in February 2001 to replace Air National Guard C-22s (Boeing 727s) from August 2003 onwards. Initial aircraft (01-0005) completed late 2002; two more followed by December 2002 and fourth (02-0042) in June 2004, serving with 1 AS of 89 AW at Andrews AFB, Washington DC (two), and 65 AS of 15 Airlift Wing, Hickham, Hawaii. C-40C designation assigned to two former Ford Air Services BBJs acquired in mid-2002 as (02-0201 and -0202) and issued to 201 AS of District Columbia Air National Guard at Andrews AFB, although nominally operated by BCC Equipment Leasing Corp. Third (02-0203) added in June 2004 from BCC. C-40B has typical accommodation for 11 crew and 26 passengers.

CUSTOMERS: Peak of 85 firm orders by December 2001, at which time 67 'green' airframes had been delivered and 43 completed aircraft were in customer service. By mid-2003 total of 73 BBJs and nine BBJ 2s on order including two BBJs in first half 2003. By October 2003, 83 BBJ/BBJ2s sold and same number delivered, of which 68 outfitted and in service. Launch customer General Electric ordered two in July 1996, of which first (N366G) flew 23 October 1993 and delivered 23 November 1998 for outfitting. First delivery of a completed aircraft to Dubai Air Wing Royal Flight 4 September 1999. Table of customers and deliveries accompanies this entry.

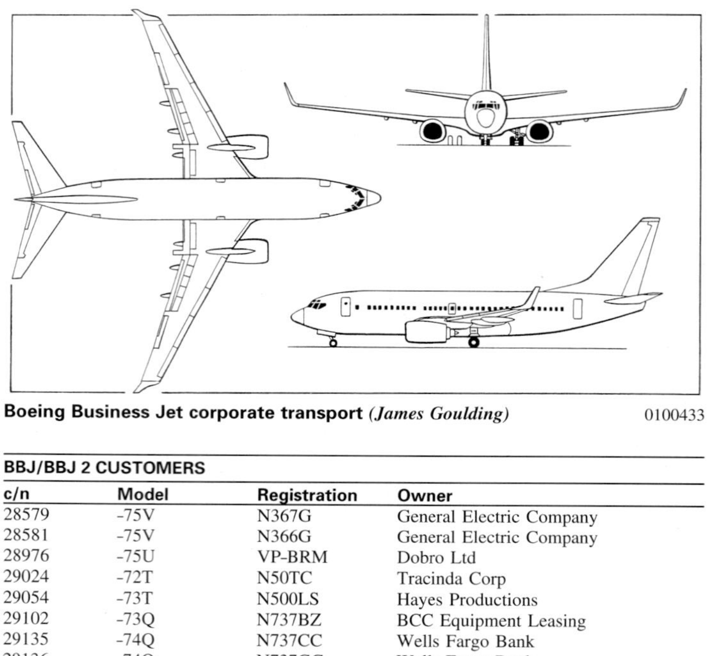

Boeing Business Jet corporate transport *(James Goulding)* 0100433

BBJ/BBJ 2 CUSTOMERS

c/n	Model	Registration	Owner
28579	-75V	N367G	General Electric Company
28581	-75V	N366G	General Electric Company
28976	-75U	VP-BRM	Dobro Ltd
29024	-72T	N50TC	Tracinda Corp
29054	-73T	N500LS	Hayes Productions
29102	-73Q	N737BZ	BCC Equipment Leasing
29135	-74Q	N737CC	Wells Fargo Bank
29136	-74Q	N737GG	Wells Fargo Bank
29139	-74T	VP-BEL	Magenta Aviation
29142	-75T	N737WH	Southern Air
29149	-7H3	TS-IOO	Government of Tunisia
29188	-7P3	HZ-TAA	HRH Talal bin AbdulAziz
29200	-73U	N742PB	Chartwell Partners

Boeing C-40C military version of BBJ *(Paul Jackson)* ***NEW***/1047786

BBJ/BBJ 2 CUSTOMERS

c/n	Model	Registration	Owner
29233	-74U	N4AS	BA Leasing & Capital
29251	-7EO	A6-HRS	Dubai Air Wing
29268	-7Z5	A6-AIN	Government of Abu Dhabi
29269	-7Z5	A6-RJZ	Government of Abu Dhabi
29272	-74V	N7378P	Raytheon Company; for Colombian Government
29273	-72U	VP-BBJ	Picton II (Switzerland)
29274	-7H6	M53-01	Malaysian Air Force
29317	-79T	VP-BWR	BEL Air
29441	-79U	N1011N	Aircraft Holdings
29749	-7AH	N134AR	BB Five Inc
29791	-7BH	P4-TBN	Tango Holding
29857	-7Z5	A6-LIW	Government of Abu Dhabi
29858	-7Z5	A6-DAS	Government of Abu Dhabi
29865	-7AK	HB-IIO	Privat Air
29866	-7AK	HB-IIP	Privat Air
29971	-7DM	01-0040	US Air Force (1 AS/89 AW)
29972	-7AN	VP-BYA	Eastern Skies
30031	-7AW	VP-CEC	BB Aviation
30070	-7AV	N889NC	News America
30076	-7BJ	VP-BBW	GAMA Aviation
30327	-7BC	N127QS	Wells Fargo Bank
30328	-7BC	N164RJ	Wells Fargo Bank
30329	-7BC	N129QS	Wells Fargo Bank
30330	-7BC	VP-BJB	Sigair
30496	-7BF	N737AG	Fun Air Corp
30547	-7BQ	HZ-DG5	Dallah al Bakara
30572	-7BC	N171QS	Wells Fargo Bank
30751	-7CG	P4-GJC	GKW Aviation
30752	-7CN	D-ABPA	Privat Air
30753	-7CP	02-0202	US Air National Guard (201 AS/DCANG
30754	-7CJ	N737ER	BBJ One Inc
30755	-7CP	02-0201	US Air National Guard (201) AS/DCANG)
30756	-7BC	N156QS	Wells Fargo Bank
30772	-7CU	N315TS	State Street Bank & Trust
30782	-7BC	N515GM	Kevin Air
30789	-73Q	N349BA	Boeing Aircraft Holding Inc
30790	-7DT	A36-002	Royal Australian Air Force
30791	-7BC	N191QS	BNJ Sales
30829	-7DT	A36-001	Royal Australian Air Force
30884	-7BC	VP-BFA	Jordanian Government
32438	-8AN	VP-BHN[1]	Eastern Skies
32450	-8EC	A6-MRM[1]	Government of Dubai Air Wing
32451	-8DP	HZ-101[1]	Saudi Air Force
32575	-7BC	PP-BSS	Grupo Safra
32627	-7ED	ZS-RSA	South African Government
32628	-7BC	N7600K	SAS Institute
32774	-7EJ	XA-AEX	Omniflys
32775	-7EL	N90R	Swiftlite Aircraft Corp
32777	-8DR	G-OBBJ[1]	Multi Flight Ltd
32805	-7DP	HZ-102	Saudi Air Force
32806	-8AW	VP-CBB[1]	Bosco Aviation
32807	-7EG	HL-7770	Samsung Aerospace Industries
32825	-8AJ	n/k[1]	Boeing
32915	-8DV	VP-BZL[1]	BZL Ltd
32916	-7DM	01-0015	US Air Force (65 AS/15 ABW)
32970	-7BC	N707BZ	Boeing
32971	-8EF	N371BC[1]	Executive Jets
33010	-7ET	N313P	Premier Executive Transport Services
33036	-7BC	N888YV	Aviation Air
33079	-8EV	EW-001PA[1]	Belarus Government
33080	-7DM	01-0005	US Air Force (1 AS/89 AW)
33102	-7BC	N108MS	Yona Aviation
33361	-8EQ	N7375P[1]	Wells Fargo Bank
33367	-7FB	3C-EGE	Equatorial Guinea Government
33405	7FG	HZ-MF1	Saudi Ministry of Finance
33434	-7BC	02-0203	US Air Force
33473	-8EX	A6-AUH[1]	Amiri Flight Line
33474	-7ES	A30-001	Australia Air Force (AEW & C)
33499	-7AJ	HZ-MF2	Saudi Ministry of Finance
33500	-7FD	02-0042	US Air Force
33542	-7AJ	A30-002	Australia Air Force (AEW & C)

Notes: Some aircraft have subsequently been sold
The BBJ has been proposed for USAF's Commander-in-Chief (CINC) support aircraft requirement.

[1] BBJ 2

BBJ/BBJ 2 ORDERS AND DELIVERIES

Year	BBJ orders	BBJ deliveries	BBJ 2 orders	BBJ 2 deliveries
1996	3			
1997	25		1	
1998	17	7		
1999		25		
2000	6	10	3	6
2001	14	13	4	
2002	6	4		1
Totals	**71**	**59**	**8**	**7**

COSTS: US$43 million 'green', estimated US$51 million to 55 million typically equipped (2005). Direct operating cost estimated at US$1,700 per hour based on operation within the USA and utilisation of 900 hours per year.

DESIGN FEATURES: Combines fuselage of 737-700, strengthened in aft section, with centre-section, wing and landing gear of 737-800. Aviation Partners Inc winglets standard, affording 5 to 7 per cent reduction in cruise drag, resulting in four to five per cent increase in range; winglets evaluated in mid-1998 by 737-800 (N737BX), first flown on BBJ prototype (N737BZ) on 20 February 1999, received FAA approval on 6 September 2000 and fitted as standard.

POWER PLANT: Two CFM International CFM56-7 turbofans, each rated at 121.4 kN (27,300 lb st). Standard N-G 737 fuel of 26,025 litres (6,875 US gallons; 5,725 Imp gallons) contained in wing, plus between three and nine belly tanks; maximum combined capacity 40,582 litres (10,721 US gallons; 8,927 Imp gallons).

ACCOMMODATION: To customer's choice; operating weights based on allowance of 5,624 kg (12,400 lb). Typical configuration includes forward lounge and private suite with double bed; mid-section conference room; 12 first class sleeper seats at 152 cm (60 in) pitch in two rows with centre aisle, and galley, lavatory and service area at rear, with crew rest area, galley and lavatory aft of flight deck. Alternative arrangements provide for exercise room/gymnasium, office, 24 first-class sleeper seats or high-density seating for up to 63 passengers, three abreast in two rows. Maximum 149 passengers in airline configuration. Cabin equivalent altitude reduced to 1,980 m (6,500 ft) from 2004, with retrofit available.

AVIONICS: Rockwell Collins Series 90 as core system.

Comms: Triple VHF comm with 8.33 kHz channel spacing; dual HF comm; L-3 Communications 120-minute CVR and Coltech Selcal.

Flight: Dual Rockwell Collins multimode GPS/ILS/VOR/DME receivers; dual ADF; TCAS II; predictive windshear; dual Smiths Industries flight management computers; dual Honeywell ADIRU; Honeywell EGPWS; L-3 Communications FDR and CVR; Flight Dynamics HGS 4000 head-up guidance system; Teledyne airborne navigation data recorder, digital flight data acquisition unit and quick-access recorder; Teledyne Navlink, including two additional navigation computers and electronic standby artificial horizon.

Instrumentation: Honeywell flat-panel LCD displays.

Mission: C-40B equipment includes fibre-optic communications management system; worldwide secure voice and data transmission (UHF, VHF, HF, UHF and commercial satcom, Magnastar airphones, UHF/AM, Boeing Connexion data/video Broadband, secure conferencing and Wideband data and LAN.

DIMENSIONS, EXTERNAL AND AREAS: As for 737-700 except:
Wing span, incl winglets ..... 35.79 m (117 ft 5 in)

DIMENSIONS, INTERNAL:
Cabin: Length ..... 24.13 m (79 ft 2 in)
Height ..... 2.16 m (7 ft 1 in)
Width ..... 3.53 m (11 ft 7 in)

Floor area ... 75.0 m^2 (807 sq ft)
Volume ... 148.7 m^3 (5,250 cu ft)

WEIGHTS AND LOADINGS:
Operating weight empty, typically equipped ... 43,082 kg (94,980 lb)
Interior completion allowance ... 5,625 kg (12,400 lb)
Max fuel weight (incl supplementary tanks) ... 32,825 kg (72,367 lb)
Max T-O weight ... 77,565 kg (171,000 lb)
Max ramp weight ... 77,790 kg (171,500 lb)
Max landing weight ... 60,780 kg (134,000 lb)
Max zero-fuel weight ... 57,155 kg (126,000 lb)
Max wing loading ... 620.5 kg/m^2 (127.09 lb/sq ft)
Max power loading ... 320 kg/kN (3.14 lb/lb st)

PERFORMANCE:
Max operating Mach No. (MMO) ... 0.82
Cruising speed: normal ... M0.80
long range ... M0.79
Approach speed ... 132 kt (244 km/h; 152 mph)
Max rate of climb ... 980 m (3,215 ft)/min
Initial cruising altitude ... 11,580 m (38,000 ft)
Max certified altitude ... 12,500 m (41,000 ft)
Service ceiling, OEI ... 7,070 m (23,200 ft)
T-O field length, S/L:
fuel for range of 4,000 n miles (7,408 km; 4,603 miles) ... 1,369 m (4,490 ft)
fuel for range of 5,000 n miles (9,260 km; 5,754 miles) ... 1,515 m (4,970 ft)
fuel for range of 6,000 n miles (11,112 km; 6,905 miles) ... 1,765 m (5,790 ft)
Landing run at typical landing weight ... 706 m (2,315 ft)
Range (nine belly tanks):
with 8 passengers ... 6,200 n miles (11,482 km; 7,134 miles)
with 25 passengers ... 5,935 n miles (10,991 km; 6,829 miles)
with 50 passengers ... 5,365 n miles (9,936 km; 6,173 miles)

OPERATIONAL NOISE LEVELS (FAR Pt 36 Stage 3):
T-O ... 85.6 EPNdB
Approach ... 95.9 EPNdB
Sideline ... 95.2 EPNdB

UPDATED

BOEING BUSINESS JET 2 (BBJ 2)

TYPE: Large business jet.

PROGRAMME: Launched 11 October 1999; rolled out 30 January 2001; first flight 9 February 2001; first 'green' delivery 3 March 2001 to an undisclosed customer; first service entry was in early 2002; forecast production rate eight per year. Certified by FAA 13 March 1998 and by JAA (EASA) 9 April 1998 (Boeing 737-800).

CUSTOMERS: BBJ 2 expected to account for some 25 per cent of all BBJ sales. Total of nine firm orders received by mid-2003, including one in first half of that year. Six in service and further two outfitting by October 2003. Refer to table in BBJ entry.

COSTS: US$53 million, 'green'; estimated US$63 million to US$68 million, typically equipped (2005).

DESIGN FEATURES: Based on 737-800 airframe (which see), affording 25 per cent more cabin volume and twice the cargo volume of the BBJ; Aviation Partners Inc winglets standard.

POWER PLANT: Two CFM56-7 turbofans, each 121.4 kN (27,300 lb st). Standard 737 'Next-Generation' fuel of 26,025 litres (6,875 US gallons; 5,725 Imp gallons) contained in wing, plus between three and seven belly tanks; maximum combined capacity 39,466 litres (10,426 US gallons; 8,681 Imp gallons).

ACCOMMODATION: Up to 78 passengers, with executive lounge and private suite. Maximum 189 in airliner configuration. Operating weights based on completion allowance of 7,257 kg (16,000 lb).

DIMENSIONS, EXTERNAL: As for BBJ, except:
Length overall ... 39.47 m (129 ft 6 in)

DIMENSIONS, INTERNAL:
Cabin: Length ... 29.97 m (98 ft 4 in)
Floor area ... 93.27 m^2 (1,004 sq ft)
Max cargo volume ... 34.7 m^3 (1,224 cu ft)

WEIGHTS AND LOADINGS:
Operating weight empty, typically equipped ... 46,226 kg (101,910 lb)
Max fuel weight (incl supplementary tanks) ... 31,922 kg (70,376 lb)
Max T-O weight ... 79,015 kg (174,200 lb)
Max ramp weight ... 79,245 kg (174,700 lb)
Max landing weight ... 66,360 kg (146,300 lb)
Max zero-fuel weight ... 62,730 kg (138,300 lb)

PERFORMANCE:
Max cruising speed ... M0.82
Long-range cruising speed ... M0.79
Max rate of climb at S/L ... 948 m (3,110 ft)/min
Initial cruising altitude ... 11,505 m (37,750 ft)
Max certified altitude ... 12,500 m (41,000 ft)
Service ceiling, OEI ... 6,090 m (22,600 ft)
T-O field length, S/L:
fuel for range of 4,000 n miles (7,408 km; 4,603 miles) ... 1,655 m (5,430 ft)
fuel for range of 5,000 n miles (9,260 km; 5,753 miles) ... 1,915 m (6,280 ft)
fuel for range of 5,650 n miles (10,464 km; 6,502 miles) ... 2,118 m (6,950 ft)
Landing run at typical landing weight ... 758 m (2,485 ft)
Range (seven belly tanks):
with 8 passengers ... 5,650 n miles (10,463 km; 6,501 miles)
with 25 passengers ... 5,315 n miles (9,843 km; 6,116 miles)
with 50 passengers ... 4,780 n miles (8,852 km; 5,500 miles)

OPERATIONAL NOISE LEVELS (FAR Pt 36 Stage 3):
T-O ... 86.0 EPNdB
Approach ... 96.3 EPNdB
Sideline ... 94.7 EPNdB

UPDATED

Interior of BBJ2 completed by Lufthansa Technik *(Gregor Schläger)* ***NEW***/0567228

BOEING SIKORSKY

BOEING COMPANY AND SIKORSKY AIRCRAFT

PARTICIPATING COMPANIES:
The Boeing Company (USA)
Sikorsky Aircraft (USA)
Web: http://www.rah66comanche.com
VICE-PRESIDENT AND RAH-66 PROGRAMME DIRECTOR: Charles 'Chuck' Allen
ARMY PROGRAMME MANAGER: Colonel Michael Cantor
COMMUNICATIONS DIRECTOR: Ed Steadham

Boeing and Sikorsky began collaboration on what became the RAH-66 in June 1985 and received a contract for the demonstration/validation programme in 1991 In early 2002, Boeing Sikorsky announced selection of Bridgeport, Connecticut, as final assembly location for the production RAH-66; also in 2002, the Joint Program Office moved from Huntsville, Alabama, to Bridgeport. Subsequently, on 23 February 2004, the Pentagon announced it was cancelling the project on grounds of rising cost and concerns over performance.

UPDATED

YRAH-66 Comanche with modified fin and added dummy radar housing
0110063

BOEING SIKORSKY RAH-66 COMANCHE

TYPE: Attack helicopter.

DEVELOPMENT MILESTONES

Requirement issued	1981
Request for proposals	21 Jun 88
Programme launched	5 Apr 91
First prototype material cut	Sept 93
Rolled out	25 May 95
First flight	4 Jan 96
Programme cancelled	23 Feb 04

PROGRAMME: Light Helicopter Experimental (LHX) design concepts requested by US Army 1981; numerous changes throughout life of programme; original plan for 5,000 to replace UH-1, AH-1, OH-58 and OH-6; reduced in 1987 to 2,096 scout/attack only, replacing 3,000 existing helicopters; further cut to 1,292 in 1990 (with another 389 possible) and then to 1,096 in 1999, although subsequently raised to 1,213 (including eight for US Army operational evaluation) by mid-2000. Under latter procurement plan, initial production examples were to have been to so-called Block I standard for armed reconnaissance; definitive Block II standard was to follow with heavy attack capability; Block III was expected to introduce added mission capabilities, including sensor fusion and AAMs. Final restructuring of programme resulted in further cut in planned procurement to 650 over 12 year period, according to defence acquisition review completed on 7 October 2002; Army then planned to field Comanche in Blocks incorporating incremental improvements with low-rate initial production to begin in 2007. Three LRIP batches, totalling 73 aircraft, were to be built to Block I standard, followed by 108 Block II helicopters, whereupon Block III was to be introduced. However, on 23 February 2004, US Department of Defense revealed intention to cancel programme in entirety; this action necessitated writing off approximately US$6.9 billion already invested in development and probable payment of up to US$4 billion more in cancellation fees to Boeing and Sikorsky.

LHX request for proposals issued 21 June 1988; 23 month demonstration/validation contracts to Boeing Sikorsky and Bell/McDonnell Douglas. Boeing Sikorsky selected 5 April 1991; to build four YRAH-66 demonstration/validation prototypes in 78 month programme, plus static test article (STA) and propulsion system testbed (PSTB).

LHTEC T800 engine specified October 1988. LHX designation changed to LH early 1990, then US Army designation RAH-66 Comanche in April 1991. Several versions of programme timetable were considered prior to that currently implemented.

Prototype critical design review, completed in December 1993, authorised production of three YRAH-66 prototypes (first item for which manufactured in September 1993). At same time, however, further R&D economies under study; December 1994 decision reduced dem/val phase to two prototypes (lacking Longbow/Hellfire capability).

Prototype construction began 29 November 1993 with forward fuselage at Sikorsky, Stratford; Boeing built aft fuselage in Philadelphia. STA airframe delivered to Stratford 1994, at which time PSTB under construction there. PSTB trials commenced in 1995 at West Palm Beach, with 100 per cent torque from both engines achieved during first 10 hours of running. PSTB subsequently suffered failure of left input bevel gear, which disintegrated and punched hole in main gearbox housing during 110 per cent power test; resonance blamed for failure.

Front and rear sections of prototype joined at Stratford 25 January 1995; completed helicopter (94-0327) rolled out 25 May 1995. Following transfer to Sikorsky's Development Flight Test Center in West Palm Beach, Florida, during June 1995, first flight accomplished on 4 January 1996. Prototype retired from flight test duty on 30 January 2002, by which time it had accumulated 387.1 flight hours in 318 sorties.

Aft fuselage section of second prototype (95-0001) delivered by Boeing to Stratford in early December 1996 for mating with forward fuselage; completed helicopter exhibited at Army Aviation Association's annual meeting in April 1998 and then to West Palm Beach. Made international debut when displayed statically at Farnborough Air Show in September

Boeing Sikorsky YRAH-66 Comanche combat helicopter 0525826

1998; flew for first time on 30 March 1999. Completed initial test schedule in April 1999, recording 4.9 hours in five sorties before temporary lay-up, due to funding constraints; also used for vertical rate of climb demonstration later in year and was scheduled to test integrated mission equipment package (MEP), including digital avionics, communications, navigation and target acquisition systems. By mid-December 2000, had logged almost 53 flight hours in 50 sorties; these figures had risen to 93 and 103.5 respectively in May 2001 when it was removed from flight status to be prepared for flight testing and validation of MEP. This phase of development began on 23 May 2002, when second prototype made first flight with MEP and new engines. Near-term objectives to be achieved by the second prototype included flight with the night vision pilotage system by October 2002, as well as completion of total weapon system critical design review in May 2003, including the Lockheed Martin Electro-Optical Sensor System (EOSS); this accomplished in early June 2003, although some discrepancies remained to be resolved later in the year.

Engineering and Manufacturing Development (EMD) officially began 1 June 2000, following RAH-66 meeting (on 4 April 2000) seven key Defense Acquisition Board Milestone 2 criteria, including a 107 m (350 ft)/min vertical climb rate, a specified detection range for the FLIR sensors, a radar cross-section specification, ballistic vulnerability and tolerance specifications and tower-testing of the selected FCR. Weight reduction effort under way in late 2000, to reduce from about 4,310 kg (9,500 lb) to target weight of 4,218 kg (9,300 lb).

Under original plan, EMD expected to take six years and include production of five RAH-66s specifically for EMD testing, followed by further eight for initial operational test and evaluation (IOT&E) by the US Army. EMD contract and plan restructured in mid-2002, at which time IOC expected in September 2009. The new plan included a start of low-rate initial production (LRIP) in 2007, with full-rate production of 60 per year from 2011–12 onwards. In meantime, second YRAH-66 would assume increasing burden of test duty. Production of components for first EMD RAH-66 began at Boeing's Philadelphia factory in early 2003, with work on assembling the first empennage starting on 21 April 2003; on completion, this shipped to Bridgeport and mated with Sikorsky-produced elements. Final assembly of first systems development and demonstration (SDD) helicopter began in August 2003; this then expected to fly in June 2005.

First -801 growth version of T800 turboshaft began bench runs in March 1994; -801 preliminary design review completed May 1993; critical design review March 1995; prototypes originally fitted with less powerful T800-LHT-800 engines, but first flight with definitive -801 engine made on 1 June 2001 by first prototype. Same engine subsequently installed on second prototype in time for resumption of flight test duty in May 2002.

CUSTOMERS: Before cancellation, programme expected to involve two prototypes, plus five SDD helicopters and four initial operational test and evaluation (IOT&E) aircraft for test duties, operational trials and training tasks, followed by planned procurement of 650 production aircraft by US Army.

COSTS: US$34,000 million programme, including US$1,960 million dem/val and US$900 million FSD but reduced to US$2,240 million dem/val/FSD between 1993 and 1997 by cancellation of three of six planned prototypes; US$8.9 million flyaway unit cost (1988 values), increased to US$13 million by early 1993. By 1993 (in 1994 dollars, estimated), procurement unit cost US$21 million, programme unit cost US$27 million. Appropriations for R&D in FY97-99 comprised US$338.6 million, US$282.0 million and US$367.8 million, with US$467 million in FY00 and US$614 million for FY01. EMD phase then projected to cost US$3.2 billion, with initial contract worth US$147.5 million awarded to Boeing Sikorsky on 1 June 2000. Under most recent programme plan (mid-2002), development costs expected to rise to about US$6.5 billion; unit cost then forecast to exceed US$32 million, although by late 2003 unit cost had reportedly risen to US$58 million.

DESIGN FEATURES: First combat helicopter designed from outset to have 'stealth' features and target acquisition radar. Embodied low-observables (LO) attributes and stated to have radar cross-section (RCS) lower than that of Hellfire missile; frontal RCS reportedly 360 times smaller than AH-64, 250 times smaller than OH-58D and 32 times smaller than OH-58D with mast-mounted sight. Also had quarter of AH-64D's IR emissions and was six times quieter, head-on.

RAH-66 lighter, but only slightly smaller, than AH-64 Apache; specified empty weight of 3,402 kg (7,500 lb) increased to 3,522 kg (7,765 lb) by early 1992, as result of Army add-ons, including allowance for Longbow radar. Design had faceted appearance for radar reflection; downward-angled engine exhausts; T tail with endplates; eight-blade fan-in-fin shrouded tail rotor; and five-blade all-composites bearingless main rotor system, with latter to be increased in diameter by 0.3 m (1 ft 0 in) and gain noise-reducing anhedral tips on EMD aircraft. New rotor incorporating anhedral tips flown for first time on first prototype on 20 July 2001. RAH-66 also featured internal weapon stowage.

Split torque transmission, obviating need for planetary gearing. Upper part of T tail folded down for air transportation. Detachable stub-wings for additional weapon carriage and/or auxiliary fuel tanks (EFAMS: external fuel and armament management system). Radar, infra-red, acoustic and visual signature requirements were set to defeat threats postulated by US Army.

Subcontractors to Boeing Sikorsky included BAE Systems (flight control computer, controller grips), Boeing (flight control computer), Hamilton Sundstrand (environmental control system, air vehicle interface controller, air data system, electrical power generation, control and distribution system), Harris Corporation (3-D digital map, controls and displays, super high-speed fibre optic databusses), Hughes-Link (operator training systems), ITT Industries (radar and laser warning receivers and point chemical detector), Kaiser Electronics (helmet integrated display sight system), Litton Guidance and Control Systems (PAGAN GPS navigation), Lockheed Martin Armament Systems (turreted gun system and ammunition feed system), Lockheed Martin Electronics and Missiles (target acquisition/designation system and night vision pilotage system), Moog (flight control

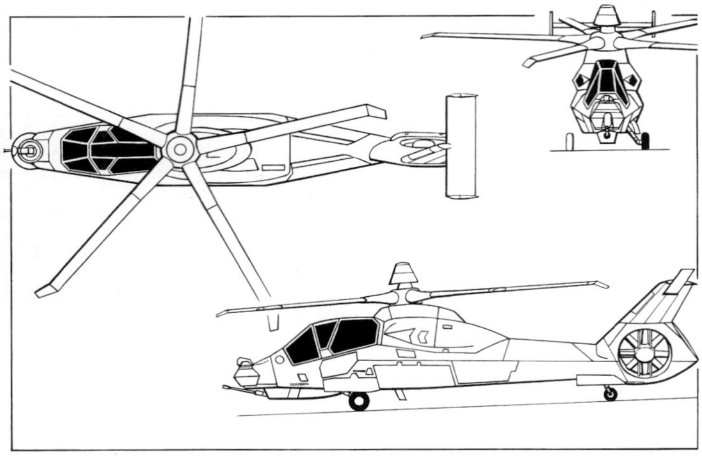

Boeing Sikorsky RAH-66 Comanche *(James Goulding)* 0525828

actuators), Northrop Grumman (signal and data processors, aided target detection/classification systems), Smiths Industries (crash survivable memory unit), TRW Military Electronics and Avionics (communication, navigation and identification system and survivability systems) and Williams (subsystem power unit).

FLYING CONTROLS: Dual triplex fly-by-wire, with sidestick cyclic pitch controllers and normal collective levers. Main rotor blades removable without disconnecting control system.

STRUCTURE: Largely composites airframe and rotor system. Fuselage built around composites internal box-beam; non-load-bearing skin panels, more than half of which hinged or removable for access to interior. Eight-blade Fantail rear rotor designed to be operable with 12.7 mm calibre bullet hits; or for 30 minutes with one blade missing. Main rotor blades and tail section by Boeing, forward fuselage and final assembly by Sikorsky.

LANDING GEAR: Tailwheel type, retractable, with single wheel on each unit; main units retracted aft, with tailwheel retracting forward. Main units could 'kneel' for air transportability.

POWER PLANT: Two LHTEC T800-LHT-801 turboshafts, each rated at 1,165 kW (1,563 shp). Transmission rating 1,639 kW (2,198 shp). Internal fuel capacity 1,142 litres (301.6 US gallons; 251.1 Imp gallons). Two external tanks totalling 3,407 litres (900 US gallons; 749.4 Imp gallons) for self-deployment; total fuel capacity 4,548 litres (1,201.6 US gallons; 1,000.5 Imp gallons). Additional fuel contained in two 424 litre (112 US gallon; 93.3 Imp gallon) tanks in side weapon bays. Main rotor tip speed (100 per cent N_r) 221 m (725 ft)/s; 355 rpm; normal operation from 95 per cent (quiet mode) to 107 per cent (load factor enhancement).

ACCOMMODATION: Pilot (in front) and WSO in identical stepped cockpits, pressurised for chemical/biological warfare protection. Crew seats designed to tolerate 11.6 m (38 ft/s) vertical crash landing.

SYSTEMS: Williams International subsystem power unit (SPU), mounted aft of starboard engine, drove hydraulic pump for actuation of landing gear and weapons bay doors; further two hydraulic pumps provided pressure for flight control systems.

AVIONICS: *Comms:* Package known as Comanche Integrated Communications, Navigation and Identification (ICNIA) systems was to have comprised dual anti-jam VHF-FM and UHF-AM Have Quick tactical communications, VHF-AM, anti-jam HF-SSB; IFF.

Radar: Miniaturised version of Longbow fire-control radar was planned to be installed in one-third of fleet although all were expected to have carriage provision.

Flight: Litton PAGAN jam-resistant four-channel GPS and radar altimeter. Litton AHRS, with two fibre optic LN-210C and one LN-100C gyro platforms.

Instrumentation: Kaiser Electronics Helmet Integrated Display Sighting System (HIDSS); NVG-compatible lighting; integrated cockpit, night vision pilotage systems, and digital map display. Two 152 × 203 mm (6 × 8 in) multifunction flat-screen LCDs in each cockpit (left in monochrome for FLIR/TV, right in colour for moving map, tactical situation and night operations), plus two 89 × 185 mm (3½ × 7¼ in) multipurpose display (MPD) flat-screen monochrome LCDs for fuel, armament and communications information in each cockpit. Microvision awarded contract by US Army in 1998 to provide a prototype helmet-mounted display incorporating virtual retinal-display technology as potential alternative to conventional helmet-mounted displays. Three redundant databusses: one low-speed (MIL-STD-1553B), one high-speed and one very high-speed (fibre optic-based) for signal data distribution.

Mission: Airborne target handover system. Laser range-finder/designator. Nose-mounted Lockheed Martin electro-optical target acquisition system with two second-generation FLIR sensors and a solid-state TV camera. Northrop Grumman TASS (Target Acquisition System Software).

Self-defence: Goodrich Electro-Optical Systems AN/AVR-2A(V) Advanced Laser Warning Receiver, chemical warning and AN/ALQ-211(V)3 radar warning receivers; RF and IR jammers.

EQUIPMENT: Side armour for cockpit fitted as standard; optional armour kit planned for floor of cockpit.

ARMAMENT: General Dynamics stowable XM-301 three-barrel 20 mm cannon in Giat undernose turret, with up to 500 rounds (320 rounds normal for primary mission) and either 750 or 1,500 rounds per minute firing rates. Aiming coverage of gun +15 to −45° in elevation and ±120° in azimuth. Integrated retractable aircraft munitions system (IRAMS) featured side-opening weapons bay door in each side of fuselage, on each of which could be mounted up to three Hellfire or six Stinger missiles or other weapons. Four more Hellfires or eight Stingers were to be deployed from multiple carriers under tip of each optional stub-wing, or auxiliary fuel tank for self-deployment. Planned to be compatible with Starstreak and Mistral air-to-air missiles. Maximum of 56 Hydra 70 2.75 in FFARs or Sura D or Snora 81 mm equivalents. All weapons fired, and targets designated, from push-buttons positioned on collective and sidestick controllers.

DIMENSIONS, EXTERNAL:
Main rotor diameter: YRAH-66 ... 11.90 m (39 ft 0½ in)
RAH-66 ... 12.19 m (40 ft 0 in)
Fantail diameter ... 1.37 m (4 ft 6 in)
Fantail blade chord ... 0.17 m (6¾ in)
Length: overall, rotor turning ... 14.28 m (46 ft 10¼ in)
fuselage (excl gun barrel) ... 13.20 m (43 ft 3¾ in)
Fuselage max width ... 2.04 m (6 ft 8¼ in)
Width over mainwheels ... 2.31 m (7 ft 7 in)
Height over tailplane ... 3.37 m (11 ft 0¾ in)
Tailplane span ... 2.82 m (9 ft 3 in)

AREAS:
Main rotor disc: YRAH-66 ... 111.22 m^2 (1,197.1 sq ft)
RAH-66 ... 116.74 m^2 (1,256.6 sq ft)
Fantail disc ... 1.48 m^2 (15.90 sq ft)

WEIGHTS AND LOADINGS (estimated):
Weight empty ... 4,218 kg (9,300 lb)
Max useful load ... 2,296 kg (5,062 lb)
Max internal fuel weight ... 870 kg (1,918 lb)
T-O weight: primary mission ... 5,601 kg (12,349 lb)[1]
max alternative ... 5,850 kg (12,897 lb)
max (self-deployment) ... 7,896 kg (17,408 lb)
Max disc loading: YRAH-66 ... 71.0 kg/m^2 (14.54 lb/sq ft)
Transmission loading at max T-O weight and power ... 4.82 kg/kW (7.92 lb/shp)

[1] *with two crew, full internal fuel, 320 rds gun ammunition, radar, four Hellfires and two Stingers*

PERFORMANCE (at 1,220 m; 4,000 ft and 35°C; 95°F, estimated):
Max level (dash) speed:
without radar ... 175 kt (324 km/h; 201 mph)
with radar ... 166 kt (307 km/h; 191 mph)
Cruising speed:
without radar ... 165 kt (306 km/h; 190 mph)
with radar ... 149 kt (276 km/h; 171 mph)
Vertical rate of climb: without radar ... 273 m (895 ft)/min
with radar ... 152 m (500 ft)/min
Masking ... 1.6 s
180° hover turn to target ... 4.7 s
Snap turn to target at 80 kt (148 km/h; 92 mph) ... 4.5 s
Operational radius, internal fuel ... 150 n miles (278 km; 173 miles)
Ferry range with external tanks ... 1,200 n miles (2,222 km; 1,380 miles)
Endurance (standard fuel) ... 2 h 30 min
g limits ... +3.5/−1

UPDATED

BRANTLY

BRANTLY INTERNATIONAL INC

12399 Airport Drive, Wilbarger County Airport, Vernon, Texas 76384
Tel: (+1 940) 552 54 51
Fax: (+1 940) 552 27 03
e-mail: sales@brantly.com
Web: http://www.brantly.com
PRESIDENT AND CEO: Henry Yao
VICE-PRESIDENT: Cy A Russum

On 23 December 1994, Brantly International obtained the type certificates for the Brantly B-2B and 305 helicopters from Japanese-American businessman James T Kimura's Brantly Helicopter Industries, which had acquired them in May 1989. In 2002, Brantly employed 40 in its 2,790 m^2 (30,000 sq ft) facility.

VERIFIED

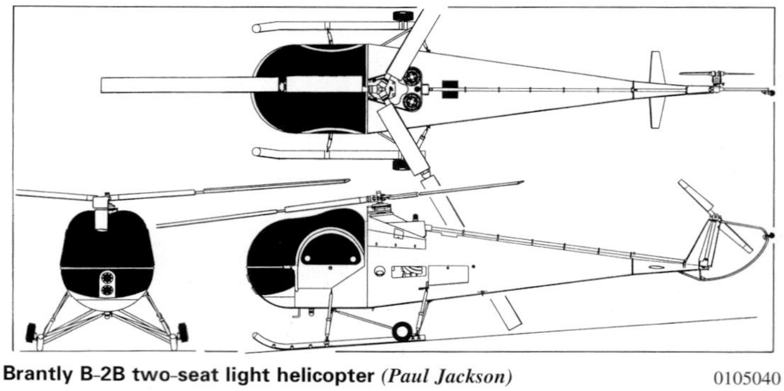

Brantly B-2B two-seat light helicopter *(Paul Jackson)* 0105040

BRANTLY B-2B

TYPE: Two-seat helicopter.

PROGRAMME: Developed from coaxial twin-rotor B-1 by Newby O Brantly. First flight (B-2) 21 February 1953; FAA certification 27 April 1959. Total of 194 B-2s and 18 B-2As (with additional headroom) produced between 1960 and 1963. Improved Model B-2B with metal rotor blades and fuel-injected Lycoming IVO-360-A1A engine certified 1 July 1963; total of 165 built between 1963 and 1967 (company owned by Gates Learjet from 1966) and a further one (as H-2) in 1975 by Brantly-Hynes Helicopter. Brantly Helicopter Industries (BHI) took over manufacturing and marketing rights and production facilities in 1989. First new-build B-2B (N25411 c/n 2001) flew 12 April 1991; three built under this name. Production continues under Brantly International, which received FAA production certificate on 19 July 1996. By 2004, Brantly had developed a new flight instrument layout incorporating GPS and was anticipating FAA approval of a tinted canopy with new bulged doors to improve shoulder room and downward vision.

CUSTOMERS: Total of 27 of current series (including BHI) registered by mid-2003; none further by mid-2004. Notified deliveries were two in 1998, none in 1999, six (to China) in 2000, two in 2001, two in 2002, including one to China and one in (first quarter of) 2003. Total of 14 remained registered in USA at May 2004, and one in Australia. (Some 80 from earlier production remain registered in USA.)

Brantly B-2B from the current production series *(Paul Jackson)* ***NEW***/0554461

COSTS: US$170,000 basic equipped (2003). Direct operating cost US$80 per hour (2003).

DESIGN FEATURES: Simple design, with blown main transparency and constant-taper fuselage. Double-articulated three-blade main rotor with pitch-change and flapping hinges close to hub and flap/lag hinges at 40 per cent blade span; symmetrical, rigid, inboard blade section with 29 per cent thickness/chord ratio, outboard section NACA 0012; outer blades quickly removable for compact storage; rotor brake standard; two-blade tail rotor mounted on starboard side, with guard. Transmission through automatic centrifugal clutch and planetary reduction gear. Bevel gear take-off from main transmission, with flexible coupling to tail rotor drive-shaft. Main rotor/engine rpm ratio 1:6.158; tail rotor ratio 1:1. Main rotor minimum speed 400 rpm; maximum 472 rpm.

FLYING CONTROLS: Conventional and manual; small fixed tailplanes on port and starboard sides of tailcone.

STRUCTURE: Semi-monocoque fuselage with alloy-stressed skin. Inboard rotor blades have stainless steel leading-edge spar; outboard blades have extruded aluminium spar; polyurethane core with bonded aluminium envelope riveted to spar. All-metal tail rotor blades.

LANDING GEAR: Fixed skid type with oleo-pneumatic shock-absorbers; small retractable ground handling wheels, size 10×3.5, pressure 4.12 bar (60 lb/sq in); fixed tailskid. Optional inflatable pontoons attach to standard skids for over-water operation.

POWER PLANT: One 134 kW (180 hp) Textron Lycoming IVO-360-A1A flat-four air-cooled piston engine, mounted vertically. Fuel contained in two interconnected bladder tanks behind cabin, total capacity 117 litres (31.0 US gallons; 25.8 Imp gallons) of which 116 litres (30.6 US gallons; 25.5 Imp gallons) are usable. Oil capacity 6.9 litres (1.83 US gallons; 1.52 Imp gallons).

ACCOMMODATION: Two, side by side in enclosed cabin; forward-hinged door on each side. Dual controls and cabin heater standard. Ground accessible baggage compartment, maximum capacity 22.7 kg (50 lb) in forward end of tailcone.

SYSTEMS: 60A alternator.

AVIONICS: To customer choice; GPS is standard.

EQUIPMENT: Twin landing lights in nose, navigation and anti-collision lights are standard.

DIMENSIONS, EXTERNAL:	
Main rotor diameter	7.24 m (23 ft 9 in)
Main rotor blade chord: inboard	0.22 m (8¾ in)
outboard	0.20 m (8 in)
Tail rotor diameter	1.30 m (4 ft 3 in)
Length: fuselage	6.43 m (21 ft 1 in)
overall, rotors turning	8.53 m (28 ft 0 in)
Height overall	2.11 m (6 ft 11 in)
Skid track	1.73 m (5 ft 8¼ in)
Track over ground handling wheels	2.08 m (6 ft 10 in)
Passenger doors (each): Height	0.79 m (2 ft 7 in)
Width	0.86 m (2 ft 9¾ in)
Baggage door: Height	0.23 m (9 in)
Width	0.53 m (1 ft 9 in)
DIMENSIONS, INTERNAL:	
Cabin: Length	1.83 m (6 ft 0 in)
Max width	1.19 m (3 ft 11 in)
Max height	0.99 m (3 ft 3 in)
Floor area	2.60 m^2 (28.0 sq ft)
Volume	2.78 m^3 (98.0 cu ft)
Baggage compartment volume	0.17 m^3 (6.0 cu ft)
AREAS:	
Main rotor blades (each)	0.69 m^2 (7.42 sq ft)
Main rotor disc	41.16 m^2 (443.0 sq ft)
Tail rotor blades (each)	0.05 m^2 (0.50 sq ft)
Tail rotor disc	1.32 m^2 (14.19 sq ft)
WEIGHTS AND LOADINGS:	
Weight empty: with skids	485 kg (1,070 lb)
with pontoons	494 kg (1,090 lb)
Baggage capacity	22.7 kg (50 lb)
Max fuel weight	82 kg (180 lb)
Max T-O weight	757 kg (1,670 lb)
Max disc loading	18.4 kg/m^2 (3.77 lb/sq ft)
Max power loading	5.65 kg/kW (9.28 lb/hp)
PERFORMANCE:	
Never-exceed speed (VNE)	87 kt (161 km/h; 100 mph)
Max cruising speed at 75% power	78 kt (144 km/h; 90 mph)
Max rate of climb at S/L	427 m (1,400 ft)/min
Service ceiling	1,829 m (6,000 ft)
Hovering ceiling IGE	1,074 m (3,525 ft)
Range: max fuel, no reserves	191 n miles (353 km; 219 miles)
with reserves	173 n miles (321 km; 200 miles)

UPDATED

CADCOR

6107 Southwest Murray Boulevard, No. 283, Beaverton, Oregon 97008
Tel: (+1 503) 577 40 67
e-mail: sales@cadcor.com
Web: http://www.cadcor.com
SALES DIRECTOR: Sean Doyle

The company's Chanute sport aircraft first flew in 2004 and is being marketed in the Experimental category.

UPDATED

CADCOR CHANUTE

TYPE: Tandem-seat sportplane kitbuilt.

PROGRAMME: Designed by Greg Cole Consulting of Bend, Oregon; airframe construction by Composites Unlimited of Scappoose, Oregon. Named for US aviation pioneer Octave Chanute. First flight (N1903N) 23 April 2004. Announced 25 May 2004, by which time 50 hours flown.

DESIGN FEATURES: High-strength structure capable of ±10 *g* with two crew and full fuel; unlimited aerobatic performance including roll rate in excess of 360°/s, more than 10 s of vertical penetration, power/weight ratio 3.39 kg/kW (5.57 lb/hp); good ground handling and superior control. Innovations include varaible chord, constant hinge-point, spadeless ailerons using leading-edge alteration form to adjust control feel to pilot's preference; efficient, 'cool face' engine cowling; and disc-damped tailwheel.

Low-set wing with 15° leading-edge sweepback introduced at one-third span and trailing-edges swept forward 5°. Mid-positioned, moderately tapered tailplane; sweptback fin with trailing-edge swept forward 15°. Roll rate 385°/s at 180 kt (333 km/h; 207 mph).

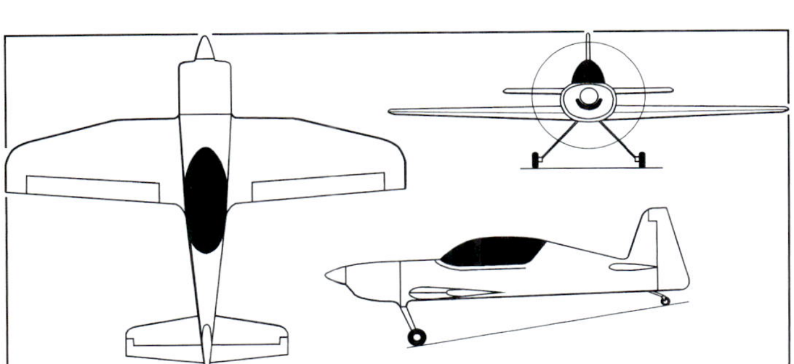

Cadcor Chanute aerobatic two-seater *(Paul Jackson)* ***NEW***/0589199

Cadcor Chanute prototype ***NEW***/1047701

FLYING CONTROLS: Conventional and manual. Horn-balanced rudder. Spadeless, variable hinge-point ailerons.

STRUCTURE: Generally of Toray 12K carbon fibre.

LANDING GEAR: Tailwheel type; fixed. Fuselage-mounted, spring aluminium cantilever mainwheel legs. Mainwheels 5.00-5. Wheel fairings. Castoring tailwheel.

POWER PLANT: One 261 kW (350 hp) Lycon Aircraft Engines (modified Textron Lycoming) AEIO-540-EXP flat-six driving an MT-Propeller MTV-9-B-C/C203-20d three-blade, constant-speed metal propeller. Fuel in two wing tanks with combined capacity of 182 litres (48.0 US gallons; 40.0 Imp gallons) and inverted capability fuselage tank of 114 litres (30.0 US gallons; 25.0 Imp gallons). Total capacity 295 litres (78.0 US gallons; 65.0 Imp gallons) of which 265 litres (70.0 US gallons; 58.3 Imp gallons) usable.

ACCOMMODATION: Two in tandem under single-piece, acrylic canopy on 20 *g* seats; Oregon Aero Inc seats reclined rearwards at 37° 30' AcroBelt five-point restraints.

AVIONICS: *Instrumentation:* Blue Mountain Inc 'glass cockpit'.

EQUIPMENT: Smoke generator; oil capacity 27 litres (7.0 US gallons; 5.8 Imp gallons).

DIMENSIONS, EXTERNAL:	
Wing span	7.32 m (24 ft 0 in)
Length overall	6.55 m (21 ft 6 in)
DIMENSIONS, INTERNAL:	
Cockpit max width: front	0.66 m (2 ft 2 in)
rear	0.58 m (1 ft 11 in)
AREAS:	
Wings, gross	9.75 m^2 (105.0 sq ft)

WEIGHTS AND LOADINGS:
Weight empty 601 kg (1,325 lb)
Max T-O weight: Utility 884 kg (1,950 lb)
Aerobatic 794 kg (1,750 lb)
Max wing loading 90.7 kg/m² (18.57 lb/sq ft)
Max power loading 3.39 kg/kN (5.57 lb/hp)

PERFORMANCE:
Never-exceed speed (VNE) 230 kt (426 km/h; 264 mph)
Cruising speed at 75% power at FL80 210 kt (389 km/h; 242 mph)
Stalling speed at Aerobatic weight: power off 58 kt (108 km/h; 67 mph) IAS
power on 40 kt (75 km/h; 46 mph) IAS
Initial rate of climb at S/L at Aerobatic weight 762 m (2,500 ft)/min

UPDATED

CAMERON

CAMERON & SONS AIRCRAFT

2772 Cessna Avenue, Coeur d'Alene Airport, Coeur d'Alene, Idaho 83814
Tel: (+1 208) 765 92 95
Fax: (+1 208) 765 94 15
e-mail: MustangMurdo@cs.com
Web: http://www.CameronAircraft.com
CEO: Murdo Cameron

Cameron Aircraft was formed in 1978. Its P-51 Mustang derivative is the company's first product; parts for the North American AT-6 Texan are also marketed. The 4,180 m² (45,000 sq ft) factory is situated on Coeur d'Alene Airport; in 2002 the workforce numbered 22.

Murdo Cameron is reported to be working on a lightweight composites air racer variant of the P-51.

UPDATED

Cameron P-51G (Textron Lycoming T53-L-701A) 0533396

CAMERON P-51G

TYPE: Tandem-seat turboprop sportplane/kitbuilt.

PROGRAMME: Full-size, turbine-powered representation of North American P-51 Mustang. Design began 1988; publicly announced end 1997; first flight 1998 and displayed at Oshkosh in July 1998 as Grand 51; later renamed P-51G.

CUSTOMERS: Two aircraft flying; two production aircraft under construction. Factory capacity for 12 per year.

COSTS: Development costs US$4.5 million; kit price US$150,000; completed airframes US$615,000 (2003).

DESIGN FEATURES: Wing span matches racing P-51s; option to extend to 'normal' 11.38 m (37 ft 4 in) available; aircraft kit comprises 12 major components. Quoted build time 1,500 to 2,000 hours.

STRUCTURE: Airframe and flight control surfaces of carbon fibre epoxy; 85 per cent is composites. One-piece 'wet' wing.

FLYING CONTROLS: Conventional and manual. No flaps. Electric three-axis trim.

LANDING GEAR: Tailwheel configuration; hydraulically operated mainwheels retract inward; tailwheel rearward. Bridgestone 24×7.7 mainwheel tyres; 12.5 in tailwheel type. Cleveland wheels and brakes.

Cameron P-51G instrument panel. Note four stick-top buttons 0121106

POWER PLANT: One 1,081 kW (1,450 shp) Textron Lycoming T53-L-701A turboprop, driving a Hamilton Sundstrand three-blade, fully feathering and reversible propeller. Rolls-Royce Merlin can be fitted. Fuel capacity 1,703 litres (450 US gallons; 375 Imp gallons) in prototype; kitbuilts have 946 litres (250 US gallons; 208 Imp gallons) with optional 757 litre (200 US gallon; 167 Imp gallon) auxiliary tank.

ACCOMMODATION: Pilot and passenger in tandem under one-piece hinged canopy. Optional P-51D canopy style available. Dual controls. Impact Dynamics leather seating. Dummy belly scoop provides baggage compartment with starboard side access door.

AVIONICS: To customer's specification; IFR is standard.

DIMENSIONS, EXTERNAL:
Wing span 11.28 m (37 ft 0 in)
Wing chord at tip 1.52 m (5 ft 0 in)
Wing aspect ratio 5.9
Length overall 10.97 m (36 ft 0 in)
Height overall 3.28 m (10 ft 9 in)
Propeller diameter 3.05 m (10 ft 0 in)
Baggage door width 0.91 m (3 ft 0 in)

DIMENSIONS, INTERNAL:
Cockpit max width 0.86 m (2 ft 10 in)

AREAS:
Wings, gross 21.65 m² (233.0 sq ft)

WEIGHTS AND LOADINGS:
Weight empty 2,041 kg (4,500 lb)
Max T-O weight 3,628 kg (8,000 lb)
Max wing loading 167.6 kg/m² (34.33 lb/sq ft)
Max power loading 3.36 kg/kW (5.52 lb/shp)

PERFORMANCE:
Max operating speed 391 kt (724 km/h; 450 mph)
Normal cruising speed 313 kt (579 km/h; 360 mph)
Stalling speed 84 kt (155 km/h; 96 mph)
Max rate of climb at S/L 1,280 m (4,200 ft)/min
Service ceiling 9,144 m (30,000 ft)
T-O run 381 m (1,250 ft)
Landing run 533 m (1,750 ft)
Range:
standard fuel 1,086 n miles (2,011 km; 1,250 miles)
auxiliary fuel 1,955 n miles (3,621 km; 2,250 miles)
g limits (prototype) ±8

UPDATED

CARLSON

CARLSON AIRCRAFT INC

PO Box 88, 50643 SR 14, East Palestine, Ohio 44413-0088
Tel: (+1 330) 426 39 34
Fax: (+1 330) 426 11 44
e-mail: mlc@sky-tek.com
Web: http://www.sky-tek.com
OWNER: Mary L Carlson

In addition to the Skycycle and newly added Criquet, Carlson also markets the Sparrow series of ultralight aircraft and a replacement wing for Baby Great Lakes and Piper Cub and Vagabond lightplanes. An ultralight version of the Cub was under construction in 2000.

Founder Ernest W Carlson was killed in the crash of the prototype Criquet on 24 May 2000; the company is now run by his widow, Mary.

UPDATED

CARLSON CA6 CRIQUET

TYPE: Tandem-seat kitbuilt.

PROGRAMME: Prototype N22CA under trial in 1999; seriously damaged in accident 24 May 2000. Reconstructed; reflown July 2003

COSTS: Kit US$24,325 excluding engine and instruments (2004).

DESIGN FEATURES: Three-quarter scale replica of Morane-Saulnier MS.500 Criquet, itself a copy of Fieseler Fi 156 Storch; uses Carlson wing, however. Quoted build time 800 hours.

POWER PLANT: One 103 kW (138 hp) LOM M332A four-cylinder piston engine driving a two-blade propeller. Fuel capacity 159 litres (42.0 US gallons; 35.0 Imp gallons) of which 152 litres (40.0 US gallons; 33.3 Imp gallons) usable plus 7.6 litre (2.0 US gallon; 1.7 Imp gallon) header tank. Optional 119 kW (160 hp) fuel-injected and supercharged M332B for higher gross weight version (998 kg; 2,200 lb) with 197 litre (52.0 US gallon; 43.3 Imp gallon) tankage.

DIMENSIONS, EXTERNAL:
Wing span 10.97 m (36 ft 0 in)
Wing chord, constant 1.63 m (5 ft 4 in)

Length: overall .... 7.49 m (24 ft 7 in)
Width, wings folded .... 3.00 m (9 ft 10 in)
Height overall .... 2.57 m (8 ft 5 in)
AREAS:
Wing, gross .... 17.84 m² (192.0 sq ft)
WEIGHTS AND LOADINGS:
Weight empty .... 476 kg (1,050 lb)
Max T-O weight .... 884 kg (1,950 lb)
PERFORMANCE:
Never-exceed speed (VNE) .... 117 kt (217 km/h; 135 mph)
Normal cruising speed .... 83 kt (153 km/h; 95 mph)
Stalling speed, power off .... 19 kt (34 km/h; 21 mph)
Max rate of climb at S/L .... 335 m (1,100 ft)/min
Service ceiling .... 3,960 m (13,000 ft)
T-O and landing run .... 23 m (75 ft)
Range .... 150 n miles (277 km; 172 miles)

UPDATED

CARTER

CARTERCOPTERS LLC

5720 Seymour Highway, Wichita Falls, Texas 76310
Tel: (+1 940) 691 08 19
Fax: (+1 940) 691 59 77
e-mail: carter@wf.net
Web: http://www.cartercopters.com
CEO: Jay Carter
VICE-PRESIDENT: Paul Redding

Company founded in 1994 by Jay Carter, who previously worked on design of the Bell XV-15 tiltrotor. CarterCopters has produced a flying prototype of what it anticipates will be a family of convertiplanes, although its ultimate purpose is to market the design, not establish a manufacturing programme itself.

In March 2004, company announced that it had been awarded a US$1 million contract from the US Army Research, Development and Engineering Command for research into Slowed-Rotor/Compound (SR/C) technology.

CarterCopters is also working with The Butterfly LLC autogyro manufacturer on a new 'smart' landing gear through a patented mechanical valve system that predicts impact energy and adjusts pressure levels accordingly.

Principle of the CarterCopter: Slowed Rotor/Compound principle attempts to overcome high cost (both initial and maintenance) and performance deficiencies of conventional helicopter. An autorotating rotor provides lift for vertical and slow-speed flight, slowing in angular velocity as the wing assumes the lift in forward flight. This angular velocity change improves lift-to-drag ratio, while at aeroplane speeds the rotor is, essentially, unloaded.

Three modes of operation are evident:

Autorotation. Air flows upwards through the rotor, the resultant thrust providing lift. Rotor disc is tilted rearwards by some 10°.

Transition. Progressive transfer of lift from rotor to wing. Rotor disc angle decreases, approaching that of the airstream, the inflow of air through the rotor being reduced, with consequent slowing of rotor angular velocity and shedding of drag. Furthermore, slowed rotation obviates the phenomenon of most other compound helicopters (except Fairey Jet Gyrodyne, Rotodyne and McDonnell XV-1 — all of 1950s vintage) of retreating blade stall and advancing blade compressibility.

High speed. Fixed wing, being optimised for high-speed flight, provides the vast majority of lift. However, the rotor retains a small angle of attack, allowing sufficient through airflow to maintain a minimal angular velocity and so ensure maintenance of the rotor's dynamic stability.

Carter principle relies upon achievement of μ (relationship of aircraft forward speed to blade speed) in excess of 1.0 during the last-mentioned phase. In this hitherto unexplored region, aerodynamic forces vary over the rotor disc. Assuming anticlockwise rotation, forces on the starboard/advancing side drive the rotor in the usual manner.

On the port/retreating side, however, the reverse flow region becomes a dominant factor affecting rotor dynamics. This is a region on the retreating side in which air velocity relative to the blade moves from the trailing- to the leading edge. Traditional helicopters and gyrocopters have a small reverse flow region (typically μ=0.3) within the total disc area. However, at μ=1.0 the full blade length is within the reverse flow area at the moment it passes through the 270° radial (at right angles to the fuselage).

As μ is further increased, the reverse flow region expands. At μ=5.0, for example, the blade is within the reverse flow area throughout almost the entire cycle of retreat.

Other aerodynamic factors requiring investigation by the Carter prototype include unstable blade flapping during slow rotation; and effects of a centre of pressure shift due to the retreating blade being in reverse flow. The latter can cause blade pitching, with consequent rotor destabilisation, vibration and increased control linkage forces.

UPDATED

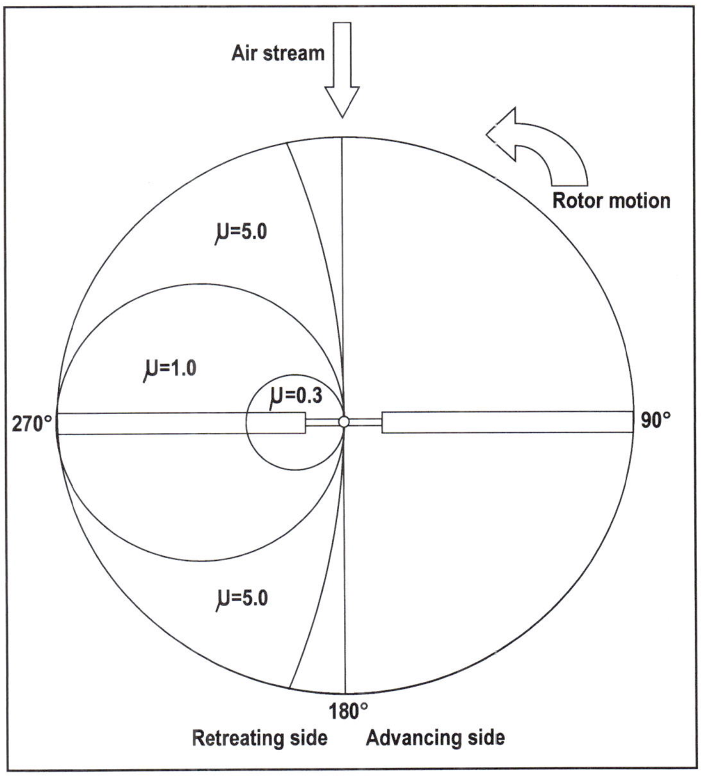

Reverse flow region within rotor disc at increasing μ values ***NEW***/0587520

Prototype CarterCopter Technology Demonstrator 0526459

CARTER CARTERCOPTER CC1

TYPE: Convertiplane.

PROGRAMME: Developed with US$670,000 funding from NASA, the CarterCopter has been conceived as the first in a series of convertiplanes. First flight (N121CC) September 1998; development delayed following damage to prototype during aborted rolling take-off 16 December 1999; flight testing resumed late October 2000 and was interrupted by minor accidents in April 2001. Attempts are planned on rotary-wing world records. On 22 March 2002, the CCTD reached μ 0.87 (relationship between forward speed and blade speed); company hoped to attain μ 1 by early 2003 but prototype again damaged in landing accident on 8 April 2003. Following repair, a further μ 1 attempt was scheduled for November 2003 but later postponed.

CURRENT VERSIONS **CC1, (Carter-Copter Technology Demonstrator):** also referred to as

CCTD: One prototype version for proof-of-concept; *as described.*

CC Heliplane Transport: See separate entry.

CC-UCAR: Smaller, unmanned version of CC1 submitted for US Navy consideration in early 2000.

CC Heliplane Trainer/Liaison/Utility (CCH-S): Passenger version for between five and nine occupants, with same fuselage width as CC1; intended to compete with medium-sized helicopters in the liaison and medical roles. Two TSX-1 turboshaft engines driving 2.44 m (8 ft 0 in) diameter wing-mounted pusher propellers; range estimated as 4,779 n miles (8,851 km; 5,500 miles) with maximum fuel or 608 n miles (1,126 km; 700 miles) with payload of 1,134 kg (2,500 lb).

CC-AEH: Heavily armed escort version of CCH-S.

CC-UBAV: Unmanned bomber version of CCH-S.

CC-KC: Aerial tanker version of CCH-S.

CC Heliplane Air-Express: Three-engined development planned for carrier on-board delivery role in competition with C-2 Greyhound and V-22 Osprey. Cruising speed 326 kt (604 km/h; 375 mph) at 9,144 m (30,000 ft); range 5,735 n miles (10,621 km; 6,600 miles) with maximum fuel and 1,303 n miles (2,414 km; 1,500 miles) with 9,072 kg (20,000 lb) payload.

CC Heliplane Hydra-Blaster: Proposed firefighting version with 34,070 litre (9,000 US gallon; 7,500 Imp gallon) foam capacity, delivering load via cannon system.

Kitplane: Two-seat kitbuilt version being marketed.

NxCC: Next-generation model announced at Sun 'n' Fun 2003. Improvement include fuselage extension of more than 90 cm (3 ft), longer wingspan, wider booms and on additional vertical stabiliser. Max T-O weight in STOL mode would be 2,721 kg (6,000 lb) and in VTOL configuration, 2,268 kg (5,000 lb). Power is expected to be provided by a turboprop engine in the 1,491 kW (2,000 shp) class.

CUSTOMERS: Prototype only by end 2001.

COSTS: Production versions expected to sell for US$250,000 to US$300,000 (1997; latest data).

DESIGN FEATURES: Twin-boom configuration with high-aspect ratio sweptback wings. Pylon-mounted two-blade rotor is powered for take-off and landing, free-rotating in forward flight, when all power is transmitted to a pusher propeller.

Maximum prerotation and flight speed 425 rpm; minimum in-flight rotor speed 80 rpm.

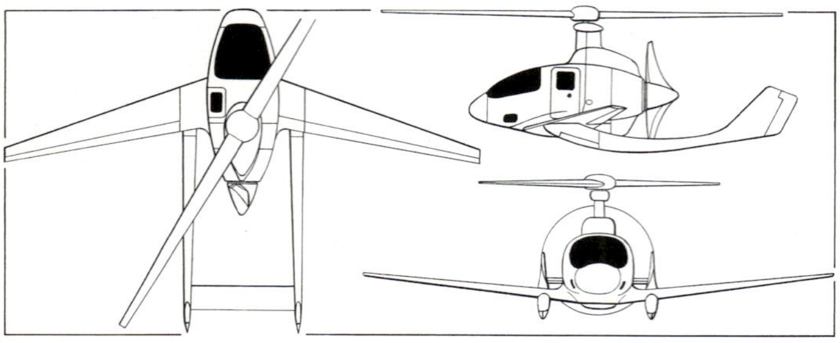

General arrangement of the Carter CarterCopter Technology Demonstrator *(James Goulding)* 0087603

Computer generated image of Next Generation CarterCopter ***NEW***/0561749

FLYING CONTROLS: All-moving elevator; twin rudders.

STRUCTURE: Fuselage of seamless moulded composites. Wing sections outboard of tailbooms removable. Tailbooms extend below propeller tips to protect latter in wheels-up landing.

Bearingless rotor system. Rotor blades of carbon composites with twistable carbon spar; 29.5 kg (65 lb) weights in each blade tip.

LANDING GEAR: Retractable tricycle type. Mainwheels retract into tailbooms.

POWER PLANT: One Corvette LS1 piston engine, with two turbochargers in series, driving a two-blade carbon composites propeller with twistable carbon spar. Engine power 447 kW (600 hp) for short duration; 298 kW (400 hp) at high altitude. Total fuel capacity 606 litres (160 US gallons; 133 Imp gallons), comprising 227 litres (60.0 US gallons; 50.0 Imp gallons) in fuselage tank; 189 litres (50.0 US gallons; 41.6 Imp gallons) in centre wing (underfuselage) tank; and two 95 litre (25.0 US gallon; 20.8 Imp gallon) tanks in wing outboard sections.

ACCOMMODATION: Pilot and passenger in front, plus three further passengers on rear bench seat.

SYSTEMS: Cabin pressurised; maximum differential 0.69 bar (10.00 lb/sq in).

DIMENSIONS, EXTERNAL:

Rotor diameter ... 13.26 m (43 ft 6 in)
Rotor blade chord: at root ... 0.43 m (1 ft 5 in)
at tip ... 0.18 cm (7 in)
Wing span ... 9.75 m (32 ft 0 in)
Fuselage length ... 6.81 m (22 ft 4 in)
Height to top of rotor head ... 3.40 m (11 ft 2 in)
Propeller diameter ... 2.44 m (8 ft 0 in)

DIMENSIONS, INTERNAL:

Cabin max width ... 1.52 m (5 ft 0 in)

AREAS:

Rotor disc ... 138.10 m^2 (1,486.7 sq ft)
Wings, gross ... 7.15 m^2 (77.0 sq ft)

WEIGHTS AND LOADINGS:

Weight empty ... 1,134 kg (2,500 lb)
Max T-O weight: STOL ... 2,268 kg (5,000 lb)
VTOL ... 1,905 kg (4,200 lb)

PERFORMANCE (two occupants):

Cruising speed: at S/L ... 191 kt (354 km/h; 220 mph)
at FL450 ... 304 kt (563 km/h; 350 mph)
Range with max fuel, 45 min reserves ... 2,085 n miles (3,862 km; 2,400 miles)

UPDATED

CESSNA

CESSNA AIRCRAFT COMPANY (SUBSIDIARY OF TEXTRON INC)

PO Box 7706, Wichita, Kansas 67277-7706
Tel: (+1 316) 517 60 00
Fax: (+1 316) 517 78 12
Web: http://www.cessna.textron.com
CHAIRMAN: Russell W Meyer Jr
PRESIDENT AND CEO: Jack J Pelton
SENIOR VICE-PRESIDENT, OPERATIONS: Ron Alberti
SENIOR VICE-PRESIDENT, PRODUCT ENGINEERING: David Brant
SENIOR VICE-PRESIDENT, SALES AND MARKETING: Roger Whyte
SENIOR VICE-PRESIDENT, FINANCE AND CHIEF FINANCIAL OFFICER: Michael Shonka
VICE-PRESIDENT, ADMINISTRATION: Michael Houeskeland
VICE-PRESIDENT, MARKETING: Phil Michel
VICE-PRESIDENT, CITATION SALES: Mark Paolucci
VICE-PRESIDENT, CORPORATE COMMUNICATIONS: Marilyn Richwine
VICE-PRESIDENT, QUALITY: Brad Thress

SINGLE-ENGINE PISTON AIRCRAFT PRODUCTION PLANT:
One Cessna Boulevard, PO Box 1996, Independence, Kansas 67301
Tel: (+1 316) 332 00 00
Fax: (+1 316) 332 03 88
GENERAL MANAGER: Terry Clark

Founded by late Clyde V Cessna 1911; incorporated 7 September 1927; acquired by General Dynamics as wholly owned subsidiary 1985; acquisition by Textron announced February 1992.

Suspended manufacture of some piston-engined light aircraft types in mid-1980s; plans for resumed production of Cessna 172 Skyhawk, 182 Skylane, 206 Stationair and T206 Turbo Stationair announced 1994; 46,450 m^2 (500,000 sq ft) new factory at Independence Municipal Airport, Montgomery County, Kansas, opened 3 July 1996 to house final assembly, painting, flight test, engineering, finance, marketing and human resource operations for single-engined light aircraft range. First new model 172 rolled out 6 November 1996. On February 2004 Cessna announced that final assembly of the Citation Mustang will take place at Independence; workforce there then stood at 350, and is expected to increase to approximately 1,000 when the Mustang is in full production.

Owned subsidiaries include Cessna Finance Corporation in Wichita and Cessna-Columbus Division in Columbus, Georgia. Sold 49 per cent interest in Reims Aviation of France to Compagnie Française Chaufour Investissement (CFCI) February 1989; Reims continues manufacturing Cessna F406 Caravan II (which see in French section).

Total 185,707 aircraft produced by 30 June 2004. Total of 842 delivered in 2002, comprising 588 single-engined piston aircraft, 57 Caravans and 197 Citations. Business jet production 305 in 2002, falling to 197 in 2003. 3,000th piston single from resumed production, a 182T Skylane, delivered in early February 2001; 5,000th following March 2004; 4,000th Citation executive jet delivered 7 October 2003; 150,000th single-engine piston aircraft, a Skyhawk SP, delivered 27 July 2004. Workforce about 9,900 in May 2003. Order backlog stood at US$4.4 billion at the end of 2003.

UPDATED

CESSNA 172 SKYHAWK

TYPE: Four-seat lightplane.

PROGRAMME: Development of previous models described in 1985–86 and earlier editions of *Jane's*; total of 35,643 commercial aircraft in the earlier Model 172/Skyhawk series built between 1955 (first flight of prototype, 12 June) and 1985, when production was suspended. Prototype 'Restart 172' (N6786R), modified from 1978 Model 172N with IO-360 engine, made its first flight at Wichita 19 April 1995; certification flight testing began 10 January 1996; three preproduction 172Rs (N172KS, N172SE and N172NU) assembled at East Pawnee, Wichita, plant under Single-Engine Pilot Program, of which the first flew on 16 April 1996; FAA FAR Pt 23 certification achieved 21 June 1996 after 145 flight test hours; first production aircraft (c/n 80004; N172FN) rolled out at Independence 6 November 1996 and delivered 18 January 1997. First export delivery began on 30 January 1997 when second Independence-built aircraft (c/n 80005; VH-NMN) left US for air ferry to Australia.

CURRENT VERSIONS: **172R Skyhawk:** Standard version. Compared with pre-1986 (172Q) versions, features fuel-injected engine; specially developed McCauley propeller optimised for reduced rpm operation; new Honeywell avionics; metal instrument panel with backlit, non-glare instruments; all-electric engine gauges, dual vacuum system, digital clock, EGT

Cessna 172R Skyhawk four-seat lightplane *(Paul Jackson)* ***NEW***/0580755

Cessna 172S Skyhawk SP *(Paul Jackson)* ***NEW***/0580751

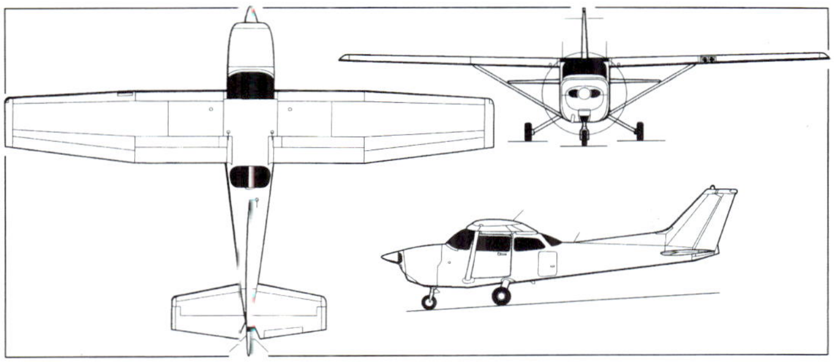

Cessna 172R Skyhawk *(Paul Jackson)* 0580697

gauge, hour meter and centrally mounted annunciator panel; stainless steel control cables; epoxy corrosion-proofing; cabin improvements as detailed below; steps for visual checking of fuel tanks, which have quick-reference quantity tabs, and consolidation of all primary electrical components into a single junction box on the firewall for simplified servicing.

New standard features announced at the NBAA Convention at New Orleans on 9 October 2000, and introduced on 172R and 172S from 2001 model year, include redesigned wingtips, restyled and relocked main landing gear fairings, improved nose gear fairing, low-drag wing strut fairings, streamlined refuelling and entrance steps, and improved interiors featuring sculptured composite side panels with integrated cupholders and cushioned armrests, removable overhead panels for simplified maintenance, Rosen sun visors, lightweight door handles and a 12 V electrical port for use with a portable GPS, CD player or laptop computer. New optional avionics include a 12.7 cm (5 in) Honeywell KMD 550 LCD colour MFD with Stormscope that interfaces with the previously available KLN 94 colour IFR GPS.

Garmin G1000 integrated glass cockpit announced 27 July 2004 as NAV III package for 172R Skyhawk GA aimed principally at flight training market, and for 172S Skyhawk SP, available in early 2005.

Detailed description applies to 172R.

172S Skyhawk SP: Special performance version, announced 19 March 1998; features Textron Lycoming IO-360-L2A engine rated at 134 kW (180 hp) driving a higher-performance McCauley 1A170E/JHA7660 fixed-pitch, 1.93 m (6 ft 4 in) diameter propeller; avionics and equipment as for baseline version except leather seats standard. Standard empty weight 746 kg (1,644 lb); maximum T-O and landing weight: normal 1,157 kg (2,550 lb), utility 998 kg (2,200 lb); maximum ramp weight: normal 1,160 kg (2,558 lb), utility 1,002 kg (2,208 lb); maximum level speed at S/L 126 kt (233 km/h; 145 mph); cruising speed, 75 per cent power at 2,590 m (8,500 ft) 124 kt (230 km/h; 143 mph); stalling speed, power off: flaps up 53 kt (99 km/h; 61 mph), flaps down 48 kt (89 km/h; 56 mph); maximum rate of climb at S/L 222 m (730 ft)/min; service ceiling 4,270 m (14,000 ft); T-O run 293 m (960 ft); T-O to 15 m (50 ft) 497 m (1,630 ft); landing from 15 m (50 ft) 407 m (1,335 ft); landing run 175 m (575 ft); range with maximum fuel: at 75 per cent power at 2,590 m (8,500 ft), 518 n miles (959 km; 596 miles); at 45 per cent power at 3,050 m (10,000 ft), 638 n miles (1,181 km; 734 miles); endurance 6 h 43 min. First delivery 31 July 1998. Further improvements introduced in October 2000, as noted.

CUSTOMERS: Total of 180 172Rs and 279 172Ss delivered in 1999, 150 and 340, respectively in 2000, 107 and 341, respectively in 2001, 57 and 258 respectively in 2002, 58 and 291 respectively in 2003, and 20 and 87 respectively in the first six months of 2004. The 1,000th 172R/SP was delivered in October 1999 to North Florida Medical. Customers include Embry-Riddle Aeronautical University (initial batch of 73 172Rs out of 300 piston-engined Cessnas of all models to be ordered over a 12 year period), TAM Training (10), Western Michigan University (38), US AFB Flight Training Centre (22 delivered from 30 March 1998), Civil Air Patrol (42 172Rs/SPs ordered by August 2000, Daniel Webster College of Nashua, New Hampshire (20 172Rs delivered by August 2000, Central Missouri State University (16 172Rs delivered by March 2000), Kansas State University (15 172Rs delivered), Singapore Flying College at Jandakot, Perth, Australia (six 172Rs ordered in February 2000 in addition to five delivered in early 1998), the University of Dubuque (three 172Rs delivered in 2003, bringing fleet total to nine), American Flyers (five 172Rs delivered in 2003, bringing fleet total to 25), Southern Illinois University Carbondale (seven 172Rs delivered in April 2003), and Theorie Training Centre of Bonn, Germany, which took delivery of five 172Rs in May 2004. 150,000th Cessna single-engine piston aircraft, 9 172S, delivered 27 July 2004 at EAA AirVenture Oshkosh, to Sporty's Pilot Shop for its 2005 Super Sweepstakes draw.

COSTS: 172R Skyhawk US$158,000 standard equipped; Skyhawk SP US$165,000. Price including NAV I US$168,150 (172R) and US$175,150 (172S); NAV II US$181,100 (172R) and US$188,100 (172S); NAV III US$199,750 (172R GA) and US$229,750 (172S) (all 2004).

DESIGN FEATURES: Classic, all-metal, braced high-wing, touring lightplane. Tapered, mid-mounted tailplane and wing outer panels; sweptback fin.

NACA 2412 wing section. Dihedral 1° 44'. Incidence 1° 30' at root, –1° 30' at tip; fin sweepback 35° at quarter-chord.

FLYING CONTROLS: Conventional and manual. Actuation by stainless steel cables; modified Frise all-metal ailerons, electrically actuated single-slotted Para Lift flaps; trim tab on elevator.

STRUCTURE: Conventional light alloy, with composites for some non-structural components such as nose-bowl, wing-, tailplane- and fin-caps, and fairings. Epoxy corrosion proofing and polyurethane exterior paint standard.

LANDING GEAR: Tricycle type; fixed. Fuselage-mounted Cessna Land-O-Matic spring cantilever tapered steel tube mainwheel legs and oleo-pneumatic nose leg; steerable nosewheel, tyre size 5.00×5, mainwheel tyres size 6.00×6; hydraulic brakes. Wheel fairings optional, but included free when Nav I and II packages are purchased.

POWER PLANT: One 119 kW (160 hp) Textron Lycoming IO-360-L2A flat-four piston engine driving a two-blade, fixed-pitch McCauley 1C235/LFA7570 metal propeller. Fuel contained in two integral wing tanks, combined capacity 212 litres (56.0 US gallons; 46.6 Imp gallons) of which 201 litres (53.0 US gallons; 44.1 Imp gallons) are usable. Refuelling points on upper surface of each wing. Oil capacity 7.6 litres (2.0 US gallons; 1.7 Imp gallons).

ACCOMMODATION: Four persons in two pairs on contoured, energy-absorbing seats dynamically tested to 26 *g*, those for pilot and front seat passenger being vertically adjustable, reclining and mounted on enlarged seat rails with dual locking pins; dual controls standard; rear bench seat with reclining back; door each side of cabin, hinged forward; inertia reel harnesses for all occupants; tinted windows with hinged side windows; ambient noise reduction soundproofing; composites headliner, and double-pin latching system for cabin doors. Baggage compartment to rear of cabin, with access door on port side. Cabin is heated and ventilated. Keith Products air conditioning system optional.

SYSTEMS: Electrical system comprising 28 V 60 A alternator and 24 V 12.75 Ah battery.

AVIONICS: *Comms:* Bendix/King KX 155A nav/com; KI 208 VOR/LOC indicator; KMA 28 audio panel. KT 76C Mode C transponder, and Pointer 3000-11 ELT, all standard.

Flight: Bendix/King KI 208 VOR/LOC standard. Optional NAV I package comprises KLN 94 GPS-IFR, second KX 155A with glideslope, KI 209A VOR/LOC/GS with GPS interface, and MD 41-231 GPS-Nav selector; optional NAV II package adds KAP 140 two-axis autopilot. Optional additions to NAV II include KCS 55 compass, KR 87 ADF, KMD 550 MFD and KDR 510 FIS weather datalink.

NAV III package for 172R GA features Garmin G1000 fully-integrated system providing all primary flight, navigation, communications, terrain, traffic, weather and engine sensor data on two 10.4 in XGA-resolution PFD/MFD displays, and includes solid-state AHRS; dual VOR/LOC/ILS/GPS; solid-state digital air-data computer; engine-monitoring display; Mode S transponder with Traffic Information Service; dual radio modules that provide WAAS-upgradeable IFR oceanic-approved GPS; VHF navigation with ILS; and VHF communication with 16-watt transceivers and 8.33-kHz channel spacing; EICAS; terrain mapping including topographic and terrain and obstruction clearance mapping; digital audio panel/intercom with ATC playback recording; and three-axis magnetometer that provides full magnetic field information in all flight attitudes. NAV III package for 172S adds to the above an autopilot; high-bandwidth datalink weather system including NEXRAD, METARS, FIS, lightning and up to 14 other weather services (Continental US only); and XM satellite radio entertainment system.

DIMENSIONS, EXTERNAL:

Wing span	11.00 m (36 ft 1 in)
Wing chord: at root	1.63 m (5 ft 4 in)
at tip	1.12 m (3 ft 8½ in)
Wing aspect ratio	7.5
Length overall	8.28 m (27 ft 2 in)
Height overall	2.72 m (8 ft 11 in)
Wheel track	2.53 m (8 ft 3½ in)
Wheelbase	1.63 m (5 ft 4 in)
Propeller diameter	1.90 m (6 ft 3 in)
Cabin doors: Max height	1.02 m (3 ft 4½ in)
Max width	0.94 m (3 ft 1 in)
Baggage door: Max height	0.56 m (1 ft 10 in)
Max width	0.39 m (1 ft 3¼ in)

DIMENSIONS, INTERNAL:

Cabin (firewall to baggage area):	
Length	3.61 m (11 ft 10 in)
Max width	1.00 m (3 ft 3½ in)
Max height	1.22 m (4 ft 0 in)

AREAS:

Wings, gross	16.17 m² (174.0 sq ft)
Ailerons (total)	1.70 m² (18.3 sq ft)
Trailing-edge flaps (total)	1.98 m² (21.26 sq ft)
Fin	1.04 m² (11.24 sq ft)
Rudder, incl tab	0.69 m² (7.43 sq ft)
Tailplane	2.00 m² (21.56 sq ft)
Elevators, incl tab	1.35 m² (14.53 sq ft)

WEIGHTS AND LOADINGS:

Weight empty, standard	735 kg (1,620 lb)
Baggage capacity	54 kg (120 lb)
Max T-O and landing weight:	
Normal	1,111 kg (2,450 lb)
Utility	952 kg (2,100 lb)
Max ramp weight: Normal	1,114 kg (2,457 lb)
Utility	956 kg (2,107 lb)
Max wing loading	68.7 kg/m² (14.08 lb/sq ft)
Max power loading	9.32 kg/kW (15.31 lb/hp)

PERFORMANCE:

Max level speed at S/L	123 kt (227 km/h; 141 mph)
Cruising speed at 80% power at FL80	122 kt (226 km/h; 140 mph)
Stalling speed, power off:	
flaps up	51 kt (95 km/h; 59 mph)
flaps down	47 kt (87 km/h; 54 mph)
Max rate of climb at S/L	219 m (720 ft)/min
Service ceiling	4,115 m (13,500 ft)
T-O run	288 m (945 ft)
T-O to 15 m (50 ft)	514 m (1,685 ft)
Landing from 15 m (50 ft)	395 m (1,295 ft)
Landing run	168 m (550 ft)
Range with max fuel, 45 min reserves:	
at 80% power at FL80	580 n miles (1,074 km; 667 miles)
at 60% power at FL100	687 n miles (1,272 km; 790 miles)
Endurance	6 h 36 min

UPDATED

CESSNA 182 SKYLANE

TYPE: Four-seat lightplane.

PROGRAMME: Development of previous versions, first of which was certified on 2 March 1956; total of 19,773 earlier Model 182/Skylanes were built between 1956 and 1985, when production was suspended. Manufacture resumed when three preproduction 182Ss assembled at East Pawnee, Wichita, plant under Single-Engine Pilot Program, of which the first (N182NU) flew on 16 July 1996; FAA FAR Pt 23 certification achieved by Model 182S on 3 October 1996 after 164 flight test hours; assembly of first new production aircraft (N182FN) began at Independence 25 September 1996, with delivery 24 April 1997. Cessna

Cessna 182T Skylane *(Paul Jackson)* ***NEW***/0580812

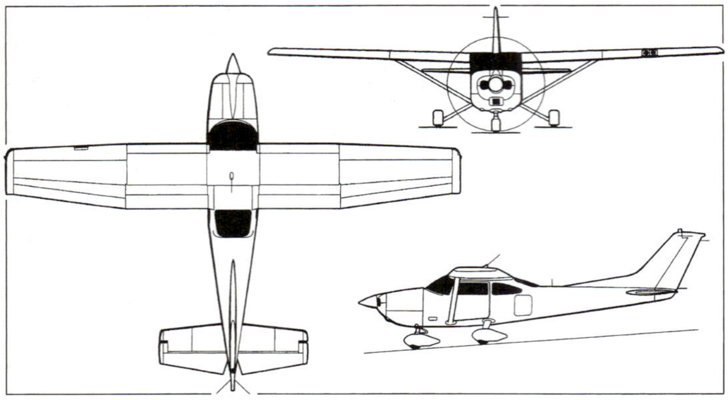

Cessna 182T Skylane *(Paul Jackson)* ***NEW***/0568986

182S supplanted when delivery of improved 182T and T182T versions began in May 2001. Garmin G1000 'glass cockpit' offered from October 2003 and demonstrator shown at Sun 'n' Fun in April 2004.

CURRENT VERSIONS: **182T Skylane:** *As described.* Standard version from 2001 model year, announced at the NBAA Convention at New Orleans on 9 October 2000; FAA certification 23 February 2001. Compared to 182S features redesigned wingtips, restyled and reclocked main landing gear fairings, improved nose gear fairing, low-drag wing strut fairings, streamlined refuelling and entrance steps, and improved interiors featuring sculptured composite side panels with integrated cupholders and cushioned armrests, removable overhead panels for simplified maintenance, Rosen sun visors, lightweight door handles, and a 12 V electrical port for use with a portable GPS, CD player or laptop computer. New optional avionics include a 12.7 cm (5 in) Honeywell KMD 550 LCD colour MFD with Stormscope that interfaces with the previously available KLN 94 colour IFR GPS. Standard empty weight 860 kg (1,897 lb); maximum level speed at S/L 150 kt (278 km/h; 173 mph); cruising speed at 80 per cent power at 2,135 m (7,000 ft), 145 kt (269 km/h; 167 mph); range with maximum fuel, 45 min reserves at 75 per cent power at 2,440 m (8,000 ft), 813 n miles (1,505 km; 935 miles).

T182T Turbo Skylane: Turbocharged version. Described separately.

CUSTOMERS: Total 1,250 normally aspirated Model 182s of current series built by mid-2003, comprising 944 182Ss and 306 182Ts.

Recent customers include the Mexican Air Force, which had taken delivery of 76 Skylanes by June 2000, the Louisiana State Department of Agriculture and Forestry, which ordered 18 for delivery in early 2003, and the US Civil Air Patrol, which ordered 15 for delivery in 2003, and six for delivery in 2004. Total of 248 delivered in 1999, 267 in 2000, 142 in 2001, 109 in 2002, 118 in 2003, and 25 in the first six months of 2004.

COSTS: US$250,000 standard equipped; with NAV 1 package, US$260,000; with NAV II package, US$278,100; with NAV III package, US$297,500 (all 2004).

DESIGN FEATURES: Generally as for Cessna 172.

Wing section NACA 2412 (modified); incidence 0° 47' at root, −2° 50' at tip; dihedral 1° 44'.

FLYING CONTROLS: Conventional and manual. Actuation by stainless steel control cables; modified Frise ailerons; electrically actuated single-slotted Para Lift flaps; trim tabs on elevator and rudder. Control surface movements: ailerons +20°/−15°; elevator +28°/−21°; rudder ±24°; flaps −38°.

STRUCTURE: Conventional light alloy, with composites for some non-structural components such as nose-bowl, wing-, tailplane- and fin-caps, and fairings. Epoxy corrosion proofing and polyurethane exterior paint standard.

LANDING GEAR: Tricycle type; fixed. Fuselage-mounted Cessna Land-O-Matic spring cantilever tapered steel tube mainwheel legs and oleo-pneumatic nose leg; steerable nosewheel, tyre size 5.00×5, mainwheel tyre size 6.00×6; hydraulic brakes. Wheel fairings optional.

POWER PLANT: One 172 kW (230 hp) Textron Lycoming IO-540-AB1A5 flat-six piston engine driving a three-blade, constant-speed McCauley B3D36C431/80VSA-1 metal propeller. Fuel contained in two integral wing tanks, combined capacity 348 litres (92.0 US gallons; 76.6 Imp gallons) of which 329 litres (87.0 US gallons; 72.4 Imp gallons) are usable. Refuelling points on upper surface of each wing. Oil capacity 8.5 litres (2.25 US gallons; 1.9 Imp gallons).

ACCOMMODATION: Four persons in two pairs on contoured, energy-absorbing seats, dynamically tested to 26 *g*, those for pilot and front seat passenger being vertically adjustable, reclining and mounted on enlarged seat rails with dual locking pins; dual controls standard; rear bench seat with reclining back; inertia reel harnesses for all occupants; tinted windows with hinged side windows; ambient noise reduction soundproofing; composites headliner, and double-pin latching system for cabin doors. Baggage compartment to rear of cabin, with access door on port side. Cabin is heated and ventilated. Keith Products air conditioning optional.

SYSTEMS: Electrical system comprising 28 V 60 A alternator and 24 V 12.75 Ah battery.

AVIONICS: *Comms:* Dual Bendix/King KX 155A nav/com glideslope; KMA 28 audio panel with marker beacon receiver, four-position stereo/com voice-activated intercom, ELT and KT 76C Mode C transponder all standard.

Flight: Bendix/King KAP 140 two-axis autopilot and KI 209A VOR/LOC/GS standard. NAV I package adds KI 209A VOR/LOC/GS with GPS interface, KLN 94 colour moving map, KMD 550 MFD and CSE GPS-NAV selector/annunciator KCS 55A HSI, Goodrich WX 500 Stormscope and CSE GPS-NAV/HSI selector/annunciator. Options to NAV II include KR 87 ADF (also with NAV I) KMH 880 traffic and terrain awareness system, and KDR 510 F1S sensor.

NAV III package, announced at the NBAA Convention in Orlando on 7 October 2003 and available as option from second quarter 2004, features Garmin G1000 all-glass integrated cockpit comprising two 264 mm (10.4 in) high-resolution, wide viewing angle XGA screens for PFD and MFD functions; solid-state AHRS and digital air data computer; dual VHF com radios; dual VOR/LOC/ILS receivers; dual GPS receivers with future WAAS upgrade capability; Mode S transponder and Traffic Information System; WX-500 Stormscope displayed on EFIS screens; digital audio panel/intercom with ATC playback recording; XM satellite Radio entertainment system; turbine-class switch panels with electro-luminescent lighting; and simplified line-replaceable units to minimise down time.

DIMENSIONS, EXTERNAL:
Wing span ..... 10.97 m (36 ft 0 in)
Wing chord: at root ..... 1.63 m (5 ft 4 in)
at tip ..... 1.09 m (3 ft 7 in)
Wing aspect ratio 7.4
Length overall ..... 8.84 m (29 ft 0 in)
Height overall ..... 2.84 m (9 ft 4 in)
Wheel track ..... 2.74 m (9 ft 0 in)
Wheelbase ..... 1.69 m (5 ft 6½ in)
Propeller diameter ..... 2.01 m (6 ft 7 in)
Cabin doors: Height ..... 1.04 m (3 ft 5 in)
Max width ..... 0.93 m (3 ft 0½ in)
Baggage door: Height ..... 0.56 m (1 ft 10 in)
Width ..... 0.40 m (1 ft 3¾ in)

DIMENSIONS, INTERNAL:
Cabin (firewall to baggage compartment):
Length ..... 3.40 m (11 ft 2 in)
Max width ..... 1.07 m (3 ft 6 in)
Max height ..... 1.23 m (4 ft 0½ in)

AREAS:
Wings, gross ..... 16.30 m^2 (175.50 sq ft)
Ailerons (total) ..... 1.70 m^2 (18.3 sq ft)
Trailing-edge flaps (total) ..... 1.97 m^2 (21.20 sq ft)
Fin ..... 1.08 m^2 (11.62 sq ft)
Rudder ..... 0.65 m^2 (6.95 sq ft)
Tailplane ..... 2.13 m^2 (22.96 sq ft)
Elevators ..... 1.54 m^2 (16.61 sq ft)

WEIGHTS AND LOADINGS:
Weight empty ..... 873 kg (1,924 lb)
Baggage capacity ..... 91 kg (200 lb)
Max T-O weight ..... 1,406 kg (3,100 lb)
Max ramp weight ..... 1,411 kg (3,110 lb)
Max landing weight ..... 1,338 kg (2,950 lb)
Max wing loading ..... 86.2 kg/m^2 (17.66 lb/sq ft)
Max power loading ..... 8.20 kg/kW (13.48 lb/hp)

PERFORMANCE:
Never-exceed speed (VNE) ..... 175 kt (324 km/h; 201 mph)
Max level speed at S/L ..... 150 kt (278 km/h; 173 mph)
Cruising speed at 80% power at FL70 ..... 145 kt (269 km/h; 167 mph)
Stalling speed, power off:
flaps up ..... 54 kt (100 km/h; 63 mph)
flaps down ..... 49 kt (91 km/h; 57 mph)
Max rate of climb at S/L ..... 282 m (924 ft)/min
Service ceiling ..... 5,515 m (18,100 ft)
T-O run ..... 242 m (795 ft)
T-O to 15 m (50 ft) ..... 462 m (1,514 ft)
Landing from 15 m (50 ft) ..... 412 m (1,350 ft)
Landing run ..... 180 m (590 ft)
Range with max fuel, 45 min reserves:
at 75% power at FL80 ..... 813 n miles (1,505 km; 935 miles)
at 55% power at FL100 ..... 930 n miles (1,722 km; 1,070 miles)

UPDATED

CESSNA T182T TURBO SKYLANE

TYPE: Four-seat lightplane.

PROGRAMME: Announced at NBAA Convention at New Orleans on 9 October 2000. Certified 23 February 2001; deliveries began in May 2001. First four Garmin G1000-equipped Turbo Skylanes delivered 29 June 2004, following FAA certification of new avionics on 7 June.

Description for Cessna 182S and 182T Skylanes applies also to the T182T Turbo Skylane, except as follows.

CUSTOMERS: Total of 96 delivered in 2001, 79 in 2002, 47 in 2003, and eight in the first six months of 2004. Recent customers include the Alabama Department of conservation, wildlife and freshwater Fisheries, which took delivery of one in July 2003 and the Civil Air Patrol, which took delivery of five on 18 November 2003.

COSTS: Standard equipped US$275,000; with NAV I package US$285,000; with NAV II package US$303,100; with NAV III package US$322,500 (all 2004).

POWER PLANT: One 175.2 kW (235 hp) Textron Lycoming TIO-540-AK1A turbocharged flat-six engine, driving a three-blade, constant-speed McCauley B3D76C442-C/80VSB-1 metal propeller.

AVIONICS: As for Skylane.

WEIGHTS AND LOADINGS:
Standard weight empty ..... 918 kg (2,023 lb)
Max power loading ..... 8.03 kg/kN (13.19 lb/hp)

PERFORMANCE:
Max level speed at FL200 ..... 176 kt (326 km/h; 203 mph)
Cruising speed at 88% power at FL125 ..... 159 kt (294 km/h; 183 mph)
Max rate of climb at S/L ..... 317 m (1,040 ft)/min
Max operating altitude ..... 6,100 m (20,000 ft)
T-O run ..... 236 m (775 ft)
T-O to 15 m (50 ft) ..... 422 m (1,385 ft)
Max range ..... 971 n miles (1,798 km; 1,117 miles)

UPDATED

Cessna T182T Turbo Skylane *(Paul Jackson)* ***NEW***/0580810

CESSNA 206 STATIONAIR AND T206 TURBO STATIONAIR

TYPE: Six-seat utility transport.

PROGRAMME: Development of previous versions described in 1984–85 and earlier editions of *Jane's*; beginning in 1964, total of 7,556 earlier Model 206/Skywagons/Super

Cessna T206H Stationair TC *(Paul Jackson)* ***NEW***/0580764

Skylanes/Stationairs/Turbo Stationairs had been built when production was suspended in 1985. New model T206H prototype (N732CP) first flown from Wichita 28 August 1996; new model 206H prototype first flown 6 November 1996; assembly of first Pilot Program T206H began 3 December 1996.

FAA certification of 206H achieved 9 September 1998, followed by approval for T206H on 1 October 1998. First delivery, of 10 206Hs to Uruguayan Air Force, began 8 December 1998. Garmin G1000 'glass cockpit' offered from October 2003 and demonstrator shown at Sun 'n' Fun in April 2004.

CURRENT VERSIONS: **206H Stationair** and **T206H Turbo Stationair (Stationair TC)**: Standard versions. New features introduced on 206H and T206H comprise fuel-injected engines; new Honeywell avionics; metal instrument panel with backlit, non-glare instruments plus back-up cold fluorescent light system; all-electric engine gauges; stainless steel control cables; epoxy corrosion-proofing; cabin improvements as detailed below; strobe lights; handles and steps to check fuel levels; static wicks; GPU receptacle; and fire extinguisher.

New standard features announced at the NBAA Convention at New Orleans on 9 October 2000, and introduced on 206H and T206H from 2001 model year, include redesigned wingtips, restyled and relocked main landing gear fairings, improved nose gear fairing, low-drag wing strut fairings, streamlined refuelling and entrance steps, and improved interiors featuring sculptured composite side panels with integrated cupholders and cushioned armrests, removable overhead panels for simplified maintenance, Rosen sun visors, lightweight door handles, and a 12 V electrical port for use with a portable GPS, CD player or laptop computer. New optional avionics include a 12.7 cm (5 in) Honeywell KMD 550 LCD colour MFD with Stormscope that interfaces with the previously available KLN 94 colour IFR GPS.

Description applies to the above version.

CUSTOMERS: Total of 218 Stationairs and 448 Turbo Stationairs delivered by 30 June 2004, including 16 and 58 respectively in 2003, and 12 and 15, respectively, in the first six months of 2004. Recent customers include Beijing Sport Aviation School of Beijing, China, which took delivery of one T206H in November 2003, and A&P Light Aircraft Service Company (China) Ltd, which ordered two for delivery in November 2004.

COSTS: Stationair US$351,900 standard equipped; with NAV I package US$370,800; with NAV II package US$394,300; with NAV III package US$413,000; Turbo Stationair US$381,900 standard equipped; with NAV I package US$400,800; with NAV II package US$424,300; with NAV III package US$443,000 (all 2004).

DESIGN FEATURES: Generally as for Cessna 172.

Wing section NACA 2412 (modified). Dihedral 2° 14'; incidence 1° 30' at root, −1° 30' at tip.

FLYING CONTROLS: As Cessna 182.

LANDING GEAR: As Cessna 182. Wipline 3450 floatplane conversion certified April 2000.

POWER PLANT: Stationair has one 224 kW (300 hp) Textron Lycoming IO-540-AC1A flat-six piston engine; Turbo Stationair has one 231 kW (310 hp) Textron Lycoming TIO-540-AJ1A turbocharged flat-six piston engine; each drives a three-blade, constant-speed McCauley B3D36C432/80VSA-1 metal propeller. Total fuel capacity 348 litres (92.0 US gallons; 76.6 Imp gallons), of which 329 litres (87.0 US gallons; 72.4 Imp gallons) are usable. Oil capacity 10.4 litres (2.75 US gallons; 2.3 Imp gallons).

ACCOMMODATION: Forward-hinged door, port side; double cargo doors starboard side, each opening away from centre. Six persons in three pairs on contoured, energy-absorbing seats, dynamically tested to 26 *g*, those for pilot and front seat passenger being vertically adjustable, reclining and mounted on enlarged seat rails with dual locking pins; other seats have reclining backs; inertia reel harnesses for all occupants; tinted windows; hinged front side windows; composites headliner; multilevel ventilation system; ambient noise reduction soundproofing and double-pin latching system on pilot's door, all standard. Utility interior and Keith Products air conditioning system optional.

AVIONICS: *Comms:* Bendix/King KX 155A nav/com/glideslope, KMA 28 audio panel, with marker beacon receiver, stereo/com splitter and voice-activated intercom. KT 76C Mode C transponder and ELT.

Flight: Bendix/King KI 209 VOR/LOC/GS and KAP 140 two-axis autopilot. Optional NAV I package adds second KX 155A, second KI 209A, plus KLN 94 colour moving map GPS-IFR, KMD 550 MFD and CSE GPS-NAV selector/annunciator. NAV II adds Bendix/King KCS 55A HSI, Goodrich WX 500 Stormscope and CSE GPS-NAV/HSI selector annunciator. KMH 880; traffic and terrain awareness system optional; NAV III package optional, as described in 182 Skylane entry.

EQUIPMENT: Optional equipment includes electric propeller de-icing, 0.45 m³ (16 cu ft) cargo pod, floatplane provisions kit, oversized rough terrain tyres and wheel fairings and horizontal stabiliser abrasion boots.

DIMENSIONS, EXTERNAL:

Wing span	10.97 m (36 ft 0 in)
Wing chord: at root	1.63 m (5 ft 4 in)
at tip	1.13 m (3 ft 8½ in)
Wing aspect ratio	7.4
Length overall	8.61 m (28 ft 3 in)
Height overall	2.83 m (9 ft 3½ in)
Wheel track	2.46 m (8 ft 1 in)
Wheelbase	1.76 m (5 ft 9¼ in)
Pilot's door (port): Max height	1.04 m (3 ft 5 in)
Max width	0.94 m (3 ft 1 in)
Cargo double door (stbd):	
Max height	1.00 m (3 ft 3¼ in)
Max width	1.09 m (3 ft 7 in)
Height to sill	0.64 m (2 ft 1 in)

DIMENSIONS, INTERNAL:

Cabin: Length	3.68 m (12 ft 1 in)
Max width	1.12 m (3 ft 8 in)
Max height	1.26 m (4 ft 1½ in)
Volume available for payload	2.87 m³ (101.2 cu ft)

AREAS:

Wings, gross	16.30 m² (175.5 sq ft)
Ailerons (total)	1.61 m² (17.35 sq ft)
Trailing-edge flaps (total)	2.68 m² (28.85 sq ft)
Fin	1.08 m² (11.62 sq ft)
Rudder, incl tab	0.65 m² (6.95 sq ft)
Tailplane	2.31 m² (24.84 sq ft)
Elevators, incl tab	1.86 m² (20.08 sq ft)

WEIGHTS AND LOADINGS:

Weight empty, standard: 206H	987 kg (2,176 lb)
T206H	1,034 kg (2,279 lb)
Max baggage	82 kg (180 lb)
Max fuel: 206H, T206H	239 kg (528 lb)
Max T-O and landing weight, 206H, T206H	1,632 kg (3,600 lb)
Max ramp weight: 206H	1,639 kg (3,614 lb)
T206H	1,641 kg (3,617 lb)
Max wing loading:	
206H, T206H	100.2 kg/m² (20.51 lb/sq ft)
Max power loading: 206H	7.30 kg/kW (12.00 lb/hp)
T206H	7.07 kg/kW (11.61 lb/hp)

PERFORMANCE:

Max level speed: 206H at S/L	151 kt (280 km/h; 174 mph)
T206H at FL170	178 kt (330 km/h; 205 mph)
Cruising speed at 75% power:	
206H at FL62	142 kt (263 km/h; 163 mph)
T206H at FL200	164 kt (304 km; 189 mph)
Stalling speed:	
206H, T206H: flaps up	62 kt (115 km/h; 72 mph)
flaps down	54 kt (100 km/h; 63 mph)
Max rate of climb at S/L: 206H	301 m (988 ft)/min
T206H	320 m (1,050 ft)/min
Service ceiling: 206H	4,785 m (15,700 ft)
T206H	8,230 m (27,000 ft)
T-O run: 206H, T206H	277 m (910 ft)
T-O to 15 m (50 ft): 206H	567 m (1,860 ft)
T206H	530 m (1,740 ft)

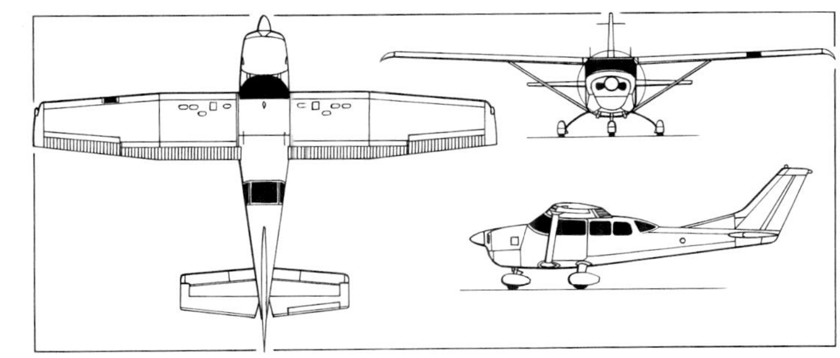

General arrangement of the Cessna 206H Stationair *(James Goulding)* ***NEW***/0580695

Landing from 15 m (50 ft): 206H, T206H ..... 425 m (1,395 ft)
Landing run: 206H, T206H ..... 224 m (735 ft)
Range with max fuel, 45 min reserves:
206H: at 75% power at FL65 ..... 605 n miles (1,120 km; 696 miles)
max range at FL65 ..... 730 n miles (1,352 km; 840 miles)
T206H: at 75% power at FL100 ..... 541 n miles (1,002 km; 622 miles)
at 75% power at FL200 ..... 568 n miles (1,052 km; 653 miles)
max range at FL200 ..... 692 n miles (1,281 km; 796 miles)

UPDATED

CESSNA 208 CARAVAN

Brazilian Air Force designation: C-98
US DoD designation: U-27A

TYPE: Light utility turboprop.

PROGRAMME: First flight of engineering prototype (N208LP) 9 December 1982; first production Caravan rolled out August 1984; FAA certification 23 October 1984; full production started 1985; wheeled float version certified March 1986. Cessna 208A certified 11 February 1985. First single-engined aircraft to achieve FAA certification for ILS in Cat. II conditions (Federal Express aircraft equipped); approval for IFR cargo operations 1989 made France and Ireland first European countries to allow single-engined public transport day/night IFR operation; also since approved in Canada, Denmark and Sweden. Russian certification of 208 and 208B achieved 4 September 1998.

The 1,000th of the 208B version was delivered on 12 December 2002 to Supap Puranitee of Thailand. 100th Caravan equipped with Wipaire 8000 floats delivered 16 May 2000. Total fleet operating time exceeded 6.1 million hours by December 2002, at which time Caravans were in service in 67 countries, with average fleet utilisation rate of 30,000 hours per month worldwide. Quoted dispatch reliability rate is 99.9 per cent. Basic utility model for passengers or cargo was Cessna 208. Engine flat rated at 447 kW (600 shp) to 3,800 m (12,500 ft). Federal Express Corporation version was 208A Cargomaster freighter with special features including T-O weight 3,629 kg (8,000 lb), Honeywell avionics, no cabin windows or starboard rear door, more cargo tiedowns, additional cargo net, underfuselage cargo pannier of composites materials, 15.2 cm (6 in) vertical extension of fin/rudder, jetpipe deflected to carry exhaust clear of pannier. These early versions now discontinued; military U-27A no longer promoted.

CURRENT VERSIONS: **Caravan 675:** Combines airframe of 208 and 208A with fully rated engine of 208B. Announced at NBAA Convention in Dallas, Texas, September 1997; FAA certification achieved in April 1998 with first delivery (N900RG; c/n 00277) 15 April to Riversville Aviation Company of New York in amphibious configuration; production follows on from 208/208A, beginning at c/n 00277.

Detailed description applies to Caravan 675, except where indicated.

208B: Stretched version, lacking cabin windows, and with ventral cargo pod as standard, developed at request of Federal Express. Commissioned by FedEx as **Super Cargomaster**; first flight (N9767F) 3 March 1986; certified 9 October 1986; first delivery to FedEx 31 October 1986. Features include 1.22 m (4 ft) fuselage plug aft of wing, payload of 1,587 kg (3,500 lb) and 12.7 m^3 (450 cu ft) of cargo volume; 503 kW (675 shp) P&WC PT6A-114A from 1991.

Grand Caravan: Announced at NBAA Convention 1990; passenger version of 208B, with cabin windows, accommodating up to 14 in quick-change interior.

CUSTOMERS: Federal Express Corporation received 40 208As and 260 208Bs. Recent customers include Ben Air Inc of Stauning, Denmark, which will take delivery of five Grand Caravans between 2000–2004 for onward sale to Scandinavian customers; Shandong Airlines of the People's Republic of China, which ordered one Grand Caravan and two Caravan 675s on amphibious floats for delivery in first quarter 2001, plus 37 options, of which two were exercised in 2002; China Northern Airlines of Shenyang, which took delivery of three Grand Caravans in 2002; Van Nuys Flight Center of California, which ordered 10 Grand Caravans for delivery between September 2001 and September 2005, four delivered to undisclosed Russian customers in the first half of 2002, four delivered to three cargo operators in Spain in the first half of 2002; Pudjiastuti of Indonesia, and Corporate Air of the Philippines, each of which ordered a Grand Caravan for delivery in 2004. Other customers include Royal Canadian Mounted Police (first amphibious version), Brazilian Air Force (eight), Chilean Army (three delivered from 7 February 1998 onwards), Liberian Army (one) and Malaysian Police (six delivered in 1994).

Total of 1,483 Caravans (all versions) delivered by 30 June 2004, including 102 in 1998; 87 in 1999, 92 in 2000, 75 in 2001, 80 in 2002, 57 in 2003, and 21 in the first six months of 2004. Out-of-production variants comprised two prototypes, 239 Model 208s and 37 Model 208As

COSTS: Caravan 675 US$1,600,000 (amphibian US$1,800,000); Super Cargomaster US$1,450,000; Grand Caravan US$1,750,000 (all 2004, typically equipped). Direct operating costs per hour: Caravan 675 US$209.95; Grand Caravan US$210.55; Super Cargomaster US$211.30; Amphibian US$226.00 (all 2004).

Cessna Caravan 675 floatplane *(Paul Jackson)* ***NEW***/0580762

DESIGN FEATURES: Launched as first all-new single-engined turboprop general aviation aircraft; intention was to replace de Havilland Canada Beavers and Otters, Cessna 180s, 185s and 206s in worldwide utility role. Main qualities advertised are high speed with heavy load, compatibility with unprepared strips, economy and reliability with minimum maintenance; can also carry weather radar, air conditioning and oxygen systems; optional packs for firefighting, photography, spraying, ambulance/hearse, border patrol, parachuting and supply dropping, surveillance and government utility missions; optional wheel or float landing gear.

Braced, high-wing design; tapered wings and tailplane; sweptback fin with dorsal fillet; auxiliary fins on floatplane.

Wing aerofoil NACA 23017.424 at root, 23012 at tip; dihedral 3° from root; incidence 2° 37' at root, –0° 36' at tip.

FLYING CONTROLS: Conventional and manual. Lateral control by small ailerons and slot-lip spoilers ahead of outer section of flaps; aileron trim standard; all tail control surfaces horn balanced; fixed tailplane with upper surface vortex generators ahead of elevator; elevator trim tabs; electrically actuated single-slotted flaps occupy more than 70 per cent of trailing-edge and deflect to maximum 30°. Control surface movements: ailerons +25/–16°; elevator +25/–20°; rudder ±25°.

STRUCTURE: All-metal. Fail-safe two-spar wing; conventional fuselage.

LANDING GEAR: Tricycle type; fixed. Fuselage-mounted tubular spring cantilever mainwheel units; oil-damped steerable nosewheel. Mainwheel tyres size 6.50–10 (8 ply); nosewheel 6.50–8 (8 ply). Oversize tyres, optional, mainwheels 8.50–10, nosewheel 22×8.00–8, and extended nosewheel fork, optional. Hydraulically actuated single-disc brake on each mainwheel. Certified with Wipline 8000 floats (with or without retractable land wheels).

POWER PLANT: One 503 kW (675 shp) P&WC PT6A-114A turboprop. McCauley 3GFR34C703/106GA-0 three-blade, constant-speed, reversible-pitch and feathering metal propeller. Alternatively, Hartzell HC-B3MN3/M10083. Integral fuel tanks in wings, total capacity 1,268 litres (335 US gallons; 279 Imp gallons), of which 1,257 litres (332 US gallons; 276.5 Imp gallons) are usable.

ACCOMMODATION: Pilot and up to nine passengers or 1,360 kg (3,000 lb) of cargo. Maximum seating capacity with FAR Pt 23 waiver is 14. Cabin has a flat floor with cargo track attachments for a combination of two- and three-abreast seating, with an aisle between seats. Forward-hinged door for pilot, with direct vision window, on each side of forward fuselage. Airstair door for passengers at rear of cabin on starboard side. Cabin is heated and ventilated. Optional air conditioning. Two-section horizontally split cargo door at rear of cabin on port side, flush with floor at bottom and with square corners. Upper portion hinges upward, lower portion forward 180°. Optional electrically operated, flight openable tambour roll-up door with airflow deflecting spoiler. In a cargo role, cabin will accommodate, typically, two D-size cargo containers or up to ten 208 litre (55 US gallon; 45.8 Imp gallon) drums.

Upgraded interior with new colour-co-ordinated fabrics and leather, and new seat design introduced in February 2001.

SYSTEMS: Electrical system is powered by 28 V 200 A starter/generator and 24 V 45 Ah lead-acid battery (24 V 40 Ah Ni/Cd battery optional). Standby electrical system, with 75 A alternator, optional. Hydraulic system for brakes only. Oxygen system, capacity 3,315 litres (117 cu ft), optional. Vacuum system standard. Cabin air conditioning system optional from c/n 208-00030 onwards. De-icing system, comprising electric propeller de-icing boots, pneumatic wing, wing strut and tail surface boots, electrically heated windscreen panel, heated pitot/static probe, ice detector light and standby electrical system, all optional.

AVIONICS: Customers' choice of Bendix/King and Garmin packages introduced in 2001; typical equipment listed below.

Comms: Dual Bendix/King KX 165 nav/com/glideslope, KT70 Mode S transponder, KHF 950 HF transceiver and KMA 24H audio control.

Radar: Honeywell RDR 200 weather radar with vertical profile.

Cessna 208B Grand Caravan *(Paul Jackson)* ***NEW***/0580760

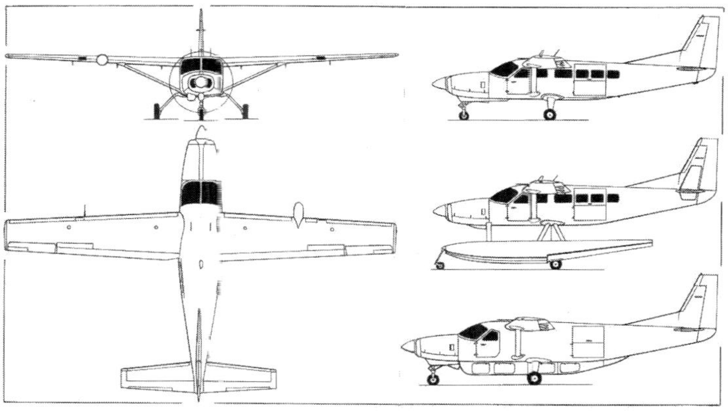

Cessna Caravan 675 with side views, top to bottom, of 208 Caravan 675, Caravan Amphibian and 208B Super Cargomaster *(Jane's/Dennis Punnett)* 0507256

Flight: Bendix/King KFC 225 flight director/autopilot, KLN 90 GPS, dual KR 87 ADF, KN 63 DME and KR 21 MKR. Garmin alternatives include GNS 340 com/nav/GPS; GNS 530 WAAS-upgradable IFR com/nav/GPS, with LOC and glideslope receiver, GMA 240 or GMA 340 audio panel and GTX 327 Mode C digital transponder with LCD display. Bendix/King KDR 510 flight information service and KMH 880 multi-hazard awareness system optional from 2003.

Instrumentation: Sensitive altimeter, electric clock, magnetic compass, attitude and directional gyros, true airspeed indicator, turn and bank indicator, vertical speed indicator, ammeter/voltmeter, fuel flow indicator, ITT indicator, oil pressure and temperature indicator.

EQUIPMENT: Standard equipment includes windscreen defrost, ground service plug receptacle, variable intensity instrument post lighting, map light, overhead courtesy lights (three) and overhead floodlights (pilot and co-pilot), approach plate holder, cargo tiedowns, internal corrosion proofing, vinyl floor covering, emergency locator beacon, partial plumbing for oxygen system, pilot's and co-pilot's adjustable fore/aft/vertical/reclining seats with armrest and five-point restraint harness, tinted windows, control surface bonding straps, heated pitot and stall warning systems, courtesy lights on wing underside, passenger reading lights, retractable crew steps (port side), rudder gust lock, tiedowns and towbar.

Optional equipment includes passenger seats, stowable folding utility seats, digital clock, fuel totaliser, turn co-ordinator, flight hour recorder, fire extinguisher, dual controls, co-pilot flight instruments, floatplane kit (from c/n 208-00030 onwards), hoisting rings (for floatplane), inboard fuel filling provisions (included in floatplane kit), ice detection light, flashing beacon, retractable crew step for starboard side, oversized tyres, electric trim system, oil quick drain valve and fan-driven ventilation system.

DIMENSIONS, EXTERNAL (L: landplane, A: amphibian):
Wing span ... 15.88 m (52 ft 1 in)
Wing chord: at root ... 1.98 m (6 ft 6 in)
at tip ... 1.22 m (4 ft 0 in)
Wing aspect ratio ... 9.7
Length overall: L ... 11.46 m (37 ft 7 in)
Height overall: L ... 4.52 m (14 ft 10 in)
A (on land) ... 4.98 m (16 ft 4 in)
Tailplane span ... 6.25 m (20 ft 6 in)
Wheel track: L ... 3.56 m (11 ft 8 in)
A ... 3.25 m (10 ft 8 in)
Wheelbase: L ... 3.54 m (11 ft 7½ in)
A ... 4.44 m (14 ft 7 in)
Propeller diameter ... 2.69 m (8 ft 10 in)
Airstair door: Height ... 1.27 m (4 ft 2 in)
Width ... 0.61 m (2 ft 0 in)
Cargo door: Height ... 1.27 m (4 ft 2 in)
Width ... 1.24 m (4 ft 1 in)

DIMENSIONS, INTERNAL (208):
Cabin: Length, excl baggage area ... 4.57 m (15 ft 0 in)
Max width ... 1.57 m (5 ft 2 in)
Max height ... 1.30 m (4 ft 3 in)
Volume ... 9.7 m³ (341 cu ft)
Cargo pannier: Length ... 5.58 m (18 ft 3½ in)
Width ... 1.28 m (4 ft 2½ in)
Depth ... 0.50 m (1 ft 7½ in)
Volume ... 2.4 m³ (83 cu ft)

AREAS:
Wings, gross ... 25.96 m² (279.4 sq ft)
Vertical tail surfaces (total, incl dorsal fin) ... 3.57 m² (38.41 sq ft)
Horizontal tail surfaces (total) ... 6.51 m² (70.04 sq ft)

WEIGHTS AND LOADINGS (A: 208-675 Caravan; B: Super Cargomaster; C: Grand Caravan; D: Caravan Amphibian):
Weight empty: A ... 1,832 kg (4,039 lb)
D ... 2,259 kg (4,980 lb)
Baggage capacity ... 147 kg (325 lb)
Cargo pannier capacity ... 372 kg (820 lb)
Max fuel weight ... 1,009 kg (2,224 lb)
Max T-O weight: A, D ... 3,629 kg (8,000 lb)
B, C ... 3,969 kg (8,750 lb)
Max ramp weight: A, D ... 3,645 kg (8,035 lb)
B, C ... 3,985 kg (8,785 lb)
Max landing weight: A, D ... 3,538 kg (7,800 lb)
B, C ... 3,855 kg (8,500 lb)
Max wing loading: A, D ... 139.8 kg/m² (28.63 lb/sq ft)
B, C ... 152.9 kg/m² (31.32 lb/sq ft)
Max power loading: A, D ... 7.21 kg/kW (11.85 lb/shp)
B, C ... 7.89 kg/kW (12.96 lb/shp)

PERFORMANCE (A: 208-675 landplane; D: Caravan Amphibian):
Max operating speed (VMO) ... 175 kt (325 km/h; 202 mph) IAS
Max cruising speed at FL100:
A ... 186 kt (344 km/h; 214 mph)
D ... 163 kt (302 km/h; 188 mph)
Stalling speed, power off:
flaps up: A ... 75 kt (139 km/h; 87 mph) CAS
D ... 74 kt (137 km/h; 86 mph) CAS
flaps down: A ... 61 kt (113 km/h; 71 mph) CAS
D ... 59 kt (110 km/h; 68 mph) CAS
Max rate of climb at S/L: A ... 376 m (1,234 ft)/min
D ... 338 m (1,110 ft)/min
Max certified altitude: A ... 7,620 m (25,000 ft)
D ... 6,100 m (20,000 ft)
T-O run: A ... 354 m (1,160 ft)
D: on land ... 335 m (1,100 ft)
on water ... 585 m (1,920 ft)
T-O to 15 m (50 ft): A ... 626 m (2,053 ft)
Landing from 15 m (50 ft) at S/L, without propeller reversal: A ... 505 m (1,655 ft)
D: on land ... 454 m (1,490 ft)
on water ... 590 m (1,935 ft)
Landing run at S/L, without propeller reversal:
A ... 227 m (745 ft)
D: on land ... 224 m (735 ft)
on water ... 319 m (1,045 ft)
Range with max fuel, at max cruise power, allowances for start, taxi and reserves stated:
A at FL100 45 min ... 932 n miles (1,726 km; 1,072 miles)
A at FL200 45 min ... 1,220 n miles (2,259 km; 1,404 miles)
D at FL100 30 min ... 990 n miles (1,833 km; 1,139 miles)
Range with max fuel at max range power, allowances as above:
A at FL100 ... 1,085 n miles (2,009 km; 1,248 miles)
A at FL200 ... 1,295 n miles (2,398 km; 1,490 miles)
g limits ... +3.8/–1.52

PERFORMANCE (B: Super Cargomaster; C: Grand Caravan):
Max cruising speed:
at FL100: B ... 175 kt (324 km/h; 201 mph)
C ... 184 kt (341 km/h; 212 mph)
Stalling speed, power off:
flaps up: B, C ... 78 kt (145 km/h; 90 mph)
flaps down: B, C ... 61 kt (113 km/h; 71 mph)
Max rate of climb at S/L: B ... 282 m (925 ft)/min
C ... 297 m (975 ft)/min
Service ceiling: B ... 6,950 m (22,800 ft)
C ... 7,220 m (23,700 ft)
T-O run: B ... 428 m (1,405 ft)
C ... 416 m (1,365 ft)
T-O to 15 m (50 ft): B ... 762 m (2,500 ft)
C ... 738 m (2,420 ft)
Landing from 15 m (50 ft), without propeller reversal:
B ... 530 m (1,740 ft)
C ... 547 m (1,795 ft)
Landing run, without propeller reversal:
B ... 279 m (915 ft)
C ... 290 m (950 ft)
Range with max fuel, max cruise power, allowances for start, taxi, climb and descent, plus 45 min reserves: at FL100:
B ... 862 n miles (1,596 km; 992 miles)
C ... 907 n miles (1,679 km; 1,043 miles)
at FL180, conditions as above:
B ... 1,044 n miles (1,933 km; 1,201 miles)
C ... 1,109 n miles (2,053 km; 1,276 miles)
Range with max fuel, max range power, allowances for start, taxi, climb and descent, plus 45 min reserves: at FL100:
B ... 963 n miles (1,783 km; 1,108 miles)
C ... 1,026 n miles (1,900 km; 1,180 miles)
at FL180, conditions as above:
B ... 1,076 n miles (1,992 km; 1,238 miles)
C ... 1,163 n miles (2,153 km; 1,338 miles)

UPDATED

CESSNA 510 CITATION MUSTANG

TYPE: Light business jet.

PROGRAMME: Initial design studies, then as turboprop, began in 1996-97; twin-jet configuration chosen in early 2001; announced at NBAA Convention, Orlando, Florida, 10 September 2002, when cabin mockup unveiled. New design but with wing scaled down from Citation Sovereign; smallest of Cessna jet family. More than 3,600 detail parts for the prototype and ground test articles had been completed by late July 2004. First flight anticipated in May 2005; certification to FAR Pt 23 (day, night, VFR, IFR, single pilot, known icing and RVSM) expected by mid-2006, with deliveries starting in fourth quarter.

Artist's impression of Cessna 510 Citation Mustang ***NEW***/0567788

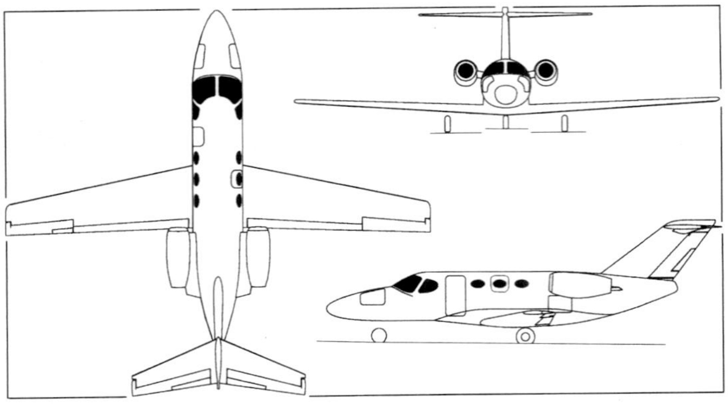

Cessna 510 Citation Mustang light business jet *(Paul Jackson)* ***NEW***/0580701

Final assembly will be at Cessna's Independence facility. Target production rate 150 to 200 per year.

CUSTOMERS: Some 200 firm orders by February 2004, of which approximately half were for US customers. Announced customers include three-times Formula One World Drivers' Champion Nelson Piquet, who ordered one on 11 September 2002 and Delta Aero-Taxi of Florence, Italy, which ordered seven on 21 October 2002 for its fractional ownership programme.

COSTS: US$2.295 million (2002).

DESIGN FEATURES: Low-wing, T tail and twin, podded, rear-mounted turbofans in usual Cessna business jet configuration, but second type to employ new Sovereign wing planform.

FLYING CONTROLS: Conventional and manual. Ailerons and elevator horn-balanced. Flight-adjustable tabs in port aileron, rudder and both elevators. Slotted flaps.

STRUCTURE: Generally of aluminium alloy; riveted and bonded joints, as required. Three-spar wing manufactured at Cessna's Columbus, Georgia, parts fabrication plant; centre-section, including fuel tank, mounted below fuselage to avoid penetration of pressure vessel. Semi-monocoque fuselage of frames and stringers with sheet metal skin; forward and aft pressure bulkheads.

LANDING GEAR: Nosewheel type; retractable. Single wheel on each leg. Trailing-link mainwheels retract inwards, nosewheel forwards; hydraulic actuation. Manually steerable nosewheel. Disc brakes with anti-skid on mainwheels.

POWER PLANT: Two pod-mounted Pratt & Whitney Canada PW615F turbofans, each rated at 6.00 kN (1,350 lb st) at ISA+15°C, with dual-channel FADEC. First PW615F ground run took place on 9 April 2004, and first flight, installed on Cessna-operated CitationJet testbed, on 27 April 2004 and had logged more than 80 hours of flight time by late July 2004. Engine certification scheduled for fourth quarter of 2005.

ACCOMMODATION: Two pilots on flight deck; four passengers in club configuration in cabin; rearmost pair of seats have centre console with single 12 V DC power outlet, dual cup holders and storage area. Door port side, ahead of wing; emergency exit, starboard, over wing. Chemical lavatory with privacy curtain opposite door. Recessed central aisle; storage and refreshments cabinets behind each pilot's seat, facing cabin. Nose and aft baggage compartments both unpressurised, with external, port-side access only. Cabin pressurised to 0.57 bar (8.3 lb/sq in), giving 2,440 m (8,000 ft) environment at 12,500 m (41,000 ft). Air conditioning vents in cabin sidewalls.

SYSTEMS: Electrical system supplied by two starter/generators via left and right busses; separate emergency bus. Hydraulic system for landing gear actuation; icing protection for wing, tailplane, windscreen and engine air intakes; certified for flight into known icing. Vapour cycle air conditioning system with distribution via sidewalls and overhead panels.

AVIONICS: Garmin G1000 as core system.

Comms: Dual VHF/COM; Mode S transponder; dual audio controls.

Flight: Dual VHF/NAV/LOC/GS; GPS; ADF; DME; marker beacon receiver; flight director; digital AHRS three-axis autopilot with yaw damper; air data computer; traffic/terrain awareness systems.

Instrumentation: Dual 25.4 cm (10 in) active matrix LCD screens for MFD functions; single 38.1 cm (15 in) MFD for display of navigation, engine indication, weather, traffic and terrain data. Standby instruments.

DIMENSIONS, EXTERNAL:
- Wing span ..... 12.88 m (42 ft 3 in)
- Length overall ..... 12.15 m (39 ft 10¼ in)
- Height overall ..... 4.19 m (13 ft 9 in)

DIMENSIONS, INTERNAL:
- Cabin: Length between pressure bulkheads ..... 4.42 m (14 ft 6 in)
  - Length excl flight deck ..... 2.74 m (9 ft 0 in)
  - Max width ..... 1.40 m (4 ft 7 in)
  - Max height ..... 1.37 m (4 ft 6 in)
  - Door width ..... 0.61 m (2 ft 0 in)
- Baggage compartment volume:
  - nose ..... 0.57 m^3(20.0 cu ft)
  - tail ..... 0.71 m^3(25.0 cu ft)

WEIGHTS AND LOADINGS:
- Basic operating weight ..... 2,336 kg (5,150 lb)
- Max fuel ..... 1,170 kg (2,580 lb)
- Payload with max fuel, one pilot ..... 272 kg (600 lb)

PERFORMANCE (estimated):
- Cruising speed at FL350 ..... 340 kt (630 km/h; 391 mph)
- T-O balanced field length ..... 950 m (3,120 ft)
- Certified ceiling ..... 12,500 m (41,000 ft)
- Range, VFR, 45 min reserves ..... 1,300 n miles (2,407 km; 1,496 miles)

UPDATED

CESSNA 525 CITATION CJ1

TYPE: Light business jet.

PROGRAMME: Original CitationJet announced at NBAA Convention 1989 as replacement for Citation 500 and I (production of which stopped 1985); first flight of FJ44 turbofans in Citation 500 April 1990; first flight of CitationJet (N525CJ) 29 April 1991; first flight of second (preproduction) prototype 20 November 1991; FAA certification for single-pilot operation received 16 October 1992; first customer delivery 31 March 1993. Russian certification achieved 4 September 1998. RVSM Group approval granted by FAA in January 2000.

Replacement Citation CJ1 announced at NBAA Convention 19 October 1998; first aircraft (N31CJ) completed late 1999; FAA certification 16 February 2000; first customer delivery 31 March 2000 to the Commercial Envelope Company of Deer Park, New York; 500th Cessna 525 series, a CJ1, rolled out 7 June 2002. By June 2002 a total of 493 Model 525s were in service, at which time their flight time totalled more than 550,000 hours.

CUSTOMERS: Total 359 CitationJets, plus three prototypes, delivered up to early 2000, when manufacture turned to CJ1. Total of 210 CJ1s delivered by 30 June 2004, including 30 delivered in 2002, 22 in 2003 and 10 in the first six months of 2004. Recent customers include Avemex SA of Toluca, Mexico (one); Taxi Marilia (TAM) of Brazil (three, including c/n 0500); Atlas Air Service GmbH of Germany (four), and Cessna's TAG Aviation USA's fractional ownership operation, CitationShares, which has received one, and the Chilean Air Force, which took delivery of three in November 2001. Total fleet time (CJ and CJ1) 697,488 hours in August 2003.

COSTS: US$4.024 million, typically equipped (2003).

DESIGN FEATURES: Small business jet of conventional appearance; wings attached below fuselage; T tail; tapered wings and tailplane; podded engines mounted clear of upper rear fuselage. Natural Laminar Flow (NLF) aerofoil section. CJ1 is generally identical to CitationJet; increased maximum T-O, ramp and landing weights to provide improved range/payload; Pro Line 21 avionics.

FLYING CONTROLS: Conventional and manual. Ailerons and elevators horn balanced; trim tab on port aileron; trim tabs on rudder and both elevators; hydraulically actuated single-slotted flaps, deflections 15, 35 and 60°, last-named for ground use only, and selectable to any intermediate position between 0 and 35°. Speed brake in upper and lower surface of each wing.

STRUCTURE: All-metal. Three-spar wing.

LANDING GEAR: Hydraulically retractable tricycle type, with single wheel on each unit. Trailing-link main units retract inward into wing; nose gear retracts forward. Main tyres 22×7.75-10 (8 ply) tubeless; nose 18×4.4 (6 ply) tubeless.

POWER PLANT: Two 8.45 kN (1,900 lb st) Williams FJ44-1A turbofans pod-mounted, with thrust attenuators. Two independent fuel systems, with fuel contained in integral wing tanks, and single filler port in each wing; see under Weights and Loadings.

ACCOMMODATION: Crew of two on flight deck. Main cabin with standard seating for five passengers, one on sideways-facing seat at front of cabin, with four in club arrangement; tray tables which stow in elbow rails, and individual reading lights and ventilation ducts, standard.

SYSTEMS: Digitally controlled pressurisation system, maximum differential 0.58 bar (8.5 lb/sq in).

AVIONICS: Rockwell Collins Pro Line 21 as core system.

Comms: Honeywell CNI5000 radios and radio altimeter.

Radar: Rockwell Collins RTA 800 colour weather radar.

Flight: Bendix/King KLN-900 FMS. Dual VHF nav receivers, ADF, DME, RMI, Bendix/King KLN-900 GPS.

Instrumentation: Two-tube EFIS with 203 × 253 mm (8 × 10 in) adaptive flight displays (third optional), LCD attitude director indicator (ADI) and horizontal situation indicator (HSI) for co-pilot.

DIMENSIONS, EXTERNAL:
- Wing span ..... 14.26 m (46 ft 9½ in)
- Wing aspect ratio 9.1

Cessna 525 Citation CJ1 six/seven-seat business jet *(Paul Jackson)* ***NEW***/0580772

Cessna 525 Citation CJ1 *(Paul Jackson)* ***NEW***/0580776

Cessna 525 Citation CJ1 *(Paul Jackson)* ***NEW***/0580775

Length overall ... 12.98 m (42 ft 7 in)
Height overall ... 4.20 m (13 ft 9¼ in)
Tailplane span ... 5.64 m (18 ft 6 in)
Wheel track ... 3.96 m (13 ft 0 in)
Wheelbase ... 4.68 m (15 ft 4¼ in)
Crew/passenger door: Height ... 1.29 m (4 ft 2¾ in)
  Width ... 0.60 m (1 ft 11½ in)

DIMENSIONS, INTERNAL:
Cabin:
  Length: between pressure bulkheads ... 4.80 m (15 ft 9 in)
    excl cockpit ... 3.35 m (11 ft 0 in)
  Max width ... 1.47 m (4 ft 10 in)
  Max height ... 1.45 m (4 ft 9 in)
Baggage compartment volume:
  nose ... 0.58 m³ (20.4 cu ft)
  cabin ... 0.11 m³ (4.0 cu ft)
  tailcone ... 0.86 m³ (30.2 cu ft)
  total ... 1.55 m³ (54.6 cu ft)

AREAS:
Wings, gross ... 22.30 m² (240.0 sq ft)
Vertical tail surfaces (total, incl tab) ... 4.35 m² (46.8 sq ft)
Horizontal tail surfaces (total, incl tabs) ... 5.64 m² (60.7 sq ft)

WEIGHTS AND LOADINGS:
Weight empty, typically equipped ... 3,025 kg (6,670 lb)
Basic operating weight, typical ... 3,112 kg (6,860 lb)
Max usable fuel weight ... 1,460 kg (3,220 lb)
Payload with max fuel, single pilot ... 281 kg (620 lb)
Max T-O weight ... 4,808 kg (10,600 lb)
Max landing weight ... 4,445 kg (9,800 lb)
Max ramp weight ... 4,853 kg (10,700 lb)
Max wing loading ... 215.6 kg/m² (44.17 lb/sq ft)
Max power loading ... 284 kg/kN (2.79 lb/lb st)

PERFORMANCE:
Max operating speed (V_{MO}):
  S/L to FL305 ... 263 kt (487 km/h; 302 mph) IAS
  above FL305 ... M0.71
Max cruising speed at 3,992 kg (8,800 lb) AUW at FL350 .. 381 kt (706 km/h; 438 mph)

Cessna 525 Citation CJ1 *(Paul Jackson)* ***NEW***/0580774

Cessna 525 Citation CJ1 *(Paul Jackson)* ***NEW***/0580773

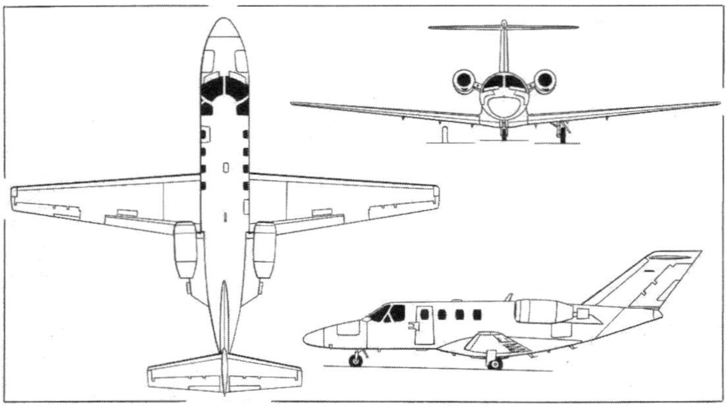

Cessna 525 Citation CJ1 (two Williams FJ44 turbofans) *(Jane's/Dennis Punnett)* 0507257

Stalling speed, landing configuration at max landing weight 83 kt (154 km/h; 96 mph) CAS
Max rate of climb at S/L 985 m (3,230 ft)/min
Rate of climb at S/L, OEI 259 m (850 ft)/min
Max certified altitude 12,500 m (41,000 ft)
Service ceiling, OEI 6,462 m (21,200 ft)
FAR Pt 25 T-O field length: 1,000 m (3,280 ft)
Landing field length 841 m (2,760 ft)
IFR range with 100 n mile (185 km; 115 mile) alternate and 45 min reserves 1,248 n miles (2,311 km; 1,436 miles)
VFR range with reserves 1,475 n miles (2,731 km; 1,697 miles)

OPERATIONAL NOISE LEVELS:
T-O 73.6 EPNdB
Approach 89.7 EPNdB
Sideline 83.6 EPNdB

UPDATED

CESSNA 525A CITATION CJ2

TYPE: Light business jet.

PROGRAMME: Design began 1 May 1998; construction of prototype started August 1998; announced at NBAA Convention at Las Vegas, Nevada, 18 October 1998; first flight of prototype (N2CJ, a rebuilt CitationJet) 27 April 1999, followed by first preproduction aircraft (N525AZ) on 15 October and second (N765CT) in mid-December 1999; total of 1,008 flight test hours accumulated by three aircraft by 9 April 2000; first production aircraft c/n 525A-0003/(N132CJ) rolled out 14 January 2000; FAA FAR Pt 23 certification achieved 21 June 2000; public debut at EAA AirVenture Oshkosh in July 2000; first customer delivery, 30 November 2000, was N200KP to King Pharmaceuticals of Tennessee (second production aircraft). By April 2004 the 200 CJ2s then in service had logged a total of more than 80,000 flight hours.

CUSTOMERS: Total of 76 firm orders held at time of launch; by December 2000 order backlog accounted for planned production until first quarter of 2004. Recent customers include WD40 Something LLC (second preproduction aircraft, re-registered N400WD), Avemex SA of Toluca, Mexico (one); Taxi Aereo Marilia (TAM) of Brazil (five); and Atlas Air Service GmbH of Germany (three). First delivery to overseas customer was D-IBBB, which arrived at Bremen for Atlas on 21 June 2001. 100th production CJ2 (N170TM) delivered to TitleMax Aviation Inc of Savannah, Georgia, on 27 August 2002. 200th production CJ2 delivered to Unijet of France 8 April 2004. Total of 86 delivered in 2002, 56 in 2003, and 14 in the first six months of 2004.

Cessna 525A Citation CJ2 *(Paul Jackson)* ***NEW***/0580769

COSTS: US$5.214 million, typically equipped (2003).

DESIGN FEATURES: Development of CitationJet/CJ1 (which see) with fuselage stretched by 1.30 m (4 ft 3 in) in cabin and tailcone areas (extra two cabin windows), wing span increased by 0.84 m (2 ft 9 in), swept tailplane of greater span, Williams FJ44-2C turbofans and Rockwell Collins Pro Line 21 avionics. Design goals included improvements in cabin room and comfort, speed and range over CitationJet. Common type rating with CJ1.

Description for the Citation CJ1 applies also to the Citation CJ2, except as follows:

FLYING CONTROLS: Spoilers, 60° flap deflection and engine nacelle thrust attenuators provide lift dump function during landing roll.

STRUCTURE: Primarily metal, with composites in non-critical areas such as fairings, wing and tailplane tips and exhaust nozzles. Three-spar wing.

POWER PLANT: Two pod-mounted Williams FJ44-2C turbofans, each flat rated at 10.68 kN (2,400 lb st) at 22°C (72°F), with thrust attenuators. Integral fuel tank in each wing, total capacity 2,245 litres (593 US gallons; 494 Imp gallons), of which 2,222 litres (587 US gallons; 489 Imp gallons) are usable; overwing gravity refuelling.

ACCOMMODATION: One or two crew on flight deck; six passengers in standard centre club layout with refreshment centre on starboard side; customised interiors to customer's choice. Flight accessible baggage area at rear of cabin; other baggage compartments in nose and tailcone. Cabin is pressurised, heated and air conditioned. Six cabin windows per side. Main door on port side forward of wing; emergency exit starboard aft.

SYSTEMS: Pressurisation system, maximum pressure differential 0.61 bar (8.9 lb/sq in). Open-centre 103.4 bar (1,500 lb/sq in) hydraulic system for operation of landing gear, flaps, speedbrakes and thrust attenuators. Vapour cycle air conditioning system. Electrical system supplied by battery and two engine-driven starter/generators. Oxygen system, capacity 623 litres (22.0 cu ft) standard, 1,417 litres (50.0 cu ft) optional.

AVIONICS: Rockwell Collins Pro Line 21 suite as core system.

Comms: Honeywell CNI-5000 panel-mounted radios; dual VHF com; dual Mode S transponders; Artex 110-4 ELT; dual Telex microphones and headsets, all standard. CVR optional.

Radar: Rockwell Collins RTA-800 solid-state colour weather radar.

Flight: Honeywell KLN 900 GPS FMS, VOR, DME, ADF, Rockwell Collins digital autopilot, ADC-3000 digital air data computer AHC-3000 AHRS and ALT-55B radar altimeter standard.

Instrumentation: Two-tube EFIS with 203 × 254 mm (8 × 10 in) PFD and MFD active matrix LCDs; co-pilot's PFD optional.

DIMENSIONS, EXTERNAL:
Wing span 15.19 m (49 ft 10 in)
Wing aspect ratio 9.4
Length overall 14.53 m (47 ft 8 in)
Height overall 4.27 m (14 ft 0 in)
Tailplane span 6.32 m (20 ft 9 in)
Wheel track 4.85 m (15 ft 11 in)
Wheelbase 5.44 m (17 ft 10 in)
Passenger door: Height 1.29 m (4 ft 2¾ in)
Width 0.60 m (1 ft 11½ in)

DIMENSIONS, INTERNAL:
Cabin: Length: overall 5.74 m (18 ft 10 in)
excl cockpit 4.19 m (13 ft 9 in)

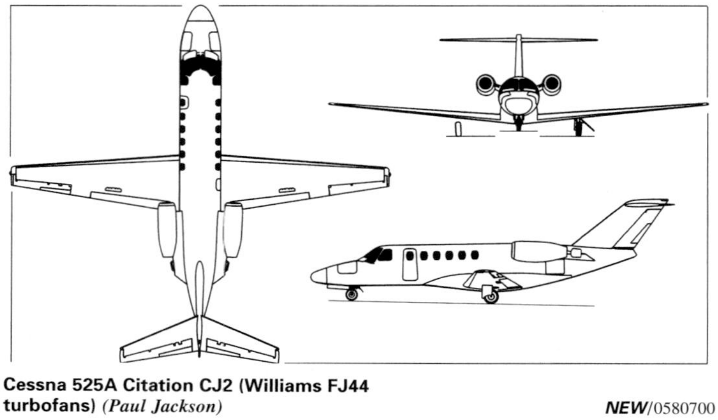

Cessna 525A Citation CJ2 (Williams FJ44 turbofans) *(Paul Jackson)* ***NEW***/0580700

Max width ... 1.47 m (4 ft 10 in)
Max height ... 1.45 m (4 ft 9 in)
Baggage volume: nose ... 0.58 m^3 (20.4 cu ft)
cabin ... 0.11 m^3 (4.0 cu ft)
tailcone ... 1.42 m^3 (50.0 cu ft)
total ... 2.11 m^3 (74.4 cu ft)

AREAS:
Wings, gross ... 24.53 m^2 (264.0 sq ft)
Vertical tail surfaces ... 4.35 m^2 (46.8 sq ft)
Horizontal tail surfaces ... 6.57 m^2 (70.7 sq ft)

WEIGHTS AND LOADINGS:
Weight empty, standard ... 3,361 kg (7,410 lb)
Max fuel ... 1,784 kg (3,932 lb)
Payload with max fuel ... 345 kg (760 lb)
Baggage capacity: nose ... 181 kg (400 lb)
cabin ... 45 kg (100 lb)
tailcone ... 272 kg (600 lb)
total ... 499 kg (1,100 lb)
Max T-O weight ... 5,613 kg (12,375 lb)
Max landing weight ... 5,216 kg (11,500 lb)
Max ramp weight ... 5,670 kg (12,500 lb)
Max zero-fuel weight ... 4,218 kg (9,300 lb)
Max wing loading ... 228.9 kg/m^2 (46.88 lb/sq ft)
Max power loading ... 263 kg/kN (2.58 lb/lb st)

PERFORMANCE (estimated):
Max operating Mach No. (MMO) ... 0.72
Max operating speed (VMO):
S/L to FL80 ... 260 kt (481 km/h; 299 mph)
FL80 to FL293 ... 275 kt (509 km/h; 316 mph)
above FL293 ... M0.72
Max cruising speed at FL330
... 410 kt (759 km/h; 472 mph)
Econ cruising speed at FL410 ... 332 kt (615 km/h; 382 mph)
Stalling speed in landing configuration at MLW ... 86 kt (160 km/h; 99 mph)
Max rate of climb at S/L ... 1,180 m (3,870 ft)/min
Rate of climb at S/L, OEI ... 342 m (1,123 ft)/min
Time to: FL370 ... 17 min
FL430 ... 36 min
Max certified altitude ... 13,715 m (45,000 ft)
FAR Pt 25 T-O balanced field length ... 1,042 m (3,420 ft)
Landing field length at max landing weight ... 908 m (2,980 ft)
Range: NBAA IFR reserves ... 1,550 n miles (2,870 km; 1,783 miles)
VFR reserves ... 1,738 n miles (3,218 km; 2,000 miles)

OPERATIONAL NOISE LEVELS
T-O ... 74.5 EPNdB
Approach ... 91.4 EPNdB
Sideline ... 88.8 EPNdB

UPDATED

CESSNA 525B CITATION CJ3

TYPE: Light business jet.

PROGRAMME: Announced on eve of NBAA Convention, Orlando, Florida, 9 September 2002; fuselage mockup shown following day. First flight (N3CJ, c/n 711) 17 April 2003, followed by first pre-production aircraft (N753CJ, c/n 525B-0001) 11 August 2003; public debut (N753CJ) at NBAA Convention, Orlando, Florida 7 October 2003; second pre-production aircraft (N763CJ, c/n 525B-0002) flew 6 November 2003; total of 900 hours in 540 sorties completed by all three aircraft by 24 May 2004, at which time the static test article had completed more than 200 test conditions, including 105,000 hours, equivalent to five life cycles, of landing gear testing; first production aircraft rolled-out 13 February 2004; engine certification achieved 31 July 2004; aircraft certification to FAR Pt 23, including single pilot operation and RVSM approval, mid-2004; deliveries from late 2004.

Cessna Citation CJ3 pre-production aircraft *(Paul Jackson)* ***NEW***/0580833

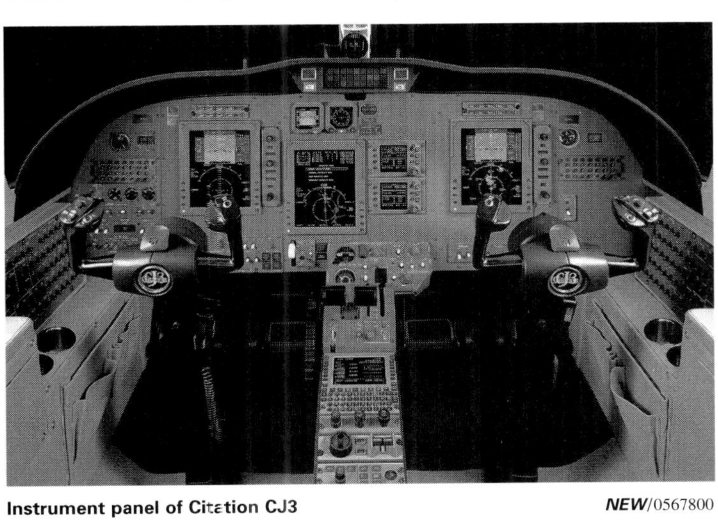
Instrument panel of Citation CJ3 ***NEW***/0567800

Citation CJ3 cabin interior ***NEW***/0567801

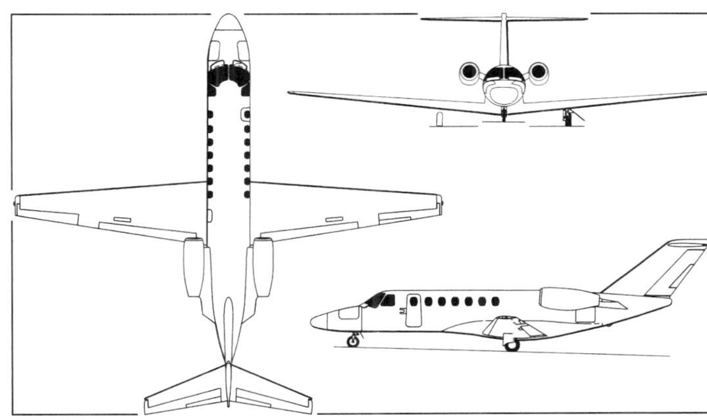
Cessna 525B Citation CJ3 *(Paul Jackson)* ***NEW***/0580698

DESIGN FEATURES: Further development of CJ1/CJ2 family. Fuselage stretched by 1.00 m (3 ft 31/2 in) (seven windows each side), giving 0.61 m (2 ft 0 in) extra in cabin, and wing span extended by 1.03 m (3 ft 41/4 in). Uprated engines with 14 per cent more T-O thrust and 12 per cent more in cruise. Improved range and passenger comfort.

Description for Citation CJ1/CJ2 applies to CJ3, except as follows:

CUSTOMERS: Orders for more than 100 held by May 2004.

COSTS: US$5,995,000 typically equipped (2003).

DESIGN FEATURES: Wing dihedral 5° 0'; sweepback 0° at 31 per cent chord; taper ratio 0.30. Tailplane sweepback 20° 0' at 25 per cent chord; no dihedral; taper ratio 0.43. Fin sweepback 49° 0' at 25 per cent chord; taper ratio 0.56.

FLYING CONTROLS: Flaps fitted (deflection 55° on ground).

POWER PLANT: Two 12.37 kN (2,780 lb st) Williams FJ44-3A turbofans with FADEC.

ACCOMMODATION: As CJ2, but chemical lavatory starboard, rear, opposite baggage area.

SYSTEMS: Hydraulic system with two engine-driven pumps, pressure 103.5 bar (1,500 lb/sq in). Secondary hydraulic system for mainwheel brakes. Cabin pressurisation as CJ2. Emergency oxygen system reservoir 1,134 litres (40.0 cu ft). Electrical system 28 V DC includes two engine-driven generators and one 44 Ah Ni/Cd battery.

Engine bleed air for anti-icing of wing leading-edges engine air intakes and windscreen, and inflation of de-icing boots on horizontal leading-deges. Backup windscreen alcohol de-icing.

AVIONICS: Rockwell Collins Pro Line 21 suite as core system.

Comms: Dual VHF-4000 transceivers. Dual TDR-94 Mode S transponders. Optional CVR. Artex C-406-2 three-frequency ELT. Provision for CVR.

Radar: As CJ2.

Flight: Collins NAV-4000/NAV-4500 dual VOR, LOC, glideslope and MKR; single (optional second) ADF. DME-4000 DME. AHC-3000 AHRS, ADC-3000 ADC, FMS-3000 FMS with performance database, GPS-4000A GPS, ALT-4000 radio altimeter, Goodrich Skywatch HP TCAS, Goodrich TAWS 8000 Landmark terrain-avoidance, dual airdata computers, standard, and FAA-approved electronic chart system. Maintenance diagnostic system.

Instrumentation: Three-tube EFIS with 203 × 254 mm (8 × 10 in) PFD and MFD active matrix LCDs. Goodrich GH-3000 standby instruments; Smiths standby EHSI.

DIMENSIONS, EXTERNAL:
Wing span ... 16.13 m (52 ft 11 in)
Wing mean chord ... 1.90 m (6 ft 23/4 in)
Wing aspect ratio ... 9.5
Length: overall ... 15.29 m (50 ft 2 in)
fuselage ... 13.79 m (45 ft 23/4 in)
Height overall ... 4.62 m (15 ft 2 in)
Tailplane span ... 6.35 m (20 ft 10 in)
Tailplane mean chord ... 1.09 m (3 ft 7 in)
Wheel track ... 4.88 m (16 ft 0 in)
Wheelbase ... 6.10 m (20 ft 0 in)
Door:
Height ... 1.29 m (4 ft 3 in)
Max width ... 0.60 m (1 ft 111/2 in)
Emergency exit width ... 0.51 m (1 ft 8 in)

DIMENSIONS, INTERNAL:
Cabin: Length overall ... 6.35 m (20 ft 10 in)
Max width ... 1.45 m (4 ft 91/4 in)
Max height ... 1.45 m (4 ft 9 in)
Baggage volume ... as CJ2

AREAS:
Wings, gross .... 27.32 m^2 (294.1 sq ft)
Vertical tail surfaces (total) .... 5.23 m^2 (56.30 sq ft)
Horizontal tail surfaces (total) .... 6.57 m^2 (70.68 sq ft)

WEIGHTS AND LOADINGS:
Standard weight empty .... 3,701 kg (8,160 lb)
Max fuel .... 2,136 kg (4,710 lb)
Baggage capacity .... as CJ2
Max ramp weight .... 6,382 kg (14,070 lb)
Max T-O weight .... 6,291 kg (13,870 lb)
Max landing weight .... 5,783 kg (12,750 lb)
Max zero-fuel weight .... 4,767 kg (10,510 lb)
Payload with max fuel .... 395 kg (870 lb)
Max wing loading .... 230.3 kg/m^2 (47.16 lb/sq ft)
Max power loading .... 254 kg/kN (2.49 lb/lb st)

PERFORMANCE (estimated):
Max operating speed (VMO): S/L to FL80 .... 260 kt (482 km/h; 299 mph) CAS
FL80 to FL293 .... 275 kt (509 km/h; 316 mph) CAS
above FL 293 .... M0.72
Max cruising speed, at mid-cruise weight: at FL350 .... 414 kt (767 km/h; 476 mph)
at FL390 .... 408 kt (756 km/h; 496 mph)
Stalling speed, landing configuration at MLW .... 86 kt (159 km/h; 99 mph) CAS
Time to: FL410 .... 22 min
FL450 .... 35 min
Max certified altitude .... 13,715 m (45,000 ft)
T-O balanced field length .... 1,052 m (3,450 ft)
Landing runway length .... 936 m (3,070 ft)
Range: with max fuel, VFR reserves .... 1,900 n miles (3,518 km; 2,186 miles)
with four passengers, NBAA IFR reserves .... 1,771 n miles (3,280 km; 2,038 miles)

UPDATED

CESSNA 550 CITATION BRAVO

TYPE: Business jet.

PROGRAMME: Cessna 550 Citation II first flew 31 January 1977; Bravo announced at Farnborough Air Show September 1994; replaced Citation II; prototype (N550BB, c/n 0734) first flight 19 April 1995; initial production aircraft (N801BB, c/n 0801) flew mid-1996; FAA certification in January 1997; first delivery, to Firebond Corporation of Minden, Louisiana, on 25 February 1997. Russian certification achieved 4 September 1998. Certification for operation into London City Airport achieved in May 2002.

CUSTOMERS: First 18 months' production had been sold by time of initial delivery. Earlier production in this series accounted for 621 Model 550 Citation IIs, 97 Model 551 Citation II SPs and 15 Model 552 T-47A Citations, or 733 in all. Total of 252 Bravos delivered by 30 September 2003, comprising 28 in 1997, 34 in 1998, 36 in 1999, 54 in 2000, 48 in 2001, 41 in 2002, 31 in 2003, and 12 in the first six months of 2004. Recent customers include Avemex SA of Toluca, Mexico (two, for delivery from May 2002); Taxi Aereo Marilia (TAM) of Brazil (three, for delivery from April 2002); Flying Partners CV of Antwerp, Belgium, which has ordered two for its shared ownership programme, for delivery from January 2002; Cessna's and TAG Aviation USA's fractional ownership operation, CitationShares, which placed an initial order for six, for delivery before the end of 2000; and The Company Jet of Grand Rapids, Michigan, which has ordered five for delivery in 2003 for its fractional ownership operation. By August 2003 some 944 Citation IIs, II SPs and Bravos had been delivered, and total fleet time stood at 4,584,470 hours.

COSTS: US$5.708 million, typically equipped (2003).

DESIGN FEATURES: Small/mid-size business jet. Based on Citation II airframe; tapered wing; tapered, mid-set tailplane; sweptback fin, sweepback at quarter-chord 33° 0'; podded engines mounted clear of upper rear fuselage; wide track landing gear.

Wing aerofoil NACA 23014 (modified) at centreline, NACA 23012 at wing station 247.95; dihedral 4°; tailplane dihedral 9°.

FLYING CONTROLS: Conventional and manual. Trim tab on port aileron is manually operated; manual rudder trim; electric elevator trim tab with manual standby; electrically actuated single-slotted flaps, maximum deflection 40°; hydraulically actuated airbrake.

STRUCTURE: Two primary, one auxiliary metal wing spars; three fuselage attachment points; conventional ribs and stringers. All-metal pressurised fuselage with fail-safe design providing multiple load paths.

LANDING GEAR: Hydraulically retractable tricycle type with single wheel on each unit. Trailing-link main units retract inward into the wing, nose gear forward. Free-fall and pneumatic emergency extension systems. Steerable nosewheel maximum delfection ±20°, or ±95° for ground handling and towing. Mainwheel tyres H22.0×8.25-10 (14 ply); steerable nosewheel with tyre size 18×4.4DD (10 ply); all tubeless. Toe-actuated multiple disc brakes on mainwheels. Parking brake and pneumatic emergency brake system. Anti-skid system standard.

POWER PLANT: Two 12.84 kN (2,887 lb st) Pratt & Whitney Canada PW530A turbofans; Nordam target-type thrust reversers standard. Two independent fuel systems with integral tank in each wing, with combined usable capacity of 2,725 litres (720 US gallons; 600 Imp gallons).

ACCOMMODATION: Crew of two on separate flight deck, on fully adjustable seats, with seat belts and inertia reel shoulder harness. Sun visors standard. Fully carpeted main cabin equipped with seats for seven to 10 passengers (seven standard, on pedestal-mounted seats). Main baggage areas in nose and tailcone; flight accessible baggage area at rear of cabin.

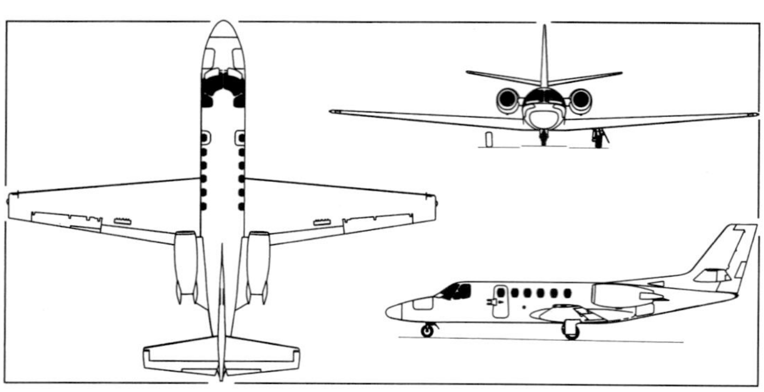

Cessna 550 Citation Bravo *(Paul Jackson)* ***NEW***/0580699

Cessna 550 Citation Bravo *(Paul Jackson)* ***NEW***/0580766

Refreshment centre standard. Cabin is pressurised, heated and air conditioned. Individual reading lights and air inlets for each passenger. Dropout constant-flow oxygen system for emergency use. Cabin door with integral airstair at front on port side and one emergency exit on starboard side. Doors on each side of nose baggage compartment. Tinted windows, each with curtains. Pilot's storm window, birdproof windscreen with de-fog system, anti-icing, standby alcohol anti-icing and bleed air rain removal system.

SYSTEMS: Pressurisation system supplied with engine bleed air, maximum pressure differential 0.61 bar (8.9 lb/sq in), maintaining a sea level cabin altitude to 7,189 m (23,586 ft), or a 2,440 m (8,000 ft) cabin altitude to 13,715 m (45,000 ft). Hydraulic system, pressure 103.5 bar (1,500 lb/sq in), with two pumps to operate landing gear and speed brakes. Separate hydraulic system for wheel brakes. Electrical system supplied by two 28 V 400 A DC starter/generators, with two 350 VA inverters and 24 V 44 Ah Ni/Cd battery. Oxygen system of 1,814 litre (64 cu ft) capacity includes two crew demand masks and five dropout constant-flow masks for passengers. Engine fire detection and extinguishing systems. Wing leading-edges electrically de-iced ahead of engines; pneumatic de-icing boots on outer wings and on leading-edges of tailplane and fins; engine inlets and windscreen have anti-ice protection via engine bleed air, with back-up alcohol system for windscreen; pitot tubes and static ports have electric anti-icing.

AVIONICS: Honeywell Primus 1000 as core system.

Comms: Dual Honeywell KY-196B VHF transceivers with 25 kHz and 8.33 kHz channel spacing; dual Avtech audio amplifiers; Dual Telex hand microphones, headsets and cockpit speackers; dual Honeywell Mode S transponders; L3 Communications FA2100 CVR with underwater locator beacon; Artex 110-4 ELT.

Radar: Honeywell Primus 660 colour weather radar.

Flight: Honeywell GNS-X_{LS} FMS; three-axis fail-passive autopilot; dual Honeywell navigation systems including VOR/LOC/GS/MKR, DME, digitally tuned ADF. Dual integrated computers combine EFIS display functions with flight guidance function; dual Honeywell VG-14A vertical gyros and C-14D compass systems supply primary heading and attitude information to cockpit displays.

Instrumentation: EFIS with two 178 × 203 mm (7 × 8 in) screens for primary flight display (PFD) and additional multifunction display (MFD) of same size; standby 3 in HSI; Davtron clock.

DIMENSIONS, EXTERNAL:
Wing span .... 15.75 m (51 ft 8 in)
Wing aspect ratio 8.4
Length overall .... 14.40 m (47 ft 3 in)
Height overall .... 4.57 m (15 ft 0 in)
Tailplane span .... 5.79 m (19 ft 0 in)
Wheel track .... 3.99 m (13 ft 1 in)
Wheelbase .... 5.64 m (18 ft 6 in)
Crew/passenger door: Height .... 1.21 m (3 ft 11½ in)
Width .... 0.60 m (1 ft 11½ in)

DIMENSIONS, INTERNAL:
Cabin: Length (between pressure bulkheads) .... 6.37 m (20 ft 11 in)
excl cockpit .... 4.75 m (15 ft 7 in)
Max width .... 1.48 m (4 ft 10¼ in)
Max height .... 1.43 m (4 ft 8¼ in)
Baggage volume: nose .... 0.44 m^3 (15.6 cu ft)
cabin .... 0.31 m^3 (10.8 cu ft)
tailcone .... 0.79 m^3 (28.0 cu ft)

AREAS:
Wings, gross .... 30.00 m^2 (322.9 sq ft)
Vertical tail surfaces (total) .... 4.73 m^2 (50.9 sq ft)
Horizontal tail surfaces (total, incl tab) .... 6.48 m^2 (69.8 sq ft)

WEIGHTS AND LOADINGS:
Standard empty weight .... 3,992 kg (8,800 lb)
Weight empty, typically equipped .... 4,073 kg (8,980 lb)
Max fuel weight (usable) .... 2,204 kg (4,860 lb)
Baggage capacity
nose .... 159 kg (350 lb)
cabin .... 136 kg (300 lb)
tailcone .... 227 kg (500 lb)
Max T-O weight .... 6,713 kg (14,800 lb)
Max ramp weight .... 6,804 kg (15,000 lb)
Max landing weight .... 6,123 kg (13,500 lb)
Max zero-fuel weight .... 5,126 kg (11,300 lb)
Payload with max fuel .... 370 kg (816 lb)
Max wing loading .... 223.8 kg/m^2 (45.83 lb/sq ft)
Max power loading .... 262 kg/kN (2.56 lb/lb st)

PERFORMANCE:
Max operating speed (VMO):
S/L to FL80 .... 260 kt (481 km/h; 299 mph)
FL80 to FL279 .... 275 kt (509 km/h; 316 mph)
at FL279 and above M0.70
Max cruising speed at FL330
.... 402 kt (744 km/h; 463 mph)
Stalling speed in landing configuration at max landing weight ....
86 kt (160 km/h; 99 mph) CAS
Max rate of climb at S/L .... 972 m (3,190 ft)/min
Rate of climb at S/L, OEI .... 345 m (1,133 ft)/min
Max certified altitude .... 13,715 m (45,000 ft)
Service ceiling, OEI .... 8,458 m (27,750 ft)
T-O balanced field length .... 1,098 m (3,600 ft)
FAR Pt 25 landing field length at max landing weight .... 970 m (3,180 ft)
NBAA IFR range, 100 n mile (185 km; 115 mile) alternate, plus 45 min reserves ....
1,744 n miles (3,230 km; 2,007 miles)
Range with max fuel, VFR reserves .... 2,000 n miles (3,704 km; 2,302 miles)

OPERATIONAL NOISE LEVELS:
T-O ... 73.7 EPNdB
Approach ... 91.2 EPNdB
Sideline ... 85.2 EPNdB

UPDATED

CESSNA 560 CITATION ENCORE

US Army designation: UC-35B
US Marine Corps designation: UC-35D

TYPE: Business jet.

PROGRAMME: Original Cessna 560 was stretched, higher-performance version of Citation S/II, first flown August 1987 and marketed as Citation V; certified 9 December 1988 as extension of Cessna 500/550 certificate; Citation Ultra introduced from c/n 560-0260 (no recertification), but replaced by current Encore.

Prototype Encore, based on rebuilt Citation Ultra, first flown (N560VU) 9 July 1998; announced at NBAA Convention at Las Vegas, Nevada 18 October 1998; first production aircraft (c/n 560-0539) rolled out in early March 2000. FAA certification achieved 26 April 2000; first delivery (N539CE/N5108G) on 29 September 2000 to J R Simplot Company of Boise, Idaho.

CURRENT VERSIONS: **Citation Encore:** Civilian business jet; *as described.*

UC-35B: In January 1996, US Army selected Ultra for its C-XX medium-range transport aircraft programme; equipped with 0.90 × 1.15 m (2 ft 11½ in × 3 ft 9 in) upward-opening clamshell cargo door on port forward fuselage; first order was for two UC-35As, of which one (95-0123) was delivered in last quarter of 1996; first two serve with 207 AvnCo at Heidelberg, Germany. Subsequent orders for five in each of FY96, FY97, FY98 and FY99; (last two aircraft of this batch are first UC-35Bs), three in 2000, one in 2001 (delivered December 2001) and one in 2003 (delivered March 2004). Total seven UC-35Bs, plus 20 earlier UC-35As.

UC-35D: Equivalent version to UC-35B for US Marine Corps. Following two UC-35C equivalents of Citation V Ultra with Reserve's MWHS-4 based at New Orleans), USMC purchased five UC-35Ds; first, 165938, delivered to Miramar in August 2001 (subsequently lost on 10 March 2004); others to Washington for VIP transport, Futenma, Japan, and elsewhere.

Citation Excel: Described separately.

CUSTOMERS: Deliveries of Ultra began July 1994; deliveries comprised 24 in 1994, 56 in 1995, 52 in 1996, 47 in 1997, 41 in 1998, 32 in 1999, six in 2000, 37 in 2001, 36 in 2002, 21 in 2003, and five in the first six months of 2004. Announced customers for Encore include J R Simplot Co of Boise, Idaho, which has ordered two; Taxi Aereo Marilia (TAM) of Brazil (one); Flying Partner CV of Antwerp, Belgium (one). By August 2003 a total of 647 Citation Vs and Encores had been delivered, and total fleet time stood at 2,117,373 flight hours.

COSTS: US$7.576 million, typically equipped (2003).

DESIGN FEATURES: Stretched version of Citation S/II for full eight-seat cabin and with fully enclosed toilet/vanity area; seventh cabin window each side; two baggage compartments outside main cabin. Rear-engined executive jet with low-mounted wing of tapered planform with root gloves; tapered horizontal tail with sweptback fin and fillet.

Encore generally as Citation Ultra and Citation Bravo (which see). Compared to Ultra, has increased wing span; bleed air anti-icing for wing leading-edges; boundary layer energisers and stall fences to improve stall characteristics; trailing-link main landing gear with narrower track; decreased fuel capacity; Pratt & Whitney Canada PW535 turbofans offering 10 per cent increase in thrust and 15 per cent improvement in specific fuel consumption; reduced fuel capacity; fuel heaters obviating the need for additives; digital pressurisation system; improved braking system; single level access electrical junction box; and redesigned interior with increased headroom, new passenger service units, pin-mounted seats for easy removal/replacement and increased serviceability.

Description for Citation Ultra applies also to Citation Encore except as follows:

FLYING CONTROLS: Conventional and manual. Trim tab on port aileron manually actuated; manual rudder trim; electric elevator trim tab with manual standby; hydraulically actuated Fowler flaps, max deflection 35°; hydraulically actuated airbrake on wing upper surfaces, each side, deflection 58°.

STRUCTURE: Two primary, one auxiliary metal wing spars; four fuselage attachment points, conventional ribs and stringers. All-metal pressurised fuselage with fail-safe design providing multiple load paths.

LANDING GEAR: Hydraulically retractable trailing-link tricycle type with single wheel on each unit. Main units retract inward into the wing, nose gear forward. Free-fall and pneumatic emergency extension systems. Goodyear mainwheels with tubeless tyres size 22×8.0-10 (12 ply), pressure 6.90 bar (100 lb/sq in). Steerable nosewheel (±20°) with Goodyear wheel and tyre size 18×4.4DD (10 ply), pressure 8.27 bar (120 lb/sq in). Goodyear hydraulic brakes. Parking brake and pneumatic emergency brake system. Anti-skid system standard. Minimum ground turning radius about nosewheel 8.38 m (27 ft 6 in).

POWER PLANT: Two Pratt & Whitney Canada PW535A turbofans, each rated at 15.12 kN (3,400 lb st) at 27°C (80°F). Integral fuel tank in each wing, combined usable capacity 3,047 litres (805 US gallons; 670 Imp gallons).

ACCOMMODATION: Standard seating for seven passengers in four forward-facing and three aft-facing seats, or eight passengers in double-club arrangement, on swivelling and fore/aft/inboard-tracking pedestal seats; refreshment centre in forward cabin area; lavatory/vanity centre with sliding doors to rear, metallic plating on cabin fittings, veneer overlay on armrests, optional pleated window shades and cabin divider mirror standard; space in aft section of cabin for 272 kg (600 lb) of baggage, in addition to baggage compartments in nose and rear fuselage.

SYSTEMS: Pressurisation system supplied with engine bleed air, maximum pressure differential 0.69 bar (8.9 lb/sq in), maintaining a sea level cabin altitude to 7,189 m (23,586 ft), or a 2,286 m (7,500 ft) cabin altitude to 13,716 m (45,000 ft). Hydraulic system, pressure 103.5 bar (1,500 lb/sq in), with two pumps to operate flaps landing gear, speed brakes and thrust reversers. Separate hydraulic system for wheel brakes. Electrical system supplied by two 28 V 300 A DC starter/generators, with two 375 VA inverters and 24 V 30 Ah Ni/Cd battery. Oxygen system of 1.90 m^3 (67 cu ft) capacity includes two crew demand masks and dropout constant flow masks for passengers. Engine fire detection and extinguishing system. Bleed air anti-icing system for wing leading-edges and engine inlets; pneumatic de-icing boots on tailplane leading-edges.

AVIONICS: Standard avionics package based on Honeywell Primus 1000 digital flight control system with integrated avionics computer.

Comms: Dual Honeywell TR-833 transceivers, dual TDR-94 transponders, dual altitude reporting systems L3 Communications FA2100 cockpit voice recorder, Artex ELT-110-4, and dual Telex microphones and headsets, all standard.

Radar: Honeywell Primus 660 colour weather radar standard.

Cessna 560 Citation Encore *(Paul Jackson)* ***NEW***/0580765

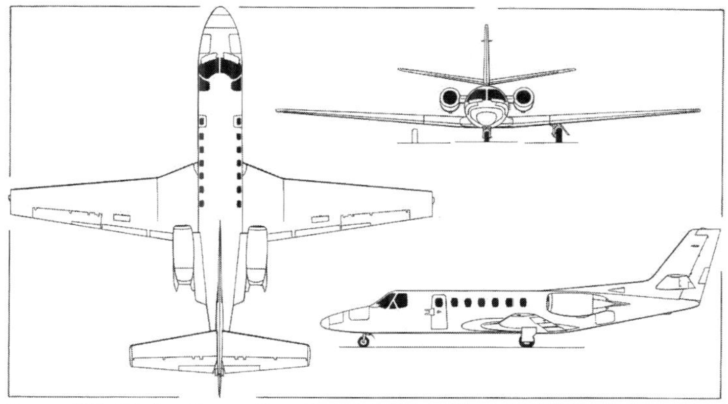

Cessna 560 Citation Encore (two Pratt & Whitney Canada JT15D-5D turbofans) *(Jane's/Dennis Punnett)* 0507258

Flight: Dual Honeywell Primus II, single Honeywell ADF, AA-300 radio altimeter, coupled vertical navigation system. Universal Avionics UNS-1Csp navigation management system, including GPS.

Instrumentation: Three-tube EFIS with 203 × 178 mm (8 × 7 in) CRTs comprising pilot's and co-pilot's primary flight displays (PFDs) and centrally mounted multifunction display (MFD); PFDs integrate functions of five flight instruments and several sources of navigation data, and provide trend data for airspeed, altitude and rate of climb. L3 Communications GH-3000 ESIS standby indicator.

DIMENSIONS, EXTERNAL:
Wing span ... 16.48 m (54 ft 1 in)
Wing aspect ratio ... 9
Length overall ... 14.91 m (48 ft 11 in)
Height overall ... 4.62 m (15 ft 2 in)
Tailplane span ... 6.55 m (21 ft 6 in)
Wheelbase ... 6.12 m (20 ft 1 in)
Wheel track ... 4.06 m (13 ft 4 in)

DIMENSIONS, INTERNAL:
Cabin:
Length: between pressure bulkheads ... 6.88 m (22 ft 7 in)
excl cockpit ... 5.28 m (17 ft 4 in)
Max width ... 1.47 m (4 ft 10 in)
Max height ... 1.43 m (4 ft 8½ in)
Baggage compartment volume ... 1.95 m^3 (69 cu ft)

AREAS:
Wings, gross ... 29.94 m^2 (322.3 sq ft)
Vertical tail surfaces (total, incl tab) ... 4.73 m^2 (50.90 sq ft)
Horizontal tail surfaces (total) ... 7.88 m^2 (84.80 sq ft)

WEIGHTS AND LOADINGS:
Standard weight empty ... 4,525 kg (9,977 lb)
Max fuel ... 2,468 kg (5,440 lb)
Baggage capacity:
nose ... 141 kg (310 lb)
aft cabin ... 272 kg (600 lb)
tailcone ... 227 kg (500 lb)
Max T-O weight ... 7,543 kg (16,630 lb)
Max ramp weight ... 7,634 kg (16,830 lb)
Max landing weight ... 6,895 kg (15,200 lb)
Max zero-fuel weight ... 5,715 kg (12,600 lb)
Fuel with max payload ... 413 kg (910 lb)
Max wing loading ... 251.9 kg/m^2 (51.60 lb/sq ft)
Max power loading ... 249 kg/kN (2.45 lb/lb st)

PERFORMANCE:
Max cruising speed at FL350 ... 429 kt (795 km/h; 494 mph)
Stalling speed, flaps down ... 83 kt (154 km/h; 96 mph)
Max rate of climb at S/L ... 1,445 m (4,740 ft)/min
Time to: FL350 ... 12 min
FL450 ... 28 min
Rate of climb at S/L, OEI ... 439 m (1,440 ft)/min
Max certified altitude ... 13,715 m (45,000 ft)
T-O balanced field length ... 1,064 m (3,490 ft)
Landing field length at max landing weight ... 844 m (2,770 ft)
NBAA IFR range, 100 n mile (185 km; 115 mile) alternate, plus 45 min reserves ... 1,178 n miles (2,182 km; 1,356 miles)
Range with max fuel, VFR reserves ... 1,970 n miles (3,648 km; 2,267 miles)

OPERATIONAL NOISE LEVELS:
T-O ... 70.0 EPNdB
Approach ... 90.5 EPNdB
Sideline ... 89.8 EPNdB

UPDATED

CESSNA 560XL CITATION EXCEL AND XLS

TYPE: Business jet.

PROGRAMME: Announced at National Business Aircraft Association Convention in New Orleans, October 1994; derivation of Model 560 Citation V/Ultra. Construction of prototype (N560XL) began February 1995; first flight 29 February 1996; public debut at NBAA convention at Orlando, Florida, November 1996, by which time prototype and preproduction aircraft (N561XL, c/n 560-5001) had completed 350 hours of flight testing in 400 sorties; first production aircraft rolled out 21 November 1997; FAA certification achieved 22 April 1998 (on Cessna 500/550/560 certificate); first delivery 2 July 1998 to Swift Transportation Inc of Phoenix, Arizona. First export, September 1998, to Automobilvertriebs AG, Austria; 100th production Excel rolled out 17 April 2000 and delivered in August 2000. Cessna delivered its 4,000th Citation, an Excel, on 7 October 2003.

CURRENT VERSION: **Citation Excel:** *As described.*

Citation XLS: Upgraded version, announced at NBAA Convention in Orlando 7 October 2003, where development aircraft (N562XL, c/n 560-5313) was displayed statically. Features 17.75 kN (3,991 lb st) Pratt & Whitney Canada PW545B turbofans providing 4.9 per cent more take-off thrust; and revised flight deck and cabin interior featuring Goodrich Aerospace crew seats, BE Aerospace passenger seats, LED lightning and restyled and upgraded furnishings. Empty weight 5,520 kg (12,170 lb); basic operating weight 5,779 kg (12,740 lb); max zero-fuel weight 6,849 kg (15,100 lb); max usable fuel 3,057 kg (6,740 lb); max T-O weight 9,163 kg (20,200 lb); max ramp weight 9,253 kg (20,400 lb); cruising speed at FL430 431 kt (798 km/h; 496 mph); max rate of climb at S/L 1,064 m (3,490 ft)/min; time to climb to FL450 29 min; IFR range, with two crew and five passengers, 1,945 n miles (3,602 km/h; 2,238 miles); VFR range with max fuel 2,110 n miles (3,907 km; 2,428 miles). FAA certification achieved 23 March 2004; first production aircraft (c/n 560-5501/N633RP) rolled out 17 April 2004; deliveries scheduled to begin mid-2004, replacing Excel from c/n 560-5501. Unit cost $9.895 million (2003).

CUSTOMERS: Total market expected to exceed 1,000. Total of 15 delivered in 1998, 39 in 1999, 79 in 2000, 85 in 2001, 81 in 2002, 48 in 2003, and 20 in the first six months of 2004. Recent customers include the Bureau of General Aviation of Civil Aviation of China, which ordered two XLSs in May 2004 for delivery in the second quarter of 2005 for operation by the CAAC flight inspection centre at Beijing Capital City Airport. By August 2003 some 340 Excels were in service and had flown a total of 311,151 hours.

COSTS: US$9.451 million (2002).

DESIGN FEATURES: Combines systems and wing and tail surfaces of Citation Ultra (Encore) with shortened version of Citation X's fuselage, providing eight-seat cabin with stand-up headroom; dual ventral strakes.

FLYING CONTROLS: As Citation Encore.

LANDING GEAR: Hydraulically retractable tricycle type with single wheel on each unit; trailing-link suspension on main legs. Mainwheel tyre size 23.5 × 8.0–10 (12 ply) tubeless, pressure 14.48 bar (210 lb/sq in); mechanically steerable nosewheel (±20°) with chined tyre, size 18×4.4DD (10 ply) tubeless, pressure 8.96 bar (130 lb/sq in). Hydraulic multiple disc carbon brakes with anti-skid and pneumatic emergency system.

POWER PLANT: Two 16.92 kN (3,800 lb st) Pratt & Whitney Canada PW545A turbofans. Nordam clamshell-type thrust reversers standard. Integral fuel tank in each wing, usable capacity 3,808 litres (1,006 US gallons; 838 Imp gallons) single-point pressure refuelling.

ACCOMMODATION: Choice of four standard seating configurations for up to 10 passengers in various layouts; seats recline, swivel and track forward and aft and laterally; forward refreshment centre and cupboard; aft lavatory and centreline cupboard. Airstair door on port side aft of flight deck. Baggage compartment in rear fuselage, capacity 317.5 kg (700 lb), with external access door incorporating integral step.

SYSTEMS: Honeywell RE-100(XL) APU optional from c/n 5021 (mid-1999) onwards, and retrofittable to earlier aircraft. Pressurisation system maximum differential 0.64 bar (9.3 lb/sq in), maintaining a sea level cabin altitude to 7,690 m (25,230 ft) or a 2,070 m (6,800 ft) cabin altitude to 13,715 m (45,000 ft). Hydraulic system, pressure 103.5 bar (1,500 lb/sq in), with two engine-driven pumps to operate landing gear, flaps, horizontal stabiliser, speed brakes and thrust reversers; separate hydraulic system for wheel brakes and anti-skid. Electrical system supplied by two 28 V 300 A DC starter/generators, with 24 V 44 Ah Ni/Cd battery. Vapour cycle air conditioning system standard; APU optional. Oxygen system, capacity 2.15 m^3(76 cu ft) with pressure demand masks for crew and dropout constant-flow masks for passengers. Engine inlets and wing leading-edges supplied by engine bleed air for anti-icing; tailplane has de-icer boots; fin unprotected; electrically heated windscreen and cockpit side windows with PPG SurfaceSeal coating for rain dispersal, with electric blower assistance.

AVIONICS: Standard Honeywell Primus 1000 integrated digital avionics suite with IC-600 avionics computer as core system. Rockwell Collins system optional.

Comms: Dual Honeywell SRZ-850 transceivers, dual XS-852B Mode S transponders, Artex 110-4 ELT, L3 communications FA2100 CVR, airborne telephone system.

Radar: Honeywell Primus 880 colour weather radar.

Flight: Dual Honeywell NV-850, dual DM-850 DME, single DF-850 ADF, long-range navigation management system incorporating GPS, and Honeywell AA-300 radio altimeter.

Instrumentation: Three-tube EFIS with 178 × 203 mm (7 × 8 in) CRT screens comprising dual primary flight displays (PFDs) showing attitude/heading and all air data information, and single multifunction display (MFD) for map/plan, weather and checklist data.

DIMENSIONS, EXTERNAL:

Wing span ... 17.16 m (56 ft 3¾ in)

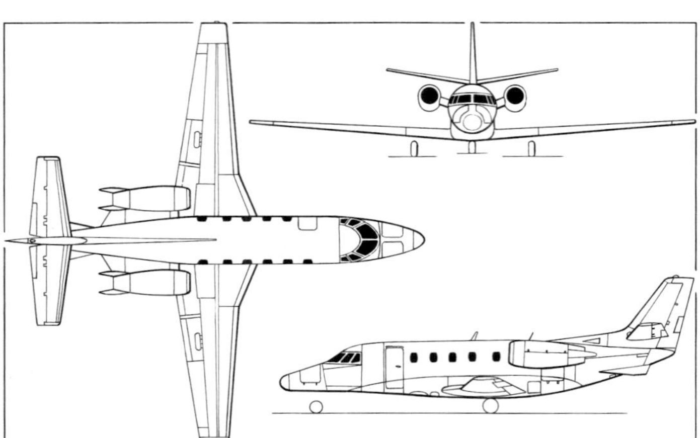

Cessna Citation Excel (PW545 turbofans) *(James Goulding)* ***NEW***/0580709

Cessna 560 Citation Excel *(Paul Jackson)* ***NEW***/0580823

Cessna 560 Citation XLS prototype *(Paul Jackson)* ***NEW***/0580820

Wing aspect ratio ... 8.4
Length: overall ... 15.79 m (51 ft 9½ in)
Height overall ... 5.30 m (17 ft 4½ in)
Tailplane span ... 6.55 m (21 ft 6 in)
Wheelbase ... 6.67 m (21 ft 10¾ in)
Wheel track ... 4.54 m (14 ft 10¾ in)

DIMENSIONS, INTERNAL:

Cabin: Length:
between pressure bulkheads ... 7.39 m (24 ft 3 in)
excl cockpit ... 5.69 m (18 ft 8 in)
Max width ... 1.70 m (5 ft 7 in)
Max height ... 1.73 m (5 ft 8 in)
Baggage capacity (aft) ... 2.26 m^3 (80 cu ft)

AREAS:

Wings, gross ... 34.35 m^2 (369.7 sq ft)
Vertical tail surfaces (total, incl tab) ... 4.73 m^2 (50.9 sq ft)
Horizontal tail surfaces (total, incl tab) ... 7.88 m^2 (84.8 sq ft)

WEIGHTS AND LOADINGS:

Weight empty, typically equipped ... 5,579 kg (12,300 lb)
Max fuel weight (usable) ... 3,057 kg (6,740 lb)
Max T-O weight ... 9,071 kg (20,000 lb)
Max ramp weight ... 9,163 kg (20,200 lb)
Max landing weight ... 8,482 kg (18,700 lb)
Max zero-fuel weight ... 6,804 kg (15,000 lb)
Max wing loading ... 264.1 kg/m^2 (54.10 lb/sq ft)
Max power loading ... 269 kg/kN (2.64 lb/lb st)

PERFORMANCE:

Max operating speed (VMO):
S/L to FL80 ... 260 kt (481 km/h; 299 mph)
FL80 to FL265 ... 305 kt (564 km/h; 351 mph) CAS
above FL265 ... M0.75
Max cruising speed at FL350
... 429 kt (795 km/h; 494 mph)
Stalling speed in landing configuration, at max landing weight ...
90 kt (167 km/h; 104 mph)
Max rate of climb at S/L ... 1,155 m (3,790 ft)/min
Rate of climb at S/L, OEI ... 259 m (850 ft)/min
Max certified altitude ... 13,715 m (45,000 ft)
Service ceiling, OEI ... 8,717 m (28,600 ft)
T-O balanced field length ... 1,095 m (3,590 ft)
Landing field length at max weight ... 969 m (3,180 ft)
NBAA IFR range ... 1,847 n miles (3,421 km; 2,157 miles)
VFR range with max fuel ... 2,080 n miles (3,852 km; 2,394 miles)

OPERATIONAL NOISE LEVELS:

T-O ... 72.4 EPNdB
Approach ... 93.1 EPNdB
Sideline ... 85.3 EPNdB

UPDATED

CESSNA 680 CITATION SOVEREIGN

TYPE: Business jet.

PROGRAMME: Design started mid-1998; announced at NBAA Convention at Las Vegas, Nevada, 18 October 1998; critical design review completed in late 1999 and Type Certificate application lodged 24 November 1999. Structural testing of fatigue test fuselage began in late 1999; official launch 3 January 2000; construction of cyclic fatigue test airframe started October 2000; manufacture of prototype started in 2001; first flight (prototype N680CS, c/n 000P, N680CS, c/n 000P) 27 February 2002, followed by first preproduction aircraft (N681CS/c/n 0001) on 27 June 2002. Public debut (N681CS) at NBAA Convention in Orlando, Florida, 10 September 2002. These aircraft, plus N682CS, c/n 0002, which flew in fourth quarter 2002, undertook a 2,000-hour, 19-month test programme. By 27 August 2003 the three aircraft had logged more than 1,600 flight hours. Two static test airframes have also been built, one of which will undertake a 36,000-cycle fatigue test programme.

After completion of initial flight tests and envelope expansion in June 2002, 000P was dedicated to stability and control checks, high altitude stall tests, cruise performance validation, systems and APU testing and airfield performance; c/n 0001 was the systems verification and certification article; and c/n 0002/N682CS dedicated to avionics and autopilot testing. Provisional FAA certification to FAR Pt 25 Amendment 98 and JAR 25 Change 15, both including RVSM compliance, granted 24 December 2003; full type certification achieved 2 June 2004; c/n 0002 was also to perform a 150-hour function and reliability programme and, with c/n 0001, undertake 300 hours of post-certification in-service testing before first customer delivery; both pre-production aircraft will then be refurbished and brought up to final production configuration before sale to customers. First production aircraft (N52114, c/n 0003, later N103SV) rolled out 27 August 2003. International debut at the European Business Aviation convention and Exhibition, Geneva 25 May 2004. First deliveries to customers in Mexico and USA, 29 September 2004. Total of nine delivered in 2004. Delivery targets are 40 in 2005 and 58 in 2006.

CUSTOMERS: Launch customer Swift Air of Phoenix, Arizona, ordered six on day of launch announcement; Executive Jet Aviation ordered 50, with 50 options, on 20 October for its NetJets fractional ownership scheme. Other announced customers include Atlas Air Service GmbH of Germany, which has ordered one. Firm orders totalled more than 100 by June 2004.

COSTS: US$13.523 million, typically equipped (2003).

DESIGN FEATURES: Design goals included large cabin, good short-field performance and US coast-to-coast range. Low wing with sweptback leading-edge; mid-mounted tailplane with leading-edge sweepback; podded engines on rear fuselage shoulders.

New wing design; sweepback 16° 18' at leading-edge, 12° 42' at quarter-chord, dihedral 3°, taper ratio 0.31. Mid-mounted tailplane, sweepback, dihedral 0° 22° 36' at quarter-chord, taper ratio 0.47; fin sweepback 38° 18' at quarter-chord taper ratio 0.41.

FLYING CONTROLS: Conventional. Five hydraulically actuated spoiler panels per wing; three centre panels on each wing function as roll spoilers, variable position speedbrakes and ground spoilers; inner (Nos. 5 and 6) and outboard (No. 1 and 10) panels function as variable speedbrakes and ground spoilers; ; roll spoilers comprise Nos. 2, 3, 4, 7, 8 and 9. Trim tabs in ailerons and rudder; trimmable tailplane maximum deflections −8°/+2°; yaw damper. Control surface deflections: ailerons +18° 48'/−13° 24' port, +18° 30'/−13° 36' starboard; elevator +16° 36'/−9° 24'; rudder ±30° 30'; brakes spoilers (Nos. 1, 2, 3, 8, 9 and 10) +35°; (Nos. 4, 5, 6 and 7) +24° flap deflections 7, 15 and 35°.

STRUCTURE: Primarily metal. Fokker Aerostructures of the Netherlands selected in 2000 to manufacture the tail surfaces of production aircraft.

LANDING GEAR: Hydraulically retractable tricycle type with twin wheels on trailing-link main units and nosewheel; main units retract inboard, nosewheel forwards. Carbon brakes; anti-skid standard; hydraulically boosted nosewheel steering.

POWER PLANT: Two Pratt & Whitney Canada PW306C turbofans with FADEC, each developing 25.7 kN (5,770 lb st) flat rated to ISA + 15°C; target-type thrust reversers. Integral fuel tanks in wings total usable capacity 6,340 litres (1,675 US gallons; 1,395 Imp gallons); further 47 litres (12.5 US gallons; 10.4 Imp gallons) unusable; single-point pressure fuelling. Oil capacity 40 litres (10.5 US gallons; 8.8 Imp gallons) per engine.

ACCOMMODATION: Crew of two on flight deck and up to 12 passengers in cabin; standard accommodation for eight passengers in double club arrangement, each with one worktable between each pair of seats, reading light and 28 V AC power outlet, with forward galley and aft lavatory. Fully dimmable LED cabin lighting. Baggage compartment in tailcone with external access. Cabin is pressurised, heated and air conditioned. Airstair door at front on port side; one emergency exit on starboard side, above the wing.

SYSTEMS: Pressurisation system, maximum differential 0.64 bar (9.3 lb/sq in), and air-conditioning system supplied by engine bleed air; pressurisation system maintains sea level cabin to 7,690 m (25,230 ft) and 2,210 m (7,250 ft) cabin to 14,325 m (47,000 ft); closed-centre hydraulic system, pressure 207 bar (3,000 lb/sq in), for operation of landing gear, nosewheel steering, braking system, spoilers and thrust reversers. 28 V DC split-bus electrical system, supplied by two 400 A starter/generators and two 44 Ah ni-cad batteries. Honeywell RE100 APU certified for in-flight operation up to 9,145 m (30,000 ft); and Honeywell environmental control and cabin pressure control systems. Oxygen system, 2.15 m^3 (76.0 cu ft), standard. Wing and tailplane leading-edges and engine inlets anti-iced by engine bleed air; electrically anti-iced windscreens and air data and AoA probes.

AVIONICS: Honeywell Primus Epic as core system.

Comms: L-3 Communications FA2100 CVR with underwater locator beacon standard.

Radar: Honeywell Primus 880 colour weather radar.

Flight: VOR, ILS, ADF, GPS, dual NZ-2000 FMS, dual Rockwell Collins AHC-3000 AHRS, TAWS, CAS 67A TCASII and three-axis autopilot standard.

Instrumentation: Four-tube EFIS with 203 × 254 mm (8 × 10 in) active matrix quartz PFD, MFD and EICAS displays, L-3 Communications GH-3000 electronic standby instrument system, Goodrich standby EHSI, and Honeywell AA-300 radio altimeter. MaxViz EVS2000 enhanced visibility system optional.

DIMENSIONS, EXTERNAL:

Wing span	19.25 m (63 ft 2 in)
Wing aspect ratio	7.7
Length overall	19.35 m (63 ft 6 in)
Height overall	6.20 m (20 ft 4 in)
Tailplane span	8.41 m (27 ft 7 in)
Wheel track	3.05 m (10 ft 0 in)
Wheelbase	8.48 m (27 ft 10 in)
Crew-passenger door: Height	1.42 m (4 ft 8 in)
Width	0.71 m (2 ft 4 in)

DIMENSIONS, INTERNAL:

Cabin: Length between pressure bulkheads	8.94 m (29 ft 4 in)
Length excluding cockpit	7.38 m (24 ft 2½ in)
Max width	1.70 m (5 ft 6¾ in)
Max height	1.73 m (5 ft 8 in)
Baggage compartment volume	2.83 m^3 (100 cu ft)

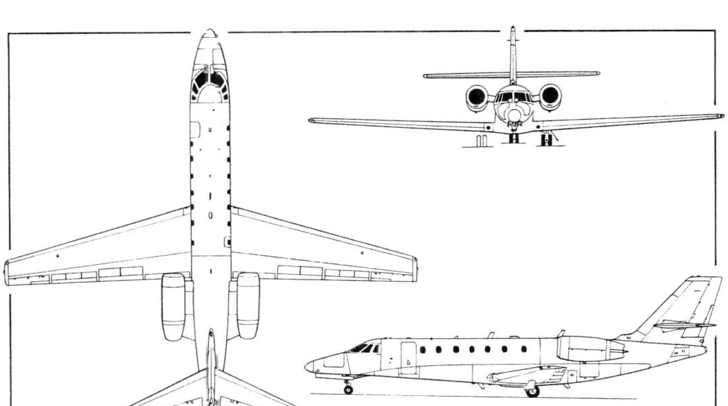

Citation Sovereign eight/11-passenger business jet ***(Michael Badrocke)*** ***NEW***/0526865

Second preproduction Cessna 680 Citation Sovereign ***NEW***/0572388

AREAS:

Wings, gross	47.93 m^3 (515.9 sq ft)
Vertical tail surfaces	8.85 m^2 (95.3 sq ft)
Horizontal tail surfaces	12.87 m^2 (138.5 sq ft)

WEIGHTS AND LOADINGS:

Empty weight, typically equipped	7,892 kg (17,400 lb)
Max fuel	4,921 kg (10,850 lb)
Max ramp weight	13,721 kg (30,250 lb)
Max T-O weight	13,607 kg (30,000 lb)
Baggage capacity	454 kg (1,000 lb)
Max landing weight	12,292 kg (27,100 lb)
Max zero-fuel weight	9,208 kg (20,300 ft)
Payload with max fuel	726 kg (1,600 lb)
Max wing loading	283.9 kg/m^2 (58.15 lb/sq ft)
Max power loading	265 kg/kN (2.60 lb/lb st)

PERFORMANCE:

Max operating speed (VMO)	M0.80
Max cruising speed:	
at FL350	459 kt (850 km/h; 528 mph)
at FL410	438 kt (811 km/h; 504 mph)
Time to FL410	18 min
Time to FL430	23 min
Max certified altitude	14,325 m (47,000 ft)
Service ceiling	13,105 m (43,000 ft)
T-O balanced field length	1,126 m (3,695 ft)
Landing run	808 m (2,650 ft)
NBAA IFR range, 100 n mile (185 km; 115 mile) alternate, 45 min reserves	2,730 n miles (5,056 km; 3,141 miles)
VFR range with max fuel	3,040 n miles (5,630 km; 3,498 miles)

UPDATED

CESSNA 750 CITATION X

TYPE: Business jet.

PROGRAMME: Announced at NBAA Convention in New Orleans in October 1990; engine flew on Citation VII testbed (N650) 21 August 1992; first flight (N750CX) 21 December 1993; two preproduction aircraft to aid integration of production systems; first of these (N751CX) flown 27 September 1994, second (N752CX) flown 11 January 1995; FAA FAR Pt 25, Amendment 74 certification 31 May 1996 after flight test programme totalling more than 3,000 hours; JAA certification achieved 26 May 1998. First customer delivery July 1996. Citation X design team awarded the National Aeronautic Association's Robert J. Collier Trophy in February 1997. Cessna delivered its 3,000th Citation, a Citation X, on 19 November 1999.

In October 2000, Cessna announced improvements to the Citation X aimed at boosting range/payload performance and enabling the aircraft to operate from shorter runways. Improvements, to be incorporated on all aircraft delivered after 1 January 2002, beginning with c/n 0173 include uprated 30.01 kN (6,764 lb st) Rolls-Royce AE 3007C-1 turbofans; 181 kg (400 lb) increase in maximum take-off weight to 16,374 kg (36,100 lb), enabling a typically equipped aircraft to carry seven passengers with maximum fuel; and take-off balanced field length at MTOW of 1,567 m (5,140 ft). Several optional items of avionics became standard, including Honeywell TCAS II and EGPWS CVR, satcom, VHF/AFIS, provisions for an FDR, and second HF transceiver, plus Teledyne angle of attack indicator/indexer, tail floodlights, red strobe light, pulse lights, Litton ELT, and 2,154 litre (76 cu ft) oxygen bottle. First delivery of an upgraded Citation X took place on 5 February 2002 to golfer Arnold Palmer (c/n 0176/N1AP).

Upgraded 2004 model year version announced at the NBAA Convention in Orlando, Florida in October 2003 features revised interior with new passenger seats, Goodrich Aerospace articulating crew seats, pin-in/pin-out cabin hardware, and LED controls for lighting, cabin temperature, window shades and audio/video equipment. New options include MaxViz EVS and Aircell satellite telephones. The upgrades were incorporated from c/n 0225 for deliveries from early 2004.

CUSTOMERS: First delivery (0003/N1AP) to Arnold Palmer July 1996; 100th Citation X delivered 23 December 1999 to Townsend Engineering of Des Moines, Iowa and 200th delivered 14 October 2002 to NetJets Inc; total 203 by 31 December 2002, comprising seven, 28, 30, 36 in 1996–99, 37 in 2000, 34 in 2001, 31 in 2002, and 18 in 2003 and seven in first six months of 2004; most to US operators, but others exported to Canada, Finland, Germany, Mexico, South Africa and UK. Executive Jet Aviation (EJA) ordered 31 for delivery beginning in 1997 and extending beyond 2000, for its NetJets fractional ownership operation. Other early recipients included General Motors (five), Honeywell (two) and Williams Companies (three). Recent customers include former World Motor Racing Champions Nigel Mansell, who took delivery of one in February 2002, and Nelson Piquet, who ordered one at the NBAA Convention at Orlando, Florida, in September 2002, and the Air Traffic Management Bureau of the Civil Aviation Authority of China, which ordered one on 1 October 2001. By February 2004 more than 220 Citation Xs had been delivered, and had accumulated more than 460,000 flight hours.

Cessna 750 Citation X *(Paul Jackson)* ***NEW***/0580815

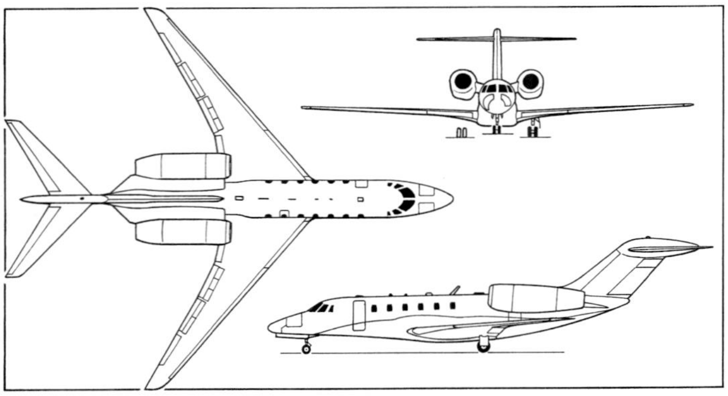

Cessna 750 Citation X *(Jane's/Dennis Punnett)* 0075950

COSTS: US$19,394 million, typically equipped. Direct operating cost, based on 1,000 n mile (1,852 km; 1,151 mile) stage length, US$1,294.92 per hour (both 2003).

DESIGN FEATURES: Optimised for high maximum operating Mach number; US transcontinental and transatlantic range. Design generally as for Citation VII (which see), but with greater angles of sweepback on all flying surfaces and of increased size and weight.

Wing sweepback at quarter-chord 37°; dihedral 2°.

FLYING CONTROLS: Dual hydraulically powered controls with manual reversion. One-piece all-moving tailplane; two-piece rudder, lower portion hydraulically powered, upper portion electrically powered; speed brakes/spoilers with manual back-up. Five spoiler panels per wing, operating in combination as aileron augmentors, airbrakes and lift dumpers.

STRUCTURE: Alloy fuselage. Thick wing skins, milled from solid; all control surfaces, spoilers, speedbrakes, wing fairings and flaps are of composites construction.

LANDING GEAR: Trailing-link main units, each with twin wheels; powered anti-skid carbon brakes; hydraulically steerable nose unit with twin wheels. Main tyres 26×6.6R14 (14 ply) tubeless; nose tyres 16×4.4D (6 or 10 ply) tubeless.

POWER PLANT: Two Rolls-Royce AE 3007C-1 turbofans, each rated at 30.01 kN (6,764 lb st) for take-off, pod-mounted on sides of rear fuselage; FADEC. Hydraulically operated target-type thrust reversers standard. Fuel contained in three separate tanks, one in each outer wing and one in centre-section/forward fairing, combined usable capacity 1,306 litres (1,930 US gallons; 1,607 Imp gallons). Two independent fuel supply systems; fuel is fed from centre tank to wing tanks; single point and over-wing refuelling.

ACCOMMODATION: Crew of two on separate flight deck, and up to 12 passengers; interior custom designed; cabin is pressurised, heated and air conditioned; heated and pressurised baggage compartment in rear fuselage with external door. Windscreen electrically heated and demisted.

SYSTEMS: Pressurisation system, maximum pressure differential 0.64 bar (9.3 lb/sq in), maintains 2,440 m (8,000 ft) cabin altitude at 15,545 m (51,000 ft). Dual isolated hydraulic systems, pressure 207 bar (3,000 lb/sq in), maintained by pressure-compensated pumps. Split-bus electrical system is powered by two engine-driven 400 A DC generators, plus an APU-driven 400 A DC generator usable to FL310, with two 24 V 44 Ah Ni/Cd batteries; wiring designed to minimise susceptibility of critical systems to HIRF interference. Wing and tail leading-edges and engine inlets heated by engine bleed air for ice protection; wing cuffs, pitot/static system, AoA system and windscreen electrically heated. Oxygen system, capacity 2.15 m^3 (76 cu ft), with pressure demand masks for crew and dropout mask for each passenger.

AVIONICS: Honeywell Primus 2000 dual digital autopilot/flight director/EICAS as core system.

Comms: Dual Honeywell RCZ-833 communication units with 25 kHz and 8.33 kHz channel spacing; dual RM-855 radio management units; dual AV-850 audio control panels; single Honeywell KHF-950 HF with provision for second; dual Mode S transponders; Coltech CSD-714 Selcal; Honeywell airborne traffic information system (AFIS); L-3 Communications FA2100 CVR; three-frequency ELT.

Radar: Honeywell Primus 880 colour weather radar.

Flight: Dual Honeywell FMSes; dual Honeywell navigation systems including VOR/LOC/GS/MKR and DME; dual Laseref IV IRS; single ADF; Honeywell AA-300 radio altimeter. Air data is provided by dual AZ-840 micro air data computers. MaxViz EVS optional.

Instrumentation: Five-tube EFIS with 178 × 203 mm (7 × 8 in) screens for pilot's and co-pilot's primary flight displays (PFD) and multifunction displays (MFD), with central EICAS display.

DIMENSIONS, EXTERNAL:
Wing span ... 19.38 m (63 ft 7 in)
Wing aspect ratio ... 7.8
Length overall ... 22.05 m (72 ft 4 in)
Height overall ... 5.87 m (19 ft 3 in)
Tailplane span ... 7.95 m (26 ft 1 in)
Wheel track ... 3.23 m (10 ft 7 in)
Wheelbase ... 8.74 m (28 ft 8 in)

DIMENSIONS, INTERNAL:
Cabin (front to mid-pressure bulkhead):
Length: between pressure bulkheads ... 8.64 m (28 ft 4 in)
excl flight deck ... 7.16 m (23 ft 6 in)
Max width ... 1.70 m (5 ft 7 in)
Max height ... 1.73 m (5 ft 8 in)
Baggage compartment volume (aft, including ski compartment) ... 2.03 m^3 (72 cu ft)

AREAS:
Wings, gross ... 48.96 m^2 (527.0 sq ft)
Vertical tail surfaces (total) ... 10.31 m^2 (111.0 sq ft)
Horizontal tail surfaces (total) ... 11.15 m^2 (120.0 sq ft)

WEIGHTS AND LOADINGS:
Weight empty, typically equipped ... 9,809 kg (21,625 lb)
Basic operating weight ... 9,988 kg (22,020 lb)
Max fuel weight ... 5,865 kg (12,931 lb)
Max T-O weight ... 16,374 kg (36,100 lb)
Max ramp weight ... 16,511 kg (36,400 lb)
Max landing weight ... 14,424 kg (31,800 lb)
Max zero-fuel weight ... 11,068 kg (24,400 lb)
Payload with max fuel ... 657 kg (1,449 lb)
Max wing loading ... 334.5 kg/m^2 (68.50 lb/sq ft)
Max power loading ... 272 kg/kN (2.67 lb/lb st)

PERFORMANCE:
Max operating Mach No. (MMO) ... 0.92
Max operating speed (VMO):
S/L to FL800 ... 270 kt (500 km/h; 310 mph)
FL80 to FL306 ... 350 kt (648 km/h; 403 mph)
above FL306 ... M0.92
Max cruising speed, mid-cruise weight at FL370 ... M0.91
Max cruising speed at FL350
... 525 kt (972 km/h; 604 mph)
Max rate of climb at S/L ... 1,113 m (3,650 ft)/min
Max certified altitude ... 15,545 m (51,000 ft)
T-O balanced field length (FAR Pt 25) ... 1,567 m (5,140 ft)
FAR Pt 25 landing field length ... 1,036 m (3,400 ft)
VFR range with two crew, M0.82 at FL490, 100 n mile (185 km; 115 mile alternate and 45 min reserves ... 3,390 n miles (6,278 km; 3,901 miles)
NBAA IFR range ... 3,070 n miles (5,686 km; 3,533 miles)

OPERATIONAL NOISE LEVELS (FAR Pt 36 Amendment 20):
T-O ... 73.2 EPNdB
Approach ... 90.3 EPNdB
Sideline ... 83.8 EPNdB

UPDATED

CIRRUS

CIRRUS DESIGN CORPORATION

4515 Taylor Circle, Duluth International Airport, Duluth, Minnesota 55811
Tel: (+1 218) 727 27 37
Fax: (+1 218) 727 21 48
Web: http://www.cirrusdesign.com
PRESIDENT: Alan Klapmeier
EXECUTIVE VICE-PRESIDENT: Dale Klapmeier
EXECUTIVE VICE-PRESIDENT AND CFO: Peter McDermott
EXECUTIVE VICE-PRESIDENT, OPERATIONS: David Coleal
EXECUTIVE VICE-PRESIDENT, SALES AND MARKETING: John M Bingham
VICE-PRESIDENT, ENGINEERING: Patrick Waddick
VICE-PRESIDENT, RESEARCH AND TECHNOLOGY: Dean Vogel
VICE-PRESIDENT, SALES AND MARKETING: Thomas Shea

Founded 1984 by Klapmeier brothers. Previously engaged in production of kits, (ST-50 and VK-30), Cirrus is concentrating on fully certified factory- built aircraft.

Current products are SR20 and SR22. Cirrus purchased 20 per cent of parachute systems company BRS in September 1999. Workforce was 961 in July 2004, based at Duluth and

CIRRUS DELIVERIES

Year	SR20	SRV	SR22	Total
1999	9			9
2000	95			95
2001	59		124	183
2002	105		292	397
2003	112	2	355	469
Totals	**380**	**2**	**771**	**1,153**

Hibbing, Minnesota, and Grand Forks, North Dakota. In August 2001, Crescent Capital acquired 58 per cent share in Cirrus for US$100 million. The 1,000th SR series aircraft, an SR22 (N1000V), was delivered in August 2003. Production rate was nine per week in 2003, increasing to 12 in mid-2004.

UPDATED

CIRRUS SR20 AND SRV

TYPE: Four-seat lightplane.

PROGRAMME: Development began 1990; mockup revealed at Oshkosh 1994; first flight (N200SR) 31 March 1995; second prototype (N202CD) flown November 1995; FAR Pt 23 certification aircraft (N203FT), designated C-1, made first flight 28 January 1998 after completion of wing redesign to lower stall speed and improve lateral control. By end 1997, the two prototypes had accumulated 1,500 hours of test flying. A second production-standard aircraft (N204CD; C-2) joined the flight test programme on 3 June 1998, committed to trials of fuel and electrical systems and avionics. Recovery parachute trials involved eight deployments, including three for FAA. Late 1998 start of deliveries delayed by decision of avionics supplier, Trimble, to withdraw from general aviation. FAR Pt 23 certification Amdt 47 received 23 October 1998; Transport Canada certification granted in March 2002; first aircraft manufactured to Canadian certification standard delivered 2 April 2002 to Tim Harpell of Toronto.

Initial production aircraft first flew (N115CD) 22 March 1999, but lost on following day. First delivery (c/n 1005, N415WM) 20 July 1999, at which time 325 on order (although many since upgraded to SR22 orders). Production certificate, to enable Cirrus to carry out its own inspections, awarded 12 June 2000. EASA certification announced 1 June 2004.

CURRENT VERSIONS: **SR20**: Total 143 built (c/n 1005 to 1147) by April 2001, when production suspended. MTOW 1,315 kg (2,900 lb).

SR20A: In production from July 2001 (c/n 1148 upwards). Additional 45 kg (100 lb) MTOW; landing light repositioned lower on engine cowling; and optional Goodrich Skywatch and Sandel SN3308 EHSI for traffic information. Upgrades (including MTOW) retrofittable. *As described.*

SR20 Version 2.0, Version 2.1 and Version 2.2: Upgraded versions available from 2003; launched at EAA Sun 'n' Fun in April 2002. Differ from SR20A principally in having no vacuum instruments and featuring electrical system and avionics packages based on those of the SR22. *Version 2.0* has a single-alternator, dual-battery, dual-bus electrical system and features Avidyne FlightMax MFD, single Garmin GNS 430 GPS/com/nav and GNC 250XL GPS/com and S-Tec System 20 autopilot as standard. First delivery in January 2003 to 907th Flight Squadron Aero club at Jameson, Pennsylvania. *Version 2.1* adds a dual-alternator electrical system and GNS 430/420 package. *Version 2.2* has dual GNS 430s, S-Tec Fifty FiveX autopilot with altitude preselect and Sandel EHSI.

SR20-G2: Announced 28 July 2004. Enhancements generally as for SR22-G2.

SRV: Basic VFR-only version, introduced at EAA AirVenture, Oshkosh, 29 July 2003, when demonstrator N936CD (c/n 1337) shown. Specification and performance generally as for SR20 except: weight empty 930 kg (2,050 lb), cruising speed at 75 per cent power 150 kt (278 km/h; 173 mph); range with reserves 634 n miles (1,174 km; 729 miles); max range 865 n miles (1,602 km; 995 miles).

SRV-G2: Announced 28 July 2004. Enhancements generally as for SR22-G2.

CUSTOMERS: Refer to table in Cirrus company description. Total 382 delivered by January 2004 and 40 (two SRV and 38 SR20s) in the first six months of 2004.

COSTS: SRV US$189,900; SR20 standard US$236,700, fully equipped US$309,650 (all 2004).

DESIGN FEATURES: Low-wing, composites construction monoplane with upturned wingtips; mid-set tailplane with horn-balanced elevators; horn-balanced rudder; fixed tab on starboard aileron. First certified aircraft with ballistic parachute as standard equipment (Cirrus Airplane Parachute System or CAPS), fitted just aft of baggage compartment. Wing dihedral 4° 30'.

STRUCTURE: Composites monocoque; one-piece main spar attached to fuselage on two locations under front seats; spar carry-through attached to fuselage side-walls. Wing upper and lower surfaces bonded to spars and ribs, forming torsion box. All flying control surfaces are flush-riveted aluminium. Fin integral with fuselage; one-piece tailplane. Cabin incorporates composites roll-cage; acrylic windscreen and windows. Airframe life limit 12,000 hours.

FLYING CONTROLS: Conventional and manual. Actuation by pushrods, cables and bellcranks. Electric trim tabs for roll and pitch activated through springed centring devices; ground-adjustable rudder tab. Control surface deflections: ailerons ±12.5°; elevator +25°/–15°; rudder 20° either side of neutral. Three-position flaps, deflections 0, 16 and 32°.

LANDING GEAR: Tricycle type; fixed. Wing-mounted composites cantilever sprung main legs with speed fairings. Main tyres 15×6.00–6, nosewheel tyre 5.00–5. Hydraulic calliper disk brakes on mainwheels. Castoring nosewheel.

POWER PLANT: One 149 kW (200 hp) Teledyne Continental IO-360-ES flat-six engine driving a two-blade Hartzell BHC-J2YF-1BF/F7694 propeller (three-blade 1.88 m; 6 ft 2 in PHC-J3YF-1MF/F77392-1 propeller optional). Single fuel tank in each wing, each of 113.5 litres (30.0 US gallons; 25.0 Imp gallons) capacity, of which 212 litres (56.0 US gallons; 46.7 Imp gallons) usable. Oil capacity 7.6 litres (2.0 US gallons; 1.7 Imp gallons).

ACCOMMODATION: Four 26 g energy attenuating seats in pairs; two forward-hinged passenger doors. Dual controls. Baggage compartment door aft of cabin on port side of fuselage. Interior reflects modern motorcar design. Leather interior optional.

Cirrus SR20 four-seat lightplane ***(Paul Jackson)*** ***NEW***/0580793

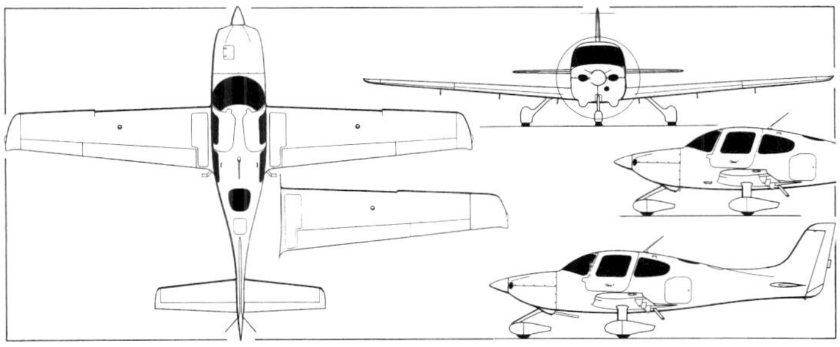

Cirrus SR20, with scrap views of SR22 wing and forward fuselage ***(Paul Jackson)*** 0110908

SYSTEMS: Single-axis S-Tec System 20 autopilot with GPS; dual-axis S-Tec System 30 and Fifty FiveX optional; 24 V DC power system with 28 V 75 A alternator and 24 V 10 Ah battery. 12 V power port.

AVIONICS: Garmin integrated package.

Comms: GNS 430 colour GPS/com/nav IFR approach-certified GPS plus back-up, GNS 250XL GPS/com, GMA 340 audio panel and GTX 327 transponder. Options include GNS 420 colour GPS/com and/or additional GNS 430 (back-up IFR GPS).

Flight: Garmin GI 106 GPS/VOR/LOC/GSI. S-Tec System FortyX autopilot. Options include Goodrich Stormscope WX-500, S-Tec System FiftyX or Fifty FiveX, Goodrich Skywatch system, GI 102 or second GI 106.

Instrumentation: Avidyne Entegra PFD and FlightMax EX3000C MFD standard, EX5000C optional.

EQUIPMENT: Integral ballistic recovery parachute (Cirrus Airplane Parachute System); vertical descent rate 7.3 m (24 ft)/s at S/L. Navigation, landing and strobe lights standard.

DIMENSIONS, EXTERNAL:

Wing span .... 10.85 m (35 ft 7 in)
Wing aspect ratio 9.1
Length overall .... 7.92 m (26 ft 0 in)
Height overall .... 2.59 m (8 ft 6 in)
Tailplane span .... 3.93 m (12 ft 10¾ in)
Wheel track .... 3.38 m (11 ft 1 in)
Wheelbase .... 2.29 m (7 ft 6 in)
Propeller diameter (two blade) .... 1.93 m (6 ft 4 in)
Cabin doors (each): Height .... 0.94 m (3 ft 1 in)
Width .... 0.86 m (2 ft 10 in)

DIMENSIONS, INTERNAL:

Cabin: Length .... 3.30 m (10 ft 10 in)
Max width .... 1.24 m (4 ft 1 in)
Max height .... 1.27 m (4 ft 2 in)

AREAS:

Wings, gross .... 12.56 m^2 (135.2 sq ft)
Horizontal tail .... 3.48 m^2 (37.50 sq ft)

WEIGHTS AND LOADINGS:

Weight empty .... 939 kg (2,070 lb)
Max fuel weight .... 163 kg (360 lb)
Max T-O weight .... 1,360 kg (3,000 lb)
Max landing weight .... 1,315 kg (2,900 lb)
Max baggage weight .... 59 kg (130 lb)
Max wing loading .... 108.3 kg/m^2 (22.19 lb/sq ft)
Max power loading .... 9.12 kg/kW (15.00 lb/hp)

PERFORMANCE:

Never-exceed speed (VNE) .... 200 kt (370 km/h; 230 mph)
Max cruising speed .... 165 kt (306 km/h; 190 mph)

Cruising speed at FL80 at 75% power ..... 156 kt (289 km/h; 180 mph)
Manoeuvring speed ..... 135 kt (250 km/h; 155 mph) IAS
Stalling speed: flaps up ..... 64 kt (119 km/h; 74 mph) IAS
flaps down ..... 54 kt (100 km/h; 63 mph) IAS
Max rate of climb at S/L ..... 274 m (900 ft)/min
Service ceiling ..... 5,335 m (17,500 ft)
T-O run ..... 409 m (1,341 ft)
T-O and landing to 15 m (50 ft) ..... 597 m (1,958 ft)
Landing from 15 m (50 ft) ..... 622 m (2.040 ft)
Landing run ..... 309 m (1,014 ft)
Range with reserves:
at 75% power ..... 733 n miles (1,357 km; 843 miles)
at econ cruise ..... 882 n miles (1,633 km; 1,015 miles)
g limits ..... +3.8/–1.9

UPDATED

Cirrus SR22-G2 four-seat tourer ***NEW***/0580806

CIRRUS SR22

TYPE: Four-seat lightplane.

PROGRAMME: Modifications to the SR20 (which see) resulted in the SR22, which comprised an SR20 airframe with modified wing, more powerful engine and all-electric instrumentation (no vacuum system); prototype (N140CD) displayed at AOPA convention in October 2000; FAA certification awarded 30 November 2000. Delivery of first production aircraft (N415PJ) 6 February 2001.

CURRENT VERSIONS: **SR22:** Initial version. Conventional Hartzell three-blade propeller.

SR22-G2: Announced 9 March 2004, replacing SR22 as standard production version. Features improvements in comfort, serviceability and performance, and a new fuselage design that reduces empty weight, increasing payload/useful load. Key features include six-point engine mount as originally offered on the limited production SR22 Centennial Edition, with optional Teledyne Continental Platinum IO-550-N engine with custom exhaust; Hartzell Scimitar Select three-blade propeller; new cowling offering improved airflow management and performance by refining the shape and diminishing the size of the exhaust contours; redesigned streamlined oil access door; two-piece lower cowling simplifying removal and installation for servicing and incorporating a high-intensity landing light, 'Secure Latch' automobile-style oversized cabin doors closed and latched at two points by a single movement, and opened by pressing a button; 'Secure Latch' baggage door; new interior inspired by luxury cars, featuring perforated-leather seats in contemporary colours with French stitching, and leather-wrapped side-yoke, throttle lever and grab handles; 'Soft Touch' composites material for cabin trim; lower wing fairing attached with fasteners rather than by bonding, for easier access to spar; door on right side of tail section for greater access and flexibility for future options; and improved access for routine inspection of the CAPS emergency parachute system. *As described.*

SR21 tdi: Announced at Friedrichshafen in April 2001; comprises SR22 airframe powered by 169 kW (227 hp) Snecma-Renault SR-305 turbocharged diesel engine; aimed initially at European market. Will offer lower fuel consumption. Initial deliveries were expected during 2004.

CUSTOMERS: Orders for 234 SR22s received by June 2001. Launch of SR22 saw approximately 20 per cent of SR20 customers upgrade to new model. Total of 124 delivered in 2001, 292 in 2002, 355 in 2003, and 193 in the first six months of 2004. Refer also to table in Cirrus company entry. Recent customers include AirShares Elite of Atlanta, Georgia, which ordered 25, plus 25 options, on 8 April 2002, for its fractional ownership programme. SR22s have been delivered to customers in Australia, Canada, Europe, Mexico, New Zealand, South Africa and South America.

COSTS: Standard US$328,700; fully equipped US$436,650 (2004).

Frontal aspect of Cirrus SR22, showing Hartzell Scimitar Select propeller of G2 version ***NEW***/0580805

DESIGN FEATURES: As for SR20, but with fuselage-mounted vortex generator ahead of wingroots and wingtip extensions.

STRUCTURE: As for SR20.

FLYING CONTROLS: As for SR20, but flight-adjustable rudder tab and ground-adjustable tab on port elevator.

LANDING GEAR: As for SR20.

POWER PLANT: One 231 kW (310 hp) Teledyne Continental IO-550-N flat-six engine driving a three-blade Hartzell Scimitar Select constant-speed propeller; Platinum IO-550-N with custom exhaust system optional. Single fuel tank in each wing, total capacity 318 litres (84.0 US gallons; 69.9 Imp gallons) of which 307 litres (81.0 US gallons; 67.4 Imp gallons) are usable. Oil capacity as SR20.

ACCOMMODATION: As for SR20.

SYSTEMS: Dual 24 V DC power system with 28 V 75 A alternator and 24 V 10 Ah battery. 12 V power port. Optional Aerospace Systems and Technologies fluid de-icing system for wing and tailplane leading-edges and propeller.

AVIONICS: *Comms:* Configuration A package includes single Garmin GNS 430 colour GPS/com/nav, GNS 420 GPS/com, GMA 340 audio panel and GTX 327 transponder as standard. Configuration B package substitutes second GNS 430 for GNS 420.

Flight: S-Tec Thirty autopilot standard in package A, replaced by S-Tec Fifty FiveX with altitude preselect in package B, which also adds Sandel SN3308 EHSI. Goodrich WX-500 Stormscope and SkyWatch traffic information system optional.

Instrumentation: Avidyne FlightMax Entegra EX5000C 10.4 in landscape format PFD and MFD standard.

EQUIPMENT: Cirrus Airframe Parachute System standard; air conditioning, TKS fluid ice protection system, and Precise Flight 0.62 m^3(22 cu ft) semi-portable oxygen system optional.

DIMENSIONS, EXTERNAL:
Wing span ..... 11.73 m (38 ft 6 in)
Wing aspect ratio 10.2

Cirrus SR22-G2 'second generation' variant *(Paul Jackson)* ***NEW***/0580802

Length overall	7.92 m (26 ft 0 in)
Height overall	2.62 m (8 ft 7 in)
Tailplane span	3.93 m (12 ft 10¾ in)
Wheel track	3.23 m (10 ft 7¼ in)
Wheelbase	2.20 m (7 ft 2½ in)
Propeller diameter	1.98 m (6 ft 6 in)
Cabin doors (each): Height	0.99 m (3 ft 3 in)
Width	0.86 m (2 ft 10 in)
DIMENSIONS, INTERNAL:	
Cabin: Length	3.30 m (10 ft 10 in)
Max width	1.24 m (4 ft 1 in)
Max height	1.27 m (4 ft 2 in)
Baggage compartment volume	0.91 m^3 (32.0 cu ft)
AREAS:	
Wings, gross	13.46 m^2 (144.9 sq ft)
Horizontal tail	3.48 m^2 (37.50 sq ft)
WEIGHTS AND LOADINGS:	
Weight empty	1,021 kg (2,250 lb)
Baggage weight	59 kg (130 lb)
Max fuel weight	218 kg (480 lb)
Max T-O weight	1,542 kg (3,400 lb)
Max wing loading	114.6 kg/m^2 (23.46 lb/sq ft)
Max power loading	6.68 kg/kW (10.97 lb/hp)
PERFORMANCE:	
Never-exceed speed (VNE)	204 kt (377 km/h; 234 mph)
Cruising speed at FL80, at 75% power	185 kt (343 km/h; 213 mph)
Manoeuvring speed	142 kt (263 km/h; 163 mph) IAS
Stalling speed: flaps up	70 kt (130 km/h; 81 mph) IAS
flaps down	59 kt (110 km/h; 68 mph) IAS
Max rate of climb at S/L	427 m (1,400 ft)/min
Service ceiling	5,335 m (17,500 ft)
T-O run	311 m (1,020 ft)
T-O to 15 m (50 ft)	480 m (1,575 ft)
Landing from 15 m (50 ft)	709 m (2,325 ft)
Landing run	348 m (1,140 ft)
Range with reserves: at 75% power	700 n miles (1,296 km; 805 miles)
at econ cruise	more than 1,000 n miles (1,852 km; 1,150 miles)
g limits	+3.8/–1.9

Cirrus SR22-G2 instrument panel and cockpit interior ***NEW***/0580807

UPDATED

COMMANDER

COMMANDER AIRCRAFT COMPANY

Wiley Post Airport, 7200 North-West 63rd Street, Bethany, Oklahoma 73008
Tel: (+1 405) 495 80 80
Fax: (+1 405) 495 83 83
e-mail: cacsales@telepath.com
Web: http://www.commanderair.com
CEO: Wirt Walker
PRESIDENT AND COO: Mat Goodman
EXECUTIVE VICE-PRESIDENT: Carl Gull

Company (a division of Aviation General Inc from 1998) acquired manufacturing, marketing and support rights for Rockwell Commander 112 and 114 from Gulfstream Aerospace Corporation in 1988; spares and support services for existing aircraft and manufacturing based in Oklahoma; 120 employees in 1999. Commander 114 maintained in production, including TC version introduced in 1995. Deliveries in 1997–2002 comprised 14, 13, 13, 21, 10 and seven (none in final quarter). In March 2000, Commander announced the introduction of the 115, an updated and refined replacement for the 114.

Factory floor area 9,638 m^2 (103,742 sq ft). Commander has over 140 authorised service centres worldwide.

Aviation General Inc filed for Chapter 11 bankruptcy protection on 27 December 2002, citing US$1 million cash flow shortfall and US$3.7 million total net indebtedness. Reorganisaion plan presented on 5 July 2003, under which Tiger aircraft LLC to acquire 80 per cent of Aviation General for US$2.8 million; this formalised on 1 November 2003 and accepted by bankruptcy court on 10 December 2003, but in July 2004 Tiger elected not to proceed with the purchase.

UPDATED

CUB CRAFTERS

CUB CRAFTERS INC

1918 South 16th Avenue, PO Box 9823, Yakima, Washington 98903
Tel: (+1 509) 248 94 91
Fax: (+1 509) 248 14 21
Web: http://www.cubcrafters.com
SALES MANAGERS:
Jim Richmond
Nathan Richmond

Having rebuilt and restored Piper PA-18 Super Cubs since 1980, Cub Crafters began production of new aircraft in 1999. It currently holds 51 Supplementary type certificates for various improvements to the original design and continues to offer engine upgrades of existing Super Cubs to 112 kW (150 hp), 119 kW (160 hp) or 134 kW (180 hp). The conversion is not approved by New Piper Aircraft.

In 2004, Cub Crafters increased the standard MTOW of its Top Cub and was preparing to launch the Cub Light upon official approval of the new LSA certification category.

UPDATED

CUB CRAFTERS PA-18-180 TOP CUB

TYPE: Two-seat lightplane.

PROGRAMME: Development of L-4 Cub, via, PA-11/US Army L-18B; first certified in PA-19/L-18C guise 1 April 1949, followed by civil PA-18 Super Cub equivalent

Floatplane version of Cub Crafters Top Cub ***(Paul Jackson)*** ***NEW***/0580831

18 November 1949 (67 kW; 90 hp Continental C90-12F); further variants included PA-18-125 (L-21A), PA-18-135 (L-21B) and PA-18A agricultural version and PA-18S seaplane. Remained in production until final 24 Super Cubs built in 1994; marketing from mid-1980s was by WTA Inc of Lubbock, Texas, but aircraft supplied by Piper.

Current production began in 1999; prototype N901CC. Type certificate retained by New Piper Aircraft Inc, although that company does not participate in present aircraft.

CURRENT VERSIONS: **PA-18-180 Top Cub:** *As described.* Not previously available in this engine power from Piper; initial MTOW 793 kg (1,750 lb); previously optional MTOW of 907 kg (2,000 lb) standard from 2004. Custom options as described.

Cub Light: *Described separately.*

CUSTOMERS: Total 10,346 (plus three PA-19s) built by Piper. Cub Crafters registered 74th from current production in April 2004. Six employed by US Air Force Academy, Colorado Springs, for glider towing. Others to US Border Patrol.

COSTS: US$149,950 (2004). Extras include floats at US$45,500.

DESIGN FEATURES: Braced high-wing monoplane, with V struts each side. Lightweight aluminium oil cooler below engine cowling is recognition feature of Cub Crafters' aircraft (but is optional Super Cub retrofit), as are optional drooped wingtips, anti-spin strakes ahead of tailplane and 36 vortex generators on each wing leading-edge, all for enhanced STOL performance.

Wing section USA 35B. Thickness/chord ratio12 per cent. Dihedral 1°. No incidence at mean aerodynamic chord. Total washout of 3° 18'.

FLYING CONTROLS: Plain ailerons and trailing edge flaps of light-alloy construction and skinning. Tailplane incidence variable for trimming. Balanced rudder and elevators. Optional extended ailerons; optional drooping ailerons (18°).

STRUCTURE: Wing has aluminium spars and stamped aluminium ribs, aluminium sheet leading-edge and aileron false spar, with Ceconite 7600 covering polyurethane finished, pop-riveted to structure. The fuselage is a rectangular welded 4130 chrome molybdenum steel tube structure covered with Stits Polyfiber; glass fibre engine cowling. Tail unit is a wire-braced structure of welded steel tubes and channels, covered with Ceconite 7600; optional metal belly skin extending 1.07 m (3 ft 6 in) forward of sternpost.

LANDING GEAR: Tailwheel type; fixed. Two side Vs and half axles hinged to bottom of fuselage. Rubber chord shock absorption. Mainwheel tyres size 8.50-6; optional Goodyear 26×10.50-6; (6 ply) or 79 cm (31 in) Alaska Bush wheels; Scott 3200 steerable leaf spring tailwheel size 2.60/2.50-4. Dual Cleveland wheels and brakes; parking brake standard. Wipline 2100 or EDO 2000 floats optional.

POWER PLANT: One 134 kW (180 hp) Textron Lycoming O-360-C4P flat-four engine, driving a Sensenich 76EM8-0-56 two-blade fixed-pitch metal propeller. Optional O-360-C1G with constant-speed propeller; optional McCauley climb propeller. From 2004, standard fit of one 87 litre (23.0 US gallon; 19.2 Imp gallon) metal fuel tank in each wing. Total fuel capacity 174 litres (46.0 US gallons; 38.3 Imp gallons); no header tank. Refuelling points on top of wing. Fuel options comprise 68 litre (18.0 US gallon; 15.0 Imp gallon) belly tank with combined cargo pod for total 242 litre (64.0 US gallon; 53.3 Imp gallon) capacity; or 121 litre (32.0 US gallon; 26.7 Imp gallon) belly tank for total 295 litre (78.0 US gallon; 65.0 Imp gallon) capacity.

ACCOMMODATION: Two persons in tandem with dual controls. Adjustable front seat. Heater and adjustable fresh air control. Downward-hinged door on starboard side, and upward-hinged window above, can be opened in flight. Sliding window on port side. Baggage compartment aft of rear seat can be enlarged by removal of latter; additional 23 kg (50 lb) of baggage in optional cargo/fuel belly tank; optional 64 kg (140 lb) baggage (only) pod; baggage door on starboard side.

SYSTEMS: Electrical system comprising 50 A alternator, 12 V engine starter and 17 A battery, standard.

AVIONICS: Standard package includes an altimeter, airspeed indicator, magnetic compass, tachometer, oil pressure gauge, ELT.

Optional avionics packages comprise: *Standard:* Garmin GNC 250XL moving map GPS/com, Garmin GTX 327 transponder and PM1000 II intercom; *Deluxe VFR:* Garmin 420 colour moving map GPS/com, GTX 327 and PM1000; and *Deluxe IFR:* Garmin 430, GI 106A course deviation indicator, GTX 327, GMA 340 audio panel, vacuum system and heated pitot.

EQUIPMENT: Navigation lights and tail strobe warning light standard.

DIMENSIONS, EXTERNAL:

Wing span	10.76 m (35 ft 3½ in)
Wing chord, constant	1.60 m (5 ft 3 in)
Wing aspect ratio	7.0
Length overall	6.88 m (22 ft 7 in)
Height overall	2.04 m (6 ft 8½ in)
Tailplane span	3.20 m (10 ft 6 in)
Wheel track	1.84 m (6 ft 0½ in)

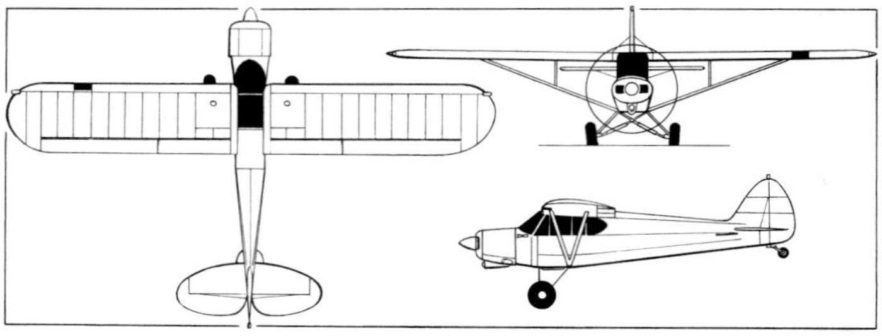

Cub Crafters Top Cub *(James Goulding)* 0143322

Propeller diameter: standard	1.93 m (6 ft 4 in)
optional (climb)	2.08 m (6 ft 10 in)

DIMENSIONS, INTERNAL:

Cargo volume, max	1.0 m³ (36 cu ft)

AREAS:

Wings, gross	16.58 m² (178.5 sq ft)
Ailerons (total)	1.75 m² (18.80 sq ft)
Trailing-edge flaps (total)	1.07 m² (11.50 sq ft)
Fin	0.43 m² (4.66 sq ft)
Rudder	0.63 m² (6.76 sq ft)
Tailplane	1.40 m² (15.10 sq ft)
Elevators (total)	1.09 m² (11.70 sq ft)

WEIGHTS AND LOADINGS:

Weight empty	544 kg (1,200 lb)
Baggage capacity	91 kg (200 lb)
T-O weight	907 kg (2,000 lb)
Max wing loading	54.7 kg/m² (11.20 lb/sq ft)
Max power loading	6.76 kg/kW (11.11 lb/hp)

PERFORMANCE:

Max level speed at S/L	122 kt (225 km/h; 140 mph)
Cruising speed:	
at 75% power	110 kt (204 km/h; 127 mph)
at 55% power	96 kt (179 km/h; 111 mph)
Stalling speed, flaps down	44 kt (81 km/h; 50 mph)
Max rate of climb at S/L	420 m (1,380 ft)/min
T-O run	77 m (250 ft)
Landing run	122 m (400 ft)

UPDATED

CUB CRAFTERS CUB LIGHT

TYPE: Two-seat lightplane; tandem-seat kitbuilt.

PROGRAMME: Developed from Top Cub to fall within limits of projected FAA Sport Pilot licence and Light Sport Aircraft certification category. Promoted at AirVenture, Oshkosh, Wisconsin, July 2003. Reduced weight and engine power.

Generally as for Top Cub, except the following.

COSTS: To be announced. Available factory-built LSA; kitbuilt LSA; or kitbuilt Experimental category.

POWER PLANT: One 74.6 kW (100 hp) Teledyne Continental O-200 flat-four driving a two-blade, fixed pitch propeller. Fuel capacity 68 litres (18.0 US gallons; 15.0; Imp gallons).

WEIGHTS AND LOADINGS:

Weight empty	386 kg (850 lb)
Max T-O weight	558 kg (1,232 lb)

PERFORMANCE:

Cruising speed	87 kt (161 km/h; 100 mph)
Stalling speed	32 kt (58 km/h; 36 mph)
Max rate of climb at S/L	290 m (950 ft)/min
T-O run	99 m (325 ft)
Landing run	89 m (290 ft)

UPDATED

CULP'S

CULP'S SPECIALTIES

PO Box 7542, Shreveport, Louisiana 71137
Tel/Fax: (+1 318) 222 08 50
e-mail: culpspecial@yahoo.com
Web: http://www.culpsspecialties.com
PRESIDENT: Steve Culp

In addition to producing the Special biplane, Culp's Specialties restores and rebuilds classic aircraft. In 1998, Culp's began building six Monocoupes, and in 2000 announced a Sopwith Pup replica.

VERIFIED

CULP'S SPECIAL

TYPE: Aerobatic three-seat biplane/kitbuilt.

PROGRAMME: Design started September 1993; work on first prototype began in following November and first flight achieved (N367LS) on 15 April 1996. Made public debut at Sun 'n' Fun 1996; initially had 670 kg (1,478 lb) empty weight and 930 kg (2,050 lb) MTOW. Manufacture of first production aircraft began in August 1994. FAA Experimental certification granted April 1996 and aircraft embarked on intensive demonstration programme.

CUSTOMERS: By January 2004, five aircraft were flying and at least seven were under construction. 200 sets of plans and six kits sold.

COSTS: Factory-built US$145,000; kit US$51,000; plans US$280 (2004).

Prototype Culp's Special *(Paul Jackson)* 0092078

DESIGN FEATURES: Fully aerobatic, single-bay biplane with 1930s styling, but offering modern performance. Loosely based on the Steen Skybolt. Cabane-mounted upper wing with single interplane strut and wire bracing; wire-braced empennage. Quoted build time 3,500 hours. Roll rate 260°/s.

FLYING CONTROLS: Conventional and manual. Actuation by pushrods; electric rudder trim and manual elevator trim. Aileron design influenced by Curtis Pitts. Horn-balanced elevator and rudder.

STRUCTURE: Fuselage is 4130 steel tubing covered with spruce; metal-skinned forward of cockpit, with fabric to rear. Wings are of fabric-covered wooden construction.

LANDING GEAR: Tailwheel type; fixed. Two faired-in side Vs of 4130 steel tubing hinged to lower longerons, plus half-axles sprung with bungee cord. Goodyear tyres: mainwheels 10.00-6, pressure 1.86 bar (27 lb/sq in); tailwheel 8.00-4, pressure 3.10 bar (45 lb/sq in). Cleveland brakes. Minimum ground turning radius 7.92 m (26 ft 0 in).

POWER PLANT: One 265 kW (355 hp) VOKBM M-14PT radial engine driving a constant-speed two-blade wooden propeller. Main fuel tank in fuselage contains 121 litres (32.0 US gallons; 26.6 Imp gallons); wing tank contains 38 litres (10.0 US gallons; 8.3 Imp gallons) and ferry tank a further 151 litres (40.0 US gallons; 33.3 Imp gallons); total usable fuel 299 litres (79.0 US gallons; 65.8 Imp gallons). Oil capacity 15 litres (4.0 US gallons; 3.3 Imp gallons).

ACCOMMODATION: Pilot and either one or two passengers, depending on model, in open cockpit. Baggage compartment behind headrest.

SYSTEMS: 10 A electrical system for radio and lights; pneumatic engine starting system.

AVIONICS: *Comms:* Honeywell radio, transponder and GPS.

DIMENSIONS, EXTERNAL:
Wing span: upper ... 7.32 m (24 ft 0 in)
lower ... 7.01 m (23 ft 0 in)
Wing chord, constant ... 1.07 m (3 ft 6 in)
Length overall ... 6.40 m (21 ft 0 in)
Max diameter of fuselage ... 1.05 m (3 ft 5½ in)
Height overall ... 2.44 m (8 ft 0 in)
Tailplane span ... 2.29 m (7 ft 6 in)
Wheel track ... 1.98 m (6 ft 6 in)
Wheelbase ... 5.18 m (17 ft 0 in)
Propeller diameter ... 2.39 m (7 ft 10 in)
Propeller ground clearance ... 0.91 m (3 ft 0 in)

DIMENSIONS, INTERNAL:
Cabin: Length ... 2.44 m (8 ft 0 in)
Max width ... 0.81 m (2 ft 8 in)
Max height ... 0.81 m (2 ft 8 in)
Floor area ... 0.56 m^2 (6.0 sq ft)
Baggage hold volume ... 0.11 m^3 (4.0 cu ft)

AREAS:
Wings, gross ... 14.96 m^2 (161.0 sq ft)
Ailerons (total) ... 2.60 m^2 (28.00 sq ft)
Rudder, incl tab ... 1.02 m^2 (11.00 sq ft)
Tailplane ... 2.97 m^2 (32.00 sq ft)
Elevators, incl tab ... 1.49 m^2 (16.00 sq ft)

WEIGHTS AND LOADINGS:
Weight empty ... 689 kg (1,520 lb)
Max fuel weight ... 218 kg (480 lb)
Max T-O and landing weight ... 1,043 kg (2,300 lb)
Max zero-fuel weight ... 848 kg (1,870 lb)
Max wing loading ... 69.7 kg/m^2 (14.29 lb/sq ft)
Max power loading ... 3.94 kg/kW (6.48 lb/hp)

PERFORMANCE:
Never-exceed speed (VNE) at S/L ... 208 kt (386 km/h; 240 mph)
Max operating speed (VMO) ... 147 kt (273 km/h; 170 mph)
Econ cruising speed at 2,590 m (8,500 ft) ... 148 kt (274 km/h; 170 mph)
Stalling speed, power off ... 59 kt (108 km/h; 67 mph)
Max rate of climb at S/L ... 1,371 m (4,500 ft)/min
Service ceiling ... 3,660 m (12,000 ft)
T-O run ... 91 m (300 ft)
T-O to 15 m (50 ft) ... 122 m (400 ft)
Landing from 15 m (50 ft) ... 213 m (700 ft)
Landing run ... 244 m (800 ft)
Range: with max fuel at econ cruising speed ... 955 n miles (1,770 km; 1,100 miles)
with max payload ... 487 n miles (902 km; 561 miles)
g limits ... ±10

UPDATED

CULP'S MONOCULP

TYPE: Four-seat kitbuilt.

PROGRAMME: Announced at Sun 'n' Fun, April 2000, when prototype under construction.

CUSTOMERS: Two kits and three sets of plans sold by May 2000.

COSTS: Kit price US$47,000 with wings completed (2004).

DESIGN FEATURES: Scaled-up (125 per cent) version of 1930s Monocoupe 90A. Quoted build time 2,500 hours.

FLYING CONTROLS: Conventional and manual.

STRUCTURE: Strut-braced high wing; wire-braced tailplane.

LANDING GEAR: Tailwheel type; fixed. Fuselage-mounted, streamlined, cantilever main legs with integral shock-absorbers. Cantilever tailwheel. Speed fairings on mainwheels.

POWER PLANT: One 294 kW (394 hp) VOKBM M-14PF nine-cylinder radial engine. Fuel capacity 318 litres (84.0 US gallons; 70.0 Imp gallons).

Original Mono Aircraft Monocoupe 110, built in 1933, on which the MonoCulp is based ***(Paul Jackson)*** ***NEW***/1047137

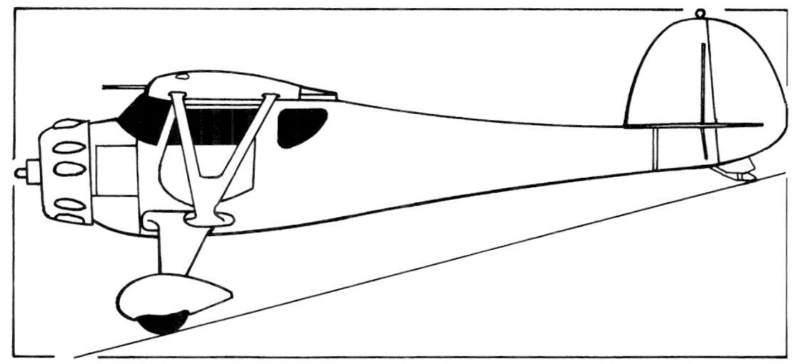

Culp's MonoCulp ***(Paul Jackson)*** 0092144

ACCOMMODATION: Pilot and three passengers in two pairs of seats; cabin door on port side.

DIMENSIONS, EXTERNAL:
Wing span ... 9.75 m (32 ft 0 in)
Length overall ... 7.47 m (24 ft 6 in)
Height overall ... 2.54 m (8 ft 4 in)

DIMENSIONS, INTERNAL:
Cabin max width ... 1.22 m (4 ft 0 in)

AREAS:
Wings, gross ... 17.84 m^2 (192.0 sq ft)

WEIGHTS AND LOADINGS:
Weight empty ... 771 kg (1,700 lb)
Max T-O weight ... 1,315 kg (2,900 lb)
Max wing loading ... 73.7 kg/m^2 (15.10 lb/sq ft)
Max power loading ... 4.48 kg/kW (7.36 lb/hp)

PERFORMANCE:
Never-exceed speed (VNE) ... 234 kt (434 km/h; 270 mph)
Normal cruising speed ... 148 kt (274 km/h; 170 mph)
Stalling speed, power off ... 54 kt (100 km/h; 62 mph)
Max rate of climb at S/L ... 914 m (3,000 ft)/min
Service ceiling ... 3,660 m (12,000 ft)
T-O run ... 122 m (400 ft)
Landing run ... 274 m (900 ft)
Range with max fuel ... 827 kt (1,532 km; 952 miles)
Endurance ... 4 h 0 min

UPDATED

CULP'S SOPWITH PUP

TYPE: Tandem-seat biplane kitbuilt.

PROGRAMME: Announced at Sun 'n' Fun, April 2000, prototype then under construction. First aircraft (N329CC), owned by John S Culp, displayed at AirVenture, Oshkosh, July 2004, immediately following completion.

CUSTOMERS: Orders for 12 kits and 18 sets of plans received by May 2000. Three completed by September 2003.

COSTS: Plans US$200; kit US$47,000 (2004).

DESIGN FEATURES: Based on First World War Sopwith Pup fighter, but using all-new materials. Single-bay biplane with wood strut and wire bracing. Quoted build time 3,500 hours.

STRUCTURE: Wood, fabric and metal tube construction.

LANDING GEAR: Tailwheel type; fixed. Two steel tube side Vs and axle.

POWER PLANT: One VOKBM M-14P nine-cylinder radial derated to 179 kW (240 hp); engines in range 164 to 268 kW (220 to 360 hp) can be fitted. Fuel capacity 189 litres (50.0 US gallons; 41.7 Imp gallons).

ACCOMMODATION: Pilot and passenger in tandem in open cockpit.

DIMENSIONS, EXTERNAL:
Wing span ... 8.08 m (26 ft 6 in)
Length overall ... 5.49 m (18 ft 0 in)
Height overall ... 2.74 m (9 ft 0 in)

DIMENSIONS, INTERNAL:
Cabin max width ... 0.71 m (2 ft 4 in)

AREAS:
Wings, gross ... 24.62 m^2 (265.0 sq ft)

WEIGHTS AND LOADINGS:
Weight empty ... 590 kg (1,300 lb)
Max T-O weight ... 1,088 kg (2,400 lb)

First Culp's Sopwith Pup displaying above EAA AirVenture at Oshkosh in July 2004 ***(Jim Koepnick/EAA)*** ***NEW***/1043415

First Culp's Sopwith Pup displaying above EAA AirVenture at Oshkosh in July 2004 *(Jim Koepnick/EAA)* ***NEW***/1043414

Max wing loading .... 44.2 kg/m² (9.06 lb/sq ft)
Max power loading .... 6.09 kg/kW (10.00 lb/hp)

PERFORMANCE:
Never-exceed speed (VNE) .... 191 kt (354 km/h; 220 mph)
Normal cruising speed .... 130 kt (241 km/h; 150 mph)
Stalling speed .... 33 kt (62 km/h; 38 mph)
Max rate of climb at S/L .... 1,372 m (4,500 ft)/min
Service ceiling .... 3,660 m (12,000 ft)
T-O run .... 76 m (250 ft)
Landing run .... 152 m (500 ft)
Range with max fuel .... 521 n miles (965 km; 600 miles)

UPDATED

DARPA

3701 North Fairfax Drive, Arlington, Virginia 22203-1714
Tel: (+1 703) 696 24 04
Web: http://www.darpa.mil
WALRUS PROGRAMME MANAGER: Preston H Carter
PUBLIC INFORMATION OFFICER: Ms Jan Walker

In 2004, DARPA launched a feasibility study for an outsize future airlift vehicle under the programme name Walrus.

NEW ENTRY

DARPA WALRUS

TYPE: Hybrid air vehicle.

PROGRAMME: Revealed on 26 April 2004 by DARPA's Tactical Technology Office, the Walrus programme is intended to investigate the feasibility of an outsize hybrid airlift vehicle generating its lift through a combination of aerodynamics and gas buoyancy, thereby flying heavier than air, unlike a conventional airship. A 12-month Phase 1 will include system studies and development of a notional concept of the objective vehicle; it will also investigate such advanced technologies as vacuum/air buoyancy compensator tanks and electrostatic atmospheric ion propulsion.

Based on Phase 1 results, and concept viability, a competitive Phase 2 will then lead to development, design, manufacture, and initial flight test in 2008, of an Advanced Technology Demonstration (ATD) vehicle having an airlift capability comparable with that of the C-130 Hercules. Requirements include VTOL capability from land or sea; maintaining hover during loading and unloading; and operation from both unprepared land sites and ships' decks. The ATD vehicle will also be used by the US armed forces to explore system concepts of operations (conops).

Depending upon the outcome of Phase 2, an eventual Walrus vehicle is foreseen that would be able to lift more than 500 US tons over intercontinental ranges; or transport a 'unit of action' from 'fort-to-fight' as a complete, integrated, action-ready contingent of personnel and equipment. It may also meet multi-agency needs for extended-range airborne patrol, persistence and intra-theatre support and resupply.

An RFP for Phase 1 was issued on 19 May 2004; two teams are to be selected from bidders, from which one will be down-selected in 2007 to build and fly the ATD demonstrator. The vacuum/air buoyancy compensation vessel will be developed by Hunt Aviation Corporation of Pass Christian, Mississippi (which see). Potential Phase 1 candidates could include such types as the ATG SkyCat, Aeros-ML, AHA Behemuth and Ohio Dynalifter.

NEW ENTRY

EAGLE

EAGLE R&D LTD CO

2512 Caldwell Boulevard, Nampa, Idaho 83651-1518
Tel: (+1 208) 461 25 67
Fax: (+1 208) 454 37 52
Web: http://www.helicycle.com

Company founded 23 April 1998 by Buford J Schramm, originator of the RotorWay kit helicopter programme, who was killed in a flying accident to an Eagle on 27 April 2004. Eagle's sales policy for its Helicycle includes mandatory test flying of customer-completed aircraft by a factory pilot before handover of final key components.

UPDATED

EAGLE HELICYCLE

TYPE: Single-seat ultralight helicopter kitbuilt.

PROGRAMME: Prototype (N3275Q) registered in April 1995; crashed 27 April 2004, with loss of B J Schramm. First customer-built aircraft (N9019U) assembled by Douglas Swochert of Burlington, Wisconsin; initially intended with two-stroke Rotax, but completed with T62 turboshaft and registered as a Swochert A/W 95.

CUSTOMERS: At least 30 kits sold by 2002, of which 16 had been registered with the FAA by mid-2004. One obtained by South African customer in 2004.

DESIGN FEATURES: Embodies best features from Mr Schramm's experience of rotary-wing design; comparable in capabilities to industry-standard Robinson R22, but single-seat. Conventional appearance; small streamlined, enclosed cabin with power plant, gearbox and fuel immediately to rear; tail rotor and empennage mounted on open boom; dorsal and ventral fins, latter incorporating tail bumper. Two-blade main and tail rotors.

Handling assisted by fully harmonised rotor; modulated collective system; elastomeric thrust bearings; generous flapping angle for low *g* and slope landing conditions; control friction devices; and electronic throttle control.

STRUCTURE: Metal tube structure with composites cabin.

LANDING GEAR: Twin-skid type.

POWER PLANT: One piston or turbine engine. Two-stroke Rotax 6/8 initially proposed, but manufacturer discontinued production. Interest changed in 2002 to Hirth 3701 rated at 64.9 kW (87 hp) at 5,000 rpm, although not pursued. Alternative 59.6 kW (79.9 hp) Rotax 912 flat-four installation under development in 2002, but proved too expensive.

First production helicopter flown with Solar T62 turbine ground power plant derated to 78.3 kW (105 hp); modified T62 installation offered by Eagle as standard power plant. Fuel capacity 79.0 litres (21.0 US gallons; 17.5 Imp gallons).

Data for T62 engine installation:

DIMENSIONS, EXTERNAL:
Main rotor diameter .... 6.35 m (20 ft 10 in)
Tail rotor diameter .... 1.12 m (3 ft 8 in)
Length overall .... 6.05 m (19 ft 10 in)
Height overall .... 2.18 m (7 ft 2 in)

Turbine-engined Helicycle built by Homer K Bell of Hillsboro, Ohio, from the 30th Eagle kit *(Paul Jackson)* ***NEW***/1047172

WEIGHTS AND LOADINGS:
Weight empty ... 229 kg (505 lb)
Max T-O weight ... 372 kg (820 lb)
PERFORMANCE:
Max cruising speed ... 87 kt (161 km/h; 100 mph)
Max rate of climb at S/L ... 274 m (900 ft)/min
Hovering ceiling IGE ... 2,895 m (9,500 ft)
Max range ... 156 n miles (289 km; 180 miles)

UPDATED

ECLIPSE

ECLIPSE AVIATION CORPORATION

2503 Clark Carr Loop SE, Albuquerque, New Mexico 87106
Tel: (+1 505) 245 75 55
Fax: (+1 505) 245 78 88
e-mail: info@eclipseaviation.com
Web: http://www.eclipseaviation.com
CHAIRMAN: Harold A Poling
PRESIDENT AND CEO: Vern Raburn
VICE-PRESIDENT, AVIONICS AND ELECTRONICS: Don Burtis
VICE-PRESIDENT, BUSINESS AFFAIRS: Jack Harrington
VICE-PRESIDENT, SALES CUSTOMER AND PRODUCT SUPPORT: Michael McConnell
VICE-PRESIDENT, ENGINEERING: Oliver Masefield
VICE-PRESIDENT, MANUFACTURING: Rod Holter
VICE-PRESIDENT, SUPPLY MANAGEMENT: Chris Herzog
VICE-PRESIDENT, MARKETING: Dottie Hall
VICE-PRESIDENT, PROGRAMME MANAGEMENT: Gene Games
VICE-PRESIDENT, TRAINING AND FLIGHT OPERATIONS: Don Taylor
VICE-PRESIDENT, FINANCE AND ADMINISTRATION/CFO: Peter C Reed
DIRECTOR OF PUBLIC RELATIONS: Cory Canada

Eclipse Aviation formed May 1998 (as Pronto Aircraft) to market the Eclipse 500, which supersedes the Williams V-Jet II (1999–2000 *Jane's*). Eclipse had funding of US$325 million in place by July 2003. At that time the company had more than 200 employees working on the project.

UPDATED

ECLIPSE ECLIPSE 500

TYPE: Light business jet.

PROGRAMME: Development began June 1999 and announced March 2000; 22 per cent scale model wind tunnel testing undertaken March to April 2000; preliminary design review completed September 2000; full-size mockup displayed at EAA AirVenture 2000 and NBAA 2000. Third round of wind tunnel tests completed January 2001 and critical design review began May 2001 transfer of engineering and certification responsibility from Williams completed July 2001. Designated 3.43 kN (770 lb st) Williams EJ-22 engine first flown in Rockwell Sabreliner 60 testbed (N15HF) 30 May 2002.

Prototype (c/n 500-100/N500EA) rolled out 13 July 2002; first flight 26 August 2002, but no further flights reported by mid-November when Eclipse announced abandonment of Williams engine; selection of Pratt & Whitney Canada announced 19 February 2003. Six flying aircraft used in the development and certification programme, as described under Current Versions. Pending availability of Pratt & Whitney Canada PW610F, the prototype resumed flight testing on 15 May 2003 with two 4.48 kN (1,000 lb st) Teledyne CAE F408-CA-400 turbofans and completed its flying career in October 2003.

First ground run of engine completed 4 May 2004; flight trials (on Pratt & Whitney Canada's Boeing 720 testbed) began 16 December 2004; first engine deliveries to Eclipse on 15 December 2004, followed by first flight of Eclipse 500 with PW610Fs. This was N503EA (third airframe, but second to fly) which achieved two sorties on 31 December 2004. The first four aircraft will accumulate some 3,500 flight hours during the test programme, leading to FAR Pt 23 plus take-off and landing to Pt 25 standard and RVSM certification, with target date of first quarter 2006, followed by JAA JAR 23 approval at the end of the year.

First conforming Eclipse 500 flying with PW610F engines ***NEW***/1043188

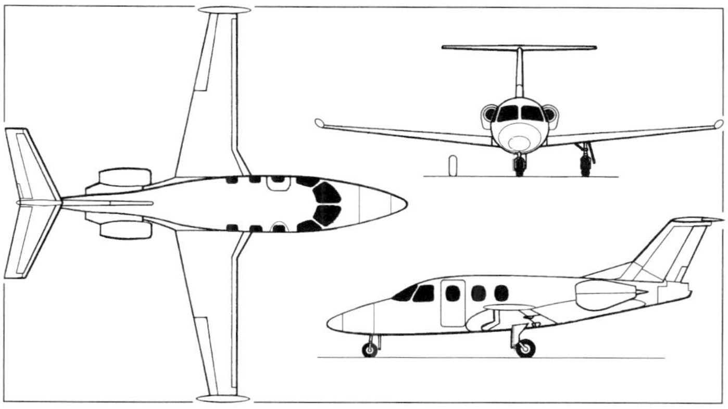
Eclipse 500 with PW610 engines ***(James Goulding)*** ***NEW***/1043199

CURRENT VERSIONS (SPECIFIC): **N500EA:** Airframe 500-100; build order 1. Prototype. First flight 26 August 2002; final flight 22 October 2003 (55 sorties; 54 hours); lightning survivability trials December 2003. (Note that N501EA is an unrelated Cessna 501 Citation).

N502EA: Build order changed from 2 to 3 in 2004. Structures and aerodynamics testing (ground vibration, loads, strain measurements, air data calibration for RVSM, flutter, stability and stall, field performance and icing).

N503EA: Airframe 500-108; build order 2. Second to fly; first production-conforming airframe. First flight 31 December 2004. FAA certification testing for mechanical systems (climate control, landing gear, fuel system, power plant, oxygen system and hot/cold weather).

N504EA: Build order 4. FAA certification testing for avionics and electrical systems (electrical navigation, communication systems, primary flight display, MFDs, autopilot, lighting).

N505EA: Build order 5. 'Beta' test airframe to full production configuration for accelerated, 1,000-hour test programme to prove reliability.

Static test: Build order 6. Airframe loads testing.

N506EA: Build order 7. 'Beta' test airframe No. 2 plus high intensity radiated fields and lightning tests.

Fatigue test: Build order 8. To undergo three 'lifetimes' of fatigue testing.

CUSTOMERS: Total of 1,357 firm orders and 715 options held by 9 September 2002, of which 530 were from individuals, from customers in the USA, Europe and South America. Announced customers include Aviace AG of Switzerland, which ordered 112 on 28 May 2002 for its private jet club fractional ownership scheme. A provisional order for 1,000 aircraft placed by Nimbus Group on 31 August 2001 subsequently lapsed in July 2002.

COSTS: Development programme US$350 million (estimated). Unit cost US$950,000-US$975,000 for pre-2003 orders, US$1.175 million thereafter (2003).

DESIGN FEATURES: Designed to fulfil goals of NASA's Small Air Transportation System programme. Friction stir welding to be used for first time on high-volume production, obviating need for large numbers of rivets and reducing weight of finished article. Unswept, low-mounted wing and T tail. Redesign work undertaken during 2001 included increasing the size of the horizontal engine pylons, modifying the rear fuselage to improve engine intake airflow and modifications to wingroot shape, wing placement (moved aft 9 cm (3½ in)) and redesigned tail.

Eclipse 500 N503EA during its maiden flight on 31 December 2004 ***NEW***/1043189

Eclipse 500 business jet prototype flying with interim F408 engines in May 2003 ***NEW***/0561747

Airframe modifications that will result from installation of the PW610Fs will include moving the engine mounts forward by 0.24 m (9 ½ in), strengthening the wing spar, moving the vapour-cycle machine from the wingroot fairing to the nose, modifying the flaps to increase Fowler effect, and addition of small tip tanks to accommodate the additional fuel needed to meet range guarantees.

FLYING CONTROLS: Conventional and manual. Sidestick controller. Electrically actuated Fowler flaps and split tailcone speedbrakes.

STRUCTURE: Principally of aluminium; sharply tapered rear fuselage means distance between engine centres will be 1.04 m (3 ft 5 in). Suppliers include: Autronics (aircraft computer system); Crossbow Technology (AHRS); FreeFlight Systems (GPS); Fuji Heavy Industries (wing); Hampson Industries PLC Aerospace Fabrications & Assemblies Division (tail unit); Harco Laboratories (pitot/static system and air data computer); Hispano-Suiza Canada (FADEC); and Meggitt Avionics (autopilot).

LANDING GEAR: Retractable tricycle type with trailing-link main gear designed by Cal-Draulics. Cleveland wheels and brakes with Michelin radial tyres fitted for improved rough-field operations.

POWER PLANT: Two pod-mounted Pratt & Whitney Canada PW610F turbofans, each rated at 4.00 kN (900 lb st) at ISA+10°C, with FADEC. Full capacity 871 litres (230 US gallons; 191.5 Imp gallons).

ACCOMMODATION: Pilot and up to five passengers; dual sidestick controllers for pilot(s); cabin laid out in club format. Polarised windows throughout; door on port side of cabin behind flight deck. Optional lavatory. Horizontally split door on port side, behind pilots.

SYSTEMS: Pressurisation system maintained sea level cabin environment to 6,553 m (21,500 ft) and 2,440 m (8,000 ft) cabin at 12,500 m (41,000 ft) altitude. Pneumatic de-icing system for wing and tailplane leading-edges, bleed anti-icing for engine inlet, and electrically heated windscreen and air system probes. Lighting system by DeVore Aviation.

AVIONICS: Avio intelligent flight system, developed in conjunction with Avidyne and General Dynamics; features all-glass, three-screen, fully IFR cockpit with two PFDs and one MFD. Avionics suite includes dual three-axis autopilots; FMS; autothrottle; aircraft performance computer; colour weather radar; traffic information services; ADS-B; TAWS-B; dual VHF nav/com/LOC/GS, Mode S transponders, GPS and AHRS with air data computer and pitot static systems; active route moving map display; flight path predictor; VNAV; LNAV; ELT; data loader, health monitoring, and RVSM capability.

All data are provisional.

DIMENSIONS, EXTERNAL:	
Wing span	11.40 m (37 ft 4¾ in)
Length overall	10.08 m (33 ft 1 in)
Height overall	3.35 m (11 ft 0 in)
DIMENSIONS INTERNAL:	
Cabin: Length	3.76 m (12 ft 4 in)
Max width	1.42 m (4 ft 8 in)
Max height	1.27 m (4 ft 2 in)
Volume	5.4 m^3 (191 cu ft)
Baggage compartment volume	1.20 m^3 (42.5 cu ft)
WEIGHTS AND LOADINGS:	
Weight empty	1,538 kg (3,390 lb)
Max fuel weight	699 kg (1,540 lb)
Max T-O weight	2,558 kg (5,640 lb)
Max ramp weight	2,576 kg (5,680 lb)
Max landing weight	2,431 kg (5,360 lb)
Max power loading	320 kg/kN (3.13 lb/lb st)
PERFORMANCE:	
Max operating speed (MMO/VMO)	M0.68 (285 kt; 528 km/h; 328 mph)
Max cruising speed	375 kt (695 km/h; 432 mph)
Stalling speed	67 kt (125 km/h; 78 mph)
Max rate of climb at S/L	911 m (2,990 ft)/min
Rate of climb at S/L, OEI	271 m (888 ft)/min
Time to climb to FL350	19 min
Service ceiling	12,495 m (41,000 ft)
Service ceiling, OEI	7,620 m (25,000 ft)
T-O run	657 m (2,155 ft)
Landing run	622 m (2,040 ft)
Range with four occupants, NBAA IFR alternate	1,280 n miles (2,370 km; 1,473 miles)

UPDATED

ENSTROM

THE ENSTROM HELICOPTER CORPORATION

PO Box 490, 2209 22nd Street, Menominee, Michigan 49858-0490
Tel: (+1 906) 863 12 00
Fax: (+1 906) 863 68 21
e-mail: sales@enstromhelicopter.com
Web: http://www.enstromhelicopter.com
PRESIDENT: Jerry M Mullins
DIRECTOR, ENGINEERING: William E Taylor Jnr
DIRECTOR, SALES: P Bayard duPont
DIRECTOR, MANUFACTURING: Billy Selmon
INFORMATION CONTACT: Marian Lambrecht

Enstrom Helicopter Corporation was established in 1959 to develop and produce light helicopters, based on Rudy Enstrom's initial design work. Extensive development and certification efforts led to start of production of the F28 helicopter in April 1965; development of turbine-powered model was initiated in 1988, culminating in certification of the TH28 in September 1992 and the 480, to FAR Pt 27 standards, in December 1994.

Initially publicly owned, the company has passed through a period as a subsidiary of a large corporation and is now privately owned by a Swiss national. By the beginning of 2000, Enstrom had produced over 1,000 helicopters of which over 700 remain in operation; deliveries have comprised eight in 1999, seven in 2000, eight in 2001, nine in 2002 and 17 in 2003; 1,000th delivery made at Heli-Expo 2002 on 14 February 2002. Total of 32 scheduled for production in 2004. The fleet has amassed over three million flying hours. Factory floor area 7,897 m^2 (85,000 sq ft); workforce 90.

UPDATED

ENSTROM F28F AND 280FX

TYPE: Three-seat helicopter.

PROGRAMME: Prototype F28 flew 12 November 1960. Basic F28A and 280 replaced by turbocharged F28C and 280C, certified by FAA 8 December 1975; production of these models ceased November 1981; succeeded by F28F and 280F Shark.

CURRENT VERSIONS: **F28F Falcon:** Basic model certified to FAR Pt 6 on 31 December 1980.

The two major changes incorporated into the F28F and 280F/280FX over the earlier F28C/280C series were an increase in power from 153 kW (205 shp) to 168 kW (225 shp) and the addition of a throttle correlator to reduce pilot workload. Most recent changes to the F28F and 280FX are the redesigned main gearbox with a heavy wall main rotor shaft (standard equipment on all new aircraft and retrofittable to all existing F models); optional lightweight exhaust silencer, which reduces noise in the hover by 40 per cent and gives a 30 per cent reduction when flying at 152 m (500 ft) (can also be retrofitted to F28F, 280F and 280FX); and a lightweight starter motor.

Enstrom 280FX three-seat helicopter *(Paul Jackson)* ***NEW***/0583418

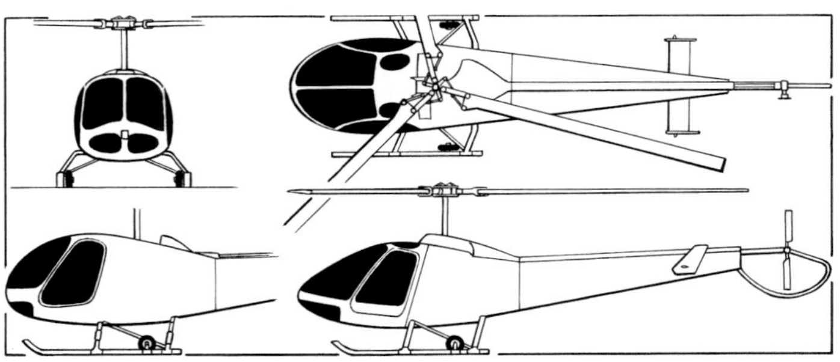

Enstrom 280FX, with scrap side view of F28F *(James Goulding)* 0085773

F28F-P Sentinel: Dedicated police patrol version; first delivery October 1986; can be fitted with a PA and siren system, Spectrolab SX-5 or Carter searchlight, specialised police equipment, FLIR and datalink. F28F-P has same specification and performance as F28F. No recent production.

280FX Shark: Certified to FAR Pt 6 on 14 January 1985. Prototype N280FX. Improvements over previous 280 series (see also F28F entry) include new seats with lumbar support and energy-absorbing NASA foam, new tailplane with endplate fins, tail rotor guard, covered tail rotor drive shaft, redesigned air inlet system, and completely faired landing gear. Optional internal auxiliary fuel tank extends range to 339 n miles (627 km; 390 miles).

CUSTOMERS: Total of 735 of earlier versions (14 F28, 315 F28A, 121 F28C, 56 F28C-2, 21 280, 206 280C and two 280L). Last of 135 F28Fs delivered 1999 to Fresno Police Department, California, but further two produced in 2002 and 2003. Total of 101 280FXs built up to mid-2004, increasing Enstrom piston-engined helicopter production to 973, including further three 280s registered in 2003. Chilean Army operates 15 280FXs for primary and instrument training; Peruvian Army has 10 F28Fs for flight training; Colombian Air Force operates 12 F28Fs for primary and instrument training. Numerous US police departments operate F28F-P for patrol and surveillance missions. Most recent military customer is the Venezuelan National Guard, which took delivery of four 280FXs for training in January 2002.

COSTS: Basic 2004 price US$320,000 for F28F Falcon and 280FX Shark.

DESIGN FEATURES: Conventional light helicopter with skid landing gear and tubular metal tail rotor protector; horizontal stabiliser with fins at tips. High inertia, three-blade fully articulated rotor head with blades attached by retention pin and drag link; control rods pass inside tubular rotor shaft to swashplate inside fuselage; no rotor brake; blade section NACA 0013.5; blades do not fold; two-blade teetering tail rotor. Thirty-groove belt drive from horizontally mounted engine to transmission.

FLYING CONTROLS: Conventional and manual. Trim system absorbs feedback from rotor and repositions stick datum as required by pilot.

STRUCTURE: Bonded light alloy blades. Fuselage has glass fibre and light alloy cabin section, steel tube centre-section frame, and stressed skin aluminium tailboom.

LANDING GEAR: Skids carried on Enstrom oleo-pneumatic shock-absorbers. Air Cruiser inflatable floats available optionally.

POWER PLANT: One 168 kW (225 hp) Textron Lycoming HIO-360-F1AD flat-four engine with Rotomaster 3BT5EE10J2 turbocharger. Two fuel tanks, each of 79.5 litres (21.0 US gallons; 17.5 Imp gallons). Total standard fuel capacity 159 litres (42.0 US gallons; 35.0 Imp gallons), of which 151 litres (40.0 US gallons; 33.3 Imp gallons) are usable. Auxiliary tank, capacity 49 litres (13.0 US gallons; 10.8 Imp gallons), can be installed in the baggage compartment. Oil capacity 9.5 litres (2.5 US gallons; 2.1 Imp gallons).

ACCOMMODATION: Pilot and two passengers, side by side on bench seat; centre place removable. Removable door on each side of cabin. Baggage space aft of engine compartment, with external door. Cabin heated and ventilated.

SYSTEMS: Electrical power provided by 24 V 70 A engine-driven alternator; 12 V 70 A system optional. No hydraulic system.

AVIONICS: Variety of fits from Honeywell and other avionics suppliers.

Instrumentation: Standard equipment includes airspeed indicator, sensitive altimeter, compass, outside air temperature gauge, turn and bank indicator, rotor/engine tachometer, manifold pressure/fuel flow gauge, EGT gauge, oil pressure gauge, gearbox and oil temperature gauge, ammeter, cylinder head temperature gauge, and fuel quantity gauge. Eight-light annunciator panel consisting of low rotor rpm, chip detectors (main and tail rotor transmissions), overboost, clutch not fully engaged, low fuel pressure, starter and low-voltage warning lights.

EQUIPMENT: Shoulder harnesses for three seats. Night lighting is optional for F28F and standard on 280FX. Night lighting includes instrument lighting with dimmer control, position light on each horizontal stabiliser tip, anti-collision light and nose-mounted landing light. Optional equipment for both F28F and 280FX includes fixed float kit, wet or dry agricultural spray kit and cargo hook for utility missions. Wide instrument panel available for IFR training.

DIMENSIONS, EXTERNAL (F28 and 280, except where specified):
Main rotor diameter ... 9.75 m (32 ft 0 in)
Tail rotor diameter ... 1.42 m (4 ft 8 in)
Distance between rotor centres ... 5.56 m (18 ft 3 in)
Main rotor blade chord ... 0.24 m (9½ in)
Tail rotor blade chord ... 0.11 m (4½ in)
Length overall, rotors stationary ... 8.92 m (29 ft 3 in)
Fuselage: Length: F28F ... 8.56 m (28 ft 1 in)
280FX ... 8.75 m (28 ft 8½ in)
Max width: F28F ... 1.55 m (5 ft 1 in)
280FX ... 1.52 m (5 ft 0 in)
Height:
to top of cabin: F28F ... 1.85 m (6 ft 1 in)
280FX ... 1.83 m (6 ft 0 in)
to top of rotor head ... 2.74 m (9 ft 0 in)
Tailplane span ... 1.60 m (5 ft 3 in)
Skid track ... 2.21 m (7 ft 3 in)
Cabin doors (each): Height ... 1.04 m (3 ft 5 in)
Width ... 0.84 m (2 ft 9 in)
Height to sill ... 0.64 m (2 ft 1 in)
Baggage door: Height ... 0.55 m (1 ft 9½ in)
Width ... 0.39 m (1 ft 3½ in)
Height to sill ... 0.86 m (2 ft 10 in)

DIMENSIONS, INTERNAL:
Cabin max width: F28F ... 1.55 m (5 ft 1 in)
280FX ... 1.50 m (4 ft 11 in)
Baggage compartment volume ... 0.18 m^3 (6.3 cu ft)

AREAS:
Main rotor disc ... 74.72 m^2 (804.25 sq ft)
Tail rotor disc ... 1.66 m^2 (17.88 sq ft)

WEIGHTS AND LOADINGS:
Weight empty, equipped: F28F ... 712 kg (1,570 lb)
280FX ... 719 kg (1,585 lb)
Baggage capacity: tailboom ... 49 kg (108 lb)
Max T-O weight: F28F, 280FX ... 1,179 kg (2,600 lb)
Max disc loading:
F28F, 280FX ... 15.8 kg/m^2 (3.23 lb/sq ft)

PERFORMANCE (both versions at AUW of 1,066 kg; 2,350 lb, except where indicated):
Never-exceed speed (V_{NE}):
F28F ... 97 kt (180 km/h; 112 mph)
280FX ... 102 kt (189 km/h; 117 mph)
Max level speed, S/L to FL30
F28F ... 97 kt (180 km/h; 112 mph) IAS
280FX ... 102 kt (189 km/h; 117 mph) IAS
Econ cruising speed: F28F ... 89 kt (165 km/h; 102 mph)
280FX ... 93 kt (172 km/h; 107 mph)
Max rate of climb at S/L:
at AUW of 1,066 kg (2,350 lb) ... 442 m (1,450 ft)/min
at AUW of 1,179 kg (2,600 lb) ... 350 m (1,150 ft)/min
Max certified altitude ... 3,660 m (12,000 ft)
Hovering ceiling:
IGE: at 1,066 kg (2,350 lb) ... 4,020 m (13,200 ft)
at 1,179 kg (2,600 lb) ... 2,345 m (7,700 ft)
OGE: at 1,066 kg (2,350 lb) ... 2,650 m (8,700 ft)
at 1,179 kg (2,600 lb) ... 1,707 m (5,600 ft)
Max range, standard fuel, no reserves:
F28F ... 229 n miles (423 km; 263 miles)
280FX ... 260 n miles (483 km; 300 miles)
Max endurance ... 3 h 30 min

UPDATED

ENSTROM 480 AND TH-28

TYPE: Light utility helicopter.

PROGRAMME: Proof-of-concept 280FX, powered by Allison 250-C20W turbine engine, flown December 1988; first flight of wide-cabin 480/TH-28 prototype (N8631E) in October 1989; second 480/TH-28 (production prototype) flew December 1991; TH-28 FAA certified September 1992 (three-seat), 480 FAA certified in June 1993 and recertified to FAR Pt 27 in December 1994.

CURRENT VERSIONS: **480:** Initial version. Five-seat (two plus three) configuration with staggered seating for forward visibility from all seats. Easily reconfigured to three seats for training or executive transport; or pilot's seat only for transporting light cargo. Instrument panel centrally located above avionics console.

480B: Certification received 8 February 2001; UK CAA certification achieved August 2003; compared to 480 has increased gross weight and improved transmission, incorporates cyclic control and airframe vibration damping, providing significant improvement in comfort. Prototype, N480EN exhibited at NBAA Convention, New Orleans, October 2000. FAA approval granted in February 2004 for increase in maximum take-off altitude at maximum take-off weight from 915 m (3,000 ft) to 3,050 m (10,000 ft) density altitude.

Guardian: Dedicated police patrol/security version.

TH-28: Prototype (N8631E) first flew 7 October 1989. Specially configured as a training/light patrol helicopter; unique features include three crashworthy seats, crashworthy fuel system, and a large instrument panel which will accommodate dual instrumentation for either VFR or IFR training. Seating configuration allows training of two students simultaneously.

CUSTOMERS: Production began in 1994 with initial deliveries to Europe. By June 2002, 43 480s, 10 480Bs and four TH-28s had been registered. Final TH-28 produced in 1993 and delivered to China in 1993. Twenty 480Bs registered by mid-2004. Certified and operating in 14 countries: Australia, Belgium, Brazil, Canada, China, France, Germany, Japan, South Africa, Sweden, Switzerland, Thailand, UK and USA. In addition, 480s are also operating in Burkina Faso and Russian Federation. Recent customers include the Indonesian Police which ordered 18 480Bs in 2003, of which two were delivered in 2003 and 14 were scheduled for delivery in 2004.

COSTS: US$639,500 for 480Bs basically equipped (2003); TH-28 price on application.

Description generally as for F28/280, except that below (for 480B).

DESIGN FEATURES: High-inertia, three-blade, fully articulated main rotor system with upgraded and sturdier main rotor gearbox than piston models; larger tail rotor; unboosted flight controls and high skid gear are standard equipment. Rotor speed 372 rpm nominal; 385 rpm maximum permissible; anti-nodal vibrating cantilever beam to damp vibration at low speeds. Extensive crashworthiness and safety features incorporated into basic design. The 480B cabin layout can be quickly reconfigured to seat one, three or five persons.

POWER PLANT: One 313 kW (420 shp) Rolls-Royce 250-C20W turboshaft engine derated to 227 kW (305 shp) for take-off and 207 kW (277 shp) for continuous operation. Engine air inlet particle separator is standard. Fuel capacity 340 litres (90.0 US gallons; 74.9 Imp gallons) in two interconnected tanks.

ACCOMMODATION: See Current Versions. Options include air conditioning (two or three evaporator system). Baggage compartment in tailboom.

EQUIPMENT: Optional equipment include emergency pop-out floats, air conditioning and cargo hook rated for external loads up to 454 kg (1,000 lb); 480B can also be fitted with police

Enstrom 480B five-seat helicopter *(Paul Jackson)* ***NEW***/0583411

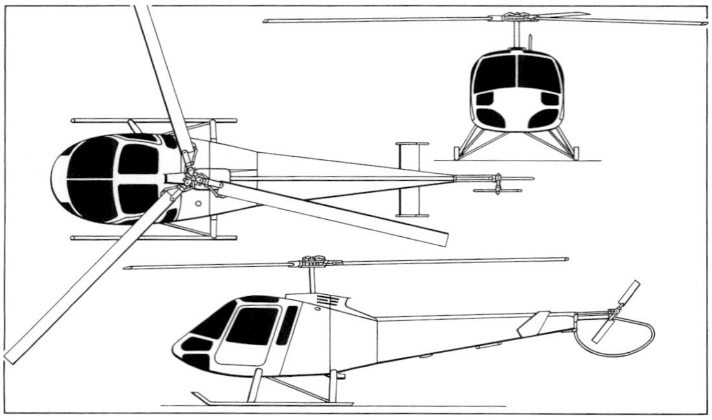

Enstrom 480B light helicopter (Rolls-Royce 250-C20W turboshaft) ***(Jane's/Mike Keep)*** 0016199

patrol equipment including special police radios, searchlight, siren/PA system, FLIR and data recording facility.

DIMENSIONS, EXTERNAL:

Main rotor diameter ... 9.75 m (32 ft 0 in)
Tail rotor diameter ... 1.54 m (5 ft 0½ in)
Distance between rotor centres ... 5.64 m (18 ft 6 in)
Fuselage: Length ... 9.09 m (29 ft 10 in)
  Max width ... 1.79 m (5 ft 10½ in)
Height: to top of cabin ... 2.11 m (6 ft 11¼ in)
  to top of rotor head ... 3.00 m (9 ft 10 in)
Skid track ... 2.50 m (8 ft 2½ in)

DIMENSIONS, INTERNAL:

Cabin max width ... 1.80 m (5 ft 10¾ in)

AREAS:

Main rotor disc ... 74.72 m^2 (804.25 sq ft)
Tail rotor disc ... 1.85 m^2 (19.96 sq ft)

WEIGHTS AND LOADINGS:

Weight empty ... 769 kg (1,695 lb)
Baggage capacity: cabin ... 23 kg (50 lb)
  tailboom ... 70 kg (155 lb)
Max T-O weight ... 1,360 kg (3,000 lb)
Max disc loading ... 18.2 kg/m^2 (3.73 lb/sq ft)

PERFORMANCE (A: at 1,360 kg; 3,000 lb AUW, B: at 1,134 kg; 2,500 lb):

Never-exceed speed (VNE) ... 125 kt (231 km/h; 144 mph)
Cruising speed: A ... 109 kt (202 km/h; 125 mph)
  B ... 115 kt (213 km/h; 132 mph)
Max rate of climb at S/L: A ... 360 m (1,180 ft)/min
  B ... 457 m (1,500 ft)/min
Service ceiling ... 3,965 m (13,000 ft)
Hovering ceiling:
  IGE: A ... 3,350 m (11,000 ft)
    B ... 4,265 m (14,000 ft)
  OGE: A ... 1,370 m (4,500 ft)
    B ... 3,655 m (12,000 ft)
Max range (no reserves) at FL30:
  A ... 350 n miles (648 km; 402 miles)
  B ... 435 n miles (806 km; 501 miles)
Endurance (no reserves): A ... 4 h 24 min
  B ... 4 h 36 min

UPDATED

EVATION

200 Timber Creek Drive, Ames, Iowa 50010-9116

PRINCIPALS:
Matt Eller
Joel Goodman

Following the bankruptcy of VisionAire, Iowa property developer, Matt Eller acquired the company's intellectual rights in the VA-10 Vantage (alias Scaled Composites 247-4) business mono-jet, together with the prototype (N247VA), moulds and jigs, for US$441,000. Funding sought to follow the US$110,000 already expended on the aircraft, which first flew on 16 November 1996, but proved to require considerable redesign, nominally becoming the VA-10A in this configuration.

Changes proposed by Evation include replacement of single Pratt & Whitney JT15D-5D turbofan by two Williams FJ33s, although it is likely that the original shortcomings in the design will be remedied as an alternative to progressing to the VA-10A configuration, estimated at the time to require a further US$120 million investment. Manufacture could take place in a plant built at Ames for VisionAire's production, but never occupied.

VisionAire VA-10 Vantage prototype, now acquired by Evation ***NEW***/0075974

NEW ENTRY

EXPLORER

EXPLORER AIRCRAFT INC

One Great Place, Jasper, Texas 75951
Tel: (+1 409) 489 15 00
Fax: (+1 409) 489 17 00
e-mail: info@exploreaircraft.com
Web: http://www.exploreaircraft.com

CEO: Graham Swannell
PRESIDENT: Donald G Joseph
DIRECTOR, OPERATIONS, RESEARCH AND DEVELOPMENT: Geoffrey P Danes

Explorer Aircraft Inc was established in Colorado in 1998 to develop and market the Australian-designed Explorer range of single-engined utility aircraft. The company moved to Texas in July 2001, and has established a new 1,300 m^2 (14,000 sq ft) office and hangar facility there. Development and promotion were continuing in 2004, but the start of production has been delayed.

UPDATED

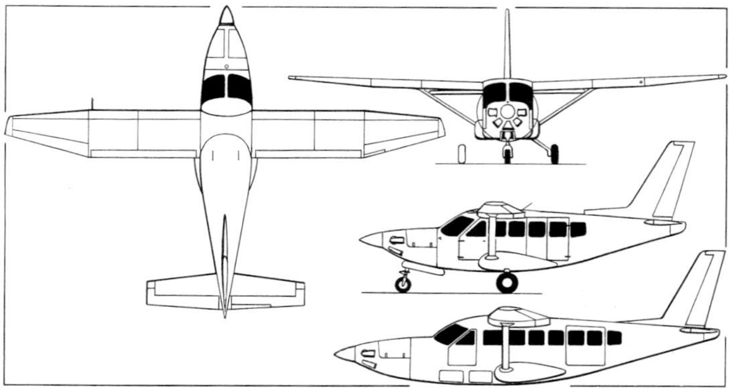

Explorer Aircraft Explorer 500T, with additional side view (lower) of stretched Explorer 750T ***(James Goulding)*** 0105701

EXPLORER EXPLORER

TYPE: Light utility turboprop.

PROGRAMME: Designed by Aeronautical Engineers Australia (AEA), beginning in 1993, with assistance from US aerodynamicist John Roncz; company renamed Explorer Aircraft Corporation. Proof-of-concept (POC) prototype (VH-OKA), designated Explorer 350R, made first flight 23 January 1998 and had flown more than 70 hours up to award of Developmental C of A on 27 April 1998. Prototype (re-registered VH-ONA in May 1998) and given Experimental C of A on 27 April 1999; IFR certification 6 May 1999; departed for USA 20 May 1999.

Explorer 500T multirole utility transport ***(Paul Jackson)*** ***NEW***/0583376

Re-engined with 447 kW (600 shp) Pratt & Whitney Canada PT6-135B turboprop as prototype Explorer 500T, in which form it first flew on 9 June 2000; simultaneously fitted with six-screen EFIS; public debut at EAA AirVenture Oshkosh in August 2000; had flown 500 hours in all configurations by mid-2003. Additional funding being sought in mid-2003 to pursue certification, which would be achieved some 30 months after funding in place, with first deliveries in 2008, allowing for 2004 launch.

CURRENT VERSIONS: **350R:** Proof-of-concept prototype, powered by one 261 kW (350 hp) Teledyne Continental TSIO-55-E3B piston engine. Not to be marketed.

500T: Initial production version employing airframe similar to that of 350R but with 447 kW (600 shp) Pratt & Whitney Canada PT6-135B turboprop. *As described.*

750T: Stretched version with plugs fore and aft of wing; up to 16 passengers; PT6-60A engine flat rated to 597 kW (800 shp).

COSTS: 500T US$1.035 million; 750T US$1.16 million (2003).

DESIGN FEATURES: High-mounted, strut-braced wing with Roncz advanced aerofoil profile and constant chord, thick centre-section; outer panels (approximately 40 per cent of each

FISHER

FISHER FLYING PRODUCTS

PO Box 468, Industrial Park, Edgeley, North Dakota 58433
Tel: (+1 701) 493 22 86
Fax: (+1 701) 493 25 39
e-mail: ffp@fisherflying.com
Web: http://www.fisherflying.com
PRESIDENT: Darlene Jackson-Hanson
VICE-PRESIDENT: Gene Hanson

Formed in 1982, Fisher Flying Products markets a range of 16 homebuilt light aircraft. These comprise the FP-202 Koala, FP-303 microlight, FP-404, FP-505 Skeeter, FP-606 Sky Baby microlight, Avenger, Avenger V and Youngster single-seat aircraft; and Super Koala, Classic, Dakota Hawk, Tiger Moth (two versions), Celebrity and Horizon 1 and 2 two-seaters. Only the more recent are described here. In 1998, Fisher took over Fisher Aero of Portsmouth, Ohio, which added the Celebrity, Horizon 1 and Horizon 2 to the product line. In late 2003, Fisher's workforce numbered seven in its 930 m^2(10,000 sq ft) factory.

Fisher Celebrity remains available, this example built in Sweden with a 1920s aura *(Paul Jackson)* ***NEW***/1047078

Fisher Koala, an earlier product still marketed in kit form *(Paul Jackson)* ***NEW***/1047077

UPDATED

FISHER AVENGER

TYPE: Single-seat ultralight kitbuilt.
PROGRAMME: First flown June 1994.
CURRENT VERSIONS: **Avenger:** Baseline version.
Avenger V: Powered by one 37.3 kW (50 hp) Volkswagen four-cylinder engine.
CUSTOMERS: Total of 191 sold by 2004, of which around 70 were flying.
COSTS: Avenger standard kit US$4,800; quick-build kit US$6,200 (2004); Avenger V standard kit US$5,200; quick-build kit US$6,600 (2004). All exclude engine (2004).
DESIGN FEATURES: Low-wing monoplane; wings braced by V-struts to mainwheel hubs; mutually braced tail surfaces. Wire-braced tailplane. Quoted build time 400 hours, reducing to 300 with quick-build kit.
FLYING CONTROLS: Conventional and manual.
STRUCTURE: Wood and fabric.
LANDING GEAR: Tailwheel type; fixed. Tyre size 4.00–8.
POWER PLANT: One 37.0 kW (49.6 hp) two-cylinder Rotax 503 UL-2V; other engines from 20.9 to 48.5 kW (28 to 65 hp) are optional. Fuel capacity 19 litres (5.0 US gallons; 4.2 Imp gallons).
DIMENSIONS, EXTERNAL:
Wing span .... 8.23 m (27 ft 0 in)
Length overall .... 4.95 m (16 ft 3 in)
Height overall .... 1.52 m (5 ft 0 in)
AREAS:
Wings, gross .... 11.24 m^2 (121.0 sq ft)
WEIGHTS AND LOADINGS:
Weight empty: Avenger .... 113 kg (250 lb)
Avenger V .... 159 kg (350 lb)
Max T-O weight: Avenger .... 272 kg (600 lb)
Avenger V .... 294 kg (650 lb)
PERFORMANCE:
Max level speed: Avenger .... 82 kt (152 km/h; 95 mph)
Avenger V .... 86 kt (160 km/h; 100 mph)
Normal cruising speed .... 52 kt (97 km/h; 60 mph)
Stalling speed: Avenger .... 23 kt (42 km/h; 26 mph)
Avenger V .... 28 kt (52 km/h; 32 mph)

Fisher Avenger single-seat kitbuilt 0110843

Max rate of climb at S/L: Avenger .... 274 m (900 ft)/min
Avenger V .... 259 m (850 ft)/min
T-O run: Avenger .... 30 m (100 ft)
Avenger V .... 61 m (200 ft)
Landing run: Avenger .... 61 m (200 ft)
Avenger V .... 76 m (250 ft)
Range: Avenger .... 150 n miles (277 km; 172 miles)
Avenger V .... 250 n miles (463 km; 287 miles)

UPDATED

FISHER DAKOTA HAWK

TYPE: Side-by-side ultralight kitbuilt.
PROGRAMME: First flight September 1994.
CUSTOMERS: Total of 80 kits sold by mid-2003 of which 15 were flying by January 2004.
COSTS: Standard kit US$11,500; quick-build kit US$13,500 (2004). Both exclude engine.
DESIGN FEATURES: Geodetic wing structure built upon I-beam spar and braced with V struts. Wire-braced tailplane. Quoted build time 700 hours for standard kit and 600 for quick-build kit. Wings fold for storage.
FLYING CONTROLS: Conventional and manual.
STRUCTURE: Precut, shaped and slotted wooden pieces; wooden truss fuselage with plywood skin; glass fibre cowling.
LANDING GEAR: Tailwheel type; fixed. Two faired-in side Vs hinged to lower longerons, plus half-axles, each with bungee cord shock-absorber, attached to compression frame. Steerable tailwheel. Matco wheels and hydraulic brakes. Steerable tailwheel; shock-absorbers and Matco speed fairings on main legs.
POWER PLANT: One 59.6 kW (79.9 hp) Rotax 912 UL four-stroke engine driving a two-blade propeller. Optional alternatives include Teledyne Continental engines between 48.4 and 63 kW (65 and 85 hp) or 74.6 kW (100 hp) Subaru. Fuel capacity 45.4 litres (12.0 US gallons; 10.0 Imp gallons).
DIMENSIONS, EXTERNAL:
Wing span .... 8.69 m (28 ft 6 in)
Length overall .... 6.02 m (19 ft 9 in)
Height overall .... 1.85 m (6 ft 1 in)
AREAS:
Wings, gross .... 11.89 m^2 (128.0 sq ft)
WEIGHTS AND LOADINGS:
Weight empty .... 272 kg (600 lb)
Max T-O weight .... 522 kg (1,150 lb)
PERFORMANCE:
Max cruising speed .... 87 kt (161 km/h; 100 mph)
Normal cruising speed .... 78 kt (145 km/h; 90 mph)
Stalling speed .... 31 kt (57 km/h; 35 mph)
Max rate of climb at S/L .... 244 m (800 ft)/min
T-O run .... 107 m (350 ft)
Landing run .... 122 m (400 ft)
Range with max fuel .... 217 n miles (402 km; 250 miles)

Fisher Dakota Hawk two-seat kitbuilt *(Paul Jackson)* ***NEW***/1047098

UPDATED

FISHER YOUNGSTER AND YOUNGSTER V

TYPE: Single-seat ultralight biplane kitbuilt.
PROGRAMME: First flown June 1994.
CURRENT VERSIONS: **Youngster:** *As described.*
Youngster V: Uprated version with 48.5 kW (65 hp) Volkswagen motorcar engine.
CUSTOMERS: Total of 35 flying by mid-2004.

Fisher Youngster (37.3 kW; 50 hp Great Plains) 0110842

COSTS: Youngster standard kit US$6,250, quick-build kit US$8,400; Youngster V standard kit US$6,550, quick-build kit US$8,750, all excluding engine (2004).
DESIGN FEATURES: Open cockpit, strut-braced biplane with detachable cockpit transparency for winter operations. Constant-chord wings of geodetic construction. Wire-braced low-mounted tailplane. Stainless steel control wires. Additional central strut between nose and upper wing forward of cockpit. Aerofoil type 2315. Quoted build time 400 hours, reducing to 300 with quick-build kit.
FLYING CONTROLS: Conventional and manual.
STRUCTURE: Wood and fabric.
LANDING GEAR: Tailwheel type; fixed. Tyres 8.00-6. Bungee cord shock absorption.
POWER PLANT: One 37.3 kW (50 hp) Great Plains four-cylinder engine; Rotax engines up to 47.8 kW (64.1 hp) are optional. Fuel capacity 30.3 litres (8.0 US gallons; 6.7 Imp gallons).
DIMENSIONS, EXTERNAL:
Wing span ... 5.49 m (18 ft 0 in)
Length overall ... 4.72 m (15 ft 6 in)
Height overall ... 1.85 m (6 ft 1 in)
AREAS:
Wings, gross ... 11.71 m^2 (126.0 sq ft)
WEIGHTS AND LOADINGS:
Weight empty ... 181 kg (400 lb)
Max T-O weight ... 294 kg (650 lb)
PERFORMANCE:
Max level speed ... 95 kt (177 km/h; 110 mph)
Normal cruising speed ... 74 kt (137 km/h; 85 mph)
Stalling speed ... 30 kt (55 km/h; 34 mph)
Max rate of climb at S/L ... 244 m (800 ft)/min
T-O run ... 61 m (200 ft)
Landing run ... 76 m (250 ft)
Range ... 225 n miles (416 km; 259 miles)

UPDATED

FISHER R-80 AND RS-80 TIGER MOTH

TYPE: Tandem-seat ultralight biplane kitbuilt.
PROGRAMME: Scale (80 per cent) replica of de Havilland D.H.82A Tiger Moth. First flight June 1994.
CURRENT VERSIONS: **R-80 Tiger Moth:** *As described.*
RS-80 Tiger Moth: Introduced 1999; has steel tube fuselage and 89.5 kW (120 hp) LOM M132 engine.
CUSTOMERS: Total of 21 flying by mid-2004, comprising 15 R-80 and six RS-80 models. Four under construction in Zimbabwe for use on aerial safaris.

Replica Tiger Moth produced by Fisher Flying Products powered by Metro 80700 engine and GSC Systems propeller *(Paul Jackson)* ***NEW***/1047097

COSTS: R.80 standard airframe kit US$12,250, quick-build kit US$15,500; RS-80 standard airframe kit US$18,000, quick-build kit US$20,000, all excluding engine (2004).
DESIGN FEATURES: Can be built using the contents of a basic tool kit. Quoted build time 550 hours.
Single-bay, staggered biplane with N-type interplane struts and slight sweepback to constant-chord wings.
FLYING CONTROLS: Conventional and manual, with horn-balanced rudder.
STRUCTURE: Wood construction using epoxy adhesives on preformed components.
LANDING GEAR: Tailwheel type; fixed. Two tubular steel side Vs hinged to lower longerons, plus half-axles, each with bungee cord shock-absorber, attached to compression frame. Matco wheels with hydraulic brakes.
POWER PLANT: Prototype has one 70.8 kW (95 hp) Mid West AE 100 rotary, but any engine between 55.9 and 89 kW (75 and 120 hp) can be fitted, including Rotax 912, Jabiru 3300, Subaru EA81-100, Norton rotary and LOM M132. One 72 litre (19.0 US gallon; 15.8 Imp gallon) centre-section fuel tank.
DIMENSIONS, EXTERNAL:
Wing span ... 7.01 m (23 ft 0 in)
Length overall ... 5.79 m (19 ft 0 in)
Height overall ... 2.24 m (7 ft 4 in)
DIMENSIONS, INTERNAL:
Cockpit max width ... 0.66 m (2 ft 2 in)
AREAS:
Wings, gross ... 15.79 m^2 (170.0 sq ft)
WEIGHTS AND LOADINGS:
Weight empty: R-80 ... 254 kg (560 lb)
RS-80 ... 363 kg (800 lb)
Max T-O weight: R-80 ... 522 kg (1,150 lb)
RS-80 ... 612 kg (1,350 lb)
PERFORMANCE:
Max level speed: R-80 ... 95 kt (177 km/h; 110 mph)
RS-80 ... 104 kt (193 km/h; 120 mph)
Cruising speed: R-80 ... 69 kt (129 km/h; 80 mph)
RS-80 ... 78 kt (145 km/h; 90 mph)
Stalling speed: R-80 ... 31 kt (57 km/h; 35 mph)
RS-80 ... 35 kt (65 km/h; 40 mph)
Max rate of climb: both ... 244 m (800 ft)/min
T-O run: R-80 ... 91 m (300 ft)
RS-80 ... 76 m (250 ft)
Landing run: R-80 ... 122 m (400 ft)
RS-80 ... 91 m (300 ft)
Range with max fuel: both ... 250 n miles (463 km; 287 miles)

UPDATED

FLIGHTSTAR

FLIGHTSTAR SPORTSPLANES

PO Box 760, Ellington, Connecticut 06029
Tel: (+1 860) 875 81 85
Fax: (+1 860) 870 54 99
e-mail: fstar@rcn.com
Web: http://www.flyflightstar.com
PRESIDENT: Thomas A Peghiny
TECHNICAL DIRECTOR: Dana Porsiani

Pioneer International formed as a division of Pioneer Parachute Company in 1981 to produce the initial Flightstar; design was sold to Pampa's Bull of Argentina as the Aviastar I (see 1992–93 *Jane's*) when Pioneer closed in 1985. Pioneer Flightstar (later Flightstar) formed in 1992 initially to import the Aviastar design, and has since developed it into the current range of ultralights. Production of the kits is done under contract by Leza-Aircam. The initial Flightstar Classic is no longer available.

In August 2004, the company employed 14 people; its Sebring production facility covered 4,180 m^2(45,000 sq ft).

UPDATED

FLIGHTSTAR IISL AND IISC

TYPE: Side-by-side ultralight kitbuilt.
PROGRAMME: Originally designed by Pioneer International; construction of prototype Flightstar started 1992; first flight 1993. Certified to German BFU-95 standards 1997; also certified in Australia. SL version introduced in 1995.
CURRENT VERSIONS: **IISL:** (SL = Sport Light) Standard version, *as described.*
IISC: (SC = Sport Cabin) Introduced 1999; cabin version with doors for cold-weather operations.

Flightstar IISC powered by an HKS 700E engine *(Paul Jackson)* ***NEW***/1047158

CUSTOMERS: Over 1,000 of original Flightstar II built. 90 IISCs and 180 IISLs completed by August 2004.
COSTS: IISL US$13,695; IISC US$14,795, both without engines (2004).
DESIGN FEATURES: Wings fold for storage and transportation. Quoted build time 100 hours for IISL; 130 hours for IISC.
Flightstar FS-II-2 aerofoil section; dihedral 2° 48'; incidence 4°, twist 2°.
FLYING CONTROLS: Conventional and manual. Push-pull cables; optional flaps on IISL.
STRUCTURE: 6061-T6 aluminium tube spar, strut-braced wing with Dacron covering; Five-ply Mylar covering optional. Composites-covered chromoly 4130 steel fuselage cage with stainless steel brackets; engine attached to heavy-duty aluminium tube running length of aircraft; all metal parts anodised to obviate painting. Strut-braced, low-mounted tailplane.
LANDING GEAR: Tricycle type; fixed. Cantilever mainwheel suspension comprising two radius arms and damped leg attached each side of pod at rear. Steerable nosewheel. Azusa wheels

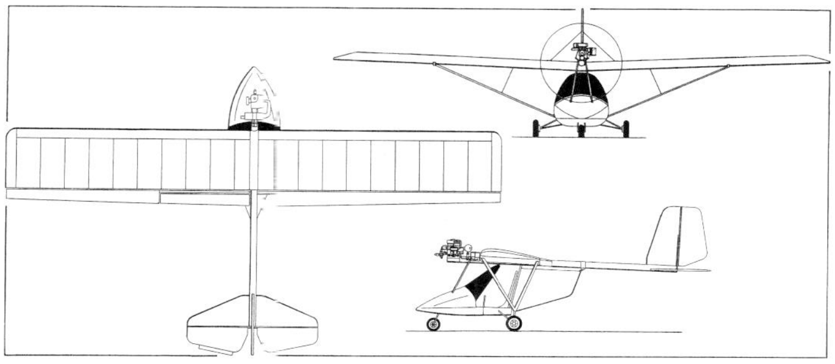

Open cockpit Flightstar IISL *(Paul Jackson)* 0105029

and optional drum brakes. Tyres size 13 in nose; 15 in main. Shock cord absorption. Optional Full Lotus floats.

POWER PLANT: One 37.0 kW (49.6 hp) Rotax 503 UL-2V driving three-blade Powerfin propeller; optional engines include 47.8 kW (64.1 hp) Rotax 582 UL and 44.7 kW (60.0 hp) four-stroke HKS 700E; latter is standard on IISC and has 3.47:1 gearbox reduction. Fuel capacity 37.9 litres (10.0 US gallons; 8.3 Imp gallons), of which 36.0 litres (9.5 US gallons; 7.9 Imp gallons) usable.

ACCOMMODATION: Fastback fabric rear fairing fitted to IISL. Cabin doors closed by zips.

EQUIPMENT: Optional BRS parachute.

DIMENSIONS, EXTERNAL:
- Wing span .... 9.75 m (32 ft 0 in)
- Length overall .... 5.79 m (19 ft 0 in)
- Height overall .... 2.39 m (7 ft 10 in)

AREAS:
- Wings, gross .... 14.59 m^2 (157.0 sq ft)

WEIGHTS AND LOADINGS:
- Weight empty .... 166 kg (365 lb)
- Max T-O weight .... 431 kg (950 lb)

PERFORMANCE (with 37.0 kW; 49.6 hp engine):
- Never-exceed speed (VNE) .... 83 kt (154 km/h; 96 mph)
- Max operating speed .... 74 kt; (137 km/h; 85 mph)
- Normal cruising speed .... 56 kt; (105 km/h; 65 mph)
- Stalling speed .... 32 kt (58 km/h: 36 mph)
- Max rate of climb at S/L .... 213 m (700 ft)/min
- T-O run .... 61 m (200 ft)
- Landing run .... 91 m (300 ft)
- Range with max fuel .... 156 n miles (289 km; 180 miles)

UPDATED

FLIGHTSTAR SPYDER

TYPE: Single-seat ultralight kitbuilt.

PROGRAMME: Derived from Pioneer Flightstar II. Can conform to FAR Pt 103.

CURRENT VERSIONS: **Spyder:** Standard version, *as described.*

Formula: High-performance upgraded model with wider windscreen and cockpit doors for cold weather flying; now discontinued.

CUSTOMERS: Over 700 kits sold by August 2004.

COSTS: Spyder US$11,495 without engine (2004).

DESIGN FEATURES: Flightstar FS-I-2 aerofoil. Wings fold for storage and transportation. Optional BRS parachute recovery system.

FLYING CONTROLS: As Flightstar IISL.

STRUCTURE: As Flightstar IISL.

LANDING GEAR: As Flightstar IISL.

POWER PLANT: One 31.0 kW (41.6 hp) Rotax 447 UL-2V driving Tennessee two-blade or Ivo three-blade propeller. Options include 37.0 kW (49.6 hp) Rotax 503 UL-2V and 47.8 kW

Flightstar Spyder single-seat ultralight 0110699

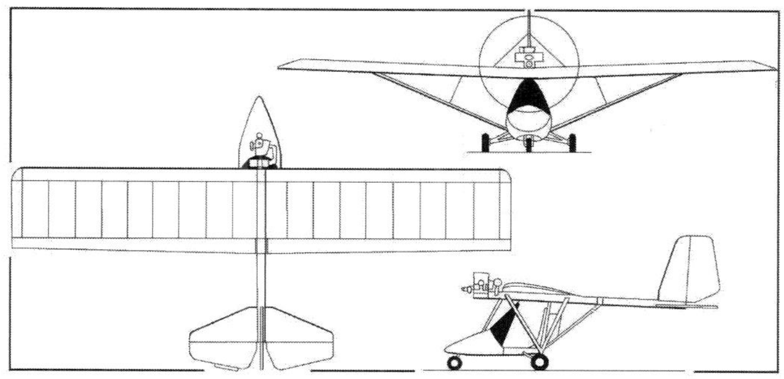

Flightstar Spyder (Rotax 447 two-cylinder engine) *(Paul Jackson)* 0533719

(64.1 hp) Rotax 582 UL. Standard fuel capacity 18.9 litres (5.0 US gallons; 4.2 Imp gallons); optionally 37.9 litres (10.0 US gallons; 8.3 Imp gallons).

DIMENSIONS, EXTERNAL:
- Wing span .... 9.14 m (30 ft 0 in)
- Length overall .... 5.03 m (16 ft 6 in)
- Height overall .... 2.39 m (7 ft 10 in)

AREAS:
- Wings, gross .... 13.38 m^2 (144.0 sq ft)

WEIGHTS AND LOADINGS:
- Weight empty .... 127 kg (280 lb)
- Max T-O weight .... 294 kg (650 lb)

PERFORMANCE:
- Never-exceed speed (VNE) .... 83 kt (154 km/h; 96 mph)
- Normal cruising speed .... 48 kt (89 km/h; 55 mph)
- Stalling speed .... 28 kt (52 km/h: 32 mph)
- Max rate of climb at S/L .... 201 m (660 ft)/min
- T-O run .... 62 m (205 ft)
- Landing run .... 61 m (200 ft)
- Range with standard fuel .... 86 n miles (161 km; 100 miles)

UPDATED

FOUR WINDS

FOUR WINDS AIRCRAFT

1501 Airway Circle, New Smyrna Beach, Florida 32168
Tel: (+1 386) 426 77 95
Fax: (+1 386) 426 53 39
Web: http://www.fourwindsaircraft.com
PRESIDENT: Jeff Rahm

The company's 6,300 m^2 (68,000 sq ft) plant employed 20 persons in early 2002. During 1990s, completed 22 Lancairs, two GlaStars, two Stallions, two Glass Gooses and one BD-5 on behalf of private owners. Marketing of new four-seat aircraft delayed by loss of prototype, but deliveries eventually began in April 2004.

UPDATED

FOUR WINDS 210 AND 250T

TYPE: Four-seat kitbuilt.

PROGRAMME: Stretched version of GlaStar (currently produced by New Glastar LLC); partly completed prototype exhibited at Sun 'n' Fun, Lakeland, Florida, April 2002; first kit deliveries then scheduled for September 2002, but date revised to September 2003. Prototype (N192FW), a Model 192, registered in July 2002, but lost in accident on 6 December 2002. By early 2004, company had decided not to proceed with 194 kW (200 hp) Model 192. Manufacturer officially notified as AAA Aircraft Leasing.

Second prototype, N250FW, a Model 210C, first flew 27 February 2004 and exhibited at Sun 'n' Fun, April 2004. Kit deliveries began later same month, at which time most orders were for Model 210.

CURRENT VERSIONS: See Power Plant.

CUSTOMERS: Total 32 kits sold and 10 delivered by April 2004.

COSTS: Kit US$68,000, less engine, avionics and interior (2004); estimated US$120,000 to flyable, equipped configuration, including builder assistance programme of 14 days in-works support.

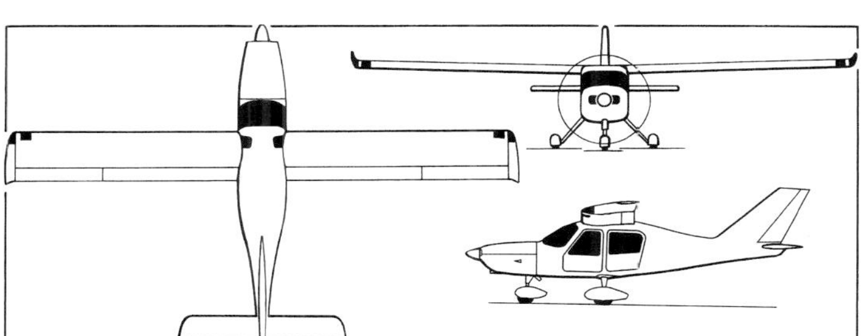

Four Winds 210 four-seat kitbuilt *(James Goulding)* ***NEW***/0589973

DESIGN FEATURES: Four-seat version of GlaStar, but with cabin reinforcement cage similar to OMF Symphony (built in Germany); however, wing of composites, with no bracing struts. Quoted build time 1,000 hours, or 200 hours with builder assistance programme.

Unswept wing and tailplane; High aspect ratio wing. Aerofoil C-65-215; dihedral 1° 30'; fin sweepback 48° 50'.

FLYING CONTROLS: Conventional and manual. Mass-balanced (lead) control surfaces. Actuation by pushrods (elevator and ailerons) and cables (rudder). Electrically actuated flaps; max deflection 30°.

STRUCTURE: Generally of fireproof 7781 E-Glass, but 4130 steel (1.5 mm; 0.058 in thickness rectangular tube) structural cage surround to cabin, twin carbon-graphite wing spars and 7075 aluminium main landing gear legs.

LANDING GEAR: Tricycle type, fixed. Fuselage-mounted spring cantilever mainwheel legs with wheel fairings; trailing-link, steerable nosewheel. Grove Gear aluminium mainwheels 15×6.00–6; nosewheel 11×4.00–5. Hydraulic brakes. Optional twin-float version.

POWER PLANT: Choice of engines to drive single propeller, as under.

210: One 224 kW (300 hp) Teledyne Continental IO-520 (Model 210C) or Textron Lycoming flat-six. Fuel capacity 341 litres (90.0 US gallons; 75.0 Imp gallons).

Four Winds 210 prototype *(Paul Jackson)* ***NEW***/0583546

250T: One 313 kW (420 shp) Rolls-Royce 250 turboprop. Fuel capacity 492 litres (130 US gallons; 108 Imp gallons) normal; 662 litres (175 US gallons; 146 Imp gallons) optional.

'Wet' wing, with fuel between spars.

ACCOMMODATION: Four persons, including one or two pilots, in two side-by-side pairs. Four doors; forward pair front-hinged; rear pair upward-hinged. Baggage space behind rear seats; latter tip up, or rapidly removable, for additional stowage.

SYSTEMS: Air conditioning optional. No pressurisation or pneumatics.

AVIONICS: To customer's choice, including Bendix/King, Chelton and Garmin.

EQUIPMENT: Optional BRS ballistic parachute.

DIMENSIONS, EXTERNAL:
Wing span ... 11.19 m (36 ft 8½ in)
Wing chord, mean ... 1.14 m (3 ft 9 in)
Wing aspect ratio ... 9.8
Length overall ... 7.92 m (26 ft 0 in)
Tailplane span ... 3.73 m (12 ft 3 in)
Tailplane chord, constant ... 0.91 m (3 ft 0 in)
Propeller diameter ... 2.08 m (6 ft 10 in)

DIMENSIONS, INTERNAL:
Cabin: Length ... 2.90 m (9 ft 6 in)
Max width ... 1.24 m (4 ft 1 in)
Max height ... 1.40 m (4 ft 7 in)

AREAS:
Wings, gross ... 12.79 m² (137.7 sq ft)
Vertical tail surfaces (total) ... 1.77 m² (19.03 sq ft)
Horizontal tail surfaces (total) ... 3.41 m² (36.75 sq ft)

WEIGHTS AND LOADINGS:
Weight empty: 210 ... 953 kg (2,100 lb)
250T ... 930 kg (2,050 lb)
Baggage capacity:
210 ... 68 kg (150 lb)
250T ... 91 kg (200 lb)
Max T-O weight: 210, 250T ... 1,542 kg (3,400 lb)
Max wing loading: 210, 250T ... 120.6 kg/m² (24.69 lb/sq ft)
Max power loading:
210 ... 6.90 kg/kW (11.33 lb/hp)
250T ... 4.93 kg/kW (8.10 lb/shp)

PERFORMANCE:
Never-exceed speed (VNE) ... 250 kt (463 km/h; 287 mph) IAS
Cruising speed at 70% power:
210 at FL80 ... 187 kt (346 km/h; 215 mph)
250T at FL150 ... 217 kt (402 km/h; 250 mph)
Stalling speed: 210, 250T ... 56 kt (104 km/h; 65 mph)
Service ceiling ... 8,535 m (28,000 ft)
T-O run: 210 ... 275 m (900 ft)
250T ... 183 m (600 ft)
Landing run: 210, 250T ... 397 m (1,300 ft)
Max range:
210 ... 1,121 n miles (2,076 km; 1,290 miles)
250T ... 1,129 n miles (2,092 km; 1,300 miles)
g limits: 210, 250T ... +9.5/–4.0

UPDATED

FREE BIRD

PO Box 904, Detroit Lakes, Minnesota 56502
Tel: (+1 218) 844 59 14
e-mail: sales@flyfbi.com
Web: www.sportlite103.com
OWNERS Eric Grina
Keith LeCleir

Free Bird Innovations purchased the Freebird and Sportlite 103 ultralights from Pro Sport Aviation in early 2004.

NEW ENTRY

FREE BIRD FREEBIRD CLASSIC

TYPE: Side-by-side kitbuilt.

PROGRAMME: Launched 1993 by earlier Freebird company. Meets FAR Pt 103 requirements.

CUSTOMERS: Over 100 under construction and over 50 flying by mid-2003.

COSTS: US$9,995 excluding engine; ready to fly models from US$19,795 (2004).

DESIGN FEATURES: Wings removable for storage. Quoted build time 125 hours.

FLYING CONTROLS: Conventional and manual; flaps. Large horn balance on rudder.

STRUCTURE: Fabric-covered aluminium tubing and composites nose skin. Strut-braced wings and tailplane. Wing-to-tailplane bracing struts augment slim rear fuselage.

LANDING GEAR: Tricycle type; fixed. Semi-recessed nosewheel.

POWER PLANT: One 47.8 kW (64.1 hp) Rotax 582 UL-2V two-cylinder engine driving a wooden fixed-pitch pusher propeller. Fuel capacity 37.8 litres (10.0 US gallons; 8.3 Imp gallons).

ACCOMMODATION: Optional enclosed cockpit with provision for heater.

SYSTEMS: Optional electric starter.

DIMENSIONS, EXTERNAL:
Wing span ... 8.84 m (29 ft 0 in)
Length overall ... 5.33 m (17 ft 6 in)
Height overall ... 1.57 m (5 ft 2 in)

AREAS:
Wings, gross ... 12.26 m² (132.0 sq ft)

WEIGHTS AND LOADINGS:
Weight empty ... 175 kg (385 lb)
Max T-O weight ... 394 kg (870 lb)

PERFORMANCE (enclosed cockpit):
Max level speed ... 70 kt (129 km/h; 80 mph)
Normal cruising speed at 75% power ... 61 kt (113 km/h; 70 mph)

Free Bird Freebird II *(Geoffrey P Jones)* 0051787

Stalling speed, flaps down ... 28 kt (52 km/h; 32 mph)
Max rate of climb at S/L ... 244 m (800 ft)/min
T-O run ... 38 m (125 ft)
Landing run ... 46 m (150 ft)
Range at 75% power ... 173 n miles (321 km; 200 miles)
g limits ... +6/–3

UPDATED

FREE BIRD SPORTLITE 103

TYPE: Single-seat ultralight kitbuilt; tandem-seat ultralight kitbuilt.

PROGRAMME: Based on Freebird Classic.

CUSTOMERS: Over 350 flying by mid-2004.

COSTS: Quick-build plus kit US$11,650 including 29.8 kW (40 hp) Hirth 2702 engine and propeller; ready-to-fly US$14,995 (2004).

DESIGN FEATURES: Intended for use under FAA Pt 103 ultralight regulations; can be converted to tandem-seater. Wings detach for storage and transportation. Quoted build time 125 hours.

FLYING CONTROLS: Conventional and manual.

STRUCTURE: As Freebird Classic. 6061-T6 aluminium tube throughout with fabric covering. Strut-braced wings. Cruciform tailplane.

LANDING GEAR: Tricycle type, fixed. Steerable nosewheel. Optional floats.
POWER PLANT: Engines in the range 16.4 to 38.8 kW (22 to 52 hp) can be fitted, including 29.5 kW (39.6 hp) Rotax 447 or 34.0 kW (45.6 hp) Rotax 503 in two-seat version. Fuel capacity 18.9 litres (5.0 US gallons; 4.2 Imp gallons) in single-seat; 32.2 litres (8.5 US gallons; 7.1 Imp gallons) in two-seat.
ACCOMMODATION: Pilot and passenger in open seat(s).
DIMENSIONS, EXTERNAL:
Wing span ... 8.08 m (26 ft 6 in)
Length overall: single-seat ... 5.05 m (16 ft 7 in)
two-seat ... 5.21 m (17 ft 1 in)
Height overall ... 2.03 m (6 ft 8 in)
AREAS:
Wings, gross ... 11.24 m^2 (121.0 sq ft)
WEIGHTS AND LOADINGS:
Weight empty ... 113 kg (250 lb)
Max T-O weight: single-seat ... 226 kg (500 lb)
two-seat ... 408 kg (900 lb)
PERFORMANCE:
Max operating speed ... 54 kt (100 km/h; 62 mph)
Normal cruising speed ... 48 kt (89 km/h; 55 mph)
Stalling speed: single seat ... 20 kt (36 km/h; 22 mph)
two seat ... 28 kt (52 km/h; 32 mph)
Max rate of climb at S/L ... 259 m (850 ft)/min
T-O run ... 23 m (75 ft)

NEW ENTRY

GLASAIR

NEW GLASAIR LLC

18810 59th Avenue Northeast, Arlington, Washington 98223
Tel: (+1 360) 435 85 33
Fax: (+1 360) 435 95 25
e-mail: info@newglasair.com
Web: http://www.newglasair.com
PRESIDENT: Mikael Via

Stoddard-Hamilton company formed in 1979. Glasair I, II and II-S were superseded by Super II and III. Details of Glasair I in 1989–90 *Jane's*. Two-seat GlaStar added to product range in 1995; float production rights sold to Aero Marine Inc in 1996. Company had delivered more than 2,400 kits by April 1999; by 2004 over 900 Glasairs and 300 GlaStars were flying.

Work began on a six-seater named Aurora, in early 2000, but Stoddard-Hamilton ceased trading in May 2000 and filed for Chapter 11 bankruptcy on 17 July. At a bankruptcy hearing on 18 August 2000, the bulk of the assets was sold to Anduril Private Investment Fund, but this agreement was set aside on 27 November 2000 and bidding for the company was re-opened.

Stoddard-Hamilton then received offer from Phoenix Composites for Glasair rights but failed to complete. Rights to both aircraft lines were purchased from Arlington Advanced Development by Thomas Wathen on 15 June 2001. Separate marketing companies were established by Mr Wathen for the two main product lines, namely New Glasair (this entry) and New GlaStar.

In July 2004, the company announced a 1,115 m^2 (12,000 sq ft) extension to its facilities to cope with increased demand.

UPDATED

GLASAIR SUPER II

TYPE: Side-by-side kitbuilt.
PROGRAMME: Derived from 1980 Glasair TD via Glasair II and stretched II-S (by which name it was initially known). Prototype first flown in 1990. Jump-start kit options, to reduce quoted build time by around 750 hours, offered from 1999 onwards. Over 3,100 hours flown by factory demonstrator N902S by August 2003.
CURRENT VERSIONS: Available in **RG** (retractable landing gear), **FT** (non-retractable tricycle gear) and **TD** (taildragger) forms introduced in 1992 to supersede earlier Glasair series; TD no longer marketed in 2002, but remains available on a special order basis.
CUSTOMERS: 560 kits sold and 200 flying by March 2004.
COSTS: Kits: US$35,950 for RG; US$29,950 for FT; US$29,950 for TD; all without engine, propeller, instruments and avionics; engine US$30,043 for 180 hp version; 'jumpstart' (quick build) fuselage extra US$6,950; 'jumpstart' wing extra US$7,850; fixed-pitch propeller US$2,000, constant-speed propeller US$5,226 (2004).
DESIGN FEATURES: Low-wing monoplane with moderately tapered wing and sweptback tail surfaces; optimised for high performance consistent with amateur assembly. Quoted build time 1,200 hours for fixed landing gear versions and 1,800 hours for RG (both with 'jumpstart' options.

Wing section NASA LS(I)-0413; dihedral 3°; incidence 2° 20'. Upswept Hoerner style wingtips. Optional 61 cm (2 ft 0 in) extension on each wingtip.
FLYING CONTROLS: Conventional and manual. Horn-balanced rudder and elevators; ailerons with trim tab; plain flaps (slotted flaps optional).
STRUCTURE: Glass fibre and foam composites construction. One-piece wing spar; stressed for aerobatic flight loads.
LANDING GEAR: Fixed or retractable. Nosewheel or tailwheel options for fixed gear. Fuselage-mounted spring cantilever mainwheel legs; trailing link nosewheel. RG version has inward-retracting mainwheels and rearward-retracting, free-castoring nosewheel; oleo shock-absorbers. Brakes fitted. Wheel size 5.00-5 in. TD has full-swivel, locking tailwheel.
POWER PLANT: One 119 to 149 kW (160 to 200 hp) Textron Lycoming O-320 or IO-360 flat-four engine driving a Hartzell constant-speed metal propeller; standard engine for RG and FT is 134 kW (180 hp) Textron Lycoming IO-360. Fuel capacity 235 litres (62.0 US gallons; 51.6 Imp gallons) in main wing tanks plus 30 litres (8.0 US gallons; 6.7 Imp gallons) in header tank. Auxiliary tanks of 41.6 litres (11.0 US gallons; 9.2 Imp gallons) capacity in optional wingtip extensions.
ACCOMMODATION: Two persons, side by side. Enclosed cockpit with gull-wing doors. NACA air vents provide cabin ventilation.

Glasair Super II-RG ***(Paul Jackson)*** ***NEW***/1047124

DIMENSIONS, EXTERNAL (all versions):
Wing span: standard ... 6.81 m (22 ft 4 in)
optional ... 8.33 m (27 ft 4 in)
Wing aspect ratio: standard ... 6.1
optional ... 8.2
Length overall ... 6.31 m (20 ft 8½ in)
Height overall ... 2.07 m (6 ft 9½ in)
Propeller diameter ... 1.78 m (5 ft 10 in)
DIMENSIONS, INTERNAL:
Cabin max width ... 1.09 m (3 ft 7 in)
Baggage volume ... 0.34 m^3 (12.0 cu ft)
AREAS:
Wings, gross: standard ... 7.55 m^2 (81.3 sq ft)
optional ... 8.50 m^2 (91.5 sq ft)
WEIGHTS AND LOADINGS:
Weight empty: RG ... 601 kg (1,325 lb)
FT ... 567 kg (1,250 lb)
TD ... 544 kg (1,200 lb)
Baggage capacity: all versions ... 45 kg (100 lb)
Max T-O weight:
standard wing:
RG, FT ... 953 kg (2,100 lb)
TD ... 907 kg (2,000 lb)
extended wing: RG, FT ... 998 kg (2,200 lb)
TD ... 952 kg (2,100 lb)
Max wing loading:
standard wing:
RG, FT ... 126.1 kg/m^2 (25.83 lb/sq ft)
TD ... 120.1 kg/m^2 (24.60 lb/sq ft)
extended wing: RG, FT ... 117.4 kg/m^2 (24.04 lb/sq ft)
TD ... 112.1 kg/m^2 (22.95 lb/sq ft)
Max power loading:
standard wing, 119 kW (160 hp) engine: FT ... 7.99 kg/kW (13.13 lb/hp)
extended wing: FT ... 8.37 kg/kW (13.75 lb/hp)
standard wing, 134 kW (180 hp) engine:
RG ... 7.10 kg/kW (11.67 lb/hp)
TD ... 6.76 kg/kW (11.11 lb/hp)
extended wing: RG ... 7.44 kg/kW (12.22 lb/hp)
TD ... 7.10 kg/kW (11.67 lb/hp)
PERFORMANCE (180 hp engine):
Never-exceed speed (VNE) ... 226 kt (418 km/h; 260 mph)
Max level speed at S/L:
RG ... 207 kt (383 km/h; 238 mph)
FT, TD ... 198 kt (367 km/h; 228 mph)
Econ cruising speed at 75% at FL80:
RG ... 192 kt (356 km/h; 221 mph)
FT, TD ... 182 kt (338 km/h; 210 mph)
Stalling speed, power off:
plain flaps down: RG ... 57 kt (105 km/h; 65 mph)
FT, TD ... 55 kt (102 km/h; 63 mph)
slotted flaps down:
RG ... 52 kt (95 km/h; 59 mph)
FT, TD ... 50 kt (92 km/h; 57 mph)
Max rate of climb at S/L: all ... 823 m (2,700 ft)/min
Max rate of roll:
standard wing: FT, TD ... 130°/s
RG ... 140°/s
extended wing: FT, TD ... 85°/s
RG ... 90°/s
Service ceiling: all ... approx 5,790 m (19,000 ft)
T-O and landing run: TD, FT ... 213 m (700 ft)
RG ... 244 m (800 ft)
Range:
RG, standard fuel ... 1,382 n miles (2,559 km; 1,590 miles)
RG, with auxiliary fuel ... 1,605 n miles (2,972 km; 1,847 miles)
FT, TD, standard fuel ... 1,315 n miles (2,435 km; 1,513 miles)
FT, TD, with auxiliary fuel ... 1,527 n miles (2,828 km; 1,757 miles)
g limits at AUW of 708 kg (1,560 lb): all ... +6/–4
... (+9/–6 ultimate)

UPDATED

GLASAIR III

TYPE: Side-by-side kitbuilt.
PROGRAMME: First flight July 1986.
CURRENT VERSIONS: **Glasair III:** *As described.*

Glasair III Turbo: Powered by 224 kW (300 hp) Textron Lycoming TIO-540. Maximum level speed 284 kt (526 km/h; 327 mph); maximum rate of climb 1,143 m (3,750 ft)/min; withdrawn in 1998 and superseded by Glasair Super III.

Glasair Super III: Prototype (N401KT) flying by third quarter 1998, when demonstrated 322 kt (596 km/h; 370 mph) at 10,060 m (33,000 ft), powered by IO-540-AA1B5 uprated

Glasair III side-by-side kitbuilt *(Paul Jackson)* **NEW**/1047150

with Honeywell turbocharger to 261 kW (350 hp) at 11,275 m (37,000 ft) and with Hartzell four-blade propeller. Predicted maximum speed of 350 kt (648 km/h; 403 mph). Has enlarged rudder, extra fuel capacity and highly modified cowling. Public debut at Sun 'n' Fun 1999.

CUSTOMERS: 385 kits delivered by May 2004; 250 flying by May 2004.

COSTS: Kit: US$35,982 without engine, propeller, instruments and avionics; engine US$42,950; constant-speed propeller US$7,003 (2004). Jumpstart options for wing (US$8,995) and fuselage (US$8,995) offered.

DESIGN FEATURES: Similar configuration to earlier Glasairs, but designed to offer better performance, constructional simplicity and economical kit price. Larger fuselage for increased baggage space, payload capacity and comfort, also improving the longitudinal and directional stability for better cross-country and IFR performance; windows to rear of cockpit doors. Thicker windscreen to improve protection against bird strikes at higher speeds, and additional glass fibre laminates, integral longerons, and lay-up schedule which provides structurally stronger and torsionally stiffer fuselage. Quoted build time 1,800 working hours with 'jumpstart' options.

Wing section LS(I)-0413; wing incidence 2° 18', dihedral 3°; strengthened; carries more fuel than previous models. Optional wingtip extensions for 8.32 m (27 ft 3½ in) or 7.86 m (25 ft 9½ in) spans.

FLYING CONTROLS: As Glasair Super II. Slotted flaps optional.

LANDING GEAR: Retractable tricycle only, otherwise as Super II-RG but with 13 cm (5 in) longer landing gear legs to accommodate larger diameter propeller; oleo strut shock absorption. Mainwheels size 5.00-5; nosewheel 4.00-5. Differential Cleveland disc brakes on mainwheels.

POWER PLANT: One 224 kW (300 hp) Textron Lycoming IO-540-K1H5 flat-six piston engine driving a two-blade constant-speed propeller. Fuel capacity in wings 235 litres (62.0 US gallons; 51.6 Imp gallons). Fuselage header tank, capacity 30 litres (8.0 US gallons; 6.7 Imp gallons). Optional tanks in wingtip extensions, combined capacity 41.6 litres (11.0 US gallons; 9.2 Imp gallons).

ACCOMMODATION: As for Super II.

Data below refer to standard wing span Glasair III.

DIMENSIONS, EXTERNAL:
- Wing span: standard ... 7.11 m (23 ft 4 in)
  - optional ... 8.33 m (27 ft 4 in)
- Wing chord: at root ... 1.28 m (4 ft 2½ in)
  - at tip ... 0.84 m (2 ft 9 in)
- Wing aspect ratio: standard ... 6.7
  - optional ... 8.2
- Length overall ... 6.52 m (21 ft 4 in)
- Height overall ... 2.29 m (7 ft 6 in)

Glasair III specially fitted with a 336 kW (450 hp) Textron Lycoming EIO-720 flat-eight and MT-Propeller MTV-16-1-B-C/C197-109 constant-speed, four-blade propeller. The aircraft crashed on landing at Lakeland during Sun 'n' Fun, April 2004 *(Paul Jackson)* **NEW**/0567030

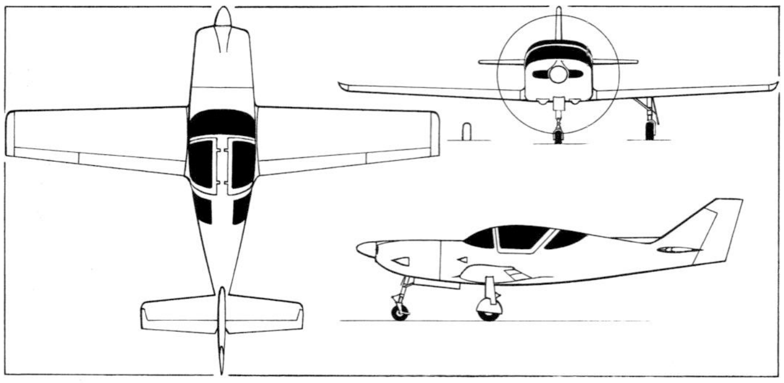

General arrangement of the Glasair III *(James Goulding)* 0093599

- Tailplane span ... 2.64 m (8 ft 8 in)
- Wheel track ... 3.10 m (10 ft 2 in)

DIMENSIONS, INTERNAL: As for Glasair Super II

AREAS:
- Wings, gross: standard ... 7.55 m² (81.3 sq ft)
  - optional ... 8.50 m² (91.5 sq ft)
- Fin ... 1.06 m² (11.40 sq ft)
- Rudder ... 0.57 m² (6.10 sq ft)
- Tailplane ... 1.51 m² (16.30 sq ft)
- Elevator ... 0.53 m² (5.70 sq ft)

WEIGHTS AND LOADINGS:
- Weight empty ... 737 kg (1,625 lb)
- Baggage capacity ... 45 kg (100 lb)
- Max T-O weight: without wingtip extensions ... 1,089 kg (2,400 lb)
  - with wingtip extensions ... 1,134 kg (2,500 lb)
- Max wing loading: standard ... 144.1 kg/m² (29.52 lb/sq ft)
  - optional ... 133.4 kg/m² (27.32 lb/sq ft)
- Max power loading: standard ... 4.86 kg/kW (8.00 lb/hp)
  - optional ... 5.07 kg/kW (8.33 lb/hp)

PERFORMANCE (standard wings):
- Never-exceed speed (VNE) ... 291 kt (539 km/h; 335 mph)
- Max level speed at S/L ... 252 kt (466 km/h; 290 mph)
- Cruising speed at FL80:
  - at 75% power ... 242 kt (448 km/h; 278 mph)
  - at 65% power ... 230 kt (426 km/h; 265 mph)
- Stalling speed: pilot only, plain flaps down ... 68 kt (126 km/h; 79 mph)
  - at max T-O weight, slotted flaps down ... 63 kt (117 km/h; 73 mph)
- Max rate of climb at S/L ... 1,143 m (3,750 ft)/min
- Max rate of roll: standard wing ... 140°/s
  - extended wing ... 90°/s
- Service ceiling: Glasair III ... approx 7,315 m (24,000 ft)
  - Glasair Super III ... 9,144 m (30,000 ft)
- T-O run ... 274 m (900 ft)
- Landing run ... 305 m (1,000 ft)
- Range at 55% power:
  - standard fuel ... 1,128 n miles (2,089 km; 1,298 miles)
  - with tip tanks ... 1,316 n miles (2,437 km; 1,514 miles)
- *g* limits at AUW of 962 kg (2,120 lb): all ... +6/–4
  - ... +9/–6 ultimate

UPDATED

GLASTAR

NEW GLASTAR LLC

Web: http://www.newglastar.com

A partial successor to Stoddard-Hamilton, New GlaStar shares accommodation, contact details and executives with New Glasair.

VERIFIED

GLASTAR GLASTAR

TYPE: Side-by-side kitbuilt.

PROGRAMME: Designed by Arlington Aircraft Developments and produced until 2000 by Stoddard-Hamilton. Prototype (N824G) first flew 29 November 1994; deliveries of kits started 1995. Over 1,400 hours flown by December 1998. Prototype converted to tailwheel configuration and then back to tricycle. Chosen by EAA as flagship aircraft for Young Eagle junior pilot scheme; first customer-built aircraft completed in July 1996. Early examples had a 93 kW (125 hp) Teledyne Continental IO-240. 'Jumpstart' kits introduced in 2003, reducing build time by estimated 400 hours.

FAA certified version being built in Germany as OMF Symphony. Four-seat, Four Winds versions produced in US.

CURRENT VERSIONS: **Glastar:** Basic model; *as described.*

Sportsman 2+2: Introduced mid-2003; strengthened wings, safety cage and landing gear. First kit delivered October 2003 to builder in New Hampshire.

CUSTOMERS: Over 900 kits sold in 16 countries up to May 2004; over 300 flying by that time. Total of 205 registered in USA alone by December 2001 increasing to 229 in October 2002.

COSTS: Glastar US$27,950: Sportsman 2+2 US$31,950 per kit, excluding engine, instruments, upholstery and paint (2004). 'Jumpstart' fuselage adds US$3,995 (Glastar) or US$4,250 (Sportsman 2+2); wing adds US$7,995 (Glastar) or US$8,750 (Sportsman 2+2) to these prices; tail kit adds US$2,895 (both).

DESIGN FEATURES: Conceived as inexpensive two-seat personal lightplane with wide performance envelope. Convertible landing gear, float compatible. Wings fold (one-person, three-minute task) and horizontal tail surfaces removable for compact hangarage or for trailer mounting. Quoted build time 1,500 to 2,000 hours. 'Jumpstart' options introduced in November 1998 to reduce time by half; wing kits for this option are assembled in the Philippines.

Braced, high wing of constant chord, unswept. Tailplane has root extensions; tall, sweptback fin. Wing section LS(I)-0413.

FLYING CONTROLS: Conventional and manual. Frise ailerons; Fowler flaps. Elevator tab for pitch trim. Aileron servo tab and trim. Two-position flaps: 15 and 35°. Vortex generators added to wings in front of ailerons and at wingroots, and root strakes to horizontal stabiliser. Optional electric elevator trim.

STRUCTURE: Fuselage construction of GFRP with 4130 steel tube frame surrounding cockpit section. Wings and tail surfaces of 2024-T3 aluminium. Two-spar wings have six full ribs per side, plus stiffeners.

LANDING GEAR: Fixed. Fuselage-mounted spring cantilever mainwheel legs with wheel fairings; option of nosewheel or steerable tailwheel. Both Glastar and Sportsman have option of Aerocet 2200, Wipline, Bilmar and Montana float; no differences in airframe among landing gear alternatives. Tailwheel (15 cm; 6 in tyre) conversion effected in 3 hours; choice of three mainwheel sizes: 5.00-5 (with speed fairings), 6.00-6 and 8.00-6. Optional Scott 3200 20 cm (8 in) tailwheel with larger mainwheels.

POWER PLANT: *Glastar:* One 119 kW (160 hp) Textron Lycoming O-320-D1F flat-four, driving a Sensenich fixed-pitch or Hartzell constant-speed two-blade propeller; or 112 kW (150 hp) 0-320-E1A and 134 kW (180 hp) 0-320-A1A; power plants in the range 93–168 kW (125–225 hp) are in use with around 100 aircraft fitted with Subaru 2.2 and 2.5 litre motorcar engine conversions. Usable fuel capacity 104 litres (27.6 US gallons; 23.0 Imp gallons). Optional 66 litre (17.5 US gallon; 14.6 Imp gallon) auxiliary tank.

Sportsman 2+2: Prototype has one 134 kW (180 hp) Textron Lycoming 0-360 engine driving two-blade Hartzell constant-speed propeller. Usable fuel capacity 114 litres (30.0 US gallons; 25.0 Imp gallons); can be increased by optional auxiliary tanks to 189 litres (50.0 US gallons; 41.6 Imp gallons).

ACCOMMODATION: Two persons side by side; seats adjust for pilot heights between 1.52 and 1.98 m (5 ft 0 in and 6 ft 6 in). Door each side. Baggage compartment behind crew with separate baggage door on port side; larger on Sportsman 2+2.

SYSTEMS: Electrical system: 12 V 30 Ah battery; alternator fit appropriate to engine.

AVIONICS: To customer's specification.

DIMENSIONS, EXTERNAL (A: Glastar; B: Sportsman):

Wing span: A, B ... 10.67 m (35 ft 0 in)
Wing chord, constant A, B ... 1.12 m (3 ft 8 in)
Wing aspect ratio: A ... 9.6
B ... 9.4
Length: fuselage: A ... 6.78 m (22 ft 3 in)
overall, Lycoming O-320 ... 6.95 m (22 ft 9½ in)
wings folded: A ... 7.47 m (24 ft 6 in)

Tailwheel GlaStar assembled in the UK *(Paul Jackson)* ***NEW***/1047147

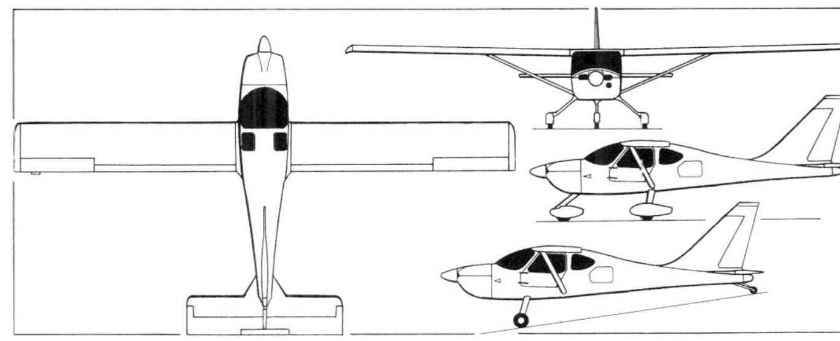
Tricycle version of GlaStar two-seat kitplane, with additional side view of tailwheel version *(Paul Jackson)* 0093597

B ... 7.01 m (23 ft 0 in)
wings folded: B ... 7.52 m (24 ft 8 in)
Height overall:
tricycle: A ... 2.77 m (9 ft 1 in)
B ... 2.84 m (9 ft 4 in)
tailwheel: A, B ... 2.11 m (6 ft 11 in)
Tailplane span ... 3.28 m (10 ft 9 in)
Propeller diameter ... 1.83 m (6 ft 0 in)
Doors (each): Height ... 0.80 m (2 ft 7½ in)
Width ... 0.94 m (3 ft 1 in)
Height to sill ... 0.84 m (2 ft 9 in)

DIMENSIONS, INTERNAL:

Cabin max width: A, B ... 1.17 m (3 ft 10 in)
Baggage compartment volume: A ... 0.91 m^3 (32.0 cu ft)
B ... 1.05 m^3(37.0 cu ft)

AREAS:

Wings, gross: A ... 11.89 m^2 (128.0 sq ft)
B ... 12.17 m^2(131.0 sq ft)
Ailerons (total) ... 0.37 m^2 (3.97 sq ft)
Trailing-edge flaps (total) ... 0.65 m^2 (6.95 sq ft)
Fin ... 1.29 m^2 (13.92 sq ft)
Rudder ... 0.47 m^2 (5.08 sq ft)
Tailplane ... 1.11 m^2 (12.00 sq ft)
Elevators (total) ... 0.99 m^2 (10.67 sq ft)

WEIGHTS AND LOADINGS:

Weight empty: A ... 544 kg (1,200 lb)
B ... 635 kg (1,400 lb)
Fuel weight ... 92 kg (203 lb)
Baggage capacity ... 113 kg (250 lb)

GlaStar Sportsman 2+2 *(Paul Jackson)* ***NEW***/1047146

Max T-O weight:
landplane: A ... 889 kg (1,960 lb)
B ... 1,043 kg (2,300 lb)
floatplane: A ... 952 kg (2,100 lb)
B ... 1,111 kg (2,450 lb)
Max wing loading:
landplane: A ... 74.8 kg/m² (15.31 lb/sq ft)
B ... 85.7 kg/m²(17.56 lb/sq ft)
floatplane: A ... 80.1 kg/m² (16.41 lb/sq ft)
B ... 91.3 kg/m² (18.70 lb/sq ft)
Max power loading: landplane:
A ... 7.46 kg/kW (12.25 lb/hp)
B ... 7.78 kg/kW (12.78 lb/hp)
floatplane: A ... 7.99 kg/kW (13.13 lb/hp)
B ... 8.28 kg/kW (13.61 lb/hp)

PERFORMANCE (119 kW, 160 hp, engine, constant-speed propeller B: 134 kW, 180 hp engine, contant-speed propeller):
Never-exceed speed (VNE):
A, B ... 162 kt (300 km/h; 186 mph)
Max level speed: A ... 145 kt (269 km/h; 167 mph)
B ... 140 kt (259 km/h; 161 mph)
Max cruising speed at 75% power:
A ... 145 kt (269 km/h; 167 mph)
B ... 140 kt (259 km/h; 161 mph)
Econ cruising speed at 65% power:
A ... 133 kt (246 km/h; 153 mph)
B ... 129 kt (240 km/h; 149 mph)
Stalling speed, power off:
flaps up: A ... 49 kt (91 km/h; 57 mph)
B ... 51 kt (94 km/h; 58 mph)
flaps down: A ... 43 kt (79 km/h; 49 mph)
B ... 42 kt (78 km/h; 48 mph)
Max rate of climb at S/L: A ... 632 m (2,075 ft)/min
B ... 594 m (1,950 ft)/min
Service ceiling: A, B ... 6,096 m (20,000 ft)
T-O run: A ... 88 m (290 ft)
B ... 107 m (350 ft)
T-O to 15 m (50 ft) ... 155 m (510 ft)
Landing from 15 m (50 ft) ... 168 m (550 ft)
Landing run: A, B ... 79 m (260 ft)
Range:
with standard fuel: A ... 529 n miles (979 km; 608 miles)
B ... 720 n miles (1,333 km; 828 miles)
with max fuel: A ... 926 n miles (1,715 km; 1,065 miles)
B ... 770 n miles (1,426 km; 886 miles)
Endurance with max fuel ... 6 h
g limits: A, B ... +3.8/–1.5

UPDATED

GREAT PLAINS

GREAT PLAINS AIRCRAFT SUPPLY CO INC

PO Box 545, Boys Town, Nebraska 68010
Tel: (+1 402) 493 65 07
Fax: (+1 402) 493 38 46
e-mail: gpasc@earthlink.net
Web: http://www.greatplainsas.com
PRESIDENT: Steve Bennett

Formed in 1982 and specialising in supply of Volkswagen engines and components for homebuilt aircraft, Great Plains took over production of kits for the Grosso Easy Eagle I in 2004; the Easy Eagle II is no longer available. The company also supplies kits and plans for the Sonerai, last described in the 2001-02 edition.

NEW ENTRY

GROEN

GROEN BROTHERS AVIATION INC

2640 West California Avenue, Suite A, Salt Lake City, Utah 84104
Tel: (+1 801) 973 01 77
Fax: (+1 801) 973 40 27
e-mail: mktg@gbagyros.com
Web: http://www.gbagyros.com
PRESIDENT AND CEO: David Groen
CHAIRMAN: Jay Groen
VICE-PRESIDENT AND COO: James P Mayfield III
CHIEF FINANCIAL OFFICER: Robin H H Wilson
DIRECTOR OF MARKETING: Scott Muir

Groen Brothers has been working on its Hawk Gyroplane family since 1986. Prototypes leading up to the Hawk 4 include the Hawk I and H2X, as described in previous editions of *Jane's*. The Hawk 4 Gyroplane is the first in a series of near VTOL-capable aircraft and is initially being developed and marketed to government agencies, with FAA certification not expected for at least two years. The RevCon 6G technology demonstrator uses the fuselage of a Cessna 337 as the basis of its fuselage and will be progressively modified into a gyrodyne by the addition of rotor blade mounted tipjets; Groen has offered its GyroLifter gyrodyne to the US Department of Defense. The company has a flight test facility at Buckeye, Arizona, where, on 12 July 2000, the prototype Jet Hawk 4T/Hawk 4 made its initial flight.

In July 2001, Groen announced plans to move to a new 18,580 m² (200,000 sq ft) facility at Phoenix, Arizona. The plant, which will cost US$14 million and employ 425, was intended to become operational by end 2002 and have the capacity to produce four aircraft per day, however this has lapsed.

In August 2001, the company concluded a joint venture with Al-Obayya Corporation to produce and market gyroplanes in Saudi Arabia. However, economic downturn of late 2001 resulted in 85 of 130-strong workforce being laid off. Earlier plans for Chinese assembly also appear to have lapsed.

Groen is also the parent of American Autogyro (AAI), whose SparrowHawk is described elsewhere.

GROEN HAWK 4

TYPE: Four-seat autogyro.

PROGRAMME: Based on proof-of-concept Hawk One (N4379X) first flown 26 September 1992; design started April 1996 and prototype two-seat H2X (N4412X) first flew 4 February 1997; later converted to three-seat Hawk III standard. First deliveries had been due June 1998, but design changes to H2X and later Hawk III in October 1998 resulted in Hawk 4. Initial aircraft (N402GB) first flew 29 September 1999, powered by a Continental piston engine, and made first vertical take-off on 9 December 1999; had flown 120 hours in 200 sorties by early April 2000. In September 2000 company switched certification effort to turbine-powered Hawk 4T (N403GB), which was renamed Hawk 4 at this time following abandonment of piston-engined version; the following October Groen changed its focus to seek government contracts for Hawk 4, slowing certification process for both piston- and turbine-powered versions until it sees market upturn.

On 28 December 2001, Groen announced contract with Utah Olympic Public Safety Command for lease of Hawk 4, beginning 20 January 2002, for security patrols at Salt Lake International Airport, equipped with video downlink system, Spectrolab SX-5 searchlight and additional radios; aircraft flew 75 hours in 67 operations.

CURRENT VERSIONS: **Hawk 4:** Currently powered by 313 kW (420 shp) Rolls-Royce 250-B17C turboprop; first flew (N403GB) 12 July 2000; originally intended for certification late 2001. Other changes include addition of underfins and taller landing gear. Two further prototypes under construction.

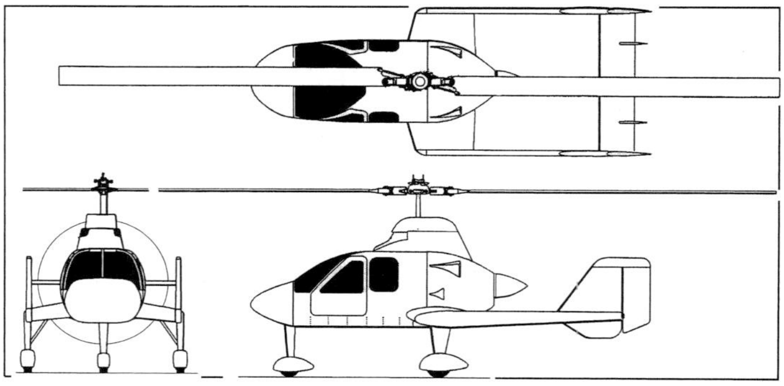

Groen Hawk 4 four-seat autogyro in its definitive configuration *(James Goulding)* ***NEW***/0567293

Applications include airborne law enforcement, electronic news gathering, aerial surveillance, utility/passenger transport and aerial application.

Jet Hawk 4T: Original designation for turbine-powered version. Renamed Hawk 4 by 2002.

Hawk 8: Proposed eight-seat variant with four-blade rotor and 559 kW (750 shp) turboprop engine.

CUSTOMERS: By May 2003, deposits on 148 aircraft had been taken, via 12 dealerships. Fractional ownership programme announced July 2001 but later dropped.

COSTS: Around US$749,000 (2003).

DESIGN FEATURES: Twin tailbooms supported by stub-wings which also house main landing gear; large twin stabilisers and rudders; with fixed horizontal tail surface mounted between the vertical tails. Two-blade, semi-rigid aluminium teetering rotor with swashplate. Rotor speed 270 rpm.

FLYING CONTROLS: Patented collective pitch-controlled rotor head allows smooth vertical take-off (zero ground roll) and enhanced flight performance. Rotor brake is standard. Actuation by pushrods. Patented dual-control stack cyclic flight controls.

STRUCTURE: Steel mast and engine mounts; stressed skin aluminium semi-monocoque fuselage, tail unit, hub structure and propeller; composites nose, engine cowling and wingtips; acrylic windscreen and doors; glass fibre nosecone and engine cowling.

LANDING GEAR: Initially fixed tricycle type; with electrohydraulically operated retractable gear to be offered later. Mainwheel tyres 6.00×6; nosewheel 5.00×5; Cleveland hydraulic brakes. Twin safety wheels at rear of tailbooms.

POWER PLANT: Production prototype for Hawk series was powered by a 134 kW (180 hp) Textron Lycoming O-360-A4M flat-four. Hawk H2X had one 335 kW (450 hp) Geschwinder V-8 aluminium liquid-cooled engine, derated to 261 kW (350 hp) at 2,500 rpm, driving a Hartzell three-blade constant-speed propeller.

Hawk 4 piston-powered version has air-cooled, six-cylinder Teledyne Continental TSIO-550 rated at 261 kW (350 hp) at 2,700 rpm; prototype had four-blade MTV propeller but production models will have Hartzell three-blade constant-speed propeller. Engine provides power to rotor for prerotation to provide for short and vertical take-off capability; power to rotor system never engaged during flight. Hawk 4 turbine-powered version has one 313 kW (420 shp) gas turbine engine driving three-blade constant-speed propeller.

Fuel capacity 284 litres (75.0 US gallons; 62.5 Imp gallons) in single tank at rear of fuselage; refuelling point at top of fuselage. Oil capacity 11.4 litres (3.0 US gallons; 2.5 Imp gallons).

ACCOMMODATION: Pilot and up to three passengers in enclosed cabin in two pairs of seats; rear seats fold to provide baggage space.

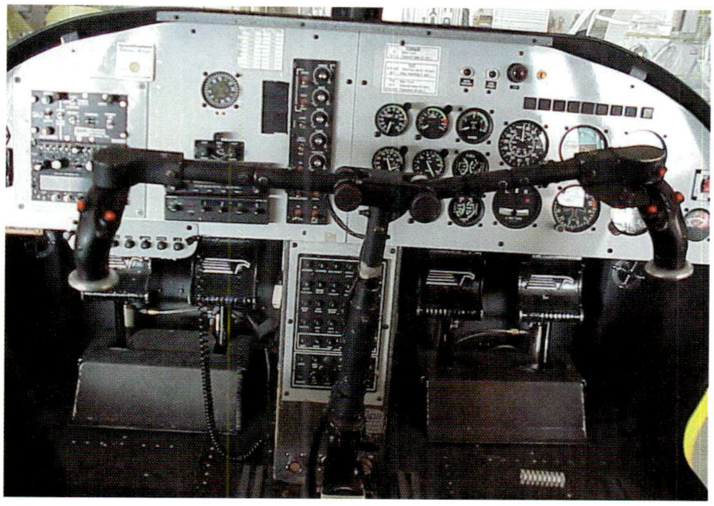

Dual cyclic control fitted to Hawk 4 0134513

Groen Hawk 4 *(Paul Jackson)* ***NEW***/0568981

SYSTEMS: Electrical system 28 V DC.

AVIONICS: *Comms:* Honeywell KLX 135A GPS/COM, KT 76A transponder with Mode C, PM 3000 intercom.

Flight: United Instruments suite.

DIMENSIONS, EXTERNAL:

- Rotor diameter .... 12.80 m (42 ft 0 in)
- Rotor blade chord .... 0.34 m (1 ft 1½ in)
- Propeller diameter .... 1.93 m (6 ft 4 in)
- Length of fuselage .... 7.31 m (24 ft 0 in)
- Height: overall .... 4.11 m (13 ft 6 in)
- to top of fuselage .... 3.35 m (11 ft 0 in)
- Wheel track .... 2.72 m (8 ft 11 in)
- Wheelbase .... 2.47 m (8 ft 1¼ in)
- Passenger doors: Width .... 0.94 m (3 ft 1 in)
- Height to sill .... 0.89 m (2 ft 11 in)

DIMENSIONS, INTERNAL:

- Cabin: Length .... 1.91 m (6 ft 3 in)
- Max width .... 1.37 m (4 ft 6 in)

AREAS:

- Main rotor disc .... 128.71 m² (1,385.4 sq ft)
- Vertical tails .... 1.88 m² (20.2 sq ft)
- Horizontal stabiliser .... 1.49 m² (16.0 sq ft)

WEIGHTS AND LOADINGS:

- Weight empty .... 835 kg (1,840 lb)
- Max T-O weight .... 1,587 kg (3,500 lb)
- Max disc loading .... 12.33 kg/m² (2.53 lb/sq ft)
- Max power loading .... 6.09 kg/kW (10.00 lb/hp)

PERFORMANCE:

- Never-exceed speed (VNE) .... 128 kt (238 km/h; 148 mph)
- Cruising speed at 75% power .... 115 kt (212 km/h; 132 mph)
- Max rate of climb at S/L .... 457 m (1,500 ft)/min
- Service ceiling .... 4,875 m (16,000 ft)
- Take-off run .... 8 m (25 ft)
- T-O to 15 m (50 ft) .... 76 m (250 ft)
- Landing from 15 m (50 ft) .... 46 m (150 ft)
- Range with max fuel at 75% power .... 315 n miles (584 km; 363 miles)

GROEN REVCON 6

TYPE: Utility autogyro.

PROGRAMME: Technology demonstrator development of Hawk 4 (see separate entry for Groen Hawk 4). Will not be marketed. Announced late 2000 as Hawk 6G with an expected first flight (prototype N9112A) in January 2001; first flight eventually made 22 September 2001.

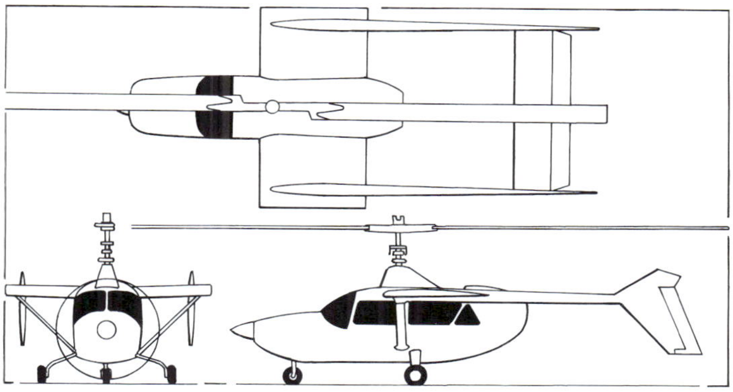

Groen RevCon 6G, based on Cessna 337 airframe *(Paul Jackson)* 0105705

Based around Cessna 337 fuselage with single 335 kW (420 shp) Rolls-Royce 250-B17F2 turboprop in nose for propulsion, driving FC9684C-6RX three-blade propeller; Rolls-Royce 250-C18 above fuselage for rotation; inverted Cessna 337 tail fin and shortened wing with spoilers and two-blade rotor system from Hawk 4. Estimated cruising speed 113 to 130 kt (209 to 241 km/h; 130 to 150 mph); useful load 907 kg (2,000 lb).

COSTS: US$950,000 (2003).

DESIGN FEATURES: Basically similar to Hawk 4.

FLYING CONTROLS: Broadly similar to Hawk 4.

STRUCTURE: As Hawk 4.

POWER PLANT: Powered by a 335 kW (420 shp) Rolls-Royce 250-C18 turboprop engine. Fuel capacity 416 litres (110 US gallons; 91.6 Imp gallons). External tanks mounted at wingtips are modified Piper PA-24 Comanche wingtip units.

AVIONICS: *Instrumentation:* Full IFR.

DIMENSIONS, EXTERNAL:

- Rotor diameter .... 13.41 m (44 ft 0 in)
- Fuselage length .... 9.07 m (29 ft 9 in)
- Height .... 4.11 m (13 ft 6 in)

AREAS:

- Rotor disc .... 168.11 m² (1,809.6 sq ft)

WEIGHTS AND LOADINGS (estimated):

- Weight empty .... 964 kg (2,125 lb)
- Max T-O weight .... 2,041 kg (4,500 lb)

Proof of concept Groen RevCon 6G *(Paul Jackson)* ***NEW***/0568982

Max disc loading .... 12.14 kg/m² (2.49 lb/sq ft)
Max power loading .... 6.52 kg/kW (10.71 lb/shp)

PERFORMANCE (estimated):
Never-exceed speed (VNE) .... 147 kt (273 km/h; 170 mph)
Cruising speed .... 104 kt (193 km/h; 120 mph)
Service ceiling .... 4,875 m (16,000 ft)
Range at 50% power .... 360 n miles (667 km; 415 miles)
Endurance .... 2 h 58 min

UPDATED

GROEN GYROLIFTER

TYPE: Convertiplane.

PROGRAMME: Response to request of US Defense Advanced Research Projects Agency (DARPA) for proposals to meet requirement for Advanced Maneuver Transport (AMT) with VTOL and heavy lift capability. Proposal is based upon the integration of a large-diameter rotor powered by tipjets with an existing fixed-wing turboprop freighter.

Groen is proposing to DARPA that it be designated to undertake the necessary research and development to produce an intermediate scale technology demonstrator leading to full-scale prototype based on the Lockheed C-130 aircraft.

DESIGN FEATURES: Heavy lift gyrodyne with tipjets for rotor drive; large fuselage with passenger accommodation and rear freight doors; VTOL capability; four turboprop engines; stub-wings to share flight load, to reduce rotor speed and increase aircraft airspeed; and mostly composites construction with metal engine attachments. Complete cargo handling system. Tipjets provide a highly efficient power/weight ratio to enable vertical take-off and landing with a 18,144 kg (40,000 lb) payload, but are shut down in other flight modes, permitting the aircraft to achieve a range of up to 1,000 n miles (1,850 km; 1,150 mile) using the efficiencies of turboprop operation in cruise. Rotor system incorporates five mission-adaptive blades incorporating Westland BERP (British Experimental Rotor Programme) paddle blade tips, which optimise rotor performance during all flight regimes, thereby increasing airspeeds above that of any contemporary helicopter, while reducing fuel consumption.

FLYING CONTROLS: Fly-by-wire, fully integrated flight, engine tipjet and rotor management system.

STRUCTURE: Mainly composites structure for fuselage and wings, and metal inserts for landing gear and engine mounting points. Composites rotor blades and unique rotor head design.

LANDING GEAR: Tricycle retractable landing gear with steerable dual-wheel nose landing gear and bogie-type main gear.

POWER PLANT: One Roll-Royce AD2100 or Roll-Royce AE 1107 C upgraded from 4,586 kW (6,150 shp) to about 6,711 kW (9,000 shp). Fuel capacity 11,340 kg (25,000 lb).

ACCOMMODATION: Pilot and navigator; or two pilots plus loadmaster.

SYSTEMS: Electrical and hydraulic systems, including APU.

AVIONICS: Fully integrated battlefield systems.

All data are provisional.

DIMENSIONS, EXTERNAL:
Rotor diameter .... 34.14 m (112 ft 0 in)
Rotor blade chord: at root .... 1.82 m (5 ft 11½ in)
at tip before BERP .... 1.21 m (3 ft 11½ in)
Length of fuselage .... 33.53 m (110 ft 0 in)
Height: overall .... 7.62 m (25 ft 0 in)
to top of fuselage .... 5.64 m (18 ft 6 in)
Wheel track .... 5.18 m (17 ft 0 in)
Wheelbase .... 12.80 m (42 ft 0 in)
Passenger doors (each):
Width .... 1.22 m (4 ft 0 in)
Height to sill .... 1.07 m (3 ft 6 in)

DIMENSIONS, INTERNAL:
Cabin; Length .... 12.50 m (41 ft 0 in)
Max width .... 4.05 m (13 ft 3½ in)

AREAS:
Rotor disc .... 914.8 m² (9,847.0 sq ft)
Vertical tails (total) .... 40.13 m²(432.0 sq ft)
Horizontal tail .... 57.6 m²(620.0 sq ft)

WEIGHTS AND LOADINGS:
Weight empty .... 24,948 kg (55,000 lb)
Max T-O weight .... 59,875 kg (132,000 lb)

PERFORMANCE (estimated):
Never-exceed speed (VNE) .... 270/330 kt (500/611 km/h; 310/379 mph)
Cruising speed at 75% power .... 250/300 kt (463/556 km/h; 288/345 mph)
Max rate of climb at S/L .... 366 m (1,200 ft)/min
Service ceiling .... 5,486 m (18,000 ft)
Take-off and landing run (powered rotor in gyroplane mode) .... zero
Range at 75% power .... 500–1,000 n miles (926–1,852 km; 575–1,150 miles)

Artist's impression of Groen GyroLifter ***NEW***/1022977

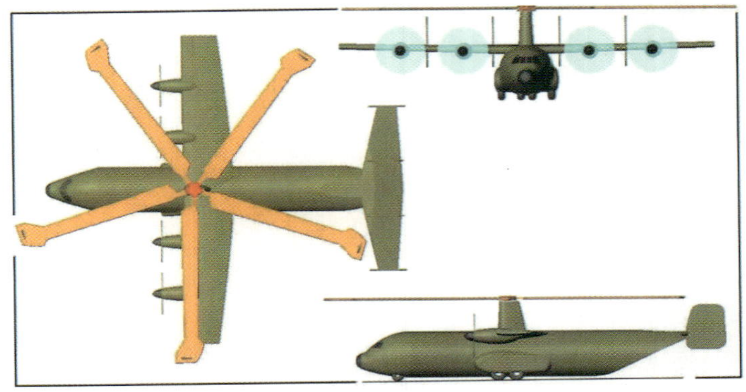

General arrangement of Groen GyroLifter, based on fuselage of Lockheed Martin C-130 Hercules ***NEW***/0569572

GROEN GYROLINER

TYPE: Convertiplane.

PROGRAMME: Groen proposes using same technology projected for its GyroLifter as the bases for a 35-seat, VTOL commuter airliner. Plan calls for mounting rotor system, with tipjets, on an existing high-wing turboprop, such as the Polish-built M-28 variant of the Antonov An-28. The resulting 19-passenger gyrodyne would provide rapid point-to-point transit and be the initial member of a family of civil runway and ATC independent gyrodynes designed to address the congestion problem at many major airports.

Artist's impression of Groen GyroLiner based on PZL M-28 ***NEW***/1022978

GROSSO

GROSSO AIRCRAFT INC

Production of the Grosso Easy Eagle has been transferred to Great Plains Aircraft Supply.

UPDATED

GREAT PLAINS EASY EAGLE

TYPE: Single-seat ultralight biplane kitbuilt.

PROGRAMME: Construction began 1996; Single-seat Easy Eagle first displayed at Oshkosh 1998. Two-seat Easy Eagle II introduced in 1999 but discontinued in 2004.

CURRENT VERSIONS: **Easy Eagle:** Single-seat version.

CUSTOMERS: Three Easy Eagles flying by mid-2004.

COSTS: Kit US$9,900 (Easy Eagle) (2004).

DESIGN FEATURES: Classic open-cockpit, single-bay biplane with steel tube cabane and flying struts and steel landing wires. Modified Clark Y aerofoil. No aerofoil on tailplane. Quoted build time 300 to 500 hours.

FLYING CONTROLS: Conventional and manual. Large ailerons on lower wings only.

STRUCTURE: Fabric-covered 4130 steel tube fuselage with metal-covered fuselage forward of cockpit and two-spar wooden wings with aluminium leading-edge.

Great Plains (Grosso-built) Easy Eagle I *(Paul Jackson)* 0126879

LANDING GEAR: Tailwheel type; fixed. Fuselage-mounted spring aluminium cantilever mainwheel legs. Azusa wheels and cable brakes; tyres 5.00-5. Solid tailwheel.

POWER PLANT: One 1,914 cc Volkswagen four-cylinder air-cooled motorcar engine; design accepts engines to 63.4 kW (85.0 hp). Wooden propeller. Fuel capacity 45.4 litres (12.0 US gallons; 10.0 Imp gallons).

DIMENSIONS, EXTERNAL:
- Wing span ... 5.49 m (18 ft 0 in)
- Length overall ... 4.37 m (14 ft 4 in)
- Height overall ... 1.78 m (5 ft 10 in)

AREAS:
- Wings, gross ... 9.57 m^2 (103.0 sq ft)

WEIGHTS AND LOADINGS:
- Weight empty ... 193 kg (425 lb)
- Max T-O weight ... 328 kg (725 lb)

PERFORMANCE:
- Max operating speed ... 95 kt (177 km/h; 110 mph)
- Normal cruising speed ... 78 kt (145 km/h; 90 mph)
- Stalling speed ... 40 kt (73 km/h; 45 mph)
- Max rate of climb at S/L ... 274 m (900 ft)/min
- T-O and landing run ... 92 m (300 ft)
- Range with max fuel ... 260 n miles (482 km; 300 miles)

UPDATED

GULFSTREAM

GULFSTREAM AEROSPACE CORPORATION

500 Gulfstream Road, Savannah, Georgia 31408
Tel: (+1 912) 965 30 00
Fax: (+1 912) 965 37 75/41 71
Web: http://www.gulfstream.com
PRESIDENT: Bryan T Moss
CHIEF OPERATING OFFICER: Joe Lombardo
SENIOR VICE-PRESIDENT, MARKETING AND SALES: Raynor Reavis
SENIOR VICE-PRESIDENT, PRODUCT SUPPORT: Larry R Flynn
SENIOR VICE-PRESIDENT, PROGRAMS, ENGINEERING AND TEST: Preston Henne
SENIOR VICE-PRESIDENT, GOVERNMENT SALES AND MARKETING: Buddy Sams
SENIOR VICE-PRESIDENT, FINANCE AND PLANNING: Dan Clare
SENIOR VICE-PRESIDENT, ADMINISTRATION: Ira Berman
VICE-PRESIDENT, ENGINEERING: Richard L Johnson
VICE-PRESIDENT, G100, G150 AND G200 PROGRAMMES: Dr Dan Nale
VICE-PRESIDENT, MARKETING AND COMMUNICATIONS: Stephanie Snyder
DIRECTOR OF CORPORATE COMMUNICATIONS: Robert Baugniet

Originally an offshoot of Grumman Corporation; passed through several owners before purchase by General Dynamics for US$4.8 billion in May 1999; deal completed 31 July 1999.

Corporate HQ and production centre at Savannah, Georgia, has 4,430 employees. Subassembly and support locations are Oklahoma City, Oklahoma (350) and Mexicali, Mexico (600); however, the company announced that its Oklahoma City subsidiary would close by end 2002. Completions and service undertaken at Long Beach, California (865), Dallas, Texas (705), Appleton, Wisconsin (400) and Brunswick, Georgia (215). Service and refurbishment centre at Westfield, Massachusetts (150); a 4,985 m (53,650 sq ft) refurbishment support facility was added at Savannah in April 2000, at which time the total direct workforce was 8,200. In early 2001, Gulfstream exchanged its engine overhaul service in Dallas with units of BBA Aviation, acquiring four regional maintenance centres in Dallas, Las Vegas, Minneapolis and West Palm Beach; the service operation was rebranded as General Dynamics Aviation Services as a result.

The 1,000th Gulfstream, a Gulfstream V, was delivered to the company's Long Beach completion centre in September 1997. During 1998, Gulfstream and Lockheed Martin's Skunk Works were revealed to be undertaking preliminary studies into the potential of a supersonic business jet; however Skunk Works withdrew from the project in 2000, following financial difficulties.

Gulfstream delivered 51 aircraft (22 Gulfstream IV-SPs and 29 Vs) in 1997 and 61 (32 Gulfstream IV-SPs and 29 Vs) in 1998. Gulfstream simultaneously handed over the 400th Gulfstream IV and 100th Gulfstream V at Savannah, Georgia, on 25 April 2000. Sales for 2000 were 82, comprising 27 new-generation V-SPs, 35 IV-SPs and 20 Vs; deliveries for same period were 70; for 2001 35 IV-SPs and 35 Vs, plus five Gulfstream 100s and 24 200s; for 2002 85 (all models); and five Gulfstream 100/200s and 14 Gulfstream 300/400/500/550 in the first nine months of 2003.

Following the purchase of Galaxy Aerospace of Israel by General Dynamics on 5 June 2001, the former's Astra and Galaxy business jets were incorporated into the Gulfstream product line as the Gulfstream 100 and 200 respectively. Production continues at Tel Aviv with 'green' aircraft flown to Lincoln, Nebraska (100) and Alliance Airport, Texas (200) for outfitting by Gulfstream; the Alliance facility closed in mid-2002 as a completion centre and became a sales and design centre.

On 9 September 2002, Gulfstream announced revised aircraft designations, the G IV becoming G300 and G400; G V becoming G500 and G550; and a widened G100 being launched as the G150. At the October 2003 NBAA Convention, the G450 was revealed following its previously unannounced maiden flight.

GULFSTREAM G100

US Air Force designation: C-38A

TYPE: Business jet.

PROGRAMME: Descendant of US Aero Commander 1121 Jet Commander (first flight 27 January 1963), acquired by IAI 1967 and developed successively as 1121 Commodore Jet, 1123 and 1124 Westwind and 1124A Westwind 2; sales of 1121/23/24/24A models by Aero Commander and IAI totalled 441 by end of 1987; 1125 model launched at NBAA Convention October 1979, renamed Astra 1981.

Construction of two prototypes and one static/fatigue test aircraft started April 1982; roll-out 1 September 1983; first flight (4X-WIN, c/n 001) 19 March 1984; first flight 4X-WIA (c/n 002) August 1984; first flight production Astra (4X-CUA) 20 March 1985; FAR Pts 25 and 36 certification 29 August 1985; first delivery 30 June 1986. Astra SP introduced at NBAA Convention October 1989; first delivery of this version (N60AJ) late 1990.

Gulfstream G100 twin-turbofan business transport *(Paul Jackson)* ***NEW***/0567036

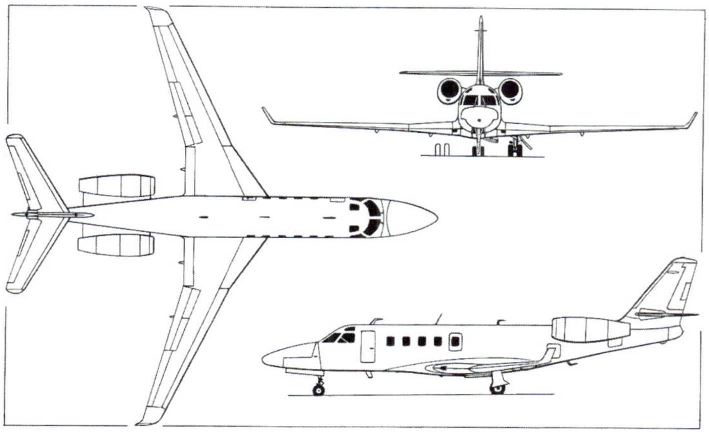

Gulfstream G100 business transport (two Honeywell TFE731-40R-200G turbofans) *(Jane's/Dennis Punnett)* 0507611

Astra SPX announced at NBAA Convention 1994 and certified by FAA 8 January 1996; type certificate acquired by Gulfstream Aerospace in June 2001, renamed Gulfstream 100 and formally unveiled as such at Paris on 17 June 2001; Gulfstream G100 name adopted from September 2002. Production rate 12 to 15 per year.

CURRENT VERSIONS: **Astra:** Initial version; 32 production aircraft; see 1992–93 and earlier *Jane's All the World's Aircraft.*

Astra SP: Introduced 1989; superseded by SPX; 36 built; details in 1997–98 *Jane's*

Gulfstream G100: First flight (as Astra SPX, 4X-WIX) 18 August 1994; FAA certification 8 January 1996; production deliveries from early 1996. Winglets and Rockwell Collins Pro Line 4 avionics. Change to more powerful TFE731-40R-200G turbofans with FADEC, hydromechanical fuel control back-up and Dee Howard thrust reversers; increased weights and payload/range performance and shorter T-O run. Type Certificate amended with G100 designation 9 August 2002.

Detailed description applies to Gulfstream 100 from c/n 138 onwards.

C-38A: Two Astra SPXs, outfitted by Tracor Flight Systems for transport and medevac duties; delivered as C-38As to 201st Airlift Squadron of the US Air National Guard at Andrews AFB, Maryland, in April and May 1998, replacing C-21A Learjets.

Gulfstream G150: Wide-cabin version; *described separately.*

Gulfstream G200: Improved and redesigned version; *described separately.*

CUSTOMERS: Combined production of Astra and Astra SP totalled 68 (plus two prototypes), of which approximately 80 per cent to US customers.

Total of 52 SPXs and 18 G100s produced up to end of 2002, including customer deliveries of 14 in 1998, 12 in 1999, 11 in 2000 and five in 2001; no figures released for 2002. Only three further aircraft registered in first nine months of 2003, increasing G100 total to 21 (although only final five of these were officially registered as G100s, beginning with c/n 146).

ASTRA/G100 PRODUCTION
(to late 2003)

Variant	Qty	Remarks
Prototypes (flying)	2	
Astra	32	
Astra SP	36	
Astra SPX	52	
Astra SPX/'G100'[1]	16	Registered as SPX
G100	5	Registered as G100
Total	143	

[1] Produced after G100 name announcement; eligible for official change of designation

DESIGN FEATURES: Typical swept wing, rear-engined business jet configuration. Meets FAR Pt 36 Stage 3 (noise), SFAR Pt 27 (fuel venting and exhaust emission) and RVSM requirements.

Wing section high-efficiency IAI Sigma 2; leading-edge sweep 34° inboard, 25° outboard; dihedral 2°; trailing-edge sweep on outer panels; winglets.

FLYING CONTROLS: Conventional, with hydraulically powered ailerons, manually actuated elevators and rudder. Control surfaces operated by pushrods; tailplane incidence controlled by three motors running together to protect against runaway or elevator disconnect; ailerons can be separated in case of jam; spoiler/lift dumper panels ahead of Fowler flaps; elevators and horn-balanced rudder each have geared tab; flaps interconnected with outboard leading-edge slats, both electrically actuated.

STRUCTURE: One-piece, two-spar wing with machined ribs and skin panels, attached by four main and five secondary frames; wing/fuselage fairings, elevator and fin tips and tailcone of GFRP; ailerons, spoilers, inboard leading-edges and wingtips of Kevlar and Nomex honeycomb; nose avionics bay door and nosewheel doors of Kevlar; Kevlar-reinforced nacelle doors and panels; chemically milled fuselage skins; some titanium fittings; heated windscreens of laminated polycarbonate with external glass layer to resist scratching.

LANDING GEAR: SHL hydraulically retractable tricycle type, with oleo-pneumatic shock-absorber and twin wheels on each unit. Trailing-link main units retract inward, nosewheels forward. Tyre sizes 23×7.00–12 (10 ply) (main), 16×4.4–10 (6 ply) deflector type (nose), pressures 8.55 bar; 124 lb/sq in and 6.90 bar; 100 lb/sq in respectively. Hydraulic extension, retraction and nosewheel steering (±58°); hydraulic multidisc anti-skid mainwheel brakes. Compressed nitrogen cylinder provides additional power source for emergency extension.

POWER PLANT: Two 18.90 kN (4,250 lb st) Honeywell TFE731-40R-200G turbofans, with Dee Howard hydraulically actuated target-type thrust reversers and FADEC, pylon-mounted in nacelle on each side of rear fuselage. Standard fuel in left and right outer wing tanks, each 1,094 litres (289 US gallons; 241 Imp gallons) and integral tank in wing centre-section and upper and lower tanks in centre-fuselage (combined capacity 2,744 litres; 725 US gallons; 604 Imp gallons). Total normal fuel 4,932 litres (1,303 US gallons; 1,085 Imp gallons), of which 4,910 litres (1,297 US gallons; 1,080 Imp gallons) are usable. Additional fuel can be carried in 378.5 litre (100 US gallon; 83.3 Imp gallon) removable auxiliary tank in forward area of baggage compartment (with baggage weight limilation imposed), all usable. Single pressure refuelling point in lower starboard side of fuselage aft of wing, or single gravity point in upper fuselage, allow refuelling of all tanks from one position. Fuel sequencing automatic.

ACCOMMODATION: Crew of two on flight deck. Dual controls standard. Sliding door between flight deck and cabin. Standard accommodation in pressurised cabin for six persons, two in forward-facing seats at front and four in club layout; galley (port or starboard) at front of cabin, coat closet forward (starboard), lavatory at rear. All six seats individually adjustable fore and aft, laterally, and can be swivelled or reclined; all fitted with armrests and headrests. Two wall-mounted foldaway tables between club seat pairs. Coat closet houses stereo tape deck. Maximum accommodation for nine passengers.

New cabin interior (first redesign since 1986), introduced in September 1999, has less intrusive (curved) cabinetry and galley units; restyled headliner and sidewall panels; new seats, tables and lighting. Forward club seating maximises recline angle and personal space for each occupant; lavatory has brighter lighting; more storage space for coats and carry-on baggage.

Plug-type airstair door at front on port side; emergency exit over wing on each side. Heated baggage compartment aft of passenger cabin, with external access and service ladder. Service compartment in rear fuselage houses aircraft batteries (or optional APU), electrical relay boxes, inverters and miscellaneous equipment.

SYSTEMS: Honeywell environmental control system, using engine bleed air, with normal pressure differential of 0.61 bar (8.8 lb/sq in). Honeywell 36–150(W) APU available optionally. Two independent hydraulic systems, each at pressure of 207 bar (3,000 lb/sq in). Primary system operated by two engine-driven pumps for actuation of anti-skid brakes, landing gear, nosewheel steering, spoilers/lift dumpers and primary control surfaces. Back-up system, operated by electrically driven pump, provides power for emergency/parking brake, primary control surfaces and thrust reversers.

Electrical system comprises two Lucas Aerospace 300 A 30 V DC engine-driven starter/generators, with two 1 kVA single-phase solid-state inverters operating in unison to supply single-phase 115 V AC power at 400 Hz and 26 V AC power for aircraft instruments. Two 24 V 24 Ah Ni/Cd batteries for engine starting and to permit operation of essential flight instruments and emergency equipment. 28 V DC external power receptacle standard.

Pneumatic de-icing of wing leading-edge slats and tailplane leading-edges; thermal anti-icing of engine intakes. Oxygen system for crew (pressure demand) and passengers (drop-down masks) supplied by 2.18 m^3 (77 cu ft) cylinder, with second cylinder of same capacity optional. Two-bottle Freon-type engine fire extinguishing system standard.

AVIONICS: Rockwell Collins Pro Line 4 suite standard.

Comms: Dual VHF-422C radios, RTU-4220 radio tuners, TDR-94D transponders, Baker B12135 audio systems and Magnastar Flightphones, Rockwell Collins HF 9000 HF, Motorola NA-1335 Selcal, Artex ELT and Universal CVR-30B CVR.

Radar: TWR-850 colour weather radar and WXP-4220 control panel.

Flight: Dual Universal UNS-1C FMS with embedded GPS, Rockwell Collins FCC-4005 autopilots, AHC-85E AHRS, ADC-850C air data systems, VIR-432 nav/GS/markers and DME-442; single ADF-462, ALT-55B radio altimeter, TCAS-94 and Honeywell EGPWS.

Instrumentation: Rockwell Collins EFD-4077 EFIS displays all flight information on four 18.4 cm (7¼ in) screens; dual Grimes engine displays and Davtron M850A digital clocks; single Flight Line 8047-10 standby altimeter, 8059-2B standby ASI, Jet AI-804CE standby AI, Precision PAI-700-02 standby compass and Hobbs 15007 hour meter.

EQUIPMENT: Standard equipment includes electric windscreen wipers, electric (warm air) windscreen demisting, wing ice inspection lights, landing light in each wingroot, taxying light inboard of each mainwheel door, navigation and Precise Flight pulse lights at wingtips and tailcone, rotating beacons under fuselage and on top of fin, provision for DeVore logo light and wing/tailplane static wicks.

DIMENSIONS, EXTERNAL:
Wing span over winglets 16.64 m (54 ft 7 in)
Wing mean aerodynamic chord 2.19 m (7 ft 2¼ in)
Wing aspect ratio 8.8
Length overall 16.94 m (55 ft 7 in)
Fuselage: Max width 1.57 m (5 ft 2 in)
Max depth 1.91 m (6 ft 3 in)
Height overall 5.54 m (18 ft 2 in)
Tailplane span 6.40 m (21 ft 0 in)
Wheel track (c/l of shock-struts) 2.77 m (9 ft 1 in)
Wheelbase 7.34 m (24 ft 1 in)
Passenger door (fwd, port): Height 1.37 m (4 ft 6 in)
Width 0.66 m (2 ft 2 in)
Overwing emergency exits (each):
Height 0.69 m (2 ft 3 in)
Width 0.48 m (1 ft 7 in)

DIMENSIONS, INTERNAL:
Cabin: Length: incl flight deck 6.86 m (22 ft 6 in)
excl flight deck 5.21 m (17 ft 1 in)
Max width 1.45 m (4 ft 9 in)
Max height 1.70 m (5 ft 7 in)
Volume, incl flight deck 12.03 m^3 (425 cu ft)
Baggage compartment volume (A: with, B: without fuel tank extension):
A 1.2 m^3 (42 cu ft)
B 1.6 m^3 (55 cu ft)

AREAS:
Wings, gross (excl winglets) 29.41 m^2 (316.6 sq ft)
Ailerons (total) 1.20 m^2 (12.92 sq ft)
Trailing-edge flaps (total) 4.20 m^2 (45.18 sq ft)
Leading-edge slats (total) 2.39 m^2 (25.76 sq ft)
Winglets (total) 0.52 m^2 (5.57 sq ft)
Fin 3.34 m^2 (35.97 sq ft)
Rudder (incl tab) 1.17 m^2 (12.63 sq ft)
Tailplane 4.86 m^2 (52.28 sq ft)
Elevators (total, incl tabs) 2.01 m^2 (21.66 sq ft)

WEIGHTS AND LOADINGS:
Basic operating weight empty 6,214 kg (13,700 lb)
Max usable fuel: standard 3,942 kg (8,692 lb)
with long-range tank 4,248 kg (9,365 lb)
Max payload 1,073 kg (2,365 lb)
Payload with max fuel 363 kg (800 lb)
Baggage compartment capacity (A: with, B: without fuel tank extension):
A 168 kg (370 lb)
B 499 kg (1,100 lb)
Max ramp weight 11,249 kg (24,800 lb)
Max T-O weight 11,181 kg (24,650 lb)
Max landing weight 9,389 kg (20,700 lb)
Max zero-fuel weight 7,711 kg (17,000 lb)
Max wing loading 380.3 kg/m^2 (77.89 lb/sq ft)
Max power loading 296 kg/kN (2.90 lb/lb st)

PERFORMANCE:
Max operating Mach No. (MMO) 0.875
Max operating speed (VMO), S/L to FL270 353 kt (653 km/h; 406 mph) IAS
Max cruising speed at FL330, 8,618 kg (19,000 lb) mid-cruise weight
484 kt (896 km/h; 557 mph)
Normal cruising speed M0.80 (459 kt; 850 km/h; 528 mph)
Long-range cruising speed 430 kt (796 km/h; 495 mph)
Max rate of climb at S/L 1,160 m (3,805 ft)/min
Rate of climb at S/L, OEI 411 m (1,348 ft)/min
Initial cruise altitude 12,497 m (41,000 ft)
Max certified altitude 13,715 m (45,000 ft)
FAR Pt 25 T-O balanced field length 1,645 m (5,395 ft)
FAR Pt 25 landing distance at MLW 890 m (2,920 ft)
Range: with eight passengers 2,286 n miles (4,235 km; 2,631 miles)
with four passengers, max fuel and NBAA IFR reserves
2,949 n miles (5,461 km; 3,393 miles)
with max fuel, VFR reserves 3,256 n miles (6,030 km; 3,747 miles)

OPERATING NOISE LEVELS (FAR Pt 36, Stage 3):
T-O 79.1 EPNdB
Sideline 89.5 EPNdB
Approach 91.9 EPNdB

GULFSTREAM G150

TYPE: Business jet.

PROGRAMME: Announced on eve of NBAA Convention at Orlando, Florida, 8 September 2002. Deliveries to launch customer NetJets were scheduled to begin in second quarter of 2005, but by early 2003 programme amended to include May 2005 first flight, certification in Januarty 2006 and first deliveries in August 2006.

CUSTOMERS: Launch customer NetJets Inc ordered 50, plus 50 options, on 9 September 2002, for delivery between 2005 and 2010.

COSTS: US$14.5 million (2003).

DESIGN FEATURES: Wide-cabin version of G100; max internal dimensicns increased by 31 cm (12¼ in) in width and 5 cm (2 in) in height. Uprated engines, increased weights, but same

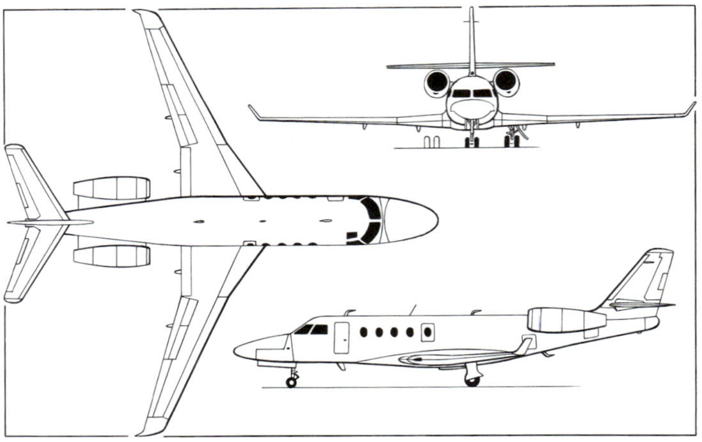

Gulfstream G150, widened derivative of G100 *(James Goulding)* ***NEW***/0568985

Model of projected Gulfstream G150 *(Paul Jackson)* ***NEW***/0567041

range and cruising speed. G200-style windscreen and cabin windows, latter reduced in number from six to five.

Differences from G100 summarised below:

POWER PLANT: Two 19.6 kN (4,400 lb st) Honeywell TFE731-40R turbofans.

ACCOMMODATION: Two pilots and between six and eight passengers in three choices of interior.

DIMENSIONS, EXTERNAL:
- Wing span ..... 16.94 m (55 ft 7 in)
- Length overall ..... 17.25 m (56 ft 7¼ in)
- Height overall ..... 5.54 m (18 ft 2 in)

DIMENSIONS, INTERNAL:
- Cabin: Length ..... 5.41 m (17 ft 9 in)
  - Max width ..... 1.75 m (5 ft 9 in)
  - Max height ..... 1.75 m (5 ft 9 in)
  - Volume ..... 13.17 m^3 (465 cu ft)
- Baggage volume:
  - main compartment ..... 1.56 m^3 (55 cu ft)
  - overhead bins ..... 0.25 m^3 (8.8 cu ft)

WEIGHTS AND LOADINGS:
- Basic operating weight (including two pilots) ..... 6,849 kg (15,100 lb)
- Payload: max ..... 1,089 kg (2,400 lb)
  - with max fuel ..... 363 kg (800 lb)
- Max fuel weight ..... 4,581 kg (10,100 lb)
- Max T-O weight ..... 11,725 kg (25,850 lb)
- Max landing weight ..... 9,843 kg (21,700 lb)
- Max zero-fuel weight ..... 7,938 kg (17,500 lb)
- Max power loading ..... 300 kg/kN (2.94 lb/lb st)

PERFORMANCE:
- Max operating Mach No (MMO) 0.85
- Normal cruising speed ..... 459 kt; (850 km/h; 528 mph)
- Long-range cruising speed ..... 430 kt (796 km/h; 495 mph)
- FAR Pt 25 T-O balanced field length ..... 1,777 m (5,830 ft)
- FAR Pt 25 landing distance at MLW ..... 1,052 m (3,450 ft)

GULFSTREAM G200

TYPE: Business jet.

PROGRAMME: Initiated as IAI 1126 derivative of Astra SP; design (then called Astra IV) finalised late 1992 in anticipation of 1993 launch; co-production with Yakovlev of Russia discussed during early part of 1993; formal announcement of launch as Galaxy, with minor design changes, announced 20 September 1993 just before NBAA Convention in Atlanta, Georgia, USA, followed next day by news that Yakovlev to be risk-sharing partner; other partners to be Rockwell Collins (avionics supplier) and eventual engine manufacturer (P&WC since selected). Replacement for Yakovlev sought in 1995, and Sogerma (France) contracted August 1996 to build production aircraft fuselages and tail units, but latter contract terminated mid-2001.

Four prototypes (003 and 004 flying plus 001 static and 002 fatigue test); first flight rescheduled for fourth quarter 1996 (later changed to second quarter 1997 and later still to fourth quarter 1997); 003 rolled out 4 September 1997 and made first flight (4X-IGA) 25 December 1997. Static testing of 001 completed September 1998. Second flight test

Flight deck of Gulfstream G200 with Rockwell Collins Pro Line 4 avionics 0100870

aircraft (4X-IGO, c/n 004) made first flight 21 May 1998; 005 (company demonstrator 4X-IGB/N505GA) first flew on 23 September 1998; combined total of 750 hours in 260 flights by three aircraft by December 1998. Israeli CAA and US FAA certification to FAR Pt 25 Amendment 25 and FAR Pts 34 and 36 awarded 16 December 1998; European JAA certification process started mid-1999. First customer delivery, to TTI of Fort Worth, Texas, made in January 2000. Production rate two per month in mid-2000. Type certificate acquired by Gulfstream Aerospace in June 2001, aircraft renamed Gulfstream 200 and formally unveiled as such at Paris on 17 June 2001; name Gulfstream G200 adopted in September 2002 although Type Certificate had been amended with new designation on 16 January 2002. G200 applies to c/n 057 and subsequent aircraft, although retrospective redesignation is permissible under US rules. Civil Aviation Administration of China certification achieved 10 September 2002.

CURRENT VERSIONS: Seen as four/eight-passenger business/executive (standard model), with option of alternative interior seating up to 18 passengers for regional transport operation.

CUSTOMERS: Between 45 and 50 commitments by October 1999, of which about 35 per cent from non-US operators in Canada, Europe, Israel, Mexico and South America. First bulk order, for seven, plus option on 15 more, placed mid-1999 by charter operator ILI Aviation of Zurich, Switzerland. IAI estimates break even at 100 sales and potential market for 200.

Totals of one delivered in 1999, six in 2000, 24 in 2001, and 12 in the first six months of 2002. Orders by September 2002 included 50 firm and 50 options by Executive Jet Inc, three by Hainan Airlines of Beijing, China, and one by MetroJet of Hong Kong. Total 82 (including protoypes) supplied to completion centres or customers by July 2003.

GALAXY/G200 PRODUCTION
(at October 2003)

Variant	Qty	Remarks
Prototypes (flying)	2	
Galaxy	48	Eligible for change of designation
G200	32	
Total	**82**	

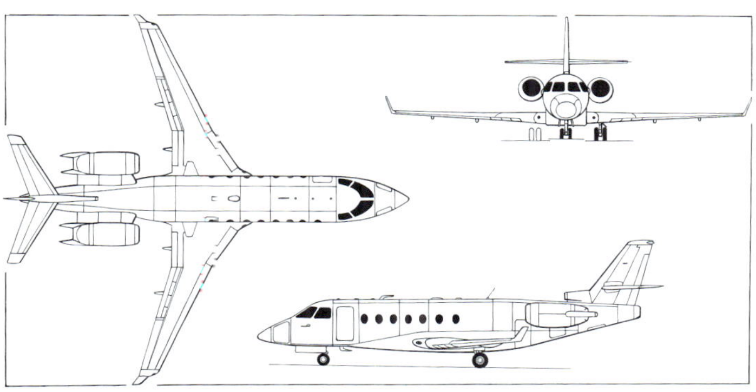

Gulfstream G200 twin-turbofan business and commuter transport *(James Goulding)* 0062974

Typical Gulfstream G200 business jet interior 0099570

COSTS: Approximately US$18 million flyaway (July 2000); total development programme cost forecast at approximately US$152 million (1993).

Detailed description applies to Gulfstream G200 from c/n 40 onwards.

DESIGN FEATURES: Designed for transatlantic range (non-stop Paris to New York). Essentially same wing as Gulfstream 100, except for 34° 30' inboard leading-edge sweep and addition of Krueger flaps; new wide-body fuselage is longer and has more headroom.

FLYING CONTROLS: All-hydraulic dual actuation except for rudder (manual); aileron movement 10° up/15° down, elevators 27° up/20° down, rudder 20° left/right; tailplane and rudder tab have electric trim. Wings fitted with outboard leading-edge slats (25° fully out), inboard Krueger leading-edge flaps (110°), four-segment upper surface airbrakes/lift dumpers (45° fully up) and Fowler single-slotted inboard and outboard trailing-edge flaps (settings 0, 12, 20 and 40°). Ailerons and elevators can be operated manually in event of hydraulic failure.

STRUCTURE: Generally similar to that described for G100. Sogerma (France) produced tail units, doors, access panels, tailcones and wing/fuselage fairings for prototypes. Flight Environments cabin flame protection and sound attenuation system.

LANDING GEAR: Hydraulically retractable tricycle type; twin wheels and oleo-pneumatic shock-absorbers on each unit. Nose unit has electrohydraulic steering (± 60°) and retracts forward. Trailing-link mainwheel units retract inward and are equipped with Honeywell multidisc anti-skid carbon brakes.

POWER PLANT: Two FADEC-equipped Pratt & Whitney Canada PW306A turbofans, each flat rated at 26.9 kN (6,040 lb st), pylon-mounted on sides of rear fuselage. Nordam nacelles and thrust reversers. Fuel in seven tanks: right and left wing, each 1,334 litres (352.5 US gallon;

Gulfstream G200 (formerly Galaxy) business jet (two PW306A turbofans) *(Paul Jackson)* ***NEW***/0567035

293-5 Imp gallons); right and left feed, each 102 litres (26.9 US gallons; 22.4 Imp gallons); centre 1,533 litres (405 US gallons; 337 Imp gallons); forward 1,011 litres (267 US gallons; 222 Imp gallons); and fuselage 3,115 litres (823 US gallons; 685 Imp gallons). Total fuel 8,532 litres (2,254 US gallons; 1,877 Imp gallons), of which 8,479 litres (2,240 US gallons 1,865 Imp gallons) are usable.

ACCOMMODATION: One or two pilots; provision for jump-seat for third crew member. Standard club-type seating for four to eight persons in business/executive version; legroom between facing seats allows enough space for full reclining and berthing; large galley with room for refrigerator, microwave oven, coffee maker and storage. Alternative 10-passenger layout has four club seats at front, with a four-place conference group and two on a divan to the rear, plus optional lavatory and additional baggage space aft. Three-abreast seating for up to 18 passengers, with single aisle, can be provided in corporate shuttle configuration. Generous baggage compartment in rear fuselage, accessed by external airstair door, can accommodate baggage for all 18 passengers. Entire accommodation, including baggage compartment, is pressurised. Airshow in-flight cabin entertainment system optional.

SYSTEMS: Dual hydraulic systems, each 207 bar (3,000 lb/sq in). Electrical system comprises three (including one on APU) Lucas Aerospace 28 V 400 Ah engine-driven starter/generators, two 24 V 43 Ah Ni/Cd batteries and a Honeywell GTCP36-150 APU. Third (24 V 27 Ah) battery for back-up powering of essential flight instruments and emergency systems. Pneumatic system for emergency extension of landing gear, actuation of wheel brakes and thrust reversers, and de-icing of wing leading-edges. Cabin pressurisation and air conditioning system, differential 0.61 bar (8.8 lb/sq in). One (optionally two) oxygen bottles.

AVIONICS: Rockwell Collins Pro Line 4 suite standard.

Comms: Dual VHF-422C radios, RTU-4220 radio tuners, TDR-94D transponders, Bendix/King KHF 950 HF and Baker B1045-F512 audio systems; triple Magnastar Flightphones; single Avtech Selcal, Artex ELT and Universal CVR-30B CVR.

Radar: TWR-850 colour weather radar with turbulence detection and WXP-4220 control panel.

Flight: Dual Universal UNS-1C FMS with embedded GPS, Rockwell Collins FCC-4005 autopilots, AHS-3000 AHRS, ADC-850C air data systems, VIR-432 VOR/ILS/GS/markers and DME-442; single ADF-462, ALT-4000 radio altimeter, TCAS-4000, Honeywell Laseref IV IRS, EGPWS Mk V and FDR.

Instrumentation: Rockwell Collins EFD-4077 EFIS displays all flight and EICAS information on five 18.4 cm (7¼ in) screens; dual Davtron M850A digital clocks; Flight Line 8047-10 standby altimeter, 8059-2B standby ASI, Jet AI-804CE standby AI, Precision PAI-700-04 standby compass and Hobbs 15007 hour meter.

DIMENSIONS, EXTERNAL:

Wing span	17.70 m (58 ft 1 in)
Length overall	18.97 m (62 ft 3 in)
Wing aspect ratio	9.1
Height overall	6.53 m (21 ft 5 in)
Tailplane span	6.86 m (22 ft 6 in)
Wheel track	3.30 m (10 ft 10 in)
Wheelbase	7.39 m (24 ft 3 in)
Passenger door: Height	1.83 m (6 ft 0 in)
Width	0.84 m (2 ft 9 in)
Baggage compartment door:	
Height	1.14 m (3 ft 9 in)
Width	0.89 m (2 ft 11 in)

DIMENSIONS, INTERNAL:

Cabin: Length: incl flight deck	9.30 m (30 ft 6 in)
excl flight deck	7.44 m (24 ft 5 in)
Max width	2.18 m (7 ft 2 in)
Max height	1.91 m (6 ft 3 in)
Volume, excl flight deck	24.6 m^3 (868 cu ft)
Baggage compartment volume	3.5 m^3 (125 cu ft)

AREAS:

Wings, gross	34.28 m^2 (369.0 sq ft)

WEIGHTS AND LOADINGS:

Basic operating weight empty	8,981 kg (19,800 lb)
Max usable fuel weight	6,804 kg (15,000 lb)
Max payload	1,905 kg (4,200 lb)
Baggage compartment capacity	898 kg (1,980 lb)
Payload with max fuel	363 kg (800 lb)
Max ramp weight	16,148 kg (35,600 lb)
Max T-O weight	16,079 kg (35,450 lb)
Max landing weight	13,608 kg (30,000 lb)
Max zero-fuel weight	10,886 kg (24,000 lb)
Max wing loading	469.1 kg/m^2 (96.07 lb/sq ft)
Max power loading	299 kg/kN (2.93 lb/lb st)

PERFORMANCE (estimated):

Max operating Mach No. (MMO)	0.85
Max operating speed (VMO):	
S/L to FL100	310 kt (574 km/h; 356 mph) IAS
FL100-FL200	310–330 kt (574–611 km/h; 356–379 mph) IAS
FL200-FL250	360 kt (667 km/h; 414 mph) IAS
Max cruising speed (VMC) at FL310, mid-cruise weight of 12,247 kg (27,000 lb)	494 kt (915 km/h; 568 mph)
Typical cruising speed at FL390	475 kt (880 km/h; 547 mph)
Long-range cruising speed	430 kt (797 km/h; 495 mph)
Stalling speed, flaps and gear down, at MLW	112 kt (208 km/h; 129 mph) IAS
Max certified altitude	13,715 m (45,000 ft)
T-O distance (S/L, ISA)	1,855 m (6,080 ft)
Landing distance at MLW (S/L, ISA)	1,001 m (3,285 ft)
Range with four passengers, NBAA IFR reserves	3,600 n miles (6,667 km; 4,142 miles)
g limits, flaps and gear up	+2.63/−1.0

OPERATING NOISE LEVELS (FAR Pt 36 Stage 3):

T-O	81.4 EPNdB
Sideline	85.8 EPNdB
Approach	90.9 EPNdB

GULFSTREAM G300 AND G400

Swedish Air Force designations: S 102B Korpen and Tp 102
US military designations: C-20F/G/H
JASDF designation: U-4
Engineering designation: G-1159C

Gulfstream IV-SP (Special Performance) 0130573

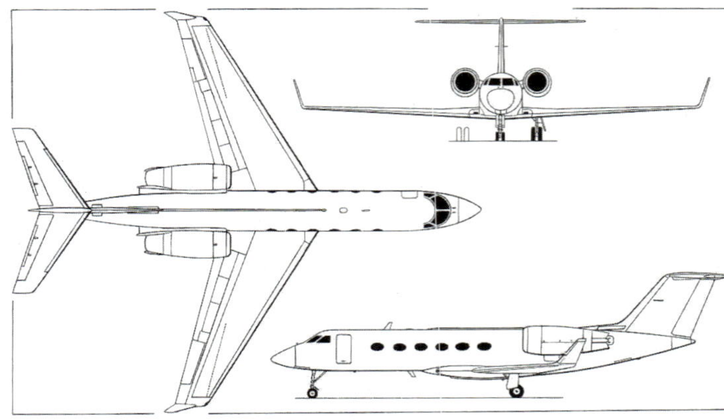

Gulfstream IV twin-turbofan business transport *(Jane's/Dennis Punnett)* 0507264

TYPE: Long-range business jet.

PROGRAMME: Design of Gulfstream IV started March 1983; manufacture of four production prototypes (one for static testing) began 1985; first aircraft (N404GA) rolled out 11 September 1985; first flight 19 September 1985; first flight of second prototype 11 June 1986 and third prototype August 1986; FAA certification 22 April 1987 after 1,412 hours' flight testing; certificate is extension of original Gulfsteam II, first approved on 19 October 1967. Westbound round-the-world flight from Le Bourget Airport, Paris, on 12 June 1987, covering 19,887.9 n miles (36,832.44 km; 22,886.6 miles), took 45 hours 25 minutes at average speed of 437.86 kt (811.44 km/h; 504.2 mph) and set 22 world records; eastbound round-the-world flight in N400GA from Houston, Texas, on 26 and 27 February 1988 covered 20,028.68 n miles (37,093.1 km; 23,048.6 miles) in 36 hours 8 minutes 34 seconds at average speed of 554.15 kt (1,026.29 km/h; 637.71 mph), setting 11 records.

In March 1993, Gulfstream IV-SP N485GA set new world speed and distance records in class, at 503.57 kt (933.21 km/h; 579.87 mph) and 5,139 n miles (9,524 km; 5,918 miles) respectively, on routine business flight from Tokyo, Japan, to Albuquerque, USA. Russian Federation and Associated States (CIS) certification achieved in June 1996. FAA approval for RVSM operation of Gulfstream IV series granted 11 August 1997. European JAA validation of FAA certification of Gulfstream IV and IV-SP achieved 16 October 2001. Gulfstream IV-SP holds 75 flight records; by September 2003 fleet had accumulated more than two million flying hours, with a 99.7 per cent despatch reliability rate.

G300/G400 introduced from January 2003 on common production line; G400 completed to full specification; G300 as baseline aircraft, with only those avionics and fittings specified by customer and lavatory location only at rear. G400 assembly includes 360 items installed on production line which previously were part of outfitting. JAA validation of FAA certification achieved for G300 and G400 in July 2003.

CURRENT VERSIONS: **Gulfstream IV:** Built until 1992. MTOW 33,203 kg (73,200 lb).

Gulfstream IV-SP: Improved (Special Performance), higher-weight version announced at NBAA Convention, Houston, in October 1991. Prototype (N476GA, converted from standard) first flown 24 June 1992; designation applied to all new IVs sold after 6 September 1992 (c/n 1214 and upwards); maximum payload increased by 1,134 kg (2,500 lb) and maximum landing weight increased by 3,402 kg (7,500 lb), with no increase in guaranteed manufacturer's empty weight. Payload/range envelope extended; expanded capability Honeywell SPZ-8400 flight guidance and control system. Production ended 3 December 2002 when 500th and final IV/IV-SP (N499GA: c/n 1499) rolled out.

Detailed description applies to Gulfstream IV-SP, except where otherwise indicated.

Gulfstream IV-MPA: Multipurpose Aircraft, announced September 1994; derived from US Navy C-20G Operational Support Aircraft (which see, below) to provide commercial operators with quick-change interior for up to 26 passengers in high-density shuttle layout, low-density executive configuration, 2,177 kg (4,800 lb) cargo capacity, or combination; large cargo door and larger/additional emergency exits standard.

Gulfstream G300: Mid-range (basic specification) version, announced on the eve of the NBAA Convention at Orlando, Florida, 8 September 2002; First aircraft A6-RJA (c/n 1503) delivered 6 August 2003 to Royal Jet of Abu Dhabi, which has ordered two, in miltipurpose aircraft (MPA) configuration.

Gulfstream G400: New (full specification) version, announced on the eve of the NBAA Convention at Orlando, Florida, 8 September 2002. Generally as IV-SP except HUD standard. First aircraft N520GA (c/n 1500) exhibited at EBACE, Geneva, 7 to 9 May 2003 (as N400GA).

Gulfstream G450: *Described separately.*

CUSTOMERS: Total 500 Gulfstream IV and IV-SPs built by December 2002 (c/n 1000 to 1499), when production switched to G300/G400. Deliveries totalled 22 in 1997, 32 in 1998, 39 in 1999, 37 in 2000, 36 in 2001 and 17 in the first six months of 2002 (no further announcements).

COSTS: US$31 million 'green' G400 (2003).

DESIGN FEATURES: Rear-engined, T-tail configuration, with all flying surfaces sweptback; wing mounted below cabin. Differences from Gulfstream III (1987–88 and earlier *Jane's All the World's Aircraft*) include aerodynamically redesigned wing, with winglets, contributing to lower cruise drag; wing also structurally redesigned with 30 per cent fewer parts, 395 kg (870 lb) lighter and carrying 544 kg (1,200 lb) more fuel; increased tailplane span; fuselage

1.37 m (4 ft 6 in) longer, with sixth window each side; Rolls-Royce Tay turbofans; flight deck with electronic displays; digital avionics; and fully integrated flight management and autothrottle systems.

Advanced sonic rooftop aerofoil; sweepback at quarter-chord 27° 40'; thickness/chord ratio 10 per cent at wing station 50, 8.6 per cent at station 414; dihedral 3°; incidence 3° 30' at root, −2° at tip; NASA (Whitcomb) winglets.

FLYING CONTROLS: Conventional. Hydraulically powered flying controls with manual reversion; trim tab in port aileron and both elevators; two spoilers on each wing act differentially to assist aileron and, with third spoiler each side, act collectively as airbrakes and lift dumpers; single-slotted Fowler flaps; four vortillons and a single 'tripper' strip under leading-edge of each wing ensure inboard part of wing stalls before outboard section; variable incidence tailplane.

STRUCTURE: Light alloy airframe except for carbon composites ailerons, spoilers, rudder and elevators, some tailplane parts, some cabin floor structure, and parts of flight deck; winglets of aluminium honeycomb. Wing box manufactured by Aerostructures Corporation.

LANDING GEAR: Retractable tricycle type with twin wheels on each unit. Main units retract inward, steerable nose unit forward. Mainwheel tyres size 34×9.25–16 (18 ply) tubeless; pressure 12.07 bar (175 lb/sq in). Nosewheel tyres size 21×7.25–10 (10 ply) tubeless, pressure 7.93 bar (115 lb/sq in); maximum steering angle ±82°. Dunlop air-cooled carbon brakes; Aircraft Braking Systems anti-skid units and digital electronic brake-by-wire system. Dowty electronic steer-by-wire system. Turning circle about wingtip 14.43 m (47 ft 4 in); about nosewheel 12.04 m (39 ft 6 in).

POWER PLANT: Two Rolls-Royce Tay Mk 611-8 turbofans, each flat rated at 61.6 kN (13,850 lb st) to ISA +15°C. Target-type thrust reversers. Fuel in two integral wing tanks, with total capacity of 16,542 litres (4,370 US gallons; 3,639 Imp gallons) in GIV/G400 and 15.141 litres (4,000 US gallons; 3,331 Imp gallons) in G300. Single pressure fuelling point in leading-edge of starboard wing. In 1999 Gulfstream considered re-engining of IV-SP as part of product improvement policy, possibilities including Rolls-Royce Deutschland BR710, and General Electric CF34; in May 2000 it placed an order for US$1.4 billion with Rolls-Royce for improved Tays.

ACCOMMODATION: Crew of two plus cabin attendant. Standard seating for up to 19 passengers (typically 12 to 14 in corporate configuration) in pressurised and air conditioned cabin. 'Quick Change' cargo/passenger version, certified for up to 26 passengers, announced 22 December 1993. Galley, lavatory and large baggage compartment, capacity 907 kg (2,000 lb), at rear of cabin. Integral airstair door at front of cabin on port side. Baggage compartment door on port side. Electrically heated wraparound windscreen. Six cabin windows, including two overwing emergency exits, on each side.

SYSTEMS: Cabin pressurisation system maximum differential 0.65 bar (9.45 lb/sq in) maintains 1,980 m (6,500 ft) cabin altitude at 13,715 m (45,000 ft); dual air conditioning systems. Two independent hydraulic systems, each 207 bar (3,000 lb/sq in). Maximum flow rate 83.3 litres (22 US gallons; 18.3 Imp gallons)/min. Two bootstrap-type hydraulic reservoirs, pressurised to 4.14 bar (60 lb/sq in). Honeywell GTCP36-100G APU in tail compartment, flight rated to 12,500 m (41,000 ft) since s/n 1156. Electrical system includes two 36 kVA alternators with two solid-state 30 kVA converters to provide 23 kVA 115/200 V 400 Hz AC power and 250 A of regulated 28 V DC power; two 24 V 40 Ah Ni/Cd storage batteries and external power socket. Wing leading-edges and engine inlets anti-iced.

AVIONICS: *Comms:* Dual VHF/HF transceivers, transponders and cockpit audio systems; cockpit voice recorder; Calquest CD-400 satellite communications equipment optional.

Radar: Digital colour weather radar.

Nav: Dual VOR/LOC/GS with marker beacon receivers; dual DME; dual ADF; dual radio altimeters; optional MLS, GPS and VLF Omega. Optional Northstar Technologies CT-1000 flight deck organiser.

Flight: Honeywell SPZ-8400 digital AFCS; Honeywell SPZ-8000 flight management system (FMS); dual fail-operational flight guidance systems including autothrottles; dual air data systems; dual flight guidance and performance computers; dual laser IRS; AHRS; VNAV; flight data recorder. System integration is accomplished through a Honeywell avionics standard communications bus (ASCB). Optional TCAS and Honeywell EGPWS.

Instrumentation: Six 203 × 203 mm (8 × 8 in) colour CRT EFIS screens, two each for primary flight display (PFD), navigation display (ND) and engine instrument and crew alerting system (EICAS). Honeywell/BAE HUD 2020 head-up display received FAA approval for Cat. II operations in early 1997, with first installation in Gulfstream IV-SP completed in May 1997. Northstar Technologies CT-1000G flight deck organiser system optional. EVS system used on Gulfstream V received FAA certification for installation on IV-SP on 19 December 2002.

Self-defence: Optional BAE Systems AN/ALQ-204 Matador IRCM available at cost of US$3.5 million.

Data applicable to both versions, except where stated.

DIMENSIONS, EXTERNAL:

Wing span over winglets	23.72 m (77 ft 10 in)
Wing chord: at root (fuselage c/l)	5.94 m (19 ft 5¾ in)
at tip	1.85 m (6 ft 0¾ in)
Wing aspect ratio	6.4
Length overall	26.92 m (88 ft 4 in)
Fuselage: Length	24.03 m (78 ft 10 in)
Max diameter	2.39 m (7 ft 10 in)
Height overall	7.44 m (24 ft 5 in)
Tailplane span	9.75 m (32 ft 0 in)
Wheel track	4.17 m (13 ft 8 in)
Wheelbase	11.61 m (38 ft 1¼ in)
Passenger door (fwd, port): Height	1.57 m (5 ft 2 in)
Width	0.91 m (3 ft 0 in)
Baggage door (rear): Height	0.90 m (2 ft 11¾ in)
Width	0.72 m (2 ft 4½ in)

DIMENSIONS, INTERNAL:

Cabin:	
Length, incl galley, lavatory and baggage compartment	13.74 m (45 ft 1 in)
Max width	2.24 m (7 ft 4 in)
Max height	1.88 m (6 ft 2 in)
Floor area	22.9 m^2 (247 sq ft)
Volume	43.2 m^3 (1,525 cu ft)
Flight deck volume	3.5 m^3 (124 cu ft)
Rear baggage compartment volume	4.8 m^3 (169 cu ft)

AREAS:

Wings, gross	88.29 m^2 (950.4 sq ft)
Ailerons (total, incl tab)	2.68 m^2 (28.86 sq ft)
Trailing-edge flaps (total)	11.97 m^2 (128.84 sq ft)
Spoilers (total)	7.46 m^2 (80.27 sq ft)
Winglets (total)	2.38 m^2 (25.60 sq ft)
Fin	10.92 m^2 (117.53 sq ft)
Rudder, incl tab	4.16 m^2 (44.75 sq ft)
Horizontal tail surfaces (total)	18.83 m^2 (202.67 sq ft)
Elevators (total, incl tabs)	5.22 m^2 (56.22 sq ft)

WEIGHTS AND LOADINGS:

Manufacturer's weight empty	16,102 kg (35,500 lb)
Allowance for outfitting	3,175 kg (7,000 lb)
Typical operating weight empty (incl crew):	
G300	19,504 kg (43,000 lb)
G400	19,912 kg (43,900 lb)
Max payload: G300	2,722 kg (6,000 lb)
G400	2,313 kg (5,100 lb)
Payload with max fuel: G300	1,134 kg (2,500 lb)
G400	726 kg (1,600 lb)
Fuel : total: G300	12,249 kg (27,005 lb)
G400	13,429 kg (29,605 lb)
max usable: G300	12,202 kg (26,900 lb)
G400	13,381 kg (29,500 lb)
Max T-O weight: G300	32,658 kg (72,000 lb)
G400	33,838 kg (74,600 lb)
Max ramp weight: G400	34,019 kg (75,000 lb)
Max landing weight	29,937 kg (66,000 lb)
Max zero-fuel weight	22,226 kg (49,000 lb)
Max wing loading: G300	369.9 kg/m^2 (75.76 lb/sq ft)
G400	383.2 kg/m^2 (78.49 lb/sq ft)
Max power loading: G300	265 kg/kN (2.60 lb/lb st)
G400	275 kg/kN (2.69 lb/lb st)

PERFORMANCE:

Max operating speed (VMO/MMO)	340 kt (629 km/h; 391 mph) CAS or M0.88
Max cruising speed at FL310	505 kt (936 km/h; 582 mph) or M0.85
Normal cruising speed at FL450	M0.80 (459 kt; 850 km/h; 528 mph)
Approach speed at max landing weight	149 kt (276 km/h; 172 mph)
Stalling speed at max landing weight:	
wheels and flaps up	130 kt (241 km/h; 150 mph)
wheels and flaps down	115 kt (213 km/h; 133 mph)
Max rate of climb at S/L	1,256 m (4,122 ft)/min
Rate of climb at S/L, OEI	314 m (1,030 ft)/min
Initial cruising altitude	12,500 m (41,000 ft)
Max certified altitude	13,715 m (45,000 ft)
Runway PCN	25
T-O run: G300	1,554 m (5,100 ft)
G400	1,384 m (5,450 ft)
FAA balanced T-O field length at S/L	1,662 m (5,450 ft)
Landing run	973 m (3,190 ft)
Range: with eight passengers and NBAA IFR reserves:	
G300	3,600 n miles (6,667 km; 4,142 miles)
G400	4,100 n miles (7,593 km; 4,718 miles)

OPERATIONAL NOISE LEVELS (FAR Pt 36):

T-O	77.5 EPNdB
Approach	92.0 EPNdB
Sideline	86.6 EPNdB

UPDATED

GULFSTREAM G450

Proposed US military designation: RC-20

TYPE: Long-range business jet.

PROGRAMME: Development, then as GIV-X and unannounced, started 2001; first of four development aircraft (c/n 4001/N401SR) flew 30 April 2003, followed by c/n 4002/N442SR on 21 June, c/n 2003/N403SR on 22 June and c/n 4004/N450GA on 18 September; formal announcement and public debut (N450GA) at NBAA Convention in Orlando, Florida 6 October 2003; more than 400 hours of flight testing in 117 sorties completed by early 8 December 2003; FAA certification scheduled for third quarter 2004, followed by EASA validation in fourth quarter, and first customer deliveries in second quarter 2005. Selected 7 October 2003 as aerial platform for Northrop Grumman's submission for the US Army's Aerial Common Sensor (ACS) programme, with requirement for up to 38 aircraft and potential US Navy requirement for up to 20 aircraft.

COSTS: US$33 million (2003).

DESIGN FEATURES: Combines wing and tail unit of G400 with a fuselage stretched by 0.30 m (1 ft 0 in), cabin door repositioned, G550 flight deck and nose section, and Rolls-Royce Tay Mk 611-8C turbofans housed in G550-type nacelles.

POWER PLANT: Two Rolls-Royce Tay Mk 611-8C turbofans, each flat rated at 61.6 kN (13,850 lb st), with FADEC. Thrust reversers standard.

SYSTEMS: Honeywell 36-150GIV APU, with air-start and run capability up to 11,280 m (37,000 ft). Pressurisation system maintains a 1,830 m (6,000 ft) cabin altitude to 12,500 m (41,000 ft).

AVIONICS: Honeywell Primus Epic PlaneView cockpit, as described for G550, with HUD and EVS standard.

DIMENSIONS, EXTERNAL:

Wing span	23.72 m (77 ft 10 in)
Length overall	27.23 m (89 ft 4 in)
Height	7.67 m (25 ft 2 in)

Gulfstream G450 has a longer fuselage and repositioned door, compared to G400 ***NEW***/0567044

Gulfstream G450 fourth prototype *(Paul Jackson)* ***NEW***/1047835

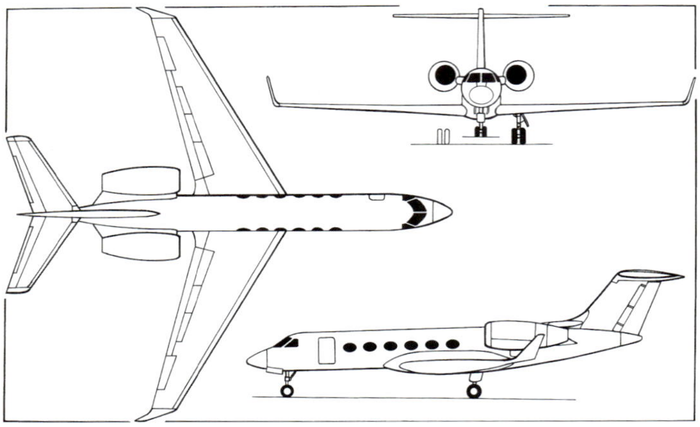

Gulfstream G350/G450 business jet *(James Goulding)* ***NEW***/1047808

DIMENSIONS, INTERNAL:
- Cabin: Length ... 13.74 m (45 ft 1 in)
  - Max width ... 2.23 m (7 ft 4 in)
  - Max height ... 1.88 m (6 ft 2 in)
  - Volume ... 43.2 m^3 (1,525 cu ft)
- Baggage compartment volume ... 4.8 m^3 (169 cu ft)

WEIGHTS AND LOADINGS:
- Basic operating weight ... 19,504 kg (43,000 lb)
- Payload: max ... 2,722 kg (6,000 lb)
  - with max fuel ... 816 kg (1,800 lb)
- Max fuel weight ... 13,381 kg (29,500 lb)
- Max T-O weight ... 33,520 kg (73,900 lb)
- Max landing weight ... 29,937 kg (66,000 lb)
- Max zero-fuel weight ... 22,226 kg (49,000 lb)
- Max wing loading ... 379.6 kg/m^2 (77.76 lb/sq ft)
- Max power loading ... 272 kg/kN (2.67 lb/lb st)

PERFORMANCE:
- Max operating Mach No. (MMO) ... M0.88
- Normal cruising speed ... M0.80 (459 kt; 850 km/h; 528 mph)
- Time to climb to FL410 ... 23 min
- Initial cruising altitude ... 12,500 m (41,000 ft)
- Maximum certified altitude ... 13,716 m (45,000 ft)
- T-O run ... 1,661 m (5,450 ft)
- Landing run ... 972 m (3,190 ft)
- Range with eight passengers at Mach 0.80, NBAA IFR reserves ... 4,350 n miles (8,056 km; 5,006 miles)

UPDATED

GULFSTREAM G500 AND G550

US military designation: C-37

TYPE: Long-range business jet.

PROGRAMME: Study announced at NBAA Convention, Houston, in October 1991. Go-ahead commitment and engine selection (BR710) announced at Farnborough Air Show in September 1992. Risk-sharing agreement with wing designers/manufacturers (Vought and ShinMaywa) announced at Paris Air Show in June 1993, and with tail (and later floor panel) manufacturer (Fokker) at NBAA Convention in September 1993. Initially named Gulfstream V.

Prototype (N501GV) rolled out 22 September 1995; first flight 28 November 1995; second aircraft, c/n 502 (N502GV) was structural test article before completion as company demonstrator; c/n 503 (N503GV), first flown 10 March 1996, used for systems testing; c/n 504 (N504GV), first flown May 1996, for engine, flight loads and environmental trials and JAA certification testing; c/n 505 (N505GV), first flown August 1996, outfitted in standard production configuration for operational testing and HIRF evaluation. Public debut (N502GV) at NBAA Convention at Orlando, Florida, in November 1996.

Gulfstream G550 company demonstrator, N550GA *(Paul Jackson)* ***NEW***/1047821

Provisional FAA certification achieved 16 December 1996 after more than 1,100 hours of flight testing in 550 sorties. Full FAA type certification (extension of Gulfstream II) granted on 11 April 1997, and FAA production certificate awarded on 11 June. First fully completed aircraft for a customer (c/n 507) delivered to Walter Annenberg, former US Ambassador to UK, on 1 July 1997. RVSM approval January 2000. JAA certification eventually granted on 31 October 2002.

From 2003, original versions discontinued and replaced by Gulfstream G500 and G550. Latter to full specification; former is baseline version with customer-specified additions only.

By 17 November 1997 company demonstrators had achieved 39 world records in time to climb and maximum altitude with payload, and city pair categories, including the first ever non-stop flights by a business jet between Los Angeles and London, London and Hong Kong, Tokyo and New York, and Washington, DC and Dubai; last-named 6,330 n mile (11,723 km; 7,284 mile) journey completed on 13 November 1997 in 12 hours 40 minutes 48 seconds with four crew and seven passengers. On 19 February 1998, N502GV became first business jet to fly non-stop from New York to Hawaii. On 4 March 2002 a Gulfstream V broke a 44-year-old record for speed over a recognised course, flying from Tokyo to Washington, DC, in 11 hours 54 minutes covering the 5,543 n miles (10,845 km; 6,739 miles) at an average speed of 492 kt (911 km/h; 556 mph). This record was previously held by a USAF KC-135 Stratotanker, which flew Tokyo to Washington, DC, in 13 hours 46 minutes in April 1958. By March 2002, the Gulfstream V held 70 world and national records. By that time the in-service fleet of 140 aircraft had flown more than 120,000 hours, with 99 per cent despatch reliability.

CURRENT VERSIONS: **Gulfstream V:** *As described.*

Gulfstream V-SP: Announced on eve of NBAA Convention, 8 October 2000; larger cabin, enhanced performance and better range by reason of aerodynamic refinement of existing V. First flight of GV-SP test article (N5SP; modified from 132nd GV) 31 August 2001, followed by first production GV-SP (c/n 5001, also N5SP) rolled out 19 June 2002 and first flown 18 July 2002, at which time the test article airframe had completed 100 flights totalling 185 hours. Redesigned as G550 (which see).

Gulfstream G500: Reduced-range version of Gulfstream V, announced on the eve of the NBAA Convention at Orlando, Florida, 8 September 2002; provisional type certificate granted by FAA 11 December 2002; followed by type certification 14 August 2003.

Gulfstream G550: New version of Gulfstream V-SP, announced on the eve of the NBAA Convention at Orlando, Florida, 8 September 2002; FAA certification 8 December 2003. First aircraft N702GA (c/n 5001); initial public appearance at EBACE, Geneva, 7 to 9 May 2003 by N812GA (c/n 5012).

Flight deck uses Honeywell Primus Epic suite with Gulfstream PlaneView cockpit, comprising four 360 mm (14 in) LCDs, including HUD and EVS, latter was certified in August 2001 and first became operational, on a USAF C-37A, on 15 May 2002. First customer is Executive Jets, which ordered 20 for delivery up to 2008, at cost of US$800 million; unit cost with most popular options US$24.9 million. Other customers include GATX Leasing (two).

Special Missions: Lockheed Martin, teamed with (then) GEC-Marconi Defence, Logica, Marshall Aerospace, MSI, Racal and CAE Electronics Montréal , chose the Gulfstream V as its platform for the UK's Airborne Stand-Off Radar (ASTOR) requirement, involving a Racal dual-mode, electronically scanned surveillance radar in a ventral fairing, with satcom antennas above and below the fuselage for real-time data transfer to ground stations, and provision for in-flight refuelling.

Additionally, when an upgraded version of the Joint STARS system was readmitted to the ASTOR competition in January 1998, the Gulfstream V was selected by Northrop Grumman as platform in preference to Boeing E-8 (707). Neither application was successful. Currently competing for NATO Ground Surveillance System programme.

Israeli Ministry of Defence ordered four G550s, plus two options, on 28 August for use as Compact Airborne early warning (CAEW) platforms, with deliveries beginning in 2005.

Gulfstream will supply airframes with interior and exterior modifications for Elta Phalcon phased array radar and other mission equipment installation and system integration to be carried out by Israel Aircraft Industries.

C-37A: Military version; two (plus four options, one of which converted to firm order in 1998, another in April 1999 and a third subsequently, with fourth converted in 2001) ordered by USAF as part of VCX requirement to replace Boeing VC-137s of 89th Airlift Wing, Andrews AFB, Maryland. Announced 5 May 1997. First (97-0400) delivered 14 October 1998; second in January 1999 and third (for C-in-C USAF) 21 February 2000. US Army's Priority Air Transport Squadron operates one example. Five C-37As leased by US Air Force for delivery between August 2001 and September 2003, first aircraft handed over 28 August 2001. Further contract awarded 13 March 2002 for up to 20 for delivery between 2002 and 2012; first firm order under this contract is for one C-37A for the USAF. US Navy requires five VC-37As as replacements for VP-3A Orions; deliveries began 2002 and extend to 2009. In December 2000 US Coast Guard ordered a C-37A which was delivered in May 2002.

EC-37A: Proposed airborne combat support aircraft; Gulfstream announced at RIAT 2000, Cottesmore, UK, that it was looking for funding during late 2000 and partner to develop systems. Aimed at two roles: standoff electronic jammer and intelligence gathering, using interchangeable underwing and underfuselage pods and in-flight workstation reconfiguration.

RQ-37A: Designation ascribed to Gulfstream candidate as surrogate UAV in US Navy Broad Area Maritime Surveillance (BAMS) competition; Gulfstream V N916GA undergoing conversion as testbed in mid-2003 to be capable of remote control from ground, though carrying an onboard safety pilot.

CUSTOMERS: Three produced in 1996, 29 in 1997, 25 in 1998, 31 in 1999, 34 in 2000, 35 in 2001 and 16 in the first six months of 2002. Executive Jet International took options on two aircraft in January 1995, for delivery to its Gulfstream Shares fleet in 1998–99. Recent customers include Nigerian government; Kuwait Airways, (three, first aircraft delivered November 1999); Brunei government (one); Time Warner (two); Chrysler Corporation (two) and Executive Jet International (10, for delivery by 2004, plus 12 options); Saudi Arabian Ministry of Defence and Aviation (first of two in medevac role, handed over 15 May 2000), and the US National Center for Atmospheric Research, which ordered one in January 2002 for service entry in 2005 as a high-performance instrumented airborne platform for environmental research (HIAPER). Japanese coast guard ordered two Gulfstream Vs on 14 November 2001 at cost of approximately US$100 million.

Gulfstream delivered 100th aircraft in April 2000. Manufacture of Gulfstream V completed in 2003 with 193 aircraft (c/n 501 to 665, 667 to 693, and 699); G500/G550 followed immediately.

COSTS: US$29.5 million (fixed price) for first 24 aircraft, then US$30.5 million (fixed price) up to 39th aircraft; typically equipped price after outfitting, US$35 million; aircraft sold on Internet during December 1999 in deal valued at US$40 million. Cost of EVS system estimated at US$1 million.

Contract for one C-37A valued at US$43.4 million (2002).

DESIGN FEATURES: Gulfstream IV fuselage re-engineered to increase length by 2.13 m (7 ft 0 in); larger wing of same basic shape and interior structure, but 10 per cent more efficient than Gulfstream IV's; larger vertical and horizontal tail surfaces; flight deck volume increased by moving bulkhead 0.30 m (1 ft) aft to provide more space for pilots and to accommodate full-size jump seat; cockpit layout and instrumentation generally similar to Gulfstream IV-SP, but redesigned to incorporate human engineering changes in system control functions; airstair door moved aft by 1.52 m (5 ft); avionics bay relocated. Computational fluid dynamics and CATIA design system used extensively in development.

FLYING CONTROLS: As G300/G400.

LANDING GEAR: As G300/G400. Turning circle about wingtip 17.07 m (56 ft 0 in); about nosewheel 14.15 m (46 ft 5 in).

POWER PLANT: Two 68.4 kN (15,385 lb st) Rolls-Royce Deutschland BR710-48 turbofans with FADEC. Fuel capacity of G550 23,417 litres (6,186 US gallons; 5,151 Imp gallons) in integral wing tanks, of which 22,993 litres 6,074 US gallons; 5,058 Imp gallons) are usable. G500 capacity 19,957 litres (5,272 US gallons; 4,390 Imp gallons) usable.

ACCOMMODATION: Crew of two/three plus cabin attendant. Standard seating for 15 to 19 passengers in pressurised and air conditioned cabin. Rear windows, each side, are emergency exits. Customised interiors according to requirements.

SYSTEMS: Digitally controlled automatic cabin pressurisation system; maximum pressure differential 0.703 bar (10.2 lb/sq in). Hamilton Sundstrand electrical power generating system profile integrated with the flight management system (FMS), and will maintain equivalent of 1,830 m (6,000 ft). Honeywell RE220 APU, designed specifically for Gulfstream V, provides engine-starting capability up to 13,110 m (43,000 ft), 40 kVA of electrical power for ground and flight use up to 13,715 m (45,000 ft), and ground air conditioning, with almost twice the cooling airflow rate of the Gulfstream IV's APU.

AVIONICS: Honeywell SPZ-8500 as core system.

Flight: Three independent IRS integrated into FMS; GPS. Honeywell enhanced ground proximity warning system (EGPWS); TCAS; turbulence-detecting Doppler radar; Hamilton Sundstrand maintenance data acquisition unit; optional Northstar Technologies CT-1000 flight deck organiser system.

Instrumentation: Honeywell SPZ-8500 digital AFCS/FMS with six 20.3 × 20.3 cm (8 × 8 in) colour LCD EFIS displays with EICAS; Honeywell/BAE Model 2020 HUD. Kollsman All Weather Window IR sensor flight testing began September 1999 to provide enhanced vision system (EVS) in conjunction with Honeywell 2020 HUD, facilitating operations in Cat. III weather on Cat. I runway, with decision height of 30 m (100 ft) and RVR of 220 m (722 ft). IR sensor flight is mounted beneath radome. Certification was due early 2001.

Mission: IBM satellite-based international communications system, including airborne voice, data, networking, fax and teleconferencing facilities, provides Gulfstream V with 'office in the sky' capability; optional Sanders AN/ALQ-204 Matador IRCM system.

Gulfstream G550 typical interior ***NEW***/1047820

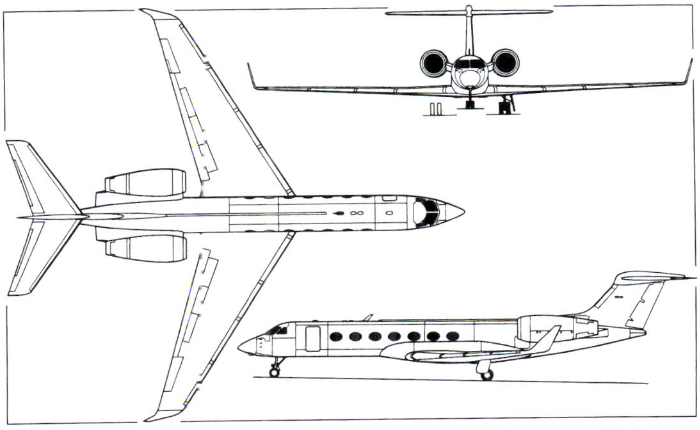

Gulfstream G550 long-range business transport ***(Paul Jackson)*** ***NEW***/1047807

Data applicable to both versions, except where stated.

DIMENSIONS, EXTERNAL:
Wing span: basic 27.69 m (90 ft 10 in)
over winglets 28.50 m (93 ft 6 in)
Length overall 29.39 m (96 ft 5 in)
Height overall 7.87 m (25 ft 10 in)
Tailplane span 10.72 m (35 ft 2 in)
Wheel track (c/l shock-absorbers) 4.37 m (14 ft 4 in)
Wheelbase 13.72 m (45 ft 0 in)

DIMENSIONS, INTERNAL:
Cabin: Length, aft of flight deck 15.57 m (50 ft 1 in)
Max width 2.24 m (7 ft 4 in)
Max height 1.88 m (6 ft 2 in)
Volume 47.3 m^3 (1,669 cu ft)
Baggage compartment volume 6.4 m^3 (226 cu ft)

AREAS:
Wings, gross 105.63 m^2 (1,137.0 sq ft)

WEIGHTS AND LOADINGS:
Weight empty, 'green' 17,917 kg (39,500 lb)
Operating weight empty, incl crew:
G500 21,682 kg (47,800 lb)
G550 21,909 kg (48,300 lb)
Allowance for outfitting 3,856 kg (8,500 lb)
Baggage capacity 1,134 kg (2,500 lb)
Payload: max: G500 3,039 kg (6,700 lb)
G550 2,812 kg (6,200 lb)
with max fuel: G500 1,134 kg (2,500 lb)
G550 816 kg (1,800 lb)
Fuel weight: max: G550 18,819 kg (41,489 lb)
usable: G500 15,966 kg (35,200 lb)
G550 18,733 kg (41,300 lb)
Payload with max fuel: G500 1,134 kg (2,500 lb)
G550 816 kg (1,800 lb)
Max T-O weight: G500 38,600 kg (85,100 lb)
G550 41,277 kg (91,000 lb)
Max landing weight 34,155 kg (75,300 lb)
Max wing loading: G500 365.4 kg/m^2 (74.85 lb/sq ft)
G550 390.8 kg/m^2 (80.04 lb/sq ft)
Max power loading: G500 282 kg/kN (2.77 lb/lb st)
G550 302 kg/kN (2.96 lb/lb st)

PERFORMANCE (at max T-O weight, except where indicated):
Max operating Mach No. (VMO) M0.885
Cruising speed: max 499 kt (924 km/h; 574 mph)
normal 488 kt (904 km/h; 562 mph) or M0.85
long-range 459 kt (850 km/h; 528 mph) or M0.80
Initial cruising altitude: G500 13,105 m (43,000 ft)
G550 12,500 m (41,000 ft)
Max certified altitude 15,545 m (51,000 ft)
T-O run: G500 1,570 m (5,150 ft)
G550 1,801 m (5,910 ft)
Landing run 844 m (2,770 ft)
T-O balanced field length 1,862 m (6,110 ft)
Landing distance, S/L, max landing weight 841 m (2,760 ft)
Range, eight passengers and four crew, M0.80, NBAA IFR reserves:
G500 5,800 n miles (10,741 km; 6,774 miles)
G550 6,750 n miles (12,501 km; 7,767 miles)

UPDATED

GULFSTREAM SBJ

TYPE: Supersonic business jet.

PROGRAMME: Gulfstream Aerospace and Lockheed Martin (which see) revealed at Farnborough on 7 September 1998 that they were jointly conducting an 18 to 24 month feasibility study into an SBJ. As then envisaged, the aircraft, similar in size to the Gulfstream II, would have featured a stand-up-headroom cabin accommodating eight passengers, and would cruise at M1.6 to M2.0 over a range of more than 4,000 n miles (7,408 km; 4,603 miles). Key design goals were the ability to operate out of existing business aviation airfields; take-off noise compatible with anticipated future emissions regulations; fuel-efficient operation at subsonic speeds; and an initial cruising altitude above that of subsonic traffic.

The feasibility phase of the project, completed in mid-2000, was followed by a further two years of wind-tunnel testing and project definition; studies carried out by the company confirmed that there is a market, but sonic boom suppression, engine emissions and noise present technological challenges. Gulfstream and Lockheed Martin, through its Advanced Development Projects division, defined the SBJ, and lobbied the US government for support via the Defense Advanced Research Projects Agency which received US$15 million for its quiet supersonic aircraft technology (QSAT) budget. However, Lockheed Martin has since withdrawn from the programme. In 2000, patents were filed with different layouts including tailplane mounted at tail fintip and drooping to attach to engine nacelles; further patents included one with large delta wing shape plus nose canard; in October Gulfstream displayed, at NBAA, the Quiet Supersonic Jet (QSJ) which had swept wing and wingtip-mounted swept tail.

By late 2003, the QSJ had evolved into a variable-geometry wing, T-tail aircraft with two pod-mounted engines on the rear fuselage, each side of the fin. Design goals include maximum take-off weight of 45,359 kg (100,000 lb), cruising speed of Mach 1.8, take-off field length of 1,981 m (6,500 ft), range of 4,800 n miles (8,889 km; 5,523 miles) and airport noise level 10 EPNdB quieter than Stage 4 requirements.

Estimated unit cost is in the region of US$70 million to US$80 million (2002), based on 200 produced. The proposed aircraft is not related to the abortive Gulfstream-Sukhoi SSBJ project from which Gulfstream withdrew in 1992. Executive Jet Aviation, which operates the NetJets fractional ownership programme, is reportedly interested in the concept of supersonic business jets.

VERIFIED

HAWKER

RAYTHEON AIRCRAFT COMPANY

At the September 2002 NBAA Convention, held in Orlando, Florida, Raytheon Aircraft (which see for further details) announced its intention to revert to separate marketing of its Beech and Hawker lines of private and executive aircraft, reducing the emphasis on its corporate name.

Raytheon had acquired British Aerospace's Corporate Jets division for US$372 million on 6 August 1993; founded Raytheon Corporate Jets Inc at Little Rock, Arkansas, with responsibility for design, development, production and support of renamed Hawker family of corporate jets. Hawker name derived by Raytheon from Hawker Siddeley, parent company of de Havilland at conception of DH/HS/BAe 125 twin-jet, which forms basis of the current Hawker line. Raytheon Corporate Jets included in Raytheon Aircraft Company upon its foundation on 15 September 1994. Former Beechjet 400A added to Hawker line as Hawker 400XP in May 2003.

UPDATED

Hawker 400XP business twin-jet ***NEW***/0583519

HAWKER 400

US Air Force designation: T-1A Jayhawk
JASDF designation: T-400

TYPE: Business jet.

PROGRAMME: Conceived as Mitsubishi MU-300 Diamond; first flight 29 August 1978; two prototypes; FAR Pt 25 certification awarded 6 November 1981; production aircraft fabricated in Japan and assembled at San Angelo, Texas; deliveries totalled 63 Diamond Is (JT15D-4 engines), 27 Diamond IAs (JT15D-4D) and one Diamond II (JT15D-5).

Beech acquired rights to Diamond II from Mitsubishi Heavy Industries and Mitsubishi Aircraft International, December 1985; made improvements to aircraft and renamed it Beechjet 400. First Beech-assembled Beechjet rolled out 19 May 1986; initial 64 used Japanese components. During 1989, Beech moved entire manufacturing operation to Wichita. Announced new Beechjet 400A November 1989, featuring certification to 13,715 m (45,000 ft), larger and more comfortable cabin, all Collins avionics with digital EFIS; customer deliveries began November 1990. Transferred to Raytheon's Hawker marque in May 2003.

CURRENT VERSIONS: **Beechjet 400:** Initial production version (65 built; superseded by 400A.

Beechjet 400A: Announced at 1989 NBAA show; production 400A first flight 22 September 1989; FAA certification received 20 June 1990; deliveries began November 1990. Also certified by July 1993 in Australia, Canada, France, Germany, Italy and UK; Brazilian and Pakistani type approval April 1994; Civil Aviation Authority of China certification achieved in second quarter of 1999; superseded by.Hawker 400XP.

Hawker 400XP: Current version (XP=eXtra Payload), announced at European Business Aviation Convention and Exhibition at Geneva 6 May 2003; first aircraft thus named was N400XP (c/n RK-356). Features 90.7 kg (200 lb increase in payload and thrust reversers, vapour-cycle air-conditioning, TCAS II and electronic ELT as standard; FAA certification granted in April 2003.

Description applies to Hawker 400XP, except where indicated.

Beechjet T-1A Jayhawk: US Air Force selected McDonnell Douglas, Beech and Quintron to supply Tanker Transport Training System (TTTS) on 21 February 1990, including requirement for 180 Beechjet 400Ts, valued at US$755 million and designated T-1A Jayhawk; represents missionised version of 400A, sharing many components and characteristics with commercial counterpart; differences include cabin-mounted avionics, increased air conditioning capability, greater fuel capacity with single-point refuelling, and strengthened windscreen and leading-edges for low-level birdstrike protection. First production aircraft (90-0400) delivered 17 January 1992; deliveries at approximately three per month; final delivery 23 July 1997. By then total fleet time exceeded 182,000 flying hours, with 90 per cent operational availability, and more than 680 pilots had been trained on the Jayhawk.

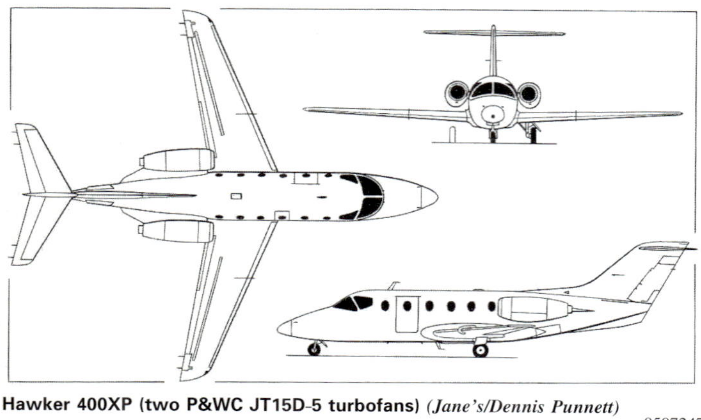

Hawker 400XP (two P&WC JT15D-5 turbofans) *(Jane's/Dennis Punnett)* 0507247

IOC for USAF Jayhawks January 1993, for Air Education and Training Command Specialised Undergraduate Pilot Training (SUPT) programme at Reese AFB (52nd FTS/64th FTW) where establishment of 41 received by October 1993; Reese closed in 1997. Second recipient was 99th FTS/12th FTW at Randolph AFB, Texas, where 16 delivered for instructor training in 1993; third unit was 86th FTS/47th FTW at Laughlin AFB, Texas, from late 1993 with training courses beginning May 1994; fourth was 71st FTW at Vance AFB, Oklahoma (first aircraft December 1994); fifth was 14th FTW at Columbus AFB, Mississippi (early 1996). T-1A used for training crews for KC-10, KC-135, C-5 and C-17, with total fleet experience of more than 376,000 hours and more than 733,000 landings by October 1999.

In February 1997 Raytheon Aircraft and its subsidiary Raytheon Aerospace were awarded a contract, valued at US$6.2 million, to retrofit 62 Jayhawks with GPS; two further options to retrofit the entire fleet would bring the total value of the GPS upgrade to about US$25.3 million.

Beechjet 400T: JASDF T-400 version, featuring thrust reversers, long-range inertial navigation and direction-finding systems; interior changes. Meets TC-X trainer requirement; three, three, two and one ordered in 1992–95, plus one in 1998; first (41-5051) delivered 31 January 1994; 10th in 2000.

CUSTOMERS: Refer to deliveries table. Total 19 new aircraft produced in 2003 (compared to 24 billings), and nine in the first six months of 2004. Total of 29 were scheduled for delivery in 2004. Recent customers include Hainan Airlines of China, which took delivery of one aircraft in February 1999, and NetJets Inc, which ordered 50 in December 2003 for delivery between 2004 and 2009, and a further 20 ordered in June 2004 for delivery in 2005–07 to NetJets Europe. US Air Force 180 T-1As ordered, of which delivery completed 23 July 1997. JASDF had received 10 by December 1999 and ordered further two in 2000; operated by 41 Hikotai at Miho. By May 2004, 376 civil 400As had been registered, in addition to 65 Model 400s, 180 Jayhawks, 13 400Ts and 93 Diamonds, or 726 in all.

COSTS: Jayhawk programme cost US$1.3 billion; Beech contracts for 180 aircraft, US$755 million. Hawker 400XP US$6.741,890 (2004).

DESIGN FEATURES: Typical low-wing, T tail, rear-engined small business jet, with sweptback wings and empennage, plus small underfin. Compared to Diamond, Beechjet has increased payload and certified ceiling, greater cabin volume achieved by moving rear-fuselage fuel tank forward under floor (balanced by moving lavatory to rear of cabin), improved soundproofing, and emergency door moved one window forward to facilitate forward club seating.

Wing has computer-designed three-dimensional Mitsubishi MAC510 aerofoil; thickness/chord ratio 13.2 per cent at root, 11.3 per cent at tip; dihedral 2° 30'; incidence 3° at root, –3° 30' at tip; sweepback 20° at quarter-chord.

T-1A Jayhawk features include student pilot in left seat, instructor on right and pupil/observer behind instructor; more bird-resistant windscreen and leading-edges; fewer cabin windows; strengthened wing carry-through structure and engine attachment points to meet low-level flight stresses; rails for four passenger seats in cabin for personnel transport; avionics relocated from nose to rack in cabin to facilitate nose installation of air conditioning; emergency door moved forward to position opposite main cabin door to allow straight-through egress; improved brakes; additional fuel tank; single-point pressure refuelling; Rockwell Collins five-tube EFIS; digital autopilot; weather radar; central diagnostic and maintenance system; Tacan with air-to-air capability.

FLYING CONTROLS: Conventional and manual. Variable incidence tailplane and elevators for pitch axis; lateral control by small ailerons and almost full semi-span, narrow chord spoilers used also as airbrakes and lift dumpers; rudder with trim tab; narrow chord Fowler-type flaps, double-slotted inboard and single-slotted outboard, occupy most of trailing-edges and are hydraulically actuated; mid-span leading-edge fences on wing; small horizontal strakes on fuselage at base of fin; small ventral fin.

BEECH AND HAWKER 400 CIVIL PRODUCTION

	400	400A	400XP
1986	16		
1987	18		
1988	19		
1989	12		
1990		1	
1991		30	
1992		31	
1993		20	
1994		14	
1995		10	
1996		26	
1997		38	
1998		41	
1999		49	
2000		41	
2001		27	
2002		16	
2003		11	8
Totals	**65**	**355**	**8**

Notes: Not including 180 USAF T-1A Jayhawks (1991–97) and 13 JASDF 400Ts. Further aircraft built by Mitsubishi in Japan.

STRUCTURE: Wings include integrally machined metal upper and lower skins joined to two box spars forming integral fuel tank; tailplane and fin similar. Wing, fuselage and tail unit certified fail-safe for unlimited life (with periodic inspections and maintenance).

LANDING GEAR: Retractable tricycle type, with single wheel and oleo-pneumatic shock-absorber on each unit. Hydraulic actuation, controlled electrically. Emergency free-fall extension. Main tyres 24×7.7 (16 ply) tubeless; nose tyre 18×4.4 (10 ply) tubeless. Nosewheel, which is steerable by rudder pedals, retracts forward; mainwheels retract inward into fuselage. Goodyear wheels and tyres; Aircraft Braking Systems brakes.

POWER PLANT: Two Pratt & Whitney Canada JT15D-5 turbofans, each rated at 13.19 kN (2,965 lb st) for take-off. Nordam thrust reversers standard on 400XP, but not fitted to T-1A. Total usable fuel capacity 400XP 2,775 litres (733 US gallons; 610 Imp gallons); 400T 2,998 litres (792 US gallons; 656 Imp gallons). One refuelling point in top of each wing, and one in rear fuselage for fuselage tank, capacity 1,158 litres (306 US gallons; 255 Imp gallons). (T-1A, single-point refuelling.) Oil capacity 7.7 litres (2.0 US gallons; 1.7 Imp gallons).

ACCOMMODATION: Crew of two on flight deck of 400XP on vertically and horizontally adjustable reclining seats with five-point safety harnesses; T-1A has seats for trainee pilot, co-pilot/instructor and observer. Improved interior introduced 1996, featuring redesigned trim panels, enhanced acoustic panels and vibration-damping engine mounts.

Standard 'centre club' layout of 400A seats eight passengers in pressurised cabin. Of these, seven are on tracking, 360° swivelling, reclining seats: four in facing pairs, two forward-facing and one aft-facing; each with integral headrest, armrest and shoulder harness. Fold-out writing table between each pair of seats. Private flushing lavatory at rear with sliding doors and optional illuminated vanity unit and hot water supply. With seat belts, this compartment can serve as eighth passenger seat.

Interior options for up to nine passengers; these include substitution of carry-on baggage compartment for one of the forward centre seats, and hot and cold service refreshment centre with integral stereo entertainment system. Independent temperature control for flight deck and cabin heating systems standard. In-flight telephone optional. Tailcone baggage compartment with external access. Optional four passenger seats in main cabin of T-1A. The 400T has an aft club arrangement with swivel chairs.

SYSTEMS: Pressurisation system, with normal differential of 0.63 bar (9.1 lb/sq in) maintaining sea level cabin environment to 7,315 m (24,000 ft) and 2,286 m (7,500 ft) cabin environment to 13,715 m (45,000 ft). Back-up pressurisation system, using engine bleed air, for use in emergency. Hydraulic system, pressure 103.5 bar (1,500 lb/sq in), for actuation of flaps, landing gear and other services. Each variable volume output engine-driven pump has a maximum flow rate of 14.76 litres (3.9 US gallons; 3.25 Imp gallons)/min, and one pump can actuate all hydraulic systems. Reservoirs, capacity 4.16 litres (1.1 US gallons; 0.9 Imp gallon), pressurised by filtered engine bleed air at 1.03 bar (15 lb/sq in). All systems are, wherever possible, of modular conception: for example, entire hydraulic installation can be removed as a single unit. Stick shaker as back-up stall warning device.

AVIONICS: *Flight:* GPS retrofitted to some T-1As.

Instrumentation: Standard avionics include pilot's integrated Rockwell Collins Pro Line 4 EFIS featuring three-tube (optional four-tube) colour CRT primary flight display (PFD) and multifunction display (MFD) units mounted side by side, and control/display unit. PFD displays airspeed, altitude, vertical speed, flight director, attitude and horizontal situation information, while MFD displays navigation, radar, map, checklist and fault annunciation information. Smaller, single or dual CRTs mounted on central console function as independent navigation sensor displays or back-up displays for main CRTs. EFIS installation features strapdown attitude/heading referencing system, electronic map navigation display, airspeed trend information and V-speeds on Mach airspeed display, TCAS II, ELT and solid-state Doppler turbulence detection radar.

DIMENSIONS, EXTERNAL:
- Wing span .... 13.25 m (43 ft 6 in)
- Wing aspect ratio 7.8
- Length overall .... 14.75 m (48 ft 5 in)
- Fuselage: Length .... 13.15 m (43 ft 2 in)
  - Max width .... 1.68 m (5 ft 6 in)
  - Max depth .... 1.85 m (6 ft 1 in)
- Height overall .... 4.24 m (13 ft 11 in)
- Tailplane span .... 5.00 m (16 ft 5 in)
- Wheel track .... 2.84 m (9 ft 4 in)
- Wheelbase .... 5.86 m (19 ft 3 in)
- Crew/passenger door: Height .... 1.27 m (4 ft 2 in)
  - Width .... 0.71 m (2 ft 4 in)

DIMENSIONS, INTERNAL:
- Cabin:
  - Length: incl flight deck .... 6.32 m (20 ft 9 in)
    - excl flight deck .... 4.72 m (15 ft 6 in)
  - Max width .... 1.50 m (4 ft 11 in)
  - Max height .... 1.45 m (4 ft 9 in)
  - Volume: incl flight deck .... 11.3 m³ (400 cu ft)
    - excl flight deck .... 8.6 m³ (305 cu ft)
- Baggage compartment volume:
  - cabin .... 0.57 m³ (20.0 cu ft)
  - tailcone .... 0.75 m³ (26.4 cu ft)

AREAS:
- Wings, net .... 22.43 m² (241.4 sq ft)
- Trailing-edge flaps (total) .... 4.22 m² (45.40 sq ft)
- Spoilers (total) .... 0.57 m² (6.20 sq ft)
- Fin, incl dorsal fin .... 5.91 m² (63.60 sq ft)
- Rudder, incl yaw damper .... 0.99 m² (10.70 sq ft)
- Tailplane .... 5.25 m² (56.50 sq ft)
- Elevators, incl tab .... 1.55 m² (16.70 sq ft)

WEIGHTS AND LOADINGS:
- Weight empty .... 4,785 kg (10,550 lb)
- Basic operating weight, incl crew, avionics and interior fittings .... 4,967 kg (10,950 lb)
- Baggage capacity: total .... 431 kg (950 lb)
  - tailcone .... 204 kg (450 lb)
- Max fuel weight .... 2,228 kg (4,912 lb)
- Max T-O weight .... 7,393 kg (16,300 lb)
- Max ramp weight .... 7,484 kg (16,500 lb)
- Max landing weight .... 7,121 kg (15,700 lb)
- Max zero-fuel weight .... 5,896 kg (13,000 lb)
- Fuel with max payload .... 1,588 kg (3,500 lb)
- Payload with max fuel .... 289 kg (638 lb)
- Max wing loading .... 329.67 kg/m² (67.52 lb/sq ft)
- Max power loading .... 280 kg/kN (2.75 lb/lb st)

PERFORMANCE:
- Max limiting Mach No. 0.78
- Max level speed at FL270 .... 468 kt (867 km/h; 539 mph)
- Cruising speed:
  - high, at FL390 .... 450 kt (834 km/h; 518 mph)
  - intermediate, at FL410 .... 420 kt (778 km/h; 483 mph)
  - long-range, at FL 430 .... 414 kt (767 km/h; 476 mph)
- Stalling speed, flaps down, idling power .... 93 kt (173 km/h; 107 mph) CAS
- Max rate of climb at S/L .... 1,149 m (3,770 ft)/min
- Rate of climb at S/L, OEI .... 218 m (715 ft)/min
- Service ceiling .... 13,243 m (43,450 ft)
- Time to FL370 .... 18 min
- Service ceiling, OEI .... 6,279 m (20,600 ft)
- Max certified altitude .... 13,715 m (45,000 ft)
- FAA (FAR Pt 25) T-O field length at S/L, ISA .... 1,190 m (3,906 ft)
- FAA landing distance at max landing weight .... 1,071 m (3,514 ft)
- Range: with 100 n mile (185 km; 115 mile) reserves:
  - with max fuel .... 1,557 n miles (2,883 km; 1,781 miles)
  - with max payload .... 874 n miles (1,618 km; 1,005 miles)
  - ferry .... 1,687 n miles (3,124 km; 1,941 miles)

OPERATIONAL NOISE LEVELS:
- T-O .... 89.0 EPNdB

UPDATED

HAWKER 800

JASDF designation: U-125

TYPE: Business jet.

PROGRAMME: Derived from the de Havilland/Hawker Siddeley/British Aerospace 125, which was built in the UK from 1962 onwards, progressing through Srs 1 to 3 and 400 to 700. Prototype Srs 800 first flew (G-BKTF) 26 May 1983; type certificate gained 4 May 1984 and Public Transport Category C of A on 30 May 1984; FAA certification 7 June 1984; Russian certification May 1993; Canadian certification (800XP) awarded in third quarter of 1997. Adopted Hawker nomenclature when programme purchased by Raytheon in 1993, at which time some 850 aircraft had been sold in 44 countries.

Final assembly of Hawker 800XP gradually transferred to Wichita; first US-assembled aircraft flew on 5 November 1996, being N297XP, the 297th Series 800; second (N1105Z; No. 301) followed on 24 November 1996; transition complete with flight of last UK-assembled aircraft, No. 337, 29 April 1997; FAA production certificate awarded to Raytheon in May 1997. One thousandth 125/Hawker series aircraft, an 800XP, was delivered as the 'Millennium Hawker' to Gainey Corporation of Grand Rapids, Michigan (N984GC), October 1998. Winglets available as option on new aircraft, or as retrofit, from 2003.

CURRENT VERSIONS: **800:** Original version superseded by 800XP in late 1995; 275 built, of which last delivered December 1995.

800XP (XP=eXtended Performance): Announced March 1995, when prototype (G-BVYW, modified from 800) completed; this and preproduction 800XP used in development programme, culminating in CAA and FAA certification in July 1995; first delivery (to Green Tree Financial of St Paul, Minnesota) October 1995 after public debut

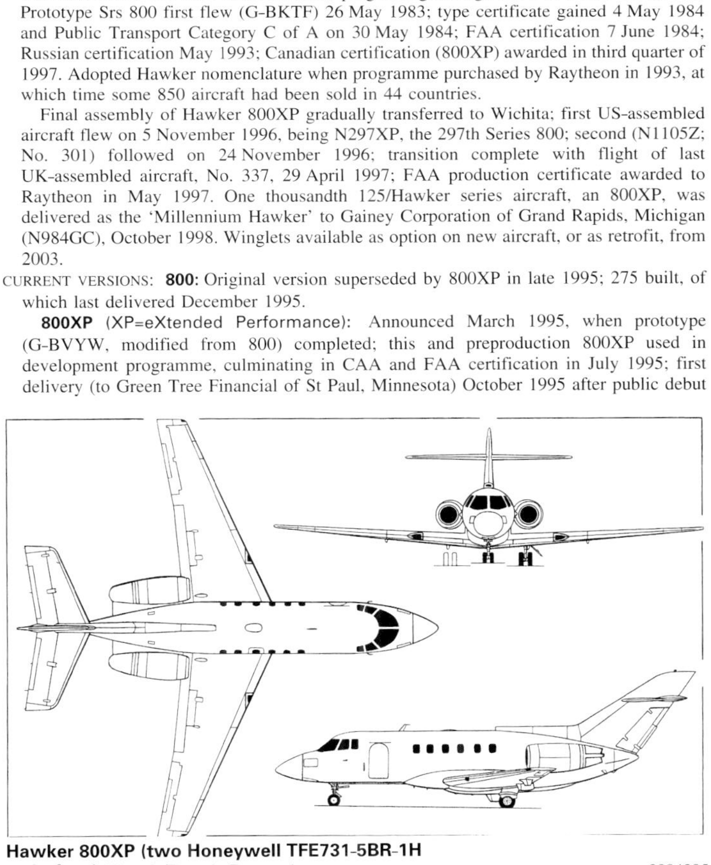

Hawker 800XP (two Honeywell TFE731-5BR-1H turbofans) ***(Jane's/Dennis Punnett)*** 0084005

Hawker 800XP twin-turbofan business jet *(Paul Jackson)* ***NEW***/0583448

at National Business Aviation Association Convention in Las Vegas during previous month. First seven delivered in 1995, followed by 26, 33, 48 and 55 in 1996–99.

Initially fitted with Honeywell SPZ 8000 avionics as standard. Collins Pro Line 21 fitted in two aircraft (c/n 258541 and 258556) in 2001 and standard fit from 2002 (258567 and onwards).

800SP: Winglet modification of 800/800XP devised by Aviation Partners Inc and shown at NBAA Convention, Orlando, Florida, September 2002; available from early 2003. Height 107 cm (3 ft 6 in); 7 per cent consequential drag reduction translates into M0.03 speed increase (18 kt; 33 km/h; 21 mph) and range extension of 180 n miles (333 km; 207 miles).

Detailed description applies to this version.

800FI, SM, RA and SIG: Special missions versions; produced as required.

U-125A: Refer to KAC entry in Japanese section.

CUSTOMERS: Refer to table. By June 2004, 682 Series 800/800XPs registered, including 55 delivered in 2001, 46 in 2002, 45 in 2003, and 20 in the first six months of 2004. Total of 49 scheduled for delivery in 2004. Largest contract for 125/Hawker placed in May 1997 when Executive Jet Inc ordered 20 800XPs, followed in September 1998 by order for a further 20, plus 16 options for NetJets fractional ownership scheme; deliveries between 1997 and 2004; further eight ordered in December 2003 for delivery from that month until 2005, and a further 20 ordered in June 2004 for delivery 2005–07 to NetJets Europe. Recent customers include National Air Service (NAS) of Jeddah, Saudi Arabia, which ordered 14 in November 1999 for its NetJets Middle East fractional ownership programme, five of these being delivered in 2000, including the first (HZ-KSRA) at the Farnborough International Air Show on 24 July 2000, and three per year thereafter until 2003; and Hainan Airlines of China, which took delivery of one aircraft on 19 July 1999. Contender for US Navy Undergraduate Military Flying Officer Training System (UMFOTS) requirement, for which 16 aircraft are required initially.

COSTS: 800XP: US$12.850,600 (2004).

DESIGN FEATURES: Classic small business jet; sweptback wing mounted below cabin floor; podded engines on rear fuselage sides; and high tailplane.

Improvements of baseline Series 800, compared with earlier 700 variant, include curved windscreen, sequenced nosewheel doors, extended fin leading-edge, larger ventral fuel tank, and increased wing span which reduces induced drag, enhances aerodynamic efficiency and carries extra fuel; outboard 3.05 m (10 ft) of each wing redesigned.

HS 125 AND HAWKER 800 PRODUCTION

	125-1	125-3	125-400	125-600	125-700	800	800XP	800XP/PL
1962	2							
1963	2							
1964	8	1						
1965	33							
1966	41	7						
1967	1	21						
1968		26	4					
1969		9	24					
1970			38					
1971		1	25	2				
1972			17	4				
1973			7	22				
1974			1	15				
1975				12				
1976				13				
1977				4	17			
1978				1	34			
1979					28			
1980					40			
1981					35			
1982					33			
1983					17	3		
1984					9	17		
1985					2	28		
1986						26		
1987						32		
1988						29		
1989						26		
1990						40		
1991						11		
1992						15		
1993						17		
1994						10		
1995						19	6	
1996						1	21	
1997						2	45	
1998						3	39	
1999						4	48	
2000						3	52	
2001						2	51	2
2002						2		44
2003						2		43
Totals	**87**	**65**	**116**	**73**	**215**	**292**	**262**	**89**

Notes: Data do not include 20 Dominie T. Mk 1 navigation trainers (125 Srs 2) delivered to the RAF in 1965 (10) and 1966 (10).

In XP version, TFE731-5BR-1H turbofans boost performance, including 14 kt (26 km/h; 16 mph) increase in cruising speed at 11,800 kg (26,015 lb) at 11,890 m (39,000 ft); 225 kg (496 lb) payload increase with eight passengers; 15 to 23 per cent reduction in time to cruising altitude, to reach 11,280 m (37,000 ft) in 23 minutes at maximum take-off weight in ISA + 10°C conditions; and enhanced take-off performance. Other improvements include installation of vortillons in place of wing fences, permitting lower V-speeds and reducing drag; enhanced TKS de-icing system with increased fluid capacity; improved high-energy brakes; restyled cabin interior to maximise use of available volume; and improved environmental control system; and redesigned interior, with increased headroom by relocating oxygen dropout units to sidewall panels, and 12.2 cm (4.8 in) extra width at shoulder level by sculpturing sidewall panels around fuselage frame. Interior further improved in 1998 — see Accommodation paragraph.

Wing thickness/chord ratio 14 per cent at root, 8.35 per cent at tip; dihedral 2°; incidence 2° 5' 42" at root, −3° 5' 49" at tip; sweepback 20° at quarter-chord; small fairings on tailplane undersurface eliminate turbulence around elevator hinge cutouts.

FLYING CONTROLS: Conventional and manual. Each control surface with geared tab; port aileron tab trimmed manually via screw-jack. Hydraulically actuated four-position double-slotted flaps; mechanically operated hydraulic cutout prevents asymmetric flap operation; upper and lower airbrakes, with interconnected controls to prevent asymmetric operation, form part of flap shrouds and provide lift dumping. Fixed incidence tailplane.

STRUCTURE: All-metal. One-piece wings, dished to pass under fuselage and attached by four vertical links, side link and drag spigot; two-spar fail-safe wings, with partial centre spar of approximately two-thirds span, to form integral fuel tankage; single-piece skins on each upper and lower wing semi-spans; detachable leading-edges; fail-safe fuselage structure of mainly circular cross-section, incorporating Redux bonding.

LANDING GEAR: Retractable tricycle type, with twin wheels on each unit. Hydraulic retraction: nosewheels forward, mainwheels inward into wings. Oleo-pneumatic shock-absorbers. Fully castoring nose unit, steerable ±45°. Dunlop mainwheels size 23×7–12 (12 ply) tubeless tyres. Dunlop nosewheels size 18×4.25–10 (6 ply) tubeless tyres. Dunlop triple-disc hydraulic brakes with Maxaret anti-skid units on all mainwheels. Minimum ground turning radius about nosewheel 9.14 m (30 ft 0 in).

POWER PLANT: Two 20.73 kN (4,660 lb st) Honeywell TFE731-5BR-1H turbofans, mounted on sides of rear fuselage in pods designed and manufactured by Northrop Grumman. Thrust reversers developed by Dee Howard fitted as standard. Integral fuel tanks in wings, with combined capacity of 4,818 litres (1,273 US gallons; 1,060 Imp gallons). Rear underfuselage tank of 882 litres (233 US gallons; 194 Imp gallons) capacity, giving total capacity of 5,700 litres (1,506 US gallons; 1,254 Imp gallons). Single pressure refuelling point at rear of ventral tank. Overwing refuelling point near each wingtip.

ACCOMMODATION: Flight deck crew of two. Dual controls standard. Seat for third crew member. Executive layout has forward baggage compartment, forward galley comprising automatic coffee maker, microwave oven, and miscellaneous storage. Seats swivel through 360°. Seating for eight passengers, with club four seating at the front of the cabin, three-place settee on the right side rear cabin and single seat opposite. Airliner style lavatory at rear with external servicing as standard. Maximum seating for 14. Interior options include differing seating layouts; microwave oven; entertainment system including CD player and video LCD screen. New interior introduced at National Business Aviation Association Convention at Las Vegas, Nevada, in October 1998 features oval internal window frames, additional sidewall lighting and restyled side panels and work tables.

SYSTEMS: Honeywell air conditioning and pressurisation system. Maximum cabin differential 0.59 bar (8.60 lb/sq in), maintains sea level cabin environment to 6,767 m (22,200 ft). Oxygen system standard, with dropout masks for passengers. Hydraulic system, pressure 186 to 207 bar (2,700 to 3,000 lb/sq in), for operation of landing gear, mainwheel doors, flaps, spoilers, nosewheel steering, mainwheel brakes and anti-skid units. Two accumulators, pressurised by engine bleed air, one for main system pressure, other providing emergency hydraulic power for wheel brakes in case of main system failure. Independent auxiliary system for lowering landing gear and flaps in event of main system failure.

DC electrical system utilises two 30 V 12 kW engine-driven starter/generators and two 24 V 23 Ah Ni/Cd batteries. A 24 V 4 Ah battery provides separate power for standby instruments. AC electrical system includes two 1.25 kVA static inverters, providing 115 V 400 Hz single-phase supplies, one 250 VA standby static inverter for avionics, and two engine-driven 208 V 7.4 kVA frequency-wild alternators for windscreen anti-icing. Ground power receptacle on starboard side at rear of fuselage for 28 V external DC supply. Honeywell 36-150 W APU. TKS liquid system de-icing/anti-icing on leading-edges of wings and tailplane. Engine ice protection system supplied by engine bleed air. Kidde-Graviner triple FD Firewire fire warning system and two BCF engine fire extinguishers. Stall warning and stick pusher system fitted.

AVIONICS: Collins Pro Line 21 as core system.

Comms: Dual Collins VHF-422C com with 8.33 kHz frequency spacing; dual TFR-94D transponders; Collins HF-9000 HF com with Coltech CSD-714 Selcal; dB Systems Dual Model 700 cockpit audio system; Universal Avionics CVR-120 cockpit voice recorder; and Artex C406-2 satellite-capable ELT, all standard. Second HF-9000; Honeywell Airborne Flight Information System, and Magnastar C2000 terrestrial-based flight phone or Universal Avionics Aero M or Aero I satcom phone, all optional.

Radar: Collins TWR-850 colour weather radar with turbulence detection as standard.

Flight: Dual Collins FMS-6000 flight management systems; dual AHC-3000 quartz AHRS; dual ADC-3000 digital air data systems; FGC-3000 AFCS; dual GPS-4000A; dual VIR-432 VOR/ILS/marker receivers; ADF-462; dual DME-442; ALT-4000 radio altimeter; TCAS-4000 TCAS II (Change 7 compliant); MDC-3000 maintenance diagnostic system; and SafeFlight AoA system, all standard. Second ADF-462, and L3 Communications F1000 FDR or JAR-OPS compliant Honeywell FDR with Teledyne FDAU, optional.

Instrumentation: Four 203 × 254 mm (8 × 10 in) active matrix LCDs for PFD, MFD and EICAS functions; and Meggitt Avionics Mark II secondary flight display system, standard. Options include 3-D mapping for MFDs; Goodrich WX-1000 lightning detection system with display on AMLCDs; Collins CDU-6200 graphics capable display; and Airshow 400 or Airshow Network cabin display systems.

DIMENSIONS, EXTERNAL:

Wing span	15.66 m (51 ft 4½ in)
Wing chord, mean	2.29 m (7 ft 6¼ in)
Wing aspect ratio	7.1
Length overall	15.60 m (51 ft 2 in)
Height overall	5.51 m (18 ft 1 in)
Fuselage max diameter	1.93 m (6 ft 4 in)
Tailplane span	6.10 m (20 ft 0 in)
Wheel track (c/l of shock-absorbers)	2.79 m (9 ft 2 in)
Wheelbase	6.41 m (21 ft 0½ in)
Passenger door (fwd, port): Height	1.30 m (4 ft 3 in)
Width	0.69 m (2 ft 3 in)
Height to sill	1.07 m (3 ft 6 in)

Emergency exit (overwing, stbd):
  Height ... 0.91 m (3 ft 0 in)
  Width ... 0.51 m (1 ft 8 in)

DIMENSIONS, INTERNAL:
Cabin (excl flight deck): Length ... 6.50 m (21 ft 4 in)
  Max width ... 1.83 m (6 ft 0 in)
  Max height ... 1.75 m (5 ft 9 in)
  Floor area ... 5.11 m² (55.0 sq ft)
  Volume ... 17.1 m³ (604 cu ft)
Baggage compartments: forward ... 0.93 m³ (33.0 cu ft)
  rear ... 0.42 m³ (15.0 cu ft)
  pannier (optional) ... 0.79 m³ (28.0 cu ft)

AREAS:
Wings, gross ... 34.75 m² (374.0 sq ft)
Ailerons (total) ... 2.05 m² (22.10 sq ft)
Airbrakes: upper (total) ... 0.74 m² (8.00 sq ft)
  lower (total) ... 0.46 m² (5.00 sq ft)
Trailing-edge flaps (total) ... 4.83 m² (52.00 sq ft)
Fin (excl dorsal fin) ... 6.43 m² (69.20 sq ft)
Rudder ... 1.32 m² (14.20 sq ft)
Horizontal tail surfaces (total) ... 9.29 m² (100.00 sq ft)

WEIGHTS AND LOADINGS:
Weight empty ... 7,029 kg (15,497 lb)
Basic operating weight ... 7,333 kg (16,166 lb)
Max payload ... 1,000 kg (2,205 lb)
Payload with max fuel ... 850 kg (1,875 lb)
Max fuel ... 4,536 kg (10,000 lb)
Fuel with max payload ... 4,386 kg (9,670 lb)
Max ramp weight ... 12,755 kg (28,120 lb)
Max T-O weight ... 12,701 kg (28,000 lb)
Max zero-fuel weight ... 8,369 kg (18,450 lb)
Max landing weight ... 10,591 kg (23,350 lb)
Max wing loading ... 365.5 kg/m² (74.86 lb/sq ft)
Max power loading ... 306 kg/kN (3.00 lb/lb st)

PERFORMANCE:
Max limiting Mach No. ... 0.87
High cruising speed at FL370 ... 447 kt (828 km/h; 514 mph)
Intermediate cruising speed at FL410 ... 428 kt (793 km/h; 493 mph)
Long-range cruising speed at FL390 ... 402 kt (745 km/h; 463 mph)
Stalling speed in landing configuration at typical landing weight ... 92 kt (170 km/h; 106 mph)
Max rate of climb at S/L ... 945 m (3,100 ft)/min
Rate of climb at S/L, OEI ... 143 m (470 ft)/min
Time to FL390 at max T-O weight ... 20 min
Max certified altitude ... 12,497 m (41,000 ft)
Service ceiling ... 11,887 m (39,000 ft)
Service ceiling, OEI ... 5,730 m (18,800 ft)
T-O balanced field length at MTOW ... 1,533 m (5,030 ft)
Landing run at MLW ... 808 m (2,650 ft)
Range, NBAA IFR reserves
  with max payload ... 2,285 n miles (4,231 km; 2,629 miles)
  with max fuel ... 2,407 n miles (4,457 km; 2,769 miles)
  with four passengers ... 2,540 n miles (4,704 km; 2,923 miles)
  ferry range ... 2,621 n miles (4,854 km; 3,016 miles)

OPERATIONAL NOISE LEVELS:
T-O ... 79.3 EPNdB

UPDATED

HAWKER HORIZON

TYPE: Business jet.

PROGRAMME: Design started in 1993; initially thought to have had preliminary designation PD1000Y, but now known to have been PD376; briefly identified as Horizon 1000; announced at NBAA Convention, Orlando, Florida, 19 November 1996; full-size cabin mockup exhibited at NBAA Convention, Dallas, Texas, September 1997. First production wing delivered to Wichita from Fuji Heavy Industries in December 1998; first three fuselage sections mated in October 2000 and mated to wing on 16 January 2001. Prototype (c/n RC-1/N4000R) rolled out 17 April 2001; first flight 11 August 2001; second aircraft (c/n RC-2/N802HH) first flew 10 May 2002; public debut (c/n RC-3/N803HH) at NBAA Convention, Orlando, Florida, 10 September 2002; three aircraft had completed 1,400 hours of test flying by August 2004, fourth aircraft (c/n RC-4/N804HH) first flew 28 April 2004, and following certification will undertake a six-month operational evaluation to validate and 'de-bug' systems. FAA certification scheduled for December 2004; production target 36 per year.

CUSTOMERS: More than 30 ordered by August 2004, spilt approximately equally between US and International customers, including 27 for Raytheon Travel Air, and one for the Jordan Grand Prix racing team.

COSTS: US$18.2 million (2004).

DESIGN FEATURES: All-new design sharing only slight family resemblance with the BAe 125-derived Hawker 800 and (discontinued) 1000. An 'advisory council' of business jet operators assisted RAC in defining the aircraft, a major requirement being a flat-floor stand-up cabin. Conventional swept-wing, T-tail design with small 'overfin' housing antenna for optional satellite communications system. Wing of supercritical, aft-loaded aerofoil section designed using computational fluid dynamics (CFD) techniques; CFD also used to design area-ruled aft fuselage to minimise engine nacelle drag and to reprofile Hawker 1000-based nose section.

Wing sweepback 28° 22' at 25 per cent chord; dihedral 4°; tailplane sweepback 33° 30' at 25 per cent chord.

FLYING CONTROLS: Supercritical section ailerons and elevators, manually operated via pushrods and cables. Pitch trim via electrically actuated, variable incidence tailplane; geared trim tabs on ailerons, fly-by-wire hydraulically powered rudder with boost for asymmetric thrust. Electrically signalled, hydraulically powered, three-segment spoilers on upper surface of each wing augment aileron roll control; outboard and middle panels provide roll and speed brake functions, with maximum deflection 35°; lift dump function provided by all spoilers at 60° deflection. Four-segment, electrically controlled and powered double-slotted flaps, deflections 0, 12, 20 and 35°. Dual controls standard.

STRUCTURE: Fuselage of graphite/epoxy laminate and honeycomb core sandwich; 25 mm (1 in) thick shell is formed by Cincinnati Milacron Viper automatic fibre-placement machines over aluminium mandrel in three sections: nose, centrebody and tailcone, including dorsal fairing and engine pylons mated using aluminium splice plates in a Nova-Tech Engineering Inc fuselage automated splice tool which automatically seals the joints and installs some 1,800 hi-shear fasteners. Wing, designed using CFD, has a supercritical aerofoil and is manufactured by Fuji Heavy Industries at Utsunomiya, Japan, as a complete unit including all systems, integral fuel tanks and leading-edge bleed-air anti-icing, and shipped to Wichita for final assembly. Horizontal stabiliser is of light alloy, two-spar construction with graphite/epoxy composites sandwich skins; vertical stabiliser has three alloy spars and graphite/epoxy composites sandwich skins.

Risk-sharing partners include: Eaton Corporation (hydraulic system); Fokker Elmo BV (wiring harness); Fuji Heavy Industries (wing); Honeywell (avionics integration); Meggitt Avionics (fire and overheat protection and standby instrumentation); Messier-Dowty (landing gear); Moog (flight controls and flap actuators); Pratt & Whitney Canada (propulsion system); Smiths Industries (fuel system); and Hamilton Sundstrand Corporation (utility systems management integration).

LANDING GEAR: Retractable tricycle type by Messier-Dowty; landing gear electrically signalled and hydraulically actuated, with free-fall emergency extension system; twin wheels on each unit. Trailing-link suspension on main units, which retract inwards; nosewheel forwards.

Hawker Horizon second (lower) and third prototypes during flight testing ***NEW***/0583444

Raytheon Hawker Horizon, cutaway drawing key

1 Forward-hinged radome with lightning protection strips
2 Weather radar scanner
3 Scanner mounting bulkhead, machined aluminium alloy
4 ILS glidescope antenna
5 Batteries
6 Nosewheel housing
7 Microwave Landing System antenna
8 Front pressure bulkhead, aluminium alloy
9 Forward electrical power distribution units
10 Radome hinge
11 Nosewheel doors, closed after cycling of landing gear
12 Nosewheel hydraulic steering jacks
13 Twin nosewheels, forward retracting
14 Torque scissor links
15 Nosewheel leg door
16 Pitot head, port and starboard
17 Incidence vane
18 Rudder pedals
19 Control column with ram's horn yoke
20 Instrument panel, five-tube EFIS displays
21 Instrument panel shroud
22 Electrically heated windscreen panels
23 Overhead switch panel
24 Audio warning system speaker
25 Co-pilot's seat
26 Captain's seat
27 Cockpit bulkhead
28 Side console panel with nosewheel steering tiller and cursor controller
29 Underfloor oxygen bottles (four)
30 TCAS antenna
31 Static ports
32 Door external handle
33 Door and airstairs
34 Folding handrail
35 Circuit breaker panels
36 Flight crew wardrobe/closet
37 Starboard side avionics equipment rack
38 No. 1 GPS antenna
39 Galley compartment
40 Upper TCAS antenna
41 Passenger wardrobe/closet
42 Cabin audio/video individual control unit
43 VHF 2 antenna
44 Landing/taxying lights, two per side
45 Position of pressure refuelling connection on starboard side
46 Hydraulic brake accumulator
47 Wing inspection light
48 Wing front spar/fuselage attachment joint
49 Composites cabin floor panels
50 Primary fuselage structure, graphite epoxy composites skins with honeycomb core
51 Cabin window panels
52 Stowage locker
53 Swivelling and reclining individual seats, eight-seat double club layout
54 Folding table
55 ATC 1 and 2 antennas
56 No. 2 GPS antenna
57 Wing front spar aluminium alloy attachment main frame/fuselage splice reinforcement
58 Starboard emergency exit
59 Starboard wing integral fuel tank
60 Fuel venting channels/stringers
61 Overwing filler cap
62 Leading-edge bleed-air de-icing
63 Wingtip vent tank
64 Starboard navigation and strobe lights
65 Starboard aileron
66 Geared tab
67 Aileron hinge control
68 Outboard double-slotted flap segment
69 Outboard spoiler panels, roll control, speed brake and lift dumping functions
70 Flap screw jack actuator
71 Ventral flap hinge fairings
72 Spoiler hydraulic actuators
73 Inboard spoiler/lift dumper
74 Flap drive torque shaft
75 Inboard double-slotted flap segment
76 VHF 1 antenna
77 Individual window blinds
78 Rear cabin seating
79 Cabin floor beams, aluminium alloy
80 Centre wing carry-through structure beneath fuselage pressure shell
81 Wing rear spar attachment joints
82 Rear spar attachment main frame/fuselage splice reinforcement
83 Cabin rear bulkhead with sliding door
84 Hand basin
85 Cabin LCD screen
86 Lavatory
87 Lavatory/baggage compartment bulkhead with flight-accessible door
88 ADF antenna
89 Baggage compartment
90 Up-and-over baggage door
91 Rear pressure bulkhead, graphite/epoxy composites
92 Starboard engine installation
93 Exhaust mixing duct
94 Rear fuselage equipment bay cooling air scoops
95 Dual air conditioning packs
96 HF controller
97 Primary heat exchangers
98 Hydraulic reservoir, dual system, port and starboard
99 Heat exchanger exhaust duct
100 Rear fuselage avionics equipment racks
101 Position of ELT antenna on starboard side
102 Fin spar attachment sloping bulkheads, aluminium alloy
103 Heat exchanger air intakes
104 Starboard target-type thrust reverser, deployed position
105 Fin leading-edge with aluminium alloy skin
106 Three-spar and rib fin torsion box structure, aluminium alloy
107 Carbon fibre fin skin panels
108 Rudder hydraulic actuators
109 VOR localiser antenna
110 Tailplane control unit
111 Trimming tailplane screw jack actuator
112 Scissor linkage
113 Trimming tailplane sealing plate
114 Tailplane pivot mounting
115 Elevator hinge control linkage
116 Upper anti-collision beacon
117 Starboard tailplane
118 Tailplane horn balance
119 Starboard elevator
120 Elevator servo tab
121 Optional antenna
122 Tail navigation light
123 Port elevator carbon fibre composites skin and substructure
124 Static dischargers
125 Tailplane two-spar and rib torsion box structure, aluminium alloy, with carbon fibre skin panels
126 Tailplane leading-edge electromagnetic expulsive de-icing system
127 Rudder
128 Rudder aluminium alloy torsion box structure with carbon fibre skins
129 APU exhaust
130 APU bay firewall
131 Honeywell 36-150(HH) APU
132 APU starboard side intake duct
133 Port engine thrust-reverser, closed
134 Engine pylon
135 Rear engine mounting
136 Bleed-air connections and pylon-mounted pre-cooler
137 Main engine mounting yoke
138 Aft fuselage hinged ventral access hatch with integral steps
139 Hinged engine cowling panels
140 Pratt & Whitney Canada PW308A turbofan
141 Core engine exhaust and multilobe mixer
142 Fan-air exhaust duct
143 Engine accessory equipment
144 FADEC engine controller
145 Engine drain mast
146 Engine air intake lip, bleed-air de-iced
147 Composites wingroot trailing-edge fairing
148 Baggage door
149 Aft electrical power distribution units
150 Mainwheel bay and pre-closing door
151 Main landing gear leg pivot mounting
152 Flap shroud structure
153 All-composites flap structure, initial production aircraft with mixed construction
154 Port inboard double-slotted flap segment
155 Inboard spoiler/lift dumper
156 Spoiler panel, composites construction
157 Dual outboard multifunction spoiler panels
158 Port outboard double-slotted flap segment
159 Aileron servo tab
160 Port aileron all-composites construction
161 Static dischargers
162 Wingtip fairing
163 Port navigation and strobe lights
164 Wingtip vent tank
165 Venting air intake
166 Wing two-spar and rib torsion box structure, machined aluminium alloy
167 Overwing fuel filler
168 Wing bottom skin/stringer panel with access hatches
169 Port wing integral fuel tank
170 Leading-edge de-icing piccolo tube
171 Leading-edge rib structure
172 Mainwheel leg door
173 Twin mainwheels
174 Trailing link mainwheel suspension
175 Shock-absorber strut
176 Hydraulic retraction jack
177 Port wing fuel venting channels
178 Machined aluminium alloy wing ribs
179 Wingroot rib
180 Leading-edge de-icing air baffle

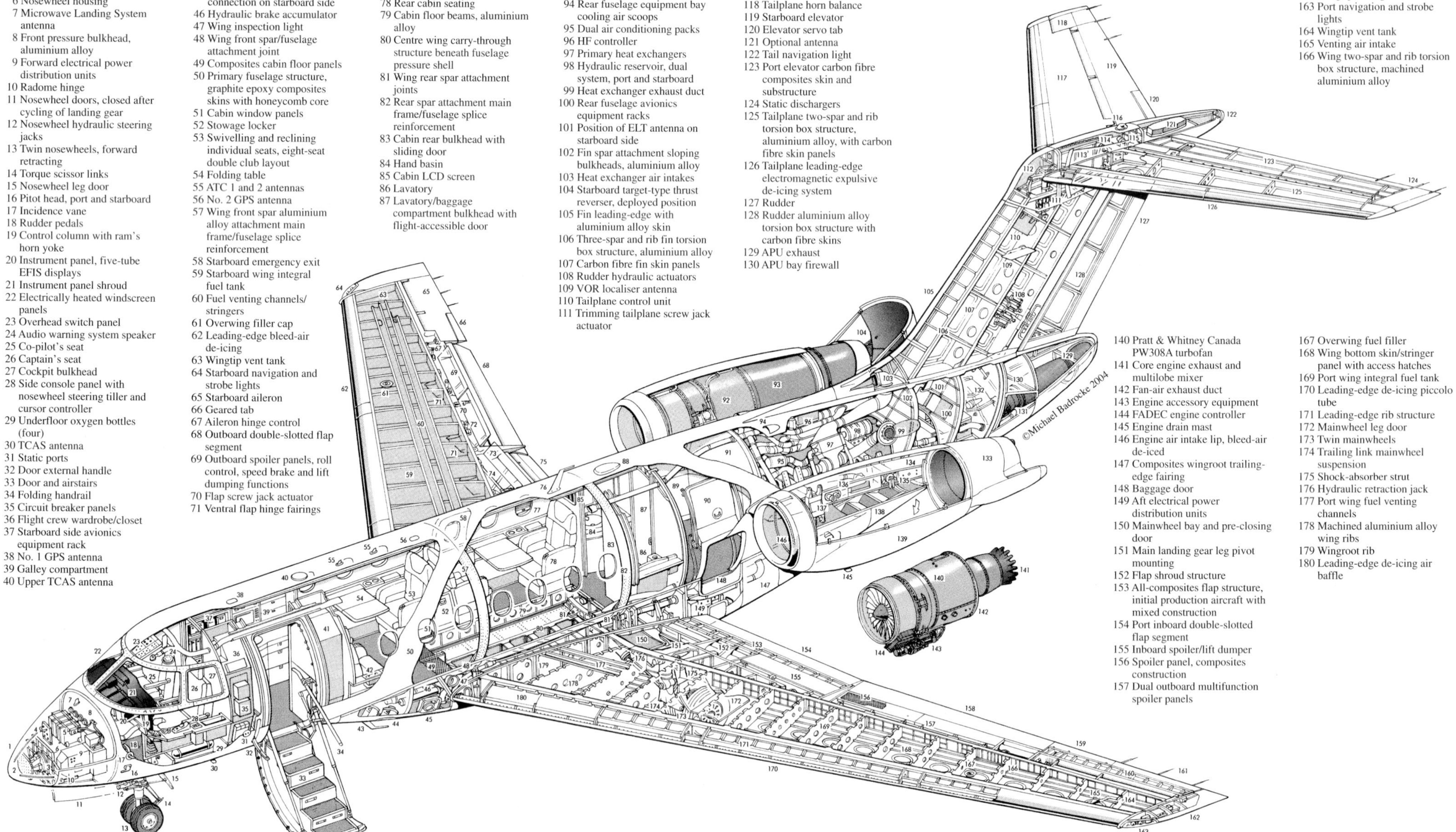

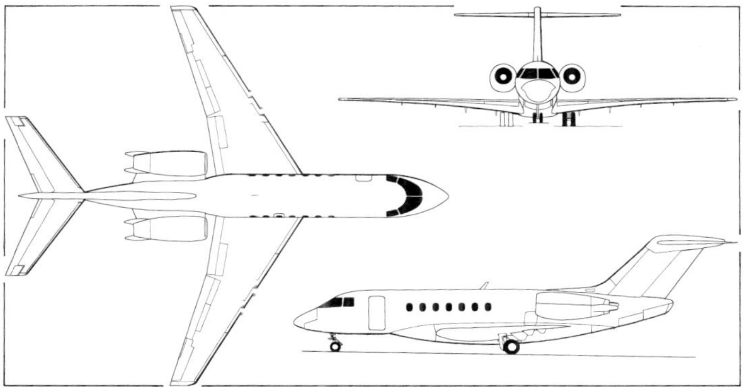

Hawker Horizon business aircraft (two P&WC PW308A turbofans) *(Paul Jackson)* ***NEW***/0568984

Steer-by-wire nosewheel, maximum deflection 70°, with disconnect for towing. Mainwheel size 26×6.6 (14 ply), nosewheel size 18×4.4 (10 ply). Hydraulic carbon disc brakes with digital brake-by-wire and electric anti-skid systems.

POWER PLANT: Two pod-mounted Pratt & Whitney Canada PW308A turbofans with FADEC, each flat-rated at 30.7 kN (6,900 lb st) at ISA + 20°C. Nordam target-type thrust reversers. Fuel in two integral wing tanks, maximum capacity 8,078 litres (2,134 US gallons; 1,777 Imp gallons). Single point fuelling/defuelling; gravity filler ports in top of each wing.

ACCOMMODATION: Crew of two, plus eight passengers in double 'club four' arrangement on reclining and side tracking seats, each pair having stowable writing table. Observer's seat on flight deck optional. Two closets and a galley immediately aft of flight deck; lavatory and flight accessible (via secondary pressurised bulkhead up to FL410) baggage compartment at rear of cabin with external door on port side aft of wing. Accommodation is air conditioned and pressurised (FL60 equivalent altitude) and includes Airshow entertainment system. Door on port side aft of flight deck; single plug-type emergency exit on starboard side over wing; Four-panel windscreen; seven cabin windows per side.

SYSTEMS: Digitally controlled pressurisation system, maximum pressure differential 0.68 bar (9.80 lb/sq in), maintains a sea level cabin environment to 7,693 m (25,240 ft) and a 1,830 m (6,000 ft) cabin environment to 13,715 m (45,000 ft). Engine bleed air anti-icing for wing leading-edges and engine inlet lips; electromagnetic expulsive de-icing of tailplane, none on fin leading-edge; electrically heated pitot and static masts, AoA and TAT probes; windscreen anti-icing and demisting and cabin window demisting via electrically conductive transparent film embedded between panels. Automatically controlled electrical power generation and distribution system (EPGDS) comprises two high-speed, variable frequency, engine-driven AC generators, one APU-driven AC generator and two 43 Ah lead acid batteries. AC/DC external power receptacles standard. Two independent hydraulic systems, operating pressure 206.85 bar (3,000 lb/sq in), for normal and emergency landing gear operation, braking, thrust reversers, spoilers, rudder, nosewheel steering and emergency electrical generation. Two-bottle oxygen system, total capacity 4,080 litres (144 cu ft), with quick-donning diluter-demand masks for crew and auto-deploy constant flow masks in overhead boxes for passengers. Tailcone-mounted Honeywell AE-36-150(HH) APU approved for in-flight operation from sea level to 10,670 m (35,000 ft) and for main engine starting from sea level to 7,925 m (26,000 ft).

AVIONICS: Honeywell Primus Epic lightweight modular avionics system based on Virtual Backplane Network architecture, which combines the cabinet-based modular capabilities of Honeywell's 777 AIMS system with the aircraft-wide network capabilities of its Primus 2000 system and built-in maintenance recording with portable access terminal. The system has 'point and click' capability via two Cursor Control Devices (CCDs) which include touchpad, numeric control and on-screen 'soft keys'. Voice Command will be a future option for some Epic functions.

The Horizon's avionics system provides functions which Raytheon states have never previously been offered in a super-mid-size business jet, including digital circuit breaker control, automatic refuelling control, and full-authority autothrottle. All avionics boxes are installed in a cabinet behind the co-pilot's seat, in an environmentally controlled, pressurised area with easy access for maintenance.

Comms: Dual VHF comms; single Collins HF-9000 HF with Selcal; dual Mode S Diversity transponders; dual audiophone/interphone/PA systems; airborne telephone.

Radar: Primus 880 colour weather radar.

Flight: Primus Epic AFCS and FMS with integrated performance computer; dual VHF nav VOR/LOC/GS/Markers; dual DME; dual IRU; dual GPS; EGPWS; TCAS II; solid-state FDR and CVR.

Instrumentation: Five 203 × 254 mm (8 × 10 in) colour, active, flat panel LCD screens comprising two PFDs, two MFDs and one EICAS, plus two smaller multifunction control and display units (MCDUs).

DIMENSIONS, EXTERNAL:

Wing span	18.82 m (61 ft 9 in)
Mean aerodynamic chord	2.92 m (9 ft 7 in)
Wing aspect ratio	7.2
Length overall	21.08 m (69 ft 2 in)
Height overall	5.97 m (19 ft 7 in)
Tailplane span	7.90 m (25 ft 11 in)
Wheel track	2.79 m (9 ft 2 in)
Wheelbase	8.46 m (27 ft 9 in)
Passenger door: Height	1.58 m (5 ft 6 in)
width	0.76 m (2 ft 6 in)

DIMENSIONS, INTERNAL:

Cabin (excl flight deck):	
Length: excl baggage compartment	7.62 m (25 ft 0 in)
incl baggage compartment	8.99 m (29 ft 6 in)
Max Width	1.97 m (6 ft 5½ in)
Width at floor	1.27 m (4 ft 2 in)
Max height	1.83 m (6 ft 0 in)
Volume (excl flight deck)	21.6 m^3 (762 cu ft)
Baggage volume: forward cabin	0.55 m^3 (19.5 cu ft)
aft cabin	2.69 m^3 (95 cu ft)

AREAS:

Wings, gross	49.3 m^2 (531.0 sq ft)
Horizontal tail surfaces	13.01 m^2 (140.00 sq ft)
Vertical tail surfaces	10.26 m^2 (110.40 sq ft)
Ailerons (total)	1.35 m^2 (14.52 sq ft)
Horizontal surfaces (total)	13.01 m^2 (140.0 sq ft)
Fin, excl dorsal fairing	10.26 m^2 (110.40 sq ft)
Rudder	2.40 m^2 (25.87 sq ft)

WEIGHTS AND LOADINGS:

Weight empty	9,596 kg (21,155 lb)
Basic operating weight (incl crew)	9,777 kg (21,555 lb)
Max fuel weight	6,486 kg (14,300 lb)
Max zero fuel weight	11,340 kg (25,000 lb)
Payload with max fuel	837 kg (1,845 lb)
Fuel with max payload	5,761 kg (12,700 lb)
Max T-O weight	17,010 kg (37,500 lb)
Max ramp weight	17,100 kg (37,700 lb)
Max landing weight	15,195 kg (33,500 lb)
Max wing loading	344.8 kg/m^2 (70.62 lb/sq ft)
Max power loading	277 kg/kN (2.72 lb/lb st)

PERFORMANCE (provisional):

Max operating Mach No. at FL350	M0.84
High cruising speed at FL410	M0.82/470 kt (870 km/h; 541 mph)
Intermediate cruising speed at FL410	M0.78/447 kt (828 km/h; 514 mph)
Long-range cruising speed at FL450	M0.75/430 kt (796 km/h; 495 mph)
Approach speed	132 kt (244 km/h; 152 mph)
Time to climb to FL370	13 min
Max rate of climb at S/L, OEI	176 m (576 ft)/min
Max certified ceiling	13,716 m (45,000 ft)
Service ceiling	13,076 m (42,900 ft)
Service ceiling, OEI	8,534 m (28,000 ft)
T-O field length	1,444 m (4,736 ft)
Landing run	750 m (2,460 ft)
Range, NBAA IFR reserves:	
with max payload	2,763 n miles (5,117 km; 3,179 miles)
with max fuel	3,294 n miles (6,100 km; 3,790 miles)
with four passengers	3,400 n miles (6,296 km; 3,912 miles)
Ferry range	3,477 n miles (6,439 km; 4,001 miles)

UPDATED

HENSLEY

HENSLEY AIRCRAFT INC

2440 Winding Creek Boulevard, Suite 204, Clearwater, Florida 33761-2556
Corporate Address
18300 Minnetonka Boulevard, Suite 204, Minnesota 55391
Tel: (+1 727) 294 69 96
e-mail: info@hensleyaircraft.com
Web: http://www.hensleyaircraft.com
CEO: Robert Hensley

The H-1 Wolf is Hensley's initial offering with final funding for construction and testing of the prototype put in place during 2003. The company's manufacturing plant is under construction at Grand Forks, North Dakota, and is scheduled to be completed by end 2005. In July 2004, the company had four employees and was planning a further increase by early 2005.

UPDATED

HENSLEY H-1 WOLF

TYPE: Four-seat kitbuilt twin.

PROGRAMME: Design began in 1991, initially as part-time venture. Construction of prototype started in 1998; exhibited, unflown, at AirVenture, Oshkosh, July 2001; maiden flight then scheduled for December 2001 and kit deliveries for second quarter of 2002; this has slipped and first flight had not taken place by October 2003. In June 2004 the first Centurion power plants were delivered and a first flight before the end of 2004 was expected, although not, apparently, achieved. Registrations N447HC, N447HM and N447HN have been reserved for first three aircraft.

CURRENT VERSIONS: **Wolf:** Baseline version.

Wolf GT: Possible future higher-powered, long-range version, with two cabin doors and increased MTOW.

CUSTOMERS: First aircraft sold October 2001.

COSTS: Wolf US$159,900 including engines (2004).

DESIGN FEATURES: Intended as manoeuvrable twin, able to operate into remote, short airstrips; affordable; easy to build and maintain.

Low-drag fuselage; constant-chord, low-mounted wings, with winglets; pusher engines located close inboard and above wings.

Wing section VH-1 Twin Turbo, purpose developed by Jeff Viken for high lift, low drag laminar flow.

FLYING CONTROLS: Conventional and manual. Fowler flaps immediately outboard of engines. Horn-balanced rudder and elevator.

STRUCTURE: Composites throughout.

LANDING GEAR: Tricycle type; fixed. Mainwheels, 6.00-6, on aluminium sprung legs; trailing-link nosewheel, 11×4.00–5, with rubber-in-compression suspension. Hydraulic brakes on mainwheels. Speed fairings on all wheels.

Wolf GT will have retractable landing gear.

POWER PLANT: Initially, prototype was due to be powered by two Quantum Powermills engines (based on Honda VTEC) with pusher propellers and dual fuel system with baseline aircraft having 101 kW (135 hp) Q-135s; this changed in second quarter 2003 to two FADEC-equipped 101 kW (135 hp) Thielert Centurion turbocharged Diesel engines driving three-blade constant-speed MT propellers. Fuel capacity 189 litres (50.0 US gallons; 41.6 Imp gallons). Standard, optional 303 litre (80.0 US gallon; 66.6 Imp gallon) tank can be added.

ACCOMMODATION: Four persons in two pairs of seats. Door, port side. Dual sidestick controls. Two doors on GT version.

Hensley Wolf under construction ***NEW***/1047404

Hensley Wolf cabin ***NEW***/1047403

SYSTEMS: Dual electrical system.
AVIONICS: Chelton Electronics EFIS system coupled with Garmin radios and TruTrak autopilot under consideration.

DIMENSIONS, EXTERNAL:
Wing span ... 10.34 m (33 ft 11¼ in)
Wing chord: at root ... 1.34 m (4 ft 4¾ in)
at tip ... 0.82 m (2 ft 8¼ in)
Wing aspect ratio ... 9.6
Length overall ... 6.78 m (22 ft 3 in)
Height overall ... 2.68 m (8 ft 9½ in)

DIMENSIONS, INTERNAL:
Cabin max width ... 1.22 m (4 ft 0 in)

AREAS:
Wings, gross ... 11.15 m² (120.0 sq ft)

Hensley Wolf nose view ***NEW***/1047402

Horizontal tail surfaces (total) ... 2.87 m² (30.86 sq ft)
Vertical tail surfaces (total) ... 2.04 m² (21.98 sq ft)

WEIGHTS AND LOADINGS:
Weight empty ... 907 kg (2,000 lb)
Max T-O weight: standard fuel ... 1,360 kg (3,000 lb)
optional fuel ... 1,587 kg (3,500 lb)
Wing loading: standard ... 122.1 kg/m² (25.00 lb/sq ft)
optional fuel ... 142.4 kg/m² (29.17 lb/sq ft)
Power loading: standard ... 6.76 kg/kW (11.11 lb/hp)
optional fuel ... 7.89 kg/kW (12.96 lb/hp)

PERFORMANCE (estimated):
Normal cruising speed at FL80 at 75% power
A ... 161 kt (298 km/h; 185 mph)
Stalling speed ... 57 kt (105 km/h; 65 mph)
Max rate of climb at S/L ... 427 m (1,400 ft)/min
single engine ... 91 m (300 ft)/min
T-O run ... 270 m (885 ft)
Landing run ... 206 m (675 ft)
Range: standard fuel ... 869 n miles (1,609 km; 1,000 miles)
optional fuel ... 2,259 n miles (4,184 km; 2,600 miles)
g limits ... +6.0/–3.0

UPDATED

HILLBERG

HILLBERG HELICOPTERS

PO Box 8974, Fountain Valley, California 92728-8974
Tel: (+1 909) 279 56 78
e-mail: Rotermouse@earthlink.com
Web: http://www.helis.com/Since80s/h_rtmous.php
PRESIDENT: Don Hillberg

Company founded 1990 to provide maintenance and modification services to helicopter operators, and moved into kit and experimental fields to help builders.

Hillberg has also developed the four-seat 1-04, which awaits funding for production. Produces a retrofit kit for the RotorWay Exec (which see), upgrading it to turbine power with a Solar T62-T32 of 112 kW (150 shp) and reducing empty weight to 340 kg (750 lb); at least 13 delivered by October 2003.

Work on a proposed four-seat EH 1-04 Baby Huey appears to have been suspended.

UPDATED

HILLBERG EH 1-01 ROTERMOUSE AND EH 1-02 SHARKMOUSE

TYPE: Single-seat helicopter kitbuilt; two-seat helicopter kitbuilt.

PROGRAMME: Design and construction of prototype began in August 1990; first flew (N10TE) August 1992. Construction of first production aircraft began in September 1996. Type is currently certified as experimental by FAA.

CURRENT VERSIONS: **EH 1-01 RoterMouse**: Single-seat version.

Description applies to EH 1-01 except where indicated.

EH 1-02 TandemMouse: Tandem two-seat version; under development; was due to fly 1999; this date slipped, but this version is believed to have flown by early 2001. Powered by 112 kW (150 shp) Solar T62-T32 turbine engine. Provisional figures appear below. Renamed **SharkMouse** 2004.

EH-1-02A: Development of EH 1-02; has 313 kW (420 shp) Rolls-Royce 250-C20B with 681 litre (180 US gallon; 150 Imp gallon) fuel tank.

Hillberg EH 1-01 RoterMouse (one Honeywell 36-55-C) 0507665

CombatMouse: Proposed military version.

CUSTOMERS: By June 2004, 29 kits had been ordered, of which at least four had been built.

COSTS: Kit: EH-1-01 US$86,000 (2004) ex-works US$100,000; EH 1-02 US$125,000; all prices encompass everything bar radio equipment (2004).

DESIGN FEATURES: Forward fuselage resembles Bell AH-1 HueyCobra. Stub-wing behind cockpit for carriage of external loads. Main rotor speed 510 rpm; tail rotor 3,250 rpm.

FLYING CONTROLS: Auto-governed hydromechanical flying controls. Optional rotor brake.

STRUCTURE: Aluminium alloy (2024-T3) monocoque fuselage, bulkheads and keel; stainless steel engine deck and firewall. Rotor has Gyrodyne QH-50 modified blades of NACA 1200 aerofoil section with 8° twist and block tip caps. Modified Robinson R22 hub and transmission. Conventional tail rotor using aluminium-skinned blades.

LANDING GEAR: Fixed steel cross-tube skid type; floats optional.

POWER PLANT: One 108 kW (145 shp) Honeywell 36-55-C turboshaft engine driving a non-folding, two-blade, semi-rigid wood-laminated main rotor. Engines in range 108–186 kW (145–250 hp) are suitable. Single 125 litre (32.9 US gallon; 27.4 Imp gallon) crash-resistant bladder fuel tank around transmission; refuelling point behind main mast. Optional 76 litre (20.0 US gallon; 16.7 Imp gallon) fuel tank on stub-wing. EH 1-02 has 170 litre (45.0 US gallon; 37.5 Imp gallon) fuel capacity.

ACCOMMODATION: Single pilot in enclosed cockpit. Entry through upward-hinged port and starboard side windows. Baggage compartment behind seat.

SYSTEMS: 24 V 60 A electrical system; Electrocraft alternator.

EQUIPMENT: Optional cargo hook.

DIMENSIONS, EXTERNAL:
Main rotor diameter: EH 1-01 ... 6.10 m (20 ft 0 in)
EH 1-02 ... 7.62 m (25 ft 0 in)
Tail rotor diameter ... 0.91 m (3 ft 0 in)
Length:
overall, rotors turning:
EH 1-01 ... 7.32 m (24 ft 0 in)
fuselage: EH 1-01 ... 6.35 m (20 ft 10 in)
EH1-02 ... 8.53 m (28 ft 0 in)
Height overall ... 2.26 m (7 ft 5 in)

DIMENSIONS, INTERNAL:
Cabin max width: EH 1-01 ... 0.61 m (2 ft 0 in)
EH 1-02 ... 0.89 m (2 ft 11 in)

AREAS:
Main rotor disc: EH 1-01 ... 29.20 m² (314.3 sq ft)
EH 1-02 ... 45.62 m² (491.1 sq ft)

WEIGHTS AND LOADINGS:
Weight empty: EH 1-01 ... 295 kg (650 lb)
EH 1-02 ... 408 kg (900 lb)
EH 1-02A ... more than 499 kg (1,100 lb)
Max T-O and landing weight:
EH 1-01 ... 621 kg (1,370 lb)
EH 1-02 ... 816 kg (1,800 lb)
EH 1-02A ... 1,451 kg (3,200 lb)

PERFORMANCE:
Never-exceed speed (VNE) ... 184 kt (341 km/h; 212 mph)
Max level speed: EH 1-01 ... 139 kt (257 km/h; 160 mph)
EH 1-02 ... 148 kt (274 mph; 170 mph)
Cruising speed: both ... 113 kt (209 km/h; 130 mph)
Max rate of climb at S/L:
EH 1-01 ... 1,433 m (4,700 ft)/min
Vertical rate of climb at S/L:
EH 1-01 ... 823 m (2,700 ft)/min
EH 1-02 ... 457 m (1,500 ft)/min
Service ceiling: EH 1-01 ... 4,495 m (14,750 ft)
EH 1-02 ... 4,115 m (13,500 ft)
Range, 20 min reserves:
EH 1-01:
with max payload ... 300 n miles (555 km; 345 miles)
with max internal and external fuel ... 695 n miles (1,287 km; 800 miles)
EH 1-02: with max payload ... 390 n miles (722 km; 448 miles)

UPDATED

HONDA

HONDA R & D AMERICAS INC

c/o Atlantic Aero Inc, Piedmont Triad International Airport, Greensboro, North Carolina 27425
Web: http://www.honda.com
PRESIDENT: Takeo Fukui
SENIOR CHIEF ENGINEER: Osamu Kubota
COMMUNICATIONS: Jeffrey Smith

Honda's first business jet was the MH02, designed and built in the late 1980s/early 1990s with the aid of Mississippi State University. This aircraft (N3079N) first flew on 5 March 1993, but the programme lapsed some three years later without entering production, the aircraft having flown some 170 hours.

More recently, Honda R & D has developed a successor, known simply as the HondaJet, together with the HF-118 engines which power it. This took place at a 2,787 m² (30,000 sq ft) facility on the Greensboro leasehold site of Atlantic Aero Inc, with whom Honda R & D teamed in 2000. The HondaJet prototype flew in December 2003.

UPDATED

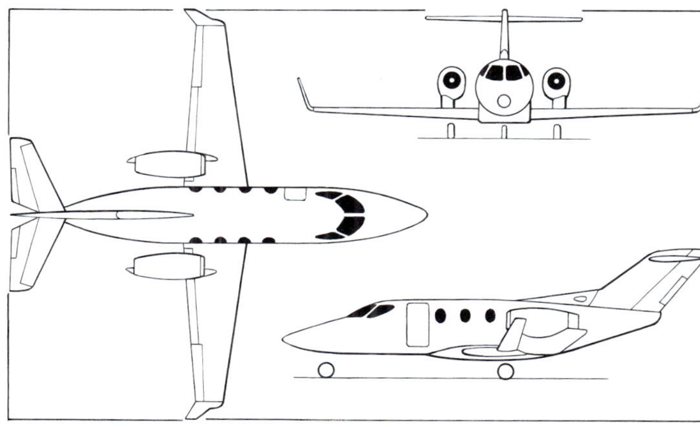

Honda R & D HondaJet (two Honda HF-118 turbofans) *(James Goulding)* ***NEW***/0567104

HONDA HA-420 HONDAJET

TYPE: Light business jet.

PROGRAMME: Development began in 2000; first brief details revealed in mid-2000; HF-118 engine made first flight 10 June 2002 in Cessna CitationJet testbed; wing aerofoil tested 2002 as 'glove' section on T-33 testbed; wind tunnel tests conducted at Greensboro facility, the University of Washington, Boeing's transonic wind tunnel and Japanese National Aeronautics Laboratory's transonic flutter wind tunnel. Static and fatigue testing of prototype (N420HA/c/n P001) under way by mid-2003; first flight undertaken in secrecy at Piedmont Triad Airport North Carolina, 3 December 2003 and announced on 16 December 2003.

DESIGN FEATURES: Low-mounted, moderately tapered wings of natural laminar flow section, with cabane-mounted engines and small winglets. T tail with sharply swept fin and rudder, plus dorsal fin; tailplane has sweptback leading-edges and elevator cut-outs for rudder movement. Primary design considerations were large cabin size and range/fuel consumption economics. Engine position allows cabin size increase of 30 per cent without penalty. Nose shape optimised for generation of fuselage laminar flow, which also achieved on wings despite their relative thickness.

HondaJet prototype N420HA ***NEW***/0572391

First flight of HondaJet, 3 December 2003 ***NEW***/0572392

Laminar flow Honda SMH-1 wing section.

FLYING CONTROLS: Double-slotted, 30 per cent chord flaps with mechanical linkage (settings 15.7° for T-O, 50° for landing). Stall characteristics said to be 'docile' in both cruise and climb.

STRUCTURE: Fuselage of co-cured carbon fibre-reinforced prepreg glass fibre and epoxy, integrally stiffened in main barrel section, with GFRP honeycomb sandwich for nosecone and tailcone. Aluminium alloy for three-spar wings and for tail assembly.

LANDING GEAR: Sumitomo hydraulically retractable tricycle type. Trailing-link main units, retracting into underside of wing centre-section; nosewheel is steer-by-wire (±10° for take-off, ±50° for taxying).

POWER PLANT: Two 7.43 kN (1,670 lb st) Honda HF118 podded turbofans, mounted on sweptback overwing pylons and equipped with Honda-developed FADEC.

Fuel in four tanks (one in each outer wing, one in wing centre-section, and one rear fuselage bladder tank). Single-point refuelling at rear on starboard side.

ACCOMMODATION: Standard seating for two crew and four passengers in pressurised cabin, with lavatory at rear; or six passengers with lavatory omitted. Four cabin windows per side; cabin door on port side immediately aft of flight deck.

SYSTEMS: Cabin pressurised to 0.60 bar (8.7 lb/sq in), giving 2,440 m (8,000 ft) equivalent environment at altitude of 13,400 m (44,000 ft). Wing leading-edge anti-icing by engine bleed air; windscreen electrically heated.

AVIONICS: Garmin 1000 integrated modular suite with three-screen EFIS.
Instrumentation: EFIS, comprising PFD for each crew member and single central MFD.
Preliminary details as follows:

DIMENSIONS, EXTERNAL:
Wing span over winglets ... 12.15 m (39 ft 10½)
Length overall ... 12.54 m (41 ft 1¾ in)
Height overall ... 4.03 m (13 ft 2½ in)

DIMENSIONS, INTERNAL:
Cabin: Length ... 4.60 m (15 ft 1 in)
Max height ... 4.50 m (4 ft 10¾ in)

WEIGHTS AND LOADINGS:
Max T-O weight ... 4,173 kg (9,200 lb)
Max power loading ... 281 kg/kN (2.75 lb/lb st)

PERFORMANCE (estimated):
Max level speed at FL300 ... 420 kt (777 km/h; 483 mph)
Max cruising speed ... 389 kt (720 km/h; 448 mph)
Optimum cruising altitude ... 12,500 m (41,000 ft)
Max certified altitude ... 13,400 m (44,000 ft)
T-O run ... 808m (2,650 ft)
Landing run ... 695 m (2,280 ft)
IFR range at FL410 ... 1,100 n miles (2,037 km; 1,265 miles)

UPDATED

HPA

INTERNATIONAL HIGH PERFORMANCE AIRCRAFT, INC

PO Box 926, Venice, Florida 34284
Tel: (+1 941) 480 98 81
Fax: (+1 941) 480 93 81
e-mail: info@international-hpa.com
Web: http://www.international-hpa.com
CHIEF DESIGNER: Petr Varadi

The M1 SpeedCruiser is being developed by International High Performance Aircraft of Lobecek 732, CZ-728 01 Kralupy nad Vlatava, Czech Republic (http://www.hpai.com), and Venice, Florida. M1 is related to Team Rocket F1 and F1EVO.

UPDATED

HPA M1 SPEEDCRUISER

TYPE: Side-by-side kitbuilt.

PROGRAMME: Announced at Oshkosh in July 2002 as the Team Rocket F2, then based on major components from the F1evo, including wings, tail and landing gear. By July 2003, had become M1.

COSTS: Estimated kit price US$25,000; fast-build kit US$35,000 (2003).

DESIGN FEATURES: Intended to combine high performance and low price. Low-wing monoplane with closely cowled engine. Tapered wings and tailplane. Meets FAA's '51 per cent' rule. Quoted build time 1,000 hours for fast-build kit.

FLYING CONTROLS: Conventional and manual. Flaps. Horn-balanced rudder and elevators.

STRUCTURE: Principally of aluminium, primed with epoxy chromate for corrosion resistance. Titanium mainwheel legs; glass fibre engine cowling and wingtips.

LANDING GEAR: Fixed; option of tailwheel or nosewheel. Fuselage-mounted spring titanium cantilever mainwheel legs. Mainwheels 5.00-5.

POWER PLANT: One 224 kW (300 hp) Textron Lycoming IO-540 flat-six and MT-Propeller MTV-9-B-C/C193-109 three-blade, constant-speed propeller. Fuel capacity 227 litres (60.0 US gallons; 50.0 Imp gallons).

ACCOMMODATION: Two persons side by side beneath rearward-sliding canopy. Fixed windscreen.

DIMENSIONS, EXTERNAL:
Wing span ... 8.20 m (26 ft 10¾ in)
Wing aspect ratio ... 6.6
Length overall ... 6.40 m (21 ft 0 in)

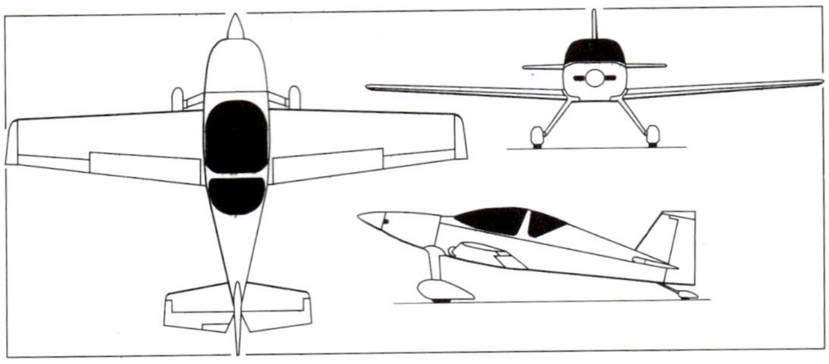

International High Performance M1 Speedcruiser *(Paul Jackson)* ***NEW***/1047841

DIMENSIONS, INTERNAL:
Cabin: Length ... 1.28 m (4 ft 2½ in)
Max width ... 1.14 m (3 ft 9 in)
Max height ... 1.12 m (3 ft 8 in)
Baggage hold volume ... 0.42 m^3 (15.0 cu ft)

AREAS:
Wings, gross ... 10.22 m^2 (110.0 sq ft)

WEIGHTS AND LOADINGS:
Weight empty ... 590 kg (1,300 lb)
Max T-O weight ... 952 kg (2,100 lb)
Max wing loading ... 93.2 kg/m^2 (19.09 lb/sq ft)
Max power loading ... 4.26 kg/kW (7.00 lb/hp)

PERFORMANCE:
Max operating speed ... 230 kt (426 km/h; 265 mph)
Normal cruising speed at 75% power ... 209 kt (386 km/h; 240 mph)
Stalling speed, power off, flaps down ... 44 kt (81 km/h; 50 mph)
Max rate of climb at S/L ... 1,067 m (3,500 ft)/min
T-O run ... 91 m (300 ft)
Landing run ... 213 m (700 ft)
Range at 55% power ... 1,099 n miles (2,035 km; 1,265 miles)

UPDATED

HUNT

HUNT AVIATION CORPORATION

101 East Beach Boulevard, Suite D, Pass Christian, Mississippi 39571
Tel: (+1 228) 452 08 08
Web: http://www.fuellessflight.com
FOUNDER: Robert D Hunt
PRESIDENT: Gene Cox
SENIOR VP AND CHIEF FINANCIAL OFFICER: Joseph A Chomko

Mr Hunt, an inventor and former nuclear engineer, has designed a hybrid aircraft relying on already commonplace effects of the force of gravity for its propulsion.

VERIFIED

HUNT FUEL-LESS AIRCRAFT

TYPE: Technology demonstrator.

PROGRAMME: Announced August 2003; promotion began at NBAA Convention, Orlando, Florida, October 2003, at which time early work had begun on construction of a prototype.

DESIGN FEATURES: Fuel-less propulsion system employing two gravity-related phenomena: buoyancy, as utilised by lighter than air craft; and gravitational attraction utilised by gliders.

Twin pontoons (booms) with cabin between and variable geometry (90° backward sweeping) wings outboard; rear fins and tailplane, latter with two differentially capable 'allerons' for roll control in addition to conventional pitch control. Reversible (dual function) wind turbine on outboard side of each pontoon (principally for steering) and behind cabin (main). Rigid pontoons are stressed to contain a vacuum for optimum buoyant lift, although early aircraft expected to rely on several helium-filled bags within pontoons (this also being safety measure in the event of vacuum failure).

Long-distance flight is accomplished by repetition of the flight cycle in which the aircraft rises almost vertically and then glides.

(1) Aircraft at horizontal rest; compressed air tanks charged from previous cycle; heavier than air because of large amount of compressed air aboard.

(2) While at rest on the ground, interior bladders are fully inflated with helium due to vacuum in cells from previous cycle, maximising bladder displacement (equal to max altitude) and increasing lift dramatically due to extreme low pressure of expanded helium because of the vacuum, but aircraft is still heavier than air because of large supply of compressed air.

(3) Take-off. Compressed air powers pneumatic turbines (on sides of pontoons) that are pointed downward to provide ground effect lift, and compressed air ballast weight is rapidly discharged, making craft lighter than air. Approximately 50 per cent of compressed air is used to power turbines and to discharge ballast weight in take-off process. Attitude of aircraft becomes almost vertical as it climbs and turbines on sides of pontoons revert to the horizontal. In part, this is accomplished by using compressed air from nose of aircraft first, making nose lighter than rear so it pitches upward after ground has been cleared.

(4) Aircraft ascends to 15,240 m (50,000 ft) and, as it rises in almost vertical nose-upward flight, a small amount of compressed air is used to maintain flight attitude and to make aircraft lighter at the same time. This process provides little propulsion, as aircraft is rising by being lighter than air.

Artist's impression of Hunt Fuel-Less Aircraft with wings fully extended ***NEW***/0568980

(5) At full height, compressed air is injected into exterior ridged cell chamber, reducing helium bladder volume; descent begins. Compressed air from very-high-pressure storage cylinders drives pneumatic motors (by expansion from higher pressure to lower pressure) that power air compressors to bring new air in from surrounding environment. New incoming air, and now-expanded lower pressure compressed air, after driving the pneumatic motors, both go into cells to compress helium within gasbags due to higher pressure than bagged helium. In this process, new weight is added to aircraft and weight of compressed air from cylinders is conserved (kept aboard aircraft); otherwise, there would be no increase in weight of aircraft if new air is not brought aboard. This is a key point to understanding the ballast process.

(6) Wings sweep back to proper angle to accomplish steep dive to pick up speed quickly as aircraft noses downward sharply.

(7) Once a downward and forward motion is achieved, wind turbines begin operation in order to bring in compressed air from atmosphere to continue to make aircraft heavier, so that it does not reach equilibrium and stop descending at a lower altitude as surrounding air's density and lifting capacity increases.

(8) Wings reduce sweep as aircraft falls into thicker air (40:1 glide ratio).

(9) Exposed wind turbines recover energy from slipstream and continue to recharge compressed air tanks until filled.

(10) After compressed air is recharged and aircraft is as heavy as it was when on ground, power from wind turbines is used to pull vacuum back into cells to inflate helium bags fully to provide lift to stabilise aircraft at low altitude; and exterior ridged cell chamber is drawn down to vacuum.

(11) Cycle repeats, with discharge of 50 per cent compressed air to provide propulsion and to make aircraft lighter than air again to rise to high altitude.

(12) Landing technique is similar to take off as compressed air is used to provide downward thrust to set craft down vertically. However, minimal compressed air consumed to conserve weight to keep craft heavier than air.

No weights or dimensions have been disclosed.

UPDATED

INDUS

INDUS AVIATION INC

5681 Apollo Drive, Dallas, Texas 75237
Tel: (+1 214) 337 63 87
Fax: (+1 214) 337 63 88
e-mail: sales@indusav.com
Web: http://www.indusav.com
PRESIDENT: Dr Ram Pattisapu
DIRECTOR, SALES AND MARKETING, NORTH AMERICA: Scott Severen
MANAGER, INDIA OPERATIONS: Kiran Kota

IndUS was formed in 1994 and in 2002 bought the type certificate, production certificate and tooling for the Thorp T-211 previously produced by Venture Light Aircraft Resources, renaming the aircraft as the Sport E in the process. IndUS has chosen TAAL of India to build the aircraft, and also to handle non-US marketing; the tooling shipped there to facilitate this.

UPDATED

INDUS SPORT E

TYPE: Side-by-side kitbuilt.

PROGRAMME: Originated in T-11 Sky Skooter, designed by John Thorp and first flown 15 August 1946; certified 22 December 1948, certificate A-791. T-211 certified 20 April 1964 by Tubular Aircraft Products of Los Angeles, which produced parts for 100 aircraft; rights passed to Adams Industries of Detroit, then Thorp Aero Inc at Sturgis, Kentucky, which acquired components in stock; production rights passed to Phoenix Aircraft at Mesa, Arizona, in 1992, but that company declared bankrupt in mid-1994. Acquired by AD Aircraft in the UK; market launch (and first sale, to Ken Fowler) of kits at PFA International Air Rally & Exhibition at Cranfield, Bedfordshire, in July 1998. AD sold six homebuild kits supplied from the USA by Venture Light Aircraft Resources (the previous owner); Venture production began with c/n 104. UK production of complete aircraft planned, but cancelled in 1999.

IndUS took over rights in 2002 (and parts stock sufficient for some 20 aircraft, all 'production' since 1970s having been from original stock of 100, from which requests for spare parts were also deducted by subsequent owners); plans to produce Sport E at rate of one per month, rising to three per month in second half of 2004. In December 2003, IndUS announced that TAAL of India had started work on assembling the first two aircraft, components coming from the aforementioned parts stock; the first aircraft was due for delivery in May 2004.

CURRENT VERSIONS: **Sport E:** Factory-built version.

Sport Experimental: Continental-engined kitbuilt version, supplied complete, or in eight sub-kits for stage-by-stage assembly.

Data refer to Continental-powered version.

CUSTOMERS: Five kits sold during PFA International Air Rally in July 1999, increasing total to six. (At least nine built in USA by Tubular Aircraft, one; Aircraft Engineering Associates, one; and Venture.) Two further kits sold after Sun 'n' Fun, April 2004 for completion under Light Sport Aircraft regulations plus one Part 23 certified example for flight training.

COSTS: Basic kit US$24,995, excluding engine, propeller and avionics; Builder assistance available; LSA compliant aircraft US$69,995, Pt23 US$99,995 (2004).

DESIGN FEATURES: Economic, simple design for recreational market.

FLYING CONTROLS: Manual; pushrod operated controls except for rudder and trims which are cable-operated. Plain ailerons; all-moving tailplane with anti-balance tab occupying 90 per cent span; wide-span manually operated three-position trailing-edge flaps with maximum deflection of 20°.

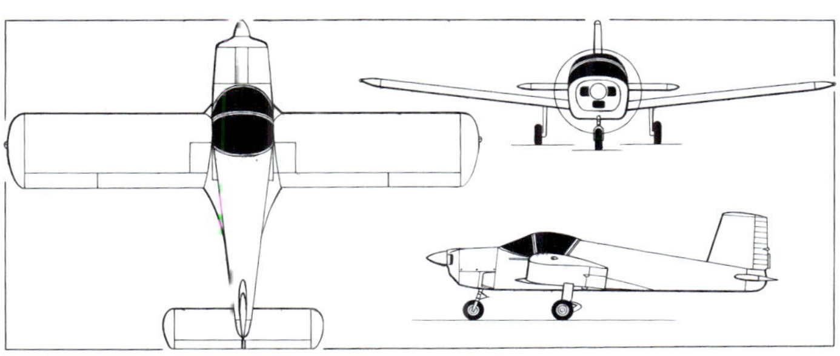

IndUS Sport Experimental two-seat kitbuilt *(James Goulding)* 0084388

Pair of Thorp T-211s overflying Florida ***NEW***/1047230

STRUCTURE: All aluminium fuselage; wing with main spar, false spar and externally stiffened (ribbed) skin; ailerons, flaps, tailplane and rudder have externally stiffened skin. IndUS investigating possibility of fitting 'smooth' wing, subject to satisfactory resolving of structural issues.

LANDING GEAR: Tricycle type; fixed. Wing-mounted, faired, cantilever mainwheel legs with integral oleo-pneumatic shock-absorbers; all three wheels size 5.00-5; steerable nosewheel; Cleveland wheels; Cleveland 30-9 brakes on main units.

POWER PLANT: One 74.6 kW (100 hp) Teledyne Continental O-200A flat-four, driving a Sensenich 69CK two-blade fixed-pitch propeller; IndUS fitting 89 kW (120 hp) Jabiru 3300 flat-six to machines in Light Sport Aircraft category. Fuel tank aft of cabin, capacity 91 litres (24.0 US gallons; 20.0 Imp gallons), of which 79.5 litres (21.0 US gallons; 17.5 Imp gallons) are usable; oil capacity 5.7 litres (1.5 US gallons; 1.25 Imp gallons).

ACCOMMODATION: Two seats side by side beneath rearward-sliding transparent canopy; fixed windscreen. Baggage compartment behind seats, capacity 8.1 kg (17.8 lb).

SYSTEMS: 12 V 60A alternator.

DIMENSIONS, EXTERNAL:

Wing span	7.62 m (25 ft 0 in)
Wing aspect ratio	6.0
Length overall	5.50 m (18 ft 0½ in)
Height overall	1.90 m (6 ft 2¾ in)
Propeller diameter	1.70 m (5 ft 7 in)

DIMENSIONS, INTERNAL:

Cabin max width	0.94 m (3 ft 1 in)

AREAS:

Wings, gross	9.72 m² (104.6 sq ft)

WEIGHTS AND LOADINGS:

Weight empty	352 kg (775 lb)
Baggage capacity	18 kg (40 lb)
Max T-O weight	576 kg (1,270 lb)
Max wing loading	59.3 kg/m² (12.14 lb/sq ft)
Max power loading	7.73 kg/kW (12.70 lb/hp)

PERFORMANCE:

Never-exceed speed (VNE)	138 kt (255 km/h; 158 mph)
Max level speed	109 kt (202 km/h; 125 mph)
Cruising speed at 75% power	107 kt (198 km/h; 123 mph)
Stalling speed, flaps down, power on	39 kt (73 km/h; 45 mph)
Max rate of climb at S/L	more than 229 m (750 ft)/min
Service ceiling	3,810 m (12,500 ft)
T-O run	137 m (450 ft)
Landing run	151 m (495 ft)
Max range	317 n miles (587 km; 365 miles)

UPDATED

JAG

JAG HELICOPTER GROUP LLC

15866 Sturgeon Street, Roseville, Michigan 48066-1819
Tel: (+1 586) 775 11 84
Fax: (+1 586) 775 13 38
e-mail: radial@bignet.net
Web: http://www.rinke-aerospace.com

Jag is promoter of the world's first turbine-powered, multi-bladed kit helicopter. Company manufacturing plant at Roseville covers 5,110 m² (55,000 sq ft) and employed 32 people in September 2002.

The company has also announced that it intends to develop a four-seat version under the designation JAG 455 (four seats, five main rotor blades, five tail rotor blades) as well as a UAV version. However, by August 2004, the JAG helicopter had been withdrawn while the company concentrates on its UAV version.

UPDATED

JOPLIN

JOPLIN LIGHT AIRCRAFT

PO Box 3805, 531 North Schifferdecker, Joplin, Missouri 64801
Tel: (+1 417) 623 29 59
Fax: (+1 417) 623 76 06
e-mail: jla@compnmore.com
Web: http://www.powerchutes.com/jla.asp
SALES MANAGER: John Lukey

Joplin Light Aircraft manufactures the Tundra and ½ TUN previously produced by Laron Aviation Technologies; it also markets the Suzi-Air engine.

UPDATED

JOPLIN TUNDRA

TYPE: Tandem-seat ultralight kitbuilt.

PROGRAMME: Originally produced by Laron; meets Transport Canada TP 10141 requirements.

CUSTOMERS: Over 30 sold and 25 flying by mid-2002 (latest information received).

COSTS: US$9,500 (2001).

DESIGN FEATURES: High-wing monoplane with high-life wing and pusher engine; cockpit enclosure optional. Quoted build time 250 hours.

FLYING CONTROLS: Manual. Full-span Junkers ailerons.

STRUCTURE: Welded metal tube fuselage. Preformed glass fibre nosecone.

LANDING GEAR: Tricycle type; fixed. Mainwheels attached to sides of pod, mounted on tubular tripod. Ski and float options. Steerable nosewheel.

POWER PLANT: One 47.8 kW (64.1 hp) Rotax 582 two-cylinder two-stroke engine; options include 34.0 kW (45.6 hp) Rotax 503 and 47.7 kW (64 hp) Suzi Air. Fuel capacity 38 litres (10.0 US gallons; 8.3 Imp gallons).

Tundra kitbuilt, previously produced by Laron 0507558

DIMENSIONS, EXTERNAL:
- Wing span ... 9.75 m (32 ft 0 in)
- Length overall ... 6.40 m (21 ft 0 in)
- Height overall ... 1.98 m (6 ft 6 in)

AREAS:
- Wings, gross ... 15.79 m^2 (170.0 sq ft)

WEIGHTS AND LOADINGS:
- Weight empty ... 181 kg (400 lb)
- Max T-O weight ... 408 kg (900 lb)

PERFORMANCE:
- Max level speed ... 87 kt (161 km/h; 100 mph)
- Normal cruising speed at 80% power ... 74 kt (137 km/h; 85 mph)
- Stalling speed ... 31 kt (57 km/h; 35 mph)
- Max rate of climb at S/L ... 366 m (1,200 ft)/min
- T-O run ... 30 m (100 ft)
- Landing run ... 61 m (200 ft)
- Range at max level speed ... 202 n miles (374 km; 232 miles)

UPDATED

JOPLIN ½ TUN

TYPE: Single-seat ultralight kitbuilt.

CUSTOMERS: Three flying by mid-2002 (latest information received).

COSTS: US$5,995 (2001).

DESIGN FEATURES: As Tundra, but meets FAR Pt 103 weight rules. Quoted build time 90 hours.

FLYING CONTROLS: As Tundra.

STRUCTURE: As Tundra.

LANDING GEAR: Tailwheel type.

POWER PLANT: One 29.5 kW (39.6 hp) two-cylinder, liquid-cooled Rotax 447; options include 2 SI 460 F35 and Suzi Air. Fuel capacity 18.9 litres (5.0 US gallons; 4.2 Imp gallons).

DIMENSIONS, EXTERNAL:
- Wing span ... 7.92 m (26 ft 0 in)
- Length overall ... 6.40 m (21 ft 0 in)
- Height overall ... 1.68 m (5 ft 6 in)

AREAS:
- Wings, gross ... 10.87 m^2 (117.0 sq ft)

WEIGHTS AND LOADINGS:
- Weight empty ... 113 kg (250 lb)
- Max T-O weight ... 249 kg (550 lb)

PERFORMANCE:
- Max level speed ... 54 kt (101 km/h; 63 mph)
- Normal cruising speed at 80% power ... 48 kt (89 km/h; 55 mph)
- Stalling speed ... 22 kt (41 km/h; 25 mph)
- Max rate of climb at S/L ... 305 m (1,000 ft)/min
- T-O run ... 46 m (150 ft)
- Landing run ... 31 m (100 ft)

UPDATED

JUST

1904 East Chicago Street, Suite G, Caldwell, Idaho 83605
Tel: (+1 208) 454 34 46
Fax: (+1 208) 454 55 80
e-mail: justaircraft@slowflight.com
Web: http://www.slowflight.com
PRESIDENT: Troy Zehr
MARKETING DIRECTOR: Kathi Jo Zehr

Just has worked in collaboration with Reality Aircraft of the UK to market the Easy Raider and Escapade lightplanes in both countries. In 2004, it launched the Summit, which type is not promoted in the UK.

NEW ENTRY

JUST SUMMIT

TYPE: Tandem-seat ultralight kitbuilt.

PROGRAMME: Marketing launch in 2004.

DESIGN FEATURES: High, constant-chord, foldable wing supported by V-struts and intermediate struts. Quoted build time 400 hours.

FLYING CONTROLS: Conventional and manual. Ailerons and elevator actuated by pushrods; rudder by cables. Three-position flaps.

STRUCTURE: Chrome-moly 4130 welded steel tube fuselage and empennage with fabric covering. Wing built on 6061 aluminium spar with wooden ribs and fabric covering. Glass fibre engine cowling.

LANDING GEAR: Tailwheel type; fixed. Two faired-in side Vs hinged to lower longerons, plus half-axles with bungee cord shock-absorbers attached to compression frame. Brakes. Matco steerable tailwheel. Optional spring aluminium cantilever main legs.

POWER PLANT: One piston engine between 37 and 89 kW (50 and 120 hp). Rotax 503, 582, 912 and 912S or Jabiru 2200 and 3300 all suitable. Fuel in wings, capacity 38 litres (10.0 US gallons; 8.3 Imp gallons) standard; 76 litres (20.0 US gallons; 16.7 Imp gallons) optional.

DIMENSIONS, EXTERNAL:
- Wing span ... 9.83 m (32 ft 3 in)
- Length: overall ... 5.79 m (19 ft 0 in)
  - wings folded ... 6.71 m (22 ft 0 in)
- Height overall ... 1.73 m (5 ft 8 in)

AREAS:
- Wings, gross ... 11.23 m^2 (120.9 sq ft)

WEIGHTS AND LOADINGS:
- Weight empty ... 204 kg (450 lb)
- Max T-O weight ... 449 kg (990 lb)

PERFORMANCE:
- No details available

NEW ENTRY

KAMAN

KAMAN AEROSPACE CORPORATION (SUBSIDIARY OF KAMAN CORPORATION)

Old Windsor Road, PO Box No. 2, Bloomfield, Connecticut 06002
Tel: (+1 860) 243 71 00 or 243 44 61
Fax: (+1 860) 243 75 14
e-mail: kmax-kac@kaman.com
Web: www.kamanaero.com
CHAIRMAN EMERITUS: Charles H Kaman
CHAIRMAN, PRESIDENT AND CEO, KAMAN CORPORATION: Paul H Kuhn
PRESIDENT, KAMAN AEROSPACE CORPORATION, HELICOPTER DIVISION: Salvatore S Bardonaro
VICE-PRESIDENT, ENGINEERING, KAMAN AEROSPACE CORPORATION, HELICOPTER DIVISION: Michael S Bowes
DIRECTOR OF MARKETING AND BUSINESS DEVELOPMENT, KAMAN AEROSPACE CORPORATION, HELICOPTER DIVISION: Melvin P French

Founded 1945 by Charles H Kaman. Developed servo-flap control of helicopter main rotor, initially in contrarotating two-blade main rotors, and still used in SH-2G Super Seasprite four-blade main rotor and on the K-MAX. R&D programmes sponsored by US Army, Air Force, Navy and NASA include advanced design of helicopter rotor systems, blades and rotor control concepts, component fatigue life determination and structural dynamic analysis and testing. In late 2001, company was said to be in early conceptual stages for a multirole helicopter of similar size to the NH 90, with a cabin wide enough to admit stretchers stowed widthways; no further details of this had emerged by early 2005. Kaman has also undertaken helicopter drone programmes since 1953; is continuing advanced research in rotary-wing unmanned aerial vehicles (UAVs).

Kaman's intermeshing rotor system design was a key ingredient in the combination put together by the Northrop Grumman team in a successful bid to become one of two competitors to continue into the next phase of the (subsequently cancelled) DARPA/US Army Unmanned Combat Armed Rotorcraft (UCAR) programme. Kaman is however pursuing US Army interest in an unmanned logistics UAV. Following a USMC demonstration of the K-MAX/BURRO (Broad-area Unmanned Responsive Resupply Operations) in 2001, a demonstration to the US Army was completed in June 2003; a second, for the aviation community, took place in September 2003, and BURRO development was continuing throughout 2004. See *Jane's Unmanned Aerial Vehicles and Targets* for further details.

Kaman is major subcontractor on many aircraft and space programmes, including design, tooling and fabrication of components in metal, metal honeycomb, bonded and composites construction, using techniques such as filament winding, braiding and RTM. Participates in programmes including Northrop Grumman E-2C, Bell Boeing V-22, Boeing 737, 747, 757, 767, 777 wing trailing-edge, KC-135 and C-17, Sikorsky UH-60 and SH-60, and NASA Space Shuttle Orbiter main fuel tank. It also designs and produces fuzes for a range of weaponry.

Kaman designed and, since 1977, has been producing all-composites rotor blades for Bell AH-1 Cobras for US and foreign forces. Appointed in April 2000 to build fuselages for MD Helicopters' single-engine line (MDH 500, 520, 530 and 600) under 10-year contract; a further contract to build MDH Explorer rotor blade systems was awarded in July 2000, but work was suspended in the second quarter of 2003 due to falling sales and a US$21 million MDHI debt written off in the third quarter of 2004. However, Kaman stated its intention of maintaining a business relationship with MDHI if the latter's economic situation improves.

KAC's activities were reorganised in early 2002 into three new business units, of which Helicopter Programs has responsibility for the Super Seasprite and K-MAX described in the following entries. Aerostructures Subcontracting in Bloomfield and Jacksonville, focuses on non-rotary-wing aircraft programmes including commercial airliner wing structures and components; major structural assemblies for military transports; aircraft thrust reversers; and business jet subassembly components. Helicopter Subcontracting, also with operations in Bloomfield and Jacksonville, undertakes work on helicopter airframes, composites rotor blades and components, including manufacture (also suspended in 2003) of MD 500 and MD 600 fuselages for MD Helicopters (which see). In early 2002, Kaman acquired Plastic Fabrication Company of Wichita, Kansas, a manufacturer of composites parts and assemblies for aerospace applications.

Kaman Aerospace International Corporation was established in 1995 as the international affiliate of Kaman Aerospace with offices in Canberra and Nowra, Australia; Kuala Lumpur, Malaysia; Cairo, Egypt; and Bloomfield, Connecticut.

Net sales of Kaman's Aerospace business were US$382.7 million in 1998, US$371.8 million in 1999, US$381.9 million in 2000, US$301.6 million in 2001 and US$275.9 million in 2002. The company employs 1,200 people.

UPDATED

KAMAN (K894) SH-2G SUPER SEASPRITE

TYPE: Naval combat helicopter.

PROGRAMME: Available as either SH-2F remanufactured airframe or as new-build helicopter. Sales promotion continues, with a new (2004) campaign based on BAE Systems Avionics' mission system and sensor suite, aimed at potential NATO customers such as Portugal and Romania.

CURRENT VERSIONS: **SH-2G(E):** Ten for Egypt, ordered 22 February 1995 and delivered between October 1997 and November 1998. Remanufactured from SH-2F.

SH-2G(A): Eleven for Royal Australian Navy, selected in January 1997. Deliveries began on a pre-acceptance basis in March 2001; 10 delivered by July 2004. Remanufactured from SH-2F.

SH-2G(NZ): Five new-build aircraft for Royal New Zealand Navy, delivered between August 2001 and February 2003.

SH-2G: Four stored US Navy aircraft reactivated, with minor modifications, and delivered to Polish Navy between September 2002 and August/September 2003.

CUSTOMERS: Australia (11), Egypt (10), New Zealand (five) and Poland (four), as described above.

UPDATED

Royal Australian Navy SH-2G(A) armed with Penguin missiles *(Kaman)* ***NEW***/1108690

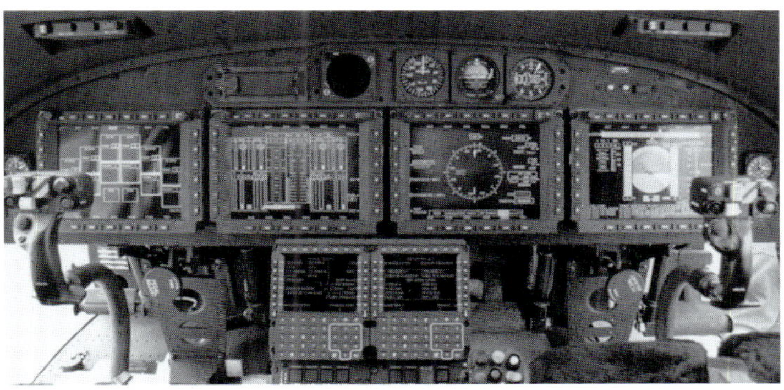

Northrop Grumman Integrated Tactical Avionics System 'glass cockpit' in the SH-2G(A) *(Kaman)* ***NEW***/1023039

KAMAN K-1200 K-MAX

TYPE: Light lift helicopter.

PROGRAMME: First flight (N3182T) 23 December 1991; first public showing 22 March 1992; first flights of second prototype (N131KA) 18 September 1993 and first production aircraft (N132KA) 12 January 1994; third prototype is static and drop test aircraft to prove 20 year life at 1,000 hours per year and with 30 return logging sorties per hour. Certification to FAR Pts 27 and 133 achieved 30 August 1994 after 800 hour/32 month programme. N133KA of Scott Paper (now Kimberly-Clark) achieved 1,000 hours in eight months, 22 June 1995, as first K-MAX to reach this total. Canadian certification awarded 23 November 1994; now certified in Australia, Austria, Canada, Ecuador, Germany, Japan, New Zealand, Switzerland, Taiwan and USA. Uses include logging, firefighting, agricultural spraying, constructing and surveying.

Kaman K-MAX operated by MacKenzie Heli Services of British Columbia *(Kaman)* ***NEW***/1023040

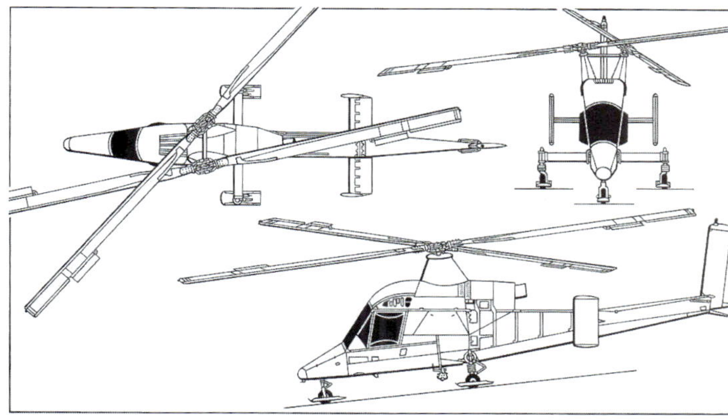

Production version of Kaman K-MAX intermeshing rotor helicopter *(Jane's/Mike Keep)* 0507267

A US$690,000 contract to demonstrate Vertrep (Vertical Replenishment) to US Navy awarded August 1995: two month demonstration period saw two K-MAXs (c/n 0010 and 0013) lift 453,592 kg (1,000,000 lb) with follow-on assessment in Guam during May 1996 during which 142 hours were flown and 2,449,400 kg (5,400,000 lb) was lifted. Two helicopters then began US$5.7 million six-month deployment in Arabian Gulf on 3 June 1996 aboard USS *Niagara Falls*. During late August 1998, the prototype K-MAX was involved in trials for Magic Lantern mine detection systems. FAA Pt 27 IFR certification received 14 May 1999 as part of bid for US Navy Vertrep contract. Kaman received certification for an external seat in June 1999; up to two may be fitted. On 14 July 1999 USMC placed contract worth US$4.2 million for remote piloting package and another worth US$2.7 million to May 2000: for further details see *Jane's Unmanned Aerial Vehicles and Targets*. In November 2001, on behalf of the US Naval Undersea Warfare Center at Newport, Rhode Island, the K-MAX recovered more than two dozen torpedoes used during exercises in local waters.

K-MAX production suspended after completion of 38th aircraft in 2003; seven sales or leases by the end of July 2004, including delivery of the last 'white tail' during that month to Grizzly Mountain Aviation. Marketing continues, but no further production or orders announced as of early 2005.

CUSTOMERS: Kaman moved from its original helicopter lease programme to a sales programme, but reinstated leasing arrangements in 2002, with Eagle Helicopters (Switzerland) and Superior Helicopters (Oregon) becoming lease customers shortly afterwards. Current operators (at July 2003) include Mountain West, Woody Contracting, Superior Helicopter (four), Petroleum Helicopters, Rainier Heli-Lift and Midwest Helicopters in USA; Cariboo Chilcotin and MacKenzie Heli Services in Canada; Helog of Switzerland; Japan Royal Helicopters (two); Rotex Helicopters AG of Lichtenstein; and Wucher Helikopter of Austria.

A K-MAX operated by Woody Contracting was the first to exceed 10,000 flying hours (increased to more than 14,000 hours by mid-2003). In December 2000, US State Department ordered five; these delivered in 2002, and in mid-2003 were performing logistics support for anti-drug operations in Colombia. Total of 37, plus one static test airframe. Of these, seven lost in accidents by mid-2004, at which time 27 were said to be in operation.

UPDATED

KIMBALL

JIM KIMBALL ENTERPRISES INC

PO Box 849, 5354 Cemetary Road, Zellwood, Florida 32798-0849
Tel: (+1 407) 889 34 51
Fax: (+1 407) 889 71 58
e-mail: jwkimball@aol.com
Web(1): http://www.jimkimballenterprises.com
Web(2): http://www.pittsmodel12.com
Web(3): http://www.dr109.com
PRESIDENT: Jim Kimball
VICE-PRESIDENT: Kevin Kimball

Kimball Enterprises was founded in 1980 and offers kits of the Pitts Model 12 Super Stinker after purchasing design rights from Mid-America Aircraft. Plans are marketed through Model 12 Inc at same address.

Recent additions to the Kimball range are the Raptor, McCullocoupe and DR 109 Rhino.

UPDATED

KIMBALL DR 109 RHINO

TYPE: Tandem-seat sportplane kitbuilt.

PROGRAMME: Designed by Dan Rihn as two-seat version of his DR 107 One Design. Added to Kimball product line in 2003. Prototype N109DR first flown 1996.

CUSTOMERS: At least nine on US Register by September 2004.

COSTS: Kit US$50,000 (2004).

DESIGN FEATURES: Low-wing monoplane with a specially designed 16% symmetrical aerofoil, conic nose and flat sides. Roll rate 360°/s.

FLYING CONTROLS: Conventional and manual. Pushrod-actuated ailerons and elevator, cable-actuated rudder. Large, symmetrical, constant-chord ailerons extend to 75 per cent of wing length and are mass balanced and aerodynamically counterbalanced; each fitted with spade ground-adjustable balances. Trim tab on each elevator also acts as a servo tab to reduce stick forces.

STRUCTURE: Fuselage of welded steel tube, sheet metal-covered from firewall aft to cockpit; fabric-covered from rear cockpit aft to empennage. Aluminium sheet metal turtledeck. Fabric-covered, wire-braced tail surfaces have steel tube spars and bent-up steel sheet metal ribs. One-piece wing has Douglas fir main and subsidiary box spars and plywood covering. All-wood ailerons. Two-piece glass fibre engine cowling.

LANDING GEAR: Tailwheel type; fixed. Fuselage-mounted spring aluminium cantilever mainwheel legs; steerable swivel-mounted aluminium spring tail unit. Mainwheels 5.00-5; hydraulic disc brakes.

POWER PLANT: One Textron Lycoming piston engine. Option of 149 kW (200 hp) AEIO-360-AIE6 flat-four; or 194 kW (260 hp) AEIO-540-D4A5; or 224 kW (300 hp) AEIO-540-L1B5 flat-six; all drive an MTV-9 three-blade variable-pitch constant-speed propeller; prototype has AEIO-540-D4A5. Fuel capacity 114 litres (30.0 US gallons; 25.0 Imp gallons) in main tank ahead of cockpit; total auxiliary fuel capacity 61 litres (16.0 US gallons; 13.3 Imp gallons) in two wing tanks.

ACCOMMODATION: Pilot and passenger in tandem cockpits under single-piece side-opening canopy; each seat has adjustable back nominally reclined to 30° and five-point aerobatic seat belt system. Can be flown solo from rear seat. Dual controls.

SYSTEMS: 12 V electrical system; electric engine starter.

AVIONICS: *Comms:* Terra VHF radio, transponder with encoder; Sigtronics voice-activated intercom.

Flight: Grand Rapids Technology engine information system.

DIMENSIONS, EXTERNAL:
- Wing span ... 7.32 m (24 ft 0 in)
- Wing aspect ratio ... 5.1
- Length overall ... 6.71 m (22 ft 0 in)
- Height overall ... 1.88 m (6 ft 2 in)

DIMENSIONS, INTERNAL:
- Cabin max width ... 0.61 m (2 ft 0 in)

AREAS:
- Wings, gross ... 10.59 m^2 (114.0 sq ft)

WEIGHTS AND LOADINGS (A: AEIO-360-A1E6, B: AEIO-540-D4A5, C: AEIO-540-L1B5):
- Weight empty ... 612 kg (1,350 lb)

Kimball (Rihn) DR 109 Rhino **NEW**/1047444

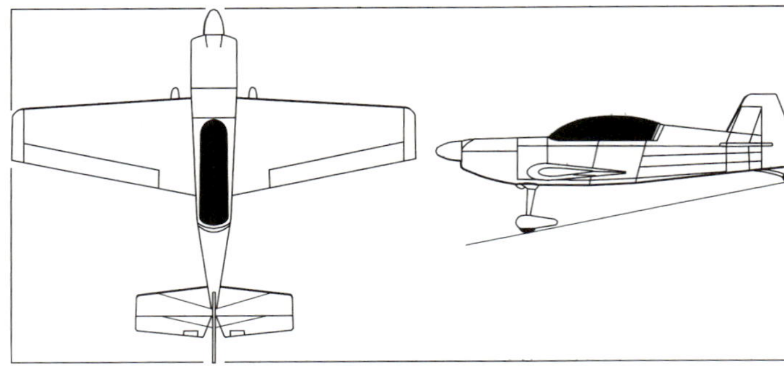

Two aspects of the Kimball (Rihn) DR 109 Rhino *(Paul Jackson)* **NEW**/1047847

- Max T-O weight ... 1,020 kg (2,250 lb)
- Max wing loading ... 96.4 kg/m^2 (19.74 lb/sq ft)
- Max power loading: A ... 6.85 kg/kW (11.25 lb/hp)
- B ... 5.27 kg/kW (8.65 lb/hp)
- C ... 4.56 kg/kW (7.50 lb/hp)

PERFORMANCE (A, B, C as above):
- Max operating speed: A ... 167 kt (309 km/h; 192 mph)
- B ... 178 kt (330 km/h; 205 mph)
- C ... 187 kt (346 km/h; 215 mph)
- Normal cruising speed: A ... 146 kt (270 km/h; 168 mph)
- B ... 155 kt (286 km/h; 178 mph)
- C ... 161 kt (298 km/h; 185 mph)
- Stalling speed: A, B, C ... 53 kt (97 km/h; 60 mph)
- Max rate of climb at S/L ... 853 m (2,800 ft)/min
- T-O run ... 183 m (600 ft)
- Landing from 15 m (50 ft) ... 549 m (1,800 ft)
- Range (45 min reserves): A ... 351 n miles (650 km; 404 miles)
- B ... 391 n miles (724 km; 450 miles)
- C ... 373 n miles (692 km; 430 miles)
- *g* limits ... ±10

NEW ENTRY

KIMBALL MCCULLOCOUPE

TYPE: Side-by-side sportplane kitbuilt.

PROGRAMME: Design initiated late 2000. Construction of prototype began in early 2001, with intended completion date of end 2002. First flight (N511) eventually achieved 1 March 2004 and aircraft made public debut at Sun 'n' Fun in April 2004.

All data are provisional.

DESIGN FEATURES: Similar to clip-wing Monocoupe designs of 1930s, incorporating some Kimball (Pitts) Model 12 items, including engine cowling and landing gear legs.

FLYING CONTROLS: Conventional and manual. Pushrod-actuated ailerons and elevators; cable-operated rudder and tailwheel steering. Mechanisms fully contained within wings to eliminate drag.

STRUCTURE: Fabric-covered tubular steel fuselage. Strut-braced, two-piece wood-covered wing with two Douglas fir main spars and mahogany plywood ribs. Compared to original Monocoupe 110 Special, has larger tail surfaces to improve handling. Strut-braced tailplane. Airfoil Clark Y.

LANDING GEAR: Fuselage-mounted spring aluminium cantilever mainwheel legs (as Kimball-built Pitts Model 12) encased in two-part, telescoping trouser, lower section of which is protected from friction damage when compressed within upper portion. Glass fibre wheel fairings. Tyres 6.00-6.

POWER PLANT: One 294 kW (394 hp) VOKBM M-14PF nine-cylinder radial engine driving a specially designed three-blade constant-speed MT propeller. Fuel capacity 288 litres (76.0 US gallons; 63.3 Imp gallons) in two 98 litre (26.0 US gallon; 21.6 Imp gallon) wing tanks and one 91 litre (24.0 US gallons; 20.0 Imp gallon) tank positioned under seats.

ACCOMMODATION: Pilot and passenger side by side.

DIMENSIONS, EXTERNAL:
- Wing span ... 7.92 m (26 ft 0 in)
- Wing chord at root ... 1.52 m (5 ft 0 in)
- Wing aspect ratio ... 5.4
- Length overall ... 6.86 m (22 ft 6 in)
- Height to top of wing ... 2.44 m (8 ft 0 in)
- Propeller diameter ... 2.49 m (8 ft 2 in)

AREAS:
- Wings, gross ... 11.57 m^2 (124.6 sq ft)

WEIGHTS AND LOADINGS:
- Weight empty ... 635 kg (1,400 lb)
- Baggage capacity ... 36 kg (80 lb)
- Max T-O weight ... 1,134 kg (2,500 lb)
- Max wing loading ... 98.0 kg/m^2 (20.06 lb/sq ft)
- Max power loading ... 3.86 kg/kW (6.35 lb/hp)

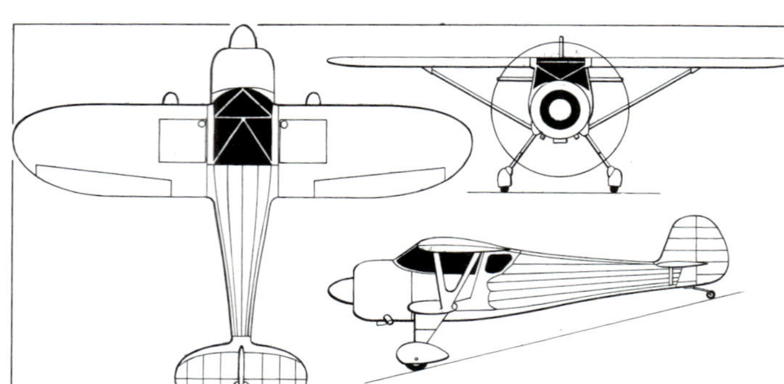

Kimball McCullocoupe general arrangement *(James Goulding)* **NEW**/1047844

Kimball McCullocoupe prototype at its public debut, April 2004 *(Paul Jackson)* ***NEW***/1047401

PERFORMANCE:
Never-exceed speed (VNE) .... 234 kt (434 km/h; 270 mph)
Normal cruising speed .... 182 kt (338 km/h; 210 mph)
Stalling speed .... 57 kt (105 km/h; 65 mph)
Endurance .... 6 h 0 min

UPDATED

KIMBALL MODEL 15 RAPTOR A S

TYPE: Tandem-seat sportplane kitbuilt.
PROGRAMME: Work began late 1999 and project details revealed early 2001. Development took second place to McCullocoupe; construction of prototype began mid-2003.
All data are provisional.
COSTS: Kit US$48,250 (2004).
DESIGN FEATURES: Based on the Kimball (Pitts) Model 12, using tail, landing gear, firewall forward and canopy. Rate of roll in excess of 360°/s.
FLYING CONTROLS: As Kimball (Pitts) Model 12. Horn-balanced rudder.
STRUCTURE: Generally as Kimball (Pitts) Model 12 with new wooden aerobatic wing; composites sports wing will be available later, if market demands.
LANDING GEAR: As Kimball (Pitts) Model 12.
POWER PLANT: Choice of 265 kW (355 hp) VOKBM M-14P or 294 kW (394 hp) M-14PF nine-cylinder radial engine driving a three-blade MT propeller; option to use Textron Lycoming power plant will be added later. Standard fuel capacity as Kimball (Pitts) Model 12; optional capacity 280 litres (74.0 US gallons; 61.7 Imp gallons).
ACCOMMODATION: As Kimball (Pitts) Model 12.
DIMENSIONS, EXTERNAL:
Wing span .... 7.92 m (26 ft 0 in)
Wing aspect ratio 5.8
Length overall .... 5.99 m (19 ft 8 in)
Height overall .... 2.36 m (7 ft 9 in)
AREAS:
Wings, gross .... 10.82 m² (116.5 sq ft)
WEIGHTS AND LOADINGS:
Weight empty .... 637 kg (1,405 lb)
Max T-O weight .... 1,043 kg (2,300 lb)
Max wing loading .... 96.4 kg/m² (19.74 lb/sq ft)
Max power loading: M-14P .... 3.94 kg/kW (6.48 lb/hp)
M-14PT .... 3.56 kg/kW (5.84 lb/hp)
PERFORMANCE:
Normal cruising speed .... 161 kt (298 km/h; 185 mph)
Max rate of climb at S/L .... 1,219 m (4,000 ft)/min
Endurance: with standard fuel .... 3 h 30 min
with max fuel .... 5 h 0 min

UPDATED

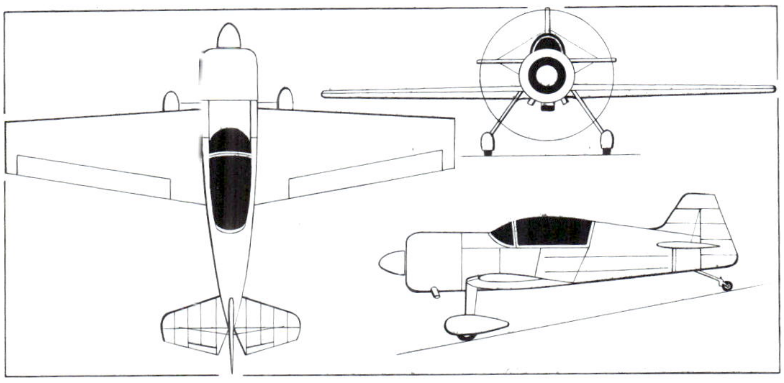

Kimball Model 15 Raptor general arrangement *(James Goulding)* 0121110

KIMBALL (PITTS) MODEL 12

TYPE: Tandem-seat biplane/kitbuilt.
PROGRAMME: Designed by Curtis Pitts and originally named Macho Stinker and, later, Pitts Monster. Design started in 1994; construction of prototype began 1996; first flight March 1996; first delivery April 1997. A 'high-performance' version has been added to the range.
CURRENT VERSIONS: **Model 12:** Baseline version, *as described.*
Model 12-540: As Model 12, but with 224 kW (300 hp) Textron Lycoming AEIO-540 or 246 kW (330 hp) Textron Lycoming AEIO-580 engine. General dimensions as Model 12.
High Performance: Strengthened version with shorter fuselage and wing spans. As standard Model 12, except where noted.
CUSTOMERS: Total of 75 under construction in nine countries: Australia, Canada, Finland, Germany, Iceland, Lithuania, South Africa, Sweden and the USA; plus 220 sets of plans sold in a further six countries. Total of 25 flying by September 2004, including one each in Australia, Canada and South Africa and two in Germany.
COSTS: Model 12 Kit US$47,995; complete aircraft US$185,000. Model 12-540 kit US$46,715; complete aircraft US$195,000 (2004).
DESIGN FEATURES: Traditional Pitts design, optimised for aerobatics; contours dictated by M-14P radial engine. Wire- and strut-braced wings of unequal span; upper wing mounted

Kimball (Pitts) Model 12 *(Paul Jackson)* ***NEW***/1047246

Artist's impression of Lycoming-engine Pitts 12-540 ***NEW***/1047846

on cabane; single interplane strut each side; wire-braced tailplane and fin. Quoted build time 2,500 hours.

FLYING CONTROLS: Conventional and manual. Rod-connected ailerons on upper and lower wings. Flight-adjustable trim tab in each elevator.

STRUCTURE: 4130 steel tube fuselage; wooden spars and ribs; fabric covering with aluminium cockpit sides. Metal interplane struts and cabane, composites engine cowling.

LANDING GEAR: Tailwheel type; non-retractable. Optional composites mainwheel fairings. Aluminium sprung main legs; mainwheels 6.00-6 in; Cleveland disc brakes.

POWER PLANT: One 265 kW (355 hp) VOKBM M-14P or 294 kW (394 hp) P.14PF nine-cylinder radial engine driving an MT MTV-9-K-C/CL260-9 three-blade constant-speed propeller. In 2003, the option of fitting a 224 kW (300 hp) Textron Lycoming AEIO-540 or 246 kW (330 hp) Textron Lycoming AEIO-580 flat-six was added. Fuel capacity 204 litres (54.0 US gallons; 45.0 Imp gallons) in 68 litre (18.0 US gallon; 15.0 Imp gallon) wing tank and 136 litre (36.0 US gallon; 30.0 Imp gallon) fuselage tank. Oil capacity 15 litres (4.0 US gallons; 3.3 Imp gallons).

ACCOMMODATION: Pilot and passenger in enclosed cockpit under one-piece Perspex canopy hinged on right side. Fixed windscreen.

SYSTEMS: 14 V or 28 V battery.

DIMENSIONS, EXTERNAL (A: Model 12; B: High Performance; C: Model 12-540):

Wing span: upper: A, C	6.71 m (22 ft 0 in)
B	7.01 m (23 ft 0 in)
lower: A, C	6.40 m (21 ft 0 in)
B	6.71 m (22 ft 0 in)
Wing chord at tip	1.07 m (3 ft 6 in)
Length overall: A, C	5.99 m (19 ft 8 in)
B	6.25 m (20 ft 6 in)
Max width of fuselage	1.07 m (3 ft 6 in)
Height overall	2.97 m (9 ft 9 in)
to top of upper wing	2.51 m (8 ft 3 in)
Tailplane span	2.44 m (8 ft 0 in)
Wheel track	2.29 m (7 ft 6 in)
Wheelbase	4.42 m (14 ft 6 in)
Propeller diameter	2.49 m (8 ft 2 in)
Propeller ground clearance	0.36 m (1 ft 2 in)

AREAS:

Wings, gross: A, C	14.01 m² (150.8 sq ft)
B	13.94 m² (150.1 sq ft)
Ailerons (total)	1.88 m² (20.21 sq ft)
Fin	0.34 m² (3.69 sq ft)
Rudder, incl tab	0.63 m² (6.82 sq ft)
Elevators, incl tab	0.68 m² (7.30 sq ft)

WEIGHTS AND LOADINGS:

Weight empty: A	703 kg (1,550 lb)
B	669 kg (1,475 lb)
C	624 kg (1,375 lb)
Baggage capacity	18 kg (40 lb)
Max T-O weight: A, B	1,043 kg (2,300 lb)
C	1,020 kg (2,250 lb)
Max wing loading: A	74.5 kg/m² (15.25 lb/sq ft)
B	74.8 kg/m² (15.32 lb/sq ft)
C	72.8 kg/m² (14.92 lb/sq ft)
Max power loading: A, B	3.94 kg/kW (6.47 lb/hp)
C AEIO-540	4.56 kg/kW (7.50 lb/hp)
AEIO-580	4.15 kg/kW (6.82 lb/hp)

PERFORMANCE:

Never-exceed and max operating speed (VNE, VMO)	207 kt (384 km/h; 239 mph)
Max cruising speed	178 kt (330 km/h; 205 mph)
Econ cruising speed, 75% power	152 kt (282 km/h; 175 mph)
Stalling speed, power off	56 kt (103 km/h; 64 mph)
Roll rate: A	220°/s
B	300°/s
Max rate of climb at S/L	More than 914 m (3,000 ft)/min
T-O run	76 m (250 ft)
T-O to 15 m (50 ft)	91 m (300 ft)
Landing from 15 m (50 ft)	274 m (900 ft)
Range with max fuel	525 n miles (972 km; 604 miles)

UPDATED

KINETIC

KINETIC AVIATION

Renamed Montagne Aviation.

KLS

KLS COMPOSITES

15425 Drayton Pike, Sale Creek, Tennessee 37373-9762
Tel: (+1 423) 451 02 09
Fax: (+1 423) 451 02 08
e-mail: sales@klscomposites.com
Web: http://www.klscomposites.com
PRESIDENT: Stanley Montgomery

KLS produces kits of, and provides support for, the former Glassic Composites SQ-2000 canard four-seater, 40 of which had been sold (and three were flying) by 2003.

A seven-seat utility kitbuilt, the SUA-7, was shown as an apparently unflown prototype at the 2003 Sun 'n' Fun, but it did not appear at the following year's event, nor have there been any further reports of progress.

UPDATED

Prototype KLS SUA-7 Sport Utility Airplane *(Geoffrey P Jones)* ***NEW***/0567226

KOLB

NEW KOLB AIRCRAFT COMPANY INC

8375 Russell Dyche Highway, London, Kentucky 40741
Tel: (+1 606) 862 96 92
Fax: (+1 606) 862 96 22
e-mail: customersupport@trikolbaircraft.com
Web: http://www.tnkolbaircraft.com
PLANT MANAGER: Ray Brown

The New Kolb Aircraft Company purchased design rights to Kolb series of light aircraft in 1999. FireStar continues to be marketed; the Laser has not been put into production, while in 2004, Kolb discontinued production of the Sling Shot. Kolbra launched at EAA convention, Oshkosh, July 2000. Total sales of all products exceed 1,500. In its 1,347 m² (14,500 sq ft) factory, Kolb has 18 employees and also produces assemblies for other kit manufacturers, including Rihn Aircraft.

In April 2002, Kolb announced that it had been appointed US assembly centre and distributor for the Ultravia Pelican Sport (which see) which is marketed as the Kolb Sport 600.

UPDATED

KOLB FIREFLY

TYPE: Single-seat ultralight kitbuilt.

PROGRAMME: Introduced at Sun 'n' Fun 1995.

CUSTOMERS: Over 100 flying by mid-2002.

COSTS: US$10,729 including engine (2004).

DESIGN FEATURES: High, constant-chord wing and boom-mounted empennage, based on FireStar; meets FAR Pt 103.7. Folding wings and tail for easy storage. Quoted build time 400 hours.

Modified NACA 4412 series aerofoil.

FLYING CONTROLS: Full-span flaperons.

STRUCTURE: Welded 4130 chromoly steel fuselage cage; 127 mm (5 in) rear fuselage tube and wing spar. Glass fibre nosecone; aluminium tube, fabric-covered wings.

Kolb FireFly single-seat ultralight kitbuilt *(Paul Jackson)* ***NEW***/1047276

LANDING GEAR: Tailwheel type; fixed. Fuselage-mounted spring cantilever mainwheel legs; speed fairings and brakes optional.

POWER PLANT: One 29.5 kW (39.6 hp) Rotax 447 UL two-cylinder piston engine with Rotax 2.58:1 reduction unit, driving WarpDrive three-blade ground adjustable wooden propeller (three-blade IVO propeller optional). Fuel capacity 18.9 litres (5.0 US gallons; 4.2 Imp gallons).

ACCOMMODATION: Optional full enclosure cockpit.

DIMENSIONS, EXTERNAL:

Wing span	6.71 m (22 ft 0 in)
Width, wings folded	1.75 m (5 ft 9 in)
Length overall	5.94 m (19 ft 6 in)
Height overall	1.75 m (5 ft 9 in)

AREAS:

Wings, gross	10.87 m² (117.0 sq ft)

WEIGHTS AND LOADINGS:
Weight empty .... 115 kg (253 lb)
Max T-O weight .... 226 kg (500 lb)
PERFORMANCE:
Never-exceed speed (VNE) .... 69 kt (128 km/h; 80 mph)
Max operating speed .... 60 kt (112 km/h; 70 mph)
Stalling speed .... 24 kt (44 km/h; 27 mph) CAS
Max rate of climb at S/L .... 305 m (1,000 ft)/min
T-O and landing run .... 46 m (150 ft)
g limits .... +4/–2

UPDATED

KOLB KOLBRA

TYPE: Tandem-seat ultralight kitbuilt.
PROGRAMME: First flown 7 May 2000 and publicly announced July 2000. Debut at Oshkosh that month. Prototype N713KA.
CUSTOMERS: Over 50 kits sold and 12 aircraft flying by mid-2002.
COSTS: Kolbra kit US$24,323 including engine and covering (2004).
DESIGN FEATURES: Comprises Mark III wings and tail, FireStar nose and Sling Shot-style landing gear; resulting hybrid then optimised for training role. Quoted build time 450 hours.
FLYING CONTROLS: Manual. Full-span flaperons.
STRUCTURE: Welded 4130 chromoly steel fuselage cage; 152 mm (6 in) diameter rear fuselage tube and wing spar. Glass fibre nosecone; aluminium tube wings, fabric-covered.
LANDING GEAR: Tailwheel type; fixed. Fuselage-mounted spring chromoly cantilever mainwheel legs with Matco wheels and high flotation or tundra tyres. Optional hydraulic disc brakes and wheel fairings.
POWER PLANT: One 47.8 kW (64.1 hp) Rotax 582 UL two-cylinder piston engine with 'B' type 2.58:1 ratio gearbox, or 'C' or 'E' type 3.47:1 ratio gearbox, driving a two- or three-blade IVO or three-blade Warp Drive propellers. Alternatively, a 59.6 kW (79.9 hp) Rotax 912 UL or 73.5 kW (98.6 hp) Rotax 912 ULS can be fitted driving 1.83 m (6 ft 0 in) two- or 1.63 m (5 ft 4 in) three-blade IVO or Warp Drive propeller.
Fuel capacity (both) 37.9 litres (10.0 US gallons; 8.3 Imp gallons).
SYSTEMS: Optional electric starter and dual Magnum strobes.
EQUIPMENT: Optional BRS ballistic parachute.
DIMENSIONS, EXTERNAL:
Wing span .... 8.64 m (28 ft 4 in)
Length overall .... 7.32 m (24 ft 0 in)
Height overall, three-blade propeller .... 2.03 m (6 ft 8 in)
AREAS:
Wings, gross .... 14.31 m^2 (154.0 sq ft)
WEIGHTS AND LOADINGS (A: Kolbra Trainer with Rotax 582, B: King Kolbra with Jabiru):
Weight empty: A .... 225 kg (496 lb)
B .... 249 kg (550 lb)
Max T-O weight: A, B .... 454 kg (1,000 lb)
PERFORMANCE:
Never-exceed speed (VNE):
A, B .... 95 kt (177 km/h; 110 mph)

Kolb Kolbra tandem-seat ultralight *(Paul Jackson)* ***NEW***/1047413

Normal cruising speed: A .... 65 kt (121 km/h; 75 mph)
B .... 74 kt (137 km/h; 85 mph)
Stalling speed, flaps down: A .... 31 kt (57 km/h; 35 mph)
B .... 40 kt (73 km/h; 45 mph)
Max rate of climb at S/L: A, B .... 305 m (1,000 ft)/min
T-O and landing run: A, B .... 76 m (250 ft)
Range: A .... 160 n miles (296 km; 184 miles)
B .... 250 n miles (463 km; 287 miles)
g limits .... +4/–2

UPDATED

KOLB MARK III EXTRA

TYPE: Side-by-side ultralight kitbuilt.
PROGRAMME: Based on the TwinStar of 1984 and its 1990 Mark III Classic derivative.
CUSTOMERS: Over 50 kits sold by June 2004 (plus over 300 of Classic).
COSTS: US$11,058 excluding engine, propeller, instruments, upholstery and covering (2004).
DESIGN FEATURES: Aerodynamic refinements to fuselage, including chisel nose, devised by Barnaby Wainfan. Wings and tail fold for transportation and storage. Quoted build time 450 hours.
FLYING CONTROLS: Manual. Half-span ailerons.
STRUCTURE: Welded 4130 chromoly steel fuselage cage. Glass fibre nosecone; fabric-covered aluminium tube wings.
LANDING GEAR: Tailwheel type; fixed. As Kolb Kolbra.
POWER PLANT: Choice of either 47.8 kW (64.1 hp) Rotax 582 or 59.6 kW (79.9 hp) Rotax 912 UL driving a three-blade propeller. Fuel capacity 37.9 litres (10.0 US gallons; 8.3 Imp gallons).
ACCOMMODATION: Pilot and passenger in side-by-side seating. Doors open forward and are removable.
EQUIPMENT: Optional BRS ballistic parachute.
DIMENSIONS, EXTERNAL:
Wing span .... 9.14 m (30 ft 0 in)
Length overall .... 7.32 m (24 ft 0 in)
Height overall .... 1.93 m (6 ft 4 in)
AREAS:
Wings, gross .... 14.87 m^2 (160.0 sq ft)
WEIGHTS AND LOADINGS:
Weight empty .... 204 kg (450 lb)
Max T-O weight .... 453 kg (1,000 lb)
PERFORMANCE (A: Rotax 582; B: Rotax 912):
Never-exceed speed (VNE) .... 86 kt (160 km/h; 100 mph)
Normal cruising speed: A .... 70 kt (129 km/h; 80 mph)
B .... 78 kt (145 km/h; 90 mph)
Stalling speed: A .... 33 kt (62 km/h; 38 mph)
B .... 36 kt (66 km/h; 41 mph)
Max rate of climb at S/L: A .... 244 m (800 ft)/min
B .... 335 m (1,100 ft)/min
T-O and landing run: A .... 61 m (200 ft)
B .... 53 m (175 ft)
Range: A .... 200 n miles (370 km; 230 miles)
g limits .... +4/–2

UPDATED

Kolb Mark III Extra two-seat ultralight *(Paul Jackson)* ***NEW***/1047443

LANCAIR

LANCAIR GROUP INC

Formed in 1984, the Lancair Group comprises Lancair International, producing kit aircraft; and Lancair Company, marketing factory-built aircraft. On 31 January 2003, the Lancair kit company was sold to Joseph C Bartels, but remains at the same site and with the same staff.
Lancair is also US dealer for the Italian-built HeliSport CH-7 Kompress.

VERIFIED

LANCAIR

LANCAIR INTERNATIONAL

2244 Airport Way, Redmond, Oregon 97756
Tel: (+1 541) 923 22 44
Fax: (+1 541) 923 22 55
e-mail: tech@lancair-kits.com
Web: http://www.lancair-kits.com
CEO: Joseph Bartels
SALES MANAGER: Kim Lorentzen
ENGINEERING MANAGER: Tim Ong

More than 2,000 Lancairs (including out-of-production Model 235) sold to customers in 40 countries; around 800 are currently flying.
In July 1999 Lancair International announced the Legacy 2000, deliveries of which began in August 2000, superseding the earlier 320/360 series, which last appeared in the 1999–2000 *Jane's*. Lancair International sold to Joseph Bartels on 30 January 2003 but remains at Redmond with same employees.

UPDATED

LANCAIR LANCAIR IV, IV-P AND SENTRY

TYPE: Four-seat kitbuilt.
PROGRAMME: Kit deliveries began 1990. On 20 February 1991, prototype set NAA world speed record between San Francisco and Denver of 314.7 kt (583.2 km/h; 362.4 mph). First kit completed July 1991.

Lancair IV-P four-seat, high-performance kitbuilt *(Paul Jackson)* ***NEW***/1047261

Lancair Propjet IV-P turboprop *(Paul Jackson)* ***NEW***/1047260

CURRENT VERSIONS: **Lancair IV:** Standard version.

Description mainly applies to Lancair IV.

Lancair IV-P: Pressurised version; first flight 1 November 1993. Provides 0.34 bar (5.0 lb/sq in) differential; equipment from Dukes Research of California.

Lancair Propjet IV-P: Powered by 560 kW (751 shp) Walter M 601E turboprop; prototype (N750TL, registered as a **IV-TP**) first flown 9 July 2001. Available as option to standard IV-P at cost of US$7,500; completed aircraft can be retrospectively converted; main structural alteration is repositioning of firewall.

On 11 September 2002, a Lancair IV-P Propjet flown by Wesley E Behel Jr claimed a Class C-1c Group 2 speed record over a 3 km course at restricted altitude by flying at 307 kt (568 km/h; 352 mph).

Lancair Sentry: Tandem-seat version based on Lancair IV and retaining wing and lower fuselage, with redesigned upper fuselage and vertical tail, latter of 10 per cent increased area. Prototype (N806EY) first flown 7 September 2002. Power supplied by same engine as Propjet IV-P, but with larger fuel tank. Intended for military training and civil air racing markets.

CUSTOMERS: Total of 500 kits sold; more than 250 flying by mid-2003. The Mexican Navy is to assess the Sentry, to augment its Aermacchi Redigo fleet.

COSTS: Fastbuild US$79,900. Pressurised version: Fastbuild US$104,900. Firewall fastbuild options and two-week builder workshop programmes available. Turbine IV-P US$115,900; Walter engine US$84,000. Sentry US$119,900 (2004).

DESIGN FEATURES: Conventional seating for four persons in airframe of new design, but following Lancair formula for high cruising speeds. Optional winglets. Quoted fastbuild construction time 1,000 hours.

FLYING CONTROLS: Conventional and manual. Fowler flaps. Sidestick controllers for both pilots. Trim tabs in port aileron and port elevator. Optional speed brakes.

STRUCTURE: Carbon fibre/epoxy airframe, with Nomex honeycomb cores.

LANDING GEAR: Tricycle layout; retractable. Mainwheels 15×6.00-6; nosewheel is 5.00-5. Cleveland dual-piston caliper brakes. Hydraulic actuation.

POWER PLANT: One 261 kW (350 hp) Teledyne Continental TSIO-550-E1B, -E2B or -E3B twin-turbocharged flat-six engine driving a three- or four-bladed MT or Hartzell constant-speed propeller. Walter M 601E turboprop can also be fitted; as described above, with three-blade Avia V-508E/2134 propeller. Two versions of PT-6A turboprop tested during 1999, but apparently not proceeded with. Lancair IV-P N540LM flew September 2000 as testbed for 485 kW (650 hp) Eagle 540 piston engine. Fuel in integral wing tanks; total standard usable capacity in IV-P 341 litres (90.0 US gallons; 74.9 Imp gallons); long-range tanks, total capacity 416 litres (110 US gallons; 91.6 Imp gallons); in Propjet IV-P 473 litres (125 US gallons; 104 Imp gallons), optional long-range tanks increasing this to 568 litres (150 US gallons; 125 Imp gallons). Oil capacity 11.4 litres (3.0 US gallons; 2.5 Imp gallons).

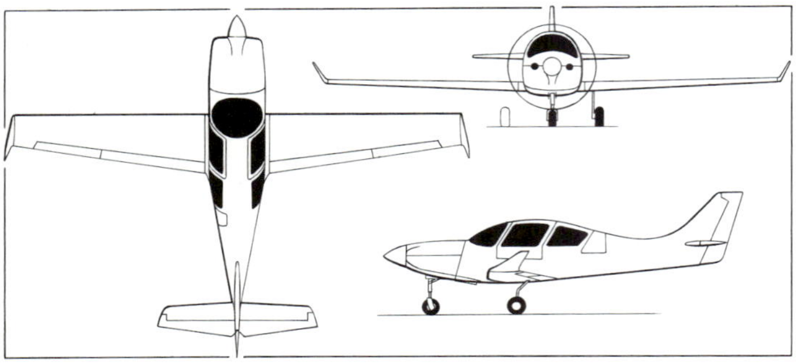

Lancair IV four-seat light aircraft *(James Goulding)* 0106496

SYSTEMS: Thermoelectric de-icing system tested during 2001.

DIMENSIONS, EXTERNAL:

Wing span: plain tips	9.19 m (30 ft 2 in)
winglets	9.93 m (32 ft 7 in)
Wing chord: at root	1.18 m (3 ft 10½ in)
at tip	0.77 m (2 ft 6¼ in)
Wing aspect ratio	9.3
Width, wings removed	2.41 m (7 ft 11 in)
Length overall: IV-P	7.62 m (25 ft 0 in)
Propjet IV-P	7.92 m (26 ft 0 in)
Height overall	2.44 m (8 ft 0 in)
Tailplane span	3.35 m (11 ft 0 in)
Wheel track	2.13 m (7 ft 0 in)
Wheelbase	1.88 m (6 ft 2 in)
Propeller diameter	1.93 m (6 ft 4 in)
Propeller ground clearance	0.25 m (10 in)

DIMENSIONS, INTERNAL:

Cabin: Length	3.20 m (10 ft 6 in)
Max width	1.17 m (3 ft 10 in)
Max height	1.22 m (4 ft 0 in)

AREAS:

Wings, gross: plain tips	9.10 m² (98.0 sq ft)
winglets	10.03 m² (108.0 sq ft)
Winglets, total	1.30 m² (14.00 sq ft)
Flaps, total	1.12 m² (12.07 sq ft)

WEIGHTS AND LOADINGS:

Weight empty: IV	907 kg (2,000 lb)
IV-P	998 kg (2,200 lb)
Baggage capacity	68 kg (150 lb)
Max T-O weight	1,610 kg (3,550 lb)
Max landing weight	1,451 kg (3,200 lb)
Max wing loading	176.9 kg/m² (36.22 lb/sq ft)
Max power loading: IV-P	6.17 kg/kW (10.14 lb/hp)
Propjet IV-P	2.88 kg/kW (4.72 lb/shp)

PERFORMANCE:

Never-exceed speed (VNE)	274 kt (507 km/h; 315 mph)
Cruising speed at 75% power at 7,315 m (24,000 ft):	
IV-P	291 kt (539 km/h; 335 mph)
Propjet IV-P	322 kt (595 km/h; 370 mph)
Stalling speed:	
IV-P standard wing	66 kt (121 km/h; 75 mph)
IV-P with winglets	61 kt (113 km/h; 70 mph)
Propjet IV-P	65 kt (120 km/h; 74 mph)
Max rate of climb at S/L: IV-P	792 m (2,600 ft)/min
Propjet IV-P	1,158 m (3,800 ft)/min
Service ceiling	8,840 m (29,000 ft)
T-O run	457 m (1,500 ft)
T-O to 15 m (50 ft)	549 m (1,800 ft)
Landing run	518 m (1,700 ft)
Range, with max fuel, no reserves:	
IV-P	1,347 n miles (2,494 km; 1,550 miles)
Propjet IV-P	999 n miles (1,850 km; 1,150 miles)
Endurance	4 h 48 min
g limits: IV-P	+4.4/–2.3 (allowable), +8/–4 ultimate
Propjet IV-P	+3.8/–1.9

UPDATED

LANCAIR LANCAIR ES AND SUPER ES

TYPE: Four-seat kitbuilt.

PROGRAMME: Prototype unveiled at Oshkosh in 1992; flight testing completed 1995. Certification of Fastbuild kit granted early 1996 under simplified version of FAR Pt 23 governing light aircraft. Further improvements to Fastbuild kit introduced in early 2002.

CURRENT VERSIONS: **ES:** Baseline version.

ES-P: Pressurised version; introduced 3 March 2004; carbon fibre wings and fuselage.

Super ES: Higher-powered version.

Description applies mainly to this version.

Lancair ES four-seat, fixed-gear lightplane *(Paul Jackson)* ***NEW***/1047221

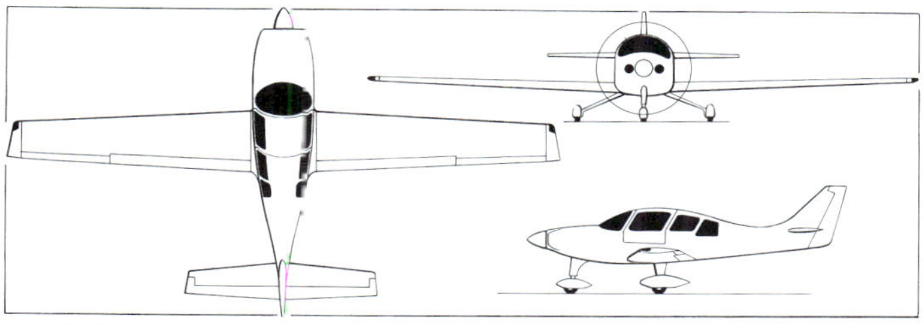

Lancair Super ES general arrangement *(James Goulding)* 0126951

CUSTOMERS: Total of 163 kits sold by early 2003, of which 50 flying by September 2004. Production continues.

COSTS: Standard Fastbuild US$58,900; New Fastbuild kit US$68,900; ES-P US$94,900 (2004).

DESIGN FEATURES: Generally as for earlier Lancairs; hardtop cabin with paired side-by-side seats. Sidestick controls. Quoted Fastbuild kit construction time 1,800 hours; with new 2002 improvements reduced to quoted time of 700 hours.

FLYING CONTROLS: Generally as for Lancair IV.

STRUCTURE: All-composites.

LANDING GEAR: Tricycle type; fixed. Fuselage-mounted spring cantilever mainwheel legs with wheel fairings. Nosewheel steerable through 90°. Main tyres 6.00–15×6; nose 5.00-5.

POWER PLANT: *ES:* One 149 kW (200 hp) Teledyne Continental IO-360-ES flat-six engine; three-blade MTV-12-D/188-17 constant-speed propeller.

Super ES: One 209 kW (280 hp) Teledyne Continental IO-550-G or 231 kW (310 hp) IO-550-N with choice of three-blade Hartzell PHC-J3YF-1RF/F7693DF or MTV-9-D/198-58 constant-speed propeller. Standard fuel capacity 360 litres (95.0 US gallons; 79.1 Imp gallons) in two wing tanks. Oil capacity 7.6 litres (2.0 US gallons; 1.7 Imp gallons).

ES-P: One 231 kW (310 hp) Teledyne Continental TSIO-550C. Fuel capacity 397 litres (105 US gallons; 87.4 Imp gallons).

ACCOMMODATION: Pilot and three passengers in two pairs of seats. Optional second door introduced 2004.

DIMENSIONS, EXTERNAL:
Wing span 10.82 m (35 ft 6 in)
Wing chord: at root 1.50 m (4 ft 11 in)
at tip 0.91 m (3 ft 0 in)
Wing aspect ratio 9.0
Width, wings removed 2.41 m (7 ft 11 in)
Length overall 7.62 m (25 ft 0 in)
Height overall 2.34 m (7 ft 8 in)
Tailplane span 4.01 m (13 ft 2 in)
Wheel track 2.36 m (7 ft 9 in)
Wheelbase 1.88 m (6 ft 2 in)
Propeller diameter 1.93 m (6 ft 4 in)

DIMENSIONS, INTERNAL:
Cabin: as Lancair IV
Baggage hold volume 1.0 m^3 (35 cu ft)

AREAS:
Wings, gross 13.01 m^2 (140.0 sq ft)

WEIGHTS AND LOADINGS:
Weight empty:
ES, Super ES with IO-550-G 862 kg (1,900 lb)
Super ES with IO-550N 907 kg (2,000 lb)
Baggage capacity 79 kg (175 lb)
Max T-O weight: Normal 1,451 kg (3,200 lb)
Utility 1,361 kg (3,000 lb)
ES-P 1,587 kg (3,500 lb)
Max wing loading: Normal 111.60 kg/m^2 (22.86 lb/sq ft)
Utility 104.6 kg/m^2 (21.43 lb/sq ft)
ES-P 122.1 kg/m^2 (25.00 lb/sq ft)
Max power loading: Normal:
ES 9.74 kg/kW (16.00 lb/hp)
Super ES 6.96 kg/kW (11.43 lb/hp)
Utility: ES 9.13 kg/kW (15.00 lb/hp)
Super ES 6.52 kg/kW (10.71 lb/hp)
ES-P 6.87 kg/kW (11.29 lb/hp)

PERFORMANCE:
Never-exceed speed (VNE) 220 kt (467 km/h; 253 mph)
Cruising speed at 75% power at FL100:
ES 174 kt (332 km/h; 200 mph)
Super ES 196 kt (362 km/h; 225 mph)
Cruising speed at 75% power at FL240: ES-P 255 kt (472 km/h; 293 mph)
Stalling speed: ES 50 kt (92 km/h; 57 mph)
Super ES 57 kt (105 km/h; 65 mph)
Rate of climb at S/L: ES 381 m (1,250 ft)/min
Super ES 610 m (2,000 ft)/min
Service ceiling: ES, Super ES 5,490 m (18,000 ft)
ES-P 7,315 m (24,000 ft)
T-O run 183 m (600 ft)
Landing run 244 m (800 ft)
Range, max fuel, no reserves:
ES 1,260 n miles (2,333 km; 1,450 miles)
Super ES 1,277 n miles (2,365 km; 1,470 miles)
ES-P 1,042 n miles (1,931 km; 1,200 miles)
Endurance: ES 7 h
Super ES 6 h
g limits: Normal +4.4/–2.3
Utility +9.0/–4.5

UPDATED

LANCAIR LEGACY

TYPE: Side-by-side kitbuilt.

PROGRAMME: Prototype (N199L) first flown 16 July 1999, publicly displayed at Oshkosh 28 July 1999. First kitbuilt example (N540L) first flown 15 May 2001. Early aircraft were marked "Legacy 2000".

Lancair Legacy side-by-side kitbuilt lightplane *(Paul Jackson)* ***NEW***/1047372

At Reno, Nevada, on 13 September 2002, a Legacy flown by Frederick E Schrameck claimed a Class C-1c Group 1 record for speed, over a 3 km course at restricted altitude, of 297 kt (550 km/h; 342 mph).

CURRENT VERSIONS: **Legacy:** Retractable landing gear version, *as described*.

Legacy FG: Fixed landing gear version introduced at Sun 'n' Fun 2003; as standard Legacy, but built mainly of glass fibre, rather than carbon fibre. Prototype N359L.

CUSTOMERS: Ten sold during Oshkosh debut; 16 by end of 1999, 150 by May 2001; by September 2002, over 100 had been delivered. Sales exceeded 262 kits by September 2004, by which time 24 aircraft were flying.

COSTS: Legacy US$49,900; Legacy FG US$32,900 (2003).

DESIGN FEATURES: Improved version of Lancair series of two-seat kitbuilts, offering increased cabin area, new wing and better performance; only part common to Lancair 360 is horizontal stabiliser. Quoted build time under 1,000 hours.

Streamlined design. Sweptforward wing trailing-edge, tapered tailplane and sweptback fin; low wing and mid-mounted horizontal tail.

Wing section NLF(1)-0215F; incidence 1°; dihedral 3°; twist 1° 30'; thickness/chord ratio 15 per cent.

FLYING CONTROLS: Conventional and manual. Fowler flaps. Aileron and elevator electric trim system included. Optional rudder trim and speedbrakes.

STRUCTURE: Carbon fibre airframe.

LANDING GEAR: Tricycle type; retractable. Mainwheels retract inwards; nosewheel rearwards. Hydraulic actuation. Cleveland 5.00-5 in wheels and brakes.

POWER PLANT: *Legacy:* Initially offered with one 149 kW (200 hp) Textron Lycoming IO-360-C1D6 flat-four driving a two-blade Hartzell or three-blade MTV-12B constant-speed propeller. Hartzell introduced 7694-4T two-blade propeller specifically for Legacy in February 2002. By 2003 this had been upgraded to a 231 kW (310 hp) Teledyne Continental IO-550-N driving a three-blade constant-speed propeller. Optional engines include 134 kW (180 hp) Lycoming IO-360-B1F and -M1A and 194 kW (260 hp) Lycoming IO-540. Standard fuel capacity 250 litres (66.0 US gallons; 55.0 Imp gallons).

Experimental example with supercharger displayed at Sun 'n' Fun 2002, offering 295 kt (547 km/h; 340 mph) at 4,570 m (15,000 ft).

Legacy FG: One 149 kW (200 hp) Textron Lycoming IO-360-A1B6 flat-four driving a Hartzell HC-F2YR-1FX/F7496X Scimitar two-blade propeller. Alternatively a 134kW (180 hp) Superior TNIO-360 Turbo Normalised XP-360 can be fitted. Standard fuel capacity 246 litres (65.0 US gallons; 54.1 Imp gallons).

ACCOMMODATION: Pilot and passenger in 25° reclined side-by-side seats beneath forward-hinging, one-piece canopy.

DIMENSIONS, EXTERNAL:
Wing span 7.77 m (25 ft 6 in)
Wing aspect ratio 7.9
Length overall 6.71 m (22 ft 0 in)
Height overall 2.54 m (8 ft 4 in)
Tailplane span 2.54 m (8 ft 4 in)
Wheelbase 1.37 m (4 ft 6 in)

DIMENSIONS, INTERNAL:
Cabin: Length 1.60 m (5 ft 3 in)
Max width 1.10 m (3 ft 7½ in)
Max height 1.17 m (3 ft 10 in)

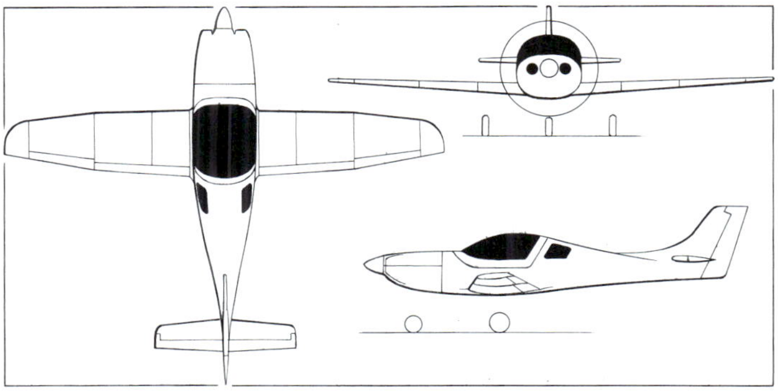

Lancair Legacy general arrangement *(James Goulding)* 0126950

Lancair Legacy two-seat kitplane *(Paul Jackson)* ***NEW***/0567777

Baggage area .... 1.49 m³ (16.0 cu ft)
AREAS:
Wings, gross .... 7.66 m² (82.5 sq ft)
WEIGHTS AND LOADINGS (A: Legacy; B: Legacy FG):
Weight empty: A .... 680 kg (1,500 lb)
B .... 590 kg (1,300 lb)
Baggage capacity: A, B .... 41 kg (90 lb)
Max T-O weight:
Normal category: A, B .... 997 kg (2,200 lb)
Utility category: A, B .... 861 kg (1,900 lb)
Max wing loading:
Normal category: A, B .... 130.2 kg/m² (26.67 lb/sq ft)
Utility category: A, B .... 112.4 kg/m² (23.03 lb/sq ft)
Max power loading:
Normal category: A .... 4.32 kg/kW (7.10 lb/hp)
B .... 6.70 kg/kW (11.00 lb/hp)
Utility category: A .... 3.73 kg/kW (6.13 lb/hp)
B .... 5.78 kg/kW (9.50 lb/hp)

PERFORMANCE (A: Legacy; B: Legacy FG):
Max cruising speed at FL180:
A .... 261 kt (483 km/h; 300 mph)
Normal cruising speed at FL80:
A .... 240 kt (444 km/h; 276 mph)
B .... 182 kt (338 km/h; 210 mph)
at FL120: A .... 217 kt (402 km/h; 250 mph)
Stalling speed, flaps down: A .... 59 kt (108 km/h; 67 mph)
B .... 57 kt (105 km/h; 65 mph)
Max rate of climb at S/L: A .... 914 m (3,000 ft)/min
B .... 792 m (2,600 ft)/min
Service ceiling: A, B .... 5,485 m (18,000 ft)
T-O run: A .... 244 m (800 ft)
B .... 305 m (1,000 ft)
Landing run: A, B .... 275 m (900 ft)
Range with max fuel, no reserves:
A .... 1,042 n miles (1,931 km; 1,200 miles)
B .... 1,173 n miles (2,172 km; 1,350 miles)
g limits: Normal: A, B .... +3.8/–1.7
Utility: A, B .... +4.4/–2.2

UPDATED

LANCAIR

LANCAIR COMPANY

22550 Nelson Road, Bend, Oregon 97701
Tel: (+1 541) 318 11 44
Fax: (+1 541) 318 11 77
e-mail: fly@lancair.com/certified
Web: http://www.lancair.com/certified
PRESIDENT: Bing Lantis
CHIEF TECHNOLOGY OFFICER: Lance Neibauer
VICE-PRESIDENT, MARKETING AND SALES: Mark Cahill

The Lancair company was formed as an autonomous firm in 1994 to complement the kit producing division of the Lancair Group. Its first certified product is the four-seat Columbia 300, which was unveiled at Oshkosh in August 1997 and assembled at a new 13,100 m² (141,000 sq ft) facility at Bend Municipal Airport, Bend, Oregon; a 3,710 m² (40,000 sq ft) extension was started in October 2003 with completion and utilisation planned by end 2004.

The company temporarily ceased production in August 2002 following financing difficulties arising from downturn in financial markets that year; production recommenced in January 2003. Factory manpower was 400 in May 2004 and was expected to reach 500 by January 2005.

In May 2003, the Columbia 350 was certified and production started immediately; this was followed by the Columbia 400 in April 2004, deliveries of which began the following month.

UPDATED

LANCAIR COLUMBIA

TYPE: Four-seat lightplane.

PROGRAMME: Announced 1996 as LC-40; aerodynamic prototype flown July 1996; first flight of certification prototype (N140LC), early 1997. Second prototype (N141LC) first flew 14 April 1997; public debut at Oshkosh August 1997; modified for FAA testing to confirm spin-resistance. Certification completed 18 September 1998; RLD certification completed March 2001.

First production aircraft (N424CH) delivered 24 February 2000.

CURRENT VERSIONS: **Columbia 300:** *As described*. Engineering designation **LC40-550FG**. Version discontinued in July 2003 following certification of Columbia 350.

Columbia 350: Engineering designation **LC42-550FG**. Announced and publicly displayed at Sun 'n' Fun 2002; prototype (N70090) first flown early 2002 with deliveries due to start late 2002; however, certification delayed until 30 March 2003 and deliveries started in late October 2003. Dual independent all-electric systems with dual busses, dual batteries and dual alternators replacing Columbia 300's vacuum pumps; fully redundant electrical system enables each bus to handle entire electrical load. Powered by 231 kW (310 hp) Teledyne Continental IO-550-N driving Hartzell three-blade propeller. Empty weight 1,043 kg (2,300 lb); performance data generally similar to Columbia 300.

Columbia 400: Engineering designation **LC41-550FG**. Turbocharged version of Columbia 350 with 231 kW (310 hp) Continental TSIO-550C. Announced 9 April 2000; prototype (N143LC converted from Columbia 300) first flew June 2000; first production aircraft (N48PD) registered April 2004; certification had been due in 2002 but was delayed until 8 April 2004; claimed to be fastest piston-engined aircraft in mass production. Certified as spin recoverable, while retaining Lancair spin-resistant design features. Cockpit includes electronic primary flight display projection system; prototype equipped as demonstrator for Highways in the Sky (HITS) instrumentation. Planned 1,632 kg (3,600 lb) MTOW not yet certified.

Lancair Columbia 300 *(Paul Jackson)* ***NEW***/1047219

CUSTOMERS: Deposits for 295 aircraft taken by end 1998; firm orders stood at 170 by late 2002. May be offered in fractional ownership schemes. Total 18 built (and nine registered in US) during 2000; 27 in 2001; 24 in 2002 and 51 in 2003; by June 2004, 141 had been delivered. Deliveries of Columbia 400 began May 2004; eight delivered by June 2004. One example delivered to NASA at Langley 10 January 2001 for use as testbed for general aviation-based AGATE and SATS programmes; designated **Columbia 300X**.

COSTS: Columbia 350 US$399,900 fully IFR-equipped (2004); Columbia 400 US$475,000 (2004).

DESIGN FEATURES: Low-wing monoplane of typical Lancair appearance and design. Integral roll-over cage to protect cockpit area. Proprietary wing section, dihedral, and twist for spin resistant features. Airframe life 12,000 hours; extendable to unlimited, subject to NDT examination.

FLYING CONTROLS: **Columbia 300/350:** Conventional and manual. Fowler flaps droop 12° for T-O; 40° maximum. Aileron travel limits 22° up and 18° down; elevator travel up 13°, down 12°; rudder travel ±17°. Optional Precise Flight SpeedBrakes.

Columbia 400: Aileron travel limits 22° up and 18° down; elevator up 23°, down 14°; rudder travel ±30°. Optional Precise Speed Brakes.

STRUCTURE: All composites construction, including flight surfaces. One-piece dual-spar wing made of a combination of carbon fibre and glass fibre. Fuselage of composites construction with carbon fibre roll-over cage to protect cockpit area. Flight control surfaces constructed of carbon fibre materials.

LANDING GEAR: Fixed tubular steel tricycle type. Mainwheels 6.00-6, nosewheel 5.00-5. Cleveland brakes. Speed fairings on all wheels. Minimum ground turning radius 4.72 m (15 ft 6 in); nosewheel steerable ±60°.

Lancair Columbia 350 *(Paul Jackson)* ***NEW***/1047218

Lancair Turbo Columbia 400 *(Paul Bowen)* ***NEW***/1067154

POWER PLANT: *Columbia 300 and 350:* One 231 kW (310 hp) Teledyne Continental IO-550-N flat-six engine, driving three-blade constant-speed Hartzell PHC-J3YF-1RF/F7691D-1 propeller. Total fuel capacity 402 litres (106 US gallons; 88.3 Imp gallons) in two integral wing tanks; usable fuel 371 litres (98.0 US gallons; 81.6 Imp gallons). Oil capacity 7.6 litres (2.0 US gallons; 1.7 Imp gallons).

Columbia 400: One 231 kW (310 hp) Teledyne Continental TSIO-550-C flat-six engine with two turbochargers and two intercoolers. Hertzell HC-H3YF-1RF/FL693DF propeller. Fuel and oil capacities as per Columbia 300 and 350.

ACCOMMODATION: Pilot and three passengers in two pairs of seats in enclosed cabin. Seating certified to 26 *g*. Double gull-wing doors hinge upwards. Emergency exit through large baggage door behind rear passenger seats with outside access by emergency hinge release pin mechanism on pilot's door.

SYSTEMS: Single 12 V battery; 14 V 60 A alternator; in Columbia 300; dual independent all-electric system with dual busses, dual 14V batteries, dual 14V 60A alternators in Columbia 350/400. Air conditioning optional. Thermal Wing de-icing on wing and tailplane leading edges optional from January 2005.

AVIONICS: *Comms:* Dual Garmin G430 GPS/nav/com, Garmin GMA-340 stereo audio panel with remote marker beacon indicators, G327 Mode A, C transponder (Mode S optional); Trans-Cal SSD120 Blind Encoder.

Flight: Avidyne Flight Max colour 267 mm (10½ in) flat panel moving map display, with optional XM WX Satellite Weather and EMax engine/fuel monitoring system. Dual Garmin 430 GPS; S-Tec 55X autopilot with altitude preselect, autotrim and GPSS.

Instrumentation: Avidyne Entegra Primary Flight Display with ADAHRS; includes altitude, attitude, heading, airspeed, vertical speed and HSI. Back-up Stack includes electric artificial horizon, airspeed and altimeter.

EQUIPMENT: Fire extinguisher and crash axe standard.

DIMENSIONS, EXTERNAL:

Wing span	10.97 m (36 ft 0 in)
Wing chord: at root	1.40 m (4 ft 7 in)
at tip	1.02 m (3 ft 4 in)
Wing aspect ratio	9.2
Length overall: 300/350	7.67 m (25 ft 2 in)
400	7.72 m (25 ft 4 in)
Fuselage max width	1.27 m (4 ft 2 in)
Height overall	2.74 m (9 ft 0 in)
Tailplane span	4.17 m (13 ft 8 in)
Wheel track	2.24 m (7 ft 4 in)
Wheelbase	2.04 m (6 ft 8¼ in)
Propeller diameter: 300/350	1.96 m (6 ft 5 in)
400	1.98 m (6 ft 6 in)

DIMENSIONS, INTERNAL:

Cabin: Length	3.54 m (11 ft 7½ in)
Max width	1.24 m (4 ft 1 in)
Max height	1.30 m (4 ft 3 in)
Baggage compartment: volume	0.76 m^3 (27.0 cu ft)

AREAS:

Wings, gross	13.12 m^2 (141.2 sq ft)

WEIGHTS AND LOADINGS:

Weight empty: 300	1,021 kg (2,250 lb)
350	1,043 kg (2,300 lb)
400	1,134 kg (2,500 lb)
Baggage capacity	54 kg (120 lb)
Max fuel weight	288 kg (636 lb)
Max T-O weight	1,542 kg (3,400 lb)
Max landing weight	1,465 kg (3,230 lb)
Max wing loading	117.6 kg/m^2 (24.08 lb/sq ft)
Max power loading: 300/350	6.90 kg/kW (11.33 lb/hp)
400	6.68 kg/kW (10.97 lb/hp)

PERFORMANCE:

Never-exceed speed (VNE): 300/350	235 kt (435 km/h; 270 mph)
400	194 kt (359 km/h; 223 mph)
Normal cruising speed at 75% power: 300/350	190 kt (352 km/h; 219 mph)
400	
Stalling speed: flaps up: 300/350	71 kt (132 km/h; 82 mph)
400	69 kt (128 km/h; 80 mph)
flaps down: 300/350	57 kt (106 km/h; 66 mph)
400	59 kt (110 km/h; 68 mph)
Max rate of climb at S/L: 300/350	427 m (1,400 ft)/min
400	457 m (1,500 ft)/min
Max certified altitude: without oxygen	4,265 m (14,000 ft)
with oxygen	5,485 m (18,000 ft)
Service ceiling: 300/350	5,480 m (18,000 ft)
400	7,620 m (25,000 ft)
T-O run: 300/350	213 m (700 ft)
400	397 m (1,300 ft)
T-O to 15 m (50 ft): 300/350	381 m (1,250 ft)
400	579 m (1,900 ft)
Landing from 15 m (50 ft): 300/350	716 m (2,350 ft)
400	792 m (2,600 ft)
Landing run: 300/350	472 m (1,550 ft)
400	381 m (1,250 ft)
Range with max fuel, economy cruise: 300/350	1,385 n miles (2,565 km; 1,593 miles)
400	1,020 n miles (1,889 km; 1,173 miles)
g limits: all	+4.4/–1.76

UPDATED

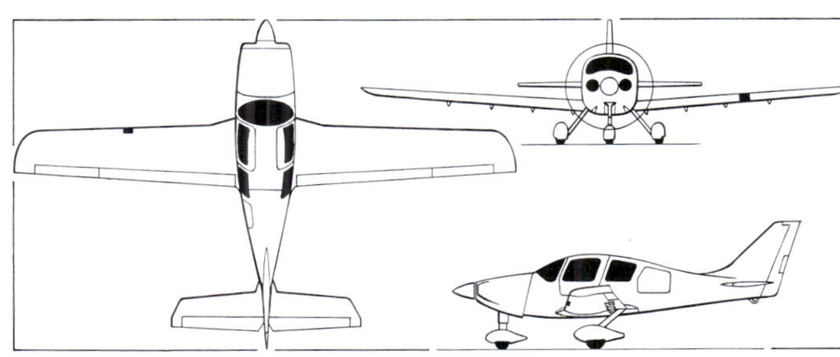

Lancair Columbia 300 (Teledyne Continental IO-550 flat-six) *(James Goulding)* 0089517

LANSHE

LANSHE AEROSPACE LLC

3100 Airman's Drive, Fort Pierce, Florida 34946
Tel: (+1 772) 465 99 96
Fax: (+1 772) 465 99 97
e-mail: sales@lansheaerospace.com
Web: http://www.lansheaercspace.com
PRESIDENT: Wadi Rahim

LanShe Aerospace LLC was founded in 2002 by former Martin-Marietta and Lockheed engineer and pilot Wadi Rahim. In late 2002, LanShe acquired the assets of Lake Aircraft Company from its former owner, Armand Rivard, and in March 2003 bought the Micco Aircraft Company. It occupies a 4,645 m^2 (50,000 sq ft) facility at Fort Pierce, Florida, where it is restarting production of the former SP20 and SP26 sport aircraft and Renegade, Seafury and Seawolf amphibians.

LANSHE LAKE LA-250

TYPE: Six-seat amphibian.

PROGRAMME: Colonial XC-1 prototype first flew 17 July 1948; entered production as three-seat C-1 Skimmer certified 19 September 1955; slightly modified, four-seat C-2 Skimmer IV approved 24 December 1957. Lake Aircraft Corporation took over Consolidated in 1959; thoroughly redesigned and increased span LA-4 produced following 26 July 1960 certification. Consolidated Aeronautics bought Lake in 1962 (production by Aerofab at Sanford, Maine; marketing by Lake Aircraft Division from Laconia, New Hampshire and Kissimmee, Florida); LA-4A and -4P similar; uprated LA-4-200 Buccaneer (certified 26 May 1970) followed, later aircraft designated LA-200 (including turbocharged, extended propeller shaft and reversible propeller subvariants); company and Type Certificate to Armand Rivard (Revo Inc) in September 1979; LA-250 Renegade stretched (0.97 m; 3 ft 2 in) version certified 30 June 1983 with four seats and 12 April 1984 with six seats; various subvariants produced until late 1980s, when manufacture almost ceased, only some 15 being built between 1990 and 1995.

Following acquisition of Lake by LanShe, Type Certificate transferred on 28 October 2003 to sister company Global Amphibians of Kissimmee, which licences LanShe manufacture; production transferred to other sister company, Mannaero, at St Lucie County Airport, Fort Pierce, Florida, which loans production tooling to LanShe and sells it all parts and non-redundant equipment.

Aerofab-built Lake LA-250 Renegade six-seat amphibian *(Paul Jackson)* ***NEW***/0567742

CURRENT VERSIONS: **Renegade 2:** Relaunched production or Renegade from 2003. *As described.*

Renegade 2T: Relaunched production of turbocharged Turbo Renegade 270, which set small amphibians world record altitude of 7,465 m (24,500 ft) in August 1983.

SeaFury: Renegade variant with salt water corrosion protection.

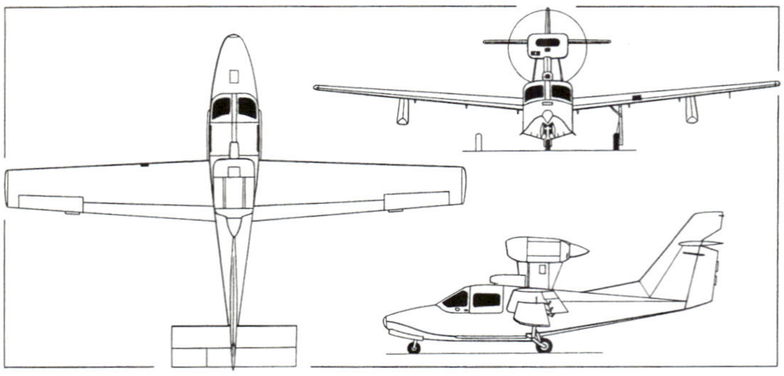

LanShe Lake LA-250 (Textron Lycoming IO-540 flat-six) *(Dennis Punnet)* ***NEW***/0572222

SeaWolf: Military version; strengthened wing with four hardpoints; forward-facing radar on engine pod. Two LA-250s converted to SeaWolf for US Department of Commerce in mid-1990s.

CUSTOMERS: Total 1,056 built before takeover by LanShe, comprising one XC-1, 22 C-1s, 19 C-2s, 185 LA-4s, two LA-4As, one LA-4P, 688 LA-4-200s, 131 LA-250s and seven military SeaWolfs.

COSTS: Renegade 2 US$449,000; Renegade 2T US$545,000; SeaFury US$749,000.

DESIGN FEATURES: Single-step all-metal double-sealed boat hull, 1.05 m (3 ft 5 in) longer than LA4-200 with deeper V bottom and additional strakes; retractable water rudder in base of aerodynaimic rudder. Tapered wings attached directly to hull sides; operable with max wave height of 46 cm (18 in).

Wing section NACA 4415 at root, 4409 at tip; dihedral 5° 30'; incidence 3° 15'.

FLYING CONTROLS: Conventional and manual. Ground adjustable aileron trim tabs; outer portion of port elevator separate from inboard and operated hydraulically as trimmer; hydraulically operated slotted flaps. Control surface movements: ailerons +29/–15°; elevators +25/–27°; rudder ±33°; flaps 20°. Optional electronic trim for rudder and ailerons.

STRUCTURE: Wing has duralumin leading/trailing-edge torsion boxes separated by single main spar; light alloy monocoque wing floats; hull alodined and zinc chromated inside and out; polyurethane external paint; metal ailerons and tail surfaces.

LANDING GEAR: Hydraulically retractable tricycle type. Oleo-pneumatic shock absorbers in main gear, which retracts inward into wings. Nosewheel retracts forward. Gerdes mainwheels with Goodyear tyres, size 6.00-6, pressure 2.41 bars (35 lb/sq in). Gerdes nosewheel with Goodyear tyre size 5.00-5, pressure 1.38 bars (20 lb/sq in). Gerdes disc brakes; parking brake. Nosewheel free to swivel 30° left/right.

POWER PLANT: *Renegade 2:* One 186 kW (250 hp) Textron Lycoming IO-540-C4B5 flat-six engine in Renegade and Sea Fury driving a Hartzell HC-E3YR-1RLF/FL7663D-2Q three-blade, constant-speed (hydraulic), Q-tip, metal pusher propeller. Oil capacity 15 litres (4.0 US gallons; 3.3 Imp gallons). Fuel in fuselage tank of 153 litres (40.5 US gallons; 33.8 Imp gallons) and two wing tanks, each of 64 litres (17.0 US gallons; 14.2 Imp gallons); total 282 litres (74.5 US gallons; 62.1 Imp gallons).

Renegade 2T and Sea Fury: One 201 kW (270 hp) Lycoming TIO-540-AA1AD. Otherwise as Renegade, except optional float tanks with combined total of 57 litres (15.0 US gallons; 12.5 Imp gallons).

Seawolf: As Renegade 2T but 333 litres (88.0 US gallons; 73.3 Imp gallons) internal fuel, plus option for up to 568 litres (150 US gallons; 123 Imp gallons) external fuel.

ACCOMMODATION: Enclosed cabin seating pilot and five passengers. Front and rear seats removable. Front seats have inertia reel harnesses as standard. Dual controls standard; dual brakes for co-pilot optional. Entry via two front-hinged windscreen sections; upward-hinged gull-wing cargo door standard. Baggage compartment (larger than in LA4-200) aft of cabin. Windscreen defrosting.

SYSTEMS: S-Tec 55X AFCS with altitude hold, altitude trim and turn co-ordinator. Vacuum system for flight instruments. Hydraulic system, pressure 86.2 bars (1,250 lb/sq in), for flaps, horizontal trim and landing gear actuation; hand pump provided for emergency operation. Engine-driven 12 V 60 A alternator and 12 V 30 Ah battery. Janitrol 30,000 BTU heater optional.

AVIONICS: *Comms:* Apollo SL40 VHF, PS Engineering PMA 7000B audio panel, Garmin GTX 327 Mode C transponder.

Radar: Optional weather radar on engine cowling.

Flight: Garmin GNS 430 nav/com/GPS and GI-106 indicator. Optional Goodrich WX-500 Stormscope, HSI electric compass, Garmin DGL 49 weather datalink and Garmin GNS 530 nav/com/GPS.

Instrumentation: JPI EDM-800 digital engine monitor.

DIMENSIONS, EXTERNAL:
- Wing span ... 11.68 m (38 ft 4 in)
- Wing chord, mean ... 1.35 m (4 ft 5 in)
- Wing aspect ratio ... 8.6
- Length overall ... 8.64 m (28 ft 4 in)
- Height overall ... 3.04 m (9 ft 11½ in)
- Tailplane span ... 3.05 m (10 ft 0 in)
- Wheel track ... 3.40 m (11 ft 2 in)
- Wheelbase ... 3.13 m (10 ft 3 in)
- Propeller diameter ... 1.93 m (6 ft 4 in)

DIMENSIONS, INTERNAL:
- Cabin: Length ... 3.15 m (10 ft 4 in)
  - Max width ... 1.04 m (3 ft 5 in)
  - Max height ... 1.09 m (3 ft 7 in)

AREAS:
- Wings, gross ... 15.24 m (164.0 sq ft)
- Ailerons (total) ... 1.16 m^2 (12.5 sq ft)
- Trailing-edge flaps (total) ... 2.28 m^2 (24.5 sq ft)
- Fin ... 1.25 m^2 (13.5 sq ft)
- Rudder ... 0.79 m^2 (8.5 sq ft)
- Tailplane ... 1.45 m^2 (15.6 sq ft)
- Elevators (total) ... 0.78 m^2 (8.4 sq ft)

WEIGHTS AND LOADINGS (all, except R2: Renegade 2, R2T: Renegade 2T, SF: SeaFury, SW: SeaWolf):
- Weight empty: R2 ... 896 kg (1,975 lb)
  - R2T ... 941 kg (2,075 lb)
  - SF ... 928 kg (2,045 lb)
  - SW ... 1,034 kg (2,280 lb)
- Baggage capacity ... 91 kg (200 lb)
- Max T-O weight: except SW ... 1,424 kg (3,140 lb)
  - SW ... 1,655 kg (3,650 lb)
- Max landing weight: except SW ... 1,383 kg (3,050 lb)
  - SW ... 1,564 kg (3,450 lb)
- Max wing loading: except SW ... 93.5 kg/m^2 (19.14 lb/sq ft)
  - SW ... 102.7 kg/m^2 (21.03 lb/sq ft)
- Max power loading: R2 ... 7.64 kg/kW (12.56 lb/hp)
  - R2T, SF ... 7.08 kg/kW (11.63 lb/hp)
  - SW ... 7.78 kg/kW (12.78 lb/hp)

PERFORMANCE (as above, all except SeaWolf):
- Max cruising speed at 75% power: R2 at FL60 ... 132 kt (245 km/h; 152 mph)
  - R2T, SF at FL200 ... 155 kt (287 km/h; 178 mph)
- Stalling speed, power off:
  - flaps and gear up ... 55 kt (102 km/h; 64 mph)
  - flaps and gear down ... 49 kt (91 km/h; 57 mph)
- Max rate of climb at S/L ... 274 m (900 ft)/min
- Max certified altitude (R2T, SF) ... 6,095 m (20,000 ft)
- Service ceiling: R2 ... 4,480 m (14,700 ft)
  - R2T, SF ... 7,255 m (23,800 ft)
- T-O run: land ... 268 m (880 ft)
  - water ... 381 m (1,250 ft)
- Landing run: land ... 145 m (475 ft)
  - water ... 183 m (600 ft)
- Range with max standard fuel: R2 ... 1,040 n miles (1,926 km; 1,196 miles)
  - R2T, SF ... 1,120 n miles (2,074 km; 1,288 miles)
  - SW with 30 min reserves ... 780 n miles (1,444 km; 897 miles)
- Endurance, SW with max external fuel ... 14 h 30 min

LANSHE SP20 AND SP26

TYPE: Aerobatic two-seat lightplane.

PROGRAMME: Based on original 1947 Meyers MAC-145A, certified 2 November 1948, and 22 built (including two MAC-125s), but with increased power. Redesign started by Micco Aircraft March 1994, using converted MAC-145 (N520SP); construction started June 1995 and prototype (N720SP) first flown 17 December 1997. First production aircraft (N820SP) construction started January 1999. Certification to FAR Pt 23 issued 5 January 2000, with deliveries soon after. Second production aircraft shown at Sun 'n' Fun, Lakeland, Florida, April 2000, before delivery to TE Aero, Texas; third production aircraft to Clipper Aviation of Oneonta, New York, June 2000. Aircraft registered using MAC-145A (SP20) and MAC-145B (SP26) type certificates, which transferred to LanShe on 22 February 2003..

CURRENT VERSIONS: **SP20:** Baseline version.

SP26: Updated and uprated aerobatic version. Six-cylinder engine in longer cowling, but firewall moved rearwards by identical distance (30 cm; 12 in), resulting in no overall change of fuselage length. Prominent strakes ahead of tailplane on prototype have now been removed. Prototype converted from N820SP; debut at Sun 'n' Fun, April 2000. Certification in Utility category granted on 19 October 2000 for 113 kg (250 lb) extension of MTOW over SP20. First production aircraft (N201MA) flew 15 November 2000 and delivered on 26 December 2000 to Andrew Spaulding of Cumberland, Maine. Aerobatic certification achieved 4 March 2002.

CUSTOMERS: Over 50 on order by end 2000, including 24 SP26; deliveries in 2000 totalled five SP20s and one SP26, and in 2001, 10 SP26s. When production suspended in October 2001, status was eight SP20s completed (two prototypes, including one to SP26 prototype; one static test; and single aircraft to SoCal Aviation, P T Esposito, Sportsman Air Park, Douglas Taylor and Dean Maggio), 11 complete production SP26s (Andrew Spaulding, Robert A Csognor, Grant Air Servicing, SoCal Aviation, Jim B Sayers, Cordes Air, Sportsman Air Park, Westminster College, Robert and Brittan William, Robert G Boston and Micco) first delivery following LanShe acquisition took place on 3 May 2003, with SP26 prototype N820SP (c/n 260001) handed over to Debra Cunningham of Gonzalez, Texas..

COSTS: SP20 US$199,000, SP26 US$256,000, standard equipped (2003).

DESIGN FEATURES: Low-wing, high-performance, metal sport aircraft; low development cost achieved by employment of existing design. Major changes from MAC-145A include replacement of the 108 kW (145 hp) Continental engine; substitution of a Meyers 200 fin and rudder; wings and flaps from the Meyers 400 Interceptor; and provision of a sliding canopy in place of the fixed canopy and side doors.

FLYING CONTROLS: Conventional and manual. Horn-balanced cable-operated rudder with ground-adjustable trim tab; horn-balanced push/pull tube-controlled elevators with flight-adjustable tabs in each. Control movements: elevator +28/–20°; ailerons +22/–12°; rudder ±22°. Electrically operated Fowler flaps, max deflection 30°.

STRUCTURE: 4130 steel tube centre section and all-aluminium wings.

LANDING GEAR: Tailwheel type; main units retract inward into wings, tailwheel rearward; mainwheel doors cover legs only when retracted, wheel wells remaining exposed. Electrohydraulic Parker Hannifin hydraulic brakes and leaf spring shock-absorbers. 6.00-6 tyres on mainwheels; 10 cm (4 in) solid tailwheel. Ground turning radius about wingtip 6.07 m (19 ft 11 in).

POWER PLANT: *SP20:* One 149 kW (200 hp) Textron Lycoming IO-360-C1E6 flat-four driving a three-blade, constant-speed McCauley QZP (B3D36C424/74SA-0) propeller.

SP26: One 194 kW (260 hp) Textron Lycoming IO-540-T4B5 flat-six driving a three-blade, constant-speed Hartzell HC-C3YR-1RF/F7693F Compact propeller.

Fuel (both versions) in integral wing tanks, combined capacity 273 litres (72.0 US gallons; 60.0 Imp gallons), of which 257 litres (68.0 US gallons; 56.6 Imp gallons) are usable. Three drain points. Oil capacity 7.6 litres (2.0 US gallons; 1.67 Imp gallons) in SP20, 9.5 litres (2.5 US gallons; 2.1 Imp gallons) in SP26. Aerobatic versions have two additional 57 litre (15.0 US gallon; 12.5 Imp gallon) tanks located in the centre-section outboard of cabin area.

LanShe SP26 originally built by Micco *(Paul Jackson)* ***NEW***/0561589

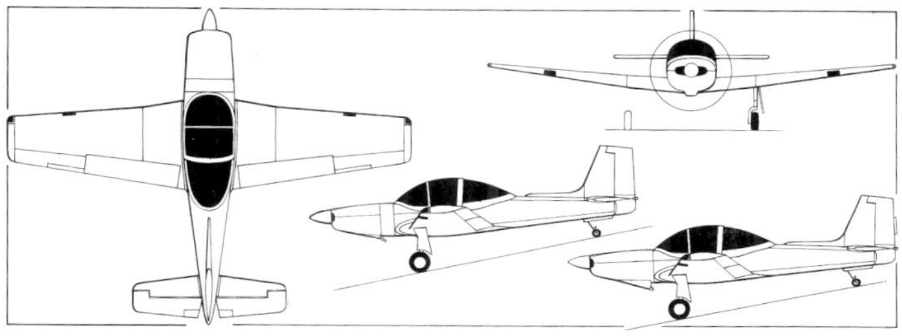

LanShe SP20 two-seat light aircraft, with additional side view (lower) of SP26 *(James Goulding)* 0059968

ACCOMMODATION: Two, side by side, under rearward-sliding canopy. Baggage area behind seats.
SYSTEMS: 28 V 70 A alternator and battery; landing lights.
AVIONICS: VFR or IFR standards available, recommended Trimble Terra suite; S-Tec 30 two-axis autopilot and Apollo MX20 multifunction display optional.
EQUIPMENT: Landing/taxying light in each wing leading-edge.

Data apply to both versions, except where stated.

DIMENSIONS, EXTERNAL:

Wing span	9.25 m (30 ft 4 in)
Wing chord: at root	2.29 m (7 ft 6 in)
at tip	1.02 m (3 ft 4 in)
Wing aspect ratio	5.9
Length overall	7.34 m (24 ft 1 in)
Max diameter of fuselage	1.22 m (4 ft 0 in)
Height overall	1.98 m (6 ft 6 in)
Tailplane span	3.76 m (12 ft 4 in)
Wheel track	3.15 m (10 ft 4 in)
Wheelbase	4.02 m (13 ft 2¼ in)
Propeller diameter: SP20	1.88 m (6 ft 2 in)
SP26	1.98 m (6 ft 6 in)
Propeller ground clearance: SP20	0.76 m (2 ft 6 in)

DIMENSIONS, INTERNAL:

Cabin: Length	1.65 m (5 ft 5 in)
Max width	1.09 m (3 ft 7 in)
Max height	0.99 m (3 ft 3 in)
Baggage volume	0.40 m³ (14.0 cu ft)

AREAS:

Wings, gross	14.55 m² (156.6 sq ft)

WEIGHTS AND LOADINGS:

Weight empty: SP20	839 kg (1,850 lb)
SP26	925 kg (2,040 lb)
Baggage capacity	45 kg (100 lb)
Max aerobatic weight: SP26	1,202 kg (2,650 lb)
Max T-O weight: SP20	1,179 kg (2,600 lb)
SP26	1,292 kg (2,850 lb)
Max landing weight: SP20	1,130 kg (2,492 lb)
SP26	1,243 kg (2,742 lb)
Max ramp weight: SP20	1,193 kg (2,630 lb)
Max wing loading: SP20	81.1 kg/m² (16.60 lb/sq ft)
SP26	88.9 kg/m² (18.20 lb/sq ft)
Max power loading: SP20	7.91 kg/kW (13.00 lb/hp)
SP26	6.67 kg/kW (10.96 lb/hp)

PERFORMANCE:

Never-exceed speed (VNE):	
SP20	180 kt (333 km/h; 207 mph)
SP26	195 kt (361 km/h; 224 mph)
Max level speed: SP20	145 kt (269 km/h; 167 mph)
SP26	165 kt (306 km/h; 190 mph)
Max cruising speed at 75% power:	
SP20	135 kt (250 km/h; 155 mph)
SP26	155 kt (287 km/h; 178 mph)
Normal cruising speed at 65% power:	
SP20	128 kt (237 km/h; 147 mph)
SP26	148 kt (274 km/h; 170 mph)
Econ cruising speed at 55% power:	
SP20	120 kt (222 km/h; 138 mph)
SP26	135 kt (250 km/h; 155 mph)
Stalling speed, power off:	
flaps and landing gear up	55 kt (102 km/h; 64 mph)
flaps and landing gear down	49 kt (91 km/h; 57 mph)
Max rate of climb at S/L: SP20	290 m (950 ft)/min
SP26	458 m (1,500 ft)/min
Service ceiling: SP20	3,660 m (12,000 ft)
SP26	5,486 m (18,000 ft)
T-O run: SP20	460 m (1,510 ft)
SP26	244 m (800 ft)
T-O to 15 m (50 ft): SP20	618 m (2,025 ft)
SP26	335 m (1,100 ft)
Landing from 15 m (50 ft): SP20, SP26	610 m (2,000 ft)
Landing run: SP20, SP26	213 m (700 ft)
Range with 30 min reserves: at max level speed:	
SP20, SP26	750 n miles (1,389 km; 863 miles)
at max cruising speed:	
SP20	850 n miles (1,574 km; 978 miles)
SP26	800 n miles (1,481 km; 920 miles)
at econ cruising speed:	
SP20	1,004 n miles (1,859 km; 1,155 miles)
SP26	1,040 n miles (1,926 km; 1,196 miles)
g limits (SP26)	+6/–3

LEARJET

BOMBARDIER AEROSPACE LEARJET (SUBSIDIARY OF BOMBARDIER INC)

One Learjet Way, PO Box 7707, Wichita, Kansas 67277
Tel: (+1 316) 946 20 00
Fax: (+1 316) 946 22 20
VICE-PRESIDENT AND GENERAL MANAGER: Jim Ziegler
VICE-PRESIDENT, OPERATIONS: Ghislain Bourque
VICE-PRESIDENT, FINANCE: Chris Chrashaw
VICE-PRESIDENT, ENGINEERING: Keith Miller
DIRECTOR, PUBLIC RELATIONS: Dave Franson

Company originally founded 1960 by Bill Lear Snr as Swiss American Aviation Corporation (SAAC); transferred to Kansas 1962 and renamed Lear Jet Corporation; prototype Lear Jet 23 (N801L) first flew 7 October 1963; Gates Rubber Company bought about 60 per cent share in 1967 and renamed it Gates Learjet Corporation, moving much of manufacturing to Tucson, Arizona; 64.8 per cent acquired by Integrated Acquisition Inc September 1987 and renamed Learjet Corporation; all manufacturing moved from Tucson to Wichita during 1988, leaving customer service completion and modification centre in Tucson.

Acquisition of Learjet by Canada's Bombardier announced April 1990 and concluded 22 June 1990 for US$75 million; name changed to Learjet Inc; Bombardier assumed responsibility for Learjet's line of credit; now part of Bombardier Inc. Learjet 31A, 40, 45 and 60 assembled in Wichita and flown to Tucson for completion and delivery.

Some 2,350 Learjets built, including 105 LJ23s, 259 LJ24s, 368 LJ25s, five LJ28s, four LJ29s, 225 LJ31s, 673 LJ35s, 62 LJ36s, 235 LJ45s, 147 LJ55s and 265 LJ60s. 2,000th Learjet, Model 45 N158PH, delivered to Parker Hannifin Corporation in August 1999.

Learjet bought manufacturing and marketing rights and tooling of Aeronca thrust reversers, for application to Learjet and other aircraft, March 1989.

Total of 36 deliveries in 1994; 43 in 1995; 34 in 1996; 45 in 1997; 61 in 1998; 99 in 1999, 134 in 2000; 182 in 2001, 60 in 2002, 31 in 2003, and 24 in the first six months of 2004.

UPDATED

LEARJET 31

TYPE: Business jet.
PROGRAMME: Learjet 31 introduced September 1987 following first flight of aerodynamic prototype (modified Learjet 35A) on 11 May 1987; first production aircraft (N311DF) used as systems testbed; FAA certification 12 August 1988. Learjet 31A and 31A/ER announced October 1990 to replace Learjet 31. Certified in 21 countries, including Argentina, Australia, Bermuda, Brazil, Canada, Czech Republic, Denmark, Germany, Grand Cayman, Guatemala, Indonesia, Italy, Japan, Luxembourg, Mexico, Namibia, Pakistan, Philippines, South Africa, Switzerland and USA. Total fleet time 400,000 hours by 1 October 2000. Superseded by Learjet 40 from 2004.
CURRENT VERSIONS: **Learjet 31:** Original version. No longer produced.

Learjet 31A: Improved version, introduced in 1990.

Description applies to Learjet 31A, except where indicated.

Learjet 31A/ER: Optional extended-range version with 2,627 litres (694 US gallons; 578 Imp gallons) of fuel.
CUSTOMERS: Last two Learjet 31s delivered to customers in 2003, final aircraft (N40043, ultimately N600AW of Pine Holdings) having been delivered to Wichita for fitting out on 27 January 2003. Production totalled 38 Learjet 31s and 208 Learjet 31As, or 246 in all.

UPDATED

LEARJET 45

TYPE: Business jet.
PROGRAMME: Design started September 1992; unveiled at NBAA Convention 20 September 1992. Other members of Bombardier group are involved; Learjet is responsible for project co-ordination, final assembly, testing and certification.

Engine testing began January 1995 with TFE731-20 installed in one nacelle of Learjet 31A testbed. Wing and fuselage of first production aircraft (N45XL) mated at Wichita 4 November 1994; first flight 7 October 1995; second prototype (N452LJ) first flown 6 April 1996, assigned to flutter testing; third aircraft (N453LJ), first flown 24 April 1996, assigned to avionics testing; fourth aircraft assigned to HIRF and lightning-strike testing, engine testing and fuel system operation; fifth aircraft fitted with production interior and assigned to function and reliability testing, including interior noise measurement.

Initial FAA certification granted 22 September 1997 followed by full approval in May 1998; JAA certification achieved 3 July 1998; FAA RVSM approval granted 25 July 2000. First customer delivery (N903HC) 28 July 1998, to Hytrol Conveyor of Jonesboro, Arkansas. First delivery to Europe (N459LJ) 8 September 1998 for Eifel Holdings Ltd of Jersey. More than 210 delivered by December 2002, including 2,000th Learjet, N158PH, to Parker Hannifin Corporation in August 1999. 100th Learjet 45 delivered in October 2000, initially as Bombardier company demonstrator, but destined for a Philippine operator; 200th delivered in April 2002.
CURRENT VERSIONS: **Learjet 45:** Current version, *as described*.

Learjet 45^{XR}: Improved version offering enhanced payload/range capability. Announced at Farnborough International Air Show, 21 July 2002 and formally launched at NBAA Convention, Orlando, Florida, 10 September 2002; certification and service entry scheduled for second quarter 2003; not achieved and prototype demonstrated at NBAA Convention, Orlando, 7 to 9 October 2003 (N45XR c/n 232). Features include a 454 kg (1,000 lb) increase in maximum T-O weight; 15.57 kN (3,500 lb st) Honeywell TFE731-20-BR turbofans flat rated at ISA +25°C (104°F) with 0.67 kN (150 lb st) automatic power reserve at T-O; and redesigned cabin interior with seats that are 5 cm (2 in) wider and increase legroom by 15 cm (6 in), increased galley storage, LED lighting system that reduces heat emissions, improved access to systems behind the aft lavatory, and quick (20 min) removal of interior components. Learjet 45^{XR} maximum T-O weight increase and engine upgrade will be retrofittable to existing Learjet 45s.
CUSTOMERS: Total of 234 delivered by September 2003; deliveries began with seven in 1998, followed by 43 in 1999, 71 in 2000, 63 in 2001, 17 in 2003, and nine in the first six months of 2004. Major orders include 40, with 10 options, by JetSolutions for its FlexJets fractional ownership scheme. Recent customers include Singapore Airlines, which ordered two in April 2001, supplementing four aircraft delivered in 1998, Cathay Pacific Airways, which

Learjet 45^{XR} twin-turbofan business jet ***NEW***/0583566

has ordered two for crew training, operated by BAE Systems Flight Training (Australia) Pty Ltd at Adelaide, South Australia, of which the first was delivered on 24 August 2001 with the second scheduled for delivery in 2002, Gold Air International which ordered five 45^{XR}s on 5 February 2003, Ocean Air Aero Taxi Ltd of Brazil, which ordered one 45^{XR} on 12 March 2003, Hughes Air Corporation of Calgary, Canada, which took delivery of a 45 on 10 July 2003, and the Irish Government, which ordered one for delivery in December 2003.

COSTS: Learjet 45 US$9.848 million, Learjet 45^{XR} US$10.423 million (both 2002).

DESIGN FEATURES: Generally as Model 31; Learjet 45 designed to combine docile handling characteristics of 31/31A and 60 with exceptional fuel efficiency and good overall performance, and offer increased maintainability and reliability; new larger fuselage, wing and tail unit; increased head and shoulder room; wing carry-through spar recessed beneath floor; latest technology systems. Wing designed with NASA; winglets; sweepback 13° 25' at 25 per cent chord.

Performance enhancement package announced at the Paris Air Show in June 1999 includes maximum T-O weight increased by 136 kg (300 lb); reductions in T-O speeds; improved nosewheel steering and removal of 40 kt (74 km/h; 46 mph) steering system limitation; improved brake-by-wire effectiveness; reconfiguration of flap selector module to permit use of 8° flap setting, instead of 20°, for approach climb in the event of a go-around; improved climb performance with bleed air anti-icing systems operating; and updated Honeywell avionics software.

FLYING CONTROLS: Conventional and manual. Two spoiler/lift dumper panels in each wing; deflections 0°, 14° 30', 28° 30' and 60°. Horn-balanced elevators. Trim tab in rudder; two in each aileron. Flaps; deflections 0, 8, 20 and 40°.

Further improvements introduced in late 2000 include restyled seats providing greater freedom of movement in the cabin, a 10 to 12 dB reduction in cabin noise levels and improvements in the cabin air distribution system.

STRUCTURE: Unigraphics, CATIA and Computervision digital design systems adopted by Learjet, de Havilland of Canada and Shorts for engineering design. Short Brothers of UK manufactures the fuselage and de Havilland Canada the wings.

LANDING GEAR: Retractable tricycle type; semi-articulated trailing-link main legs retract inward, nose leg forward; twin-wheel main units, size 22×5.75-12 (10 ply) with brake-by-wire anti-skid carbon multidisc brakes; nosewheel has dual-chine tyre, size 18×4.4 (10 ply), with steer-by-wire.

POWER PLANT: Two Honeywell TFE731-20-1B turbofans (progressively 20R, 20AR and 20BR), each flat rated at 15.57 kN (3,500 lb st) at ISA +16°C (61° F), or 16.24 kN (3,650 lb st) with APR. Dee Howard target-type thrust reversers; digital electronic engine control. Fuel capacity 3,426 litres (905 US gallons; 754 Imp gallons) usable, plus 61 litres (16.0 US gallons; 13.4 Imp gallons) unusable. Oil capacity 7 litres (1.9 US gallons; 1.6 Imp gallons) per engine.

ACCOMMODATION: Two crew and up to nine passengers; eight-passenger cabin typically with PMP fully adjustable swivelling and reclining seats in double club arrangement, galley/refreshment centre and storage cabinet at front of cabin; lavatory, doubling as optional ninth seat, at rear; cabin pressurised, maximum differential 0.65 bar (9.43 lb/sq in); clamshell door with integral steps on port side at front of cabin, upper part serves as emergency exit; eight cabin windows per side, one forward of starboard wing leading-edge serving as emergency exit; externally accessible heated and lined baggage compartment, capacity 227 kg (500 lb), in aft fuselage accessed via door on port side beneath engine nacelle.

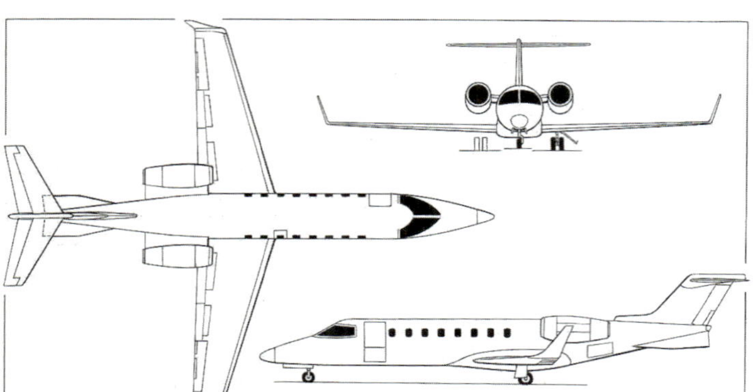

Learjet 45 business jet (two Honeywell TFE731-20-AR turbofans) *(Jane's/Dennis Punnett)* 0507269

SYSTEMS: Honeywell air conditioning and pressurisation system, maximum differential 0.65 bar (9.4 lb/sq in), with dual independent digital control system and pneumatic redundancy; dual-zone automatic temperature control. Gaseous oxygen system, pressure 127.6 bar (1,850 lb/sq in). Main and auxiliary back-up hydraulic systems, pressure 207 bar (3,000 lb/sq in). Dual independent anti-icing and de-icing systems comprising bleed air anti-icing on wing and tailplane leading-edges and engine inlets, electric de-icing on pitot-static probes and electric anti-icing and de-fogging on windscreen. Honeywell RE 100 APU optional.

AVIONICS: Honeywell Primus 1000 integrated avionics system.

Comms: Dual Primus II nav/ident radios.

Radar: Primus 660 weather radar standard.

Flight: Dual Primus II nav radios. Primus 1000 digital autopilot/flight director standard; Honeywell traffic-alert and collision-avoidance (TCAS II) optional.

Instrumentation: Primus 1000 with EICAS; dual PFDs and MFDs; flight and navigation information displayed on four 203 × 178 mm (8 × 7 in) EFIS screeens; heart of system is IC-600 integrated avionics computer, which combines EFIS and EICAS processor.

DIMENSIONS, EXTERNAL:
- Wing span ..... 14.56 m (47 ft 9¼ in)
- Wing aspect ratio 7.3
- Length overall ..... 17.81 m (58 ft 5 in)
- Max diameter of fuselage ..... 1.75 m (5 ft 9 in)
- Height overall ..... 4.31 m (14 ft 1½ in)
- Wheel track ..... 2.84 m (9 ft 4 in)
- Wheelbase ..... 7.87 m (25 ft 9¾ in)

DIMENSIONS, INTERNAL:
- Cabin:
  - Length: incl flight deck ..... 7.52 m (24 ft 8¼ in)
    - excl flight deck ..... 6.02 m (19 ft 9 in)
  - Max width ..... 1.55 m (5 ft 1 in)
  - Max height ..... 1.50 m (4 ft 11 in)
  - Floor area, excl flight deck ..... 5.76 m² (62.0 sq ft)
  - Volume, excl flight deck ..... 11.6 m³ (410 cu ft)
- Baggage compartment volume ..... 1.4 m³ (50 cu ft)

AREAS:
- Wings, basic ..... 28.95 m² (311.6 sq ft)

WEIGHTS AND LOADINGS (A: Learjet 45, B: Learjet 45^{XR}):
- Weight empty: A ..... 5,797 kg (12,780 lb)
  - B ..... 5,869 kg (12,939 lb)
- Basic operating weight, typical:
  - A, B ..... 6,227 kg (13,729 lb)
- Max baggage weight: cabin ..... 77 kg (170 lb)
  - aft fuselage ..... 227 kg (500 lb)
- Max payload: A, B ..... 1,030 kg (2,271 lb)
- Payload with max fuel: A ..... 435 kg (959 lb)
  - B ..... 889 kg (1,959 lb)
- Max fuel weight: A, B ..... 2,750 kg (6,062 lb)
- Fuel with max payload: A ..... 2,154 kg (4,750 lb)
  - B ..... 2,608 kg (5,750 lb)
- Max T-O weight: A ..... 9,298 kg (20,500 lb)
  - B ..... 9,752 kg (21,500 lb)
- Max landing weight: A, B ..... 8,709 kg (19,200 lb)
- Max ramp weight: A ..... 9,412 kg (20,750 lb)
  - B ..... 9,866 kg (21,750 lb)
- Max zero-fuel weight: A, B ..... 7,257 kg (16,000 lb)
- Max wing loading: A ..... 321.2 kg/m² (65.79 lb/sq ft)
  - B ..... 336.9 kg/m² (69.00 lb/sq ft)
- Max power loading: A ..... 286 kg/kN (2.81 lb/lb st)
  - B ..... 300 kg/kN (2.95 lb/lb st)

Learjet 40, cutaway drawing key

1 Glass fibre radome with lightning diverter strips
2 Weather radar scanner
3 Dual ILS glideslope antennas
4 Radar scanner mounting bulkhead
5 Nose avionics equipment bay
6 Nosewheel bay doors
7 Access doors port and starboard
8 Forward retracting steerable nosewheel
9 Nosewheel leg pivot mounting
10 Hydraulic retraction jack
11 Windscreen rain dispersal air duct
12 Machined front pressure bulkhead
13 Pitot heads
14 Incidence vane
15 Rudder pedals
16 Instrument panel
17 Control column handwheel
18 Honeywell Primus four-screen full-colour EFIS displays
19 Instrument panel shroud
20 Electrically-heated windscreen panels
21 Sidewall mounted circuit-breaker panel
22 First Officer's seat
23 Flight deck bulkhead and curtained doorway
24 Centre control pedestal
25 Captain's seat
26 Cockpit section frame structure
27 Bulkhead stowage locker
28 Entry door lower segment with integral airstairs
29 Fold-out step
30 Internal door latch
31 Entry lobby
32 Fold-down seat for bar service
33 External door latch
34 Upper door segment
35 Bar/galley unit
36 Starboard side wardrobe
37 TCAS antenna
38 Optional individual video monitor, stowable
39 Cabin wall trim panelling
40 Fold-out table
41 Cabin window panels
42 Six-seat passenger cabin, four 'club' seats and two singles
43 Lower VHF antenna
44 Fuselage frames and stringers
45 Individual swivelling and reclining seats
46 Composites cabin floor panels with seat mounting rails
47 Wing centre-section U-beam spar structure beneath fuselage pressure cabin
48 Starboard main landing gear, stowed position
49 Pull-down window blinds
50 Starboard side emergency escape hatch
51 Upper VHF antenna
52 Starboard wing integral fuel tank
53 Fuel capacitance probe
54 Fuel system vent and feed piping
55 Fuel scavenge pump
56 Leading-edge thermal de-icing
57 Starboard navigation and recognition lights
58 Winglet
59 Starboard aileron
60 Geared tab
61 Aileron cable-operated hinge control
62 Starboard 'spoileron', spoiler, lift-dumper and aileron coupled functions
63 Starboard single-slotted flap, extended
64 Starboard engine intake
65 ATC antenna
66 ADF antenna
67 Lavatory bulkhead with sliding door
68 Lavatory unit, optional seventh passenger seat
69 Hand basin
70 Port side cabin baggage stowage space
71 Honeycomb panel and machined frame rear pressure bulkhead structure
72 Fuselage fuel tank
73 Engine mounting cross-beam
74 Starboard engine installation
75 Cabin air conditioning pack
76 Baggage compartment
77 Starboard nacelle pylon
78 Engine multilobe exhaust mixer
79 Target-type thrust reverser doors, open
80 Heat exchanger ram-air intake
81 ELT antenna
82 Composites fin root fairing
83 Tailplane de-icing air duct
84 Three-spar fin torsion box structure
85 VOR antenna
86 Elevator control rods
87 Fin ribs, skins and stringer panels
88 Tailplane trim electric actuator
89 Trimming tailplane sealing plate
90 Starboard tailplane segment
91 Elevator horn balance
92 Starboard elevator
93 Elevator hinge control linkage
94 Anti-collision beacon
95 Tail navigation light
96 Port elevator rib structure
97 Static dischargers
98 Three-spar and rib tailplane torsion box structure
99 Leading edge de-icing air duct
100 Tailplane pivot mounting
101 Rudder rib structure
102 Rudder trim tab
103 Glass fibre tailcone
104 Cable-operated rudder control quadrant
105 Autopilot servos
106 Port 'delta' fin
107 Elevator cable quadrant
108 Machined sloping fin spar attachment bulkheads
109 Engine fire suppression bottles
110 Rear avionics equipment rack, ventral access hatch on starboard side
111 Dual batteries
112 Port thrust reverser doors, closed position
113 Reverser actuator fairing
114 Rear engine mounting
115 Main engine mounting yoke and drag strut
116 Baggage door, open
117 Honeywell TFE-731-20 turbofan
118 Cold stream exhaust
119 Hot stream exhaust
120 Accessory equipment gearbox
121 Engine oil tank
122 Hinged engine cowling panels
123 Intake lip, bleed-air de-iced
124 Composites wing root trailing edge fairing
125 Wing main and rear spar/ fuselage attachment joints
126 Port mainwheel bay
127 Hydraulic retraction jack
128 Mainwheel leg pivot mounting
129 Flap operating screw jacks and guides, central drive unit
130 Flap shroud ribs
131 Flap rib structure
132 Port single-slotted flap, extended
133 Port 'spoileron'
134 'Spoileron' hydraulic jack
135 Aileron trim tab, port only
136 Aileron geared tab
137 Port aileron rib structure
138 Port winglet
139 Static dischargers
140 Two-spar winglet mounting ribs
141 Port navigation and recognition lights
142 Wing tank bottom skin access panels
143 Three spar wing torsion box structure
144 Wing rib structure
145 Port wing integral fuel tank
146 Fixed leading edge
147 Leading edge de-icing 'picolo' air duct
148 Twin mainwheels
149 Mainwheel leg door
150 Trailing axle mainwheel suspension
151 Shock absorber leg strut
152 Leg-mounted landing light
153 Fuel collector box and pump bay
154 Wing tank end rib
155 Wing/fuselage mounting drag link
156 Front spar/fuselage attachment joint
157 Composites leading edge fairing.

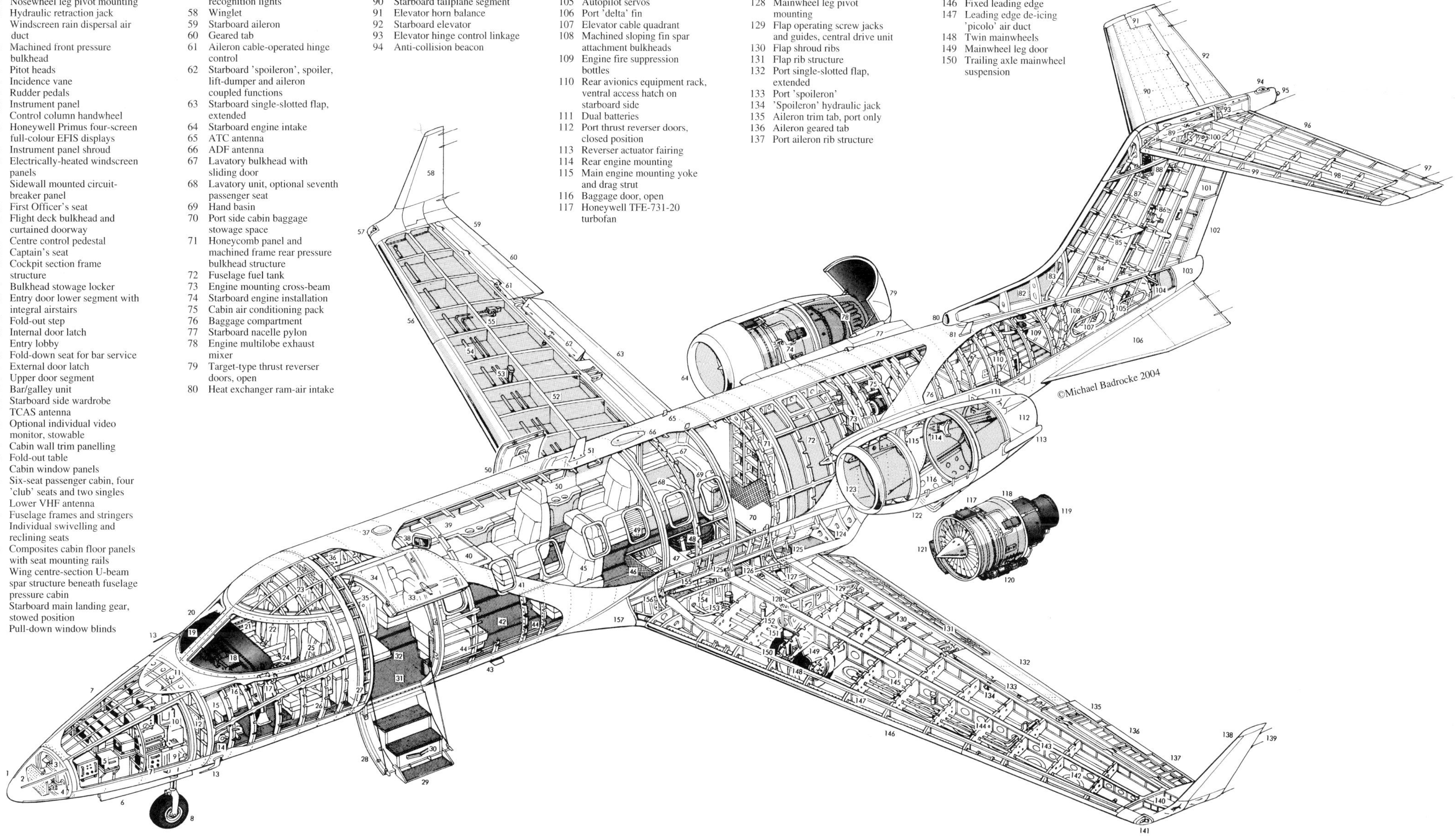

PERFORMANCE:
Max operating speed (VMO) to FL267 ... 330 kt (611 km/h; 379 mph)
Max operating Mach No. (MMO) above FL267 ... 0.81
High cruising speed: A, B ... 464 kt (859 km/h; 534 mph)
Normal cruising speed: A, B ... 457 kt (846 km/h; 526 mph)
Long-range cruising speed:
A, B ... 430 kt (796 km/h; 495 mph)
Minimum control speed, landing, flaps 40° ... 97 kt (180 km/h; 112 mph) IAS
Time to FL430 after MTOW departure:
A ... 23 min 6 s
B ... 24 min 54 s
Max certified altitude: A, B ... 15,545 m (51,000 ft)
T-O field length: A ... 1,326 m (4,350 ft)
B ... 1,542 m (5,060 ft)
FAR Pt 91 landing distance: A, B ... 811 m (2,660 ft)
Range with two crew and four passengers, NBAA IFR reserves:
A, B ... 2,098 n miles (3,885 km; 2,414 miles)
Range with two crew and eight passengers, IFR reserves:
B ... 2,007 n miles (3,717 km; 2,309 miles)

OPERATIONAL NOISE LEVELS:
T-O: A, B ... 74.4 EPNdB
Sideline: A ... 85.2 EPNdB
B ... 85.1 EPNdB
Approach: A, B ... 93.4 EPNdB

UPDATED

LEARJET 40

TYPE: Business jet.

PROGRAMME: Announced at Farnborough International Air Show 21 July 2002 as replacement for Learjet 31A. Prototype (N40LX, converted from Learjet 45 prototype) first flight 31 August 2002, immediately followed by first production aircraft (45-2001/N40LJ) 5 September. Public debut (N40LJ) and formal launch at NBAA Convention, Orlando, Florida, 10 September 2002; prototype and first production aircraft undertook flight test programme, leading to FAA certification on 11 July 2003, (and award of first airworthiness certificate to N40LJ on 18 September 2003), followed by EASA certification on 9 January 2004. Later that month, the first five aircraft were delivered, two to Bombardier's Flexjet fractional ownership programme, two to US operators, and one to a German customer.

CUSTOMERS: Total of nine delivered in the first six months of 2004. Recent customers include Cirrus Aviation of Germany, which ordered two, of which the first entered service in February 2004.

COSTS: Approximately US$6.915 (2004).

DESIGN FEATURES: Generally as for Learjet 45, but fuselage shortened by 0.62 m (2 ft 0½ in) forward of wing, three cabin windows removed and capacity of fuselage fuel tank reduced.

POWER PLANT: As Learjet 45; TFE731-20AR-1B engines at outset. Fuel 3,036 litres (802 US gallons; 668 Imp gallons) usable, plus 61 litres (16.0 US gallons; 13.4 Imp gallons) unusable.

ACCOMMODATION: Two crew and up to seven passengers in forward club arrangement with galley on starboard side at front of cabin and full-width lavatory at rear. Compared with Learjet 45, cabin features redesigned seats that are 5 cm (2 in) wider and increase legroom by 15 cm (6 in), and an LED lighting system. The first 40 aircraft will feature an optional racing car-inspired red and black custom interior reflecting a sponsorship agreement between Bombardier and Indianapolis Motor Speedway Corporation.

DIMENSIONS, EXTERNAL:
Length overall ... 16.92 m (55 ft 6¼ in)
Wheelbase ... 7.86 m (25 ft 9½ in)

DIMENSIONS, INTERNAL:
Cabin: Length: incl flight deck ... 6.92 m (22 ft 8¼ in)
excl flight deck ... 5.38 m (17 ft 8 in)
Floor area, excl flight deck ... 5.17 m² (55.7 sq ft)
Volume, excl flight deck ... 10.28 m³ (363 cu ft)

WEIGHTS AND LOADINGS:
Weight empty ... 5,779 kg (12,740 lb)
Basic operating weight ... 6,091 kg (13,428 lb)
Payload: max ... 1,167 kg (2,572 lb)
with max fuel ... 815 kg (1,797 lb)
Fuel weight: max ... 2,438 kg (5,375 lb)
with max payload ... 2,087 kg (4,600 lb)
Max ramp weight: initial ... 9,344 kg (20,600 lb)
SB 40-11-1[1] ... 9,639 kg (21,250 lb)
Max T-O weight: initial ... 9,230 kg (20,350 lb)
SB 40-11-1[1] ... 9,525 kg (21,000 lb)
Max landing weight ... as Learjet 45
Max zero-fuel weight ... as Learjet 45
Max wing loading ... 329.0 kg/m² (67.39 lb/sq ft)

First production Learjet 40 *(Paul Jackson)* ***NEW***/0567042

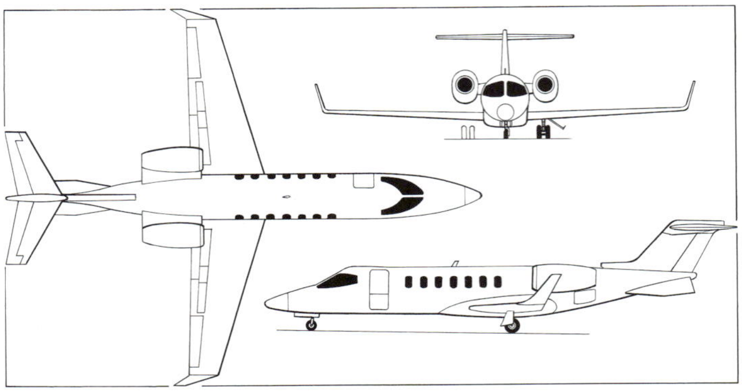

Learjet 40 business jet *(James Goulding)* 0526908

Max power loading ... 293 kg/kN (2.88 lb/lb st)

[1] *Certified September 2004 (Service Bulletin)*

PERFORMANCE:
Max operating Mach No. (MMO) ... as Learjet 45
Cruising speed: high ... M0.81 (464 kt; 839 km/h; 534 mph)
normal ... M0.79 (457 kt; 846 km/h; 526 mph)
long-range ... M0.75 (480 kt; 796 km/h; 494 mph)
Minimum control speed ... as Learjet 45
Time to FL430 at MTOW ... 22 min 42 s
Max certified altitude ... as Learjet 45
T-O balanced field length ... 1,306 m (4,285 ft)
FAR Pt 91 landing distance ... as Learjet 45
Range with two crew and four passengers, IFR reserves
... 1,803 n miles (3,339 km; 2,074 miles)

OPERATIONAL NOISE LEVELS: As Learjet 45.

UPDATED

LEARJET 60

TYPE: Business jet.

PROGRAMME: Announced 3 October 1990 as Learjet 55C successor; first flight of proof-of-concept aircraft with one PW305 turbofan 18 October 1990; flight testing resumed 13 June 1991 with two PW305s and stretched fuselage (more than 300 hours flown by May 1992); first production aircraft first flight (N601LJ) 15 June 1992; certification awarded 15 January 1993; deliveries started immediately. Certified in Argentina, Austria, Bermuda, Brazil, Canada, China, Denmark, Germany, Grand Cayman, Italy, Luxembourg, Malaysia, Mexico, the Philippines, South Africa, Switzerland, Turkey, the United Arab Emirates and USA.

CUSTOMERS: Total of 265 built by June 2003, including 22 delivered in 1994, 24 in 1995, 23 in 1996, 24 in 1997, 32 in 1998, 32 in 1999, 35 in 2000, 29 in 2001, 18 in 2002, 12 in 2003, and six in the first six months of 2004. Recent customers include Publix Super Markets Inc, Canadian National Railway company, Sears Roebuck & Co, Krispy Kreme Doughnut Corporation, and Subic International Air Charter, which took delivery of two in 2003.

COSTS: US$12.45 million (2002).

DESIGN FEATURES: Largest Learjet; otherwise generally as for earlier versions.

FLYING CONTROLS: As Learjet 31A. Spoilers can be partially extended to adjust descent rates.

LANDING GEAR: Retractable tricycle type, hydraulically actuated and electrically controlled; single wheel on nose unit and twin wheels on main units; nosewheel retracts forward, mainwheels inward; steerable nosewheel (±60°). Mainwheel tyre size 17.5×5.75–8, pressure 14.76 bar (214 lb/sq in); nosewheel tyre size 18×4.4 (10 ply), pressure 7.24 bar (105 lb/sq in).

Learjet 60 operated by BBJS/Flexjets of Addison, Texas *(Paul Jackson)* ***NEW***/0583436

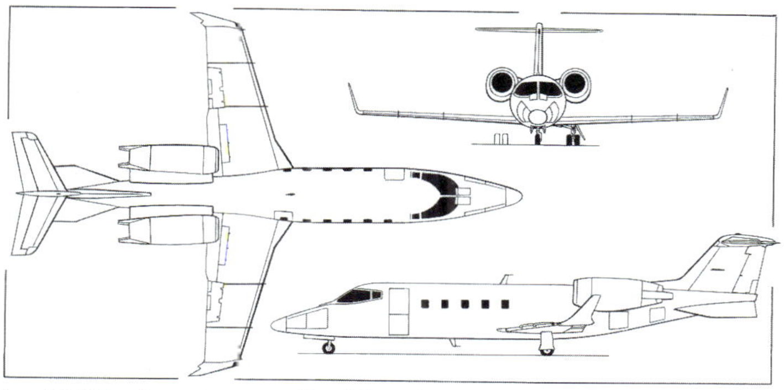

Learjet 60 business transport *(Jane's/Dennis Punnett)* 0507270

POWER PLANT: Two Pratt & Whitney Canada PW305A turbofans with FADEC, each flat rated at 20.46 kN (4,600 lb st) at up to 27°C (80°F). Fuel details under Weights and Loadings below. Single-point pressure refuelling system standard; gravity-feed fuel filler ports in each wing.

ACCOMMODATION: Two crew and up to 10 passengers; gross pressure cabin volume 15.57 m^3 (550 cu ft); compared with 55C, main cabin is 0.71 m (2 ft 4 in) longer and rear baggage hold section 0.38 m (1 ft 3 in) longer; full-across aft lavatory has flat floor, large mirror, wardrobe and external servicing; total 1.67 m^3 (59 cu ft) baggage capacity divided between an externally accessible hold (larger than that of Learjet 55C) and internal pressurised, heated compartment that is accessible in flight; galley cabinet has warming oven, cold liquid dispensers, ice compartment and storage for dinnerware; entertainment centre; 10-way adjusting seating is standard. New interior introduced in 1998 (from c/n 115), featuring redesigned passenger service unit, wider headliner, revised beverage cabinet, optional television monitor and wider choice of upholstery fabrics and leather trim.

SYSTEMS: Environmental control system uses engine bleed air for air conditioning and pressurisation, maximum differential 0.65 bar (9.4 lb/sq in), maintaining sea level cabin altitude to 7,835 m (25,700 ft) and 2,440 m (8,000 ft) cabin altitude to 15,545 m (51,000 ft). Hydraulic system, pressure 103.4 bar (1,500 lb/sq in), provided by two engine-driven, variable volume pumps, with electrically driven auxiliary pump. Electrical system comprises two 30 V DC 400 A engine-driven starter-generators and two 24 V DC batteries. Oxygen system, capacity 2.18 m^3 (77.0 cu ft), with demand-type masks for crew and drop-down constant-flow masks for passengers. Bleed air anti-icing on wing leading-edges, engine inlets, engine spinners and inlet guide vanes, electric anti-icing on windscreen. Hamilton Sundstrand T-20G-10C3A APU.

AVIONICS: Standard fully integrated all-digital Rockwell Collins Pro Line 4.

Radar: Rockwell Collins WXR840 colour weather radar.

Flight: Four-tube (152 × 178 mm; 6 × 7 in) EFIS, dual digital air data computers, dual navigation and communications radios, UNS-1C FMS, dual automatic AHRS, Rockwell Collins AMS-850 avionics management system, advanced autopilot and long-range navaid as standard; circuit breaker and control panels redistributed, as in Learjet 31A.

DIMENSIONS, EXTERNAL:

- Wing span ... 13.34 m (43 ft 9 in)
- Wing chord: at root ... 2.74 m (9 ft 0 in)
  - at tip ... 1.12 m (3 ft 8 in)
- Wing aspect ratio ... 7.2
- Length: overall ... 17.89 m (58 ft 8¼ in)
  - fuselage ... 17.02 m (55 ft 10 in)
- Fuselage max diameter ... 1.96 m (6 ft 5 in)
- Height overall ... 4.44 m (14 ft 6¾ in)
- Tailplane span ... 4.48 m (14 ft 8½ in)
- Wheel track ... 2.51 m (8 ft 3 in)
- Wheelbase ... 7.73 m (25 ft 4½ in)
- Cabin door: Width ... 0.64 m (2 ft 1 in)
  - Height to sill ... 0.69 m (2 ft 3 in)

DIMENSIONS, INTERNAL:

- Cabin: Length: incl flight deck ... 7.07 m (23 ft 2½ in)
  - excl flight deck ... 5.38 m (17 ft 8 in)
  - Max width ... 1.80 m (5 ft 11 in)
  - Max height ... 1.73 m (5 ft 8 in)
  - Floor area, excl flight deck ... 6.40 m^2 (68.9 sq ft)
  - Volume, excl flight deck ... 12.8 m^3 (453 cu ft)

AREAS:

- Wings, gross ... 24.57 m^2 (264.5 sq ft)
- Horizontal tail surfaces (total) ... 5.02 m^2 (54.00 sq ft)
- Vertical tail surfaces (total) ... 4.79 m^2 (51.53 sq ft)

WEIGHTS AND LOADINGS:

- Weight empty ... 6,364 kg (14,030 lb)
- Basic operating weight empty ... 6,641 kg (14,640 lb)
- Max payload ... 1,070 kg (2,360 lb)
- Payload with max fuel ... 544 kg (1,200 lb)
- Max usable fuel weight ... 3,588 kg (7,910 lb)
- Max T-O weight ... 10,659 kg (23,500 lb)
- Max ramp weight ... 10,773 kg (23,750 lb)
- Max landing weight ... 8,845 kg (19,500 lb)
- Max wing loading ... 433.8 kg/m^2 (88.85 lb/sq ft)
- Max power loading ... 260 kg/kN (2.55 lb/lb st)

PERFORMANCE:

- Max operating speed (VMO):
  - S/L-FL80 ... 300 kt (555 km/h; 345 mph) IAS
  - FL80-FL200 ... 340 kt (629 km/h; 391 mph) IAS
  - FL200-FL230 ... 340–330 kt (630–611 km/h; 391–378 mph) IAS
  - FL230-FL267.5 ... 330 kt (611 km/h; 378 mph) IAS
- Max operating Mach No. (MMO):
  - FL265-FL370 ... 0.81
  - FL370-FL430 ... 0.81-0.78
  - above FL430 ... 0.78
- Cruising speed: high ... 466 kt (863 km/h; 536 mph)
  - normal ... 453 kt (839 km/h; 521 mph)
  - long-range ... 422 kt (782 km/h; 486 mph)
- Stalling speed, flaps and landing gear down ... 106 kt (197 km/h; 122 mph) CAS
- Approach speed ... 139 kt (257 km/h; 160 mph) IAS
- Max rate of climb at S/L ... 1,371 m (4,500 ft)/min
- Rate of climb at S/L, OEI ... 378 m (1,240 ft)/min
- Time to FL410 after MTOW departure ... 18 min 30 s
- Max certified altitude ... 15,545 m (51,000 ft)
- Service ceiling, OEI ... 7,195 m (23,600 ft)
- T-O balanced field length ... 1,661 m (5,450 ft)
- FAR Pt 91 landing distance ... 1,043 m (3,420 ft)
- Range with two crew and four passengers:
  - VFR ... 2,685 n miles (4,972 km; 3,089 miles)
  - NBAA IFR ... 2,510 n miles (4,648 km; 2,888 miles)

OPERATIONAL NOISE LEVELS:

- T-O ... 78.9 EPNdB
- Sideline ... 83.2 EPNdB
- Approach ... 87.7 EPNdB

UPDATED

LEGEND

LEGEND AIRCRAFT INC

PO Box 11, 1015 Airport Drive, Winnsboro, Louisiana 71295
Tel: (+1 318) 435 44 01
e-mail: lrundell@3g.quik.com
Web: http://www.turbinelegend.com
PRESIDENT: Lanny Rundell

Lanny Rundell of Southern Air purchased the assets of Performance Aircraft in February 2002.

UPDATED

LEGEND LEGEND AND LEGEND TURBINE LEGEND

TYPE: Tandem-seat sportplane kitbuilt; tandem-seat turboprop sportplane kitbuilt.

PROGRAMME: Prototype (N620L) first flown by Performance Aircraft Inc in 1996 with Chevrolet V-8 engine; subsequently converted to Turbine Legend with Walter M 601E turboprop; public debut in turboprop form at EAA Sun 'n' Fun at Lakeland, Florida, in April 1999.

Performance Turbine Legend built in Canada *(Paul Jackson)* ***NEW***/1047381

CURRENT VERSIONS: **Legend:** Piston-engined version, powered by one 429 kW (575 hp) liquid-cooled Chevrolet V-8 driving a three-blade Hartzell propeller via a Geschwender 2:1 reduction gearbox. Wing span 8.22 m (27 ft 0 in). Ventral (P-51-type) air scoop).

Turbine Legend: Turboprop version, *as described.*

JC 100: Designation of Turbine Legend built in 2000 by Toys 4 Boys at Deland, Florida.

CUSTOMERS: By second quarter of 2004, more than 37 kits had been sold and at least 14 aircraft were flying, including one in Canada.

COSTS: Fast-build kit, less engine US$119,700 (2004). Engine (new) US$104,000 (1999).

DESIGN FEATURES: Highly streamlined, low-wing monoplane with sweptback tail surfaces and steeply raked windscreen. Tapered wings and mid-mounted tailplane. Optional winglets. Quoted build time: 2,500 hours, standard; 1,900 hours fast-build. Roll rate 190°/s.

FLYING CONTROLS: Conventional and mechanical; horn-balanced elevators and rudder; three-axis electric trim; electrically actuated slotted flaps, maximum deflection 38°.

STRUCTURE: Mostly CFRP.

LANDING GEAR: Retractable tricycle type; mainwheel size 15×6.0-6, nosewheel size 5.00×5; mainwheels retract inwards, nosewheel rearwards; steering by differential braking; dual disc brakes standard.

POWER PLANT: One Walter M 601 turboprop, rated at 540 kW (724 shp) for take-off and 490 kW (657 shp) maximum continuous, driving an Avia V 508E/84 three-blade, constant-speed, reversible-pitch feathering propeller. Fuel contained in integral tanks in outer wings 379 litres (100 US gallons; 83.3 Imp gallons); optional tip tanks, combined capacity 95 litres (25.0 US gallons; 20.8 Imp gallons); three optional extra tanks in fuselage, total capacity 167 litres (44.0 US gallons; 36.7 Imp gallons).

ACCOMMODATION: Two persons in tandem under rear-hinged, upward-opening canopy with fixed single-piece windscreen. Dual controls standard.

DIMENSIONS, EXTERNAL:

- Wing span ... 8.97 m (29 ft 5 in)
- Wing aspect ratio ... 8.6
- Length overall ... 7.84 m (25 ft 8½ in)
- Height overall ... 2.93 m (9 ft 7¼ in)
- Propeller diameter ... 2.13 m (7 ft 0 in)

DIMENSIONS, INTERNAL:

- Cabin: max width ... 0.75 m (2 ft 5½ in)

AREAS:

- Wings, gross ... 9.38 m^2 (101.0 sq ft)

WEIGHTS AND LOADINGS:
Weight empty: 930 kg (2,050 lb)
Baggage capacity: 54 kg (120 lb)
Max T-O weight: 1,496 kg (3,300 lb)
Max wing loading: 159.5 kg/m^2 (32.67 lb/sq ft)
Max power loading: 3.06 kg/kW (5.02 lb/shp)
PERFORMANCE:
Never-exceed speed (VNE): 347 kt (643 km/h; 400 mph)
Max level speed at S/L: 274 kt (507 km/h; 315 mph)
Max cruising speed at 7,620 m (25,000 ft): 290 kt (537 km/h; 334 mph)
Econ cruising speed at 7,620 m (25,000 ft): 245 kt (454 km/h; 282 mph)
Manoeuvring speed: 261 kt (482 km/h; 300 mph)
Stalling speed: flaps up: 66 kt (123 km/h; 76 mph)
landing configuration: 58 kt (107 km/h; 66 mph)
Max rate of climb at S/L: 1,981 m (6,500 ft)/min
Service ceiling: 10,670 m (35,000 ft)
T-O and landing run: 244 m (800 ft)
Range: with standard fuel and reserves:
at max cruising speed: 821 n miles (1,520 km; 944 miles)
at econ cruising speed: 993 n miles (1,839 km; 1,142 miles)
with optional fuel: 1,207 n miles (2,237 km; 1,390 miles)
g limits: normal: +6/–4
ultimate: +9/–6

UPDATED

LEZA

LEZA AIRCAM CORPORATION

1 Leza Drive, Sebring, Florida 33870
Tel: (+1 863) 655 42 42
Fax: (+1 863) 655 03 10
e-mail: aircam@ct.net
Web: http://www.leza-aircam.com
CEO AND PRESIDENT: Antonio Leza

Leza AirCam has a 4,090 m^2 (44,000 sq ft) factory at Sebring Airport and also produces the former Maxair Drifter and Super Drifter series of single- and two-seat ultralights.

In 2004, company announced that it was planning a 3,440 m^2 (37,000 sq ft) extension to its factory.

UPDATED

LEZA AIRCAM

TYPE: Two-seat kitbuilt twin.

PROGRAMME: Initially designed as camera platform for *National Geographic* magazine and first flew (N5084T) 1994. Redesigned as a kit and made public debut at Sun 'n' Fun 1996. First float-equipped version made initial flight 9 June 1999.

CUSTOMERS: More than 130 kits sold by mid-2004; over 40 flying by early 2001. Export destinations include Australia, Namibia and Zimbabwe.

COSTS: US$42,263 without engines; US$78,840 with Rotax 912 S engines (2004).

DESIGN FEATURES: Configured for slow flight. Strut- and wire-braced wings and mid-mounted tailplane wire-braced to tall, narrow-chord fin. Quoted build time 800 hours.

FLYING CONTROLS: Conventional and manual. Full-width in-flight-adjustable trim tab on elevator, Pushrod-actuated ailerons taper at tips. Electric flaps.

STRUCTURE: All-metal monocoque fuselage with glass fibre cockpit enclosure. Constant chord, swept-tip wings of 6061-T6 aluminium tubing with aluminium spars and ribs, and fabric covering. Low-drag wing struts. Tailfin has glass fibre fillet and fabric-covered rudder and elevator with metal trim tab.

LANDING GEAR: Tailwheel type; fixed. Fuselage-mounted spring steel cantilever mainwheel legs; Hegar wheels; differential hydraulic disc brakes and 6.00-6 in tyres. Maule steerable tailwheel. Full Lotus floats optional.

POWER PLANT: Two 47.8 kW (64.1 hp) Rotax 582s or two 73.5 kW (98.6 hp) Rotax 912 ULSs or mounted above wings, driving three-blade ground-adjustable Warp Drive pusher propellers. Optionally two 85.8 kW (115 hp) turbocharged Rotax 914s can be fitted. Fuel in two wing-mounted aluminium tanks, total capacity 106 litres (28.0 US gallons; 23.3 Imp gallons).

ACCOMMODATION: Pilot and passenger/observer in tandem seats within open cockpit. Lexan windshield. Dual controls. Open baggage compartment behind second seat.

AVIONICS: To customer's specification. Options include Terra T X 760D COM, TRT 250D transponder.

EQUIPMENT: Fire extinguisher standard.

DIMENSIONS, EXTERNAL:
Wing span: 11.07 m (36 ft 4 in)
Wing aspect ratio: 6.5

Prototype Leza AirCam N5084T *Ndoli* in 2004, following its recovery from Africa and refurbishment in USA ***(Paul Jackson)*** ***NEW***/1047396

Leza AirCam equipped with floats ***(W J Bushell)*** ***NEW***/1047843

Length overall: 8.23 m (27 ft 0 in)
Height overall: 2.54 m (8 ft 4 in)
Tailplane span: 3.96 m (13 ft 0 in)
Wheel track: 2.59 m (8 ft 6 in)
Min distance between propeller blade tips: 0.15 m (6 in)
DIMENSIONS, INTERNAL:
Baggage compartment: Length: 1.07 m (3 ft 6 in)
Max width: 0.63 m (2 ft 0¾ in)
AREAS:
Wings, gross: 18.95 m^2 (204.0 sq ft)
WEIGHTS AND LOADINGS (Rotax 912):
Weight empty: 472 kg (1,040 lb)
Max T-O weight: 762 kg (1,680 lb)
Max wing loading: 40.21 kg/m^2 (8.24 lb/sq ft)
Max power loading: 5.19 kg/kW (8.52 lb/hp)
PERFORMANCE (Rotax 912):
Never-exceed speed (VNE): 95 kt (177 km/h; 110 mph)
Max operating speed: 87 kt (161 km/h; 100 mph)
Normal cruising speed: 43–87 kt (80–161 km/h; 50–100 mph)
Stalling speed, power off, flaps down: 34 kt (63 km/h; 39 mph)
Max rate of climb at S/L: 610 m (2,000 ft)/min
Rate of climb at S/L, OEI: 91 m (300 ft)/min
Service ceiling: 5,485 m (18,000 ft)
T-O run: less than 61 m (200 ft)
Landing run: 91 m (200 ft)
Range at 57 kt (106 km/h; 70 mph): 295 n miles (547 km; 340 miles)
Endurance: 6 h
g limits: ±6

UPDATED

LEZA DRIFTER AND SUPER DRIFTER

TYPE: Tandem-seat ultralight kitbuilt.

PROGRAMME: Originated in 15 kW (20 hp) Maxair Hummer V-tail ultralight of early 1980s; Maxair Drifter marketed from 1984 by Lockwood Aviation Supply; also supplied in quantity to Antipodes as Austflight Drifter; production resumed by Leza-Lockwood. Previously sold with Rotax XP505, 618 or 447 engine.

CURRENT VERSIONS: **Drifter MU582** : Base model, *as described.*

Super Drifter SD912XL : Higher-powered model with 59.6 kW (79.9 hp) Rotax 912 engine, electric start, flaps and hydraulic brakes.

CUSTOMERS: More than 1,000 of both models flying by September 2004.

COSTS: MU582 US$18.960; SD912XL; US$27,730, both including engine and propeller (2004).

DESIGN FEATURES: Circular tube fuselage; wings supported by cabane and V struts; wire-braced empennage. Noise fairing/windscreen for partial protection of pilot. Single-piece wing is removable for storage and transportation. Quoted assembly time 150 hours for Drifter and 200 hours for Super Drifter.

FLYING CONTROLS: Conventional and manual. Full-span ailerons on Drifter; flaps on Super Drifter.

STRUCTURE: Metal airframe with fabric-covered wings and empennage; composites nose fairing.

LANDING GEAR: Generally as AirCam; Drifter has drum brakes and Super Drifter has hydraulic brakes. Steerable tailwheel. Optional floats.

POWER PLANT: *Drifter:* One 48.0 kW (64.4 hp) Rotax 582 driving a 1.52 m (5 ft 0 in) Warp Drive three-blade propeller. Fuel capacity 37.9 litres (10.0 US gallons; 8.3 Imp gallons).

Super Drifter: One 59.6 kW (79.9 hp) Rotax 912 driving a 1.73 m (5 ft 8 in) Warp Drive three-blade propeller. Fuel capacity as Drifter.

ACCOMMODATION: Pilot and passenger in tandem open cockpit.

SYSTEMS: Super Drifter has electric starter.

DIMENSIONS, EXTERNAL:
Wing span: 9.14 m (30 ft 0 in)
Length overall: 6.55 m (21 ft 6 in)

Austflight (Leza) Drifter tandem-seat kitplane ***(Paul Jackson)*** ***NEW***/1047463

Height overall: Drifter .... 2.82 m (9 ft 3 in)
Super Drifter .... 2.87 m (9 ft 5 in)

AREAS:
Wings, gross .... 13.94 m² (150.0 sq ft)

WEIGHTS AND LOADINGS:
Weight empty: Drifter .... 193 kg (425 lb)
Super Drifter .... 225 kg (495 lb)
Max T-O weight: Drifter .... 408 kg (900 lb)
Super Drifter .... 453 kg (1,000 lb)

PERFORMANCE:
Never-exceed speed (VNE) .... 69 kt (128 km/h; 80 mph)
Normal cruising speed .... 61 kt (113 km/h; 70 mph)
Stalling speed: Drifter .... 34 kt (63 km/h; 39 mph)
Super Drifter .... 35 kt (65 km/h; 40 mph)
Max rate of climb at S/L .... 305 m (1,000 ft)/min
T-O run .... 61 m (200 ft)
Landing run .... 91 m (300 ft)
Range .... 200 n miles (370 km; 230 miles)

NEW ENTRY

LIBERTY

LIBERTY AEROSPACE INC

1383 General Aviation Drive, Melbourne, Florida 32935
Tel: (+1 321) 752 03 32
Fax: (+1 321) 752 03 77
e-mail: sales@libertyaircraft.com
Web: http://www:libertyaircraft.com
PRESIDENT AND CEO: Anthony J P Tiarks
VICE-PRESIDENT: Ivan Shaw
VICE PRESIDENT, SALES AND MARKETING: Robert Stangarone
PUBLIC RELATIONS AND MARKETING: Russell Greenberg

Liberty was established specifically to produce the XL-2 lightplane. It received refinancing when KFH Bahrain acquired 75 per cent of the stock of Liberty's parent company in October 2004.

UPDATED

LIBERTY XL-2

TYPE: Two-seat lightplane

PROGRAMME: Design, by the team that created the Europa kitbuilt (which see in UK section), began in 1997. Announced 26 May 2000; mockup displayed at NBAA Convention in New Orleans and AOPA-USA Convention in Long Beach during October 2000, when Scaled Technology Works (STW) announced as manufacturing partner; first three aircraft registered as built by STW Composites, but partnership ended in 2003 when STW closed.

'Official' first flight of prototype (N202XL) 2 April 2001, followed by public debut at EAA Sun 'n' Fun at Lakeland, Florida later that month; second prototype and sales demonstrator (N203XL) appeared at 2002 Sun 'n' Fun fitted with a redesigned engine cowling; production conforming prototype (N204XL) entered the flight test programme in September 2003; FAA FAR Pt 23 certification for day VFR achieved 19 February 2004; will be certified for IFR. Target production of 10 in 2003, rising to 140 in 2004, with potential for up to 400 per year. However, by late 2004, only second production aircraft (N505XL; fourth XL-2 overall) had been registered.

CURRENT VERSIONS: **XL-2** Standard version, adoption of Teledyne Continental IOF-240 engine announced 21 July 2001. *As described.*

XL-2R: Proposed future version with Rotax 912 S flat-four.

XL-2 Surveillance Flatform: Version under consideration in early 2004 would feature FLIR Systems Ultra 8500 FW long-range FLIR sensor; endurance 5 hours; estimated cost US$400,000.

CUSTOMERS: Launch customer Civil Flying School at Moorabbin Airport, Melbourne, Australia, ordered two in July 2000; Bill Crokaris of Melbourne has ordered one. First 50 aircraft (for Founders Club members, at special price of US$85,000); 60 sold by April 2004. Target production four per week by 2005.

COSTS: Basic US$129,500 with Continental engine; typically equipped US$139,500 (2004).

DESIGN FEATURES: Based on Europa, but with modifications to optimise airframe for mass production; cabin is 10 cm (4 in) wider than that of Europa. Design goals included low price, efficient high-speed cruise with STOL performance, economy and ease of maintenance, advanced structural materials offering strength and durability, excellent handling with positive stability, and ability to store at home on a purpose-designed transporter.

FLYING CONTROLS: Conventional and manual, via pushrods. Dual controls and adjustable rudder pedals standard. All-moving mass-balanced tailplane with electrically actuated 1.3:1 geared anti-balance/trim tab maximum deflection +13/5°; mass-balanced ailerons with differential action, maximum deflection +25/−19°; electrically actuated slotted flaps occupying 70 per cent span and 30 per cent chord, maximum deflection 30°; rudder maximum deflection ±30°.

STRUCTURE: Welded 4130 steel tube forward fuselage and centre-section to carry engine, landing gear and wing attachment loads, with modular carbon fibre skin; flying surfaces comprise riveted subassembly with bonded aluminium skin; single-spar wings have optional folding facility, quick connect attachment to centre-section and self-connecting flap and aileron controls; de-riggable tailplane optional. Matrix Composites of Rockledge, Florida,

Liberty XL-2 instrument panel *(Paul Jackson)* ***NEW***/0583537

First production Liberty XL-2 *(Paul Jackson)* ***NEW***/0583538

manufactures composites fuselage components, and Heart of Georgia Metal Crafters supplies wings; final assembly by Liberty at Melbourne, Florida.

LANDING GEAR: Non-retractable tricycle type; aluminium main legs; tubular steel nose leg; steering via castoring nosewheel and differential brakes. Cleveland wheels and hydraulic disc brakes. Tyre size 5.00×5.

POWER PLANT: One 93 kW (125 hp) Teledyne Continental IOF-240-B flat-four with Aerosance PowerLink FADEC driving a Sensenich W69EK7-63G two-blade, fixed-pitch, metal propeller, in standard version. Alternatively, for later certification, one 73.5 kW (98.6 hp) Rotax 912 S flat-four, driving a purpose-designed two-blade, fixed-pitch carbon fibre Dowty propeller via 2.43:1 reduction gearing. Fuel capacity 112 litres (29.5 US gallons; 24.6 Imp gallons) of which 106 litres (28 US gallons; 23.3 gallons) are usable.

ACCOMMODATION: Two persons, side by side. Baggage compartment, with cargo net, behind seats, maximum capacity 45 kg (100 lb). Upward-hinged door on port side. Cabin heating.

SYSTEMS: Electrical system powered by 14 V 60 A engine-driven alternator. 14 V battery and 14 V auxiliary power outlet standard.

AVIONICS: Optional Garmin and UPS (Apollo) packages.

Basic VFR: Garmin GNC 250XL VFR GPS/com and GTX 327 transponder/altitude encoder; or UPS Apollo GX 65 GPS/com and SL 70 transponder/altitude encoder.

Basic IFR: Garmin GNS 430 GPS/com/nav/map, GNC 300XL IFR GPS/com, GTX 327 transponder/altitude encoder, GMA 340 audio panel, and GI 102A and GI 106A indicators; or UPS (Apollo) GX 60 GPS/com, SL 30 nav/com/GPS, SL 70 transponder/altitude encoder, SL 15M audio panel, ACU-14V annunciator and Mid-Continent MD200-306 indicator.

Deluxe IFR: as Basic IFR with the addition of second Garmin GNS 430 and substitution of GI 105A indicator for GI 102A; or with addition of UPS (Apollo) MX20 MFD. ELT standard on aircraft for US market, optional elsewhere.

Instrumentation: Vision Microsystems VM1000FX digital engine information display. Sensitive altimeter, magnetic compass, airspeed indicator, vertical speed indicator, electric tachometer, trim position indicator, flap position indicator, electric stall warning, annunciator panel, fuel contents gauge, fuel pressure gauge, manifold pressure gauge, percentage power gauge, OAT gauge, CHT gauge, EGT gauge, oil temperature and pressure gauges, voltmeter, ammeter and quartz clock.

EQUIPMENT: Standard equipment includes tinted windows, windscreen defrost, four-point safety harnesses, cloth upholstery, map/storage pockets, map shelf, baggage restraint net, static port and alternate static port, wingtip navigation lights, anti-collision beacon, tiedown points, fuel quick-drain and white paint finish. Optional equipment includes electric gyro package comprising artificial horizon, direction indicator and turn co-ordinator; pitot heat; speed kit comprising nose- and mainwheel fairings and flap hinge fairings; landing/taxi/instrument/ cabin flights; deluxe leather interior; pearlescent exterior paint; choice of two exterior decal designs; and towbar.

DIMENSIONS, EXTERNAL:
Wing span .... 8.53 m (28 ft 0 in)
Wing aspect ratio 7.0
Length overall .... 6.10 m (20 ft 0in)
Height overall .... 2.24 m (7 ft 4 in)
Propeller diameter .... 1.75m (5 ft 9in)

DIMENSIONS, INTERNAL:
Cabin max width .... 1.22 m (4 ft 0 in)
Baggage volume .... 0.68 m³ (24.0 cu ft)

AREAS:
Wings, gross .... 10.41 m² (112.0 sq ft)

WEIGHTS AND LOADINGS (Continental engine):
Weight empty, VFR .... 483 kg (1,065 lb)
Baggage capacity .... 45 kg (100 lb)
Max T-O weight .... 749 kg (1,653 lb)
Max wing loading .... 72.1 kg/m² (14.76 lb/sq ft)
Max power loading .... 8.05 kg/kW (13.22 lb/hp)

PERFORMANCE:
Never-exceed speed (VNE) .... 162 kt (300 km/h; 186 mph)
Max level speed .... 125 kt (232 km/h; 144 mph)
Manoeuvring speed .... 100 kt (185 km/h; 115 mph)
Stalling speed: flaps up .... 50 kt (93 km/h; 58 mph)
flaps down .... 43 kt (80 km/h; 50 mph)
Max rate of climb at S/L .... 208 m (682 ft)/min
T-O run .... 251 m (822 ft)
T-O to 15 m (50 ft) .... 456 m (1,496 ft)
Landing run .... 256 m (841 ft)
Range with max fuel and 55% power with reserves .... 500 n miles (926 km; 575 miles)
g limits .... +3.8/−1.9

UPDATED

LITTLE WING

LITTLE WING AUTOGYROS INC

746 Highway 89 North, Mayflower, Arkansas 72106
Tel: (+1 501) 470 74 44
Fax: (+1 501) 470 74 87
e-mail: ron@littlewingautogyro.com
Web: http://www.littlewingautogyro.com
PRESIDENT: Ronald Herron

Founded in 1995, company is developing a series of autogyros; the original Roto-Pup was to be joined by a new design, the tricycle landing gear, open cockpit **Skylark;** this is in abeyance while company works on other projects, including a radial-engined aircraft. Workforce one; manufacturing area 55 m^2 (600 sq ft).

By April 2005, the Little Wing LW-5 held six class E-3a world records: East to West Coast USA; West to East Coast USA; Round trip US transcontinental; distance without landing (573 n miles; 1,061 km; 659 miles); time to 6,000 m (19,685 ft) (24 min 28 s); and altitude (8,049 m; 26,408 ft).

UPDATED

LITTLE WING ROTO-PUP

TYPE: Single-seat autogyro ultralight kitbuilt; two-seat autogyro ultralight kitbuilt.

PROGRAMME: Initial design work began in 1980 and construction of LW-1 prototype (N45LW), using a Piper PA-11 fuselage, started December 1990; this made its first flight 21 October 1994. LW-2, scaled-down version of LW-1, first flew April 1995; LW-3, controlled by full universal tilting of rotor head, first flew (N46LW) 22 September 1996. Certified by FAA in Experimental category. In 2003, LW-3 was fitted with a seven-cylinder radial engine which has increased empty weight by 45 kg (100 lb); testing continues.

CURRENT VERSIONS: **Roto-Pup LW-2**: Controlled by elevator, rudder and laterally tilting rotor head.

Roto-Pup LW-3: Controlled by rudder and universally tilting rotor head. Single seat.

Roto-Pup LW-3+2: Tandem two-seat version. Two flying by end 2000.

Description applies to LW-3, unless otherwise stated.

Roto-Pup LW-5: Tandem-seat, shorter version (N100MK). Landing gear moved 24 cm (9½ in) forward to compensate for 61 cm (2 ft 0 in) reduction in length. Flown solo from rear seat.

Roto-Pup Ultralight: First flew April 1995. McCulloch O-100 engine; fuselage uncovered to remain within 115 kg (254 lb) ultralight limit. Controlled by elevator, lateral tilting rotor head and rudder. Offered as plans only.

CUSTOMERS: Four prototypes undertook flight test programme. Total of 165 sets of plans also sold. Three customer kit-built examples have also flown, including Andrew Keech's LW-5 (N100MK) with Rotax 914F-engine which set world records in 2004 for altitude, distance and time to 6,000 m (19,685 ft). Also flew 1,000 mile cross-country in June 2002. Others nearing completion in Belgium, France and Japan, as well as four in Michigan.

COSTS: LW-3 kit approximately US$10,000; LW-3+2 US$15,000, both without engine (2004).

DESIGN FEATURES: Fuselage design based on Piper PA-11 Cub Special, giving classic autogyro appearance, but scaled-down and using Pratt trusses in place of Cub's Warren trusses. Two-blade Rotor Flight Dynamics rotor; each blade is aluminium bonded and has positive twist. Rotor speed 390 rpm, blade tip speed 140 m (459 ft)/s. (On LW-2, pitch link to airframe is isolated by elastomeric dampers, determining rotor axis inclination relative to airframe). Rotor mast dampened by large rubber bushes at aft pylon braces. Rotor brake optional, but recommended on all versions.

FLYING CONTROLS: Floor-mounted control stick; cable and pulley actuated elevator in LW-2; LW-3 has no elevators, but floor-mounted control column, pushrods and universally tilting rotor head for pitch and roll, plus cable-actuated rudder. Ground-adjustable horizontal stabiliser for airframe pitch; tip shields at ends of horizontal stabilisers. Push-pull cables used on two-seat versions.

Little Wing LW-5 **NEW**/1037887

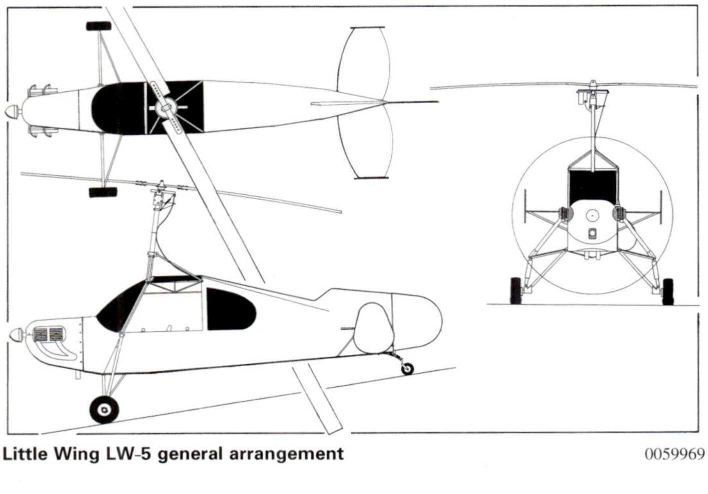
Little Wing LW-5 general arrangement 0059969

Little Wing LW-3 fitted with Rotec R2800 radial engine ***NEW***/0553233

STRUCTURE: Full chromoly 4130 steel tube welded fuselage and rotor pylon; fuselage covered with Dacron fabric and polyurethane finish. Hub structure of 2024-T3 aluminium bar with bolt attachments, pivoted at rotor attachment point for ground adjustment of blade pitch.

LANDING GEAR: Non-retractable type with tailwheel. Mainwheels Hegar 6.50-6 in with 1.03 bar (15 lb/sq in) tyre pressure. Matco 3.00-2 steerable tailwheel of solid rubber. Internal expanding go-kart style brakes by Leaf. Optional shock-absorbing gear. Floats optional.

POWER PLANT: *LW-3:* One 52.2 kW (70 hp) TEC converted Volkswagen four-stroke engine with dual ignition; options include 2si 52.2 kW (70 hp) two-stroke water-cooled engine and various Rotax, Subaru and Hirth engines. Fuel capacity 38 litres (10.0 US gallons; 8.3 Imp gallons) in seat tank. Refuelling position on starboard side of cabin. Oil capacity 2.4 litres (5.0 US pints; 4.2 Imp pints).

LW-3+2: One 82.0 kW (110 hp) Hirth F30 or Rotax 912; or 84.6 kW (113.4 hp) Rotax 914 optional power plants in the range 67.1 to 112 kW (90 to 150 hp) can be fitted. Fuel capacity 49.2 litres (13.0 US gallons; 10.8 Imp gallons). In 2003, company fitted an 82 kW (110 hp) Rotec R2800 seven-cylinder radial engine driving a two-blade propeller.

ACCOMMODATION: One or two occupants, according to version, in enclosed cockpit; forward-opening door on starboard side. Baggage compartment behind single seat. Lexan polycarbonate windscreen and side windows.

SYSTEMS: 1,000 A 12 V battery for ignition back-up and rotor pre-spin. Hydraulic pre-spin optional.

AVIONICS: Micronair 760 nav/com and EIS engine electronic monitoring system are recommended.

DIMENSIONS, EXTERNAL:	
Rotor diameter: LW-2/3	7.62 m (25 ft 0 in)
LW-3+2	8.53 m (28 ft 0 in)
Rotor blade chord	0.18 m (7 in)
Fuselage length: standard	5.49 m (18 ft 0 in)
LW-5	4.88 m (16 ft 0 in)
Height to top of rotor head	2.59 m (8 ft 6 in)
Tail unit span	2.13 m (7 ft 0 in)
Wheel track	2.13 m (7 ft 0 in)
Wheelbase: standard	3.96 m (13 ft 0 in)
LW-5	3.51 m (11 ft 6 in)
Propeller diameter: VW engine	1.57 m (5 ft 2 in)
2si engine	1.73 m (5 ft 8 in)
Rotec engine	1.88 m (6 ft 2 in)
DIMENSIONS, INTERNAL:	
Cabin: Length: standard	1.24 m (4 ft 1 in)
LW-5	1.90 m (6 ft 3 in)
Max width: standard	0.57 m (1 ft 10½ in)
LW-5	0.66 m (2 ft 2 in)
Max height	1.07 m (3 ft 6 in)
AREAS:	
Rotor blades (each)	0.54 m^2 (5.83 sq ft)
Rotor disc: LW-2/3	45.62 m^2 (491.1 sq ft)
LW-3+2	57.21 m^2 (615.8 sq ft)
Dorsal fin	0.42 m^2 (4.52 sq ft)
Auxiliary tip plates (each)	0.15 m^2 (1.60 sq ft)
Rudder	0.28 m^2 (3.03 sq ft)
Tailplane	0.81 m^2 (8.80 sq ft)
Elevators (total)	0.84 m^2 (9.00 sq ft)
WEIGHTS AND LOADINGS:	
Weight empty: LW-2	160 kg (352 lb)
LW-3+2	215 kg (475 lb)
LW-3	204 kg (450 lb)
Baggage capacity	11 kg (25 lb)
Max fuel weight	24 kg (54 lb)
Max T-O weight: LW-2	340 kg (750 lb)
LW-3+2	499 kg (1,100 lb)
Max disc loading: LW-3+2	8.72 kg/m^2 (1.79 lb/sq ft)
PERFORMANCE:	
Never-exceed speed (VNE)	104 kt (193 km/h; 120 mph)
Max operating speed at S/L:	
LW-3, LW-3+2	87 kt (161 km/h; 100 mph)
Econ cruising speed at FL10:	
LW-3	52 kt (97 km/h; 60 mph)
LW-3+2	65 kt (121 km/h; 75 mph)
Touchdown speed for power-off landing	9–13 kt (16–24 km/h; 10–15 mph)

Max rate of climb at S/L: LW-3 ..... 518 m (1,700 ft)/min
LW-3+2 ..... 183 m (600 ft)/min
Service ceiling: standard ..... 3,050 m (10,000 ft)
with Rotax 914 ..... 7,925 m (26,000 ft)
T-O run: LW-3 ..... 61 m (200 ft)
LW-3+2 ..... 92 m (300 ft)
Landing run: LW-3, LW-3+2 ..... 3–6 m (10–20 ft)
Range: LW-3 ..... 100 n miles (185 km; 115 miles)
LW-3+2 ..... 150 n miles (277 km; 172 miles)

UPDATED

LOCKHEED MARTIN

LOCKHEED MARTIN CORPORATION

6801 Rockledge Drive, Bethesda, Maryland 20817
Tel: (+1 301) 897 61 21
Fax: (+1 301) 897 62 52
Web: http://www.lockheedmartin.com
CHAIRMAN AND CEO: Vance D Coffman
PRESIDENT AND COO: Robert J Stevens
VICE-PRESIDENT AND CHIEF FINANCIAL OFFICER: Christopher E Kubasik
EXECUTIVE VICE-PRESIDENT, AERONAUTICS COMPANY: Dain M Hancock
EXECUTIVE VICE-PRESIDENT, SPACE SYSTEMS COMPANY: G Thomas Marsh
EXECUTIVE VICE-PRESIDENT, ELECTRONIC SYSTEMS: Robert B Coutts
EXECUTIVE VICE-PRESIDENT, INFORMATION AND TECHNOLOGY SERVICES: Michael F Camardo
EXECUTIVE VICE-PRESIDENT, INTEGRATED SYSTEMS AND SOLUTIONS: Albert E Smith
SENIOR VICE-PRESIDENT, CORPORATE COMMUNICATIONS: Dennis Boxx

Former Lockheed Aircraft Corporation renamed Lockheed Corporation in September 1977. Merger with Martin Marietta announced 30 August 1994 and completed 15 March 1995. Further expansion resulted from the acquisition of Loral Corporation's defence electronics and systems integration businesses in April 1996 for approximately US$9.1 billion and completed purchase of Comsat in August 2000 for US$2.6 billion. Workforce total of about 132,500 at beginning of 2004. Net sales in 2003 reported as US$31.8 billion. Activities include design and production of aircraft, electronics, satellites, space systems, missiles, ocean systems, information systems, and systems for strategic defence and for command, control, communications and intelligence.

Major business groups of Lockheed Martin are:

Aeronautics Company
Space Systems Company, comprising:
- Lockheed Martin Space Systems — Astronautics Operations
- Lockheed Martin Space Systems — Michoud Operations
- Lockheed Martin Space Systems — Missiles & Space Operations
- Lockheed Martin Commercial Space Systems
- International Launch Services

Electronic Systems, comprising:
- Lockheed Martin Air Traffic Management
- Lockheed Martin Systems Integration — Owego
- Lockheed Martin Canada
- Lockheed Martin Distribution Technologies
- Lockheed Martin Information Systems
- Lockheed Martin Tactical Systems
- Lockheed Martin Maritime Systems & Sensors
- Lockheed Martin Missiles and Fire Control
- Lockheed Martin UK

Information and Technology Services, comprising:
- Knolls Atomic Power Laboratory
- Lockheed Martin Aircraft & Logistics Centers
- Lockheed Martin Information Support Services
- Lockheed Martin Space Operations
- Lockheed Martin Technical Operations
- Sandia National Laboratories
- Technology Ventures

Integrated Systems and Solutions, comprising:
- Lockheed Martin Management & Data Systems
- Lockheed Martin Mission Systems
- Lockheed Martin Advanced Concepts

UPDATED

LOCKHEED MARTIN AERONAUTICS COMPANY (LM AERO)

1 Lockheed Boulevard, Fort Worth, Texas 76108
Tel: (+1 817) 777 20 00
Fax: (+1 817) 777 21 15
PRESIDENT: Dain M Hancock
EXECUTIVE VICE-PRESIDENT AND GENERAL MANAGER, F-16 PROGRAMME: John Bean
EXECUTIVE VICE-PRESIDENT AND GENERAL MANAGER, F-35 PROGRAMME: Tom Burbage
EXECUTIVE VICE-PRESIDENT AND GENERAL MANAGER, F/A-22 PROGRAMME: Ralph D Heath
EXECUTIVE VICE-PRESIDENT FOR PROGRAMS: Robert T Elrod
EXECUTIVE VICE-PRESIDENT FOR FINANCE: John C McCarthy

LM Aero includes the following operating units:
Lockheed Martin Aeronautics Company - Marietta
Lockheed Martin Aeronautics Company - Palmdale
Lockheed Martin Aeronautics Company - Fort Worth

Lockheed Martin Aeronautics Company established in January 2000 through consolidation of aeronautics businesses, with headquarters at Fort Worth, Texas. Sales in 2002 were valued at US$6.5 billion, rising to US$10.2 billion in 2003.

Teaming arrangement agreed with Northrop Grumman to collaborate in development and marketing of AEW aircraft; also with KAI of South Korea to develop and market the T-50 Golden Eagle advanced trainer/lead-in fighter trainer.

UPDATED

LOCKHEED MARTIN AERONAUTICS — MARIETTA

86 South Cobb Drive, Marietta, Georgia 30063-0264
Tel: (+1 770) 494 44 11
Fax: (+1 770) 494 76 56
MANAGER, MEDIA RELATIONS: Peter Simmons

In April 1991, Lockheed won competition to produce F-22 (now F/A-22) with General Dynamics (now Lockheed Martin Tactical Aircraft Systems) and (then) Boeing Military Airplanes. Lockheed Martin Aeronautics Company-Marietta also involved in studies into advanced mobility aircraft for 21st century strategic transport; was subcontractor to Boeing on NASA High-Speed Civil Transport (HSCT) programme; working with NASA on advanced subsonic technology transport programme; and engaged in other, classified projects.

Long-term activities at Marietta include production of C-130J Hercules and F/A-22 Raptor aircraft. LMAS signed joint venture agreement with Alenia of Italy in September 1996 concerning development and marketing of the C-27J tactical transport that incorporates systems developed for the C-130J Hercules. However, lack of sales resulted in Lockheed Martin seeking to scale down involvement in closing months of 2002, although it continues marketing C-27J as joint effort with Alenia.

Lockheed Martin Aeronautics Company-Marietta working to develop standard avionics suite for future versions (upgrades) of P-3 Orion and AEW aircraft; open architecture design for maximum flexibility and system growth; elements include AN/APS-145 radar, GPS and communications/navigation system. Also interest from several countries on possible new-build aircraft.

Aeronautical Systems Support, with headquarters in Smyrna, Georgia, is subordinate element, with responsibility for after sales support, including provision of spare parts and technical back-up for variety of aircraft types.

UPDATED

LOCKHEED MARTIN (645) F/A-22 RAPTOR

TYPE: Multirole fighter.

DEVELOPMENT MILESTONES

Requirement issued	1981
Request for proposals	Sep 85
Design submitted	Jul 86
Programme launched	31 Oct 86
First flight	29 Sep 90
First flight, pre-production	7 Sep 97
Production go-ahead	15 Aug 01
First flight, production	Sep 03
First delivery	26 Sep 03

PROGRAMME: US Air Force Advanced Tactical Fighter (ATF) requirement called for 750 McDonnell Douglas F-15 Eagle replacements incorporating low observables technology and supercruise (supersonic cruise without afterburning); parallel assessment of two new power plants; request for information issued 1981; concept definition studies awarded September 1983 to Boeing, General Dynamics, Grumman, McDonnell Douglas, Northrop and Rockwell; requests for proposals issued September 1985; submissions received by 28 July 1986; USAF selection announced 31 October 1986 of demonstration/validation phase contractors: Lockheed YF-22 and Northrop YF-23; each produced two prototypes and ground-based avionics testbed. Competing engine demonstration/validation programmes launched September 1983; ground testing began 1986–87; flight-capable Pratt & Whitney YF119s and General Electric YF120s ordered early 1988; all four aircraft/engine combinations flown.

Decision of 11 October 1989 extended evaluation phase by six months; draft request for engineering and manufacturing development (EMD) proposals issued April 1990; first artists' impressions released May 1990; Lockheed teamed with General Dynamics (Fort Worth) and Boeing Military Airplanes to produce two YF-22 prototypes, which first flew on 29 September and 30 October 1990.

Acquisition of former General Dynamics gave Lockheed Martin control of 67.5 per cent of programme. Final engineering and manufacturing development (EMD) requests issued for both weapon system and engine 1 November 1990; proposals submitted 2 January 1991; F-22 and F119 power plant announced by USAF as winning combination, 23 April 1991; EMD contract given 2 August 1991 for 11 (later reduced to nine) flying prototypes, plus one static test article and one fatigue test airframe; design underwent several detail refinements through early 1990s.

Combat roles reassessment of May 1993 added air-to-ground attack with precision-guided munitions (PGMs) to F-22's roles. Under US$6.5 million contract addition on 25 May 1993, main weapon bay and avionics adapted for delivery of AIM-9X missile and 454 kg (1,000 lb) GBU-32 Joint Direct Attack Munition (JDAM). Addition of ground attack capability resulted in redesignation as the **F/A-22**, which was announced by USAF Chief of Staff General Jumper on 17 September 2002.

Preliminary design review, covering all aspects of the design, completed 30 April 1993; critical design review completed February 1995; preproduction verification (PPV) batch of four aircraft was scheduled to be ordered 1997 but these deleted from overall programme at start of that year; long lead items for first production batch (two aircraft), subsequently reclassified as Production Representative Test Vehicles (PRTV), ordered in 1998, with full funding in FY99. Minor design changes for production aircraft announced July 1991. Suggested name of SuperStar rejected in 1991 and it remained unnamed until occasion of roll-out in April 1997 when it was announced that the name Raptor had been chosen.

Fabrication of first component for first EMD aircraft (91-4001, c/n 4001) began 8 December 1993 at Boeing's facility in Kent, Washington; assembly of forward fuselage launched at Marietta on 2 November 1995 with start of work on nose landing gear well; assembly work also begun at Fort Worth in mid-1995 with mating of three assemblies that comprise the mid-fuselage of first EMD aircraft taking place in early 1996, followed by road transfer of entire section to Marietta in September 1996 for start of final assembly. Delivery by Boeing to Lockheed Martin at Marietta of subassemblies, including wing and aft fuselage, for first EMD aircraft took place on schedule in third quarter of 1996. Pratt &

Whitney also delivered first F119 flight test power plant in September. Prototype rolled out 9 April 1997; planned May 1997 first flight delayed by fuel leaks and hardware-related anomalies. First flight accomplished 7 September 1997, with aircraft airborne for 58 minutes during which initial handling evaluation undertaken before landing gear was retracted and further handling assessment performed at speeds of up to 250 kt (463 km/h; 288 mph); maiden flight also included simulated powered approach at medium altitude before landing at Marietta. Second sortie of 35 minutes took place on 14 September 1997, after which aircraft underwent minor structural modifications and was then placed in structural test fixture for load ground tests and strain gauge calibration.

Low-rate initial production (LRIP) decision originally dependent upon accumulating 183 hours of flight testing; this milestone passed on 23 November 1998 and cleared way for release of US$195.5 million in late December 1998 for advance procurement (long lead items) for six Lot 1 LRIP aircraft, which subsequently reclassified as PRTV 2 aircraft after wrangle over funding in mid-1999. Earlier, in December 1998, Lockheed Martin received contract worth US$503 million for two PRTVs and associated programme support.

Wind tunnel testing occupied 19,195 hours up to YF-22 stage and a further 16,930 hours up to mid-1995, when Configuration 645 was finalised; six major categories were investigated (aerodynamic loads and weapons bay acoustics, inlet and engine compatibility, mission/manoeuvre performance, inlet icing, stability and control flying qualities, weapons and stores separation), with last-named comprising majority of 900 hours that remained to be done in 1995–97.

Avionics trials from November 1997 in a Boeing 757 (N757A) Flying Test Bed (FTB) with AN/APG-77 radar in F-22-type nosecone; block 1 software, permitting basic radar operation including simultaneous search-and-track modes, delivered to Boeing, April 1998 and subsequently tested on 757; block 2 software delivered on 7 December 1998, with block 3S beginning flight testing on 24 April 2000; this is early version of block 3.0 software, which delivered to Seattle on 11 August 2000 in readiness for airborne trials programme that began in mid-September. Communications/navigation/identification (CNI) system and EW suite being tower-tested on full-scale model of forward fuselage at Fort Worth, Texas, during 1998–99; subsequently tested on Boeing 757 FTB, which is fitted with three common integrated processors (CIPs) for 1,400 hour software flight test programme. Further modification of Boeing 757 to mount representative wing section above forward fuselage completed in late 1998, and flight testing of conformal antennas began on 11 March 1999; by December 2000, 757 FTB had accumulated 641.9 flight test hours in 126 sorties. Radar testing also accomplished using T-39 Sabreliner as target aircraft and involved a Lockheed T-33 for calibrated airborne trials from December 1998.

Milestones in 1999 included delivery of first AN/ALR-94 EW system on 15 February by Sanders to the Avionics Integration Laboratory in Seattle; 100th sortie in May; successful completion of design load limit testing of article 3999 on 25 September; and compliance with all five major 1999 flight test objectives (including flight at altitude of 50,000 ft; opening of side and main weapon bay doors in flight; supercruise demonstration; and flutter envelope expansion) by 24 September.

Flight test programme interrupted at least twice during 2000, with most serious occurrence arising in May after discovery of hairline cracks in canopies of first two aircraft; grounding order lifted on 5 June, when second EMD F-22 resumed testing with some restrictions while awaiting replacement unit. Major event of year was expected to be Pentagon Defense Acquisition Board review to culminate in award of contract for initial batch of LRIP aircraft; this was scheduled for 21 December, but slipped to 3 January 2001 and was further delayed by poor weather that prevented three of 11 critical test objectives being achieved; funding release dependent upon compliance with several 'exit criteria', including first flight of F-22 with block 3.0 software, first AMRAAM launch and first flights of aircraft 4004, 4005 and 4006. First AIM-120C AMRAAM launch (of 60 planned) achieved 24 October 2000; static tests completed 28 December 2000. Remaining three objectives satisfied by 6 February 2001, with first flight of aircraft 4006. However, decision to proceed with LRIP phase not taken until 15 August 2001, when initial batch of 10 aircraft approved with FY01 funds; decision on moving to full-rate production reprogrammed for December 2004. If authorised, this should continue until at least FY13.

Four Lockheed Martin F/A-22A Raptors formate during test missions from their trials base. Nearest are 4008 and 4007 ***(Kevin Robertson/Edwards AFB)*** ***NEW***/0583433

Objectives for 2002 comprised first flight of first Production Representative Test Vehicle, which achieved on 12 October; first supercruise launch of AIM-120, including successful interception of aerial target, which accomplished on 5 November; expansion of flight envelope to permit start of USAF pilot training in anticipation of initial operational test and evaluation (IOT&E); and supercruise launch of, and interception by, heat-seeking AIM-9M, which undertaken on 22 November. Block 3.1 software package delivered 5 February 2002 and flown for first time (on 91-4006) on 25 April at Edwards AFB.

In February 2003, the fifth and sixth EMD aircraft successfully demonstrated the Intraflight Datalink (IFDL) facility for the first time. Subsequently, flight operations were briefly interrupted when nose gear of one aircraft retracted as engines were being shut down; this incident occurred on 18 March 2003 and resulted in the F/A-22 being grounded until 22 March.

Notable milestone passed on 18 April 2001, when 1,000th flight test hour recorded; 2,000th hour completed on 7 June 2002; 3,000th hour on 26 February 2003; 4,000th on 2 September 2003 and 5,000th on 9 February 2004. Nevertheless, delays in production and delivery of test aircraft meant programme was behind schedule, with knock-on effect on start of IOT&E; this was scheduled to start in August 2002, but was delayed and did not begin until 29 April 2004. At that time, 24 aircraft were flying and IOT&E was due to be

F/A-22A Raptor 4012 accompanied by an F-15C ***(USAF)*** ***NEW***/0583426

completed in September 2004. One year earlier, the first delivery to an operational unit was effected 26 September 2003, when Raptor 01-4018 was flown from Marietta to Tyndall AFB, Florida, for 43rd FS of 325th FW. Overall development programme was to be completed in March 2004, but will now continue until at least November 2005; this delay largely attributed to problems experienced with avionics software, such as tendency for elements of the avionics suite to shut down every 3 to 4 hours. By late April 2004, avionics package had demonstrated increased stability by completing five hours of continuous operation between malfunctions.

CURRENT VERSIONS: (SPECIFIC) : Test programme involves total of nine EMD aircraft, plus two non-flying test articles. Airframe 3999 for static loads testing built between second and third flying F-22s, and airframe 4000 for fatigue testing built between third and fourth flying F-22s. Final assembly of 3999 began in July 1998; following completion in January 1999, it was transferred to the structural test facility at Marietta and began load testing in March 1999. All planned static testing completed by mid-May 2002, with 'first service-life' fatigue testing satisfactorily concluded on 17 May 2002, at which time second cycle of full lifetime testing began.

Clear division of test assignments resulted in three aircraft (4001-03) being allocated to airframe structure evaluations, with remaining six concentrating on avionics test taskings. Use of separate instrumentation configurations for airframe and avionics-dedicated test articles provides back-up for almost every aircraft and offers potential to switch missions between test fleet if necessary. Tasks allocated to EMD aircraft are:

4001/91-4001: 'Spirit of America'. Rolled out on 9 April 1997 at Lockheed Martin's Marietta, Georgia, facility and made first flight on 7 September 1997. Following completion of structural ground tests, disassembled and airlifted to Edwards AFB on 5 February 1998; used for evaluation of flying qualities, flutter and loads characteristics. Flown to Wright-Patterson AFB, Ohio, on 2 November 2000 and formally retired from flight test duty; 175 flights and 372.7 hours; stripped of useful components and used for live-fire testing, involving exploding shells and missile fragments, during 2001.

Raptor 4003 flying the Lockheed Martin F/A-22A's 5,000th hour on 9 February 2004 *(Bill Krause/Edwards AFB)* ***NEW***/0583432

F/A-22A Raptor 4006 overflying the USAF Academy at Colorado Springs *(Kevin Robertson/Edwards AFB)* ***NEW***/0583475

4002/91-4002: 'Old Reliable'. Rolled out 10 February 1998; maiden flight on 29 June 1998 at Marietta; flew to Edwards AFB on 26 August and by end October 1998 had made 27 flights (66.1 hours), expanding flutter and handling qualities envelope, including 26° AoA. Used to launch first AIM-9M Sidewinder on 25 July 2000, followed by first AIM-120C AMRAAM on 24 October and had previously been employed for fit checks and captive-carry trials with AGM-88 HARM in April 1999. Tasks include performance assessment (propulsion; high AoA) plus stores separation and jettison as well as some electronic warfare and IR signature evaluations. Undertook six-month test programme of M61A2 cannon in 2002–03, expending more than 5,400 20 mm rounds in 12 ground and 38 airborne firings; final clearance of envelope then expected in second quarter of 2004.

4003/91-4003: First Block 2 aircraft, with internal structure fully representative of production version; rolled out at Marietta on 25 May 1999; first engine runs in October 1999; taxi trials completed at beginning of March 2000, with first flight following on 6 March and delivery to Edwards AFB (fourth flight) on 15 March; regained flight status on 19 September 2000, being previously engaged on planned ground testing and upgrade programme. Assigned to envelope expansion, replacing 4001, including loads testing, crosswind landings, validation of arrester hook and weapons bay environment work. Mid-fuselage section used for fit checks of Sidewinder and AMRAAM weapons at Fort Worth in July 1998 shortly before delivery to Marietta for final assembly. Accomplished first AIM-120 AMRAAM launch at supersonic speed on 21 August 2002, when single missile fired while flying at M1.2 at 3,650 m (12,000 ft). Will also test M61A2 cannon and JDAM integration.

4004/91-4004: First EMD aircraft with Hughes CIP software, including AN/APG-77 radar and ILS. Allocated to avionics development, but also to be used for low observables evaluation including radar cross-section and IR signature assessments; and comms/nav/ident (CNI) testing. Mid-fuselage section delivered from Fort Worth to Marietta on 28 December 1998. Block 1.1 avionics installed at Marietta, with electrical power applied for first time on 31 August 1999; Block 1.2 avionics then incorporated to support taxi trials and initial flight testing at Dobbins Air Reserve Base, from where first flight took place on 15 November 2000. Aircraft ferried to Edwards AFB at end January 2001.

Lockheed Martin F/A-22A Raptor trials behind a 'rainmaking' Boeing KC-135 *(Derk Blanset/Edwards AFB)* ***NEW***/0583434

F/A-22A Raptor 4005 ***NEW***/0583427

F/A-22A Raptor aerobatics ***NEW***/0583429

4005/91-4005: Radar, CNI and armament development tasks. Assigned to initial testing of Block 3.0 software. Primary fire-control evaluation aircraft, with first flight made on 5 January 2001; made first guided launch of AIM-120C against target drone over Point Mugu test range on 21 September 2001.

4006/91-4006: Avionics development tasks. Primarily for integrated avionics testing, RCS testing and, eventually, for systems effectiveness/military utility evaluation. Expected to fly for first time in December 2000, but did not do so until 5 February 2001.

4007/91-4007: Was to have been initial two-seat F-22B but completed as single-seat F-22A and assigned to integrated avionics testing. First flight made 15 October 2001, thereafter joining test fleet at Edwards AFB on 5 January 2002. Assigned to IOT&E programme as back-up aircraft.

4008/91-4008: Allocated to avionics development tasks and also destined to validate observability specification data. Subsequently to be used for IOT&E programme, beginning in August 2003. Flown for first time on 8 February 2002 and ferried to Edwards AFB on 31 May 2002. On 2 July 2002, aircraft flown to Palmdale for modifications and upgrades necessary to prepare it for DIOT&E; this was completed at start of October 2002, when aircraft resumed developmental flight test duty at Edwards.

4009/91-4009: Was to have been the second F-22B but completed as F-22A. Avionics development and observability trials. First flight originally planned for 1 June 2001, but retained for ground testing at Marietta, where formally delivered for maintenance trials 15 April 2002; flown to Edwards to join test fleet on 31 March 2003 (last of four — 4008 to 4011 — modified at Palmdale to production standard) and will participate in IOT&E programme from August 2003.

4010/99-4010 and **4011/99-4011:** PRTV aircraft; both to participate in IOT&E and also used for additional service testing at Nellis AFB, Nevada from the third quarter of 2003. First flight of a PRTV aircraft (99-4011, the second example) at Marietta on 16 September 2002; formally accepted by USAF on 26 November 2002, completing IOT&E fleet. First PRTV aircraft (99-4010) made maiden flight on 12 October 2002; officially delivered to USAF at Marietta on 23 October; and transferred to Palmdale on 30 October 2002 for modifications before joining Air Force Operational Test and Evaluation Center (AFOTEC) Detachment 6 at Edwards AFB.

4012/00-4012: Initial delivery to Nellis AFB, Nevada, for 422nd Test and Evaluation Squadron, from 14 January 2003. This aircraft to be joined by seven more in developing F/A-22 tactics, techniques and operating procedures at the USAF Air Warfare Center; **4013/00-4013** and **4014/00-4014** both at Nellis by 22 May 2003, with three more following later in year. By beginning of May 2004, aircraft **4024/01-4024** was latest Raptor to have flown.

Trials by 4001 to 4009 originally to occupy 4,337 hours in 2,409 sorties. Of these totals, 2,110 hours and 1,200 sorties dedicated to airframe and systems testing, with balance allocated to mission avionics testing; however, as part of an effort to make up for delays in test programme, amount of time allocated to avionics flight testing was cut to 1,530 hours in third quarter of 2001.

CURRENT VERSIONS (GENERAL): **F/A-22A:** Originally designated F-22A. Single-seat production version for USAF. Planned in-service date is December 2005.

F-22B: Projected two-seat version for USAF; development terminated 10 July 1996 to reduce costs.

FB-22: Proposed single- or two-seat long-range strike-dedicated derivative under study by Lockheed Martin at USAF request in 2002; larger than F/A-22, with combat radius of up to 1,565 n miles (2,897 km; 1,800 miles) and significantly bigger payload of 30 small diameter bombs rather than eight as on F/A-22. If proceeded with, likely to have 90 per cent avionics and 30 per cent structural commonality with F/A-22, but may feature different power plant, with General Electric F110 and Pratt & Whitney F135 also under consideration. Other probable changes could include modified delta wing planform, insertion of fuselage plug to increase weapons bay size and carriage characteristics and compromised air-to-air combat capability, with *g* limit of 5. Secretary of the Air Force James Roche informed US Congress in late February 2003 that his optimum bomber force would include at least 150 FB-22s; however, USAF unlikely to require new aircraft for many years.

NATF: Projected US Navy variant to replace Grumman F-14 Tomcat; development abandoned.

CUSTOMERS: US Air Force: two YF-22 demonstrators; nine EMD aircraft plus one static and one fatigue test airframes; original 648 production aircraft programme of 1991 reduced to 442 in January 1994; latter originally to be funded from 1997 (long lead), beginning with four preproduction verification (PPV) aircraft in FY98, followed by series production of 438 (but PPV aircraft cancelled in early 1997). Quadrennial Defense Review (QDR) report in May 1997 resulted in planned procurement falling to 339 (including eight PRTV aircraft); further cuts followed, first to 333 in 1999 and then to 295 in 2001, although in early 2003, it appeared that as few as 276 could be acquired, with peak production rate of 32 per year during FY07 to FY10. Contracts for two PRTV 1 aircraft and long lead items for batch of six PRTV 2 aircraft signed December 1998. Decision to authorise LRIP expected in December 2000 but delayed until 15 August 2001; full-rate production (FRP) decision due to be taken in December 2004. Pilot training at Tyndall AFB, Florida, by 43rd Fighter Squadron, 325th Fighter Wing from 2003, with IOC of first squadron following in December 2005, this being 27th Fighter Squadron of 1st Fighter Wing at Langley AFB, Virginia.

F/A-22 PROCUREMENT
(at June 2003)

FY	Lot	Quantity
99	PRTV 1	2
00	PRTV 2	6
01	LRIP 1	10
02	LRIP 2	13
03	LRIP 3	21
04	LRIP 4	22
05	LRIP 5	24
06	FRP1	28
Total		**126**

Notes: Excludes two prototypes and nine EMD aircraft. FY05 and FY06 totals subject to approval and could change.

COSTS: US$818 million contracts to both ATF teams, October 1986, for 54-month studies; each airframe team investing own funds (Lockheed/Boeing/GD team investment totalled US$675 million in addition to DoD funding); each engine contractor, about US$50 million; total US$3,800 million spent by USAF on both ATFs up to April 1991; programme cost for 648 aircraft was US$13 billion for development (1991 base year) and US$52.5 billion for production (1994 then-year); flyaway cost US$61.2 million at 1991 prices.

EMD contract, 2 August 1991, comprised US$9,550 million for 11 (subsequently nine) airframes, plus US$1,375 million to P&W for 33 (later amended to 27) engines. FY94 US Congressional appropriation of US$2,100 million was US$163 million below expectations, resulting in slippage of critical design review and first flight. Similarly, FY95 appropriation of US$2,300 million was US$110 million below expected figure, leading to further delay in maiden flight; and on 9 December 1994, Defense Secretary William Perry announced 10 per cent cut (approximately US$210 million) in FY96 budget, necessitating a third restructuring of the programme.

On 30 December 1999, Lockheed Martin awarded contract for approximately US$1.3 billion, covering procurement of six PRTV 2 aircraft, augmenting earlier US$195.5 million contract for long-lead items; at same time, Pratt & Whitney received separate US$180 million award for 12 F119 engines. Additional appropriation of US$277.1 million allocated to Lockheed Martin for long-lead items for 10 Lot 1 LRIP aircraft; further US$862 million contract awarded 19 September 2001 towards remaining cost of producing these 10 aircraft. Total cost of LRIP 2 batch of 13 aircraft quoted as US$3.03 billion; subsequent appropriations for production aircraft have been US$4.46 billion in FY03 (21 aircraft); US$4.11 billion in FY04 (22) and US$4.16 billion in FY05 (24); RDT&E funding over same period totalled US$2.40 billion. Congressional cap on production budget currently US$37.6 billion although, in third quarter of 2001, US DoD was considering asking Congress to increase this to US$45 billion; most recent (March 2003) production cost estimate is US$42.2 billion. At same time, total development cost estimated to be US$21.9 billion. By April 2004, total programme cost being reported as US$71.8 billion, of which development process said to account for US$28.7 billion, although US General Accounting Office report of 15 March 2004 claimed that development costs are of the order of US$40 billion. Same GAO report said USAF would only be able to afford 218 aircraft, rather than 275 to 277 estimated by the Air Force.

DESIGN FEATURES: Low-observables configuration and construction; stealth/agility trade-off decided by design team. Antennas located in leading- or trailing-edges of wings and fins, or flush with surfaces, to minimise radar signature. Target thrust/weight ratio 1.4 (achieved

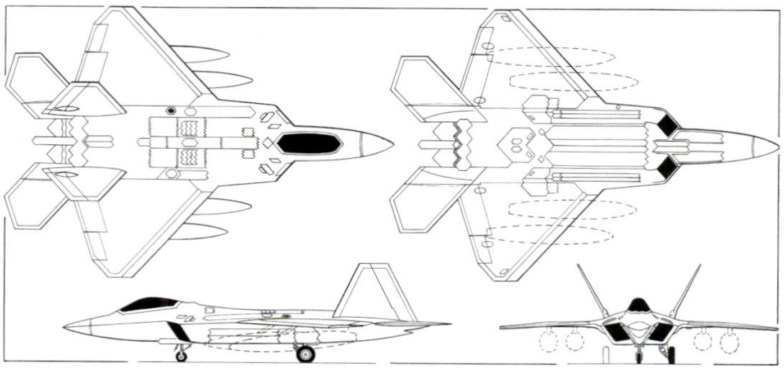

Lockheed Martin F/A-22A ***(James Goulding)*** 0526924

ratio 1.2 at T-O weight); greatly improved reliability and maintainability for high sortie-generation rates, including under 20 minutes combat turnround time; enhanced survivability through 'first-look, first-shot, first-kill' capability; short T-O and landing distances; supersonic cruise and manoeuvring (supercruise) in region of M1.5 without afterburning; internal weapons storage and generous internal fuel; conformal sensors.

Highly integrated avionics for single-pilot operation and rapid reaction. Radar, RWR and comms/ident managed by single system presenting relevant data only, and with emissions controlled (passive to fully active) in stages, according to tactical situation. CIP handles all avionics functions, including self-protection and radio, and automatically reconfigures to compensate for faults and failures. Two CIPs, with space for third, linked by 400 Mbits/s fibre optic network (see Avionics). Computer architecture and avionics processes will need to be upgraded, in five to six years, following Intel decision to phase out 32-bit, 25 MHz i960MX chip which is obsolete; this chip has insufficient power to support intended air-to-ground applications after 2013, but is being installed on first 155 Raptors.

Wing and horizontal tail leading-edge sweep 42°; trailing-edge 17° forward, increased to 42° outboard of ailerons; all-moving five-edged horizontal tail. Vertical tail surfaces canted outwards at 28°; leading- and trailing-edge sweep 22.9°; biconvex aerofoil. Wing taper ratio 0.169; leading-edge anhedral 3.25°; root twist 0.5°; tip twist −3.1°; thickness/chord ratio 5.92 per cent at root, 4.29 per cent at tip; custom-designed aerofoil. Horizontal tails have no dihedral or twist.

Diamond-shaped cheek air intakes with highly contoured air ducts; single-axis thrust vectoring included on F119, but most specified performance achievable without.

Production aircraft to be coated with new, Boeing-developed, stealthy paint intended to enhance low-visibility attributes; this first applied to second EMD aircraft in March 2000.

FLYING CONTROLS: Triplex, digital, fly-by-wire system with GEC sidestick control, using line-replaceable electronic modules to enhance maintainability; thrust vectoring utilised to augment aerodynamic pitch control power and provide firm control even at low speeds and high angles of attack. Technology and control concepts demonstrated throughout the flight envelope during prototype air vehicle test programme, including flight at AoA greater than 60°; wind tunnel testing with models of production aircraft successfully attained AoAs greater than 85°.

Ailerons and flaperons occupy almost entire wing trailing-edge; full-span leading-edge flaps; conventional rudders in vertical tail surfaces; slab taileron surfaces; airbrake not included (differential rudder and wing trailing-edge surfaces for speed control). Control surface authorities: leading-edge flaps 0° up/35° down (2°/37° overtravel); trailing-edge flaperons 20° up/35° down; ailerons ±25°; horizontal tail leading-edges 30° up/25° down; rudders ±30°; speedbrake (rudder) 30° out.

STRUCTURE: Lockheed Martin at Marietta constructs forward fuselage, including cockpit (with avionics architecture, displays, controls, air data system), apertures, edges, tail assembly, landing gear and environmental control system and undertakes final assembly. Lockheed Martin's Fort Worth plant builds the centre-fuselage; Palmdale, the radome. Boeing responsible for wings, fuselage aft sections, power plant installation, auxiliary power generation system, radar, arresting gear system and avionics integration laboratory.

Total airframe weight comprises approximately 36 per cent titanium 64, three per cent titanium 62222, 24 per cent thermoset composites (both epoxy resin and bismaleimide), one per cent thermoplastic composites, 16 per cent aluminium, 6 per cent steel and 14 per cent other materials. Forward fuselage substructure is aluminium and composites; centre-fuselage includes titanium, aluminium and composites; both forward and centre-fuselage skins primarily graphite bismaleimide. Four main mid-fuselage section bulkheads are single-piece, closed-die, titanium forgings. Rear fuselage approximately 67 per cent titanium, 22 per cent aluminium and 11 per cent composites. Engine bay doors titanium honeycomb, produced by liquid-interface diffusion bonding. Tailbooms assembled by electron beam welding. Titanium vent screens (to reduce radar reflectivity) contain thousands of precisely shaped holes in special alignment, each cut by abrasive water-jet. Thermoplastics used in areas requiring high tolerance to damage, examples including doors for landing gear and weapon bay.

Wing skins are monolithic graphite bismaleimide. Main (front) wing spars are machined titanium forgings; intermediate spars are mix of resin transfer moulded (RTM) sine-wave composites and titanium for strength to meet vulnerability requirements in wing fuel tank; rear spars composites and titanium. Wingroot and control surface actuator attachment fittings are titanium HIP castings. Horizontal stabiliser incorporates 'tow placed' composites pivot shaft in addition to aluminium honeycomb core and graphite bismaleimide skins produced by GKN Aerospace Services; vertical stabilisers use solid graphite bismaleimide skins over graphite epoxy RTM spars. Wing control surfaces are combination of co-cured composites skins/substructure and non-metallic honeycomb core construction.

LANDING GEAR: Menasco retractable tricycle type, stressed for no-flare landings of up to 3.05 m (10 ft)/s. Honeywell wheels, brakes and anti-skid system. Nosewheel tyre Goodyear or Michelin 23.5×7.5-10 (22 ply) tubeless; mainwheel tyres Goodyear or Michelin 37×11.50-18 (30 ply) tubeless. Kaiser airfield arrester hook in enclosed fairing between engines.

POWER PLANT: Two 156 kN (35,000 lb st) class Pratt & Whitney F119-PW-100 advanced technology reheated turbofans, each fitted with a two-dimensional, convergent/divergent thrust vectoring (±20° in vertical plane) exhaust nozzle for enhanced performance and manoeuvrability.

Fuel in eight tanks in forward fuselage, mid-fuselage, wings and tailbooms; provision for later addition of fuel in saddle and fin tanks. Additionally, up to four external fuel tanks, each 2,271 litres (600 US gallons; 500 Imp gallons), on underwing hardpoints. Dorsal Xar Industries aerial refuelling receptacle covered by doors, except when required. Fuel grade JP-8.

ACCOMMODATION: Pilot only on zero/zero modified Boeing ACES II ejection seat and wearing tactical life support system with improved *g*-suits, pressure breathing and arm restraint. Pilot's view over nose is −15°. Canopy manufactured by Sierracin Sylmar Corp as single-piece unit, hinged at rear.

SYSTEMS: Twin 276 bar (4,000 lb/sq in) hydraulic systems with four pumps (each 273 litres; 72.0 US gallons; 60.0 Imp gallons per minute), but single Parker Bertea actuator on each control surface to save weight and cost. Curtiss Wright actuators on leading-edge flaps. Two 65 kW engine-driven generators. Honeywell G250 335 kW (450 hp) APU driving a Hamilton Sundstrand 27 kW generator and 100 litre (26.5 US gallon; 22.0 Imp gallon)/min pump. Smiths 270 V DC electrical distribution system. Honeywell environmental control system for life support system and flight avionics (open loop air cycle), mission avionics (closed loop vapour cycle) and fuel cooling (thermal management). OBIGGS and Normalair-Garrett OBOGS.

AVIONICS: Final integration, as well as integration of entire suite with non-avionics systems, undertaken at Avionics Integration Laboratory, Seattle, Washington; airborne integration supported by Boeing 757 flying testbed and the Air Vehicle Integration Facility at Marietta.

Comms: TRW AN/ASQ-220 communication/navigation/identification (CNI) system includes Mk 12 IFF and UHF/VHF. CNI system uses modules contained in two integrated CNI racks, CIP assets and integrated display/control panels. Intra-Flight Data Link (IFDL) encrypted radio and wireless communications modem allows all Raptors in a flight to share target and system data without fear of being overheard. Supplier team, responsible for electronics and software, also includes Rockwell Collins, ITT and GEC.

Radar: Northrop Grumman/Raytheon AN/APG-77 multimode radar incorporates active electronically scanned array (AESA), capable of interleaving air-to-air search and multitarget track functions. Also has weather mapping mode and provisions for air-to-ground modes and side arrays. Radar reported to be capable of detecting 1 m^2 (10.76 sq ft) target at range of approximately 109 n miles (201 km; 125 miles). AN/APG-77 to be replaced starting with Lot 5 production by new AESA radar that will eventually incorporate advanced air-to-ground computer software.

Flight: Vehicle management system (VMS) combines flight and propulsion controls; integrated vehicle subsystem control (IVSC) operates utilities via digital databus. Total of 18 Raytheon 1750A common processor modules used for VMS, IVSC and stores management system. F/A-22 is first fighter with triplex digital flight control computers and no electrical or mechanical back-up; control reconfiguration modes provide safe flying after actuator or hydraulic failures. VMS controls 14 surfaces (horizontal tail, ailerons, flaperons, rudders, leading-edge flaps and inlet bleed and bypass doors). No AoA limitation, but overstressing made impossible by restrictions to roll rate and load factor, according to fuel state, stores carriage and flight condition. Lear Astronics VMS integrated with Rosemount low-observable air data system, including two AoA probes and four sideslip plates on nose. CNI system includes GPS, Tacan and ILS. Twin Litton LN-100F INS. Throttle and stick contain 20 controls with 63 functions.

Software Block 0 for initial flight tests. Block 1.1 installed on aircraft 4004 in 1999, was primarily for radar, but included more than half of the avionics suite's source lines of code. Block 2, installed in the 757 FTB in October 1999, began sensor fusion, including radio frequency co-ordination and some electronic warfare functions. Block 3S added CNI and ECCM; Block 3.0, tested in the 757 FTB in fourth quarter of 2000 and flown for first time in F/A-22 on 5 January 2001, provided full sensor fusion and weapon delivery function; Block 3.1.3 adds GBU-32 JDAM, JTIDS receive and GPS and will be available to IOC aircraft. Block 4 will incorporate helmet cueing and AIM-9X. Work is under way on a Block 5 upgrade that should provide JTIDS transmit as well as enhanced air-ground capability from around 2006, including compatibility with forthcoming Small Diameter Bomb (SDB).

Instrumentation: Fused situational awareness information is displayed to pilot via four Lockheed Martin colour liquid crystal multifunction displays (MFD); MFD bezel buttons provide pilot format control. Centre screen measures 203 × 203 mm (8 × 8 in) and typically will function as situation display; right and left screens measure 152 × 152 mm (6 × 6 in) and function as attack and defensive displays respectively; fourth screen (directly below situation display) can be used to provide global side-view depiction of tactical situation and may also present other data such as fuel status, engine parameters, stores data, BIT reports and electronic checklists. Additionally, 76 × 102 mm (3 × 4 in) upfront display screens on each side of the integrated control panel, immediately below the BAE Systems HUD. Illumination is NVG-compatible.

Mission: Two Hughes common integrated processors (CIP); CIP also contains mission software that uses tailorable mission planning data for sensor emitter management and multisensor fusion; mission-specific information delivered to system through Fairchild data transfer equipment/mass memory (DTE/MM) system that contains mass storage for default data and air vehicle operational flight programme; stores management system. General purpose processing capacity of CIP is rated at more than 700 million instructions per second (Mips) with growth to 2,000 Mips; signal processing capacity greater than 20 billion operations per second (Bops) with expansion capability to 50 Bops; CIP contains more than 300 Mbytes of memory with growth potential to 650 Mbytes. Of 132 slots available in CIPs 1 and 2, 41 are initially vacant and thus available for growth. Intra-flight datalink automatically shares tactical information between two or more F/A-22s. CNI system includes JTIDS (receive-only terminal). Lockheed Martin airborne video recorder. Lockheed Martin stores management system. Airframe contains provisions for IRST and side-mounted phased-array radar.

Self-defence: BAE Systems AN/ALR-94 electronic warfare (RF warning and countermeasures and missile launch detection functions) subsystem. AN/ALE-52 flare dispenser.

ARMAMENT: Internal long-barrel General Dynamics M61A2 20 mm cannon with hinged muzzle cover and 480-round magazine capacity (production aircraft). Three internal bays for AIM-9M Sidewinder or next-generation AIM-9X (one in each side bay on Hughes LAU-141/A trapeze-type launcher) and six AIM-120C AMRAAM AAMs and/or 460 kg (1,015 lb) GBU-32 JDAM PGMs on Edo LAU-142/A hydraulic ejection launchers in main weapons bay. Four underwing stores stations at 317 mm (125 in) and 442 mm (174 in) from centreline of fuselage capable of carrying 2,268 kg (5,000 lb) each.

Typical weapon loads include six AIM-120s and two AIM-9s carried internally for air combat; two JDAMs, two AIM-120s and two AIM-9s carried internally for air-to-ground attack; and six AIM-120s and two AIM-9s internally, plus two external fuel tanks and further four AAMs (AIM-120 or AIM-9) underwing for long-range air combat.

Other weaponry envisaged for use by the F/A-22 includes the BLU-109 Penetrator, the wind-corrected munitions dispenser (WCMD), AGM-88 HARM, GBU-22 Paveway 3 guidance unit (with 500 lb bomb), new Small Diameter Bomb (SDB) and the low-cost autonomous attack system (LOCAAS) submunitions dispenser package.

DIMENSIONS, EXTERNAL:

Wing span	13.56 m (44 ft 6 in)
Wing chord: at root (theoretical)	9.85 m (32 ft 3½ in)
at tip (reference)	1.66 m (5 ft 5½ in)
at tip (actual)	1.14 m (3 ft 9 in)
Wing aspect ratio	2.4
Length overall	18.92 m (62 ft 1 in)
Height overall	5.08 m (16 ft 8 in)
Tail span: horizontal surfaces	8.84 m (29 ft 0 in)
vertical surfaces	5.97 m (19 ft 7 in)
Wheelbase	6.04 m (19 ft 9¾ in)
Weapon bay ground clearance	0.94 m (3 ft 1 in)

AREAS:

Wings, gross	78.0 m^2 (840.0 sq ft)
Leading-edge flaps (total)	4.76 m^2 (51.20 sq ft)
Flaperons (total)	5.10 m^2 (55.00 sq ft)
Ailerons (total)	1.98 m^2 (21.40 sq ft)
Vertical tails (total)	16.54 m^2 (178.00 sq ft)
Rudders/speedbrakes (total)	5.09 m^2 (54.80 sq ft)
Stabilators (total)	12.63 m^2 (136.00 sq ft)

WEIGHTS AND LOADINGS (estimated):

Weight empty (target)	14,365 kg (31,670 lb)
Max T-O weight	almost 27,216 kg (60,000 lb)
Max wing loading	348.7 kg/m^2 (71.43 lb/sq ft)
Max power loading	87 kg/kN (0.86 lb/lb st)

PERFORMANCE (YF-22, demonstrated):

Max level speed: supercruise	M1.58
with afterburning	M1.7 at 9,150 m (30,000 ft)

Ceiling .... 15,240 m (50,000 ft)
g limit .... +7.9
PERFORMANCE (F/A-22A, design target, estimated):
Max level speed at S/L .... 800 kt (1,482 km/h; 921 mph)
g limit .... +9

UPDATED

LOCKHEED MARTIN 382U/V HERCULES

US Air Force designations: C-, EC-, WC-130J
US Coast Guard designation: HC-130J
US Marine Corps designation: KC-130J
RAF designations: Hercules C. Mk 4 (C-130J-30); Hercules C. Mk 5 (C-130J)
TYPE: Medium transport/multirole.

DEVELOPMENT MILESTONES

Programme launched	1991
First order (RAF)	Dec 94
Rolled out	18 Oct 95
First flight	5 Apr 96
First delivery (RAF)	24 Aug 98
Entered service (RAF)	14 Nov 00

PROGRAMME: US Air Force specification issued 1951, leading to first-generation Hercules (Allison T56 turboprops); first production contract for C-130A to Lockheed September 1952; first flight 23 August 1954; two YC-130 prototypes, 231 C-130As, 230 C-130Bs, 491 C-130Es, 1,089 C-130Hs and 113 L-100s manufactured before introduction of C-130J and commercial L-100J equivalent. Official total of 2,156 (including prototypes) delivered by January 1998, when final C-130H handed over.

Privately funded development and flight test programme for next-generation version began in 1991 as Hercules II, but now known as C-130J Hercules, or L-100J in equivalent civilian form.

Initial British delivery accomplished on 24 August 1998; aircraft involved was ZH865, which arrived at Boscombe Down on 26 August for start of clearance trials by UK Defence Evaluation Research Agency (DERA). First aircraft ferried to RAF transport force at Lyneham was ZH878 on 21 November 1999. Final RAF aircraft flown from Marietta to Cambridge in late May 2000; final delivery to RAF on 21 June 2001.

First 'operational' mission accomplished by Lockheed Martin test crew in late November 1998, when C-130J completed three sorties from Marietta and airlifted 37,650 kg (83,000 lb) of hurricane relief supplies to Tegucigalpa, Honduras. Subsequently, 1999 witnessed start of deliveries to US Air Force Reserve Command and Air National Guard, as well as to operating units in Australia and the UK. Entire fleet had accumulated 30,000 flight hours by February 2002; by 24 July 2002, this had risen to 50,000 hours. Roll-out of the 100th C-130J — an aircraft for the US Coast Guard — took place on 17 February 2003.

Block 5.3 software configuration available from third quarter 2001 is standard equipment for latest deliveries and was retrofitted to existing aircraft by end of 2002; Block 5.3 offers ability to fly integrated precision radar approaches, gives enhanced navigation capabilities and permits fully automatic formation flying using co-ordinated aircraft positioning system (CAPS). Next version will be Block 5.4, which will be introduced in about FY05-06; new features include an AN/APX-119 IFF transponder and 8.33 kHz VHF radios. A Block 6.0 configuration will follow and is expected to incorporate improvements originating with the Co-operative Systems and Software Upgrade Requirements Management (COSSURM); latter centres around a three-year, US$20 million contract awarded to Lockheed Martin in November 2003 by C-130J operators in Australia, Italy, the UK and the USA with the object of facilitating major avionics improvements. Upgrades that could feature in the Block 6.0 configuration include integration of the civil global air traffic management system as well as infrared countermeasures, inert gas generation system, Link 16 and improved logistic systems.

CURRENT VERSIONS: **C-130J**: Baseline version. Dimensionally similar to preceding C-130H, but incorporating new equipment and features as subsequently described. Subject of initial order for two from USAF in FY94, with subsequent contract for two in FY96; these initially earmarked for trials and eventually joined Air Force Reserve Command (AFRC), while further eight funded in FY97 and FY98 assigned to Air National Guard's 135th Airlift Squadron at Warfield ANGB, Martin State Airport, Baltimore, Maryland, which fully equipped by mid-July 2000. Initial aircraft delivered to AFRC 403rd Wing at Keesler AFB, Mississippi, on 31 March 1999 for training.

Description applies mainly to baseline C-130J except where indicated.

Royal Australian Air Force Lockheed Martin C-130J-30 Hercules *(Paul Jackson)* ***NEW***/0583474

Royal Air Force Lockheed Martin C-130J Hercules C. Mk 5 *(Paul Jackson)* ***NEW***/0583425

C-130J-30: Stretched version of current production C-130; fuselage lengthened by 4.57 m (15 ft 0 in), offering increases in capability of between 31 and 50 per cent, dependent upon mission and configuration (see accompanying diagram). Orders received from Australia, Denmark, Italy, UK and USA.

CC-130J: Designation allocated to C-130J-30 aircraft in USAF service during 2003 but subsequently dropped in 2004, when basic C-130J designation adopted by USAF. First three funded in FY99, with two more in FY00 and five in FY02. Initial deliveries to ANG units in Rhode Island and California; FY02 aircraft to be assigned to 146th AW (two aircraft), 143rd AW (one), 403rd Wing (one) and a new USAF C-130J training unit (48th Airlift Squadron) that was established within the 314th Airlift Wing at Little Rock AFB, Arkansas on 5 December 2003 (one); all five to be delivered in 2004. In March 2003, USAF concluded contract for multiyear procurement of 40 CC-130Js, at planned rate of eight per year from FY04 up to and including FY08, although only four funded in FY04, rising to 11 in FY05.

EC-130J: 'Commando Solo' psychological warfare version; first two funded in FY98 budget, with first (99-1933) handed over on 17 October 1999; after flight testing, it moved to Palmdale, California, in July 2000 for fitting out before being delivered to the 193rd Special Operations Squadron, Air National Guard, at Harrisburg IAP, Pennsylvania. IOC was expected by the end of 2003, but slipped into 2004 because of problems encountered in integrating a new switchable 60/90 kVA generator with the broadcast system; this in turn caused delay in delivery of fully equipped aircraft to the 193rd SOS. Additional procurement comprises third example in FY99 budget, fourth in FY00 and fifth in FY01.

HC-130J: Replacement for earlier US Coast Guard HC-130s; funding for initial six contained in FY01 budget. First example accepted at Marietta in late 2002 and used for crew training before being formally delivered to Coast Guard at Elizabeth City, North Carolina, on 31 October 2003; these initially being used in logistics role but will eventually be fitted with full mission kit, including inverse synthetic aperture sea search radar. All six delivered by end of 2003.

KC-130J: Tanker/transport version for US Marine Corps, which has requirement for 51 to replace KC-130F, KC-130R and KC-130T variants. Originally to have two Flight Refuelling Mk32B-901E hose-and-drogue wing-mounted refuelling pods; three aircraft funded in FY97 budget, two in FY98, two in FY99, one in FY00, three in FY01 and two in FY02. Further 20 aircraft for USMC covered by multiyear procurement contract concluded in March 2003; this comprises initial batch of four in FY03, with additional aircraft at rate of four per year during FY05 to FY08. First contract for five USMC

Italian Air Force Lockheed Martin C-130J-30 Hercules *(Paul Jackson)* ***NEW***/0583472

Royal Air Force Lockheed Martin C-130J Hercules C. Mk 5 *(Paul Jackson)* ***NEW***/0583423

US Coast Guard HC-130J Hercules ***NEW***/0583342

conversions (to be accomplished by end of 2000) announced 21 July 1998. Final assembly of first KC-130J (165735) began 22 March 1999, with first flight on 9 June 2000; total of three aircraft assigned to test programme at Patuxent River in latter half of 2000; first drogue engagement accomplished by Navy F/A-18 Hornet on 30 August 2000; initial trials revealed that aft fairing of pod was unsatisfactory, with cracks appearing in hose/drogue coupling; redesign of fairing in 2000–01 cleared way for further testing in second quarter of 2001, which confirmed much improved performance in areas of flying quality and durability. On completion of trials, all three test aircraft assigned to USMC tanker/transport squadron VMGR-252 at MCAS Cherry Point, North Carolina, which took delivery at beginning of September 2001; total of seven handed over by end 2001. Two aircraft subsequently assigned to intensive 270-hour operational evaluation from October 2003, which paved way for operational introduction by USMC in January 2004. Further difficulties experienced with Mk 32B pods during flight testing (including 'sine wave' oscillation of hose) eventually culminated in Marine Corps deciding to abandon Flight Refuelling system and switch to proven Sargent Fletcher pods taken from existing KC-130F/R aircraft; new-build pods, with a higher fuel flow rate, are to be acquired by the Marines and Lockheed Martin will continue to work with Flight Refuelling on finding a solution that will allow the Mk 32B to be offered to export customers. Provisions for installation of refuelling probe incorporated in basic aircraft. Additional fuel in underwing and cargo hold tanks; see Power Plant.

MC-X Combat Talon 3: Potential new variant for USAF Special Operations Command, which has requirement for up to 54 aircraft, with procurement likely to begin in 2006. This will almost certainly be based on the C-130J. The same organisation is also contemplating acquisition of a new **AC-X** gunship that could utilise the C-130J airframe.

WC-130J: Weather reconnaissance version with aerial reconnaissance weather officer console, dropsonde system operator console and dropsonde launch tube; initial four in FY96 budget, with three more in FY97, two in FY98 and one in FY99; for 53rd WRS, Air Force Reserve Command at Keesler AFB, Mississippi. US$46.9 million contract for modification of first six, with option for further four, signed 18 September 1998; deliveries began on 30 September 1999; formal acceptance 12 October 1999.

CUSTOMERS: See tables. Total of 179 ordered by March 2004. First customer was RAF, which ordered 25 in December 1994; of these, 15 are stretched C-130J-30, designated C. Mk 4, with final 10 as standard C-130Js, designated Hercules C. Mk 5. Delivery of first example to the trials unit at Boscombe Down was due in November 1996, but was delayed until 26 August 1998. First service recipient was J Conversion Flight of No. 57 (Reserve) Squadron, followed by No 24 and then No. 30 Squadrons at RAF Lyneham. Deliveries to Lyneham began on 21 November 1999, when C. Mk 4 ZH878 arrived for duty as temporary ground procedures trainer; two days later, on 23 November, C. Mk 4 ZH875 was formally handed over in official ceremony at Lyneham. Completion of deliveries occurred on 21 June 2001; operational service with RAF began 14 November 2000, when No.24 Squadron flew scheduled mission to Puerto Rico. No.30 Squadron attained operational status in June 2002.

Second order covered two C-130Js for evaluation by the USAF and was finalised on 13 October 1995, one week before the first was rolled out at Marietta. Further two funded in FY96, four in FY97 and four in FY98 for Air Force Reserve Command and Air National Guard units; first ANG squadron was 135th AS at Martin State Airport, Baltimore. Further orders began FY99 with contract for three C-130J-30 aircraft, with FY01 procurement including two more C-130J-30s for USAF and initial batch of six C-130Js for long-range SAR duties with US Coast Guard; five more C-130J-30s in FY02 budget. First C-130J-30 rolled out 25 January 2001; following testing, this delivered to 143rd AW, Rhode Island ANG, on 2 December 2001. Two more C-130J-30s to 143rd AW by end 2001, with next two for 146th AW, California ANG, which received its first example on 2 June 2002. USAF also now purchasing C-130J-30 version and has identified requirement for more than 160, of which initial 40 covered by multiyear procurement contract signed in March 2003.

Second order received in 1995 was from Australia, for 12 C-130J-30s to replace C-130Es of No. 37 Squadron at Richmond, at total cost of US$660 million. Order placed on 21 December 1995 and included options for an additional 24 aircraft, plus eight options for New Zealand, which was guaranteed pricing based on Australia's larger order; options have not been converted to firm orders. First Australian C-130J (A97-440/N130JQ, c/n 5440) flew on 15 February 1997; delivery began 7 September 1999, when A97-464 arrived at Richmond; last aircraft handed over on 1 June 2000, with operational status achieved in December 2001.

Third overseas customer was Italy, which initially ordered 18 C-130Js, subsequently adding two more in January 2000 and another two in March 2000; last four were stretched C-130J-30 version. At same time as placing final order, Italy revised overall procurement plan and will now receive 12 C-130Js and 10 C-130J-30s. First aircraft rolled out at Marietta on 11 July 2000. Italian military certification also awarded in July 2000, with first C-130J delivery (actually second aircraft, MM 62176, c/n 5497) to 2° Gruppo, 46° Aerobrigata at Pisa departing USA on 16 August; formal acceptance 21 September 2000. All will be tanker-capable and configured to receive fuel, although only six likely to be operated as such at any one time. Two of the original aircraft are to be adapted as sigint platforms by 2006, with antennae fitted to removable panels; reconfiguration from basic transport role to specialised sigint mission would require maximum of 24 hours; equipment is to include 12 workstations, a communications suite and secure datalinks. C-130J-30 version assigned to 50° Gruppo, 46° Aerobrigata, from third quarter of 2002. Most recent export customer is Denmark, which ordered three C-130J-30s, with option on fourth, in December 2000; all three formally accepted at Marietta on 21 October 2003, with delivery to Denmark following in March and April of 2004.

Other potential customers reported to be in discussion with Lockheed Martin include Bahrain, Canada, Egypt, Israel, Kuwait and Portugal. Norway has expressed interest in obtaining six as replacements for its current fleet of six C-130Hs, while Saudi Arabia reported to have requirement for up to 24 aircraft. However, no recent export orders have been received

MILITARY/GOVERNMENT C-130J SALES
(To March 2004, excluding USA)

Country	J	J-30	Total
Australia		12	12
Denmark		3	3
Italy	12	10	22
UK	10	15	25
Total	**22**	**40**	**62**

Notes: Options not included

COSTS: US$55 million programme unit cost (Australia) (1995). Italian order of November 1997 valued at US$1.2 billion. Baseline price of C-130J-30 quoted as US$67 million in early 2002. Multiyear procurement of 40 C-130J-30s for USAF and 20 KC-130Js for US Marine Corps valued at US$4.05 billion (unit price US$67.5 million). FY03 contract for one C-130J-30 for Air Force Reserve Command worth US$70.5 million.

DESIGN FEATURES: Archetypal tactical transport: upswept rear fuselage for ramp access; high wing for propeller ground clearance despite low floor height; latter provided by pannier-mounted landing gear, which obviates long mainwheel legs stowed in wings. Can deliver loads and parachutists over open ramp and parachutists through side doors; cargo hold pressurised.

Significant changes introduced with C-130J, which optimised for economical operation, justifying customers' substitution for earlier C-130s on 30 year lifetime savings alone. Entirely revised flight deck reduces LRUs by half and wire assemblies by 53 per cent, with wire terminations cut by 81 per cent; has four MFDs, plus HUD for both pilots; lighting compatible with NVGs. Most systems have digital interfaces with the main mission computer to include unmodified mechanical systems like the hydraulics, which are largely unaltered from those of C-130H; provision for integrated self-defence suite (RWR, MAW, chaff/flare dispensers and IR jammers). Propulsion system provides 29 per cent more take-off thrust and is 15 per cent more efficient; fuel efficiencies obviate requirement for external tanks on most types of mission; propeller has 50 per cent fewer parts and weighs 15 per cent less. Manpower requirements of typical 16-aircraft squadron cut by 38 per cent compared with earlier versions of C-130, as result of reduced flight crew and 50 per cent better maintainability. Comprehensive computerised maintenance system employs a hand-held data module as interface between aircraft's BITE and operating base's central technical records.

Wing section NACA 64A318 at root and NACA 64A412 at tip; dihedral 2° 30'; incidence 3° at root, 0° at tip.

FLYING CONTROLS: All flying controls integrated with digital autopilot/flight director and comprise control surfaces boosted by dual hydraulic units; trim tabs on ailerons, both elevators and rudder; elevator tabs have AC main supply and DC standby; Lockheed-Fowler composites trailing-edge flaps.

STRUCTURE: All-metal two-spar wing with integrally stiffened taper-machined skin panels up to 14.63 m (48 ft 0 in) long. Incorporates carbon fibre composites materials for flaps and 32 graphite-epoxy trailing-edge panels.

LANDING GEAR: Hydraulically retractable tricycle type. Each main unit has two wheels in tandem, retracting into fairing built on to fuselage side. Nose unit has twin wheels and is steerable ±60°. Mainwheels 20.00-20 (26 ply) tubeless; nose 12.50-16 (12 ply) tubed or tubeless. Oleo shock-absorbers. Minimum ground turning radius: C-130J, 11.28 m (37 ft) about nosewheel and 25.91 m (85 ft) about wingtip; C-130J-30 14.33 m (47 ft) about nosewheel and 27.43 m (90 ft) about wingtip.

POWER PLANT: Four Rolls-Royce AE 2100D3 turboprops, flat rated to 3,424 kW (4,591 shp) (manufacturer's rating 3,458 kW; 4,637 shp at ISA −25°C), fitted with Dowty Aerospace R391 six-blade composites propellers and Lucas Aerospace FADEC. Automatic thrust control system (ATCs) and autofeather systems, plus engine monitoring system (EMS) which is incorporated into aircraft's integrated diagnostic system (IDS).

Total internal fuel capacity of 25,552 litres (6,750 US gallons; 5,621 Imp gallons) without foam and 24,363 litres (6,436 US gallons; 5,359 Imp gallons) with foam. Provisions only for two optional underwing pylon tanks, each with capacity of 5,220 litres (1,379 US gallons; 1,148 Imp gallons) without foam and 4,883 litres (1,290 US gallons; 1,074 Imp gallons) with foam. Total fuel capacity 35,992 litres (9,508 US gallons; 7,917 Imp gallons) without foam and 34,129 litres (9,016 US gallons; 7,507 Imp gallons) with foam. Tanker versions can carry cargo hold tank with additional 13,578 litres (3,587 US gallons; 2,987 Imp gallons). Single pressure refuelling point and overwing gravity fuelling. In-flight refuelling probe fitted as standard on port side of RAF aircraft; optional for all others, with Italian aircraft currently unique in being configured as receiver/tankers.

ACCOMMODATION: Crew of two on flight deck, comprising pilot and co-pilot, with provisions for optional third workstation. Ergonomic problems suffered by short pilots necessitated a number of alterations to cockpit, including redesign of seat and seat track, HUD and main control yoke, with throttle quadrant also modified. Two crew bunks, with lower incorporating three additional seats and harnesses for relief crew/flight deck passengers. Galley. Separate loadmaster's station in cargo hold, including folding desk, is standard equipment on RAF aircraft; optional for all others. Flight deck and main cabin pressurised and air conditioned.

Standard complements for C-130J as follows: 92 troops, 64 paratroopers, 74 litter patients plus two attendants, 54 passengers on palletised airline seating. Corresponding data for C-130J-30 are 128 troops, 92 paratroopers, 97 litter patients plus four attendants and 79 passengers on palletised airline seating. Airdrop loads comparable to C-130H/H-30 and

US C-130J PROCUREMENT
(To March 2004)

FY	C-130J	C-130J-30	EC-130J	HC-130J	KC-130J	WC-130J	Total
1994	2						2
1996	2					4	6
1997	4				3	3	10
1998	4		2		2	2	10
1999		3	1		2	1	7
2000			1		1		2
2001		2	1	6	3		12
2002		5			2		7
2003		1			4		5
2004		4					4
2005		11			4		15
2006–08		25			12		37
Total	**12**	**51**	**5**	**6**	**33**	**10**	**117**

Notes: FY03-08 procurement for USAF and US Marine Corps covered by multiyear contract, with initial tranche of four KC-130Js funded in FY03. USAF planned to acquire eight C-130J-30s per year from FY04 to FY08, with Marine Corps receiving four KC-130Js per year from FY05 to FY08. However, changes in USAF procurement from planned figures are apparent in above table, with funds appropriated for only four in FY04, followed by 11 in FY05. Additional aircraft are being purchased for the Air National Guard and Air Force Reserve Command, including one for latter organisation in FY03

include light armoured vehicles. Light and medium towed artillery pieces, wheeled and tracked vehicles and 463L palletised loads (five in C-130J and seven in C-130J-30, plus one on ramp in each model) are transportable. Hydraulically operated (with dual actuators) main loading door and ramp at rear of hold; can be opened in flight at up to 250 kt (463 km/h; 288 mph). Crew door on port forward fuselage side. Paratroop door on each side aft of landing gear fairing. Two emergency exit doors standard. Optional cargo handling system ordered by the USAF includes flush-mounted winch; in-ramp towplate for airdrop operations; low-profile rails and electric locks; flip-over rollers with covers; container delivery system centre vertical restraint rails. Capable of automatic preprogrammed cargo drops.

SYSTEMS: Lucas generator. Environmental control system in starboard undercarriage fairing is similar to that of the C-130H, incorporating dual air cycle machines, but with 30 per cent greater cooling capacity and a digital electronic control system. Honeywell GTCP85-185L(A) auxiliary power unit in port undercarriage fairing furnishes ground electrical power and bleed air for environmental control system. MIL-STD-1553B digital databus architecture. Integrated diagnostic system (IDS) incorporating fault detection/isolation subsystem with BIT (built-in test) facility. Goodrich pneumatic fin anti-icing system.

AVIONICS: *Comms:* Honeywell com/nav/ident management system, with Intel 80960 processor. AN/ARC-222 VHF, HF/UHF radio, intercom and IFF, with provisions for satcom system. All communication radios have secure features. AN/APX-119 IFF to be included on Block 5.4 and subsequent aircraft.

Radar: Northrop Grumman AN/APN-241 low-power colour radar incorporates Doppler beam-sharpening ground mapping mode, air-to-air skin paint mode and protective windshear mode as well as conventional colour weather radar.

Flight: HG-9550 radar altimeter, AN/ARN-153(V) Tacan, digital autopilot/flight director (DA/FD), dual Honeywell laser INS with embedded GPS receivers, Doppler velocity sensor, VOR, ILS, marker beacon receiver, UHF/VHF DF, ADF, E-TCAS, ground collision avoidance system (GCAS), global digital map display units and provision for microwave landing system.

Instrumentation: 'Dark cockpit' concept. Flight Dynamics HUD as certified primary flight instrument at pilot and co-pilot positions, four 152 × 203 mm (6 × 8 in) Avionics Display Corporation active matrix liquid-crystal display (AMLCD) colour multipurpose display systems (CMDSs) which are NVG-compatible for flight instrumentation, navigation and engine information and five Avionics Display Corporation 58 × 76 mm (2.3 × 3 in) monochrome AMLCDs for digital selector panels.

Mission: DRS Technologies AN/APN-243(V) station-keeping equipment. Provision for secure voice communication system.

Self-defence: Provisions for Lockheed Martin AN/AAR-47 missile warning system, Sanders AN/ALQ-157 IR countermeasures system, BAE Systems AN/ALE-47 chaff/flare dispensing systems and AN/ALR-56M radar warning receiver or AN/ALR-69 enhanced radar warning system. Testing of RWR, countermeasures dispensing system (CMDS) and missile warning system undertaken in September and October 2003; an aircraft from the 135th AS was used, spending a week in the Benefield Anechoic Facility at Edwards AFB to evaluate RWR and CMDS operation, before moving to the Naval Weapons Center at China Lake to test the missile warning system in flight. Northrop Grumman Large Aircraft Infra-Red Countermeasures (LAIRCM) system to be installed on Air Mobility Command C-130J, following successful conclusion of EMD testing in 2002–04.

EQUIPMENT: USMC KC-130Js were intended to use Flight Refuelling Mk 32B-901E hose pods underwing, but problems encountered during flight testing have resulted in existing Sargent Fletcher refuelling pods being installed instead. Italy also expected to fit the Mk 32B-901E pod to some aircraft and may also seek alternative solution. Latest C-130J-30s for Air National Guard's 146th AW are equipped with Aero Union Systems Airborne Fire Fighting System (AFFS), which allows them to drop up to 15,142 litres (4,000 US gallons; 3,331 Imp gallons) of fire retardant in a single pass. Italian C-130J configured to operate with Special Avionics Mission Strap-on-Now (SAMSON) C-130 Open Skies System (COPS) in support of international arms control verification efforts.

DIMENSIONS, EXTERNAL:
Wing span ... 40.41 m (132 ft 7 in)
Wing aspect ratio ... 10.1
Length overall: C-130J ... 29.79 m (97 ft 9 in)
C-130J-30 ... 34.37 m (112 ft 9 in)
Height overall: C-130J ... 11.84 m (38 ft 10 in)
C-130J-30 ... 11.81 m (38 ft 9 in)
Tailplane span ... 16.05 m (52 ft 8 in)
Wheel track ... 4.34 m (14 ft 3 in)
Propeller diameter ... 4.11 m (13 ft 6 in)
Main cargo door (rear of cabin):
Height ... 2.77 m (9 ft 1 in)
Width ... 3.12 m (10 ft 3 in)
Height to sill ... 1.03 m (3 ft 5 in)
Paratroop doors (each): Height ... 1.83 m (6 ft 0 in)
Width ... 0.91 m (3 ft 0 in)
Height to sill ... 1.03 m (3 ft 5 in)
Emergency exits (each): Height ... 1.22 m (4 ft 0 in)
Width ... 0.71 m (2 ft 4 in)

DIMENSIONS, INTERNAL:
Cabin, excl flight deck:
Length excl ramp: C-130J ... 12.19 m (40 ft 0 in)
C-130J-30 ... 16.76 m (55 ft 0 in)
Length incl ramp: C-130J ... 15.32 m (50 ft 3 in)
C-130J-30 ... 19.89 m (65 ft 3 in)
Max width ... 3.12 m (10 ft 3 in)
Max height ... 2.74 m (9 ft 0 in)
Total usable volume: C-130J ... 128.9 m^3 (4,551 cu ft)
C-130J-30 ... 170.5 m^3 (6,022 cu ft)

AREAS:
Wings, gross ... 162.12 m^2 (1,745.0 sq ft)
Ailerons (total) ... 10.22 m^2 (110.00 sq ft)
Trailing-edge flaps (total) ... 31.77 m^2 (342.00 sq ft)
Fin ... 20.90 m^2 (225.00 sq ft)
Rudder, incl tab ... 6.97 m^2 (75.00 sq ft)
Tailplane ... 35.40 m^2 (381.00 sq ft)
Elevators, incl tabs ... 14.40 m^2 (155.00 sq ft)

WEIGHTS AND LOADINGS (internal fuel only, except where specified):
Operating weight empty:
C-130J ... 34,274 kg (75,562 lb)
C-130J-30 ... 35,966 kg (79,291 lb)
Max fuel weight: internal ... 20,819 kg (45,900 lb)
external (optional) ... 8,506 kg (18,754 lb)
Max payload, 2.5 *g*: C-130J ... 18,955 kg (41,790 lb)
C-130J-30 ... 21,687 kg (47,812 lb)
Max T-O weight: C-130J-30 ... 74,389 kg (164,000 lb)
Max overload T-O weight:
C-130J, C. Mk 4 ... 79,380 kg (175,000 lb)
Max normal landing weight ... 58,965 kg (130,000 lb)
Max overload landing weight ... 70,305 kg (155,000 lb)
Max zero-fuel weight, 2.5 *g* ... 53,230 kg (117,350 lb)
Max wing loading: C-130J-30 ... 458.9 kg/m^2 (93.98 lb/sq ft)
Max power loading: C-130J-30 ... 5.44 kg/kW (8.93 lb/shp)

PERFORMANCE (C-130J except where indicated):
Never-exceed speed (VNE) at 3,050 m (10,000 ft)
C. Mk 4 ... 378 kt (700 km/h; 435 mph)
Max cruising speed:
C-130J ... 348 kt (645 km/h; 400 mph)
C. Mk 4 at 7,620 m (25,000 ft) ... 356 kt (659 km/h; 410 mph)
Econ cruising speed ... 339 kt (628 km/h; 390 mph)
Stalling speed ... 100 kt (185 km/h; 115 mph)
Max rate of climb at S/L ... 640 m (2,100 ft)/min
Time to 6,100 m (20,000 ft) ... 14 min
Cruising altitude ... 8,535 m (28,000 ft)
Service ceiling at 66,680 kg (147,000 lb) AUW ... 9,315 m (30,560 ft)
Service ceiling, OEI, at 66,680 kg (147,000 lb) AUW ... 6,955 m (22,820 ft)
T-O run ... 930 m (3,050 ft)
T-O to 15 m (50 ft) ... 1,433 m (4,700 ft)
T-O run using max effort procedures ... 549 m (1,800 ft)
Landing from 15 m (50 ft) at 58,967 kg (130,000 lb) AUW ... 777 m (2,550 ft)
Landing run at 58,967 kg (130,000 lb) AUW ... 427 m (1,400 ft)
Runway LCN: asphalt ... 37
concrete ... 42
Range with 15,876 kg (35,000 lb) payload: C-130J-30 ...
2,832 n miles (5,244 km; 3,259 miles)

UPDATED

LOCKHEED MARTIN ADVANCED MOBILITY AIRCRAFT (AMA)

TYPE: Medium transport/multirole.

PROGRAMME: Studies under way since early 1995; initially known as 'World Airlifter' and New Strategic Aircraft (NSA). Lockheed Martin's primary objective was to develop a replacement for Boeing KC-135 Stratotanker in-flight refuelling aircraft, Lockheed C-141 StarLifter strategic transport and tanker/transport types such as the Lockheed L-1011 TriStar and McDonnell Douglas KC-10 Extender, although it is also intended to offer commercial freighter versions as well as special mission aircraft configured for AEW and battlefield surveillance tasks.

Lockheed Martin is proposing to develop the AMA as a private venture and has sought two or three risk-sharing international partners to form a consortium; past discussions took place with several potential partners, including Aerospatiale Matra, DaimlerChrysler Aerospace and BAE Systems.

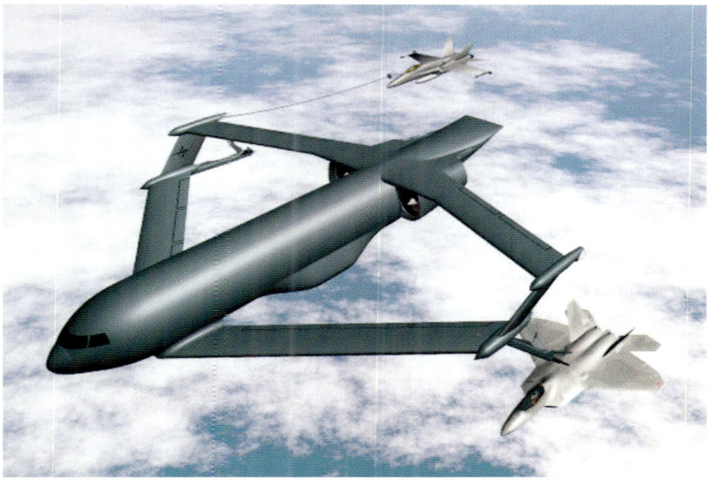

Computer-generated image of Lockheed Martin box-wing tanker/transport design 0121681

Recent studies envisage a twin-engined design with high bypass turbofans in the 267 to 311 kN (60,000 to 70,000 lb st) class, an M0.85 cruise speed, with 30 per cent greater fuel offload than the KC-135R/T Stratotanker. In the cargo role, AMA will carry a payload of 45,360 to 54,430 kg (100,000 to 120,000 lb) over 4,000 n miles (7,408 km; 4,603 miles).

Over 40 advanced aircraft designs examined, leading to further study of four basic concepts, all of which feature modular design using common basic structure and systems to reduce initial manufacturing costs and facilitate airframe upgrades during service life. Modular systems and avionics bus architecture will easily accommodate mission-orientated equipment for specific roles. Configurations studied include a conventional high-wing aircraft; a blended wing/body aircraft; a box-wing aircraft with two refuelling booms and two hose-and-drogue assemblies; and a global transport with an unrefuelled range of 12,000 n miles (22,220 km; 13,810 miles).

Design effort directed to the box-wing aircraft concept during 1997–2000, by virtue of aerodynamic and structural efficiency, combined with greatly reduced aircraft size. As recently envisaged, it will have two flight deck crew, plus advanced refuelling and loadmaster workstations; incorporate roll-on/roll-off cargo handling capability and be compatible with 20 and 40 ft ISO containers; and embody fly-by-light/power-by-wire flight control systems. Testing of a radio-controlled scale model began on 7 March 1997, this exhibiting excellent flight characteristics and meeting, or surpassing, test objectives during a total of 18 sorties.

Current planning expects AMA development effort to reach a peak during 2004–13, with resultant production aircraft ready for delivery from 2013; however, USAF purchase seems less likely in view of decision to acquire Boeing KC-767 tanker. No recent news received.

UPDATED

LOCKHEED MARTIN AERONAUTICS — PALMDALE

1001 Lockheed Way, Palmdale, California 93599
Tel: (+1 661) 572 41 53
VICE-PRESIDENT AND SITE GENERAL MANAGER: Richard S Baker
DIRECTOR, COMMUNICATIONS: Dianne Knippel

Palmdale performs upgrades and modifications to Lockheed U-2, F-117 and C-130 aircraft. Is also headquarters for Advanced Development Programs (ADP) initiatives across the Lockheed Martin Aeronautics Company, this facility also being known as the legendary Skunk Works. Responsibilities include derivatives or upgrades of current aircraft systems, new development systems, critical technology development and integration, and operational effectiveness and system analysis. Palmdale performs rapid prototyping — examples including the X-35 Joint Strike Fighter concept demonstrators. Next-generation platforms such as Unmanned Air Vehicle concepts are currently undergoing research and test at Palmdale.

UPDATED

LOCKHEED MARTIN AERONAUTICS — FORT WORTH

1 Lockheed Boulevard, Fort Worth, Texas 76108
Tel: (+1 817) 777 20 00
Fax: (+1 817) 763 47 97
Web: http://www.lmaeronautics.com
VICE-PRESIDENT, F-16 PROGRAMMES: John L Bean
EXECUTIVE VICE-PRESIDENT, JSF PROGRAMME: C T 'Tom' Burbage

General Dynamics' Fort Worth Division sold to Lockheed; became Lockheed Fort Worth Company on 1 March 1993; renamed Lockheed Martin Tactical Aircraft Systems following merger between Lockheed and Martin Marietta in 1995. Adopted current title in January 2000 with consolidation of all Lockheed Martin aeronautical operations into single company with headquarters in Fort Worth, Texas.

Activities include production, development and support of F-16 Fighting Falcon; one-third share of F/A-22 Raptor development and initial production; and leadership of F-35 Joint Strike Fighter development team (with Northrop Grumman and BAE Systems). Is principal subcontractor to Mitsubishi for production of F-2 in Japan and to KAI for development of T-50 in South Korea.

Fort Worth work force numbered 16,200 in August 2003; factory and associated office facilities occupied 687,500 m² (7.4 million sq ft).

UPDATED

LOCKHEED MARTIN F-35 JOINT STRIKE FIGHTER

TYPE: Multirole fighter.

DEVELOPMENT MILESTONES

Requirement issued	Nov 94
Request for proposals	Dec 95
Programme launched	16 Nov 96
First prototype material cut	Sep 97
First flight (prototype)	24 Oct 00
Development go-ahead	26 Oct 01
First production material cut	10 Nov 03

PROGRAMME: Origins of Joint Strike Fighter (JSF) programme vested in separate USAF/USN Joint Advanced Strike Technology (JAST) and Defense Advanced Research Project Agency (DARPA) Common Affordable Lightweight Fighter (CALF) projects of early 1990s.

Projects merged in November 1994, as JAST, after Congressional directive in mid-1994; programme renamed JSF in latter half of 1995. Previously, formal request for proposals (RFP) for preliminary research contracts released on 2 September 1994, stipulating industry response by 4 November and issue of contract awards by 16 December.

Some elements of US industry joined forces to win JAST/JSF work, with international collaboration in evidence. McDonnell Douglas led one team after signing October 1994 Memorandum of Understanding (MoU) with Northrop Grumman and British Aerospace; each company submitted individual bids, but all three would participate in event of securing contract. Boeing allied with Dassault of France on aspects of subsystem design effort.

Subsequent research contracts worth US$99.8 million were distributed between four companies: Boeing (US$27.6 million), Lockheed Martin (US$19.9 million), McDonnell Douglas (US$28.2 million) and Northrop Grumman (US$24.1 million). Further US$28 million allocated for associated avionics, propulsion systems, structures and materials, and modelling and simulation.

Merger of JAST and CALF resulted in expanded flight test programme, involving two finalists; each to build two demonstrators, one with ASTOVL capability and the other to use conventional take-off and landing (CTOL).

Draft RFP issued December 1995, with USA and UK signing MoU on 20 December 1995, which committed UK to participate in four year weapons system concept demonstration (WSCD) phase. MoU also stipulated that UK must contribute some 10 per cent (approximately US$200 million) of demonstration phase costs as full collaborative partner.

Formal release of the final RFP for JSF was expected on 7 March 1996, but was delayed to June 1996, with contract award date in November 1996.

All WSCD contenders chose Pratt & Whitney's F119 (later modified and redesignated F135) engine for their proposals, although a General Electric/Allison/Rolls-Royce team secured a US$7 million contract in March 1996 to examine alternative power plants. These were based on the General Electric F110 and YF120 engines, with the latter being chosen in May 1996 following Congressional directive aimed at fostering competition and also overcoming possible impact of developmental or operational problems with the F135. Further US$96 million multiyear contract awarded in February 1997 to cover technology maturation and core engine development of alternate engine over four-year period. Original intent was to begin full-scale EMD of F136 (as it was known) in about 2004, however, budget cut in 2003 resulted in delay, with F136 SDD to begin in August 2005, with first flight of CTOL version targeted for October 2009, followed by STOVL version in April 2010. Production engines to be available in early 2012.

On 16 November 1996, US Secretary of Defense, William J Perry announced that Boeing and Lockheed Martin had been chosen to participate in WSCD. Simultaneously, Boeing was awarded a US$661.8 million contract for the X-32, while Lockheed Martin received US$718.8 million for the X-35; in addition, Pratt & Whitney secured a contract worth US$804 million for the associated Engine Ground and Flight Demonstration Program. Subsequently, Northrop Grumman and British Aerospace joined Lockheed Martin team.

Australia, Canada, France, Germany, Greece, Israel, Singapore, Spain and Sweden were all briefed on JSF programme. For System Development and Demonstration (SDD) phase, four partnership options were available. Most costly is Level 1, with responsibility for 10 per cent of cost; UK is only partner at this level. Italy and the Netherlands are Level 2 partners, each contributing about 5 per cent of cost. Level 3 involves payment of 1 to 2 per cent, with Denmark and Norway having teamed up to share burden, while Australia, Canada and Turkey meet the cost alone. Finally, Security Co-operation Participant (SCP) Level involves smaller contribution of about US$50 million; Israel and Singapore were first two SCP subscribers, signing letters of intent in February 2003 and letters of agreement thereafter. Greece, Poland and South Korea are potential future additions, while Spain and Taiwan may also acquire STOVL version in due course. Levels 1 and 2 entitled to receive contracts; Level 3 may receive subcontracts from Levels 1/2 participants; and SCP receive data only.

Lockheed Martin development included production of 91 per cent scale powered model of JAST demonstrator for wind-tunnel tests. Model JAST, with F100 engine, began trials at Pratt & Whitney's West Palm Beach, Florida, facility in February 1995; subsequently to outdoor hover test rig at NASA Ames and then installed in 24 × 36 m wind tunnel at Mountain View, California, for series of powered hover and transition tests which ran from December 1995 to 5 March 1996; total of 196 hours accumulated, representative of

Manufactured image of Lockheed Martin F-35B in vertical landing configuration *NEW*/1048033

Lockheed Martin X-35 Article 301 in its STO/VL incarnation ***NEW***/1047747

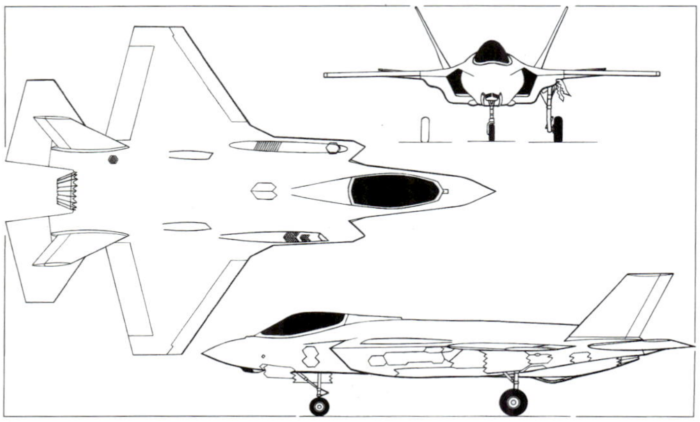

Lockheed Martin F-35A as planned in 2004 *(James Goulding)* ***NEW***/1047715

approximately 2,400 take-offs and landings with the vertical lift system. Midway through outdoor hover tests, design was reconfigured to eliminate canards.

Lockheed Martin design, development, construction and flight testing of full-scale demonstrator aircraft, required one initially to be flown as CTOL version (X-35A) to demonstrate land-based USAF model, before being reconfigured to serve as STOVL version (X-35B) for US Marine Corps, Royal Air Force and Royal Navy; other aircraft representative of carrier-capable US Navy model (X-35C). Although Fort Worth is team leader, both X-35 aircraft built at Palmdale, California, using rapid prototyping techniques.

Design of X-35 frozen as **Configuration 220** 13 June 1997 after Initial Design Review, at which time 11,000 hours of model testing accumulated. Development team joined by Northrop Grumman on 8 May 1997 and BAe (now BAE Systems) on 18 June 1997.

Release of engineering drawings in early September 1997 heralded start of parts production at Palmdale for demonstrator aircraft; by end of 1997, Lockheed Martin had completed about 70 per cent of required tooling and conducted second interim programme review. X-35 final design review completed in September 1998 and coincided with roll-out of full-scale mockup at Fort Worth. Assembly of first aircraft began in April 1998, with main wing carry-through bulkhead installed in early August 1998, by which time manufacturing of large composites skins for upper and lower wing surfaces also complete; aircraft moved from assembly testing to factory floor on 18 September 1999, in readiness for installation of flight control surfaces and landing gear as well as systems checks.

Flight control system software tested on NF-16D VISTA in 1998 as part of integrated subsystem technology demonstration. Further trials using AFTI/F-16 in 1999–2000 involved all-electric flight control system and modular electric power system planned for JSF. JSF avionics also tested on Northrop Grumman's BAe One-Eleven Co-operative Avionics Test Bed (CATB), which was fitted with sensors, processors and software in 1999; trials begun of Northrop Grumman radar and distributed infra-red sensor system, Kaiser helmet-mounted display and Lockheed Martin core processor in first quarter 2000.

More than 100 hours of flight time accumulated by early September 2000 using CATB, when avionics development and integration process completed; successful demonstrations included automatic target cueing (ATC), whereby sensors acquire targets rapidly and automatically; electro-optical targeting system (EOTS); electronic warfare suite and electronically scanned radar. Subsequent testing included co-operative engagement between CATB and E-8C Joint STARS, demonstrating all-weather precision targeting and combat identification techniques for fixed and moving targets.

Design of JSF continued to be refined after selection of Configuration 230-1. By 1998, third version of **Configuration 230 (230-3)** had reduced area of USAF/USMC JSF wings by seven per cent, but increased USN variant's wing area by 11 per cent. Further redesign occurred in 1999, culminating in September with Configuration 230-5, which has enlarged wing to satisfy sustained turn performance requirement and strengthened to meet 9 *g* stress requirement for the CTOL variant; further main change involved redesign of lift-fan nozzle from D-shape extendible box to venetian blind-type box, offering dual benefits of simpler design and reduced weight for the STOVL configuration.

Redesign process continued into 2000, with Configuration 230-5 adding further refinements aimed at reducing weight and increasing payload bringback capability; resultant structural modification entailed weight reduction in several areas such as weapon bay and landing gear door assemblies. Configuration 230-5 finally submitted as Preferred Weapons System Concept design in mid-2000.

Significant programme events in latter part of 1998 and 1999 included first test run of basic Pratt & Whitney JSF119 engine (designated FX661) at West Palm Beach, Florida, facility on 11 June 1998; over 330 hours of developmental testing performed by end August 1999, validating complete X-35 flight envelope. Final altitude flight qualification testing undertaken with another engine (designated FX663) in fourth quarter of 1999. National Aerospace Laboratory low-speed wind tunnel in the Netherlands used for testing of scale models of JSF starting in June 1998; initial trials of lift fan system's clutch, fan and gearbox rigs at Indianapolis, Indiana, in May and June 1998; and installation of engine inlet duct in assembly tool at Palmdale at beginning of July 1998. Subsequently, also in July, over 50 hours of preliminary engine testing were completed at West Palm Beach, including vibration surveys, fan and core running-in, operating performance calibration and engine control stability assessment.

Testing of first STOVL engine (designated FX662) undertaken initially at West Palm Beach; first run with lift fan engaged accomplished on 10 November 1998, with operation to 100 per cent speed following on 22 November; first series of tests included stress surveys and performance calibration and occupied about six weeks. Validation of STOVL

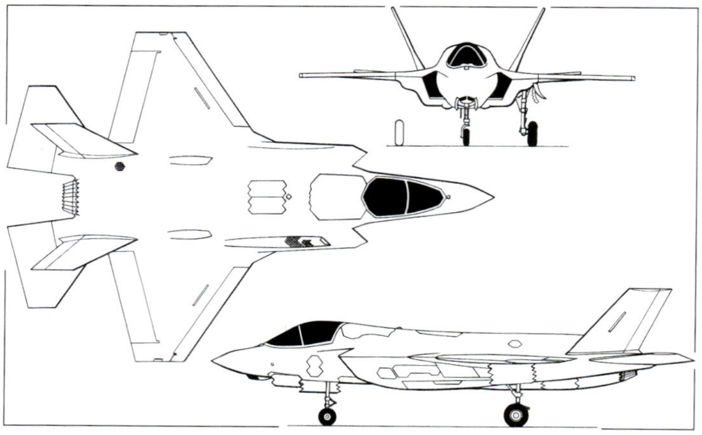

Lockheed Martin F-35B as planned in 2004 *(James Goulding)* ***NEW***/1047714

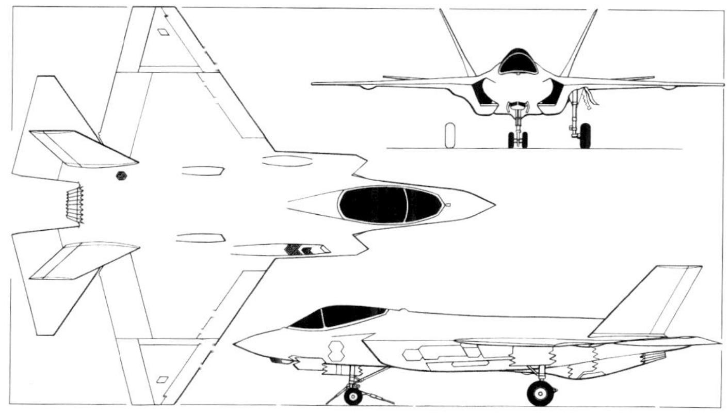

Lockheed Martin F-35C as planned in 2004 *(James Goulding)* ***NEW***/1047713

propulsion system achieved in late August 1999, with high-power clutch engagement of shaft-driven lift fan, simulating seamless conversion from conventional wing-borne flight configuration to jet-borne flight configuration for STOVL approach ending with vertical landing.

Following this, on 9 December 1999, flight engine YF001 successfully installed on the X-35A demonstrator at Palmdale, with integration tests, including plumbing connections, data communications and electrical checks, beginning immediately thereafter.

In UK, DERA vectored-thrust advanced aircraft control (VAAC) Harrier T. Mk 4 completed 20 hour, 36 sortie, flight test programme in November 1998, during which a sidestick control column was evaluated by civilian and military pilots. Two-phase programme began with initial calibration to validate STOVL control laws and stick characteristics; subsequent evaluation included pattern work, approach and transition to hover and precision and aggressive hover tasks, resulting in confirmation that side-stick provides satisfactory control of STOVL aircraft at low speeds.

Lockheed Martin also reached preliminary agreement with partners over work-sharing arrangements to be implemented for large-scale production. Lockheed Martin has responsibility for forward fuselage, cockpit, wing edges and final assembly; Northrop Grumman to fabricate mid-fuselage and wing box, with BAE Systems producing tails and aft fuselage section.

Radar signature testing of full-scale pole model began in late 1999 at Helendale, California; this included measurement of radar cross-section, assessment of antenna performance and demonstration of the robustness of supportable low-observable materials.

Significant events in 2000 included X-35A flight readiness review in March. Subsequently, in April, testing associated with development and flight qualification of the JSF119-PW-611 engine for the X-35A and X-35C was completed after 193 hours of operating time. Final assembly and painting of X-35A in late May was followed by lengthy series of ground tests, including systems checkout, engine running at full military power and with full afterburner augmentation, and low- and medium-speed taxi tests. These extended into October and culminated in a successful first flight by X-35A Article 301 from Palmdale to Edwards on 24 October; by 6 November, further four flights had been made, including first by USAF pilot, and envelope had been expanded to 390 kt (722 km/h; 449 mph); first aerial refuelling (from KC-135) on 7 November; maiden supersonic flight 21 November, when 25 hours had been flown in 25 sorties. X-35A test programme completed on 22 November, whereupon Article 301 returned to Palmdale for conversion to X-35B. X-35C (Article 300) first flight took place on 16 December 2000 from Palmdale to Edwards AFB. Initial X-35C testing at Edwards was completed in early February 2001, with aircraft making transcontinental ferry flight to Navy test centre at Patuxent River, Maryland on 9–10 February for specialised trials.

Following installation of shaft-driven lift fan in late December 2000, X-35B STOVL version began series of hover-pit trials on 22 February 2001; these concluded 16 March and included 26 lift-fan clutch engagements from CTOL to STOVL mode at high rpm settings. Accelerated mission testing followed and was concluded on 6 April, with X-35B then being readied for flight trials; this process included installation of flight-ready lift fan. Taxi tests began 12 June 2001, followed by first brief vertical take-off and landing on 23 June, with sustained hover accomplished on next day. Initial testing at Palmdale completed successfully by 3 July, when X-35B flown to Edwards AFB for remainder of trials programme; notable events included first airborne transition from STOVL to conventional mode on 9 July; first vertical landing from wingborne flight on 16 July and 'Mission X' demonstrations involving short take-off, level supersonic dash and vertical landing on 20 and 26 July, before STOVL testing concluded on 30 July.

Both X-35 aircraft subsequently allocated to museums, with Article 300 going to the naval air museum at Patuxent River and Article 301 joining the Smithsonian collection in Washington, DC.

After study of test results and company proposals, the US Department of Defense announced on 26 October 2001 that Lockheed Martin would be awarded a US$19 billion contract to cover SDD of what now became known as the F-35. Originally scheduled to occupy 126 months (but since extended by one year), SDD began immediately. Simultaneously, it was revealed that Pratt & Whitney would receive a contract worth US$4 billion for development

X-35 FLIGHT TESTS

	X-35A	X-35B	X-35C
First flight	24 Oct 00	23 Jun 01	16 Dec 00
Last flight	22 Nov 00	6 Aug 01	10 Mar 01
Sorties	27	39	73
Hours	27.4	21.5	58.0
Max *g*	5.0	5.0	4.8
Max speed	M1.05	M1.2	M1.22
Max AoA	20°	20°	20°
Max alt (ft)	34,000	34,000	34,000
Dummy deck landings			252
VTOs		17	
STOs		14	
Short landings		6	
Vertical landings		27	

and production of the associated F135 engine. As a major partner, the UK will contribute US$2 billion towards total SDD expenditures of approximately US$30 billion. Majority of preliminary design review successfully concluded on 27 March 2003, with a few items concerning aircraft weight and weapons integration not finally resolved until end of June; next major hurdle is critical design review, scheduled for 2005.

A total of 14 fully instrumented flying aircraft will be built at Fort Worth and assigned to SDD. Five will emerge as **F-35As** (USAF CTOL version); five will be **F-35Cs** (US Navy CV version); and the remaining four will be **F-35Bs** (US Marine Corps/UK STOVL version). Further eight non-flying airframes: two of each version as static test articles, plus another F-35C for drop testing and one pole-test airframe for radar signature evaluation. Production of SDD aircraft began in 2004, with fabrication of the first composites upper wing starting on 6 February; assembly of first centre fuselage section by Northrop Grumman began in May, followed by start of assembly of first forward fuselage by Lockheed Martin on 12 July and first aft fuselage by BAE Systems on 24 August.

F-35A first flight planned for August 2006. Just over 40 per cent of total planned flight testing by all JSF variants to be undertaken by F-35A. F-35B to fly about one year later, in 2007, with second F-35A following in late 2007. Second F-35B and first F-35C both due to fly for first time in 2008. Flight trials will be undertaken by Lockheed Martin at Fort Worth as well as joint industry/service teams at Edwards AFB, California (USAF) and NAS Patuxent River, Maryland (US Navy). Programme originally expected to involve almost 15,000 flight hours, although this reduced to 10,185 (about 5,700 sorties) according to most recent estimates.

Avionics testing to be accomplished initially by Boeing 737-300 that will feature a modified nose housing an F-35 radar as well as aerodynamically neutral canard surfaces on both sides of forward fuselage; later will carry antennae and sensors for electro-optical distributed aperture system. F-35 E-O targeting system also likely to be installed on this aircraft, which is set to make first flight with mission system in May 2005. Full funding for Low-Rate Initial Production (LRIP) was scheduled to begin in FY06 but has been delayed by about a year, with consequent effect on delivery of production aircraft. Projected IOC dates have also slipped, with the F-35B still expected to be first, in 2012, followed by both the F-35A and F-35C in 2013. Initial aircraft will be to Block 1 standard, with basic warfighting capability (including compatibility with JDAM and AIM-120 AMRAAM); additional weaponry for enhanced air-to-air, SEAD, close air support and interdiction will be added on Block 2 standard aircraft, while Block 3 will introduce more missiles and bombs and add deep strike capability. Lockheed Martin anticipates production rate of 17 aircraft per month at Fort Worth from 2011, but would like to achieve rate of 22 per month. With peak demand likely to be for about 30 a month, this could only be accomplished with second production centre. In early 2003, UK government commissioned Rand to perform feasibility study into creating second assembly line in UK; one key finding was that creating second assembly line in UK would increase direct assembly bill for UK aircraft by almost UK£47 million and indirectly raise US government costs by a further UK£30 million, principally arising from reduced efficiencies.

CURRENT VERSIONS: Three variants of basic design, optimised for the mission requirements of different armed services. As currently envisaged, basic company designation for operational JSF is Configuration 240-3, with following variants expected to enter service with US armed forces.

F-35A: Land-based **CTOL** derivative for USAF. Evaluated by X-35A between 24 October and 22 November 2000.

F-35B: STOVL version for US Marine Corps, Royal Air Force and Royal Navy, with USAF revealing intention in February 2004 also to acquire some F-35Bs. At present, it is not clear if these will be identical to the Marine Corps version. Engine-driven lifting fan behind cockpit replaces some fuel. Wing folding originally specified on Royal Navy version only, but requirement since dropped. Broader and higher spine behind cockpit to accommodate lifting fan and air intake; shortened cockpit canopy replaces blister-type glazing of F-35A and B. Evaluated by X-35B between 23 June and 30 July 2001; final flight on 6 August 2001 was from Edwards AFB to Lockheed Martin facility at Palmdale.

In fourth quarter of 2003, concerns emerged over weight of the F-35B version. Reviews of structural design eliminated some of the excess, but in early 2004, (by which time design had progressed to **configuration 240-3**), it was evident that further weight reduction initiatives would be necessary. These concerns were sufficiently serious in nature to warrant postponing the critical design review (CDR), with that for the F-35B now set to occur in mid-2006; equivalent CDRs for the F-35A and F-35C versions are scheduled to take place in November 2005 and early 2007 respectively.

Another unwelcome consequence was that it became necessary to delay the start of SDD testing, in order to give Lockheed Martin additional time to solve the weight problem. In May 2004, the extent of this became known when it was revealed that the STOVL version was about 1,500 kg (3,300 lb) overweight. A special task force known as SWAT (STOVL Weight Attack Team) was created by Lockheed Martin to investigate and, by September 2004, this had succeeded in removing about 1,225 kg (2,700 lb) of the unwanted weight.

This was accomplished through a mixture of airframe and engine design improvements as well as some compromise of customer requirements. Most of the extra weight was eradicated by a redesign that involved moving a number of internal elements, but additional benefit stemmed from a change in methods used for mating of fuselage sections and by positioning wing spars more closely together so as to permit use of thinner wing skin. Revisions to the air inlet and engine nozzle designs provided further benefit in the form of extra effect from installed thrust of the F13 engine.

It also proved necessary to modify the weapons bays, which, on the F-35B, have reverted to the size originally specified. Thus, internal carriage of weaponry will be limited to a pair of weapons in the 454 kg (1,000 lb) class (plus two AIM-120 AMRAAMs), rather than the two 900 kg (2,000 lb) weapons that may be carried internally by both the F-35A and F-35C.

In November 2004, Lockheed Martin received go-ahead to alter design schedule and build **F-35B1** version, which is 2,087 kg (4,600 lb) lighter than original conception.

F-35C: Carrier-based CTOL (CV). Wing, fin and elevator areas increased by chord extension; ailerons in addition to flaperons on wing; enlarged control surfaces and modified control system; strengthened landing gear with catapult launch bar on twin-wheel nose leg; concealed arrester hook; and folding wing. Evaluated by X-35C between 16 December 2000 and 10 March 2001 from test centres at Edwards AFB and Patuxent River, Maryland (from 10 February 2001).

EA-35B: Unofficial designation relating to Lockheed Martin studies into potential two-seat electronic attack version that could replace Marine Corps EA-6B from about 2015. Would retain internal weapons carriage capability, with sensors and jamming equipment embedded in fuselage and wings.

CUSTOMERS: Two X-35 demonstrators (Article 300 was X-35C; Article 301 first flew as X-35A and was subsequently modified to X-35B). Initial planning called for some 3,000 F-35s for USA and UK, but US Navy and Marine Corps requirement cut significantly in 2003 (see table). UK confirmed on 30 September 2002 that F-35B (STOVL) version will be procured for both RAF and Royal Navy, with service introduction then anticipated during 2012; however, in September 2004, it became apparent that first UK delivery has been delayed until 2011, with service introduction also likely to be deferred by at least one year.

Orders also expected from Australia (up to 100, possibly including some configured for long-range surveillance and reconnaissance), Canada, Denmark, Israel, Italy, the Netherlands, Norway, Singapore and Turkey.

US AND UK F-35 REQUIREMENTS
(as at October 2003)

Service	Qty	Remarks
US Air Force	1,763	Replaces A-10 and F-16; complements F/A-22. Primarily F-35A but expected to include some F-35Bs
US Navy/Marine Corps	680	To include F-35B and F-35C versions
Royal Navy (UK)	60	Replaces Sea Harrier
Royal Air Force (UK)	90	Replaces Harrier
Total	**2,593**	

COSTS: At time of downselect for SDD phase, unit cost of USAF F-35A quoted as US$37.3 million, with CV/STOVL versions costing under US$50 million each. US$3.27 billion SDD funding for FY03, following US$1.52 billion appropriation in FY02; US$4.25 billion authorised for FY04 with US$4.57 billion requested for FY05. Fly away costs mentioned in connection with Italian involvement quoted as US$36.6 million for CTOL version, US$47.4 million for CV version and US$45.3 million for STOVL version. UK purchase of 150 F-35Bs estimated to cost around UK £10 billion. Total programme cost (excluding procurement) estimated at US$33 billion at start of SDD; by mid-2004, this figure had risen to US£40.5 billion. In September 2004, unit cost estimates reported as US$45 million for the F-35A and US$55 to US$60 million for the F-35B and F-35C versions.

DESIGN FEATURES: Trapezoidal mid-wing configuration, optimised for low observability. Twin tailfins; internal weapon bays. Wing and tailplane leading-edges swept back approximately 33°; trailing-edges swept forward approximately 14°; fins swept back approximately 42° and canted outward at tips by approximately 25°. Twin 'divertless' fibre-placed graphite/epoxy composites engine air intakes with no moving parts produced by ATK. All-electric flight control system.

STOVL version employs a lifting fan behind the cockpit, driven by a shaft from the single engine; inlet and outlet are covered by doors, except when in use; original side-hinged bi-fold hatch of X-35B replaced in early 2003 by single-piece cantilever inlet door hinged at rear. Other changes related to thermal management also incorporated at this time, to satisfy requirement to operate on 49°C (120°F) day. The resultant cold air barrier prevents hot air from being reingested when on or near the ground.

FLYING CONTROLS: All-electric flight control system for movement of primary flight control surfaces (flaps, tailerons and rudder) incorporating Parker Hannifin electrohydrostatic actuators. Moog leading-edge flap drive system. Flight control computer originally to be supplied by Honeywell, but replaced by advanced Lockheed Martin unit in third quarter of 1998 to eliminate anticipated throughput problems arising from growth in flight control software.

LANDING GEAR: Retractable tricycle type; mainwheels retract inwards; nosewheel(s) forward. Single wheels on each unit, except twin nosewheels and catapult towbar on F-35C, which also has reinforced gear for deck landings. Tyres will feature embedded transponder including integrated circuit and capacitive pressure sensor to facilitate monitoring of pressure and condition. Dunlop selected to provide 34 × 11.OR16 radial main tyres for SDD phase of STOVL F-35B version; Honeywell to develop wheels and brakes, with Crane Hydro-Aire supplying brake control and anti-skid system. Arresting gear by SP Aerospace.

POWER PLANT: One 178 kN (40,000 lb) class (111 kN; 25,000 lb st dry thrust) Pratt & Whitney F135 (formerly JSF119-PW-611; F119 derivative) turbofan. Rolls-Royce three-bearing swivel-duct nozzle on -611S version to deflect thrust downwards for STOVL, plus a Rolls-Royce engine-driven fan behind cockpit and a bleed air reaction control valve in each wingroot to provide stability at low speeds. For F-35B, total vertical lift of 177 kN (39,700 lb st) comprises some 40 per cent from main nozzle, 48 per cent from fan and 12 per cent from reaction control valves. Ground testing of first of seven F135 engines (including four STOVL powerplants) began at West Palm Beach, Florida, on 11 October 2003; first run of STOVL engine followed on 14 April 2004, although testing of latter halted in June after discovery of erosion damage in second-stage vanes of low-pressure turbine. F-35A has in-flight refuelling receptacle on spine; US Navy and Marine Corps require retractable probe on starboard side. JP-5 or JP-8 fuel. General Electric/Rolls-Royce F136 turbofan with same performance characteristics under development as alternative interchangeable powerplant. Ground testing of F-136 began at Evendale, Ohio, on 22 July 2004; SDD of this powerplant due to begin August 2005, with first flight targeted for October 2009. Production standard F136 engines should be available with effect from LRIP Lot 4 in 2011-12.

ACCOMMODATION: Pilot only; canopy by Sierracin; canopy frame assembly by Smiths Aerospace. STOVL versions have canopy of reduced length. Martin-Baker Mk 16E lightweight ejection seat in X-35s; F-35 expected to have new seat resulting from Joint Ejection Seat Program (JESP).

SYSTEMS: Hamilton Sundstrand 80 kW engine-driven switched-reluctance starter/generator providing two independent channels of 270 V DC electrical power; electric distribution units, power panels, power distribution centres, batteries and battery charger equipment provided by Smiths Industries; electrical wiring on SDD aircraft by Stork Aerospace of the Netherlands; Honeywell thermal- and energy-management module (T/EMM) combining functions of auxiliary and emergency power units and environmental control system; 270 V DC lithium ion emergency battery. Weapons bay door drive system by TRW Aeronautical Systems. 276 bar (4,000 lb/sq in) high pressure hydraulic power generation and distribution system by Eaten Corporation.

AVIONICS: *Comms:* Northrop Grumman Radio Systems to provide next-generation CNI (communications, navigation and identification) avionics for F-35; package will include VHF/UHF radio, Have Quick I/II, SINCGARS/SIP, SatCom, IFF/SIF transponder, ILS, MLS, ACLS, Tacan, intra-flight data link, Link 4A, Link 16/JTIDS and weapons data link.

Radar: Northrop Grumman MIRFS/MFA (multifunction integrated RF system/multifunction nose array) combines AN/APG-81 active electronically scanned array (AESA) radar, electronic warfare and communications functions planned for production F-35.

Flight: Lockheed Martin Tactical Defense Systems Integrated Core Processor (ICP) incorporating open system architecture.

Instrumentation: Two side-by-side 200 × 250 mm (8 × 10 in) flat panel MFDs. Meggitt secondary flight display system. Vision Systems International selected in third quarter of 1999 to provide advanced integrated helmet-mounted display (HMD) system; alternative HMD being developed by BAE Systems, which delivered first helmet to Lockheed Martin at end of September 2004. UK design based on binocular helmet developed for Typhoon.

Mission: Electro-optical sensor system (EOSS) being developed jointly by Lockheed Martin and Northrop Grumman; latter responsible for electro-optical distributed aperture system (EODAS) with functions that include air-to-air search-and-track, target cueing and missile warning, and air-to-ground surface-target tracking; EODAS uses six CMC Electronics conformal compact lightweight imaging IR sensors around airframe and seventh in targeting system pod, with combined data providing all-aspect multifunction imaging to pilot via wide-angle helmet-mounted display, overlaid with target and threat data information. Second key element of EOSS is Lockheed Martin internal Electro-Optical Targeting System (EOTS); incorporating BAE Systems Avionics laser precision ranger/designator, this can be used when engaging targets on the ground and in the air.

Self-defence: EW capability incorporated into MIRFS/MFA. BAE Systems is prime contractor for EW equipment and systems integration; Litton Amecom to supply low-cost ESM equipment.

ARMAMENT: Internal cannon (in port engine air intake trunk upper surface) specified by USAF and under consideration by US Navy; Boeing/Mauser BK 27 originally selected, but replaced by General Dynamics GAU-12 25 mm weapon in fourth quarter of 2002. US Marine Corps and UK variants to have 'missionised' 25 mm cannon in low-observables pod. Internal weapon bays, incorporating pneumatic weapon suspension and release equipment by EDO; F-35A and F-35C internal weapons fit is two AIM-120C AMRAAMs and two GBU-31 JDAM bombs. Other weapons expected to be used by the F-35 include CBU-105 WCMD Sensor Fuzed Weapon, GBU-12 Paveway II, AGM-154 JSOW, AIM-132 ASRAAM and GBU-32 JDAM, all for internal carriage; externally carried weapons are expected to include AGM-158 JASSM, Storm Shadow cruise missile and the AIM-9X Sidewinder. Optional external stores on six hardpoints, which can accommodate fuel tanks or up to 6,800 kg (15,000 lb) additional ordnance. Station capacities, according to October 2003 briefing document, will be as follows; 1 and 11 (outer wing) 136 kg (250 lb); 2 and 10 (mid-wing) 1,134 KG (2,500 lb); 3 and 9 (inner wing) 2,268 kg (5,000 lb); 4 and 8 (internal) 1,134 kg (2,500 lb); 5 and 7 (internal) 159 kg (350 lb); and 6 (centreline) 454 kg (1,000 lb). By February 2004, company briefings were quoting 136 kg (300 lb); 907 kg (2,000 lb), 2,268 kg (5,000 lb), 907 kg (2,000 lb); 159 kg (350 lb); and 454 kg (1,000 lb), respectively. Internal capacity of F-35B will be lower, following reduction in size of weapons bays, with stations 4 and 8 limited to maximum of 454 kg (1,000 lb).

Data are provisional.

DIMENSIONS, EXTERNAL (estimated):
Wing span: F-35A, F-35B ... 10.67 m (35 ft 0 in)
F-35C (wings spread) ... 13.11 m (43 ft 0 in)
F-35C (wings folded) ... 9.47 m (31 ft 1 in)
Length overall: F-35A, F-35B ... 15.57 m (51 ft 1 in)
F-35C ... 15.67 m (51 ft 5 in)
Height overall: F-35A, F-35B ... 4.57 m (15 ft 0 in)
F-35C ... 4.72 m (15 ft 6 in)
Tailplane span: F-35A, F-35B ... 7.29 m (23 ft 11 in)
F-35C ... 8.64 m (28 ft 4 in)
Wheel track: F-35A, F-35B ... 4.34 m (14 ft 3 in)
F-35C ... 4.22 m (13 ft 10 in)
Wheelbase: F-35A, F-35B ... 6.02 m (19 ft 9 in)
F-35C ... 6.17 m (20 ft 3 in)

AREAS:
Wings, gross: F-35A, F-35B ... 42.7 m^2 (460 sq ft)
F-35C ... 58.3 m^2 (628 sq ft)

WEIGHTS AND LOADINGS (approx):
Weight empty: F-35A ... 12,426 kg (27,395 lb)
F-35C ... 13,888 kg (30,618 lb)
Max weapon load ... more than 9,072 kg (20,000 lb)
Max internal fuel weight:
F-35A ... 8,391 kg (18,498 lb)
F-35C ... 8,901 kg (19,624 lb)
Max T-O weight: F-35A ... 29,710 kg (65,500 lb)
F-35C ... 30,322 kg (66,850 lb)
Max wing loading: F-35A ... 695.2 kg/m^2 (142.39 lb/sq ft)
F-35C ... 519.7 kg/m^2 (106.45 lb/sq ft)

PERFORMANCE (approx):
Max level speed: all ... M1.6
Combat radius: F-35A ... more than 600 n miles (1,111 km; 690 miles)
F-35B ... more than 450 n miles (833 km; 517 miles)
F-35C ... more than 700 n miles (1,296 km; 805 miles)

UPDATED

LOCKHEED MARTIN F-16 FIGHTING FALCON

Israel Defence Force names: F-16C Barak (Lightning), F-16D Brakeet (Thunderbolt) and F-16I Suefa (Storm)
Turkish Air Force name: Savaşan Şahin (Fighting Falcon)
United Arab Emirates Air Force name: Desert Falcon

TYPE: Multirole fighter.

DEVELOPMENT MILESTONES

Request for proposals	6 Jan 1972
Design submitted	Feb 1972
Programme launched	Apr 1972
Rolled out	13 Dec 1973
First flight (YF-16)	20 Jan 1974
Official first flight (YF-16)	2 Feb 1974
First flight, pre-production	8 Dec 1976
First flight, production	7 Aug 1978
First delivery (USAF)	Aug 1978
Entered service (USAF)	6 Jan 1979

PROGRAMME: Emerged from General Dynamics YF-16 of US Air Force Lightweight Fighter prototype programme in 1972; first flight of prototype YF-16 (72-01567) 2 February 1974; first flight of second prototype (72-01568) 9 May 1974; selected for full-scale development (FSD) 13 January 1975; day fighter requirement extended to add air-to-ground capability with multimode radar and all-weather navigation; production of FSD aircraft began July 1975; first flight of FSD aircraft 8 December 1976. F-16 achieved 5 millionth flying hour

Israel's first Lockheed Martin F-16I Suefa ***NEW***/0560894

late in 1993; 10 millionth flying hour passed in March 2002. 4,000th aircraft delivered (to Egyptian Air Force) on 28 April 2000; total of 4,152 delivered as at 1 May 2004.

Total sales 4,347 at end of 2001. Further orders from Chile (10 aircraft), Oman (12) and Poland (48), increased sales to 4,417 at start of 2003. No further orders have since been forthcoming.

CURRENT VERSIONS: **F-16A:** First version for air-to-air and air-to-ground missions; not currently in production but aircraft in storage available for export customers.

F-16B: Standard tandem two-seat version of F-16A; fully operational both cockpits; fuselage length unaltered; reduced fuel.

F-16C/D: Single-seat and two-seat USAF Multinational Staged Improvement Program (**MSIP**) aircraft respectively, implemented February 1980. MSIP expands growth capability to allow for precision ground attack and BVR missiles, and all-weather, night and day missions; **Stage I** applied to Block 15 F-16A/Bs delivered from November 1981; **Stage II** applied to Block 25 F-16C/Ds from July 1984 included core avionics, cockpit and airframe changes. Block 25 aircraft originally delivered with F100-PW-200 engine, but all surviving examples now have F100-PW-220E following retrofit, as well as other improvements as detailed under Stage III of MSIP. **Stage III** involved installation of systems as they became available, beginning 1988 and extending to Block 50/52, including selected retrofits back to Block 25. Changes included adoption of Northrop Grumman AN/APG-68 multimode radar, with improved range, resolution, more operating modes and better ECCM than AN/APG-66; advanced cockpit with multifunction displays and upfront controls, BAE Systems wide-angle HUD, Fairchild mission data transfer equipment and radar altimeter; expanded base of fin giving space for proposed later fitment of AN/ALQ-165 Airborne Self-Protection Jamming system (since cancelled, though now installed in Korean aircraft); increased electrical power and cooling capacity; structural provision for increased take-off weight and manoeuvring limits; and smart weapons such as AIM-120A AMRAAM and AGM-65D Maverick.

Common engine bay introduced at **Block 30/32** (FMS deliveries from December 1985 and USAF deliveries from July 1986) to allow fitting of either P&W F100-PW-220 (Block 32) or GE F110-GE-100 (Block 30) Alternate Fighter Engine. Other changes include computer memory expansion and seal-bonded fuselage fuel tanks. Additions in 1987 included voice message unit, doubled chaff/flare capacity, repositioning of RWR antennas to provide better coverage in forward hemisphere, Shrike anti-radiation missiles (from August 1987), crash survivable flight data recorder and modular common inlet duct ('large-mouth') allowing full thrust from F110 at low airspeeds.

Software upgraded for full Level IV multitarget compatibility with AMRAAM early 1988. Industry-sponsored development of radar missile capability for several air forces resulted in firing of AIM-7F and AIM-7M missiles from F-16C in May 1988; capability introduced mid-1991; missiles guided using pulse Doppler illumination while tracking targets in a high PRF mode of the AN/APG-68 radar.

Block 40/42 (deliveries from December 1988) upgrades include AN/APG-68(V)1 radar allowing 100 hour operation before maintenance, full compatibility with Lockheed Martin low-altitude navigation and targeting infra-red for night (LANTIRN) pods, four-channel digital flight control system, expanded capacity core computers, diffractive optics HUD, enhanced envelope gunsight, GPS, improved leading-edge flap drive system, improved cockpit ergonomics, high gross weight landing gear, structural strengthening, increased performance battery and provision for improved EW equipment, including advanced interference blanker. LANTIRN targeting pod gives day/night standoff target identification, automatic target handoff for multiple launch of Mavericks, autonomous laser-guided bomb delivery and precision air-to-ground laser ranging. LANTIRN navigation pod provides real-world IR view through HUD for night flight plus automatic/manual terrain following with dedicated radar sensor. Combat Edge pressure breathing system installed 1991 for higher pilot *g* tolerance.

A total of 39 Block 40 F-16C/Ds of the 31st FW was involved in a quick response capability (QRC) modification effort, known as 'Sure Strike', to install Improved Data Modem (IDM) equipment. Work was undertaken by a joint USAF/LMTAS team at Aviano AB, Italy, and was completed in December 1995, these being the first Block 40 aircraft to receive the IDM which was standard equipment on current production Block 50 F-16s. In mid-1998, a demonstration programme was conducted to adapt existing IDM with Lockheed Martin kit to provide 'Gold Strike' system capable of two-way transmission of digitised video imagery of targets and thus enhance pilot's situational awareness.

First Block 40 F-16C/Ds issued in late 1990 to 363rd FW (Shaw AFB, South Carolina); first LANTIRN pods to 36th FS/51st FW at Osan, South Korea, in 1992. USAF Block 40/42 F-16Cs unofficially designated **F-16CG**. Following retirement of F-111F, Block 40/42 F-16Cs and F-16Ds with LANTIRN comprise more than 50 per cent of USAF night/precision strike force. All USAF F-16Cs and F-16Ds to receive FLIR targeting pod capability.

Block 50/52 (deliveries began with F-16C 90-0801 in October 1991 for operational testing) upgrades included F110-GE-129 and F100-PW-229 increased performance engines (IPE), AN/APG-68(V)5 radar with advanced programmable signal processor employing VHSIC technology, Have Quick IIA UHF radio and AN/ALR-56M advanced RWR.

Lockheed Martin F-16C Fighting Falcon leading a pair of North American P-51 Mustangs at Sun 'n' Fun 2004, Lakeland, Florida *(Paul Jackson)* ***NEW***/0583469

Lockheed Martin F-16C of USAF's 22 FS, Spangdahlem, Germany *(Paul Jackson)* ***NEW***/0583471

Changes initiated at **Block 50D/52D** in 1993 included full integration of AGM-88 HARM (high-speed anti-radiation missile) via HARM aircraft launcher interface computer (ALIC), improved data modem (IDM), upgraded programmable display generator with growth potential for colour and map, expanded data transfer cartridge, ring laser INS (Honeywell and Litton units both used) and improved VHF/FM antenna. AN/ALE-47 advanced chaff/flare dispenser fitted to all FMS Block 50 aircraft delivered after mid-1996 and incorporated as standard on USAF aircraft with effect from FY97 purchase; also retrofitted to earlier USAF Block 40/50 aircraft.

First Block 50D/52D (91-0360) delivered to USAF on 7 May 1993; optimised for defence suppression missions, having software for horizontal situation display on existing two MFDs and provision for AN/ASQ-213 HARM targeting system (HTS); sensor in pod on starboard side of engine inlet; HTS has capability similar to F-4G 'Wild Weasel' which it replaced in SEAD role. USAF has increased HTS inventory and incorporated software and hardware improvements enabling more targets to be tracked with enhanced ambiguity resolution and speedier reaction time.

Deliveries of Block 50/52 began to 4th FS of 388th FW at Hill AFB, Utah, from October 1992. Block 50D/52D aircraft initially to 309th FS (now 79th FS) of 363rd (now 20th) FW at Shaw AFB, South Carolina in 1993. Production for USAF was due to terminate with FY94 batch, but six additional F-16Cs funded in each of FY96 and FY97, plus three in FY98, one in FY99, 10 in FY00 and four in FY01; FY96 and subsequent aircraft to Block 50D standard. F-16s delivered from mid-2000 (FY97 and later) to improved configuration, incorporating modular mission computer by Raytheon (replacing three core avionics processors) that was developed for F-16A/B MLU programme, Honeywell colour liquid crystal multifunction displays (replacing monochrome CRT MFDs), Honeywell colour programmable display generator, Teac colour airborne videotape recorder, colour cockpit TV sensor and Litton onboard oxygen generating system (OBOGS). FY00 and subsequent aircraft have AN/APX-113 and associated avionics improvements. Majority of FY00 and FY01 aircraft delivered to USAF between April and December 2002, but final example to receive full CCIP upgrade before being handed over in early 2005. USAF Block 50/52 F-16Cs unofficially designated **F-16CJ**.

First Samsung-assembled F-16C Block 52D rolled out in South Korea on 7 November 1995. First flight of initial TAI-built F-16C Block 50D in late May 1996, with delivery to Turkish Air Force following on 29 July; last TAI-produced F-16 delivered October 1999. First Block 50D delivered to Greece 29 January 1997. First Block 52D delivered to Singapore 30 January 1998. First production lease (PL) Block 52D aircraft for Singapore was delivered 28 May 1998; this was under terms of commercial contract, with delivery achieved in less than 24 months from placing of order.

First delivery of F-16C Block 52 in Korea Fighter Program II in June 2003; aircraft produced under license by KAI, which delivered last aircraft in August 2004.

Advanced Block 50/52: Latest production versions, originally referred to as Block **50+/52+**. Basic configuration includes upgraded AN/APG-68(V)9 radar with 30 per cent greater air-to-air detection range and synthetic aperture radar (SAR) mode for high-resolution mapping and target detection/recognition. Is also compatible with latest FLIR navigation and targeting pod systems and has upgraded core avionics, including an improved modular mission computer, two 102 × 102 mm (4 × 4 in) colour cockpit displays, cockpit and exterior lighting compatible with night vision goggles, helmet-mounted cueing system, a digital terrain system, IFF interrogator/transponder, high off-boresight missile compatibility. Link 16 datalink and OBOGS. Available with a choice of internal electronic countermeasures equipment and able to take various customer-unique systems. Maximum take-off gross weight increased to 21,772 kg (48,000 lb). Can be fitted with low-drag conformal fuel tanks, with combined capacity of 1,705 litres (450 US gallons; 375 Imp gallons). Additional fuel in optional 2,271 litre (600 US gallon; 500 Imp gallon) auxiliary wing tanks. Two-seat aircraft have a rear cockpit configured for either a weapon system operator or instructor pilot (converted with a single switch), plus a dorsal avionics compartment that accommodates all of the systems of the single-seat aircraft, plus additional chaff/flare dispensers and specialised mission equipment.

First customer for Advanced Block 52 version was Greece, which revealed intention on 30 April 1999 to buy as many as 70 although subsequent contract covered only 50 aircraft, with fixed price option for up to 10 more that was converted to firm order on 14 September 2001. Deliveries began 31 October 2002, with formal inauguration ceremony at Forth Worth on 2 April 2003, when first five aircraft handed over. Initial aircraft ferried to Greece in May 2003. Other customers for Advanced Block 50/52 aircraft are Chile (Block 50), Oman (Block 50), Poland (Block 52) and Singapore (Block 52).

USAF F-16C/D Retrofit Programmes: F-16 originally designed to fly 8,000 hours based on specified usage spectrum, but actual usage has in most cases been more severe, with

Lockheed Martin F-16C of New York Air National Guard *(USAF)* ***NEW***/0583497

US Air Force Lockheed Martin F-16C Fighting Falcons ***NEW***/0583567

aircraft regularly flying at higher operational weights than predicted. USAF F-16C/D aircraft now undergoing 'Falcon STAR' (structural augmentation roadmap). Modifications accomplished under this programme will ensure that aircraft achieve 8,000 hour service life, without depot inspection. USAF 'Falcon STAR' retrofit kit proofing conducted in 2003, with pilot production and installation beginning in 2004; first Falcon STAR aircraft delivered to 148th FW, Minnesota ANG, in February 2004. At least 10 other countries are involved in this programme.

ANG/AFRC Block 25/30/32 F-16C/Ds subject to combat upgrade plan integration details (CUPID) which completed by mid-2003. CUPID brought approximately 620 older F-16s to a standard close to that of the early Block 50/52 aircraft. Among the improvements are situation awareness datalink (SADL), improved airborne videotape recorder, colour camera, initial NVG-compatible cockpit lighting, LANTIRN and Rafael Litening II FLIR targeting pod capability, AN/ALQ-213 countermeasures control system and provisions for GPS/laser gyro INS. Future enhancements to include expanded central computer, joint helmet-mounted cueing system, AIM-9X missile, follow-on NVIS capability, PIDS-3 pylon upgrade for smart weapon compatibility, ACES II ejection seat improvements, enhanced main battery and software upgrades.

Block 40/42 F-16C/Ds are currently being upgraded to include NVG compatibility, MD-1295/A improved data modem, digital terrain system, AN/ALE-47 chaff dispenser, AN/ALE-50 towed decoy and smart weapons compatibility (including the GBU-27, JDAM, JSOW and WCMD). Block 50/52 F-16C/Ds already have these capabilities. Three squadrons of ANG Block 42 aircraft currently receiving F100-PW-229 engine as replacement for original F100-PW-220; first engines installed in mid-2002.

In June 1998, USAF launched an upgrade effort known as common configuration implementation program (CCIP), which is intended to provide common hardware and software capability to 648 Block 40/42/50/52 aircraft. CCIP is a multiphase effort and is being implemented in stages, based on availability of subsystems. Work is being undertaken at the Ogden Air Logistics Center, Hill AFB, Utah, to where the first eight modification kits were shipped on 29 June 2001; these Phase 1 kits included the modular mission computer and colour MFDs applicable to 107 older aircraft of the Block 50/52 versions. The first aircraft to undergo CCIP was completed ahead of schedule and delivered to the 20th FW on 11 January 2002.

Phase 1A Block 50/52 kits include the AN/APX-113 combined electronic interrogator/transponder which gives autonomous BVR intercept capability. These aircraft

Colorado Air National Guard Lockheed Martin F-16C turning tightly in a vapour-laden atmosphere *(Paul Jackson)* ***NEW***/0583470

also capable of alternative carriage of an advanced Lockheed Martin Sniper XR FLIR targeting pod in addition to the HARM targeting system pod. The first of 251 aircraft to receive this capability was delivered in October 2002; first operational unit with Phase 1A aircraft was 389th Fighter Squadron at Mountain Home AFB, Idaho.

Phase 2, fielded in July 2003, adds Link 16 multifunctional information distribution system (MIDS) datalink, the Vision Systems International joint helmet-mounted cueing system (JHMCS) and an electronic horizontal situation indicator to 251 Block 50/52 aircraft. Starting in 2005, 397 Block 40/42 Fighting Falcons are to receive the full package of modifications detailed above, in Phase 3 of CCIP upgrade. Several other F-16 operators are considering CCIP upgrade of their Block 40/50 aircraft.

F-16E/F: Originally known as Block 60 and redesignated as F-16E (single-seat) and F-16F (two-seat) in late 2003, this is the most advanced version yet and has Northrop Grumman AN/APG-80 multimode agile beam radar with active electronically scanned array (AESA) antenna, offering numerous advantages, including mode interleaving; also has internal Northrop Grumman AN/AAQ-32 FLIR navigation and targeting system, plus advanced cockpit layout, with three 127 × 178 mm (5 × 7 in) colour liquid crystal displays having picture-in-picture and moving map capability. New core avionics suite based on advanced mission computer utilising commercial hardware and software and a high-speed fibre optic databus. Other improvements include digital fuel management system, higher-capacity environmental control system, new air data system (eliminating probe on tip of nose) and expanded digital flight control system with additional automatic modes, such as terrain following. Specialised equipment includes Northrop Grumman Falcon Edge advanced internal ECM system and Thales secure radio and datalink. Power plant is General Electric F110-GE-132 engine in the 144.65 kN (32,500 lb st) class. High-capacity tyres and brakes, allowing maximum take-off gross weight to rise to 22,679 kg (50,000 lb). Many of the features introduced by Advanced Block 50/52 aircraft also adopted as standard on the F-16E/F, including conformal fuel tanks, dorsal avionics compartment and with rear cockpit configured for weapon system operator.

United Arab Emirates announced selection of the F-16E/F **Desert Falcon** on 12 May 1998, although signature of contract for 80 aircraft delayed until first quarter of 2000. First flight (by F-16F) made 6 December 2003, with first three two-seaters to be used for flight testing. F-16E first flight accomplished on 6 August 2004. Delivery to UAE was expected to begin in third quarter of 2004; first three aircraft to go to Tucson for training, with further five to be in UAE by end of 2004; final delivery set for 2007.

Initial deliveries to UAE will be to so-called Standard 0 configuration to allow pilot training in USA to begin; first operational aircraft will be in Standard 1 configuration which marries Block 50 capability to new EW suite, advanced cockpit and new radar. Standard 2 will become available in early 2006 and feature advanced EW modes as well as terrain-following radar capability and additional weaponry; definitive Standard 3 version in 2007 will include BAE Systems TERPROM digital terrain avoidance and navigation system and, possibly, a helmet-mounted cueing system.

NF-16D: Variable stability in-flight simulator test aircraft (**VISTA**) modified from Block 30 F-16D (86-0048) ordered December 1988 to replace NT-33A testbed. Features include Calspan variable stability flight control system, fully programmable cockpit controls and displays, additional computer suite, permanent flight test data recording system, variable feel centrestick and sidestick in front cockpit, with latter also available for use in F-16 mode; safety pilot in rear cockpit. Internal gun, RWR and chaff/flare equipment removed, providing space for Phase II and III growth including additional computer, reprogrammable display generator and customer hardware allowance. Dorsal avionics compartment in bulged spine. Aircraft transferred to USAF Test Pilot School at Edwards AFB in 2001 to support TPS missions and various research and development activities. Aircraft participated in USAF Auto Air Collision Avoidance System demonstration in mid-2003.

F-16I: Two-seat version, basically similar to Advanced Block 52, for Israel, featuring Pratt & Whitney F100-PW-229 engine, conformal fuel tanks and SAR. Incorporates Northrop Grumman AN/APG-68(V)9 radar and significant amount of Israeli avionics, including EW suite, cockpit displays, helmet-mounted sight and advanced self-protection system (ASPS). EW suite, by Elisra, will include radar and missile approach warning systems plus jammers, while Elbit Systems to provide head-up display system, central mission computer, advanced display processor, DASH IV display and sight helmet system and stores management systems. RADA teamed with Smiths Aerospace to supply data acquisition system; Elop to provide HUD unit. Aircraft also to utilise Rafael Litening II targeting pod. Total of 50 initially purchased in US$2.5 billion deal; further 60 aircraft then on option, of which 52 converted to firm order on 19 December 2001. First aircraft completed in June 2003, thereafter being modified to accommodate flight test instrumentation. Roll-out and formal acceptance of first F-16I (401, marked as '253' to indicate the operating squadron) at Fort Worth 14 November 2003, with first flight (by fourth aircraft, also marked as 253) on 23 December 2003. First two aircraft delivered to Israel (for 253 Squadron) on 19 February 2004; balance to follow at rate of about two per

Maiden flight of the first F-16F Desert Falcon for the United Arab Emirates, 6 December 2003 ***NEW***/0563220

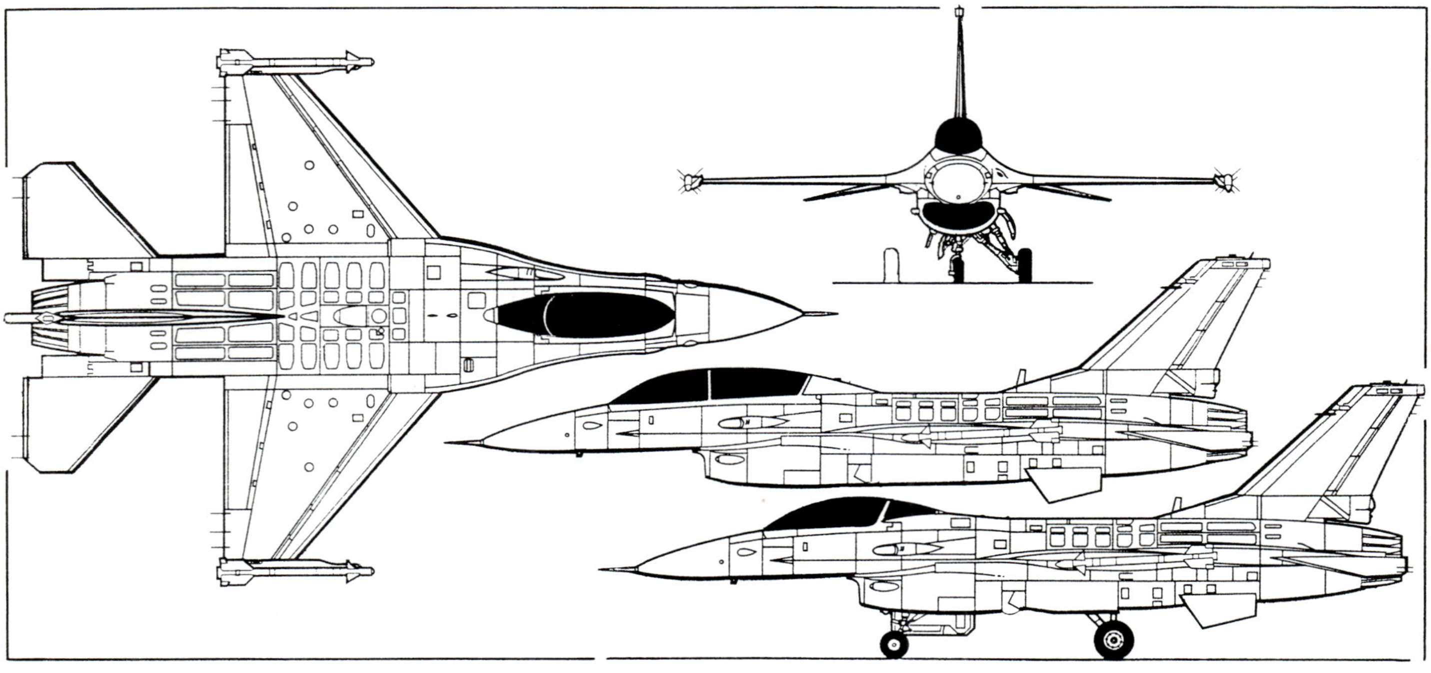

F-16C (GE F110 turbofan) with extra side view (top) of two-seat F-16D (P&W F100 turbofan) *(Paul Jackson)* 0016166

month. In early 2004, however, it emerged that Israel was 'highly dissatisfied' with synthetic aperture mode of AN/APG-68 radar and that additional trials would be undertaken in Israel.

F-16N: US Navy supersonic adversary aircraft (SAA) modified from F-16C/D Block 30. Four of 26 were two-seat **TF-16N**. Entire fleet retired during 1994–95.

FS-X and TFS-X: F-16 design selected by Japan Defence Agency as basis for its FS-X (now F-2) requirement 19 October 1987; details under Mitsubishi in Japanese section.

F-16 Recce: Following trials with a prototype system in mid-1995, an LMTAS-designed reconnaissance pod for the US Air National Guard was flown for the first time on 29 September 1995 and subsequently tested on an F-16C (86-227) of the 149th Fighter Squadron, Virginia ANG. Four pods were built and then deployed with the 149th FS to Aviano AB, Italy, in May 1996 for operational validation in support of NATO missions over Bosnia. This successful trial culminated in decision to procure a total of 20 podded systems for service with F-16C Block 30 aircraft of five ANG squadrons, each of which will have four pods and one ground exploitation system. The core of the programme, known as the BAE Systems Theater Airborne Reconnaissance System (TARS), is the Per Udsen MRP with USAF-supplied KS-87 cameras incorporating E-O video back instead of wet film; the Lockheed Martin Fairchild Systems medium-altitude E-O (MAEO) camera; an Ampex DCRsi-240 digital data recorder and a TERMA Elektronik cockpit control device. The TARS pods were delivered in 1999 and certified for operational use by the F-16 in first half of 2000. Egypt ordered six TARS pods and two ground stations at beginning of 2003.

Elta EL/M-2060P pod also cleared for use by F-16 in 1999. System is contained in standard 300 US gallon drop tank and comprises autonomous, all-weather, day and night high-resolution reconnaissance synthetic aperture radar sensor with ability to transmit imagery to ground station via bidirectional datalink.

Most F-16s delivered since the early 1990s have provisions for a reconnaissance pod. Final 20 aircraft for South Korea (15 F-16C and 5 F-16D) will be configured for reconnaissance mission, although decision on system is not expected until first quarter of 2005; delivery of these aircraft began in mid-2003. Omani F-16s to use BAE Systems F-9120 Advanced Airborne Reconnaissance System (AARS) and will receive two sensor pods and a ground exploitation station at a cost of about US$27.5 million. Poland has selected the Goodrich DB-110 EO reconnaissance pod and is understood to require seven pod systems and two ground exploitation stations (one of which will be mobile) as well as associated data recording, datalink and mission control systems.

F-16ES: Enhanced Strategic two-seat, long-range interdictor F-16 proposal; now defunct but provided basis for Advanced Block 50/52 and Israeli F-16I.

F-16U: Proposed two-seat version unsuccessfully offered to United Arab Emirates.

F-16 'Falcon 2000': Private venture design of early 1990s, similar to F-16U, which Lockheed Martin aimed at USAF as a follow-on to the F-16 before introduction of JSF. Also known as the **F-16X**; now defunct.

CUSTOMERS: See tables. Total 4,417 production aircraft ordered or requested to date, including planned USAF procurement of 2,230 and 28 embargoed Pakistan Air Force examples that have been distributed equally between the USAF and US Navy. Total of 64 delivered in 2003, (including 10 from KAI in South Korea), leaving backlog of about 300 at start of 2004. Backlog had fallen to 265 as at 1 May 2004.

COSTS: Approximately US$40 million, flyaway, depending on configuration. Sale of 48 aircraft to Poland valued at US$3.5 billion. F-16E/F version unit cost reported to be in region of US$50–55 million.

DESIGN FEATURES: Conceived as 'lo' complement to Boeing F-15 Eagle in hi-lo fighter mix; optimised for high agility in air combat. Cropped delta wings blended with fuselage, with highly swept vortex control strakes along fuselage forebody and joining wings to increase lift and improve directional stability at high angles of attack; wing section NACA 64A-204; leading-edge sweepback 40°; relaxed stability (rearward CG) to increase manoeuvrability; deep wingroots increase rigidity, save 113 kg (250 lb) structure weight and increase fuel volume; fixed geometry engine intake; pilot's ejection seat inclined 30° rearwards; single-piece birdproof forward canopy section; two ventral fins below wing trailing-edge.

Baseline F-16 airframe life planned as 8,000 hours with average usage of 55.5 per cent in air combat training, 20 per cent ground attack and 24.5 per cent general flying; structural strengthening programme for pre-Block 50 aircraft required during 1990s.

FLYING CONTROLS: Four-channel digital fly-by-wire (analogue in earlier variants); pitch/lateral control by pivoting monobloc tailerons and wing-mounted flaperons; maximum rate of flaperon movement 52°/s; automatic wing leading-edge manoeuvring flaps programmed for Mach number and angle of attack; flaperons and tailerons interchangeable left and right; sidestick control column with force feel replacing almost all stick movement.

STRUCTURE: Wing, mainly of light alloy, has 11 spars, five ribs and single-piece upper and lower skins; attached to fuselage by machined aluminium fittings; leading-edge flaps are one-piece bonded aluminium honeycomb and driven by rotary actuators; fin is multispar, multirib with graphite epoxy skins; brake parachute or ECM housed in fairing aft of fin root; tailerons have graphite epoxy laminate skins, attached to corrugated aluminium pivot shaft and removable full-depth aluminium honeycomb leading-edge; ventral fins have aluminium honeycomb and skins; split speedbrakes in fuselage extensions inboard of tailerons open to 60°. Nose radome by Brunswick Corporation.

LANDING GEAR: Goodrich (formerly Menasco) hydraulically retractable type, nose unit retracting rearward and main units forward into fuselage. Nosewheel is located aft of intake to reduce the risk of foreign objects being thrown into the engine during ground operation, and rotates 90° during retraction to lie horizontally under engine air intake duct. Oleo-pneumatic struts in all units. Aircraft Braking Systems mainwheels and brakes; Goodyear or Goodrich tubeless mainwheel tyres, size 27.75×8.75–14.5 (or 27.75×8.75R14.5) (24 ply), pressure 14.48 to 15.17 bar (210 to 220 lb/sq in) at T-O weights less than 13,608 kg (30,000 lb). Steerable nosewheel with Goodyear, Goodrich or Dunlop tubeless tyre, size 18×5.7–8 (18 ply), pressure 20.68 to 21.37 bar (300 to 310 lb/sq in) at T-O weights less than 13,608 kg (30,000 lb). All but two main unit components interchangeable. Brake-by-wire system on main gear, with Aircraft Braking Systems anti-skid units. Runway arresting hook under rear fuselage; Irvin 7.01 m (23 ft 0 in) diameter braking parachute fitted in Greek and Turkish (Block 30/40 only) F-16s. Israeli (F-16C/D models only) and Singaporean (F-16Ds) aircraft have braking parachute compartment configured for electronic equipment. Landing/taxying lights on nose landing gear door.

POWER PLANT: One 131.6 kN (29,588 lb st) General Electric F110-GE-129, or one 129.4 kN (29,100 lb st) Pratt & Whitney F100-PW-229 afterburning turbofan as alternative standard. These Increased Performance Engines (IPE) installed from late 1991 in Block 50 and Block 52 aircraft and retrofitted to about 50 Air National Guard Block 42 aircraft. Pratt & Whitney has proposed F100-PW-229A version, with new fan module among other radical improvements that will raise airflow by more than 10 per cent, lower turbine temperatures by almost 50°C (122°F) and permit inspection intervals to rise from 4,300 cycles to 6,000. New version offers potential to increase maximum augmented thrust rating to about 142 kN (31,860 lb st), although this would require larger inlet on F-16. General Electric also engaged in improvement efforts, using company funding to begin development of F110-GE-129 EFE (Enhanced Fighter Engine) in October 1997; EFE initially to be rated at up to 151.0 kN (33,945 lb st), with further growth potential to 160.0 kN (35,970 lb st); alternatively, improved thrust levels can be sacrificed for up to a 50 per cent increase in TBO and servicing intervals. Production derivative known as F110-GE-132 rated at 144.6 kN (32,500 lb st) installed in F-16E/F aircraft for UAE. Immediately prior standard was 128.9 kN (28,984 lb st) F110-GE-100 or 105.7 kN (23,770 lb st) F100-PW-220 in Blocks 40/42.

Of 1,446 F-16Cs and F-16Ds ordered by USAF, 556 with F100 and 890 with F110. Fixed geometry intake, with boundary layer splitter plate, beneath fuselage. Apart from first few, F110-powered aircraft have intake widened by 30 cm (1 ft 0 in) from 368th F-16C (86-0262); Israeli second-batch F-16D-30s have power plants locally modified by Bet-Shemesh Engines to F110-GE-110A with provision for up to 50 per cent emergency thrust at low level.

Standard fuel contained in wing and five seal-bonded fuselage cells which function as two tanks; 3,986 litres (1,053 US gallons; 876 Imp gallons) in single-seat aircraft; 3,297 litres (871 US gallons; 726 Imp gallons) in two-seat aircraft. Halon inerting system. In-flight refuelling receptacle in top of centre-fuselage, aft of cockpit. Auxiliary fuel can be carried in drop tanks: one 1,136 litre (300 US gallon; 250 Imp gallon) under fuselage; 1,402 litre (370 US gallon; 308 Imp gallon) under each wing. Optional Israel Military Industries 2,271 litre (600 US gallon; 500 Imp gallon) underwing tanks initially adopted only by Israel, but have since been selected by one or two other operators; also adopted for F-16E/F versions. Latter have conformal fuel tanks (CFTs) with a combined capacity of 1,703 litres (450 US gallons; 375 Imp gallons). CFTs available as optional equipment on Advanced Block 50/52.

ACCOMMODATION: Pilot only in F-16C, in pressurised and air conditioned cockpit. Boeing (formerly McDonnell Douglas) ACES II zero/zero ejection seat. Bubble canopy made of polycarbonate advanced plastics material. Inside of USAF F-16C/D canopy coated with gold film to dissipate radar energy. In conjunction with radar-absorbing materials in air intake, this reduces frontal radar signature by 40 per cent. Windscreen and forward canopy are an integral unit without a forward bow frame, and are separated from the aft canopy by a simple support structure which serves also as the breakpoint where the forward section pivots upward and aft to give access to the cockpit. A redundant safety lock feature prevents canopy loss. Windscreen/canopy design provides 360° all-round view, 195° fore and aft, 40° down over the side, and 15° down over the nose.

To enable the pilot to sustain high *g* forces, and for pilot comfort, the seat is inclined 30° aft and the heel line is raised. In normal operation the canopy is pivoted upward and aft by electrical power; the pilot is also able to unlatch the canopy manually and open it with a back-up handcrank. Emergency jettison is provided by explosive unlatching devices and two rockets. A limited displacement, force-sensing control stick is provided on the right-hand console, with a suitable armrest, to provide precise control inputs during combat manoeuvres.

The F-16D has two cockpits in tandem, equipped with all controls, displays, instruments, avionics and life support systems required to perform both training and combat missions. The layout of the F-16D second station is similar to the F-16C, and is fully systems-operational. A single-enclosure polycarbonate transparency, made in two pieces and spliced aft of the forward seat with a metal bow frame and lateral support member, provides outstanding view from both cockpits.

Advanced Block 50/52 F-16Ds, F-16Fs and F-16Is are configured with weapon system operator station in rear cockpit, plus a large dorsal equipment compartment extending from the rear of the canopy to the leading edge of the fin; compartment houses avionics unique to each operator, plus additional chaff/flare dispensers and an in-flight refuelling receptacle.

SYSTEMS: Regenerative 12 kW environmental control system, with digital electronic control, uses engine bleed air for pressurisation and cooling of crew station and avionics compartments. Two separate and independent hydraulic systems supply power for operation of the primary flight control surfaces and the utility functions. System pressure (each) 207 bar (3,000 lb/sq in), rated at 161 litres (42.5 US gallons; 35.4 Imp gallons)/min. Bootstrap-type reservoirs, rated at 5.79 bar (84 lb/sq in).

Electrical system powered by engine-driven 60 kVA main generator and 10 kVA standby generator (including ground annunciator panel for total electrical system fault reporting), with Hamilton Sundstrand constant speed drive and powered by a Hamilton Sundstrand accessory drive gearbox. 17 Ah battery. Four dedicated, sealed cell batteries provide transient electrical power protection for the fly-by-wire flight control system.

An onboard Hamilton Sundstrand/Solar jet fuel starter is provided for engine self-start capability. Simmonds fuel measuring system. AlliedSignal emergency power unit automatically drives a 5 kVA emergency generator and emergency pump to provide uninterrupted electrical and hydraulic power for control in the event of the engine or primary power systems becoming inoperative.

AVIONICS: *Comms:* Magnavox AN/ARC-164 UHF transceiver (AN/URC-126 Have Quick IIA in Block 50/52); provision for Magnavox KY-58 secure voice system; Rockwell Collins AN/ARC-186 VHF AM/FM transceiver, ARC-190 HF radio, government-furnished AN/AIC-18/25 intercom and SCI advanced interference blanker, Teledyne Electronics AN/APX-101 IFF transponder with government-furnished IFF control, government-furnished National Security Agency KIT-1A/TSEC cryptographic equipment. F-16C/D Block 52 aircraft of Singapore and South Korea have Litton AN/APX-109+ advanced interrogator/transponder. AN/APX-113 advanced interrogator/transponder installed in Greek Block 50, Taiwanese Block 20 and Turkish Block 50 aircraft and is standard equipment on USAF aircraft procured in FY00 and FY01; is being retrofitted to earlier USAF aircraft as part of CCIP upgrade.

Radar: Northrop Grumman AN/APG-68(V) pulse Doppler range and angle track radar, with mechanically scanned planar array in nose. Provides air-to-air modes for range-while-search, uplook search, velocity search with ranging, air combat, track-while-scan (10 targets), raid cluster resolution, single target track and pulse Doppler track to provide target illumination for AIM-7 missiles, plus air-to-surface modes for ground-mapping, Doppler beam-sharpening, ground moving target, sea target, fixed target track, target freeze after pop-up, beacon, and air-to-ground ranging. Improved AN/APG-68(V)9 radar installed on Advanced Block 50/52 aircraft and Northrop Grumman plans to offer an upgrade kit enabling existing radars to be brought to latest standard, which has synthetic aperture radar (SAR) mapping and terrain following (TF) modes, plus interleaving of all modes; if internal FLIR targeting system is selected, this could share processor with ABR. F-16E/F aircraft for UAE to have Northrop Grumman AN/APG-80 agile beam radar with active electronically scanned array (AESA).

Flight: Litton LN-39 standard inertial navigation system (ring laser Litton LN-93 or Honeywell H-423 in Block 50/52: LN-93 for Egypt, Indonesia, Israel, South Korea, Pakistan, Portugal and Taiwan, plus Netherlands retrofit and Greek second batch); Rockwell Collins AN/ARN-108 ILS, Rockwell Collins AN/ARN-118 Tacan, Rockwell Collins GPS, Honeywell central air data computer, Elbit Fort Worth enhanced stores management computer, Gould AN/APN-232 radar altimeter. Fairchild digital terrain system (incorporating BAE Systems Terprom algorithms) to be installed in all new USAF F-16s and USAF Reserve F-16C/Ds. Optional equipment includes Rockwell Collins VIR-130 VOR/ILS.

Instrumentation: Marconi wide-angle holographic electronic HUD with raster video capability (for LANTIRN) and integrated keyboard; data entry/cockpit interface and dedicated fault display by Litton Canada and Elbit Fort Worth; Astronautics cockpit/TV set. Cockpit lighting and external strip lighting compatible with night imaging systems.

Mission: Honeywell multifunction displays. Lockheed Martin LANTIRN package comprises AN/AAQ-13 (navigation) and AN/AAQ-14 (targeting) pods. Turkish aircraft (150+ modified by 1996) to share 60 LANTIRN pod systems; LANTIRN also purchased by Greece, South Korea and Singapore, although Singapore now seeking a replacement system. Sharpshooter pod (down-rated export version of AAQ-14 LANTIRN targeting system) acquired by Bahrain and Israel, but latter obtained indigenous Rafael Litening IR targeting and navigation pod as replacement. Total of 168 Litening II navigation/targeting pods ordered from Rafael and Northrop Grumman to equip Block 25/30/32/40/42 F-16C/Ds of the Air National Guard (136) and Air Force Reserve Command (32); programme called Precision Attack Targeting System, with first AFRC unit (457th FS at Fort Worth, Texas) receiving initial batch of four pods in February 2000; this and three more AFRC squadrons equipped by end of 2000. ANG accepted first pods during 2000; goal is to allocate eight pods to each 15-aircraft squadron. Litening ER (Extended Range) targeting and navigation pod system also being supplied to ANG, which received first eight (of 16) in fourth quarter of 2002; all delivered by end of year. Under CCIP modification programme, Block 40/42/50/52 aircraft of USAF adopting new Lockheed Martin Sniper XR advanced FLIR targeting pod system from late 2002 onwards; more than 500 pods could eventually be acquired, with total of 17 received by USAF by February 2004; deliveries of export equivalent to Norway have also begun. Export version known as Pantera (Precision Attack Navigation and Targeting with Extended Range Acquisition) also selected by Oman (seven pod systems) and Poland (22) by late 2003. Pantera one of several systems in contention to satisfy South Korea's requirement for new targeting pod system, which is expected to go to tender in about 2007.

Raytheon AN/ASQ-213 HARM Targeting System (HTS) pod introduced on Block 50D/52D aircraft and subsequently retrofitted to entire Block 50/52 fleet. Entered service 1994 and currently deployed by USAF units in USA, Japan and Germany.

Self-defence: Dalmo Victor AN/ALR-69 radar warning system replaced in USAF Block 50/52 by BAE Systems AN/ALR-56M advanced RWR, which also ordered for USAF Block 40/42 retrofit and (first export) Korean Block 52s. Korean aircraft retrofitted with the ITT Avionics/Northrop Grumman AN/ALQ-165 Airborne Self-Protection Jammer (ASPJ) from mid-2000. Provision for Northrop Grumman AN/ALQ-131 or Raytheon AN/ALQ-184 jamming pods. AN/ALQ-131 supplied to Bahrain and Egypt. Israeli Air Force F-16s extensively modified with locally designed and manufactured equipment, as well as optional US equipment to tailor them to the IAF defence role. This includes Elisra SPS 3000 self-protection jamming equipment in enlarged spines of F-16D-30s and Elta EL/L-8240 ECM in third batch of F-16C/Ds, replacing AN/ALQ-178(V)1 Rapport ECM in Israeli F-16As. Elisra Passive Airborne Warning System 2 (PAWS-2) IR-based missile defence system is being installed on F-16I following delivery to Israel. Chilean aircraft will have ITT Industries Advanced Integrated Defensive Electronic Warfare Suite (AIDEWS), incorporating radar warning and RF countermeasures. Northrop Grumman Falcon Edge integrated EW system to be installed on UAE F-16E/F.

BAE Systems AN/ALQ-178(V)3 Rapport III integral self-protection system in Turkish F-16C/Ds will almost certainly be replaced by improved AN/ALQ-178(V)5 system. In March 1993, Greece ordered Raytheon ASPIS (Advanced Self-Protection Integrated Suite) self-defence EW system, comprising Northrop Grumman AN/ALR-93 RWR, BAE Systems AN/ALE-47 chaff/flare dispensers and Raytheon AN/ALQ-187 I-DIAS jammer; enhanced version known as ASPIS II is being installed on Advanced Block 52 aircraft.

USAF Air National Guard procured Terma PIDS wing weapon pylon with additional chaff/flare dispensers.

BAE Systems AN/ALE-40(V)-4 chaff/flare dispensers (AN/ALE-47 in FY97 Block 50, FMS Block 20/50 since mid-1996 and for retrofit to Block 40/42 and 50/52 of USAF). Raytheon AN/ALE-50(V)2 towed decoy installed in AMRAAM missile pylons and adopted by USAF for all Block 40/42/50/52 aircraft was widely fielded in 1999; USAF to buy total of 961. USAF Air National Guard participating in wing weapon pylon upgrade programme (to be fielded in 2004) that adds MIL-STD-1760 interface to existing PIDS pylons. This programme also includes provisions for future incorporation of a passive missile approach warning system.

ARMAMENT: General Dynamics M61A1 20 mm multibarrel cannon in the port side wing/body fairing, equipped with a General Dynamics ammunition handling system and an enhanced envelope gunsight (part of the head-up display system) and 511 rounds of ammunition. There is a mounting for an air-to-air missile at each wingtip, one underfuselage centreline hardpoint, and six underwing hardpoints for additional stores. For manoeuvring flight at 5.5 *g* the underfuselage station is stressed for a load of up to 1,000 kg (2,200 lb), the two inboard underwing stations for 2,041 kg (4,500 lb) each, the two centre underwing stations for 1,587 kg (3,500 lb) each, the two outboard underwing stations for 318 kg (700 lb) each, and the two wingtip stations for 193 kg (425 lb) each. For manoeuvring flight at 9 *g* the underfuselage station is stressed for a load of up to 544 kg (1,200 lb), the two inboard underwing stations for 1,134 kg (2,500 lb) each, the two centre underwing stations for 907 kg (2,000 lb) each, the two outboard underwing stations for 204 kg (450 lb) each, and the two wingtip stations for 193 kg (425 lb) each. There are mounting provisions on each side of the inlet shoulder for the specific carriage of sensor pods (electro-optical, FLIR and so on); each of these stations is stressed for 408 kg (900 lb) at 5.5 *g*, and 250 kg (550 lb) at 9 *g*.

Typical stores loads can include two wingtip-mounted AIM-9L/M/P Sidewinders, with up to four more on the outer underwing stations; Rafael Python 3 on Israeli F-16s from early 1991 and Python 4 from mid-1997; centreline GPU-5/A 30 mm cannon; drop tanks on the inboard underwing and underfuselage stations; HARM targeting system pod along the starboard side of the nacelle; and bombs, air-to-surface missiles or flare pods on the four inner underwing stations. Stores can be launched from Aircraft Hydro-Forming MAU-12C/A bomb ejector racks, Hughes LAU-88 launchers, Orgen triple or multiple ejector racks and Lucas Aerospace Flight Structures Twin Store Carrier (TSC). New BRU-57 dual-station 'smart' bomb racks fitted to Block 50/52 aircraft of USAF from fourth quarter of 2001; installed at mid-span hardpoint, each BRU-57 will be able to carry munitions such as JSOW, JDAM and WCMD; in April 2004, USAF certified Block 40/50 F-16s to carry AGM-158 Joint Air-to-Surface Standoff Missile (JASSM) and WCMD on BRU-57.

Weapons launched successfully from F-16s, in addition to AIM-9 Sidewinder and AIM-120A AMRAAM, include radar-guided AIM-7 Sparrow, Rafael Derby and Sky Flash BVR air-to-air missiles, AIM-132 ASRAAM and Magic 2 IR homing air-to-air missiles, AGM-65A/B/D/G Maverick air-to-surface missiles, AGM-88 HARM and AGM-45 Shrike anti-radiation missiles, AGM-84 Harpoon anti-ship missiles (clearance trials 1993–94) and, in Royal Norwegian Air Force service, the Penguin Mk 3 anti-ship missile. Israeli F-16s will soon have capability to operate with Rafael Spice PGM, which marries guidance kit (including control surfaces, EO sensor, GPS SatNav and datalink) to a 900 kg (2,000 lb) bomb; up to four weapons can be carried by F-16. LGBs include GBU-10, GBU-12, GBU-22, GBU-24 and GBU-27; F-16 can also deliver GBU-15 glide bomb, which used in conjunction with datalink pod. Israeli IMI STAR-1 anti-radiation weapon has also begun carriage trials on F-16D, although full-scale development is dependent upon receipt of a firm order; IMI runway attack munition (RAM) introduced into IDF/AF service in about 2000. CMS Defense Systems Autonomous Free-flight Dispenser System (AFDS) was tested at Eglin AFB, Florida, during 1992–93 and can be loaded with a variety of submunitions, including cratering bombs, shaped charge bomblets, anti-tank mines, area denial submunitions and general purpose bomblets.

Newest capability, introduced on Block 50/52 aircraft of USAF, incorporates 50T5 software upgrade, allowing F-16 to carry and deliver latest family of precision munitions; release to service occurred in mid-2000. New weapons comprise GBU-31 Joint Direct Attack Munition (JDAM), AGM-154 Joint StandOff Weapon (JSOW) and CBU-103, CBU-104 and CBU-105 wind-corrected munitions dispensers (WCMDs). First operational unit with JSOW and WCMD was 20th FW at Shaw AFB, South Carolina.

Data applicable to Block 50/52 versions:

DIMENSIONS, EXTERNAL (F-16C, D):
Wing span: over missile launchers ... 9.45 m (31 ft 0 in)
over missiles ... 10.00 m (32 ft 9¾ in)
Wing aspect ratio ... 3.2
Length overall ... 15.03 m (49 ft 4 in)
Height overall ... 5.09 m (16 ft 8½ in)
Tailplane span ... 5.58 m (18 ft 3¾ in)
Wheel track ... 2.36 m (7 ft 9 in)
Wheelbase ... 4.00 m (13 ft 1½ in)

AREAS (F-16C, D):
Wings, gross ... 27.87 m^2 (300.0 sq ft)
Flaperons (total) ... 2.91 m^2 (31.32 sq ft)
Leading-edge flaps (total) ... 3.41 m^2 (36.72 sq ft)
Fin, incl dorsal fin ... 4.00 m^2 (43.10 sq ft)
Rudder ... 1.08 m^2 (11.65 sq ft)
Horizontal tail surfaces (total) ... 5.92 m^2 (63.70 sq ft)

WEIGHTS AND LOADINGS:
Weight empty
F-16C: F100-PW-229:
with CFTs ... 9,358 kg (20,631 lb)
without CFTs ... 8,910 kg (19,643 lb)
F110-GE-129:
with CFTs ... 9,466 kg (20,868 lb)
without CFTs ... 9,017 kg (19,880 lb)
F-16D: F100-PW-229
with CFTs ... 9,760 kg (21,517 lb)
without CFTs ... 9,312 kg (20,529 lb)
F110-GE-129:
with CFTs ... 9,867 kg (21,754 lb)
without CFTs ... 9,419 kg (20,766 lb)
Max internal fuel (JP-8): F-16C ... 3,228 kg (7,116 lb)
Max external fuel (JP-8), F-16C/D (with 300 US gallon centreline tank):
normal: with CFTs ... 4,569 kg (10,072 lb)
without CFTs ... 3,208 kg (7,072 lb)
optional: with CFTs ... 5,879 kg (12,962 lb)
without CFTs ... 4,519 kg (9,962 lb)
Max external load (full internal fuel):
F-16C: F100-PW-229:
with CFTs ... 9,635 kg (21,241 lb)
without CFTs ... 8,855 kg (19,522 lb)
F110-GE-129:
with CFTs ... 9,190 kg (20,260 lb)
without CFTs ... 8,742 kg (19,272 lb)
Typical combat weight (two AAMs, 50% fuel):
F-16C: F100-PW-229:
with CFTs ... 12,254 kg (27,015 lb)
without CFTs ... 11,125 kg (24,527 lb)
F110-GE-129:
with CFTs ... 12,367 kg (27,265 lb)
without CFTs ... 11,239 kg (24,777 lb)
Max T-O weight with two AAMs, no tanks:
F-16C: F100-PW-229:
with CFTs ... 14,548 kg (32,073 lb)
without CFTs ... 12,723 kg (28,050 lb)
F110-GE-129:
with CFTs ... 14,661 kg (32,323 lb)
without CFTs ... 12,852 kg (28,335 lb)
with full external load:
F-16C/D Block 50/52 ... 21,772 kg (48,000 lb)
Wing loading: at 12,927 kg (28,500 lb) AUW ... 463.8 kg/m^2 (95.00 lb/sq ft)
at 19,187 kg (42,300 lb) AUW ... 688.4 kg/m^2 (141.00 lb/sq ft)
at 21,772 kg (48,000 lb) AUW ... 781.2 kg/m^2 (160.00 lb/sq ft)
Thrust/weight ratio (clean, without CFTs) ... 1.03 to 1

PERFORMANCE:
Max level speed at 12,200 m (40,000 ft) ... above M2.0
Service ceiling ... more than 15,240 m (50,000 ft)
Radius of action:
F-16C Block 50, with CFTs, two 907 kg (2,000 lb) bombs, two Sidewinders, 3,940 litres (1,040 US gallons; 867 Imp gallons) external fuel, tanks retained, hi-lo-lo-hi ... 735 n miles (1,361 km; 845 miles)
F-16C Block 50, with CFTs, armament as above, 5,542 litres (1,464 US gallons; 1,219 Imp gallons) external fuel, tanks retained, hi-lo-lo-hi ... 845 n miles (1,565 km; 972 miles)
F-16C Block 50, two BVR missiles, two Sidewinders, 3,940 litres (1,040 US gallons; 867 Imp gallons) external fuel, not including CFTs, tanks dropped when empty, combat air patrol mission ... 950 n miles (1,759 km; 1,093 miles)
Ferry range:
F-16C Block 50, with CFTs and 3,940 litres (1,040 US gallons; 867 Imp gallons) external fuel ... 2,150 n miles (3,981 km; 2,474 miles)
F-16C Block 50, with 5,542 litres (1,464 US gallons; 1,219 Imp gallons) external fuel, not including CFTs ... 2,415 n miles (4,472 km; 2,779 miles)
Symmetrical *g* limit with full internal fuel ... +9

F-16 TABLES: Table 1 provides a rapid reference to customers and quantities; more detailed information on F-16C/D production block numbers appears in Table 2; F-16E/F/I production in Table 3.

TABLE 1: F-16 CUSTOMERS

Operator	Total	Single-seat	Qty	Two-seat	Qty	Power plant	First aircraft A/C	First aircraft B/D	First delivery	Squadrons (or base)
Bahrain	22	F-16C-40	18	F-16D-40	4	F110-GE-100	101	150	March 1990	1, 2
Belgium	160[1]	F-16A-10	55	F-16B-10	12	F100-PW-200	FA01	FB01	January 1979	
		F-16A-15	41	F-16B-15	8	F100-PW-200	FA56	FB13	October 1982	1, 2, 23, 31, 349, 350
		F-16A-15OCU	40	F-16B-15OCU	4	F100-PW-220	FA97	FB21	January 1988	
Chile	10	F-16C-50 (Adv)	6	F-16D-50 (Adv)	4	F100-GE-129			2006	
Denmark	70	F-16A-10	30[1]	F-16B-10	8[1]	F100-PW-200	E-174	ET-204	January 1980	723, 727, 730
		F-16A-15	16[1]	F-16B-15	4[1]	F100-PW-200	E-596	ET-613	May 1982	723, 727, 730
		F-16A-15OCU	8[3]	F-16B-15OCU	4[3]	F100-PW-220	E-004	ET-197	December 1987	726
Egypt	220	F-16A-15	34	F-16B-15	8[2]	F100-PW-200	9301	9201	March 1982	72, 74
		F-16C-32	34	F-16D-32	6	F100-PW-220	9501	9401	August 1986	68, 70
		F-16C-40	34	F-16D-40	7	F110-GE-100	9901	9801	October 1991	60, 64
		F-16C-40	1	F-16D-40	5	F110-GE-100	9935	9808	1994	
		F-16C-40	34[7]	F-16D-40	12[7]	F110-GE-100	9951	9851	March 1994	75, 77
		F-16C-40	21			F110-GE-100B	(96-0086)	—	May 1999	71, 73
		F-16C-40	12	F-16D-40	12	F110-GE-100B	(99-0105)	(99-0117)	June 2001	
Greece	140	F-16C-30	34	F-16D-30	6	F110-GE-100	110	144	November 1988	330, 346
		F-16C-50	32	F-16D-50	8	F110-GE-129	045	077	January 1997	341, 347
		F-16C-52 (Adv)	40	F-16D-52 (Adv)	20	F110-PW-229	500	534	Late 2002	
Indonesia	12	F-16A-15OCU	8	F-16B-15OCU	4	F100-PW-220	S-1605	S-1601	December 1989	3
Israel	312	F-16A-10	67	F-16B-10	8	F100-PW-200	100	001	January 1980	140, 147, 253
		F-16C-30	51	F-16D-30	24	F110-GE-100A	301	020	December 1986	101, 105, 109, 110, 117
		F-16C-40	30	F-16D-40	30	F110-GE-100A	502	601	July 1991	101, 105, 109, 110, 117
				F-16I	102	F100-PW-229		401	February 2004	253
Korea, South	180	F-16C-32	30	F-16D-32	10	F100-PW-220	85-574	84-370	March 1986	161, 162, 155
		F-16C-52	95[8]	F-16D-52	45[8]	F100-PW-229	92-000	92-028	December 1994	
Lockheed Martin	12[11]	F-16C-52	4	F-16D-52	8	F100-PW-229	96-5033	96-5025	June 1998	Singapore (lease in USA)
Netherlands	213[3]	F-16A-10	46	F-16B-10	13	F100-PW-200	J-212	J-259	June 1979	306, 311, 312, 313,
		F-16A-15	84	F-16B-15	18	F100-PW-200	J-258	J-649	May 1982	315, 322, 323
		F-16A-15OCU	47	F-16B-15OCU	5	F100-PW-220	J-141	J-065	1988	
Norway	74	F-16A-10	28[3]	F-16B-10	7[3]	F100-PW-200	272	301	January 1980	332, 338
		F-16A-15	32[3]	F-16B-15	5[3]	F100-PW-200	300	690	June 1982	331, 334
				F-16B-15OCU	2	F100-PW-220	—	711	July 1989	331
Oman	12	F-16C-50 (Adv)	8	F-16D-50 (Adv)	4	TBD			2005	
Pakistan	68	F-16A-15	28	F-16B-15	12	F100-PW-200	82-701	82-601	January 1983	9, 11, 14
		F-16A-15OCU	13	F-16B-15OCU	15	F100-PW-220	91-729	91-613	—	See note 10
Portugal	20	F-16A-15OCU	17	F-16B-15OCU	3	F100-PW-220E	15101	15118	February 1994	201
Poland	48	F-16C-52 (Adv)	36	F-16D-52 (Adv)	12	F100-PW-229			2006	
Singapore	58	F-16A-15OCU	4	F-16B-15OCU	4	F100-PW-220	880	884	February 1988	140
		F-16C-52	18	F-16D-52	12	F100-PW-229	608	624	January 1998	143
				F-16D-52 (Adv)	20	F100-PW-229	—	661	Early 2004	
Taiwan	150	F-16A-20	120	F-16B-20	30	F100-PW-220	6601	6801	July 1996	12, 14, 17, 21, 22, 23, 26, 27
Thailand	36	F-16A-15OCU	14	F-16B-15OCU	4	F100-PW-220	10305	10301	June 1988	103
		F-16A-15OCU	12	F-16B-15OCU	6	F100-PW-220	07020	07032	September 1995	403
Turkey	240[4]	F-16C-30	34	F-16D-30	9	F110-GE-100	86-0066	86-0191	May 1987	142, Oncel Flight
		F-16C-40	102	F-16D-40	15	F110-GE-100	88-0033	88-0014	July 1990	141, 161, 162, 191, 192, 181, 182
		F-16C-50	60	F-16D-50	20	F110-GE-129	93-0657	93-0691	July 1996	141, 151, 152
UAE	80	F-16E	55	F-16F	25	F110-GE-132	00-6001	3001	2004	
USAF	2,230[9]	F-16A-10	255	F-16B-10	74	F100-PW-200	78-0001	78-0077	August 1978	See note A
		F-16A-15	409[5]	F-16B-15	46[6]	F100-PW-200	80-0541	80-0635	September 1981	See note A
		F-16C-25	209	F-16D-25	35	F100-PW-200	83-1118	83-1174	July 1984	See note B
		F-16C-30/32	360/56	F-16D-30/32	48/5	both (100/220)	85-1398	85-1509	July 1986	See note B
		F-16C-40/42	234/150	F-16D-40/42	31/47	both (100/220)	87-0350	87-0391	December 1988	See note B
		F-16C-50/52	189/42	F-16D-50/52	28/12	both (129/229)	90-0801	90-0834	October 1991	See note B

Operator	Total	Single-seat	Qty	Two-seat	Qty	Power plant	First aircraft		First delivery	Squadrons (or base)
							A/C	B/D		
US Navy	26	F-16N-30	22	TF-16N-30	4	F110-GE-100	163268	163278	June 1987	withdrawn
Venezuela	24	F-16A-15	18	F-16B-15	6	F100-PW-200	1041	1715	September 1983	161, 162
Totals	**4,417**		**3,477**		**940**					

[1] Built by Sabca (Belgium); 222nd and last Sabca F-16 (BAF FA-136) delivered 22 October 1991
[2] One built by Fokker (Netherlands)
[3] Built by Fokker; 300th and last Fokker F-16 (RNethAF J-021) delivered 27 February 1992
[4] Two F-16Cs and six F-16Ds built by GD; remainder by TAI
[5] Two built by Fokker
[6] Four built by Sabca
[7] TAI production
[8] 12 built by GD, 36 CKD kits and 92 produced locally by Korea Aerospace Industries (formerly Samsung Aerospace)
[9] Deliveries completed December 2002, except for last Block 50 which will be handed over in December 2004 with full CCIP modifications. Table excludes full-scale development (FSD) aircraft (six F-16A and two F-16B)
[10] 28 aircraft embargoed, placed in storage at Davis-Monthan AFB, Arizona and are being distributed between US Navy Strike and Air Warfare Center (10 F-16A and four F-16B) as adversary aircraft for tactics training and US Air Force (three F-16A and 11 F-16B) for test support duties
[11] New production aircraft leased to Singapore for training in the USA

Notes: **Note A:** Currently operated by US Air Force Flight Test Center/412th Test Wing, Edwards AFB, California; US Air Force Air Armament Center/46th Test Wing, Eglin AFB, Florida; 56th Fighter Wing (FW), Luke AFB, Arizona (Taiwan AF training squadron); Air National Guard.

Note B: Currently operated by US Air Force Flight Test Center/412th Test Wing, Edwards AFB, California; US Air Force Air Armament Center/46th Test Wing, Eglin AFB, Florida; 8th FW, Kunsan AB, South Korea; 20th FW, Shaw AFB, South Carolina; 27th FW, Cannon AFB, New Mexico; 31st FW, Aviano AB, Italy; 35th FW, Misawa AB, Japan; 51st FW, Osan AB, South Korea; 52nd FW, Spangdahlem AB, Germany; 53rd W, Eglin AFB, Florida; 56th FW, Luke AFB, Arizona; 57th W, Nellis AFB, Nevada; 354th FW, Eielson AFB, Alaska; 366th W, Mountain Home AFB, Idaho; 388th FW, Hill AFB, Utah; and also five fighter squadrons of the Air Force Reserve Command and more than 25 Air National Guard fighter squadrons.

Blocks 1 and 5 retrofitted to Block 10 standard 1982–84; Block 15 retrofitted to Block 15OCU (avionics standard only) from 1987. New-build F-16As are Block 15OCU from November 1987.

Export programme codenames are: Bahrain — Peace Crown I-II; Chile — Peace Puma; Egypt — Peace Vector I, II, III, IIIA, IV, V and VI; Greece — Peace Xenia I-III; Indonesia — Peace Bima-Sena; Israel — Peace Marble I-V; Italy — Peace Caesar; Jordan — Peace Falcon I-II; South Korea — Peace Bridge I-II and Korean Fighter Program I-II; Oman — Peace A'sama A'safiya; Pakistan — Peace Gate I-IV; Poland — Peace Sky; Portugal — Peace Atlantis I-II; Singapore — Peace Carvin I-IV; Taiwan — Peace Fenghuang; Thailand — Peace Naresuan I-III; Turkey — Peace Onyx I-II (and CAE-Link simulator, Peace Onyx III); and Venezuela — Peace Delta.

Recent orders: Greek option for 10 aircraft (six F-16C and four F-16D) converted to firm order in 2001; Israel also converted option in 2001, when further 52 F-16I added to outstanding order. USAF resumed procurement of F-16C with batch of six aircraft in FY96 plus six in FY97, three in FY98, one in FY99, 10 in FY00 and four in FY01; further orders unlikely. Chile placed order for 10 aircraft (six F-16C and four F-16D) in March 2002; Oman also placed order in 2002, for 12 aircraft (eight F-16C and four F-16D). Poland selected F-16 at beginning of 2003 and will receive 48 aircraft (36 F-16C and 12 F-16D) in 2006–08.

Transfers: Israel received 36 F-16As and 14 F-16Bs from surplus USAF stocks August 1994 to March 1995 for 140, 144 and 253 Squadrons. **Denmark** received three ex-USAF F-16As in July 1994, plus three F-16As and one F-16B in 1997. **Jordan** accepted initial batch of ex-USAF ADF aircraft (12 F-16As and four F-16Bs) on lease basis (subsequently permanent transfer) in December 1997/April 1998 and took delivery of eight more ex-USAF aircraft (seven F-16As and one F-16B) in January and February 2003, with further nine F-16As due for delivery by end of 2003. **Portugal** received 25 second-hand aircraft (21 F-16As and four F-16Bs, of which five F-16As to be broken-down for spares, with the rest receiving F-16 Mid-Life Update improvements); first upgraded aircraft rolled out at Alverca 26 June 2003; **Thailand** request for surplus USAF aircraft approved by US Congress at beginning of 2000, with delivery of 15 F-16As and one F-16B drawn from storage beginning in August 2002; **Italy** receiving 30 F-16A and four F-16B ADF aircraft in 2003–04 on five-year lease arrangement, with five-year follow-on option; first aircraft rolled out at Hill AFB, Utah on 9 May 2003, with initial batch of five aircraft ferried to Trapani 17 July 2003; deliveries due for completion in November 2004 (deal includes extra four aircraft for use as spares source).

TABLE 2: F-16C/D PRODUCTION

Block	USAF		USN		Bahrain		Egypt		Greece		Israel		Korea		Singapore		Turkey		Lockheed Martin		Chile		Oman		Poland	
	F-16C	F-16D	F-16N	TF-16N	C	D	C	D	C	D	C	D	C	D	C	D	C	D	C	D	C	D	C	D	C	D
25	7	4																								
25A	16	3																								
25B	25	5																								
25C	35	5																								
25D	40	4																								
25E	47	4																								
25F	39	10																								
Subtotal	**209**	**35**																								
30	16	2									4	—														
32							1	4					2	6												
30A	40	2									15	1														
32A							13	2					4	—												
30B	47	3	4	—							15	—					17	6								
32B							20	—					4	—												
30C	35	2	8	—							16	—														
32C	20	3											4	—												
30D	42	4	8	—							1	4														
32D	13	—											4	—												
30E	55	5	2	4							—	10					17	3								
32E													4	—												
30F	51	6									—	9														
32F	2	1											4	—												
30H	34	12							2	4																
32H	11	1											4	—												
30J	25	8							12	2																
32J	10	—																								
30K	15	3							16	—																
30L									4	—																
32Q													—	4												
30VISTA	—	1																								
Subtotal	**625**	**88**	**22**	**4**			**34**	**6**	**34**	**6**	**51**	**24**	**30**	**10**			**34**	**9**								
40	2	—																								
40A	6	3															17	3								
42A	5	3																								
40B	22	—																								
42B	7	13																								

Block	USAF		USN		Bahrain		Egypt		Greece		Israel		Korea		Singapore		Turkey		Lockheed Martin		Chile		Oman		Poland	
	F-16C	F-16D	F-16N	TF-16N	C	D	C	D	C	D	C	D	C	D	C	D	C	D	C	D	C	D	C	D	C	D
40C	38	3																								
42C	17	2																								
40D	38	2			2	4											17	3								
42D	17	3																								
40E	37	—			6	—																				
42E	18	5																								
40F	33	4															17	3								
42F	15	8																								
40G	31	5					2	—																		
42G	21	3																								
40H	14	5					—	4			6	4														
42H	27	6																								
40J	7	5					9	2			8	8					16	3								
42J	21	4																								
40K	6	4					15	1			8	8														
42K	2	—																								
40L							8				8	10					17	3								
40M							1	1																		
40N							—	8																		
40P																	18	—								
40Q							3	3																		
40R							12	5																		
40S							19																			
40T							5																			
40U							8																			
40V							8																			
40W					10																					
40							12	12																		
Subtotal	**1,009**	**166**	**22**	**4**	**18**	**4**	**136**	**42**			**81**	**54**					**136**	**24**								
50	4	—																								
50A	7	7																								
52A	1	1																								
50B	24	8																								
50C	21	4																								
50D	133	9							32	8							60	20								
52D	41	11											95	45	18	12			4	8						
50 (Adv)																					6	4	8	4		
52 (Adv)									40	20						20									36	12
Totals	**1,240**	**206**	**22**	**4**	**18**	**4**	**136**	**42**	**106**	**34**	**81**	**54**	**125**	**55**	**18**	**32**	**196**	**44**	**4**	**8**	**6**	**4**	**8**	**4**	**36**	**12**
Grand total												**2,499**														

UPDATED

TABLE 3: OTHER VERSIONS PRODUCTION

Model	Israel	UAE	Total
F-16E		55	55
F-16F		25	25
F-16I	102		102
Total	**102**	**80**	**182**

LUSCOMBE

LUSCOMBE AIRCRAFT CORPORATION

5333 North Main Street, Altus, Oklahoma 73521
Tel: (+1 580) 477 33 55
Fax: (+1 580) 477 33 68
e-mail: sales@luscombeaircraft.com
Web: http://www.luscombeaircraft.com
FOUNDER AND CEO: Bill McKown
PRESIDENT: John Daniel
SENIOR VICE-PRESIDENT: Charles Gibson Jr
SALES MANAGER: Miles Hoover

Original Luscombe company formed in 1934, specialising in all-metal private light aircraft; ownership passed to Temco in 1949, Silvaire Aircraft in 1955, and to Luscombe Aircraft Corporation on termination of aircraft production in 1960. Type certificate of Model 11 transferred to Classic Air of Lansing, Michigan; relaunch of modified version undertaken from 1998 in newly built 11,150 m² (120,000 sq ft) factory at Altus.

A second design of the original Luscombe company, the Model 8, is marketed in the US by Renaissance.

UPDATED

LUSCOMBE 11E

TYPE: Four-seat lightplane.

PROGRAMME: Model 11A Sedan first flew (NX74202) 11 September 1946; certified 4 October 1948; total of 198 built with tailwheel and 123 kW (165 hp) Continental E-165 engine; production ended in 1949; 41 remained registered in US during 2002. Planned 1970 relaunch as Alpha Aviation Alpha IID failed to materialise.

Considerably redesigned Luscombe 185-11E being produced at Altus, Oklahoma, following conversion of Sedan N1674B as proof-of-concept followed by construction of preproduction prototype (N747BM, c/n 00995; later NX11E) for certification. Initially marketed under name of Spartan, which had been dropped by 2002. First production aircraft/demonstrator N11XA.

Structural test fuselage completed December 1997; production of new aircraft due to have started April 2000, but this slipped to early 2004. Certification (Certificate No. A804) for sales and marketing purposes awarded November 1998, with full type certification (Certificate No. A804) received 17 December 2002.

Changes from original include engine, 24 V electrical system, nosewheel, composites engine cowling, modified windscreen, electric (formerly hydraulic) flaps, electric (formerly manual) elevator trim and three-point seatbelts.

CUSTOMERS: Two (including prototype) registered by October 2001. Announced customers include the US Civil Air Patrol, which has a requirement for eight.

COSTS: US$155,900 basic (2003).

DESIGN FEATURES: High-wing braced monoplane; original 1946 design modified for nosewheel and 0.30 m (1 ft 0 in) increase in constant-chord inner wing, with sweptforward outer trailing-edge and Hoemer-style wingtips; tapered tailplane; slightly sweptback fin with ventral fillet. Wing section NACA 43012A.

FLYING CONTROLS: Conventional and manual. Stainless steel control cables. All control surfaces aluminium skinned with external stiffeners. Horn-balanced elevators with flight-adjustable, electric trim tab on port side; ground-adjustable tab on starboard aileron. Control movements: ailerons +15/−9°, elevators +29/−13° and rudder ±15°. Three-position electric flaps; maximum deflection 25°.

STRUCTURE: Steel tube semi-monocoque corrosion-proofed fuselage and strut-braced two-spar wings, all covered with aluminium skins, except composites engine cowling. Dorsal tail fillet.

LANDING GEAR: Alloy spring leaf main legs with 6.00-6 tyres, pressure 2.4 bar (35 lb/sq in). Air/oil shock-absorbing strut nose gear with 5.00-5 tyre, pressure 1.7 bar (25 lb/sq in), steerable ±10°. Hydraulic disc brakes and parking brake on mainwheels; turning radius approximately 8.23 m (27 ft 0 in).

POWER PLANT: One 157 kW (210 hp) Teledyne Continental IO-360-ES4 six-cylinder fuel-injected piston engine derated to 138 kW (185 hp) and driving a McCaulley 1B235/EFC7667 two-blade fixed-pitch metal propeller with D4245 spinner. Fuel capacity 159 litres (42.0 US gallons; 35.0 Imp gallons) in two bladder wing tanks, of which 152 litres (40.0 US gallons; 33.3 Imp gallons) are usable. Filler port in each wing. Oil capacity

Production Luscombe 11E, N11XE *(Paul Jackson)* ***NEW***/0583485

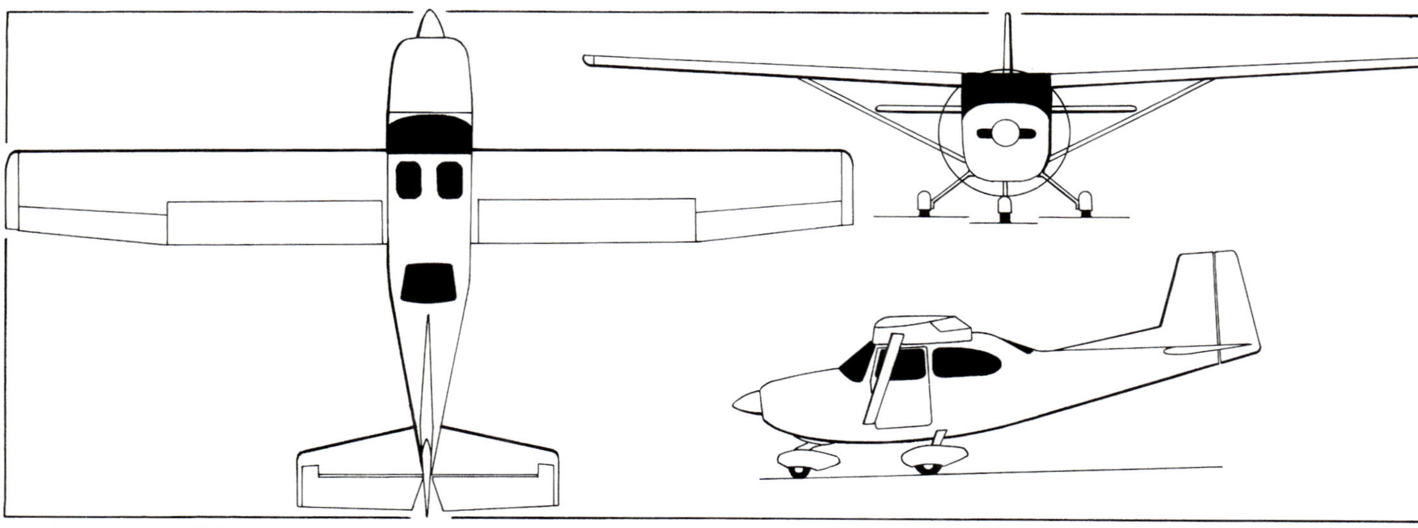

General arrangement of the Luscombe 185-11E *(Paul Jackson)* 0016820

7.5 litres (2.0 US gallons; 1.6 Imp gallons). Proof-of-concept aircraft has three-blade propeller, which is optional on production airframes.

ACCOMMODATION: Pilot and three passengers in side-by-side pairs; door each side; front seats fully adjustable; dual controls. Inertia reel harnesses for all seats. Cabin windows are tinted, as are two vision ports in cabin roof.

SYSTEMS: 28 V 70 A alternator; 24 V 12.75 Ah battery; dual vacuum pumps.

AVIONICS: *Comms:* Garmin GNC 250XL GPS/com, GTX 327 transponder and GMA 340 audio panel standard. GTX 330 transponder and Senheiser HMEC 300/BP-03 or HME 100 headsets optional.

Flight: Garmin GNS 430 or GNS 530 GPS/com/nav, GI 106A CDI and flight data recorder optional.

Instrumentation: Altimeter, compass, ASI, VSI, turn co-ordinator, ammeter, OAT gauge, vacuum gauge, fuel gauge, tachometer, EGT gauge and oil pressure gauge standard.

EQUIPMENT: Position lights, strobe lights and white paint with accent colour all standard. Options include leather seats, ergonomic seating package, Rosen sun visors, wheel fairings, metallic paint accent strip and polished propeller spinner.

DIMENSIONS, EXTERNAL:
- Wing span ... 11.73 m (38 ft 6 in)
- Wing aspect ratio ... 8.9
- Length overall ... 7.32 m (24 ft 0 in)
- Height overall ... 2.69 m (8 ft 10 in)
- Tailplane span ... 3.58 m (11 ft 9 in)
- Wheel track ... 2.64 m (8 ft 8 in)
- Wheelbase ... 2.24 m (7 ft 4 in)
- Passenger door: Max height ... 1.19 m (3 ft 11 in)
  - Max width ... 0.81 m (2 ft 8 in)

DIMENSIONS, INTERNAL:
- Cabin: Length ... 2.70 m (8 ft 10¼ in)
  - Max width ... 1.16 m (3 ft 9½ in)
  - Max height ... 1.30 m (4 ft 3¼ in)

AREAS:
- Wings, gross ... 15.51 m² (167.0 sq ft)

WEIGHTS AND LOADINGS:
- Weight empty ... 658 kg (1,450 lb)
- Baggage capacity ... 45 kg (100 lb)
- Max T-O and landing weight ... 1,034 kg (2,280 lb)
- Max ramp weight ... 1,037 kg (2,286 lb)
- Max wing loading ... 66.7 kg/m² (13.65 lb/sq ft)
- Max power loading ... 7.50 kg/kW (12.32 lb/hp)

PERFORMANCE:
- Never-exceed speed (VNE) ... 157 kt (290 km/h; 180 mph)
- Max level speed ... 130 kt (241 km/h; 150 mph)
- Cruising speed at FL75:
  - 70% power ... 117 kt (217 km/h; 135 mph)
  - 60% power ... 110 kt (204 km/h; 127 mph)
- T-O speed ... 60 kt (111 km/h; 69 mph)
- Stalling speed, power off:
  - flaps up ... 47 kt (87 km/h; 54 mph)
  - flaps down ... 43 kt (79 km/h; 49 mph)
- Max rate of climb at S/L ... 267 m (876 ft)/min
- Service ceiling ... 4,877 m (16,000 ft)
- T-O run ... 274 m (900 ft)
- T-O to 15 m (50 ft) ... 419 m (1,375 ft)
- Landing from 15 m (50 ft) ... 579 m (1,900 ft)
- Landing run ... 264 m (866 ft)
- Range with reserves:
  - at 70% power cruising speed ... 500 n miles (926 km; 575 miles)
  - at 60% power cruising speed ... 550 n miles (1,018 km; 632 miles)
- Endurance ... 5 h 42 min
- *g* limits ... +3.8/−1.52
- Noise level ... 72.4 dB(A)

UPDATED

MAULE

MAULE AIR INC

2099 GA Highway 133 South, Lake Maule, Moultrie, Georgia 31768
Tel: (+1 229) 985 20 45
Fax: (+1 229) 890 24 02
e-mail: sales@mauleairinc.com
Web: http://www.mauleairinc.com
CHAIRMAN: June D Maule
DIRECTOR: Ray Maule
VICE-PRESIDENT: David Maule
ENGINEER: David Wright
SALES MANAGER: Brent Maule

Original Maule Aircraft Corporation formed in Michigan by the late Belford D Maule to produce M-4, extending existing business of manufacturing aircraft accessories. Company transferred to Moultrie, Georgia, 1968; Maule Air Inc formed 1984 to produce uprated M-5 Lunar Rocket and M-7 Super Rocket; Lunar Rocket discontinued, but modified M-4 being prepared for return to production. Delivered 32 aircraft in 2003.

UPDATED

MAULE M-4

TYPE: Two-seat lightplane.

PROGRAMME: First post Second World War design by Belford D Maule. Four-seat M-4 Bee Dee first flew February 1957 and received FAA certification on 10 August 1961. Produced in −145, −180, −210 and −220 versions (indicating power of Continental or Franklin engines), being named Jetasen and Rocket in later versions; some 471 manufactured up to replacement by M-5 from 1975.

In preparation for Light Sport Airplane licensing category, prototype **M-4-100 Sport 100** (c/n 54001C/N5505C) first flew 18 July 2003 with 73.5 kW (98.6 hp) Rotax 912S flat-four and made public debut at AirVenture, Oshkosh, on 29 July. Changes from original include lower-powered engine and deletion of flaps. This version not pursued, however.

CURRENT VERSIONS: **M-4-180V:** Announced August 2004, month following first flight of initial aircraft (c/n 530C1, N799ZZ). Designation indicates 'Vintage'. Flaps reinstated; higher-powered engine. Available from 2005.

COSTS: US$95,999 fixed-pitch propeller; US$105,999 variable-pitch propeller (2005).

DESIGN FEATURES: High wing attached to upper longerons and braced to lower longerons by streamline V struts. Vortex generators along entire wing leading-edge. Curved fin and tailplane tips. Wire-braced tailplane and fin.

Wing section USA 35B (modified); dihedral 1°; incidence 30'.

FLYING CONTROLS: Conventional and manual. Cable actuation. Mass-balanced rudder; horn-balanced elevators. Flight-adjustable, metal trim tab in port elevator. Flap settings include −7° for high-speed cruise.

STRUCTURE: Welded steel tube fuselage and integral fin with fabric covering, except aluminium doors and composites engine cowling; all-metal wing (including ailerons and flaps) built on two spars, but with composites tips.

LANDING GEAR: Tailwheel type; fixed. Two faired side Vs with half-axles sprung under fuselage with compressed rubber shock-absorbers. Maule steerable tailwheel. Speed Mainwheels 6.00-6; tailwheel 3.80/2.50-4. Hydraulic mainwheel brakes. 'Half' speed fairings to rear of tyres only.

POWER PLANT: One 134 kW (180 hp) Textron Lycoming O-360 flat-four driving a two-blade fixed- or variable-pitch propeller.

ACCOMMODATION: Two seats, side by side. Baggage space to rear. Door each side.

AVIONICS: *Comms:* Garmin GTX 327 VHF and Narco Mode C transponder.

Flight: Garmin GNC 250XL GPS.

DIMENSIONS, EXTERNAL:
- Wing span: 9.40 m (30 ft 10 in)
- Wing chord, constant: 1.60 m (5 ft 3 in)
- Wing aspect ratio: 6.2
- Length overall: 6.71 m (22 ft 0 in)
- Height overall: 1.89 m (6 ft 2½ in)
- Tailplane span: 2.97 m (9 ft 8¾ in)
- Wheel track: 1.83 m (6 ft 0 in)
- Wheelbase: 4.82 m (15 ft 10 in)

DIMENSIONS, INTERNAL:
- Cabin max width: 1.08 m (3 ft 6½ in)

AREAS:
- Wings, gross: 14.17 m^2 (152.6 sq ft)

WEIGHTS AND LOADINGS:
- Useful load: 408 kg (900 lb)

DIMENSIONS, EXTERNAL:
- Wing span: 9.40 m (30 ft 10 in)
- Wing chord, constant: 1.60 m (5 ft 3 in)
- Wing aspect ratio: 6.2
- Length overall: 6.71 m (22 ft 0 in)
- Height overall: 1.89 m (6 ft 2½ in)
- Tailplane span: 2.97 m (9 ft 8¾ in)
- Wheel track: 1.83 m (6 ft 0 in)
- Wheelbase: 4.82 m (15 ft 10 in)

DIMENSIONS, INTERNAL:
- Cabin max width: 1.08 m (3 ft 6½ in)

AREAS:
- Wings, gross: 14.17 m^2 (152.6 sq ft)

PERFORMANCE(estimated):
- Max cruising speed: 138 kt (256 km/h; 159 mph)

UPDATED

MAULE M-7 AND M-9

TYPE: Light utility transport; light utility turboprop.

PROGRAMME: First certified 9 November 1983 as M-7-235 Super Rocket (prototype N5656A: based on M-6-235 but with extended cabin and additional windows). In May 2000, Maule received first of an initial batch of 10 SMA SR305 turbocharged diesel engines for installation in M-9s.

CURRENT VERSIONS: Produced with short (M-5), mid-length (M-6) or long (A Model) wing span and engine/landing gear options as detailed below; short option deleted on 1999 versions, while piston-engined and turboprop/nosewheel aircraft have only M-6 wing from that year, the turboprop/tailwheel variants employing A Model wings. Designation prefixes are T = tricycle and X = short span.

MX-7-160 Sportplane: Engineering designation 19. Certified 13 November 1992. Four-seater, 119 kW (160 hp) Textron Lycoming O-320-B2D engine, Sensenich two-blade fixed-pitch metal propeller, and four-position flaps. Prototype N9232G. Last delivery in 1999.

MXT-7-160 Comet: Engineering designation 17. Certified 13 November 1992. As MX-7-160, but tricycle landing gear; two seats standard, four seats optional. Previously known as Maule Trainer. Last deliveries in 1998 (five to Henderson State University).

MX-7-160C Sportplane: Engineering designation 34. Certified 12 March 1999. As M-7-160, but spring aluminium landing gear. Prototype only (N41510); no reported deliveries by June 2004.

MX-7-180: Engineering designation 11. Certified 18 December 1984. Last built in 1994.

MX-7-180A Sportplane: Engineering designation 20. Certified 3 June 1993. As MX-7-160, but 134 kW (180 hp) Textron Lycoming O-360-C4F engine. Prototype N9225G. Two delivered in 2002 and one in the first nine months of 2003.

MX-7-180AC Sportplane: Engineering designation 33. Certified 4 May 2000. As MX-7-180A, but spring aluminium landing gear. Three delivered in 2003, and one in the first six months of 2004.

MXT-7-180A Comet: Engineering designation 21. Certified 3 June 1993. As MX-7-180, but tricycle landing gear and four seats standard. Introduced and first production aircraft (N1002N) flown 1996; optimised for flight training schools. Total of 18 delivered in 1999, six in 2000, one in 2001 and three in 2002, one in 2003, and one in the first six months of 2004.

MX-7-180B Star Rocket: Engineering designation 22. Certified 12 July 1993. 134 kW (180 hp) Textron Lycoming O-360-C1F engine, Hartzell two-blade constant-speed metal propeller and five-position flaps. Two delivered in 2002, none in 2003, and one in the first six months of 2004.

MX-7-180C: Engineering designation 28. Certified 27 August 1996. As MX-7-180B, but with spring aluminium main landing gear. Two delivered in 1999, two in 2000, one in 2001, two in 2002, three in 2003, and one in the first six months of 2004.

MXT-7-180 Star Rocket: Engineering designation 14. Certified 9 November 1990. As MX-7-180B, but with tricycle landing gear. Previously known as Star Craft. One delivered in 1999, seven in 2000, and two in 2001; none delivered in 2002; one in 2003, and five in the first six months of 2004.

Maule M-4-180V prototype ***NEW***/1047438

Maule M-7-235 floatplane participating in a spot-landing competition *(Paul Jackson)* ***NEW***/0583483

MX-7-205C Rocket: Engineering designation 31. One 153 kW (205 hp) PZL-Franklin 6A-350-C1R engine, McCauley two-blade propeller. Prototype (N205MX) was being test flown by second quarter of 1999. Engine may be offered as an option on Textron Lycoming O-360-powered models. Not certified by 2004.

M-7-235: Engineering designation 4. Certified 9 November 1983. Last built in 1994.

M-7-235B Super Rocket: Engineering designation 23. Certified 14 October 1993. Five-seater; choice of 175 kW (235 hp) carburetted Textron Lycoming O-540-J1A5, low-compression Mogas approved O-540-B4B5 or fuel-injected IO-540-W1A5 engines; McCauley constant-speed propeller, five-position flaps; fuselage raised 7.6 cm (3 in) at trailing-edge of wing and baggage area moved aft 12.7 cm (5 in) to accommodate fifth seat; recommended for high gross weight short-field operation, and for floatplane operation. Eight delivered in 1999, seven in 2000, five in 2001, six in 2002, three in 2003, and one in the first six months of 2004.

M-7-235C Orion: Engineering designation 25. Certified 10 October 1995. As for M-7-235B but with spring aluminium main landing gear. Sixteen delivered in 1999, 18 in 2000, 14 in 2001, 15 in 2002, nine in 2003, and two in the first six months of 2004.

MT-7-235 Super Rocket: Engineering designation 18. Certified 20 March 1992. As M-7-235B, but tricycle landing gear, four-position flaps and IO-540-W1A5 engine only. Prototype N9226Y. Four delivered in 1999, five in 2000, 16 in 2001 and 12 in 2002, seven in 2003, and one in the first six months of 2004. Civil Air Patrol approved purchase of 15 in glider towing configuration mid-2000.

MX-7-235: Engineering designation 10. Certified 18 October 1984. Last built in 1995.

M-7-260 Super Rocket: Engineering designation 26. Certified 17 April 1998. As M-7-235B, but 194 kW (260 hp) Textron Lycoming IO-540-V4A5 engine; McCauley two-blade constant-speed propeller standard, Hartzell and MT propellers optional. Seven delivered in 1999, one in 2000, four in 2001, one in 2002, and one in 2003.

M-7-260C Orion: Engineering designation 30. Certified 19 October 1998. As M-7-235C, but 194 kW (260 hp) Textron Lycoming IO-540-V4A5 engine; McCauley two-blade constant-speed propeller standard, Hartzell and MT propellers optional. Eight delivered in 1999, seven each in 2000 and 2001, two in 2002, three in 2003, and two in the first six months of 2004.

MT-7-260 Super Rocket: Engineering designation 27. Certified 17 April 1998. As MT-7-235, but 194 kW (260 hp) Textron Lycoming IO-540-V4A5 engine; McCauley two-blade constant-speed propeller standard, Hartzell and MT propellers optional. Three delivered in 1999, two in 2000, four in 2001, and one in 2002.

Maule M-9-230 prototype *(Paul Jackson)* ***NEW***/0583512

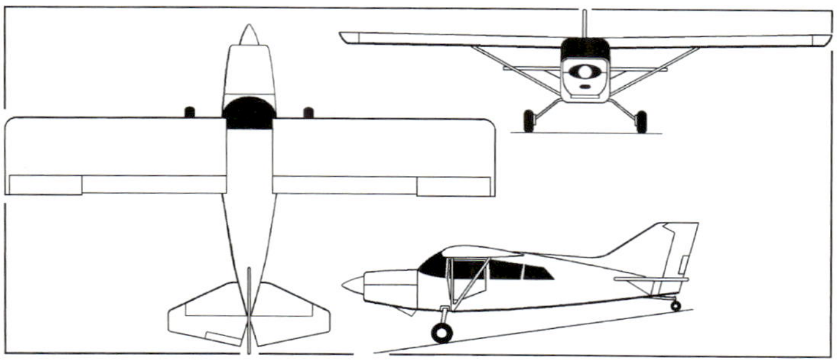

Typical Maule M-7 tailwheel configuration *(Paul Jackson)* ***NEW***/0567240

M-7-420: Engineering designation 12. Prototype N450X completed 1990. One other known.

M-7-420AC: Engineering designation 29. Certified 17 December 1998. M-7 fuselage with long-span wings, 313 kW (420 shp) Rolls-Royce 250-B17C turboprop, spring aluminium tailwheel landing gear. One delivered in 1999, none in 2000, one each in 2001 and 2002, and one in 2003.

MT-7-420: Engineering designation 51. Certified 6 January 2003. As MT-7-260 but with 313 kW (420 shp) Rolls-Royce 250-B17C turboprop. First aircraft N683CE. One delivered in the first six months of 2004.

MX-7-420: Engineering designation 13. Prototype N98MT completed 1986. Certified 1 June 1989. Last built in 1994.

M-9-230: Engineering designation 44. M-7 with 169 kW (227 hp) SMA SR305-230 Diesel engine and LoPresti cowling; max take-off weight 1,270 kg (2,800 lb). Fuel capacity 322 litres (85.0 US gallons; 70.8 Imp gallons); estimated range more than 869 n miles (1,609 km; 1,000 miles). Prototype (c/n 44001/N305SR) first flown on 18 July 2003; public debut at EAA AirVenture Oshkosh 29 July 2003 at which time it had completed 18 hours' flying time. Estimated cost US$200,000 (2003).

CUSTOMERS: Total of 990 produced by 31 March 2004, plus 908 of earlier M-5 and M-6 series. At 17 April 2001 Maule's order backlog stood at 100 aircraft, of which 10 were for SMA SR305-powered MX-9s.

COSTS: MX-7-160 US$106,575; MX-7-160C US$112,100; MX-7-160A US$112,520; MXT-7-160 US$116,320; MX-7-180AC US$117,900; MXT-7-180A US$122,200; MX-7-180B US$125,350; MX-7-180C US$130,730; MXT-7-180 US$136,620; M-7-235B (O-540) US$141,700 (IO-540) US$149,675; M-7-235C (O-540) US$147,990 (IO-540) US$156,260; M-7-260 US$160,780; MT-7-235 US$162,600; M-7-260C US$167,680; MT-7-260 US$173,420; M-7-420AC US$450,000; MT-7-420 US$470,000. All 2004, standard equipment.

DESIGN FEATURES: Rugged, STOL utility aircraft. High, constant chord, wing braced by V-struts; mid-mounted tailplane; large sweptback fin.

USA 35B (modified) wing section; dihedral 1°; incidence 0° 30'; cambered wingtips standard.

FLYING CONTROLS: Conventional and manual. Ailerons linked to rudder servo tab to reduce adverse yaw; trim tab in port elevator; rudder trim by spring to starboard rudder pedal; flap deflection 40° down for slow flight (further setting of 48° down on all except-420), 24° down, 0 and 7° up for improved cruise performance; underfin on floatplane and amphibious versions.

STRUCTURE: All-metal two-spar wing with dual struts and glass fibre tips; fuselage frame of welded 4130 steel tube with Ceconite covering aft of cabin and metal doors and skin forward of cabin; glass fibre engine cowling.

LANDING GEAR: Fixed; tailwheel or nosewheel type, with spring cantilever or oleo shock-absorbtion. Refer to table for current version application. Narrow-track versions have two faired-in side Vs hinged to lower longerons plus half-axles incorporating. Maule oleo-pneumatic shock-absorbers joined under centreline. Wide-track versions have fuselage mounted spring aluminium cfantilever mainwheel legs. Maule steerable tailwheel or cantilever, steerable nosewheel. Cleveland mainwheels with Goodyear or McCreary tyres size 7.00-6, pressure 1.79 bar (26 lb/sq in). Tailwheel tyre size 8×3.5-4, pressure 1.03 to 1.38 bar (15 to 20 lb/sq in). Cleveland hydraulic disc brakes. Parking brake. Oversize tyres, size 20×8.5-6 (pressure 1.24 bar; 18 lb/sq in) optional

Provisions for fitting optional Aqua 2200 and 2400, Baumann 2720 or Wipline 2350 and 3000 floats, or Baumann 2750A or Wipline 2350A and 3000A amphibious floats (also available on some tricycle models). Float option available for MX-7-160, MX-7-180A/B/C, M-7-235B/C, M-7-260 and M-7-420AC. Ski option available for M-7-235B.

POWER PLANT: One flat-four or flat-six engine as described under Current Versions, driving a Sensenich two-blade fixed-pitch or Hartzell two-blade constant-speed propeller (three-blade McCauley propeller optional on 175 kW; 235 hp and 194 kW; 260 hp models); or Rolls-Royce 250-B17C turboprop with three-blade, Hartzell HC-B3TF-7A/T10173F-21R constant-speed feathering and reversing propeller in -420 models. Piston versions have two fuel tanks in wings with total usable capacity of 163 litres (43.0 US gallons; 35.8 Imp gallons). Auxiliary fuel tanks in outer wings (standard on all 1999 models), to provide total capacity of 276 litres (73.0 US gallons; 60.8 Imp gallons). Turboprop versions have total fuel capacity of 322 litres (85.0 US gallons; 70.8 Imp gallons). Refuelling points on wing upper surface.

ACCOMMODATION: Four or five seats according to model, as described under Current Versions; individual, adjustable front seats; non-adjustable, rear seats. Dual controls and three-point shoulder harnesses standard. Baggage compartment, capacity 113 kg (250 lb), aft of seats; cargo capacity with passenger seats removed 349 kg (770 lb). One front-hinged door on port side; three doors on starboard side, forward and centre doors hinged at front edge, rear baggage door hinged at rear edge to form double cargo door providing an opening 1.30 m

(4 ft 3 in) wide to facilitate loading of bulky cargo; aircraft may be flown with doors removed. Accommodation heated and ventilated.

SYSTEMS: Hydraulic system for brakes only; electrical system powered by 60 A engine-driven alternator; 12 V battery (24 V battery on turboprops).

AVIONICS: Standard SP-3 VFR package comprises Garmin GNC 250XL GPS/com (GNC 420 upgrade optional), GTX 327 digital transponder, PS Engineering PM 1000II four-place intercom. IFR-4 package comprises GNS 430 GPS/nav/com, GNC 250XL GPS/com, GI-106A CDI with GS, GTX 327 digital transponder, PS Engineering PMA 7000B stereo audio panel. IFR-4KX package deletes GNC 250XL and adds Bendix/King KX-155-42 nav/com/GS and KI-209-01 VPR/GS indicator. IFR-5 package adds second GNS 430 to IFR-4 package. GNS 530 replacing one GNS 430 optional in IFR packages. Optional avionics upgrades include 28 V system for IFR packages, Century NSD 360A-15 or 360A-26 non-slaved/slaved HSIs replacing one GI-106A in IFR packages; S-TEC System 20, System 30 or System 50 autopilots and heading bug directional gyro; Insight Strikefinder SF-2000 Bose Headset X; Light Speed Technologies Twenty 3G ANR headset; and fifth-seat microphone and headset jack. All avionics packages include Narco AR-850 remote altitude encoder, Bendix/King KA 33-00 cooling blower (KA 33-01 optional), ELT, pilot and co-pilot PTT switches, radio master switch, and broadband antenna; IFR packages also include second broadband antenna, marker antenna and triplexer.

Instrumentation: Standard equipment includes full gyro panel/vacuum system comprising attitude and directional gyros, electric turn co-ordinator, VSI, OAT gauge and suction gauge; ASI; altimeter; compass; audio/visual stall warning indicator; CHT gauge; tachometer; electric fuel gauge, manifold pressure gauge (constant-speed propeller models only); fuel pressure gauge; oil pressure and temperature gauges; ammeter, and clock. Options include TAS indicator, carburettor air temperature gauge, hour meter, and KS Mix-Mizer EGT probe.

EQUIPMENT: Standard equipment Includes instrument and dome lights, auxiliary cabin heater, auxiliary fuel pump, auxiliary power plug, heated pitot tube, vinyl seats with velour inserts, vinyl cabin side panels, cloth velour cabin headliner, cabin soundproofing, cabin overhead speaker, window vents, cabin steps, cargo tiedowns, smoke-tinted windscreen, windscreen defroster, landing light in port wing, navigation lights, wingtip strobe lights, seven fuel drains, tiedown rings, airframe powder coating, wing corrosion proofing, and standard external paint scheme in Glacier White or Dune White base colour with single-colour trim. Options include dual-caliper brakes, dynamically balanced propeller, engine heater, leather seats and interior trim, four-point shoulder harnesses for pilot and co-pilot, Oregon Aero seat cushion system for pilot and co-pilot, two-position middle seat, colour co-ordinated instrument panel, landing light in starboard wing, Precise Flight pulselight, Micro Dynamics Inc vortex generators, dual tailplane struts (for glider towing), ground service plug, fuselage grab handles, Halon fire extinguisher, Schweizer or Tost glider tow/release kit, aircraft towbar, observation door, camera port, observation window, skylight, co-pilot's swing-out window, smoke-tinted cabin windows, and additional external trim colours.

DIMENSIONS, EXTERNAL:
Wing span: all except M-7-420AC ... 10.03 m (32 ft 11 in)
 M-7-420AC ... 10.26 m (33 ft 8 in)
Wing chord, constant: all ... 1.60 m (5 ft 3 in)
Wing aspect ratio: all versions except M-7-420AC ... 6.5
 M-7-420AC ... 6.7
Length overall:
 piston-engined versions ... 7.21 m (23 ft 8 in)
 M-7-420AC ... 7.32 m (24 ft 0 in)
 MT-7-420 ... 7.47 m (24 ft 6 in)
Height overall: tailwheel versions ... 1.93 m (6 ft 4 in)
 tricycle versions ... 2.54 m (8 ft 4 in)
 floatplane ... 3.05 m (10 ft 0 in)
 amphibian ... 3.20 m (10 ft 6 in)
Wheel track:
 MX-7-160, MX-7-180A, MX-7-180B, M-7-235B, M-7-260 ... 1.83 m (6 ft 0 in)
 MXT-7-160, MXT-7-180, MXT-7-180A, MT-7-235, MT-7-260, MT-7-420 ... 2.34 m (7 ft 8 in)
 MX-7-160, MX-7-180AC, MX-7-180C, M-7-235C, M-7-260C, M-7-420AC ... 2.39 m (7 ft 10 in)
Propeller diameter:
 MX-7-160, MX-7-160C, MXT-160 ... 1.88 m (6 ft 2 in)
 MX-7-180A, MX-7-180AC, MX-7-180C, MXT-7-180A, MX-7-180B, MXT-7-180 ... 1.93 m (6 ft 4 in)
 M-7-260, M-7-260C, MT-7-260 ... 1.98 m (6 ft 6 in)
 M-7-235B, M-7-235C, MT-7-235, M-7-260, M-7-260C, MT-7-260 three-blade:
 option 1 ... 1.98 m (6 ft 6 in)
 option 2 ... 2.03 m (6 ft 8 in)
 M-7-235B, M-7-235C, MT-7-235 ... 2.06 m (6 ft 9 in)
 M-7-420, MT-7-420 ... 2.03 m (6 ft 8 in)

DIMENSIONS, INTERNAL:
Cabin: max width ... 1.07 m (3 ft 6 in)

AREAS:
Wings, gross: all except M-7-420AC ... 15.38 m^2 (165.6 sq ft)
 M-7-420AC ... 15.72 m^2 (169.2 sq ft)

MAULE M-7 OPTIONS

Version	Power Plant							Gear		
	B2D	C4F	C1F	B4B5	W1A5	V4A5	B17	TW/S	TW/A	TR/A
MX-7-160	•							•		
MXT-7-160	•									•
MX-7-160C	•								•	
MX-7-180A		•						•		
MX-7-180AC		•							•	
MX-7-180B			•					•		
MX-7-180C			•						•	
MXT-7-180			•							•
MXT-7-180A		•								•
M-7-235B				•				•		
M-7-235C				•					•	
MT-7-235					•					•
M-7-260						•		•		
M-7-260C						•			•	
MT-7-260						•			•	
M-7-420AC							•		•	
MT-7-420							•			•

Notes: Power plants are Lycoming O-320-B2D, O-360-C4F, O-360-C1F, O-540-B4B5, IO-540-W1A5, IO-540-V4A5 (or earlier equivalents) and Rolls-Royce (Allison) 250-B17-C. Landing gears are tailwheel/oleo main gear, tailwheel/aluminium and tricycle/aluminium.

MAULE M-7 PRODUCTION

Version	Annual Deliveries																					Totals
	1983	1984	1985	1986	1987	1988	1989	1990	1991	1992	1993	1994	1995	1996	1997	1998	1999	2000	2001	2002	2003	
MX-7-160											15	16	9	3								**43**
MXT-7-160											1	1	1			5						**8**
MX-7-180			20	12	8	17	6	7	12	9	4	1										**96**
MX-7-180A											9	17	22	5	7	1						**61**
MX-7-180B												1	4	8		4				2	2	**19**
MX-7-180C														1	2	6	2	2	1	2	3	**19**
MXT-7-180								3	21	7	16	4	4	18	3	4	1	7	2	2	1	**93**
MXT-7-180A											1	1	3	6	22	24	18	6	1	3	1	**86**
M-7-235	2	17	13	11	16	15	13	9	13	3	13	7										**132**
M-7-235B												5	17	11	7	6	8	7	5	6	3	**75**
M-7-235C														7	11	5	16	17	14	15	9	**95**
MT-7-235										2	6	9	6	4	2	6	4	5	16	12	7	**79**
MX-7-180AC																		1			3	**4**
MX-7-160C																	1					**1**
MX-7-235		2	23	25	18	14	5	3	16	9	2		1									**118**
M-7-260																1	7	1	4	1	1	**15**
M-7-260C																1	8	9	7	2	3	**28**
MT-7-260																	3	1	4	1	1	**10**
M-7-420A																			1		1	**1**
M-7-420AC																			1	2	1	**4**
MX-7-420								2	1			1										**4**
Yearly totals	**2**	**19**	**56**	**48**	**42**	**46**	**24**	**24**	**63**	**30**	**67**	**63**	**67**	**63**	**54**	**63**	**69**	**56**	**57**	**46**	**32**	**991**

Notes: Versions not achieving series production have included M-7-235A, MX-7-205C, M-7-420, MT-7-420 and MXT-7-420

WEIGHTS AND LOADINGS:
Weight empty: MX-7-160 .... 621 kg (1,370 lb)
MX-7-160C .... 670 kg (1,477 lb)
MXT-7-160 .... 660 kg (1,456 lb)
MX-7-180A .... 636 kg (1,402 lb)
MX-7-180AC .... 654 kg (1,442 lb)
MXT-7-180A .... 683 kg (1,505 lb)
MX-7-180B .... 652 kg (1,438 lb)
MX-7-180C .... 673 kg (1,483 lb)
MXT-7-180 .... 693 kg (1,528 lb)
M-7-235B .... 728 kg (1,605 lb)
M-7-235C .... 750 kg (1,653 lb)
MT-7-235 .... 751 kg (1,655 lb)
M-7-260 .... 730 kg (1,610 lb)
M-7-260C .... 758 kg (1,671 lb)
MT-7-260 .... 769 kg (1,696 lb)
M-7-420AC .... 712 kg (1,570 lb)
MT-7-420 (estimated) .... 726 kg (1,600 lb)
Max T-O weight:
MX-7-160, MX-7-160C, MXT-7-160 .... 997 kg (2,200 lb)
MX-7-180A, MX-7-180AC, MXT-7-180A .... 1,088 kg (2,400 lb)
MX-7-180B, MX-7-180C, MXT-7-180, M-7-235B, M-7-235C, MT-7-235, M-7-260, M-7-260C, MT-7-260, M-7-420AC, MT-7-420 .... 1,134 kg (2,500 lb)
Max wing loading:
MX-7-160, MX-7-160C, MXT-7-160 .... 64.86 kg/m² (13.28 lb/sq ft)
MX-7-180A, MX-7-180AC, MXT-7-180A .... 76.76 kg/m² (14.49 lb/sq ft)
MX-7-180B, MX-7-180C, MXT-7-180, M-7-235B, M-7-235C, MT-7-235, M-7-260, M-7-260C, MT-7-260, MT-7-420 .... 73.71 kg/m² (15.10 lb/sq ft)
M-7-420AC .... 72.14 kg/m² (14.78 lb/sq ft)
Max power loading:
MX-7-160, MX-7-160C, MXT-7-160 .... 8.37 kg/kW (13.75 lb/hp)
MX-7-180A, MX-7-180AC, MXT-7-180A .... 8.12 kg/kW (13.33 lb/hp)
MX-7-180B, MX-7-180C, MXT-7-180 .... 8.45 kg/kW (13.89 lb/hp)
M-7-235B, M-7-235C, MX-7-235 .... 6.47 kg/kW (10.64 lb/hp)
M-7-260, MJ-7-260 .... 5.85 kg/kW (9.61 lb/hp)
MX-7-420AC, MJ-7-420 .... 3.62 kg/kW (5.95 lb/shp)

PERFORMANCE:
Cruising speed at 75% power at optimum altitude:
MX-7-160, MX-7-160C, MXT-7-160 .... 104 kt (193 km/h; 120 mph)
MX-7-180A, MX-7-180AC, MXT-7-180A .... 117 kt (217 km/h; 135 mph)
MX-7-180B, MX-7-180C, MXT-7-180 .... 126 kt (233 km/h; 145 mph)
M-7-235B, M-7-235C, MT-7-235 .... 139 kt (257 km/h; 160 mph)
M-7-260, M-7-260C, MT-7-260 .... 142 kt (264 km/h; 164 mph)
M-7-420AC, MT-7-420 .... 165 kt (306 km/h; 190 mph)
Stalling speed, flaps down, power off:
MX-7-160, MX-7-160C, MX-7-180A, MX-7-180AC, MX-7-180C, MXT-7-180A, MX-7-180B, MXT-7-180, M-7-235B, M-7-235C, MT-7-235, M-7-260, M-7-260C, MT-7-260 .... 35 kt (64 km/h; 40 mph)
MX-7-420AC, MT-7-420 .... 44 kt (81 km/h; 50 mph)
Max rate of climb at S/L:
MX-7-160, MX-7-160C, MXT-7-160 .... 251 m (825 ft)/min
MX-7-180A, MX-7-180AC, MXT-7-180A .... 280 m (920 ft)/min
MX-7-180B, MX-7-180C, MXT-7-180 .... 365 m (1,200 ft)/min
M-7-235B, M-7-235C, MT-7-235 .... 457 m (1,500 ft)/min
M-7-260, M-7-260C, MT-7-260 .... 503 m (1,650 ft)/min
MX-7-420AC, MT-7-420 .... 853 m (2,800 ft)/min
Service ceiling:
MX-7-160, MX-7-160C, MXT-7-160 .... 3,965 m (13,000 ft)
MX-7-180A, MX-7-180AC, MX-7-180C, MXT-7-180A, MX-7-180B, MX-7-180C, MXT-7-180 .... 4,575 m (15,000 ft)
M-7-235B, MT-7-235, M-7-235C, M-7-260, M-7-260C, MT-7-260, M-7-420AC, MT-7-420 .... 6,100 m (20,000 ft)
T-O run (solo, half fuel):
MX-7-160, MX-7-160C, MXT-7-160 .... 183 m (600 ft)
MX-7-180C, MX-7-180B, MXT-7-180 .... 92 m (300 ft)
MX-7-420AC, MT-7-420 .... 61 m (200 ft)
MX-7-180A, MX-7-180AC, MXT-7-180A .... 168 m (550 ft)
M-7-235B, M-7-235C, MT-7-235, M-7-260, M-7-260C, MT-7-260 .... 76 m (250 ft)
M-7-420AC, MT-7-420 .... 61 m (200 ft)
T-O to 15 m (50 ft):
MX-7-160, MX-7-160C, MXT-7-160 .... 360 m (1,180 ft)
MX-7-180A, MX-7-180AC, MXT-7-180A .... 350 m (1,150 ft)
MX-7-180B, MX-7-180C, MXT-7-180, M-7-235B, M-7-235C, MT-7-235, M-7-260, MT-7-260, M-7-420AC, MT-7-420 .... 183 m (600 ft)
Landing from 15 m (50 ft):
all landplane models .... 152 m (500 ft)
Range with main tank fuel only, optimum altitude, 30 min reserves:
MX-7-160, MXT-7-160 .... 469 n miles (869 km; 540 miles)
MX-7-180A, MXT-7-180A .... 434 n miles (804 km; 500 miles)
MXT-7-180, all tanks .... 825 n miles (1,528 km; 950 miles)
Range with max fuel, no reserves:
MX-7-180C .... 951 n miles (1,762 km; 1,095 miles)
M-7-235B:
O-540 engine .... 792 n miles (1,467 km; 912 miles)
IO-540 engine .... 829 n miles (1,537 km; 955 miles)
M-7-235C:
O-540 engine .... 792 n miles (1,467 km; 912 miles)
IO-540 engine .... 864 n miles (1,601 km; 995 miles)
M-7-235B floatplane:
O-540 engine .... 708 n miles (1,311 km; 815 miles)
IO-540 engine .... 772 n miles (1,430 km; 889 miles)
MX-7-420:
at 75% power .... 551 n miles (1,022 km; 635 miles)
at 50% power .... 782 n miles (1,448 km; 900 miles)

UPDATED

MAVERICK

MAVERICK JETS INC

1371 General Aviation Drive, Melbourne, Florida 32935
Tel: (+1 321) 752 41 11
Fax: (+1 321) 752 44 55
e-mail: info@maverickjets.com
Web: http://www.maverickjets.com
PRESIDENT AND CEO: Jim McCotter
DIRECTOR, AIRCRAFT SYSTEMS: Thomas R Sevier

The company moved into new premises in May 1998 and immediately began manufacture of the first Twinjet. In early 2001, Colorado businessman Jim McCotter acquired a majority share in Maverick Air Inc, renaming it Maverick Jets, and announced plans to establish a new company, known as McCotter Aviation, to develop a six-seat, factory-built, FAA-certified version of the Leader, designated MC2400. Development has been delayed by loss of the prototype.

UPDATED

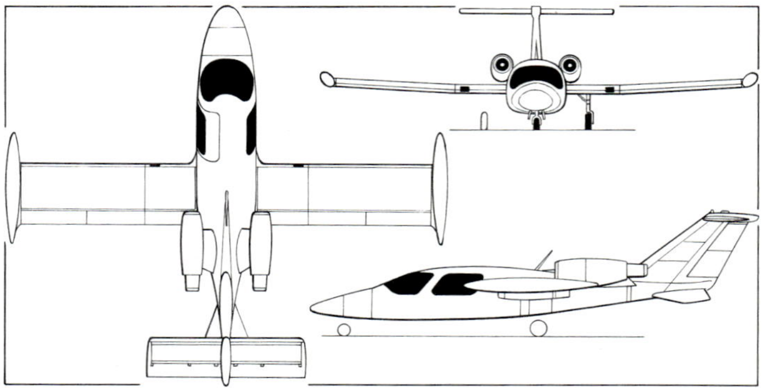

General arrangement of Maverick Leader *(James Goulding)* 0105706

MAVERICK LEADER

TYPE: Light business jet.

PROGRAMME: Launched at EAA Convention at Oshkosh, Wisconsin, July 1997; then known as Twinjet 1200; construction of proof-of-concept prototype, undertaken by Maverick Air of Pueblo, Colorado started September 1999; first flight (N750TJ) 4 August 1999; designated TJ-1500 Twinjet 1500; public debut at EAA AirVenture, Oshkosh, August 2000. Became Maverick Leader in 2002. Prototype destroyed in fatal crash 24 January 2003. FAA/JAA certification planned for 2004.

Prototype Maverick Twinjet 1500, now known as Maverick Leader *(Paul Jackson)* ***NEW***/0554466

Planned production by Tbilaviamsheni in Georgia terminated by TAM in early 2004.

COSTS: Unit cost US$1.25 million; estimated operating cost US$170 per hour or 55 cents per mile (both 2003).

DESIGN FEATURES: Designed using advanced CAD and three-dimensional modelling techniques; design goals include comfort and performance of a traditional business jet in a light piston twin configuration and good short-field performance; unswept, mid-mounted, constant-chord wing with tip tanks; T tail; engines pod-mounted on rear fuselage shoulders.

FLYING CONTROLS: Conventional and manual. Electrically-actuated flaps, approximately two-thirds span; speedbrakes in upper wing; dual (stick) controls; electric trim.

STRUCTURE: Composites, employing prepreg glass/carbon fibre and nomex.

LANDING GEAR: Retractable tricycle type with single wheel on each unit; main units retract inwards; nosewheel, forwards; Cleveland brakes.

POWER PLANT: Two 4.89 kN (1,100 lb st) Williams International FJ33-4 turbofans with FADEC, pod-mounted on rear fuselage. Fuel contained in two integral wing tanks and single fuselage tank, combined capacity 1,249 litres (330 US gallons; 275 Imp gallons).

ACCOMMODATION: Five persons in pressurised cabin; single upward-opening gullwing door on port side; baggage compartments in nose and to rear of cabin. Leather interior standard.

SYSTEMS: Pressurisation system, maximum differential 0.39 bar (5.6 lb/sq in) maintains 3,050 m (10,000 ft) cabin altitude to 9,150 m (30,000 ft). Split-bus electrical system supplied by two 24 V batteries and two alternators. Airframe anti-icing and oil-heated engine inlets standard.

AVIONICS: Integrated all-glass flight deck.

DIMENSIONS, EXTERNAL:
Wing span: standard .... 10.13 m (33 ft 3 in)
with tip tanks .... 10.52 m (34 ft 6 in)
Wing aspect ratio 7.6
Length overall .... 8.69 m (28 ft 6 in)
Height overall .... 2.74 m (9 ft 0 in)

DIMENSIONS, INTERNAL:

Cabin: Length	2.64 m (8 ft 8 in)
Max width	1.32 m (4 ft 4 in)
Max height	1.09 m (3 ft 7 in)
Baggage volume	0.42 m³ (15.0 cu ft)

AREAS:

Wings, gross	13.43 m² (144.6 sq ft)

WEIGHTS AND LOADINGS:

Weight empty	1,315 kg (2,900 lb)
Max T-O weight	2,630 kg (5,800 lb)
Max wing loading	195.8 kg/m² (40.11 lb/sq ft)
Max power loading	394 kg/kN (3.87 lb/lb st)

PERFORMANCE (estimated):

Never-exceed speed (VNE) and max level speed	391 kt (724 km/h; 450 mph)
Max cruising speed	350 kt (648 km/h; 403 mph)
Econ cruising speed	330 kt (611 km/h; 380 mph)
Stalling speed, flaps down	78 kt (145 km/h; 90 mph)
Max rate of climb at S/L	914 m (3,000 ft)/min
Rate of climb at S/L, OEI	305 m (1,000 ft)/min
Max certified altitude	9,450 m (31,000 ft)
Service ceiling, OEI	6,095 m (20,000 ft)
T-O run	610 m (2,000 ft)
Landing run	607 m (1,990 ft)
Range: with max internal fuel	1,500 n miles (2,775 km; 1,726 miles)
with tip tanks	1,700 n miles (3,148 km; 1,956 miles)

UPDATED

MD

MD HELICOPTERS INC

4555 East McDowell Road, Mesa, Arizona 85215-9797
Tel: (+1 480) 346 63 00
Fax: (+1 480) 346 68 07
e-mail: helicoptersales@mdhelicopters.com
Web: http://www.mdhelicopters.com
CHAIRMAN AND CEO: Henk Schaeken
COO: Albert Halder
VICE-PRESIDENT OF SALES AND MARKETING: Colin Whicher
MARKETING MANAGER: Mike McNabb

McDonnell Douglas' line of light helicopters, derived from the Hughes company's products, was acquired by Boeing as part of its purchase of the former company in August 1997. Having no part in the Boeing business strategy, all except the AH-64 Apache were offered for sale and in January 1999 it was announced that Netherlands-based holding company MD Helicopters, owned by the Rotterdam Dockyard Company, had been successful in its bid.

MD now owns all production jigs and tooling for the MD 500, 530F, 520N, 600N and Explorer, as well as a licence to employ the NOTAR system in future helicopters (the technology remaining in Boeing ownership) and in August 2000 acquired the former Boeing facility at Falcon Field, Mesa, Arizona, and began a 6,975 m² (75,000 sq ft) expansion to include a spares warehouse, completion and delivery centre and customer training facility.

Production Certificate 714NM issued to MDHI on 18 February 1999, covering interim manufacture of two MD 500Es (c/n 0542 and 0543), four MD 530Fs (c/n 0134 to 0137), four MD 520Ns (c/n LN086 to LN089) and three MD 600Ns (c/n RN047, 052 and 054. PC 715NM issued 5 November 1999 for all subsequent production (including MD 600s c/n RN053 and 055 upwards).

In April 2000, Kaman was contracted to build fuselages for MD's single-engine product range at its Moosup and Jacksonville plants, deliveries beginning June 2000 and, in July 2000, was contracted also to supply rotor blades for the MD Explorer. Composite Solutions of Auburn, Washington, supplies tailbooms for the NOTAR range of helicopters. TAI of Turkey is building Explorer fuselages at Akinci.

Hongdu MD Helicopters established in early 2003 as 60:40 venture between Chinese manufacturer and MD's parent (RDM Holdings); RDM investing US$10 million for Chinese assembly line for MD 500 and MD 600 at Nanchang.

MD Helicopters delivered 33 aircraft in 1999 (five MD 500Es, six MD 530Fs, five MD 520Ns, six MD 600Ns and 11 Explorers), 41 in 2000 (11 MD 500Es, three MD 530Fs, four MD 520Ns, seven MD 600Ns and 16 Explorers), 28 in 2001 (four MD 500Es, two MD 520Ns, two MD 600Ns and 20 Explorers), 15 in 2002 (five MD 500Es, four MD 520Ns, two MD 600Ns and four Explorers), and 15 in 2003 (three MD 500Es, three MD 530Fs, one MD 600N, and eight Explorers).

Total fleet time of all MD NOTAR helicopters exceeded 500,000 hours by December 2003. High-time airframe was an MD 520N operated by the Phoenix Police Department in Arizona.

UPDATED

MD 500 AND MD 530

TYPE: Light utility helicopter.

PROGRAMME: Derived from Hughes OH-6A/civil Model 500 first flown in 1963; US certification 30 June 1964; earlier versions produced for military and civil (engineering designations 369H, 369HM, 369HS, 369HE and 369D) use. First flight MD 500E (N5294A) 28 January 1982; first flight MD 530F, 22 October 1982.

CURRENT VERSIONS: **MD 500E:** Replaced MD 500D in production 1982; deliveries started following issue of Type Certificate on 15 December 1982; Rolls-Royce 250-C20R became optional replacement for standard 250-C20B in late 1988; window area of forward canopy increased in 1991 model. MD 500E introduced many cabin improvements including more space for front and rear seat occupants, lower bulkhead between front and rear seats, T tail and optional four-blade Quiet Knight tail rotor.

MD 530F Lifter: Powered by Rolls-Royce 250-C30; transmission rating increased from 280 kW (375 shp) to 317 kW (425 shp) from 11 July 1985; diameter of main rotor increased by 0.3 m (1 ft 0 in) and of tail rotor by 5 cm (2 in); cargo hook kit for 907 kg (2,000 lb) external load available; certified 29 July 1983; first delivery 20 January 1984. Upgraded drive system in **369FF**, certified 11 July 1985.

Detailed description applies to MD 500E and 530F, except where indicated.

MD 500/530 Defender: No longer in production.

MD 500N: Produced as MD 520N.

CUSTOMERS: Some 4,669 OH-6/MD 500/530 series (excluding licensed manufacture) produced by late 2003. Total of 567 MD 500Es registered by early 2004, plus 145 MD 530Fs by April 2004. Eleven MD 500s and four MD 530s delivered in 2000, four MD 500Es in 2001, five in 2002, and three each MD 500Es and MD 530Fs in 2003. The Police Departments of Mesa, Arizona, Columbus, Ohio, Wichita, Kansas, and Atlanta, Georgia; and the Xiongying Aero Club of Zhongshan, China, each took delivery of an MD500E during 2002; the Metropolitan Police Department of Las Vegas, Nevada, has three MD 530Fs. Mesa fleet rose to three in December 2002 while Oklahoma City's fleet simultaneously increased to two.

COSTS: Typical MD 500E US$860,000; MD 530F US$1.1 million (2002). Direct operating cost: MD 500E US$211, MD 530F US$237 per hour (both 2002).

DESIGN FEATURES: Fully articulated five-blade main rotor with blades retained by stack of laminated steel straps; blade aerofoil section NACA 015; blades can be folded after removing retention pins; two-blade tail rotor with optional X-pattern four-blade Quiet Knight tail rotor to reduce external noise; optional high-skid landing gear to protect tail rotor in rough country; protective skid on base of lower fin; narrow chord fin with high-set tailplane and endplate fins introduced with MD 500D. Main rotor rpm (500E/530F) 492/477 normal; main rotor tip speed 207 to 208 m (680 to 684 ft)/s; tail rotor rpm, 2,933/2,848.

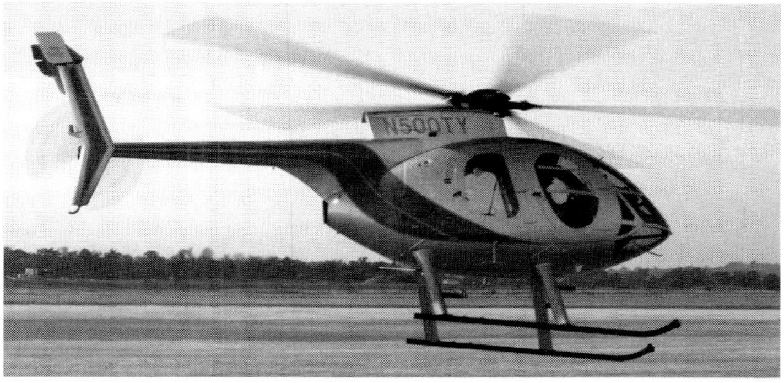

MD Helicopters MD 500E light utility helicopter *(Paul Jackson)* ***NEW***/0583505

FLYING CONTROLS: Plain mechanical without hydraulic boost.

STRUCTURE: Airframe based on two A frames from rotor head to landing gear legs, enclosing rear-seat occupants; front-seat occupants protected within straight line joining rotor hub and forward tips of landing skids; engine mounted inclined in rear of fuselage pod, with access through clamshell doors; main rotor blades have extruded aluminium spar hot-bonded to wraparound aluminium skin; tail rotor blades have swaged tubular spar and metal skin. Thicker fuselage skins, to reduce surface rippling, introduced during 2001.

LANDING GEAR: Tubular skids carried on oleo-pneumatic shock-absorbers. Utility floats, snow skis and emergency inflatable floats optional.

POWER PLANT: *MD 500E:* One 313 kW (420 shp) Rolls-Royce 250-C20B or 335.5 kW (450 shp) 250-C20R turboshaft, derated in both cases to 280 kW (375 shp) for T-O (5 minutes); maximum continuous rating 261 kW (350 shp).

MD 530F: One 485 kW (650 shp) Rolls-Royce 250-C30 turboshaft, derated to 317 kW (425 shp) up to 50 kt (92 km/h; 57 mph) and 280 kW (375 shp) above 50 kt (92 km/h; 57 mph) and for maximum continuous power (MCP).

MCP transmission rating 261 kW (350 shp); improved, heavy-duty transmission, rating 447 kW (600 shp), derated to 280 kW (375 shp) optional on production aircraft or for retrofit from June 1995.

Two interconnected bladder fuel tanks with combined usable capacity of 242 litres (64.0 US gallons; 53.3 Imp gallons). Self-sealing fuel tank optional. Refuelling point on starboard side of fuselage. Auxiliary fuel system, with 79.5 litre (21.0 US gallon; 17.5 Imp gallon) internal tank, available optionally. Oil capacity 5.7 litres (1.5 US gallons; 1.2 Imp gallons). Greater capacity internal fuel tanks also available.

ACCOMMODATION: Forward bench seat for pilot and two passengers, with two or four passengers, or two litter patients and one medical attendant, in rear portion of cabin. Pilot sits on left instead of normal right-hand seating. Low-back front seats and individual rear seats, with fabric or leather upholstery, optional. Baggage space, capacity 0.31 m³ (11 cu ft), under and behind rear seat in five-seat form. Clear space for 1.2 m³ (42 cu ft) of cargo or baggage with only three front seats in place. Two doors on each side. Interior soundproofing optional.

SYSTEMS: Aero Engineering Corporation air conditioning system or Fargo pod-mounted air conditioner optional.

AVIONICS (MD 500E): Optional avionics listed below.

Comms: Dual Bendix/King KY 195 or Rockwell Collins VHF-251 transceivers; Bendix/King KT 76 or Rockwell Collins TDR-950 transponder; intercom system, headsets, microphones; and optional public address system.

Radar: Optional installation.

Flight: Dual Bendix/King KX 175 or Rockwell Collins VHF-251/231 nav receivers, latter with IND-350 nav indicator; Bendix/King KR 85 or Rockwell Collins ADF-650 ADF.

Instrumentation: Basic VFR instruments and night flying lighting, attitude and directional gyros and rate of climb indicator.

Mission: Optional, FLIR and 30 Mcd Spectrolab SX-16 Nightsun searchlight.

EQUIPMENT: Optional equipment includes shatterproof glass, heating/demisting system, nylon mesh seats, dual controls, cargo hook, cargo racks, underfuselage cargo pod, heated pitot tube, extended landing gear, blade storage rack, litter kit, emergency inflatable floats and inflated utility floats.

DIMENSIONS, EXTERNAL:

Main rotor diameter: 500E	8.05 m (26 ft 5 in)
530F	8.33 m (27 ft 4 in)
Main rotor blade chord	0.171 m (6¾ in)
Tail rotor diameter: 500E	1.37 m (4 ft 6 in)
530F	1.42 m (4 ft 8 in)
Distance between rotor centres:	
500E	4.67 m (15 ft 4 in)
530F	4.88 m (16 ft 0 in)
Length overall, rotors turning:	
500E	8.61 m (28 ft 3 in)
530F	9.94 m (32 ft 7¼ in)
Length of fuselage	7.49 m (24 ft 7 in)

Height to top of rotor head:
standard skids ..... 2.67 m (8 ft 9 in)
extended skids ..... 2.97 m (9 ft 8¾ in)
Tailplane span ..... 1.65 m (5 ft 5 in)
Skid track (standard) ..... 1.91 m (6 ft 3 in)
Cabin doors (each): Height ..... 1.13 m (3 ft 8½ in)
Max width ..... 0.76 m (2 ft 6 in)
Height to sill: 500E ..... 0.79 m (2 ft 7 in)
530F ..... 0.76 m (2 ft 6 in)
Cargo compartment doors (each):
Height ..... 1.12 m (3 ft 8¼ in)
Width ..... 0.88 m (2 ft 10½ in)
Height to sill: 500E ..... 0.71 m (2 ft 4 in)
530F ..... 0.66 m (2 ft 2 in)

DIMENSIONS, INTERNAL:
Cabin: Length ..... 2.44 m (8 ft 0 in)
Max width ..... 1.31 m (4 ft 3½ in)
Max height ..... 1.52 m (5 ft 0 in)

AREAS:
Main rotor blades (each): 500E ..... 0.62 m² (6.67 sq ft)
530F ..... 0.65 m² (6.96 sq ft)
Tail rotor blades (each): 500E ..... 0.063 m² (0.675 sq ft)
530F ..... 0.066 m² (0.711 sq ft)
Main rotor disc: 500E ..... 50.89 m² (547.8 sq ft)
530F ..... 54.58 m² (587.5 sq ft)
Tail rotor disc: 500E ..... 1.53 m² (16.47 sq ft)
530F ..... 1.65 m² (17.72 sq ft)
Fin ..... 0.56 m² (6.05 sq ft)
Tailplane ..... 0.76 m² (8.18 sq ft)

WEIGHTS AND LOADINGS:
Weight empty: 500E ..... 672 kg (1,481 lb)
530F ..... 722 kg (1,591 lb)
Max normal T-O weight: 500E ..... 1,361 kg (3,000 lb)
530F ..... 1,406 kg (3,100 lb)
Max overload T-O weight:
500E, 530F ..... 1,610 kg (3,550 lb)
Max T-O weight, external load:
530F ..... 1,701 kg (3,750 lb)
Max hook capacity: 530F ..... 907 kg (2,000 lb)
Disc loading at max normal T-O weight:
500E ..... 26.8 kg/m² (5.48 lb/sq ft)
530F ..... 25.8 kg/m² (5.28 lb/sq ft)
Transmission loading at max T-O weight and power:
500E standard ..... 5.22 kg/kW (8.57 lb/shp)
500E optional ..... 4.87 kg/kW (8.00 lb/shp)
530F standard ..... 5.39 kg/kW (8.86 lb/shp)
530F optional ..... 5.03 kg/kW (8.27 lb/shp)

PERFORMANCE (at max normal T-O weight, except where indicated):
Never-exceed speed (VNE) at S/L:
500E, 530F ..... 152 kt (282 km/h; 175 mph)
Max cruising speed at S/L:
500E ..... 134 kt (248 km/h; 154 mph)
530F ..... 133 kt (247 km/h; 154 mph)
Max cruising speed at FL50:
500E ..... 132 kt (245 km/h; 152 mph)
530F ..... 135 kt (249 km/h; 155 mph)
Econ cruising speed at S/L:
500E ..... 129 kt (239 km/h; 149 mph)
530F ..... 131 kt (243 km/h; 151 mph)
Econ cruising speed at FL50:
500E, 530F ..... 123 kt (228 km/h; 142 mph)
Max rate of climb at S/L, ISA:
500E ..... 536 m (1,760 ft)/min
530F ..... 631 m (2,070 ft)/min
Max rate of climb at S/L, ISA+20°C:
530F ..... 628 m (2,061 ft)/min
Vertical rate of climb at S/L: 500E ..... 248 m (813 ft)/min
530F ..... 446 m (1,462 ft)/min
Service ceiling: 500E ..... 4,575 m (15,000 ft)
530F ..... 5,700 m (18,700 ft)
Hovering ceiling IGE: ISA: 500E ..... 2,590 m (8,500 ft)
530F ..... 4,875 m (16,000 ft)
ISA+20°C: 500E ..... 1,830 m (6,000 ft)
530F ..... 4,358 m (14,300 ft)
Hovering ceiling OGE: ISA: 500E ..... 1,830 m (6,000 ft)
530F ..... 4,389 m (14,400 ft)
ISA+20°C: 500E ..... 975 m (3,200 ft)
530F ..... 3,535 m (11,600 ft)
Range, 2 min warm-up, standard fuel, no reserves:
500E at S/L ..... 233 n miles (431 km; 268 miles)
530F at S/L ..... 200 n miles (371 km; 231 miles)
500E at FL50 ..... 264 n miles (488 km; 303 miles)
530F at FL50 ..... 232 n miles (429 km; 267 miles)
Endurance, 530F at S/L ..... 2 h 0 min

UPDATED

MD 520N

TYPE: Light utility helicopter.

PROGRAMME: First flight OH-6A NOTAR (no tail rotor) testbed 17 December 1981; extensive modifications during 1985 with second blowing slot, new fan, 250-C20B engine and MD 500E nose; flight testing resumed 12 March 1986 and completed in June; retired to US Army Aviation Museum, Fort Rucker, Alabama, October 1990. Commercial MD 520N and uprated (485 kW; 650 shp Rolls Royce 250-C30) MD 530N NOTAR helicopters announced February 1988 and officially launched January 1989; first flight of MD 530N (N530NT) 29 December 1989, but this variant not pursued; first flight 520N (N520NT) 1 May 1990 and first production 520N 28 June 1991; 520N certified 12 September 1991 on MD500 type certificate; first production 520N (N521FB) delivered to Phoenix Police Department 31 October 1991. MD 520N set new Paris to London speed record in September 1992, at 1 hour 22 minutes 29 seconds. Now certified in 21 countries.

MD 520N wearing imaginative Belgian registration OO-SZO *(Paul Jackson)* ***NEW***/0561657

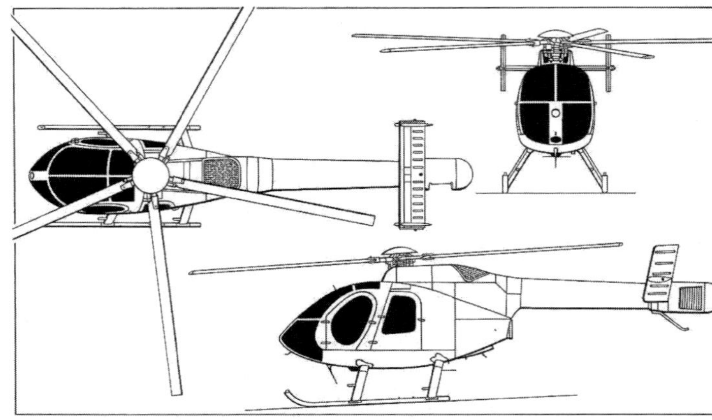

MD 520N five-seat NOTAR helicopter *(Jane's/Mike Keep)* 0507284

CURRENT VERSIONS: **MD 520N:** NOTAR version of MD 500, offering more power, higher operating altitude and greater maximum T-O weight than MD 500E. Engineering designation is **MD 500N**.

Description applies to 520N.

MD 600N: Stretched version, described separately.

MD 520N Defender: Military variant, being developed.

CUSTOMERS: Total of 99 registered by mid-2002; none further known by June 2004. Four delivered in 2000, two in 2001 and four in 2002. Law enforcement agencies flying MD 520Ns include Phoenix, Arizona (first operator; two more delivered in early 2002); Burbank, Glendale, Huntington Beach, Los Angeles, Ontario and San Jose, California; Hernando County, Florida; Orange County, Florida; Jefferson County, Kentucky, which took delivery of one in March 2002; Prince George County, Clinton, Maryland, which took delivery of two on 4 October 2000; Hamilton County, Ohio, which has seven 520Ns; El Salvador, whose Policia Nacional Civil took delivery of two in 1996; San Juan, Puerto Rico; Honolulu, Hawaii; and Calgary, Alberta, Canada. Other operators include the Tata Group of Mumbai, India, Weetabix Ltd, UK, and Belgian Gendarmerie (two).

COSTS: US$980,000 (2002). Direct operating cost US$224.71 per hour (2002).

DESIGN FEATURES: NOTAR system provides anti-torque and steering control without an external tail rotor, thus eliminating the danger of tail strikes; air emerging through Coanda slots and steering louvres is cool and at low velocity. Believed to be currently the quietest turbine helicopter, based on FAA certification test noise figures. Main rotor rpm 477; main rotor tip speed 208 m (684 ft)/s; NOTAR system fan rpm 5,388. Emergency floats among options. Redesigned diffuser in NOTAR tailboom and revised fan rigging introduced from early 2000 and available for retrofit; combined with uprated Rolls-Royce 250-C20R+ engine (see below); these improvements increase the aircraft's payload capability.

FLYING CONTROLS: Unboosted mechanical, as in earlier versions. Rotor downwash over tailboom deflected to port by two Coanda-type slots fed with low-pressure air from engine-driven variable-pitch fan in root of tailboom; this counters normal rotor torque; some fan air is also vented at tail through variable-aperture louvres controlled by pilot's foot pedals, giving steering control in hover and forward flight. Port moving fin on tailplane connected to foot pedals, primarily to increase directional control during autorotation and allow touchdown at under 20 kt (37 km/h; 23 mph); starboard fin operated independently by yaw damper.

STRUCTURE: Same as for MD 500E/530F, except graphite composites tailboom; metal tailplane and fins; new high-efficiency fan with composites blades fitted in production aircraft. NOTAR system components now have twice the lifespan of conventional tail rotor system assemblies. During 1993, NOTAR system components' warranty increased from two to three years. Thicker fuselage skins, to reduce surface rippling, will be introduced during 2001.

POWER PLANT: One Rolls-Royce 250-C20R turboshaft, derated to 317 kW (425 shp) for T-O (5 minutes) and 280 kW (375 shp) maximum continuous. Improved, heavy-duty transmission, rating 447 kW (600 shp), derated to 317 kW (425 shp) for T-O, 280 kW (375 hp) maximum continuous, on production aircraft from June 1995. Rolls-Royce 250-C20R+ engine, providing improved hot weather performance and 3 to 5 per cent more power, introduced from early 2000. Fuel capacity 235 litres (62.0 US gallons; 51.6 Imp gallons).

SYSTEMS: Electrical system includes 87 Ah starter-generator. Aero Aire air conditioning system optional.

EQUIPMENT: Phoenix Police Department pioneered use of MD 520Ns for firefighting, using 341 litre (90.0 US gallon; 75.0 Imp gallon) 'Bambi' buckets to drop 151,416 litres (40,000 US gallons; 33,307 Imp gallons) of water on fires in remote desert and mountain areas.

DIMENSIONS, EXTERNAL:
Rotor diameter ..... 8.33 m (27 ft 4 in)
Length: overall, rotor turning ..... 9.78 m (32 ft 1¼ in)
fuselage ..... 7.77 m (25 ft 6 in)

Height to top of rotor head:
standard skids ... 2.74 m (9 ft 0 in)
extended skids ... 3.01 m (9 ft 10¾ in)
Height to top of fins ... 2.83 m (9 ft 3½ in)
Tailplane span ... 2.01 m (6 ft 7¼ in)
Skid track ... 1.92 m (6 ft 3½ in)

AREAS:
Rotor disc ... 54.51 m² (586.8 sq ft)

WEIGHTS AND LOADINGS:
Weight empty: standard ... 719 kg (1,586 lb)
Max fuel weight ... 183 kg (404 lb)
Max hook capacity ... 1,004 kg (2,214 lb)
Max T-O weight: normal ... 1,519 kg (3,350 lb)
with external load ... 1,746 kg (3,850 lb)
Max normal disc loading ... 27.9 kg/m² (5.71 lb/sq ft)
Transmission loading at max T-O weight and power ... 4.80 kg/kW (7.88 lb/shp)

PERFORMANCE (at normal max T-O weight, ISA, except where indicated):
Never-exceed speed (VNE) ... 152 kt (281 km/h; 175 mph)
Max cruising speed at S/L ... 135 kt (249 km/h; 155 mph)
Max rate of climb at S/L: ISA ... 564 m (1,850 ft)/min
ISA + 20°C ... 480 m (1,575 ft)/min
Service ceiling ... 4,320 m (14,175 ft)
Hovering ceiling IGE ... 3,414 m (11,200 ft)
Hovering ceiling OGE: ISA ... 1,830 m (6,000 ft)
ISA +20°C ... 1,292 m (4,240 ft)
Range at S/L ... 229 n miles (424 km; 263 miles)
Endurance at S/L ... 2 h 24 min

OPERATIONAL NOISE LEVELS:
T-O ... 85.4 EPNdB
Approach ... 87.9 EPNdB
Flyover ... 80.2 EPNdB

UPDATED

MD 600N

TYPE: Light utility helicopter.

PROGRAMME: Stretched version of MD 520N. Announced as 'concept', 8 November 1994; prototype, then known as MD 630N (N630N, converted from MD 530F demonstrator), first flight 22 November 1994, 138 days after project approval; public debut at Heli-Expo in Las Vegas January 1995; production go-head 28 March 1995, at which time designation changed to MD 600N prototype first flown with production standard engine and rotor system 6 November 1995; production prototype (N600RN) first flown 15 December 1995, and became certification test vehicle leading to FAR Pt 27 certification, but was destroyed in ground fire 28 May 1996, following emergency landing after rotor/tailboom strike during abrupt control reversal tests. This resulted in changes to tailboom/rotor clearance; third prototype (N605AS) flown 9 August 1996; further accidents to N630N on 4 and 21 November 1996 and on 18 January 1997, all during autorotational descents, culminated in total loss and delayed certification and first delivery, originally scheduled for 18 December 1996, to 15 May and 6 June 1997, respectively. MD 600N certified on MD500 certificate.

In July 1998, Boeing completed a year-long envelope expansion programme for the MD 600N leading to FAA approval for operation at a density altitude of 2,135 m (7,000 ft) at a T-O weight of 1,746 kg (3,850 lb) and at a density altitude of 1,220 m (4,000 ft) at a T-O weight of 1,860 kg (4,100 lb). Other performance enhancements approved by the FAA included provision for doors-off operation at speeds up to 115 kt (213 km/h; 132 mph), operation at temperatures -40°C/52°C (-40°F/126°F), lifting up to 968 kg (2,134 lb) on the external cargo hook, making slope landings up to 10° in any direction, operation with emergency floats and for installation of a movable landing light and additional wire strike protection on the fuselage. The MD 600N also completed HIRF trials at NAS Patuxent River, Maryland. Yaw-stability augmentation system (Y-SAS) under development during 2000, aimed at reducing pilot workload during extended flights and in turbulent conditions; Y-SAS was expected to be available from March 2001, following FAA certification, and will be field-installable.

CUSTOMERS: Total of 68 registered by May 2003; none further by June 2004. Launch customer AirStar Helicopter of Arizona (two, of which first delivered 6 June 1997); Saab Helikopter AB of Sweden and Ro air Limited of Hong Kong ordered one each in June 1995; other customers include Guangdong General Aviation Company (GGAC) of the People's Republic of China, which took delivery of one MD 600N in November 2000, during Airshow China 2000 in Zhuhai, Los Angeles County Sheriff's Department Aero Bureau (three), Orange County, California, Sheriff's Department, Indianapolis, Indiana, Police Department (one), Presta Services of France (one), Turkish National Police, which ordered 10 in December 2000 for delivery during 2002 (subsequently postponed to 2003), UND (University of North Dakota) Aerospace (one), West Virginia State Police (one) and the US Border Patrol (45, of which 11 delivered by end of 1998, when procurement halted pending evaluation of UAVs for border patrol role). Deliveries totalled 15 in 1997, 21 in 1998, six in 1999, seven in 2000, two in 2001, two in 2002 and one in 2003.

COSTS: US$1.315 million (2002). Direct operating cost US$269 per hour (2002).

DESIGN FEATURES: Stretched MD 520N airframe (less than 1 per cent new parts) by means of 0.76 m (2 ft 6 in) plug aft of cockpit/cabin bulkhead and 0.71 m (2 ft 4 in) plug in tailboom, combined with more powerful engine, uprated transmission and six-blade main rotor. Cabin has flat floor to assist cargo handling, and will feature quick-change interior configurations to suit multiple-use operators. Intended for civil, utility, offshore, executive transport, medevac, aerial news gathering, touring, law enforcement and other noise-sensitive operations; also adaptable for armed scout, utility and other military missions.

Description generally as for MD 520N except as follows:

POWER PLANT: One 603 kW (808 shp) Rolls-Royce 250-C47M turboshaft, derated to 447 kW (600 shp) for T-O (5 minutes) and 395 kW (530 shp) maximum continuous, with FADEC. Transmission manufactured from WE43A magnesium alloy for lower weight, greater strength and enhanced corrosion resistance, rating 447 kW (600 shp). Fuel contained in two crashworthy bladder tanks in lower fuselage, total capacity 440 litres (116 US gallons; 96.8 Imp gallons), of which 434 litres (115 US gallons; 95.5 Imp gallons) are usable.

ACCOMMODATION: Standard seating for eight, including pilot, in 3+3+2 configuration with centre bench; alternative six-seat club and coach layouts; utility with main cabin seats removed; EMS with accommodation for pilot, one stretcher case and two medical attendants; or electronic news gathering, with accommodation for pilot and passenger in cockpit and operator with cabin and monitor in cabin. All facilitated by quick-release seat mechanisms. Two removable centre-opening doors on each side of cabin; door to cockpit on each side. Bubble 'comfort' windows and sliding windows for cabin, custom soundproofing and Integrated Flight Systems Inc air conditioning optional.

MD 600N (Rolls-Royce 250 turboshaft) *(Paul Jackson)* ***NEW***/0583504

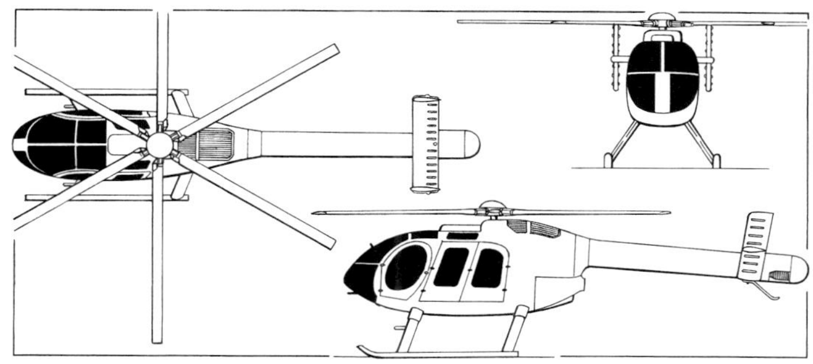

MD 600N eight-seat NOTAR helicopter *(James Goulding)* 0051783

SYSTEMS: Electrical system comprises 28 V 200 Ah starter-generator and 28 V 17 Ah Ni/Cd battery. 24 V auxiliary power receptacle inside starboard cockpit door standard.

AVIONICS: To customer's choice, including VHF/UHF/FM/AM transceivers, GPS, transponder, ELT, intercom, stereo tape system and PA system.

EQUIPMENT: Optional equipment includes dual controls, particle separator, heated pitot tube, rotor brake, cargo hook, wire strike kit and emergency floats.

DIMENSIONS, EXTERNAL:
Rotor diameter ... 8.38 m (27 ft 6 in)
Rotor ground clearance ... 2.65 m (8 ft 8½ in)
Length: overall, rotor turning ... 10.79 m (35 ft 4¾ in)
fuselage ... 8.99 m (29 ft 6 in)
Fuselage width at cabin ... 1.40 m (4 ft 7 in)
Height to top of rotor head:
standard skids ... 2.65 m (8 ft 8½ in)
extended skids ... 2.74 m (9 ft 0 in)
Height to top of fins:
standard skids ... 2.68 m (8 ft 9½ in)
extended skids ... 2.83 m (9 ft 3½ in)
Fuselage ground clearance, extended skids ... 0.58 m (1 ft 10¾ in)
Skid track ... 2.47 m (8 ft 1¼ in)
Tailplane span, over fins ... 2.01 m (6 ft 7¼ in)

DIMENSIONS, INTERNAL:
Cabin, excl cockpit: Length ... 1.83 m (6 ft 0 in)
Width ... 1.22 m (4 ft 0 in)
Height ... 1.22 m (4 ft 0 in)

AREAS:
Rotor disc ... 55.18 m² (594.0 sq ft)

WEIGHTS AND LOADINGS:
Weight empty, standard ... 952 kg (2,100 lb)
Useful load: internal ... 907 kg (2,000 lb)
external ... 1,179 kg (2,600 lb)
Max hook capacity ... 1,361 kg (3,000 lb)
Max T-O weight: internal load ... 1,859 kg (4,100 lb)
slung load ... 2,131 kg (4,700 lb)
Max disc loading:
internal load ... 33.7 kg/m² (6.90 lb/sq ft)
slung load ... 38.6 kg/m² (7.91 lb/sq ft)
Transmission loading at max T-O weight and power:
internal load ... 4.16 kg/kW (6.83 lb/shp)
slung load ... 4.77 kg/kW (7.83 lb/shp)

PERFORMANCE:
Never-exceed speed (VNE) ... 135 kt (250 km/h; 155 mph)
Max cruising speed, S/L to FL50, ISA ... 134 kt (248 km/h; 154 mph)
Max rate of climb at S/L, ISA ... 411 m (1,350 ft)/min
Max operating altitude ... 6,100 m (20,000 ft)
Hovering ceiling, ISA: IGE ... 3,383 m (11,100 ft)
OGE ... 1,829 m (6,000 ft)
Max range:
at S/L ... 342 n miles (633 km; 393 miles)
at FL50, ISA ... 382 n miles (707 km; 439 miles)
Endurance: at S/L ... 3 h 36 min
at FL50 ... 3 h 54 min

UPDATED

MD EXPLORER

US Coast Guard designation: MH-90 Enforcer

TYPE: Light utility helicopter.

PROGRAMME: Initially known as MDX, then MD 900 (proposed MD 901 with Turbomeca engines was not pursued); McDonnell Douglas design; announced February 1988; launched January 1989; Hawker de Havilland of Australia designed and manufactures airframe; Canadian Marconi tested initial version of integrated instrumentation display system (IIDS) early 1992; Kawasaki completed 50 hour test of transmission early 1992. Other partners include Aim Aviation (interior), IAI (cowling and seats) and Lucas Aerospace (actuators).

MD Explorer light utility helicopter *(Paul Jackson)* ***NEW***/0583503

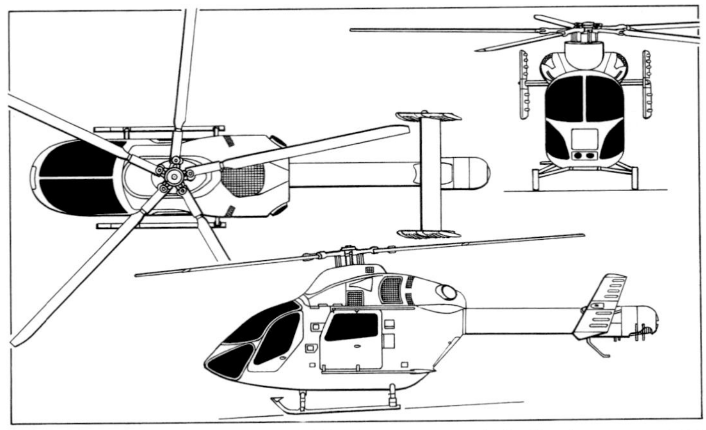

MD Explorer eight-seat commercial helicopter *(Jane's/Mike Keep)* 0100457

Ten prototypes and trials aircraft, of which seven (Nos. 1, 3-7 and 9) for static tests; first flight (No.2/N900MD) 18 December 1992, followed by No.8/N900MH 17 September 1993 and No.10/N9208V 16 December 1993; first production/demonstrator Explorer (No.11/N92011) flown 3 August 1994. FAA certification 2 December 1994; first delivery 16 December 1994; JAA certification July 1996; FAA certification for single-pilot IFR operation achieved January 1997. Type certificate transferred to MDHI on 18 February 1999.

FAA certification of uprated PW207E engine achieved in July 2000, providing 11 per cent more power for take-off and 610 m (2,000 ft) increase in hovering capability OEI in hot-and-high conditions; first delivery of PW207E-engined Explorer to Police Aviation Services, UK, 27 September 2000. "100th production" Explorer (actually 89th overall, including prototypes) delivered 1 March 2002 to Tomen Aerospace Corporation of Japan for ENG operations by Aero Asahi of Hiroshima. Total fleet time stood at more than 120,000 hours by December 2002.

CURRENT VERSIONS: **MD Explorer:** Initial civilian utility version, as described.

Details apply to civilian version except where indicated.

MD Enhanced Explorer: Improved version, announced September 1996; originally MD 902, but now known as "902 Configuration". Main features include Pratt & Whitney Canada PW206E engines with increased OEI ratings; transmission approved for dry running for 30 minutes at 50 per cent power; improved engine air inlets, NOTAR inlet design and engine fire suppression system, and more powerful stabiliser control system, resulting in 7 per cent increase in range, 4 per cent increase in endurance and 113 kg (250 lb) increase in payload over Explorer. First flight (N9224U; c/n 900-0051, 41st Explorer) 5 September 1997. FAA certification to Category A performance standards (including continued take-off with one failed engine) and single-pilot IFR operation achieved 11 February 1998; JAA certification for Category A performance achieved July 1998. Retrofit kits to convert Explorers to Category A standard. First Enhanced Explorer delivery in May 1998 to Tomen Aerospace of Japan. PW206E replaced by PW207E from late 2000, beginning at c/n 900-0077, allowing further MTOW increase to 2,948 kg (6,500 lb).

MH-90 Enforcer: Beginning March 1999, under a programme code-named Operation New Frontier, the US Coast Guard used two leased MD 900 Explorers for shipboard anti-drug smuggling operations. Armed with a pintle-mounted M240 7.62 mm minigun at the door station. In September 1999 the MD900s were exchanged for two leased MD 902 Enhanced Explorers. These subsequently replaced by Agusta A 109s. Six delivered to Mexican Navy at Acapulco (two each respectively in May and December 1999 and April 2000) for anti-drug operations, equipped with 0.50 in General Dynamics GAU-19/A Gatling guns, and 70 mm rocket pods; further four in process of delivery. Weapons qualification trials were completed at Fort Bliss, Texas in November 2000.

CUSTOMERS: Market estimated at 800 to 1,000 in first decade; however, only 103 registered and built by end of 2003. First delivery 16 December 1994 to Petroleum Helicopters Inc (PHI) which ordered five; second delivery (N901CF) December 1994 to Rocky Mountain Helicopters for EMS duties with affiliate Care Flight unit of Regional Emergency Medical Services Authority (REMSA) in Reno, Nevada. Total of two delivered in 1994, 12 in 1995, 15 in 1996, one in 1997, four in 1998, 11 in 1999, 16 in 2000, 20 in 2001, four in 2002, and eight in 2003; initial (MD 900) series comprised 40 aircraft, including three flying prototypes; PW207E engine from 64th production (67th overall) aircraft.

The US Drug Enforcement Administration took delivery of one on 21 January 2004.

COSTS: US$2.285 million (2002); direct operating cost US$408.11 (2002) per hour.

DESIGN FEATURES: NOTAR anti-torque system; all-composites five-blade rotor of tapered thickness with parabolic swept outer tip with bearingless flexbeam retention and pitch case; tuned fixed rotor mast and mounting truss for vibration reduction; replaceable rotor tips; maximum rotor speed 392 rpm; modified A-frame construction from rotor mounting to landing skids protects passenger cabin; energy-absorbing seats absorb 20 *g* vertically and 16 *g* fore and aft; onboard health monitoring, exceedance recording and blade track/balance.

FLYING CONTROLS: NOTAR tailboom (see details under MD 520N); mechanical engine control from collective pitch lever is back-up for electronic FADEC. Automatic stabilisation and autopilot available for IFR operation.

STRUCTURE: Cockpit, cabin and tail largely carbon fibre; top fairings Kevlar composites; no magnesium; lightning strike protection embedded in composites skin. Transmission overhaul life 5,000 hours; glass fibre blades have titanium leading-edge abrasion strip and are attached to bearingless hub by carbon fibre encased glass fibre flexbeams; rotor blades and hub on condition.

LANDING GEAR: Fixed skids with replaceable abrasion pads; emergency floats optional.

POWER PLANT: Baseline MD 900 powered by two Pratt & Whitney Canada PW206E turboshafts with FADEC, each rated at 463 kW (621 shp) for 5 minutes for T-O, 489 kW (656 shp) for 2½ minutes OEI and 410 kW (550 shp) maximum continuous. Transmission rating 820 kW (1,100 shp) for T-O, 746 kW (1,000 shp) maximum continuous, 507 kW (680 shp) for 2½ minutes OEI and 462 kW (620 shp) maximum continuous OEI.

Fuel contained in single tank under passenger cabin, capacity 564 litres (149 US gallons; 124 Imp gallons), of which 553 litres (146 US gallons; 122 Imp gallons) are usable. Single-point refuelling; self-sealing fuel lines.

ACCOMMODATION: Two pilots or pilot/passenger in front on energy-absorbing adjustable crew seats with five-point shoulder harnesses/seat belts; six passengers in club-type energy-absorbing seating with three-point restraints; rear baggage compartment accessible through rear door; cabin can accept long loads reaching from flight deck to rear door; hinged, jettisonable door to cockpit on each side; sliding door to cabin on each side.

SYSTEMS: Hydraulic system, operating pressure 34.475 bar (500 lb/sq in).

AVIONICS: Full IFR capability for single- or two-pilot operation.

Comms: Two headsets standard. Bendix/King Silver Crown VHF transceiver, audio control panel, ELT, cockpit voice recorder and Wulfsberg Flexcomm II optional.

Radar: Honeywell RDR 2000 vertical profile colour weather radar optional with IFR package.

Flight: Optional equipment includes Bendix/King Silver Crown VOR/ILS, HSI, ADF, DME, marker beacon receiver, radar altimeter, Loran C and KLN 90B GPS. Coupled three-axis autopilot optional.

Instrumentation: Single- or two-pilot instrument panels incorporate Canadian Marconi integrated instrumentation display system (IIDS) with high-resolution sunlight-readable LCD screen displaying engine and system information including engine condition trend monitoring, exceedance recording, caution annunciators, onboard track and balance of rotor and fan, weight on cargo hook, outside air temperature, digital clock, running time meter and RS-232 download modem interface for personal computer. Other standard instrumentation includes airspeed indicator, encoding altimeter, vertical speed indicator, turn and slip indicator, wet compass and clock. EFIS 40 electronic flight information system and parallel IIDS monitor for long-line hook operations from left seat optional.

Mission: Law enforcement panel with space for FLIR screen available. Avionics and equipment options for law enforcement conversions are described in *Jane's Aircraft Upgrades*.

EQUIPMENT: Standard equipment includes magnetic chip detectors on engines, tiedown fittings, flush-mounted cargo tiedowns, rotor blade tiedowns, right side passenger step, utility beige colour carpet, trim, wall and ceiling panels, soundproofing, tinted windows, map case, recessed hover and approach light, wander and white dome lights in cockpit, white dome light in cabin, utility light in baggage compartment, single 28 V DC power outlet each in cockpit and cabin, single colour external paint with two-colour accent stripes, FOD covers, pitot tube cover and cockpit fire extinguisher.

Optional equipment includes dual controls, heated pitot head, rotor brake, pilot-activated engine fire extinguisher, engine air particle separator, maintenance hand pump for hydraulics, external cargo hook with 1,361 kg (3,000 lb) capacity 272 kg (600 lb) personnel hoist, wire strike kit, emergency floats for skids, retractable landing light, port side cabin step, landing gear and rotor fairing canopy cover, heater/defogger, vapour-cycle air conditioner, upgraded soundproofing, passenger service unit with air gaspers and reading lights, window reveal panels, matching close-out panel for aft baggage area, upgraded passenger seats, smoke detector in baggage compartment, jack-point fittings and ground handling wheels.

DIMENSIONS, EXTERNAL:
Rotor diameter .... 10.31 m (33 ft 10 in)
Length: overall, rotor turning .... 11.83 m (38 ft 10 in)
  fuselage .... 9.85 m (32 ft 4 in)
Fuselage width at cabin .... 1.63 m (5 ft 4 in)
Height: to top of rotor head .... 3.66 m (12 ft 0 in)
  to top of fins .... 2.79 m (9 ft 2 in)
Min fuselage ground clearance .... 0.38 m (1 ft 3 in)
Tailplane span .... 2.84 m (9 ft 4 in)
Skid track .... 2.24 m (7 ft 4 in)
Cabin door width .... 1.27 m (4 ft 2 in)

DIMENSIONS, INTERNAL:
Cabin:
  Length: overall, incl baggage compartment .... 3.93 m (12 ft 10¾ in)
    passenger compartment only .... 1.91 m (6 ft 3 in)
  Max height .... 1.24 m (4 ft 1 in)
  Max width .... 1.45 m (4 ft 9 in)
  Volume .... 4.9 m^3 (173 cu ft)
  Baggage (if closed off) .... 1.39 m^3 (49 cu ft)

AREAS:
Rotor disc .... 83.52 m^2 (899.0 sq ft)

WEIGHTS AND LOADINGS:
Weight empty, standard configuration .... 1,531 kg (3,375 lb)
Standard fuel load .... 489 kg (1,078 lb)
Max internal payload .... 1,292 kg (2,848 lb)
Max slung load .... 1,361 kg (3,000 lb)
Max T-O weight: internal load .... 2,835 kg (6,250 lb)
  external load .... 3,129 kg (6,900 lb)
Max disc loading:
  internal load .... 33.9 kg/m^2 (6.95 lb/sq ft)
  external load .... 37.5 kg/m^2 (7.68 lb/sq ft)
Transmission loading at max T-O weight and power:
  internal load .... 3.80 kg/kW (6.25 lb/shp)
  external load .... 3.82 kg/kW (6.27 lb/shp)

PERFORMANCE (at max internal load T-O weight, ISA, except where indicated):
Never-exceed speed (VNE) at S/L, ISA .... 140 kt (259 km/h; 161 mph)
Max cruising speed at S/L, 38°C (100°F) .... 134 kt (248 km/h; 154 mph)
Max rate of climb at S/L .... 579 m (1,900 ft)/min
Vertical rate of climb at S/L .... 411 m (1,350 ft)/min
Rate of climb at S/L, OEI .... 305 m (1,000 ft)/min
Service ceiling: both engines .... 5,335 m (17,500 ft)
  OEI .... 3,200 m (10,500 ft)
Hovering ceiling:
  IGE, ISA .... 3,353 m (11,000 ft)

IGE, ISA + 20°C .... 2,042 m (6,700 ft)
OGE, ISA .... 2,743 m (9,000 ft)
OGE, ISA + 20°C .... 1,433 m (4,700 ft)
Hovering ceiling, OGE:
IGE, ISA at 87% max T-O weight .... 1,220 m (4,000 ft)
Max range: at S/L .... 257 n miles (476 km; 295 miles)
at FL50, ISA .... 293 n miles (542 km; 337 miles)
Max endurance: at S/L .... 2 h 54 min
at FL50, ISA .... 3 h 12 min
OPERATIONAL NOISE LEVELS:
T-O .... 84.1 EPNdB
Approach .... 88.9 EPNdB
Flyover .... 83.1 EPNdB

UPDATED

ME 262 PROJECT

ME 262 PROJECT

Paine Field, Building 221, 10727 36th Avenue West, Everett, Washington 98204
Tel: (+1 425) 290 78 78
Fax: (+1 425) 355 13 23
e-mail: me262project@juno.com
Web: http://www.stormbirds.com
PROJECT MANAGER: Bob Hammer

Founded by the late Stephen L Snyder, Classic Fighter Industries Inc launched a programme in 1993 to build five exact flying replicas of a Second World War German jet fighter, the Messerschmitt Me 262. Work was originally contracted to the Texas Airplane Factory, but halted during 1997 because of a legal dispute. The production line was relocated to Paine Field (Snohomish County Airport) in early 1999; renamed Me 262 Project in 2001; first aircraft flew in late 2002 but was damaged soon after; it has since been repaired.

UPDATED

MESSERSCHMITT ME 262

TYPE: Two-seat jet sportplane
PROGRAMME: Original prototype first flew 25 March 1942; entered operational service June 1944; flying terminated May 1945, except for limited number assigned to Allied technical evaluation or flown by Czechoslovak Air Force. Total of 1,443 built by Messerschmitt.

Classic programme based on reverse-engineering of WkNr 110639, former US Me 262B-1a evaluation aircraft, refurbished by Classic from 1993 onwards and redelivered to US Navy at Willow Grove, Pennsylvania, on 8 September 2000. Completion of first new-build aircraft rescheduled to 2002, due to company restructuring. First taxi testing undertaken 25 June 2002, following engine tests conducted from February 2002. Maiden flight (following earlier short 'hops') (N262AZ) 20 December 2002, but damaged in landing accident on 13 January 2003; repairs undertaken during 2003 and made its second 'first' flight (actually third individual flight) on 29 June 2004. Second aircraft was in final assembly in April 2004.

CURRENT VERSIONS: **Me 262A-1c:** Single-seat version; one (501245/'White 3') under construction by Classic; 55 per cent complete in mid-2001, but suspended, pending customer interest.

Me 262B-1c: Two-seat trainer; two (501241/N262AZ/'White 1' and 501243/'White ') under construction; first airframe for private owner in US; second 55 per cent complete, but suspended.

'Me 262A/B-1c': Further two aircraft (501242/'White 8' and 501244/N262MS/'Red 13', or 'Tango Tango'); can be converted to either of above variants; second machine earmarked for private owner in Germany in final assembly in late 2004; will be second aircraft to fly.

Current production aircraft assigned 'c' type suffix by Messerschmitt Foundation to indicate employment of CJ610 engine; maker's serial numbers follow on from German 1945 production.

Descriptions of Me 262 appear in 1945-46 Jane's *and in various historical references. Differences from original are noted below.*

COSTS: Around US$2 million for flying example and US$1.1 million for static display example, excluding engines and avionics.

DESIGN FEATURES: Modifications minimal, except where required to enhance safety. Replacement of Jumo 004 engine involves smaller, lighter and more powerful CJ610 being located in rear of original-sized nacelles; interior air duct and false Jumo engine exterior shape below access panels restore weight to original for aerodynamic and balance purposes. Landing gear mountings strengthened to obviate original weakness; modern disc brakes within mainwheel hubs; nosewheel brake deleted.

First reproduction Messerschmitt Me 262, N262AZ *(Nick Cirelli)* ***NEW***/0522814

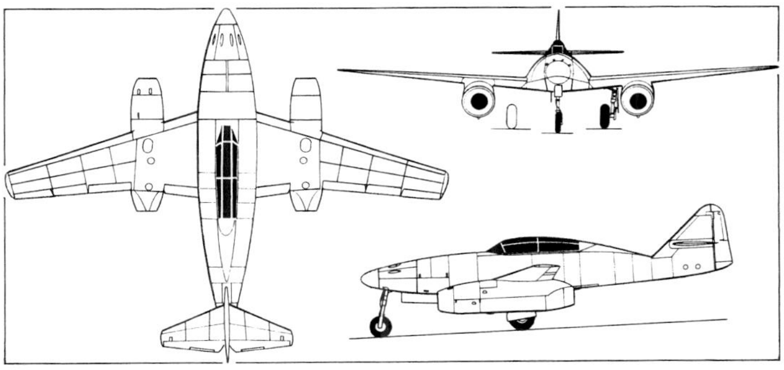

Me 262 Project Messerschmitt Me 262B-1c two-seat trainer version of Second World War jet fighter *(Paul Jackson)* 0103513

POWER PLANT: Two 12.7 kN (2,850 lb st) General Electric CJ610 turbojets; soft throttle stop at 8.0 kN (1,800 lb st) to emulate Jumo engine max output; full estimated installed thrust rating of up to 11.1 kN (2,500 lb st) employed for initiating (only) take-off roll and may be re-engaged above 226 kt (418 km/h; 260 mph) asymmetric safety speed.

PERFORMANCE:
Never-exceed speed (VNE) .... 434 kt (804 km/h; 500 mph)[1]
Max level speed (estimated) .... 539 kt (998 km/h; 620 mph)
Range with max fuel .... more than 869 n miles (1,609 km; 1,000 miles)

[1] *voluntary limitation*

UPDATED

MIDWEST AEROSPORT

MIDWEST AEROSPORT INC

By late 2004, the prototype Midwest Aerosport Formula GT was for sale, the company having ceased trading and reassigned the Formula GT design to New Century Aerosport, which also manufactures the Radial Rocket.

UPDATED

MINI-IMP

MINI-IMP AIRCRAFT COMPANY

PO Box 2011, Weatherford, Texas 76086-2011
Tel: (+1 817) 596 32 78
e-mail: info@mini-imp.com
Web: http://www.mini-imp.com
PRESIDENT: Gary S James

Company formed to produce kit version of Mini-IMP high-performance lightplane, designed and marketed in 1970s by Moulton B 'Molt' Taylor, who gained earlier prominence with the FAA-certified (1956) Aerocar flying car. Also holds rights to Micro-IMP.

VERIFIED

MINI-IMP MINI-IMP

TYPE: Single-seat sportplane kitbuilt.
PROGRAMME: Single-seat version of four-place Aerocar IMP ('Independently Made Plane'); first flown in mid-1970s (ahead of Imp); fitted in early 1976 with folding wing. Marketed as minimal kit or plans-only homebuilt by Aerocar Inc. Last described in 1992-93 *Jane's*.

Relaunched as partial kit by current owner in April 2002. Available in original configuration, or as Model C, with lengthened nose and additional vertical fin for higher-powered engines.

CUSTOMERS: Some 25 of original series sold as minimal kits, of which between six and 10 have flown, including one in Brazil.

COSTS: Kit US$12,500; 'professional' kit US$6,450 (2004), excluding engine.

DESIGN FEATURES: Compact, high-speed lightplane of high aerodynamic efficiency; mid-mounted, engine and propeller at extreme rear to achieve low cross-sectional area and minimal airflow interruption. 'Inverted butterfly' tail surfaces, optionally augmented by vertical fin above fuselage. Drive-shaft resonance obviated by Flexodyne coupling.

High, constant-chord wing, with downturned tips; wings foldable for storage in 10 minutes; flaperons deflect +10° in cruising flight to unload tail surfaces and increase speed. Aerofoil GA-PC(1).

FLYING CONTROLS: Manual. Almost full-span flaperons. Inverted V ruddervators with 40° anhedral.

STRUCTURE: Aluminium overall, with glass fibre for some non-structural parts.

LANDING GEAR: Tricycle type. Electric retraction. Mainwheels hinge upwards, legs stowed in fuselage side, wheels in wing; nosewheel retracts forward on separate electrical circuit. Optional spring aluminium mainwheel legs. Typical tyre sizes 5.00-5 main; 10×3.50-4 nose.

POWER PLANT: Mid-mounted power pant of 60 to 75 kW (80 to 100 hp), including Teledyne Continental O-200, Rotax 912 or Jabiru 3300. Internal cooling fan.

Centre-section fuel tank, capacity 45 litres (12.0 US gallons; 10.0 Imp gallons). Optional external wingtip tanks, each 34 litres (9.0 US gallons; 7.5 Imp gallons).

DIMENSIONS, EXTERNAL:
Wing span .... 7.57 m (24 ft 10 in)

Mini-IMP of original series *(Paul Jackson)* 0137384

Wing chord, constant .... 0.91 m (3 ft 0 in)
Length overall: normal .... 4.88 m (16 ft 0 in)
  elongated nose .... 5.18 m (17 ft 0 in)
Height: over fuselage .... 1.22 m (4 ft 0 in)
  propeller turning .... 1.89 m (6 ft 2½ in)

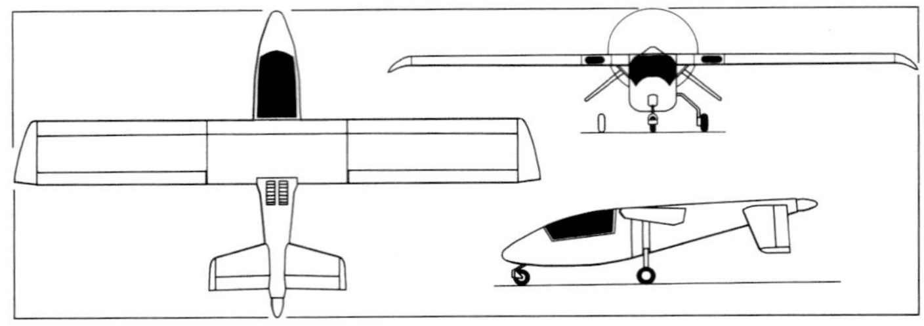

Mini-Imp single-seat sportplane *(Paul Jackson)* ***NEW***/0587519

WEIGHTS AND LOADINGS:
Weight empty: normal .... 236 kg (520 lb)
  elongated nose .... 354 kg (780 lb)
Max T-O weight: normal .... 362 kg (800 lb)
  elongated nose .... 453 kg (1,000 lb)
PERFORMANCE (75 kW; 100 hp engine):
Max level speed .... more than 174 kt (322 km/h; 200 mph)
Cruising speed .... more than 156 kt (290 km/h; 180 mph)
Stalling speed: .... 44 kt (81 km/h; 50 mph)
Max rate of climb at S/L .... 457 m (1,500 ft)/min
T-O run .... 244 m (800 ft)
g limits .... +6/–4

UPDATED

MONTAGNE

MONTAGNE AVIATION LLC

6330 East Mountain Goat Circle, Wasilla, Alaska 99654
Tel: (+1 907) 745 75 97
e-mail: mtgoatacft@ak.net
Web: http://www.bushplanes.com
PRESIDENT: Bill Montagne

Montagne Aviation (formerly Kinetic) has moved to Wasilla in Alaska to start production of the Mountain Goat and its proposed Big Horn development. Few details of the latter have been revealed.

VERIFIED

MONTAGNE MOUNTAIN GOAT

TYPE: Two-seat lightplane.
PROGRAMME: Fourth pre-production (but definitive) aircraft registered N101MG in February 1997. Company moved to Alaska in 2003 to start production and develop new variants.
CURRENT VERSIONS: **Mountain Goat:** Standard version, *as described.*
Mountain Goat UAV: Proposed UAV version. Major advantages stated to be all-weather operations and ability to use rougher terrain for take-off and landing; quiet technology propeller and exhaust systems.
CUSTOMERS: Orders being taken; first completed kit is N93GW, built by Goat Works Inc and registered in September 2001. Company intends construction of 50 aircraft in first full year of production, rising to 500 per year in fifth. Two titanium-framed aircraft under construction late 2003.
COSTS: US$145,000, equipped (2003), US$160,000 with titanium frame.
DESIGN FEATURES: Similar in appearance to Piper PA-18 Super Cub, but with 10 cm (4 in) wider fuselage. High, single strut-braced wings; empennage strut-braced. Sweptback fin. Heavy-duty construction and refined aerodynamics for inhospitable terrain.
FLYING CONTROLS: Conventional and manual. Actuation bellcrank and pushrod. Fowler flaps and flaperons both reflex up for faster cruise speeds. Horn-balanced rudder and elevators; flight-adjustable trim tab on port elevator; ground-adjustable tab on rudder.
STRUCTURE: Fuselage and empennage of fabric-covered 4130 steel tube. All-metal flush riveted high wing has single bracing struts and flush riveted flaps and flaperons. Strut-braced tailplane. Optional titanium 3-2.5 and 6-4 replacing steel.
LANDING GEAR: Tailwheel type; fixed. Suspension of bungee cord with hinged side Vs and half-axles or spring steel cantilever. Mainwheels have Goodyear 26×10.5-6 tyres (tyres in the range 6.00-6 to 31-6 can be fitted) and toe-operated dual hydraulic caliper brakes; dual-piston brakes optional. Scott 3200 steerable tailwheel with heavy-duty tail spring. Float and ski mountings optional.
POWER PLANT: One 134 kW (180 hp) Textron Lycoming IO-360-B2E flat-four driving a McCauley 76-66 two-blade, ground-adjustable pitch propeller; Hartzell and MT propellers can optionally be fitted. Fuel capacity 246 litres (65.0 US gallons; 54.1 Imp gallons) of which 235 litres (62.0 US gallons; 51.6 Imp gallons) usable.

Montagne Mountain Goat fourth pre-production aircraft built by Bill Montagne with appropriate registration and decoration ***NEW***/0533718

ACCOMMODATION: Two seats in tandem with optional jump seat. One-piece full Plexiglas door on each side of fuselage. Dual controls. Four-point seat belts. Baggage space behind rear seat.
SYSTEMS: Skytec lightweight starter; B&C 40 A alternator; electric auxiliary fuel pump.
AVIONICS: *Comms:* moving map GPS/COM, Mode C transponder. Vision Microsystems engine management system.
EQUIPMENT: Wingtip strobe lighting; navigation lights, landing light and anti-collision beacon.
DIMENSIONS, EXTERNAL:
Wing span .... 10.82 m (35 ft 6 in)
Wing aspect ratio 6.7
Length overall .... 7.42 m (24 ft 4 in)
Height overall .... 2.06 m (6 ft 9 in)
Propeller diameter .... 1.93 m (6 ft 4 in)
DIMENSIONS, INTERNAL:
Baggage hold volume .... 0.91 m^3 (32.0 cu ft)
AREAS:
Wings, gross .... 17.47 m^2 (188.0 sq ft)
WEIGHTS AND LOADINGS:
Weight empty: standard .... 567 kg (1,250 lb)
  titanium version .... 531 kg (1,170 lb)
Baggage capacity: standard .... 159 kg (350 lb)
  titanium version .... 195 kg (430 lb)
Max T-O weight .... 1,134 kg (2,500 lb)
Max zero-fuel weight .... 943 kg (2,079 lb)
Max wing loading .... 64.9 kg/m^2 (13.30 lb/sq ft)
Max power loading .... 8.45 kg/kW (13.89 lb/hp)
PERFORMANCE:
Never-exceed speed (VNE) .... 185 kt (342 km/h; 213 mph)
Max cruising speed at 75% power .... 138 kt (256 km/h; 159 mph)
Econ cruising speed at 55% power .... 115 kt (212 km/h; 132 mph)
Stalling speed, power off:
  flaps up .... 48 kt (89 km/h; 55 mph)
  flaps down .... 33 kt (62 km/h; 38 mph)
Max rate of climb at S/L .... 671 m (2,200 ft)/min
T-O and landing run .... 107 m (350 ft)
Range with max fuel, 45 min reserves .... 659 n miles (1,287 km; 800 miles)

UPDATED

MOONEY

MOONEY AIRPLANE COMPANY (MAC)

Louis Schreiner Field, Kerrville, Texas 78028
Tel: (+1 830) 896 60 00 or 792 29 10
Fax: (+1 830) 896 81 80
e-mail: sales@mooney.com
Web: http://www.mooney.com
PRESIDENT: J Nelson Happy
EXECUTIVE VICE-PRESIDENT: Nicholas Chabbert
VICE-PRESIDENT, SALES AND MARKETING: David J Copeland
MANAGING DIRECTOR: Tom Gray

Original Mooney Aircraft Corporation formed in Wichita, Kansas, 1948; produced single-seat M-18 Mite until 1952 (total 282); M20 family in production since then. Alexandre Couvelaire, President of Euralair/Avialair Paris, France, and Michel Seydoux, President of MSC, jointly acquired Mooney in 1985; Mooney and Aerospatiale (Socata) announced joint development of TBM 700 June 1987 (see under Socata in French section), but Mooney withdrew in 1991. Controlled by AVAQ Group Inc since 1998. By early 2001, Mooney had produced 10,320 aircraft, including 9,940 of M20 family. New Mooney Encore announced October 1996; Eagle in 1998. Company decided in 1999 not to proceed with a turboprop, but was considering reintroducing a pressurised aircraft with 240 kt (444 km/h; 276 mph) cruising speed. Mooney delivered 100 aircraft in 2000, compared with 97 in 1999, 93 in 1998, 86 in 1997 and 73 in 1996. Personnel totalled 220 in 2001, including those engaged on Lockheed Martin F-16 and C-130, Boeing, Air Tractor, Aviat Pitts Special and Commander subcontracts; by early February 2002, this had been reduced to 11.

Mooney Aircraft Corporation filed for Chapter 11 bankruptcy protection on 27 July 2001, continuing production while restructuring the company; delivered 29 aircraft in 2001 against year's target of 70. Announced on 16 November 2001 that extended financing had been received to continue operations, including spares support, pending sale of the company. On 8 February 2002 AASI announced it had purchased Congress Financial Corporation's position as senior creditor and intended restarting production as well as moving its own production facility to Kerrville.

AASI subsequently reorganised as Mooney Aerospace Group, Ltd, with Mooney Airplane Company as wholly owned subsidiary. MAC received FAA production certificate on 26 June 2002 for Eagle 2, Ovation 2 and Bravo 2. Employment at Kerville totalled 186 in May 2004. Total of 10 aircraft delivered in 2002, all in final quarter, 36 in 2003, and 13 in the first six months of 2004. Target production for 2004 is 70 aircraft.

On 1 August 2003 Mooney signed an agreement with BAE Systems, Foxton Investments, the Kaskol Group of Russia, and Venture Industries, to form an international partnership for the development and marketing of a range of new-generation aircraft under the Mooney name.

In late May 2004, Allen Holding Finance Ltd bought the stock of Mooney Airplane Company from its parent Mooney Aerospace Group Ltd for US$4 million. Under the terms of the agreement Allen Holding assumed more than US$21 million of MAC's debt owed to the secured debenture holders and also agreed to provide $4 million of new capital within 30 days.

UPDATED

MOONEY M20M BRAVO

TYPE: Four-seat lightplane.

PROGRAMME: Wood-winged M20 prototype first flew 10 August 1953; some 700 of early series built before design translated into all-metal with M20B, introduced in 1961. Progressed through several further subvariants, of which M20M announced 2 February 1989 as TLS (Turbo Lycoming Sabre, with 201 kW; 270 hp TIO-540-AF1A); Bravo introduced in 1996, followed by Bravo DX on 2 April 2003 and Bravo GX in February 2004.

CURRENT VERSIONS: **Bravo DX:** Introduced at EAA Sun 'n' Fun at Lakeland 2 April 2003. Features Garmin avionics suite including GNS 530/430, GDL 49 Nexrad weather uplink, GTX 330 Mode S transponder and Bendix/King KFC 225 autopilot/flight director, 3.28 m^3 (115.7 cu ft) oxygen system, Precise Flight speedbrakes, and six-way adjustable leather seats as standard; Honeywell K1825 electronic HSI, TKS anti-icing system and electric 17,000 BTU air conditioning system optional.

Bravo GX: Announced 10 February 2004. Features Garmin G1000 all-glass integrated avionics comprising dual 10.4 in XGA high-resolution colour LCD primary flight (PFD) and multifunction (MFD) displays; digital audio control panel with integrated marker beacon and clearance playback capability; dual integrated radio modules that provide IFR oceanic-approved GPS, VHF navigation with ILS and VHF communication, with 16-watt transceivers and 8.33 kHz channel spacing; Mode S transponder with Traffic Information Service; advanced solid-state Attitude and Heading Reference System (AHRS) with rapid alignment whether on the ground or in the air; digital air data computer; Mid-Continent 3400 backup attitude indicator; worldwide terrain database depicting topographic information and relative terrain; and (US specification aircraft only) XM satellite radio receiver featuring full-colour, graphical weather information via digital datalink and 101 channels of CD-quality audio. The G1000 installation adds 16 kg (36 lb) to the aircraft's empty weight. Deliveries from third quarter 2004.

Bravo GX Freedom: Limited edition; announced at NBAA Convention, Las Vegas, October 2004. Standard features include Goodrich WX500 Stormscope and increase in usable fuel capacity to 385 litres (102 US gallons; 84.7 Imp gallons).

CUSTOMERS: Total of 315 TLS/Bravos delivered by June 2001 halt of production. Deliveries resumed in July 2002. Da Kine, Hawaii Inc of Hood River, Oregon, became the first customer for a Mooney Airplane Company-built aircraft (a Bravo), with an order announced

Mooney M20M Bravo DX Garmin avionics version *(Paul Jackson)* ***NEW***/0583498

on 25 July 2002. Five delivered in 2003, and nine in 2004, including one to German customer.

COSTS: Bravo DX US$439,950; Bravo GX US$459,950, both standard equipped (2004).

DESIGN FEATURES: High-efficiency touring aircraft, originally designed by Mooney brothers. Low wing and mid-mounted tailplane; flying surfaces of characteristic Mooney 'reversed' configuration with unswept leading-edges and sharply forward-swept trailing-edges.

Wing section NACA $63_2$182-215 at root, $64_1$181-412 at tip; dihedral 5° 30'; incidence 2° 30' at root, 1° at tip; wing swept forward 2° 29' at quarter chord.

FLYING CONTROLS: Conventional and manual. Sealed gap, differential ailerons; fin and tailplane integral so that both tilt, varying tailplane incidence for trimming; no trim tabs; electrically actuated single-slotted flaps; electric rudder trim; Precise Flight speed brakes. Control surface movements: ailerons +14° 30'/-8°, elevator ±22°, rudder ±24°, flaps -10° T-O and -33° landing.

STRUCTURE: Single-spar wing with auxiliary spar out to mid-position of flaps; wing and tail surfaces covered with stretch-formed wraparound skins. Steel tube cabin section covered with light alloy skin; semi-monocoque rear fuse-lage with extruded stringers and sheet metal frames.

LANDING GEAR: Electrically retractable levered suspension tricycle type with airspeed safety switch bypass. Nosewheel retracts rearward, main units inward into wings. Rubber disc shock-absorbers in main units. Cleveland mainwheels, size 6.00-6 (6-ply), and steerable nosewheel, size 5.00-5 (6-ply). Tyre pressure, mainwheels 2.07 bar (30 lb/sq in), nosewheel 3.38 bar (49 lb/sq in). Cleveland hydraulic single-disc (double optional) brakes on mainwheels. Parking brake.

POWER PLANT: One 201 kW (270 hp) turbocharged Textron Lycoming TIO-540-AF1B flat-six engine, driving a McCauley B3D32C417/82NRD-7 three-blade, metal propeller. Two integral fuel tanks in inboard wing leading-edges, with a combined capacity of 363 litres (96.0 US gallons; 79.9 Imp gallons), of which 337 litres (89.0 US gallons; 74.1 Imp gallons) usable. Two-piece nose cowling is of glass fibre/graphite composites construction.

ACCOMMODATION: Cabin accommodates four persons in pairs on individual vertically adjusting seats with reclining back, armrests, lumbar support and headrests; side armrests removable in rear seats. Front seats have inertia reel shoulder harnesses. Dual controls standard. Overhead ventilation system. Cabin heating and cooling system, with adjustable outlets and illuminated control. One-piece wraparound windscreen. Tinted Plexiglas windows. Rear seats removable for freight stowage. Rear seats fold forward for carrying cargo. Single door on starboard side. Compartment for 54 kg (120 lb) baggage behind cabin, with access from cabin or through door on starboard side. Amsafe inflatable restraint seatbelts optional from third quarter 2004.

SYSTEMS: Dual 70 Ah 28 V alternators, dual 24 V 10 Ah batteries, voltage regulator and warning lights, together with protective circuit breakers. Oxygen system, capacity

MOONEY M20 PRODUCTION (TO MID-2004)

Model	Name	Eng desig	Built	Certification
M20			200	24 Aug 55
M20A			501	13 Feb 58
M20B	Mk 21		224	14 Dec 60
M20C	Mk 21		2,222	20 Oct 61
	Ranger			
	Aerostar 200			
M20D	Master		160	15 Oct 62
M20E	Super 21	21	1,485	4 Sep 63
	Chapparal	21		
	Aerostar 201	21		
M20F	Executive	22	1,251	25 Jul 65
	Aerostar 220	22		
M20G	Statesman		190	13 Nov 67
M20J	201	24	2,135	27 Sep 76
	201LM			
	201SE			
	205			
	ATS			
	MSE			
	Allegro			
MT20	TX-1		2	nil
M20K	231	25	1,151	16 Nov 78
	231SE	25		
	252 TSE	25		
	Encore	25		
M20L	PFM	26	41	25 Feb 88
M20M	TLS	27	315	28 Jun 89
	Bravo/Bravo DX/GX	27	19	
M20R	Ovation	29	280	30 Jun 94
	Ovation 2	29	54	
M20S	Eagle	30	65	7 Feb 99
M20T	EFS	28	1	nil
Total			**10,298**	

Notes: Individual models have had several marketing names. Engineering designation forms part of aircraft's serial number. Certification date is for FAA.

3,259 litres (115 cu ft), with masks and overhead outlets, standard. TKS known ice protection system and air conditioning optional.

AVIONICS: Standard equipment (Bravo) comprises Garmin GNS 530 nav/com/GPS, GNS 430 with GI 106A indicator, transponder/encoder, two-axis autopilot, and slaved HSI. Goodrich WX-500 Stormscope and SkyWatch optional. Bravo DX avionics as under Current Versions, above.

EQUIPMENT: Standard equipment includes attitude indicator, IFR directional gyro, two electric fuel quantity gauges, electric OAT gauge, CHT and EGT gauges, Precise Flight speed brakes, push-pull tube-actuated flight control system, annunciator panel; navigation lights, wing-mounted dual landing/taxying lights, internally lit instruments with rheostat; sound-damping composites interior, vertically adjusting front seats, wing jackpoints and external tiedowns, fuel tank quick drains, auxiliary power plug, heated pitot tube, polished propeller spinner, epoxy polyimide and conversion anti-corrosion treatment, external polyurethane paint finish, dual static ports with drains, and alternate static air source.

DIMENSIONS, EXTERNAL:
Wing span ... 11.00 m (36 ft 1 in)
Wing chord, mean ... 1.50 m (4 ft 11¼ in)
Wing aspect ratio ... 7.4
Length overall ... 8.15 m (26 ft 9 in)
Height overall ... 2.54 m (8 ft 4 in)
Tailplane span ... 3.58 m (11 ft 9 in)
Wheel track ... 2.79 m (9 ft 2 in)
Wheelbase ... 2.02 m (6 ft 7½ in)
Propeller diameter ... 1.91 m (6 ft 3 in)
Baggage door: Width ... 0.53 m (1 ft 9 in)
Height ... 0.43 m (1 ft 5 in)

DIMENSIONS, INTERNAL:
Cabin: Length ... 3.20 m (10 ft 6 in)
Max width ... 1.10 m (3 ft 7½ in)
Max height ... 1.13 m (3 ft 8½ in)
Floor area ... 3.53 m^2 (38.0 sq ft)
Volume ... 3.9 m^3 (137 cu ft)
Baggage compartment volume ... 0.64 m^3 (22.6 cu ft)

AREAS:
Wings, gross ... 16.26 m^2 (175.0 sq ft)
Ailerons (total) ... 1.06 m^2 (11.40 sq ft)
Trailing-edge flaps (total) ... 1.66 m^2 (17.90 sq ft)
Fin ... 0.73 m^2 (7.92 sq ft)
Rudder ... 0.58 m^2 (6.23 sq ft)
Tailplane ... 1.99 m^2 (21.45 sq ft)
Elevators (total) ... 1.11 m^2 (12.05 sq ft)

WEIGHTS AND LOADINGS:
Weight empty ... 1,029 kg (2,268 lb)
Max T-O weight ... 1,528 kg (3,368 lb)
Max landing weight ... 1,452 kg (3,200 lb)
Max wing loading ... 94.0 kg/m^2 (19.25 lb/sq ft)
Max power loading ... 7.59 kg/kW (12.47 lb/hp)

PERFORMANCE:
Max cruising speed:
at FL130 ... 195 kt (361 km/h; 224 mph)
at FL250 ... 220 kt (407 km/h; 253 mph)
Stalling speed:
flaps and wheels up ... 66 kt (122 km/h; 76 mph)
flaps and wheels down ... 59 kt (109 km/h; 68 mph)
Max rate of climb at S/L ... 344 m (1,130 ft)/min
Max certified altitude ... 7,620 m (25,000 ft)
Range with reserves, at FL130 ... 1,070 n miles (1,982 km; 1,231 miles)
Endurance ... 6 h 42 min

UPDATED

MOONEY M20R OVATION2

TYPE: Four-seat lightplane.

PROGRAMME: Prototype (N20XR) rolled out April 1994; first flight May 1994; FAA certification June 1994. Similar M20S Eagle certified February 1999, but discontinued in 2003.

CURRENT VERSIONS: **Mooney M20R Ovation2:** Baseline version, superseded Ovation; unveiled at the AOPA Convention on 22 October 1999. A limited edition Mooney Ovation2 Platinum Edition, with advanced avionics and instrumentation, plus airframe and cabin upgrades, has been introduced. An Ovation2 which made its first flight on 18 June 2002 was the first new production Mooney following the formation of Mooney Aerospace Group, Ltd.

Ovation2 DX: Introduced at EAA Sun 'n' Fun at Lakeland 2 April 2003. Features thin-blade, McCauley two-blade propeller providing a 6 kt (11 km/h; 7 mph) increase in cruising speed at FL80; Garmin avionics suite including GNS 530/430, GDL 49 Nexrad weather uplink, GTX 330 Mode S transponder and Bendix/King KFC 225 autopilot flight director; six-way adjustable leather seats with adjustable lumbar support and Precise Flight speedbrakes as standard; 3.28 m^3 (115.7 cu ft) oxygen system, TKS anti-icing system and 25,000 BTU air conditioning system optional.

Ovation2 GX: Announced 10 February 2004. Features Garmin G1000 all-glass integrated avionics comprising dual 264 mm (10.4 in) XGA high-resolution colour LCD primary flight (PFD) and multifunction (MFD) displays; digital audio control panel with integrated marker beacon and clearance playback capability; dual integrated radio modules that provide IFR oceanic-approved GPS, VHF navigation with ILS and VHF communication, with 16-watt transceivers and 8.33 kHz channel spacing; Mode S transponder with Traffic Information Service; advanced solid-state attitude and heading reference system (AHRS) with rapid alignment whether on the ground or in the air; digital air data computer; worldwide terrain database depicting topographic information and relative terrain; (US specification aircraft only) XM satellite radio receiver featuring full-colour, graphical weather information via digital datalink and 101 channels of CD-quality audio; and Mid-Continent 3400 backup electric attitude indicator. The G1000 installation adds 16 kg (36 lb). First Ovation2 GX is c/n 29-0333.

Mooney M20R Ovation2 DX *(Paul Jackson)* ***NEW***/0583527

Mooney M20R Ovation2 *(James Goulding)* 0092656

Ovation2 GX Freedom: Limited edition announced at NBAA Convention, Las Vegas, October 2004. Standard features include Goodrich WX500 Stormscope and increase in usable fuel capacity to 385 litres (102 US gallons; 84.7 Imp gallons).

CUSTOMERS: Most sold in USA, but others exported to Brazil, France, Germany, Israel, Italy, Paraguay, South Africa, Switzerland, Thailand and UK; orders include three for Western Michigan University. Total of 280 Ovations and 62 Eagles had been registered before Mooney reconstituted in February 2002; 65th and final Eagle completed in 2003. Ten delivered in 2002, comprising eight Ovations and two Eagles, 19 Ovation2s, 11 Ovation2 DXs, and one Eagle 2 in 2003, and 28 Ovation2 DXs in 2004.

COSTS: Ovation US$319,950; Ovation2 DX US$389,950; Ovation2 GX US$409,950, all standard equipped (2004).

DESIGN FEATURES: Combines airframe of M20M Bravo (which see) with normally aspirated flat-six engine. Completely restyled instrument panel, seats and cabin interior with sandwich-core rigid trim/soundproofing panels and leather and wool fabric upholstery.

POWER PLANT: M20R Ovation2 has one 224 kW (300 hp) Teledyne Continental IO-550-G5B flat-six engine, derated to 209 kW (280 hp), driving a McCauley 3A32C418-G/82NRC-9 three-blade, constant-speed metal propeller; Ovation2 has redesigned propeller giving slight performance improvement. Usable fuel 337 litres (89.0 US gallons; 74.1 Imp gallons).

AVIONICS: Generally as for Mooney M20M Bravo.

EQUIPMENT: Generally as for M20M Bravo, except as detailed under Current Versions.

DIMENSIONS, EXTERNAL:
Propeller diameter ... 1.85 m (6 ft 1 in)

WEIGHTS AND LOADINGS:
Weight empty ... 1,009 kg (2,225 lb)
Baggage capacity ... 54 kg (120 lb)
Max usable fuel ... 242 kg (534 lb)
Max T-O weight ... 1,528 kg (3,368 lb)
Max landing weight ... 1,451 kg (3,200 lb)
Max wing loading ... 94.0 kg/m^2 (19.25 lb/sq ft)
Max power loading ... 7.32 kg/kW (12.03 lb/hp)

PERFORMANCE :
Max cruising speed ... 192 kt (356 km/h; 221 mph)
Stalling speed:
flaps and wheels up ... 66 kt (123 km/h; 76 mph)
flaps and wheels down ... 59 kt (110 km/h; 68 mph)
Max rate of climb at S/L ... 381 m (1,250 ft)/min
Service ceiling ... 6,100 m (20,000 ft)
Max range, economy cruise, with reserves, at FL90 ... 1,280 n miles (2,370 km; 1,473 miles)

UPDATED

MOONEY SPEEDSTER (TOXO)

This programme was abandoned in early 2004.

UPDATED

NAGY

IMRE NAGY

32148 Shirey Road #A, Escondido, San Diego, California 92026-4716
e-mail: nagyairship@aol.com
Web: http://www.nagyairship.com

Mr Nagy has applied for an international patent for an airship of innovative design, a small prototype of which was nearing completion in September 2004.

NEW ENTRY

NAGY HIGH-SPEED AIRSHIP

TYPE: Helium rigid airship.

PROGRAMME: First reported in mid-2003. Prototype of 2-ton version (c/n 001) registered as N241H on 1 June 2004 and said to be nearing completion in September 2004. All R & D completed; all versions said to be production-ready on receipt of order.

CURRENT VERSIONS: **2-ton version::** 20-passenger prototype/POC vehicle for projected larger versions. Potential as air patrol, rescue, ambulance or media vehicle, or as 'flying home'.

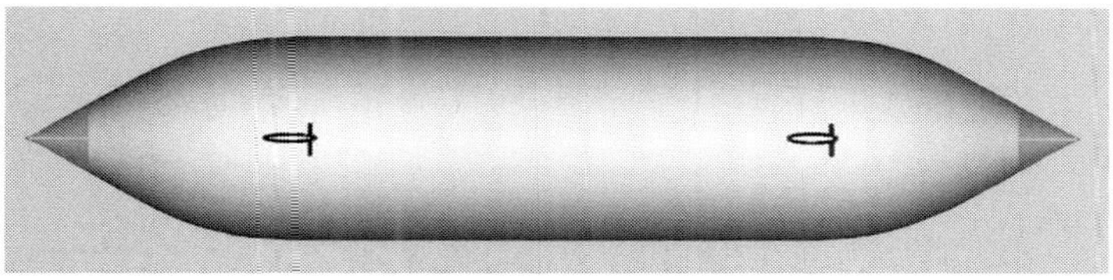

Artist's impression of NAGY High-Speed Airship ***NEW***/1047198

10-ton version: : Projected 100-passenger version. Additional applications could include airborne hospital, coast guard, military or homeland security vehicle.

100-ton version: : Projected 1,000-passenger version. Possible high-density cargo or military transport or luxury travel vehicle.

COSTS: Basic unit price US$4 million for 2-ton, US$8 million for 10-ton, US$12 million for 100-ton (2004).

DESIGN FEATURES: Mainly cylindrical hull, tapering to pointed cone at each end. Crew and passenger accommodation entirely internal. Vectored-thrust propulsion units (360° travel) mounted on external outriggers each side, fore and aft. Payload capability quoted as up to 85 per cent of total body volume. Requires no ground handling crew.

FLYING CONTROLS: Computer-controlled, with GPS-based navigation. Helium recovery system eliminates need for ballast.

STRUCTURE: Described only as 'space-age stronger-than-steel fabrics' and 'stronger than rigid' frame body. Helium contained in multiple independent cells ("can not crash, even if cut in half").

Following data all provisional.

POWER PLANT: Four 559 kW (750 hp) engines (type unspecified) in 2-ton, eight in 10-ton and 12 in 100-ton versions. Fuel capacity (all versions) adequate for non-stop global circumnavigation.

EQUIPMENT: Hoist to upload or download passengers or cargo where landings are not possible.

DIMENSIONS, EXTERNAL (A: 2-ton, B: 10-ton, C: 100-ton):
Length overall: A ... 45.72 m (150 ft)
B ... 82.30 m (270 ft)
C ... 167.64 m (550 ft)
Diameter: A ... 10.67 m (35 ft)
B ... 16.76 m (55 ft)
C ... 33.53 m (110 ft)

DIMENSIONS, INTERNAL:
Total volume: A ... 2,720 m^3 (96,000 cu ft)
B ... 11,600 m^3 (410,000 cu ft)
C ... 116,100 m^3 (4,100,000 cu ft)
Passenger compartment:
Length: A ... 42.67 m (140 ft)
B ... 76.20 m (250 ft)
C ... 152.40 m (500 ft)
Width: A ... 3.05 m (10 ft)
B ... 6.10 m (20 ft)
C ... 9.14 m (30 ft)
Height: A ... 3.05 m (10 ft)
B ... 6.10 m (20 ft)
C ... 9.14 m (30 ft)
Floor area: A ... 130.1 m^2 (1,400 sq ft)
B (2 floors, total) ... 929.0 m^2 (10,000 sq ft)
C (3 floors, total) ... 4,180.6 m^2 (45,000 sq ft)

WEIGHTS AND LOADINGS:
Max T-O weight: A ... 1,814 kg (4,000 lb)
B ... 9,071 kg (20,000 lb)
C ... 90,718 kg (200,000 lb)

PERFORMANCE (estimated):
Max level speed: A, B, C ... more than 173 kt (321 km/h; 200 mph)
Absolute ceiling: A, B, C ... 21,340 m (70,000 ft)

NEW ENTRY

NASA

NATIONAL AERONAUTICS AND SPACE ADMINISTRATION

(Office of Aeronautics)
600 Independence Avenue SW, Washington, DC 20546
Tel: (+1 202) 453 26 93
Fax: (+1 202) 426 42 56
Web: http://www.nasa.gov

ADMINISTRATOR: Sean O Keefe
CHIEF ENGINEER: Daniel R Mulville
ASSOCIATE ADMINISTRATOR, OFFICE OF AERONAUTICS: J Victor Lebacqz
CHIEF FINANCIAL OFFICER: Arnold G Holz

NASA continues to undertake research and development allied to aerospace technology but space science and space flight activity now receives the greater share of the annual budget. For FY04, total of US$15.47 billion requested, including US$959 million for aeronautics technology. Personnel totalled just over 18,000 at beginning of June 2003. Prime NASA concerns have traditionally been, and continue to be, R&D, rather than operationally driven. In addition to programmes described here in detail, NASA funds are also allocated to a variety of projects and concept studies, including research into 'quiet' sonic booms (with Northrop Grumman), air traffic management and UAVs. However, in March 2004, NASA announced intent to cancel future hypersonic research, although it will complete third and final X-43A flight, which should take place by the end of 2004.

UPDATED

NASA AIRCRAFT TRIALS PROGRAMMES

Of some 100 aircraft operated by NASA, most are standard types assigned to specific research programmes. The fleet comprises about 35 dedicated to research and 65 to programme support; most carry civilian registrations between N400NA and N999NA, assigned according to base. Main operating bases of NASA's Aerospace Technology division are:

Ames Research Center, Moffett Federal Airfield, California (N700-799NA)
Dryden Flight Research Center, Edwards AFB, California (N800-899NA)
Glenn Research Center, Cleveland-Hopkins IAP, Ohio (N600-699NA)
Langley Research Center Langley AFB, Hampton, Virginia (N500-599NA)
Other significant aircraft operations bases are:
Johnson Space Center, Ellington ANGB, Houston, Texas (N900-999NA)
Wallops Flight Facility, Wallops Island, Virginia (N400-499NA)
Details follow of some significant research projects.

X-43 HYPER-X: Second of NASA's 'Waverider' projects, X-43 objective was to allow testing of subscale hydrogen-fuelled GASL Inc scramjet engine at speeds of M7 to M10. Forebody shaped to generate shockwaves which compress air entering scramjet intake; hydrogen-based fuel combusts spontaneously on contact with oxygen contained in the compressed air.

Request for proposals issued on 13 October 1996, with contract, worth US$33.4 million over 55 month period, following in March 1997 to MicroCraft Inc of Tullahoma, Tennessee, although total project cost, including flight testing, subsequently rose to around US$250 million. MicroCraft headed a consortium that included Accurate Automation, Boeing North American and GASL, and produced three (originally to have been four) test vehicles, each approximately 3.7 m (12 ft) long, with a span of about 1.5 m (5 ft) and gross weight of 1,270 kg (2,800 lb). Preliminary design review completed in June 1997, with fabrication commencing in third quarter of 1997. First flight vehicle delivered to NASA's Dryden Flight Research Center in October 1999.

Three flights were made with first originally due to take place in May 2000, after captive-carry flight on NASA's NB-52B launch platform; however, development delays resulted in postponement to 2001. Initial 'flight' was a captive-carry test on 28 April 2001. The first launch was planned to reach M7, followed by second flight to M7 and final flight to M10. After drop launch at altitude of 6,100 m (20,000 ft), an Orbital Sciences' Pegasus booster rocket accelerated test vehicle to specified speed at approximately 30,500 m (100,000 ft), whereupon X-43 separated to fly under own power. Flight path was entirely over water and no attempt was made to recover test vehicles from Pacific Ocean. First flight, on 2 June 2001, terminated prematurely when X-43A suffered structural failure and deviated from planned flight path following booster ignition; investigations identified loss of control of Pegasus booster as cause of failure.

Boeing F-15B Eagle NASA 836 ***(NASA/Jim Ross)*** ***NEW***/0583524

Unmanned work by NASA includes the X-45 combat vehicle ***(NASA/Lori Losey)*** ***NEW***/0589980

Second attempt to reach speed of M7 successfully accomplished on 27 March 2004, following redesign of Pegasus booster rocket and modification of flight plan to include increased launch altitude from NB-52B of 12,200 m (40,000 ft) to reduce dynamic pressure on booster. X-43A scramjet ignited satisfactorily after separation from booster at M7 and burned for 11 sec, during which time the X-43A continued to accelerate. The third and final mission took place on 16 November 2004 and achieved a speed of M9.8.

Three more versions of X-43 were planned, but all have been cancelled in wake of March 2004 decision to terminate research into hypersonic flight. **X-43B** was to be scaled-up version of X-43A with objective of testing reusable combined-cycle propulsion system. **X-43C** was expected to fly in 2007–08, and would have been powered by a HyTech hydrocarbon-fuelled dual-mode scramjet that was under development for the USAF by Pratt & Whitney. Maximum speeds of the order of M7 were expected to be achieved by the X-43C, which was a 2,270 kg (5,000 lb), 5.03 m (16 ft 6 in) long lifting-body design. Finally, there was the **X-43D**, a potential successor to the X-43A that would have used a cooled hydrogen-fuelled dual-mode scramjet capable of propelling the vehicle to M15.

BOEING F-15B: Original F-15B prototype (NASA837/71-0290) modified for ACTIVE (Advanced Control Technology for Integrated Vehicles) project involving thrust vectoring nozzles developed by Pratt & Whitney. Series of trials, in non-reversionary mode, undertaken in 1999, included take-offs and landings using vectoring only as well as flight to unlimited angles of attack and speeds of M2. New nozzles can turn up to 20° in any direction, permitting thrust control in pitch and yaw. ACTIVE programme is a joint venture involving NASA, the USAF's Wright Laboratory, Boeing (formerly McDonnell Douglas) and Pratt & Whitney.

ACTIVE F-15B also used to evaluate 'neural network'-based controller as opening part of intelligent flight control system (IFCS) programme; total of 15 flights accomplished during March and April 1999. Additional flights conducted by third quarter of 2003, when evaluation completed; latter phase included revolutionary 'self-learning' FCS with potential to allow damaged aircraft to be safely recovered by automatic adjustment of FCS to take into account damaged or defective surfaces and systems.

NASA FLEET 2004

Regn	Type	Base
N1NA	Gulfstream III	Washington, DC
N2NA	Gulfstream I	Houston, Texas
N3NA	Gulfstream I	Huntsville, Alabama
N7NA	Beech B200 King Air	Edwards AFB, California
N8NA	Beech B200 King Air	Wallops Island, Virginia
N145UA	Boeing 747SP-21	Washington, DC
N417NA	Bell UH-1H Iroquois	Kennedy Space Center, Florida
N418NA	Bell UH-1H Iroquois	Kennedy Space Center, Florida
N419NA	Bell UH-1H Iroquois	Kennedy Space Center, Florida
N420NA	Bell UH-1H Iroquois	Kennedy Space Center, Florida
N426NA	Lockheed P-3B Orion	Wallops Island, Virginia
N427NA	Lockheed EC-130Q Hercules	Wallops Island, Virginia
N501NA	Cirrus SR22	Langley, Virginia
N504NA	Cessna 206H	Langley, Virginia
N507NA	Lancair Columbia 300X	Langley, Virginia
N511NA	Northrop T-38A Talon	Houston, Texas
N515NA	Boeing 737-100	Langley, Virginia
N519NA	Dassault HU-25 Guardian	Langley, Virginia
N523NA	Dassault Falcon 20	Langley, Virginia
N524NA	NA OV-10A Bronco	Langley, Virginia
N529NA	Beech 200 King Air	Langley, Virginia
N535NA	Bell UH-1H Iroquois	Langley, Virginia
N557NA	Boeing 757-225	Langley, Virginia
N564NA	Northrop T-38B Talon	Langley, Virginia
N601NA	Lockheed S-3B Viking	Cleveland, Ohio
N607NA	DHC-6 Twin Otter	Cleveland, Ohio
N616NA	Learjet 25	Cleveland, Ohio
N703NA	Bell 301	Moffett Field, California
N710NA	Convair 990	Edwards AFB, California
N714NA	Lockheed C-141A StarLifter	Moffett Field, California
N801NA	Beech 200 King Air	Edwards AFB, California
N803NA	Molino PIK-20E	Edwards AFB, California
N806NA	Lockheed ER-2	Edwards AFB, California
N809NA	Lockheed ER-2	Edwards AFB, California
N816NA	LM F-16A Fighting Falcon	Edwards AFB, California
N817NA	DC-8-72	Edwards AFB, California
N818NA	Lockheed YO-3A	Edwards AFB, California
N819NA	Beech T-34C Turbo-Mentor	Edwards AFB, California
N836NA	Boeing F-15B Eagle	Edwards AFB, California
N837NA	Boeing F-15B Eagle	Edwards AFB, California
N840NA	Boeing F/A-18A Hornet	Edwards AFB, California
N843NA	Boeing F/A-18A Hornet	Edwards AFB, California
N845NA	Boeing F/A-18B Hornet	Edwards AFB, California
N846NA	Boeing F/A-18B Hornet	Edwards AFB, California
N847NA	Boeing F/A-18A Hornet	Edwards AFB, California
N850NA	Boeing F/A-18A Hornet	Edwards AFB, California
N851NA	Boeing F/A-18A Hornet	Edwards AFB, California
N852NA	Boeing F/A-18B Hornet	Edwards AFB, California
N853NA	Boeing F/A-18A Hornet	Edwards AFB, California
N901NA	Northrop T-38A Talon	Houston, Texas
N902NA	Northrop T-38A Talon	Houston, Texas
N903NA	Northrop T-38A Talon	Houston, Texas
N904NA	Northrop T-38A Talon	Houston, Texas
N905NA	Boeing 747-123	Houston, Texas
N907NA	Northrop T-38A Talon	Houston, Texas
N908NA	Northrop T-38A Talon	Houston, Texas
N909NA	Northrop T-38A Talon	Houston, Texas
N910NA	Northrop T-38A Talon	Houston, Texas
N911NA	Northrop T-38A Talon	Houston, Texas
N912NA	Northrop T-38A Talon	Houston, Texas
N913NA	Northrop T-38A Talon	Houston, Texas
N914NA	Northrop T-38A Talon	Houston, Texas
N915NA	Northrop T-38A Talon	Houston, Texas
N916NA	Northrop T-38A Talon	Houston, Texas
N917NA	Northrop T-38A Talon	Houston, Texas
N918NA	Northrop T-38A Talon	Houston, Texas
N919NA	Northrop T-38A Talon	Houston, Texas
N920NA	Northrop T-38A Talon	Houston, Texas
N921NA	Northrop T-38A Talon	Houston, Texas
N923NA	Northrop T-38A Talon	Houston, Texas
N924NA	Northrop T-38A Talon	Houston, Texas
N926NA	General Dynamics RB-57F	Houston, Texas
N931NA	Boeing KC-135A Stratotanker	Houston, Texas
N933NA	Learjet 23	Stennis Space Center, Mississippi
N940NS	Boeing 377 Super Guppy	Houston, Texas
N941NA	Airbus 377SGT-F Super Guppy	Houston, Texas
N944NA	Gulfstream II	Houston, Texas
N945NA	Gulfstream II	Houston, Texas
N946NA	Gulfstream II	Houston, Texas
N947NA	Gulfstream II	Houston, Texas
N948NA	Gulfstream II	Houston, Texas
N949NA	Gulfstream II	Houston, Texas
N950NA	Gulfstream II	Houston, Texas
N955NA	Northrop T-38A Talon	Houston, Texas
N956NA	Northrop T-38A Talon	Houston, Texas
N959NA	Northrop T-38A Talon	Houston, Texas
N960NA	Northrop T-38A Talon	Houston, Texas
N961NA	Northrop T-38A Talon	Houston, Texas
N962NA	Northrop T-38A Talon	Houston, Texas
N963NA	Northrop T-38A Talon	Houston, Texas
N965NA	Northrop T-38A Talon	Houston, Texas
N966NA	Northrop T-38A Talon	Houston, Texas
N967NA	Northrop T-38A Talon	Edwards AFB, California

Notes: Aircraft currently registered to NASA, not necessarily all airworthy at any one time. N714NA and N940NS are in museums. Does not include some temporary military assignments or B-52H 61-0025

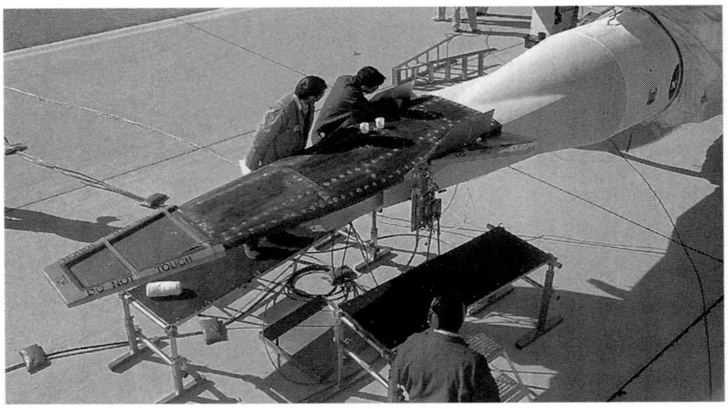

Second X-43A hypersonic research vehicle being mounted under the wing of a B-52 *(NASA/Tony Landis)* ***NEW***/0583560

A second F-15B (NASA836/74-0141), known as the Aerodynamic Flight Facility and equipped with a flight test fixture (airborne test-bench) on the centreline pylon, flew seven sorties between 26 February and 1 April 1996 to test materials for the X-33's thermal protection system; used during 1999 for investigation of supersonic natural laminar flow of advanced wing design, carrying 0.91 by 1.22 m (3 by 4 ft) test article on centreline pylon during four test flights at speeds approaching M2 and altitudes of up to 13,715 m (45,000 ft). This aircraft recently involved in experiments into gust monitoring and aeroelasticity, with 0.45 m (1 ft 6 in) long, flexible, test wing mounted on the flight test fixture. Future projects include trials of Pulse Detonation Engine; these will involve an experimental engine suspended from a new centreline Propulsion Flight Test Fixture (PFTF) that was flown for the first time in late 2001.

BOEING F/A-18: F/A-18A NASA853 refurbished to serve as testbed for Active Aeroelastic Wing (AAW) programme that examines use of 'flexible wings' to research airflow over the wing and assess how this affects controllability. Following modification, aircraft rolled out at Dryden on 27 March 2002 and made initial flight on 15 November 2002, this being one of 30 to 40 sorties planned for the first phase of testing. Second series of trials incorporating wing-warping was intended for early 2004, after development of flight control laws; this phase planned to last for three to four months and include about 30 sorties. However, flying did not resume until December 2004.

Two NASA F/A-18As (NASA843 and NASA847) used for demonstration of autonomous close formation flight in 2001. Achieved automatic 61 m (200 ft) horizontal and 15 m (50 ft)

A pair of F/A-18 Hornets participates in a NASA automated refuelling trial *(NASA/Carla Thomas)* ***NEW***/0583521

The NASA Boeing B-52H *(NASA/Tony Landis)* ***NEW***/0583520

lateral separation on 21 February 2001 on seventh of planned 12 sorties. This test effort also included study of wingtip vortex effect, confirming that close formation flying could offer significant fuel economies; in one test, one F/A-18 flew in wingtip vortex of another F/A-18, with subsequent analysis revealing that trailing aircraft consumed 270 kg (600 lb) less fuel than a third F/A-18 which was not part of the formation. This equates to about a 14 per cent reduction in fuel burn.

Same pair of aircraft subsequently used in early 2003 for initial phase of test project intended to ultimately develop viable automated air-to-air refuelling system for use by unmanned air vehicles.

BOEING 747SP: Engineering work under way to build Stratospheric Observatory For Infra-red Astronomy (SOFIA) with 2.69 m (8 ft 10 in) telescope housed in unpressurised cavity in aft fuselage and looking to port; partner is German Ministry of Science (BMFT), which delivered telescope to USA on 4 September 2002; cruise altitude 12,500 to 13,700 m (41,000 to 45,000 ft); to fly up to 160 sorties (8 hours each) per year for 20 years. US$484 million, 10 year contract secured in December 1996 by a team led by Universities Space Research Association (USRA) to design, assemble, test and operate SOFIA. Former United Airlines Boeing 747SP (N145UA) chosen from several in storage at Las Vegas, Nevada, and ferried to San Francisco in February 1997 for attention by United before being handed over to NASA; following acceptance, 747SP delivered to Waco, Texas, in May 1997 for modification and fitting out by L-3 Communications Integrated Systems; suspension assembly for telescope successfully installed by end of first quarter 2003. Several months of testing of telescope and associated systems will presage delivery to NASA Ames, from where observation missions are expected to begin in September 2004, with full scientific operations set to commence at the end of 2004.

DC-8 airborne science research aircraft *(NASA/Tony Landis)* ***NEW***/0583522

BOEING 757: The ARIES (Airborne Research Integrated Experiments System) aircraft (N557NA) undertook research into turbulence associated with convective storms in early 2002, using an experimental radar designed to detect atmospheric disturbance. Total of 13 missions undertaken from Langley with object of encountering turbulence and comparing it with what radar predicted.

BOEING NB-52B STRATOFORTRESS: Used for aerial launch of test vehicles, with most recent activity involving X-43A Hyper-X project. After more than 40 years of service as launch platform, NB-52B has been replaced by a B-52H (61-0025) which was delivered on 30 July 2001 and then flown to Tinker AFB, Oklahoma for programmed depot maintenance, returning on 17 April 2002; it will have a universal pylon able to accept different adaptors and initially accommodate loads of up to 11,340 kg (25,000 lb), although it is eventually planned to increase payload capability to in excess of 31,750 kg (70,000 lb). The first project involving the B-52H is the X-37 Approach and Landing Test Vehicle (ALTV); trials were due to begin during 2004 and entail air launch from altitudes of up to 12,200 m (40,000 ft).

MCDONNELL DOUGLAS DC-8-72: Assigned to Airborne Science Program at Dryden, following outfitting with Jet Propulsion Laboratory (Pasadena, California) advanced airborne synthetic aperture radar and other sensor systems including imaging spectroradiometer and thermal emission and reflection radiometer. Used for convection and moisture experiments, Earth science studies and also supporting NASA investigation into ozone depletion. In first quarter of 2002, used for ice and snow studies, employing synthetic aperture radar and other sensors to gather data that will give scientists a better understanding of how the snow process works. Previously, the DC-8 (N817NA) had been deployed to the Pacific Rim to conduct studies of volcanic activity and gather data to assist in planning for, and responding to, natural disasters. Most recent task, in March 2004, involved visits to Chile, Costa Rica and Antarctica to map environmental effects of global warming.

UPDATED

NORTHROP GRUMMAN

NORTHROP GRUMMAN CORPORATION

1840 Century Park East, Los Angeles, California 90067-2199
Tel: (+1 310) 553 62 62
Fax: (+1 310) 201 30 23
Web: http://www.northropgrumman.com
CHAIRMAN AND CEO: Dr Ronald D Sugar
CORPORATE VICE-PRESIDENT AND CHIEF FINANCIAL OFFICER: Richard B Waugh Jr
CORPORATE VICE-PRESIDENT, COMMUNICATIONS: Rosanne O'Brien

Northrop company formed 1939 to produce military aircraft; activities extended to missiles, target drones, electronics, space technology, communications, support services and commercial products. Grumman Aircraft Engineering Corporation incorporated 6 December 1929; became major supplier of carrierborne aircraft to US Navy.

Acquisition of Grumman by Northrop completed 1 May 1994 and new corporation formed 18 May. Northrop Grumman subsequently completed acquisition of Vought Aircraft Company, with US$130 million purchase of Carlyle Group's 51 per cent interest, in August 1994. Further expansion announced on 3 January 1996, when Northrop Grumman revealed agreement to acquire the defence and electronic systems businesses of the Westinghouse Electric Corporation for US$3 billion. Finalisation of this purchase in March 1996 resulted in significant increase in products and technologies offered by Northrop Grumman. A major addition, finalised in July 1999, was that of Teledyne Ryan Aeronautical at cost of approximately US$140 million. Subsequently, in early April 2001, Northrop Grumman completed acquisition of Litton Industries Inc; value of this acquisition estimated at US$5.1 billion (including US$1.3 billion net debt). Another acquisition, announced on 20 April 2001, involved purchase of Electronic and Information Systems Group of Aerojet-General Corporation for US$315 million. Further growth followed on 30 November 2001, with acquisition of Newport News Shipbuilding, manufacturer of nuclear-powered submarines and aircraft carriers. In first quarter of 2002, Northrop Grumman announced intention of merging with TRW Inc; deal was finalised by 1 July, when Northrop Grumman revealed details of US$7.8 billion purchase. TRW Systems renamed Northrop Grumman Mission Systems and TRW Space and Electronics renamed Northrop Grumman Space Technology. Workforce totalled about 39,000 in November 2000, rising to about 80,000 after purchase of Litton and further increasing to approximately 120,000 with addition of TRW. Net sales of US$26,206 million were achieved in 2003, compared with US$17,206 million in 2002 and US$13,012 million in 2001.

Collaboration with Lockheed Martin announced in September 1996 concerning development and marketing of airborne early warning and control system aircraft. Further co-operation occurred in May 1997, when Northrop Grumman joined Lockheed Martin Joint Strike Fighter team, having previously been associated with unsuccessful McDonnell Douglas/British Aerospace JSF submission. Closer collaboration with DASA on surveillance and reconnaissance systems to result from MoU signed on 19 April 2000. In late 2000, Northrop Grumman received a Phase I study contract from US Defense Advanced Research Projects Agency for the Quiet Supersonic Platform (QSP) programme, intended to develop technology required for any future long-range, supersonic, military and/or civil aircraft; follow-on contracts were awarded in 2002 for continued development.

Northrop Grumman now organised into seven operating sectors as follows:

ELECTRONIC SYSTEMS Responsibilities include the development and production of radar and electronic systems for installation in the F-16 Fighting Falcon, F-22 Raptor, B-1B Lancer, AH-64D Apache Longbow, C-130 Hercules, E-3B/C Sentry and Boeing 737 AWACS and E-8C Joint STARS. Teamed with Rafael of Israel for sale and production of Litening II target-designation and navigation pod. Revenues for FY02 were US$5,339 million and workforce totalled 22,000 in December 2002. Development, integration and manufacturing expertise also embraces military airborne, space and undersea radar, surveillance and reconnaissance systems, air defence systems, tactical communications equipment, anti-submarine warfare sensors and systems, submersibles, mine countermeasures equipment, marine systems and shipboard instrumentation. Radar systems are produced for civil air traffic control agencies in the USA, Europe, Africa, the Middle and Far East, Asia and Latin America.

INFORMATION TECHNOLOGY Wholly owned subsidiary, now including much of Litton, provides technical engineering, project management and support resource services for information technology systems, plus technical and professional services in operations and maintenance.

MISSION SYSTEMS: At Reston, Virginia.

INTEGRATED SYSTEMS: *As described.*

SHIP SYSTEMS: At Pascagoula, Mississippi.

NEWPORT NEWS Production and assembly of nuclear-powered submarines, warships and aircraft carriers.

SPACE TECHNOLOGY At Redondo Beach, California.

UPDATED

INTEGRATED SYSTEMS

1 Hornet Way, El Segundo, California 90450
Tel: (+1 310) 332 10 00
Fax: (+1 310) 331 12 00
Web: http://www.is.northropgrumman.com
CORPORATE VICE-PRESIDENT AND PRESIDENT, INTEGRATED SYSTEMS: Scott J Seymour

Having divested itself of the aerostructures business, Integrated Systems now functions as prime contractor for the B-2A Spirit stealth bomber and the E-8C Joint STARS airborne targeting and battlefield management system, while continuing production of the E-2C Hawkeye for the US Navy and export customers. Serves as principal subcontractor for the F/A-18 (see Boeing) and is teamed with Lockheed Martin on the F-35 JSF programme. Other work includes modification and support of the EA-6B Prowler, F-5 Tiger II, F-14 Tomcat and Fairchild A-10 Thunderbolt II. Also engaged in production and support of UAVs for reconnaissance, surveillance and deception as well as aerial target systems.

Revenues for 2003 were US$3,803 million and workforce totalled 13,400 in early 2004.

Operating elements of Integrated Systems are organised into business areas, with responsibilities as described.

Air Combat Systems
One Hornet Way, El Segundo, California 90450
Tel: (+1 310) 331 36 16
MANAGER, COMMUNICATIONS: James F Hart

B-2 Spirit, F/A-18 Hornet/Super Hornet, F-35 Joint Strike Fighter, F-5 Tiger II/T-38 Talon and UAV systems (including RQ-8A Fire Scout unmanned version of Schweizer 330 helicopter. Has sites at Palmdale and San Diego, California; New Town, North Dakota; Whiteman AFB, Missouri; Tinker AFB, Oklahoma; and Hill AFB, Utah.

Airborne Early Warning And Electronic Warfare Systems
South Oyster Bay Road, Bethpage, New York 11714
Tel: (+1 516) 575 51 19
MANAGER, COMMUNICATIONS: John Vosilla

Production of E-2C Hawkeye 2000, plus modification, repair and support of other types, including EA-6B Prowler, A-10 Thunderbolt II, C-2 Greyhound, F-5 Tiger II, F-14 Tomcat and S-2T Turbo-Tracker. Has sites at St Augustine and Cecil Field, Florida; Point Mugu, California, plus field support services at many other locations.

Airborne Ground Surveillance And Battle Management Systems
2000 NASA Boulevard, Melbourne, Florida 32902
Tel: (+1 321) 726 75 26
MANAGER, COMMUNICATIONS: James Stratford

Production and upgrading of E-8C Joint STARS, plus associated advanced surveillance and battle management systems; currently developing new surveillance systems as part of Multi-Platform Radar Technology Insertion Program (MP-RTIP) for USAF E-10A multisensor command and control aircraft. Has additional site at Lake Charles, Louisiana.

UPDATED

NORTHROP GRUMMAN E-2C HAWKEYE

TYPE: Airborne early warning and control system.

DEVELOPMENT MILESTONES

E-2A	
First flight	21 Oct 60
First delivery	19 Jan 64
Subsequent versions	
E-2C	
First flight	20 Jan 71
First flight, production	23 Sep 72
Entered service (US Navy)	Nov 73

PROGRAMME: First flight of first of three prototypes 21 October 1960; total 59 production E-2As, of which 51 updated to E-2B by end 1971 apart from two TE-2A trainers and two converted to E-2C prototypes; first flight of E-2C prototype 20 January 1971; production started mid-1971; first flight production aircraft 23 September 1972; 219 of all E-2C versions ordered, of which about 200 had been delivered by end of first quarter of 2004.

CURRENT VERSIONS: **E-2C:** Current service and production version (*as detailed*). Baseline aircraft (Group '0') had AN/APS-120, but remaining examples to be phased out of US Navy inventory by 2004.

AN/APS-139 and Allison T56-A-427 engines form **Group I** update; first operational aircraft (163538) delivered to VAW-112 on 8 August 1989; 18 built; AN/APS-139 can detect cruise missiles at ranges exceeding 100 n miles (185 km; 115 miles); also monitors maritime traffic; radar coverage extended by AN/ALR-73 passive detection system (PDS), detecting electronic emitters at twice radar detection range. AN/APS-145 in **Group II** aircraft from December 1991; other enhancements give Group II 96 per cent expansion in radar volume, 400 per cent extra target tracking capability, 40 per cent more radar and identification range and 960 per cent increase in numbers of targets displayed. Group II added JTIDS in 1993–94; also has GPS. Final Group II aircraft for US Navy delivered in mid-2001. Retrofit with eight-blade propellers was due to start in 2001, with first operational example expected in fourth quarter; however, vibration problems encountered in testing caused delay and new propeller did not enter service until mid-2004. Group II aircraft phase-out started in 2002–03, when improved Hawkeye 2000 began to join fleet.

TE-2C: Training model, based on E-2C; two conversions (158639, 158648) originally undertaken, of which one (158639) was later assigned to JTIDS development with Northrop Grumman. At least three further conversions (159105, 163029, 163848) subsequently made for training purposes with VAW-120 at Norfolk, Virginia. Five new-build examples to be acquired in FY04–07, with these including wiring and hardware systems to facilitate future conversion.

E-2T: Originally reported to be conversion of E-2B for Taiwan, but new-build aircraft equivalent to E-2C Group I actually supplied; AN/APS-138 radar and electronic warfare upgrades. Delivery began in 1995, with further two to be obtained following agreement in July 1999; these will be to Hawkeye 2000E standard, with AN/APS-145 radar, but lacking CEC and satcom equipment. Delivery of first accomplished 10 August 2004, with second to follow by end of year.

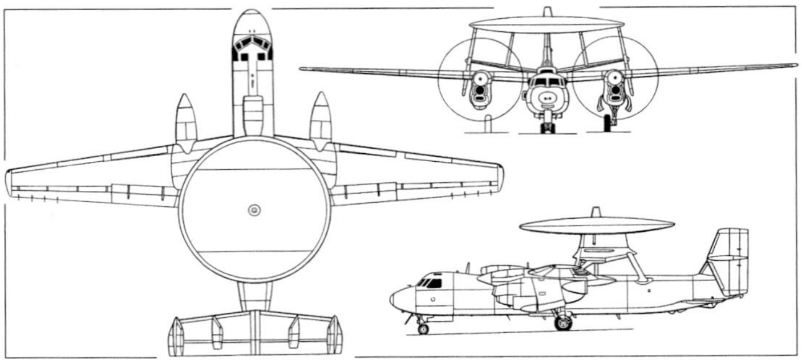

Northrop Grumman E-2C Hawkeye (Allison T56 turboprop engines) *(Jane's/Dennis Punnett)* ***NEW***/0558329

Hawkeye 2000: In December 1994, company received US$155 million contract to redefine E-2C as the Hawkeye 2000. Key element is mission computer upgrade (MCU), with new equipment based on Raytheon's Model 940.

Initial trials of upgraded mission computer installed on second Group II aircraft (164109) began with first flight on 24 January 1997 and were completed in July1997, at which time authorisation given for low-rate initial production of new mission computer. However, early flight trials revealed software problems that delayed production of new mission computer by about a year. More ambitious technical and operational evaluations undertaken with five modified aircraft in 1999–2001. All were Group II aircraft fitted with MCU and ACIS (see below) elements of proposed Hawkeye 2000; first two delivered to Patuxent River for initial evaluation by May 1999. At least four to Point Mugu from August 1999, joining VAW-117 for operational evaluation from October, with latter phase including deployed duty aboard a carrier for full battlegroup operations. In meantime, another E-2C (163849) used as a testbed for satcom, vapour-cycle cooling upgrade and Navy's USG-3 co-operative engagement capability (CEC) package, following first flight in April 1998; latter results in addition of 1.37 m (4 ft 6 in) antenna dish under belly containing omnidirectional transceiver that connects with command centres on parent aircraft carrier and surface combatant warships; provision of satcom evident through addition of fairing on top of rotodome.

New mission computer is less than half the weight of L-304, one-third of volume, and offers 15 times the processing power; other improvements for Hawkeye 2000 include government-furnished advanced control indicator set (ACIS), satellite-based voice and data communications capability, a new Honeywell vapour-cycle cooling system, air-to-air refuelling capability (if required) and inclusion of equipment and systems that form part of Navy CEC package. MCU and ACIS make use of commercial off-the-shelf technology incorporating open architecture.

In April 1999, contract awarded for 24 aircraft for US Navy (21 Hawkeye 2000s) as five-year procurement package, Taiwan (two Hawkeye 2000Es) and France (one Hawkeye 2000); in early 2003, US Navy announced intention to award second multiyear procurement contract, for total of eight aircraft to be acquired at rate of two per year during FY04 to FY07. Five will be configured as TE-2C trainers, lacking mission system, while final pair will be employed as development aircraft for the Advanced Hawkeye. First deployment of Hawkeye 2000 with US Navy by VAW-117 squadron in 2003. Export-configured Hawkeye 2000s lack CEC and satcom facilities.

Advanced Hawkeye: Full-scale development of this enhanced derivative began in early August 2003, following award of system development and demonstration (SDD) contract to Northrop Grumman. Total of US$1.9 billion to be invested in new version by 2012, of which US$413.5 million will go towards developing new space-time adaptive UHF radar; Lockheed Martin is prime contractor for radar, with Northrop Grumman Electronic Systems furnishing multiple solid-state transmitters and Raytheon providing receivers. The new radar will feature an electronically scanned array in the existing rotodome and is expected to have twice the range of the current AN/APS-145 radar, as well as superior performance in eliminating clutter and ability to track small targets like cruise missiles. New antenna and rotary coupler developed by L-3 Communications.

Other features include surveillance infra-red search and track (SIRST) system, modular communications equipment, multisensor integration and tactical ('glass') cockpit featuring flat-panel displays, with latter enabling co-pilot to augment mission system operators by

Second of France's three Northrop Grumman E-2C Hawkeyes *(Paul Jackson)* ***NEW***/0589328

performing some tactical functions. US Navy is prime customer, but another potential buyer is UK Royal Navy, which has emerging requirement for new AEW platform to operate from future aircraft carrier.

Communication suite to be greatly expanded and include additional satellite communication systems, plus multiple AN/ARC-210 radios with MIDS tactical datalink and Block 2 CEC sensor datalink. Advanced Hawkeye also planned to be first US Navy aircraft to have Joint Tactical Radio System (replacing AN/ARC-210 units). Upgraded IFF to be produced by BAE Systems.

Powerplant unchanged from current E-2C, but Advanced Hawkeye will use eight-blade Hamilton Sundstrand NP2000 propellers from outset. MTOW will exceed 25,850 kg (57,000 lb).

Northrop Grumman received US$49 million pre-systems development and demonstration contract for E-2C RMP (Radar Modernization Program) in late 2001; work done under this 12-month contract concerned definition of physical architecture of future mission system. Northrop Grumman also conducted initial flight testing of new radar's transmitter, receiver, antenna and rotary coupler on an NC-130H Hercules.

Two aircraft (one E-2C Hawkeye 2000 and one TE-2C trainer) being acquired under the FY04-07 MYP will be modified for SDD. Subsequent deployment planning anticipates starting pilot production in FY08, with an initial batch of four aircraft for operational evaluation in 2011. Low-rate initial production (LRIP) should begin in FY09, when five Advanced Hawkeyes are likely to be purchased by US Navy; thereafter, at least 75 aircraft will eventually be needed to sustain carrier battle group operations at present levels, with peak production rate expected to be eight aircraft per year. Full operational capability due between 2015 and 2017.

CUSTOMERS: US Navy orders for E-2C had totalled 139 by FY92, of which all delivered by March 1994. Further procurement authorised December 1994, when Northrop Grumman awarded US$122.5 million contract for start-up of new assembly line at St Augustine, Florida. Initial orders for seven Group II aircraft (four with FY95 funds, three with FY96 funds); first of these new aircraft (165293) was rolled out on 24 February 1997, flew on 22 March, and was delivered to VAW-120 at Norfolk, Virginia. Further contract in December 1996 to cover advance acquisition costs of four Group II aircraft with FY97 funds; four more ordered before April 1999 award of multiyear contract for 24 aircraft for US Navy (21), France (one) and Taiwan (two). Delivery of first new-build Hawkeye 2000 to US Navy accomplished in October 2001, following handover of single upgraded aircraft converted from existing E-2C under US$17.8 million contract awarded in April 2000; four Hawkeye 2000s to be delivered in 2002, then five per year in 2003–05, with last in February 2006; additional eight aircraft (three E-2C and five TE-2C) to follow in 2006–09 at rate of two per year. Hawkeye 2000 for France ordered by US Navy on 28 March 2001, for delivery in first quarter of 2004; two Hawkeye 2000E for Taiwan to follow in latter half of 2004 2004. Current US Navy programme status in adjacent table:

Major retrofit programme for USN aircraft to Group II standard was planned from FY95, but 1994 defence review established new E-2Cs more economical than retrofit; however, 16 Group I E-2Cs were upgraded to Group II at St Augustine, Florida; first returned to service on 21 December 1995. First operational squadron with upgraded Group II E-2C was VAW-123 which accepted its fourth and final aircraft on 29 April 1996.

E-2C entered service with VAW-123 at NAS Norfolk, Virginia, November 1973 and went to sea on board USS *Saratoga* late 1974; E-2C issued to 19 other squadrons, including three of Naval Reserve current squadrons are VAW-112, 113, 116 and 117 at Point Mugu, California; VAW-115 at Atsugi, Japan; VAW-120, 121, 123 to 126 with VAW-78 of Reserves at Norfolk, Virginia, plus VAW-77 of Reserves at Atlanta, Georgia, for anti-drug smuggling surveillance duty. VAW-120 is training unit. Miramar was base of first Group II squadron, VAW-113, June 1992. VAW-113 first operational evaluation cruise in USS *Carl Vinson,* 1993. New Build Group II aircraft also issued to VAW-110 (disbanded September 1994), 112, 116 and 117. Hawkeye 2000 in process of delivery to US Navy; first operational squadron was VAW-117. VAW-116 and VAW-125 both also using Hawkeye 2000 by March 2004, when about 12 aircraft had been accepted.

See table for exports. Singaporean aircraft with AN/APS-138 radar; Taiwan received E-2T with AN/APS-138 radar. Israeli aircraft in storage by 1994, with three being sold to Mexico in 2002; these refurbished by Israel Aircraft Industries, with first aircraft (AMP-100) being handed over in Israel on 21 January 2004. Following training of Mexican personnel, this and second aircraft ferried to Mexico in June 2004; third example delivered via UK in November 2004. France signed Letter of Offer and Acceptance (LOA) in June 1995 for two aircraft; first flight of first Aéronavale aircraft on 12 March 1998, with formal roll-out ceremony at St Augustine, Florida, on 28 April 1998. Both aircraft initially retained

Variant	Quantity	First aircraft
E-2C 'Group 0'	100	158638
E-2C Group I	18[1]	163029
E-2C Group II	36	164108
Hawkeye 2000	29[2]	165648
Total	**183**	

[1] Two aircraft later converted to TE-2C and 16 upgraded to Group II standard
[2] Includes eight aircraft for procurement in FY04-07 to sustain production until Advanced Hawkeye becomes available; five of these aircraft will be completed as TE-2C trainers

in USA for crew training, with first E-2C (actually No 2/165456) delivered to Lann-Bihoué, France, on 14 December 1998; second in April 1999. Operating unit, 4 Flottille, formed at Lann-Bihoué January 1999.

French Navy given authorisation in 1998 to purchase third Hawkeye; delivered to Lann-Bihoué in February 2004, this was completed as a Hawkeye 2000, to which configuration both earlier examples will be upgraded. Italy a potential customer in longer term, with requirement for at least two AEW aircraft by 2007. Egypt and Singapore said to be considering acquisition of surplus US Navy aircraft in mid-2000, with former subsequently signing contract for single example as attrition replacement; this delivered in February 2003, following upgrade to Hawkeye 2000 standard.

United Arab Emirates expected to obtain five refurbished Hawkeye 2000 aircraft from surplus US Navy stocks; possibility of sale at estimated cost of US$400 million notified to Congress on 4 September 2002. Contract was expected to be signed in first quarter of 2004, but there has been no confirmation of this; aircraft will probably be fitted with in-flight refuelling probe above cockpit. India reported in early 2004 to be evaluating Hawkeye.

COSTS: Procurement of 21 new Hawkeye 2000s for US Navy will cost US$1.47 billion. Upgrading of 13 Japanese aircraft to Hawkeye 2000 standard expected to cost approximately US$400 million, with cost of Egyptian upgrade of six aircraft valued at US$174 million. System Development and Demonstration (SDD) of Advanced Hawkeye predicted to cost US$2.1 billion between 2009 and 2012.

DESIGN FEATURES: E-2C can cover naval task force in all weathers flying at 9,150 m (30,000 ft) and can detect and assess approaching aircraft at in excess of 300 n miles (556 km; 345 miles); AN/APS-145 has total radiation aperture control antenna (TRAC-A) to reduce sidelobes to offset jamming; radar sweeps 6 million cu mile envelope and simultaneously monitors surface ships; long-range, automatic target track initiation and high-speed processing enable each E-2C to track more than 2,000 targets simultaneously and automatically, and control more than 40 intercepts; Randtron Systems AN/APA-171 antenna housed in 7.32 m (24 ft) diameter radome, rotating at 5 to 6 rpm above rear fuselage; antenna arrays in rotodome provide radar sum and difference signals and IFF.

High-mounted tapered wing; moderately sweptback tailplane with pronounced dihedral; twin fins and twin auxiliary fins, all with sweptback leading-edges and at right angles to tailplane. Rotating, circular radar housing mounted on cabane above rear fuselage; Hawkeye 2000 has satcom antenna housing on top of radome. Nose-tow catapult attachment, arrester hook and tail bumper; parts of tail made of composites to reduce radar reflection; wings fold hydraulically on skewed hinges to lie parallel to fuselage.

Wing incidence 4° at root, 1° at tip.

FLYING CONTROLS: Conventional and fully powered, with artificial feel; tailplane has 11° dihedral; four fins and three double-hinged rudders; long-span ailerons droop automatically when hydraulically operated Fowler flaps are extended; autopilot provides autostabilisation or full flight control. Empennage (including horizontal stabilisers, elevators, rudders, vertical fins and tabs) produced by Potez Aéronautique of France, following award of contract in mid-1997 for initial batch of five assemblies; first completed assembly delivered in second quarter of 1999.

STRUCTURE: Wing centre-section has three beams, ribs and machined skins; hinged leading-edge provides access to flying and engine controls. Fuselage conventional light metal. Composites used in parts of tail.

LANDING GEAR: Hydraulically retractable tricycle type. Pneumatic emergency extension. Steerable nosewheel unit retracts rearward. Mainwheels retract forward and rotate to lie flat in bottom of nacelles. Twin wheels on nose unit only. Oleo-pneumatic shock-absorbers. Mainwheel tyres size 36×11 (24 ply) tubeless, pressure 17.93 bar (260 lb/sq in) on ship,

First Mexican E-2C, pictured on its delivery flight in June 2004 *(Peter R Foster)* ***NEW***/1047702

E-2C EXPORTS

Customer	Quantit	Group	First aircraft	Delivery	Unit
Egypt[4]	6	0	162791	1987–88(5)	87 Squadron
		0	164626	1993(1)	
France[5]	2	II	1(165455)	1998	4 Flottille
	1	2000	3(166417)	2004	
Israel[1]	4	0	160771	1978	192 Sqdn
Japan[2]	13	0	34-3451	1982(4)	601 Sqdn
		0	54-3455	1984(4)	
		0	34-3459	1992–93(3)	
		0	44-3462	1993(2)	
Singapore	4	0	011(162793)	1987	111 Sqdn
Taiwan[3]	4		2501	1995	78 Sqdn
	2	2000		2004	78 Sqdn
Total	**36**				

[1] Israeli aircraft withdrawn from use in 1994; three sold to Mexico in 2002 refurbished by IAI before delivery in 2004
[2] Japanese aircraft being upgraded to Hawkeye 2000 standard
[3] Taiwanese aircraft known as E-2Ts; total includes two on order as E-2T Hawkeye 2000E
[4] Five Egyptian aircraft to be upgraded to Hawkeye 2000 standard; a sixth, former US Navy aircraft, was first to be upgraded and was delivered to Egypt in February 2003 (not included in table); second upgraded aircraft delivered in February 2004, with remaining four to follow by April 2007
[5] Original two French aircraft to be upgraded to Hawkeye 2000 configuration

14.48 bar (210 lb/sq in) ashore. Nosewheel tyres 20×5.5 (12/14 ply) tubeless. Hydraulic brakes. Hydraulically operated retractable tailskid. A-frame arrester hook under tail.

POWER PLANT: Two 3,803 kW (5,100 ehp) Allison T56-A-427 turboprops, driving Hamilton Sundstrand Type 54460-1 four-blade fully feathering reversible-pitch constant-speed propellers. These have foam-filled blades which have a steel spar and glass fibre shell. T56-A-427 engines provide 15 per cent improvement in efficiency, compared with -425 installed before 1989. US Navy and Northrop Grumman seeking to extend mission duration and will undertake in-flight refuelling trials in October 2004, using F/A-18 Hornet as tanker; previously, only Israel had configured Hawkeye to be refuelled in flight. Hamilton Sundstrand/Ratier Figeac NP2000 eight-blade, all-composites propeller flight tested on E-2C 163535 starting on 19 April 2001, following successful ground running on T56 and completion of critical design review in September 1998, with carrier suitability evaluation in USS *John F Kennedy* in November 2003. Fleet introduction began in 2004, with VAW-124 being one of first operational squadrons to receive new propeller. In 2002, US Navy requested Northrop Grumman to undertake studies into possibility of adopting a new engine, although current plans do not include a change of engine.

ACCOMMODATION: Normal crew of five, consisting of pilot and co-pilot on flight deck, plus ATDS Combat Information Center (CIC) staff of combat information centre officer, air control officer and radar operator. Downward-hinged door, with built-in steps, on port side of centre-fuselage and three overhead escape hatches.

SYSTEMS: Pneumatic boot de-icing on wings, tailplane and fins. Spinners and blades incorporate electric anti-icing.

AVIONICS: *Comms:* AN/AIC-14A intercom. AN/ARC-210 wideband/narrowband radio to be installed on Hawkeye 2000.

Radar: Lockheed Martin AN/APS-145 advanced radar processing system (ARPS) with fully automatic overland/overwater detection capability, Randtron AN/APA-171 rotodome (radar and IFF antennas). New Lockheed Martin radar now being developed for installation on Advanced Hawkeye.

Flight: Lockheed Martin AN/ASN-92 CAINS carrier aircraft inertial navigation system, GPS, AN/ASN-50 heading and attitude reference system, Rockwell Collins AN/ARA-50 UHF ADF, AN/ASW-25B ACLS, BAE Systems standard central air data computer, ASM-400 in-flight performance monitor, Honeywell AN/APN-171(V) radar altimeter.

Mission: BAE Systems/Hazeltine AN/APA-172 control indicator group with Lockheed Martin enhanced (colour) main display units (EMDU), which being replaced by L-3 Communications flat panel display screen during 2000–07; Litton OL-77/ASQ computer programmer (L-304) with Lockheed Martin enhanced high-speed processor, BAE Systems/Hazeltine OL-483/AP airborne interrogator system, Litton AN/ALR-73 passive detection system, AN/ARC-158 UHF datalink, AN/ARQ-34 HF datalink and JTIDS Class 2 HP terminal. Barco Display Systems graphics controllers with radar display capability and colour flat panel displays on Hawkeye 2000. Hawkeye 2000 also has AN/ALQ-217 ESM in place of AN/ALR-73 and Multi-Mission Advanced Tactical Terminal (MATT) for data communications.

DIMENSIONS, EXTERNAL:
Wing span .... 24.56 m (80 ft 7 in)
Wing chord: at root .... 3.96 m (13 ft 0 in)
at tip .... 1.32 m (4 ft 4 in)
Wing aspect ratio 9.3
Width, wings folded .... 8.94 m (29 ft 4 in)
Length overall .... 17.53 m (57 ft 6 in)
Height overall .... 5.58 m (18 ft 3¾ in)
Diameter of rotodome .... 7.32 m (24 ft 0 in)
Tailplane span .... 7.99 m (26 ft 2½ in)
Wheel track .... 5.93 m (19 ft 5¾ in)
Wheelbase .... 7.06 m (23 ft 2 in)
Propeller diameter .... 4.11 m (13 ft 6 in)

AREAS:
Wings, gross .... 65.03 m^2 (700.0 sq ft)
Ailerons (total) .... 5.76 m^2 (62.00 sq ft)
Trailing-edge flaps (total) .... 11.03 m^2 (118.75 sq ft)
Fins, incl rudders and tabs:
outboard (total) .... 10.25 m^2 (110.36 sq ft)
inboard (total) .... 4.76 m^2 (51.26 sq ft)
Tailplane .... 11.62 m^2 (125.07 sq ft)
Elevators (total) .... 3.72 m^2 (40.06 sq ft)

WEIGHTS AND LOADINGS:
Weight empty .... 18,363 kg (40,484 lb)
Max fuel (internal, usable) .... 5,624 kg (12,400 lb)
Max T-O weight .... 24,687 kg (54,426 lb)
Max wing loading .... 379.6 kg/m^2 (77.75 lb/sq ft)
Max power loading .... 3.25 kg/kW (5.34 lb/ehp)

PERFORMANCE:
Max level speed .... 338 kt (626 km/h; 389 mph)
Max cruising speed .... 325 kt (602 km/h; 374 mph)
Cruising speed (ferry) .... 268 kt (496 km/h; 308 mph)
Approach speed .... 103 kt (191 km/h; 119 mph)
Stalling speed (landing configuration) .... 75 kt (138 km/h; 86 mph)
Service ceiling .... 11,275 m (37,000 ft)
Min T-O run .... 564 m (1,850 ft)
T-O to 15 m (50 ft) .... 793 m (2,600 ft)
Min landing run .... 439 m (1,440 ft)
Ferry range .... 1,541 n miles (2,854 km; 1,773 miles)
Time on station, 175 n miles (320 km; 200 miles) from base .... 4 h 24 min
Endurance with max fuel .... 6 h 15 min

UPDATED

NORTHROP GRUMMAN E-8 JOINT STARS

TYPE: Airborne ground surveillance system.

DEVELOPMENT MILESTONES

Programme launched	27 Sep 1985
First flight, production	17 Aug 1995
First delivery	4 March 1996
Entered service (IOC)	18 Dec 1997
Final delivery (planned)	31 March 2005

PROGRAMME: Full-scale development contract for Joint Surveillance Target Attack Radar System (Joint STARS) programme awarded to Grumman (later Northrop Grumman), 27 September 1985; two Boeing 707-328Cs used as EC-18C testbeds, later redesignated E-8A.

Airframe modifications done at company's Lake Charles, Louisiana, plant following Boeing's withdrawal from programme; avionics installation at Melbourne.

Northrop Grumman awarded US$132 million (two contracts) in June 1997 for computer replacement programme taking advantage of commercial off-the-shelf (COTS) technology and intended to reduce costs; upgrade entails integration of new, more powerful central computers significantly faster than those originally installed. It also involves replacement of programmable signal processors and substitution of high-capacity switch and fibre-optic cable for existing copper-wired workstation network. This expected to produce savings of about US$19.3 million per aircraft, with original five central computers (three operating and two 'hot' spares) giving way to just two Compaq Clippers (one operating and one 'hot' spare). Testing of the new computers was completed in early August 2000; installation in production aircraft began in 2001.

Major radar upgrade programme begun in December 1998, with award of US$14.5 million contract to Northrop Grumman for pre-Engineering and Manufacturing Development (EMD) phase of radar technology insertion program (RTIP) to replace existing AN/APY-3 with new 2-D electronically scanned array radar, offering upgraded signal processing and operation and control performance. Scope of programme subsequently broadened and renamed Multi-Platform Radar Technology Insertion Programme (MP-RTIP), with Northrop Grumman securing US$303 million contract at end of 2000; resultant system destined for installation in new E-10A Multi-sensor Command and Control Aircraft (MC2A). This will be a modified Boeing 767-400ER. MP-RTIP EMD began in fourth quarter of 2003 with full-rate production decision due in third quarter of 2007. Potential applications include installation in RQ-4B Global Hawk UAV in addition to manned platform, with latter expected to attain operational capability in about 2012; some

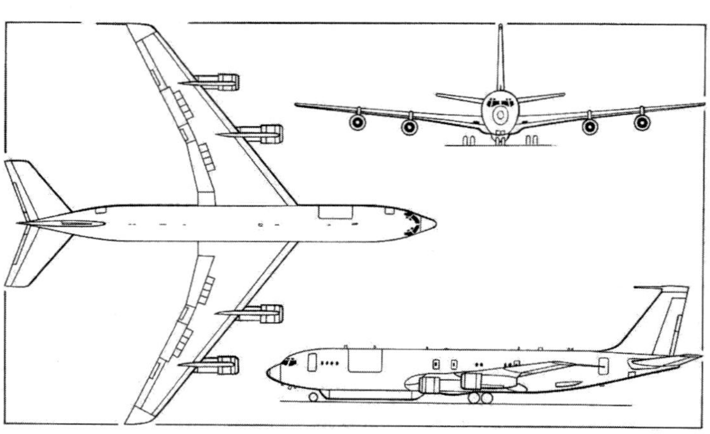

Northrop Grumman E-8C Joint STARS surveillance aircraft *(James Goulding)*
0507565

Northrop Grumman E-8C Joint STARS *(Paul Jackson)* ***NEW***/0576515

E-8 JOINT STARS PROCUREMENT

Lot	FY Advance Acquisition	FY Full Funding	Type	Identity	Delivered
FSD	85[1]		E-8A	86-0416/T1	
FSD	85[1]		E-8A	86-0417/T2	
FSD	90[2]		E-8C	90-0175/T3	
1	92	93	E-8C	92-3289/P1	4 March 1996
1	92	93	E-8C	92-3290/P2	11 December 1996
2	93	94	E-8C	93-0597/P3	25 November 1997
2	93	94	E-8C	93-1097/P4	18 August 1998
3	94	95	E-8C	94-0284/P5	13 August 1999
3	94	95	E-8C	94-0285/P6	27 November 2000
4	95	96	E-8C	95-0122/P7	1 December 1999
4	95	96	E-8C	95-0121/P8	30 March 2001
5	95	97	E-8C	96-0042/P9	6 March 2000
5	95	97	E-8C	96-0043/P10	27 July 2000
6	97	98	E-8C	97-0100/P11	6 August 2001
7	98	99	E-8C	97-0200/P12	5 November 2001
7	98	99	E-8C	97-0201/P13	25 April 2002
8	99	00	E-8C	99-0006/P14	19 August 2002
9	00	01	E-8C	00-2000/P15	25 February 2003
10	01	02	E-8C	01-2005/P16	(26 February 2004)
11	02	03	E-8C	02-0911/P17	(31 March 2005[3])

[1] Year of full-scale development (FSD) contract award
[2] Year of follow-on FSD contract award; aircraft serves as permanent testbed
[3] Scheduled delivery date

E-8Cs may eventually be retrofitted with MP-RTIP as the E-8D. Research and development effort divided between Northrop Grumman (prime contractor) and Raytheon (subcontractor), with former taking overall responsibility for radar and latter working on antenna development.

CURRENT VERSIONS: **E-8A:** Two development aircraft. Fitted with consoles for 10 operators; Pratt & Whitney JT3D-3B engines. Returned to Grumman after participation in Gulf War for completion of performance testing in 100 sorties and several thousand hours of ground testing. Both accepted by USAF, December 1993, for trials at Edwards AFB. First currently serving as **TE-8A** crew trainer.

E-8B: Originally proposed production version, based on new-built airframe; F108 turbofan engines; 15 operator consoles. One prototype YE-8B, 88-0322, flown 12 June 1990 in 'green' state; avionics not installed; overtaken by decision to use remanufactured Boeing 707 airframes; delivered to USAF 3 October 1991 for storage at Davis-Monthan AFB, Arizona; bartered with Omega Air, 1993, for five used 707s.

E-8C: Production version; 18 operator consoles (17 operations, one navigation/self-defence). Initial two aircraft to Lake Charles for conversion, May-June 1992. First E-8C (T3/90-0175) is permanent testbed, not for delivery to USAF; completed 500th mission on 3 September 2003 and may serve as Universal Testbed (UTB) for Multi-Platform Radar Technology Insertion Program (MP-RTIP). First operational aircraft (P1/92-3289) commenced flight testing on 17 August 1995 and was delivered to USAF in March 1996, making operational debut with mission from Frankfurt over former Yugoslavia on 15 November 1996. Total of four production aircraft completed to so-called **'Block 0'** configuration and delivered to USAF by August 1998; all four are scheduled to be upgraded to Block 20 standard by March 2005.

Following description applies to E-8C, except where otherwise indicated.

E-8C Block 10: Baseline USAF aircraft, featuring interoperability with other US and allied aircraft; specific enhancements include TADIL-J secure communications, a digital autopilot and Y2K computer compliance. First example was P5/94-0284, in service 1999; same standard also applicable to P6-P10. These being upgraded to Block 20 configuration; first aircraft to receive upgrade was P7/95-0122, which was redelivered on 11 February 2002; second was P9/96-0042, redelivered on 25 September 2002; third was P6/94-0285, redelivered on 21 April 2003; fourth was P8/95-0121, redelivered on 15 October 2003. Remaining two aircraft (P5/94-0284 and P10/96-0043) both expected to complete upgrade by October 2004.

E-8C Block 20: Modified to accept COTS hardware, with computer replacement programme tied to networked communications; introduces US Army improved data modem, basic SINCGARS radio and radar advanced signal processing. Planned for 2000, but first aircraft to this configuration was P11/97-0100, which was delivered in August 2001. Nine more delivered by February 2004, including first four of 10 aircraft due to be upgraded by March 2005.

E-8C Block 30: Was expected to be baseline attack control platform with UHF satcom, broadcast intelligence, TADIL-J attack support upgrade (ASU) and phase one of global air traffic management (GATM), but some of these features already incorporated in late production Block 20 aircraft. Block 30 designation may not be adopted until new engines are fitted.

CUSTOMERS: US Air Force and US Army (aircraft operated by USAF). See table for details of USAF procurement; total 20 originally expected, including two upgraded E-8As and one E-8C permanent testbed, but proposed fleet reduced to 13 by 1997 Quadrennial Defense Review, although subsequent purchases raised eventual total to 17 (excluding two E-8As and E-8C testbed).

Operating unit has been 93rd Air Control Wing (ACW) which activated at Robins AFB, Georgia, in January 1996, although all aircraft reassigned to 116th ACW of Georgia ANG at same base in fourth quarter of 2002, when 93rd ACW inactivated. IOC achieved with delivery of third aircraft (93-0597) to Robins AFB on 18 December 1997.

COSTS: E-8A development costs US$1,006.2 million; E-8C development costs US$868.7 million (FY91). Procurement of Lot 1 to Lot 4 (eight aircraft) totalled US$2,141 million. Total of US$221.8 million appropriated for 16th aircraft in FY01-02.

DESIGN FEATURES: Boeing 707-320C airliner converted with 7.32 m (24 ft) antenna covered by a 12.19 m (40 ft) 'canoe' radome and fairing under forward fuselage, for phased-array SLAR.

FLYING CONTROLS, STRUCTURE, LANDING GEAR: As, or similar to, commercial Boeing 707.

POWER PLANT: Four 80.1 kN (18,000 lb st) Pratt & Whitney TF33-P-102B turbofans initially installed; now being upgraded to 85.4 kN (19,200 lb st) TF33-P-102C standard following award of US$10.5 million contract in December 1998 to United Technologies for 42 modification kits. See Weights and Loadings for fuel. In late June 2002, it was announced that re-engining with Pratt & Whitney JT8D-219 turbofans is to be undertaken so as to reduce need for aerial refuelling, increase time on station and raise operating ceiling to 12,800 m (42,000 ft) to increase radar coverage significantly. Funding still awaited in March 2004, but re-engining is scheduled to take 60 days per aircraft and will be accomplished during periodic depot maintenance.

ACCOMMODATION: Standard mission crew of 21, comprising pilot, co-pilot, flight engineer, navigator/self-defence suite operator and 17 Air Force/Army operators; on longer missions replacement flight deck crew and system operators can be carried up to maximum of 34.

SYSTEMS: As Boeing 707; additional electrical generating capacity. World Auxiliary Power Company APU.

AVIONICS: *Comms:* Telephonics multiple intercom net control system; Raytheon UHF communications system.

Radar: Northrop Grumman (Norden Systems) AN/APY-3 multimode side-looking phased-array I-band radar, scanned electronically in azimuth and steered mechanically in elevation from either side of aircraft to provide 120° field of view. Synthetic aperture (SAR) mode used to detect stationary objects, such as parked tanks. Can interleaf Doppler mode to detect moving targets in moving target indicator/wide-area search (MTI/WAS), which is primary operating mode. Coverage is approximately 50,000 km^2 (19,305 sq miles) per minute, cruising at 9,150 to 12,200 m (30,000 to 40,000 ft), with ability to detect targets at range of 50 km (31 miles) to 250 km (155 miles) from aircraft.

Flight: Litton INS, Rockwell Collins flight management system.

Mission: Five Raytheon Model 920/866 supermini computers per aircraft (being replaced by two Compaq Clippers from 2001), Computing Devices International programmable signal processors (three), Interstate Electronics graphic displays, 18 Raytheon Model 920 workstations, Orbit International workstation keyboards, Miltope message printers (replaced by Data Metrics printers on fifth and subsequent aircraft), Cubic Defense Systems surveillance and control datalink, JTIDS for TADIL-J generation and processing. Satellite communications link, Magnavox encrypted UHF radios (12), two encrypted HF radios, three encrypted VHF radios with single channel ground and airborne radio system

(SINCGARS) provision. Radar data can be transmitted instantaneously to ground stations; or attacks by aircraft and ground forces directed via JTIDS datalink.

Self-defence: Unspecified, but includes chaff/flare dispensers and probably also incorporates threat-sensing equipment.

ARMAMENT: Nil.

DIMENSIONS, EXTERNAL (abbreviated):

Wing span	44.42 m (145 ft 9 in)
Length overall	46.61 m (152 ft 11 in)
Height overall	12.95 m (42 ft 6 in)

WEIGHTS AND LOADINGS:

Weight empty	77,564 kg (171,000 lb)
Max fuel weight	70,307 kg (155,000 lb)
Max T-O weight	152,407 kg (336,000 lb)

PERFORMANCE:

Max operating Mach No. (MMO)	0.84
Service ceiling	12,800 m (42,000 ft)
Endurance: internal fuel	11 h
with one in-flight refuelling	20 h

UPDATED

NORTHROP GRUMMAN E-10

TYPE: Airborne multisensor command and control system.

PROGRAMME: Boeing 767-400ER selected by USAF in second quarter of 2002 to serve as testbed for next-generation multisensor command and control aircraft (MC2A) and may also be chosen as platform for production system, for which decision expected in about July 2005 as part of Milestone B approval process. On 14 May 2003, Northrop Grumman and Raytheon received a pre-system development and demonstration contract, worth US$215 million, for weapon system integration. An initial batch of four operational E-10 Increment 1 MC2As is planned to start the process of replacing the E-8C Joint STARS and will be equipped with Northrop Grumman Multi-Platform Radar Technology Insertion Program (MP-RTIP) J-band radar; this merges moving target indicator over synthetic aperture radar, resulting in enhanced ground moving target indicator capability.

Trials intended to validate this new concept of intelligence gathering undertaken with a Boeing 707 (N404PA) that flew for the first time in modified form on 18 April 2002; known as *Paul Revere*, this aircraft performed a series of operational test flights in mid-2002, including some as part of the Joint Expeditionary Forces experiment at Nellis AFB, Nevada. Testing continues, but key objectives to be addressed by *Paul Revere* comprise directing activities of manned and unmanned air and space sensor systems used for surveillance and reconnaissance; integration and monitoring of all-source sensor data; location, identification, designation and tracking of targets using multisensor integration software techniques; undertaking weapon-target matching; establishing tactical priorities and objectives when engaging time-sensitive targets; and accomplishment of rapid and timely bomb damage assessment following an attack.

Total procurement of E-10 is dependent upon success of first four aircraft and also whether USAF chooses to add AMTI (airborne moving target indicator) capability or deploy another type of platform to replace the E-3 Sentry from 2035 onwards. At present, notional Increment 2 E-10 aircraft will not feature AMTI sensor (radar), but may have the potential for integrating AMTI data obtained from external source.

The 767-400ER testbed was the subject of a US$126 million contract awarded to Boeing in August 2003. Delivery of the 'green' aircraft is expected in December 2005, at which time it will be flown to Northrop Grumman's facility at Lake Charles, Louisiana, for modification to E-10 configuration. Due to fly for the first time in modified form in 2009, it will be fitted with prototype systems tasked with battle management, command, control, communications and intelligence. Development of this subsystem is expected to take five years and cost US$400 million.

The system development and demonstration (SDD) phase was originally due to begin in mid-2004, but has been delayed by a year, until June 2005. In the interval, work has continued on the key MT-RTIP sensor, with Northrop Grumman receiving a US$888 million contract in April 2004 for Phase 2 SDD work that is expected to result in application to the E-10 as well as the RQ-4 Global Hawk UAV.

Solicitation for the E-10 battle management command and control (BMC2) system began in June 2003. Three teams, headed by Boeing, Lockheed Martin and Northrop Grumman, were in competition, with all three invited in April 2004 to bid for the SDD contract. Formal submission of proposals followed in May 2004, with Northrop Grumman's system being selected in September 2004, at which time development cost was projected to be slightly more that US$300 million.

CURRENT VERSIONS: Three basic versions in prospect, although only the **E-10A** is now funded; this will feature the primary MP-RTIP sensor and is likely to have cruise missile defence as a cornerstone capability. Under original schedule, system testbed, with MP-RTIP, expected to make maiden flight in 2009; followed by first two production aircraft in 2012 and remaining two in 2013. However, budget cut of US$115 million for FY05 appropriation imposed by Congress, with probable consequent delay in schedule.

E-10B perceived as a replacement for E-3 Sentry and will have enhanced air moving target indicator capability if it comes to fruition, while **E-10C** viewed as a replacement for RC-135 Rivet Joint, with enhanced sigint/comint gathering capability.

COSTS: Development and production of a system testbed and four operational E-10 aircraft estimated to be US$5.3 billion.

ACCOMMODATION: BMC2 system of basic E-10A expected to require about 25 operators in addition to flight deck crew. Additional personnel will be required on E-10B and E-10C versions.

SYSTEMS: Supplemental electric power system by Hamilton Sundstrand.

AVIONICS: Northrop Grumman MP-RTIP J-band radar sensor accommodated in 6.09 m (20 ft 0 in) ventral housing beneath forward fuselage section.

UPDATED

Manufactured image of Northrop Grumman E-10A ***NEW***/1044848

NORTHROP GRUMMAN QSP

TYPE: Technology demonstrator.

PROGRAMME: In 2002, the Defense Advanced Research Projects Agency awarded Northrop Grumman two contracts for continuation of conceptual studies for a future Quiet Supersonic Platform (QSP). The first contract, valued at US$2.7 million, enabled Northrop Grumman to validate earlier studies and undertake wind-tunnel testing of a scale model of its preferred QSP design. At least a dozen different concepts were considered after work began in 2000, culminating in selection of the so-called 'dual relevant' approach, whereby the 'joined wing' or 'strut-braced wing' configuration could be adapted into either a strike/attack platform or an executive jet.

In its current guise, the QSP utilises a slender fuselage that can accommodate side-by-side weapons bays some 8.3 m (27 ft) in length or a 6.9 m (23 ft) long cabin. The main wing is shoulder-mounted and has a sharply swept cranked arrow planform with a straight trailing-edge; inboard leading-edge sweep is 84°, reducing to 71.5° outboard. The QSP is approximately 47.5 m (156 ft) long with a span of 17.6 m (57 ft 9 in); max T-O weight will be about 45,360 kg (100,000 lb), including payload of approximately 9,072 kg (20,000 lb). Project objectives include a very low sonic boom signature, range of 6,000 n miles (11,112 km; 6,904 miles) and maximum cruise speed of M2.4. Two General Electric GE449 engines are located at the rear of the aircraft, with a single vertical tail protruding above the engine nacelles. Northrop Grumman to decide in mid-2004 whether or not to go ahead with a QSP test aircraft as a precursor to development of a long-range strike aircraft. News of the outcome was still awaited in November.

The second contract received by Northrop Grumman was worth US$3.4 million and related to modification and flight trials of a Northrop F-5E Tiger II; these were expected to take place in the latter half of 2002, but did not begin until 24 July 2003. This aircraft featured a revised forward fuselage that produced a shaped sonic boom, which is significantly quieter. Modification was accomplished at St Augustine, Florida; following its first flight, it was transferred to Edwards AFB, California, from where it undertook a series of at least eight supersonic test flights. These confirmed predictions.

UPDATED

Northrop Grumman F-5E modified with new forward fuselage for QSP development ***(NASA/Carla Thomas)*** ***NEW***/0561598

Northrop Grumman design for a quieter supersonic aircraft 0533360

NUVENTURE

NUVENTURE AIRCRAFT

4184 West Kelley, Fresno, California 93722-9798
Tel: (+1 559) 447 11 12
e-mail: alantolle@cs.com
Web: http://www.nuventureaircraft.com
PRESIDENT: Alan E Tolle

NuVenture was established to maintain availability of components for the former Questair Venture. All jigs and stocks of spares were bought from Henry Bouley of North Windham, Connecticut by Richard Clawson and transferred to Visalia, California. In February 2001, ownership of NuVenture was purchased by Alan Tolle, manufacture remaining at Visalia.

VERIFIED

NUVENTURE VENTURE

TYPE: Side-by-side kitbuilt.

PROGRAMME: Questair M20 Venture designed by Jim Griswald and associates. Prototype (N62V) first flew 1 July 1987; made public debut at Oshkosh in same month; several improvements introduced in following year. Marketed in kit form; some 100 sold, of which about half completed during early 1990s and 36 remained on US civil register in 2002. Further five of fixed landing gear version, known as Spirit, also produced, first having flown in February 1991; last described in 1992–93 *Jane's*. Venture established nine speed and climb records in FAI Class C1b during 1989 and 1990; inaugurated Sport Class in Reno air races.

Kit production resumed by NuVenture; Spirit also available.

COSTS: US$49,900 in kit form, excluding engine and shipping charges (2004).

DESIGN FEATURES: Optimised for high-speed flight, but with comfortably large cockpit in spite of overall small size. 'Egg shape' fuselage; high-aspect ratio wings; narrow-chord fin and tapered tailplane.

Wing section NACA 23017 at root; 23010 at tip; twist 3°. Leading-edge cuff on each wing, from 60 per cent semi-span to tip, adds droop; leading-edge slots at cuff end and 40 per cent span generate vortices at high AoA to retard stalled airflow spreading from root to tip.

FLYING CONTROLS: Conventional and manual. No flaps, but ailerons droop when landing gear extended. Electrically operated trim tabs on ailerons and elevators.

STRUCTURE: All-aluminium.

LANDING GEAR: Retractable tricycle type. Single, small wheel on each leg. Electric retraction; separate systems for mainwheels and nosewheel. Steerable nosewheel. All wheels retract rearwards. Hydraulic brakes. Option of hybrid configuration with retractable nosewheel and fixed main gear with spring cantilever legs.

POWER PLANT: One 231 kW (310 hp) Teledyne Continental IO-550-N6B flat-six, derated to 209 kW (280 hp), driving a McCauley two-blade, constant-speed propeller. Fuel capacity 212 litres (56.0 US gallons; 46.6 Imp gallons).

DIMENSIONS, EXTERNAL:
- Wing span ... 8.38 m (27 ft 6 in)
- Length overall ... 4.95 m (16 ft 3 in)
- Height overall ... 2.35 m (7 ft 8½ in)
- Propeller diameter ... 1.73 m (5 ft 8 in)

AREAS:
- Wings, gross ... 6.74 m^2 (72.50 sq ft)

WEIGHTS:
- Weight empty ... 553 kg (1,220 lb)
- Max T-O weight ... 907 kg (2,000 lb)
- Max wing loading ... 134.7 kg/m^2 (27.59 lb/sq ft)
- Max power loading ... 4.35 kg/kW (7.14 lb/hp)

PERFORMANCE:
- Never-exceed speed (VNE) ... 300 kt (555 km/h; 345 mph)
- Max level speed ... 265 kt (491 km/h; 305 mph)
- Cruising speed at FL120 ... 240 kt (144 km/h; 276 mph)
- Stalling speed: flaps up ... 68 kt (126 km/h; 79 mph)
  - flaps down ... 61 kt (113 km/h; 71 mph)
- Max rate of climb at S/L ... 762 m (2,500 ft)/min
- Service ceiling ... 8,840 m (29,000 ft)
- T-O to 15 m (50 ft) ... 305 m (1,000 ft)
- Landing from 15 m (50 ft) ... 488 m (1,600 ft)
- Range, cruising at FL120:
  - VFR reserves ... 1,000 n miles (1,853 km; 1,151 miles)
  - IFR reserves ... 833 n miles (1,542 km; 958 miles)
- Endurance ... 4 h
- *g* limits: solo ... +6/–3
  - at max T-O weight ... +5/–2.5

UPDATED

Questair Venture with fixed mainwheels and retractable nosewheel *(Paul Jackson)* ***NEW***/1047153

OHIO

OHIO AIRSHIPS INC

11177 Bowen Road, PO Box 70, Mantua, Ohio 44255
Tel: (+1 866) 568 06 52
Web (1): http://www.ohio-airships.com
Web (2): http://www.dynalifter.com

PRESIDENT: Robert Rist
VICE-PRESIDENT AND CORPORATION STRATEGIST: Brian Martin
VICE-PRESIDENT AND CHIEF FINANCIAL OFFICER: Scott Gindlesberg
ENGINEERING ADVISER: Don VanFossen

This company, founded in 1999, has a hybrid semi-rigid in the development stage. Design assistance is being given by Conceptual Research Corporation, Analytical Methods Inc and Composite Engineering Inc.

UPDATED

OHIO DYNALIFTER

TYPE: Hybrid rigid air vehicle.

PROGRAMME: Announced in 2000, following taxi tests of 14.16 m^3 (500 cu ft) subscale prototype which began in March of that year. US Patent for concept awarded in 2001. Design concept is scalable, with four different sized versions undergoing design development in 2002–03. CFD analysis completed in 2003. Pentagon interest reported in late 2002 as potential addition to US Civil Reserve Air Fleet; interest from Department of Homeland Security in 2004.

Barber Airport in Alliance, Ohio, selected in July 2003 as site for 30.5 m (100 ft) long, 4,645 m^2 (50,000 sq ft) shelter where, in 2004, 36.58 m (120 ft) long Dynalifter 1 began construction; this expected to fly late 2004 or early 2005.

CURRENT VERSIONS (PROJECTED): **PSC-1 Freighter:** Outsize, 500 US ton (446.4 UK ton) carrier, half of which is payload. Potential applications include time-definite shipping market; parcels shipping; sealed intermodal containers; discounted international shipping across oceans or undeveloped areas; light, high-volume goods; or long-loiter patrol and surveillance.

PSC-2 Sightseer/Transport: Smaller version (one-fifth size of Freighter) for advertising; coastal patrol; tundra transportation; tourist transport and passenger ferry; or other civil applications.

PSC-3: International-range version with detachable 1,699 m^3 (60,000 cu ft) cargo bay; MTOW 158,757 kg (350,000 lb); payload in 65 US ton (58 UK ton) range.

PSC-4 Dynalifter 1: Small (two-person) prototype/PoC aircraft; patroller with 454 kg (1,000 lb) payload; aimed at most tasks currently performed by helicopters.

DESIGN FEATURES: Flattened oval hull shape, of approximately aerofoil shape in side elevation, with detachable, container-loaded ventral cargo bay; option for ramp-loaded bay on PSC-1. Low/mid-mounted wings, with winglets, just aft of CG; low-mounted foreplanes

Artist's impression of Ohio Dynalifter ***NEW***/1022774

at nose; engines (see Power Plant paragraph) mounted on wings and foreplanes. Conventional tail surfaces, with twin fins above and below tailplane halves.

Design allows concentrated wing lift, concentrated cargo lift, and high *g* loading from landing gear. Power/weight ratio said to be similar to that of powered sailplane and about one-fifth that of typical turboprop cargo aircraft. Ground handling weight is 75 per cent of maximum in-flight weight.

Company designates Dynalifter concept not as an airship but as a new category of helium-assisted cargo aircraft, combining static (helium) and dynamic (wings/foreplanes) lift in equal proportions, permitting it to be smaller than traditional airship for given weight of payload and to offer capability for higher speeds. It also has STOL capabilities, using medium-sized runways, and detachable nature of cargo bay can shorten turnaround times.

FLYING CONTROLS: Honeywell FBW actuators. Hull provides about one-third of total dynamic lift.

STRUCTURE: Ohio Airships (US patented) cable stay-bridge suspension.

LANDING GEAR: *PSC-1:* Quadricycle type, comprising four six-wheel (46×16) bogies mounted equidistant from CG; complete steering rotation by 100° in each direction.

PSC-2: As PSC-1 except bogies are two-wheel (36×11).

POWER PLANT: *PSC-1:* Eight 3,803 kW (5,100 shp) Rolls-Royce (Allison) T56-A-427 turboprops (three on each wing and one on each canard), each driving a four-blade propeller.

PSC-2: Four 1,864 kW (2,500 shp) Pratt & Whitney Canada PW127 turboprops (one on each wing and one on each canard).

ACCOMMODATION: Minimum flight crew of four pilots in PSC-1 (two for take-off and landing, two for in-flight duties); two-person ground crew for container operation. Take-off and landing pilots only in PSC-2, and no ground crew.

SYSTEMS: Dual 207 bar (3,000 lb/sq in) engine-driven hydraulic systems, one in each wing; third (back-up) system driven by pair of redundant APUs in rear of hull. AC electrical power from four 90 kVA generators.

AVIONICS: Commercial off-the-shelf.

DIMENSIONS, EXTERNAL:
Hull:
Length: PSC-1 ... 302.6 m (992.8 ft)
PSC-2 ... 176.8 m (580.0 ft)
Max width: PSC-1 ... 51.2 m (168.0 ft)
PSC-2 ... 29.9 m (98.0 ft)
Max depth: PSC-1 ... 36.9 m (121.0 ft)
PSC-2 ... 21.6 m (71.0 ft)
Fineness ratio: PSC-1, PSC-2 ... 7
Wing span: PSC-1 ... 163.2 m (535.5 ft)
PSC-2 ... 73.2 m (240.0 ft)

DIMENSIONS, INTERNAL:
Hull volume: PSC-1 ... 311,485 m^3 (11 million cu ft)
PSC-2 ... 63,430 m^3 (2.24 million cu ft)
Cargo hold:
Length: PSC-1 ... 45.72 m (150.0 ft)
PSC-2 ... 25.91 m (85.0 ft)
Width: PSC-1 ... 12.19 m (40.0 ft)
PSC-2 ... 6.10 m (20.0 ft)

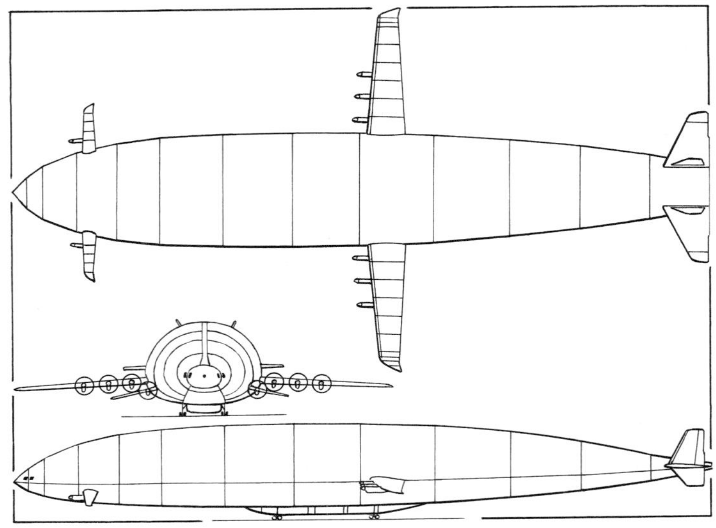

General appearance of the Dynalifter Freighter and Sightseer/Transport *(James Goulding)* ***NEW***/0567103

Height: PSC-1 ... 4.57 m (15.0 ft)
PSC-2 ... 4.27 m (14.0 ft)
Volume: PSC-1 ... 2,548.5 m^3 (90,000 cu ft)
PSC-2 ... 673.9 m^3 (23,800 cu ft)

WEIGHTS AND LOADINGS:
Weight empty: PSC-1 ... 188,618 kg (415,832 lb)
PSC-2 ... 48,693 kg (107,350 lb)
Fuel weight: PSC-1 ... 105,430 kg (232,432 lb)
PSC-2 ... 21,144 kg (46,615 lb)
Payload weight:
PSC-1: for 5,000 n mile range ... 102,058 kg (225,000 lb)
max ... 158,757 kg (350,000 lb)
PSC-2: for 3,200 n mile range ... 6,350 kg (14,000 lb)
max ... 20,412 kg (45,000 lb)
Static displacement lift: PSC-1 ... 226,796 kg (500,000 lb)
PSC-2 ... 45,359 kg (100,000 lb)
Dynamic lift at designed airspeed:
PSC-1 ... 226,796 kg (500,000 lb)
PSC-2 ... 45,359 kg (100,000 lb)
Max T-O weight: PSC-1 ... 453,592 kg (1,000,000 lb)
PSC-2 ... 90,718 kg (200,000 lb)

PERFORMANCE:
Max level speed:
PSC-1, PSC-2 ... 120 kt (222 km/h; 138 mph)
Cruising speed: PSC-2 ... 90 kt (167 km/h; 104 mph)
Stalling speed: PSC-1, PSC-2 ... 40 kt (75 km/h; 47 mph)
Cruise altitude/pressure ceiling:
PSC-2 ... 3,050 m (10,000 ft)
Range with max payload:
PSC-1 ... 3,200 n miles (5,926 km; 3,682 miles)
PSC-2 ... 1,800 n miles (3,333 km; 2,071 miles)
Range with reduced payload:
PSC-1 with 102,058 kg (225,000 lb) ... 5,000 n miles (9,260 km; 5,754 miles)
PSC-2 with 6,350 kg (14,000 lb) ... 3,200 n miles (5,926 km; 3,682 miles)
Endurance:
PSC-1 with 11,340 kg (25,000 lb) payload ... more than 2.5 days
PSC-2 at 3,050 m (10,000 ft) with max payload ... 18 h

UPDATED

PAM

PERFORMANCE AVIATION MANUFACTURING GROUP

PO Box 80, Williamsburg, Virginia 23187
Tel: (+1 757) 229 03 67
Fax: (+1 757) 229 73 80
e-mail: pamgroup@cox.net
Web: http://www.flying-platform.com
CEO: Clement Makowski
PRESIDENT: Robert J Pegg

PAM Group was founded in 1998 and is developing an individual lifting vehicle.

UPDATED

PAM 100B INDIVIDUAL LIFTING VEHICLE

TYPE: Lifting vehicle.

PROGRAMME: Project initiated October 1989; construction of prototype started January 1993; first flight (N6172N) June 1994, powered by single engine; rebuilt with two Hirth engines and reflown May 1998; FAA certification April 1999. A total of 50 hours (tethered and free) flown between 1994 and mid-1999; over 140 hours had been accumulated by February 2004. There are no reports of sales or series production.

UPDATED

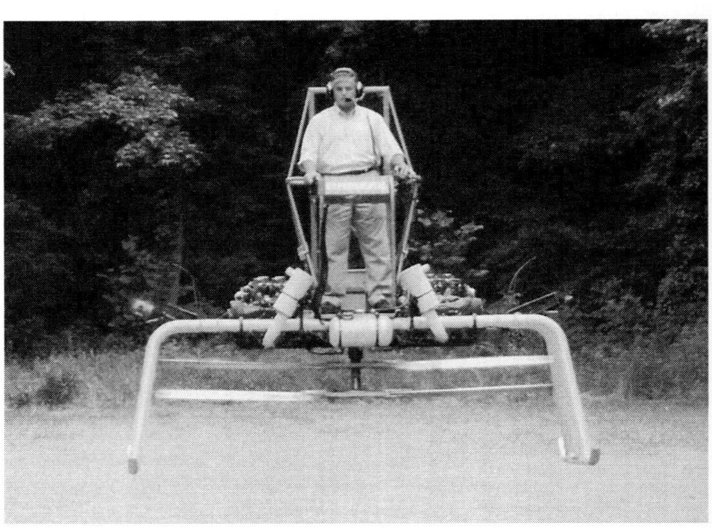

PAM 100B Individual Lifting Vehicle in hovering flight ***NEW***/0553236

PARABOUNCE

ONE GIANT LEAP LLC

PO Box 461845, Los Angeles, California 90046
Tel: (+1 323) 876 67 72
Fax: (+1 323) 882 61 72
e-mail: AGiantLeap@aol.com
Web: http://www.parabounce.com
PRESIDENT AND CEO: Stephen Meadows

Parabounce manufactures recreational balloons filled with sufficient helium to permit their users to leap up to 38 m (125 ft) into the air before floating back down to the ground. In late 2002, it introduced the Propbike, a small, human-powered airship.

UPDATED

PARABOUNCE PROPBIKE

TYPE: Helium non-rigid airship.

PROGRAMME: Announced in late 2002. US airship manufacturer Advanced Hybrid Aircraft was contracted to design, build and test prototype in 1999.

DESIGN FEATURES: Spherical helium-filled balloon, from which pilot module is suspended by means of a load ring. Pedal power turns ducted pusher propeller behind pilot's seat, geared to turn at 300, 600 (optimum) or 900 rpm depending on pilot's fitness standard; rudder and elevators situated aft of propeller duct. Balloon is fitted with two 30.5 m (100 ft) tethers, to the ends of which lead weights can be attached to counterbalance pilot weight; two slow-deflation ports are provided for use if balloon escapes its tethers. Advertising banners can be attached to envelope, and options include internal illumination for nocturnal operation. Electric motor may also be offered as customer option.

STRUCTURE: Envelope of polyurethane coated fabric.

DIMENSIONS, EXTERNAL:
Balloon diameter .... 6.86 m (22 ft 6 in)

WEIGHTS AND LOADINGS:
Pilot weight range .... 27-118 kg (60-260 lb)

PERFORMANCE (estimated):
Typical speed at 600 propeller rpm .... 4 kt (7.5 km/h; 4.5 mph)

UPDATED

Parabounce Propbike man-powered airship ***NEW***/0568974

PIASECKI

PIASECKI AIRCRAFT CORPORATION

Second Street West, Essington, Pennsylvania 19029-0360
Tel: (+1 610) 521 57 00
Fax: (+1 610) 521 59 35
e-mail: piasecki_jw@piasecki.com
PRESIDENT: Frank N Piasecki
VICE-PRESIDENTS:
Frederick W Piasecki (Technology)
John W Piasecki (Contracts)
DIRECTOR, MILITARY REQUIREMENTS: Jimmy R Hayes
DIRECTOR, PROGRAM REQUIREMENTS: Joseph P Cosgrove

Frank Piasecki developed and flew the USA's second successful helicopter on 11 April 1943 and the world's first tandem rotor helicopter on 7 March 1945; created Piasecki Helicopter Corporation in 1946, this forming subsidiary Piasecki Aircraft Corporation in 1955, following which the original company was sold to Boeing.

Piasecki assisted PZL Swidnik of Poland (which see) in FAA certification of the W-3A helicopter, which was awarded in May 1993. Piasecki has exclusive sales agreement for W-3A in the Americas and Pacific Rim countries. W-3A Sokół complies with FAR Pt 29, is certified for full IFR operations and has US instrumentation.

UPDATED

PIASECKI VECTORED THRUST DUCTED PROPELLER (VTDP)

US military designation: X-49A

TYPE: New-concept rotorcraft.

PROGRAMME: Began with US Army contract to develop a compound helicopter incorporating the Piasecki VTDP concept for the AH-64 Apache and AH-1W SuperCobra. Programme objectives were met or exceeded by both the AH-64 VTCAD (Vectored Thrust Combat Agility Demonstrator) and AH-1W VTCAD configurations, resulting in increased maximum level flight speed; 50 per cent improvement in longitudinal acceleration and deceleration capability in level flight; 50 per cent decrease in turn and pull-up radii at speeds in excess of 95 kt (176 km/h; 109 mph); and handling qualities at least as good as those of the baseline AH-64A and AH-1W. In addition, tactical simulations confirmed superiority of the VTCADs over the standard Apache and AH-1W SuperCobra. Is currently a US Army programme directed towards Sikorsky UH-60 improvements.

A separate US Navy contract, awarded in 1995 and valued at US$16.1 million, involved investigation into application of VTDP technology to the AH-1W(4BW)/AH-1Z four-blade rotor configuration. The Navy contract included ground testing of the full-scale VTDP and additional flight controls simulation and testing of the 4BW/VTDP configuration.

Piasecki then proposed flight demonstration of this technology to the Navy on an AH-1W(4BW), but instead was awarded a four-year US$26.1 million Advanced Technology Demonstration (ATD) contract on 28 September 2000 for integration, testing and flight demonstration of VTDP on a modified Sikorsky YSH-60F. This flight test programme, to begin in 2005, will be conducted by Piasecki under an FAA experimental

Computer-generated image of H-60/VTDP compound helicopter 0563225

flight certificate. The VTDP concept is being investigated by the DoD as an affordable means of upgrading the capabilities extending the service life of existing single main rotor helicopters such as the UH-1, UH/SH-60 and AH-64, until the follow-on Joint Replacement Aircraft is fielded some time after 2030. Most recently, the US Air Force selected the H-60/VTDP concept unofficially known as SpeedHawk, as one of a number of alternatives being considered as an upgrade or replacement, to be fielded by 2010, for its ageing HH-60G combat search and rescue helicopters.

By mid-2000, Piasecki had constructed and completed ground testing of a full-scale 2.44 m (8 ft 0 in) diameter duct and integrated five-blade propeller. In addition to installation of this unit, modifications to the YSH-60F will include addition of lifting wings (from an Aerostar business twin) with flaperons and integration of VTDP controls into the helicopter's existing mechanical control system, all of which is expected to add some 726 kg (1,600 lb) to the demonstrator's empty weight. The increase in empty weight will be lower in production versions, which will employ a supplementary power unit (SPU) that will provide an increase in hover payload to offset the increased empty weight. By May 2004 the drive system, including the SPU, has successfully completed endurance testing, all critical structural tests had been completed, and modification of the test aircraft was 70 per cent complete.

DESIGN FEATURES: The VTDP comprises a ducted propeller with integral vanes and spherical sectors to vector propeller thrust. It provides lateral thrust for anti-torque and lateral control, in lieu of a tail rotor, as well as forward thrust for auxiliary propulsion. Combined with the lifting wing, it provides for increasing speed to more than 200 kt (370 km/h; 230 mph); greater manoeuvrability; reduced vibration and fatigue loads; and consequent improvement in component lives and reduction in maintenance requirements and life cycle costs.

Piasecki estimated that a VTDP-based H-60 compound helicopter would have an unrefuelled combat radius of 762 n miles (1,411 km; 876 miles) following a rolling take-off, and 520 n miles (963 km; 598 miles) after a vertical take-off, representing a three-fold increase over the standard H-60.

UPDATED

PIPER

THE NEW PIPER AIRCRAFT INC

2926 Piper Drive, Vero Beach, Florida 32960
Tel: (+1 561) 567 43 61
Fax: (+1 561) 778 21 44
Web: http://www.newpiper.com
PRESIDENT AND CEO: Charles (Chuck) M Suma
VICE-PRESIDENT, PRODUCT SERVICES: Werner Hartlieb
MARKETING AND SALES DIRECTOR: Larry Bardon
CORPORATE COMMUNICATIONS DIRECTOR: Mark S Miller

In July 1995, after Piper had stabilised its financial position during four years of Chapter 11 protection, the US Bankruptcy Court approved a new reorganisation plan under which Piper's assets were bought for US$95 million by Newco Pac Inc, a new company jointly owned by Philadelphia-based investment firm Dimeling, Schreiber and Park, Teledyne Continental Motors (which was Piper's largest creditor), and the remaining creditors. Under this ownership, The New Piper Aircraft Inc was established. Earlier history of Piper last appeared in the 1995–96 *Jane's*.

During 1999, Piper delivered 329 aircraft, 395 in 2000; 441 in 2001, 290 in 2002, and 229 in 2003. Workforce 680 in March 2003. Sales revenue totalled US$243 million in 2001, up from US$157 million in 2000.

In July 2003, American Capital Strategies Ltd bought US$57 million of Piper's debt to acquire 94 per cent of the company; the remaining 6 per cent was acquired by affiliates of Exeter Capital Partners, a long-time investor in New Piper.

Kits and parts for classic Pipers are available from other outlets, not necessarily with Piper's approval. These include:

J3 CUB: Wag-Aero Sport trainer. Replacement fuselages (standard and with additional 10 cm; 4 in cockpit width) from
Airframes Inc,
Big Lake, Alaska 99652; *tel* (+1 907) 892 82 44; http://www.supercubs.com.

PA-12 SUPER CRUISER: Fuselages manufactured by
Dakota Air Frame Inc,
1372 Cleveland Avenue, Larchwood, Iowa 51241; *tel/fax* (+1 712) 477 20 84.

PA-14 FAMILY CRUISER: Wag-Aero Sportsman.

PA-15 VAGABOND: Wag-Aero Wag-A-Bond.

PA-18 SUPER CUB: Fuselages from Airframes Inc (as for J-3 Cub); complete aircraft from Cub Crafters.

UPDATED

Piper PA-28 Warrior III four-seat lightplane *(Paul Jackson)* ***NEW***/0589372

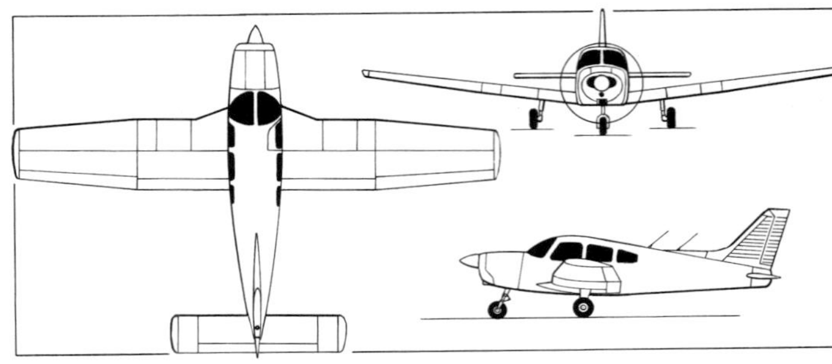

Piper PA-28-161 Warrior III (Lycoming O-320 flat-four) *(James Goulding)* 0100458

PIPER PA-28-161 WARRIOR III

TYPE: Four-seat lightplane.

PROGRAMME: As replacement for Cherokee 140 series, Piper redesigned airframe with several refinements, including fuselage stretch and new wing; first flight of prototype PA-28-151 17 October 1972; FAA certification 9 August 1973; PA-28-161 Warrior II first flown 27 August 1976; two/four-seat Cadet trainer version introduced April 1988, but no longer in production; Warrior III introduced late 1994.

CUSTOMERS: Five delivered in 1996, seven in 1997, 20 in 1998, 25 in 1999, 43 in 2000, 32 in 2001, 29 in 2002, 31 in 2003, and 17 in the first nine months of 2004. Piper has built some 30,000 of PA-28 series since prototype Cherokee flew on 10 January 1960. Recent customers include the University of North Dakota, which ordered eight in April 2003.

COSTS: Standard equipped price US$180,600 (2004).

DESIGN FEATURES: Classic tourer and trainer. Low, moderately tapered wing (with root glove) and low-set tailplane; sweptback fin.

NACA 65_2-415 wing section on inboard panels, Mod No. 5 of NACA TN 2228 on leading-edge of outboard panels; dihedral 7°, incidence 2° at root, −1° at tip; sweepback at quarter-chord 5°.

FLYING CONTROLS: Manual. Conventional ailerons and rudder, plus all-moving tailplane with combined anti-servo and trim tab. Four-position manually operated flaps.

STRUCTURE: Conventional light alloy, with semi-monocoque fuselage, single-spar wings and ribbed light alloy skins on fin and tailplane; glass fibre nose cowl and wing/tailplane/fintips.

LANDING GEAR: Tricycle type; fixed. Wing-mounted, trousered, cantilever main legs with Piper oleo-pneumatic shock-absorbers; single wheel, with glass fibre speed fairing, on each unit. Cleveland disc brakes and wheels with tyres size 6.00–6 (4 ply) on main units, pressure 1.65 bar (24 lb/sq in). Cleveland steerable nosewheel and tyre size 5.00–5 (4 ply), pressure 2.06 bar (30 lb/sq in). Parking brake.

POWER PLANT: One 119 kW (160 hp) Textron Lycoming O-320-D3G flat-four engine, driving a Sensenich 74DM6-0-60 two-blade fixed-pitch metal propeller. Fuel in two wing tanks, with total capacity of 189 litres (50.0 US gallons; 41.6 Imp gallons), of which 181.5 litres (48.0 US gallons; 40.0 Imp gallons) are usable. Refuelling point on upper surface of each wing. Oil capacity 7.5 litres (2.0 US gallons; 1.7 Imp gallons).

ACCOMMODATION: Four persons in pairs in enclosed cabin. Individual horizontally adjustable front seats with seat belts and shoulder harnesses; bench-type rear seat with seat belts and shoulder harnesses. Dual controls standard. Large door on starboard side. Baggage compartment at rear of cabin, with external baggage door on starboard side. Heating, ventilation and windscreen defrosting standard.

SYSTEMS: Hydraulic system for brakes only. Electrical system powered by 28 V 60 A engine-driven alternator. 24 V 10 Ah battery standard.

AVIONICS: Standard and optional Advanced packages by Garmin/Meggitt, as described. Avidyne FlightMax Entegra integrated flight deck available as option from late 2004.

Comms: Standard: Garmin GNS-430 com/nav/IFR GPS; GMA-340 audio panel with marker beacon lights and four-position intercom; GTX-330 transponder; ELT; avionics master switch; cabin speaker; Telex 100T microphone and Airman 760 headset. Advanced Training Group: as above but with dual GNS-430.

Flight: Standard: GI-106A VOR/LOC/GS/GPS indicator; Narco AR-850 altitude reporter. Advanced package as above, plus dual GNS-430 and GI-106A and single RCR 650D ADF with IND 650A indicator. ST-180 HSI with slaved compass system, and DME 450, optional.

Instrumentation: Piper true airspeed indicator, magnetic compass, sensitive altimeter, gyro horizon, directional gyro, rate of turn indicator, rate of climb indicator, OAT gauge, digital ammeter, electric clock, annunciator panel with push-to-test, recording tachometer, three-way oil temperature/pressure and fuel-pressure gauge, dual fuel quantity gauges and vacuum gauge.

EQUIPMENT: Standard equipment includes engine hour recorder, alternate static source, heated pitot head, colour co-ordinated control wheels, internally lit rocker switches, external power receptacle, electrical engine primer system, instrument panel lighting package, cabin dome light, navigation lights, landing/taxying light, wingtip Comet strobe lights, avionics dimming switch, pilot's and co-pilot's vertically adjustable seats in fabric and vinyl with magazine storage pockets on backs, rear bench seat, four floor-mounted cabin fresh-air vents, vinyl cabin side panels, wall-to-wall carpet and headliner, crew armrests, pilot's storm window, two sun visors, two map pockets, 'Quietised' soundproofing, carpeted baggage compartment with security straps, halon fire extinguisher, tiedown points, jack pads, static discharge wicks, and Du Pont Imron polyurethane exterior paint in base colour with two contrasting trim stripes. S-Tec electric trim, carburettor ice detector, second altimeter, wingtip recognition lights, low profile tail strobe light, two-tone base colour exterior paint, metallic paint, and foreign certification gross weight kit optional.

DIMENSIONS, EXTERNAL:
Wing span ... 10.67 m (35 ft 0 in)
Wing chord: at root, excl glove ... 1.60 m (5 ft 3 in)
at tip ... 1.07 m (3 ft 6¼ in)
Wing aspect ratio ... 7.2
Length overall ... 7.25 m (23 ft 9½ in)
Height overall ... 2.22 m (7 ft 3½ in)
Tailplane span ... 3.96 m (12 ft 11¾ in)
Wheel track ... 3.05 m (10 ft 0 in)
Wheelbase ... 2.03 m (6 ft 8 in)
Propeller diameter ... 1.88 m (6 ft 2 in)
Propeller ground clearance ... 0.21 m (8¼ in)
Cabin door: Height ... 0.89 m (2 ft 11 in)
Width ... 0.91 m (3 ft 0 in)
Baggage door: Height ... 0.51 m (1 ft 8 in)
Max width ... 0.56 m (1 ft 10 in)
Height to sill ... 0.71 m (2 ft 4 in)

DIMENSIONS, INTERNAL:
Cabin (instrument panel to rear bulkhead):
Length ... 2.49 m (8 ft 2 in)
Max width ... 1.05 m (3 ft 5¼ in)
Max height ... 1.14 m (3 ft 8¾ in)
Floor area ... 2.28 m² (24.5 sq ft)
Volume (incl baggage) ... 3.0 m³ (106 cu ft)
Baggage compartment volume ... 0.68 m³ (24.0 cu ft)

AREAS:
Wings, gross ... 15.79 m² (170.0 sq ft)
Ailerons (total) ... 1.23 m² (13.20 sq ft)
Trailing-edge flaps (total) ... 1.36 m² (14.60 sq ft)
Fin ... 0.69 m² (7.40 sq ft)
Rudder ... 0.38 m² (4.10 sq ft)
Tailplane, incl tab ... 2.76 m² (29.70 sq ft)

WEIGHTS AND LOADINGS:
Weight empty, equipped ... 697 kg (1,536 lb)
Baggage capacity ... 91 kg (200 lb)
Max T-O weight ... 1,106 kg (2,440 lb)
Max wing loading ... 70.1 kg/m² (14.35 lb/sq ft)
Max power loading ... 9.28 kg/kW (15.25 lb/hp)

PERFORMANCE:
Max level speed ... 117 kt (216 km/h; 134 mph)
Normal cruising speed at 75% power ... 115 kt (213 km/h; 132 mph)
Stalling speed: flaps up ... 50 kt (93 km/h; 58 mph)
flaps down ... 44 kt (82 km/h; 51 mph)
Max rate of climb at S/L ... 196 m (644 ft)/min
Service ceiling ... 3,355 m (11,000 ft)
T-O to 15 m (50 ft) ... 494 m (1,620 ft)
Landing from 15 m (50 ft) ... 354 m (1,160 ft)
Range, 45 min reserves:
at 75% power ... 426 n miles (789 km; 490 miles)
at 55% power at FL100 ... 513 n miles (950 km; 590 miles)

UPDATED

PIPER PA-28-181 ARCHER III

TYPE: Four-seat lightplane.

PROGRAMME: Introduced (as Cherokee Challenger 180) 9 October 1972 as successor to Cherokee 180; Archer 180, featuring minor changes, available from 1974; PA-28-181 Archer II launched 1976, and since 1978 has featured the tapered wings of Warrior II; Archer III from 1994, with axisymmetric engine inlets, redesigned windscreen and cabin side windows, plus interior restyling and improvements.

CUSTOMERS: 37 Archer IIIs delivered in 1995, and 45 each in 1996 and 1997; 91 in 1998; 100 in 1999, 102 in 2000, 88 in 2001, 38 in 2002, 49 in 2003, and 12 in the first nine months of 2004.

Description of the Warrior III applies also to Archer III except as follows:

COSTS: Standard, equipped US$211,600 (2004).

LANDING GEAR: Tricycle type. Tyres size 6.00–6 (4 ply), on all three wheels. Mainwheel tyre pressure 1.65 bar (24 lb/sq in), nosewheel 1.24 bar (18 lb/sq in). Cleveland high-capacity disc brakes. Parking brake. Wheel speed fairings standard.

POWER PLANT: One 134 kW (180 hp) Textron Lycoming O-360-A4M flat-four engine, driving a Sensenich 76EM8S14-O-62 two-blade fixed-pitch metal propeller. Fuel in two tanks in

Piper PA-28-181 Archer III *(Paul Jackson)* ***NEW***/0589370

wing leading-edges, with total capacity of 189 litres (50.0 US gallons; 41.6 Imp gallons), of which 181.5 litres (48.0 US gallons; 40.0 Imp gallons) are usable. Oil capacity 7.5 litres (2.0 US gallons; 1.7 Imp gallons).

ACCOMMODATION: Four persons in pairs in enclosed cabin. Individual adjustable front seats, with dual controls; individual rear seats. Large door on starboard side. Baggage compartment at rear of cabin; door on starboard side. Rear seats removable to provide 1.3 m³ (44 cu ft) cargo space. Accommodation heated and ventilated. Windscreen defrosting.

SYSTEMS: 28 V 70 A alternator and 24 V 10 Ah battery.

AVIONICS: Standard Garmin International and Meggitt package, as for Warrior III. Optional Premium Select package adds second GNS-430 and GI-106A VOR/LOC/GS/GPS indicator, S-Tec System 55X dual-axis autopilot with automatic electric trim, DG with heading bug and trim indicator and Meggitt ST-360 altitude preselect. Avidyne FlightMax Entegra integrated flight deck available as option from late 2004.

Comms: Garmin GMA-340 audio panel with marker beacon receiver and stereo intercom; GTX-327 solid-state transponder; Telex 100T microphone and Airman 760 headset.

Flight: GI-106A VOR/LOC/GS/COM/GPS indicator standard. ADI/HSI, RCR 650D ADF and DME 450, optional.

Instrumentation: Standard GNS-430 high-resolution colour LCD MFD with GI-106A VOR/LOC/GS/GPS indicator. WX500 Stormscop, Sky 497 Skywatch with GNS interface, and ST-360 altitude select/alerter optional.

EQUIPMENT: Standard equipment generally as for Warrior III except: polished propeller spinner, metal instrument panel, illuminated side-mounted OAT gauge, overhead switch panel, EGT gauge, flush locking fuel caps, overhead vent fan system, restyled windscreen and window lines with tinted transparencies, full chemical corrosion protection, Aztec Silver interior decor, DuPont Imron 6000 clearcoat/basecoat and exterior paint, wool carpeting and Hobnail side panels, passenger armrests and headrests. Additional options include Piper Aire air conditioning, carburettor ice detector, stainless steel cowling fasteners; Meggitt electric trim, low-noise exhaust system and leather seats in choice of three colours.

Piper PA-28-181 Archer II *(Paul Jackson)* ***NEW***/0568429

Instrument panel of baseline-equipped Piper PA-28-181 Archer III *(Paul Jackson)* 0110707

DIMENSIONS, EXTERNAL: As for Warrior III except:

Length overall	7.32 m (24 ft 0 in)
Tailplane span	3.92 m (12 ft 10½ in)
Wheelbase	2.00 m (6 ft 7 in)
Propeller diameter	1.93 m (6 ft 4 in)

DIMENSIONS, INTERNAL: As for Warrior III except:

Cabin: Width	1.06 m (3 ft 5¾ in)
Height	1.14 m (3 ft 9 in)
Baggage compartment volume, incl hatshelf	0.74 m³ (26.0 cu ft)

AREAS: As for Warrior III

WEIGHTS AND LOADINGS:

Weight empty, equipped	763 kg (1,682 lb)
Baggage capacity	90 kg (200 lb)
Max T-O weight	1,156 kg (2,550 lb)
Max ramp weight	1,160 kg (2,558 lb)
Max wing loading	73.2 kg/m² (15.00 lb/sq ft)
Max power loading	8.62 kg/kW (14.17 lb/hp)

PERFORMANCE:

Max level speed	133 kt (246 km/h; 153 mph)
Normal cruising speed at 75% power at FL79	128 kt (237 km/h; 147 mph)
Stalling speed, flaps down	45 kt (84 km/h; 52 mph)
Max rate of climb at S/L	211 m (692 ft)/min
Service ceiling	4,300 m (14,100 ft)
T-O run	346 m (1,135 ft)
T-O to 15 m (50 ft)	490 m (1,608 ft)
Landing from 15 m (50 ft)	427 m (1,400 ft)
Landing run	280 m (920 ft)
Range at S/L, allowances for taxi, T-O, climb and descent and 45 min reserves, at FL60:	
at 75% power	444 n miles (822 km; 511 miles)
at 65% power	487 n miles (901 km; 560 miles)
at 55% power	522 n miles (966 km; 600 miles)

UPDATED

PIPER PA-28R-201 ARROW

TYPE: Four-seat lightplane.

PROGRAMME: Derived from Cherokee Archer II, but with retractable landing gear, more powerful engine and untapered wings of 1975 PA-28-180 Archer; replaced original retractable gear Cherokee, the PA-28R Arrow 180, which had first flown 1 February 1967; PA-28R-201 Arrow III with increased span, tapered wings first flown 16 September 1975; first production aircraft 7 January 1977; new Arrow IV and Turbo Arrow IV models introduced 1979 with all-moving T tails; these subsequently discontinued and earlier low-tail **Arrow III** version restored to production.

CUSTOMERS: Seven delivered in 1996, two in 1997, two in 1998, six in 1999, 18 in 2000, 23 in 2001, 26 in 2002, 16 in 2003, and 11 in the first nine months of 2004. Recent customers include the University of North Dakota, which ordered one in April 2003.

Description of Archer III applies also to Arrow except as follows:

COSTS: Standard, equipped US$279,800 (2004).

LANDING GEAR: Tricycle type; retractable. Hydraulic actuation by an electrically operated pump. Main units retract inward into wings, nose unit rearward. All units fitted with oleo-pneumatic shock-absorbers. Mainwheels and tyres size 6.00-6 (6 ply), pressure

Piper PA-28R-201 Arrow III four-seat lightplane *(Paul Jackson)* ***NEW***/0589366

2.07 bar (30 lb/sq in). Nosewheel and tyre size 5.00–5 (4 ply), pressure 1.86 bar (27 lb/sq in). High-capacity dual hydraulic disc brakes and parking brake.

POWER PLANT: One 149 kW (200 hp) Textron Lycoming IO-360-C1C6 flat-four engine, driving a McCauley two-blade constant-speed metal propeller. Fuel tanks in wing leading-edges with total capacity of 291 litres (77.0 US gallons; 64.1 Imp gallons), of which 273 litres (72.0 US gallons; 60.0 Imp gallons) are usable. Oil capacity 7.5 litres (2.0 US gallons; 1.7 Imp gallons).

SYSTEMS: Generally as for Archer III and Warrior III except for electrohydraulic system for landing gear actuation.

AVIONICS: Standard and Deluxe avionics packages by Garmin, as described. Avidyne FlightMax Entegra integrated flight deck available as option from late 2004.

Comms: Standard package comprises Garmin GNS 430 com/nav/GPS, GMA 340 audio with marker beacon receiver and intercom, GTX-327, Telex 100T noise cancelling microphone and Telex Airman 760 headset. DeLuxe package adds second GNS 430.

Flight: Standard package comprises single GI-106A VOR/LOC/GSA/GPS indicator. DeLuxe package adds second GNS 430, GI-106A and VOR/LOC/GPS/HSI indicator with slaved compass system, replacing standard directional gyro. Optional equipment includes S-Tec System 55X dual-axis autopilot with automatic electric trim, turn indicator and DG with heading bug; ST-361 ADI; DME 40; and RCR 650D ADF with IND-650A indicator.

EQUIPMENT: Generally as for Warrior III and Archer III.

DIMENSIONS, EXTERNAL:

Wing span	10.80 m (35 ft 5 in)
Wing chord: at root, excl glove	1.60 m (5 ft 3 in)
at tip	1.07 m (3 ft 6¼ in)
Wing aspect ratio	7.4
Length overall	7.52 m (24 ft 8¼ in)
Height overall	2.39 m (7 ft 10¼ in)
Tailplane span	3.92 m (12 ft 10½ in)
Wheel track	3.19 m (10 ft 5½ in)
Wheelbase	2.39 m (7 ft 10¼ in)

AREAS:

Wings, gross	15.79 m^2 (170.0 sq ft)

WEIGHTS AND LOADINGS:

Weight empty, equipped	811 kg (1,789 lb)
Max T-O weight	1,247 kg (2,750 lb)
Max wing loading	79.0 kg/m^2 (16.18 lb/sq ft)
Max power loading	8.37 kg/kW (13.75 lb/hp)

PERFORMANCE:

Max level speed	145 kt (268 km/h; 166 mph)
Normal cruising speed	137 kt (253 km/h; 157 mph)
Stalling speed: flaps up	60 kt (112 km/h; 69 mph)
flaps down	55 kt (102 km/h; 64 mph)
Max rate of climb at S/L	253 m (831 ft)/min
Service ceiling	4,935 m (16,200 ft)
T-O to 15 m (50 ft)	488 m (1,600 ft)
Landing from 15 m (50 ft)	463 m (1,520 ft)
Cruising range at 55% power, at FL90, 45 min reserves	880 n miles (1,630 km; 1,013 miles)

UPDATED

PIPER PA-32R-301 SARATOGA II HP

TYPE: Six-seat utility transport.

PROGRAMME: Saratoga family of fixed and retractable landing gear light aircraft announced 17 December 1979 to replace earlier PA-32 Cherokee SIX 300 and T-tail Lance series (using PA-32 Type Certificate issued 4 March 1965); fixed-gear versions were discontinued, but reinstated in 2003, as described separately.

Saratoga SP certified 7 November 1979; fixed-gear Saratoga followed on 9 January 1980. Saratoga II HP certified 26 May 1993, featuring axisymmetric engine inlets, aerodynamic clean-up and interior improvements and restyling.

CURRENT VERSIONS: **Saratoga II HP:** Standard version, *as described.*

Saratoga II TC: Turbocharged version with Textron Lycoming TIO-540 engine; *described separately.*

6X: Fixed-gear version; *described separately.*

CUSTOMERS: Total of 38 HPs delivered in 1997, 28 in 1998, 30 in 1999, 28 in 2000, 22 in 2001, five in 2002, nine in 2003, and nine in the first nine months of 2004.

Piper has built over 7,200 of PA-32 family since prototype first flew on 6 December 1963.

COSTS: Standard, equipped US$454,000 (2004).

DESIGN FEATURES: Larger counterpart to PA-28 series, featuring stretched fuselage. Saratoga versions have 'semi-tapered' wing, with root glove, unswept inboard panel and tapered outer panel with small fence inboard of tip.

Wing section NACA 66$_2$-415, dihedral 7°, thickness/chord ratio 15 per cent, twist 2°, incidence 2°, washout 0°.

FLYING CONTROLS: Conventional. Ailerons, rudder and all-moving tailplane with combined anti-servo and trim tab. Rudder trim; four-position electrically actuated flaps with preselect. Control surface max deflections: ailerons +28/–22°; tailplane +14° 30'/–5° 30'; rudder ±28°; flaps 40°.

STRUCTURE: Conventional light alloy, with semi-monocoque fuselage, single-spar wings and ribbed light alloy skins on fin and tailplane; glass fibre nose cowl and wing/tailplane/fintips.

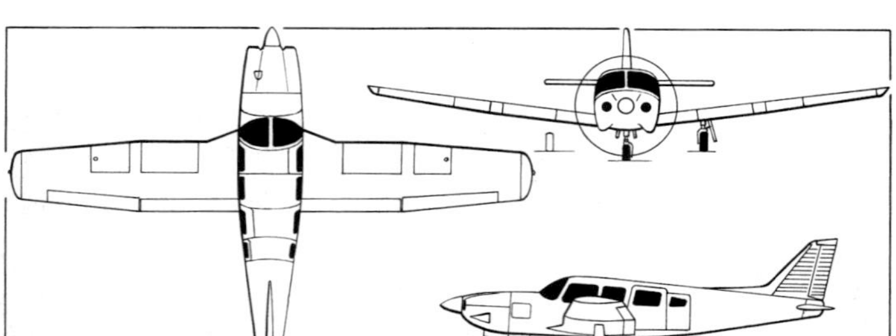

General arrangement of the Piper PA-32R-301 Saratoga II HP *(James Goulding)* 0100459

Piper PA-32R-301 Saratoga II HP *(Paul Jackson)* ***NEW***/1047096

LANDING GEAR: Hydraulically retractable tricycle type with single wheel on each unit. Main units retract inward, nosewheel aft. Emergency free-fall extension system. Piper oleo-pneumatic shock-absorbers. Steerable nosewheel. Mainwheels and tyres size 6.00–6 (8 ply), pressure 2.62 bar (38 lb/sq in). Nosewheel and tyre size 5.00–5 (6 ply), pressure 2.41 bar (35 lb/sq in).

POWER PLANT: One 224 kW (300 hp) Textron Lycoming IO-540-K1G5 flat-six engine, driving a Hartzell HC-I3YR-1RF/F7663DR three-blade, constant-speed, metal propeller with C3575-1 spinner. Two fuel tanks in each wing with combined capacity of 405 litres (107 US gallons; 89.1 Imp gallons), of which 386 litres (102 US gallons; 84.9 Imp gallons) are usable. Refuelling points on wing upper surface. Oil capacity 11.5 litres (3.0 US gallons; 2.5 Imp gallons).

ACCOMMODATION: Enclosed cabin seating six people, rear four in club arrangement; dual controls standard. Two forward-hinged doors, one on starboard side forward, overwing, and one on port side at rear end of cabin. Space for 45 kg (100 lb) baggage at rear of cabin, with external, lockable baggage/utility door on port side. Additional baggage space, capacity 45 kg (100 lb), between engine fireproof bulkhead and instrument panel, with external door on starboard side. Pilot's storm window. Accommodation heated and ventilated. Piper Aire air conditioning system optional. Windscreen defroster standard.

SYSTEMS: Electrically driven hydraulic pump for landing gear actuation. Electrical system includes 28 V 90 A engine-driven alternator and 24 V 10 Ah battery. Standby vacuum system standard. Optional oxygen system and Piper Aire air conditioning.

AVIONICS: Standard Garmin International and Meggitt package.

Comms: Garmin GMA-340 audio panel with marker beacon receiver and stereo intercom; GTX-330 solid-state transponder with traffic information system; Telex 100T microphone, and Telex Airman headset; ELT.

Flight: Dual Garmin GNS-430 and GNS-530 integrated VOR/LOC/GS/COM and GPS, GI-106A VOR/LOC/GD/GPS indicator, Meggitt ST-180 HSI, DME 450 and S-TEC-System 55x two-axis autopilot with automatic electric trim, DG with heading bug and trim indicator, ST-361 ADI with flight director features. BFGoodrich Skywatch and WX-500 Stormscope optional.

Instrumentation: Dual GNS-430/530 high-resolution colour LCD MFD; Piper true airspeed indicator, illuminated magnetic compass, sensitive altimeter (with second optional) ADI, HSI, rate of turn indicator, rate of climb indicator, OAT gauge, digital ammeter, annunciator panel with push-to-test, recording tachometer, manifold pressure/fuel flow gauge, oil pressure gauge, oil temperature gauge, dual fuel quantity gauges, fuel quantity sight gauges, CHT gauge, EGT gauge and vacuum gauge. Co-pilot's 76 mm (3 in) instrument panel with electric attitude indicator optional.

Optional from 2004 is the Avidyne FlightMax Entegra integrated flight deck, comprising two 264 mm (10.4 in) high-resolution, sunlight-readable displays for PFD and MFD functions, and EMax engine instrumentation system. EXP5000 PFD with solid-state air data and attitude/heading reference system presents standard flight instrumentation including EADI, EHSI, altitude, primary engine instruments, airspeed, and vertical speed; EX5000 MFD provides situational awareness data on an integrated display that includes moving map, lightning detection, terrain, traffic, engine information and optional Narrowcast Datalink with Nexrad weather information. EMax displays manifold pressure, rpm, fuel flow and oil pressure on the PFD, fuel totaliser, automatic leaning function, percentage horsepower, fuel remaining, six-cylinder CHT, EGT, oil temperature, OAT and electrical bus voltages, and computes nautical miles per gallon, fuel-to-waypoint and fuel-to-destination on the MFD, plus data logging of critical engine parameters which are downloadable via a dataport in a standard spreadsheet format. Entegra provides fully integrated autopilot control and mode annunciations, with altitude capture, vertical speed selection and heading control.

EQUIPMENT: Standard equipment includes engine hour recorder, alternate static source, electric clock, internally lit rocker switches, heated pitot head, resettable circuit breakers in CB panel, standby electric vacuum pump, electric emergency fuel pump, external power receptacle, instrument panel lighting package, avionics dimming, map lights, navigation lights, landing/taxying light, wingtip strobe lights, pulsating recognition lights, pilot's and co-pilot's vertically adjustable all-leather seats, pilot's vent window, sun visors, four all-leather passenger seats with headrests, shoulder harnesses, seat belts, and quick-release facility, fold-down armrests (fifth and sixth seats), refreshment console, executive writing table, Hobnail fabric side panels, super 'Quietised' soundproofing, utility door for rear baggage access and cargo loading, static discharge wicks, and Du Pont Imron 6000 clearcoat/basecoat exterior paint. Optional equipment includes entertainment/executive console; provision for AM/FM radio and CD player, multimedia entertainment system and laptop computer workstation; and metallic paint.

DIMENSIONS, EXTERNAL:

Wing span	11.02 m (36 ft 2 in)
Wing aspect ratio	7.3
Length overall	8.43 m (27 ft 8 in)
Height overall	2.59 m (8 ft 6 in)
Tailplane span	3.94 m (12 ft 11 in)
Wheel track	3.38 m (11 ft 1 in)
Wheelbase	2.41 m (7 ft 11 in)
Cabin door (fwd, stbd): Height	0.89 m (2 ft 11 in)
Width	0.91 m (3 ft 0 in)
Cabin door (rear, port): Height	0.72 m (2 ft 4½ in)
Width	0.71 m (2 ft 4 in)
Baggage door (fwd): Height	0.41 m (1 ft 4 in)
Width	0.56 m (1 ft 10 in)
Baggage/utility door (fwd): Height	0.52 m (1 ft 8½ in)
Width	0.66 m (2 ft 2 in)

DIMENSIONS, INTERNAL:

Cabin: Length (instrument panel to rear bulkhead)	3.16 m (10 ft 4¼ in)
Max width	1.24 m (4 ft 0¾ in)

Max height .... 1.07 m (3 ft 6 in)
Volume (incl rear baggage area) .... 5.5 m^3 (195 cu ft)
Baggage compartment volume:
forward .... 0.20 m^3 (7.0 cu ft)
rear .... 0.49 m^3 (17.3 cu ft)

AREAS:
Wings, gross .... 16.56 m^2 (178.3 sq ft)
Ailerons (total) .... 0.98 m^2 (10.60 sq ft)
Trailing-edge flaps (total) .... 1.36 m^2 (14.60 sq ft)
Fin .... 0.70 m^2 (7.50 sq ft)
Rudder, incl tab .... 0.40 m^2 (4.30 sq ft)
Horizontal tail surfaces (total) .... 2.94 m^2 (31.60 sq ft)

WEIGHTS AND LOADINGS:
Weight empty, equipped .... 1,088 kg (2,398 lb)
Baggage capacity .... 91 kg (200 lb)
Max T-O and landing weight .... 1,633 kg (3,600 lb)
Max ramp weight .... 1,639 kg (3,615 lb)
Max wing loading .... 98.6 kg/m^2 (20.19 lb/sq ft)
Max power loading .... 7.30 kg/kW (12.00 lb/hp)

PERFORMANCE:
Never-exceed speed (VNE) .... 191 kt (353 km/h; 219 mph)
Max level speed .... 175 kt (324 km/h; 201 mph)
Normal cruising speed .... 166 kt (307 km/h; 191 mph)
Stalling speed: flaps up .... 67 kt (124 km/h; 78 mph) CAS
flaps down .... 63 kt (117 km/h; 73 mph)
Max rate of climb at S/L .... 398 m (1,305 ft)/min
Service ceiling .... 4,751 m (15,588 ft)
T-O run .... 366 m (1,200 ft)
T-O to 15 m (50 ft) .... 540 m (1,770 ft)
Landing from 15 m (50 ft) .... 464 m (1,520 ft)
Landing run .... 195 m (640 ft)
Range, long-range cruise power, allowances for start, taxi, T-O, climb and descent, and 45 min reserves .... 859 n miles (1,590 km; 988 miles)

UPDATED

PIPER PA-32-301FT 6X

TYPE: Six-seat utility transport.

PROGRAMME: First flight (N326X) 24 February 2003. Announced at EAA Sun 'n' Fun at Lakeland, Florida 2 April 2003. FAA certification 22 July 2003.

CURRENT VERSIONS: **6X:** Standard version, *as described.*

6XT: Turbocharged version; *described separately.*

CUSTOMERS: 10 delivered in 2003 and 23 in the first nine months of 2004.

COSTS: Standard, equipped, VFR, US$365,500 (2004).

The description for the Saratoga II HP applies also to the Piper 6X except:

DESIGN FEATURES: Generally similar to Saratoga II HP except for fixed landing gear, deletion of rearmost cabin window on each side and NASA air intake port side of nose.

POWER PLANT: One 224 kW (300 hp) Textron Lycoming IO-540-K1G5 flat-six engine, driving a Hartzell F7663D three-blade, constant-speed, metal propeller. Two interconnected fuel tanks in each wing, combined usable capacity 386 litres (102 US gallons; 84.9 Imp gallons).

LANDING GEAR: Non-retractable tricycle type. Speed fairings standard.

AVIONICS: Standard Garmin International package. Avidyne FlightMax Entegra integrated flight deck optional, as described under Saratoga II HP.

Comms: Garmin GMA-340 audio panel with marker beacon receiver and intercom; GTX-327 digital altitude encoding transponder; Telex 100T microphone; Telex Airman headset; four PS Engineering headsets; and cabin speaker with microphone jacks.

Flight: Dual Garmin GNS-430 com/nav/GS/GPS; GI-106 GPS/VOR/LOC/GS indicator. Optional Premium avionics package adds GNS-530, Meggitt System 55x autopilot with electric trim and altitude preselect and HSI with compass system. DME 450, RCR 650D ADF with 650A indicator and Goodrich Stormscope and Skywatch situational awareness package, optional.

Instrumentation: Garmin GNS-430 MFD.

EQUIPMENT: Options include Premium interior package comprising leather seats, carpet upgrade and Piper Aire air conditioning; super soundproofing; UK lighting package; exterior paint scheme with solid bottom base; and metallic paint.

WEIGHTS AND LOADINGS:
Weight empty, equipped .... 995 kg (2,194 lb)
Standard useful load .... 645 kg (1,421 lb)

PERFORMANCE:
Never-exceed speed (VNE) .... 189 kt (350 km/h; 217 mph)
Max level speed .... 155 kt (287 km/h; 178 mph)
Normal cruising speed at FL72 .... 148 kt (274 km/h; 170 mph)
Stalling speed, flaps down .... 59 kt (109 km/h; 68 mph)
Service ceiling .... 5,242 m (17,200 ft)
T-O run .... 391 m (1,284 ft)
T-O to 15 m (50 ft) .... 619 m (2,030 ft)
Landing from 15 m (50 ft) .... 557 m (1,825 ft)
Landing run .... 278 m (911 ft)
Range, long-range cruise power, at FL110, 45 min reserves .... 804 n miles (1,489 km; 925 miles)

UPDATED

Prototype Piper 6X fixed landing gear utility transport *(Paul Jackson)* ***NEW***/1047197

PIPER PA-32R-301T SARATOGA II TC

TYPE: Six-seat utility transport.

PROGRAMME: Turbocharged parallel to Saratoga II HP; PA-32RT-300T Turbo Lance II certified 20 April 1978, followed by PA-32R-301T Turbo Saratoga SP on 7 November 1979 and fixed-gear Turbo Saratoga on 9 January 1980.

Current Saratoga II TC certified 9 July 1997; first customer deliveries August 1997.

CURRENT VERSIONS: **Saratoga IITC:** Standard version, *as described.*

6XT: Fixed-gear, *described separately.*

CUSTOMERS: Total of 26 delivered in 1997, 48 in 1998, 46 in 1999, 70 in 2000, 68 in 2001, 45 in 2002, 28 in 2003, and 28 in the first nine months of 2004.

COSTS: Standard, equipped US$486,400 (2004).

The description for the Saratoga II HP applies also to the Saratoga II TC except:

DESIGN FEATURES: New interior and relocated battery and ground power receptacle.

POWER PLANT: One 224 kW (300 hp) Textron Lycoming TIO-540-AH1A turbocharged flat-six engine, with automatic wastegate, driving a Hartzell HC-I3YR-1RF/F7663DR three-blade, constant-speed, metal propeller. Fuel as for Saratoga II HP.

WEIGHTS AND LOADINGS:
Weight empty, equipped .... 1,122 kg (2,473 lb)

PERFORMANCE:
Never-exceed speed (VNE) .... 191 kt (353 km/h; 219 mph)
Max level speed .... 187 kt (346 km/h; 215 mph)
Cruising speed:
at FL80 .... 172 kt (318 km/h; 198 mph)
at FL100 .... 175 kt (324 km/h; 201 mph)
at FL150 .... 185 kt (343 km/h; 213 mph)
Max certified altitude .... 6,100 m (20,000 ft)
T-O run .... 338 m (1,110 ft)
T-O to 15 m (50 ft) .... 552 m (1,810 ft)
Landing from 15 m (50 ft) .... 519 m (1,700 ft)
Landing run .... 269 m (880 ft)
Range, 45 min reserves and normal allowances, normal (N) or long-range (LR) cruise power:
at FL80:
N .... 842 n miles (1,559 km; 969 miles)
LR .... 950 n miles (1,759 km; 1,093 miles)
at FL100:
N .... 844 n miles (1,563 km; 971 miles)
LR .... 948 n miles (1,756 km; 1,091 miles)
at FL150:
N .... 844 n miles (1,563 km; 971 miles)
LR .... 945 n miles (1,750 km; 1,087 miles)

UPDATED

Piper PA-32R-301T Saratoga II TC *(Paul Jackson)* ***NEW***/1047095

Piper PA-32R-301T Saratoga II TC *(Paul Jackson)* ***NEW***/0568426

PIPER PA-32-301XTC 6XT

TYPE: Six-seat utility transport.

PROGRAMME: Announced at EAA Sun 'n' Fun at Lakeland, Florida, 2 April 2003. Prototype (N326XT) flew shortly thereafter. FAA certification awarded 28 August 2003.

CUSTOMERS: 11 delivered in 2003 and 12 in the first nine months of 2004.

COSTS: Standard, equipped, VFR, US$387,100 (2004).

Piper PA-32-301XTC 6XT turbocharged, fixed-gear six-seater *(Paul Jackson)* ***NEW***/1047196

Instrument panel of Piper 6XT *(Paul Jackson)* ***NEW***/1047194

The description for the Piper 6X applies also to the Piper 6XT except:

POWER PLANT: One 224 kW (300 hp) Textron Lycoming TIO-540-AH1A turbocharged flat-six engine, driving a Hartzell three-blade, constant-speed, metal propeller.

EQUIPMENT: Oxygen system optional.

WEIGHTS AND LOADINGS:

Weight empty, equipped	1,026 kg (2,261 lb)
Standard useful load	614 kg (1,354 lb)

PERFORMANCE:

Never-exceed speed (VNE)	189 kt (350 km/h; 217 mph)
Max level speed at mid-cruise weight	172 kt (319 km/h; 198 mph)
High cruising speed at FL100	161 kt (298 km/h; 185 mph)
Service ceiling	6,100 m (20,000 ft)
T-O run	426 m (1,397 ft)
T-O to 15 m (50 ft)	576 m (1,890 ft)
Landing from 15 m (50 ft)	557 m (1,825 ft)
Landing run	278 m (911 ft)
Range, long-range cruise power, 45 min reserves	931 n miles (1,724 km; 1,071 miles)

UPDATED

PIPER PA-34-220T SENECA V

TYPE: Six-seat utility twin.

PROGRAMME: Original PA-34 Seneca certified 7 May 1971 and announced 23 September 1971, having been preceded by prototypes of fixed-gear PA-34-180 Twin Six (flown 25 April 1967) and retractable gear version in following year; 149 kW (200 hp) engines; redesignated Seneca II with increased MTOW, certified 18 July 1974 and introduced for 1975; improved Seneca III with more powerful counter-rotating (C/R) engines certified 17 December 1980 and introduced 15 February 1981; Seneca IV with axisymmetric engine inlets, aerodynamic refinements and interior improvements and restyling certified 17 November 1993 and introduced in 1994. Seneca V with L/TSIO-360-RB engines unveiled 16 January 1997 (having been secretly certified on 11 December 1996); engines, equipped with intercoolers and density control system (automatic wastegate), maintain S/L rated power to 5,945 m (19,500 ft). Other new features include redesigned instrument panel with digital display monitoring panel (DDMP), overhead switch/dome light/speaker panel, entertainment/executive console with extendable work table and in-flight phone/fax option replacing the second row seat on the starboard side. Senecas have also been built in Brazil and Poland.

CUSTOMERS: Total of 28 Seneca IVs delivered in 1995, and 18 in 1996; 39 Seneca Vs delivered in 1997, 55 each in 1998 and 1999, 42 in 2000, 38 in 2001, 43 in 2002, 28 in 2003, and nine in the first nine months of 2004. Widely used as twin-conversion trainer; recent customers include Sabena Airlines, which took delivery of two in 1999 for its training centre at Scottsdale, Arizona.

Production of the PA-34 series in the USA totals some 4,700; the prototype first flew on 30 August 1968.

COSTS: Standard equipped price US$643,500 (2004).

DESIGN FEATURES: Twin-engined version of Cherokee SIX/Saratoga. Low-mounted, constant-chord wings and tailplane; leading-edge gloves inboard of engines; constant-chord tailplane.

Wing section NACA 65_2-415; dihedral 7°; thickness/chord ratio 15 per cent; washout 0° 41'; twist 2° 41'.

FLYING CONTROLS: Manual. Frise ailerons, rudder and one-piece all-moving tailplane with combined anti-balance and trim tab; anti-servo tab in rudder; wide-span electrically operated slotted flaps. Control surface movements: ailerons +35/−20°; tailplane +12° 3'/−7° 30'; rudder ±35°. Maximum flap deflection 40°.

STRUCTURE: Conventional light alloy, with semi-monocoque fuselage; single-spar wings; glass fibre wingtips.

LANDING GEAR: Hydraulically retractable tricycle type. Main units retract inward, nose unit forward. Oleo-pneumatic shock-absorbers. Steerable nosewheel. Emergency free-fall extension system. Mainwheels and tyres size 6.00-6 (8 ply), pressure 3.79 bar (55 lb/sq in); nosewheel and tyre size 6.00-6 (6 ply), pressure 2.76 bar (40 lb/sq in). Nosewheel safety mirror. High-capacity disc brakes. Parking brake.

POWER PLANT: One 164 kW (220 hp) Teledyne Continental TSIO-360-RB (port) and one 164 kW (220 hp) LTSIO-360-RB (starboard) flat-six turbocharged counter-rotating engine, each driving a Hartzell BHC-C2YF-2/FC8459(B)-8R (port) or /FJC8459(B)-8R (starboard) two-blade, constant-speed, fully feathering metal propeller; McCauley 82NJA-6 and L82NJA-6 three-blade propellers optional, but mandatory when de-icing equipment installed. Propeller synchrophasers standard. Fuel in four tanks in wings, with a total capacity of 485 litres (128 US gallons; 107 Imp gallons), of which 462 litres (122 US gallons; 102 Imp gallons) are usable. Oil capacity 7.5 litres (2.0 US gallons; 1.7 Imp gallons). Glass fibre engine cowlings.

ACCOMMODATION: Enclosed cabin, seating five people on individual seats with 0.25 m (10 in) centre aisle; sixth seat (second row, starboard side) optional, replacing entertainment/executive console. Dual controls standard. Pilot's storm window. Two forward-hinged doors, one on starboard side at front, the other on port side at rear. Large

Piper PA-34-220 Seneca V *(Paul Jackson)* ***NEW***/1047092

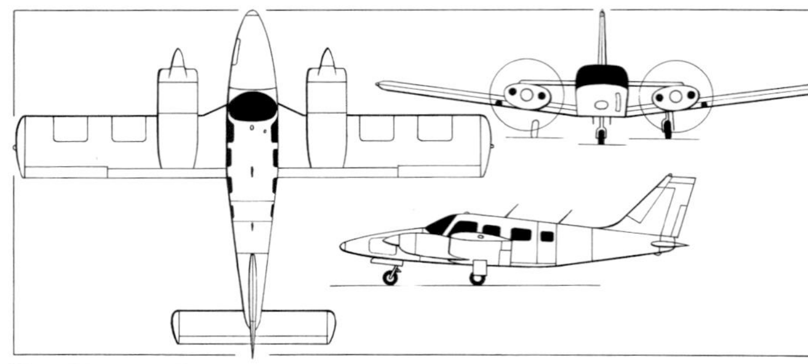

Piper PA-34-220T Seneca V six-seat twin *(James Goulding)* 0016212

utility door adjacent rear cabin door provides an extra-wide opening for loading bulky items. Passenger seats easily removable to provide different seating/baggage/cargo combinations. Space for 39 kg (85 lb) baggage at rear of cabin, and for 45 kg (100 lb) in nose compartment with external door on port side. Cabin heated and ventilated. Windscreen defrosters standard.

SYSTEMS: Electrohydraulic system for landing gear actuation. Electrical system powered by two 28 V 85 A alternators; 24 V 19 Ah battery. Optional pneumatic de-icing boots on wing, tailplane and fin leading-edges. Oxygen system optional.

AVIONICS: *Comms:* Dual Garmin GNS-430/530 com/nav/GS/GPS; GMA-340 audio panel with marker beacon receiver and intercom; GTX-327 transponder; ground clearance system; Telex 100T microphone and Airman 760 headset.

Radar: Provision in nose and on instrument panel for weather radar installation.

Flight: Dual Garmin GNS-430/530; Meggitt ST-361 ADI; ST-180 slaved HIS; dual GI-106A VOR/LOC/GS/GPS indicators; AR-850 altitude encoder; Meggitt System 55 dual-axis FCS.

Instrumentation: Piper true airspeed indicator, magnetic compass, sensitive altimeter, ADI, HIS, rate of turn indicator, rate of climb indicator, illuminated OAT gauge, digital ammeter, annunciator panel with push-to-test, recording tachometer, dual manifold pressure gauges, three-way oil pressure, temperature and CHT gauge, dual fuel quantity gauges and vacuum gauge with warning indicator. Co-pilot's 75 mm (3 in) instrument and United 5035-P40 encoding altimeter panel optional.

EQUIPMENT: Standard equipment as listed for Saratoga II HP, plus emergency landing gear extension system, cabin dome light, pilot's storm window, sun visors with power setting table and checklist, chemical corrosion protection, flush fuel caps and nose gear safety mirror. Piper Aire air conditioning system, built-in oxygen system and foreign certification gross weight kit optional.

DIMENSIONS, EXTERNAL:

Wing span	11.85 m (38 ft 10¾ in)
Wing chord, constant	1.60 m (5 ft 3 in)
Wing aspect ratio	7.3
Length overall	8.72 m (28 ft 7½ in)
Height overall	3.02 m (9 ft 10¾ in)
Tailplane span	4.14 m (13 ft 6¾ in)
Wheel track	3.38 m (11 ft 1 in)
Wheelbase	2.13 m (7 ft 0 in)
Propeller diameter	1.93 m (6 ft 4 in)
Distance between propeller centres	3.80 m (12 ft 5½ in)
Cabin door (stbd, fwd): Height	0.89 m (2 ft 11 in)
Width	0.91 m (3 ft 0 in)
Cabin door (port, rear): Height	0.72 m (2 ft 4½ in)
Width	0.71 m (2 ft 4 in)
Baggage door (stbd, rear): Height	0.52 m (1 ft 8½ in)
Width	0.66 m (2 ft 2 in)
Baggage door (port, fwd): Height	0.53 m (1 ft 9 in)
Width	0.61 m (2 ft 0 in)

DIMENSIONS, INTERNAL:

Cabin (incl flight deck): Length	3.15 m (10 ft 4¼ in)
Max width	1.24 m (4 ft 0¾ in)
Max height	1.07 m (3 ft 6 in)
Volume	5.5 m^3 (195 cu ft)
Baggage compartment volume:	
forward	0.43 m^3 (15.3 cu ft)
rear	0.49 m^3 (17.3 cu ft)

AREAS:

Wings, gross	19.39 m^2 (208.7 sq ft)
Ailerons, incl tab (total)	1.17 m^2 (12.60 sq ft)
Trailing-edge flaps (total)	1.94 m^2 (20.84 sq ft)
Fin	1.14 m^2 (12.32 sq ft)
Rudder, incl tab	0.71 m^2 (7.62 sq ft)
Horizontal tail surfaces (total)	3.60 m^2 (38.74 sq ft)

WEIGHTS AND LOADINGS:

Weight empty, equipped	1,534 kg (3,382 lb)
Baggage capacity	83 kg (185 lb)
Max T-O weight	2,154 kg (4,750 lb)
Max ramp weight	2,165 kg (4,773 lb)
Max zero-fuel weight	2,031 kg (4,479 lb)

Max landing weight .... 2,047 kg (4,513 lb)
Max wing loading .... 111.1 kg/m² (22.76 lb/sq ft)
Max power loading .... 6.57 kg/kW (10.80 lb/hp)

PERFORMANCE:
Max level speed .... 218 kt (404 km/h; 251 mph)
Cruising speed:
at FL100:
at max cruise power .... 182 kt (337 km/h; 209 mph)
at normal cruise power .... 176 kt (326 km/h; 202 mph)
at FL185:
at max cruise power .... 201 kt (372 km/h; 231 mph)
at normal cruise power .... 190 kt (352 km/h; 219 mph)
Stalling speed, flaps down .... 61 kt (113 km/h; 71 mph)
Max rate of climb at S/L .... 446 m (1,462 ft)/min
Rate of climb at S/L, OEI .... 76 m (250 ft)/min
Max certified altitude .... 7,620 m (25,000 ft)
Service ceiling, OEI .... 5,030 m (16,500 ft)
T-O run .... 349 m (1,145 ft)
T-O to 15 m (50 ft) .... 521 m (1,710 ft)
Landing from 15 m (50 ft) .... 665 m (2,180 ft)
Landing run .... 427 m (1,400 ft)
Range at long-range cruise power, incl allowance for taxi, T-O, climb and 45 min reserves:
at FL100 .... 812 n miles (1,503 km; 934 miles)
at FL150 .... 828 n miles (1,533 km; 952 miles)
at FL185 .... 819 n miles (1,516 km; 942 miles)

UPDATED

PIPER PA-44-180 SEMINOLE

TYPE: Four-seat utility twin.

PROGRAMME: Prototype first flown May 1976; production version announced 21 February 1978; FAA certification 10 March 1978; two versions, Seminole, and Turbo Seminole (latter certified 29 November 1979), produced until 1982; normally aspirated Seminole restored to production in 1988, suspended in 1990 and restored again in 1995.

CUSTOMERS: Four delivered in 1995, eight in 1996, six in 1997, four in 1998, six in 1999, 11 in 2000, 62 in 2001, 60 in 2002, 16 in 2003, and 11 in the first nine months of 2004. Recent customers include the University of North Dakota, which ordered one in April 2003, and the Royal Jordanian Air Academy, which took delivery of three in 2003.

Manufacture of all PA-44 variants reached 600 in 2002. Many used in twin-conversion and IFR training.

COSTS: Standard, equipped US$437,700 (2004).

DESIGN FEATURES: Twin-engined derivation of PA-28 Arrow (which see). Constant-chord low wing and T tail; leading-edge gloves inboard of engines; sweptback fin with small fillet.

Wing section NACA 65_2-415; thickness/chord ratio 15 per cent; dihedral 7°; incidence 2°; twist 3°; washout −1°.

FLYING CONTROLS: Manual. Ailerons, rudder and all-moving tailplane with full-span anti-servo tab; rudder tab; four-position manually operated flaps. Control surface movements: ailerons +23/−17°; tailplane +15/−3°; rudder ±37°; maximum flap deflection 40°.

STRUCTURE: Conventional light alloy, with semi-monocoque fuselage; single-spar wings.

LANDING GEAR: Hydraulically retractable tricycle type. Free-fall emergency extension system. Piper oleo-pneumatic shock-absorbers. Mainwheels and tyres size 6.00-6 (8 ply), with tubes. Steerable nosewheel with tyre size 5.00-5 (6 ply), with tube. Dual toe-operated high-capacity disc brakes. Heavy-duty brakes and tyres optional.

POWER PLANT: Two 134 kW (180 hp) Textron Lycoming flat-four counter-rotating engines (one O-360-A1H6 to port and one LO-360-A1H6), each driving a Hartzell two-blade constant-speed fully feathering metal propeller: HC-C2Y(K,R)-2CEUF/FC7666A-2R and /FJC7666A-2R, respectively. One bladder-type fuel tank in each engine nacelle, with total capacity of 416 litres (110 US gallons; 91.6 Imp gallons), of which 409 litres (108 US gallons; 89.9 Imp gallons) are usable. Refuelling point on upper surface of each nacelle. Oil capacity 11.5 litres (3.0 US gallons; 2.5 Imp gallons).

ACCOMMODATION: Cabin seats four in two pairs of individual seats. Dual controls standard. Emergency exit on port side. Pilot's storm window. Baggage compartment at rear of cabin, capacity 91 kg (200 lb). Accommodation heated and ventilated. Windscreen defrosters.

SYSTEMS: Electrohydraulic system for landing gear actuation and brakes. Electrical system includes two engine-driven 14 V 60 A alternators and 12 V 35 Ah battery. Janitrol combustion heater of 45,000 BTU capacity. Dual vacuum systems standard.

AVIONICS: Standard and DeLuxe packages by Garmin, as described.

Comms: Standard: Garmin GNS-430 com/nav/GPS; GI-106A VOR/LOC/GS/CPS indicator; GTX-327 transponder; GMA-340 audio panel with marker beacon receiver and four-position intercom; Telex 100T noise-cancelling microphone and Airman 760 headset.

Flight: Standard: Garmin GNS-430 com/nav/IFR GPS; Narco AR-850 altitude reporter.

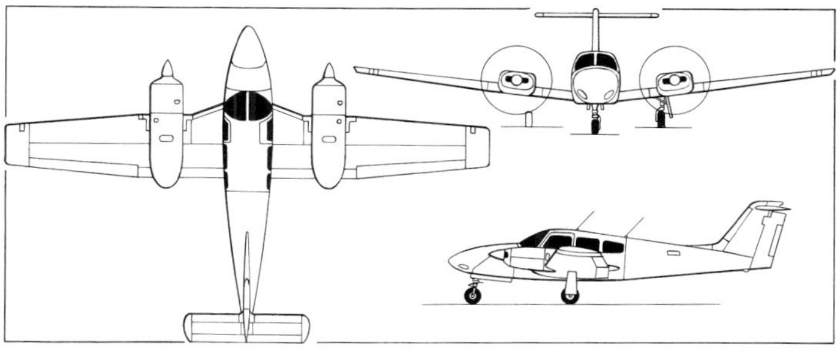

Piper PA-44-180 Seminole four-seat light twin *(James Goulding)* 0100460

DeLuxe package as above, plus second GNS 430 and GI-106A; VOR/LOC/GPS/HSI with slaved compass system, replacing standard directional gyro; and RCR-650D ADF with IND-650A indicator. Optional equipment includes ST-361 ADI; ST-360 altitude select/alerter; DME-450; Meggitt System 55 dual-axis autopilot; and Meggitt electric trim (standard with autopilot).

EQUIPMENT: Metal instrument panel, engine hour recorders, alternate static source, heated pitot head, instrument panel lights and overhead blue lighting, avionics dimming, dome light, navigation lights, landing/taxying light, wingtip strobe lights, pilot's and co-pilot's vertically adjustable seats in fabric and vinyl, with optional lumbar support, two reclining rear passenger seats, vinyl cabin side panels, crew armrests, tinted windscreen and windows, pilot's storm window, 'Quietised' soundproofing, tiedown points, jack pads, nose gear safety mirror, external power receptacle, static discharge wicks, and Du Pont Imron polyurethane exterior paint in white base colour with two contrasting trim stripes. Metallic paint optional.

DIMENSIONS, EXTERNAL:
Wing span .... 11.75 m (38 ft 6½ in)
Wing aspect ratio 8.1
Length overall .... 8.41 m (27 ft 7¼ in)
Height overall .... 2.59 m (8 ft 6 in)
Tailplane span .... 3.05 m (10 ft 0 in)
Wheel track .... 3.21 m (10 ft 6½ in)
Wheelbase .... 2.56 m (8 ft 4¾ in)
Propeller diameter .... 1.88 m (6 ft 2 in)
Distance between propeller centres .... 3.86 m (12 ft 7¼ in)
Cabin door (stbd): Height .... 0.89 m (2 ft 11 in)
Width .... 0.91 m (3 ft 0 in)
Baggage door: Height .... 0.51 m (1 ft 8 in)
Width .... 0.56 m (1 ft 10 in)

DIMENSIONS, INTERNAL:
Cabin (instrument panel to rear bulkhead):
Length .... 2.46 m (8 ft 1 in)
Max width .... 1.05 m (3 ft 5½ in)
Max height .... 1.25 m (4 ft 1 in)
Volume .... 3.0 m³ (106 cu ft)
Baggage compartment volume .... 0.74 m³ (26.0 cu ft)

AREAS:
Wings, gross .... 17.08 m² (183.8 sq ft)
Ailerons (total) .... 1.13 m² (12.12 sq ft)
Trailing-edge flaps (total) .... 1.36 m² (14.60 sq ft)
Fin .... 1.26 m² (13.56 sq ft)
Rudder, incl tab .... 0.76 m² (8.21 sq ft)
Horizontal tail surfaces (total) .... 2.27 m² (24.40 sq ft)

WEIGHTS AND LOADINGS:
Weight empty, equipped .... 1,173 kg (2,587 lb)
Max T-O weight .... 1,723 kg (3,800 lb)
Max wing loading .... 100.9 kg/m² (20.67 lb/sq ft)
Max power loading .... 6.42 kg/kW (10.55 lb/hp)

PERFORMANCE:
Max level speed .... 168 kt (311 km/h; 193 mph)
Cruising speed:
at 75% power .... 162 kt (300 km/h; 186 mph)
at 65% power .... 157 kt (291 km/h; 181 mph)
Stalling speed: flaps up .... 57 kt (106 km/h; 66 mph) IAS
flaps down .... 55 kt (102 km/h; 63 mph) IAS
Max rate of climb at S/L .... 408 m (1,340 ft)/min

Piper PA-44-180 Seminole *(Paul Jackson)* ***NEW***/1047119

Rate of climb at S/L, OEI ... 65 m (212 ft)/min
Service ceiling ... 4,575 m (15,000 ft)
Service ceiling, OEI ... 1,155 m (3,800 ft)
T-O to 15 m (50 ft) ... 671 m (2,200 ft)
Landing from 15 m (50 ft) ... 454 m (1,490 ft)
Cruising range at 55% power, 45 min reserves ... 770 n miles (1,426 km; 886 miles)

UPDATED

PIPER PA-46-350P MALIBU MIRAGE

TYPE: Business prop.

PROGRAMME: Prototype Malibu first flew 30 November 1979; FAA certification of original PA-46-310P Malibu (TSIO-520 engine and two-blade propeller) received 27 September 1983; production deliveries began December 1983; 404 built before replaced by PA-46-350P Malibu Mirage October 1988, FAA certification having been received on 30 August 1988. Production temporarily suspended in 2000 to enable the company to concentrate on launching the turboprop Malibu Meridian; resumption then planned for late 2001.

CURRENT VERSIONS: **Malibu Mirage:** *As described.* Improvements introduced on 1995 model include pilot's heated glass windscreen, inflatable lumbar support on pilot/co-pilot's seats, colour co-ordinated control wheels and restyled interior trim, cabinetry and seats. Strengthened wing structure of Malibu Meridian (which see) introduced as standard from 1999, affording 18 kg (40 lb) increase in maximum take-off weight.

20th Anniversary Special Edition: Limited production version introduced in 2004 to celebrate the 20th anniversary of service entry of the original PA-46-310P Malibu. Features include 20th anniversary logos on the external paint finish, leather seats and threshold kick-plate, and special anniversary option packages for the most popular avionics suites, typically for US customers comprising Meggitt Magic EFIS System, Honeywell IHAS 8000 Integrated Hazard Aviodance System, Flight Information System weather datalink and a second Garmin GTX 330 transponder.

Malibu Meridian: Described separately.

JetPROP DLX: Supplemental Type Certificate awarded in 1998 to JetPROP LLC and Rocket Engineering, both of Spokane, Washington, for conversion of Malibu and Malibu Mirage with Pratt & Whitney PT6A-35 turboprop and Hartzell four-blade propeller.

CUSTOMERS: Total of 40 delivered in 1995, 57 in 1996, 54 in 1997, 55 in 1998, 61 in 1999, 63 in 2000, 10 in 2001, 19 in 2002, seven in 2003, and 12 in the first nine months of 2004.

COSTS: Standard, equipped US$970,000 (2004).

DESIGN FEATURES: High-speed, long-range Piper single, with streamlined appearance and moderately high-aspect ratio wing, which is low mounted and tapered; mid-tailplane, also tapered; sweptback fin; wide track landing gear.

Wing section NASA 23016 at root, 23009 at tip; dihedral 4° 30'; thickness/chord ratio 16 per cent; twist 2° 57'; incidence 3° 38'.

FLYING CONTROLS: Conventional and manual. Horn-balanced elevators and rudder; stainless steel control cables. Electronic trim tab in elevator; electrically operated trailing-edge flaps. Precise Flight speed brakes standard. Control surface movements: ailerons ±18°; elevator ±23° 30'/−14° 30'; rudder 26° port/30° starboard; elevators +19/−14° 30'; maximum flap deflection 35°.

STRUCTURE: Cantilever high-aspect ratio all-metal wings; light alloy fuselage, fail-safe construction in pressurised area; light alloy tail surfaces.

LANDING GEAR: Hydraulically retractable tricycle type with single wheel on each unit; main units retract inward into wingroots, nosewheel rearward, rotating 90° to lie flat under baggage compartment. Mainwheel size 6.00-6 (8 ply), pressure 3.80 bar (55 lb/sq in); nosewheel 5.00-6 (6 ply) pressure 3.45 bar (50 lb/sq in). Toe-operated brakes.

POWER PLANT: One 261 kW (350 hp) Textron Lycoming TIO-540-AE2A turbocharged and intercooled flat-six engine, driving a Hartzell HC-13YR-1E/7890K three-blade constant-speed propeller with polished spinner. Composites (Kevlar) propeller replaced metal from 1998. Fuel system capacity 462 litres (122 US gallons; 102 Imp gallons), of which 454 litres (120 US gallons; 100 Imp gallons) are usable. Oil capacity 11.5 litres (3.0 US gallons; 2.5 Imp gallons).

ACCOMMODATION: Pilot and five passengers in pressurised, heated and air conditioned cabin; dual controls standard; front two occupants have vertical, fore-and-aft adjusting and reclining leather seats with inflatable lumbar supports, stowaway armrests, inertia reel shoulder harnesses and map-holders. Leather reclining passenger seats in club arrangement with stowaway armrests and inertia reel shoulder harnesses; unpressurised baggage compartment in nose, and pressurised space at rear of cabin. Door with integral steps on port side aft of wing.

SYSTEMS: Pressurisation, maximum differential 0.38 bar (5.5 lb/sq in), to provide a cabin altitude of 2,400 m (7,900 ft) to a height of 7,620 m (25,000 ft). Hydraulic system pressure 107 bar (1,550 lb/sq in). Dual engine-driven vacuum pumps standard. Split bus electrical system has two 28 V/70 A alternators; 24 V 10 Ah battery; full icing protection standard. Optional electrically heated pilot's windscreen and fuel management system.

AVIONICS: Standard Garmin New Generation IFR avionics package, as detailed.

Comms: Dual GNS-530 com/nav/IFR GPS with MFDs; GTX-327 transponder (second optional); GMA-340 audio panel with marker beacon receiver and intercom; Telex 100T noise-cancelling microphone and Airman 760 headset.

Radar: RDR-2000 vertical profile colour weather radar in wing-mounted pod. Goodrich WS-1000+ Stormscope optional.

Flight: Garmin GNS-430 com/nav/IFR GPS; ST-361 ADI; ST-180 HSI with slaved compass system; ST-360 altitude select/alerter; United 5035P-P40 altitude encoder; Meggitt System 55X three-axis autopilot with electric trim, turn indicator and yaw damper; co-pilot's longitudinal electric trim button. Options include single or dual RCR-650A ADF with IND-650A indicators; DME 450; Honeywell KI 229 RMI (exchange for IND-650A): and KRA-10A radar altimeter.

Instrumentation: Dual MFDs with GNS-430 installation; dual GI-106A VOR/LOC/GS/GPS indicators.

EQUIPMENT: Standard equipment includes heated lift detector; stall warning computer and horn; digital ammeter/voltmeter; six-channel CHT monitoring system with selectable cylinder readout; gyro air filter; alternate static source; heated pitot head; nosewheel light; wingtip taxying light; navigation lights, wingtip strobe lights; cockpit dome lights; solid-state dimming landing gear position and instrument panel lights; seven cabin overhead lights; PPG pilot's heated windscreen; windscreen defrosters; pilot's opening storm window; pilot's relief tube; supplemental electric heater; emergency oxygen system; leather sidewalls; stowaway executive writing table; forward refreshment/entertainment centres; super 'Quietised' soundproofing with inner passenger windows; halon fire extinguisher; Truax static discharge wicks; chemical corrosion protection; and Du Pont Imron polyurethane exterior paint in single- or two-tone base colour with graphics striping in choice of three trim colours. Optional equipment includes 'Infinity' paint scheme and stainless steel cowling fasteners.

Piper PA-46-350P Malibu Mirage *(Paul Jackson)* ***NEW***/1047110

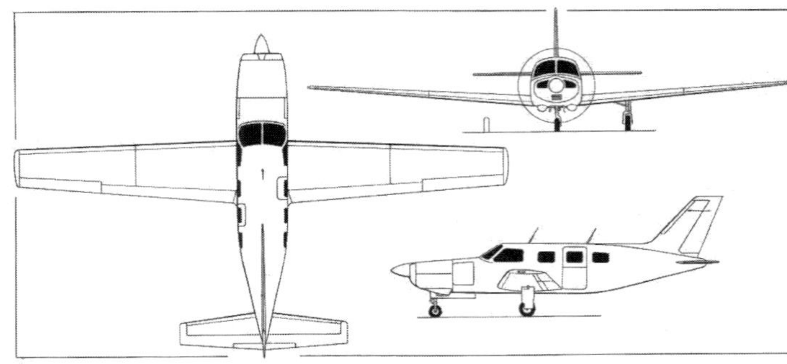

Piper PA-46-350P Malibu Mirage(Textron Lycoming TIO-540-AE2A) *(Jane's/Dennis Punnett)* 0507288

DIMENSIONS, EXTERNAL:
Wing span ... 13.11 m (43 ft 0 in)
Wing aspect ratio ... 10.6
Length overall ... 8.81 m (28 ft 10¾ in)
Height overall ... 3.44 m (11 ft 3½ in)
Tailplane span ... 4.42 m (14 ft 6 in)
Wheel track ... 3.75 m (12 ft 3½ in)
Wheelbase ... 2.44 m (8 ft 0 in)
Propeller diameter ... 2.03 m (6 ft 8 in)
Passenger door (port, rear): Height ... 1.17 m (3 ft 10 in)
Width ... 0.61 m (2 ft 0 in)
Baggage door (port, nose): Height ... 0.58 m (1 ft 11 in)
Width ... 0.48 m (1 ft 7 in)

DIMENSIONS, INTERNAL:
Cabin (instrument panel to rear pressure bulkhead):
Length ... 3.76 m (12 ft 4 in)
Max width ... 1.26 m (4 ft 1½ in)
Max height ... 1.19 m (3 ft 11 in)
Baggage compartment volume:
nose ... 0.37 m³ (13.0 cu ft)
rear cabin ... 0.57 m³ (20.0 cu ft)

AREAS:
Wings, gross ... 16.26 m² (175.0 sq ft)

WEIGHTS AND LOADINGS:
Weight empty, equipped ... 1,416 kg (3,121 lb)
Baggage capacity: nose ... 45 kg (100 lb)
rear cabin ... 45 kg (100 lb)
Max T-O weight ... 1,968 kg (4,340 lb)
Max ramp weight ... 1,976 kg (4,358 lb)
Max landing and max zero-fuel weight ... 1,870 kg (4,123 lb)
Max wing loading ... 121.1 kg/m² (24.80 lb/sq ft)
Max power loading ... 7.55 kg/kW (12.40 lb/hp)

PERFORMANCE:
Max level speed at mid-cruise weight ... 220 kt (407 km/h; 253 mph)
Cruising speed at optimum altitude, mid-cruise weight, high-speed cruise power ... 213 kt (394 km/h; 245 mph)
Stalling speed, flaps and wheels down ... 58 kt (108 km/h; 67 mph)
Max rate of climb at S/L ... 372 m (1,220 ft)/min
Max certified altitude ... 7,620 m (25,000 ft)
T-O run ... 332 m (1,090 ft)
T-O to 15 m (50 ft) ... 637 m (2,090 ft)
Landing from 15 m (50 ft) ... 597 m (1,960 ft)
Landing run ... 311 m (1,020 ft)
Range with max fuel, allowances for start, T-O, climb and descent, plus 45 min reserves, at optimum altitude, long-range cruise power .... 1,345 n miles (2,490 km; 1,547 miles)

UPDATED

PIPER PA-46-500TP MALIBU MERIDIAN

TYPE: Business turboprop.

PROGRAMME: Launched at 1997 National Business Aviation Association Convention in Dallas, Texas, where a full-scale fuselage mockup was displayed; prototype (N400PT, converted from second Malibu Mirage) rolled out 13 August 1998; first flight 21 August 1998. Static test airframe and three further certification flight test aircraft (N403MM, first flown in July 1999, N402MM first flown 27 August 1999 and N401MM first flown in September 1999) built on production tooling. N401MM was dedicated to stability and autopilot testing and ice shape testing; N402MM, first with production standard interior and exterior paint finish, conducted performance and avionics testing and certification for flight into known icing, as well as serving as marketing demonstrator, N403MM was responsible for systems and power plant testing, high-speed flight testing and flutter testing. Public debut (N402MM) at the NBAA Convention in Atlanta, Georgia, in October 1999. First

Piper PA-46-500TP Malibu Meridian turboprop *(Paul Jackson)* ***NEW***/1047109

flight of production aircraft (c/n 003/N375RD) 30 June 2000. FAR Pt 23 certification achieved 27 September 2000, followed by UK CAA approval on 21 June 2001.

Increase in maximum take-off weight to 2,310 kg (5,092 lb), resulting in 15 per cent increase in useful load, certified at end of 2002; modifications include strengthening of airframe, modified stall strips and addition of vortex generators to the wings and undersurface of tailplane.

CURRENT VERSIONS: **Malibu Meridian:** *As described.*

20th Anniversary Special Edition: As described under Malibu Mirage entry.

CUSTOMERS: Total of 239 orders held by October 2000. First delivery (N375RD) to Richard Dumais of Richardson, Texas, in November 2000. Total of 98 delivered in 2001, 25 in 2002, 24 in 2003, and 21 in the first nine months of 2004.

COSTS: US$1.81 million typically equipped (2004).

Description generally as for Malibu Mirage, except that below.

DESIGN FEATURES: Strengthened wing, incorporating wingroot leading-edge gloves to increase area and reduce stalling speed; strengthened tail surfaces; and 37 per cent increase in area of horizontal stabiliser.

FLYING CONTROLS: Deflections as for Malibu Mirage except flaps 36°. Flight-adjustable trim tab in rudder. Ground-adjustable tab on starboard aileron.

POWER PLANT: One Pratt & Whitney Canada PT6A-42A turboprop, thermodynamic rating 767 kW (1,029 shp), flat rated at 373 kW (500 shp) maximum continuous, driving a Hartzell HC-E4N-3Q/E8501B-3.5 four-blade constant-speed reversible propeller. Fuel capacity 655 litres (173 US gallons; 144 Imp gallons), of which 644 litres (170 US gallons; 141.5 Imp gallons) are usable. Oil capacity 11 litres (3.0 US gallons; 2.5 Imp gallons).

ACCOMMODATION: Pilot and five passengers in standard club layout; passenger capacity reduced to four with optional entertainment centre comprising beverage cooler, storage cabinets, AM/FM/CD stereo and provision for VCR with flat-panel monitor. Fully automatic bleed-air conditioning and temperature control systems.

SYSTEMS: Upgraded electrical system, with 200 A starter/generator and 130 A standby alternator.

AVIONICS: Standard Garmin International and Meggitt package with dual Garmin GNS 530 nav/com/GPS as core system.

Comms: Garmin transceiver with 8.33 kHz spacing; GMA-340 audio panel; GTX-327 transponder.

Radar: Honeywell RDR 2000 vertical profile colour weather radar with KMD 850 MFD.

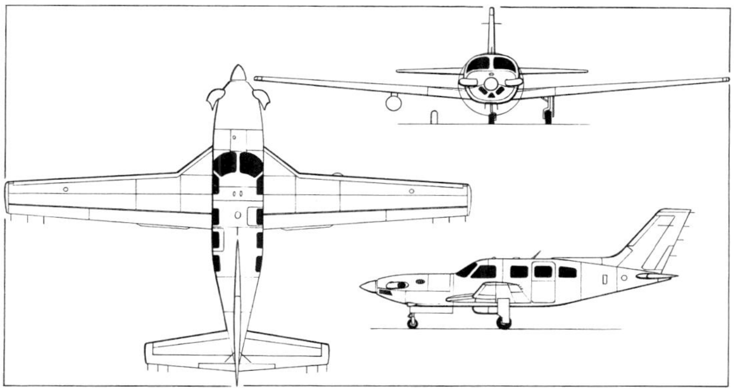

Piper Malibu Meridian (Pratt & Whitney Canada PT6A turboprop) *(James Goulding)* 0075961

Flight: S-Tec Magic 1500 AFCS with integrated yaw damper; dual Meggitt air data, attitude, heading and reference systems (ADAHRS) provides digital readout of pitch and roll attitudes, roll and yaw rates, altitude, rate of altitude change, airspeed and heading.

Instrumentation: Meggitt Avionics MAGIC EFDS fit comprising pilot's and co-pilot's colour flat panel LCD primary flight display and navigation display driven by dual air data attitude heading reference systems (ADAHRS).

DIMENSIONS, EXTERNAL: As for Malibu Mirage except:
- Length overall ... 9.02 m (29 ft 7¼ in)
- Height overall ... 3.45 m (11 ft 4 in)
- Propeller diameter ... 2.095 m (6 ft 10½ in)

DIMENSIONS, INTERNAL: As for Malibu Mirage

AREAS:
- Wings, gross ... 17.00 m² (183.0 sq ft)

WEIGHTS AND LOADINGS:
- Weight empty, equipped ... 1,540 kg (3,395 lb)
- Baggage capacity ... 45 kg (100 lb)
- Max T-O weight ... 2,310 kg (5,092 lb)
- Max ramp weight ... 2,329 kg (5,134 lb)
- Max wing loading ... 135.8 kg/m² (27.82 lb/sq ft)
- Max power loading ... 6.20 kg/kW (10.2 lb/shp)

PERFORMANCE:
- Max cruising speed at FL300 at mid-cruise weight ... 260 kt (481 km/h; 299 mph)
- Max rate of climb at S/L ... 474 m (1,556 ft)/min
- Max certified altitude ... 9,150 m (30,000 ft)
- T-O run ... 518 m (1,698 ft)
- T-O to 15 m (50 ft) ... 776 m (2,545 ft)
- Landing from 15 m (50 ft) ... 644 m (2,114 ft)
- Landing run ... 308 m (1,009 ft)
- Max range at FL300, max cruise power, mid-cruise weight, 45 min reserves ... more than 1,000 n miles (1,852 km; 1,151 miles)
- Endurance ... 4 h 12 min

UPDATED

PRECISION TECH

PRECISION TECH AIRCRAFT INC

155E Hangar 33, Highway 61 SE, Cartersville, Georgia 30120
Tel: (+1 770) 607 40 09
Fax: (+1 770) 386 63 55
e-mail: info@fergy.net
Web: http://www.fergy.net
PRESIDENT: Lance McAfee

Precision Tech took over manufacture and distribution rights for the Ferguson F-II from its designer during 1999.

VERIFIED

PRECISION TECH F-IIB FERGY

TYPE: Side-by-side ultralight kitbuilt.

PROGRAMME: Designed by Bill Ferguson; awarded Best New Design at Sun 'n' Fun 91.

CURRENT VERSIONS: **Ferguson F-II:** Initial version, no longer produced.

F-IIB Fergy: Current production model, *as described.*

CUSTOMERS: At least 40 flying by October 2003.

COSTS: Kit US$11,500 (2004).

DESIGN FEATURES: Pod-and-boom configuration; high wing and low tailplane. Wings and tail fold for transportation. Gull wing cabin doors for easy access. Quoted build time 300–350 hours.

FLYING CONTROLS: Conventional and manual. Dual controls. Full-span flaperons deflect to −15°. Deflections: aileron +27/−23°, rudder ±35°.

STRUCTURE: Fuselage of 4130 chromoly steel tubing with glass fibre cowling; 6063-T6 aluminium tube tailboom and wing spars; wing ribs and tail 6061-T6 aluminium tubing with chromoly steel brackets. Fabric covering; aluminium wing struts. Wing dihedral 0°45'.

LANDING GEAR: Tailwheel type; fixed. Fuselage-mounted spring cantilever mainwheel legs. Brakes on mainwheels; optional speed fairings. Main wheels 6.00–6; larger tyres can be fitted. Options include skis and straight or amphibious floats.

POWER PLANT: One 37.0 kW (49.6 hp) Rotax 503UL-2V engine mounted on rear of wing driving two-blade (optional three-blade) Warp Drive propeller mounted above wing; engines in the range 33.6 to 48.5 kW (45 to 65 hp) recommended. Fuel capacity 47.3 litres (12.5 US gallons; 10.4 Imp gallons).

EQUIPMENT: Optional BRS parachute.

DIMENSIONS, EXTERNAL:
- Wing span ... 8.99 m (29 ft 6 in)
- Length overall ... 6.71 m (22 ft 0 in)
- Height overall ... 1.73 m (5 ft 8 in)

PrecisionTech F-IIB Fergy II kitplane *(Paul Jackson)* 0089508

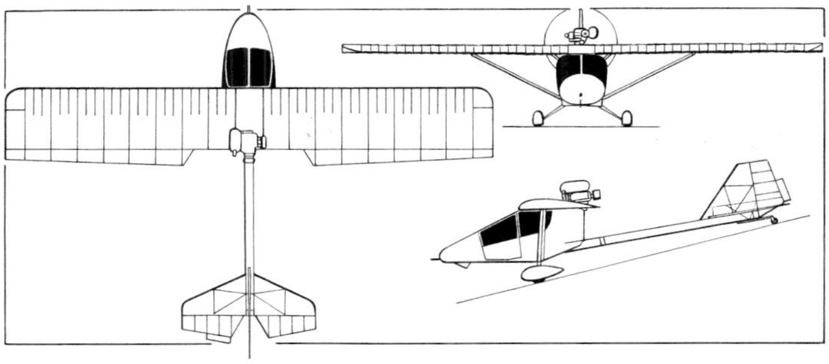

General arrangement of PrecisionTech Fergy *(James Goulding)* 0110973

AREAS:
- Wings, gross ... 13.01 m² (140.0 sq ft)

WEIGHTS AND LOADINGS:
- Weight empty ... 181 kg (400 lb)
- Max T-O weight ... 453 kg (1,000 lb)

PERFORMANCE(48.5 kW; 65 hp engine):
- Never-exceed speed (VNE): ... 86 kt (161 km/h; 100 mph)
- Normal cruising speed range ... 43–70 kt (80–129 km/h; 50–80 mph)

Stalling speed, power off:
- flaperons up ..... 30 kt (55 km/h; 34 mph)
- flaperons down ..... 25 kt (45 km/h; 28 mph)

Max rate of climb at S/L ..... 366 m (1,200 ft)/min
T-O run ..... 53 m (175 ft)
Landing run ..... 61 m (200 ft)
Range ..... 191 n miles (354 km; 220 miles)

UPDATED

PROSPORT

PROSPORT AVIATION

Rights to the Freebird and Sportlite 103 have been acquired by Free Bird Innovations.

UPDATED

PULSAR

PULSAR AIRCRAFT CORPORATION

4233 North Santa Anita Avenue, Suite 5, El Monte Airport, El Monte, California 91731
Tel: (+1 626) 443 10 19
Fax: (+1 626) 443 13 11
e-mail: info@pulsaraircraft.com
Web: http://www.pulsaraircraft.com
PRESIDENT Solly Melyon

Pulsar Aircraft Corporation acquired rights to Pulsar series of light aircraft from SkyStar in July 1999 and announced the Super Pulsar 100, based on the Pulsar III, at that time. It has also obtained rights to the former Tri-R KIS and KIS Cruiser, marketing of which began in 2001.

In 2004, the Pulsar XP-LSA and Super Pulsar-LSA were added to the range in anticipation of Light Sport Aircraft regulations introduced 1 September 2004.

UPDATED

PULSAR SUPER PULSAR 100

TYPE: Side-by-side kitbuilt.

PROGRAMME: Designed by Mark Brown. Prototype Pulsar II first flew 3 April 1988 as two-seat version of Aero Designs Star-Lite (itself first flown in 1983). Originally with 47.8 kW (64.1 hp) Rotax 582 engine. Until suspended in 1999, in favour of Super Pulsar, parallel main production versions were Pulsar XP, with 59.6 kW (79.9 hp) Rotax 912 UL and Pulsar III with 84.6 kW (113.4 hp) turbocharged Rotax 914, giving improved performance. Larger canopy and redesigned cowling with chin intake; 100 per cent mass-balanced controls. Nose— and tailwheel versions of both were available. Programme acquired by SkyStar in late 1996 and by Pulsar in 1999.

CURRENT VERSIONS: **Super Pulsar 100:** Refined version of Pulsar III with repositioned tail unit for improved stability and higher, wider cabin. Prototype (N601SP) first flew August 2000 but damaged 20 August 2000; flight testing resumed 2 April 2001 and aircraft made public debut at Sun 'n' Fun 13 April 2001; by April 2002, 228 hours had been flown by prototype. Initial customer-built aircraft first flown 3 September 2002 in Poland. Tailwheel version under consideration.

Description applies to Super Pulsar 100.

Super Pulsar-LSA: Inreoduced 2004, powered by 59.7 kW (80 hp) Jabiru 2200. Offered as kit or, from January 2005, as ready-to-fly; latter cost US$79,950.

Wega: Version of Pulsar II-914 Turbo, certified to JAR-VLA with 600 kg (1,323 lb) maximum T-O weight, marketed by HK Aircraft Technology in Germany, which see.

Pulsar XP: Returned to production in 2001 as kit; cost US$22,950 (2004); described in 1997–98 *Jane's.*

Pulsar XP-LSA: As Super Pulsar-LSA; ready-to-fly US$69,950 (2004).

CUSTOMERS: More than 250 of all Pulsar variants flying by mid-2003, with a further 300 then under construction.

COSTS: Basic Super Pulsar kit (no engine) US$26,950 (2003); fast-build options add US$7,500.

DESIGN FEATURES: Highly streamlined low-wing configuration, with sweptback fin. Designed for ease of home construction with premoulded components where possible. Wings removable in about 15 minutes for mounting on trailer. Quoted build time 1,000 hours, or 600 hours for 'Super' kit.

FLYING CONTROLS: Conventional and manual; 60 per cent mass-balanced surfaces operated by rods and cables.

STRUCTURE: GFRP with carbon fibre and foam cores. Wing skins of composites.

LANDING GEAR: Tricycle type; fixed. Fuselage-mounted spring cantilever mainwheel legs with wheel fairings.

POWER PLANT: One 88.0 kW (118 hp) Pulsar Aeromaxx flat-four piston engine recommended; alternatively 93 kW (125 hp) Teledyne Continental IO-240, 59.7 kW (80 hp) Jabiru 2200 or 89 kW (120 hp) Jabiru 3300. Fuel capacity 144 litres (38.0 US gallons; 31.6 Imp gallons).

DIMENSIONS, EXTERNAL:
- Wing span ..... 7.62 m (25 ft 0 in)
- Wing aspect ratio 7.8
- Length overall ..... 6.10 m (20 ft 0 in)
- Height overall ..... 1.80 m (5 ft 11 in)

DIMENSIONS, INTERNAL:
- Cabin: Max width ..... 1.09 m (3 ft 7 in)
  - Max height ..... 1.14 m (3 ft 9 in)

AREAS:
- Wings, gross ..... 7.43 m² (80.0 sq ft)

WEIGHTS AND LOADINGS (Aeromaxx):
- Weight empty, equipped ..... 340 kg (750 lb)
- Baggage capacity ..... 27 kg (60 lb)
- Max T-O weight ..... 635 kg (1,400 lb)
- Max wing loading ..... 85.4 kg/m² (17.50 lb/sq ft)
- Max power loading ..... 7.22 kg/kW (11.86 lb/hp)

PERFORMANCE (Aeromaxx, estimated):
- Never-exceed speed (VNE) ..... 191 kt (354 km/h; 220 mph)
- Max level speed at S/L ..... 165 kt (305 km/h; 190 mph)
- Stalling speed, power off ..... 48 kt (89 km/h; 55 mph)
- Max rate of climb at S/L, solo ..... 671 m (2,200 ft)/min
- T-O run ..... 213 m (700 ft)
- Landing run ..... 244 m (800 ft)
- Range ..... 1,086 n miles (2,011 km; 1,250 miles)
- *g* limits ..... +9/–6

UPDATED

PULSAR SPORT 150

TYPE: Side-by-side kitbuilt.

PROGRAMME: Designed by Rich Trickle and Vance Jaqua. First flown 1991 and initially known as Tri-R KIS; name derived from 'keep it simple'. Programme sold to Pulsar in 1999; new owner modifying design with revised canopy (rear-hinged or sliding) and empennage.

CURRENT VERSIONS: **TR-1:** Tricycle landing gear version, *as described.*

TD: Tailwheel version, introduced 1992; powered by 88.0 kW (118 hp) Textron Lycoming O-235-C1B; seven flying by end of 1996.

CUSTOMERS: At least 31 (24 tricycle and seven tailwheel) flying by 2004.

COSTS: Basic kit US$24,950, without engine, avionics, propeller, spinner, upholstery, battery, instruments and paint (2004).

DESIGN FEATURES: Conventional low-wing monoplane of composites construction. Upturned wingtips and sharply tapered fin. Quoted build time 1,000 hours.

FLYING CONTROLS: Conventional and manual. Horn-balanced rudder and elevators, latter with trim tab.

STRUCTURE: Constructed of high-temperature epoxy pre-impregnated GFRP/CFRP premoulded components, with either Divinycell or honeycomb core. All metal components prewelded or premachined. Aerofoil NACA 63215 slightly modified.

LANDING GEAR: Fixed. Fuselage-mounted spring cantilever mainwheel legs with optional speed fairings in tricycle or tailwheel layout. Matco wheels and brakes; McCreary 5.00-5 tyres. Tricycle version has steerable nosewheel.

POWER PLANT: One piston engine driving a two— or three-blade propeller: typically 88 kW (118 hp) Textron Lycoming O-235-C1B; 93 kW (125 hp) Teledyne Continental IO-240; and 119 kW (160 hp) Textron Lycoming O-320. Alternatives include 88.0 kW (118 hp) Aeromaxx 100, 59.7 kW (80 hp) Limbach L 2000 and 74.6 kW (100 hp) Fire Wall Forward CAM 100 four-cylinder converted Honda motorcar engine. Fuel capacity 95 litres (25.0 US gallons; 20.8 Imp gallons); optionally 129 litres (34.0 US gallons; 28.3 Imp gallons) for Lycoming— and Continental-powered versions.

ACCOMMODATION: Two, side by side. Upward-hinged door each side.

DIMENSIONS, EXTERNAL:
- Wing span ..... 7.01 m (23 ft 0 in)
- Wing chord ..... 1.17 m (3 ft 10 in)
- Wing aspect ratio 6.0
- Length overall ..... 6.71 m (22 ft 0 in)
- Height overall: TR-1 ..... 1.83 m (6 ft 0 in)
  - TD ..... 1.89 m (6 ft 2½ in)
- Propeller diameter ..... 1.42 m (4 ft 8 in)

DIMENSIONS, INTERNAL:
- Cabin: Length ..... 1.65 m (5 ft 5 in)
  - Max width ..... 1.07 m (3 ft 6 in)
  - Max height ..... 1.02 m (3 ft 4 in)

AREAS:
- Wings, gross ..... 8.18 m² (88.0 sq ft)

WEIGHTS AND LOADINGS (L: Lycoming O-320 engine, C: Continental engine):
- Weight empty: L ..... 386 kg (850 lb)
  - C ..... 367 kg (810 lb)
- Max T-O weight: L, C ..... 680 kg (1,500 lb)
- Max wing loading: L, C ..... 83.2 kg/m² (17.05 lb/sq ft)

Pulsar XP side-by-side kitbuilt *(Paul Jackson)* ***NEW***/1047383

Pulsar Sport 150 (Tri-R KIS) nosewheel version *(Paul Jackson)* ***NEW***/1047262

Max power loading: L .... 5.71 kg/kW (9.38 lb/hp)
C .... 7.30 kg/kW (12.00 lb/hp)

PERFORMANCE (L, C, as above):
Never-exceed speed (VNE) .... 191 kt (354 km/h; 220 mph)
Max level speed: L .... 174 kt (322 km/h; 200 mph)
C .... 165 kt (306 km/h; 190 mph)
Cruising speed at 75% power:
L .... 156 kt (290 km/h; 180 mph)
C .... 152 kt (282 km/h; 175 mph)
Stalling speed: L, C .... 48 kt (89 km/h; 55 mph)
Max rate of climb at S/L: L .... 610 m (2,000 ft)/min
C .... 457 m (1,500 ft)/min
Service ceiling: C .... 5,180 m (17,000 ft)
T-O run: L, C .... 229 m (750 ft)
Landing run: L, C .... 244 m (800 ft)
Range at cruising speed standard tankage:
L .... 521 n miles (965 km; 600 miles)
C .... 651 n miles (1,207 km; 750 miles)
g limits .... +4.4/–2.2

UPDATED

PULSAR CRUISER AND SUPER CRUISER

TYPE: Four-seat kitbuilt.

PROGRAMME: Designed by Rich Trickel and Vance Jaque. First flown 1994, when known as Tri-R Cruiser TR-4. Design purchased by Pulsar in 1999.

CURRENT VERSIONS: **Cruiser:** Baseline model; being replaced by Super Cruiser.

Super Cruiser: Higher-powered version with minor redesign to dorsal fin to give larger area. First aircraft to this standard (N98WG) flew 8 March 1999.

CUSTOMERS: More than 60 kits sold and 18 flying by mid-2002 (latest data supplied).

COSTS: Kit US$34,950 (2004); fast-build options add US$7,000.

DESIGN FEATURES: Larger version of the KIS TR-1. Quoted build time 1,500 hours.

Description generally as for Sport 150 except that below.

LANDING GEAR: Tricycle type only.

POWER PLANT: *Cruiser:* One 134 kW (180 hp) Textron Lycoming O-360 flat-six, driving a two-blade, fixed-pitch propeller.

Super Cruiser: One 157 kW (210 hp) six-cylinder Teledyne Continental IO-360, driving a two-blade, constant-speed propeller.

Usable fuel capacity 189 litres (50.0 US gallons; 41.6 Imp gallons).

DIMENSIONS, EXTERNAL:
Wing span .... 8.83 m (29 ft 0 in)
Wing aspect ratio 6.2
Length overall .... 7.16 m (23 ft 6 in)
Height overall .... 2.29 m (7 ft 6 in)

DIMENSIONS, INTERNAL:
Cabin: Length .... 1.98 m (6 ft 6 in)
Max width .... 1.14 m (3 ft 9 in)
Max height .... 1.17 m (3 ft 10 in)

Pulsar Super Cruiser four-seat kitbuilt *(Paul Jackson)* ***NEW***/1047355

AREAS:
Wings, gross .... 12.54 m² (135.0 sq ft)

WEIGHTS AND LOADINGS:
Weight empty: Cruiser .... 544 kg (1,200 lb)
Super Cruiser .... 590 kg (1,300 lb)
Baggage capacity .... 29 kg (65 lb)
Max T-O weight:
Cruiser, Super Cruiser .... 1,088 kg (2,400 lb)
Max wing loading:
Cruiser, Super Cruiser .... 86.8 kg/m² (17.78 lb/sq ft)
Max power loading:
Cruiser .... 8.12 kg/kW (13.33 lb/hp)
Super Cruiser .... 6.96 kg/kW (11.43 lb/hp)

PERFORMANCE:
Never-exceed speed (VNE) .... 186 kt (346 km/h; 215 mph)
Cruising speed: at 75% power:
Cruiser .... 161 kt (298 km/h; 185 mph)
Super Cruiser .... 174 kt (322 km/h; 200 mph)
at 60% power: Cruiser .... 156 kt (290 km/h; 180 mph)
Super Cruiser .... 165 kt (306 km/h; 190 mph)
Stalling speed .... 48 kt (89 km/h; 55 mph)
Max rate of climb at S/L: Cruiser .... 396 m (1,300 ft)/min
Super Cruiser .... 457 m (1,500 ft)/min
Service ceiling .... 5,180 m (17,000 ft)
T-O run: Cruiser .... 305 m (1,000 ft)
Super Cruiser .... 274 m (900 ft)
T-O to 15 m (50 ft): Cruiser .... 366 m (1,200 ft)
Super Cruiser .... 305 m (1,000 ft)
Landing run: Cruiser .... 395 m (1,300 ft)
Super Cruiser .... 411 m (1,350 ft)
Range:
Cruiser .... more than 782 n miles (1,448 km; 900 miles)
Super Cruiser .... more than 695 n miles (1,287 km; 799 miles)

UPDATED

QUIKKIT

RAINBOW FLYERS INC, QUIKKIT DIVISION

9002 Summer Glen, Dallas, Texas 75243-7445
Tel/Fax: (+1 214) 349 04 62
e-mail: quikkit@glassgoose.com
Web: http://www.glassgoose.com
PRESIDENT: Thomas W Scott
PRODUCTION MANAGER: Robert Giddens

WORKS:
Lakeview Airport, Lake Dallas, Texas 75065

Quikkit was initially involved in the production of parts and modification kits for the Aero Composites Sea Hawker; the Glass Goose, while visually similar, is a new design.

VERIFIED

QUIKKIT GLASS GOOSE

TYPE: Two-seat amphibian kitbuilt.

PROGRAMME: Extensively redesigned Aero Composite Sea Hawker, obviating deficiencies of that discontinued aircraft. Modifications initially applied to Tom Scott's Sea Hawker and later marketed for incorporation by other Sea Hawker operators, although no longer available.

Prototype new-build Glass Goose flew 1996, but lost to flaperon flutter. Second prototype, in definitive configuration, first flew (N96GG) 20 March 1999. Production undertaken on Sea Hawker tools and jigs, acquired by Quikkit in 1993.

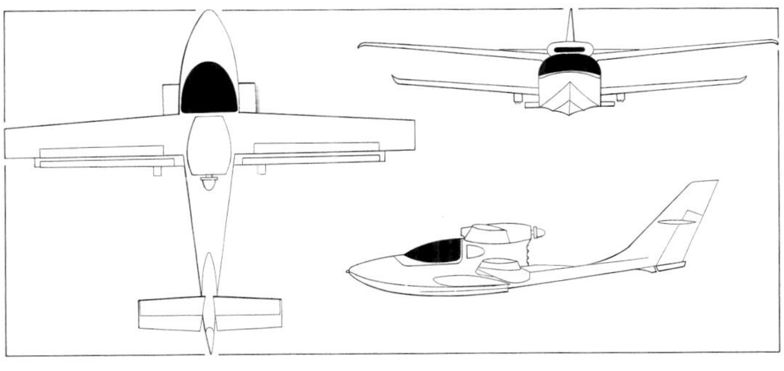

Quikkit Glass Goose two-seat amphibian *(Paul Jackson)* 0084616

Quikkit Glass Goose (Textron Lycoming flat-four) *(Paul Jackson)* ***NEW***/1047388

CUSTOMERS: Total 42 under construction, of which seven flying, by October 2003. At least six others extant by 2000, reflecting Sea Hawker conversions; eight on US Register, September 2004.

COSTS: US$29,500, excluding engine, avionics and equipment; quick-build options add US$13,000 (2004).

DESIGN FEATURES: Pusher, cantilever biplane, with single-step hull, sponsons and underwing pods for retractable main landing gear. Dished upper rear fuselage; mid-mounted tailplane. Wing positive stagger of 20 cm (8 in); upswept lower wingtips; downswept upper wingtips.

Modifications to Sea Hawker design include extended (91 cm; 3 ft 0 in) upper wing span; sponsons of aerofoil section (replacing wingtip floats); and extended pylon chord, with associated fillets, vortex generators and augmentation slots in reshaped cowling to eliminate propeller vibration.

FLYING CONTROLS: Conventional and manual, with drooping ailerons on all wings, max deflection +10/–40°. Actuation by pushrods and cables.

STRUCTURE: Glass fibre, with Kevlar reinforcement. Forward wing spars of glass fibre beam type, with carbon caps, at 27 per cent chord; glass fibre C-type rear spar at 85 per cent chord; glass fibre/foam core sandwich ribs. Pre-moulded flaperons aid construction. Intersecting fuselage bulkheads for additional strength. Quoted build time 1,000 hours.

LANDING GEAR: Retractable tricycle type for land operation. Two, side-by-side Cheng Shin 11×4.00–5 tyres on each main unit, underside of landing gear stowage pod also being forward-hinged leg/door. Mainwheel brakes.

Rearward retracting, stainless steel, fully castoring nose leg, with 4.00–5 tyre, extends against spring-loaded single door. Electrohydraulic actuation. Ventral rudder, cable operated, for water manoeuvring.

POWER PLANT: One 119 kW (160 hp) Textron Lycoming O-320-B2B flat-four driving a Warp Drive, four-blade, ground-adjustable pitch propeller. Four equal-size, integral fuel tanks in wings (upper pair main; lower pair for long range), total capacity 265 litres (70.0 US gallons; 58.3 Imp gallons). No crossfeed. Optional provision for total of 455 litres (120 US gallons; 100 Imp gallons) involves fuselage tank.

ACCOMMODATION: Two, side by side, with baggage stowage areas behind seats and between lower wings.

DIMENSIONS, EXTERNAL:
Wing span: upper ... 8.23 m (27 ft 0 in)
lower ... 7.62 m (25 ft 0 in)
Wing chord: at root ... 0.84 m (2 ft 9 in)
at tip ... 0.65 m (2 ft 1½ in)
Length overall ... 5.94 m (19 ft 6 in)
Height overall ... 2.29 m (7 ft 6 in)
Tailplane span ... 2.44 m (8 ft 0 in)

DIMENSIONS, INTERNAL:
Cockpit max width ... 1.07 m (3 ft 6 in)
Baggage volume: cockpit ... 0.24 m^3 (8.5 cu ft)
between wings ... 0.45 m^3 (16.0 cu ft)

AREAS:
Wings, gross ... 12.17 m^2 (131.0 sq ft)
Sponsons (lifting) ... 1.44 m^2 (15.50 sq ft)
Flaperons (total) ... 1.86 m^2 (20.0 sq ft)

WEIGHTS AND LOADINGS:
Weight empty ... 454 kg (1,000 lb)
Max T-O weight ... 816 kg (1,800 lb)
Max wing loading ... 67.1 kg/m^2 (13.74 lb/sq ft)
Max power loading ... 6.85 kg/kW (11.25 lb/hp)

PERFORMANCE:
Max level speed ... 139 kt (257 km/h; 160 mph)
Max cruising speed at 75% power ... 122 kt (225 km/h; 140 mph)
Stalling speed: flaps up ... 44 kt (81 km/h; 50 mph)
flaps down ... 37 kt (68 km/h; 42 mph)
Max rate of climb at S/L ... 366 m (1,200 ft)/min
Service ceiling ... 3,660 m (12,000 ft)
T-O run ... 244 m (800 ft)
Landing run ... 274 m (900 ft)
Max range, 30 min reserves ... 955 n miles (1,770 km; 1,100 miles)
g limits ... ±9

UPDATED

RANS

RANS INC

4600 Highway 183 Alternate, Hays, Kansas 67601
Tel: (+1 785) 625 63 46
Fax: (+1 785) 625 27 95
e-mail: rans@media-net.net
Web: http://www.rans.com
PRESIDENT AND CEO: Randy J Schlitter
VICE-PRESIDENT: Paula Schlitter

Formed in 1974 to build sand yachts Rans' earlier aircraft designs include S-4/5 Coyote, S-6 Coyote II, S-9 Chaos, S-10 Sakota, S-11 Pursuit and S-14 Airaile. Rans delivered its 3,000th kitplane during 1998 and in 1999 achieved FAA '51 per cent rule' certification of the S-7, S-12 and S-16 kits. Latest products are the related S-17 and S-18 single- and two-seat ultralights. At Sun 'n' Fun 2004, preliminary details of the S-19 were released.

In October 2003, Rans employed 45 at its 4,645 m^2 (50,000 sq ft) factory. By June 2004, 3,884 kits had been delivered.

UPDATED

Rans S-7C Courier *(Paul Jackson)* ***NEW***/1047269

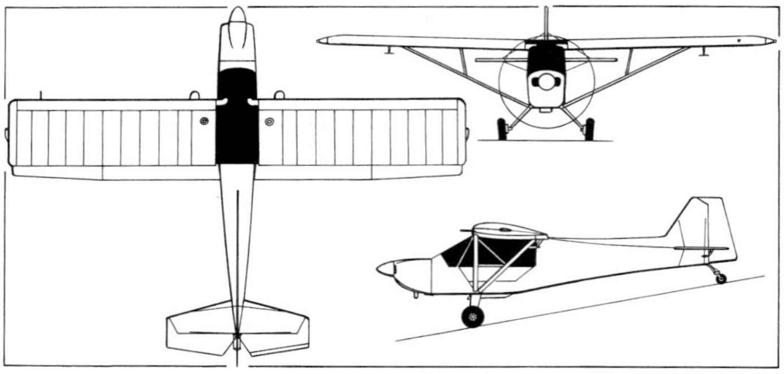

Rans S-7C Courier two-seat factory-built light aircraft *(James Goulding)* 0089516

RANS S-7 COURIER

TYPE: Tandem-seat kitbuilt.

PROGRAMME: Prototype first flew November 1985; in production from January 1986.

CURRENT VERSIONS: **S-7:** Original kitbuilt version, superseded by S-7S.

S-7C: Factory-built version; changes include additional fuselage side stringers, smaller ailerons, larger flaps, taller (15 cm; 6 in) fin and rudder, extended dorsal strake, control surface gap seals, improved low-friction control cable runs, tab for pitch trim, tapered spring steel main landing gear legs with improved fairings, dual calliper brakes, larger baggage compartment and corrosion-proofing as standard. First flight (N11632; second airframe) 20 December 1996. Further aircraft, N2506A for trials, built 2000. Type certificate awarded 14 September 2001, permitting start to deliveries; however, production was on hold until January 2003, pending definitive Sport Pilot regulations.

S-7S: Kitbuilt version of S-7C, *as described.* Kit conforms to FAA 51 per cent rule.

CUSTOMERS: Total of 367 5–7 kits sold, of which 190 completed by September 2004; 31 S-7Cs built for stock by this date.

COSTS: S-7S kit US$18,265 without engine, US$34,180 with Rotax 912ULS (2004). Production version US$65,000.

DESIGN FEATURES: High-wing, mid-tailplane configuration with sweptback fin, constant-chord wings and tapered tailplane. Wings braced by A-struts; empennage mutually wire-braced. Wings fold rearwards and tailplanes upwards for stowage; usual rigging/de-rigging time 30 to 45 minutes. Quoted build time for kit version is 500 to 700 hours.

NACA 2412 wing section, thickness/chord ratio 14 per cent; dihedral 1°; twist 25°.

FLYING CONTROLS: Conventional and manual. Cable-operated rudder and ailerons, the last-named spade-assisted, and torque tube-operated elevators; trim tab on starboard elevator; cable-operated four-position flaps (0, 8, 16 and 26°).

STRUCTURE: Wings have two tubular 6061-T6 alloy spars, preformed aluminium tube ribs riveted and clipped to spars and sheet alloy leading-edges; fuselage and tail surfaces of welded 4130 steel; glass fibre forward decking and engine cowling. Airframe is fabric-covered.

LANDING GEAR: Tailwheel type; fixed. Fuselage-mounted spring tubular steel mainwheel legs with streamlined trouser; steerable, full-swivelling tailwheel; mainwheel size 16×5 (6.00–6 on S-7C), maximum pressure 2.41 bar (35 lb/sq in); tailwheel with solid tyre, size 8×2; Matco/Cleveland wheels with hydraulic brakes. Optional speed fairings on main legs increase cruising speed by 2 to 4 kt (5 to 8 km/h; 3 to 5 mph). Tundra tyres, skis, floats and Aerocet amphibious floats optional.

POWER PLANT: One 59.6 kW (79.9 hp) Rotax 912 UL or one 73.5 kW (98.6 hp) Rotax 912 ULS four-cylinder piston engine, each with 2.27:1 reduction gear, driving a two-blade, ground-adjustable propeller (Tennessee on 912 UL and Sensenich on 912 ULS). S-7C and S-7S have Rotax 912 ULS with 2.43:1 reduction gear; driving Sensenich two-blade propeller. Fuel contained in two wing tanks, combined capacity 68 litres (18.0 US gallons; 15.0 Imp gallons), of which 58 litres (15.2 US gallons; 12.7 Imp gallons) are usable. Oil capacity 2.9 litres (0.75 US gallon; 0.6 Imp gallon).

ACCOMMODATION: Pilot and passenger in tandem on fore- and aft-adjustable seats; lap and shoulder harness standard; optional deluxe seats. Dual controls standard. Cabin is heated and ventilated. Upward-hinged door on each side is flight-openable with uplocks; quick-release hinge pins optional. Baggage compartment at rear of cabin.

AVIONICS: To customer's choice; optional GPS, VHF com and intercom on S-7C.

DIMENSIONS, EXTERNAL(all versions, except: A: S-7 with Rotax 912 UL, B: S-7 with Rotax 912 ULS, C: S-7C, D: S-7S):
Wing span ... 8.92 m (29 ft 3 in)
Wing chord, constant ... 1.52 m (5 ft 0 in)
Wing aspect ratio ... 5.8
Length overall: A, B ... 6.96 m (22 ft 10 in)
C, D ... 7.09 m (23 ft 3 in)
Length overall, wings folded: A, B ... 7.11 m (23 ft 4 in)
Height overall: A, B, D ... 1.90 m (6 ft 3 in)
C ... 2.01 m (6 ft 7 in)
Tailplane span ... 2.44 m (8 ft 0 in)
Wheel track ... 1.93 m (6 ft 4 in)
Wheelbase ... 5.18 m (17 ft 0 in)
Propeller diameter:
A, B ... 1.73 m to 1.83 m (5 ft 8 in to 6 ft 0 in)
C, D ... 1.83 m (6 ft 0 in)
Cabin door: Max height ... 0.97 m (3 ft 2 in)
Max width ... 1.52 m (5 ft 0 in)
Height to sill ... 0.81 m (2 ft 8 in)

DIMENSIONS, INTERNAL (all versions):
Cabin: Length ... 1.93 m (6 ft 4 in)
Max width ... 0.76 m (2 ft 6 in)

Selection of current Rans products comprising (l to r) S-6S Coyote II tailwheel, Coyote II nosewheel, S-7C Courier and S-6ES Coyote II nosewheel *(Paul Jackson)* ***NEW***/1047234

Max height .... 1.07 m (3 ft 6 in)
Floor area .... 1.11 m² (12.00 sq ft)
Volume .... 1.4 m³ (48 cu ft)
Baggage compartment
Volume .... 0.28 m³ (10.0 cu ft)

AREAS (A, B, C and D as above):
Wings, gross: all .... 13.67 m² (147.1 sq ft)
Ailerons (total): A, B .... 1.11 m² (12.00 sq ft)
Trailing-edge flaps (total): A, B .... 1.11 m² (12.00 sq ft)
Fin: A, B .... 0.74 m² (8.00 sq ft)
Rudder: A, B .... 0.56 m² (6.00 sq ft)
Tailplane: A, B .... 1.11 m² (12.00 sq ft)
Elevators, incl tabs (total): A, B .... 0.93 m² (10.00 sq ft)

WEIGHTS AND LOADINGS (A, B, C and D as above):
Weight empty: A, B .... 306 kg (675 lb)
C, D .... 318 kg (700 lb)
Max T-O weight: all .... 544 kg (1,200 lb)
Baggage capacity .... 23 kg (50 lb)
Max wing loading: all .... 39.8 kg/m² (8.16 lb/sq ft)
Max wing loading: A .... 9.14 kg/kW (15.02 lb/hp)
B, C .... 7.41 kg/kW (12.17 lb/hp)

PERFORMANCE (A, B, C and D as above):
Never-exceed speed (VNE) .... 113 kt (209 km/h; 130 mph)
Cruising speed: A, C .... 96 kt (177 km/h; 110 mph)
B .... 103 kt (190 km/h; 118 mph)
Stalling speed: flaps up .... 40 kt (74 km/h; 46 mph)
flaps down .... 36 kt (66 km/h; 41 mph)
Max rate of climb at S/L: A .... 244 m (800 ft)/min
B .... 335 m (1,100 ft)/min
C, D .... 305 m (1,000 ft)/min
Service ceiling: A .... 4,270 m (14,000 ft)
B, C, D .... 4,420 m (14,500 ft)
T-O run: A .... 72 m (235 ft)
B .... 69 m (225 ft)
C, D .... 99 m (325 ft)
Landing run: A, B .... 99 m (325 ft)
C, D .... 114 m (375 ft)
Range: A .... 420 n miles (779 km; 484 miles)
B .... 410 n miles (759 km; 472 miles)
C .... 417 n miles (772 km; 480 miles)
D .... 296 n miles (548 km; 341 miles)
Endurance: A .... 4 h 24 min
B, C .... 4 h 0 min
D .... 3 h 6 min
g limits .... +4/−2

UPDATED

RANS S-12XL AIRAILE AND S-12S SUPER AIRAILE

TYPE: Side-by-side ultralight kitbuilt.

PROGRAMME: First flown March 1990. Improved, production versions of S-12 Airaile kitbuilt. R&D prototype of S-12XL (N8045X) first flown March 1995; made its debut at Sun 'n' Fun in Lakeland, Florida, April 1995.

CURRENT VERSIONS: **S-12XL Airaile:** Standard version.

Description applies to S-12XL, except where indicated.

S-12S Super Airaile: Improved version, first flown (N80887; modification of an earlier S-12XL) January 1999. 59.6 kW (79.9 hp) Rotax 912 or 73.5 kW (98.6 hp) Rotax 912 ULS engines, with thrust line lowered 25 mm (1 in) and driving ground-adjustable, composites Warp Drive propeller new wing with improved aerofoil section, identical in rib construction to that of S-7C Courier; stamped aluminium ribs in tail surfaces; reduced tailplane incidence; cabin height raised 50 mm (2 in), with more sharply raked windscreen incorporating a new flow fence; increased range of seat tilt to improve headroom and comfort; hydraulic brakes standard; pre-sewn Dacron skins replaced by traditional doped fabric secured by pop rivets.

Performance improvements include 9 kt (16 km/h; 10 mph) increase in cruising speed. Marketed in parallel with S-12XL.

CUSTOMERS: 987 S-12s (all models) ordered by June 2004, of which 850 S-12XLs and 20 S-12Ss flying.

COSTS: S-12XL kits, with engine: Rotax 503DC, US$16,393; Rotax 582, US$18,294; Rotax 912, US$26,243; Rotax 912S, US$27,430 (2004).

S-12S kits, with engine: Rotax 912, US$30,343; Rotax 912S, US$31,330 (2004).

DESIGN FEATURES: Pod and boom fuselage; V-strut-braced wings; wire-braced tail surfaces; wings fold rearwards and tailplanes upwards for storage. Quoted build time 175 to 225 hours; or 325 to 500 hours with full enclosure cockpit. (Extra 200 hours for S-12S.)

FLYING CONTROLS: Conventional and manual. Cable-operated; ground-adjustable flaps primarily used for setting best trim, but can also be lowered for landing; tailplane incidence is ground-adjustable. Dual controls standard.

STRUCTURE: Welded steel tube cockpit cage; aluminium alloy tubular tailboom; wings have two tubular alloy spars and preformed ribs; ailerons, flaps and tail surfaces of alloy tube with stamped alloy or moulded ABS ribs; fabric covering.

LANDING GEAR: Tricycle type; fixed. Fuselage-mounted tubular spring steel cantilever mainwheel legs (with streamlined trouser on S-12S); with small tailwheel/bumper; spring-loaded telescoping nose leg; pedal-operated hydraulic brakes on mainwheels; steerable nosewheel. All wheels 15 cm (6 in) aluminium on S-12XL and 13 cm (5 in) on S-12S. Optional spats, tundra tyres and Puddle Jumper floats.

Rans S-12S Super Airaile *(Paul Jackson)* ***NEW***/1047266

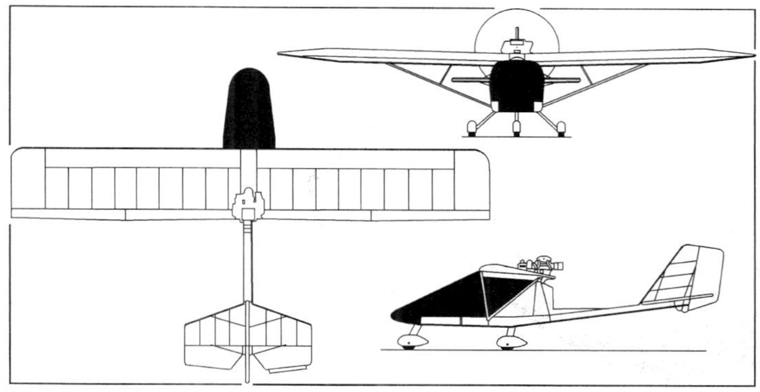

Rans S-12S Super Airaile general arrangement ***NEW***/0567241

POWER PLANT: One air-cooled 34.0 kW (45.6 hp) Rotax 503 or liquid-cooled 47.8 kW (64.1 hp) Rotax 582 two-cylinder engine with 2.58:1 reduction gear, or 59.6 kW (79.9 hp) Rotax 912 or 73.5 kW (98.6 hp) Rotax 912 ULS flat-fours with 2.27:1 reduction gear, driving a two-blade Tennessee wooden propeller; three-blade Warp Drive propeller optional; electric start optional on 503 and 582, standard on 912. Version with 48.5 kW (65 hp) Hirth 2706 two-cylinder two-stroke is available in Germany. Fuel in single wing tank, maximum capacity 34 litres (9.0 US gallons; 7.5 Imp gallons); dual tanks optional, (standard in S-12S) combined capacity 68 litres (18.0 US gallons; 15.0 Imp gallons).

ACCOMMODATION: Pilot and passenger side by side in open cockpit; transparent mini pod fairing standard; optional partial- and full-enclosure cockpits provide increased leg, shoulder and head room and lower noise levels and, combined with flow fences attached to fuselage pod and boom, raise cruising speed.

DIMENSIONS, EXTERNAL:
Wing span .... 9.45 m (31 ft 0 in)
Length overall .... 6.25 m (20 ft 6 in)
Height overall .... 2.36 m (7 ft 9 in)
Propeller diameter:
S-12XL .... 1.73 to 1.83 m (5 ft 8 in to 6 ft 0 in)
S-12S .... 1.73 m (5 ft 8 in)

AREAS:
Wings, gross .... 14.12 m² (152.0 sq ft)

WEIGHTS AND LOADINGS (A: S-12XL with Rotax 912 UL, B: S-12S with Rotax 912 ULS):
Weight empty: A .... 261 kg (575 lb)
B .... 295 kg (650 lb)
Max T-O weight: A .... 499 kg (1,100 lb)
B .... 521 kg (1,150 lb)

PERFORMANCE (A and B as above):
Never-exceed speed (VNE): A .... 87 kt (161 km/h; 100 mph)
B .... 104 kt (193 km/h; 120 mph)
Cruising speed: A .... 74 kt (137 km/h; 85 mph)
B .... 87 kt (161 km/h; 100 mph)[1]
Stalling speed, flaps down: A, B .... 31 kt (57 km/h; 35 mph)
Max rate of climb at S/L: A .... 274 m (900 ft)/min
B .... 366 m (1,200 ft)/min
T-O run: A .... 87 m (285 ft)
B .... 81 m (265 ft)
Landing run: A, B .... 61 m (200 ft)
Range: A .... 325 n miles (601 km; 374 miles)
B .... 312 n miles (579 km; 360 miles)
Endurance: A .... 4 h 24 min
B .... 3 h 54 min

[1] *With cockpit and wheel fairings*

UPDATED

RANS S-16 SHEKARI

TYPE: Side-by-side kitbuilt.

PROGRAMME: Prototype (N8072U), powered by 59.6 kW (79.9 hp) Rotax 912 UL engine, first flew 25 March 1997 in tricycle landing gear configuration, and won its class in the 1997 Sun 'n' Fun Race at Lakeland, Florida, averaging 113 kt (209 km/h; 130 mph) over a 52 n mile (97 km; 60 mile) course; re-engined with Teledyne Continental IO-240B and flown April 1998; subsequently converted to tailwheel configuration, although this configuration not marketed. Further prototype, N8073U, (plus one static test airframe) completed for trial installation of 119.3 kW (160 hp) Textron Lycoming O-320; FAA approval under homebuilt rules achieved 16 December 1998; first kit delivered to customer two days later; production capacity 50 kits per year.

CUSTOMERS: Total 26 sold and 20 flying by July 2003.

COSTS: Quick-build kit US$29,000, not including engine (2004).

DESIGN FEATURES: Conventional configuration low-wing monoplane with tailwheel or nosewheel landing gear. Quoted build time 500 to 1,000 hours.

FLYING CONTROLS: Conventional and manual. Pushrod-operated elevators, cable-operated rudder; half-span, five-position flaps. Triangular trim tab at base of rudder.

STRUCTURE: Composites fuselage, made up of four moulded shells, with powder-coated welded chromoly steel cockpit cage; aluminium wings, elevators and rudder with composites tips.

LANDING GEAR: Fixed. Fuselage-mounted spring cantilever mainwheel legs with streamlined trousers and option of steerable nosewheel or tailwheel. Hydraulic toe brakes standard; wheel speed fairings on all units.

POWER PLANT: One 93.2 kW (125 hp) Teledyne Continental IO-240B flat-four or one 119.3 kW (160 hp) Textron Lycoming O-320 flat-four, driving a three-blade ground-adjustable Warp Drive wooden propeller. Alternative engines include Subaru EA81. Fuel in two fuselage tanks, total capacity 121 litres (32.0 US gallons; 26.6 Imp gallons); oil capacity 4.7 litres (1.25 US gallons; 1.04 Imp gallons).

ACCOMMODATION: Two persons, side by side under one-piece forward-hinged canopy supported by gas struts; dual controls, central throttle, shoulder harnesses, lap belts and roll-over protection standard.

Rans S-16 Shekari with nosewheel landing gear *(Paul Jackson)* 0526969

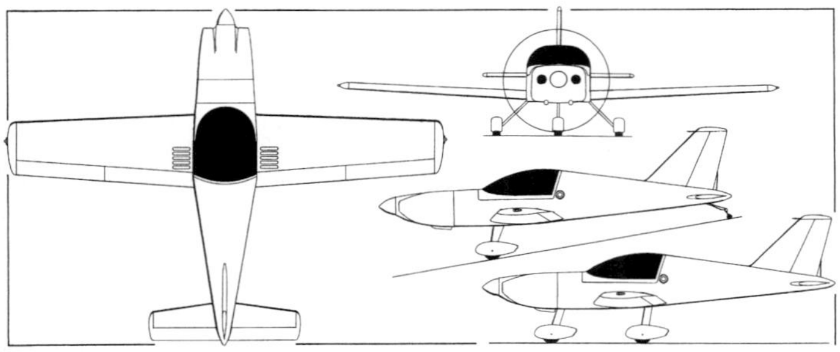

Rans S-16 Shekari, with additional side view of optional nosewheel version *(Paul Jackson)* 0064496

DIMENSIONS, EXTERNAL:
Wing span 7.32 m (24 ft 0 in)
Wing aspect ratio 6.6
Wing chord, mean 1.12 m (3 ft 8 in)
Length overall 5.89 m (19 ft 4 in)
Height overall 2.24 m (7 ft 4 in)
Tailplane span 2.54 m (8 ft 4 in)
Propeller diameter 1.78 m (5 ft 10 in)
DIMENSIONS, INTERNAL:
Cabin: Max width 1.02 m (3 ft 4 in)
Max height 1.00 m (3 ft 3½ in)
Baggage volume 0.20 m^3 (7.0 cu ft)
AREAS:
Wings, gross 8.08 m^2 (87.0 sq ft)
WEIGHTS AND LOADINGS (IO-240 engine):
Weight empty 422 kg (930 lb)
Baggage capacity 23 kg (50 lb)
Max T-O weight: Normal 657 kg (1,450 lb)
Aerobatic 590 kg (1,300 lb)
Max wing loading 81.4 kg/m^2 (16.67 lb/sq ft)
Max power loading 7.06 kg/kW (11.60 lb/hp)
PERFORMANCE (IO-240 engine):
Never-exceed speed (VNE) 191 kt (354 km/h; 220 mph)
Max cruising speed 139 kt (257 km/h; 160 mph)
Stalling speed: flaps up 54 kt (100 km/h; 62 mph)
flaps down 51 kt (94 km/h; 58 mph)
Max rate of climb at S/L 305 m (1,000 ft)/min
Service ceiling 4,420 m (14,500 ft)
T-O run 152 m (500 ft)
Landing run 160 m (525 ft)
Range 741 n miles (1,372 km; 853 miles)
Endurance 5 h 18 min
g limits +4/–4

UPDATED

RANS S-17 STINGER

TYPE: Single-seat ultralight/kitbuilt.

PROGRAMME: First flight April 1996; development continuing towards FAA FAR Pt 103 certification. Had undergone fundamental redesign by February 2000, when production version made first flight, changing from tractor to pusher propeller and relocating tailboom to low position, complete with tailwheel; by 2003, certification still continuing.

CURRENT VERSIONS: **S-17:** *As described.*

S-18: Two-seat; described separately.

CUSTOMERS: 45 sold and 35 completed by February 2004.

COSTS: Kit, with engine: Rotax 447, US$11,200; Rotax 503 DC, US$12,193 (all 2004). Flyaway US$3,000 to US$4,000 extra.

DESIGN FEATURES: Pod and boom fuselage; V-strut-braced wings; quoted building time 150 hours.

FLYING CONTROLS: Conventional and manual.

STRUCTURE: Light alloy fuselage, pre-assembled with all controls in situ; wing has tubular spar. Fuselage is 12.7 cm (5 in) aluminium tube with 4130 welded steel cockpit cage; optional composites nosecone. Dacron skin; spring steel landing gear.

LANDING GEAR: Tailwheel type; fixed. Fuselage-mounted spring steel cantilever mainwheel legs; hand-lever operated brakes on mainwheels.

POWER PLANT: Prototype has one 34.0 kW (45.6 hp) Rotax 503 two-cylinder piston engine driving a three-blade IVO propeller via 2.58:1 reduction gearing. Alternatives include 31.0 kW (41.6 hp) Rotax 447 UL-2V. Fuel capacity 19 litres (5.0 US gallons; 4.2 Imp gallons) standard; 34 litres (9.0 US gallons; 7.5 Imp gallons) optional.

ACCOMMODATION: One person in open cockpit with Lexan windscreen/nosecone; four-point Hooker harness.

Rans S-17 Stinger *(Paul Jackson)* 0089510

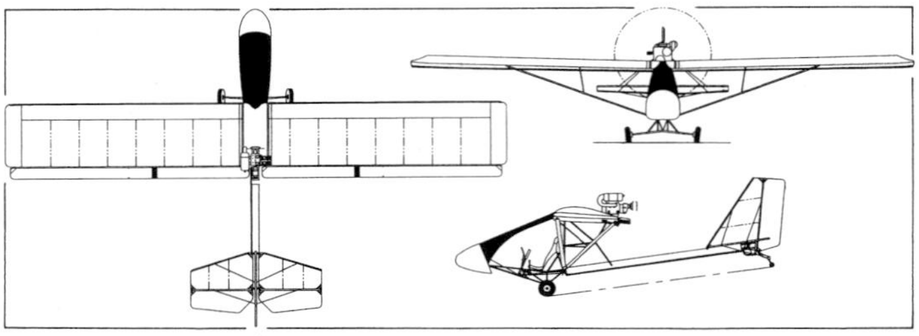

Rans S-17 general arrangement 0089513

DIMENSIONS, EXTERNAL:
Wing span 8.99 m (29 ft 6 in)
Length overall 5.28 m (17 ft 4 in)
Height overall 2.13 m (7 ft 0 in)
Propeller diameter 1.73 m (5 ft 8 in)
AREAS:
Wings, gross 11.80 m^2 (127.0 sq ft)
WEIGHTS AND LOADINGS (Rotax 503):
Weight empty 118 kg (261 lb)
Max T-O weight 243 kg (536 lb)
PERFORMANCE (Rotax 503):
Never-exceed speed (VNE) 82 kt (152 km/h; 95 mph)
Cruising speed 52 kt (97 km/h; 60 mph)
Stalling speed, power off, flaps up 25 kt (45 km/h; 28 mph)
Max rate of climb at S/L 335 m (1,100 ft)/min
Service ceiling 3,965 m (13,000 ft)
T-O run 24 m (80 ft)
Landing run 30 m (100 ft)
Range 52 n miles (97 km; 60 miles)
g limits +4/–2

UPDATED

RANS S-18 STINGER II

TYPE: Tandem-seat ultralight kitbuilt.

PROGRAMME: Developed from single-seat S-17 Stinger. First flight of prototype S-18 (N24052) 9 September 2000; production from December 2000; kit deliveries from March 2001. When canopied version made first flight.

CUSTOMERS: 35 sold and 28 flying by February 2004.

COSTS: Kit, including engine and exhaust: Rotax 503 DC US$17,183; Rotax 582 US$19,094; Rotax 912 US$27,033 (2004).

DESIGN FEATURES: Pod-and-boom fuselage; V-strut-braced wings are identical to those of S-12XL Airaile and fold rearwards for transport or storage; wire-braced tail surfaces. Quoted build time under 200 hours.

FLYING CONTROLS: Conventional and manual. Flaps.

STRUCTURE: Welded 4130 steel tube fuselage cage and 15 cm (6 in) light alloy tailboom, pre-assembled and painted; wing has tubular spar; all parts prefabricated. Wing and empennage covered by fabric; control gaps sealed by Velcro.

LANDING GEAR: Tailwheel type; fixed. Fuselage-mounted spring aluminium cantilever mainwheel legs; main tyre size 8.00–6; manually operated brakes on mainwheels.

POWER PLANT: Prototype has one 37.0 kW (49.6 hp) Rotax 503 DC air-cooled, two-cylinder piston engine with 1:2.58 reduction gearing, driving a three-blade Ivo pusher propeller; alternatives are 47.8 kW (64.1 hp) Rotax 582 and 73.5 kW (98.6 hp) Rotax 912 liquid-cooled engines with angled radiator on wing upper surface, cooling airflow optimised by louvres. Fuel contained in single tank, capacity 34 litres (9.0 US gallons, 7.5 Imp gallons); second tank of similar capacity optional.

ACCOMMODATION: Two persons in tandem. Options of open cockpit with Lexan windscreen/nosecone; full windscreen; and fully enclosed cabin. Baggage in optional 'saddlebags' each side of cabane on open versions.

DIMENSIONS, EXTERNAL:
Wing span 9.45 m (31 ft 0 in)
Width, wings folded 1.75 m (5 ft 9 in)
Length overall 6.88 m (22 ft 7 in)

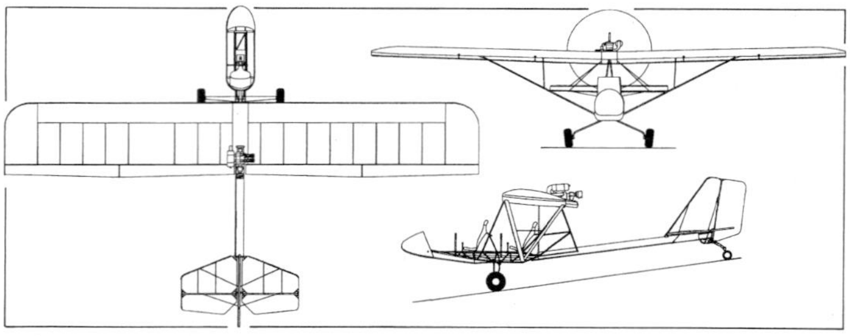

Two-seat Rans S-18 Stinger II open cockpit version *(Paul Jackson)* 0105043

Fully enclosed version of Rans S-18 Stinger II ultralight kitbuilt *(Paul Jackson)* ***NEW***/0567261

Height overall .... 2.06 m (6 ft 9 in)
Propeller diameter .... 1.68 m (5 ft 6 in)
AREAS:
Wings, gross .... 14.12 m² (152.0 sq ft)
WEIGHTS AND LOADINGS:
Weight empty .... 217 kg (478 lb)
Max T-O weight .... 417 kg (920 lb)
PERFORMANCE (Rotax 503 DC):
Never-exceed speed (VNE) .... 78 kt (144 km/h; 90 mph)
Cruising speed .... 52 kt (97 km/h; 60 mph)
Stalling speed, power off .... 32 kt (58 km/h; 36 mph)
Max rate of climb at S/L .... 152 m (500 ft)/min
T-O run .... 101 m (330 ft)
Landing run .... 61 m (200 ft)
Endurance with max fuel, with reserves .... 3 h 0 in

UPDATED

RANS S-19

At Sun 'n' Fun, April 2004, Rans released preliminary information about the S-19, a two-seat (side-by-side) low-wing ultralight intended to be constructed of sheet metal and powered by a 73.5 kW (98.6 hp) Rotax 912S. First flight was due in 2005.

NEW ENTRY

RAYTHEON

RAYTHEON AIRCRAFT COMPANY

Among the divisions and subsidiaries of this major aerospace/defence contractor are Raytheon Aircraft Company and Raytheon Systems Limited.
9709 East Central, Wichita, Kansas 67201-0085
Tel: (+1 316) 676 71 11
Fax: (+1 316) 676 82 86
Web: http://www.raytheonaircraft.com
BRANCH DIVISIONS: Salina, Kansas and Little Rock, Arkansas.
CHAIRMAN AND CEO: James E Schuster
PRESIDENT AND GENERAL MANAGER, BEECHCRAFT BUSINESS: Brad Hatt
PRESIDENT AND GENERAL MANAGER, HAWKER BUSINESS: Randy Groom
VICE-PRESIDENTS:
Jackie Berger (Communications and Corporate Affairs)
John Brauneis (Contracts)
Doug Debrecht (Information Technology and Infrastructure)
Ed Dolanski (Customer Support/RAPID)
Sherry Grady (Government Business)
Daisy Jenkins (Human Resources)
Toby O'Brien (CFO)
Glenn Oka (Quality Assurance)
William Patterson (Director, Special Projects)
David Riemer (Product Development and Engineering)
Paul Schumacher (Manufacturing Operations)
Wayne Wallace (General Counsel & Corporate Secretary, Legal)
Brad Widmann (Integrated Supply Chain)
MANAGER, PUBLIC RELATIONS: Tim Travis

Refer to **Beech** and **Hawker** entries for aircraft descriptions.

Raytheon Aircraft Company (RAC) formed 15 September 1994, combining Raytheon Company subsidiaries Beech Aircraft Corporation and Raytheon Corporate Jets Inc. Raytheon acquired British Aerospace's Corporate Jets division 6 August 1993. RAC builds civil and military aircraft, and components; Salina division supplies wings for most models, non-metallic interior components, ventral fins, nosecones and tailcones.

Final assembly of Hawkers transferred from UK to new 41,800 m² (450,000 sq ft) plant at Wichita beginning last quarter of 1995 and finalised by April 1997, when last UK-built Hawker was completed; 17,745 m² (191,000 sq ft) Training Systems Division plant added for Texan II assembly line. Little Rock continues to provide custom interiors, avionics, sales and customer support services for Hawker models and aircraft painting services for Hawkers. Beech and Hawker product names are retained and separate Beechcraft and Hawker divisions were established in mid-2002 for marketing purposes, Raytheon title having been dropped from aircraft designations.

Wholly owned subsidiaries include Raytheon Aircraft Credit Corporation Inc (business aircraft retail financing and leasing); Travel Air Insurance Company Ltd (aircraft liability insurance); Raytheon Aircraft Services network of fixed-base operations; Raytheon Travel Air, formed in 1997, which offers fractional ownership; and Raytheon Aircraft Charter and Management, which offers aircraft charter and full-service management.

In November 2001 RAC had 11,700 employees worldwide, of whom 8,800 were in Kansas, and occupied 380,900 m² (4,100,000 sq ft) of plant area in Wichita.

Total production by RAC more than 54,417 by 30 September 2004.

RAYTHEON SYSTEMS LIMITED: SENTINEL CONVERSION PLANT:
Hawarden Airport, Broughton, North Wales, CH4 0BA, UK
Tel: (+44 1244) 52 38 88
Fax: (+44 1244) 52 35 13 Co-located with Raytheon Aircraft Services Ltd, Broughton plant of Raytheon Systems is responsible for conversion of four of the RAF's five Sentinels from Bombardier Global Express business jets. Purpose-built, £3.5 million facility for Sentinel programme officially opened on 7 March 2003, including 5,575 m² (60,000 sq ft) hangar.

UPDATED

BEECH/HAWKER FIVE-YEAR DELIVERIES

Type	1999	2000	2001	2002	2003
T-6A	1	49	47	60	
Bonanza	96	103	89	56	55
Baron	49	50	47	27	27
King Air	128[1]	151[2]	119[3]	71[4]	80[5]
Beech 1900	24	54	11	11	1
Beechjet/Hawker 400XP (civil)	48	51	25	19	24
Premier			18	29	29
Hawker 800	55	67	55	46	47
Annual totals	**401**	**525**	**411**	**319**	**337**

[1] 41 King Air C90Bs; 44 King Air B200s and 43 King Air 350s
[2] 46 King Air C90Bs; 59 King Air B200s and 46 King Air 350s
[3] 41 King Air C90Bs, 46 King Air B200s and 32 King Air 350s
[4] 21 King Air C90Bs, 26 King Air B200s and 24 King Air 350s
[5] 18 King Air C90Bs, 38 King Air B200s and 24 King Air 350s

RAYTHEON SENTINEL

RAF designation: Sentinel R. Mk 1

TYPE: Airborne ground surveillance system.

PROGRAMME: On 15 June 1999 Raytheon Systems Limited was chosen as the preferred bidder for the UK Ministry of Defence's Airborne Stand-Off Radar (ASTOR) programme, for which five Global Express airframes are being modified by Bombardier's Belfast division to provide the airborne platform for radar and communications systems. Contract value £800 million (1999). Preliminary Design Review (PDR) scheduled for completion in March 2001; 200-hour flight test programme of an aerodynamically representative airframe, modified from Global Express prototype C-FBGX, began on 3 August 2001, leading to Critical Design Review (CDR) and design freeze in fourth quarter 2002. Sentinel height is 7.95 m (26 ft 1 in) due to extension of fintip, other dimensions being as for Global Express.

First production airframe (C-GJR6/ZJ690) delivered 7 February 2002 to Raytheon Systems Limited at Greenville, Texas, for airframe modification and systems integration, being followed in 2003 by flight tests and operational acceptance trials before delivery to UK Ministry of Defence in 2004. Second production airframe (C-FZVM/ZJ691) delivered to Hawarden, UK, on 29 January 2003. Known as ASTOR until 'Sentinel' name assigned June 2003. First flight of fully-equipped aircraft took place at Greenville on 26 May 2004. In-service date is 2005, with aircraft based at RAF Waddington, Lincolnshire, and operated by No. 5 Squadron, which reformed on 1 April 2004. Final delivery to RAF due in 2007.

UPDATED

Raytheon Sentinel ZJ690's first flight in fully equipped configuration ***NEW***/0589423

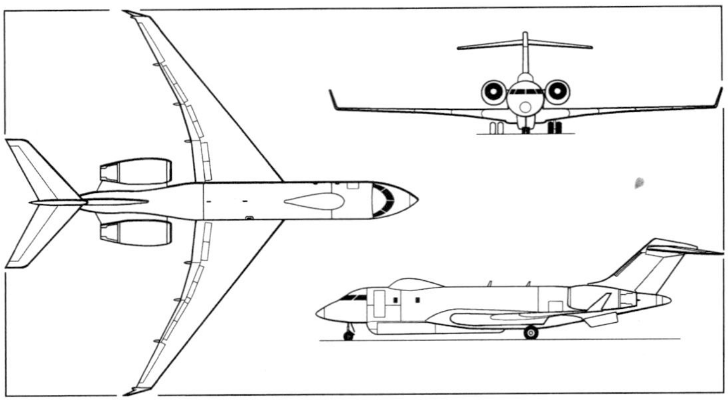

Raytheon Sentinel R. Mk 1 airborne surveillance platform *(Paul Jackson)* ***NEW***/1047692

RENAISSANCE

RENAISSANCE AIRCRAFT LLC

PO Box 596, Cape Girardeau, Missouri 63702
Tel: (+1 573) 651 39 33
Fax: (+1 573) 651 39 23
Web: http://www.renaissanceaircraft.com
PRESIDENT: John Dearden

Renaissance Aircraft has relaunched production of the classic Luscombe 8F light aircraft, assembling airframes, manufactured on original tooling, by subcontractors in the Czech Republic at its 4,459 m^2 (48,000 sq ft) factory in Missouri.

Launch of production was seriously delayed by a dispute over ownership of the Luscombe 8 Type Certificate and manufacturing tools, all of which were acquired through a legal judgment of 14 May 2004. However, Renaissance then began search for investors to finance manufacture of new aircraft.

UPDATED

RENAISSANCE 8F

TYPE: Two-seat lightplane.

PROGRAMME: Luscombe Airplane Corp Model 8 Silvaire first flew 18 December 1938; progressed through several subvariants, built by Luscombe, Temco Engineering and Silvaire Aircraft until 1960; total of some 5,867 produced. In 1999, Renaissance refurbished an original Luscombe 8F (N999RA) as prototype for relaunched production. First deliveries were due late 2000, but not effected. However, promotion was continuing in mid-2001, and assembly of first aircraft (N722J) began late July 2001. However, by mid-2004 it had not appeared.

CUSTOMERS: More than 10 deposits placed by July 2001.

COSTS: US$77,500, basic VFR; US$86,000 Mode C VFR; US$97,400 IFR (all 2004).

DESIGN FEATURES: High-wing, all-metal touring lightplane of late 1930s technology, but incorporating modifications and some modern features. Changes include considerable increase in power from original 67.1 kW (90 hp) Continental C-90; 60 per cent more baggage area or optional seating for two children; additional fuel; and new cockpit interior.

FLYING CONTROLS: Conventional and manual. Flight-adjustable tab in port elevator; ground-adjustable tabs on both ailerons. No horn balances. Plain flaps; deflections 0, 10, 20 and 30°.

STRUCTURE: Alclad aluminium monocoque fuselage; oval section duralumin bulkhead stampings and riveted Alclad skin. Wings constructed on two I-type spars of extruded aluminium and ribs of riveted T-section extrusions with Alclad skin. Ailerons covered with beaded Alclad sheet riveted to a single duralumin spar. Wings attached to upper sides of cabin and braced by single streamlined strut each side. Wing struts and landing gear attached to aluminium forgings riveted to forward section of metal seat bottom on each side. Duralumin/Alclad tailplane bolted to fuselage. Steel tube main landing gear legs hinged at fuselage sides and with single oleo-spring unit within fuselage.

Prototype Renaissance 8F *(Paul Jackson)* ***NEW***/0561576

LANDING GEAR: Tailwheel type; fixed. Fuselage-mounted, trousered, spring cantilever mainwheel legs with wheel fairings. Mainwheel tyres 6.00–6; toe-operated hydraulic brakes; optional speed fairings. Steerable tailwheel, size 2.30/2.50–4. Optional skis, floats or tundra tyres.

POWER PLANT: One 112 kW (150 hp) Textron Lycoming O-320 flat-four driving a Sensenich 74–62, two-blade, fixed-pitch aluminium propeller with spinner. Optionally, 103 kW (138 hp) LOM M332A straight-four. Fuel tank in each wing, total 117 litres (30.0 US gallons; 25.0 Imp gallons).

ACCOMMODATION: Two, on individual leather seats, with dual controls. Forward-hinged door each side. Baggage compartment behind seats may be replaced by two child seats.

SYSTEMS: Electrical system with starter, ammeter and circuit breakers

AVIONICS: *Comms:* ELT standard. Optional Mode C VFR package includes Bendix/King Silver Crown KX 155 nav/com/glideslope with KI 209 VOR/LOC/GS head; KT 76C-10 transponder; KMA 12801 audio panel with intercom and push-to-talk switches on control columns, and Ameriking AK 350 encoder. Optional IFR package includes Mode C items as above and adds a second KX 155 with KI 1209 VOR/LOC/GS; KT 76C-10X transponder in place of KT 76C-10, and KA 33-00 cooling blower.

Flight: Optional Mode C VFR and IFR packages each include Bendix/King KMD 150-05 GPS with colour moving map display.

Instrumentation: Optional full gyro package includes engine-driven vacuum system and gauge; vacuum-driven directional gyro and attitude indicator; electric turn-and-bank indicator; ASI, OAT gauge and precision electric clock.

EQUIPMENT: Navigation, landing and wingtip strobe lights; vinyl and fabric cabin upholstery, and carved wood control column grips and fuel escutcheons, all standard. Combination EGT/CHT gauge, custom exterior paint; leather interior; thermal acoustic cabin insulation; custom wood instrument panel overlays with integral lighting, and interior and panel lights, all optional. Heated pitot tube included in optional IFR avionics package.

DIMENSIONS, EXTERNAL:	
Wing span	6.71 m (35 ft 0 in)
Wing aspect ratio	8.8
Length overall	6.70 m (22 ft 0 in)
Height overall	2.13 m (7 ft 0 in)
Propeller diameter	1.88 m (6 ft 2 in)
DIMENSIONS, INTERNAL:	
Cabin max width	0.99 m (3 ft 3 in)
AREAS:	
Wings, gross	13.01 m^2 (140.0 sq ft)
WEIGHTS AND LOADINGS:	
Weight empty	449 kg (990 lb)
Max T-O weight: interim	635 kg (1,400 lb)
planned	793 kg (1,750 lb)
Max wing loading: interim	48.8 kg/m^2 (10.00 lb/sq ft)
planned	61.0 kg/m^2 (12.50 lb/sq ft)
Max power loading: interim	5.68 kg/kW (9.33 lb/hp)
planned	7.10 kg/kW (11.67 lb/hp)
PERFORMANCE:	
Max level speed	130 kt (241 km/h; 150 mph)
Max cruising speed at 75% power:	
at S/L	122 kg (225 km/h; 140 mph)
at FL80	126 kt (233 km/h; 145 mph)
Econ cruising speed at 65% power	119 kt (220 km/h; 137 mph)
Stalling speed: flaps up	41 kt (76 km/h; 47 mph)
flaps down	38 kt (70 km/h; 43 mph)
Max rate of climb at S/L	457 m (1,500 ft)/min
Service ceiling	6,400 m (21,000 ft)
g limits	+4.6/–2.2

UPDATED

ROBINSON

ROBINSON HELICOPTER COMPANY

2901 Airport Drive, Torrance, California 90505
Tel: (+1 310) 539 05 08
Fax: (+1 310) 539 51 98
Web: http://www.robinsonheli.com
PRESIDENT: Franklin D Robinson
VICE-PRESIDENTS:
Kurt Robinson (Product Support)
Wayne Walden (Production)
MARKETING DIRECTOR: Tim Goetz

By March 2004, Robinson had produced 5,283 helicopters, including 390 in 2000, 328 in 2001, 255 in 2002 and 422 in 2003. 5,000th production helicopter, an R44 Raven II delivered in September 2003. Production rate 15 helicopters per week in 2004. Factory floor area 24,150 m^2 (260,000 sq ft), but scheduled to increase when new building was completed during 2004. Workforce more than 1,000. Company is ISO 9001 certified.

UPDATED

ROBINSON R22 BETA II

TYPE: Two-seat helicopter

PROGRAMME: Design began 1973; first flight 28 August 1975; first flight of second R22 early 1977; FAA certification 16 March 1979; UK certification June 1981; Japanese certification 18 November 2002; deliveries began October 1979; early versions were R22 (O-320-A2B engine; 199 built) and R22HP (O-320-B2C; 151 built), both with 621 kg (1,300 lb) max T-O weight. R22 Alpha (O-320-B2C and 621 kg; 1,370 lb MTOW) certified October 1983 (further 151 built); R22 Beta certified 5 August 1985. Current Beta II introduced in 1995. More powerful Textron Lycoming O-360 engine provides better high-level hover performance and allows take-off power to be sustained up to 2,285 m (7,500 ft). Previously optional, tinted windscreen and door windows fitted as standard. Production began at c/n 2571 in 1995.

CUSTOMERS: Total production more than 3,400 by end of 2004. 3,000th handed over 15 October 1999; deliveries in 1999 totalled 128, followed by 126 in 2000, 134 in 2001, 107 in 2002, 128 in 2003, and 115 in the first six months of 2004.

COSTS: US$176,000 for basic R22 Beta II (2004). Direct operating cost US$27.30 per hour; total cost (500 hours annually) US$77.48 per hour (2004).

DESIGN FEATURES: Simple, pod-and-boom light helicopter; horizontal stabiliser, starboard side only; vertical stabiliser above and below boom; offset to starboard; tall rotor mast. Horizontally mounted piston engine drives transmission through multiple V belts and sprag-type overrunning clutch; main and tail gearboxes use spiral bevel gears; maintenance-free flexible couplings of proprietary manufacture used in both main and tail rotor drives. Two-blade semi-articulated main rotor, with tri-hinged underslung rotor head to reduce blade flexing, rotor vibration and control force feedback, and an elastic teeter hinge stop to prevent blade-boom contact when starting or stopping rotor in high winds; blade section NACA 63-015 (modified); two-blade tail rotor on port side; rotor brake standard.

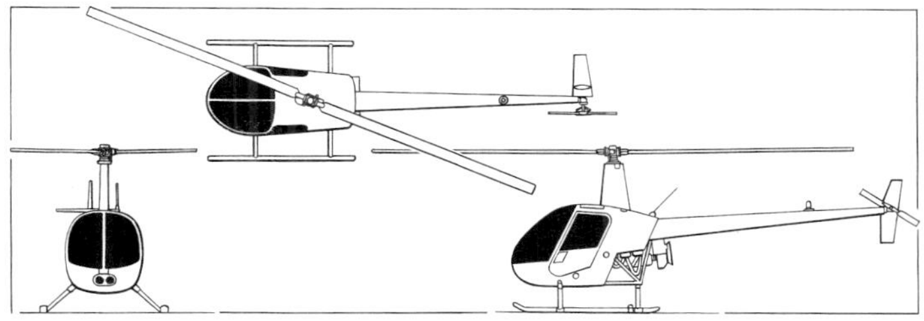

Robinson R22 Beta II two-seat helicopter *(James Goulding)* 0051888

FLYING CONTROLS: Manual. Removable dual controls standard.

STRUCTURE: All-metal bonded blades with stainless steel spar and leading-edge, light alloy skin and light alloy honeycomb filling; frame section of steel tube with light alloy skinning; full monocoque light alloy tailboom.

LANDING GEAR: Welded steel tube and light alloy skid landing gear, with energy-absorbing crosstubes.

POWER PLANT: One Textron Lycoming O-360 flat-four engine, derated to 97.5 kW (131 hp), mounted in the lower rear section of the main fuselage, with cooling fan. Light alloy main fuel tank in upper rear section of the fuselage on port side, usable capacity 72.5 litres (19.2 US gallons; 16.0 Imp gallons). Optional auxiliary fuel tank, capacity 39.75 litres (10.5 US gallons; 8.7 Imp gallons). Oil capacity 5.7 litres (1.5 US gallons; 1.25 Imp gallons). Transmission overhaul interval 2,200 hours or 12 years.

ACCOMMODATION: Two seats side by side in enclosed cabin, with inertia reel shoulder harness. Curved two-panel, tinted windscreen. Removable door, with tinted window, on each side. Baggage space beneath each seat. Cabin heated and ventilated.

SYSTEMS: Electrical system, powered by 12 V DC alternator, includes navigation, panel and map lights, dual landing lights, anti-collision light and battery.

AVIONICS: *Comms:* Bendix/King KY 197A VHF com radio; KT 76C transponder with Mode C altitude encoder; and NAT AA 80 intercom system with floor and hand switches. Options include Garmin 430 GPS/COM/VOR/LOC/GS with 106A CDI, replacing KY 197A; Pointer 3000-10 or 4000-10 (Canadian spec) ELT; NAT AA 12 intercom controller; and David Clark H10-13H or Bose Series X headsets.

Flight: Garmin 150XL, 250XL and 400 GPS optional.

Instrumentation: Standard equipment includes ASI; VSI; rotor/engine dual tachometer; sensitive altimeter; magnetic compass; digital OAT gauge; CHT gauge; oil temperature and pressure gauges; ammeter; hours meter; manifold pressure gauge; quartz clock; and warning lights for low voltage, low oil pressure, low fuel, low rotor rpm and horn, main gearbox temperature, main and tail rotor gearbox chips, starter engaged, and rotor brake engaged.

Optional instruments include BFG (AIM) 1100 artificial horizon with/without slip indicator; BFG (AIM) 205-1A DG; turn co-ordinator; PAI-700 vertical compass (replaces standard compass); United Instruments IVSI (replaces standard VSI); LC-2 digital clock (replaces standard clock); and Millibar altimeter.

Robinson R22 Beta II two-seat helicopter *(Paul Jackson)* ***NEW***/0589463

EQUIPMENT: Standard equipment includes rotor brake; tinted windscreen and windows; belly hardpoint; dual landing lights; navigation, panel and map lights; anti-collision light; ground handling wheels; rotor blade tiedowns; and windscreen cover.

Optional equipment includes three-cylinder engine priming system; RHC oil filter; cabin heater/defogger; metallic base or trim exterior colours; and leather seats.

DIMENSIONS, EXTERNAL:
Main rotor diameter 7.67 m (25 ft 2 in)
Tail rotor diameter 1.07 m (3 ft 6 in)
Main rotor blade chord 0.18 m (7¼ in)
Distance between rotor centres 4.39 m (14 ft 5 in)
Length overall, rotors turning 8.76 m (28 ft 9 in)
Fuselage: Length 6.30 m (20 ft 8 in)
Max width 1.12 m (3 ft 8 in)
Height: overall 2.72 m (8 ft 11 in)
to top of cabin 1.75 m (5 ft 9 in)
Skid track 1.93 m (6 ft 4 in)

DIMENSIONS, INTERNAL:
Cabin max width 1.12 m (3 ft 8 in)

AREAS:
Main rotor blades (each) 0.70 m^2 (7.55 sq ft)
Tail rotor blades (each) 0.04 m^2 (0.40 sq ft)
Main rotor disc 46.21 m^2 (497.4 sq ft)
Tail rotor disc 0.89 m^2 (9.63 sq ft)
Fin 0.21 m^2 (2.28 sq ft)
Stabiliser 0.14 m^2 (1.53 sq ft)

WEIGHTS AND LOADINGS:
Weight empty (without auxiliary fuel tank) 388 kg (855 lb)
Fuel weight: standard 52 kg (115 lb)
auxiliary 28.6 kg (63 lb)
Max T-O and landing weight 621 kg (1,370 lb)
Max disc loading 13.4 kg/m^2 (2.75 lb/sq ft)
Max power loading at T-O 6.37 kg/kW (10.46 lb/hp)

PERFORMANCE:
Never-exceed speed (VNE):
without sling load 102 kt (190 km/h; 118 mph)
with sling load 75 kt (139 km/h; 86 mph)
Max level speed 97 kt (180 km/h; 112 mph)
Cruising speed at 70% power at FL80 96 kt (177 km/h; 110 mph)
Econ cruising speed 82 kt (153 km/h; 95 mph)
Max rate of climb at S/L more than 305 m (1,000 ft)/min
Rate of climb at FL100 more than 183 m (600 ft)/min
Service ceiling 4,265 m (14,000 ft)
Hovering ceiling IGE 2,865 m (9,400 ft)
Range with max payload, no reserves:
normal fuel more than 173 n miles (321 km; 200 miles)
max fuel more than 260 n miles (482 km; 300 miles)
Endurance at 65% power, auxiliary fuel, no reserves 3 h 20 min

UPDATED

ROBINSON R44

TYPE: Four-seat helicopter.

PROGRAMME: Development began 1986; first flight (N44RH) 31 March 1990; first flight of third R44 March 1992; marketing began 22 March 1992; certification completed 10 December 1992. In August 1997 an R44 became first piston helicopter flown around the world. In June 2002 an R44 Raven (G-NUDE) became the first piston-engined helicopter to land at the North Pole. Japanese certification achieved 18 November 2002.

CURRENT VERSIONS: **R44 Astro:** Initial version; no longer produced.

R44 Raven I: Basic version, introduced in April 2000 to replace Astro; features 194 kW (260 hp) Textron Lycoming O-540 engine derated to 168 kW (225 hp) at T-O, 153 kW (205 hp) continuous; hydraulic control system standard; plus elastomeric tail rotor bearings and adjustable pedals for pilot.

R44 Raven II: Upgraded version introduced in June 2002. FAA certification 10 October 2002; first deliveries November 2002. Features a Textron Lycoming IO-540 engine, 28 V 70 A electrical system, increased lifting area on main rotor blades, and aerodynamic tip caps on main and tail rotor blades. Total of 580 ordered and 390 delivered by May 2004, including two ENG variants. *As described.*

R44 Clipper II: Float-equipped version, initially certified (as R44 Clipper) 17 July 1996. Available with twin utility floats or pop-out floats.Total of 141 Clippers sold by 1 June 2001.

R44 Police Helicopter: Specially modified Raven II for law enforcement, with Inframetrics 445G Mk II IR sensor, TV camera and ×7 zoom lens mounted in gyrostabilised nose turret; LCD video monitor; Spectrolab SX-5E 15 to 20 million candlepower searchlight; FM radio package; bubble door windows; Bendix/King KFM 985 dual-band transceiver, KY 196A VHF, II Morrow Apollo SL-60 standby comms, KT 76C transponder, 28 V electrical system and extended landing gear struts. Empty weight, equipped, approx 715 kg (1,570 lb). Certified July 1997; first customer El Monte Police Agency of Los Angeles. Recent customers include the Estonian Air Force, which ordered two in February 2002 for delivery commencing May 2002. China's Zheng Zhou Public Security Bureau, which took delivery of one in June 2002 following its appearance at the China Expo 2002 exhibition, Fontana Police Department, California, which took delivery of one in October 2003, and Merced County Sheriff's Department, California, which took delivery of one in November 2003, and the National Police of El Salvador, which took delivery of one on 25 June 2004.

R44 Newscopter: Digital electronic news gathering (ENG) version of Raven II intended for media companies and fitted with Ikegami HL-59WNA digital camera with ×21 lens and 360° continuous rotation with five-axis gyro stabilisation; and microwave transmitter; four video monitors, two micro cameras, KY 196A VHF, KT 76C transponder (Mode C), GPS, two FM radios and bubble window in port door. Deliveries began January 1998; 29 sold by 1 June 2002. First digital ENG Raven II delivered on 9 February 2003 to Sky Helicopters of Garland, Texas, and was the 30th ENG Newscopter delivered. Other customers include flying TV Ltd of the UK, which took delivery of one (G-PIXX) in May 2004.

R44 IFR: Equipped for IFR helicopter training IFR package adds US$27,200 to standard cost. 18 Sold by 1 June 2001.

CUSTOMERS: Total of 1,632 delivered by 31 December 2003. Deliveries totalled 150 in 1999, 264 in 2000, 194 in 2001, and 294 in 2003, comprising 75 Raven Is and 219 Raven IIs, and 228 in the first six months of 2004. 1,000th delivery took place in February 2001. 5,000th

Robinson R44 Raven II *(Paul Jackson)* ***NEW***/0589361

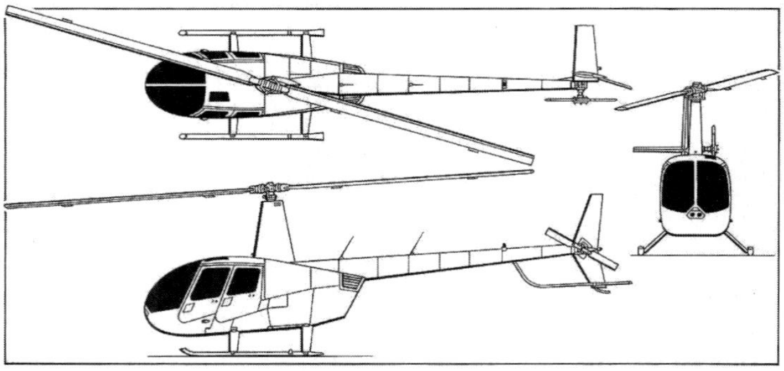

Robinson R44 Raven II light helicopter *(Jane's/Mike Keep)* 0507289

Robinson helicopter delivered was an R44 Raven II for National Airways Corporation of Lanseria, South Africa, in September 2003.

COSTS: Raven I US$307,000, Raven II US$346,000. Clipper II with utility floats US$362,000, Clipper II with pop-out floats US$368,000, R44 Police Helicopter US$509,000, R44 Newscopter US$549,000 (all 2004). Direct operating cost US$43.30 per hour; total (500 hours annually) US$131.87 per hour (both 2004).

DESIGN FEATURES: New design, incorporating general configuration (but larger cabin) and some proven concepts of R22; designed to requirements of FAR Pt 27; features for comfort and safety include electronic throttle governor to reduce pilot workload by controlling rotor and engine rpm during normal operations, rotor brake, advanced warning devices, automatic clutch engagement to simplify and reduce start-up procedure and reduce chance of overspeed, T-bar pistol grip cyclic control, and crashworthy features including energy-absorbing landing gear and lap/shoulder strap restraints designed for high forward *g* loads. High reliability and low maintenance; patented rotor design with tri-hinge (see R22); low noise levels.

FLYING CONTROLS: Conventional, with Robinson central cyclic stick; rpm governor; rotor brake standard; left-hand collective lever and pedals removable.

LANDING GEAR: Fixed skids; or, on R44 Clipper II, twin utility floats, or pop-out helium floats, which inflate in 2 seconds.

POWER PLANT: One 224 kW (300 hp) Textron Lycoming IO-540 AE1A5 flat-six engine, derated to 183 kW (245 hp) for 5 minutes, 153 kW (205 hp) maximum continuous. Standard fuel capacity 116 litres (30.6 US gallons; 25.5 Imp gallons) in tank on port side; optional 69 litres (18.3 US gallons; 15.2 Imp gallons) in starboard tank. No transmission limit; transmission TBO 2,200 hours or 12 years.

ACCOMMODATION: Four persons seated 2 + 2. Baggage stowage beneath each seat. Dual controls. Cabin heated and ventilated. Tinted windscreen and door windows.

AVIONICS: *Comms:* Bendix/King KY 196A VHF radio; KT 76C transponder with Mode C altitude encoder; and NAT AA 80 intercom system with floor and hand switches. Options include Garmin 420 GPS/COM; Garmin 430 GPS/COM/VOR/LOC/GS with 106A CDI; Pointer 3000–10 or 4000–10 (Canadian spec) ELT; NAT AA 12 intercom controller; and David Clark H10-13H or Bose Series X headsets.

Flight: Garmin 150XL, and 400 GPS optional.

Instrumentation: Standard equipment includes ASI; VSI; rotor/engine dual tachometer; sensitive altimeter; magnetic compass; digital OAT gauge; CHT gauge; oil temperature and pressure gauges; ammeter; hours meter; manifold pressure gauge; quartz clock; and warning lights for low voltage, low oil pressure, low fuel, low rotor rpm and horn, main gearbox temperature, main and tail rotor gearbox chips, starter engaged, and rotor brake engaged.

Optional instruments include BFG (AIM) 1100 artificial horizon with/without slip indicator; BFG (AIM) 205-1A DG; turn co-ordinator; PAI-700 vertical compass (replaces standard compass); United Instruments IVSI (replaces standard VSI); LC-2 digital clock (replaces standard clock); and millibar altimeter.

EQUIPMENT: Standard equipment includes rotor brake; tinted windscreen and windows; belly hardpoint; dual landing lights; navigation, panel and map lights; anti-collision light; ground handling wheels; tow cart adapter, rotor blade tie-downs; and windscreen cover. Optional equipment includes three-cylinder engine priming system; RHC oil filter; observation bubble windows; cabin heater/defogger; metallic base or trim exterior colours; and leather seats.

DIMENSIONS, EXTERNAL:
Main rotor diameter 10.06 m (33 ft 0 in)
Tail rotor diameter 1.47 m (4 ft 10 in)
Length overall, rotors turning 11.76 m (38 ft 7 in)
Fuselage: Length 9.07 m (29 ft 9 in)
Max width 1.28 m (4 ft 2½ in)
Height: overall 3.28 m (10 ft 9 in)
to top of cabin 1.83 m (6 ft 0 in)
Skid track 2.18 m (7 ft 2 in)

DIMENSIONS, INTERNAL:
Cabin max width 1.19 m (3 ft 11 in)

AREAS:
Main rotor disc 79.46 m^2 (855.3 sq ft)
Tail rotor disc 1.70 m^2 (18.35 sq ft)

WEIGHTS AND LOADINGS:
Weight empty: standard ... 683 kg (1,506 lb)
Fuel weight: standard ... 83 kg (184 lb)
auxiliary ... 50 kg (110 lb)
Max T-O and landing weight ... 1,134 kg (2,500 lb)
Max disc loading ... 14.27 kg/m² (2.92 lb/sq ft)
Max power loading at T-O ... 6.21 kg/kW (10.20 lb/hp)
PERFORMANCE:
Cruising speed at 75% power ... 117 kt (217 km/h; 135 mph)
Max rate of climb at 1,830 m (6,000 ft) ... over 305 m (1,000 ft)/min
Service ceiling ... 4,270 m (14,000 ft)
Hovering ceiling:
IGE at 1,134 kg (2,500 lb) ... 2,728 m (8,950 ft)
OGE: at 1,043 kg (2,300 lb) ... 2,286 m (7,500 ft)
at 1,134 kg (2,500 lb) ... 1,372 m (4,500 ft)
Max range, at 105 kt (194 km/m; 121 mph) no reserves ... approx 347 n miles (643 km; 400 miles)

UPDATED

ROCKY MOUNTAIN

ROCKY MOUNTAIN WINGS INC

PO Box 1188, Nampa, Idaho 83653-1188
Tel/Fax: (+1 208) 466 66 99
e-mail: rmwwings@aol.com
Web: http://www.realflying.com
PRESIDENT: Stace E Schrader

Rocky Mountain was formed in 1999 to market the Ridge Runner. In August 2004, the Bushwhacker (Ridge Runner IV) was introduced.

UPDATED

ROCKY MOUNTAIN RIDGE RUNNER

TYPE: Tandem-seat ultralight kitbuilt.
PROGRAMME: Ridge Runner I awarded Grand Champion at AirVenture 2001 and Arlington 2001; Ridge Runner II awarded outstanding lightplane award at Sun 'n' Fun 2004.
CURRENT VERSIONS: **Ridge Runner I:** Single-seat ultralight model.
Ridge Runner II: As Ridge Runner I but 1+1-seat version; *as described.*
Ridge Runner III: Slightly larger, fully two-seat version of Ridge Runner II; wider fuselage.
Bushwhacker/Ridge Runner IV: Two-seat version meeting LSA criteria; fuselage frame displayed at AirVenture 04. To be powered by Rotec radial engine; max T-O weight 598 kg (1,320 lb).
CUSTOMERS: Total of 22 of all versions sold by August 2003.
COSTS: Ridge Runner I US$9,900; Ridge Runner II US$10,900; Ridge Runner III US$11,900 (2004).
DESIGN FEATURES: Wings fold for transportation and storage with no need to disconnect controls. Quoted build time 600 hours.
FLYING CONTROLS: Conventional and manual. Full-span ailerons and slotted flaps. Trim tab on port elevator. Horn-balanced fin and elevators.
STRUCTURE: Fabric-covered 4130 chromoly steel tube fuselage and wooden wings with tubular steel V-struts. Composites engine cowling. Tailplane has V-strut supports below.
LANDING GEAR: Tailwheel type; fixed. Two side Vs hinged to lower longerons, plus half-axles, each with bungee cord shock-absorber, attached to compression frame with fabric-covered side panels and half axle. Mainwheels 8.00–6. Maule solid tailwheel. Cable operated brakes.
POWER PLANT: Choice of 19.0 kW (25.5 hp) Rotax 277; 29.5 kW (39.6 hp) Rotax 477, 37.0 kW (49.6 hp) Rotax 503, 47.8 kW (64.1 hp) Rotax 582, or 20.9 kW (28.0 hp) Hirth F33 engines. Fuel tank(s) in inboard section of wings, capacity 18.9 litres (5.0 US gallons; 4.2 Imp gallons) in Ridge Runner I; 38 litres (10.0 US gallons; 8.3 Imp gallons) in Ridge Runner II and III.
ACCOMMODATION: Pilot and passenger in tandem. Fully enclosed cockpit.
DIMENSIONS, EXTERNAL:
Wing span ... 7.99 m (26 ft 2½ in)

Rocky Mountain Ridge Runner II *(Paul Jackson)* ***NEW***/1047435

Length overall ... 5.18 m (17 ft 0 in)
Height overall ... 1.57 m (5 ft 2 in)
DIMENSIONS, INTERNAL (A: Ridge Runner I, B: Ridge Runner II, C: Ridge Runner III):
Cabin max width: A, B ... 0.61 m (2 ft 0 in)
C ... 0.74 m (2 ft 5 in)
AREAS:
Wings, gross ... 9.23 m² (99.4 sq ft)
WEIGHTS AND LOADINGS (A, B and C as above):
Weight empty: A ... 113 kg (250 lb)
B ... 159 kg (350 lb)
C ... 163 kg (360 lb)
Max T-O weight: A ... 249 kg (550 lb)
B, C ... 408 kg (900 lb)
PERFORMANCE (A, B and C as above, Rotax 582):
Never-exceed speed (VNE) ... 95 kt (177 km/h; 110 mph)
Normal cruising speed ... 70 kt (129 km/h; 80 mph)
Stalling speed: flaps up ... 28 kt (52 km/h; 32 mph)
flaps down ... 25 kt (45 km/h; 28 mph)
Max rate of climb at S/L: A ... 244 m (800 ft)/min
B ... 366 m (1,200 ft)/min
C ... 488 m (1,600 ft)/min
T-O run ... 23 m (75 ft)
Landing run ... 31 m (100 ft)
Range with max fuel: A ... 152 n miles (281 km; 175 miles)
B, C ... 282 n miles (523 km; 325 miles)

UPDATED

ROTORWAY

ROTORWAY INTERNATIONAL

4140 West Mercury Way, Stellar Air Park, Chandler, Arizona 85226
Tel: (+1 480) 961 10 01
Fax: (+1 480) 961 15 14
e-mail: rotorway@rotorway.com
Web: http://www.rotorway.com
VICE-PRESIDENT, PRODUCTION: Rusty Mueller
VICE-PRESIDENT, SALES: Brent Marshall
MARKETING DIRECTOR: Susie Bell

Initial RotorWay Company set up by the late B J Schramm in early 1960s; successively produced the Javelin, Scorpion I and Scorpion II before launching Exec in 1980. Company purchased and re-established on 1 June 1990 by the late John Netherwood; assets acquired in 1996 by company's employees, who are current owners.

RotorWay currently manufactures Exec 162F helicopter kits for final assembly by amateur builders and operators. The Exec is powered by the company's own engine, which has been in production for more than 20 years and has recently been upgraded to improve performance at altitude.

Some foreign distributors also sell helicopters; approximately 40 per cent of sales are outside the USA. RotorWay which had 40 employees in mid-2004, offers flight orientation and maintenance training, and customer service programme; 3,440 m² (37,000 sq ft) facility near Phoenix, Arizona, incorporates all departments, including manufacturing, sales and flight school.

UPDATED

ROTORWAY EXEC 162F

TYPE: Two-seat helicopter kitbuilt.
PROGRAMME: Development of first RotorWay helicopter (Scorpion) began in 1967; followed by Scorpion Too in 1971 and Exec in 1980. With some 23 modifications, became Exec 90, certified in UK in 1990 and in Poland in 1993. Exec 162 introduced in 1994, although latter's improvements can be retrofitted.
Kits, which lack only avionics and paint, are currently marketed in USA and 50 other countries; 40 per cent of production is shipped abroad. French certification received in October 2001 following approval of modifications to meet DGAC requirements, including improved fire protection, repositioned instrumentation and foot controls.
CURRENT VERSIONS: **Exec 162F:** *As described.* Marketed in Australia as *Exec 162FA.*
Exec 162 Pro: Not approved by RotorWay. Optional improvement package from Pro-Drive Inc, comprising modified main rotor drive system, 'E Z Start' clutch and carbon fibre tail rotor.
JetExec: Not approved by RotorWay. Turbine conversion offered by Kiss Aviation of Perris, California; powered by 112 kW (150 hp) Solar T-62T-32 Titan. Fuel capacity 151 litres (40.0 US gallons; 33.3 Imp gallons); max rate of climb 548 m (1,800 ft)/min; range 260 n miles (482 km; 300 miles).
CUSTOMERS: Over 250 Exec 90 and 750 Exec 162F kits sold and over 600 flying; 500th 162F delivered to Mexican Navy June 2000, 600th to Heli Diffusion of France in June 2002 and 700th to Australia in August 2003. International uses include police surveillance, forestry observation, powerline inspection, ranch work, crop-spraying, recreation and flight training.
COSTS: Kit: US$64,350 including engine (2003). ACIS US$4,675 extra. Also available in four batches of parts to spread cost.
DESIGN FEATURES: Asymmetrical RotorWay-designed aerofoil section two-blade main rotor. All-metal aluminium alloy blades attached to aluminium alloy teetering rotor hub by

RotorWay Exec 162F two-seat kitbuilt helicopter ***NEW***/1048020

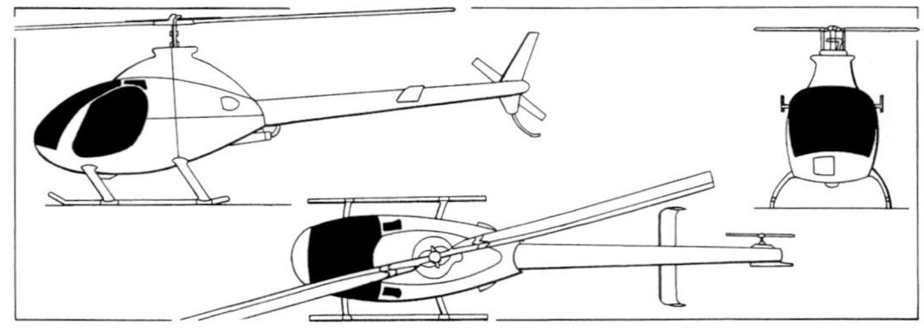

RotorWay 162F homebuilt (RotorWay International RI 162 engine) *(Paul Jackson)* 0051838

retention straps. Teetering tail rotor, with two blades each comprising steel spar and aluminium alloy skin. Elastomeric bearing rotor hub system with dual push/pull cable-controlled swashplate for cyclic pitch control. Quoted build time 450 hours. Company offers on-line diagnostic testing by modem.

STRUCTURE: Blades as detailed under Design Features. Basic 4130 steel tube airframe structure, with wraparound glass fibre fuselage/cabin enclosure. Aluminium alloy monocoque tailboom.

LANDING GEAR: Twin-skid type. Floats optional (from outside source).

POWER PLANT: One 113 kW (152 hp) RotorWay International RI 162F 2.66 litre (162 cu in) liquid-cooled engine with FADEC. Optional Altitude Compensation Induction System (ACIS) supercharger supplies pressurised air to engine at altitude, obviating need for turbocharging and prolonging engine life. Standard fuel capacity 64.4 litres (17.0 US gallons; 14.2 Imp gallons).

ACCOMMODATION: Two persons side by side; optional external HeliPak baggage pod fits between skids, capacity 0.21 m^3 (7.4 cu ft).

AVIONICS: Extra navigational equipment optional on US market helicopters.

EQUIPMENT: Optional crop-spraying kit, including two 42 litre (11.0 US gallon; 9.2 Imp gallon) tanks.

DIMENSIONS, EXTERNAL:
Main rotor diameter ... 7.62 m (25 ft 0 in)
Tail rotor diameter ... 1.28 m (4 ft 2½ in)
Length of fuselage ... 6.71 m (22 ft 0 in)
Length overall, rotors turning ... 8.99 m (29 ft 6 in)
Height to top of main rotor ... 2.44 m (8 ft 0 in)
Skid track ... 1.65 m (5 ft 5 in)

DIMENSIONS, INTERNAL:
Cabin max width ... 1.12 m (3 ft 8 in)

AREAS:
Main rotor disc ... 45.60 m^2 (490.9 sq ft)
Tail rotor disc ... 1.17 m^2 (12.57 sq ft)

WEIGHTS AND LOADINGS:
Weight empty ... 442 kg (975 lb)
Crew weight ... 193 kg (425 lb)
Max T-O weight ... 680 kg (1,500 lb)
Max disc loading ... 14.9 kg/m^2 (3.06 lb/sq ft)
Max power loading at T-O ... 6.01 kg/kW (9.87 lb/hp)

PERFORMANCE:
Never-exceed (VNE) and max level speed ... 100 kt (185 km/h; 115 mph)
Normal cruising speed ... 82 kt (153 km/h; 95 mph)
Max rate of climb at S/L ... 305 m (1,000 ft)/min
Service ceiling ... 3,050 m (10,000 ft)
Hovering ceiling, with two persons:
IGE ... 2,135 m (7,000 ft)
OGE ... 1,525 m (5,000 ft)
Range with max fuel at optimum cruising power ... 156 n miles (289 km; 180 miles)
Endurance with max fuel at optimum cruising power ... 2 h

UPDATED

SAFIRE

SAFIRE AIRCRAFT COMPANY

Opa Locka Airport, 15001 NW 42nd Avenue, Building 47, Miami Florida 33054
Tel: (+1 305) 779 40 40
Fax: (+1 305) 779 40 50
e-mail: info@safireaircraft.com
Web: http://www.safireaircraft.com
CHAIRMAN: Michael Margaritoff
PRESIDENT AND CEO: Camilo A Salomon
SENIOR VICE-PRESIDENT, ENGINEERING AND PROGRAMME MANAGEMENT: Joseph Furnish
VICE-PRESIDENT, OPERATIONS AND MANUFACTURING: Dr Joseph Cox
VICE-PRESIDENT, SUPPLY MANAGEMENT: Alan E Schwartz
VICE-PRESIDENT, STRATEGIC PLANNING AND SPECIAL PROJECTS: Miguel Vasquez

Safire, formed in September 1998, was developing the S-26 six-seat business jet, but in April 2003 announced the larger, all metal Safire Jet, which will be the first of a series of light jets the company is planning. On 7 March 2003, Safire moved from its former premises in West Palm Beach to a 4,645 m^2 (50,000 sq ft) facility at Miami's Opa Locka Airport. Employment stood at 100 in April 2004. Safire secured Italian financial booking in 2004 to carry the programme into the manufacturing phase, but progress was halted by lengthy Honeland Security checks on all foreign investment.

UPDATED

SAFIRE JET

Artist's impression of Safire Jet ***NEW***/0589444

TYPE: Light business jet.

PROGRAMME: Announced 25 April 2003, as replacement for S-26; first metal cut June 2003; maiden flight of first of two flying prototypes (plus one static test airframe) was scheduled for September 2004, but delayed pending additional funding; certification to FAR Pt23 single-pilot qualification and flight into known icing conditions and first customer deliveries were scheduled for 2006. Major structural components will be supplied by subcontractors, with Safire undertaking final assembly (on computer-controlled, eight-station production line), flight test, completion and delivery. Target assembly time 16 days per airframe; maximum production rate estimated at more than 500 per year by 2009.

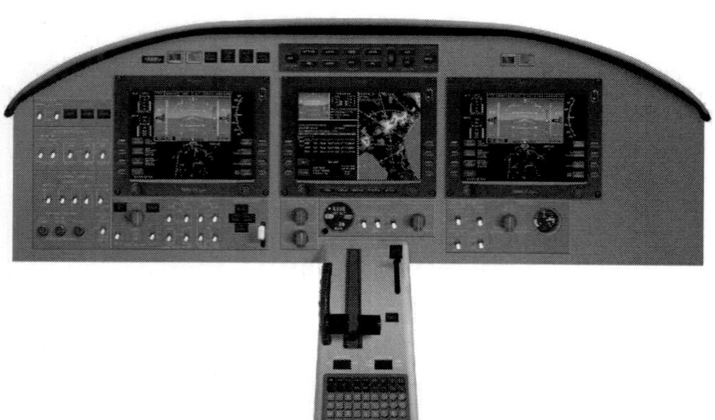

Safire Jet instrument panel ***NEW***/0589442

CUSTOMERS: Firm orders for 396 held by April 2004.

COSTS: US$1.295 million for holders of deposits for S-26, US$1.395 million for customers placing orders before first flight, US$1.495 million thereafter; direct operating costs US$360 to 30 390 per hour (all 2004).

DESIGN FEATURES: Conventional low-wing monoplane with rear-mounted engines and sweptback cruciform tail, Supercritical laminar-flow wing aerofoil.

FLYING CONTROLS: Conventional and manual. Dual control yokes standard

STRUCTURE: Light alloy throughout. Suppliers include Apex Engineering International/Excel Manufacturing (wing, tail unit and engine nacelles). Avidyne (avionics), Metalcraft Technologies (fuselage), Kollsman (pressurisation), Keith Products (vapour cycle air conditioning), Lee Aerospace (transparencies), Barry Controls Aerospace (engine vibration isolators and mounting structure), and Williams (power plant).

LANDING GEAR: Hydraulically retractable tricycle type with single wheel on each unit.

POWER PLANT: Two pod-mounted Williams FJ33-4 turbofans, each rated at 4.89 kN (1,100 lb st). Usable fuel 939 litres (248 US gallons; 207 Imp gallons) in integral wing tanks.

ACCOMMODATION: Pilot and co-pilot or passenger in cockpit. Seating for four in club arrangement in main cabin. Accommodation is pressurised and air conditioned. Lavatory and in-flight accessible baggage compartment aft of cabin.

SYSTEMS: Pressurisation supplied by engine bleed air, maximum differential 0.54 bar (7.8 lb/sq in), maintaining 2,440 m (8,000 ft) cabin altitude to certified ceiling of 12,500 m (41,000 ft).

AVIONICS: Avidyne FlightMax Entegra integrated flight deck, designed to support FAA/NASA Highways in the Sky and Synthetic Vision Programmes.

Instrumentation: EFIS with three 264 mm (10.4 in) flat panel displays, comprising two MFDs each with its own solid-state airdata and attitude/heading reference system (ADAHRS), and a PFD incorporating EADI and EHSI functions.

Following data are provisional:

DIMENSIONS, EXTERNAL:
Wing span ... 12.01 m (39 ft 4¾ in)
Length overall ... 11.06 m (36 ft 3½ in)
Height overall ... 4.57 m (15 ft 0 in)
Cacin door: Width ... 0.61 m (2 ft 0 in)

DIMENSIONS, INTERNAL:
Cabin: Length ... 4.24 m (13 ft 10¾ in)
Max width ... 1.43 m (4 ft 8½ in)
Max height ... 1.40 m (4 ft 7 in)

Aisle width .... 0.23 m (0 ft 9 in)
Baggage compartment volume .... 0.40 m³ (14.0 cu ft)

WEIGHTS AND LOADINGS:
Weight empty .... 1,721 kg (3,795 lb)
Max payload .... 635 kg (1,400 lb)
Max T-O weight .... 2,835 kg (6,250 lb)
Max power loading .... 290 kg/kN (2.84 lb/lb st)

PERFORMANCE(estimated):
Max cruising speed .... 380 kt (704 km/h; 437 mph)
Long-range cruising speed .... 300 kt (556 km/h; 345 mph)
Approach speed .... 90 kt (167 km/h; 104 mph)
Stalling speed in landing configuration .... 69 kt (128 km/h; 80 mph)
Max certified operating altitude .... 12,500 m (41,000 ft)
Max rate of climb at S/L .... 884 m (2,900 ft)/min
Rate of climb at S/L, OEI .... 268 m (880 ft)/min
Max range, at 45 min reserves .... 1,300 n miles (2,407 km; 1,496 miles)
Range with 363 kg (800 lb) payload, long-range cruising speed at FL410, 45 min reserves .... 1,150 n miles (2,129 km; 1,323 miles)

UPDATED

SCALED

SCALED COMPOSITES LLC

1624 Flight Line, Mojave California 93501-1663
Tel: (+1 661) 824 45 41
Fax: (+1 661) 824 41 74
e-mail: info@scaled.com
Web: http://www.scaled.com
PRESIDENT: Burt (Elbert L) Rutan
VICE-PRESIDENT AND GENERAL MANAGER: Michael W Melvill

Scaled Composites founded 1982; bought by Beech Aircraft Corporation (now part of Raytheon) June 1985; sold back to Burt Rutan November 1988 and integrated in joint venture with Wyman-Gordon Company of North Grafton, Massachusetts; is now owned by private investors following buy-out by Burt Rutan and 10 other investors; continues to provide R&D facilities to individuals and companies; several projects developed for Beech retained by Scaled.

Scaled Composites is responsible for research and development and prototype construction; company also active in UAV field. Current programmes include proprietary products; those emerging in 2002–03 included White Knight, SpaceShipOne and Capricorn, known details of which follow.

Scaled produced the prototype and undertook early test flights of the Visionaire Vantage business jet. Similarly, Scaled was involved in the Adam A500, which it designed, built and flight tested.

The company's V-Jet II testbed for the Williams FJX-2 turbofan was replaced by the Eclipse Eclipse 500.

Company currently has 130 employees and annual revenue of around US$22 million.

UPDATED

SCALED COMPOSITES 311 CAPRICORN

TYPE: Technology demonstrator.

PROGRAMME: Announced 23 October 2003 and unveiled at roll-out on 8 January 2004; named **Virgin Atlantic GlobalFlyer**. Construction began in late 2002. Vehicle for attempt at first solo, non-stop, eastbound aeroplane global circumnavigation (minimum 19,863.692 n miles; 36,787.559 km; 22,858.739 miles between 66° 33' South and 66° 33' North, as stipulated by FAI) by celebrated balloon pilot Steve Fossett, with Sir Richard Branson of sponsor Virgin Atlantic Airways as 'reserve pilot'. First flight (N277SF), lasting 1½ hours, took place at Mojave 5 March 2004; four more flights by early June 2004; record attempt planned for 2004/05 winter, to capture optimum jetstream assistance), launching from central USA and flying round the world in less than 80 hours.

DESIGN FEATURES: Typical, innovative Scaled Composites configuration. Single-piece, high aspect ratio wing carrying central accommodation module and dorsal power plant pod, plus two separate pontoons, each with fin, underfin and two-piece tailplane.

FLYING CONTROLS: Conventional and manual (cables and pushrods). Six-segment, narrow-chord aileron on each outer wing; rudder and elevators on tail unit of each boom. No flaps, slats or spoilers.

STRUCTURE: Composites throughout, with graphite/epoxy skins and aramid honeycomb core. Single graphite/epoxy I-beam wing spar.

LANDING GEAR: Tricycle type; retractable. Single mainwheels, on aluminium oleos with steel sliders and hydraulic disc brakes; fully castoring twin nosewheels. Due to large wing span, main gear can be rotated 90° to allow aircraft to exit hangar sideways. Paratech braking parachute (diameter 1.83 m; 6 ft) in canister on each wing inboard trailing-edge.

POWER PLANT: One 10.23 kN (2,300 lb st) Williams FJ44-3 ATW turbofan. Fuel (JP-4) in 2,161.5 litre (571 US gallon; 475.5 Imp gallon) forward tank and 2,055.5 litre (543 US gallon; 452 Imp gallon) aft tank in each boom; further 1,241.5 litres (328 US gallons; 273 Imp gallons) in four outboard tanks in each wing; and 117.3 litres (31.0 US gallons; 25.8 Imp gallons) in fuselage tank, giving total capacity (13 tanks) of 11,034 litres (2,915 US gallons; 2,427 Imp gallons). Fuel load represents more than 82 per cent of total T-O weight.

ACCOMMODATION: Pressurised and soundproofed cockpit for one occupant, including horizontal sleeping space and stowage for food and personal equipment. Cockpit pressure differential 0.55 bar (8.0 lb/sq in) gives 3,050 m (10,000 ft) environment at FL450.

AVIONICS: Chelton FlightLogic EFIS; Wulfsberg Series III VHF radios; NAT audio control panel. Dual TruTrak Flight Systems digital autupilots. Inmarsat C satellite datalink.

DIMENSIONS, EXTERNAL:
Wing span .... 34.75 m (114 ft 0 in)
Wing aspect ratio 32.5
Length overall .... 13.44 m (44 ft 1¼ in)
Height overall .... 4.05 m (13 ft 3½ in)

DIMENSIONS, INTERNAL:
Cabin length .... 2.13 m (7 ft 0 in)

AREAS:
Wings, gross .... 37.16 m² (400.0 sq ft)

WEIGHTS AND LOADINGS:
Weight empty .... 1,520 kg (3,350 lb)
Max fuel weight .... 8,165 kg (18,000 lb)
Max T-O weight .... 9,981 kg (22,006 lb)
Landing weight (all fuel expended) .... less than 1,814 kg (4,000 lb)
Max wing loading .... 268.6 kg/m² (55.01 lb/sq ft)
Max power loading .... 976 kg/kN (9.57 lb/lb st)

The Scaled 311 Virgin Atlantic GlobalFlyer deploying its drag-chutes during its maiden flight on 5 March 2004 ***NEW***/1047184

Scaled Composites 311 Capricorn Virgin Atlantic GlobalFlyer during testing at Edwards AFB *(Kristie Hogbin/USAF)* 1043413

PERFORMANCE (estimated):
- Normal cruising speed .... 250 kt (463 km/h; 288 mph)
- Normal cruising altitude .... 13,720 m (45,000 ft)
- Service ceiling .... 15,850 m (52,000 ft)
- T-O field length .... 3,660 m (12,000 ft)
- Range:
  - in still air .... 18,250 n miles (33,799 km; 21,001 miles)
  - with optimum jetstream assistance .... 20,000 n miles (37,040 km; 23,015 miles)
- Endurance .... 80 h

UPDATED

SCALED COMPOSITES 318 WHITE KNIGHT

TYPE: High-altitude platform.

PROGRAMME: Design conception started 1996, but existence unknown until shortly before first flight (N318SL) on 1 August 2002; officially unveiled 18 April 2003 as launch platform for suborbital SpaceShipOne and, possibly, other payloads. First captive-carry flight with SpaceShipOne attached, 20 May 2003; this programme completed in October 2004.

DESIGN FEATURES: Designed to provide high-altitude airborne launch of company's SpaceShipOne manned, suborbital spacecraft, and equipped to flight-qualify all of latter's systems except rocket propulsion. Objective is to demonstrate possibility of relatively inexpensive spaceflight. Cockpit avionics, ECS, pneumatics, trim servos, data system and electrical system components are identical to those in SpaceShipOne. High thrust/weight ratio and large speed brakes allow trainee astronauts to practise such spaceflight manoeuvres as boost, approach and landing in a realistic environment, thus effectively serving as a flying simulator for SpaceShipOne pilot training.

Twin-boom configuration with separate T tails and short fuselage pod. High-aspect ratio wings have marked anhedral inboard of booms and approximately equal dihedral outboard, raising fuselage high above ground. Podded jet engine pylon-mounted to fuselage above each wingroot. Tailbooms sharply tapered to narrow point at front. Forward section of fuselage pod bulged, in similar fashion to Proteus but with 16 circular windows and forward-opening circular crew door.

Although intended primarily for high-altitude missions such as launch of SpaceShipOne, other roles could include reconnaissance, surveillance, atmospheric research, data relay, telecommunications, imaging, and booster-launch for microsatellites.

FLYING CONTROLS: Manual, with three-axis electric trim; ailerons outboard of tailbooms, rudders and elevators on T tails. Pneumatic speed brakes inboard of tailbooms can be operated separately as inboard and outboard pairs to allow steep descent with lift/drag ratio of less than 4.5:1.

STRUCTURE: All carbon fibre epoxy materials, primarily using sandwich construction with honeycomb and PVC foam cores.

LANDING GEAR: Wide-track, four-wheel type. Pneumatically retracting (rearward) mainwheel with compressed rubber shock absorption in each tailboom, in line with wing anhedral/dihedral junction. Small, non-retractable, 'spatted' nosewheel at front of each boom; starboard one is steerable.

POWER PLANT: Two General Electric J85-GE-5 turbojets (each 17.13 kN; 3,850 lb st with afterburning). Fuel tanks (see under Weights and Loadings for capacity) in wing leading-edges and forward part of tailbooms.

ACCOMMODATION: Three-place cabin, qualified for unlimited altitude. Two crew doors (one on port side, one in nosecone), with dual-pane windows. Cockpit allows single-pilot operation (VMC day conditions only).

SYSTEMS: Pneumatic system for main landing gear extension/retraction. Hydraulic system for wheel brakes and nose gear steering. Dual-bus electrical system.

White Knight/SpaceShipOne combination during a captive-carry flight, with Beech Starship 1 acting as chase aircraft ***NEW***/1047171

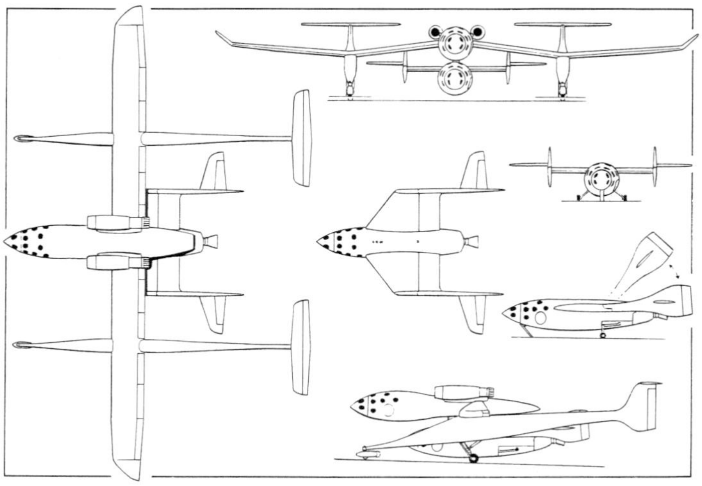

Provisional drawings of White Knight and SpaceShipOne, with scrap view of latter in 'feathered' configuration *(Michael Badrocke)* ***NEW***/0567229

AVIONICS: *Flight:* Include INS-GPS nav, flight director, flight test data (recording and telemetry), air data, vehicle health monitoring and video systems

Instrumentation: Back-up flight instruments.

DIMENSIONS, EXTERNAL:
- Wing span .... 24.99 m (82 ft 0 in)
- Cabin diameter .... 1.52 m (5 ft 0 in)
- Crew door diameter (2, each) .... 0.66 m (2 ft 2 in)
- Window diameter (16, each) .... 0.23 m (9 in)

DIMENSIONS, INTERNAL:
- Cabin max width .... 1.50 m (4 ft 11 in)

AREAS:
- Wings, gross .... 43.48 m^2 (468.0 sq ft)

WEIGHTS AND LOADINGS:
- Max fuel weight .... 2,903 kg (6,400 lb)
- Max payload .... approx 3,629–4,082 kg (8,000–9,000 lb)
- Estimated max T-O weight with SpaceShipOne .... 8,164–8,618 kg (18,000–19,000 lb)

PERFORMANCE:
- Never-exceed speed (VNE) .... M0.6 (160 kt; 296 km/h; 184 mph EAS)

UPDATED

SCALED COMPOSITES 316 SPACESHIPONE

TYPE: Technology demonstrator.

PROGRAMME: Design concepts for 'Tier One' suborbital manned spacecraft inaugurated 1996; SpaceShipOne development programme began with award of customer contract in April 2001; aircraft registered N328KF (signifying non-metric equivalent of 100,000 m) on 20 March 2003 and existence revealed 18 April 2003. Initial two sorties (first unmanned, second manned) on 20 May and 29 July 2003 were captive-carry flights under company's White Knight aircraft; followed by first manned glide 7 August 2003; first 'feather' conversion made on fourth flight, 27 August 2003. Pitch-up problem encountered during third gliding flight on 23 September; rectified, and flying resumed, 17 October 2003. First supersonic flight was made on the 17 December 2003 Wright brothers centennial anniversary, after release from White Knight at 14,630 m (48,000 ft), aircraft suffering minor damage when port mainwheel leg collapsed on landing. On same date, telecommunications and entertainments businessman Paul G Allen identified as project's sponsor.

On 1 April 2004, FAA Office of Commercial Space Transportation issued SpaceShipOne with licence as suborbital manned rocket (first of its kind), paving way for attempt on Ansari X-Prize. During the course of flight 15P (see table) on 21 June 2004, N328KF achieved an altitude under rocket power of 86,106 m (282,500 ft), which was submitted to the FAI for ratification in Classes C-1 and C-1d as a record for a landplane launched from a mother-ship.

Flights on 29 September and 4 October 2004 successfully qualified for the US$10,000 X-Prize, considerably exceeding in the process both the official and unofficial absolute records for altitude achieved in 1962 and 1963 by the North American X-15A. Shortly before this, on 27 September, Sir Richard Branson's Virgin Group announced its entry into agreement to license the SpaceShipOne technology (currently owned by sponsor Paul

SPACESHIPONE FLIGHTS

Flight*	Date	Duration h min s	Speed at burnout	Apogee m (ft)	Remarks
01C	20 May 03	1 48 0			Unmanned
02C	29 Jul 03	2 6 0			Manned
03G	7 Aug 03	- 19			First glide
04GC	27 Aug 03	1 6 0			Launch aborted
05G	27 Aug 03	- 10 30			Second glide
06G	23 Sep 03	- 12 15			Pitch-up problems
07G	17 Oct 03	- 17 49			Tail strake/fence mods
08G	14 Nov 03	- 19 55			Enlarged tailplanes/elevators
09G	19 Nov 03	- 12 25			First to emergency aft CG limit
10G	4 Dec 03	- 13 14			
11P	17 Dec 03	- 18 10	M1.2	20,665 (67,800)	First powered/first supersonic; minor damage on landing
12G	11 Mar 04	- 18 30			First with thermal protection coating
13P	8 Apr 04	- 16 27	M1.6	>32,004(105,000)	
14P	13 May 04	- 20 44	M2.5	64,435(211,400)	
15P	21 Jun 04	- 24 5	>M2.9	100,124(328,491)	Asymmetric thrust caused trim system failure
16P	29 Sep 04	- 24 0	>M2.92	102,870(337,500)	First for X-Prize
17P	4 Oct 04	- 24 0	>M3.09	111,996(367,442)	Second for X-Prize; exceeds X-15 official record by more than 15,850 m (52,000 ft) and the unofficial record by more than 3,960 m (13,000 ft)

Notes: * C: Captive carry, G: glide, P: powered

Allen's Mojave Aerospace Ventures) to develop the world's first privately funded spaceships to carry commercial passengers. Separately, it formed new company Virgin Galactic to become the world's first commercial space tourism operator. Initial discussions with Scaled concerned suborbital spacecraft with accommodation for up to nine passengers, with configuration to be finalised in 2005. Anticipated design changes may include an attempt to eliminate bouts of uncommanded roll encountered on both of the X-Prize flights.

DESIGN FEATURES: Rocket-powered research aircraft, designed primarily using CFD and intended for suborbital flights to 100 km (62 miles) altitude following air launch from White Knight. Unique configuration enables aircraft-like qualities for boost, glide and landing phases. Craft converts, via a pneumatically actuated 'feather' of control surfaces, to a stable, high-drag shape to re-enter atmosphere, allowing 'hands off' re-entry and greatly reduced aerodynamic and thermal loads.

Wing sweepback 30° 48' on leading-edge, 24° 6' at quarter-chord; dihedral, root incidence and washout all 0°. Wings, tailbooms and horizontal stabilisers all have Scaled-designed blunt trailing-edge aerofoil sections.

FLYING CONTROLS: Three flight control systems: manual subsonic, electric supersonic, and cold-gas remote control system. Each vertical tail surface has a manual upper rudder for subsonic flight and an electrically actuated lower rudder for trim and supersonic control. Horizontal tail surfaces, outboard of each vertical fin, comprise an all-flying stabiliser for trim and supersonic control, with a manual elevon for subsonic pitch and roll. Boom strake ahead of tailplanes, and small fence on outer tailplane leading-edges, added in 2003 following pitch-up problems. Beyond Earth atmosphere, roll, pitch and yaw are controlled by redundant pneumatic attitude thrusters at outer ends of wing leading-edges (roll) and at front of cabin (pitch and yaw).

Wings have no control surfaces as such; instead, entire one-third of wing trailing-edge, plus adjacent booms and tail surfaces, are hinged to 'feather' 65° upwards by pneumatic actuation, creating a high-drag configuration which decelerates craft for re-entry and permits a high AoA steep descent before 'defeathering' the configuration for a glide landing.

Craft enters shallow glide-climb after release from White Knight before igniting rocket and rotating into near-vertical climb for 65 seconds until rocket burnout, when it continues in zero-*g* climb for about 3½ minutes to reach 100 km altitude before starting free fall back towards Earth.

Cables from rudder pedals in cockpit run entire length of aircraft, terminating at bellcranks forward of rudders. These bellcranks drive pushrods that actuate the top rudders at their root ribs. Lower (trim) rudders are driven by single electric servo, attached by continuous loop cable to similar system of bellcranks and pushrods. Manual top rudder deflections are 0° inboard and 24° outboard, and can be operated independently; trim rudders deflect -3.5 and +17° respectively. Horizontal stabilisers can trim via electric servos from +7 to -13°. Elevons are 100 per cent mass balanced, and are actuated by pushrods through a root torque tube from the centre stick in the cockpit.

STRUCTURE: All structures are carbon fibre semi-monocoques, with foam and/or honeycomb sandwich panels, except for rear quarter of vertical booms and rudders, which are of glass fibre for comms and telemetry antenna 'transparency'. Fuselage is a circular cross-section shell, sized to house the nitrous oxide tank, on the front of which is a self-contained carbon fibre pressure vessel for pilot and passengers. Rear lower section of fuselage aft of tank is removable, together with landing gear, to allow easy access to hybrid motor and feather systems before and after each flight. Wings are centred around two load-bearing carbon fibre spars which attach to fuselage fore and aft of NO_2 tank. Tailbooms are vertical to aircraft axis and canted outwards 0° 30' from front to rear. Booms and rear one-third of wing comprise the 'feather', which is attached to top of rear wing spar with four custom-built carbon fibre and steel piano hinges. Rudders are attached to ends of tailbooms with large custom-built glass fibre piano hinges. Main structural element of horizontal stabilisers is a central spar or spindle that cantilevers each tailplane from two large bearings within the boom; elevons each hinge about three points at quarter-chord.

SpaceShipOne gliding back to Earth at the end of flight 15P, 21 June 2004 *NEW*/1047183

LANDING GEAR: Rearward-retracting 'tricycle', comprising mainwheel units, on cantilever, self-sprung glass fibre struts, and nose skid. Main gear tyres are 11X4.00-5 Lamb-type on custom-built Matco rims, axles and brakes; extension and retraction is spring-actuated with pneumatic back-up. Spring-actuated nose skid comprises glass fibre strut, a crushable foam damper for shock absorption, and a maple shoe (replaced after each flight) for abrasion resistance and braking.

POWER PLANT: Specially developed 75.6 kN (17,000 lb thrust) hybrid rocket propulsion system, using non-toxic liquid nitrous oxide (NO_2) and hydroxy-terminated polybutadiene (HTPB) fuel. Development of fuel, bulkhead, controller, valve, injector, igniter and ground test programme was completed by Environmental Aeroscience Corporation of Miami and SpaceDev of San Diego; first firing November 2002; latter's system selected in September 2003. Composites NO_2 tank and case, throat and nozzle components developed by Scaled; tank's filament-wound carbon fibre over-wrap provided by Thiokol; ablative nozzle supplied by AAE Aerospace.

ACCOMMODATION: Three-place cabin, with space-qualified environmental control system; otherwise as described for White Knight. NO_2 tank and some systems can be heated by engine bleed air while attached to White Knight. Pressure cabin has two emergency egress systems. A large locking ring around the nose can be disengaged, releasing entire nose and opening a 0.91 m (3 ft) diameter escape path for pilot. Plug door on port side is normal primary mode of entry, but can be used as secondary emergency exit if necessary.

SYSTEMS: Pneumatic system, pressure 310 bar (4,500 lb/sq in) for all systems except remote control system, which is 414 bar (6,000 lb/sq in), actuates wing/tail 'feather', spaceflight attitude control thrusters, cabin pressurisation and window demisting. Hydraulic system for wheel brakes only. Lithium batteries for electrical power. Bottled oxygen system for emergency use.

AVIONICS: *Flight:* New INS-GPS nav/flight director developed jointly by Scaled Composites and Fundamental Technology Systems of Orlando, Florida, provides pilot with precise guidance information needed to fly SpaceShipOne manually during boost and re-entry, and for approach and landing; it also monitors vehicle health. Unit stores flight test data and telemeters data to mission control station.

DIMENSIONS, EXTERNAL:

Wing span .... 5.00 m (16 ft 4¾ in)
Wing mean chord (geometric reference) .... 3.08 m (10 ft 1¼ in)
Wing aspect ratio 1.64
Length overall .... 8.50 m (27 ft 10¾ in)
Height overall .... 2.71 m (8 ft 10¾ in)
Fuselage diameter .... 1.52 m (5 ft 0 in)
Feather surface span .... 5.00 m (16 ft 4¾ in)
Feather surface chord .... 1.16 m (3 ft 9½ in)
Rudder span (both) .... 0.73 m (2 ft 4¾ in)
Rudder chord (both) .... 0.335 m (1 ft 1¼ in)

A befogged Los Angeles basin as seen from space by N328KF's onboard video camera *NEW*/1047182

Tailplane span (each side) ... 1.22 m (4 ft 0 in)
Span over horizontal tails ... 8.17 m (26 ft 9½ in)
Elevon span (each side) ... 1.04 m (3 ft 4¾ in)
Elevon chord ... 28% of local chord
Wheel track ... 3.05 m (10 ft 0 in)
Wheelbase ... 1.95 m (6 ft 4¾ in)

DIMENSIONS, INTERNAL:
Cabin volume ... 36.13 m^3 (1,276 cu ft)

AREAS:
Wings (reference) ... 15.22 m^2 (163.8 sq ft)
Tailplanes (total reference) ... 2.25 m^2 (24.2 sq ft)
Tailbooms (total reference, excl rudders) ... 2.76 m^2 (29.7 sq ft)
Rudders (both, total reference) ... 0.23 m^2 (2.5 sq ft)

WEIGHTS AND LOADINGS (ESTIMATED):
Propellant weight ... 1,633 kg (3,600 lb)
Launch weight ... 2,994 kg (6,600 lb)
Landing weight ... 1,361 kg (3,000 lb)
Max applied shear force at VNE ... 1,162.02 kg/m^2 (238.00 lb/sq ft)
Max wing loading at max AUW ... 179.43 kg/m^2 (36.75 lb/sq ft)
Thrust/weight ratio at altitude at max AUW ... 2.3 : 1

PERFORMANCE:
Never-exceed (VNE) and max level speed ... 265 kt (490 km/h; 305 mph) EAS
Max Mach No. ... more than 3.0
Max speed with landing gear extended ... 130 kt (241 km/h; 150 mph) EAS
Design manoeuvring speed ... 124 kt (230 km/h; 143 mph) EAS
Max speed with feather extended ... 110 kt (204 km/h; 127 mph) EAS
Stalling speed:
solo, no fuel or oxidiser ... 70 kt (130 km/h; 81 mph)
at max AUW ... 103 kt (191 km/h; 119 mph)
Max operating altitude ... more than 100,000 m (328,000 ft)
g limit at max AUW ... +3.5

OPERATIONAL NOISE LEVELS:
In cabin, rocket motor lit ... 116 dB

UPDATED

SCHWEIZER

SCHWEIZER AIRCRAFT CORPORATION

PO Box 147, Elmira, New York 14902
Tel: (+1 607) 739 38 21
Fax: (+1 607) 796 24 88
e-mail: marketing@sacusa.com
Web: http://www.schweizer-aircraft.com
PRESIDENT: W Stuart Schweizer
EXECUTIVE VICE-PRESIDENTS: Leslie E Schweizer
VICE-PRESIDENT: Michael D Oakley
GENERAL MANAGER: Randy Simpson
MARKETING CO-ORDINATOR: Peter S Schweizer
MARKETING DIRECTORS:
David K Savage
Barbara J Tweedt

Established 1939 by Ernest Schweizer (1912–2000) to produce sailplanes; from mid-1957 to 1995 made Grumman (later Gulfstream American) Ag-Cat, initially under subcontract, but as Schweizer-owned product from January 1981; manufacturing rights sold to Ag-Cat Corporation of Malden, Missouri, but production terminated soon thereafter.

Schweizer acquired rights for sole US manufacture of Hughes 300 light helicopter 13 July 1983; continues to support earlier Hughes 300s; first Elmira-built 300C completed June 1984; Schweizer purchased US rights for whole 300C programme from McDonnell Douglas Helicopter Company (formerly Hughes Helicopters) 21 November 1986. Deliveries totalled 56 helicopters in 1996, 44 in 1997, 41 in 1998, 43 in 1999, 36 in 2000, 33 in 2001, 32 in 2002 and 38 in 2003, by which time Schweizer had delivered nearly 930 helicopters.

Schweizer also manufactures the Model 300C, 300CB, 300CB*i* and 333 helicopters and the 2–37 surveillance aircraft. An unmanned version of the 300CB, known as the RoboCopter 300, has been developed in collaboration with Kawada Industries Inc of Japan and is described in *Jane's Unmanned Aerial Vehicles and Targets*.

At the Heli-Expo show at Orlando, Florida, in February 2002 Schweizer, Sikorsky Aircraft and Shanghai Little Eagle Science and Technology (SLEC) of the People's Republic of China announced an agreement whereby a new joint-venture company, Shanghai Sikorsky Aircraft Co (SSAC), in which SLEC and Sikorsky hold 51 per cent and 49 per cent shares respectively, will produce Schweizer Model 300 series helicopters under licence. Initially, Schweizer 300C/CBCBis and 333s are being assembled from Schweizer-supplied kits (as Shen3A, Shen2B and Shen4T respectively), leading to manufacture of detail parts and eventually to full manufacture in China. First three kits arrived in Shanghai on 26 August 2002, followed by two more in September. Target production in China is 24 helicopters per year.

On 26 August 2004 Sikorsky Aircraft announced that it had reached agreement to acquire Schweizer as a wholly owned subsidiary. Completion of the acquisition was announced in the following month.

Subcontracts include aircraft structural components and assemblies for Northrop Grumman, Sikorsky and Boeing. The company employed 421 in 2004 in its 18,580 m^2 (200,000 sq ft) facility.

UPDATED

SCHWEIZER SA 2-37

US military designation: RG-8A

TYPE: Multisensor surveillance lightplane.

PROGRAMME: First flight 1985; prototype fitted with Hughes AN/AAQ-16 and other manufacturers' thermal imaging systems.

CURRENT VERSIONS: **SA 2-37A:** Baseline version.

SA 2-37B: Powered by 186 kW (250 hp) Textron Lycoming TIO-540-AB1AD engines. First aircraft (N6150U, c/n 015) registered February 2001; all subsequent production is to this standard. *As described.*

CUSTOMERS: Details in table. Three for US Army, of which one lost in accident; remaining two transferred in 1987 to US Coast Guard at Opa Locka for anti-narcotics operations; one converted to SA 2-38A.

Fifth, sixth and seventh 2-37As built for Central Intelligence Agency late 1989/early 1990; others to Colombian (CACOM 2), Jordanian (14 Squadron) and Mexican (EA 502) air forces, possibly to assist US drugs interdiction efforts as transfers not announced. Colombian purchase of one aircraft being followed by US government provision of five more.

CIA aircraft employed as airborne communications relay platforms for long-range reconnaissance UAVs; believed used over former Yugoslavia in 1994, supporting General Atomics Gnat 750 vehicles; also fitted with surface surveillance equipment to identify sites for NATO air strikes in same operational theatre. Participated in hostage rescue at Japanese embassy in Peru, 1997.

DESIGN FEATURES: Modification of Schweizer SGM 2-37 motor glider, but with slightly greater wing span, drooped leading-edge and leading-edge fences on outer wing panels to improve stall, much more powerful engine with large exhaust silencers on fuselage sides, three-blade quiet propeller, fuselage modified to accept bulged canopy and larger engine, streamlined fairings and hydraulic parking brake on mainwheels; more than treble standard fuel capacity, but optional extra tank also available; removable underfuselage skin and hatches give access to 1.84 m^3 (65 cu ft) payload bay behind cockpit, in which pallets holding LLLTV, FLIR or camera payloads can be quickly removed and installed; other engines and larger payloads available for surveillance, basic and advanced training, operator training, glider and banner towing, and priority cargo delivery. Certified to FAR Pt 23 for day and night IFR; inaudible when overflying at 'quiet mode' speed at about 610 m (2,000 ft) over land and 183 m (600 ft) over water with propeller turning at 1,100 rpm to 1,300 rpm and engine producing 48.5 kW (65 hp) of its maximum 186 kW (250 hp).

Wing section Wortmann FX-61-163 at root and FX-60-126 (modified) at tip; outer wing panels and horizontal tail can be removed for transport.

FLYING CONTROLS: Manual. All-moving tailplane with anti-balance tab; externally mass-balanced ailerons; rudder.

LANDING GEAR: Tailwheel type; fixed. Wing-mounted, cantilever main legs. Cantilever tailwheel. Cleveland disc brakes.

POWER PLANT: Current production SA 2-37B has one 186 kW (250 hp) 175 kW (235 hp) Textron Lycoming TIO-540-AB1AD turbocharged flat-six engine, driving a McCauley three-blade constant-speed propeller. Usable fuel capacity 375 litres (99 US gallons; 82.4 Imp gallons). Shadin fuel flow system.

ACCOMMODATION: Seats for two persons side by side under two-piece upward-opening canopy, hinged on centreline. Dual controls, seat belts and inertia reel harnesses standard. Compartment aft of seats enlarged to accommodate pallet containing up to 340 kg (750 lb) of sensors or other equipment.

SCHWEIZER SA 2-37 PRODUCTION

c/n	Reg/serial	Date	Operator
1	N3623C	1985	Schweizer; to
	85–0047	1986	US Army; to
	8101	1987	USCG; to SA 2-38A
2	N3623F	1985	Schweizer; to
	85–0048	1986	US Army; lost 22 Sep 86
3	N9237A		Schweizer;
	OES-2251	1994	Mexican AF
	5051	2000	Mexican AF
4	86–0404	1987	US Army; to
	8102	1987	USCG; lost 1996
5	N7508U	Jul 89	CIA; to
	N701AN	Dec 96	CIA
6	N7508W	Nov 89	CIA; to
	N87260	Feb 97	CIA
7	N7508Y	Jan 90	CIA; to
	N4122L	Dec 96	CIA
8	N7508Z	Feb 91	Schweizer; to
	FAC 2046		Colombian AF; to
	N12139	Mar 97	CIA
9	N7510U	May 94	Schweizer; to
	OES-2252	Jul 94	Mexican AF; to
	EAOE-2252		Mexican AF; to
	5052		Mexican AF; crashed 20 Mar 03
010	N6147U	Oct 97	LAIRD International; to
	1450	Dec 01	Royal Jordanian AF
011	N61474	Nov 97	Crashed 19 Feb 98
012	N61499	Oct 98	Schweizer; to
	FAC 5751	May 02	Colombian AF
014	N20086	May 99	Schweizer; to
	1451	May 00	Royal Jordanian AF
015	N6150U	Jun 00	Schweizer; to
	N2061L	Feb 01	Schweizer; to
	FAC	May 02	Colombian AF
016	N40623	Sep 01[1]	Schweizer; to
	FAC	02	Colombian AF
017	N10673	Feb 02[1]	Schweizer; to
	FAC	02	Colombian AF
018	N5206W	Aug 03*	Schweizer
019	N30745	May 02[1]	Schweizer; to
	FAC	02	Colombian AF
020	N2067F	Aug 02[1]	Schweizer; to
	FAC	02	Colombian AF

Notes: CIA aircraft registered to Vantage Leasing Inc of Dover, Delaware.
Nos. 1 and 2 remain currently registered (2004) as N3623C and N3623F, apparently due to administrative oversight. Similarly, Nos 010, 012, 014, 016, 017, 019 and 020 remain on the US civil aircraft register.
No. 010 LAIRD International at Marietta, Georgia.
No. 013: This serial number not allocated by Schweizer company.
Nos. 015 to 017 and 019 upwards are SA 2-37Bs.
Nos. 015 to 017 and 019 to 020 believed to be Colombian FAC5752 onwards.

[1] Issue of Airworthiness Certificate.

Schweizer RG-8A of the Colombian Air Force 0085838

DIMENSIONS, EXTERNAL:
- Wing span .... 21.70 m (71 ft 2½ in)
- Wing aspect ratio 25
- Fuselage length .... 8.79 m (28 ft 10 in)
- Height overall .... 2.36 m (7 ft 9 in)

AREAS:
- Wings, gross .... 18.68 m^2 (201.1 sq ft)

WEIGHTS AND LOADINGS:
- Weight empty .... 1,157 kg (2,550 lb)
- Payload .... 322 kg (710 lb)
- Max mission payload .... 227 kg (500 lb)
- Max fuel .... 272 kg (600 lb)
- Max T-O weight .... 1,950 kg (4,300 lb)
- Max wing loading .... 104.4 kg/m^2 (21.38 lb/sq ft)
- Max power loading .... 10.47 kg/kW (17.2 lb/hp)

PERFORMANCE:
- Never-exceed speed (VNE) .... 165 kt (305 km/h; 189 mph)
- Mission speed .... 85 kt (157 km/h; 98 mph)
- Optimum climbing speed .... 85 kt (157 km/h; 98 mph) CAS
- Stalling speed .... 67 kt (125 km/h; 78 mph)
- Service ceiling .... 7,315 m (24,000 ft)
- T-O run .... 449 m (1,473 ft)
- T-O to 15 m (50 ft) .... 742 m (2,433 ft)
- Landing from 15 m (50 ft) .... 726 m (2,383 ft)
- Landing run .... 375 m (1,230 ft)
- Endurance: standard, 200 n mile (370 km; 230 mile) mission radius .... 7 h
  - quiet mode .... up to 12 h

UPDATED

SCHWEIZER SA 2-38

US military designations: RU-38A/B Twin Condor

TYPE: Multisensor surveillance lightplane.

PROGRAMME: Development started 1993; substantial redesign of single-engined SA 2-37A/RG-8A; first US Coast Guard 2-37A returned to Schweizer 24 January 1994 for conversion beginning in April; first flight (N61428) 31 May 1995; second aircraft was due to fly in mid-1996, but was destroyed in a crash earlier the same year while still an RG-8; third was to be first 'new' build, but with wing taken from stock; all three for USCG. Type publicly revealed 20 July 1995. First delivery due to US Coast Guard in late 1996; however, programme suspended following RG-8 crash and not resumed until December 1996 when contract amended to two aircraft with engines uprated with turbochargers. Modifications introduced in 1997 comprised a redesigned tailplane and improved inlet ducts for the rear engine. Second (new-build) aircraft registered N61449 in July 1997. Flight testing (aircraft reserialled 8103) for USCG acceptance began at Edwards AFB on 10 July 1998 by 445th Flight Test Squadron, USAF programme was 100 sorties in three/four months; however test programme was terminated in the following year. Both aircraft remain registered to Schweizer.

Development of SA 2-38B first revealed in June 2003, when three aircraft were registered (N2101J, c/n 001; N2082C, c/n 002; and N2083N, c/n 003), all destined for undisclosed 'US government customer'. First flight was expected in early 2004; initial aircraft declared airworthy on 4 March 2004 and certified 1 April 2004.

CURRENT VERSIONS: **SA 2-38A:** Two-seat version, *as described except where otherwise indicated.*

SA 2-38B: Two/four-seat version powered by two 336 kW (450 shp) Rolls-Royce 250-B17F turboprops. Compared with SA 2-38A has increased wing span, overall length and MTOW.

CUSTOMERS: US Coast Guard (two); second US customer, and one overseas, also reported. However, by late 1999, only two SA 2-38As known to have been built.

COSTS: US$450,000 (1993) for initial redesign, US$35 million (1994) USAF contract to convert existing two aircraft; approximately US$1 million each (excluding sensors) for any additional procurement. Renegotiated contract (December 1996) covered two aircraft for US$53 million.

DESIGN FEATURES: Main objectives of SA 2-38A were to increase night patrol capability and reduce engine coking problems compared with SA 2-37A; additional engine also increases safety factor for overwater operation, though normal mode is single-engine cruise with second engine shut down.

Utilises wings and cockpit forward section of SA 2-37A; wing section Wortmann FX-61-163 at root and FX-60-126 (modified) at tip; principal design changes are adoption of pod and twin tailboom configuration with two engines in push-pull layout; cabin slightly widened.

Schweizer RU-38A Twin Condor surveillance aircraft ***NEW***/0077028

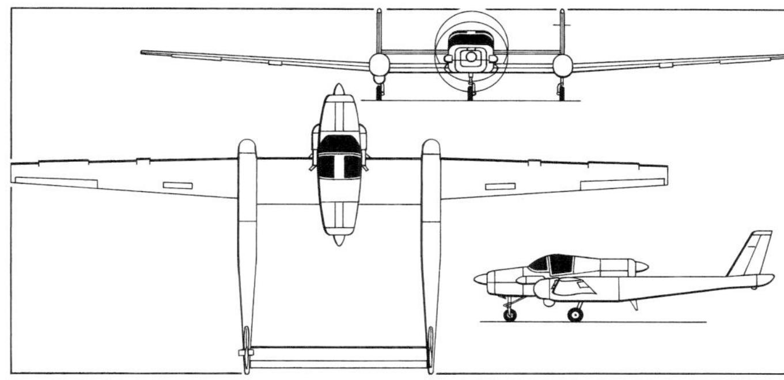

Schweizer SA 2-38A twin-engined surveillance aircraft *(Jane's/Mike Keep)* ***NEW***/0064927

LANDING GEAR: Tricycle type; fixed. Boom-mounted cantilever main legs; strut-braced nosewheel leg; single wheel on each unit. Main units Cleveland 6.50–10 wheel assemblies with Cleveland 30-95A brakes; nosewheel is Cleveland 6.00–8.

POWER PLANT: Prototype SA 2-38A originally with two heavily muffled Teledyne Continental GIO-550A flat-six engines, each rated at 261 kW (350 hp) at 3,400 rpm, with 3:2 reduction gear to limit propeller speed to 2,267 rpm. One engine in nose and one in rear of fuselage pod, respectively driving a tractor and pusher constant-speed, fully feathering three-blade propeller. Turbocharged engines in service version. Usable fuel capacity 375 litres (99.0 US gallons; 82.5 Imp gallons).

SA 2-38B has two 336 kW (450 shp) Rolls-Royce 250-B17F turboprops, geared to drive propellers at 800-1,000 rpm at mission speed. Rear engine is essentially redundant, to reduce risk of engine failure and to provide higher cruising speeds en route to and from missions. Front engine drives four-blade, constant-speed propeller; rear propeller is three-blade, constant-speed and full-feathering to enable rear engine to be shut down in flight in quiet surveillance mode, when aircraft noise is said to be undetectable from the ground during overflight at 610 m (2,000 ft); engine exhaust is directed over wings to reduce infra-red signature. Usable fuel capacity 1,060 litres (280 US gallons; 233 Imp gallons).

ACCOMMODATION: SA 2-38A accommodates pilot and sensor operator, side by side. SA 2-38B accommodates up to four, with seats for sensor operators behind the pilots' seats, but will operate with two crew with initial customer.

SYSTEMS: Three independent 28 V power sources.

AVIONICS: *Comms:* Rockwell Collins AN/ARC-182 VHF/UHF and Honeywell HF 990 radios, Bendix/King KY 196 VHF, KMA 24H-65 audio selector (two) and KFS 594 HF control; Wolfsburg RT9600 marine band radio.

Radar: Honeywell AN/APN-215(V) colour weather radar with search and mapping modes in nose of port tailboom.

Flight: Bendix/King KI 229 RMI, KI 256 flight director, KI 525A HSI, KI 250 radar altimeter, KDI 572 DME, KC 192 autopilot, KR 87 ADF and KNS 81 VOR/LOC/GS/RNAV. BAE Systems 7880 Omega/GPS.

Mission: Surveillance radar, as above. FLIR/LLLTV and dual recorder in nose of starboard boom; Bendix/King KY 58 and KY 75 communications encryption devices. Fully integrated sensor suite under development.

DIMENSIONS, EXTERNALA: SA 2-38A; B: SA 2-38B:
- Wing span: A .... 19.51 m (64 ft 0 in)
  - B .... 25.64 m (84 ft 1½ in)
- Wing aspect ratio: A 18.1
  - B 21.2
- Length overall: A .... 9.19 m (30 ft 2 in)
  - B .... 10.70 m (35 ft 1¼ in)
- Height overall: A .... 2.90 m (9 ft 6 in)
- Wheelbase: A .... 5.79 m (19 ft 0 in)

AREAS:
- Wings, gross: A .... 20.98 m^2 (225.9 sq ft)
  - B .... 31.05 m^2 (334.2 sq ft)
- Fins (total): A .... 2.26 m^2 (24.30 sq ft)
- Tailplane: A .... 3.26 m^2 (35.10 sq ft)

WEIGHTS AND LOADINGS:
- Weight empty: A .... 1,524 kg (3,360 lb)
  - B .... 1,935 kg (4,265 lb)
- Max payload: A .... 408 kg (900 lb)
  - B .... 363 kg (800 lb)
- Max fuel weight: A .... 272 kg (600 lb)
  - B .... 669 kg (1,475 lb)
- Max T-O weight: A .... 2,404 kg (5,300 lb)
  - B .... 3,265 kg (7,200 lb)
- Max wing loading: A .... 114.6 kg/m^2 (23.46 lb/sq ft)
  - B .... 105.2 kg/m^2 (21.54 lb/sq ft)
- Max power loading: A .... 4.61 kg/kW (7.57 lb/hp)
  - B .... 4.87 kg/kW (8.00 lb/hp)

Image purporting to show a Schweizer RU-38A with modifications including winglets and improved cooling for rear engine in form of twin side intakes and smaller, fixed, spine intake ***NEW***/1047186

PERFORMANCE :
Never-exceed speed (VNE): A ... 165 kt (305 km/h; 189 mph)
B ... 168 kt (311 km/h; 193 mph)
Cruising speed at 1,525 m (5,000 ft): A
at 75% power ... 136 kt (252 km/h; 157 mph)
at 63% power ... 127 kt (235 km/h; 146 mph)
Mission speed: A ... 90 kt (167 km/h; 104 mph)
B ... 83 kt (154 km/h; 96 mph)
Best climbing speed: A ... 87 kt (161 km/h; 100 mph)
B ... 83.5 kt (155 km/h; 96 mph)
Stalling speed: A ... 75 kt (139 km/h; 87 mph)
B ... 62 kt (115 km/h; 72 mph)
Max rate of climb at S/L: A ... 670 m (2,200 ft)/min
Rate of climb at S/L, nose engine only: A ... 350 m (1,150 ft)/min
Service ceiling: A ... 7,315 m (24,000 ft)
B ... 9,144 m (30,000 ft)
T-O run: A ... 295 m (960 ft)
B ... 294 m (965 ft)
T-O to 15 m (50 ft): A ... 430 m (1,400 ft)
B ... 311 m (1,020 ft)
Landing from 15 m (50 ft): A ... 755 m (2,475 ft)
B ... 945 m (3,100 ft)
Landing run: A ... 410 m (1,350 ft)
B ... 600 m (1,970 ft)
Endurance with no reserves: typical mission: A ... 6 h
max: A ... 10 h

UPDATED

SCHWEIZER 300C

Engineering designation: Model 269

TYPE: Three-seat helicopter.

PROGRAMME: Hughes 269 first flew 2 October 1956; production deliveries began October 1961; marketed later as Model 300. Developed 300C first flew August 1969; first flight Hughes production model December 1969; FAA certification May 1970. Production of 300C transferred from Hughes Helicopters to Schweizer July 1983; first flight Schweizer 300C (1,166th 300C) June 1984; Schweizer bought entire programme November 1986. Deliveries of 300CB began in August 1995 with initial three of 10 ordered by launch customer Helicopter Adventures, of Concord, California. More than 3,600 Hughes/Schweizer 300/330s have been produced. Production rate stood at one per week in 2004.

CURRENT VERSIONS: **300C**: Standard civil version.

Main description applies to 300C, unless otherwise indicated.

300C Sky Knight: Special police version; options include safety mesh seats with inertia reel shoulder harnesses, public address/siren system, searchlight, integrated communications, infra-red sensor, heavy-duty 28 V 100 A electrical system, cabin heater, night lights with strobe beacons, cabin utility light, fire extinguisher, first aid kit and map case. Chinese designation **Shen3A**.

TH-300C: Military training version.

300CB: 'Basic' version (otherwise designated 269C-1) for training role; 134 kW (180 hp) Textron Lycoming HO-360-C1A engine, 132 litre (35 US gallon; 29.1 Imp gallon) fuel tank and 2,000 hour TBO; stated to be cheaper to operate than 300C. First flight (N6002X) 28 May 1993; type certificate awarded August 1995. Changes from standard model include one-piece upper pulley hub; improved air filtration system; lightweight exhaust; and improved oil filter. 200,000 hours flown by October 2002, at which time 11 had been involved in accidents without fatalities. Chinese designation **Shen2B**.

300CB*i*: Enhanced version of 300CB, which it was scheduled to replace from 2002; introduced at Heli-Expo 2002 at Orlando, Florida, in February 2002. Features fuel-injected engine, new splined (replacing bolted) main rotor driveshaft and hub with 4,000-hour life, and electronic automatic engagement system (AES). First delivery (G-OCBI, c/n 0139) to CSE Aviation at Oxford, UK, in August 2002. One delivered to Great Slave Helicopters at Yellowknife, NWT, October 2002.

RoboCopter: Unmanned version developed by Kawada Industries in Japan for cropspraying, pipeline patrol, aerial crane and reconnaissance; first flight October 1996; Schweizer teamed with Northrop Grumman to offer helicopter for US Navy VTOL UAV requirement (subsequently met by RQ-8A Fire Scout version of Schweizer 333, which see); RoboCopter still being promoted in late 2002 under US marketing name **Argus**; further details in *Jane's Unmanned Aerial Vehicles and Targets*.

Schweizer 300CB*i* three-seat helicopter *(Paul Jackson)* ***NEW***/1047133

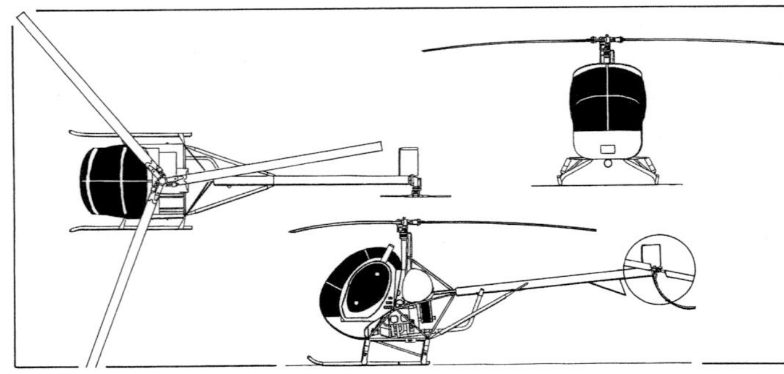

General arrangement of the Schweizer 300 *(Paul Jackson)* 0064928

CUSTOMERS: Hughes produced 2,775 of all versions, including TH-55A Osage for US Army, before transfer to Schweizer. Royal Thai Army received 58 TH-300C helicopters between March 1986 and mid-1989; 500th Schweizer-manufactured 300C delivered at Heli Expo 1994. Sky Knight customers include police departments of Baltimore, Maryland; Lakewood, California; and Kansas City, Missouri. Recent customers include Helicopter Adventures Inc of California, which ordered 23 300CB*is*, plus 10 options, for its rotary-wing training schools in California and Florida; High Desert Helicopters of West Jordan, Utah, which ordered two 300Cs in March 2004; and the Civil Aviation University of China, which took delivery of two 300Cs in early 2004, bringing to nine the total of Schweizer 300Cs delivered to China in 2003-04. By late 2004, Schweizer had built some 705 300Cs and 176 300CBs; during 2000, 13 300Cs and 17 300CBs were delivered, followed by 17 and 12 in 2001; 13 and 17 in 2002; and 20 and 15 in 2003. Argentine Coast Guard took delivery of two 300Cs in August 2001 to supplement two supplied in 1996 and two in 1999.

COSTS: 300C: US$235,650 (2001); 300CB US$204,750 (2001).

DESIGN FEATURES: Simple, pod-and-boom configuration, with exposed power plant and fuel tanks. Aerofoil-section stabiliser, starboard side, rear, at approx 45° dihedral; triangular underfin. Fully articulated three-blade main rotor, turning at 471 rpm; fully interchangeable blades; blade section NACA 0015; elastomeric dampers; two-blade teetering tail rotor; limited blade folding; no rotor brake; multiple V-belt and pulley reduction gear/drive system between horizontally mounted engine and transmission, with electrically controlled belt-tensioning system instead of clutch; braced tubular tailboom; separate dihedral tailplane and fin.

FLYING CONTROLS: Manual. Electric two-axis cyclic trim, cyclic and collective friction control.

STRUCTURE: Main rotor blades bonded with constant-section extruded aluminium spar, wraparound skin and trailing-edge; tail rotor blades have steel tube spar and glass fibre skin; steel tube cabin section with light alloy and stainless steel skin and Plexiglas transparencies.

LANDING GEAR: Tubular skids carried on booted oleo-pneumatic shock-absorbers. Replaceable heavy-duty skid shoes. Two ground handling wheels with 0.25 m (10 in) balloon tyres, pressure 4.14 to 5.17 bar (60 to 75 lb/sq in). Available optionally on floats made of polyurethane-coated nylon fabric, 4.70 m (15 ft 5 in) long and with a total installed weight of 27.2 kg (60 lb). Heavy-duty skid plates optional.

POWER PLANT: One 168 kW (225 hp) Textron Lycoming HIO-360-D1A flat-four engine, derated to 142 kW (190 hp), mounted horizontally aft of seats. Normal fuel capacity of 300C and 300CB is 123 litres (32.5 US gallons; 27.1 Imp gallons) of which 114 litres (30.0 US gallons; 25.0 Imp gallons) are usable; auxiliary tank of 133 litres (35.0 US gallons; 29.2 Imp gallons) can be fitted to 300C and tank of 114 litres (30.0 US gallons; 25.0 Imp gallons) to 300CB. Oil capacity 9.5 litres (2.5 US gallons; 2.1 Imp gallons).

ACCOMMODATION: Three persons side by side on sculptured and cushioned bench seat, with shoulder harness, in Plexiglas enclosed cabin. Accommodation ventilated. Carpet and tinted canopy standard. Forward-hinged, removable door on each side. Dual controls optional in 300C, standard in 300CB; captain sits on left in 300C and on right in 300CB. Offset anti-torque pedals. Exhaust muff heating kit available.

SYSTEMS: Standard electrical system includes 24 V 70 A alternator, 24 V battery, lightweight high-torque starter and external power socket. Night lights with strobe beacons standard on 300CB. Automatic engagement system (AES) introduced as an option from February 2002 incorporates a start-up overspeed governor, computer-controlled rotor engagement device and low rotor rpm warning.

AVIONICS: *Comms:* Avionics include Honeywell KY 96A com transceiver and headsets, KR 86 ADF and GTX 320 transponders and ACK A-30 blind encoder.

EQUIPMENT: Standard for all models are fire extinguisher, first aid kit and engine hour meter. Model 300C has external power plug and fuel pressure indicator; Model 300CB has night lights with strobes. Optional equipment includes stretcher kits, cargo racks with combined capacity of 91 kg (200 lb), external load sling of 408 kg (900 lb) capacity, Simplex Model 5200 agricultural spray or dry powder dispersal kits, instrument training package, throttle governor, start-up overspeed control unit, night flying kit, single or dual exhaust mufflers, door lock and dual oil coolers.

DIMENSIONS, EXTERNAL:
Main rotor diameter ... 8.18 m (26 ft 10 in)
Main rotor blade chord ... 0.17 m (6¾ in)
Tail rotor diameter ... 1.30 m (4 ft 3 in)
Distance between rotor centres ... 4.66 m (15 ft 3½ in)
Length: fuselage ... 6.77 m (22 ft 2½ in)
overall, rotors turning ... 9.40 m (30 ft 10 in)

Height: to top of rotor head .... 2.66 m (8 ft 8¾ in)
to top of cabin .... 2.19 m (7 ft 2 in)
Width: rotor partially folded .... 2.44 m (8 ft 0 in)
cabin .... 1.30 m (4 ft 3 in)
Skid track .... 1.99 m (6 ft 6½ in)
Length of skids .... 2.51 m (8 ft 3 in)
Passenger doors (each): Height .... 1.09 m (3 ft 7 in)
Width .... 0.97 m (3 ft 2 in)
Height to sill .... 0.91 m (3 ft 0 in)

DIMENSIONS, INTERNAL:
Cabin: Length .... 1.50 m (4 ft 11 in)
Max width .... 1.45 m (4 ft 9 in)

AREAS:
Main rotor blades (each) .... 0.70 m^2 (7.55 sq ft)
Tail rotor blades (each) .... 0.08 m^2 (0.86 sq ft)
Main rotor disc .... 52.54 m^2 (565.5 sq ft)
Tail rotor disc .... 1.32 m^2 (14.20 sq ft)
Fin .... 0.23 m^2 (2.50 sq ft)
Horizontal stabiliser .... 0.246 m^2 (2.65 sq ft)

WEIGHTS AND LOADINGS:
Weight empty: 300C .... 499 kg (1,100 lb)
300CB .... 494 kg (1,088 lb)
Baggage capacity .... 45 kg (100 lb)
Max T-O weight:
Normal category: 300C .... 930 kg (2,050 lb)
300CB .... 794 kg (1,750 lb)
external load .... 975 kg (2,150 lb)
Max disc loading:
Normal category: 300C .... 17.7 kg/m^2 (3.63 lb/sq ft)
300CB .... 15.1 kg/m^2 (3.09 lb/sq ft)
external load .... 18.6 kg/m^2 (3.80 lb/sq ft)
Max power loading at T-O:
Normal category: 300C .... 6.57 kg/kW (10.79 lb/hp)
300CB .... 5.61 kg/kW (9.21 lb/hp)
external load .... 6.89 kg/kW (11.32 lb/hp)

PERFORMANCE (at max Normal T-O weight, ISA):
Never-exceed speed (VNE) at S/L:
300C .... 95 kt (176 km/h; 109 mph) IAS
300CB .... 94 kt (174 km/h; 108 mph) IAS
Max cruising speed: 300C .... 86 kt (159 km/h; 99 mph)
300CB .... 80 kt (148 km/h; 92 mph)
Speed for max range, at FL40 .... 67 kt (124 km/h; 77 mph)
Max rate of climb at S/L: 300C .... 229 m (750 ft)/min
300CB .... 381 m (1,250 ft)/min
Hovering ceiling: IGE: 300C .... 3,290 m (10,800 ft)
300CB .... 2,135 m (7,000 ft)
OGE: 300C .... 2,620 m (8,600 ft)
300CB .... 1,465 m (4,800 ft)
Range at FL40, 2 min warm-up, max normal fuel, no reserves .... 208 n miles (386 km; 240 miles)
Max endurance at S/L: 300C .... 3 h 48 min
300CB .... 3 h 18 min

UPDATED

SCHWEIZER 330 AND 333

TYPE: Four-seat helicopter.

PROGRAMME: Announced 1987; first flight in public (N330ST; converted from a 300C) 14 June 1988; FAA certification September 1992. Italian certification April 2002. Deliveries started mid-1993. Three sets of controls available for student training. Model 330SP is marketing designation; type is 269D. Production rate one per month in 2004.

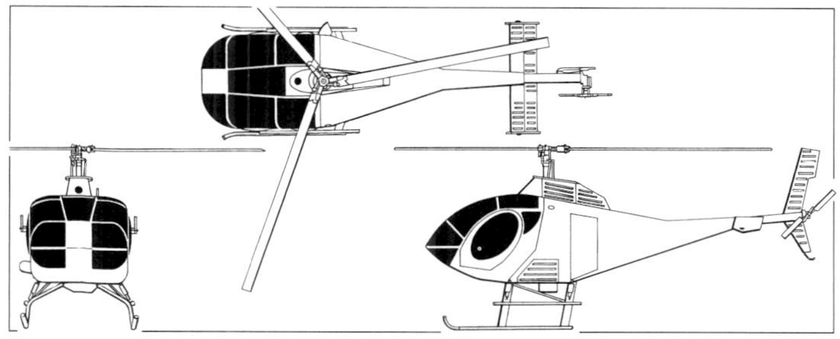

Schweizer 330SP three/four-seat light helicopter *(Jane's/Mike Keep)* 0130925

CURRENT VERSIONS: **330**: Original version delivered from 1993.

330SP: Announced May 1997, incorporates larger main rotor hub, increased blade chord and raised landing gear. Modifications can be retrofitted to existing 330s.

333: Announced early 2000 as an upgraded version of the 330SP; lead customer was San Antonio Police Department (two aircraft). FAA certification awarded 28 September 2000, at which time first three already delivered. 330SP can be upgraded to 333 standard. Differences include new cambered wider chord aerofoil main rotor blades, larger diameter rotor system, increased height landing gear and increased transmission rating. Chinese designation **Shen 4T**.

Five-seat version under study in 2003 for Indonesian Police Force requirement for up to 35 helicopters, with initial procurement expected to total 16 to 18.

Details apply to 333 unless otherwise stated.

RQ-8A Fire Scout: Schweizer 333 selected as airframe basis for Northrop Grumman Model 379 VTOL tactical UAV under development for US Navy (see *Jane's Unmanned Aerial Vehicles and Targets*) for which a US$93.7 million contract was awarded on 10 February 2000. In late March 2003, Schweizer began flight tests of a 333 (original prototype N330ST) equipped with a four-blade main rotor destined for the RQ-8A.

MQ-8A: Proposed armed version of RQ-8A Fire Scout for US Marine Corps.

CUSTOMERS: Recent customers include Chesapeake Bay Helicopters of Virginia and Aero Optics Inc of Wisconsin, which have each taken delivery of a 333 for powerline/pipeline inspection and other aerial work duties; the Dominican Air Force, which took delivery of four 333s in 2004 for pilot training and airborne patrol; and the Mexican Procuraduria General de la Republica's Agencia General de Investigacion, which ordered five 333s for delivery during 2004 for airborne patrol and law enforcement duties.

Three aircraft delivered in 1993; five in 1994, plus one prototype and one demonstrator; five in 1996, five in 1997, three in 1998, one in 1999, six in 2000, four in 2001; two in 2002; and three in 2003, for a total of 39, excluding two Fire Scout prototypes.

COSTS: US$607,500 for 333 (2001).

DESIGN FEATURES: Developed from Model 300C, with turbine power, enclosed engine and tapered tailboom. Civil roles include law enforcement, scout/observation, aerial photography, light utility, agricultural spraying and personal transport; uses turbine fuel rather than scarcer Avgas; extremely low flat rating of engine for good hot-and-high performance; streamlined external envelope and large tailplane with endplate fins added during development; rotor rpm 471.

FLYING CONTROLS: Similar to 300C; see Design Features.

STRUCTURE: Generally similar to 300C.

POWER PLANT: *330/330SP:* One 313.2 kW (420 shp) Rolls-Royce 250-C20W turboshaft derated to 175 kW (235 shp). Transmission rating 175 kW (235 shp) for T-O; 164 kW (220 hp) maximum continuous. Maximum usable fuel capacity 276 litres (73.0 US gallons; 60.8 Imp gallons).

Schweizer 333 four-seat helicopter *(Paul Jackson)* ***NEW***/1047132

333: One 313.2 kW (420 shp) Rolls-Royce 250-C20W turboshaft derated to 173 kW (232 shp). Transmission rating 209 kW (280 shp) for T-O; 173 kW (232 shp) maximum continuous. Fuel capacity as 330SP.

ACCOMMODATION: Three persons side by side (central occupant slightly forward), with up to three sets of controls; four point inertia reel seat belts. Fourth occupant in 'high density' configuration (central seat replaced by double seat). Cabin ventilated.

SYSTEMS: Engine inlet anti-ice system; 24 V lead-acid battery; 150 A starter/generator; external APU plug; circuit breaker protection.

AVIONICS: *Instrumentation:* Tachometer, engine torquemeter, engine oil temperature and pressure indicators; electric digital OAT, turbine outlet temperature indicator; annunciator warning panel.

Mission: Law enforcement options include FLIR, video recorder, Spectrolab SX-5 Starburst searchlight, Pronet and Litton computer system.

DIMENSIONS, EXTERNAL:
Main rotor diameter ... 8.38 m (27 ft 6 in)
Tail rotor diameter ... 1.30 m (4 ft 3 in)
Length overall, rotors turning ... 9.50 m (31 ft 2 in)
Fuselage: Length ... 6.82 m (22 ft 4½ in)
Max width ... 1.72 m (5 ft 7¾ in)
Height overall ... 3.35 m (11 ft 0 in)
Width overall ... 2.13 m (7 ft 0 in)
Tailplane span ... 2.04 m (6 ft 8½ in)
Skid track ... 1.92 m (6 ft 3½ in)

DIMENSIONS, INTERNAL:
Cabin: Max width ... 1.84 m (6 ft 0½ in)
Length ... 1.98 m (6 ft 6 in)
Height at seat ... 1.35 m (4 ft 5¼ in)

AREAS:
Main rotor disc area ... 55.20 m^2 (594.2 sq ft)
Tail rotor disc area ... 1.32 m^2 (14.20 sq ft)

WEIGHTS AND LOADINGS:
Weight empty ... 567 kg (1,250 lb)
Max T-O weight ... 1,156 kg (2,550 lb)
Max disc loading ... 20.95 kg/m^2 (4.29 lb/sq ft)
Transmission loading at max T-O weight and power ... 5.54 kg/kW (9.11 lb/shp)

PERFORMANCE:
Never-exceed speed (VNE) ... 120 kt (222 km/h; 138 mph)
Normal cruising speed ... 105 kt (194 km/h; 121 mph)
Econ cruising speed ... 94 kt (174 km/h; 108 mph)
Max rate of climb at S/L ... 420 m (1,380 ft)/min
Hovering ceiling at MTOW: IGE ... 2,651 m (8,700 ft)
OGE ... 1,554 m (5,100 ft)
Max range at FL40, no reserves ... 319 n miles (590 km; 367 miles)
Max endurance, no reserves ... 4 h 10 min

UPDATED

SEASTAR

SEASTAR AIRCRAFT INC

5409 Overseas Highway 333, Marathon, Florida 33050
Tel: (+1 305) 744 99 86 or (+36 20) 471 77 90
e-mail: info@seastarplane.com
Web: http://www.seastarplane.com
PRESIDENT: János Dósa
DESIGNER: Craig Easter

WORKS:
Box 107, Szigetszentmiklós, H-2311, Hungary
Tel: (+36 24) 45 10 00
Fax: (+36 24) 45 20 00

Seastar has undertaken a fundamental redesign of the Seawind amphibian which it now offers complete or as a kit. The next product will be the Adventurer turbo-powered version; other versions planned include an armed version for patrol work and a search and rescue equipped model.

The company is establishing a 5,000 m^2 (53,820 sq ft) facility in Hungary where certified aircraft will be built before being transferred to the USA for completion.

UPDATED

SeaStar SeaStar multipurpose amphibian *(Susan Bushell)* 0121672

SEASTAR SEASTAR

TYPE: Six-seat amphibian kitbuilt.

PROGRAMME: Production prototype Seawind first flew 23 August 1982. Progressed through several versions, last of which built by Seawind International Inc of Canada.

Compared to Seawind, the completely new design SeaStar offers a changed aerofoil, single-piece Fowler flaps, gull-wing cabin doors and stainless steel landing gear. Adapted for optional pressurisation and alternative diesel and turboprop engines. Prototype (N601SS) exhibited, partly complete, at Sun 'n' Fun, April 2000; first flown 12 November 2000; by April 2002, 111 hours had been flown and certification programme begun with certification expected 2005–06; however, by April 2004 certification was no longer planned. By this date, prototype reported to have flown 601 hours.

CURRENT VERSIONS: **SeaStar (initially marketed as SeaStar 7T and later SeaStar 200):** piston-engined version, *as described*.

SeaStar 300 (initially marketed as SeaStar 7P): Intended pressurised version, envisaged also for law enforcement duties. Will have two pop-in doors and flight controls for pilot only. Prototype due for construction 'in near future'.

SeaStar 400: Proposed paramilitary version with four underwing hardpoints.

SeaStar 600: Proposed medevac version, capable of taking one stretcher plus attendant.

COSTS: Basic kit, less engine and avionics, piston engined US$119,900, turbine engined US$129,900; estimated cost by completion US$250,000 to US$270,000. Factory-built, M 601 turboprop-powered law enforcement version, but without avionics, US$590,000 (2004).

CUSTOMERS: Several kits under construction by end 2001; company assisting with completion of first ten; over 30 deposits taken by April 2002; however, by October 2004 none appeared to be complete.

DESIGN FEATURES: Unconventional configuration; shoulder wing with stabilising floats at tips; long-chord fin mounts engine and tailplane at mid-position, shielding propeller from water spray. Optional longer-span wing for high-altitude operations.

FLYING CONTROLS: Conventional and manual. Full-span elevator with two flight-adjustable tabs; ground-adjustable tab on rudder; flight-adjustable tab in port aileron. Electrically operated two-section slotted flaps, max deflection 40°.

STRUCTURE: Carbon fibre throughout; stainless steel landing gear. Wing aerofoil GA37A315.

POWER PLANT: *Piston version:* One 261 kW (350 hp) Textron Lycoming TIO-540-J2B flat-six driving a four-blade, reversible pitch MT propeller or three-blade Avia propeller, with standard fuel capacity of 379 litres (100 US gallons; 83.3 Imp gallons). Auxiliary fuel tank optional, capacity 151 litres (40.0 US gallons; 33.3 Imp gallons).

Turbine version: One 490 kW (657 shp) derated Walter M 601D turboprop driving a three-blade, reversible pitch Avia propeller. Standard fuel capacity 530 litres (140 US gallons; 117 Imp gallons); auxiliary fuel tank capacity 227 litres (60.0 US gallons; 50.0 Imp gallons).

LANDING GEAR: Tricycle type; electrically retractable; spring steel legs. Mainwheels are 7.00–6, have hydraulic brakes and retract upwards into wing; steerable 6.00–6 nosewheel retracts forwards.

ACCOMMODATION: Two front seats and bench for three passengers in rear. Four centreline-hinged, upward-opening doors in piston-engined version.

ARMAMENT: Military version has provision for total of four underwing pylons.

SeaStar SeaStar multipurpose amphibian *(Paul Jackson)* ***NEW***/1047387

DIMENSIONS, EXTERNAL:
Wing span: standard ..... 11.58 m (38 ft 0 in)
optional ..... 12.80 m (42 ft 0 in)
Length overall ..... 9.19 m (30 ft 2 in)
Height overall ..... 3.35 m (11 ft 0 in)
Baggage door: Width ..... 0.76 m (2 ft 6 in)
Height ..... 0.69 m (2 ft 3 in)
DIMENSIONS, INTERNAL:
Cabin: Length ..... 3.25 m (10 ft 8 in)
Width: at front ..... 1.35 m (4 ft 5 in)
at rear ..... 1.52 m (5 ft 0 in)
Height ..... 1.08 m (3 ft 6½ in)
Baggage compartment: Length ..... 1.83 m (6 ft 0 in)
Volume ..... 0.76 m^3 (27.0 cu ft)
AREAS:
Wings, gross ..... 15.89 m^2 (171.0 sq ft)
WEIGHT AND LOADINGS:
Weight empty ..... 1,134 kg (2,500 lb)
Max T-O weight: piston ..... 1,814 kg (4,000 lb)
turboprop ..... 1,886 kg (4,600 lb)
Max wing loading: piston ..... 114.2 kg/m^2 (23.39 lb/sq ft)
turboprop ..... 131.3 kg/m^2 (26.90 lb/sq ft)
Max power loading: Lycoming ..... 6.96 kg/kW (11.43 lb/hp)
Walter ..... 4.26 kg/kW (7.00 lb/shp)
PERFORMANCE:
Max level speed: Lycoming ..... 181 kt (335 km/h; 208 mph)
Walter ..... 217 kt (402 km/h; 250 mph)
Max cruising speed:
Lycoming at 75% power:
at FL100 ..... 175 kt (323 km/h; 201 mph)
at FL200 ..... 185 kt (343 km/h; 213 mph)
Walter at 90% power:
at FL100 ..... 216 kt (400 km/h; 249 mph)
at FL200 ..... 237 kt (439 km/h; 273 mph)
Stalling speed ..... 52 kt (97 km/h; 60 mph)
Service ceiling ..... 7,010 m (23,000 ft)
Max rate of climb at S/L: Lycoming ..... 427 m (1,400 ft)/min
Walter ..... 610 m (2,000 ft)/min
T-O run on land: Lycoming ..... 274 m (900 ft)
Walter ..... 235 m (770 ft)
Max range, no reserves:
Lycoming ..... 1,215 n miles (2,251 km; 1,399 miles)
Walter ..... 1,203 n miles (2,228 km; 1,385 miles)

UPDATED

SEAWIND

SEAWIND INC

Box 1041, Kimberton, Pennsylvania 19442-1041
Tel: (+1 610) 917 11 20
Fax: (+1 610) 933 33 35
e-mail: seawind@seawind.net
Web: http://www.seawind.biz
PRESIDENT: Richard Silva
SALES DIRECTOR: Vincent A Parisi Jr

Seawind has recently announced that it is pursuing certification for its Seawind 300C amphibian, a derivative of the earlier Seawind 2000 and 3000. Production will be in a new 7,620 m^2 (82,000 sq ft) plant at St Jean-sur-Richelieu Canada.

In third quarter of 2004, Seawind announced an eight-month delay in operations because of a 20 per cent shortfall in funding.

UPDATED

Trials vehicle Seawind 300C, N46SW, at Sun 'n' Fun in April 2004 *(Paul Jackson)* ***NEW***/1047432

SEAWIND SEAWIND 300C

TYPE: Five-seat amphibian.

PROGRAMME: Developed from Seawind 2000 through Seawind 3000, which first flew 23 August 1982; both versions sold as kits; now discontinued.

Factory-constructed Seawind 300C is being prepared for certification in late 2004 by FAA and Transport Canada, with deliveries to start in mid-2005. Trials aircraft N46SW had flown 1,640 hours by July 2003. Certification flying by two preproduction aircraft, first having been due to fly in March 2004 but failed to do so.

CUSTOMERS: Total of 150 kits sold and 63 flying by June 2004; orders currently being taken for certified examples.

COSTS: US$294,700 (2004).

DESIGN FEATURES: Intended to be first-ever composites amphibian to receive certification.

FLYING CONTROLS: Conventional and manual. Slotted flaps and flaperons.

STRUCTURE: Entire aircraft constructed from glass fibre, vinylester resin and foam composites.

LANDING GEAR: Tricycle type; retractable. Trailing-link-main gear, nosewheel retracts rearwards. Single-step hull with stabilising floats in downturned wingtips. Honda outboard motor mounted internally for manoeuvring on water.

POWER PLANT: One 224 kW (300 hp) Textron Lycoming IO-540-K1H5 flat-six, driving Hartzell Scimitar three-blade, constant-speed propeller. Fuel capacity 280 litres (74.0 US gallons; 61.6 Imp gallons) in main fuel tanks with option to increase to 416 litres (110 US gallons; 91.6 Imp gallons).

ACCOMMODATION: Pilot and three or four passengers in enclosed cabin. Front bucket seats move forward and aft; rear seats hinge forward for access to cabin baggage area and are easily removable. Retractable cargo net.

AVIONICS: Three packages available, based around UPS products. Optional IFR layouts include full 'glass cockpit' from Avidyne systems with full moving map and EFIS displays.
Comms: UPS Apollo audio intercom, GPS/com and Mode C transponder.
Radar: Goodrich Stormscope.
Flight: Optional Century Trident autopilot.

EQUIPMENT: Automatic bilge pump.

DIMENSIONS, EXTERNAL:
Wing span ..... 10.67 m (35 ft 0 in)
Wing aspect ratio 7.4
Length overall ..... 8.28 m (27 ft 2 in)
Height overall: ..... 3.10 m (10 ft 2 in)
Propeller diameter ..... 1.93 m (6 ft 4 in)
Baggage door: Height ..... 0.48 m (1 ft 7 in)
Width ..... 1.02 m (3 ft 4 in)
DIMENSIONS, INTERNAL:
Cabin: Length ..... 2.67 m (8 ft 9 in)
Max width ..... 1.37 m (4 ft 6 in)
Max height ..... 1.09 m (3 ft 7 in)
Baggage volume: cabin ..... 0.28 m^3 (10.0 cu ft)
nose ..... 0.08 m^3(3.0 cu ft)
Freight hold: Length ..... 1.80 m (5 ft 11 in)
Volume ..... 0.85 m^3(30.0 cu ft)
AREAS:
Wings, gross ..... 15.42 m^2 (166.0 sq ft)
WEIGHTS AND LOADINGS:
Weight empty ..... 1,021 kg (2,250 lb)
Max T-O weight ..... 1,542 kg (3,400 lb)
Max wing loading ..... 100.0 kg/m^2 (20.48 lb/sq ft)
Max power loading ..... 6.90 kg/kW (11.33 lb/hp)
PERFORMANCE:
Max operating speed ..... 173 kt (321 km/h; 200 mph)
Max cruising speed at 75% power ..... 166 kt (307 km/h; 191 mph)
Normal cruising speed at 65% power ..... 156 kt (290 km/h; 180 mph)
Econ cruising speed at 55% power ..... 147 kt (272 km/h; 169 mph)
Stalling speed, power off:
flaps up ..... 63 kt (117 km/h; 73 mph)
flaps and wheels down ..... 50 kt (93 km/h; 58 mph)
Max rate of climb at S/L ..... 381 m (1,250 ft)/min
Service ceiling ..... 6,100 m (20,000 ft)
T-O run: land ..... 265 m (870 ft)
water ..... 335 m (1,100 ft)
T-O to 15 m (50 ft): land ..... 358 m (1,175 ft)
water ..... 442 m (1,450 ft)
Landing from 15 m (50 ft): land ..... 396 m (1,300 ft)
water ..... 351 m (1,150 ft)
Landing run: land ..... 235 m (770 ft)
water ..... 189 m (620 ft)
Range, no reserves:
with normal fuel ..... 856 n miles (1,585 km; 985 miles)
with max fuel ..... 1,270 n miles (2,352 km; 1,461 miles)

UPDATED

SEQUOIA

SEQUOIA AIRCRAFT CORPORATION

2000 Tomlynn Street, PO Box 6861, Richmond, Virginia 23230
Tel: (+1 804) 353 17 13
Fax: (+1 804) 359 26 18
e-mail: seqair@aol.com
Web: http://www.SeqAir.com
PRESIDENT: Alfred P Scott

Sequoia has marketed the F.8L Falco since 1979.

VERIFIED

SEQUOIA FALCO F.8L

TYPE: Side-by-side lightplane/kitbuilt.

PROGRAMME: Sequoia Aircraft markets plans and kits to build improved version of Falco F.8L high-performance monoplane, designed in Italy by Ing Stelio Frati and first flown on 15 June 1955. Italian production totalled 76; kit version first flown 1974.

CUSTOMERS: Total of 78 Sequoia aircraft flown by August 2004, in Australia, Brazil, Canada, Chile, France, Germany, Italy, Netherlands, New Zealand, Norway, South Africa, UK and USA.

COSTS: Kit: US$96,000 (2001) including propeller but excluding engine. Plans: US$400. Extra costs to complete estimated by manufacturer as US$13,000.

DESIGN FEATURES: Streamlined low-wing, retractable-gear monoplane. Quoted build time 2,000 hours.

Sequoia F.8L two-seat kitbuilt *(Paul Jackson)* ***NEW***/1047128

NACA $64_2212.5$ wing section at root, NACA 64_2210 at tip. Wing dihedral 5°, washout 3°. Tailplane section NACA 65009.

FLYING CONTROLS: Conventional and manual.

STRUCTURE: Entire airframe has plywood-covered wood structure, with overall fabric covering and glass fibre wing fillets. Optional metal control surfaces.

LANDING GEAR: Tricycle type; retractable. Steerable nosewheel. Cleveland single calliper disc brakes. 5.00–5 mainwheels and 4.00–5 nosewheel.

POWER PLANT: One 119 kW (160 hp) Textron Lycoming IO-320-B1A is standard for kitbuilt aircraft, driving Hartzell HCC-ZYL-1BF/F7663-4 two-blade constant-speed metal propeller. Optional engines for which Sequoia offers installation kits are the 112 kW (150 hp) IO-320-A1A and 134 kW (180 hp) IO-360-B1E. Fuel capacity 151 litres (40.0 US gallons; 33.3 Imp gallons). Optional 11.4 litre (3.0 US gallon; 2.5 Imp gallon) header tank to permit inverted flight.

ACCOMMODATION: Two seats, side by side; child's seat or baggage area to rear. Dual controls.

DIMENSIONS, EXTERNAL:

Wing span	8.00 m (26 ft 3 in)
Wing chord: at root	1.65 m (5 ft 5 in)
at tip	0.83 m (2 ft 8½ in)
Wing aspect ratio	6.4
Length overall	6.63 m (21 ft 9 in)
Height overall	2.29 m (7 ft 6 in)
Tailplane span	2.71 m (8 ft 10½ in)
Wheel track	2.08 m (6 ft 10 in)
Wheelbase	1.50 m (4 ft 11 in)
Propeller diameter	1.83 m (6 ft 0 in)

DIMENSIONS, INTERNAL:

Cabin max width	1.07 m (3 ft 6 in)

AREAS:

Wings, gross	10.00 m^2 (107.6 sq ft)
Rudder	0.48 m^2 (5.20 sq ft)
Tailplane	2.17 m^2 (23.40 sq ft)
Elevator	0.83 m^2 (9.00 sq ft)

WEIGHTS AND LOADINGS (119 kW; 160 hp engine):

Weight empty	550 kg (1,212 lb)
Payload with max fuel	194 kg (428 lb)
Baggage capacity	41 kg (90 lb)
Max aerobatic weight	748 kg (1,650 lb)
Max T-O weight	852 kg (1,880 lb)
Max wing loading	85.3 kg/m^2 (17.47 lb/sq ft)
Max power loading	7.15 kg/kW (11.75 lb/hp)

PERFORMANCE (119 kW; 160 hp engine):

Never-exceed speed (VNE)	208 kt (386 km/h; 240 mph)
Max level speed at S/L	184 kt (341 km/h; 212 mph)
Cruising speed at 75% power	165 kt (306 km/h; 190 mph)
Stalling speed: clean	66 kt (121 km/h; 75 mph)
flaps and wheels down	54 kt (100 km/h; 62 mph)
Max rate of climb at S/L	347 m (1,140 ft)/min
Service ceiling	5,790 m (19,000 ft)
T-O run	174 m (570 ft)
T-O to 15 m (50 ft)	351 m (1,150 ft)
Landing from 15 m (50 ft)	351 m (1,150 ft)
Landing run	229 m (750 ft)
Range at econ cruising speed	869 n miles (1,609 km; 1,000 miles)
Endurance	4 h 24 min
g limits	+6/–3

UPDATED

SHERPA

SHERPA AIRCRAFT MANUFACTURING CO

34100 Skyway Drive, Scappoose, Oregon 97056
Tel: (+1 503) 543 40 04
Fax: (+1 503) 543 40 05
e-mail: sherpaworldwide@att.net
Web: http://www.sherpaaircraft.com
DIRECTORS:
Byron Root
Glen Gordon

Sherpa Aircraft is in the process of certifying the Sherpa 300 derivative of its earlier six-seat aircraft version. In 2001, the company announced that it had entered into a joint venture agreement with Khrunichev (of Russia) which would result in parts being manufactured in Moscow and Sherpa marketing the T-411 Aist and Aviotechnica SL-90. Sherpa is expanding its manufacturing facility at Scappoose, Oregon, at present 1,400 m^2 (15,000 sq ft), to accommodate increased production.

VERIFIED

SHERPA SHERPA

TYPE: Light utility transport.

PROGRAMME: Development of the Sherpa series began in 1991, with first prototype (N1415B) making debut at Oshkosh in July 1994. Second prototype (N711SA) first flown 1995; both prototypes five-seat versions. Third prototype (N712SA), shown at AirVenture 2001, is the eight-seat Sherpa C-400 version. All versions use same airframe, except for modified wings on K-200, K-300 and C-400. By August 2003, construction of the first K-200 was underway.

In August 2004, the first two aircraft from Russia were delivered to Sherpa for assembly.

CURRENT VERSIONS: **Sherpa:** Original six-place version; kit production under consideration.

Sherpa K-200: Five/six-seat version; under development. Reduced wing span and area.

Second prototype Sherpa N711SA, basis of K-200 version *(Paul Jackson)* ***NEW***/1047415

Sherpa K-300: Six-seat version; main differences from K-200 are in cabin and CG envelope.

Sherpa C-400: Eight-seat version*; as described unless otherwise stated.*

Sherpa C-500T: Proposed certified eight/10-seat version.

Sherpa K-500T: Turbine-powered eight-seater; same airframe as C-400; wing displayed at AirVenture '04, at which time it was hoped that the first K-500T would be flying by April 2005.

Sherpa C-700T: Proposed 10-seat version.

CUSTOMERS: Three flying by end 2001.

COSTS: Sherpa K-200: US$119,000; Sherpa K-300: US$139,000, Sherpa K-500T US$225,000 Sherpa C-400 US$425,000, Sherpa C-500T US$750,000; C-700T US$895,000 (2004).

DESIGN FEATURES: Optimised for use in inhospitable terrain, with STOL performance, strengthened landing gear, option for large tyres and easy maintenance access.

Constant-chord, high-mounted wing with V bracing struts and auxiliary struts; wire-braced empennage. Tail surfaces have rounded tips and horn balances. Fully glazed flight deck door.

FLYING CONTROLS: Conventional and manual. Two-track Fowler flaps increase wing area by 15 per cent when fully deployed at 40°. Inboard trim tab each side above flap. Tabs on rudder and starboard elevator adjustable in flight, as is tailplane incidence.

STRUCTURE: Steel 4130 fuselage frame has fabric covering aft and metal sheet covering forward of cabin. Fabric-covered 4130 steel tube wing has metal leading-edges and control surfaces. Fabric-covered steel tube empennage. Graphite composites nose cowlings. Single-piece 6 mm (¼ in) thick wrap-round windscreen.

LANDING GEAR: Tailwheel type; fixed. Standard tyre size 29×11.0–10; optional 1.07 m (42 in) Tundra tyres can be fitted. Cleveland hydraulic brakes. Optional floats, wheeled floats and skis; conversion time quoted as 90 minutes.

POWER PLANT: *Sherpa K-200:* One 265 kW (355 hp) VOKBM M-14X nine-cylinder radial. Fuel capacity 530 litres (140 US gallons; 117 Imp gallons).

Sherpa K-300T: , C-500T: and C-700T: One 560 kW (751 shp) Walter M 601 turboprop driving a three-blade propeller. Fuel capacity 852 litres (225 US gallons; 187 Imp gallons).

Sherpa K-300 : and C-400: One 347 kW (465 hp) Textron Lycoming TIO-720-X152 flat-eight driving a three-blade Hartzell F8483 propeller. Normal fuel capacity 568 litres (150 US gallons; 125 Imp gallons) in six individually removable tanks.

ACCOMMODATION: Pilot and up to seven passengers on individual seats with central aisle; seats can be removed individually to allow carriage of freight (up to 10 empty 208 litre (55.0 US gallon; 45.8 Imp gallon) drums) or two casualties plus two attendants. Double door on starboard side of fuselage for cabin; individual doors for crew positions. Additional baggage door on starboard side aft of cabin area.

SYSTEMS: 12 V electrical system.

AVIONICS: Honeywell avionics suite plus S-Tek autopilot.

DIMENSIONS, EXTERNAL (all, except where indicated):

Wing span: C-500T, K-500T	13.72 m (45 ft 0 in)
K-200, K-300, C-400	13.41 m (44 ft 0 in)
Wing chord, constant	1.85 m (6 ft 1 in)
Wing aspect ratio	7.3
Length overall	9.96 m (32 ft 8 in)
Height overall	2.87 m (9 ft 5 in)
Tailplane span	5.03 m (16 ft 6 in)
Wheel track	2.69 m (8 ft 10 in)
Propeller diameter	2.54 m (8 ft 4 in)
Propeller ground clearance	0.69 m (2 ft 3 in)
Cabin door: Height	1.12 m (3 ft 8 in)
Width	1.55 m (5 ft 1 in)
Baggage door: Height	0.51 m (1 ft 8 in)
Width	0.76 m (2 ft 6 in)

DIMENSIONS, INTERNAL (all):

Cabin volume	4.42 m^3 (156 cu ft)
Baggage hold	0.36 m^3 (12.7 cu ft)

AREAS (all, except where indicated):
- Wings, gross: C-500T, K-500T ... 25.55 m^2 (275.0 sq ft)
  - K-200, K-300, C-400 ... 24.71 m^2 (266.0 sq ft)
- Ailerons (total) ... 1.83 m^2 (19.7 sq ft)
- Trailing-edge flaps (total) ... 4.51 m^2 (48.5 sq ft)
- Spoilerons (total) ... 0.39 m^2 (4.2 sq ft)
- Fin ... 1.24 m^2 (13.4 sq ft)
- Rudder, incl tab ... 1.37 m^2 (14.7 sq ft)
- Tailplane ... 2.81 m^2 (30.3 sq ft)
- Elevators, incl tab ... 2.39 m^2 (25.7 sq ft)

WEIGHTS AND LOADINGS:
- Weight empty: K-200 ... 1,252 kg (2,760 lb)
  - K-300 landplane ... 1,390 kg (3,065 lb)
  - K-300T ... 1,393 kg (3,070 lb)
  - C-400 ... 1,445 kg (3,185 lb)
  - C-500T ... 1,427 kg (3,146 lb)
- Max T-O weight: K-200 ... 1,846 kg (4,070 lb)
  - K-300: landplane ... 2,268 kg (5,000 lb)
    - floatplane ... 2,404 kg (5,300 lb)
  - K-300T ... 2,494 kg (5,500 lb)
  - C-500T ... 2,844 kg (6,270 lb)
- Max wing loading: K-200 ... 74.7 kg/m^2 (15.30 lb/sq ft)
  - K-300: landplane ... 87.8 kg/m^2 (17.99 lb/sq ft)
    - floatplane ... 93.1 kg/m^2 (19.06 lb/sq ft)
  - K-300T ... 96.6 kg/m^2 (19.78 lb/sq ft)
  - C-500T ... 110.1 kg/m^2 (22.55 lb/sq ft)
- Max power loading: K-200 ... 6.98 kg/kW (11.46 lb/hp)
  - K-300: landplane ... 6.54 kg/kW (10.75 lb/hp)
    - floatplane ... 6.94 kg/kW (11.40 lb/hp)
  - K-300T ... 4.46 kg/kW (7.33 lb/hp)
  - C-500T ... 5.09 kg/kW (8.36 lb/hp)

PERFORMANCE:
- Max operating speed:
  - K-200 ... 123 kt (229 km/h; 142 mph)
  - K-300T ... 182 kt (338 km/h; 210 mph)
- Max cruising speed: K-300 ... 132 kt (245 km/h; 152 mph)
  - K-300T ... 162 kt (301 km/h; 187 mph)
  - C-400 ... 140 kt (259 km/h; 161 mph)
- Normal cruising speed at 75% power: K-200 ... 105 kt (195 km/h; 121 mph)
  - C-400 ... 144 kt (267 km/h; 166 mph)
  - C-500T ... 162 kt (301 km/h; 187 mph)
- Econ cruising speed: K-300 ... 120 kt (222 km/h; 138 mph)
  - K-300T ... 146 kt (270 km/h; 168 mph)
  - C-400 ... 129 kt (238 km/h; 148 mph)
- Stalling speed:
  - flaps up: K-200 ... 51 kt (94 km/h; 58 mph)
    - K-300, K-300T, C-400 ... 54 kt (100 km/h; 62 mph)
  - flaps down: K-200 ... 35 kt (65 km/h; 40 mph)
    - K-300 ... 40 kt (73 km/h; 45 mph)
    - K-300T, C-400 ... 42 kt (78 km/h; 48 mph)
    - C-500T ... 46 kt (84 km/h; 52 mph)
- Max rate of climb at S/L:
  - K-200, C-400 ... 300 m (985 ft)/min
  - K-300 ... 274 m (900 ft)/min
  - K-300T ... 509 m (1,670 ft)/min
- Service ceiling: K-200 ... 5,030 m (16,500 ft)
  - K-300 ... 5,640 m (18,500 ft)
  - K-300T, C-400 ... 7,620 m (25,000 ft)
- T-O run: K-200 ... 119 m (390 ft)
  - K-300 ... 159 m (522 ft)
  - K-300T ... 108 m (355 ft)
  - C-400 ... 151 m (494 ft)
- Landing run: K-200 ... 104 m (340 ft)
  - K-300 ... 122 m (400 ft)
  - K-300T ... 81 m (265 ft)
  - C-400 ... 86 m (283 ft)
- Range at max cruising speed:
  - K-300 ... 938 n miles (1,738 km; 1,080 miles)
  - K-300T ... 1,147 n miles (2,124 km; 1,320 miles)
  - C-400 ... 833 n miles (1,543 km; 959 miles)
- Range at econ cruising speed:
  - K-300T ... 1,364 n miles (2,526 km; 1,570 miles)
  - C-400 ... 943 n miles (1,747 km; 1,086 miles)

UPDATED

SIKORSKY

SIKORSKY AIRCRAFT (SUBSIDIARY OF UNITED TECHNOLOGIES CORPORATION)

6900 Main Street, Stratford, Connecticut 06615-9129
Tel: (+1 203) 386 40 00
Fax: (+1 203) 386 73 00
Web: http://www.sikorsky.com

OTHER WORKS: Troy, Alabama; South Avenue, Bridgeport, Connecticut; Shelton, Connecticut; West Haven, Connecticut; Development Flight Test Center, West Palm Beach, Florida

PRESIDENT: Stephen Finger
SENIOR VICE-PRESIDENT, PRODUCTION OPERATIONS: Robert R E Moore
SENIOR VICE-PRESIDENT, GOVERNMENT AND ADVANCED DESIGN PROGRAMS: Paul E Martin
SENIOR VICE-PRESIDENT, MARKETING AND COMMERCIAL PROGRAMS: Jeffrey P Pino
VICE-PRESIDENT, DEVELOPMENT: David Adler
VICE-PRESIDENT, PRODUCTION ENGINEERING: Mark Miller
VICE-PRESIDENT, FINANCE AND CFO: Peter Longo
VICE-PRESIDENT, GOVERNMENT BUSINESS DEVELOPMENT: Joseph Haddock
MANAGER OF PUBLIC RELATIONS: William S Tuttle

Founded as Sikorsky Aero Engineering Corporation in 1923 by late Igor I Sikorsky; has been division of United Technologies Corporation since 1929, but established as a subsidiary with effect 1 January 1995; began helicopter production in 1940s.

Headquarters and main plant at Stratford, Connecticut; also has manufacturing facilities elsewhere in Connecticut; other smaller facilities in Alabama and Florida. Workforce in October 2002 was about 8,000 worldwide. Main current programmes include UH-60 Black Hawk and derivatives, S-76 series and, in co-operation with international partners, development of S-92 Helibus. In 2003, Sikorsky delivered total of 69 helicopters, comprising 28 Black Hawks (20 UH-60L, two remanufactured UH-60M and six S-70As), 18 MH-60S KnightHawks and 23 S-76C+s.

Sikorsky licensees include Agusta of Italy, Eurocopter of France and Germany, Korean Air Lines of South Korea, Mitsubishi of Japan, Pratt & Whitney Canada Ltd and UK element of AgustaWestland. Sikorsky and Embraer of Brazil signed agreement in mid-1983 to transfer technology covering design and manufacture of composites components. Sikorsky and CASA of Spain signed MoU in June 1984 covering long-term helicopter industrial co-operation programme; CASA builds tail rotor pylon, tailcone and stabiliser components for H-60 and S-70, with first CASA S-70 components delivered to Sikorsky January 1986. Most recent overseas venture, in collaboration with the Alpata Group of Turkey, concerns creation of Alp Aviation, which manufactures high-technology, precision-machined aerospace and defence components in 6,500 m^2 (70,000 sq ft) facility in Eskisehir.

In October 1998, Sikorsky purchased Helicopter Support Inc, so as to offer enhanced support to helicopter operators on worldwide basis. In latter half of 1998, Sikorsky also secured US$150 million deal with US Navy covering contractor maintenance of jet trainer aircraft as well as HH-1N Iroquois and UH-3H Sea King helicopters. Overall responsibility for latter allocated to Sikorsky Support Services Inc, with the work undertaken at Meridian, Mississippi; Pensacola, Florida; and Corpus Christi, Texas. Joint venture Shanghai Sikorsky Aircraft Company markets Schweizer light helicopters in China and is also planning to undertake indigenous production of light helicopters.

UPDATED

SIKORSKY CH-53X

TYPE: Medium lift helicopter.

PROGRAMME: Variant of proven CH-53E Super Stallion (refer to *Jane's Aircraft Upgrades*) to meet US Marine Corps' need for up to 154 new-build helicopters for delivery between 2009 and 2020. Operational requirements document prepared by Marine Corps by first quarter of 2004 and submitted to US Department of Defense for final authorisation, which should take place before end of 2004, allowing contract to be awarded to Sikorsky for system development and demonstration (SDD) phase.

Proposed CH-53X will feature new powerplant in the 4,474 kW (6,000 shp) class, with Marine Corps reportedly favouring Rolls-Royce T406 turboshaft on grounds of commonality with Bell Boeing MV-22 Osprey tiltrotor; other candidates include Honeywell's T55-715 and a growth version of the General Electric T64-GE-416 that currently powers the CH-53E. Further improvements will include low maintenance elastomeric rotor head, high efficiency rotor blades with anhedral tips, 'glass cockpit' with multifunction displays, improved internal and external cargo handling systems and enhanced survivability features. In Marine Corps service, it will almost certainly adopt the designation CH-53F.

Sikorsky CH-53 Super Stallion from earlier production *(US Navy)* ***NEW***/0106783

UPDATED

SIKORSKY S-70A

US Army designations: UH-60A, UH-60C, UH-60L, UH-60M and UH-60Q Black Hawk, AH-60L, EH-60A, HH-60L, HH-60M, MH-60K, MH-60L and MH-60M
US Air Force designation: HH-60G Pave Hawk
US Navy designation: MH-60S KnightHawk
US Marine Corps designation: VH-60N
Israel Defence Force name: Yanshuf (Owl)
Japan Self-Defence Forces designations: UH-60J and UH-60JA
South Korean Army designation: UH-60P
Turkish Armed Forces name: Yarasa (Bat)

TYPE: Multirole medium helicopter.

DEVELOPMENT MILESTONES

Request for proposals	Jan 72
Requirement issued	Aug 72
Selected by US Army	23 Dec 76
First flight	17 Oct 74
First flight, production	17 Oct 78
First delivery (US Army)	31 Oct 78
Entered service (101st AD)	Jun 79

Sikorsky UH-60L Black Hawks of the US Army ***NEW***/0589360

PROGRAMME: UH-60A declared winner of US Army Utility Tactical Transport Aircraft System (UTTAS) competition against Boeing Vertol YUH-61A 23 December 1976; first flight of first of three YUH-60A competitive prototypes 17 October 1974; 2,000th H-60 delivered May 1994; 2,500th followed at end of 2001.

AgustaWestland in UK, Korean Air Lines of South Korea and Mitsubishi of Japan have licences to build the S-70 series, although UK programme not activated.

CURRENT VERSIONS: **UH-60A Black Hawk:** Initial production version, designed to carry crew of three and 11 troops; also can be used without modification for medevac, reconnaissance, command and control, and troop supply; cargo hook capacity 4,082 kg (9,000 lb); one UH-60A can be carried in C-130, two in C-141 and six in C-5.

Medevac kits delivered from 1981; missile qualification completed June 1987, with day and night firing of Hellfire in various flight conditions; airborne target handover system (ATHS) qualified; cockpit lighting suitable for night vision goggles fitted to production UH-60s since November 1985 and retrofitted to those built earlier; US Army began testing Alliant Techsystems Volcano mine dispensing system July 1987; modular Volcano container is disposable and dispenses 960 Gator anti-tank and anti-personnel mines; deployment of 2,940 kg (6,482 lb) system began in FY95; usage monitor to measure certain rotor loads installed in 30 UH-60As; wire strike protection added to UH-60s and EH-60s during 1987; accident data recorders also fitted. Total 1,049 built for US Army (including 66 conversions to EH-60A) before production of UH-60L began in 1989. UH-60A 85-24441, delivered in 1985, became first Black Hawk to complete 10,000 flight hours, with milestone passed in December 2002. Of original UH-60A, majority will be upgraded to UH-60M, but 193 to be renovated in near term, to sustain non-remanufactured fleet until UH-60M programme is completed.

Detailed description applies to UH-60A/L except where indicated.

Enhanced Black Hawk: Incorporates active and passive self-defence systems, retrofitted by Corpus Christi Army Depot, Texas, to new-build UH-60A/Ls; first 15 delivered to US Army in South Korea November 1989. Equipment includes BAE Systems AN/ARN-148 Omega navigation receiver, Motorola AN/LST-5B satellite UHF communications transceiver, Honeywell AN/ARC-199 HF-SSB, and AEL AN/APR-44(V)3 specific threat RWR complementing existing AN/APR-39 general threat RWR; M134 Minigun can be fitted on each of two pintle mounts, replacing M60 machine gun.

JUH-60A: At least seven used temporarily for trials.

GUH-60A: At least 20 grounded airframes in use as technical training aids.

HH-60D Night Hawk: One prototype (82-23718) completed for abandoned USAF combat rescue variant; subsequently became an HH-60G.

EH-60A: (Designation **EH-60C** reserved, but not adopted by US Army.) Prototype YEH-60A (79-23301) ordered in October 1980 to carry 816 kg (1,800 lb) Quick Fix IIB battlefield ECM detection and jamming system. TRW Electronic Systems Laboratories was prime contractor for AN/ALQ-151(V)2 ECM kit, with installation by Tracor Aerospace; four dipole antennas on fuselage and deployable whip antenna; hover IR suppressor system (HIRSS) standard. YEH-60A first flight 24 September 1981; order for Tracor Aerospace to modify 40 UH-60As to EH-60A standard under US$51 million contract placed October 1984; first delivery July 1987 as part of US Army Special Electronics Mission Aircraft (SEMA) programme; 66 funded by FY87 excluding prototype; programme completed 1989. Under current Army planning, EH-60A to be phased out in 2005.

Intercepts/locates AM, FM, CW and SSB radio emissions from upper HF to mid-VHF ranges over bandwidths of 8, 30 or 50 kHz; jams VHF communications. Protective systems of UH-60A/L (M-130 chaff/flare and AN/ALQ-144 IR jammer) augmented by Sanders AN/ALQ-156(V) missile approach warning system, ITT AN/ALQ-136(V) pulsed transmitter, Northrop Grumman AN/ALQ-162(V) CW transmitter and Litton AN/APR-39(V) RWR.

AN/ALQ-151(V)3 Advanced Quick Fix mission system tasked with providing ESM capability to forward ground units at division level; at least six EH-60As (84-24027, 85-24468, 85-24473, 87-24657, 87-24662 and 87-24670) adapted to **EH-60L** configuration. As fielded for Task Force XXI trials (1997), the EH-60L utilised the following equipment: sensor subsystem comprised Sanders TACJAM-A ESM for detection, direction-finding, identification and tracking of communications signals in HF, UHF, VHF and SHF frequency bands; Lockheed Martin Federal Systems communications high-accuracy locating system — exploitable (CHALS-X) for direction-finding in HF, UHF and SHF frequency bands; and signal location subsystem (SILO) for direction-finding in VHF frequency band. Navigation and timing subsystem featured INS, GPS and control display unit in cockpit. Mission control and interface subsystem comprised control, navigation, workstation and graphics processors, mass storage unit, keyboards with trackball and 483 × 483 mm (19 × 19 in) colour monitors. Communications subsystem comprised modified PRC-118 wideband control datalink, two AN/ARC-201A SINCGARS radios, AN/ARC-164 tasking and reporting datalink, intercom and AN/UYH-15 digital temporary storage recorder/reproducer set. Antenna subsystem. Airborne survivability subsystem comprised AN/ALQ-144(V)1 IR jammer, AN/ALQ-156(V)2 missile approach warning system (MAWS), AN/ALQ-162(V)2 CW transmitter and AN/APR-39(V)2 RWR.

Improvements to airframe include installation of UH-60L engines and gearbox for increase in maximum weight from 7,845 kg (17,295 lb) to 10,206 kg (22,500 lb).

MH-60A: About 30 modified for Army 160th Special Operations Aviation Regiment (SOAR); fitted with Raytheon Systems AN/AAQ-16 FLIR, BAE Systems AN/ARN-148 Omega/VLF navigation, M-130 chaff/flare dispensers, AN/ALQ-144 IR jammer, night vision equipment, multifunction displays, auxiliary fuel tanks and door-mounted Minigun; fitted with -701C engines; interim equipment, pending MH-60K. Replaced by MH-60L in late 1990 and reverted to UH-60A configuration.

UH-60C: Designation for command and control (C2) version; was under development at the US Naval Research Laboratory on behalf of the US Army. This version subsequently abandoned but UH-60C project continues, with Raytheon as prime contractor for C2 system. This to be installed on UH-60L, with first three systems delivered in first half of 2003. First five modified helicopters will serve as prototypes to validate performance in advance of initial operational test and evaluation, which is currently scheduled for late 2004. Initial version to be known as **Block I**, with more sophisticated **Block II** expected to enter development in 2005; latter will be configured for UH-60M, which should facilitate system integration, while Block II also expected to use Joint Tactical Radio system. Contract worth US$110 million for development and production awarded in late 2001. Original plan was to obtain 207 helicopters, although this reduced to 121 by start of 2001. Rockwell Collins AN/ASC-15B/C consoles have been fitted to more than 50 US Army aircraft as interim **UH-60A(C)**. Communications suite to be finalised, but expected to include Have Quick II, SINCGARS/SIP and JTIDS; other capabilities also to be incorporated, including FLIR, NVG compatibility, digital map display, mission planner facility, improved ECCM and storage space for ground power generators and antennas. Five workstations will be provided for system operators.

HH-60G Pave Hawk: Replaced US Air Force HH-60D Night Hawk rescue helicopters, which were not funded; converted from UH-60A/L, including 10 originally delivered to 55th Aerospace Rescue and Recovery Squadron (later Special Operations Squadron) at Eglin AFB, Florida, in 1982–83, initially remaining as UH-60As; all progressively fitted by Sikorsky Support Services at Troy, Alabama, with aerial refuelling probe, 443 litre (117 US gallon; 97.5 Imp gallon) internal auxiliary fuel tank and fuel management panel; then to Pensacola NAD for mission avionics and modified instrument panel; some retrofitted with replacement internal tank of 700 litres (185 US gallons; 154 Imp gallons) capacity; -701C engines fitted to 10 special operations examples and later production aircraft (FY89 onwards); recent retrofit programme, begun in November 1999, entailed installation of -701C engine on older HH-60Gs.

Further procurement began with batch of nine in FY87, followed by purchases of 16, 18, 22, 15 and 13 in FY88-92; eight more funded in FY97 and delivered in 1998. All designated MH-60G until 1 January 1992, when 82 in combat rescue role redesignated HH-60G, with balance of 16 remaining as MH-60G for special operations units; by fourth quarter of 1998, only nine still in MH-60G configuration and at start of 2000, all were using HH-60G designation. All have rescue hoist, Doppler/INS, electronic map display, Tacan, Honeywell

AN/APN-239 lightweight weather/ground-mapping radar, secure HF, and satcom; MH-60G had ESSS (see Armament paragraph) for weapons and additional fuel carrying capability, plus door-mounted 0.50 in machine guns and Raytheon AN/AAQ-16 Pave Low III FLIR.

Upgraded version of HH-60G, known as **Block 152**, made debut at Stratford, Connecticut, on 29 April 1999, when first of 49 planned aircraft rolled out in Upgraded Communication, Navigation/Integrated Electronic Warfare (UCN/IEW) configuration; new features include enhanced com/nav system and EW suite integrated into MIL-STD-1553 databus to reduce crew workload. Contractor trials in May and June 1999, after which modified HH-60G (possibly 92-26460) delivered to Nellis AFB, Nevada, for operational test and evaluation with 422nd TES. Retrofit included installation of revised, externally mounted, armament system with 0.50 in machine guns, expendable chaff/flare defensive countermeasures and repositioned nose radar.

Further extensive upgrade under consideration by USAF for HH-60G fleet from about 2003, with various options being examined; these include remanufacture to new **Block 162** standard with 'glass cockpit', new defensive aids and other changes; and less ambitious structural life extension programme.

MH-60K: US Army special operations aircraft (SOA); prototype (89-26194) ordered in January 1988; first flight 10 August 1990. US Army funded two batches of 11 with options for another 38, which not taken up; first production aircraft (91-26368) completed, February 1992; trials at Patuxent River and Edwards AFB before intended first deliveries in June 1992 to 160th Special Operations Aviation Group (part of 160 SOA Regiment). Deliveries delayed by software problems with special operations equipment; first 10 accepted in 1992 in non-operational state; remaining 12 initially stored, then delivered with new software installed, October to December 1993, to permit start of training by 160 SOA Group, February 1994.

Features include provision for additional 3,141 litres (829 US gallons; 691 Imp gallons) of internal and external fuel (see Power Plant), plus flight refuelling capability, integrated avionics system with electronic displays, Raytheon AN/AAQ-16 FLIR and AN/APQ-174B terrain-following, ground-mapping and air-to-ground ranging radar, T700-GE-701C engines and uprated transmission, external hoist, wire-strike protection, rotor brake, tiedown points, folding tailplane, AFCS similar to that of SH-60B, strengthened pintle mounts for 0.50 in machine guns, provision for Stinger missiles, missile warning receiver, pulse radio frequency jammer, CW radio jammer, laser detector, chaff/flare dispensers, and IR jammer.

Expected to undergo service life extension programme and emerge as MH-60Ms.

SH-60B Seahawk: US Navy ASW/ASST helicopter, *described separately*.

SH-60F Seahawk: US Navy carrierborne inner-zone ASW helicopter to replace SH-3D Sea King. See Seahawk entry.

HH-60H and HH-60J Jayhawk: Search and rescue/special warfare helicopters; see Seahawk entry.

UH-60J: Designation of Japanese-built S-70A-12 for Air and Maritime Self-Defence Forces; procurement details under Mitsubishi in Japanese section.

UH-60JA: Japanese Ground Self-Defence Force version; procurement began in 1995. Manufactured by Mitsubishi.

UH-60L: Replaced UH-60A in production for US Army from October 1989 (aircraft 89-26179 onwards); prototype (84-23953) first flight 22 March 1988; two pre-series aircraft (89-26149 and 26154); first delivery 7 November 1989 to Texas ArNG. Powered by T700-GE-701C engines with uprated 2,535 kW (3,400 shp) transmission. More than 600 delivered to US Army, with production expected to continue until replaced by new-build UH-60M version. Current production aircraft fitted with hover IR suppression system (HIRSS) to cool exhaust in hover as well as forward flight; older UH-60s retrofitted. Composites wide-chord main rotor blades of improved design flight tested at West Palm Beach, beginning 8 December 1993; new blade 16 per cent wider than current titanium rotor and has anhedral tip angled down at 20°; testing revealed much lower vibration plus anticipated benefits in payload, speed and manoeuvrability; projected retrofit from 1997 did not occur, but new rotor will be fitted to forthcoming UH-60M. Underslung load capability increased to 4,082 kg (9,000 lb).

HH-60L: Alternative designation for new-build medical evacuation version based on UH-60L airframe but incorporating UH-60Q specialised mission equipment. Sikorsky awarded US$11 million contract on 22 February 2000 for design definition and conversion of four UH-60Ls to this standard. Initial delivery expected in 2000, but delayed until March 2001, when three helicopters assigned to 507th Medical Company (Air Ambulance) at Fort Hood, Texas. Procurement continues, with additional 18 aircraft ordered between FY00 and FY04.

MH-60L: Similar to MH-60A; approximately 40 helicopters modified for 160th SOAR, US Army; some helicopters upgraded to MH-60L 'Defensive Armed Penetrator' configuration in 1990 with FLIR, radar and standard UH-60 external stores support system; two Black Hawk companies of 160th SOAR each have MH-60K platoon and MH-60L platoon. Armament includes multiple 30 mm Chain Gun, racks of four Hellfires and 2.75 in rocket pods, 40 mm grenade launcher or trainable 7.62 mm Gatling guns. Expected to undergo service life extension programme and emerge as MH-60Ms.

HH-60M: Medical evacuation version based on upgraded UH-60M; as many as 357 could be acquired for service with first- and second-line medical evacuation units.

MH-60M: Special Forces derivative, intended to replace existing MH-60 fleet with 160th SOAR. Few details available, but majority of existing MH-60Ks and MH-60Ls will be subjected to service life extension programme and emerge as MH-60Ms to common configuration. Development contract due to be awarded in second quarter of FY04; Milestone C go-ahead for refurbishment scheduled for final quarter of FY06. At least 62 helicopters will be acquired, although US Army is hopeful of increasing fleet to total of 96 by 2010. Remanufacture expected to provide new 20-year airframe life; improved avionics, electrical and dynamic systems also anticipated, with replacement engine to be developed, tested and procured. RFP for latter issued in mid-2004, with Rolls-Royce and General Electric both expected to respond to requirement to re-engine about 100 helicopters from 2006 onwards.

UH-60M: First use of designation was for proposed enhanced version for US Army. Cancelled early 1989 in favour of UH-60L. Designation subsequently re-used in 2000; see immediately below for details.

UH-60M: Improved Black Hawk version (originally known as the **UH-60L+**) for service with US Army; involves avionics and power plant modernisation effort to extend operational life by 25 to 30 years, with benefits in terms of payload (up to 907 kg; 2,000 lb advantage over UH-60A) and performance (up to 15 kt; 28 km/h; 17 mph faster). Sikorsky awarded US$7.45 million contract in August 2000 for preliminary risk reduction effort. US Army envisages initial procurement via major upgrade programme, whereby approximately 906 UH-60As will be brought to UH-60M standard, while another 311 UH-60Ls are probably to be similarly upgraded. New-build examples also to be produced from 2007, with US Army having requirement for up to 300.

Improvements for the UH-60M include a wide-chord, composite-spar main rotor, a digitised 'glass cockpit' based on the MIL-STD-1553 databus and new avionics, Stormscope weather mapping system, an advanced flight control computer, new diagnostic monitoring systems, a strengthened centre fuselage, larger fuel tanks and advanced infra-red suppression.

First Sikorsky UH-60M conversion ***NEW***/0567012

Upgrade programme will also allow entire US Army Black Hawk fleet to standardise on General Electric T700-GE-701D engine and improved durability main gearbox. General Electric and US Army announced joint effort to remanufacture existing engines into more reliable -701D form at start of 2001, benefits including longer engine life and lower operating costs, as well as offering about 4 per cent more power.

On 30 March 2001, Defense Acquisition Board approved Milestone B system development and demonstration phase; this followed in May by US$219.7 million contract for research, development, test and evaluation. Total of four UH-60M prototypes planned, comprising three conversions of existing Black Hawks (first, second and fourth aircraft) and one new-build helicopter (third aircraft; serial number 02-26969). Upgrade programme began in November 2001, with arrival of three UH-60s at Troy, Alabama, for dismantling and evaluation. Some structural reconditioning will be undertaken at Troy, before components and assemblies are shipped to Stratford for reconditioning of dynamic components and reassembly. First aircraft to be upgraded is UH-60A 85-24432; second is conversion of UH-60L 89-26217; third (actually the fourth UH-60M) is conversion of UH-60A 77-22716 to **HH-60M** medical evacuation configuration. Maiden flight of first UH-60M prototype (now serialled 02-26976) took place at Sikorsky's West Palm Beach, Florida, facility on 17 September 2003; 75-minute sortie included flight at 120 kt (222 km/h; 138 mph) and 45 degree turns, as well as verifying performance of systems and instrumentation. Second aircraft (now serialled 02-26977) flew in October 2003, with remaining two to follow suit in January (02-26969) and April 2005 (02-26978, the former 77-22716).

First pair of aircraft primarily concerned with flight testing, while third used for electromagnetic interference chamber testing and fourth to evaluate specialised medical equipment. Four pre-production UH-60Ms (04-26998 to 04-27001, all previously UH-60As) also to be produced for US Army operational evaluation, which should begin at end of FY06, coincident with delivery of first LRIP helicopters.

Development phase to take four years, leading to low-rate initial production in FY04, FY05 and FY06, with full-rate production decision due to be taken in FY05. Total cost of upgrade estimated to be about US$11.6 billion.

VH-60N: Nine for US Marine Corps Executive Flight Detachment of squadron HMX-1 at Quantico, Virginia, to replace UH-1Ns; deliveries started November 1988; known as VH-60A until redesignated 3 November 1989. Name **White Hawk** adopted by Marine Corps.

Additional equipment includes more durable gearbox, weather radar, SH-60B-type flight control system and ASI, -401 engines as in SH-60B, cabin soundproofing, VIP interior, cabin radio operator station, EMP hardening, 473 litre (125 US gallon; 104 Imp gallon) internal fuel tank and extensive avionics upgrading. SPAR (Special Progressive Aircraft Rework) undertaken on VH-60N fleet from 1998.

UH-60P: South Korean Army version of UH-60L (S-70A-18) with minor avionics modifications to meet local requirements; first (KA-1602) of three UH-60Ls delivered by Sikorsky 10 December 1990; balance of 81 UH-60Ps on initial contract assembled locally by Korean Air Lines with increasing indigenous content, in US$500 million, five year programme. Deliveries from follow-on batch of 57 since completed, but acquisition of third batch, to replace remaining UH-1 Iroquois, has not taken place. South Korea has requirement for about 12 medical evacuation helicopters and could obtain UH-60Q or equivalent.

UH-60Q: 'Dustoff' (Dedicated Unhesitating Service To Our Fighting Forces) medical evacuation/search and rescue version for US Army. Development began in early 1990s, after Gulf War, when it was realised that a requirement existed for a longer-range medevac helicopter to replace the UH-1V Iroquois. A proof-of-principle conversion of a UH-60A (86-24560) was undertaken by Serv-Air Inc of Richmond, Kentucky, and flown for the first time as the **YUH-60A(Q)** on 31 January 1993. This aircraft was subsequently delivered to the Tennessee Army National Guard at Lovell Field, Chattanooga, on 12 March 1993, where it underwent a 12-month evaluation programme beginning in September 1993, with an organisation known as CECAT (Combat Enhancing Capable Aviation Team).

Sikorsky eventually selected as prime integrator for production and was awarded an initial contract on 9 February 1996 for two Phase 2 conversions in FY96, with second contract for further two in FY97; YUH-60A(Q) designation also applied to these aircraft, which participated in two-year qualification programme, with initial flight tests successfully completed in second quarter of 1997; formal operational test followed at Fort Campbell, Kentucky, between July and September 1998, using three YUH-60A(Q)s, with fourth delivered to CECAT in early 1999. Major subcontractors are Air Methods (medical interiors); Breeze-Eastern (HS-29900 external electric rescue hoist); BAE Systems Canada (mission management system); FLIR Systems Inc (SAFIRE thermal imaging system); Litton (LITOX onboard oxygen generating system); Simula (medical attendant seats) and Telephonics (intercom).

Definitive UH-60Q configuration includes a medical interior able to accommodate six stretcher patients, with integrated suction and oxygen systems plus defibrillation, ventilation and intubation equipment, as well as apparatus for monitoring of vital signs. It also has a 'glass cockpit' incorporating Litton smart multifunction displays (SMFDs); Doppler 128C with embedded GPS; NVG-compatible lighting; AN/ARS-6(V)2 personnel locator system; HIRSS; chaff/flare dispensers; ESSS; -701C engines; plus an improved data modem and SINCGARS radios, which allow it to transmit and receive digital data.

MH-60R Seahawk: US Navy ASW/ASST helicopter, *described separately*.

MH-60S KnightHawk: Shipboard transport helicopter. Key element in 1996 US Navy Helicopter Master Plan, entailing retirement of CH/HH-46Ds, HH-60Hs and SH-3s by FY12 and their replacement by navalised MH-60S (previously designated CH-60S until 6 February 2001). Design is fundamentally baseline UH-60L Black Hawk with T700-GE-401C engines and dynamics of SH-60 Seahawk, plus automatic rotor blade folding system, folding tail pylon, improved durability gearbox, rotor brake, automatic flight control system (AFCS), HIFR capability and rescue hoist for SAR/CSAR missions. Also has 'glass cockpit', with active matrix liquid crystal displays (AMLCDs); common cockpit on MH-60S and MH-60R, with Lockheed Martin securing US$61 million contract in August 1998 to develop and produce two prototypes for flight testing from late 1999. Equipment includes two integrated inertial navigation/GPS units, mass memory unit,

Sikorsky UH-60A, previously used by US Army and now in black and gold colour scheme of US Department of Homeland Security *(Paul Jackson)* ***NEW***/0589358

mission computer, flight management computer, operational software and four Litton flat-panel displays replacing all but standby instruments; MH-60S cockpit is 'scaled-down' version, with potential for upgrading if combat SAR mission is added at later date. MH-60S also has provision for external stores support system (ESSS), allowing carriage of additional fuel and forward-firing weapons.

Design includes a convertible cargo handling system; when configured for pure cargo operations, MH-60S can carry two 1.02 × 1.22 × 1.02 m (40 × 48 × 40 in) tri-wall pallets with total weight of 1,588 to 1,814 kg (3,500 to 4,000 lb); as a personnel transport, is able to accommodate a crew of four, plus 13 passengers. Underslung loads up to 4,082 kg (9,000 lb) may also be carried, with total payload capacity about 4,536 kg (10,000 lb).

MH-60S potential demonstrated by modified UH-60L in June 1995, when internal and external cargo-carrying capability studied by Sikorsky and the US Navy. Detail design began October 1996, with award in April 1997 of US$5.75 million contract to Sikorsky for demonstrator. This was hybrid vehicle, based around UH-60L (96-26673) borrowed from the US Army, married to components from SH-60F furnished by US Navy. Resulting YCH-60S (Navy identity 966673) made first flight on 6 October 1997 and was used for joint Navy/Sikorsky 35 hour flight test programme that ended on 10 January 1998. First shipboard demonstration on 19 November 1997, with YCH-60S completing 17 landings aboard combat store ship USS *Saturn*; initial trial included 12 vertical replenishment lifts with 680 kg (1,500 lb) slung load and three hot refuellings.

YCH-60S evaluated in 1999–2000 as potential airborne mine countermeasures (AMCM) platform to replace MH-53E. Initial trials at Stratford in third quarter of 1999, followed by transfer to Patuxent River in fourth quarter for tow tests, plus carriage, winch deployment and recovery of AN/AQS-20/X mine detection sonar. Thicker frames for helicopter's rear-cabin structure will permit cable loads up to 2,722 kgf (6,000 lbf). Raytheon Airborne Mine Neutralisation System (AMNS) selected in first quarter of 2003, with first AMCM-configured MH-60S making initial flight on 20 July 2003. Trials with Northrop Grumman Airborne Laser Mine Detection System (ALMDS) began in April 2004, this being one of five sensors that will be utilised by the MH-60S. AMCM version is basic MH-60S with operator station in cabin, plus additional internal fuel and ability to tow detection equipment; neutralisation accomplished by BAE Systems Archerfish expendable underwater vehicle, which guided into position and then detonated to destroy mine. System Development and Demonstration (SDD) phase subject of US$18 million contract awarded in first quarter of 2003. MH-60S AMCM platform to achieve IOC in 2005, with ultimate total of 66 MH-60Ss planned for AMCM mission.

Features not embodied in demonstrator include fuel dump vents, flotation gear, HIFR and navalised T700 engines. Avionics also unrepresentative, although featuring a databus and four 127 × 127 mm (5 × 5 in) Rockwell Collins AMLCDs for pilot and co-pilot vertical and horizontal situation data, plus two control display/navigation units and an LN-100G embedded GPS/INS.

Decision to proceed with MH-60S low-rate initial production (LRIP) taken in early 1998, although firm fixed-price contract for first lot not awarded to Sikorsky until September 1999; valued at US$67.4 million, this was for initial five aircraft, plus option for one more (taken up in November 1999) and associated engineering and logistic services. Maiden flight of initial production MH-60S (165742) at Stratford, Connecticut on 27 January 2000; delivered to Patuxent River, Maryland, on 15 May to begin US Navy development testing and operational evaluation. At least four helicopters involved in trials, with initial technical evaluation completed by November 2000; three-month Opeval began November 2001 and yielded disappointing result, with MH-60S considered neither operationally "effective" nor "suitable"; further evaluations concluded in mid-2003 by VX-1 at Patuxent River resulted in MH-60S being described as "operationally effective and operationally suitable" in October 2003. Changes to software and operational requirements ensued, allowing approval to be given for start of full-rate production at end of August 2002.

Navy has requirement for 271 MH-60Ss. First example funded in FY98, with five more (Lot 1) in FY99, 16 (Lot 2) in FY00, 15 (Lot 3) for FY01, 13 (Lot 4) for FY02, 15 (Lot 5) for FY03, 13 (Lot 6) for FY04 and 21 and 26 respectively requested for FY05 and FY06. First squadron is HC-3 at North Island, California, which received initial example (165745) for use as ground maintenance trainer in early 2001 and which assumed responsibility for flight training in February 2002; second squadron is HC-5 at Andersen AFB, Guam, which began converting from H-46 in mid-2002. HC-6 at Norfolk, Virginia, also accepted its first MH-60S in second quarter of 2002. USN to undertake major reorganisation of fleet helicopter squadrons in next few years, which will result in creation of Helicopter Sea Combat (HSC) squadrons; see table for details of implementation plan and future assignments. Approximately 40 delivered by end of 2002, with first operational deployment, by HC-5 in USS *Essex*, completed at end of January 2003; 50th MH-60S delivered in June 2003. By end April 2004, US Navy had received 64 helicopters and completed 50,000 flight hours. Turkey also planning to acquire initial batch of four to six MH-60Ss.

UH-60X: Designation allocated to potential follow-on to UH-60M, but now known as Future Utility Rotorcraft (FUR), for which UH-60M may provide basis. Total of 256 required for service with 'first-to-fight' units, but currently unfunded and unlikely to enter inventory until at least 2025.

Firehawk: Trials of specialist firefighting version began in July 1998, using modified UH-60L (96-26728) with extended landing gear and removable 3,785 litre (1,000 US gallon; 833 Imp gallon) ventral watertank manufactured by Aero Union of Chico, California. Replenishment of tank accomplished in two ways: by landing next to water source for water to be pumped into tank via side connector, or by hovering over source and using snorkel hose and pump assembly to suck up water. Subsequent three month demonstration of firefighting capability by Los Angeles County Fire Department proved validity of system. Demonstrator then returned to US Army for further testing, before delivery to Oregon Army National Guard in 1999. Two examples delivered to Los Angeles County Fire Department in 2001, with Army National Guard units in California and Florida also receiving single examples during 2002–03.

Maple Hawk: Unsuccessful contender in contest to supply Canadian Forces with new SAR helicopter; offer reportedly priced at C$300 million for 15 helicopters, but EHI EH 101 Cormorant selected in December 1997.

Battle Hawk: Offered unsuccessfully to Australian Army for Project Air 87 armed reconnaissance requirement; based on MH-60K.

Exports comprise: S-70A-1: FMS deal for Royal Saudi Land Forces Army Aviation Command; 12 delivered January to April 1990 to squadron based at King Khaled Military City; modified to **Desert Hawk** and one (delivered December 1990) fitted with VIP interior; Desert Hawk has 15 troop seats, blade erosion protection using polyurethane tape and spray-on coating, Racal Jaguar 5 frequency-hopping radio, provision for searchlights, internal auxiliary fuel tanks, and external hoist.

S-70A-1L: Medical evacuation version for Saudi Arabia; IR filtered searchlight, rescue hoist, improved AN/ARC-217 HF com, AN/ARN-147 VOR/ILS, AN/ARN-149 ADF, air conditioning and provision for six stretchers; eight delivered from December 1991; further eight required.

S-70A-5: Two for Philippine Air Force, delivered March 1984.

S-70A-9: Royal Australian Air Force; 39 replaced Bell UH-1s; deliveries from October 1987 to 1 February 1991; first completed by Sikorsky, remainder assembled by Hawker de Havilland in Australia; aircraft transferred to Australian Army in February 1989, but RAAF continues to maintain them.

S-70A-11: Three to Jordan in 1986-87.

S-70A-12: Japan Self-Defence Forces acquiring **UH-60J/JA** versions of Mitsubishi SH-60J for search and rescue. Sikorsky-built prototype (N7267D), plus two CKD kits, delivered late 1990. Further production by Mitsubishi (which see).

S-70A-16: Reserved for Westland Helicopters.

S-70A-17: Turkish Jandarma ordered six in September 1988; deliveries completed December 1988; further six (including two VIP) delivered from late 1990 to Turkish National Police. See also S-70A-28.

S-70A-18: Korea (see UH-60P).

S-70A-19: Reserved for GKN Westland of UK.

S-70A-21: Two VIP versions to Egypt, 1990. Two VIP-configured UH-60L ordered by Egypt in third quarter of 1999; further two VIP-configured UH-60Ls requested in September 2002. Latter pair were delivered in November and December 2003.

S-70A-22: Korean VIP version. Three aircraft built by Sikorsky.

S-70A-24: Two UH-60Ls for Mexico. Delivered 1991. Further four received in about 1996.

S-70A-25/26: Moroccan Gendarmerie ordered two Black Hawks with different seating arrangements in 1991; delivered October 1992; began operations 11 November 1992; fitted with colour weather radar.

S-70A-27: Hong Kong. Two delivered 16 December 1992 to Royal Hong Kong Auxiliary Air Force; unit became Government Flying Service 1 April 1993. Fitted with FLIR and searchlight. Requirement for further four reportedly existed in 1995, but only one additional aircraft delivered.

S-70A-28: Turkish follow-on batch; 90 ordered 8 December 1992, of which first five to Jandarma on 4 January 1993, followed by 40 to armed forces during 1993–94; balance of 45 to have been co-produced in Turkey by TAI (which see) but programme suspended. However, fresh negotiations for 50 additional Black Hawks concluded in latter half of 1998,

MH-60S PLANNED ASSIGNMENT

New Identity	Current Identity	Planned Date	Fleet Assignment	Type of Deployment
HSC-2	HC-2	Jan 06	Atlantic	Training
HSC-3	HC-3	Mar 05	Pacific	Training
HSC-4	HS-4	Feb 07	Pacific	Carrier Force
HSC-5	HS-5	Feb 09	Atlantic	Carrier Force
HSC-6	HS-6	Apr 10	Pacific	Carrier Force
HSC-7	HS-7	Jan 08	Atlantic	Carrier Force
HSC-8	HS-8	Mar 08	Pacific	Carrier Force
HSC-9	HS-3	Feb 10	Atlantic	Carrier Force
HSC-11	HS-11	Feb 11	Atlantic	Carrier Force
HSC-12	HS-2	Apr 09	Pacific	Carrier Force
HSC-14	HS-14	Apr 11	Pacific	Carrier Force
HSC-15	HS-15	Feb 12	Atlantic	Carrier Force
HSC-21	HC-11	Mar 05	Pacific	Expeditionary
HSC-22	(new)	Jan 07	Atlantic	Expeditionary
HSC-23	(new)	Jan 07	Pacific	Expeditionary
HSC-25	HC-5	Mar 05	Pacific	Expeditionary
HSC-26	HC-6	Mar 05	Atlantic	Expeditionary
HSC-28	HC-8	Mar 05	Atlantic	Expeditionary

Notes:
HSC-3, HSC-21, HSC-25, HSC-26 and HSC-28 already equipped with MH-60S; date given refers to change of unit identity; all other dates refer to planned conversion to MH-60S

with first five airlifted to Turkey by An-124 in mid-June 1999 and deliveries completed in 2001. Final 30 produced with 'glass cockpit', which to be retrofitted to earlier machines; first flight of **S-70A-28D** in this guise on 29 March 2000. Installation includes four Rockwell Collins MFDs dual flight management system and LN-100G INS/GPS; four to incorporate Tadiran Spectralink ASR-700 airborne search and rescue system (ASARS) for use by Turkish Army Special Forces in combat search and rescue and covert operations.

S-70A-30: One VIP transport ordered for Argentine Air Force, January 1994; delivered 4 September 1994.

S-70A-33: Four ordered by Brunei in 1995; delivered 1997–98. Equipment includes radar, AN/AAQ-21 FLIR and external stores support system.

S-70A-34: Malaysia ordered two S-70A Black Hawks in 1996 as replacements for AS 332 Super Puma in VIP transport role; first of pair delivered by end of 1997, with second following in February 1998.

S-70A-36: Brazil received four S-70As in August 1997 for use in Peru/Ecuador peacekeeping support mission; equipment includes GPS, HF radio, internal rescue hoist and weather radar.

S-70A-37: Version of Firehawk; two to Sultan of Brunei, 2000, replacing S-70Cs.

S-70A-39: Chilean order for one, announced in March 1998; delivered in July 1998, with further purchases expected to replace UH-1H Iroquois over next few years.

S-70A-41: Colombia. Believed to refer to 22 aircraft delivered during 1994 (two), 1995 (two), 1997 (seven) and 1998 (11); these are UH-60L derivative, unlike initial delivery of 1988–89 which was baseline UH-60A. Further helicopters requested in November 1999; two contracts received by Sikorsky in mid-December 2000 cover purchase of 21 helicopters for Colombian Army, plus four for Air Force and six for National Police. Ultimately, as many as 60 in prospect as part of US drive to combat increased narcotics trafficking in this region.

S-70A-42: Austria signed contract in December 2000 for nine aircraft to begin replacement of Agusta-Bell 204/212; total cost of deal put at US$184 million. First example handed over in USA on 10 June 2002.

S-70A-43: Thailand. Initial three helicopters delivered in December 2001 (two) and January 2002 (one); further two due to be handed over in September 2004. Further procurement seems likely since Thailand said to require at least 30 Black Hawks.

S-70A-50: Israel. Request for 15 UH-60Ls revealed by US DoD in April 1997; first so-called 'Peace Hawk' handed over at Stratford, Connecticut, on 23 March 1998, being airlifted (with four others) to Israel by C-5 on 27 May 1998. Deliveries completed by end of 1998. Additional 35 requested in September 2000 at estimated cost of US$525 million,

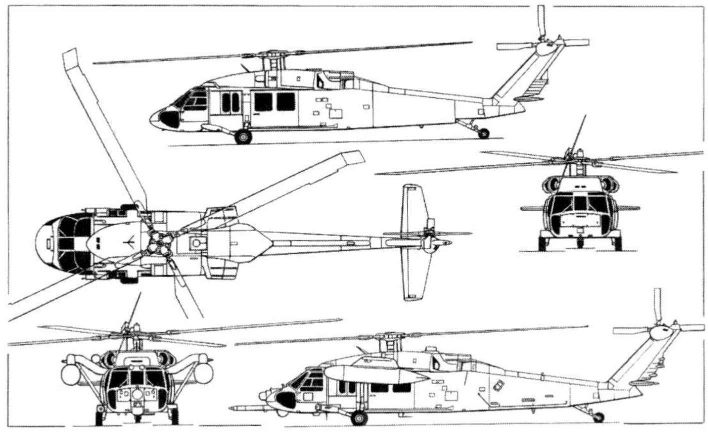

Sikorsky UH-60L Black Hawk combat assault helicopter, with additional lower side view and lower front view of MH-60K special operations variant *(Dennis Punnett)* 0507463

but subsequent contract received on 31 January 2001 covered supply of 24 **S-70A-55** aircraft at total cost of US$211.8 million. Delivery of these was accomplished in second half of 2002. Israeli Black Hawks subject of major upgrade that began in 2002; accomplished indigenously, much of the work undertaken on IDF/AF UH-60As and UH-60Ls is classified, but does include addition of in-flight refuelling probes and external fuel tanks as well as more powerful environmental conditioning systems.

Other recent customers include Los Angeles County Fire Department accepted two S-70A Firehawks in the first half of 2001 and ordered a third in 2003, for delivery in December 2004.

Direct transfers include one UH-60L to Bahrain, early 1991; five UH-60As delivered to Colombian Air Force in July 1988 for anti-narcotics operations; five more sold February

US BLACK HAWK PROCUREMENT

Fiscal Year	Lot	UH-60A	EH-60A	HH-60G	UH-60L	MH-60K	HH-60L	UH-60M	MH-60S	FMS	Year Total	Cum Total
FY73	N/A	6									6	6
FY77	1	15									15	21
FY78	2	56									56	77
FY79	3	92									92	169
FY80	4	94									94	263
FY81	5	80		5							85	348
FY82	6	96		6							102	450
FY83	7	96									96	546
FY84	8	84	12								96	642
FY85	9	86	18								104	746
FY86	10	78	18								96	842
FY87	11	84	18	9							111	953
FY88	12	72		16						21[1]	108	1,061
FY89	13	49		18	23					5[2]	95	1,156
FY90	14			22	76					1[3]	99	1,255
FY91	15			15	45	22				8[4]	90	1,345
FY92	16			13	52						65	1,410
FY93	17				60						60	1,470
FY94	18				63						63	1,533
FY95	19				68						68	1,601
FY96	20				65					9[5]	74	1,675
FY97	21			8	30		4			15[6]	57	1,732
FY98	22				34				1		35	1,767
FY99	23				18				5		23	1,790
FY00	24				19		3		16	9[7]	47	1,837
FY01	25				16		2		15	34[8]	67	1,904
FY02	26				19		2	4[9]	13	24[10]	62	1,966
FY03	27				12		7		15		34	2,000
FY04	28				13		4	4[13]	13	4[11]	38	2,038
FY05	29				44				21		65	2,103
FY06[12]	30				28				26		54	2,157
FY07[12]	31				22						22	2,179
FY08[12]	32				5						5	2,184
FY09[12]	33				4						4	2,188
Totals		**988**	**66**	**112**	**716**	**22**	**22**	**8**	**125**	**129**		**2,188**

[1] For Columbia (5), Egypt (2) and Saudi Arabia (13)
[2] For Columbia
[3] For Bahrain
[4] For Saudi Arabia
[5] For Columbia (7?) and Egypt (2)
[6] For Israel
[7] For Columbia (7) and Egypt (2)
[8] For Columbia (31) and Thailand (3)
[9] Prototypes. One new-build example plus three remanufactured from UH-60A (2) and UH-60L (1)
[10] For Israel
[11] For Egypt (2) and Thailand (2)
[12] Totals for FY06 to FY09 not complete. Additional procurement for US Army is expected and MH-60S purchases still to be determined. Further FMS deliveries also likely to occur
[13] All remanufactured from UH-60A

1989. Israel received 10 former US Army UH-60As in August 1994, for 124 Squadron at Palmachim, under local name of Yanshuf (Owl).

Taiwan requires new utility helicopter as replacement for existing UH-1H Iroquois; S-70A in contention and reportedly preferred option over competing Bell 412; up to 80 of chosen type to be obtained, with local manufacture expected to be a key factor in selection. Decision originally anticipated in 1998 but has been deferred, with initial contract for 20 to 25 helicopters likely. Jordan also to acquire UH-60L version, with initial batch of eight helicopters likely to be provided under FMS programme.

Follow-on purchase of Black Hawk by Australia is possible, since UH-60M version is being offered to satisfy the Air 9000 requirement for up to 50 utility helicopters between 2006 and 2011.

S-70C: Commercial version; no recent news to report.

CUSTOMERS: By 1999, over 2,400 H-60s of all variants had flown more than 3,600,000 hours. US Army total includes EH-60As and diversions to USAF, Bahrain, Colombia, Egypt and Saudi Arabia; 1,000th of S-70 series accepted 17 October 1988 and 2,000th May 1994; US Army Black Hawks in service in Germany, Hawaii and South Korea and with Army National Guard and Army Reserve.

US Army UH-60As loaned to the US Drug Enforcement Agency, augmenting five bought direct from Sikorsky. Fleet in 1999 totalled 13 (all with military serial numbers), plus three in storage and further two lost in accidents.

See Current Versions for export models and details.

COSTS: UH-60L US$8.6 million (1997) US Army unit cost; MH-60G US$10.2 million. Two VIP aircraft for Egypt cost about US$47 million (2002), with spare engines, spare parts, tools and support equipment and other logistical support. MYP6 purchase of 80 UH-60L and 82 MH-60S estimated to be worth US$1.5 billion during FY02-06. Cost of single S-70A Firehawk for Los Angeles County Fire Department quoted as US$14.5 million (2003).

DESIGN FEATURES: Represented new generation in technology for performance, survivability and ease of operation when introduced to replace UH-1 as US Army's main squad-carrying helicopter; adapted to wide variety of other roles, including several maritime applications. Four-blade main rotor; one-piece forged titanium rotor head with elastomeric blade retention bearings providing all movement and requiring no lubrication; hydraulic drag dampers; bifilar self-tuning vibration absorber above head; blades have 18° twist, and tips swept at 20°; thickness and camber vary over the length of blades, based on Sikorsky SC-1095 aerofoil; blades tolerant up to 23 mm hits and spar tubes pressurised with gauges to indicate loss of pressure following structural degradation.

Two pairs of tail rotor blades fastened in cross-beam arrangement, mounted to starboard; tail rotor pylon tilted to port to produce lift as well as anti-torque thrust and to extend permissible CG range; fixed fin large enough to allow controlled run-on landing following loss of tail rotor.

FLYING CONTROLS: Rotor pitch control powered by two independent hydraulic systems; Hamilton Sundstrand AFCS with digital three-axis autopilot provides speed and height control and coupled modes. Full-time autostabilisation includes feet-off heading hold cancelling torque-induced yaw at all airspeeds and during hover; positive fuselage attitude control provided by electrically driven variable incidence tailplane moving from +34° in hover to −6° during autorotation; angle is controlled by combined sensing of airspeed, collective-lever position, pitch attitude rate and lateral acceleration.

STRUCTURE: Main blade spar is formed and welded into oval titanium tube, with Nomex core, graphite trailing-edge, and covered by glass fibre/epoxy skin; titanium leading-edge abrasion strip and Kevlar tip. New main blades, with modified tips and 16 per cent increase in chord, under development for UH-60L; available for retrofit from 1997. Cross-beam composites tail rotor, eliminating all rotor head bearings. Light alloy airframe designed to retain 85 per cent of its flight deck and passenger space intact after vertical impact at 11.5 m (38 ft)/s, lateral impact at 9.1 m (30 ft)/s, and longitudinal impact at 12.2 m (40 ft)/s; also withstands simultaneous 20 *g* forward and 10 *g* downward impact; glass fibre and Kevlar used for cockpit doors, canopy, fairings and engine cowlings; glass fibre/Nomex floors; tailboom folds to starboard and main rotor mast can be lowered for transport/storage.

LANDING GEAR: Non-retractable tailwheel type with single wheel on each unit. Energy-absorbing main gear with a tailwheel which gives protection for the tail rotor in taxying over rough terrain or during a high-flare landing. Axle assembly and main gear oleo shock-absorbers by General Mechatronics. Mainwheel tyres size 26×10.00–11, pressure 8.96 to 9.65 bar (130 to 140 lb/sq in); tailwheel tyre size 15×6.00–6, pressure 6.21 to 6.55 bar (90 to 95 lb/sq in). Alaskan-based H-60s have Airglass Engineering ski landing gear.

Vertical replenishment by a Sikorsky MH-60S KnightHawk ***NEW***/0589359

UH-60A transferred from US Army to Customs Service, but retaining its military serial number ***(Paul Jackson)*** 0526994

POWER PLANT: Two 1,163 kW (1,560 shp) General Electric T700-GE-700 turboshafts on UH-60A. From late 1989 (UH-60L), two T700-GE-701C engines, each developing 1,402 kW (1,880 shp). UH-60M will have two T700-GE-701D, each rated at 1,652 kW (2,215 shp). (T700-GE-701A engines with maximum T-O rating of 1,285 kW; 1,723 shp optional in export models.) Transmission rating 2,109 kW (2,828 shp) in UH-60A, uprated to 2,535 kW (3,400 shp) in models with T700-GE-701C engines.

Two crashworthy, bulletproof fuel cells, with combined usable capacity of 1,361 litres (360 US gallons; 300 Imp gallons), aft of cabin. Single-point pressure refuelling, or gravity refuelling via point on each tank. Auxiliary fuel can be carried internally in one of several optional arrangements, or externally by the ESSS system. Two external tanks each of 871 litres (230 US gallons; 192 Imp gallons); up to two internal tanks, each of 700 litres (185 US gallons; 154 Imp gallons).

ACCOMMODATION: Two-man flight deck, with pilot and co-pilot on armour-protected seats. A third crew member is stationed in the cabin at the gunner's position adjacent forward cabin windows. Forward-hinged jettisonable door on each side for access to flight deck area. Main cabin open to cockpit to provide good communication with flight crew and forward view for squad commander. Accommodation for 11 fully equipped troops, or 14 in high-density configuration; 20 minimally armed personnel in optional configuration. Eight troop seats can be removed and replaced by four litters for medevac missions, or to make room for internal cargo. An optional layout is available to accommodate a maximum of six litter patients. Executive interiors for seven to 12 passengers available for the S-70A. Cabin heated and ventilated. Simula Safety Systems Inc received US$7.1 million contract in April 1999 covering supply of 290 cockpit airbag systems for installation in US Army UH-60A/L; this is low-rate initial production phase and total comprises 275 aircraft units and 15 spares.

External cargo hook, having a 3,630 kg (8,000 lb) lift capability, enables UH-60A to transport a 105 mm howitzer, its crew of five and 50 rounds of ammunition. Rescue hoist of 272 kg (600 lb) capacity optional. Large rearward-sliding door on each side of fuselage for rapid entry and exit.

SYSTEMS: Solar 67 kW (90 hp) T-62T-40-1, Honeywell or Hamilton Sundstrand APU. An optional winterisation kit provides a second hydraulic accumulator installed in parallel with the APU hydraulic start accumulator, maintaining engine start capability at low ambient temperatures; Honeywell 30 to 40 kVA and 20 to 30 kVA electrical power generators; 17 Ah Ni/Cd battery. Engine fire extinguishing system. Rotor blade de-icing system standard on US Army aircraft, optional for export. Electric windscreen de-icing.

AVIONICS: Configurations vary between aircraft. Additional avionics and self-protection equipment installed in Enhanced Black Hawk, as described under Current Versions. Improvement options offered from 1996 for new-build and retrofit on S-70 series include 'glass cockpit' and digital avionics; equipment available includes EFIS and digital automated flight computer system (AFCS).

Comms: Raytheon AN/ARC-186 VHF-FM, Raytheon AN/ARC-115 VHF-AM, Raytheon AN/ARC-164 UHF-AM, Rockwell Collins AN/ARC-186(V) VHF-AM/FM, Honeywell AN/APX-100 IFF transponder, Raytheon TSEC/KT-28 voice security set, and intercom. HH-60G has AN/URC-108 satcom and is being upgraded with Rockwell Collins AN/ARC-210 integrated communications system; Rockwell Collins AN/ARC-220 nap of the earth digital radio and AN/ARC-222 installed on Block 152 Upgrade HH-60G. UH-60M will have Telephonics secure digital intercom system.

Radar: MH-60K has Raytheon AN/APQ-147A terrain-following/terrain-avoidance radar, HH-60G has Honeywell AN/APN-239 (RDR-1400C) radar. MH-60L and some export S-70s also equipped with radar.

Flight: Hamilton Sundstrand AFCS with digital three-axis autopilot, Honeywell AN/ARN-123(V)1 VOR/marker beacon/glideslope receiver, Emerson AN/ARN-89 ADF, BAE Systems AN/ASN-128 Doppler, AN/ASN-43 gyrocompass, Honeywell AN/APN-209(V)2 radar altimeter. HH-60G has BAE Systems AN/ASN-137 Doppler, Rockwell Collins AN/ASN-149 GPS and Litton ring laser gyro INS (replacing Carousel IV). Northrop Grumman LISA-200 AHRS to be retrofitted to US Army helicopters. UH-60M to have CMC Electronics CMA-2082M flight management system and CMA-2088 emergency control panel.

Instrumentation: HH-60G has Teldix KG-10 map display. US Army MH-60 special operations versions to receive Elbit ANVIS 7 NVG/HUD system as retrofit; system already fitted in UH-60A and L. Elbit ANVIS/HUD system ordered by Australia in August 1999; 12 units acquired for installation by Raytheon Australia on S-70A-9. NVG-compatible version of Lockheed Martin GH-3000 electronic standby instrument system to be installed on MH-60K. UH-60M will have four Rockwell Collins 152 × 203 mm (6 × 8 in) landscape colour AMLCDs.

Mission: HH-60G has Raytheon AN/AAQ-16 FLIR. UH-60Q and HH-60L have (and HH-60M will have), FLIR Systems Inc AN/AAQ-22 SAFIRE thermal imaging system. Raytheon ZSQ-1(V) electro-optic sensor system to be installed on US Army MH-60s; two versions exist, with suite of equipment including laser rangefinders or laser spot trackers/designators, laser pointers, FLIR sensors, image-intensifying TV and associated processing and control electronics.

Self-defence: Baseline UH-60 Black Hawk has Raytheon AN/APR-39(V)1 RWR, Sanders AN/ALQ-144 IR countermeasures set and BAE Systems M-130 chaff/flare dispenser. MH-60K has BAE Systems AN/AAR-47 missile warning system, Northrop Grumman AN/ALQ-136 pulse radio frequency jammer, Northrop Grumman AN/ALQ-162 CW radio jammer, Raytheon AN/APR-39A and AN/APR-44 pulse/CW warning receivers, Raytheon AN/AVR-2 laser detector, BAE Systems M-130 chaff/flare dispenser and Sanders AN/ALQ-144 IR countermeasures set. HH-60G has chaff/flare dispenser (BAE Systems M-130 being replaced by AN/ALE-47 since 1998) and Sanders AN/ALQ-144 IR countermeasures set. Development testing of Sanders AN/ALQ-212 Advanced Threat IR Countermeasures (ATIRCM) system on EH/MH-60s of US Army to begin in 1999.

AN/AAR-47 missile warning system, AN/ALR-69(V) RWR, AN/ALE-47 chaff/flare dispensers and AN/ALQ-213 EW management system installed on Block 152 Upgrade HH-60G.

EQUIPMENT: HH-60G has Lucas Aerospace internal rescue hoist with 76 m (250 ft) of cable. Lucas awarded contract in mid-2000 to supply initial batch of 21 hoists for installation on MH-60S and other S-70 variants, with potential options for additional 70 in 2001 and 2002 plus 34 in 2003. UH-60Q has external Breeze-Eastern HS-29900 electric rescue hoist. MH-60S to use Raytheon Airborne Mine Neutralisation System (AMNS) beginning in 2005.

ARMAMENT: New production UH-60As and Ls from c/n 431 onward incorporate hardpoints for an external stores support system (ESSS). This consists of a combination of fixed provisions built into the airframe and four removable external pylons from which fuel tanks and a variety of weapons can be suspended. Able to carry more than 2,268 kg (5,000 lb) on each side of the helicopter, the ESSS can accommodate two 871 litre (230 US gallon; 192 Imp gallon) fuel tanks outboard, and two 1,703 litre (450 US gallon; 375 Imp gallon) tanks inboard. This allows the UH-60A to self-deploy 1,200 n miles (2,222 km; 1,381 miles) without refuelling. The ESSS also enables the Black Hawk to carry Hellfire laser-guided anti-armour missiles, gun or M56 mine dispensing pods, FIM-92 Stinger AAMs, ECM packs, rockets and motorcycles. Up to 16 Hellfires can be carried externally on the ESSS, with another 16 in the cabin to provide capability to land and reload. Two pintle mounts in cabin can each accommodate a 0.50 in calibre General Electric GECAL 50 or 7.62 mm six-barrel Minigun.

DIMENSIONS, EXTERNAL:
Main rotor diameter ... 16.36 m (53 ft 8 in)
Main rotor blade chord ... 0.53 m (1 ft 8¾ in)
Tail rotor diameter ... 3.35 m (11 ft 0 in)
Length overall: rotors turning ... 19.76 m (64 ft 10 in)
rotors and tail pylon folded ... 12.60 m (41 ft 4 in)
Length of fuselage:
UH-60A/HH-60G, excl flight refuelling probe ... 15.26 m (50 ft 0¾ in)
HH-60G, incl retracted refuelling probe ... 17.38 m (57 ft 0¼ in)
Fuselage max width: UH-60A ... 2.36 m (7 ft 9 in)
Max depth of fuselage ... 1.75 m (5 ft 9 in)
Height: overall, tail rotor turning ... 5.13 m (16 ft 10 in)
to top of rotor head ... 3.76 m (12 ft 4 in)
in air-transportable configuration ... 2.67 m (8 ft 9 in)
Tailplane span ... 4.38 m (14 ft 4½ in)
Tailplane chord ... 0.88 m (2 ft 10½ in)
Wheel track ... 2.705 m (8 ft 10½ in)
Wheelbase ... 8.83 m (28 ft 11¾ in)
Tail rotor ground clearance ... 1.98 m (6 ft 6 in)
Cabin doors (each): Height ... 1.37 m (4 ft 6 in)
Width ... 1.75 m (5 ft 9 in)

DIMENSIONS, INTERNAL:
Cabin: Volume ... 11.6 m^3 (410 cu ft)

AREAS:
Main rotor blades (each) ... 4.34 m^2 (46.70 sq ft)
Tail rotor blades (each) ... 0.41 m^2 (4.45 sq ft)
Main rotor disc ... 210.15 m^2 (2,262.0 sq ft)
Tail rotor disc ... 8.83 m^2 (95.03 sq ft)
Fin ... 3.00 m^2 (32.30 sq ft)
Tailplane ... 4.18 m^2 (45.00 sq ft)

WEIGHTS AND LOADINGS:
Weight empty: UH-60A ... 5,118 kg (11,284 lb)
UH-60L ... 5,224 kg (11,516 lb)
Payload: internal, UH-60A/L ... 1,197 kg (2,640 lb)
underslung, UH-60A ... 3,629 kg (8,000 lb)
underslung, UH-60L/Q and MH-60S ... 4,082 kg (9,000 lb)
Mission T-O weight: UH-60A ... 7,708 kg (16,994 lb)
UH-60L ... 7,907 kg (17,432 lb)
HH-60G ... 8,119 kg (17,900 lb)
MH-60K ... 11,113 kg (24,500 lb)
Max alternative T-O weight (ferry mission), UH-60A/L:
... 11,113 kg (24,500 lb)
Max disc loading:
UH-60L at mission T-O weight ... 36.7 kg/m^2 (7.52 lb/sq ft)
UH-60L at max alternative T-O weight ... 52.9 kg/m^2 (10.83 lb/sq ft)
Transmission loading:
UH-60L at mission T-O weight and max power ... 3.04 kg/kW (5.00 lb/shp)
UH-60L at max alternative T-O weight and power ... 4.20 kg/kW (6.91 lb/shp)

PERFORMANCE (UH-60A at mission T-O weight, except where indicated):
Never-exceed speed (VNE):
UH-60A/L/Q ... 195 kt (361 km/h; 224 mph)
MH-60S ... 180 kt (333 km/h; 207 mph)
Max level speed at S/L ... 160 kt (296 km/h; 184 mph)
Max level speed at max T-O weight ... 158 kt (293 km/h; 182 mph)
Max cruising speed at 1,220 m (4,000 ft) and 35°C (95°F):
UH-60A ... 139 kt (257 km/h; 160 mph)
UH-60L ... 152 kt (282 km/h; 175 mph)
Single-engine cruising speed at 1,220 m (4,000 ft) and 35°C (95°F) ...
105 kt (195 km/h; 121 mph)
Vertical rate of climb at 1,220 m (4,000 ft) and 35°C (95°F): UH-60A ...
119 m (390 ft)/min
UH-60L ... 472 m (1,550 ft)/min
Service ceiling: UH-60A ... 5,700 m (18,700 ft)
UH-60L ... 5,835 m (19,150 ft)
Hovering ceiling:
OGE at 35°C: UH-60A ... 1,645 m (5,400 ft)
UH-60L ... 2,330 m (7,650 ft)
Range with max internal fuel at max T-O weight, 30 min reserves: UH-60A ...
319 n miles (592 km; 368 miles)
UH-60L ... 315 n miles (584 km; 363 miles)
Range with external fuel tanks on ESSS pylons:
with two 870 litre (230 US gallon; 191.5 Imp gallon) tanks ...
880 n miles (1,630 km; 1,012 miles)
with two 870 litre (230 US gallon; 191.5 Imp gallon) and two 1,703 litre (450 US gallon; 375 Imp gallon) tanks ... 1,200 n miles (2,222 km; 1,381 miles)
Endurance: UH-60A ... 2 h 18 min
UH-60L ... 2 h 6 min

UPDATED

SIKORSKY S-70B

US Navy designations: SH-60B and MH-60R Seahawk, SH-60F and HH-60H
US Coast Guard designation: HH-60J Jayhawk
Japan Maritime Self-Defence Force designation: SH-60J and SH-60K
Spanish Navy designation: HS.23
Republic of China Navy designation: S-70C(M)-1 and S-70C(M)-2 Thunderhawk

TYPE: Naval combat helicopter.

DEVELOPMENT MILESTONES

Programme launched	1977
First order (US Navy)	FY82
First flight	12 Dec 79
First flight, production	11 Feb 83
Entered service (US Navy)	1984

PROGRAMME: Naval development of Sikorsky UTTAS (UH-60A Black Hawk) utility helicopter; won US Navy LAMPS Mk III competition for shipboard helicopter in 1977; first flight of first of five YSH-60B prototypes (161169) 12 December 1979; first 18 SH-60Bs authorised FY82. Changed USN planning in 1993 resulted in premature end to SH-60B/F production; original intent was to remanufacture SH-60B/Fs and HH-60Hs as SH-60R (redesignated MH-60R in mid-2001), but acquisition strategy changed in 2001 and most MH-60Rs will be new-build helicopters.

Most recent development, revealed in model form at Asian Aerospace, Singapore, February 2000, involves incorporation of external stores support system (ESSS) of UH-60/S-70A series, as well as MH-60R sensor suite. According to Sikorsky, this proposed version evolved in response to requests from potential customers in and around Pacific Rim, such as Indonesia, Malaysia and Singapore. If proceeded with, new version would have maximum take-off weight of around 11,340 kg (25,000 lb).

CURRENT VERSIONS: **SH-60B:** Initial production version for ASW/ASST; 181 built for US Navy, excluding prototypes.

Detailed description applies to SH-60B, unless otherwise stated.

NSH-60B: Designation applied to two SH-60Bs (162337 and 162974) assigned to permanent test duties at Patuxent River, Maryland.

SH-60F: CV Inner Zone ASW helicopter, known as CV-Helo, for close-in ASW protection of aircraft carrier groups; US$50.9 million initial US Navy contract for full-scale development and production options placed 6 March 1985; replacing SH-3H Sea King; Seahawk prototype modified as SH-60F test aircraft; first flight 19 March 1987; initial fleet deployment by HS-2 aboard USS *Nimitz* in 1991. Currently assigned to 10 deployable squadrons (HS-2 to HS-8, HS-11, HS-14 and HS-15) plus one training unit (HS-10) and one Reserve Force squadron (HS-75). Production terminated with delivery of 82nd example 1 December 1994.

SH-60F has all LAMPS Mk III avionics, fairings and equipment removed, including cargo hook and RAST system main and tail probes, but installation provisions retained. Replaced by integrated ASW mission avionics including Honeywell AN/AQS-13F dipping sonar, MIL-STD-1553B databus, dual Litton AN/ASN-150 tactical navigation computers and AN/ASM-614 avionics support equipment, automatic flight control system with quicker automatic transition and both cable and Doppler autohover, tactical datalink with other aircraft, communications control system, multifunction keypads and displays for each of four crew members; internal/external fuel system and extra weapon station to port allowing carriage of three Mk 50 homing torpedoes; provision for surface search radar, FLIR, night vision equipment, passive ECM, MAD, air-to-surface missile capability, sonobuoy datalink, chaff/sonobuoy dispenser, attitude and heading reference system (AHRS), Navstar GPS, fatigue monitoring system and increase of maximum T-O weight to 10,659 kg (23,500 lb); secondary missions include SAR and plane guard.

YSH-60F: Designation applied to second production SH-60F (163283) which served as 'prototype' on test duties at Patuxent River, Maryland. To be fitted with lifting wing with flaperons and 2.44 m (8 ft) diameter vectored thrust ducted propeller ('ring tail') by Piasecki Aircraft Corporation (which see) as **X-49A** for trials project as part of advanced technology

Initial aircraft of Spain's second S-70B order ***NEW***/0567014

Sikorsky HH-60H fitted with nose-mounted AN/AAS-44(V) FLIR and laser designator *(Jane's/Patrick Allen)* ***NEW***/0567015

Royal Australian Navy Sikorsky S-70B-2 Seahawk *(Paul Jackson)* ***NEW***/0589374

demonstration programme. Known as SpeedHawk, this programme originated with Naval Air Systems Command, but was transferred to US Army control in late 2003; a 14-month flight test effort is expected to begin in about July 2005, but this target date may slip in view of transfer to Army.

HH-60H: US Navy procurement of 42 completed in 1996; used for strike-rescue/special warfare support (HCS); designated HH-60H in September 1986; first flight (163783) 17 August 1988; accepted by USN 30 March 1989; in service with HCS-4 at Norfolk, Virginia, January 1990; initial procurement ended with 18th delivery July 1991, completing HCS-5 at Point Mugu, California; both squadrons are part of Navy Reserve. Regular SH-60F squadrons later added pairs of HH-60Hs for deployed duty when embarked aboard aircraft carriers; missions are to recover four-man crew at 250 n miles (463 km; 288 miles) from launch point or fly 200 n miles (371 km; 230 miles) and drop eight SEALs from 915 m (3,000 ft).

Close derivative of SH-60F, with T700-GE-401C engines and HIRSS as SH-60B/F; equipment includes Litton AN/APR-39A(XE)2 RWR, Raytheon AN/AVR-2A(V) laser warning receiver, Honeywell AN/AAR-47 missile plume detector, Lockheed Martin AN/ALE-47 chaff/flare dispenser, Sanders AN/ALQ-144 IR jammer, Elbit ANVIS 7 NVG/HUD system and two cabin-mounted M60D 7.62 mm machine guns; provision for weapon pylons; required to operate from decks of FFG-7, DD-963, CG-47 and larger vessels, as well as unprepared sites. Cubic AN/ARS-6 personnel locator system installed from FY91. Some equipped with Indal RAST (recovery assist, secure and traverse) equipment. Armament development authorised October 1991 for installation of Hellfire ASM, 70 mm (2.75 in) rockets and forward-firing guns. Some HH-60H now fitted with nose-mounted Raytheon AN/AAS-44(V) FLIR/laser designator system for use with Hellfire missile.

HH-60J Jayhawk: Ordered in parallel with HH-60H; adapted for US Coast Guard medium-range recovery (MRR) role; last of 42 delivered in 1996. First flight (USCG 6001) 8 August 1989; first delivery to USCG (6002 at Elizabeth City CGAS) 16 June 1990; subsequently to Mobile, Traverse City, San Diego, Astoria, San Francisco, Cape Cod, Sitka, Kodiak and Clearwater CGAS. When carrying three 455 litre (120 US gallon; 100 Imp gallon) external tanks, HH-60J can fly out 300 n miles (556 km; 345 miles) and return with six survivors in addition to four-man crew, or loiter for 1 hour 30 minutes when investigating possible smugglers; other duties include law enforcement, drug interdiction, logistics, aids to navigation, environmental protection and military readiness; compatible with decks of 'Hamilton' and 'Bear' class USCG cutters. Equipment includes Honeywell RDR-1300C search/weather radar, AN/ARN-147 VOR/ILS, KDF 806 direction-finder, GPS, Tacan, VHF/UHF-DF, TacNav, dual U/VHF-FM radios, HF radio, IFF, V/U/HF IFF crypto computers, NVG-compatible cockpit, rescue hoist and external cargo hook.

XSH-60J: Japan Maritime Self-Defence Force (JMSDF) placed US$27 million order for two **S-70B-3** s for installation of Japanese avionics and mission equipment; first flights

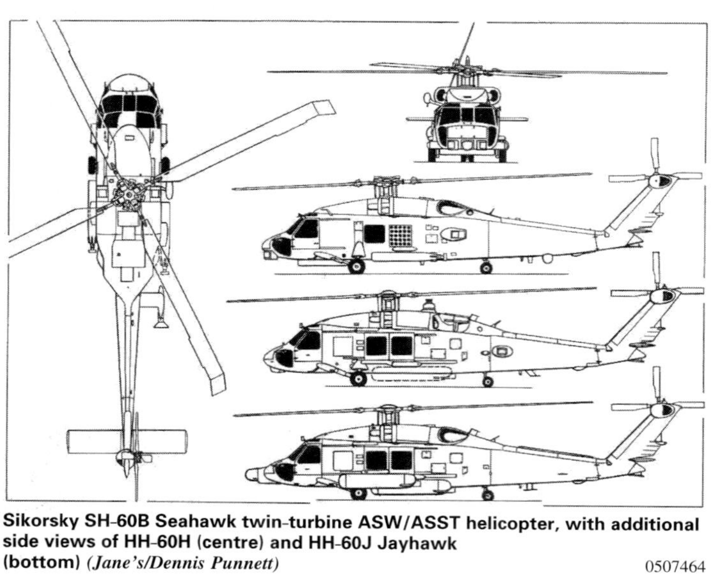

Sikorsky SH-60B Seahawk twin-turbine ASW/ASST helicopter, with additional side views of HH-60H (centre) and HH-60J Jayhawk (bottom) *(Jane's/Dennis Punnett)* 0507464

31 August and early October 1987; 1,007 hour test programme by Japan Defence Agency Technical Research and Development Institute between 1 June 1989 and 7 April 1991 to evaluate largely Japanese avionics for SH-60J, but AN/APS-124 radar.

SH-60J: Mitsubishi is manufacturing more than 100 SH-60J Seahawks for JMSDF.

SH-60K: Improved version of Seahawk for JMSDF; prototype rolled out 8 August 2001 and made first hover flight on 9 September 2001. Production was expected to begin in FY01 but has been delayed; JMSDF reportedly seeking initial batch of 50, with second batch of 50 to follow, as well as upgrade project involving SH-60J version.

MH-60R: Originally designated SH-60R and also known as LAMPS Block II; combines SH-60B capabilities with dipping sonar of SH-60F; original plan was for rebuild of existing fleet; first two conversions were to be funded in FY98; 15 in FY99 and more thereafter; however, concerns over cost led to one year delay in launch of remanufacture programme, which began in FY00 with batch of four helicopters for test duties (ordered 25 April 2000) and was followed by five low-rate initial production (LRIP1) helicopters in FY01, also converted from existing airframes. Training unit HSL-41 at North Island, California, planned to be first squadron under new identity as HSM-41.

Development delays and cost concerns in 2000–01 prompted Navy to restructure programme with mostly new-build MH-60Rs, which will be purchased for US$1 million to US$3 million more than remanufactured examples and also allow the Navy to implement measures to improve power to weight performance. Of total 243 MH-60Rs required, only two prototypes and first nine 'production' aircraft are remanufactured airframes.

MH-60R systems orientated towards littoral warfare operations, with ability to process and prosecute large number of air and sea contacts in a comparatively confined space, the latter in relatively shallow water. New systems added to enhance countermeasures and passive and active detection capability. Initial upgrade package abandoned on grounds of high cost in 1998, when less expensive programme, making extensive use of COTS technology, was adopted.

Lockheed Martin secured US$61 million contract in third quarter of 1998 for development of common cockpit prototype applicable to MH-60R and MH-60S variants. Under terms of contract, Lockheed Martin provided flight instrument displays, two MFDs, two operator keysets and digital communications suite as well as Litton integrated INS/GPS, mass memory unit, mission and flight management computers and applicable operational software for both versions. New 'glass cockpit' centred around Lockheed Martin-developed computer systems, but using commercial PowerPC processors, with data presented to pilots via electronic flight instrument display and multifunction mission display.

Other changes on MH-60R include deletion of MAD, addition of AGM-114 Hellfire anti-armour missile and two more stores stations, databus, Telephonics AN/APS-147 multimode radar, Raytheon/Thomson-Marconi Sonar AN/AQS-22 (FLASH) advanced airborne low-frequency dipping sonar, AN/AYK-14 mission processor and AN/UYS-2A enhanced modular signal processor, Lockheed Martin AN/ALQ-210 ESM, Raytheon AN/AAS-44 FLIR/laser ranger and NVG compatibility. MTOW expected to rise to 10,659 kg (23,500 lb). May eventually be fitted with new, more powerful engine.

Two SH-60Bs (162976 and 162977) selected as prototypes; conversion undertaken by Lockheed Martin Systems Integration at Owego, New York, where first (then designated SH-60R) was rolled out on 5 August 1999. First flight scheduled for October 1999, following electronic systems functional test and checkout on the ground, but delayed until 11 December; first prototype half analogue/half 'glass cockpit' for initial testing, with full 'glass cockpit' installed after about three months. After initial trials at Owego, first prototype delivered to Patuxent River, Maryland, in early May 2000 for start of two-year Navy/contractor developmental test programme.

Initial test aircraft (166402), remanufactured by Sikorsky, first flew 19 July 2001 and formally accepted by Navy (still in manufacturer's hands) later in same month. Subsequently to Patuxent River, on 10 August 2001 for installation of flight test instrumentation and then to Lockheed Martin at Owego for fitting of new mission systems. First flight with 'total weapon system' made on 4 April 2002.

All four test aircraft (166402 to' 405) had been delivered to the Navy by the beginning of February 2002, with first of five LRIP1 MH-60Rs (166406 to '410) flying on 9 July and being delivered to US Navy by end of that month. Further two LRIP1 machines were accepted by Navy in third quarter of 2002 and then allocated to Lockheed Martin at Owego for fitment of mission systems. LRIP2 batch of four helicopters procured with FY04 funding; total cost of these stated to be US$183.2 million, with delivery due in 2006. Final hurdle before start of full-rate production is US Navy OpEval (operational evaluation), which is due to start with VX-1 Squadron at Patuxent River in May 2005. Successful conclusion of OpEval will clear way for Milestone III approval of full-rate production starting in FY05. Total of eight helicopters planned for FY05, but rate will ultimately build to peak of 30 per year. USN to reorganise fleet and redesignate units as Helicopter Maritime Strike (HMS) squadrons; see table for details of implementation plan and future assignments.

MH-60R PLANNED ASSIGNMENT

New Identity	Current Identity	Planned Date	Fleet Assignment	Type of Deployment
HSM-37	HSL-37	Jan 14	Pacific	Expedetionary
HSM-40	HSL-40	Jan 10	Atlantic	Training
HSM-41	HSL-41	Mar 05	Pacific	Training
HSM-42	HSL-42	Mar 14	Atlantic	Expedetionary
HSM-48	HSL-48	Feb 15	Atlantic	Expedetionary
HSM-49	HSL-49	Apr 14	Pacific	Expedetionary
HSM-51	HSL-51	Mar 13	Pacific	Expedetionary
HSM-60	HSL-60	Mar 14	Reserve	Reserve
HSM-70	(new)	Jan 08	Atlantic	Carrier Force
HSM-71	(new)	Apr 09	Pacific	Carrier Force
HSM-72	(new)	Feb 09	Atlantic	Carrier Force
HSM-73	HSL-43	Feb 07	Pacific	Carrier Force
HSM-74	HSL-44	Feb 10	Atlantic	Carrier Force
HSM-75	HSL-45	Mar 08	Pacific	Carrier Force
HSM-76	HSL-46	Feb 11	Atlantic	Carrier Force
HSM-77	HSL-47	Apr 10	Pacific	Carrier Force
HSM-78	(new)	Feb 12	Atlantic	Carrier Force
HSM-79	(new)	Apr 11	Pacific	Carrier Force

Notes:
Dates refer to planned conversion to MH-60R

Exports comprise: S-70B-1: Spanish Navy received six from December 1988 (designated **HS.23**) for operation from four FFG-7 frigates by Escuadrilla 010 at Rota; similar to USN SH-60B, but with Honeywell AN/AQS-13F dipping sonar. Spanish government approval to order additional six granted in December 1998, with order placed in third quarter 2000; five of them delivered to Rota, in October 2002, with final aircraft retained at Owego, New York, for additional trials before being handed over in early 2003. Deal for new helicopters also included funds to upgrade original six to same standard, including armament kits and compatibility with AGM-114 Hellfire and AGM-119 Penguin ASMs. First two upgraded helicopters completed modernisation process in July and October 2003.

S-70B-2: Royal Australian Navy (RAN) selected Seahawk for role adaptable weapon system (RAWS) full-spectrum ASW helicopter with autonomous operating capability; order for eight confirmed 9 October 1984; eight more ordered May 1986. S-70B-2 has substantially different avionics from USN version: Racal Super Searcher radar (capable of tracking 32 surface targets) and Rockwell Collins advanced integrated avionics including cockpit controls and displays, navigation receivers, communications radios, airborne target handoff datalink and tactical data system (TDS). Upgrade of Australian Seahawks, known as Project Sea 1405, includes installation of Raytheon AN/AAQ-27 FLIR and an electronic warfare support measures package based on Elisra's AES-210 system; also installation of Smiths NVG-compatible aircraft standby attitude indicators and Northrop Grumman AN/AAR-54(V) passive MAWS. All 16 Seahawks upgraded by third quarter of 2003; first helicopter handed over to Tenix Defence Systems in first quarter 2000. Mid-life upgrade (MLU) expected to follow in due course, with project definition study to begin in 2003–04; MLU is likely to involve provision of dipping sonar and integration of ASM, with Penguin Mk2 anti-ship missile a strong possibility as this already purchased for use by RAN Seasprite helicopters.

S-70B-6: Hybrid SH-60B/F for Greece, unofficially known as **Aegean Hawk**; selected December 1991 and initial quantity of five ordered 17 August 1992 for MEKO 200 frigates. Option for three more subsequently converted to firm order and contract for further two (later increased to three) signed on 12 June 2000. Armament includes NFT Penguin Mk2 ASMs; avionics include AN/AQS-18(V)-3 dipping sonar, AN/APS-143(V3) radar and AN/ALR-66(V)-2 ESM; towed MAD and sonobuoy launcher omitted. First two delivered fourth quarter of 1994, with three more in 1995, one in 1997 and two in 1998; two helicopters from latest order were due for delivery to Greece in June 2004, with final machine following by end of 2005. Original eight aircraft being modified to operate with AN/AAQ-22Q Star SAFIRE FLIR sensor; three final aircraft have Raytheon AN/AAS-44 FLIR/laser rangefinder.

S-70B-7: Six Seahawks ordered by Royal Thai Navy in October 1993; equipped for coastal surveillance, maritime patrol and SAR from aircraft carrier HTMS *Chakri Naruebet*; first handed over at Stratford on 6 March 1997, with all six delivered by June.

S-70B-28: Initial batch of four ordered by Turkish Navy on 14 February 1997, with option on another four subsequently converted to firm order; the first example made its maiden flight on 18 January 2001 and all eight were delivered in 2002 for service aboard frigates in ASW and surveillance roles. They are first export Seahawks with a Rockwell Collins 'glass cockpit' and also have L-3 Communications Ocean Systems HELRAS long-range active dipping sonar and Telephonics AN/APS-143(V) radar installed. Original order includes supply of AGM-114 Hellfire II ASM. Turkey has ultimate requirement for up to 28 S-70Bs, of which further eight ordered in 2002.

S-70C(M)-1/2 Thunderhawk: S-70C designation used for H-60 purchases not qualifying for FMS. Principally assigned to aircraft delivered to Taiwan. Production complete.

CUSTOMERS: Total US Navy requirement originally 260 SH-60Bs; 186 on order, including five prototypes, when procurement terminated in FY94. First flight production Seahawk 11 February 1983; last SH-60B delivered to US Navy on 25 September 1996; first squadron was HSL-41 at NAS North Island, San Diego, California; operational deployment began 1984; 10 US Navy squadrons operating by March 1991 (HSLs 41, 43, 45, 47 and 49 at NAS North Island; 40, 42, 44, 46 and 48 at NAS Mayport, Florida); subsequently HSL-51 formed at Atsugi, Japan, 1 October 1991, and HSL-37 at NAS Barbers Point, Hawaii, began converting from SH-2Fs on 6 February 1992; most recent unit to equip is HSL-60 of the Reserve Force, also at Mayport. SH-60Bs deployed in 'Oliver Hazard Perry' (FFG-7) class frigates, 'Spruance' class and Aegis-equipped destroyers and 'Ticonderoga' class guided

Sikorsky MH-60R at the hover *NEW*/0589373

missile cruisers. US Navy originally required 150 SH-60Fs; total 82 completed, comprising seven pre-series plus 18 each in FY88, 89 and 91, 12 in FY92 and nine in FY93; procurement then halted; in West Coast service with HS-2, 4, 6, 8, 10 and 14 squadrons at NAS North Island, California; HS-3 at Jacksonville, Florida, equipped from 27 August 1991 as first East Coast squadron, followed by HS-1, 5, 7, 11 and 15, of which training squadron HS-1 since disestablished, leaving HS-10 of Pacific Fleet to conduct all US Navy SH-60F instruction. Reserve Force unit HS-75 at Jacksonville now also has SH-60F.

Exported to Australia, Greece, Japan, Spain, Taiwan, Thailand and Turkey (see Current Versions). S-70B-4 and -5 are derivatives of SH-60F and HH-60H, respectively; not taken up. In contention for order from Singapore for six naval helicopters; decision expected by end of 2004.

COSTS: US$20.25 million (1992) USN programme unit cost. Flyaway cost of final USN SH-60B about US$16 million; total MH-60R development programme costs expected to be around US$400 million, with unit flyaway cost quoted as US$16 million to US$18 million (FY96 dollars) for remanufactured aircraft and slightly more for new-build examples.

DESIGN FEATURES: SH-60B Seahawk designed to provide all-weather detection, classification, localisation and interdiction of surface ships and submarines, either controlled through datalink from parent ship or operated independently; secondary missions include SAR, vertical replenishment, medevac, fleet support and communications relay.

Revised features, compared with UH-60A, include more powerful navalised GE T700-GE-401 engines, additional fuel, sensor operator's station, port-side internal launchers for 25 sonobuoys, pylon on starboard side of tailboom for MAD bird, lateral pylons for two torpedoes or external tanks, chin-mounted ESM pods, sliding cabin door, rescue hoist, electrically actuated blade folding, rotor brake, folding tail, short-wheelbase tailwheel landing gear with twin tailwheels stressed for lower crash impact, DAF Indal RAST recovery assist, secure and traversing for haul-down landings on small decks and moving into hangar, hovering in-flight refuelling system, and emergency flotation system; pilots' seats not armoured. SH-60B gives 57 minutes' more listening time on station and 45 minutes' more ship surveillance and targeting time than SH-2F Seasprite LAMPS Mk I.

For operation in Gulf during mid-1980s Iran-Iraq war, 25 SH-60Bs fitted with upper and lower Sanders AN/ALQ-144 IR jammers, BAE Systems AN/ALE-39 chaff/flare dispensers, Honeywell AN/AAR-47 electro-optical missile warning, and a single 7.62 mm machine gun in door, for a weight penalty of 169 kg (369.5 lb); seven Seahawks fitted with Raytheon AN/AAS-38 FLIR on root weapon pylon with instantaneous relay to parent ship.

First Block I SH-60B update, introduced in production Lot 9, delivered from October 1991, includes provision for NFT AGM-119 Penguin anti-ship missile, Mk 50 advanced lightweight torpedo, Flightline AN/ARR-84 99-channel sonobuoy receiver (replacing ARR-75), Rockwell Collins AN/ARC-182 V/UHF FM radio and Rockwell Collins Class 3A Navstar GPS; before production cutbacks, 115 Penguin-capable Seahawks to come from retrofitting back to Lot 5, but only 28 launch kits (delivered 1997) so far ordered.

FLYING CONTROLS: As for UH-60.

STRUCTURE: Basically as for UH-60 plus marine corrosion protection; single cabin door, starboard side, narrower than on UH-60.

LANDING GEAR: Generally as for UH-60, but with twin tailwheel positioned further forward to facilitate operation from landing platforms on warships.

POWER PLANT: Two 1,260 kW (1,690 shp) intermediate rating General Electric T700-GE-401 turboshafts in early aircraft; 1,342 kW (1,800 shp) T700-GE-401C turboshafts introduced in 1988 and on HH-60H/J. Transmission rating 2,535 kW (3,400 shp). Internal fuel capacity 2,233 litres (590 US gallons; 491 Imp gallons). Hovering in-flight refuelling capability. Two 455 litre (120 US gallon; 100 Imp gallon) auxiliary fuel tanks on fuselage pylons optional (three on HH-60J). Hover IR suppressor subsystem (HIRSS) exhaust cowling fitted to HH-60H.

ACCOMMODATION: Pilot and airborne tactical officer/back-up pilot in cockpit, sensor operator in specially equipped station in cabin. Dual controls standard. Sliding door with jettisonable window on starboard side. Accommodation heated, ventilated and air conditioned.

SYSTEMS: Generally as for UH-60A.

AVIONICS: Refer also to Current Versions for variants other than SH-60B.

Comms: Rockwell Collins AN/ARC-159(V)2 UHF, Rockwell Collins AN/ARC-174(V)2 HF, Hazeltine AN/APX-76A(V) and Honeywell AN/APX-100(V)1 IFF transponders, TSEC/KG-45(E-1) communications security set, TSEC/KY-75 voice security set, Telephonics OK-374/ASC communications system control group. Satellite communications planned for MH-60R.

Radar: Raytheon AN/APS-124 search radar on SH-60B and Telephonics AN/APS-147 on MH-60R (Racal Super Searcher for Australia; Telephonics AN/APS-128PC for Taiwan; Telephonics AN/APS-143(V) for Turkey).

Flight: Rockwell Collins AN/ARN-118(V) Tacan, Northrop Grumman AN/APN-127 Doppler, Rockwell Collins AN/ARA-50 UHF DF, Honeywell AN/APN-194(V) radar altimeter. US Navy began testing version of Raytheon Joint Precision Approach and Landing System (JPALS) on an SH-60 in 2002; this system was expected to begin replacing existing radar-based shipboard precision approach system from 2005–06, but funding shortages have resulted in it being indefinitely delayed.

Mission: Sikorsky sonobuoy launcher, Flightline AN/ARR-75 and R-1651/ARA sonobuoy receiving sets (AN/ARR-84 receiver in Australian Seahawks and for USN Block 1 upgrade), Raytheon AN/ASQ-81(V)2 towed MAD (CAE AN/ASQ-504(V) internal MAD in Australian Seahawks), Raymond MU-670/ASQ magnetic tape memory unit, Astronautics IO-2177/ASQ altitude indicator, Fairchild AN/ASQ-164 control indicator set, Fairchild AN/ASQ-165 armament control indicator set, IBM AN/UYS-1(V)2 Proteus acoustic processor (Computing Devices UYS-503 for Australia) and CV-3252/A converter display, GD Information AN/AYK-14 (XN-1A) digital computer, Raytheon AN/ALQ-142 ESM, Sierra Research AN/ARQ-44 datalink and telemetry (Rockwell Collins DHS-901 in Australian Seahawks). SH-60F has Honeywell AN/AQS-13F dipping sonar (AN/AQS-18 in Taiwanese S-70s). During 1991 Gulf War, pod-mounted Hughes (now Raytheon) AN/AAQ-16 FLIR fitted to five SH-60Bs and Raytheon AN/AAQ-17 FLIR deployed on one SH-60B; BAE Systems Sea Owl IR turret evaluated later in 1991. Raytheon AN/AAS-44 FLIR/laser ranger on MH-60R. Australian examples also acquired AN/AAQ-16 FLIR for 1991 Gulf War. Raytheon AN/AAQ-27 FLIR system adopted as part of upgrade of Australian Seahawks to be undertaken by Tenix Defence Systems. US Navy to acquire airborne laser mine detection system for SH-60B and MH-60R; Northrop Grumman system selected in mid-2000, with 36 month engineering and manufacturing development (EMD) from 2001, with production decision due to be taken before the end of 2004. Greek aircraft retrofitted with AN/AAQ-22Q Star SAFIRE FLIR sensor between November 2002 and June 2003; Raytheon AN/AAS-44 installed on final three Aegean Hawks.

Self-defence: ESM systems include Raytheon AN/APR-39 RWR on HH-60H; none on SH-60F. MH-60R has Lockheed Martin AN/ALQ-210 ESM. Australian Seahawks fitted with AN/ALE-47 chaff/flare dispensers and AN/AAR-47 missile detectors for 1991 Gulf War. Upgrade of Australian Seahawks to include ESM based on Elisra AES-210 system.

EQUIPMENT: External cargo hook (capacity 2,722 kg; 6,000 lb) and rescue hoist (272 kg; 600 lb) standard. SH-60B/F and MH-60R have provision for eight sonobuoys.

ARMAMENT: US Navy armament includes up to three Mk 46 torpedoes and (IOC 1993) NFT AGM-119B Penguin Mk 2 Mod 7 anti-shipping missiles. Block I upgrade integrated Penguin and Honeywell Mk 50 Advanced Lightweight Torpedo from 1993. HH-60H has two pintle-mounted M240G 7.62 mm machine guns and was cleared in 1996 to operate with AGM-114 Hellfire ASM. HH-60H can operate with 70 mm (2.75 in) rocket pods and GAU-17/A 7.62 mm forward-firing guns. Hellfire to be included in SH-60B and MH-60R armament for attacking small ships.

DIMENSIONS, EXTERNAL(As UH-60A except:):

Length overall, rotors and tail pylon folded:
- SH-60B ... 12.47 m (40 ft 11 in)
- HH-60H ... 12.51 m (41 ft 0⅝ in)
- HH-60J ... 13.13 m (43 ft 0⅞ in)

Length of fuselage: HH-60J ... 15.87 m (52 ft 1 in)
Width, rotors folded ... 3.26 m (10 ft 8½ in)
Height:
- overall, tail rotor turning ... 5.18 m (17 ft 0 in)
- overall, pylon folded ... 4.04 m (13 ft 3¼ in)
- to top of rotor head ... 3.79 m (12 ft 5⅜ in)

Wheelbase ... 4.83 m (15 ft 10 in)
Tail rotor ground clearance ... 1.83 m (6 ft 0 in)
Main/tail rotor clearance ... 6.6 cm (2⅝ in)

AREAS: As UH-60A

WEIGHTS AND LOADINGS:

Weight empty: SH-60B ASW ... 6,191 kg (13,648 lb)
- HH-60H ... 6,114 kg (13,480 lb)
- HH-60J ... 6,086 kg (13,417 lb)

Useful load: HH-60J ... 3,551 kg (7,829 lb)
Internal payload: HH-60H ... 1,860 kg (4,100 lb)
Mission gross weight:
- SH-60B ... 9,575 kg (21,110 lb)

Max T-O weight:
- SH-60B Utility, HH-60H ... 9,926 kg (21,884 lb)
- SH-60F, MH-60R ... 10, 659 kg (23,500 lb)
- HH-60J ... 9,637 kg (21,246 lb)

Max disc loading:
- SH-60B Utility, HH-60H ... 47.2 kg/m^2 (9.67 lb/sq ft)
- SH-60F, MH-60R ... 50.7 kg/m^2 (10.39 lb/sq ft)
- HH-60J ... 45.9 kg/m^2 (9.39 lb/sq ft)

Transmission loading at max T-O weight and power:
- SH-60B Utility, HH-60H ... 3.92 kg/kW (6.44 lb/shp)
- SH-60F, MH-60R ... 4.21 kg/kW (6.91 lb/shp)
- HH-60J ... 3.80 kg/kW (6.25 lb/shp)

PERFORMANCE:

Cruising speed at S/L:
- HH-60H ... 147 kt (272 km/h; 169 mph)
- HH-60J ... 146 kt (271 km/h; 168 mph)

Dash speed at 1,525 m (5,000 ft), tropical day:
- SH-60B ... 126 kt (234 km/h; 145 mph)

Vertical rate of climb at S/L, 32.2°C (90°F):
- SH-60B ... 213 m (700 ft)/min

Vertical rate of climb at S/L, 32.2°C (90°F), OEI: SH-60B ... 137 m (450 ft)/min

UPDATED

SIKORSKY S-76

TYPE: Multirole medium helicopter.

PROGRAMME: S-76A announced 19 January 1975; first flight (N762SA) 13 March 1977; certification to FAR Pt 29, Category B, on 21 November 1978; Category A on 9 January 1979; deliveries started early 1979; delivery of Mk II began 1 March 1982; S-76B programme initiated October 1983; first flight (N3123U) 22 June 1984; certified in category B on 31 October 1985 and in Category A on 3 February 1987; final S-76B delivered December 1997; see 1997–98 and previous *Jane's All the World's Aircraft* for details of these early models.

S-76C announced June 1989; replaced S-76B's P&WC PT6B turboshafts with Arriel engines; first flight 18 May 1990; FAA certification in Category B on 15 March 1991 and in Category A on 12 April 1991, deliveries beginning immediately thereafter.

Following manufacture of c/n 760514 in mid-2000, production of S-76C+ fuselages progressively transferred to Aero Vodochody of the Czech Republic under a US$200 million contract, beginning with Sikorsky-built c/n 760515, which was shipped to

Sikorsky S-76 instrument panel 0527145

Sikorsky S-76C+ executive helicopter ***NEW***/1047362

the Czech Republic for final assembly and scheduled to return to the USA in January 2001. Three further shipsets to be sent from Sikorsky, with first wholly Czech-built fuselage expected to be completed in late 2001. Fiescher Advanced Composites of Austria supplies composites components to Aero Vodochody. Keystone Helicopter Corporation of West Chester, Pennsylvania, selected in April 2000 as principal completion centre for S-76C+s, following flight testing and certification at Sikorsky's Stratford, Connecticut, facility, and completed its 50th S-76C+ on 8 September 2004.

By March 2004 more than 550 S-76s of all models were in service with 192 operators in 44 countries, and had accumulated more than three million flight hours.

CURRENT VERSIONS: **S-76C:** Initial production version with Arriel 1S1 engines; now superseded by S-76C+.

S-76C+: FAA and CAA certification and first delivery, mid-1996; features Arriel 2S1 turboshafts with FADEC for improved single-engine performance and fuel efficiency. *As described.*

CUSTOMERS: See table. The 580th production S-76 was registered in July 2004. Recent customers include Offshore Logistics Inc (OLOG), which ordered 15 S-76C+s on 10 February 2003, plus 24 options, for delivery over a five-year period, Senior Taxi Aerea of Porto Allegre, Brazil which ordered four on 3 March 2003 for delivery in 2003, the Fisheries and Maritime Matters Department of the Xunta de Galicia in Spain, which ordered two in SAR configuration for delivery in early 2005, and Air Harrods of the UK, which ordered one in VIP configuration for delivery in July 2005.

COSTS: Unit cost approximately US$8.5 million in offshore configuration (2004).

DESIGN FEATURES: Meets FAR Pt 29 with Category A IFR; intended for offshore support, business transport, medical evacuation and general utility use; technology and aerodynamics based on those of UH-60 Black Hawk.

Four-blade main rotor with high twist and varying section and camber based on Sikorsky SC-1095; tapered blade tip has 30° leading-edge sweep; fully articulated rotor head with single elastomeric bearings; hydraulic drag dampers; dual bifilar vibration absorber assemblies above rotor head; four-blade cross-beam tail rotor on port side; rotor brake optional. Max rotor speed, power on, 316 rpm (108 per cent). Main rotor blade life of 28,000 hours; tail rotor life unlimited. Optional manual blade folding.

Emergency medical service installation includes multiple-position pivoting primary patient litter, a second litter, and track-mounted seats for four attendants, forward and rear oxygen systems, and dual access to external power on ground; cabin volume 4.0 m^3 (141 cu ft).

FLYING CONTROLS: Dual powered hydraulic controls with autostabilisation and autopilot; releasable spring-centring trim system for both cyclic and collective controls.

STRUCTURE: Main rotor blades have formed and welded titanium oval-section tubular main spar with Nomex honeycomb aerofoil core, glass fibre composites outer skin and titanium/nickel leading-edge abrasion strips. Fuselage has bonded aluminium/honeycomb skin panels to permit flush riveting; extensive use made of carbon fibre (replacing Kevlar and glass fibre in earlier S-76s). Aerospace Industrial Development Corporation (AIDC) of Taiwan signed five-year contract in December 1998 for supply of composites crew and passenger doors.

LANDING GEAR: Hydraulically retractable tricycle type, with single wheel on each unit. Nosewheel retracts rearward, main units inward into rear fuselage; all three wheels are enclosed by doors when retracted. Mainwheel tyres size 14.5×5.5–6 (14 ply) tubeless, pressure 11.38 bar (165 lb/sq in); nosewheel tyre size 5.00–4 (12 ply) tubeless, pressure 9.31 bar (135 lb/sq in). Hydraulic brakes; hydraulic mainwheel parking brake. Non-retractable tricycle gear, with low-pressure tyres and skid landing gear both optional.

POWER PLANT: Two Turbomeca Arriel 2S1 turboshafts with FADEC, each rated at 638 kW (856 shp) for take-off, 731 kW (980 shp) OEI for 30 seconds and 592 kW (794 shp) maximum continuous; transmission rating 1,193 kW (1,600 shp) for take-off and maximum continuous. Standard fuel capacity 1,083 litres (286 US gallons; 238 Imp gallons) of which 1,064 litres (281 US gallons; 234 Imp gallons) are usable; optional auxiliary tank capacity 405 litres (107 US gallons; 89.1 Imp gallons). Self-sealing tanks optional.

ACCOMMODATION: One or two pilots and 12 to 13 passengers. Three four-abreast rows of seats, floor-mounted at a pitch of 79 cm (31 in). A number of executive layouts are available, including a four-passenger 'office in the sky' configuration. Executive versions have luxurious interior trim, full carpeting, special soundproofing, radiotelephone and co-ordinated furniture. Dual controls optional. Two large doors on each side of fuselage, hinged at forward edges; sliding doors available optionally. Baggage hold aft of cabin, with external door each side of fuselage. Cabin heated and ventilated; air conditioning optional. Windscreen demisting and dual windscreen wipers. Windscreen heating and external cargo hook optional.

SYSTEMS: Hydraulic system at pressure of 207 bar (3,000 lb/sq in) supplied by two pumps driven from main gearbox. Hydraulic system maximum flow rate 15.9 litres (4.2 US gallons; 3.5 Imp gallons)/min. Bootstrap reservoir. Pump head pressure 3.45 bar (50 lb/sq in). In VFR configuration, electrical system comprises two 200 A DC starter/generators and a 24 V 17 Ah Ni/Cd battery. In IFR configuration, system comprises gearbox-driven 7.5 kVA generator, and a 115 V 600 VA 400 Hz static inverter for AC power. 34 Ah battery optional. Engine fire detection and extinguishing system.

S-76 DELIVERIES

Year	S-76A	S-76A+	S-76B	S-76C	S-76C+	Total	Cum
1979	35					35	35
1980	85					86	121
1981	50					50	171
1982	29					29	200
1983	36					36	236
1984	30					30	266
1985	8		11			19	285
1986	3		13			16	301
1987	3		7			10	311
1988	2		9			11	322
1989		3	14			17	339
1990	[illegible]	13	5			19	358
1991		1	8	10		19	377
1992			4	8		12	389
1993	[illegible]		3	7		11	400
1994			6	7		13	413
1995			6	10		16	429
1996			11	1	7	19	448
1997			4		14	18	466
1998					16	16	482
1999					7	7	489
2000					7	7	496
2001					8	8	510
2002					6	6	
2003					23		533
Totals	**284**	**17**	**101**	**43**	**89**	**533**	

Notes: Some S-76As retrofitted to A+. Excludes seven (six flying) prototypes and pre-series

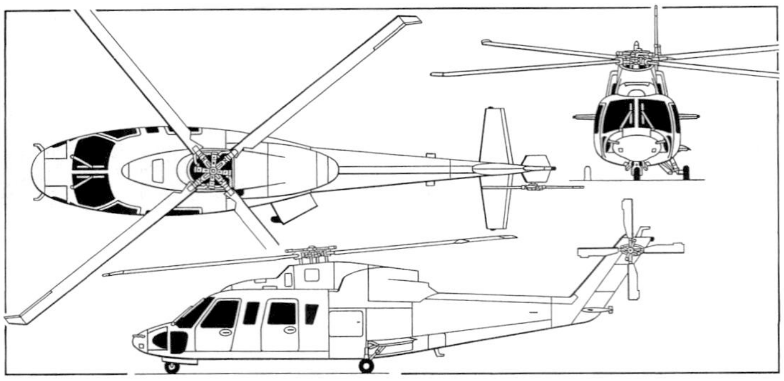

Sikorsky S-76C+ commercial transport helicopter *(Jane's/Mike Keep)* 0016738

AVIONICS: *Comms:* Honeywell Primus II integrated radios.

Radar: Honeywell Primus 440 colour weather radar.

Flight: Honeywell SPZ-7600 dual digital AFCS. Optional Universal UNS-1D navigation management system, L-3 Com/Sextant TCAS and Argus moving map display.

Instrumentation: Four-tube Honeywell colour EFIS; Parker Hannifin Gull Electronics active colour matrix LCD three-screen integrated instrument display system (IIDS).

EQUIPMENT: Standard equipment includes cabin fire extinguishers; cockpit, cabin, instrument, navigation and anti-collision lights; landing light; external power socket; and utility soundproofing. Optional equipment includes 1,497 kg (3,000 lb) capacity cargo hook, 272 kg (600 lb) capacity rescue hoist, emergency flotation gear, engine air particle separators and litter installation.

DIMENSIONS, EXTERNAL:
- Main rotor diameter ... 13.41 m (44 ft 0 in)
- Main rotor blade chord ... 0.39 m (1 ft 3½ in)
- Tail rotor diameter ... 2.44 m (8 ft 0 in)
- Tail rotor blade chord ... 0.16 m (6½ in)
- Length: overall, rotors turning ... 16.00 m (52 ft 6 in)
  - fuselage ... 13.21 m (43 ft 4 in)
- Height overall, tail rotor turning ... 4.42 m (14 ft 6 in)
- Tailplane span ... 3.05 m (10 ft 0 in)
- Width of fuselage ... 2.13 m (7 ft 0 in)
- Wheel track ... 2.44 m (8 ft 0 in)
- Wheelbase ... 5.00 m (16 ft 5 in)
- Tail rotor ground clearance ... 1.97 m (6 ft 5¾ in)

DIMENSIONS, INTERNAL:
- Passenger cabin: Length ... 2.46 m (8 ft 1 in)
  - Max width ... 1.93 m (6 ft 4 in)
  - Max height ... 1.35 m (4 ft 5 in)
  - Floor area ... 4.18 m^2 (45.0 sq ft)
  - Volume ... 5.8 m^3 (204 cu ft)
- Baggage compartment volume ... 1.1 m^3 (38 cu ft)

AREAS:
- Main rotor disc ... 141.26 m^2 (1,520.5 sq ft)
- Tail rotor disc ... 4.67 m^2 (50.27 sq ft)
- Tailplane ... 2.00 m^2 (21.5 sq ft)

WEIGHTS AND LOADINGS:
- Operating weight empty ... 2,545 kg (5,611 lb)
- Weight empty, executive configuration ... 3,691 kg (8,138 lb)
- Max baggage ... 272 kg (600 lb)
- Useful load ... 1,616 kg (3,562 lb)
- Max fuel ... 3,030 kg (6,680 lb)
- Max T-O weight ... 5,307 kg (11,700 lb)
- Max disc loading ... 37.6 kg/m^2 (7.69 lb/sq ft)
- Transmission loading at max T-O weight and power ... 4.45 kg/kW (7.31 lb/shp)

PERFORMANCE:
- Never-exceed speed (VNE) and max level speed ... 155 kt (287 km/h; 178 mph)
- Long-range cruising speed ... 140 kt (259 km/h; 161 mph)
- Max rate of climb at S/L ... 495 m (1,625 ft)/min
- Service ceiling ... 3,871 m (12,700 ft)
- Service ceiling, OEI ... 975 m (3,200 ft)
- Hovering ceiling: IGE ... 1,722 m (5,650 ft)
  - OGE ... 549 m (1,800 ft)
- Range at 140 kt (259 km/h; 161 mph) at FL40, standard fuel:
  - no reserves ... 439 n miles (813 km; 505 miles)
  - 30 min reserves ... 385 n miles (713 km; 443 miles)

UPDATED

VIP interior of Sikorsky S-76 0527142

SIKORSKY S-92 AND H-92 SUPERHAWK

TYPE: Medium transport helicopter.

DEVELOPMENT MILESTONES

Programme launched	Jun 95
Announced	Mar 92
First flight	23 Dec 98
First flight, production representative	5 Oct 01
First flight, production	14 Jun 04
Certification	17 Dec 02

PROGRAMME: Announced March 1992; originally envisaged as S-92C 'Growth Hawk' development of S-70; market evaluation co-ordinated with Mitsubishi Corporation and Mitsubishi Heavy Industries. Launched at Paris Air Show, June 1995 as S-92A Helibus and military S-92IU (international utility). Risk-sharing partners Mitsubishi Heavy Industries (7.5 per cent), Jingdezhen Helicopter Group of China (2 per cent) and Gamesa of Spain (7 per cent), with Taiwan Aerospace (6.5 per cent) and Embraer (4 per cent) as additional fixed-price supplier/partners; Russia's Mil is associated with programme, but not yet a full partner; other suppliers include Aerazur (fuel cells), Goodrich (health and usage monitoring system), Dunlop (engine inlet), Eaton (emergency flotation bag electro-optical sensors); Hamilton Sundstrand (automatic flight control system and active vibration computers), Honeywell (radar and APU), Lucas-Western (rescue hoist), Martin-Baker (crew seats), Messier-Bugatti (wheels and brakes), Moog (active vibration controls), Parker Bertea (servos), Rockwell Collins (multifunction displays and nav/com suite); Universal (FMS) and Vickers (hydraulic pumps). Programme rationalisation resulted in discontinuation of S-92IU and adoption of standard civil/military configuration with new designation S/H-92 applied in 2002; designations S-92 and H-92 Superhawk adopted in 2003 for civilian and military variants respectively.

Design changes announced in July 2000 in response to customer requests include a 0.41 m (1 ft 4 in) increase in cabin length to permit installation of a 1.27 m (4 ft 2 in) wide cabin door to improve hoisting capability and to accommodate a Stokes litter during SAR operations; reduction in height of tail rotor pylon by about 1.02 m (3 ft 4 in) to offset additional weight of cabin extension; and relocation of the horizontal stabiliser. These changes will provide additional benefits in creating an improved fold configuration for shipboard operations, increased birdstrike protection deriving from the relocation of tail rotor drive shaft and controls aft of the tail spar, and a flatter hover attitude arising from a forward shift of the helicopter's centre of gravity, improving visibility for confined space and shipboard landing, and increasing aft fuselage ground clearance. The revised configuration has been incorporated from the third prototype and on all production aircraft.

Test programme comprises one ground test vehicle (PA1/GTV; first airframe built, first ground run 4 September 1998 and completed 200-hour FAA certification endurance run in September 1999) and four flying prototypes, of which first (PA2/N292SA) completed maiden flight at Sikorsky's Development Flight Center in Florida on 23 December 1998 and was subsequently modified to the revised configuration described above. GTV and first flying prototype have 1,305 kW (1,750 shp) CT7-6D engines; remaining three flying prototypes have production-standard CT7-8s and APUs. PA3/N392SA (second flying) joined the test programme in May 1999, flew in October, and is devoted principally to engine and AFCS development; subsequently upgraded to production configuration with extended main rotor shaft; new main rotor servos; redesigned electrical system to maintain rotor ice protection after a generator failure; new tail rotor pylon; and state-of-the-art cockpit displays. Flown again in this configuration on 22 March 2002. PA4/N492SA delayed during assembly to incorporate airframe changes described above and final avionics updates and first flew in definitive production configuration on 5 October 2001. PA5/N592SA, exhibited unflown at the Paris Air Show in June 1999, is in utility configuration with MIL-STD-1553 databus, rear loading ramp, folding seats for 22 troops, sliding windows, cargo hook and provision for 7.62 mm machine gun pintle mounts; was later modified to revised configuration as described above, and made its first flight on 8 February 2001.

By February 2002, PA2 had flown more than 380 hours, and PA3 221 hours, including flights at maximum take-off weight of 13,998 kg (30,860 lb), demonstrating a range of

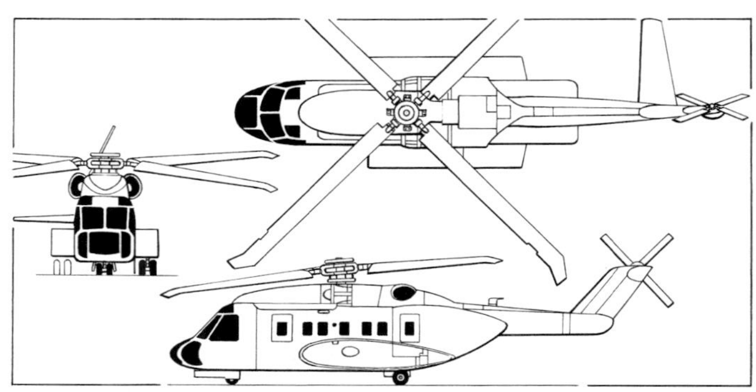

Sikorsky S-92 in envisaged production configuration *(James Goulding)* 0103496

First Sikorsky S-92A customer delivery, March 2004, to PHI ***NEW***/0576465

Sikorsky S-92 second flying prototype ***NEW***/1047293

Sikorsky S-92 first flying prototype ***NEW***/1047292

800 n miles (1,482 km; 921 miles) with mid-mission hover; PA4 had flown 60 hours; and PA5 more than 260 hours. PA2 was withdrawn from use in May 2002.

FAA FAR Pt 29 Category A certification achieved 17 December 2002 after 1,570-hour flight test programme; Category B full release awarded 7 May 2004. EASA certification achieved 8 June 2004 as the first helicopter certified by the Agency. Initial production aircraft (c/n 920006/N192PH) made its first flight on 14 June 2004. Total of 12 scheduled for delivery in 2004.

CURRENT VERSIONS: **S-92A:** Civilian variant, *as described*.

H-92 Superhawk: Military variant. Name announced at the Paris Air Show 16 June 2003. Features uprated 2,289 kW (3,070 shp) CT7-8C engines providing some 25 per cent more power than the S-92's CT7-8As; folding main and tail rotors; BAE Systems fly-by-wire controls; self-sealing fuel tanks; armour protection for critical components; self-defence systems including chaff/flare dispensers and IR suppression; and Rockwell Collins mission avionics. Programme timetable dependent on engine development, with first flight tests of CT7-8C targeted for 2006 and certification on H-92 in 2007.

Sikorsky is teamed with Bombardier Aerospace Defence Services to offer the H-92 Superhawk for the Canadian Maritime Helicopter Project to replace Canadian Forces Sea King helicopters. If the H-92 is selected, General Dynamics Canada will be responsible for mission system integration, and Bombardier will perform interior completion, installation and checking of mission equipment, exterior painting and final acceptance and delivery, in addition to in-service support and fleet management. The H-92 Superhawk is also competing for the USAF combat search and rescue helicopter programme to replace HH-60G Pave Hawks, with a requirement for 100 to 150 helicopters, and in a proposed **VH-92** variant which was the unsuccessful contender for the US Marine Corps VX requirement to replace VH-3Ds for Presidential transport missions.

H-92 Cyclone: Canadian military version. Selection announced 22 July 2004; 28 required for service in maritime role aboard Halifax-class vessels at cost of C$1.8 billion (including ship adaptation). Deliveries from early 2007.

CUSTOMERS: Total of 22 firm orders and 17 options held by May 2004. Launch customer Cougar Helicopter of Eastern Canada announced intent to order up to five for delivery in 2002; following programme delays, reaffirmed requirement for three at February 2003 Heli-Expo, Dallas, Texas. Other announced civil customers include Norsk Helikopter (two for Stratoil contract, agreed February 2003), Petroleum Helicopters Inc (PHI), which ordered two in August 2003 for offshore support operations in the Gulf of Mexico, of which the first was "handed over" in unflown condition at Heli-Expo 2004 at Dallas in March 2004, and the Turkmenistan Government, which ordered two in July 2004 for delivery in 2005-06 for Presidential transport. First VIP configured S-92 (c/n 920007/N908W) delivered to Harrods Aviation at London-Stansted Airport, UK, on 22 September 2004, for operation on behalf of Laws Helicopter LLC.

COSTS: Programme US$600 million; unit cost US$13 million to US$14 million. Direct operating cost US$2,38 per hour, comprising $1,194 fixed and $1,175 variable costs (2002).

DESIGN FEATURES: Design, manufacture and assembly use CATIA database system; modular design simplifies customised configurations. Standard medium helicopter configuration, with engines above cabin; raised tailboom to permit rear loading via ramp; high-set, strut-braced tailplane to port. Prominent sponsons accommodate main landing gear and fuel.

Dynamic components based on proven Sikorsky technology to reduce development risks, including: titanium yoke-type infinite-life main rotor head with elastomeric bearings; four-blade, quick-release, all-composites main rotor with swept, tapered anhedral tips based on scaled-up version of blades tested on Black Hawk in 1995–96; blade chord increased 12 per cent, compared with Black Hawk; damping masses at mid-span of each blade; normal rotor speed 246 rpm; new transmission based on upgraded four-stage version of Black Hawk's three-stage main gearbox; rotor brake; new intermediate tail rotor gearbox; and new four-blade, fully articulated, birdstrike-tolerant tail rotor with individually removable blades to starboard, meeting FAR/JAR 29 birdstrike requirements. Manual rotor blade folding (two forward; two aft). New-style bifiliar vibration absorber, with Hamilton Sundstrand computer and Moog actuators, on top of rotor hub comprises metal drum enclosing five composites springs allowing absorber to move in opposition to in-plane forces. Service life initially set (2004) at 30,000 hours. Target TBO for rotor head is 50,000 hours. Gearbox TBO 6,000 hours.

FLYING CONTROLS: Similar, but not identical, to Black Hawk; dual digital AFCS with autopilot and dual independent, triple-axis stability augmentation features Hamilton Sundstrand primary processor based on that of RAH-66 Comanche, and is expected to have 8,000 hour MTBF. BAE Systems contracted in 2003 to design FBW control system.

STRUCTURE: Modular structure of aluminium and composites (about 40 per cent of structure is of composites, though mostly non-structural to reduce costs) designed to be highly crack-resistant, with extensive lightning/HIRF protection; composites main rotor blades (including spars); structure optimised for minimum parts count.

Sikorsky responsible for rotor and transmission systems, final assembly and flight test. Airframe largely designed and manufactured by 'Team S-92' partners, as follows: Mitsubishi (main cabin), AIDC of Taiwan (flight deck), Embraer (sponson front halves, landing gear and incorporation of Aerazur/Intertechnique fuel system), Gamesa (cabin interior, aft fuselage, tail boom and upper fuselage transmission housing), and Jingdezhen (vertical tail including horizontal stabiliser).

LANDING GEAR: Retractable tricycle; main units retract rearwards into sponsons; nosewheel retracts forwards under flight deck. Wheels and brakes supplied by Messier-Bugatti. Optional inflatable emergency flotation bags, usable up to Sea State 5, activated by Eaton electro-optical sensors.

POWER PLANT: Two GE CT7-8A turboshafts, each rated at 1,879 kW (2,520 shp) for T-O, 1,742 kW (2,336 shp) for 30 minutes, and 1,523 kW (2,043 shp) maximum continuous; 1,939 kW (2,600 shp) OEI 30 seconds; 1,879 kW (2,520 shp) OEI 2 minutes; 1,863 kW (2,498 shp) OEI 30 minutes; and 1,523 kW (2,043 shp) OEI continuous. Hamilton Sundstrand dual-channel FADEC with autostart, power assurance and OEI training modes. Rolls-Royce Turbomeca RTM322 will be offered as alternative engine if demand is forthcoming. Transmission via modular compound planetary gearbox with 30 minute run-dry capability and 140 per cent overtorque certification, rating 3,110 kW (4,170 shp).

Single-point pressure refuelling/defuelling. Standard fuel in sponsons, each with optional gravity fuelling port, combined capacity 2,896 litres (765 US gallons; 637 Imp gallons) of which 2,877 litres (760 US gallons; 633 Imp gallons) are usable. Standard auxiliary fuel option of 1,401 litres (370 US gallons; 308 Imp gallons, comprising two 700 litre (185 US gallon; 154 Imp gallon) tanks in cabin. Other fuel options include two external tanks, each 871 litres (230 US gallons; 192 Imp gallons); or single or dual 322 litre (85.0 US gallon; 70.8 Imp gallon) bench-type tanks in cabin. Crash-resistant fuel system standard. In-flight refuelling probe optional.

ACCOMMODATION: Two-pilot crew on separate flight deck on FAA/JAA 16 *g* crashworthy seats; provision for one flight deck observer; 19 (FAA certification limit) to 24 passengers, with baggage on rear ramp; 10 executive passengers; or up to three LD-3 cargo containers in civil version. Main door, starboard side, front in three options: (a) split horizontally in upward- and downward-hinged sections, latter including steps; (b) lower airstair section with inward/backward-sliding upper portion (allowing installation of external SAR winch) and (c) full-height sliding door incorporating Type IV emergency exit. Further three FAA/JAA Type IV cabin emergency exits, plus pop-out windows at each seat row, for rapid emergency evacuation. Total of 22 combat-ready troops on sideways seats in military version; both versions have rear-loading ramp and 366 kg/m^2 (75 lb/sq ft) minimum floor rating (optionally 976 kg/m^2; 200 lb/sq ft). Military version has sliding cabin windows and weapons mounts. Martin-Baker crew and passenger seats. Accommodation is heated and ventilated with independent cockpit and cabin zones; air conditioning optional. Active noise suppression system under consideration.

SYSTEMS: Goodrich IMD-HUMS health and usage monitoring system standard, with cockpit displays and downloading facility to enable ground crew to access system via hand-held diagnostic equipment; active noise control system reduces cabin noise by 3 to 4 dB; active vibration control may be employed in key airframe areas.

Honeywell 36-150 (S92) APU with in-flight start and continuous run capability. Electrical system comprising two 75 kVA, 115 V, 400 Hz, three-phase main gearbox-driven AC generators, two 400 A, AC/28 V DC converters and one 35 kVA APU-driven back-up generator with 100 A AC/DC backup converter. Battery 28 V, 15 Ah. Three hydraulic systems, supplied by main gearbox-driven pumps, two serving main and tail rotor and

VIP interior of S-92 0527135

S-92 flight deck 0103498

stability augmentation system, while third serves utilities and acts as back-up; all 276 bar (4,000 lb/sq in). Anti-icing system for engine inlets, windscreen and pitot head; main and tail rotor de-icing optional. Automatically deployed emergency flotation system meets or exceeds FAR/JAR stability requirements and has demonstrated Sea State 5 capability; externally mounted 14-person liferaft in forward end of each sponson, with 50 per cent overload capability.

AVIONICS: Open architecture avionics system accommodating ARINC 429 and MIL-STD-1553 interfaces with Collins Pro Line 4 as core system.

Comms: Dual VHF, radio management and audio controls. Mode S transponder. BAE Systems CVR/FDR. Optional deployable emergency beacon.

Radar: Provision for radar in nose compartment.

Flight: UNS-1C FMS; Hamilton Sundstrand automatic flight control system featuring three-axis stability augmentation system and fully coupled dual digital autopilots with automatic approach-to-hover option. Independent standby instruments. Optional UNS-1ESP with GPS; TCAS I; and EGPWS.

Instrumentation: Rockwell Collins EFIS cockpit with four 203 × 152 mm (8 × 6 in) MFD-268EP active matrix liquid crystal displays (AMLCD) for PFD, EICAS, health monitoring, navigational and weather radar functions; fifth AMLCD, 152 × 203 mm (6 × 8 in) centre-mounted, optional for displaying sensor information such as moving map or FLIR data.

All avionics housed in removable mission equipment rack behind co-pilot station with wiring routed through conduits in fuselage frames for added protection.

Mission: Optional satcom, FLIR and loudhailer.

EQUIPMENT: Optional, single or dual, hydraulically powered, electrically controlled SAR rescue hoist, maximum capacity 272 kg (600 lb), with cable viewing window and spotlight. Windscreen wiper/washer system. Optional cargo handling system, which is compatible with 1.07 × 1.22 m (3 ft 6 in × 4 ft 0 in) pallets, includes 1,814 kg (4,000 lb) capacity centreline winch with 28 V DC winch motor, easy on/off cabin rollers, and 16 *g* cargo tiedowns at all major frames.

DIMENSIONS, EXTERNAL:
Main rotor diameter ... 17.71 m (56 ft 4 in)
Tail rotor diameter ... 3.35 m (11 ft 0 in)
Length overall, rotors turning ... 20.88 m (68 ft 6 in)
Fuselage: Length: airframe ... 17.12 m (56 ft 2 in)
incl tail rotor ... 18.47 m (60 ft 7 in)
Max width: over sponsons ... 3.89 m (12 ft 9 in)
incl horizontal stabiliser ... 5.26 m (17 ft 3 in)
Max width, main rotor in X position ... 12.19 m (40 ft 0 in)
Height: overall, tail rotor turning ... 5.46 m (17 ft 11 in)
tail rotor in X position ... 5.13 m (16 ft 10 in)
to top of main rotor head ... 4.70 m (15 ft 5 in)
to top of fin ... 4.32 m (14 ft 2 in)
Wheel track (c/l shock-absorbers) ... 3.17 m (10 ft 5 in)
Wheelbase ... 6.20 m (20 ft 4 in)
Forward door (stbd): Height ... 1.85 m (5 ft 11 in)
Width ... 1.02 m (3 ft 4 in)

DIMENSIONS, INTERNAL:
Cabin: Length (incl bulkhead) ... 6.10 m (20 ft 0 in)
Max width ... 2.01 m (6 ft 7 in)
Max height ... 1.83 m (6 ft 0 in)
Volume ... 16.9 m^3 (596 cu ft)
Baggage volume (civil) ... 3.97 m^3 (140 cu ft)

AREAS:
Main rotor disc ... 231.55 m^2 (2,492.4 sq ft)
Tail rotor disc ... 8.83 m^2 (95.03 sq ft)

WEIGHTS AND LOADINGS:
Weight empty: military utility ... 6,895 kg (15,200 lb)
offshore oil ... 7,212 kg (15,900 lb)
airline ... 7,076 kg (15,600 lb)
executive ... 7,802 kg (17,200 lb)
SAR ... 7,348 kg (16,200 lb)
Max baggage weight ... 454 kg (1,000 lb)
Max internal fuel ... 2,327 kg (5,130 lb)
Max external load ... 4,536 kg (10,000 lb)
Max T-O weight (civil):
internal load ... 11,861 kg (26,150 lb)
external load* ... 12,836 kg (28,300 lb)
Max disc loading:
internal load ... 51.2 kg/m^2 (10.49 lb/sq ft)
external load* ... 55.4 kg/m^2 (11.35 lb/sq ft)
Transmission loading at max T-O weight and power:
internal load ... 3.81 kg/kW (6.26 lb/shp)
external load* ... 4.12 kg/kW (6.77 lb/shp)

* *Not yet certified*

PERFORMANCE (at max internal load T-O weight):
Never-exceed speed (VNE): power on ... 165 kt (305 km/h; 189 mph) IAS
power off ... 120 kt (222 km/h; 138 mph) IAS
Max cruising speed ... 153 kt (283 km/h; 176 mph)
Econ cruising speed ... 139 kt (257 km/h; 160 mph)
Max certified altitude ... 4,575 m (15,000 ft)
Hovering ceiling, A: IGE ... 3,450 m (11,320 ft)
OGE ... 2,170 m (7,120 ft)
Range, offshore configuration, normal fuel, 19 passengers 915 m (3,000 ft), ISA+10°C,
30 min reserve +10% ... 444 n miles (822 km; 510 miles)
no reserve ... 544 n miles (1,007 km; 626 miles)
Range with standard internal auxiliary fuel ... 726 n miles (1,344 km; 835 miles)
Radius of action, SAR: normal fuel ... 235 n miles (435 km; 270 miles)
auxiliary fuel ... 345 n miles (638 km; 397 miles)

UPDATED

SINO SWEARINGEN

SINO SWEARINGEN AIRCRAFT COMPANY

1770 Skyplace Boulevard, San Antonio, Texas 78216
Tel: (+1 210) 258 39 00
Fax: (+1 210) 258 39 73
e-mail: gcomfort@sj30jet.com
Web: http://www.sj30jet.com
CHAIRMAN: Chuen-Huei Tsai
PRESIDENT AND CEO: Dr Carl Chen
CHIEF FINANCIAL OFFICER AND SENIOR VICE-PRESIDENT, FINANCE AND ADMINISTRATION: David Turner
SENIOR VICE-PRESIDENT, OPERATIONS: Alfred Baumbusch
DEPUTY SENIOR VICE-PRESIDENT, OPERATIONS: T H Tsiang
VICE-PRESIDENTS:
Gene Comfort (Sales and Marketing)
John W Dieker (Manufacturing)
David Gee (Engineering)
Robert E Homan (Quality Assurance)
Chester J Schickling (Sales and Marketing)
Jim L Wooley (Human Resources)
Carol Zuniga (Accounting)
SENIOR ADVISOR: Ed Swearingen

Ed Swearingen is well known for designing Fairchild Merlin and Metro commuter and business aircraft, and for engineering such aircraft as Piper Twin Comanche and Lockheed JetStar II. Current project is SJ30-2 business jet.

Swearingen Aircraft Inc re-formed on 1 April 1995 as joint venture company with Sino Aerospace of Taiwan, which provides financial support; will certify and produce SJ30-2 at new 8,130 m^2 (87,500 sq ft) facility at Martinsburg, West Virginia, where the workforce totalled 85 in September 2002, and is expected to rise to about 350 by 2006. San Antonio workforce stood at 400 in late 2002.

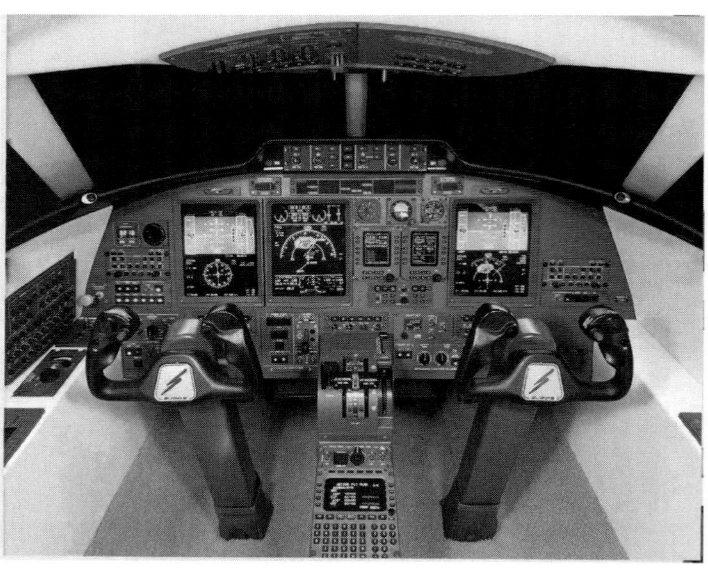

Instrument panel of the SJ30-2 ***NEW***/1022931

SINO SWEARINGEN SJ30-2

TYPE: Light business jet.

PROGRAMME: Announced 30 October 1986 as SA-30 Fanjet; Gulfstream Aerospace, Williams International and Rolls-Royce announced they were joining programme in October 1988 and aircraft renamed Gulfstream SA-30 Gulfjet; Gulfstream withdrew from programme 1 September 1989; place taken by Jaffe Group of San Antonio, Texas, and aircraft renamed Swearingen/Jaffe SJ30; now a Sino Swearingen project. First flight of prototype (N30SJ) 13 February 1991 (with FJ44-1 engines).

Certification originally intended in 1995 but delayed pending development of increased performance SJ30-2, *as described*, which features fuselage stretched by 1.32 m (4 ft 4 in), wing span increased by 1.83 m (6 ft 0 in), increased wing dihedral, revised wing/fuselage fairing, increased fuel capacity and more powerful FJ44-2A engines. Prototype (N30SJ, modified from SJ30 prototype) first flown (original engines) 8 November 1996.

FJ44-2A engines installed and flown for first time on 4 September 1997; total of 270 flight hours accumulated with these engines (or 370 hours overall in SJ30-2 configuration) by late April 1999, including operation at speeds up to M0.83 and maximum altitude of 13,105 m (43,000 ft), 1.8 *g* turns at 12,495 m (41,000 ft), verification of stall and engine-out minimum control speeds, stability and control tests, engine operating tests throughout the flight envelope, simulated icing tests, yaw/damper/rudder bias system testing and handling trials with simulated flight control failures including trim runaways and asymmetric speed brake deployment. N30SJ withdrawn from use in mid-1999.

Static and fatigue tests on a representative section of primary wing joint completed February 1999, to 150 per cent of limit load in 230,000 loading cycles. Risk-sharing partner Gamesa Aeronautica of Vitoria, Spain, manufactured fuselages and wings of five certification test airframes; first of three production conforming flying prototypes (c/n 002/N138BF) rolled out at San Antonio 17 July 2000 and first flown 30 November 2000; public debut at the NBAA Convention in Orlando, Florida, 9 September 2002, but destroyed in crash near Del Rio, Texas on 26 April 2003; 003/N30SJ (second use of registration), first flown 6 March 2003 and 004/N404SJ, first flown 17 October 2003, were each scheduled to complete 400 flight hours during the test and certification programme; No. 005 added to test programme to replace 003, having previously been earmarked as initial customer delivery aircraft; and will serve as autopilot, interior and function/reliability certification airframe c/n TF-2 and TF-3 are static test and fatigue airframes respectively, last-named being the final airframe to be assembled in Texas; TF-2 transferred to Sino Swearingen static test facility at San Antonio 8 March 2001, with static tests scheduled for completion by mid-2002. The

Second of the three production-confirming prototypes SJ30-2 No. 003 *(Paul Jackson)* ***NEW***/0567967

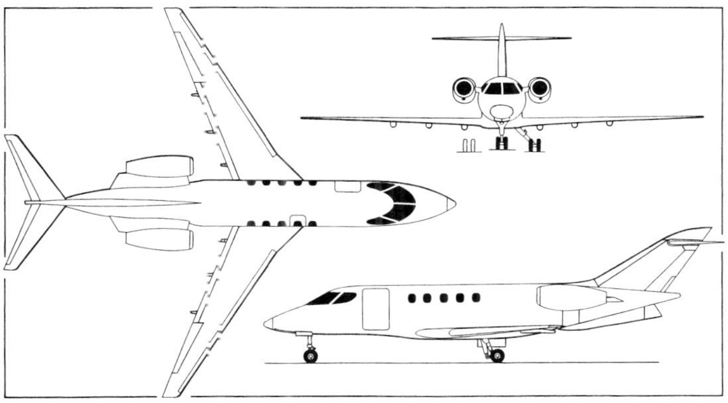

Swearingen SJ30-2 business jet *(Paul Jackson)* ***NEW***/0567227

1,500 hour flight test programme was scheduled to culminate in FAA certification in FAR Pt 23 Commuter category and first customer deliveries in early 2004 prior to the loss of the first conforming prototype.

CUSTOMERS: More than 150 orders placed by customers and distributors in Brazil, Canada, France, Germany, Mexico, South Africa, Sweden, United Kingdom and USA by September 2002, representing planned production to mid-2005.

COSTS: US$5.495 million (2003). Estimated direct operating cost, based on 1,000 n mile (1,852 km; 1,150 mile) stage length, US$601.15 per hour (2003).

DESIGN FEATURES: High-speed and high-altitude cruise with long-range capabilities, plus good slow speed handling; highly efficient swept wing planform.

Proprietary, computer-designed aerofoil sections. Wing sweep 32° 36', dihedral 2° 30'.

FLYING CONTROLS: Conventional and manual. Electrically actuated variable incidence tailplane, trim tab on rudder, aileron force bias spring for trimming in roll axis; single ventral fin under tail for yaw damping; slotted Fowler trailing-edge flaps actuated electrically; hydraulically actuated full-span leading-edge slats; single electrohydraulically actuated spoiler/lift dumper panel on each wing ahead of flap.

STRUCTURE: All-metal with chemically milled skins on fuselage.

LANDING GEAR: Retractable tricycle type, manufactured by Integrated Aerospace of Santa Ana, California, with twin wheels on each unit. Trailing-link oleo-pneumatic suspension on main units. Hydraulic actuation, main units retracting inward and rearward into fuselage, nose unit forward. Hydraulically steerable (±60°) nose unit. All wheels 41 cm (16 in) diameter. Main unit tyre pressure 10.76 bar (155 lb/sq in), nosewheel tyre pressure 2.14 bar (31 lb/sq in). Standard air-cooled power braking; dual ABS anti-skid.

POWER PLANT: Two 10.23 kN (2,300 lb st) Williams FJ44-2A turbofans flat rated to 72°C, pod-mounted on pylons on sides of rear fuselage. Six fuel tanks: one in each outer wing (total 1,055 litres; 278.8 US gallons; 232.1 Imp gallons); two wing hopper tanks (total 103 litres; 27.2 US gallons; 22.6 Imp gallons); centre wing tank (199 litres; 52.7 US gallons; 43.9 Imp gallons); and fuselage (torsion box) tank (1,440 litres; 380.3 US gallons; 316.7 Imp gallons) for total of 2,797 litres (739 US gallons; 615.3 Imp gallons) under FAR Pt 23 certification, or 2,487 litres (657 US gallons; 547 Imp gallons) under JAR 23 certification. Fuel is burned sequentially to maintain aircraft centre of gravity. Single refuelling point in fuselage tank. Oil capacity 3.4 litres (0.9 US gallon; 0.75 Imp gallon) per engine.

ACCOMMODATION: Pilot and one passenger (or co-pilot) on flight deck. Main cabin separated by a bulkhead; six passengers seats comprising club four and two forward-facing seats; aft-facing seats recline and have lateral tracking capability; forward-facing seats have full berthing capability; lavatory, separated from main cabin by a half-cabin height partition with fan close-out, has belted, side-facing seat that is approved for occupation during take-off and landing. Full-height refreshment centre and foldaway tables, dual cupholders and cabin environment controls standard; recessed storage area aft of lavatory, with foldaway coat hooks, optional. Airstair passenger door with light at front on port side. Emergency exit over starboard wing. Baggage compartment aft of main cabin, with external access via port side door under engine nacelle. Two-piece birdproof electrically heated wraparound windscreen.

SYSTEMS: Cabin pressurised to 0.83 bar (12.0 lb/sq in), maintaining sea level pressure to 12,500 m (41,000 ft); cabin heated by engine bleed air; cooled by a vapour-cycle system. Hydraulic system (207 bar; 3,000 lb/sq in) for actuation of leading-edge slats, airbrake/lift dumpers and landing gear extension/retraction. Two 300 A 28 V DC engine-driven starter/generators and static inverters. Redundant frequency-wild alternators provide power for windscreen heating. Wing and engine inlets anti-iced by engine bleed air; tailplane has pneumatic de-icer boots, de-icing system. Oxygen system has 646 litre (22.8 cu ft) capacity.

AVIONICS: Honeywell Primus Epic CDS as core system.

Comms: Honeywell Primus II radio system; optional 8.33 kHz spacing.

Radar: Honeywell Primus 331 colour weather radar.

Flight: Dual IC-615 integrated avionics computers with combined flight director, autopilot and FMS/GPS functions; dual RVSM-compatible Honeywell AZ-850 micro air data computers; ADF, DME. Options include TCAS 2000, second FMS/GPS integrated with IC-615, second ADF, second DME and lightning sensor system.

Instrumentation: EFIS cockpit comprising two Honeywell DU-180 flat panel 203 × 254 mm (8 × 10 in) active matrix colour LCDs, for PFD and MFD functions, with third optional.

SJ30-2 cabin ***NEW***/1022932

DIMENSIONS, EXTERNAL:
Wing span ... 12.90 m (42 ft 4 in)
Wing chord: at root ... 2.41 m (7 ft 11 in)
at tip ... 0.50 m (1 ft 7½ in)
Wing aspect ratio ... 9.4
Length overall ... 14.26 m (46 ft 9½ in)
Fuselage: Length ... 12.79 m (41 ft 11½ in)
Max diameter ... 1.52 m (5 ft 0 in)
Height overall ... 4.34 m (14 ft 3 in)
Tailplane span ... 4.44 m (14 ft 7 in)
Wheel track (mean) ... 2.09 m (6 ft 10¼ in)
Wheelbase ... 5.70 m (18 ft 8½ in)
Passenger door: Height ... 1.42 m (4 ft 8 in)
Width ... 0.81 m (2 ft 8 in)
Baggage door: Height ... 0.65 m (2 ft 1½ in)
Width ... 0.47 m (1 ft 6½ in)
Emergency door: Height ... 0.56 m (1 ft 10 in)
Width ... 0.48 m (1 ft 7 in)

DIMENSIONS, INTERNAL:
Cabin: Length:
between pressure bulkheads ... 5.36 m (17 ft 7 in)
passenger section ... 3.81 m (12 ft 6 in)
Max width ... 1.43 m (4 ft 8½ in)
Max height ... 1.31 m (4 ft 3½ in)
Floor area ... 5.46 m^2 (58.8 sq ft)
Volume ... 8.7 m^3 (307.0 cu ft)
Baggage compartment volume ... 1.50 m^3 (53 cu ft)

AREAS:
Wings, gross ... 17.72 m^2 (190.7 sq ft)
Ailerons (total) ... 0.76 m^2 (8.20 sq ft)
Trailing-edge flaps (total) ... 3.31 m^2 (35.63 sq ft)
Leading-edge flaps (total) ... 2.83 m^2 (30.48 sq ft)
Fin ... 1.06 m^2 (11.40 sq ft)
Rudder, incl tab ... 0.69 m^2 (7.40 sq ft)
Tailplane ... 3.41 m^2 (36.70 sq ft)
Elevator, incl tabs ... 0.79 m^2 (8.53 sq ft)

WEIGHTS AND LOADINGS (A: FAR Pt 23 commuter certification; B: JAR23 certification):
Weight empty, equipped: A, B ... 3,538 kg (7,800 lb)
Operating weight empty: A, B ... 3,674 kg (8,100 lb)
Max baggage weight: A, B ... 227 kg (500 lb)
Max payload: A, B ... 635 kg (1,400 lb)
Max fuel weight: A ... 2,245 kg (4,950 lb)
B ... 1,996 kg (4,400 lb)
Max T-O weight: A ... 6,123 kg (13,500 lb)
B ... 5,670 kg (12,500 lb)
Max ramp weight: A ... 6,169 kg (13,600 lb)
B ... 5,715 kg (12,600 lb)
Max landing weight: A ... 5,688 kg (12,540 lb)
B ... 5,670 kg (12,500 lb)
Max zero-fuel weight: A, B ... 4,536 kg (10,000 lb)
Max wing loading: A ... 345.6 kg/m^2 (70.79 lb/sq ft)
B ... 320.0 kg/m^2 (65.55 lb/sq ft)
Max power loading: A ... 299 kg/kN (2.93 lb/lb st)
B ... 277 kg/kN (2.72 lb/lb st)

PERFORMANCE (estimated):
Max operating Mach No. (MMO) ... 0.83
High-speed cruising speed ... M0.80 (459 kt; 850 km/h; 528 mph)
Long-range cruising speed ... M0.78 (447 kt; 828 km/h; 514 mph)
Stalling speed, flaps down, flight idle power ... 94 kt (174 km/h; 108 mph)
Max rate of climb at S/L: A ... 1,189 m (3,900 ft)/min
B ... 1,250 m (4,100 ft)/min
Rate of climb at S/L, OEI: A ... 305 m (1,000 ft)/min
B ... 335 m (1,100 ft)/min
Service ceiling ... 14,935 m (49,000 ft)
Service ceiling, OEI: A ... 6,126 m (20,100 ft)
B ... 6,340 m (20,800 ft)
FAA T-O balanced field length: A ... 1,217 m (3,993 ft)
B ... 1,103 m (3,620 ft)

FAA landing distance:
at max landing weight: A, B .... 1,042 m (3,420 ft)
at typical mission completed weight, pilot and three passengers: A, B .... 847 m (2,780 ft)
Range, pilot and three passengers, NBAA IFR reserves, 100 n mile (185 km; 115 mile) alternate, at M0.80:
A, B .... 2,363 n miles (4,376 km; 2,719 miles)
Range, pilot and three passengers, NBAA IFR reserves, 100 n mile (185 km; 115 mile) alternate at M0.78:
A .... 2,500 n miles (4,630 km; 2,877 miles)
B .... 1,840 n miles (3,407 km; 2,117 miles)
OPERATIONAL NOISE LEVELS (estimated):
T-O .... 74.7 EPNdB
Approach .... 93.6 EPNdB
Sideline .... 85.6 EPNdB

SKYCRAFT

SKYCRAFT INTERNATIONAL INC

5000 West Airport Drive, Canton/Akron Regional Airport, North Canton, Ohio 44720
Tel: (+1 330) 966 69 66
Fax: (+1 330) 862 62 62
e-mail: arvsuper2@aol.com
PRESIDENT: John Hall

SkyCraft purchased rights to Super2 from Highlander Aircraft Corporation of Golden Valley, Minnesota, in 1999. Previously, Highlander had obtained the design from Aviation Scotland in UK and became sole source when co-licensee, ASL Hagfors Aero of Sweden, ceased trading in September 1995.

No production has been reported since 1999.

UPDATED

SKYCRUISER

SKYCRUISER CORPORATION

Two Soundview Drive, Greenwich, Connecticut 06830
Tel: (+1 203) 625 00 71
Fax: (+1 203) 625 00 65
e-mail: airship@airshipman.com
Web (1): http://www.skycruisergroup.com
Web (2): http://www.airshipman.com
PRESIDENT: George A Spyrou

Global Skyship Industries, then the US subsidiary of Aviation Support Group Ltd of the UK, acquired the assets of the former Westinghouse Airships Inc (WAI) in December 1996. As a result, GSI became holder of the type certificates for the UK-designed Skyship 500 HL and Skyship 600 and the WAI Sentinel series. In late 1997, GSI announced that it had re-opened the production line and had these airships available for immediate delivery.

Formation of Skycruiser Corporation was announced on 18 November 2002, as a joint venture between Airship Management Services Inc (AMS) of Greenwich, Connecticut, and Skyship Cruise Ltd of Lindau, Switzerland. New holder of the Type Certificates for the Skyship and Sentinel series of airships, it is part of the Skycruiser group of companies that include AMS, GSI (now co-located with AMS), Skyship Cruise Ltd and Skycruise Switzerland. The companies also operate a maintenance facility and FAA repair station at Skycruiser Corporation's base in Elizabeth City, North Carolina.

UPDATED

SKYCRUISER SKYSHIP 500HL

TYPE: Helium non-rigid airship.
PROGRAMME: First flight (G-SKSB) (converted Skyship 500) 30 July 1987.
CUSTOMERS: Three conversions completed in UK; one (N601LP) in service in USA with Airship Operations Inc in 1998, but destroyed by severe storm on 7 September 1998. Type remains available from GSI which, in mid-2001, still had three (including 07/N503LP and 09/N504LP) ready for assembly on receipt of orders; however, no new sales reported by mid-2004.
DESIGN FEATURES: Combines modified Skyship 500 gondola with larger envelope of Skyship 600, providing ability to operate with greater payload in hotter climates and at higher altitudes than Skyship 500.
POWER PLANT: Two 153 kW (205 hp) Porsche 930/01 non-turbocharged engines; fuel capacity 545 litres (144 US gallons; 120 Imp gallons), of which 530 litres (140 US gallons; 117 Imp gallons are usable).
ACCOMMODATION: Pilot, co-pilot and up to nine passengers. Ground crew of 18.
DIMENSIONS, EXTERNAL: As for Skyship 600 except:
Length overall .... 59.13 m (194 ft 0 in)
Gondola: Length .... 9.75 m (32 ft 0 in)
Max width .... 2.13 m (7 ft 0 in)
Max height .... 1.93 m (6 ft 4 in)
DIMENSIONS, INTERNAL: As for Skyship 600 except:
Envelope volume .... 6,666 m^3 (235,400 cu ft)
Gondola cabin: Length .... 4.20 m (13 ft 9½ in)
Height .... 1.96 m (6 ft 5 in)
WEIGHTS AND LOADINGS:
Max usable fuel .... 382 kg (842 lb)
Gross disposable load .... 2,190 kg (4,829 lb)
PERFORMANCE:
Max level speed .... 53 kt (99 km/h; 62 mph)
Cruising speed .... 30 kt (56 km/h; 35 mph)
Pressure ceiling .... 2,135 m (7,000 ft)
Max range .... 347 n miles (643 km; 400 miles)
Endurance at 35 kt (65 km/h; 40 mph) .... 17 h

UPDATED

SKYCRUISER SKYSHIP 600

TYPE: Helium non-rigid airship.
PROGRAMME: First flight 6 March 1984; special category C of A awarded by UK CAA 1 September 1984; aerial work certification received second quarter 1986; full passenger-carrying C of A 8 January 1987, initiating Skycruise aerial sightseeing service over London, San Francisco, Munich and Sydney in 1987, and over Paris in 1988. First US FAA type certificate awarded to an airship for civil use was issued to Skyship 600 on 9 May 1989. Certificate passed to Slingsby Aviation on 24 September 1990; Westinghouse Airships Inc on 24 February 1994; Global Skyship Industries on 7 August 1997; and Skycruiser Corporation in November 2002. During 1990 Farnborough Air Show a Skyship 600 (G-SKSC) with two 227 litre (60 US gallon; 50 Imp gallon) long-endurance fuel tanks made unrefuelled flight of 50 hours 15 minutes; sufficient fuel for a further 20 hours remained at end of flight.
CURRENT VERSIONS: **Skyship 600:** Initital (UK-built) version (nine completed).

Description applies to Skyship 600 except where indicated.

Skyship 600B: With larger envelope, using Tedlar fabric, manufactured by TCOM LP. First example, converted from Skyship 600 c/n 07 (N602SK) and operated for Fuji Film by Airship Management Services, flew for first time 25 July 1998. Second example (c/n 10, N610SK) completed for Airship Operations Inc (Hilfiger contract) to replace destroyed Skyship 500HL.
CUSTOMERS: Total of nine built in UK, of which 01 (G-SKSC) purchased by Global Skyships from UK MoD and 07 (N602SK) converted to 600B in 1998. Two more 600Bs began construction in 1999, of which 02/B-04 (N602CL) was delivered to CargoLifter, Germany, in May 2000, certified by LBA 3 August 2001 and re-registered as D-LCLA *Charly*; for use as a crew trainer. In 2001, N606SA had its Porsche engines replaced by Lycoming IO-540s, making its maiden flight in this form on 18 January 2002 before being redelivered to Airship Management Services Europe in Paris in the second quarter of 2002. In 2003 it was being operated by STI Aviation on behalf of the US Office of Naval Research.
COSTS: Skyship 600B approximately US$6 million (2000).
DESIGN FEATURES: Non-rigid envelope, with one ballonet forward and one aft (together 26 per cent of total volume); ballonet air intake aft of each propulsor unit. Cruciform tail unit.
FLYING CONTROLS: Vectored thrust propulsion (see Power Plant); differential inflation of ballonets for static fore and aft trim; cable-operated rudders and elevators, each with spring tab; ballast in box below crew seats; disposable water ballast in tanks at rear.
STRUCTURE: Envelope manufactured by TCOM LP from single ply polyester fabric, coated with titanium dioxide-loaded polyurethane to reduce ultraviolet degradation; polyvinylidene chloride film bonded on to inner coating of polyurethane on inside of envelope minimises loss of helium gas. Four parabolic arch load curtains, carrying multiple Kevlar 29 gondola suspension cables. Nose structure is domed disc, moulded from GFRP and carrying fitting by which airship is moored to its mast. Each tail surface attached to envelope at root and braced by wires on each side; all four surfaces constructed from interlocking ribs and spars of Fibrelam with GFRP skins.

One-piece moulded gondola of Kevlar-reinforced plastics, with flooring and bulkheads of Fibrelam panels; those forming engine compartment at rear are faced with titanium for fire protection.

Skyship 600B in operation with a US customer ***NEW***/0561599

SKYSHIP 500 PRODUCTION

c/n	Registrations	Remarks
1214/01	G-BECE	AD500; first flight 3 Feb 79; destroyed Mar 79
1214/02	G-BIHN, N502LP	Now 500HL, Trans-Continental Leasing
1214/03	G-SKSA	Deregistered 1996
1214/04	G-SKSB, N602LP, N501LP	Trans-Continental Leasing
1214/05	G-SKSE, JA1003	
1214/06	G-SKSH, N601LP	500HL; destroyed 7 Sep 98
1214/07	N503LP	Now 500HL, Trans-Continental Leasing
1214/09	N504LP	Trans-Continental Leasing

SKYSHIP 600 PRODUCTION

c/n	Registrations	Remarks
1215/01	G-SKSC, ZH762	Stored
1215/02	G-SKSD, VH-HAA, N602CL, D-CLCA, N601SK	Now 600B, Wells-Fargo (trustee)
1215/03	G-SKSG	Destroyed; cancelled 1996
1215/04	G-SKSF, N601SK, N430KK	Sentinel 1000; destroyed
1215/05	G-SKSJ, N600LP, N605SK	Wells-Fargo (trustee)
1215/06	G-SKSK, VH-HAN, JA1004, F-GLSD, N606SA	Wilmington Trust
1215/07	G-SKSL, N602SK	Now 600B, Airship Management Services
1215/08	G-SKSM, HL-	
1215/09	G-SKSN, JA1006	
1215/10	N610SK	Wells-Fargo (trustee)

LANDING GEAR: Single two-wheel assembly with double tyres, mounted beneath rear gondola.

POWER PLANT: Two 190 kW (255 hp) Porsche 930/67/AI/3 six-cylinder air-cooled and turbocharged piston engines mounted in rear of gondola. Each drives a ducted propulsor consisting of a Hoffmann HO-V155A-R/137 five-blade reversible-pitch propeller within an annular duct of carbon fibre-reinforced GFRP. Each propulsor can be rotated about its pylon attachment to gondola through an arc of 210°, 90° upward and 120° downward, vectored thrust thus providing both V/STOL and in-flight hovering ability.

Fuel tank, capacity 682 litres (180 US gallons; 150 Imp gallons), at rear of engine compartment; 668 litres (176.5 US gallons; 147 Imp gallons) usable. Auxiliary fuel tanks optional. Engine modifications include provision of automatic mixture control, fuel injection and electronic ignition.

ACCOMMODATION: Seats for pilot and co-pilot, with dual controls. Maximum capacity 12 passengers in addition to pilots. Ground crew of 18.

SYSTEMS: 28 V electrical system, supplied by engine-driven alternators.

AVIONICS: Include Honeywell Silver Crown series dual nav/com, ADF, Omega, VOR/ILS and weather radar.

EQUIPMENT: Night signs (30.48 m; 100 ft long and 9.14 m; 30 ft high) are full-colour aerial boards comprising 8,400 primary-coloured lamps computerised to enable the display of signs, animation and graphic logos in 16 colours.

DIMENSIONS, EXTERNAL:
- Envelope:
  - Length overall: 600 .... 59.00 m (193 ft 7 in)
  - 600B .... 60.96 m (200 ft 0 in)
  - Max diameter .... 15.20 m (49 ft 10½ in)
- Fineness ratio: 600 .... 3.88
  - 600B 3.91
- Height overall, incl gondola and landing gear .... 20.30 m (66 ft 7¼ in)
- Width over tailfins .... 19.20 m (63 ft 0 in)
- Gondola: Length overall .... 11.67 m (38 ft 3½ in)
  - Max width .... 2.56 m (8 ft 4¾ in)
  - Max height .... 1.93 m (6 ft 4 in)
- Propeller diameter .... 1.37 m (4 ft 6 in)

DIMENSIONS, INTERNAL:
- Envelope volume: 600 .... 6,666 m^3 (235,400 cu ft)
  - 600B .... 7,192.5 m^3 (254,000 cu ft)
- Ballonet volume (two, total) .... 1,733 m^3 (61,200 cu ft)
- Gondola cabin: Length .... 6.89 m (22 ft 7¼ in)
  - Height .... 1.92 m (6 ft 3½ in)
  - Floor area (usable) .... 12.00 m^2 (130.0 sq ft)

WEIGHTS AND LOADINGS:
- Max usable fuel .... 484 kg (1,067 lb)
- Gross disposable load .... 2,343 kg (5,165 lb)

PERFORMANCE:
- Max level speed .... 56 kt (104 km/h; 65 mph)
- Cruising speed .... 35 kt (64 km/h; 40 mph)
- Pressure ceiling .... 2,135 m (7,000 ft)
- Max range (standard fuel) .... 347 n miles (643 km; 400 miles)

UPDATED

SKYMEDIA

SKYMEDIA AIRSHIPS INC

PRESIDENT: Paul K Cohen
CHAIRMAN: Paul M Thomas

SkyMedia Airships was formed in 2001 with equity holding, management and marketing assistance from VentureNet Capital Group of Temecula, California. It designed the SMA, a low-cost hot-air airship for advertising purposes, but by mid-2004 the company appeared to be no longer extant, its website having expired on 19 August. There is no evidence that the SMA had been built or flown by that time.

UPDATED

SKY RAIDER

SKY RAIDER LLC

8668 Linden Road, Caldwell, Idaho 83687
Tel: (+1 208) 465 71 16
Fax: (+1 208) 465 79 37
e-mail: skyraiderllc@earthlink.net
Web: http://www.skyraiderllc.com
MANAGER: Fred Parker

Company, previously named Flying K Enterprises, was established by former SkyStar employee, the late Kenny Schrader. A version of its Sky Raider is produced in the UK as Reality Easy Raider.

In 2003, the company added a third version, the Super Sky Raider, to its range.

UPDATED

Sky Raider Sky Raider built by Brian Raeder and Grand Champion Ultralight at EAA AirVenture 2003, fitted with Kitfox wheels, modified tail and redesigned cowling ***NEW***/1047223

SKY RAIDER SKY RAIDER

TYPE: Single-seat ultralight kitbuilt; tandem-seat ultralight kitbuilt.

PROGRAMME: First flown in 1996.

CURRENT VERSIONS: **Sky Raider:** Single-seat model, available as certified aeroplane or under FAR Pt 103 ultralight regulations.

Sky Raider II: Tandem-seat version; introduced in 1999.

Super Sky Raider: Slightly larger version with dual controls, introduced 2003.

CUSTOMERS: Total of 162 SkyRaiders and 26 SkyRaider IIs sold by 4 June 2001 (latest information supplied).

COSTS: Sky Raider kit US$9,350; Sky Raider II kit US$10,945; Super Sky Raider kit US$13,995 (2003).

DESIGN FEATURES: High wing with tubular V-struts, and two auxiliary struts to each wing. Wire-braced fin and V-strut braced tailplane. Wings foldable for storage without disconnection of control runs. Quoted build time 300 hours for Sky Raider and 350 hours for Sky Raider II.

FLYING CONTROLS: Conventional and manual. Frise ailerons, cable actuated. Slotted flaps.

STRUCTURE: Fabric-covered steel tube. Optional wingtip extensions to Sky Raider and Sky Raider II add 55 cm (21½ in) to span; choice of rounded or square fin on Sky Raider. Heavier wing spar (6 cm; 2½ in diameter) and taller fin on Sky Raider II. Horn-balanced rudder on Sky Raider II.

LANDING GEAR: Tailwheel type: fixed. Two side Vs hinged to lower longerons, plus half-axles, each with bungee cord shock-absorber, attached to compression frame. Mainwheels 8.00–6. Optional floats and skis.

POWER PLANT: *Sky Raider:* One 20.1 kW (27 hp) Rotax 277 two-cylinder two-stroke driving a two-blade propeller for Pt 103; 31.0 kW (41.6 hp) Rotax 447 for normal operations. Optionally, engines up to 37.3 kW (50 hp) can be fitted. Fuel capacity 18.9 litres (5.0 US gallons; 4.2 Imp gallons).

Sky Raider II: One 44.7 kW (60 hp) HKS flat-two, four-stroke engine driving a GSC ground-adjustable three-blade or Powerfin carbon fibre propeller. Optional engines include the 41.8 kW (56.0 hp) Zanzottera MZ 202. Fuel capacity 37.9 litres (10.0 US gallons; 8.3 Imp gallons) due to extra wing fuel tank.

Super Sky Raider: One 59.7 kW (80.0 hp) Jabiru 2200 driving a GSC ground-adjustable two-blade propeller; optional engines include the 68.6 kW (92 hp) Zanzottera MZ 301. Fuel capacity 53 litres (14.0 US gallons; 11.7 Imp gallons).

ACCOMMODATION: *Sky Raider:* Pilot only, in open-sided cabin.

Sky Raider II: Two seats in tandem with single controls within fully enclosed cabin.

Super Sky Raider: Two seats in tandem with dual controls; taller and longer cabin.

DIMENSIONS, EXTERNAL (A: Sky Raider, B: Sky Raider II, C: Super Sky Raider):
- Wing span: A, B .... 7.99 m (26 ft 2½ in)
  - C .... 8.53 m (28 ft 0 in)
- Length overall: A, B .... 5.18 m (17 ft 0 in)
  - C .... 5.49 m (18 ft 0 in)
- Height overall: A .... 1.57 m (5 ft 2 in)
  - B .... 1.73 m (5 ft 8 in)
  - C .... 1.83 m (6 ft 0 in)

AREAS:
- Wings, gross, A, B:
  - without extensions .... 9.10 m^2 (98.0 sq ft)
  - with extensions; and C .... 9.94 m^2 (107.0 sq ft)

WEIGHTS AND LOADINGS:
Weight empty: A with Rotax 447 ... 115 kg (253 lb)
B with HKS ... 184 kg (405 lb)
C with Jabiru ... 204 kg (450 lb)
Max T-O weight: A ... 249 kg (550 lb)
B ... 431 kg (950 lb)
C ... 476 kg (1,050 lb)
PERFORMANCE (A with Rotax 447; B with HKS, C with Jabiru 2200):
Never-exceed speed (VNE):
all ... 86 kt (160 km/h; 100 mph)
Normal cruising speed: A ... 56 kt (105 km/h; 65 mph)
B ... 65 kt (121 km/h; 75 mph)
C ... 70 kt (129 km/h; 80 mph)
Stalling speed, flaps down: A ... 20 kt (37 km/h; 23 mph)
B ... 26 kt (49 km/h; 30 mph)
C ... 28 kt (52 km/h; 32 mph)
Max rate of climb at S/L: A ... 320 m (1,050 ft)/min
B ... 396 m (1,300 ft)/min
C ... 427 m (1,400 ft)/min
T-O and landing run: all ... 23 m (75 ft)
Range: A ... 120 n miles (222 km; 138 miles)
B ... 250 n miles (463 km; 287 miles)

UPDATED

SKYSTAR

SKYSTAR AIRCRAFT CORPORATION

3901 Aviation Way, Caldwell, Idaho 83605
Tel: (+1 800) 554 83 69 and (+1 208) 454 24 44
Fax: (+1 800) 454 64 64
e-mail: info@skystar.com
Web: http://www.skystar.com
PRESIDENT AND CEO: Frank Miller

SkyStar Aircraft Corporation took over Kitfox programme from former Denney Aerocraft Company in 1993. Kitfox production continues. By 2000, more than 4,000 Kitfox kits had been sold. In 1998, the company added the Kitfox Lite and in the following year streamlined its range. The Series 6 Kitfox and Kitfox Lite^2 were added in January 2000; the Series 7 was introduced in early 2002 to replace the Series 6; the Kitfox Lite^2 was completed in 2003.

Rights to Aero Designs Pulsar purchased in late 1996, but in July 1999 the design was sold to the Pulsar Aircraft Corporation (which see).

In early 2001, Skystar opened a spare parts shop at Caldwell and launched an Internet retailing operation. In January 2004, in advance of the Light Sport Aircraft category, SkyStar reorganised into two separate divisions: Sport Plane and Experimental/Manufacturing. The former is headed by former President and CEO Ed Downs; the latter by Frank Miller.

UPDATED

SKYSTAR KITFOX

TYPE: Side-by-side kitbuilt.

PROGRAMME: Prototype Kitfox first flown 7 May 1984. Initial variants, now discontinued, had MTOW below US 544 kg (1,200 lb) ultralight limit; kit production totalled 257 Model 1s, 491 Model 2s, and 467 Model 3s. Series IV (retrospectively renamed Classic IV) introduced 1991; 323 kits of 476 kg (1,050 lb) MTOW version produced before introduction of current version. First certified version was Series 5, introduced April 1994, but replaced by Series 6 in January 2000; itself replaced by Series 7 in early 2002.

CURRENT VERSIONS: **Classic IV:** Ultralight version, introduced in 1992. Engines up to 74.6 kW (100 hp) can be fitted under helmeted engine cowling. MTOW 544 kg (1,200 lb). **Speedster** option includes short-span wings and improved streamlining for up to 17 kt (32 km/h; 20 mph) additional speed. Still available; quoted build time 900 hours.

Kitfox 4B: Biplane version. One only built (N95FX).

Kitfox 6 Convertible: Introduced 2000; tailwheel and nosewheel subvariants. Considerably refined design, with only 5 per cent parts commonality with Kitfox V. Now discontinued.

Kitfox 7: Public debut at Sun 'n' Fun, April 2002; modifications include different struts and redesigned cowling, plus wing technology developed for Kitfox Lite.

Detailed description applies to above version except where indicated.

Kitfox Sport: Intended to meet Sport Pilot category; empty weight of 318 kg (700 lb). Prototype (N702KS) built 2002.

Kitfox Lite: *Described separately.*

CUSTOMERS: Some 4,400 Kitfox aircraft kits delivered to 42 countries, and over 2,500 built.

COSTS: Classic IV US$12,795; Series 7 US$15,995, excluding engine and avionics; Sport US$29,995 including engine and instruments (2003).

DESIGN FEATURES: Designed to have good short-field performance. Constant-chord, high wing with V strut bracing each side; braced tailplane, mid-mounted. With Kitfox IV, introduced in 1991, new wing with laminar flow section and all-metal hinge brackets for full-span flaperons became standard; windscreen material thickened and given increased slope and, with other changes, resulted in increase in cruising speed; stalling and landing speeds decreased, former due to use of flaperons; wing folding standard. Optional agricultural spray pods. Quoted build time 1,000 hours.

FLYING CONTROLS: Manual. Junkers-type differential flaperons have been redesigned for Series 7; rudder and elevators with mass balance. Variable incidence tailplane.

STRUCTURE: Wings of aluminium alloy, plywood ribs and glass fibre tips, with fabric covering overall. Steel tube fuselage and tail unit, with fabric covering.

LANDING GEAR: Fixed. Fuselage-mounted spring aluminium cantilever mainwheel legs with optional speed fairings. Option of tailwheel or trailing link, castoring nosewheel; can be converted after completion; hydraulic disc brakes; optional floats, amphibious floats and skis.

POWER PLANT: Option include 73.5 kW (98.6 hp) Rotax 912 ULS or 84.6 kW (113.4 hp) Rotax 914 driving a Hoffmann HO-V352F two-blade constant-speed propeller; 74.6 kW (100 hp) Teledyne Continental O-200, 93.2 kW (125 hp) Teledyne Continental IO-240B and 86.5 kW (116 hp) Textron Lycoming O-235. For Sport Pilot compliance, a 47.8 kW (64.1 hp) Rotax 582, 59.6 kW (79.9 hp) Rotax 912 or 73.5 kW (98.6 hp) Rotax 912S are recommended in the Classic IV. Fuel in two wing tanks, total capacity 102 litres (27.0 US gallons; 22.5 Imp gallons) of which 98 litres (26.0 US gallons; 21.7 Imp gallons) are usable.

ACCOMMODATION: Two, side by side, can have 0.24 m^3 (8.5 cu ft) storage space behind seat. Optional drop tank style cargo pod.

DIMENSIONS, EXTERNAL:
Wing span: normal ... 9.75 m (32 ft 0 in)
optional (Speedster and Sport) ... 8.84 m (29 ft 0 in)
Wing aspect ratio: normal ... 7.8
Width, wings folded ... 2.44 m (8 ft 0 in)
optional (Speedster and Sport) ... 7.0
Length: overall ... 5.99 m (19 ft 8 in)
wings folded: Series 7 ... 6.88 m (22 ft 7 in)
Sport ... 6.71 m (22 ft 0 in)
Max width of fuselage ... 1.08 m (3 ft 6½ in)

Ultralight Kitfox Classic IV *(Paul Jackson)* ***NEW***/1047357

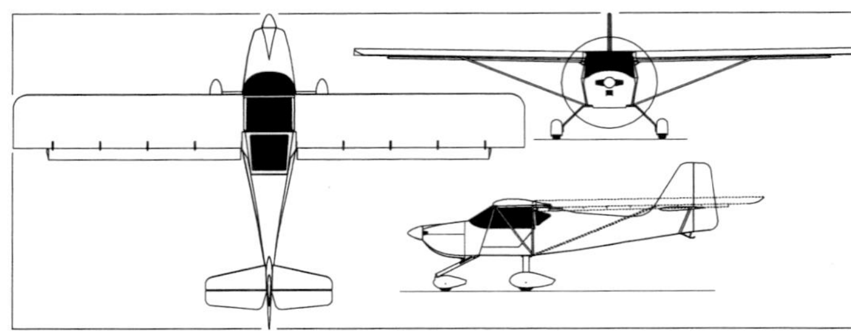

Skystar Kitfox Series 7, showing folded wing position in profile *(Paul Jackson)* ***NEW***/1047848

Height overall: tailwheel ... 1.73 m (5 ft 8 in)
nosewheel ... 2.44 m (8 ft 0 in)
Tailplane span ... 2.41 m (7 ft 11 in)
Wheel track ... 2.03 m (6 ft 8 in)
Propeller diameter ... 1.83 m (6 ft 0 in)
DIMENSIONS, INTERNAL:
Cabin max width ... 1.09 m (3 ft 7 in)
AREAS:
Wings, gross (normal): Series 7 ... 12.26 m^2 (132.0 sq ft)
Speedster and Sport ... 11.15 m^2 (120.0 sq ft)
WEIGHTS AND LOADINGS (Rotax 912 ULS):
Weight empty: Series 7 ... 363 kg (800 lb)
Sport ... 318 kg (700 lb)
Baggage capacity ... 68 kg (150 lb)
Max T-O weight: Series 7 ... 703 kg (1,550 lb)
Sport ... 558 kg (1,232 lb)
Max wing loading: Series 7 ... 57.3 kg/m^2 (11.74 lb/sq ft)
Sport ... 50.1 kg/m^2 (10.27 lb/sq ft)
Max power loading: Series 7 ... 9.57 kg/kW (15.72 lb/hp)
Sport ... 7.60 kg/kW (12.49 lb/hp)
PERFORMANCE (two crew, Rotax 912 ULS, unless stated otherwise):
Max level speed ... 111 kt (205 km/h; 125 mph)
Cruising speed ... 104 kt (193 km/h; 120 mph)
Stalling speed ... 36 kt (66 km/h; 41 mph)
Max rate of climb at S/L ... 366 m (1,200 ft)/min
T-O and landing run ... 92 m (300 ft)
Range, with reserves:
Series 7 ... 664 n miles (1,231 km; 765 miles)
Sport ... 573 n miles (1,062 km; 660 miles)
g limits, pilot only ... +3.8/–1.52

UPDATED

SKYSTAR KITFOX LITE AND KITFOX LITE^2

TYPE: Single-seat ultralight/kitbuilt; side-by-side ultralight/kitbuilt.

PROGRAMME: First displayed at Sun 'n' Fun in April 1998. Planned to be certified in Europe under BCAR standards.

CURRENT VERSIONS: **Kitfox Lite:** Single-seat version.

Description applies to the above.

Kitfox Lite^2: Two-seat training version introduced in 2000. Has 37.0 kW (49.6 hp) Rotax 503 UL-2V engine. Tricycle version first flown 15 February 2001. Production was halted in 2003 due to delays with the FAA Sport Pilot rule; 10 aircraft were due to be built in last batch.

CUSTOMERS: 200 kits and 10 factory-built aircraft delivered by January 2001.

COSTS: Kitfox Lite kit US$18,350 including engine, factory built US$21,995 (2004). Kitfox Lite^{2}kit at close of production was US$21,995 including engine, factory built US$28,995 (2003).

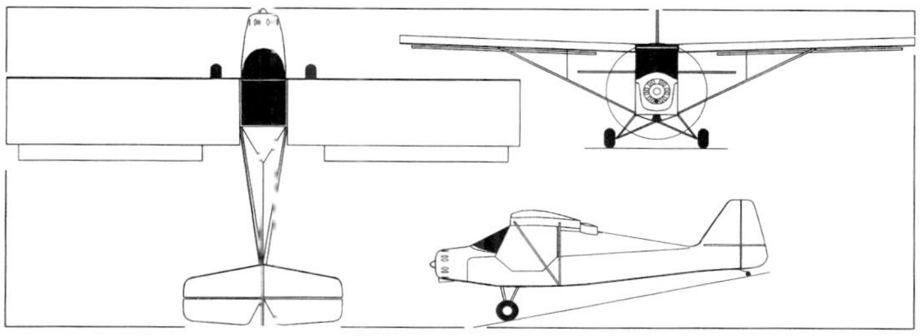

General arrangement of Skystar Kitfox Lite *(Paul Jackson)* **NEW**/0567242

DESIGN FEATURES: Generally as for Kitfox. Meets FAR Pt 103 standards. Wings fold for storage. Quoted build time 200 hours; wing pre-built by factory.

FLYING CONTROLS: Manual, including flaperons.

STRUCTURE: Welded 4130 steel fuselage; aluminium wing spar; aluminium and wood wing, all fabric-covered.

LANDING GEAR: Fixed. Two faired-in side Vs hinged to lower longerons, plus half-axis, each with bungee cord shock-absorber, attached to compression frame. Option of tricycle or tailwheel configuration; free-castoring 6 cm (2½ in) tailwheel. 4.00-6 main tyres. Individual wheel brakes. Options also include floats and skis.

POWER PLANT: One 26.1 kW (35 hp) Skystar/2SI Series 460 engine driving Tennessee two-bladed propeller through 2.5:1 reduction gear. Alternatively, a 37.0 kW (49.6 hp) Rotax 503 can be fitted. Fuel capacity 19 litres (5.0 US gallons; 4.2 Imp gallons) in Lite; 38 litres (10.0 US gallons; 8.3 Imp gallons) in Lite^2.

EQUIPMENT: Kitfox Lite option includes BRS parachute (standard on factory-built examples).

DIMENSIONS, EXTERNAL:

Wing span: Lite	7.67 m (25 ft 2 in)
Lite^2	8.89 m (29 ft 2 in)
Length overall: Lite	5.00 m (16 ft 5 in)
Lite^2	5.61 m (18 ft 5 in)
Length: wings folded: Lite	5.36 m (17 ft 7 in)
Height in level flight: Lite	1.88 m (6 ft 2 in)

AREAS:

Wings, gross: Lite	9.38 m^2 (101.0 sq ft)
Lite^2	11.61 m^2 (125.0 sq ft)

Skystar Kitfox Lite single-seat version *(Paul Jackson)* **NEW**/1047392

WEIGHTS AND LOADINGS:

Weight empty: Lite	115 kg (254 lb)
Lite^2	223 kg (492 lb)
Max T-O weight: Lite	249 kg (550 lb)
Lite^2	476 kg (1,050 lb)

PERFORMANCE (with 2SI):

Max level speed: Lite	55 kt (101 km/h; 63 mph)
Lite^2	72 kt (133 km/h; 83 mph)
Normal cruising speed: Lite	48 kt (89 km/h; 55 mph)
Lite^2	65 kt (121 km/h; 75 mph)
Stalling speed: Lite	24 kt (44 km/h; 27 mph)
Lite^2	32 kt (58 km/h; 36 mph)
Max rate of climb at S/L	229 m (750 ft)/min
T-O run: Lite	31 m (100 ft)
Lite^2	77 m (250 ft)
Landing run: Lite	23 m (75 ft)
Lite^2	67 m (220 ft)
Range: Lite	95 n miles (177 km; 110 miles)
Lite^2	199 n miles (370 km; 230 miles)
Endurance: Lite	2 h 0 min
Lite^2	3 h 30 min
g limits	+3.8/-1.52

UPDATED

SLIPSTREAM

SLIPSTREAM INDUSTRIES INC

W 8407 Cottonville Drive, Wautoma, Wisconsin 54982
Tel: (+1 920) 787 58 86
Fax: (+1 920) 787 57 15
e-mail: mail@slipstream.bz
Web: http://www.slipstream.bz
PRESIDENT: Michael Puhl

Formed on 1 April 1999, SlipStream produces a range of ultralight aircraft and accessories, and distributes the Verner range of engines. Rights to the Viking Dragonfly were sold to Dart Industries of South Africa in March 2004. The Skyquest was also withdrawn from the product line at this time. The company also manufactures the Shark range of amphibious floats at its 1,394 m^2 (15,000 sq ft) factory. In April 2004, the workforce numbered six.

UPDATED

SLIPSTREAM DRAGONFLY

Rights to the Dragonfly were sold to Dart Industries of South Africa in March 2004.

NEW ENTRY

SLIPSTREAM GENESIS AND REVELATION

TYPE: Side-by-side kitbuilt; side-by-side ultralight kitbuilt.

PROGRAMME: Prototype Genesis first flew 1992; Revelation developed as a lower-powered 'recreational' version.

CURRENT VERSIONS: **Genesis:** Baseline version. Experimental category.

Revelation: Lower-powered version; can be upgraded to Genesis standard by replacement only of windscreen and engine and addition of glass fibre enclosure.

CUSTOMERS: Total of 142 Genesis and 127 Revelation aircraft flying by end 2003.

COSTS: Kit including engine (2004): Genesis US$18,995-US$34,525; Revelation US$13,995-US$32,525, depending on engine.

DESIGN FEATURES: Quoted build time for Genesis 400 hours; for Revelation 300 hours (deduct 90 when opting for sailcloth-covered wing).

FLYING CONTROLS: Conventional and manual. Horn-balanced elevator.

STRUCTURE: Preformed glass fibre nose; fabric-covered strut-braced wings; tubular steel rear fuselage; main landing gear legs of 4 cm (1½ in) glass fibre.

LANDING GEAR: Tricycle type; fixed. Fuselage-mounted spring glass fibre cantilever mainwheel legs with optional composites wheel fairings. Cable brakes; hydraulic brakes optional. Optional floats.

SlipStream Genesis two-seat kitbuilt *(Paul Jackson)* **NEW**/0567797

POWER PLANT: *Genesis:* One 47.8 kW (64.1 hp) Rotax 582 driving Tennessee two-blade wooden propeller. Fuel capacity 76 litres (20.0 US gallons; 16.7 Imp gallons). Other engines including Verner 133M, Rotax 912 UL, 912 ULS and 914 can be fitted.

Revelation: One 39.3 kW (52.7 hp) Rotax 503B. Fuel capacity 38 litres (10.0 US gallons; 8.3 Imp gallons). Same engine options as Genesis.

Data apply to both versions, except where otherwise stated.

DIMENSIONS, EXTERNAL:

Wing span	9.35 m (30 ft 8 in)
Length overall	5.89 m (19 ft 4 in)
Height overall	1.92 m (6 ft 3½ in)

AREAS:

Wings, gross	16.62 m^2 (179.0 sq ft)

WEIGHTS AND LOADINGS (A: Genesis, B: Revelation):

Weight empty: A	295 kg (650 lb)
B	211 kg (465 lb)
Max T-O weight: A	589 kg (1,300 lb)
B	499 kg (1,100 lb)

PERFORMANCE (A and B as above):

Never-exceed speed (VNE):	
A	104 kt (193 km/h; 120 mph)

Slipstream Revelation ultralight *(Paul Jackson)* **NEW**/1047409

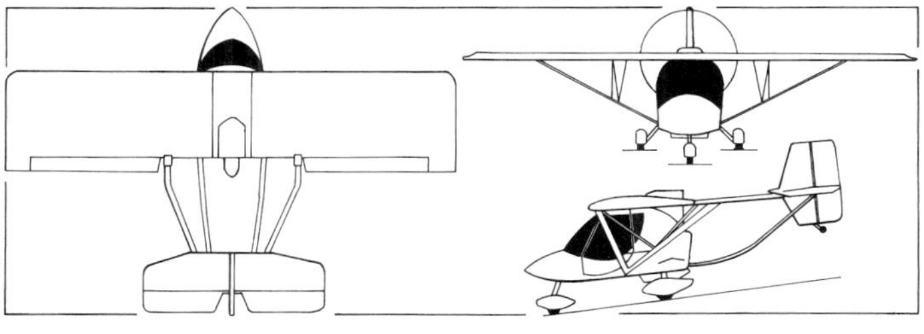

SlipStream Revelation two-seat ultralight *(Paul Jackson)* 0100466

Slipstream Genesis as a floatplane *(Paul Jackson)* ***NEW***/1047407

SlipStream Revelation ultralight derivative of Genesis *(Paul Jackson)* ***NEW***/0567796

Normal cruising speed: A .... 70 kt (129 km/h; 80 mph)
B .... 57 kt (106 km/h; 66 mph)
Stalling speed: A .... 39 kt (72 km/h; 45 mph)
B .... 38 kt (70 km/h; 43 mph)
Max rate of climb at S/L: A .... 305 m (1,000 ft)/min
B .... 244 m (800 ft)/min
T-O run: A, B .... 122 m (400 ft)
Landing run: A .... 77 m (250 ft)
B .... 46 m (150 ft)
Range with max fuel: A .... 252 n miles (466 km; 290 miles)
B .... 130 n miles (241 km; 150 miles)

UPDATED

SLIPSTREAM SCEPTER

TYPE: Single-seat ultralight kitbuilt.
PROGRAMME: Development of Genesis, introduced in 1999.
CUSTOMERS: Total of 14 flying by end 2003.
COSTS: US$11,495 to US$13,295 (2004) including engine.
DESIGN FEATURES: Quoted build time 120 hours.

Data as Genesis, except that below.

POWER PLANT: One 39.3 kW (52.7 hp) Rotax 503B, 29.5 kW (39.6 hp) Rotax 447 UL-1V or 47.8 kW (64.1 hp) Rotax 582. Fuel capacity 19 litres (5.0 US gallons; 4.2 Imp gallons); options to increase to 37.9 litres (10.0 US gallons; 8.3 Imp gallons) and 75.7 litres (20.0 US gallons; 16.7 Imp gallons) available.
WEIGHTS AND LOADINGS (Rotax 503):
Weight empty .... 199 kg (438 lb)
Max T-O weight .... 544 kg (1,200 lb)
PERFORMANCE (Rotax 447):
Normal cruising speed .... 48 kt (89 km/h; 55 mph)
Stalling speed .... 24 kt (44 km/h; 27 mph)
Max rate of climb at S/L .... 229 m (750 ft)/min

SlipStream Scepter single-seat ultralight *(Paul Jackson)* 0526985

T-O and landing run .... 31 m (100 ft)
Range with max fuel .... 60 n miles (112 km; 70 miles)

UPDATED

SLIPSTREAM DEFIANCE

TYPE: Two-seat kitbuilt twin.
PROGRAMME: Introduced in 2000 as the Skyblaster; based on Genesis/Revelation with 98 per cent commonality of parts; withdrawn from product line in 2003 for redesign of cowling and front-mounted engine; reintroduced in late-2004 as the Defiance.
CUSTOMERS: Four flying by mid-2002 (latest data supplied).
COSTS: Estimated US$37,995 with Verner engines.
DESIGN FEATURES: Two in-line engines, with counter-rotating propellers, mounted on front and back of wing; layout simplifies single-engine handling. Forward engine to be cowled within nose cone.

Data as Genesis, except that below.

POWER PLANT: Two 61.1 kW (82 hp) Verner four-stroke engines in push/pull layout. Fuel capacity 151 litres (40.0 US gallons; 33.3 Imp gallons).
DIMENSIONS EXTERNAL:
Height overall .... 2.42 m (7 ft 11¼ in)
WEIGHTS AND LOADINGS:
Weight empty .... 363 kg (800 lb)
Max T-O weight .... 635 kg (1,400 lb)
PERFORMANCE:
Normal cruising speed .... 70 kt (129 km/h; 80 mph)
Stalling speed .... 44 kt (81 km/h; 50 mph)
Max rate of climb at S/L .... 366 m (1,200 ft)/min
T-O run .... 23 m (75 ft)

UPDATED

SLIPSTREAM SKYQUEST

Skyquest production was terminated in 2004 due to difficulty in obtaining a suitable power plant.

UPDATED

SlipStream SkyQuest twin-engined kitbuilt *(Jane's/Susan Bushell)* 0110685

SOLARIS

SOLARIS AVIATION INC

Palm Beach North County Airport, 11250-4 Aviation Boulevard, West Palm Beach, Florida 33412
Tel: (+1 954) 415 93 37
Fax: (+1 954) 757 51 82
Web: http://www.solarisaviation.com
PRESIDENT: Nick Schneider
VICE-PRESIDENTS:
George Rodgers
J Cook
CHAIRMAN: David Schuldt

Solaris Aviation was established in 1999 and purchased the assets of Ruschmeyer Luftfahrttechnik GmbH of Germany including Type Certificates, moulds and tooling. Ruschmeyer had filed for bankruptcy in January 1996, having built a small number of R 90 four-seat touring aircraft. Solaris has also acquired a 10,220 m^2 (110,000 sq ft) manufacturing facility in Bielsko-Biała, Poland, where labour-intensive parts of the relaunched aircraft are produced before being shipped to Florida for final assembly. The Sigma, as the R 90 is now known, was first shown publicly at AirVenture, Oshkosh, in July 2001, the project having been formally announced on 5 July 2001. No production had ensued by late 2004, although promotion was then continuing.

UPDATED

SOLARIS SIGMA AND ALPHA

TYPE: Four-seat lightplane.
PROGRAMME: Development of original Ruschmeyer MF-85 started in Germany in 1985; first prototype MF-85P-RG (V0001) flew with Porsche engine; changed to R 90-230 RG using flat-rated Textron Lycoming IO-540 after cessation of Porsche production in March 1990; first flight of prototype V001 (D-EEHE) 8 August 1988; first flight of V002 (D-EERO) 25 September 1990; V003 (D-EERH) first flight 12 February 1992; LBA certification June 1992; FAA 24 June 1994. Main production variant was R 90-230 RG, of which 25 built; Ruschmeyer also produced prototype fixed landing gear R 90-230 FG; lower and higher-powered versions were overtaken by company's bankruptcy. Programme transferred to Solaris in USA, 1999; German-built R 90-230 RG registered N230S in June 2001 as demonstrator. Production deliveries due to begin in third quarter of 2002, but no production reported by mid-2004.

US-registered Ruschmeyer R 90-230 RG employed by Solaris as a Sigma demonstrator *(Paul Jackson)* ***NEW***/1047155

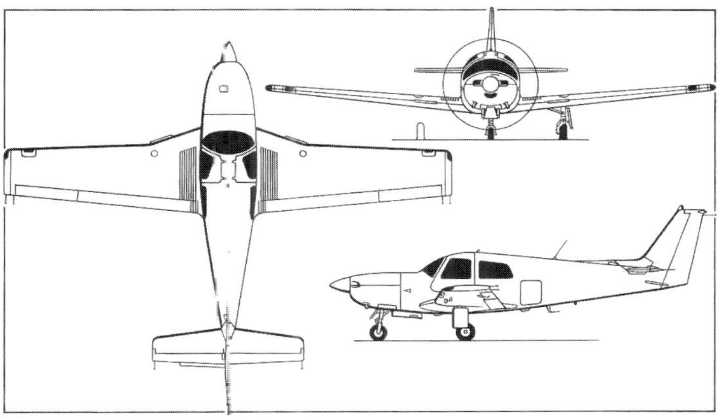

Solaris Sigma four-seat tourer *(Jane's/Mike Keep)* 0507144

CURRENT VERSIONS: **Sigma 230:** One Textron Lycoming IO-540-C4D5 flat-six, derated to 172 kW (230 hp). Retractable landing gear; intended for European market.

Alpha 230: As above. Fixed gear.

Sigma 250: IO-540-C4D5 engine, fully rated at 186 kW (250 hp). Retractable gear. Baseline US version.

Alpha 230: As immediately above. Fixed gear.

Sigma 310: Teledyne Continental IO-550N flat-six, rated at 231 kW (310 hp). Retractable gear. Uprated US version.

Alpha 310: As immediately above. Fixed gear.

COSTS: Sigma 250 US$244,900; Sigma 310 US$279,900 (2002).

DESIGN FEATURES: Objectives were FAR Pt 23 compliance; high performance by low-drag all-composites airframe, use of derated engine for European version, requiring less cooling, reduced noise by lower engine rpm, special silencer and, originally, matched four-blade constant-speed propeller. Consequently, reduced fuel consumption and emissions; noise level under German regulations demonstrated 66 dBA, 8 dBA below limit; also 10.2 dBA below ICAO Chapter 10 limit. Laminar flow aerofoil. Certified 18,000 hour life.

FLYING CONTROLS: Conventional and manual; all push/pull rod operated to minimise friction; dual controls with sticks standard; fixed tailplane; combined trim and trim tab on port elevator; stall strips on inboard wing leading-edge; three-position Fowler flaps operated by electrohydraulic system are linked to elevator to avoid trim changes; max deflection 30°. Aileron deflections +16/–11°; elevator +18/–15°; rudder starboard 240 mm (9.45 in), port 220 mm (8.66 in).

STRUCTURE: All-composites moulded airframe with 70 per cent fewer parts than equivalent metal structure. Ruschmeyer qualified BASF Palatal A430 resin fibre composites material for aviation; advantages are durability at up to 72°C ambient, improved strength, field repair without special tools, negligible allergic risks during manufacture, and material is recyclable. GFRP with Rohacell foam cores and CFRP reinforcement round doors; steel tube running from firewall, over roof and into tail, supports doors and protects against roll-over; components earth bonded to German and FAA standards. Meets 1.75 safety factor for composites airframes.

LANDING GEAR: Tricycle type, retractable. Electrohydraulic actuation; trailing-link levered mainwheel suspension retracts inward and covered by mechanically linked doors; foot-powered hydraulic disc brakes; steerable nosewheel retracts rearward, also covered by doors.

POWER PLANT: As for Current Versions. Hartzell three-blade, constant-speed propeller. Adjustable cowl flaps. Standard Sigma 250/Alpha 250 fuel 311 litres (82.0 US gallons; 68.3 Imp gallons) in integral tanks in inner wings; Sigma 310/Alpha 310 349 litres (92.0 US gallons; 76.7 Imp gallons). Oil capacity 11.4 litres (3.0 US gallons; 2.5 Imp gallons) for Sigma 250/Alpha 250; 12 litres (3.25 US gallons; 2.7 Imp gallons) for Sigma 310/Alpha 310.

ACCOMMODATION: Four-seat interior with ergonomically designed seats; three-point harness; upward-opening gull wing doors; heating and ventilation system with electric blower; baggage compartment accessible from inside and outside.

SYSTEMS: S-Tec 40 (optional 55X) single-axis autopilot. Electrohydraulic actuation for flaps; pitot heater and night lighting. Electrical system 28 V DC; 70 Ah alternator; 24 V 18 Ah battery.

AVIONICS: *Comms:* Garmin GNS-430 transceiver; GTX-327 transponder; GMA-340 audio system; ELT. Optional GNS-530 nav/com with GI-106 moving map indicator.

Flight: Garmin GNS-420 GPS. Optional second GNS-430, S-Tec ST-180 slaved compass, ST-361 flight director, ST-360 altitude selector, Goodrich WX-500 Stormscope and Shadin fuel/air data computer.

DIMENSIONS, EXTERNAL:
Wing span ... 9.50 m (31 ft 2 in)
Wing aspect ratio ... 7.0
Length overall ... 7.98 m (26 ft 2 in)
Height overall ... 2.74 m (9 ft 0 in)
Wheelbase ... 1.93 m (6 ft 4 in)
Propeller diameter ... 1.88 m (6 ft 2 in)

DIMENSIONS, INTERNAL:
Cabin: Length ... 2.86 m (9 ft 4½ in)
Width ... 1.14 m (3 ft 9 in)
Height ... 1.24 m (4 ft 0¼ in)
Baggage compartment volume ... 0.80 m^3 (28.3 cu ft)

AREAS:
Wings, gross ... 12.94 m^2 (139.3 sq ft)

WEIGHTS AND LOADINGS (Sigma 230, 250 and 310):
Weight empty: S230 ... 850 kg (1,874 lb)
S250 ... 880 kg (1,940 lb)
S310 ... 1,051 kg (2,317 lb)
Baggage capacity: all ... 50 kg (110 lb)
Max T-O weight: S230 ... 1,352 kg (2,980 lb)
S250 ... 1,372 kg (3,025 lb)
S310 ... 1,587 kg (3,500 lb)
Max wing loading: S230 ... 104.4 kg/m^2(21.39 lb/sq ft)
S250 ... 106.0 kg/m^2 (21.72 lb/sq ft)
S310 ... 122.7 kg/m^2 (25.13 lb/sq ft)
Max power loading: S230 ... 7.89 kg/kW (12.96 lb/hp)
S250 ... 7.36 kg/kW (12.10 lb/hp)
S310 ... 6.87 kg/kW (11.29 lb/hp)

PERFORMANCE (Sigma 230, 250 and 310):
Max level speed: S230 ... 175 kt (324 km/h; 201 mph)
S250 ... 184 kt (340 km/h; 211 mph)
S310 ... 215 kt (398 km/h; 247 mph)
Cruising speed: S250 ... 173 kt (320 km/h; 199 mph)
S310 ... 205 kt (380 km/h; 236 mph)
Econ cruising speed: S250 ... 125 kt (232 km/h; 144 mph)
S310 ... 150 kt (278 km/h; 173 mph)
Stalling speed:
flaps up:
S230, S250 ... 67 kt (125 km/h; 78 mph)
S310 ... 71 kt (132 km/h; 82 mph)
flaps down: S230, S250 ... 57 kt (106 km/h; 66 mph)
S310 ... 60 kt (112 km/h; 69 mph)
Max rate of climb at S/L: S250 ... 399 m (1,310 ft)/min
S310 ... 408 m (1,340 ft)/min
Service ceiling: S250, S310 ... 6,100 m (20,000 ft)
T-O run: S230 ... 261 m (855 ft)
S250 ... 249 m (815 ft)
S310 ... 250 m (820 ft)
T-O to 15 m (50 ft): S250 ... 427 m (1,400 ft)
S310 ... 503 m (1,650 ft)
Landing from 15 m (50 ft): S250 ... 412 m (1,350 ft)
S310 ... 488 m (1,600 ft)
Landing run: S230, S250 ... 328 m (1,075 ft)
S310 ... 406 m (1,330 ft)
Range:
normal cruise:
S250 ... 840 n miles (1,555 km; 966 miles)
S310 ... 1,150 n miles (2,129 km; 1,323 miles)
long-range cruise:
S230 ... 870 n miles (1,611 km; 1,001 miles)
S250 ... 1,160 n miles (2,148 km; 1,334 miles)
S310 ... 1,488 n miles (2,755 km; 1,712 miles)

UPDATED

SONEX

SONEX LTD

PO Box 2521, Oshkosh, Wisconsin 54903-2521
Tel: (+1 920) 231 82 97
Fax: (+1 920) 426 83 33
e-mail: info@sonex-ltd.com
Web: http://www.sonex-ltd.com
PRESIDENT: John Monnett
GENERAL MANAGER: Jeremy Monnett

EUROPEAN AGENT:
ARES Srl, Via Papiria 54a, I-00175 Roma, Italy
Tel: (+39 06) 76 90 01 32
Fax: (+39 06) 76 90 00 89
e-mail: arescomp@tin.it
Web: http://www.rce.it/sonex

Company formed by John Monnett, designer of the Sonerai, Monerai and Moni light aircraft, to produce kits of his latest design, the Sonex, its Y-tailed Waiex and motor glider Xenos derivatives. In late 2002 the company had a workforce of four in its 968 m^2 (10,424 sq ft) factory.

SONEX SONEX

TYPE: Side-by-side ultralight kitbuilt.
PROGRAMME: Announced March 1996 in response to request for ultralight to meet Italian requirements; first flight of prototype (N12SX) with Jabiru 2200 engine, 28 February 1998. Five prototypes. First customer-built aircraft flown 12 June 2001.
CURRENT VERSIONS: **Sonex:** Kitbuilt, *as described.*

Waiex: Y-tailed derivative (name indicates Y-tailed experimental). Prototype (N12YX) first flown 19 July 2003 and made public debut at AirVenture 2003 following week. Data as per Sonex, except length 5.51 m (18 ft 1 in), empty weight 283 kg (625 lb), never-exceed speed 166 kt (309 km/h; 192 mph), stalling speed, flaps down 37 kt (68 km/h; 42 mph).

Xenos: Motor glider, based around Sonex front fuselage; *described separately.*
CUSTOMERS: Over 500 sets of plans and kits sold and 50 flying by August 2003.
COSTS: Sonex Airframe kit US$11,985 tailwheel, US$12,265 tricycle both excluding engine and wheels and brakes; plans US$600 (including two-day workshop) (2002). Waiex estimated completed cost US$25,000 (2003).
DESIGN FEATURES: Simple to assemble; 198 parts to easy-build kit. Wings removable for transport and storage. Quoted build time 700 hours.

NACA 64-415 aerofoil.
FLYING CONTROLS: Conventional and manual. Flaps.
STRUCTURE: Of 6061-T6 aluminium throughout.
LANDING GEAR: Choice of tailwheel or tricycle; both fixed. Titanium mainwheel legs and glass fibre speed fairings. Azusa wheels and brakes. Mainwheels 5.00×5, tailwheel (where fitted) 4×1 in.
POWER PLANT: Choice of power plants between 59.7 and 89.5 kW (80 and 120 hp), including Volkswagen 2180 and Jabiru 2200 and 3300. Two-blade, wooden, fixed-pitch Sensenich or Sterba propeller. Fuel capacity 61 litres (16.0 US gallons; 13.3 Imp gallons) in single fuel tank.
DIMENSIONS, EXTERNAL:
Wing span .... 6.71 m (22 ft 0 in)
Length overall .... 5.36 m (17 ft 7 in)
Height overall: tailwheel .... 1.63 m (5 ft 4 in)
nosewheel .... 1.93 m (6 ft 4 in)
AREAS:
Wings, gross .... 9.10 m^2 (98.0 sq ft)
WEIGHTS AND LOADINGS (59.7 kW; 80 hp Jabiru 2200):
Weight empty .... 263 kg (580 lb)
Baggage capacity .... 18 kg (40 lb)

Third Prototype Sonex Sonex in nosewheel configuration *(Paul Jackson)* ***NEW***/0567769

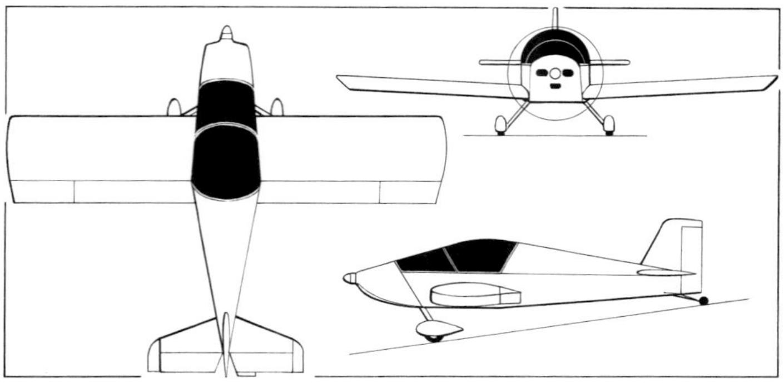

Production version of Sonex Sonex *(James Goulding)* 0530165

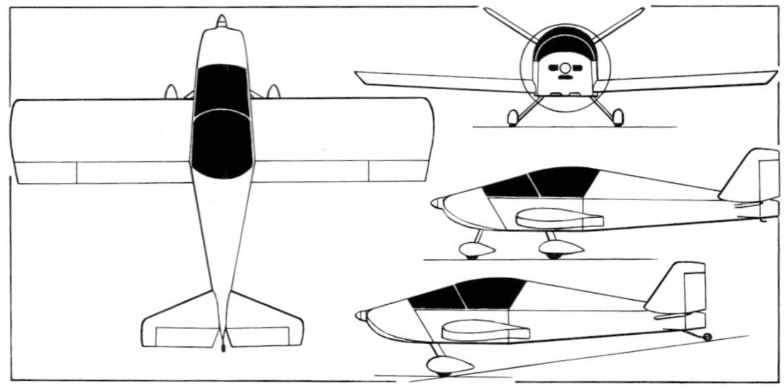

Y-tailed Sonex Waiex *(James Goulding)* ***NEW***/0561623

Prototype Sonex Waiex at Oshkosh in July 2003 *(Paul Jackson)* ***NEW***/0567798

Max T-O weight: Utility .... 499 kg (1,100 lb)
Aerobatic .... 385 kg (850 lb)
PERFORMANCE (engine as above):
Never-exceed speed (VNE) .... 171 kt (317 km/h; 197 mph)
Max level speed .... 130 kt (241 km/h; 150 mph)
Normal cruising speed .... 113 kt (209 km/h; 130 mph)
Stalling speed: flaps up .... 40 kt (74 km/h; 46 mph)
flaps down .... 35 kt (65 km/h; 40 mph)
Max rate of climb at S/L .... 305 m (1,000 ft)/min
T-O run .... 91 m (300 ft)
Landing run .... 153 m (500 ft)
Range with max fuel .... 412 n miles (764 km; 475 miles)
g limits: Utility .... +4.4/–2.2
Aerobatic .... +6/–3

SONEX XENOS

TYPE: Side-by-side ultralight motor glider kitbuilt.
PROGRAMME: Announced in 2000; wing static load testing completed March 2001; Jabiru 3300 prototype (N212XS) first flown 19 July 2003, followed same day by Aero-Vee 2002 VW prototype (N112XS).
COSTS: Estimated completed cost US$28,000 (2003).
DESIGN FEATURES: Based on Sonex fuselage (which see). Quoted build time 1,200 hours.
FLYING CONTROLS: Conventional and manual. Speed brakes.
STRUCTURE: Primarily of 6061 aluminium. Wing has Wortmann FX-61 aerofoil. V tail similar to designer's earlier Moni motor glider.
LANDING GEAR: Tailwheel type; fixed.
POWER PLANT: Choice of engines including 2,180 cc Aero-Vee 2002 VW conversion and 2200 and 3300 Jabiru. Fuel capacity 61 litres (16.0 US gallons; 13.3 Imp gallons) in single fuel tank.

All data are provisional.
DIMENSIONS, EXTERNAL:
Wing span: Utility .... 13.91 m (45 ft 7½ in)
Aerobatic .... 11.99 m (39 ft 4 in)
Length overall .... 6.02 m (19 ft 9 in)

Prototype Xenos N212XS on public display at Oshkosh a week after its maiden flight *(Paul Jackson)* ***NEW***/0567802

AREAS:
Wings, gross: Utility .... 14.68 m^2 (158.0 sq ft)
Aerobatic .... 13.38 m^2 (144.0 sq ft)
WEIGHTS AND LOADINGS:
Weight empty: Utility .... 345 kg (760 lb)
Aerobatic .... 340 kg (750 lb)
Max T-O weight: Utility .... 578 kg (1,275 lb)
Aerobatic .... 476 kg (1,050 lb)
PERFORMANCE, POWERED:
Never-exceed speed (VNE) .... 130 kt (241 km/h; 150 mph)
Stalling speed: utility .... 39 kt (71 km/h; 44 mph)
aerobatic .... 36 kt (66 km/h; 41 mph)
g limits: utility .... +4.4/–2.2
aerobatic .... +6.0/–3.0
PERFORMANCE, UNPOWERED:
Best glide ratio: Utility 24
Aerobatic 21

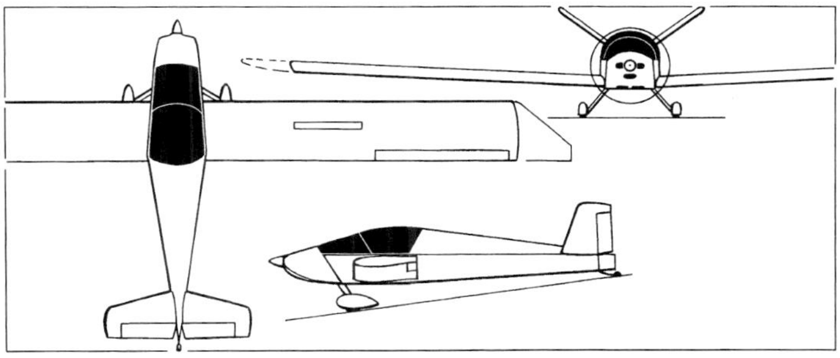

Sonex Xenos motor glider kitplane *(James Goulding)* ***NEW***/0561625

STEEN

STEEN AERO LAB LLC

1451 Clearmont Street Northeast, Palm Bay, Florida 32905
Tel: (+1 321) 725 41 60
Fax: (+1 321) 725 30 58
Web: http://www.steenaero.com
PRESIDENT: David Stone

Steen is a long-established company which provides plans, materials and components to aircraft builders. Its latest product is a new design from the renowned Curtis Pitts.

STEEN PITTS 14

TYPE: Tandem-seat biplane kitbuilt.

PROGRAMME: Frame of prototype revealed at AirVenture, Oshkosh, July 2003; first flight due before July 2004.

COSTS: Not priced as a complete kit, but all components available from Steen Aero Lab.

DESIGN FEATURES: Continues Pitts' 60-year history of aerobatic biplane development. Differs from earlier designs in concentrating bracing wire loads in a single plane between the interplane struts. Designed for simplicity of structure, low weight, ease of construction and strength. Suitable as airshow aerobatic performer.

Choice of semi-symmetric NACA 23012 aerofoil or symmetric SS. Former on prototype. Straight or swept upper wing, to builder's choice; range of empennage shapes, including classic Samson style.

Single-bay biplane with broad-chord interplane strut each side and supporting cabane for upper wing. Wire-braced empennage.

LANDING GEAR: Tricycle type; fixed. Oleo shock absorbers.

POWER PLANT: One 294 kW (394 hp) VOKBM M-14PF nine-cylinder radial.

ACCOMMODATION: Two in tandem beneath single-piece blown canopy.

EQUIPMENT: Optional 'smoke tank' in lower fuselage with external filler.

Provisional data follow:

Fuselage of prototype Pitts 14 in July 2003 *(Paul Jackson)* ***NEW***/0561716

DIMENSIONS, EXTERNAL:
Wing span .... 6.55 m (21 ft 6 in)
Wing aspect ratio 5.7
AREAS:
Wings, gross .... 14.59 m^2 (157.0 sq ft)
WEIGHTS AND LOADINGS:
Weight empty .... 680 kg (1,500 lb)
Airshow weight .... 848 kg (1,871 lb)
Max aerobatic weight .... 907 kg (2,000 lb)
Max T-O weight .... 1,021 kg (2,250 lb)
Wing loading: aerobatic .... 62.2 kg/m^2 (12.74 lb/sq ft)
max .... 70.0 kg/m^2 (14.33 lb/sq ft)
Power loading: aerobatic .... 3.08 kg/kW (5.06 lb/hp)
max .... 3.47 kg/kW (5.70 lb/hp)

Artist's impression of Steen Pitts 14 ***NEW***/0561653

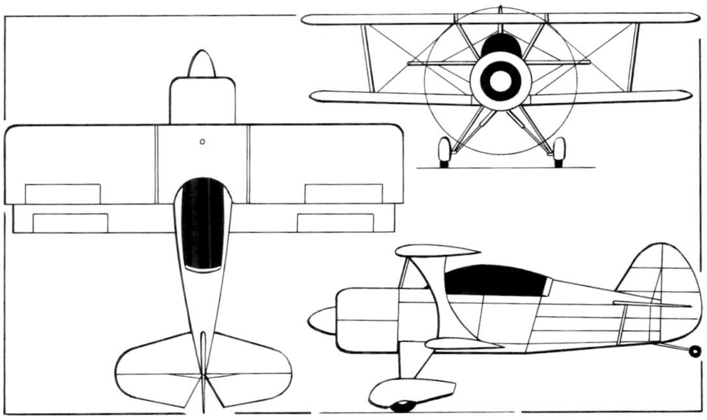

Steen Pitts 14 aerobatic biplane *(James Goulding)* ***NEW***/0567292

TAYLORCRAFT

TAYLORCRAFT AVIATION INC

4495 West State Highway 71, La Grange, Texas 78945-5150
Tel: (+1 979) 242 3695
Fax: (+1 979) 242 3693
e-mail: sales@taylorcraft.com
Web: http://www.taylorcraft.com
PRESIDENT: Harry Ingram

Present Taylorcraft Aviation inherits traditions of Taylor-Young Airplane Company, formed 1935 on C G Taylor's departure from Piper. Taylor produced A, BC, BF, DC, 15, 16, 18, 19 and 20 until late 1950s; Taylorcraft Aviation Corporation established 1971 at Alliance, Ohio, to build F-19 and, from 1971 F21 until 1985, when transferred to Lock Haven, Pennsylvania, where 16 more built before bankruptcy. Taylorcraft Aircraft Company re-established in 1989 to manufacture F22, of which only 17 produced up to 1992 closure. Type Certificate passed through several hands, including Taylorcraft Aerospace and Taylorcraft 2000, none of which initiated production.

Taylorcraft Aviation established by Harry Ingram in 5,020 m^2 (54,000 sq ft) plant to deliver new F22s from 2004 onwards; total of 20 full-time staff by mid-2004. Obtained F22 Type Certificate 5 March 2003.

UPDATED

TAYLORCRAFT F22

TYPE: Two-seat lightplane.

PROGRAMME: Taylorcraft 19 awarded FAA certification 20 June 1951, followed by F19 on 3 July 1973, F21 on 2 July 1980, F21A (fuel tanks transferred to wings) 15 November 1982, F21B (793 kg; 1,750 lb MTOW) 6 September 1985 and F22 (prototype N180GT) on 1 August 1988. Production re-launched in 2004, although none registered by the end of that year.

CURRENT VERSIONS: **F22 Classic:** Derived from F21B with addition of flaps, wider doors, up-hinged windows and adjustable individual seats. Lycoming O-235 engine. Tailwheel.

F22A Tracker: F22 with tricycle landing gear. Prototype N14KV; certified 20 February 1991.

Taylorcraft F22A from the first production series 1036874

F22B Ranger: F22 with Lycoming O-360 engine. Prototype N24UK; certified 14 July 1992.

F22C Trooper: F22A (tricycle gear) with O-360 engine. Prototype N223UK; certified 20 February 1992.

CUSTOMERS: Previous series comprised four F22s; 11 F22As; one F22B; and one F22C up to 1992.

COSTS: F22 US$59,995; F22A US$ 63,995; F22B US$69,995; F22C US$73,995 (all 2004).

DESIGN FEATURES: Progressive development of Taylorcraft 19 series, incorporating classic features of earlier designs. High, constant-chord wing with rounded tips and V-type braces with intermediate struts. Wire-braced empennage.

Wing section NACA 23012.

FLYING CONTROLS: Conventional and manual. Flaps. Flight-adjustable tab in elevator. Control movements: ailerons ±23°; elevators +27/ −25°; rudder ±26°. Max flap deflection 30°.

STRUCTURE: Welded steel tube airframe with fabric covering.

LANDING GEAR: Nosewheel or tailwheel type; fixed. Split type with bent half-axles incorporating bungee shock absorbers. Mainwheels 6.00–6 (4 ply). Optional nosewheel 5.00–5. Cleveland mainwheel brakes.

POWER PLANT: One 88.0 kW (118 hp) Textron Lycoming O-235-L2C flat-four or 134 kW (180 hp) O-360-A4M driving a Sensenich (72CK-0-50 or 76EM8S-0-60, respectively) two-blade, metal, fixed-pitch propeller. Fuel tank in each wing; total capacity 159 litres (42.0 US gallons; 35.0 Imp gallons), of which 151 litres (40.0 US gallons; 33.3 Imp gallons) are usable.

ACCOMMODATION: Two, side by side; door each side. Baggage area behind seats. Cabin heated and ventilated.

SYSTEMS: Electrical system includes 14 V 60 A alternator and 25 Ah battery.

AVIONICS: Provision for IFR avionics.

Data apply to all versions, except where otherwise stated.

DIMENSIONS, EXTERNAL:

Wing span	10.97 m (36 ft 0 in)
Length overall	6.78 m (22 ft 2¾ in)
Height overall	1.88 m (6 ft 2 in)
Propeller diameter: O-235	1.83 m (6 ft 0 in)
O-360	1.93 m (6 ft 4 in)

AREAS:

Wings, gross	17.07 m^2 (183.7 sq ft)

WEIGHTS AND LOADINGS:

Weight empty: F22, F22A	494 kg (1,090 lb)
F22B, F22C	538 kg (1,185 lb)
Baggage capacity	91 kg (200 lb)
Max T-O weight	793 kg (1,750 lb)
Max wing loading	46.5 kg/m^2 (9.53 lb/sq ft)
Max power loading: F22, F22A	9.03 kg/kW (14.83 lb/hp)
F22B, F22C	5.92 kg/kW (9.72 lb/hp)

Taylorcraft F22A from the first production series 1036873

PERFORMANCE:

Never-exceed speed (VNE)	128 kt (238 km/h; 148 mph)
Max level speed: F22, F22A	104 kt (193 km/h; 120 mph)
F22B, F22C	117 kt (217 km/h; 135 mph)
Cruising speed at 75% power:	
F22, F22A	96 kt (177 km/h; 110 mph)
F22B, F22C	109 kt (201 km/h; 125 mph)
Stalling speed, power off: flaps up	42 kt (78 km/h; 48 mph)
flaps down	37 kt (68 mph; 42 mph)
Max rate of climb at S/L: F22, F22A	229 m (750 ft)/min
F22B, F22C	457 m (1,500 ft)/min
Service ceiling	5,485 m (18,000 ft)
T-O run: F22, F22A	153 m (500 ft)
F22B, F22C	122 m (400 ft)
T-O to 15 m (50 ft): F22, F22A	243 m (795 ft)
F22B, F22C	199 m (650 ft)
Landing from 15 m (50 ft)	176 m (575 ft)
Landing run	107 m (350 ft)
Range at 75% power, 30 min reserves:	
F22, F22A	530 n miles (981 km; 610 miles)
F22B, F22C	398 n miles (737 km; 458 miles)

UPDATED

Taylorcraft F22A from the first production series ***NEW***/0567732

TBG

THUNDER BUILDERS' GROUP LLC

Tel: (+1 719) 591 93 73
Fax: (+1 509) 472 61 08
e-mail: sales@thundermustang.com
Web: http://www.thundermustang.com
MANAGING MEMBER: John McCartney

Scale replica of the classic P-51 Mustang fighter was originally produced by Papa 51 company which ceased trading on 1 December 2000. Group of 25 kitbuilders (six of them with almost complete aircraft) then formed Thunder Builder's Group LLC and acquired company's intellectual rights and production moulds with a view to sustaining their own projects and resuming manufacture. This seemed to have been achieved on 13 September 2001, when GUT Works reached agreement with Group; however, the two partners parted on 25 July 2002 when TBG again put the assets up for sale.

TEAM ROCKET

TEAM ROCKET

80 CR 406, Taylor, Texas 76574
Tel: (+1 512) 365 81 31
Fax: (+1 512) 352 50 80
e-mail: mark@teamrocketaircraft.com
Web: http://www.teamrocketaircraft.com
SALES MANAGER: Mark Frederick

Team Rocket was formed in 1998. Its F1 kitbuilt was developed by International High Performance Aircraft of Lobeček 732, CZ-728 01 Kralupy nad Vlatava, Czech Republic (http://www.hpai.cz), and Venice, Florida (http://www.ihpaircraft.com).

A tapered-wing version, the F1EVO, was introduced in 2003, at which time Team Rocket had a workforce of three and a 697 m^2 (7,500 sq ft) factory.

TEAM ROCKET F1

TYPE: Tandem-seat sportplane kitbuilt.

PROGRAMME: Developed from Van's RV-4 by way of Harmon Rocket. First flight (in USA) November 2000 (N121JC). Second flew in Australia, July 2001.

CURRENT VERSIONS: **Team Rocket F1:** Baseline version.

Description applies to this version, unless otherwise indicated.

Team Rocket F1EVO: Tapered-wing version announced as under development in 2003. As F1, except where otherwise indicated.

Team Rocket F2: Planned side-by-side two-seat version, announced at Oshkosh in July 2002 but discontinued; project has been moved to High Performance Aircraft of USA.

CUSTOMERS: Sales totalled 130 by August 2003, of which 20 were flying.

COSTS: Kit, excluding engine, propeller and avionics, US$37,400 (2003).

DESIGN FEATURES: Intended to combine high performance and low price. Low-wing monoplane with closely cowled engine. Constant-chord wings, tapered tailplane and moderately sweptback fin.

Meets '51 per cent' rule. Quoted build time 2,000 hours (1,200 for fast-build kit).

FLYING CONTROL: Conventional and manual. Flaps. Horn-balanced rudder and elevators.

STRUCTURE: Principally of aluminium, primed with epoxy chromate for corrosion resistance. Titanium mainwheel legs; glass fibre engine cowling and wingtips.

LANDING GEAR: Tailwheel type; fixed. Spring titanium mainwheel legs. Mainwheels 5.00–5. F1EVO will be offered with either tailwheel or nosewheel.

POWER PLANT: One flat-six piston engine rated between 175 and 224 kW (235 and 300 hp), driving a two- or three-blade propeller recommended for both versions; typically prototypes have Textron Lycoming IO-540 and three-blade, constant-speed MTV-9-B-C/C198-52 propeller. Fuel capacity 197 litres (52.0 US gallons; 43.3 Imp gallons).

Team Rocket F1 sportplane *(Paul Jackson)* ***NEW***/0567794

ACCOMMODATION: Two persons in tandem, beneath rearwards-sliding canopy. Fixed windscreen.

DIMENSIONS, EXTERNAL:
- Wing span: F1 .... 6.56 m (22 ft 6 in)
  - F1EVO .... 7.75 m (25 ft 5 in)
- Wing aspect ratio: F1 .... 4.8
  - F1EVO 5.9
- Length overall .... 6.40 m (21 ft 0 in)

AREAS:
- Wings, gross: F1 .... 9.85 m² (106.0 sq ft)
  - F1EVO .... 10.22 m² (110.0 sq ft)

WEIGHTS AND LOADINGS:
- Weight empty .... 544 kg (1,200 lb)
- Max T-O weight .... 907 kg (2,000 lb)
- Max wing loading: F4 .... 92.1 kg/m² (18.87 lb/sq ft)
  - F1EVO .... 88.8 kg/m² (18.18 lb/sq ft)
- Max power loading .... 4.06 kg/kW (6.67 lb/hp)

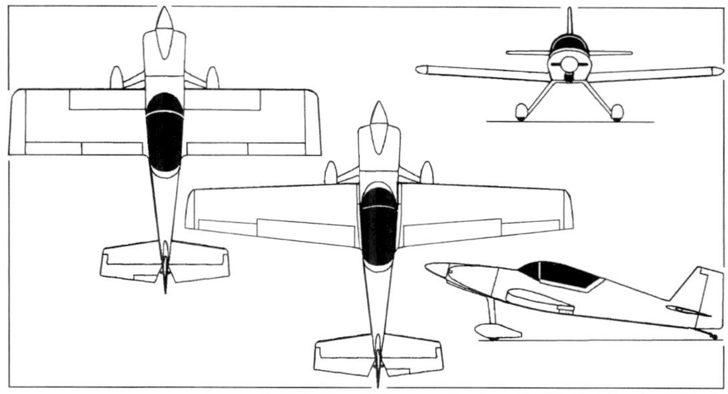

Team Rocket F1, with additional plan view (lower) of F1EVO ***NEW***/0567251

PERFORMANCE (224 kW; 300 hp engine):
- Max level speed: F1 .... 220 kt (407 km/h; 253 mph)
  - F1EVO .... 230 kt (426 km/h; 265 mph)
- Cruising speed at 75% power
  - F1 .... 200 kt (370 km/h; 230 mph)
  - F1EVO .... 209 kt (386 km/h; 240 mph)
- Stalling speed: F1 .... 47 kt (87 km/h; 54 mph)
  - F1EVO .... 44 kt (81 km/h; 50 mph)
- Max rate of climb at S/L .... 1,067 m (3,500 ft)/min
- T-O run .... 92 m (300 ft)
- Landing run .... 214 m (700 ft)
- Range at 55% power .... 1,000 n miles (1,850 km; 1,150 miles)

TEAM TANGO

TEAM TANGO INC

1990 SW 19th Avenue, Williston, Florida 32696
Tel: (+1 352) 528 09 82
e-mail: info@teamtango.com
Web: http://www.teamtango.com
PRESIDENT AND CEO: Tom Miller

The former DFL was purchased by a consortium of constructors of its projects in June 2002 and now a subsidiary of Revolution Aviation. A new 557 m² (6,000 sq ft) Builder's Center has been opened at Williston Municipal Airport, where the company continues to develop and market the Tango and Foxtrot.

TEAM TANGO FOXTROT

TYPE: Four-seat kitbuilt.

PROGRAMME: Development began in 1999; first engine runs of prototype (N747F) completed 27 August 2001; first flight was due before end of 2001 and achieved 23 December 2002. Two further aircraft had flown by mid-2003. By June 2003, prototype had completed 35 hours of flying. An MTOW upgrade to 1,361 kg (3,000 lb) is planned.

COSTS: Basic kit US$34,995 (2003).

CUSTOMERS: Several under construction by August 2003.

DESIGN FEATURES: Conventional design similar in appearance to Tango. Vinylester resin used in place of epoxy to reduce cure time of composites.

FLYING CONTROLS: Conventional and manual. Flaps (electric actuation) optional.

STRUCTURE: Extensive use of composites throughout.

LANDING GEAR: Tricycle type; fixed. Speed fairings on all wheels.

POWER PLANT: Prototype fitted with 224 kW (300 hp) Textron Lycoming IO-540; engines in the range 149 to 224 kW (200 to 300 hp) are recommended. Standard fuel capacity 379 litres (100 US gallons; 83.3 Imp gallons).

ACCOMMODATION: Pilot and three passengers in two pairs of seats. Upward-opening, centreline-hinged door each side; fixed windscreen. Baggage stowage behind seats.

DFL Foxtrot prototype displayed incomplete at Sun 'n' Fun, 2001 *(Jane's/Susan Bushell)* 0121673

DIMENSIONS, EXTERNAL:
- Wing span .... 9.75 m (32 ft 0 in)
- Wing aspect ratio 8
- Length overall .... 7.56 m (24 ft 9½ in)
- Height overall .... 2.21 m (7 ft 3 in)

DIMENSIONS, INTERNAL:
- Cabin: Max width .... 1.17 m (3 ft 10 in)
  - Max height .... 1.09 m (3 ft 7 in)
- Baggage volume .... 0.40 m³ (14.0 cu ft)

AREAS:
- Wings, gross .... 11.89 m² (128.0 sq ft)

WEIGHTS AND LOADINGS (224 kW; 300 hp engine):
- Weight empty .... 774 kg (1,706 lb)
- Baggage capacity .... 68 kg (150 lb)
- Max payload .... 250 kg (552 lb)
- Max T-O weight .... 1,270 kg (2,800 lb)
- Max wing loading .... 106.8 kg/m² (21.88 lb/sq ft)
- Max power loading .... 5.68 kg/kW (9.33 lb/hp)

PERFORMANCE (224 kW; 300 hp engine):
- Never-exceed speed (VNE) .... 212 kt (394 km/h; 245 mph)
- Max operating speed .... 200 kt (370 km/h; 230 mph)
- Normal cruising speed .... 189 kt (351 km/h; 218 mph)
- Stalling speed, flaps down .... 54 kt (100 km/h; 62 mph)
- Max rate of climb at S/L .... 914 m (3,000 ft)/min
- Service ceiling .... 7,315 m (24,000 ft)
- T-O run .... 107 m (350 ft)
- Landing run .... 244 m (800 ft)
- Range with max fuel .... 1,564 n miles (2,896 km; 1,800 miles)
- *g* limits .... +6/–4

TEAM TANGO TANGO II

TYPE: Side-by-side kitbuilt.

PROGRAMME: Developed from the Aero Mirage TC-2 of 1983. Prototype flew November 1996. First two production aircraft displayed at Oshkosh in July 1999; third flew March 2000. Production delayed from mid-2001 to mid-2002 whilst buyer sought for company.

CUSTOMERS: Five flying and seven under construction by mid-2003.

COSTS: Basic kit US$24,995, US$36,500 O-360 engine kit and US$11,995 VFR avionics kit (2003).

DESIGN FEATURES: Conventional design optimised for amateur assembly; low-mounted, constant-chord wings and mid-positioned tailplane. Steeply raked windscreen. Quoted build time 570 hours for quick-build kit version; company has building facility to enable purchasers to complete in 30 to 45 days. Roll rate 120°/s.

FLYING CONTROLS: Conventional and manual. Flaps deflect 10/20/35°.

STRUCTURE: Generally of composites. Two-spar wing has modified NACA64415 aerofoil and 3° dihedral.

POWER PLANT: One 134 kW (180 hp) Textron Lycoming O-360 flat-four driving a two-blade Hartzell HCC2YR-7666A constant-speed propeller. Fuel capacity 152 litres (40.0 US gallons; 33.3 Imp gallons) in two wing tanks; two optional extra 38 litre (10.0 US gallon; 8.3 Imp gallon) tanks can be fitted. Optionally, 149 kW (200 hp) IO-360 or 194 kW (260 hp) IO-540.

LANDING GEAR: Tricycle type; fixed. Cantilever mainwheel legs are fibre glass; nose leg is 4130 steel tube; speed fairings on all wheels. Main units have Cleveland wheels and brakes with Goodyear 5.00–5 tyres; nosewheel is steerable and has 4.50–5 Lamb tyre.

ACCOMMODATION: Two, side by side. Upward-opening, centreline hinged door each side; fixed windscreen. Baggage stowage behind seats.

AVIONICS: To customer's specification. Standard VFR option includes Honeywell KLX 135 GPS/com, KT 76A transponder and Sigtronics intercom; IFR version adds KX 155 nav/com and KI 209 VOR/LOC.

DIMENSIONS, EXTERNAL:
Wing span ... 7.62 m (25 ft 0 in)
Wing aspect ratio ... 8.3
Length overall ... 6.31 m (20 ft 8½ in)
Height overall ... 2.06 m (6 ft 9 in)
Wheel track ... 2.54 m (8 ft 4 in)
Propeller diameter ... 1.93 m (6 ft 4 in)
AREAS:
Wings, gross ... 6.97 m² (75.0 sq ft)
DIMENSIONS, INTERNAL:
Cockpit: Max width ... 1.12 m (3 ft 8 in)
Max height ... 0.94 m (3 ft 1 in)
Baggage compartment volume ... 0.34 m³ (12.0 cu ft)
WEIGHTS AND LOADINGS:
Weight empty ... 488 kg (1,075 lb)
Baggage capacity ... 45 kg (100 lb)
Max T-O weight ... 839 kg (1,850 lb)
Max wing loading ... 120.4 kg/m² (24.67 lb/sq ft)
Max power loading ... 6.26 kg/kW (10.28 lb/hp)
AREAS:
Wings, gross ... 6.97 m² (75.0 sq ft)
PERFORMANCE:
Never-exceed speed (VNE) ... 212 kt (394 km/h; 245 mph)
Max level speed ... 191 kt (354 km/h; 220 mph)
Max cruising speed ... 174 kt (322 km/h; 200 mph)
Stalling speed ... 53 kt (97 km/h; 60 mph) IAS

Team Tango Tango2 built from the 15th kit *(Paul Jackson)* ***NEW***/0567027

Max rate of climb at S/L ... 549 m (1,800 ft)/min
Service ceiling ... 7,315 m (24,000 ft)
T-O run ... 107 m (350 ft)
Landing run ... 244 m (800 ft)
Max range ... 869 n miles (1,609 km; 1,000 miles)
g limits ... +6/–4

THRUSH

THRUSH AIRCRAFT INC

PO Box 3149, Albany, Georgia 31706-3149
Tel: (+1 229) 883 14 40
Fax: (+1 229) 439 97 90
e-mail: mhumphries@thrushaircraft.com
Web: http://www.thrushaircraft.com
PRESIDENT: Larry Bays
CHIEF EXECUTIVE OFFICER: K Payne Hughes
GENERAL MANAGER: Milton R Humphries
PRODUCTION MANAGER: Ken Gum

Former Ayres Corporation bought Albany factory and manufacturing and world marketing rights to Thrush Commander-600 and -800 from Rockwell International General Aviation Division in November 1977. Acquired 93 per cent shareholding in Let Kunovice of Czech Republic in 1998, but stopped providing finance in July 2000 after Czech bank cancelled a debt repayment freeze; Let declared bankrupt in October 2000.

Ayres filed for Chapter 11 bankruptcy protection in late November 2000 and was seeking further investment of US$80 million; work restarted on the Loadmaster programme after a financing agreement with GATX, as debtor-in-possession, was concluded. However, GATX foreclosed on the company on 7 August 2001, securing all assets in return for a debt reduction of US$10.3 million.

Quality Aerospace established, also on 7 August 2001, to take over almost all of Ayres's assets at 21,100 m² (227,000 sq ft) Albany plant and continues to produce aerostructures, support existing Thrush fleet and promote Turbo-Thrush. Thrush Aircraft Inc purchased Ayres assets on 30 June 2003, and began the process of restarting the production line, with intention to manufacture two new aircraft by November 2003. Employment stood at about 300 in 2004.

UPDATED

Ayres-built TPE331-engined Thrush S2R-G10 ***NEW***/0559124

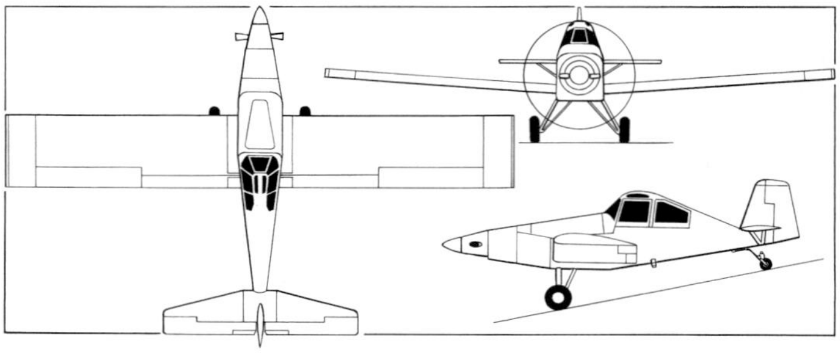

Turbo-Thrush S2R-T34 *(James Goulding)* ***NEW***/0569767

THRUSH TURBO-THRUSH S2R

TYPE: Agricultural sprayer.

PROGRAMME: Leland Snow designed S-1 prototype, first flown with radial engine on 17 August 1953; entered production as S-2; Type Certificate for original S2D issued 1 November 1965; SR2 followed 21 March 1968; over 1,800 built by Snow and (following 18 February 1970 transfer of Type Certificate) Rockwell. Type Certificate obtained by Ayres on 28 November 1977; Ayres introduced turboprop version, having previously specialised in turbine conversions. PT6-engined S2R-T34 certified 28 April 1977; Garrett TPE331 versions certified from 5 March 1992 (S2R-G6) onwards. Quality Aerospace obtained Type Certificate on 26 November 2001 and this transferred to Thrush Aircraft on 2 September 2003. New company standardised version designations, adopting marketing names of Model 400, 510 and 660.

CURRENT VERSIONS: **Thrush 400:** Light airframe version with 1,514 litre (400 US gallon; 333 Imp gallon) hopper. Undergoing redesign in 2004, prior to relaunch of marketing.

S2RHG Thrush 510: Baseline version with 1,931 litre (510 US gallon; 425 Imp gallon) hopper and option of PT6A (designation S2RHG-T34) or TPE331 turboprop. *As described.*

Thrush 660: Described separately.

S2R-T34: 559 kW (750 shp) P&WC PT6A-34AG turboprop; standard or optional hoppers. High Gross (4,763 kg; 10,500 lb) MTOW version certified 5 November 1997.

CUSTOMERS: By July 2001, 15 G1s, 55 G6s, 68 G10s, 39 T15s, 272 T34s and 15 T45s had been produced and over 2,500 of all versions delivered to 80 countries.

Production by new Thrush company began with 273rd T34 N550AG, registered December 2003; this initially had c/n T34-273, which changed to T34HG-103. Four registered by late 2004.

COSTS: About US$600,000, depending on engine and customer fit.

DESIGN FEATURES: Conventional crop-sprayer configuration, with hopper ahead of cockpit; cantilever low wing of constant chord; braced empennage with sweptback leading-edges; robust landing gear.

Advantages over piston-engined versions include much improved take-off and climb, 454 kg (1,000 lb) higher payload because of lower engine weight, operation on aviation turbine fuel or diesel, 3,500 hour TBO, quieter operation and ability to feather propeller without stopping engine while refuelling and reloading.

FLYING CONTROLS: Conventional and manual. Plain ailerons; servo tab in each elevator; electrically actuated flaps.

STRUCTURE: Two-spar light alloy wing with 4130 chrome molybdenum steel spar caps; welded chrome molybdenum steel tube fuselage structure covered with quickly removable light alloy skin panels; underfuselage skin of stainless steel; all-metal tail surfaces and strut-braced tailplane; metal ailerons and flaps. Extended wing, originally optional, is now standard, replacing earlier 13.54 m (44 ft 5 in), 30.34 m² (326.6 sq ft) version.

LANDING GEAR: Tailwheel type; fixed. Fuselage-mounted spring steel cantilever mainwheel legs standard from 2004. Ayres-built aircraft have narrower track (2.51 m; 8 ft 3 in) gear comprising two side Vs hinged to lower longerons, plus half-axles sprung under centreline with rubber-in-compression shock-absorbers; 8.50–10 (10 ply) tyres. Hydraulically operated Cleveland dual calliper disc brakes. Parking brakes. Wire cutters on main gear. Steerable, spring steel locking tailwheel, size 5.00-5.

POWER PLANT: One turboprop rated at 559 kW (750 shp): P&WC PT6A-34G standard, Honeywell TPE331-6 optional. Hartzell three-blade, constant-speed, feathering and reversing propeller; usable fuel capacity 515 litres (136 US gallons; 113 Imp gallons) in 400 gallon hopper version and 863 litres (228 US gallons; 190 Imp gallons) in 510 gallon hopper version.

ACCOMMODATION: Single adjustable mesh seat in 'safety pod' sealed cockpit enclosure, with steel tube overturn structure; aft fuselage pressurised. Tandem seating optional, with forward-facing second seat. Dual controls optional with forward-facing rear seat, for pilot training; these versions have 'DC' suffix to c/n; available since 1985. Adjustable rudder pedals. Downward-hinged door on each side. Tempered safety glass windscreen. Cockpit wire cutter. Dual inertia reel safety harness with optional second seat. Baggage compartment standard on single-seat aircraft.

SYSTEMS: Electrical system powered by a 24 V 50 A alternator. Dual lightweight 24 V 35 Ah batteries.

AVIONICS: To customer's requirements.

EQUIPMENT: GFRP hopper forward of cockpit can hold 1,931 litres (510 US gallons; 425 Imp gallons) of liquid, equivalent to 1.87 m³ (66.0 cu ft). Standard equipment includes Universal spray system with external 50 mm (2 in) stainless steel plumbing, 50 mm pump with wooden fan, Transland gate, 50 mm valve, quick-disconnect pump mount and strainer. Streamlined spraybooms with outlets for 68 nozzles. Micro-adjust valve control (spray) and calibrator (dry). A 51 mm (2 in) side-loading system is installed on the port side. Stainless steel rudder cables. Navigation lights, instrument lights and two strobe lights. Windscreen washer/wiper.

Optional items are a Transland high-volume spreader; agitator installation; 10-unit AU5000 Micronair installation in lieu of standard booms and nozzles; Transland gatebox with stiffener casting; quick-disconnect flange and kit; night working lights including landing light and wingtip turn lights; cockpit fire extinguisher; and water-bomber configuration.

DIMENSIONS, EXTERNAL:
- Wing span ... 14.48 m (47 ft 6 in)
- Wing chord, constant ... 2.29 m (7 ft 6in)
- Wing aspect ratio ... 6.4
- Length overall ... 10.06 m (33 ft 0 in)
- Height overall ... 2.79 m (9 ft 2 in)
- Tailplane span ... 5.18 m (17 ft 0 in)
- Wheel track ... 2.76 m (9 ft 0¾ in)
- Wheelbase ... 5.88 m (19 ft 3½ in)

AREAS:
- Wings, gross ... 32.52 m² (350.0 sq ft)

WEIGHTS AND LOADINGS (Thrush 510)::
- Weight empty ... 1,950 kg (4,300 lb)
- Typical operating weight (CAM 8) ... 4,400 kg (9,700 lb)
- Max wing loading ... 135.3 kg/m² (27.71 lb/sq ft)
- Max power loading ... 7.87 kg/kW (12.93 lb/shp)

PERFORMANCE (Thrush 510)::
- Never-exceed speed (VNE) ... 138 kt (256 km/h; 159 mph)
- Max level speed with spray equipment ... 138 kt (256 km/h; 159 mph)
- Cruising speed at 55% power ... 130 kt (241 km/h; 150 mph)
- Working speed at 30–50% power ... 78–130 kt (145–241 km/h; 90–150 mph)
- Stalling speed: flaps up ... 61 kt (113 km/h; 70 mph)
  - flaps down ... 57 kt (106 km/h; 66 mph)
- Stalling speed at normal landing weight:
  - flaps up ... 51 kt (95 km/h; 59 mph)
  - flaps down ... 50 kt (92 km/h; 57 mph)
- Max rate of climb at S/L ... 267 m (875 ft)/min
- T-O run ... 366 m (1,200 ft)
- Landing run ... 183 m (600 ft)
- Landing run with propeller reversal ... 122 m (400 ft)
- Ferry range at 45% power ... 669 n miles (1,239 km; 770 miles)

UPDATED

THRUSH 660 TURBO-THRUSH

TYPE: Agricultural sprayer.

PROGRAMME: Prototype produced in 1997 to verify new hopper and landing gear designs, but was otherwise as S2R. Certification achieved 13 March 2000 with first delivery same day.

CUSTOMERS: Eight delivered by October 2000; ninth was first of type registered to Thrush company (c/n T660-109, N204PE), December 2003. Further two by mid-2004.

DESIGN FEATURES: Based on S2R (which see) but with enlarged hopper, made possible by change to cantilever main landing gear. Enlarged wing centre-section extends span. Wing section NACA 4412 modified; incidence root 3.5°, tip 1.72°, twist 1.784°. Tailplane section NACA 0008 modified. Rear fuselage is pressurised.

Data as for Turbo-Thrush S2R, except as below.

FLYING CONTROLS: Aileron droop system, plus enlarged flaps. Control surface deflections: ailerons +21°, −17°; elevators +27°, −17°; rudder 19°; flaps −15°.

LANDING GEAR: Tailwheel type; fixed. Fuselage mounted spring steel cantilever mainwheel legs with heavy-duty 29 in tyres (31 in tyres optional) and Cleveland dual calliper hydraulic disc brakes. Wire cutters on main undercarriage legs. Tailwheel 6.00×6.

POWER PLANT: One 917 kW (1,230 shp) Pratt & Whitney Canada PT6A-65AG turboprop. Optionally, one 783 kW (1,050 shp) P&WC PT6A-60AG turboprop. Hartzell HC-B65MP-3C/M10876 five-blade, constant-speed feathering and reversing propeller. Fuel capacity 871 litres (230 US gallons; 191.5 Imp gallons), of which 854 litres (225.6 US gallons; 187.8 Imp gallons) are usable.

ACCOMMODATION: Optional air conditioning.

EQUIPMENT: GFRP hopper can hold 2,500 litres (660 US gallons; 550 Imp gallons) of liquid. Total capacity, including spray system, is 2,687 litres (710 US gallons; 591 Imp gallons); total capacity with fire-bomber equipment 2,971 litres (785 US gallons; 654 Imp gallons). Side loader size 76 mm (3 in) on port side. Standard equipment includes Universal spray

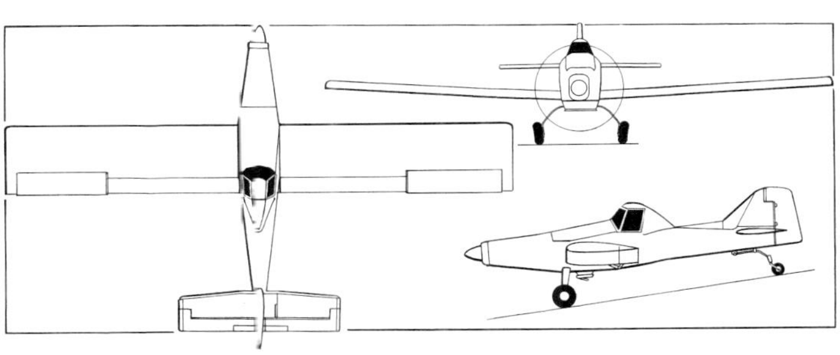

Thrush 660 Turbo-Thrush *(James Goulding)* ***NEW***/0569764

Thrush 660 Turbo-Thrush ***NEW***/1047087

Thrush 660 spray gear ***NEW***/1047086

system with external 76 mm (3 in) stainless steel plumbing, five-blade Weath-Aerofan, 76 mm Transland pump, 68-nozzle sprayboom extending over 70 per cent of span. Optional equipment as per Turbo-Thrush S2R.

DIMENSIONS, EXTERNAL (PT6A engine):
- Wing span ... 16.46 m (54 ft 0 in)
- Wing aspect ratio ... 7.2
- Wing chord, constant ... 2.29 m (7 ft 6 in)
- Length overall ... 10.89 m (35 ft 8¾ in)
- Height overall ... 3.08 m (10 ft 1¼ in)
- Tailplane: span ... 5.18 m (17 ft 0 in)
  - chord: at root ... 1.36 m (4 ft 5½ in)
    - at tip ... 1.13 m (3 ft 8½ in)
- Wheel track ... 2.90 m (9 ft 6 in)
- Wheelbase ... 2.76 m (9 ft 0¾ in)
- Propeller diameter ... 2.82 m (9 ft 3 in)
- Propeller ground clearance:
  - fuselage horizontal ... 0.42 m (1 ft 4½ in)
  - tail down ... 0.80 m (2 ft 7½ in)

DIMENSIONS, INTERNAL:
- Hopper volume ... 2.5 m³ (88 cu ft)

AREAS:
- Wings, gross ... 37.63 m² (405.0 sq ft)
- Flaps, total ... 4.26 m² (45.87 sq ft)
- Ailerons, total ... 4.34 m² (46.71 sq ft)
- Tailplane ... 6.58 m² (70.78 sq ft)
- Elevators, total ... 2.29 m² (24.61 sq ft)
- Fin ... 3.15 m² (34.0 sq ft)
- Rudder ... 1.39 m² (15.0 sq ft)

WEIGHTS AND LOADINGS (PT6A-65AG):
- Weight empty ... 2,880 kg (6,350 lb)
- Max T-O weight ... 6,418 kg (14,150 lb)
- Max landing weight ... 5,670 kg (12,500 lb)
- Max wing loading ... 170.6 kg/m² (34.94 lb/sq ft)
- Max power loading (PT6A-65AG) ... 7.00 kg/kW (11.50 lb/shp)

PERFORMANCE (PT6A-65AG):
- Never-exceed speed (VNE) ... 191 kt (354 km/h; 220 mph)
- Cruising speed at 55% power ... 161 kt (298 km/h; 185 mph)
- Working speed ... 86–152 kt (185–282 km/h; 100–175 mph)
- Stalling speed, power off, flaps down ... 50 kt (92 km/h; 57 mph)
- Max rate of climb at S/L:
  - at 5,670 kg (12,500 lb) ... 312 m (1,025 ft)/min
  - at 6,418 kg (14,150 lb) ... 213 m (700 ft)/min
- T-O run ... 335 m (1,100 ft)
- Landing run ... 183 m (600 ft)
- Ferry range at 50% power ... 521 n miles (966 km; 600 miles)

UPDATED

First Thrush 660 built by Thrush Aircraft Inc ***NEW***/1047085

TIGER

TIGER AIRCRAFT LLC

226 Pilot Way, Martinsburg, West Virginia 25401
Tel: (+1 304) 267 10 00
Fax: (+1 304) 262 00 69
e-mail: billcrum@tigeraircraft.com
Web: http://www.tigeraircraft.com
CEO AND PRESIDENT: Gene Criss
VICE-CHAIRMAN: Peter Lo
CFO: Ron Shade
MARKETING: Bill Crum

On 12 May 2000, Tiger Aircraft acquired the Type Certificate for the AG-5B Tiger, last employed by American General Aircraft Corporation. It also holds the AA-1 Type Certificate and intends to produce new designs. The company employed 63 at the end of 2001 with plans to increase workforce to 85 over following 12 months. In late 2002, it was owned jointly by Teleflex (30 per cent) and Taiwanese investors. Production certification for Tiger granted by FAA on 3 September 2002.

UPDATED

Tiger Aircraft AG-5B Tiger *(Paul Jackson)* ***NEW***/1047178

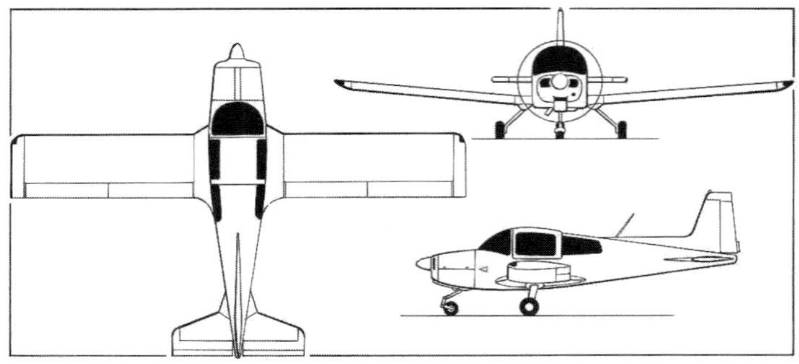

Tiger Aircraft AG-5B Tiger general arrangement *(James Goulding)* 0131346

TIGER AG-5B TIGER

TYPE: Four-seat lightplane.

PROGRAMME: Original two-seat BD-1 designed by Jim Bede; prototype first flown 11 July 1963; developed into AA-1 by American Aviation (later bought by Grumman and then by Gulfstream American), which built 461 AA-1s, 470 AA-1As, 680 AA-1Bs and 211 AA-1Cs. Four-seat AA-5 certified 12 November 1971 and 3,054 built before designs sold to American General Aircraft in 1989. Updated AG-5B Tiger first flown 21 April 1990; certified 21 September 1990; company sold to Teleflex in early 1992 but later filed for bankruptcy in January 1994 after building 89 examples.

First Tiger Aircraft AG-5B (N999TE) made maiden flight 9 July 2001. Symbolically delivered to new owner, Herb Hortman, 26 July 2001 at AirVenture 2001. Certification completed 3 December 2001, when first three aircraft handed over. In late 2002 Tiger Aircraft was negotiating with Chengdu Aircraft Industrial Corporation (CAC) of China with a view to licensed assembly of the AG-5B for Chinese commercial pilot training schools.

CUSTOMERS: 41 orders by July 2001. Total of 14 delivered in 2002, 18 in 2003, and 15 in the first nine months of 2004.

COSTS: Basic trainer version US$186,800; standard US$235,800 (2004).

DESIGN FEATURES: Conventional low-wing monoplane with constant-chord wings and tailplane and moderately sweptback fin with fillet. Extensive use of honeycomb and adhesive bonding throughout the structure.

FLYING CONTROLS: Conventional and manual with adjustable tab on each aileron and rudder. Electrically operated flaps, maximum deflection 45°.

STRUCTURE: Semi-monocoque fuselage of aluminium honeycomb. Cantilever low wing with tubular single spar, section NACA $64_2$415 (modified), clad in aluminium skin. Dihedral 5°, incidence 1° 25'.

LANDING GEAR: Tricycle type; fixed. Mainwheels 6.00–6, tyre pressure 2.34 bar (34 lb/sq in); nosewheel 5.00–5, tyre pressure 1.72 bar (25 lb/sq in). Cleveland hydraulic brakes.

POWER PLANT: One 134 kW (180 hp) Textron Lycoming O-360-A4K flat-four, driving a two-blade fixed-pitch Sensenich 76EM8S-10-0 propeller. Fuel in two wing tanks, combined capacity 199 litres (52.6 US gallons; 43.8 Imp gallons), of which 193 litres (51.0 US gallons; 42.5 Imp gallons) are usable. Oil capacity 7.6 litres (2.0 US gallons; 1.7 Imp gallons).

ACCOMMODATION: Pilot and three passengers in individual pairs of leather seats; baggage area behind rear seats. Fixed wraparound windscreen; entry to cockpit via sliding canopy.

SYSTEMS: 24 V electrical system.

AVIONICS: Full Garmin IFR panel standard.

Comms: Dual GNS 430 com/nav with VOR/LOC/GPS/GPS-LOC; GMA 340 audio panel with marker beacon and four-place intercom system; GTX 327 digital transponder, and ground clearance system

Flight: Dual Garmin 106A glideslope indicators; S-Tec 30 two-axis autopilot with altitude hold.

Instrumentation: Sigma/Tek electronic instrument cluster comprising vacuum, load amp, oil pressure, oil temperature, fuel pressure and CHT gauges; Davtron M800 chronometer and A301C digital OAT gauge, and Datcom Hobbs meter.

EQUIPMENT: Standard equipment includes leather interior; leather-wrapped control yokes, machined aluminium instrument panel with backlit instruments and panel/instrument light dimmers, cabin dome light, map light, and flush-mounted anti-collision beacons and landing lights.

DIMENSIONS, EXTERNAL:
Wing span .... 9.60 m (31 ft 6 in)
Wing chord, constant .... 1.35 m (4 ft 5¼ in)
Wing aspect ratio 7.1
Length overall .... 6.71 m (22 ft 0 in)
Height overall .... 2.44 m (8 ft 0 in)
Tailplane span .... 2.65 m (8 ft 8½ in)
Wheel track .... 2.51 m (8 ft 3 in)
Wheelbase .... 1.64 m (5 ft 4½ in)
Propeller diameter .... 1.98 m (6 ft 6 in)

DIMENSIONS, INTERNAL:
Cabin: Length .... 1.98 m (6 ft 6 in)
Max width .... 1.04 m (3 ft 5 in)
Max height .... 1.23 m (4 ft 0¼ in)
Floor area .... 2.18 m² (23.5 sq ft)
Baggage hold volume .... 0.34 m³ (12.0 cu ft)

AREAS:
Wings, gross .... 13.02 m² (140.1 sq ft)

WEIGHTS AND LOADINGS:
Weight empty .... 680 kg (1,500 lb)
Baggage capacity: four-seat .... 54 kg (120 lb)
two-seat .... 154 kg (340 lb)
Max T-O weight .... 1,089 kg (2,400 lb)
Max wing loading .... 83.6 kg/m² (17.13 lb/sq ft)
Max power loading .... 8.12 kg/kW (13.33 lb/hp)

PERFORMANCE:
Never-exceed speed (VNE) .... 173 kt (93 km/h; 200 mph)
Max operating speed (VMO) .... 148 kt (274 km/h; 170 mph)
Max cruising speed at 75% power at FL85 .... 143 kt (265 km/h; 165 mph)
Stalling speed: flaps up .... 56 kt (104 km/h; 65 mph)
flaps down .... 53 kt (99 km/h; 61 mph)
Max rate of climb at S/L .... 259 m (850 ft)/min
Service ceiling .... 4,205 m (13,800 ft)
T-O run .... 260 m (854 ft)
T-O to 15 m (50 ft) .... 472 m (1,550 ft)
Landing from 15 m (50 ft) .... 341 m (1,120 ft)
Landing run .... 125 m (410 ft)
Range: no reserves .... 681 n miles (1,261 km; 783 miles)
with 45 min reserves .... 572 n miles (1,059 km; 658 miles)
Endurance, no reserves .... 4 h 42 min

UPDATED

TITAN

TITAN AIRCRAFT

1419 State Route 45 South, PO Box 190, Austinburg, Ohio 44010
Tel: (+1 440) 275 32 05
Fax: (+1 440) 275 31 92
e-mail: sales@titanaircraft.com
Web: http://www.titanaircraft.com
PRESIDENT: John Williams

While continuing to produce the Tornado ultralight, Titan has expanded its range with a scaled-down representation of the Second World War North American P-51 Mustang.

Titan's Tornado continues to be produced *(Paul Jackson)* ***NEW***/0561715

TITAN T51 MUSTANG

TYPE: Tandem-seat sportplane kitbuilt.

PROGRAMME: Construction started September 2001 and prototype (N151TD) flew 17 June 2002. By late 2003, a turbocharged version with a Rotax 914 was nearing completion.

COSTS: Kit US$37,500, less engine, propeller and instruments (2003).

CUSTOMERS: More than 20 sold and two flying by mid-2003.

DESIGN FEATURES: Configuration approximates to 75% scaled-down NA P-51 of 1940s, Laminar flow wing is 72% scale. Quoted build time 600 hours.

FLYING CONTROLS: Conventional and manual. Horn-balanced elevators and rudder; flight-adjustable tab in starboard elevator; elevator airflow energised by vortex generators above and below tailplane. Fowler flaps; maximum deflection 30°.

STRUCTURE: Welded 4130 steel tube fuselage. Aluminium wing has foam-filled D section. Composites wing-, fin- and elevator tips.

LANDING GEAR: Tricycle type; retractable. Mainwheels 120/90-10; tailwheel 2.8/2.5-4. Hydraulic retraction; hydraulic brakes.

POWER PLANT: One 73.5 kW (98.6 hp) Rotax 912S flat-four, driving a three-blade IVO, ground-adjustable pitch propeller. Fuel capacity 91 litres (24.0 US gallons; 20.0 Imp gallons) of which 87 litres (23.0 US gallons; 19.2 Imp gallons) are usable.

ACCOMMODATION: Canopy slides rearward; baggage compartments fore and aft of cabin accessed by doors on port side of fuselage.

DIMENSIONS, EXTERNAL:
Wing span ... 7.32 m (24 ft 0 in)
Length overall ... 7.16 m (23 ft 6 in)
Height overall ... 2.79 m (9 ft 2 in)

DIMENSIONS, INTERNAL:
Cabin max width ... 0.61 m (2 ft 0 in)

AREAS:
Wings, gross ... 10.96 m² (118.0 sq ft)

Titan T51 Mustang prototype *(Paul Jackson)* ***NEW***/0561714

WEIGHTS AND LOADINGS:
Weight empty ... 340 kg (750 lb)
Max T-O weight ... 558 kg (1,232 lb)

PERFORMANCE:
Never-exceed speed (VNE) ... 171 kt (317 km/h; 197 mph)
Max cruising speed ... 130 kt (241 km/h; 150 mph)
Stalling speed, flaps down ... 34 kt (63 km/h; 39 mph)
Max rate of climb at S/L ... 365 m (1,200 ft)/min
T-O and landing run ... 92 m (300 ft)
Range ... 625 n miles (1,158 km; 720 miles)
g limits ... +6/−4

TREK

PO Box 3700, Los Altos, California 94024
Tel: (+1 209) 368 5800
Fax: (+1 209) 368 5802
e-mail: info@trekaerospace.com
Web: http://www.trekaerospace.com
PRESIDENT Harry Falk
VICE PRESIDENTS Timothy Worley
Robert Bulaga

Company founded in 1995 to conduct research on powered-lift, ducted rotor aircraft. US Defense Advanced Research Projects Agency (DARPA) awarded a 36 month, US$4.7 million development contract in December 2000. NASA Ames assisting with development of stability augmentation system. Initial proof-of-concept vehicle, SoloTrek XFV (see *Jane's All the World's Aircraft* 2003-04) conducted first controlled, tethered flight on 18 December 2001. Proof-of-concept vehicle retired in March 2002 after 39 hours of tethered flight testing.

Pre-series vehicle (001) first tethered flight on 13 December 2002. This was seriously damaged four days later when the rotors ingested the overhead safety tether and the craft fell from a height of 1.3 m (4 ft). Additional design and safety enhancements were made throughout 2003. Activities temporarily suspended on 31 December 2002, but resumed with further funding in following year. Pre-series vehicle (002) was constructed in July/August 2003 and successfully completed the DARPA contract in February 2004.

Trek's core technology can be applied to several vehicles.

UPDATED

TREK SPRINGTAIL EFV-4

TYPE: Exoskelitor flying vehicle.

PROGRAMME: Follow-on aircraft to previously tested SoloTrek XFV. Designed in early 2003. First tethered flight (N401XV) 2 October 2003. First free (untethered) hover 20 October 2003. First transitional, forward flight 5 November 2003.

CURRENT VERSIONS: **EFV-4A**: Prototype N401XV; employed for pure research.

EFV-4B: Non-research, envelope-expansion aircraft. Redesigned fans with variable pitch blades. Under development in 2004.

COSTS: Target unit cost for production aircraft is approximately US$100,000 to US$125,000.

DESIGN FEATURES: Proprietary, core technology consists of twin, counter-rotating, variable-pitch ducted rotors which have the ability to generate 4.8 kg/kW (8 lb/hp) of thrust, which Trek states is superior to other designs. Use of the twin counter-rotating, variable-pitch ducted rotors also has the advantage of eliminating torque problems associated with some VTOL vehicles. EFV-4 uses an advanced fly-by-wire stability augmentation system (SAS) which allows it to be operated either manned, remotely piloted, or autonomously. SAS significantly reduces the pilot workload; reduces RPV operator-training time; and can be programmed to limit vehicle performance to operator's capabilities (low, medium or high performance).

Twin, counter-rotating, variable-pitch ducted rotors are tilted in conjunction with power variation to achieve vertical take-off, climb, forward flight and vertical landing. Maximum duct forward tilt angle 35°. Windmilling, variable-pitch rotors limit power-out descent rate to 5 m (17 ft)/s.

Airframe configuration supports pilot in standing position: engine behind pilot. Power transmitted by a short driveshaft with universal joints to self-lubricating main gearbox above and behind the pilot's head. Twin driveshafts to fans, each with satellite gearboxes. Small cooling fan for main gearbox. Projected time between overhauls 1,000 hours.

FLYING CONTROLS: No conventional aerodynamic surfaces. Ducted rotors tilted by electromechanical actuators; small vanes, electromechanically actuated, behind each fan for precise control at slow speed and hover. Three-axis trim.

Hovering trials of the Trek Springtail EFV-4A ***NEW***/1047809

STRUCTURE: Aluminium honeycomb airframe with carbon fibre, Kevlar, stainless steel and titanium components. Welded, aluminium tube landing skids. Composites ducts.

LANDING GEAR: Twin skids incorporate crush technologies that reduce damage in the event of a hard landing. Pneumatic forward mainwheels; solid, castoring rear wheels for ground handling and taxying.

POWER PLANT: One 88 kW (118 hp) Wankel-type rotary engine driving pair of three-blade, variable-pitch rotors. Fuel tank below engine; capacity 46 litres (12.3 US gallons; 10.2 Imp gallons). Other power plant options are possible.

ACCOMMODATION: One person in standing position, supported by safety harness. Two hand levers for engine and flight control; engine/vertical axis, left; longitudinal and lateral axes, plus trim, right. Small area between landing gear for a limited cargo.

SYSTEMS: Fully FBW stability control system. 24 V DC electrical system.

AVIONICS: *Instrumentation:* Production aircraft utilise a helmet-mounted display for all necessary flight data. A small touch-screen 'maintenance' panel to monitor system health.

EQUIPMENT: Ballistic recovery parachute as backup to power-out descent capability.

DIMENSIONS, EXTERNAL:
Height over rotor hub ... 2.53 m (8 ft 4 in)
Width over ducts ... 2.95 m (9 ft 8 in)
Width at landing gear ... 1.07 m (3 ft 6 in)
Length (front to rear) ... 1.63 m (5 ft 4 in)

WEIGHTS AND LOADINGS:
Weight empty ... 170 kg (375 lb)
Payload with max fuel ... 162 kg (358 lb)
T-O weight: normal ... 281 kg (622 lb)
max ... 378 kg (834 lb)

PERFORMANCE (estimated):
Max level speed ... 98 kt (181 km/h; 113 mph)
Cruising speed ... 82 kt (152 km/h; 94 mph)
Loiter speed ... 56 kt (104 km/h; 64 mph)
Max rate of climb at S/L ... 1,676 m (5,500 ft)/min
Hovering ceiling, OGE ... 3,475 m (11,400 ft)
Range ... 160 n miles (296 km; 184 miles)
Endurance ... 2 h 10 min

UPDATED

TREK DRAGONFLY UMR-1

TYPE: Lifting vehicle.

PROGRAMME: Under development.

COSTS: Target US$150,000 to US$200,000.

DESIGN FEATURES: Conceived as unmanned utility cargo vehicle with over 1.2 m^3 (41 cu ft) capacity; equally capable of manned or remote operation, however. Contains same core technology, advanced FBW stability augmentation and control systems as Springtail, but enclosed configuration supports pilot in seated position in the forward bay of the fuselage. Engine in the aft fuselage bay. Power transmitted by short driveshaft with universal joints to self-lubricating main gearbox slightly above and behind pilot's head. Twin driveshafts to fans, each with satellite gearboxes. Small cooling fan for main gearbox. Intended time between overhauls 1,000 hours.

FLYING CONTROLS: Ducted rotors tilted by electromechanical actuators; small vanes, electromechanically actuated, behind each fan for precise control at slow speed and hover. Inverted V-tail provides directional and pitch control in forward flight. Three-axis trim.

STRUCTURE: Aluminium honeycomb airframe with carbon fibre, Kevlar, stainless steel and titanium components. Welded, aluminium tube landing skids. Composites ducts.

LANDING GEAR: Twin skids incorporate crush technologies that reduce damage in the event of a hard landing. Pneumatic forward mainwheels; solid, castoring rear wheels for ground handling and taxying.

POWER PLANT: One 132 kW (177 hp), Wankel-type rotary engine driving two three-blade, variable-pitch rotors. Fuel tank in central bay of fuselage, capacity 59 litres (15.5 US gallons; 12.9 Imp gallons). Other power plant options possible.

ACCOMMODATION: One person seated in the forward bay of the fuselage. Cargo capacity of approximately 0.37 m^3 (13 cu ft) in the centre bay. Two hand levers for engine and flight control; engine and vertical axis, left; longitudinal and lateral axes plus trim, right.

SYSTEMS: Fully FBW stability control system. 24 V DC electrical system.

AVIONICS: *Instrumentation:* Two touch-screen monitors mounted in instrument panel; left screen displays flight data, right screen displays engine and systems data. Military variants have helmet-mounted display.

Trek Dragonfly UMR-1 enclosed lifting vehicle ***NEW***/1047813

EQUIPMENT: Ballistic recovery parachute as backup to power-out glide and descent capability.

DIMENSIONS, EXTERNAL:
Height over rotor hub ... 1.88 m (6 ft 2 in)
Width: over ducts ... 2.95 m (9 ft 8 in)
at landing gear ... 1.14 m (3 ft 9 in)
Length ... 3.99 m (13 ft 1 in)

WEIGHTS AND LOADINGS:
Weight empty ... 224 kg (493 lb)
Payload with max fuel ... 205 kg (450 lb)
T-O weight ... 367 kg (808 lb)

PERFORMANCE(estimated):
Max level speed ... 204 kt (378 km/h; 235 mph)
Cruising speed ... 145 kt (269 km/h; 167 mph)
Loiter speed ... 74 kt (137 km/h; 85 mph)
Max rate of climb at S/L ... 1,676 m (5,500 ft)/min
Hovering ceiling, OGE ... 3,930 m (12,900 ft)
Range ... 500 n miles (926 km; 575 miles)
Endurance ... more than 3 h

UPDATED

TSI

750 Hammond Drive, Building 10, Suite 100, Atlanta, Georgia 30328
Tel: (+1 404) 446 22 06
e-mail: info@techsphere.us
Web: http://www.techspheresystems.com
CHAIRMAN AND COO: Keith Vierela
PRESIDENT AND CEO: Mike Lawson
EXECUTIVE VICE-PRESIDENTS: Ed Pickett (Government Applications)
John Youngbeck
COMMUNICATIONS SPECIALIST: Josh Vierela

Techsphere Systems was formed in mid-2002 to manufacture, own, operate, lease and sell spherical airship platforms for manned and unmanned use in commercial, research and government applications. To this end, in November 2002 it entered an agreement giving it worldwide exclusive rights to manufacture and sell certain types of airship of this configuration designed by the Canadian company 21st Century Airships. That company's 19.05 m (62.5 ft) diameter SPAS-70-1A, redesignated **AeroSphere SA-60** and registered N8041X, was acquired as a low-altitude prototype for larger new products, the first of which is a 28.65 m (94 ft) diameter AeroSphere, due for completion in February or March 2005 and capable of reaching altitudes up to 6,100 m (20,000 ft).

Marketing rights are delegated to Cyber Aerospace Corporation of West Palm Beach, Florida, to whom the SA-60 was sold by TSI in early 2004. This prototype was evaluated by US Naval Air Systems Command for a surveillance role in June 2004, being sold on in the following month to Sierra Nevada Corporation of Sparks, Nevada, for systems integration to convert it to unmanned operation. Also in 2004, Cyber took delivery of a 23.2 m (76 ft) **SA-76**, a five-engined variant intended as an optionally piloted vehicle with the ability to operate at up to 4,880 m (16,000 ft). Georgia Tech Research Institute (GTRI) is assisting in developing some TSI platforms for unmanned (UAV) applications.

Immediate TSI production is at an initial 3,902 m^2 (42,000 sq ft) manufacturing facility on Phillips Drive, Columbus, Georgia, opened on 10 November 2004. At a second Columbus location, TSI plans to complete a 10,035 m^2 (108,000 sq ft) modern factory during the first quarter of 2005, as well as a hangar able to house airships of up to 91.4 m (300 ft) diameter.

NEW ENTRY

AeroSphere SA-60 N8041X during early trials 1043438

TURBINE DESIGN

TURBINE DESIGN INC

1335 Saratoga Street, Deland, Florida 32724
Tel: (+1 386) 736 82 62
Fax: (+1 386) 738 05 10

Company activities have included US sales of a kit version of Eurospace F-15-F Excalibur using some of the 10 uncompleted airframes imported into the USA* from Sokol of Russia, a turboprop Rolls-Royce 250-engined F-15-F having first flown on 10 May 2002; conversions of ADI Stallion, Beech Bonanza Ryan Navion, Piper Comanche, Cessna 341 and others with Walter M 601; and engineering of Maxima conversion of Piper Malibu and Mirage with Walter M 601 turboprop, although this activity was later discontinued.

There have been no recent reports of the TD-2 Tempest turboprop sportplane, at least two examples of which had been fabricated (but not, apparently, registered) by 2003.

* Unconnected with Turbine Design, one of these became a Douglas K& (N24VK) in 2002.

UPDATED

Turboprop conversion of F-15-F by Turbine Design ***NEW***/0567037

ULLMANN

ULLMANN AIRCRAFT COMPANY

5132 Willow Point Road, Wichita, Kansas 67220
Tel/Fax: (+1 316) 744 11 45
e-mail: brian.ullmann@ullmannaircraft.com
Web: http://www.ullmannaircraft.com
PRESIDENT AND CEO: Bill Ullmann
CHIEF ENGINEER: Brian Ullmann

The Ullmann Aircraft Company's first aircraft is the Panther.

ULLMANN 2000 PANTHER

TYPE: Four-seat kitbuilt.

PROGRAMME: Design began 15 April 1997 and construction of first prototype (N202KT) on 2 June 1998. Formally announced at AirVenture '99; mockup displayed at Sun 'n' Fun 2000. Part-completed prototype displayed at Sun 'n' Fun 2001 and again at AirVenture 01, with first flight then expected before end 2001; this later slipped to June 2002 but had not taken place by November 2002, when it was said to be expected in the following month; achieved 29 March 2003 and by August 2003 had completed 60 hours of flying. Kit deliveries planned to begin December 2003.

COSTS: Development costs estimated at US$500,000 in 1997. Kit US$49,000, excluding engine, propeller, avionics, instruments, interior or paint; four sub-kits are available (2003 prices).

DESIGN FEATURES: Cantilever high-wing monoplane of rectangular fuselage cross-section. Constant-chord wings and tailplane; sweptback fin with fillet. High-performance kitbuilt meeting FAR Pt 23 certification and 51 per cent homebuilt rule. Many parts supplied pre-formed, pre-drilled or pre-assembled. Quoted build time 1,200 to 1,600 hours.
Aerofoil section NACA 64_2A215, dihedral 1° 30'; incidence 1°; no twist.

FLYING CONTROLS: Conventional and manual. Horn-balanced elevator and rudder, both with electric trim. Single-slotted Fowler flaps. Dorsal fillet and drooped outboard cuff assists spin resistance.

STRUCTURE: All-metal wing with square tips. Fuselage of 4130 steel tube covered with aluminium sheet.

LANDING GEAR: Tricycle type; fixed, Wittmann sprung steel main legs. Cleveland wheels; disc brakes on mainwheels. Nosewheel 5.00-5, steerable ±60°. Mainwheels 6.00-6, all with 2.07 bar (30.0 lb/sq in) tyre pressure. Pressure Recovery wheel fairings.

POWER PLANT: Prototype has one 224 kW (300 hp) Teledyne Continental IO-550-L driving a three-blade Hartzell constant-speed propeller; engine mount also compatible with IO-520-L. Mounts and cowlings for 194 kW (260 hp) Textron Lycoming IO-540 and 224 kW (300 hp) Zehrbach engines under consideration. Fuel in two integral wing leading-edge tanks, combined capacity 277 litres (73.0 US gallons; 60.8 Imp gallons); or alternative long-range tanks combined capacity 390 litres (103 US gallons; 85.8 Imp gallons), of which 386 litres (101 US gallons; 84.9 Imp gallons) are usable. Gravity refuelling point on wing. Oil capacity 7.6 litres (2.0 US gallons; 1.7 Imp gallons).

ACCOMMODATION: Pilot and three passengers in two pairs of seats. Split doors on each side of fuselage; single baggage door at rear of cabin.

SYSTEMS: 12 V 24 Ah Concorde RG-25XC battery. Electric fuel pump.

AVIONICS: Prototype has two Bendix/King KX 155A com radios, Bendix/King KLN 94 VOR/DME and GPS.

DIMENSIONS, EXTERNAL:

Wing span	10.44 m (34 ft 3 in)
Wing chord, constant	1.16 m (3 ft 9½ in)
Wing aspect ratio	9.0
Length overall	7.87 m (25 ft 10 in)
Fuselage max width	1.29 m (4 ft 2¼ in)
Height overall	2.90 m (9 ft 6 in)
Tailplane span	3.43 m (11 ft 3 in)
Wheel track	2.84 m (9 ft 4 in)
Wheelbase	2.34 m (7 ft 8 in)

Ullmann Panther prototype *(Paul Jackson)* ***NEW***/0567789

Propeller diameter	1.98 m (6 ft 6 in)
Propeller ground clearance	0.43 m (1 ft 5 in)
Passenger doors (each): Height	1.07 m (3 ft 6 in)
Width	1.05 m (3 ft 5¼ in)
Height to sill	0.88 m (2 ft 10½ in)
Baggage door: Height	0.45 m (1 ft 5¾ in)
Width	0.54 m (1 ft 9¼ in)
Height to sill	0.89 m (2 ft 11 in)

DIMENSIONS, INTERNAL:

Cabin: Length	2.32 m (7 ft 7½ in)
Max width	1.27 m (4 ft 2 in)
Max height	1.22 m (4 ft 0 in)
Volume	2.97 m^3 (105.0 cu ft)
Baggage hold volume	0.61 m^3 (21.5 cu ft)

AREAS:

Wings, gross: flaps in	12.08 m^2 (130.0 sq ft)
flaps out	13.19 m^2 (142.0 sq ft)
Ailerons (total)	1.48 m^2 (15.88 sq ft)
Trailing-edge flaps (total)	2.37 m^2 (25.52 sq ft)
Fin	1.32 m^2 (14.25 sq ft)
Rudder, incl tab	0.47 m^2 (5.08 sq ft)
Tailplane	2.35 m^2 (25.31 sq ft)
Elevators, (total incl tab)	0.84 m^2 (9.00 sq ft)

WEIGHTS AND LOADINGS (estimated):

Weight empty	925 kg (2,040 lb)
Baggage capacity	54 kg (120 lb)
Max payload	363 kg (800 lb)
Max fuel weight	281 kg (620 lb)
Max T-O and landing weight	1,383 kg (3,050 lb)
Max wing loading	114.5 kg/m^2 (23.46 lb/sq ft)
Max power loading	6.19 kg/kW (10.17 lb/hp)

PERFORMANCE (estimated):

Never-exceed speed (VNE)	250 kt (463 km/h; 287 mph)
Max operating speed	200 kt (370 km/h; 230 mph)
Normal cruising speed at 75% power	200 kt (370 km/h; 230 mph)
Stalling speed, power off, flaps down	59 kt (110 km/h; 68 mph)
Range, 45 min reserves: with max fuel	1,227 n miles (2,272 km; 1,412 miles)
with max payload	823 n miles (1,524 km; 947 miles)
Endurance	6 h 8 min
g limits	+3.8/−1.52

UNITED STATES AIR FORCE

AIR FORCE MATERIEL COMMAND

3753 Chidlaw Road, Wright-Patterson AFB, Ohio 45433-5006
Tel: (+1 937) 257 63 06
Fax: (+1 937) 257 25 58
e-mail: HQAFMC.PA@wpafb.af.mil
Web: http://www.afmc.wpafb.af.mil
COMMANDER, AFMC: General Gregory S Martin
DIRECTOR, PUBLIC AFFAIRS: Col Donna Pastor

Air Force Materiel Command (AFMC) is composed of four product centres, each responsible for developing and acquiring specific weapon systems with the assistance of the Air Force Research Laboratory at Wright-Patterson AFB, Ohio and the Air Force Office of Scientific Research at Bolling AFB, Washington, DC. These comprise the Aeronautical Systems Center, Wright-Patterson AFB, Ohio; the Air Armament Center, Eglin AFB, Florida; the Electronic Systems Center, Hanscom AFB, Massachusetts; and the Materiel Systems Group, also at Wright-Patterson AFB.

AFMC also has two test establishments, namely Air Force Flight Test Center at Edwards AFB, California; and Arnold Engineering Development Center at Arnold AFB, Tennessee. These are responsible for evaluation of systems under development and in the field.

Air Logistics Centers (Ogden ALC at Hill AFB, Utah; Oklahoma City ALC at Tinker AFB, Oklahoma and Warner Robins ALC at Robins AFB, Georgia) provide weapon system sustainment and maintenance services for just under 50 types of aircraft, with a total inventory of about 6,200 airframes and 20,000 engines. Finally, two specialist organisations (Aerospace Maintenance and Regeneration Center at Davis-Monthan AFB, Arizona, and Air Force Security Assistance Center at Wright-Patterson AFB, Ohio) provide for storage, regeneration, reclamation and disposal of aircraft and associated aerospace equipment.

UPDATED

REPLACEMENT INTERDICTOR AIRCRAFT (RIA)

Exploratory studies, for an aircraft to replace the Lockheed Martin F-117A Nighthawk and Boeing F-15E Eagle in the interdiction role and also augment the bomber force, have been launched by the US Air Force, with Boeing, Lockheed Martin and Northrop Grumman undertaking studies into a potential RIA. Service entry approximately 2015–20. Manned and unmanned concepts have been and are being considered, with bomber version of F/A-22 Raptor emerging as possible option in 2003-04.

UPDATED

UNITED STATES NAVY

NAVAL AIR SYSTEMS COMMAND

47123 Buse Road, Unit Moffett Building, Patuxent River, Maryland 20650-1547
Tel: (+1 301) 757 78 25
Web: http://www.navair.navy.mil

NAVAIR has over 31,500 military and civilian employees and provides total life cycle support of all naval aviation weapon systems. Subordinate elements include test organisations at China Lake, California; Patuxent River, Maryland; and Point Mugu, California, as well as aircraft repair and overhaul depots at Cherry Point, North Carolina; Jacksonville, Florida; and North Island, California. It is prime authority for the USAF/USN Lockheed Martin F-35 Joint Strike Fighter, via the Program Executive Office (PEO); other PEOs have responsibility for tactical aircraft programmes; air ASW, assault and special mission programmes; and strike weapons and unmanned aviation.

VERIFIED

MULTIMISSION MARITIME AIRCRAFT (MMA)

In late 1997, US Navy initiated two-year requirements study into replacement for Lockheed Martin P-3C Orion maritime patrol/anti-submarine warfare aircraft, as well as intelligence-gathering EP-3E, submarine-communications Boeing E-6 Mercury and tanker/transport Lockheed Martin KC-130 Hercules of US Marine Corps. Last-mentioned type now being replaced by new KC-130J aircraft, but maritime requirement remains.

A Defense Acquisition Board review in March 2000 presaged an analysis of alternatives and paved the way for exploratory concept studies by industry between June 2000 and January 2002. Concepts that emerged at this time included separate and distinct remanufacture programmes involving the P-3C by Lockheed Martin and Raytheon; adoption of a 737 derivative by Boeing; and studies by Northrop Grumman into using the RQ-4A Global Hawk UAV as an adjunct to the manned system.

Navy assessment of the results took place in the first half of 2002 and resulted in selection of two proposals to proceed to the Component Advanced Development (CAD) stage and the award on 10 September 2002 of US$7 million contracts to Boeing and Lockheed Martin. These covered CAD Phase 1, which concluded five months later, in February 2003. Phase 1 provided the US Navy with a variety of concepts, from which two were selected to progress to 14-month CAD Phase 2 (February 2003 to 30 April 2004). Boeing's proposal was based on the 737 airliner, while Lockheed Martin offered a new version of the P-3 known as the Orion 21.

A draft System Development and Demonstration (SDD) request for proposals was released to industry on 15 July 2003, with the definitive SDD RFP following in late September 2003. Milestone B source selection of the winning proposal was announced on 14 June 2004, when Boeing received a US$3.89 billion contract award. SDD began in the second quarter of FY04 and will continue until the fourth quarter of FY08.

First flight is unlikely to take place before 2009, but the Milestone C decision to go ahead with low-rate initial production is due in the first quarter of FY09, with IOC anticipated at the end of FY12. Ultimately, the Navy expects to purchase 108 MMAs, augmenting them with UAVs for certain surveillance missions.

UPDATED

US LIGHT AIRCRAFT

US LIGHT AIRCRAFT CORPORATION

27080 Rancho Ballena Lane, Ramona, California 92065
Tel/Fax: (+1 760) 789 86 07
e-mail: j.millett@gte.net or sales@flyhornet.com
Web: http://www.flyhornet.com
PRESIDENT: James Millet

US Light Aircraft formed from partnership between Millett Engineering and Forespar Company in 1993.

US LIGHT AIRCRAFT HORNET

TYPE: Tandem-seat ultralight kitbuilt.
PROGRAMME: Design began in 1991 and construction of prototype in 1992; first flight June 1993. Winner of 15 awards at Copperstate, Oshkosh and Sun 'n' Fun between 1993 and 1999. Qualifies under FAR Pt 21.191 as 51 per cent experimental kit.
CUSTOMERS: About 65 sold and 45 flying by end 2002.
COSTS: Kit US$20,400, including engine; US$13,750 without engine (2003).
DESIGN FEATURES: 400 parts in kits include 7,500 factory-installed rivets. Quoted build time 300 hours.
FLYING CONTROLS: Conventional and manual. Electrically operated flaps with maximum deflection of 30°. Horizontal stabiliser trimmed electrically. Mass-balanced elevator.
STRUCTURE: Fabric-covered fuselage of 3 cm (1.25 in) 6063-T8 aluminium tube with tapered-base tubular longitudinal structure. Fabric-covered dual-spar wings with single bracing struts.
LANDING GEAR: Fixed tricycle type with 15 cm (6 in) Goodyear Airspring pneumatic suspension on mainwheels and 7.5 cm (3 in) Airspring on nosewheel; hydraulic brakes on mainwheels with aluminium Hegar wheels and 15×6-6 Cheng Shin tyres; nose tyre 13×5-6.
POWER PLANT: One 41.0 kW (55 hp) Hirth 2703 engine with electric start is standard, driving Tennessee two-blade, wooden pusher propeller of diameter up to 1.52 m (5 ft 0 in); 48.5 kW (65 hp) Hirth 2706 optional. Fuel capacity 68 litres (18.0 US gallons; 15.0 Imp gallons) in two tanks between spars.
EQUIPMENT: Optional BRS VLS 1050 ballistic parachute.
DIMENSIONS, EXTERNAL:

US Light Aircraft Corporation Hornet *(Geoffrey P Jones)* 0100462

Wing span .... 8.38 m (27 ft 6 in)
Length overall .... 6.10 m (20 ft 0 in)
Height overall .... 1.83 m (6 ft 0 in)
AREAS:
Wings, gross .... 12.82 m² (138.0 sq ft)
WEIGHTS AND LOADINGS:
Weight empty .... 215 kg (475 lb)
Max T-O weight .... 454 kg (1,000 lb)
PERFORMANCE (41.0 kW; 55 hp engine):
Never-exceed speed (VNE) .... 108 kt (201 km/h; 125 mph)
Max level speed .... 104 kt (193 km/h; 120 mph)
Normal cruising speed at 75% power .... 61 kt (113 km/h 70 mph)
Stalling speed, flaps down .... 35 kt (65 km/h; 40 mph)
Max rate of climb at S/L .... 305 m (1,000 ft)/min
T-O run .... 114 m (375 ft)
Landing run .... 76 m (250 ft)
Range, 10% reserves .... 304 n miles (563 km; 350 miles)

USA

ULTRALIGHT SOARING AVIATION

1830 East Virginia Street, Anna, Illinois 62906
Tel/Fax: (+1 618) 833 31 87
e-mail: ekstrom@midwest.net
Web: http://www.members.aol.com/cumulusman
PRESIDENT: Dave Ekstrom

Marketing of the Cumulus motor glider began in 2003.

NEW ENTRY

USA CUMULUS

TYPE: Single-seat motor glider/kitbuilt.
PROGRAMME: Proof-of-concept Cloud Dancer prototype flew 1996; marketing in considerably modified present form began 2003; exhibited at Sun 'n' Fun, Lakeland, Florida, April 2004.
CUSTOMERS: Total 11 sold by April 2004.
COSTS: Kit less instruments US$7,995; power plant US$3,835 (2004).
DESIGN FEATURES: Low-wing, open cockpit (optionally enclosed) pusher of pod-and-boom configuration with strut-braced cruciform tailplane.
Quoted build time 150 hours.
Optional short-span wing.
FLYING CONTROLS: Conventional and manual. Flaps with upward reflex for improved cruising speed. Door-type spoilers in wing upper surface. Horn-balanced rudder.
STRUCTURE: Spec 6061-T6 aluminium (optional 4130 steel) boom and cockpit frame with glass fibre skin. Tubular metal wing with single spar; aluminium skin with fabric covered ailerons and flaps.
LANDING GEAR: Tailwheel type; fixed. Fuselage-mounted spring steel cantilever mainwheel legs with optional speed fairings. Steerable, solid tailwheel.
POWER PLANT: One 31.0 kW (41.6 hp) Rotax 447 two-cylinder piston engine driving a Tennessee Propellers Inc two-blade wooden propeller. Fuel capacity 19 litres (5.0 US gallons; 4.2 Imp gallons).
EQUIPMENT: Optional BRS ballistic recovery parachute.
DIMENSIONS, EXTERNAL:
Wing span .... 13.11 m (43 ft 0 in)

Partly complete USA Cumulus on display at Lakeland, April 2004 *(Paul Jackson)* ***NEW***/0576836

Length overall .... 6.10 m (20 ft 0 in)
Height overall .... 1.35 m (4 ft 5 in)
AREAS:
Wings, gross .... 13.01 m² (140.0 sq ft)
WEIGHTS AND LOADINGS:
Weight empty .... 163 kg (360 lb)
Max T-O weight .... 290 kg (640 lb)
PERFORMANCE, POWERED:
Max level speed .... 78 kt (144 km/h; 90 mph)
Max cruising speed .... 65 kt (121 km/h; 75 mph)
Max rate of climb at S/L .... 304 m (1,000 ft)/min
T-O run .... 52 m (170 ft)
Range .... 173 n miles (321 km; 200 miles)
PERFORMANCE, UNPOWERED:
Stalling speed, flaps down .... 28 kt (52 km/h; 32 mph)
Best glide ratio at 37 kt (69 km/h; 43 mph) 20
Min rate of sink at 30 kt (55 km/h; 34 mph) .... 0.97 m (3.2 ft)/s

NEW ENTRY

VAN'S

VAN'S AIRCRAFT INC

14401 NE Keil Road, Aurora, Oregon 97002
Tel: (+1 503) 678 65 45
e-mail: info@vansaircraft.com
Web: http://www.vansaircraft.com
PRESIDENT: Richard VanGrunsven

Company derives its name from its president's nickname and began in 1973 with the sale of plans (and, later, kits) for the RV-3 sporting homebuilt. More than 10,000 Van's kits sold by October 2003, of which 3,427 were then flying. The four-seat RV-10 was announced in October 2001 and made its first flight on 29 May 2003.

The company constructed a new 5,110 m^2 (55,000 sq ft) facility at Aurora, Oregon, to which it moved during 2000; by June 2002 the workforce numbered 60. All Van's kits conform to the FAA's 51 per cent rule.

The company is now working on the **RV-11** motor glider; by early 2004 the forward fuselage was under construction.

VAN'S RV-4

TYPE: Tandem-seat kitbuilt

PROGRAMME: First flight of prototype 21 August 1979. Plans and kits available to homebuilders.

CURRENT VERSIONS: **RV-4:** *As described.*

CUSTOMERS: Over 4,600 sets of plans and kits sold, with about 1,800 aircraft under construction and 1,087 RV-4s flying by 9 October 2003.

COSTS: Kit: US$13,115 (2003). Information pack: US$8.

DESIGN FEATURES: Cantilever low-wing monoplane with constant-chord, low aspect ratio wings, tapered horizontal tail surfaces and sweptback fin.

Wing section Van's Aircraft 135; dihedral 3° 30'; incidence 1°.

FLYING CONTROLS: Conventional and manual. Horn-balanced elevators; Frise ailerons; plain flaps. No tabs.

STRUCTURE: Generally of light alloy. I-beam main wing spar and pressed ribs; moulded glass fibre wingtips and engine cowling.

LANDING GEAR: Tailwheel type; fixed cantilever tapered steel spring main gear struts. Glass fibre speed fairings on mainwheels, tyre size 5.00-5; steerable tailwheel, tyre size 6 in.

POWER PLANT: One 112 kW (150 hp) Textron Lycoming O-320-E1F flat-four engine; two-blade fixed-pitch propeller. Options for engine up to 134 kW (180 hp). Fuel capacity 121 litres (32.0 US gallons; 26.6 Imp gallons).

ACCOMMODATION: Two persons in tandem under starboard-hinged canopy. Baggage compartments forward of instrument panel and aft of rear seat.

DIMENSIONS, EXTERNAL:	
Wing span	7.01 m (23 ft 0 in)
Wing aspect ratio	4.8
Length overall	6.20 m (20 ft 4 in)
Height overall	1.65 m (5 ft 5 in)
Wheel track	1.88 m (6 ft 2 in)
Propeller diameter	1.73 m (5 ft 8 in)
DIMENSIONS, INTERNAL:	
Cabin: Length	1.91 m (6 ft 3 in)
Max width	0.72 m (2 ft 4 in)
Max height	1.02 m (3 ft 4 in)
Baggage volume	0.21 m^3 (7.3 cu ft)
AREAS:	
Wings, gross	10.22 m^2 (110.0 sq ft)
WEIGHTS AND LOADINGS:	
Weight empty	411 kg (905 lb)
Baggage capacity	23 kg (50 lb)
Max T-O weight	680 kg (1,500 lb)
Max wing loading	66.6 kg/m^2 (13.64 lb/sq ft)
Max power loading	6.09 kg/kW (10.00 lb/hp)
PERFORMANCE (112 kW; 150 hp engine; MTOW):	
Max level speed at S/L	174 kt (322 km/h; 200 mph)
Cruising speed at FL80	
at 75% power	163 kt (303 km/h; 188 mph)
at 55% power	148 kt (274 km/h; 170 mph)
Stalling speed	47 kt (87 km/h; 54 mph)
Max rate of climb at S/L	457 m (1,500 ft)/min
Service ceiling	5,485 m (18,000 ft)
T-O run	145 m (475 ft)
Landing run	130 m (425 ft)
Range with max fuel at 55% power	686 n miles (1,271 km; 790 miles)

Van's RV-4 kitbuilt *(Paul Jackson)* **NEW**/0567755

VAN'S RV-7 AND RV-7A

TYPE: Side-by-side kitbuilt.

PROGRAMME: Replacement for the RV-6 and RV-6A. Design work undertaken during 2000; prototype (N137RV) first flown 19 March 2001. Public debut at Sun'n'Fun in April 2001. RV-6/6A kits currently under construction can be completed to RV-7/7A standard by using optional parts.

Van's RV-7 side-by-side kitbuilt *(Paul Jackson)* **NEW**/0567793

Non-standard Van's RV-7SS has a Subaru powerplant *(Paul Jackson)* **NEW**/0567792

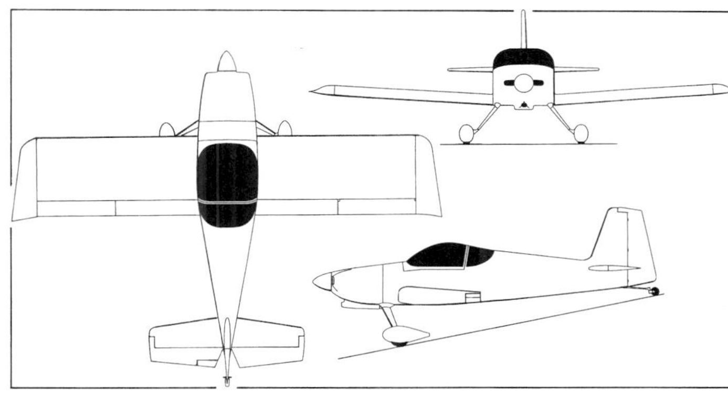

Van's RV-7 general arrangement *(Paul Jackson)* 0110689

CURRENT VERSIONS: **RV-7:** Tailwheel version.

RV-7A: Nosewheel version.

CUSTOMERS: Total of 935 kits sold by July 2002 and 1,000 by August 2002, of which 41 were flying by 9 October 2003.

COSTS: RV-7 kit US$16,450, quick-build kit US$24,375; RV-7A kit US$17,125, quick-build kit US$25,050 (2003).

DESIGN FEATURES: Shares commonality of parts with the RV-8/8A and RV-9A. Quoted build time reduced to 1,400 hours compared to RV-6 family due to improvements made in kit production.

FLYING CONTROLS: Conventional and manual. Horn-balanced rudder and elevators, flight-adjustable tab in port tailplane. Plain electric flaps.

STRUCTURE: Fuselage and wings of metal, with composites wingtips, tailplane tip, fintip, fin-root fairing, cowling, landing gear legs, fairings and spats.

LANDING GEAR: Fixed; choice of tricycle or tailwheel; Wittman leaf-spring main legs. Speed fairing on main legs. Solid tailwheel.

POWER PLANT: Prototype has 149 kW (200 hp) Textron Lycoming IO-360 driving a two-blade Hartzell constant-speed propeller; the design accepts engines in the 112 to 149 kW (150 to 200 hp) range. Fuel capacity 159 litres (42.0 US gallons; 35.0 Imp gallons) in two wing tanks.

ACCOMMODATION: Pilot and passenger side by side under forward-hinged one-piece canopy; optional rearward-sliding canopy with fixed windscreen. Compared to RV-6, cabin room is increased.

DIMENSIONS, EXTERNAL:	
Wing span	7.62 m (25 ft 0 in)
Wing aspect ratio	5.2
Length overall	6.20 m (20 ft 4 in)
Height overall: RV-7	1.78 m (5 ft 10 in)
RV-7A	2.39 m (7 ft 10 in)
DIMENSIONS, INTERNAL:	
Cabin max width	1.09 m (3 ft 7 in)
Baggage volume	0.34 m^3 (12.0 cu ft)
AREAS:	
Wings gross	11.15 m^2 (120.0 sq ft)
WEIGHTS AND LOADINGS (149 kW; 200 hp engine):	
Weight empty	505 kg (1,114 lb)
Baggage capacity	45 kg (100 lb)
Max T-O weight	816 kg (1,800 lb)
Max wing loading	73.2 kg/m^2 (15.00 lb/sq ft)
Max power loading	5.48 kg/kW (9.00 lb/hp)
PERFORMANCE (149 kW; 200 hp engine):	
Never-exceed speed (VNE)	199 kt (370 km/h; 230 mph)
Max operating speed: RV-7	188 kt (348 km/h; 216 mph)
RV-7A	185 kt (343 km/h; 213 mph)

Nosewheel Van's RV-7A *(Paul Jackson)* ***NEW***/0567791

Cruising speed:
at 75% power:
RV-7 .... 179 kt (332 km/h; 206 mph)
RV-7A .... 177 kt (328 km/h; 204 mph)
at 55% power: RV-7 .... 162 kt (299 km/h; 186 mph)
RV-7A .... 159 kt (295 km/h; 183 mph)

Stalling speed, flaps down:
RV-7, RV-7A .... 51 kt (94 km/h; 58 mph)
Max rate of climb at S/L: RV-7 .... 579 m (1,900 ft)/min
RV-7A .... 564 m (1,850 ft)/min
Service ceiling: RV-7 .... 6,860 m (22,500 ft)
RV-7A .... 6,555 m (21,500 ft)
T-O and landing run .... 152 m (500 ft)
Range with max fuel, 55% power:
RV-7 .... 812 n miles (1,504 km; 935 miles)
RV-7A .... 803 n miles (1,488 km; 925 miles)

VAN'S RV-8 AND RV-8A

TYPE: Tandem-seat kitbuilt.

PROGRAMME: First flown (N118RV) 22 July 1995; launched at Oshkosh five days later. Second prototype (N58RV) flew early 1997 and accumulated 325 hours in seven months.

CURRENT VERSIONS: **RV-8:** Tailwheel version.

RV-8A: Tricycle version; first flown (N58VA) April 1998; public debut at Sun 'n' Fun, same month.

CUSTOMERS: By mid-2002, 1,300 sets of plans and kits had been sold. By 9 October 2003, 339 were flying.

COSTS: RV-8 kit US$16,895, quick-build kit US$24,820; RV-8A kit US$17,295, quick-build kit US$25,220 (2003 prices).

DESIGN FEATURES: Similar in appearance to the RV-4, but with wider, longer fuselage; rounded engine cowling; non-sweptback main landing gear legs; larger canopy; however, very little parts commonality with RV-4.

Description generally as for RV-4/RV-6.

LANDING GEAR: Wittman leaf-spring main legs; nose- or tailwheel. Speed fairings on mainwheels and (when fitted) nosewheel.

POWER PLANT: One 149 kW (200 hp) Textron Lycoming IO-360-A16D four-cylinder piston engine driving two-blade Hartzell HC-C2YK-BF/F7666A4 constant-speed propeller. Fuel capacity 159 litres (42.0 US gallons; 35.0 Imp gallons). Oil capacity 7.6 litres (2.0 US gallons; 1.7 Imp gallons).

ACCOMMODATION: Tandem seating for pilot and passenger under sliding canopy. Separate windscreen.

Specification for RV-8/RV-8A, except where noted.

DIMENSIONS, EXTERNAL:
Wing span .... 7.32 m (24 ft 0 in)
Wing chord, constant .... 1.47 m (4 ft 10 in)
Wing aspect ratio 4.8
Length overall: RV-8 .... 6.40 m (21 ft 0 in)

Van's RV-8A nosewheel kitbuilt *(Paul Jackson)* ***NEW***/0567783

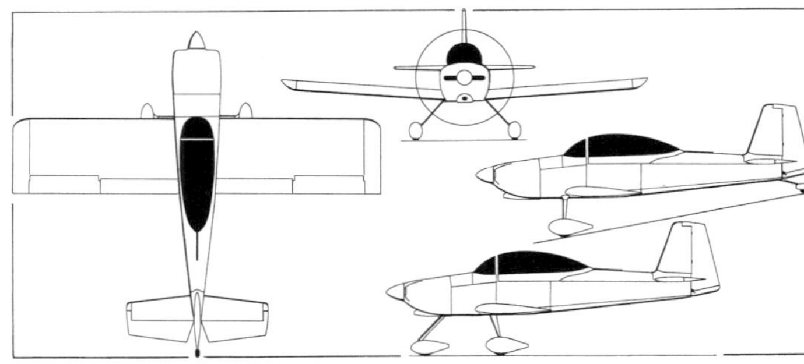

Van's RV-8 two-seat light aircraft, with additional side view of RV-8A *(Paul Jackson)* 0051856

RV-8A .... 6.35 m (20 ft 10 in)
Height overall: RV-8 .... 1.70 m (5 ft 7 in)
RV-8A .... 2.24 m (7 ft 4 in)
Wheel track .... 1.90 m (6 ft 3 in)
Propeller diameter .... 1.83 m (6 ft 0 in)

DIMENSIONS, INTERNAL:
Cockpit max width .... 0.91 m (3 ft 0 in)
Baggage volume .... 0.21 m³ (7.5 cu ft)

AREAS:
Wings, gross .... 10.78 m² (116.0 sq ft)

WEIGHTS AND LOADINGS:
Weight empty .... 508 kg (1,120 lb)
Baggage capacity .... 57 kg (125 lb)
Max T-O weight .... 816 kg (1,800 lb)
Max wing loading .... 79.9 kg/m² (16.36 lb/sq ft)
Max power loading .... 5.48 kg/kW (9.00 lb/hp)

PERFORMANCE (two persons, 149 kW; 200 hp engine):
Max operating speed: RV-8 .... 191 kt (354 km/h; 220 mph)
RV-8A .... 189 kt (351 km/h; 218 mph)
Cruising speed at 55% power at FL80:
RV-8 .... 162 kt (301 km/h; 187 mph)
RV-8A .... 161 kt (298 km/h; 185 mph)
Stalling speed, power off:
one occupant: both .... 45 kt (83 km/h; 51 mph)
two occupants: both .... 51 kt (94 km/h; 58 mph)
Max rate of climb at S/L: RV-8 .... 579 m (1,900 ft)/min
RV-8A .... 549 m (1,800 ft)/min
Service ceiling: RV-8 .... 6,860 m (22,500 ft)
RV-8A .... 6,550 m (21,500 ft)
T-O and landing run: both .... 153 m (500 ft)
Range at 55% power:
RV-8 .... 816 n miles (1,512 km; 940 miles)
RV-8A .... 808 n miles (1,496 km; 930 miles)

Van's RV-8 tailwheel version *(Paul Jackson)* 0567754

VAN'S RV-9 AND RV-9A

TYPE: Side-by-side kitbuilt.

PROGRAMME: Prototype RV-9A (N96VA) first flown late 1997; was hybrid proof-of-concept aircraft using unique RV-6T fuselage plus new wings; had flown more than 200 hours when lost in accident 2 April 2000; second prototype, and first 'true' RV-9A (N129RV), flown 15 June 2000. Empennage kits became available in late 1999, followed by wing and fuselage/finishing kits in mid-2000 and quick-build kits in late 2000; first customer-built aircraft flown 17 June 2001. RV-9 prototype (N179RV) first flown 4 March 2002.

CURRENT VERSIONS: **RV-9:** Tailwheel version.

RV-9A: Tricycle version.

CUSTOMERS: 400 kits sold by April 2001; 55 flying by 9 October 2003.

COSTS: Basic kit RV-9 US$16,425, RV-9A US$17,100; quick-built RV-9 US$24,350, RV-9A US$25,025 (2003).

DESIGN FEATURES: Based upon, and similar in appearance to, RV-6/6A, but with redesigned wing (Roncz aerofoil and sheared wingtips) and enlarged, constant-chord horizontal tail surfaces.

FLYING CONTROLS: Conventional and manual. Horn-balanced rudder. Slotted electric flaps extend approximately two-thirds of wing span, compared with one-half on other RV designs.

LANDING GEAR: Fixed tricycle type on RV-9A; tailwheel on RV-9. Brakes.

POWER PLANT: First prototype flown with 88.0 kW (118 hp) Textron Lycoming O-235-L2C flat-four driving two-blade Sensenich 72 × 66 fixed-pitch metal propeller. Second prototype has 119 kW (160 hp) Textron Lycoming O-320-D3G driving three-blade MT propeller. Engines between 80.5 and 119 kW (108 and 160 hp) can be fitted. Fuel capacity 136 litres (36.0 US gallons; 30.0 Imp gallons) in two wing tanks. Wilksch reportedly developing conversion during 2002, to enable use of its 90 kW (120 hp) WAM-120 diesel engine.

ACCOMMODATION: Pilot and passenger side by side under sliding canopy. Dual controls.

DIMENSIONS, EXTERNAL:

Wing span: both	8.53 m (28 ft 0 in)
Length overall: both	6.22 m (20 ft 5 in)
Height overall: RV-9	1.83 m (6 ft 0 in)
RV-9A	2.39 m (7 ft 10 in)
Propeller diameter	1.83 m (6 ft 0 in)

DIMENSIONS, INTERNAL: As RV-7

AREAS:

Wings, gross	11.52 m² (124.0 sq ft)

WEIGHTS AND LOADINGS (119 kW; 160 hp engine):

Weight empty	476 kg (1,050 lb)
Baggage capacity	45 kg (100 lb)
Max T-O weight	793 kg (1,750 lb)
Max wing loading	68.9 kg/m² (14.11 lb/sq ft)
Max power loading	6.66 kg/kW (10.94 lb/hp)

PERFORMANCE (119 kW; 160 hp engine):

Max operating speed: RV-9	170 kt (315 km/h; 196 mph)
RV-9A	169 kt (312 km/h; 194 mph)
Cruising speed at 75% power:	
RV-9	163 kt (303 km/h; 188 mph)
RV-9A	162 kt (299 km/h; 186 mph)
Stalling speed, flaps down: both	44 kt (81 km/h; 50 mph)
Max rate of climb at S/L: both	427 m (1,400 ft)/min
Service ceiling: RV-9	5,790 m (19,000 ft)
RV-9A	5,640 m (18,500 ft)
T-O run: both	145 m (475 ft)
Landing run: both	137 m (450 ft)
Range at 75% power:	
RV-9	617 n miles (1,142 km; 710 miles)
RV-9A	608 n miles (1,126 km; 700 miles)

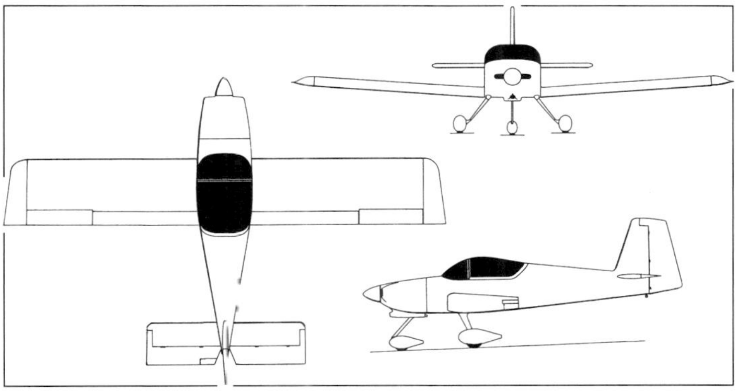

Van's RV-9A general arrangement *(Paul Jackson)* 0110690

Van's RV-9A kitbuilt *(Paul Jackson)* ***NEW***/0567775

First UK-built Van's RV-9 tailwheel version with 'jester' registration and decoration *(Paul Jackson)* ***NEW***/0567774

VAN'S RV-10

TYPE: Four-seat kitbuilt.

PROGRAMME: Programme announced in late 2001 and wooden mockup of cabin shown when first flight expected second quarter of 2003; this achieved by prototype (N410RV) on 29 May 2003. Exhibited at AirVenture, Oshkosh, July 2003. Kit deliveries started 25 September 2003.

CUSTOMERS: By October 2003, 85 kits had been sold.

COSTS: Standard kit US$35,000; quick-build kit US$45,000 (2003).

DESIGN FEATURES: Similar to other members of RV series; designed to carry four 1.83 m (6 ft 0 in) tall adults.

FLYING CONTROLS: Conventional and manual. Horn-balanced rudder and elevator; actuation by cables (rudder) and pushrods. Fowler electric flaps. Tabs in both elevators.

STRUCTURE: Similar to RV-7 and RV-9; aluminium fuselage with composites cabin top, wingtips and wingroot. Two-spar wing.

LANDING GEAR: Tricycle type; fixed. 6.00-6 tyres on all wheels; Cleveland disc brakes on mainwheels. Optional speed fairings.

POWER PLANT: Prototype fitted with one 194 kW (260 hp) Textron Lycoming IO-540 flat-six driving a Hartzell F8477-4 two-blade, constant-speed propeller. Engines in the range 149 to 194 kW (200 to 260 hp) can be fitted. Total fuel capacity 227 litres (60.0 US gallons; 50.0 Imp gallons) in two wing leading-edge tanks.

ACCOMMODATION: Pilot and three passengers in two pairs of seats; rear seatbacks remove to increase baggage space. Cabin has gull-wing door on each side. Baggage door on port fuselage side behind cockpit.

DIMENSIONS, EXTERNAL:

Wing span	9.68 m (31 ft 9 in)
Wing aspect ratio	6.8
Length overall	7.44 m (24 ft 5 in)
Height overall	2.72 m (8 ft 11 in)

DIMENSIONS, INTERNAL:

Cabin: Max width	1.23 m (4 ft 0¼ in)
Max height	1.03 m (3 ft 4½ in)
Baggage volume	0.37 m³ (13.0 cu ft)

AREAS:

Wings, gross	13.75 m² (148.0 sq ft)

WEIGHT AND LOADINGS:

Weight empty	726 kg (1,600 lb)
Baggage capacity	45 kg (100 lb)
Max T-O weight	1,224 kg (2,700 lb)
Max wing loading	89.1 kg/m² (18.24 lb/sq ft)
Max power loading	6.32 kg/kW (10.38 lb/hp)

PERFORMANCE (194 kW; 260 hp engine):

Max operating speed (VMO)	180 kt (334 km/h; 208 mph)
Normal cruising speed at 75% power	171 kt (317 km/h; 197 mph)
Econ cruising speed at 55% power	152 kt (282 km/h; 175 mph)
Stalling speed, flaps down	55 kt (102 km/h; 63 mph)
Max rate of climb at S/L	442 m (1,450 ft)/min
Service ceiling	6,096 m (20,000 ft)
T-O run	152 m (500 ft)
Landing run	198 m (650 ft)
Range at 75% power at FL80	716 n miles (1,327 km; 825 miles)

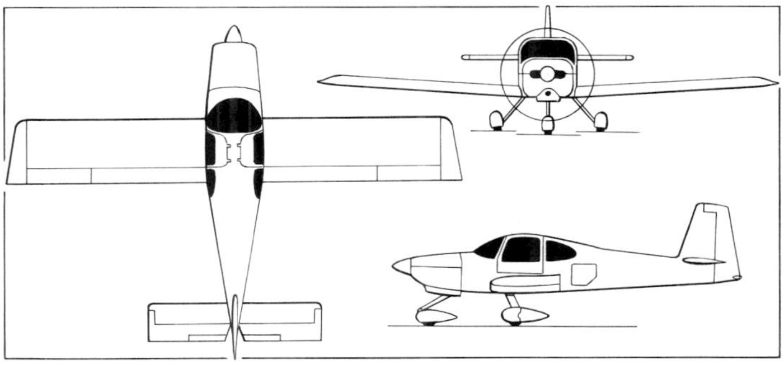

Van's RV-10 general arrangement *(James Goulding)* ***NEW***/0567244

Van's RV-10 four-seat kitbuilt ***NEW***/0553234

VAT

VERTICAL AVIATION TECHNOLOGIES INC

1609 Hangar Road, Sanford, Florida 32773
Tel: (+1 407) 322 94 88
Fax: (+1 407) 330 26 47
e-mail: sales@vertical-aviation.com
Web: http://www.vertical-aviation.com
PRESIDENT: Bradley G Clark

Vertical Aviation Technologies Inc is FAA-approved repair facility for various Sikorsky helicopters. The company remanufactures Sikorsky S-55/H-19s, S-58/H-34s and S-62s. In 1984, began to modify four-seat Sikorsky S-52-3 former production helicopter (of which 95 originally produced, mainly for US Marine Corps as HO5S) into kit for assembly by individuals, corporations or military; all tooling and fixtures completed by 1988. Promotion and manufacture was interrupted by S-55 upgrade effort, but resumed at Sun 'n' Fun in April 2002, when upgraded S-52 was first shown.

VAT HUMMINGBIRD

TYPE: Four-seat helicopter kitbuilt.

PROGRAMME: Based on previously type-certified Sikorsky S-52-3 and first flown 1990. (S-52 prototype flew 12 February 1947; total 95 built). First batch of sales between 1989 and 1993; promotion lapsed between 1993 and 2001, when VAT efforts directed towards S-55QT Whisper Jet conversion. Modified Hummingbird shown at Sun 'n' Fun in April 2002; twin anhedral tail strakes replaced by more conventional horizontal stabilisers with endplates; Lycoming engine replacing 194 kW (260 hp) Ford Taurus V-8; doors and engine cowling of composites.

CUSTOMERS: Fourteen Hummingbird kits sold by mid-1990s, of which 12 flying. Customers in China, Hungary, Slovak Republic, USA and UK.

COSTS: Kit US$169,900 (2003), excluding engine, avionics and upholstery.

DESIGN FEATURES: Newly manufactured airframes and tailcones identical to S-52-3, except for nose section. Quoted build time 800 hours; kit includes original Sikorsky components.

FLYING CONTROLS: Conventional and manual. Two anhedral stabilisers of original S-52 replaced from 2002 by horizontal unit with endplates.

STRUCTURE: Mainly aluminium, with GFRP nose (from Bell JetRanger), doors, engine cowling, tailplane endplates and tailcone. Monocoque cabin and tailboom; metal tube centre-section.

LANDING GEAR: Four-wheel type, fixed; oleo-pneumatic shock-absorbes on each leg; rear wheels 5.00-5; forward wheels 4.10/3.50. Hydraulic brakes on rear.

POWER PLANT: One 194 kW (260 hp) Textron Lycoming VO-435-A1F derated to 186 kW (250 hp) driving three-blade fully articulated main rotor and two-blade tail rotor. Unlimited transmission. Fuel capacity 216 litres (57.0 US gallons; 47.5 Imp gallons). Oil capacity 11 litres (3.0 US gallons; 2.5 Imp gallons).

ACCOMMODATION: Four persons, including pilot(s); door each side at front; rear door port only.

DIMENSIONS, EXTERNAL:
Main rotor diameter .... 10.06 m (33 ft 0 in)
Tail rotor diameter .... 1.75 m (5 ft 9 in)
Length: overall, rotors turning .... 12.11 m (39 ft 9 in)
fuselage .... 9.30 m (30 ft 6 in)
Distance between rotor centres .... 6.07 m (19 ft 11 in)
Height: to top of rotor head .... 2.62 m (8 ft 7 in)
overall .... 2.95 m (9 ft 8 in)
Fuselage max width .... 1.52 m (5 ft 0 in)
Wheel track: forward .... 1.83 m (6 ft 0 in)
rear .... 2.49 m (8 ft 2 in)
Wheelbase .... 1.88 m (6 ft 2 in)

Current version of Vertical Aviation Technologies Hummingbird *(Paul Jackson)* 0137387

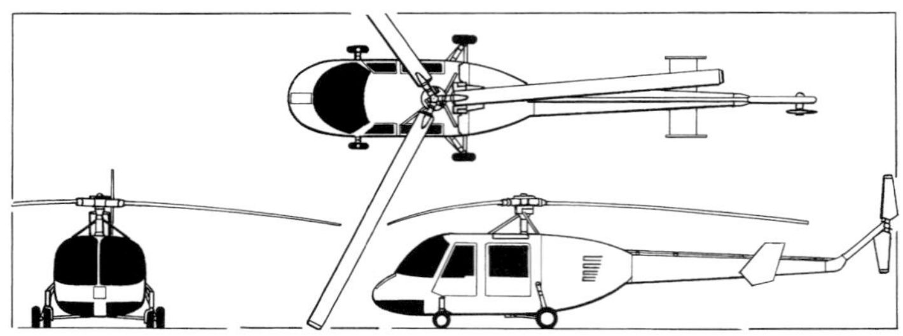

VAT Hummingbird four-seat helicopter *(James Goulding)* 0137335

AREAS:
Main rotor disc .... 79.46 m² (855.3 sq ft)
Tail rotor disc .... 2.41 m² (25.97 sq ft)

WEIGHTS AND LOADINGS:
Weight empty .... 816 kg (1,800 lb)
Max T-O weight .... 1,225 kg (2,700 lb)
Max disc loading .... 15.4 kg/m² (3.16 lb/sq ft)
Max power loading .... 6.57 kg/kW (10.80 lb/hp)

PERFORMANCE (three-blade rotor):
Never-exceed speed (VNE) .... 95 kt (177 km/h; 110 mph)
Cruising speed .... 78 kt (145 km/h; 90 mph)
Max rate of climb at S/L at 1,043 kg (2,300 lb) .... 381 m (1,250 ft)/min
Hovering ceiling: IGE .... 2,745 m (9,000 ft)
OGE .... 1,220 m (4,000 ft)
Service ceiling .... 3,660 m (12,000 ft)
Range .... 304 n miles (563 km; 350 miles)

VIPER

VIPER AIRCRAFT CORPORATION

3908 Stearman Avenue, Pasco, Washington 99301
Tel: (+1 509) 543 35 70
Fax: (+1 509) 547 31 42
e-mail: info@viper-aircraft.com
Web: http://www.viper-aircraft.com
PRESIDENT: Scott Hanchette
VICE-PRESIDENT: Dan Hanchette

Viper Aircraft's first product is the ViperJet. Series production is now under way.

VIPER VIPERJET

TYPE: Two-seat jet sportplane kitbuilt.

PROGRAMME: Development began in 1988; engineering by AirBoss Aerospace; tooling began in 1996; prototype was initially designed around a Teledyne Continental TSIO-520 and named ViperFan; in early 1999 a switch to jet propulsion was revealed and aircraft renamed ViperJet; development of ViperFan will continue as a lower priority. Prototype (N520VF), officially registered as Hanchette ViperJet) first flown 22 October 1999 and displayed publicly at Oshkosh in July 2000; underwent further modifications; seriously damaged in crash at St George, Utah, on 13 March 2002, but expected to return to flying in early 2004.

CURRENT VERSIONS: **ViperFan:** Initial piston-engined version.

ViperJet: Jet-powered prototype.

ViperJet Mk II: Production version, *as described.*

CUSTOMERS: Ten sold and one flown by end 2003; three customer aircraft then under construction.

COSTS: Kit US$158,400; engine extra. Pressurisation upgrade US$39,000; builder-assistance programme US$75,000 (2003).

DESIGN FEATURES: High-performance personal jet at self-build price. Low wing with sweptback leading-edge and initially with winglets; sweptback tail surfaces with fin fillet and ventral strakes.

Wing aerofoil section NACA 65-218 at root, NACA 65-212 at tip; horizontal tail NACA 66-012; vertical tail NACA 63-015.

Quoted build time 2,000 hours using builder-assistance programme.

Prototype Viper ViperJet in 2001 ***NEW***/0572343

FLYING CONTROLS: Conventional and manual. Electric double-slotted flaps, span 1.65 m (5 ft 5 in) deflecting between 0 and 40°; aileron span 1.22 m (4 ft 0 in), deflection 27° up and 18° down; rudder deflection ±25°; elevator deflection ±30°. Dual controls.

STRUCTURE: Extensive use of composites; much of this provided by Unlimited Composites of Scappoose, Oregon. Preformed skins and wing spars.

LANDING GEAR: Retractable tricycle type; nosewheel retracts rearwards, mainwheels inwards. Production aircraft will not have landing gear doors.

POWER PLANT: Prototype initially powered by one refurbished Turbomeca Marboré 2 turbojet rated at 3.91 kN (880 lb st); later changed to converted General Electric T58-8F rated at 4.44 kN (1,000 lb st). Other engine options include versions of Marboré up to 4.89 kN (1,100 lb st). Also recommended is 13.12 kN (2,950 lb st) General Electric J85/CJ-610. Fuel capacity 511 litres (135 US gallons; 112 Imp gallons) in prototype; 1,136 litres (300 US gallons; 250 Imp gallons).

ACCOMMODATION: Pilot and passenger in tandem in unpressurised cockpit; cockpit pressurisation upgrade is available.

DIMENSIONS, EXTERNAL:
Wing span .... 7.87 m (25 ft 10 in)
Wing chord: at root .... 1.78 m (5 ft 10 in)
at tip .... 0.86 m (2 ft 10 in)
Wing aspect ratio 7.7
Length overall .... 7.57 m (24 ft 10 in)
Height overall .... 3.25 m (10 ft 8 in)
Tailplane span .... 4.06 m (13 ft 4 in)
DIMENSIONS, INTERNAL:
Cockpit: Length .... 2.92 m (9 ft 7 in)
Max width .... 0.76 m (2 ft 6 in)
Max height .... 1.24 m (4 ft 1 in)
AREAS:
Wings, gross .... 10.50 m² (113.0 sq ft)
WEIGHTS AND LOADINGS (J85 engine):
Weight empty .... 998 kg (2,200 lb)
Max T-O weight .... 1,950 kg (4,300 lb)
Baggage capacity .... 45 kg (100 lb)
Max wing loading .... 185.8 kg/m² (38.05 lb/sq ft)
Max power loading .... 149 kg/kN (1.46 lb/lb st)
PERFORMANCE (estimated):
Never-exceed speed (VNE) .... 434 kt (805 km/h; 500 mph)
Max level speed at FL250 .... 391 kt (724 km/h; 450 mph)
Stalling speed, flaps down .... 73 kt (136 km/h; 84 mph)
Max rate of climb at S/L .... 3,050 m (10,000 ft)/min

Third production ViperJet under construction *NEW*/0572342

Service ceiling .... 7,620 m (25,000 ft)
T-O run .... 366 m (1,200 ft)
Landing run .... 610 m (2,000 ft)
Range with max fuel, 45 min reserve .... 869 n miles (1,609 km/h; 1,000 miles)
Endurance .... 2 h 0 min
g limits .... ±12

VERIFIED

VSTOL

VSTOL AIRCRAFT CORPORATION

Suite 100, 8619 Northwest Brostrom Road, Kansas City, Missouri 64152
Tel: (+1 816) 746 59 37 and (+1 941) 218 54 68
Fax: (+1 941) 931 54 68
e-mail: info@vstolaircraft.com
Web: http://www.vstolaircraft.com
CEO: Dick Turner

VSTOL is responsible for marketing the SST 2000 Pairadigm and SS 2000 Super Solution series of kitbuilt aircraft. Kits are produced in Venezuela.

VSTOL SST 2000 PAIRADIGM

TYPE: Two-seat kitbuilt twin.
PROGRAMME: Developed from the Star Flight XC 280 Stiletto, which first flew January 1997. A single-engined version is known as the **Super Solution 2000**.
CUSTOMERS: Over 400 of both single- and twin-engined versions are flying. Some used for crop-spraying.
COSTS: Kit US$44,995, or US$85,000 with full spray equipment (2002).
DESIGN FEATURES: Two pusher engines mounted side by side at rear of fuselage; propellers partly overlap and rotate in same direction. Wings have full length leading-edge slats and are removable for storage. Quoted build time 250 hours.
FLYING CONTROLS: Conventional and manual. Actuation by stainless steel cables. Leading-edge slats, Frise ailerons, slotted flaps and horn-balanced rudder.
STRUCTURE: 4130 steel and aluminium tubing throughout; fabric-covered V-braced wings and composites fuselage skin; braced tailplane. Extended wings are optional. Cabin enclosure removable.
LANDING GEAR: Fixed; choice of nose- or tailwheel versions; no brakes; wheel size 152 mm (6 in). Steerable nosewheel on tricycle model.
POWER PLANT: Two engines, each between 37.3 and 74.6 kW (50 and 100 hp); 48.5 kW (65 hp) Hirth 2706 engines are recommended. Fuel capacity 75.7 litres (20.0 US gallons; 16.7 Imp gallons).
EQUIPMENT: Crop-spraying kit includes 189 litre (50.0 US gallon; 41.6 Imp gallon) chemical tank and 8.23 m (27 ft) sprayboom with 16 nozzles. Swath width 16.76 m (55 ft).
DIMENSIONS, EXTERNAL:
Wing span .... 10.52 m (34 ft 6 in)

VSTOL Pairadigm *(Geoffrey P Jones)* 0016518

Wing aspect ratio 6.3
Length overall .... 7.16 m (23 ft 6 in)
Height overall .... 2.59 m (8 ft 6 in)
AREAS:
Wings, gross .... 17.65 m² (190.0 sq ft)
WEIGHTS AND LOADINGS:
Weight empty .... 408 kg (900 lb)
Max T-O weight .... 771 kg (1,700 lb)
Max wing loading .... 43.7 kg/m² (8.95 lb/sq ft)
PERFORMANCE:
Never-exceed speed (VNE) .... 108 kt (201 km/h; 125 mph)
Max level speed .... 78 kt (145 km/h; 90 mph)
Normal cruising speed .... 61 kt (113 km/h; 70 mph)
Stalling speed .... 26 kt (47 km/h; 29 mph)
Max rate of climb at S/L .... 244 m (800 ft)/min
Rate of climb at S/L, OEI .... 61 m (200 ft)/min
Service ceiling .... 3,660 m (12,000 ft)
T-O run .... 46 m (150 ft)
Landing run .... 31 m (100 ft)
Range at cruising speed .... 250 n miles (463 km; 287 miles)
g limits .... +4.4/−2.2

WACO

WACO AIRCRAFT COMPANY OHIO INC

174 Barklow, Portsmouth, Ohio 45662
Tel/Fax: (+1 740) 820 51 00
e-mail: wacoohio@falcon1.net
Web: http://www.wacokit.com
DIRECTORS:
Mike Fisher
John Kennard
Ben Bagnall

Waco produces kit replicas of the UMF design built in the early 1930s by the US company of the same name and certified under ATC 546 on 11 July 1934. Mike Fisher is founder of Fisher Flying Products.

(Weaver Aircraft Company formed 1920 and traded as WACO; renamed Waco Aircraft Company at Troy, Ohio, in 1928 and ceased aircraft production in 1947.)

WACO UMF

TYPE: Three-seat biplane/kitbuilt.
PROGRAMME: UMF indicates Continental R-670 or W-670 engine; 1934 open cockpit; tandem cockpits. Originals built 1934 and 1935. Current series prototype (N2021N) built 1999. Demonstrator N30056, built 2002, registered as Waco YMF, having Jacobs R755-B2 engine (for which ZMF more accurate designation).
COSTS: Kit US$80,000, excluding engine, propeller and instruments.

Waco 'UMF' (YMF) reproduction kitplane *(Paul Jackson)* *NEW*/0561713

DESIGN FEATURES: Single-bay biplane; upper plane mounted on cabane and N-type interplane struts and braced by flying and drag wires. Wire-braced empennage.
Wing section Clarke Y.
FLYING CONTROLS: Conventional and manual. Horn-balanced elevators and variable incidence tailplane. Ground-adjustable rudder trim tab. Ailerons on all wings, with vertical connecting rods.

STRUCTURE: Fuselage of pre-welded 4130 steel square-section tube and wooden longerons; aluminium wings. Fabric covering, but metal cladding forward of cockpits on spine and on wing leading edges.

LANDING GEAR: Tricycle type; fixed. Mainwheels size 7.50–10; Maule steerable tailwheel 2.80/2.50-4. Hydraulic brakes. Optional floats. Rubber-in-tension suspension.

POWER PLANT: One 205 kW (275 hp) Jacobs R755-B2 or 164 kW (220 hp) Continental W-670 seven-cylinder radial, driving a Sensenich two-blade wooden propeller. Fuel in two foam-filled upper wing tanks, total 189 litres (50.0 US gallons; 41.6 Imp gallons) of which 182 litres (48.0 US gallons; 40.0 Imp gallons) are usable. Oil capacity 13 litres (3.5 US gallons; 2.9 Imp gallons).

ACCOMMODATION: Pilot in rear cockpit; two passengers side by side in front cockpit, with single set of controls. Option of windscreen (only) ahead of each cockpit or fully enclosed cockpits, with rear-sliding front canopy and forward-sliding rear canopy.

DIMENSIONS, EXTERNAL:
Wing span: upper ... 9.14 m (30 ft 0 in)
lower ... 8.18 m (26 ft 10 in)
Wing chord (constant) ... 1.45 m (4 ft 9 in)
Wing stagger ... 0.71 m (2 ft 4 in)
Length overall ... 6.86 m (22 ft 6 in)
Height overall ... 2.64 m (8 ft 8 in)
Wheel track ... 2.11 m (6 ft 11 in)

WEIGHTS AND LOADINGS:
Weight empty ... 726 kg (1,600 lb)
Max T-O weight ... 1,134 kg (2,500 lb)
Max power loading: Continental engine ... 6.92 kg/kW (11.36 lb/hp)
Jacobs engine ... 5.53 kg/kW (9.09 lb/hp)

PERFORMANCE (continental engine):
Max level speed ... 109 kt (201 km/h; 125 mph)
Cruising speed ... 96 kt (177 km/h; 110 mph)
Max rate of climb at S/L ... 366 m (1,200 ft)/min
Service ceiling ... 4,875 m (16,000 ft)
Landing run ... 92 m (300 ft)

WACO CLASSIC

WACO CLASSIC AIRCRAFT CORPORATION

PO Box 1229, Battle Creek, Michigan 49016-1229
Tel: (+1 296) 565 10 00
Fax: (+1 296) 565 11 00
e-mail: flywaco@wacoclassic.com
Web: http://www.wacoclassic.com
PRESIDENT: Mitchell Lampert
GENERAL MANAGER: Patrick Horgan

Waco Classic (known until 1997 as Classic Aircraft) delivers some five or six aircraft per year. It originated in 1983 at Lansing, Michigan, and acquired the Waco YMF-5 type certificate from the FAA; manufacturing authority was assigned in February 1984. The company moved into a purpose-built, 2,800 m² (30,000 sq ft) factory at W K Kellogg Airport, Battle Creek, in April 2000; 90th aircraft handed over at official opening of factory on 20 May 2000; 100th followed in 2002.

UPDATED

Waco Classic YMF Super three-seat biplane *(Paul Jackson)* ***NEW***/0554429

WACO CLASSIC YMF SUPER

TYPE: Three-seat biplane.

PROGRAMME: Construction began March 1984 under type certificate of original, pre-Second World War Waco YMF-5; first flight of prototype (N1935B) 20 November 1985; FAA certification 10 March 1986; marketed initially as F-5, differing from original in having 300 modifications, including new throttle control, wheels, brakes, electrical system, starting system, engine and propeller, plus steerable tailwheel; modern constructional materials and methods; MTOW 1,202 kg (2,650 lb). Recertified 28 January 1988 at 1,256 kg (2,770 lb) from 10th aircraft onwards.

YMF Super certified 24 June 1991 with stretched fuselage and further increase in MTOW; balanced rudder; designation F5C; introduced at 40th aircraft.

CURRENT VERSIONS: **YMF Super**: Fuselage stretched 15 cm (6 in); larger front cockpit than previous standard F-5 version for commercial pleasure flying; enlarged forward door; front and rear cockpits 10 cm (4 in) longer; front cockpit 6.3 cm (2½ in) wider.

CUSTOMERS: Total 106 (including F-5s) delivered by November 2004. Exported to Australia, Canada, Germany, Japan, Kenya, Mexico, South Africa, Switzerland and the Caribbean. Two scheduled for delivery to Germany in February and June 2005, and one to the UK in April 2005.

COSTS: Standard equipped aircraft US$289,000 (2001).

DESIGN FEATURES: Single-bay biplane of 1930s vintage. Wings braced with N-type interplane struts, plus flying and landing wires, with double-N cabane. Elliptical wingtips. Cross-braced landing gear. Helmeted engine cowling.

Wing section Clark Y; dihedral 2°, incidence 0° on upper and lower wings.

FLYING CONTROLS: Conventional and manual. Interconnected ailerons on upper and lower wings; no tabs. Horn-balanced tail surfaces. Tailplane trim by screw-jack actuator; ground-adjustable trim tab on rudder. Control surface movements; ailerons ±25°; elevator +33°/−27°; rudder ±30°.

STRUCTURE: Modern construction techniques, tolerances and materials applied to original design. Steel interplane struts; streamlined stainless steel flying and landing wires; all-wood wing with Dacron covering; aluminium ailerons with external chordwise stiffening. Fuselage of 4130 welded steel tubes with internal oiling for corrosion protection; wooden bulkheads; Dacron covering. Braced welded steel tube tail surfaces with Dacron covering. Helmeted aluminium engine cowling.

LANDING GEAR: Tailwheel type; fixed. Shock-absorption by oil and spring shock-struts. Steerable tailwheel. Cleveland 30-67F hydraulic brakes on mainwheels only. Cleveland 40-101A mainwheels, tyre size 7.50–10; Cleveland 40-199A tailwheel, tyre size 3.50–10. Fairings standard on main legs and wheels; tailwheel fairing optional. DeVore Aviation PK3000 floats optional.

The 100th Waco Classic YMF Super *(Paul Jackson)* ***NEW***/1047139

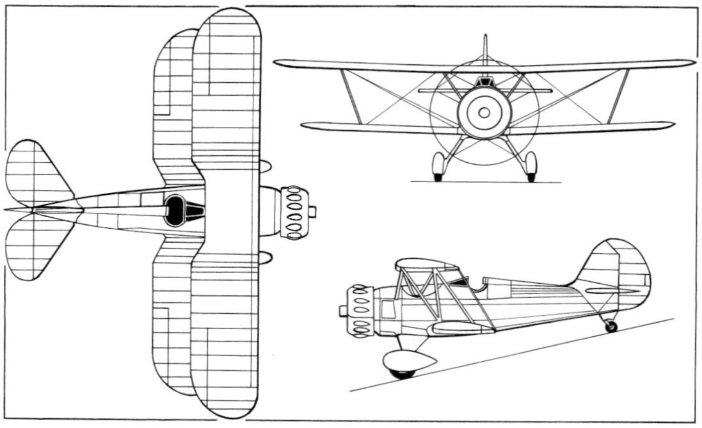

Waco Classic YMF Super three-seat biplane *(James Goulding)* ***NEW***/0587518

POWER PLANT: One 205 kW (275 hp) Jacobs R-755-B2M or −B2 air-cooled radial engine (remanufactured), driving a Sensenich two-blade, fixed-pitch, wooden propeller. Hamilton Sundstrand 2B20 constant-speed propeller with spinner optional. Fuel in two aluminium tanks in upper wing centre-section, usable capacity 182 litres (48.0 US gallons; 40.0 Imp gallons). Refuelling point for each tank in upper wing surface. Auxiliary tanks, usable capacity 45.4 litres (12.0 US gallons; 10.0 Imp gallons) each, optional in either or both inboard upper wing panels. Oil capacity 15 litres (4.0 US gallons; 3.3 Imp gallons) standard;18.9 litres (5.0 US gallons; 4.2 Imp gallons) when additional fuel tanks installed. Fuel injection system optional.

ACCOMMODATION: Three seats in tandem open cockpits, two side by side in front position, single seat at rear. Controls in both cockpits; toe brakes in rear only. Seat belts with shoulder harness, and pilot's adjustable seat, standard. Lockable front and rear baggage compartments. Front windscreen removable.

SYSTEMS: 24 V electrical system with two batteries, alternator and starter for electrical supply to navigation, strobe and rear cockpit lights. Hydraulic system for brakes only.

AVIONICS: Customer specified. Bendix/King and Garmin VFR and IFR packages available. S-Tec 30 autopilot optional.

Instrumentation: Compass, airspeed indicator, turn and bank indicator, rate of climb indicator, sensitive altimeter, recording tachometer, cylinder head temperature gauge and oil pressure and oil temperature gauges standard in rear cockpit. Front cockpit instruments optional. Other options include exhaust gas temperature gauge, carburettor temperature gauge, *g* meter, vacuum or electrically driven gyro system, Hobbs meter (engine-time recorder), outside air temperature gauge and manifold gauge.

EQUIPMENT: Instrument post lighting, heated pitot, three-colour paint scheme with choice of two designs, landing and taxying lights, front and rear cockpit heaters, also standard. Optional equipment includes ground service plug, flight-approved metal front cockpit cover, glider tow hook, deluxe interior with carpet, leather sidewalls and interior trim, and special exterior paint designs.

DIMENSIONS, EXTERNAL:
Wing span: upper ... 9.14 m (30 ft 0 in)
lower ... 8.18 m (26 ft 10 in)
Length overall ... 7.26 m (23 ft 10 in)
Height overall ... 2.59 m (8 ft 6 in)
Wheelbase ... 1.95 m (6 ft 5 in)
Propeller diameter ... 2.44 m (7 ft 8 in)

DIMENSIONS, INTERNAL:
Rear baggage compartment volume ..... 0.21 m³ (7.5 cu ft)
AREAS:
Wings, gross ..... 21.69 m² (233.5 sq ft)
WEIGHTS AND LOADINGS:
Basic weight empty ..... 900 kg (1,985 lb)
Baggage capacity: front ..... 11 kg (25 lb)
rear ..... 34 kg (75 lb)
Max T-O and landing weight ..... 1,338 kg (2,950 lb)
Max wing loading ..... 61.7 kg/m² (12.63 lb/sq ft)
Max power loading ..... 6.53 kg/kW (10.73 lb/hp)

PERFORMANCE:
Never-exceed speed (VNE) ..... 186 kt (344 km/h; 214 mph)
Max level speed at S/L ..... 117 kt (217 km/h; 135 mph)
Max cruising speed at S/L ..... 104 kt (193 km/h; 120 mph)
Econ cruising speed at FL80 ..... 95 kt (177 km/h; 110 mph)
Stalling speed, power off ..... 51 kt (95 km/h; 59 mph)
Max rate of climb at S/L ..... 235 m (770 ft)/min
T-O run ..... 152 m (500 ft)
Range, standard fuel, 30 min reserves ..... 286 n miles (531 km; 330 miles)
g limits ..... +5.5/−2.2

UPDATED

WAG-AERO

WAG-AERO INC

PO Box 181, 1216 North Road, Lyons, Wisconsin 53148
Tel: (+1 262) 763 95 86
Fax: (+1 262) 763 75 95
e-mail: wagaero-sales@wagaero.com
Web: http://www.wagaero.com
PRESIDENT: Mary M Myers
DIRECTOR, SALES: Sandy Hana
DIRECTOR, TECHNICAL: Tom O'Neill
MARKETING SUPERVISOR: Mary Pat Henningfield

Company founded in early 1960s, initially to supply light aircraft spares. Three aircraft types currently offered, based on Piper designs. Founders Dick and Bobbie Wagner sold Wag-Aero to Bill Read and Mary Myers on 1 September 1995. Some 7,000 kits or sets of plans have been sold.

WAG-AERO SPORT TRAINER

TYPE: Tandem-seat kitbuilt.
PROGRAMME: Modernised kitbuilt version of 1937 Piper J3. Sport Trainer first flew 12 March 1975.
CURRENT VERSIONS: **Sport Trainer:** Basic two-seat design. Quoted build time of 1,200 hours; conforms to FAA 51 per cent rule.
Description applies to Sport Trainer.
Acro Trainer: Differs from standard version by having strengthened fuselage, shortened wings (8.23 m; 27 ft 0 in), modified lift struts, improved wing fittings and rib spacing, and new leading-edge.
Observer: Replica of Piper L-4 military liaison aircraft.
Super Sport: Structural modifications to accept engines of up to 112 kW (150 hp), making it suitable for glider towing, bush operations, or as floatplane.
CUSTOMERS: 4,193 sets of plans and kits sold by mid-2001.
COSTS: Plans: US$69.50; kit US$17,000 (2003).
DESIGN FEATURES: High-wing cabin monoplane with braced wings and tailplane.
FLYING CONTROLS: Conventional and manual. Horn-balanced rudder.
STRUCTURE: Fuselage and empennage of welded 4130 chrome molybdenum steel tube with fabric covering; wooden wing with light alloy leading-edge and fabric covering.
LANDING GEAR: Tailwheel type; fixed. Welded steel tube side Vs and half-axles. Bungee shock-absorbers. Speed fairings.
POWER PLANT: Can be powered by any flat-four Teledyne Continental, Franklin or Textron Lycoming engine between 48.5 and 93 kW (65 and 125 hp). Standard fuel capacity 45.4 litres (12.0 US gallons; 10.0 Imp gallons); auxiliary fuel capacity 98.4 litres (26.0 US gallons; 21.7 Imp gallons).
ACCOMMODATION: Two persons in tandem. Horizontally split door on starboard side.
DIMENSIONS, EXTERNAL:
Wing span ..... 10.73 m (35 ft 2½ in)
Wing chord, constant ..... 1.60 m (5 ft 3 in)

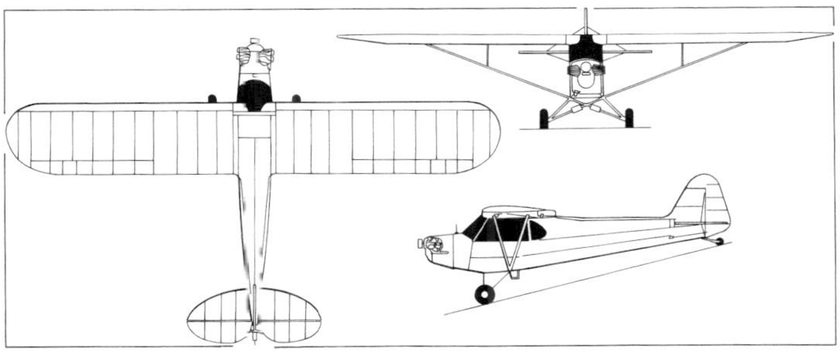
General arrangement of the Wag-Aero Sport Trainer *(Paul Jackson)* 0526902

Wag-Aero Sport Trainer wearing Piper Cub markings *(Paul Jackson)* ***NEW***/0567782

Wing aspect ratio 6.9
Length overall ..... 6.82 m (22 ft 4½ in)
Height overall ..... 2.03 m (6 ft 8 in)
Wheel track ..... 1.80 m (5 ft 11 in)
AREAS:
Wings, gross ..... 16.58 m² (178.5 sq ft)
WEIGHTS AND LOADINGS:
Weight empty ..... 327 kg (720 lb)
Max T-O weight ..... 635 kg (1,400 lb)
Max wing loading ..... 38.3 kg/m² (7.84 lb/sq ft)
Max power loading ..... 6.82 kg/kW (11.20 lb/hp)
PERFORMANCE:
Never-exceed speed (VNE) ..... 104 kt (193 km/h; 120 mph)
Max level speed at S/L ..... 89 kt (164 km/h; 102 mph)
Cruising speed ..... 82 kt (151 km/h; 94 mph)
Stalling speed ..... 34 kt (63 km/h; 39 mph)
Max rate of climb at S/L ..... 149 m (490 ft)/min
Service ceiling ..... 3,660 m (12,000 ft)
T-O run ..... 122 m (400 ft)
Landing run ..... 130 m (425 ft)
Range:
at cruising speed with standard fuel ..... 270 n miles (500 km; 310 miles)
with auxiliary fuel ..... 395 n miles (732 km; 455 miles)

WAG-AERO WAG-A-BOND

TYPE: Side-by-side kitbuilt.
PROGRAMME: Replica of 1948 Piper PA-17 Vagabond. Prototype Wag-A-Bond completed May 1978; plans and kits available. Conforms to FAA 51 per cent rule.
CUSTOMERS: 1,519 plans and kits sold by mid-2001.
COSTS: Plans US$66 (2003). Kit US$18,000 (2003).
DESIGN FEATURES: Two versions (**Classic** and **Traveler**); Traveler is modified and updated version of Vagabond with port and starboard doors, overhead skylight window, extended sleeping deck (conversion from aircraft to camper interior taking about 2 minutes and accommodating two persons), extended baggage area, engine of up to 85.7 kW (115 hp), and provision for full electrical system.
Wing section USA 35B.
FLYING CONTROLS: Conventional and manual. Horn-balanced rudder. Tab on port elevator.
STRUCTURE: All-wood wing and aluminium alloy aileron structures. Welded 4130 steel tube and flat plate fuselage structure, and steel tube tail unit. Complete airframe is fabric covered.
LANDING GEAR: Non-retractable, tailwheel type; as Sport Trainer; mainwheel tyres 7.00–6; Cleveland brakes. Optional skis.
POWER PLANT: Traveler can be powered by a Textron Lycoming engine of 80.5 to 85.7 kW (108 to 115 hp). Classic can be powered by a Teledyne Continental engine of 48.5 to 74.5 kW (65 to 100 hp). Fuel capacity: Traveler 98.4 litres (26.0 US gallons; 21.6 Imp gallons), Classic 45.4 litres (12.0 US gallons; 10.0 Imp gallons).
ACCOMMODATION: Two persons side by side on adjustable bench seat; cabin includes baggage stowage. Forward-opening cabin door on each side of fuselage.
SYSTEMS: Optional electrical system in Traveler.
DIMENSIONS, EXTERNAL:
Wing span ..... 8.32 m (29 ft 3½ in)
Wing chord, constant ..... 1.60 m (5 ft 3 in)
Wing aspect ratio 5.0
Length overall ..... 5.70 m (18 ft 8½ in)
Height overall ..... 1.83 m (6 ft 0 in)
Wheel track ..... 1.83 m (6 ft 0 in)
DIMENSIONS, INTERNAL:
Cabin max width ..... 1.02 m (3 ft 4 in)
AREAS:
Wings, gross ..... 13.70 m² (147.5 sq ft)
WEIGHTS AND LOADINGS:
Weight empty: Classic ..... 318 kg (700 lb)
Traveler ..... 363 kg (800 lb)
Baggage capacity: Classic ..... 18 kg (40 lb)
Traveler ..... 27 kg (60 lb)

Wag-Aero Wag-A-Bond 0016516

Max T-O weight: Classic .... 567 kg (1,250 lb)
Traveler .... 658 kg (1,450 lb)
Max wing loading: Classic .... 41.4 kg/m^2 (8.47 lb/sq ft)
Traveler .... 48.0 kg/m^2 (9.83 lb/sq ft)
Max power loading: Classic .... 7.61 kg/kW (12.50 lb/hp)
Traveler .... 7.67 kg/kW (12.61 lb/hp)

PERFORMANCE:
Max level speed: Classic .... 91 kt (169 km/h; 105 mph)
Traveler .... 118 kt (219 km/h; 136 mph)
Cruising speed: Classic .... 83 kt (153 km/h; 95 mph)
Traveler .... 104 kt (193 km/h; 120 mph)
Stalling speed:
Classic, Traveler .... 39 kt (71 km/h; 44 mph)
Max rate of climb at S/L: Classic .... 190 m (625 ft)/min
Traveler .... 259 m (850 ft)/min
Service ceiling: both .... 4,260 m (14,000 ft)
T-O run: both .... 119 m (390 ft)
Landing run: both .... 232 m (760 ft)
Range: both .... 620 n miles (1,148 km; 713 miles)

Wag-Aero Sportsman *(Paul Jackson)* 0093689

WAG-AERO 2+2 SPORTSMAN

TYPE: Four-seat kitbuilt.

PROGRAMME: Version of the Piper PA-14 Family Cruiser. Plans and material kits available. Conforms to FAA 51 per cent rule.

CUSTOMERS: 1,971 sets of plans and kits sold by May 2001.

COSTS: Plans: US$89; kit US$18,000 (2003).

DESIGN FEATURES: True four-seater, with option of hinged rear fuselage decking to provide access to baggage and rear seat areas. Conventional, high-wing configuration with V-type bracing struts, wire-braced empennage, mid-mounted tailplane and classic fin and rudder.
Wing section USA 35B.

FLYING CONTROLS: Conventional and manual. Horn-balanced rudder. Upper and lower spoilers; trim tabs on both elevators. Aileron limits ±17°; elevator limits 34° up, 29° down.

STRUCTURE: Similar construction to Wag-A-Bond, with glass fibre wingtips. Alternatively, drawings and materials provided to modify standard PA-12, PA-14 or PA-18 wings. Prewelded fuselage structure also available.

LANDING GEAR: As Wag-A-Bond; main units use 8.00-6 or 7.00-6 tyres. Optional floats.

POWER PLANT: Engine of 93 to 149 kW (125 to 200 hp). Usable fuel capacity 148 litres (39.0 US gallons; 32.5 Imp gallons).

ACCOMMODATION: Four persons; two separate front seats, plus rear bench seat which is removable for stretcher or cargo carrying. Baggage area to rear of bench seat. Forward-opening cabin door on each side.

DIMENSIONS, EXTERNAL:
Wing span .... 10.90 m (35 ft 9 in)
Wing chord, constant .... 1.60 m (5 ft 3 in)
Wing aspect ratio 7.3
Length overall .... 7.12 m (23 ft 4½ in)
Height overall .... 2.02 m (6 ft 7½ in)
Wheel track .... 1.98 m (6 ft 6 in)

DIMENSIONS, INTERNAL:
Cabin max width .... 0.99 m (3 ft 3 in)

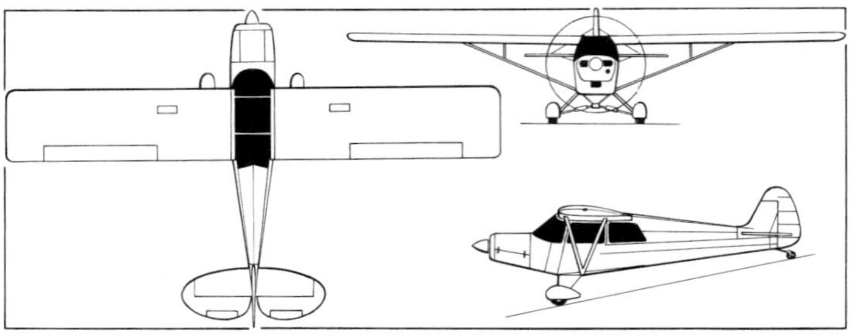
Wag-Aero 2+2 Sportsman *(James Goulding)* 0526903

AREAS:
Wings, gross .... 16.18 m^2 (174.1 sq ft)

WEIGHTS AND LOADINGS:
Weight empty .... 490 kg (1,080 lb)
Max T-O weight .... 998 kg (2,200 lb)
Max wing loading .... 61.7 kg/m^2 (12.63 lb/sq ft)
Max power loading .... 6.70 kg/kW (11.00 lb/hp)

PERFORMANCE (typical; actual data depend on engine fitted):
Max level speed .... 112 kt (207 km/h; 129 mph)
Cruising speed .... 108 kt (200 km/h; 124 mph)
Stalling speed .... 33 kt (62 km/h; 38 mph)
Max rate of climb at S/L .... 244 m (800 ft)/min
Service ceiling .... 4,510 m (14,800 ft)
T-O run .... 70 m (230 ft)
Landing run .... 104 m (340 ft)
Range at cruising speed .... 670 n miles (1,240 km; 771 miles)

WARNER

WARNER AEROCRAFT COMPANY

9415 Laura Court, Seminole, Florida 33776-1625
Tel: (+1 727) 595 23 82
Fax: (+1 617) 598 10 14
e-mail: info@warnerair.com
Web: http://www.warnerair.com
PRESIDENT: Dana Axelrod
SALES MANAGER: Mike Dwyer

Warner Aerocraft markets the Warner Space Walker, Revolution and Sportster.

Warner Space Walker tandem-seat sportplane *(Paul Jackson)* ***NEW***/0567771

WARNER SPORTSTER

TYPE: Tandem-seat sportplane kitbuilt.

PROGRAMME: Development of Country Air (later Warner) Space Walker, of which 46 built. Two-seat versions of Space Walker developed as Revolution I (48.5 kW; 65 hp Teledyne Continental) and Revolution II (74.6 kW; 100 hp Textron Lycoming); further power increase produced Sportster; prototype (N269U) first flown 18 April 1998. New version with revised canopy introduced 2002 and has replaced original model aircraft; length reduced to 6.20 m (20 ft 4 in). Prototype registered N69DA.

CUSTOMERS: 125 Space Walkers and Revolutions built, plus six of original Sportster; two of new Sportsters flying by October 2003.

COSTS: Basic kit US$16,495; fast-build kit US$19,995 (2003).

Single-seat, enclosed cockpit option for Warner Sportster *(Paul Jackson)* 0527005

DESIGN FEATURES: Quoted build time 700 hours. Roll rate 45°/s.

FLYING CONTROLS: Conventional and manual. Horn-balanced rudder and elevators; trim tab in port elevator. Dual controls.

STRUCTURE: 4130 steel tube and fabric fuselage with wooden turtle deck; fabric-covered two-spar wooden cantilever wings. Glass fibre cowling and wingtips. Tail unit identical to Revolution/Space Walker. Compared to early version, new Sportster has lower bottom longeron and lowered bulkheads forward of cockpit.

LANDING GEAR: Tailwheel type; fixed. Tubular welded V legs with tubular bracing and spring shock-absorbers. 6.00-6 tyres, Cleveland wheels with disc brakes. Steerable 15 cm (6 in) Maule tailwheel. Speed fairings on mainwheels.

POWER PLANT: Engines from 63.4 to 119 kW (85 to 160 hp) can be fitted; prototype has 85.6 kW (115 hp) Textron Lycoming O-290-D. Fuel capacity 76 litres (20.0 US gallons; 16.7 Imp gallons).

ACCOMMODATION: Pilot and passenger in tandem in individual open cockpits; choice of three styles of windscreen; alternatively, rear cockpit can be covered with a bubble canopy and front canopy temporarily faired over. Compared to early Sportster, new version has slightly more curved lips to accommodate canopy curvature, and instrument panel moved forward.

DIMENSIONS, EXTERNAL:
Wing span .... 7.92 m (26 ft 0 in)
Wing chord, constant .... 1.37 m (4 ft 6 in)
Wing aspect ratio 6.0
Length overall .... 6.20 m (20 ft 4 in)
Max diameter of fuselage .... 0.76 m (2 ft 6 in)
Height overall .... 1.68 m (5 ft 7 in)
Propeller ground clearance .... 0.46 m (1 ft 6 in)
Tailplane span .... 2.41 m (7 ft 11 in)
Wheel track .... 1.83 m (6 ft 0 in)
Wheelbase .... 4.41 m (14 ft 5½ in)

DIMENSIONS, INTERNAL:
Cockpit max width .... 0.76 m (2 ft 6 in)
AREAS:
Wings, gross .... 10.41 m^2 (115.0 sq ft)
WEIGHTS AND LOADINGS (O-290 engine):
Weight empty .... 386 kg (850 lb)
Baggage capacity .... 11 kg (25 lb)
Max T-O weight .... 680 kg (1,400 lb)
Max wing loading .... 65.4 kg/m^2 (13.39 lb/sq ft)
Max power loading .... 7.94 kg/kW (13.04 lb/hp)
PERFORMANCE (O-290 engine):
Never-exceed speed (VNE) .... 130 kt (241 km/h; 140 mph)
Max level speed .... 115 kt (212 km/h; 132 mph)
Normal cruising speed .... 104 kt (193 km/h; 120 mph)
Stalling speed .... 40 kt (73 km/h; 43 mph)
Max rate of climb at S/L .... 358 m (1,200 ft)/min
Service ceiling .... 3,660 m (12,000 ft)
T-O and landing run .... 130 m (350 ft)
Range with max fuel .... 356 n miles (659 km; 410 miles)
g limits .... +6/–4

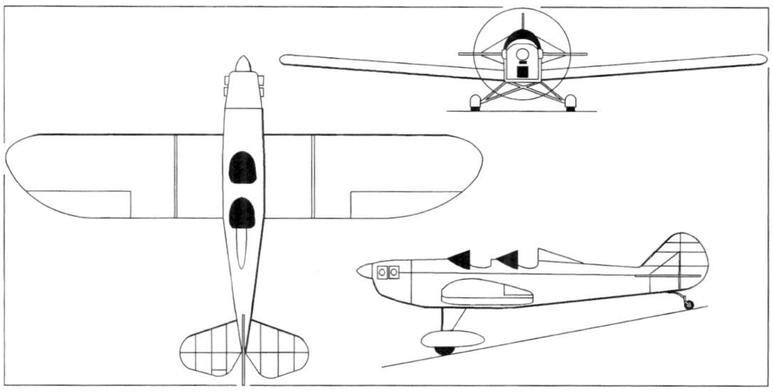

Warner Sportster tandem-seat, open-cockpit version *(Paul Jackson)* **NEW**/0567254

WEATHERLY

WEATHERLY AIRCRAFT COMPANY

5000 Bailey Loop, McClellan, California 95652
Tel: (+1 916) 640 01 20
Fax: (+1 916) 640 01 16
e-mail: fly@weatherlyaircraft.com
Web: http://www.weatherlyaircraft.com
PRESIDENT: Gary Beck
MANUFACTURING MANAGER: Carl Lange

Weatherly developed conversion of Fairchild M-62 in the early 1960s for agricultural role, and later designed and built the Model 201. Further refinement of the design resulted in the Weatherly 620, production of which continued at a low rate but ceased in 1997 when the company was declared bankrupt. Weatherly was purchased by GBECK Inc in 2000. The company has moved to a 6,782 m^2 (73,000 sq ft) facility at McClellan Park, California, the former McClellan AFB, which closed in July 2001.

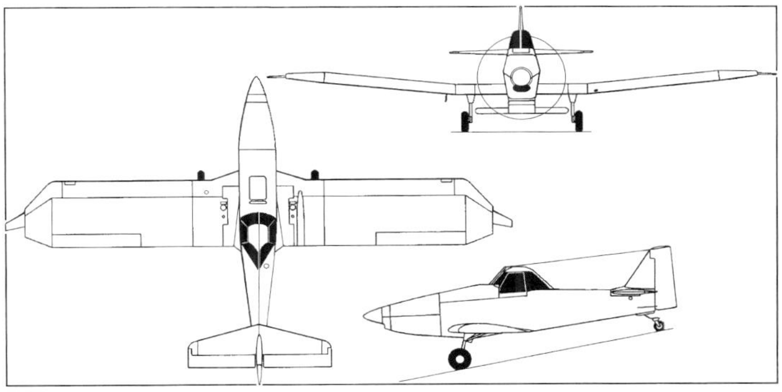

Weatherly 620-BTG general arrangement **NEW**/0567253

WEATHERLY 620

TYPE: Agricultural sprayer.

PROGRAMME: Developed from Weatherly 201; prototype (N9256W) first flown 1979.

CURRENT VERSIONS: **620:** Initial version; no longer available.

620-B: Current piston-engined version, powered by Pratt & Whitney radial engine. Length 8.15 m (26 ft 9 in); height 2.44 m (8 ft 0 in). Slightly different spray equipment but same sized hopper.

620-TP: Initial turboprop version, with P&WC PT6A-11AG; no longer available.

620-BTG: Current turboprop version; engine can be retrofitted to earlier versions.

Description applies to Weatherly 620-BTG.

CUSTOMERS: By 1999, 155 Weatherly 620s of all versions had been built. The 156th was registered in 2001; two more built by October 2003.

COSTS: 620B US$225,000; 620BTG US$360,000 (2003).

DESIGN FEATURES: High lift drag ratio wings, with tip vanes for vortex diffusion and increase of swath width; vanes fold downwards for ground handling. Hinged wing leading-edges aid inspection.

FLYING CONTROLS: Conventional and manual. Stainless steel rudder cables.

STRUCTURE: All-metal monocoque fuselage; metal wings and control surfaces; glass fibre tail fairings. Rudder deflects ±20°; elevator 27° up and 15° down; ailerons 29° up and 11° 30' down.

LANDING GEAR: Tailwheel type; fixed. Mainwheel oleos connected to spar to disperse stresses. Spring tailwheel assembly with fully castoring and locking wheel. Three-piston Cleveland brakes. Mainwheel tyre size 8.5–10; tailwheel 12.5-4.5.

POWER PLANT: *620-B:* One Pratt & Whitney 336 kW (450 hp) R-985-AN radial engine driving a Hartzell HC-B3R30-4 or -4B three-blade propeller. Fuel capacity 369 litres (97.5 US gallons; 81.1 Imp gallons), of which 341 litres (90.1 US gallons; 75.1 Imp gallons) is usable. Oil capacity 25.3 litres (6.7 US gallons; 5.6 Imp gallons).

620-BTG: One Honeywell TPE331-1 turboprop, derated to 373 kW (500 shp), driving a McCauley 3GFR34C602/100LA-2 three-blade propeller. Fuel capacity 492 litres (131 US gallons; 108 Imp gallons) in seven tanks (two in each wing and three in wing centre-section) of which 439 litres (116 US gallons; 96.5 Imp gallons) usable. Oil capacity 0.6 litre (1.6 US gallons; 1.3 Imp gallons).

ACCOMMODATION: Pilot only. Four-way adjustable mesh seat. Baggage compartment behind seat.

SYSTEMS: 24 V electrical system; 50 A Jasco starter/generator in 620-B, 250A Jasco model in 620-BTG. Optional air conditioning and windshield washer/wiper.

EQUIPMENT: Agricultural dispersal system comprises a 1,344 litre (355 US gallon; 296 Imp gallon) hopper in forward fuselage, size 1.35 m^3 (47.5 cu ft); Weath-Aero fan; 5 cm (2 in) SS bottom load pump; 5 cm (2 in) Agrinautics spray pump with three-way valve; 6.10 m (20 ft) pylon-mounted drop boom with nozzles every 15 cm (6 in); 63.5 cm (2 ft 1 in) Transland gate box. Main gear and cockpit wire deflectors, plus deflector cable from cabin to fin. Rinse-out system optional, as are stainless steel spray system and flow meters.

DIMENSIONS, EXTERNAL:
Wing span (with vanes extended) .... 14.22 m (46 ft 8 in)
Wing aspect ratio 7.9
Length overall .... 9.04 m (29 ft 8 in)
Height overall .... 2.90 m (9 ft 6 in)
Wheel track .... 3.20 m (10 ft 6 in)
Propeller diameter: 620-B .... 2.43 m (7 ft 11½ in)
620-BTG .... 2.49 m (8 ft 2 in)
AREAS:
Wings, gross .... 25.73 m^2 (277.0 sq ft)

Weatherly 620-B (one Pratt & Whitney R-985) *(W J Taylor)* 0084556

WEIGHTS AND LOADINGS:
- Weight empty .... 1,374 kg (3,030 lb)
- Max T-O weight under CAM 8 .... 2,721 kg (6,000 lb)
- Baggage shelf .... 11 kg (25 lb)
- Max wing loading .... 105.8 kg/m² (21.66 lb/sq ft)
- Max power loading .... 8.12 kg/kW (13.33 lb/shp)

PERFORMANCE:
- Never-exceed speed (VNE) .... 153 kt (283 km/h; 176 mph)
- Normal operating speed .... 122 kt (226 km/h; 140 mph)
- Stalling speed .... 62 kt (115 km/h; 72 mph)
- Max rate of climb at S/L .... 427 m (1,400 ft)/min
- Service ceiling .... 4,572 m (15,000 ft)
- T-O run .... 287 m (940 ft)

WINDCRAFTER

430 West Enterprise Road, Building 3, Shelton, Washington 98584
Tel/Fax: (+1 360) 462 94 63
e-mail: windcrft@hctc.com
Web: http://www.windcrafterinc.com
OWNERS/FOUNDERS: Steve Swearingen
Therin Laney
PILOT: Aaron Shell
SECRETARY/TREASURER: Rick Marecle

Windcrafter manufactures a range of airships of unorthodox design, both manned (ultralight) and unmanned. Basic manned version is the CA 108.

NEW ENTRY

WINDCRAFTER CA 108

TYPE: Helium non-rigid airship.
PROGRAMME: No programme history known. A larger model, the CA 2010 (6.10 × 3.05 m; 20 × 10 ft advertising area) is also available.
COSTS: From US$16,000 including training and accessories (2004).
DESIGN FEATURES: Upright rectangular envelope, of carangifoil (fish-shaped) aerofoil section, with low or negative internal pressure. Engine at base.
POWER PLANT: Electric brushless motor, powered by lithium polymer battery.
DIMENSIONS, EXTERNAL:
- Height .... 3.53 m (11 ft 7 in)
- Advertising banner area (each side):
  - Height .... 3.05 m (10 ft 0 in)
  - Width .... 2.44 m (8 ft 0 in)

DIMENSIONS, INTERNAL:
- Volume .... more than 2.83 m³ (100 cu ft)

NEW ENTRY

One of Windcrafter's radio-controlled unmanned carangifoil airships ***NEW***/1047199

WINGTIP TO WINGTIP

WINGTIP TO WINGTIP LLC

7453 M-50, Onsted, Michigan 49265
Tel: (+1 517) 467 74 20
email: wingtiptowingtip@aol.com
PRESIDENT: Greg Panzl

Since 1990, some 34 aerobatic aircraft in the S-300 family have been built and have gained high placings in competitions in the US and abroad. The Staudacher company was purchased in 1999 by Greg Panzl, brother of S-300 aerobatic pilot Chris Panzl, and has launched an improved version under the new name of Wingtip to Wingtip. The company has its own private airstrip at Onsted. Staudachers have proved popular subjects for scale radio-controlled aeromodellers.

WINGTIP TO WINGTIP PANZL S-330

TYPE: Aerobatic two-seat sportplane.
PROGRAMME: Designed by Jon A Staudacher. First Staudacher S-300 built for Mike Goulian in 1990; design progressively refined throughout 1990s. Seven Staudachers participated in 1993 US national championships.
CURRENT VERSIONS: **S-300:** Baseline version; *as described*. Several subvariants, up to S-300E.

S-300RG: Retractable landing gear; cantilever tailplane; IO-540-K engine and two-blade Hartzell propeller. One only (N126RG), first flown 22 February 1998.

Staudacher S-300D flown by Chris Panzl, brother of company president, Greg Panzl *(Paul Jackson)* ***NEW***/0567257

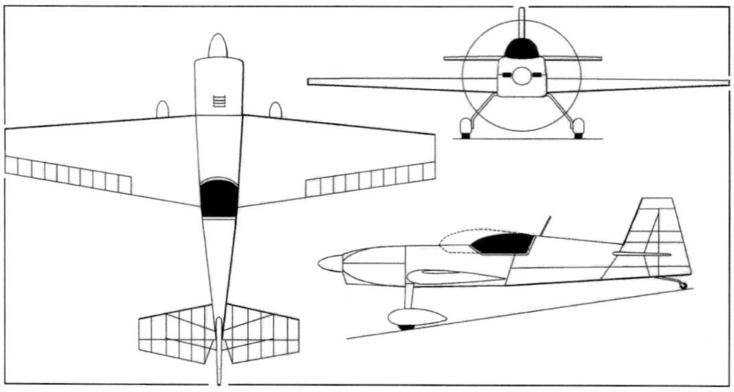

Staudacher S-300 single-seat version, with two-seat canopy shown by broken line *(Paul Jackson)* ***NEW***/0567243

S-600: Tandem two-seat version; cockpit enlarged forward. Four built.

S-600F: Tandem two-seat derivative of S-300 and S-600 intended for cross-country and aerobatic training. Full dual controls and second baggage space between firewall and cockpit; composites shell over tubular fuselage. Prototype (N603SA) under construction during 2003.

Panzl S-330 P: Prototype (N330LS) delivered to Loren Smith 24 July 2000; second built 2001; third registered in July 2002; and fourth in July 2003. Derived from S-300, but with improved canopy and carbon fibre panels throughout, resulting in weight saving. Empty weight, equipped, 585 kg (1,289 lb); 246 kW (330 hp) Textron Lycoming IO-540-G1D5 and MTV-9-B-C propeller. Aerobatic fuel 87 litres (23.0 US gallons; 19.2 Imp gallons); ferry tank in each wingroot, combined capacity 159 litres (42.0 US gallons; 35.0 Imp gallons).

Additionally, several custom versions of Staudacher, comprising **S-341** (N341SA), built 1992 with Textron Lycoming GO-480; **S-1000** (N6195V), built 1996 for Jon Staudacher; **DC 11** (N327MS), with Textron Lycoming O-320 built 1992; and **GU 11** (N11GU), with Teledyne Continental O-200 built for Jon Staudacher in 2000.

CUSTOMERS: Total 34 aircraft of Staudacher design built by end of 2003, including four Panzl S-330s.
DESIGN FEATURES: Created to rival Sukhoi and Extra Unlimited class aerobatic aircraft; roll rate in excess of 420°/s. Low-wing monoplane with wire-braced tail surfaces. Moderately tapered wing; ailerons occupy two-thirds of trailing-edges; all subvariants have same aerofoil.
FLYING CONTROLS: Conventional and manual. Spade-type aileron servos. Trim tab in each elevator. Twin arrowhead-shape horn balance at fin tip.
STRUCTURE: Sitka spruce (wooden) wing built on box spar reinforced with carbon fibre and covered with 3 mm (⅛ in) plywood. Metal (4130 steel) tube fuselage and fabric-covered empennage; carbon fibre front panels, cowling and turtledeck. Fabric-covered ailerons.
LANDING GEAR: Tailwheel type; fixed. Aluminium cantilever mainwheel legs.
POWER PLANT: One 186 kW (250 hp) Textron Lycoming IO-540-B1A5 or 224 kW (300 hp) IO-540-K1B5 flat-six driving an MTV-9-B-C/C200-15 three-blade, constant-speed propeller. Fuel tank ahead of pilot, capacity 129 litres (34.0 US gallons; 28.3 Imp gallons).
ACCOMMODATION: One or two (in tandem) pilots, according to version.
DIMENSIONS, EXTERNAL:
Wing span ... 7.72 m (25 ft 4 in)
Length overall ... 6.64 m (21 ft 9½ in)
Height overall ... 1.80 m (5 ft 11 in)
AREAS:
Wings, gross ... 10.29 m² (110.8 sq ft)
WEIGHTS AND LOADINGS:
Weight empty ... 567 kg (1,250 lb)
Max T-O weight (Aerobatic) ... 725 kg (1,600 lb)
Max wing loading ... 70.3 kg/m² (14.44 lb/sq ft)
Max power loading (IO-540-K) ... 3.25 kg/kW (5.33 lb/hp)
PERFORMANCE:
Never-exceed speed (VNE) ... 217 kt (402 km/h; 250 mph)
Stalling speed ... 52 kt (95 km/h; 59 mph)
Max rate of climb at S/L:
S-330 ... more than 1,219 m (4,000 ft)/min

YESTERYEAR

BIPLANES OF YESTERYEAR LLC

1455 Crestway, Ontario, Oregon 97914
Tel: (+1 541) 889 04 06
Fax: (+1 541) 881 04 06
e-mail: mifyter@fmtc.com or maxadx@fmtc.com
Web: http://www.mifyter.com
DIRECTOR AND DESIGNER: Rod Cowgill

Having won several awards and mentions for his evocation of a First World War German biplane fighter, Mr Cowgill launched a production kit version in 2003.

YESTERYEAR MIFYTER

TYPE: Single-seat ultralight biplane kitbuilt.
PROGRAMME: Prototype (N666RC) first flown 12 August 1996. Nominated Champion at Arlington 1997 and Grand Champion two years later; further awards at AirVenture, Oshkosh, in 1997 and 2002. Kit available to order from July 2003 and will ship early 2004. Pronounced 'my fighter'.
COSTS: Kit US$14,995, excluding engine, propeller and instruments (2003). Spoked wheels, dummy bombs, flowing white scarf and machine guns optional.
DESIGN FEATURES: Based on Fokker biplane, with third lifting surface between wheels but with same aerofoil as Avid series. Two-bay wings with N-type interplane struts. Quoted build time 650 to 850 hours. Cross-axle lifting surface has Clark 23012 airfoil and is free-floating to accommodate landing gear movement during take-off and landing.
FLYING CONTROLS: Conventional and manual. Large, horn-balanced rudder. Ailerons on both mainplanes.
STRUCTURE: Fuselage of 4130 welded steel tube; wood and 6061-T6 steel wing. Both covered in Stits Polyfiber. Glass fibre nose cowling and wingtips.
POWER PLANT: One piston engine of between 48.5 and 55.9 kW (65 and 75 hp); prototype has inverted 47.0 kW (63.1 hp) Rotax 532 with radiator installed vertically under louvered top cowling.
LANDING GEAR: Tailwheel type; fixed. Main wheels are 35 cm (14 in) with heavy-duty spokes; 51 cm (20 in) Cheng Shin tyres 17 cm (7 in) Matro tailwheel on prototype; kit aircraft will have 13 cm (5 in) vers on.
EQUIPMENT: Two fake 7.92 mm Spandau machine guns mounted forward of cockpit.
DIMENSIONS, EXTERNAL:
Wing span: upper ... 6.25 m (20 ft 6 in)
lower ... 5.64 m (18 ft 6 in)

Prototype Biplanes of Yesteryear Mifyter *(Paul Jackson)* ***NEW***/0561712

Length overall ... 5.03 m (16 ft 6 in)
Height overall ... 2.01 m (6 ft 7 in)
AREAS:
Wings, gross ... 13.56 m² (146.0 sq ft)
WEIGHTS AND LOADINGS:
Weight empty ... 195 kg (430 lb)
Max T-O weight ... 342 kg (755 lb)
PERFORMANCE:
Max level speed ... 83 kt (153 km/h; 95 mph)
Cruising speed ... 65 kt (121 km/h; 75 mph)
Stalling speed ... 33 kt (60 km/h; 37 mph)
Max rate of climb at S/L ... 411 m (1,350 ft)/min
T-O run ... 23 m (75 ft)
Landing run ... 46 m (150 ft)
Range ... 304 n miles (563 km; 350 miles)

ZENITH

ZENITH AIRCRAFT COMPANY

Mexico Memorial Airport, PO Box 650, Mexico, Missouri 65265-0650
Tel: (+1 573) 581 90 00
Fax: (+1 573) 581 00 11
e-mail: info@zenithair.com
Web: http://www.zenithair.com
PRESIDENT: Sebastien C Heintz
MANAGER, OFFICE: Joyce Fort
MANAGER, PRODUCTION: Nicholas M Heintz

Zenith Aircraft Company has acquired all rights to manufacture and sell the Canadian Zenair range of light aircraft kits detailed below. For details of the CH 2000, see AMD in this section. By the end of 2002, over 2,700 kits had been produced.

In 1997, Zenith announced a joint venture with Czech Aircraft Works (CZAW, which see) to assemble the CH 601 and CH 701 for the European market. Additionally these Zenith aircraft are marketed both as complete aircraft and in kit form in Germany and Austria by Roland Aircraft; they are distinguished by the suffix 'D' in model numbers.

ZENITH ZODIAC CH 601HD

TYPE: Side-by-side ultralight kitbuilt.
PROGRAMME: First flight of CH 600 prototype June 1984. Plans and kits of parts (45 per cent premanufactured) for improved model followed. New **CH 601UL** developed to meet TP 10141 advanced ultralight (AULA) category in Canada. Construction of prototype CH 601 (59.7 kW; 80 hp Rotax 912) began September 1990 (first flight October that year). Kit production started January 1991. UL version (2001 cost US$12,380) remains available.
CURRENT VERSIONS: **CH 601HD:** Offers wider cockpit and extra power for increased performance and true cross-country capability.

CH 601HDS: First flown August 1991. Has shorter, tapered 'speed' wing of 7.01 m (23 ft 0 in) span. HD and UL are available as kits and as 85 per cent pre-assembled kits requiring only quoted 120 working hours to complete, compared to quoted 400 hours for unassembled kits.

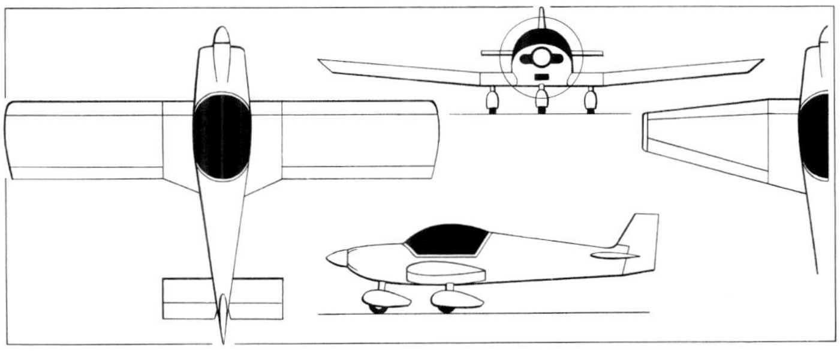

Zenith Zodiac CH 601HD, with half-plan view of tapered CH 601HDS wing *(James Goulding)* 0016514

Description applies to CH 601HD.
CUSTOMERS: Over 600 of all versions flying by end 2002.
COSTS: Plans: US$315. Kit: US$12,620. Information pack: US$15 (2003). Materials and component parts are available.
DESIGN FEATURES: Conventional low-wing monoplane with all-moving vertical tail. Conforms to Canadian TP 10141 airworthiness standards.

Wing section NACA 65018 (modified); dihedral 6°, incidence 8°.
FLYING CONTROLS: Manual. Long-span ailerons; elevators; and Zenith all-moving vertical fin.
STRUCTURE: All-metal monocoque. Single-piece, starboard-hinged canopy.
LANDING GEAR: Non-retractable tricycle or tailwheel type. Skis and floats optional.
POWER PLANT: One engine of 47.7 to 85.75 kW (64 to 115 hp), including 59.6 kW (79.9 hp) Rotax 912 UL, with reduction gear, driving three-blade ground adjustable propeller, 47.8 kW (64.1 hp) Rotax 582 UL, 48.5 to 74.5 kW (65 to 100 hp) Teledyne Continental or 52 kW (70 hp) Volkswagen. 74.6 kW (100 hp) four-cylinder four-stroke liquid-cooled Subaru EA81 is available from Stratos for the CH 601HD. Fuel capacity 60 litres (16.0 US gallons; 13.3 Imp gallons); wing baggage lockers may be fitted with two 26.5 litre (7.0 US gallon; 5.8 Imp gallon) auxiliary tanks.

Zenith CH 601 built by an Australian owner *(Paul Jackson)* ***NEW***/0567781

DIMENSIONS, EXTERNAL:
- Wing span .... 8.23 m (27 ft 0 in)
- Wing chord (constant) .... 1.24 m (4 ft 1 in)
- Wing aspect ratio 5.6
- Length overall .... 5.79 m (19 ft 0 in)
- Height overall: tailwheel .... 1.90 m (6 ft 2¾ in)
  - nosewheel .... 1.98 m (6 ft 6 in)
- Tailplane span .... 2.31 m (7 ft 7 in)
- Wheel track .... 1.88 m (6 ft 2 in)
- Wheelbase .... 1.30 m (4 ft 3¼ in)
- Propeller diameter .... 1.73 m (5 ft 8 in)

DIMENSIONS, INTERNAL:
- Cockpit max width .... 1.12 m (3 ft 8 in)

AREAS:
- Wings, gross .... 12.08 m^2 (130.0 sq ft)

WEIGHTS AND LOADINGS (Rotax 912 for power loading):
- Weight empty .... 290 kg (640 lb)
- Baggage capacity .... 36 kg (80 lb)
- Max T-O weight .... 544 kg (1,200 lb)
- Max wing loading .... 45.1 kg/m^2 (9.23 lb/sq ft)
- Max power loading .... 9.11 kg/kW (15.00 lb/hp)

PERFORMANCE (Rotax 912):
- Never-exceed speed (VNE) .... 130 kt (241 km/h; 150 mph)
- Max level speed at S/L .... 117 kt (217 km/h; 135 mph)
- Econ cruising speed .... 104 kt (193 km/h; 120 mph)
- Stalling speed .... 41 kt (76 km/h; 47 mph)
- Max rate of climb at S/L .... 351 m (1,150 ft)/min
- Service ceiling .... 4,875 m (16,000 ft)
- T-O run .... 131 m (430 ft)
- Landing run .... 159 m (520 ft)
- Range with max fuel .... 677 n miles (1,255 km; 780 miles)
- Endurance .... 4 h
- *g* limits .... ±6

ZENITH SUPER ZODIAC CH 601XL

TYPE: Side-by-side kitbuilt.

PROGRAMME: Certifiable version of CH 601 (which see), with new wing planform and 589 kg (1,300 lb) MTOW. Introduced at Oshkosh in July 1998.

Description generally as for CH 601, except that below.

CURRENT VERSIONS: **CH 601XL**: US version.

CH 601DX: German designation.

COSTS: Plans US$395; Kit US$15,890 (2003).

DESIGN FEATURES: Wing section NACA 23018 at root, NACA 23015 at tip.

FLYING CONTROLS: Flaps added.

POWER PLANT: One 89 kW (120 hp) Jabiru 3300 flat-six or 86.5 kW (116 hp) Textron Lycoming O-235. Options include 73.5 kW (98.6 hp) Rotax 912 ULS and conversions of Subaru motorcar engines. Fuel capacity 91 litres (24.0 US gallons; 20.0 Imp gallons) in two wing tanks.

DIMENSIONS, EXTERNAL:
- Wing span .... 8.23 m (27 ft 0 in)
- Wing chord: at main panel root .... 1.60 m (5 ft 3 in)
  - at tip .... 1.33 m (4 ft 4¼ in)
- Wing aspect ratio 5.5

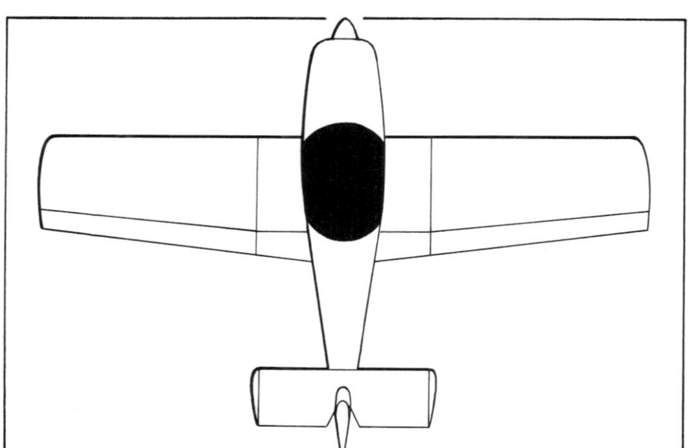

Plan view of Zenith Zodiac CH 601XL, showing revised wing shape of this version *(James Goulding)* 0062649

Zenith CH 601 XL registered in the USA *(Paul Jackson)* ***NEW***/0567779

AREAS:
- Wings, gross .... 12.26 m^2 (132.0 sq ft)

WEIGHTS AND LOADINGS (O-235):
- Weight empty .... 363 kg (800 lb)
- Baggage capacity .... 18 kg (40 lb)
- Max T-O weight .... 589 kg (1,300 lb)
- Max wing loading .... 47.5 kg/m^2 (9.85 lb/sq ft)
- Max power loading .... 6.82 kg/kW (11.21 lb/hp)

PERFORMANCE (O-235):
- Never-exceed speed (VNE) .... 156 kt (289 km/h; 180 mph)
- Max level speed .... 129 kt (238 km/h; 148 mph)
- Normal cruising speed .... 120 kt (222 km/h; 138 mph)
- Stalling speed, flaps down .... 39 kt (71 km/h; 44 mph)
- Max rate of climb at S/L .... 283 m (930 ft)/min
- Service ceiling .... 4,267 m (14,000 ft)
- T-O and landing run .... 152 m (500 ft)
- Max range .... 521 n miles (965 km; 600 miles)
- Endurance, no reserves .... 4 h 30 min
- *g* limits .... ±6

ZENITH STOL CH 701

TYPE: Side-by-side ultralight kitbuilt.

PROGRAMME: Experimental category prototype made first flight mid-1986. Plans, 49 or 85 per cent kits available. Meets Canadian TP 10141 standards for advanced ultralight trainers. German promotion as **CH-701D**. Unauthorised copies are available in Czech Republic as **Kappa-1**, with slight performance differences. Ekoflug of Slovak Republic builds version designated MXP-740. CH-701-AG agricultural version ceased US production after 55 completed, but revived by Czech Aircraft Works (CZAW). Marketed in Italy as **ICP Savannah** and previously in France as **Aerotrophy TT 2000**.

In July 2001 design was upgraded with redesigned wing spar to accommodate increased MTOW, increased fuel capacity, top-hinged cabin doors and redesigned windscreen. Aircraft from serial number 7–4512 are affected.

CUSTOMERS: Over 700 flying by end 2002. Indian National Cadet Corps, part of the Indian Air Force, ordered 85 plus 48 options in 2000; these assembled by CZAW.

COSTS: Plans: US$380. Kit: US$12,980. Information pack: US$35. Amphibian kit: US$17,730 (2003). Materials and component parts available.

DESIGN FEATURES: Strut-braced, high-wing, STOL cabin monoplane with foldable wings and fixed leading-edge slats. In most countries, CH 701 can be registered either as advanced ultralight (AUL) or as experimental homebuilt. Quoted build time of standard CH 701 400 hours.

FLYING CONTROLS: Manual. Junkers-type flaperons; elevators; and Zenith all-moving fin. Full-span leading-edge slats for STOL performance.

STRUCTURE: All-metal semi-monocoque.

LANDING GEAR: Tricycle or tailwheel gear; optional floats, amphibious or ski gears.

POWER PLANT: One 37.0 kW (49.6 hp) Rotax 503 UL-2V engine in prototype, but 59.6 kW (79.9 hp) Rotax 912 UL, 47.8 kW (64.1 hp) Rotax 582 UL, 44.7 to 52.2 kW (60 to 70 hp) VW or 48.5 to 67.1 kW (65 to 90 hp) Teledyne Continental engine optional. Two- or three-blade propeller. Standard fuel capacity 75.7 litres (20.0 US gallons; 16.7 Imp gallons) in two wing tanks.

DIMENSIONS, EXTERNAL:
- Wing span .... 8.23 m (27 ft 0 in)
- Length overall .... 6.38 m (20 ft 11 in)
- Width, wings folded .... 2.06 m (6 ft 9 in)
- Height overall .... 2.62 m (8 ft 7 in)

DIMENSIONS, INTERNAL:
- Cabin max width .... 1.07 m (3 ft 6 in)

AREAS:
- Wings, gross .... 11.33 m^2 (122.0 sq ft)

Zenith STOL CH 701 displaying its short-field capability *(Paul Jackson)* ***NEW***/0567780

Zenith CH 701 STOL floatplane *(Paul Jackson)* 0527001

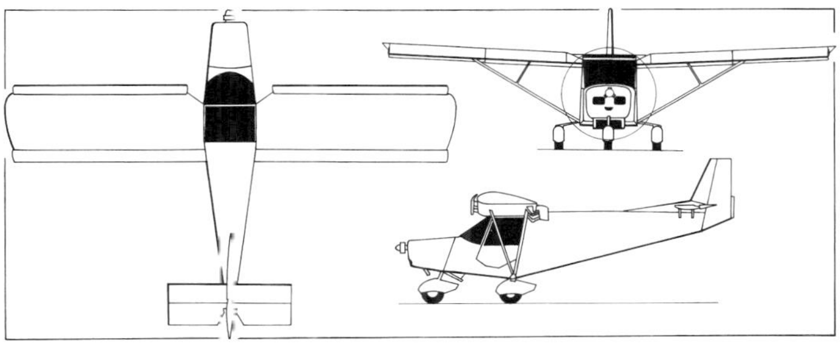

Zenith STOL CH 701 light aircraft *(Paul Jackson)* 0051860

WEIGHTS AND LOADINGS:

Weight empty .... 218 kg (480 lb)
Max T-O weight .... 499 kg (1,100 lb)

PERFORMANCE (47.8 kW; 64.1 hp engine):

Max level speed at S/L .... 82 kt (152 km/h; 95 mph)
Max cruising speed .... 70 kt (129 km/h; 80 mph)
Stalling speed .... 27 kt (49 km/h; 30 mph)
Max rate of climb at S/L .... 427 m (1,400 ft)/min
T-O run .... 28 m (90 ft)
Landing run .... 24 m (80 ft)
Range: standard fuel .... 182 n miles (338 km; 210 miles)
with wing tanks .... 360 n miles (667 km; 415 miles)

ZENITH SUPER STOL CH 801

TYPE: Four-seat kitbuilt.

PROGRAMME: Based on STOL CH 701 design; work started in 1988 for an undisclosed customer, but was suspended when order cancelled. Subsequently revived; prototype built by Flypass Ltd in Ontario, Canada, and first flown March 1998. Launched at Oshkosh, July 1998, with kit deliveries following.

In 2003, Jordan Aerospace Industries of Amman announced the Hawk-I CH801, a locally marketed and assembled CH 801.

CUSTOMERS: 150 kits sold and 20 flying by end 2002. Marketing by CZAW (Czech Republic), which offers builder's assistance programme.

COSTS: Kit excluding engine US$20,950 (2003).

DESIGN FEATURES: High-wing, strut-braced monoplane. Although visually similar to the CH 701, no commonality of parts. Quoted build time 700 hours.

Wing aerofoil NACA 65015.

FLYING CONTROLS: Manual. Full-span Junkers flaperons with outboard section offset 2° higher; deflection to 35° elevator; all-moving, horn-balanced fin. Fixed full-span leading-edge wing slats.

STRUCTURE: All-metal semi-monocoque construction, making extensive use of blind rivets. Single-spar 6061-T6 aluminium wing. Fuselage welded 4130 tubular steel frame.

LANDING GEAR: Tricycle type; cantilever mainwheel legs optimised for rough-field operations with Matco or Cleveland wheels and brakes. Nosewheel steerable ±15°. Tundra tyres on all wheels. Optional flotation gear.

POWER PLANT: One 134 kW (180 hp) Textron Lycoming O-360-A driving two-blade, fixed-pitch Sensenich 76-EMB-O-54 metal propeller in production prototype; engines in the range 112 to 179 kW (150 to 240 hp) can be fitted, including four-stroke LOM M337B of 173 kW (232 hp). Standard fuel capacity 151 litres (40.0 US gallons; 33.3 Imp gallons); optional fuel capacity 303 litres (80.0 US gallons; 66.6 Imp gallons).

ACCOMMODATION: Pilot and three passengers in enclosed cabin. Single door on port side. Underfuselage cargo pod optional.

DIMENSIONS, EXTERNAL:

Wing span .... 9.45 m (31 ft 0 in)
Wing chord, constant .... 1.60 m (5 ft 3 in)
Wing aspect ratio 5.8
Length overall .... 7.47 m (24 ft 6 in)
Height overall .... 2.67 m (8 ft 9 in)
Tailplane span .... 2.54 m (8 ft 4 in)
Tailplane chord .... 0.91 m (3 ft 0 in)
Wheel track .... 2.03 m (6 ft 8 in)
Wheelbase .... 1.65 m (5 ft 5 in)
Passenger door: Height .... 0.97 m (3 ft 2 in)
Width .... 0.91 m (3 ft 0 in)

DIMENSIONS, INTERNAL:

Cabin: Length .... 1.98 m (6 ft 6 in)
Max width .... 1.12 m (3 ft 8 in)
Max height .... 1.17 m (3 ft 10 in)

AREAS:

Wings, gross .... 15.51 m² (167.0 sq ft)
Tailplane .... 2.32 m² (25.00 sq ft)

Zenith Super STOL CH 801 (Textron Lycoming O-360A flat-four) 0527136

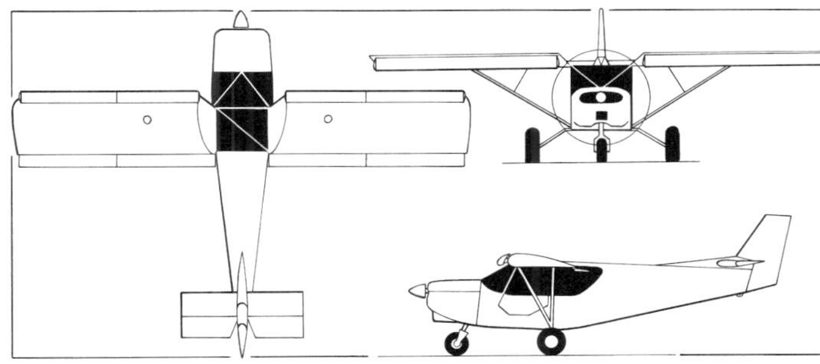

Zenith STOL CH 801 *(James Goulding)* 0051861

WEIGHTS AND LOADINGS:

Weight empty .... 522 kg (1,150 lb)
Max T-O weight .... 975 kg (2,150 lb)
Max wing loading .... 62.9 kg/m² (12.87 lb/sq ft)
Max power loading .... 7.27 kg/kW (11.94 lb/hp)

PERFORMANCE:

Never-exceed speed (VNE) .... 130 kt (241 km/h; 150 mph)
Max level speed .... 96 kt (177 km/h; 110 mph)
Max cruising speed at 75% power .... 91 kt (169 km/h; 105 mph)
Stalling speed: flaps up .... 48 kt (89 km/h; 56 mph)
flaps down .... 34 kt (63 km/h; 39 mph)
Max rate of climb at S/L .... 219 m (720 ft)/min
Service ceiling .... 4,267 m (14,000 ft)
T-O run .... 119 m (390 ft)
T-O to 15 m (50 ft) .... 122 m (400 ft)
Landing from 15 m (50 ft) .... 91 m (300 ft)
Landing run .... 46 m (150 ft)
Range:
with standard fuel .... 273 n miles (507 km; 315 miles)
with optional fuel .... 547 n miles (1,013 km; 630 miles)
Endurance: with standard fuel .... 3 h 0 min
with optional fuel .... 6 h 0 min
g limits .... +6/−3

ZENITH ZODIAC CH 640

TYPE: Four-seat kitbuilt.

PROGRAMME: Design work started 2 January 2001; prototype (N640Z) first flown 1 March 2001. Public debut at Sun 'n' Fun, April 2001.

CUSTOMERS: Three completed by August 2004.

COSTS: Standard kit US$24,800; quick-build kit introductory price US$35,190, both without engine, propeller, instruments, electrics and upholstery (2001). Can be purchased in smaller kit packages.

DESIGN FEATURES: Fuselage based on AMD CH 2000 with cabin rear wall moved to allow extra seating; 90 per cent commonality of parts. Influenced by other Zenith designs: area forward of firewall is based on similarly engined Zenith CH 801. However, wings are of new design, incorporating high-lift aerofoil. Quick-build kit includes almost-complete fuselage, wings, ailerons, flaps and tailplane; quoted build time 1,500 hours. Standard kit build time quoted as 1,500 hours. Meets FAA 51 per cent rule.

Wing section LS(1)0417 (mod).

FLYING CONTROLS: Manual. Single-piece all-flying tailplane and rudder. Electrically operated split flaps with maximum deflection of 30°. Anti-balance tab on both ailerons.

Zenith Zodiac CH 640 prototype *(Paul Jackson)* ***NEW***/1047157

STRUCTURE: Stressed-skin semi-monocoque all-metal fuselage of 6061-T6 aluminium sheet riveted to aluminium extrusions. All-metal stressed-skin wings consist of two panels bolted to fuselage spar-box assembly; Hoerner wingtips.

LANDING GEAR: Tricycle type; fixed. Cantilever main legs comprise single piece of 19 mm (¾ in) aluminium. Steerable nosewheel with heavy-duty bungee shock absorption. Cleveland 5.00×5 wheels standard throughout, with option to fit 6.00×6. Cleveland hydraulic single-disc brakes on mainwheels. Optional wheel fairings. Tailskid protects empennage in hard landings.

POWER PLANT: One 134 kW (180 hp) Lycoming IO-360-A1A driving a Sensenich 76-EM8-0-63 two-blade, fixed-pitch propeller; engines between 112 and 149 kW (150 and 200 hp) can be fitted. Standard fuel capacity 144 litres (38.0 US gallons; 31.6 Imp gallons) in two wing tanks forward of wing spar; optional capacity 174 litres (46.0 US gallons; 38.3 Imp gallons).

ACCOMMODATION: Two pairs of seats, designed to withstand loads of 27 *g* . Dual controls. Gull-wing cabin door on each side of fuselage. Roll-over protection bar of 4130N tubing between two forward seat backs is bolted to top of wing main spar.

SYSTEMS: 12 V battery and 14 V 70 A alternator standard. Auxiliary electric fuel pump.

AVIONICS: To customer's choice.

DIMENSIONS, EXTERNAL:

Wing span	9.60 m (31 ft 6 in)
Wing chord at root	1.46 m (4 ft 9½ in)
Wing aspect ratio	6.6
Length overall	7.01 m (23 ft 0 in)
Height overall	2.24 m (7 ft 4 in)
Tailplane span	3.00 m (9 ft 10 in)

DIMENSIONS, INTERNAL:

Cabin: Length	1.88 m (6 ft 2 in)
Max width	1.17 m (3 ft 10 in)
Height, front seat to roof	0.96 m (3 ft 2 in)

AREAS:

Wings, gross	13.94 m² (150.0 sq ft)
Tailplane	2.79 m² (30.0 sq ft)

WEIGHTS AND LOADINGS:

Weight empty	520 kg (1,147 lb)
Max T-O weight	998 kg (2,200 lb)
Max wing loading	71.6 kg/m² (14.66 lb/sq ft)
Max power loading	7.44 kg/kW (12.22 lb/hp)

PERFORMANCE:

Never-exceed speed (VNE)	139 kt (257 km/h; 160 mph)
Max operating speed	136 kt (253 km/h; 157 mph)
Normal cruising speed at 75% power at FL70	130 kt (241 km/h; 150 mph)
Stalling speed: flaps up	51 kt (94 km/h; 58 mph)
flaps down	41 kt (76 km/h; 47 mph)
Max rate of climb at S/L	290 m (950 ft)/min
Service ceiling	3,900 m (12,800 ft)
T-O run	290 m (950 ft)
Landing run	229 m (1,750 ft)
Range with standard fuel	443 n miles (820 km; 510 miles)
Endurance	3 h 45 min

UPDATED

ZIVKO

ZIVKO AERONAUTICS INC

502 Airport Road, Building 11, Guthrie, Oklahoma 73044
Tel: (+1 405) 282 13 30
Fax: (+1 405) 282 13 39
Web: http://www.zivko.com
PRESIDENT: Bill Zivko

Founded in 1987, Zivko manufactures the Edge 540 series of aerobatic monoplanes from a 1,858 m² (20,000 sq ft) factory on the airport at Guthrie. In August 2000, it received FAA certification under Pt 21.191(g). The improved Edge 540A was introduced in April 2001. In mid-2002 the workforce numbered 20.

ZIVKO EDGE 540

TYPE: Aerobatic single-seat sportplane; aerobatic two-seat sportplane.

PROGRAMME: Initial version was Zivko Edge 360. Design work on 540 began in 1992 with first flight in 1993; 540T followed in 1999 with first flight 2000.

CURRENT VERSIONS: **540:** Single-seat version.

540A: Improved version introduced at Sun 'n' Fun, April 2001 (N24KC); has larger rudder and elevator for improved handling. 540AK version may have IO-540-K engine.

540T: Two-seat model introduced March 2000; two are TK versions.

CUSTOMERS: Totals of 23 540s and seven 540Ts built by end of 2003, including examples to Mexico, South Africa, Switzerland and Australia. Three 540Ts registered in 2003; no 540/As.

COSTS: Programme cost US$1.1 million by 2003. Edge 540 US$231,655; Edge 540T US$245,632 (2003).

DESIGN FEATURES: Optimised for aerobatics; customised John Roncz aerodynamic design; rate of roll 420°/s. Mid-positioned, tapered wing with straight leading-edge and large, horn-balanced ailerons and unique aerofoil. Transparencies in lower sides and floor of cockpit.

FLYING CONTROLS: Conventional and manual. Ailerons and elevator actuated by pushrods; horn-balanced rudder by cables. Flight adjustable tabs in port elevator. Suspended, spade-type tab on each aileron; electric trim.

STRUCTURE: 4130 Steel tube fuselage; upper fuselage and fairings of composites; lower aft fuselage fabric covered, stressed to ultimate ±15 *g*; all-composites fully cantilevered two-spar wing. Fully composites empennage with no wire bracing.

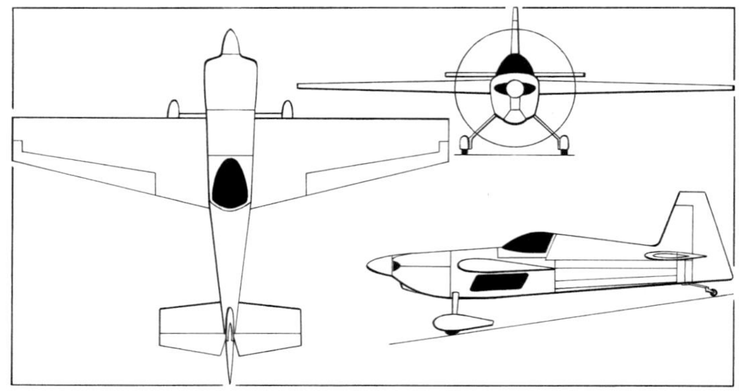

Single-seat version of Zivko Edge 540 *(James Goulding)* 0530164

LANDING GEAR: Tailwheel type; fixed. Main legs aluminium sprung steel; Cleveland 5.00–5 wheels and hydraulic brakes; optional speed fairings. Tyre pressure 3.45 bar (50.0 lb/sq in). Spring steel tail gear, with chrome-plated, solid 4.00–4 tailwheel, choice of fixed or steerable.

POWER PLANT: One Textron Lycoming IO-540 modified to produce 254 kW (340 hp), driving Hartzell HC-C3YR-4AX or optional Hartzell HC-C3YR-1AX1 three-blade, composites propeller. Edge 540 has single 72 litre (19.0 US gallon; 15.8 Imp gallon) forward fuselage fuel tank, plus optional total of 159 litres (42.0 US gallons; 35.0 Imp gallons) in wing tanks; Edge 540T has single 76 litre (20.0 US gallon; 16.6 Imp gallon) fuselage tank and two optional 83 litre (22.0 US gallon; 18.3 Imp gallon) wing tanks. Oil capacity 11.4 litres (3.0 US gallons; 2.5 Imp gallons). Inverted oil (Christen) and fuel (ZAI) systems.

ACCOMMODATION: Pilot (pilot and passenger in tandem in 540T) under one-piece perspex canopy hinged on starboard side; carbon fibre contoured seats. Baggage compartment forward of front seat.

SYSTEMS: 12 V DC system with light weight B&C 8 A alternator and regulator. Bendix fuel injection system.

AVIONICS: *Comms:* Becker com and Mode C transponder.
Flight: Optional Sierra Flight Systems EFIS.

Zivko Edge 540A operated in Australia ***(Paul Jackson)*** ***NEW***/0567770

Zivko Edge 540T two-seat aerobatic lightplane *(Paul Jackson)* ***NEW***/0567745

DIMENSIONS, EXTERNAL:
- Wing span: 540 ... 7.42 m (24 ft 4 in)
  - 540T ... 7.67 m (25 ft 2 in)
- Wing chord:
  - at root: 540 ... 1.57 m (5 ft 2 in)
    - 540T ... 1.68 m (5 ft 6 in)
  - at tip: 540 ... 0.74 m (2 ft 5 in)
    - 540T ... 0.84 m (2 ft 9 in)
- Length overall: 540 ... 6.27 m (20 ft 7 in)
  - 540T ... 6.91 m (22 ft 8 in)
- Height overall ... 2.36 m (7 ft 9 in)
- Tailplane span ... 2.46 m (8 ft 1 in)

AREAS:
- Wings, gross: 540 ... 9.10 m^2 (98.0 sq ft)
  - 540T ... 9.87 m^2 (106.2 sq ft)

WEIGHTS AND LOADINGS:
- Weight empty: 540 ... 531 kg (1,170 lb)
  - 540T ... 562 kg (1,240 lb)
- Baggage capacity ... 5 kg (10 lb)
- Max T-O weight:
  - 540: Aerobatic ... 703 kg (1,550 lb)
    - Utility ... 816 kg (1,800 lb)
  - 540T: Aerobatic ... 726 kg (1,600 lb)
    - Utility ... 885 kg (1,950 lb)
- Max wing loading:
  - 540: Aerobatic ... 77.2 kg/m^2 (15.82 lb/sq ft)
    - Utility ... 89.7 kg/m^2 (18.37 lb/sq ft)
  - 540T: Aerobatic ... 73.6 kg/m^2 (15.07 lb/sq ft)
    - Utility ... 89.6 kg/m^2 (18.36 lb/sq ft)
- Max power loading: 540 ... 3.22 kg/kW (5.29 lb/hp)
  - 540T: Aerobatic ... 2.86 kg/kW (4.71 lb/hp)
    - Utility ... 3.49 kg/kW (5.74 lb/hp)

PERFORMANCE:
- Never-exceed speed (VNE) ... 230 kt (426 km/h; 265 mph)
- Max cruising speed at 75% power:
  - 540 ... 180 kt (333 km/h; 207 mph)

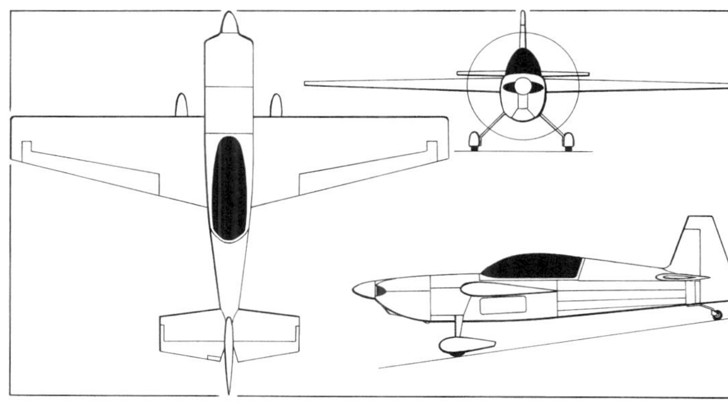

Two-seat Zivko Edge 540T *(James Goulding)* 0572340

  - 540T ... 185 kt (343 km/h; 213 mph)
- Stalling speed:
  - 540: Aerobatic ... 51 kt (95 km/h; 57 mph)
    - Utility ... 61 kt (113 km/h; 71 mph)
  - 540T: Aerobatic ... 53 kt (99 km/h; 61 mph)
    - Utility ... 61 kt (113 km/h; 71 mph)
- Max rate of climb at S/L: 540 ... 1,128 m (3,700 ft)/min
  - 540T ... 1,189 m (3,900 ft)/min
- Service ceiling ... 3,810 m (12,500 ft)
- T-O run ... 61 m (199 ft)
- T-O to 15 m (50 ft) ... 90 m (293 ft)
- Landing from 15 m (50 ft) ... 300 m (982 ft)
- Range with max fuel ... 456 n miles (844 km; 525 miles)
- *g* limits: Aerobatic ... ±10
  - Utility ... ±8

Uzbekistan

TAPO

TASHKENTSKOYE AVIATSIONNOYE PROIZDSTVENNOYE OBEDINENIE IMENI V P CHKALOVA (TASHKENT AIRCRAFT PRODUCTION ENTERPRISE NAMED FOR VALERY P CHKALOV)

ulitsa Elbek 61, 700016 Tashkent
Tel: (+810 99871) 136 11 67
Fax: (+810 99871) 268 03 18
e-mail: oms@tapo.tashkent.ru
GENERAL DIRECTOR: Vladimir P Kucherov
DEPUTY GENERAL DIRECTOR: Nabil A Artykov
TECHNICAL DIRECTOR: Nikolai A Kuznetsov
MARKETING DIRECTOR: Zafar M Islamov

Tashkent plant established 20 November 1941 on arrival of first elements of GAZ 84 (formed 1932 and incorporating Polikarpov from 1936) evacuated from Khimki, Moscow. Enterprise, dating from 1972 in current form, is said to be the largest aircraft manufacturing centre on the Asian continent. It comprises five plants at Tashkent, Andizhan and Fergana. Total of over 7,000 aircraft built, including I-15s, I-16s, I-153s, Li-2s (total 2,258), Il-14s, An-8s, Ka-22s, An-12s, An-22s and Il-76s. It has manufactured wing centre-sections for the An-124 and An-225 transports and two An-70 prototypes, and has repaired MiG and Sukhoi fighters and Tu-22M bombers. It developed the Il-78 tanker version and collaborated with Beriev on the A-50 AEW derivative of the Il-76. Low-rate production is under way of the Ilyushin Il-114 twin-turboprop transport and Il-76, for which TAPO is the sole source, although airframes are stored partly complete until orders are received. Final report of State Tests of Il-76MF stretched version issued 24 April 2003, paving way for manufacture, although Russia plans own assembly line. Il-114s retrofitted with US engines, as required by demand.

TAPO was registered as a joint stock company in May 1996; the state holds 82.7 per cent, National Bank of Foreign Economic Activity 6.7 per cent and employees 10.6 per cent. An attempt in 1997 to sell 14 per cent of the company was unsuccessful; by mid-1998, offers were being invited for a 25 per cent share, but no acceptances have been reported. Company began airbrake manufacture for Avro RJ in 1997; this now discontinued. Following April 2000 foundation of Ilyushin International company, TAPO expected to join forces with design bureau responsible for its current production aircraft. Gained ISO 9002 certification in 1998 and Boeing D1-9000 subcontractor approval in 2000.

Subsidiary, TAPO-Avia operates commercial freight charter flights with fleet of four An-12s, one An-24 and five Il-76MDs. TAPO aircraft are marketed via Uzavialeasing, established December 1997 in collaboration with local government and Russian and Uzbek banks.

Ilyushin Il-76 awaiting completion at TAPO ***NEW***/0589375

UPDATED

Vietnam

VMA

VIETNAM MECHANICS ASSOCIATION

Ho Chi Minh City
PRESIDENT: Nguyen Van Dao
PROJECT MANAGER: Nguyen Xuan Hung

At a press conference on 23 August 2003, Professor Nguyen Van Dao revealed a model of an ultralight aeroplane then under construction by the VAM. Reports published in the West initially described the aircraft as a helicopter.

VAM expected to begin construction of another lightplane in December 2003. This is understood to be the VAM-, a VAM-1 variant with local content increased from 20 per cent to 70 per cent. A four-seat aircraft, believed designated VAM-3, is planned to have 90 per cent local content.

UPDATED

Model of VMA Ultralight ***NEW***/0561759

VAM VAM-1

TYPE: Tandem-seat ultralight.
PROGRAMME: Vietnamese adaptation of Beaver RX550, marketed in Canada by Aircraft Sales and Parts (http://www.ultralight.ca). Construction of prototype began April 2003, following granting of government approval on 18 April and donation of private financial resources. Announced August 2003, when first flight predicted in following month; however maiden sortie delayed aircraft made first fast taxying runs at Phuoc Long Airport, Binh Phuoc province, on 28 March 2004, following which an application for permission to fly was lodged. Actual first flight date not known.
CUSTOMERS: Requirement for 120 to be imported into the USA by expatriate Vietnamese partners Ngyuen Sang and Tim Trung Tran, who are investing US$5 million in the project.
COSTS: US$45,000 (2004).
DESIGN FEATURES: Pod-and-boom configuration with high-mounted, V strut-braced, constant-chord, slightly sweptback wings.
FLYING CONTROLS: Conventional and manual. Full-span ailerons. No aerodynamic balances on tail surfaces. Boom-mounted empennage with sweptback leading-edges.
STRUCTURE: Aluminium 6061-T6 tube with fabric covering.
POWER PLANT: Unspecified piston engine, believed 48.0 kW (64.4 hp) Rotax 582, driving a two-blade, pusher propeller.
LANDING GEAR: Tricycle type; fixed. Mainwheels mounted on hinged horizontal, metal tube Vs with damped compression arm attached at wing strut junction with central boom.
DIMENSIONS, EXTERNAL:
Wing span .... 9.75 m (32 ft 0 in)
Length overall .... 6.30 m (20 ft 8 in)
Height overall .... 2.03 m (6 ft 8 in)
WEIGHTS AND LOADINGS:
Weight empty .... 150 kg (330 lb)

Beaver RX550, basis of VAM-1 *(Capital Air Sports)* ***NEW***/1047705

Max T-O weight .... 470 kg (1,035 lb)
PERFORMANCE (estimated):
Max level speed .... 71 kt (132 km/h; 82 mph)
T-O run .... 70 m (230 ft)
Service ceiling .... 2,500 m (8,200 ft)

UPDATED

AIR-LAUNCHED MISSILES

Raytheon Maverick launched from an AV-8B ***NEW***/0567252

The following pages support the Armament paragraphs of aircraft descriptions in this book by explaining in brief how the potential of an individual aircraft is enhanced by its missile armament. Coverage is restricted to missiles carried by, or applicable to, aircraft in the current edition; for this reason, some older missiles are excluded, as are future projects still in the early stages of definition. Contents do include certain anti-tank and shoulder-launched anti-aircraft missiles which have airborne applications, mostly on helicopters.

To expedite retrieval of data, missiles are listed in alphabetical order of name or designation, with full cross-references to alternative epithets. In many instances, the 'manufacturer' of Chinese and Russian missiles is actually the export sales agency. More detailed information is contained in *Jane's Air-Launched Weapons*.

KEY

Roles

AAM	Air-to-air missile
ARM	Anti-radiation missile
ASM	Air-to-surface missile
AshM	Anti-ship missile
ATM	Anti-tank missile
LGB	Laser-guided bomb
SOM	Standoff missile

Guidance

AL	Active laser	LR	Laser radar
AP	Autopilot	MMW	Millimetric-wave
A/P	Active/passive radar	PR	Passive radar
ARH	Active radar homing	RC	Radio command
GPS	Global Positioning System	RF	Radio frequency
I	Inertial	SAL	Semi-active laser
Im	Imaging	SARH	Semi-active radar homing
IR	Infra-red	T	Terrain reference
L	Laser	TV	Television

Name/designation	Role	Manufacturer/country	Length m (in)	Diameter m (in)	Weight kg (lb)	Guidance	Range n miles (km)
3M55 see 'Yakhont'							
3M60 see 'Kayak'							
3M80 see Moskit							
9M14 see 'Sagger'							
9M17 see 'Swatter'							
9M32 see 'Grail'							
9M36 see 'Gremlin'							
9M39 see 'Grouse'							
9M114 see 'Spiral'							
9M120 see 'Spiral' and AT-X-16							
9M121 see AT-X-16							
9M313 see 'Gimlet'							
AA-6 see 'Acrid'							
AA-7 see 'Apex'							
AA-8 see 'Aphid'							
AA-9 see 'Amos'							
AA-10 see 'Alamo'							
AA-11 see 'Archer'							
AA-12 see 'Adder'							
AA-X-13 see R-37							
AAM-3, Type 90	AAM	Mitsubishi/Japan	3.00 (118)	0.13 (5)	91 (201)	IR	4.3 (8)
AAM-4	AAM	Mitsubishi/Japan				ARH	medium
AAM-5	AAM	not assigned/Japan				ImIR	
AAM-L, KS-172	AAM	Novator/Russia	7.40 (291)		750 (1,653)	I/ARH	216 (400)
AARGM	ARM	SAT & NAWC/USA				GPS/MMW	100 (185)

Name/designation	Role	Manufacturer/country	Length m (in)	Diameter m (in)	Weight kg (lb)	Guidance	Range n miles (km)
ACM, AGM-129	ASM	Raytheon/USA	6.35 (250)	0.70 (28)	1,250 (2,756)	I/LR	1,620 (3,000)
'Acrid', AA-6/R-46TD	AAM	Vympel/Russia	6.20 (244)	0.36 (14)	467 (1,030)	I/IR	27 (50)
A-Darter, V3E	AAM	Denel/South Africa	2.98 (117)	0.17 (7)	89 (196)	AL	11 (20)
'Adder', AA-12/R-77	AAM	Vympel/Russia	3.60 (142)	2.00 (79)	175 (386)	I/ARH	40 (75)
'Adder', R-77M-PD	AAM	Vympel/Russia	3.60 (142)	2.00 (79)	225 (496)	I/ARH	81 (150)
AFDS (unpowered)	SOM	LFK/Germany	3.47 (137)	0.63 (25)	660 (1,455)	GPS/I	
AGM-65 see Maverick							
AGM-78 see Standard							
AGM-84 see Harpoon/SLAM							
AGM-88 see HARM							
AGM-114 see Hellfire							
AGM-119 see Penguin							
AGM-122 see Sidearm							
AGM-123 see Skipper							
AGM-129 see ACM							
AGM-130A	LGB	Boeing/USA	3.94 (155)	0.46 (18)	1,323 (2,917)	TV, ImIR	24 (45)
AGM-130C	LGB	Boeing/USA	3.95 (156)	0.46 (18)	1,353 (2,983)	TV/IIR	24 (45)
AGM-142 see Popeye							
AGM-154C JSOW	SOM	Raytheon/USA	4.26 (168)		475 (1,047)/ 680 (1,499)	I/GPS	32 (60)
AGM-158 see JASSM							
AIM-7 see Sparrow							
AIM-9 see Sidewinder							
AIM-120 see AMRAAM							
AIM-132 see ASRAAM							
'Alamo', AA-10/R-27AE	AAM	Vympel/Russia	4.78 (188)	0.26 (10)	350 (772)	I/ARH	43 (80)
'Alamo', AA-10/R-27EM	AAM	Vympel/Russia	4.78 (188)	0.26 (10)	350 (772)	I/SARH	59 (110)
'Alamo', AA-10/R-27ER	AAM	Vympel/Russia	4.70 (185)	0.26 (10)	350 (772)	I/SARH	40 (75)
'Alamo', AA-10/R-27ET	AAM	Vympel/Russia	4.50 (177)	0.26 (10)	343 (756)	I/IR	38 (70)
'Alamo', AA-10/R-27R	AAM	Vympel/Russia	4.00 (157)	0.23 (9)	253 (558)	I/SARH	27 (50)
'Alamo', AA-10/R-27T	AAM	Vympel/Russia	3.70 (146)	0.23 (9)	254 (560)	I/IR	22 (40)
ALARM	ARM	MBDA/International	4.30 (169)	0.22 (9)	265 (584)	PR	24 (45)
AM 39 see Exocet							
'Amos' AA-9/R-33	AAM	Vympel/Russia	4.15 (163)	0.38 (15)	490 (1,080)	I/SARH	65 (120)
AMRAAM, AIM-120A/B/C	AAM	Raytheon/USA	3.65 (144)	0.18 (7)	157 (346)	I/ARH	27 (50)
APACHE AP	SOM	MBDA/International	5.10 (201)	0.63 (25)	1,230 (2,712)	I/ARH	76 (140)
'Apex', AA-7/R-24R	AAM	Vympel/Russia	4.46 (176)	0.20 (8)	235 (518)	SARH	27 (50)
'Apex', AA-7/R-24T	AAM	Vympel/Russia	4.16 (164)	0.20 (8)	235 (518)	IR	27 (50)
'Aphid', AA-8/R-60	AAM	Vympel/Russia	2.08 (82)	0.13 (5)	65 (143)	IR	1.6 (3)
'Aphid', AA-8/R-60M	AAM	Vympel/Russia	2.08 (82)	0.13 (5)	65 (143)	AL	2.7 (5)
'Archer', AA-11/R-73M1	AAM	Vympel/Russia	2.90 (114)	0.17 (7)	105 (231)	I/IR	11 (20)
'Archer', AA-11/R-73M2	AAM	Vympel/Russia	2.90 (114)	0.17 (7)	110 (243)	I/IR	16 (30)
ARMAT	ARM	MBDA/International	4.15 (163)	0.40 (16)	550 (1,213)	I/PR	49 (90)
Armiger	ARM	BGT/Germany	3.90 (153)	0.20 (8)	230 (507)	PR/ImIR	108 (200)
AS-7 see 'Kerry'							
AS-9 see 'Kyle'							
AS-10 see 'Karen'							
AS-11 see 'Kilter'							
AS-12 see 'Kegler'							
AS-13 see 'Kingbolt'							
AS-14 see 'Kedge'							
AS-15 see 'Kent'							
AS 15TT	AShM	MBDA/International	2.30 (91)	0.18 (7)	96 (212)	Radio	8.1 (15)
AS-16 see 'Kickback'							
AS-17 see 'Krypton'							
AS-18 see 'Kazoo'							
AS-19 see 'Koala'							
AS-20 see 'Kayak'							
AS-30L	ASM	MBDA/International	3.65 (144)	0.34 (13)	520 (1,146)	I/SAL	5.4 (10)
ASM-1, Type 80	AShM	Mitsubishi/Japan	4.00 (157)	0.34 (13)	600 (1,323)	I/ARH	27 (50)
ASM-1C, Type 91	AShM	Mitsubishi/Japan	4.00 (157)	0.35 (14)	510 (1,124)	I/ARH	35 (65)
ASM-2, Type 93	AShM	Mitsubishi/Japan	4.10 (161)	0.35 (14)	520 (1,146)	I/ImIR	54 (100)
ASMP	SOM	MBDA/International	5.38 (212)	0.38 (15)	860 (1,896)	I/T	135 (250)
ASMP Plus (ASMP-A)	SOM	MBDA/International					270 (500)
Aspide 1	AAS	MBDA/International	3.70 (146)	0.20 (8)	220 (485)	SARH	19 (35)
ASRAAM (AIM-132)	AAM	MBDA/International	2.90 (114)	0.17 (7)	87 (192)	ImIR	11 (20)
AT-2 see 'Swatter'							
AT-3 see 'Sagger'							
AT-6 see 'Spiral'							
AT-9 see 'Spiral 2'							
AT-12 see 'Swinger'							

AA-12 'Adder' AAM *(Paul Jackson)* 0121705

MBDA ALARM anti-radar weapon *(Paul Jackson)* 0121684

Name/designation	Role	Manufacturer/country	Length m (in)	Diameter m (in)	Weight kg (lb)	Guidance	Range n miles (km)
AT-16, 9M120M/9M121 Vikhr M	ATM	Shipunov/Russia	2.80 (110)	0.13 (5)	45 (99)	SAL	5.4 (10)
Ataka see 'Swinger'							
ATAM see Mistral							
ATASK see Helstreak							
BGM-71 see TOW							
Black Shaheen see SCALF EG							
Brimstone	ASM	MBDA/International	1.63 (64)	0.18 (7)	50 (110)	MMW/I	4.3 (8)
Burya see 'Kitchen'							
C-101, YJ-16	AShM	CPMIEC/China	7.50 (295)	0.54 (21)	1,850 (4,079)	I/ARH	24 (45)
C-201, HY-4	AShM	CPMIEC/China	7.36 (290)	0.76 (30)	1,740 (3,836)	AP/ARH	73 (135)
C-601, CAS-1 'Kraken'/YJ-6	AShM	CPMIEC/China	7.36 (290)	0.76 (30)	2,440 (5,379)	AP/ARH	54 (100)
C-701	AShM	CPMIEC/China	2.51 (99)	0.54 (21)	100 (220)	I/TV	8.1 (15)
C-801, YJ-1	AShM	CPMIEC/China	4.65 (183)	0.36 (14)	655 (1,444)	I/ARH	27 (50)
C-802, YJ-2	AShM	CPMIEC/China	6.40 (252)	0.36 (14)	715 (1,576)	I/ARH	70 (130)
CAS-1 see C-601							
Dandy see NT-D							
Darter, V-3C (see U-Darter)	AAM	Denel/South Africa	2.75 (108)	0.16 (6)	90 (198)	IR	2.7 (5)
Derby	AAM	Rafael/Israel	3.62 (143)	0.16 (6)	118 (260)	ARH	63 (34)
DWS24/DWS39 (unpowered)	SOM	LFK/Germany	3.50 (138)		600 (1,323)	I	5.4 (10)
Exocet, AM 39	AShM	MBDA/International	4.70 (185)	0.35 (14)	670 (1,477)	I/ARH	27 (50)
FIM-92 see Stinger							
Gabriel 3AS	AShM	IAI/Israel	3.85 (152)	0.34 (13)	560 (1,235)	I/ARH	19 (35)
Gabriel 4LR	AShM	IAI/Israel	4.70 (185)	0.44 (17)	960 (2,116)	I/ARH	108 (200)
'Gimlet', SA-16/9M313 Igla 1	AAM	Kolomna/Russia	1.69 (67)	0.07 (3)	11 (24)	IR	2.7 (5)
'Grail' SA-7/9M32 Strela 2	AAM	Turopov/Russia	1.22 (48)	0.07 (3)	10 (22)	IR	2.7 (5)
'Gremlin', SA-14/9M36 Strela 3	AAM	Turopov/Russia	1.47 (58)	0.07 (3)	11 (24)	IR	2.7 (5)
'Grouse', SA-18/9M39 Igla	AAM	Kolomna/Russia	1.69 (67)	0.07 (3)	11 (24)	IR	2.7 (5)

SA-18 Grouse AAM *(Paul Jackson)* 0105006

BGT IRIS-T AAM ***NEW***/0527138

Hakeem see PGM							
HARM, AGM-88A/B/B+/C/D	ARM	Raytheon/USA	4.17 (164)	0.25 (10)	361 (796)	PR	43 (80)
Harpoon, AGM-84A	AShM	Boeing/USA	3.90 (154)	0.34 (13)	530 (1,168)	I/ARH	65 (120)
Have Lite see Popeye 2							
Have Nap see Popeye 1							
Hellfire, AGM-114A	ATM	Hellfire Systems LLC/ USA	1.63 (64)	0.18 (7)	46 (101)	SAL	4.3 (8)
Hellfire, AGM-114B/C	ATM	Hellfire Systems LLC/ USA	1.73 (68)	0.18 (7)	48 (106)	SAL, ImIR or RF+IR	4.3 (8)
Hellfire, AGM-114F	ATM	Hellfire Systems LLC/ USA	1.80 (71)	0.18 (7)	49 (108)	SAL	4.3 (8)
Hellfire 2, AGM-114K	ATM	Hellfire Systems LLC/ USA	1.63 (64)	0.18 (7)	46 (101)	SAL	4.9 (9)
Hellfire 2, AGM-114 Longbow	ATM	Hellfire Systems LLC/ USA	1.78 (70)	0.18 (7)	50 (110)	MMW/I	4.3 (8)
Helstreak/ATASK (Starstreak)	AAM	Shorts/UK	1.40 (55)	0.13 (5)	16 (35)	RC	3.2 (6)
HJ-8A	ATM	Norinco/China	0.88 (35)	0.12 (5)	11 (24)	Wire	1.6 (3)
HJ-8B	ATM	Norinco/China	1.00 (39)	0.12 (5)	13 (29)	Wire	2.2 (4)
HOT 1	ATM	MBDA/International	1.27 (50)	0.14 (6)	24 (53)	Wire	2.2 (4)
HOT 2	ATM	MBDA/International	1.30 (51)	0.15 (6)	24 (53)	Wire	2.2 (4)
HOT 3	ATM	MBDA/International	1.30 (51)	0.15 (6)	33 (73)	Wire	2.2 (4)
Hsiung Feng 2	AShM	Chung Shan/Taiwan	3.90 (154)	0.34 (13)	520 (1,146)	I/ARH+ImIR	43 (80)
HY-4 see C-201							
Igla see 'Grouse' and 'Gimlet'							
Ingwe, ZT35	ATM	Denel/South Africa	1.75 (69)	0.40 (16)	29 (64)	L	2.7 (5)
IRIS-T	AAM	BGT/Germany	3.00 (118)	0.13 (5)	87 (192)	ImIR	6.5 (12)
JASSM, AGM-158	ASM	Lockheed Martin/USA	4.26 (168)	0.55 (22)	1,024 (2,257)	ImIR/GPS/INS	200 (370)
JSOW see AGM-154							
'Karen', AS-10/Kh-25MR	ASM	Zvezda/Russia	4.04 (159)	0.28 (11)	300 (661)	RC	5.4 (10)
'Karen', AS-10/Kh-25ML	ASM	Zvezda/Russia	4.04 (159)	0.28 (11)	300 (661)	SAL	11 (20)
'Kayak', AS-20/Kh-35/3M50 Uran	AShM	Zvezda/Russia	3.75 (148)	0.42 (17)	480 (1,058)	I/ARH	70 (130)
'Kazoo', AS-18/Kh-59M Ovod M	ASM	Raduga/Russia	5.37 (211)	0.38 (15)	930 (2,050)	I/TV	62 (115)
'Kedge', AS-14/Kh-29L	ASM	Vympel/Russia	3.87 (152)	0.38 (15)	657 (1,448)	SAL	5.4 (10)

Name/designation	Role	Manufacturer/country	Length m (in)	Diameter m (in)	Weight kg (lb)	Guidance	Range n miles (km)
'Kedge', AS-14/Kh-29T	ASM	Vympel/Russia	3.87 (152)	0.38 (15)	670 (1,477)	TV	6.5 (12)
'Kedge', AS-14/Kh-29TE	ASM	Vympel/Russia	3.87 (152)	0.38 (15)	700 (1,543)	TV	16 (30)

Raytheon AGM-154 JSOW
NEW/0527130

Kh-29L/AS-14 'Kedge' *(Paul Jackson)* *NEW*/0567248

Name/designation	Role	Manufacturer/country	Length m (in)	Diameter m (in)	Weight kg (lb)	Guidance	Range n miles (km)
'Kegler', AS-12/Kh-25MP/Kh-27	ASM	Zvezda/Russia	4.36 (172)	0.28 (11)	310 (683)	I/PR	22 (40)
'Kent', AS-15A/Kh-55/RKV-500	SOM	Raduga/Russia	6.04 (238)	0.51 (20)	1,400 (3,086)	I/T	1,296 (2,400)
'Kent', AS-15B/Kh-55SM/ RKV-500M	SOM	Raduga/Russia	7.10 (280)	0.77 (30)	1,700 (3,748)	I/T	1,620 (3,000)
KEPD 150 (PDWS 200)	SOM	Taurus Systems/ International	4.50 (177)		1,060 (2,337)	I/GPS/T/ImIR	81 (150)
KEPD 350 (Taurus)	SOM	Taurus Systems/ International	5.00 (197)		1,400 (3,086)	I/GPS/T/ImIR	189 (350)
'Kerry', AS-7/Kh-23	ASM	Zvezda/Russia	3.53 (139)	0.28 (11)	287 (633)	SAL or RC	2.7 (5)
Kh-15 see 'Kickback'							
Kh-22 see 'Kitchen'							
Kh-23 see 'Kerry'							
Kh-25 see 'Karen' and 'Kegler'							
Kh-27 see 'Kegler'							
Kh-28 see 'Kyle'							
Kh-29 see 'Kedge'							
Kh-31 see 'Krypton'							
Kh-35 see 'Kayak'							
Kh-41 see Moskit							
Kh-55/65 see 'Kent'							
Kh-58 see 'Kilter'							
Kh-59 see 'Kingbolt'							
Kh-101	SOM	Raduga/Russia	7.45 (293)		2,400 (5,291)	I/Im	2,700 (5,000)
'Kickback', AS-16/Kh-15/RKV-500B	ASM	Raduga/Russia	4.78 (188)	0.46 (18)	1,200 (2,646)	I/PR or I/ARH	81 (150)
'Kilter' AS-11/Kh-58E	ARM	Raduga/Russia	5.00 (197)	0.38 (15)	650 (1,433)	I/PR	86 (160)
'Kingbolt', AS-13/Kh-59 Ovod	ASM	Raduga/Russia	5.40 (213)	0.38 (15)	850 (1,874)	TV	86 (160)
Kokon see 'Spiral'							
'Kraken' see C-601							
Krypton', AS-17/Kh-31A-1	ASM	Zvezda/Russia	4.70 (185)	0.36 (14)	600 (1,323)	I/ARH	27 (50)
'Krypton', AS-17/Kh-31A-2	ASM	Zvezda/Russia	5.23 (206)	0.36 (14)	600 (1,323)	I/ARH	38 (70)
'Krypton', AS-17/Kh-31P-1	ASM	Zvezda/Russia	4.70 (185)	0.36 (14)	600 (1,323)	I/PR	54 (100)
'Krypton', AS-17/Kh-31P-2	ASM	Zvezda/Russia	5.23 (206)	0.36 (14)	600 (1,323)	I/PR	108 (200)
Kukri, V-3B	AAM	Denel/South Africa	2.94 (116)	0.13 (5)	73 (161)	IR	2.2 (4)
'Kyle', AS-9/Kh-28	ARM	Zvezda/Russia	6.00 (236)	0.43 (17)	715 (1,576)	PR	49 (90)
LY-60 see PL-11							
MAA-1 Piranha/Mol	AAM	Orbita/Brazil	2.82 (111)	0.15 (6)	90 (198)	IR	2.7 (5)
Magic 1, R 550	AAM	MBDA/International	2.72 (107)	0.16 (6)	89 (196)	IR	1.6 (3)
Magic 2, R 550	AAM	MBDA/International	2.75 (108)	0.16 (6)	89 (196)	IR	11 (20)
Marte 2	AShM	MBDA/International	4.80 (189)	0.32 (13)	340 (750)	I/ARH	11 (20)
Marte 2A	AShM	MBDA/International	3.90 (154)	0.32 (13)	269 (593)	I/AR	16 (30)
Marte 2B	ARM	MBDA/International	3.90 (154)	0.32 (13)	269 (593)	PR	32 (60)

MBDA Marte on NH 90 *NEW*/0567760

Mokopa anti-tank missile *(Paul Jackson)* 0121685

Name/designation	Role	Manufacturer/country	Length m (in)	Diameter m (in)	Weight kg (lb)	Guidance	Range n miles (km)
Maverick, AGM-65A	ASM	Raytheon/USA	2.49 (98)	0.31 (12)	210 (463)	TV	1.6 (3)
Maverick, AGM-65B	ASM	Raytheon/USA	2.49 (98)	0.31 (12)	210 (463)	TV	4.3 (8)
Maverick, AGM-65D	ASM	Raytheon/USA	2.49 (98)	0.31 (12)	220 (485)	ImIR	11 (20)
Maverick, AGM-65E	ASM	Raytheon/USA	2.49 (98)	0.31 (12)	293 (646)	SAL	11 (20)
Maverick, AGM-65F/G/J/K	ASM	Raytheon/USA	2.49 (98)	0.31 (12)	307 (677)	ImIR	13 (25)
Maverick, AGM-65H	ASM	Raytheon/USA	2.60 (102)	0.31 (12)	305 (672)	ARH	13 (25)
Meteor	AAM	MBDA/International	3.65 (144)		185 (408)	ARH	54 (100)
MICA	AAM	MBDA/International	3.10 (122)	0.16 (6)	110 (243)	I/ARH or IR	27 (50)
Mistral, ATAM	AAM	MBDA/International	1.80 (71)	0.09 (4)	18 (40)	IR	2.7 (5)
Mokopa, ZT6	ATM	Denel/South Africa	1.80 (71)	0.18 (7)	52 (115)	MMW or SAL	4.3 (8)
Moskit, Kh-41/3M80	ASM	Raduga/Russia	9.74 (383)	0.76 (30)	4,500 (9,921)	I/ARH or I/PR	135 (250)
Nag	ATM	DRDO/India			43 (95)	RC + ImIR or MMW	2.2 (4)
Nimrod	ASM	IAI/Israel	2.84 (112)	0.21 (8)	100 (220)	I/SAL	13 (25)
NTD Dandy	ATM	Rafael/Israel	1.20 (47)	0.14 (6)	11 (24)	I/TV/IR	2.2 (4)
NT-S Spike	ATM	Rafael/Israel	1.90 (75)	0.14 (6)	27 (60)	I//TV/IR	6.2 (10)
Ovod see 'Kazoo' and 'Kingbolt'							
Penguin 2, AGM-119B	AShM	Kongsberg/Norway	2.96 (117)	0.28 (11)	385 (849)	I/IR	19 (35)
Penguin 3, AGM-119A	AShM	Kongsberg/Norway	3.18 (125)	0.28 (11)	370 (816)	I/IR	30 (55)
PGM-A (1A, 2A, 3A) Hakeem	ASM	MBDA/International	3.60 (142)	0.35 (14)	300 (661)	I+SAL/IR/TV	11 (20)
PGM-B (1B, 2B, 3B) Hakeem	ASM	MBDA/International	4.00 (157)	0.43 (17)	1,115 (2,458)	I+SAL/IR/TV	11 (20)
Piranha see MAA-1							
PL-2/PL-3	AAM	CATIC/China	2.99 (118)	0.13 (5)	76 (168)	IR	1.6 (3)
PL-5	AAM	CATIC/China	2.89 (114)	0.13 (5)	85 (187)	IR	1.6 (3)
PL-7	AAM	CATIC/China	2.75 (108)	0.16 (6)	90 (198)	IR	1.6 (3)
PL-8	AAM	CATIC/China	3.00 (118)	0.16 (6)	120 (265)	IR	2.7 (5)
PL-9	AAM	Luoyang/China	2.99 (118)	0.16 (6)	115 (254)	IR	2.7 (5)
PL-10	AAM	CATIC/China	3.99 (157)	0.29 (11)	300 (661)	SARH	8.1 (15)
PL-11, LY-60	AAM	CATIC/China	3.89 (153)	0.20 (8)	220 (485)	SARH	13 (25)
PL-12	AAM	CATIC/China	3.85 (152)	0.20 (8)	180 (396)	I/ARH	38 (70)
Popeye 1, AGM-142A Have Nap	ASM	Rafael/Israel	4.82 (190)	0.53 (21)	1,360 (2,998)	I/TV or ImIR	43 (80)
Popeye 2, AGM-142B Have Lite	ASM	Rafael/Israel	4.00 (157)	0.53 (21)	1,115 (2,458)	I/TV or ImIR	40 (75)
Popeye, AGM-142C	ASM	Rafael/Israel		0.53 (21)		I/TV	
Popeye, AGM-142D	ASM	Rafael/Israel		0.53 (21)		I/ImIR	

Kongsberg Penguin anti-ship weapon *(Paul Jackson)* ***NEW***/0567247

Israel's Python 4 AAM 0073040

Name/designation	Role	Manufacturer/country	Length m (in)	Diameter m (in)	Weight kg (lb)	Guidance	Range n miles (km)
Python 3	AAM	Rafael/Israel	3.00 (118)	0.16 (6)	120 (265)	IR	8.1 (15)
Python 4	AAM	Rafael/Israel	3.00 (118)	0.16 (6)	105 (231)	IR	8.1 (15)
QW-1 Vanguard	AAM	CPMIEC/China	1.53 (60)	0.07 (3)	16.5 (36)	IR	2.7 (5)
QW-2 Vanguard	AAM	CPMIEC/China	1.59 (63)	0.07 (3)	11.3 (25)	AL	3.2 (6)
R-24 see 'Apex'							
R-27 see 'Alamo'							
R-33 see 'Amos'							
R-37, AA-X-13	AAM	Vympel/Russia	4.20 (165)	0.38 (15)	600 (1,323)	I/ARH	81 (150)
R-46 see 'Acrid'							
R-60 see 'Aphid'							
R-73 see 'Archer'							
R-77 see 'Adder'							
R 530 see Super 530							
R 550 see Magic							
RB 04	AShM	Saab/Sweden	4.25 (167)	0.50 (20)	600 (1,323)	I/AR	16 (30)
RB 05	ASM	Saab/Sweden	3.60 (142)	0.30 (12)	305 (672)	RC	4.3 (8)
RB 15F Mk 1	AShM	Saab/Sweden	4.35 (171)	0.50 (20)	598 (1,318)	I/AR	49 (90)
RB 15F Mk 2/3	AShM	Saab/Sweden	4.33 (170)	0.50 (20)	630 (1,389)	I/AR	81 (150)/108 (200)
RB 24J Swedish AIM-9J							
RB 71 Swedish Sky Flash							
RB 74 Swedish AIM-9L							
RB 75 Swedish Maverick							
R-Darter, V4	AAM	Denel/South Arfica	2.87 (113)	0.13 (5)	90 (198)	IR/AL	4.3 (8)
S-25LD	ATM	PMED/Russia	4.10 (161)	0.34 (13)	400 (882)	SAL	6.2 (10)
SA-7 see 'Grail'							
SA-14 see 'Gremlin'							
SA-16 see 'Gimlet'							
SA-18 see 'Grouse'							
'Sagger', AT-3/9M14 Malyutka	ATM	Kolomna/Russia	0.86 (34)	0.13 (5)	113 (249)	Wire	1.6 (3)
SCALP EG (Storm Shadow)	SOM	MBDA/International	5.10 (201)	0.63 (25)	1,300 (2,866)	I/MMW/ImIR	135 (250)
Sea Eagle	AShM	MBDA/International	4.14 (163)	0.40 (16)	600 (1,323)	I/ARH	59 (110)
Sea Skua	AShM	MBDA/International	2.50 (98)	0.25 (10)	147 (324)	SARH	8.1 (15)
Shafrir 2	AAM	Rafael/Israel	2.60 (102)	0.16 (6)	95 (209)	IR	1.6 (3)

Name/designation	Role	Manufacturer/country	Length m (in)	Diameter m (in)	Weight kg (lb)	Guidance	Range n miles (km)
Shturm see 'Spiral 2'							
Sidearm, AGM-122	ARM	Motorola/USA	3.00 (118)	0.13 (5)	91 (200)	PR	4.3 (8)
Sidewinder, AIM-9L/M/S	AAM	several/USA and Europe	2.87 (113)	0.13 (5)	87 (192)	IR	4.3 (8)
Sidewinder, AIM-9P	AAM	several/USA	3.07 (121)	0.13 (5)	87 (192)	IR	4.3 (8)
Sidewinder, AIM-9R	AAM	several/USA	2.87 (113)	0.13 (5)	87 (192)	Visual band CCD	4.3 (8)
Sidewinder, AIM-9S	AAM	Raytheon/USA	2.87 (113)	0.13 (5)	86 (190)	IR	4.3 (8)
Sidewinder, AIM-9X	AAM	Raytheon/USA	2.90 (114)	0.13 (5)	85 (187)	ImIR	5.4 (10)
Skipper, AGM-123	LGB	ESC/USA	4.33 (170)	0.36 (14)	582 (1,283)	SAL	3.8 (7)
Sky Flash	AAM	MBDA/International	3.66 (144)	0.20 (8)	195 (430)	SARH	22 (40)
Sky Sword (Tien Chien) 1	AAM	Chung Shan/Taiwan	2.87 (113)	0.13 (5)	90 (198)	IR	2.7 (5)
Sky Sword (Tien Chien) 2	AAM	Chung Shan/Taiwan	3.60 (142)	0.20 (8)	190 (419)	SARH	22 (40)
SLAM, AGM-84E	ASM	Boeing/USA	4.50 (177)	0.34 (13)	630 (1,389)	I/GPS/ImIR	51 (95)
SLAM-ER, AGM-84H	ASM	Boeing/USA	4.37 (172)	0.34 (13)	726 (1,600)	I/GPS/ImIR	151 (280)
Sparrow, AIM-7F	AAM	Raytheon/USA	3.66 (144)	0.20 (8)	227 (500)	SARH	22 (40)
Sparrow, AIM-7M	AAM	Raytheon/USA	3.66 (144)	0.20 (8)	230 (507)	SARH	24 (45)
Sparrow, AIM-7P	AAM	Raytheon/USA	3.66 (144)	0.20 (8)	230 (507)	RC/SARH	24 (45)
Sparrow, AIM-7R	AAM	Raytheon/USA	3.66 (144)	0.20 (8)	230 (507)	RC/SARH/IR	24 (45)
Spike see NT-S							
'Spiral', AT-6/9M114 Kokon	ATM	Kolomna/Russia	1.83 (72)	0.13 (5)	34 (75)	RC	3.2 (6)
'Spiral 2', AT-9/9M/14 Shturm	ATM	Kolomna/Russia	1.83 (72)	0.13 (5)	40 (88)	RC	4.3 (8)
Standard, AGM-78	ARM	Raytheon/USA	4.57 (180)	0.34 (13)	615 (1,356)	PR	30 (55)
Starstreak see Helstreak							
Stinger, FIM-92	AAM	Raytheon/USA	1.52 (60)	0.07 (3)	16 (35)	IR	1.6 (3)

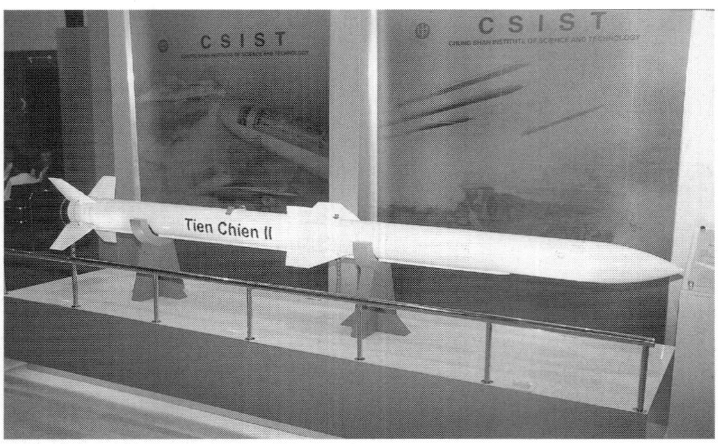

Chung Shan Tien Chen 2 ***(Paul Jackson)*** ***NEW***/0567255

MBDA Storm Shadow ***NEW***/0527139

Name/designation	Role	Manufacturer/country	Length m (in)	Diameter m (in)	Weight kg (lb)	Guidance	Range n miles (km)
Storm Shadow see SCALP EG							
Strela see 'Grail' and 'Gremlin'							
Super 530D	AAM	MBDA/International	3.80 (150)	0.26 (10)	270 (595)	SARH	22 (40)
Super 530F1	AAM	MBDA/International	3.54 (139)	0.26 (10)	245 (540)	SARH	13 (25)
'Swatter', AT-2C/9M17 Skorpion	ATM	Nudelman/Russia	1.16 (46)	0.13 (5)	30 (66)	RC	2.2 (4)
Swift, ZT3	ASM	Denel/South Africa	1.35 (53)/	0.13 (5)	19 (42)	L	2.2 (4)
'Swinger', AT-12/9M120 Vikhr/Ataka	ATM	Shipunov/Russia	1.70 (67)	0.13 (5)	43 (95)	RC	4.3 (8)
Taurus see KEPD 350							
Tien Chien see Sky Sword							
Torgos	SOM	Kentron/South Africa	4.86 (191)		980 (2,161)	I/GPS/ImIR	162 (300)
TOW, BGM-71A/B	ATM	Raytheon/USA	1.17 (46)	0.15 (6)	19 (42)	Wire	2.2 (4)
TOW, BGM-71C I-TOW	ATM	Raytheon/USA	1.45 (57)	0.15 (6)	19 (42)	Wire	2.2 (4)
TOW, BGM-71D/E TOW 2/2A	ATM	Raytheon/USA	1.55 (61)	0.15 (6)	22 (49)	Wire	2.2 (4)
TOW, BGM-71F TOW 2B	ATM	Raytheon/USA	1.17 (46)	0.15 (6)	23 (51)	Wire	2.2 (4)
TRIGAT, ATGW-3LR	ATM	consortium/Europe	1.57 (62)	0.15 (6)	49 (108)	ImIR	2.7 (5)
TY-90	AAM	Luoyang/China	1.9 (75)	0.09 (4)	20 (44)	IR	3.7 (7)
Type 80 see ASM-1							
Type 88 see ASM-2							
Type 90 see AAM-3							
U-Darter (see Darter)	AAM	Denel/South Africa	2.75 (108)	0.16 (6)	96 (212)	IR	4.3 (8)
Uran see 'Kayak'							
V3B see Kukri							
V3C see Darter							
V3E see A-Darter							
V4 see R-Darter							
Vikhr see 'Swinger' and AT-16							
X- see Kh-							
Yakhont (3M55)	AShM	Strela/Russia	8.30 (327)	0.67 (26)	2,550 (5,622)		
YJ-1 see C-801							
YJ-2 see C-802							
YJ-6 see C-601							
YJ-16 see C-101							
ZT3 see Swift							
ZT35 see Ingwe							
ZT6 see Mokopa							

AERO-ENGINES

Introduction

The following pages summarise the vital statistics of power plants mentioned in the main body of this book. They are divided into piston engines, turboprops/turboshafts and jet engines, and listed in alphabetical order of the manufacturer's name. Readers requiring further data on the two last-mentioned categories are referred to *Jane's Aero-Engines*. Some very small engines, employed by UAVs (and ultralights), are described in *Jane's Unmanned Aerial Vehicles and Targets* ; those for large missiles and spacecraft, in *Jane's Space Directory* ; and turboshafts, in *Jane's Helicopter Markets and Systems.* Space precludes an entry on each subvariant of more widely produced power plants, and therefore the aero-engine or helicopter publications should be consulted for these data.

Note that engine power ratings given below have been supplied, principally, by their manufacturers and are usually uninstalled output; data may thus vary from that quoted in the Aircraft section of this book, in which information is mainly supplied by the producer of the airframe. In most cases, the engine manufacturer's data reflect take-off power, generally under ISA sea-level conditions. Many gas-turbine engines, especially those for helicopters, are cleared to higher powers for brief periods in an emergency. Conversely, piston (and rotary) engine outputs are normally the maximum possible, which may be time-limited.

UPDATED

Not quite a Merlin or Griffon, the Jabiru 5100 eight-cylinder piston engine nevertheless convincingly fills the forward bay of a Supermarine Spitfire Mk 26 scale version of Second World War fighter *(Paul Jackson)* ***NEW***/1043446

Piston Engines

Arrangement: 4-O (2) means an engine with four cylinders, horizontally opposed, two-stroke; 8-IV (4) means eight cylinders in inverted-vee form, four-stroke; 4-X (2D) means four cylinders in X configuration, two-stroke diesel; 9-R is a nine-cylinder radial; L indicates in line. Wankel-type engines (W) are alternatively known as rotary (or, more accurately, rotating-piston) engines; 2 × 2-O is two horizontally opposed pistons within single cylinder.Cooling: A = air cooling, L = liquidIn many cases the engines exist in numerous variants, for example with a geared drive or a supercharger (mechanically driven or turbo). Horsepower ratings to one place of decimals are based on metric (CV/PS) original data.

Manufacturer	Country	Designation/ name	Arrangement	Cooling	Bore (mm)	Cylinder Stroke (mm)	Cylinder capacity (cc)	Dry weight (kg)	Max power (kW)	Remarks
Aero-Conversions	USA	Aero-Vee 2002	4-O (4)	A	92.0	82.0	2,180	71.7	59.7	
AlfaPrag	Czech Republic	TP-422	4-O (4)	A	95.0	77.0	2,193	79.0	63	
Alturair	USA	A-650	1-rotor (W)	L	—	—	650	62.1	74.6	
American Eagle	USA	540	8-V (4)	L	—	—	8,827	316	484.9	
Arrow	Italy	AE 530AC	2-O (2)	A	74.6	61.0	533	50.0	50.7	
Arrow	Italy	AE 1070AC	4-O (2)	A	74.6	61.0	1,066	65.0	90	
Arrow	Italy	GP 1000	4-O (2)	A	74.6	57.0	996	65.0	90	
Arrow	Italy	GP 1500	6-O (2)	A	74.6	57.0	1,495	87.5	134	
ATG	UK	A-Tech 100	2-O (2D)	L	—	—	1,810	100	74.6	
ATG	UK	A-Tech 600	4-O (2D)	L	—	—	9,162	225	447	
BMW	Germany	R1100RS	2-O (4)	A+oil	—	—	1,085	69.3	66	
BMW	Germany	R1100S	2-O (4)	A+oil	—	—	1,085	73.2	70	
BMW	Germany	R115ORS	2-O (4)	A+oil	—	—	1,130	76.3	72	
Bombardier (Rotax)	Canada	V220	6-V (4)	L	97.0	70.0	3,104	190	164	
Bombardier (Rotax)	Canada	V300	6-V (4)	L	97.0	70.0	3,104	210	223	
CAM	Canada	100	4-L (4)	L	74.0	86.5	1,488	92.1	74.6	
										Continental — see Teledyne Continental
CRM	Italy	18D/SS	18-W (4D)	L	150.0	180.0	57,260	1,700	1,380	
DaimlerChrysler	Germany	Suprex ('Smart')	3-L(4)	L	—	—	599	66	40.0	
Diamond	Austria	IAE 50R	1-rotor (W)	L	—	—	294	33.0	40.4	
Diamond	Austria	GIAE 110R	2-rotor (W)	L	—	—	588	54.0	78.0	
Diesel Air GmbH	Germany	D 280	2 × 2-O (2D)	A	74.0	2 × 78.0	1,400	78	58.8	
Diesel Air Ltd	UK	Dair-100	2 × 2-O (2D)	L	—	—	1,810	93	74.6	

Diesel Air 280 *(Paul Jackson)* ***NEW***/1043461

Diesel Air 100 *(Paul Jackson)* ***NEW***/0561740

Manufacturer	Country	Designation/ name	Arrangement	Cooling	Bore (mm)	Cylinder Stroke (mm)	Cylinder capacity (cc)	Dry weight (kg)	Max power (kW)	Remarks
FAM	France	200	6-V (4)	L	—	—	3,000	179	136	
FDM	Germany	100c	4-L (4)	L	76.0	71.4	1,206	84.0	73.6	
FDM	Germany	125C	4-L (4)	L	76.0	87.6	1,590	90.0	91.9	
FDM	Germany	140i	4-L (4)	L	76.0	87.6	1,590	92.0	103.0	
FDM	Germany	160Ti	4-L (4)	L	76.0	87.6	1,590	104.0	117.7	
HCI	USA	R180	5-R (4)	A	85.5	87.5	2,520	55.3	55.9	
Hirth	Germany	F23A	2-O (2)	A	72.0	64.0	521	24.0	29.8	
Hirth	Germany	F30	4-O (2)	A	72.0	64.0	1,042	36.0	70.8	
Hirth	Germany	F30A	4-O (2)	A	72.0	64.0	1,042	39.0	77.2	
Hirth	Germany	F30A36	4-O (2)	A	72.0	64.0	1,042	39.0	88.0	
Hirth	Germany	F30E	4-O (2)	A	72.0	64.0	1,042	42.0	61.0	
Hirth	Germany	F30ES	4-O (2)	A	72.0	64.0	1,042	42.0	75.0	
Hirth	Germany	F31	2-L (2)	A	76.0	69.0	625	26.5	29.1	
Hirth	Germany	F33A	1 (2)	A	76.0	69.0	313	12.7	18.1	
Hirth	Germany	F33B	1 (2)	A	76.0	69.0	313	13.0	18.1	
Hirth	Germany	2701	2-L (2)	A	70.0	64.0	493	32.8	32.1	
Hirth	Germany	2703	2-L (2)	A	72.0	64.0	521	32.8	40.4	
Hirth	Germany	2704	2-L (2)	A	76.0	69.0	625	31.0	39.0	
Hirth	Germany	2706	2-L (2)	A	76.0	69.0	625	31	48.5	
Hirth	Germany	3503E	2-L (2)	L	76.0	69.0	625	36.0	51.5	
Hirth	Germany	3701	3-L (2)	A	76.0	69.0	939	45.0	59.0	
Hirth	Germany	3701ES	3-L (2)	A	76.0	69.0	939	45.0	74.0	
HKS	Japan	700E	2-O (4)	A	85.0	60.0	680	55.0	44.0	
Honda	Japan		4-O (4)	L			—	—	—	teamed with Teledyne continental (USA)
HPower	USA	HKS 700E	(4)						44.7	
Jabiru	Australia	1600	4-O (4)	A	88.0	66.0	1,606	54.0	44.7	
Jabiru	Australia	2200	4-O (4)	A	97.5	74.0	2,200	55.8	59.7	
Jabiru	Australia	3300	6-O (4)	A	97.5	74.0	3,300	73.0	89	
Jabiru	Australia	5100	8-O (4)	A	97.5	85.0	5,077	117.0	241.4	
Jabiru	Australia	6000	8-O (4)	A	97.5	100.0	5,973	108.4	149.2	
JPX	France	4T60/A	4-O (4)	A	93.0	75.4	2,050	73.0	47.8	
JPX	France	4TX75/A	4-O (4)	A	95.0	82.0	2,325	78.0	59.5	
JPX	France	4TX75/M	4-O (4)	A	95.0	82.0	2,325	71.0	59.5	
KMPO	Russia	P-800	4-O (2)	A	62.0	66.0	797	48.0	48.5	
KMPO	Russia	P-1000	4-O (2)	A	72.0	66.0	925	55.0	59.7	
König	Germany	SD 750	4-R (2)	A	66.0	42.0	570	18.5	20.8	
König	Germany	SF 930	4-R (2)	A	70.0	60.0	930	36.0	35.8	
Limbach	Germany	L 550	4-O (2)	A	66.0	40.0	548	15.5	32.0–33.6	
Limbach	Germany	L 550E	4-O (2)	A	66.0	40.0	548	16.0	37.0	
Limbach	Germany	SL 1700, L 1700EC	4-O (4)	A	88.0	69.0	1,680	73.0	50.7	
Limbach	Germany	L 1700EA	4-O (4)	A	88.0	69.0	1,680	69.0	44.0	
Limbach	Germany	L 2000	4-O (4)	A	90.0	78.4	1,994	70.0	59.0	
Limbach	Germany	L 2000EO	4-O (4)	A	90.0	78.4	1,994	74.0	59.0	
Limbach	Germany	L 2000EB	4-O (4)	A	90.0	78.4	1,994	76.5	59.0	
Limbach	Germany	L 2400EB	4-O (4)	A	97.0	82.0	2,424	82.0	64.9	
Limbach	Germany	L 2400EF	4-O (4)	A	97.0	82.0	2,424	82.0	73.5	
Limbach	Germany	L 2400DFi/EFi	4-O (4)	A+L	97.0	82.0	2,424	77.0	74.0	
Limbach	Germany	L 2400DWFIG	4-O (4)	A	97.0	82.0	2,424	100	96	
Limbach	Germany	L 2400DT/ET	4-O (4)	A+L	97.0	82.0	2,424	86.0	96.0	
Limbach	Germany	L 2400DT.X	4-O (4)	A+L	97.0	82.0	2,424	105	118	
LOM	Czech Republic	M132A, AK, AR	4-L (4)	A	105.0	115.0	3,980	102	90	
LOM	Czech Republic	M132B	4-L (4)	A	105.0	115.0	3,980	105	97.0	
LOM	Czech Republic	M137A, AZ, AR	6-L (4)	A	105.0	115.0	5,970	141	134	
LOM	Czech Republic	M137B	6-L (4)	A	105.0	115.0	5,970	141	143	
LOM	Czech Republic	M332A	4-L (4)	A	105.0	115.0	3,980	102	103	
LOM	Czech Republic	M332B	4-L (4)	A	105.0	115.0	3,980	113	118	

Aero-Conversions Aero-Vee *(Paul Jackson)* 1043465

Jabiru 5100 *(Paul Jackson)* ***NEW***/0561738

Manufacturer	Country	Designation/ name	Arrangement	Cooling	Bore (mm)	Cylinder Stroke (mm)	Cylinder capacity (cc)	Dry weight (kg)	Max power (kW)	Remarks
LOM	Czech Republic	M332C	4-L (4)	A	105.0	115.0	3,980	113	124.0	
LOM	Czech Republic	M337A	6-L (4)	A	105.0	115.0	5,970	153	154	
LOM	Czech Republic	M337B	6-L (4)	A	105.0	115.0	5,970	153	173	
LOM	Czech Republic	M337C	6-L (4)	A	105.0	115.0	5,970	153	185	
LPE	USA	IVG-600	8-IV (4)	L	—	—	9,832	255	448	
										Lycoming — see Textron Lycoming
Modena Avio	Italy	323	3-R (4)	L	100.0	95.0	2,238	82.0	96.9	
										Morane Renault — see Société de Motorisations Aéronautiques
Motorstar	Romania	R-263	5-R (4)	A	107	94.0	4,310	100	93.2	
Nelson	USA	H-63CP	4-O (2)	A	68.3	70.0	1,030	30.8	35.8	
Novikov (RKBM)	Russian Federation	DN-200	3 × 2-O (2D)	L	72.0	72.0	4,440	105	110	
Orenda	Canada	OE600	8-V(4)	L	112.6	101.6	8,112	340	447	Production halted because of market downturn
Orenda	Canada	OE600 Turbo	8-V (4)	L	112.6	101.6	8,112		559	Production halted because of market downturn
Parma-Technik (ex-Walter)	Czech Republic	Mikron IIIAE	4-L (4)	A	90.0	96.0	2,440	70.0	48.5	
Parma-Technik (ex-Walter)	Czech Republic	Mikron IIIB	4-L (4)	A	90.0	96.0	2,440	69.0	55.0	
Pulsar Aircraft	USA	Aeromaxx	4-O (4)	A	—	—	1,998	74.8	88.0	
PZL	Poland	PZL-3S	7-R (4)	A	155.5	155.0	20,600	411	442	
PZL	Poland	PZL AI-14RA	9-R (4)	A	105.0	130.0	10,160	200	191	
PZL	Poland	PZL ASz-62	9-R (4)	A	155.0	174.0	29,870	580	735	
PZL	Poland	PZL K-9	9-R (4)	A	155.0	174.0	29,870	580	860	
PZL	Poland	PZL-F 2A-120-C1	2-O (4)	A	117.48	88.9	1,916	58.5	44.7	
PZL	Poland	PZL-F 4A-235-B31	4-O (4)	A	117.48	88.9	3,850	103	86.5	
PZL	Poland	PZL-F 6A6350-C1	6-O (4)	A	117.48	88.9	5,735	150	153	
Q-Drive	Czech Republic	QD-1400	4-O (4)	A	94.0	100.0	1,387	75.0	59.0	
Rotax	Austria	447 UL-1V	2-L (2)	A	67.5	61.0	436.5	26.8	29.5	
Rotax	Austria	447 UL-2V	2-L (2)	A	67.5	61.0	436.5	26.8	31.0	
Rotax	Austria	462	2-L (2)		—	—	—	—	38.8	
Rotax	Austria	503 UL-1V	2-L (2)	A	72.0	61.0	496.7	31.4	34.0	
Rotax	Austria	503 UL-2V	2-L (2)	A	72.0	61.0	496.7	31.4	37.0	
Rotax	Austria	582 UL	2-L (2)	L	76.0	64.0	580.7	27.4	48.0	
Rotax	Austria	618 UL-2V	2-L (2)	L	76.0	68.0	617.0	31.0	55.0	
Rotax	Austria	912 UL, 912A, 912F	4-O (4)	A+L	79.5	61.0	1,211.2	59.0	59.6	
Rotax	Austria	912 UL DCDI	4-O (4)	A+L	84.0	61.0	1,352	56.6	73.5	
Rotax	Austria	912 ULS, 912S	4-O (4)	A+L	84.0	61.0	1,352	56.6	73.5	
Rotax	Austria	914 UL DCDI, 914F	4-O (4)	A+L	79.5	61.0	1,211.2	64.0	84.5	
Rotax	Austria	936	6-V	—	—	—	—	—	162	
Rotec	Australia	R2800 Fireball 7	7-R (4)	A	—	—	2,800	95	82.0	
RotorWay	USA	RI 162	4-O (4)	L	—	—	2,660	77.1	113	
SAEC	China									HS5: ASh-62IR made under licence, see PZL ASz-62 (Poland)
Samara MKB	Russia	P-032	2-O (2)	A	72.0	54.0	440	12.5	24.6	
Samara MKB	Russia	P-033	2-O (2)	A	72.0	54.0	440	—	28.3	
Sauer	Germany	UL 1800	4-O (4)	A	90.0	69.0	1,745	56.5	35.0	
Sauer	Germany	UL 2100	4-O (4)	A	90.0	84.0	2,135	60.5	62.0	
Sauer	Germany	SE 1800	4-O (4)	A	90.0	69.0	1,745	63.3	40.0	
Sauer	Germany	SA 2100	4-O (4)	A	90.0	84.0	2,135	69.0	59.0	
Sauer	Germany	SD 2500	4-O (4)	A	97.0	84.0	2,481	79.0	68.0	
Sauer	Germany	SM 2700	4-O (4)	A	97.0	90	2,660	82.0	75.0	
Société de Motorisations Aéronautiques	France	SR305	4-O (4D)	A	undisclosed	undisclosed	4,988	192	169	
Stol Techno	Japan	HKS 700E	2-O (4)	A+L	85.0	60.0	680	47.0	44.7	

Limbach L2400DT/ET ***NEW***/1043464

Limbach L1700EO/EC and L2000EO/EC ***NEW***/1043472

Manufacturer	Country	Designation/ name	Arrangement	Cooling	Bore (mm)	Cylinder Stroke (mm)	Cylinder capacity (cc)	Dry weight (kg)	Max power (kW)	Remarks
Subaru	Japan	EJ22	4-L (4)	L	96.9	75.0	2,212	119	119	
Subaru	Japan	EA81–100	4-O (4)	L	92.0	67.0	1,781	97.1	74.6	
Subaru	Japan	EA81–140	4-O (4)	L	92.0	67.0	1,781	100	104	
Superior	USA	Vantage SV-360	4-O (4)	A	130.175	111.125	5,915	130.6	134	
Teledyne Continental Motors	USA	O-200-A & B	4-O (4)	A	103.1	98.4	3,280	77.2	74.6	
Teledyne Continental Motors	USA	IO-240-A & B	4-O (4)	A	112.7	98.4	3,940	93	93.0	
Teledyne Continental Motors	USA	IOF-240	4-O (4)	A	112.7	98.4	3,940			
Teledyne Continental Motors	USA	O-300-A, C & D	6-O (4)	A	103.1	98.4	4,916	112.8	108.0	
Teledyne Continental Motors	USA	IO-300-CCB, D.DB, G.GB, H, HB	6-O (4)	A	103.1	98.4	4,916	148.4	156.7	
Teledyne Continental Motors	USA	IO-300-ES	6-O (4)	A	103.1	98.4	4,916	138.3	156.7	
Teledyne Continental Motors	USA	IO-300-J, JB, K, KB	6-O (4)	A	103.1	98.4	4,916	148.4	145.5	
Teledyne Continental Motors	USA	TSIO-300-A, AB	6-O (4)	A	103.1	98.4	4,916	128.7	156.7	
Teledyne Continental Motors	USA	TSIO-300-C, CB	6-O (4)	A	103.1	98.4	4,916	136.5	167.8	
Teledyne Continental Motors	USA	TSIO-300-D, DB	6-O (4)	A	103.1	98.4	4,916	128.7	167.8	
Teledyne Continental Motors	USA	L/TSIO-360-E, EB, F, -FB	6-O (4)	A	112.7	98.4	5,900	145.8	149.1	
Teledyne Continental Motors	USA	TSIO-360-H, -HB	6-O (4)	A	112.7	98.4	5,900	134.6	156.7	
Teledyne Continental Motors	USA	TSIO-360-JB	6-O (4)	A	112.7	98.4	5,900	134.6	167.8	
Teledyne Continental Motors	USA	L/TSIO-360-KB	6-O (4)	A	112.7	98.4	5,900	148.9	164.1	
Teledyne Continental Motors	USA	TSIO-360-LB	6-O (4)	A	112.7	98.4	5,900	155.7	156.7	
Teledyne Continental Motors	USA	TSIO-360-MB	6-O (4)	A	112.7	98.4	5,900	148.6	156.7	
Teledyne Continental Motors	USA	L/TSIO-360-RB	6-O (4)	A	112.7	98.4	5,900	148.6	164.1	
Teledyne Continental Motors	USA	TSIO-360-SB	6-O (4)	A	112.7	98.4	5,900	148.9	164.1	
Teledyne Continental Motors	USA	O-470-GCI	6-O (4)	A	127.0	101.6	7,702	195.9	179.0	
Teledyne Continental Motors	USA	O-470-J	6-O (4)	A	127.0	101.6	7,702	160.6	167.8	
Teledyne Continental Motors	USA	O-470-K, L & M	6-O (4)	A	127.0	101.6	7,702	172.4	171.5	
Teledyne Continental Motors	USA	O-470-R & S	6-O (4)	A	127.0	101.6	7,702	172.2	171.5	
Teledyne Continental Motors	USA	O-470-U	6-O (4)	A	127.0	101.6	7,702	176.4	171.5	
Teledyne Continental Motors	USA	IO-470-C	6-O (4)	A	127.0	101.6	7,702	182.6	186.4	
Teledyne Continental Motors	USA	IO-470-D & E	6-O (4)	A	127.0	101.6	7,702	193.3	193.9	
Teledyne Continental Motors	USA	IO-470-F	6-O (4)	A	127.0	101.6	7,702	181.0	193.9	
Teledyne Continental Motors	USA	IO-470-H	6-O (4)	A	127.0	101.6	7,702	182.6	193.9	
Teledyne Continental Motors	USA	IO-470-J & K	6-O (4)	A	127.0	101.6	7,702	182.3	167.8	
Teledyne Continental Motors	USA	IO-470-L & M	6-O (4)	A	127.0	101.6	7,702	186.6	193.9	
Teledyne Continental Motors	USA	IO-470-N	6-O (4)	A	127.0	101.6	7,702	188.0	193.9	
Teledyne Continental Motors	USA	IO-470-S	6-O (4)	A	127.0	101.6	7,702	185.7	193.9	
Teledyne Continental Motors	USA	IO-470-U, V & VO	6-O (4)	A	127.0	101.6	7,702	184.2	193.9	

PZL-F4A 0526914

Rotec Fireball 7 *(Paul Jackson)* ***NEW***/1043463

Manufacturer	Country	Designation/ name	Arrangement	Cooling	Bore (mm)	Cylinder Stroke (mm)	Cylinder capacity (cc)	Dry weight (kg)	Max power (kW)	Remarks
Teledyne Continental Motors	USA	IO-520-A & J	6-O (4)	A	133.0	101.6	8,500	187.1	212.5	
Teledyne Continental Motors	USA	IO-520-B, BA, BB	6-O (4)	A	133.0	101.6	8,500	184.5	212.5	
Teledyne Continental Motors	USA	IO-520-C, CB	6-O (4)	A	133.0	101.6	8,500	180.9	212.5	
Teledyne Continental Motors	USA	IO-520-D & E	6-O (4)	A	133.0	101.6	8,500	186.7	212.5	
Teledyne Continental Motors	USA	IO-520-F	6-O (4)	A	133.0	101.6	8,500	186.7	212.5	
Teledyne Continental Motors	USA	IO-520-K	6-O (4)	A	133.0	101.6	8,500	186.7	212.5	
Teledyne Continental Motors	USA	IO-520-L	6-O (4)	A	133.0	101.6	8,500	187.2	212.5	
Teledyne Continental Motors	USA	IO-520-M, MB	6-O (4)	A	133.0	101.6	8,500	186.5	212.5	
Teledyne Continental Motors	USA	L/IO-520-P	6-O (4)	A	133.0	101.6	8,500	174.4	186.4	
Teledyne Continental Motors	USA	L/TSIO-520-AE	6-O (4)	A	133.0	101.6	8,500	165.5	186.4	
Teledyne Continental Motors	USA	TSIO-520-AF	6-O (4)	A	133.0	101.6	8,500	197.5	212.5	
Teledyne Continental Motors	USA	TSIO-520-B, BB	6-O (4)	A	133.0	101.6	8,500	184.7	212.5	
Teledyne Continental Motors	USA	TSIO-520-BE	6-O (4)	A	133.0	101.6	8,500	200.5	231.2	
Teledyne Continental Motors	USA	TSIO-520-C & H	6-O (4)	A	133.0	101.6	8,500	193.4	212.5	
Teledyne Continental Motors	USA	TSIO-520-CE	6-O (4)	A	133.0	101.6	8,500	197.5	242.4	
Teledyne Continental Motors	USA	TSIO-520-D, DB	6-O (4)	A	133.0	101.6	8,500	184.5	212.5	
Teledyne Continental Motors	USA	TSIO-520-E, EB	6-O (4)	A	133.0	101.6	8,500	184.0	223.7	
Teledyne Continental Motors	USA	TSIO-520-G	6-O (4)	A	133.0	101.6	8,500	188.2	212.5	
Teledyne Continental Motors	USA	TSIO-520-J, JB, N, NB	6-O (4)	A	133.0	101.6	8,500	179.8	231.2	
Teledyne Continental Motors	USA	TSIO-520-K, KB	6-O (4)	A	133.0	101.6	8,500	179.8	212.5	
Teledyne Continental Motors	USA	TSIO-520-L, LB	6-O (4)	A	133.0	101.6	8,500	188.7	231.2	
Teledyne Continental Motors	USA	TSIO-520-(M) P & R	6-O (4)	A	133.0	101.6	8,500	189.4	212.5	
Teledyne Continental Motors	USA	TSIO-520-T	6-O (4)	A	133.0	101.6	8,500	193.4	231.2	
Teledyne Continental Motors	USA	TSIO-520-U & UB	6-O (4)	A	133.0	101.6	8,500	191.5	223.7	
Teledyne Continental Motors	USA	TSIO-520-VB	6-O (4)	A	133.0	101.6	8,500	184.4	242.4	
Teledyne Continental Motors	USA	TSIO-520-WB	6-O (4)	A	133.0	101.6	8,500	216.0	242.4	
Teledyne Continental Motors	USA	GTSIO-520-D & H	6-O (4)	A	133.0	101.6	8,500	219.5	275.9	
Teledyne Continental Motors	USA	GTSIO-520-L	6-O (4)	A	133.0	101.6	8,500	220.0	275.9	
Teledyne Continental Motors	USA	GTSIO-520-M	6-O (4)	A	133.0	101.6	8,500	219.1	275.9	

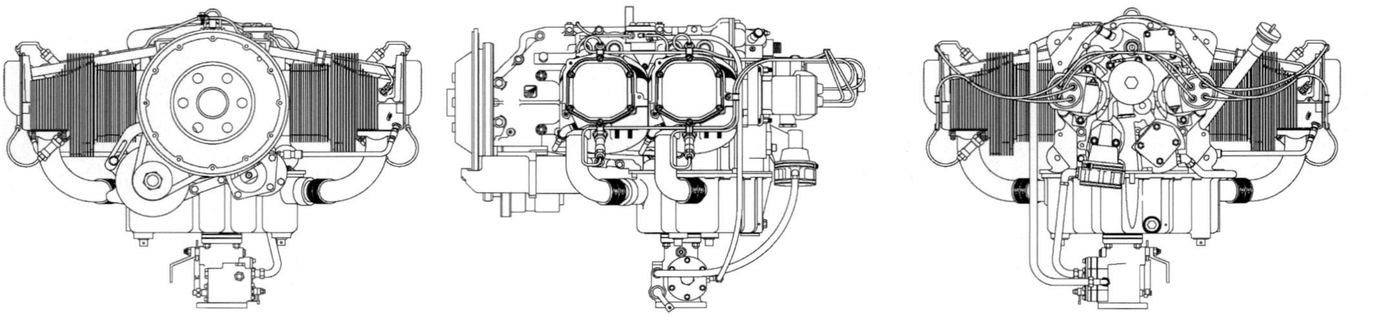

Superior Vantage ***NEW***/1043462

Manufacturer	Country	Designation/ name	Arrangement	Cooling	Bore (mm)	Cylinder Stroke (mm)	Cylinder capacity (cc)	Dry weight (kg)	Max power (kW)	Remarks
Teledyne Continental Motors	USA	GTSIO-520-N	6–O (4)	A	133.0	101.6	8,500	220.4	275.9	
Teledyne Continental Motors	USA	IO-550-A	6–0 (4)	A	133.0	108.0	9,000	188.0	223.7	
Teledyne Continental Motors	USA	IO-550-B	6–0 (4)	A	133.0	108.0	9,000	184.5	223.7	
Teledyne Continental Motors	USA	IO-550-C	6–0 (4)	A	133.0	108.0	9,000	188.3	223.7	
Teledyne Continental Motors	USA	IO-550-D, E, F & (L)	6–0 (4)	A	133.0	108.0	9,000	192.8	223.7	
Teledyne Continental Motors	USA	IO-550-G	6–0 (4)	A	133.0	108.0	9,000	186.9	208.8	
Teledyne Continental Motors	USA	IO-550-N	6–0 (4)	A	133.0	108.0	9,000	186.9	231.2	
Teledyne Continental Motors	USA	TSIO-550-B & E	6–0 (4)	A	133.0	108.0	9,000	200.5	261.0	
Teledyne Continental Motors	USA	TSIO-550-C	6–0 (4)	A	133.0	108.0	9,000	200.5	231.2	
Teledyne Continental Motors	USA	TSIOL-550-A	6–0 (4)	L	133.0	108.0	9,000	182.1	261.0	
Teledyne Continental Motors	USA	TSIOL-550-B	6–0 (4)	L	133.0	108.0	9,000	188.4	242.4	
Teledyne Continental Motors	USA	TSIOL-550-C	6–0 (4)	L	133.0	108.0	9,000	188.4	261.0	
Textron Lycoming	USA	O-235-C	4-O (4)	A	111.0	98.4	3,850	96.6	85.8	
Textron Lycoming	USA	O-235-I, M	4-O (4)	A	111.0	98.4	3,850	98.0	88.0	
Textron Lycoming	USA	O-235N, P	4-O (4)	A	111.0	98.4	3,850	98.0	86.5	
Textron Lycoming	USA	O-320-A, E	4-O (4)	A	130.0	98.4	5,200	110.7	112	
Textron Lycoming	USA	(H)O-320-B2C	4-O (4)	A	130.0	98.4	5,200	112	119	
Textron Lycoming	USA	O-320-D	4-O (4)	A	130.0	98.4	5,200	114	119	
Textron Lycoming	USA	AEIO-320-D	4-O (4)	A	130.0	98.4	5,200	123	119	
Textron Lycoming	USA	AEIO-320-E	4-O (4)	A	130.0	98.4	5,200	117	112	
Textron Lycoming	USA	(L)IO-320-B, C	4-O (4)	A	130.0	98.4	5,200	117.5	119	
Textron Lycoming	USA	(L)O-360-A	4-O (4)	A	130.0	111.0	5,920	120	134	
Textron Lycoming	USA	O-360-F	4-O (4)	A	130.0	111.0	5,920	122	134	
Textron Lycoming	USA	TO-360-C, F	4-O (4)	A	130.0	111.0	5,920	154	157	
Textron Lycoming	USA	IO-360-A, C	4-O (4)	A	130.0	111.0	5,920	133	149	
Textron Lycoming	USA	IO-360-B	4-O (4)	A	130.0	111.0	5,920	122.5	134	
Textron Lycoming	USA	LIO-360-C	4-O (4)	A	130.0	111.0	5,920	139	149	
Textron Lycoming	USA	TIO-360-C	4-O (4)	A	130.0	111.0	5,920	158	157	
Textron Lycoming	USA	HIO-360-B1A	4-O (4)	A	130.0	111.0	5,920	118.4	134.3	
Textron Lycoming	USA	HO-360-C1A	4-O (4)	A	130.0	111.0	5,920	116.6	134.3	
Textron Lycoming	USA	HIO-360-D1A	4-O (4)	A	130.0	111.0	5,920	132	142	
Textron Lycoming	USA	(L)HIO-360-F1AD	4-O (4)	A	130.0	111.0	5,920	133	142	
Textron Lycoming	USA	AEIO-360-A	4-O (4)	A	130.0	111.0	5,920	139	149	
Textron Lycoming	USA	AEIO-360-B	4-O (4)	A	130.0	111.0	5,920	125	134	
Textron Lycoming	USA	O-540-A	6-O (4)	A	130.0	111.0	8,860	170	186	
Textron Lycoming	USA	IO-540-A1A5	6-O (4)	A	130.0	111.0	8,860	203	201	
Textron Lycoming	USA	O-540-B	6-O (4)	A	130.0	111.0	8,860	166	175	
Textron Lycoming	USA	IO-540-C	6-O (4)	A	130.0	111.0	8,860	170	186	
Textron Lycoming	USA	IO-540-D	6-O (4)	A	130.0	111.0	8,860	173	194	
Textron Lycoming	USA	O-540-E	6-O (4)	A	130.0	111.0	8,860	167	194	
Textron Lycoming	USA	O-540-F1B5	6-O (4)	A	130.0	111.0	8,860	167	194	
Textron Lycoming	USA	O-540-J	6-O (4)	A	130.0	111.0	8,860	165	175	
Textron Lycoming	USA	IO-540-K	6-O (4)	A	130.0	111.0	8,860	199	224	
Textron Lycoming	USA	IO-540-S	6-O (4)	A	130.0	111.0	8,860	200	224	
Textron Lycoming	USA	IO-540-T4B5	6-O (4)	A	130.0	111.0	8,860	176	196	
Textron Lycoming	USA	IO-540-W1A5	6-O (4)	A	130.0	111.0	8,860	167	175	
Textron Lycoming	USA	AEIO-540-D	6-O (4)	A	130.0	111.0	8,860	174	194	
Textron Lycoming	USA	AEIO-540-L	6-O (4)	A	130.0	111.0	8,860	204	224	
Textron Lycoming	USA	TIO-540-A	6-O (4)	A	130.0	111.0	8,860	231.8	231.2	
Textron Lycoming	USA	TIO-540-AB1AD	6-O (4)	A	130.0	111.0	8,860	205	186	
Textron Lycoming	USA	TIO-540-AF1A, AF1B	6-O (4)	A	130.0	111.0	8,860	211	201	
Textron Lycoming	USA	TIO-540-AE2A	6-O (4)	A	130.0	111.0	8,860	246	261	
Textron Lycoming	USA	TIO-540-C	6-O (4)	A	130.0	111.0	8,860	205	186	
Textron Lycoming	USA	(L)TIO-540-F	6-O (4)	A	130.0	111.0	8,860	233	242	
Textron Lycoming	USA	(L)TIO-540-J	6-O (4)	A	130.0	111.0	8,860	239	261	
Textron Lycoming	USA	TIO-540-S	6-O (4)	A	130.0	111.0	8,860	228	224	
Textron Lycoming	USA	(L)TIO-540-U	6-O (4)	A	130.0	111.0	8,860	248	261	
Textron Lycoming	USA	(L)TIO-540-V	6-O (4)	A	130.0	111.0	8,860	248	269	
Textron Lycoming	USA	(L)TIO-540-W	6-O (4)	A	130.0	111.0	8,860	243	261 kN (350 hp)	
Textron Lycoming	USA	TIO-541-E	6-O (4)	A	130.0	111.0	8,860	270	283	
Textron Lycoming	USA	TIGO-541-E	6-O (4)	A	130.0	111.0	8,860	319	317	
Textron Lycoming	USA	AEIO-580	6-O (4)	A					246	

TAE Centurion 1.7 *(Paul Jackson)* ***NEW***/0561736

Zanzottera B2000 *(Paul Jackson)* ***NEW***/0561735

Manufacturer	Country	Designation/ name	Arrangement	Cooling	Bore (mm)	Cylinder Stroke (mm)	Cylinder capacity (cc)	Dry weight (kg)	Max power (kW)	Remarks
Textron Lycoming	USA	IO-720-A, B, D	8-O (4)	A	130.0	111.0	11,840	258	298	
Thielert	Germany	Centurion 1.7	4-L (4D)	L	80.0	84.0	1,689	134	99	also called TAE
Thielert	Germany	Centurion 4.0	8-V (4D)	L	80.0	84.0	3,378	c188	228	
UFA	Russian Federation	Turbo-Diesel							331	
VAZ	Russian Federation	VAZ-416	2-rotor (W)	L	—	—	1,308	125	134	
VAZ	Russian Federation	VAZ-4161	2-rotor (W)	L	—	—	1,308	125	132	
VAZ	Russian Federation	VAZ-4162	2-rotor (W)	L	—	—	1,308	125	147	
VAZ	Russian Federation	VAZ-426	3-rotor (W)	L	—	—	1,962	155	201	
VAZ	Russian Federation	VAZ-4261	3-rotor (W)	L	—	—	1,962	155	177	
VAZ	Russian Federation	VAZ-4262	3-rotor (W)	L	—	—	1,962	155	198	
VAZ	Russian Federation	VAZ-4263	3-rotor (W)	L	—	—	1,962	145	221	
VAZ	Russian Federation	VAZ-4265	3-rotor (W)	L	—	—	1,962	130	201	
VAZ	Russian Federation	VAZ-526	4-rotor (W)	L	—	—	2,616	175	298	
Verner	Czech Republic	SVS 1400	2-O (4)	A	94.0	100.0	1,400	62.0	58.8	
Verner	Czech Republic	VM 133 M	2-O (4)	A	97.0	90.0	1,329	62.0	58.2	
VM	Italy	1304HF	4-O (4D)	A	130.0	110.0	5,840	185	154	
VM	Italy	1306HF	6-O (4D)	A	130.0	110.0	8,760	243	235	
VM	Italy	1308HF	8-O (4D)	A	130.0	110.0	11,680	298	316	
VOKBM	Russian Federation	M-3	3-R (4)	A	105.0	130.0	3,387	119	77.2	
VOKBM	Russian Federation	M-5	5-R (4)	A	105.0	130.0	5,644	115	118	
VOKBM	Russian Federation	M-6	6-O (4)	L	105.0	82.0	4,260	270	254	
VOKBM	Russian Federation	M-7	7-R (4)	A	105.0	130.0	7,902	156	191	
VOKBM	Russian Federation	M-9F	9-R (4)	A	105.0	130.0	10,160	214	294	
VOKBM	Russian Federation	M-14NTK	9-R (4)	A	105.0	130.0	10,160		316	
VOKBM	Russian Federation	M-14PF	9-R (4)	A	105.0	130.0	10,160	214	294	
VOKBM	Russian Federation	M-14PT	9-R (4)	A	105.0	130.0	10,160	217	265	
VOKBM	Russian Federation	M-16	8-X (4)	A	—	—	—	150	224	
VOKBM	Russian Federation	M-17	4-O (4)	A	—	—	—	118	130.5	
Walter	Czech Republic	M202	2-O (2)	A	82.0	64.0	676	36.0	48.5	
Wankel Rotary	Germany	LCR-407 SGti	1-rotor (W, D)	L	—	—	407	25.0	27.2	
Wankel Rotary	Germany	LOCR-407 SD	1-rotor (W, D)	L	—	—	407	38.0	33.1	
Wankel Rotary	Germany	LCR-814 TG ti	2-rotor (W, D)	L	—	—	814	35.0	55.1	
Wankel Rotary	Germany	LOCR-814 TD	2-rotor (W, D)	L	—	—	814	50.0	66.2	
Wankel Rotary	Germany	Twinpack	4-rotor (W, D)	L	—	—	—	119	110	
Wilksch	UK	WAM 120	3-L(2D)	L	—	—	—	100	89.5	
Wilksch	UK	WAM 160	4-L(2D)	L	—	—	—	120	119.3	
Zanzottera	Italy	MZ202	2-L (2)	A	76.0	69.0	626	39.0	48.5	
Zanzottera	Italy	498ia	2-O (2)	A	75.0	56.0	498	17.0	28.3	
Zanzottera	Italy	B1000I	4-O (2)	A	75.0	56.0	996	29.0	59.7	
Zanzottera	Italy	B2000I	6-O (2)	A	76.0	69.0	1,880	51.0	112	
Zoche	Germany	ZO 01A	4-X (2D)	A	95.0	94.0	2,665	84.0	110	
Zoche	Germany	ZO 02A	8-X (2D)	A	95.0	94.0	5,330	118	220	
Zoche	Germany	ZO 03A	2-IV (2D)	A	95.0	94.0	1,332.5	55.0	51.0	
Zöllner	Germany	DZ	4-O (4)	A	—	—	1,998	74.0	86.8	

UPDATED

Turboprop Engines

Arrangement: A = axial stages, C = centrifugal stages, C+C= two centrifugal stages on the same shaft, C, C two stages on different shafts.Prop drive: FT = free turbine, SS = single shaft, 2S = two-shaft engine but no independent power turbineFor turboshaft engines, T-O rating is the maximum except contingency, usually a 30 minute, or maximum continuous power.

Manufacturer	Country	Designation/ name	Arrangement	Air flow (kg/s)	Prop drive	Length (mm)	Width (mm)	Weight, dry (kg)	T-O rating kW	Remarks
AlliedSignal										see Honeywell
Allison										see Rolls-Royce (USA)
DEMC	China								2,080	WJ5A I: AI-24A made under licence, details as AI-24T under Progress (ZMKB) (Ukraine)
DEMC	China	WJ5E	10A	14.6	SS	2,381	770	720	2,130 720[1]	
EPI (Europrop International)	International	TP400-D6	5A, 6A	—	FT	3,500	—	1,800	8,203	
General Electric	USA	CT7–5A	5A+C	4.41	FT	2,438	737	355	1,294	
General Electric	USA	CT7–9	5A+C	5.20	FT	2,438	737	365	<1,447	
General Electric	USA	T64-P4D	14A	12.2	FT	2,793	683	538	2,535	
Honeywell	USA	TPE331–3	C+C	3.54	SS	1,092	533	161	626	
Honeywell	USA	TPE331–12	C+C	3.49	SS	1,168	533	181	834	
Honeywell	USA	TPE331–14GR	C+C	5.26	SS	1,333	533	281	1,462	
Honeywell	USA	LTP 101–700	A+C	2.31	FT	949	592	147	522	
Innodyn	USA	165TE	—	—	—	—	—	85.3	123	
Innodyn	USA	255TE	—	—	—	—	—	85.3	190	
KKBM	Russian Federation	NK-12MV	14A	65.0	FT	4,785	1,150	2,900	11,033	
Klimov	Russian Federation	TV3–117VMA-SB2	12A	9.0	FT	2,055	650	560	1,864	
Klimov	Russian Federation	TV7–117S	5A+C	7.95	FT	2,143	886	520	1,839	
Klimov	Russian Federation	TV7–117S Series 2	5A+C		FT	2,136	940	430	2,610	
LHTEC	USA	CTP800–4T Twin	2×(C+C)	2×3.27	2×FT	1,769	1,252	517	2,013	
OMKB	Russian Federation	TVD-20	7A+C	5.40	FT	1,770	850	240	1,081	
Pratt & Whitney Canada	Canada	PT6A-27	3A+C	3.08	FT	1,575	483	149	507	
Pratt & Whitney Canada	Canada	PT6A-41	3A+C	3.40	FT	1,701	483	183	634	
Pratt & Whitney Canada	Canada	PT6A-65R	4A+C	4.31	FT	1,905	483	218	1,026	
Pratt & Whitney Canada	Canada	PT6A-68	4A+C	4.5	FT	1,930	483	259.5	1,193	
Pratt & Whitney Canada	Canada	PW120	C, C	—	FT	2,134	635	418	1,491	
Pratt & Whitney Canada	Canada	PW121	C, C	—	FT	2,134	635	425	1,603	
Pratt & Whitney Canada	Canada	PW123 C/D	C, C	—	FT	2,134	635	450	1,603	
Pratt & Whitney Canada	Canada	PW127	C, C	—	FT	2,134	660	481	2,051	
Pratt & Whitney Canada	Canada	PW150A	3A+C	—	FT	2,423	767	690	3,781	
Progress (ZMKB)	Ukraine	AI-20M	10A	20.7	SS	3,096	842	1,040	2,940	
Progress (ZMKB)	Ukraine	AI-24T	10A	14.4	SS	2,346	677	600	1,880	
PZL	Poland	TWD-10B	6A+C	4.58	FT	2,060	555	230	754	
RKBM	Russian Federation	TVD-1500S	3A+C	4.0	FT	1,965	620	240	1,044	
Rolls-Royce	UK	Dart Mk 536	C+C	10.7	SS	2,480	963	569	1,700	
Rolls-Royce	UK	Tyne 21	6A, 9A	21.1	2S	2,762	1,400	1,085	4,226	
Rolls-Royce	USA	250-B17	6A+C	1.56	FT	1,143	483	88.4	313	
Rolls-Royce	USA	T56–15	14A	14.7	SS	3,708	686	828	3,424	
Rolls-Royce	USA	T56–427	14A	15.2	SS	3,711	686	880	3,910	
Rolls-Royce	USA	AE 2100A	14A	16.3	FT	—	—	715.8	3,095	
Rolls-Royce	USA	AE 2100C	14A	—	FT	—	—	715.8	2,685	
Rolls-Royce	USA	AE 2100D2, D3	14A	16.3	FT	2,743	1,151	702	3,424	

Walter M 601B turboprop *(Paul Jackson)* 0126836

Experimental ATP turboprop for light aircraft, first flown in USA in 2002 *(Paul Jackson)* 0526932

P&W Canada PT6A-67B *(Paul Jackson)* 0126837

Honeywell TPE331 *(Paul Jackson)* ***NEW***/1043466

Manufacturer	Country	Designation/ name	Arrangement	Air flow (kg/s)	Prop drive	Length (mm)	Width (mm)	Weight, dry (kg)	T-O rating kW	Remarks
Rybinsk	Russia	TVD-1500V (RD-600V)	3A+C	4.0	FT	1,250	620	220	1,156	
SAEC	China									WJ6: AI-20M made under licence; details as under Progress (ZMKB) (Ukraine)
Saturn	Russian Federation	AL-34–1	C	—	FT	1,700	640	178	809	
Turbomeca	France	Arrius 1D	C	not disclosed	FT	826	476	111	313	
Turbomeca	France	Arrius 2F	C	not disclosed	FT	945	459	103	376	
Walter	Czech Republic	M 601E	2A+C	3.60	FT	1,675	590	200	560	
Walter	Czech Republic	M 601F	2A+C	3.60	FT	1,675	590	202	580	
Walter	Czech Republic	M 602A	C+C	7.33	FT	2,565	753	570	1,360	
Walter	Czech Republic	M 602B	C+C		FT	2,285	753	480	1,500	

Residual thrust

UPDATED

Turboshaft Engines

See Turboprop engines

Manufacturer	Country	Designation/ name	Arrangement	Air flow (lbs/s)	Drive	Length (mm)	Width (mm)	Weight, dry (kg)	T-O rating kW	Remarks
AlliedSignal										see Honeywell
Allison										see Rolls-Royce
Aviadvigatel	Russian Federation	D-25V	9A	26.2	FT	2,737	1,086	1,325	4,050	
CLXMW	China	WZ6							1,104	Turbomeca Turmo IIIC derivative made under licence
General Electric	USA	CT7–2	5A+C	4.5	FT	1,194	635	212	1,212	
General Electric	USA	CT7–6	5A+C	5.9	FT	1,194	660	220	1,491	
General Electric	USA	CT7–6D	5A+C	6.1	FT	1,194	660	229	1,514	
General Electric	USA	CT7–8	5A+C	5.9	FT	1,194	660	—	1,879	
General Electric	USA	T64–419	14A	13.3	FT	2,006	660	343	3,542	
General Electric	USA	T700–401	5A+C	4.5	FT	1,168	635	197	1,260	
General Electric	USA	T700–700	5A+C	4.5	FT	1,168	635	198	1,210	
General Electric	USA	T700/T6	5A+C	5.9	FT	1,205	660	220	1,652	
Granit	Russia	TVD-400	A+C	2.15	FT	820	380	96.0	298	
Honeywell	USA	LTS 101–600A	1A+C	2.31	FT	785	599	115	459	(formerly AlliedSignal)
Honeywell	USA	LTS 101–750B	1A+C	2.30	FT	795	470	123	515	
Honeywell	USA	T53–703	5A+C	5.90	FT	1,209	584	247	1,343	
Honeywell	USA	T5317A	5A+C	5.53	FT	1,209	584	256	1,119	
Honeywell	USA	T55–714	7A+C	13.19	FT	1,181	610	377	3,630	
Klimov	Russian Federation	GTD-350								see under PZL (Poland)
Klimov	Russian Federation	TV2–117	10A	8.4	FT	2,842	547	338	1,118	
Klimov	Russian Federation	TV3–117V	12A	8.7	FT	2,085	640	285	<1,633	
Klimov	Russian Federation	TV3–117VMA-SB3	12A	9.0	FT	2,055	650	310	1,864	
Klimov	Russian Federation	TV7–117V	5A+C	9.2	FT	1,780	635	360	2,796	
Klimov	Russian Federation	VK-1500	12A	7.3	FT	1,714	708	340	1,118	
Klimov	Russian Federation	VK-2500	12A	9.3	FT	2,055	660	295	1,790	
LHTEC	USA	T800–800	C+C	3.27	FT	843.3	550.1	143	995	
LHTEC	USA	T800–801	C+C	4.43	FT	843.3	550.1	149.7	1,165	
LHTEC	USA	CTS800–1G	C+C	4.01	FT	856.0	550	149.7	1,312	
LHTEC	USA	CTS800–4	C+C	3.54	FT	1,047.2	561.6	173.7	1,016	
Mitsubishi	Japan	MG5–100	—	—	FT	—	—	—	597	
Mitsubishi	Japan	XTS1–10	—	—	FT	—	—	—	659	
MTR	International	390	C+C	3.20	FT	1,078	442	169	958	
Omsk (OMKB)	Russia	TVD-20V	C+C	5.7	FT	1,850	855	210	1,119	
Omsk (OMKB)	Russia	TV-O-100	2A+C	2.66	FT	1,275	780	125	537	
Pratt & Whitney Canada	Canada	PT6B-36	3A+C	3.08	FT	1,504	495	169	732	
Pratt & Whitney Canada	Canada	PT6T-3B Twin-Pac	2×(3A+C)	2×3.08	2×FT	1,702	1,118	299	1,342	
Pratt & Whitney Canada	Canada	PT6C-67	4A+C	4.3	FT	1,930	483	—	1,252	
Pratt & Whitney Canada	Canada	PW206A	C	—	FT	912	500	108	477	
Pratt & Whitney Canada	Canada	PW207D	C	—	FT	—	—	110	529	
Progress (ZMKB)	Ukraine	AI-450	C	1.72	FT	1,085	515	110	345	
Progress (ZMKB)	Ukraine	D-136	6A, 7A	35.6	FT	3,715	1,382	1,077	7,457	
Progress (ZMKB)	Ukraine	D-127	5A, 2A+C	27.4	FT	3,950	1,400	—	10,700	
PZL	Poland	PZL-10W	6A+C	4.6	FT	1,875	740	141	662	
PZL	Poland	PZL GTD-350	7A+C	2.19	FT	1,385	626	140	294	
RKBM	Russian Federation	TVD-1500V (RD-600V)	3A+C	4.0	FT	1,250	620	220	969	
Rolls-Royce	UK	Gem 42	4A, C	3.41	FT	1,099	575	183	746	
Rolls-Royce	UK	Gnome	10A	6.3	FT	1,392	577	148	1,145	
Rolls-Royce	USA	250-C20B	6A+C	1.56	FT	985	483	71.5	313	(formerly Allison)
Rolls-Royce	USA	250-C28	C	2.02	FT	1,021	557	106	373	
Rolls-Royce	USA	250-C30	C	2.54	FT	1,041	557	114	485	
Rolls-Royce	USA	250-C40B	C	2.77	FT	1,041	557	127	533	
Rolls-Royce	USA	T406	14A	16.1	FT	1,958	671	440	4,586	
RRTI	International	RTM 322–01/8	3A+C	5.75	FT	1,171	604	240	1,566	
RRTI	International	RTM 322–01/9	3A+C	5.9	FT	1,171	659	233	1,785	

Turbomeca Arriel turboshaft ***NEW***/0561744

General Electric T700 turboshaft ***(Paul Jackson)*** ***NEW***/0561734

International MTR 390 turboshaft ***NEW***/0561743

Progress AI-450 *(Paul Jackson)* ***NEW***/1043467

Manufacturer	Country	Designation/ name	Arrangement	Air flow (kg/s)	Prop drive	Length (mm)	Width (mm)	Weight, dry (kg)	T-O rating kW	Remarks
RRTI	International	RTM322–01/12	3A+C	5.75	FT	1,171	736	249	1,566	
Rybinsk	Russia	RD-600V (TVD-1500V)	3A+C	4.0	FT	1,250	620	220	1,156	
Soyuz	Russia	TV-O-100	2A+C	2.66	FT	1,275	780	125	537	
Turbomeca	France	Ardiden 1	C+C	not disclosed	FT	—	—	198	1,255	
Turbomeca	France	Ardiden 2K	C+C	not disclosed	FT	—	—	190	1,165	
Turbomeca	France	Arriel 1	1A+C	not disclosed	SS	1,090	430	120	551	
Turbomeca	France	Arriel 2	1A+C	not disclosed	SS	—	—	—	632	
Turbomeca	France	Arrius 1A	C	not disclosed	FT	782	360	87.0	380	
Turbomeca	France	Arrius 1A/1M	C	not disclosed	FT	793	367	101	357	
Turbomeca	France	Arrius 2B	C	not disclosed	FT	782	360	87.0	519	
Turbomeca	France	Arrius 2B1	C	not disclosed	FT	1,158	518	114	559	
Turbomeca	France	Arrius 2K1	C	not disclosed	—	968	470	115	500	
Turbomeca	France	Artouste III	A+C	4.31	SS	1,815	507	178	440	
Turbomeca	France	Astazou XIVM	2A+C	2.49	SS	1,470	460	160	640	
Turbomeca	France	Makila 1A1	3A+C	5.5	FT	2,103	528	241	1,357	
Turbomeca	France	Makila 1A2	3A+C	5.5	FT	2,117	498	247	1,376	
Turbomeca	France	Makila 1A4	3A+C	5.7	FT	1,836	498	247	1,567	
Turbomeca	France	TM 333	2A+C	—	FT	1,045	454	156	747	
Turbomeca	France	Turmo IIIC	1A+C	5.9	FT	2,184	637	225	1,104	
ZEF	China								522	WZ8: Turbomeca Arriel 1C made under licence

UPDATED

Jet Engines

Arrangement: A = axial stages, C = centrifugal stages, F = fan stages, a/b = afterburner or augmentor
Air flow: Flow is through fan, where applicable, not core; BPR = bypass ratio; PF = propfan
<: Various ratings up to this maximum

without reverser (other NK-8 family include it)
measured over propfan blades
has vectored nozzle(s)

W: Afterburning or augmented ('wet') thrust
D: Unaugmented ('dry') thrust of afterburning engine

Manufacturer	Country	Designation /name	Arrangement	Air flow (lbs/s	BPR	Length (mm)	Diameter (mm)	Weight, dry (kg)	T-O rating kW	Remarks
Aerosud-Marvol	South Africa/Russia	SMR-95	4F, 9A, a/b	75.5	0.49	4,229	1,080	1,235	81.4 W	
Aerosud-Marvol	South Africa/Russia								49.4 D	
Agilis	USA	turbofan	1F, 1A	n/a	5.0	1,420	580	104	4.45	
Aviadvigatel	Russian Federation	D-20P	3F, 8A	113	1.0	3,304	976	1,468	53.0	
Aviadvigatel	Russian Federation	D-21A1	5F, 10A	153	0.83	4,837	1,020	2,100	52.3	
Aviadvigatel	Russian Federation	D-30 III	5F, 10A	127	1.0	3,983	1,050	1,550	66.7	
Aviadvigatel	Russian Federation	D-30F6	5F, 10A, a/b	150	0.57	7,040	1,020	2,416	186.1 W	
Aviadvigatel	Russian Federation								93.2 D	
Aviadvigatel	Russian Federation	D-30KU	3F, 11A	269	2.42	5,700	1,560	2,668	107.9	
Aviadvigatel	Russian Federation	D-30KU-90	3F, 13A	265	2.44	5,700	1,560	2,400	117.7	
Aviadvigatel	Russian Federation	PS-7	1F+2A, 5A	n/a	n/a	n/a	n/a	n/a	68.7	
Aviadvigatel	Russian Federation	PS-9	1F+3A, 5A	n/a	n/a	n/a	n/a	n/a	100.0	
Aviadvigatel	Russian Federation	PS-14	1F+3A, 5A	n/a	n/a	n/a	n/a	n/a	137.3	
Aviadvigatel	Russian Federation	PS-18R	1F+4A, 5A	n/a	n/a	n/a	n/a	n/a	176.5	
Aviadvigatel	Russian Federation	PS-90A	2F+2A, 13A	470	4.6	4,964	1,900	2,950	156.9	
Aviadvigatel	Russian Federation	PS-90A10	1F, 12A	264	3.76	4,280	1,400	1,900	90.2	
Aviadvigatel	Russian Federation	PS-90A12	1F, 13A	370	5.05	4,795	1,670	2,300	117.7	
CFE	USA	CFE-738	1F, 5A+C	95.3	5.3	2,514	1,092	601	26.5	
CFM	International	CFM56–2B	1F+3A, 9A	370	6.0	2,430	1,735	2,119	97.9	
CFM	International	CFM56–3C	1F+3A, 9A	312	5.0	2,360	1,524	1,951	<104.5	
CFM	International	CFM56–5A	1F+3A, 9A	386	6.0	2,422	1,735	2,257	<117.9	
CFM	International	CFM56–5C	1F+4A, 9A	466	6.6	2,616	1,836	2,644	<151.3	
CFM	International	CFM56–7B	1F+4A, 9A	<355	5.1	2,507	1,549	2,384	<117.4	
CFM	International	CFM56–9	1F+2A, 9A	<282	5.08	2,329	1,422	n/a	>82.3	
Engine Alliance	USA	GP7270	IF+5A, 9A	1,361	8.7	4,750	3,150	6,085	311.4	
Eurojet	International	EJ200	3F, 5A, a/b	<77.1	c0.4	4,000	740	990	c89.0 W	
Eurojet	International								c60 D	
General Electric	USA	GE90–76B	1F+3A, 10A	1,361	9.0	5,182	3,404	7,559	340	
General Electric	USA	GE90–85B	1F+3A, 10A	1,415	8.7	5,182	3,404	7,559	377	
General Electric	USA	GE90–90B	1F+3A, 10A	1,449	8.4	5,182	3,404	7,559	401	
General Electric	USA	GE90–92B	1F+3A, 9A	1,461	8.3	5,182	3,404	7,559	409	
General Electric	USA	GE90–115B	1F+4A, 9A	1,641	8.9	c5,400	3,556	c8,618	511.6	
General Electric	USA	GEnx	IF+4A, 10A		9.2	4,623	2,822	5,642	283.76	
General Electric	USA	CF6–50C2/E2	1F+3A, 14A	658	4.3	4,394	2,195	3,956	234	
General Electric	USA	CF6–80A2/A3	1F+3A, 14A	663	4.6	3,998	2,195	3,819	222.4	
General Electric	USA	CF6–80C2	1F+3A, 14A	802	5.05	4,267	2,692	4,309	<270	
General Electric	USA	CF6–80E1	1F+3A, 14A	874	5.3	4,405	2,794	5,075	<298	
General Electric	USA	CF34–3A	1F, 14A	151	6.3	2,616	1,245	739	<41.0	
General Electric	USA	CF34–8C1	1F, 10A	c181	n/a	3,264	1,321	1,059	56.38	
General Electric	USA	CF34–8D3	1F, 10A	c181	n/a	3,264	1,321	1,120	59.17	
General Electric	USA	CF34–8E	1F, 10A	n/a	n/a	n/a	n/a	1,120	62.28	
General Electric	USA	CF34–10E	1F, 10A	n/a	5.3	n/a	n/a	n/a	82.26	
General Electric	USA	F101–102	2F, 9A, a/b	159.7	2.01	4,590	1,402	2,023	136.9 W	
General Electric	USA	F110–100	3F, 9A, a/b	113	0.87	4,630	1,181	1,492	124.6 W	
General Electric	USA								78.0 D	
General Electric	USA	F110–129	3F, 9A, a/b	122	0.76	4,620	1,181	1,791	129.0 W	
General Electric	USA								75.7 D	
General Electric	USA	F110–400	3F, 9A, a/b	113	0.87	4,620	1,181	1,599	119.2 W	
General Electric	USA								71.6 D	
General Electric	USA	F118–100	3F, 9A	130+		2,553	1,181	1,452	84.5	
General Electric	USA	F404–400	3F, 7A, a/b	64.4	0.34	4,030	880	989	71.2 W	
General Electric	USA								48.9 D	
General Electric	USA	F404–402	3F, 7A, a/b	66.2	0.27	4,030	880	1,035	78.7 W	
General Electric	USA								53.2 D	
General Electric	USA	F404–F1D2	3F, 7A	65	n/a	n/a	880	n/a	46.8	
General Electric	USA	F414–400	3F, 7A, a/b	76.8	n/a	n/a		n/a	97.9 W	
General Electric	USA								55.6 D	
General Electric/Rolls-Royce	International	F136	1F+3A, 5A	n/a	0.8	n/a	n/a	n/a	180	
GTRE	India	Kaveri	3F, 6A, a/b	n/a	n/a	n/a	n/a	n/a	81.0 W	
GTRE	India								52.0 D	

RRTI Adour *(Paul Jackson)* ***NEW***/1043457

Pratt & Whitney PW4000 series *(Paul Jackson)* ***NEW***/1043458

Pratt & Whitney PW600 *(Paul Jackson)* ***NEW***/1043460

Pratt & Whitney PW610F *(Paul Jackson)* ***NEW***/1043459

Manufacturer	Country	Designation /name	Arrangement	Air flow (lbs/s	BPR	Length (mm)	Diameter (mm)	Weight, dry (kg)	T-O rating kW	Remarks
Honda	Japan	F118	1A+2A		4.43		1,224		7.418	
Honeywell	USA	ALF502R-6	1F, 7A+C	87.0	5.6	1,487	1,059	624	33.4	
Honeywell	USA	AS907	1F, 4A+C	n/a	n/a	2,290	868.7	619	28.9	
Honeywell	USA	ATF3-6A	1F, 5A, C	73.5	2.8	2,591	853	510	24.2	
Honeywell	USA	LF507	1F, 7A+C	87.0	5.7	1,487	1,059	624	31.1	
Honeywell	USA	TFE731-2	1F, 4A+C	51.3	2.66	1,520	716	337	15.57	
Honeywell	USA	TFE731-3	1F, 4A+C	53.7	2.8	1,520	716	342	16.46	
Honeywell	USA	TFE731-5B	1F, 4A+C	64.9	3.48	1,665	754	408	21.13	
Honeywell	USA	TFE731-20	1F, 4A+C	66.2	3.1	1,547	716	406	15.57	
Honeywell	USA	TFE731-40	1F, 4A+C	65.8	2.9	1,547	716	406	18.91	
Honeywell	USA	TFE731-60	1F, 4A+C	84.8	3.9	2.083	781	448	22.24	
Honeywell	USA	TFE1042(F124)	3F, 4A+C	42.7	0.4	1,925	591	499	28.0	
Honeywell	USA	TFE1042-70(F125)	3F, 4A+C, a/b	43.3	0.4	3,561	591	617	41.1 W	
Honeywell	USA								26.8 D	
IAE	International	V2522-A5	1F+4A, 10A	384	4.9	3,200	1,600	2,359	97.9	
IAE	International	V2527-A5	1F+4A, 10A	384	4.75	3,200	1,600	2,500	117.9	
IAE	International	V2528-D5	1F+4A, 10A	384	4.8	3,200	1,600	2,540	124.6	
IAE	International	V2533-A5	1F+4A, 10A	384	4.4	3,200	1,600	2,500	146.8	
IHI	Japan	F3-30	2F, 5A	34.0	0.9	1,340	560	340	16.37	
IL	Poland	D-18A	2F, 5A	38.4	0.7	1,940	750	380	17.65	
IL	Poland	K-15	6A	23.5	0	1,560	725	320	14.71	
IL	Poland	SO-3	7A	n/a	0	2,151	707	321	10.79	
KKBM	Russian Federation	NK-8-2U	2F+2A, 6A	228	1.049	5,288	1,442	2,350	103.0	
KKBM	Russian Federation	NK-8-4	2F+2A, 6A	222	1.042	5,101	1,442	2,440	103.0	
KKBM	Russian Federation	NK-86	2F+2A, 6A	292	1.6	3,638[1]	1,600	2,450	127.5	
Klimov	Russian Federation	RD-33	4F, 9A, a/b	75.5	0.49	4,229	1,040	1,055	81.4 W	
Klimov	Russian Federation								49.4 D	
Klimov	Russian Federation									RD-35 — see PS/ZMK DV-2
Klimov	Russian Federation	RD-133	4F, 9A	77	0.49	4,230	1,040	1,145	88.25	
Klimov	Russian Federation	RD-43	4F, 9A	85	n/a	4,230	n/a	c1,000	98.01	
LM	China	WP6	9A, a/b	46.3	0	5,483	668	725	39.72 W	
LM	China								29.42 D	
LM	China	WP6A	9A, a/b	46.3	0	5,483	668	725	36.78 W	
LM	China	WS6	3F, 11A, a/b	155	1.0	4,654	1,370	2,100	122.1 W	
LM	China								71.1 D	
LMC	China								138.3 W	WS6 uprated for JH-7
LMC	China								71.3 D	
LMC	China	WP7	3A+5A, a/b	65.0	0	4,600	825	1,053	59.8 W	
LMC	China								43.2 D	
LMC	China	WP7F	3A+5A, a/b	n/a	0	n/a	n/a	n/a	63.7 W	
LMC	China								44.1 D	
LMC	China	WP13	3A+5A, a/b	65.6	0	4,600	907	1,211	64.7 W	
LMC	China								40.2 D	
LMC	China	WP13A II	3A+5A, a/b	65.6	0	5,150	907	1,201	65.9 W	
LMC	China								47.1 D	
LMC	China	WP13B	3A+5A, a/b	n/a	0	n/a	n/a	n/a	68.7 W	
LMC	China								47.1 D	
LMC	China	WP13F	3A+5A, a/b	n/a	0	n/a	n/a	n/a	64.7 W	
LMC	China								44.1 D	
Powerjet	International, France/Russia	SM146	1F, 6A	n/a	4.43	n/a	1,224	n/a	70.82	
Pratt & Whitney	USA	F100-220	3F, 10A, a/b	n/a	0.7	5,280	1,181	1,451	106.0 W	
Pratt & Whitney	USA								65.3 D	
Pratt & Whitney	USA	F100-220P	3F, 10A, a/b	n/a	0.6	5,280	1,181	1,526	120.1 W	
Pratt & Whitney	USA								74.3 D	
Pratt & Whitney	USA	F100-229	3F, 10A, a/b	n/a	0.36	4,855	1,181	1,681	129.5 W	
Pratt & Whitney	USA								79.2 D	
Pratt & Whitney	USA	F117-100	1F+4A, 12A	608	6.0	3,729	2,154	3,220	181.0	
Pratt & Whitney	USA	F119-100	3F, 6A, a/b		0.45				c155.6 W	
Pratt & Whitney	USA	F135	3F, 9A, a/b	n/a	n/a	n/a	n/a	n/a	c177.9 k WN	
Pratt & Whitney	USA	J52-408	5A, 7A	64.9	0	3,020	814	1,052	49.8	
Pratt & Whitney	USA	JT8D-219	1F+6A, 7A	221	1.77	3,911	1,250	2,092	93.4	
Pratt & Whitney	USA	JT9D-7R4H	1F+4A, 11A	769	4.8	3,371	2,463	4,029	249	
Pratt & Whitney	USA	PW2040	1F+4A, 12A	608	6.0	3,729	2,154	3,311	185.5	
Pratt & Whitney	USA	PW4050	1F+4A, 11A	798	5.1	3,371	2,463	4,179	222	
Pratt & Whitney	USA	PW4060	1F+4A, 11A	816	4.8	3,371	2,463	4,179	267	

Williams FJ44-1 *(Paul Jackson)* ***NEW***/1043454

Williams FJ33 *(Paul Jackson)* ***NEW***/1043455

Manufacturer	Country	Designation /name	Arrangement	Air flow (lbs/s	BPR	Length (mm)	Diameter (mm)	Weight, dry (kg)	T-O rating kW	Remarks
Pratt & Whitney	USA	PW4168	1F+5A, 11A	903	5.1	4,143	2,715	6,396	305	
Pratt & Whitney	USA	PW4084	1F+6A, 11A	1,089	6.4	4,869	3,010	6,768	386	
Pratt & Whitney	USA	PW4098	1F+7A, 11A	1,293	5.8	4,945	3,035	7,484	436	
Pratt & Whitney	USA	PW6000	1F+4A, 6A	n/a	4.9	2,728	1,435	2,359	100.24	
Pratt & Whitney	USA	PW8000	1F+3A, 5A	n/a	10.0	3,150	1,930	3,629	<155.7	
Pratt & Whitney	USA	PW7000 (XTE-66)	n/a	n/a	n/a	n/a	n/a	n/a	<155.7	
Pratt & Whitney Canada	Canada	JT15D-1B	1F+C	34.0	3.3	1,506	691	235	9.79	
Pratt & Whitney Canada	Canada	JT15D-5	1F+1A, C	42.0	n/a	1,549	711	288	12.9	
Pratt & Whitney Canada	Canada	PW305	1F, 4A+C	n/a	4.3	2,070	970	450	23.4	
Pratt & Whitney Canada	Canada	PW306	1F, 4A+C	n/a	4.5	1,920	970	473	25.4	
Pratt & Whitney Canada	Canada	PW308	1F, 4A+C	n/a	4.5	n/a	n/a	n/a	35.6	
Pratt & Whitney Canada	Canada	PW308A	1F, 4A+C	n/a	3.8	n/a	n/a	n/a	28.9	
Pratt & Whitney Canada	Canada	PW530A	1F, 2A+C	n/a	3.9	1,524	701	278	13.34	
Pratt & Whitney Canada	Canada	PW535A	1F, 2A+C	n/a	3.9	1,524	701	n/a	14.94	
Pratt & Whitney Canada	Canada	PW600	1F, 1A+C	n/a	n/a	673	356	c150	6.0	
Pratt & Whitney Canada	Canada	PW800	1F, 6A	n/a	n/a	n/a	n/a	n/a	44.5+	
Progress (ZMKB)		AI-22	1F+5A, 7A	45.6	4.57	2,400		750	37.95	
Progress (ZMKB)	Ukraine	AI-222–25F	2F, 8A	50.3	1.19	3,010	810	540	41.1 W	
Progress (ZMKB)	Ukraine								24.5 D	
Progress (ZMKB)	Ukraine	AI-222–25KFK	2F, 8A	50.3	1.19	2,460	810	520	30.07 W	
Progress (ZMKB)	Ukraine								24.5 D	
Progress (ZMKB)	Ukraine	AI-222-28	2A, 8A	49.4	1.18	1,960			29.42	
Progress (ZMKB)	Ukraine	AI-25TL	3F, 7A	46.8	2.0	1,993	820	350	16.87	
Progress (ZMKB)	Ukraine	AI-25TLSH	3A, 8A	46.8	1.98					
Progress (ZMKB)	Ukraine	D-18T	1F, 7A, 7A	765	5.6	5,400	2,330	4,100	230	
Progress (ZMKB)	Ukraine	D-18TM	1F, 7A, 7A	770	5.5	5,700	2,330	4,750	248	
Progress (ZMKB)	Ukraine	D-27 propfan	2PF, 5A, 2A+C	27.4	29.95	n/a	n/a	1,650	109.8	
Progress (ZMKB)	Ukraine	D-36	1F, 6A, 7A	255	5.6	3,470	1,373	1,109	63.7	
Progress (ZMKB)	Ukraine	D-436T	1F, 6A, 7A	265	4.98	3,470	1,373	1,250	73.5	
Progress (ZMKB)	Ukraine	D-436T2	1F+A, 6A, 7A	265	4.9	4,157	1,655	1,450	80.4	
PS/ZMK	International	DV-2	1F+2A, 7A	49.5	1.46	1,721	994	440	21.58	
PW/MTU	International	MTFE	1F+2A, 6A	c370	5.2	2,515	1,397	1,769	89.0	
RKBM	Russian Federation	RD-41	7A	53.5	0	1,594	635	290	40.2	
Rolls-Royce	Germany	BR710	1F, 10A	197	4.2	3,409	1,321	1,633	<68.9	
Rolls-Royce	Germany	BR715	1F+2A, 10A	288	4.5	3,599	1,575	2,114	<102.3	
Rolls-Royce	UK	535C	1F, 6A, 6A	518	4.4	3,010	1,877	3,309	166.4	
Rolls-Royce	UK	535E4B	1F, 6A, 6A	522	4.3	2,995	1,892	3,261	<191.7	
Rolls-Royce	UK	Pegasus 11–61	3F, 8A[3]	208	1.2	3,485	1,222	1,932	105.9	
Rolls-Royce	UK	RB211–22B	1F, 7A, 6A	626	5.0	3,033	2,154	4,171	186.8	
Rolls-Royce	UK	RB211–524B	1F, 7A, 6A	671	4.4	3,106	2,180	4,452	222	
Rolls-Royce	UK	RB211–524G/H	1F, 7A, 6A	728	4.3	3,175	2,192	4,387	273	
Rolls-Royce	UK	Spey 512	5A, 12A	94.3	0.71	2,911	942	1,168	55.8	
Rolls-Royce	UK	Spey 807	5A, 12A	91.6	0.96	2,456	825	1,096	49.1	
Rolls-Royce	UK	Tay 611, 620	1F+3A, 12A	176	3.18	2,405	1,118	1,422	61.6	
Rolls-Royce	UK	Tay 651	1F+3A, 12A	193	3.10	2,405	1,138	1,533	68.5	
Rolls-Royce	UK	Trent 556	1F, 8A, 6A	858	8.5	3,912	2,474	4,719	252	
Rolls-Royce	UK	Trent 600	1F, 8A, 6A	922	8.0	3,912	2,474	4,719	306	
Rolls-Royce	UK	Trent 772	1F, 8A, 6A	897	4.9	3,912	2,474	4,785	320	
Rolls-Royce	UK	Trent 892	1F, 8A, 6A	1,200	5.7	4,369	2,794	5,957	411	
Rolls-Royce	UK	Trent 895	1F, 8A, 6A	1,217	5.79	4,369	2,794	5,981	425	
Rolls-Royce	UK	Trent 977	1F, 8A, 6A	1,225	8.7	4,546	2,946	6,181	340.3	
Rolls-Royce	UK	Trent 1000	1F, 8A, 6A	1,247	11	3,861	2,845		311.4	
Rolls-Royce	UK	Viper 535	8A	23.9	0	1,806	740	358	14.94	
Rolls-Royce	UK	Viper 632	8A	26.5	0	1,806	740	376.5	17.66	
Rolls-Royce	UK	Viper 680	8A	27.2	0	1,806	740	379	19.39	
Rolls-Royce	USA	AE 3007A	1F, 14A	96.6	5.0	2,705	1,105	717.1	33.7	
Rolls-Royce	USA	AE 3007H	1F, 14A	118	5.0	2,705	1,105	717.1	36.9	
RRTI	International	Adour 811/815	2F, 5A, a/b	43.1	0.75	2,970	559	738	37.4 W	
RRTI	International								24.6 D	
RRTI	International	Adour 871	2F, 5A	44.0	0.80	1,956	559	603	26.6	
Samara	Russian Federation	NK-25	3F, 9A	339	0.9	5,200	1,348	2,850	245	
Samara	Russian Federation	NK-44	n/a	n/a	6.0	n/a	3,100	n/a	392	

Williams FJ44-2 *(Paul Jackson)* ***NEW***/1043453

Internal details of Pratt & Whitney PW610F ***NEW***/1043456

Manufacturer	Country	Designation /name	Arrangement	Air flow (lbs/s)	BPR	Length (mm)	Diameter (mm)	Weight, dry (kg)	T-O rating kW	Remarks
Samara	Russian Federation	NK-93	2PF, 7A, 8A	1,000	17.0	5,972	2,900[2]	3,650	176.5	
Samara	Russian Federation	NK-321	3F, 5A, 7A, a/b	365	1.4	c6,000	1,460	3,442	245 W	
Samara	Russian Federation								137.2 D	
SARI	China	Kunlun	3A, 5A	n/a	n/a	907	n/a	1,200	76.53	
Saturn	Russian Federation	AL-21F3	14A, a/b	104	n/a	5,340	1,030	1,800	110.5 W	
Saturn	Russian Federation								76.5 D	
Saturn	Russian Federation	AL-31FM	4F, 9A, a/b	112	0.57	4,950	1,277	1,488	122.6 W	
Saturn	Russian Federation								79.3 D	
Saturn	Russian Federation	AL-37FU	4F, 9A, a/b	n/a	0.65	c5,000	1,277	1,660	142.2 W	
Saturn	Russian Federation								83.4 D	
Saturn	Russian Federation	AL-41 (SAT-41)	2F, 6A, a/b	c150	n/a	n/a	n/a	c1,850	c196.1 W	
Saturn	Russian Federation								c113.8 D	
Saturn	Russian Federation	AL-55	4F, 6A	29.8	0.6	1,210	590	315	21.58	
Saturn	Russian Federation	AL-55F	4F, 6A, a/b	29.8	0.6	2,520	590	385	34.34	
SNECMA	France	Atar 9K50	9A, a/b	73.0	0	5,944	1,020	1,582	70.6 W	
SNECMA	France								49.2 D	
SNECMA	France	M53 P2	3F, 5A, a/b	86.0	0.35	5,070	1,055	1,500	95.0 W	
SNECMA	France								64.3 D	
SNECMA	France	M88-2	3F, 6A, a/b	67.0	0.25	3,540	660	909	75.0 W	
SNECMA	France								50.0 D	
Soyuz	Russian Federation	R-195	3A	67.0	0	3,300	914	990	44.13	
Soyuz	Russian Federation	R-29-300	5A, 6A, a/b	110	0	4,960	912	1,880	122.3 W	
Soyuz	Russian Federation								78.5 D	
Soyuz	Russian Federation	R-35-300	5A, 6A, a/b	110	0	4,950	912	1,765	127.5 W	
Soyuz	Russian Federation								83.9 D	
Soyuz	Russian Federation	R-79	5F, 6A, a/b[3]	120	1.0	5,229	1,100	2,750	152.0 W	
Soyuz	Russian Federation								107.6 D	
Soyuz	Russian Federation	RD-1700	2F, 4A	30	0.78	n/a	n/a	297.58	19.61 W	
Soyuz	Russian Federation								16.67 D	
Turbomeca-SNECMA	France	Larzac 04-20	2F, 4A	28.6	1.04	1,179	602	302	14.12	
Turbo-Union	International	RB199 Mk 105	3F, 3A, 6A, a/b	74.6	c1	3,302	752	980	74.7 W	
Turbo-Union	International								42.5 D	
Volvo/GE	Sweden	RM12	3F, 7A, a/b	68.0	0.28	4,100	880	1,050	80.5 W	
Volvo/GE	Sweden								54.0 D	
Williams and Williams Rolls	USA	FJ33-4A	1F+2A, C	n/a	n/a	978	534.7	<136	<6.7	
Williams and Williams Rolls	USA	FJ44-1A	1F+1A, C	29.575	3.28	1,356	530.9	205	8.45	
Williams and Williams Rolls	USA	FJ44-1C	1F+1A, C	26.49	3.4	1,356	530.9	208	6.67	
Williams and Williams Rolls	USA	FJ44-2A	1F+3A, C	35.8	4.1	1,507	553.7	n/a	10.23	
Williams and Williams Rolls	USA	FJ44-2C	1F+3A, C	n/a	n/a	1,507	553.7	n/a	10.68	
Williams and Williams Rolls	USA	FJ44-3A	1F+3A, C			2,091	573	238	10.23	
Williams and Williams Rolls	USA	FJ44-4A	1F+3A, C						15.575	
XAE	China									WS9: Rolls-Royce Spey Mk 202 made under licence; data presumed identical
XAE	China	WP8	8A	150	n/a	5,380	1,400	3,133	93.2 W	

UPDATED

Auxiliary Power Units (APUs)

Manufacturer	Country	Designation/name	Start up ceiling(m)	Weight, dry (kg)	S/L compression(bar)	Ground power(kVA)	Rating kW	Remarks
Aerosila	Russia	TA-4FE	<2,500	180	2.4	18	124	
Aerosila	Russia	TA-6A	<3,000	245	4.4	40	235	
Aerosila	Russia	TA-6V	<3,000	255	4.4	40	235	
Aerosila	Russia	TA-8	<3,000	165	3.2	—	107	
Aerosila	Russia	TA-8V	<5,000	185	3.2	40	107	
Aerosila	Russia	TA-8K	<3,500	178	3.1	60	55	
Aerosila	Russia	TA-12	<7,000	290	4.8	40	287	
Aerosila	Russia	TA-12-60	<7,000	335	4.8	60	287	
Aerosila	Russia	TA-14	<8,000	50	3.6	20	79	
Aerosila	Russia	TA-18-100	<11,000	140	4.4	60	214	
Honeywell	USA	36-100	6,095	53	3.52	30	127	
Honeywell	USA	36-150W	6,095	52-55	3.24	10	142	
Honeywell	USA	36-150CX/IAI	6,095	57	3.38	10	142	
Honeywell	USA	36-150F2	6,095	55	3.17	10	142	
Honeywell	USA	36-150F	6,095	53	3.24	10	142	
Honeywell	USA	36-150DD	6,095	53	3.24	10	145	
Honeywell	USA	36-150 M	6,095	65	3.59	23	145	
Honeywell	USA	36-150RJ	11,280	73	3.11	30	179	
Honeywell	USA	RE50		25				
Honeywell	USA	RE100	6,095	40	3.45	10	101	
Honeywell	USA	RE220	13,105	109	3.86	45	261	
Honeywell	USA	36-280	10,670	159	3.24	90	291	
Honeywell	USA	36-300	10,670	136	3.24	90	291	
Honeywell	USA	131-9A	12,495	159	3.52	90	343	
Honeywell	USA	131-9B	12,495	186	3.52	96	90	
Honeywell	USA	331-200	13,105	227	3.45	90	432	
Honeywell	USA	331-250	13,105	227	3.04	90	447	
Honeywell	USA	331-350	12,495	254	3.45	115	618	
Honeywell	USA	331-500	13,105	331	3.65	120	969	
Honeywell	USA	700	7,620	295	3.05	100	648	
Microturbo	France	Rubis 3	7,600	56	5.00	6	125	
Microturbo	France	Saphir 20	6,000	38		10	70	
Microturbo	France	Saphir 100	4,000	74	4.5		135	
Omsk	Russian Federation	VSU-10						
Pratt & Whitney	Canada	PW901	7,620	379	3.72	180	1,145	
ZMKB Progress (design), **Motor Sich** (manufacture)	Ukraine	AI-8	7,620	145	—	60	60	
ZMKB Progress (design), **Motor Sich** (manufacture)	Ukraine	AI-9V	—	70	2.89			
ZMKB Progress (design), **Motor Sich** (manufacture)	Ukraine	AI-9-3B	6,000	128	3.25	—	16	
Stupino	Russian Federation	TA-6						
Stupino	Russian Federation	TA-12						
Stupino	Russian Federation	TA-14						
Sundstrand	USA	APS500	<9,145	46-52	c3.6	19	200	
Sundstrand	USA	APS1000	<7,620	70-82	4.14	45	261	
Sundstrand	USA	APS2000	11,280	132	3.65	60	376	
Sundstrand	USA	APS2100	11,280	122	3.65	60	376	
Sundstrand	USA	APS3200	11,885	138	4.14	90	450	
Ufa	Russian Federation	VD-100						

Honeywell RE50 *(Paul Jackson)* ***NEW***/1043450

ZMKB Progress AI-9-3B *(Paul Jackson)* ***NEW***/1043447

UPDATED

PROPELLERS

Introduction

In the table will be found explanations of various manufacturers' designation systems, to be read in conjunction with data supplied in the 'Engines' paragraph of aircraft descriptions.
First and simplest of the propeller types is **fixed-pitch**, constructed of carved wood (normally glued multiple-ply for warp resistance), wood with composites skin, or aluminium. Pitch is blade angle in respect to a flat plane; proportional to the distance the propeller will advance through the air on each rotation. Fine pitch — a small angle — allows a low-power engine to develop a high rotational speed, but delivers low forward airspeed. Coarse pitch is optimised for higher speeds, but requires proportional engine power.
While diameter and blade area are related factors in performance, as is the subtle change of pitch along the blade's length, fixed-pitch is, by definition, a compromise made at the time of manufacture. Light aircraft options are normally restricted to the operator's choice of a fine-pitch 'climb propeller' or coarse-pitch 'cruise propeller', according to operating environment and preference.
Increasingly, light aircraft have a **ground-adjustable pitch** propeller. Blades are mounted in sockets in the hub and clamped in place by a bolt. In theory, the pitch can be re-set to meet the specifics of each flight, but it is more normal for the owner to experiment with blade angles until the optimum for the aircraft is reached. Wood and wood/composites are used in construction, as is carbon fibre.
For special applications, such as motor gliders, a fixed-pitch, but **folding propeller** offers the simplest means of eliminating propeller drag in engine-off models. The propeller is at rest in a folded position (blades either parallel to the axis of motion or at right-angles to it, held in place by spring or bungee rubber) and extends when engine torque is applied
Variable pitch enables fine pitch to be used to enhance take-off and initial climb performance (maximum engine power at low airspeed), before a change to coarse pitch for fastest cruising speed without overspeeding the engine. Before and during landing, fine pitch and a low throttle setting will have the effect of increasing drag. Several methods of variable pitch are available:
Two-position: The propeller is pre-set in fine pitch for take-off and the pilot provided with a once-only means of reverting to coarse pitch. Rare in current usage.
Automatic pitch: Automatic alteration of pitch angle according to airspeed, often indicated by a vaned spinner ahead of the propeller. Rare in current usage.
Flight-adjustable pitch: Change controlled manually by the pilot and constantly monitored for each stage of flight, in concert with selected engine power. Electric or hydraulic actuation. The term 'variable pitch' is often used, loosely, to describe flight-adjustable pitch.
Constant-speed: A governor (constant-speed unit) is set by the pilot to the desired engine (and propeller) speed and pitch is constantly and automatically altered to maintain the desired value. Throttle changes (in fact, manifold pressure variations) can then be made by the pilot without reference to the propeller, which will adjust to the correct setting. However, the setting is changed manually for each stage of the flight (take-off, climb and cruise).
The availability of flight-adjustable and constant-speed mechanisms permits two further options:
Feathering: Blade pitch can be extended to neutral (edge-on to the airflow), reducing drag upon engine failure, or when deliberately gliding. The propeller may 'windmill' (turn slowly on account of residual pitch).
Reverse pitch: Change of pitch beyond feathering angle to produce forward thrust for more rapid stopping or ground and water manoeuvring.

UPDATED

Model of the immense, 5.33 m (17 ft 6 in) Ratier-Figeac propeller being produced for the Airbus A400M airlifter *(Paul Jackson)* ***NEW***/1044625

From craftsman-carved to computer-controlled, the techniques for propeller manufacture have changed beyond recognition since a hand-crafted, wooden propeller was attached to the Armstrong Siddeley Genet Major of this Southern Martlet in 1931. The aircraft still flies regularly at Old Warden, England, with the Shuttleworth Trust *(Paul Jackson)* ***NEW***/1044626

Propellers

Manufacturer	Example designation	Explanation
Aerosila Aerosila NPP OAO ulitsa Metallistov 6 Stupino 142800 Moscovskoya oblast Russian Federation *Tel:* (+7 095) 333 22 72 *Fax:* (+7 095) 333 51 47 *e-mail:* vint@aerosila.ru *Web:* http://www.aerosila.ru	AV-17 AV-24AN SV-27 SV-34 AV-36 SV-92 AV-101 AV-110 AV-140	Three-blade, variable pitch; diameter 3.60 m Three-blade, variable pitch, Antonov An-28; diameter 2.80 m 8 + 8 contrarotating propfan; diameter 4.50 m Six-blade, constant-speed; diameter 3.60 m Six-blade, constant-speed; diameter 2.65 m 8 + 10 ducted fan; diameter 2.90 m Two/three/four-blade, variable pitch; diameter 1.20 m Four-blade, variable pitch; diameter 1.40 m Six-blade, constant-speed; diameter 3.72 m
Airmaster Airmaster Propellers Ltd Unit B 144 Central Park Drive PO Box 21220 Henderson New Zealand *Tel:* (+64 9) 836 00 65 *Fax:* (+64 9) 836 00 69 *e-mail:* sales@propellor.com *Web:* http://www.propellor.com	AP308 AP332	Electric variable pitch (constant speed); three-blade Electric variable pitch (contant speed); feathering; three-blade; max diameter 72 in/1,830 mm
Avia Avia Propeller s.r.o. Beranových 666 CZ-199 02 Prague 9 Czech Republic *Tel:* (+420 266) 11 20 30 *Fax:* (+420 266) 11 20 31 *e-mail:* brunclik@aviapropeller.cz *Web:* http://www.aviapropeller.com	AV 723-5-B AV 803-3 AV 842 AV 843 AV 844-5W-E AV 845	Aluminium alloy; Avia; variable pitch (G = ground adjustable, F = fixed pitch); 72 mm blade shank diameter; 3 blades; constant-speed, non-feathering, non-counterweighted (1 = constant speed, non-feathering, counterweighted, 2 = CS, feathering, 3 = CS, feathering, reversing, 4 = constant speed, reversing, 6 = autonomous); SAE 2 flange with ½ in bolts (A = special flange, 80 mm bolts, C = SAE 2 7/16 in bolts, D = ARP 502 flange, E = ARP 880 flange, F = SAE 1 flange, G = Walter/LOM flange, H = PW115 flange) Aluminium alloy; Avia; hydraulic variable pitch (constant-speed); three blade; diameter 99 in/2,500 mm; anticlockwise rotation (803-4J-E = clockwise) Aluminium alloy; Avia; hydraulic variable pitch (constant-speed and optional feathering); two-blade; diameter 72–80 in/1,830–2,040 mm Aluminium alloy; Avia; hydraulic variable pitch (constant-speed and optional feathering); three-blade; diameter 88 in/2,235 mm (843-5W-E = 84 in/2,080 mm diameter, reversing) Aluminium alloy; Avia; hydraulic variable pitch (constant-speed, optional feathering); four blade; diameter 82 in/2,080 mm Aluminium alloy; Avia hydraulic variable pitch (constant speed); five blade; diameter 82 in/2,080 mm
	V-500A V 508D	2,000 mm diameter (/1690 = 1,690 mm diameter, /1905 = 1,905 mm diameter). Two-blade 2,500 mm diameter. Three-blade. Walter M601D engine. (508Z = Walter M601Z, 508E/84 = 84 in diameter, 508E-AG/99B = 99 in diameter, agricultural, 508E 106/2690 = 106 in/2,690 mm diameter, 508E 91/2300 = 91 in/2,300 mm diamter, 508H = 106 in/2,690 mm diameter, 633 kW/850 hp rating)
	V 510 V 520	2,300 mm diameter. Five-blade. Walter M601E-F. (510T = Walter M601F-T, 510AG agricultural.) 2,700 mm diameter. Two-blade
Avtek Avtek Srl Via A Gramsci 18 I-20050 Zoccorino (MI) Italy *Tel:* (+39 0362) 60 20 77 *Fax:* (+39 0362) 80 26 20 *e-mail:* lucianomerati@tin.it		Carbon fibre, two- or three-blade, flight-adjustable
Baoding Baoding Propeller Plant Box 609, PO Box 818 Baoding 072152 Hebei People's Republic of China *Tel:* (+86 312) 705 15 82 *Fax:* (+86 312) 705 14 54	J17-G13	Four-blade, constant-speed, fully feathering, metal; diameter 4.50 m
Bolly The Propeller Company Pty Ltd Unit 8 100 Hewittson Road Elizabeth West, SA 5113 Australia *Tel:* (+61 08) 82 55 96 88 *Fax:* (+61 08) 82 55 96 66 *e-mail:* bolly@bolly.com.au *Web:* http://www.bolly.com.au	BOS1-54-24-N-L BOS2-58-33-S-R BOS3-62-52-S-R BOS4-54-22-N-L	Bolly Optima Series 1; 54 in diameter (60 = 60 in) ; 24 in nominal pitch (22 = 22 in); narrow aspect ratio blade (S = Standard, W = wide); left-hand (anticlockwise) rotation viewed from behind (R = clockwise). Composites, fixed pitch; two-blade Series 2; 58 (55½, 59½, 62) in diameter; 33 (31, 34, 36, 38, 39, 42) in pitch; standard aspect ratio; clockwise rotation. Composites; individual, bolted, fixed pitch blades; two to five blades Series 3; 62 (68, 72) in diameter; 52 (58, 60) in pitch; S, R as above. Composites, ground-adjustable pitch; two-, three- or four-blade Series 4 (folding blades); 54 in diameter; 22 in pitch; N, L as above. Composites, ground-adjustable pitch; two-blade
Dowty Dowty Propellers Anson Business Park Cheltenham Road East Staverton Gloucester GL2 9QN United Kingdom *Tel:* (+44 1452) 71 60 00 *Fax:* (+44 1452) 71 61 01 *Web:* http://www.dowty.com	R352/6-123-F/1 R381 R391 R408 R414	Rotol type 352; six-blade; blade shank size 123; engine flange mounted; first subvariant Six-blade, constant-speed, feathering/reversing; Saab 2000 diameter 12 ft 6 in Six-blade, constant-speed, feathering/reversing; composites; diameter 13 ft 6 in; Lockheed C-130J, Alenia C-27J Six-blade, constant-speed, feathering/reversing; composites; diameter 13 ft 6 in; Dash 8 Q400 Six-blade, constant-speed, feathering/reversing; composites; ShinMaywa US-1A Kai
EVRA EVRA 4 avenue de la Forêt d'Halatte F-60100 Creil France *Tel:* (+33 3) 44 25 50 31 *Fax:* (+33 3) 44 25 90 72	160.150.136 D9.28 D11.29 6006 Merville DH.5220 180.170 Evolution	Wood or composites; various specific designations according to aircraft type

Manufacturer	Example designation	Explanation
GT GT Propellers Viale del Commercio I-47838 Riccione (RN) Italy *Tel:* (+39 0541) 69 33 99 *Fax:* (+39 0541) 69 33 3[illegible] *e-mail:* info@gt-propellers.com *Web:* http://www.gt-propellers.com		Wood, two- and three-blade, fixed pitch Wood, two-, three- and four-blade, fixed pitch, foldable Wood, two-, three- and four-blade, ground-adjustable pitch Wood, two-, three- and four-blade, electric or hydraulic variable pitch
Halter Helices Halter SARL F-05130 Tallard France *Tel:* (+33 4) 92 54 05 41 *Fax:* (+33 4) 92 54 05 42 *e-mail:* halter@uav-propellers.com *Web:* http://www.silent-propellers.com		Produced to customer's specification
Hamilton Sundstrand Hamilton Sundstrand Propulsion Systems One Hamilton Road Windsor Locks Connecticut 06096-1010 *Tel:* (+1 860) 654 27 72 *e-mail:* props.info@hs.utc.com *Web:* http://www.hamiltonsundstrandcorp.com	14SF-5 247F-1 568F-1	Four-blade, 13 ft diameter; aluminium spar with glass fibre blade shell; hydromechanical constant-speed; ATR 42 (7 = Dash 8 Q100, 11 = ATR 72, 11E = ATR 72 electronic pitch control, 15 = Dash 8 Q100/Q200/Q300, 17 = Canadair 215T, 19 = Canadair 215T/415, 23 = Dash 8 Q100/Q200/Q300) Four-blade, 13 ft diameter, graphite spar and Kevlar shell; hydromechanical constant-speed; ATR 72 Srs 211/212 (1E = ATR 72 Srs 211/212 electronic pitch control; 3 = XAC MA-60) Six-blade, 13 ft diameter, graphite spar and Kevlar shell; electromechanical constant-speed; ATR 42 Srs 400/500 (5 = CASA C-295, 7 = Ilyushin Il-114, 11 = previously assigned to Ayres Loadmaster)
Hartzell Hartzell Propeller Inc One Propeller Place Piqua Ohio 45356-2634 USA *Tel:* (+1 937) 778 42 00 *Fax:* (+1 937) 778 43 91 *Web:* http://www.hartzellprop.com	BHC-J2YF-1BF/F7694	Hartzell hub series B (blank = A, C = C, D = D); two-blade (3 = three, 4 = four); /; blades diameter 76 in; subtype 94 (7068 = 70 in, 7280 = 72/72½ in, 7663 = 76 in, 7666 = 76 in, 7681 = 76 in, 7692 = 76 in, 8052 = 80 in, 8459 = 84 in)
Helix Helix Carbon GmbH Düserhofstraße 20 D-52074 Aachen *Tel:* (+49 241) 931 92 17 *Fax:* (+49 241) 931 92 18 *e-mail:* mail@helix-propeller.de *Web:* http://www.helix-propeller.de	H50V1, 45mL-S2 H50F1, 75mR-S-16-2	Carbon Fibre; fixed pitch 1.45 m diameter (1.60 m, 1.75 m), 2.20 m); left-hand rotation; S (N) blade design; two-blade (3 = three, 4 = four) Carbon fibre; fixed pitch; 1.75 m diameter; right-hand rotation; S(c) blade design; 16° pitch (9°); two-blade (3 = three)
Hoffmann Hoffmann Propeller GmbH Küpferlingstrasse 9 D-83022 Rosenheim Germany *Tel:* (+49 8031) 187 80 *Fax:* (+49 8031) 18 78 78 *e-mail:* info@hoffmann-prop.com *Web:* http://www.hoffmann-prop.com	HO-V343K-V/S180GY-B+10 HO27HM-180/138	Hoffmann; Variable pitch; Type 34; three blades; AS 127D flange, SAE 2, ½ in bolts (F = ARP 502 Type 1, L= AS 127D SAE 2, $\frac{7}{16}$ in bolts, B = AS 127D SAE 1, G = ⅜ in bolts); hydraulic pitch change, oil pressure to decrease/counterweights to increase (blank = oil pressure increase/no counterweights, F = oil pressure increase/counterweights, S = oil pressure decrease/counterweights to increase up to feathering); /; feathering blades (blank = right hand, tractor, D = right hand, pusher, L = left hand, tractor, LD = left hand, pusher, V = pitch change pin position, altered to allow oil pressure to decrease pitch); 1,800 mm diameter; blade design GY; electrical de-icing; 10 cm over basic size Fixed pitch Hoffmann, 1,800 mm diameter
Kasparaero Kaspar a synove — strojirna Kalmar sro Nadrazni 76 CZ-150 00 Praha 5-Smichov Czech Republic *Tel:* (+420) 257 32 01 39 *Fax:* (+420) 257 32 22 26	KA-1/2-PA	Flight adjustable (hydraulic); diameter 1,610 mm; Kasparaero; Type 1; two-blade (3 = three-blade); -; right-hand rotation (L = left-hand); tractor (T = pusher)
KievProp c/o ULM Europe Ch du Rié 4 CH-1053 Bretigny Switzerland *Tel:* (+41 21) 732 29 87 *e-mail:* prop@kievprop.com.ua *Web:* http//www.kievprop.com.ua	163,173 PV-173	Composites; ground-adjustable pitch; three-blade; left-hand rotation; 1,710 mm diameter (183=1,810 mm; 263/273/283 = right-hand rotation; 165/175/185 = five-blade; 265/275/285 = five-blade, right-hand rotation) Variable-pitch (mechanical), as 173 (273 = as 273), but 1,800 mm diameter
LOM Praha LOM Praha SP Černokostelcká 270 CZ-100 38 Praha 10 Czech Republic *Tel:* (+420 270) 36 38 *Fax:* (+420 270) 65 23 *e-mail:* lompraha@mbox.vol.cz *Web:* http://www.lom.cz	V-231 V-341 V-532 V-541 V-546	Two-blade, wood, fixed-pitch, 1,650 to 1,800 mm diameter Two-blade, metal, ground-adjustable pitch, 1,600 to 1,900 mm diameter Three-blade, wood, hydraulically variable pitch, 1,850 mm diameter Two-blade, metal, hydraulically constant-speed, 1,750 to 1,900 mm diameter Three-blade, metal, hydraulically constant-speed, reversing/feathering, 1,750 to 1,900 mm diameter
McCauley McCauley Propeller Systems PO Box 5053 Vandalia Ohio 45377-5053 *Fax:* (+1 937) 890 60 01 *Web:* http://www.mccauley.textron.com	D3AF32C72-A/S-82NC-0 1A100/MCM6950	Dowel (indexing hole) location 90° and 270° (blank = 60°/240°, B = 0°/180°, C = 30°/210°); three-blade; 4 in flange (D = SAE 2 flange, F = 4½ in flange); feathering; shank size 32; constant-speed; version change 72; minor change; / ; blade minor change; 82 in diameter; blade design characteristics; zero reduction from basic diameter Basic design (fixed pitch); / ; SAE 1 flange (SLM = SAE 2); 69 in diameter; 50 in pitch at ¾ radius

Manufacturer	Example designation	Explanation
MT-Propeller MT-Propeller Entwicklung GmbH Flugplatzstrasse 1 D-94348 Atting Germany *Tel:* (+49 9429) 940 90 *Fax:* (+49 9429) 84 32 *e-mail:* sales@mt-propeller.com *Web:* http://www.mt-propeller.com	MTV-9-B-C/CL160-03 MT186R140-3D	MT-Propeller; variable pitch; Type 9; SAE 2½ in bolts (A-F as for Avia propellers); counterweight moment towards high pitch position (blank = none or small counterweights with low pitch moment); blade pitch change pin moment towards high pitch (blank = low pitch); left hand rotation, tractor (blank = right tractor; RD = right, pusher, LD = left, pusher); 160 cm diameter; Type 3 MT-Propeller; two-blade, fixed pitch; 1,860 mm diameter
Neuform Neuform Composites GmbH Ameke 63 D-48317 Dreinsteinfurt Germany *Tel:* (+49 2387) 940 26 *Fax:* (+49 2387) 940 28 *e-mail:* info@neuform-propeller.de *Web:* http://www.neuform-propeller.de	T2 C3-V	Composites, T-type blade (C, TX), 1,730 to 1,750 mm diameter, fixed pitch, two-blade (3 = three blade, 4 = four blade) Composites, C-type blade (T, TX), 1,730 to 1,750 mm diameter, variable pitch (2 = two blade, 4 = four blade)
Powerfin Powerfin Inc 4700, 188th Street NE Arlington Washington 98223 *Tel:* (+1 360) 403 06 35 *Fax:* (+1 360) 403 05 99 *e-mail:* info@powerfin.com *Web:* http://www.powerfin.com	Model A Model B Model C Model D Model E Model F	64 to 72 in diameter, three-blade, composites 60 to 64 in diameter, three-blade, composites 64 to 69 in diameter, three-blade, composites 69 to 72 in diameter, three-blade, composites 51 in diameter, three-blade, composites 66 to 70 in diameter, three-blade, composites
PZL Warszawa PZL Warszawa-Okecie SA aleja Krakowska 110/114 PL-00-971 Warszawa Poland *Tel:* (+48 22) 846 61 52 *Fax:* (+48 22) 846 27 01 *e-mail:* wasiucionek@pzl-okecie.pl *Web:* http://www.pzl-okecie.com.pl	AW-2-30	Four-blade, constant-speed; metal; Antonov An-2; 30 cm reduction from full AW-2 diameter of 2.60 m (M-18 Dromader)
Quinti Avio Via Cassia Zona PIP n 21 I-52040 Marciano della Chiana (AR) Italy *Tel:* (+39 0575) 84 21 29 *Fax:* (+39 0575) 84 23 74 *e-mail:* quintiavio@quintiavio.com *Web:* http://www.quintiavio.com	QA2WD00.R QA3WD00.R QA3WD00.S QA4WD000.S	Constant-speed (electric) hub; two-blade; Rotax engine; blades as required, 1,000 to 1,800 mm Constant-speed (electric) hub; three-blade; Rotax engine; blades as required, 1,000 to 1,800 mm Constant-speed (electric) hub; three-blade; SAE1 fitting; blades as required, 1,000 to 1,800 mm Constant-speed (electric) hub; four-blade; SAE1 fitting; blades as required, 1,000 to 1,800 mm
Ratier-Figeac BP2 F-46101 Figeac Cedex France *Tel:* (+33 5) 65 50 50 50 *Fax:* (+33 5) 65 34 23 63	FH386	Composites; eight-blade; 5.33 m (17 ft 6 in) diameter; Airbus A400M
Sensenich Sensenich Wood Propeller Co Inc 2008 Wood Court Plant City Florida 33567 USA *Tel:* (+1 813) 752 37 11 *Fax:* (+1 813) 742 28 18 *e-mail:* donr@sensenichprop.com *Web:* http://www.sensenichprop.com	W70L-43 W72CK-42	Wood; 70 in diameter; Lycoming engine (A = Aeronca, C = Continental, F = Franklin, J = Jacobs, R = Ranger, W = Warner); pitch angle Wood; 72 in diameter; Continental engine (AC/BR/BE/BS/BRCK/DK/EY/FK/GK/JR/RK Continental, AF/CF = Franklin, JA = Jacobs, AA/FE/FM/JB/LY/RM = Lycoming, BB/BS/RB = Ranger, CA/CB = Warner); pitch angle
Sensenich Propeller Manufacturing Co 14 Citation Lane Lititz Pennsylvania 17543 USA *Tel:* (+1 717) 569 04 35 *Fax:* (+1 717) 560 37 25 *e-mail:* denisen@sensenich.com *Web:* http://www.sensenich.com	74DM6S5-2-60L	Metal; 74 in diameter; blade design D; hub design M6 for SAE 2 flange, ⅜ in bolts (C = ARP 502 flange, K = SAE 1, M7 = SAE 2, $\frac{7}{16}$ in bolts, M8 = SAE 2, ½ in bolts, R = SAE 3); integral dowled spacer 5 × ¼ in; 2 in reduction from basic diameter; blade pitch 60 in at 75% radius; design change
VPERED ooo VPERED pr Entuziastov 15 111024 Moskva Russian Federation *Tel:* (+7 095) 273 49 28 *Fax:* (+7 095) 273 49 28 *e-mail:* vpered@online.ru	VM-3	Wood, two-blade, fixed pitch 1,850 mm diameter
VZLU VZLU Beranových 130 Prague-Letňany CZ-199 05 Czech Republic *Tel:* (+420 272) 11 53 04 *Fax:* (+420 286) 92 05 18 *e-mail:* vrtule@vzlu.cz *Web:* http://www.vzlu.cz	V230B V234 V237 V331 V534AD	Wood, two-blade, fixed pitch; 1,625 mm diameter; Rotax 447/503 (C = Rotax 912/582/618, E = Rotax 912F) two-blade Wood, two-blade, fixed pitch; 1,800 mm diameter Wood, two-blade, fixed pitch; 1,600 mm diameter (237A 1,600 mm; 237DF 1,450 mm) Wood, three-blade, ground-adjustable pitch; 1,650 mm diameter Wood, two-blade, hydraulically controlled constant-speed; 1,613 mm diameter (BD = 1,650 mm)

Manufacturer	Example designation	Explanation
Warp Drive Warp Drive Inc 1207 Highway 18 East Ventura Iowa 50482 USA *Tel:* (+1 641) 357 60 00 *Fax:* (+1 641) 357 75 92		Carbon fibre, ground-adjustable pitch; two-, three-, four-, five-blade
Woodcomp Woodcomp sro Vodoská 4 CZ-250 70 Czech Republic *Tel:* (+420) 283 97 13 09 *Fax:* (+420) 283 97 02 86 *e-mail:* info@woodcomp.cz *Web:* http://www.woodcomp.cz	Junkers JU 160	Carbon fibre; two-, three- or four-blade; 1,600 mm diameter (165 = 1,650 mm; 170 = 1,700 mm); ground-adjustable pitch
	Junkers Vario 165	Carbon fibre; 1,650 mm diameter (170 = 1,700 mm); mechanical flight-adjustable pitch, two-blade
	Kremen SR 116	Wood; two- or three- blade; 1,600 mm diameter (117 = 1,450 mm); ground-adjustable pitch
	Kremen SR 200 A	Wood; 1,450 mm diameter (B/C = 1,680 mm)
	Kremen SR 2000 XA	Wood; two- or-three-blade; electrically variable pitch design; 1,700 mm diameter (XB = 1,500 mm)
	Kremen SR 3000	As SR 2000, optional reverse and feather
	SportProp Winglet 136	Glass fibre; ground-adjustable pitch (Classic 150R, Winglet 165, Speedy 165, Klassisch 160, Speedy 175, Para 110, Para 137)
	SportProp Varia 165/2R	Glass fibre; mechanical flight-adjustable pitch (Varia 170); two-blade

UPDATED

INDEXES

AIRCRAFT

To help users of this title evaluate the published data, *Jane's Information Group* has divided entries into three categories.
N NEW ENTRY Information on new equipment and/or systems appearing for the first time in the title.
V VERIFIED The editor had made a detailed examination of the entry's content and checked its relevance and accuracy for publication in the new edition to the best of his ability.
U UPDATED During the verification process, significant changes to content have been made to reflect the latest position known to *Jane's* at the time of publication.
Items in italics refer to entries which have been deleted from this edition. These entries are still available in the online version of this product-jawa.janes.com.

AIRCRAFT APPEARING IN THIS EDITION ONLY. For those appearing in previous 10 editions, please see following index on tinted paper.

Index begins with aircraft generally identified only by numbers. Most were previously listed as 'Model 707' and so on under M

B

C

D

E

F

G

H

I

J

K

L

M

N

O

P

Q

R

S

T

U

V

W

X

Y

Z

AIRCRAFT–PREVIOUS TEN EDITIONS

Index begins with aircraft generally identified only by numbers. Most were previously listed as 'Model 707' and so on under M

D

E

F

G

H

I

J

K

L

M

Note: Aircraft previously identified as 'Model xxx' are now listed as numbers only at the beginning of the index.

N

O

P

Q

R

S

T

U

AERO-ENGINES–PREVIOUS TEN EDITIONS

Detailed descriptions of aero-engines last appeared in the 1995-96 edition. For current data refer to table in this edition, or to full details in Jane's Aero-Engines

U

V

W

X

Y

Z

NOTES

NOTES

NOTES

NOTES

NOTES